BK 621.3 H236
HANDBOOK OF MODERN ELECTRONICS AND ELECTRIC
ENGINEERING
C1986 85.00 FV
3000 706001 30018
St. Louis Community College
C0-AUQ-416

WITHDRAWN

St. Louis Community
College

HANDBOOK OF MODERN ELECTRONICS AND ELECTRICAL ENGINEERING

HANDBOOK OF MODERN ELECTRONICS AND ELECTRICAL ENGINEERING

Editor-in-Chief

CHARLES BELOVE

Department of Electrical and Computer Engineering
Florida Atlantic University
Boca Raton, Florida

A Wiley-Interscience Publication

JOHN WILEY & SONS

New York · Chichester · Brisbane · Toronto · Singapore

Library of Congress Cataloging in Publication Data:

Main entry under title:

Handbook of modern electronics and electrical engineering.

"A Wiley-Interscience publication."

Includes index.

1. Electronics—Handbooks, manuals, etc.
2. Electric engineering—Handbooks, manuals, etc.
I. Belove, Charles. II. Hopkins, Phillip.

TK7825.H38 1986 621.3 85-29450

ISBN 0-471-09754-3

Printed in the United States of America

10 9 8 7 6 5 4 3 2 1

CONTRIBUTORS

Peter G. Anderson
School of Computer Science
Rochester Institute of Technology
Rochester, New York

J. Ronald Bailey
Systems Products Division
IBM Corporation
Boca Raton, Florida

Michael Baltrush
New Jersey Institute of Technology
Wayne, New Jersey

Mark B. Barron
Calma Company
Sunnyvale, California

Eleanor Baum
Pratt Institute
Brooklyn, New York

Paul M. Baum
Queens College
Queens, New York

Don Berlincourt
President, Channel Products, Inc.
Chesterland, Ohio

C. F. Bieber
Hughes Aircraft Company
Tucson, Arizona

John A. Biles
School of Computer Science
Rochester Institute of Technology
Rochester, New York

Frank T. Boesch
Department of Electrical Engineering
Stevens Institute of Technology
Hoboken, New Jersey

Clifford Bogen
Professor, Life Science
New York Institute of Technology
Old Westbury, New York

James R. Carbin
School of Computer Science
Rochester Institute of Technology
Rochester, New York

Warren R. Carithers
School of Computer Science
Rochester Institute of Technology
Rochester, New York

Frank R. Castella
Applied Physics Laboratory
The Johns Hopkins University
Baltimore, Maryland

James A. Chmura
School of Computer Science
Rochester Institute of Technology
Rochester, New York

K. K. Chow
Lockheed Missiles and Space Company
Palo Alto Research Laboratory
Palo Alto, California

Chris Comte
School of Computer Science
Rochester Institute of Technology
Rochester, New York

Lawrence A. Coon
School of Computer Science
Rochester Institute of Technology
Rochester, New York

Richard G. Costello
Professor, The Cooper Union for the Advancement of Science and Art
New York, New York

Josefa Cubina
Assistant Professor, Life Science
New York Institute of Technology
Old Westbury, New York

Thomas J. Cutler
Lieutenant Commander, Department of Seamanship and Navigation
United States Naval Academy
Annapolis, Maryland

John Dawson, M.D.
Chief Resident in Surgery
Department of Surgery
New York Medical College
Valhalla, New York

Louis R. M. Del Guercio, M.D.
Professor and Chairman
Department of Surgery
New York Medical College
Valhalla, New York

E. T. Dickerson
University of Houston–Clear Lake
Houston, Texas

Daniel F. DiFonzo
COMSAT Laboratories
Clarksburg, Maryland

Daniel B. Diner
Assistant Professor, Computer Science Department
New York Institute of Technology
Old Westbury, New York

Henry Domingos
Professor, Electrical Engineering Department
Clarkson College of Technology
Potsdam, New York

Melvyn M. Drossman
New York Institute of Technology
Old Bethpage, New York

Mary Ann Dvonch
School of Computer Science
Rochester Institute of Technology
Rochester, New York

George H. Ebel
The Singer Company
Wayne, New Jersey

Yusuf Z. Efe
Professor, The Cooper Union for the Advancement of Science and Art
New York, New York

Electronic Defense Laboratories
Sylvania Electric Products, Inc.
Mountainview, California

Steven G. Epstein, M.D.
Resident in Surgery
Department of Surgery
New York Medical College
Valhalla, New York

Henry A. Etlinger
School of Computer Science
Rochester Institute of Technology
Rochester, New York

Roger D. Fruechte
Assistant Head, Electronics Department
General Motors Research Laboratories
Warren, Michigan

Gustave R. Gaschnig
Missile Systems Division
Raytheon Company
Bedford, Massachusetts

Martin Gazourian
Associate Professor, Department of Electrical and Computer Engineering
Florida Atlantic University
Boca Raton, Florida

Stanley A. Gelfand
Chief, Audiology and Speech Pathology Service
Veterans Administration Medical Center
East Orange, New Jersey

Sorab K. Ghandhi
Professor of Electrophysics
Electrical and Systems Engineering Department
Rensselaer Polytechnic Institute
Troy, New York

Amrit L. Goel
Professor, Industrial Engineering and Operations Research
Syracuse University
Syracuse, New York

Alex Goldman
Magnetics Division of Spang Industries
Butler, Pennsylvania

Sam Goldwasser
The Moore School of Electrical Engineering
University of Pennsylvania
Philadelphia, Pennsylvania

Peter Graham
Department of Electrical Engineering
Florida Atlantic University
Boca Raton, Florida

Albert B. Grundy
Audio Research Institute
New York, New York

Joel Halpert
Grumman
Hauppage, New York

James Hammerton
School of Computer Science
Rochester Institute of Technology
Rochester, New York

Jeffrey Hantgan
Assistant Professor, Department of Electrical Engineering
State University of New York
Stony Brook, New York

Harold Z. Haut
Biomedical Engineer
Avery Laboratories
Farmingdale, New York

Jack Hollingsworth
School of Computer Science
Rochester Institute of Technology
Rochester, New York

Hing-Loi A. Hung
Manager, Advanced Microwave Techniques Department
COMSAT Laboratories
Clarksburg, Maryland

Guy Johnson
Professor, School of Computer Science
Rochester Institute of Technology
Rochester, New York

Benedict Kingsley
National Foundation for
Non-Invasive Diagnostics
Princeton, New Jersey

Yeng S. Kuo
Lockheed Engineering and
Management Services
Company, Inc.
Houston, Texas

Edward J. Lancevich
Manhattan College
Brooklyn, New York

William Lembeck
Research Scientist,
New York University
Post Graduate Medical School
New York, New York

Harry Levitt
Director of the Communication
Sciences Laboratory
State University of New York
New York, New York

Shin-R Lin
New York Institute of Technology
Old Westbury, New York

Jefferson F. Lindsey III
Southern Illinois University
at Carbondale
Carbondale, Illinois

Daniel D. Lingelbach
Professor, School of Electrical
Engineering
Oklahoma State University
Stillwater, Oklahoma

Peter H. Lutz
School of Computer Science
Rochester Institute of Technology
Rochester, New York

James A. M. McHugh
Computer and Information Science
Department
New Jersey Institute of Technology
Newark, New Jersey

Manu Malek-Zavarei
Bell Laboratories
Holmdel, New Jersey

Alan B. Marcovitz
Florida Atlantic University
Boca Raton, Florida

Thomas E. Marlin
Engineering Associate,
Exxon Research and
Engineering Company
Florham Park, New Jersey

Stephen C. Martin
Palo Alto Research Laboratory
Lockheed Missiles and
Space Company
Palo Alto, California

Jay Michlin
Exxon Research and Engineering
Company
Morristown, New Jersey

Peter Monsen
SIGNATRON, Inc.
Lexington, Massachusetts

J. Keith Nelson
Professor, Center for Electric Power
Engineering
Rensselaer Polytechnic Institute
Troy, New York

Ray W. Nettleton
Department of Electrical Engineering
and Systems Science
Michigan State University
East Lansing, Michigan

Robert B. Newman
Senior Vice President and Principal Consultant, Bolt Beranek and Newman, Inc.
Cambridge, Massachusetts

Edward H. Nicollian
Professor,
University of North Carolina
Charlotte, North Carolina

Rayno D. Niemi
School of Computer Science
Rochester Institute of Technology
Rochester, New York

John H. Painter
Altair Corporation
College Station, Texas

Jerald D. Parker
Professor, School of Mechanical and Aerospace Engineering
Oklahoma State University
Stillwater, Oklahoma

Thomas W. Parsons
Hofstra University
Hempstead, New York

Peter Pleshko
Manager, Plasma Display Technology
IBM Corporation
Kingston, New York

Frederick B. Pogust
Eaton Corporation
AIL Division
Farmingdale, New York

Icarius E. Pyros
Associate Professor,
Electrical Engineering Department
New York Institute of Technology
Old Westbury, New York

Matthew J. Quinn, Jr.
College of Technology
University of Houston
Houston, Texas

George W. Raffoul
Lockheed Engineering and Management Services Company, Inc.
Houston, Texas

RCA Corporation
Consumer Electronics Technical Training
Indianapolis, Indiana

Milton Rosenstein
Associate Professor,
Electrical Engineering Department
New York Institute of Technology
Old Westbury, New York

Mansoor A. Saifi
Belcom Research
Murray Hill, New Jersey

Joseph Scaturro
Associate Professor,
Electrical Engineering Department
Clarkson College of Technology
Potsdam, New York

Emil R. Schiesser
NASA Johnson Space Center
Houston, Texas

Krishna Seshan
Department of Metallurgical Engineering
University of Arizona
Tucson, Arizona

Jack W. Seyl
NASA Johnson Space Center
Houston, Texas

Yacov Shamash
Professor, Department of Electrical and Computer Engineering
Florida Atlantic University
Boca Raton, Florida

Julius Simon
Laboratory for Chromatography
Flushing, New York

Bernard Sklar
Aerospace Corporation
Los Angeles, California

Gordon R. Slemon
University of Toronto
Toronto, Canada

Bernard Smith
Commander, ERADCOM, U.S. Army
Fort Monmouth, New Jersey

Charles Suffel
Mathematics Department
Stevens Institute of Technology
Hoboken, New Jersey

Frederick L. Swern
Bendix
Succasunna, New Jersey

Jeff Tosk
Department of Microbiology
Loma Linda University
Loma Linda, California

William H. Tranter
Department of Electrical Engineering
University of Missouri
Rolla, Missouri

Katsuaki Tsurushima
Manager, Hi-Fi Audio Division and Video Group
Sony Corporation of America
Park Ridge, New Jersey

Kwei Tu
Lockheed Engineering and Management Services Company, Inc.
Houston, Texas

Wayne C. Turner
Professor, Montana State University
Boseman, Montana

Thomas C. Upson
Mathematics Department
Rochester Institute of Technology
Rochester, New York

Richard VanSlyke
Department of Electrical Engineering
Stevens Institute of Technology
Hoboken, New Jersey

Surya V. Varanasi
Lockheed Engineering and Management Services Company Inc.
Houston, Texas

L. S. Watkins
Research Leader, Engineering Research Center
Western Electric Company
Princeton, New Jersey

Gerald Weiss
Professor, Polytechnic Institute of New York
Brooklyn, New York

Martin Wolf
Professor, Department of Electrical Engineering and Science
University of Pennsylvania
Philadelphia, Pennsylvania

Rodger E. Ziemer
Department of Electrical Engineering
University of Missouri
Rolla, Missouri

PREFACE

Spurred by military requirements during World War II, advances in electronics and information technology from about 1940 to 1960 caused a virtual revolution in the practice of engineering. In the years since then, semiconductor technology, the advent of the integrated circuit, and the resulting digital revolution have further led electrical engineering to proceed with new developments at a breakneck pace.

For example, circuit design is rapidly becoming a process of selecting an appropriate combination of integrated circuit blocks based on their input–output characteristics. Many types of design that were not feasible in the past because of computational difficulty are now routinely done by computer. The design of VLSI (Very Large Scale Integration) circuits, each of which contains over 1000 transistors, would be virtually impossible without the computer to aid in the design and determine optimum placement for the many elements involved in each VLSI chip.

However, engineers are still required in all phases of design, development, and manufacture. During the course of an assignment, an engineer may come in contact with many different areas of engineering. It is impossible for an individual to be expert in all of these areas. For those in which he or she does not have immediate expertise, the engineer must rely on others or spend time studying the literature.

This has led to a constantly increasing need for up-to-date state-of-the-art information in the form of textbooks and handbooks. This is especially true in electronics, a field which crosses all boundaries and is used extensively in all branches of engineering. For example, most instrumentation, measurement, and control systems involve electronic circuits, usually in integrated circuit form. In addition, most engineers today have electronic computer service available at their desks in the form of an individual PC (personal computer) which may or may not be part of a network, or a terminal connected to a large "mainframe" computer.

This handbook adopts a fresh approach. It is designed to provide service to engineers in fields other than electronics, to practicing electrical and electronics engineers, to management personnel, and to anyone else needing this kind of specialized information. The 69 chapters in the handbook provide an overview of each subject area in reasonable depth along with basic theory and design information where appropriate. An extensive bibliography is included with each article so that up-to-date references are available if additional information is required.

The handbook is divided into three parts in accordance with the natural flow of subject matter in the field of electronics and electrical engineering. The first part contains sections on mathematics, properties of materials, and components. The second part is devoted to electronic and electric circuits. It contains separate sections on passive circuits, active circuits, and digital circuits. All of this material is brought together in the part on systems. Here we have a comprehensive treatment of systems engineering and automatic control, including chapters on the new sciences of robotics and reliability. There follows an extensive treatment of medical applications of electronics. Other sections cover sound and video recording and reproduction; telecommunications; computers, including new material on microprocessors; and energy engineering.

The authors who have contributed articles to the handbook have done a superb job of condensing mountains of material into readable accounts of their respective areas. The emphasis throughout has been to provide a practical introduction and overview of each subject.

Finally, I would like to thank the associate editors who labored long and diligently with infinite patience, and Thurman R. Poston and George Telecki, our Wiley editors, for their untiring efforts to bring this project to fruition.

CHARLES BELOVE

Boynton Beach, Florida
February 1986

CONTENTS

Mathematics, Materials, and Components

PART ONE MATHEMATICS

Circuits: Passive, Active, and Digital

PART FOUR PASSIVE CIRCUITS

PART FIVE ACTIVE CIRCUITS

PART SIX DIGITAL CIRCUITS

Systems

PART SEVEN SYSTEMS ENGINEERING, AUTOMATIC CONTROL, AND MEASUREMENTS

PART EIGHT MEDICAL APPLICATIONS OF ELECTRONICS

PART NINE SOUND AND VIDEO RECORDING AND REPRODUCTION

PART THIRTEEN ENERGY ENGINEERING

HANDBOOK OF MODERN ELECTRONICS AND ELECTRICAL ENGINEERING

PART 1
MATHEMATICS

CHAPTER 1
UNITS AND CONSTANTS

SHIN-R LIN

New York Institute of Technology
Old Westbury, New York

1.1 SI BASE UNITS

In physical or engineering measurements, many different quantities are involved. To compare magnitudes of quantities, it is necessary to have a basis for comparison. This basis for a quantity is called the unit of magnitude of that quantity.

The International System of Units (abbreviated SI) was adopted by the Eleventh General Conference on Weights and Measures (abbreviated CGPM from the official French name Conférence Générale des Poids et Measures) in 1960. SI is intended to be a basis for worldwide standardization of units of measurement and is being adopted universally. It represents the modernized form of the metric system.

SI units are divided into three classes: base units, derived units, and supplementary units. The SI base units are listed in Table 1.1.

Authorized translations of the original French definitions of the base units are as follows:

Meter (m). The meter is the length equal to 1 650 763.73 wavelengths in vacuum of the radiation corresponding to the transition between the levels $2p_{10}$ and $5d_5$ of the krypton-86 atom. [Eleventh CGPM (1960).]

Kilogram (kg). The kilogram is the unit of mass; it is equal to the mass of the international prototype of the kilogram. [First CGPM (1889); Third CGPM (1901).]

Second (s). The second is the duration of 9 192 631 770 periods of the radiation corresponding to the transition between the two hyperfine levels of the ground state of the cesium-133 atom. [Thirteenth CGPM (1967).]

Ampere (A). The ampere is that constant current which, if maintained in two straight parallel conductors of infinite length, of negligible circular cross section, and placed 1 meter apart in vacuum, would produce between these conductors a force equal to 2×10^{-7} newton per meter of length. [Ninth CGPM (1948).]

Kelvin (K). The kelvin, the unit of thermodynamic temperature, is the fraction 1/273.16 of the thermodynamic temperature of the triple point of water. [Thirteenth CGPM (1967).]

Mole (mol) **(1)** The mole is the amount of substance of a system that contains as many elementary entities as there are atoms in 0.012 kilogram of carbon-12. **(2)** When the mole is

TABLE 1.1. SI BASE UNITS

Quantity	SI Unit Name	SI Unit Symbol
Length	meter	m
Mass	kilogram	kg
Time	second	s
Electric current	ampere	A
Thermodynamic temperature	kelvin	K
Amount of substance	mole	mol
Luminous intensity	candela	cd

used, the elementary entities must be specified and may be atoms, molecules, ions, electrons, other particles, or groups of such particles. [Fourteenth CGPM (1971).]

Candela (cd). The candela is the luminous intensity, in a given direction, of a source that emits monochromatic radiation of frequency 540×10^{12} hertz and that has a radiant intensity in that direction of 1/683 watt per steradian. [Sixteenth CGPM (1979).]

1.2 SI DERIVED UNITS

Derived units are expressed algebraically in terms of base units by means of the mathematical symbols of multiplication and division according to the definitions or physical laws.

TABLE 1.2. SI DERIVED UNITS WITH SPECIAL NAMES

Quantity	SI Unit Name	SI Unit Symbol	Expression in Terms of Other Units
Frequency	hertz	Hz	
Force	newton	N	
Pressure, stress	pascal	Pa	N/m^2
Energy, work, quantity of heat	joule	J	$N \cdot m$
Power, radiant flux	watt	W	J/s
Electric charge, quantity of electricity	coulomb	C	
Electric potential, potential difference, electromotive force	volt	V	W/A
Capacitance	farad	F	C/V
Electric resistance	ohm	Ω	
Electric conductance	siemens	S	A/V
Magnetic flux	weber	Wb	$V \cdot s$
Magnetic flux density	tesla	T	Wb/m^2
Inductance	henry	H	Wb/A
Celsius temperature	degree Celsius	°C	
Luminous flux	lumen	lm	
Illuminance	lux	lx	lm/m^2

TABLE 1.3. SI DERIVED UNITS WITH SPECIAL NAMES ADMITTED FOR REASONS OF SAFEGUARDING HUMAN HEALTH

	SI Unit		
Quantity	Name	Symbol	Expression in Terms of Other Units
Activity (of a radionuclide)	becquerel	Bq	
Absorbed dose, specific energy imparted, kerma, absorbed dose index	gray	Gy	J/kg
Dose equivalent, does equivalent index	sievert	Sv	J/kg

Certain derived units have been given special names and symbols. These names and symbols are given in Table 1.2. They may themselves be used to express other derived units. These are shown in Table 1.3.

1.3 SI SUPPLEMENTARY UNITS

Supplementary units contain two units: the radian, for SI unit of plane angle, and the steradian, for SI unit of solid angle. [Eleventh CGPM (1960).] Refer to Table 1.4.

In 1980, the CIPM (International Committee for Weights and Measures) specified that in SI units the quantities plane angle and solid angle should be considered dimensionless derived quantities. Therefore, the supplementary units radian and steradian are to be regarded as dimensionless derived units that may be used or omitted in the expressions for derived unit.

TABLE 1.4. SI SUPPLEMENTARY UNITS

	SI Unit	
Quantity	Name	Symbol
Plane angle	radian	rad
Solid angle	steradian	sr

1.4 SYMBOLS FOR PHYSICAL QUANTITIES

Table 1.5 lists symbols and measurement units for physical quantities.

TABLE 1.5. SYMBOLS FOR PHYSICAL QUANTITIES

Quantity	Symbol	Measurement Name	Measurement Unit
Space and Time			
Area	A, S	square meter	m^2
Volume	V	cubic meter	m^3
Distance, displacement	d, r	meter	m
Angular displacement	θ	radian	rad

TABLE 1.5. (*Continued*)

Quantity	Symbol	Measurement Name	Measurement Unit
Space and Time			
Speed, velocity	v	meter per second	m/s
Angular velocity	ω	radian per second	rad/s
Linear acceleration	a	meter per second square	m/s^2
Angular acceleration	α	radian per second square	rad/s^2
Period	T	second	s
Frequency	f, ν	hertz	Hz
Angular frequency	ω	revolution per second	r/s
Rotational frequency	n	radian per second	rad/s
Wavelength	λ	meter	m
Wave number	$\tilde{\nu}$	1 per meter	m^{-1}
Angular wave number	κ	radian per meter	rad/m
Mechanics			
Density, mass density	ρ	kilogram per cubic meter	kg/m^3
Momentum	p	kilogram meter per second	$kg \cdot m/s$
Angular momentum	L	kilogram square meter per second	$kg \cdot m^2/s$
Moment of inertia	I	kilogram meter square	$kg \cdot m^2$
Force	F	newton	N
Weight	W	newton	N
Weight density	γ	newton per cubic meter	N/m^3
Moment of force	M	newton meter	$N \cdot m$
Torque	T, M	newton meter	$N \cdot m$
Pressure stress	p	pascal	Pa
Young's modulus	E	pascal	Pa
Shear modulus	G	pascal	Pa
Bulk modulus	K	pascal	Pa
Energy	J	joule	J
Kinetic energy	K	joule	J
Potential energy	U	joule	J
Work	W	joule	J
Power	P	watt	W
Efficiency	η	(numeric only)	
Electricity and Magnetism			
Electric charge, quantity of electricity	Q, q	coulomb	C
Linear charge density	λ	coulomb per meter	C/m
Surface charge density	σ	coulomb per square meter	C/m^2
Volume charge density	ρ	coulomb per cubic meter	C/m^3

TABLE 1.5. (*Continued*)

Quantity	Symbol	Measurement Name	Measurement Unit
Electricity and Magnetism			
Electric field strength	E	volt per meter	V/m
Electric potential, potential difference, electromotive force	V	volt	V
Electric flux	Ψ	coulomb	C
Electric flux density	D	coulomb per square meter	C/m^2
Capacitance	C	farad	F
Permittivity	ϵ	farad per meter	F/m
Current density	J	ampere per square meter	A/m^2
Resistance	R	ohm	Ω
Resistivity	ρ	ohm meter	Ω · m
Conductance	G	siemens	S
Conductivity	σ	siemens per meter	S/m
Magnetic field strength	H	ampere per meter	A/m
Magnetic flux	Φ	weber	Wb
Magnetic flux linkage	Λ	weber	Wb
Magnetic flux density	B	tesla	T
Permeability of free space	μ_0	henry per meter	H/m
Permeability	μ	henry per meter	H/m
Magnetic susceptibility	χ_m	(numeric only)	
Inductance	L	henry	H
Reluctance	R	reciprocal henry	H^{-1}
Impedance	Z	ohm	Ω
Reactance	X	ohm	Ω
Capacitive reactance	X_c	ohm	Ω
Power	P	watt	W
Frequency	f	hertz	Hz
Angular frequency	ω	radian per second	rad/s
Wavelength	λ	meter	m
Heat and Thermodynamics			
Thermodynamic temperature (absolute)	T, Θ	Kelvin	K
Celsius temperature	t, θ	degree Celsius	°C
Quantity of heat	Q	joule	J
Internal energy	U	joule	J
Free energy	F	joule	J
Enthalpy	H	joule	J
Entropy	S	joule per Kelvin	J/K
Heat capacity	C	joule per Kelvin	J/K
Thermal conductivity	λ, κ	watt per meter Kelvin	W/m · K
Specific heat capacity	c	joule per kilogram Kelvin	J/kg · K

TABLE 1.5. *(Continued)*

Quantity	Symbol	Measurement Name	Measurement Unit
Light and Radiation			
Radiant intensity	I	watt per steradian	W/sr
Radiant power, radiant flux	P	watt	W
Radiant energy	W	joule	J
Radiance	L	watt per steradian square meter	$W/sr \cdot m^2$
Radiant exitance	M	watt per square meter	W/m^2
Irradiance	E	watt per square meter	W/m^2
Luminous intensity	I	candela	cd
Luminous flux	ϕ	lumen	lm
Quantity of light	Q	lumen-second	$lm \cdot s$
Luminance	L	candela per square meter	cd/m^2
Luminous efficacy	$\kappa(\lambda)$	lumen per watt	lm/W
Refractive index, index of refraction	n	(numeric only)	
Emissivity	$\epsilon(\lambda)$	(numeric only)	
Absorptance	$\alpha(\lambda)$	(numeric only)	
Transmittance	$\tau(\lambda)$	(numeric only)	
Reflectance	$\rho(\lambda)$	(numeric only)	
Sound and Acoustics			
Velocity of sound	v	meter per second	m/s
Sound energy flux	P	watt	W
Sound intensity	I	watt per meter	W/m
Acoustic impedance	Z	Newton second per meter cubed	$N \cdot s/m^3$
Acoustic absorption factor	α	(numeric only)	
Reflection factor	ρ	(numeric only)	
Transmission factor	τ	(numeric only)	
Dissipation factor	δ	(numeric only)	
Loudness level	T	phon	

Source. "Letter Symbols for Quantities Used in Electrical Science and Electrical Engineering," IEEE Standards 280, 1968, with permission.

1.5 SYMBOLS AND VALUES FOR PHYSICAL CONSTANTS

Table 1.6 lists symbols, values, and measurement units for physical constants.

TABLE 1.6. PHYSICAL CONSTANTS

Quantity	Symbol	Value	Error (ppm)	Prefix	Measurement Unit
Speed of light in vacuum	c	2.997 924 58	0.004	10^8	m/s
Gravitational constant	G	6.672 000	615	10^{-11}	Nm^2/kg^2
Avogadro constant	N_A	6.022 045	5.1	10^{26}	$kmol^{-1}$
Boltzmann constant	k	1.380 662	32	10^{-23}	J/K
Planck constant	$\hbar$	1.054 588 7	5.4	10^{-34}	J.s
Volume of ideal gas at STP	V	2.241 383	31	10^1	$m^3/kmol$
Electronic charge	e	1.602 189 2	2.9	10^{-19}	C
Million electron volt	Mev	1.602 189 2	2.9	10^{-13}	J
Atomic mass unit	amu	$1/12\ M_{c12}$			
		= 1.660 566	5.1	10^{-27}	kg
		= 9.315 016	2.8	10^2	MeV/c^2
Electron rest mass	m_e	9.109 534	5.1	10^{-31}	kg
		0.511 003 4	2.8		MeV/c^2
Proton rest mass	m_p	1.672 648	5.5	10^{-27}	kg
		9.382 796	2.8	10^2	MeV/c^2
Stefan Boltzmann constant	σ	5.670 32	125	10^{-8}	W/m^2K^4
Rydberg constant	R_∞	1.097 373 177	0.075	10^7	m^{-1}
Fine structure constant	α	1/137.03604	0.82		
Bohr radius	a_0	5.291 770 6	0.82	10^{-11}	m
Classical electron radius	r_e	2.817 938 0	2.5	10^{-15}	m
Compton wavelength of electron	λ_e	3.861 590 5	1.6	10^{-13}	m
Bohr magneton	μ_β	9.274 077 5	4.5	10^{-24}	J/T
Nuclear magneton	μ_N	5.050 823 7	4.6	10^{-27}	J/T
Electron magnetic moment	μ_c	9.284 851	7.0	10^{-24}	J/T
Proton magnetic moment	μ_p	1.410 617 0	4.5	10^{-26}	J/T

Source. Particle Data Group, Rev. Mod. Phys., 52(2), Part II, 1980, and CODATA, Bull. 11, Dec. 1973.

1.6 CONVERSION FACTORS

Table 1.7 lists conversion factors for quantities that are used frequently. For a more complete listing see Mechtly, 1973.

TABLE 1.7. CONVERSION FACTORS FOR THE INTERNATIONAL SYSTEM OF SI UNITS

To Convert From	Multiply By	To Obtain
Length		
Angstrom	1.000 000 E − 10	meter
Inch	2.540 000 E − 02	meter
Foot	3.048 000 E − 01	meter
Light-year	9.46055 E 15	meter
Micron	1.000 000 E − 06	meter
Mile (international nautical)	1.852 000 E 03	meter
Mile (international)	1.609 344 E 03	meter
Area		
Acre (US survey)	4.046 873 E − 03	square meter
Inch, square	6.451 600 E − 04	square meter
Foot, square	9.290 304 E − 02	square meter
Mile, square (international)	2.589 988 E 06	square meter
Volume		
Fluid ounce (US)	2.957 353 E − 05	cubic meter
Gallon (US liquid)	3.785 412 E − 03	cubic meter
Inch, cubic	1.638 706 E − 05	cubic meter
Pint (US liquid)	4.731 765 E − 04	cubic meter
Quart (US liquid)	9.463 529 E − 04	cubic meter
Angle		
Degree	1.745 329 E − 02	radian
Minute	2.908 882 E − 04	radian
Second	4.848 137 E − 06	radian
Velocity		
Feet per second	3.048 000 E − 01	meter per second
Miles per hour	4.470 400 E − 01	meter per second
Velocity		
Miles per hour	1.609 344 E 03	kilometer per hour
Kilometers per hour	2.777 778 E − 01	meter per second
Knot (international)	5.144 444 E − 01	meter per second
Acceleration, feet per square second	3.048 000 E − 01	meter per square second

TABLE 1.7. *(Continued)*

To Convert From	Multiply By	To Obtain
Mass		
Ounce (avoirdupois)	2.834 952 E − 02	kilogram
Ounce (troy)	3.110 348 E − 02	kilogram
Pound (avoirdupois)	4.535 924 E − 01	kilogram
Pound (troy)	3.732 417 E − 01	kilogram
Slug	1.459 390 E 01	kilogram
Ton (long, 2240 lb)	1.016 047 E 03	kilogram
Ton (metric)	1.000 000 E 03	kilogram
Ton (short, 2000 lb)	9.071 847 E 02	kilogram
Density		
Pound per cubic foot	1.601 846 E 01	kilogram per cubic meter
Slug per cubic foot	5.153 788 E 02	kilogram per cubic meter
Force		
Dyne	1.000 000 E − 05	newton
Ounce-force	2.780 139 E − 01	newton
Pound-force	4.448 222 E 00	newton
Ton-force	8.896 444 E 03	newton
Work-Energy		
British thermal unit (thermochemical)	1.054 350 E 03	joule
Calorie (thermochemical)	4.184 000 E 00	joule
Electronvolt	1.602 189 E − 19	joule
Erg	1.000 000 E − 07	joule
Foot-pound-force	1.355 818 E 00	joule
Power		
Horsepower (550 ft · lb/s)	7.456 999 E 02	watt
Foot-pound-force per second	1.355 818 E 00	watt
Pressure or Stress (Force per Unit Area)		
Atmosphere (standard)	1.013 250 E 05	pascal
Bar	1.000 000 E 05	pascal
Millimeter of mercury (at 0°C)	1.333 224 E 02	pascal
Temperature		
Degree Celsius	$t_k = t°C + 273.15$	Kelvin
Degree Fahrenheit	$t_k = (t°F + 459.67)/1.8$	Kelvin
Degree Rankine	$t_k = t°R/1.8$	Kelvin
Kelvin	$t°C = t_k - 273.15$	degree Celsius
Light		
Footcandle	1.076 391 E − 01	lux
Foot lambert	3.426 259 E − 02	candela per square meter
Lambert	3.183 099 E − 03	candela per square meter

Bibliography

Goldman, D. and R. J. Bell, eds., *The International System of Units (SI)*, NBS Special Publication 330, 1981.

IEEE, *Letter Symbols for Quantities Used in Electrical Science and Electrical Engineering* (IEEE No. 280), ASME, New York, 1968.

Mechtly, E., *The International System of Units*, NASA sp-7012, 1973.

Particle Data Group, *Review of Particle Properties, Rev. Mod. Phys.* **52**(2), April 1980.

Report of the CODATA Bulletin No. 11, Dec. 1973, *Recommended Consistent Values of the Fundamental Physical Constants 1973*, Pergamon Press, Oxford, 1973.

CHAPTER 2
MATHEMATICS USED IN ENGINEERING

PAUL BAUM

Queens College
Flushing, New York

ELEANOR BAUM

Pratt Institute
Brooklyn, New York

2.1 ALGEBRA

Following are some basic formulas and concepts that are useful to electrical engineers.

Exponents, Radicals, and Logarithms

$$x^0 = 1 \quad (x \neq 0) \qquad x^{-n} = \frac{1}{x^n} \qquad x^n x^m = x^{n+m}$$

$$\frac{x^n}{x^m} = x^{n-m} \qquad (x^n)^m = x^{nm} \qquad (xy)^n = x^n y^n \qquad \left(\frac{x}{y}\right)^n = \frac{x^n}{y^n}$$

$$\sqrt{x} = x^{1/2} \qquad \sqrt[n]{x} = x^{1/n} \qquad \sqrt[n]{x^m} = x^{m/n} \qquad \sqrt[n]{xy} = x^{1/n} y^{1/n}$$

$$\log x^n = n \log x \qquad \log xy = \log x + \log y \qquad \log \frac{x}{y} = \log x - \log y$$

Binomial Theorem

$$(x \pm y)^n = \sum_{k=0}^{n} \frac{n!}{k!(n-k)!} x^k (\pm y)^{n-k}$$

where $n! = n$ factorial $= n(n-1)\ldots(3)(2)(1)$. For example:

$$(x-y)^4 = x^4 - 4yx^3 + 6y^2x^2 - 4y^3x + y^4$$

Polynomials

$$p(x) = a_n x^n + a_{n-1}x^{n-1} + \cdots + a_1 x + a_0$$

b is a "zero" of the polynomial and a root of the equation $p(x) = 0$ if $p(b)$ gives a value of zero. In fact, $(x - b)$ is a factor of the polynomial.

An nth degree polynomial has n zeros. These may be real or complex, and may be repeated values. If the coefficients (a's) of the polynomial are real numbers, complex roots appear in complex conjugate pairs.

It is sometimes useful to see that:

$$\frac{a_{n-1}}{a_n} = -\sum \text{roots} \quad \text{and} \quad \frac{a_0}{a_n} = (-1)^n \prod \text{roots}$$

Quadratic Equation

The solution of $ax^2 + bx + c = 0$ is

$$x_{1,2} = \frac{-b \pm \sqrt{b^2 - 4ac}}{2a}$$

If a, b, and c are real, then the roots x_1 and x_2 are:

$$x_1 \neq x_2, \quad \text{and } x_1 \text{ and } x_2 \text{ are real if } \sqrt{b^2 - 4ac} > 0$$

$$x_1 = x_2 \text{ if } \sqrt{b^2 - 4ac} = 0$$

$$x_1 \text{ and } x_2 \text{ are complex conjugates if } \sqrt{b^2 - 4ac} < 0$$

Determinants

The square array that follows is known as a determinant of order n:

$$A = \begin{vmatrix} a_{11} & a_{12} & a_{13} & \cdots & a_{1n} \\ a_{21} & a_{22} & a_{23} & \cdots & a_{2n} \\ \cdots & \cdots & \cdots & \cdots & \cdots \\ a_{n1} & a_{n2} & a_{n3} & \cdots & a_{nn} \end{vmatrix}$$

Coefficients in horizontal lines form rows, and those in vertical lines form columns. Elements are identified by their position in this array by a double subscript, the first subscript indicating the row and the second indicating the column. The main diagonal consists of elements along the line from a_{11} to a_{nn}.

The minor of any element a_{jk} is the determinant that remains when the row and column containing that element are removed.

For example, for the determinant A

$$A = \begin{vmatrix} a_{11} & a_{12} & a_{13} \\ a_{21} & a_{22} & a_{23} \\ a_{31} & a_{32} & a_{33} \end{vmatrix}$$

the minor for a_{22} is

$$M_{22} = \begin{vmatrix} a_{11} & a_{13} \\ a_{31} & a_{33} \end{vmatrix}$$

The cofactor of an element a_{jk} is defined:

$$\text{cofactor} = (-1)^{j+k} \quad (\text{minor})_{jk}$$

where j is the element row and k the element column.

The determinant of order n is evaluated by successive reduction. It is equal to the sum of the product of the elements of any row or column multiplied by the corresponding element cofactor. For the third order determinant already given,

$$A = a_{11}M_{11} - a_{21}M_{21} + a_{31}M_{31}$$

$$A = a_{11}\begin{vmatrix} a_{22} & a_{23} \\ a_{32} & a_{33} \end{vmatrix} - a_{21}\begin{vmatrix} a_{12} & a_{13} \\ a_{32} & a_{33} \end{vmatrix} + a_{31}\begin{vmatrix} a_{12} & a_{13} \\ a_{22} & a_{23} \end{vmatrix}$$

This process is continued

$$A = a_{11}(a_{22}a_{33} - a_{32}a_{23}) - a_{21}(a_{12}a_{33} - a_{32}a_{13}) + a_{31}(a_{12}a_{23} - a_{22}a_{13})$$

Simultaneous Equations

Determinants are used to solve sets of simultaneous equations of the form

$$\begin{array}{c} a_{11}X_1 + a_{12}X_2 + a_{13}X_3 + \cdots + a_{1n}X_n = y_1 \\ \cdots \\ a_{n1}X_1 + a_{n2}X_2 + a_{n3}X_3 + \cdots + a_{nn}X_n = y_n \end{array}$$

Cramer's rule gives the solution to such a set as

$$X_1 = \frac{D_1}{\Delta} \qquad X_2 = \frac{D_2}{\Delta} \qquad \cdots \qquad X_n = \frac{D_n}{\Delta}$$

Here

$$\Delta = \text{system determinant} = \begin{vmatrix} a_{11} & a_{12} & \cdots & a_{1n} \\ a_{12} & a_{22} & \cdots & a_{2n} \\ \vdots & \vdots & \vdots & \vdots \\ a_{n1} & a_{n2} & \cdots & a_{nn} \end{vmatrix}$$

and D_j is the determinant formed by replacing the jth column of coefficients in Δ by the column $y_1, y_2, \ldots, y_n$.
For example:

$$\begin{aligned} 5x_1 - 2x_2 - 3x_3 &= 10 \\ -2x_1 + 4x_2 - x_3 &= 0 \\ -3x_1 - x_2 + 6x_3 &= 0 \end{aligned}$$

Using Cramer's rule:

$$x_1 = \frac{D_1}{\Delta} = \frac{\begin{vmatrix} 10 & -2 & -3 \\ 0 & 4 & -1 \\ 0 & -1 & 6 \end{vmatrix}}{\begin{vmatrix} 5 & -2 & -3 \\ -2 & 4 & -1 \\ -3 & -1 & +6 \end{vmatrix}}$$

$$X_1 = \frac{10\begin{vmatrix} 4 & -1 \\ -1 & 6 \end{vmatrix} - 0\begin{vmatrix} -2 & -3 \\ -1 & 6 \end{vmatrix} + 0\begin{vmatrix} -2 & -3 \\ 4 & -1 \end{vmatrix}}{5\begin{vmatrix} 4 & -1 \\ -1 & 6 \end{vmatrix} - (-2)\begin{vmatrix} -2 & -3 \\ -1 & 6 \end{vmatrix} + (-3)\begin{vmatrix} -2 & -3 \\ 4 & -1 \end{vmatrix}}$$

$$X_1 = \frac{10(24 - 1)}{5(24 - 1) + 2(-12 - 3) - 3(2 + 12)} = \frac{230}{43}$$

2.2 TRIGONOMETRIC AND HYPERBOLIC FUNCTIONS AND COMPLEX NUMBERS

Trigonometric Definitions: (Fig. 2.1)

$$\sin\theta = \frac{\text{opposite}}{\text{hypotenuse}} = \frac{1}{\csc\theta}$$

$$\cos\theta = \frac{\text{adjacent}}{\text{hypotenuse}} = \frac{1}{\sec\theta}$$

$$\tan\theta = \frac{\text{opposite}}{\text{adjacent}} = \frac{\sin\theta}{\cos\theta}$$

$$\cot\theta = \frac{\text{adjacent}}{\text{opposite}} = \frac{\cos\theta}{\sin\theta}$$

$$\sec\theta = \frac{\text{hypotenuse}}{\text{adjacent}} = \frac{1}{\cos\theta}$$

$$\csc\theta = \frac{\text{hypotenuse}}{\text{opposite}} = \frac{1}{\sin\theta}$$

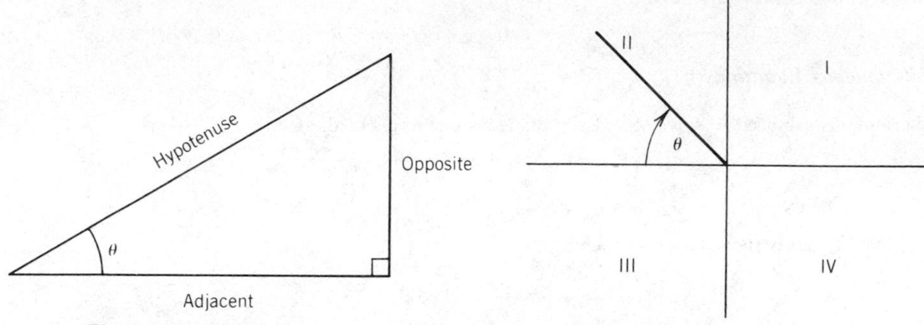

Fig. 2.1 Right triangle.

Fig. 2.2 Quadrants.

Referring to the rectangular coordinate system and the quadrants as labeled in Fig. 2.2, angle θ is in quadrant II. It should also be noted that angles repeat themselves by adding 360° or 2π radians. The algebraic signs of the function are as follows:

Quadrant	Sine	Cosine	Tangent
I	+	+	+
II	+	−	−
III	−	−	+
IV	−	+	−

Some commonly used values for trigonometric functions are:

Angle in Degrees and Radians	Sin θ	Cos θ	Tan θ
0°, 360°, 2π	0	1	0
30°, $\pi/6$	$1/2$	$\sqrt{3}/2$	$1/\sqrt{3}$
45°, $\pi/4$	$1/\sqrt{2}$	$1/\sqrt{2}$	1
60°, $\pi/3$	$\sqrt{3}/2$	$1/2$	$\sqrt{3}$
90°, $\pi/2$	1	0	$\pm\infty$
180°, π	0	-1	0
270°, $3\pi/2$	-1	0	$\pm\infty$

Reduction Formulas

Angle \ Function	Sine	Cosine	Tangent
$-\theta$	$-\sin\theta$	$\cos\theta$	$-\tan\theta$
$90° - \theta$	$\cos\theta$	$\sin\theta$	$\cot\theta$
$90° + \theta$	$\cos\theta$	$-\sin\theta$	$-\cot\theta$
$180° - \theta$	$\sin\theta$	$-\cos\theta$	$-\tan\theta$
$180° + \theta$	$-\sin\theta$	$-\cos\theta$	$\tan\theta$
$270° - \theta$	$-\cos\theta$	$-\sin\theta$	$\cot\theta$
$270° + \theta$	$-\cos\theta$	$\sin\theta$	$-\cot\theta$
$360° - \theta$	$-\sin\theta$	$\cos\theta$	$-\tan\theta$

Inverse Trigonometric Functions

Since $\sin\theta = a$, we can say that θ is the angle whose sine is a. This is expressed by the symbol $\sin^{-1}a = \theta$ and can also be read as "the inverse sine of a is θ."

For example:

$$\sin^{-1}0.319 = 18.6°, \quad \text{or} \quad 180° - 18.6° = 161.4°$$

The angle may lie in either the first or the second quadrant.

Formulas for Reference†

Functions of a sum or difference

(a) $\sin(x + y) = \sin x \cos y + \cos x \sin y$.

(b) $\sin(x - y) = \sin x \cos y - \cos x \sin y$.

(c) $\cos(x + y) = \cos x \cos y - \sin x \sin y$.

(d) $\cos(x - y) = \cos x \cos y + \sin x \sin y$.

(e) $\tan(x + y) = \dfrac{\tan x + \tan y}{1 - \tan x \tan y}$.

(f) $\tan(x - y) = \dfrac{\tan x - \tan y}{1 + \tan x \tan y}$.

Sum or difference of functions into products

(a) $\sin x + \sin y = 2\sin\frac{1}{2}(x + y)\cos\frac{1}{2}(x - y)$.

(b) $\sin x - \sin y = 2\cos\frac{1}{2}(x + y)\sin\frac{1}{2}(x - y)$.

(c) $\cos x + \cos y = 2\cos\frac{1}{2}(x + y)\cos\frac{1}{2}(x - y)$.

(d) $\cos x - \cos y = -2\sin\frac{1}{2}(x + y)\sin\frac{1}{2}(x - y)$.

Products of functions into sums or differences

(a) $2\sin x \cos y = \sin(x + y) + \sin(x - y)$.

(b) $2\cos x \sin y = \sin(x + y) - \sin(x - y)$.

(c) $2\cos x \cos y = \cos(x + y) + \cos(x - y)$.

(d) $2\sin x \sin y = -\cos(x + y) + \cos(x - y)$.

Functions of double and half angles

(a) $\sin 2x = 2\sin x \cos x$.

(b) $\cos 2x = \cos^2 x - \sin^2 x, \quad = 2\cos^2 x - 1, \quad = 1 - 2\sin^2 x$.

(c) $\tan 2x = \dfrac{2\tan x}{1 - \tan^2 x}$.

(d) $\sin\frac{1}{2}x = \pm\sqrt{\dfrac{1 - \cos x}{2}}$ or $2\sin^2\theta = 1 - \cos 2\theta$.

(e) $\cos\frac{1}{2}x = \pm\sqrt{\dfrac{1 + \cos x}{2}}$ or $2\cos^2\theta = 1 + \cos 2\theta$.

(f) $\tan\frac{1}{2}x = \pm\sqrt{\dfrac{1 - \cos x}{1 + \cos x}} = \dfrac{\sin x}{1 + \cos x} = \dfrac{1 - \cos x}{\sin x}$.

An important transformation

(a) $a\cos x + b\sin x = \sqrt{a^2 + b^2}\cos(x - \alpha)$, where $\cos\alpha = \dfrac{a}{\sqrt{a^2 + b^2}}$, $\sin\alpha = \dfrac{b}{\sqrt{a^2 + b^2}}$.

† From Smail, L., *Calculus*, Appleton-Century-Crofts, New York, 1948, with permission.

(b) $a\cos x + b\sin x = \sqrt{a^2+b^2}\,\sin(x+\beta)$, where $\sin\beta = \dfrac{a}{\sqrt{a^2+b^2}}$, $\cos\beta = \dfrac{b}{\sqrt{a^2+b^2}}$.

Relations Among Sides and Angles of Any Plane Triangle (Fig. 2.3)

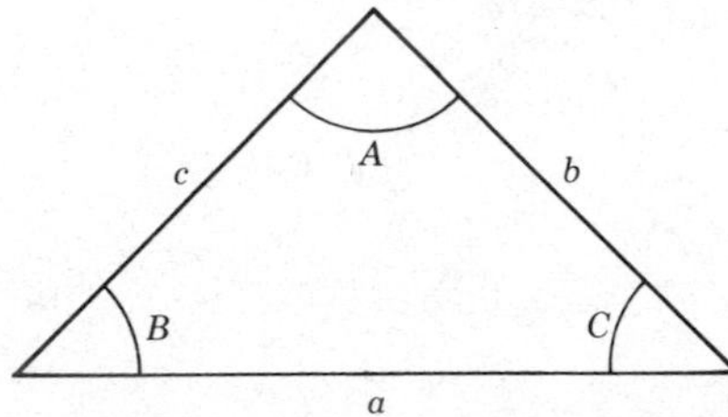

Fig. 2.3 Plane triangle.

Law of Sines.

$$\frac{a}{\sin A} = \frac{b}{\sin B} = \frac{c}{\sin C}$$

Law of Cosines.

$$a^2 = b^2 + c^2 - 2bc\cos A$$
$$b^2 = a^2 + c^2 - 2ac\cos B$$
$$c^2 = a^2 + b^2 - 2ab\cos C$$

Law of Tangents.

$$\frac{a-b}{a+b} = \frac{\tan\frac{1}{2}(A-B)}{\tan\frac{1}{2}(A+B)}$$

$$\frac{a-c}{a+c} = \frac{\tan\frac{1}{2}(A-C)}{\tan\frac{1}{2}(A+C)}$$

$$\frac{b-c}{b+c} = \frac{\tan\frac{1}{2}(B-C)}{\tan\frac{1}{2}(B+C)}$$

Exponential and Hyperbolic Functions

In electrical engineering problems the functions e^x and e^{-x} often occur. Here e is the base of the natural system of logarithms and $e = 2.7183\ldots$. Definitions of hyperbolic functions are:

$$\sinh x = \frac{e^x - e^{-x}}{2}$$

$$\cosh x = \frac{e^x + e^{-x}}{2}$$

$$\tanh x = \frac{\sinh x}{\cosh x}$$

$$\coth x = \frac{\cosh x}{\sinh x}$$

$$\operatorname{sech} x = \frac{1}{\cosh x}$$

$$\operatorname{csch} x = \frac{1}{\sinh x}$$

Common relationships between hyperbolic functions are:

$$\cosh^2 x - \sinh^2 x = 1$$

$$1 - \tanh^2 x = \frac{1}{\cosh^2 x}$$

$$\sinh(-x) = -\sinh x$$

$$\cosh(-x) = \cosh x$$

$$\tanh(-x) = -\tanh x$$

Sums of angle relationships are:

$$\sinh(x \pm y) = \sinh x \cosh y \pm \cosh x \sinh y$$

$$\cosh(x \pm y) = \cosh x \cosh y \pm \sinh x \sinh y$$

$$\tanh(x \pm y) = \frac{\tanh x \pm \tanh y}{1 \pm \tanh x \tanh y}$$

Products of the functions of two angles are:

$$\sinh x \sinh y = \tfrac{1}{2}[\cosh(x + y) - \cosh(x - y)]$$

$$\sinh x \cosh y = \tfrac{1}{2}[\sinh(x + y) + \sinh(x - y)]$$

Double angles and half angles are:

$$\sinh 2x = 2 \sinh x \cosh x$$

$$\cosh 2x = \sinh^2 x + \cosh^2 x$$
$$= 2 \sinh^2 x + 1$$

$$\tanh 2x = \frac{2 \tanh x}{1 + \tanh^2 x}$$

$$\sinh \frac{x}{2} = \sqrt{\frac{\cosh x - 1}{2}}$$

$$\cosh \frac{x}{2} = \sqrt{\frac{\cosh x + 1}{2}}$$

$$\tanh \frac{x}{2} = \sqrt{\frac{\cosh x - 1}{\cosh x + 1}}$$

Complex Numbers

The concept of an imaginary number arises when one considers the square root of -1.

$$j = \sqrt{-1}$$

$$j^2 = -1$$

$$j^3 = -j$$

The complex number $\underline{A}$ is shown in the complex plane in Fig. 2.4*a*. $\underline{A}$ is defined by its projections on the real and the imaginary axes, where

$$\underline{A} = a + jb$$

$$a = \mathrm{Re}[\underline{A}]$$

$$b = \mathrm{Im}[\underline{A}]$$

The number $\underline{A}$ can also be thought of as a vector from the origin to point $\underline{A}$, as in Fig. 2.4*b*.

In polar coordinates

$$\underline{A} = A \angle \theta$$

when A is the magnitude of the vector $\underline{A}$ and θ is the angle the vector makes with the positive real axis. Clearly,

$$A = \sqrt{a^2 + b^2} \qquad \theta = \tan^{-1}\frac{b}{a}$$

$$a = A \cos\theta \qquad b = A \sin\theta$$

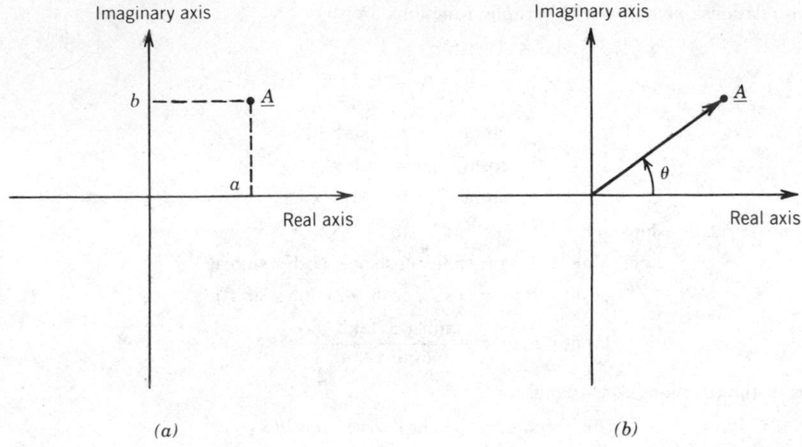

Fig. 2.4 (a) Rectangular form of complex number. (b) Vector form of complex number.

It is now useful to remember Euler's equation:

$$e^{j\theta} = \cos\theta + j\sin\theta$$

Then

$$\underline{A} = a + jb = A(\cos\theta + j\sin\theta) = Ae^{j\theta} = A\,\underline{/\theta}$$

The above are all equivalent ways of expressing $\underline{A}$. Note that Euler's equation leads to some interesting trigonometric results useful in AC circuit analysis.

$$\cos\theta = \frac{e^{j\theta} + e^{-j\theta}}{2} = \mathrm{Re}[e^{j\theta}]$$

$$\sin\theta = \frac{e^{j\theta} - e^{-j\theta}}{2j} = \mathrm{Im}[e^{j\theta}]$$

Addition and Subtraction

If $\underline{A} = a + jb$ and $\underline{B} = c + jd$, then

$$\underline{A} + \underline{B} = (a + c) + j(b + d)$$

$$\underline{A} - \underline{B} = (a - c) + j(b - d)$$

Multiplication

In rectangular form

$$\underline{A}\,\underline{B} = (a + jb)(c + jd) = (ac - bd) + j(ad + bc)$$

Looking at the exponential form

$$\underline{A}\,\underline{B} = (A\,\underline{/\theta})(B\,\underline{/\phi}) = Ae^{j\theta}Be^{j\phi}$$

$$= ABe^{j(\theta+\phi)}$$

$$\underline{A}\,\underline{B} = AB\,\underline{/(\theta+\phi)}$$

Complex Conjugate

The complex conjugate of $\underline{A}$ is $\underline{A}^*$ and is shown in Fig. 2.5.

$$\underline{A}^* = a - jb = A\,\underline{/-\theta}$$

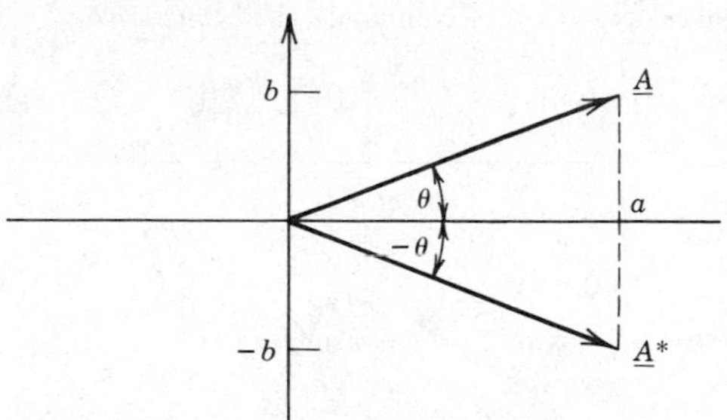

Fig. 2.5 Complex conjugate.

The sum of a complex number and its complex conjugate is

$$\underline{A} + \underline{A}^* = a + jb + a - jb = 2a$$

Multiplication of a complex number by its complex conjugate results in

$$\underline{A}\,\underline{A}^* = (a + jb)(a - jb) = a^2 + b^2$$

or

$$\underline{A}\,\underline{A}^* = (Ae^{j\theta})(Ae^{-j\theta}) = A^2$$

Division

If $\underline{A}/\underline{B} = \underline{C}$, then

$$\frac{\underline{A}}{\underline{B}} = \frac{\underline{A}}{\underline{B}} \cdot \frac{\underline{B}^*}{\underline{B}^*} = \frac{\underline{A}\,\underline{B}^*}{B^2}$$

The multiplication by $\underline{B}^*/\underline{B}^*$ is called rationalization and is performed so that the denominator of the fraction becomes real:

$$\frac{\underline{A}}{\underline{B}} = \frac{(a + jb)(c - jd)}{(c + jd)(c - jd)} = \frac{ac + bd}{c^2 + d^2} + j\frac{bc - ad}{c^2 + d^2}$$

In polar or exponential forms

$$\frac{\underline{A}}{\underline{B}} = \frac{A\angle\theta}{B\angle\phi} = \frac{Ae^{j\theta}}{Be^{j\phi}} = \frac{A}{B}e^{j(\theta - \phi)} \quad \text{or} \quad \frac{\underline{A}}{\underline{B}} = \frac{A}{B}\angle(\theta - \phi)$$

Powers and roots of complex numbers are found using laws of exponents. Thus

$$(\underline{A})^N = (Ae^{j\theta})^N = A^N e^{jN\theta} = A^N \angle N\theta$$

Also

$$(\underline{A}^{1/N}) = [Ae^{j(\theta + 2\pi k)}]^{1/N} = \sqrt[N]{A}\ \angle\frac{\theta + k2\pi}{N}$$

2.3 CALCULUS

Differentiation

If the function $y = f(t)$ is differentiable at t, then its derivative

$$\frac{dy}{dt} = f'(t) = \lim_{\Delta t \to 0} \frac{f(t + \Delta t) - f(t)}{\Delta t}$$

denotes both:

(1) the slope of the graph of $f(t)$ versus t at time t
(2) the instantaneous rate of change in y with respect to t

The following chart gives derivatives of commonly used functions.

$f(t)$	$\frac{df}{dt} = f'(t)$
t^n	nt^{n-1}
e^{at}	ae^{at}
a^t	$a^t \ln a$
$\cos at$	$-a \sin at$
$\sin at$	$a \cos at$
$\ln t$	$1/t$
$\tan at$	$a \sec^2 at$
$\cot at$	$-a \csc^2 at$
$\sec at$	$a(\sec at)(\tan at)$
$\csc at$	$-a(\csc at)(\cot at)$
$\sinh at$	$a \cosh at$
$\cosh at$	$-a \sinh at$
$\tanh at$	$a \operatorname{sech}^2 at$
$\coth at$	$-a \operatorname{csch}^2 at$
$\operatorname{sech} at$	$-a \operatorname{sech} at(\tanh at)$
$\sin^{-1} t$	$1/\sqrt{1-t^2}$
$\cos^{-1} t$	$-1/\sqrt{1-t^2}$
$\tan^{-1} t$	$1/(1+t^2)$
$\cot^{-1} t$	$-1/(1+t^2)$

Differentiation Rules. Suppose that u and v are functions of t.

Constant Rule.

$$\frac{d}{dt}(\text{const}) = 0$$

Scalar Multiplication Rule.

$$\frac{d}{dt}(cu) = c\frac{du}{dt}$$

Sum Rule.

$$\frac{d}{dt}(u \pm v) = \frac{du}{dt} \pm \frac{dv}{dt}$$

Product Rule.

$$\frac{d}{dt}(uv) = u\frac{dv}{dt} + v\frac{du}{dt}$$

Quotient Rule.

$$\frac{d}{dt}\left(\frac{u}{v}\right) = \frac{v\frac{du}{dt} - u\frac{dv}{dt}}{v^2}$$

Chain Rule. (y is a function of u)

$$\frac{d}{dt}(y) = \frac{dy}{du} \cdot \frac{du}{dt}$$

Power Rule.

$$\frac{d}{dt}(u^n) = nu^{n-1}\frac{du}{dt}$$

$$\frac{d}{dt}(t^n) = nt^{n-1}$$

Absolute Value Rule.

$$\frac{d}{dt}(|u|) = \frac{u}{|u|}\frac{du}{dt}$$

Maxima and Minima. Let x be a function of t, namely $x(t)$. To find the value of t that will minimize or maximize $x(t)$, set

$$\frac{dx(t)}{dt} = 0$$

The corresponding value of t, say $t = t_m$, that satisfies this equation will be a maximum provided that

$$\frac{d^2x(t)}{dt^2} < 0 \qquad \text{at} \quad t = t_m$$

$t = t_m$ will be a minimum if

$$\frac{d^2x(t)}{dt^2} > 0 \qquad \text{at} \quad t = t_m$$

The maxima and minima are often called the extreme values of $f(t)$.

Differential of a Function. If $y = f(t)$, then the differential of $f(t)$ is denoted by $df(x)$ and is defined by

$$dy = df(t) = f'(t)\,dt$$

Example. If $y = f(t) = t^2$, then the derivative

$$f'(t) = \frac{dy}{dt} = 2t$$

and the differential

$$dy = 2t\,dt$$

Partial Derivatives. Let $u = f(t, x)$ be a function of the two independent variables t and x. If x is kept fixed, then u becomes temporarily a function of the single variable t. From this point of view, the function has a derivative with respect to t called the partial derivative of u with respect to t and denoted by

$$\frac{\partial u}{\partial t} \quad \text{or} \quad u'_t \quad \text{or} \quad f'_t(t, x)$$

The partial derivative with respect to x is defined in a like manner. In general, for a function of several variables, there is a partial derivative with respect to each independent variable.

Example. Let $f(t, x) = t^3x^2$. Then

$$\frac{\partial f}{\partial t} = 3t^2x^2 \qquad \frac{\partial f}{\partial x} = 2t^3x \qquad \frac{\partial^2 f}{\partial t\,\partial x} = 6t^2x \qquad \frac{\partial^2 f}{\partial x^2} = 2t^3$$

Integration

Integration is the opposite of differentiation. Thus, if

$$\frac{d}{dt}\left[\int f(t)\,dt\right] = f(t)$$

then

$$\int f'(t)\,dt = f(t) + G \qquad \text{where} \quad G = \text{any constant}$$

Performing the process of integration is simply learning to use the differentiation formulas in reverse. The arbitrary added constant G is called the constant of integration.

Definite Integrals. The fundamental theorem of calculus states that if $f(t)$ is a function continuous in the interval $[a, b]$, then

$$\int_a^b f(t)\,dt = F(t)\Big|_a^b = F(b) - F(a)$$

where $F(t)$ is any function such that $F'(t) = f(t)$ for all t in $[a, b]$.

It is important to realize that this theorem describes a method for evaluating a definite integral rather than a procedure for finding antiderivatives.

Properties of Definite Integrals.

$$\int_a^b Kf(t)\,dt = K\int_a^b f(t)\,dt \qquad \text{where} \quad K = \text{constant}$$

$$\int_a^b f(t)\,dt = \int_a^c f(t)\,dt + \int_c^b f(t)\,dt \qquad \text{where} \quad a < c < b$$

$$\int_a^b [f(t) \pm g(t)]\,dt = \int_a^b f(t)\,dt \pm \int_a^b g(t)\,dt$$

Variable Limit of Integration. If the integral of $f(t)$ is considered over a variable interval (a, t), then the definite integral is a function of the upper limit of integration:

$$F(t) = \int_a^t f(x)\,dx$$

where x is a dummy variable. Note also that the derivative of a definite integral with respect to the upper limit of integration is equal to the value of the integrand at this upper limit:

$$\frac{d}{dt}\int_0^t f(x)\,dx = f(t)$$

Table of Integrals. The following table of standard integration formulas is useful. The variable u is used instead of t for broader range of application.

TABLE 2.1 A BRIEF TABLE OF INTEGRALS

Some Fundamental Forms

$$\int du = u + C$$

$$\int c\,du = c\int du$$

$$\int (f + g + \cdots)\,du = \int f\,du + \int g\,du + \cdots$$

$$\int u\,dv = uv - \int v\,du$$

$$\int u^n\,du = \frac{u^{n+1}}{n+1} + C \qquad (n \neq -1)$$

$$\int \frac{du}{u} = \ln u + C$$

Rational Forms Involving $a + bu$

$$\int \frac{u\,du}{a + bu} = \frac{1}{b^2}[a + bu - a\ln(a + bu)] + C$$

$$\int \frac{u^2\,du}{a + bu} = \frac{1}{b^3}[\tfrac{1}{2}(a + bu)^2 - 2a(a + bu) + a^2\ln(a + bu)] + C$$

$$\int \frac{u\,du}{(a + bu)^2} = \frac{1}{b^2}\left[\frac{a}{a + bu} + \ln(a + bu)\right] + C$$

$$\int \frac{u^2\,du}{(a + bu)^2} = \frac{1}{b^3}\left[a + bu - \frac{a^2}{a + bu} - 2a\ln(a + bu)\right] + C$$

TABLE 2.1. *(Continued)*

Rational Forms Involving a + bu

$$\int \frac{du}{u(a+bu)} = -\frac{1}{a}\ln\frac{a+bu}{u} + C$$

$$\int \frac{du}{u^2(a+bu)} = -\frac{1}{au} + \frac{b}{a^2}\ln\frac{a+bu}{u} + C$$

$$\int \frac{du}{u(a+bu)^2} = \frac{1}{a(a+bu)} - \frac{1}{a^2}\ln\frac{a+bu}{u} + C$$

Forms Involving $\sqrt{a+bu}$

$$\int u\sqrt{a+bu}\,du = \frac{2(3bu-2a)}{15b^2}(a+bu)^{\frac{3}{2}} + C$$

$$\int u^2\sqrt{a+bu}\,du = \frac{2(15b^2u^2-12abu+8a^2)}{105b^3}(a+bu)^{\frac{3}{2}} + C$$

$$\int \frac{u\,du}{\sqrt{a+bu}} = \frac{2(bu-2a)}{3b^2}\sqrt{a+bu} + C$$

$$\int \frac{u^2\,du}{\sqrt{a+bu}} = \frac{2(3b^2u^2-4abu+8a^2)}{15b^3}\sqrt{a+bu} + C$$

$$\int \frac{du}{u\sqrt{a+bu}} = \frac{1}{\sqrt{a}}\ln\frac{\sqrt{a+bu}-\sqrt{a}}{\sqrt{a+bu}+\sqrt{a}} + C \qquad (a>0)$$

$$\int \frac{du}{u\sqrt{a+bu}} = \frac{2}{\sqrt{-a}}\arctan\sqrt{\frac{a+bu}{-a}} + C \qquad (a<0)$$

$$\int \frac{du}{u^2\sqrt{a+bu}} = -\frac{\sqrt{a+bu}}{au} - \frac{b}{2a}\int\frac{du}{u\sqrt{a+bu}}$$

$$\int \frac{\sqrt{a+bu}}{u}\,du = 2\sqrt{a+bu} + a\int\frac{du}{u\sqrt{a+bu}}$$

$$\int \frac{\sqrt{a+bu}}{u^2}\,du = -\frac{\sqrt{a+bu}}{u} + \frac{b}{2}\int\frac{du}{u\sqrt{a+bu}}$$

Forms Involving $a^2 \pm u^2$ and $u^2 - a^2$

$$\int \frac{du}{a^2+u^2} = \frac{1}{a}\arctan\frac{u}{a} + C, \qquad \text{if} \quad a>0$$

$$\int \frac{du}{a^2-u^2} = \frac{1}{2a}\ln\frac{a+u}{a-u} + C = \frac{1}{a}\tanh^{-1}\frac{u}{a} + C, \qquad \text{if} \quad u^2<a^2$$

$$\int \frac{du}{u^2-a^2} = \frac{1}{2a}\ln\frac{u-a}{u+a} + C = -\frac{1}{a}\coth^{-1}\frac{u}{a} + C, \qquad \text{if} \quad u^2>a^2$$

Forms Involving $\sqrt{a^2-u^2}$

$$\int \frac{du}{\sqrt{a^2-u^2}} = \arcsin\frac{u}{a} + C, \qquad \text{if} \quad u^2<a^2, \quad a>0$$

$$\int \sqrt{a^2-u^2}\,du = \frac{u}{2}\sqrt{a^2-u^2} + \frac{a^2}{2}\arcsin\frac{u}{a} + C$$

$$\int u^2\sqrt{a^2-u^2}\,du = -\frac{u}{4}(a^2-u^2)^{\frac{3}{2}} + \frac{a^2}{8}u\sqrt{a^2-u^2} + \frac{a^4}{8}\arcsin\frac{u}{a} + C$$

TABLE 2.1. (*Continued*)

Forms Involving $\sqrt{a^2 - u^2}$

$$\int \frac{\sqrt{a^2-u^2}}{u}\,du = \sqrt{a^2-u^2} - a\ln\left(\frac{a+\sqrt{a^2-u^2}}{u}\right) + C$$

$$\int \frac{\sqrt{a^2-u^2}}{u^2}\,du = -\frac{\sqrt{a^2-u^2}}{u} - \arcsin\frac{u}{a} + C$$

$$\int \frac{u^2\,du}{\sqrt{a^2-u^2}} = -\frac{u}{2}\sqrt{a^2-u^2} + \frac{a^2}{2}\arcsin\frac{u}{a} + C$$

$$\int \frac{du}{u\sqrt{a^2-u^2}} = -\frac{1}{a}\ln\left(\frac{a+\sqrt{a^2-u^2}}{u}\right) + C$$

$$\int \frac{du}{u^2\sqrt{a^2-u^2}} = -\frac{\sqrt{a^2-u^2}}{a^2u} + C$$

$$\int (a^2-u^2)^{\frac{3}{2}}\,du = -\frac{u}{8}(2u^2-5a^2)\sqrt{a^2-u^2} + \frac{3a^4}{8}\arcsin\frac{u}{a} + C$$

$$\int \frac{du}{(a^2-u^2)^{\frac{3}{2}}} = \frac{u}{a^2\sqrt{a^2-u^2}} + C$$

Forms Involving $\sqrt{a^2 + u^2}$

$$\int \frac{du}{\sqrt{a^2+u^2}} = \ln(u+\sqrt{a^2+u^2}) + C = \sinh^{-1}\frac{u}{a} + C$$

$$\int \sqrt{a^2+u^2}\,du = \frac{u}{2}\sqrt{a^2+u^2} + \frac{a^2}{2}\ln(u+\sqrt{a^2+u^2}) + C = \frac{u}{2}\sqrt{a^2+u^2} + \frac{a^2}{2}\sinh^{-1}\frac{u}{a} + C$$

$$\begin{aligned}\int u^2\sqrt{a^2+u^2}\,du &= \frac{u}{8}(2u^2+a^2)\sqrt{a^2+u^2} - \frac{a^4}{8}\ln(u+\sqrt{a^2+u^2}) + C\\ &= \frac{u}{8}(2u^2+a^2)\sqrt{a^2+u^2} - \frac{a^4}{8}\sinh^{-1}\frac{u}{a} + C\end{aligned}$$

$$\int \frac{\sqrt{a^2+u^2}}{u}\,du = \sqrt{a^2+u^2} - a\ln\left(\frac{a+\sqrt{a^2+u^2}}{u}\right) + C$$

$$\int \frac{\sqrt{a^2+u^2}}{u^2}\,du = -\frac{\sqrt{a^2+u^2}}{u} + \ln(u+\sqrt{a^2+u^2}) + C = -\frac{\sqrt{a^2+u^2}}{u} + \sinh^{-1}\frac{u}{a} + C$$

$$\int \frac{u^2\,du}{\sqrt{a^2+u^2}} = \frac{u}{2}\sqrt{a^2+u^2} - \frac{a^2}{2}\ln(u+\sqrt{a^2+u^2}) + C = \frac{u}{2}\sqrt{a^2+u^2} - \frac{a^2}{2}\sinh^{-1}\frac{u}{a} + C$$

$$\int \frac{du}{u\sqrt{a^2+u^2}} = -\frac{1}{a}\ln\left(\frac{\sqrt{a^2+u^2}+a}{u}\right) + C$$

$$\int \frac{du}{u^2\sqrt{a^2+u^2}} = -\frac{\sqrt{a^2+u^2}}{a^2u} + C$$

$$\begin{aligned}\int (a^2+u^2)^{\frac{3}{2}}\,du &= \frac{u}{8}(2u^2+5a^2)\sqrt{a^2+u^2} + \frac{3a^4}{8}\ln(u+\sqrt{a^2+u^2}) + C\\ &= \frac{u}{8}(2u^2+5a^2)\sqrt{a^2+u^2} + \frac{3a^4}{8}\sinh^{-1}\frac{u}{a} + C\end{aligned}$$

$$\int \frac{du}{(a^2+u^2)^{\frac{3}{2}}} = \frac{u}{a^2\sqrt{a^2+u^2}} + C$$

Forms Involving $\sqrt{u^2 - a^2}$

$$\int \frac{du}{\sqrt{u^2-a^2}} = \ln(u+\sqrt{u^2-a^2}) + C = \cosh^{-1}\frac{u}{a} + C$$

TABLE 2.1. (*Continued*)

Forms Involving $\sqrt{u^2 - a^2}$

$$\int \sqrt{u^2 - a^2}\, du = \frac{u}{2}\sqrt{u^2 - a^2} - \frac{a^2}{2}\ln(u + \sqrt{u^2 - a^2}) + C = \frac{u}{2}\sqrt{u^2 - a^2} - \frac{a^2}{2}\cosh^{-1}\frac{u}{a} + C$$

$$\int u^2\sqrt{u^2 - a^2}\, du = \frac{u}{8}(2u^2 - a^2)\sqrt{u^2 - a^2} - \frac{a^4}{8}\ln(u + \sqrt{u^2 - a^2}) + C$$

$$= \frac{u}{8}(2u^2 - a^2)\sqrt{u^2 - a^2} - \frac{a^4}{8}\cosh^{-1}\frac{u}{a} + C$$

$$\int \frac{\sqrt{u^2 - a^2}}{u}\, du = \sqrt{u^2 - a^2} - a \operatorname{arc\,cos}\frac{a}{u} + C = \sqrt{u^2 - a^2} - a \operatorname{arc\,sec}\frac{u}{a} + C$$

$$\int \frac{\sqrt{u^2 - a^2}}{u^2}\, du = -\frac{\sqrt{u^2 - a^2}}{u} + \ln(u + \sqrt{u^2 - a^2}) + C = -\frac{\sqrt{u^2 - a^2}}{u} + \cosh^{-1}\frac{u}{a} + C$$

$$\int \frac{u^2\, du}{\sqrt{u^2 - a^2}} = \frac{u}{2}\sqrt{u^2 - a^2} + \frac{a^2}{2}\ln(u + \sqrt{u^2 - a^2}) + C = \frac{u}{2}\sqrt{u^2 - a^2} + \frac{a^2}{2}\cosh^{-1}\frac{u}{a} + C$$

$$\int \frac{du}{u\sqrt{u^2 - a^2}} = \frac{1}{a}\operatorname{arc\,cos}\frac{a}{u} + C = \frac{1}{a}\operatorname{arc\,sec}\frac{u}{a} + C$$

$$\int \frac{du}{u^2\sqrt{u^2 - a^2}} = \frac{\sqrt{u^2 - a^2}}{a^2 u} + C$$

$$\int (u^2 - a^2)^{\frac{3}{2}}\, du = \frac{u}{8}(2u^2 - 5a^2)\sqrt{u^2 - a^2} + \frac{3a^4}{8}\ln(u + \sqrt{u^2 - a^2}) + C$$

$$= \frac{u}{8}(2u^2 - 5a^2)\sqrt{u^2 - a^2} + \frac{3a^4}{8}\cosh^{-1}\frac{u}{a} + C$$

$$\int \frac{du}{(u^2 - a^2)^{\frac{3}{2}}} = -\frac{u}{a^2\sqrt{u^2 - a^2}} + C$$

Forms Involving $\sqrt{2au - u^2}$

$$\int \sqrt{2au - u^2}\, du = \frac{u - a}{2}\sqrt{au - u^2} + \frac{a^2}{2}\operatorname{arc\,cos}\left(1 - \frac{u}{a}\right) + C$$

$$\int u\sqrt{2au - u^2}\, du = \frac{2u^2 - au - 3a^2}{6}\sqrt{2au - u^2} + \frac{a^3}{2}\operatorname{arc\,cos}\left(1 - \frac{u}{a}\right) + C$$

$$\int \frac{\sqrt{2au - u^2}}{u}\, du = \sqrt{2au - u^2} + a\operatorname{arc\,cos}\left(1 - \frac{u}{a}\right) + C$$

$$\int \frac{\sqrt{2au - u^2}}{u^2}\, du = -\frac{2\sqrt{2au - u^2}}{u} - \operatorname{arc\,cos}\left(1 - \frac{u}{a}\right) + C$$

$$\int \frac{du}{\sqrt{2au - u^2}} = 2\operatorname{arc\,sin}\sqrt{\frac{u}{2a}} + C = \operatorname{arc\,cos}\left(1 - \frac{u}{a}\right) + C$$

$$\int \frac{u\, du}{\sqrt{2au - u^2}} = -\sqrt{2au - u^2} + a\operatorname{arc\,cos}\left(1 - \frac{u}{a}\right) + C$$

$$\int \frac{u^2\, du}{\sqrt{2au - u^2}} = -\frac{(u + 3a)}{2}\sqrt{2au - u^2} + \frac{3a^2}{2}\operatorname{arc\,cos}\left(1 - \frac{u}{a}\right) + C$$

$$\int \frac{du}{u\sqrt{2au - u^2}} = -\frac{\sqrt{2au - u^2}}{au} + C$$

$$\int \frac{du}{(2au - u^2)^{\frac{3}{2}}} = \frac{u - a}{a^2\sqrt{2au - u^2}} + C$$

Trigonometric Forms

$$\int \sin u\, du = -\cos u + C$$

TABLE 2.1. *(Continued)*

Trigonometric Forms

$$\int \cos u \, du = \sin u + C$$

$$\int \tan u \, du = -\ln \cos u + C = \ln \sec u + C$$

$$\int \cot u \, du = \ln \sin u + C = -\ln \csc u + C$$

$$\int \sec u \, du = \ln(\sec u + \tan u) + C = \ln \tan\left(\frac{u}{2} + \frac{\pi}{4}\right) + C$$

$$\int \csc u \, du = -\ln(\csc u + \cot u) + C = \ln \tan \frac{u}{2} + C$$

$$\int \sec^2 u \, du = \tan u + C$$

$$\int \csc^2 u \, du = -\cot u + C$$

$$\int \sec u \tan u \, du = \sec u + C$$

$$\int \csc u \cot u \, du = -\csc u + C$$

$$\int \sin^2 u \, du = \tfrac{1}{2}(u - \sin u \cos u) + C = \tfrac{1}{2}u - \tfrac{1}{4}\sin 2u + C$$

$$\int \cos^2 u \, du = \tfrac{1}{2}(u + \sin u \cos u) + C = \tfrac{1}{2}u + \tfrac{1}{4}\sin 2u + C$$

$$\int \tan^2 u \, du = \tan u - u + C$$

$$\int \sec^3 u \, du = \tfrac{1}{2}\sec u \tan u + \tfrac{1}{2}\ln(\sec u + \tan u) + C$$

$$\int \sin mu \sin nu \, du = \frac{\sin(m-n)u}{2(m-n)} - \frac{\sin(m+n)u}{2(m+n)} + C$$

$$\int \sin mu \cos nu \, du = -\frac{\cos(m-n)u}{2(m-n)} - \frac{\cos(m+n)u}{2(m+n)} + C$$

$$\int \cos mu \cos nu \, du = \frac{\sin(m-n)u}{2(m-n)} + \frac{\sin(m+n)u}{2(m+n)} + C$$

$$\int u \sin u \, du = \sin u - u \cos u + C$$

$$\int u \cos u \, du = \cos u + u \sin u + C$$

$$\int u^2 \sin u \, du = (2 - u^2)\cos u + 2u \sin u + C$$

$$\int u^2 \cos u \, du = (u^2 - 2)\sin u + 2u \cos u + C$$

$$\int \sin^m u \cos^n u \, du = -\frac{\sin^{m-1} u \cos^{n+1} u}{m+n} + \frac{m-1}{m+n}\int \sin^{m-2} u \cos^n u \, du$$

$$\int \sin^m u \cos^n u \, du = \frac{\sin^{m+1} u \cos^{n-1} u}{m+n} + \frac{n-1}{m+n}\int \sin^m u \cos^{n-2} u \, du$$

$$\int \frac{du}{a + b\cos u} = \frac{2}{\sqrt{a^2 - b^2}} \arctan\left(\frac{\sqrt{a^2 - b^2}\tan \frac{u}{2}}{a + b}\right) + C, \quad \text{if} \quad a^2 > b^2$$

$$\int \frac{du}{a + b\cos u} = \frac{1}{\sqrt{b^2 - a^2}} \ln\left[\frac{a + b + \sqrt{b^2 - a^2}\tan \frac{u}{2}}{a + b - \sqrt{b^2 - a^2}\tan \frac{u}{2}}\right] + C, \quad \text{if} \quad b^2 > a^2$$

TABLE 2.1. (*Continued*)

Trigonometric Forms

$$\int \frac{du}{a + b\sin u} = \frac{2}{\sqrt{a^2 - b^2}} \arctan\left(\frac{a \tan \frac{u}{2} + b}{\sqrt{a^2 - b^2}} \right) + C, \qquad \text{if} \quad a^2 > b^2$$

$$\int \frac{du}{a + b\sin u} = \frac{1}{\sqrt{b^2 - a^2}} \ln \left[\frac{a \tan \frac{u}{2} + b - \sqrt{b^2 - a^2}}{a \tan \frac{u}{2} + b + \sqrt{b^2 - a^2}} \right] + C, \qquad \text{if} \quad b^2 > a^2$$

Inverse Trigonometric Forms

$$\int \arcsin u \, du = u \arcsin u + \sqrt{1 - u^2} + C$$

$$\int \arccos u \, du = u \arccos u - \sqrt{1 - u^2} + C$$

$$\int \arctan u \, du = u \arctan u - \tfrac{1}{2}\ln(1 + u^2) + C$$

Exponential and Logarithmic Forms

$$\int e^u \, du = e^u + C$$

$$\int a^u \, du = \frac{a^u}{\ln a} + C$$

$$\int u\, e^u \, du = e^u(u - 1) + C$$

$$\int u^n e^u \, du = u^n e^u - n \int u^{n-1} e^u \, du$$

$$\int \frac{e^u}{u^n} du = -\frac{e^u}{(n-1)u^{n-1}} + \frac{1}{n-1} \int \frac{e^u \, du}{u^{n-1}}$$

$$\int \ln u \, du = u \ln u - u + C$$

$$\int u^n \ln u \, du = u^{n+1} \left[\frac{\ln u}{n+1} - \frac{1}{(n+1)^2} \right] + C$$

$$\int \frac{du}{u \ln u} = \ln(\ln u) + C$$

$$\int e^{au} \sin nu \, du = \frac{e^{au}(a \sin nu - n \cos nu)}{a^2 + n^2} + C$$

$$\int e^{au} \cos nu \, du = \frac{e^{au}(a \cos nu + n \sin nu)}{a^2 + n^2} + C$$

Hyperbolic Forms

$$\int \sinh u \, du = \cosh u + C$$

$$\int \cosh u \, du = \sinh u + C$$

$$\int \tanh u \, du = \ln \cosh u + C$$

$$\int \coth u \, du = \ln \sinh u + C$$

$$\int \operatorname{sech} u \, du = \arctan(\sinh u) + C = \operatorname{gd} u + C$$

$$\int \operatorname{csch} u \, du = \ln \tanh \tfrac{1}{2} u + C$$

$$\int \operatorname{sech}^2 u \, du = \tanh u + C$$

TABLE 2.1. (*Continued*)

Hyperbolic Forms

$$\int \operatorname{csch}^2 u\, du = -\coth u + C$$

$$\int \operatorname{sech} u \tanh u\, du = -\operatorname{sech} u + C$$

$$\int \operatorname{csch} u \coth u\, du = -\operatorname{csch} u + C$$

$$\int \sinh^2 u\, du = \tfrac{1}{4}\sinh 2u - \tfrac{1}{2}u + C$$

$$\int \cosh^2 u\, du = \tfrac{1}{4}\sinh 2u + \tfrac{1}{2}u + C$$

$$\int u \sinh u\, du = u \cosh u - \sinh u + C$$

$$\int u \cosh u\, du = u \sinh u - \cosh u + C$$

$$\int e^{au} \sinh nu\, du = \frac{e^{au}(a \sinh nu - n \cosh nu)}{a^2 - n^2} + C$$

$$\int e^{au} \cosh nu\, du = \frac{e^{au}(a \cosh nu - n \sinh nu)}{a^2 - n^2} + C$$

Wallis' Formulas

$$\int_0^{\frac{\pi}{2}} \sin^n u\, du = \int_0^{\frac{\pi}{2}} \cos^n u\, du = \begin{cases} \dfrac{(n-1)(n-3)\cdots 4\cdot 2}{n(n-2)\cdots 5\cdot 3\cdot 1}, & \text{if } n \text{ is an odd integer} > 1 \\[2ex] \dfrac{(n-1)(n-3)\cdots 3\cdot 1}{n(n-2)\cdots 4\cdot 2}\cdot\dfrac{\pi}{2}, & \text{if } n \text{ is a positive even integer} \end{cases}$$

$$\int_0^{\frac{\pi}{2}} \sin^m u \cos^n u\, du = \begin{cases} \dfrac{(n-1)(n-3)\cdots 4\cdot 2}{(m+n)(m+n-2)\cdots(m+5)(m+3)(m+1)}, & \text{if } n \text{ is an odd integer} > 1 \\[2ex] \dfrac{(m-1)(m-3)\cdots 4\cdot 2}{(n+m)(n+m-2)\cdots(n+5)(n+3)(n+1)}, & \text{if } m \text{ is an odd integer} > 1 \\[2ex] \dfrac{(m-1)(m-3)\cdots 3\cdot 1\cdot(n-1)(n-3)\cdots 3\cdot 1}{(m+n)(m+n-2)\cdots 4\cdot 2}\cdot\dfrac{\pi}{2}, & \text{if } m \text{ and } n \text{ are both positive even integers} \end{cases}$$

Source. Smail, L., *Calculus*, Appleton-Century-Crofts, New York, 1948, with permission.

2.4 MATRICES

The rectangular array of numbers or functions

$$A = \begin{bmatrix} a_{11} & a_{12} & \cdots & a_{1n} \\ a_{21} & a_{22} & \cdots & a_{2n} \\ a_{m1} & a_{m2} & \cdots & a_{mn} \end{bmatrix}$$

is called a matrix of order (m, n) or an $m \times n$ matrix. The numbers or functions a_{ij} are called the elements of the matrix, with the first subscript indicating the row position and the second subscript the column position. A matrix of order $m \times n$ has m rows and n columns. Two matrices are equal if they have the same order and have equal corresponding elements. For example, if two matrices of the same order A and B are equal, then $a_{ij} = b_{ij}$ for all i and j.

Special Matrices

Column matrix. A matrix of one column but any number of rows is known as a column matrix or a vector.

Square matrix. A matrix of order (n, n) is a square matrix of order n. The main or principal diagonal of a square matrix consists of the elements $a_{11}, a_{22}, a_{33}, \ldots, a_{nn}$. A square matrix in which all elements except those on the principal diagonal are zeros is known as a *diagonal* matrix. Further, if all elements of a diagonal matrix have the value 1, the matrix is known as a *unit* or *identity* matrix.

Zero matrix. If all elements of a matrix are zero, $a_{ij} = 0$, the matrix is known as a zero matrix, 0.

Symmetrical matrix. If $a_{ij} = a_{ji}$ in a matrix, the matrix is known as a symmetrical matrix.

Matrix Operations

Addition and Subtraction. The sum of two matrices of the same order is found by adding the corresponding elements. If the elements of A are a_{ij} and of B are b_{ij} and if $C = A + B$, then $C_{ij} = a_{ij} + b_{ij}$ and clearly $A + B = B + A$ for all matrices.

If a matrix A is multiplied by a constant α, then every element of A is multiplied by α:

$$\alpha A = [\alpha a_{ij}]$$

In particular, if $\alpha = -1$, then $-A = [-a_{ij}]$. From this, we see that to subtract B from A we multipy all elements of B by -1 and add the resulting matrix to A.

If $C = A - B$, then $C_{ij} = a_{ij} - b_{ij}$.

Multiplication. The multiplication of matrices A and B is defined only if the number of columns of A equals the number of rows of B. If A is of order $m \times n$ and B is of order $n \times p$, then the product AB is a matrix C of order $m \times p$.

$$C_{m \times p} = A_{m \times n} B_{n \times p}$$

The elements of C are found from the elements of A and B by multiplying the ith row of A and the jth column of B and summing these products to give C_{ij},

$$C_{ij} = a_{i1}b_{1j} + a_{i2}b_{2j} + \cdots + a_{ip}b_{pj} = \sum_{k=1}^{p} a_{ik}b_{kj}$$

In general, if AB is defined, BA need not be, and even when BA is defined, $AB \neq BA$.

Finally, if U (unit matrix) and A are both square and of order n, $UA = AU = A$.

Other Definitions

Transpose. The *transpose* of a matrix A is A^T and is formed by interchanging the rows and columns of A. For example, if

$$A = \begin{bmatrix} 1 & 2 & -1 \\ 0 & -1 & 1 \end{bmatrix} \quad \text{then} \quad A^T = \begin{bmatrix} 1 & 0 \\ 2 & -1 \\ -1 & 1 \end{bmatrix}$$

Determinant. The *determinant* of a square matrix has elements that are the elements of the matrix.

$$\det A = \det[a_{ij}] = |a_{ij}|$$

Cofactor. The *cofactor* A_{ij} of the element of a matrix a_{ij} is defined for a square matrix A. It has the value $(-1)^{i+j}$ times the determinant formed by deleting the ith row and the jth column in $\det A$.

Adjoint matrix. The *adjoint matrix* of a square matrix A is formed by replacing each element a_{ij} by the cofactor A_{ij} and transposing. Thus

$$\text{adjoint of } A = [A_{ij}]^T$$

Inverse matrix. The *inverse matrix* of A, A^{-1}, is defined as the adjoint matrix divided by the determinant of A.

$$A^{-1} = \frac{\text{adjoint of } A}{\det A}, \qquad \det A \neq 0$$

Example. Given

$$A = \begin{bmatrix} 1 & 2 \\ -1 & 1 \end{bmatrix}, \qquad \det A = 3$$

then

$$[A_{ij}] = \begin{bmatrix} 1 & 1 \\ -2 & 1 \end{bmatrix}, \quad [A_{ij}]^T = \begin{bmatrix} 1 & -2 \\ 1 & 1 \end{bmatrix} \text{ and } A^{-1} = \frac{1}{3}\begin{bmatrix} 1 & -2 \\ 1 & 1 \end{bmatrix} = \begin{bmatrix} \frac{1}{3} & -\frac{2}{3} \\ \frac{1}{3} & \frac{1}{3} \end{bmatrix}$$

The inverse matrix A^{-1} has the property that

$$AA^{-1} = A^{-1}A = 1$$

In the example given,

$$AA^{-1} = \begin{bmatrix} 1 & 2 \\ -1 & 1 \end{bmatrix}\begin{bmatrix} \frac{1}{3} & -\frac{2}{3} \\ \frac{1}{3} & \frac{1}{3} \end{bmatrix} = \begin{bmatrix} 1 & 0 \\ 0 & 1 \end{bmatrix} = 1 \text{ and } A^{-1}A = \begin{bmatrix} \frac{1}{3} & -\frac{2}{3} \\ \frac{1}{3} & \frac{1}{3} \end{bmatrix}\begin{bmatrix} 1 & 2 \\ -1 & 1 \end{bmatrix} = \begin{bmatrix} 1 & 0 \\ 0 & 1 \end{bmatrix} = 1$$

Matrix Solution of Simultaneous Linear Equations

As an example of a set of simultaneous linear equations, consider Kirchhoff's voltage equations which have the form

$$ZI = V$$

or

$$\sum_{j=1}^{n} Z_{ij}I_j = V_i$$

In matrix notation, Z is a square matrix of order n, and I and V are column matrices.

$$\begin{bmatrix} Z_{11} & \cdot & \cdots & Z_{1n} \\ Z_{21} & \cdot & \cdots & Z_{2n} \\ \vdots & \vdots & \vdots & \vdots \\ Z_{n1} & \cdot & \cdot\cdot & Z_{nn} \end{bmatrix}\begin{bmatrix} I_1 \\ I_2 \\ \vdots \\ I_n \end{bmatrix} = \begin{bmatrix} V_1 \\ V_2 \\ \vdots \\ V_n \end{bmatrix}$$

To solve these equations using matrices $ZI = V$, multiplying by Z^{-1} gives

$$Z^{-1}ZI = Z^{-1}V$$

but

$$Z^{-1}Z = U \quad \text{and} \quad UI = I$$

so that

$$I = Z^{-1}V = \frac{\text{adjoint of } Z}{\det Z} V$$

which is equivalent to Cramer's rule.

2.5 VECTORS

Vector Algebra

Components and Unit Vectors. A vector **A** can be specified by its components (projections) along any three mutually perpendicular axes, A_x, A_y, A_z. The vector can be uniquely expressed in terms of its components through the use of unit vectors $\hat{i}, \hat{j}, \hat{k}$, which are defined as vectors of unit magnitude in the positive x, y, and z directions.

$$\mathbf{A} = A_x\hat{i} + A_y\hat{j} + A_z\hat{k}$$

Addition and Subtraction. Given two vectors

$$\mathbf{A} = A_x\hat{i} + A_y\hat{j} + A_z\hat{k} \quad \text{and} \quad \mathbf{B} = B_x\hat{i} + B_y\hat{j} + B_z\hat{k},$$

the sum of **A** and **B** is

$$\mathbf{A} + \mathbf{B} = (A_x + B_x)\hat{i} + (A_y + B_y)\hat{j} + (A_z + B_z)\hat{k}$$

Note that

$$\mathbf{A} + \mathbf{B} = \mathbf{B} + \mathbf{A}$$

and that

$$\mathbf{A} + (\mathbf{B} + \mathbf{C}) = (\mathbf{A} + \mathbf{B}) + \mathbf{C} = \mathbf{A} + \mathbf{B} + \mathbf{C}$$

The negative of a vector **B** is defined as

$$-\mathbf{B} = -B_x\hat{i} - B_y\hat{j} - B_z\hat{k}$$

The vector **B** subtracted from the vector **A** is

$$\mathbf{A} - \mathbf{B} = (A_x - B_x)\hat{i} + (A_y - B_y)\hat{j} + (A_z - B_z)\hat{k}$$

Scalar Multiplication. If a is a scalar, then $a\mathbf{A}$ is a vector in the direction of **A** if $a > 0$ and in the direction of $-\mathbf{A}$ if $a < 0$.

$$a\mathbf{A} = aA_x\hat{i} + aA_y\hat{j} + aA_z\hat{k}$$

Scalar (Dot) Product of Vectors. If **A** and **B** are vectors and θ is $\measuredangle \mathbf{A}, \mathbf{B}$, then the scalar product is defined as the scalar

$$\mathbf{A} \cdot \mathbf{B} = |\mathbf{A}|\,|\mathbf{B}|\cos\theta$$

From the definition of the scalar product

$$\mathbf{B} \cdot \mathbf{A} = \mathbf{A} \cdot \mathbf{B}$$
$$\mathbf{A} \cdot (\mathbf{B} + \mathbf{C}) = \mathbf{A} \cdot \mathbf{B} + \mathbf{A} \cdot \mathbf{C}$$
$$a\mathbf{A} \cdot \mathbf{B} = a(\mathbf{A} \cdot \mathbf{B})$$
$$\hat{i} \cdot \hat{i} = \hat{j} \cdot \hat{j} = \hat{k} \cdot \hat{k} = 1$$
$$\hat{i} \cdot \hat{j} = \hat{j} \cdot \hat{k} = \hat{k} \cdot \hat{i} = 0$$
$$\mathbf{A} \cdot \mathbf{B} = \left(A_x\hat{i} + A_y\hat{j} + A_z\hat{k}\right) \cdot \left(B_x\hat{i} + B_y\hat{j} + B_z\hat{k}\right) = A_xB_x + A_yB_y + A_zB_z$$

Vector (Cross) Product of Vectors. The vector product of two vectors **A** and **B** is denoted by $\mathbf{A} \times \mathbf{B}$ and is defined as a vector perpendicular to the plane containing **A** and **B** and is in the direction of the axial motion of a right-handed screw turning **A** into **B**.

The magnitude of $\mathbf{A} \times \mathbf{B}$ is

$$|\mathbf{A} \times \mathbf{B}| = |\mathbf{A}|\,|\mathbf{B}|\sin\theta$$

From the definition of the vector product

$$\mathbf{A} \times \mathbf{B} = -(\mathbf{B} \times \mathbf{A})$$
$$\mathbf{A} \times (\mathbf{B} + \mathbf{C}) = (\mathbf{A} \times \mathbf{B}) + (\mathbf{A} \times \mathbf{C})$$
$$\hat{i} \times \hat{i} = \hat{j} \times \hat{j} = \hat{k} \times \hat{k} = 0$$
$$\hat{i} \times \hat{j} = \hat{k}, \qquad \hat{j} \times \hat{k} = \hat{i}, \qquad \hat{k} \times \hat{i} = \hat{j}, \qquad \hat{j} \times \hat{i} = -\hat{k}, \qquad \text{etc.}$$

In terms of components

$$\begin{aligned}\mathbf{A} \times \mathbf{B} &= \left(A_x\hat{i} + A_y\hat{j} + A_z\hat{k}\right) \times \left(B_x\hat{i} + B_y\hat{j} + B_z\hat{k}\right) \\ &= \hat{i}(A_yB_z - A_zB_y) + \hat{j}(A_zB_x - A_xB_z) + k(A_xB_y - A_yB_x) \\ &= \begin{vmatrix} \hat{i} & \hat{j} & \hat{k} \\ A_x & A_y & A_z \\ B_x & B_y & B_z \end{vmatrix}\end{aligned}$$

Scalar Triple Product.

$$\begin{aligned}\mathbf{A} \cdot (\mathbf{B} \times \mathbf{C}) = \mathbf{B} \cdot (\mathbf{C} \times \mathbf{A}) = \mathbf{C} \cdot (\mathbf{A} \times \mathbf{B}) &= \mathbf{ABC} \\ &= \begin{vmatrix} A_x & B_x & C_x \\ A_y & B_y & C_y \\ A_z & B_z & C_z \end{vmatrix}\end{aligned}$$

and

$$(\mathbf{ABC})(\mathbf{DEF}) = \begin{vmatrix} \mathbf{A} \cdot \mathbf{D} & \mathbf{A} \cdot \mathbf{E} & \mathbf{A} \cdot \mathbf{F} \\ \mathbf{B} \cdot \mathbf{D} & \mathbf{B} \cdot \mathbf{E} & \mathbf{B} \cdot \mathbf{F} \\ \mathbf{C} \cdot \mathbf{D} & \mathbf{C} \cdot \mathbf{E} & \mathbf{C} \cdot \mathbf{F} \end{vmatrix}$$

Other useful identities are

$$\mathbf{A} \times (\mathbf{B} \times \mathbf{C}) = (\mathbf{A} \cdot \mathbf{C})\mathbf{B} - (\mathbf{A} \cdot \mathbf{B})\mathbf{C}$$

$$(\mathbf{A} \times \mathbf{B}) \cdot (\mathbf{C} \times \mathbf{D}) = \mathbf{A} \cdot [\mathbf{B} \times (\mathbf{C} \times \mathbf{D})]$$

$$= (\mathbf{A} \cdot \mathbf{C})(\mathbf{B} \cdot \mathbf{D}) - (\mathbf{A} \cdot \mathbf{D})(\mathbf{B} \cdot \mathbf{C})$$

$$(\mathbf{A} \times \mathbf{B}) \times (\mathbf{C} \times \mathbf{D}) = [(\mathbf{A} \times \mathbf{B}) \cdot \mathbf{D}]\mathbf{C} - [(\mathbf{A} \times \mathbf{B}) \cdot \mathbf{C}]\mathbf{D}$$

$$= (\mathbf{ABD})\mathbf{C} - (\mathbf{ABC})\mathbf{D}$$

$$\mathbf{A} \times [\mathbf{B} \times (\mathbf{C} \times \mathbf{D})] = (\mathbf{B} \cdot \mathbf{D})(\mathbf{A} \times \mathbf{C}) - (\mathbf{B} \cdot \mathbf{C})(\mathbf{A} \times \mathbf{D})$$

$$(\mathbf{A} \times \mathbf{B}) \cdot [(\mathbf{B} \times \mathbf{C}) \times (\mathbf{C} \times \mathbf{A})] = [\mathbf{A} \cdot (\mathbf{B} \times \mathbf{C})]^2 = (\mathbf{ABC})^2$$

Vector Analysis

Differential Operators. The gradient of a scalar function $\Phi(x, y, z)$ is defined by

$$\text{grad}\,\Phi \equiv \nabla\Phi = \hat{i}\,\frac{\partial\Phi}{\partial x} + \hat{j}\,\frac{\partial\Phi}{\partial y} + \hat{k}\,\frac{\partial\Phi}{\partial z}$$

The divergence of a vector **A** is defined by

$$\text{div}\,\mathbf{A} \equiv \nabla \cdot \mathbf{A} = \frac{\partial A_x}{\partial x} + \frac{\partial A_y}{\partial y} + \frac{\partial A_z}{\partial z}$$

The curl of a vector **A** is defined by

$$\text{curl}\,\mathbf{A} \equiv \nabla \times \mathbf{A} = \begin{vmatrix} \hat{i} & \dfrac{\partial}{\partial x} & A_x \\ \hat{j} & \dfrac{\partial}{\partial y} & A_y \\ \hat{k} & \dfrac{\partial}{\partial z} & A_z \end{vmatrix}$$

$$= \hat{i}\left(\frac{\partial A_z}{\partial y} - \frac{\partial A_y}{\partial z}\right) + \hat{j}\left(\frac{\partial A_x}{\partial z} - \frac{\partial A_z}{\partial x}\right) + \hat{k}\left(\frac{\partial A_y}{\partial x} - \frac{\partial A_x}{\partial y}\right)$$

The Laplacian of a scalar function Φ is defined by

$$\nabla^2\Phi \equiv \nabla \cdot \nabla\Phi = \text{div grad}\,\Phi = \frac{\partial^2\Phi}{\partial x^2} + \frac{\partial^2\Phi}{\partial y^2} + \frac{\partial^2\Phi}{\partial z^2}$$

The Laplacian of a vector function $A(x, y, z)$ is defined by

$$\nabla^2\mathbf{A} \equiv (\text{div grad})\mathbf{A} = \hat{i}\nabla^2 A_x + \hat{j}\nabla^2 A_y + \hat{k}\nabla^2 A_z$$

Useful differential operations are:

$$\text{grad}\,(\Phi + \psi) = \text{grad}\,\Phi + \text{grad}\,\psi$$

where Φ and ψ are scalar functions.

$$\text{grad}\,(\Phi\psi) = \Phi\,\text{grad}\,\psi + \psi\,\text{grad}\,\Phi$$

$$\text{div}\,(\mathbf{A} + \mathbf{B}) = \text{div}\,\mathbf{A} + \text{div}\,\beta$$

$$\text{curl}\,(\mathbf{A} + \mathbf{B}) = \text{curl}\,\mathbf{A} + \text{curl}\,\mathbf{B}$$

$$\text{div}\,(\Phi\mathbf{A}) = \mathbf{A} \cdot \text{grad}\,\Phi + \Phi\,\text{div}\,\mathbf{A}$$

$$\text{curl}\,(\Phi\mathbf{A}) = \Phi\,\text{curl}\,\mathbf{A} - \mathbf{A}_x\,\text{grad}\,\Phi$$

$$(\mathbf{A} \cdot \text{grad})\mathbf{B} = A_x\frac{\partial\mathbf{B}}{\partial x} + A_y\frac{\partial\mathbf{B}}{\partial y} + A_z\frac{\partial\mathbf{B}}{\partial z} = \hat{i}\,(\mathbf{A} \cdot \nabla B_x) + \hat{j}\,(\mathbf{A} \cdot \nabla B_y) + \hat{k}\,(\mathbf{A} \cdot \nabla B_z)$$

$$\text{grad}\,(\mathbf{A} \cdot \mathbf{B}) = (\mathbf{A} \cdot \text{grad})\mathbf{B} + (\mathbf{B} \cdot \text{grad})\mathbf{A} + \mathbf{A} \times \text{curl}\,\mathbf{B} + \mathbf{B} \times \text{curl}\,\mathbf{A}$$

$$\text{div}\,(\mathbf{A} \times \mathbf{B}) = \mathbf{B} \cdot \text{curl}\,\mathbf{A} - \mathbf{A} \cdot \text{curl}\,\mathbf{B}$$

$$\text{curl}\,(\mathbf{A} \times \mathbf{B}) = \mathbf{A}\,\text{div}\,\mathbf{B} - \mathbf{B}\,\text{div}\,\mathbf{A} + (\mathbf{B} \cdot \text{grad})\mathbf{A} - (\mathbf{A} \cdot \text{grad})\mathbf{B}$$

$$\text{curl curl}\,\mathbf{A} = \text{grad div}\,\mathbf{A} - \nabla^2\mathbf{A}$$

$$\text{curl grad}\,\Phi = 0$$

$$\text{div curl}\,\mathbf{A} = 0$$

2.6 LINEAR DIFFERENTIAL EQUATIONS

Definitions

The linear differential equation with constant coefficients will be considered. It may be written in the form:

$$a_n \frac{d^n y}{dt^n} + a_{n-1} \frac{d^{n-1} y}{dt^{n-1}} + \cdots + a_1 \frac{dy}{dt} + a_0 y = b_m \frac{d^m x}{dt^{m-1}} + \cdots + b_1 \frac{dx}{dt} + b_0 x$$

where $x(t)$ is the input and $y(t)$ is the output. Since $x(t)$ is generally known, the previous equation may be written:

$$a_n \frac{d^n y}{dt^n} + a_{n-1} \frac{d^{n-1} y}{dt^{n-1}} + \cdots + a_1 \frac{dy}{dt} + a_0 y = F(t)$$

$F(t)$ is often called the forcing function. When $F(t)$ is not equal to zero, the differential equation is called a nonhomogeneous differential equation. When $F(t)$ is equal to zero, the equation is called a homogeneous differential equation.

Solution of the Homogeneous Differential Equation

$$a_n \frac{d^n y}{dt^n} + a_n \frac{d^{n-1} y}{dt^{n-1}} + \cdots + a_1 \frac{dy}{dt} + a_0 y = 0$$

The homogeneous equation has n linearly independent solutions $y_1(t), y_2(t), \ldots, y_n(t)$. The general solution is the linear combination of these solutions

$$y_h(t) = A_1 y_1(t) + A_2 y_2(t) + \cdots + A_n y_n(t)$$

where the A's are arbitrary constants. The solution must satisfy n initial conditions, which are generally known, namely:

$$y(0), \frac{dy}{dt}(0), \ldots, \frac{d^{n-1}y}{dt^{n-1}}(0)$$

The specific values of the A's are to be found later so that the solution $y_h(t)$ satisfies the initial conditions. It is assumed that the solutions of the homogeneous equation have the form

$$y(t) = e^{pt}$$

Here p is a constant to be determined. Substitution of this assumed solution into the differential equation leads to the conclusion that the assumption is correct if p is a root of the algebraic equation

$$a_n p^n + a_{n-1} p^{n-1} + \cdots + a_1 p + a_0 = 0$$

This algebraic equation, called the characteristic equation, yields n values of p, say $p_1, p_2, \ldots, p_n$. Suppose that these are real values and all different from one another. The solutions to the homogeneous differential equation are then

$$y_1 = e^{p_1 t}, \qquad y_2 = e^{p_2 t}, \ldots, \qquad y_n = e^{p_n (t)}$$

The homogeneous solution is then

$$y_h(t) = A_1 e^{p_1 t} + A_2 e^{p_2 (t)} + \cdots + A_n e^{p_n (t)}$$

If the roots of the characteristic equation are complex, they must occur in complex conjugate pairs (if the coefficients of the original differential equation are real numbers). So if $p_1 = -\alpha + jw$, then $p_2 = -\alpha - jw$ and $y_1 = e^{-\alpha t}(C_1 \cos wt + C_2 \sin wt)$.

If the characteristic equation yields repeated roots, say p_1 occurs q times, then

$$y_h(t) = (A_1 + A_2 t + \cdots + A_q t^{q-1}) e^{p_1 t} + \ldots + A_n e^{p_n t}$$

It should be noted that other terms for the homogeneous solution are "source free solution," "natural solution," and "transient response."

Solution of Nonhomogeneous Differential Equations

The solution to the nonhomogeneous equation $F(t) \neq 0$ consists of the sum of two parts:

$$y(t) = y_h(t) + y_p(t)$$

$y_h(t)$ is the homogeneous solution containing n arbitrary constants. $y_p(t)$ is called the particular solution. It is also referred to as the forced solution, the complementary solution, or the solution due to the source.

The complete solution $y(t)$ must satisfy the differential equation and must contain n arbitrary constants so that the initial conditions can be satisfied. Because $y_h(t)$ contains these n arbitrary constants, $y_p(t)$ must have all of its constants determined (they will have specific values).

Method of Undetermined Coefficients. The procedure outlined as follows is used when the driving function $F(t)$ is one of the functions ordinarily encountered in electrical engineering. The procedure consists of assuming a particular solution of a form that satisfies the form for the solution of a nonhomogeneous differential equation. This generally means that one assumes $y_p(t)$ to be the linear combination of functions of the form of $F(t)$ and all of its successive derivatives. If a term in $F(t)$ is a sum of a form shown in the chart that follows, then $y_p(t)$ is assumed to be the sum of the corresponding choice (combining similar time functions). If a term in $F(t)$ already appears as a term in the homogeneous solution, the $y_p(t)$ form shown in the chart is multiplied by t.

$F(t)$	Assume for $y_p(t)$
Constant	Constant
At	$K_1t + K_0$
At^2	$K_2t^2 + K_1t + K_0$
At^n	$K_nt^n + K_{n-1}t^{n-1} + \cdots + K_1t + K_0$
Ae^{mt}	Ke^{mt}
At^ne^{mt}	$e^{mt}(K_nt^n + \ldots K_1t + K_0)$
$A \cos wt$	$K_1\cos wt + K_2\sin wt$
$A \sin wt$	$K_1\cos wt + K_2\sin wt$
$At^n\cos wt$	$(K_nt^n + \cdots + K_1t + K_0)\cos wt$ $+(C_nt^n + \cdots + C_1t + C_0)\sin wt$
$Ae^{mt}\cos wt$	$e^{mt}(K_1\cos wt + K_2\sin wt)$

Example. If

$$y_h(t) = A + Be^{-t} + Ce^{-2t} \quad \text{and} \quad F(t) = 6 + 3e^{-t}$$

then let

$$y_p(t) = K_1t + K_2te^{-t}$$

The procedure for determining the constants that $y_p(t)$ contains is:

1. Assume a solution for $y_p(t)$ from the chart.
2. Substitute this solution into the differential equation
$$a_n\frac{d^ny}{dt^n} + \cdots + a_1\frac{dy}{dt} + a_0y = F(t)$$
3. Equate the coefficients of like functions of time on the left side of the equation to those on the right side.
4. Solve for the constants (K's) appearing in the assumed $y_p(t)$.

Complete Solution to Nonhomogeneous Differential Equations

Procedure.

1. Solve for $y_h(t)$, the solution to the homogeneous equation.
2. Solve for $y_p(t)$, the particular solution.
3. $y(t) = y_h(t) + y_p(t)$.
4. Since the complete solution must satisfy the initial conditions, now solve for the arbitrary constants so that the initial conditions are indeed satisfied.

2.7 LINEAR DIFFERENCE EQUATIONS

Difference Equations

Difference equations occur when functions do not continuously vary with time, but rather have meaning only at discrete time intervals. Certainly, any system containing digital components involves the processing of discrete numbers or signals according to a specified rule. This rule is a recursion formula or a difference equation. It relates the value of a function or signal at some time k to its value at other specified times. It can also be thought of as a relationship between a sample of a signal or an output with other samples of the output.

An nth order linear difference equation with constant coefficients is

$$a_n y_{k+n} + a_{n-1} y_{k+n-1} + \cdots + a_1 y_{k+1} + a_0 y_k = U_k$$

Here U_k is the input sequence and acts as the forcing function does in differential equations. The similarity of the difference equation solution to the differential equation solution will be utilized frequently.

The general solution to the general difference equation must satisfy n known sample values (or initial conditions). One may specify values for

$$y_0, y_1, y_2, \ldots, y_{n-1}$$

The complete solution y_k is again the sum of the homogeneous solution and the particular solution $y_k = y_k$ homogeneous + y_k particular.

Solution of Homogeneous Difference Equations

Homogeneous equations occur when $u_k = 0$

$$a_n y_{k+n} + a_{n-1} y_{k+n-1} + \cdots a_1 y_{k+1} + a_0 y_k = 0$$

Again assume a solution in the form of

$$y_k = e^{pk} = b^k \qquad \text{where} \quad b = e^p$$

Substitution of this result into the difference equation results in the conclusion that the assumption is correct for the n values of b, which are solutions of the algebraic equation

$$a_n b^n + a_{n-1} b^{n-1} + \cdots + a_1 b + a_0 = 0$$

If the solutions to this "characteristic equation" are all real and not repeated values, then

$$y_k = b_1^k, y_k = b_2^k, \ldots, y_k = b_n^k$$

The linear combination of these solutions is the complete homogeneous solution

$$y_k = A_1 b_1^k + A_2 b_2^k + \cdots + A_n b_n^k$$

Note that the values of b are numbers. If the characteristic equation yields a value for b repeated q times, then

$$y_k = \left(A_1 + A_2 k + A_3 k^2 + \ldots + A_q k^{q-1}\right) b_1^k + \cdots + A_k b_n^k$$

If a complex value for b arises, and if the coefficients of the difference equation are real, then the complex conjugate value of b will also be a root.

If $b_1 = \alpha + j\beta$ then $b_2 = \alpha - j\beta$

then $y_k = r^k (C_1 \cos \theta k + C_2 \sin \theta k) + \cdots + A_k b_n^k$

where $r^k = \sqrt{\alpha^2 + \beta^2}$ and $\theta = \tan^{-1} \dfrac{\beta}{\alpha}$

Multiple complex roots are treated in the usual way.

Solution of Nonhomogeneous Difference Equations

$$y_k = y_k \text{ homogeneous} + y_k \text{ particular}$$

To find y_k particular, the method of undetermined coefficients will be used.

U_k	Assume for y_k Particular
a^k or e^{rk}	Ca^k
k^p	$C_p k^p + C_{p-1} k^{p-1} + \cdots + C_1 k + C_0$
$\cos wk$	$C_1 \cos wk + C_2 \sin wk$
$\sin wk$	$C_1 \cos wk + C_2 \sin wk$
$a^k \cos wk$	$a^k(C_1 \cos wk + C_2 \sin wk)$
$k^p a^k$	$a^k(C_p k^p + \cdots + C_1 k + C_0)$
$k^p \cos wk$	$(C_p k^p + \cdots + C_1 k + C_0)\cos wk$ $+(D_p k^p + \ldots + D_1 k + D_0)\sin wk$

If U_k contains a sum of functions, the solution is assumed to be the sum of the individual assumptions (combining like terms, of course). If a term in U_k is of the same functional form as a term in the homogeneous solution, the assumed solution due to that term is multiplied by k.

For terms of the form $\cos wk$ and $\sin wk$, it is important to remember that

$$\cos w(k+m) = (\cos wm)\cos wk - (\sin wm)\sin wk$$

$$\sin w(k+m) = (\sin wm)\cos wk + (\cos wm)\sin wk$$

The procedure for determining the coefficients (C's) for y_k particular is:

1. Assume a solution for y_k particular.
2. Substitute this solution into the difference equation

$$a_n y_{k+n} + a_{n-1} y_{k+n-1} + \cdots + a_1 y_{k+1} + a_0 y_k = U_k$$

Note that this substitution will involve finding the $k+1, k+2, \ldots, k+n$ value of the particular solution involved. For example: If we assume y_k particular $= C_1 3^k$, then

$$y_{k+1} = C_1 3^{k+1} = 3C_1 3^k$$

$$y_{k+2} = C_1 3^{k+2} = 9C_1 3^k$$

$$y_{k+3} = C_1 3^{k+3} = 27C_1 3^k \quad \text{etc.}$$

3. Equate the coefficients of like terms in k on the left side of the equation to those on the right side.
4. Solve for the C's (the constants) appearing in y_k particular.

Complete Solution to Difference Equations

Procedure.

1. Solve for y_k homogeneous.
2. Solve for y_k particular.
3. $y_k = y_k$ homogeneous $+ y_k$ particular.
4. Now solve for the arbitrary constants in y_k homogeneous so that the initial sample values are satisfied.

2.8 POWER SERIES

The closed-form expressions for some commonly encountered series are as follows.

Arithmetic Progression

$$a + (a+d) + (a+2d) + \ldots n \text{ terms} = \frac{n}{2}[2a + (n-1)d]$$

Geometric Progression

$$a + ar + ar^2 + \ldots n \text{ terms} = a\left[\frac{1-r^n}{1-r}\right]$$

$$a + ar + ar^2 + \ldots \infty \text{ terms} = \frac{a}{1-r} \qquad \text{where} \quad r < 1$$

$$a + ax + a^2x^2 + \ldots \infty \text{ terms} = \frac{1}{1-ax}$$

Powers of Natural Numbers

$$\sum_{1}^{\infty} x^n = \frac{x}{1-x} \qquad \text{where} \quad x < 1$$

$$\sum_{1}^{\infty} nx^n = \frac{x}{(1-x)^2} \qquad \text{where} \quad x < 1$$

Power Series for Common Functions

$$e^{ax} = 1 + ax + \frac{a^2x^2}{2!} + \frac{a^3x^3}{3!} + \ldots$$

$$\sin ax = ax - \frac{a^3x^3}{3!} + \frac{a^5x^5}{5!} - \ldots$$

$$\cos ax = 1 - \frac{a^2x^2}{2!} + \frac{a^4x^4}{4!} - \ldots$$

$$\frac{1}{x} = 1 - (x-1) + (x-1)^2 - (x-1)^3 + (x-1)^4 - \ldots$$

Taylor Series Expansion

$$f(x) = f(a) + f'(a)(x-a) + \frac{f''(a)}{2!}(x-a)^2 + \frac{f'''(a)}{3!}(x-a)^3 + \cdots + \frac{f^{(n)}(a)}{n!}(x-a)^n + \ldots$$

This formula holds for points close to $x = a$ and for functions whose derivatives at the point $x = a$ exist.

If $a = 0$, the series is called the Maclaurin series and takes the form

$$f(x) = f(0) + f'(0)x + \frac{f''(0)x^2}{2!} + \ldots + \frac{f^{(n)}(0)}{n!}x^n + \ldots$$

If the Taylor series is terminated after n terms the remainder is

$$R_n = \frac{f^{(n+1)}(z)}{(n+1)!}(x-c)^{n+1}$$

where z is some number between c and x.

2.9 FOURIER SERIES

Trigonometric Fourier Series

Periodic functions are those for which $f(t) = f(t+T)$. If, over a period, a periodic function is single valued, has a finite number of minima and maxima, has a finite number of discontinuities, and satisfies

$$\int_{t_0}^{t_0+T} |x(t)|\, dt < \infty$$

then it can be expanded into a Fourier series of the form

$$f(t) = A_0 + \sum_{k=1}^{\infty} [A_k \cos k\omega_0 t + B_k \sin k\omega_0 t]$$

Here $\omega_0 = 2\pi/T = 2\pi f$ is called the fundamental frequency and $k2\pi/T$ is the kth harmonic. The coefficients must be evaluated using:

$$A_0 = \frac{1}{T}\int_{t_0}^{t_0+T} f(t)\, dt$$

Note that this can also be interpreted as the average value over a period of $f(t)$.

$$A_k = \frac{2}{T}\int_{t_0}^{t_0+T} f(t) \cos k\omega_0 t\, dt$$

$$B_k = \frac{2}{T}\int_{t_0}^{t_0+T} f(t) \sin k\omega_0 t\, dt$$

TABLE 2.2. FOURIER COEFFICIENTS OF SYMMETRICAL PERIODIC WAVEFORMS

Type of Symmetry	Conditions	Form of Fourier Series	Formulas for Coefficients
Odd function	$f(t) = -f(-t)$	$f(t) = \sum_{k=1}^{\infty} B_k \sin k \frac{2\pi}{T} t$	$A_k = 0;\ B_k = \frac{4}{T}\int_0^{T/2} f(t)\sin k \frac{2\pi}{T} t\, dt$
Even function	$f(t) = f(-t)$	$f(t) = \sum_{k=0}^{\infty} A_k \cos k \frac{2\pi}{T} t$	$B_k = 0;\ A_k = \frac{4}{T}\int_0^{T/2} f(t)\cos k \frac{2\pi}{T} t\, dt;\ A_0 = \frac{2}{T}\int_0^{T/2} f(t)\, dt$
Half-wave	$f(t) = -f\left(t + \frac{T}{2}\right)$	$f(t) = \sum_{i=1}^{\infty}\left[A_{2k-1}\cos(2k-1)\frac{2\pi}{T} t + B_{2k-1}\sin(2k-1)\frac{2\pi}{T} t\right]$	$\left.\begin{matrix} A_{2k-1} \\ B_{2k-1} \end{matrix}\right\} = \frac{4}{T}\int_0^{T/2} f(t) \begin{matrix}\cos \\ \sin\end{matrix} (2k-1)\frac{2\pi}{T} t\, dt$
Odd quarter-wave	$f(t) = -f(-t)$ and $f(t) = -f\left(t + \frac{T}{2}\right)$	$f(t) = \sum_{k=1}^{\infty} B_{2k-1}\sin(2k-1)\frac{2\pi}{T} t$	$A_k = 0;\ B_{2k-1} = \frac{8}{T}\int_0^{T/4} f(t)\sin(2k-1)\frac{2\pi}{T} t\, dt$
Even quarter-wave	$f(t) = f(-t)$ and $f(t) = -f\left(t + \frac{T}{2}\right)$	$f(t) = \sum_{k=1}^{\infty} A_{2k-1}\cos(2k-1)\frac{2\pi}{T} t$	$B_k = 0;\ A_{2k-1} = \frac{8}{T}\int_0^{T/4} f(t)\cos(2k-1)\frac{2\pi}{T} t\, dt$

Source. Javid and Brenner, 1963, with permission.

TABLE 2.3. THE THREE FORMS OF A FOURIER SERIES

Complex Form	Trigonometric Form 1	Trigonometric Form 2
$f(t) = \sum_{k=-\infty}^{+\infty} \mathbf{D}_k e^{jk(2\pi/T)t}$	$f(t) = C_0 + \sum_{k=1}^{\infty} C_k \cos\left(k \frac{2\pi}{T} t + \varphi_k\right)$ $= F_0 + \sum_{k=1}^{\infty} \sqrt{2}\, F_k \cos\left(k \frac{2\pi}{T} t + \varphi_k\right)$ $= D_0 + \sum_{k=1}^{\infty} 2\,\mathrm{Re}\left[\mathbf{D}_k e^{jk(2\pi/T)t}\right]$	$f(t) = A_0 + \sum_{k=1}^{\infty} \left(A_k \cos k \frac{2\pi}{T} t + B_k \sin k \frac{2\pi}{T} t\right)$

Conversion formulas for $k = 1, 2, 3, \ldots$

$$\mathbf{D}_k = \tfrac{1}{2} C_k \angle \varphi_k \qquad \mathbf{D}_{-k} = \mathbf{D}_k^* \qquad \mathbf{D}_k = \tfrac{1}{2} A_k - j\tfrac{1}{2} B_k$$

$$C_k = 2D_k \qquad \varphi_k = \text{angle } \mathbf{D}_k \qquad C_k = \sqrt{A_k^2 + B_k^2} \qquad \varphi_k = -\tan^{-1} \frac{B_k}{A_k} \qquad F_k = \frac{C_k}{\sqrt{2}}$$

$$A_k = 2\,\mathrm{Re}[\mathbf{D}_k] \qquad B_k = -2\,\mathrm{Im}[\mathbf{D}_k] \qquad A_k = C_k \cos \varphi_k \qquad B_k = -C_k \sin \varphi_k$$

For $k = 0$,

$$D_0 = C_0 = A_0 = F_0 \qquad B_0 \equiv 0$$

Source. Javid and Brenner, 1963, with permission.

The integration may start at any place that is convenient, but must extend over a period. If $t_0 = 0$, then

$$\int_0^T$$

but if $t_0 = -T/2$, then the form is

$$\int_{-T/2}^{T/2}$$

Table 2.2 shows the form of these equations for situations in which $f(t)$ exhibits certain defined symmetry properties.

An alternate form for the Fourier series is

$$f(t) = \sum_{k=0}^{\infty} C_k \cos\left(k \frac{2\pi}{T} t + \phi_k\right)$$

Exponential Fourier Series

Using Euler's equation

$$e^{j\theta} = \cos\theta + j\sin\theta$$

An exponential form of the Fourier series can be developed

$$f(t) = \sum_{-\infty}^{\infty} \mathbf{D}_k e^{jk(2\pi/T)t}$$

where

$$\mathbf{D}_k = \frac{1}{T}\int_{t_0}^{t_0+T} f(t) e^{-jk(2\pi/T)t}\, dt$$

Table 2.3 gives relationships between the coefficients of the various forms of the Fourier series.

Theorems Involving Fourier Coefficients

Superposition. If a periodic function can be expressed as the sum of several simpler periodic functions, then the Fourier series of the sum is equal to the sum of the individual Fourier series.

Time Displacement. If a periodic function $f(t)$ has the Fourier series

$$f(t) = \sum_{-\infty}^{\infty} \mathbf{D}_k e^{jk(2\pi/T)t}$$

then the delayed function $f(t-a)$ has as its coefficient

$$[\mathbf{D}_k]_{\text{delayed function}} = [D_k]_{\text{original function}} \cdot e^{-jk(2\pi/T)a}$$

Differentiation. If $f(t)$ is differentiable

$$f(t) = \sum_{-\infty}^{\infty} \mathbf{D}_k e^{jk(2\pi/T)t}$$

then

$$f'(t) = \sum_{-\infty}^{\infty} \left[jk \frac{2\pi}{T} \mathbf{D}_k \right] e^{jk(2\pi/T)t}$$

Fourier Series of Some Common Functions

Example 1. Rectangular Waveform (Fig. 2.6).

$$f(t) = V \sum_{-\infty}^{\infty} \frac{1 - e^{-jk(2\pi/T)T_1}}{jk2\pi} e^{jk(2\pi/T)t}$$

For $T_1 = T/2$:

$$f(t) = V\left(\frac{1}{2} + \frac{2}{\pi}\sin\frac{2\pi}{T}t + \frac{2}{3\pi}\sin\frac{6\pi}{T}t + \dots\right)$$

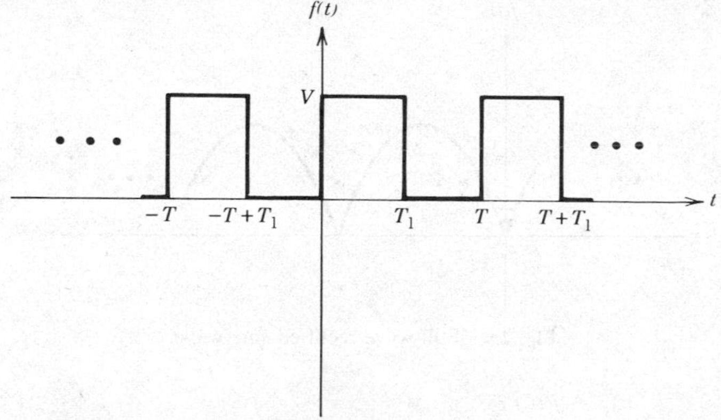

Fig. 2.6 Rectangular waveform.

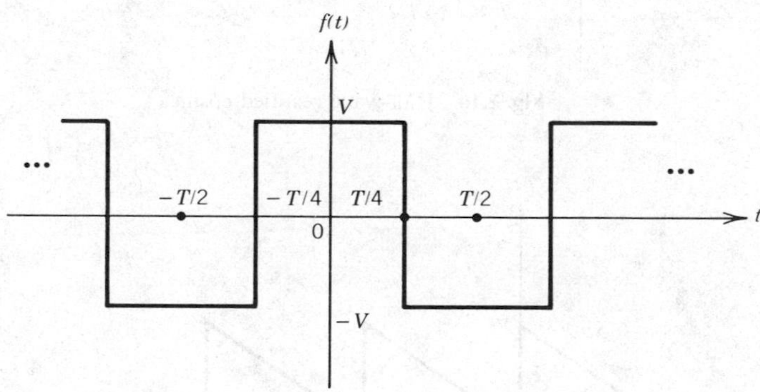

Fig. 2.7 Square wave.

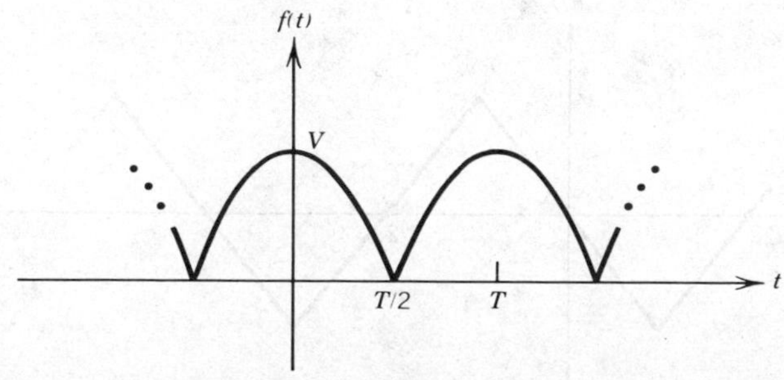

Fig. 2.8 Full-wave rectified cosine wave.

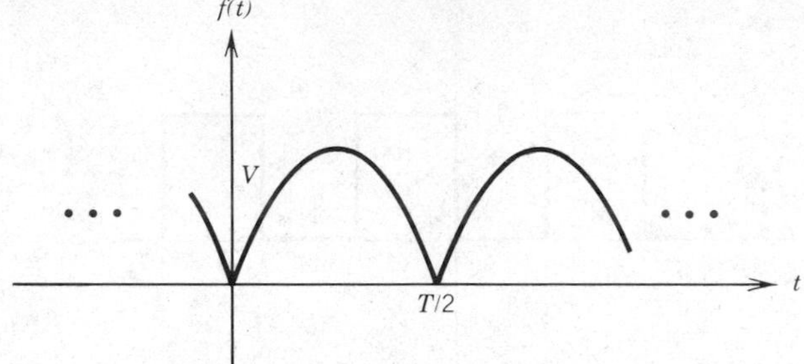

Fig. 2.9 Full-wave rectified sine wave.

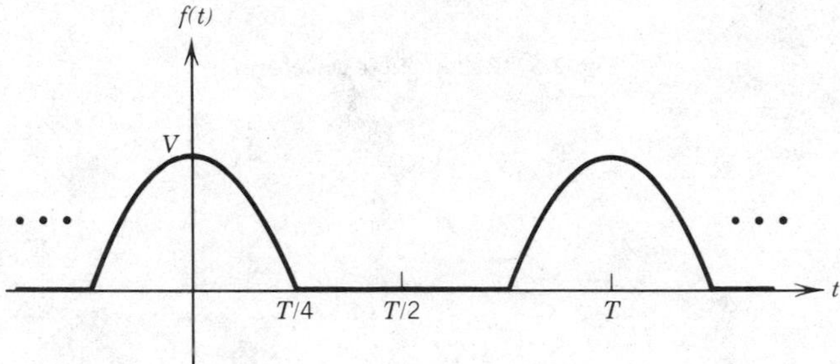

Fig. 2.10 Half-wave rectified cosine.

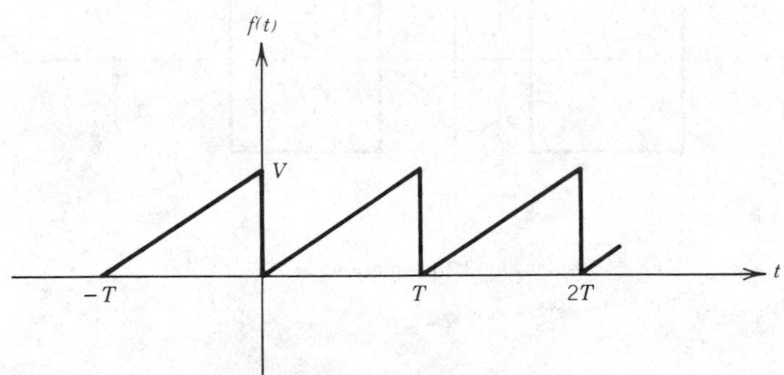

Fig. 2.11 Sawtooth waveform.

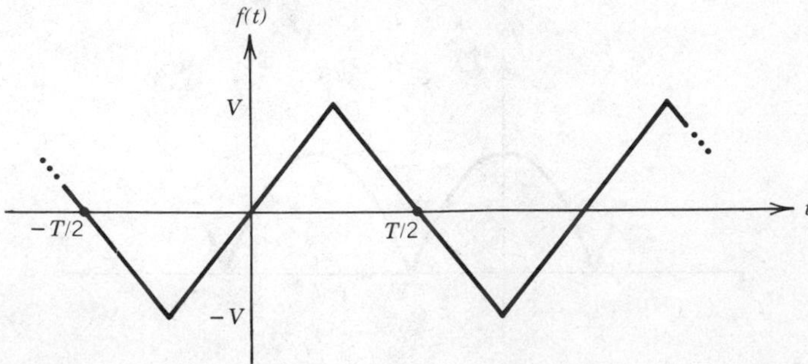

Fig. 2.12 Triangular waveform.

For $T_1 = T/4$:

$$f(t) = V\left(\frac{3}{4} - \frac{1}{\pi}\cos\frac{2\pi}{T}t + \frac{1}{\pi}\sin\frac{4\pi}{T}t + \frac{1}{3\pi}\cos\frac{6\pi}{T}t + \frac{1}{3\pi}\sin\frac{6\pi}{T}t - \frac{1}{5\pi}\cos\frac{10\pi}{T}t + \frac{1}{5\pi}\sin\frac{10\pi}{T}t + \ldots\right)$$

Example 2. Square Wave (Fig. 2.7).

$$f(t) = \frac{4V}{\pi}\left(\cos\frac{2\pi}{T}t - \frac{1}{3}\cos\frac{6\pi}{T}t + \frac{1}{5}\cos\frac{10\pi}{T}t - \frac{1}{7}\cos\frac{14\pi}{T}t + \ldots\right)$$

Example 3. Full-Wave Rectified Cosine Wave (Fig. 2.8).

$$f(t) = \frac{4V}{\pi}\left(\frac{1}{2} + \frac{1}{3}\cos\frac{2\pi}{T}t - \frac{1}{15}\cos\frac{4\pi}{T}t + \frac{1}{35}\cos\frac{6\pi}{T}t - + \ldots\right)$$

Example 4. Full-Wave Rectified Sine Wave (Fig. 2.9).

$$f(t) = \frac{2V}{\pi}\sum_{-\infty}^{\infty}\frac{1}{1-4k^2}e^{jk2t}$$

and if $T/2 = \pi$, then

$$f(t) = \frac{2V}{\pi} + \frac{4V}{\pi}\sum_{k=1}^{\infty}\frac{1}{1-4k^2}\cos 2kt$$

Example 5. Half-Wave Rectified Cosine Wave (Fig. 2.10).

$$f(t) = \frac{V}{\pi}\left(1 + \frac{\pi}{2}\cos\frac{2\pi}{T}t + \frac{2}{3}\cos\frac{4\pi}{T}t - \frac{2}{15}\cos\frac{8\pi}{T}t + - \ldots\right)$$

Example 6. Sawtooth Waveform (Fig. 2.11).

$$f(t) = V\left[\frac{1}{2} - \frac{1}{\pi}\left(\sin\frac{2\pi}{T}t + \frac{1}{2}\sin\frac{4\pi}{T}t + \frac{1}{3}\sin\frac{6\pi}{T}t + \ldots\right)\right]$$

Example 7. Triangular Waveform (Fig. 2.12).

$$f(t) = \frac{8V}{\pi^2}\left(\sin\frac{2\pi}{T}t - \frac{1}{3^2}\sin\frac{6\pi}{T}t + \frac{1}{5^2}\sin\frac{10\pi}{T}t - + \ldots\right)$$

Bibliography

Ayres, F., *Matrices*, Schaum Outline Series, McGraw-Hill, New York, 1962.

Cadzow, J., *Discrete Time Systems*, Prentice-Hall, Englewood Cliffs, NJ, 1973.

Cooper, G. R. and C. D. McGillem, *Methods of Signal and System Analysis*, Holt, Rinehart & Winston, New York, 1974.

Fraser, M., *College Algebra and Trigonometry*, Benjamin/Cummings, Menlo Park, CA, 1978.

Gabel, R. A. and R. A. Roberts, *Signals and Linear Systems*, 2nd ed., Wiley, New York, 1980.

Goodman, A. W., *Analytic Geometry with Calculus*, Macmillan, New York, 1980.

Hildebrand, F. B., *Methods of Applied Mathematics*, Prentice-Hall, Englewood Cliffs, NJ, 1958.

Jahnke, E. and F. Emde, *Table of Functions*, Dover, New York, 1945.

Javid, M. and E. Brenner, *Analysis, Transmission and Filtering of Signals*, McGraw-Hill, New York, 1963.

Jolley, L. B. W., *Summation of Series*, Dover, New York, 1961.

Larson, R. A. and R. P. Hostetler, *Calculus*, D. C. Heath, Lexington, MA, 1981.

Lathi, B. P., *Signals, Systems and Communications*, Wiley, New York, 1965.

Lipschutz, S., *Finite Mathematics*, Schaum Outline Series, McGraw-Hill, New York, 1966.

Marion, J., *Classical Electromagnetic Radiation*, Academic, New York, 1965.
Robson, J., *Basic Tables in Physics*, McGraw-Hill, New York, 1967.
Ross, S., *Introduction to Ordinary Differential Equations*, Wiley, New York, 1980.
Smail, L., *Calculus*, Appleton-Century-Crofts, New York, 1948.
Spiegel, M., *Vector Analysis*, Schaum Outline Series, McGraw-Hill, New York, 1959.
Van Valkenberg, M. E., *Network Analysis*, 3rd ed., Prentice-Hall, Englewood Cliffs, NJ, 1974.

CHAPTER 3
COMPUTATION

THOMAS C. UPSON

Rochester Institute of Technology
Rochester, New York

3.1 COMPUTATIONAL TOOLS AVAILABLE

Little, if any, computation is performed manually today. Machines have made possible the solution of problems that only a short time ago were unsolved because of the computation time required.

Computational tools available range from pocket adding machines and minicomputers to large time-sharing systems. It is inappropriate to comment in detail on the tools available to the engineer, since advances are so rapid as to make comparisons obsolete before publication. Suffice it to say that the present level of sophistication of even the most basic equipment now calls for the exclusion from engineering and scientific handbooks of the mathematical tables traditionally found in them.

In addition to computational machines (hardware), a variety of software "packages" are available. Again the spectrum of complexity and sophistication is broad. Table 3.1 describes a few of the scientific software packages available, their language, and the number of subroutines they contain. Further details are given in the Engineering Systems Software Referral Catalog.[1] For example, the IMSL (International Mathematical and Statistical Library) package contains over 400 subroutines programmed in Fortran that are capable of matrix inversion, evaluation of Bessel functions, solution of linear programs, and a variety of tasks in data analysis. The IMSL package is capable of supporting many computer-compiler environments.†

3.2 SOURCES OF ERROR IN COMPUTATION[2]

The user of any numerical scheme should be aware of sources of error. The error in an approximation has several components.

1. The accuracy of the data used. In a laboratory or other testing environment in which data are being collected, the measurements are only an approximation of the actual value of the variable. Thus the measurement device itself introduces error. It is important for the user to know the level at which the error is introduced. For example, five-place accuracy cannot be expected from a numerical approximation scheme when the data collected have only three-place accuracy.

† For a complete list of the computer-compiler environments supported by IMSL, see IMSL Library, 1982. Available from IMSL Inc., 7500 Bellaire Blvd., 6th Floor, NBC Building, Houston, TX 72036.

TABLE 3.1. REPRESENTATIVE SOFTWARE PACKAGES AVAILABLE†

Software Title	Software Capabilities	Languages	Source of Further Information
NAG (Numerical Algorithms Group) Library	Numerical and statistical analysis (466 subroutines)	FORTRAN	Numerical Algorithms Group Inc. Attn: Company Secretary 1250 Grace Ct. Downers Grove, IL 60516 312 971-2337
IMSL (International Mathematical and Statistical Library)	Numerical and statistical analysis (500 subroutines)	FORTRAN IV, FORTRAN IV-Plus	IMSL, Inc. 7500 Bellaire Blvd. 6th Floor, NBC Building Houston, Texas 72036 713 772-1927
SPSS	Statistical analysis and data management	FORTRAN	SPSS Inc. Suite 3000 444 North Michigan Ave. Chicago, IL 60611 312 329-2400
Microsolve	Linear programming system for use on microcomputer	FORTRAN IV	Advanced Management Techniques Box 1597 Houston, TX 77057 713 373-1905

† See Engineering Systems Software Referral Catalog[1] for details.

2. The computation facility. Each machine has the capability for a maximum level of accuracy. The user should be aware of this level and how it affects the results of lengthy computation. For example, in a calculator with eight-place accuracy, computations using *e*, the base of the natural logarithms, introduce an error, called the roundoff error. This is because *e*, being an irrational number, has infinitely many digits, only the first eight of which can be accommodated. In lengthy computations, roundoff errors can accumulate and affect the result.

3. The algorithm. Each computational method provides an approximate numerical value for the desired actual value. Thus there is an inherent error in the approximation. Significant increases in accuracy can usually be gained only at the expense of increased algorithm complexity and/or computation time. For example, the various numerical techniques for approximating $y(x')$, given: $y' = f(x, y)$, $y(x_0) = y_0$ (described later in this chapter) start at the point (x_0, y_0) and proceed by n steps, each of length h, to compute an approximate value of the solution at $x' = x_0 + nh[h = (x' - x_0)/n]$. Even if the initial data are accurate and no roundoff error is introduced during computation, an error is still introduced at each step, called the local error. Local errors accumulate from step to step, to create a "global error" at x'. Global and local errors may differ by as much as an order of magnitude. For example, the local error in the Euler method (discussed later) is proportional to the square of the step size, h^2, while the global error is proportional to h.

Another form of error is introduced into an algorithm when only a finite number of terms of an infinite series is used. This is called truncation error. For example, the formula of evaluation of

$$\int_a^b f(x)\,dx$$

by Simpson's rule uses a truncated Taylor's series for $f(x)$. Both Simpson's rule and Taylor's series are discussed later in this chapter.

Needless to say, the exact error introduced in these ways is not known. The best that can be determined is the maximum error that may be present. For instance, when the first three nonzero terms of the series

$$x - \frac{x^3}{3!} + \frac{x^5}{5!} - \frac{x^7}{7!} + \cdots$$

are used to represent the sine function, it is known that the error will be no larger than the first term omitted, in this case

$$E_5(x) \leqslant \left| \frac{x^7}{7!} \right|$$

since the terms of the series have alternating signs.

It may be possible for the user to reduce somewhat the influence of roundoff error by taking into account the order in which operations are performed. For example, to compute $(\sin x + \cos x)^2$ it would be wise to multiply first:

$$(\sin x + \cos x)^2 = \sin^2 x + 2 \sin x \cos x + \cos^2 x = 1 + \sin 2x$$

Taking advantage of trigonometric identities reduces the number of operations in which roundoff is involved from three to one.

3.3 NUMERICAL TECHNIQUES IN COMPUTATION

Newton–Raphson Root-Finding Technique[2]

The algorithm will be presented for differentiable functions of one variable, $y = f(x)$.

To solve an equation of the form $f(x) = 0$, the following iteration scheme is used. (See Fig. 3.1.)

A starting value, x_1, is chosen which is presumed to lie in the neighborhood of the actual solution.† x_2 is computed as $x_1 - f(x_1)/f'(x_1)$, which represents the intersection of the tangent line (to the graph at x_1) with the x-axis.

Similarly after the nth iteration, x_n, has been computed the $(n + 1)$st may be computed as

$$x_{n+1} = x_n - \frac{f(x_n)}{f'(x_n)}$$

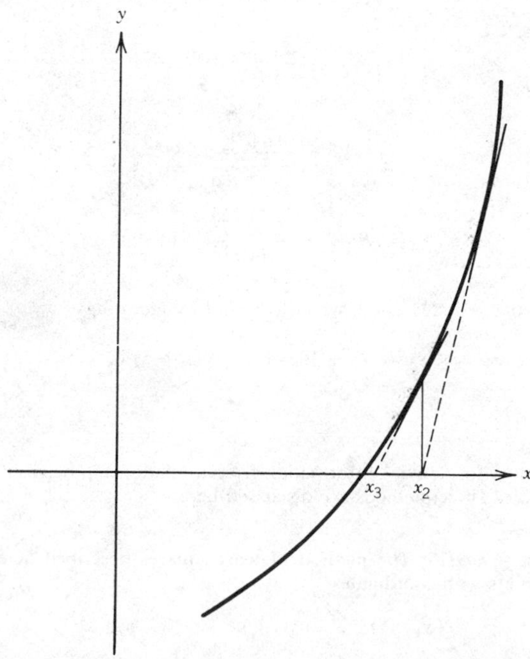

Fig. 3.1 Iteration scheme in the solution of $f(x) = 0$ using Newton–Raphson root-finding technique.

† x_1 may be determined by evaluation of the function until a change of sign is detected. (Cases of multiple roots are discussed in Ref. 2.)

Convergence of the Newton–Raphson method to the root is rapid, provided the ratio of the second derivative to the first derivative of the function is not large. In fact, it is known that the error (see Ref. 2, pp. 43–50) at the nth iteration is given by

$$\epsilon_{n+1} = \frac{1}{2}(\epsilon_n)^2 \frac{f''(a)}{f'(a)}$$

where a is the actual root of the equation, that is, $f(a) = 0$. As a is not known, it is only necessary to determine an upper bound for

$$\left| \frac{f''(a)}{f'(a)} \right|$$

Example. Find a positive root of $x^2 - 7x + 10$ near $x = 1$.

$$x_{n+1} = x_n - \frac{f(x_n)}{f'(x_n)}$$

Setting

$$f(x) = x^2 - 7x + 10$$

then

$$f'(x) = 2x - 7$$

so that

$$x_{n+1} = x_n - \frac{x_n^2 - 7x_n + 10}{2x_n - 7}$$

$x_1 = 1$:

$$x_2 = x_1 - \frac{x_1^2 - 7x_1 + 10}{2x_1 - 7}$$

$$x_2 = 1 - \frac{4}{-5} = 1.8$$

$x_2 = 1.8$:

$$x_3 = 1.8 - \frac{0.64}{-3.4} = 1.9882$$

$x_3 = 1.9882$:

$$x_4 = 1.9882 - \frac{0.0355}{-3.0236} = 1.9999$$

$x_4 = 1.9999$.
For comparison, the actual value is $x = 2$, which is found by factoring:

$$x^2 - 7x + 10 = (x - 2)(x - 5)$$

Curve Fittings

There are many techniques for deriving information from a relationship between variables when only limited data are available. Three methods are discussed here.

1. Method of least squares.[3] The method of least squares described here fits a straight line, $y = a + bx$, to the n points with coordinates

$$(x_1, y_1), \ldots, (x_i, y_i), \ldots, (x_n, y_n)$$

so as to minimize the sum of the squares of the error, S:

$$S = \sum_{i=1}^{n} [y_i - (a + bx_i)]^2$$

The coefficients a and b are given by

$$a = \frac{\left(\sum_{i=1}^{n} y_i\right)\left(\sum_{i=1}^{n} x_i^2\right) - \left(\sum_{i=1}^{n} x_i\right)\left(\sum_{i=1}^{n} x_i\, y_i\right)}{n\left(\sum_{i=1}^{n} x_i^2\right) - \left(\sum_{i=1}^{n} x_l\right)^2}$$

$$b = \frac{n\left(\sum_{i=1}^{n} x_i y_i\right) - \left(\sum_{i=1}^{n} x_i\right)\left(\sum_{i=1}^{n} y_i\right)}{n\left(\sum_{i=1}^{n} x_i^2\right) - \left(\sum_{i=1}^{n} x_i\right)^2}$$

Other least squares methods using higher order polynomial and exponentially related data are discussed in Ref. 3, pp. 136–145.

Example. Find the least squares linear approximation for the points in Fig. 3.2.

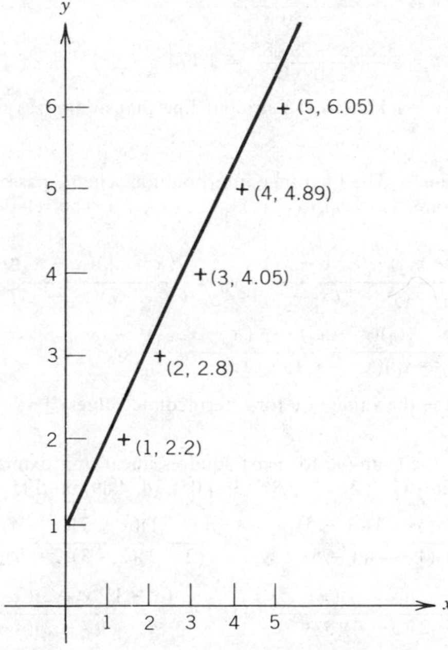

Fig. 3.2 Least squares linear approximation.

TABLE 3.2. DATA COLLECTED FOR THE LEAST SQUARES LINEAR APPROXIMATION TO THE POINTS SHOWN IN FIG. 3.2

i	x_i	y_i	x_i^2	$x_i y_i$
1	1	2.20	1	2.20
2	2	2.80	4	5.60
3	3	4.05	9	12.15
4	4	4.89	16	19.56
5	5	6.05	25	30.25
$\sum_{i=1}^{5}$	15	19.99	55	69.76

The data are organized in Table 3.2. From this table one obtains

$$a = \frac{\left(\sum_{i=1}^{5} y_i\right)\left(\sum_{i=1}^{5} x_i^2\right) - \left(\sum_{i=1}^{5} x_i\right)\left(\sum_{i=1}^{5} x_i y_i\right)}{5\left(\sum_{i=1}^{5} x_i^2\right) - \left(\sum_{i=1}^{5} x_i\right)^2}$$

$$a = \frac{(19.99)(55) - (15)(21.21)}{(5)(55) - (15)^2}$$

$$a = \frac{1099.45 - 1046.40}{275 - 225} = \frac{53.05}{50} = 1.061$$

$$b = \frac{5\left(\sum_{i=1}^{5} x_i y_i\right) - \left(\sum_{i=1}^{5} x_i\right)\left(\sum_{i=1}^{5} y_i\right)}{50}$$

$$b = \frac{5(71.71) - (15)(19.99)}{50}$$

$$b = \frac{358.55 - 299.85}{50} = 1.174$$

$y = a + bx$. Thus, $y = 1.061 + 1.174x$ is the straight line that minimizes the least squares error with the data points.

2. Lagrange interpolation.[4] The Lagrange interpolation scheme passes a polynomial of degree n through the $(n + 1)$ data points $(x_0, y_0), (x_1, y_1), \ldots, (x_n, y_n)$. The relationship between x and y is given by

$$y = \frac{(x - x_1)(x - x_2) \cdots (x - x_n)}{(x_0 - x_1)(x_0 - x_2) \cdots (x_0 - x_n)} y_0 + \frac{(x - x_0)(x - x_2) \cdots (x - x_n)}{(x_1 - x_0)(x_1 - x_2) \cdots (x_1 - x_n)} y_1 + \cdots + \frac{(x - x_0)(x - x_1) \cdots (x - x_{n-1})}{(x_n - x_0)(x_n - x_1) \cdots (x_n - x_{n-1})} y_n$$

This is then used to determine the value of y for intermediate values of x.†

Example. Use the data in the example for least squares linear approximation to fit a fourth-degree polynomial through the points (1, 2.2), (2, 2.8), (3, 4.05), (4, 4.89), and (5, 6.05).

$$y = \frac{(x-2)(x-3)(x-4)(x-5)}{(1-2)(1-3)(1-4)(1-5)}(2.2) + \frac{(x-1)(x-3)(x-4)(x-5)}{(2-1)(2-3)(2-4)(2-5)}(2.8) + \frac{(x-1)(x-2)(x-4)(x-5)}{(3-1)(3-2)(3-4)(3-5)}(4.05) + \frac{(x-1)(x-2)(x-3)(x-5)}{(4-1)(4-2)(4-3)(4-5)}(4.89) + \frac{(x-1)(x-2)(x-3)(x-4)}{(5-1)(5-2)(5-3)(5-4)}(6.05)$$

After the necessary calculations

$$y = 0.0329x^4 - 0.5058x^3 + 2.5371x^2 - 3.9641x + 4.1$$

3. Taylor's series.[5] The Taylor polynomial approximates a known function with a polynomial of degree n and is useful when a good local approximation is sought. This is in contrast to the previous two techniques in which approximations provide a good fit throughout some interval. The Taylor's polynomial of degree n will agree with the function and its first n derivatives at $x = a$. If $y = f(x)$ and its derivatives are known at $x = a$, then the Taylor series for $f(x)$ is

$$f(x) = f(a) + \sum_{i=1}^{\infty} \frac{f^{(i)}(a)}{i!}(x - a)^n$$

If the series is truncated after n terms or if only the first n derivatives are known, a polynomial

† Other interpolations, polynomials, and associated error are given in Ref. 4, pp. 281–296.

TABLE 3.3. FREQUENTLY USED TAYLOR POLYNOMIALS

$f(x)$	Taylor Polynomial	Interval of Convergence
$\frac{1}{1-x}$	$\sum_{k=0}^{n} x^k$	$\lvert x\rvert < 1$
$\sin x$	$\sum_{k=0}^{n} (-1)^k \frac{x^{2k+1}}{(2k+1)!}$	$\lvert x\rvert < \infty$
$\cos x$	$\sum_{k=0}^{n} (-1)^k \frac{x^{2k}}{(2k)!}$	$\lvert x\rvert < \infty$
e^x	$\sum_{k=0}^{n} \frac{x^k}{k!}$	$\lvert x\rvert < \infty$
$\ln(1+x)$	$\sum_{k=0}^{n} (-1)^n \frac{x^{k+1}}{k+1}$	$\lvert x\rvert < 1$
$\sqrt{1+x}$	$1 + \frac{x}{2} - \frac{x^2}{22} + \frac{3}{23}x^3 - \frac{3.5}{24}x^4 + \cdots$	$\lvert x\rvert < 1$

approximation is obtained with the remainder, $R_n(x)$:

$$R_n(x) = \frac{f^{(n+1)}(t)}{(n+1)!}(t-a)^{n+1}$$

for some t:

$$a \leqslant t \leqslant x.$$

Table 3.3 contains Taylor's series for commonly used functions.

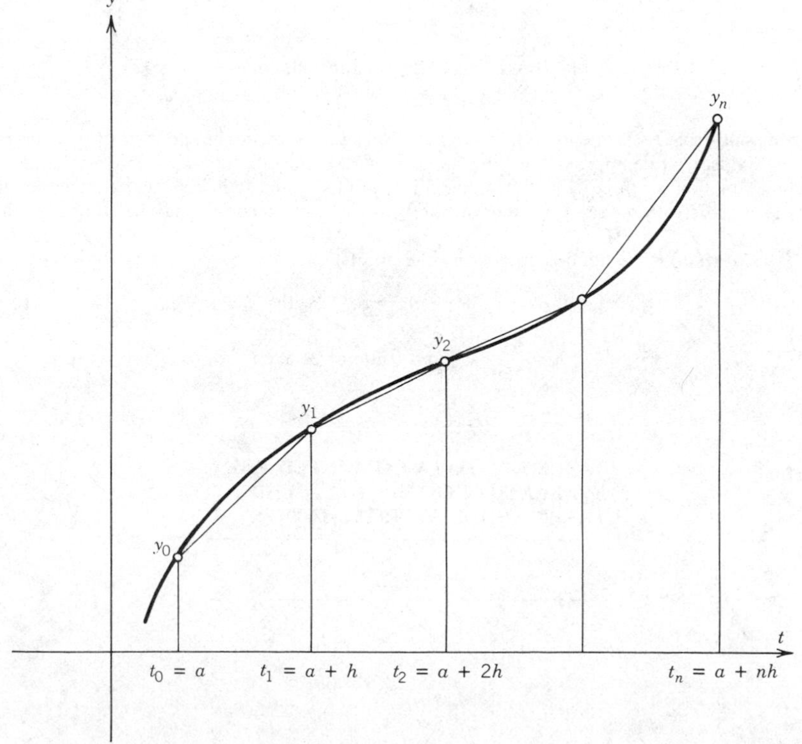

Fig. 3.3 Trapezoidal approximation to $y = f(t)$ for $a \leqslant t \leqslant b$.

3.4 NUMERICAL INTEGRATION

It is possible to obtain an approximate value of $\int_a^b f(t)\,dt$ when either $f(t)$ has no antiderivative in closed form or when $f(t)$ is defined by a set of data points. Three techniques are described here. The first two are used when $f(t)$ has no closed-form antiderivative, the third when only f and its first n derivatives are available at a.

1. Trapezoidal rule.[6] The trapezoidal rule uses a series of linear approximations to the function $y = f(t)$ to determine an approximate value of the definite integral of f for $a \leqslant x \leqslant b$ as follows. (See Fig. 3.3.)

The interval (a, b) is partitioned equally by the points

$$a = t_0 < t_1 < t_2 < \cdots < t_n = b \qquad \text{with} \quad t_i = a + ih \quad \text{and} \quad h = \frac{b-a}{n}$$

Setting $y_i = f(t_i)$ yields the approximation

$$\int_a^b f(t)\,dt = \frac{h}{2}(y_0 + 2y_1 + 2y_2 + \cdots + 2y_{n-1} + y_n)$$

Example. Determine an approximate value for

$$\int_0^1 \sqrt{1+t^3}\,dt$$

First select the number, n, of subintervals, say, $n = 4$. Next compute

$$h = \frac{b-a}{n} = \frac{1-0}{4} = \frac{1}{4}$$

so that $t_0 = 0$, $t_1 = \frac{1}{4}$, $t_2 = \frac{1}{2}$, $t_3 = \frac{3}{4}$, and $t_4 = \frac{4}{4} = 1$. Next compute the corresponding function values to arrive at the last column in Table 3.4. Using the trapezoidal rule,

$$\int_0^1 \sqrt{1+t^3}\,dt = \frac{\frac{1}{4}}{2}[1 + 2(1.008) + 2(1.061) + 2(1.192) + 1.414]$$

or

$$\int_0^1 \sqrt{1+t^3}\,dt = 1.117, \quad \text{approximately}$$

2. Simpson's rule.[6] While the trapezoidal rule uses linear approximations to the functions to be integrated, Simpson's rule approximates the function with quadratic functions designed to agree with the original function at three consecutive partition points. Thus it is expected that either greater accuracy is possible with the same number of points or the same accuracy may be attained with fewer points.

The interval (a, b) is partitioned equally by the points

$$a = t_0 < t_1 < t_2 < \cdots < t_n = b \qquad \text{with} \quad t_i = a + ih$$

$$h = \frac{b-a}{n} \quad \text{and} \quad n \textit{ must } \text{be even}$$

TABLE 3.4. DATA COMPUTED FOR EVALUATION OF $\int_0^1 \sqrt{1+t^3}\,dt$ USING TRAPEZOIDAL APPROXIMATION

i	t_i	$y_i = f(t_i) = \sqrt{1+t_i^3}$
0	0	1
1	$\frac{1}{4}$	1.008
2	$\frac{1}{2}$	1.061
3	$\frac{3}{4}$	1.192
4	1	1.414

Setting $y_i = f(t_i)$ yields Simpson's integral approximation rule:

$$\int_a^b f(t)\,dt = \frac{h}{3}(y_0 + 4y_1 + 2y_2 + 4y_3 + 2y_4 + \cdots + 2y_{n-2} + 4y_{n-1} + y_n)$$

Example. Using the data from Table 3.4, Simpson's rule gives

$$\int_0^1 \sqrt{1+t^3}\,dt = \frac{h}{3}(y_0 + 4y_1 + 2y_2 + 4y_3 + y_4)$$

or

$$\int_0^1 \sqrt{1+t^3}\,dt = \frac{\frac{1}{4}}{3}[1.000 + (4)(1.008) + (2)(1.061) + 4(1.192) + 1.414]$$

$$\int_0^1 \sqrt{1+t^3}\,dt = 1.111$$

For comparison, this same accuracy for the trapezoidal rule would require 100 subintervals.

3. Taylor's series. When a function and its first n derivatives are known at a, then from

$$f(t) = f(a) + \sum_{i=1}^{n} \frac{f^{(i)}(a)}{i!}(t-a)^i$$

the integral may be approximated by

$$\int_a^b f(t)\,dt = f(a)(b-a) + \sum_{i=1}^{n} \frac{f^{(i)}(a)(b-a)^{i+1}}{(i+1)!}$$

Example. Using the basic Taylor's series for

$$\sqrt{1+t^3} = 1 + \frac{1}{2}t^3 + \frac{t^6}{2^2 \cdot 2!} + \frac{3t^9}{3^3 \cdot 3!} \cdots$$

and using only the first three nonzero terms leads to

$$\int_0^1 \sqrt{1+t^3}\,dt = \int_0^1 \left(1 + \frac{1}{2}t^3 - \frac{t^6}{2^2 \cdot 2!} + \frac{t^9}{2^4}\right) dt = 1.113392$$

3.5 NUMERICAL SOLUTION OF FIRST ORDER ORDINARY DIFFERENTIAL EQUATIONS[7]

Three methods of determining a solution for the first order ordinary differential equation $y' = f(x, y)$, which passes through the point (x_0, y_0), are described. (This equation is referred to as an initial value problem.) The same example is used to illustrate all three. Note that the more sophisticated techniques decrease the error at the expense of increased computation time.

1. Euler's method. Euler's method approximates the actual solution to the initial value problem by lines parallel to the tangent to the solution curve. (See Fig. 3.4 for an arbitrary example.) The procedure is to start at the initial point on the curve (x_0, y_0) and proceed to the point (x_1, y_1) ($x_1 = x_0 + h$, h is the step size) along the tangent to the curve at (x_0, y_0). At the next step the process starts at (x_1, y_1) and proceeds along a line parallel to the tangent line to the point (x_2, y_2) ($x_2 = x_1 + h$). The process continues in this manner until the desired value for x is reached and the approximate solution is obtained. The relationship that computes the approximations is

$$y_{i+1} = y_i + hf(x_i, y_i),$$

where h is the step size, y_{i+1} is the new y value, y_i is the approximation after i steps, and $x_i = x_{i-1} + h = x_0 + ih$. In particular,

$$y_1 = y_0 + hf(x_0, y_0)$$

Example. To solve the initial value problem

$$y' = x - y, \qquad y(0) = 1$$

for $y(1)$, choose the step size $h = 0.25$. Thus $x_0 = 0.00$, $x_1 = 0.25$, $x_2 = 0.50$, $x_3 = 0.75$, and $x_4 = 1.00$. It is convenient to collect the data as in Table 3.5.

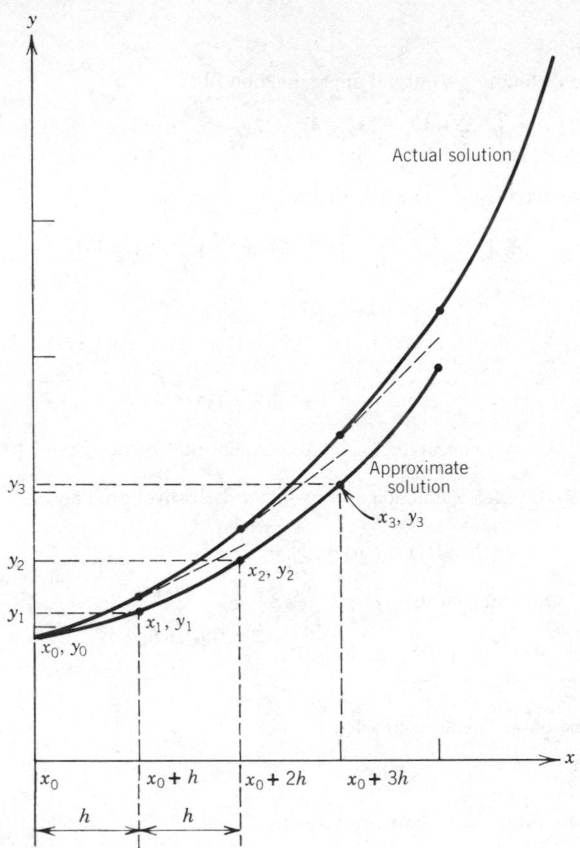

Fig. 3.4 Euler's approximation.

TABLE 3.5. DATA FOR EULER APPROXIMATION TO THE SOLUTION OF $y' = x - y$, $y(0) = 1$

i	x_i	y_i	$y_i' = x_i - y_i$	y_{i+1} (new y_i)
0	0.00	1.00	− 1.00	0.75
1	0.25	0.75	− 0.50	0.625
2	0.50	0.625	− 0.425	0.59375
3	0.75	0.59375	0.15625	0.7890625
4	1.00	0.7890625	—	—

TABLE 3.6. COMPARISON OF THE ACTUAL SOLUTION AND EULER APPROXIMATION OF THE SOLUTION FOR $y' = x - y$, $y(0) = 1$

x	y	Error
0.00	1.0000	0.0000
0.25	0.8076	0.0576
0.50	0.7131	0.1197
0.75	0.6947	0.1010
1.00	0.7358	0.0533

For comparison, the closed form of the solution is

$$y = (x - 1) + 2e^{-x}$$

The values of this function along with the error from the previous approximation are given in Table 3.6.

2. The improved Euler method. The improved Euler method utilizes an average of the slopes at the two end points rather than the slope at one end point only to correct for the heavy weight assigned to the behavior at the single point. The first order differential equation is

$$y' = f(x, y) \qquad \text{subject to} \quad y(x_0) = y_0$$

The relationship between the approximation for x_i and that for x_{i+1} is

$$y_{i+1} = y_i + h\left\{\frac{f(x_i, y_i) + f[x_i + h, y_i + hf(x_i, y_i)]}{2}\right\}$$

This method is also referred to as the modified Euler method or the average slope method.

Example. Solve $y' = x - y$, subject to $y(0) = 1$, using the improved Euler method with $h = 0.25$ to find an approximate value for $y(1)$. To start:

$$y_1 = y_0 + (0.25)\left\{\frac{f(x_i, y_i) + f[0.25, y_i + 0.25f(x_i, y_i)]}{2}\right\}$$

$$x_0 = 0, \qquad y_0 = 1$$

so that

$$y_1 = 1 + (0.25)\left\{\frac{f(0, 1) + f[0.25, 1 + 0.25f(0, 1)]}{2}\right\}$$

with

$$f(x, y) = x - y$$

$$f(0, 1) = -1$$

and

$$y_1 = 1 + 0.25\left[\frac{-1 + f(0.25, 0.75)}{2}\right]$$

$$y_1 = 1 + 0.25\left[\frac{-1 - 0.50}{2}\right]$$

$$y_1 = 1 + (0.25)(-0.75)$$

$$y_1 = 0.8125$$

The algorithm is continued and the data are collected in Table 3.7 adopting the notation

$$k_i = x_i + h \quad \text{and} \quad l_i = y_i + hf(x_i, y_i)$$

3. Fourth order Runge–Kutta method. The (fourth order) Runge–Kutta method uses a weighted average of the slope at several places along the interval and is based on a five-term Taylor's series approximation. The algorithm for the solution at $x = x'$ of the differential equation $y' = f(x,$

TABLE 3.7. IMPROVED EULER APPROXIMATION FOR $y' = x - y$ AND $y(0) = 1$

i	x_i	y_i	$f(x_i, y_i)$	$x_i + h = k_i$	$y_i + hf(x_i, y_i) = l_i$	$f(k_i, l_i)$	y_{i+1}
0	0.00	1.00	−1.00	0.25	0.75	−0.50	0.8125
1	0.25	0.8125	−0.5625	0.50	0.6719	−0.1719	0.7207
2	0.50	0.7207	−0.2207	0.75	0.6655	0.0845	0.7036
3	0.75	0.7036	0.0463	1.00	0.7152	0.2848	0.7451
4	1.00	0.7451					

y), subject to $y(x_0) = y_0$ is

$$y_{i+1} = y_i + \frac{h}{6}(k_{i1} + 2k_{i2} + 2k_{i3} + k_{i4}) \qquad 0 \leqslant i \leqslant \eta$$

where

$$k_{i1} = f(x_i, y_i)$$

$$k_{i2} = f\left(x_i + \frac{h}{2}, y_i + \frac{hk_{i1}}{2}\right)$$

$$k_{i3} = f\left(x_i + \frac{h}{2}, y_i + \frac{hk_{i2}}{2}\right)$$

$$K_{i4} = f(x_i + h, y_i + hk_{i3})$$

and

$$h = \frac{x' - x_0}{n}$$

Example. The solution of $y' = x - y$, subject to $y(0) = 1$, using the method of Runge–Kutta to determine $y(1)$ with $h = 0.25$ gives the data collected in Table 3.8. Solutions of $y' = x - y$, $y(0) = 1$ at $x = 1$ for various step sizes and computational techniques are given in Table 3.9. Table 3.10 gives a

TABLE 3.8. SOLUTION OF $y' = x - y$, $y(0) = 1$ BY THE RUNGE–KUTTA METHOD

i	x_i	y_i
0	0.00	1.0000
1	0.25	0.8076
2	0.50	0.7131
3	0.75	0.6948
4	1.00	0.7358

TABLE 3.9. SOLUTIONS TO $\hat{y} = x - y$, $y(0) = 1$, FOR VARIOUS STEP SIZES USING EULER, MODIFIED EULER, AND RUNGE–KUTTA COMPUTATIONAL METHODS

Step Size	Euler	Improved Euler	Runge–Kutta	Actual
0.25	0.6328	0.7451	0.7358	0.7358
0.2	0.6554	0.7414	0.7358	0.7358
0.1	0.6974	0.7371	0.7358	0.7358
0.05	0.7170	0.7360	0.7358	0.7358
0.025	0.7265	0.7358	0.7358	0.7358
0.01	0.7321	0.7358	0.7358	0.7358

TABLE 3.10. ORDER OF MAGNITUDE OF ERROR FOR EULER, IMPROVED EULER, AND RUNGE–KUTTA METHODS

Type of Error	Euler	Improved Euler	Runge–Kutta
Local	h^2	h^3	h^5
Global	h	h^2	h^4

comparison of the orders of magnitude of the local and global errors for the various computational techniques discussed.

Note. In the event f does not depend on y, the Runge–Kutta approximation agrees with Simpson's rule to evaluate the integral

$$y_{n+1} - y_n = \int_{x_n}^{x_n+h} f(x)\,dx$$

References

1 Engineering Systems Software Referral Catalog, 7th ed., Digital Equipment Corp., Marlboro, MA, 1981.

2 A. J. Bajpai, I. M. Calus, and J. A. Fairley, *Numerical Methods for Scientists and Engineers*, Wiley, New York, 1977.

3 R. L. Burden, J. D. Faires, and A. C. Reynolds, *Numerical Analysis*, Prindle, Weber & Schmidt, Boston, 1978.

4 P. A. Stark, *Introduction to Numerical Methods*, Macmillan, New York, 1970.

5 M. Kline, *Calculus/An Intuitive and Physical Approach*, 2nd ed., Wiley, New York, 1978.

6 F. B. Hildebrand, *An Introduction to Numerical Analysis*, 2nd ed., McGraw-Hill, New York, 1974.

7 W. E. Boyce and R. C. DiPrima, *Elementary Differential Equations and Boundary Value Problems*, 3rd ed., Wiley, New York, 1977.

CHAPTER 4
TRANSFORMS

MANU MALEK-ZAVAREI

Bell Communications Research
Holmdel, New Jersey

4.1 INTRODUCTION

Linear transforms, especially the Fourier transform, the Laplace transform, and the z-transform, provide techniques for solving problems in linear systems. They are used as mathematical tools that change difficult problems into solvable ones. This chapter provides an introduction to these transforms.

We will begin with the basic concept of the Fourier transform. We will introduce the Fourier series and the discrete Fourier transform. Also, we will present the fast Fourier transform method. Next we will discuss basic definitions, properties, and applications of the Laplace transform. A similar section will follow on the z-transform. Finally, the concept of the transfer function in continuous-time and discrete-time linear time-invariant systems will be discussed.

4.2 THE FOURIER TRANSFORM

In his study of vibrating strings, Daniel Bernoulli (1700–1782) first used the idea that any periodic function can be represented by a series of harmonically related sinusoidal components. However, this analysis technique was widely accepted only after Jean B. J. Fourier's publication in 1822 of his systematic study of such representations.[1–4]

To understand this basic idea, let us first consider an arbitrary (not necessarily periodic) function $g(t)$. The "Fourier transform" $G(f)$, or $G_1(\omega)$, of $g(t)$, is defined as

$$G(f) = \int_{-\infty}^{+\infty} g(t)e^{-2\pi jft}\,dt \tag{4.1a}$$

$$G_1(\omega) = \int_{-\infty}^{+\infty} g(t)e^{-j\omega t}\,dt \tag{4.1b}$$

where $j = \sqrt{-1}$ and $\omega = 2\pi f$. The function $G(f)$ or $G_1(\omega)$ is also known as the frequency spectrum of $g(t)$. The inverse Fourier transform of $G(f)$, or $G_1(\omega)$, is given by

$$g(t) = \int_{-\infty}^{+\infty} G(f)e^{2\pi jft}\,df = \frac{1}{2\pi}\int_{-\infty}^{+\infty} G_1(\omega)e^{j\omega t}\,d\omega \tag{4.2}$$

Eq. (4.1) is also known as the Fourier integral. A function is said to have a Fourier transform if the

TABLE 4.1. CORRESPONDENCE BETWEEN A FUNCTION AND ITS FOURIER TRANSFORM

Function	Fourier Transform
Real and even	Real and even
Real and odd	Imaginary and odd
Imaginary and even	Imaginary and even
Complex and even	Complex and even
Complex and odd	Complex and odd
Even	Even
Odd	Odd

TABLE 4.2. SOME COMMON FOURIER TRANSFORMS

$g(t)$	$G(f)$	$G_1(\omega)$
1	$\delta(f)$	$2\pi\delta(\omega)$
$\delta(t-T)$	$\cos 2\pi fT - j\sin 2\pi fT$	$e^{-j\omega T}$
$u(t)$ (unit step function)	$\dfrac{\delta(f)}{2} - \dfrac{j}{2\pi f}$	$\pi\delta(\omega) - \dfrac{j}{\omega}$
$\text{sgn}(t)$	$\dfrac{-j}{\pi f}$	$\dfrac{-2j}{\omega}$
$\lvert t\rvert^{-1/2}$	$\lvert f\rvert^{-1/2}$	$2\pi\lvert\omega\rvert^{-1/2}$
$e^{-a\lvert t\rvert}$	$\dfrac{2a}{a^2+(2\pi f)^2}$	$\dfrac{2a}{a^2+\omega^2}$
e^{-at^2}	$\sqrt{\dfrac{\pi}{a}}\, e^{-\pi^2 f^2/a}$	$\sqrt{\dfrac{\pi}{a}}\, e^{-\omega^2/4a}$
$\sin at$	$\dfrac{j}{2}\left[\delta\left(f+\dfrac{a}{2\pi}\right) - \delta\left(f-\dfrac{a}{2\pi}\right)\right]$	$j\pi[\delta(\omega+a) - \delta(\omega-a)]$
$\cos at$	$\dfrac{1}{2}\left[\delta\left(f+\dfrac{a}{2\pi}\right) + \delta\left(f-\dfrac{a}{2\pi}\right)\right]$	$\pi[\delta(\omega+a) + \delta(\omega-a)]$

corresponding Fourier integral converges. From the Fourier integral, Eq. (4.1), the correspondence in Table 4.1 between a function and its Fourier transform can be deduced.

Fourier transform methods are used in signal processing. Table 4.2 presents some common Fourier transform pairs. In this table $g(t)$ is defined in the interval $(-\infty, +\infty)$.

4.3 DISCRETE FOURIER TRANSFORM

Now let $g(t)$ be a continuous periodic function with period T; that is, let $g(t+T) = g(t)$ for all t, where T is the smallest positive number with this property. The Fourier transform $G(f)$ of $g(t)$ is then a discrete function given by

$$G(f_n) = \int_T g(t)e^{-2\pi j f_n t}\,dt \tag{4.3}$$

where

$$f_n = \frac{n}{T} = n\,\Delta f \tag{4.4}$$

and the integral is evaluated over one period of $g(t)$. The inverse Fourier transform of $G(f_n)$ in Eq. (4.3) is

$$g(t) = \sum_{n=-\infty}^{+\infty} G(f_n)e^{2\pi j f_n t} \tag{4.5}$$

and is often referred to as the *Fourier series* of $g(t)$. Note that Eq. (4.5) corresponds to the discretized version of Eq. (4.2) when the increments Δf are very small. In fact, using Eq. (4.4), we have

$$\int_{-\infty}^{+\infty} G(f)e^{2\pi jft}\,df = \lim_{\Delta f\to 0}\sum_{n=-\infty}^{+\infty} G(f_n)e^{2\pi jnt\,\Delta f} = \lim_{\Delta f\to 0}\sum_{n=-\infty}^{+\infty} G(f_n)e^{2\pi jf_nt} \tag{4.6}$$

However, in the transform pair of Eqs. (4.1) and (4.2), the function $g(t)$ is nonperiodic, while in the transform pair Eqs. (4.3) and (4.5), it is periodic with period T.

Consider the case where both the time and the frequency variables are discrete. In this case both the time function and its Fourier transform will be periodic. Let $g(t_k)$ represent the discrete-time signal in which $t_k = k\,\Delta t$ and let the time increment Δt be given by

$$\Delta t = \frac{1}{f_s} \tag{4.7}$$

where f_s is the sampling frequency at which the function $g(t)$ is sampled. Then the corresponding Fourier transform pair will be

$$G(f_n) = \Delta t \sum_k g(t_k)e^{-2\pi jf_nt_k} \tag{4.8}$$

$$g(t_k) = \sum_n G(f_n)e^{2\pi jf_nt_k} \tag{4.9}$$

where the summation in Eq. (4.8) is evaluated over one period of $g(t_k)$ and the summation in Eq. (4.9) is evaluated over one period of $G(f_n)$. Equations (4.8) and (4.9) are of primary interest in digital signal processing. They are referred to as the *discrete Fourier transform* (DFT) pair. Note that Eqs. (4.8) and (4.9) can be viewed as the discretized versions of Eqs. (4.1) and (4.2). However, in the transform pair of Eqs. (4.1) and (4.2) the functions $g(t)$ and $G(f)$ are not periodic, while in the DFT pair of Eqs. (4.8) and (4.9), these functions are both periodic with periods T and $2\pi f_s$, respectively. Some of the properties of the DFT pair are shown in Fig. 4.1.

If the time domain interval is shifted to the right so that a period of $g(t_k)$ starts at $t_k = 0$, then by Eq. (4.7),

$$K\,\Delta t = \frac{K}{f_s} = T \tag{4.10}$$

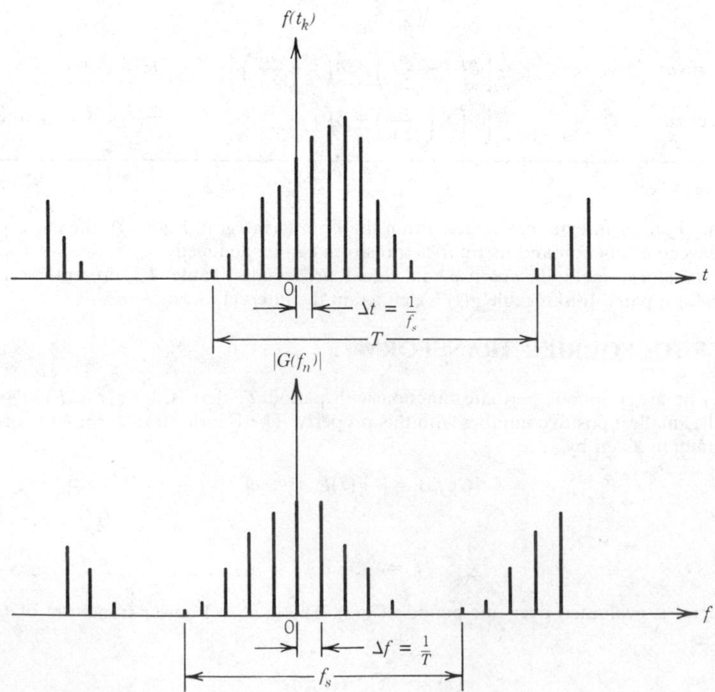

Fig. 4.1 Illustration of some properties of the DFT pair. (Functions do not necessarily represent any actual transform pair.)

TABLE 4.3. PROPERTIES OF A DISCRETE FOURIER TRANSFORM (DFT) PAIR

$g(k)$	$G(n)$
Real	Real part even, imaginary part odd
Real and even	Real and even
Real and odd	Imaginary and odd

where K is the number of samples in the range $0 \leqslant t \leqslant T$, and by Eq. (4.4),

$$K\,\Delta\omega = K\frac{2\pi}{T} = 2\pi f_s \tag{4.11}$$

If the frequency domain interval of interest is shifted to the right so that $n = 0$ corresponds to the beginning of a period, then from Eq. (4.11) it is evident that K is also the number of samples in the frequency range $0 \leqslant f \leqslant f_s$. Note that $n = K$ corresponds to the beginning of a new period.

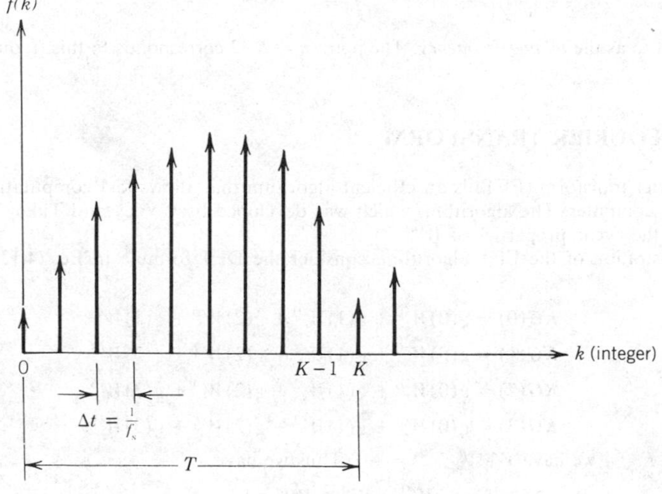

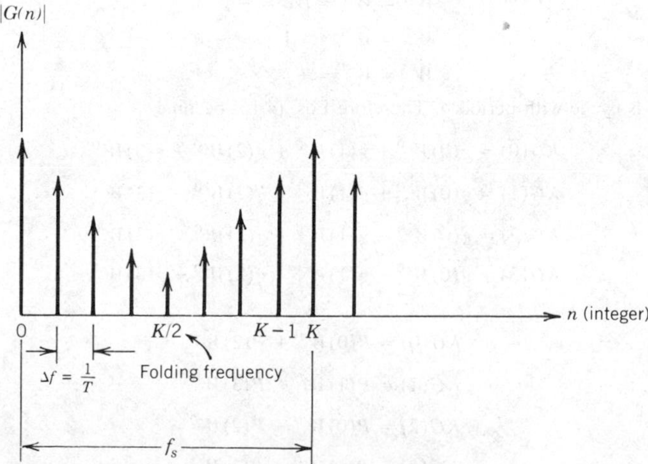

Fig. 4.2 Illustration of some properties of the DFT pair when the time function $f(k)$ is real. (Functions do not necessarily represent any actual transform pair.)

Let $T = 1$, $W = e^{-j(2\pi/K)}$, $G(n) = G(f_n)$, and $g(k) = g(t_k)$. Eqs. (4.8) and (4.9) become

$$G(n) = \frac{1}{K}\sum_{k=0}^{K-1} g(k)W^{nk}, \qquad n = 0, 1, \ldots, K-1 \tag{4.12}$$

$$g(k) = \sum_{n=0}^{K-1} G(n)W^{-nk}, \qquad k = 0, 1, \ldots, K-1 \tag{4.13}$$

The above equations form the DFT pair that is commonly used. In practice, K is usually taken to be a power of 2 for computational convenience; that is,

$$K = 2^{\gamma}, \qquad \gamma \text{ integer} > 1 \tag{4.14}$$

From Eqs. (4.12) and (4.13) it can be readily shown that the behavior of both of the functions $g(k)$ and $G(n)$ in the ranges $K/2$ to K and $-K/2$ to 0 is identical. Thus, if $g(t_k)$ is an even function, we have $g(K-k) = g(k)$, and if it is an odd function, we have $g(K-k) = -g(k)$ for any integer k. Similar properties hold for the function $G(f_n)$. Thus, one can verify the properties of the DFT pair in Table 4.3 by using Eqs. (4.12) and (4.13). As seen from Table 4.3, when $g(k)$ is a real time function, which is most often the case, the magnitude $|G(n)|$ over half of the interval is identical to that over the other half of the interval. Thus, the maximum unambiguous frequency is

$$f_0 = \frac{f_s}{2} \tag{4.15}$$

and is referred to as the *folding frequency*. The point $n = K/2$ corresponds to this frequency. (See Fig. 4.2.)

4.4 FAST FOURIER TRANSFORM

The fast Fourier transform (FFT) is an efficient algorithm that allows fast computation of the DFT on the digital computer. The algorithm, which was developed by Cooley and Tukey[3] in 1965, takes advantage of the cyclic properties of W^{nk}.

As an illustration of the FFT algorithm, consider the DFT formula in Eq. (4.12). For $K = 4$ it yields

$$KG(0) = g(0)W^0 + g(1)W^0 + g(2)W^0 + g(3)W^0 \tag{4.16a}$$

$$KG(1) = g(0)W^0 + g(1)W^1 + g(2)W^2 + g(3)W^3 \tag{4.16b}$$

$$KG(2) = g(0)W^0 + g(1)W^2 + g(2)W^4 + g(3)W^6 \tag{4.16c}$$

$$KG(3) = g(0)W^0 + g(1)W^3 + g(2)W^6 + g(3)W^9 \tag{4.16d}$$

Note that for $K = 4$ we have $We^{-j(2\pi/4)} = -j$. Thus we have

$$W^0 = W^4 = W^8 = 1 \tag{4.17a}$$

$$W^1 = W^5 = W^9 = -j \tag{4.17b}$$

$$W^2 = W^6 = -1 \tag{4.17c}$$

$$W^3 = W^7 = +j \tag{4.17d}$$

Note that W^{nk} is cyclic with period 4. Therefore Eqs. (4.16) become

$$KG(0) = g(0)W^0 + g(1)W^0 + g(2)W^0 + g(3)W^0 \tag{4.18a}$$

$$KG(1) = g(0)W^0 + g(1)W^1 - g(2)W^0 - g(3)W^1 \tag{4.18b}$$

$$KG(2) = g(0)W^0 - g(1)W^0 + g(2)W^0 - g(3)W^0 \tag{4.18c}$$

$$KG(3) = g(0)W^0 - g(1)W^1 - g(2)W^0 + g(3)W^1 \tag{4.18d}$$

or

$$KG(0) = P(0)W^0 + P(2)W^0 \tag{4.19a}$$

$$KG(1) = P(1)W^0 + P(3)W^1 \tag{4.19b}$$

$$KG(2) = P(0)W^0 - P(2)W^0 \tag{4.19c}$$

$$KG(3) = P(1)W^0 - P(3)W^1 \tag{4.19d}$$

where

$$P(0) = g(0) + g(2) \tag{4.20a}$$
$$P(1) = g(0) - g(2) \tag{4.20b}$$
$$P(2) = g(1) + g(3) \tag{4.20c}$$
$$P(3) = g(1) - g(3) \tag{4.20d}$$

Comparison of Eqs. (4.16) and (4.19) shows that the FFT algorithm considerably reduces the amount of computation required to evaluate $G(n)$, $n = 0, 1, 2, 3$. In general there are K^2 multiplications in the form similar to Eqs. (4.16) but there are only $2K$ multiplications per factor in the form similar to Eqs. (4.19).

The FFT method has applications in signal processing, especially where real-time processing is essential (e.g., in radar signal processing).

4.5 THE LAPLACE TRANSFORM

Determining the Laplace Transform

The Laplace transform is useful in transforming functions in the time domain to those in the complex frequency domain. It can be used to transform a constant coefficient linear differential equation to an algebraic equation, thus considerably reducing the computational effort required for its solution.

Let $f(t)$ be an integrable function of time defined on the interval $[0, \infty]$. Then the (unilateral) Laplace transform of $f(t)$, denoted by $L[f(t)] \triangleq F(s)$, is defined by the integral

$$F(s) = \int_{0^-}^{\infty} f(t)e^{-st}\,dt \tag{4.21}$$

for all values of the complex frequency s for which the integral exists. The lower limit of integration is 0^- so that if the time function f includes an impulse at $t = 0$, the integral includes the impulse.

It is clear from Eq. (4.21) that if $F(s)$ exists for some $s = \sigma_c + j\omega_c$, then it will also exist for all s such that $\mathrm{Re}(s) > \sigma_c$. The greatest lower bound on σ_c for which an integral in Eq. (4.21) exists is called the *abscissa of convergence* for the function $f(t)$. The region on the complex plane to the right of the abscissa of convergence is called the *region* or the *domain* of convergence of $F(s)$. (See Fig. 4.3.)

The Laplace integral, Eq. (4.21), can be extended to other values of $s = \sigma + j\omega$. That is, $F(s)$ can be considered a well-defined function for all values of s except for those that cause the integral in Eq. (4.21) to diverge.

Example 1. Consider the function $f(t) = e^{\alpha t}$ where α is a real or a complex number. Then

$$L[f(t)] \triangleq F(s) = \int_{0^-}^{\infty} e^{\alpha t}e^{-st}\alpha t = \left.\frac{e^{-(s-\alpha)t}}{-(s-\alpha)}\right|_{0^-}^{\infty} \tag{4.22}$$

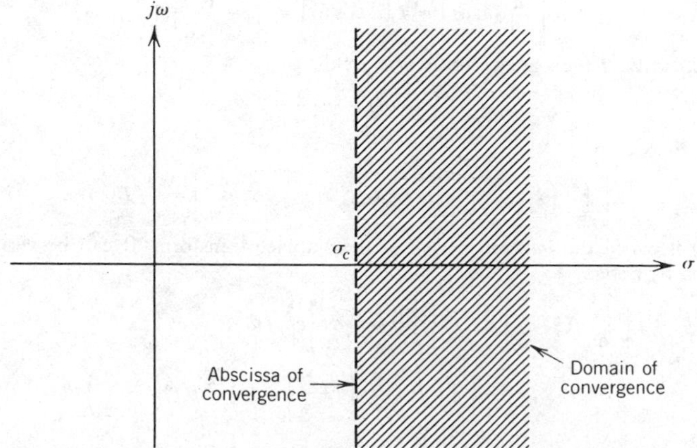

Fig. 4.3 Domain of convergence of the Laplace transform in the s-plane.

Thus $F(s)$ exists; that is, it is finite, when $\mathrm{Re}(s-\alpha)>0$, and it is infinite otherwise. That is,

$$F(s)=L[e^{\alpha t}]=\frac{1}{s-\alpha} \qquad \text{for} \quad \mathrm{Re}(s)>\mathrm{Re}(\alpha) \tag{4.23}$$

and $\mathrm{Re}(\alpha)$ is the abscissa of convergence for $e^{\alpha t}$. However, we will regard the Laplace transform of $e^{\alpha t}$ to be $1/(s-\alpha)$, which is well-defined for all s except for $s=\alpha$.

Uniqueness is a fundamental property of the Laplace transform. That is, if two time functions $f_1(t)$ and $f_2(t)$ have the same function of the complex frequency s, say $F(s)$, as their Laplace transform, then $f_1(t)$ and $f_2(t)$ can differ only trivially. Therefore, for a time function $f(t)$, a "unique" function $F(s)$ exists as its Laplace transform. Conversely, given a Laplace transform $F(s)$, there is a unique time function $f(t)$ over the interval $[0,\infty]$ (except for trivialities) such that $F(s)=L[f(t)]$. This is written as

$$f(t)=L^{-1}[F(s)] \tag{4.24}$$

meaning that $f(t)$ is the *inverse Laplace transform* of $F(s)$.

Another important property of the Laplace transform is linearity. That is, if $F_1(s)$ and $F_2(s)$ are the Laplace transforms of the time functions $f_1(t)$ and $f_2(t)$, respectively, then

$$L[\alpha_1 f_1(t)+\alpha_2 f_2(t)]=\alpha_1 F_1(s)+\alpha_2 F_2(s) \tag{4.25}$$

for arbitrary constants α_1 and α_2. The linearity property of the Laplace transform follows immediately from the linearity of the Laplace integral, Eq. (4.21).

Example 2. Consider

$$f(t)=\sin\alpha t=\frac{1}{2j}\left(e^{j\alpha t}-e^{-j\alpha t}\right) \tag{4.26}$$

Then using Eq. (4.23) and the linearity property of the Laplace transform yields

$$L[\sin\alpha t]=\frac{1}{2j}\left[\frac{1}{s-j\alpha}-\frac{1}{s+j\alpha}\right]=\frac{\alpha}{s^2+\alpha^2} \tag{4.27}$$

Let $F(s)$ be the Laplace transform of $f(t)$. Then

$$L\left[\frac{d^n}{dt^n}f(t)\right]=s^nF(s)-s^{n-1}f(0^-)-s^{n-2}f^{(1)}(0^-)-\cdots-sf^{(n-2)}(0^-)-f^{(n-1)}(0^-) \tag{4.28}$$

where $f^{(i)}$ indicates the ith derivative of f with respect to t. This is referred to as the *differentiation rule* the Laplace transform. It can be verified by evaluating the Laplace integral, Eq. (4.21), for $(d/dt)f(t)$ by parts:

$$L\left[\frac{d}{dt}f(t)\right]=f(t)e^{-st}\Big|_{0^-}^{\infty}-\int_{0^-}^{\infty}f(t)(-se^{-st})\,dt=-f(0^-)+s\int_{0^-}^{\infty}f(t)e^{-st}\,dt=sF(s)-f(0^-) \tag{4.29}$$

Application of Eq. (4.29) $n-1$ times yields Eq. (4.28).

Example 3. Consider $(d/dt)\sin\alpha t=\alpha\cos\alpha t$. Thus, using Eqs. (4.27) and (4.28) results in

$$L\left[\frac{d}{dt}\sin\alpha t\right]=L[\alpha\cos\alpha t]=s\left[\frac{\alpha}{s^2+\alpha^2}\right] \tag{4.30}$$

and using the linearity of the Laplace transform yields

$$L[\cos\alpha t]=\frac{s}{s^2+\alpha^2} \tag{4.31}$$

If $L[f(t)]=F(s)$, then

$$L\left[\int_{0^-}^{t}\int_{0^-}^{\tau_1}\cdots\int_{0^-}^{\tau_{n-1}}f(\tau_n)\,d\tau_n\,d\tau_{n-1}\cdots d\tau_1\right]=\frac{1}{s^n}F(s) \tag{4.32}$$

Equation (4.32) is called the *integration rule* of the Laplace transform. It can be verified again by using integration by parts:

$$\begin{aligned}L\left[\int_{0^-}^{t}f(\tau)\,d\tau\right]&\triangleq\int_{0^-}^{\infty}\left[\int_{0^-}^{t}f(\tau)\,d\tau\right]e^{-st}\,dt\\&=\left[\int_{0^-}^{t}f(\tau)\,d\tau\right]\frac{e^{-st}}{-s}\Big|_{0^-}^{\infty}-\int_{0^-}^{\infty}f(t)\left(\frac{e^{-st}}{-s}\right)dt\\&=0+\frac{1}{s}\int_{0^-}^{\infty}f(t)e^{-st}\,dt=\frac{1}{s}F(s)\end{aligned} \tag{4.33}$$

Repeated application of Eq. (4.33) results in Eq. (4.32).

The "time function translation" property of the Laplace transform implies that if $L[f(t)] = F(S)$, then for any positive scalar α,

$$L[f(t-\alpha)] = e^{-\alpha s}F(s) \tag{4.34}$$

Further, the scalar α may be negative if $f(t-\alpha)$ vanishes for $t < 0$. This property follows easily from the Laplace transform defining integral:

$$L[f(t-\alpha)] \triangleq \int_{0^-}^{\infty} f(t-\alpha)e^{-st}\,dt = \int_{0^-}^{\infty} f(t-\alpha)e^{-st}\,dt$$

$$\int_{0^-}^{\infty} f(\tau)e^{-s(\tau+\alpha)}\,d\tau = e^{-s\alpha}\int_{0^-}^{\infty} f(\tau)e^{-s\tau}\,d\tau = e^{-s\alpha}F(s) \tag{4.35}$$

where $\tau = t - \alpha$. Note that $f(t-\alpha)$ is the original function delayed by α.

The "Laplace transform translation" property implies that if $L[f(t)] = F(s)$, then

$$L[e^{\alpha t}f(t)] = F(s-\alpha) \tag{4.36}$$

where α is any complex scalar. Two other properties of the Laplace transform are the "initial value" and the "final value" properties. The former implies that if $f(t)$ approaches a limit as t approaches zero from the right, then

$$f(0+) = \lim_{s\to\infty} sL[f(t)] \tag{4.37}$$

Note that the limit on the right side of Eq. (4.37) may exist without the existence of $f(0+)$. The latter property implies that if $f(t)$ approaches a limit as t approaches ∞, then

$$\lim_{t\to\infty} f(t) = \lim_{s\to 0} sL[f(t)] \tag{4.38}$$

Again, note that the limit on the right side of Eq. (4.38) may exist without the existence of the limit on the left side.

Example 4. Consider $f(t) = e^{-\alpha t}\cos\omega t$ where α is a real positive number. Using Eqs. (4.31) and (4.36) we have

$$L[f(t)] \triangleq F(s) = \frac{s+\alpha}{(s+\alpha)^2+\omega^2} \tag{4.39}$$

Now $\lim_{s\to\infty} sF(s) = 1$, which is equal to $f(0+)$. Also $\lim_{s\to 0} sF(s) = 0$, which is equal to $\lim_{t\to\infty} f(t)$.

The last property of the Laplace transform to be discussed is the "convolution" property. If time functions $f_1(t)$ and $f_2(t)$ are defined on the interval $[0, \infty]$, the convolution of f_1 and f_2, denoted $f_1 * f_2$, is a time function defined by

$$f_1 * f_2(t) = \int_0^t f_1(\tau)f_2(t-\tau)\,d\tau \tag{4.40}$$

provided that the integral exists for all t. If $F_1(s)$ and $F_2(s)$ are the Laplace transforms of $f_1(t)$ and $f_2(t)$, respectively, then according to the convolution property of the Laplace transform,

$$L[f_1 * f_2(t)] = F_1(s)F_2(s) \tag{4.41}$$

This is an important property of the Laplace transform often used in the study of linear time-invariant continuous-time systems.

Inversion of the Laplace Transform

The importance of the Laplace transform lies in the fact that there is a unique time function $f(t)$ for each Laplace transform $F(s)$. Let $F(s) = L[f(t)]$ and let σ_c be the abscissa of convergence for the function $f(t)$. Then $f(t)$ can be recovered as follows:

$$f(t) \triangleq L^{-1}[F(s)] = \frac{1}{2\pi j}\lim_{\omega\to\infty}\int_{\sigma-j\omega}^{\sigma+j\omega} F(s)e^{st}\,ds \tag{4.42}$$

for all $\sigma > \sigma_c$. The proof of this fact may be found in Ref. 4. Note that by assumption $f(t) = 0$ for $t < 0$; thus the integral, Eq. (4.42), yields $f(t)$ for $t > 0$. Also, $f(\cdot)$ could be discontinuous at t; in fact the left side of Eq. (4.42) may be replaced by $\frac{1}{2}[f(t+0) + f(t-0)]$.

The integral in Eq. (4.42) is, in general, difficult to compute and is rarely used directly for the inversion of the Laplace transform. Rather, the inverse Laplace transform is usually found by expanding the function into easily invertible components and applying the linearity property of the Laplace transform. This procedure can be formalized as follows. Let $F(s)$ be a rational function, that is, a ratio of two polynomials in s. Further, let $F(s)$ be strictly proper, that is, if the degree of its numerator is m and that of its denominator is n, then $m < n$. Thus, $F(s)$ can be written as

$$F(s) = \frac{Q(s)}{P(s)} = \frac{q_m s^m + q_{m-1}s^{m-1} + \cdots + q_1 s + q_0}{p_n s^n + p_{n-1}s^{n-1} + \cdots + p_1 s + p_0} = \frac{q_m}{p_n}\frac{\prod_{i=1}^{m}(s - z_j)}{\prod_{i=1}^{n}(s - p_i)} \tag{4.43}$$

where the coefficients $q_j, j = 0, 1, \ldots, m$ and $p_i, i = 0, 1, \ldots, n$ are real. Then complex numbers K_{ij} exist such that $F(s)$ can be written as

$$F(s) = \sum_{i=1}^{p}\sum_{j=1}^{m_i}\frac{K_{ij}}{(s - p_i)^j} \tag{4.44}$$

where m_i is the multiplicity of root p_i of polynomial $P(s)$ and p is the number of distinct roots of $P(s)$. The complex numbers K_{ij} are determined as follows:

$$K_{i,m_i-j+1} = \frac{1}{(j-1)!}\frac{d^{j-1}}{ds^{j-1}}(s - p_i)^{m_i}F(s)|_{s=p_i}, \qquad j = 1, 2, \ldots, m_i; \quad i = 1, 2, \ldots, p \tag{4.45}$$

Then the inverse Laplace transform of $F(s)$ is

$$f(t) = \sum_{i=1}^{p}\sum_{j=1}^{m_i}\frac{K_{ij}t^{j-1}}{(j-1)!}e^{\lambda_i t} \tag{4.46}$$

The proof of this theorem can be found in Ref. 4.

The roots $z_j, j = 1, 2, \ldots, m$ of polynomial $Q(s)$ are called the *zeros* of $F(s)$ and the roots p_i, $i = 1, 2, \ldots, n$ of polynomial $P(s)$ are called the *poles* of $F(s)$. Clearly, the abscissa of convergence for $f(t)$ is $\max_i\{\mathrm{Re}(p_i)\}$. The complex numbers $K_{ij}, j = 1, 2, \ldots, m_i$ are called the *residues* corresponding to the pole p_i. The poles, the zeros, and the residues either are real or occur in complex conjugate pairs because the coefficients p_i and q_j are assumed to be real. Note that $\sum_{i=1}^{p} m_i = n$. Also note that in the case where all the poles of $F(s)$ are distinct, that is, where $m_i = 1, i = 1, 2, \ldots, p$, then $p = n$ and Eq. (4.44) becomes

$$F(s) = \sum_{i=1}^{n}\frac{K_i}{s - p_i} \tag{4.47}$$

where

$$K_i = (s - p_i)F(s)|_{s=p_i}, \qquad i = 1, 2, \ldots, n \tag{4.48}$$

and Eq. (4.46) becomes

$$f(t) = \sum_{i=1}^{n} K_i e^{\lambda_i t} \tag{4.49}$$

Example 5. Consider

$$F(s) = \frac{s}{(s+1)^3(s+2)} \tag{4.50}$$

It can be written as

$$F(s) = \frac{K_{11}}{s+1} + \frac{K_{12}}{(s+1)^2} + \frac{K_{13}}{(s+1)^3} + \frac{K_{21}}{s+2} \tag{4.51}$$

where

$$K_{13} = (s+1)^3F(s)|_{s=-1} = -1 \tag{4.52}$$

$$K_{12} = \frac{d}{ds}(s+1)^3F(s)\Big|_{s=-1} = \frac{2}{(s+2)^2}\Big|_{s=-1} = 2 \tag{4.53}$$

$$K_{11} = \frac{1}{2}\frac{d^2}{ds^2}(s+1)^3F(s)\Big|_{s=-1} = \frac{-2}{(s+1)^3}\Big|_{s=-1} = -2 \tag{4.54}$$

$$K_{21} = (s+2)F(s)|_{s=-2} = 2 \tag{4.55}$$

Thus

$$F(s) = \frac{-2}{s+1} + \frac{2}{(s+1)^2} - \frac{1}{(s+1)^3} + \frac{2}{s+2} \tag{4.56}$$

and

$$L^{-1}[F(s)] = -2e^{-t} + 2te^{-t} - \frac{t^2}{2}e^{-t} + 2e^{-2t} \qquad t \geqslant 0 \tag{4.57}$$

Table of Laplace Transforms

The Laplace transforms of some common functions of time as well as the inverse Laplace transforms of some common functions of s are collected in Table 4.4 for convenience. Note that the time function $f(t)$ is assumed to vanish for $t < 0$.

TABLE 4.4. SOME COMMON FUNCTIONS AND THEIR LAPLACE TRANSFORMS

No.	$f(t)$	$F(s)$
1	$\delta(t)$	1
2	$u(t)$ (unit step function)	$\frac{1}{s}$
3	$u(t-\alpha)$	$\frac{e^{-\alpha s}}{s}$
4	$r(t) = tu(t)$ (unit ramp)	$\frac{1}{s^2}$
5	t^n	$\frac{n!}{s^n + 1}$
6	$t^n e^{\alpha t}$	$\frac{n!}{(s-\alpha)^{n+1}}$
7	$-1 - \alpha t + e^{\alpha t}$	$\frac{\alpha^2}{s^2(s-\alpha)^2}$
8	$1 - e^{\alpha t} + \alpha t e^{\alpha t}$	$\frac{\alpha^2}{s(s-\alpha)^2}$
9	$\frac{\beta}{\alpha^2} + \frac{1 + \alpha t - \beta t}{\alpha} e^{\alpha t}$	$\frac{s-\beta}{s(s-\alpha)^2}$
10	$\sin \omega t$	$\frac{\omega}{s^2+\omega^2}$
11	$\cos \omega t$	$\frac{s}{s^2+\omega^2}$
12	$e^{\alpha t} \sin \omega t$	$\frac{\omega}{(s-\alpha)^2+\omega^2}$
13	$e^{\alpha t} \cos \omega t$	$\frac{s-\alpha}{(s-\alpha)^2+\omega^2}$
14	$\frac{1}{\omega^2}(1 - \cos \omega t)$	$\frac{1}{s(s^2+\omega^2)}$
15	$e^{-\zeta\omega_n t} \sin\left(\omega_n\sqrt{1-\zeta^2}\, t\right)$	$\frac{\omega_n\sqrt{1-\zeta^2}}{s^2 + 2\zeta\omega_n s + \omega_n^2}$
16	$\frac{e^{\alpha t}}{\alpha^2+\omega^2} + \frac{\sin(\omega t - \phi)}{\omega\sqrt{\alpha^2+\omega^2}}$ where $\phi = -\tan^{-1}\frac{\omega}{\alpha}$	$\frac{1}{(s-\alpha)(s^2+\omega^2)}$
17	$e^{\alpha t}(A \cos \omega t + B \sin \omega t)$	$\frac{A(s-\alpha) + B\omega}{(s-\alpha)^2+\omega^2}$
18	$e^{\alpha t}\left(A \cos \omega t + \frac{B + \alpha A}{\omega} \sin \omega t\right)$	$\frac{As + B}{(s-\alpha)^2+\omega^2}$
19	$\frac{\sqrt{(\alpha+\beta)^2+\omega^2}}{\omega} e^{\alpha t} \sin(\omega t + \delta)$ where $\delta = \tan^{-1}\frac{\omega}{\alpha+\beta}$	$\frac{s+\beta}{(s-\alpha)^2+\omega^2}$
20	$2\lvert K\rvert e^{\alpha t} \cos(\omega t + \angle K)$	$\frac{K}{s-\alpha-j\omega} + \frac{\overline{K}}{s-\alpha+j\omega}$

TABLE 4.4. (*Continued*)

No.	$f(t)$	$F(s)$
21	$\dfrac{1}{\alpha^2+\omega^2}+\dfrac{e^{\alpha t}}{\omega\sqrt{\alpha^2+\omega^2}}\sin(\omega t-\delta)$ where $\delta=\tan^{-1}\dfrac{\omega}{\alpha}$	$\dfrac{1}{s[(s-\alpha)^2+\omega^2]}$
22	$\dfrac{1}{\omega_n^2}-\dfrac{e^{-\zeta\omega_n t}}{\omega_n\sqrt{1-\zeta^2}}\sin\left(\omega_n\sqrt{1-\zeta^2}t+\delta\right)$ where $\delta=\cos^{-1}\zeta$	$\dfrac{1}{s(s^2+2\zeta\omega_n s+\omega_n^2)}$
23	$\dfrac{\beta}{\alpha^2+\omega^2}+\dfrac{e^{\alpha t}\sqrt{(\alpha+\beta)^2+\omega^2}}{\omega\sqrt{\alpha^2+\omega^2}}\sin(\omega t+\delta)$ where $\delta=\tan^{-1}\dfrac{\omega}{\alpha+\beta}-\tan^{-1}\dfrac{\omega}{\alpha}$	$\dfrac{s+\beta}{s[(s-\alpha)^2+\omega^2]}$

Solution of Differential Equations Using the Laplace Transform

An important application of the Laplace transform is in the solution of linear constant-coefficient differential equations. By taking the Laplace transform of both sides of such a differential equation, the Laplace transform of the unknown function can be expressed as a rational function of s. The inverse Laplace transform of this function then provides the solution. Thus the Laplace transform converts the solution of a linear constant-coefficient differential equation to that of an algebraic equation. The following example illustrates the method.

Example 6. Consider the differential equation

$$\dddot{f}(t)+10\ddot{f}(t)+37\dot{f}(t)+52f(t)=-3u(t)+\dot{\delta}(t),\qquad t\geqslant 0 \tag{4.58}$$

with initial conditions

$$f(0^-)=\dot{f}(0^-)=1,\qquad \ddot{f}(0^-)=0 \tag{4.59}$$

Taking the Laplace transform of both sides of Eq. (4.58) yields

$$[s^3F(s)-s^2f(0^-)-s\dot{f}(0^-)-\ddot{f}(0^-)]+10[s^2F(s)-sf(0^-)-\dot{f}(0^-)]$$
$$+37[sF(s)^-f(0-)]+52F(s)=\frac{3}{s}+s \tag{4.60}$$

or

$$F(s)=\frac{s^3+12s^2+47s-3}{s(s^3+10s^2+37s+52)} \tag{4.61}$$

which can be written in partial fraction form as

$$F(s)=\frac{-3/52}{s}+\frac{63/20}{s+4}-\frac{(136/65)(s+3)+541/130}{(s+3)^2+4} \tag{4.62}$$

Thus the solution is

$$f(t)=L^{-1}[F(s)]=\frac{3}{52}+\frac{63}{20}e^{-4t}-\frac{136}{65}e^{-3t}\cos 2t+\frac{541}{130}e^{-3t}\sin 2t,\qquad t\geqslant 0 \tag{4.63}$$

4.6 THE z-TRANSFORM

Determining the z-Transform

The concept of the z-transform is closely related to the idea of sampling a time function. Consider a function of continuous time $f(t)$ where $f(t)=0$ for $t<0$. An "ideal sampler" is one that takes samples of infinitesimal width of $f(t)$ at regular intervals of time (see Fig. 4.4). Thus an ideal sampling $f^*(t)$ of $f(t)$ can be viewed as a sequence of equally spaced impulses with magnitudes equal to the

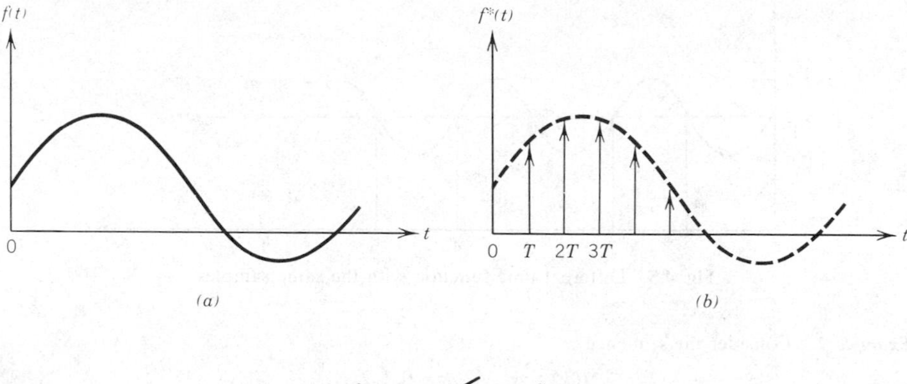

Fig. 4.4 Ideal sampling: (a) a function of continuous time; (b) its ideal samples; (c) symbol of ideal sampler.

values of the function $f(\cdot)$ at the corresponding discrete times. That is,

$$f^*(t) = \sum_{k=0}^{\infty} f(kT)\delta(t - kT) \tag{4.64}$$

where T is called the *sampling period*. Taking the Laplace transform of both sides of Eq. (4.64) yields

$$L[f^*(t)] \triangleq F^*(s) = \sum_{k=0}^{\infty} f(kT)L[\delta(t - kT)] = \sum_{k=0}^{\infty} f(kT)e^{-kTs} \tag{4.65}$$

Now define the complex variable z as

$$z = e^{Ts} \tag{4.66}$$

Thus Eq. (4.65) becomes

$$F^*(s) = \sum_{k=0}^{\infty} f(kT)z^{-k} \tag{4.67}$$

which is defined as the (*one-sided*) *z-transform* of $f(t)$ and is indicated by $F(z)$. That is,

$$Z[f(k)] \triangleq F(z) = \sum_{k=0}^{\infty} f(kT)z^{-k} \tag{4.68}$$

Also, the z-transform of a sequence of scalars $f(kT)$, $k = 0, 1, 2, \ldots$ is defined as in Eq. (4.68).

Note that the z-transform exists if the infinite sum in Eq. (4.68) converges. The *radius of convergence* r_c of the infinite series in Eq. (4.68) is defined as

$$r_c = \lim_{k\to\infty} |f(k)|^{1/k} \tag{4.69}$$

The series in Eq. (4.68) is analytic for $|z| > r_c$. Thus it converges absolutely for all z in the domain $|z| > r_c$. For notational convenience, T is usually taken to be unity.

Example 1. Consider the unit step function $u(t)$. Then

$$Z[u(t)] \triangleq U(z) = \sum_{k=0}^{\infty} z^{-k} \tag{4.70}$$

which converges to

$$\frac{1}{1 - 1/z} = \frac{z}{z - 1} \tag{4.71}$$

if $|z| > 1$. Thus the radius of convergence of the series in Eq. (4.70) is 1.

Note that Eq. (4.71) is also the z-transform of the sequence $u^*(t) = 1, 1, 1, \ldots$. Although Eq. (4.71) is the z-transform of this unique sequence, it does not correspond to a unique function of continuous time since there are many functions whose samples $f(k)$ at $k = 0, 1, 2, \ldots$ are 1 (see Fig. 4.5).

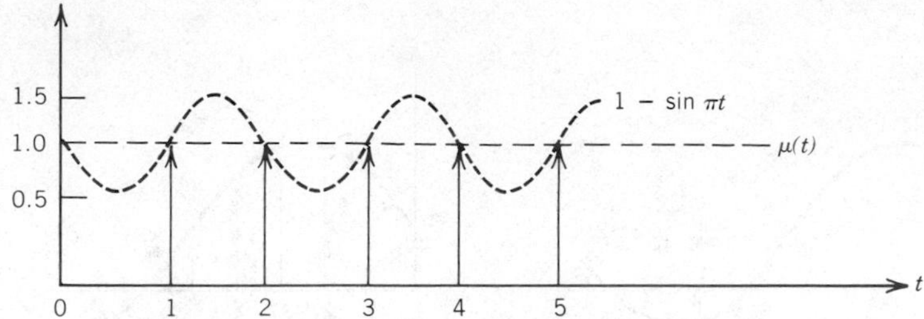

Fig. 4.5 Different time function with the same samples.

Example 2. Consider the sequence

$$f(k) = \alpha^k, \qquad k = 0, 1, 2, \ldots \tag{4.72}$$

Then

$$Z[f(k)] \triangleq \sum_{k=0}^{\infty} f(k) z^{-k} = \sum_{k=0}^{\infty} \alpha^k z^{-k} = \frac{1}{1 - \alpha/z} = \frac{z}{z - \alpha} \tag{4.73}$$

provided that $|\alpha/z| < 1$ or $|z| > |\alpha|$. Thus the radius of convergence of the infinite series in Eq. (4.73) is $|\alpha|$.

Since the z-transform is defined through the Laplace transform, the properties of the z-transform can be deduced from those of the Laplace transform. The uniqueness property holds for the z-transform in the sense that a given function $F(z)$ corresponds to a unique "sequence" $f(k)$. It does not, however, correspond to a unique function of continuous time $f(t)$, as has been discussed. Similar to the Laplace transform, the z-transform is linear, that is,

$$Z[\alpha_1 f_1(t) + \alpha_2 f_2(t)] = \alpha_1 Z[f_1(t)] + \alpha_2 Z[f_2(t)] \tag{4.74}$$

for arbitrary constants α_1 and α_2.

The "advance" property of the z-transform is analogous to the differentiation rule of the Laplace transform:

$$Z[f(k+m)] = z^m F(z) - \sum_{z=0}^{m-1} f(k) z^{m-k} \tag{4.75}$$

for any positive integer m. Similarly, the "delay" property of the z-transform is analogous to the integration rule of the Laplace transform:

$$Z[f(k-m)] = z^{-m} F(z) \tag{4.76}$$

for any positive integer m.

Two other properties of the z-transform are the "initial value" and the "final value" properties. The former implies that

$$f(0+) = \lim_{z \to \infty} F(z) \tag{4.77}$$

provided that the limit exists. The latter implies that if a sequence $f(k)$ tends to a limit as k tends to infinity, then

$$\lim_{k \to \infty} f(k) = \lim_{z \to 1} (z - 1) F(z) \tag{4.78}$$

provided that $F(z)$ is analytic for $|z| > 1$. Finally, the "convolution" property of the z-transform implies that if $F_1(z)$ and $F_2(z)$ are the z-transforms of $f_1(t)$ and $f_2(t)$, respectively, then

$$Z\left[\sum_{j=0}^{k} f_1(j) f_2(k-j)\right] = F_1(z) F_2(z) \tag{4.79}$$

Inversion of the z-Transform

Given a function $F(z)$ of the complex variable z, one can find a function $f(\cdot)$ in the time domain such that

$$Z[f(t)] = F(z) \tag{4.80}$$

or equivalently,

$$f(t) = Z^{-1}[F(z)] \tag{4.81}$$

Note, however, that unlike the Laplace transform, the inverse z-transform $f(t)$ is not unique. In fact an infinite number of functions $f(t)$ exist for which $F(z)$ is the z-transform. These functions have the same values only at the sampling instants, that is, they can all be represented by the same ideal sampling $f^*(t)$. The methods of determining the inverse z-transform can be grouped into three: the power series method, the residue method, and the partial fraction expansion method. In the power series method, the defining infinite series, Eq. (4.68), is used to determine the inverse z-transform of any given function $F(z)$. This is done by expanding the rational function $F(z)$ into an infinite series in z^{-k}. Then the coefficients of z^{-k} will be equal to $f(k)$. However, this procedure is, in general, tedious and does not yield a closed-form solution.

Example 3. Consider

$$F(z) = \frac{z}{z-2} \tag{4.82}$$

Long division results in

$$F(z) = 1 + 2z^{-1} + 4z^{-2} + 8z^{-3} + \cdots \tag{4.83}$$

Thus, by comparison with Eq. (4.68), we have

$$f(0) = 1, \qquad f(1) = 2, \qquad f(2) = 4, \qquad f(3) = 8, \ldots \tag{4.84}$$

The residue method uses the fact that if $F(z) = Z[f(k)]$, then

$$f(k) = \frac{1}{2\pi j} \oint_{\Gamma} F(z) z^{k-1}\, dz, \qquad k = 0, 1, 2, \ldots \tag{4.85}$$

where integration is performed in the counterclockwise direction and Γ is any closed curve enclosing the origin and lying outside the circle $|z| > r_c$ [the radius of convergence of $F(z)$], or, equivalently, Γ is any closed curve enclosing the poles of $zF(z)$, $k = 0, 1, 2, \ldots$. The contour integral in Eq. (4.85) can be evaluated by using Cauchy's residue theorem which states that if $F(z)$ is an analytic function in a region R except at a finite number of singularities, and if Γ is a closed curve in R not passing through these singularities, then

$$\frac{1}{2\pi j} \oint_{\Gamma} F(z)\, dz = \text{sum of the residues of } F(z) \tag{4.86}$$

where integration is performed in the counterclockwise direction and the residue of $F(z)$ at a pole $z = p_i$ with multiplicity m_i is found from Eq. (4.45). From this discussion it is evident that

$$f(k) = Z^{-1}[F(z)] = \text{sum of the residues of } F(z) z^{k-1} \tag{4.87}$$

Example 4. Again consider

$$F(z) = \frac{z}{z-\alpha} \tag{4.88}$$

$F(z)$ has a simple pole at $z = \alpha$. Thus,

$$f(k) = Z^{-1}[F(z)] = \text{Res}\left[\frac{z}{z-\alpha} z^{k-1}\right] = z^k\big|_{z=\alpha} = \alpha^k \tag{4.89}$$

The partial expansion method of determining the inverse z-transform is similar to the method used in determining the inverse Laplace transform of a rational function. Instead of $F(s)$ in the case of the Laplace transform, here we expand $F(z)/z$ in partial fractions. The reason for this is that it is usually more convenient to find the inverse z-transform of a fraction with a free z in its numerator. The following example illustrates this method.

Example 5. Consider

$$F(z) = \frac{z(-2z+1)}{z^2 - 3z + 2} \tag{4.90}$$

The partial fraction expansion of $F(z)/z$ yields

$$\frac{F(z)}{z} = \frac{1}{z-1} - \frac{3}{z-2} \tag{4.91}$$

Thus

$$F(z) = \frac{z}{z-1} - 3\frac{z}{z-2} \tag{4.92}$$

and using Eq. (4.89), the inverse z-transform of $F(z)$ is

$$f(k) = 1 - 3(2^k) \tag{4.93}$$

Table of z-Transforms

The z-transforms of some common functions and sequences are collected in Table 4.5. Note that $f(k) = 0$ for $k < 0$.

TABLE 4.5. SOME COMMON FUNCTIONS AND SEQUENCES AND THEIR z-TRANSFORMS

No.	$f(k)$	$F(z)$
1	$\delta(k)^a$	1
2	1	$\dfrac{z}{z-1}$
3	α^k	$\dfrac{z}{z-\alpha}$
4	k	$\dfrac{z}{(z-1)^2}$
5	k^2	$\dfrac{z(z+1)}{(z-1)^3}$
6	k^3	$\dfrac{z(z^2+4z+1)}{(z-1)^4}$
7	k^n	$(-1)^n z^n \dfrac{d^n}{dz^n}\left(\dfrac{z}{z-1}\right)$
8	$\dfrac{1}{k}, \quad k > 0$	$\ln \dfrac{z}{z-1}$
9	$e^{-\alpha k}$	$\dfrac{z}{z-e^{-\alpha}}$
10	$ke^{-\alpha}k$	$\dfrac{ze^{-\alpha}}{(z-e^{-\alpha})^2}$
11	$k^{\alpha k}$	$\dfrac{\alpha z}{(z-\alpha)^2}$
12	$k^2 \alpha k$	$\dfrac{\alpha z(z+\alpha)}{(z-\alpha)^3}$
13	$\dfrac{\alpha^k}{k!}$	$e^{\alpha/z}$
14	$\sin \alpha k$	$\dfrac{z \sin \alpha}{z^2 - 2z\cos\alpha + 1}$
15	$\cos \alpha k$	$\dfrac{z(z-\cos\alpha)}{z^2 - 2z\cos\alpha + 1}$
16	$e^{-\alpha k}\sin \beta k$	$\dfrac{ze^{-\alpha}\sin\beta}{z^2 - 2ze^{-\alpha}\cos\beta + e^{-2\alpha}}$
17	$e^{-\alpha k}\cos \beta k$	$\dfrac{z(z-e^{-\alpha}\cos\beta)}{z^2 - 2ze^{-\alpha}\cos\beta + e^{-2\alpha}}$
18	$\dfrac{\alpha^k + (-\alpha^k)}{2\alpha^2}$	$\dfrac{1}{z^2-\alpha^2}$
19	$\dfrac{\alpha^k - \beta^k}{\alpha - \beta}$	$\dfrac{z}{(z-\alpha)(z-\beta)}$

$^a\delta(k) = \begin{cases} 1 & k = 0 \\ 0 & k \neq 0 \end{cases}$ is the Kronecker delta.

Solution of Difference Equations Using the z-Transform

The z-transform can be used to solve linear difference equations with constant coefficients. By taking the z-transform of both sides of such a difference equation, the z-transform of the unknown sequence can be expressed as a rational function of z. The inverse z-transform of this function then provides the solution. Thus the z-transform converts the solution of a linear constant-coefficient difference equation to that of an algebraic equation. The following example illustrates the method.

Example 6. Consider the difference equation

$$f(k+2)+3f(k+1)+2f(k)=0, \qquad k=0,1,2,\ldots \tag{4.94}$$

subject to

$$3f(0)+f(1)=0 \tag{4.95}$$

Taking the z-transform of both sides of Eq. (4.94) yields

$$[z^2F(z)-z^2f(0)-zf(1)]+3[zF(z)-zf(0)]+2F(z)=0 \tag{4.96}$$

or using Eq. (4.95)

$$F(z)=\frac{z^2f(0)}{z^2+3z+2} \tag{4.97}$$

which can be written as

$$\frac{F(z)}{z}=f(0)\left[\frac{-1}{z+1}+\frac{2}{z+2}\right] \tag{4.98}$$

Thus the solution is

$$f(k)=z^{-1}[F(z)]=f(0)[-(-1)^k+2(-2)^k] \tag{4.99}$$

4.7 TRANSFER FUNCTIONS

The Laplace transform and the z-transform are fundamental tools for studying linear time-invariant systems. One of the methods of describing such systems in the input-output form is use of the transfer function, which relates the Laplace or the z-transform of the output to the Laplace or the z-transform of the input. For a linear time-invariant system with r inputs and m outputs the transfer function is an $m\times r$ matrix.

For the purpose of discussion, let $m=r=1$. Then the system is referred to as a *scalar* system. The transfer function of such a system in the continuous case is defined as the ratio of the Laplace transform of the output and the Laplace transform of the input when all the initial conditions are zero. Consider a scalar linear time-invariant system described by the constant-coefficient differential equation

$$\begin{aligned}p_ny^{(n)}(t)+p_{n-1}y^{(n-1)}(t)+\cdots+p_1y^{(1)}(t)+p_0y(t)\\=q_mu^{(m)}(t)+q_{m-1}u^{(m-1)}(t)+\cdots+q_1u^{(1)}(t)+q_0u(t)\end{aligned} \tag{4.100}$$

where $u(t)$ indicates input, $y(t)$ indicates output, and superscript i indicates ith-order derivative with respect to time. The transfer function for this system is

$$H(s)\triangleq\frac{Y(s)}{U(s)}=\frac{q_ms^m+q_{m-1}s^{m-1}+\cdots+q_1s+q_0}{p_ns^n+p_{n-1}s^{n-1}+\cdots+p_1s+p_0} \tag{4.101}$$

Note that $H(s)$ in Eq. (4.101) is a scalar-valued rational function of s. Equation (4.101) can also be written as

$$H(s)=\frac{q_m(s-z_1)(s-z_2)\cdots(s-z_m)}{p_n(s-\lambda_1)(s-\lambda_2)\cdots(s-\lambda_n)} \tag{4.102}$$

where $z_i (i=1,2,\ldots,m)$ are called the *zeros* and λ_j $(j=1,2,\ldots,n)$ are called the *poles* of the system. There may be common factors in the numerator and the denominator of Eq. (4.102). If these are canceled, the remaining factors will then be referred to as the poles and zeros of the transfer function.

Equation (4.101) implies

$$Y(s)=H(s)U(s) \tag{4.103}$$

If the input is a unit impulse, that is, if $u(t) = \delta(t)$, then $U(s) = 1$ and $Y(s) = H(s)$. That is

$$y(t) \triangleq h(t) = L^{-1}[H(s)] \tag{4.104}$$

where $h(t)$ is known as the *impulse response* of the system. Therefore the transfer function of a scalar linear time-invariant continuous system is the Laplace transform of its impulse response.

For multiinput, multioutput (multivariable), linear time-invariant systems, the Laplace transform of the output vector, $Y(s)$, and the Laplace transform of the input vector, $U(s)$, are related through the transfer function as

$$Y(s) = H(s)U(s) \tag{4.105}$$

which is similar to Eq. (4.103). Note that here $H(s)$ is an $m \times r$ matrix where r is the number of inputs and m is the number of outputs.

The development of a transfer function for discrete-time systems is analogous to that for continuous-time systems. The transfer function of a linear time-invariant discrete-time scalar system is defined as the ratio of the z-transform of the output and the z-transform of the input when all the initial conditions are zero. Consider a scalar linear time-invariant system described by the constant-coefficient difference equation

$$\begin{aligned} p_n y(k+n) + p_{n-1} y(k+n-1) + \cdots + p_1 y(k+1) + p_0 y(k) \\ = q_m u(k+m) + q_{m-1} u(k+m-1) + \cdots + q_1 u(k+1) + q_0 u(k) \end{aligned} \tag{4.106}$$

where $u(k)$ and $y(k)$ indicate input and output, respectively, at time k. The transfer function for this system is

$$H(z) \triangleq \frac{Y(z)}{U(z)} = \frac{q_m z^m + q_{m-1} z^{m-1} + \cdots + q_1 z + q_0}{p_n z^n + p_{n-1} z^{n-1} + \cdots + p_1 z + p_0} \tag{4.107}$$

which is a scalar-valued rational function of z. The poles and zeros of the system and of the transfer function are defined similar to the continuous-time case. Equation (4.107) implies

$$Y(z) = H(z)U(z) \tag{4.108}$$

which is the defining equation for the transfer function of linear time-invariant discrete-time systems in both scalar and multivariable cases. It has been shown[5] that

$$H(z) = Z[h(k-1)] \tag{4.109}$$

where $h(k)$ is the pulse response of the scalar system. Equation (4.109) also holds in the case of multivariable discrete systems. In such a case $h(k)$ will be an $m \times r$ matrix where r and m are the number of inputs and outputs, respectively. The ith column of $h(k)$ is then the zero-state response to a pulse applied at the ith input when all the other inputs are zero.

References

1 A. Beurling, "Sur les integrales de Fourier absolument convergentes et leur application à une transformation fonctionnelle," 9ème Congrès des Mathematiciens Scandinaves, Helsingfors, Finland, 1939.

2 S. Bochner and K. Chandrasekharan, *Fourier Transforms*, Princeton University Press, Princeton, NJ, 1949.

3 J. W. Cooley and J. W. Tukey, "An algorithm for machine calculation of complex Fourier Series," *Mathematics of Computation* **19**(90):297–301 (1965).

4 W. R. LaPage, *Complex Variables and the Laplace Transform for Engineers*, McGraw-Hill, New York, 1961.

5 E. I. Jury, *Theory and Application of the z-Transform Method*, Wiley, New York, 1964.

Bibliography

Bellman, R. E., E. Kalaba, and J. A. Lockett, *Numerical Inversion of the Laplace Transform*, American Elsevier, New York, 1966.

Blackman, R. B. and J. W. Tukey, *The Measurement of Power Spectra*, Dover, New York, 1968.

Bracewell, R. N., *The Fourier Transform and its Applications*, 2nd ed., McGraw-Hill, New York, 1978.

Campbell, G. A. and R. M. Foster, *Fourier Integrals for Practical Applications*, D. Van Nostrand, N.Y., 1948.

Carslaw, H. S., *Introduction to the Theory of Fourier's Series and Integrals*, Dover, New York, 1930.

Doetsch, G., *Guide to the Applications of the Laplace and z-Transforms*, Van Nostrand Reinhold, New York, 1971.

Doetsch, G., *Introduction to the Theory and Application of the Laplace Transformation*, Springer-Verlag, New York, 1974.

Erdelyi, A., *Tables of Integal Transforms*, Vol. 1, McGraw-Hill, New York, 1954.

Freeman, H. and O. Lowenschuss, *I.R.E. Trans. Automatic Control (PGAC)*, pp. 28–30, March 1958.

Helm, H. A., *Bell System Technical Journal* **38**(1):177–196 (1956).

Jury, E. I. and C. A. Galtieri, *I.R.E. Trans. Circuit Theory* **CT-9**:371–374 (1961).

Lago, G. V., *Trans. A.I.E.E.* **74**(part II):403–408 (1955).

Lighthill, M. J., *An Introduction to Fourier Analysis and Generalized Functions*, Cambridge University Press, Cambridge, England, 1958.

Nussbaumer, H. J., *Fast Fourier Transforms and Convolution Algorithms*, Springer-Verlag, New York, 1981.

Paley, R. E. and N. Wiener, "Fourier transforms in the complex domain," *American Mathematical Society, Colloquium Publications*, Vol. 19, New York, 1934.

Papoulis, A., *The Fourier Integral and Its Applications*, McGraw-Hill, New York, 1962.

Piessens, R., *J. Comput. Appl. Math.* **1**(2):115–128 (1975).

Piessens, R. and N. D. P. Dang, *J. Comput. Appl. Math.* **2**(3):225–230 (1976).

Talbot, A., *J. Inst. Math. Appl.* **23**:97–120 (1979).

CHAPTER 5
PROBABILITY

JOEL HALPERT

Grumman Aerospace Corp.
Bethpage, New York

5.1 DEFINITIONS AND PROBABILITY THEOREMS[1]

Probability is that part of mathematics that deals with the outcome of experiments involving uncertainties. The laws of probability can be developed from concepts that deal with what is known as set theory. Following are ten basic definitions describing set theory and four definitions relating to experimental outcomes. Finally, the basic probability definition is given along with some probability theorems.

Set. A set is a well-defined collection of objects.

Element. An element (or member) of a set is an object within the set.

Equality of Sets. Two sets are equal if they each have exactly the same elements.

Null Set. A null set (or empty set) is a set that contains no elements. A null set is usually denoted by the symbol Ø.

Universal Set. A universal set is a set that contains all elements. A universal set is usually denoted by the symbol U.

Subset. If every element of a set A is also an element of a set B, then A is called a subset of B.

It should be noted that a universal set depends on the nature of the experiment. For example, the set of numbers representing the outcome of the tossing of a single die is (1, 2, 3, 4, 5, 6), which is the universal set for that procedure. The number (2) is a subset of this universal set.

The relationship between sets can be defined and associated with logical AND, OR, and NOT functions as well as pictorially represented in what are known as Venn diagrams.

Intersection. The intersection of two sets A and B is the set of elements that are common to A and B. This is the logical AND function. The intersection of A and B is written $A \cap B$. See Fig. 5.1.

Disjoint Sets. $A \cap B = \emptyset$. Sets A and B have no elements in common.

Union. The union of two sets A and B is the set of elements that belong to A or to B or to both. This is the logical INCLUSIVE OR function. The union of A and B is written as $A \cup B$. See Fig. 5.2.

Complement. If A is a subset of the universal set U, then the complement of A with respect to U is

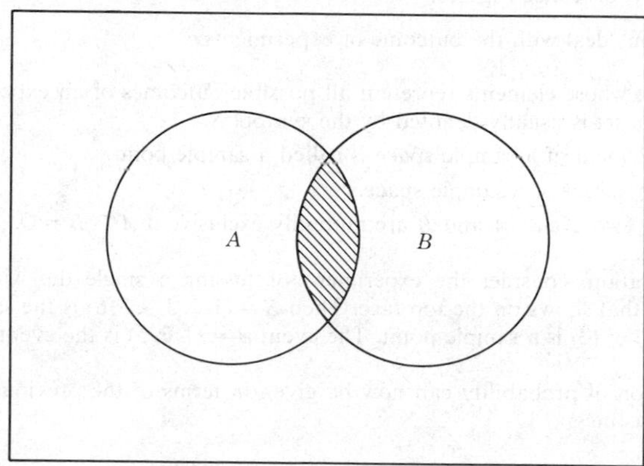

Fig. 5.1 The intersection of two sets, $A \cap B$, shown in the hatched region.

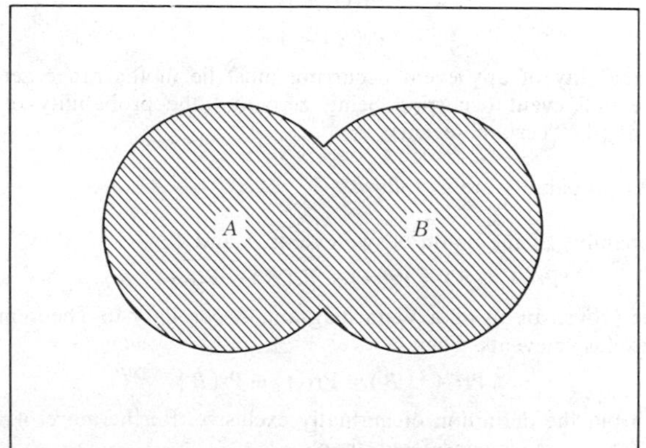

Fig. 5.2 The union of two sets, $A \cup B$, shown in the hatched region.

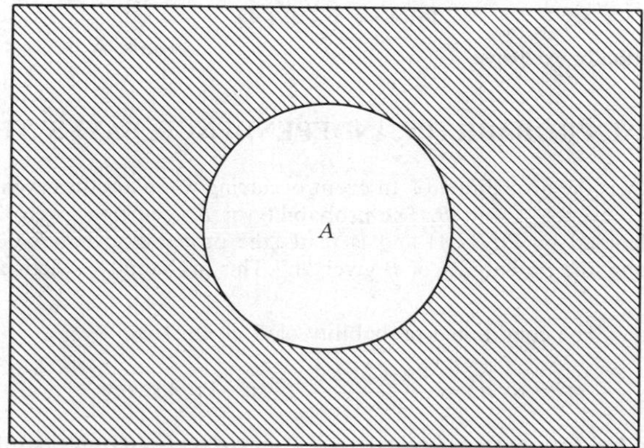

Fig. 5.3 The complement of a set, $\overline{A}$ (or A'), shown in the hatched region.

the set of all elements of U that are not in A. This is the logical NOT function. The complement of A is written as A' or $\bar{A}$. See Fig. 5.3.

The next definitions deal with the outcome of experiments.

Sample Space. A set whose elements represent all possible outcomes of an experiment is called the sample space. This set is usually denoted by the symbol S.

Sample Point. An element of a sample space is called a sample point.

Event. An event is a subset of a sample space.

Mutually Exclusive. Two events A and B are mutually exclusive if $A \cap B = \emptyset$.

By way of illustration, consider the experiment of tossing a single die with the objective of studying the number that shows on the top face. Then $S = (1, 2, 3, 4, 5, 6)$ is the sample space for the experiment. The number (2) is a sample point. The event $A = (1, 3, 5)$ is the event that the number is odd.

The basic definition of probability can now be given in terms of the previous definitions of sets and experimental outcomes:

Probability. The probability of any event A, denoted by $\Pr(A)$, is the sum of the weights of all sample points in A

$$0 \leqslant \Pr(A) \leqslant 1$$

$$\Pr(\emptyset) = 0$$

$$\Pr(S) = 1$$

In words, the probability of any event occurring must lie in the range zero to one, with the probability of the null event occurring being zero and the probability of the sample space occurring being one (the "certain" event).

Some basic laws of probability are as follows:

Theorem 1. The probability of the union of two events A and B:

$$\Pr(A \cup B) = \Pr(A) + \Pr(B) - \Pr(A \cap B) \tag{5.1}$$

Equation (5.1) can be proven by using a Venn diagram. A corollary to Theorem 1 is obtained if A and B are mutually exclusive events. Then

$$\Pr(A \cup B) = \Pr(A) + \Pr(B) \tag{5.2}$$

since $\Pr(A \cap B) = 0$ from the definition of mutually exclusive. Furthermore, if a set of n mutually exclusive events $A_1, A_2, \ldots, A_n$ is considered, then

$$\Pr(A_1 \cup A_2 \cup \cdots \cup A_n) = \Pr(A_1) + \Pr(A_2) + \cdots + \Pr(A_n) \tag{5.3}$$

Theorem 2. The probability of the complementary event A

$$\Pr(\bar{A}) = 1 - \Pr(A) \tag{5.4}$$

since A and $\bar{A}$ are mutually exclusive.

5.2 CONDITIONAL PROBABILITY, INDEPENDENCE, BAYES' THEOREM[1]

Conditional probability is the probability of an event occurring when another event is known to have occurred. Consider two events, A and B. The probability of B occurring given the fact that A has already occurred is denoted by $\Pr(B/A)$ and is read "the probability that B occurs given that A occurs," or more simply "the probability of B given A." This probability is defined as follows:

Conditional Probability. The conditional probability of B given A is

$$\Pr(B/A) = \frac{\Pr(A \cap B)}{\Pr(A)} \quad \text{if} \quad \Pr(A) > 0 \tag{5.5}$$

Equation (5.5) can be rearranged for calculation of any quantity given the other two quantities.

Independent Events. The events A and B are independent if and only if

$$\Pr(A \cap B) = \Pr(A)\Pr(B) \tag{5.6}$$

Note that if A and B are independent events, then Eq. (5.5) becomes

$$\Pr(B/A) = \Pr(B) \tag{5.7}$$

and the conditional probability degenerates to a simple probability.

Bayes' theorem provides a means of calculating the conditional probability of an event that is a subset of a set of events that forms a partition of the sample space.

Theorem 3. Bayes' Theorem. Let $(B_1, B_2, \ldots, B_n)$ be a set of n events forming a partition of the sample space S, with $\Pr(B_i) \neq 0$ for $i = 1, 2, \ldots, n$. Let A be any event of S, with $\Pr(A) \neq 0$. Then, for $k = 1, 2, \ldots, n$:

$$\Pr(B_k/A) = \frac{\Pr(B_K \cap A)}{\Pr(B_1 \cap A) + \Pr(B_2 \cap A) + \cdots + \Pr(B_n \cap A)} \tag{5.8}$$

5.3 RANDOM VARIABLES, PROBABILITY DISTRIBUTIONS, DENSITIES[1,2]

In this section, a number is assigned to every outcome of an experiment. This in turn defines a function whose independent variable is not a number but an element of the sample space. All values and functions are taken to be real.

Random Variable. A random variable is a function, X, whose value x, is a real number determined by each element in the sample space.

Consider the event $X < x$. Its probability, $\Pr(X < x)$, is a function of x. This function, denoted by $F(x)$, is called the distribution function of the random variable X.

Distribution Function. The distribution function of the random variable X is the function

$$F(x) = \Pr(X < x) \tag{5.9}$$

defined for any number x from $-\infty$ to ∞. It is a nondecreasing function of x, that is, $F(x)$ does not decrease as x increases, and has $F(-\infty) = 0$ and $F(+\infty) = 1$.

To define the probability density function (pdf), a distinction is usually made between discrete random variables and continuous random variables. Discrete random variables have distribution functions of the "staircase" type (the interval between discrete values is normally filled with a

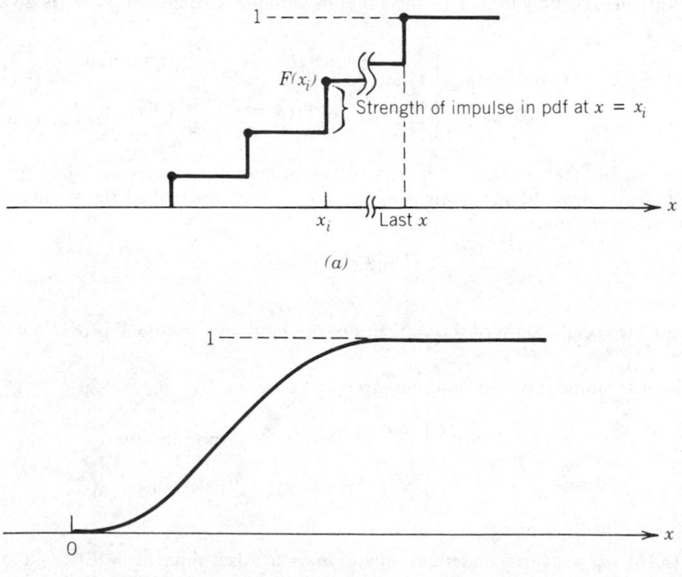

Fig. 5.4 Examples of probability distributions: (a) discrete, (b) continuous.

horizontal line creating the "staircase"). Continuous random variables usually have "smooth," continuous distribution functions. See Fig. 5.4. In either case, the pdf can be rigorously defined as follows:

Probability Density Function (pdf). The density function, $f(x)$, of the random variable X is given by

$$f(x) = \frac{dF(x)}{dx} \tag{5.10}$$

For discrete random variables, the pdf consists of impulse functions at the discrete points. For continuous random variables, the density function is given by Eq. (5.10) at all points where the derivative exists. Furthermore, the total area under the $f(x)$ curve is one, that is

$$\int_{-\infty}^{\infty} f(x)\,dx = 1 \tag{5.11}$$

and the area under the $f(x)$ curve from $x = a$ to $x = b$ gives the probability that the random variable X lies between a and b, that is

$$\int_a^b f(x)\,dx = \Pr(a < X < b) \tag{5.12}$$

For a continuous random variable X, $\Pr(X = \text{any one value}) = 0$, thus the strict inequality in Eq. (5.12).

5.4 EXPECTED VALUE, DISPERSION, MOMENTS[2]

This section deals with several important parameters of random variables. Definitions are given for continuous random variables and discrete random variables.

Expected Value or Mean. The expected value of the random variable X, denoted by $E(X)$, is also the mean value, denoted by μ, and is given by

$$\mu = E(X) = \begin{cases} \int_{-\infty}^{\infty} xf(x)\,dx & \text{(continuous)} \\ \sum_n x_n \Pr(x = x_n) & \text{(discrete)} \end{cases} \tag{5.13}$$

This defines the center of gravity of $f(x)$.

Dispersion or Variance. The variance of the random variable X, denoted by σ^2, is given by

$$\sigma^2 = E\left[(X-\mu)^2\right] = \begin{cases} \int_{-\infty}^{\infty} (x-\mu)^2 f(x)\,dx & \text{(continuous)} \\ \sum_n (x_n-\mu)^2 \Pr(x = x_n) & \text{(discrete)} \end{cases} \tag{5.14}$$

The variance equals the moment of inertia of the probability masses and gives some notion of their concentration near the mean. Its positive square root, σ, is called the standard deviation. It can be further shown that

$$\sigma^2 = E(X^2) - \mu^2 \tag{5.15}$$

A more complete specification of the statistics of the random variable X is possible if its moments are known.

Moments. The kth moment of the random variable X is given by

$$m_k = E(X^k) = \begin{cases} \int_{-\infty}^{\infty} x^k f(x)\,dx & \text{(continuous)} \\ \sum_n x_n^k \Pr(x = x_n) & \text{(discrete)} \end{cases} \tag{5.16}$$

From Eq. (5.16), $m_0 = 1$ (area under curve) and $m_1 = \mu$. Also, since $m_2 = E(X^2)$, the variance can be given as

$$\sigma^2 = m_2 - m_1^2 \tag{5.17}$$

5.5 SOME USEFUL DENSITIES AND DISTRIBUTIONS

Presented in this section are some of the more commonly used discrete and continuous probability densities and distributions.

Uniform[1,2]

A discrete uniformly distributed random variable, X, is one in which the random variable assumes all its values with equal probability. Consider k such values, that is, $x_1, x_2, \ldots, x_k$. The distribution function can then be written as

$$F(x) = \Pr(X = x) = \frac{1}{k} \qquad x = x_1, x_2, \ldots, x_k \tag{5.18}$$

A continuous uniformly distributed random variable has a density function that looks like a rectangular pulse. Consider the range of values from x_1 to x_2. The pdf can then be written as

$$f(x) = \begin{cases} \dfrac{1}{x_2 - x_1} & \text{for } x_1 < x < x_2 \\ 0 & \text{elsewhere} \end{cases} \tag{5.19}$$

Binomial[1]

A binomial experiment consists of a set of repeated trials, each trial having two possible outcomes. These outcomes are usually labeled success or failure. A simple example of such a trial is the tossing of a coin, for which the two possible outcomes are heads and tails. Either outcome can be taken as the "success," the other outcome as the "failure." In general, a binomial experiment is one that possesses the following properties:

1. The experiment consists of n repeated trials.
2. Each trial results in an outcome that may be classified as a success or a failure.
3. The probability of a success, denoted by p, remains constant from trial to trial. The probability of a failure, denoted by q, is therefore $1 - p$.
4. The repeated trials are independent.

For a binomial experiment, one can define a binomial random variable X as the number of successes in n trials of a binomial experiment. The distribution function for this random variable is called a binomial distribution. For x successes in n independent trials, the binomial distribution function is given by

$$F(x) = \Pr(X = x) = \frac{n!}{x!(n-x)!} p^x q^{n-x} \qquad x = 0, 1, 2, \ldots, n \tag{5.20}$$

In Eq. (5.20) the symbol ! means factorial. For the binomial distribution, the mean and variance are given, respectively, as in Eqs. (5.21) and (5.22)

$$\mu = np \tag{5.21}$$

$$\sigma^2 = npq \tag{5.22}$$

Poisson[1,3]

A Poisson experiment is one that yields numerical values of a random variable, X, during a given time interval or in a specified region. A Poisson random variable X is the number of successes in a Poisson experiment. In general, a Poisson experiment is one that possesses the following properties:

1. The average number of successes, μ, occurring in the given time interval or specified region is known.
2. The probability that a single success will occur during a very short time interval or in a small region is proportional to the length of the time interval or the size of the region and does not depend on the number of successes occurring outside this time interval or region.
3. The probability that more than one success will occur in such a short time interval or falling in such a small region is negligible.

The distribution function for a Poisson random variable is called a Poisson distribution and is given by

$$F(x) = \Pr(X = x) = \frac{e^{-\mu}\mu^x}{x!} \qquad x = 0, 1, 2, \ldots \tag{5.23}$$

In Eq. (5.23), μ is the average number of successes occurring in the given time interval or region. The mean and the variance both have the value μ.

Normal[1,2]

A continuous random variable is normally distributed if its pdf is a Gaussian curve, named in honor of the scientist Gauss who derived its equation from a study of errors in repeated measurements of the same quantity. The Gaussian curve is often called the bell-shaped curve because its trace most closely resembles that of a bell. The curve describes so many sets of data that occur in nature, industry, and research that it is easily the most important distribution in statistics. Some areas of statistics that make extensive use of the normal distribution are sampling theory (where the chi-square, F, and t-distributions are developed from the statistics of samples from normally distributed populations), estimation theory, and hypothesis testing.

The parameters of the Gaussian curve are the mean, μ, and the standard deviation, σ (or the variance, σ^2). The curve is shown in Fig. 5.5a. It has the following properties:

1. The maximum occurs at the mean, where $x = \mu$.
2. The curve is symmetric about a vertical axis through the mean.

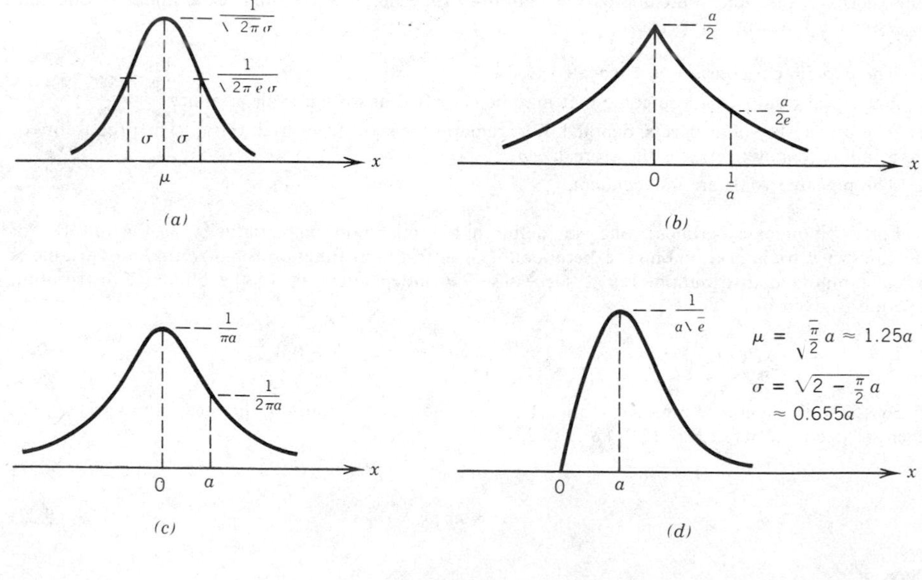

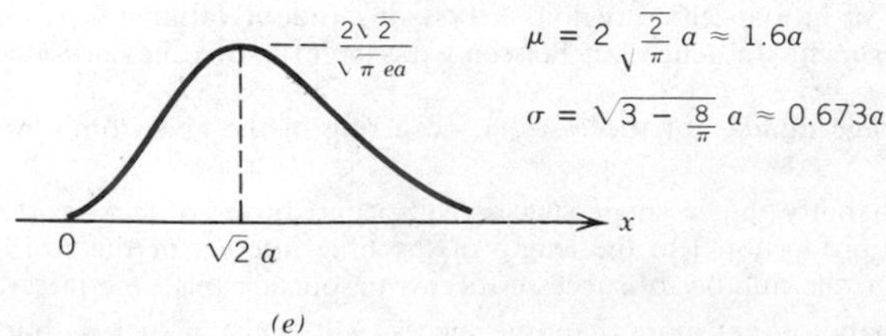

Fig. 5.5 Some probability densities: (a) normal; (b) Laplace; (c) Cauchy; (d) Rayleigh; (e) Maxwell.

3. The curve approaches the horizontal axis asymptotically as we proceed away from the mean in either direction.
4. Since it is a pdf, the area under the curve equals 1.

A random variable, X, described by the Gaussian curve is called a normal random variable. A normal random variable with mean μ and standard deviation σ has a pdf given by

$$f(x) = \frac{1}{\sqrt{2\pi}\,\sigma} \exp\left[-\frac{1}{2}\left(\frac{x-\mu}{\sigma}\right)^2 \right] \tag{5.24}$$

The corresponding distribution function is given by

$$F(x) = \Pr(X < x) = \frac{1}{2} + \operatorname{erf}\left(\frac{x-\mu}{\sigma}\right) \tag{5.25}$$

In Eq. (5.25), erf is the "error function" and is defined by

$$\operatorname{erf} z = \frac{1}{\sqrt{2\pi}} \int_0^z e^{-y^2/2}\, dy \tag{5.26}$$

Two properties of the error function are given in Eqs. (5.27) and (5.28)

$$\operatorname{erf}(-z) = -\operatorname{erf} z \tag{5.27}$$

and

$$\operatorname{erf}(\infty) = 0.5 \tag{5.28}$$

It would be very difficult to set up tables of values for the distribution function in Eq. (5.25) since this function is parametric in μ and σ. Instead, a new normal random variable Z is defined as

$$Z = \frac{X-\mu}{\sigma} \tag{5.29}$$

It can be shown that Z has a mean of zero and a variance of one. The random variable Z is called a standard normal random variable. A single table can now be generated for the standard normal distribution, and the corresponding values for normal random variables with parametric values other than $\mu = 0$ and $\sigma = 1$ can be found from Eq. (5.29). Table 5.1 shows the area under the standard normal curve, corresponding to $\Pr(Z < z)$, for values of z from -3.49 to $+3.49$. Equation (5.29) is used to convert the given normal random variable to the standard before using the table. The table is an evaluation of Eq. (5.25) with $\mu = 0$ and $\sigma = 1$ and can be generated on most scientific high-speed computers which contain an error function subroutine in their math library. This is precisely how Table 5.1 was generated, with each value carried to four decimal places. As an example of how to use the table, consider a normal distribution with a mean of 50 and a standard deviation of 4. To find the probability that $x < 45$, Eq. (5.29) is used to convert to the standard normal random variable, giving $z = -1.25$. Hence to find $\Pr(X < 45)$, Table 5.1 is used to find $\Pr(Z < -1.25)$. This value is 0.1056.

Gamma[2,3]

The gamma density is given by

$$f(x) = \frac{c^{b+1}}{\Gamma(b+1)} x^b e^{-cx} \qquad x \geqslant 0, \qquad b > 0, \qquad c > 0 \tag{5.30}$$

where

$$\Gamma(b+1) = \int_0^\infty y^b e^{-y}\, dy \tag{5.31}$$

is called the gamma function. An example of a random variable described by the gamma density is the waiting time required to observe the $b + 1$ occurrence of an event of a specified type when events of this type are occurring randomly at a mean rate c per unit time. The chi-square density and F-density both contain gamma functions in their analytic descriptions. A random variable is said to have a gamma distribution if its density is given by Eq. (5.30). The mean and variance are given by

$$\mu = \frac{b+1}{c} \tag{5.32}$$

$$\sigma^2 = \frac{b+1}{c^2} \tag{5.33}$$

TABLE 5.1. AREAS UNDER THE NORMAL CURVE

Z	0.00	0.01	0.02	0.03	0.04	0.05	0.06	0.07	0.08	0.09
-3.4	.0003	.0003	.0003	.0003	.0003	.0003	.0003	.0003	.0003	.0002
-3.3	.0005	.0005	.0005	.0004	.0004	.0004	.0004	.0004	.0004	.0003
-3.2	.0007	.0007	.0006	.0006	.0006	.0006	.0006	.0005	.0005	.0005
-3.1	.0010	.0009	.0009	.0009	.0008	.0008	.0008	.0008	.0007	.0007
-3.0	.0013	.0013	.0013	.0012	.0012	.0011	.0011	.0011	.0010	.0010
-2.9	.0019	.0018	.0018	.0017	.0016	.0016	.0015	.0015	.0014	.0014
-2.8	.0026	.0025	.0024	.0023	.0023	.0022	.0021	.0021	.0020	.0019
-2.7	.0035	.0034	.0033	.0032	.0031	.0030	.0029	.0028	.0027	.0026
-2.6	.0047	.0045	.0044	.0043	.0041	.0040	.0039	.0038	.0037	.0036
-2.5	.0062	.0060	.0059	.0057	.0055	.0054	.0052	.0051	.0049	.0048
-2.4	.0082	.0080	.0078	.0075	.0073	.0071	.0069	.0068	.0066	.0064
-2.3	.0107	.0104	.0102	.0099	.0096	.0094	.0091	.0089	.0087	.0084
-2.2	.0139	.0136	.0132	.0129	.0125	.0122	.0119	.0116	.0113	.0110
-2.1	.0179	.0174	.0170	.0166	.0162	.0158	.0154	.0150	.0146	.0143
-2.0	.0228	.0222	.0217	.0212	.0207	.0202	.0197	.0192	.0188	.0183
-1.9	.0287	.0281	.0274	.0268	.0262	.0256	.0250	.0244	.0239	.0233
-1.8	.0359	.0351	.0344	.0336	.0329	.0322	.0314	.0307	.0301	.0294
-1.7	.0446	.0436	.0427	.0418	.0409	.0401	.0392	.0384	.0375	.0367
-1.6	.0548	.0537	.0526	.0516	.0505	.0495	.0485	.0475	.0465	.0455
-1.5	.0668	.0655	.0643	.0630	.0618	.0606	.0594	.0582	.0571	.0559
-1.4	.0808	.0793	.0778	.0764	.0749	.0735	.0721	.0708	.0694	.0681
-1.3	.0968	.0951	.0934	.0918	.0901	.0885	.0869	.0853	.0838	.0823
-1.2	.1151	.1131	.1112	.1093	.1075	.1056	.1038	.1020	.1003	.0985
-1.1	.1357	.1335	.1314	.1292	.1271	.1251	.1230	.1210	.1190	.1170
-1.0	.1587	.1562	.1539	.1515	.1492	.1469	.1446	.1423	.1401	.1379
-.9	.1841	.1814	.1788	.1762	.1736	.1711	.1685	.1660	.1635	.1611
-.8	.2119	.2090	.2061	.2033	.2005	.1977	.1949	.1922	.1894	.1867
-.7	.2420	.2389	.2358	.2327	.2296	.2266	.2236	.2206	.2177	.2148
-.6	.2743	.2709	.2676	.2643	.2611	.2578	.2546	.2514	.2483	.2451
-.5	.3085	.3050	.3015	.2981	.2946	.2912	.2877	.2843	.2810	.2776
-.4	.3446	.3409	.3372	.3336	.3300	.3264	.3228	.3192	.3156	.3121
-.3	.3821	.3783	.3745	.3707	.3669	.3632	.3594	.3557	.3520	.3483
-.2	.4207	.4168	.4129	.4090	.4052	.4013	.3974	.3936	.3897	.3859
-.1	.4602	.4562	.4522	.4483	.4443	.4404	.4364	.4325	.4286	.4247
-.0	.5000	.4960	.4920	.4880	.4840	.4801	.4761	.4721	.4681	.4641
0.0	.5000	.5040	.5080	.5120	.5160	.5199	.5239	.5279	.5319	.5359
.1	.5398	.5438	.5478	.5517	.5557	.5596	.5636	.5675	.5714	.5753
.2	.5793	.5832	.5871	.5910	.5948	.5987	.6026	.6064	.6103	.6141
.3	.6179	.6217	.6255	.6293	.6331	.6368	.6406	.6443	.6480	.6517
.4	.6554	.6591	.6628	.6664	.6700	.6736	.6772	.6808	.6844	.6879
.5	.6915	.6950	.6985	.7019	.7054	.7088	.7123	.7157	.7190	.7224
.6	.7257	.7291	.7324	.7357	.7389	.7422	.7454	.7486	.7517	.7549
.7	.7580	.7611	.7642	.7673	.7704	.7734	.7764	.7794	.7823	.7852
.8	.7881	.7910	.7939	.7967	.7995	.8023	.8051	.8078	.8106	.8133
.9	.8159	.8186	.8212	.8238	.8264	.8289	.8315	.8340	.8365	.8389
1.0	.8413	.8438	.8461	.8485	.8508	.8531	.8554	.8577	.8599	.8621
1.1	.8643	.8665	.8686	.8708	.8729	.8749	.8770	.8790	.8810	.8830
1.2	.8849	.8869	.8888	.8907	.8925	.8944	.8962	.8980	.8997	.9015
1.3	.9032	.9049	.9066	.9082	.9099	.9115	.9131	.9147	.9162	.9177
1.4	.9192	.9207	.9222	.9236	.9251	.9265	.9279	.9292	.9306	.9319
1.5	.9332	.9345	.9357	.9370	.9382	.9394	.9406	.9418	.9429	.9441
1.6	.9452	.9463	.9474	.9484	.9495	.9505	.9515	.9525	.9535	.9545
1.7	.9554	.9564	.9573	.9582	.9591	.9599	.9608	.9616	.9625	.9633
1.8	.9641	.9649	.9656	.9664	.9671	.9678	.9686	.9693	.9699	.9706
1.9	.9713	.9719	.9726	.9732	.9738	.9744	.9750	.9756	.9761	.9767
2.0	.9772	.9778	.9783	.9788	.9793	.9798	.9803	.9808	.9812	.9817
2.1	.9821	.9826	.9830	.9834	.9838	.9842	.9846	.9850	.9854	.9857
2.2	.9861	.9864	.9868	.9871	.9875	.9878	.9881	.9884	.9887	.9890
2.3	.9893	.9896	.9898	.9901	.9904	.9906	.9909	.9911	.9913	.9916
2.4	.9918	.9920	.9922	.9925	.9927	.9929	.9931	.9932	.9934	.9936
2.5	.9938	.9940	.9941	.9943	.9945	.9946	.9948	.9949	.9951	.9952
2.6	.9953	.9955	.9956	.9957	.9959	.9960	.9961	.9962	.9963	.9964
2.7	.9965	.9966	.9967	.9968	.9969	.9970	.9971	.9972	.9973	.9974
2.8	.9974	.9975	.9976	.9977	.9977	.9978	.9979	.9979	.9980	.9981
2.9	.9981	.9982	.9982	.9983	.9984	.9984	.9985	.9985	.9986	.9986
3.0	.9987	.9987	.9987	.9988	.9988	.9989	.9989	.9989	.9990	.9990
3.1	.9990	.9991	.9991	.9991	.9992	.9992	.9992	.9992	.9993	.9993
3.2	.9993	.9993	.9994	.9994	.9994	.9994	.9994	.9995	.9995	.9995
3.3	.9995	.9995	.9995	.9996	.9996	.9996	.9996	.9996	.9996	.9997
3.4	.9997	.9997	.9997	.9997	.9997	.9997	.9997	.9997	.9997	.9998

TABLE 5.1. (*Continued*)

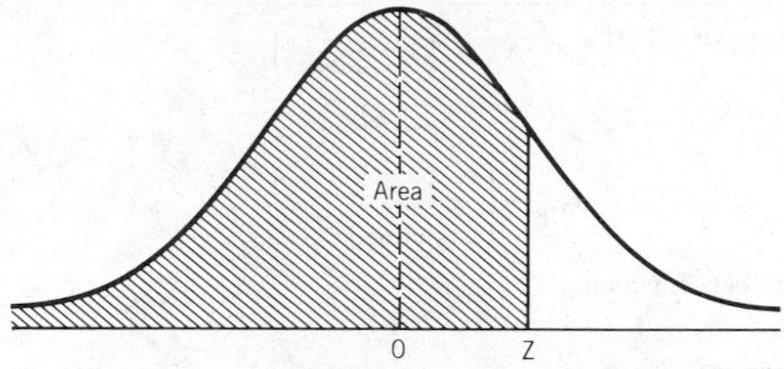

Area under normal curve in the hatched region.

Beta[2,3]

The beta density is given by

$$f(x) = Ax^b(1-x)^c \qquad 0 \leqslant x \leqslant 1, \qquad b > 0, \qquad c > 0 \tag{5.34}$$

where

$$A = \frac{\Gamma(b+c+2)}{\Gamma(b+1)\Gamma(c+1)} \tag{5.35}$$

The beta function can be expressed in terms of the gamma function and is given by

$$B(b,c) = \frac{\Gamma(b)\Gamma(c)}{\Gamma(b+c)} \tag{5.36}$$

Thus the constant A in Eq. (5.34) can be expressed in terms of the beta function as

$$A = \frac{1}{B(b+1, c+1)} \tag{5.37}$$

A random variable is said to have a beta distribution if its density is given by Eq. (5.34). The mean and variance are given by

$$\mu = \frac{b+1}{b+c+2} \tag{5.38}$$

$$\sigma^2 = \frac{(b+1)(c+1)}{(b+c+2)^2(b+c+3)} \tag{5.39}$$

t-Distribution[3,4]

The t-distribution deals with the statistics of a sample drawn from a normal population. Specifically, suppose a random sample of size n is drawn from a normal population. The population statistics have a mean of μ but an unknown variance. The sample statistics have a mean of $\bar{x}$ and a variance of s^2, that is,

$$\bar{x} = \frac{1}{n}\sum_{i=1}^{n} x_i \tag{5.40}$$

$$s^2 = \frac{1}{n}\sum_{i=1}^{n} (x_i - \bar{x})^2 \tag{5.41}$$

Then a value t can be defined such that

$$t = \frac{\bar{x} - \mu}{s/\sqrt{n-1}} = \sqrt{n-1}\left(\frac{\bar{x} - \mu}{s}\right) \tag{5.42}$$

t is a value of a new random variable T having the t-distribution with $n-1$ degrees of freedom. Notice that t is defined as a function of the known statistics of the sample and the population from which it is drawn. This distribution is widely used in sampling theory as a test on the mean. Its density function, mean, and variance are given by

$$f(t) = \frac{1}{\sqrt{n-1}\,B\left(\frac{1}{2}, \frac{n-1}{2}\right)} \frac{1}{\left(1 + \frac{t^2}{n-1}\right)^{n/2}} \qquad -\infty < t < \infty \tag{5.43}$$

$$\mu_{f(t)} = 0 \tag{5.44}$$

$$\sigma^2_{f(t)} = \begin{cases} \dfrac{n}{n-2} & n > 2 \\ \text{undefined} & n \leqslant 2 \end{cases} \tag{5.45}$$

In Eq. (5.43), B is the beta function.

Chi-Square Distribution[3]

The chi-square distribution deals with the variance statistic of a sample drawn from a normal population. Specifically, suppose a random sample of size n is drawn from a normal population. The population has a known variance of σ^2. Then a value χ^2 can be defined such that

$$\chi^2 = \sum_{i=1}^{n} x_i^2 \tag{5.46}$$

χ^2 is a value of a new random variable having the chi-square distribution with n degrees of freedom. This distribution is widely used in sampling theory as a test on the variance. Its density function, mean, and variance are given by

$$f(y) = \frac{1}{2^{n/2}\sigma^n\Gamma\left(\frac{n}{2}\right)} y^{n/2-1}e^{-y/2\sigma^2} \qquad y > 0 \tag{5.47}$$

$$\mu_{f(y)} = n\sigma^2 \tag{5.48}$$

$$\sigma^2_{f(y)} = 2n\sigma^4 \tag{5.49}$$

F-Distribution[4,5]

The F-distribution essentially deals with the ratio of two independent chi-square distributions, each divided by their degrees of freedom. Let y_1 be a value of the chi-square variable Y_1 with n degrees of freedom and y_2 be a value of the chi-square random variable Y_2 with m degrees of freedom, Y_1 and Y_2 independent. Then a value f can be defined such that

$$f = \frac{y_1/n}{y_2/m} \tag{5.50}$$

f is a value of a new random variable F having the F-distribution with n and m degrees of freedom. There are two parameters, the two different degrees of freedom. While the f statistic can be defined as the ratio given in Eq. (5.50) or its inverse, it is important that the chi-square variable in the numerator be specified so that the f curves can be properly defined. This implies that the order in which the degrees of freedom are stated is important; the values of the F-distribution are not the same when the degrees of freedom are interchanged, that is, in general, $F(n, m) \neq F(m, n)$. The F-distribution is applied primarily in the analysis of variance. Its density function, mean, and variance are given by

$$g(f) = \frac{\Gamma\left(\frac{n+m}{2}\right)\left(\frac{n}{m}\right)^{n/2}}{\Gamma\left(\frac{n}{2}\right)\Gamma\left(\frac{m}{2}\right)} \frac{f^{n/2-1}}{\left(1 + \frac{nf}{m}\right)^{(n+m)/2}} \qquad 0 < f < \infty \tag{5.51}$$

$$\mu_{g(f)} = \frac{m}{m-2} \qquad m > 2 \tag{5.52}$$

$$\sigma^2_{g(f)} = \frac{2m^2(n+m-2)}{n(m-2)^2(m-4)} \qquad m > 4 \tag{5.53}$$

$F(n, m)$ has no mean for $m \leqslant 2$ and no variance for $m \leqslant 4$.

Laplace[2]

The Laplace density is given by (see Fig. 5.5*b*)

$$f(x) = \frac{a}{2} e^{-a|x|} \tag{5.54}$$

Cauchy[2]

The Cauchy density is given by (see Fig. 5.5*c*)

$$f(x) = \frac{a/\pi}{a^2 + x^2} \tag{5.55}$$

Rayleigh[2]

The Rayleigh density is given by (see Fig. 5.5*d*)

$$f(x) = \frac{x}{a^2} e^{-x^2/2a^2} \qquad x \geqslant 0 \tag{5.56}$$

Maxwell[2]

The Maxwell density is given by (see Fig. 5.5*e*)

$$f(x) = \frac{\sqrt{2}}{(a^3)\sqrt{\pi}} x^2 e^{-x^2/2a^2} \qquad x \geqslant 0 \tag{5.57}$$

5.6 MULTIVARIATE DISTRIBUTIONS[2,5]

Multivariate distributions deal with distributions of more than one random variable. This implies that the sample space is now greater than one dimension: The one-dimensional sample space is described by one random variable. Thus, two distinct random variables can be described by a two-dimensional sample space. Consider two random variables, X and Y, with individual means μ_1 and μ_2, respectively. A number called the covariance of X and Y can be defined in a manner similar to that in Eq. (5.14) as

$$\sigma_{12} = E[(X - \mu_1)(Y - \mu_2)] = E(XY) - \mu_1 \mu_2 \tag{5.58}$$

If X and Y are independent random variables, then their covariance is zero. Furthermore, if random variables X and Y have positive standard deviations σ_1 and σ_2, respectively, then a number called the correlation coefficient of X and Y can be defined as

$$\rho_{12} = \frac{\sigma_{12}}{\sigma_1 \sigma_2} \tag{5.59}$$

Again if X and Y are independent random variables, then their correlation coefficient is zero.

A joint probability density function (pdf) and a joint probability distribution function can be defined for two random variables, as was done for one random variable in Eqs. (5.9) and (5.10). Thus the joint distribution function for (continuous) random variables is given by

$$F(x, y) = \Pr(X \leqslant x, Y \leqslant y) \tag{5.60}$$

and the joint density function is given by

$$f(x, y) = \frac{\partial^2 F(x, y)}{\partial x \partial y} \tag{5.61}$$

A function called the marginal distribution can be defined as the one-dimensional distribution of either random variable alone. A similar statement can be made for the marginal density. For example, the marginal distribution for X is given by

$$F_x(x) = F(x, \infty) \tag{5.62}$$

and the marginal density for X is given by

$$f_x(x) = \int_{-\infty}^{\infty} f(x, y)\, dy \tag{5.63}$$

Bivariate Normal Distribution[5]

Consider random variables X and Y with means μ_1 and μ_2 and standard deviations σ_1 and σ_2, respectively. These random variables are said to have a bivariate normal distribution if they have a joint pdf given by

$$f(x, y) = \frac{1}{2\pi\sigma_1\sigma_2\sqrt{1-\rho^2}}$$

$$\times \exp\left\{ -\frac{1}{2(1-\rho^2)}\left[\left(\frac{x-\mu_1}{\sigma_1}\right)^2 + \left(\frac{y-\mu_2}{\sigma_2}\right)^2 - 2\rho\left(\frac{x-\mu_1}{\sigma_1}\right)\left(\frac{y-\mu_2}{\sigma_2}\right)\right]\right\} \tag{5.64}$$

In Eq. (5.64), ρ is the correlation coefficient. The joint moments of X and Y with covariance σ_{12} is given by†

$$m_{kn} = E(X^kY^n) = kn\int_0^{\sigma_{12}} E(X^{k-1}Y^{n-1})\,d\sigma_{12} + E(X^k)E(Y^n) \tag{5.65}$$

Two Functions of Two Random Variables[2]

Given two functions $g(x, y)$ and $h(x, y)$ of the real variables x and y, and two random variables X and Y, two other random variables are formed as $V = g(X, Y)$ and $W = h(X, Y)$. These random variables have a joint distribution $F(v, w)$ and a joint density $f(v, w)$. With v and w as two real numbers, R denotes the region of the xy plane such that $g(x, y) \leqslant v$ and $h(x, y) \leqslant w$. Then

$$F(v, w) = \int\int_R f(x, y)\,dx\,dy \tag{5.66}$$

is the joint distribution function for V and W. To determine the joint density function, the equations $v = g(x, y)$ and $w = h(x, y)$ must be solved for x and y in terms of v and w. Assume that there are n pairs (x, y) that are all real solutions of these equations. By noting that the Jacobian of the variable transformation is given by

$$J(x, y) = \begin{vmatrix} \dfrac{\partial g(x, y)}{\partial x} & \dfrac{\partial g(x, y)}{\partial y} \\ \dfrac{\partial h(x, y)}{\partial x} & \dfrac{\partial h(x, y)}{\partial y} \end{vmatrix} \tag{5.67}$$

it can be shown that

$$f(v, w) = \frac{f(x_1, y_1)}{J(x_1, y_1)} + \cdots + \frac{f(x_n, y_n)}{J(x_n, y_n)} + \cdots \tag{5.68}$$

If for certain values of (v, w) there are no real solutions, then $f(v, w) = 0$.

5.7 THE CENTRAL LIMIT THEOREM[2,5]

A very general statement of the central limit theorem is that given a set of independent random variables of any type distribution, the distribution of their sum asymptotically approaches a normal distribution as the number of variables increases. This theorem is useful not only in the study of probability but in other areas; for example, the output of a linear system tends to be normally distributed even though the random inputs may not be. It can also be shown to represent the convolution of a large number of positive functions, and can therefore be applied to linear systems in cascade.

Consider n independent continuous random variables $X_1, X_2, \ldots, X_n$ such that each has mean μ_i and variance σ_i^2 (in general n distinct values for each parameter). Define a new random variable equal to the sum, that is, $X = X_1 + X_2 + \cdots + X_n$. Then the mean of X is the sum of the component means and the variance of X is the sum of the component variances. It can be shown that the density function for X is given by

$$f(x) = f_i(x)*f_2(x)*\cdots*f_n(x) \tag{5.69}$$

where the * denotes the convolution. Thus the density of X equals the convolution of the densities of $X_1, X_2, \ldots, X_n$. The central limit theorem states that under certain general conditions $f(x)$ approaches a normal curve as n increases.

†Equation (7-121) in Ref. 2.

A special case of the central limit theorem deals with a sample of size n drawn from any population having a known mean μ and a known variance σ^2, both of which are finite. Thus consider that $X_1, X_2, \ldots, X_n$ denotes the items of a random sample of size n drawn from any distribution having a mean μ and variance σ^2. Define a random variable called the sample mean, $\bar{X} = (X_1 + X_2 + \cdots + X_n)/n$. A new random variable, Y, can now be defined as a function of $\bar{X}$ and the population parameters μ and σ, and is given by

$$Y = \frac{\bar{X} - \mu}{\sigma/\sqrt{n}} \tag{5.70}$$

The central limit theorem states that Y has a limiting normal distribution with a mean of zero and a variance of one. Thus the distribution of Y is approximately normal with a mean of zero and a variance of one. It is then possible to use this approximate normal distribution to compute approximate probabilities concerning $\bar{X}$, to find an approximate confidence interval for μ, and to test certain statistical hypotheses without even knowing the exact pdf of $\bar{X}$ in every case.

5.8 MARKOV PROCESS[2]

This section defines a Markov process and provides an example of one which is of first order.

Before a Markov process can be defined, it is necessary to define what is meant by a stochastic process. We are given an experiment specified by its outcomes ζ forming the sample space S, by certain subsets of S called events, and by the probabilities of these events. To every outcome ζ is now assigned, according to a certain rule, a time function $x(t, \zeta)$ real or complex. A family of functions has now been created, one for each ζ. This family is called a stochastic process. It can be viewed as a function of the two variables t and ζ. The domain of ζ is the set S, and the domain of t is a set of real numbers assumed here to be the entire time axis. The notation $x(t)$ is used to represent a stochastic process, omitting the dependence on ζ. It should be noted, however, that $x(t)$ represents four different things:

1. A family of time functions (t and ζ variables).
2. A single time function (t variable, ζ fixed). This signifies a specific outcome, ζ_i.
3. A random variable (t fixed, ζ variable).
4. A single number (t fixed, ζ fixed).

A stochastic process $x(t)$ is called Markov if for every n and $t_1 < t_2 \cdots < t_n$ we have

$$\Pr[x(t_n) \leqslant x_n / x(t_{n-1}), \ldots, x(t_1)] = \Pr[x(t_n) \leqslant x_n / x(t_{n-1})] \tag{5.71}$$

Equation (5.71) is equivalent to

$$\Pr[x(t_n) \leqslant x_n / x(t) \quad \text{for all} \quad t \leqslant t_{n-1}] = \Pr[x(t_n) \leqslant x_n / x(t_{n-1})] \tag{5.72}$$

The following are some properties of a Markov process:

1. A process $x(t)$ is Markov if the past has no influence on the statistics of the future under the condition that the present is known.
2. A Markov process is Markov in reverse, that is, if $t_1 < t_2$, then
$$\Pr[x(t_1) \leqslant x_1 / x(t) \quad \text{for all} \quad t \geqslant t_2] = \Pr[x(t_1) \leqslant x_1 / x(t_2)] \tag{5.73}$$
3. A process $x(t)$ is Markov if, with $t_1 < t_2$, $x(t_2) - x(t_1)$ is independent of $x(t)$ for every $t \leqslant t_1$.
4. If $x(t)$ is a normal Markov process with zero mean, then its autocorrelation satisfies
$$R(t_3, t_2) R(t_2, t_1) = R(t_3, t_1) R(t_2, t_2) \tag{5.74}$$
for every $t_3 > t_2 > t_1$.
5. A Markov process $x(t)$ can be associated with a first order differential equation
$$\frac{dx(t)}{dt} - h[x(t), t] = g(t) \tag{5.75}$$
where $g(t)$ is such that the random variables $g(t_1), \ldots, g(t_n)$ are independent for any n, $t_1, \ldots, t_n$. With
$$g(t) = \frac{dw(t)}{dt} \tag{5.76}$$
the process $w(t)$ has independent increments. A process $w(t)$ has independent increments if

$w(t_i) - w(t_{i+1})$, $i = \ldots, -1, 0, 1, \ldots$ is a sequence of uncorrelated random variables, where the intervals (t_i, t_{i+1}) are nonoverlapping but otherwise arbitrary.

An example of a first order Markov process is the velocity $v(t)$ of a free particle in Brownian motion. $v(t)$ satisfies the equation of motion

$$\frac{dv(t)}{dt} + \beta v(t) = \frac{dw(t)}{dt} \tag{5.77}$$

where $w(t)$ is a process with independent increments such that $E[dw(t)] = 0$ and $E[|dw(t)|^2] = \alpha\, dt$. The solution of Eq. (5.77) is a Markov process $v(t)$ with statistics $\mu(v, t) = -\beta v$ and $\sigma^2(v, t) = \alpha$.

6. If $x(t)$ is continuous, a conditional density can be defined as

$$p(x, t; x_0, t_0) = f[x/x(t_0) = x_0] \tag{5.78}$$

assuming $x(t_0) = x_0$. Then

$$p(x, t; x_0, t_0) = \int_{-\infty}^{\infty} p(x, t; x_1, t_1) p(x_1, t_1; x_0, t_0)\, dx_1 \tag{5.79}$$

Equation (5.79) is the Chapman–Kolmogorov equation for Markov processes. If p depends only on $t - t_0$, the process is called homogeneous. The conditional density $p(x, t; x_0, t_0)$, $t > t_0$, also satisfies the following equations:

$$\frac{\partial p}{\partial t_0} + \mu(x_0, t_0)\frac{\partial p}{\partial x_0} + \tfrac{1}{2}\sigma^2(x_0, t_0)\frac{\partial^2 p}{\partial x_0^2} = 0 \tag{5.80}$$

$$\frac{\partial p}{\partial t} + \frac{\partial}{\partial x}[\mu(x, t)p] - \tfrac{1}{2}\frac{\partial^2}{\partial x^2}[\sigma^2(x, t)p] = 0 \tag{5.81}$$

These are the backward and forward diffusion equations derived by Kolmogorov. Equation (5.81) is called the Fokker–Planck equation. For the problem of the velocity of a free particle in Brownian motion, using the resultant values of the statistics of $v(t)$ in Eq. (5.81), the conditional density p satisfies the equation

$$\frac{\partial p}{\partial t} = \beta\frac{\partial(vp)}{\partial v} + \frac{\alpha}{2}\frac{\partial^2 p}{\partial v^2} \tag{5.82}$$

References

1 R. E. Walpole, *Introduction to Statistics*, 2nd ed., Macmillan, New York, 1974.
2 A. Papoulis, *Probability, Random Variables, and Stochastic Processes*, McGraw-Hill, New York, 1965.
3 M. Fisz, *Probability and Mathematical Statistics*, S. Chand, New Delhi, 1961.
4 F. L. Wolf, *Elements of Probability and Statistics*, McGraw-Hill, New York, 1962.
5 R. V. Hogg and A. T. Craig, *Introduction to Mathematical Statistics*, Macmillan, New York, 1959.

Bibliography

Hastings, N. A. and J. B. Peacock, *Statistical Distributions*, Halsted Press, Wiley, New York, 1975.

CHAPTER 6
BOOLEAN ALGEBRA

FRANK T. BOESCH
CHARLES SUFFEL

Stevens Institute of Technology
Hoboken, New Jersey

6.1 HUNTINGTON POSTULATES FOR A BOOLEAN ALGEBRA[1]

Boolean, or simply switching, algebra constitutes the mathematical base for the analysis and design of switching circuits. There are several equivalent ways of defining a Boolean algebra. Here we give the Huntington postulates, a set of axioms for a Boolean algebra that are not only consistent (i.e., not self-contradictory) but also independent (i.e., no axiom of the set can be deduced from the others).

The set B, together with two binary operations, denoted $+$ and $\cdot$, is called a Boolean algebra if

1. **Closure.** For each pair $a, b \in B$, $a + b \in B$, and $a \cdot b \in B$.
2. **Identities.** There exist distinct elements 0 and 1 such that for all $b \in B$

$$b + 0 = b \quad \text{and} \quad b \cdot 1 = b$$

3. **Commutativity.** For each pair $a, b \in B$

$$a + b = b + a \quad \text{and} \quad a \cdot b = b \cdot a$$

4. **Distributivity.** For each triple $a, b, c \in B$

$$a + b \cdot c = (a + b) \cdot (a + c) \quad \text{and} \quad a \cdot (b + c) = a \cdot b + a \cdot c$$

5. **Complements.** For each $b \in B$ there is an element $\bar{b}$, called a complement of b, such that

$$b + \bar{b} = 1 \quad \text{and} \quad b \cdot \bar{b} = 0$$

To illustrate, we give two examples, the first being the simplest Boolean algebra possible.

Example 1. The Binary Boolean Algebra. Let $B = \{0, 1\}$ and define $+$ and $\cdot$ as shown in Fig. 6.1. It is readily verifiable that this satisfies the Huntington postulates.[1]

Example 2. The Power Set.[1] Let U be a set and $P(U)$ be the collection of all subsets of U. Define

$$X + Y = X \cup Y \quad \text{and} \quad X \cdot Y = X \cap Y$$

for each pair $X, Y \in P(U)$. Here 0 is the empty set and 1 is the universal set U. The well known set-theoretic properties of union and intersection generalize to arbitrary Boolean algebras. Some of the important ones follow.

+	0	1
0	0	1
1	1	1

·	0	1
0	0	0
1	0	1

Fig. 6.1 A two-element Boolean algebra.

6.2 PROPERTIES OF A BOOLEAN ALGEBRA

Let B be a Boolean algebra. Then the following properties are valid.[1]

1. **Uniqueness of identities.** The elements 0 and 1 are unique.
2. **Universal bounds.** The identities 0 and 1 are universal bounds. For each $b \in B$
$$b \cdot 0 = 0 \quad \text{and} \quad b + 1 = 1$$
3. **Idempotency.** For all $b \in B$
$$b + b = b \cdot b = b$$
4. **Uniqueness of complement.** For each $b \in B$ there is only one complement $\bar{b}$. In addition
$$\bar{0} = 1 \quad \text{and} \quad \bar{1} = 0$$
5. **Involution.** For each $b \in B$
$$(\bar{\bar{b}}) = b$$
6. **Associativity.** For each triple $a, b, c \in B$,
$$(a + b) + c = a + (b + c) \quad \text{and} \quad (a \cdot b) \cdot c = a \cdot (b \cdot c)$$
7. **DeMorgan's laws.** For each pair $a, b \in B$
$$(\overline{a + b}) = \bar{a} \cdot \bar{b} \quad \text{and} \quad \overline{a \cdot b} = \bar{a} + \bar{b}$$
8. **Absorption.** For each pair $a, b \in B$
$$a + a \cdot b = a \cdot (a + b) = a$$

6.3 SWITCHING FUNCTIONS

In preparation for the application of Boolean algebra to switching circuits, we introduce the definition of switching variable and switching polynomial. Let $\{0, 1\}$ be denoted by Q.

Switching variable. A switching variable x is a variable over the set Q.

Switching polynomial. A switching polynomial $p(x_1, \ldots, x_n)$ in the independent switching variables $x_1, \ldots, x_n$ is any expression obtained recursively by repeated application of the operations $+, \cdot$ and complementation to 0, 1 and the variables $x_1, \ldots, x_n$.

By substitution of values for the switching variables $x_1, \ldots, x_n$ in $p(x_1, \ldots, x_n)$ one obtains an n-variable function on Q taking values in Q. We refer to this function as the n-variable function associated with $p(x_1, \ldots, x_n)$.

In the application to switching theory, a switching variable represents a "gate," a two-state device that either permits or blocks the flow of a signal through it. When the variable assumes the value 0 the gate is open; it is closed when the variable is 1. A physical interconnection of gates is called a switching circuit. The circuit assumes an open or closed state depending on the individual states of the gates. Thus associated with a switching circuit is an n-variable function on Q with values in Q, often referred to as the transmission function of the circuit.

If x and y represent gates, then $x + y$ represents the parallel combination of the gates while $x \cdot y$ represents the series combination. The variable $\bar{x}$ represents the dependent gate whose state is the opposite of the gate represented by x. For convenience we allow 0 to represent an open circuit gate and 1 to represent a closed circuit gate. Thus the n-variable function associated with a switching polynomial is the transmission function of a series-parallel switching circuit. The import of the next theorem is that any given transmission function can be synthesized in a series-parallel configuration.

Synthesis Theorem. Any n-variable function (i.e., transmission function) is the n-variable function associated with some switching polynomial.

TABLE 6.1. A THREE-VARIABLE TRANSMISSION FUNCTION

x	0	0	0	0	1	1	1	1
y	0	0	1	1	0	0	1	1
z	0	1	0	1	0	1	0	1
f	0	1	0	1	1	0	1	0

We now present an algorithm for finding a switching polynomial for a given transmission function. Let an n-variable function be specified in an $(n + 1) \times 2^n$ table, where the first n entries of a column contain a particular choice of values for the switching variables $x_1, \ldots, x_n$ and the last entry gives the corresponding function value. Each of the following two methods yields a switching polynomial with associated n-variable function equal to the given transmission function.[2]

Method 1. For each column with 1 in the last entry, form a product of n-variables in which x_i appears if the ith entry is 1 and $\bar{x}_i$ appears if the ith entry is 0. Set $p(x_1, \ldots, x_n)$ equal to the sum of these products.

Method 2. For each column with 0 in the last entry, form a sum of n-variables in which x_i appears if the ith entry is 0 and $\bar{x}_i$ appears if the ith entry is 1. Set $p'(x_1, \ldots, x_n)$ equal to the product of these sums.

Example 3. A Transmission Function. Consider the transmission function defined in Table 6.1. Using Method 1 from the algorithm previously given we obtain the switching polynomial

$$p(x, y, z) = \bar{x} \cdot \bar{y} \cdot z + \bar{x} \cdot y \cdot z + x \cdot \bar{y} \cdot \bar{z} + x \cdot y \cdot \bar{z}$$

The application of Method 2 from the algorithm yields

$$p'(x, y, z) = (x + y + z) \cdot (x + \bar{y} + z) \cdot (\bar{x} + y + \bar{z}) \cdot (\bar{x} + \bar{y} + \bar{z})$$

Physical realizations of these polynomials are given in Fig. 6.2. Note that p has the same associated n-variable function as the simpler polynomial

$$\hat{p}(x, y, z) = \bar{x} \cdot z + x \cdot \bar{z}$$

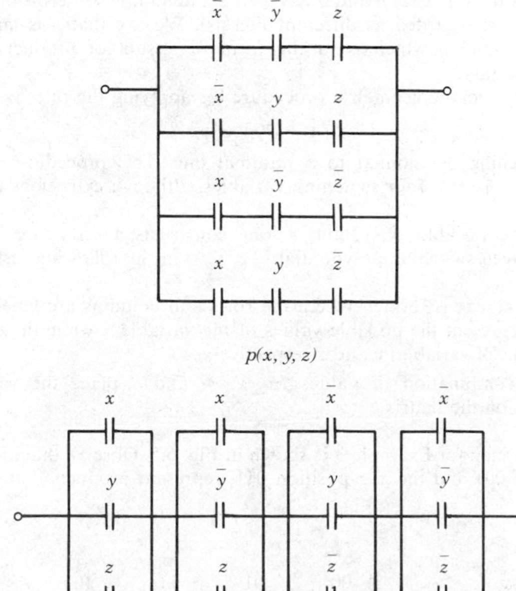

Fig. 6.2 Switching circuits for Table 6.1.

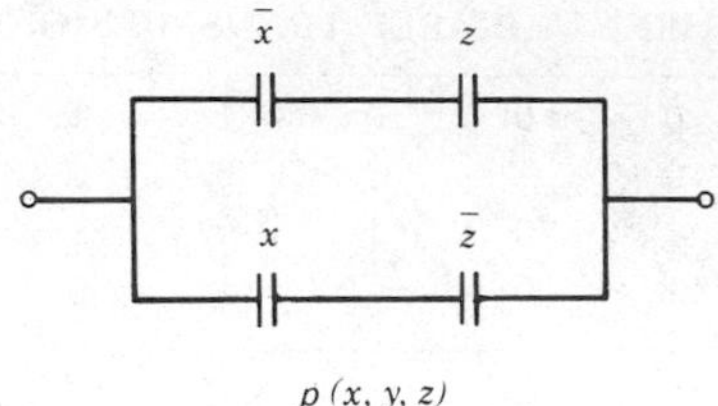

Fig. 6.3 Another realization of Table 6.1.

since for all choices of particular values of x, y, and z

$$p(x, y, z) = \bar{x} \cdot \bar{y} \cdot z + \bar{x} \cdot y \cdot z + x \cdot \bar{y} \cdot \bar{z} + x \cdot y \cdot \bar{z}$$

but

$$\bar{x} \cdot \bar{y} \cdot z + \bar{x} \cdot y \cdot z = \bar{x} \cdot (\bar{y} + y) \cdot z = \bar{x} \cdot 1 \cdot z = \bar{x} \cdot z$$

and

$$x \cdot \bar{y} \cdot \bar{z} + x \cdot y \cdot \bar{z} = x \cdot (\bar{y} + y) \cdot \bar{z} = x \cdot \bar{z}$$

Thus we have the more economical realization shown in Fig. 6.3.

6.4 SIMPLIFICATION OF BOOLEAN FUNCTIONS

In the remainder of this section we describe a useful method for the simplification of switching polynomials of the type obtained in Method 1 of the algorithm. First, however, we shall describe what we mean by a "simpler" expression. Note that in the preceding example we required fewer parallel banks to realize $\hat{p}(x, y, z)$ than to realize $p(x, y, z)$. In the design of a physical switching circuit, the actual measure of economy depends on the type of implementation, for example, diodes, relays, or logic gates. Rather than develop a separate theory for each of these types of implementation, we have one common measure of minimality. Let $p(x_1, \ldots, x_n)$ and $q(x_1, \ldots, x_n)$ be switching polynomials associated with the same n-variable function, both in the form of sums of products, wherein each product term (either x_i or $\bar{x}_i$) but not both may appear. Then p is *simpler* than q if (1) p has fewer terms than q or (2) p and q have the same number of terms but p has fewer literals than q (where x and $\bar{x}$ are regarded as different literals). We say that p is minimal if there is no simpler switching polynomial q which is in the form of a sum of products and has the same associated n-variable function.

We now describe a systematic geometric procedure for applying the rule

$$x \cdot y + \bar{x} \cdot y = y$$

which simplifies a switching polynomial to a minimal one. This procedure is most effective for switching polynomials in three or four switching variables, although extensions to the case of five or six variables do exist.[1,2]

For the case of three variables x, y, and z, one constructs a geometric figure known as the Karnaugh map of the given switching polynomial $p(x, y, z)$ in the following fashion:

1. First a two-by-four matrix is formed where the rows and columns are labeled as shown in Fig. 6.4. The two rows represent the possible values of the variable x while the columns indicate the possible combinations of variables y and z respectively.
2. For each possible combination of values for x, y, and z place the value of $p(x, y, z)$ in appropriate position on the matrix.

Hence the Karnaugh map for Example 3 is shown in Fig. 6.5. Observe that any two adjacent cells, such as the xyz position 001 and the xyz position 011, represent products that differ in exactly one

x \ yz	00	01	11	10
0				
1				

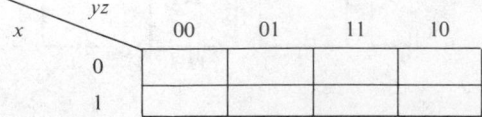

Fig. 6.4 General Karnaugh map for three variables.

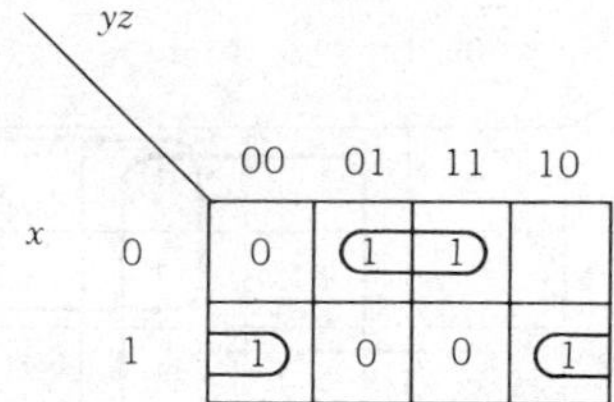

Fig. 6.5 Karnaugh map corresponding to Example 6.3.

literal. More precisely, the exceptional literal appears in uncomplemented form in one product and complemented form in the other. Thus the terms in a sum of products expression corresponding to adjacent cells containing 1's may be combined with use of the law:

$$A \cdot a + A \cdot \bar{a} = A$$

where A represents some combination of literals and a is any other literal. It should also be noticed that cells such as the 100 position and the 110 position must be considered adjacent cells. Therefore it may be helpful to visualize the map wrapped into a cylindrical shape. Two adjacent cells containing 1's form a "two cell." A square of four 1's on the cylinder form a "four cell." In this instance the four corresponding terms in the sum of products expression may be combined so as to eliminate two literals.

At this point it should be clear that the way in which the columns and rows are labeled provides a geometric visualization via "k cells" of every possible simplification obtainable by the use of the law

$$A \cdot a + A \cdot \bar{a} = A$$

It remains to describe an algorithmic procedure applied to the map which yields a minimal sum of products expression having the same three-variable function as the original expression used in creating the map. To this end consider the Karnaugh map for a three-variable expression that does not contain all 1's.

1. Find all one cells that are not contained in two cells.
2. Find all two cells that are not contained in four cells.
3. Find all four cells.
4. Write a minimal expression consisting of products corresponding to the simplifications related to the one, two, and four cells already found.[1,2]

Applying this algorithm to our example, we obtain a pair of two cells as shown in Fig. 6.5, namely 001-011 and 100-110; the corresponding minimal expression is

$$\hat{p}(x, y, z) = \bar{x} \cdot z + x \cdot \bar{z}$$

The procedure naturally extends to four-variable expressions as described hereafter. First form the four × four matrix with rows and columns labeled as shown in Fig. 6.6. As in the three-variable case, the values of the four-variable function associated with a sum of products expression $p(w, x, y, z)$ are entered into the matrix. To obtain the geometric visualization of adjacency we now imagine the map to be in the form of a torus, that is, the top and bottom of the matrix are joined, as are the left and right sides. Using the resulting adjacencies, one, two, and four cells are defined as they were in the three-variable case. In addition we introduce the notion of an "eight cell," that is, a two-by-four rectangle of eight 1's on the torus. Again these "k-cells" lead to simplifications obtained from application of the law $A \cdot a + A \cdot \bar{a} = A$. The algorithm for writing a minimal expression using the map is the immediate extension of the one stated for three variables.

yz \ wx	00	01	11	10
00				
01				
11				
10				

Fig. 6.6 General Karnaugh map for four variables.

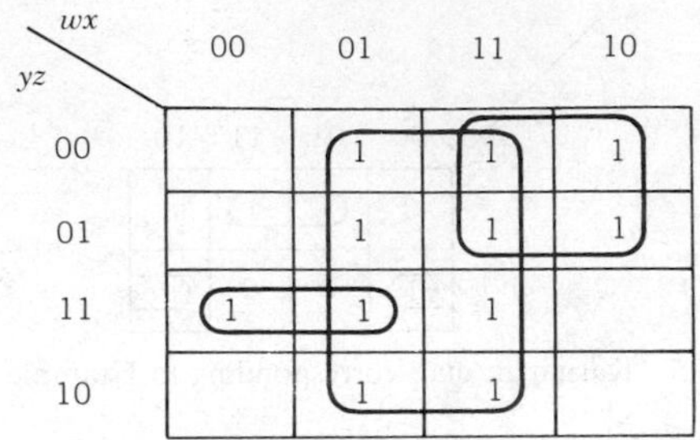

Fig. 6.7 An example of a 4-variable Karnaugh map.

We conclude by illustrating the algorithm for the Karnaugh map shown in Fig. 6.7. Application of the algorithm yields the two cell, four cell, and eight cell indicated in the figure. Now in the two cell, $\bar{x}$ and x both appear, hence they are absorbed and eliminated to yield $\bar{w} \cdot y \cdot z$. In the four cell, x and z are both eliminated, yielding $w \cdot \bar{y}$. In the eight cell, all variables but x are eliminated. Thus the corresponding minimal expression is

$$\hat{p}(w, x, y, z) = \bar{w} \cdot y \cdot z + w \cdot \bar{y} + x$$

References

1 F. J. Hill and G. R. Peterson, *Introduction to Switching Theory and Logical Design*, 3rd ed., Wiley, New York, 1981.

2 Z. Kohavi, *Switching and Finite Automata Theory*, 2nd ed., McGraw-Hill, New York, 1978.

PART 2

PROPERTIES OF MATERIALS

CHAPTER 7

PROPERTIES OF MATERIALS

KRISHNA SESHAN*

University of Arizona
Tucson, Arizona

C. F. BIEBER

Hughes Aircraft Company
Tucson, Arizona

7.1 UNITS AND CLASSIFICATION OF MATERIALS

Although the use of SI units is strongly recommended,[1] several useful and practical units are in common use. Where appropriate, these practical units are shown. Listed below are these practical units and their SI equivalents.

	Practical Unit	Equivalent SI	Table
Coefficient of expansion	ppm/°C	10^{-6}/K	7.1
Weight/1000 ft	lb/1000 ft	kg/m	7.7
Electrical resistivity	$\mu\Omega$-cm	S-m	7.7
Electrical conductivity	mhos/m	S/m	
Coefficient of resistance	ppm/°C	S · m/K	

This chapter deals with the properties of resistive, conductive, and packaging materials used in the electrical, microelectronic, and packaging industries. The chapter begins with the properties of resistive materials and then deals with conductive materials. Often related mechanical and thermal properties also come into play. Therefore, properties like tensile strength and temperature coefficient of expansion are also discussed.

Conductive, resistive, and packaging materials in device applications are limited by processes of creep, electromigration, and diffusion. Data on these properties are included in this chapter.

*Now with IBM, E. Fishkill, NY, 12533.

Also included is a discussion of the sparse data in a growing field: microelectronic packaging. Here severe restrictions exist on material compatibility. Normally insulating materials, like epoxy or glass, have to be made thermally conductive.

The chapter concludes with a survey of some typical materials used for conductive, resistive, and packaging applications and lists some of their limitations.

Classification of Materials

Materials can be classified as metallic conductors[2] (1 – 150 μΩ-cm), semiconductors (150–1000 μΩ-cm), or insulators (greater than 10^3 μΩ-cm). Amorphous semiconductors and insulators and glassy conductors are produced by special rapid solidification techniques and may become important to semiconducting technology in the near future. Ionic conductors, such as fused salts, are important at temperatures over 500°C. The band theory of solids provides a theoretical basis for this classification. This theory also explains the differences in the temperature dependence of resistivity exhibited by these materials.

A class of materials important to microelectronic packaging is conductive and nonconductive joining materials. These include conductive and nonconductive epoxies, polymides, soft solders, solder glasses, and brazes. These materials have specially tailored properties of thermal expansion coefficient and bondability to make them compatible with the dissimilar surfaces that they bond. Properties of such materials are described in Section 7.6.

7.2 RESISTIVE PROPERTIES

Origin of Resistivity; Matthiessen's Rule

Any disturbance of the absolute regularity of the crystal structure, any perturbation of the atoms from their ideal lattice site, will scatter electrons and thereby cause electrical resistance.[3,4] At 0 K, the phonon, or lattice vibration, contribution to resistivity drops to zero. The remaining bulk resistivity arises from impurities, grain boundaries, dislocations, vacancies, and other imperfections. In pure metals and dilute alloys, the total resistivity (ρ) is the sum of two terms: the thermal component, ρ_T, which arises from lattice vibrations, and the residual resistivity, ρ_r, caused by impurities and structural imperfections.[4] The latter is temperature independent. The total resistivity is given by

$$\rho = \rho_T + \rho_r \tag{7.1}$$

Equation (7.1) is Matthiessen's rule.[2]

Residual Resistivity Ratio (RRR)

The residual resistivity ratio (RRR) is a number used to denote the purity of a sample. The purer the sample, the higher its RRR. RRR is defined as

$$\mathrm{RRR} = \frac{\rho 298\ \mathrm{K}}{\rho 4.2\ \mathrm{K}} \tag{7.2}$$

Ratios of up to 100,000 have been obtained by zone refining techniques.

Volume Resistivity

Volume, or "bulk," resistivity, ρ, is defined as the ratio of the electric field to the resultant current density

$$\rho = \frac{E}{J} \tag{7.3}$$

where ρ = volume or bulk resistivity, Ω-m
E = electric field strength, V/m
J = current density, A/m^2

The resistance, R, is related to the resistivity

$$R = \rho \cdot \frac{L}{A} \tag{7.4}$$

where A = conductor cross section, m^2
L = conductor length, m
R = resistance, Ω

Table 7.1 lists the electrical resistivities of pure metals and Table 7.2 lists the resistivities of some selected alloys and compounds.

Bulk Resistivity

Bulk resistivity, ρ, is the reciprocal of the bulk conductivity, σ

$$\rho = \frac{1}{\sigma} \tag{7.5}$$

The units are expressed as 10^{-8} Ω-m, or as microohm-cm, $\mu\Omega$-cm. Note that the two are identical.

Mass Resistivity

Mass resistivity, δ, is defined as

$$\delta = \frac{R \cdot m}{L^2} \qquad (\Omega\text{-kg/m}) \tag{7.6}$$

where R = resistance, Ω
m = conductor mass, kg
L = conductor length, m

Sheet Resistivity

Sheet resistivity, ρ/t, is used when the conductor thickness (t) is much smaller than its width or length. It is defined as

$$\frac{\rho}{t} = R \cdot \frac{W}{L} \frac{\text{ohm}}{\text{square}} \tag{7.7}$$

where t = conductor thickness, m
L = conductor length, m
W = conductor width, m

The resistance of a sheet is its sheet resistivity multiplied by the number of squares L/W. L/W is also known as the aspect ratio.

Surface Resistivity

At high frequencies the electric current is confined to a near surface region. At a skin depth, δ, the surface resistivity, R_s, falls to $1/e$ of its value at the surface. The surface resistivity, R_s, is the DC sheet resistivity of a conductor with thickness

$$R_s = \frac{\rho}{\delta} = \frac{1}{\sigma\delta} \tag{7.8}$$

where ρ = electrical resistivity
δ = skin depth, m
σ = conductivity, S/m

Table 7.3 shows the high-frequency characteristics of some metals.

Nordheim's Rule

The addition of an impurity raises the residual resistivity according to

$$\rho_r(X) = A \times (1 - X) \tag{7.9}$$

where ρ_r = resistivity of the dilute alloy
X = concentration of the impurity
A = constant depending on the host metal and impurity; it increases with valence, atomic size, and other differences between the host and impurity

Equation (7.9) is called Nordheim's rule. Figure 7.1 shows the relationship between resistivity and the phase diagram for a two-component system. The effect of alloy additions on resistivity for selected metals is shown in Table 7.4.

Thin Film Resistivity

The resistivity of thin films increases from the bulk value as the thickness approaches the mean free path of the conduction electrons, which is typically 10 to 100 nm at room temperature.[3,17,18] Electron scattering from the surface is the dominant mechanism.

TABLE 7.1. PHYSICAL PROPERTIES OF PURE METALS[15]

Metal	Melting Point (°C)	Density at 20°C (g/cm^3)	Thermal Conductivity, 0-100°C (W/m-K)	Mean Specific Heat, 0-100°C (J/kg-K)	Resistivity at 20°C ($\mu\Omega$-cm)	Temperature Coefficient of Resistivity, 0-100°C (ppm/K)	Coefficient of Expansion, 0-100°C (10^{-6}/K)
Aluminum	660.1	2.70	238	917	2.67	4500	23.5
Antimony	630.5	6.68	23.8	209	40.1	5100	8-11
Barium	729	3.5	—	285	60(0°C)	—	18
Beryllium	1287	1.848	194	2052	3.3	9000	12
Bismuth	271	9.80	9	124.8	117	4600	13.4
Cadmium	320.9	8.64	103	233.2	7.3	4300	31
Calcium	839	1.54	125	624	3.7	4570	22
Cerium	798	6.75	11.9	188	85.4	8700	8
Cesium	28.5	1.87	36.1[a]	234	20	4800	97
Chromium	1860	7.1	91.3	461	13.2	2140	6.5
Cobalt	1492	8.9	96	427	6.34	6600	12.5
Copper	1083.4	8.96	397	386.0	1.694	4300	17.0
Gallium	29.7	5.91	41.0[a]	377	[b]	—	18.3
Germanium	937	5.32	56.4	310	$\sim 89 \times 10^3$	—	5.75
Gold	1063	19.3	315.5	130	2.20	4000	14.1
Hafnium	2227	13.1	22.9	147	32.2	4400	6.0
Indium	156.4	7.3	80.0	243	8.8	5200	24.8
Iridium	2454	22.4	146.5	130.6	5.1	4500	6.8
Iron	1536	7.87	78.2	456	10.1	6500	12.1
Lead	327.4	11.68	34.9	129.8	20.6	4200	29.0
Lithium	181	0.534	76.1	3517	9.29	4350	56
Magnesium	649	1.74	155.5	1038	4.2	4250	26.0
Manganese	1244	7.4	7.8	486	$160(\alpha)$[c]	—	23
Mercury	− 38.87	13.546	8.65	138	95.9	1000	61
Molybdenum	2615	10.2	137	251	5.7	4350	5.1
Nickel	1455	8.9	88.5	452	6.9	6800	13.3
Niobium	2467	8.6	54.1	268	16.0	2600	7.2
Osmium	3030	22.5	87.5	130	8.8	4100	4.57

Palladium	1552	12.0	75.5	247	10.8	4200	11.0
Platinum	1769	21.45	71.5	134.4	10.58	3920	9.0
Potassium	63.2	0.86	104[a]	754	6.8	5700	83
Radium	700	5	—	—	—	—	—
Rhenium	3180	21.0	47.6	138	18.7	4500	6.6
Rhodium	1966	12.4	149	243	4.7	4400	8.5
Rubidium	38.8	1.53	58.3[a]	356	12.1	4800	9.0
Ruthenium	2310	12.2	116.3	234	7.7	4100	9.6
Silicon	1412	2.34	138.5	729	10^3-10^6	—	7.6
Silver	960.8	10.5	425	234	1.63	4100	19.1
Sodium	97.8	0.97	128	1227	4.7	5500	71
Strontium	770	2.6	—	737	23(0°C)	—	100
Tantalum	2980	16.6	57.55	142	13.5	3500	6.5
Tellurium	450	6.24	3.8	134	1.6×10^5(0°C)	—	[d]
Thallium	304	11.85	45.5	130	16.6	5200	30
Thorium	1755	11.5	49.2	100	14	4000	11.2
Tin	231.9	7.3	73.2	226	12.6	4600	23.5
Titanium	1667	4.5	21.6	528	54	3800	8.9
Tungsten	3400	19.3	174	138	5.4	4800	4.5
Uranium	1132	19.05(α)[c] 18.89(β)[e]	28	117	27	3400	[c]
Vanadium	1902	6.1	31.6	498	19.6	3900	8.3
Zinc	419.5	7.14	119.5	394	5.96	4200	31
Zirconium	1852	6.49	22.6	289	44	4400	5.9

Source. Data From Ref. 5.

[a] Solid.

[b] 17.4$\|a$ axis, 8.1$\|b$ axis, 54.3$\|c$ axis.

[c] α-Uranium $\left.\begin{array}{ll} 23 & \|a \text{ axis} \\ -3.5 & \|b \text{ axis} \\ 17 & \|c \text{ axis} \end{array}\right\}$ 25–300°C.

[d] 1.7$\|c$ axis, 27.5$\perp c$ axis.

[e] β-Uranium $\left.\begin{array}{ll} 4.6 & \|c \text{ axis} \\ 23 & \perp c \text{ axis} \end{array}\right\}$ 20–720°C.

TABLE 7.2 SELECTED PROPERTIES OF SOME ALLOYS AND COMPOUNDS[a]

Alloy or Composition	Melting Range (°C)	Density (g/cm^3)	Thermal Conductivity (W/m-K)	Thermal Expansion (10^{-6}/K)	Resistivity ($\mu\Omega$-cm)
Carbon Steel					
AISI 1010	1510–1525	7.84	60.2	12.2	16.8
1042	1480–1510	7.84	50.7	11.3	17.1
1078	1450–1500	7.84	48.2	11.1	18.0
Stainless Steel					
Type 304	1400–1450	7.9	16.2	17.2	72
316	1370–1400	7.9	16.2	16.0	74
405	1480–1530	7.7	27.0	10.8	60
420	1450–1500	7.7	24.9	10.3	55
Aluminum					
Type 1100	643–657	2.71	222	23.6	2.92
2024	502–638	2.77	190	22.9	3.4
5056	568–638	2.64	120	24.1	5.9
6061	582–652	2.70	180	23.6	3.7
7075-T6	477–635	2.80	130	23.4	5.2
Copper					
Type C10100 OFHC	1083	8.94	391	17.0	1.72
C11000 ETP	1065–1083	8.89	388	16.8	1.72
C14500 0.5% Te	1051–1075	8.94	355	17.1	1.86
C17200 2.0% Be	865–980	8.25	105–130	16.7	5.7–11.5
C21000 Gilding	1050–1065	8.86	234	18.1	3.1
C23000 Red brass	990–1025	8.75	159	18.7	4.7
C27000 Yellow brass	905–932	8.47	116	20.3	6.4
C51000 Phosphor bronze	880–1050	8.86	69	17.8	11.5
Other					
Gray cast iron ASTM-A48	1175–1300	7.2	50	10.8	70
Malleable cast iron ASTM A-47	1175–1300	7.36	51	13.5	40
Maraging steel	1430–1450	8.03	21	10.1	—
Inconel 601	1370–1400	8.1	11	11.5	—

TABLE 7.2 (*Continued*)

Alloy or Composition	Melting Range (°C)	Density (g/cm^3)	Thermal Conductivity (W/m-K)	Thermal Expansion (10^{-6}/K)	Resistivity ($\mu\Omega$-cm)
80Ni-20Cr	1395–1425	8.4	13	13.2	108
Hastelloy C	1265–1345	8.9	11	11.3	130
92.5Ti-5A1-2.5Sn	1550–1650	4.48	7.8	9.5	157
Borides					
CrB_2	1900	5.2	—	6.0–6.7	21–56
HfB_2	3250	11.2	—	5.3–7.5	10–12
MoB_2	2100	7.7	—	7.7	30
NbB_2	2980	6.9	17	5.9–8.4	12–65
TaB_2	3090	12.6	13	5.2–5.9	14–68
TiB_2	2899	4.4	26	4.8–6.6	10–20
VB_2	2425	5.0	—	3.8–8.0	16–38
W_2B_5	2200	13.1	—	—	21–56
ZrB_2	3040	6.2	23	5.5–6.2	7–10
Carbides					
B_4C	2450	2.5	121	2.6–3.1	10^6
SiC	2700	2.6–3.2	95	3.9–4.3	100–200
TaC	3885	14.4	22	8.3	20–30
TiC	3127	5.0–7.2	17–20	6.6–7.4	105
ZrC	3521	6.6–6.8	21	6.6	63
Nitrides					
BN	2730	1.9–2.2	182	7.5	10^{19}
Si_3N_4	1900	3.2	14–33	2.4	10^{19}
TaN	3343	14.4	10	5.5	135
TiN	3220	5.4	17	6.3	22
VN	2320	6.1	—	—	86
Oxides					
Al_2O_3, alumina	2040	3.9	17	7.7	10^{20}
$3Al_2O_3$-$2SiO_2$, mullite	1871	2.9–3.3	2.5	4.8–5.4	10^{21}
BeO, beryllia	2550	2.9	250	7.5	10^{20}
CaO, calcia	2599	3.3–5.5	—	12.6	—
MgO, magnesia	2799	3.5–3.6	47	12.8–14	—
2mgO-SiO_2, forsterite	1888	2.8–3.2	3.3–4.1	8.5	10^{20}
SiO_2, silica	1670	1.9–2.6	1.4	0.5	—
ZrO_2, zirconia	2599	5.5–9.9	12–14	5.6	—

Source. Data from Refs. 5–10.
[a]Values approximate due to effects of method of preparation.

TABLE 7.3. CONDUCTORS AT HIGH FREQUENCIES

Metal	Resistivity (Ω-m)10^8	Relative Permeability at 0.002 Wb/m^2	$\delta\sqrt{v}$ (δ = depth of penetration, m; v = frequency, Hz)	$10^7 R_s/\sqrt{v}$ (R_s = surface resistivity, Ω/m^2)
Aluminum	2.828	1	0.085	3.33
Brass (65.8 Cu, 34.2 Zn)	6.29	1	0.126	4.99
Brass (90.9 Cu, 9.1 Zn)	3.65	1	0.096	3.79
Graphite	1,000	1	1.592	62.81
Chromium	2.6	1	0.081	3.21
Copper	1.724	1	0.066	2.61
Gold	2.22	1	0.075	2.96
Lead	22	1	0.236	9.32
Magnesium	4.6	1	0.108	4.26
Mercury	95.8	1	0.493	19.43
Nickel	7.8	100	0.014	55.71
Phosphor bronze	7.75	1	0.140	5.54
Platinum	9.83	1	0.158	6.22
Silver	1.629	1	0.064	2.55
Tin	11.5	1	0.171	6.73
Tungsten	5.51	1	0.118	4.67
Zinc	5.38	1	0.117	4.60
Magnetic iron	10	200	0.011	90.9
Permalloy (78.5 Ni, 21.5 Fe)	16	8,000	0.0022	727
Supermalloy (5 Mo, 79 Ni, 16 Fe)	60	100,000	0.0012	4,880
Mumetal (75 Ni, 2 Cr, 5 Cu, 18 Fe)	62	20,000	0.0029	2,140

Source. Data from Ref. 11.

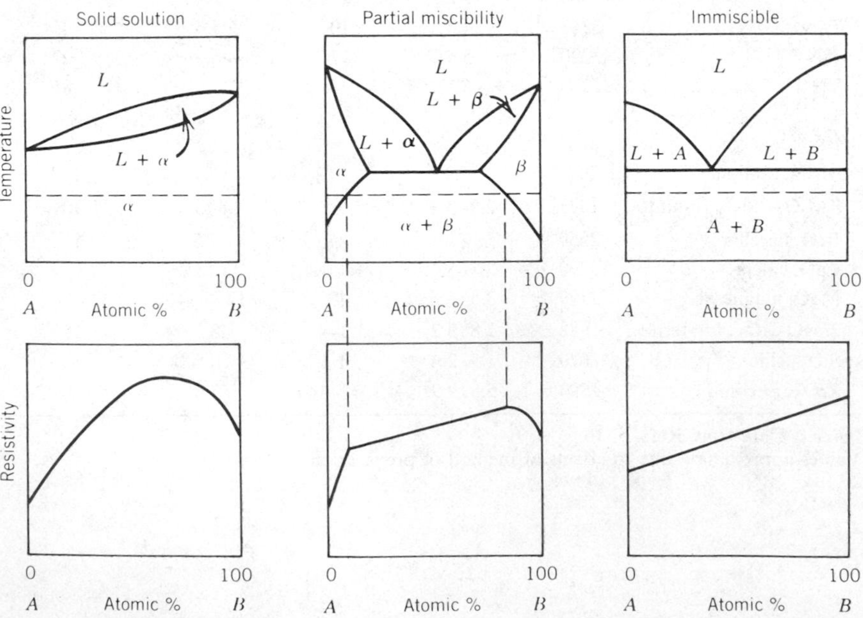

Fig. 7.1 Typical resistivity of binary alloys as a function of composition.[12,13] (Shows Nordheim's rule.)

TABLE 7.4. MAXIMUM SOLID SOLUBILITY c_{max} AND ATOMIC RESISTIVITY INCREASE, A, Ω-m/(at %) FOR COPPER, SILVER, GOLD, AND ALUMINUM

Solute	Copper		Silver		Gold		Aluminum	
	c_{max}	A	c_{max}	A	c_{max}	A	c_{max}	A
Ag	4.9	1.4	20.34	—	*	3.6	23.8	115
Al	19.6	12.5	20.34	5	6	18.7	—	—
As	—	68	8.8	85	—	80	—	—
Au	*	55	*	36				
Be	16.4	62						
Cd	—	30	42.2	3.8	32.5	6.3	0.11	0.57
Cr	0.8	36	—	—	23	42.5	0.40	77
Cu	—	—	14.1	0.77	*	4.5	24.8	7.85
Fe	4.5	93	—	—	75	79	0.025	53
Ga	19.9	14.2	18.7	23.6	13	22	—	2.5
Ge	11.8	37.9	9.6	55	3.2	52		
In	11	10.6	20	17.8	12.6	13.9		
Mg	7	6.5	29.3	19.5	25	13	16.3	4.9
Mn	*	29	47	16	—	24.1	0.90	59.7
Ni	*	12.5	—	—	*	7.9	0.023	17.7
P	3.5	6.7						
Pd	*	8.9	*	4.4	*	4.1		
Pt	*	21	40.5	16	*	10.1		
Sb	6	54	7.2	72.5	1.1	68		
Si	11.25		39.5				1.59	6.5
Sn	9.1	28.8	11.5	43.6	6.8	33.6		
Zn	38.3	3.2	40.2	6.4	31	9.5	66.4	2.11

Source. Data from Refs. 4, 6, 12–16.

The resistivity of thin films tends to be higher than the bulk, even when the films are thicker than the electron mean free path. Grain boundaries, oxides, voids, and other structural imperfections are usually responsible for the increased resistivity. Annealing, deposition at a higher temperature, and epitaxial growth are processes that cause the film resistivity to approach the bulk value.

A further increase in resistivity occurs when the films are extremely thin and physically discontinuous. Conduction in such a film proceeds by tunneling or thermionic emission, and resistance is nonohmic.

The temperature coefficient of resistivity (TCR), of very thin films tends to be negative. This occurs when tunneling and thermionic emission processes offer a significant contribution to conductivity. Both these processes are enhanced by an increase in temperature, causing a decrease in resistivity.[6]

Table 7.5 shows the correlation between grain size and resistivity of sputtered aluminum–copper–silicon films used for IC interconnections.[19] Table 7.6 shows the properties of refractory silicides. Tables 7.7 and 7.8 show the properties of thin film conductor and resistor materials. Table 7.9 shows the properties of copper and aluminum wires that are used for interconnections.

7.3 CURRENT-CARRYING CAPACITIES OF CONDUCTORS

The current-carrying capacities of electrical wires, bus bars, and cables are limited by heating effects produced by the current. A 30° rise above a 40° ambient corresponds to a current density of about 10^6 A/m^2 in copper bus. The properties and maximum current-carrying capacities of wires are shown in Tables 7.10 and 7.11.

The current-carrying capacities of thin films are limited by electromigration; the standard value of the current density in thin films is about 100 A/m^2. Electromigration is the migration of atoms under the influence of electric currents.[20,28–31]

TABLE 7.5. FILM DEPOSITION, RESISTIVITY, AND GRAIN SIZE FOR ALUMINUM ALLOY FILMS

			Planned Compositions					Specific Resistivity		Grain Size			
										As Deposited		After 450°C 15 min Anneal in N_2	
No.	Film		Cu %	Si %	Substrate Temperature	Rate (nm/s)	Thickness (nm)	As Deposited ($\mu\Omega$-cm)	After 450°C 15 min Anneal in N_2 ($\mu\Omega$-cm)	Average Size (μm)	Range (μm)	Average Size (μm)	Range (μm)
1	Al	IN · S	—	—	RT	3.3	838	2.77	2.43[a]	0.53	0.2–0.7	0.9	0.2–1.2
2	Al	IN · S	—	—	200°C	3.6	828	2.85	2.40[a]	0.96	0.2–1.5	1.4	0.2–1.8
3	Al	M · S	—	—	RT	2.1	813	2.93	2.68	0.30	0.1–1.0	0.7	0.1–1.8
4	Al + Cu	IN · S	2	—	RT	2.9	762	4.19	3.28	0.23	0.1–0.5	1.0	0.1–1.8
5	Al + Cu	IN · S	2	—	200°C	2.9	762	3.12	3.35	0.67	0.2–1.0	2.3	0.2–2.6
6	Al + Cu	M · S	2	—	RT	1.8	836	4.10	3.51	0.22	0.1–0.5	1.2	0.2–2.0
7	Al + Cu + Si	IN · S	3.85	1.0	RT	3.0	850	7.99	4.42	0.55	—	0.2	—
8	Al + Cu + Si	IN · S	3.85	1.0	200°C	2.4	813	8.02	4.07	0.07	—	0.2	—
9	Al + Cu + Si	M · S	2.0	1.0	RT	1.3	938	4.50	3.47[b]	0.07	—	0.1	—

[a] Room temperature resistivity of annealed Al film (IN · S) (Incandescent Source) is low (2.43); this is possibly due to an error in thickness measurements. All previous data show a specific resistivity of 2.71 $\mu\Omega$-cm.

[b] DC magnetron sputter deposited, Al + Cu + Si films show a lower resistivity as compared with the resistivity of films prepared from an IN · S. This difference is attributed to the difference in copper concentration.

IN · S = incandescent source; M · S = magnetron sputtering.

TABLE 7.6. PROPERTIES OF REFRACTORY SILICIDES

Silicide	Lowest Binary Eutectic Temperature (°C)	Barrier Height on n-type Silicon (eV)		Annealed Resistivity ($\mu\Omega$-cm)
$CoSi_2$	1195	0.64		18–20
$CrSi_2$	1300	0.57		~600
$FeSi_2$	1208	—		$\gg$ 1000
$HfSi_2$	1300	—		45–50
$MoSi_2$	1410	0.55		40–100
$NbSi_2$	1295	—		50
$NiSi_2$	966	0.70		~50
Pd_2Si	720	0.74		30–35
PtSi	830	0.87		28–35
$TaSi_2$	1385	0.59		35–70
$TiSi_2$	1330	0.60		13–26
VSi_2	1385	—	Solderable	50–55
WSi_2	1440	0.65	with Ni	30–100
$ZrSi_2$	1355	0.55	underlayer	35–40

Source. Data from Refs. 20–22.

TABLE 7.7. THICK AND THIN FILM CONDUCTORS

Material	Thickness (μm)	(ohm/sq.)	Properties
Thick Film			
80Ag-20Pd	12	0.015–0.020	Solderable; fair leach resistance
60Ag-40Pd	12	0.040–0.050	Solderable; moderate leach resistance
90Ag-10Pt	15	0.015–0.020	Solderable; good leach resistance
Au	12	0.003–0.004	Conventional gold
Au	7	0.003–0.005	Thin printing gold
Au-Pt	12	0.020–0.050	Solderable; good leach resistance
Cu	15	0.002–0.003	Solderable; excellent leach resistance fired in N_2 atmosphere
Ni	25	0.060–0.075	Solderable; excellent leach resistance fired in N_2 atmosphere
Thin Film	Å		
Au	100	0.27–0.30	
Au	500	~0.05	
Au	10,000	~0.004	
Al	100	~0.33	
Al + Cu	5000–1000		IC interconnect
Al	1000	~0.04	

Source. Data from Refs. 7, 14, 23, 24.

TABLE 7.8. THICK AND THIN FILM RESISTORS

Selected Thick Film Resistor Systems
(Temperature Coefficient of Resistance +25 to −55°C and +25 to +125°C ppm/K)

ohm/sq.	Dupont Series 17[a]	Cermalloy 8000 Series[b]	ESL 3900 Series[c]	EMCA 5500-1 Series[d]
10	0 ± 175	0 ± 100	+50 ± 50	—
100	0 ± 50	0 ± 50	0 ± 50	0 ± 50
1000	0 ± 50	0 ± 50	0 ± 50	0 ± 50
10K	0 ± 100	0 ± 50	0 ± 50	0 ± 50
100K	0 ± 100	0 ± 100	0 ± 100	0 ± 50
1M	0 ± 100	0 ± 100	− 100 ± 100	—

Selected Thin Film Resistors

Material	Typical Sheet Resistivity (μΩ-cm)	(ppm/K)
NiCr	50–300	0 ± 25
Ta-N	50–500	0 to −200
Cr-SiO	100–5000	0 ± 100
Diffused Si	2–200	+600 to +2500

Source. Data from Refs. 7, 9, 15, 24, 25.
[a] Electro-Science Laboratories, Inc., Pennsauken, NJ 08110.
[b] Cermalloy, Inc., West Conshohocken, PA 19428.
[c] Electro Materials Corp. of America, Mamaroneck, NY 10543.
[d] E. I. du Pont de Nemours & Co., Inc., Wilmington, DE 19898.

7.4 CONDUCTIVE PROPERTIES

Volume Conductivity

Volume, or "bulk," electrical conductivity, σ, is the electrical current density, J, carried by the conductor per unit electric field strength, E:

$$\sigma = \frac{J}{E} \tag{7.10}$$

where σ = volume or bulk electrical conductivity, S/m
J = electric current density, A/m^2
E = electric field strength, V/m

In a noncubic single crystal, the conductivity is a tensor having nine terms. Cubic isotropic materials can be represented as having one conductivity. Polycrystalline noncubic materials can be thought of as having one "average" conductivity.[20]

The conductance, G, is related to the bulk conductivity

$$G = \sigma \cdot \frac{A}{L} \tag{7.11}$$

where A = conductor cross section, m^2
L = conductor length, m
σ = conductivity, S/m in SI, mho/m in MKS

Mass Conductivity

The mass conductivity, σ_m, is defined as

$$\sigma_m = G\frac{L^2}{m} \tag{7.12}$$

where σ_m = mass conductivity, $S \cdot m^2/kg$
G = conductance, S
L = conductor length, m
m = mass, kg

7.5 RELATED MATERIAL PROPERTIES

Mechanical

Tensile Stress, Tensile Strain. The engineering tensile stress, σ, is the force, F, per unit undeformed area, A_0 (see Fig. 7.2*a*)

$$\sigma = \frac{F}{A_0} \tag{7.13}$$

where A_0 = original undeformed area, m^2
F = force, N
σ = stress, N/m^2 or Pa

Use of the SI unit of stress, Pascal, Pa, and megaPascal, MPa, is highly recommended.[1] The following are useful conversions:

$$1\ \frac{\text{lbf}}{\text{in.}^2} = 6.8948\ \text{kPa} \tag{7.14}$$

$$1\ \frac{\text{kgf}}{\text{mm}^2} = 9.80665\ \text{MPa} \tag{7.15}$$

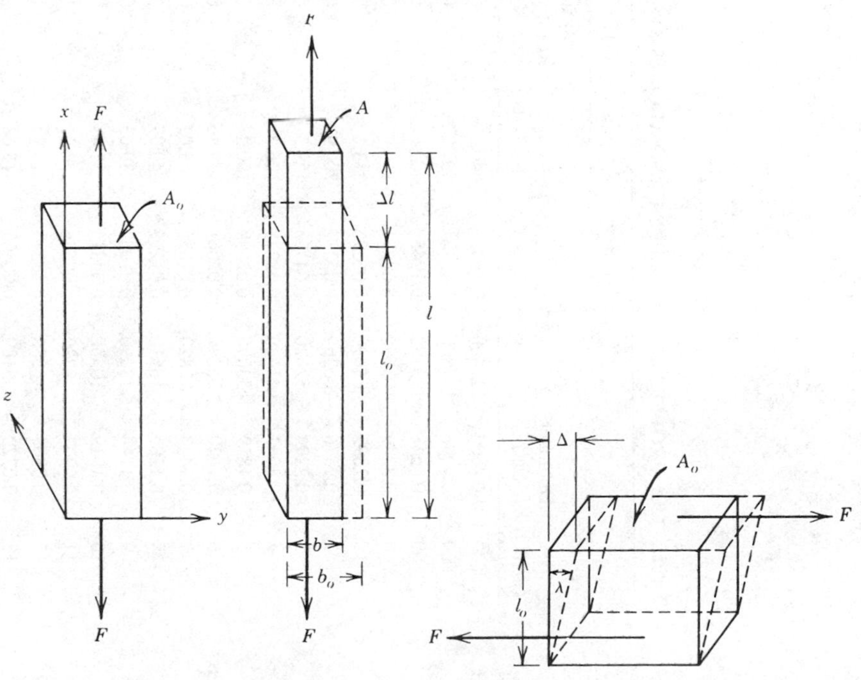

Fig. 7.2 Uniaxial tensile force before (*a*) and after (*b*) deformation ΔL. (*c*) Shear force, F, and shear strain, γ.

TABLE 7.9. PROPERTIES OF COPPER AND ALUMINUM WIRES

Size (AWG)	Diameter (in.)	Area (c mils)	Area (in.^{2})	Cu per ASTM B3-74[a] Weight (#/1000′)	Cu per ASTM B3-74[a] Maximum DC Resistivity at 20°C (Ω/1000′)	Al per ASTM B609-81[a] Weight (#/1000′)	Al per ASTM B609-81[a] Maximum DC Resistivity at 20°C (Ω/1000′)
0000	0.4600	211600.	0.1662	640.5	0.04901	194.7	0.08035
000	0.4096	167800.	0.1318	507.9	0.06180	154.4	0.1013
00	0.3648	133100.	0.1045	402.8	0.07793	122.5	0.1278
0	0.3249	105600.	0.08291	319.4	0.09827	97.15	0.1611
1	0.2893	83690.	0.06573	253.3	0.1239	77.03	0.2031
2	0.2576	66360.	0.05212	200.9	0.1563	61.07	0.2562
3	0.2294	52620.	0.04133	159.3	0.1970	48.43	0.3230
4	0.2043	41740.	0.03278	126.3	0.2485	38.41	0.4073
5	0.1819	33090.	0.02599	100.2	0.3133	30.45	0.5136
6	0.1620	26240.	0.02061	79.46	0.3951	24.15	0.6476
7	0.1443	20820.	0.01635	63.02	0.4982	19.16	0.8167
8	0.1285	16510.	0.01297	49.97	0.6282	15.20	1.030
9	0.1144	13090.	0.01028	39.63	0.7921	12.04	1.299
10	0.1019	10380.	0.00815	31.43	0.9988	9.556	1.637
11	0.0907	8230.	0.00646	24.92	1.260	7.571	2.065
12	0.0808	6530.	0.00513	19.77	1.588	6.008	2.604
13	0.0720	5180.	0.00407	15.68	2.003	4.771	3.283
14	0.0641	4110.	0.00323	12.43	2.525	3.781	4.140

15	0.0571	3260.	0.00256	9.858	3.184	3.001	5.220
16	0.0508	2580.	0.00203	7.818	4.015	2.375	6.583
17	0.0453	2050.	0.00161	6.200	5.063	1.889	8.301
18	0.0403	1620.	0.00128	4.917	6.385	1.495	10.47
19	0.0359	1290.	0.00101	3.899	8.051	1.186	13.20
20	0.0320	1020.	0.000804	3.092	10.15	0.9424	16.64
21	0.0285	812.	0.000638	2.452	12.80	0.7475	20.99
22	0.0253	640.	0.000503	1.945	16.14	0.5891	26.46
23	0.0226	511.	0.000401	1.542	20.36	0.4701	33.37
24	0.0201	404.	0.000317	1.223	25.67	0.3718	42.08
25	0.0179	320.	0.000252	0.9699	32.37	0.2949	53.06
26	0.0159	253.	0.000199	0.7692	40.81	0.2327	66.91
27	0.0142	202.	0.000158	0.6100	51.47	0.1856	84.37
28	0.0126	159.	0.000125	0.4837	64.90	0.1461	106.4
29	0.0113	128.	0.000100	0.3836	81.83	0.1175	134.2
30	0.0100	100.	0.0000785	0.3042	103.2	0.09203	169.2
31	0.0089	79.2	0.0000622	0.2413	130.1	0.07289	213.3
32	0.0080	64.0	0.0000503	0.1913	164.1	0.05890	270.0
33	0.0071	50.4	0.0000396	0.1517	206.9	0.04638	339.2
34	0.0063	39.7	0.0000312	0.1203	260.9	0.03654	427.7
35	0.0056	31.4	0.0000246	0.09542	329.0	0.02890	539.3
36	0.0050	25.0	0.0000196	0.07567	414.8	0.02301	680.1

Source. Data from Refs. 14, 24, 25.

[a]Based on ASTM Specifications B3-74, B609-81 ($\rho = 2.8264\ \mu\Omega$-cm), and B258-81.

TABLE 7.10. MAXIMUM CURRENT CAPACITY OF COPPER AND ALUMINUM WIRES (AMPERES)

	MIL-W-5088								
	Copper		Aluminum		National Electrical Code	Underwriters Laboratory		American Insurance Association	500 c mils/A
Size (AWG)	Single-wire	Wire-bundled	Single-wire	Wire-bundled		60°	80°		
30	—	—	—	—	—	0.2	0.4	—	0.20
28	—	—	—	—	—	0.4	0.6	—	0.32
26	—	—	—	—	—	0.6	1.0	—	0.51
24	—	—	—	—	—	1.0	1.6	—	0.81
22	9	5	—	—	—	1.6	2.5	—	1.28
20	11	7.5	—	—	—	2.5	4.0	3	2.04
18	16	10	—	—	6	4.0	6.0	5	3.24
16	22	13	—	—	10	6.0	10.0	7	5.16
14	32	17	—	—	20	10.0	16.0	15	8.22
12	41	23	—	—	30	16.0	26.0	20	13.05
10	55	33	—	—	35	—	—	25	20.8
8	73	46	58	36	50	—	—	35	33.0
6	101	60	86	51	70	—	—	50	52.6
4	135	80	108	64	90	—	—	70	83.4
2	181	100	149	82	125	—	—	90	132.8
1	211	125	177	105	150	—	—	100	167.5
0	245	150	204	125	200	—	—	125	212.0
00	283	175	237	146	225	—	—	150	266.0
000	328	200	—	—	275	—	—	175	336.0
0000	380	225	—	—	325	—	—	225	424.0

Source. Data from Refs. 14, 24, 25.

TABLE 7.11. ELASTIC CONSTANTS AND MECHANICAL PROPERTIES OF POLYCRYSTALLINE METALS AND ALLOYS AT ROOM TEMPERATURE

Material	Condition or Temper	Young's Modulus (GPa)	Yield Pt or 0.2% Offset (MPa)	Ultimate Tensile Strength (MPa)	Elongation in 2 in. (%)	Hardness
Carbon steel	Hot rolled	200	180	325	28	Brin 95
AISI 1010	cold drawn		305	365	20	Brin 105
Carbon steel	Hot rolled	200	305	550	16	Brin 163
AISI 1042	cold drawn		515	615	12	Brin 179
Carbon steel	Hot rolled	200	380	690	12	Brin 207
AISI 1078	cold drawn		500	650	10	Brin 192
Stainless steel	Annealed	195	205	515	40	—
Type 304	ASTM A666 Gr C		515	860	10	—

TABLE 7.11. (*Continued*)

Material	Condition or Temper	Young's Modulus (GPa)	Yield Pt or 0.2% Offset (MPa)	Ultimate Tensile Strength (MPa)	Elongation in 2 in. (%)	Hardness
Stainless steel Type 450	Annealed	200	170	415	—	R_B 88
54Fe-29Ni-17Co ASTM F15	Cold worked	140	345	515	—	R_B 100
58Fe-42Ni ASTM F30	Annealed	145	275	450	30	R_B 70
48Fe-52Ni ASTM F30	Annealed	165	275	450	35	R_B 70
Gold	Annealed	83	—	131	45	Vick 25
	cold worked		207	221	4	—
Platinum	Annealed	172	14	124	40	Vick 40
	cold worked		186	234	2.5	Vick 100
Silver	Annealed	76	55	152	48	Vick 26
	cold worked		303	345	2.5	—
Aluminum alloy 1100	– O	69	34	90	35	R_B 23[a]
	– H18		150	165	5	R_B 44[a]
Aluminum alloy 2024	– O	72	76	185	20	R_B 47[a]
	– T3		345	485	18	R_B 120[a]
Aluminum alloy 5056	– O	72	150	290	—	R_B 65[a]
	– H18		405	435	—	R_B 105[a]
Aluminun alloy 6061	– O	68	55	125	25	R_B 30[a]
	– T6		275	310	12	R_B 95[a]
Aluminum alloy 7075	– O	71	105	230	17	R_B 60[a]
	– T6		505	570	11	R_B 150[a]
Copper C10100 OFHC	Annealed	115	69	220	45	R_F 40
	– HO2		205	260	35	R_F 70
	– HO4		250	290	14	R_F 84
Copper C14500 0.5% Te	Annealed	115	69	220	50	R_F 40
	– HO2		275	290	25	R_F 42
	– HO4		305	330	20	R_F 48
Copper C17200 2.0% Be	Sol. treated & TD04	125	515	655	20	R_B 92
Copper C23000 red brass	Annealed	115	69	270	48	R_F 56
	– HO2		340	395	12	R_B 65
	– HO4		395	485	5	R_B 77

Source. Data from Refs. 5, 6, 26, 27.
[a] Rockwell B scale, 500 kg load, 10 mm ball.

The engineering strain, Σ, is the change in length per unit undistorted length, l_0

$$\Sigma = \frac{\Delta l}{l_0} = \frac{l - l_0}{l_0} \tag{7.16}$$

Table 7.11 shows the elastic and mechanical properties of some metals and alloys.

Shear Stress and Strain. The shear stress is the shear force F per unit area, A_0 (see Fig. 7.2*b*)

$$\tau = \frac{F}{A_0} \tag{7.17}$$

where τ = shear stress, Pa
F = shear force, N
A_0 = area, m^2

The shear strain is defined as the amount of deformation, Δ, in shear per unit length (see Fig. 7.2*c*)

$$\gamma = \frac{\Delta}{l_0} \tag{7.18}$$

Young's, Shear, and Bulk Modulus. Young's modulus, E, is the constant of proportionality between stress σ and strain Σ

$$\sigma = E\Sigma \tag{7.19}$$

This equation is valid for elastic strains only. Young's modulus is also referred to as Hooke's law.

Shear modulus, G, is the proportionality between shear stress, τ, and shear strain, γ

$$\tau = G\gamma \tag{7.20}$$

This is sometimes referred to as the rigidity modulus.

The bulk modulus, K, is the inverse of the compressibility, β

$$K = \frac{1}{\beta} \tag{7.21}$$

The compressibility, β, is the proportionality between the decrease in volume, $\Delta V/V$, and the pressure change, ΔP

$$-\frac{\Delta V}{V} = \beta(\Delta P) \tag{7.22}$$

Poisson's Ratio. Poisson's ratio is the ratio between the lateral contraction, ϵ_y, and the longitudinal elongation, ϵ_x, when a material is strained in uniaxial tension.

The lateral strain ϵ_y is found from the original (b_0) and contracted (b) width when the Poisson's ratio v is

$$v = -\frac{\epsilon_y}{\epsilon_x} \tag{7.23}$$

For common materials v has a value close to $\frac{1}{3}$.

Thermal

Thermal Conductivity. The flux of heat, q, flowing along a rod is proportional to the temperature gradient dT/dx and the cross section of the rod A. The constant of proportionality, λ, is called the thermal conductivity.[7,32]

$$q = -\lambda A \frac{dT}{dx} \tag{7.24}$$

where λ = thermal conductivity, $W/m \cdot K$
A = area, m^2
dT/dx = temperature gradient

Wiedemann–Franz Law. In metals the ratio of the thermal conductivity, λ, and the product of the temperature T and the electrical conductivity, σ, is found to be a constant, L.

$$L = \frac{\lambda}{\sigma T} \tag{7.25}$$

where λ = thermal conductivity, W/m · K
σ = electrical conductivity, S/m
T = temperature, K

This ratio, called the Wiedemann–Franz ratio, is constant for host pure metals and has a value of 20,000 to 30,000 $(\mu V/K)^2$.

Specific Heat. Specific heat, sometimes referred to as heat capacity, is the amount of heat required to raise the temperature of 1 kg of the material by 1 K at constant pressure; this is C_p. The same quantity at constant volume is called C_v. C_p and C_v are related by

$$C_p - C_v = \alpha^2 VTK \tag{7.26}$$

where α = coefficient of volume expansion
V = molar volume
T = temperature, K
K = bulk modulus

For solid metals, $C_p \simeq C_v$.

Temperature Coefficient of Electrical Resistivity (TCR). Above the Debye temperature, where all lattice vibration modes are excited, the thermal component of resistivity of conductors is approximately linear. The resistivity can be expressed as

$$\rho = \rho_0[1 + 2(T - \alpha T) + \cdots] \tag{7.27}$$

where ρ_0 = room temperature resistivity
α = temperature coefficient of electrical resistivity, °C^{-1}

For pure metals α is about 0.004 per degree centigrade or 4000 ppm/°C in popular units. α for alloys is usually lower. This equation does not hold for high temperatures, where other electron scattering processes set in.

Thermal Coefficient of Linear Expansion. The thermal coefficient of linear expansion, α_L, is defined as the change in length, dl, per unit length, l, per degree rise in temperature, dT.

$$l = l_0[l + \alpha_L(T - T_0)] \tag{7.28}$$

where l_0 = length, m, at temperature T_0
l = length, m, at temperature T
T = temperature, K

If the length of a rod l_0 is known at a temperature T_0, its length, l, at any other temperature, T, is

$$l = l_0[l + \alpha_L(T - T_0)] \tag{7.29}$$

Volume Coefficient of Thermal Expansion. The volume coefficient of thermal expansion, α_V, is defined as

$$\alpha_V = \frac{1}{V}\frac{dV}{dT} \tag{7.30}$$

where V = volume, m^3
T = temperature, K

Usually $\alpha_V = 3\alpha_L$.

Creep. Creep is the continued plastic deformation of materials as a function of time under a given stress. It is greatly accelerated at high temperatures approaching the melting temperature.

Electromigration

Electromigration is the transport of mass under the influence of an electric current. For a full account, consult Ref. 30. Electromigration causes failures of metal-conducting components in an integrated circuit. Table 7.12 shows the mean time to failure (MTF) for aluminum, gold, and copper films. Electromigration can proceed in thin films via grain boundaries.[29]

Diffusion

Diffusion is the transport of mass in a solid, liquid, or gas. The driving force is a concentration gradient. Fick's law states that the flux, J, of a component is proportional to its concentration

TABLE 7.12. MEDIAN FAILURE TIMES FOR ALUMINUM, GOLD, AND COPPER CONDUCTORS AS A FUNCTION OF ALLOYING ADDITIONS ($J = 2 \times 10^6$ A/cm^2)

Conductor	Alloy (wt%)	Mean Time to Failure (hr)
Al at 175° C		
Al		30–45[a]
	Si 1.8	100–200
	Cu 4	2500
	Cu 4, Si 1.7	4000[a]
	Ni 1	3000
	Cr	8300
	Mg 2	1000
	Cu 4, Mg 2	10,000[a]
	Cu 4, Ni 2, Mg 1.5[b]	32,000[a]
	Au 2	55
	Ag 2	45
Au at 300° C[c]		
	Ta	4000[a]
	Ni–Fe	800
	Mo	100[a]
	W + Ti	25
Cu at 300° C		
Cu		180
	Al 1	300
	Al 10	6000
	Be 1.7[b]	20,000[a]

Source. Data from Refs. 19, 28, 33.
[a] Extrapolated by means of appropriate values for n and ΔH.
[b] Deposited from an alloyed source by evaporation; the composition indicated is that of the source.
[c] Gold films on SiO_2 need an adhesion layer, which is made of the material indicated in the table. After annealing some degree of alloying takes place.

gradient, dC/dx

$$J = -D\frac{dC}{dx} \tag{7.31}$$

where D = diffusion coefficient, m^2/sec
J = flux, sec^{-1}cm^{-2}
C = concentration, cm^{-3}

The diffusion coefficient is temperature dependent and follows the exponential law

$$D = D_0 \exp\frac{(-Q)}{(kT)} \tag{7.32}$$

where D = diffusion coefficient at a temperature, T
D_0 = preexponential constant or diffusion coefficient
Q = activation energy for diffusion, J or eV
k = Boltzmann's constant, J/K or eV/K
T = temperature, K

Tables 7.13 and 7.14 show relevant diffusion data for various couples.[34–36]

TABLE 7.13. RESULTS OF THIN FILM DIFFUSION MEASURED BY SURFACE TECHNIQUES AND LATERAL SPREADING METHOD

Diffusant / Matrix	Method of Measurement	Temperature (°C)	Diffusivity (cm^2/s)	Activation Energy (eV)
Ag / Au	AES surface accumulation	30–260	$10^{-5}(D_b^0)$	0.63
Ag / Au	AES sputter profiling	30	$\sim 2 \times 10^{-18}$	
Ag / Au	He ion scattering	30	$\sim 10^{-14}$	
Ag / Au	AES surface accumulation	150–260	$10^{-6}(D_b^0)$	1.1
Cr / Au	AES surface accumulation	210–293	$10^{-3}(D_b^0)$	1.09
Cu / Au–0.2 Co	AES surface accumulation	100 125 150	6.4×10^{-15} 3.6×10^{-14} 1.3×10^{-13}	
Cu / Au	He ion backscattering	25	$\sim 10^{-17}$	
Co / Au–0.86 Co	AES sputter profiling	100 150	3.3×10^{-18} 2.4×10^{-16}	
Cu / Al	AES sputter profiling	175	8.6×10^{-13}	
Cu / Al	Lateral spreading	150–300	$10(D_b^0)$	0.81
Au / Ag	ESCA surface accumulation	141	4.3×10^{-17}	
Pt / Cr	AES surface accumulation	400–580	$10^{-2}(D_b^0)$	1.69
Ag / Cu	AES surface accumulation	300	$\sim 10^{-14}$	
Pb / Cu	AES surface accumulation	125–230		0.52
Ag / Ni	AES surface accumulation	300	$\sim 10^{-18}$	
Au / Pt	AES surface accumulation	250–350		0.96
Si / W	AES surface accumulation	670–850		2.6–3.2
Sn / Sn	Tracer scanning	142–213	1.8×10^{-5}	0.46

Source. Data from Ref. 34.

TABLE 7.14. RADIOACTIVE TRACER GRAIN BOUNDARY DIFFUSION DATA IN SELECTED SYSTEMS

Diffusant / Matrix[a]	Method of Measurement	δD^0(cm^3/s)	Activation Energy (eV)	Remarks[b]
^{195}Au / Au (epi. film)	Ion sputtering	1.9×10^{-10}	1.16	Diss. disl.
^{195}Au / Au (p. bulk)	Ion sputtering	3.1×10^{-10}	0.88	High-angle GBs
^{195}Au / Au (p. film)	Ion sputtering	9.0×10^{-10}	1.0	High-angle GBs
^{195}Au / Au–1.2 Ta (p. bulk)	Ion sputtering	5×10^{-7}	1.26	High-angle GBs
^{195}Au / Au–1.2 Ta (p. film)	Ion sputtering	1×10^{-9}	1.2	Low-angle GBs
^{195}Au / Au–1.2 Ta (p. bulk)	Ion sputtering	1×10^{-9}	1.2	Low-angle GBs
^{195}Au / Ni–0.5 Co (p. film)	Ion sputtering	1.4×10^{-10}	1.6	High-angle GBs
^{110}Ag / Ag (p. bulk)	Lathe sectioning	1.3×10^{-9}	0.80	High-angle GBs
Ni / Ni bicrystal	Autoradiography	7×10^{-10}	1.08	45 ⟨100⟩
^{63}Ni / Ni bicrystal	Autoradiography	2.2×10^{-8}	1.77	10° ⟨112⟩ tilt

Source. Data from Refs. 28, 34.
[a] p. = polycrystalline matrix; epi. = epitaxial.
[b] diss. disl. = dissociated dislocations.

TABLE 7.15. PROPERTIES OF STRESS RELIEVED AND ANNEALED MICROELECTRONIC BONDING WIRES

Bonding Wire	Diameter (in.)	Diameter (μm)	Elongation (%)	Tensile Strength (g)	Resistance (Ω/0.100″)	Current Capacity[a] (A)
Gold	0.0007	18	2–6	3.5	0.225	0.380
			6–10	3.5		
	0.001	25	2–27	9	0.110	0.648
			6–10	9		
	0.00125	32	3–8	14	0.071	0.906
			7–12	12		
	0.002	51	5–12	32	0.028	1.834
			12–20	30		
Aluminum, 1% silicon	0.0007	18	0.75–3	6	0.273	0.282
			0.75–3.5	4		
	0.001	25	1–3.5	15	0.133	0.481
			1–4	12		
	0.00125	32	1–4	22	0.086	0.671
			1–4	17		
	0.002	51	2–7	62	0.034	1.359
			2–8	50		

[a] Allowable per MIL-M-38510E for wires greater than 0.040″ long.

7.6 MICROELECTRONIC PACKAGING MATERIALS

Microelectronic packaging[25] is a broad term that includes all of the materials and processes by which mechanical and environmental protection, thermal management, and electrical interconnection are provided to a semiconductor device or assemblage of semiconductor devices.

In many cases, gold and/or aluminum wires[28] are utilized for the electrical connection to the semiconductor chip. Properties of these bonding wires are shown in Table 7.15. Specific packaging materials are selected for their abilities to provide a satisfactory solution to a particular problem. Various trade-offs are usually required and the chosen packaging approach represents a compromise. Major factors and related considerations affecting the choice of packaging materials include:

1. **Thermal requirements**[34]
   (a) Maximum allowable junction temperatures
   (b) Matching of expansion coefficients
   (c) Ambient operating temperature range
2. **Electrical requirements**
   (a) Frequency effects
   (b) Electrostatic or electromagnetic shielding
   (c) Voltage drops in interconnections
   (d) Cross talk or Electromagnetic Interactions (EMI)
   (e) Sensitivity to light
   (f) Insulation resistance
3. **Reliability requirements**
   (a) Hermeticity
   (b) Environmental protection
   (c) Radiation hardness
   (d) Susceptibility to alpha particles
   (e) Fatigue resistance
4. **Practical considerations**
   (a) Cost
   (b) Available equipment
   (c) Available expertise
   (d) Available materials
   (e) Schedule
   (f) Process sequencing
   (g) Ease of rework
   (h) Weight
   (i) Structure
   (j) Size
   (k) Imposed specifications

Some typical materials used in the microelectronics industry and their limitations follow. Most applications are materials limited. Some resistive materials are as follows:

Application	Material	Drawback
Zero-temperature coefficient of resistivity	Nichrome	
Precision wire resistors	Nichrome	Aging of wire creep
Precision resistors	Carbon	Incompatibility with microelectronic processing
Thick film resistors	Ruthenium based	Expensive; prone to failure with hydrocarbons in the atmosphere

TABLE 7.16. PROPERTIES OF SELECTED PACKAGING MATERIALS

Material	Melting Point (°C)		Density (g/cm^3)	Thermal Conductivity (W/m-K)	Thermal Expansion (10^{-6}/K)	Resistivity ($\mu\Omega$-cm)
54Fe-29Ni-17Co ASTM F15	1450		8.37	17.6	5.5	49
58Fe-42Ni ASTM F30	1440		8.12	10.9	4.3	72
48Fe-52Ni ASTM F30	1435		8.30	13.3	9.8	43
Stainless steel type 304	1400		7.9	16.2	17.2	72
Carbon steel AISI 1010	1510		7.84	60.2	12.2	16.8
Aluminum	660		2.70	226	23.5	2.87
Copper (OFHC)	1083		8.96	391	17.0	1.72
Molybdenum	2615		10.2	134	5.1	5.7
Nickel	1455		8.9	88	13.3	6.9
Silicon	1412		2.34	138	7.6	10^5
Tungsten	3400		19.3	159	4.5	5.5
Al_2O_3, 96%	—		3.7	30.2	6.4	10^{20}
Al_2O_3, 99.5%	2040		3.9	32.3	6.5	10^{20}
BeO, 99.5%	2550		2.9	215	7.5	10^{20}
Solders	**Solid**	**Liquid**				
52In-48Sn	118	118	7.3	34	20	14.7
97In-3Ag	143	143	7.38	73	22	7.5
62Sn-36Pb-2Ag	179	179	8.41	31	25	14.8
63Sn-37Pb	183	183	8.4	50	25	15
50In-50Pb	180	209	8.86	22	27	28.7
96.5Sn-3.5Ag	221	221	7.36	33	30	10.8
Adhesives						
Epoxy—Ag filled	—	—	—	1.6	—	200
Epoxy—unfilled	—	—	—	0.35	—	10^{13}

Source. Data from Refs. 6, 23, 27.

Some conductive materials are as follows:

Application	Material	Drawback
Interconnect wires	Al	Not strong;
	Al-Si	Si tends to precipitate;[37] prone to electromigration
	Al-Cu-Si	Cu prevents electromigration
Interconnect wire 25 μm diameter	Au pure	Expensive; soft
	Au + 3 ppm Be	Harder and bondable
High temperature, high current, thin	Al-Cu-Si	Prone to degradation[37] and electromigration
film interconnects	Mo-silicides, W-silicides	Sensitive to method of vapor deposition
Thick film interconnects	Pt-Pd-Ag films, Au doped with Be or Pt or Cu	Problems with interfacial silicide formation;[38] alloy with components; expensive

Some packaging materials are as follows:

Application	Material	Drawback
Substrates	Al_2O_3, alumina	Radioactivity in substrate causes device memory reversal
Substrates	BeO	
Bond devices to substrates and conduct heat away	Conducting epoxys	
Seal covers in microelectronic packaging; covers are gold-plated KOVAR or alloy 42	Au-Sn solder, Ag-Sn solder	Cosmetic defects; solder wets top surface
Headers to hold devices	Nickel-plated, cold-rolled steel; pins are nickel-plated KOVAR	Won't gold plate; nickel-plated pins have bonding and soldering problems

Some properties of selected packaging materials are shown in Table 7.16.

References

1 American National Standard Metric Practice IEEE Std. 268, 1976; ANSI Z210, 1979; ASTM E 380-76.

2 R. M. Rose, L. A. Shepard, and J. Wolff, "The Structure and Properties of Materials," vol. IV, *Electronic Properties*, Wiley, New York, 1966.

3 E. H. Sondheimer, *Adv. Phys.* **1**:1 (1952).

4 A. N. Gerritsen, "Metallic Conductivity, Experimental Part," vol. 19, p. 137, in S. Flugge, ed., *Handbuch der Physik*, Springer-Verlag, Berlin, 1956.

5 C. J. Smithels, *Metals Reference Book*, 5th ed., Butterworth, London, 1976.

6 B. P. Bardes, ed., *Metals Handbook*, 9th ed., American Society for Metals, Metals Park, PA, 1978.

7 C. A. Harper, ed., *Handbook of Materials and Processes for Electronics*, McGraw-Hill, New York, 1970.

8 E. S. Parker, *Material Data Book*, McGraw-Hill, New York, 1965.

9 C. T. Lynch, *Handbook of Materials Science*, CRC, Boca Raton, FL, 1975.

10 L. E. Toth, *Transition Metal Carbides and Nitrides*, Academic Press, New York, 1971.

11 O. E. Gray, ed., *AIP Handbook*, 3rd ed., McGraw-Hill, New York, 1972.

12 M. Hansen and K. Anderko, *Constitution of Binary Alloys*, McGraw-Hill, New York, 1958.

13 R. P. Elliot, *Constitution of Binary Alloys*, 1st Sppl., McGraw-Hill, New York, 1965.

14 D. G. Fink and D. Christiansen, *Electronics Engineers Handbook*, 2nd ed., McGraw-Hill, New York, 1982.

15 A. S. Berezhnoi, *Silicon and Its Binary Systems*, Consultants Bureau, New York, 1960.

16 F. A. Trumbore, "Solid Solubilities of Important Elements in Ge and Si," *Bell System Tech. J.* **39**:205 (1960).

17 D. S. Campbell, *The Use of Thin Films in Physical Investigation*, Academic, New York, p. 299, 1966.

18 C. A. Neugebauer, in B. Schwartz and N. Schwartz, eds., *Measurement Techniques for Thin Films*, The Electrochemical Society, New York p. 191, 1967.

19 P. B. Ghate and T. C. Blair, *Thin Solid Films* **55**:113 (1978).

20 F. Mohammadi, *Solid State Tech.* **24**(1):65 (1981).

21 A. K. Sinha, *J. Vac. Sci. Technol.* **19**(3):778 (1981).

22 S. P. Murarka, *J. Vac. Sci. Technol.* **17**(4):(1980).

23 C. A. Harper, ed., *Handbook of Electronic Packaging*, McGraw-Hill, New York, 1970.

24 D. G. Fink, *Electronics Engineers Handbook*, McGraw-Hill, New York, 1975.

25 D. G. Fink and H. W. Beaty, *Standard Handbook for Electrical Engineers*, McGraw-Hill, New York, 1978.

26 L. F. Mondolfo, *Aluminum Alloys, Structure and Properties*, Butterworth, London, 1976.

27 *Materials Engineering Magazine*, 1983 Materials Selector, Penton/IPC, 1983.

28 J. M. Poate, K. N. Tu, and J. W. Mayer, *Thin Films: Interdiffusion and Reactions*, Wiley, New York, 1978.

29 F. J. d'Heurle and P. S. Ho, "Electromigration in Thin Films," in Poate et al.,[28] pp. 243.

30 H. B. Huntington and A. R. Grone, *J. Phys. Chem. Solids* **20**:76 (1961).

31 J. R. Black, Electromigration—A Brief Survey and Some Recent Results, *IEEE Trans. Electron Dev.* **16**:338 (1969).

32 A. Goldsmith, et al., *Handbook of Thermophysical Properties of Solid Materials*, CRC, Boca Raton, FL, 1982.

33 G. T. Meadea, *Electrical Resistance of Metals*, Plenum, New York, 1965.

34 D. Gupta, D. R. Campbell, and P. S. Ho, "Grain Boundary Diffusion in Thin Films," in Poate et al.,[28] pp. 161.

35 D. Gupta, "Thin Film Couples," in Poate et al.[29]

36 Y. Adda and J. Philibert, *La Diffusion dans Les Solides*, 1, Vols. 1 and 2, Presse. Univ. France (Paris), 1966.

37 M. Paugh and K. Seshan, "Degradation of Mechanical Properties of Annealed 25 μm Al-1% Si Microelectronic Interconnect Wire," *Met. Trans.* **14(A)**, 921 (1983).

38 G. Ottaviani and J. W. Meyer, "Mechanisms and Interfacial Layers on Silicide Formation; Thin Film Interactions," in *Reliability and Degradation*, M. J. Howes and D. V. Morgan, eds., Wiley, New York, pp. 105–149, 1981.

CHAPTER 8
INSULATING AND DIELECTRIC MATERIALS

SORAB K. GHANDHI
J. KEITH NELSON

Rensselaer Polytechnic Institute
Troy, New York

8.1 INSULATING FILMS FOR SOLID STATE DEVICES AND INTEGRATED CIRCUITS

Sorab K. Ghandhi

Insulating films are used today at many points in the manufacture of silicon and gallium arsenide monolithic integrated circuits.[1,2] First and foremost, they serve as masks against inpurities during important fabrication processes such as diffusion and ion implantation. Thus they allow the patterned introduction of dopants into the semiconductor substrate, so that multiple devices can be made on a single chip.

Next they play an active role in device operation, as the dielectric material in insulated gate field effect transistors. They also allow electrical connections to be made between devices and serve as the insulating layer over which conductors can be placed. Often a number of layers of these films are used in multilayer interconnection schemes, which are common in modern integrated circuit technology.

Finally these films play an important role in protecting the complete microcircuit. Ideally their use can avoid expensive hermetic packaging and thus result in an important cost reduction for these circuits.

Each of the mentioned application areas requires different film properties; moreover many different preparative techniques can be used in each case.[3,4] Here space considerations wil! limit the description to the most commonly used methods.

Native insulating films are grown out of reaction with the semiconductor itself.[5] These films, usually the oxides of silicon and gallium arsenide, are relatively free from contamination; moreover they can be grown in situ as part of other processes, such as diffusion. In the case of silicon, oxide growth is accomplished conventionally by subjecting the slice to either dry oxygen or water vapor at elevated temperatures. This process is carried out in a resistively heated quartz reaction chamber in which the silicon slices are held on a carrier. Dry oxygen or water vapor is introduced at the inlet of this chamber, to result in the formation of amorphous silicon dioxide. Details of this type of system and its many variations are available in the literature.[1,2]

Native insulating films of gallium arsenide usually consist of mixtures of Ga_2O_3 and As_2O_3 and are commonly grown at room temperature by anodization. Many different anodization solutions have been used for this purpose.

A variety of deposited insulating films are also used in integrated circuit technology. The most common technique[6–8] for their preparation is chemical vapor deposition (CVD), although vacuum evaporation, sputtering, and ion deposition techniques are also used.[9,10] CVD films are grown out of the chemical reaction of volatile species in systems that are very similar in configuration to those used for thermal oxidation. Here, however, both resistance-heated hot-wall systems and RF-heated cold-wall systems are in use. In many cases growth is carried out at low pressures, typically in the 10^{-3} atm range. This greatly increases the rate of gaseous diffusion in the reactor and allows tight stacking of slices while still maintaining uniform film thickness.

Yet other CVD systems incorporate in the deposition process the use of the electronically excited species, which are created by means of a plasma. These plasma-enhanced CVD systems allow fabrication of insulating films at low temperatures and are especially useful in the final processing steps for integrated circuits.

Deposited films can also be prepared by the spinning of a liquid with subsequent bakeout. Again the ability to fabricate these films at low temperatures makes them attractive in the final phases of integrated circuit fabrication.

This section of Chapter 8 discusses the preparative aspects of insulating films and their electrical and physical properties. An extensive list of references is provided.

8.1-1 Native Oxides

The native oxide of silicon is usually grown by passing either oxygen or water vapor over silicon held at 800–1200°C. The use of dry oxygen results in dense oxides with a relatively slow growth rate.[10,11] Water vapor, obtained by bubbling oxygen or nitrogen through deionized water held at 95°C, corresponding to a vapor pressure of 640 torr (0.842 atm), results in a somewhat less dense oxide but one with a faster growth rate. Pyrogenic systems, in which hydrogen and oxygen are burned, are often used for this purpose because of their operational convenience.

Figures 8.1 to 8.4 show the oxide growth rate in wet and dry oxygen as a function of temperature

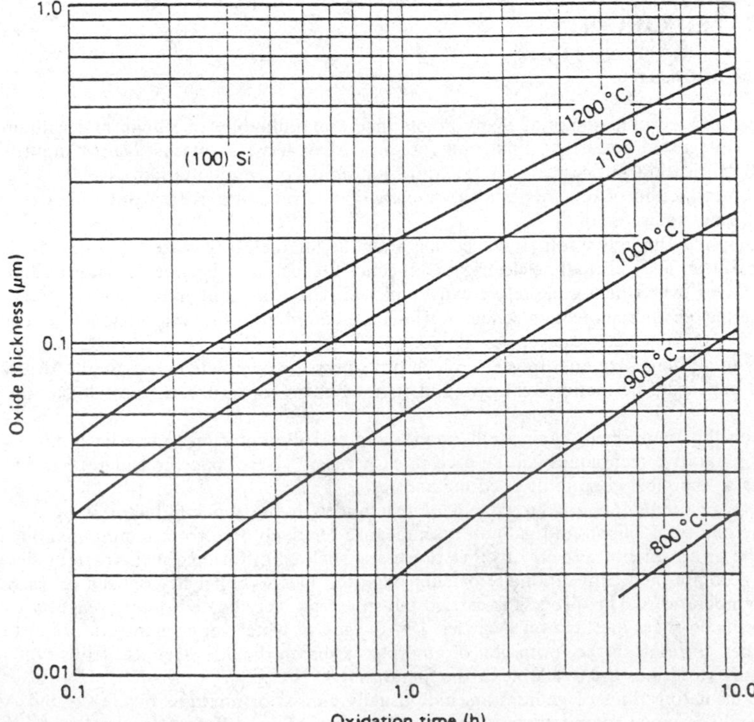

Fig. 8.1 Oxide growth rate for dry oxygen: (100) silicon.

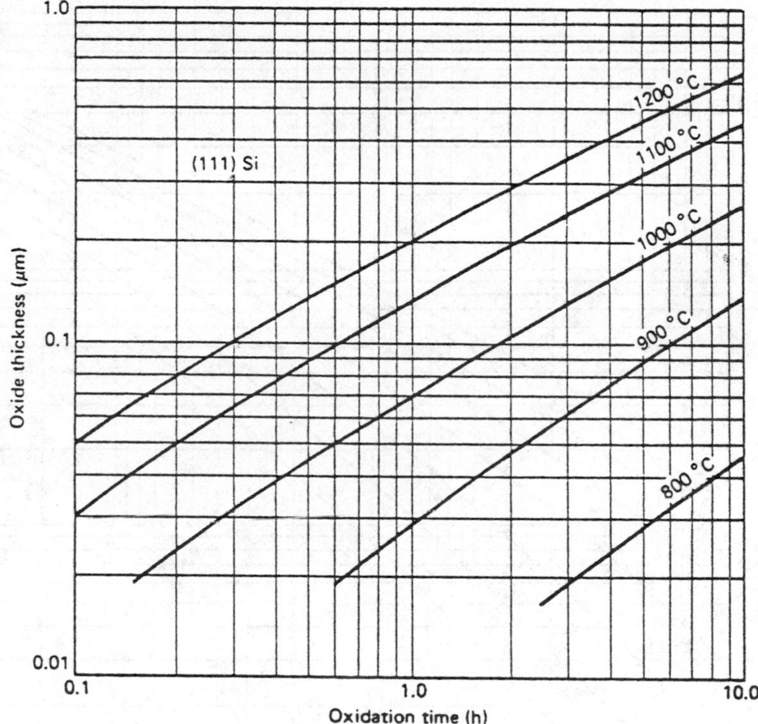

Fig. 8.2 Oxide growth rate for dry oxygen: (111) silicon.

and crystal orientation. As seen,[2] growth is much faster during wet oxidation. High-pressure oxidation systems are currently under development to allow oxide growth to be carried out at even faster rates, or alternatively, at reduced temperatures.[12,13]

Table 8.1 lists the properties of oxides grown under wet and dry conditions. Note that the oxide grown dry is of somewhat better quality, with higher density and higher dielectric strength, than its wet counterpart.

An important property of these oxides is their ability to mask against diffusing impurities. A measure of this capability is provided in Table 8.2, which lists the relative diffusivities of common dopants in silicon. With the exception of gallium, all of these are about three to four orders of magnitude below the values for impurity diffusivity in silicon. As a consequence, silicon dioxide films serve as excellent diffusion masks for boron, phosphorus, arsenic, and antimony. Figures 8.5 and 8.6 show the relative masking properties of silicon dioxide against the most commonly used dopants, boron and phosphorus. It should be emphasized, however, that these as well as other properties of silicon dioxide are a strong function of the preparative method. Thus this figure serves at best as a guideline of this masking capability.

The electronic properties of oxides created dry are superior to those made by wet oxidation. Thus dry oxides have been shown to have a lower fixed oxide charge[15] as well as a lower density of trapped charge at the oxide–silicon interface than wet oxides. The use of halogenic species during oxidation has resulted in a further reduction of these charges and an improvement in oxide quality.[16–19] Native oxides of silicon, grown under dry conditions, are thus especially useful in the fabrication of gate oxides in metal oxide semiconductor (MOS) devices. The wet oxides, on the other hand, are used in most general purpose applications because of their faster growth rate.

The best oxides of gallium arsenide are grown by wet anodization techniques, which are carried out at room temperature in an anodization cell. Here the gallium arsenide serves as the anode and a noble metal such as platinum is used as a cathode. A wide variety of electrolytes are used as cathodes and all result in a mixture of Ga_2O_3 and As_2O_3, in varying proportions.

Aqueous electrolyte systems for gallium arsenide are based on water or water-H_2O_2 mixtures[20,21] with additives for pH control. These additives may be acidic (such as H_3PO_4), basic (such as NH_4OH), or neutral [such as $(NH_4)_2HPO_4$]. Only a narrow pH range is allowed with each system in

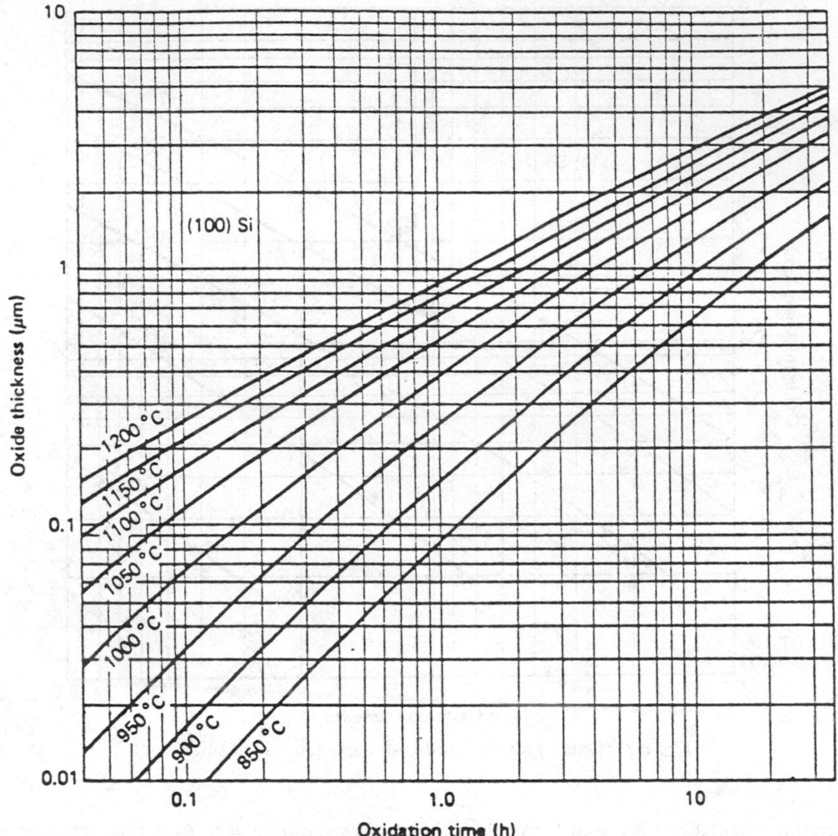

Fig. 8.3 Oxide growth rate for wet oxygen: (100) silicon.

order to prevent dissolution of the oxide during its formation. Nonaqueous systems have also been used for this purpose[22] and are similar in their characteristics.

Recently much work has centered around an anodization system of the mixed aqueous/nonaqueous type, using a solution of water and propylene glycol containing tartaric acid.[23] Here a 3% aqueous solution of tartaric acid is used, with ammonia to adjust its pH to the 5 to 7 range. To this aqueous solution is added propylene glycol in a 1:2 (parts-by-volume) ratio (aqueous/glycol). This mixture results in an extremely low dissolution rate for GaAs, typically 4 to 6×10^{-2} Å/s, compared with rates as high as 1 to 3 Å/s for aqueous electrolytes. Oxides grown in this system have been used successfully as gates in MOS devices,[24] and also as diffusion masks for gallium arsenide.

Anodic oxides of silicon can also be grown. However, these oxides are quite porous in character and are not used in device fabrication processes at the present time. Their use is confined to diagnostic applications where measurements must be made on successively removed layers of silicon.[25]

Table 8.1 provides a brief summary of the important electrical properties of native oxide films. The wide range of parameters serves to emphasize the sensitivity of these films to the preparative technique that is used.

In summary, thin insulating films must meet a number of requirements if they are to be suitable for use in microcircuits. First and foremost, their electrical resistivity must be sufficiently high so that circuit operation is not affected by their presence. Typically most insulators conduct by electronic processes, and with ohmic behavior, at electric fields below 10^5 V/cm; here a bulk resistivity of 10^{12} Ω-cm is considered adequate for most practical purposes. High field conduction becomes dominant in the 10^6 V/cm range and can occur by both electronic or ionic processes. In addition, tunneling and electron hopping can also occur, and the general conduction characteristic takes on a nonohmic

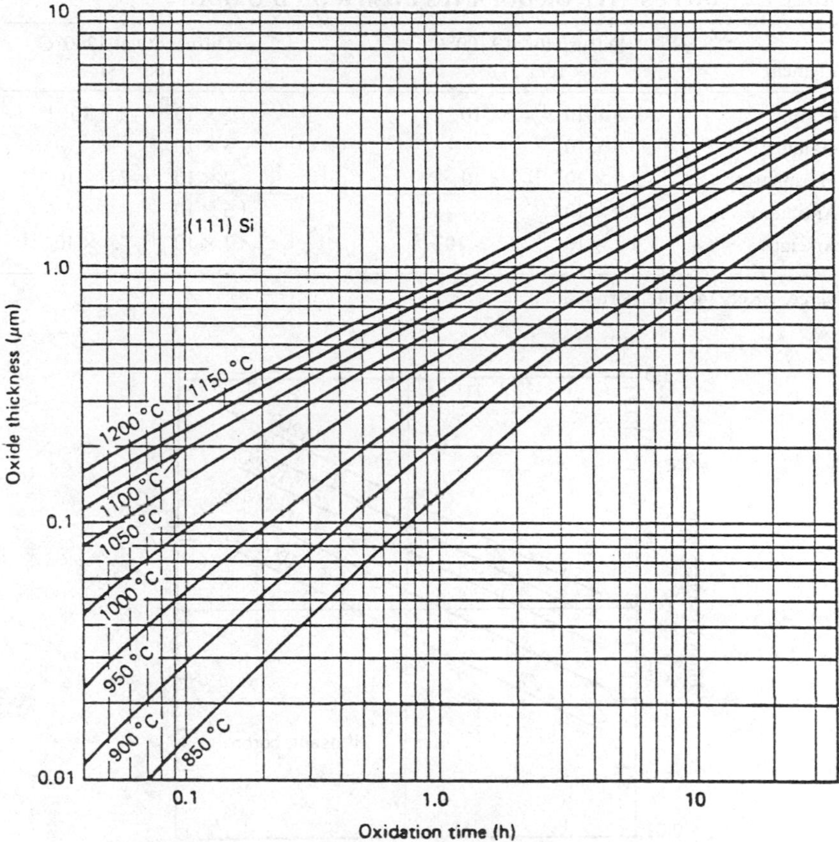

Fig. 8.4 Oxide growth rate for wet oxygen: (111) silicon.

behavior. High field conduction processes are of special importance in oxides that are used in the gate region of MOS devices, because of their extreme thinness ($\simeq 250$ Å).

Loss mechanisms become apparent at high frequencies. They are a complex function of interface and bulk trapping phenomena. Often they are directly related to the preparative technique for the oxide. Here native oxides generally outperform deposited films because of their inherently cleaner character.

TABLE 8.1. PROPERTIES OF NATIVE OXIDES OF SILICON AND GALLIUM ARSENIDE

Method of Preparation	Density (g/cm^3)	Resistivity at 300 K (Ω-cm)	Dielectric Strength (V/cm)
Fused silica	2.2	$1\text{–}5 \times 10^{16}$	$2\text{–}9 \times 10^6$
Dry thermal oxide	2.2–2.3	$10^{15}\text{–}10^{16}$	$5\text{–}9 \times 10^6$
Wet thermal oxide	2–2.2	$10^{15}\text{–}10^{16}$	$1\text{–}5 \times 10^6$
Anodic oxide of silicon	1.8	$10^{12}\text{–}10^{15}$	$1\text{–}5 \times 10^6$
Anodic oxide (gallium arsenide)	4.2	$10^{14}\text{–}10^{15}$	$1\text{–}5 \times 10^6$

TABLE 8.2. DIFFUSIVITY OF DOPANTS IN SILICON DIOXIDE

Element	Diffusivity at 1100°C (cm^2/s)	Diffusivity at 1200°C (cm^2/s)
Boron	3×10^{-17}–2×10^{-14}	2×10^{-16}–5×10^{-14}
Gallium	5.3×10^{-11}	5×10^{-8}
Phosphorus	2.9×10^{-16}–2×10^{-13}	2×10^{-15}–7.6×10^{-13}
Antimony	9.9×10^{-17}	1.5×10^{-14}
Arsenic	1.2×10^{-16}–3.5×10^{-15}	2×10^{-15}–2.4×10^{-14}

Source. Ref. 14, with permission.

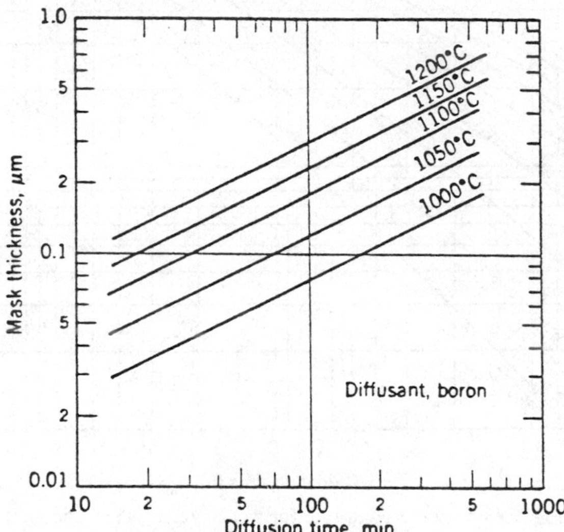

Fig. 8.5 Mask thickness for boron.

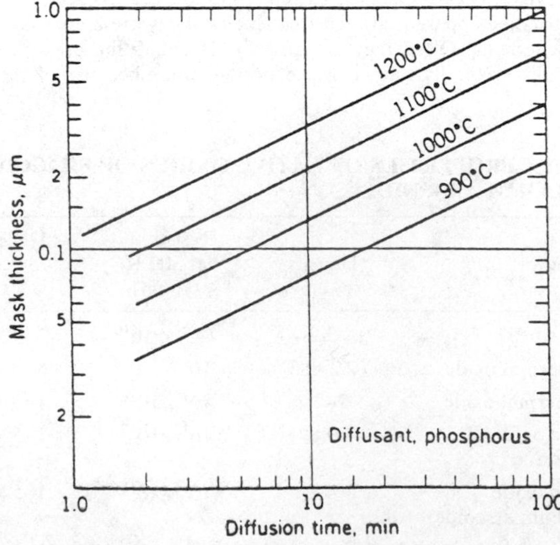

Fig. 8.6 Mask thickness for phosphorus.

Insulating films must also be metallurgically compatible with the substrate and interconnection metals. Thus they must adhere firmly to both and neither crack of themselves or damage the associated layers during thermal cycling. In some instances a pad layer of an intermediate material must be used to provide this stress relief. A case in point is the use of silicon nitride films as oxidation masks. The poor thermal expansion mismatch between silicon nitride and the underlying silicon causes a massive dislocation network to be formed in the silicon, thereby making it unsuitable for device operation. Typically a 100 to 200 Å silicon dioxide layer is used to relieve the stress created during this process.

8.1-2 Deposited Films

Many different films have been investigated for use in integrated circuit technology. However, those in most common use today are silicon dioxide, phosphosilicate glass, and silicon nitride. These are considered separately here.

Silicon Dioxide

This insulator has many unique properties which make it near ideal for semiconductor and integrated circuit technology. It is dense, hard, and inert to many chemicals that are used during device processing. It can be used as a diffusion or ion implantation mask against many common dopants, and also as a cap layer for postimplantation annealing purposes. At the same time it is readily etched with fine-line definition, so that it can be patterned for microcircuit applications. It can be deposited at temperatures ranging from 250 to 1000°C so that it can readily accommodate the thermal restrictions placed by previous processes.

Deposited silicon dioxide is an excellent insulator, with a specific resistance of about 10^{14} Ω-cm and a dielectric breakdown strength of above 10^6 V/cm. Thus it is used as an insulating layer over which metal films can be placed for connection between components. Its ability to bond firmly to aluminum (the most commonly used interconnection metal) makes it near-ideal for this application. Finally, its excellent abrasion-resistant properties make it suitable for a protective cover layer over the entire microcircuit.

The simplest deposition technique for this material is the spin-on method. Here, the slice is spin coated by holding it in a vacuum chuck which is rotated at 2500 to 5000 rpm. A drop of a silica-bearing mixture is next applied, to form a thin layer ($\simeq$ 5000 Å) across the slice by means of centrifugal force. This method can be used to form reasonably uniform layers, with appropriate attention to viscosity control. Moreover, successive applications can allow the buildup of these layers to 1 to 2 μ in thickness.

A number of proprietary formulations are available for this purpose. In general they consist of the acyloxysilanes and alkylsiloxanes, with additives such as butyl carbitol and cellulose for viscosity control during the spinning operation. Upon application, and subsequent oven drying, these films are converted to silicon dioxide by baking to about 250°C. Baking them at even higher temperatures allows them to become densified, with an overall improvement in their microstructure.

Spin-on films are useful as top coatings, since they simplify the handling of finished slices. Their lack of uniformity and their porosity restrict them to this application. Films of considerably better quality can be grown by the pyrolytic oxidation of a variety of alkoxysilanes[26] in the 700 to 800°C temperature range. One such compound is tetraethylorthosilane (TEOS), which is transported in vapor form to the reaction chamber by means of a bubbler arrangement. The oxidation reaction is

TABLE 8.3. ALKOXYSILANES FOR USE IN THE GROWTH OF SILICA FILMS

Material	Formula	Molecular Weight	Boiling Point (°C)	Vapor Decomposition Temperature (°C)
Tetraethoxysilane	$Si(OC_2H_5)_4$	208	167	728–840
Ethyltriethoxysilane	$(C_2H_5)Si(OC_2H_5)_3$	192	161	650–750
Amyltriethoxysilane	$C_5H_{11}Si(OC_2H_5)_3$	234	198	600–740
Vinyltriethoxysilane	$CH_2=CHSi(OC_2H_5)_3$	190	160	600–700
Phenyltriethoxysilane	$C_6H_5Si(OC_2H_5)_3$	240	234	610–750
Dimethyldiethoxysilane	$(CH_3)_2Si(OC_2H_5)_2$	148	111	760–900

Source: Ref. 10, with permission.

TABLE 8.4. PROPERTIES OF SILICON DIOXIDE FILMS GROWN BY CHEMICAL VAPOR DEPOSITION (CVD) AND PLASMA CVD

Property	CVD	Plasma CVD
Pinhole density	1–10 cm^{-2}	$\leqslant 1\ cm^{-2}$
Particulate density	10–100 cm^{-2}	$\leqslant 0.2\ cm^{-2}$
Step coverage	Poor	Conformal
Adhesion to aluminum	Good	Good
Crack resistance	Poor	Good
Intrinsic film stress	Tensile	Compressive
	6–8 $\times 10^8$ dynes/cm^2	2×10^9 dynes/cm^2
Breakdown voltage	8×10^6 V/cm	8×10^6 V/cm
Film resistivity	$> 10^{17}$ Ω-cm	$> 10^{17}$ Ω-cm

carried out in a cold-wall CVD system at 800°C; it proceeds as follows:

$$Si(C_2H_5O)_4 + 12\ O_2 = SiO_2 + 8\ CO_2 + 10\ H_2O$$

Table 8.3 lists the properties of a number of alkoxysilanes that can be used for this purpose.[10] In all cases oxide formation is accompanied by formation of a large amount of water as a reaction byproduct, resulting in films with somewhat impaired quality. Nevertheless, the electronic properties of these films are adequate for their use in a number of situations where a thick oxide must be deposited by a relatively low temperature process.

The growth of silica films is most commonly carried out by the pyrolytic oxidation of silane.[27,28] The silane reaction is conducted at lower temperatures (300–500°C), than the alkoxysilane reaction and proceeds as follows:

$$SiH_4 + O_2 \rightarrow SiO_2 + 2\ H_2$$

resulting in high-quality silica films. Typically a cold-wall CVD system of the resistance-heated type is used because of the low temperatures involved. The resulting film has a built-in tensile stress of about 3×10^9 dynes/cm^2 for a 450°C growth temperature.

Low-pressure CVD systems of the hot-wall type are increasingly used for this purpose[29] and result in improved film uniformity from slice to slice as well as in increased throughput. In addition, film quality is generally superior to that obtained in an atmospheric pressure system, with a reduced pinhole density. However, the growth rate is somewhat slower (100–150 Å/min as compared with 500–600 Å/min for atmospheric pressure systems).

Silicon dioxide films can also be grown at low pressure ($\simeq$0.1–0.5 torr) and low temperature (250°C) in plasma-enhanced systems[30,31] by means of reactions involving SiH_4/O_2, SiH_4/CO_2 and SiH_4/N_2O mixtures. This technique results in a built-in compressive stress in the deposited films, and greatly reduces the tendency to cracking during subsequent thermal cycling. As a result, films grown by this method can be much thicker than those grown at atmospheric pressure.

Some selected properties of deposited silica films are summarized in Table 8.4.

Phosphosilicate Glass

Phosphosilicate glass (PSG) is made by incorporating P_2O_5 into the silica film during the CVD process. This glass is more dense and void-free than silica, and it makes a better protective layer for this reason.[32] In addition, the incorporation of P_2O_5 into silica films reduces the built-in tensile stress from 3×10^9 dynes/cm^2 for the undoped film to about 2×10^9 dynes for films with 13% P_2O_5 by weight, and to zero for films with 20% P_2O_5 by weight. This, in turn, greatly improves their integrity during thermal cycling.

A further advantage comes about because the thermal expansion coefficient of silica rapidly increases with the incorporation of P_2O_5. Consequently, PSG films can be tailored to provide a more suitable thermal match to the underlying semiconductor. The thermal expansion coefficient of PSG is shown in Fig. 8.7 as a function of P_2O_5 content, over the composition range of practical interest.[33] Values for undoped silica (6×10^{-7}/°C), for silicon (2.6×10^{-6}/°C), and for gallium arsenide (5.9×10^{-6}/°C) are also shown in this figure.

PSG layers have been found to be effective in immobilizing sodium ions in MOS technology and are commonly used for this important application.[34] They are also used to enhance the stability of

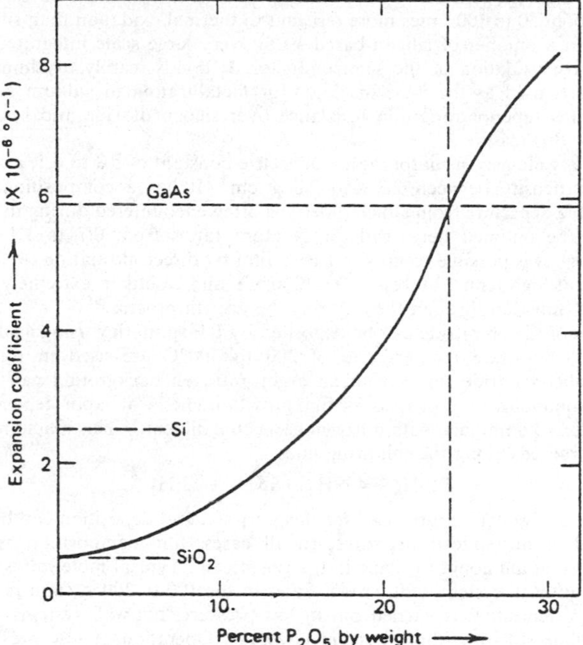

Fig. 8.7 Thermal expansion coefficient of phosphosilicate glass (PSG) films.

bipolar devices and microcircuits.[35] Application of these films followed by a heat treatment results in the gettering of metallic impurities in linear integrated circuits and in semiconductor power devices.[36,37] Finally they are often used as coatings on finished microcircuits, to protect the aluminum metallization from scratches during the final bonding operation and to provide permanent protection against alkali ion migration. These passivation layers must contain no more than 8% by weight of P_2O_5 to prevent corrosion reactions with the aluminum metallization in the presence of moisture.[38]

PSG films are grown by the simultaneous oxidation of silane and phosphine gases in the 300 to 500°C range, the same as that for the growth of silica films. Thus their deposition technique is a natural extension of the silicon dioxide growth process and can be carried out in the same system. Phosphine gas is usually provided in a 5 to 10% dilution in argon or nitrogen for this application. The reaction of phosphine gas with oxygen results in the formation of P_2O_5, which is incorporated as a network former in the resulting glass. The phosphine oxidation reaction is

$$2\ PH_3 + 4\ O_2 \rightarrow P_2O_5 + 3\ H_2O$$

so that a small amount of water is produced as a byproduct of this process.

Almost any amount of P_2O_5 can be incorporated by this technique. However, the films become increasingly hygroscopic, so that their P_2O_5 content is limited to about 2 to 8% by weight for films left permanently in place on the finished product. This content goes as high as 20% in films used only during device processing. The weight percentage of P_2O_5 in the glass is approximately equal to $1\frac{1}{2}$ times the PH_3/SiH_4 mole ratio in the gas phase, over this range of compositions.

Silicon Nitride

Films of silicon nitride are used extensively in both silicon and gallium arsenide device technology. They are more difficult to grow than films of silica or PSG so that their use is dictated in situations where they provide improved properties over these insulators. Often they are used in conjuction with silica films to obtain a combination of characteristics that neither can provide alone.

Unlike silicon dioxide, silicon nitride is an excellent barrier to alkali ion migration and is used extensively as a cover layer in MOS technology for this reason. It is superior to PSG since it is not hygroscopic and it has allowed the use of unencapsulated circuits in many consumer applications. It is also an excellent diffusion mask for gallium and is used in power device applications for this reason. Its ability to restrict the diffusion of gallium results in its use as a capping material for gallium arsenide during the high temperature (900°C) anneal process which must be carried out after ion implantation. It is about 50 to 100 times more resistant to thermal oxidation than silicon so that it can be used as a mask in a number of silicon-based VLSI (very large scale integrated circuit) schemes which require selective oxidation of the semiconductor. It bonds readily to aluminum metal upon heat treatment, and is used as the base insulator for metallization in gallium arsenide integrated circuits. Finally, it has superior radiation resistance over silicon dioxide and is used in radiation-hardened devices for this reason.[39]

Silicon nitride is a wide-gap insulator with a dielectric constant of 5.8 to 6.1, a refractive index of 1.98 to 2.05, and a density between 2.3 and 2.8 g/cm^3. It has a composition given by Si_3N_4. However, considerable departure from stoichiometry is often encountered during its deposition by the various methods to be outlined here, with Si/N atom ratios from 0.7 to 1.1 being commonly encountered. Although it is possible to grow a native film by direct nitridation of silicon, the process must be carried out at high temperatures (1100–1300°C), and results in extremely thin films due to the low diffusivity of nitrogen through them during the growth process.[40]

Amorphous films of silicon nitride can be deposited by RF sputtering using a silicon target with a nitrogen discharge.[41] Substrate temperatures of 200 to 300°C are used in this process. Direct sputtering from a silicon nitride target with an argon-nitrogen background can be accomplished. However, the most commonly used process for film growth is chemical vapor deposition involving the reaction of silane gas and ammonia, with nitrogen gas as the diluent.[42] This reaction is usually carried out at 700°C, and proceeds along the following lines:

$$3\ SiH_4 + 4\ NH_3 \rightarrow Si_3N_4 + 12\ H_2$$

Both hot- and cold-wall systems can be used for this purpose, and deposition can be accomplished at atmospheric as well as at reduced pressures. In all cases, film composition and properties are controlled by the ratio of ammonia to silane in the gas stream. Typical mole ratios of NH_3/SiH_4 are 150 or higher for a hot-wall system, with growth rates in the 100 to 200 Å/min range.

Increasingly, film deposition is carried out in low-pressure, hot-wall systems which allow close stacking of slices while still providing uniform coverage. Operation at low pressure ($\simeq$1 torr) is achieved by greatly reducing or eliminating the use of carrier gases, so that these systems can be run with reactant partial pressures (and growth rates) comparable to those achieved with atmospheric systems.

Plasma-enhanced cold-wall systems can also be used for the growth of silicon nitride.[43,44] Here an RF plasma is used to obtain one or more active species of the reactants, and films can be grown at low temperatures (275–300°C). This allows films to be deposited directly on finished microcircuits. Growth rates for these systems are in the 200 Å/min range.[43] Films grown in this manner have large quantities of hydrogen incorporated in them. This greatly affects properties such as their etch rate. Their growth at elevated temperatures, or their densification by heat treatment at 700°C, results in greatly improved film quality but obviates the initial advantage of low temperature growth for this approach.

Deposited films of silicon nitride have a large amount of built-in tensile stress (5×10^9 dynes/cm^2) when grown at 700°C. They are usually deposited in thicknesses below 1000 Å to avoid breakage, peeling, or damage to the underlying semiconductor. Typically, films with a Si/N ratio of 0.75 have been found to have minimum stress and are favored for this reason.

Films grown by plasma-enhanced techniques generally have a lower stress ($\simeq 2 \times 10^9$ dynes/cm^2) than those grown in hot-wall systems; in addition this stress may be either tensile or compressive, depending on the conditions of film growth. Relatively thick (0.5–1 μ) films can be grown with a compressive stress, with excellent adhesion to the semiconductor surface.[44]

The etching properties of silicon nitride films are highly variable and are related to their Si/N ratio as well as the manner in which they are deposited. Pure Si_3N_4 etches at about 68 Å/min in HF (49%), whereas CVD and plasma-grown films usually etch much faster (250–500 Å/min). Boiling phosphoric acid is often used to selectively etch these films in the presence of silica layers.

8.1-3 Organic Films

There are two problems with interconnections which increase in importance with increasing complexity in integrated circuit schemes. First, it becomes very difficult to make these interconnections in an orderly manner and to avoid (or minimize) crossovers; and second, the area taken up by the interconnection metal increases to the point where it becomes the dominant factor in chip area utilization. Both these problems necessitate the use of multiple levels of interconnection. Typically three such layers are commonplace, and some circuits require even more. With each level the surface of the circuit becomes increasingly irregular, and the step height encountered by the interconnection metal increases. This results in a significant yield loss during circuit fabrication due to failure of the metal at these steps.

Organic films have been used successfully to alleviate these problems. These films are put down by spin coating, in liquid form, and they smooth out step irregularities by means of surface tension. As a result, their use allows many levels of metallization without a progressive deterioration of the surface topology.[45–50]

These films must meet a number of requirements to be successful in this application. Their electrical resistivity and dielectric breakdown strength must be comparable to those of silicon dioxide. They must be capable of being formed in pinhole-free layers and must not alter dimensionally during the subsequent curing process, since this would damage the metal interconnection layer on which they are placed. They must adhere firmly to the surface on which they are laid and provide adhesion to the layer of metallization that follows. Thus they must adhere to the first insulating layer of silicon dioxide on the surface of the wafer; to aluminum metal, which is the common choice for interconnections; and also to aluminum oxide, which forms as a thin native oxide on the metal. Finally, these films must be capable of surviving the heat treatments that are required in successive fabrication processes. These include the formation of interconnections between successive layers, die bonding to the substrate, and wire bonding to the package terminals.

Polymeric organic materials, used for this purpose, include synthetic rubbers such as cyclized cis-poly isoprene, epoxyphenol resins, and silicones. However, the polyimides have been found to be the most thermally stable materials and work has concentrated on their use in microcircuit applications.

A number of polyimide formulations are commercially available. Polyimide isoindroquinazolinedione (PIQ) is one of the most stable, since it contains the thermally stable isoindroquinazolinedione ring.[46] This material exhibits dimensional stability when processed at temperatures as high as 450°C, for times as long as 1 hr. In contrast, materials such as the silicones cannot be used in processes subjected to temperatures above 250°C for this period of time.[47]

In practice polyimide resins are diluted with additional solvents such as methylpyrolidinone or dimethylformamide to allow them to be spin coated in layers of 1 to 4 μ thickness. Next the films are cured by heat treatments in the 300 to 400°C temperature range. Via holes, for making interconnections through these films, are cut by the use of wet chemical etches such as hydrazine. However, gaseous etching with use of an oxygen plasma is the preferred approach for high-density circuit applications, since it avoids problems associated with the capillary action of the liquid etchant.

Properly cured polyimide films generally have a breakdown field strength of 4 to 8×10^6 V/cm, which is comparable to that of silicon dioxide. Their electrical resistivity is, however, about two decades lower than that of silicon dioxide, but this is not a problem at the voltage levels encountered in microcircuits.[48]

Adhesion to the first insulation layer of silicon dioxide, as well as to the metallization, is perhaps the most important problem area for these films. Typically this adhesion has been enhanced by the use of a variety of coupling agents, of which hexamethyldisilazane (HMDS) has been the most popular. Recently, however, new coupling materials based on the use of aluminum organic chelates have been found to be superior.[49] Use of these materials has allowed high peel strengths (in excess of 200 g/cm) to be preserved even when circuits have been maintained under high temperature, high humidity conditions (120°C with water vapor pressure of 2 atm).

References

1 R. A. Colclaser, *Microelectronics: Processing and Device Design*, Wiley, New York, 1980.

2 S. K. Ghandhi, *VLSI Fabrication Principles: Silicon and Gallium Arsenide*, Wiley, New York, 1983.

3 L. I. Maissel and R. Glang, *Handbook of Thin Film Technology*, McGraw-Hill, New York, 1970.

4 J. L. Vossen and W. Kern, eds., *Thin Film Processes*, Academic, New York, 1978.

5 C. J. Frosh and L. Derick, "Surface Protection and Selective Masking During Diffusion in Silicon," *J. Electrochem. Soc.* **104**:547 (1957).

6 B. Mattson, "CVD Films for Interlayer Dielectrics," *Solid State Tech.*, p. 60, January 1980.

7 W. Kern and R. S. Rosler, "Advances in Deposition Processes for Passivation Films," *J. Vac. Sci. Tech.* **14**:1082 (1977).

8 W. Kern, G. L. Schnable, and A. W. Fisher, "CVD Glass Films for Passivation of Silicon Devices: Preparation, Composition, and Stress Properties," *RCA Rev.* **37**:3 (1976).

9 H. V. Schreiber and E. Froschle, "High Quality RF-Sputtered Silicon Dioxide Layers," *J. Electrochem. Soc.* **123**:30 (1976).

10 R. M. Burger and R. P. Donovan, *Fundamentals of Silicon Integrated Circuit Device Technology*, Vol. 1, Prentice-Hall, Englewood Cliffs, NJ, 1967.

11 B. E. Deal, "The Oxidation of Silicon in Dry Oxygen, Wet Oxygen, and Steam," *J. Electrochem. Soc.* **110**:527 (1963).

12 J. R. Ligenza, "Oxidation of Silicon by High Pressure Steam," *J. Electrochem. Soc.* **109**:73 (1962).

13 R. J. Zeto, C. G. Thornton, E. Hryckowian, and C. D. Bosco, "Low Temperature Thermal Oxidation of Silicon by Dry Oxygen Pressure Above 1 Atm," *J. Electrochem. Soc.* **122**:1409 (1975).

14 M. Ghezzo and D. M. Brown, "Diffusivity Summary of B, Ga, P, As, and Sb in SiO_2," *J. Electrochem. Soc.* **120**:146 (1973).

15 E. H. Nicollian and J. R. Brews, *MOS (Metal Oxide Semiconductor) Physics and Technology*, Wiley, New York, 1982.

16 A. Rohatgi, S. R. Butler, and F. J. Feigl, "Mobile Sodium Ion Passivation in HCl Oxides," *J. Electrochem. Soc.* **126**:149 (1979).

17 B. E. Deal, "Thermal Oxidation Kinetics of Silicon in Pyrogenic H_2O and 5% HCl/H_2O Mixtures," *J. Electrochem. Soc.* **125**:576 (1978).

18 B. R. Singh and P. Balk, "Thermal Oxidation of Silicon in O_2-Trichloroethylene," *J. Electrochem. Soc.* **126**:1288 (1979).

19 T. Hattori, "Elimination of Stacking Faults in Silicon by Trichloroethylene Oxidation," *J. Electrochem. Soc.* **123**:945 (1976).

20 R. A. Logan, B. Schwartz, and W. J. Sundburg, "The Anodic Oxidation of GaAs in Aqueous H_2O_2 Solution," *J. Electrochem. Soc.* **120**:1385 (1973).

21 B. Schwartz, F. Ermanis, and M. H. Brastad, "The Anodization of GaAs and GaP in Aqueous Solution," *J. Electrochem. Soc.* **123**:1089 (1976).

22 B. N. Arora and M. G. Bidnukar, "Anodic Oxidation of Gallium Arsenide," *Sol. State Electron.* **19**:657 (1976).

23 H. Hasegawa and H. L. Hartnagel, "Anodic Oxidation of GaAs in Mixed Solutions of Glycol and Water," *J. Electrochem. Soc.* **123**:713 (1976).

24 H. Tokuda, Y. Adachi, and T. Ikoma, "Microwave Capability of 1.5 Micron-Gate GaAs MOSFET," *Electron. Lett.* **13**:761 (1977).

25 H. D. Barber, H. B. Lo, and J. E. Jones, "Repeated Removal of Thin Layers of Silicon by Anodic Oxidation," *J. Electrochem. Soc.* **123**:1404 (1976).

26 C. R. Barnes and C. R. Geesner, "Pyrolytic Deposition of Silicon Dioxide for 600°C. Thin Film Capacitors," *J. Electrochem. Soc.* **110**:361 (1963).

27 N. Goldsmith and W. Kern, "The Deposition of Vitreous Silicon Dioxide from Silane," *RCA Rev.* **28**:153 (1967).

28 B. J. Baliga and S. K. Ghandhi, "Growth of Silica and Phosphosilicate Films," *J. Appl. Phys.* **44**:990 (1973).

29 R. S. Rosler, "Low Pressure CVD Production Processes for Poly, Nitride, and Oxide," *Solid State Technol.*, p. 20, April 1977.

30 R. S. Rosler and G. M. Engle, "Plasma Enhanced CVD in a Novel LPCVD-Type System," *Solid State Technol*, p. 172, April 1981.

31 P. G. Evert and T. van de Ven, "Plasma Deposition of Silicon Dioxide and Silicon Nitride Films," *Solid State Technol.*, p. 167, April 1981.

32 M. M. Schlacter, E. S. Schlegel, R. S. Kan, R. A. Lathlaen, and G. L. Schnable, "Advantages of Vapor-Plated Phosphosilicate Glass Films in Large Scale Integrated Circuit Arrays," *IEEE Trans. Electron Dev.* **ED-17**:1077 (1970).

33 B. J. Baliga and S. K. Ghandhi, "Lateral Diffusion of Zink and Tin in Gallium Arsenide," *IEEE Trans. Electron Dev.* **ED-21**:410 (1974).

34 E. Yon, W. H. Ko, and A. B. Kuper, "Sodium Distribution in Thermal Oxide by Radiochemical and MOS Analysis," *IEEE Trans. Electron Dev.* **ED-13**:276 (1966).

35 M. Yamin, "Observation of Phosphorus Stabilized SiO_2 Films," *IEEE Trans. Electron Dev.* **ED-13**:256 (1966).

36 S. P. Murarka, "A Study of the Phosphorus Gettering of Gold in Silicon by the Use of Neutron Activation Analysis," *J. Electrochem. Soc.* **123**:765 (1976).

37 S. K. Ghandhi, *Semiconductor Power Devices*, Wiley, New York, 1977.

38 R. B. Comizzoli, "Aluminum Corrosion in the Presence of Phosphosilicate Glass and Moisture," *RCA Rev.* **37**:483 (1976).

39 J. A. Appels, E. Kooi, M. M. Paffen, J. J. H. Schatorje, and W. H. C. G. Verkuylen, "Local Oxidation of Silicon and Its Application in Semiconductor Device Technology," *Philips Res. Repts.* **25**:118 (1970).

40 T. Ito, S. Hijiya, T. Nozaki, H. Arakawa, M. Shinoda, and Y. Fukukawa, "Very Thin Silicon Nitride Films Grown by Direct Thermal Reaction with Nitrogen," *J. Electrochem. Soc.* **125**:448 (1978).

41 G. J. Kominiak, "Silicon Nitride Films by Direct RF Sputter Deposition," *J. Electrochem. Soc.* **122**:1271 (1975).

42 R. Ginsburgh, D. L. Heald, and R. C. Neville, "Silicon Nitride Chemical Vapor Deposition in a Hot Wall Diffusion System," *J. Electrochem. Soc.* **125**:1557 (1978).

43 R. S. Rosler, W. C. Bensing, and J. Baldo, "A Production Reactor for Low Temperature Plasma-Enhanced Silicon-Nitride Deposition," *Solid State Technol.*, p. 45, June 1976.

44 A. K. Sinha, H. J. Levinstein, T. E. Smith, G. Quinitana, and S. E. Haszko, "Reactive Plasma Deposited Si-N Films for MOS-LSI Passivation," *J. Electrochem. Soc.* **125**:60 (1978).

45 K. Kato, S. Harada, A. Saiki, T. Kimura, and T. Okubo, "A Novel Planar Multilevel Interconnection Technology Utilizing Polyimide," *IEEE Trans. Parts, Hybrids Packag.* **PHP-9**: 176 (1973).

46 Y. Homma, H. Nozawa, and S. Harada, "Polyimide Liftoff Technology for High-Density LSI Metallization," *IEEE Trans. Electron Dev.* **ED-28**:522 (1981).

47 A. Saiki, S. Harada, T. Okubo, K. Makai, and T. Kimura, "A New Transistor with Two-Level Metal Electrodes," *J. Electrochem. Soc.* **124**:1619 (1977).

48 L. B. Rothman, "Properties of Thin Polyimide Films," *J. Electrochem. Soc.* **127**:2216 (1980).

49 A. Saiki and S. Harada, "New Coupling Method for Polyimide Adhesion to LSI Surface," *J. Electrochem. Soc.* **129**:2278 (1982).

50 L. B. Rothman, "Process for Forming Passivated Metal Interconnection System with a Planar Surface," *J. Electrochem. Soc.* **130**:1131 (1983).

8.2 DIELECTRIC MATERIALS FOR CHARGE STORAGE

J. Keith Nelson

8.2-1 Classification of Dielectrics

The introduction of a dielectric material of relative permittivity ϵ_r between the plates of a plane parallel capacitor of area A and gap d increases the capacitance and energy storage by a factor ϵ_r to a value of $\epsilon_0\epsilon_r A/d$ coulombs where ϵ_r is the permittivity of free space (8.85×10^{-12}F/m). The increase in charge storage comes about because some of the free charge will be neutralized by the action of electrostatic dipoles in the material introduced. In applied fields such dipoles, formed by the separation of charges on an atomic or molecular scale, may be considered to align themselves in the electric field. The nature of the dipoles has a profound influence on the behavior of the dielectric material and leads to several broad groups of materials:[1]

1. **Nonpolar.** Materials whose molecular structures and bonding are such that there are no dipolar chemical groups. The only source of polarization is electronic and polarization is induced when the center of charge of the electrons is displaced from the nucleus. In addition to electronic polarizability, some materials may contain dipolar groups of atoms whose net dipole moment is zero but which exhibit an *induced* ionic polarizability in an applied field. Examples of materials having only electronic polarizability include hydrogen, nitrogen, benzene, polyethylene, and polystyrene. These typically have dielectric constants in the range 1 to 4 and the dominance of electronic polarizability may be tested by observing that $\epsilon_r = n^2$, where n is the refractive index. Additional induced ionic (or atomic) polarizability is observed in materials such as carbon dioxide, carbon tetrachloride, titanium dioxide, and the alkali halide crystals.
2. **Dipolar.** In addition to having the above *induced* polarization mechanisms, substances that contain permanent *dipolar* groups which can orientate in an applied field have an additional important source of polarizability which generates dielectrics having a much higher relative permittivity. Unlike the induced forms of polarization, this polarizability is reduced with temperature due to the hindrance of thermal agitation. Dipolar liquids, such as water and nitrobenzene, can have relative dielectric constants up to about 100.
3. **Ferroelectric.** By analogy with ferromagnetism, ferroelectric materials are spontaneously polarized, having domains about 1μm across where all the permanent dipoles are oriented in the same

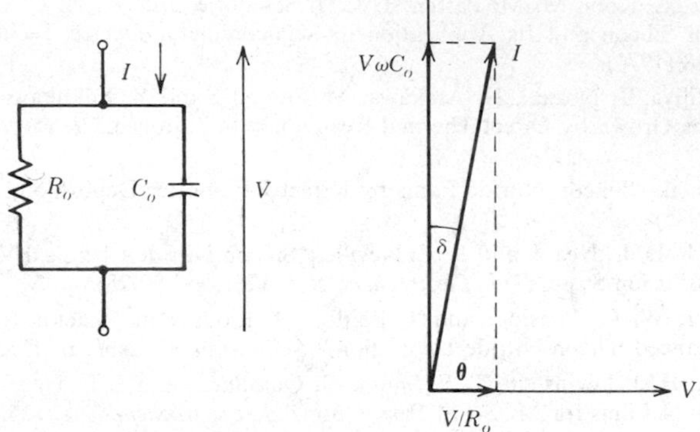

Fig. 8.8 Simple representation of a lossy dielectric. (*a*) Equivalent circuit. (*b*) Phasor diagram.

direction. This internal ordering below the Curie temperature gives rise to dielectric constants up to 20,000. Many ferroelectric formulations have been based on barium titanate ($BaTiO_3$) and potassium dihydrogen phosphate and their related materials.

8.2-2 Fundamental Aspects

Real dielectric materials are not perfect insulators but also suffer electrical losses. A lossy capacitor dielectric may thus be represented by the simplified equivalent circuit of Fig. 8.8*a*, where C_o is loss free and R_o dissipates energy to simulate the nonideal nature of the capacitor. Under DC conditions the losses are mainly accounted for by electronic and ionic conductivity. However, for time-varying electric fields, energy will also be dissipated in the process of dipole orientation (or hysteresis loss for ferroelectric materials). As a circuit element, Fig. 8.8*a* will have the phasor diagram of Fig. 8.8*b*, from which it is clear that R_o has introduced a phase angle, descriptive of the losses. Since θ is close to 90°, it is usual to express the losses in terms of the angle ($\delta = 90 - \theta$), where

$$\tan \delta = \frac{1}{\omega C_o R_o}$$

for a field having an angular frequency ω. Alternatively, the loss may be incorporated into the relative permittivity by providing both real and imaginary terms.

$$\epsilon_r^* = \epsilon_r' - j\epsilon_r''$$

so that $\tan \delta = \epsilon''/\epsilon'$ and may be regarded as the ratio of magnitudes of the in-phase and quadrature component of current.

Dipole reorientation takes place with a relaxation time, τ, characteristic of the molecular structure. Dielectric response to a sinusoidal electric field is thus frequency dependent and, for a simple system, is given by the well-known Debye equations:

$$\epsilon_r' = \epsilon_\infty + \frac{\epsilon_s - \epsilon_\infty}{1 + \omega^2\tau^2}$$

$$\epsilon_r'' = \frac{\omega\tau(\epsilon_s - \epsilon_\infty)}{1 + \omega^2\tau^2}$$

where ϵ_s is the static (DC) permittivity and ϵ_∞ that at very high frequency where dipoles are effectively "frozen" and unable to contribute to the dielectric constant.

In general, as the frequency is increased a number of loss peaks may be measured corresponding to the particular polarization mechanism operative. Concomitant reductions in dielectric constant occur until the last dispersion takes place at optical frequencies due to the induced electronic polarization, as shown in Fig. 8.9.

8.2-3 Breakdown of Solid Dielectric Materials

Since the energy density of a capacitor is proportional to $\epsilon_r E^2$, an increase of the working stress, E, is clearly an attractive way of improving energy storage. Despite popular claims to the contrary, solid materials do not have well-defined breakdown strengths. Many polar polymeric materials at low temperatures can exhibit electric strengths higher than 10 MV/cm, as shown in Fig. 8.10, although in

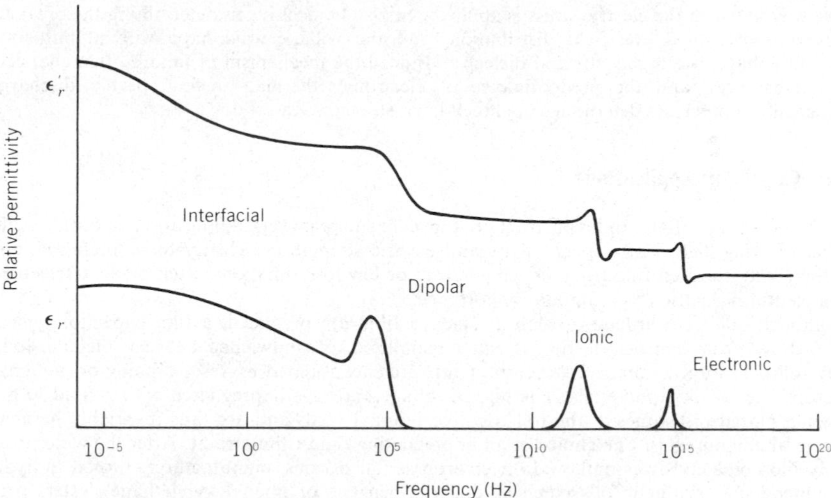

Fig. 8.9 Idealized representation of the frequency dependence of dielectric permittivity for a range of possible polarization mechanisms.

different circumstances they might fail at stresses 100 times smaller. The reasons for this involve both the material and its environment. The breakdown of most materials in industrial use is dominated by physical imperfections and chemical impurities and not by the material per se. Microcracks and fissures formed during the casting or extrusion of a material and ionic contaminants are examples of problems in this category that are very difficult to eliminate completely. Although environmental parameters such as temperature and humidity have a marked influence on the electric strength, the

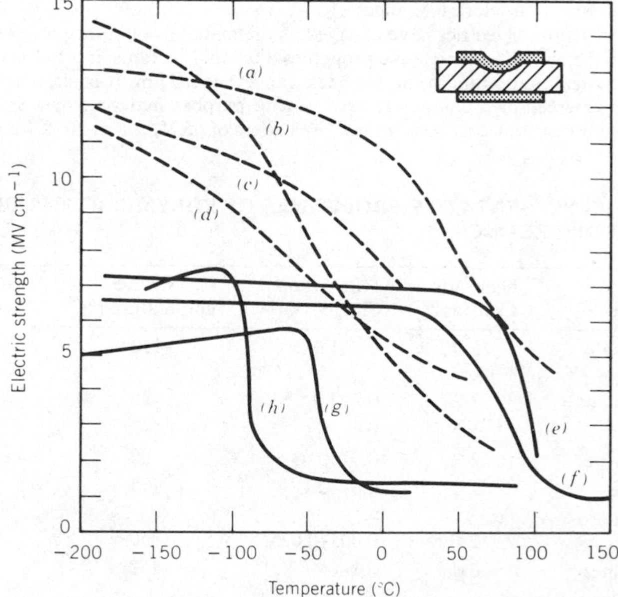

Fig. 8.10 Examples of the temperature dependence of the DC electric strength of a range of polar (broken lines) and nonpolar (solid lines) polymers: (*a*) polymethylmethacrylate, (*b*) polyvinyl alcohol, (*c*) polyvinyl chloride acetate, (*d*) 55% chlorinated polyethylene, (*e*) atactic polystyrene, (*f*) low-density polyethylene, (*g*) polyisobutylene, and (*h*) polybutadiene. Reprinted with permission from Nelson.[2]

precise way in which the electric stress is applied can also be decisive in determining the breakdown.[2] The area under stress, the field distribution, and the voltage waveshape will all influence the withstand voltage of a highly stressed dielectric. Indeed the mechanism of failure often changes with the circumstances, and the predominance of electronic, thermal, treeing, partial discharge, or electrochemical processes determines the breakdown strength of a solid.[3]

8.2-4 Capacitor Applications

The choice of materials and construction for capacitors is very application specific. Although materials having high relative permittivity and electric strength are clearly to be preferred, applications involving elevated frequency or temperature or low loss situations often make a selection of a low tangent δ dielectric of paramount importance.

Although other constructions, such as stacked film, are practiced, a film capacitor typically is manufactured from a dielectric that is either metallized or sandwiched between metallic foils and tightly rolled to form a compact structure. Units are encapsulated by resin dipping or molding with polypropylene or alkyd materials. The higher voltage units are impregnated with a fluid to prevent damaging partial discharges at the foil edges or internal voids and are thus invariably hermetically sealed in aluminum cans. The choice of an impregnating fluid is important. After the widespread use of polychlorobiphenyls was outlawed for environmental reasons, manufacturers turned to hydrocarbon mineral oils, synthetic oils (such as dodecyl benzene or phenyl xylylethane), esters (such as dioctyl phthalate), or silicone fluids.[4] Again the choice is dependent on application and on the nature of the main dielectric film being employed. A number of mixed fluids have emerged in an effort to achieve a better balance of properties. For example, an increase in aromatic content improves gas-absorbing properties, while some formulations may require the additions of chemical stabilizers.

Impregnated paper has been widely used as a capacitor dielectric in the past, but since its working stress is limited to about 20 V/μm and it has a typical loss tangent of 0.003 there has been a gradual change to polymeric materials, which can outperform paper and in some cases also tolerate service temperatures in excess of the 100°C limit for paper. For specialist applications such as radio transmitting equipment, mica offers an attractive range of properties for low-loss, high-frequency capacitors. Muscovite mica has a relative permittivity between 6 and 7, loss tangent in the range 10^{-3} to 10^{-4}, and an electric strength in excess of 100 V/μm coupled with chemical stability up to about 500°C. Precision mica capacitors can also yield 1% stability over long periods but are difficult to manufacture competitively to close tolerances.

Synthetic polymer film dielectrics have emerged as the materials of choice for many applications. Table 8.5 indicates the approximate relative properties of candidate films, most of which are available as pinhole-free electrical grade film in the thickness range 1 to 10 μm. It is clear from the table why polypropylene has emerged as a commonly used general purpose material, since it provides reasonable electrical properties at low cost and can sustain a stress of 65 V/μm at 70°C for a 20-year life. In

TABLE 8.5. REPRESENTATIVE PROPERTIES OF POLYMERIC MATERIALS OF CAPACITOR DIELECTRICS

Material	Dielectric Constant	Dissipation Factor (%)	Service Temperature (°C)	Relative[a] Cost
Polycarbonate	3.0	0.1–1.0	110	7
Polyethylene terephthalate	3.2	0.2–2.0	85	3.6
Polyimide	3.5	0.2	240	68
Polypropylene	2.2	0.01–0.03	90	1.8
Polystyrene	2.5	0.01–0.1	75	1.3
Polytetrafluoroethylene	2.0	0.003–0.025	250	21
Polysulphone	3.1	0.1–0.4	150	23
Polyvinylidene fluoride[b]	11	0.01–0.12	160	

Source. Ref. 4, with permission.

[a] Relative to low-density polyethylene = 1.

[b] Not presently available commercially as a capacitor grade film.

general, films offering a high service temperature (such as polyimide and polysulphone) carry a high cost penalty. In some cases the dielectric losses are very sensitive to temperature and frequency. Capacitor designs utilizing such films as polyvinylidene fluoride would run the risk of thermal breakdown for sustained high temperature AC operation but are attractive for intermittent discharge duty.

Where the required current densities are low, the use of 300 to 500 Å aluminum layer vapor deposited onto the polymer as an electrode can raise the volume efficiency to about 1.5 $\mu F/cm^3$. By careful control of the thickness of the metallized layer (1-5 Ω/sq.), the capacitor unit can become self-healing.[5] Local breakdown caused by a flaw in the dielectric film will vaporize the metallization and hence relieve the electric stress to permit recovery at the expense of some capacitance loss.

In addition to synthetic polymers, ceramics form the basis for a major part of the capacitor industry for signal and low-power applications. Ceramic capacitors are usually formed by sintering mixtures of inorganic materials at temperatures up to 1800°C. The resulting integral polycrystalline structure can also be fabricated as a multilayer device with in-built electrodes formed by printing with a precious metal ink or paste.

Ceramic capacitors based on low-loss materials such as magnesium silicate ($MgSiO_3$) with bonding additives have found use in resonant circuit applications, since they can exhibit high stability and a positive temperature coefficient of about 100 ppm/°C. Formulations based on titanium dioxide (TiO_2) or its derivatives can be adjusted to allow the temperature coefficients to be tailored in the range +100 to −1500 ppm/°C, but they have permittivities less than about 300. Ferroelectric ceramics can exhibit relative permittivities as high as 15,000 and are extensively used to fabricate capacitors of high value. However, ferroelectrics are inherently very lossy due to their ferroelectric hysteresis. Barium titanate ($BaTiO_3$), for example, has a dissipation factor of 0.02 and a dielectric constant of about 1500 that is sensitive to both frequency and temperature in the range of the Curie point at 120°C. More stable formulations have been developed based on substitutions into the $BaTiO_3$ structure and multiphase ceramics such as piezoelectric La-modified lead zirconate titanate $PbZrO_3$-$PbTiO_3$ (PLZT) have yielded materials with reduced dielectric constant dependence on field strength. Failure of ceramic capacitors is often due to formation cracks, voids, and electrode delamination and thus they are not recommended for pulse discharge duty. The electric strength of ceramics is ultimately limited by the porosity, although ionic migration and moisture ingress can also often cause premature failure.

For applications, such as power supply smoothing, in which large capacitance is required but losses are of secondary importance, high-energy densities may be obtained by employing electrolytic technology. Layers of Al_2O_3 ($\epsilon_r = 8.5$) or Ta_2O_5 ($\epsilon_r = 25$) typically as thin as 0.02 μm may be formed on an etched aluminum or tantalum foil by an anodizing process. The foil is wound together with a separator and suitable liquid or solid electrolyte [e.g., MnO_2 formed by in situ pyrolysis of $Mn(NO_3)_2$] to form a compact assembly. The active dielectric layer is self-healing, since flaws in the film are continually rebuilt by corrosion of the supporting metal. As a consequence such films may operate at stresses in excess of 10^8 V/m, which is close to breakdown. Nevertheless the very thin film provides a severe voltage limitation, since coherent films greater than 0.8 μm are difficult to make, and the technology imposes a fixed voltage polarity restriction.

An emerging concept for charge storage, which has some similarities with electrolytic technology, involves the utilization of a Helmholtz double layer as a capacitor. Such layers, which separate charges at the interface of a polarizable electrode and electrolyte, are only a few angstroms wide and can thus generate very large capacitances providing the voltage is limited to about 1 V dictated by the decomposition potential of the electrolyte. The first commercial devices[6] using this principle utilizing an activated carbon/sulfuric acid system offer considerable space savings over standard electrolytic devices and have attracted some interest as battery substitutes. A major problem with the Helmholtz layer devices, as presently configured, is the high internal impedance; a typical 5 V, 1 F unit might have an equivalent source resistance of 5 Ω.

8.2-5 Other Applications

The application of electric fields to dielectric materials at elevated temperatures permits movement of charges. Subsequent reduction of temperature for low-loss substances immobilizes the charges and produces a material with permanent polarization known as an electret. Materials commonly used for polymer electrets include polyethylene terephthalate, fluorinated ethylene-propylene, polytetrafluoroethylene, and polyethylene. The charged species responsible for the phenomena can be trapped carriers or orientated molecular dipoles, but an electret is capable of producing a long-lived static electric field somewhat analogous to that of a permanent magnet. This quality has found applications in electrostatic microphones, such as that shown in Fig. 8.11. The polymer foil electret poled at 10^6 to 10^7 V/m at 200°C is vibrated causing the electrostatic field across the air gap, *h*, to vary, producing a small fluctuation in voltage across the device.[7]

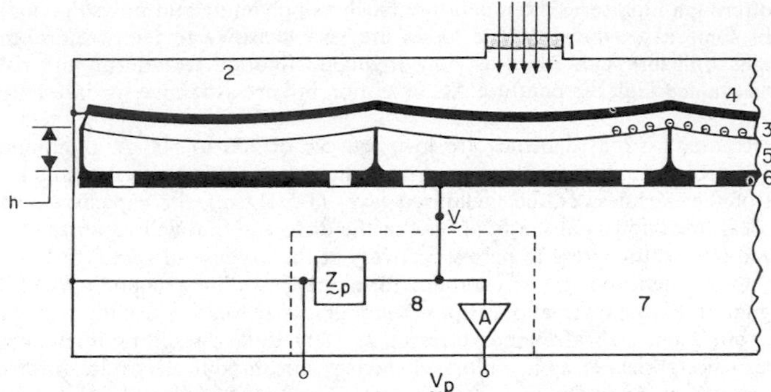

Fig. 8.11 Multicell electret microphone. 1 Sound port and impedance material, 2 Front acoustic chamber, 3 Electret film, 4 Metallization, 5 Air film, 6 Stationary electrode with holes, 7 Rear acoustic chamber, 8 Preamplifier. Adapted from reference 7 after Baumhauer and Brzezinski.[7]

Electrets are also finding application in advanced filtration schemes where the electric field can be used to electrostatically augment the collection efficiency of a filter. The ability to control charge on a dielectric material has also led to a number of useful applications such as xerographic copying, electrostatic printing, and a variety of instrumentation applications.[8]

In the context of charge storage in dielectrics, the polymer polyvinylidene fluoride (PVF_2) is of particular interest. This is a unique semicrystalline polymer with the monomer unit $CH_2 = CF_2$ which has at least two stable crystal structures,[9] one of which is polar with a dipole moment of 7.01×10^{-30} $C \cdot m$. Thin films may be poled at fields of about 10^8 V/m to form an electret that has a piezoelectric and pyroelectric response. These properties make the polymer of interest as a transducer and as an IR imaging device.

References

1 J. C. Anderson, *Dielectrics*, Chapman & Hall, London, 1964.

2 J. K. Nelson, "Breakdown Strength of Solids," in R. Bartnikas and R. M. Eichhorn, eds., *Engineering Dielectrics Vol. 2A*, ASTM, Philadelphia, 1983.

3 W. P. Baker, *Electrical Insulation Measurements*, Newnes, London, 1965.

4 J. K. Nelson, "Solid Dielectrics for Capacitors," in M. Bever, ed., *Encyclopedia of Materials Science and Engineering*, Pergamon, Oxford (in press).

5 D. G. Shaw, S. W. Cichanowski, and A. Yializis, "Changing Capacitor Technology: Failure Mechanisms and Design Innovations," *Trans. IEEE* **E1-16**:399 (1981).

6 K. Sanada and M. Hosokawa, "Electric Double Layer Capacitor, Super Capacitor," *NEC Res. Devel.* **55**:21 (1979).

7 J. C. Baumhauer, Jr. and A. M. Brzezinski, "The EL2 Electret Transmitter: Analytical Modeling, Optimization and Design," *Bell Sys. Tech. J.* **58**:1557 (1979).

8 A. D. Moore, *Electrostatics and Its Applications*, Wiley, New York, 1973.

9 D. K. Das Gupta and K. Doughty, "On the Nature and Mechanism of Piezo- and Pyro-Electricity in Polyvinylidene Fluoride," in Y. Wada, M. M. Perlman, and H. Kokado, eds., *Charge Storage, Charge Transport and Electrostatics With Their Applications*, Elsevier, Amsterdam, 1979.

8.3 DIELECTRIC MATERIALS AS INSULATORS

J. Keith Nelson

8.3-1 Background and Selection Criteria

The selection of gaseous, liquid, or solid media for duty as insulators for electrical conductors must be undertaken with care, especially when composite insulating structures are being designed. It will be clear from Section 8.2-3 that the electric strength (electric field at breakdown) is highly dependent on the circumstances under which it is measured and is usually dominated by the impurities and imperfections present. Furthermore, values quoted usually assume a uniform or well-defined electric field such as that specified in IEC Spec 243. In practice, electric fields are seldom uniform since the geometrical configuration of the conductors and asperities on surfaces usually create bulk and local field divergence. Electric field intensification can also occur due to space charges, which can build up both within a dielectric and at its surface. Assessment of such internal field distortion is difficult but can be accomplished with sophisticated techniques[1] and indicates that in specialized circumstances distortion can be significant.

The properties of relative permittivity (dielectric constant) and loss tangent also assume importance in the failure of insulating structures. In a composite insulator configuration stressed with alternating voltages, the electric field will be capacitively distributed and thus affected by the permittivity of the materials employed. For example, a small gas void trapped within an epoxy resin ($\epsilon_r = 3.8$) molding will be subjected to about four times the electric field of the host material, which can result in partial discharges, as indicated in Section 8.3-3. The dissipation factor assumes importance not only in those applications where loss must be minimized, but also because of its effect on the breakdown phenomena. If the rate of increase of energy of the electric field exceeds that at which heat can be lost by the insulation, thermal failure may ensue.

There are many instances when the selection of an electrical insulating material may be dictated by nonelectrical requirements. Considerations of mechanical loading, rigidity, or dimensional stability of solids may make the choice of some polymeric materials inappropriate. Similarly, service temperatures and/or chemical compatibility often prove to be constraints. This is especially the case in selection of impregnating fluids for use with polymeric solids (see Section 8.2-4).

Interfaces between solids and gases or liquids invariably form a weak link in any insulation system. This occurs because of field enhancements and interfacial charges and, in the case of a liquid, may be influenced by fluid motion. An example is shown in Fig. 8.12 in which the electric stress distribution at a phenolic spacer is compared with that in the liquid.[2] For high-voltage systems, difficulties may be minimized by informed geometrical design such as increasing creepage paths and shielding the triple point (electrode/solid/fluid) junctions.[3] Electrically stressed surfaces that are exposed to a hostile environment can be a severe limitation. Contamination by moisture, atmospheric pollutions, or carbonaceous or metallic particles can prejudice dielectric integrity.

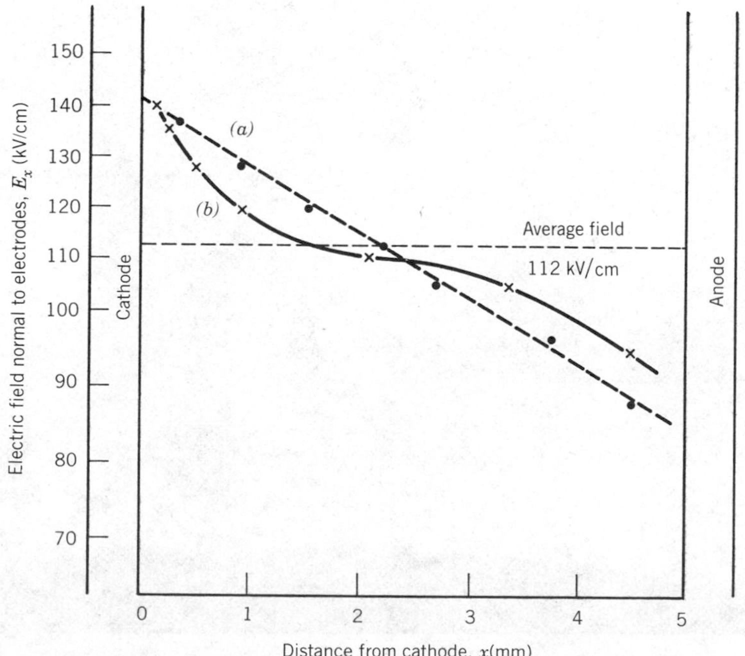

Fig. 8.12 Local field E_x plotted as a function of position x in Aroclor 1242 (a) and at a phenolic spacer interface. (b) Adapted from Cherney and Cross.[2]

8.3-2 Low Voltage Applications

The use of silicon dioxide and other materials employed as insulators in the semiconductor industry is the subject of Section 8.1, but in general the high electric fields associated with insulating films substantially less than 1 μm thick favor the operation of electron injection and ionization leading to rapid failure, as reviewed in Ref. 4. Eventual failure in such circumstances may be modified by migration by ionic impurities and may sometimes come about by a thermal mechanism,[5] but usually occurs on a submicrosecond time scale.

Other mechanisms of failure are slower. Organic materials can contribute to electrical stress aging at low voltages by electrochemical means. This form of long-term failure is particularly important for materials such as epoxy resins used as molded encapsulation for microelectronic packages. Although the migration of atmospheric ionic pollutants (e.g., salts, solder, fluxes, and moisture) is a primary cause of failure, the encapsulating package itself may contribute to the problem. Epoxy materials may contain ionic contaminants, halogens from thermal degradation, and corrosive acids from incomplete curing with an anhydride hardener. Once a corrosive species arrives at the integrated circuit surface, the corrosion kinetics leading to eventual failure will be principally dependent on the prevailing temperature and bias voltage which acts not only to drive the electrolytic corrosion but also to aid the migration of impurities.[6]

Electrochemical deterioration frequently occurs in impregnated structures such as capacitors, which contain mobile impurity ions. Movement of such ions results in electrode reactions and undesirable chemical changes in the solid or liquid dielectrics. Certain organic compounds such as quinones and aromatic azo and azoxy compounds may be used as stabilizers to retard electrochemical failure by acting as hydrogen acceptors.[7]

Surface tracking can also cause an insulating material to fail at comparatively low voltages. The combination of moisture and pollutants on a surface causes excessive surface conductivity. In turn the conduction currents cause local heating and dry band formation. Excessive stresses are thus thrown across local areas of the insulating surfaces with the resulting degradation and erosion of the underlying organic material.[8] Standard tests such as IEC Spec 112 have been devised to define a comparative tracking index to screen materials.

Fig. 8.13 Vented electrical trees growing radially outward into insulation from the semiconducting shield of a power distribution cable. Courtesy of Union Carbide Corporation.

8.3-3 High Voltage Applications

As discussed in Section 8.3-1, discharges can occur locally without immediately precipitating the complete failure of the insulation. This occurs where the electric field in an insulator is highly divergent due to occluded gas, foreign inclusions, or the geometrical design. These discharges cause cumulative erosion of the insulation and the generation of harmful chemical byproducts which cause eventual failure. Insulating structures such as epoxy-mica used for the stators of large rotating machines can survive continuous discharge magnitudes in excess of 1000 pC and still provide an acceptable service life, although such magnitudes in a capacitor might lead to failure in hours. This phenomenon has led to the widespread use of discharge detection, location, and measurement techniques in high-voltage equipment.[9]

A special case of partial discharges is that of "electrical" treeing (as distinct from "water" or "electrochemical" treeing[10]) in highly nonuniform fields, such as those obtained at sharp protrusions, where discharges may be initiated and will propagate in a series of steps like those shown in Fig. 8.13. Tree initiation is often ascribed to the formation of microcracks in the high field region. Such cracks may be the result of processing or may be formed as the consequence of the stress application. Generally the destruction of an insulating material by this channeling process is much more rapid than is discharge erosion in gaseous cavities. More details of the process may be found in Ref. 11.

8.3-4 Materials Summary

The reader may find comprehensive data on a wide range of insulating materials in Refs. 12 and 13. Among the cross-linked synthetic polymeric solids, phenolic and epoxy resins are widely used. Both groups are often used with fillers or reinforcing materials and thus their properties vary with the type and loading of the filler material. The phenol formaldehydes provide tensile strengths in the range 50 to 70 N/mm^2 and moisture absorption between 0.1 and 1%. Working temperatures up to 120°C can be accommodated and a well-cured resin has a relative permittivity of about 4.5 and a loss tangent of about 0.01. However, these latter properties are sensitive to temperature and frequency and the common use of cellulose filler makes phenolic materials susceptible to tracking. Increased track resistance may be obtained from the amino resins, such as melamine formaldehyde, but at considerable extra cost.

There are two types of epoxy resins: aromatic formed by the reaction of epichlorhydrin with bisphenol A and cycloaliphatic forming a saturated ring structure. Epoxies, when properly cured, exhibit many attractive mechanical and electrical features and are compatible with a wide range of materials. Epoxy insulation is commonly cast or molded, often under vacuum, and is cured up to temperatures of 180°C. Unfilled resin has a relative permittivity of 3.5 to 5 with a loss tangent in the range of 0.005 to 0.02 at 20°C up to frequencies of about 1 MHz. Epoxies are widely used when reinforced with glass fabrics or cellulose-based boards when tensile strengths of up to 400 N/mm^2 may be obtained and sustained temperatures of up to 130°C may be accommodated. This limit can be raised to about 150°C for special discharge-resistant epoxy-mica composites widely used in the insulation of rotating machinery. Cycloaliphatic resins offer a somewhat greater resistance to tracking and have found application as outdoor insulators.

Of the linear polymers, polyethylene in both its high- and low-density forms is widely used as an insulating material. It is characterized by low losses ($\tan \delta = 4 \times 10^{-4}$), a permittivity of about 2.2, and mechanical properties that make it an attractive candidate for cable applications, especially at high frequency. The working temperature is limited to about 80°C, although this can be increased somewhat by cross-linking. Polyethylene is track resistant but is not compatible with some solvents, especially at elevated temperatures. Although its water absorption is low, its use as a high-voltage cable insulant is complicated by the incidence of both electrochemical and electrical treeing.[11] In the cable application, maximum working stresses are limited to about 50 kV/cm, which is about 1% of the electric strength exhibited in a small sample laboratory test (see Fig. 8.10 in Section 8.2). In applications that require even higher insulation resistance, such as high-impedance electrometer circuits, polyethylene can be replaced by polytetrafluoroethylene (PTFE) having a loss tangent of 10^{-5}. PTFE has many other useful properties such as a high working temperature of 250°C, unique resistance to solvents, and nonflammability but its price restricts its use to situations for which there is no substitute.

A variety of materials are used for low-voltage cable insulation. Ethylene–propylene copolymers, when cross-linked, have properties similar to natural rubber. Polyvinyl chloride (PVC), when provided with suitable plasticizers and stabilizers to minimize high-temperature decomposition, is a polymer ideally suited for low-voltage cables on account of its good adhesion, oil resistance, and mechanical properties. Being a polar material (see Section 8.2-1), PVC is electrically a lossy material having a marked loss peak at 100°C, but this is of little consequence for low-voltage applications. PVC may by used up to about 80°C.

Wire enamels of the type common in providing the turn-to-turn insulation in rotating machines, transformers, and other electromagnetic devices are also an important class of insulating materials. The main class of materials used for this application is the aromatic polyimides which can be used up to 250°C. These are applied as a varnish prepolymer and then cured to form a coating that combines flexibility and resistance to solvents and acids. Polyimide materials are expensive and hybrid formulations have been developed (e.g., polyamide-imide) which offer a variety of properties and corresponding costs.

Although polymeric materials dominate the insulation industry, traditional materials such as cellulose paper are still the dielectrics of preference for extra-high-voltage transformers and cables. Paper is a low-cost material that offers good mechanical properties and the ability for impregnation with mineral oils. Properties[12] are very dependent on the type and purity of the paper and the extent to which moisture is removed. However, well-prepared oil-impregnated papers have a loss tangent of about 2×10^{-2} at 20°C. Properties of oil and other impregnating fluids are extensively reviewed in Ref. 14. Naphthenic and paraffinic mineral oils are widely used as coolants/insulants but are flammable with an exothermic reaction and can suffer from oxidation, especially in the presence of metals. Applications that require fire resistance were served by polychlorinated biphenyl fluids until these were outlawed because of their ecological unacceptability. Silicone fluids are used as alternatives.

Inorganic solids such as glasses and ceramics have DC resistivities that are a strong function of temperature due to the predominance of ionic conduction. However, their inorganic nature makes them resistant to thermal and electrical degradation. Electrical losses of most glasses decrease with frequency up to 10^{-6} to 10^{-7} Hz and then increase markedly in the microwave region. Lamp and vacuum tube envelopes are the biggest applications for glass as an insulator, although aluminoborosilicate materials are being employed as bases for thin film circuits.[15]

Ceramics offer a higher impact strength than do glasses and an unrivaled temperature capability. Many single or multiple oxide ceramics such as alumina (Al_2O_3) and beryllia (BeO) have melting points in excess of 2000°C. Beryllia has a thermal conductivity in the range 5×10^4 to 25×10^4 W/K, which makes it unique as an insulator with a thermal conductivity similar to that of a metal. It is used as a heat-sinking material for high-power semiconductor devices. Electrical porcelains fabricated from clay fired at 1000 to 1400°C form the basis for most outdoor high-voltage insulators. Such insulators are limited by external flashover rather than internal puncture and thus are always provided with a glaze to minimize the ability of the porous surface to collect contamination.

Except for the widespread use of atmospheric air as an insulant, the application of electronegative dielectric gases such as the chlorofluorocarbons and sulfur hexafluoride (SF_6) is mainly restricted to high-voltage power systems. At a pressure of 3 atm, SF_6 can withstand electric stresses similar to those withstood by an insulating oil and has the advantage of being light and nonflammable. Recent developments in SF_6 technology are leading to the design of all-enclosed high-voltage substations, and high-pressure gases are being applied as alternatives to liquids as impregnating agents for polymeric systems.[16] As with liquid dielectrics, the electric strength of gases is critically dependent on the presence of impurity particles which affect the electric field distribution.

8.3-5 Statistics of Breakdown

Insulation failure is not only a physical and chemical phenomenon but also a statistically distributed event. Multiple tests under nominally identical circumstances yield a distribution of breakdown values often having a wide scatter, especially for systems involving liquids. Furthermore, the probability of failure is time as well as stress dependent. It is usually (but not universally) the case that an insulating system offers a higher electric strength under impulse voltages of short duration because some time-dependent mechanisms (such as the migration of particles or space charge) may be eliminated. On the basis that failure is controlled by a "weak link" process,[17] some success has been achieved by describing dielectric failure by an extreme value distribution[18] such as the Weibull type relating the probability of failure to a stress or time. For the case of a stress E

$$\Pr(E) = 1 - \exp\left[-\left(\frac{E-\gamma}{\alpha}\right)^{\beta}\right]$$

where α, β, and γ are parameters specifying scale, shape, and threshold of the distribution, respectively. A consequence of this type of description is that most insulation systems exhibit an approximately logarithmic dependence of breakdown strength on area (or volume).[19] This results from the fact that the larger the system, the greater the probability of its having a flaw or initiating event.

The ability to predict the long-term life of a stressed structure is also important. Many mathematical aging models have been based on an empirical aging law of the form

$$E^n t = \text{constant}$$

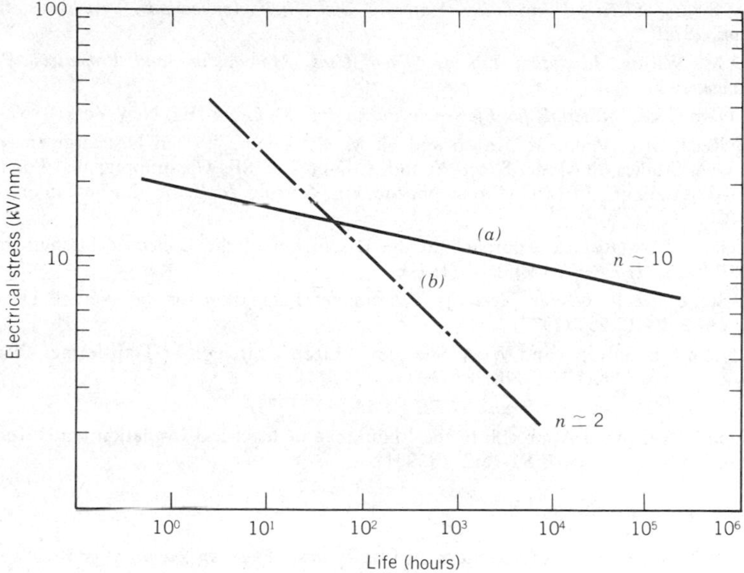

Fig. 8.14 Representative voltage endurance curves for (*a*) epoxymica composite and polyethylene (*b*) subjected to partial discharges.

where E is a constant electric stress and t is the time to failure. The exponent n for organic insulation is typically in the range 1 to 6 and for micaceous systems it is between 6 and 13. Figure 8.14 shows an example of such a voltage endurance curve for an epoxy-mica material in comparison with that for 0.25 mm polyethylene samples. The characteristics show typical behavior of materials subjected to partial discharges, but precise values are very dependent on the individual geometry.[20] In reality the endurance of insulation is dependent upon its thermal, mechanical, and chemical environment, and accurate life assessments must take into account multifactor stressing.[21]

References

1 D. W. Tong, "Electron Beam Probing of Space Charge in PET Films," *Trans. IEEE* **E1-17**:377 (1982).

2 E. A. Cherney and J. D. Cross, "Electric-Field Distortions at Solid-Liquid Dielectric Interfaces," *Trans. IEEE* **E1-9**:37 (1974).

3 K. Nakanishi, Y. Shibuya, and T. Nitta, "Experimental Study of the Breakdown Characteristics of Large Scale Gas Insulated Systems," in L. G. Christophorou, ed., *Gaseous Dielectrics II*, Pergamon, Elmsford, NY, 1980.

4 N. Klein, "A Theory of Localized Electronic Breakdown in Insulating Films," *Adv. Phys.* **21**:605 (1972).

5 N. Klein and E. Burstein, "Electrical Pulse Breakdown of Silicon Oxide Films," *J. Appl. Phys.* **40**:2728 (1969).

6 L. G. Feinstein, "Failure Mechanisms in Molded Microelectronic Packages," *Semiconductor Internat.*, p. 51, September 1979.

7 N. Parkman, "Some Properties of Solid-Liquid Composite Dielectric Systems," *Trans. IEEE* **E1-13**:289 (1978).

8 P. J. Lambeth, "Effect of Pollution on High-Voltage Outdoor Insulators," *Proc. IEE* **118R**:1107 (1971).

9 J. H. Mason, "Discharge Detection and Measurements," *Proc. IEE* **112**:1407 (1965).

10 S. L. Nunes and M. T. Shaw, "Water Treeing in Polyethylene—A Review of Mechanisms," *Trans. IEEE* **E1-15**:437 (1980).

11 R. M. Eichhorn, "Treeing in Solid Organic Dielectric Materials," in R. Bartnikas and R. M. Eichhorn, eds., *Engineering Dielectrics Vol 2A*, ASTM, Philadelphia, 1983.

12 F. M. Clark, *Insulating Materials for Design and Engineering Practice*, Wiley, New York, 1962.

13 R. W. Sillars, *Electrical Insulating Materials and Their Applications*, Peregrinus, Stevenage, England, 1973.

14 A. C. M. Wilson, *Insulating Liquids: Their Uses, Manufacture and Properties*, Peregrinus, Stevenage, 1980.

15 C. A. Harper, ed., *Materials for Electronic Packaging*, McGraw-Hill, New York, 1969.

16 C. W. Reed, S. F. Philp, M. Kawai and H. M. Schneider, "Partial Discharge Inception and Breakdown Studies on Model Sheet-Wound, Compressed SF_6 Gas-Impregnated Polymer Film-Insulated Windings," in L. G. Christophorou, ed., *Gaseous Dielectrics II*, Pergamon, Elmsford, NY, 1980.

17 E. Occhini, "A Statistical Approach to the Discussion of the Dielectric Strength in Electric Cables," *Trans. IEEE* **PAS-90**:2671 (1971).

18 G. C. Stone and R. G. van Heeswijk, "Parameter Estimation for the Weibull Distribution," *Trans. IEEE* **E1-12**:253 (1977).

19 J. K. Nelson, B. Salvage, and W. A. Sharpley, "Electric Strength of Transformer Oil for Large Electrode Areas," *Proc. IEE* **118**:388 (1971).

20 J. H. Mason, "Discharges," *Trans. IEEE* **E1-13**:211 (1978).

21 L. Simoni, "A General Approach to the Endurance of Electrical Insulation under Temperature and Voltage," *Trans. IEEE* **E1-16**:277 (1981).

Bibliography, Sections 8.2 and 8.3

Apps, L. T., "Metallised Plastics Capacitors for Electronics," *Electronics and Power* **16**:369 (1970).

Bartnikas, R. and R. M. Eichhorn, *Engineering Dielectrics Vol. 2A*, ASTM, Philadelphia, 1983.

Gumbel, E. J., *Statistics of Extremes*, Columbia Press, New York, 1938.

Harrop, P. J., *Dielectrics*, Butterworth, London, 1972.

McMahon, E. J., "A Tutorial on Treeing," *Trans. IEEE* **E1-13**:277, 1978.

Nicker, D. A., "High Voltage Ceramic Capacitors," *Electrocomp. Sci. Tech.* **1**:113 (1974).

Perlman, M. M., ed., *Electrets, Charge Storage and Transport in Dielectrics*, Electrochemical Society, Princeton, NJ, 1973.

Proceedings of the 2nd Capacitor and Resistor Technology Symposium, Components Technology Institute, Huntsville, AL, 1982.

Reed, C. W., ed., *Proceedings of a Symposium on High Energy Density Capacitors and Dielectric Materials*, National Academy of Science, Washington, DC, 1981.

Sato, K., et al., "Characteristics of Film and Oil for All-Polypropylene-Film Power Capacitors," *Trans. IEEE* **PAS-99**:1937 (1980).

Wada, Y., M. M. Perlman, and H. Kokado, eds., *Charge Storage, Charge Transport and Electrostatics with Their Applications*, Elsevier, Amsterdam, 1979.

Yoshida, Y., et al., "Evolution of Power Capacitor as a Result of New Material Development," CIGRE, Paper No. 15-01, Paris, 1980.

CHAPTER 9
MAGNETIC MATERIALS

ALEX GOLDMAN

Spang Industries
Butler, Pennsylvania

9.1 DEFINITIONS

A magnetic material is one that can be magnetically polarized, either on an atomic basis or on a massive or macroscopic basis. This polarization involves alignment (partial or complete) of the electronic magnetic moments of a material by an external applied magnetic field. As with other fields (such as gravitational or electric), the detection of the effect is evidenced by the use of a probe, in this case, a magnetic field from a permanent magnet or its equivalent, a field from a current-carrying wire.

Paramagnetism is the parallel alignment of unpaired atomic spins. Diamagnetism is caused by the antiparallel opposing reaction of the electronic orbital momentum to the external field. Paramagnetic alignment leads to a slight increase in the intensity of magnetization, while the diamagnetic effect is negative and is only observed when there are no unpaired spins. Manganese and chromium are examples of paramagnetic substances, while bismuth and aluminum are diamagnetic. In paramagnetism a large external field is necessary for the alignment of relatively few atomic spins. In ferromagnetism and ferrimagnetism the effect is many orders of magnitude greater because it involves a cooperative effect (called exchange energy) involving extremely large numbers of spins in areas called magnetic domains. All atomic spins in a domain are parallel in ferromagnetic materials. In ferrimagnetic materials (hard and soft ferrites) this is not strictly true, but the net effect of the interactions is similar in that the uncompensated spins of the ionic combination will again be aligned cooperatively in a domain. Additionally, a ferromagnetic (or ferrimagnetic) material has different values of degree of polarization for a fixed value of applied field depending on the magnetic history of the material. This effect is called magnetic hysteresis. Our discussion in this chapter concerns only ferromagnetic and ferrimagnetic materials because these are the only important ones in electrical or electronic applications. Emphasis is placed on commercial materials, although many others are important for scientific and specialized purposes. Ferromagnetic materials are classified as hard or soft, the former meaning that the material has a permanent polarization and is used for the field it generates. A soft material is one that must be energized during use by an external field, especially an alternating, or AC, field.

Most important ferromagnetic materials contain either iron, nickel, or cobalt or often combinations of these. These are the elements in which the cooperative action of exchange energy is most conducive. Some compounds of manganese (Heusler alloys) and some rare earth elements are somtimes ferromagnetic, but these are of minor importance.

In addition to their chemistry, ferromagnetic materials may be classified according to their physical form. They may be massive (of relatively equal size in all dimensions) as in permanent magnets. They may be metals rolled down to flat strips with thickness ranging from 0.000125 in. to

0.030 in. (0.0031–0.75 mm). They may also be made into wire, and finally they may be broken down into powder or flake. The latter forms may be compacted into components called powder cores or they may be used uncompacted as in magnetic inks or recording tapes. Oxide magnetic materials (ferrites) are usually formed as a powder, compacted and fired to form practically shaped component bodies. Occasionally these materials are used as single crystals (recording heads). Often the same material may be used in several forms. Permalloy 80 can be used as a strip material as well as in powder or flake form. We will treat each material in the form that it appears. The assembly of the various forms into components is the subject of another chapter.

9.2 ELECTRICAL AND MAGNETIC PROPERTIES

The units of magnetic and associated electrical quantities may be expressed in either cgs units or SI units. Both are given in this section. A conversion chart for the two systems is given in Table 9.1.

The signature of a magnetic material is the magnetic hysteresis loop (BH loop) (see Fig. 9.1). In an AC signal, one sine wave current cycle sends the magnetic material through one hysteresis loop cycle. Many of the important properties can be inferred from this figure. The abscissa is H, the

TABLE 9.1. QUANTITIES IN BOTH CENTIMETER-GRAM-SECOND (cgs) UNITS AND INTERNATIONAL SYSTEM UNITS (SI)

Quantity	cgs Units	SI Units	cgs to SI Conversion Factor
ϕ, Flux	maxwell or line	weber or V · s	10^{-8}
B, Flux density	gauss or line/cm^2	tesla or Wb/m^2 or (V · s)/m^2	10^{-4}
H, Field strength	oersted	amp/m	79.6
μ, Relative permeability	μ (gauss/oersted)	μ_r	1
$(BH)_{max}$, Energy product	mega-gauss-oersted or gauss-oersted $\times 10^6$	kwatt-sec/m^3 or (kJ)/m^3	7.96

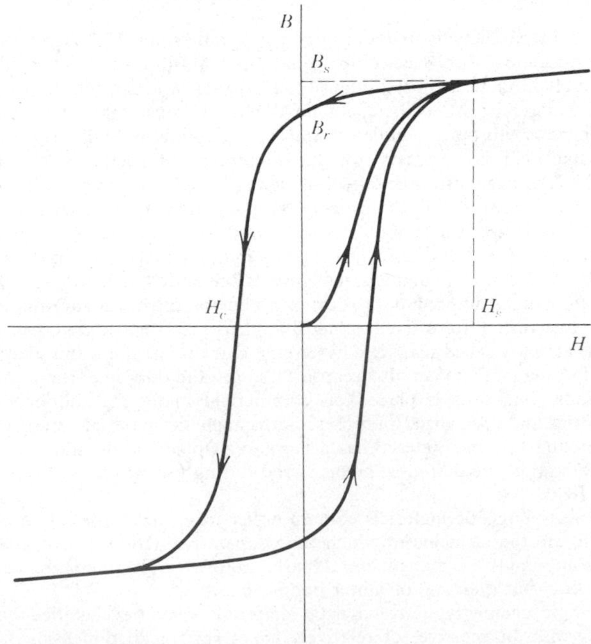

Fig. 9.1 Initial-magnetization curve and hysteresis loop.

magnetic field strength measured in oersteds or amps per meter. This is the intensity of the polarizing field. The ordinate B is the induction or flux density, which is the sum of the field strength and the $4\pi M$ [or the total polarization produced by the unit magnetic moments (or poles) M at a distance of 1 cm], thus $B = H + 4\pi M$. In soft materials where $M >> H$, $B \cong 4\pi M$; in hard materials both must be considered. The total magnetic flux is the product of the induction or flux density times the cross-sectional area, $\phi = BA$. The saturation induction, B_s, is the peak polarization obtainable per unit area. Although B may increase slightly after that point, it is entirely due to increasing H. Another major property is the remanence or residual induction B_r, which is the remaining polarization after removal of the polarizing field. B, B_s, and B_r are measured in gausses (G) or teslas (webers/meter^2). The ratio B_r/B_s is called the squareness ratio, which can range from 10 to 20% to almost 100%. The coercive force H_c is the reverse field necessary to reduce the induction to zero. It is measured in oersteds or amps per meter, as is H_s, the field strength at which B_s occurs.

A ratio of extreme importance is the relative permeability μ, which is B/H and is unitless. At low fields approaching zero, this ratio is called μ_i (or sometimes μ_0). At the steepest point of the magnetization curve it is called μ_{max}. Different conventions are used in expressing the permeability at different drive levels. In the United States, μ_{40} usually means the permeability at 40 G, in Europe, μ_5 may mean permeability at 5 mA/cm. It is important to know the convention used.

A property partly related to the hysteresis loop is the magnetic loss, that is, the energy lost in the material from the total traversing the hysteresis loop. One part of this loss, called the hysteresis loss, can be inferred by the included area swept out by the loop. Another portion, called the eddy current loss, is due to the circular current loops generated in the material caused by the oscillating magnetic field. A third component, called the residual (or anomalous) loss, represents the unexplained portion. The sum total of these losses is called the specific core loss and is expressed in watts per pound or watts per kilogram. In ferrites, conventionally the core loss is given in milliwatts per cubic centimeter.

At low drive levels, such as those used in telecommunications, the loss is usually specified as a loss resistance and a loss breakdown may be made in many different manners, again depending on the convention. In the United States, the Legg equation is used as a basis

$$\frac{R}{\mu f L} = aB_m + bf + c$$

where a, b, and c represent the loss coefficients for hysteresis, eddy current, and anomalous losses, respectively, and f is the frequency in hertz. By varying either B_m or f independently, the coefficients are determined.

Another loss criterion at low levels (especially in ferrites) is the loss factor (LF) that can be expressed as the ratio of tangent of the loss angle, tan δ, divided by the initial permeability.

$$\text{LF} = \frac{\tan\delta}{\mu_i} = \frac{1}{\mu_i Q} = \frac{R_s}{\mu_i X_L} = \frac{R_s}{2\pi\mu_i f L}$$

Sometimes the inverse of the LF or the $\mu_i Q$ product is specified.

Two parameters related to the stability of a material (especially ferrites) are the temperature factor (TF) and the disaccommodation factor (DF). TF is the relative change in inductance per degree per unit of inductance

$$\text{TF} = \frac{\Delta L}{L} \times \frac{1}{\mu_i \Delta T}$$

where L is inductance in henrys and ΔT is temperature change in degrees Celsius. The disaccommodation factor describes the fraction decrease in permeability per decade of time per unit of permeability.

$$\text{DF} = \frac{\mu_1 - \mu_2}{\mu_1^2 \log \dfrac{t_2}{t_1}}$$

A final electrical property of the magnetic material unrelated to the hysteresis loop but extremely important for power application is the resistivity, defined by $R = \rho l/A$, where R = resistance, l = length, and A = cross-sectional area. The unit for ρ is either ohm-centimeter or ohm-meter. A related property, the dielectric constant, becomes important at microwave frequencies.

Desirable Properties for Specific Applications

The choice of a magnetic material is usually made by matching the requirements of the application to material characteristics including frequency of operation, energy losses, stability, physical requirements, and last but not least cost.

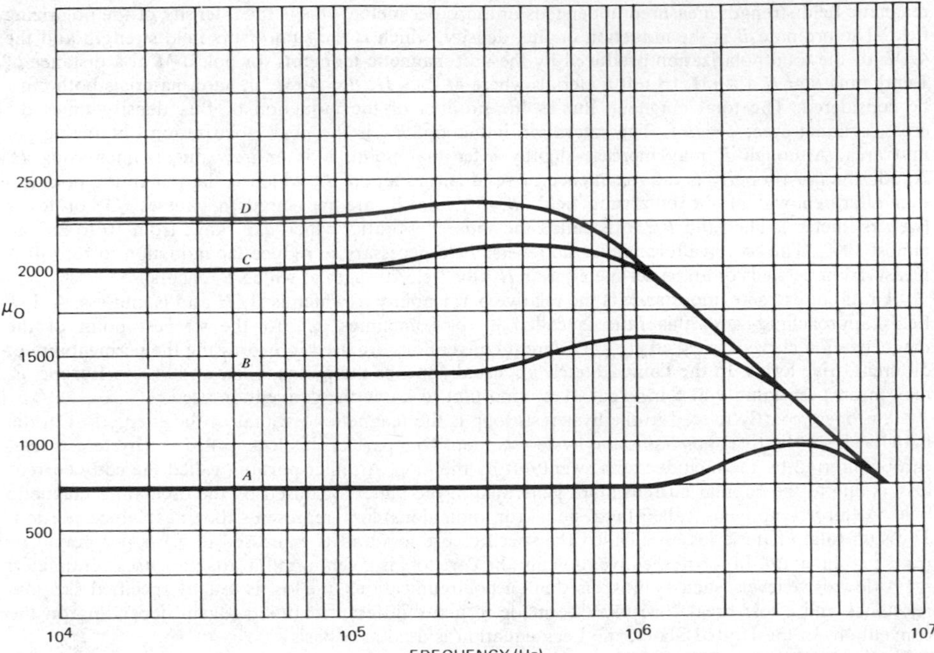

Fig. 9.2 μ_0 vs. frequency for four different ferrite materials. Note that the drop-off frequency decreases as the permeability increases.

Power Applications. At low frequencies (such as 60 Hz), high saturation, low cost, and low core loss at low frequencies are required. High saturation reduces the number of turns needed so that lower copper losses are incurred. At high frequencies, eddy current losses must be kept low either by use of thin-gage metallic strip or by increasing the resistivity of the material.

Telecommunication Applications. Here a high μ_i at the operating frequency is needed. Published μ vs. f curves (Fig. 9.2) should be consulted to avoid exceeding the critical frequency for a material (that frequency where the permeability drops off precipitously). A low loss factor, $1/\mu Q$, narrows band width and increases selectivity. In addition, high stability of μ with respect to flux density, time, and temperature prevents demodulation. If there is a superimposed DC signal, stability of μ to DC is also important. Since the frequency of operation is high, use of high-resistivity materials (ferrites) or thin-gage strip is a prime requirement.

Memory and Computer Applications. Here a high squareness ratio, (B_r/B_s), is needed. B_s should be high, and if the pulse repetition rate is high, the eddy current losses must be low (thin-gage strip or high-resistivity material).

Shielding Applications. A high initial permeability is needed. At low frequencies a low hysteresis loss is useful. Good formability will assist in the fabrication.

Magnetostrictive Materials. A high magnetostriction is of prime importance. The magnetomechanical coupling factor should be high. The mentioned precautions to prevent high eddy current losses should also be observed.

Permanent Magnet Materials. The most important feature here is a high $(BH)_{max}$ product. Coercive force and remanence should be as high as possible. Stability of these properties against time and temperature will prevent demagnetization.

Ranges of Magnetic Properties Available

Obviously the available magnetic properties of materials have some practical limits. It is important to know what these limits are and what other properties will do concurrently.

Saturation Induction. The saturation induction is an intrinsic property depending solely on the chemical composition at any specific temperature. It does vary, however, with temperature in a fashion common to all ferromagnetic materials. Thus it is not surprising that the saturation induction and Curie point (the temperature at which a material loses its ferromagnetic properties) usually parallel one another.

The saturation inductions of magnetic metals and alloys are generally quite high. The high saturation values (15,000–24,000 G or 1.5–2.4 T) are important in applications where large flux excursions are necessary for power applications (Power transformers, motors). Of the ferromagnetic elements, iron has the highest saturation (21,500 G or 2.15 T), cobalt the next highest (19,000 G or 1.9 T), and finally nickel (6000 G or 0.6 T). Mixtures of two or more of these generally result in an averaging of the individual elements. A notable exception is the 50 : 50 iron-cobalt alloy which has a higher saturation (24,500 G or 2.45 T) than either element alone. Alloys of iron, nickel, or cobalt with additions of nonmagnetic elements such as silicon, aluminum, molybdenum, and such have lower saturation values but other useful properties (e.g., resistivity, permeability) may be enhanced.

The saturations of these oxide materials are lower than those of metals for two reasons. (1) The large oxygen ions in the crystal lattice dilute the magnetic metal ion contribution since they are nonmagnetic. (2) The spins of the magnetic ions are not all aligned parallel in the crystal as described previously. As a result, maximum saturation inductions of 6000 G or 0.6 T or less are encountered. Magnetite or ferrous ferrite has one of the highest saturations but little use is made of it because of its low resistivity. Of the common useful ferrites, the ones with the largest saturations are the manganese-zinc-iron ferrites which have saturations up to 5000 G (0.5 T). Garnets, which are useful for high-frequency (microwave) applications, have saturations between 500 and 2000 G (0.05–0.2 T).

Permeability. High initial permeability depends partially on intrinsic properties such as composition by affecting basic magnetic parameters called anisotropy and magnetostriction. The first is the affinity of the magnetization for certain crystal directions. The second is the sensitivity of the magnetization to external stresses. In addition to chemistry, the permeability is sensitive to such things as crystalline texture, external stress, inclusions, dislocations, and degree of atomic ordering.

Permeability can be measured at DC or higher frequencies. For shielding, high permeability at DC or low frequency is needed. High initial permeability is most important for electronic applications at very low drive levels (below 100 G). In the power applications listed earlier, the maximum permeability or power permeability is most important. The nickel-iron alloys have the highest initial permeability of all materials, especially in the 80% nickel range and aided by the addition of 4 to 5% molybdenum. This is due to the simultaneous attainment of zero anisotropy and magnetostriction. Initial permeabilities of over 100,000 are possible with maximum permeabilities as high as 1,000,000. In silicon iron, initial permeabilities are much lower (1800 at 40 G) but maximum permeabilities (50,000) are sufficient to make this a very important power material. To emphasize the need for chemical purity, initial permeability in iron can be raised from a few thousand to 100,000 by lengthy and costly refinement processes.

Originally ferrites were only considered for very high frequencies and permeabilities were of secondary importance. In recent years, however, high-permeability ferrites have become available with μ_i of commercial material as high as 15,000. Laboratory materials have been made with μ_i up to 40,000. The materials in which these are possible are the manganese-zinc-iron ferrites which have low magnetostriction and anisotropy. The magnetostriction is lowered by use of excess iron, which directly controls the Fe^{2+} content. The proper chemistry is used to minimize the anisotropy. High purity is also extremely important. The availability of high-purity raw materials has partially been responsible for the achievement of these high permeabilities.

Resistivity. The eddy current losses are inversely proportional to the resistivity. They are also proportional to the square of the frequency. To keep these losses to a minimum, the resistivity must be raised as the frequency is raised.

The pure magnetic metals (nickel, iron, cobalt) have low resistivities ($\sim$10 $\mu\Omega$-cm). However, additions of silicon, aluminum, and molybdenum increase the resistivity significantly ($\sim$50 $\mu\Omega$-cm). In addition, metals can be powdered and insulated, retaining many of the metal characteristics such as saturation while increasing the resistivity greatly. The recently developed amorphous metal alloys have resistivities about three times those of their crystalline counterparts ($\sim$50 $\mu\Omega$-cm).

High resistivity is the main attraction of ferrites, being many orders of magnitude higher than the metals. The highest resistivities of this group are in the garnets and the iron-deficient ferrites such as $NiFe_2O_4$, $MgFe_2O_4$, and $MnFe_2O_4$, which have resistivities from 10^6 to 10^{10} Ω-cm. The ferrite with lowest resistivity is magnetite followed by those with excess iron (mostly manganese-zinc-iron ferrites) which have $\rho = \sim 10^2$ Ω-cm.

9.3 PHYSICAL PROPERTIES

Tensile and Compressive Strengths

Mechanical properties must be considered when choosing a magnetic material because of the fabrication needs of the component (e.g., drawing, punching, slitting, rolling) but also because of the effect of mechanical conditions on magnetic properties. Additionally the material may be under dynamic stresses in application (rotors in motors). 2-Vanadium Permendur must have high tensile strength in aircraft generators, so some of the magnetic properties are compromised for this need. Ferrites have very poor tensile properties (as is the case with most ceramics) but do have high compressive strengths.

Hardness

Hardness in metals is important as an indication of workability (drawing, etc.). While high magnetic softness usually corresponds to total mechanical softness, such a material may be too soft for fabrication (e.g., punching). Another important hardness concern is in the application (e.g., recording heads) where mechanical hardness usually goes along with wear resistance. Some alloys such as Sendust (iron-silicon-aluminum) or high aluminum-iron alloys (6.5 or 16% Al) have high hardnesses for this application. In addition, either single-crystal or hot-pressed ferrites are also excellent for the same reason.

Coefficient of Expansion

Many magnetic alloys are also important as expansion or temperature-compensating alloys. This is especially true of nickel-iron alloys in the 36 to 40% nickel range, some of which have a temperature coefficient of thermal expansion close to zero.

Phase Transition Temperatures

Phase transitional temperatures are important in the processing of magnetic materials. For example, the transition for iron, which normally occurs in pure iron at 930°C, is eliminated by the addition of other elements such as silicon. High-temperature heat treatments can be performed on this alloy which would not be possible in the pure material.

Corrosion Resistance

Many metallic materials are fairly corrosion resistant as the high-nickel alloys (80 Permalloy). Vanadium Permendur and Mumetal are even somewhat better. High-iron alloys such as silicon-iron corrode badly and are usually protected with some type of insulating coating such as magnesium or aluminum phosphate. Ferrites, of course, are quite resistant because of their oxide nature.

Radiation Effects

Most magnetic materials are altered to some degree by various types of radiation. In addition, materials containing cobalt are intensely radioactive. Thus it may be necessary in some applications to substitute more radiation-resistant materials at the expense of magnetic properties.

The metals that are most affected by γ or neutron radiation are the high-permeability materials such as permalloys. The most resistant are the silicon-irons and permanent magnet materials such as Alnicos. Grain-oriented materials can lose orientation on irradiation. The structure-sensitive properties such as permeability μ or coercive force H_c are mainly affected while the structure insensitive ones such as B_s are not.

Ferrites are fairly stable under room temperature irradiation but are unsuitable at high temperatures.

9.4 PRACTICAL MATERIALS

Iron

Iron is the least costly of all magnetic materials. With small amounts of purification to reduce carbon, an acceptable material can be produced with high saturation and moderate losses. High-purity iron can be made with very high permeability (about 100,000) but the purification is too lengthy to be practical. The saturation induction is about 21,000 G (2.15 T). The low mechanical strength and low resistivity prevent its use in many AC circuits. Iron is also available in compressed powder cores. The main advantage is the high resistivity and the consequent ability to operate at high frequencies extending into the high megahertz region. Of course the permeability drops dramatically to a range of about 10 to 100. In addition, flake cores are available with permeabilities of about 100.

Silicon-Iron Alloys

Silicon-iron is available in both nonoriented and oriented grades. Most practical grades contain between 0.5 and 3.5% silicon. Silicon levels above this range make the material too brittle for commercial working. Melting is done in electric or BOF furnaces using scrap charging. The ingots are hot and cold rolled to various gages from 0.001 to 0.017 in. (0.025–0.4 mm). If they are nonoriented grades, they are coated with phosphate and supplied to the customer. If they are oriented grades, they are usually coated with a MgO coating and given a special transformation anneal to develop the texture. They are then coated with phosphate like the nonoriented grades. With either the oriented or nonoriented grades, the customer, after fabricating the shape, must give the material a stress-relief anneal to develop the best magnetic properties.

The original method of specifying the quality of the material was to multiply by 10 the core loss (in watts per pound) of 29-gage material at 60 Hz and 15,000 G (1.5 T). Thus a material having a core loss of 2.2 W/lb under those conditions is called M-22. For highest permeability, low silicon levels are best. Lowest core losses are obtained in 3.25% silicon oriented material. There has been a recent breakthrough in processing of oriented silicon-iron material. The material for this process is called Hi-B or Orient-Core. Several manufacturers are now producing this type of material, which has superior orientation, lower core losses, and higher operating inductions.

Nickel-Iron Alloys

The important nickel-iron alloys fall into two categories, the ones with a nickel percentage of about 80% and the ones with a nickel percentage of about 50%. The 80% nickel alloys have higher permeabilities but lower saturations, whereas the 50% alloys have the reverse situation. The 80% alloys are made of high-purity materials with 4 to 5% molybdenum or copper additions (Mumetal). Carbon, oxygen, and sulfur contents are kept quite low and a very careful heat treatment and controlled cool rate are used to develop the proper atomic ordering. Annealing is done in high-purity hydrogen at rather high temperatures.

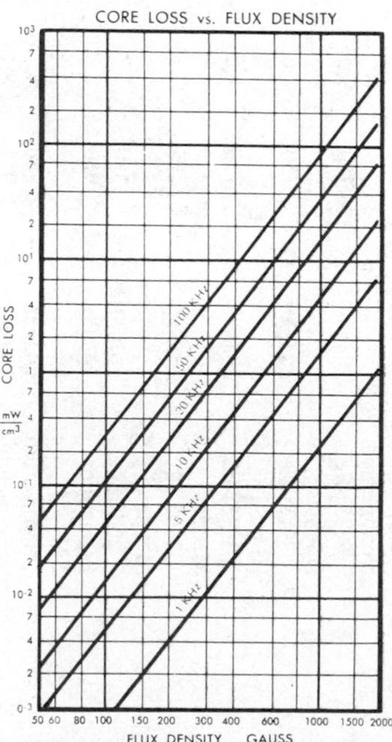

Fig. 9.3 Core loss of a power ferrite material ($\mu_0 = 3000$) as a function of flux density and frequency. Courtesy of Magnetics, Division of Spang Industries, Inc., Butler, Pa.

TABLE 9.2. PROPERTIES OF SOFT MAGNETIC MATERIALS

Materials	Saturation Flux Density (G)	Saturation Flux Density (T)	Curie Point (°C)	Density (gm/cm^3)	Resistivity ($\mu\Omega$-cm)	Initial Permeability (μ_0)	Maximum Permeability	Core Loss (W/kg)	At Frequency	At Induction (kG)
Iron	21,500	21.5	770	7.85	9.6	300	6,000			
Cobalt	19,000	1.9	1,121	8.84	9	70	250			
Nickel	6,080	0.6	358	8.89	8.7	250	2,500			
Low-Frequency Power Material										
Cold-rolled low-carbon steel	21,500	21.5	770	7.85	9.6	200	5,000	7.3	60 Hz	15 kG
Nonoriented 3% Si-Fe	20,000	2.0	750	7.65	47	280	8,000	3.2	60 Hz	15 kG
Oriented 3% Si-Fe	20,000	2.0	750	7.65	50	1,400	50,000	1.17	60 Hz	15 kG
Hi B 3% Si-F3	20,000	2.0	750	7.65	50	1,920		1.46	60 Hz	17 kG
Metglas[a] 2605 S2	15,600	1.56	415	7.18	130			0.25	60 Hz	14 kG
Metglas 2605 SC	16,100	1.6	370	7.32	125			0.3	60 Hz	14 kG
Permendur	24,500	2.45	950	8.3	7	1,000	11,000	25	400 Hz	20 kG
2-V Permendur	24,000	2.4	980	8.15	26	800	5,000	8.8	50 Hz	18 kG
Supermendur	24,000	2.4	940	8.15	26	800	70,000	12.1	400 Hz	15 kG
27% Co-Fe	24,200	2.42	970	8.02	19	650	2,800	6	60 Hz	15 kG
Inductor and Shielding Materials										
50% Ni-Fe	15,500	1.55	500	8.2	45	3,500	100,000	6.6	1 kHz	10 kG
Oriented 50% Ni-Fe	16,000	1.6	500	8.2	45		200,000	8.8	1 kHz	10 kG
Mumetal	6,500	0.65	460	8.25	55	20,000	100,000	5.5	5 kHz	5 kG
Metglas[a] 2826MB	8,800	0.88	353	8.02	160		750,000	96	50 kHz	2 kG
4-79 Permalloy	7,500	0.75	460	8.74	55	50,000	300,000	12.2	10 kHz	4 kG
Supermalloy	7,000	0.7	460	8.77	65	6,000	800,000	17.5	10 kHz	4 kG

High-Frequency Inductor Materials

Materials	Saturation Flux Density (G)	(T)	Curie Point (°C)	Resistivity (Ω-cm)	Initial Permeability (μ_0)	Frequency Range	Loss Factor ($1/\mu Q$)
Permalloy powder	8,000	0.8	460	500	14–300	1–300 kHz	300×10^{-6}
Sendust powder	10,000	1.0	500	10	80	1–100 kHz	300×10^{-6}
Carbonyl iron powder	12,000	1.2	700	100	3–40	0.1–100 MHz	20×10^{-6}
Mn-Zn ferrite	3,800	0.38	145	100	2,000	1–200 kHz	2×10^{-6}
Ni-Zn ferrite	1,500	0.15	350		80		
	3,500	0.35	400	10^7	1,500	1–10 MHz	100×10^{-6}

High-Frequency Power Materials

Materials	Saturation Flux Density (G)	(T)	Curie Point (°C)	Resistivity	Initial Permeability (μ_0)	Maximum Permeability	Core Loss (W/kg)	At Frequency	At Flux Level
Metglas[a] 2605 S3	15,800	1.5	405	125 μΩ-cm	—	20,000	10	20 kHz	2 kG
Mn-Zn ferrite	4,700	0.47	250	100 Ω-cm	2700	4,500	14	20 kHz	2 kG
Powder core 50-50 Ni-Fe	15,000	1.5	500	10 Ω-cm	125	150	88	20 kHz	2 kG (0.2 T)
Iron powder core	19,000	1.9	770	5 Ω-cm	90	250	360	20 kHz	2 kG

[a] Metglas® is a proprietary trademark of Allied Corporation.

The 50% alloys do not require quite as much care as the 80% alloys, since their permeabilities are considerably lower. However, their saturation is about twice that of the 80% alloys. Some 50% alloys can also be grain oriented to produce extremely high squareness ratios, on the order of 98 to 99%.

Both the 80% and 50% materials are available in compressed powder forms. A powder with 81% nickel material, 2% molybdenum, and the rest iron has extremely low losses, excellent stability to DC bias, and very good temperature stability. It is used in low-loss inductors and loading coils. The 50% nickel-iron powder becomes increasingly important in power applications, primarily in switch-mode power supplies. The losses are about twice as high but the saturation induction is also twice as high.

Cobalt-Iron Alloys

Cobalt-iron alloys have the highest saturation of all common magnetic materials. They are also some of the most expensive because of the high cost of cobalt. The highest saturation (24,500 G or 2.35 T) occurs at 35% cobalt content. The most important alloys are the ones with 50% cobalt content, followed by ones with lower (27%) cobalt contents. After forming, the material is often given a magnetic anneal for orientation. Most cobalt-iron alloys are used in aircraft or aerospace applications.

Soft Ferrites

Soft ferrites are becoming increasingly important as frequencies of operation of many electrical and electronic circuits increase. Unlike the materials mentioned previously, they are ceramics that are solid solutions of iron oxide, Fe_2O_3, with oxides of divalent metals such as manganese, nickel, cobalt, and magnesium. Their generic formula is $MO \cdot Fe_2O_3$ or MFe_2O_4, where M is the divalent metal ion. As previously stated, their high resistivities lead to very low losses, especially at high frequencies.

Ferrites are usually made from rather inexpensive raw materials by blending the component oxides, calcining, milling the calcined powder, drying, granulating, and molding the powder into a core. The properties are developed by a high-temperature firing process. There are other methods of forming ferrite powder, such as coprecipitation, but these have not reached commercial use. The relatively low cost of materials and the ease of formation have also been responsible for rapid growth. Because of the wide range of compositions, additives, and processing conditions, many different commercial ferrites are available for varying applications. For electronic applications such as telecommunications, the low losses and moderately high permeability permit their use in LC circuits such as filters and high frequency transformers (pulse, impedance matching). For high-frequency power applications, their low losses at moderate flux levels and lower cost make them useful materials for power transformers, fly-back transformers, and choke coils. Certain ferrites can be made with very high squareness ratios, which make them applicable for memory cores and switching cores. The important parameters for soft ferrites are as follows:

1. **Initial permeability.** The initial permeability μ_i varies from about 100 to 15,000.
2. **Loss factor.** The loss factor $1/\mu Q$ varies from a low of about 1×10^{-6} to much higher values at higher frequencies.
3. **Temperature factor.** The temperature factor varies from about 1×10^{-6} to about 10×10^{-6}.
4. **Disaccommodation factor.** The disaccommodation factor is very low, 1×10^{-6} for high-permeability ferrites to 10×10^{-6} for low-loss lower permeability ferrites.
5. **Core losses.** These are usually specified at a specific frequency and flux level. Until recently these were 16,000 Hz and 2000 G (0.2 T), but with the trend to higher frequencies and flux levels, these have changed. More useful are the log-log graphs shown in Fig. 9.3 with families of curves showing the whole range of frequencies and drive levels.

The large improvement in ferrites over the years is due to higher purity raw materials and the use of additives such as silicon, calcium, titanium, and tin. Better firing furnaces capable of good control of temperature and atmosphere are also helpful. This is true especially in the case of the most important soft ferrites, the manganese-zinc-iron ferrites. The excess iron in the form of Fe^{2+} must be carefully equilibrated with the oxygen partial pressure of the atmosphere, especially during the cooling-down period.

Square-loop ferrites for use in memory cores are mostly confined to magnesium-manganese ferrites. The squareness is achieved by an appropriate heat-treating cycle. Materials with high squareness usually have high anisotropy and low magnetostriction.

For very high-frequency applications, nickel-zinc or nickel ferrites are better than manganese-zinc, although the permeabilities are lower.

Typical values for the various materials described above are found in Table 9.2.

TABLE 9.3. PROPERTIES OF PERMANENT MAGNET MATERIALS

Material	Saturation Flux Density, B_s		Coercive Force, H_c		Maximum Energy Product, $(BH)_{max}$	
	G	T	Oe	A/m	G-Oe × 10^{-6}	(kJ)/m^3
Carbon steel	9,000	0.9	51	4,060	0.20	1.6
Tungsten steel	10,500	1.05	70	5,570	0.33	2.6
37% Co steel	10,400	1.04	230	18,300	0.98	7.8
Alnico 5	12,800	1.28	640	50,900	5.5	43.8
Alnico 9	10,500	1.05	1500	119,000	9.0	71.6
Alnico 8	8,200	0.82	1650	13,100	5.3	42
Oriented Ba ferrite	2,300	0.23	3800	30,200	3.4	27.1
Nonoriented Ba ferrite	2,300	0.23	1860	14,800	1.05	8.36
Elongated single-domain iron	7,350	0.735	940	74,800	3.5	27.9
Cunife	5,400	0.54	550	4,3800	1.5	11.9
Cunico	3,400	0.34	680	54,100	0.8	6.37
Vicalloy	9,050	0.905	415	33,000	2.3	18.3
Remalloy	8,500	0.85	345	27,500	1.2	9.55
Pt-Co	6,450	0.645	5400	43,0000	9.2	73.2
Rare earth cobalt	9,200	0.92	9000	716,000	21	167
Mn-Al-C	5,200–6,000	0.52–0.6	2000–6000	60–48	5–6	40–48
Nd-Fe-B	11,800–12,500	1.18–1.25	9,500–11,500	750,000–915,000	33–36	263–287

Amorphous Materials

In the past five years there has been a rapid development of several types of new materials called amorphous metals or metallic glasses. These are made by very rapid quench of a melted material as soon as it is formed into a thin strip. The two important categories thus far are:

1. $M_{80}Z_{20}$, where M is a metal combination involving nickel, cobalt, and iron and Z is the glass-forming nonmetal combination. These materials can be designed to have low magnetostriction and should find use in recording heads and other specialty applications.

2. $Fe_{80}Z_{20}$, where Z is the glass-former such as silicon, boron, or carbon or a combination of these. These materials may be processed either for power applications down to 60 Hz or for high-frequency applications. The losses are quite low, much lower than those of silicon-iron. These materials could produce a dramatic change in magnetic material usage if the price can be reduced to a competitive level.

Hard Ferrites

For permanent magnet materials, the first quadrant of the BH loop is no longer the important one. Since permanent magnets possess a significant remanent induction, and since the magnet is further demagnetized by external fields in actual application, it is really the second quadrant that is important. Unlike the case with soft materials, a high coercive force is helpful to prevent demagnetization; the remanence should also be high. A very useful specification for permanent magnet material is the $(BH)_{max}$ product, which is the maximum value of the product of the remanence and coercive force possible in the second quadrant. Hard ferrites, unlike soft ferrites, have high anisotropies. As stated previously, the B_s in ferrites is lower than that in most magnetic metals, but in hard ferrites the coercive force is higher. The generic formula for hard ferrites is $MO \cdot 6Fe_2O_3$, where M is either barium, strontium, or lead. They are formed by the same ceramic processes as those used for soft ferrites. Two varieties exist, oriented and nonoriented. The nonoriented materials are pressed in the same way as soft ferrites. The oriented materials are aligned in a magnetic field in either wet or dry condition before pressing. This allows for the axis of easy magnetization to be oriented parallel to a high degree, which increases the squareness and therefore the $(BH)_{max}$ product. Representative properties of hard ferrites are shown in Table 9.3.

Metallic Permanent Magnet Materials

The criteria for hard ferrites are also true for metallic magnet materials. There are massive permanent magnets such as bar or horseshoe configurations and there are thin-gage strip magnetic materials. The former are hard and brittle, while the latter are malleable. Many permanent magnet materials contain cobalt, mostly because it forms hexagonal-type crystal structures that have an easy direction or anisotropy parallel to the hexagonal axis. Prior to the recent escalation of cobalt prices, Alnico V was the most widely used permanent magnet material. It is formed by either casting or sintering processes. Before Alnico became available, certain grades of carbon, chromium tungsten, or cobalt steels were used for permanent magnets. High-grade anisotropic Alnicos with high BH products of about 9×10^6 G-Oe (72 kJ/m^3) were developed. Elongated single-domain (ESD) iron particles can be aligned and compressed to form special magnets. Platinum-cobalt is one of the best permanent magnet materials but its high price precludes large-scale usage.

In the past 15 years a new variety of metallic magnet has come along to somewhat revolutionize the phase modulation (PM) industry with its super-high $(BH)_{max}$ values [as high as 25×10^6 G-Oe (200 kJ/m^3)]. These are known as the rare-earth-cobalt magnets, and although they are somewhat expensive, their use has mushroomed recently. Some typical values of these materials are shown in Table 9.3. In the ductile category, cobalt-containing alloys such as Vicalloy and Remalloy are produced by melting, rolling, and other metallurgical techniques. A recent innovation has been the development of iron-cobalt-chromium alloys with lower cobalt contents. Within the past year, a new material composed of Neodynium, iron, and boron has been made commercially with a $(BH)_{max}$ value of 35×10^6G-Oe (280 kJ/m^3) and in the laboratory with a corresponding value of 45×10^6 G-Oe (360 kJ/m^3).

Microwave Materials

Microwave magnetic materials are only oxide (ferrite and garnet) materials because of the high resistivity needed at microwave frequencies. The ferrites are either nickel or magnesium-manganese ferrites. The newer garnet materials, which have the generic formula $3M_2O_3 \cdot 5Fe_2O_3$, have achieved large-scale usage in microwave components such as rotators, circulators, and phase shifters. This subject is discussed in another chapter.

Memory Materials

The square-loop ferrites for memory cores have already been discussed. Very thin-gage permalloy strip wound on small bobbins is also used in memory applications requiring high reliability. For recording purposes, magnetic wire and strip have been used but now are almost entirely supplanted by oxide-coated plastic tape. The oxide is of a semihard variety consisting of either γ-Fe_2O_3, Fe_3O_4 (magnetite), or chromium dioxide (CrO_2). The newest memory materials are the recently developed bubble memory materials. These are anisotropic garnet films epitaxially grown on substrates such as gadolinium-gallium garnets. The memory elements are the cylindrical domains, through the thickness of the film. These domains can be generated, stored, moved, and read. Thin-film memories of cobalt-nickel, cobalt-phosphorus and such have been made by electroplating, direct reduction from solution, vacuum deposition, or sputtering.

Magnetostrictive Materials

Alloys processed for high magnetostriction can be used as magnetostrictive transducers for cleaning or machining. Pure nickel can be used, or 50 : 50 nickel-iron, 95 : 5 nickel-cobalt, or the 2V Permendur (49Co-49Fe-2V) have high magnetostriction.

Bibliography

Bozorth, R. M., *Ferromagnetism*, Van Nostrand, New York, 1951.

Cullity, B. D., *Introduction to Magnetic Materials*, Addison-Wesley, Reading, MA, 1972.

Heck, C., *Magnetic Materials and Their Applications*, Crane Russack, New York, 1974.

Rado, G. T. and H. Suhl, *Magnetism*, Academic, New York, 1963.

Smit, J., *Magnetic Properties of Materials*, McGraw-Hill, New York, 1971.

Tebble, R. S., and D. J. Craik, *Magnetic Materials*, Wiley, New York, 1969.

Wohlfarth, E. P., *Ferromagnetic Materials*, Vol. 2, North-Holland, New York, 1980.

CHAPTER 10
SEMICONDUCTOR MATERIALS

EDWARD H. NICOLLIAN

University of North Carolina at Charlotte
Charlotte, North Carolina

10.1 INTRODUCTION

Elemental and compound semiconductors have device applications based on their material properties. The elemental semiconductor predominantly used in integrated circuits is silicon (Si). A major reason for the predominance of silicon in this application is that a thin, chemically stable, and electrically inert film of silicon dioxide can be easily grown on it. Also its bandgap of 1.12 eV ensures that free carrier[†] density is temperature independent over the temperature range (25–125°C) common during integrated circuit operation.

Silicon is emphasized in this section because it has by far the greatest economic impact of any semiconductor. However, the elemental semiconductor germanium (Ge) and the compound semiconductor gallium-arsenide (GaAs) also are discussed. Table 10.1 summarizes many of the properties of Si, Ge, and GaAs that are discussed in this section. Germanium has specialized applications and still is used in consumer applications. Compound semiconductors are used mainly in photonic and microwave applications. For example, GaAs is used in photonic devices because it has a direct gap (Si and Ge are indirect gap semiconductors) and in microwave devices because of its high electron mobility.

All of the topics in this section are treated in greater detail in Refs. 1 to 3.

10.2 ELECTRICAL, THERMAL, AND OPTICAL PROPERTIES

Energy Bands

The most important result of applying the principles of quantum mechanics to electrons in a semiconductor is that these electrons are allowed only certain energy levels grouped in bands separated by an "energy gap," also termed the bandgap. In silicon, the two allowed bands that participate in the electrical conduction process are the "valence band" (mostly filled with electrons) and the higher lying (mostly empty) "conduction band." Free carrier concentrations in the two bands depend on the density of allowed energy levels and the probability that given energy levels are occupied by electrons.

Electrons in the outermost shell of each silicon atom are called valence electrons. Each silicon atom has four valence electrons that it shares with four neighbors in the lattice by forming covalent

[†]A free carrier is an electron or a hole that is free to move throughout the semiconductor crystal.

TABLE 10.1. PROPERTIES OF GERMANIUM, SILICON, AND GALLIUM-ARSENIDE AT 300 K

Properties	Ge	Si	GaAs
Atoms/cm^3	4.42×10^{22}	5.0×10^{22}	4.42×10^{22}
Atomic weight	72.60	28.09	144.63
Breakdown field (V/cm)	$\sim 10^5$	$\sim 3 \times 10^5$	$\sim 4 \times 10^5$
Crystal structure	Diamond	Diamond	Zincblende
Density (g/cm^3)	5.3267	2.328	5.32
Dielectric constant	16.0	11.9	13.1
Effective density of states in conduction band, M_c (cm^{-3})	1.04×10^{19}	2.8×10^{19}	4.7×10^{17}
Effective density of states in valence band, M_v (cm^{-3})	6.0×10^{18}	1.04×10^{19}	7.0×10^{18}
Effective mass, m^*/m_0	$m_l^* = 1.64$	$m_l^* = 0.98$	0.067
Electrons	$m_t^* = 0.082$	$m_t^* = 0.19$	
	$m_{lh}^* = 0.044$	$m_{lh}^* = 0.16$	$m_{lh}^* = 0.082$
Holes	$m_{hh}^* = 0.28$	$m_{hh}^* = 0.49$	$m_{lh}^* = 0.45$
Energy gap (eV)	0.66	1.12	1.424
Intrinsic carrier concentration (cm^{-3})	2.4×10^{13}	1.45×10^{10}	1.79×10^6
Intrinsic Debye length (μm)	0.68	24	2250
Intrinsic resistivity (Ω-cm)	47	2.3×10^5	10^8
Lattice constant (Å)	5.64613	5.43095	5.6533
Linear coefficient of thermal expansion $\Delta L/L_0 \Delta T$(°C^{-1})	5.8×10^{-6}	2.6×10^{-6}	6.86×10^{-6}
Melting point (°C)	937	1415	1238
Minority carrier lifetime(s) (pure)	10^{-3}	2.5×10^{-3}	$\sim 10^{-8}$
Mobility (drift) (cm^2/V-s)			
Electrons	3900	1500	8500
Holes	1900	450	400
Optical-phonon energy (eV)	0.037	0.063	0.035
Phonon mean free path λ_0(Å)	105	76(electron) 55(hole)	58
Specific heat (J/g-°C)	0.31	0.7	0.35
Thermal conductivity (W/cm-°C)	0.6	1.5	0.46
Thermal diffusivity (cm^2/s)	0.36	0.9	0.44
Vapor pressure (Pa)	1 at 1330°C 10^{-6} at 760°C	1 at 1650°C 10^{-6} at 900°C	100 at 1050°C 1 at 900°C

Source. Reprinted from Sze,[1] with permission.

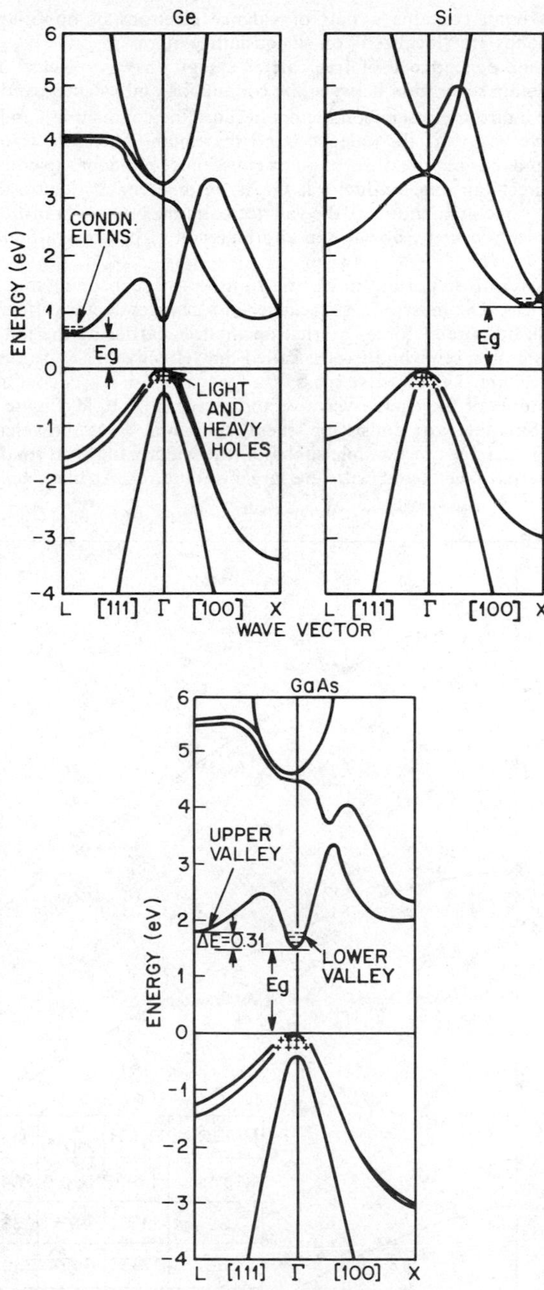

Fig. 10.1 Free carrier energy vs. momentum of germanium, silicon, and gallium-arsenide. Values of E_q are listed in Table 10.1. Plus (+) signs indicate holes in the valence bands and minus (−) signs indicate electrons in the conduction bands. After Chelikowsky and Cohen.[4]

bonds. Each covalent bond contains a pair of valence electrons of opposite spin. The valence electrons in covalent bonds are "localized" on silicon lattice atoms.

Figure 10.1 shows the dependence of free carrier energy on wave vector[†] or momentum. The bandgap E_g is the minimum separation between the conduction and valence bands.

Both Si and Ge are indirect gap semiconductors because the conduction band minimum does not occur at the same wave vector as the valence band maximum. Therefore transitions between the valence and conduction bands require absorption or emission of phonons[‡] to conserve momentum.

An example of a direct gap semiconductor is GaAs. In this semiconductor, the conduction band minimum occurs at the same momentum as the valence band maximum. Transitions between valence and conduction bands can occur by absorption and emission of photons, a fact important in many photonic devices.

The mass of free carriers is important in electric transport in semiconductors and therefore in the operation of many devices. The mass of a free carrier in a semiconductor differs from the mass of a free electron in vacuum because of forces exerted on the free carrier by the ion cores of the lattice. The mass of a free carrier in a semiconductor is called the "effective mass" to distinguish it from the free electron mass in vacuum. The effective mass of a free carrier in a semiconductor depends on the reciprocal of the curvature[§] of the energy-wave vector curves (Fig. 10.1). Figure 10.1 shows that the valence bands of Ge, Si, and GaAs consist of several subbands. Near the valence band maximum (minimum hole energy), there is a heavy hole subband (the wider band with smaller curvature) and a light hole subband (the narrower band with the larger curvature). At the valence band maximum,

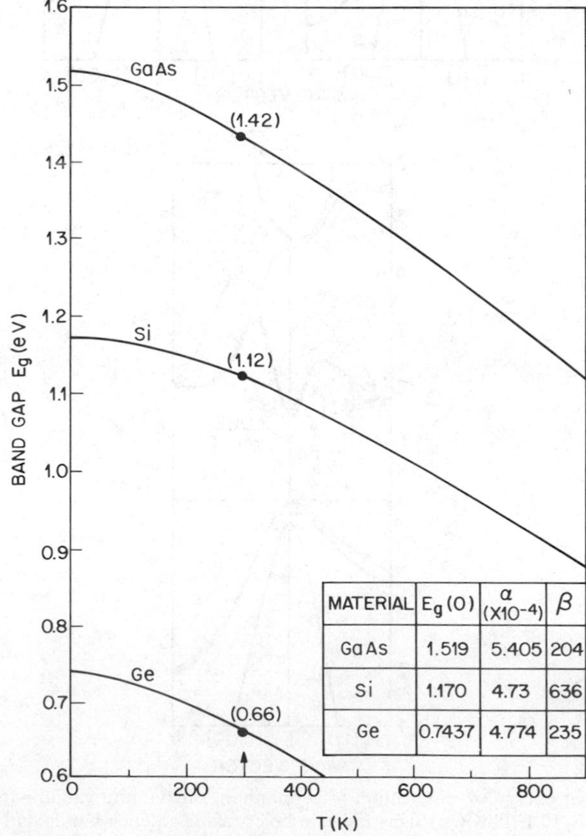

MATERIAL	$E_g(0)$	α ($\times10^{-4}$)	β
GaAs	1.519	5.405	204
Si	1.170	4.73	636
Ge	0.7437	4.774	235

Fig. 10.2 Bandgaps of germanium, silicon, and gallium-arsenide as functions of temperature. After Thurmond.[5] Reprinted by permission of the publisher, The Electrochemical Society, Inc.

[†] Wave vector is momentum divided by $h/2\pi$, where h is Planck's constant.

[‡] A phonon is the quantum of lattice vibration energy.

[§] The curvature is the second derivative of energy with respect to wave vector.

these two bands have the same curvature. In most integrated circuit applications, only holes at the valence band maximum play a role so that there is no distinction between heavy and light holes.

Figure 10.2 shows minimum bandgap energies and their variation with temperature for GaAs, Si, and Ge. The variation of bandgap $E_g(T)$ with temperature follows the approximate empirical relation[5]

$$E_g(T) = E_g(0) - \frac{\alpha T^2}{T+\beta} \text{ (eV)} \tag{10.1}$$

where $E_g(0)$ is the bandgap energy at 0 K, α and β are given in Fig. 10.2, and T is the absolute temperature. The temperature coefficient dE_g/dT, determined by measurement, is negative for all three semiconductors. Some semiconductors have positive dE_g/dT, such as lead-sulfide (PbS) for which the bandgap energy increases from 0.28 eV at 0 K to 0.41 eV at 300 K.

Measured near room temperature, the bandgaps of Ge and GaAs increase with pressure[6] $dE_g/dP = 5 \times 10^{-6}$ eV/(kg/cm^2) for Ge and about 12.6×10^{-6} eV/(kg/cm^2) for GaAs. The measured bandgap energy for Si decreases with pressure, $dE_g/dP = -2.4 \times 10^{-6}$ eV/(kg/cm^2).

Details of the energy-band structure of semiconductors are discussed in Refs. 1 and 7.

Electrons and Holes

An electron excited (by thermal agitation for example) into the conduction band is a negative charge free to move throughout the crystal under the influence of temperature or potential gradients. The valence charge deficiency produced by the excitation of an electron to the conduction band is a positive charge, called a hole, that is free to migrate throughout the crystal under the influence of temperature or potential gradients.

Intrinsic Semiconductor

In a pure semiconductor, a number of hole-electron pairs are present at a given temperature with the concentration of electrons, n, equal to the concentration of holes, p. These concentrations are intrinsic to the semiconductor and equal the intrinsic concentration, n_i (values are given in Table 10.1 and Fig. 10.3).

Fermi Level. The most probable energy distribution of an ensemble of free carriers at thermal equilibrium, subject to the Pauli exclusion principle,† is the Fermi–Dirac distribution. The probability f_n that an energy level E is occupied at thermal equilibrium by an electron is given by the Fermi–Dirac function[8]

$$f_n = \left(1 + \exp\frac{E - E_F}{kT}\right)^{-1} \tag{10.2}$$

where k is Boltzmann's constant (8.617×10^{-5} eV/K), and E_F is the Fermi level, the energy at which a state, if present, has a 50% chance of being occupied by an electron.

The probability f_p that a level E is vacant or that it is occupied by a hole is $f_p = 1 - f_n$ or

$$f_p = \left(1 + \exp\frac{E_F - E}{kT}\right)^{-1} \tag{10.3}$$

Electron and Hole Concentrations. The zero applied field electron concentration (cm^{-3}) in the conduction band is[3]

$$n = \int_{E_c}^{\infty} M_c(E) f_n(E - E_F)\, dE = M_c \exp\left[-\frac{E_c - E_F}{kT}\right] \tag{10.4}$$

where $M_c(E)$ is the density of states in the conduction band (cm^{-3}/J), E_c is the conduction band edge, and $M_c = 5.4 \times 10^{15} T^{3/2}$ (cm^{-3}) for silicon (values for Ge and GaAs at 300 K are listed in Table 10.1). Similarly, the zero applied field hole concentration (cm^{-3}) in the valence band is[3]

$$p = \int_{-\infty}^{E_v} M_v(E) f_p(E_F - E)\, dE = M_v \exp\left[-\frac{E_F - E_v}{kT}\right] \tag{10.5}$$

†The Pauli exclusion principle states that no two electrons can have the same quantum state and that all electrons are indistinguishable.

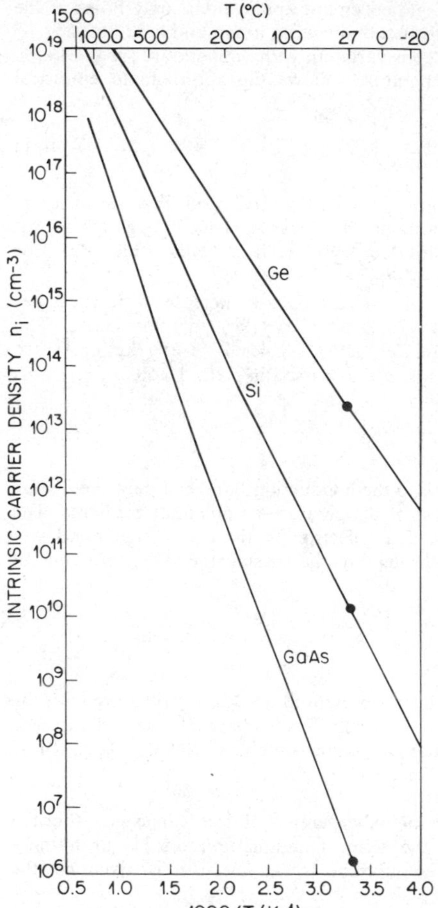

Fig. 10.3 Temperature dependence of the intrinsic carrier densities of germanium, silicon, gallium-arsenide. After Thurmond.[5] Reprinted by permission of the publisher, The Electrochemical Society, Inc.

where $M_v(E)$ is the density of states in the valence band (cm^{-3}/J), E_v is the valence band edge, and $M_v = 2.0 \times 10^{15} T^{3/2}$ (cm^{-3}) for silicon (values for Ge and GaAs at 300 K are listed in Table 10.1).

The position of the Fermi level in the bandgap for an intrinsic semiconductor is obtained from the condition of charge neutrality $n = p$ (see Ref. 3)

$$E_F = E_g/2 + \frac{kT}{2} \ln \frac{M_v}{M_c} = E_i \tag{10.6}$$

where E_i is the intrinsic Fermi level.

Because $p = n = n_i$ in an intrinsic semiconductor, $pn = n_i^2$. Thus[3]

$$n_i = (M_c M_v)^{1/2} \exp\left[-\frac{E_g}{2kT}\right] \tag{10.7}$$

Figure 10.3 shows the temperature dependence of n_i.

Extrinsic Semiconductor

Donors and Acceptors. When elemental impurities, called dopants, are added to the semiconductor, it becomes an "extrinsic" semiconductor. Dopants are added to modify the free carrier densities of the intrinsic semiconductor, controlling electrical conductivity and allowing fabrication of *pn* junctions and various unipolar and bipolar transistor structures.

Effective dopants are elemental impurities that enter the semiconductor lattice substitutionally; that is, they take the place of a silicon atom in the lattice, and have very shallow energy levels in the bandgap. Because silicon has a valence of 4, substitutional addition of elements from group V of the periodic table such as phosphorus, antimony, and arsenic results in an extra electron weakly bonded to the impurity atom. This extra electron occupies an energy level in the bandgap very close to the conduction band edge (0.05 eV). At normal temperatures this extra electron is thermally excited into the conduction band, leaving behind a positively charged impurity atom called a donor. The substitutional addition of elements to silicon from group III of the periodic table, such as boron and aluminum, results in the acceptance of an additional electron from the valence band to complete the four valence electrons shared with the surrounding silicon atoms. At ordinary temperatures, an electron is thermally excited from the valence band to the shallow energy level (0.05 eV above the valence band edge) of the impurity, creating a hole in the valence band and a negatively charged impurity atom called an acceptor. When the donor concentration N_D cm^{-3} is much larger than n_i and the acceptor concentration N_A cm^{-3} is zero, $n = N_D$ and the semiconductor is n type. When $N_A \gg n_i$ and $N_D = 0$, $p = N_A$ and the semiconductor is p type. When both donors and acceptors are present, the semiconductor is compensated, but if $N_D > N_A$ it is n type and if $N_A > N_D$ it is p type.

Free Carrier Concentrations

The position of the Fermi level in the bandgap of an n-type semiconductor is[3]

$$E_F = E_i + kT \ln\left[\frac{N_D - N_A}{n_i} + \frac{n_i}{N_D - N_A}\right] \tag{10.8}$$

and of a p-type semiconductor is[3]

$$E_F = E_i - kT \ln\left[\frac{N_A - N_D}{n_i} - \frac{n_i}{N_A - N_D}\right] \tag{10.9}$$

Figure 10.4 shows $E_F - E_i$ as a function of temperature for silicon with N_D and N_A as parameters.

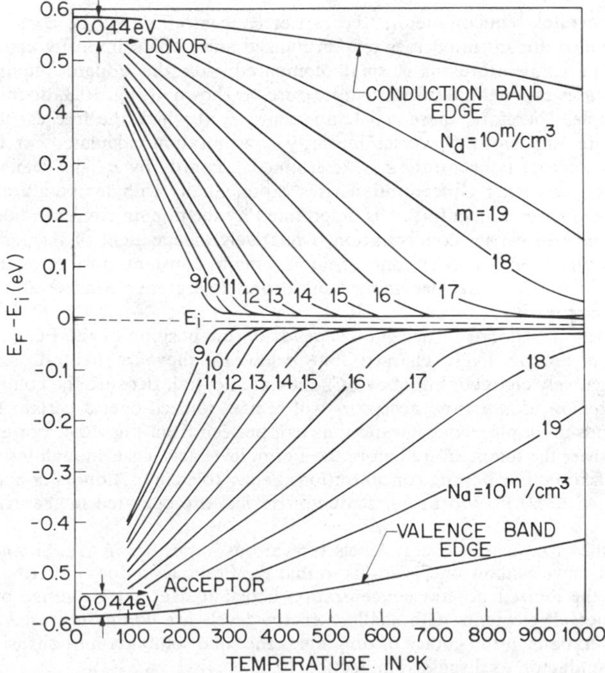

Fig. 10.4 Dependence of the Fermi level measured with respect to midgap on temperature with doping concentration as parameter. $E_g(T)$ from Eq. (10.1) is also shown. After Beadle et al.[9] Copyright (1981), American Telephone and Telegraph Company. Reprinted with permission.

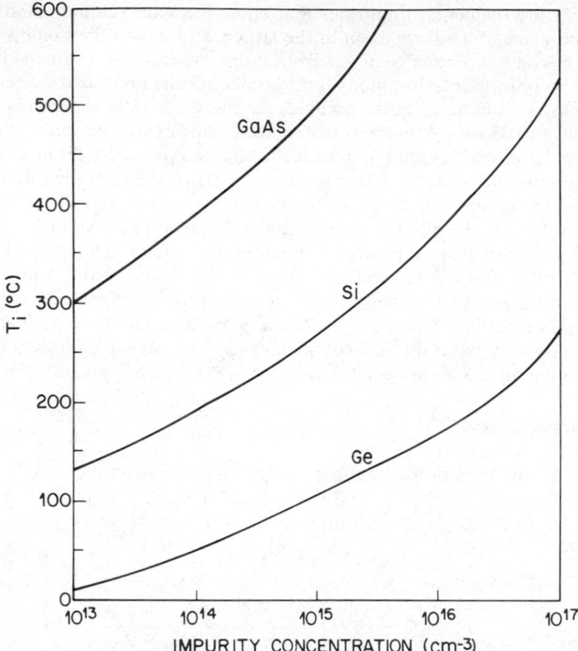

Fig. 10.5 Intrinsic temperature as a function of doping concentration. After Sze.[1]

In a nondegenerate† semiconductor, free carrier concentration will be constant independent of temperature provided dopant impurities remain ionized and dopant impurity concentration is larger than n_i. At room temperature, n_i is small compared with the dopant impurity concentration. However, n_i increases exponentially with temperature, as shown in Fig. 10.3, doubling every 11°C for silicon, for example. Therefore, there is a temperature T_i at which the free carrier concentration is equal to n_i. Figure 10.5 shows T_i versus impurity concentration calculated for GaAs, Si, and Ge. Above T_i the free carrier concentration is determined primarily by n_i, the semiconductor becomes intrinsic, and the free carrier concentration rises exponentially with temperature, as shown in Fig. 10.3. Below T_i free carrier concentration is determined by the doping concentration, the semiconductor is extrinsic, and free carrier concentration is relatively independent of temperature. As temperature is decreased further, free carrier concentration remains constant, until at relatively low temperatures the dopant impurities start becoming neutral and free carrier concentration starts decreasing with decreasing temperature.

The charge state of a donor or acceptor depends on the position of the Fermi level relative to it. Above E_F donors are positively charged, but below E_F they are neutral. Similarly, below E_F acceptors are negatively charged, but above E_F they are neutral. Because the commonly used dopant levels are very shallow, donors and acceptors will remain ionized over a certain temperature range. This range depends on doping concentration, as can be seen from Fig. 10.4. For example, the lighter the doping, the lower the temperature before the Fermi level gets close enough to the dopant level for significant deionization. For doping concentrations below 10^{15} cm^{-3}, donors or acceptors will remain ionized down to 45 to 50 K. Most semiconductor devices are operated in the extrinsic temperature range.

Figure 10.6 shows measured energy levels for various impurities in Ge, Si, and GaAs. Elements chosen as dopants have shallow energy levels so that the Fermi level does not cross the impurity level, thereby keeping the ionized dopant concentration constant over a wide range of temperature and bias. However, not all elements with shallow energy levels are commonly used as dopants. Other important considerations in the choice of dopants is the solid solubility and diffusivity of the element in the host semiconductor, as discussed in Section 10.4.

†The semiconductor is degenerate[3] if it is so highly doped that either $n \approx M_c$ or $p \approx M_v$ and the Fermi level is near either the conduction or valence band edge.

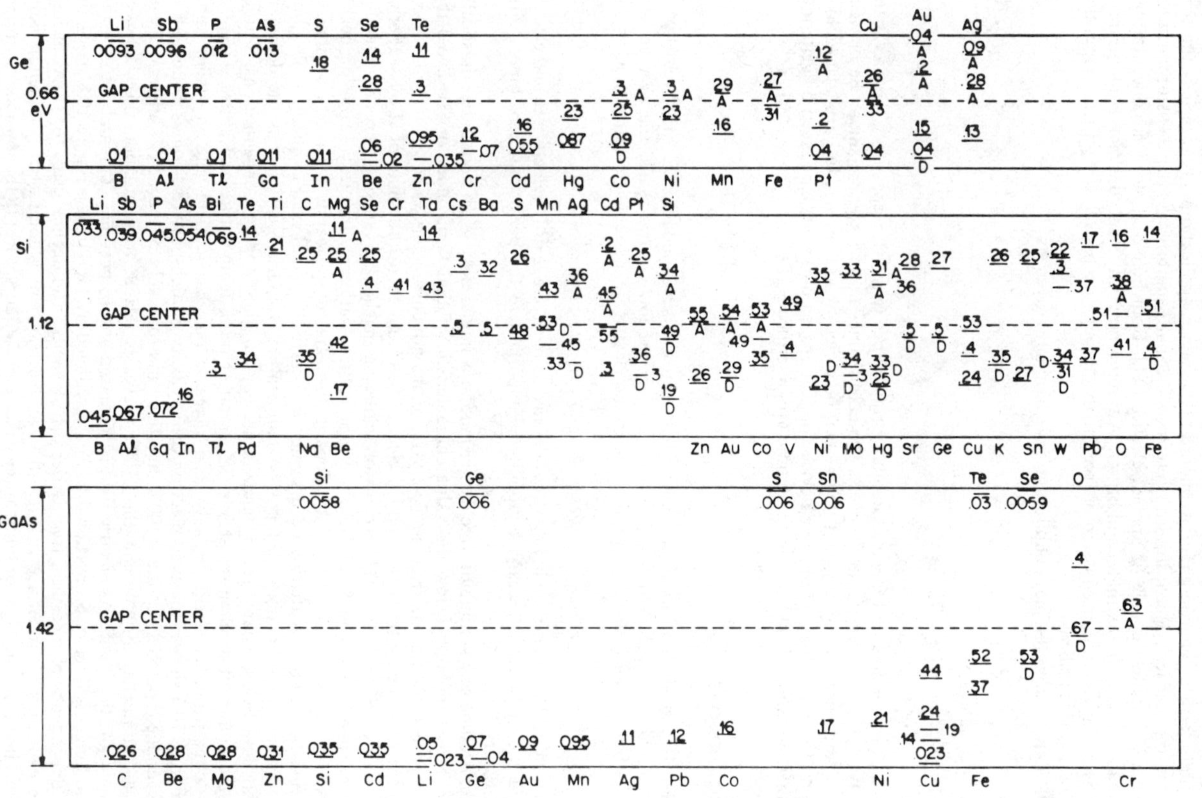

Fig. 10.6 Measured ionization energies for various impurities. Energy levels below midgap are measured from the valence band edge and are acceptor type unless marked with a D for donor type. Energy levels above midgap are measured from the conduction band edge and are donor type unless marked with an A for acceptor type. Bandgaps are at 300 K. After Sze.[1]

Deep level impurities such as the heavy metals shown in Fig. 10.6 usually are not deliberately introduced. Rather they exist as residual impurities at concentrations several orders of magnitude less than the dopant impurities in device grade semiconductors.

Lifetime

Free carrier lifetime in a semiconductor is important in device applications because it is the time required for the free carrier concentration to return to thermal equilibrium† after a disturbance has occurred, such as a sudden change of applied electric field. Both hole and electron concentrations must change to reestablish thermal equilibrium. An important way that thermal equilibrium is reestablished is by the capture or emission of free carriers through deep level heavy metal impurities, such as gold or copper in silicon.

For small changes from thermal equilibrium, hole lifetime is given by[3]

$$\tau_p = (c_p n_T)^{-1} \quad \text{(s)} \tag{10.10}$$

and electron lifetime by[3]

$$\tau_n = (c_n n_T)^{-1} \quad \text{(s)} \tag{10.11}$$

where c_p and c_n are the hole and electron capture probabilities (cm^3/s), respectively, and n_T is the volume density (cm^{-3}) of deep impurity levels. Typical lifetimes in device grade silicon lie between the millisecond and microsecond ranges. Lifetime can be varied by varying n_T. In particular, gold is added to silicon to shorten lifetime in some device applications. In others, n_T is reduced by gettering[3] to lengthen lifetime. Methods of measuring lifetime are described in Ref. 3.

Free Carrier Transport

Bulk Mobility. Upon application of an electric field across a semiconductor, a "drift velocity" is superimposed on the thermal motion of the free carriers. At low electric fields, the average drift velocity $\bar{v}_d$ of a free carrier is directly proportional to the electric field strength F (V/cm) or[1]

$$\bar{v}_d = \mu F \quad \text{(cm/s)} \tag{10.12}$$

where μ in $\text{cm}^2/\text{V-s}$, the proportionality constant defined as the average drift velocity per unit electric field, is called the mobility. Free carrier mobility is important in the performance of bipolar and unipolar transistors.

In a polycrystalline semiconductor, free carrier motion is impeded by grain boundaries drastically reducing free carrier mobility below the value characteristic of that semiconductor. To avoid this degradation, transistors are made from single-crystal semiconductor material.

Free carriers gain energy from the applied electric field and then lose some of it by scattering resulting in a finite drift velocity. The two major scattering mechanisms in single-crystal, nonpolar semiconductors such as Ge and Si are impurity and lattice scattering. Impurity scattering occurs when charged free carriers are deflected by dopant ions, and its probability depends on ionized dopant concentration. Lattice scattering occurs when free carriers are deflected by thermal vibrations of lattice atoms, and its probability depends on temperature. Impurity scattering is the dominant mechanism at low fields, below room temperature, or at high doping concentrations. Lattice scattering is the dominant mechanism at low fields and at room temperature and above, except at high doping concentrations. Drift velocity also depends on electric field and crystallographic orientation. Figure 10.7 shows measured electron drift velocity in high-purity silicon as a function of electric field applied parallel to the ⟨100⟩ direction and parallel to the ⟨111⟩ direction with temperature as parameter. Structure in the ⟨100⟩ curves is related to population shifts between valleys in the conduction band. As electric field is increased and $\bar{v}_d$ becomes comparable to thermal velocity, $\bar{v}_d$ no longer is directly proportional to F, as predicted by Eq. (10.12). Figure 10.8 shows the electric field dependence of measured free carrier drift velocity. The decline of $\bar{v}_d$ for GaAs at high fields can be explained by reference to Fig. 10.1. As electric field is increased, conduction band electrons are excited from the high mobility lower valley (smaller curvature) to the low mobility upper valley (larger curvature), increasing the electron population of the upper valley at the expense of the lower valley. At high fields, the electron population in the upper valley becomes significant, causing overall drift velocity to decline. This transferred electron effect leads to differential negative resistance, which has been exploited in GaAs and indium-phosphide (InP) to make microwave power

†A system is in thermal equilibrium when all of its macroscopic parameters are time invariant.

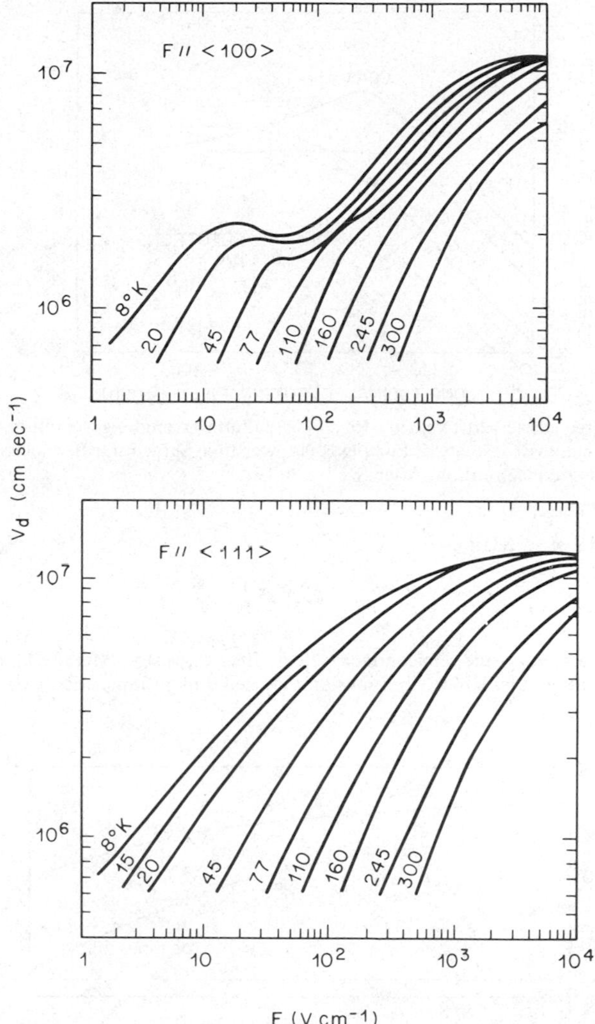

Fig. 10.7 Measured electron drift velocity as a function of electric field. After Canali et al.[10]

amplifiers and local oscillators. A more complete discussion of the transferred electron effect is given in Ref. 1.

Figure 10.9 shows electron and hole mobilities in ⟨111⟩ Ge, Si, and GaAs measured at 300 K as functions of ionized dopant concentration at low fields. Equation (10.12) applies here. Drift velocity is weakly dependent on orientation at low fields and at room temperature, so that mobility for ⟨100⟩ is about the same as for ⟨111⟩. Figure 10.10 shows the temperature dependence of electron and hole mobility in ⟨111⟩ silicon at low fields where Eq. (10.12) applies. For lower doping concentrations, the slopes of these curves deviate slightly from the slopes predicted by lattice and impurity scattering because of other scattering mechanisms.[1]

At low fields, the interval between collisions t_c is independent of the applied electric field, and the relation between mobility and effective mass m^* is[2]

$$\mu = \frac{qt_c}{2m^*} \tag{10.13}$$

where q is the electronic charge in C. Ratios m^*/m_0, where m_0 is the free electron mass in vacuum, are listed for electrons and light and heavy holes in Table 10.1.

The diffusivity D of free carriers in a concentration gradient in the semiconductor is related to

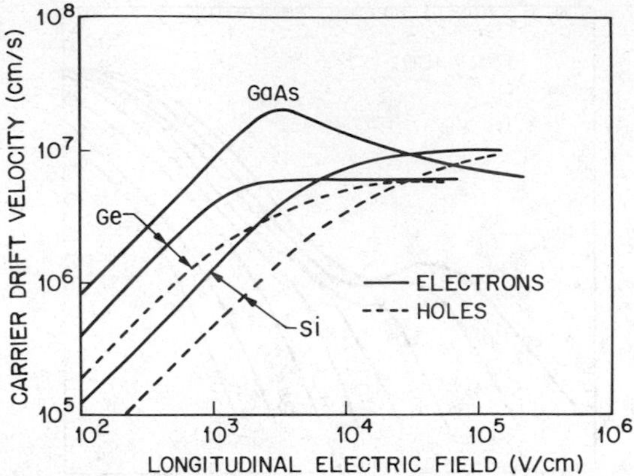

Fig. 10.8 Measured free carrier drift velocity for $\langle 111 \rangle$ gallium-arsenide, germanium, and silicon at 300 K. For heavy doping, drift velocity at low fields is lower than shown. Drift velocity at high fields is independent of doping concentration. After Sze.[1]

their mobility by the Einstein relation[1]

$$D = \frac{kT}{q} \mu \text{ (cm}^2\text{/s)} \tag{10.14}$$

Surface Mobility. In a metal oxide semiconductor field effect transistor (MOSFET), inversion layer free carriers also are scattered by lattice vibrations and ionized dopant impurities. However, inversion

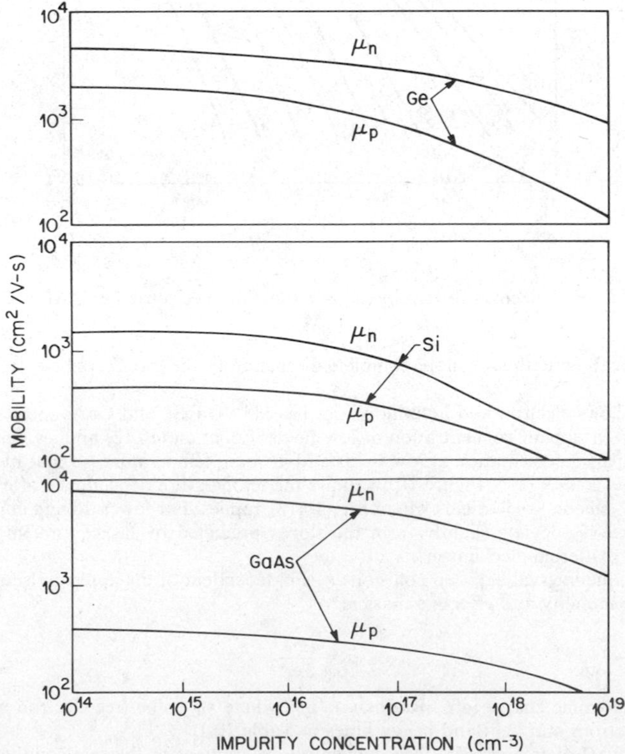

Fig. 10.9 Dependence of free carrier drift mobility on ionized impurity concentration. After Sze.[1]

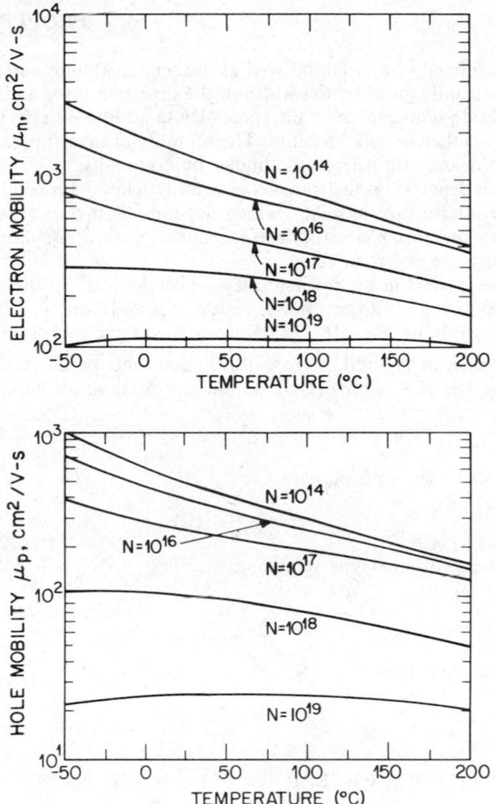

Fig. 10.10 Temperature dependence of electron and hole mobility in ⟨111⟩ silicon at low fields. After Beadle et al.[9] Copyright (1981), American Telephone and Telegraph Company. Reprinted with permission.

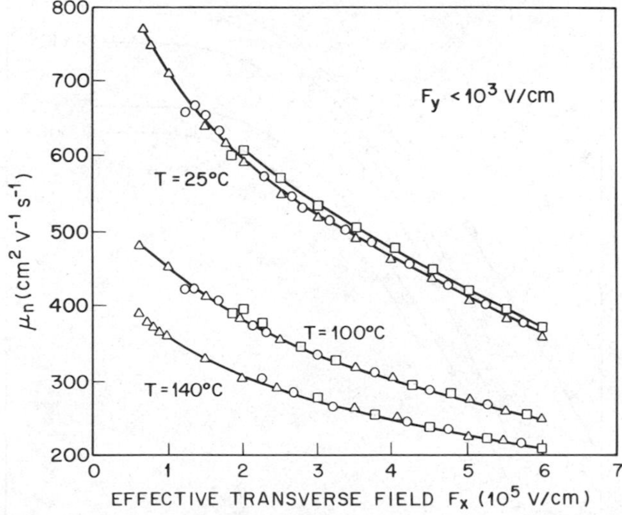

Fig. 10.11 Dependence of electron surface mobility in *p*-type ⟨100⟩ silicon on effective transverse electric field at low F_y with temperature as parameter. Effective transverse electric field is defined as the transverse field averaged over the electron distribution in the inversion layer and is given by $(F_x)_{\text{eff}} = \epsilon_s^{-1}(Q_B + \frac{1}{2}Q_n)$ where ϵ_s is the dielectric permittivity (C/V-Cm), Q_B is the depletion layer charge density (C/cm^2), and Q_n is the inversion layer charge density (C/cm^2). After Sabnis and Clemens.[11] Copyright (1979) IEEE.

layer free carriers are confined to a potential well at the semiconductor surface. The mobility of free carriers in this well also is influenced by the width of the inversion layer and by scattering by charges in the oxide near the oxide-semiconductor interface. These additional effects reduce the mobility of the free carriers in the well below bulk mobility. The mobility of inversion layer free carriers is called the surface mobility, to distinguish it from the higher bulk mobility.

Inversion layer width depends on the transverse electric field F_x applied between gate and source. This dependence makes the surface mobility, which depends on inversion layer width, a function of F_x. Figure 10.11 shows measured electron surface mobility as a function of effective transverse electric field with temperature as a parameter.

Drift velocity of free carriers in the inversion layer also depends on the longitudinal electric field F_y applied between source and drain. For a given F_x and low F_y, drift velocity is directly proportional to F_y, as given by Eq. (10.12). However, as F_y is increased, drift velocity tends to saturate at lower fields than in the bulk. Figure 10.12 shows the measured drift velocity of inversion layer electrons as a function of F_y with F_x as a parameter. A curve for bulk drift velocity is included for comparison.

Resistivity. The resistance of a semiconductor is[2]

$$R = \rho \frac{L}{A} \quad (\Omega) \tag{10.15}$$

The "resistivity" ρ in Ω-cm, of an n-type semiconductor is[2]

$$\rho = \frac{1}{q\mu_n n} \tag{10.16}$$

Similarly, for a p-type semiconductor[2]

$$\rho = \frac{1}{\mu_p p} \tag{10.17}$$

where μ_n is the electron mobility and μ_p is the hole mobility. In general, when both holes and

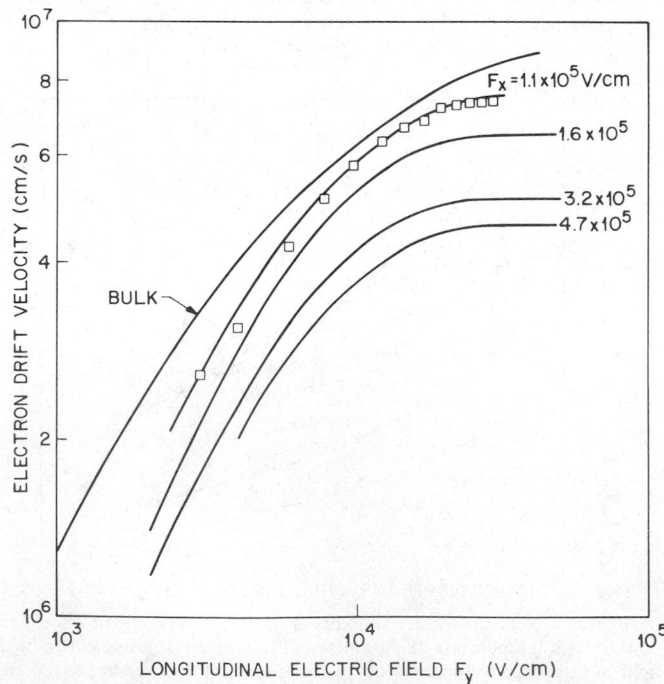

Fig. 10.12 Inversion layer electron drift velocity as a function of F_y in ⟨100⟩ p-type silicon at 300 K with F_x as a parameter. After Sze.[1]

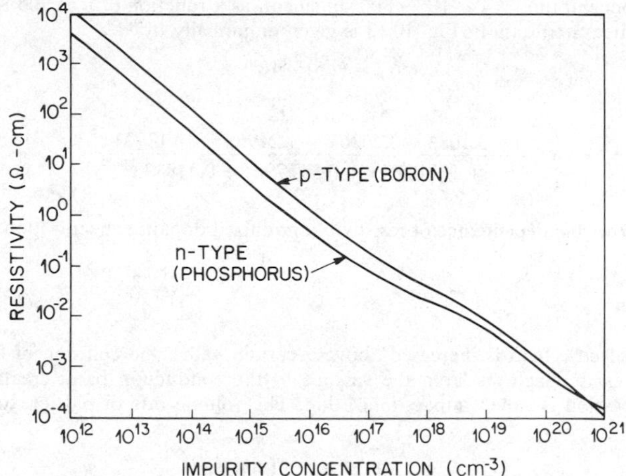

Fig. 10.13 Dependence of the resistivity of extrinsic silicon at 300 K on dopant concentration. After Beadle et al.[9] Copyright (1981), American Telephone and Telegraph Company. Reprinted with permission.

electrons have to be taken into account[2]

$$\rho = \frac{1}{q(\mu_n n + \mu_p p)} \tag{10.18}$$

When only one type of dopant impurity is present, ρ has a simple dependence on ionized dopant concentration. Figure 10.13 shows the measured resistivity of silicon as a function of ionized boron concentration (p type) and ionized phosphorus concentration (n type).

Boron concentration $N_B < 10^{20}\ \text{cm}^{-3}$ in silicon as a function of ρ at 300 K that accurately approximates the measured data in Fig. 10.13 is given empirically by[9]

$$N_B = (q\rho)^{-1}\left[2.13 \times 10^{-3} + 1.73 \times 10^{-2}(1 + 83.02\rho^{1.105})^{-1}\right] \tag{10.19}$$

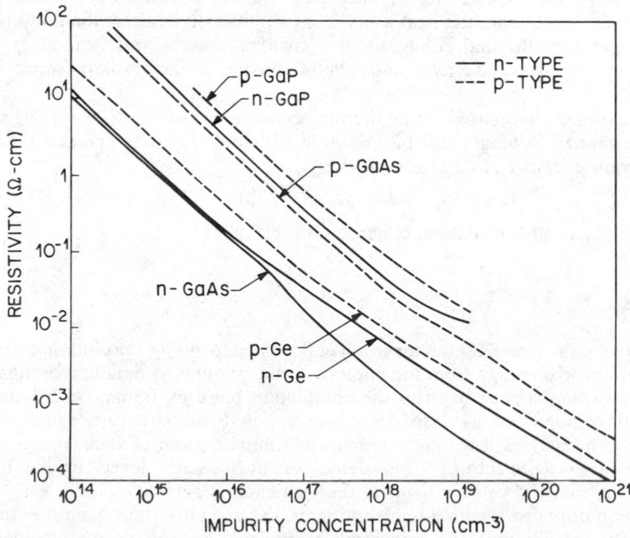

Fig. 10.14 Dependence of the resistivity of extrinsic germanium, gallium-arsenide, and gallium-phosphorus at 300 K on dopant concentration. After Beadle et al.[9] Copyright (1981), American Telephone and Telegraph Company. Reprinted with permission.

and phosphorus concentration $N_P < 10^{20}$ cm^{-3} in silicon as a function of ρ at 300 K that accurately approximates the measured data in Fig. 10.13 is given empirically by[9]

$$N_P = (q\rho)^{-1}10^{u_1} \tag{10.20}$$

where

$$u_1 = \frac{-3.1083 - 3.2626x - 1.2196x^2 - 0.13923x^3}{1 + 1.0265x + 0.38755x^2 + 0.41833x^3}$$

and $x = \log(\rho)$.

Figure 10.14 shows the dependence of resistivity on ionized dopant concentration.

Impact Ionization

When an applied electric field is increased above a certain value, the collision of free carriers with lattice atoms will excite electrons from the valence to the conduction band creating hole-electron pairs. Such pair creation is called impact ionization. The volume rate of pair creation G by impact ionization is[1]

$$G = a_n n\bar{v}_n + \alpha_p p\bar{v}_p \ (\text{cm}^{-3}/\text{s}) \tag{10.21}$$

where α_n is the electron ionization rate (cm^{-1}) defined as the number of electron-hole pairs generated by an electron per unit distance traveled, α_p is the ionization rate defined analogously for holes, and $\bar{v}_n$ and $\bar{v}_p$ are electron and hole drift velocity, respectively.

Figure 10.15 shows the measured dependence of α_n and α_p on $1/F$.

Ionization rates are useful in predicting free carrier multiplication and avalanche breakdown voltage and field in *pn* junctions. Impact ionization is discussed in greater detail in Ref. 1.

Thermal Properties

For a temperature gradient dT/dx along the axis (x-direction) of a semiconductor rod, the resulting heat flux Q_x flowing across unit area per unit time determines the thermal conductivity of the semiconductor as[12]

$$K = -Q_x/(dT/dx) \tag{10.22}$$

where the units of Q_x are watts per square centimeter and the units of K are watts per square centimeter-degree.

Figure 10.16 shows the dependence of measured thermal conductivity on lattice temperature. Copper commonly is used to conduct heat away from *pn* junction devices; diamond (type II) has the highest room temperature thermal conductivity known at present and is used as a heat sink in junction lasers and microwave devices; and silicon dioxide is the insulator most widely used in integrated circuits.

Figure 10.17 shows the measured linear thermal expansion coefficient of silicon as a function of temperature. The change in length ΔL of a sample of length L_0 at temperature T_0 caused by an increase of temperature from T_0 to T is given by[12]

$$\Delta L = aL_0(T - T_0) \tag{10.23}$$

where a is the coefficient of linear thermal expansion in Fig. 10.17.

Optical Properties

Photons can interact with a semiconductor in several ways depending on photon energy. In all cases, the semiconductor absorbs energy from the photons. First, photons at bandgap or higher energies will excite electrons from the valence band to the conduction band by impact with lattice atoms. Such photon-induced transitions are used to determine energy-band structure,[7] such as in Fig. 10.1. Second, photons with energies less than bandgap will impart some of their energy to free carriers. Third, photons can excite electrons from defect or trap energy levels in the bandgap to the conduction band or from the valence band to the trap energy level.

In the interaction of photons with a semiconductor, the two important quantities measured are the transmission coefficient Tr and the reflection coefficient R. For normal incidence, these two quantities are given by[1]

$$Tr = \frac{(1 - R^2)\exp(-4\pi x/\lambda)}{(1 - R^2)\exp(-8\pi x/\lambda)} \tag{10.24a}$$

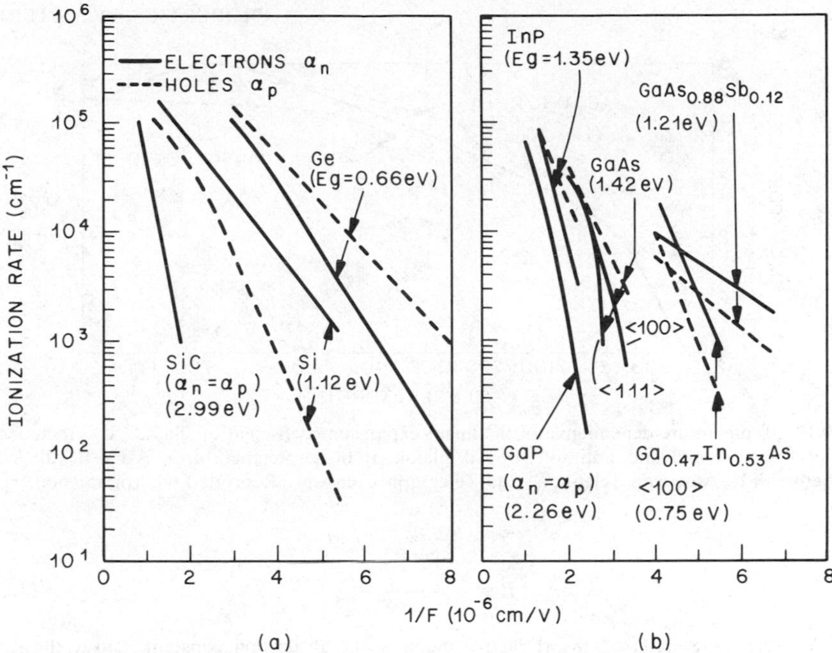

Fig. 10.15 Ionization rate vs. reciprocal electric field measured at 300 K for germanium, silicon, gallium-arsenide, and other compound semiconductors. After Sze.[1]

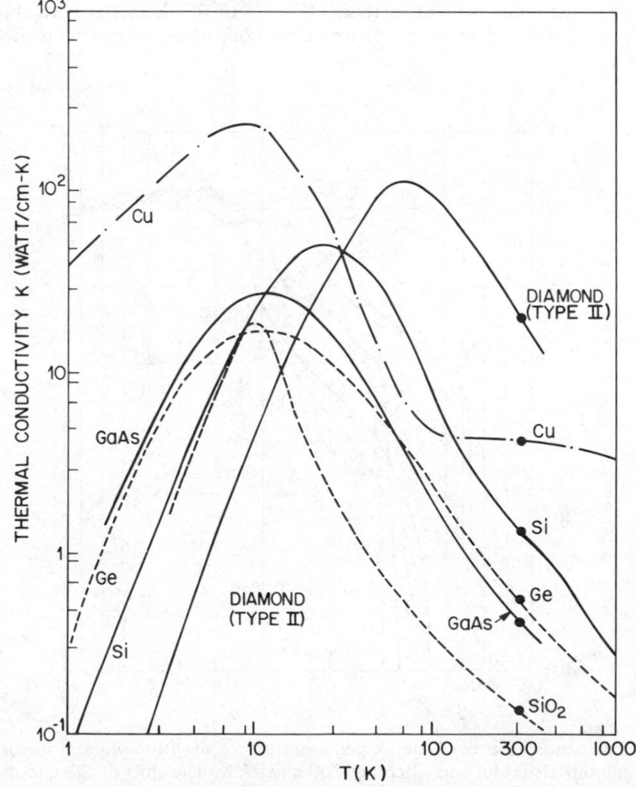

Fig. 10.16 Measured thermal conductivity as a function of temperature for pure germanium, silicon, gallium-arsenide, copper, diamond type II, and silicon dioxide. After Sze.[1]

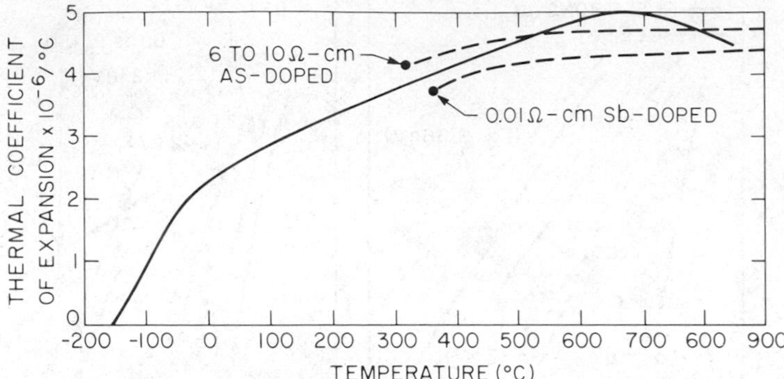

Fig. 10.17 Temperature dependence of the linear expansion coefficient of silicon. Also included are data for arsenic-doped and antimony-doped silicon at higher temperatures. After Beadle et al.[9] Copyright (1981), American Telephone and Telegraph Company. Reprinted with permission.

and[1]

$$R = \frac{(1 - n^*)^2 + \kappa^2}{(1 + n^*)^2 + \kappa^2} \tag{10.24b}$$

where λ is the wavelength, n^* the refractive index, κ the absorption constant, and x the sample thickness. The absorption coefficient per unit length A is given by[1]

$$A \equiv \frac{4\pi\kappa}{\lambda} \tag{10.25}$$

Both n^* and κ can be determined from an analysis of $Tr - \lambda$ or $R - \lambda$ data at normal incidence or by observing R or Tr at different angles of incidence. Then both n^* and κ can be related to interband transition energies.

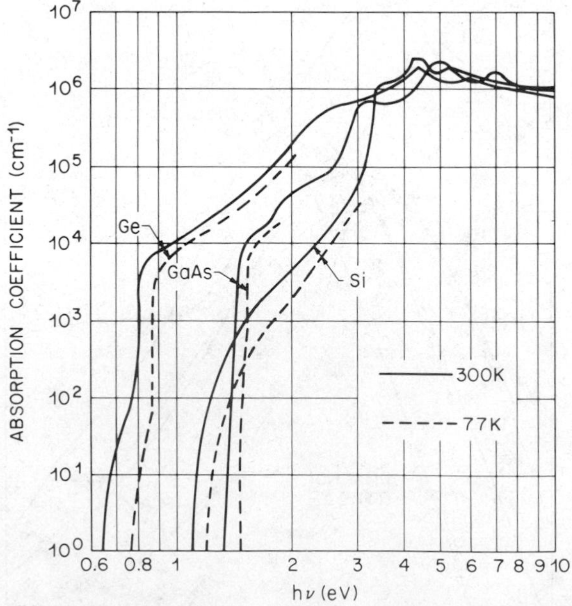

Fig. 10.18 Measured absorption coefficients near and above the fundamental absorption edge for pure germanium, gallium-arsenide, and silicon at 300 and 77 K. The shift of the curves toward higher photon energies at lower temperatures is caused by the temperature dependence of bandgap energy [see Fig. 10.2 and Eq. (10.1)]. After Sze.[1]

Figure 10.18 shows the measured dependence of κ near and above the fundamental absorption edge (band-to-band transitions) on photon energy.

10.3 OXIDATION OF SILICON

Growth of a thin layer of silicon dioxide (0.01–1 μm) on silicon is a key feature of integrated circuit technology. The chemical stability and electrical inactivity of silicon dioxide and its relatively easy preparation makes silicon the most widely used semiconductor in semiconductor electronics. Silicon dioxide layers provide surface "passivation" for bipolar transistors and junction diodes where the oxide layer ensures that the electrical characteristics of these devices are dominated by bulk rather than surface properties. A silicon dioxide layer also can serve (1) as a "diffusion mask," (2) to isolate one device from another, (3) to insulate the gate from the silicon in field effect devices, one of the most important applications, and (4) to isolate multiple levels of interconnections in an integrated circuit.

Oxides that fulfill these requirements can be made by several methods. Of most importance are vapor phase reaction,[13] plasma anodization,[14,15] wet anodization,[16] and thermal oxidation in both wet and dry environments.[17] Thermal oxidation yields oxides having the best electrical properties. Moreover, thermally grown oxides meet all the other requirements just described. Therefore, thermal oxidation is the key oxidation process in integrated circuit technology today. The other oxide preparation methods serve to increase the options available to the devicc fabricator. Details of thermal oxidation kinetics and technology for silicon are described in Refs. 2 and 3.

Interfacial Defects and Oxide Charge

Interfacial Defects. Electrically active defects are characteristic of the interface between a semiconductor and an oxide or other insulator. These defects are located at or near the interface and have densities ranging from $\leqslant 10^{10}$ cm^{-2} to $\geqslant 10^{12}$ cm^{-2} depending on oxide or insulator preparation conditions. Those defects having energy levels in the semiconductor bandgap that change occupancy by capturing free carriers from the semiconductor or by emitting carriers to the semiconductor when gate bias is varied are called interface traps. Those defect levels that do not change occupancy with variations of gate bias are called oxide fixed charge. In the Si–SiO_2 system, oxide fixed charge is usually positive. Both types of defect are undesirable at moderate to high densities ($> 10^{11}$ cm^{-2}) because they adversely affect device characteristics. For example, high interface trap and oxide fixed charge densities degrade gain and increase noise in bipolar transistors and metal-oxide-semiconductor field-effect transistors (MOSFETs), alter threshold voltage in MOSFETs, alter avalanche breakdown voltage and degrade forward characteristics of *pn* junctions, and decrease the efficiency of photonic devices. To optimize device performance and stability, interface trap and oxide fixed charge densities must be controlled during fabrication and kept invariant during device operation.

How these defects affect device characteristics and how their properties are measured and controlled during device fabrication and operation are described for the Si–SiO_2 interface in Ref. 3.

Oxide Charge. There are two types of charge in the oxide, "mobile charge" and "oxide trapped charge." The most common mobile charge in SiO_2 is the positive sodium ion. Sodium is the most important alkali metal contaminant because of its widespread presence in the environment. Sodium ion concentrations of a part per million or less in SiO_2 can cause intolerably large shifts of the electrical characteristics of devices. Sodium ions affect device characteristics by the large image charge they induce in the silicon after drifting to the Si–SiO_2 interface under an applied electric field.† The sodium story in SiO_2 is explained in Ref. 3.

Oxide trapped charge is created by exposure to ionizing radiation or by hot carrier injection. Ionizing radiation produces hole-electron pairs in the SiO_2. The major effect is the build-up of positive charge with exposure time. Positive oxide charging occurs when the holes produced drift under an applied field to the Si–SiO_2 interface where they become trapped. The image charge induced by trapped holes‡ in the silicon alters device characteristics.

Some of the electrons drifting from source to drain in a MOSFET gain sufficient energy from applied fields to surmount the energy barrier of 3.2 eV between silicon and SiO_2. Once in the SiO_2, a small fraction are trapped by water-related centers producing negative oxide charge and interface traps. Hot electron injection is avoided by operating MOSFETs at voltages below 3.2 V. Ionizing radiation, hot carrier injection, and trapping are more fully discussed in Ref. 3.

†Sodium ions present a serious stability problem becuase they do not lost their charge at an electrode.
‡Few electrons are trapped compared with holes.

10.4 EPITAXY, DIFFUSION, AND ION IMPLANTATION

Epitaxial growth of thin semiconductor films, ion implantation, diffusion of dopant impurities into semiconductors, and oxidation, discussed in Section 10.3, form the main processes of semiconductor device technology.

Epitaxial Growth

Epitaxy is the process of growing a thin layer of single-crystal semiconductor on a single-crystal semiconductor substrate. Epitaxial growth is important in semiconductor device technology for two reasons. First, thin films of relatively low dopant concentration can be grown on substrates containing the same type of dopant. Usually substrate doping concentration is much higher than in the epitaxial film to reduce series resistance associated with the substrate without otherwise altering device characteristics. Second, dopant concentration in the epitaxial film can be controlled easily and independently of the dopant concentration in the substrate by controlling the dopant impurity concentration in the phase from which the epitaxial layer grows. Thus, epitaxial growth can be used to make a *pn* junction between the epitaxial layer and the substrate.

Epitaxial layer thicknesses are typically between the micron range and atomic dimensions. Three commonly used methods of growing epitaxial layers are (1) vapor phase growth,[2] used in silicon technology; (2) liquid phase growth,[18] used extensively in compound semiconductor device technology; and (3) molecular-beam epitaxy,[19,20] which can give precise control of semiconductor compositions and layer thicknesses down to atomic dimensions. See Ref. 2 for a discussion of vapor phase growth of silicon films.

Diffusion

Semiconductor devices and integrated circuits are made by introducing dopant impurities of a given type and concentration into specific regions of a semiconductor. Solid state diffusion is one practical way of implementing this aim. If a dopant of conductivity type opposite to that initially in the semiconductor is diffused into the semiconductor, a *pn* junction is formed. A major use of diffusion in modern integrated circuit technology is doping polysilicon gates.

Diffusion usually is a two-step process. The first step, called the predeposition step, is that in which the dopant impurity is deposited, usually from a gas, on the semiconductor surface to a depth up to a few tenths of a micron. In the second step, called the drive-in, the predeposited dopant impurities are diffused deeper to form the desired concentration distribution. For silicon, drive-in is done at elevated temperatures in an oxidizing ambient to grow an oxide layer that effectively seals in the predepositied impurities.

The one-dimensional diffusion equation relating rate of change of dopant concentration to dopant concentration gradient is[1]

$$\frac{\partial C(x,t)}{\partial t} = D_I \frac{\partial^2 C(x,t)}{\partial x^2} \qquad (10.26)$$

where $C(x,t)$ is the dopant concentration in reciprocal centimeters cubed and D_I is the dopant impurity diffusion coefficient. There are two practical conditions under which Eq. (10.26) can be solved: (1) The dopant concentration at the semiconductor surface C_s (cm^{-3}) is always constant, and the solution is[1]

$$C(x,t) = C_s \operatorname{erfc}\left(\frac{x}{2\sqrt{D_I t}}\right) \qquad (10.27)$$

and (2) the initial dopant concentration on the semiconductor surface C_T (cm^{-2}) constitutes the total amount of dopant impurities available, and the solution is[1]

$$C(x,t) = \frac{C_T}{\sqrt{\pi D_I t}} \exp\left(\frac{x^2}{4D_I t}\right) \qquad (10.28)$$

The profiles of many dopants are well approximated by Eqs. (10.27) and (10.28).

The diffusion coefficient depends on temperature and dopant concentration. For low dopant concentrations, D_I becomes concentration independent. In a limited temperature range for low dopant concentrations[1]

$$D_I(T) = D_\infty \exp(-E_A/kT) \qquad (10.29)$$

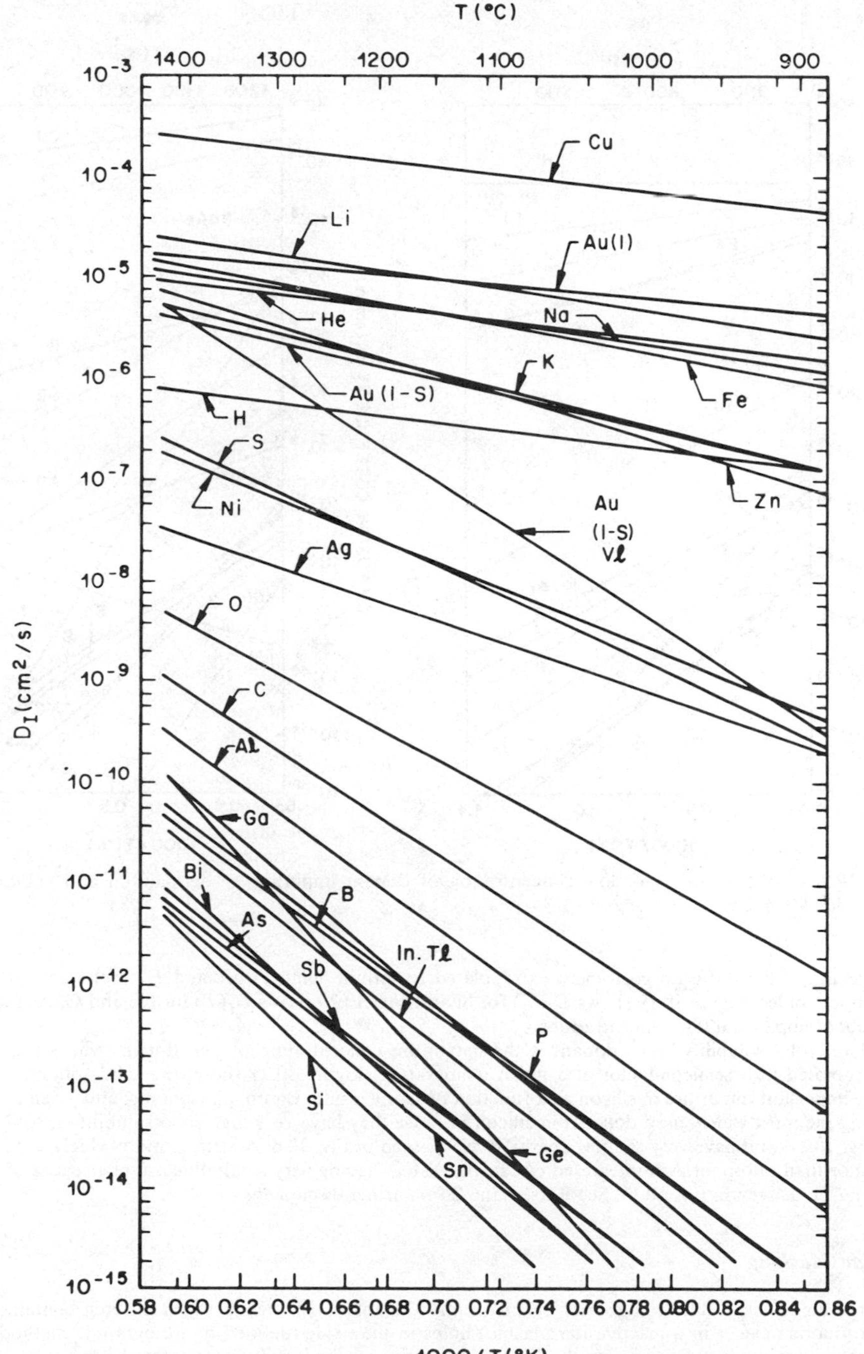

Fig. 10.19 Plot of $D_I(T)$ for low concentration of dopant impurities in silicon. After Beadle et al.[9] Copyright (1981), American Telephone and Telegraph Company. Reprinted with permission.

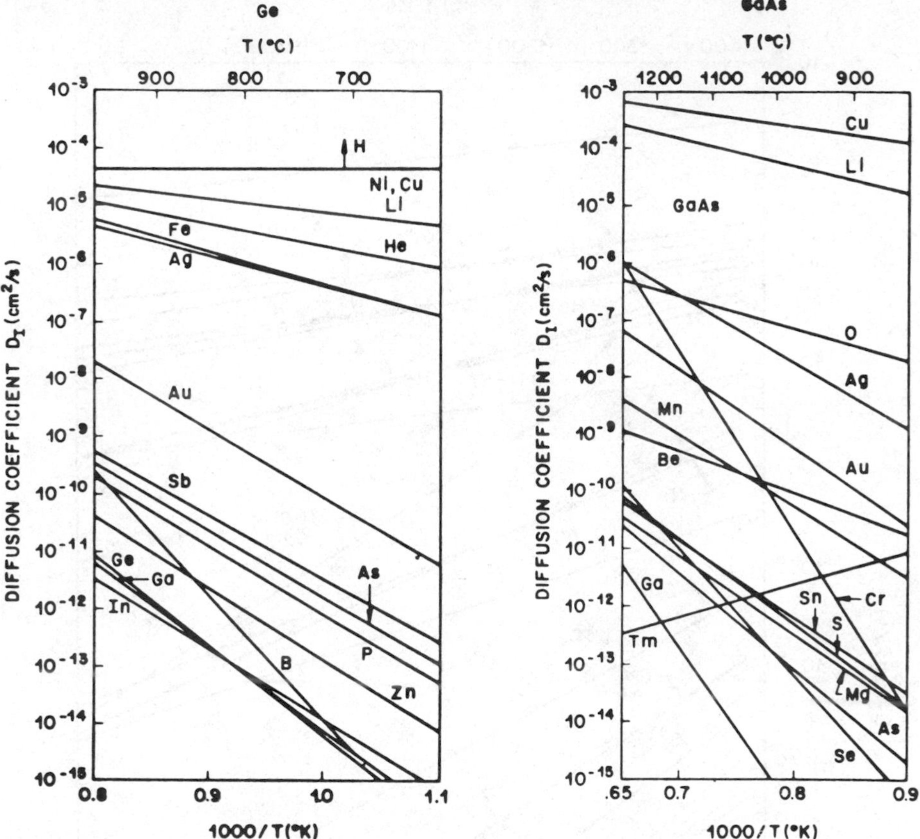

Fig. 10.20 Plot of $D_I(T)$ for low concentration of dopant impurities in germanium and gallium-arsenide. After Sze.[1]

where D_∞ is the diffusion coefficient extrapolated to infinite temperature and E_A is the activation energy in joules. Figure 10.19 shows $D_I(T)$ for Si and Fig. 10.20 shows $D_I(T)$ for Ge and GaAs for a variety of dopants at low concentrations.

The "solid solubility" of a dopant is the maximum concentration of that dopant which can be incorporated in a semiconductor at a given temperature. Figure 10.21 shows the solid solubility of some important impurities in silicon as a function of temperature. Boron, phosphorus, and arsenic are among the most widely used dopants in silicon because they have very high solid solubilities, as seen in Fig. 10.21, and have very shallow energy levels as seen in Fig. 10.6. Arsenic is more widely used as a donor than phosphorus in integrated circuit transistors having very small dimensions because of its lower D_I,† as seen in Fig. 10.19. See Refs. 1 and 2 for further discussion.

Oxide Masking

SiO_2 is an effective diffusion barrier against many dopants.[21] By etching (with an etch containing hydrofluoric acid or in a reactive ion plasma) holes in the oxide (defined by lithographic methods), the remaining oxide acting as a mask, dopant impurities can be incorporated into a silicon substrate in selected areas to make the active *pn* junction regions in an integrated circuit. To be effective as a diffusion mask, the diffusivity of the dopant in SiO_2 must be smaller than in silicon. Commonly used dopants having diffusivities smaller in SiO_2 than in silicon are phosphorus, boron, arsenic, and

†To maintain small device dimensions, dopant impurities should not diffuse far during thermal processing steps.

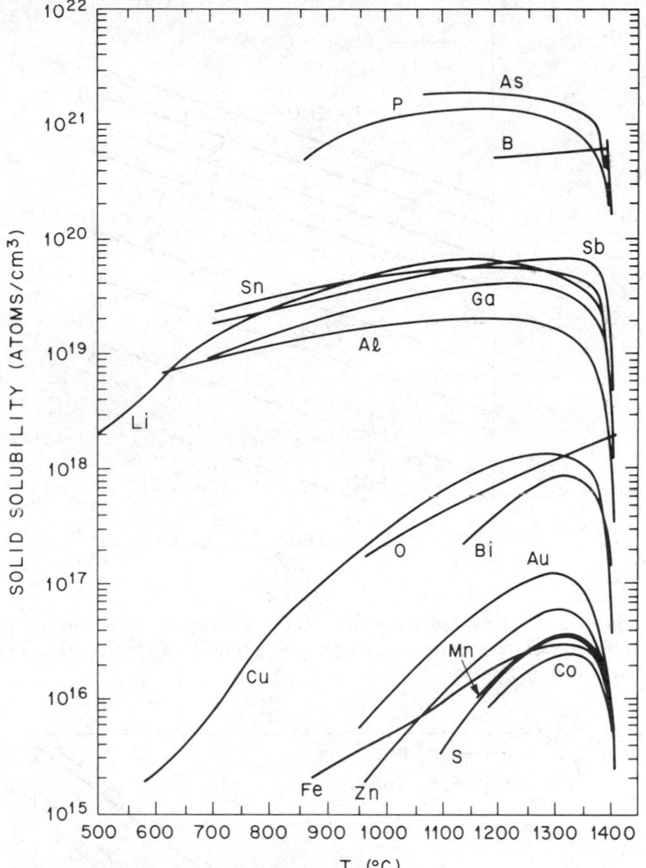

Fig. 10.21 Dependence of the solid solubility of various dopants in silicon on temperature. After Sze.[1]

antimony. The minimum thickness of SiO_2 x_m required to mask against boron diffusion is[22]

$$\frac{x_m^2}{t_m} = 4.9 \times 10^5 \exp(-2.80q/kT_m)\ (\mathrm{cm^2/s}) \tag{10.30}$$

and against phosphorus diffusion is[23]

$$\frac{x_m^2}{t_m} = 1.7 \times 10^7 \exp(-1.46q/kT_m)\ (\mathrm{cm^2/s}) \tag{10.31}$$

where t_m is the time in seconds and T_m is the diffusion temperature in degrees Kelvin.

Figure 10.22 shows the thickness x_m at which SiO_2 thermally grown in dry oxygen fails to mask against P_2O_5 and B_2O_3 vapors as a function of time measured with diffusion temperature as a parameter, and described by Eqs. (10.30) and (10.31). The diffusivities of boron and phosphorus are concentration dependent. However, Fig. 10.22 is conservative because it is based on the highest effective diffusivities of boron and phosphorus reported in the literature.

Ion Implantation

Ion implantation is a method of doping a semiconductor by bombarding it with energetic ions of the dopant impurity. This step is usually followed by heat treatment at elevated temperatures typically in

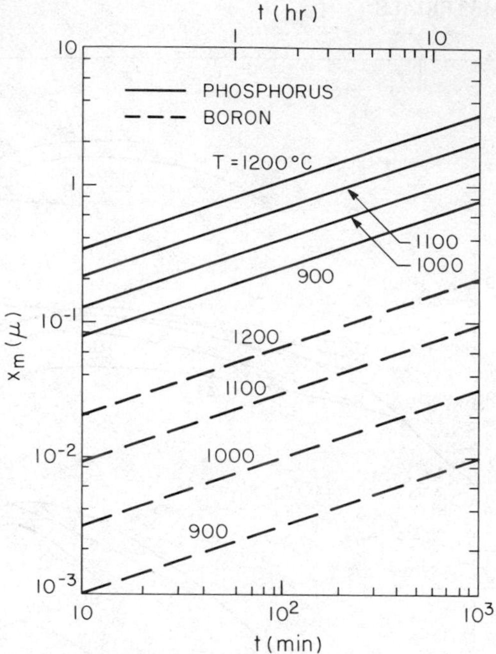

Fig. 10.22 Dependence of minimum oxide thickness required to mask against phosphorus and boron on time with diffusion temperature as parameter. From Wolfe.[24] Copyright (1975), Pergamon Press, Ltd. Reprinted with permission.

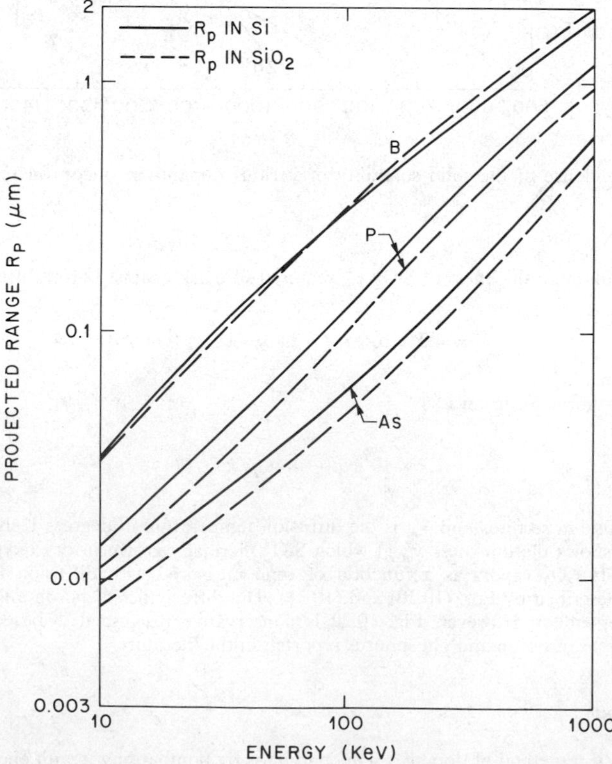

Fig. 10.23 Dependence of projected range of boron, phosphorus, and arsenic ions on incident ion beam energy. After Pickar.[25]

the 900°C range to anneal out lattice damage caused by the energetic ions,† electrically activate the implanted dopant atoms, and diffuse these atoms deeper into the substrate.‡ The main advantages over chemical predeposition are[25] (1) precise control over total dose, depth profile, and area uniformity and (2) implanted junctions can be self-aligned to the edge of the mask.

Typical ion energies are between 10 and 400 keV, and doses are between 10^{11} and 10^{16} ions/cm^2. Figure 10.23 shows the projected range of boron, phosphorus, and arsenic ions in Si and SiO_2 as a function of energy.

When the ion beam is scanned uniformly across the semiconductor surface, a common practice, the distribution of dopant atoms into the semiconductor approximates a gaussian curve. The doping profile after device fabrication is completed usually differs from the profile immediately after implantation because of subsequent heat treatments. Methods of doping profile measurement are described in Ref. 3.

Polysilicon, SiO_2, and photoresist layers can be used for masking against dopant ions in ion implantation. Figure 10.23 shows the penetration of commonly used dopant ions into SiO_2. An SiO_2 layer much thicker than the projected range for a given energy and dopant ion will effectively mask against that dopant ion. For example, a 0.1-μm-thick SiO_2 film will effectively mask 30 keV phosphorus ions. Tables of projected range statistics for a variety of commonly used dopants in semiconductors can be found in Ref. 26. Details of ion implantation are given in Refs. 25 and 27.

†Lattice damage must be annealed out to ensure that the free carrier concentrations are determined by dopant concentration rather than by lattice damage.

‡In some applications, ion implantation is a predeposition step followed by a high temperature oxidation-diffusion step that anneals out lattice damage.

References

1 S. M. Sze, *Physics of Semiconductor Devices*, 2nd ed., Wiley-Interscience, New York, 1981.

2 A. S. Grove, *Physics and Technology of Semiconductor Devices*, Wiley, New York, 1967.

3 E. H. Nicollian and J. R. Brews, *MOS (Metal Oxide Semiconductor) Physics and Technology*, Wiley-Interscience, New York, 1982.

4 J. R. Chelikowsky and M. L. Cohen, *Phys. Rev. B* **14**:556 (1976).

5 C. D. Thurmond, *J. Electrochem. Soc.* **122**:1133 (1975).

6 W. Paul and D. M. Warschau, eds., *Solids Under Pressure*, McGraw-Hill, New York, 1963.

7 J. C. Phillips, *Bands and Bonds in Semiconductors*, Academic, New York, 1973; and F. Herman, *Proc. IRE* **43**:1703 (1955).

8 Max Born, *Atomic Physics*, 5th ed., Hafner, New York, 1935, Ch. 8, p. 261.

9 W. F. Beadle, R. D. Plummer, and J. C. C. Tsai, eds., *Quick Reference Manual for Semiconductor Engineers*, Bell Laboratories, Murray Hill, NJ, 1981.

10 C. Canali, C. Jacoboni, F. Nava, G. Ottaviani, and A. Alberigi-Quaranta, *Phys. Rev. B* **12**:2265 (1975).

11 A. G. Sabnis and J. T. Clemens, *IEEE Tech. Dig.*, Int. Electron Dev. Meet., 1979, p. 18.

12 C. Kittel, *Introduction to Solid State Physics*, 2nd ed., Wiley, New York, 1956, Ch. 6.

13 E. L. Jordan, *J. Electrochem. Soc.* **108**:478 (1961).

14 J. R. Ligenza, *J. Appl. Phys* **36**:2703 (1965).

15 J. R. Ligenza and M. Kuhn, *Solid-State Technol.* **13**:33 (1970).

16 P. F. Schmidt and W. Michel, *J. Electrochem. Soc.* **104**:230 (1957).

17 M. M. Atalla, in H. Gatos, ed., *Properties of Elemental and Compound Semiconductors*, Vol. 5, Wiley-Interscience, New York, 1960, pp. 163–181.

18 H. C. Casey and M. B. Panish, *Heterostructure Lasers*, Academic, New York, 1978.

19 A. Y. Cho, *J. Vac. Sci. Technol.* **16**:275 (1979).

20 J. C. Bean, "Growth of Doped Silicon Layers by Molecular Beam Epitaxy," in F. F. Y. Wang, ed., *Impurity Doping Processes in Silicon*, North-Holland, Amsterdam, 1981.

21 C. J. Frosch and L. Derick, *J. Electrochem. Soc.* **104**:547 (1957).

22 S. Horiuchi and J. Yamaguchi, *Jap. J. Appl. Phys.* **1**:314 (1962).

23 C. T. Sah, H. Sello, and D. A. Tremere, *J. Phys. Chem. Solids* **11**:288 (1959).

24 H. F. Wolfe, *Silicon Semiconductor Data*, Pergamon, New York, 1969, p. 601.

25 K. A. Pickar, "Ion Implantation in Silicon-Physics, Processing and Microelectronic Devices," in R. Wolfe, ed., *Applied Solid State Science*, Vol. 5, Academic, New York, 1975.

26 J. F. Gibbons, W. S. Johnson, and S. W. Mylroie, *Projected Range Statistics*, 2nd ed., Dowden, Hutchinson, and Ross, Stroudsburg, PA, 1975.

27 J. W. Mayer and O. J. Marsh, "Ion Implantation in Semiconductors," in R. Wolfe, ed., *Applied Solid State Science*, Vol. 1, Academic, New York, 1969.

CHAPTER 11

EMISSIVE, OPTICAL, AND PHOTOSENSITIVE MATERIALS

BERNARD SMITH

Eradcom Delet-BS, US Army
Fort Monmouth, New Jersey

MANSOOR A. SAIFI

Belcom Research
Murray Hill, New Jersey

11.1 EMISSIVE MATERIALS

Bernard Smith

11.1-1 Introduction

Electron emission from solids has been studied with varying intensity since the last century. The heart of any electron tube, be it a receiving, microwave, or millimeter wave tube, is the source of electrons, commonly called the cathode. Cathodes available to date for use in electron devices fall into two general categories, primary and secondary emitters. The proper functioning of any of these devices is directly dependent upon the proper functioning of the cathode. Therefore, the importance of the proper choice of cathode in achieving satisfactory performance from any electron tube becomes readily apparent.

The important advances in electron emitters made in the last half century have been closely tied to the degree of perfection achieved with high-vacuum techniques and to unique fabrication techniques. The advent of solid state technology has led to the development of modern vacuum research equipment, such as auger electron spectroscopes, scanning electron microscopes, and secondary ion mass spectroscopes. This equipment has been used to identify the composition of elements on the surface of cathodes. With use of these tools, significant advances have been made in understanding electron emission from composite materials such as alkaline earth oxides and pure metal cathodes. The literature in the field of cathodes is so extensive that no attempt can be made here to cover every aspect of emission materials and technology. Therefore, this section gives an overview of the most promising emitters available and briefly discusses the developmental cathodes, which are undergoing various stages of development. References are provided for those who wish more in-depth discussion of the physics and performance of the various emitters.

A brief review of the equation that describes thermionic emission is important for understanding the differences between various emission materials. Emission from a surface at a temperature T

TABLE 11.1. PROPERTIES OF EMISSION MATERIALS

Material	Richardson Constant, A_R	Work Junction, ϕ_R (eV)	Current Density, J (A/cm^2)	Operating Temperature (K)
Tungsten (W)	70	4.50	0.25–0.7	2500–2600
Oxide-coated nickel	< 10	1.0–1.3	0.1–2	1000–1150
Impregnated tungsten (type B)	2.4	1.65	1–5	1325–1450
M-Cathode	5.67	1.67	1–5	1225–1300
Thoriated tungsten	4	2.65	0.25–3	1900–2000

depends on the value of the work function ϕ and is given by the well-known Richardson equation:[1]

$$J = AT^2(1 - r)e^{-\phi/KT}$$

Since in most cases $r \ll 1$, this equation reduces to

$$J = AT^2e^{-\phi/KT}$$

where J = current density in A/cm^2
r = reflection coefficient
T = temperature in Kelvins
ϕ = work function in eV
K = Boltzmann's constant

Because of the temperature dependence of ϕ, it can be written

$$\phi = \phi_0 + \alpha T$$

Substituting in the equation for J we get

$$J = (Ae^{-\alpha/K})T^2e^{-\phi_0/KT}$$

where $Ae^{-\alpha/K}$ is defined as the Richardson constant A_0 and J becomes

$$J = A_0T^2e^{-\phi_0/KT}$$

Listed in Table 11.1 are important properties of some emission materials.

11.1-2 Thermionic Emission Cathodes

Oxide-Coated Cathodes

This class of thermionic emitter has been the most widely investigated and extensively used of all cathode systems. The basic oxide cathode is composed of alkaline earth (barium, strontium, calcium) oxides coated on a metal base.[2,3] The most commonly used base is nickel or a nickel-alloy base metal. In these cathodes the emission comes from a monolayer of barium which forms on the surface when the oxides are dissociated. A number of variants of the standard oxide-coated cathode have been reported in the literature, including the pressed nickel,[3] the Medicus,[4] and the coated-powder cathode.[5] These cathodes are considered variants of the dispenser cathode or, more accurately, variants between the oxide cathode and the dispenser cathode. Details of the fabrication techniques for these variants of the oxide-coated cathode can be found in the references cited. In general, all techniques adhere to the basic emission theory reported for these cathodes by Wagner and Herrman,[2] Wright,[6] and Nergaard.[7] Most investigators agree that the emission of oxide cathodes comes from the monolayer of barium which forms on the surface when the oxides are decomposed. At low operating temperatures, the chemical reaction leading to the generation of free barium is slow. Residual contaminants may deactivate the cathode faster than the barium is released, a condition commonly called poisoning. At high temperatures, life is shortened due to gross evaporation of the oxide cathode coating. Thus for any tube employing oxide emitters, the design becomes a compromise between cathode life and tube performance. At sustained current densities above 1 A/cm^2, the oxide coating overheats because of coating resistance and vaporizes. This phenomenon occurs to a lesser extent in all the oxide cathode variants that were previously mentioned.

Figure 11.1 shows emission characteristics of oxide-coated cathodes. In general, oxide-coated cathodes are capable of delivering DC emission currents of 100 to 400 mA/cm^2 for thousands of

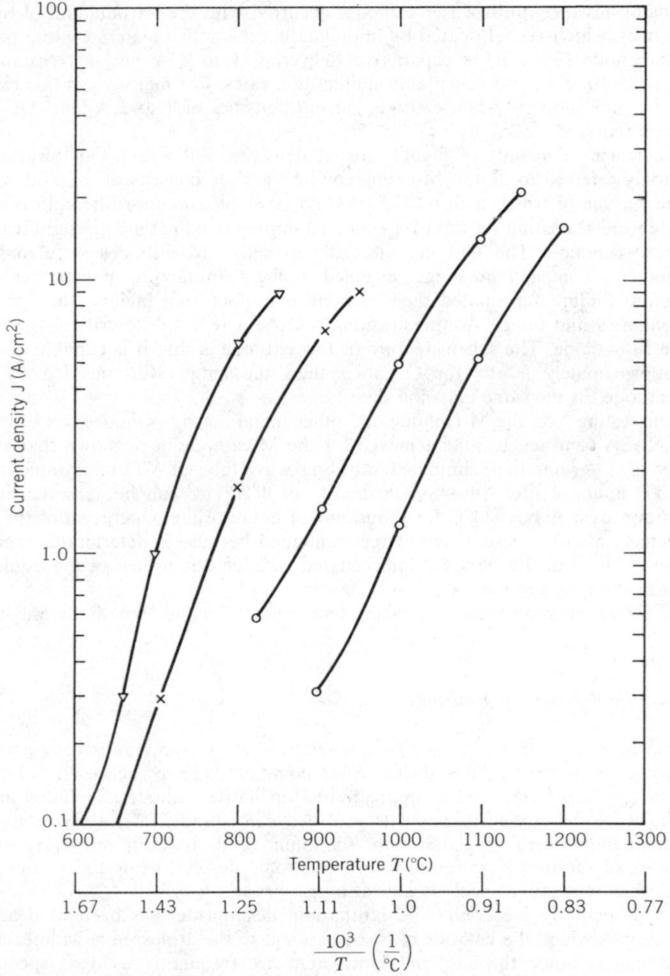

Fig. 11.1 Emission vs. temperature for commercially available cathodes.

hours of life. For very short pulses (low duty cycle applications), current densities of 1 A/cm^2 can be obtained and the cathodes still have reasonably long life. A well-activated cathode will permit drawing of current loadings up to 0.5 A/cm^2 for many thousands of hours and peak emission values of 50 A/cm^2 have been reported.[8]

Metal Matrix Emitters With Active Metal Coating

Metal matrix or barium monolayer cathodes consist of a metal base covered by an absorbed electropositive monolayer. For practical cathodes the electropositive monolayer (usually barium) is coabsorbed with oxygen to form a dipole layer which effectively lowers the work function.[9] The modern microwave or millimeter wave tube, which requires DC current densities in excess of 1 A/cm^2 for thousands of hours of life in almost all cases, is built with the barium monolayer cathode as its electron source. The modern era of the barium monolayer cathode began with the introduction of the L-cathode. This cathode consists of a porous tungsten plug attached to a vapor-tight reservoir that contains barium and strontium carbonates. During processing these carbonates are decomposed to form free barium which diffuses through the porous tungsten pellet to coat the tungsten with barium. The work function of these cathodes is about 2.2 eV, which is typical of pure barium on clean tungsten. The main difficulties encountered with these cathodes are those caused by an excessively high barium evaporation rate.

The next major advance in dispenser cathodes occurred with the introduction of barium calcium aluminate cathodes, which are fabricated by impregnating the active material into a porous tungsten plug.[10,11] This cathode (Type B) is capable of delivering 1 to 2 A/cm^2 at reasonable operating temperatures (1000–1050°C) and acceptable sublimation rates. For many years this cathode was the workhorse of the tube industry when cathode current densities of 1 to 2 A/cm^2 DC for high duty cycle pulses were required.

With the increasing demands of higher current densities, Zalm and Van Stratum[12] developed what is commonly referred to as the M-cathode. This cathode consists of a standard impregnated cathode on the surface of which a thin (< 10,000 Å) layer of osmium-ruthenium is sputtered. The effect of this thin metal coating on top of a standard impregnated cathode is significant lowering of the surface work function. The first commercially available M-cathodes were manufactured by Phillips Metalonic, a company no longer engaged in the manufacture of cathodes in the United States. Just before Phillips terminated their cathode manufacturing facility, they prepared for the National Aeronautics and Space Administration (NASA) a report[13] describing the manufacturing process for the M-cathode. The salient feature of this cathode is that it is capable of operating at a temperature approximately 50 to 100°C cooler than the temperature needed by the standard impregnated cathode for the same emission density.

Extensive life testing[14] on the M-cathode and other metal matrix cathodes has been and still is in progress on a NASA contract. Results achieved for the M-cathode have shown that at an operating current density of 2 A/cm^2 in a simulated traveling wave-tube (TWT) environment, this cathode gives over 40,000 hours of life. An emission density of 4 A/cm^2 can be achieved at an operating temperature of approximately 1020°C for thousands of hours of life. Operation of the M-cathode at an emission density above 5 A/cm^2 is not recommended because it deteriorates rapidly in performance down to the level of the standard impregnated cathode due to loss of the osmium-ruthenium coating on top of the tungsten matrix.

Figure 11.1 shows emission versus operating temperature for the M-cathode and the B-cathode.

High Temperature Refractory Emitters

The electronic properties of many refractory and semirefractory compounds have also been studied in the search for an improved cathode system. Such compounds have been used as bulk emitters, as coating on refractory substrates, and as impregnants for matrix cathodes. Included in the listing of refractory emitters are the rare earth elements and their compounds. Literature references are found for thorium, lanthanum, cerium, gadolinium, scandium, and uranium refractory emitters. Compounds and alloys of thorium have found use in electronic devices, while thoria and carbides have been used in special applications such as ionization gauges.

Lanthanum, particularly the compound lanthanum hexaboride, has been used as the emission source in applications where the cathode must be exposed to the atmosphere without degradation in emission performance. Since thorium and lanthanum are frequently used as special application emitters, only these two cathodes will be discussed in this section.

Thorium and lanthanum compound cathodes, having higher work functions than the oxide, impregnated, or refractory metal cathodes require a higher operating temperature and are therefore more susceptible to heater failure. Typically thorium and lanthanum cathodes operate in the temperature range of 1200 to 1800°C. They are rugged, not readily damaged by operation in less than high vacuum, resistant to sparking, and relatively stable. They are easy to activate, have a very low electrical resistance, and do not form high-resistance interface layers with their base metals. Extensive work has been done on the hexaboride cathodes by Lafferty[15] at the General Electric Research Laboratory.

Thoriated tungsten cathodes have been used in radio frequency transmitting tubes and are capable of high current densities at reasonably long life. Thoriated tungsten cathodes are produced by adding a small percentage of thorium oxide to tungsten powder, then sintering, swaging, or drawing the material to form wire or a filament that is heated in a hydrocarbon atmosphere to form the tungsten carbide surface layers. Thoriated tungsten filaments are more susceptible to damage by bombardment of residual gas ions than are hexaboride cathodes.

A method has been developed whereby a particular composition of thorium carbide, which is a modification of the thoriated tungsten cathode, can be fabricated by compaction and then incorporated into a vehicle for vacuum hot pressing. The cathode consists of a tantalum cup to hold the emission mix which is then covered with a tantalum foil. Cathodes containing an emission composition of thorium carbide, thorium oxide, and zirconium carbide have been operated at continuous current densities in excess of 30 A/cm^2 for periods longer than 25 hours. The typical operating temperature for the cathodes is 1740 to 1780°C. A more complete discussion of thoriated tungsten cathodes can be found in papers by Glascock[16,17] and a report by Bondley.[18]

11.1-3 Field Emission Cathodes

High-Voltage Field Emission Cathodes

The attractiveness of this class of emitters lies in the fact that current densities in excess of 100 A/cm^2 have been reported from very fine, single-point, refractory material tips operated in electric fields in excess of 10^7 V/cm. Considerable understanding and knowledge concerning the basic process of field emission has developed since the theoretical formulation of field emission cathodes by Fowler and Nordheim[19] over 50 years ago. The physical properties desirable for an emitter source are (1) high tensile strength to withstand electrostatic stress, (2) high resistance to sputtering and contamination, and (3) high melting point. Materials possessing these properties are (1) pure metals, (2) heterogeneous cathodes consisting of a surface layer on a solid substance, and (3) homogeneous metalloid compounds, or alloys. These materials have found application as both high-voltage thermionic field emitters[20] (TFE) and as field ionization (FI) sources. Very high specimen current densities can be obtained at 0.1 μm or less beam size with relatively simple beam optics from TFE sources. Current densities of 1300 A/cm^2 or power densities of 1.6×10^7 W/cm^2 have been achieved in a 0.1 μm beam spot.[21,22] Similar results have also been demonstrated by Chapman[23] and by Spindt.[24] The major barriers to the utilization of these cathodes in microwave tubes are (1) they need high operating voltage, (2) stable operation can only be achieved in vacuums of 10^{-8} torr or less, and (3) the total current is very low, that is, microamps to several milliamps.

Low-Voltage Field Emission Cathodes

The low-voltage or thin film field emission cathode[25,26] consists of a conductor/insulator/conductor sandwich that has a dielectric thickness of about 1 to 2 μm. Holes about 2 μm in diameter go through the top conductor or metal gate film, and dielectric layer. The holes have undercut cavities in the dielectric layer and metal cones within the cavities. Field emission is obtained from the tips of the cones when the tips are driven negative with respect to the gate film. Because of the field enhancement of the tips and the very close spacing between the rim of the hole in the gate film and the small tips, a potential of only 100 to 200 V across the sandwich is required for field emission.

Both high-voltage field emitters and low-voltage field emitters obey the Fowler–Nordheim equation[25,27]

$$J = \frac{AF^2}{t^2(y)\phi} \exp - \left[\frac{BV(y)\phi^{3/2}}{F} \right]$$

where J = emission current density in A/cm^2
F = field at the tip
ϕ = work function in eV

A and B are constants, and $V(y)$ and $t(y)$ are slowly varying functions of y which are tabulated in the literature.[28] The field at the tip is

$$F = BV \, (\text{V/cm}^2)$$

where V is the applied voltage and

$$B = f(\gamma, R, \theta) \text{ cm}^{-1}$$

with γ being the tip radius, R the anode-to-tip spacing, and θ the emitter cone half angle. In general it can be said that as γ, R, and θ become smaller, B, the field enhancement factor, becomes larger.

To date these emitters have demonstrated emission capabilities of 10 to 20 A/cm^2 operating in vacuums of 10^{-8} torr or better. Typical total current from the emitters has been as high as 100 mA. The potential of these emitters is very promising for future application in military fast warm-up devices, in millimeter wave tubes, and in commercial applications where high current density, small spot size, and reasonable total current (i.e., 10–100 mA) are needed. To ascertain that these requirements can be fulfilled, the feasibility of operating these cathodes in electron guns as well as their performance in practical operating devices must be established. Efforts are now in process to demonstrate the above at the Stanford Research Institute[25] and at the Georgia Institute of Technology.[26]

11.1-4 Room-Temperature Emitters

Primary Emission Cold Cathodes

The advent of the electron beam semiconductor in late 1960s generated a considerable amount of interest in the development of solid state, room-temperature cold cathode devices.[29–32] Because in

TABLE 11.2. SECONDARY EMISSION CHARACTERISTICS OF FREQUENTLY USED MATERIALS

Material	Secondary Emission Yield, δ	Average Power
Platinum	1.8	Medium–high
Al_2O_3	3–5	Low
BeO	8	Medium
MgO	8–15	Low
BaO-SrO	5–12	Medium
W-$BaCaAlO_4$	2–3	Medium
W-ThO_2	2–3	Medium
Au-MgO	3–8	Medium
GaAs	>3	Low

most cases these emitters required a coating of cesium to achieve negative electron affinity properties for reasonable current densities (i.e., 0.1–1 A/cm^2), operation in a normal tube vacuum was difficult if not impossible to achieve. Therefore, because of the difficulty of maintaining an active cesium surface, development of this type of emitter for high-power microwave/millimeter wave devices is no longer active.

Secondary Emission Cold Cathodes

Secondary electron emission has been reported in review articles by Bruining[33] and by Hachenberg and Brauer.[34] It is well known that this phenomenon may be desirable (as an emitter) or undesirable (multipactor) in the design of electron devices. This discussion is limited to its potential as an emitter.

Secondary electron emission takes place primarily at the surface of materials. Metals, semiconductors, and insulators as well as liquids exhibit secondary emission characteristics. The secondary emission yield of a target material is defined as the ratio of the total number of secondaries per primary electron. The secondary emission yield, δ, is largely dependent on the energy of the primary electron, in electron volts, and the angle of incidence of the primary beam. Table 11.2 gives a partial listing of the secondary emission characteristics of frequently used materials. For metals, as a general rule, δ ranges from 0.7 to 1.7. For alkali-halides, alkali-oxides, and the alkaline-earth oxides, δ may reach values of 20 or more but has a fairly short life. The ideal secondary emission cathode material is one that (1) has a δ greater than 2.5, (2) has a low energy first crossover (<30 V), (3) is able to survive under bombardment, (4) is insensitive to normal microwave tube environment, and (5) is chemically, mechanically, and thermally stable. Listed in Table 11.2 is an Au-MgO cermet that has

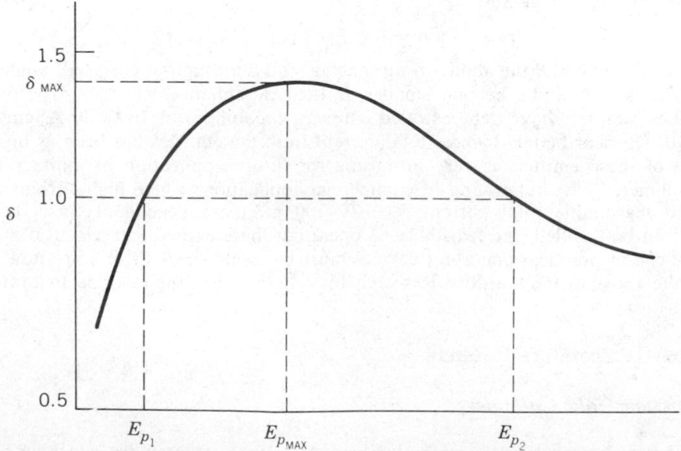

Fig. 11.2 Representative secondary emission curve.

demonstrated the potential to meet all the criteria cited. Values of δ as high as 20 have been reported[34] for MgO, but widespread use of this material in microwave tubes is dependent on improvements in its stability in a normal tube operating environment.

Figure 11.2 is a plot of a representative secondary emission curve where Ep_1 is the first crossover point, that is, δ is greater than 1, Ep_m is the point where δ is a maximum, and Ep_2 is the point on the curve where δ decreases to less than 1. In all cases, the first crossover energy, Ep_1, is lower than the second crossover energy, Ep_2. Data on other secondary emission materials can be found in Refs. 33 and 34.

11.1-5 New Cathodes Under Development

A variation of the osmium-ruthenium-coated M-cathode is under development[35] at Varian Associates. This cathode consists of a tungsten-iridium (W-Ir) matrix that is impregnated with barium calcium aluminate impregnant. Emission densities of 4 to 8 A/cm^2 have been demonstrated by this cathode operating in the temperature range of 1025 to 1100°C. This cathode has had limited application in development-type tubes which require current densities above 4 A/cm^2.

At Phillips Laboratories[36] in Eindhoven, Holland, development of a cathode based on a tungsten matrix pressed or impregnated with barium scandate (instead of the conventional barium aluminate usually used in the W-matrix) is in process. Very high emission values have been reported, but these results have yet to be repeated by investigators in this country.

Cathodes based on a tungsten matrix fabricated with $Ba_5Sr(WO_6)_2$ have also been reported. The original work on this cathode was done by General Electric, Schenectady in a pressed matrix. Work is now under way by Smith and Newman[37] to fabricate the tungstate cathode by impregnation. When fully developed the impregnated version is anticipated to be capable of emission densities of 8 A/cm^2 at 1025°C, with life in excess of 5000 hours.

TABLE 11.3. DEVELOPMENTAL CATHODES

Cathode Type	Benefits	Comments	Application Area
Tungsten-iridium	High current density	Difficult to fabricate, current density greater than 6 A/cm^2, temperature approximately 1000°C	ECM,[a] millimeter wave high-power tubes
Controlled porosity density (CPD)	Uniform current density, potential for long life	Manufacturing program planned	Space tubes, millimeter wave tubes
Tungstate	Current density 2–20 A/cm^2	Pressed version difficult to fabricate; development of impregnated version in process; needs work on impregnants to reduce impregnating temperature	ECM, millimeter wave high-power tubes
Co-deposited cathode	Current density 2–20 A/cm^2, low sublimation rate	Not commercially available, more information needed	ECM, millimeter wave tubes
Low-voltage field emitters	No heater, high current density, instant on	Total current 100 mA, potential for much higher currents; data needed on performance in tube environment	Space, ECM medium-power millimeter wave devices
Scandate	Low impregnating temperature, easy to fabricate	Needs more work to improve reliability and repeatability, no life data available	High-power microwave tubes, millimeter wave tubes

[a] ECM, electronic countermeasures.

The Naval Research Laboratory (NRL)[38] in Washington, DC, is developing a controlled porosity density matrix cathode. This cathode consists of a thin film of BaO on an iridium surface made using microprocessing techniques to fabricate a 0.001-in. iridium foil with a uniform array of holes. During operation this foil receives a very thin coating of Ba-BaO. In this manner the cathode achieves a very uniform and low work function surface. Emission densities of 6 A/cm^2 DC in a depressed collector system have been achieved at Varian from an Ir-BaO cathode.

It should be pointed out that at their present stage of development, not one of the new cathodes can be considered superior. A considerable amount of work remains to be done before a level of confidence in their reliability and repeatability of performance can be estabished. Table 11.3 lists the developmental cathodes along with information on their emission potential and possible areas of application.

References

1 Herring, Conyers, "Thermionic Emission," *Rev. Mod. Phys.* **21**(2) (Apr. 1949).

2 S. Wagner and G. Herrman, *The Oxide Coated Cathode*, Vols. 1 and 2, Chapman and Hall, London, 1951.

3 C. P. Hadley, W. G. Rudy, and A. J. Sloechert, "A Study of the Molded Nickel Cathode," *J. Electrochem. Soc.* **105**:395–398 (July 1958).

4 G. Medicus, Presentation at 1978 Tri-Services Cathode Workshop, Naval Research Laboratory, Washington, D.C.

5 D. W. Mauer and C. M. Pleass, "The CPC: A Medium Current Density, High Reliability Cathode," *Bell Syst. Tech. J.* **46**(10) (Dec. 1967).

6 D. A. Wright, "A Survey of Present Knowledge of Thermionic Emitters," *Proc. Inst. Elec. Engr.* **100**:1250142 (May 1953).

7 L. S. Nergaard, "Studies of the Oxide Cathode," *RCA Rev.* **13**:464–545 (Dec. 1952).

8 E. A. Coomes, *J. Appl. Phys.* **17**:647 (1946).

9 J. Affleck and W. T. Boyd, "Investigation of Various Activator Refractory Substrate Combinations," *Gen. Elec. Tech. Inform. Series*, R66ET 1-9, 12 (1966).

10 H. J. Lemmens, "A New High Emission Density Thermionic Emitting Cathode," *Phillips Tech. Rev.* **11**:341 (1950).

11 A. Venema, R. C. Hughes, P. P. Coppola, and R. Levi, "Dispenser Cathodes," *Phillips Tech. Rev.* **19**(6):177 (1957/58).

12 P. Zalm and A. J. Van Stratum, "Osmium Dispenser Cathode," *Phillips Tech. Rev.* **27**:69 (Oct. 1966).

13 A. Gupta, "The Manufacture of M-Type Impregnated Cathode," NASA/Lewis Research Center, NASA Order No. C-2129-D (1977).

14 R. Gorske, NASA Contract NAS 3-22335, "Design, Construction and Long Life Endurance Testing of Cathode Assemblies for Use in Microwave High Power Transmitting Tubes" (1980).

15 J. M. Lafferty, "Boride Cathodes," *J. Appl. Phys.* **22**:299 (Mar. 1951).

16 H. Glascock, *Rev. Science Instru.* **43**:698 (1972).

17 H. Glascock, *Rev. Science Instru.* **47**:90 (1976).

18 R. J. Bondley, Contract DA28-043-AMC-01719(E), ARPA ORDER No. 679, "High Current Density, Short Life Cathodes for Linear Beam Tubes" (Mar. 1968).

19 R. W. Fowler and L. W. Nordheim, "Electron Emission in Intense Electric Fields," *Proc. Roy. Soc.* **A119**:173 (1928).

20 L. W. Swanson and N. A. Martin, "Field Emission Cathode Stability Studies: Zirconium/Tungsten Thermal Field Cathode," *J. Appl. Phys.* **46**:2029 (1975).

21 W. P. Dyke and J. K. Trolan, "Field Emission: Large Current Densities, Space Charge, and the Vacuum Arc," *Phys. Rev.* **89**:799–808 (1953).

22 L. W. Swansen, J. Orloff, and A. E. Bell, "Field Electron and Ion Source Research for High Density Information Storage System," Air Force Contract AFA1-TR-79-1133 (1929).

23 A. T. Chapman, Arpa Contract NAAH01-71-C-1046 "Melt Grown Oxide Metal Composites" (1973).

24 C. A. Spindt, *J. Appl. Phys.* **39**:304 (1968).

25 C. A. Spindt, NASA Contract NAS3-21507, "Development Program on a Cold Cathode Electron Gun" (Jun. 1981).

26 Contract DAAK40-77-0096, MICOM, "Manufacturing Method, for the Production of Low Voltage Field Emitters" (1978).

27 R. H. Fowler and L. W. Nordheim, *Proc. Roy. Soc.* (London) **A-119**:73 (1928).

28 R. B. Burgess, H. Kroemer, and J. M. Houston, *Phys. Rev.* **90**:515 (1953).

29 K. R. Faulkner and J. R. Howorth, CVD Research Project RP 28-1, "Semiconductor Cold Cathode Displaying a Sharp High Energy Cut-Off," Contract N/CP16/1711/67.

30 K. R. Faulkner, et al., "Negative Electron Affinity GaAsP Cold Cathode Silicon Vidicon," 1973 International Electron Device Meeting Technical Digest, Washington, DC, 3–5 Dec. (1973).

31 H. Kressel, et al., "An Optoelectronic Cold Cathode Using an $AlxGa_{1-x}$ as Heterojunction Structure," *App. Phys. Let.* **16**:9 (1970).

32 H. Schade, et al., "Optoelectronic Electron Emitter," Contract DAA-V07-71-C-0054 (Oct. 1972).

33 H. Bruining, *Physics and Applications of Secondary Emission*, McGraw-Hill, New York, 1954.

34 O. Hachenberg and W. Brauer, "Secondary Electron Emission from Solids," in *Advances in Electronics and Electron Physics*, Vol XI, Academic, New York, 1959, p. 413.

35 L. K. Falce, Patent No 4, 165, 473, "Electron Tube with Dispenser Cathode" (21 Aug. 1979).

36 A. Van Ooostron and L. Augustus, "Activation and Early Life of a Pressed Barium Scandate Cathode," *Appl. Surface Sci.* **2**:2 (1979).

37 B. Smith and A. Newman, Patent No 4, 078, 900, "Method of Making a High Current Density Long Life Cathode" (Mar. 1978).

38 L. Falce, Contract 00173-77-C-0186, "Iridium Foil Cathode Development" (May 1978).

11.2 OPTICAL AND PHOTOSENSITIVE MATERIALS

Mansoor A. Saifi

11.2-1 Optical Materials and Indicatrix

The three states of matter—gaseous, liquid, and solid—can all be used as optical materials. Optical materials of most engineering interest are crystalline and amorphous solids. A new class of materials called liquid crystals are used in many devices as a read-out matrix; they are discussed in detail in Section 15.3.

Crystalline solids are characterized by an orderly periodic array of atoms. Solids that do not have this periodicity of structure are called amorphous. The largest group of amorphous materials used in optical applications is glasses and ceramics. An important optical property is the refractive index. When light passes from a vacuum to a denser medium such as glass, its velocity is reduced. The ratio between these velocities is known as the refractive index n of the denser medium and is given by

$$n = \frac{\text{velocity in vacuum } (c)}{\text{velocity in medium } (v)} = \sqrt{\frac{\epsilon\mu}{\epsilon_0 \mu_0}} \tag{11.1}$$

where ϵ and μ are the permittivity and permeability of the medium and ϵ_0 and μ_0 those of the vacuum, respectively. The wavelength and directional dependence of the refractive index are related to materials, electronics, and structural properties such as symmetry. In isotropic materials the refractive index is independent of the light propagation direction and polarization, whereas in anisotropic materials it depends on both. Anisotropy can also be induced by application of electric and magnetic fields or by mechanical stress and strain. Materials exhibiting large field-induced changes in their optical properties are of special interest for various device applications. A convenient way to describe the refractive index as a function of propagation direction and polarization is by the index ellipsoid, or indicatrix.[1] The generalized equation for the ellipsoidal surface is

$$B_{11}x_1^2 + B_{22}x_2^2 + B_{33}x_3^2 + B_{23}x_2x_3 + B_{13}x_1x_3 + B_{12}x_1x_2 = 1 \tag{11.2}$$

where x_i are the cartesian coordinates, and the reciprocals of the coefficients B_{ij} are related to the refractive index. The effect of applied field on refractive index is then described by suitable coefficients relating B_{ij} to applied fields. Several such coefficients are discussed in a later section. Referring to Fig. 11.3, the refractive index of a propagating wave can be obtained from the plane that is perpendicular to the propagation direction and that passes through the index ellipsoid origin. The intersection of this plane with the index ellipsoid defines an ellipse. The semimajor and semiminor axes of this ellipse, OA and OB, represent the refractive indices of light polarized along these directions. In a special case when the propagation direction coincides with an indicatrix major axis,

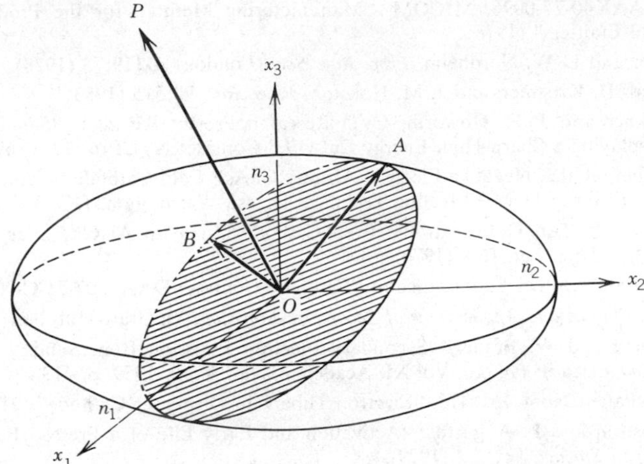

Fig. 11.3 Index ellipsoid. OP-The direction of light propagation. The crosshatched plane is perpendicular to OP.

the refractive indices are given by the lengths of the other two axes. The indices corresponding to the three major axes are called principal refractive indices, and in the principal axis coordinate system the index ellipsoid is

$$\frac{x_1^2}{n_1^2} + \frac{x_2^2}{n_2^2} + \frac{x_3^2}{n_3^2} = 1 \tag{11.3}$$

where the n_i can be obtained from the coefficients B_{ij}.

11.2-2 Optical Glasses

Glass, because of its transparency in the visible spectral region (400–800 nm wavelength) is the most commonly used optical material in applications such as lenses, prisms, and windows. When a beam of light falls on a glass surface the phenomena of reflection, refraction, and absorption occur. The proportion of light reflected from a single polished surface can be determined from Fresnel's formula

$$R = \frac{(n-1)^2}{(n+1)^2} \tag{11.4}$$

TABLE 11.4. OPTICAL GLASSES AND THEIR REFRACTIVE INDICES

Glass Type[a]	N_f	N_d	N_c
FK5-487704	1.492	1.487	1.486
PK1-504669	1.509	1.504	1.501
BK3-498651	1.504	1.498	1.496
SK1-610567	1.607	1.610	1.618
KF6-517522	1.524	1.517	1.514
SF1-717295	1.735	1.717	1.711

[a] The glass types are from "Optical Glass," Catalog No. 3050/USA, Schott Optical Glass Inc. The six-digit code is composed of two parts. The first three digits describe the nominal refractive index N_d and the last three digits describe the Abbe number.

where R is the ratio of reflected to incident light intensity and n is the glass refractive index. (See Table 11.4.) In the visible region the refractive index of most commercial glasses lies between 1.45 to 2, which corresponds to reflectance of 3.36 to 11.11%. These reflectance values can be reduced by applying suitable optical coatings (see Section 11.2-3). A major component of optical glasses is silicon dioxide (SiO_2), with addition of oxides of several other elements such as barium, calcium, boron, potassium, and lead. The refractive index and its dispersion (variation with wavelength) then depends upon the glass composition. For a single-component glass, that made of silica has the lowest refractive index. The rate of change of refractive index with wavelength λ, that is, $dn/d\lambda$, is called dispersion. For glasses used in the visible spectrum, the refractive characteristics are conventionally specified by two numbers, the index of refraction for the sodium D line (0.5893 μm) and the Abbe V-number. The V-number, often referred to as the reciprocal relative dispersion, is defined as

$$V = \frac{n_d - 1}{n_f - n_c} \tag{11.5}$$

where n_d, n_f, and n_c are the refractive indices for the sodium D line, the hydrogen F line (0.4861 μm), and the hydrogen C line (0.6563 μm), respectively. The index difference $(n_f - n_c)$ is a good measure of the dispersion. This information is useful in the design of optical elements such as lenses and more complex instruments operating in the visible spectral range. It is interesting to note that traditionally glasses in the range $n_d > 1.6$, $V_d > 50$, and $n_d < 1.6$, $V_d > 55$ are called crowns while the others are called flints.

11.2-3 Optical Coatings

Thin dielectric or metal coatings are used to modify the reflection and transmission properties of lenses, prisms, and windows. Depending upon the intended use, single or multilayer coatings on a suitable substrate are available for a variety of applications. These range from a very low to a very high reflectance in a narrow or wide spectral range. Most commonly used terminology for such structures is AR (antireflection), HR (high reflection), BBAR (broad-band antireflection), and optical filters (high transmission in a given band). It is apparent that the selection of coating and substrate material depends upon the operating wavelength. Some commonly used coating materials and their useful spectral range are given in Table 11.5. The coatings are applied by thermal evaporation, electron beam heating, or sputtering. The following briefly describes the principles characterizing the behavior of single and multilayer coatings.

From Fresnel's formula, Eq. (11.4), for normal incidence the reflectance R of a dielectric substrate is given by

$$R = \frac{(n_s - n_0)^2}{(n_2 + n_0)^2} \tag{11.6}$$

where n_s and n_0 are the substrate and medium refractive indices at the wavelength λ. The application of a single dielectric film of index n_1 and thickness $d_1 = (\lambda/4n_1)(1 + 2m)$, where $m = 0$ or any integer

TABLE 11.5. COMMONLY USED COATING MATERIALS

Material, Nominal Refractive Index	Spectral Range: Ultraviolet	Visible	Infrared (< 5 μm)	Infrared (> 5 μm)	Ref. No.
PbF_2, 1.75	←	—	—	→	2
MgF_2, 1.38	←	—	→		2
Al_2O_3, 1.62	←	—	→		2
SiO_2, 1.46	←	—	→		2
ZrO_2, 2	←	—	→		2
SiO, 1.9		←	→		2
Si, 3.5		←	—	→	2
Ge, 4.05		←	—	→	2
CdTe, 2.69		←	—	→	3
ZnSe, 2.44		←	—	→	3

value, modifies[4] the reflectance to

$$R = \frac{(n_1^2 - n_0 n_s)^2}{(n_1^2 + n_0 n_s)^2} \tag{11.7}$$

For $m = 0$ the thickness d_1 equals a quarter wavelength of light in the coating material and the integer values of m provide an additional thickness equal to one half the wavelength $(\lambda/2n_1)$. In essence the reflection properties of the coated substrate are modified by the interference of light reflected from the air-film and film-substrate interfaces. Two cases are of special interest. First, for $n_0 < n_1 < n_s$ it can be seen from Eq. (11.7) that the reflectance of a coated substrate is always less than that of the substrate itself [Eq. (11.6)]. These then form the AR coatings mentioned earlier. Indeed from Eq. (11.7) for film index $n_1 = \sqrt{n_0 n_s}$ the theoretical value of R is zero. It is often not possible to fabricate films of this exact refractive index, and hence multilayer coatings are employed. Double-layer coatings provide four adjustable parameters—two refractive indices and two thicknesses. For a

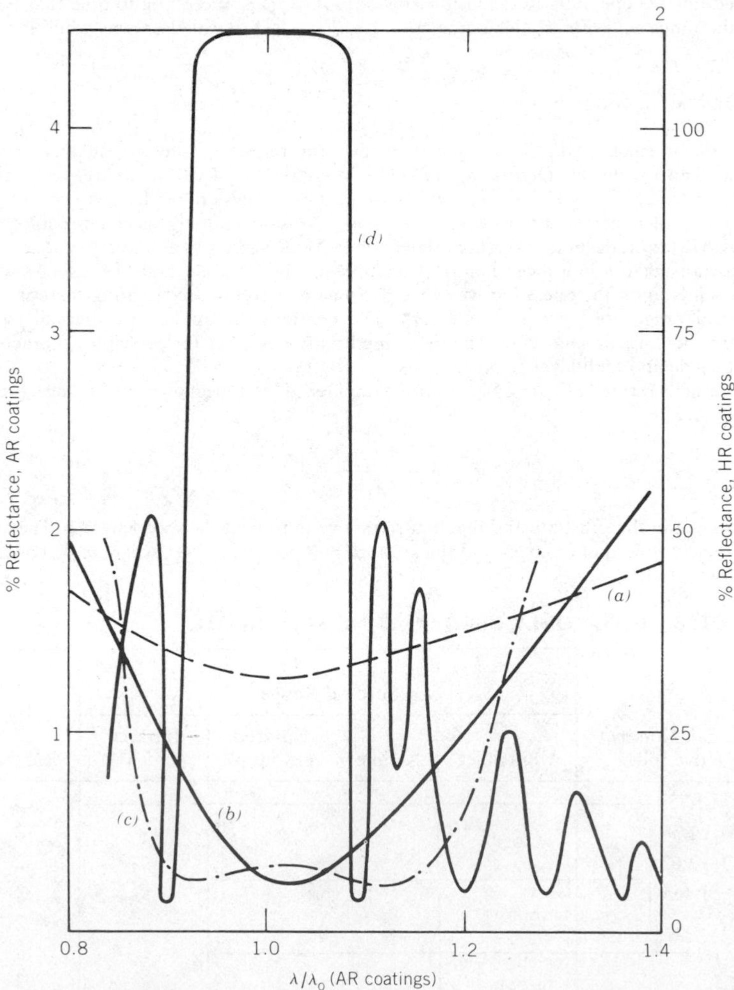

Fig. 11.4 Typical reflection spectra for various coatings: (*a*) Single-layer antireflection (AR); (*b*) two-layer AR coating, V-coat; (*c*) broad-band AR coating; (*d*) multilayer high-reflection (HR) coating. λ_0, Operating wavelength.

quarter-quarter double-layer coating it can be shown[4] that

$$R = \frac{\left(n_0 n_2^2 - n_1^2 n_s\right)^2}{\left(n_0 n_2^2 + n_1^2 n_s\right)^2} \tag{11.8}$$

Once again for $n_1/n_2 = \sqrt{n_0/n_s}$ the theoretical value of R is zero. In case this condition cannot be satisfied, coating thicknesses can be adjusted to minimize reflection.[2] Multilayer coatings designed to provide low reflectance in a narrow spectral range are sometimes called V-coatings, a typical characteristic of which is shown in Fig. 11.4.

For the second case, when $n_0 < n_1 > n_s$, reflectance of a substrate coated with a single film [Eq. (11.7)] exceeds that of the substrate itself. Reflectance of over 99% (particularly useful for laser mirrors) can be obtained by building up successive pairs of quarter-wave coatings of high (H) and low (L) refractive indices. The high reflectivity of this structure is due to the constructive interference of reflection from all interfaces. This structure can be denoted by $[a(HL)^m S]$ where m represents the number of stacks and a and S, air and substrate, respectively. Note that the film next to air is always of high index. Furthermore, if $n_H > n_L < n_S$, an additional H layer is used next to the substrate. Coatings designed for intermediate values of reflection and transmission are called beam splitters.

Long-wave pass filters exhibit high reflectivity at short wavelengths and transmit long wavelength radiation, while shortwave pass filters do the opposite. In many cases quarter-wave stacks are used to obtain high reflectance properties, with additional non-quarter-wave films near air and substrate to achieve high transmittance at the desired wavelengths. For example the structures

$$S(0.5H)L(HL)^m(0.5H)a, \qquad S(0.5L)(HL)^m H(0.5L)a$$

are typically used as long-wave and shortwave pass filters, respectively. Here $0.5H$ and $0.5L$ are eighth-wave coatings of high and low index. The additional coatings are designed to suppress the secondary reflection peaks observed around the high reflectivity region of the quarter-wave stacks (see Fig. 11.4).

It should be noted that reflectance and transmittance properties strongly depend on the angle of incidence and therefore should be clearly specified. Often the polarization of the reflected and transmitted waves is quite different.

11.2-4 Electrooptic Materials

Electrooptic effect refers to change in refractive index due to an applied electric field. The physical origin of the effect can be separated into two parts. A purely electronic part is produced by interaction of the applied electric field with the electrons, and a nuclear part is produced by coupling between the field-induced nuclear displacement and the electrons. In many cases the refractive index can be expressed in a power series of the applied field

$$n = n_0 + a_1 E + a_2 E^2 + a_3 E^3 \cdots + a_m E^m \tag{11.9}$$

where n_0 is the refractive index in the absence of the electric field E and the a_m are constants. In isotropic materials such as liquids and in crystals with a center of symmetry, the refractive index is independent of the direction of the applied field ($\pm E$), and hence the coefficients a_m for m, an odd integer, vanish. The change in the refractive index ($\Delta n = n - n_0$) with the square of the applied electric field and higher order even terms is called the quadratic electrooptic effect, or the Kerr effect. The change in refractive index (Δn) with the applied field produces birefringence, which in terms of the Kerr coefficient K is given by

$$\Gamma = \frac{(n_p - n_s)l}{\lambda} = lKE^2 \tag{11.10}$$

where n_p and n_s are refractive indices parallel and perpendicular to the applied field E at wavelength λ and l is the path length.

An important class of electrooptic materials is ferroelectrics. These materials possess a high dielectric constant near the Curie temperature and exhibit a higher order electrooptic effect of significant magnitude. The electrooptic and elastooptic coefficients for crystalline materials depend upon the crystallographic direction and are tensor quantities.[5] The optical properties are then described by a generalized index ellipsoid, Eq. (11.2).

The change in B_{ij} due to the applied electric field describes the electrooptic effect. For the quadratic electrooptic effect in ferroelectrics it is convenient to use the electric polarization P instead of the electric field E as the independent variable. The change in B_{ij} is then given by

$$\Delta B_{ij} = g_{ij,k}(P^2)_k \tag{11.11}$$

TABLE 11.6. SOME CHARACTERISTICS OF FERROELECTRIC MATERIALS

Material	Electrooptic Coefficient	Index of Refraction	Wavelength (μm)	Curie Temperature, T_c(K)	Ref. No.
$BaTiO_3$	$g_{11} = 0.12, g_{12} = -0.01$	2.4	0.6328	401	6
$KTaO_3$	$g_{11} - g_{12} = 0.16, g_{44} = 0.12$	2.24	0.6328	4	7
$KTa_{.65}Nb_{.35}O_3$	$g_{11} = 0.136, g_{12} = -0.38, g_{44} = 0.147$	2.29	0.6328	283	7
$BaTiO_3$	$r_{51} = r_{42} = 1280$, $r_c{}^{(b)} = 1.08$ } T^a	$n_1 = n_2 = 2.44$, $n_3 = 2.37$	0.546	393	6
$LiNbO_3$	$r_{13} = 8.6, r_{33} = 30.8$, $r_{51} = r_{42} = 28$, $r_c = 21$ } S^c	$n_1 = n_2 = 2.29$, $n_3 = 2.2$	0.6328	1470	8
GaP	$r_{41} = 1.06\ S$	$n = 3.315$	0.6	—	9
GaAs	$r_{41} = 0.27\text{–}1.2\ S$	$n = 3.6\text{–}3.42$	1–1.8	—	10

[a] T = constant stress.
[b] $r_c = r_{33} - (n_1/n_2)^2 r_{13}$.
[c] S = constant strain.
r in 10^{-12} m/V, g in m^4/C^2.

where k is the polarization direction. In reduced† tensor notation

$$\Delta B_1 = g_{1k}(P^2)_k \tag{11.12}$$

In device application these materials are generally kept above their Curie temperature, in which case the conventional Kerr coefficients are given by

$$K_{100} = \frac{2n^3(g_{11} - g_{12})\epsilon^2}{\lambda} \tag{11.13}$$

and

$$K_{110} = \frac{2n^3 g_{44}\epsilon^2}{\lambda} \tag{11.14}$$

where n is the refractive index and ϵ is the dielectric constant. Above the Curie temperature T_c, the dielectric constant of ferroelectrics follows the Curie–Weiss law $\epsilon = C/(T - T_c)$, where T is the operating temperature and C the Curie constant. The values of the various constants for some ferroelectric materials are given in Table 11.6.

A very important class of electrooptic materials is crystals whose molecular structures do not possess a center of symmetry. Such materials have nonzero coefficients for odd powers of electric field E in Eq. (11.9). The electrooptic effect due to the linear term in the electric field is designated the Pockell effect. Once again the electrooptic effect is described by the field-induced changes in the coefficients B_{ij}.

By convention the r coefficients describe the Pockell effect with the electric field E, and the g coefficients describe the quadratic electrooptic effect with electric polarization P. The values of the Pockell coefficient for various materials are given in Table 11.6.

Electrooptic materials are used in devices such as modulators, memories, variable retardation plates, and optical switches. These applications are based on the phenomenon of field-induced birefringence, which causes an incident plane polarized wave to become elliptically polarized upon passage through the material. In electrooptic modulators, crossed polarizers as shown in Fig. 11.5 are employed to impress information on a carrier wave by intensity modulation. A rapidly growing new area is the development of thin-film electrooptic devices such as waveguide modulators and deflectors.[10,11,12]

†The reduced notation for suffixes ij is 11 = 1, 22 = 2, 33 = 3, 23 = 32 = 4, 13 = 31 = 5, 12 = 21 = 6.

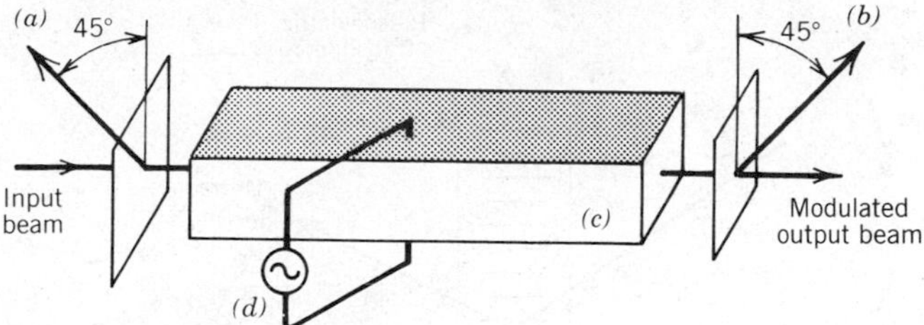

Fig. 11.5 Schematic drawing of an electrooptic modulator. (*a*), (*b*) Polarizer and analyzer. Note that their axes are at 90° to each other. (*c*) Electrooptic crystal with evaporated electrodes. (*d*) Modulating signal.

11.2-5 Elastooptic Materials

The elastooptic effect refers to a change in the refractive index due to strain or stress. The physical origin of this effect is the shift in the nuclear spacing and coupling between various crystal fields. As in the case of electrooptic materials, the effect is described by the strain- or stress-induced deformation of the index ellipsoid. The changes in the refractive index are given by a fourth rank strain-optic or stress-optic tensor.[5] The changes in the B coefficient [Eq. (11.2)] due to applied strain(s) are given in reduced notation by

$$\Delta B_i = \sum_{j=1}^{6} p_{ij} S_j \tag{11.15}$$

Thus the strain-optic tensor components B_{ij} are given by an array of 6×6 matrix elements. Depending upon the crystal structure and symmetry, many of these components are either zero or have the same magnitude. Strain-optic coefficients for several materials are listed in Table 11.7. The strain-induced birefringence Γ is found by solving the equation of the modified index ellipsoid for the desired wave propagation direction. This is generally of the form

$$\Gamma = n^3 (p_{ij} S_j - p_{lk} S_k) \gamma l \tag{11.16}$$

where n is the refractive index, l the propagation path length, and γ a material constant close to unity.

The stress-optic coefficients q_{ij} are defined in a manner similar to strain-optic coefficients. The two are related by the equation of elastic moduli

$$q_{ij} = p_{ij} C_{kj} \tag{11.17}$$

Elastooptic materials are used in optical devices such as laser Q-switches, modulators, and scanners. All these devices are based on the scattering of light by sound waves.[16,17] An acoustic wave is launched into a transparent elastooptic material by a piezoelectric transducer (Fig. 11.6). The elastooptic effect causes the acoustic wave to be accompanied by a similar wave of varying refractive index n. This periodic pattern of alternately higher and lower refractive index acts like an efficient

TABLE 11.7. ELASTOOPTIC COEFFICIENTS OF SOME MATERIALS

Material	Wavelength (μm)	Elastooptic Coefficient	$M_2{}^a$	Ref. No.
Fused silica	0.63	$p_{11} = 0.121, p_{12} = 0.27$	1	13
Lucite	0.63	$p_{11} = 0.3, p_{12} = 0.28$	33	14
GaP	0.63	$p_{11} = -0.151, p_{12} = -0.082,$ $p_{44} = 0.074$	29.5	13
GaAs	1.15	$p_{11} = -0.165, p_{12} = -0.14,$ $p_{44} = -0.074$	69	13
Ge	10.6	$p_{11} = 0.27, p_{12} = 0.235,$ $p_{44} = 0.125$	540	15

[a] M_2, Figure of merit, normalized to fused silica. Figure of merit $= n^6 p^2/\rho v^3$, where $\rho =$ density, $v =$ acoustic velocity, $n =$ refractive index.

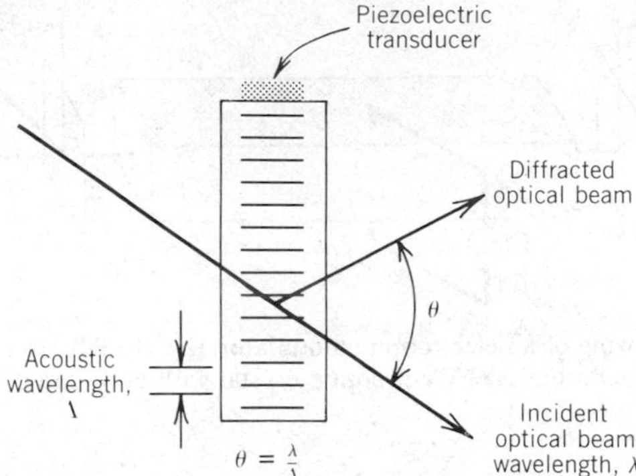

Fig. 11.6 Acoustic Bragg reflector.

diffraction grating. For the light path $l < \Lambda^2/\lambda$ (Λ and λ are acoustic and optical wavelengths in the medium) the light is split up into many orders separated by angles of about λ/Λ. This is known as Debye–Sears (liquid medium) or Raman–Nath (solid medium) scattering. For the light path $l > \Lambda^2/\lambda$ (very-high-frequency sound waves) the incident light is strongly diffracted into one (first order) spot. This occurs for a particular angle of light incidence, and because of its similarity to Bragg reflection of x-rays by the parallel planes of a crystal lattice, it is called acoustic Bragg reflection. The deflection angle with reference to the incident light beam (for small angles) is given by $\theta = \lambda/\Lambda$. The deflection angle can thus be easily controlled by controlling the acoustic frequency with a VCO (voltage-controlled oscillator).

11.2-6 Magnetooptic Materials

The most important magnetooptic effect is the Faraday effect. This is defined as the rotation ϕ_f of the direction of polarization of a linearly polarized light in passing through the material under the influence of an external magnetic field. In paramagnetic or diamagnetic materials, it is given by

$$\phi_f = VHl \tag{11.18}$$

where H is magnetic field intensity, l the path length parallel to the magnetic field, and the coefficient V is known as the Verdet constant.

In ferromagnetic materials the magnetic field is replaced by the magnetic flux density (magnetization), and in this case the Faraday rotation coefficient (K) is known as the Kundst constant. In these materials a useful quantity is specific Faraday rotation F obtained for saturation magnetization M_s.

$$F = KM_s \tag{11.19}$$

Wemple[18] has reviewed materials for magnetooptic modulators and a comprehensive table of magnetooptic constants is given by Chen.[19] Magnetooptic materials can be used in optical modulators, similar to the way electrooptic materials are used.[20,21] However, compared with electrooptic materials they suffer from high absorption at optical frequencies.

11.2-7 Nonlinear Optical Materials

The electrooptic, elastooptic, and magnetooptic effects described in the previous sections refer to changes in optical properties due to low-frequency applied fields. In this section nonlinear optical materials are discussed. The distinction being made here is that the changes in the optical properties are induced by the electromagnetic fields at optical frequencies. This gives rise to a host of new phenomena such as second harmonic generation, parametric oscillators, multiphoton absorption, self-focusing, and stimulated Raman, Brillouin, and Rayleigh scattering. The observation and device application of these phenomena is largely due to the advent of lasers, a source of intense monochromatic electromagnetic radiation. In many cases the crystalline materials used for these devices are the same as those used for the electrooptic effect. An important difference is that at optical frequencies the lattice motion can no longer be excited and the induced polarization at the sum frequency is due only to nonlinearities of the electronic motion. As in the case of the quadratic electrooptic effect, the electric polarization P due to intense electric fields at optical frequencies can be represented as a

TABLE 11.8. PHOTOGRAPHIC FILMS[a]

Film Type	Spectral Sensitivity[b] (μm)	Resolution (lines/mm)	Speed (ASA)	Contrast	Application
Kodak					
Spectrum Analysis No. 1	250–440	> 225	—	High	Emission spectroscopy
103-0	250–500	56–68	—	Medium	Spectroscopy-low intensity
649-F	—	2000	—	Extra high	Holograms
High Speed Recording					
2485	PR	20–50	800	—	High-speed cathode ray tube (CRT) recording
RAR-2495	O	32–100	320	—	CRT recording-extended blue sensitivity
LPF-7	O	—	—	Very high	Photo plotters-intermediate reduction microphotography
Polaroid					
612	P	15–18	20,000	High	CRT recording
47, 57, 107	P	22–28	3, 000	Medium	Camera
552	P	20–25	400	Medium	Scanning electron microscopy photographs
1462	—	40–50	200	High	Blue sensitivity

[a] These are representative films. Consult manufacturers' catalogues for detailed specifications.
[b] P, panchromatic; PR, P with extended red sensitivity; O, orthochromatic.

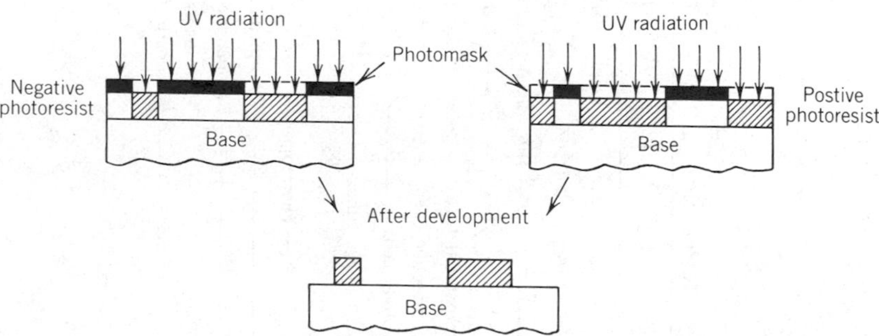

Fig. 11.7 Image formation with photoresists. Crosshatched regions are those exposed to radiation. On development both provide chemically resistant regions for further processing.

TABLE 11.9. TYPICAL PHOTORESISTS

Photoresist Type	Supplier	Trade Name
Liquid negative	Kodak	Microneg
	Hunt	Waycoat and SC
	Dynachem	CMR-5000, OMR 83-DC
	Norland	NPR 6, NPR 22, NPR 29
Liquid positive	Kodak	Kodak 809
	Hunt	Waycoat HPR
	Shipley	Micropaset 1300
Dry film	Dupont	Riston
	Dynachem	Laminar
	Hercules	Akuamer

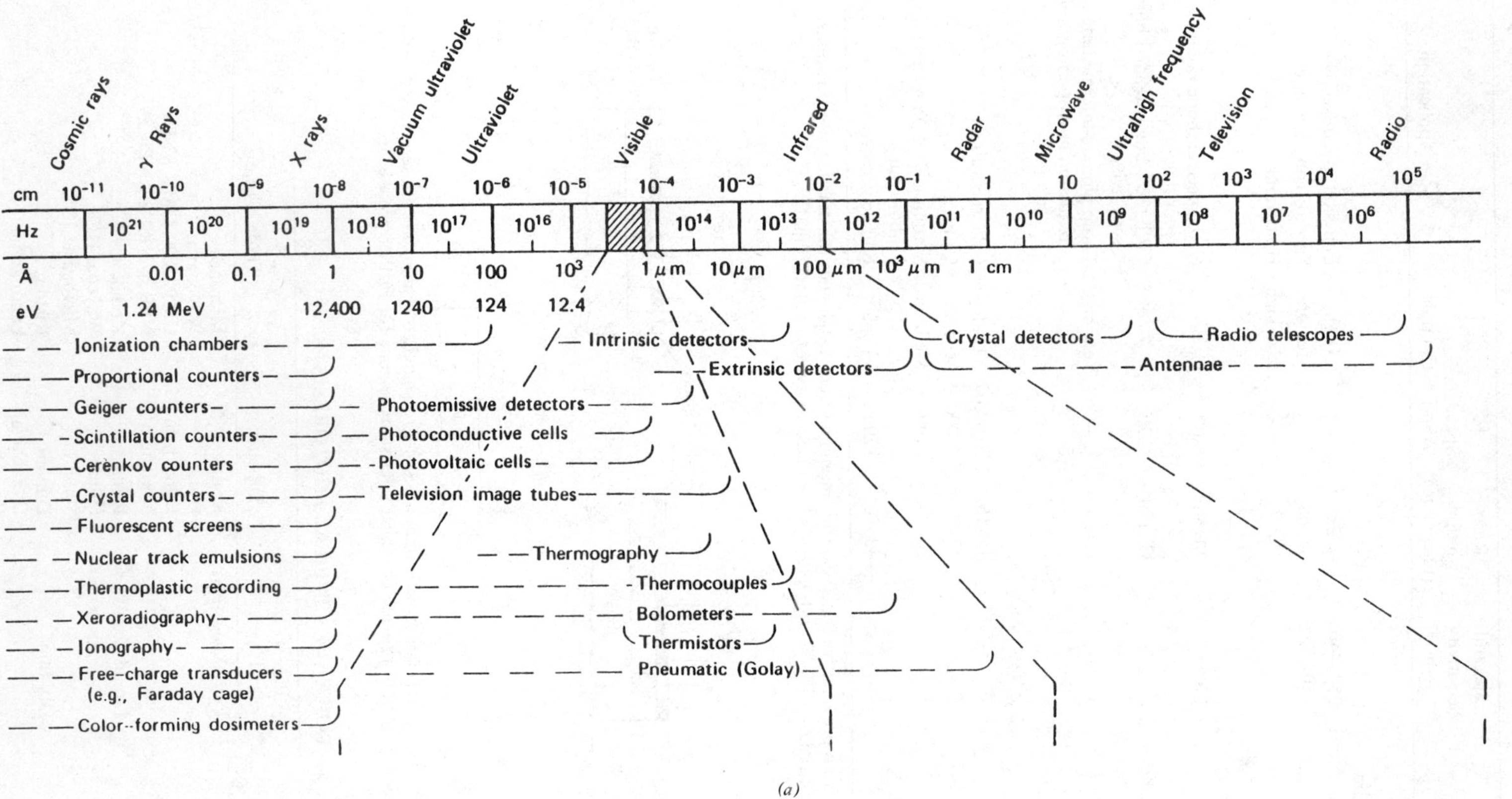
Cosmic rays
γ Rays
X rays
Vacuum ultraviolet
Ultraviolet
Visible
Infrared
Radar
Microwave
Ultrahigh frequency
Television
Radio
cm
10^-11
10^-10
10^-9
10^-8
10^-7
10^-6
10^-5
10^-4
10^-3
10^-2
10^-1
1
10
10^2
10^3
10^4
10^5
Hz
10^21
10^20
10^19
10^18
10^17
10^16
10^14
10^13
10^12
10^11
10^10
10^9
10^8
10^7
10^6
Å
0.01
0.1
1
10
100
10^3
1 μm
10 μm
100 μm
10^3 μm
1 cm
eV
1.24 MeV
12,400
1240
124
12.4
Ionization chambers
Proportional counters
Geiger counters
Scintillation counters
Cerenkov counters
Crystal counters
Fluorescent screens
Nuclear track emulsions
Thermoplastic recording
Xeroradiography
Ionography
Free-charge transducers
(e.g., Faraday cage)
Color-forming dosimeters
Photoemissive detectors
Photoconductive cells
Photovoltaic cells
Television image tubes
Thermography
Thermocouples
Bolometers
Thermistors
Pneumatic (Golay)
Intrinsic detectors
Extrinsic detectors
Crystal detectors
Radio telescopes
Antennae

(a)

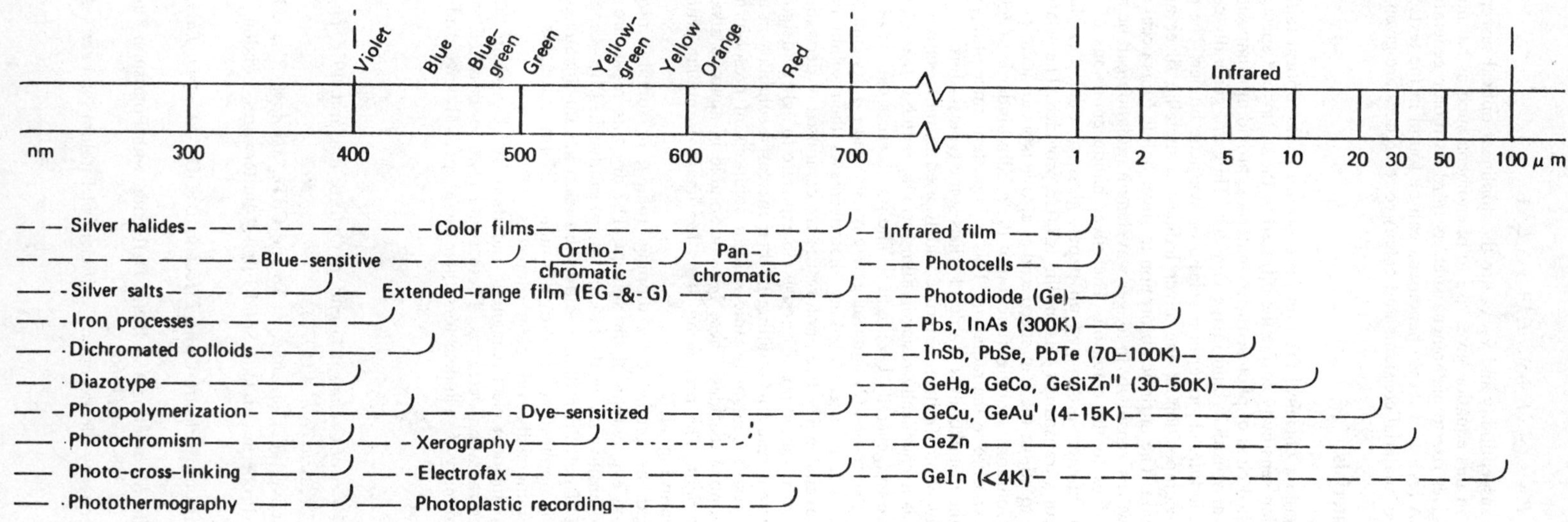

(b)

Fig. 11.8 Electromagnetic spectrum. (*a*) Spectral sensitivities of radiation detectors. (*b*) Spectral sensitivities of radiation detectors in the ultraviolet to infrared region.

series expansion

$$P = \epsilon_0(\chi_i \cdot E + \chi_2 \cdot E^2 + \chi_3 \cdot E^3 + \cdots) \tag{11.20}$$

where χ_1 is the linear optical susceptibility and χ_2, χ_3 are the nonlinear optical susceptibilities of the material. The quadratic term in this equation gives rise to phenomena such as harmonic generation. The cubic polarization term gives rise to phenomena such as third harmonic generation and several stimulated scattering effects. A review of these phenomena can be found in the texts by Yariv[22] and Nelson.[23] An extensive table of material constants and extensive references are given by Singh.[24]

11.2-8 Photosensitive Materials

Photosensitive materials undergo a chemical reaction on exposure to the optical radiation to which they are sensitive. The radiation sensitivity covers the infrared to the ultraviolet spectral range. Silver halides form the most diversified class of photosensitive materials, providing compositions for a wide range of spectral and temporal sensitivity and image contrast. They are generally colloidal suspensions of silver salts (bromide, iodide, chloride) in a gelatin or synthetic binder. The photosensitivity results from the chemical decomposition of the silver halides to metallic silver and halogen on exposure to the optical radiation. The intrinsic absorption in silver halide films occurs at blue, violet, and shorter wavelengths. However, sensitivity to longer wavelength radiation can be achieved by the addition of suitably chosen dyes to the emulsion. The silver halide emulsions of thickness from several microns to several millimeters are coated on paper, glass, or other flexible supports. For protection a thin overcoating of a transparent gelatin film is often provided. The gelatin binder in the emulsion contains sensitizers to adjust the film speed and it also acts as an ideal medium for rapid penetration of the processing solutions. Another function of the gelatin binder is to act as an acceptor for the halogen released on exposure to the radiation. This is important since otherwise the released halogen can recombine with the metallic silver. Table 11.8 lists some typical films.

Another important class of photosensitive materials widely used in the electronic industry is the photoresists. The photoresist is a chemically resistant film-forming material that undergoes marked solubility changes on exposure to (generally UV) light.[25] The photoresists can be divided into three distinct classes: (1) liquid positive resists, (2) liquid negative resists, and (3) dry film resists. Negative resists, on exposure to light, undergo a cross-linking that decreases their solubility in their prior solvents. Some commercially available negative photoresists are composed of derivatives of polyvinyl cinnamate or polyisoprene rubber. Liquid positive resists on exposure to light undergo photochemical decomposition and rearrangement reactions resulting in their increased solubility in aqueous solvent systems. For example, some resists contain organic solvent solutions of cresol-formaldehyde plus photodecomposable naphtoquinone diazides. These become soluble in aqueous alkaline developers after exposure to ultraviolet light. The processing steps for image formation with negative and positive photoresists are shown in Fig. 11.7.

The liquid photoresists can be applied by a dipping, rolling, or spinning process, the resin adhering to the substrate upon evaporation of the solvent. The dry film resists are negative resists composed of a predyed photosensitive layer (10–125 μm thick) sandwiched between a thin transparent mylar film and a backing of polyolefin film. The dry film resist has the advantage of eliminating several processing steps in film application and provides more uniform coating as compared to wet films. Table 11.9 lists some commercially available photoresists.

Materials that respond to incident electromagnetic radiation by changes in their electrical or optical properties are used as radiation detectors. These could also be classified as photosensitive materials. Fig. 11.8 gives a summary of such detectors, their useful spectral range, and applications.[26]

References

1 J. F. Nye, *Physical Properties of Crystals*, Clarendon, Oxford, England, 1976, pp. 236–240.

2 G. Haas and J. Ritter, *Vac. Sci. Tech.* **4** (1971).

3 D. T. F. Marple, "Refractive Index of ZnSe, ZnTe and CdTe," *J. Appl. Phys.* **33** (1964).

4 D. S. Heavens, "Optical Properties of Thin Solid Films," Butterworth, London, 1955.

5 Nye, *loc. cit.*, pp. 241–258.

6 A. R. Johnston, "Dispersion of Electro-Optic Effect in $BaTiO_3$," *J. Appl. Phys.* **42**:3501–3507 (1971).

7 J. E. Geusic, et al., "Light Modulation and Beam Deflection with Potassium Tantalate-Niobate Crystals," *J. Appl. Phys.* **37**:388–398 (1966).

8 E. H. Turner, "High Frequency Electro-optic Coefficient of Lithium Niobate," *Appl. Phys. Lett.* **8**:303–304 (1966).

9 D. F. Melson and E. H. Turner, "Electro-optic and Piezoelectric Coefficients and Refractive Index of Gallium Phosphide," *J. Appl. Phys.* **39**:3337–3343 (1968).

10 L. Ho and C. F. Barber, "Electro-optic Effect of Gallium Arsenide," *Appl. Opt.* **2**:647–648 (1963).

11 I. P. Kaminow, "Optical Waveguide Modulator," *IEEE Trans. Microwave Theory Tech.* **MTT-23**:57–70 (1975).

12 M. Izutu, T. Itoh, and T. Sueta, "10 GHz Bandwidth Travelling Wave $LiNbO_3$ Optical Waveguide Modulator," *IEEE J. Quantum Electr.* **QE-14**(6):394–395 (1978).

13 R. W. Dixon, "Photoelastic Properties of Selected Materials and Their Relevance for Application to Acoustic Light Modulators and Scanners," *J. Appl. Phys.* **38**:5149–5153 (1967).

14 T. M. Smith and A. Korpel, "Measurement of Light-Sound Interaction Efficiencies in Solids," *IEEE J. Quant. Electr.* **QE-1**:283–284 (1965).

15 R. L. Abrams and D. A. Pinow, "The Acousto-optic Properties of Crystalline Germanium," *J. Appl. Phys.* **41**:2765–2768 (1970).

16 C. F. Quate, C. D. W. Wilkinson, and D. K. Winslow, "Interaction of Light and Microwave Sound," *Proceed. IEEE* **53**(10):1604–1623 (1965).

17 E. I. Gordon, "A Review of Acousto-optical Deflection and Modulation Devices," *Appl. Opt.* **5**(10):1629–1639 (1966).

18 S. H. Wemple, "Materials for Magneto-optic Modulators," *J. Electr. Mater.* **3**(1):243–263 (1974).

19 D. Chen, "Magneto-optic Materials," *Handbook of Lasers with Selected Data on Optical Technology*, CRC Press, Boca Raton, FL, pp. 460–477.

20 F. S. Chen, "Modulators for Optical Communications," *Proceed. IEEE* **58**(10):1440–1457 (1970).

21 P. D. Tien, et al., "Switching and Modulation of Light in Magneto-optic Waveguides of Garnet Films," *Appl. Phys. Lett.* **21**:394–396 (1972).

22 A. Yariv, *Quantum Electronics*, Wiley, New York, 1967.

23 D. F. Nelson, *Electric, Optic and Acoustic Interactions in Dielectrics*, Wiley, Chichester, England, 1979.

24 S. Singh, "Non-Linear Optical Materials," in *Handbook of Lasers with Selected Data on Optical Technology*, Chemical Rubber Co., Cleveland, Ohio.

25 J. Pacansley, "Recent Advances in Photodecomposition Mechanisms of Diazo-Oxides," *Poly. Eng. Sci.* **20**(16) (1980).

26 Thomas Woodlief, Jr., ed., *SPSE Handbook of Photographic Science and Engineering*, Wiley, New York, 1973.

Bibliography

Arecchi, F. T. and E. O. Schulz-DuBois, eds., *Laser Handbook Vol. I*, North-Holland, Amsterdam, 1972.

Born, M. and E. Wolf, *Principles of Optics*, 4th ed., Pergamon, New York, 1970.

DeBell, G. W. and D. H. Harrison, eds., *Optical Coatings*, Proceed. SPIE, Vol. 50 (1974).

DeForest, W. S., *Photoresist Materials and Processes*, McGraw-Hill, New York, 1975.

Driscoll, W. J. and W. Vaughan, eds., *Handbook of Optics*, McGraw-Hill, New York, 1972.

Kallard, T., ed., *Acoustic Surface Wave and Acousto-optic Devices*, Optosonic, New York, 1971.

Kodak Microelectronics Seminar Proceedings, Kodak publication No. G-130, Eastman Kodak Co., Rochester, NY, 1981.

Lines, M. E. and A. M. Glass, *Principles and Applications of Ferroelectrics and Related Materials*, Clarendon, Oxford, 1977.

Photofabrication Methods with Kodak Resists, Eastman Kodak Co., Rochester, NY, 1979.

Potter, R. F., ed., *Physical Properties of Optical Materials*, Proceed. SPIE, Vol. 204 (1979).

Woodlief, Thomas, Jr., ed., *SPSE Handbook of Photographic Science and Engineering*, Wiley, New York, 1973.

PART 3

COMPONENTS

CHAPTER 12

PASSIVE LUMPED CIRCUIT ELEMENTS

HENRY DOMINGOS
JOSEPH SCATURRO

Clarkson University
Potsdam, New York

12.1 RESISTIVE COMPONENTS

Henry Domingos

12.1-1 Discrete Resistors[1-4]

Introduction

Resistors are the most commonly used electronic component, with sales volume closely paralleling the fortunes of the integrated circuit industry. The design engineer faced with choosing a resistor has many factors to consider: price, availability, tolerance, power dissipation, stability, reliability, frequency response, temperature coefficient, voltage coefficient, size, and package, to name a few. For insight into these factors the materials and construction of the various types must be considered.

Most discrete resistors fall into one of four basic categories: carbon composition, carbon film, metal film, or wirewound. Carbon composition resistors have been in use for nearly 100 years and are still popular. Part of this popularity must be attributed to the momentum of widespread use over the years and a reluctance to make design changes. However, as a general purpose resistor this type of resistor has certain distinct advantages, namely low price, wide resistance range, low inductance, excellent surge capability, good performance under temperature cycling, good reliability, and ready availability. On the other hand, it is not available in tolerances less than 5%, has poor long-term stability, is sensitive to ambient moisture levels, and has a high noise figure (in high resistance values). Sales have been gradually declining over the past few years as prices of other, higher performance resistors have become more competitive. Allen-Bradley Co. is now the sole US manufacturer of carbon composition resistors, although other companies sell imported products.

Carbon film resistors have supplanted carbon composition resistors for general purpose use because of lower cost, better tolerances, better stability, lower noise, and better high-frequency performance. Parts for electronic applications are available in power ratings from 1/10 to 2 W, with tolerances as low as 0.5%, and with load life and moisture resistance changes of 2% or less. The distinctive feature of carbon film resistors is the negative resistance coefficient, typically −200 ppm/°C for low resistance values to more than −1000 ppm/°C for larger sizes. Sales volume is now several billion units per year, with principal manufacturers being Dale Electronics, R-Ohm, Mepco/Electra, and Airco Electronics.

So-called metal film resistors may indeed consist of a thin metallic layer on an insulating core, or may instead be an oxide or other metal compound or even a metallic glaze. Manufacturers such as Corning Glass, Dale Electronics, Mepco/Electra, and TRW offer a wide variety of styles, packages, and specifications for many different applications. These resistors are manufactured in power ratings down to 1/20 W, and generally they are smaller in size than other equivalent resistors. Although their resistance range is somewhat less than that of carbon composition resistors, their tolerance is far better—0.1 to 1% generally, with some products available with tolerance down to 0.025%. In addition, their stability, shelf life, and temperature coefficient are superior. Finally, their high-frequency performance outclasses that of all other types of resistors.

Because of the wide range of specifications now available for them, metal film resistors are encroaching on traditional markets for low-cost carbon composition and carbon film resistors and also for the more costly wirewound resistors. They presently lead all other types in dollar sales volume, with a downward trend in price as their flexibility enlarges their market base.

Wirewound resistors fall into three categories; a low-cost general purpose type, a power wirewound, and a precision wirewound. Precision and power wirewound resistors are used whenever their high cost, large size, and poor frequency response must be tolerated to take advantage of their outstanding accuracy, stability, noise figure, temperature coefficient, and voltage coefficient. Precision wirewounds have large case sizes to keep the internal temperature rise small, which in turn minimizes resistance changes. Even though low-inductance configurations, such as the Ayrton-Perry winding, can be employed, the inductance and distributed capacitance generally limit their use to the audio frequency range. All wirewounds are limited in high-resistance values by the small wire diameter and long wire lengths which would be required. Except for general purpose types, the purchase tolerance is usually from 0.01 to 1%, although values as low as 0.002% are advertised. Major manufacturers include TRW, Dale Electronics, RCL, and Ohmite.

Fabrication

Carbon composition resistors employ graphite or calcined carbon black dispersed in a resin system such as phenol/formaldehyde filled with microcrystalline silica. The bulk resistivity of the conducting core is controlled by the carbon particle size and density. The insulating shell is made of the same material without the carbon, which results in a rugged assembly with minimum thermal mismatch problems. In assembly the shell is formed around one lead of tinned copper. To improve lead strength and lower contact resistance the leads may be knurled, swaged, or formed into an inverted taper or double nailhead. The core material is inserted, the second lead aligned, and the end cap added. The entire resistor is then compressed and cured. Vacuum impregnation with aromatic compounds follows, then the part is tested, sorted, and labeled.

Carbon film resistors are fabricated by pyrolytic decomposition of carbon on the surface of a ceramic substrate. The source of the carbon is an organic gas such as methane, cracked at a temperature of 1100°C. The characteristics of the resistor are sensitive to the deposition conditions. Film thickness usually ranges from less than 2 μm to about 100 μm. The thinnest films have sheet resistivities of 10,000 Ω/sq and exhibit the highest temperature coefficient, about -1000 ppm/°C. Thicker films have sheet resistivities as low as 10 Ω/sq and temperature coefficients of -100 ppm/°C.

Contact is made to each end of the film by force-fitted end caps with a layer of silver cement for better contact. At this stage the resistors are adjusted to the proper size by a procedure known as spiraling. A thin grinding wheel is used to cut a groove through the film along a helical path, increasing the resistance value by changing the configuration of the current-carrying film. The spiral cut may extend for only a turn or so on up to more than a dozen turns. The resistance is continuously measured and the spiraling automatically halted when the desired value is reached. The finished resistor is coated with a thermosetting resin, a molded epoxy, a ceramic coating, or a conformal coating of various materials.

When any type of film resistor is spiraled, the current flow and power distribution are no longer uniform over the film. Current density is a maximum at the tip of the spiral cut; this leads to a hot spot at that point and possible permanent changes in resistor value. For high values of resistance the power dissipation is too low for thermal damage but the maximum voltage may be limited by breakdown across a spiral cut. Both types of damage have been observed during high-power pulse operation.

Metal film resistors employ a variety of materials and manufacturing methods. Evaporated films of nickel-chromium alloys are widely used. The sheet resistance and temperature coefficient are determined by the deposition process, with such variables as film thickness, deposition rate, substrate temperature, annealing, and oxidation treatments being the most important. An 80% nickel alloy is the usual starting material, but the deposited film is richer in chromium. With film thicknesses from 5 to 100 nm and the inclusion of other materials in the film it is possible to obtain sheet resistivities of 10 to 10^4 Ω/sq. However, most films have a resistivity of 100 to 500 Ω/sq. Temperature coefficients are likewise dependent on deposition conditions, with values as low as 25 ppm/°C being readily obtainable.

Mixtures of metal, metal compounds, glass, and solvents are commonly used for the resistive element. The usual method of application in this case is by dipping or rolling. The films tend to be relatively thick, 5 to 50 μm. A typical sequence would begin with an alumina substrate dipped in a liquid glaze composed of tantalum, tantalum nitride, glass, and a carrier. The coated substrate is fired at 1100°C to fuse the glass particles and bind the film to the substrate. The rods are then cut to length and the ends nickel plated and tinned with hard solder. The resistor blanks are spiraled, then attached to end caps and lead assemblies. A jacket such as molded phenolic completes the assembly.

Wire for wirewound resistors is specially formulated for high stability and low temperature coefficient. The most common types are nickel chromium alloys (with trade names such as Evanohm, Karma, and Moleculoy) with a resistivity of 1.33×10^{-4} Ω-cm (800 Ω per circular mil foot) and copper-nickel alloys (such as Cupron, Advance, and Neutraloy) with resistivities near 5×10^{-5} Ω-cm (300 Ω per circular mil foot). Wire diameters range from 7 μm to 10 mm. Rods and ribbon are also available for very low resistance values. Wire is supplied bare or coated with various enamels, plastics, or fabrics in single or multiple layers.

The construction of power wirewounds and precision wirewounds provides an interesting contrast. Power wirewounds are intended to withstand higher operating temperatures, hence the materials and assembly are different. Construction starts with a ceramic core such as steatite, alumina, or beryllia. End caps and lead assemblies are attached and a bare resistance wire is welded to one end. The wire is wound in a single layer with wide spacing between turns. The resistance is continuously measured during the winding process, and when the desired value is reached the wire is cut and welded to the other end cap. A silicone, ceramic, or vitreous enamel coating is applied for insulation. All the materials are capable of operating at high temperatures, and the type of winding facilitates heat transfer to the ambient.

Precision wirewounds, on the other hand, are designed for accurate, stable resistance. Such resistance is best achieved by long lengths of wire and restricted temperature excursions. This implies greater bulk and the use of different, cheaper materials for the package. Enameled wire is cut to length, wound onto a plastic bobbin divided into several segments, and welded to the leads. Since resistance variations in the starting spool of wire may be as large as 1%, and the insulation precludes continuous resistance measurement, the resistance is trimmed to final value by sandblasting. The winding is sometimes anchored in place with epoxy or RTV (room temperature vulcanized) rubber and the assembly inserted in an epoxy case, closed at one end except for a hole for the lead. Both ends are then sealed. Because the winding operation induces bending and tensile stresses which change the resistance, the parts are thermally stabilized by a 160°C bake for 24 to 96 hours.

Characteristics

Some of the characteristics of resistors that are of concern to users are accuracy (or tolerance), stability, temperature coefficient, voltage coefficient, humidity effects, power dissipation, frequency effects, and reliability. The last three are treated separately in later sections.

Tolerance. General purpose resistors are available in standard values for tolerances of 5, 10, and 20%. Carbon composition resistors and some wirewounds use a color code with three to five bands. Resistor color codes are listed in Table 12.1.

Since the resistors may be sorted as a final step, the distribution of values may be bimodal rather than normal. Initial purchase tolerances are 5 to 20% for carbon composition resistors and typically 0.5 to 10% for carbon film resistors, 0.1 to 1% for metal film resistors, and 0.01 to 1% for precision wirewounds.

Stability. Stability refers to the change in resistance value following exposure to a specified environmental stress. Stresses include high or low temperature during storage, long shelf life, application of full rated power, moisture, soldering heat, short-time overload, and radiation exposure. Carbon composition resistors generally have the poorest stability, and relatively mild stresses over a period of time can cause resistance shifts well outside the purchase tolerance; it is not uncommon to find parts with such resistance shift in inventory. Wirewound resistors are the most stable, followed by metal film and carbon film.

Temperature Coefficient. Temperature coefficient is defined as $(1/R)(\partial R/\partial T)$, expressed in parts per million per degree Celsius (ppm/°C). Carbon composition resistors have the largest temperature

TABLE 12.1. RESISTOR COLOR CODES

Color	1st Band,[a] 1st Significant Figure	2nd Band, 2nd Significant Figure	3rd Band, Multiplier	4th Band,[b] Tolerance (%)	5th Band,[b] Failure Rate (%/1000 hr)
Black	0	0	1	—	—
Brown	1	1	10	—	1
Red	2	2	10^2	—	0.1
Orange	3	3	10^3	—	0.01
Yellow	4	4	10^4	—	0.001
Green	5	5	10^5	—	—
Blue	6	6	10^6	—	—
Violet	7	7	10^7	—	—
Gray	8	8	10^8	—	—
White	9	9	10^9	—	—
Silver	—	—	0.01	10	—
Gold	—	—	0.1	5	—
None	—	—	—	20	—

[a] The first band is the one closest to one end of the resistor. A first band wider than the others indicates a wirewound resistor.
[b] Certain MIL parts.

coefficient. It is larger for higher value resistors, and may be negative at high temperatures. Over the specified operating ambient temperature range of −55 to 130 or 150°C it may vary between +1600 and −800 ppm/°C. Hence a 5 or 10% resistor may be out of spec at extreme temperatures under no-load conditions. For carbon film resistors the temperature coefficient is characteristically negative, ranging from −200 ppm/°C at low resistance values to more than −1000 ppm/°C at high values. Wirewounds have the best temperature coefficient with values as low as ±20 ppm/°C. Metal film resistors are almost as good with values ranging from ±20 to ±200 ppm/°C.

Voltage Coefficient. When voltage is applied to a resistor there may be a slight decrease in resistance (apart from temperature-induced changes). The percent change in resistance per applied volt is called the voltage coefficient. For carbon composition resistors the change is as high as −0.05%/V; for carbon film resistors it may be an order of magnitude lower. Other types of resistors have a negligible voltage coefficient.

Humidity Effects. Moisture can produce two reversible effects. On the surface of a high-value resistor it can provide a path for leakage currents and thus lower the apparent resistance. If it is absorbed through the jacket of a carbon composition resistor the resistance value will increase as much as 10%. In either case the moisture can be removed by baking, and the effect is minimized as long as the resistor is dissipating power. Other types of resistors are susceptible to chemical reactions if moisture is allowed to penetrate to the resistive element.

Power Dissipation. Discrete resistors for electronic applications are available in power dissipation ratings from 1/20 to 2 W or more. However, for high values of resistance, voltage is the limiting factor, ranging from 150 V for smaller packages to 750 V or so for 2-W sizes. At low values of resistance the application of these voltages would exceed the rated power dissipation. The value of resistance that is the boundary between these two regimes is called the critical resistance, given by $R = V^2/P$, where V and P are the rated voltage and dissipation, respectively.

The important consideration in power dissipation is the temperature rise. For example, a 1-W carbon composition resistor can be allowed to dissipate 1 W only if it is mounted so that air can circulate freely around it and the ambient temperature is less than 70°C. Under these conditions more than half of the power is conducted away through the leads and most of the rest is carried away by convection. If the resistor is mounted so that it is close to other heat-producing components or has restricted ventilation, the power dissipation must be reduced to avoid reaching the maximum allowable temperature of 150°C. Of course, for ambient temperatures higher than 70°C the dissipation must be linearly derated to 150°C.

Under pulse operation the situation is more complex.[5] When a single pulse is applied the

TABLE 12.2. PULSE-HANDLING CAPABILITIES FOR CARBON COMPOSITION RESISTORS

Rated Dissipation (W)	Pulse Energy (J)	Equivalent Source
1/8	0.45	2 μF @ 670 V
1/4	1.8	10 μF @ 600 V
1/2	6.4	32 μF @ 630 V
1	16	32 μF @1000 V
2	44	32 μF @1650 V

Source. Allen-Bradley Co.

limitation is also usually the peak temperature, even when the voltage momentarily exceeds the rated value. Immediately after application of power the temperature rise is limited by the thermal mass of the resistive element. The temperature increase is given by

$$T = \frac{1}{\rho c V} \int_0^t P(t)\, dt$$

where $P(t)$ = applied power, W
ρ = density, kg/cm^3
V = volume, cm^3
c = specific heat, J/kg °C

If P is constant, then the temperature is directly proportional to time. Eventually the generated heat diffuses to the jacket, substrate, and leads and the temperature rise is proportional to $\sqrt{P}$ for constant applied power. The time when the temperature rise changes from mainly adiabatic to mainly diffusive can be roughly estimated from the dimensions of the resistive element. The changeover occurs when the dimensions become comparable to the heat diffusion length of the resistive material, given by $L = \sqrt{Dt}$, where D is the diffusion coefficient in square centimeters per second and t is the time in seconds. For carbon composition resistors with a rated dissipation of 1/8 W, this transition occurs at about 1 s, whereas for metal film resistors it can be as short as 100 ns.

Because the volume of the resistive core in carbon composition resistors is so large, these resistors can absorb large amounts of pulse energy before the temperature rise reaches the relatively moderate limit of 150°C for the plastic materials. Allen-Bradley Co. advertises pulse energies for carbon composition resistors as listed in Table 12.2.

The ability to absorb pulse energy depends not only on the thermal mass but also on the peak temperature that can be tolerated before permanent damage occurs. The peak temperature is lowest for plastic materials and highest for ceramics. Generally speaking, carbon composition resistors have the best pulse-handling capability (because the thermal mass is so large) followed by precision wirewounds, carbon film, and metal film resistors.

When power has been applied for several minutes, steady state conditions are reached and the temperature rise reaches a constant value given by $T = P/\theta$, where θ is the thermal resistance in watts per degree Celsius between the resistive element and the ambient temperature. θ can be approximated by the slope of the derating curve. For example, for a 1-W carbon composition resistor, $\theta = 1\text{ W}/(150° - 70°) = 12.5\text{ mW/°C}$. With a pulse train or periodic waveform the average power is often used as the steady state power, although this is not valid at low-duty cycle where the peak temperature excursion is much greater than the average temperature.

Frequency Effects. Resistance remains a constant value only for low frequencies. Frequency response is affected by distributed capacitance and inductance in the resistive path, lead inductance, capacitance from case to ground, skin effect, and dielectric losses. An exact equivalent circuit is impractical, so approximate models must be resorted to, with the complexity depending on the desired accuracy. Unfortunately, parameters for even the simplest models are seldom supplied by manufacturers. Instead typical frequency behavior is given by impedance plots such as that shown in Fig. 12.1.

For a given type of resistor the frequency response tends to improve as resistance value is lowered and as case size decreases. The best frequency performance is shown by film resistors with a minimum of spiraling. Wirewound resistors have the poorest frequency response of any type, being inductive at audio frequencies and capacitive at radio frequency (RF) frequencies.

Reliability.[6] Reliability refers to the probability that a resistor will still be within specification after a given time under certain operating conditions. This is not the same as quality, which is related to the

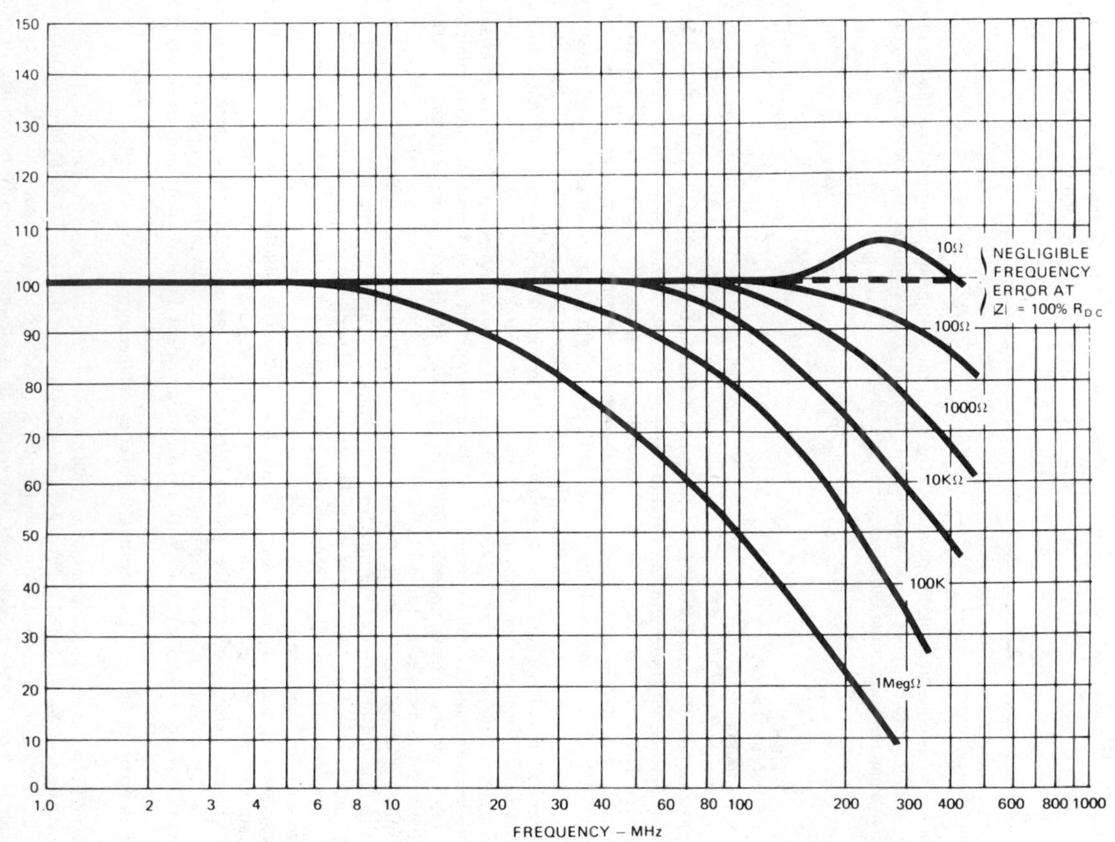

Fig. 12.1 Frequency variation of the magnitude of total impedance as a percentage of DC resistance for film resistors. Courtesy of Mepco/Electra, Inc.

TABLE 12.3. FAILURE RATES FOR RESISTORS[a]

Type	Mil. Spec.	Failure Rate (FITS)[b]
Fixed composition	MIL-R-39008	0.94
Fixed film (insulated)	MIL-R-39017	1.9
Fixed film	MIL-R-55182	2.0
Fixed wirewound (accurate)	MIL-R-39005	10

[a] Low-value resistors dissipating rated power at 25°C in ground-benign environment, qualified to failure rate level M.
[b] Failures per 10^9 part-hours of operation.

percentage of defective parts in a received shipment, although there may be some correlation. Reliability is expressed in terms of a failure rate in percent per thousand hours of operation, or in FITS (failures per 10^9 part-hours of operation). The failure rate for any resistor increases with ambient temperature and with power dissipation.

Resistors may fail catastrophically (open or short circuit) or may drift out of spec. Open circuits can be caused by cracked substrates, open welds, broken resistive elements, or chemical corrosion. Causes of short circuits include electrical breakdown, foreign objects or materials, and silver migration. Excessive drift is due to moisture penetration, corrosion, insufficient annealing and stabilization, electrostatic discharge or other transients, or improper use.

Military Handbook 217C[6] lists base failure rates for different types of resistors as a function of ambient temperature and power dissipation. Multiplicative factors are based on environmental service condition, purchase quality factor, and resistance range. As a means of comparison, failure rates are listed in Table 12.3 for several types of low-value resistors dissipating rated power at 25°C in a ground-benign environment, qualified to failure rate level M (1% per 1000 hours at 60% confidence).

Resistor Networks and Chip Resistors

Use of resistor networks and chips is a way of reducing costs through reduced component count and need for less space on printed circuit boards, although improved performance with their use in certain applications is an important advantage. Resistor networks can be either discrete components in a single package, thick-film network arrays (applied by screen printing), or thin-film networks (applied by vacuum evaporation). Most products are in a dual inline package (DIP) or single inline package (SIP), with flatpacks and cans also available. Chip resistors have single surface or wrap-around terminations that enable them to be attached with conductive epoxy or by solder reflow.

Resistor networks of the thick-film or cermet type typically have tolerances of 2% and temperature coefficients of 100 ppm/°C. Thin-film networks have tolerances as low as 0.1% and temperature coefficients of resistance of 25 ppm/°C. Significant features of both types of network are the ability to provide resistance ratios considerably tighter than the absolute tolerances and temperature tracking to ±5 ppm/°C (±25 ppm/°C for thick films). These advantages are exploited in certain applications such as ladder networks for digital/analog converters. Chip resistors range in value from 5 Ω to 5 MΩ with tolerances of 1% and TCRs of 100 ppm/°C.

Networks can be purchased with separate, isolated resistors or special resistor arrays. Important applications include voltage dividers, ladder networks pull-up/pull-down terminations, digital logic family translators, opamp networks, interfaces, and attenuators.

Summary

Table 12.4 lists the main characteristics of each family of resistors. Typical values are given, not those available under special conditions or at extra cost. Note that extreme values may not be attainable simultaneously, for example, precision wirewounds are available to 10 MΩ only in certain styles, 0.01% tolerance is not usually available for low values of resistance, and high values of resistance have low frequency response. Manufacturers' catalogs must be consulted for specific details.

12.1-2 Microelectronic Circuit Resistors[7–9]

This section considers three different processing technologies: thick film, in which the resistors are deposited by a silk screen process; thin film, where the deposition is by a vacuum evaporation or sputtering method; and silicon monolithic, in which the resistors are part of the silicon chip.

TABLE 12.4. TYPICAL RESISTOR CHARACTERISTICS

Characteristic	Carbon Composition	Carbon Film	Metal Film	Precision Wirewound	Chip
Resistance	1 Ω to 100 MΩ	1 Ω to 10 MΩ	1 Ω to 10 MΩ	0.1 Ω to 10 MΩ	5 Ω to 5 MΩ
Power dissipation	1/8 to 2 W	1/10 to 2 W	1/20 to 2 W	0.1 W & up	50 to 600 mW
Purchase tolerance	5, 10, 20%	0.5 to 10%	0.1 to 1%	0.01 to 1%	1 to 10%
Frequency limit	1 MHz	100 MHz	400 MHz	50 kHz	100 MHz
Temperature coefficient	− 800 to + 1600	− 200 to − 1000	± 20 to ± 200	± 20 to ± 200	± 100
Reliability	Best	Good	Good	Poorest	Good
Stability	Poorest	Good	Good	Best	Fair
Approximate cost	1¢	0.05 to 1¢	2 to 20¢	50–100¢	1 to 10¢
Main advantages	Low cost, high reliability, high surge capability	Lowest cost	Good accuracy and stability, small size	Highest accuracy and stability	Smallest size
Main disadvantages	Poor accuracy, poor stability		Poor surge capability	Most expensive, poorest frequency response	Limited availability

Thick-Film Resistors

Thick-film resistors are used in resistor networks or in hybrid circuits because of low cost, low equipment investment, fast turnaround time, good cost effectiveness for small quantities, good flexibility, and the ability to be combined with a mixture of active and passive components on the same substrate such as inductors, large capacitors, and linear, digital, and power devices.

Alumina is almost universally used as a substrate. Before the resistor paste is deposited, a conductor paste for terminations consisting of gold, platinum-gold, palladium-silver, or another combination is laid down. The presence of the conductor affects the properties of the resistor through chemical reactions between the two materials and because the thickness of the resistor during printing is affected as the screen moves over the conductor. Resistor pastes consist of a conducting material, glass frit, and an organic vehicle to control basic flow properties (rheology). Palladium-silver at one time was the most widely used resistor paste, but its resistance value is sensitive to firing conditions because it depends on partial oxidation of the palladium. Use of ruthenium dioxide is more common today.

Resistor inks are supplied in families in which the sheet resistivities change by decade values over at least some portion of the range from 1 Ω/sq to 100 MΩ/sq for a thickness of 25 μm. The thickness depends on the paste rheology, screen mesh size, and speed and pressure of the squeegee. The typical value is 25 μm. The preferred minimum width of the printed line is 1.25 mm. Good design dictates that conductor terminations should overlap the resistor by 0.25 mm around the ends, and that resistor aspect ratio (length/width) should lie in the range 0.2 to 5. Serpentine patterns and sharp corners should be avoided if possible. To avoid use of more than one ink, resistors are sometimes connected in series or parallel. Line width and aspect ratio are designed to assure a power density less than about 7.5 W/cm^2.

The tolerance of a thick-film resistor as deposited is seldom less than 10% and may be as much as 30%. However, final tolerances of 1% may be achieved by laser trimming or abrasive trimming. The resistors are designed for a value 70 to 80% of the desired value, then adjusted as required. Fig. 12.2 shows several examples of resistor trimming.

Thick-film resistors have temperature coefficients of 50 to 250 ppm/°C and voltage coefficients of 50 ppm/V. They are useful to at least 1 MHz and exhibit excellent aging and load life characteristics —usually less than 1% change after thousands of hours of operation.

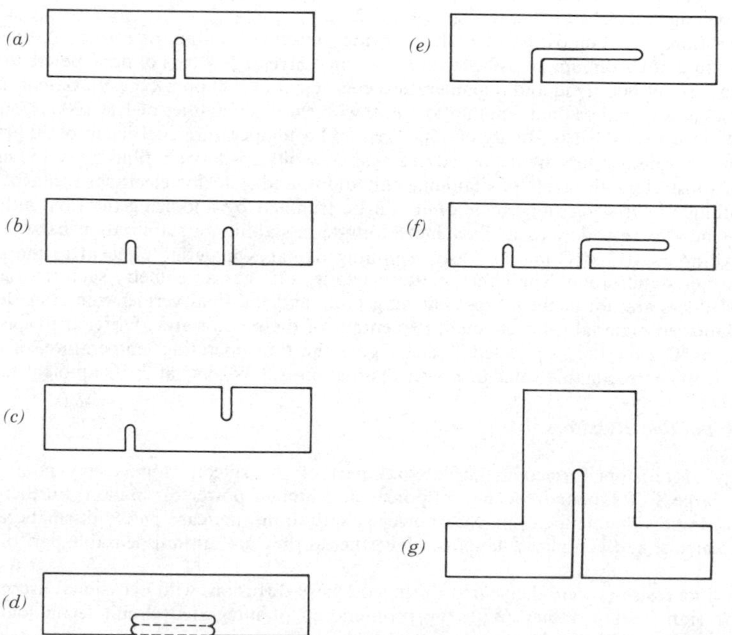

Fig. 12.2 Examples of geometries used in laser trimming of thick-film resistors: (*a*) straight cut, (*b*) double cut, (*c*) reverse double cut, (*d*) scan cut, (*e*) L cut, (*f*) L cut with vernier, and (*f*) top hat design.

Fig. 12.3 Standard pattern to obtain a wide range of resistance values. Large increments can be added by opening "loops" on the right, smaller values by opening rungs on the "ladders" in the center, with a continuous vernier adjustment by a straight laser cut into the column on the left.

Thin-Film Resistors

Thin-film resistors are used in hybrid circuits where greater accuracy, better stability, smaller size, and better high-frequency performance than those provided by thick-film resistors are needed. Thin-film technology is well developed but is generally more expensive than thick-film production.

Resistors, conductors, and dielectrics are deposited by evaporation or sputtering with use of an etching process or possibly a mask to delineate appropriate geometries. Minimum line widths and separations are 25 to 50 μm. All resistors are put down in a single step. This means that only a single value of sheet resistance is available, compared with up to three values for a thick-film process. As a consequence the resistor geometry must account for wide ranges of resistor values, and serpentine patterns are common. "Library" resistor patterns use a single layout that can be trimmed over a wide range by opening "loops" and "ladders" (see Fig. 12.3).

Resistive films are composed of tantalum nitride, nickel-chromium, or cermet. A typical cermet film is a mixture of chromium and silicon monoxide in a layer a few tens of nanometers thick, having a sheet resistivity of 500 Ω/sq and a temperature coefficient of 100 ppm/°C. Nickel-chromium films range in thickness from less than 5 nm up to 1 μm, with sheet resistivities of 1 to 1000 Ω/sq. A typical value is 10 nm with a sheet resistivity of 150 Ω/sq and a temperature coefficient of 30 ppm/°C.

Tantalum nitride resistors are deposited in a sequence of steps to yield films 20 to 150 nm thick on a prepared substrate with layers of titanium, palladium, and gold for electrical contacts. Tantalum nitride is unique in that resistors made of it can be trimmed by anodizing the film, although laser trimming is now more widely used. To achieve long-term stability, a stabilization bake is performed (say for five hours at 300°C in air). Laser trimming damages material adjacent to the kerf, which requires further stabilization. The library pattern in Fig. 12.3 has a geometry such that laser cuts in loops and ladders are not in the current carrying path, and the final vernier trim is made in such a way that damaged material is a very small percentage of the resistor area. Drifts as low as 0.1% over 20 years at 65°C have been projected.[10] The higher the film operating temperature, of course, the greater the drift. A reasonable value of power dissipation is 3 W/cm^2 at 25°C ambient temperature.

Silicon Monolithic Resistors

The quality of resistors formed as an integral part of the silicon chip is very poor, with large tolerances, large TCRs, poor frequency response, and limited power dissipation. Furthermore, such resistors consume valuable area, increase power supply drain, increase power dissipation, introduce parasitic elements, and complicate layouts. Nevertheless they are an indispensable part of integrated circuits.

At one time resistors were delineated during the base diffusion, with occasional use made of the emitter diffusion for low values. With the proliferation of integrated circuit technologies and the unusual requirements of very-large-scale integration (VLSI) design, chip designers are taking advantage of the full range of capabilities afforded during normal circuit processing. Of course the fabrication process is tailored to the requirements of the active devices, and optimization of resistor properties is only of secondary concern.

An adequate discussion of integrated circuit resistor design is not possible in the limited space

TABLE 12.5. RESISTIVE MATERIALS FOR INTEGRATED CIRCUITS

Material	Resistivity ρ (Ω-cm) or Sheet Resistivity R (Ω/sq.)	Comments
Substrate	$\rho = 0.1–100$	Almost never used
Epitaxial layer	$\rho = 0.1–50$	Occasionally used for high-value, high-voltage resistors
Base diffusion	$R = 100–300$	Most commonly used; practical resistor range 20 Ω to 100 kΩ; minimum width 5 μm
Emitter diffusion	$R = 1–5$	Used for low-value resistors
Ion-implanted regions	$R = 10^3–10^4$	Used for high resistance values
Polysilicon layer	$R = 10–200$	
Aluminum metallization	$R = 0.1–0.05$	Generally considered a conductor, although IR drop needs to be considered in I^2L designs
Nichrome and tantalum	$R = 1–1000$	High-quality resistors, but require extra processing steps

available here. Generally speaking, typical tolerances range up to 20%, temperature coefficients are several thousand parts per million per degree Celsius, frequency response is limited to a few tens of megahertz, and power dissipation is limited to roughly 10 $\mu W/\mu m^2$. However, resistance ratios can be held to less than 1%, and temperature tracking is excellent. These are important advantages that have been incorporated in sophisticated designs of digital/analog and analog/digital converters and differential amplifiers, among others. Furthermore, high-quality thin-film resistors are available by deposition on the surface of the silicon chip.

A summary of properties of potential resistive materials for use in integrated circuits is given in Table 12.5. The table lists representative values only and does not include many aspects of resistor design. For these see the reference list.

12.1-3 Thermistors[11,12]

Thermistors are thermally sensitive resistors, that is, resistors with an unusually large temperature coefficient. The resistance may change by two orders of magnitude between 0 and 100°C or by as much as seven orders of magnitude from −100 to +400°C. This makes thermistors extremely sensitive temperature sensors, to the extent that qualification testing or calibration cannot be performed in an air ambient but only in a constant temperature bath held to ±0.1°C at most.

Construction

The more common negative temperature coefficient (NTC) thermistors are made from ceramic materials such as oxides of manganese, chromium, nickel, cobalt, iron, copper, and uranium. With the addition of small amounts of certain metals these compounds become semiconductors (either p type or n type) that exhibit a decrease in resistance as additional carriers are thermally activated.

Disk-shaped thermistors are manufactured in the same way as ceramic disk capacitors. The raw materials are milled to a fine powder which is pressed into disks. The disks are heated to drive off the organic binder, then sintered at high temperatures. Silver paint is applied, leads attached, and insulation is provided. Other forms include beads (made by applying a drop of slurry on two lead wires), rods (extruded), and washers in a wide range of sizes. Fast response time requires a small thermal mass, and beads as small as 100 μm in diameter can be supplied with response times of a fraction of a second in still air or of a few milliseconds in a liquid plunge. The thermistor material is commonly encapsulated in a conformal epoxy coating, but glass packages are also used for probes and special sensor assemblies are readily available.

Positive temperature coefficient (PTC) thermistors are made from barium or strontium titanate or from silicon. The titanate mixtures produce an increase in resistance with temperature as a result of a complicated space charge effect at the surface of grain boundaries. In silicon devices the desired characteristic is produced by virtue of the decrease in mobility. In either material the PTC effect occurs only over a limited range of temperature; outside this range the temperature coefficient is small and negative.

Characteristics

The resistance-vs.-temperature characteristics for a typical family of NTC thermistors are illustrated in Fig. 12.4. The parametric resistance values refer to the resistance at an ambient temperature of

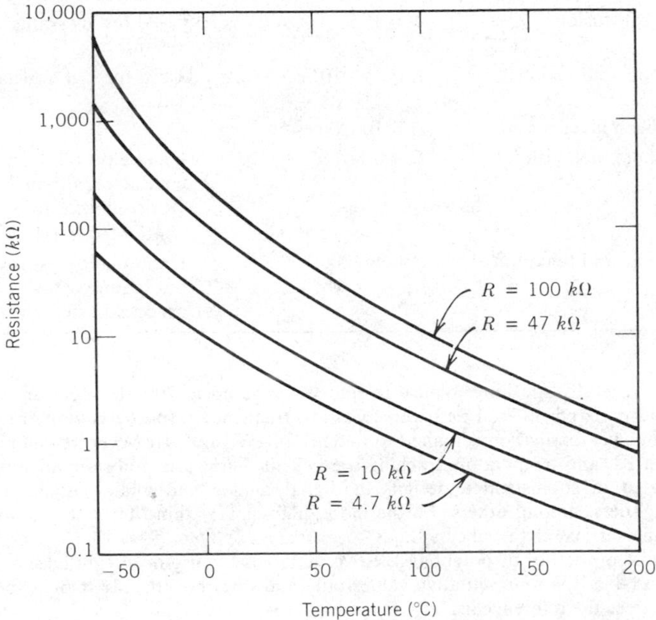

Fig. 12.4 Example of the variation of resistance with temperature for a negative temperature coefficient (NTC) thermistor. Labeled values of resistance are the nominal values at 25°C.

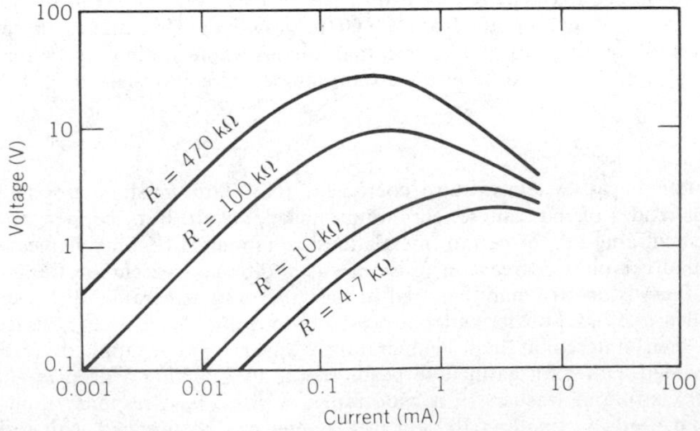

Fig. 12.5 Example of volt-ampere characteristics for a negative temperature coefficient (NTC) thermistor at an ambient temperature of 25°C. The nonlinearity is a result of internal heating at high power dissipation.

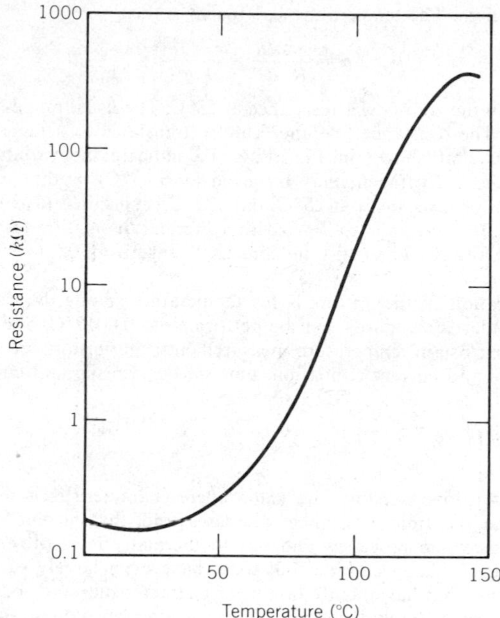

Fig. 12.6 Typical behavior of a positive temperature coefficient (PTC) thermistor.

25°C under conditions of negligible power dissipation in the thermistor. If the power dissipation is high enough to raise the initial temperature of the thermistor by even a small amount the resistance changes. The relationship between the applied voltage and the current is then highly nonlinear, as illustrated in Fig. 12.5. For a given thermistor these curves depend on the ambient temperature.

Corresponding curves for a PTC thermistor are shown in Figs. 12.6 and 12.7.

The resistance curves in Fig. 12.4 are sometimes approximated by an equation of the form

$$\frac{R_{T1}}{R_{T2}} = \exp\left\{ \beta\left(\frac{1}{T_1} - \frac{1}{T_2} \right) \right\}$$

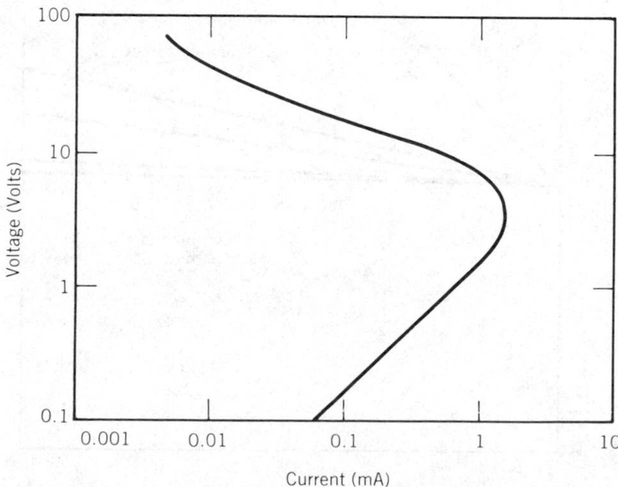

Fig. 12.7 Volt-ampere characteristic for a positive temperature coefficient (PTC) thermistor. At large values of voltage the current begins to decrease, since the internal temperature rise causes a large increase in resistance.

where T is in degrees Kelvin. The temperature coefficient is then

$$\alpha = \frac{1}{R}\frac{dR}{dT} = -\frac{\beta}{T^2}$$

Manufacturers specify the zero-power resistance at 25°C. The usual range is 10 Ω to 1 MΩ, with a tolerance of 10 or 20%. The resistance tolerance can be translated to a temperature tolerance by a resistance curve such as that shown in Fig. 12.4. To indicate the variation of resistance with temperature the temperature coefficient may be given (3–6%/°C) or the value of β given (2000–5000/K). Often the ratio of resistances at 25°C and 125°C is specified (5–50, typically). Maximum power dissipation ranges from 10 mW to 2 W, with a thermal dissipation factor (inverse of thermal resistance) of 0.1 to 25 mW/°C. Thermal time constants range from 0.5 to 150 s, depending on size and package.

An important application of thermistors is for temperature standards. Calibration, traceable to National Bureau of Standards standards, can be performed to 0.0015°C. Stability is within 0.005°C per year. Apart from their use in temperature measurements, thermistors are used for compensation in active circuits, voltage and current regulation, time delays, sensing and control, surge protection, and spark suppression.

12.1-4 Varistors[13–15]

A varistor is a voltage-sensitive resistor. The volt-ampere characteristic is highly nonlinear; as the voltage is increased the dynamic resistance decreases and the current increases rapidly. The mechanism is due to electronic processes and not to thermal effects (in contrast to thermistors). Earlier types were made of silicon carbide, but these have been largely supplanted by zinc oxide devices. Other components that have highly nonlinear characteristics and some of the same applications are selenium suppressors (now obsolete), silicon *pn* junction diodes, and gas discharge tubes. Although this discussion is largely restricted to zinc oxide varistors, comparisons are made with other devices, as in Fig. 12.8. The figure illustrates the rapid rise in current in these devices as the voltage is increased.

Characteristics

The characteristics of zinc oxide varistors are controlled by the raw materials, their processing, and the size and shape of the finished device. The starting material is zinc oxide with up to 10% of other

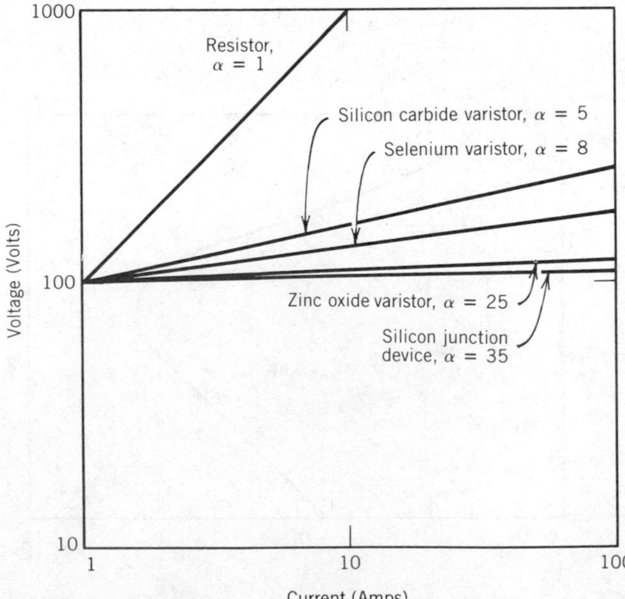

Fig. 12.8 Comparison of different kinds of varistors, α is the exponent of voltage in the equation $I = kV^{\alpha}$.

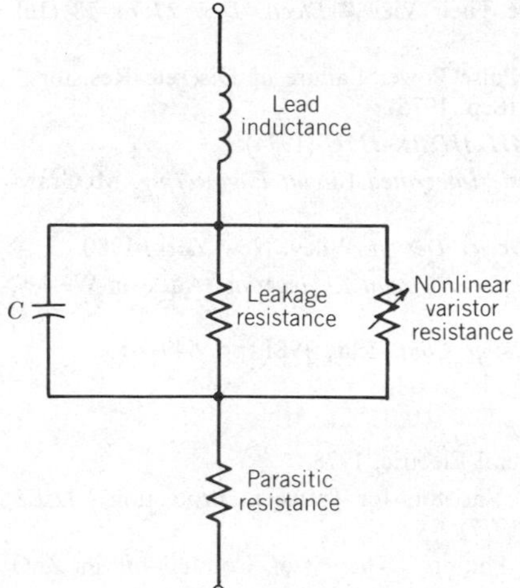

Fig. 12.9 Equivalent circuit for a ZnO varistor.

metallic oxides. The manufacturing process is exactly like that for ceramic disk capacitors or disk thermistors. The completed disk consists of conductive zinc oxide particles about 25 μm in diameter surrounded by a layer of insulating material. Varistor conduction comes about from electron tunneling across the insulating layer, initiated by excess carriers in the zinc oxide. This requires a voltage of 3 to 3.5 V per grain boundary for a current density of 1 mA/cm^2. Controlling the grain size and the thickness of the disk means that the voltage at which appreciable current begins to flow is fixed. The diameter of the disk determines the total current and power-handling capability of the varistor.

The volt-ampere characteristic is usually approximated by a relationship of the form

$$I = kV^{\alpha}$$

Values of α for different kinds of devices are indicated in Fig. 12.8. For zinc oxide varistors, α ranges from 20 to 80. At low voltage values the current predicted by the preceding equation is enhanced by leakage, whereas at high current levels the voltage drop across the finite conductivity of the zinc oxide particles is too large to be ignored. This leads to an equivalent circuit for the varistor, as shown in Fig. 12.9. The capacitance is surprisingly high, not because the dielectric constant of the constituent materials per se is so large but because the conducting grains in an insulating matrix act as an artificial dielectric and increase the effective dielectric constant to about 1000. The 1-MHz capacitance ranges from 10 pF for small, high-voltage devices to 0.02 μF for high current, low-voltage parts.

Applications

By far the most common application for varistors is in transient suppression. Under normal operation the resistance is high enough that negligible current flows. When a potentially destructive transient appears, the resistance decreases and the device conducts a much higher current, in effect clamping the voltage. Important specifications include the voltage at which 1 mA flows, the maximum voltage for a specified surge current, the energy-absorbing capability, the peak current, and the capacitance. For example, a varistor to protect an appliance from transients in a 120-V line might have a minimum voltage of 184 V at 1 mA DC, a maximum of 273 V AC peak at 1 mA, a clamping voltage of 390 V at 10 A (hence a clamping ratio of $390/120\sqrt{2} = 2.30$ at 10 A), a peak current of 500 A, and a surge energy of 10 J. In general the minimum voltage at 1 mA is determined from the nominal working voltage, and the clamping voltage, peak current, and surge capability requirements must be estimated from the characteristics of the transient environment, source impedance, and nature of the circuit to be protected.

References

1 Charles A. Harper, ed., *Handbook of Components for Electronics*, McGraw-Hill, New York, 1977.
2 Gerald L. Ginsberg, *A User's Guide to Selecting Electronic Components*, Wiley, New York, 1981.
3 "Military Standard-Resistors, Selection and Use of," *MIL-STD-199B* (1978).

4 "Choosing the Right Resistor: Three Experts Give Their View," *Electr. Des.* **27**:74–78 (Jul. 1979).

5 Henry Domingos and Donald C. Wunsch, "High Pulse Power Failure of Discrete Resistors," *IEEE Trans. Parts, Hyb. Packag.* **THP-11**:225–229 (Sep. 1975).

6 "Reliability Prediction of Electronic Equipment," *MIL-HDBK-217C* (1974).

7 Douglas A. Hamilton and William G. Howard, *Basic Integrated Circuit Engineering*, McGraw-Hill, New York, 1975.

8 Roy A. Colclaser, *Microelectronics: Processing and Device Design*, Wiley, New York, 1980.

9 Arthur B. Glaser and Gerald E. Subak-Sharpe, *Integrated Circuit Engineering*, Addison-Wesley, Reading, MA, 1977.

10 P. L. Scarff and L. J. Kiszka, in *Proc. 31st Electr. Comp. Conf.*, May 1981, pp. 449–455.

11 *Thermistor Manual*, Fenwal Electronics, 1974.

12 *PTC/NTC Thermistors*, Mepco/Electra, 1974.

13 *Transient Voltage Suppression Manual*, 2nd ed., General Electric, 1978.

14 Lionel M. Levinson and Herbert R. Philipp, "ZnO Varistors for Transient Protection," *IEEE Trans. Parts, Hyb. Packag.* **THP-13**:338–343 (Dec. 1977).

15 G. D. Mahan, Lionel M. Levinson, and H. R. Philipp, "Theory of Conduction in ZnO Varistors," *J. Appl. Phys.* **50**:2799–2812 (Apr. 1979).

12.2 CAPACITORS

Henry Domingos

12.2-1 Introduction

Capacitors are one of the most widely used electronic components, and sales have increased steadily to the point where the market in the United States alone is over one billion dollars and more than six billion units. The growth rate (unit sales have doubled in the past 10 years) is due to the explosive growth of the computer and integrated circuit industries and to the increasing use of electronics in new applications such as automobiles and home appliances. The diversified markets have led to a wide variety of products to meet a wide range of electrical requirements, and the industry is constantly improving its products to achieve smaller size, higher reliability, better performance, and lower costs. The development of new product lines has led to changes in the product mix which reflect new demands. In the United States about 50% of the capacitors sold are ceramic multilayer capacitors, whereas in Japan electrolytic capacitors and in Western Europe plastic film capacitors are dominant. Recent new technological developments attest to the vitality of the capacitor industry, and point to continued evolution of product lines in the forseeable future.

12.2-2 Definitions and General Circuit Properties

Capacitance is defined as the ratio of charge to voltage between two conductors. That is,

$$C = \frac{q}{v} \tag{12.1}$$

where C is in farads, q is in coulombs, and v is in volts. Since current is the time derivative of charge, the above equation leads to

$$i = C\frac{dv}{dt} \tag{12.2}$$

or

$$v = \frac{1}{C}\int i\,dt \tag{12.3}$$

where the current i (amperes) and voltage v may be functions of time. The instantaneous energy stored in a capacitor is given by

$$w = \tfrac{1}{2}Cv^2 \tag{12.4}$$

or

$$w = \frac{1}{2}\frac{q^2}{C} \tag{12.5}$$

where w is in joules.

Under steady-state AC conditions the current and voltage in an ideal capacitor are related by

$$I = 2\pi fCV \tag{12.6}$$

where I and V are the rms (or peak) current and voltage, respectively, and f is the frequency in hertz. The reactance is given by

$$X_c = -\frac{1}{2\pi fC} \tag{12.7}$$

and has the unit of ohms.

Figure 12.10 shows several conductor configurations of interest for electronics applications. The capacitance of two parallel plate capacitors, part a of the figure, is given by

$$C = \frac{\epsilon A}{d} \tag{12.8}$$

where $\epsilon = \epsilon_r \epsilon_0$ is the permittivity of the dielectric, ϵ_r is the relative permittivity (or dielectric constant), and ϵ_0 is the permittivity of free space with the value 8.85×10^{-12} F/m. Equation (12.8) assumes that the electric field lines are straight, parallel, uniformly spaced, and confined to the region between the plates. If fringing effects are considered the capacitance value is somewhat greater, but an exact expression cannot be obtained in the general case. Sometimes an empirical amount, $0.44d$, is added to the length of each side to approximate the effect of the fringing field.

Part b of Fig. 12.10 is important for applications such as microstrip, stripline, printed circuit boards, and interconnections on an integrated circuit chip. The approximate value of the capacitance per unit length is given by

$$C = \frac{\epsilon w}{h}\left\{1 + \frac{2h}{\pi w}\left[1 + \ln\left(\frac{\pi w}{h}\right)\right]\right\} \tag{12.9}$$

for $w/h \gg 1$. The error in this expression is less than 10% for $w = 2h$ and decreases as w/h becomes larger. This formula underestimates the capacitance between a conductor and a ground plane if the thickness of the conductor is not negligible compared with w and h. Also the formula applies only if one dielectric material is present.

The other geometries in Fig. 12.10 are exact (neglecting end effects) and are useful for calculating the capacitance of a coaxial cable, a conductor near a metal chassis, or the capacitance of a pair of wires (half that given in part d).

Any real capacitor dissipates energy as well as stores it. This energy loss, which appears as heat, is a result of several effects: the finite conductivity of the electrodes, electrode contacts, and lead wires;

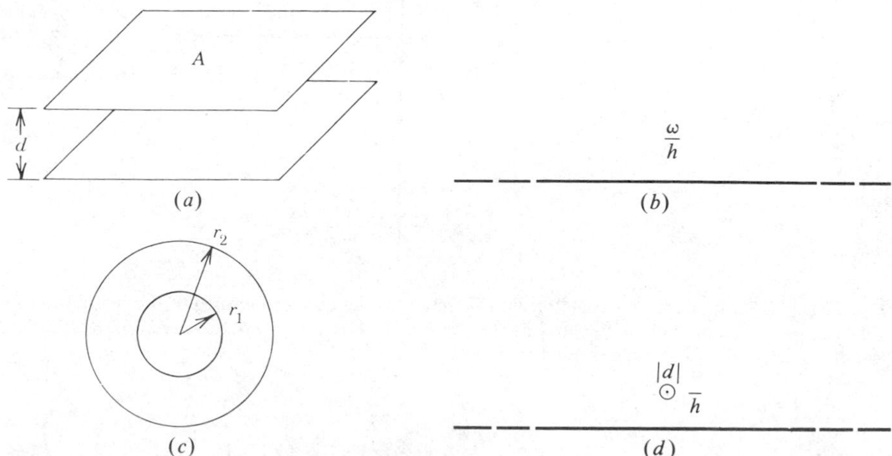

Fig. 12.10 Capacitance of several different configurations of conductors: (a) parallel plates, $C = \epsilon A/d$; (b) microstrip, $C/L = (\epsilon\omega/h)\{1 + (2h/\pi\omega)[1 + \ln(\pi\omega/h)]\}$; ($c$) concentric cylinders, $C/L = 2\pi\epsilon/\ln(r_2/r_1)$; and ($d$) cylindrical conductor above a ground plane, $C/L = 2\pi\epsilon/\ln[2h/d + (4h^2 - d^2)^{1/2}/d]$.

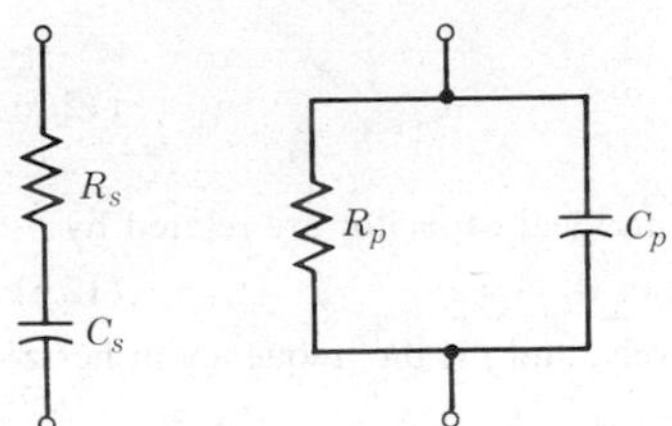

Fig. 12.11 Equivalent series and parallel circuits for capacitors with losses. $R_s = R_p/(1 + R_p^2\omega^2C_p^2)$; $C_s = (1 + R_p^2\omega^2C_p^2)/R_p^2\omega^2C_p$; $R_p = R_s + 1/R_s\omega^2C_s^2$; $C_p = C_s/(1 + R_s^2\omega^2C_s^2)$.

the finite resistivity of the dielectric, which results in a DC leakage current; AC dielectric losses, which are frequency dependent and depend on the polarization mechanism within the dielectric; corona effects, and dielectric absorption (manifested as a buildup of voltage across a capacitor after it has been discharged).

The simplest model to account for capacitor losses is a resistor in series with an ideal capacitor, as shown in Figure 12.11. This resistance is called the equivalent series resistance, or ESR. Some capacitor bridges measure the equivalent model of a parallel resistance and capacitance. The conversion formulas are given in the figure and emphasize the point that C_s is not equal to C_p and that components in either model generally depend on the frequency.

The quality factor, or Q, of a network is defined as 2π times the ratio of the maximum stored energy to the energy dissipated per cycle at a given frequency. For a capacitor

$$Q = \frac{1}{2\pi f C_s R_s} = 2\pi f C_p R_p \tag{12.10}$$

The higher the value of Q, the more nearly ideal the capacitor is.

Bridges often measure the dissipation factor, DF, which is the reciprocal of Q, often expressed in percent. Losses are sometimes specified in terms of a loss angle or power factor. Figure 12.12 shows the relationship of these various quantities.

An accurate model for a capacitor that would be valid over a wide frequency range would be difficult to specify, since any capacitor is a distributed network. A reasonably accurate representation that is useful in explaining certain characteristics is given in Fig. 12.13. Here the dielectric losses are represented by R_p and the conductor losses by R_s. The important addition is the series inductance L, which arises from lead inductance and the inductance of the capacitor structure itself. If R_p is

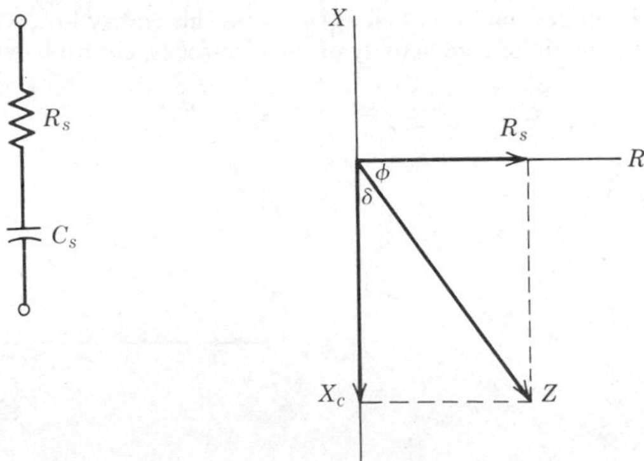

Fig. 12.12 Definitions for a capacitor with losses. Q is the quality factor, DF the dissipation factor, PF the power factor, and δ the loss angle. $X_c = -1/2\pi fC$; $Q = 1/\text{DF} = 1/2\pi f C_s R_s$; $\text{DF} = \tan\delta = \cot\phi = 2\pi f C_s R_s$; $\text{PF} = \sin\delta = \cos\phi = R_s/Z$.

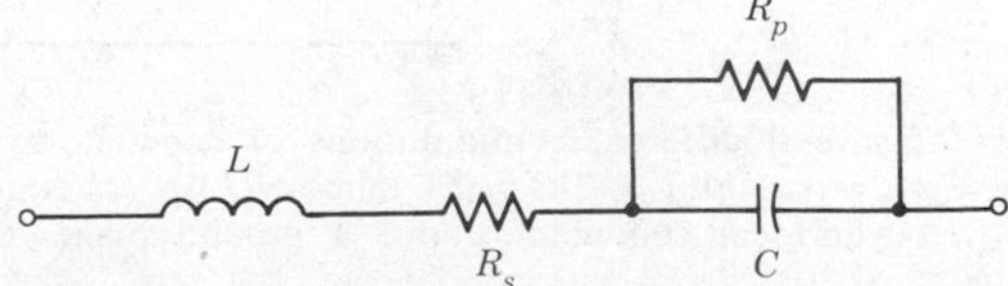

Fig. 12.13 A reasonably accurate model of a capacitor.

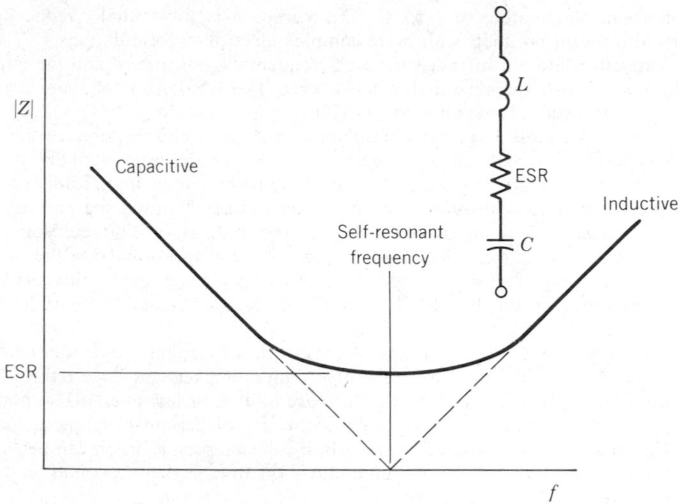

Fig. 12.14 Impedance variation of a capacitor with frequency.

neglected the capacitor behaves as a series RLC circuit whose impedance is shown in Fig. 12.14. Generally speaking, large capacitors have low self-resonant frequencies. Above the self-resonant frequency a capacitor behaves as an inductor, with impedance increasing with frequency. For this reason it is common practice to place a small capacitor, say 0.01 μF, in parallel with a large capacitor to maintain a low combined impedance at high frequencies.

12.2-3 Aluminum Electrolytic Capacitors

Aluminum electrolytics are popular because of low cost and large capacitance per unit volume. They are used in filtering, bypass, and coupling applications where large capacitance is required, but where tolerances and capacitance variation with temperature are not important. Both polarized and nonpolarized units are available.

Aluminum electrolytics are of the foil type, with an electrolyte that can be aqueous, paste, or "dry" (without water). The anode is formed from high purity (99.99%) aluminum foil from 25 to 100 μm thick. The foil can be plain but is more commonly electrochemically etched to increase the surface area and hence increase the capacitance. The increase is by a factor of 8 to 30 times, depending on the oxide thickness.

The foil is anodized to produce a layer of aluminum oxide. The voltage used during the final stage of anodizing is called the forming voltage, and determines the oxide thickness (about 1.2 nm/V). An upper limit to the forming voltage, called the scintillation voltage, is reached when the electric field causes localized momentary breakdown. The scintillation voltage is about 600 V for aluminum, depending on the resistivity of the electrolyte. The rated voltage of the capacitor is typically 60 to 70% of the forming voltage.

A second aluminum foil is the cathode connection. If the capacitor is a nonpolarized type, the cathode foil is also anodized with the same forming voltage. For polarized types the foil has either no oxide or a very thin layer. The cathode foil serves only as an electrical contact to the electrolyte, which is the actual cathode. The most common electrolyte is of the glycol-borate type, made by dissolving ammonium pentaborate in glycol or by dissolving boric acid in glycol. Water is sometimes added to improve the characteristics at low temperatures. Other electrolytes include ethylene glycol and dimethyl formamide.

The anode and cathode foils are welded to lead wires and then wound into a tight roll with porous paper separators. The paper prevents short circuits between the foils caused by jagged edges or rough surfaces, and it absorbs the electrolyte to maintain uniform and intimate contact with the surfaces of the foils. In a newer type of construction, the foils are arranged in stacks, with a tab extension on each foil. This arrangement can reduce both the ESR and inductance by an order of magnitude.

Aluminum electrolytics are available in values from 1 μF to 1 F, with rated voltages from 3 to 475 V. Higher voltages imply lower capacitance values for a given volume, since higher voltages require thicker oxides and hence less capacitance per unit area of foil. Tolerances are typically -10 to $+150\%$. The capacitance is a strong function of temperature and may decrease by an order of magnitude from room temperature to $-55°C$. This variation is substantially reduced in premium grade capacitors and recent products with more complex electrolyte formulations.

Electrolytic capacitors are not intended for high frequency applications, and the impedance may reach a minimum value at frequencies as low as 10 kHz. The ESR is low at room temperature but increases by roughly an order of magnitude at $-55°C$.

The leakage current is quite large for aluminum electrolytics and requires certain precautions. During extended periods of high storage temperature the oxide may partially dissolve in the electrolyte. If rated voltage is then applied the current may be extremely high, leading to catastrophic failure. The proper procedure is to reform the oxide by initially limiting the current with a series resistor or with reduced voltage until the oxide is restored. It is characteristic of aluminum electrolytics that the leakage current decreases during use. During normal use the leakage current increases with applied voltage and with temperature. As a very rough guide, the current doubles as the applied voltage increases from 50 to 100% of rated voltage, and it doubles for each 25° increase in temperature.

Aluminum electrolytics exhibit a gradual decrease in capacitance over an extended period because of loss of electrolyte through the seals. New kinds of packages have reduced this deterioration significantly, and capacitors now show a decrease of 10% or less over 10,000 hours.

Another problem that should be noted involves the use of certain cleaning agents on printed circuit boards. Chlorine from halogenated hydrocarbon solvents such as freon can penetrate the seals and attack the internal structure and cause failure in a short time. Xylene, alcohols, and certain types of detergents are recommended for cleaning.

12.2-4 Tantalum Electrolytic Capacitors

Tantalum capacitors are more versatile and reliable and have better characteristics than do aluminum electrolytics, but also have a much higher cost. There are three types: foil, wet slug, and solid.

Foil capacitors are made in the same way as aluminum electrolytics. Tantalum foil, 10 to 25 μm thick, is anodized to form tantalum pentoxide, which serves as the dielectric. The thickness is about 1.8 nm/V of forming voltage. The anode foil is anodized for polarized units, while both foils are anodized for nonpolarized capacitors. Tantalum lead wires are spot welded to both the anode and cathode foils, which are then wound with paper separators into a compact roll. The roll is inserted into a metallic case and a suitable electrolyte such as ethylene glycol or dimethyl formamide with ammonium nitrate, ammonium pentaborate, or polyphosphates is added for improved performance.

Tantalum foil capacitors are available in sizes ranging from 0.12 to 3500 μF, at voltages up to 450 V. They can be used over the full military temperature range from -55 to $+125°C$. The capacitance increases slowly over this temperature range, and may vary as little as $\pm 10\%$ for some units. The ESR decreases sharply as the temperature is increased from the lower limit, and the leakage current increases drastically. At 125°C the leakage current is about 40 times its room temperature value.

Most applications for this type of capacitor are in the higher voltage ranges where slug-type tantalum capacitors are not applicable and where the superior qualities compared with those of aluminum electrolytics are required in spite of the higher cost. The disadvantages, compared with other tantalum types, are the large size, high ESR, high leakage currents, and large change in capacitance with temperature.

The wet-slug tantalum capacitor has the highest volumetric efficiency of any capacitor. The key to this feature is the large surface area of the porous tantalum pellet used as an anode. To form the pellet, tantalum powder is mixed with a binder and compressed around a tantalum lead wire. The assembly is sintered at very high temperatures in a vacuum to evaporate the binder and any impurities and to weld the tantalum particles together. This results in a strong but light and porous anode assembly. The slug is immersed in an electrochemical bath and anodized to produce a uniform layer of tantalum pentoxide. The anode is then inserted into a case plated with silver and platinum which serves as the cathode. The case is filled with an electrolyte, either lithium chloride or a solution of sulphuric acid in gel form, and sealed.

The main application of wet-slug capacitors is in power supply filters. Although the maximum rated voltage is only 125 V, multiple unit packages are available with ratings of several hundred volts. Capacitance values range from a fraction of a microfarad to 2000 μF.

There are two main disadvantages to wet-slug capacitors. One is the possibility of leakage of the very corrosive liquid electrolyte. The second is the necessity of preventing the application of reverse voltage for even a short duration. When reverse current flows, silver is electroplated onto the anode and can cause a catastrophic short circuit. Tantalum case units are now available which circumvent this difficulty.

The solid tantalum capacitor is similar to the wet-slug capacitor in its initial stages of manufacture. After the pellet is anodized to form the dielectric layer, several layers of cathode material are applied. The anodized pellet is dipped into a manganous nitrate solution, which is then pyrolytically converted to manganese dioxide. This step is repeated several times to provide a good coating of the manganese dioxide, which is a reasonably good conductor and serves as the first layer of the cathode. The next step is to apply several layers of colloidal graphite, followed by silver paint, to provide a good connection to the cathode. The completed assembly is soldered into a metal can or is epoxy dipped after leads are attached.

Solid tantalums are used for low-frequency filtering, bypassing, and coupling applications where large capacitance and small case size are important. They are available in values of 0.0047 to 1000 μF at voltages up to 125 V. Their stability and reliability are unsurpassed among electrolytics. There is no liquid to evaporate, and the solid electrolyte is stable. The capacitance varies as little as $\pm 10\%$ from its room temperature value over the entire temperature range from -55 to $+125$°C. Unfortunately, the dielectric and electrolyte do not exhibit the self-healing qualities associated with other electrolytics. At small defect sites MnO_2 can be converted to Mn_2O_3 or Mn_3O_4, which is an insulator, and hence isolate any defect. This results in a decrease in leakage current with use. However, small defects can lead to catastrophic failure from localized heating, and solid tantalum capacitors must be protected from situations where high inrush currents can occur. This has led to the recommended practice of inserting a series resistor of 3 Ω/V of rated voltage. This rules out the use of solid tantalums for most power supply applications. This same failure mechanism also makes these capacitors susceptible to high voltage spikes. Transients which exceed the forming voltage will result in permanent increases in leakage current and possible catastrophic failure.

To screen capacitors from early failures due to oxide and electrolyte defects, a 100-hour burn-in at rated voltage and maximum temperature is recommended, using a low-impedance power source. It is further recommended that the operating voltage not exceed 60% of rated voltage. Under these conditions the reliability is excellent.

12.2-5 Ceramic Capacitors

Approximately three fourths of the capacitors sold in the United States are ceramic capacitors. This wide usage is due to low cost, small size, wide range of capacitance value, and general applicability for electronic applications. Ceramic capacitors are especially suited for hybrid circuits and integrated circuit filtering, bypass, and coupling applications where appreciable changes in capacitance can be tolerated.

Ceramic capacitors are manufactured in disk form, as multilayer or monolithic capacitors, or in tubular form. The dielectric material is mainly barium titanate, calcium titanate, or titanium dioxide with small amounts of other additives to give the desired characteristics. Class I capacitors utilize calcium titanate and are characterized by low dielectric constant (6–500), good control of tolerances, excellent stability, excellent aging characteristics, low dissipation, and well-controlled capacitance-vs.-temperature behavior. Class II capacitors have much higher dielectric constants (200 to over 10,000) and therefore much higher volumetric efficiency. However, the electrical characteristics are inferior.

Disk capacitors are made from carefully formulated powders, milled to produce a small particle size. The powder is compressed into a thin disk which is then fired at a high temperature to fuse the material. Electrodes are screen printed on each side of the disk and fired. Leads are attached, and the unit is encapsulated.

To fabricate multilayer ceramic capacitors, a slurry consisting of the dielectric powder, a binder, and a solvent is cast into thin sheets on a stainless-steel belt or a plastic tape. After drying, electrodes are printed on the sheets, which are then stacked and compressed. The stacks are cut into individual capacitors, heat treated, and fired at a high temperature. Finally, electrodes are attached on both ends and the parts are encapsulated.

Disk capacitors have a limited range of electrode area and of thickness. Consequently the dielectric formulation is varied over the full range to achieve a wide range in capacitance values. Multilayer capacitors, on the other hand, use a restricted set of dielectrics. The range of capacitance values is readily accommodated by changes in electrode area and the number of layers in the stack. Class I dielectrics can be used for capacitance values from 1 pF to about 1 nF, with tolerances as low as $\pm 10\%$ and voltage ratings up to kilovolts. The capacitance is intended to be stable with temperature or to achieve a specific temperature characteristic to compensate for variations in other components. The temperature coefficients range from P100 (positive 100 ppm/°C) to N750 (negative 750 ppm/°C). Extended ranges are also manufactured from N1400 to N5600. The zero temperature coefficient specification (NPO) has a tolerance of ± 30 ppm/°C. The capacitance is very stable in storage or in operation. The dissipation factor for class I dielectrics is less than 0.1% over the full temperature range, up to frequencies of several tens of megahertz. The voltage coefficient is essentially zero.

Class II dielectrics are used for general purpose applications where small size is important and where stability and tolerance are not so important. Capacitance values range from 2 pF to 2 μF, with typical ratings of 50 to 200 V for multilayer chip capacitors and up to 1000 V for other packages.

Class II materials are ferroelectrics that show an aging effect, that is, the capacitance decreases with time. The aging rate is highest for high values of the dielectric constant and ranges from 0.8 to 8% per decade of time measured from one hour after a high temperature treatment. The capacitance can be fully restored by heating the capacitor to 150°C.

The dissipation factor ranges from 5 to 10% at −55°C to 1 to 4% at room temperature, and decreases slowly to 125°C. The dissipation factor decreases with DC voltage but increases with AC voltage. It also increases with frequency at frequencies above 1 to 100 kHz.

Class II dielectrics have a pronounced voltage coefficient. The decrease in capacitance can be as high as 30% at rated voltage at low temperatures. On the other hand, the application of AC voltages increases the capacitance. The variation of capacitance with temperature is large and nonlinear. The capacitance generally decreases at both very low and very high temperatures; it can decrease by as much as 80%.

The change in capacitance with frequency also depends heavily on the type of dielectric. Class II materials with high dielectric constants show a 10% decrease in capacitance at a frequency of 10 kHz, whereas for those with low dielectric constants the corresponding frequency is greater than 10 MHz. Class I materials show similar changes at frequencies of hundreds of megahertz.

12.2-6 Paper and Plastic Capacitors

Paper, plastic, and paper-plastic combinations are used in a wide variety of applications including filtering, coupling, bypassing, timing, and noise suppression. They are capable of operation at high temperatures, have high insulation resistance, have good stability, and are available in tolerances as low as 1%. The self-healing property of metallized films is very useful in certain applications. The availability of extremely thin films and the wide variety of materials provides the versatility for a wide range of applications.

These capacitors are made by three construction techniques: wound foil, metallized film, and stacked film. A wound foil capacitor consists of two aluminum foils separated by sheets of dielectric and rolled into a compact cylinder. Contact to the foils is made by inserting tabs during winding or, in the extended foil type, by allowing the foils to extend beyond the dielectric on opposite sides. After winding, leads are attached to the tabs or exposed foil edges. If paper sheets are used, the capacitor is usually impregnated with oil, resin, wax, or a synthetic compound. This prevents the absorption of water and reduces corona and high-voltage breakdown. Plastic film capacitors are sometimes impregnated but are often dry. After leads are attached the assembly is molded, potted, dipped, or sealed in a metal can.

Metallized film capacitors are made by vacuum deposition of aluminum 0.1 μm thick directly onto the dielectric film. The pattern is basically that of an extended foil configuration. After the winding, contact to each end of the roll is made with a fine spray of molten metal, to which leads are finally soldered. This type of construction reduces the volume of low-voltage large-value capacitors and provides a voltage breakdown property known as self-clearing. If a defect should occur in the dielectric, the discharge current through the defect generates enough heat to vaporize the thin metal electrodes. This isolates the defect site and permits restoration of normal operation. In the wound foil construction the electrodes are much thicker and a breakdown results in a permanent short. Self-clearing requires a minimum energy of 10 to 50 μJ, and low values of metallized film capacitors used in low-voltage high-impedance applications may be shorted and fail instead of clearing.

Stacked film is the newest type of construction. Metallized films are wound onto a large cylinder, then cut into rectangular sections. Connections are made to alternate electrodes on opposite ends, resulting in a stack of metallized films connected in parallel. The structure is similar to that of a multilayer ceramic capacitor, except that the dielectric is much thinner. This compensates for the low dielectric constant of film dielectrics and makes the stacked film capacitor a contender for applications for which ceramic capacitors were previously used.

A wide variety of film materials are currently employed in film capacitors. Table 12.6 lists the properties of some common materials.

Paper and plastic capacitors for electronic applications are available in sizes from 1 nF to 10 μF and in voltage ratings from 50 V to several thousand volts. Tolerances are 10 and 20% for paper capacitors, down to 1% for some film types. The variation of capacitance with temperature varies from one dielectric to another, as indicated in Table 12.6.

Dissipation factors for film capacitors are less than 1% at 25°C. Films that have low DFs near room temperature generally have low DFs over the entire temperature range. DFs for paper and polyester dielectrics increase significantly at either extreme and can reach 2% for certain impregnants.

Insulation resistance decreases with temperature from the values shown in Table 12.6. The typical decrease is roughly three orders of magnitude from room temperature to 125°C.

Paper and film capacitors can be used at high frequencies, depending on size and length of leads. The self-resonant frequency of a 1 μF capacitor is about 1 MHz and increases about an order of magnitude as the capacitance decreases two orders of magnitude.

TABLE 12.6. TYPICAL PROPERTIES OF FILM CAPACITOR DIELECTRICS

Dielectric	Dielectric Constant	Dissipation Factor at 25°C (%)	Maximum Temperature (°C)	ΔC^a at − 55°C (%)	ΔC^a at Maximum Temperature (%)	Insulation Resistance at 25°C (MΩ × μF)
Paper	2.2	0.5	125	− 10	+4	40,000
Polyester	3.2	0.4	125	−5	+15	100,000
Polycarbonate	2.9	0.1	125	−2	+1	300,000
Polystyrene	2.5	0.05	85	+1	−2	500,000
Polysulfone	3.1	0.1	150	−2	+1	100,000
Polypropylene	2.1	0.05	105	−6	+2	100,000
TFE[b]	2.1	0.02	150	+2	−2	5,000,000

[a] ΔC is the change from room temperature value; it is altered by the use of impregnants.
[b] Polytetrafluorethylene.

12.2-7 Mica and Glass Capacitors

Applications of mica and glass dielectric capacitors are those that require high Q and excellent stability with respect to temperature, aging, and frequency. They are used in high-frequency coupling, bypass, and tuning circuits where outstanding performance and high reliability are required.

Mica capacitors are made by screening a silver paste onto thin mica sheets. After being fired the sheets are stacked with foils, alternately extended beyond the mica on opposite ends of the stack. The foils are folded over the ends, then compressed and clamped by the lead assembly. The capacitors are dipped repeatedly into phenolic or epoxy resins or are molded in phenolic or polyester.

Glass capacitors are made from a potash lead glass, drawn into a thin ribbon, which is then stacked with alternate layers of aluminum foil. Alternate foils are welded to lead wires, cover glass is added, and the entire assembly is sealed in vacuum at high temperature and pressure.

Mica capacitors are available in sizes from 1 pF to about 100 nF in tolerances as low as 0.5%. Voltage ratings range from 50 to 1000 V, with special construction up to 35 kV. Capacitance values for glass capacitors range from 0.5 pF to 10 nF with tolerances of 1 to 10%. Voltage ratings are either 300 or 500 V.

Both mica and glass capacitors are stable with respect to temperature. For some values of capacitance the temperature coefficient can be zero. The range extends from -20 to ± 200 ppm/°C.

The DF is typically between 0.01 and 0.1% for medium-size capacitors. For small capacitances it may reach 1%; it is generally larger for mica than for glass capacitors. The DF increases with temperature, roughly an order of magnitude over the full temperature range. It also increases with frequency. Insulation resistance at room temperature is typically 10^3 MΩ × μF for mica capacitors, and about 10^6 for glass capacitors.

Both types of capacitor are capable of high-frequency operation. The self-resonant frequency is about 10 MHz for large capacitor values and is well over 100 MHz for smaller values.

12.2-8 Selection of Capacitors

Performance is the important criterion for selecting capacitors for particular applications, but availability and price must also be considered. "Available on special request" usually means long

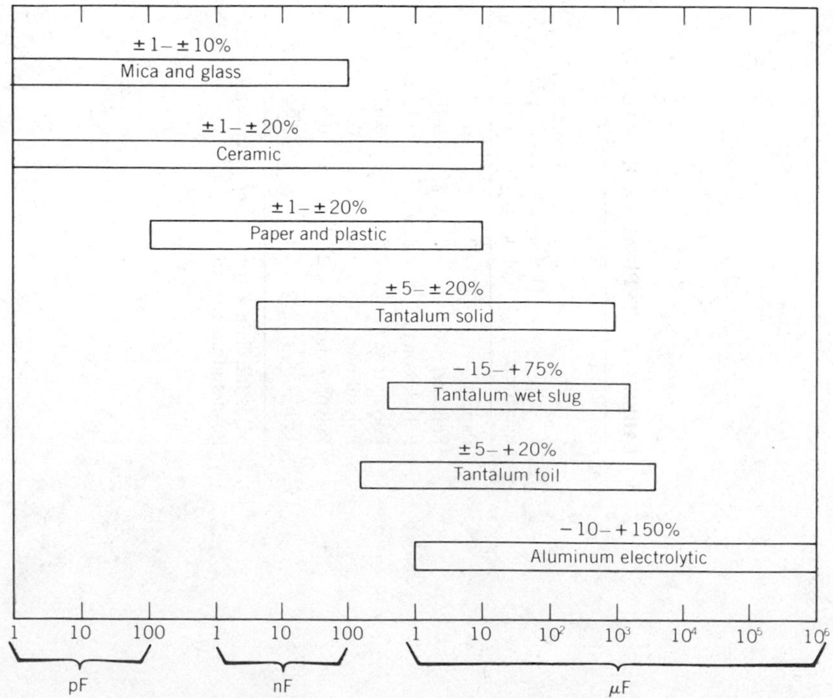

Fig. 12.15 Range of capacitance values and tolerances for different capacitor types.

delays, limited quantities, no second source, and premium prices. General performance characteristics are difficult to specify because of continued changes in product lines in response to improved technologies and changes in markets. There is also a wide overlap in specifications among the various families of capacitors. For these reasons capacitor selection is sometimes difficult.

Figure 12.15 shows the range of capacitance values and tolerances for different kinds of capacitors. In mid-range sizes there are several families to choose from. However, such factors as peak current, AC ripple current, and nonpolarized vs. polarized requirements may limit the choice.

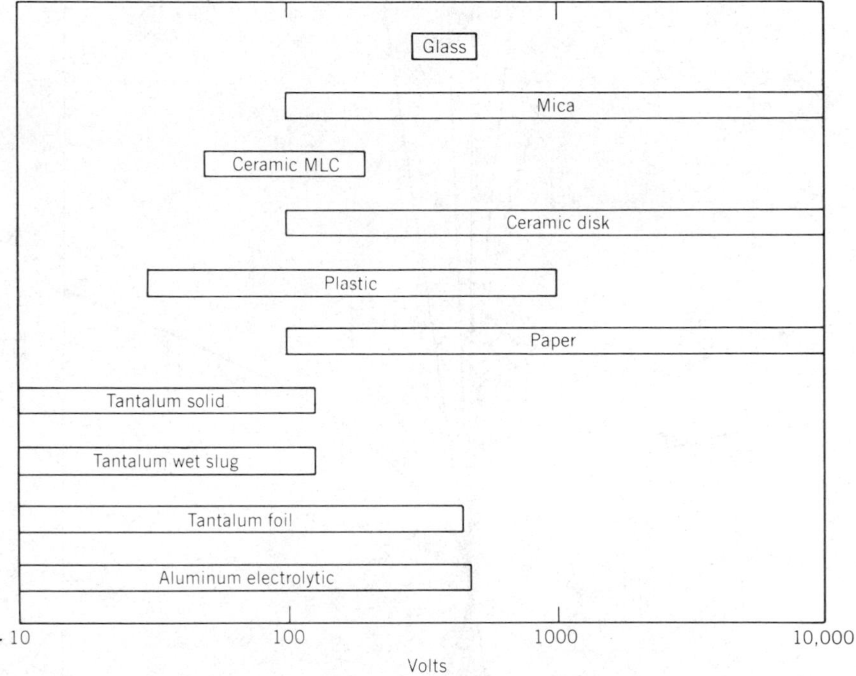

Fig. 12.16 Voltage ratings for different capacitor types.

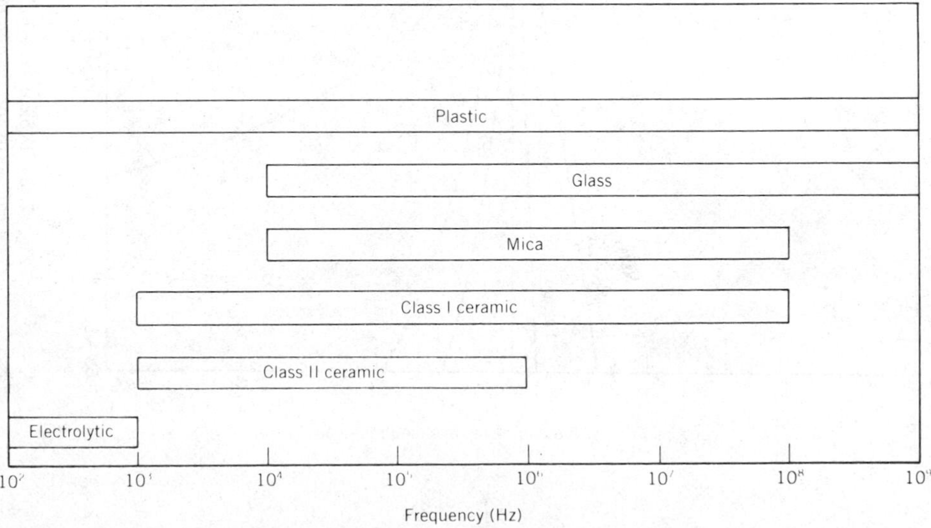

Fig. 12.17 Useful frequency range for capacitor families.

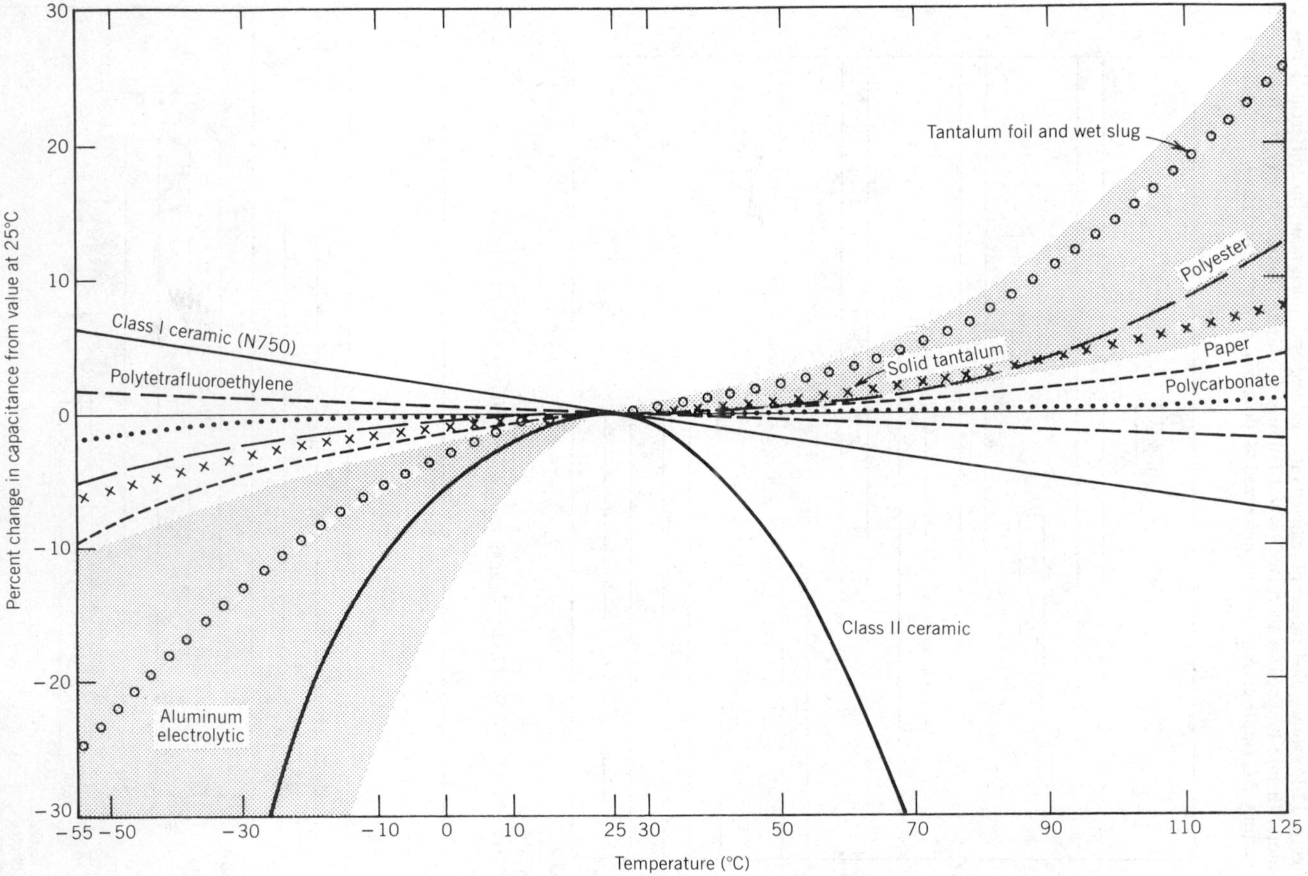

Fig. 12.18 Typical variation of capacitance with temperature. Shaded area shows the wide variation that can occur for aluminum electrolytics with different capacitance values, voltage rating, and type of construction. Some other types of capacitors exhibit similar variation.

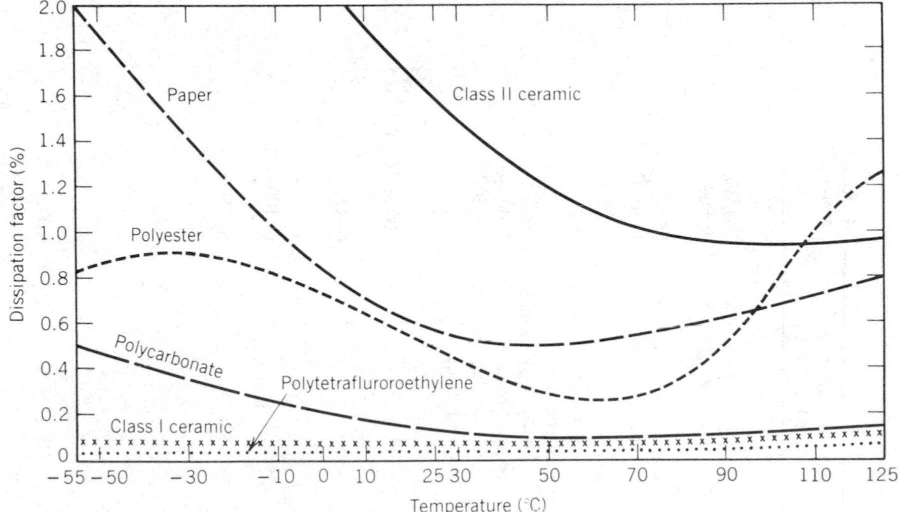

Fig. 12.19 Typical variation of dissipation factor with temperature.

Voltage ratings of different families are shown in Fig. 12.16. Large capacitance values are usually not available in the highest voltage ratings, and high voltages may involve large dimensions or different style packages. The voltage rating does not necessarily correlate with the actual breakdown voltage. Small capacitance values require impractically small electrode areas unless multiple layers of dielectric are used, but the manufacturer may specify the same working voltage as that of other capacitors in the same product line. It should also be emphasized that the reliability of a capacitor increases as the voltage is reduced.

The useful frequency range of capacitor families is given in Figure 12.17. The upper frequency is limited by the self-resonant frequency (which depends in part on lead length), equivalent series resistance, and decrease in capacitor value.

Variations in capacitance with temperature are shown in Fig. 12.18. These are representative values only since they depend on capacitance size, voltage rating, type of electrolyte or impregnant, and so on. Figure 12.19 shows the variation of dissipation factor with temperature. Again these are representative values, intended for a general comparison.

A summary of capacitor characteristics is given in Table 12.7. Reliability data are available in Ref. 3. For other detailed characteristics and application information consult the manufacturer's literature.

12.2-9 Integrated Circuit Capacitors

Silicon monolithic integrated circuits employ three types of capacitors: those made of a heavily diffused layer, a silicon dioxide layer, and an aluminum electrode; those that utilize a reverse-biased *pn* junction; and those that rely on parasitic capacitance and the gate input capacitance of MOS transistors. The last type is used in semiconductor memories and dynamic logic.

Figure 12.20 shows a cross-section view of an oxide capacitor. The *n*-type emitter diffusion forms one plate of the capacitor and the aluminum metalization forms the other. To avoid extra processing

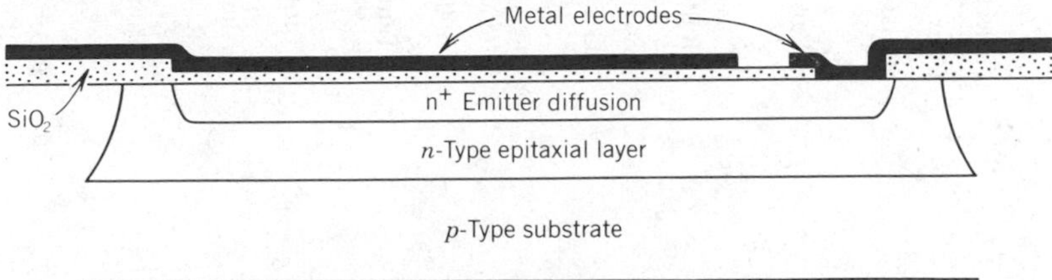

Fig. 12.20 Cross section of an oxide capacitor.

TABLE 12.7. CHARACTERISTICS OF DIFFERENT TYPES OF CAPACITORS

Dielectric	Applicable Specification	Capacitance: Range	Capacitance: Tolerance	Capacitance: Stability after 2000 Hr Life Test	DC Rated Voltage (V)
Glass					
Fixed	MIL-C-23269 (ER)	0.5–10,000 pF	0.25 pF–5%	0.5%–0.5 pF (whichever is greater)	100, 300, & 500
Variable	MIL-C-14409	0.6–1.8 pF thru 1–120 pF	. . .	Cap. change vs. rotation: $< 10\%$	250–1250
Mica					
Button style	MIL-C-10950	5–2400 pF	± 1, ± 2, ± 5, or $\pm 10\%$	$\leqslant 1\%$ or 0.5 pF (whichever is greater)	500
General purpose	MIL-C-5	47–27,000 pF	± 1, ± 2, or $\pm 5\%$	$\leqslant 5\%$ or 1 pF (whichever is greater)	300–2500
	MIL-C-39001 (ER)	1–91,000 pF	0.5 pF–$\pm 5\%$	$\leqslant 1\%$ or 1 pF (whichever is greater)	50–500
Electrolytic					
Aluminum	MIL-C-62	1–1000 μF	-10, $+50\%$	$\pm 15\%$	400 & 450
Tantalum (nonsolid)	MIL-C-39006 (ER)	0.5–1200 μF	-15: $+30$, $+50$, $+75$, $\pm 5\%$ to $\pm 20\%$	$\leqslant 15\%$	6–450
Tantalum (solid)	MIL-C-39003 (ER)	0.0023–330 μF	± 5, ± 10, or $\pm 20\%$	$\leqslant 2\%$	6–100
Aluminum oxide	MIL-C-39018	0.68–220,000 μF	-10: $+5$, $+30$, $+50$, $+75\%$	$\leqslant 15\%$	5–350
Tantalum (solid) chip	MIL-C-55365 (ER)	0.001–4700 μF	± 5, ± 10, or $\pm 20\%$	. . .	2–50
Paper					
Wax-impregnated	MIL-C-12889	0.01–0.5 μF	$+0$, -30 or $+50$, -10%	$\leqslant 10\%$	100: 250 & 500 (AC-DC)

Paper-Plastic					
Polycarbonate	MIL-C-19978 (ER)	0.001–1 μF	± 5 or ± 10%	⩽ 6%	50–600
Metallized	MIL-C-39022 (ER)	0.022–10 μF	± 5 or ± 10%	⩽ 10%	50–400
Paper & polyethylene terephthalate	MIL-C-19978 (ER)	0.001–1 μF	± 2, ± 5, or ± 10%	⩽ 6%	200–1000
Plastic or metallized plastic	MIL-C-55514 (ER)	0.001–10 μF	± 1, ± 2, ± 5, or ± 10%	⩽ 5%	50–600
Polyethylene terephthalate	MIL-C-19978 (ER)	0.001–10 μF	± 2, ± 5, or ± 10%	⩽ 6%	30–1000
Ceramic					
Fixed, general purpose	MIL-C-11015	2.2–2,200,000 pF	0.5 pF, ± 10, ± 20%	⩽ 10% or ⩽ 20%	50–1600
	MIL-C-39014 (ER)	10–1,000,000 pF	± 10, ± 20%	⩽ 20%	50–1600
Temperature compensating	MIL-C-20 (ER)	1.0–68,000 pF	0.25 pF, ± 1, ± 2, ± 5, ± 10%	± 3% or 0.5 pF (whichever is greater)	50–200
Variable	MIL-C-81	1.5–7 thru 15–60 pF	. . .	. . .	200–500
Fixed, chip	MIL-C-55681 (ER)	10–180,000 pF	± 5, ± 10, or ± 20%	. . .	50 & 100
Gas or Vacuum					
Fixed	MIL-C-23183	6–1000 pF	± 10%	. . .	30 kV
Variable	MIL-C-23183	5–3000 pF	± 10%	. . .	2 & 3 kV

TABLE 12.7. (*Continued*)

Temperature		Relative Cost for Equiv. CV Rating[a]	Relative Size		Dissipation Factor (%)		
Operating Temperature (°C)	Temperature Coefficient (% or ppm/°C)		Varies as	For Equiv. CV Rating	60 Hz	1000 Hz	1 MHz
− 55– + 125	140, 105, 0 ± 25	Medium	CV^2	Large	. . .	< 0.001	. . .
− 55– + 125 or + 150	± 50, ± 100, ± 150, and + 50, − 0	Medium high	C	Large	. . .	. . .	. . .
− 55– + 85 or + 150	± 100, − 20– + 100, and not specified	Medium high	CV^2	Large	. . .	< 0.17	< 1.2
− 55– + 125 or + 150	0– + 70, − 20– + 100 ± 100, ± 200, and not specified	Medium low	CV^2	Large	. . .	< 0.18	< 0.12
− 55– + 125 or + 150	0– + 70, − 20– + 100, ± 200, and not specified	Medium low	CV^2	Large	. . .	< 0.1	< 1000
− 40– + 85	Capacitance drops from 30–60% at − 40°C	Medium	CV	Very small	. . .	15 to 18% at 120 Hz; varies with V	. . .
− 55– + 85, derated to + 125	Capacitance drops from 12–50% at − 55°C	High	CV	Very small	. . .	10–32% at 120 Hz; varies with C and V	. . .
− 55– + 85, derated to + 125	Capacitance drops 10% max at − 55°C	Medium	CV	Very small	. . .	3–8% at 120 Hz; varies with V	. . .
− 40– + 85, derated to + 125	. . .	Medium	CV	Very small	. . .	10–35% at 120 Hz; varies with C and V	. . .
− 55– + 125	. . .	Medium	CV	Very small	. . .	0.6–15% at 120 Hz; varies with V	. . .
− 55– + 85	Capacitance drops < 30% at − 55°C	Medium	CV^2	Small	. . .	< 1	. . .
− 55– + 85 or + 125	Capacitance drops < 10% at − 55°C	Medium	CV^2	Small	. . .	< 1	. . .

−55–+125	Capacitance change ±2% at −55°C	High	CV^2	Large	< 0.1	< 0.1	Higher
−65–+125	± 10%	High	CV^2	Medium large	< 0.1	< 0.1	. . .
−55–+85 or +125	. . .	Medium	CV^2	Small	. . .	< 2.0	. . .
−65–+85	− 7 to +5%	High	CV^2	Small	< 0.6	< 0.6	. . .
−55–+85 or +125	Capacitance change < +30, − 80% at −55°C	Very low	$CV^2 + k$	Small	. . .	< 2.5	< 2.5
−55–+85 or +125	Capacitance change < +30, − 80% at −55°C	Very low	$CV^2 + k$	Small	. . .	< 2.5	< 2.5
−55–+125	0. ± 30, 0 ± 60	Very low	$CV^2 + k$	Small	. . .	0.15	< 0.10
−55–+85	Capacitance change < −4.5, + 2% at −55°C	Medium low	$CV^2 + k$	Large	. . .	. . .	0.2
−55–+125	0 ± 30 or 0 ± 15	Low	CV	Small	. . .	< 2.5	< 2.5
−55–+85	. . .	High	. . .	Large	. . .	< 0.001	. . .
−55–+85	. . .	High	. . .	Large	. . .	< 0.001	. . .

Source. Reprinted from *MIL-STD-198D*.
[a]Where "C" = capacitance and "V" = voltage.

steps the oxide layer is the same as that covering emitter diffusion regions. Occasionally a thin layer of silicon nitride (Si_3N_4) is added to improve stability and to provide a somewhat higher capacitance per unit area. The capacitance is given by Eq. (12.8).

The dielectric constant for silicon dioxide is about 3.9, depending on growth conditions; for silicon nitride it is about 7. For a typical emitter oxide thickness of 0.12 μm, the capacitance per unit area is about 3×10^{-4} pF/μm^2.

The oxide capacitor is a fairly good capacitor with a temperature coefficient of roughly 15 ppm/°C, fairly low dissipation factor, good tolerances, low voltage coefficient, and nonpolarized construction (however, one terminal is connected to the p-type substrate through a pn junction). The main disadvantage is the large chip area required, which limits sizes to about 100 pF. The main application is for frequency compensating capacitors in linear integrated circuits.

The capacitance of a reverse biased pn junction is also given by Eq. (12.8). The plate separation d is the depletion layer thickness. For an abrupt pn junction d is given by

$$d = \left\{ \frac{2\epsilon(\phi/q + V_R)(N_A + N_D)}{qN_A N_D} \right\}^{1/2} \tag{12.11}$$

where $\epsilon = 11.9\ \epsilon_0$ is the permittivity of silicon, q is the electron charge (1.6×10^{-19} C), V_R is the reverse applied voltage, and N_A and N_D are the acceptor and donor concentrations on each side of the junction. ϕ/q is the barrier potential given by

$$\phi/q = \frac{kT}{q} \ln\left(\frac{N_A N_D}{n_i^2} \right) \tag{12.12}$$

k is Boltzmann's constant, T is the absolute temperature, and n_i is the intrinsic carrier concentration (1.5×10^{10}/cm^3 at room temperature). ϕ/q is a slowly varying function of impurity concentration and is about 0.65 V at room temperature.

Equation (12.11) predicts that the capacitance is proportional to the square root of the doping concentration. Hence more heavily doped junctions yield the highest capacitance. Capacitance is also inversely proportional to the square root of reverse voltage for $V_R > \phi/q$, a feature that is utilized in voltage-controlled electronic tuning.

Typical values of capacitance per unit area range from about 1×10^{-3} to 1×10^{-4} pF/μm^2. Unfortunately these capacitors have high dissipation factors, and furthermore can only be operated with reverse bias. In addition the capacitance is limited to small values to avoid large junction areas. It should also be noted that junctions in integrated circuits are diffused junctions, whose behavior is not necessarily approximated by Eq. (12.11). For more exact expressions consult standard texts on semiconductor physics.

Capacitors for thin-film integrated circuits are made from silicon monoxide or tantalum oxide. Silicon monoxide ($\epsilon_r = 6$) is evaporated onto an aluminum layer which is the bottom electrode. A second layer of aluminum on top of the dielectric serves as the other electrode. Tantalum capacitors are made by sputtering a layer of pure tantalum onto a substrate. This layer can be used for thin-film resistors as well. The tantalum layer is then partially anodized to form Ta_2O_5 ($\epsilon_r = 22$). A second layer of metal is evaporated to form the other electrode and interconnection pattern. The capacitor so formed is polarized, and the bottom plate must have a positive polarity with respect to the top.

Thin-film capacitors have a capacitance per unit area of 10^{-3} to 10^{-4} pF/μm^2, a temperature coefficient of 100 to 200 ppm/°C, and a dissipation factor of 0.001 to 0.01. Although the initial tolerance is 10% or so, the values can be trimmed to less than 1%.

Various dielectric compositions are also available for screen-printing thick-film capacitors. However, the dielectric thickness is too great for high values of capacitance unless materials with high dielectric constants are used. It is more common to use discrete multilayer ceramic or tantalum chip capacitors.

References

1 "Selection and use of Capacitors," *MIL-STD-198D* (Nov. 8, 1976).

2 *Parts Application and Reliability Information Manual for Navy Electronic Equipment*, NAVSEA 0967-LP-597-1011 (Oct. 1980).

3 "Reliability Prediction of Electronic Equipment," *MIL-HDBK-217C* (Apr. 9, 1979).

4 Gerald L. Ginsburg, *A User's Guide to Selecting Electronic Components*, Wiley, New York, 1981.

5 Charles A. Harper, ed., *Handbook of Components for Electronics*, McGraw-Hill, New York, 1977.

6 *Component Technology and Standardization Manuals*, General Electric Co., 1978

12.3 INDUCTORS

Joseph Scaturro

12.3-1 Introduction

Inductance may be characterized, in a general way, as the property of a circuit element by which energy is capable of being stored in a magnetic field. However, it only has significance in an electric circuit when the current is changing with respect to time. When current increases or decreases, the effect that opposes this change is called inductance (L) or self-inductance (L_s). Inductance, therefore, is caused by a changing magnetic field produced by a varying current. The inductance in henrys (H) per coil turn may be expressed as

$$\frac{L_s}{N} = \frac{d\phi}{di}$$

where ϕ = magnetic flux, webers
i = current, amperes
N = number of turns

Generally, inductive components are unique compared with resistors and capacitors, which are available as standard items, in that they are usually designed for a specific application. However, a wide variety of inductors have recently become available as stock items because of the trend toward miniaturization. Inductors with low inductive values are generally wound on noninductive forms (air or phenolic cores), those with intermediate values on powdered iron cores, and those with higher values on ferrite cores.

12.3-2 Inductance of Simple Configurations

An ideal inductor is a coil wound with resistanceless (or lossless) wire. When motion of charge exists in the wire, energy is stored in the magnetic field around the coil. A direct current (DC) steady state or constant current through the coil wire results in zero voltage across the coil. A current that varies with respect to time produces an induced electromotive force, or voltage. This emf results because of the interaction of voltage, current, and magnetic field. The expression for voltage is

$$v = L\frac{di}{dt}$$

where v = voltage, volts
i = current, amperes
L = inductance, henrys
t = time, seconds.

Note that an abrupt change in current in zero time is impossible in a circuit-containing inductance, since an infinite voltage is required. The schematic representation of inductance is shown in Fig. 12.21. The unit equation for inductance is

$$1\text{H} = 1\,\frac{\text{V}\cdot\text{s}}{\text{A}}$$

Inductance may also be expressed as the slope of the curve when voltage is plotted versus the change in motion of charge (or charge acceleration).

Most inductors consist of a coil of wire with a core appropriately selected for the particular application. However, every circuit exhibits some inductance (stray inductance), since motion of charge through any conductor, including a straight wire, results in a magnetic field around it. In general, most coil shapes do not lend themselves to simple mathematical computations. Empirical formulas, tables, and nomograms are utilized as aids in determining inductance (or self-inductance). Empirical equations for some simple configurations are as follows:

1. Single layer coils or solenoids, where the length is greater than one half the diameter and the core is of nonferromagnetic material

$$L = \frac{\mu_0 N^2 A}{l + 0.45d} \text{ (henrys)}$$

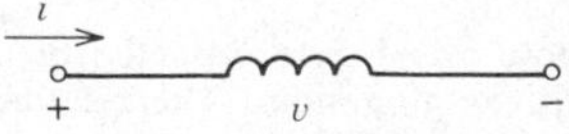

Fig. 12.21 Reference directions for inductor current and voltage.

where N is the number of turns, A is the cross-sectional area in square meters, l is the length in meters, d is the diameter of the coil in meters, and μ_0 is the constant $4\pi 10^{-7}$, known as the permeability of free space.

2. The low-frequency inductance of the toroidal coil of N turns with a winding circular cross section of diameter d in inches and a mean radius of core revolution $D/2$ in. is

$$L_s = 0.01595 N^2\left[D - (D^2 - d^2)^{1/2}\right]10^{-6}\text{ (henrys)}$$

12.3-3 Air Core Coils

Air core coils are usually single or multilayer types wound on cylindrical forms. Toroids, flat spirals, and single or multilayer windings on square hexagonal forms are occasionally used. The inductance of many types of air core coils may be calculated by empirical formulas involving the dimensions of the coil and the number of turns (as shown in Section 12.3-2). These relationships may be found in US Department of Commerce Circular of the National Bureau of Standards C74. However, the inductance of an air core coil at high frequencies cannot be accurately calculated because of skin effect and distributed capacitance of the coil.

Single-layer air core coils (or solenoids) are used for resonant circuits at broadcast (535 khz–1605 kHz) and higher frequencies and for radio frequency chokes at very high frequencies (30 MHz–300 MHz). (For large values of inductance in a small space, multilayer coils are used.)

Air core coil losses may be expressed in a number of ways: as the equivalent series resistance of the coil, as the power factor of the coil, or in terms of the quality factor (Q).

The equivalent series resistance is usually greater than the DC resistance except at extremely low frequencies. At higher frequencies the current distribution over the cross section of the conductor is modified such that an increasing current density concentration occurs in the surface layers of the conductors; it occurs because the reactance associated with the possible current paths is smallest near the conductor surface. This phenomenon is known as skin effect and is equal to the $\sqrt{\sigma\pi\mu f}$, where σ is conductivity, f is the frequency of the current, and μ is the permeability of the medium. This equation determines the depth at which current density has decreased to $1/\epsilon$ (37%) of its value at the surface of the conductor. The proximity of the conductors in the coil may also affect current distribution. Dielectric losses occur primarily in the coil form. Eddy current losses may exist in surrounding metallic objects. The effects of these losses as frequency increases result in only a small change in quality factor Q, which is the ratio of inductive reactance to effective resistance ($2\pi fL/R$) with variations in frequency.

The power factor of an inductor is the ratio of resistance to impedance of the coil

$$pf = \frac{R}{\sqrt{R^2 + (\omega L)^2}}$$

Dividing through by R

$$pf = \frac{1}{\sqrt{\left(\frac{\omega L}{R}\right)^2 + 1}}$$

But $Q = \omega L/R$ and therefore

$$pf = \frac{1}{\sqrt{Q^2 + 1}}$$

If Q is > 5, the power factor may be assumed to be equal to the reciprocal of Q.

12.3-4 Magnetic Materials (See Also Chapter 9)

In terms of magnetic characteristics, the vast majority of materials are either diamagnetic or paramagnetic. The relative permeability of diamagnetic materials such as silver and copper is slightly less than unity. Paramagnetic materials such as aluminum and platinum have a relative permeability slightly greater than unity. Relative permeability (μ_r) is the ratio of the absolute permeability (μ) with respect to the permeability of free space (μ_0). The unit for permeability is henrys per meter; in the SI system of units $\mu_0 = 4\pi \times 10^{-7}$.

The materials of most significance in electrical engineering are those called ferromagnetic; they are principally composed of iron and its alloys with nickel, cobalt, or aluminum. The relative permeability of ferromagnetic materials is many times greater than that of free space. The differences in the magnetic characteristics of materials may be attributed to the manner in which a magnetic field affects the electron spins, the proton spins, and the orbital motions of the electrons of each atom of the material. In ferromagnetic materials the spins are aligned parallel to each other. The material is divided into magnetic domains, each having a net magnetization even without an external field. An overall net magnetization, however, will not exist, since the magnetization in the various domains will cancel each other. Application of a small magnetic field causes growth of favorable domains, resulting in high magnetization. These materials become paramagnetic above a critical temperature known as the Curie temperature.

The permeability of ferromagnetic materials is not constant. Permeability is the slope of normal magnetization curve, which is a plot of magnetic flux density (B) versus magnetic intensity (H). This curve is nonlinear, since saturation conditions exist in ferromagnetic materials. The most salient losses that exist in ferromagnetic materials are hysteresis loss and eddy current loss. Many applications involve application of AC. The varying magnetic flux that results induces emfs in the material, which in turn create eddy currents. The heating due to these currents is an energy loss, called eddy current loss. Reduction of eddy current loss is brought about by utilizing laminated sheets for cores so as to confine the currents within the relatively high electrical resistivities. As the material is magnetized during each half cycle, the amount of energy stored in the magnetic field exceeds that which is released upon demagnetization. The difference between the two energy densities may be determined by integrating the areas of a hysteresis loop representing the energy stored in the magnetic field and the energy released. The difference represents the energy that is not returned to the source but dissipated as heat as the domains are realigned in response to the changing magnetic field intensity. This dissipation of energy is called hysteresis loss. The sum of all losses in a core is termed total core loss.

Magnetic materials may be designated nonretentive soft materials and retentive hard materials. The nonretentive materials are low-loss materials used in motor, generator, and transformer cores; electromagnetic devices; and memory cores. Nonalloyed iron is used for low-frequency applications, while ferrites, permalloy, silicon iron, and garnets are used for high-frequency applications. Purified iron is used for DC applications such as relays and other electromagnetic cores.

Some manufacturers of magnetic materials are Allen-Bradley Co., 1201 S. Second St., Milwaukee, WI 53204; Ceramic Magnetics, Inc., 87 Fairfield Rd., Fairfield, NJ 07006; Fair-Rite Products Corp., Wallkill, NJ 12589; Ferronics Inc., 69 North Lincoln Rd., East Rochester, NY 14445; Ferroxcube Div., Amperex Electronic Corp., 5083 Kings Hwy., Saugerties, NY 12477; Indiana General Div., EM & M Corp, Crows Mill Rd., Kensby, NJ 08832; Krystinel Corp., 126 Pennsylvania Ave., Paterson, NJ 07509; Magnetics Div., Spang Industries, PO Box 391, Butler, PA 16001; and Stackpole Carbon Co., Electronic Components Div., Stackpole St., St. Mary's, PA 15857. Brochures, application information, and magnetic design manuals are available upon request.

12.3-5 Iron-Core Inductors

Iron cores have higher permeabilities than does air, thereby concentrating magnetic flux. This effect increases the inductance of winding, which reduces the cost because the number of turns and therefore the winding time required for a given value of inductance are reduced. An increase in the value of Q also results, since the resistance is obviously less. Therefore, for AC applications where high values of inductance are required, the windings are usually placed on laminated iron cores and the turns are wound as close to the core as possible to achieve the best results.

The calculation of inductance for an iron-core coil including an air gap subjected to an AC corresponding to a maximum flux density (B_M) in the iron is

$$L = \frac{1.256 N^2}{\dfrac{l_i}{\mu a_i} + \dfrac{l_a}{a_a}} \times 10^{-8} \text{ (henrys)}$$

where N = number of turns
- l_i = mean length of magnetic circuit in the iron, cm
- l_a = mean length of magnetic circuit in the air gap, cm
- μ = permeability (H/cm) of magnetic material evaluated at appropriate maximum alternating flux density B_{max}
- a_i = cross-sectional area (cm^2) of iron (not including area of insulation between laminations; stacking factor 0.9 times actual area is sometimes used)
- a_a = effective cross-sectional area (cm^2) of air gap (for small air gaps, length of air gap is added to each dimension affecting area)

For dimensions in inches

$$L = \frac{3.19N^2}{\dfrac{l_i}{\mu a_i} + \dfrac{l_a}{a_a}} \times 10^{-8} \text{ (henrys)}$$

For an air gap 0.1 in. and a 2 × 3 in. rectangle, the effective area = 2.1 × 3.1, or 6.51 in.^{2}. This calculation corrects for fringing flux effect.

Toroidal inductors yield much higher inductance values than do axial lead or solenoid-type inductors. Since the core is round, leakage flux (flux in the air about the core) tends to be minimized and shielding requirements are reduced. In fact, the toroidal inductor is considered self-shielding. Toroids usually have high Qs, due to low core losses. They are, however, considerably more expensive than are axial lead types. For a single-layer toroid of rectangular cross section without an air gap

$$L = 0.0020N^2 b/\mu_d \ln(r_2/r_1)$$

where L = inductance, μH
b = core width, cm
μ_d = average incremental premeability
r_1 = inside radius, cm
r_2 = outside radius, cm
N = total number of turns

Some manufacturers of inductors are Dale Electronics Inc., Yankton Div., East Highway 50, Yankton, SD 57078; Bourns Inc. Trimpot Products Div., Magnetic Products, 28151 Hwy. 74, Romoland, CA 92380; Delevan Div., American Precision Industries, Inc., 270 Quaker Rd., East Aurora, NY 14052; TRW, VTC Transformers, 150 Varick St., New York, NY 10013; Litton TRIADUTRAD, 305 No. Briant St., Huntington, IN 46750; Thordarson, Electronic Center, Mt. Carmel IL 62863; Mecrotran Co. Inc., PO Box 236, Valley Stream, NY 11582.

12.3-6 Inductors for Integrated Circuit Applications

The accuracy of fabrication of integrated components is on the order of 10%. However, relative ratios between various components to approximately 3% can be maintained. For example, if two resistors are to be made and the ratio of resistance is to be 4:1, that ratio can be obtained to 3% accuracy even though there may be 10% errors in the values of resistance themselves.

Fabrication of integrated inductors has not been satisfactory. One of the limitations of the integrated circuit technology is the lack of integrated inductors. Therefore, whenever possible, inductors are avoided in the design of integrated circuits. In many instances the need for inductive elements can be eliminated through the use of a technique known as RC synthesis. If inductors with a Q greater than 5 μH are required, discrete inductors are used and connected externally to the silicon chip. The physical size of these inductors is usually much larger than the size of the chip.

The Delevan Division of American Precision Industries, Inc., fabricates microinductors. The micro "i" (Delevan trademark) inductors offer inductances of 0.010 μH through 1000 μH in various low-profile packages (0.075 in.) for use on thick-film hybrid packages. These parts have ribbon leads or reflow solder pads. Bourns Inc. Trimpot Products Division, Magnetic Products, also designs microminiature inductors in packages from 0.125 × 0.125 × 0.125 to 0.310 × 0.410 × 0.465 in. with printed circuit pins or ribbon leads.

12.3-7 Qualifying the Inductor Application

Qualifying an inductor application requires the following information:

1. The value of inductance and tolerance required.
2. The mechanical dimensions and tolerances.
3. The Q desired.
4. The current rating of the inductor.
5. The frequency at which the inductor will operate.

With reference to No. 3, the higher the Q, the better the performance of the inductor. Since high Q in most cases is desirable, Q is expressed as a minimum. With reference to No. 5, each inductor has a frequency above which it will not function in a desired manner. It appears capacitive above the self-resonant frequency (abbreviated SRF). Therefore, SRF is normally specified as a minimum. Normally in the production of an inductor, the SRF is maintained at its highest achievable level, thus broadening the usable frequency range of the inductor.

Bibliography

Angus, R. B., *Electrical Engineering Fundamentals*, 2nd ed., Wesley, Reading, MA, 1968.

Fink, D. G., et al., eds., *Electronics Engineers Handbook*, 2nd ed., McGraw-Hill, New York, 1982.

Langford-Smith, F., ed., *Radiotron Designers Handbook*, 4th ed., Wireless Press, Harrison, NJ, 1952. Reproduced and distributed by RCA Victor Division, Radio Corporation of America.

Terman, F. E., *Radius Engineers Handbook*, 3rd ed., McGraw-Hill, New York, 1943.

12.4 TRANSFORMERS

Joseph Scaturro

12.4-1 Coupled Circuits

"Coupled circuits" generally refers to magnetic coupling, that is, the effect of magnetic interaction between two parts of a circuit or between two separate and distinct circuits. In 1830 Faraday and Henry observed that if a coil of wire was placed in the vicinity of another coil carrying a varying current, a voltage was induced in it. This phenomenon may be expressed as

$$v = N\frac{d\phi}{dt}$$

The equation is known as Faraday's law and states that if a coil of wire of N turns is situated in a time-varying magnetic flux ($d\phi/dt$), a voltage (v) is induced in the coil. For example, a simple coupled circuit may consist of two coils. Assume the voltage applied to one coil is sinusoidal, $v = V_m \sin \omega t$. This voltage source results in sinusoidally varying current which, in turn, creates a time-varying magnetic flux. A portion of the flux from this coil passes through or links the other coil. Since the flux is varying with respect to time, a voltage is induced in that coil. The voltage induced is a function of the number of turns of the coil and the amount of varying flux linking the coil. It should be kept in mind that a voltage is also induced in the coil to which voltage is applied. The self-induced voltage opposes the applied or excitation voltage (Lenz's law).

The inductance of a coil due to the varying current in another coil is termed mutual inductance. Again assume a simple coupled circuit consisting of two coils, as shown in Fig. 12.22. In the figure, if

$$v = V_m \sin \omega t$$

then

$$L_1 = N_1\frac{d\phi_1}{di_1}$$

and

$$L_2 = N_2\frac{d\phi_2}{di_2}$$

where L_1 and L_2 = self-inductance of each coil, H
ϕ = flux, Wb
N = number of turns
i = current, A

In addition

$$M_{21} = N_2\frac{d\phi_{21}}{di_1}$$

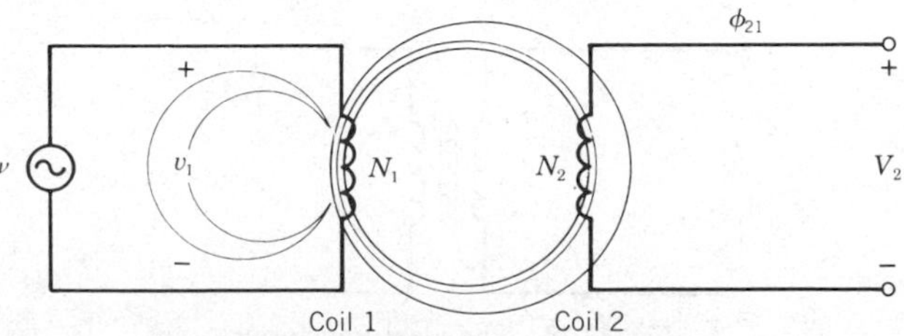

Fig. 12.22 Coupled coils.

where M_{21} is the mutual inductance of coil 2 due to the current in coil 1 and

$$M_{12} = N_1 \frac{d\phi_{12}}{di_2}$$

where M_{12} is the mutual inductance of coil 1 due to the current in coil 2.

When the permeability of the medium is constant

$$M_{21} = M_{12} = M$$

Coils may be either closely or loosely coupled. For example, two air core coils placed a few inches from each other are loosely coupled because only a small portion of the flux from coil 1 will link coil 2. If they are wound one over the other or if they are wound together on a ferromagnetic material, they are closely coupled. In commercial power transformers, the coils are so closely coupled that over 98% of the flux generated in one coil links the other. The coefficient of coupling K, which defines the magnetic proximity of two coils, is defined as

$$K = \frac{M}{\sqrt{L_1 L_2}}$$

where M is the mutual inductance of the coils in henrys and L_1 and L_2 are self-inductances. A coefficient of coupling of unity indicates maximum coupling; as previously mentioned, for power transformers $K \geqslant 0.98$. For loosely coupled coils, as may be used in radios, K might be $\leqslant 0.1$. The desired degree of coupling depends on the particular needs in a given circuit.

12.4-2 The Ideal Transformer

A transformer is a device that transfers energy from one circuit to another by electromagnetic induction. Mutual inductance, previously mentioned, describes the basic principle involved, that is, the circuits maintain their coupling in such a way that any changing current in the first coil, or primary, causes a changing flux that induces a voltage in the second coil, or secondary. When a load is connected to the second coil, this voltage causes a load current, or secondary current, which in turn creates a counterflux that causes the first coil current to increase in the attempt to supply more flux. Such action, called transformer action, causes energy to pass from the primary to the secondary through the medium of the changing magnetic field. A ferromagnetic core may be used to provide tighter coupling.

The analysis of the transformer is a difficult task unless assumptions or idealizations are initially made. The actual circuit model or equivalent circuit must include the resistance of the coils, leakage flux, core losses, and the fact that the core has a finite permeability. Therefore assumptions for the ideal transfomer are:

1. The resistance of the windings is negligible.
2. The energy lost or the core loss in the core material is negligible.
3. All the flux links every turn in each coil or the leakage flux is zero.
4. The core material has an infinite permeability or the magnetization curve (B vs. H) of the core material is the B axis (a magnetic short circuit).
5. The effects of capacitance are negligible.

The circuit representing the ideal transformer may be drawn as shown in Fig. 12.23. The secondary induced voltage is produced by the same flux as the primary induced voltage and is either in phase with or 180° out of phase with the primary voltage, depending on the manner in which the coils are wound on the core. To be specific in this regard, dots are placed near the transformer terminals in diagrams. The dots indicate that the induced voltage rises in the two windings are in phase when both are defined as rises from the unmarked to the dotted terminal.

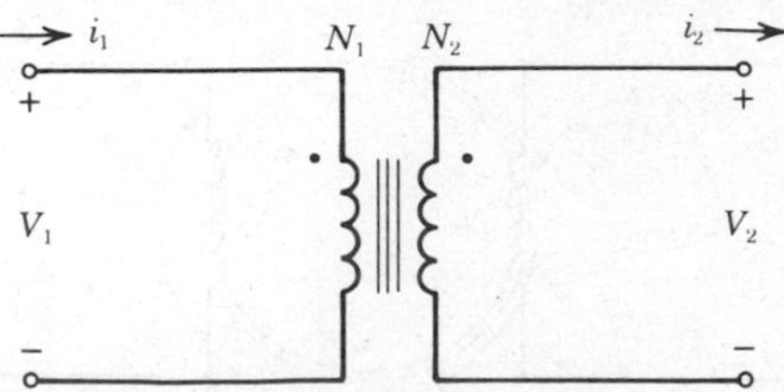

Fig. 12.23 The ideal transformer.

To analyze ideal transformer action, the law of conservation of energy is applied and is extended to include reactance as well as resistance

$$v_1 i_1 = v_2 i_2$$

$$v_1 = N_1 \frac{d\phi}{dt} \quad \text{and} \quad v_2 = N_2 \frac{d\phi}{dt}$$

$$\frac{v_1}{v_2} = \frac{N_1 (d\phi/dt)}{N_2 (d\phi/dt)}$$

or

$$\frac{v_1}{v_2} = \frac{N_1}{N_2} = a$$

where a is called the ratio of the transformation. Therefore for the ideal transformer the voltage across the windings is proportional to the number of turns in the winding.

Since

$$v_1 i_1 = v_2 i_2$$

then

$$\frac{v_1}{v_2} = \frac{i_2}{i_1} = \frac{N_1}{N_2}$$

or

$$\frac{i_1}{i_2} = \frac{N_2}{N_1} = \frac{1}{a}$$

That is, the current through the windings is inversely proportional to the number of turns in the winding.

If an impedance Z_2 is connected across the secondary terminals, then

$$Z_2 = \frac{v_2}{i_2}$$

but

$$Z_1 = \frac{v_1}{i_1}$$

$$\frac{Z_1}{Z_2} = \frac{v_1/i_1}{v_2/i_2} = \frac{v_1}{v_2}\frac{i_2}{i_1} = \frac{N_1}{N_2}\frac{N_1}{N_2}$$

or

$$\frac{Z_1}{Z_2} = \left(\frac{N_1}{N_2}\right)^2 \quad \text{or} \quad a^2$$

That is, the apparent or equivalent load impedance as seen from (or reflected through) the transformer windings is proportional to the square of the ratio of transformation (or the turns ratio).

12.4-3 Lossless Transformers

Power losses of actual transformers consist of copper or winding loss due to the resistance of each winding and core loss when the core material is ferromagnetic. Core loss results in heating of the ferromagnetic material. Therefore it may be simulated by an appropriate value of resistance in the circuit model (or equivalent circuit). Actual transformers also differ from ideal transformers in that the flux produed by one winding does not link all the turns of the second winding. That is, leakage flux exists primarily in air, which may be simulated by an appropriate value of inductance. Also cores have a finite permeability, and therefore a current is required to produce an mmf to set up a flux. The relationship between the current producing the flux and the induced primary voltage is 90°, where the current lags the voltage. Hence the element simulating this effect is an inductance in shunt with the primary winding (L_m).

If power losses are neglected, the circuit model for the lossless transformer, which includes the ideal transformer, may be as shown in Fig. 12.24.

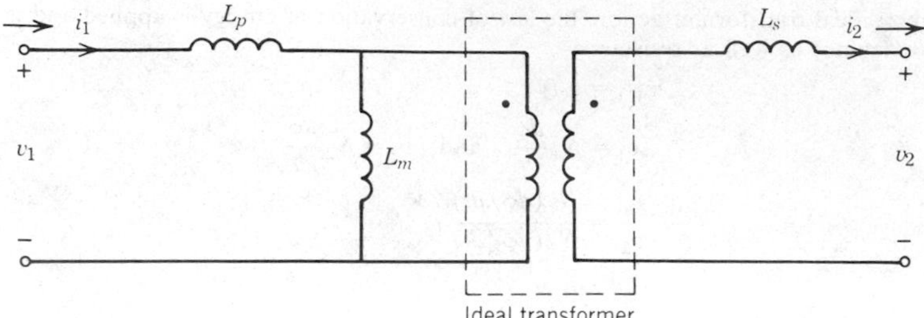

Fig. 12.24 Lossless transformer model.

12.4-4 Electronic Transformers

Power Transformers

Electronic power transformers usually operate at a single frequency. Generally the frequencies are 50, 60, or 400 Hz. In Europe 50 Hz is common; 400 Hz is the power frequency of most airborne equipment. Future airborne power transformers, however, will operate at higher frequencies to reduce weight and size.

Power transformer design considerations are controlled by the following specifications:

1. **Efficiency.** This is the ratio of power output to power input. This quality factor of power transformers is a function of core losses and copper losses
$$\%\ \text{Eff.} = \frac{P_0}{P_i}(100)$$
Typical efficiencies range from 70 to 98%. Larger transformers usually have higher efficiencies.
2. **Power factor.** The power factor is particularly important in large transformers where a great deal of power is consumed. The power factor is simply the cosine of the phase angle or the ratio of the actual or real power dissipated (watts) to the apparent power (volt amperes)
$$\text{Power factor (PF)} = \cos\theta = \frac{\text{power dissipated}}{\text{apparent power}} \qquad \left(\frac{\text{W}}{\text{VA}}\right)$$
3. **Temperature rise.** This is an important specification because it stipulates the operating temperature of the device. Power losses cause the temperature rise.
4. **Voltage regulation.** This is defined as the change in magnitude of the secondary voltage as the current changes from no load to full load, with the primary voltage held fixed. Regulation, therefore, has no significance if the transformer load is constant. However, when the load is changing the voltage across the load is expected to remain within certain limits, regulation becomes an important design consideration.
5. **Phase shift.** Phase shift is an important factor in the design of reference transformers. The phase shift allowable for specific conditions is stated in many specifications. Phase shift is a function of the DC resistance of the primary winding, the leakage inductance, and the impedance of the driving generator. Generally a minimum phase shift is desirable.
6. **In-rush current.** This is a function of voltage switch or value. Peak currents many times the normal excitation current can result. This current is usually of a transient nature and lasts for only a few cycles of the power frequency. Then it stabilizes to the normal excitation current.
7. **Leakage inductance.** As previously mentioned, leakage inductance does not represent a power loss. It consists of lines of magnetic force that do not cut or link any turns of the winding and hence do not produce a usable voltage. Leakage inductance affects voltage regulation. The greater the leakage inductance, the poorer the regulation.
8. **Core losses and corona.** These are also factors to be considered when designing transformers (see Section 12.4-5).

Audio Transformers

Audio transformers are used for voltage current and impedance matching over a nominal frequency range of 20 to 20,000 Hz. They may also be used for other auxiliary functions, such as offering a path for direct current through the primary while at the same time keeping it out of the secondary circuit.

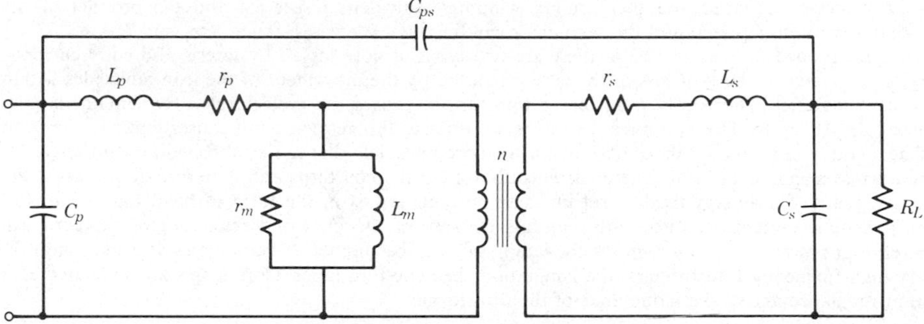

Fig. 12.25 Broadband transformer model.

The span of the frequency response is the prime consideration in the design, size, and cost of audio transformers. In wideband audio transformers, the frequency coverage may be separated into three independent ranges for the purpose of analysis. In the high-frequency range, the leakage inductance and distributed capacitance (C_p, primary shunt and distributed capacitance; C_s, secondary shunt and distributed capacitance; and C_{ps}, primary to secondary capacitance) have the most significance. In the low-frequency range, the open circuit inductance (L_m) is important. At the medium-frequency range, about 1000 Hz, the effect of the transformer on frequency response may be neglected. The equivalent circuit of the broadband transformer with a resistive load is shown in Fig. 12.25.

Miniaturized audio transformers have excellent high-frequency response because as the size of the transformer decreases, the leakage inductance and distributed capacitance of the windings also decreases. However, the small size of these transformers results in increased cost and in degradation of the low-frequency response (which is dependent on the primary open circuit inductance, L_m).

Pulse Transformers

The pulse transformer is a magnetic component designed to generate or transfer a pulse of electrical energy possessing specific electrical characteristics. Basically the pulse is one of two types: high power or low power. Low-power pulse transformers are generally used in triggering and coupling circuits. In a triggering circuit, the pulse transformer is used either to initiate another function or to form another pulse in a regeneration-type circuit. In coupling circuits, the pulse transformers are used for circuit optimization by impedance matching, phase reversal, or isolation. High-power pulse transformers (above 300 W peak) are usually used in modulators for radar systems. Their function is to provide impedance matching between the pulse-forming network and the magnetron. The prime concern is the transformation of the pulse with a minimum of distortion.

The specifications for a pulse transformer should include the amplitude, usually expressed in terms of volts into a specific load; pulse width; overshoot, expressed in voltage or percent of 100% pulse; and droop, expressed in volts or percent of 100% pulse.

Broadband Radio-Frequency (RF) Transformers

RF transformers provide a low-cost, simple, and compact means for impedance transformation at higher frequencies. Bifilar windings (a form of noninductive winding) and powdered iron or ferrite cores provide optimum coupling. The use of high permeability cores at lower frequencies reduces the number of turns needed and the distributed capacitance. At the upper frequencies the reactance increases, even though the permeability may decrease.

Double-Tuned Transformers

The double-tuned transformer is one of the most widely used circuit configurations for intermediate-frequency systems in the frequency range of 250 kHz to 50 MHz. It consists of a primary and secondary tuned to the same frequency and coupled inductively to a degree dependent on the desired shape of the selectivity curve. The coils may be undercoupled, critically coupled, or overcoupled.

12.4-5 Transformer Losses

Transformer losses consist of copper or winding losses and core losses. The winding losses are due to the resistance of the primary and secondary windings. They may be expressed as the product of the primary current squared and the primary winding equivalent resistance and the product of the secondary current squared and the secondary winding equivalent resistance, $I_1^2 r_1$ and $I_2^2 r_2$.

As mentioned in Section 12.3-4, there are two types of core losses, hysteresis and eddy currents. Hysteresis refers to the heat loss in the core generated by the movement of the iron molecules within the core material. The AC flowing in the transformer causes the molecules to reorient themselves during an AC cycle. The resistance the molecules offer to this reorientation causes losses in the form of heat. Eddy currents are those currents in the core materials that are established at right angles to the normal magnetizing field. These currents do not cut the coil turns and therefore do not assist the normal fields. The energy used to set up these currents is lost in the form of heat. Laminations of ferromagnetic materials reduce eddy currents because of the increased resistance of the core. An increase in power frequency requires the laminations to be thinner to keep losses at a minimum. In very-high-frequency transformers, the laminations become thin tapes. Core losses are simulated as a shunt resistance across the inductance of the transformer.

12.4-6 Corona

Corona is a high-voltage phenomenon in which the air surrounding a high-voltage terminal or wire is ionized. This ionized air appears to glow and makes audible crackling sounds. Ionized air or corona also causes RF noise. Corona is a destructive agent to electrical insulating materials. Prolonged periods of exposure to corona can lead to breakdown.

12.4-7 Failure Modes and Mechanisms—Transformers and Inductors

The primary causes of failure in transformers and inductors may be attributed to the grade of the insulation, the ambient temperature, and the electrical stress. Several of the failure mechanisms are discussed as follows:

1. **Excessive primary voltage.** If the primary voltage is sufficiently high, immediate puncture of the insulation may result. Moderate overvoltage (approximately 20% above rated voltage) will lead to premature breakdown of the insulation.
2. **Input frequency fluctuation.** Frequencies below the rated value will cause low reactance and higher-than-rated currents to flow. Frequencies above the rated value will cause higher core losses. Either case will result in overheating or the temperature rise of the device will be higher than the designed maximum temperature. This will eventually lead to insulation breakdown.
3. **Excessive secondary current.** Secondary currents above rated values result in overheating of the transformer, which weakens the dielectric strength of the insulation. This may cause the windings to open or short circuit. Also if the containers are of potting or filling compounds, the overheating may cause breakage or distortion of the container.
4. **Corona.** Corona occurs at points of high potential stress. It causes accelerated aging of the insulation by the liberation of ozone and by increase in temperature. Weak spots are thereby created in the insulation which will eventually lead to breakdown.

Molded and encapsulated coils tend to fail due to shorted turns, open terminations, and degradation of the quality factor Q. Although molding or encapsulating improves the coil's resistance to moisture, shock, and vibration, improper flow of the molding compound can crush the winding, which may result in open circuits or dielectric strength failures. In instances where there is a degradation of Q, there will be a higher incidence of failure in coils with powdered iron cores. Powdered iron slugs, if overheated, will degrade. This occurs at 105°C (ambient temperature plus temperature rise).

Bibliography

Del Toro, V., *Principles of Electrical Engineering*, Prentice Hall, Englewood Cliffs, NJ, 1965.

Fink, D. G., et al., eds., *Electronics Engineering Handbook*, 2nd ed., McGraw-Hill, New York, 1982.

Fitzgerald, A. E. and D. E. Higginbotham, *Basic Engineering*, 2nd ed., McGraw-Hill, New York, 1957.

"Reliability Stress and Failure Rate Data for Electronic Equipment," *MIL HDBK-217A* (Dec. 1, 1965).

CHAPTER 13
ACTIVE LUMPED CIRCUIT ELEMENTS

MARK B. BARRON

General Electric Microelectronics Center
Research Triangle Park, North Carolina

HENRY DOMINGOS

Clarkson University
Potsdam, New York

GEORGE H. EBEL

Singer Company
Wayne, New Jersey

13.1 DIODES, RECTIFIERS, VARISTORS

Mark B. Barron

13.1-1 *Pn*-Junction Diodes

The backbone of most active semiconductor devices, those that amplify, switch, or emit radiation, is the *pn* junction. Formed by placing a *p*-type semiconductor material adjacent to an *n*-type semiconductor, the *pn* junction has the property of blocking the flow of current in one direction while allowing it to pass in the other direction. While dissimilar semiconductor materials can be used for *n*- and *p*-type semiconductors, thus forming a heterojunction, most *pn* junctions are formed out of the same material, for example, silicon or germanium. The physical operation of a *pn* junction can be visualized if one remembers that *n*-type material has electrical carriers that are predominantly electrons (negatively charged) and *p*-type material has electrical carriers that are predominantly holes (positively charged). Therefore if a bias is put onto the junction such that the *p* side is positive and the *n* side is negative, as in Fig. 13.1*a*, electrons will be attracted from the *n* material to the positive side of the supply, and holes will be attracted to the negative terminal. Current will therefore flow across the junction (Fig. 13.1*c*).

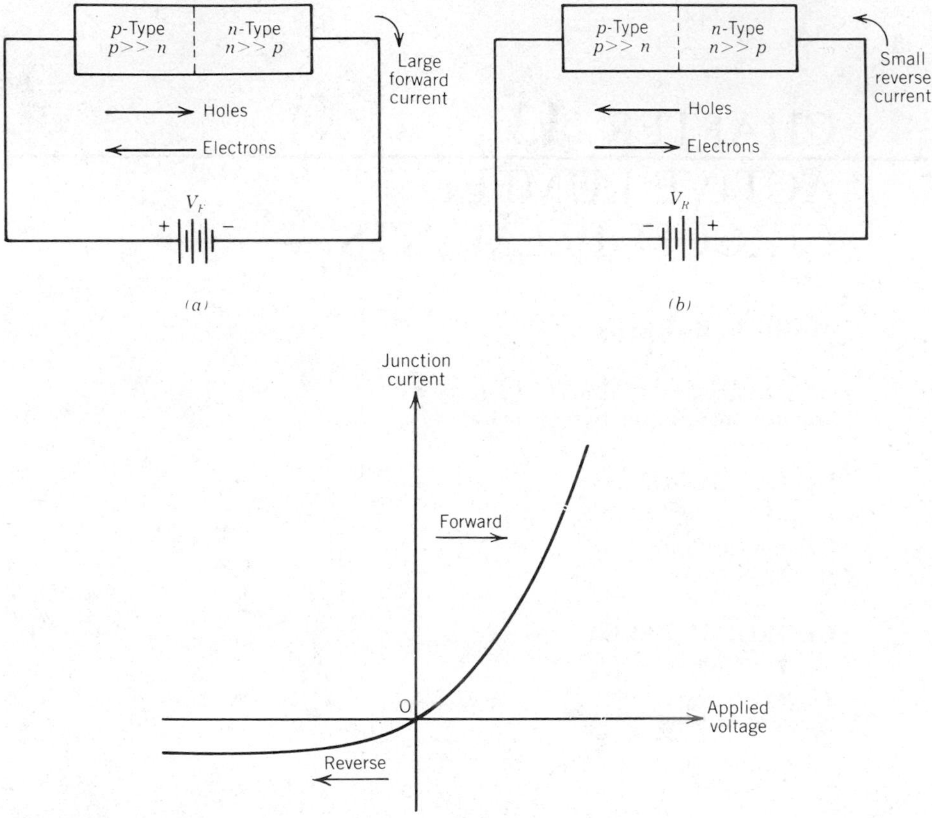

Fig. 13.1 (*a*) Forward-biased *pn* junction. (*b*) Reverse biased *pn* junction. (*c*) Current-voltage characteristic of a *pn* junction.

If the bias is reversed (Fig. 13.1*b*), and the *n* material is positively biased with respect to the *p* material, electrons will be attracted out of the *p* material by the positive side of the supply. Similarly, holes will be attracted out of the *n* material by the negative side of the supply. Since only a few electrons exist in the *p* material, the current attributed to these carriers that flows across the junction will be small.

The ideal current across a *pn* junction is given by the equation[1]

$$I = I_0(e^{qV/kT} - 1) \tag{13.1}$$

where V is the voltage across the junction, T is temperature in kelvins, and q/k is a constant of value 11.6×10^4 K/V. For real devices there will be a recombination of the holes and electrons as they flow across the junction and the current will be reduced from the ideal case. The diode equation above can be modified to include this effect by including an ideality factor, n

$$I = I_0(e^{qV/nkT} - 1) \tag{13.2}$$

where n varies between 1 and 2 depending on material, temperature, and current level. For germanium n is close to 1, while for silicon at low currents it is approximately 2. At higher current levels in silicon the value of n is approximately 1. At very high current levels there is a significant alteration of carrier concentration outside the regions of the junction, and a voltage drop appears across these regions. At high injection levels the current varies as[2]

$$I = I_0' e^{qV/2kT} \tag{13.3}$$

where I_0' is a constant somewhat larger than I_0 in Eq. (13.2). Figure 13.2*a* shows forward-biased current-voltage characteristics of a typical silicon diode.

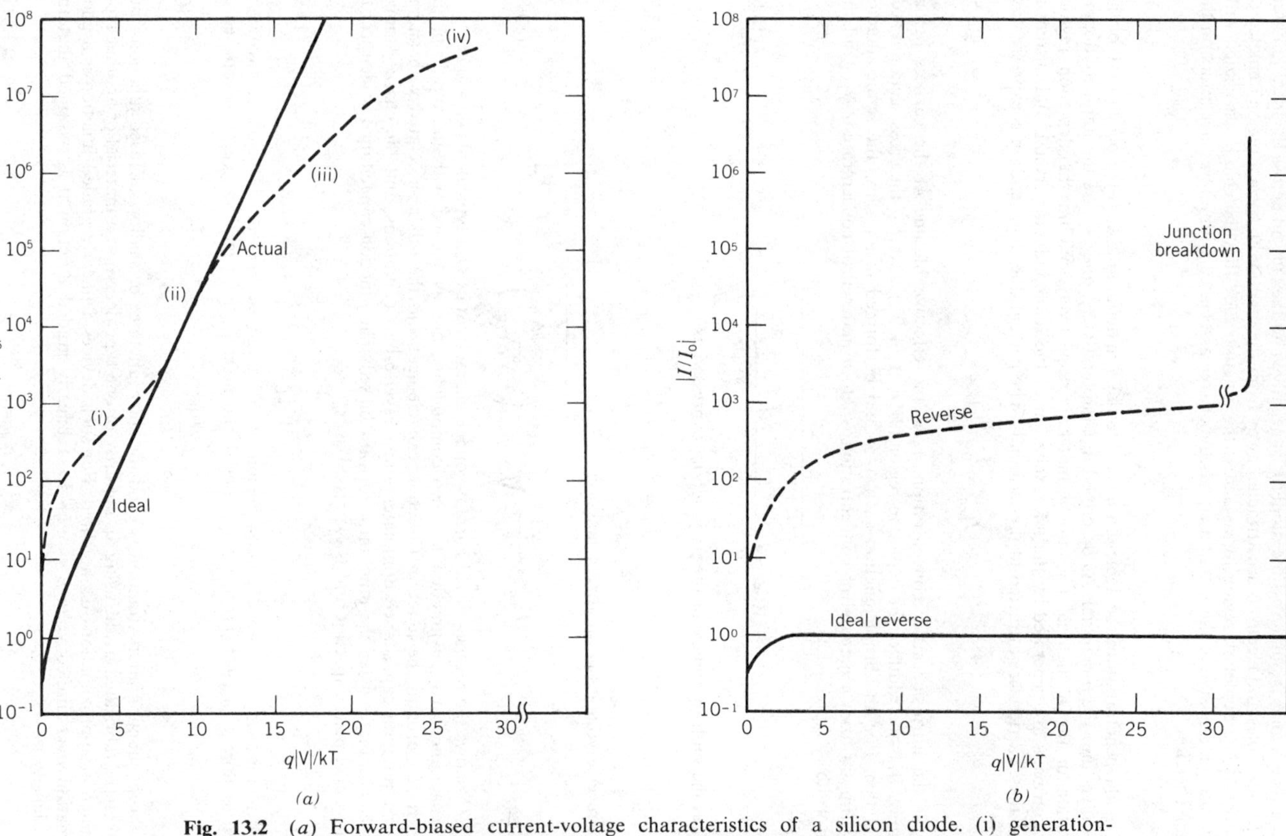

Fig. 13.2 (*a*) Forward-biased current-voltage characteristics of a silicon diode. (i) generation-recombination current, (ii) diffusion current, (iii) high-injection, (iv) series resistance. (*b*) Reverse-biased current-voltage characteristics of a silicon diode.

When a reverse bias is put across the diode, V is negative and the current across the ideal junction is $I = -I_0$. The holes and electrons are attracted away from the junction to form a region depleted of mobile carriers. The depletion region forms across the junction such that the charge associated with ionized donors on one side of the junction equals the charge associated with ionized acceptors on the other side. Therefore

$$N_D X_n = N_A X_p \tag{13.4}$$

where X_n is the penetration of the depletion layer into the n material and X_p is the penetration of the depletion layer into the p material. As described in Chapter 10, N_D and N_A are the impurity doping concentration in the n material and the p material, respectively. Within the depletion region, hole-electron pairs are generated by thermal excitation of the semiconductor crystal, and current is created as these carriers are attracted to the terminals. The current due to generation is expressed as[3]

$$I_{\text{gen}} = \frac{1}{2} q \frac{n_i}{\tau_0} WA \tag{13.5}$$

where n_i is the intrinsic carrier concentration, τ_0 is the effective lifetime of the carriers before recombining, W is the width of the depletion region ($W = X_n + X_p$), and A is the cross-sectional area of the junction. For an abrupt junction such as would be formed in an alloy, the semiconductor changes abruptly from p material of concentration N_A to n material of concentration N_D, and the depletion width is expressed as

$$W = \sqrt{\frac{2\epsilon(V_0 - V)}{q} \left(\frac{N_A + N_D}{N_A N_D} \right)} \tag{13.6}$$

where V_0 is the built-in voltage of the pn junction, expressed as

$$V_0 = \frac{kT}{q} \ln \frac{N_A N_D}{n_i^2} \tag{13.7}$$

For large reverse voltages in an abrupt junction

$$I_{\text{gen}} = \frac{n_i q A}{2\tau_0} \cdot \sqrt{\frac{2\epsilon V}{q} \cdot \left(\frac{N_A + N_D}{N_A N_D} \right)} \tag{13.8}$$

Figure 13.2*b* shows the reverse-biased current-voltage characteristics of a typical silicon diode.

The capacitance in a pn junction has two components, one associated with the dipole in the depletion region around the junction (junction capacitance) and the other associated with charge storage. The junction capacitance dominates under reverse-bias conditions, and the charge storage capacitance dominates when the junction is forward biased. The junction capacitance is identical to that of a parallel plate capacitor and is expressed as

$$C_j = \frac{\epsilon A}{W} \tag{13.9}$$

where W is the width of the depletion region. The charge storage (or diffusion) capacitance, attributed to storage of electrons in the p material and holes in the n material, can be expressed as

$$C_D = \frac{q I \tau_0}{kT} \tag{13.10}$$

The abrupt junction approximation is good for alloy junctions or shallow diffused junctions, but for pn junctions fabricated by diffusing an impurity of one type deep into a material of another type, abrupt junction approximations do not hold. For such devices a linearly graded junction is a good approximation over limited voltage ranges. In a linearly graded junction it is assumed that the impurity density is

$$N_D - N_A = ax \tag{13.11}$$

The width of the depletion region can be calculated to be[4]

$$W = \left(\frac{12\epsilon(V_0' - V)}{qa} \right)^{1/3} \tag{13.12}$$

where V_0' is the built-in potential found by solution of the equation

$$V_0' = \frac{kT}{q} \ln \left[\frac{a}{2n_i} \left(\frac{12\epsilon V_0'}{qa} \right)^{1/3} \right]^2 \tag{13.13}$$

With the depletion width as given in Eq. (13.13), the generation current, I_{gen}, and the capacitance, C_j, given in Eq. (13.9) still apply.

Breakdown Voltage

When a reverse bias is applied across a *pn* junction, most of the voltage occurs across the depletion region. The width of the depletion region varies only by $V^{\frac{1}{2}}$ or $V^{\frac{1}{3}}$, depending on whether the junction is abrupt or graded. Therefore the electric field $E = dv/dx \approx V/W$ increases with voltage. At a high enough electric field the carriers generated in the depletion region will gain enough kinetic energy from the field so that when they collide with the silicon lattice they will break loose other hole-electron pairs. This multiplication of carriers is called the avalanche process and gives rise to a current

$$I_r = MI_{r0} \tag{13.14}$$

where I_{r0} is the current without avalanche and

$$M = \frac{1}{1-(V/V_{\mathrm{Br}})^n} \tag{13.15}$$

V_{Br} is the breakdown voltage of the junction and n varies between 3 and 6 depending on the semiconductor. It can be seen from Eq. (13.15) that when $V = V_{\mathrm{Br}}$ the multiplication factor $M \to \infty$. The current is limited only by the resistance of the semiconductor outside the junction region. For abrupt junctions, V_{Br} is given as

$$V_{\mathrm{Br}} = \frac{\epsilon(N_A + N_D)E_{\mathrm{crit}}^2}{2qN_A N_D} \tag{13.16}$$

with E_{crit} being the field at which breakdown occurs. For a graded junction the breakdown voltage is given as[5]

$$V_{\mathrm{Br}} = \sqrt{\frac{32\epsilon E_{\mathrm{crit}}^3}{9qa}} \tag{13.17}$$

Figure 13.3 shows the breakdown voltages of one-sided p^+n step-junctions in silicon, germanium, and galium-arsenide as a function of impurity concentration in the lightly doped *n*-type material. Figure 13.4 shows breakdown voltages in the same materials for linearly graded junctions.

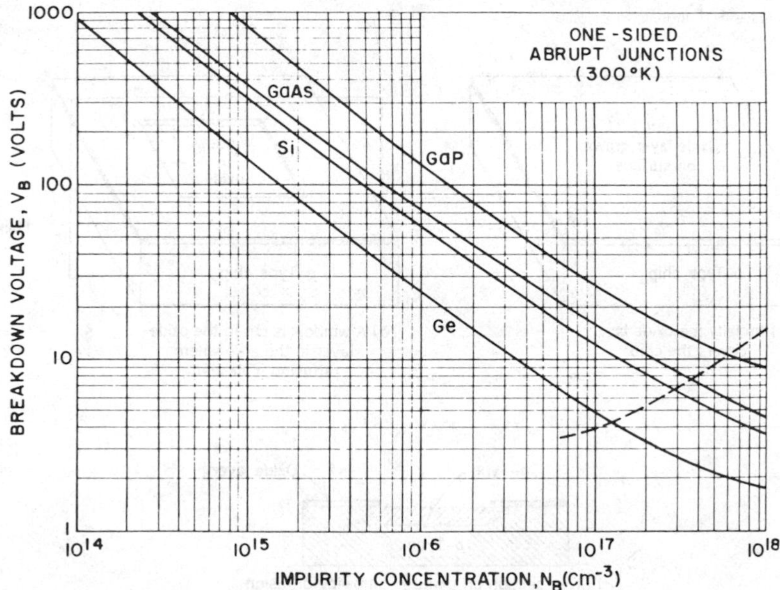

Fig. 13.3 Avalanche breakdown voltage vs. impurity concentration for one-sided abrupt junctions.

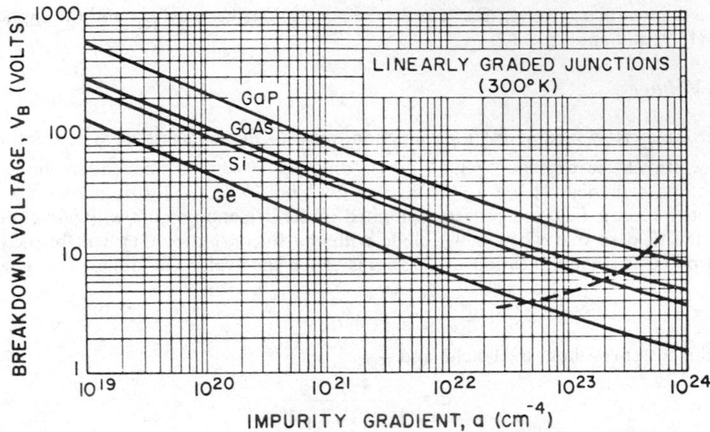

Fig. 13.4 Avalanche breakdown voltage vs. impurity concentration for *p* linearly graded junctions.

For heavily doped materials, with impurity concentration in excess of 5×10^{17} cm^{-3}, another breakdown mechanism may occur in place of avalanche breakdown; this is the Zener effect. In Zener breakdown the field is so high that the electrons are stripped from the crystal atoms without need for carrier collisions as required for avalanche breakdown. The conditions for Zener breakdown to occur can be met only with abrupt junctions having a narrow depletion region, for example, heavily doped semiconductors with low breakdown voltages. Zener breakdown also occurs more easily in narrow bandgap semiconductors such as germanium than in wider gap materials such as gallium phosphide. Experimentally it has been found that Zener breakdown occurs at fields of approximately 10^6 V/cm.[6]

Breakdown in Actual *pn*-Junction Diodes. In practical *pn* junctions fabricated with modern technology, breakdown voltages are dominated by surface effects and/or curvature of the junctions. Most *pn*-junction diodes are fabricated using diffusion processes (described in Chapter 10). Diode location is determined either by patterned oxide windows through which impurities are diffused (Fig. 13.5) or, in the case of power rectifiers, by cutting or etching islands out of a blanket diffused region (Fig. 13.6). In the first case, the breakdown voltage is lowered by the curvature at the edges of the diffused junction. In the second case, surface imperfections effectively lower E_{crit} at the edge of the junction, and $V_{Br\ surface} < V_{Br\ bulk}$.

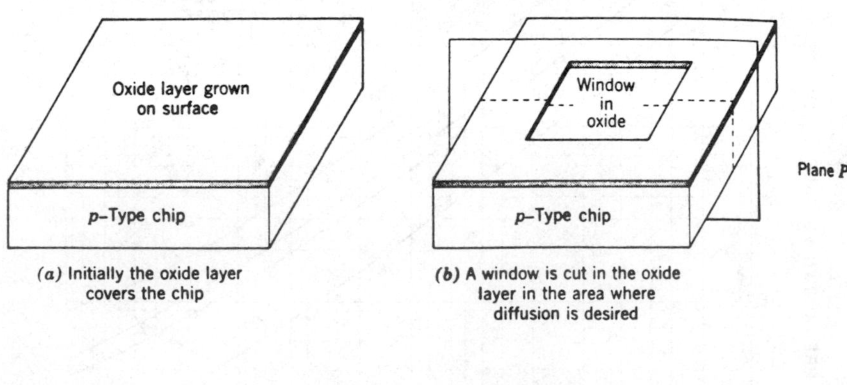

(a) Initially the oxide layer covers the chip

(b) A window is cut in the oxide layer in the area where diffusion is desired

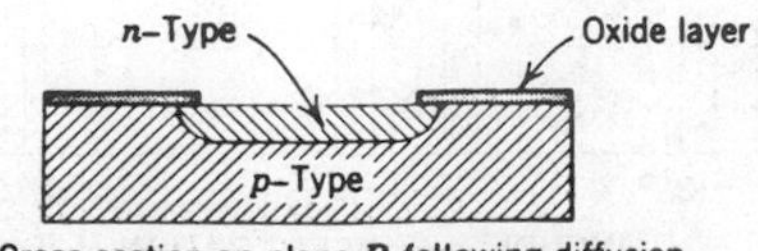

(c) Cross section on plane *P* following diffusion

Fig. 13.5 Fabrication of planar *pn* junction diode.

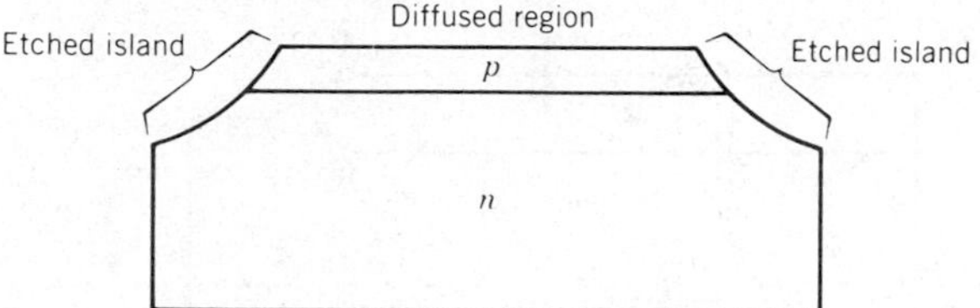

Fig. 13.6 Mesa diode resulting from etching process.

Figure 13.7 shows the breakdown voltage of diffused silicon junctions vs. diffusion depth, with substrate impurity concentration and diffusion surface concentration as parameters.

Transient Characteristics. When the voltage across a *pn* junction is suddenly switched from positive to negative, the current does not immediately go to the steady-state reverse-biased value. The reason is that there are excess minority carriers on each side of the junction under forward-bias conditions. When the voltage bias is changed from forward to reverse, the excess holes on the *n* side of the junction are attracted to the negative terminal on the *p* material, and likewise the excess electrons on the *p* side of the junction flow toward the positive terminal on the *n* material. Usually the impedance of the drive circuit limits the reverse current to some peak value, I_R, during turnoff. As seen in Fig. 13.8, the current remains at the value I_R for a storage time t_s until the excess minority carriers reach their equilibrium values. Thereafter the current falls to the steady-state value associated with the

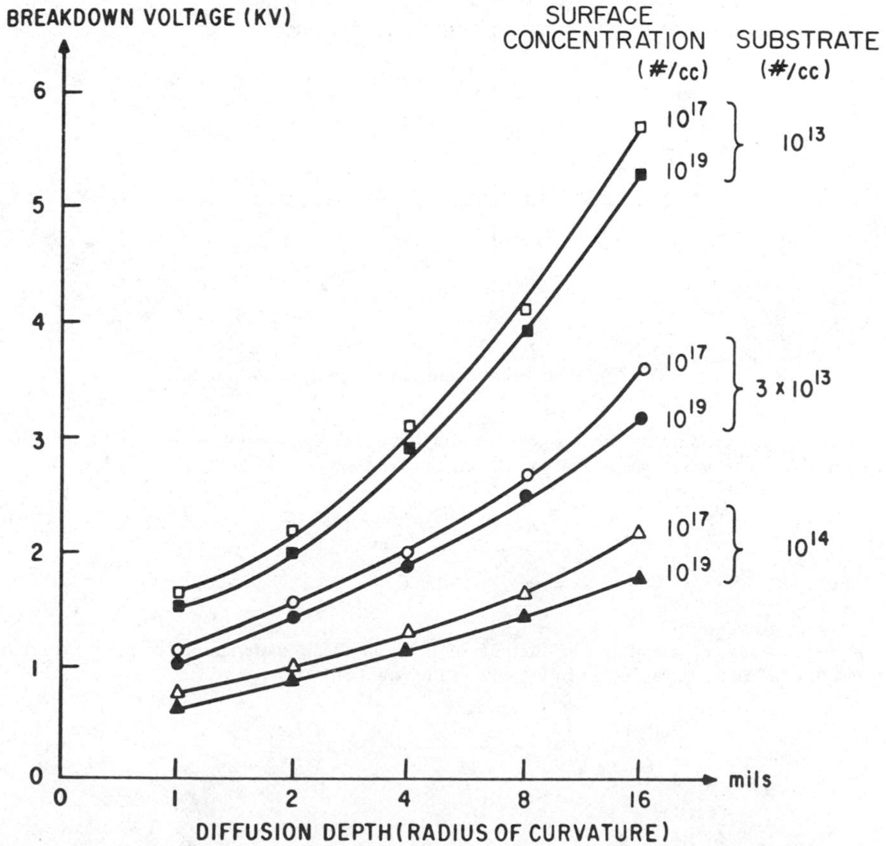

Fig. 13.7 Avalanche breakdown voltage of diffused silicon junctions vs. diffusion depth with impurity concentration and diffusion surface concentration as parameters.

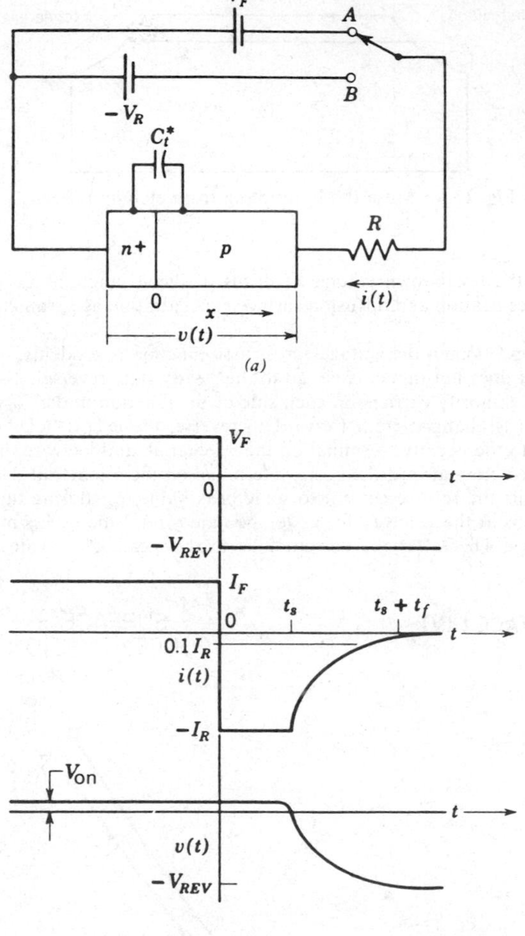

Fig. 13.8 Transient voltage and current in a junction diode.

reverse voltage as the minority carriers near the junction are depleted below their equilibrium values. An analysis of the problem[7] shows that the storage time is given as

$$t_s = \tau_0 \left[\operatorname{erf}^{-1}\left(\frac{I_F}{I_F + I_R} \right) \right]^2 \tag{13.18}$$

where τ_0 = effective lifetime
I_F = forward current
I_R = reverse current

For p^+n junctions, τ_0 would be the lifetime of holes in the n material before they recombine. Likewise the fall time, t_f, can be given by solving the equation[3]

$$\operatorname{erf}\sqrt{\frac{t_f}{\tau_0}} + \frac{e^{-t_f/\tau_0}}{\sqrt{\dfrac{\pi t_f}{\tau_0}}} = 1 + 0.1\left(\frac{I_R}{I_F}\right) \tag{13.19}$$

In some applications it is desirable to have turnoff be as fast as possible, and manufacturers reduce the lifetime, τ_0, by doping the semiconductor with gold or platinum or by irradiating it with

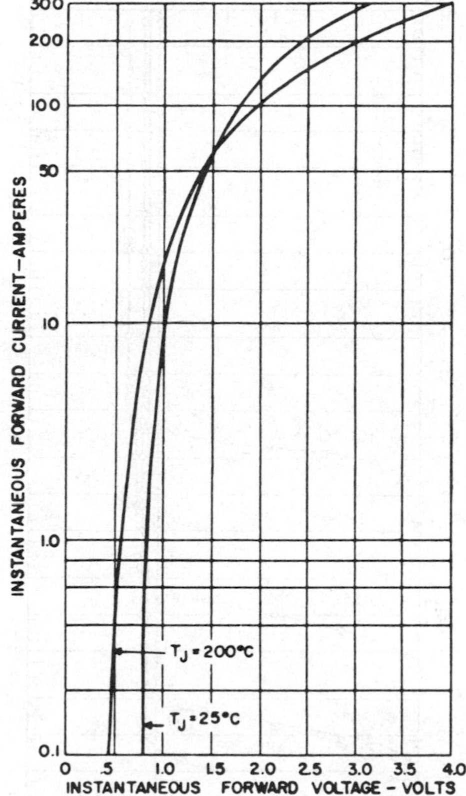

Fig. 13.9 Temperature dependence of forward-biased diode characteristics.

high-energy electrons. However, there is a tradeoff between fast switching speed and low reverse leakage current since, as indicated in Eq. (13.5), the reverse current increases with decreasing lifetime, τ_0.

Thermal Effects. As temperature increases in a *pn* junction biased at a fixed forward voltage, conduction current can either increase or decrease, depending on the level of operating current. At low and medium levels of current, the current will increase with temperature. At high-level injection the current decreases, as shown in Fig. 13.9. Under reverse-bias conditions the reverse current increases exponentially with temperature (Fig. 13.10). Avalanche breakdown voltage increases slightly with temperature (Fig. 13.11). On the other hand, because the bandgap narrows at higher temperatures, Zener diode breakdown voltage will go down with temperature. In higher voltage devices, current may become so large at high temperatures that a sustained high voltage will generate so many watts that the device will heat itself and the current will regeneratively increase until the device self-destructs. In applying diodes at high temperatures one must consider the possibility of thermal runaway and design the circuits with appropriate temperature-sensing devices.

Diode Types

Varactor Diodes. While the *pn* junction capacitance is a menace to some applications, it is a worthwhile feature for applications that require a voltage-controlled capacitance. For such applications it is desirable to have the capacitance vary at a rate more rapid than the $V^{-\frac{1}{2}}$ that is achieved with abrupt junctions. More rapid variation with voltage can be achieved by using a hyperabrupt junction, such as that shown in Fig. 13.12. With such a device, constructed using controlled epitaxial techniques, varactors can be obtained that have capacitances varying as $(V_0 - V)^{-2}$.

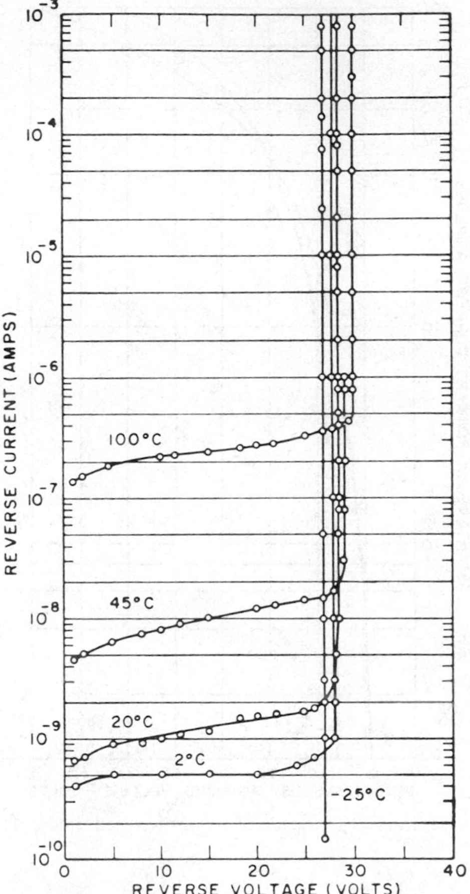

Fig. 13.10 Temperature dependence of reverse-biased diode characteristics.

***Pin* Diodes.** A *pin* diode is constructed with a high resistivity layer (*i*ntrinsic silicon) sandwiched between *p* material and *n* material (Fig. 13.13*a*). Such diodes are characterized by relatively constant depletion layer capacitance, given as

$$C = \frac{\epsilon A}{W_i} \tag{13.20}$$

where W_i is the width of the intrinsic region. The breakdown voltage is given as

$$V_B = E_{\text{crit}} W_i \tag{13.21}$$

where E_{crit} is the critical electrical field at which breakdown occurs. For a silicon *pin* structure with $W_i = 10$ μm, the breakdown voltage would be $V_B = 2 \times 10^5 \times 10 \times 10^{-4} = 200$ V. Because of the high resistance of the intrinsic region, most of the voltage drop occurs across this region. For low to moderate levels of forward current, the resistance varies inversely with current (Fig. 13.13*b*) and for this reason a *pin* diode is often used in microwave circuits as a variable attenuator.

Photodiodes. In the discussion of *pn* junctions it was noted that hole-electron pairs are thermally generated, and with reverse bias contribute to the reverse-bias current. At zero bias the generation current is in equilibrium with the recombination current, and the net current is zero. It is possible to shine light on the semiconductor and generate hole-electron pairs in excess of those generated at thermal equilibrium. The only requirement is that the energy for the light exceed the bandgap energy of the semiconductor: $h\nu > E_g$. For silicon the bandgap is approximately 1.11 eV, which says that the

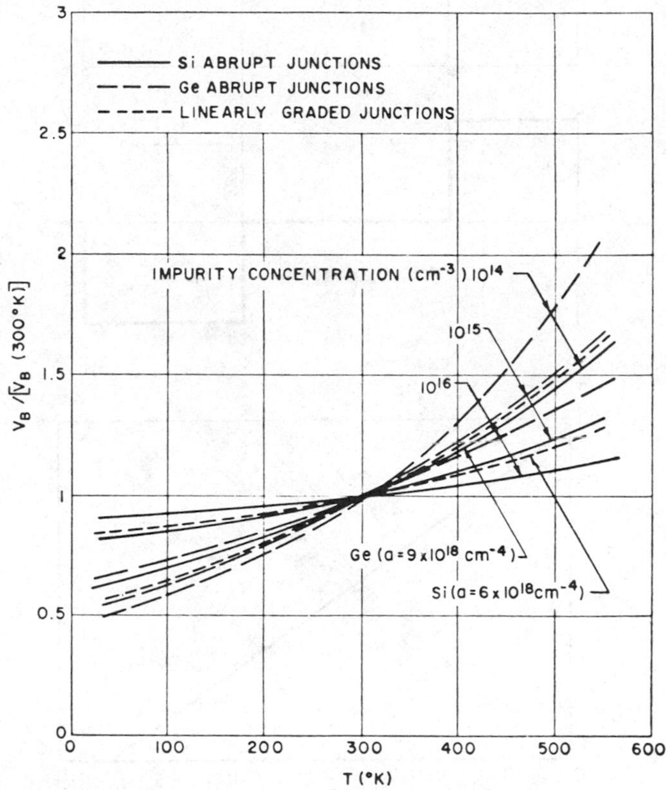

Fig. 13.11 Temperature dependence of avalanche breakdown voltage in germanium and silicon.

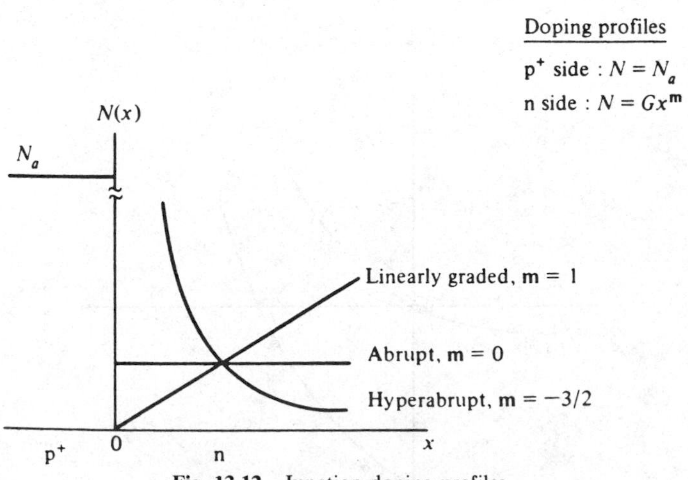

Fig. 13.12 Junction doping profiles.

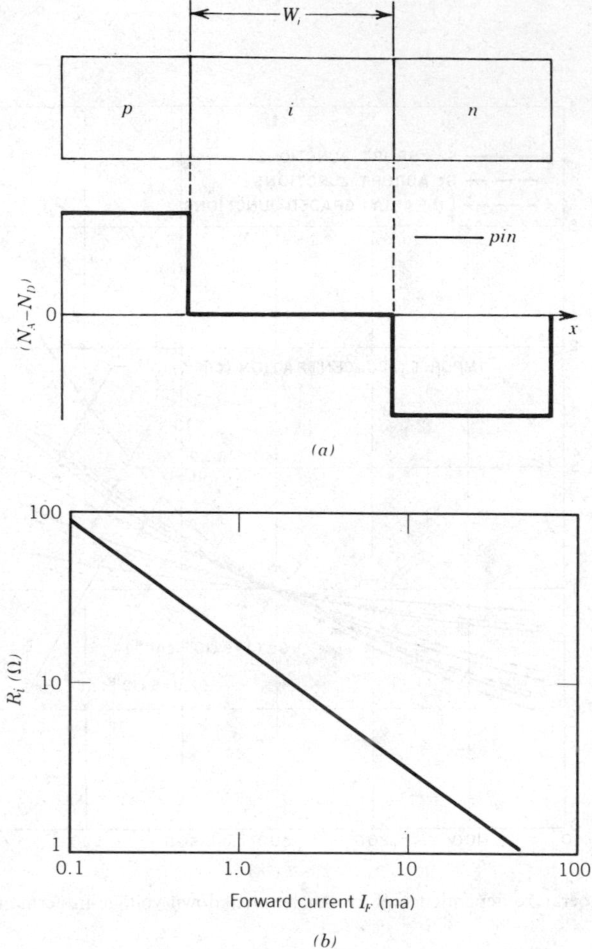

Fig. 13.13 (*a*) A *pin* diode. (*b*) Resistance versus current for forward-biased *pin* diode.

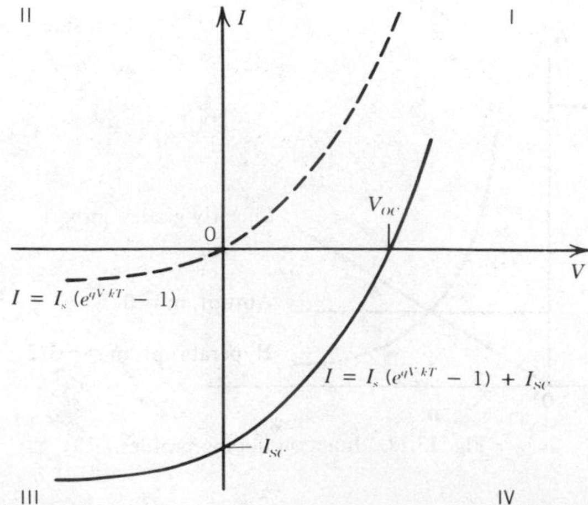

Fig. 13.14 Current-voltage characteristics of a photodiode.

wavelength of the light has to be less than $\lambda_T = hc/E_g = 1.12\ \mu m$. Silicon is therefore sensitive to light in the shallow infrared through the visible region. The *VI* characteristics of a photodiode are shown in Fig. 13.14. When the photodiode is operating in quadrant IV it is delivering power to the external circuit. In quadrants I and III, power is being delivered from the external circuit to the photodiode. The short-circuit current, I_{SC}, is determined by the flux of the light, whereas the open-circuit voltage, V_{OC}, is limited to the bandgap of the semiconductor. For silicon, $V_{OC} < 1.1$ V.

Since photodiodes can generate power in quadrant IV, they are used for generating electricity. In such applications they are referred to as solar cells. Silicon solar cells convert only about 20% of the incident light having a wavelength less than 1.1 μm, and none of the light at greater wavelengths. In the case of sunlight, some of the energy is at wavelengths greater than 1.1 μm where $h\nu < E_g$. Therefore the composite efficiency of silicon solar cells for converting sunlight to electrical power is only 10%. Since about 1 kW/m^2 of radiation falls on the earth on a bright day, about 100 W/m^2 can be generated from solar cells.

Zener Diodes. Zener diodes are a class of *pn*-junction devices that have an accurately specified breakdown voltage and are intended to be operated as a voltage clamp at that voltage. In reality the breakdown mechanism is more often avalanche than Zener, but the generic name of Zener diode is applied to these devices even though the name does not accurately describe the physical cause of the breakdown.

Zener diodes are used as voltage regulators or voltage references. The breakdown voltage is a parameter of specification, as is the dynamic resistance of the device. Zener diodes are fabricated by manufacturers for certain ranges of voltage breakdown points, and are then selected to narrower ranges (usually ±5%) by automatic test equipment.

Zener diodes can also be used for suppression of transient voltages. In this application the diode is not operated in the breakdown region until a voltage transient occurs. The voltage across the diode under normal operating conditions is usually less than 90% of the breakdown value. When there is a transient, the diode clamps the voltage and acts as a shunt element to absorb the energy of the surge current. The rating on the maximum nonrepetitive transient current is designated I_{RSM}; it is usually a pulse of a certain time duration.

Voltage breakdown ratings for Zener diodes are usually between 4 and 100 V. Suppression power ratings of up to 20 kW can be dissipated for short pulses in such diodes. Published specifications give information on the repetitive peak reverse power that can be dissipated under various conditions.

Metal Oxide Varistors. For applications having higher voltages, such as circuits operating at 120 V AC, or having potential power transients exceeding the capability of Zener diodes, a *m*etal *o*xide *v*aristor (MOV) can be a practical alternative to Zener diodes. MOVs are fabricated using zinc oxide or lead oxide powders containing various doping elements. Electrically MOVs appear as back-to-back *pn* junctions (Fig. 13.15) and are rated at maximum rms applied voltage, recurrent peak idle voltage,

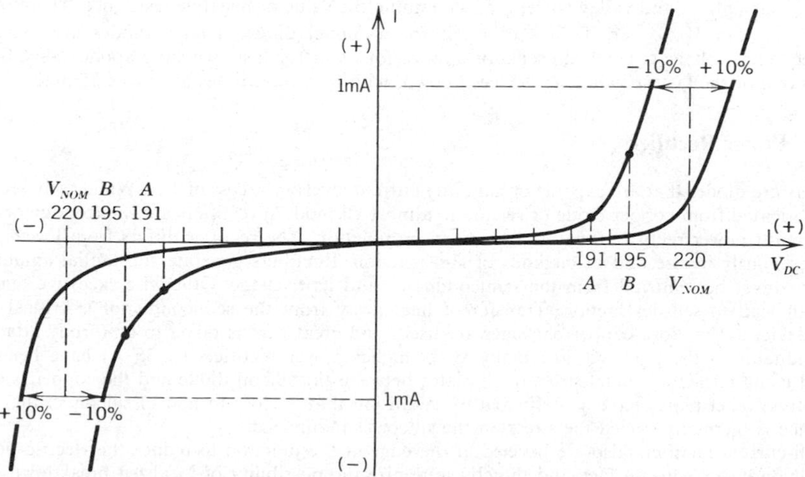

Fig. 13.15 Current-voltage characteristics of zinc oxide metal oxide varistor (MOV). A = maximum allowable steady-state DC applied voltage; B = maximum allowable steady-state recurrent peak applied voltage; V_{NOM} = nominal varistor voltage at 1 MA DC.

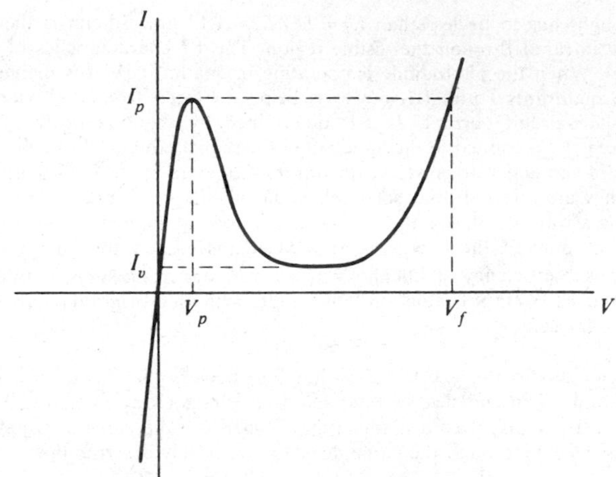

Fig. 13.16 Tunnel diode current-voltage characteristic.

maximum DC applied voltage, energy (joules), average power dissipation, and peak current for a designated pulse width. Metal oxide varistors can handle peak currents of several thousand amps, several thousand volts, and energy of hundreds of joules. As with Zener diodes, MOVs have a specified dynamic resistance as a function of current.

Tunnel Diodes. A tunnel diode is a *pn* junction with extremely heavily doped *p*-type and *n*-type regions. Generally tunnel diodes are manufactured from germanium, although they can be fabricated from gallium-arsenide. A thorough explanation of the tunnel diode requires the use of quantum mechanics, which is beyond the scope of this handbook. For those interested in a detailed physical description of the process, several good references exist.[8]

The tunnel diode is capable of conducting substantial current under reverse bias because electrons and holes can move across the junction without experiencing the normal effects of the electric field. For small voltages, positive or negative, a tunnel diode has an almost linear *V-I* characteristic (Fig. 13.16). At a forward voltage of V_p, the device begins to lose its tunneling properties and the current actually decreases with increasing voltage. This negative resistance region continues until the normal diode current becomes large enough to cause the resistance to become positive again. It is this controlled negative resistance region that gives the tunnel diode its usefulness. The values of peak tunneling current, I_p, and valley current, I_v, determine the value of negative resistance. Therefore the ratios I_p/I_v and V_p/V_v are figures of merit for a tunnel diode. Tunnel diodes are useful for high-frequency voltage-controlled oscillator applications and for fast switching applications. Such a diode is one of the fastest switching devices known, with transition times as low as 25 psec.

13.1-2 Power Rectifiers

Rectifiers are diodes that are capable of handling current levels in excess of 1 A. While early rectifiers were fabricated from copper oxide or selenium, almost all modern rectifiers are semiconductors. The predominant power rectifier today is the silicon *pn* junction. The rectifier differs from lower power diodes primarily in size and in methods of construction. Rectifiers generate substantial amounts of heat that must be removed from the semiconductor and its package. Otherwise excessive temperatures can lead to self-destruction. Transfer of heat away from the semiconductor is critical in all power devices. Therefore copper packages are used, and great care is taken to uniformly attach the semiconductor to the package. For many years higher power rectifiers (> 35 A) have been constructed using refractory metal stress relief plates between the silicon diode and the copper package. These stress relief plates have a coefficient of expansion that is close to that of silicon so that when the device is thermally cycled the stress on the silicon is minimized.

High-current rectifiers also are beveled at the edge of the junction to reduce the electric field on the semiconductor edge surface and thereby minimize the possibility of localized breakdown at the surface (Fig. 13.17). Lower current devices often have an etched groove that is passivated with a glass (Fig. 13.18*a*) or have field rings[9] that spread the field out on the surface (Fig. 13.18*b*). Power rectifier packages are either stud type (Fig. 13.19*a*), flat base (Fig. 13.19*b*), or in the case of devices less than 3

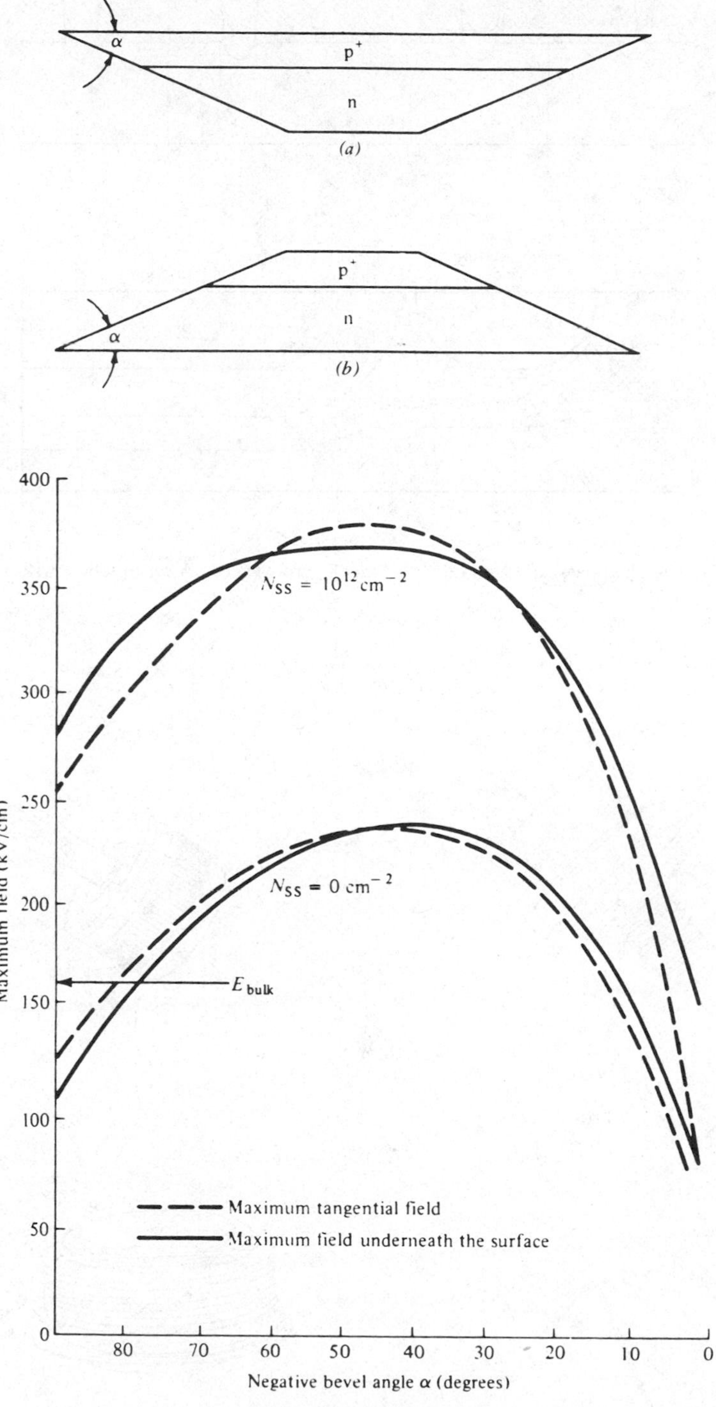

Fig. 13.17 (*a*) Positive beveled junction. Highly doped region has a larger area than does the low-doped region. (*b*) Negative beveled junction. Highly doped region has a smaller area than does the low-doped region. (*c*) Maximum electric field vs. bevel angle for two values of surface charge.

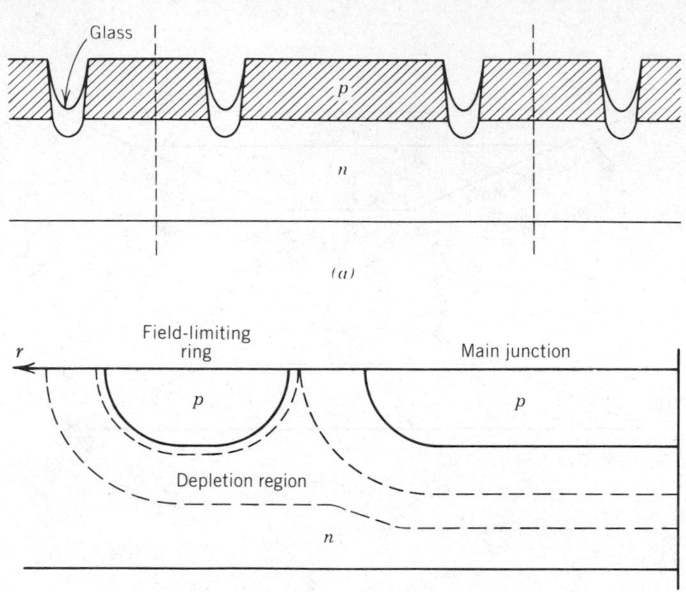

Fig. 13.18 (*a*) A glass-passivated diode with "moat" grooves. (*b*) A *pn* junction with field-limiting ring.

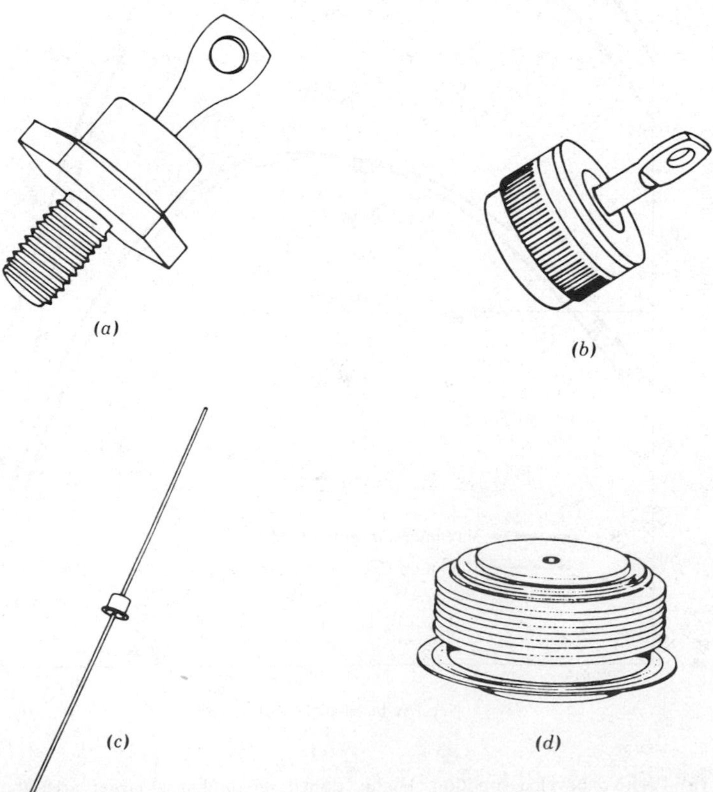

Fig. 13.19 Types of power rectifier packages. (*a*) Stud, (*b*) flat base (or press fit), (*c*) axial lead, (*d*) pressure contact.

A, axial lead (Fig. 13.19*c*). For currents above 200 A, pressure contact packages are often used (Fig. 13.19*d*). A major concern in using any package is ensuring that the thermal resistance between the package and the heat sink is minimized. Most manufacturers specify the maximum junction-to-case thermal resistance. Therefore, knowledge of the power generated in the semiconductor allows estimation of the upper temperature limit of the package. For a 5-A rated rectifier having a 1-V forward drop and a 7°C/W junction-to-case thermal resistance, the device would have an increase in package temperature of $5 \times 1 \times 7 = 35°C$. If the device is rated at 200°C maximum operating junction temperature, then the package has to be maintained at less than 165°C when 5 W of power is dissipated through the heat sink. With an ambient temperature of T_A, the thermal resistance from package to ambient has to be less than $R_{\theta CA} = (165 - T_A)/5$ W. For a maximum ambient temperature of 50°C, $R_{\theta CA} \leqq 25°C/W$.

References

1 E. S. Yang, *Fundamentals of Semiconductor Physics*, Ch. 4, McGraw-Hill, New York, 1978.

2 S. K. Ghandi, *Semiconductor Power Devices*, Ch. 2 and 3, Wiley, New York, 1977.

3 S. M. Sze, *Physics of Semiconductor Devices*, Ch. 3, Wiley, New York, 1969.

4 B. G. Streetman, *Solid State Electronic Devices*, Sec. 5.5.3, Prentice-Hall, Englewood Cliffs, NJ, 1980.

5 D. A. Fraser, *The Physics of Semiconductor Devices*, Oxford University Press, New York, 1977.

6 P. E. Gray and C. L. Searle, *Electronic Principles—Physics, Models, and Circuits*, Sec. 6.5, Wiley, New York, 1969.

7 B. G. Streetman,[4] Sec. 5.5.2.

8 S. M. Sze,[3] Ch. 4.

9 Y. C. Kao and E. D. Wolley, "High-Voltage Planar *p-n* Junctions," *Proc. IEEE* **55**(8):1409–1414 (1967).

10 *Transient Voltage Suppression Manual*, General Electric Company, Syracuse, NY, 1976.

11 J. F. Gibbons, *Semiconductor Electronics*, Chs. 5–8, McGraw-Hill, New York, 1966.

12 A. Blicher, *Thyristor Physics*, Chs. 14 and 15, Springer-Verlag, New York, 1976.

13 M. S. Adler and V. A. Temple, *Semiconductor Avalanche Breakdown Design Manual*, General Electric Company, Schenectady, NY, 1979.

14 A. S. Grove, *Physics and Technology of Semiconductor Devices*, Wiley, New York, 1967.

15 *Semiconductor Data Handbook*, General Electric Company, Syracuse, NY, 1977.

13.2 BIPOLAR TRANSISTORS

Mark B. Barron

13.2-1 Transistor Action

Current Amplification-Carrier Flow

In the sections covering *pn*-junction diodes it was noted that a diode when forward biased will conduct current according to the expression

$$I = I_0 e^{qV/nkT} \tag{13.22}$$

where I_0 = a constant
V = forward applied voltage
n = "ideality" factor usually close to 1 in value

When the forward-biased junction is placed near (within a few micrometers) a reverse-biased *pn* junction, a three-terminal device called a transistor can be obtained. Figure 13.20*a* illustrates a *p*-type semiconductor sandwiched between two *n*-type semiconductors. If region 3 is reverse biased relative to region 2, and region 2 is positive biased relative to region 1, the electrons injected from region 1 will see a more positive potential at terminal 3 than at terminal 2. Therefore the electrons emitted from region 1 into region 2 will be swept into region 3. By changing the bias voltage across the junction between regions 1 and 2, the current into region 3 can be made to change. For historical reasons region 2 is called the base, region 1 the emitter, and region 3 the collector. For this arrangement to yield an effective amplifying device, the current into the base lead should be small relative to the total current into the emitter. The number of holes injected from the base into the emitter is proportional to the inverse of the emitter doping concentration, N_E, and the number of

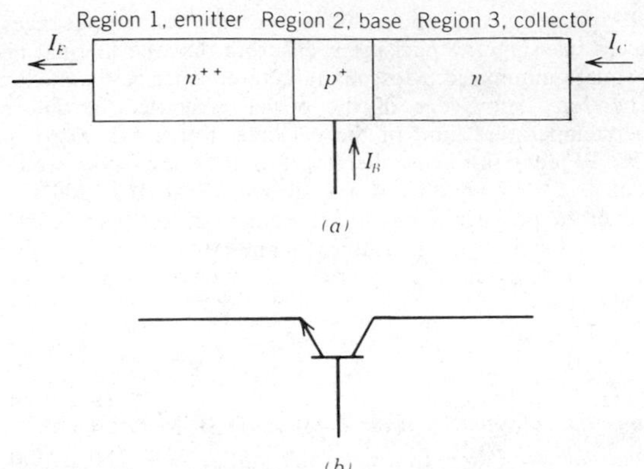

Fig. 13.20 (*a*) ***npn*** transistor. (*b*) ***npn*** transistor symbol.

electrons injected from the emitter into the base is proportional to the inverse of the base region doping concentration, N_B. Therefore for the highest injection ratio of emitter electrons to base holes, the emitter should be heavily doped and the base should be lightly doped. The ratio of number of electrons injected from the emitter to the total current across the junction is called the injection efficiency, γ.

$$\gamma = \frac{I_n}{I_p + I_n} \tag{13.23}$$

***npn* vs. *pnp*.** The previous discussion dealt with a sandwich of three layers: n, p, n. In the normal mode of operation the n-type collector is biased positive relative to the base and the emitter, and when a positive bias is applied to the base relative to the emitter, electrons flow into the emitter terminal, or equivalently, positive current flows out of the emitter. Thus the symbolic representation of an *npn* device denotes current out of the emitter (Fig. 13.20*b*). An alternate structure having a p-type emitter and collector and an n-type base can also be used for constructing a transistor. This *pnp* structure is operated in the normal mode with the collector biased negative, so that with application of a slightly negative base voltage, electrons flow out of the emitter representing positive current into the emitter terminal. The symbolic representation of a *pnp* structure is shown in Fig. 13.21.

Ideal Device Characteristics

In high carrier lifetime semiconductors with a narrow base, the electrons diffuse across the base region undisturbed. In semiconductors with a low carrier lifetime and/or a wide base, the electrons

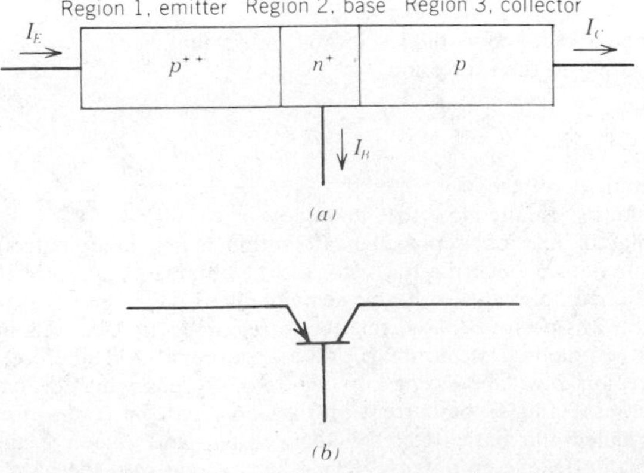

Fig. 13.21 (*a*) ***pnp*** transistor. (*b*) ***pnp*** transistor symbol.

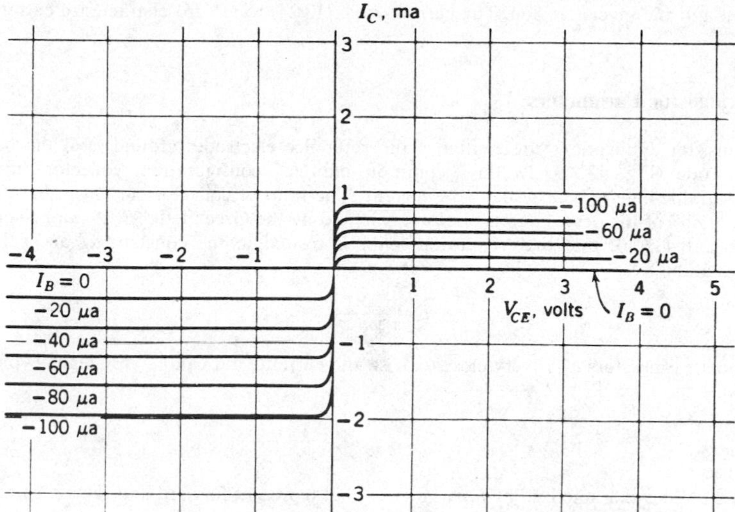

Fig. 13.22 Current-voltage characteristics for an ideal *pnp* transistor.

recombine with holes in the base region, and additional base terminal current is needed to supply these holes. Therefore, high gain transistors are best achieved with a narrow base and high carrier lifetime. For transistors of this type, the following equations describe behavior:

$$I_E = I_{ES}(e^{qV_{EB}/kT} - 1) - \alpha_R I_{CS}(e^{qV_{CB}/kT} - 1) \tag{13.24}$$

$$I_C = -\alpha_F I_{ES}(e^{qV_{EB}/kT} - 1) + I_{CS}(e^{qV_{CB}/kT} - 1) \tag{13.25}$$

$$I_B = -(I_E + I_C) \tag{13.26}$$

where $\alpha_R, \alpha_F, I_{ES}, I_{CS}$ are constants that depend on the characteristics of the semiconductor material and the dimensional properties of the device. The *VI* characteristics associated with Eqs. (13.24) and (13.25) are shown in Fig. 13.22.

Operating Regions

A transistor has four basic regions of operation, since each of the two junctions can be forward or reverse biased.

Cutoff Region. When both junctions are reverse biased, only leakage current flows and

$$I_E = -I_{ES} + \alpha_R I_{CS} \tag{13.27}$$

$$I_C = \alpha_F I_{ES} - I_{CS} \tag{13.28}$$

Normal Active Region. The emitter-base junction is forward biased and the base-collector junction is reverse biased. If the base-collector junction is biased with a large enough voltage, the currents are given by

$$I_E = I_{ES}(e^{qV_{EB}/kT} - 1) + \alpha_R I_{CS} \tag{13.29}$$

$$I_C = -\alpha_F I_E - (1 - \alpha_F \alpha_R) I_{CS} \tag{13.30}$$

Inverse Region. The base-collector junction is forward biased and the base-emitter junction is reverse biased. Under these conditions

$$I_E = -I_{ES} - \alpha_R I_{CS}(e^{qV_{CB}/kT} - 1) \tag{13.31}$$

$$I_C = -\frac{I_E}{\alpha_R} + \alpha_F I_{ES}\left(1 - \frac{1}{\alpha_R \alpha_F}\right) \tag{13.32}$$

Saturation Region. Both junctions are forward biased. As far as the terminal currents are concerned, the current in the saturation region can be thought of as a superposition of operation in the normal active region and the inverse region. The current Eqs. (13.24) to (13.26) characterize current flow in this region.

13.2-2 Transistor Parameters

In most transistor applications the emitter is the reference electrode (ground) and the base is the control electrode (Fig. 13.23). In this "common emitter" configuration, collector current is a dependent variable determined by the base current. The ratio of collector current to base current is the current "gain" of the device and is often represented by the Greek letter β. It can be shown from Eqs. (13.29) and (13.30) that the AC current gain is related to the constant α_F according to the following equation

$$\beta = \frac{\alpha_F}{1 - \alpha_F} \tag{13.33}$$

In most modern transistors α_F is very close to 1, so the current gain can be very high. Typical values are over 100.

h Parameters

Transistors are often characterized by h parameters that represent forward and reverse gain as well as input and output impedance/admittance. The first subscript denotes the parameter, large case type for DC and small case for AC parameters. The second subscript indicates common emitter, common base, or common collector. In the common emitter case $i_c/i_b = h_{\text{fe}}$ is the AC forward current gain, which is the same as β.

$$h_{\text{re}} = v_{\text{be}}/v_{\text{ce}}$$

is the reverse current gain, $v_{\text{be}}/i_b = h_{\text{ie}}$ is the input impedance, and $h_{\text{oe}} = i_c/v_{\text{ce}}$ is the output admittance.

Gain vs. Current

The ideal Eqs. (13.24) to (13.26) indicate that h_{fe} is a constant equal to $\alpha_F/(1 - \alpha_F)$. However, in actual transistors the current gain decreases at very low and very high currents (Fig. 13.24). The reason for this is that at low currents, surface imperfections in the semiconductor cause the injected emitter carriers and the base carriers to recombine at the surface, and there is no opportunity for them to be collected by the collector. At high currents, holes and electrons recombine more often in the base because of carrier-carrier collisions, which reduce the lifetime. Also the injection efficiency decreases because of the effects of conductivity modulation.

In addition, at high currents there is a resistive voltage drop beween the base lead and the middle of the emitter region (Fig. 13.25). Therefore the emitter-base junction at the middle of the emitter is not as heavily forward biased as the edge. This effect essentially reduces the effective current-carrying area of the transistor.

Leakage Currents

The equations for the ideal transistor indicated that for $V_{EB} = 0$, the collector current is equal to $-I_{CS}$ for large reverse collector voltages. For $I_B = 0, I_C$ is

$$I_C|_{I_B=0} = \frac{-I_{CS} + \alpha_F \alpha_R I_{CS}}{1 - \alpha_F} \tag{13.34}$$

In actual transistors there is excess leakage current associated with crystal imperfections at the surface region of the junctions. Such currents are usually specified on a vendor's specification sheet as I_{CBO}, collector-base current with emitter open, or I_{CEO}, collector-emitter current with base open. The relation of I_{CEO} to I_{CBO} is approximately

$$I_{\text{CEO}} = h_{\text{FE}} I_{\text{CBO}} \tag{13.35}$$

Therefore the collector-base junction leakage current is magnified substantially by the gain of the transistor. At high temperatures, especially, significant power dissipation can be associated with the leakage current.

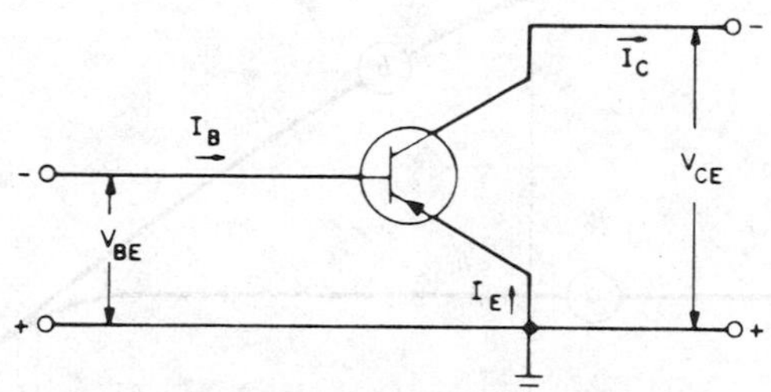

Fig. 13.23 *PNP* transistor in common-emitter configuration.

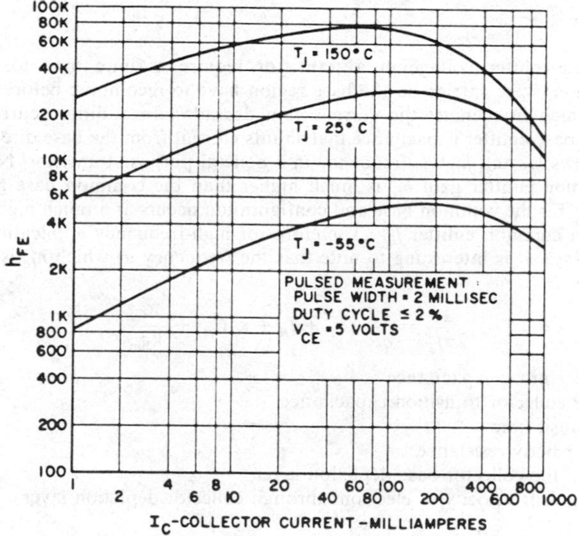

Fig. 13.24 Transistor current gain vs. collector current.

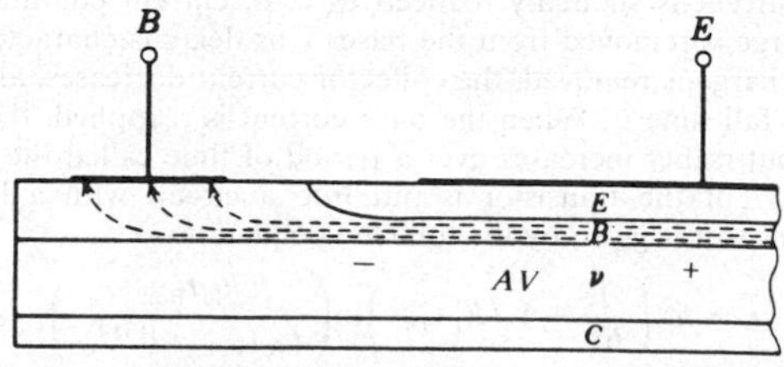

Fig. 13.25 Resistive voltage drop between center of emitter and edge of emitter due to lateral base current.

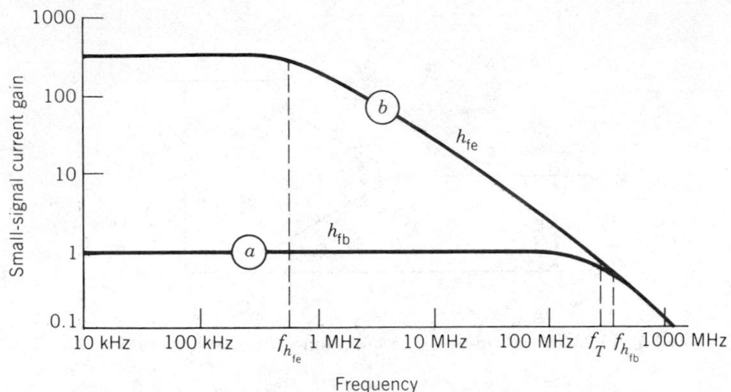

Fig. 13.26 Gain vs. frequency for common base (h_{fb}) and common emitter (h_{fe}) configurations.

Gain vs. Frequency

Changes in the base-emitter voltage in a transistor require a finite time to propagate to the base-collector junction, since carriers in the base region have to recombine before reaching a steady state. Therefore at high frequencies the current gain degrades. In addition, current gain degrades because of parasitic base-emitter capacitance that shunts current from the base directly to the emitter.

Figure 13.26 shows h_{fe} and h_{fb} vs. frequency in a general purpose transistor. Note that while the low-frequency common emitter gain h_{fe} is much higher than the common base gain h_{fb}, the 3 dB attenuation point $f_{h_{fb}}$ for the common base and configuration occurs at a much higher frequency than the 3 dB point for a common emitter $f_{h_{fe}}$. Therefore for high-frequency applications, common base offers some advantages. It is interesting to note that the frequency at which h_{fe} is equal to 1 (f_T) is very close to $f_{h_{fb}}$.

$$\frac{1}{2\pi f_T} = \frac{kT}{qI_C} C_{tE} + t_B + r_C C_{tC} + \frac{x_C}{V_{\text{lim}}} \tag{13.36}$$

where C_{tE} = average emitter capacitance
C_{tC} = average collector transition capacitance
t_B = base transit time
r_C = collector body resistance
x_C = width of the collector-base depletion layer
V_{lim} = emitting drift velocity of electrons through collector depletion layer

Transient Response

As mentioned, when base current suddenly changes a period of time is required before steady state is achieved in the collector current. In digital or power switching applications, the transistor is often in the saturation mode when turned on in order to minimize conduction losses in the devices. In the saturation mode, excess charge is built up in the base because the base-collector junction is forward biased. When the base current is suddenly reduced to zero, current continues to flow at the same value until the excess charge is removed from the base. This delay is characterized by a storage time t_s. After the excess base charge is removed, the collector current decreases, and the time that it takes for it to reach zero is the fall time t_f. When the base current is reapplied, the collector current does not rise instantaneously but rather increases over a period of time called the rise time, t_r.

Referring to Fig. 13.27, if the transistor is put into a circuit with a load resistance R_L, the switching time for the transistor is given by the following equations

$$t_r = h_{fe}\left(\frac{1}{6f_r} + 1.7R_L C_{tC}\right)\ln\left(\frac{h_{fe}I_{\text{Bon}}}{h_{fe}I_{\text{Bon}} - 0.9I_C}\right) \tag{13.37}$$

$$t_f = h_{fe}\left(\frac{1}{6f_r} + 1.7R_L C_{tC}\right)\ln\left(\frac{I_C - h_{fe}I_{\text{Boff}}}{0.1I_C - h_{fe}I_{\text{Boff}}}\right) \tag{13.38}$$

$$t_s = \tau_s hr.\ \ln\left(\frac{I_{\text{Bon}} - I_{\text{Boff}}}{I_C/h_{fe} - I_{\text{Boff}}}\right) \tag{13.39}$$

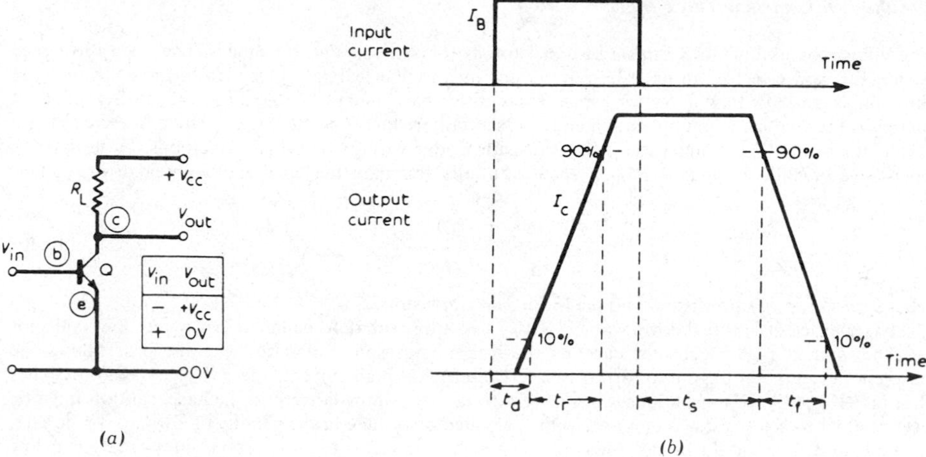

Fig. 13.27 Transistor switching. (*a*) Circuit arrangement for switching. (*b*) Transient current during switching.

where C_{tC} = collector transition capacitance
I_{Boff} = forced current out of base during turn-off
I_{Bon} = forced current into base during turn-on
τ_s = characteristic storage time

For a given transistor the switching speed can be improved by preventing the device from going into saturation, which means that $t_s \approx 0$. This can be accomplished by shunting the collector-base junction with a germanium or Schottky diode which conducts current at a lower voltage than does silicon. When the collector-base junction becomes forward biased as the transistor enters saturation, the voltage is clamped by the bypass diodes and the excess charge bypasses the base of the transistor. The disadvantage of a clamping diode is that the voltage drop of the transistor will be higher than the drop of one in true saturation.

Transistor switching speed is also improved during the manufacturing process by degrading the carrier lifetime, by diffusing heavy metals such as gold or platinum into the semiconductor, or by irradiating with high-energy electrons. The penalty for doing this, however, is that leakage currents are increased and current gain β is decreased.

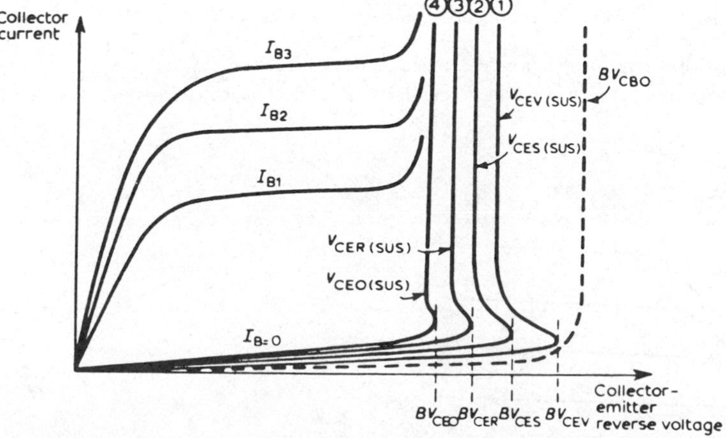

Fig. 13.28 Transistor avalanche breakdown under various base-bias conditions. (1) V_{CEV}: reverse-biased base emitter; (2) V_{CES}: base emitter short circuit; (3) V_{CER}: external resistance across base-emitter junction; (4) V_{CED}: open-circuit base.

Breakdown Characteristics

The voltage breakdown mechanism in transistors is the same as that for diodes. The main difference is that the gain effect in an open-base transistor means that leakage current from the collector-base junction is amplified as it passes across the emitter-base junction. Therefore the transistor effect increases the avalanche multiplication effect, essentially reducing the voltage at which $M \rightarrow \infty$. Figure 13.28 shows the *VI* characteristics of a transistor under various base-bias conditions. Note that the open-base breakdown voltage BV_{CEO} is substantially less than the open-emitter breakdown voltage BV_{CBO}. In fact

$$BV_{CEO} = \frac{BV_{CBO}}{(h_{fe})^{1/n}} \tag{13.40}$$

where $n \approx 4$ for *npn* transistors and $n \approx 6$ for *pnp* transistors.

The shorted-base breakdown voltage BV_{CES} is approximately equal to BV_{CBO} at low collector currents, but at higher collector currents the internal resistance in series with the base allows the base-emitter junction to go into forward bias. Therefore at high currents the BV_{CES} curve approaches that of BV_{CEO}. Since most applications have a finite drive impedance into the base, caution must be exercised when a transistor is operated with the collector voltage in the vicinity of BV_{CEO}. To be safe, normal operation should be designed to occur well below BV_{CEO}, with some safety margin below tested values.

Temperature Effects

As with a *pn*-junction diode, increasing temperature can cause leakage currents to increase. Since the reverse-biased base-collector junction leakage current I_{CBO} is amplified by the current gain of the transistor [Eq. (13.35)], a significant amount of power can be generated by the leakage current I_{CEO} falling across the potential drop of V_{CE}.

Increasing temperature also increases the current gain, up to the point at which the number of thermally generated carriers in the base region starts to equal the impurity dopant concentration in the base. The transistor then begins to lose its amplifying characteristics. Since the transistor degradation occurs at some junction temperature T_{jmax}, the power that can be dissipated in the transistor is given as

$$P_{diss} = \frac{T_{jmax} - T_{amb}}{R_{\theta jc}} \tag{13.41}$$

where T_{amb} is the ambient temperature of the transistor case and $R_{\theta jc}$ is the thermal resistance between the transistor junction and the case (see Section 13.1 for a discussion of thermal resistance).

Actual Transistor Characteristics

The ideal transistor characteristics shown in Fig. 13.22 change significantly during actual transistor operation (Fig. 13.29). The physical phenomena responsible for the changes are:

1. The resistive voltage drop across the collector when current is flowing
2. The narrowing of the effective base region as the collector-depletion region penetrates the base region

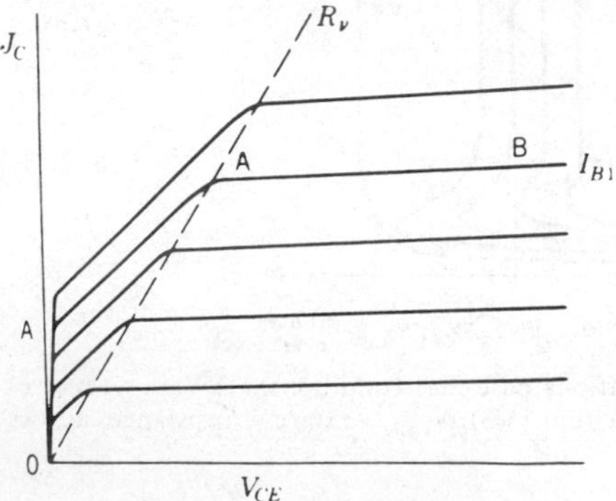

Fig. 13.29 Characteristics of actual transistor.

3. The resistive voltage drop laterally across the base, which causes the injected emitter current to crowd toward the edge of the emitter

In higher voltage transistors the collector region is lightly doped with impurities. In the operating region 0 to A′ the transistor is heavily saturated and the base-collector junction, under strong forward bias, injects many carriers into the collector region, which reduces the resistance in the collector. In the region A′ to A the base-collector junction becomes less heavily forward biased and there is less modulation of the conductivity in the collector. Therefore the collector body resistance increases and the transistor is said to be in quasi-saturation. At point A on the curve the transistor goes out of saturation and the base-collector junction becomes reverse biased. It can be seen that quasi-saturation in a transistor significantly reduces the current gain at low collector-emitter voltages.

Second Breakdown

Another phenomenon that limits transistor performance is second breakdown, the usually destructive breakdown of the transistor voltage-withstand capability that is below the value normally associated with avalanche breakdown. Two types of second breakdown are usually observed: forward-bias second breakdown, where the base current is positive, turning on the transistor which has been blocking large voltages; and reverse-bias second breakdown, where the transistor is in an "on" condition and the base-emitter junction is reverse biased to turn the transistor off. Second breakdown is usually associated with localized regions of excess heat due to anomalously high thermal resistance (e.g., over solder voids) or due to localized current crowding. Current crowding can occur during turn-off when carriers in the base are removed near the edge of the emitter first, forcing the current to squeeze under the center of the emitter. It can also occur during turn-on because the emitter periphery turns on before the rest of the emitter and large amounts of power can be dissipated at the emitter edge. The effect of second breakdown is to reduce the safe operating area (SOA) within which the transistor can be reliably operated. Figure 13.30 gives an example of the safe region of operation for a transistor.

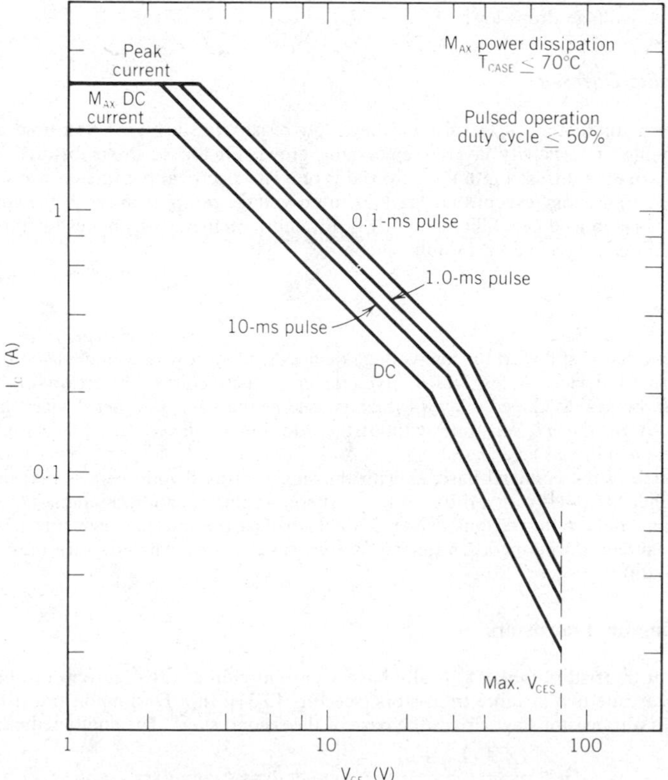

Fig. 13.30 Safe operating area (SOA) of transistor.

13.2-3 Power Transistor Types

Transistor construction varies with the age of the device type and with the applications the device is intended to fill.

Single Diffused

Probably the simplest construction is the single-diffused transistor, which is fabricated by simultaneously diffusing the emitter and collector dopants into a thin slice of base material. This process is limited to low-voltage devices, since the base-collector depletion region extends into the base region. The thick base region that is required for the fabrication of the transistor leads to a relatively low current gain in this structure. The thick base region combined with the absence of a drift field in the uniformly doped base region also makes this structure somewhat slow.

Triple Diffused

For higher voltage devices, high-resistivity n material (ν) is used as the starting material. This forms the bulk-collector region, which supports the base-collector junction voltage. The low-resistance n^+ collector is then diffused into the material, with the top surface protected. Diffusion is followed by a p^+-base diffusion into the top surface and an n^+-emitter diffusion. The depletion region associated with the reverse-biased base-collector junction is primarily into the ν region. Therefore the base region can be much narrower than that in a single-diffused transistor. The thinner base combined with the drift field associated with the diffused base makes the triple-diffused device faster than the single-diffused device. However, crystal dislocations associated with the base diffusion, as well as charge storage in the ν region during turn-off, tend to make the triple-diffused transistor electrically less rugged than the single-diffused device. That is to say, second breakdown can be more of a problem in the triple-diffused transistor than in a single-diffused one.

From a manufacturing point of view both the single-diffused and triple-diffused device structures are unattractive because the starting semiconductor material must be thin in order to keep collector and base resistive voltage drops low.

Epitaxial Double Diffused

To facilitate the manufacture of transistors, thick, low-resistivity silicon is often used as the starting material and a higher resistivity layer is epitaxially grown on top of this substrate. The base and emitter regions are then diffused into the epitaxial layer. The electrical parameters are similar to those of triple-diffused transistors, except that the maximum voltage rating is lower in the epitaxial device. The reason for this reduced capability is that it is difficult to achieve resistivities in epitaxial layers as high as those that can be achieved in bulk silicon.

Epitaxial Base

An older process that is still used for low-voltage devices is to start with a thick n^+ substrate, as with the double-diffused device. A high-resistivity n layer is then epitaxially grown on the substrate, followed by a more heavily doped p layer, which is used as the base. In older devices the p layer may be grown directly on the n^+ substrate without the additional n layer. In either case the emitter is diffused into the p epitaxial layer.

The advantage of the epitaxial base, as with the single-diffused transistor, is that the base region resistivity is higher than that with a diffused base. Therefore there is additional resistance between the base contacts and the emitter regions, which helps to ballast the internal segments of the transistor. This helps to ensure uniform operation during turn-on and turn-off, which in turn reduces the chance of second breakdown.

13.2-4 Darlington Transistors

For high-current transistors, which typically have a current gain at rated current of only about 10, it is common to cascade two or more transistors (see Fig. 13.31). In a Darlington transistor the emitter current of the first transistor stage drives the base of the second stage. The combined current gain is

$$\frac{I_C}{I_{B1}} = \frac{I_{C1} + I_{C2}}{I_{B1}} = h_{\mathrm{FE1}} + h_{\mathrm{FE2}} + h_{\mathrm{FE1}} h_{\mathrm{FE2}} \tag{13.42}$$

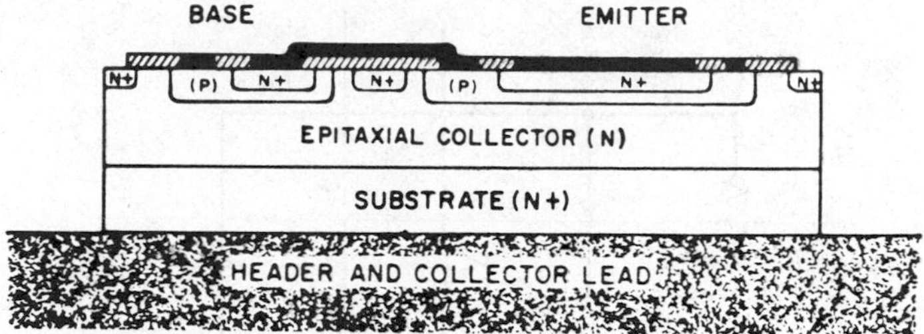

Fig. 13.31 Construction of a monolithic Darlington transistor.

Monolithic Darlington transistors are usually constructed so that resistors shunt the base-emitter junction of the transistors. There is also sometimes an antiparallel bypass diode around the base-emitter junction of the first stage in order to remove the charge from the second base during turn-off.

While the gain of a Darlington transistor is considerably higher than that of a single-stage transistor, it should be recognized that for a given level of current the Darlington transistor will dissipate more power than a single-stage device. The reason for this is that V_{CEsat} in a Darlington transistor will be at least 0.5 V because the emitter of the first stage has to be at least a diode drop above the emitter of the second stage in order to turn on the second stage. Therefore the collector of the first stage, which is common with the second stage, has to be at a voltage at least equal to the voltage of the second stage base-emitter junction, which is approximately 0.5 V or larger. A single-stage transistor, on the other hand, can have a V_{CEsat} approaching 0 V at low currents.

13.2-5 Considerations for Transistor Applications

Parallel Operation

In applications that call for currents larger than a single transistor can handle, several transistors are often connected in parallel. The problem with this is that the transistors have to be carefully matched. Otherwise the device with the lowest V_{CEsat} can "hog" all of the current. Since transistors have a positive temperature coefficient, the overloaded device can overheat and conduct even more of the current in a regenerative process that can eventually lead to destruction.

One way to balance the loading through parallel transistors, other than by testing and matching *VI* characteristics, is to put a low-value resistor in the emitter leg of each transistor (Fig. 13.32). With this configuration there is negative feedback. As the current increases in any one leg, the voltage increases across the resistor, reducing the base-emitter junction voltage which in turn reduces the current through the guilty transistor.

Serial Operation

Some applications may require that a transistor block more voltage than its rating allows. In these cases more than one transistor may be connected in series if certain precautions are taken, that is, if

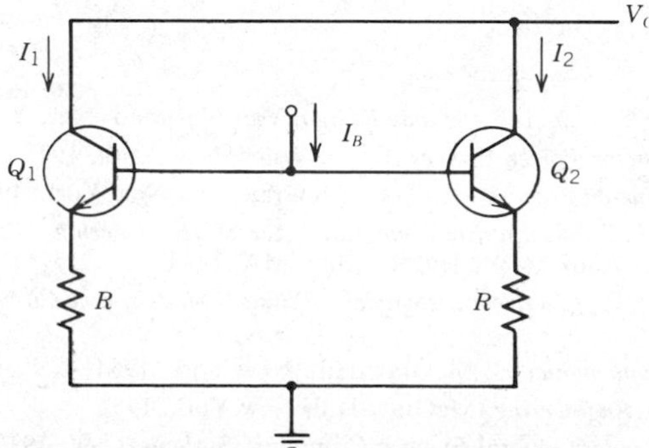

Fig. 13.32 Emitter ballasting for equalization of current in parallel transistors.

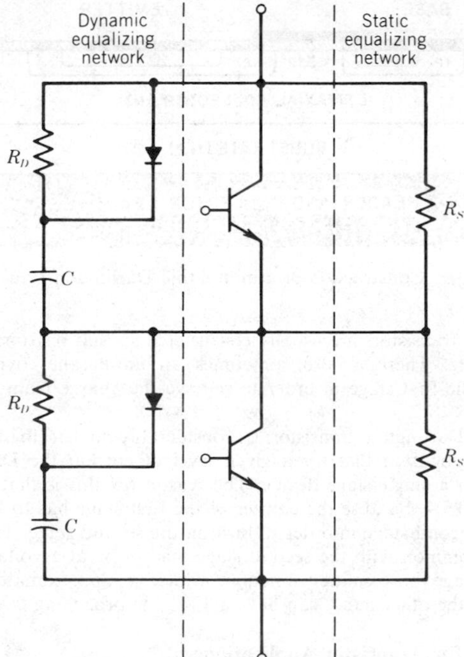

Fig. 13.33 Equalizing network for serial transistor connection.

voltage sharing between the individual transistors at steady state and transient operating conditions is equalized. Due to differences in junction capacitances, delay times, and leakage currents for individual transistors, external equalization networks and special consideration for base-drive circuits are required. One such equalizing network is shown in Fig. 13.33. The resistors, R_s, equalize the static leakage currents. The capacitors, C, equalize the voltage during turn-on and turn-off while the resistors limit the power dissipation in the RC network.

Snubber Networks

The RC network across the transistors, in the example of serial operation in the previous section, is sometimes used with single transistors in applications where voltage transients can potentially take the transistor out of its safe operating area. For example, inductive loads such as motors can create voltage transients if the current is suddenly interrupted. The values of R and C are often determined empirically to give the lowest power loss and fastest switching speed consistent with the possible transients in the circuit.

Bibliography

Fitchen, F. C., *Transistor Circuit Analysis and Design*, Van Nostrand, New York, 1960.

Ghandi, S. K., *Semiconductor Power Devices*, Ch. 4, Wiley, New York, 1977.

Gibbons, J. F., *Semiconductor Electronics*, Ch. 9, McGraw-Hill, New York, 1966.

Glaser, A. B. and G. E. Subak-Sharpe, *Integrated Circuit Engineering—Design, Fabrication, and Applications*, Ch. 2, Addison-Wesley, Reading, MA, 1979.

Gray, P. E. and C. L. Searle, *Electronic Principles—Physics, Models, and Circuits*, Ch. 7, Wiley, New York, 1969.

Moll, J. L., *Physics of Semiconductors*, McGraw-Hill, New York, 1964.

Phillips, A. B., *Transistor Engineering*, McGraw-Hill, New York, 1962.

Power Transistor Users Guide, General Electric Company, Syracuse, NY, 1975.

SCR Manual, Ch. 6, General Electric Company, Syracuse, NY, 1972.

Searle, C. L., A. R. Boothroyd, E. J. Angelo, P. E. Gray, and D. D. Pederson, *SEEC Notes 2—Elementary Circuit Properties of Transistors*, Wiley, New York, 1964.

Semiconductor Data Book, General Electric Company, Syracuse, NY, 1977.

Streetman, B. G., *Solid State Electronic Devices*, Ch. 7, Prentice-Hall, Englewood Cliffs, NJ, 1980.

Sze, S. M., *Physics of Semiconductor Devices*, Ch. 6, Wiley, New York, 1969.

Towers, T. D. and S. Libes, *Semiconductor Circuit Elements*, Chs. 5–9, Hayden, Rochelle Park, NJ, 1977.

Yang, E. S., *Fundamentals of Semiconductor Devices*, Ch. 9, McGraw-Hill, New York, 1978.

13.3 FIELD EFFECT TRANSISTORS

Mark B. Barron

There are two basic types of field effect transistors: insulated gate and junction gate. The latter can be broken down into *pn* junctions and Schottky junctions.

13.3-1 Insulated Gate Field Effect Transistors

The most prevalent transistors in today's applications are the insulated-gate variety, primarily because they are the easiest to fabricate and apply in integrated circuits. In an insulated-gate device, a metallic layer is placed on top of an insulator that is on the surface of a semiconductor. A voltage applied between the metal and the semiconductor establishes an electric field across the insulator and into the semiconductor. The electric field at the surface of the semiconductor attracts and repels holes and electrons, the carrier attracted being determined by the polarity of the field. For example, if a positive voltage is applied to the metal, electrons will be attracted to the surface of the semiconductor and holes will be repelled. If the semiconductor is *p* type, a small positive voltage will repel holes from the semiconductor surface and the electric field will be terminated by the ionized acceptor dopant atoms. As the voltage is increased, the surface becomes totally depleted of holes and mobile electrons begin to accumulate at the surface. When the surface region becomes heavily populated with electrons, the semiconductor becomes "inverted." If *n*-type regions have been diffused or implanted into the *p*-type semiconductor at the edge of the metal (Fig. 13.34*a*) and a voltage is applied between them, current can flow between these *n* regions if the semiconductor surface has inverted (Fig.

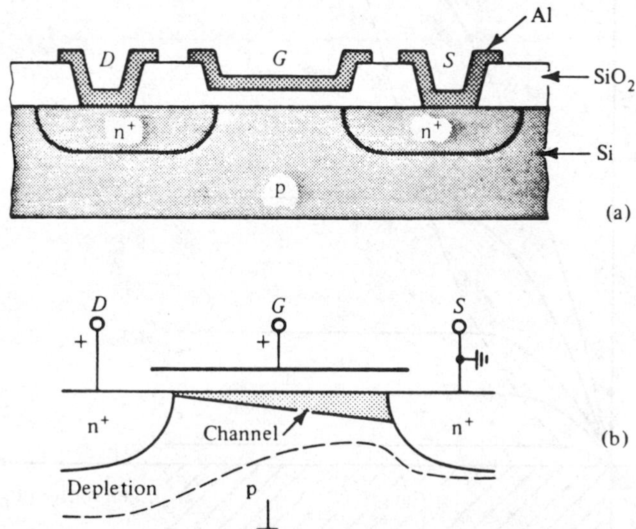

Fig. 13.34 (*a*) Construction of metal oxide semiconductor field effect transistor (MOSFET). S, source; D, drain; G, gate. (*b*) Inversion region (channel) in MOSFET under voltage bias.

13.34*b*). When the positive voltage on the metal is reduced below a threshold level, V_t, the *p*-type semiconductor will no longer be inverted and current will cease to flow between the *n* regions. The value of the threshold voltage is determined by the thickness of the insulator, the doping concentration of the semiconductor substrate, and the amount of trapped charge that may reside in the insulator.

Most insulated-gate transistors are fabricated using silicon dioxide as the insulator. Thus, the *m*etal *o*xide *s*emiconductor *f*ield *e*ffect *t*ransistor, or MOSFET, is the prevalent structure in use today. Figure 13.34 shows an "*n*-channel" MOSFET. The device consists of two *n*-type regions formed in a *p*-type substrate. One of these regions is called the *source* (because it is the source of carriers when the transistor is "turned on"), and the other region is called the *drain* (because it acts as a drain for the carriers to flow into). The metal over the oxide is called the *gate* because it is used to gate the device on and off. A common material for the gate is aluminum, but other materials such as polycrystalline silicon and refractory metals (or metal silicides) such as tungsten or molybdenum are sometimes used.

A *p*-channel MOSFET is identical to the *n*-channel device except that *n*-type material is used as the starting substrate and *p* regions are used to form the source and drain. In a *p*-channel transistor, a negative voltage is needed to invert the surface of the *n*-type substrate and thereby allow current to flow between the source and a negatively biased drain.

The charge in the inverted surface region of a MOSFET device can be easily derived by comparing the device with a simple capacitor. The charge on the capacitor is $Q = CV$. In a MOSFET device, the gate capacitance per unit area when the surface is strongly inverted is

$$C_g = \frac{\epsilon_{ox}}{x_o} \tag{13.43}$$

where x_o is the oxide thickness and ϵ_{ox} is the permittivity of oxide. The mobile charge per unit area at the surface of the MOSFET is therefore

$$q = \epsilon_{ox}(V_g - V_t) \tag{13.44}$$

where V_g is the gate-substrate voltage and V_t is the threshold voltage that must be applied to turn the

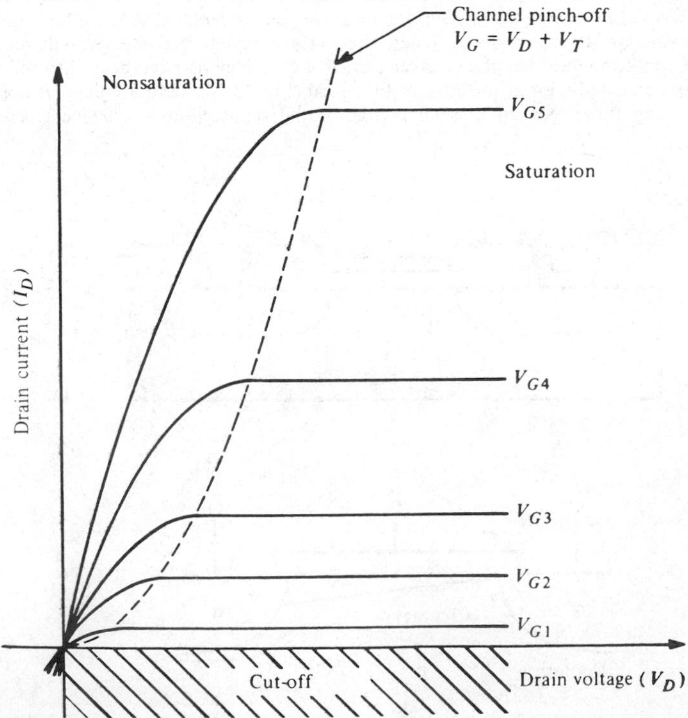

Fig. 13.35 Current-voltage characteristics of a metal oxide semiconductor field effect transistor (MOSFET).

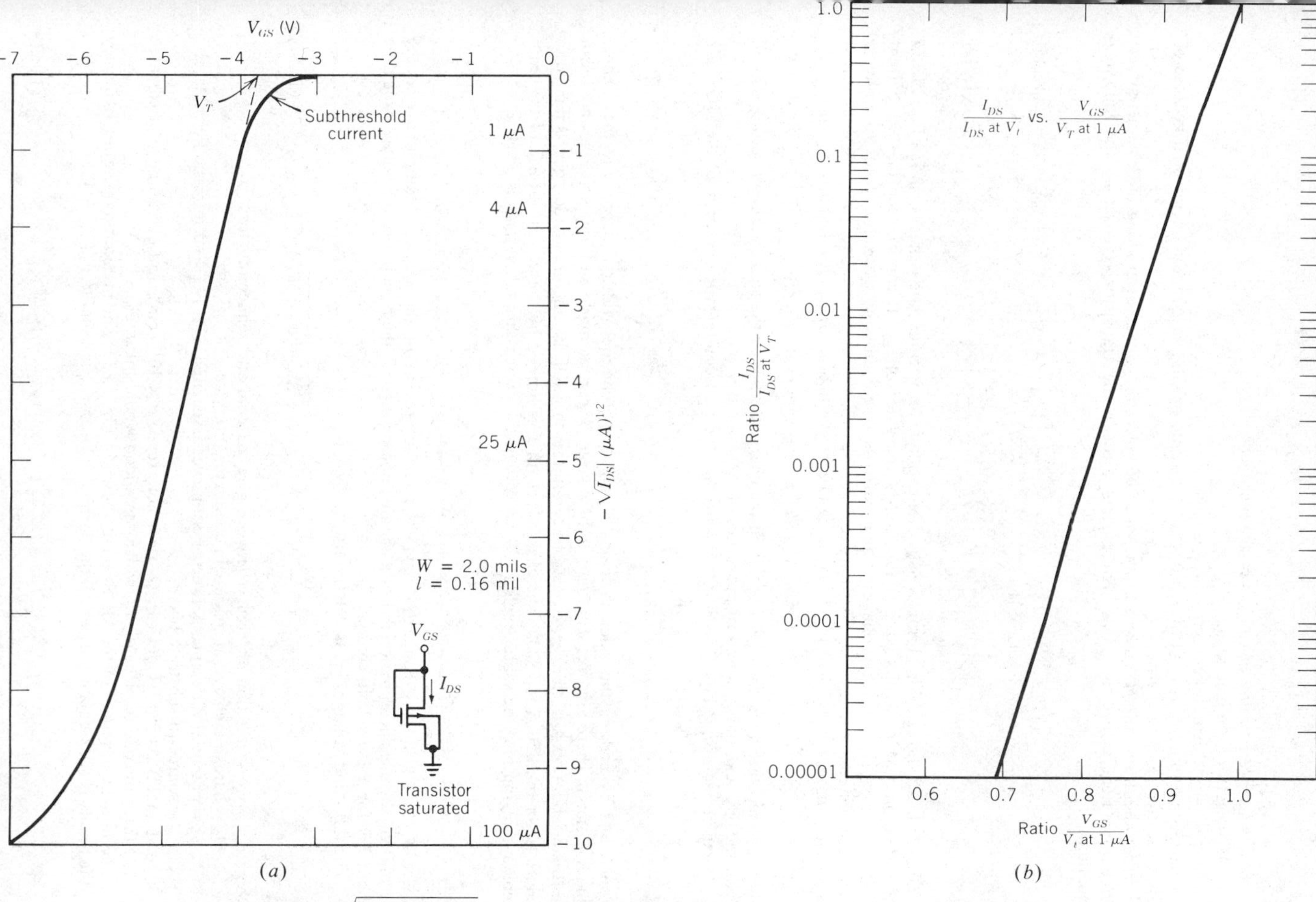

Fig. 13.36 (*a*) $\sqrt{\text{Drain current}}$ vs. gate voltage for a metal oxide semiconductor field effect transistor (MOSFET) in saturation. (*b*) Subthreshold current vs. gate voltage.

device on. For small drain-to-source voltages ($V_D \ll V_g - V_t$) it can be shown that the current through the channel of the MOSFET is

$$I_D = \frac{\mu_s \epsilon_{\text{ox}}}{x_o} \frac{W}{L} (V_g - V_t) V_D, \qquad V_D \ll (V_g - V_t) \tag{13.45}$$

where μ_s = surface mobility of carriers in channel
W = width of the gate region
L = channel length between source and drain

The surface mobility μ_s is typically about half the value in bulk semiconductor material because of surface scattering of the carriers.

When the drain voltage exceeds a few tenths of volts, the voltage drop along the channel affects the electric field across the oxide. The current through the transistor is given by the following equations

$$|V_g - V_t| > V_D \ : I_D = \frac{\mu_s \epsilon_{\text{ox}} W}{x_o L} \left[(V_g - V_t) V_D - \tfrac{1}{2} V_D^2 \right] \tag{13.46}$$

Nonsaturation region

$$|V_g - V_t| < V_D \ : I_D = \frac{\mu_s \epsilon_{\text{ox}} W}{2 x_o L} (V_g - V_t)^2 \tag{13.47}$$

Saturation region

Figure 13.35 shows drain current versus drain voltage for various gate voltages, and in Fig. 13.36*a* the saturation current is plotted as a function of the gate voltage. It can be seen that the current varies as the square of gate voltage at high currents where mobility saturation leads to a slower increase in current with voltage. Also, as seen in Fig. 13.36*b*, at gate voltages near V_t the current varies much more rapidly than V_g^2. In fact it can be shown that at low values the current is given as

$$I_D = I_o \exp(qV_g / nkT) \cdot [1 - \exp(-qV_D / kT)] \tag{13.48}$$

where I_o and n are constants that depend on the characteristics of the semiconductor surface as well as on the physical and geometrical properties of the transistor.

It can be seen from Eqs. (13.45) and (13.46) that the transconductance, or gain, of the MOSFET

$$g_m = \left. \frac{\partial I_D}{\partial V_g} \right|_{V_D > V_g - V_t} = \frac{\mu_s \epsilon_{\text{ox}}}{x_o L} W (V_g - V_t) \tag{13.49}$$

is increased by making the oxide thickness x_o small, the channel width W large, and the channel length L small. There are practical limits to how large W can be made and how small x_o and L can be. Below about 50 Å of thickness, electrons can tunnel through the silicon dioxide, so x_o is limited to values larger than 50 Å for SiO_2. The channel length L is limited by the punch-through of the drain depletion region into the source region. To first order, the punch-through voltage is given as

$$V_{\text{PT}} = \frac{L^2 q N_{\text{ch}}}{2 \epsilon_s} \tag{13.50}$$

where N_{ch} is the doping concentration in the surface region of the semiconductor and ϵ_s is the permittivity of the silicon material. For a channel doping of 10^{16}, the punch-through voltage would be about 7 V for a 1-μm-long channel.

The threshold voltage, V_t, is determined by

$$V_t = \phi_{\text{ms}} + \frac{Q_{fs}}{C_{\text{ox}}} + 2\psi_B + \frac{x_o}{\epsilon_{\text{ox}}} \cdot \sqrt{2 q \epsilon_s N_{\text{ch}} (2\psi_B + V_{\text{ss}})} \tag{13.51}$$

where ϕ_{ms} = work function difference between gate material and semiconductor
Q_{fs} = parasitic charge at oxide-semiconductor interface
ψ_B = potential difference between Fermi level associated with N_{ch} and intrinsic Fermi level
V_{ss} = source-to-substrate bias

For good-quality MOSFET devices, Q_{fs} does not exceed $q \times 10^{11}/\text{cm}^3$, and ψ_B can be approximated by half the semiconductor bandgap, which for silicon would mean $\psi_B \approx 0.5$ V.

The equivalent circuit of the MOSFET is shown in Fig. 13.37. The maximum operating frequency of the device is the frequency at which the current through C_{in} is equal to $g_m(V_G - V_t)$

$$f_{\max} = \frac{g_m}{2\pi C_{\text{in}}} \approx \frac{\mu_s}{2\pi L^2} V_D, \qquad |V_D| < |V_g - V_t| \tag{13.52}$$

$$f_{\max} = \frac{\mu_s (V_g - V_t)}{2\pi L^2}, \qquad |V_D| > |V_g - V_t| \tag{13.53}$$

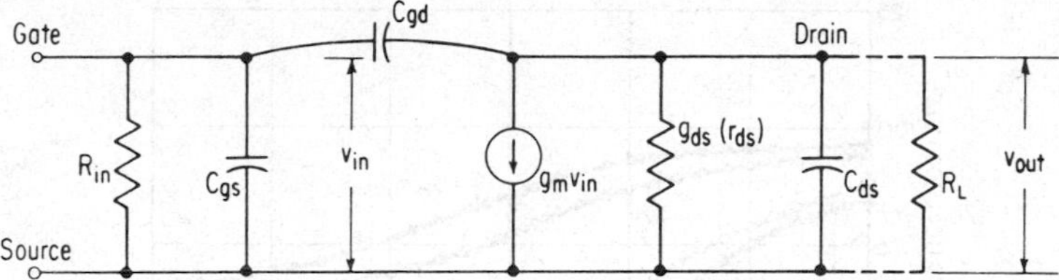

Fig. 13.37 Small signal equivalent circuit of a metal oxide semiconductor field effect transistor (MOSFET).

The drain output conductance, g_D, is associated with the modulation of the effective channel length, L, by the drain depletion region

$$\frac{1}{r_D} = g_D = \frac{\partial I_D}{\partial V_D} \approx \frac{\mu_s C_{\text{ox}}}{\pi L^2} \frac{(V_g - V_t)^2}{(V_D + V_t - V_g)}, \qquad |V_D| > 2|V_g - V_t| \tag{13.54}$$

The gain of a MOSFET device having a negative feedback source resistance is given as

$$A'_V = \frac{-g_m r_D/(G_L + g_D)}{r_D + R_L + (g_m r_D + 1)R_s} \approx \frac{-g_m R_L}{1 + g_m R_s} \tag{13.55}$$

if r_D and g_m are assumed large.

Because a MOSFET is a unipolar device (i.e., only one type of carrier is associated with current flow), the switching speed is limited only by the current the device can carry and the capacitive load to be driven. For a MOSFET in saturation, the transit time of carriers from source to drain is

$$t_T = \frac{4}{3}\left(\frac{L^2}{\mu_s}\right)\frac{1}{(V_g - V_t)} \tag{13.56}$$

For a 3-μm channel length and $V_G - V_t = 1$ V, the transit time is approximately 0.2 ns. As long as the circuit RC time constants exceed this value, the transit time can be ignored when assessing the transient performance of a MOSFET in the circuit. For a MOSFET driving a capacitive load C_L, the switching time is found by adding the time required for the voltage across C_L to decrease from V_{DD} to $V_G - V_t$, plus the time for it to decrease from $V_G - V_t$ to near zero. Referring to Fig. 13.38, the

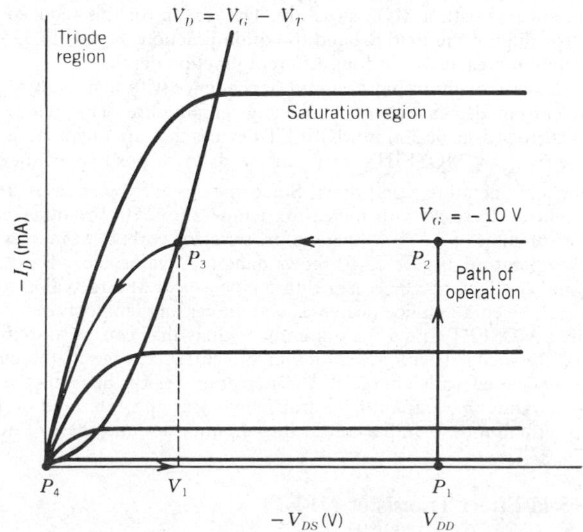

Fig. 13.38 Transient path of operation of a metal oxide semiconductor field effect transistor (MOSFET) ($P_1 \rightarrow P_4$).

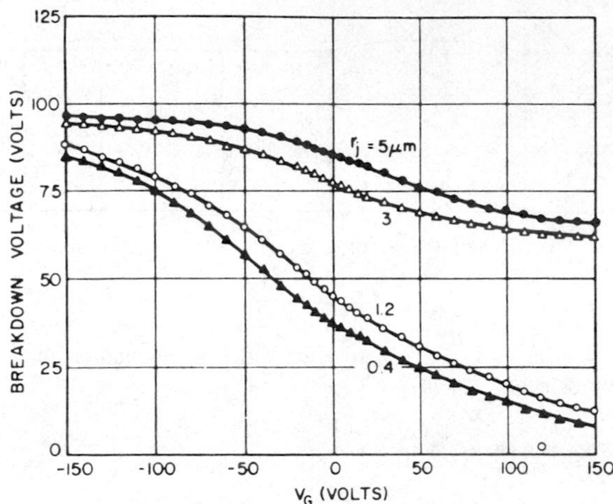

Fig. 13.39 Drain-substrate breakdown voltage vs. gate voltage for four different drain diffusion depths.

time required for the load voltage to decay from $V_{\rm DD}$ to $V_G - V_t$ is

$$t_1 = \frac{2C_L|V_{\rm DD} - V_g + V_t|}{g_m(V_g - V_t)} \tag{13.57}$$

The time, t_2, which is the time for the load voltage to decay from $(V_G - V_t)$ to zero, is found by solving for t in the equation

$$\frac{v(t)}{(V_g - V_t)} = \frac{2e^{-t/\tau}}{1 + e^{-t/\tau}} \tag{13.58}$$

where

$$\tau = C_L/g_m, \qquad g_m = \frac{\mu_s C_{\rm ox} W}{L}(V_g - V_t)$$

Junction breakdown voltages in MOSFET devices are generally lower than those in normal *pn* junctions that are not covered with a MOS structure. The reason for this reduced voltage is that the gate metal causes a crowding of the field around the drain junction. Figure 13.39 shows the effect of gate voltage on the junction breakdown of four different junction depths.

MOSFETs are normally thought to have current dependence with temperature that has a negative coefficient such that current decreases with increasing temperature. This temperature dependence helps to ensure equal current distribution in MOSFET devices that are operated in parallel. However, there are opposing effects in MOSFETs that can lead to a positive or negative temperature coefficient, depending on operating conditions. Since the mobility decreases as $T^{-\frac{3}{2}}$, there is a tendency for the current to decrease with increasing temperature. On the other hand, the threshold voltage V_t varies as a function of $1/T$, so there is pressure for current to increase since $(V_G - V_t)^2$ would increase with temperature. Figure 13.40 shows that at V_G values near V_t, the threshold voltage variation dominates and conductance increases with temperature. At large values of V_G, the mobility variation dominates and the conductance decreases with increasing temperature.

Electrical noise in a MOSFET limits the minimum signals that can be amplified. Three types of noise are thermal noise associated with variations in the carrier charge in the channel, generation-recombination noise associated with charge in the depletion region beneath the channel, and $1/f$ noise that is due to a random occupation of trap centers at the interface of the oxide and the semiconductor. The thermal noise is important at high frequencies and the $1/f$ noise is dominant at low frequencies.

13.3-2 Junction Field Effect Transistor (JFET)

A *j*unction *f*ield *e*ffect *t*ransistor is similar to a MOSFET except that the gate is replaced by a *pn* junction (Fig. 13.41*a*) and the device normally operates by depleting an already existing channel. The

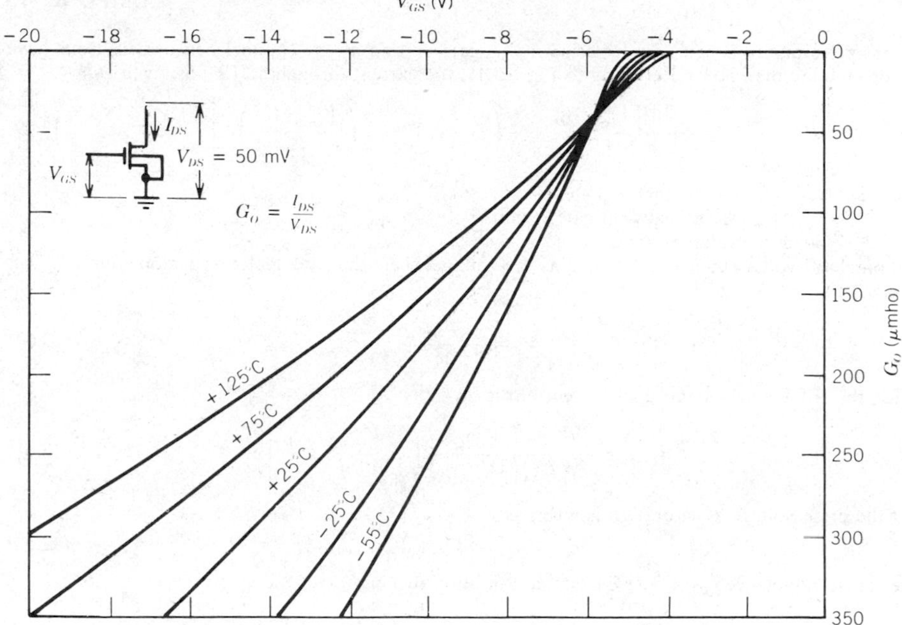

Fig. 13.40 Variation of channel conductance with temperature in a metal oxide semiconductor field effect transistor (MOSFET).

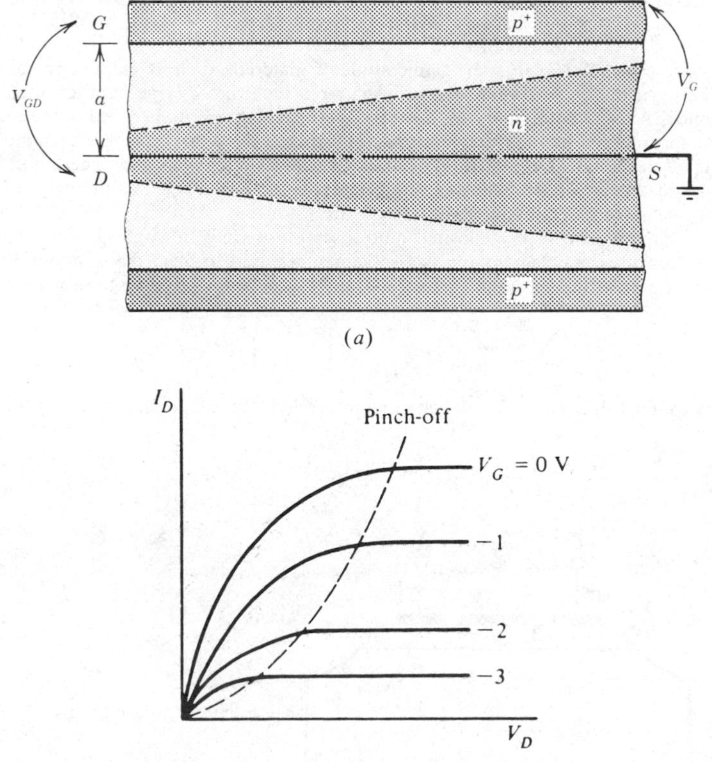

Fig. 13.41 (*a*) Construction of a junction field effect transistor (JFET). (*b*) Current-voltage characteristics in an *n*-channel JFET.

highest performance is achieved by using a $p+$ gate and an n-type channel because electrons have higher mobility than holes. Referring to Fig. 13.41, the current through a JFET is given as

$$I_D = \frac{2aWV_p}{\rho_c L}\left[\frac{V_D}{V_p} + \frac{2}{3}\left(-\frac{V_G}{V_p}\right)^{\frac{3}{2}} - \frac{2}{3}\left(\frac{V_D - V_G}{V_p}\right)^{\frac{3}{2}}\right] \tag{13.59}$$

where ρ_c = resistivity of channel
$a = 1/2$ the spacing between gate regions
V_p = pinch-off voltage

The pinch-off voltage can be calculated as the voltage where the gate depletion region equals a. This voltage is

$$V_p = \frac{-a^2 q N_D}{2\epsilon_s} \tag{13.60}$$

When the JFET enters saturation, the current is given as

$$I_{D\,\mathrm{sat}} = \frac{2aWV_p}{\rho_c L}\left[\frac{V_G}{V_p} + \frac{2}{3}\left(-\frac{V_G}{V_p}\right)^{\frac{3}{2}} + \frac{1}{3}\right] \tag{13.61}$$

and the drain voltage at saturation is given as

$$V_{D\,\mathrm{sat}} = V_G + V_p \tag{13.62}$$

The transconductance g_m of a JFET in the saturation region is

$$g_{m\,\mathrm{sat}} = \frac{2aW}{\rho_c L}\left[1 - \left(-\frac{V_G}{V_p}\right)^{1/2}\right] \tag{13.63}$$

Junction field effect transistors at one time were selected for applications requiring a high input impedance, such as the front-end amplifiers for radios and for electrometers. However, because they are difficult to manufacture and because of the improving characteristics of MOSFETs, JFETs are less widely used today than are MOSFETs.

MOSFETs are not practical in some semiconductor materials such as gallium-arsenide because the interface between the GaAs and a surface insulator is such that a large number of surface traps inhibit gate action. Also gate junctions are not easily diffused into GaAs because of the instability of the semiconductor material at high temperatures. However, since GaAs has a higher mobility than silicon, its use is desirable in applications requiring fast switching speeds. Therefore JFET-type structures are used, except that a Schottky diode replaces the pn junction as the gate. Figure 13.42 shows such a *m*etal *s*emiconductor *f*ield *e*ffect *t*ransistor, or MESFET. Often MESFETs are fabricated on a thin epitaxial layer deposited onto a semi-insulating substrate. Ion implantation can also be used to form this thin region, and is also sometimes used to form the source-drain contacts. Since only a single-sided gate is used in such structures, the current is given by the equation

$$I_D = \frac{x_c WV_p}{\rho_c L}\left[\frac{V_G}{V_p} + \frac{2}{3}\left(-\frac{V_G}{V_p}\right)^{\frac{3}{2}} + \frac{1}{3}\right] \tag{13.64}$$

where x_c is the channel thickness, ρ_c is the channel resistivity, and the pinch-off voltage V_p is given as

$$V_p = \frac{q x_c^2 N_D}{2\epsilon_s} \tag{13.65}$$

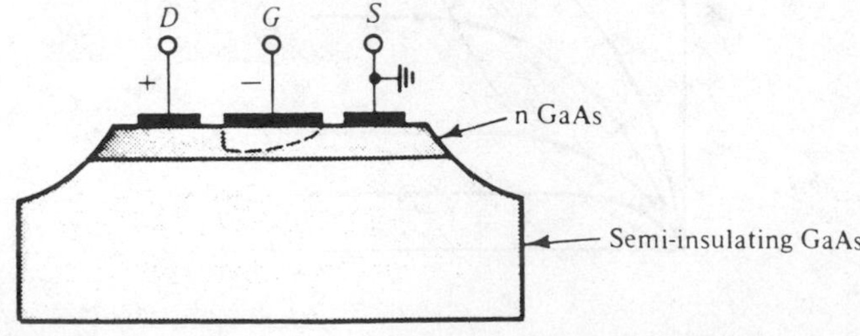

Fig. 13.42 Gallium-arsenide metal semiconductor field effect transistor (MESFET).

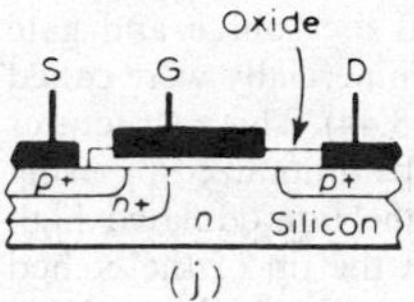

Fig. 13.43 Double-diffused metal oxide semiconductor field effect transistor (DMOSFET) construction.

13.3-3 MOSFET Variants

Dual Gate MOSFETs

For high-frequency applications where low distortion is needed, dual gate MOS structures are sometimes used. These MOSFET tetrodes operate much the same as a standard MOSFET except that the primary gate is decoupled from the drain region by the presence of the secondary gate. The benefit of the second gate is that the gate-drain capacitance is substantially reduced. Also such devices may be used in applications such as mixers and automatic gain control.

Double-Diffused MOSFETs

Another approach to obtaining high-speed MOSFETs is to fabricate the device in a high-resistivity substrate. The active channel region is formed by diffusing an n-type region followed by a $p+$ diffusion, which forms the source. The channel length is then determined by the difference in junction depths, as indicated in Fig. 13.43. Such a structure is referred to as a DMOSFET. The high-resistivity n region between the $p+$ drain and the source is depleted when a few volts are applied to the drain. Because high transconductance values can be maintained even with fairly large distances between the drain and the $n+$ region, higher breakdown voltages can be obtained than would be the case if the drain $p+$ region were immediately adjacent to the $n+$ channel region.

Power MOSFETs

The DMOS structure described above can be applied to the construction of power MOSFET devices. However, to achieve control of significant amounts of power with a MOSFET, it is necessary to have large device widths [W in Eq. (13.47)]. For maximum utilization of silicon area, power MOSFETS are

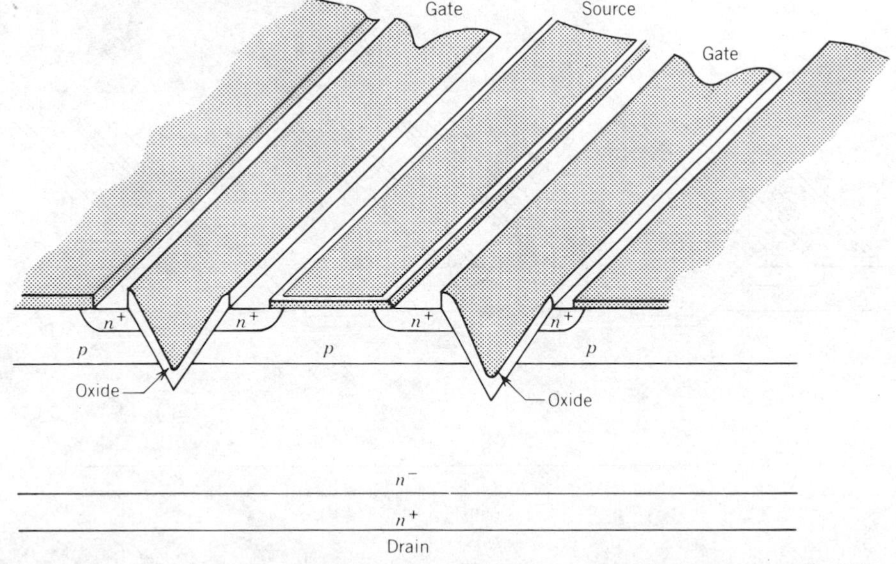

Fig. 13.44 Construction of a metal oxide semiconductor with a V-shaped groove (VMOS).

generally fabricated with the drain contact on the bottom of the silicon and the source and gate contacts on the top surface. The first power MOS devices to be introduced commercially were called VMOS because of the V-shaped groove that was etched into the silicon (Fig. 13.44). These structures are good for lower voltage applications (100 V), since the drain resistance can be minimized by using the etched "V" to reduce the distance that carriers have to travel through the unmodulated high resistivity n region. However, the V structure causes electric field crowding at the tip of the etched region, and this field concentration reduces the drain-source breakdown voltage. For higher voltage devices, a preferred structure is a planar DMOS structure (Fig. 13.45*a*) or a truncated VMOS structure (Fig. 13.45*b*). The dominant tradeoff in power MOSFETs is between on-resistance and breakdown voltage. It turns out that the ideal on-resistance varies as the 2.5 power of the ideal breakdown voltage, that is,

$$R_{\text{ideal}} \propto V_{\text{BV}}^{2.5} \tag{13.66}$$

Spreading resistance from the channel region to the drain contact increases the actual resistance. Figure 13.46 shows some actual device measurements compared with the ideal resistance as a function of breakdown voltage.

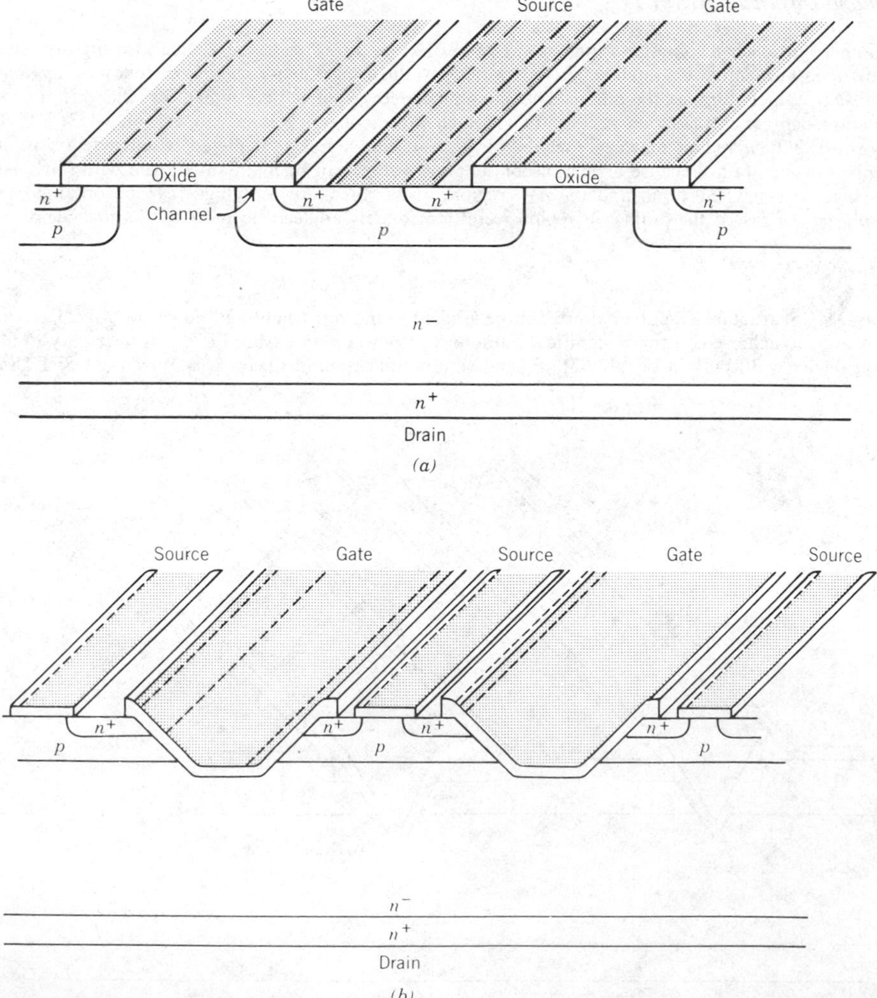

Fig. 13.45 (*a*) Power double-diffused metal oxide semiconductor field effect transistor (DMOSFET). (*b*) Truncated V-groove power MOSFET.

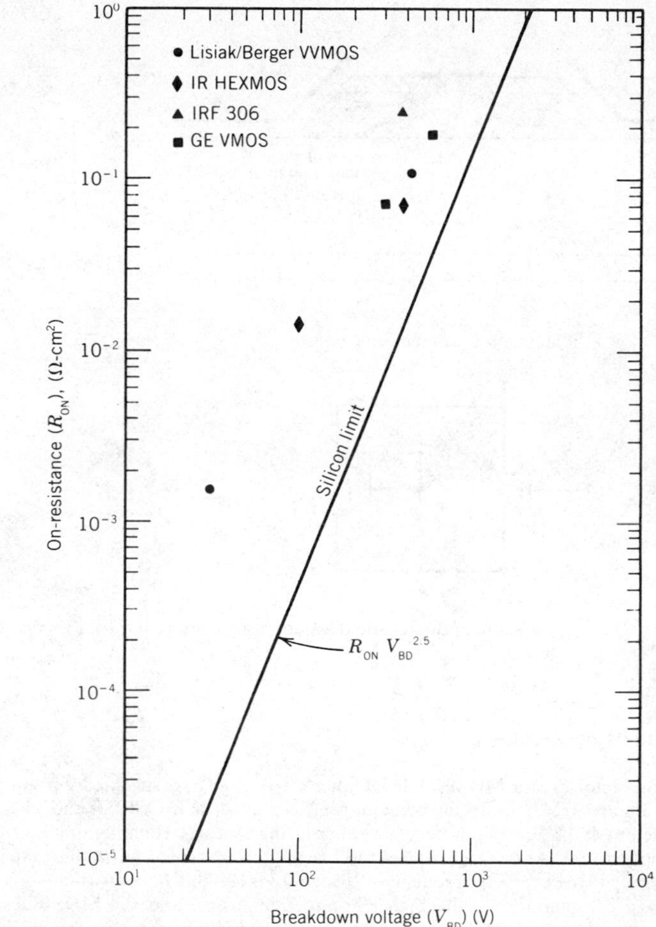

Fig. 13.46 On-resistance vs. voltage in various power metal oxide semiconductor field effect transistor (MOSFET) devices.

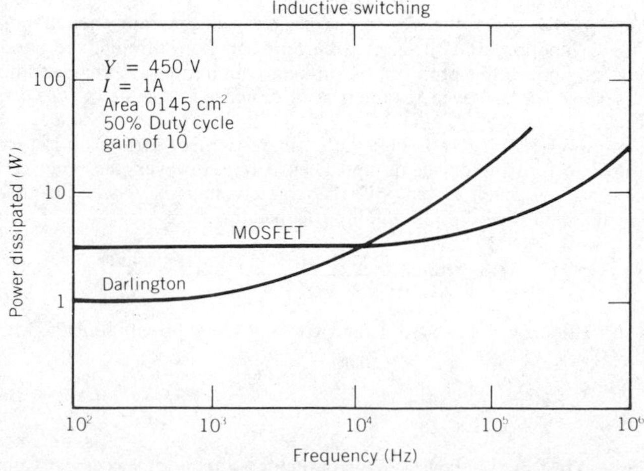

Fig. 13.47 Comparison of power metal oxide semiconductor field effect transistor (MOSFET) (IR330) with bipolar Darlington transistor of equivalent current rating (ZJ499D).

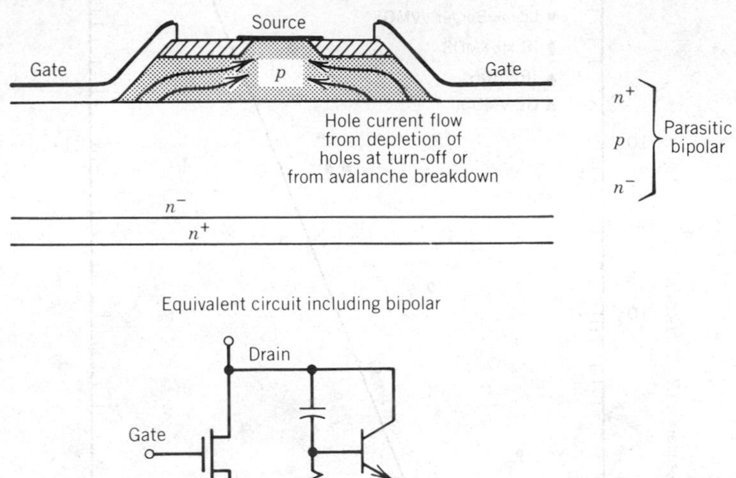

Fig. 13.48 Power metal oxide semiconductor field effect transistor (MOSFET) showing parasitic bipolar transistor.

13.3-4 Application Considerations

For a given semiconductor area a MOSFET has higher on-voltage at a specified current level than a bipolar transistor. Figure 13.47 shows the switching efficiency of a MOSFET and of a Darlington transistor, both rated at 450 V, and each being switched using a peak switching input current equal to a tenth of the output current. At low frequencies the Darlington transistor has a third the losses of the power MOS transistor. However, at frequencies above 13 kHz the lower switching losses of the MOSFET more than compensate for the higher conduction losses, and the MOSFET is the more efficient device. For low-voltage MOSFETs (< 100 V) the conduction losses are much closer to those of bipolar devices.

Some care has to be exercised when applying MOSFETs in high-voltage switching circuits because of a parasitic bipolar transistor that exists in parallel with the MOSFET (Fig. 13.48). As the drain voltage increases rapidly, the *pn*-junction capacitance can allow charge to flow into the *p*-type region, which is the base of the parasitic *npn* transistor. If the *p*-region is wide enough, the capacitive current flowing horizontally under the n^+ source region will generate enough potential drop to forward bias the $n + p$ junction, and substantial current can flow through the parasitic transistor. Modern designs have minimized this problem, but there are limitations on the maximum allowed rate of change of drain voltage (dv/dt) even for the best of designs. The value of $(dv/dt)_{max}$ is specified on application sheets.

An advantage of power MOSFETs is that the input resistance to the gate is very high, so very little power is required to turn the device on and keep it on. However, the input capacitance for a 10-A, 400-V device can be as much as 4000 pF. If the gate voltage is switched from 0 to 10 V in 10 ns, the peak current that will flow through the input capacitor is

$$I = C\frac{dv}{dt} = \frac{4000 \times 10^{-12} \times 10}{10^{-8}} = 4 \text{ A}$$

Therefore a fairly stiff gate driver is needed if the switching speed potential of the MOSFET is to be realized.

Bibliography

Adler, M. S. and S. R. Westbrook, "Power Semiconductor Switching Devices—A Comparison Based on Inductive Switching," *IEEE Trans. Electr. Dev.* **ED-29**(6):947–952 (Jun. 1982).

Baliga, B. J., "Switching Lots of Watts at High Speeds," *IEEE Spectr.* **18**(12):42–48 (Dec. 1981).

Barron, M. B., "Low Level Currents in Insulated Gate Field Effect Transistors," *Solid State Electr.* **15**:293–302 (1972).

Cobbold, R. S. C., *Theory and Applications of Field Effect Transistors*, Wiley, New York, 1970.

Collins, H. W. and B. Pelly, "HEXFET—A New Power Technology," *Electr. Des.* (Jun. 7, 1979).

Crawford, R. H., *MOSFET in Circuit Design*, McGraw-Hill, New York, 1967.

Glaser, A. B. and G. E. Subak-Sharpe, *Integrated Circuit Engineering—Design, Fabrication, and Applications*, Ch. 3, Addison-Wesley, Reading, MA, 1979.

Grebene, A. B., *Analog Integrated Circuit Design*, Van Nostrand Reinhold, New York, 1972.

Grove, A. S., *Physics and Technology of Semiconductor Devices*, Wiley, New York, 1967.

Penney, W. M. and L. Lau, *MOS Integrated Circuits*, Ch. 2, Van Nostrand Reinhold, New York, 1972.

Streetman, B. G., *Solid State Electronic Devices*, Ch. 8, Prentice-Hall, Englewood Cliffs, NJ, 1980.

Sze, S. M., *Physics of Semiconductor Devices*, Chs. 7–10, Wiley, New York, 1969.

Towers, T. D. and S. Libes, *Semiconductor Circuit Elements*, Ch. 10, Hayden, Rochelle Park, NJ, 1977.

Wallmark, J. T. and H. Johnson, *Field Effect Transistors*, Prentice-Hall, Englewood Cliffs, NJ, 1966.

Yang, E. S., *Fundamentals of Semiconductor Devices*, Chs. 7 and 8, McGraw-Hill, New York, 1978.

13.4 INTEGRATED CIRCUITS

Henry Domingos

13.4-1 Introduction

History and Development of the Integrated Circuit[1,2]

The invention of the integrated circuit in more or less the form known today was preceded by over ten years of intense activity in discrete semiconductor device development. Although schemes were suggested as early as 1952 for building many components within a single silicon block, it was not until 1960 that the first integrated circuits were fabricated. Fabrication was made possible by the invention of the planar process in 1959 to replace the alloy, mesa, and other techniques used up to that time. The planar process employed oxide masking, diffusion, and deposited metal interconnects to achieve new levels of circuit density, reliability, performance, and cost effectiveness.

The decade of the 1960s saw the development of many of the integrated circuit families and processing methods in use today. The early small-scale integration (SSI) circuits consisted of gates implemented with resistor-transistor logic (RTL), diode-transistor logic (DTL), emitter-coupled logic (ECL), complementary transistor logic (CTL), and transistor-transistor logic (TTL), among others. The typical chip was 1 to 2 mm square, with about 10 components, laid out with 25-μm minimum dimensions. Typical propagation delays were about 25 ns, although some were capable of a surprisingly low 5 ns.

By the mid-1960s, medium-scale integration (MSI, up to 200 gates per chip) had arrived. Circuits such as JK flip-flops, 10-bit counters and shift registers, and 64-bit RAMs were available. PMOS (*p*-channel metal oxide semiconductor), CMOS (complementary MOS), NMOS (*n*-channel MOS), and SOS (silicon on sapphire) rounded out the technology arsenal (leaving only integrated injection logic, I^2L, to come later) and paved the way for later spectacular advances. By the end of the decade, 256-bit RAMs represented the state of the art. The integrated circuit industry was now able to boast 2×3-mm chips, 10-μm dimensions, propagation delays of 1 ns, and component densities of 300 components per square millimeter.

The large-scale integration (LSI) era arrived in the 1970s. Calculator chips, 1- and 4-kbit memories, and microprocessors were manufactured in great quantities. Furthermore the technology array was virtually complete—suppliers had developed all the major fabrication families in use today: aluminum gate PMOS, silicon gate NMOS, CMOS, SOS, Schottky TTL, I^2L, and ECL. By the end of the decade, chips up to 7-mm square were carrying 2400 components per square millimeter.

Today the industry is capable of submicron dimensions, subnanosecond propagation times, picowatt gate dissipations, and placement of hundreds of thousands of components on a chip. Continued hectic progress is inevitable, since markets already exist in the industrial, military, and consumer sectors for even more complex circuits. However, the move into VLSI (very large-scale integration) places great demands on the designer. The sheer complexity of the new circuits requires innovative techniques in logic design, chip layout, fabrication, and testing.

Major Integrated Circuit Families

The proliferation of processing variations, circuit types, and acronyms presents a confusing picture. However, it should be noted that many of the technological developments reported in the literature never reach the production stage. Furthermore, some of the largest manufacturers of integrated circuits produce only for their own captive use, and their innovative designs and processes are not used in the commercial market. Of products that are commercially available, one should distinguish between generic logic families, modifications within a family, and processing variations. For example, consider the logic family TTL. It exists in several modifications such as low-power TTL, high-speed TTL, Schottky TTL, low-power Schottky TTL, and radiation-hardened TTL. Some of these are available in several options such as open collector, active pull-up, and tri-state outputs. Any of these can be fabricated by several techniques: standard epitaxial with buried subcollector, ion implanted, oxide isolated, triple diffused, refractory metal contacts, and so on. In this section a brief overview of the major families is given, with their relative advantages and disadvantages being pointed out.

Integrated circuits can be conveniently divided into bipolar and MOS types. The bipolar families discussed here include TTL, ECL, and I^2L. Discussion of the MOS families is limited to PMOS, NMOS, and CMOS. Generally speaking, bipolar circuits have higher power dissipation and faster switching speed.

Bipolar Families. TTL was one of the earliest logic families available and for many years was the workhorse of the industry. Its ready availability in a broad range of SSI and MSI circuits made it the designer's choice. Indeed it is considered an advantage for any logic family to be TTL compatible, that is, to operate from a single 5-V supply and with the same logic levels. Modifications of the basic logic gate have given TTL added flexibility, and the development of low-power Schottky has made it a viable LSI technology.

ECL is the fastest logic family. Its main drawbacks are high power dissipation, unusual logic levels, and nonstandard supply voltages. Although its high power dissipation would seem to preclude its use in LSI, it is available in very high speed 1-kbit RAMs.

I^2L is the newest bipolar logic and in many ways the most interesting. Its most astonishing feature is the ability to continuously trade off speed for power over an extremely wide range, even on the same chip. In a watch circuit, for example, only certain portions of the circuit are required to operate at the oscillator frequency; once this frequency is scaled down the rest of the circuitry can be designed for low power consumption to increase battery life. Another distinctive feature is the ability to combine analog and digital functions on the same chip. This is not normally done with other families because of conflicting requirements in the fabrication process. The main disadvantage of I^2L is low speed in its simplest form. Modifications to increase speed cause higher manufacturing costs. Nevertheless I^2L is used in LSI circuits, and certain advantages in layout and circuit design make it a natural choice for custom applications.

MOS Families. MOS circuits have been in production since the mid-1960s. The first LSI circuits were calculator chips that used a metal gate PMOS process. Although the performance of PMOS is inherently inferior to that of NMOS because its majority carriers (holes) have lower mobility, PMOS was initially used because of the industry's inability to make stable, high-quality NMOS products. It became the cheapest LSI technology. However, it is considered an obsolete technology and is not a serious contender for VLSI designs.

NMOS is the dominant technology in LSI and early VLSI circuits. Creative circuit design and progress in fabrication techniques have resulted in continuing improvements in speed, density, and cost effectiveness. It appears that the silicon or silicide gate, depletion load NMOS family will continue to be widely used in the foreseeable future.

CMOS is the only MOS family that is available in a large selection of SSI and MSI circuits. It is also used in LSI memories and logic circuits and shows considerable promise for the VLSI era. One of its advantages is extremely low power dissipation in the standby mode. This has led to its use in nonvolatile memories (employing battery or even capacitor backup) and in calculators with memory. Another advantage is its ability to operate over a wide range of power supply voltages, from 3 to 20 V. It also has outstanding noise immunity. Its main disadvantage is limited drive compared with that of bipolar circuits.

Cost Considerations

The cost of an integrated circuit is determined by a number of factors. A newly introduced product is expensive, costing perhaps more than $100. As the manufacturer moves down the "learning curve" and especially as competitors introduce their version of the same product, prices decline drastically. Older products continue to decline in price as the processing technology matures. SSI circuits are available for \$0.15 each in large quantities. The asymptotic price, as it were, for any circuit is strongly affected by the package cost (plastic versus hermetic), the size of the chip, the cost of processing a wafer, the processing yield (percentage of good chips on a wafer), and performance specifications such as speed and reliability.

The cost of processing a wafer depends on the complexity of the particular technology. For a mature low-cost process such as PMOS, the fabrication cost (materials, labor, capital equipment, and other direct costs) may be as low as \$30 per wafer. For state-of-the-art processes requiring 10 to 12 masking steps, the cost approaches \$100. The starting wafers themselves cost \$5 to \$15 depending on size and specifications such as flatness, oxygen content, and resistivity. Sapphire and gallium-arsenide wafers are much more costly; this high cost has been a deterrent to their use in commercial integrated circuits.

Semiconductor processing is a batch process wherein up to 50 wafers are processed simultaneously. The greater the number of chips on a wafer, the lower the manufacturing cost per chip will be. Figure 13.49 shows the number of chips on wafers of different diameters. The figure makes it abundantly clear why manufacturers attempt to squeeze circuits into smaller and smaller areas to minimize the "real estate" required.

Upon completion of wafer processing each chip is tested on a go, no-go basis before any further manufacturing steps are undertaken. This is the earliest point at which functional testing is possible; such testing eliminates further processing costs for defective chips. The wafer probe yield depends on the chip area, the processing complexity, the design rules, and the maturity of the process. For new VLSI chips the yield can be a small fraction of 1% initially, whereas for SSI circuits it approaches

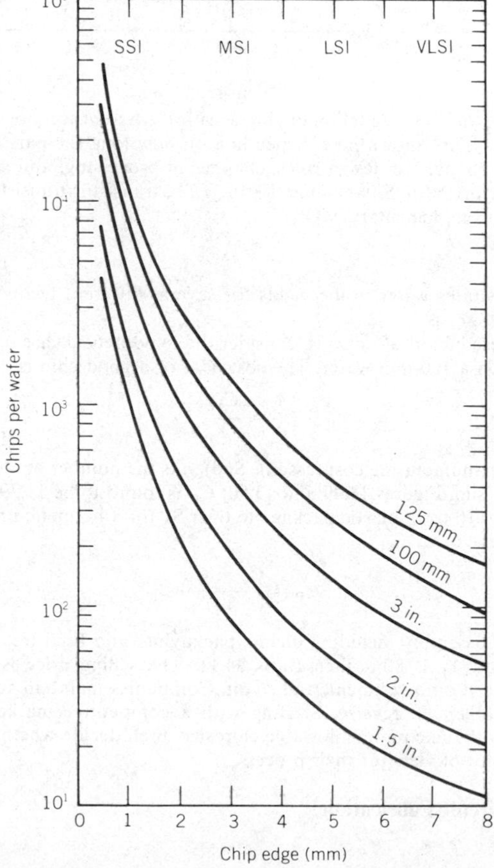

Fig. 13.49 Number of whole, square chips on a wafer as a function of the chip size (linear dimension) for different wafer diameters. SSI, small-scale integration; MSI, medium-scale integration; LSI, large-scale integration; VLSI, very-large-scale integration.

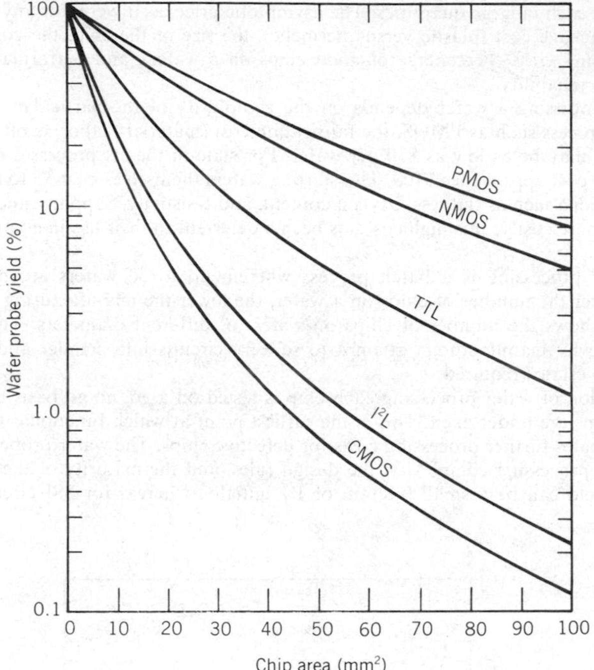

Fig. 13.50 Wafer probe yield as a function of chip area for different technologies. Actual values may differ considerably from those shown here, depending on how long the part has been in production, number of masks used, changes in design rules, changes in processing, and so on. PMOS, *p*-channel metal oxide semiconductor; NMOS, *n*-channel MOS; TTL, transistor-transistor logic; I^2L, integrated injection logic; CMOS, complementary MOS.

100%. Figure 13.50 illustrates wafer probe yields for several different technologies as a function of chip size.

To illustrate how these factors affect cost, consider the hypothetical case of a CMOS circuit with a chip size of 4×4 mm on a 100-mm wafer. The cost, C_C, of a good chip is

$$C_C = \frac{C_W}{nY_P} \tag{13.67}$$

where C_W is the wafer manufacturing cost (assume \$60), n is the number of chips per wafer, and Y_P is the wafer probe yield. Using Figures 13.49 and 13.50, C_C is found to be \$1.79. The cost of packaging can range from under \$0.10 for a plastic package to over \$1 for a hermetic unit. The final cost of the part, C_P, is

$$C_P = \frac{C_C + C_A}{Y_F} \tag{13.68}$$

where C_A is the cost of assembly including dicing, packaging, and final test and Y_F is the final test yield. If C_A is \$1.50 and Y_F is 80%, then C_P is \$4.11. The selling price is C_P plus overhead and indirect costs, taxes, return on investment, and so on. Companies maintain accurate models for costs and often work the problem in reverse. Starting with a competitive market price for a proposed product, they determine the maximum allowable chip size, then decide whether the project is feasible from an engineering point of view for their process.

13.4-2 Integrated Circuit Fabrication[3,4]

Crystal Growth

Almost all silicon crystals for integrated circuits are grown by the Czochralski or crystal-pulling method depicted in Fig. 13.51. A charge of polycrystalline silicon is placed in the crucible and melted

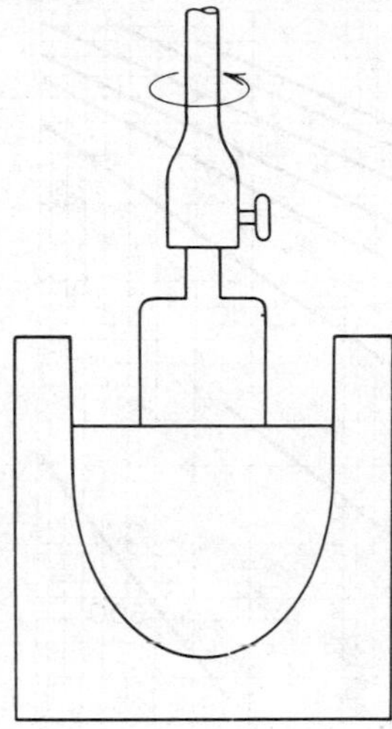

Fig. 13.51 Czochralski or crystal-pulling method for growing silicon single crystals.

with RF or resistance heaters. An exact amount of impurity is added to achieve the desired impurity concentration in the finished product. A seed crystal is lowered until it is wetted by the molten silicon, then is slowly withdrawn. The liquid silicon solidifies on the seed to form a continuous, high-quality single crystal. Relative rotation of the seed and crucible maintains a uniform distribution of impurity in the melt and uniform temperature and growth conditions. By careful programming of temperature and withdrawal rate, good uniformity of crystal resistivity and diameter can be maintained throughout the entire crystal in spite of the tendency for impurities to segregate during the solidification process. High-quality ingots 125 or 150 mm in diameter up to 1 m long can be grown in this manner. Crystals are grown with the axis either in a $\langle 100 \rangle$ or $\langle 111 \rangle$ direction.

Growth orientation is verified by an x-ray Laue technique, then heads and tails are cropped. The outer surface is machined to the required diameter and one or more flats are ground to identify orientation and resistivity type. The ingot is then ready for slicing into wafers.

Single crystals are sometimes grown by the floating zone process, whereby a molten zone is slowly moved through the crystal, sweeping impurities toward one end. This technique is especially useful for preparing crystals of very high purity. When combined with neutron transmutation doping,[5] high-resistivity material with very uniform doping can be produced for high-voltage applications.

Wafer Preparation

Ingots are cut into wafers with the inside edge of a thin disk-shaped diamond-edged saw blade. Wafers are then lapped, etched, and polished. One face is polished to a mirror finish, whereas the backside is roughened to act as a sink for precipitates and crystal imperfections. The edge may be rounded to minimize chipping and to facilitate handling in automatic processing equipment.

Oxidation

Silicon is readily oxidized at high temperatures to produce an oxide layer with excellent mechanical and electrical properties. The oxide is used as a passivating layer to protect the silicon surface from contamination, as a dielectric for capacitors and MOS gates, as a layer of insulation, as a mask for diffusion and ion implantation, and for isolation between active elements. Thicknesses range from a few tens of nanometers for gate oxides to 1 μm for MOS field oxides and bipolar circuits.

A freshly cleaved silicon surface will quickly acquire a very thin layer of oxide under standard atmospheric conditions. However, growth is usually undertaken at elevated temperatures to achieve useful thicknesses. Oxidation takes place at the interface between the oxide and the silicon surface. When the oxide layer is thin, the oxidation rate is limited by the reaction rate at the interface, and the oxide thickness is proportional to the time spent in the oxidation furnace. For thick oxides the rate is limited by the diffusion of oxygen through the oxide layer, and proceeds with the square root of time. Oxide growth is summarized in Figs. 13.52 and 13.53.

Silicon dioxide is often doped with phosphorus or boron to serve as a getter for impurities or as a source of impurities during diffusion. The introduction of chlorine has also been found to improve

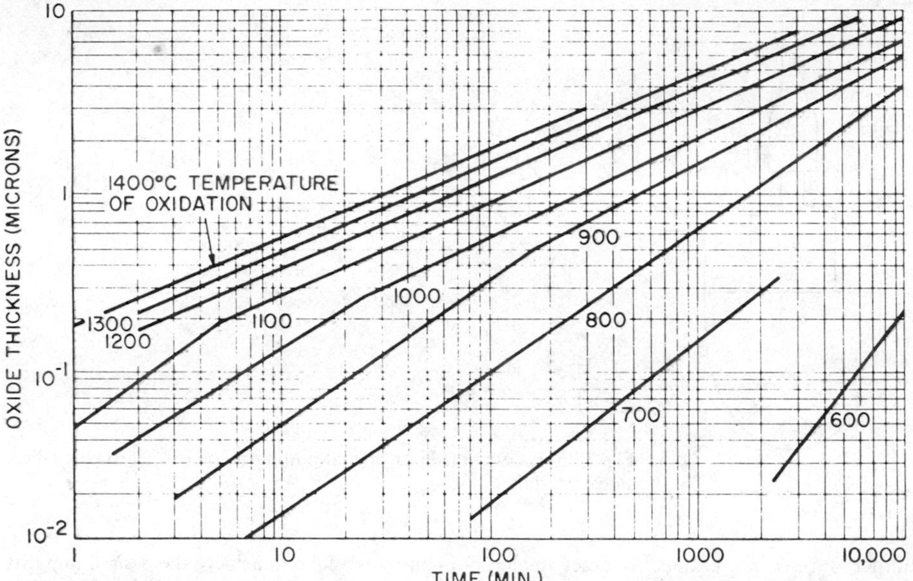

Fig. 13.52 Oxide growth in atmospheric steam. Courtesy of Integrated Circuit Engineering Corp.

the quality of the oxide layer and to reduce the occurrence of silicon dislocations. Recently oxide growth under high pressure has been found to have several benefits, chief of which is greatly reduced oxidation time.

A protective layer of glass is applied to fully processed wafers by chemical vapor deposition (CVD) or reactive sputtering. This final application must be done at temperatures below the melting point or eutectic point of any materials used in the processing. The glass serves to protect the surface during handling, dicing, and so on.

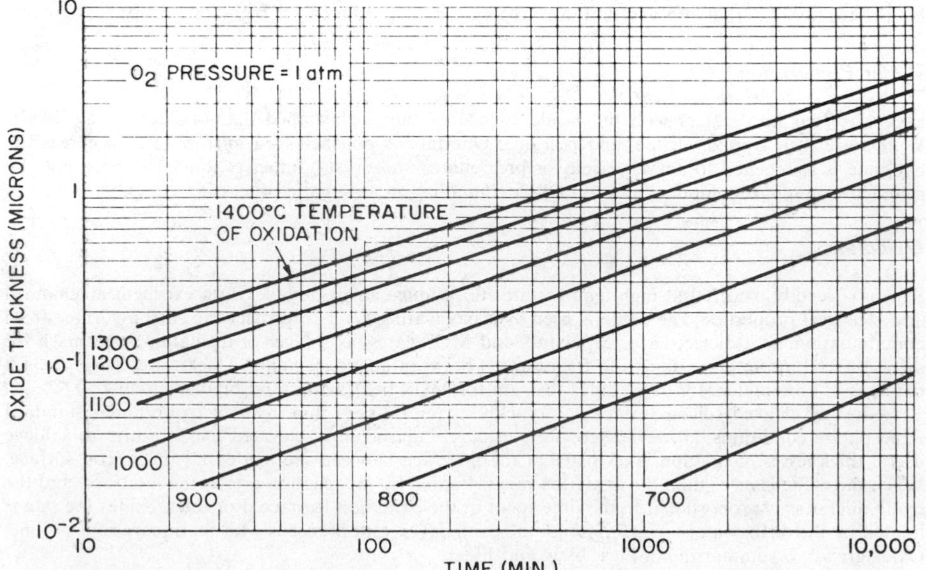

Fig. 13.53 Oxide growth in dry oxygen. Courtesy of Integrated Circuit Engineering Corp.

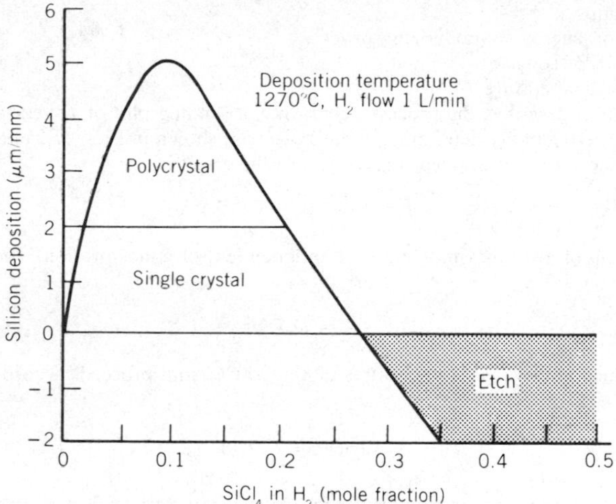

Fig. 13.54 Deposition rate in silicon epitaxy as a function of chlorine content. From Sorab K. Ghandhi, *The Theory and Practice of Microelectronics*, Wiley, New York, 1968.

Epitaxy

Epitaxy is a method of crystal growth in which atoms are deposited on a substrate in such a way as to form a continuous single crystal. The crystal structure is determined by the substrate, but the impurity concentration is independently controlled and can even be changed from one type to another within the layer. The planar epitaxial method was practically universal in the early 1960s and is common today among a wide variety of families, both bipolar and MOS.

Epitaxial growth takes place in a vessel called a reactor. Silicon wafers are placed on a graphite block called a susceptor which is heated to about 1200°C. Wafers are initially etched with water vapor or anhydrous hydrogen chloride to remove any damaged material. Several chemical reactions can be utilized to provide the source of free silicon atoms; hydrogen reduction of silicon tetrachloride is the common method. Growth conditions must be carefully controlled, as implied by Fig. 13.54.

In a typical operation, wafers are placed in the reactor and the system flushed with nitrogen. Hydrogen is then turned on at a flow rate of 30 l/min. The temperature is adjusted to 1200°C and the wafers are etched for several minutes in a 1 : 100 ratio of HCl. Deposition begins with the introduction of $SiCl_4$ in a ratio of 1 : 800, with a small quantity of phosphine (PH_3), diborane (B_2H_6), or an arsenic compound to determine the impurity concentration. The film is grown for several minutes, with typical thicknesses ranging from 1 to 10 μm.

Vapor phase epitaxy is also used to grow layers of silicon on sapphire substrates. Liquid phase epitaxy (LPE) is commonly employed in gallium-arsenide technology.

Diffusion and Ion Implantation

Diffusion is a process whereby random thermal motion results in a transport of material from a region of high concentration to one of low concentration. In a silicon crystal the diffusion of foreign atoms can occur via several mechanisms: interstitial diffusion, substitutional diffusion, field-enhanced diffusion, or structurally enhanced diffusion. Whatever the mechanism, the process can be described phenomenologically by Fick's first law of diffusion

$$\bar{f} = -D\nabla n \tag{13.69}$$

where $\bar{f}$ = flow of particles per unit area per unit time
n = particle concentration
D = diffusion coefficient with units of length squared per unit time

D depends on the diffusing species and the host lattice. Impurities such as gold that diffuse by an interstitial mechanism have a much higher diffusion coefficient than those that diffuse substitutionally. D is very temperature sensitive, and a temperature dependence is usually assumed of the form

$$D = D_0 e^{-E_A/kT} \tag{13.70}$$

where D_0 = a constant
E_A = activation energy characterizing process
k = Boltzmann's constant
T = absolute temperature

Assuming the equation describes the process accurately, a semilog plot of D versus $1/kT$ will result in a straight line. Experimentally determined values of D are shown in Figs. 13.55 and 13.56.

If the continuity equation is applied to Eq. (13.69) the result is

$$\nabla \cdot \bar{f} = -\frac{\partial n}{\partial t} = -\nabla \cdot D\,\nabla n \tag{13.71}$$

If D is not a function of position (implying D is independent of concentration), Fick's second law is the result

$$\frac{\partial n}{\partial t} = D\,\nabla^2 n \tag{13.72}$$

Diffusion is performed at elevated temperatures (900–1200°C) and proceeds according to Eq. (13.72).

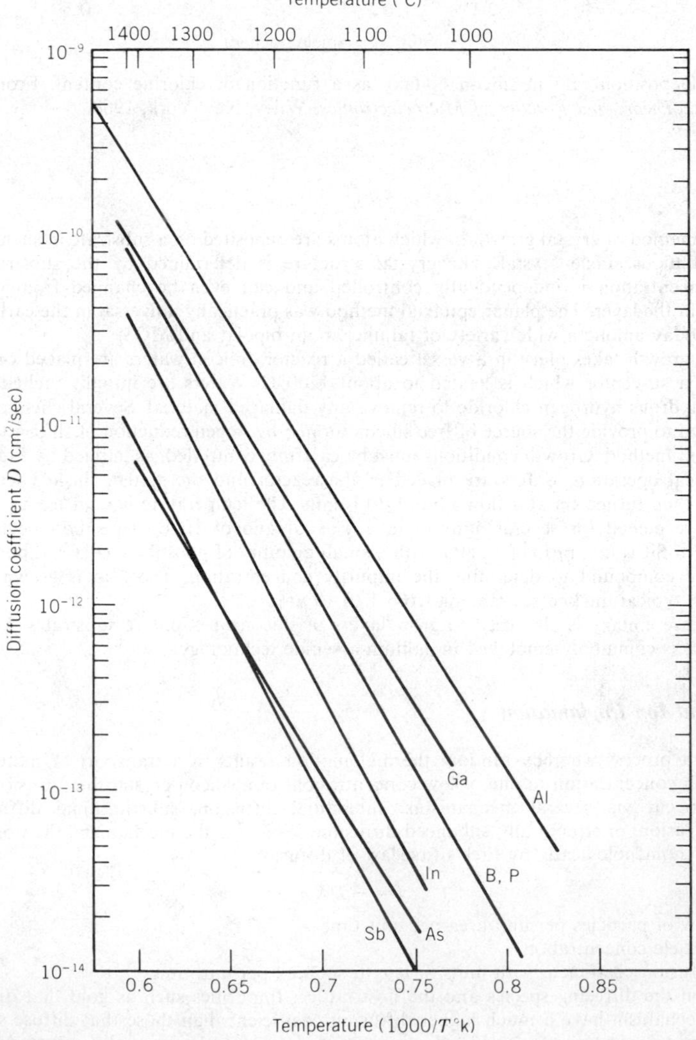

Fig. 13.55 Diffusion coefficients for commonly used n- and p-type impurities in silicon. From Sorab K. Ghandhi, *The Theory and Practice of Microelectronics*, Wiley, New York, 1968.

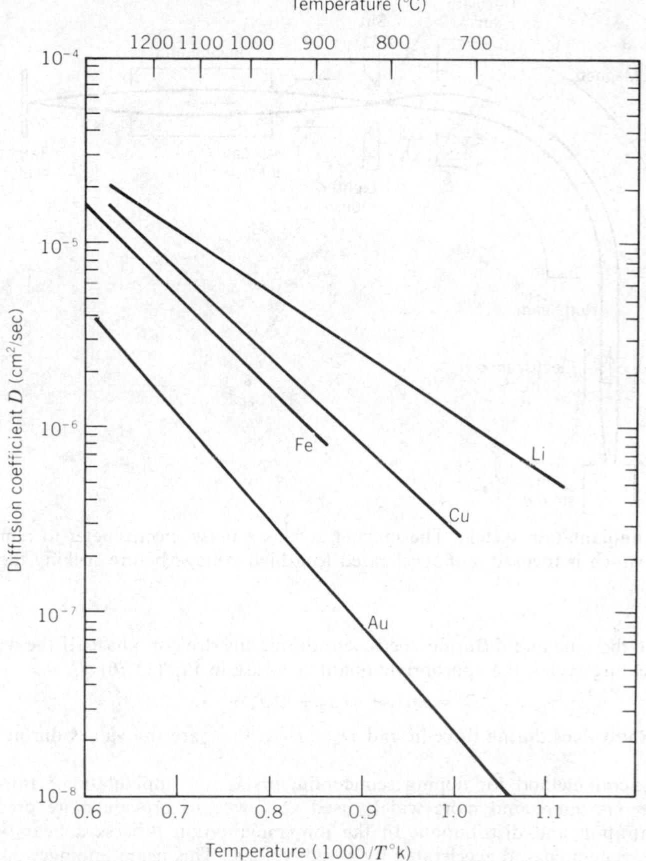

Fig. 13.56 Diffusion coefficients for interstitial diffusers in silicon. Because of the different diffusion mechanisms for these impurities, D is several orders of magnitude larger than values in Fig. 13.55. From Sorab K. Ghandhi, *The Theory and Practice of Microelectronics*, Wiley, New York, 1968.

Then the temperature is lowered and the impurities are "frozen in," since D is extremely small at ordinary temperatures.

Diffusion is commonly performed in a two-step process. In the predeposition step the impurity concentration at the surface of the wafer is held constant while the impurity diffuses into the wafer. The solution to Eq. (13.72) in this case is

$$n = n_0 \operatorname{erfc}\left(x / 2\sqrt{Dt} \right) \tag{13.73}$$

where n_0 is the surface concentration and erfc is the complementary error function defined by

$$\operatorname{erfc}(\mu) = 1 - \frac{2}{\sqrt{\pi}} \int_0^{\mu} e^{-\alpha^2}\, d\alpha \tag{13.74}$$

The predeposition step introduces a total quantity of impurity per unit area

$$Q = 2n_0 \sqrt{\frac{Dt}{\pi}} \tag{13.75}$$

in a thin surface layer. During the drive-in step the total impurity content is held constant at this value, but the impurities are allowed to diffuse further into the wafer. During this step the solution to Eq. (13.72) is given by

$$n = \frac{Q}{\sqrt{\pi D t}} e^{-x^2/4Dt} \tag{13.76}$$

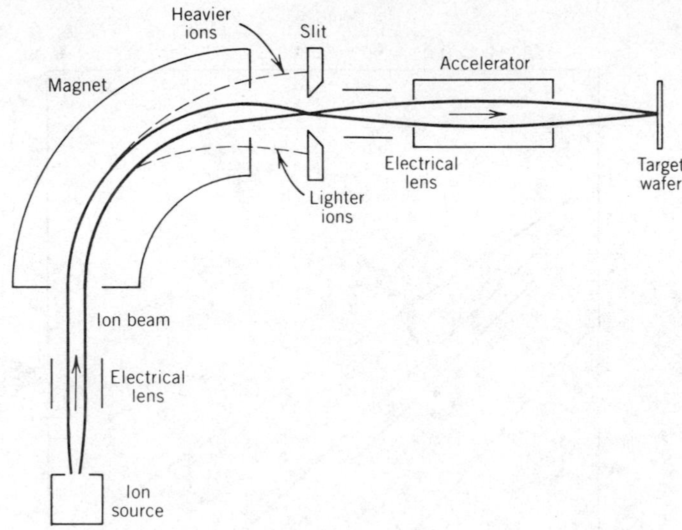

Fig. 13.57 Ion implantation system. The magnet acts as a mass spectrometer to remove impurities from the beam, which is focused and accelerated to a high voltage before striking the silicon wafer.

Here t and D are the time and diffusion coefficient during the drive-in phase. If the wafer is exposed to subsequent heating cycles, the appropriate quantity to use in Eq. (13.76) is

$$Dt = D_1t_1 + D_2t_2 + D_3t_3 + \cdots \qquad (13.77)$$

where D_1t_1 are the values during drive-in and $D_2t_2, D_3t_3, \cdots$, are the values during any following heat treatments.

A second general method for doping semiconductors is ion implantation.[6] Introduced in the 1960s, it has become more and more widely used as a way to provide more precise control of impurity concentration and distribution. In the ion implantation process a beam of ions of the desired species is generated and accelerated by a high voltage. This beam impinges on the surface of a wafer and the ions are driven into the crystal lattice to a depth up to 1 μm. The process is depicted in Fig. 13.57.

TABLE 13.1. MEAN AND STANDARD DERIVATION OF RANGE FOR IMPLANTED IONS IN A SILICON WAFER

	Boron		Phosphorus	
Energy (kV)	Mean Range (μm)	Standard Deviation (μm)	Mean Range (μm)	Standard Deviation (μm)
20	0.071	0.026	0.026	0.009
40	0.14	0.043	0.049	0.016
60	0.21	0.056	0.073	0.023
80	0.27	0.065	0.098	0.029
100	0.33	0.073	0.12	0.035
120	0.38	0.079	0.15	0.046
140	0.43	0.086	0.17	0.046
160	0.47	0.091	0.20	0.051
180	0.52	0.096	0.23	0.056
200	0.56	0.10	0.25	0.063

Source. Gibbons.[7]

The average depth of penetration, or range, depends on the accelerating voltage and the type of ion used. The mean and standard deviation of the distribution are illustrated in Table 13.1.

Ion implantation usually requires an annealing step to (1) anneal any crystalline damage caused by the high energy beam, (2) cause the implanted ions to become electrically active, and (3) redistribute (by diffusion) the implanted impurities. Depending on the beam energy, a 30-min anneal at 500°C may be sufficient for doses less than $10^{12}/\text{cm}^2$. Doses greater than $10^{14}/\text{cm}^2$ may require temperatures up to 1100°C.

Photolithography

Integrated circuits can be manufactured at a low cost because they are mass produced by a process similar to that used in the printing industry, whereby a single master can be used to make an unlimited number of copies. From 4 to 12 photomasks can be used to define the geometric patterns for each stage of the fabrication process. The number of masks required is a measure of the processing complexity, the wafer manufacturing cost, and the yield. Table 13.2 lists the number of masks needed for different processing technologies.

Each mask requires the preparation of initial artwork based on the actual chip layout. For SSI circuits, this was accomplished by manual drafting techniques, where accuracies of 1 mm on a 200X to 1000X layout were sufficient. For MSI circuits, computer-controlled drafting tables are necessary to control original tolerances to less than 25 μm. The initial reduction is to a size of 10X or 20X, then a final reduction and step-and-repeat operation replicates the image enough times to cover the entire wafer. The common technique uses a computer-controlled light beam to expose a 10X reticle, which is then reduced and stepped. The entire set of finished masks must meet critical tolerances and exact registration over the entire wafer.

More recent techniques include electron-beam-generated masks, electron-beam exposure of photosensitive material on the wafer itself, photomasks for use with x-ray exposure systems, and direct step-and-repeat on the wafer.

Working masks consist of a layer of emulsion, chromium, or iron oxide on a glass plate. Emulsion masks are the least expensive, but are used for only 20 exposures before being discarded. The life of hard masks ranges from 50 exposures on up, depending on whether contact, proximity, or projection printing is used. However, masks must be regularly cleaned and inspected for best results.

Continued reduction of the minimum feature size on photomasks has been one of the important factors in the increasing complexity of integrated circuits. Early design rules called for minimum line widths and line separations of 25 to 30 μm. This has been reduced to 2 to 5 μm in production lines today through use of better masks, improved photosensitive materials, better etching methods, and shorter wavelengths in mask aligners. To further reduce this below 1 μm will require electron-beam and x-ray photolithography. The trends in minimum feature size are shown in Fig. 13.58.

TABLE 13.2. NUMBER OF MASKS (INCLUDING SCRATCH PROTECTION) REQUIRED FOR SEVERAL DIFFERENT TECHNOLOGIES

Technology	Masks
PMOS (*p*-channel metal oxide semiconductor), aluminum gate	5
NMOS (*n*-channel MOS), silicon gate depletion loads	7
CMOS (complementary MOS)	7
SOS (silicon on sapphire), silicon gate ion implant	7–9
TTL (transistor-transistor logic), standard	7
TTL Schottky	8–9
I^2L (integrated-injection logic)	5–10
ECL (emitter-coupled logic)	7–8
Linear bipolar	7

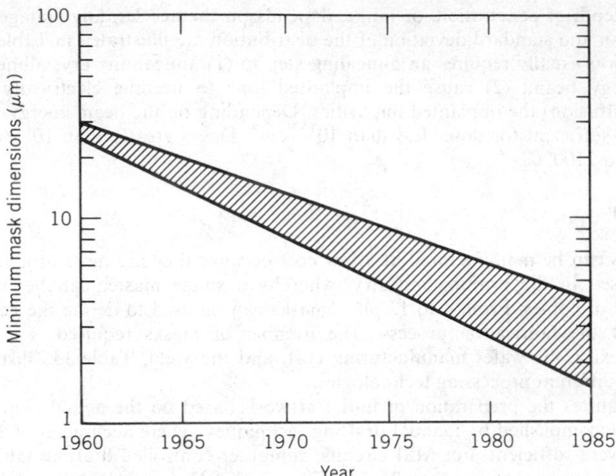

Fig. 13.58 Trends in minimum mask dimensions for integrated circuits in large-volume production.

Bipolar Circuit Fabrication

The fabrication of a typical bipolar integrated circuit begins with a ⟨111⟩ oriented *p*-type wafer, doped with boron, with a resistivity of 5 to 20 Ω-cm. The first step is thermal oxidation to a thickness of nominally 1 μm. This oxide will serve as a diffusion mask for the first diffusion, which is an *n*-type buried subcollector. To define the areas where the diffusion is to take place, a standard photolithographic sequence is performed.

After being thoroughly cleaned and dried, the wafers are coated with a thin layer of photosensitive material called photoresist. The layer is dried, and each wafer is exposed to ultraviolet light through a mask that defines the regions where the collectors of transistors will be located. If "negative" photoresist is being used these regions on the mask will be opaque. The ultraviolet light polymerizes the photoresist in areas where the mask is clear. The photoresist is then developed; unexposed areas are washed away, exposing the underlying silicon dioxide. After a hardening bake, the wafers are immersed in a solution of hydrofluoric acid. The oxide is completely removed in any areas not protected by the remaining photoresist. When this step is completed, the wafers are thoroughly cleaned to remove all traces of photoresist. At this stage the wafers have a thick layer of oxide everywhere except for the areas to be diffused and are ready for the buried subcollector diffusion.

The purpose of the buried subcollector is to reduce the collector saturation resistance by providing a low resistance path for the collector current, and also to reduce the gain of the parasitic *pnp* transistor formed by the substrate, epitaxial layer, and base diffusion. The diffusion is performed with an arsenic compound, for example, arsenic trioxide, at 1200°C for 15 hours. This produces a diffusion depth of 5 μm with a sheet resistance of 10 to 20 Ω/sq. The sequence of steps is illustrated in Fig. 13.59.

Next the wafers are completely stripped of oxide and an epitaxial layer is grown, 4 to 10 μm thick, phosphorus doped, with a resistivity of 0.1 to 0.5 Ω-cm. This layer will serve as the bulk collector material. The doping level is chosen for considerations of collector breakdown voltage, collector capacitance, collector resistance, depletion layer thickness, and requirements of subsequent processing steps.

An oxide layer is again grown to protect the surface and serve as a diffusion mask for the isolation diffusion. This diffusion is a deep *p*-type diffusion that separates transistors whose collectors must be electrically isolated from each other. After another photolithographic operation, a boron diffusion is performed at 1200°C for three hours. This is sufficient to convert the epitaxial layer in the isolation regions to *p* type all the way through to the substrate. During the drive-in phase an oxide layer is again grown to protect the surface and to serve as the diffusion mask for the next step.

The third mask level defines a *p*-type diffusion that forms the base regions within the *n*-type collector regions, as well as resistors. Boron is diffused at 1100°C for about two hours, producing a junction depth of 1 to 3 μm and a sheet resistance of 100 to 300 Ω/sq.

The final diffusion is a phosphorus diffusion to form the emitters, collector contact regions, low-value resistances, and cross-unders. The depth must be carefully controlled, since it determines

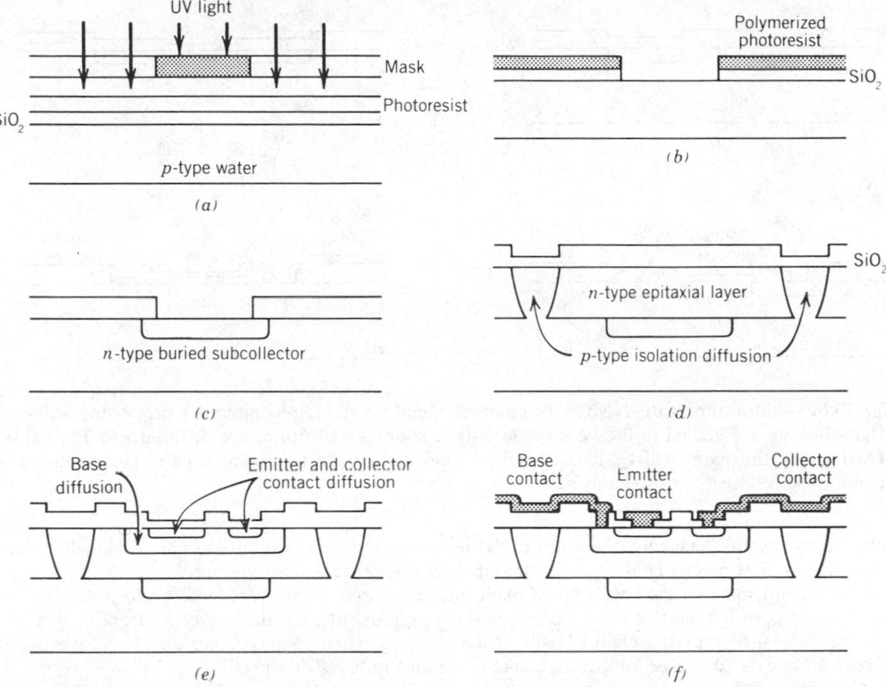

Fig. 13.59 Bipolar processing steps. (*a*) A photomask designates the areas to be exposed to the ultraviolet light. (*b*) The polymerized photoresist protects the silicon dioxide during the oxide etching step. (*c*) The buried subcollector is diffused through the oxide mask. (*d*) After the epitaxial layer is grown, an isolation diffusion is performed. (*e*) The wafer after the base and emitter diffusions. (*f*) Contact windows are cut through the oxide and metal is deposited and patterned.

the base width of the transistors, with attendant effects on the transistor gain and transient response. The emitter diffusion must be heavily doped to compensate the doping level of the base diffusion, to provide high injection efficiency at the emitter, and to allow an ohmic contact to the collector region by the aluminum metallization. The emitter diffusion is a phosphorus diffusion at 1050°C for 30 minutes to provide a sheet resistance of 10 to 30 Ω/sq. and a junction depth a fraction of a micrometer less than the base junction depth.

The wafer at this stage contains all the diffusions that define the transistors, diodes, and resistors, protected by oxide of varying thickness over the different regions. With some exceptions the remaining steps are the same for all integrated circuits.

The fifth mask designates the places where electrical contacts will be made to the underlying silicon. The procedure is exactly the same as used heretofore, except that the process is complete when the oxide is etched. The next operation is vacuum deposition of aluminum over the entire wafer to a thickness of 1 μm. The metal is patterned with the sixth mask and the aluminum etched. This is followed by an anneal at 450 to 500°C for several minutes to break down the thin layer of oxide at the interface that is inevitably present, and to form a good low-resistance contact to the underlying silicon. Finally a thick layer of silicon dioxide is deposited by CVD or sputtering to protect the surface of the wafer during handling and assembly operations.

The last, seventh, mask is used to remove the deposited glass over the bonding and test pads. This completes the wafer processing, and the wafers are now ready for probing and packaging.

MOS Circuit Processing

The simplest fabrication procedure is for a single-channel aluminum-gate MOS family. Although PMOS was the earliest and least expensive family, it is now obsolete and the processing steps will be illustrated for aluminum gate NMOS and for ion-implanted silicon-gate CMOS.

The sequence of steps for the NMOS process is shown in Fig. 13.60. The starting wafer is ⟨100⟩ or ⟨111⟩ *p* type, doped to 5 to 10 Ω-cm. The initial step is the growth of field oxide 1 μm thick. This is

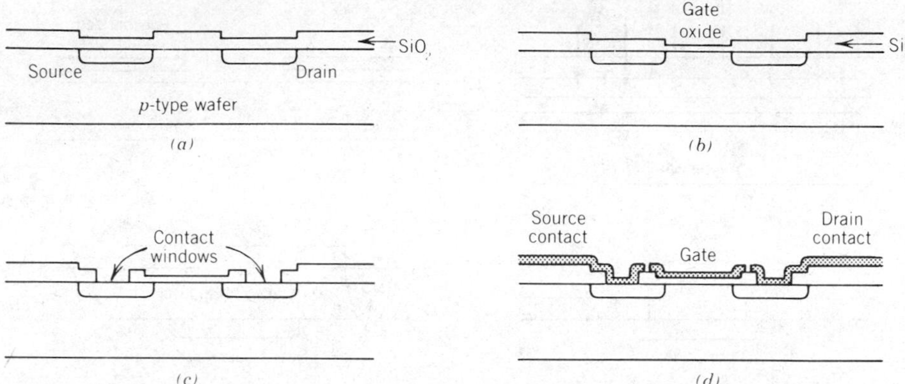

Fig. 13.60 Aluminum gate NMOS (*n*-channel metal oxide semiconductor) processing steps. (*a*) After windows are etched in the field oxide, *n*-type sources and drains are diffused. (*b*) The oxide is etched where the gates will be located, and a thin gate oxide is grown. (*c*) Contact windows are etched. (*d*) Metal is deposited and patterned.

followed by the first masking operation to define sources, drains, and cross-unders. This diffusion is not critical in that precise control of depth and sheet resistance is not required.

The second mask is used to remove oxide in the regions where gates will be formed. This step requires precise alignment of the mask, since the gate must overlap the source and drain by a small amount to ensure that the channel will be conducting when required, yet any excessive overlap increases the gate to source and drain capacitance and reduces the speed.

Gate oxide growth is also a critical step, since the oxide must be thin (25–100 nm) yet free from any pinholes and from any contaminants that might cause drifts in the threshold voltage. Otherwise the threshold voltage depends primarily on the substrate doping concentration and on the oxide thickness, as shown in Fig. 13.61 for an *n*-channel transistor and in Fig. 13.62 for a PMOS device.

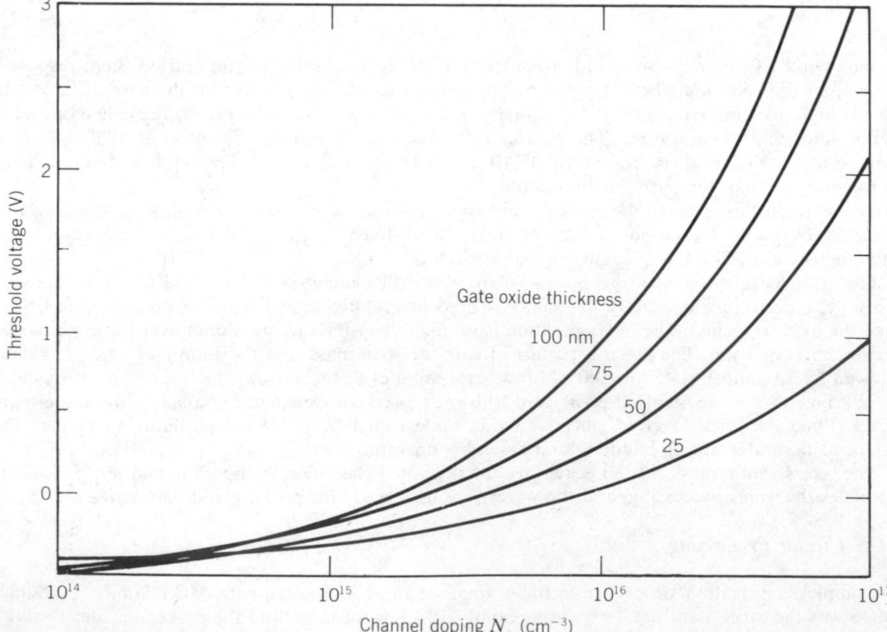

Fig. 13.61 Threshold voltage for an NMOS (*n*-channel metal oxide semiconductor) transistor. The interface charge density is assumed to be $5 \times 10^{10}/cm^2$.

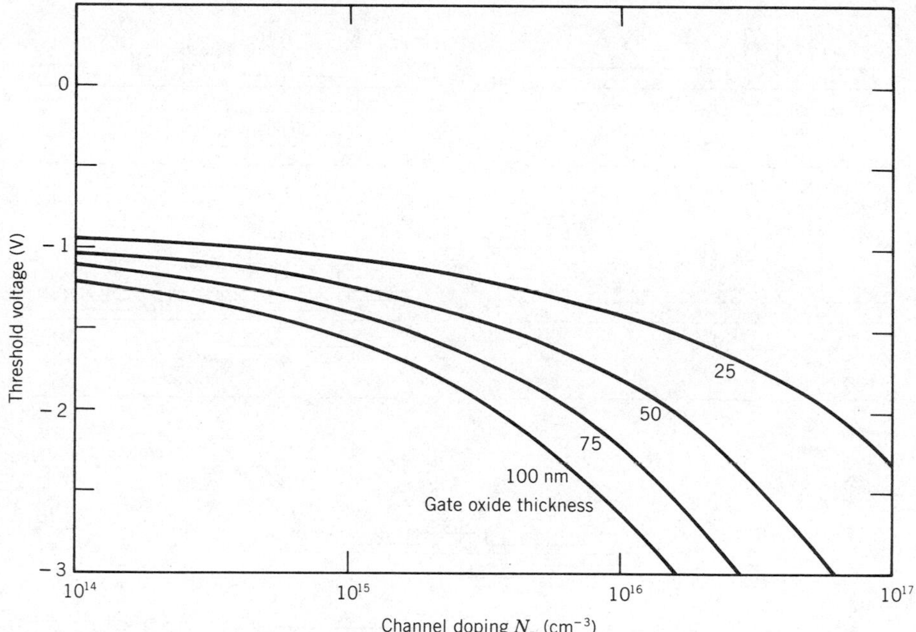

Fig. 13.62 Threshold voltage for a PMOS (p-channel metal oxide semiconductor) transistor. The interface charge density is assumed to be $5 \times 10^{10}/\text{cm}^2$.

The third and fourth masks define the contact cuts (or pre-ohmic etch) and the metallization patterns, respectively. In this case the gate electrodes and the interconnections are both aluminum, as contrasted with the silicon gate process (to be discussed next). Finally the scratch protection layer is applied, and the fifth mask is used to expose the bonding pads.

The simplicity of this process—no epitaxial layer, a single, noncritical diffusion, and only five required masks—makes it obviously the least expensive. In spite of the slower speed of NMOS, the simplicity of its gate layout and its low power dissipation made it the only candidate for LSI at one time. Today NMOS is capable of much higher performance at the penalty of more complex processing (up to 12 masks); in spite of increased competition from circuits in bipolar logic families it is still the dominant LSI process.

The benefits of implementing logic gates from complementary MOS transistors were recognized early, and the practical difficulties of building both n- and p-channel devices on the same chip were soon overcome. Only one of many possible variations in building CMOS logic will be described here; it is an ion-implanted silicon gate process.

The starting wafer is $\langle 100 \rangle$ n type with a resistivity of 2 to 10 Ω-cm. After an initial oxide growth of 1 μm, the first mask is used to define areas for p wells or tubs, which will contain the n-channel transistors. After a photo-patterning operation the oxide is removed from the p-well areas. Then the photoresist is removed and the wafers are ion implanted with boron. The oxide shields the regions outside the p-wells from the ion implant. A high-temperature heat treatment anneals any lattice damage and allows the boron to diffuse to a depth of roughly 3 μm. The sheet resistance is about 800 Ω/sq. The ion implant allows the channel doping and hence the threshold voltage to be accurately controlled.

The second mask is used to pattern p-type guard rings around the periphery of the p wells. These are diffused, heavily doped p-type regions that require a very high voltage to invert, and thus prevent unwanted channels from forming, and reduce leakage currents between transistors.

The third mask is the gate oxide mask. It patterns the entire area for each type of transistor, including source and drain as well as gate. A thin gate oxide is then grown, and polysilicon 1 μm thick is deposited over the whole wafer. The polysilicon is heavily doped with phosphorus to reduce its resistivity. A mask is then used to pattern the polysilicon.

Mask number five is the pattern for the n-channel sources and drains. The field oxide, polysilicon, photoresist, and sometimes a layer of aluminum act as a mask against the ion beam everywhere except where the thin gate oxide is exposed in the p wells. (Refer to Fig. 13.63.) A heavy dose of phosphorus ions produces heavily doped source and drain regions for the n-channel transistors that

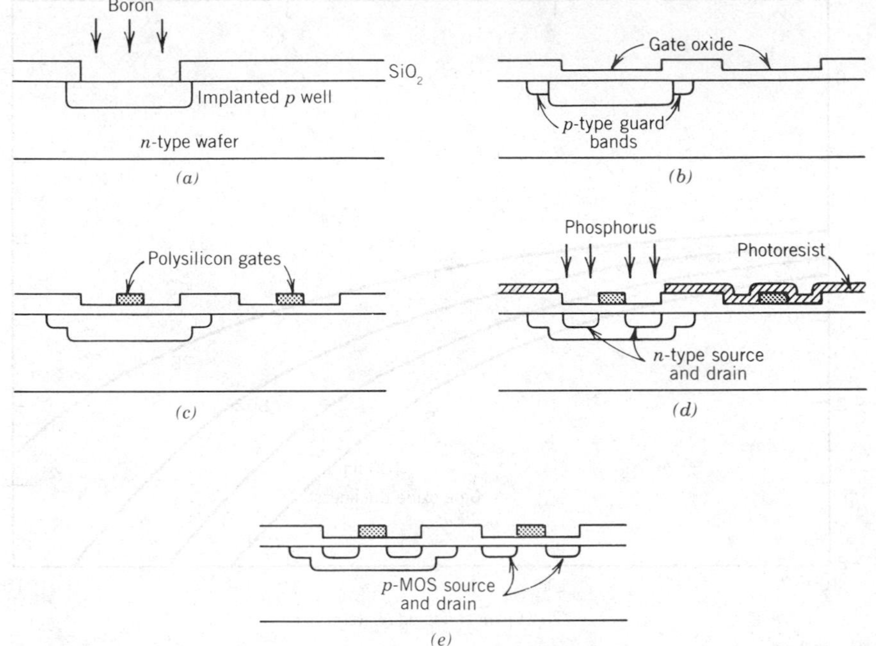

Fig. 13.63 Silicon gate, ion-implanted CMOS (complementary metal oxide semiconductor) processing steps. (*a*) The oxide is patterned to define the ion-implanted *p* wells. (*b*) After a second masking step and boron implant to form the *p*-type guard bands, the entire transistor area is defined and a thin oxide is grown. (*c*) Polysilicon is deposited and patterned. (*d*) A thick layer of photoresist masks the phosphorus implant, which forms the sources and drains of the NMOS (*n*-channel MOS) transistor. (*e*) A similar step is used for the PMOS (*p*-channel MOS) transistor. In the remaining steps, contact windows are etched, metal is deposited and patterned, and glass is deposited, then removed over the bonding pads.

are self-aligned with the gates. This assures turn-on of the channels, yet minimizes overlap capacitance. An annealing cycle removes crystalline damage and activates the phosphorus atoms.

The process is repeated to form *p*-channel sources and drains with a boron implant, patterned with the sixth mask. The fabrication is completed with the seventh, eighth, and ninth masks for contact cuts, metallization patterns, and bond pads, respectively.

The process just described is for a modern technology for high-performance CMOS. The process may have several modifications, for example, a phosphorus implant in the field regions to increase the field threshold, special oxides for radiation hardening, and so on. An important major variation involves the use of oxide isolation to improve packing density and reduce stray capacitance.

Assembly, Packaging, and Testing

After completion of wafer processing, each chip is electrically tested on a probe station. Those that fail to meet required specifications are marked to be discarded. The fraction of chips that pass the probe test is called the yield and is an indication of the quality of the manufacturing process.

The next step is the separation of the individual dice. Scribe lanes, roughly 150-μm wide, form a rectangular grid between the dice. During wafer processing all oxide is removed from the scribe lanes to facilitate die separation. The earliest method of separation, still commonly used, is a simple scribe and break. In this method a diamond-tipped scribing tool scratches the wafer surface along the scribe lanes in each direction, and gentle pressure from soft rubber rollers breaks the wafers into individual dice. Other separation methods in use today include diamond sawing, laser cutting, and occasionally chemical etching.

Good chips are die bonded in a suitable package or on a lead frame by eutectic bonding, with a gold-germanium solder preform, or with epoxy. The aim is to achieve good thermal contact with the header as well as a strong mechanical bond. Other mounting methods include beam leads and flip chip bonding. Beam leads are thick layers of aluminum or titanium, platinum, and gold which extend from the contact pad well out into a wide scribe lane. The dice are separated by etching, which leaves cantilevered beams extending out beyond the edge of the dice. The beam lead chips are spider bonded onto a suitable lead frame or substrate. In flip chip bonding, solder "bumps" are built up over the contact pads with layers of metal such as aluminum, nickel, copper, tin, and tin/lead solder. The chip is mounted upside down on a heated header so that the bumps make contact with and solder onto a suitable interconnection pattern.

If a hermetic package is being used, the lid is attached in a controlled, low-moisture atmosphere. Many dual in-line packages use a ceramic or Kovar lid attached with a low-temperature glass frit or a solder preform. Axial lead packages are sealed by welding a Kovar cap to the header flange. A plastic package is much cheaper and simplifies the inventory of package parts. The chip is often coated with a lacquer, and an epoxy or silicone plastic body is formed by transfer, injection, or compression molding.

Completed parts are electrically tested and may be subjected to a wide variety of tests, depending on customer requirements. Some of the commonly used tests include burn-in, gross and fine leak, high-temperature storage, temperature cycling, thermal shock, acceleration, mechanical shock, solderability, lead integrity, particle impact noise, functional testing, leakage current, power supply current, delay measurements, load conditions, logic levels, and noise margins. These tests can be valuable for weeding out infant mortality failures or detecting problems in processing; however, they also add cost to the final product.

13.4-3 Bipolar Integrated Circuits

Transistor-Transistor Logic

A standard TTL gate is fabricated by the process depicted earlier in Fig. 13.59. A schematic of a two-input NAND gate is given in Fig. 13.64.

The circuit functions as follows. If one or both inputs are at a low voltage, Q_1 has one or both emitter junctions forward biased. Q_2 presents a large load resistance to the collector of Q_1, hence Q_1 is in a saturated condition and its collector is at a low voltage. This puts Q_2 in a cutoff condition. The base of Q_3 is low, hence Q_3 is also nonconducting. The base of Q_4 is returned to a high voltage through the 1.6-kΩ resistor, which turns on Q_4. Hence the output voltage is high.

On the other hand, if both inputs are high, Q_1 operates in the inverse mode. The base of Q_2 is connected to V_{cc} through the forward-biased collector-base junction and the 4-kΩ resistor. Q_2 turns on, and its emitter voltage is high, turning on Q_3. With both Q_2 and Q_3 in the saturated mode, the voltage difference between the base of Q_4 and the output is not sufficient to forward bias both the diode and the emitter junction of Q_4 (this is the purpose of the diode). Hence the output is low.

The transfer characteristic of the gate is shown in Fig. 13.65. When the input voltage is low, the output voltage is at its highest level. Q_2 and Q_3 are both off, and Q_4 is operating in its active region.

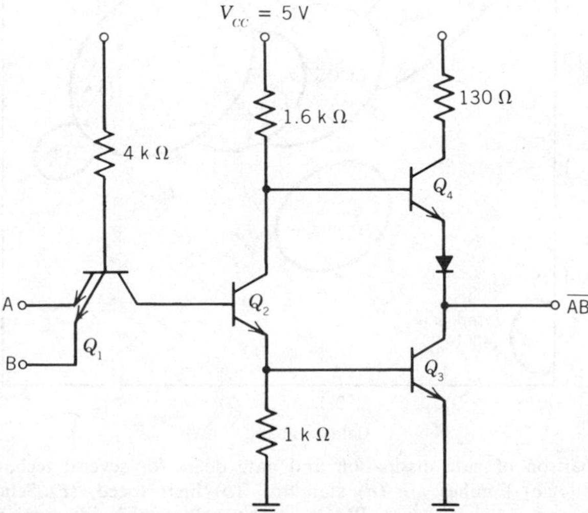

Fig. 13.64 Standard TTL (transistor-transistor logic) two-input NAND gate.

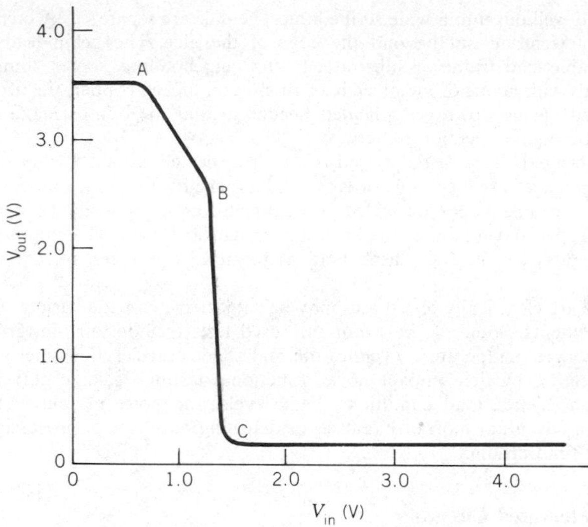

Fig. 13.65 Transfer characteristic of a TTL (transistor-transistor logic) gate.

As the input voltage is increased to point A on the transfer characteristic, current is diverted from the emitter to the collector of Q_1, and Q_2 begins to enter its active region. The gain of Q_2 is approximately 1.6, the ratio of its two resistors. As Q_2 moves further into its active region, the drive to Q_4 decreases. Since Q_4 is basically an emitter follower, the output voltage decreases at a slope of approximately 1.6. Eventually, at point B on the characteristic, Q_3 begins to conduct. Its low input impedance increases the gain of Q_2, and the output voltage drops sharply to point C, the output saturation level.

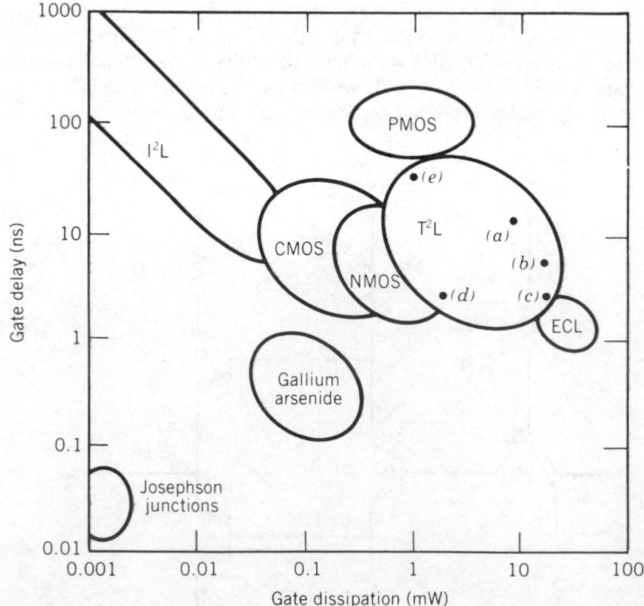

Fig. 13.66 A comparison of gate dissipation and gate delay for several technologies. The TTL (transistor-transistor logic) families are (*a*) standard, (*b*) high speed, (*c*) Schottky barrier, (*d*) low-power Schottky, and (*e*) low power. PMOS, *p*-channel metal oxide semiconductor; CMOS, complementary MOS; NMOS, *n*-channel MOS; ECL, emitter-coupled logic; T^2L, integrated injection logic.

There are several possible minor circuit modifications of the basic gate. However, the circuit shown in Fig. 13.64 optimizes such considerations as chip area, stray capacitance, noise immunity, and transfer characteristics. It is also possible to alter the gate delay and gate dissipation of 10 ns and 10 mW, respectively, for the standard gate by changing the values of resistors to tradeoff speed and power. Figure 13.66 compares delay-power levels for several variations of TTL as well as for other logic families to be discussed.

The standard TTL gate is not suitable for LSI applications because of the high gate dissipation. An important modification makes use of Schottky barrier diodes to prevent the transistors from saturating. This enables the speed to be increased, and innovative gate designs have decreased the power dissipation so that low-power Schottky TTL is a viable LSI technology.

Emitter-Coupled Logic

The emitter-coupled or current-mode logic family is the fastest logic commercially available. One series of SSI gates is characterized by propagation delays of 2 ns with 25 mW dissipation and another series by 1-ns delays and 50 mW dissipation. At these speeds, printed circuit board interconnections longer than a few centimeters require special design techniques.

ECL electrical parameters are different from those of other logic families in several ways. The power supply voltage is different, and logic levels assume unusual values. The availability of OR/NOR outputs in a single gate and the possibility of OR-tieing permits more design versatility and lower gate count. Even though the gate dissipation is high, operation at very high speeds does not increase the dissipation, and smaller switching transients simplify power supply filtering. The noise immunity is aided by the high common-mode rejection of the differential current switch at the input. Finally the emitter-follower outputs offer low output impedance and simplify transmission line driving.

The basic OR/NOR gate is shown in Fig. 13.67. Note that the gate requires a negative power supply, that the V_{cc} connection is separated between the low-current circuitry and the high-current emitter followers, and that both V_{cc} connections are grounded. These arrangements tend to reduce noise and improve performance. The logic swing is relatively small, with a voltage of -0.96 for logical "1" to -1.8 for a logical "0."

The input transistors Q_1 and Q_2 form a differential pair with transistor Q_3. If the input transistors are both supplied with a low voltage, Q_3 conducts and maintains its emitter voltage high enough to bias the input transistors off. If one or both input voltages are at the "1" level, Q_3 will be biased off and current will switch from one load resistor to the other. A detailed DC analysis would show that the transistors are never biased into deep saturation, which accounts for the high speed of ECL.

Since ECL is normally applied only where its high speed can be taken advantage of, signal paths between devices must often be treated as transmission lines. Only if the path propagation time is short compared with pulse rise and fall times can transmission line characteristics be ignored. Usual practice is to pattern the interconnections between packages to act as 50- to 100-Ω transmission lines, with matching terminations on the receiving end. Unfortunately, 50-Ω terminating resistors usually need to be returned to an intermediate voltage supply of -2 V to reduce the emitter-follower drive and reduce gate dissipation.

ECL gates are fabricated by standard bipolar techniques, with a few modifications. Gold doping, to reduce storage time in saturated logic families, is not required. An extra, deep, heavily doped, *n*-type diffusion is usually performed to provide a low resistance path to the buried subcollector. This reduces collector resistance to a minimum.

In spite of high gate dissipation, ECL has found application where high-speed computation is a necessity. To accompany high-speed logic, small ECL memories with access times of 3 to 10 ns are available. In some cases ECL logic is incorporated on the same chip with other logic families. Even though ECL was introduced more than 20 years ago, it appears that its usage will continue to increase to meet the needs of new high-performance applications.

Integrated Injection Logic

I^2L was the latest major logic family to be announced (1971). It is an outgrowth of an early logic called direct-coupled transistor logic. Because of its simple gate structure and low power dissipation it showed promise of becoming a major LSI logic family. Its most serious drawback is its slow speed. Nevertheless it has several unusual features and enough versatility to make it useful in special LSI applications.

Figure 13.68 shows the basic gate. Part (*a*) of the figure illustrates the inverting operation. The *pnp* transistor is called the injector and acts as a current source to supply a constant current to the node at the base terminal of the inverter. If the transistor in the previous stage is cut off, the injector current will forward bias the *npn* transistor into its saturation region. If the previous transistor is saturated, all of the injector current will flow through it, and the node voltage will be too low to turn the inverter

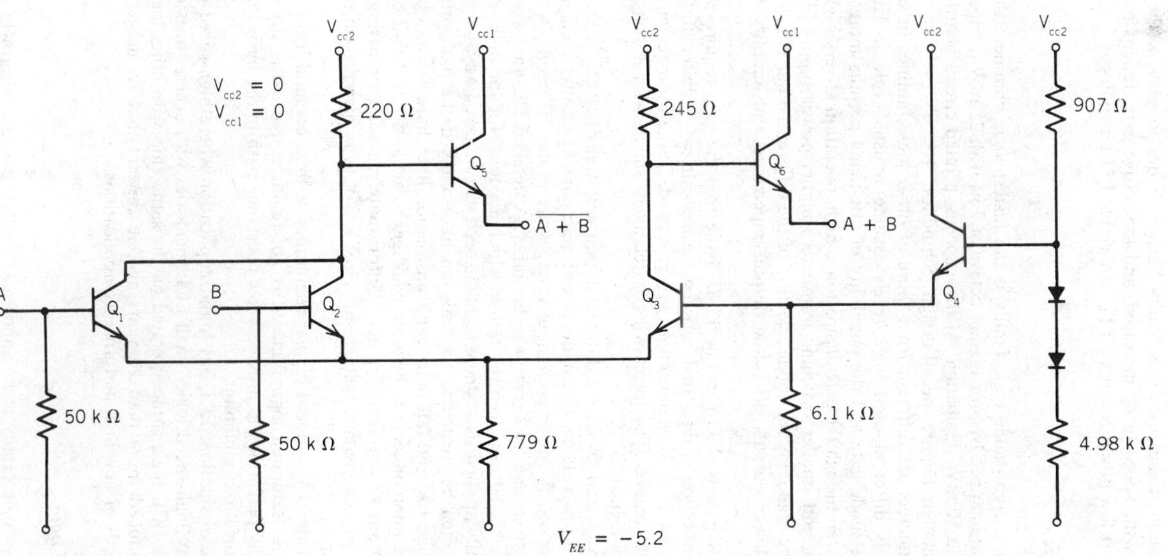

Fig. 13.67 The basic 2-ns 25-mW ECL (emitter-coupled logic) OR/NOR gate.

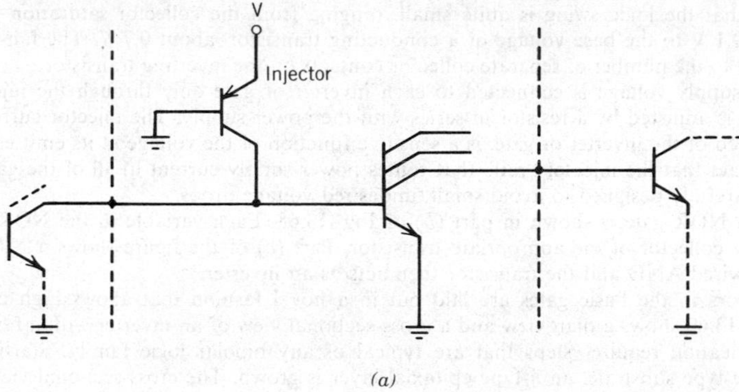

(a)

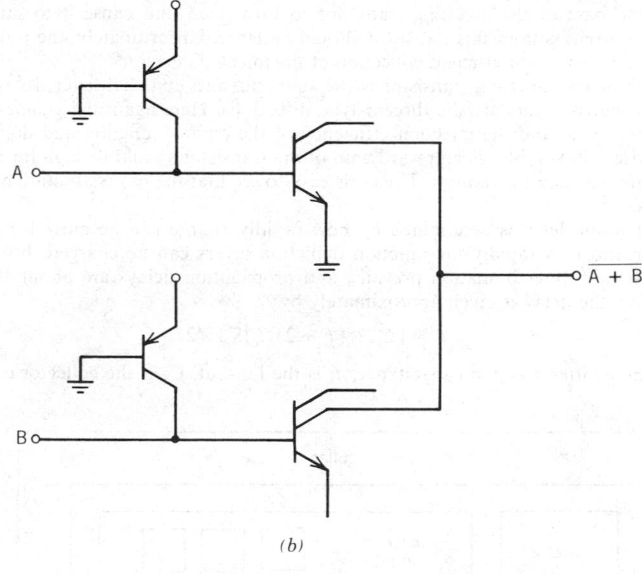

(b)

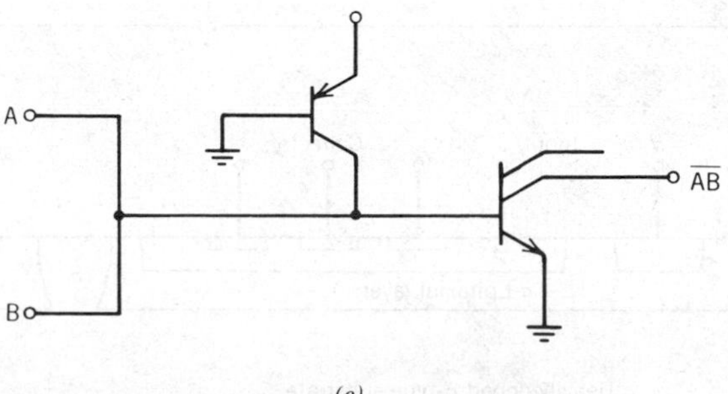

(c)

Fig. 13.68 Integrated injection logic gate. (*a*) Inverter. The injector is a *pnp* transistor that supplies current either to the collector of the previous transistor or to the base of the inverting transistor. (*b*) NOR gate. The logical NOR is achieved by wiring together outputs from inverters. (*c*) NAND gate. The wired-AND inputs are assumed to be the collectors of other transistors.

on. It is clear that the logic swing is quite small, ranging from the collector saturation voltage of approximately 0.1 V to the base voltage of a conducting transistor, about 0.7 V. The fan-out of the inverter is equal to the number of separate collector contacts on the inverting transistor.

The power supply voltage is connected to each inverter or gate only through the injector. The injector current is adjusted by a resistor in series with the power supply. The injector current, which controls the speed of the inverter or gate, is a sensitive function of the voltage at its emitter terminal; so sensitive in fact that the injector "rail" that routes power supply current to all of the gates on the chip must be carefully designed to avoid small, undesired voltage drops.

A two-input NOR gate is shown in part (*b*) of Fig. 13.68. Each variable in the NOR output is supplied from a collector of the appropriate transistor. Part (*c*) of the figure shows a NAND gate. The input is a wired AND and the transistor then acts as an inverter.

The transistors in the basic gates are laid out in a novel fashion that allows high component density. Figure 13.69 shows a plan view and a cross-sectional view of an inverter with a fan-out of 3. The wafer fabrication requires steps that are typical of any bipolar logic family. Starting with a heavily doped *n*-type substrate, an *n*-type epitaxial layer is grown. The cross-sectional view shows a deep, heavily doped *n*-type isolation collar that extends down to the substrate. This collar is an optional feature that requires an extra mask but improves performance. The next step is the *p* diffusion. One *p* region is the emitter of the injector, the other forms the collector of the injector and also the base of the inverting transistor. The injector is a lateral *pnp* transistor that must inject enough current into the base of the inverting transistor to turn it on and cause it to saturate. Hence the common base current gain of this transistor should be large. Unfortunately, the transistor has a poor geometrical arrangement for efficient collection of the injected current.

The emitter of the inverting transistor is the substrate and epitaxial layer. Its base is the large *p* diffusion, and the collectors are the three *n*-type diffusions. Here again the geometry is unfavorable for good current gain, and the injection efficiency of the emitter junction and doping profile of the base are likewise unfavorable. The upward gain of the transistor should be high for best performance, and must be greater than the fan-out. Thus it is easy to see that the basic structure has several built-in disadvantages.

The propagation delay is determined by how rapidly charge can be moved into and out of the epitaxial layer and how rapidly the junction depletion layers can be charged. For injector currents above 100 μA the former limitation prevails, and propagation delays are about 10 ns. For smaller injector currents the delay is given approximately by

$$t_d = [C_e + (F+2)C_c]\nabla V / 2I_{inj} \qquad (13.78)$$

where C_e is the emitter junction capacitance, F is the fan-out, C_c is the collector capacitance, ∇V is

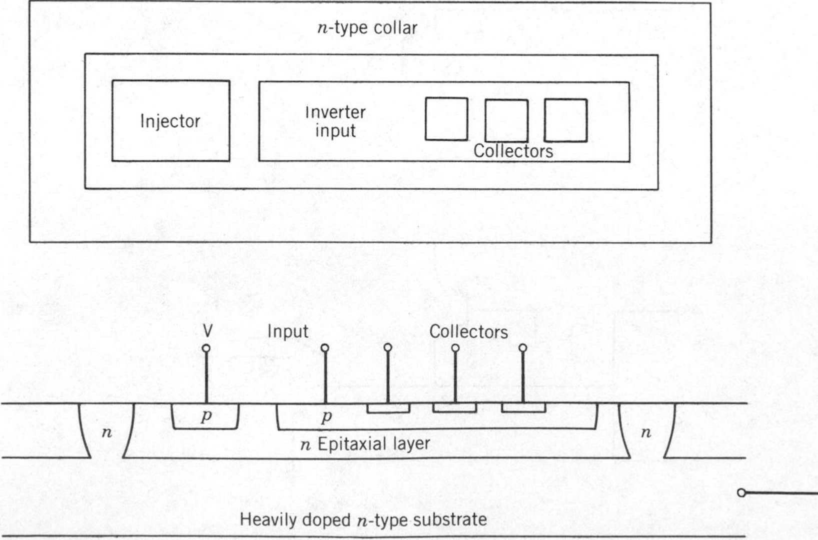

Fig. 13.69 Layout and cross-section view of an inverter. The grounded substrate is both the base of the lateral *pnp* injector and the emitter of the vertical *npn* transistor. The collectors are heavily doped *n*-type diffusions.

the logic swing, and I_{inj} is the injector current. It is possible to vary the injector current over a range of several orders of magnitude with corresponding changes in propagation delay. All of this occurs at a nearly constant power-delay product, since the product is approximately independent of injector current. No other technology is capable of such wide ranges in performance.

Another unusual feature of I^2L is the ability to combine with other bipolar logic families and even linear circuits on the same chip. This usually means extra masking steps. However, the possibility of combining operational amplifiers, I^2L, and interface circuits leads to great versatility.

Several variations of I^2L have been developed to improve performance.[8] Some of these include Schottky I^2L, Schottky transistor logic, substrate fed logic, and vertical injection logic. These variations are capable of reducing gate delay to 5 ns or less, but require extra masks.

Although I^2L has not dominated LSI integrated circuits as originally foreseen, it is widely used in a diverse array of products for consumer, telecommunications, automotive, computer, industrial, and military applications. It will continue to enjoy widespread use where its flexibility and design simplicity are advantages.

13.4-4 MOS Integrated Circuits

p-Channel MOS Logic

PMOS logic is fabricated by the same processing steps described earlier for NMOS, except that opposite impurity types are employed. A cross section of a PMOS transistor, its schematic symbol, and the electrical characteristics are shown in Fig. 13.70. Analytic expressions for the drain current are given by

$$I_D = -\frac{\mu\epsilon_{ox}}{t_{ox}}\frac{W}{L}\left[(V_G - V_T)V_D - \frac{V_D^2}{2}\right] \quad \text{for} \quad V_G < V_D + V_T \tag{13.79}$$

$$I_D = -\frac{\mu\epsilon_{ox}}{2t_{ox}}\frac{W}{L}(V_G - V_T)^2 \quad \text{for} \quad V_G > V_D + V_T \tag{13.80}$$

where μ = mobility of carriers (holes) in channel
ϵ_{ox} = permittivity of gate oxide
t_{ox} = thickness of gate oxide
W = width of channel
L = length of channel
V_G = gate voltage
V_T = threshold voltage
V_D = drain voltage

Two other constants are commonly used

$$K' = \frac{\mu\epsilon_{ox}}{t_{ox}} = \mu C_0 \tag{13.81}$$

$$\beta = K'\frac{W}{L} \tag{13.82}$$

C_0 is the capacitance per unit area of the gate. K' depends only on the properties of the processing. Typical values range from 2 to 20 $\mu A/V^2$. β depends on the width-to-length ratio of the channel and is readily adjusted by the geometry of the layout.

A PMOS inverter is shown in Fig. 13.71. Typical values of V_{DD}, V_{GG}, and V_T are -5, -17, and -2 V, respectively, and a typical value of K' is 10 $\mu A/V^2$. The transfer characteristic shown in the figure is given for two different values of width-to-length ratios of the input and load transistors. This ratio is called the beta ratio, β_R.

$$\beta_R = \frac{W/L\ (\text{input})}{W/L\ (\text{load})} \tag{13.83}$$

The larger the beta ratio, the more the transfer function approaches the ideal. In particular, it is essential that the output voltage of an inverter be able to turn off a following inverter when required. However, large beta ratios require more area for logic gates and increase gate delays. A typical value is 10, and represents a compromise among these conflicting requirements. A layout of an inverter with $\beta_R = 20$ is shown in Fig. 13.72.

The transient response of an inverter is determined by the current drive capability of the input and load transistors and by the capacitance at the output terminal. The total capacitance is the sum of the output capacitance of the inverter itself, the capacitance of connecting lines, and the gate

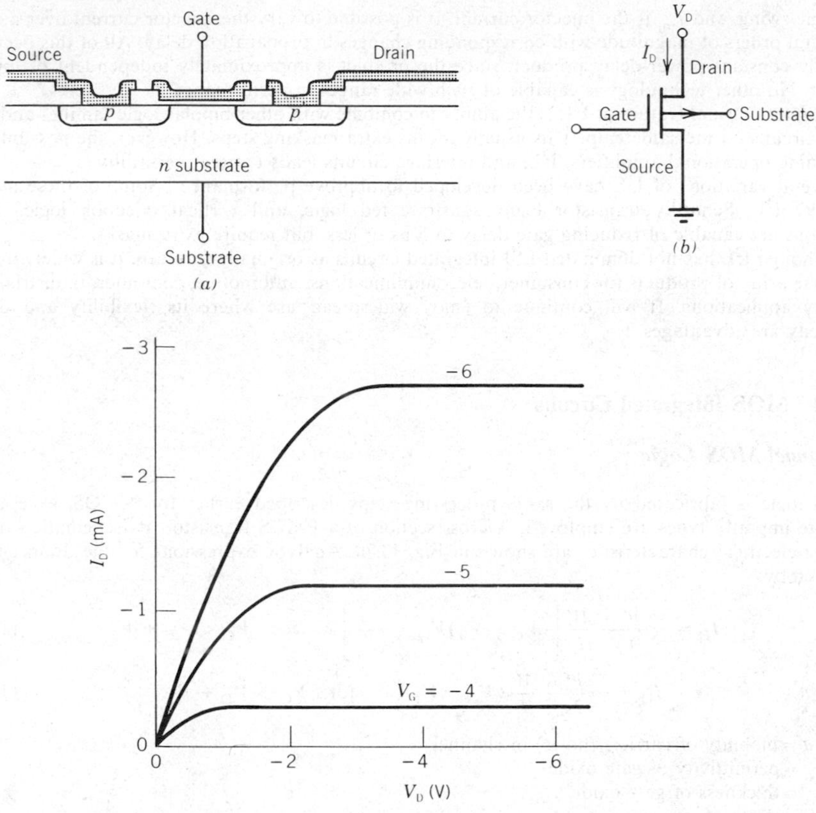

Fig. 13.70 PMOS (*p*-channel metal oxide semiconductor) transistor: (*a*) structure, (*b*) electrical symbol, and (*c*) drain characteristics for an ideal device with $\beta = 0.6\ \text{mA/V}^2$ and $V_T = -3$ V. The substrate is assumed to be grounded.

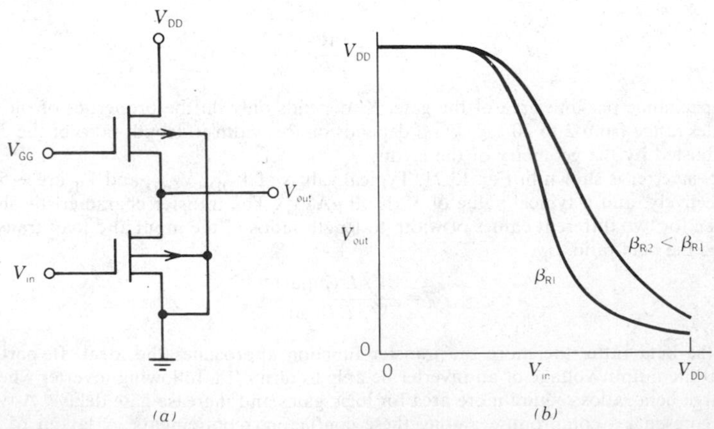

Fig. 13.71 PMOS (*p*-channel metal oxide semiconductor) inverter. (*a*) Schematic showing two power supplies. V_{GG} is chosen to be more negative than $V_{DD} + V_T$ so that the load transistor operates in the nonsaturated region. (*b*) Transfer characteristic. Large values of β_R result in better inverter characteristics.

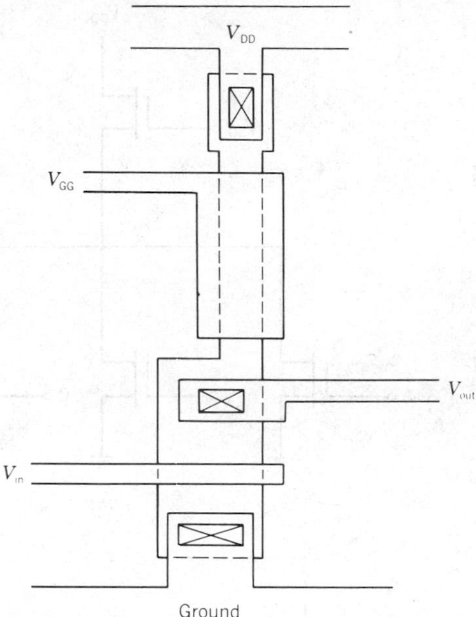

Fig. 13.72 Example of layout of an inverter with a beta ratio of 20.

capacitance of following stages. This last term usually dominates. Detailed methods for finding transient response are available in standard texts.[9,10] However, it should be pointed out that the rise and fall times differ by the beta ratio of the inverter (roughly), since the current capability of each transistor is different. Note also that the transient response of inverters depends on the fan-out unless the widths (or lengths) of the transistors are adjusted accordingly. If a series of inverters must drive a large capacitive load, it can be shown that the optimum increase in inverter drive capability should be in the ratio $e = 2.72$, to minimize overall delay. This sometimes is impractical to implement, and other circuits can be used for output buffers.

Several basic gates are illustrated in Fig. 13.73. When transistors are connected in series (as in the NAND gate), adjustments must be made in W/L to keep the overall beta ratio the desired value. For transistors or branches in parallel, the beta ratio will change as different numbers of parallel paths are turned on. Because of this, NOR circuits are more economical of chip area. Transient response is governed by the same considerations as with an inverter.

PMOS logic is rarely used in new designs but in the 1970s was almost universally employed in calculators and other consumer products. It has been overshadowed by NMOS technology, to be presented next. Note, however, that virtually all of the circuit design considerations are the same for any single-channel logic.

NMOS Logic

The main differences between NMOS and PMOS transistors are the polarity of the voltages and the higher speed of NMOS due to the higher mobility of electrons. Consequently all of the previous discussion of PMOS logic applies to NMOS as well. However, many PMOS circuit variations have been brought to maturity in NMOS and many new innovations have given NMOS a dominant role in VLSI.

For example, saturated-load devices and ion-implanted depletion mode-load devices were sometimes used in PMOS logic, but are much more common in NMOS. The differences in circuit connections and transfer functions are illustrated in Fig. 13.74. The depletion load transistor is a device that has a negative threshold voltage (for NMOS) and consequently is always conducting when used as a load device in an inverter as shown in the figure. Unfortunately an extra processing step is required to achieve the two different threshold voltages for the input and load transistors. However, the elimination of the V_{GG} power supply and power supply bus on the chip are distinct advantages. The depletion transistor is a better load device than the saturated-mode load, which

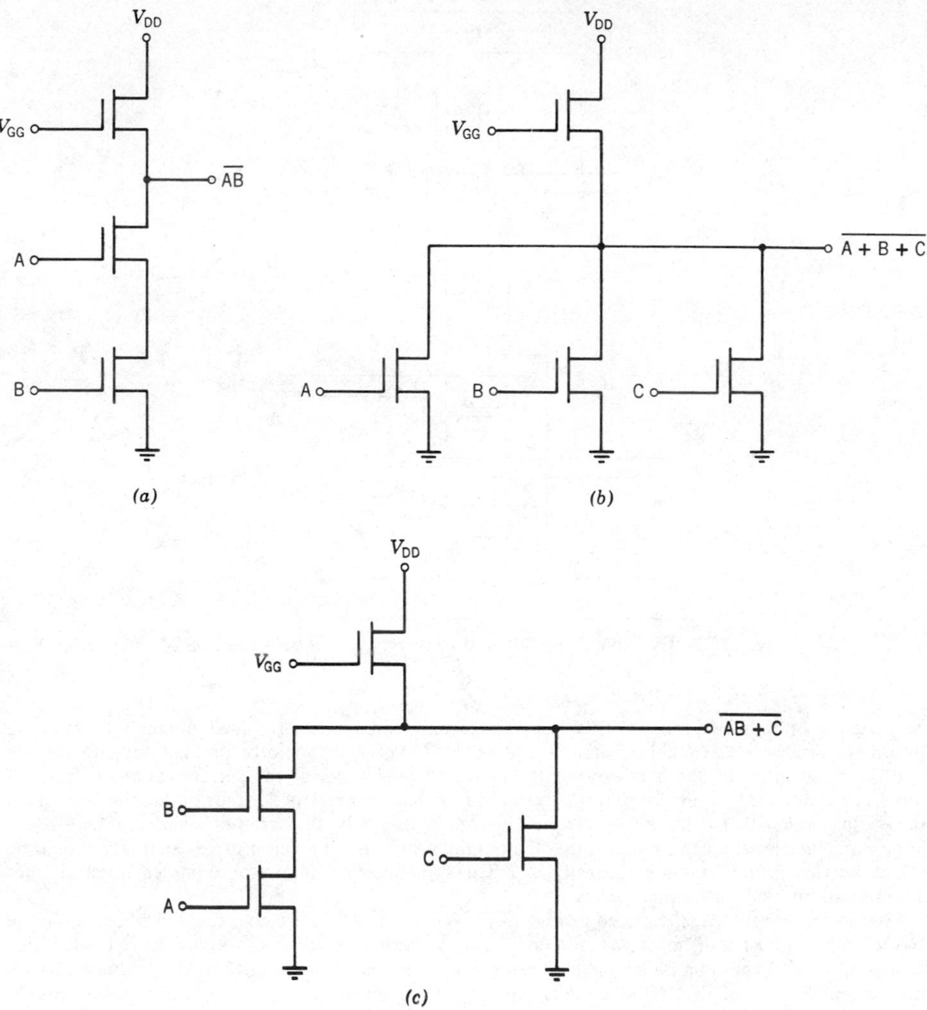

Fig. 13.73 PMOS (*p*-channel metal oxide semiconductor) logic gate. (*a*) two-input NAND gate. (*b*) three-input NOR gate. (*c*) AND-OR-INVERT gate. These examples assume that the voltage level equal to V_{DD} is a logical "1."

tends to turn off as the output voltage rises, slowing the rise time. The depletion transistor can also be designed with a much smaller beta ratio, which reduces the gate area.

NMOS technology has continued to improve dramatically in performance as designers have introduced new variations in circuit design and processing. Substrate bias, multilevel polysilicon, low-resistivity silicide gates, scaling, dry etching, and laser recrystallized poly are only a few of the improvements that have enabled NMOS to maintain its lead as the most popular VLSI technology.

Complementary MOS

Unlike single-channel MOS, CMOS is a ratioless logic because the output voltage of a gate does not depend on the ratio of betas in the transistors. However, its most unusual characteristic is the extremely low power dissipation, of the order of nanowatts per gate in the standby mode. This is achieved by connecting the transistors so that a conducting path from power supply to ground exists only when the transistors are switching. This is illustrated by the inverter shown in Fig. 13.75.

The source of the PMOS transistor, as well as its channel region, is connected to the power

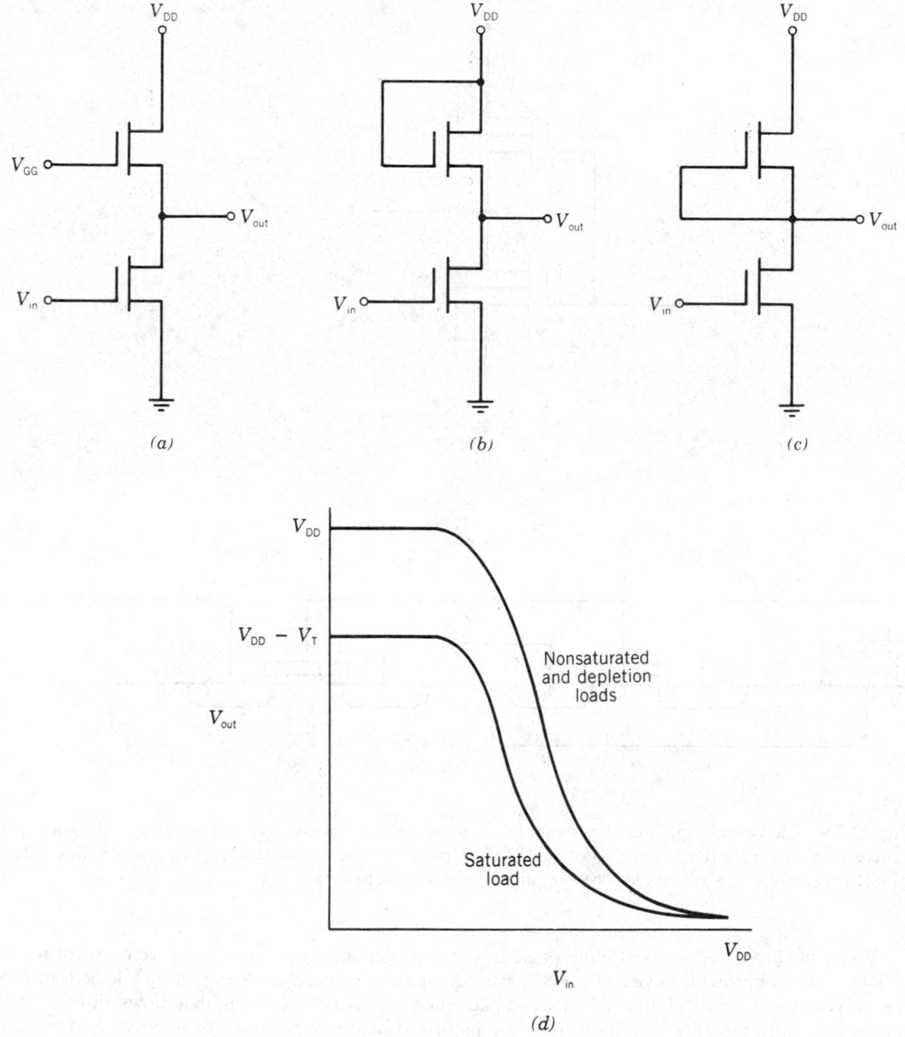

Fig. 13.74 NMOS (n-channel metal oxide semiconductor) inverters. (a) Nonsaturated load, assuming $V_{GG} > V_{DD} + V_T$. (b) Saturated load. (c) Depletion load. (d) Typical transfer characteristics.

supply. This allows the PMOS transistor, which acts as a load for the inverter, to be used in a circuit with only a single, positive voltage supply. The source of the NMOS transistor and the p well is connected to ground. The two transistors are connected in series by joining their drains together and the inverter output is taken from this common node. When the input voltage is low, the NMOS transistor is cut off. However, the gate of the PMOS device is negative with respect to its source, which turns it on. Consequently the output node is connected to V_{DD} through the load device and the capacitance at the node is charged to the power supply voltage through an active pull-up. When the input voltage is high, the load transistor is cut off and the driver is turned on, discharging the node capacitance through an active pull-down.

The transfer function of the inverter is shown in Fig. 13.76. If the threshold voltage magnitudes are equal and the transistor W/L ratios are adjusted to account for the higher mobility of carriers in the NMOS transistor, the transfer function is nearly ideal, with perfect symmetry about the midpoint, a sharp transition between logic levels, and a logic swing equal to the supply voltage. Furthermore the inverter will operate over a wide range of supply voltages and is insensitive to changes in ambient temperature.

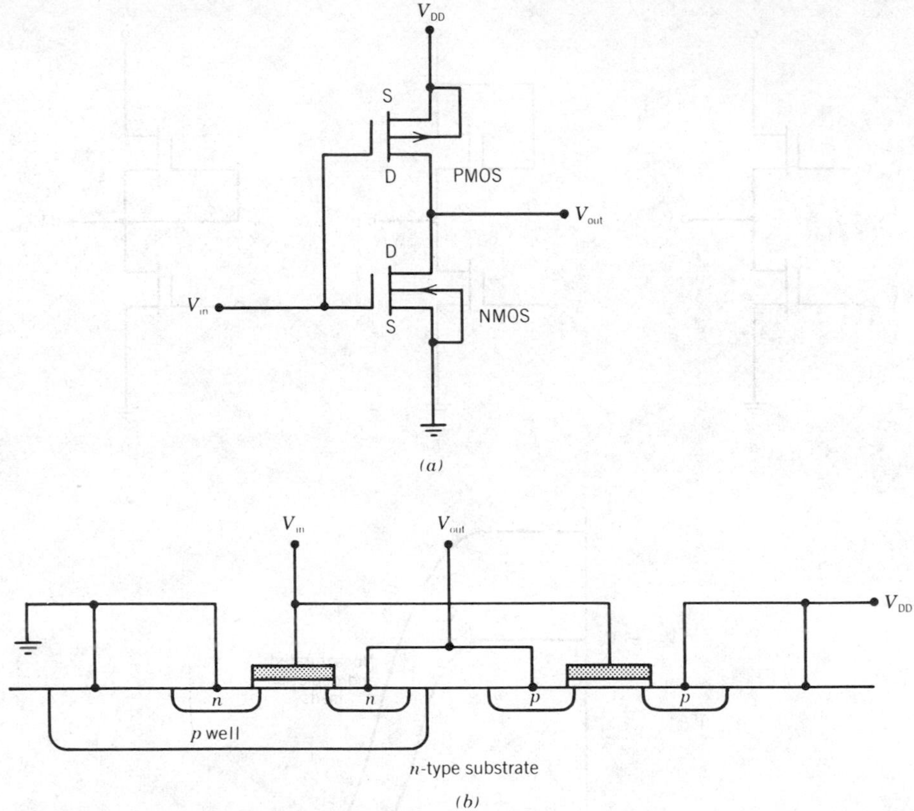

Fig. 13.75 CMOS (complementary metal oxide semiconductor) inverter. (*a*) Schematic diagram. The source and substrate (i.e., channel) of the PMOS transistor are connected to V_{DD}, while those of the NMOS transistor are grounded. (*b*) Simplified cross-section view.

When the input voltage is at either logic level the power supply current I_{DD} is zero except for the leakage current in the off device. The DC power dissipation is therefore low. During a logic transition the power supply must charge the node capacitance and supply current that flows through both transistors while they are simultaneously conducting. Usually the former component is dominant. The total inverter dissipation is then approximately

$$P_d = V_{DD} I_L + V_{DD}^2 f C_0 \tag{13.84}$$

where I_L is the leakage current, f is the operating frequency, and C_0 is the node capacitance. For a typical SSI inverter or gate, the DC term is several nanowatts and the frequency-dependent term is comparable to that of a TTL gate above 10 MHz.

In a VLSI circuit, not all of the gates are exercised at the clock frequency, hence the average gate dissipation can be substantially less than the maximum value.

The propagation delay of a CMOS inverter depends on the node capacitance, the supply voltage, and the electrical parameters of the transistors. The delay is given by[11]

$$t_p = \frac{1.8 t_{ox} L_n C_0}{\epsilon_{ox} \mu_n W_n V_{DD}} \left[\frac{1}{\left(1 - \dfrac{V_{Tn}}{V_{DD}}\right)^2} + \frac{1}{B\left(1 - \dfrac{V_{Tp}}{V_{DD}}\right)^2} \right] \tag{13.85}$$

$$B = \frac{\mu_p W_p L_n}{\mu_n W_n L_p}$$

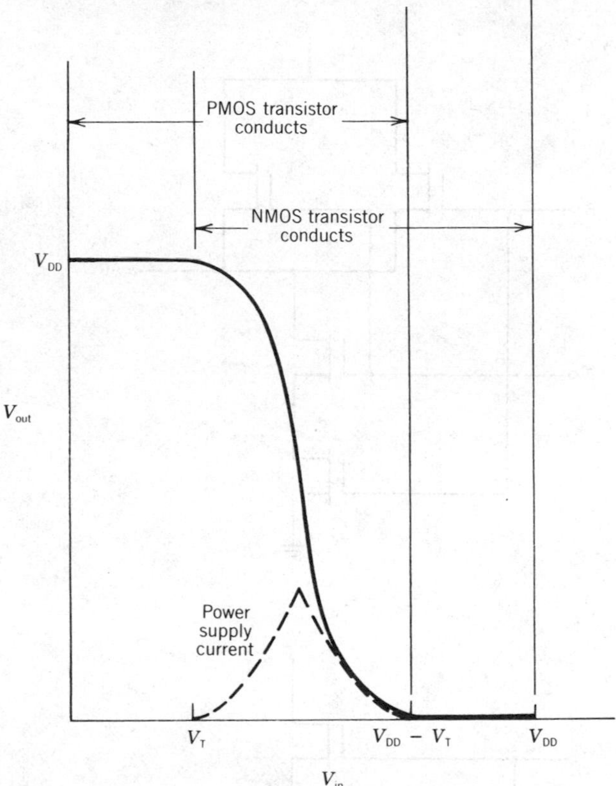

Fig. 13.76 Transfer characteristic of a CMOS (complementary metal oxide semiconductor) inverter. Current is drawn from the power supply only during transitions from one logic level to another.

where t_{ox} = thickness of gate oxide
ϵ_{ox} = permittivity of gate oxide
V_T = threshold voltage
μ = mobility
L = channel length
W = channel width

The subscripts n and p refer to the NMOS and PMOS transistors, respectively. The gate delay is typically a few tens of nanoseconds per picofarad of node capacitance. The node capacitance is determined primarily by the gate capacitance of following stages, but also includes interconnection capacitance and output capacitance. Silicon-on-sapphire and oxide-isolated CMOS minimize the latter contribution.

CMOS NAND and NOR gates are illustrated in Fig. 13.77. As in the case of single-channel gates, transistor W/L ratios should be adjusted to maintain equal rise and fall times. It is also clear that CMOS gates require more transistors and more complex interconnections and hence require more chip area. Of course, the gate dissipation is substantially lower at low frequencies.

This low power dissipation makes it possible to put thousands of gates on a chip without requiring special packaging to remove the heat. In standby the current requirements are so low that battery backup can be used for what amounts to nonvolatile memories. New design ideas that combine NMOS and CMOS on one chip or invert the structure (using n wells instead of p wells) have increased the speed and decreased the area. The continued progress of CMOS technology indicates that CMOS will become the dominant VLSI family.

References

1 Michael F. Wolff, "The Genesis of the Integrated Circuit," *IEEE Spectr*. **13**:45–53 (Aug. 1976).
2 *Electronic Design*, Special Anniversary Issue (Nov. 23, 1972).

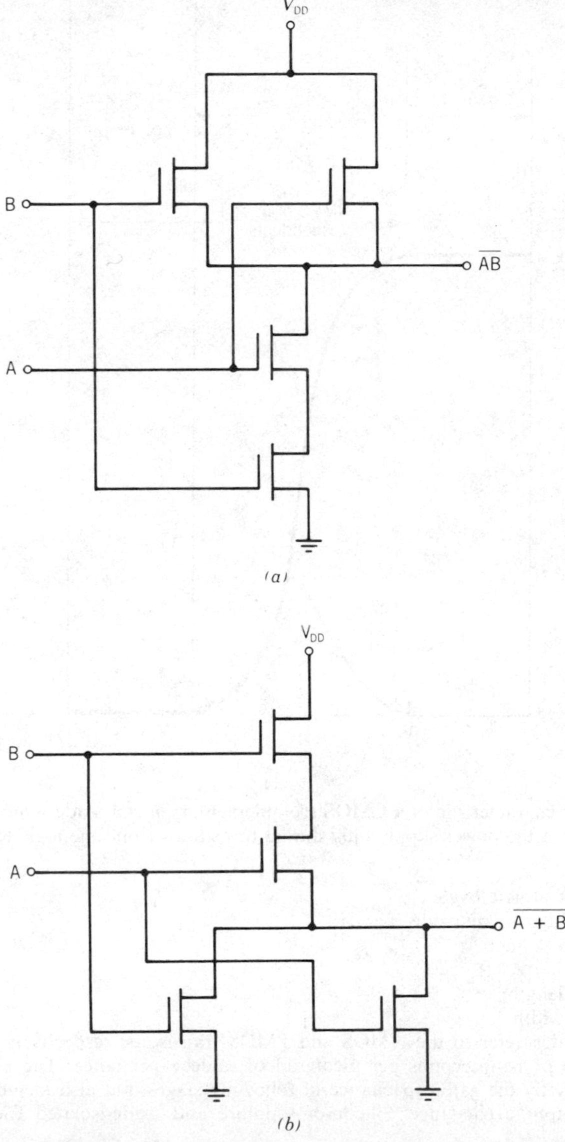

Fig. 13.77 CMOS (complementary metal oxide semiconductor) gates: (*a*) NAND gate, (*b*) NOR gate.

3 Arthur B. Glaser and Gerald E. Subak-Sharpe, *Integrated Circuit Engineering*, Addison-Wesley, Reading, MA, 1977.

4 Douglas J. Hamilton and William G. Howard, *Basic Integrated Circuit Engineering*, McGraw-Hill, New York, 1975.

5 Hans Mark Janus and Olof Malmros, "Application of Thermal Neutron Irradiation for Large Scale Production of Homogeneous Phosphorus Doping of Floatzone Silicon," *IEEE Trans. Electr. Dev.* **ED-29**:797–802 (Aug. 1976).

6 George Carter and W. A. Grant, *Ion Implantation of Semiconductors*, Wiley, New York, 1976.

7 James F. Gibbons, "Ion Implantation in Semiconductors, Part I, Range Distribution Theory and Experiments," *Proc. IEEE* **56**:295–319 (Mar. 1968).

8 J. L. Stone, "I^{2}L: A Comprehensive Review of Techniques and Technology," *Solid State Technol.* **20**:42–47 (June 1977).

9 Robert H. Crawford, *MOSFET in Circuit Design*, McGraw-Hill, New York, 1967.

10 William M. Penney and Lillian Lau, eds., *MOS Integrated Circuits*, Van Nostrand Reinhold, New York, 1972.

11 Thomas Klein, "Technology and Performance of Integrated Complementary MOS Circuits," *IEEE J. Solid State Cir.* **SC-4**:122–130 (June 1969).

13.5 HYBRID CIRCUITS—THICK AND THIN FILM

George H. Ebel

13.5-1 General Description

There is no generally acceptable definition for hybrid circuits in the microelectronic industry. Most definitions tend to classify hybrid circuits by design, fabrication, and materials; by function; or by use. The common thread in all definitions is the miniaturization of electronic circuit fabrication by means other than placement of all functions in a single monolithic semiconductor integrated circuit. A typical example of this is the replacement of a printed circuit board having discrete parts by a hybrid circuit, using one of several packaging methods to be described later. Size reduction using hybrid construction techniques range from about 5 : 1 to 20 : 1.

Some persons contend that hybrid circuits should be classified as an assembly technique rather than as a unique component. This is probably because the one item that seems to be common to all such circuits is a microminiature interconnection scheme.

The title of this section (13.5) suggests that all hybrid circuits are divided into thick-film and thin-film categories. (See Figs. 13.78 and 13.79.) Although substrate metallization is one aspect of

Fig. 13.78 Multilevel thick-film hybrid construction with a single-level substrate attached to the main substrate. The single-level portion shows abrasive trimming of thick-film resistors (dark areas). Large elements on the main substrate are ceramic chip capacitors. The hybrid is $1\frac{1}{4}$ in. square and contains 8 integrated circuits, 18 transistors, 3 diodes, 33 capacitors, 44 resistors, and 428 wirebonds.

Fig. 13.79 Single-level thin-film hybrid construction showing four large interdigitated power transistors. The hybrid is 1″ square and contains 9 integrated circuits, 25 transistors, 6 diodes, 18 capacitors, 53 resistors, and 634 wirebonds.

hybrid circuits, there are many others. An outline of three popular methods of classification follows:

A. Design, Fabrication, and Materials

- (1) Substrate metallization
  - (a) Thick film
  - (b) Thin film
  - (c) Multilevel
- (2) Substrate resistors
  - (a) Thick film
  - (b) Thin-film nichrome
  - (c) Thin-film tantalum nitride
- (3) Substrate material
  - (a) Alumina
  - (b) Beryllia
  - (c) Polymers
  - (d) Metal core
- (4) Interconnection methods
  - (a) Wirebonding
  - (b) Reflow solder
  - (c) Conductive polymers
  - (d) Welding
- (5) Component attachment
  - (a) Eutectic
  - (b) Reflow solder
  - (c) Polymers
- (6) Packaging
  - (a) Hermetic
  - (b) Nonhermetic

B. Function

- (1) Analog
- (2) Digital
- (3) Power
- (4) Microwave
- (5) Optical

C. Use/Environment

- (1) Military
  - (a) Aircraft
  - (b) Missile
  - (c) Space
  - (d) Vehicular
  - (e) Ground
- (2) Commerical
  - (a) Vehicular
  - (b) Medical implanted devices
  - (c) Office, laboratory, etc.

This outline is not intended to be exhaustive but only to show the diversity of items that can be classified as hybrid circuits. Items in the classification can be combined to describe a particular hybrid circuit in more precise detail. This diversity also leads to unique manufacturing processes at each location at which hybrid circuits are manufactured. Although as the hybrid circuit industry matures the manufacturing processes and procedures are becoming more standardized, it is still difficult to find two manufacturers that use identical materials, processes, and procedures.

13.5-2 Definition of Terms

Many documents contain definitions of terms used in the field of hybrid circuits. The most common reference is MIL-STD-883. The following definitions are in the August 25, 1983 version of test method 2017 of that document.

Active circuit area. Active circuit area includes all areas of functional circuit elements and operating metallization or any connected combinations thereof excluding beam leads.

Controlled environment. A controlled environment shall be in accordance with the requirements of Federal Standard 209, Class 100, environment for air cleanliness and humidity, except that the maximum allowable relative humidity shall not exceed 50%. The use of an inert gas environment such as nitrogen shall satisfy the requirement for a controlled environment.

Foreign material. Foreign material is any material that is displaced from its original or intended position within the microcircuit package. Conductive foreign material is any substance that appears opaque under the conditions of lighting and magnification used in routine visual inspection. Particles shall be considered embedded in the glassivation when there is evidence of color fringing around the periphery of the particle.

Glassivation. Glassivation is the top layer(s) of transparent insulating material that covers the active chip area, including metallization except bonding pads and beam leads. Crazing is the presence of minute cracks in the glassivation.

Insulating layer. Insulating layer is a dielectric layer used to isolate multilevel conductive and resistive material or to protect top-level conductive resistive material.

Multilayered metallization. Multilayered metallization (conductors) is two or more layers of metal or any other material used for interconnections that are not isolated from each other by a grown or deposited insulating material. The term "underlying metal" shall refer to any layer below the top layer of metal.

Kerf. Kerf is that portion of the resistor area from which resistor material has been removed or modified by trimming.

Multilevel metallization. Multilevel metallization (conductors) is two or more layers of metal or any other material used for interconnections that are isolated from each other by a grown or deposited insulating material.

Operating metallization. Operating metallization (conductors) is all metal or any other material used for interconnection except metallized scribe lines, test-pattern-unconnected functional circuit elements, unused bonding pads, and identification markings.

Organic polymer. Organic polyer (epoxy) vapor residue is the material that is emitted from the polymer that forms on an available surface.

Original design separation. Original design separation is the separation dimension or distance that is intended by design.

Original width. Original width is the width dimension or distance that is intended by design (i.e., original metal width, original diffusion width, original beam width, etc.).

Passivation. Passivation is the silicon oxide, nitride, or other insulating material that is grown or deposited directly on the die prior to the deposition of metal.

String. String is a filamentary run-out or whisker of organic polymer material.

Thick film. Thick film is a film deposited by screen printing processes and fired at high temperature to fuse into its final form. The basic processes of thick-film technology are screen printing and firing.

Thin film. A thin film (usually less than 10,000 Å thick) is one that is deposited on a substrate by an accretion process such as vacuum evaporation, sputtering, or pyrolytic decomposition.

Unused component or unused deposited element. Such a component or element is one that is not connected to a circuit or circuit path at only one point. A connection may be made by design or by visual anomaly.

Substrate. A substrate is the supporting structural material into and/or upon which the passivation, metallization, and circuit elements are placed.

Ground plane. A ground plane is a metallized area on a substrate that is used to separate voltage distribution layers from signals or from different classes of signals. The ground plane may or may not be on the circuit side of the substrate.

Narrowest resistor width. This width is the narrowest portion of a given resistor prior to trimming. However, the narrowest resistor width for a block resistor may be specified in the approved design documentation.

Edge metallization. Edge metallization is the metallization that electrically connects the metallization from the top surface to the opposite side of the substrate. It is also called wraparound metallization.

Interdigitated capacitor. This is a capacitor formed by spacing of parallel metallized lines.

RF tuning. RF tuning is the adjustment of the output signals from an RF circuit by alteration of lines or pads and/or changing resistance and capacitance values to meet a specific electrical specification.

Through-hold metallization. Such metallization is that which electrically connects the metallization on the top surface of the substrate to the opposite surface of the substrate.

Compound bond. A compound bond is the monometallic bonding of one bond on top of another.

13.5-3 Manufacturing Processes

There are many manufacturing processes for building hybrid circuits. These processes, from substrate technology through component attachment and interconnection techniques to packaging and testing can be combined in an almost limitless fashion.

Substrate Technology

The material most commonly used for substrates today is unglazed alumina. Beryllia is sometimes used if higher power dissipation is required. However, people working with beryllia must be protected from beryllia dust, and this generally precludes the use of beryllia for substrate resistors that require trimming. Recent work has investigated the use of metal, such as porcelainized steel, for substrate material. Another technology being studied is the use of organics such as polyimides. Other materials that have been used for substrate manufacture are glazed alumina, sapphire, glass, and oxidized silicon.

To complete a hybrid substrate, some form of interconnection pattern must be placed on the substrate material. This is done using either thick-film or thin-film technology. Thick-film technology is basically the age-old technique of screen printing. Thin-film technology is newer photolithographic techniques, which allow denser packaging with single-level interconnections. Thick-film technology, however, allows for more current-carrying capability, which is required for high-power hybrids.

Single level thick-film circuits are made by screening paste through fine-mesh stainless-steel screens that have appropriate patterns on them. Most conductor patterns are made with a gold-based paste, although other less expensive materials are continually being evaluated. Resistive inks are screened on the substrate to create most of the resistors used in the hybrid circuit. Resistors are not normally screened on multilevel thick-film circuits. The multilevel construction is used for higher density circuits (see Fig. 13.80). Since it is difficult to screen reliable resistors on levels above the substrate, discrete chip resistors or resistor networks are used. A multilevel substrate is built up by alternately screening on layers of conductors and insulation. Actually to make a reliable multilevel hybrid the following are required:

1. Screen and fire first-level conductor pattern.
2. Screen and fire first level of insulating material with via holes to connect second-level conductor pattern.
3. Screen and fire conductive material into the via holes so that the screening of the second conductive level is done on as flat a surface as possible.
4. Repeat steps 2 and 3 to eliminate pinhole short circuits between conductor levels.
5. Screen and fire second-level conductor pattern.
6. Continue until desired level of conductor patterns has been completed. The metallization of the top level is quite often a different material than the other levels. This is done to optimize the bonding reliability of items placed on the substrate and to assure the best interconnection process for the hybrid. Other ways to accomplish this are to double screen the final level or to plate the desired metallization on the final level.

Thin-film substrate technology (see Fig. 13.81) is generally categorized by the resistor material used. The two most common materials are nichrome and tantalum nitride. Nichrome is generally better for tight tolerance situations but suffers from the problem of open resistors in the presence of moisture and voltage. The characteristics of tantalum nitride have improved and this material is now competitive with nichrome.

Thin-film substrates are made by first depositing the resistor materials on the substrate material. The conductor material, usually gold, is deposited over the resistor material. Then by photolitho-

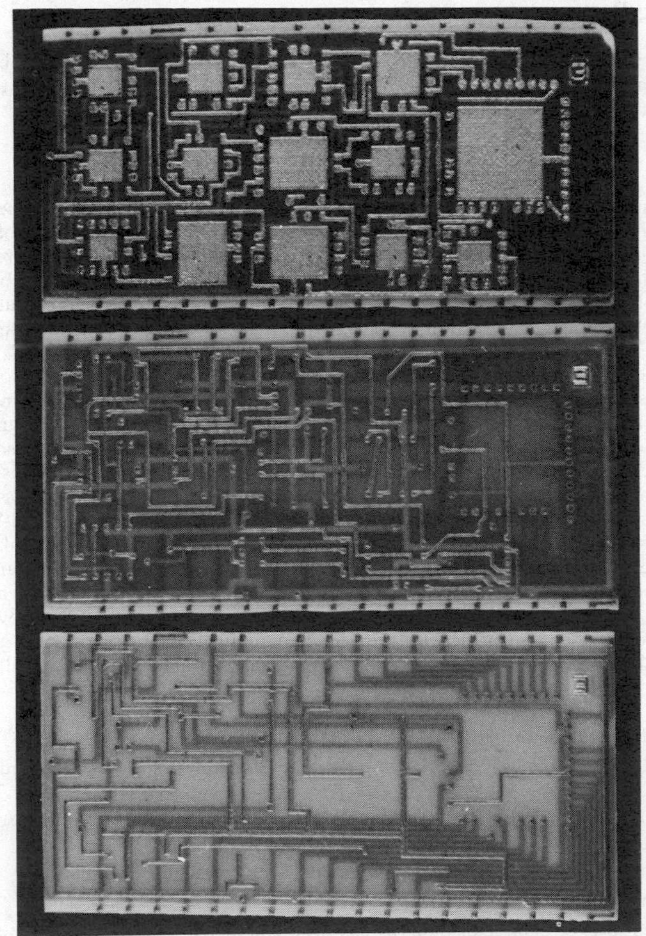

Fig. 13.80 Top three levels of a four-level digital multilevel thick-film substrate. The first level can be seen through the second-layer pattern on the left. This substrate fits into a $1 \times 1/2$-in. package. The conductors are 5 mil (minimum) wide and the minimum spacing between conductors is 5 mil. These dimensions are difficult to maintain in large-scale production. Generally the conductors and the spacings between them are 10 mils or greater to assure high yield factors.

Fig. 13.81 Section of a thin-film hybrid circuit showing a serpentine nichrome resistor, an integrated circuit, and a transistor. The dark areas on the conductor runs where the bondwires pass over them are deposits of silicon dioxide which prevent short circuits.

graphic techniques all of the resistor and conductor material that will not be used in the final circuit is etched away. The next step is removal of the conductor material wherever a resistor is desired. This leaves the final substrate containing the interconnections and resistor pattern.

For both thick- and thin-film substrates, critical resistors are usually added to the substrate later in chip form.

Attachment of Chip Element Devices

How chip element devices are attached to finished substrates varies considerably from one manufacturer to another. Both active and passive chip elements are used. The list that follows shows some of the more common elements used in hybrid circuits:

Active Devices	Passive Devices
Integrated circuits	Thin-film resistors (both discrete and networks)
Transistors	Thick-film resistors (both discrete and networks)
Diodes	Capacitors (ceramic, tantalum, and MOS)
	Inductors and transformers

The bonding of these elements onto the substrate is accomplished by metallic or polymeric means. Typical metal schemes are various solder types or eutectics of gold-silicon, gold-germanium, or gold-tin. With use of metallic bonding for items such as ceramic chip capacitors, cracking of the bond can occur owing to a thermal mismatch of the parts. Another problem with use of solder is the leaching of gold.

Conductive and nonconductive epoxies are the predominant polymeric materials used for bonding circuit elements to hybrid substrates. There are two major problems with epoxies. The scope of the problems increases directly with the amount of epoxy used. The first, which applies to both conductive and nonconductive epoxies, is one of outgassing of the epoxies. Outgassing can leave a thin organic film on the surface of a hybrid before it is sealed that prevents a reliable interconnection at some future time. For hermetically sealed hybrids, continued outgassing of, say, moisture and ammonia from the epoxies can cause field problems.

The second major problem pertains only to conductive epoxies. A phenomenon that is still not understood causes a high electrical resistance path at the epoxy bond interface. Manifestation of this in the field usually takes one to three years. This high-resistance interface has been reported for bonds to such materials as silicon, copper, and aluminum but has not been reported for materials such as gold and silver.

Interconnection Methods

Once the substrate is populated with circuit element chips, electrical connections usually have to be made from the chips to the substrate metallization. Most designers of hybrids attempt to make as many of these electrical connections as possible during mounting of the chip elements. Indeed, with use of such techniques as flip chips, bump chips, or beam-leaded semiconductor devices, a complete hybrid can be made with no additional interconnection operations. However, because of the customized nature of most hybrids, semiconductor devices and resistor networks are usually connected electrically to the substrate with fine (about 1 mil) gold or aluminum wire (see Fig. 13.82).

The energy for making the wirebond interconnections is derived from a combination of such things as pressure and heat. The strength of the silicon sets the upper limit for the pressure; since this

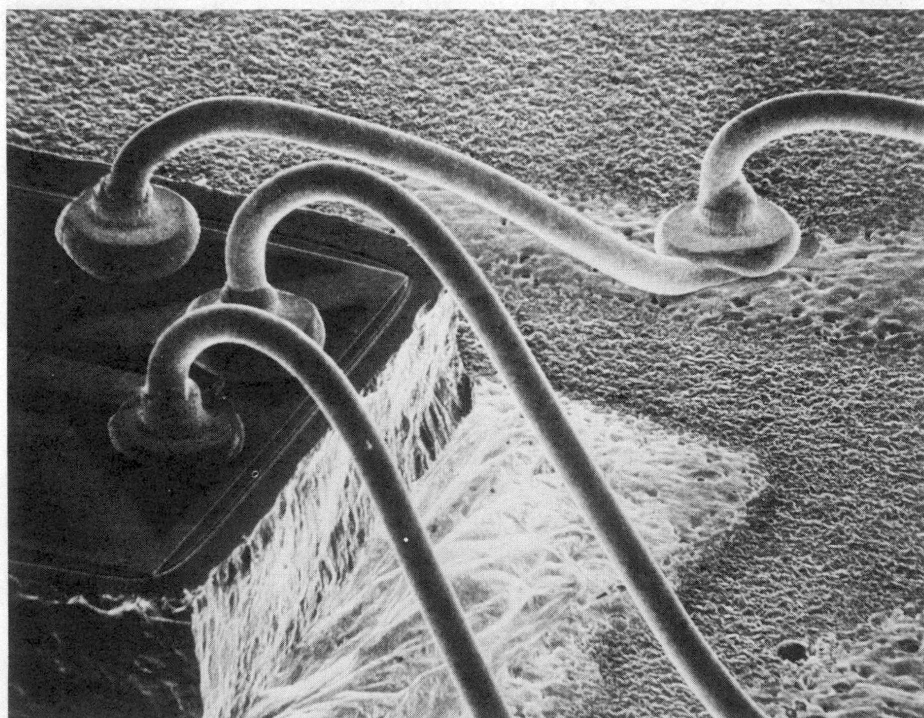

Fig. 13.82 Fine gold wires (1 mil) connecting a dual transistor chip to the thick-film substrate metallization. The transistor chip was mounted to the substrate with a gold-silicon eutectic. The wirebond in the upper left is a composite bond. In this case it is used to give added reliability to the wedge bond on the substrate and is sometimes referred to as a safety bond. The other end of the wire is attached to the same substrate metallization run with a wedge bond.

Fig. 13.83 Thick-film substrate resistor that has been laser trimmed.

is not enough to make a bond, heat must be added. This can be obtained by heating the entire hybrid, by heating the bonding tool, or by using ultrasonic energy. The most popular bonders used for hybrid construction today are the thermosonic bonders that combine pressure, a heated stage for the hybrid, and an ultrasonic bonding tool. The thermosonic bonder is used for gold wire. Gold is the predominant material used for wirebonding in hybrids, because the bonding parameters for gold wire are not as critical as those for aluminum. Ultrasonic wirebonders are used for bonding aluminum wire.

It is important to have a clean surface to make reliable wirebonds. Recently a mild plasma cleaning with argon or an oxygen-argon mixture just before the wirebonding operation has been shown to be effective.

Resistor Trimming

To build hybrid circuits with close resistor tolerances, substrate and chip resistors are trimmed to accurate values. The use of lasers to trim both thick- and thin-film resistors (see Fig. 13.83) is the most common method. Some thick-film resistors are still trimmed with an air-grit abrasion technique. The manufacturing stage at which trimming is done depends largely on how accurate the resistor value has to be. Many circuits require active trimming of resistors, that is, the circuit must be functioning during the trimming operation. Therefore the circuit must be essentially complete before final trimming is done.

Final Packaging and Sealing

Final packaging can be separated into two general categories: hermetic and nonhermetic. Hermetic hybrids are defined as being enclosed in metal, ceramic, or glass or a combination of these. Most hermetic hybrids are used by the military or other high-reliability operations and therefore the packaging is more standardized than is nonhermetic packaging. Metal packages with metal lids welded on to make the final seal constitute the majority of hybrids in the hermetic category. Ceramic packages are sometimes used, but large ceramic packages tend to crack and sealing yields are not as high as those for welded metal packages.

The substrate, with chips and interconnections made, is bonded into the package with either polymeric or metallic systems in the same manner as the chip elements were bonded to the substrate. The same problems that were discussed earlier in this section concerning chip element attachment also apply to substrate attachment. After the substrate is secure in the package, the electrical connections to the package pins are made using wirebonding techniques.

The packaging of nonhermetic hybrids is extremely diverse. The less expensive hybrids are usually molded in some polymeric material. More expensive and complex hybrids generally have "dust covers" placed over the electronics for mechanical protection. Epoxy is the most common material for attaching the dust covers. With most nonhermetic hybrids the substrate is used for the package base.

Care must be taken when polymeric molding compounds are used for final packaging of hybrid circuits. Since moisture will eventually find its way to the critical circuit elements, such as semiconductors and nichrome resistors, the surface of the hybrid must be as clean as possible when packaged. Also, both gold and silver will migrate rapidly in the presence of voltage, moisture, and any halogen ion. The other problem with molding of hybrids is one of thermal mismatch during temperature cycling. Molded hybrids should be thermally cycled over the environments of actual use for a long enough period to ensure an adequate design.

Probably the major failure mechanism for hybrids is surface contamination. Therefore the active surfaces should be made as clean as possible just prior to final packaging. Dry plasma chemistry has been extremely effective for accomplishing this.

13.5-4 Current Specifications

Although many attempts have been made to develop specifications for commercial hybrids, none have been generally accepted by the industry. Almost all hybrid specifications are patterned after three military specifications: (1) MIL-STD-883, Test Methods and Procedures for Microelectronics; (2) MIL-M-38510, General Specification for Military Specification Microcircuits; and (3) MIL-STD-1772, Certification Requirements for Hybrid Microcircuit Facilities and Lines.

MIL-STD-883 covers all microelectronics, and most of the documents cover monolithic integrated circuits. There are, however, two test methods that apply only to hybrids. The first is test method 2017, "Internal Visual (Hybrid)." This method sets standards for visual inspection for construction details of hybrids. The second "hybrid only" test method, 5008, "Test Procedures for Hybrid and Multichip Microcircuits," refers to many other test procedures in the document that are also used for monolithic integrated circuits. This test method also specifies how both active and passive chip elements used in hybrids are to be screened prior to assembly into the hybrid.

MIL-M-38510 is also a general document primarily aimed at monolithic integrated circuits. Appendix G of this document, General Requirements for Custom Hybrid Microcircuits, deals solely with hybrids and details those portions of MIL-M-38510 that are applicable to hybrids.

13.5-5 Thermal Considerations

For medium- and low-power hybrids, adequate thermal designs must be generated. In general, most design rules for semiconductor devices call for maximum junction temperatures of 125 to 130°C. A limit is also put on temperature in resistors, according to the materials used. The temperature rise for any device in a hybrid is usually calculated as a function of the case temperature. Generally, simple calculations are adequate to assure good design. These calculations almost always result in a slightly conservative design.

The temperature rise between any two points can be calculated using the simple equation

$$\Delta T = RP \tag{13.86}$$

where ΔT = temperature rise, °C
R = thermal resistance, °C/W
P = power dissipation, W

Usually temperature is reported in degrees centigrade and power in watts. The power generated in a particular device is calculated from the circuit schematic while the thermal resistance is calculated from the equation

$$R = \frac{t}{KA}$$

where t = thickness of item, m
K = thermal conductivity, W/m · °C
A = cross-sectional area, m^2

The values for K can usually be obtained from handbooks or materials specification data sheets. The cross-sectional area is easy to calculate. If the material whose thermal impedance is being calculated does not act to spread heat (e.g., unfilled nonconductive epoxy), then the measured cross-sectional area where the heat enters is used. If heat spreading occurs, the angle of heat spreading is assumed to be 45°. The area used to calculate the thermal impedance is then the average of the area where the heat enters and of the area where the heat exits. If the heat paths of two adjacent devices overlap due to heat spreading, adjustments must be made to correct the problem. A simple method is to assume no heat spreading beyond the point where the two heat paths overlap.

Thickness values are critical for materials like epoxies, and they are often the major uncertainties in the calculations. Particular care should be taken in the selection of the bond-line thickness that will actually be realized in production line hybrids.

Individual thermal impedances are calculated for each element in the thermal path and these elements are then combined in series-parallel combinations to arrive at the case-to-heat-source thermal impedance. This is similar to the combination of electrical resistors in series parallel to arrive at an equivalent series resistance.

The temperature rise from heat-source-to-case is calculated using Eq. (13.86), and the equivalent thermal resistance is determined by combining the individual thermal resistances, as described previously.

13.5-6 Frequency Considerations

Standard hybrid technology can be used for circuits operating in the range of 0 to 100 MHz. Thin-film technology, because of its narrower line widths, can be used at higher frequencies than thick-film construction. In the range from about 100 to 300 MHz, lumped circuit elements give way to distributed elements. Above 300 MHz, microwave technology takes over. This area is exclusively in the realm of thin-film technology.

In laying out substrate conductor paths even for moderate frequency ranges, care should be taken to separate paths that could cause harmful cross talk. Also long parallel runs should be eliminated from the substrate design.

13.5-7 Applications

Hybrid circuits are used for many reasons. These include (1) size reduction, (2) small production runs, (3) performance, and (4) reliability.

The size reduction is obvious. Typically a printed circuit board of about 25 in.^{2} constructed using discrete parts can be reduced to one or two hybrids of about 1 in.^{2} each. The size and weight reductions make hybrids extremely attractive for military and space applications.

Associated with size reduction is the problem of small production runs. In many cases, especially with digital circuits, even greater size and weight reductions would be possible through customization of large-scale integrated circuits. However, the cost penalty for a customized monolithic circuit often is prohibitive. An additional problem with such circuits is that of obsolescence. Reorders are also difficult. In the case of hybrids, these problems are considerably less important, since most designs are made with standard high-volume parts. If a problem does arise it can usually be solved by a minor redesign.

Critical circuits can often be produced most effectively in hybrid form. Good performance can be achieved because of the close proximity of the circuit elements, which reduces thermal gradients. This results in good tracking of matched elements. Active trimming is routinely used to produce hybrid circuits with critical requirements. Also the miniaturization obtained in a hybrid enhances high-frequency performance.

Interconnection of discrete devices is one of the major causes of failure in electronic equipment. Use of hybrid technology reduces the number of interconnections almost in half. Recent efforts in the hybrid industry to improve wirebonding have resulted in highly reliable connections: New wirebonding equipment and cleaning processes have made interconnection failures within a hybrid an extremely unlikely event.

The hybrid industry is a multibillion dollar industry that has had a steady growth rate. The future for the industry looks promising. The use of microprocessor-controlled automatic wirebonders is common in the industry, and pattern-recognition devices are being developed. These innovations, coupled with the automatic "pick and place" equipment that is being developed, will make complete automatization of hybrid assembly a reality in the near future.

The applications of hybrids are as varied as the methods of assembly and the materials used. The military and NASA have used hybrid circuits for over a decade to reduce the size and weight of airborne and space equipment. The influence of the military and NASA reaches far beyond the devices used by them. The only widely accepted specification for custom hybrids has been generated by these organizations, and most high-reliability applications such as medical device and commercial aircraft equipment use all or portions of the military specifications.

The medical industry is one of the largest users of hybrids in the construction of implantable devices such as heart pacemakers. The auto industry is using hybrid technology to miniaturize the electronics for vehicles. These hybrids are often in the form of "plug-in" modules for easy maintenance. Some other hybrid uses are:

1. Detection equipment in oil drilling bits
2. Entertainment systems on commercial aircraft
3. Computers and peripherals
4. Television
5. Microwave ovens
6. Sewing machine controls
7. Home entertainment equipment
8. Medical analysis machines
9. Electronic typewriters
10. Power tool controls
11. Telecommunications
12. Commercial microwave equipment

Some areas in which hybrids are used only minimally today but in which use could expand in the future are industrial controls, laboratory equipment, and electronic toys. The future for the hybrid industry appears to be bright.

13.5-8 General References

One of the most concentrated sources of references on hybrid circuits is the International Society for Hybrid Microelectronics, PO Box 3255, Montgomery, AL 36109. This society has published the proceedings of their annual symposium since 1967. They also have published technical journals twice a year since 1978.

Two symposiums devoted solely to hybrid microelectronics were held at Fort Monmouth, NJ, in 1976 and 1980 under the auspices of the US Army Electronics Research Development Command (published by U.S. Army ERADCOM, Ft. Monmouth, NJ 07703). The proceedings of these symposiums both contain many excellent technical papers on hybrid microcircuits.

The Proceedings of the International Reliability Physics Symposium contain several papers on hybrid technology (published by IEEE, Canterbury Press, New York). These are listed by year as follows:

1975

Libove, C., "Rectangular flat-pack lids under external pressure—formulas for screening and design," pp. 38–47.

Ebel, G. H., "Tutorial session on hybrid technology—introduction," pp. 221–223.

Redemske, R. F., "Multilevel substrate technology and epoxy component attach for hybrid fabrication," pp. 224–229.

Lane, C. H., "Packages and film resistors for hybrid microcircuits," pp. 230–241.

Bertin, A. P., Terwilliger, T. W., "Repairs to complex hybrid circuits—their effect on reliability," pp. 242–247.

Murphy, C., "Hybrid technology—loose particles and coating materials," pp. 248–252.

1977

Ebel, G. H., "Failure analysis techniques applied in resolving hybrid microcircuit reliability problems," pp. 70–81.

Dermarderosian, A., "Humidity threshold variations for dendrite growth on hybrid substrates," pp. 92–100.

Somerville, D. T., "The role of hybrid construction techniques on sealed moisture levels," pp. 107–111.

Christon, A., Wilkins, W., "Assessment of silicon encapsulation materials—screening techniques," pp. 112–119.

Kossowsky, R., Mitchell, J., "Characterization of reaction bonded gold and silver thick-film metallizations," pp. 262–271.

1978

Gedney, R. W., "Trends in packaging technology," pp. 127–129.

Christon, A., Griffith, J. R., Wilkens, W., "Reliability testing of fluorinated polymeric materials (FNP) for hybrid encapsulation," pp. 194–199.

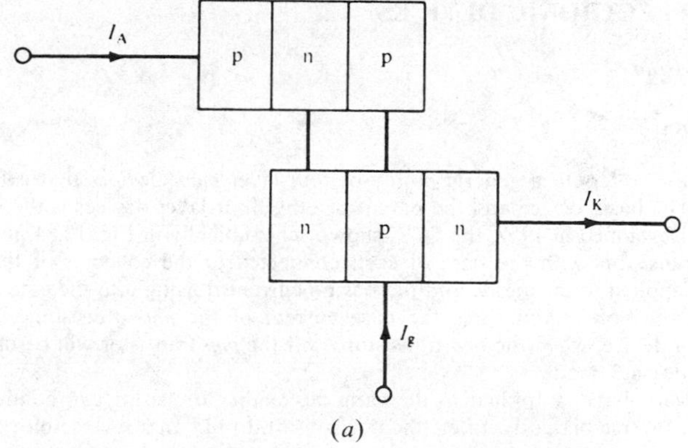

(a)

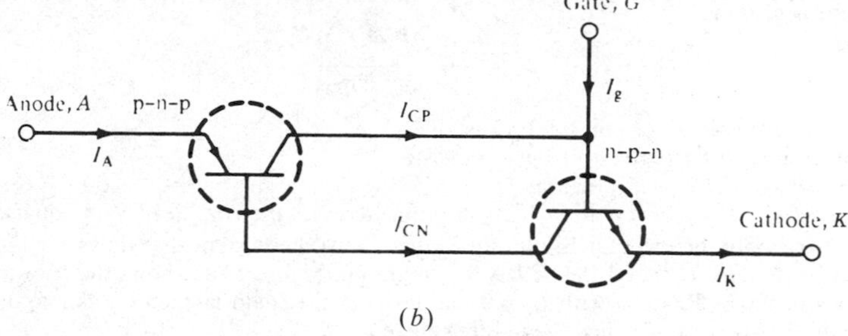

(b)

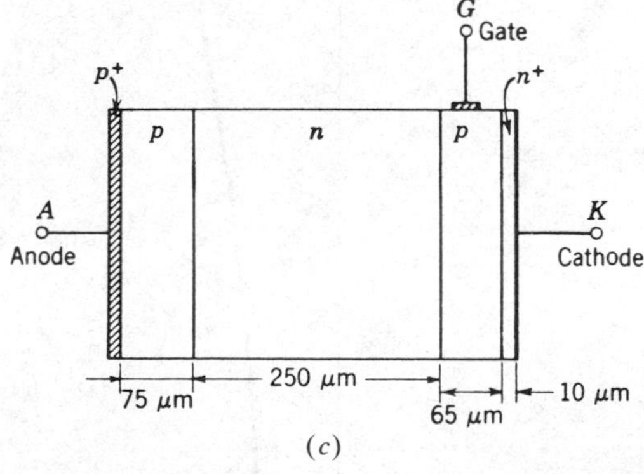

(c)

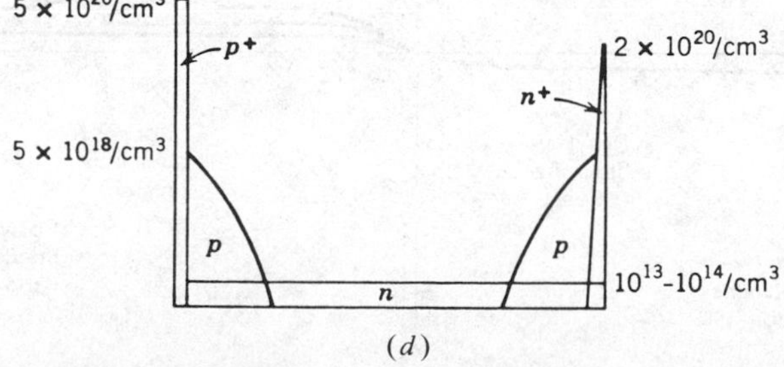

(d)

Fig. 13.84 (*a, b*) Two-transistor equivalent of a silicon controlled rectifier (SCR). (*c*) Cross-sectional dimensions of a typical SCR. (*d*) Doping profile of an SCR.

13.6 OTHER ELECTRONIC DEVICES

Mark B. Barron

13.6-1 Thyristors

The name "thyristor" refers to a generic family of four-layer *pnpn* devices that exhibit regenerative switching action. The basic device and the parent of other four-layer devices is the *s*ilicon *c*ontrolled *r*ectifier, or SCR. Developed in 1957, the SCR, shown schematically in Fig. 13.84, is essentially a pair of *npn* and *pnp* transistors with the base of each connected to the collector of the other. When a positive voltage is applied to the anode and there is no current flowing into the gate of the device, the *npn* transistor will be "off." And since the base current of the *pnp* transistor is supplied by the collector of the *npn* device, when the *npn* transistor is off the *pnp* transistor will be off too. This mode is the "forward blocking" mode.

When a positive voltage is applied to the cathode, neither transistor can conduct and the SCR simply looks like a reverse-biased rectifier; the device is said to be in a reverse-blocking mode. When a positive voltage is applied to the anode and a current is passed between the gate and the cathode, the *npn* transistor will begin to conduct current, thereby driving the base of the *pnp* transistor. The anode current is given as

$$I_A = \frac{\alpha_{npn} I_g + I_L}{1 - \alpha_{npn} - \alpha_{pnp}} \tag{13.87}$$

where α_{npn} = current gain I_C/I_E of *npn* transistor
α_{pnp} = equivalent current gain of *pnp* transistor
I_L = leakage current of junctions

At low currents $\alpha_{npn} + \alpha_{pnp}$ is less than 1. As current increases the combined gain equals 1, the SCR latches "on," and only the external circuit impedance limits the current that flows. The gate current, I_G, can then be reduced to 0 and the SCR will remain on. Figure 13.85 shows the forward operating characteristics of the SCR. Even with $I_G = 0$ the device will remain latched on as long as the anode current remains above a value called the holding current.

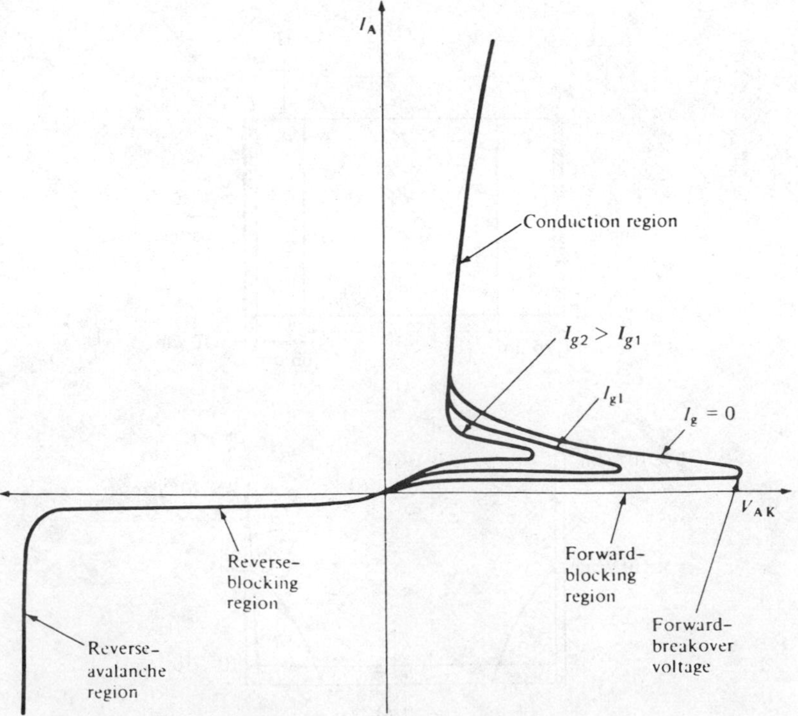

Fig. 13.85 Current-voltage characteristics of a silicon controlled rectifier (SCR).

Sometimes an SCR can be switched on even in the absence of gate current. One condition is if the temperature of the device increases and the leakage current through the SCR is enough to exceed the latching current. Another condition is if the voltage between cathode and anode is increased rapidly. The capacitive current in this case, $C(dv/dt)$, can trigger the device into a conducting state. Another situation that will trigger an SCR is when the anode voltage is increased beyond the maximum blocking value V_{BO}. The blocking junction then goes into avalanche and the breakdown current is sufficient to turn on the device.

To minimize the parasitic triggering action, SCR manufacturers usually incorporate shunt resistors (called emitter shorts) between the cathode and the gate. The effect of these is to increase the amount of current needed to trigger the device. To keep the required gate trigger currents small, amplifying gates are used, which are equivalent to pilot SCRs driving the main SCR. The cross section of an amplifying-gate SCR having a shorted emitter is shown in Fig. 13.86.

When an SCR turns on, the entire area of the device does not turn on simultaneously. Rather the portion near the gate begins to conduct, and the plasma of mobile carriers then spreads over the remaining portion of the SCR. The time required for the plasma to spread increases with increasing density of emitter-gate shunting contacts. The spreading of the conduction process determines the maximum rate (di/dt) at which current can increase through the SCR. If di/dt is too large, the current will be forced through an area of the SCR that is too small to handle the resulting power dissipation. One way to increase the allowable di/dt is to drive the gate at a current level higher than the amount needed to just initiate turn-on. Manufacturers also design interdigitated-gate fingers that distribute the gate contact over a large area so that no part of the cathode is further than some maximum allowed distance from the gate electrode. One such design is the involute pattern shown in

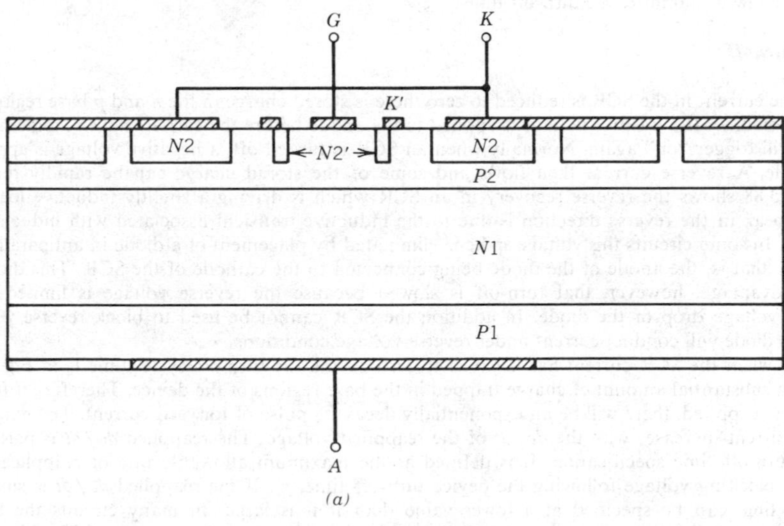

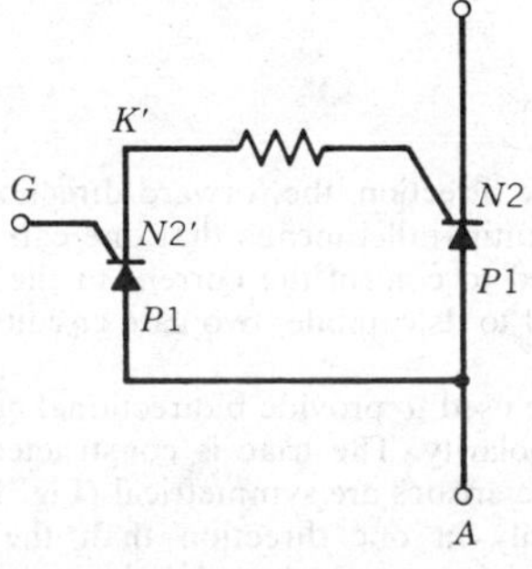

Fig. 13.86 (a, b) Amplifying-gate silicon controlled rectifier (SCR).

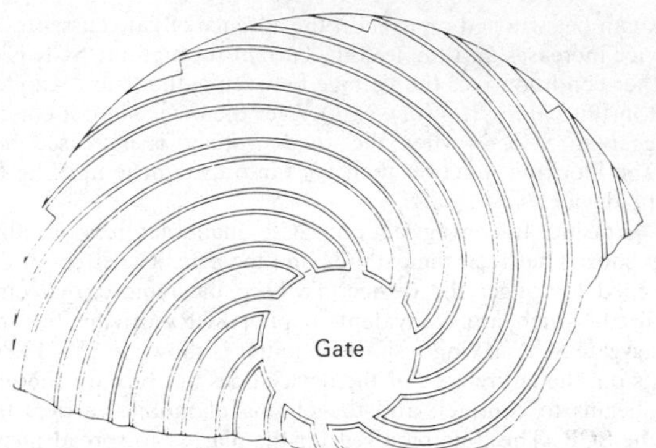

Fig. 13.87 Involute Design for interdigitated-gate silicon controlled rectifier (SCR).

Fig. 13.87. Rated *di/dt* values of 1000 A/μs can be achieved with such structures. The obvious disadvantage of a highly interdigitated structure is that some of the active emitter area is lost for conduction. The device, therefore, cannot carry as much average current as can one without the interdigitation. The allowed *di/dt* value can also be increased by increasing the carrier lifetime in the semiconductor. However, for high-frequency applications where a high *di/dt* is needed, the lifetime should be low to minimize the turn-off time.

SCR Turnoff

When the current in the SCR is reduced to zero there is stored charge in the *n* and *p* base regions that must recombine. If a positive voltage is applied to the anode before the carriers have recombined, the device will trigger "on" again. Normally when an SCR is turned off, a negative voltage is applied to the anode. A reverse current then flows and some of the stored charge can be rapidly removed. Figure 13.88 shows the reverse recovery in an SCR which is driving a slightly inductive load. The voltage peak in the reverse direction is due to the inductive transient associated with inductance in the load. In some circuits this voltage spike is eliminated by placement of a diode in antiparallel with the SCR, that is, the anode of the diode being connected to the cathode of the SCR. This diode has the disadvantage, however, that turn-off is slowed because the reverse voltage is limited to the forward voltage drop of the diode. In addition the SCR cannot be used to block reverse voltages, since the diode will conduct current under reverse-voltage conditions.

Even when the SCR current has decayed to zero during the turn-off cycle (time t_4 in Fig. 13.88) there is a substantial amount of charge trapped in the base regions of the device. Therefore if forward voltage is reapplied, there will be an exponentially decaying pulse of forward current. The magnitude of this current increases with the *dv/dt* of the reapplied voltage. The reapplied *dv/dt* is part of the SCR's turn-off time specification. It is defined as the maximum allowable rate of reapplication of off-state blocking voltage following the device turn-off time, t_q. If the reapplied *dv/dt* is small, the turn-off time can be specified at a lower value than if it is large. In many circuits the SCR is protected by placing a resistor-capacitor (RC) snubber circuit in parallel with the SCR (Fig. 13.89). The capacitor limits the value of *dv/dt* and the resistor limits the amount of current that flows through the capacitor.

Triacs

An SCR can control current in only one direction, the forward direction in which the anode voltage is positive. Placement of two SCRs in antiparallel means that one can be used to control current in one direction and the other can be used to control the current in the opposite direction. However, since the gate of each SCR is referenced to its cathode, two gate circuits would be needed in such an application.

A single device, called a triac, can be used to provide bidirectional current control with a gate that can be triggered by signals of either polarity. The triac is constructed essentially as two SCRs in antiparallel (Fig. 13.90*a*). The *VI* characteristics are symmetrical (Fig. 13.90*b*) except that the device can be triggered somewhat more easily in one direction than the other. Besides offering the convenience and economy associated with a single gate for device triggering, the triac offers the primary advantage that the output characteristics under both forward and reverse bias are almost

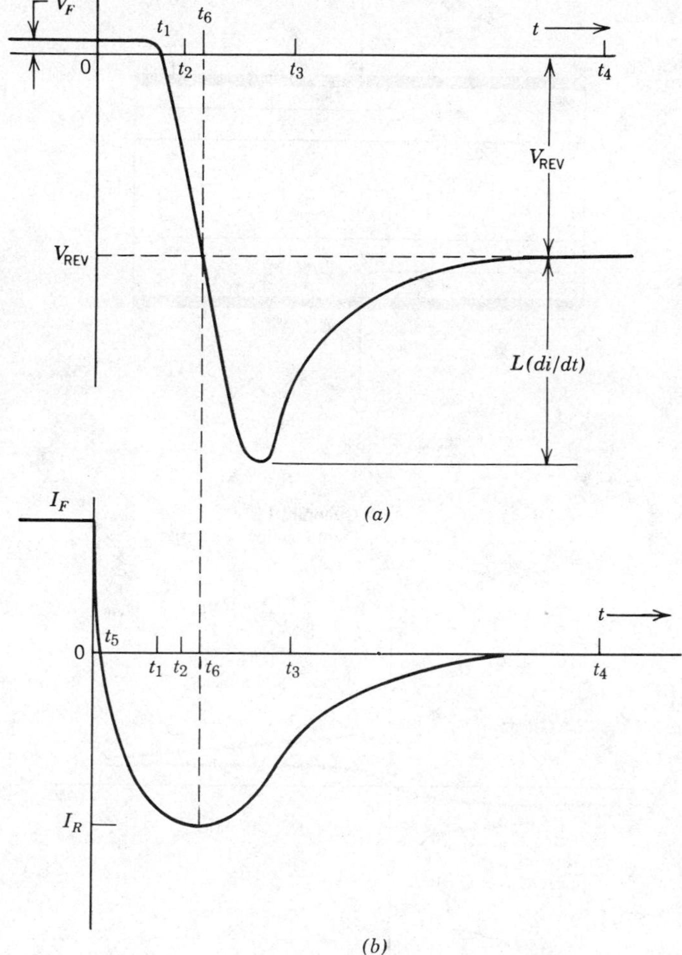

Fig. 13.88 (*a*, *b*) Turn-off wave forms of a silicon controlled rectifier (SCR) driving an inductive load.

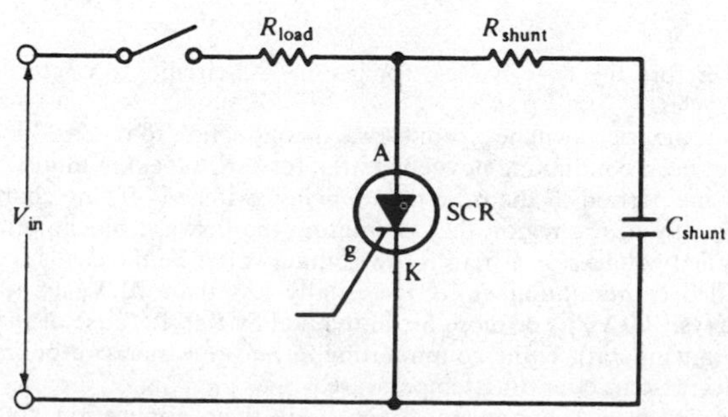

Fig. 13.89 RC snubber circuit for dv/dt limitation.

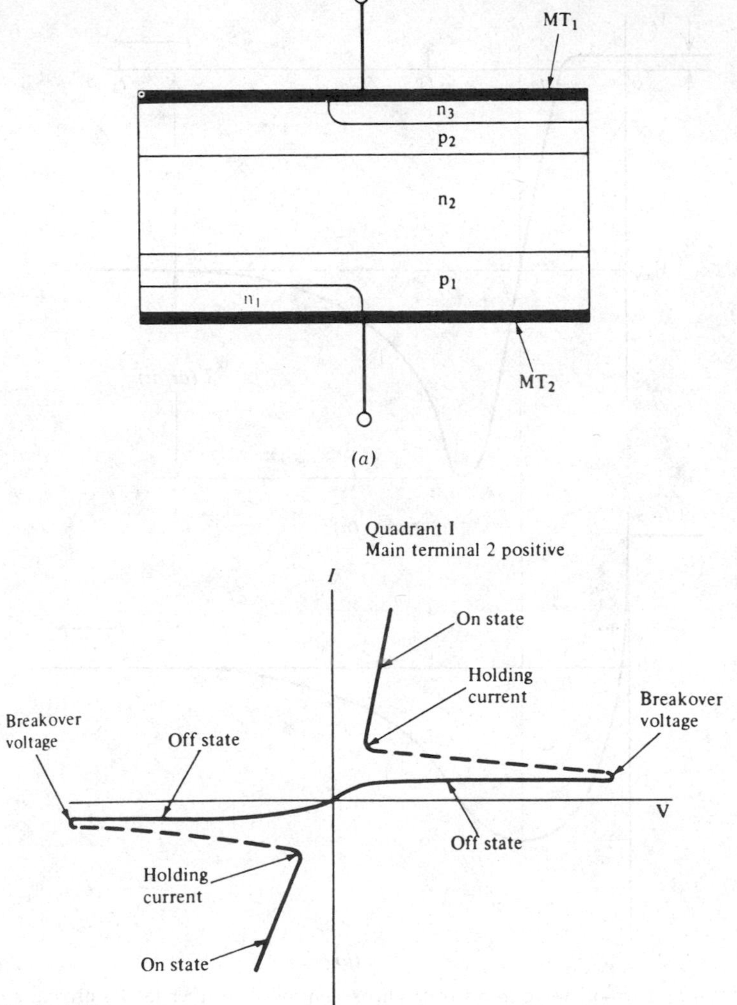

Fig. 13.90 (*a*) Cross section of a triac. (*b*) Current-voltage characteristics of a triac.

perfectly matched. Therefore the triac is ideal for use in AC circuits to control the power on both positive and negative cycles.

When the voltage on the triac switches from forward conduction to reverse-blocking state, the half of the triac that has not been conducting now enters the forward-blocking mode. Since there is stored charge in the bases of the portion of the triac that is being switched off, the charge from that region can be transported laterally to the region that is entering the forward blocking mode. Therefore the rate of increase of applied voltage for a triac is lower than would be the case for two separate SCRs. This specification, called commutating dv/dt, is usually less than 10 V/μs at maximum allowed operating temperature (vs. 100 V/μs or more for individual SCRs). Because of the decrease of carrier lifetime with decreasing temperature, the commutating dv/dt in a triac can be improved by a factor of 2 for every 40°C decrease in operating temperature.

The commutating dv/dt restrictions on triacs limit their application to circuits where the maximum frequency is around 400 Hz and the maximum voltage around 800 V. While higher voltage triacs can be built, the industrial applications for which these voltages would be typical are usually best served by separate SCRs in antiparallel.

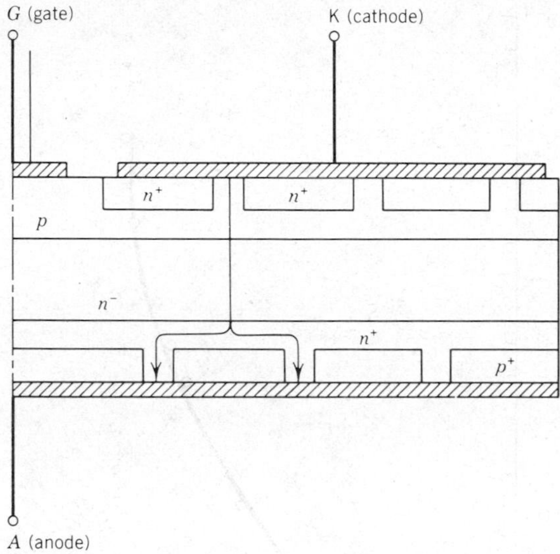

Fig. 13.91 Asymmetrical silicon controlled rectifier (SCR).

Other Thyristor Devices

Asymmetrical SCRs. As mentioned earlier in this section, an antiparallel diode is sometimes connected across the SCR and the reverse-blocking potential of the SCR is not used. If such an application is intended, the parameters of the SCR can be optimized differently than if the device is designed to accommodate reverse blocking. By adding an n^+ region adjacent to the anode (Fig. 13.91), the n^- base region can be made thinner, since the n^+ will prevent the depletion region of the forward-blocking junction from "punching through" to the anode. The thinner device structure means that the on-state voltage can be reduced. Anode shorting metallization can also be added to the anode, as is usually done on the cathode. These anode shorts not only improve the dv/dt rating but also improve the turn-off time by allowing stored charge in the n-type base to recombine more rapidly.

Reverse Conducting Thyristors (RCTs). When an asymmetrical SCR is used with an antiparallel diode, the next step of integrating the diode with the SCR is a natural one. Such a structure is shown in Fig. 13.92. Care must be taken to isolate the diode from the SCR with a region of low-lifetime material, or else the RCT can have the same reapplied dv/dt problems that a triac has. Since the antiparallel diode is often specified for optimized performance in a specific application, RCTs are often designed with certain applications in mind (e.g., transit cars).

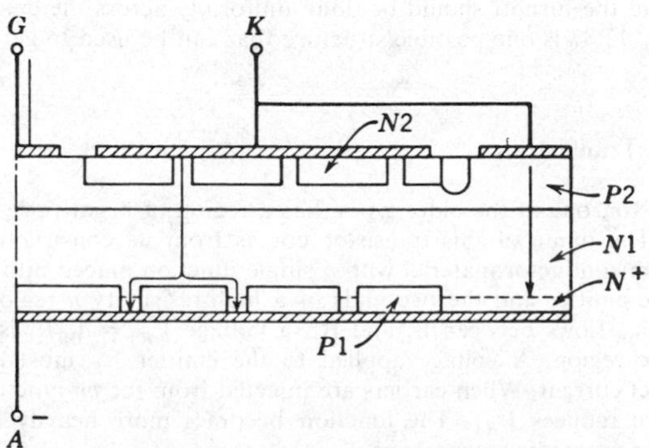

Fig. 13.92 Reverse conducting thyristor.

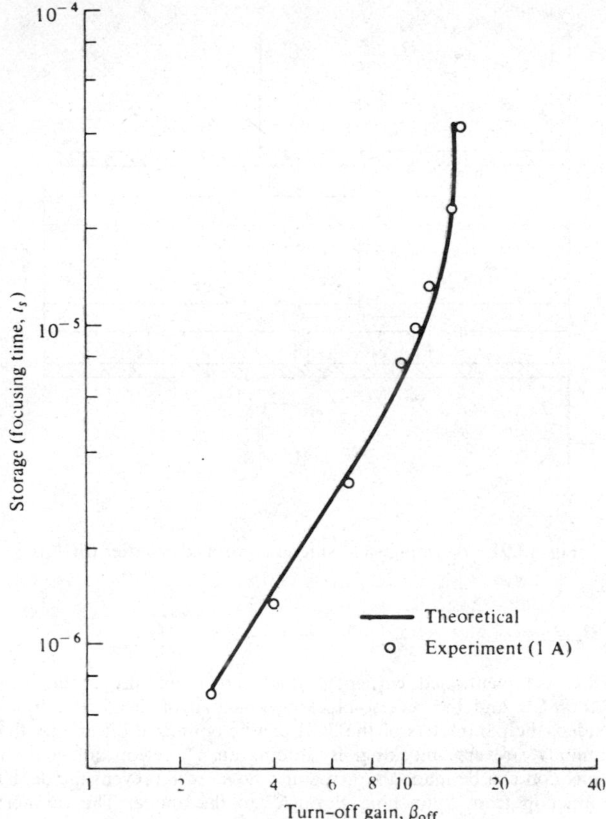

Fig. 13.93 Turn-off time versus gain for gate turn-off thyristor (GTO) with anode current of 1 A.

Gate Turn-Off Thyristors. Thyristors normally latch on with the application of a gate signal, and stay on until the forward-driving voltage and current are reversed. It is possible to turn off a thyristor with the application of a negative-gate bias. Devices designed specifically for turn-off by the gate are known as *g*ate *t*urn-*o*ff thyristors, or GTOs.

A GTO operates by reducing the main current below the holding current level. In a GTO, the turn-off gain, I_A / I_G, is a tradeoff with the turn-off time. However, even for long turnoff times, gains seldom exceed 20. Figure 13.93 shows the turn-off time vs. gain for a 1-A device. It can be seen that for a turnoff time of 1 μs the gain is about 4, compared with a maximum gain of 20. To achieve the fastest turn-off times the distance of all points of the emitter from the gate contact region should be as short as possible and the turnoff should be done uniformly across the emitter area. The involute structure shown in Fig. 13.87 is one possible structure that can be used to get uniform turnoff of the emitter.

13.6-2 Unijunction Transistors

The unijunction transistor, one of the older types, has a region of negative resistance which is useful in oscillator circuits. The name of this transistor comes from its construction (Fig. 13.94*a*); it is essentially a slab of semiconductor material with a single junction placed into the base material. The p^+ region is called the emitter and the two ends of a high-resistivity n region (B_1 and B_2 are base contacts. If a current i_{BB} flows between B_1 and B_2, a voltage $V_{R1} = i_{BB} R_1$ is established across the lower half of the base region. A voltage applied to the emitter V_E must exceed V_{R1} before the junction begins to inject current. When carriers are injected from the *pn* junction, the resistance R_1 is reduced, which in turn reduces V_{R1}. The junction becomes more heavily forward biased with a decreasing V_{R1}, and the process is regenerative.

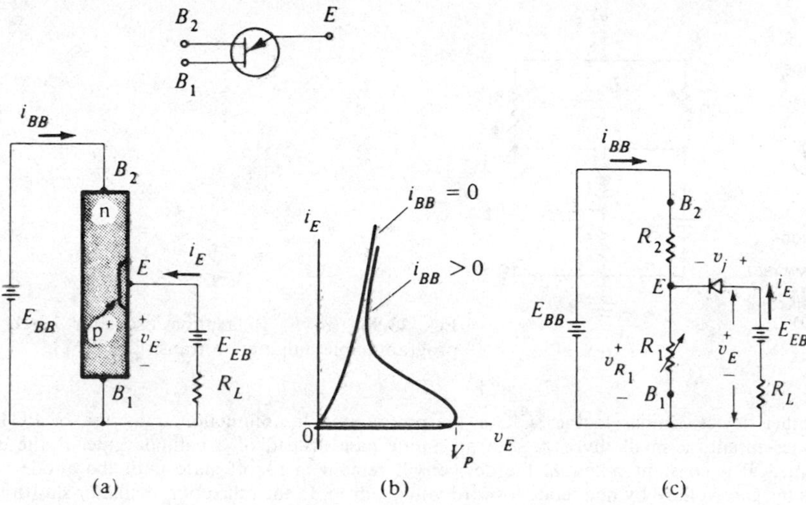

Fig. 13.94 Biasing circuit for (*a*) and current-voltage characteristic (*b*) of unijunction transistor. (*c*) Equivalent circuit.

The most imporant unijunction parameters are the intrinsic standoff ratio, $\eta = R_1/(R_1 + R_2)$ and the transit time and charge storage properties in the lower base region between E and B_1. The total n region resistance, $(R_1 + R_2)$ should be large so that i_{BB} can be small, and η should be large so that V_p can be close to E_{BB}. Referring to the VI curves in Fig. 13.94*b*, the peak voltage V_p is

$$V_p = \eta E_{\text{BB}} + \frac{kT}{q} \ln\left(\frac{I_p}{I_S}\right) \tag{13.88}$$

where I_S is the *pn*-junction diode saturation current. The valley current I_V is determined by the properties of the semiconductor material, and the difference between V_p and V_V is a measure of the switching efficiency of the transistor.

13.6-3 Other Semiconductor Devices

A host of other semiconductor elements are used in electronics circuits, too many to be exhaustively covered in a general handbook. *Semiconductor Circuit Elements* by Towers and Libes is a useful supplement that covers the broad spectrum of all semiconductor devices.

A couple of devices are worthy of mention, however, because of their use with or in lieu of devices described previously in this section. One such device is the diac. It is used in the gate circuits of triacs to provide a known, controlled triggering voltage. The diac is essentially a *pnp* transistor with an open base which exhibits negative resistance after it reaches avalanche voltage. Unlike in a normal transistor, the emitter and collector regions are uniformly and symmetrically formed in the *n* region. The symbol and *VI* characteristics of the diac are shown in Fig. 13.95. Diacs generally have avalanche breakdown voltages in the range of 25 to 40 V.

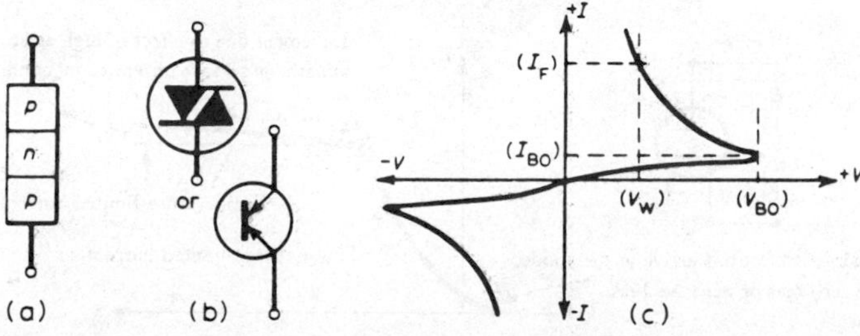

Fig. 13.95 The diac: (*a*) basic structure, (*b*) circuit symbols, and (*c*) current-voltage characteristics.

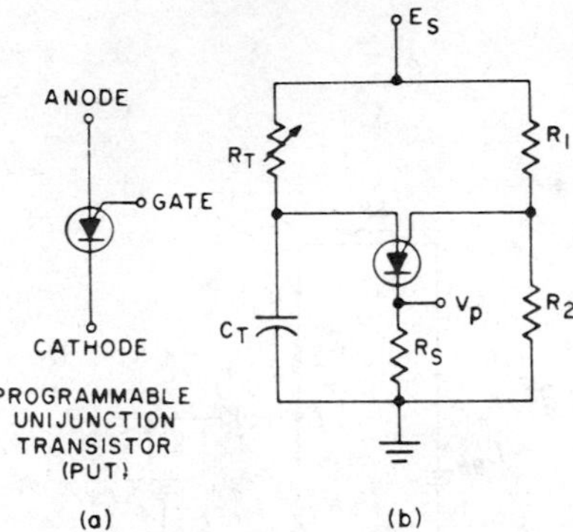

Fig. 13.96 (*a, b*) Relaxation oscillator based on a programmable unijunction transistor (PUT).

Another device, similar to the SCR, is the programmable unijunction transistor, or PUT. The PUT is essentially a small thyristor with an anode gate instead of a cathode gate. If the gate is maintained at a constant potential the device will remain in its off state until the anode voltage exceeds the gate voltage by one diode forward-voltage drop. In the relaxation oscillator shown in Fig. 13.96, the gate voltage is maintained by the supply voltage and the resistor divider network, R_1 and R_2. The voltage determines the peak-point voltage V_p. The peak current I_p and the valley current I_V both depend on the equivalent resistance on the gate, $R_1R_2/(R_1 + R_2)$ and the source voltage, E_S. The relaxation period of the oscillator is given as

$$T \approx R_T C_T \ln\left(1 + \frac{R_2}{R_1}\right) \tag{13.89}$$

13.6-4 Vacuum Tubes

Vacuum tubes are included in this section for completeness. However, except for replacements in old products and some high-frequency microwave applications, vacuum tubes are all but obsolete. The basis of tube operation is thermionic emission, that is, the emission of electrons from a solid into a vacuum if the electrons can gain enough energy to overcome the atomic forces that normally keep them in the solid. In an electronic tube this energy is provided by a heater element that forms the cathode of the tube. The heat is provided by passage of current through a filamentary wire, normally tungsten. In a simple vacuum diode an anode terminal is provided to collect the electrons emitted from the cathode. The current per unit area that can flow in such a diode is determined by the cathode temperature according to the formula

$$J_{\max} = A_e T^2 e^{-W_F/kT} \tag{13.90}$$

where A_e is a constant that depends on the emitter material and W_F is the work function, which ranges from 1 to 4 V for metals. Because the thermionic emission current is such a strong function of temperature, most vacuum diodes are operated in the space-charge-limited region (Fig. 13.97). The current in the space-charge-limited portion of the curve is given by the equation

$$I = BV_A^{\frac{3}{2}} \tag{13.91}$$

where V_A is the anode voltage and B is a constant called perveance that depends on geometry.

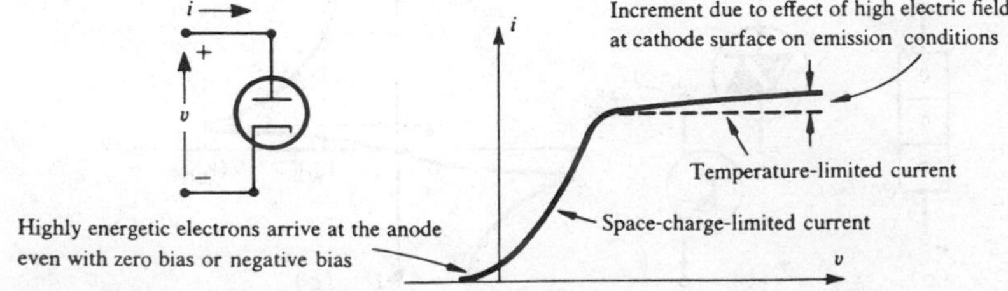

Fig. 13.97 Current-voltage characteristics of a thermionic emission vacuum diode.

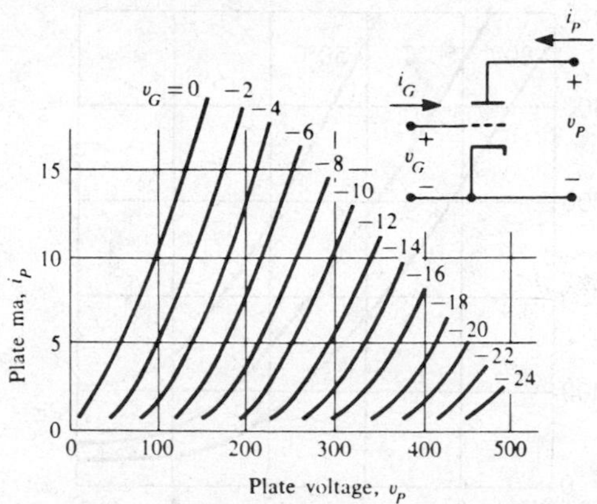

Fig. 13.98 Operating characteristics for a vacuum triode (type 6N7).

Lee deForest in 1906 discovered that the current in a vacuum diode could be controlled by inserting a grid of fine wires between the anode and the cathode. Such a device is called a triode, and the current through it is given as

$$I_A = B\left(V_G + \frac{V_A}{\mu}\right)^{\frac{3}{2}} \tag{13.92}$$

where μ = amplification factor
V_A = anode voltage
V_G = grid voltage

Operating characteristics for a triode are shown in Fig. 13.98.

A drawback of the triode is that the grid-anode capacitance is large and at high frequencies there is unwanted coupling between the anode (also called the plate) and the grid. To reduce the feedback capacitance, a device called a pentode is used for low-power high-frequency applications. The pentode is a triode with two additional grids inserted between the control grid and the anode. The grid closest to the control grid, called the screen grid, is biased at a positive potential so it appears as a permeable triode plate. To avoid distortion associated with this screen grid, another grid is inserted between it and the plate. This grid, called the suppressor, is operated at ground potential. The characteristics of the pentode are shown in Fig. 13.99.

13.6-5 Thyratrons

The vacuum tube, although providing good amplification of small signals, is inefficient as a power control device. The thyratron, a precursor of the thyristor, was developed from a hot cathode

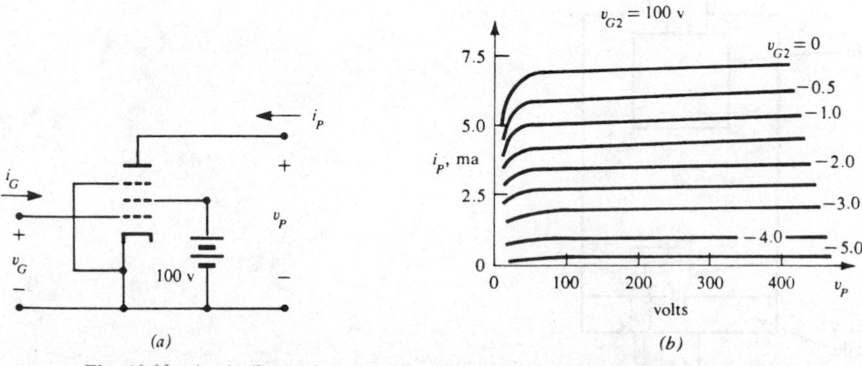

Fig. 13.99 (*a, b*) Operating characteristics of a vacuum pentode (type 6J7).

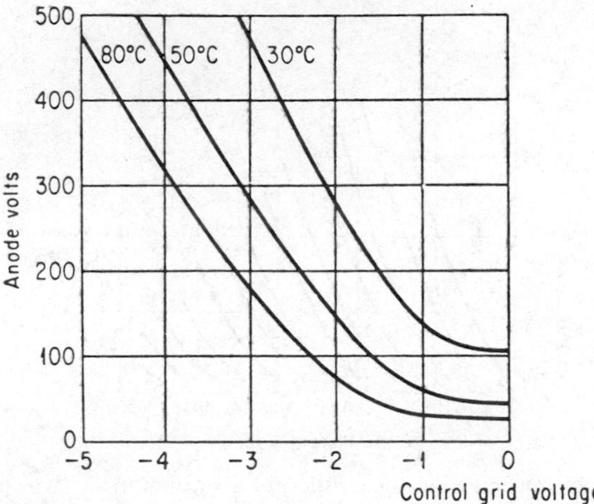

Fig. 13.100 Critical grid voltage curves for a mercury-vapor thyratron.

gas-filled diode. Gases such as argon, neon, xenon, or mercury vapor are frequently used. The gas-filled diode is capable of controlling large amounts of power, and the addition of a control grid allows current to be controlled.

With anode voltage applied and the grid held at a large negative potential, the effect of the anode voltage on the field near the cathode is negligible compared with the field associated with the grid. As the grid is made less negative, a few electrons succeed in passing through the grid. These accelerate toward the anode and receive enough energy so that upon colliding with a gas atom, ionization occurs. The resulting positive ions further reduce the negative field near the grid allowing more electrons to pass through the grid in a regenerative manner. The grid potential is instrumental in initiating the ionization process, but once started, the grid potential has no effect on the current flowing to the anode. Like a thyristor, the current can be interrupted only by removing the positive anode potential. An example of the critical grid curves for a mercury-vapor tube is shown in Fig. 13.100.

A device somewhat related to the thyratron is the ignitron. In this device a mercury pool is used as the cathode and an emission spot can be initiated by a special ignitor electrode that dips into the mercury pool (Fig. 13.101). When a spark appears between the ignitor and the mercury, a cathode spot forms that spreads in a few microseconds to handle thousands of peak amperes. In comparing an ignitron with a thyratron, the ignitor in the ignitron causes the gas discharge to form while the grid in the thyratron prevents discharge. Also, the mercury pool cathode of the ignitron is more rugged than the thermionic cathode of the thyratron. On the other hand, the control power is much smaller for a thyratron than for an ignitron.

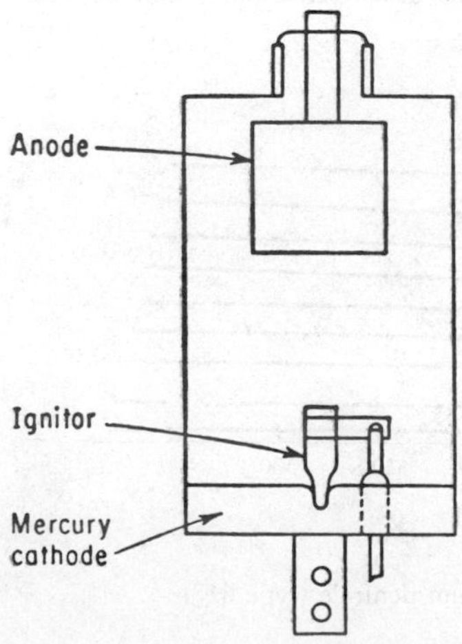

Fig. 13.101 Internal construction of an ignitron.

Bibliography

Blicher, A., *Thyristor Physics*, Springer-Verlag, New York, 1976.

Dewan, S. B. and A. Straughen, *Power Semiconductor Circuits*, Wiley, New York, 1975.

Gentry, F. E., F. W. Gutzwiller, N. Holonyak, and E. E. Von Zastrow, *Semiconductor Controlled Rectifiers*, Prentice-Hall, Englewood Cliffs, NJ, 1964.

Ghandhi, S. K., *Semiconductor Power Devices*, Ch. 5, Wiley, New York, 1977.

Gibbons, J. F., *Semiconductor Electronics*, Ch. 10, McGraw-Hill, New York, 1966.

Ryder, J. D., *Electronic Engineering Principles*, Ch. 15, Prentice-Hall, Englewood Cliffs, NJ, 1964.

SCR Manual, General Electric Company, Syracuse, NY, 1972.

Streetman, B. G., *Solid State Electronic Devices*, Ch. 11, Prentice-Hall, Englewood Cliffs, NJ, 1980.

Towers, T. D. and S. Libes, *Semiconductor Circuit Elements*, Hayden, Rochelle Park, NJ, 1977.

CHAPTER 14 FERROMAGNETIC AND PIEZOELECTRIC COMPONENTS

ALEX GOLDMAN

Spang Industries
Butler, Pennsylvania

DON BERLINCOURT

Channel Products, Inc.
Chesterland, Ohio

14.1 FERROMAGNETIC COMPONENTS

Alex Goldman

14.1-1 General Description

A magnetic component is an element composed of magnetic material, metallic or ceramic, and shaped in a particular configuration so that its magnetic properties can be used for a wide variety of functions. It may be used alone, as in a permanent magnet, or, as in most cases involving soft magnetic material, in conjunction with a winding and other accessories to form a device. Magnetic components come in many different sizes, from a 0.01 in. (0.25 mm)-diameter memory core to massive stacks of generator laminations. The functions performed by these components include:

1. Provision of DC magnetic field; for example, permanent magnets.
2. Provision of AC field in rotating electrical machinery; for example, motors, generators.
3. Provision of inductance in low-level telecommunication applications; for example, tuned circuits and channel filters.
4. Detection of changes in current in electrical circuits; for example, meters.
5. Detection of changes of magnetization in surroundings; for example, magnetometers.
6. Soaking up of sudden high currents in protective chokes.
7. Passage of DC or low-frequency signals while attenuating high frequencies; for example, bandpass filters, chokes.
8. Conversion of voltage or current in power transformers.
9. Concentration of radiated electromagnetic energy in antennas.
10. Shielding of other electrical components from electromagnetic noise and external fields.

11. Acting as transducer from magnetic to mechanical energy in magnetostrictive devices.
12. Sensing of electrical energy and temperature.
13. Control of power in magnetic amplifiers.
14. Performance of logic functions in computer memory storage and control devices.
15. Provision of circuit elements for control of microwave electromagnetic fields.

14.1-2 Metallic and Oxide Components

Metallic components are formed from metals in several different degrees of subdivision.

1. Massive, such as permanent magnets or soft pole tips for electromagnets.
2. Sheet or strip material from 0.000125 to 0.050 in. (0.003–1.25 mm)-thick, as in laminations and tape-wound cores.
3. Wire, as in semihard reed switches.
4. Powder.
5. Thin films.

Massive. Permanent magnets come in many shapes, such as bar magnets, ring magnets, horseshoe magnets, and so on. The shape is determined by the strength and configuration of the required field. Most metallic magnets have high flux densities so that smaller cross-sectional areas can be used for the given total flux.

$$\Phi = BA$$

Φ = flux, Wb or maxwell (lines)

B = flux density, Wb/m^2 (or gauss)

A = cross-sectional area, m^2 (or cm^2)

However, the coercive force is usually small (except in Pt-Co and $SmCo_5$) making these magnets prone to demagnetization. To overcome this, the poles are kept far apart, which increases the length. As a result of these two effects, most metal magnets (such as Alnicos) are long and narrow, as in the horseshoe shape.

Metal Strip. This is normally made by melting and casting an ingot, then hot and cold rolling it to a thin strip. The cold working, such as rolling, slitting, punching, and winding, requires that the material be annealed to relieve stresses in order to produce optimum magnetic properties. Powder metallurgical techniques can also be used to form the starting ingot. The new amorphous strip materials are made by melting and extrusion through a narrow orifice followed by rapid cooling on a rapidly rotating wheel. The forms into which strip material can be formed include (see Figs. 14.1 and 14.2):

1. **Laminations.** Usually punched in a press or occasionally photoetched into the desired configuration. The individual laminations are usually insulated and stacked. (See Fig. 14.2.)

(*a*)

Fig. 14.1 Typical soft magnetic components: (*a*) tape-wound cores,

(*b*)

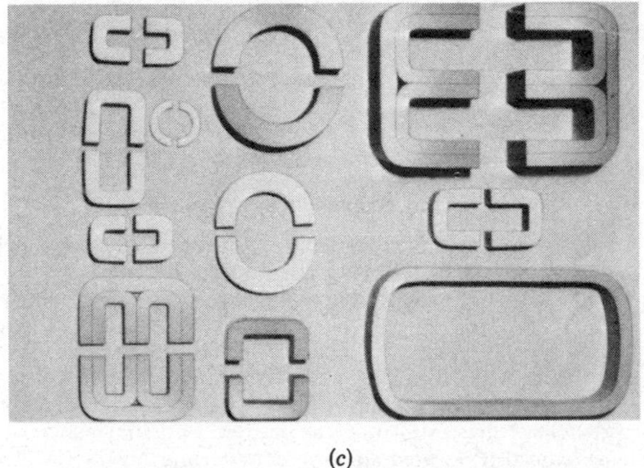

(*c*)

(*d*)

Fig. 14.1 (*Continued*) (*b*) bobbin cores, (*c*) cut cores, (*d*) laminations,

(e)

(f)

Fig. 14.1 (*Continued*) (*e*) ferrites, (*f*) powder cores.

2. **Tape wound cores.** The strip is slit into the desired width and then insulated and spirally wound into a core.
3. **Bobbin cores.** A very thin strip is wound as a tape core but onto a nonmagnetic bobbin to give the necessary mechanical support.
4. **Cut cores.** These are similar to tape-wound cores and may be rectangular or circular. The tape core is impregnated with a resin, then cut into two "C" or "U" cores, the cut surfaces ground, and the two halves banded together.
5. **Shields.** These are prepared from rather heavy-gage strip and formed into protective covers for devices such as cathode ray tubes.
6. **Thin films.** These are usually sputtered, plated, or vacuum deposited.

Metal Powder Cores. Metal powder cores are made mostly in two materials, powdered iron and powdered nickel-iron alloys. The powder is made by one of several means including mechanical disintegration, atomization, or the carbonyl process. It is then insulated and pressed into various

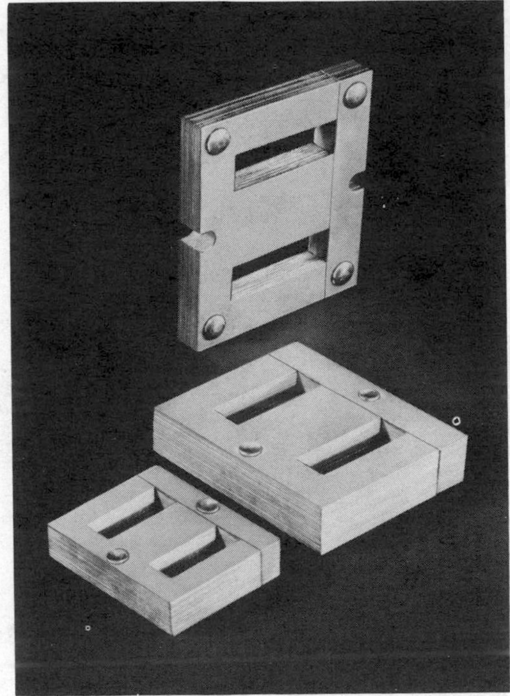

Fig. 14.2 A stack of E-I laminations. Photo courtesy of Magnetic Division of Spang Industries, Inc.

shapes including:

1. Toroids, for loading coils and linear inductors (see Fig. 14.1*f*).
2. Tuning slugs, for television sets and radios.
3. E cores and pot cores.

Oxide Components (Ferrites and Garnets). Here the powdered material is made by ceramic processes, pressed into the required shape, and fired to develop the final magnetic properties. The shapes include:

1. Pot cores.
2. E cores and U-I cores.
3. Tuning slugs.
4. Deflection yokes.
5. Toroids.
6. Permanent magnet shapes—disks, rings, and so on.
7. Microwave components.
8. Bubble memories.
9. Antenna rods.
10. Oxide-coated tape.

14.1-3 Geometrical Form

Gapless Structures

Toroid. This is the simplest type of magnetic structure, since it contains no discrete air gap and the permeability can be considered constant throughout, which simplifies analysis. For example, the inductance of this structure can be given by

$$L = 0.4\pi N^2 \mu \frac{A_e}{l_e} \times 10^{-8}$$

where L = inductance, H
N = number of turns
μ = permeability
A_e = effective cross-sectional area, cm^2
l_e = effective mean length, cm

Solid toroids of magnetic material are made of ferrites or metal powder cores. Both are made by molding of powders. The ferrite will sinter to a continuous ceramic structure. The powdered metal core (iron or nickel-iron alloy), because of the ceramic insulation between metal particles, is considered to have a very large number of uniformly distributed air gaps, which are equivalent to one localized air gap.

Sizes of these toroids can vary from small memory cores [0.010 in. (0.25 mm) or less in diameter] to large accelerator cores, which can have diameters of over 50 cm.

With regard to applications, the toroid has the advantage of having the highest inductance for the amount of material used. It also has a lower flux leakage than a gapped structure. It has the disadvantage of being difficult to wind. Special toroid winders are needed and the operation can be relatively costly. Another disadvantage results from the lower quality factor, Q, or higher losses at higher frequencies. The change of inductance with temperature, time, drive level, and DC bias is also greater than in a lower permeability gapped structure. Still another disadvantage is the lack of a mechanism to prevent saturation at high power levels.

Toroids can also be made by stacking ring laminations; some of the advantages and disadvantages just listed still apply. Stacking factors, which are lower with thinner laminations, reduce the amount of magnetic material per usable volume. Ring laminations also may not use the advantage of grain-oriented material. Here the easy direction of magnetic action is not always in the flux direction which, in a toroid, rotates at a full 360°. Toroids as beads or sleeves are sometimes slipped over conductor wires for radio frequency interference (RFI) suppression or to soak up sudden current surges.

Tape-Wound Cores (Fig. 14.1*a*). The analysis of this spirally wound structure is similar to that of a toroid in that the flux path is circular but not exactly continuous, since there are free ends. Flux overlap between adjacent turns results in almost a gapless structure. At higher frequencies a large portion of the magnetic losses are due to eddy current effects. These losses are given by

$$P_E = \frac{KB^2l^2f}{\rho}$$

where P_E = eddy current losses, W/kg
B = peak flux density, G or T
K = constant relating to shape of component
f = frequency, Hz
l = shortest dimension transverse to flux, cm
ρ = resistivity, Ω-cm

In a tape-wound core, the shortest dimension is the thickness. Reducing the thickness reduces eddy current losses and allows for higher frequency operation. Reducing the thickness reaches a point of diminishing return because of (1) the cost of rolling to a very thin gage and (2) lowering of the stacking factor as the tape gets thinner.

Tape-wound cores possess mostly the same advantages and disadvantages as listed for toroids. Most often the tape-wound cores are placed into core boxes (plastic or aluminum) to protect them from mechanical and environmental hazards and to protect the copper windings from damage. Sometimes two tape cores of different materials are combined and wound as one core. This provides a different kind of hysteresis loop that combines several useful effects.

For thicknesses less than 0.001 in. (0.025 mm), and especially for very small cores, a supporting holder or bobbin is usually used. Bobbin cores are made from material as thin as one eighth of a thousandth of an inch (0.003 mm) and as narrow as 50 mil (1.25 mm). Most bobbin cores are used for logic elements in computer circuitry. (See Fig. 14.1*b*.)

Gapped Structures

Aside from metal powder cores with distributed gap, previously mentioned, the usual gapped structures are:

1. **Stacked laminations (except rings).** These may be made in U-I, EE, or EI shapes as well as in motor or generator laminations (rotors and stators). Because of the air gap, the permeability is lower than that of toroids. However, the use of a bobbin that can easily be wound separately and then assembly of the laminations around it is a simple, inexpensive core construction method. The gap can be of the butt-gap type, as in EI and EE cores, or of the overlap type, such as in D-U laminations. (See Fig. 14.1*d*.)
2. **Cut cores.** These are essentially tape-wound cores which are vacuum impregnated with a resin and cut in half. The bobbin containing the winding is inserted on one or both legs and the two halves are banded together. The cores may be circular or rectangular for single-phase circuits but more often are EE shaped for three-phase circuits. As with other gapped structures, ease of winding and saturation protection are main features of cut cores. These cores are mainly used for power applications and are mostly of silicon-iron, nickel-iron, or cobalt-based alloys (for aircraft applications). (See Fig. 14.1*c*.)
3. **Gapped ferrites.** These are usually made in EE, EI, U-I, or pot-core shapes. In power applications all of these are used. Because of the limited heat dissipation capability of ceramics, ferrites are limited in size and thus power levels. For low-level telecommunication applications, the pot core is the design of choice. The advantages and disadvantages are the same as for other gapped cores. However, the pot cores, which can be considered circular E cores, have several further advantages. The enclosed shape of the pot core provides a built-in shield against external noise or interference, which can be critical at low signal levels. A second advantage is that of adjustability. While the inductance of a wound toroid or other ungapped structure can be trimmed only by changing the number of turns, the inductance of the pot core can be tuned by changing the air gap, most often by means of a tuning slug which can be screwed into the gap. A third advantage is the increased stability to variations in temperature (see Section 14.1-4) and to decreases in permeability over time (disaccommodation). So called ungapped pot cores used for power applications do not have an air gap ground into the center post but do in reality have small gaps at the two mating surfaces. To reduce this gap for higher effective permeability, the mating surfaces are often lapped. (See [Figs. 14-1*e* and 14-3].)
4. **Rod and slug shapes.** These may be considered a type of gapped structure. The larger the ratio of length to cross-sectional area becomes, the closer the structure approaches a toroidal design. Wound rods are used as antennas. Threaded slugs can be screwed into a wire coil (on a form) for use as variable inductors for radio or television tuning circuits.

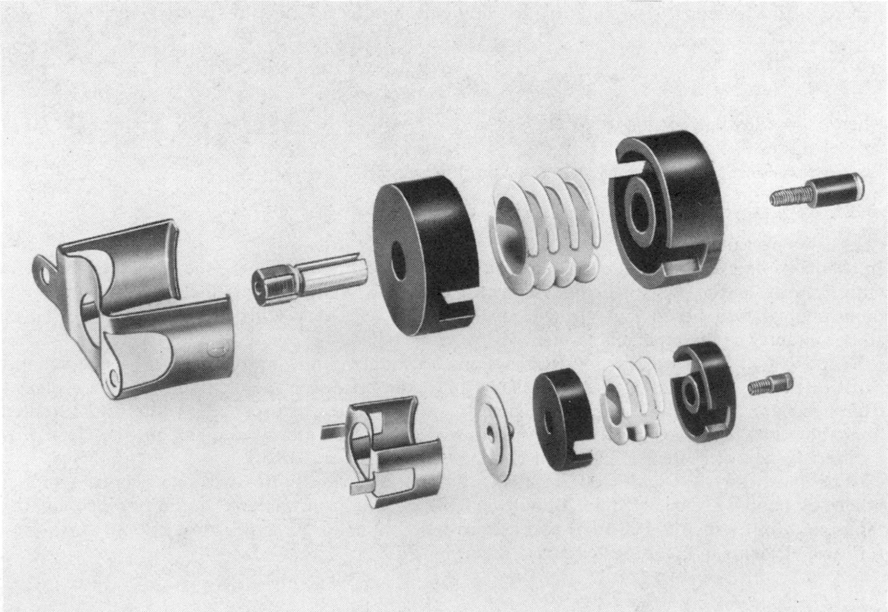

Fig. 14.3 Construction of a pot-core assembly. Photo courtesy of Magnetic Division of Spang Industries, Inc.

5. **Deflection yokes.** These are made from ferrite materials as toroids which are cracked into two C-shaped halves, wound, and then reassembled. Ease of winding is a major consideration.

Miscellaneous Structures

1. Permanent magnet shapes, as mentioned previously.
2. Wire, formerly used in magnetic recording applications and now replaced by oxide-coated tape. Also used in some semihard materials for reed switch contacts.
3. Oxide-coated tape for analog or digital storage and readout of electrical signals. The analog applications include audio or video recording, while the digital applications include computer and other memory devices.
4. Coatings for electromagnetic interference (EMI) suppression. These are usually ferrite materials that are flame sprayed.

14.1-4 Temperature Considerations

The range of temperature in which a component can function if mostly dictated by the choice of magnetic material and very little by the configuration of the component. This is discussed in Chapter 9 on magnetic materials. However, two other considerations relating to temperature effects are important.

1. The heating effect from either losses in the core or winding due to higher power levels (power transformers, etc.) must be considered. Here the ability of the structure to dissipate heat becomes a determining factor. For example, many closed structures, such as ferrite pot cores, may not be as effective (especially in large sizes) as ferrite E cores. In addition, because of the poor heat conductivity of ferrites, very large ferrite cores cannot be used for power applications. Aside from the possibility of heat buildup destroying the insulation and other organic materials in the pot core, the pot core may exceed the fairly low Curie point of ferrite and become magnetically ineffective.

2. Low-level devices, although outside the previous consideration, may be subject to variations in ambient temperatures, especially if the device is outdoors, as is the case of some public telephones or loading coils. Temperature variations can be extreme (−30 to +60°C). In this case the inductance change can vary the resonant frequency in a tuned inductance-capacitance (L-C) circuit. The frequency is given by

$$f_r = \frac{1}{2\pi\sqrt{LC}}$$

While the shape of the inductance vs. temperature curve, which compensates for the temperature variation of the capacitor, is determined by the composition of the material used, the actual inductance change of the device is decreased by the inclusion of an air gap.

The temperature coefficient is given by

$$\mathrm{TC} = \mathrm{TF} \times \mu_{\mathrm{eff}}$$

where TC = temperature coefficient of component is

$$\frac{\Delta L}{L} \times \frac{10^{-6}}{\Delta T} = \frac{\text{ppm change}}{\Delta T}$$

TF = ungapped or toroidal temperature coefficient normalized to unit permeability (related to material property)

$\mu_{\mathrm{eff}} = \mu_0 \times$ gap factor (less than 1)

If the temperature coefficient is too high for a core of a given inductance, the core size may have to be increased and gapped to a lower μ_e, which decreases the TC but allows the same inductance per turn due to increased core size. Powdered iron or nickel-iron cores provide the best temperature stability because of their lower effective permeabilities.

14.1-5 Frequency Range of Operation

As previously stated, the most critical factor for determining frequency limitations is the increase in eddy current losses at high frequencies. For metallic strip components, the lowering of these losses is brought about by reduction of the strip thickness. In the case of a ferrite component, the high-frequency operation is accomplished by its high resistivity (10^5–10^{10} times that of metals). Where high flux is necessary, use is made of laminations. Use of gapped cores also increases the frequency of operation.

14.1-6 Applications

Low-Frequency Transformers

For low frequencies (primarily 60 and 400 Hz), transformers are used for coupling AC signals and converting voltage and current.

Since the voltage and current are related to the number of turns

$$\frac{V_1}{V_2} = \frac{n_1}{n_2} \quad \text{and} \quad \frac{I_1}{I_2} = \frac{n_2}{n_1}$$

by converting the primary electrical signal to a magnetic flux and then coupling this to the secondary winding, power is transferred. The voltage induced in a winding for a sine wave input is

$$V = 4.44BNAf \times 10^{-8}$$

where V = induced voltage, V
B = maximum flux density, kgauss
N = number of turns
A = cross-sectional area, cm^2
f = frequency, Hz

For square-wave drive, the coefficient becomes 4.

To sustain the highest voltage per turn or to use the least material (related to cross-sectional area, A), the B level should be as high as possible and the permeability at high flux levels should also be as high as possible. This would mean lower primary current or lower number of primary turns. Core loss and cost are prime considerations for lower frequencies (60 and 400 Hz). In larger transformers, lamination stacks are the preferred configuration. Low-carbon steel and silicon-iron up to about 3% silicon are used extensively. The new amorphous metal alloy strip is now under development and may dictate changes in configuration and flux levels used.

Communications Transformers

Communications transformers are mainly used to match impedances between stages or between speaker coils and power amplifiers. This can be accomplished by

$$\frac{Z_1}{Z_2} = \frac{n_1^2}{n_2^2}$$

Thin-gage silicon-iron may be used up to about 10 kHz and is often compared with manganese-zinc ferrite cores. For the most demanding application, tape-wound cores or laminations of high permeability, low loss, 80% nickel-iron are the components of choice. Material for these cores can be as thin as 0.001 to 0.002 in. (0.025–0.05 mm).

For very high frequencies, only ferrites and iron powder cores remain. Pot cores of manganese-zinc and nickel-zinc ferrites extend far into the megahertz range.

Higher Frequency Power Transformers

A recent trend is to the use of more efficient high-frequency power supplies in place of the older inefficient 60-Hz units. The transformers are smaller in size and weight and also use less energy. The components of choice for these switching power supplies (especially for the 20,000-Hz and higher frequencies) are E cores made of manganese-zinc ferrite.

Radio and TV Transformers

Ferrite E or U-I cores are again the component of choice. They are used in components such as fly-back transformers, pincushion transformers, and convergence coils.

Radio-Frequency (RF) Inductors

For RF inductors, the principal consideration is stability and high Q, which is given by

$$Q = \frac{X_L}{R_s} = \frac{2\pi fL}{R_s}$$

Therefore increases in inductance and reduction of R (losses) contribute to higher Q. At low frequencies metal laminations can be used, but at higher frequencies mostly powdered metal cores or ferrites are chosen. Where the inductance must be tuned very exactly, as in channel filters, pot cores with adjusters are used exclusively.

Loading Coil Inductors

In telephone transmission lines, capacitance in the line increases with length. The increase will detune the transmitted circuits unless compensated. This is accomplished by adding the inductance at various distances in the line. This process is called loading, and the cores used to supply the inductance are called load coil cores. What is required is a very precisely controlled inductance and low losses so that the Q will not be lowered. Two components are mainly used—molybdenum-permalloy toroids and ferrite pot cores. The former has much greater stability due to the distributed air gap, especially with regard to DC shock and temperature. In the United States, where most transmission lines are above ground and subject to great temperature variations and possible lightning, the molyperm cores are used exclusively. In many parts of the world where transmission lines are buried, ferrite pot cores with 66 or 88 mH inductances are used.

Ground Fault Interrupter (GFI) Cores

A new use for magnetic cores is for detection of a ground fault in an electrical circuit coupled with the interrupter device (circuit breaker). This device, the GFI (shown in Fig. 14.4) detects the imbalance of current between the two lines when one line is momentarily shorted, such as occurs when a child places a metallic object in an outlet. The magnetic field generated by the resultant current (as low as 5 mA) is sensed and a signal is sent to the interrupting device. For great sensitivity, high-permeability ring laminations or tape cores are used (80% nickel-iron alloys are preferred). These should have low temperature variations of permeability.

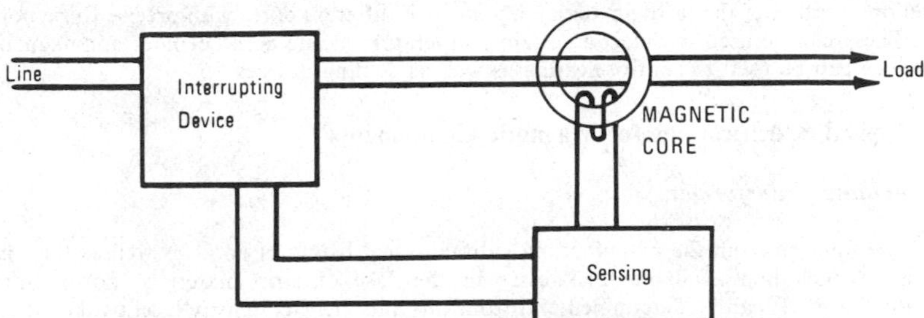

Fig. 14.4 Application of a magnetic core in a GFI (ground fault interrupter) circuit.

Theft Protection Devices

A small piece of permalloy placed in a book or an article of clothing can be detected by sensitive magnetometers, thus alerting libraries and stores to possible theft. The permalloy can be removed by bypassing the detector or by neutralizing (saturating) the permalloy with a permanent magnet (Vicalloy) strip.

Choke Coils

Inductors are used to pass DC or low-frequency AC and provide high impedance to high frequency. For operation at lower frequencies, metal tape cores or laminations of silicon-iron are used with ferrites, again at the higher frequencies. One application of a choke coil is for soaking up of large DC currents when a DC motor is started. The high current could overload silicon-controlled rectifiers (SCRs) and burn them out. A large tape-wound core of 65 permalloy (which has low remanence and high saturation) can absorb a large inrush current due to the large possible ΔB flux swing. At high frequencies, ferrite beads are used for suppression of external noise.

Magnetic Amplifiers

Magnetic amplifiers have mostly been replaced by SCRs but are still used in some applications because of their ruggedness and insensitivity to noise. High premeability nickel-iron tape-wound cores can be used, especially with high squareness.

Memory Storage Components

Magnetic memory systems are used extensively. There are several types, involving magnetic tape, toroid core memories and drums and disk memories, thin-film memories, and plated-wire memories. Magnetic tape memories use metals (iron or iron-cobalt) or oxide coatings ($\gamma - Fe_2O_3$ or CrO_2). Core memories use small ferrite toroids or bobbin cores of square permalloy threaded with wires for driving and reading. Bobbin cores are used mainly in computers or logic circuitry of extremely high reliability, such as in military and space applications. In core memories the cores in the matrix are pulsed to reset them to a "0" position, which is one extreme of remanence (close to saturation). Depending on the digital signal impressed, some cores will be switched to the "1" position (the other extreme of remanence) for storage. To read, all cores are switched back to the original 0 position. The ones that have had a 1 impressed will yield a large flux change, inducing a current in the read wire threaded through the core. Thin-film memories are made of permalloy deposited on a glass plate. Another variation of this technique is the plated-wire memory. Disk and drum memories are similar to tape in that a head or sensor passes over the magnetized regions to write or read a sense of magnetization. The magnetic head itself is a magnetic component usually made of ferrite or permalloy "head" laminations with an air gap in which the magnetic state of the area being sensed is determined. Audio tape recording is done the same way but an analog recording is produced as opposed to the digital recording.

A recent magnetic memory system that is now being developed is the "bubble" system. Bubbles are small regions (as small as 1 μm) of reverse magnetization in a biased thin sheet (film) of material. The bubbles can be generated, stored at designated locations, and then read by having them move past a sensor.

Thermal Detection Devices

By careful preparation of the ferrite material, a variety of different cores with precise Curie points can be made. These can be used to detect a certain temperature as the core becomes nonmagnetic. This type of device can be used to control heating as well as cooling devices.

14.1-7 Typical Specifications for Magnetic Components

Power Handling Transformers

Much can be inferred from the manufacturer's dimensional listing of cores as well as their material characteristics, including available thicknesses in the case of strip materials. For example, the cross-sectional area A_c can be determined and from this and the flux density level used (saturation or somewhat less), the total flux swing can be calculated

$$\Phi_t = 2BA_c$$

In addition, the size of the winding area or window can be found. From the frequency f, the turns per volt can be calculated

$$\frac{E}{N} = 4.44BA_c f \times 10^{-8}$$

and from this the number of turns required. With square-wave excitation, the coefficient in the above equation is 4.

A useful piece of data is the so-called core product, or $W_A A_c$, where A_c = area of core and W_A = window area.

It can be shown with some reasonable assumptions that in a saturating transformer, the power output is related to the core product by a simple equation of the type

$$W_A A_c = \frac{KP_{\text{out}}}{f}$$

where K is a constant that includes material efficiency, window utilization factor, and stacking factor. The values of K for inverter applications and for different materials are

Materials	K
50% cobalt	5.25
3% silicon	6.5
50% nickel	7.6
80% nickel	15.7
Ferrites	31.5

In addition to power-handling capacity of the component, it is also important to know the core loss. Curves exhibiting core losses in watts per pound (or watts per kilogram) or in milliwatts per cubic centimeter appear in manufacturers' catalogs. The former is used for metallic strip materials, while the latter is used for ferrites and some iron powder cores. The conversion from one set of units to the other is made with a knowledge of the density of the material. In addition to the core loss, the winding loss must be considered. This loss can be calculated from the number of turns, the length per turn, the core resistivity, and the current. The higher the permeability (especially impedance permeability or that at the operating flux level), the lower the required magnetomotive force (HI) and therefore the fewer turns. It is important to know the temperature characteristics of the core loss, since the core in operation may be at a higher-than-ambient temperature and it is the loss at that temperature that is relevant. New ferrite materials with minimum core loss at elevated temperatures are available commercially.

In the case of tape-wound cores, it is important to specify the type of protective box (if any) that is needed. This is determined by the winding pressures that will be used and the temperature that will be encountered. Aluminum or plastic is used for construction and a special GVB (guaranteed voltage breakdown) coating is also available. Oil or silicone rubber is used to cushion the core in the box. In some cases (silicon-iron), the material may be uncased.

Low-Level Transformers or Inductors

In the initial permeability (low drive level or Rayleigh) region where most electronic or telecommunications applications are concentrated, power considerations are not usually important. In this case the inductance is the main concern. In the case of laminations and tape cores, the inductance may be given for each lamination size in terms of the stacking factor of the material, the number of turns, and the AC permeability. The stacking factor may be listed separately for different materials and

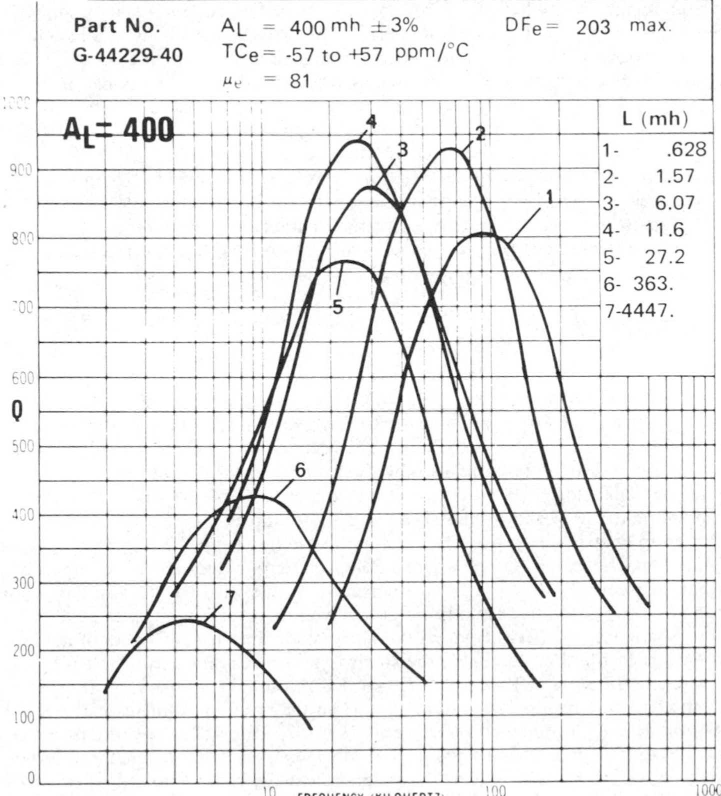

Fig. 14.5 Q vs. frequency curves for a 42 × 29-mm pot core with various windings (inductances).

thicknesses and the permeability as a function of frequency and flux density for the various materials. The flux density is given in terms of the stacking factor and turns for a specified voltage and frequency. The DC magnetizing force is given for the turns and a specific current. As the lamination size is reduced or becomes more complex (as in F lams), the greater the deviation from the calculated values.

In the case of powder cores and ferrites, the inductance per thousand turns N is listed as the A_L

$$A_L = \frac{L}{(1000)^2}\,(\text{mH}) \quad \text{and} \quad L = A_L N^2\,(\text{mH})$$

For toroidal cores (powder core, tape-wound core, ferrite toroids), changing the inductance means changing the number of turns. For ferrite pot cores, an easier way to change inductance is by varying the air gap. Thus the manufacturer can provide several different A_Ls for the same size pot core. For example, an 18 × 11-mm pot core is available in A_Ls of 160, 250, 315, and 400.

For applications such as bandpass filters, the band width is determined by the Q of the core

$$\frac{\Delta f}{f_r} = \frac{1}{Q}$$

Thus the higher the Q the narrower the band width or the greater the selectivity. Q vs. frequency curves are given for powder cores (iron or molyperm) and for ferrite pot cores. (See Fig. 14.5.) The core and winding should be chosen to have a Q maximum coinciding with the resonant frequency. If temperature considerations are important, this should also be checked either by noting the temperature coefficient of the core in the catalog or by calculating it as follows: $\text{TC} = \text{TF} \times \mu_e$.

If time stability is also important, the disaccommodation coefficient should be determined similarly to the temperature coefficient.

Logic Applications

In logic or switching applications, the component is considered nonlinear (as compared with the linear or Rayleigh region of low-level inductor applications). The components are made of materials with high squareness ratios (ferrites, "square" permalloy, or oriented 50-50 nickel-iron). The voltage output in the read process is dependent on the flux change and the switching time. The flux capacity of the core is usually given in maxwells. This is dependent on the cross-sectional area of the toroid. The switching time is dependent on the material, including the thickness and the current pulse used. Graphs showing this relationship are offered by the component manufacturer. Also available is a guide for thickness as a function of frequence or pulse repetition rate.

Core dimensions may be dictated by the relationship of the O.D. and I.D. with respect to the core area and window area. For example, there may be a need for increased window area (at the expense of smaller flux capacity) for greater number of turns for impedance matching. Usually, the smallest bobbin that supports the required flux is used. A stainless-steel bobbin can reduce that size because of the increased material strength. Ceramic bobbins are also used.

Permanent Magnet Specifications

To specify the size and shape needed for a permanent magnet application, a survey of the demagnetization curve of the material is usually made. Normally the flux (BA) needed in the gap and the total magnetomotive force ($H_g l_g$) are specified. The point of operation is somewhere between H_c and B_r, since in the presence of a gap the magnet is demagnetized by a reverse self-demagnetizing field. The slope of a line through the origin to the operating point is the ratio B_d/H_d and is known as the shear-line slope. This slope is a function of the dimensions of the magnet and air gap and is independent of the material. Usually, the shear-line slope and consequently the dimensions of the magnet are chosen to coincide or approach the slope of the line from the origin and passing through the point on the demagnetization curve where $(BH)_{\text{max}}$ contour lines are superimposed (Fig. 14.6). The relationship between B_d/H_d and the dimensions of the magnet (length and diameter in a cylindrical magnet) can be obtained from nomograms from magnet suppliers, calculated by formulas, or predicted by experience. Figure 14.6 shows the shear-slope lines for two types of magnet materials, Alnico V and barium ferrite. For Alnico V, the shear-line slope passing through $(BH)_{\text{max}}$ is about 20. For the barium ferrite it occurs at about $B_d/H_d = 1$. Corresponding L/D ratios are 4.4 : 1 for Alnico and 0.4 : 1 for the ferrite. Thus, there is a factor of 11 in the optimum dimensional ratio. The Alnico magnet works best at a B of about 10,000 G and H of 520 Oe, while for the ferrite the values are 1900 G and 1900 Oe. The B level obtained is that at the center of the magnet. Other variables that must be accounted for are stabilization of the magnet, leakage characteristics, operating conditions, and temperature and mechanical shock effects. The vendor's product information bulletin should be consulted.

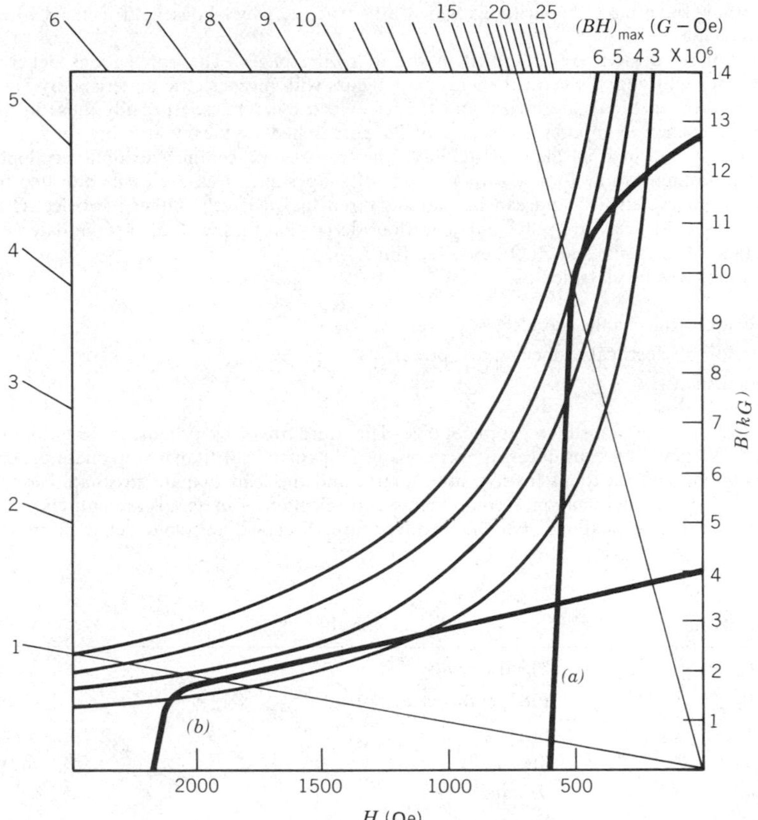

Fig. 14.6 Demagnetization curves for Alinco V (*a*) and anisotropic barium ferrite (*b*). The corresponding shear-line slopes passing through the $(BH)_{max}$ point are shown. The constant BH product contours are also shown.

Bibliography

Bourne, H. C., *Magnetic Circuits*, California Book, Berkeley, CA, 1961.

Bradley, F. N., *Materials for Magnetic Functions*, Hayden, New York, 1971.

DeMaw, M. F., *Ferromagnetic Core Design and Application Handbook*, Prentice-Hall, Englewood Cliffs, NJ, 1981.

MIT Staff, *Magnetic Circuits and Transformers*, Wiley, New York, 1943.

Olsen, E., *Applied Magnetism—A Study in Quantities*, Springer-Verlag, New York, 1966.

Parker, R. J. and R. J. Studders, *Permanent Magnets and Their Applications*, Wiley, New York, 1942.

Watson, J. K., *Applications of Magnetism*, Wiley, New York, 1980.

14.2 PIEZOELECTRIC COMPONENTS

Don Berlincourt

14.2-1 General Description

During the last decade, piezoelectric devices have come into extensive use, particularly in consumer products. Prior to about 1950 the only large-scale consumer application was in phonograph elements using Rochelle salt, with smaller related use in microphones and headphones. Piezoelectric ceramics and quartz now have many applications which affect nearly every household in the United States, Europe, and Japan.

Many of these devices are described in the following pages. The applications depend on the coupling of mechanical and electrical energy that occurs with piezoelectric materials. By far the most important of these materials are quartz and the ferroelectric ceramics, especially those in the Pb(Zr, Ti)O_3 family. Quartz is important because of its unmatched frequency stability, high mechanical quality factor Q, and high mechanical stability. The ferroelectric ceramics exhibit very high electromechanical coupling, can be formed into a variety of shapes and sizes, are easily adjusted for a very wide range of characteristics, and can be manufactured inexpensively. Other piezoelectric materials used much less extensively are polyvinylidene fluoride (PVF_2) plastic, lead metaniobate ($PbNb_2O_6$) ceramics, $LiNbO_3$ crystals, and ZnO deposited films.

Applications can be classified as:

1. Conversion of mechanical to electric power.
2. Conversion of electrical to mechanical power.
3. Circuit elements.

The first two are the transducer applications. The third involves primarily oscillator and filter applications. Virtually all transducer applications now involve lead titanate zirconate ceramics. The filter and oscillator applications involve both quartz and the lead titanate zirconate ceramics, with quartz applications predominating. Requirements and selection of materials are entirely different for the three general classifications. Factors involved are discussed in some detail in the following sections.

Notation	Variable	Unit
E_m	Electric field	V/m
D_m	Electric displacement	C/m^2
T	Stress	N/m^2
S	Strain	m/m
V	Voltage	V
δ	Displacement	m
d_{mi}	Piezoelectric constant	m/V or C/N
g_{mi}	Piezoelectric constant	Vm/N
e_{mi}	Piezoelectric constant	C/m^2
s_{ij}	Elastic compliance constant	m^2/N
c_{ij}	Elastic stiffness constant	N/m^2
ϵ_{mk}	Dielectric permittivity	C/Vm
k	Piezoelectric coupling factor	Dimensionless
f_s	Series resonance frequency	Hz
f_p	Parallel resonance frequency	Hz
f_r	Resonance frequency; lower zero Reactance frequency	Hz
f_a	Antiresonance frequency; upper zero reactance frequency	Hz
f_m	Frequency of minimum impedance	Hz
f_n	Frequency of maximum impedance	Hz
$\tan \delta_E$	Dielectric loss factor	Dimensionless
$\tan \delta_M$	Mechanical loss factor	Dimensionless
$Q_M = 1/\tan \delta_M$	Mechanical quality factor	Dimensionless
Q_L	Loaded Q of a transducer	Dimensionless
k_{mat}	Piezoelectric coupling factor for uniform electric field and mechanical stress	Dimensionless
k_{eff}	Piezoelectric coupling factor for the dynamic case with nonuniform electric field and/or mechanical stress	Dimensionless

14.2-2 Piezoelectric Materials—Characteristics

The properties of important piezoelectric materials are discussed very briefly in this section. Succeeding sections discuss some properties in more detail in relation to specific applications.

In general, a piezoelectric material is characterized by its permittivity, elastic constants, and piezoelectric constants. These coefficients are related by tensor equations of the following type (utilizing matrix notation)

$$S_i = s_{ij}^E T_j + d_{mi} E_m \tag{14.1}$$

$$T_i = c_{ij}^E S_j - e_{mi} E_m \tag{14.2}$$

These equations are equivalent, but Eq. (14.1) yields simpler solutions in static or low-frequency modes, Eq. (14.2) in high-frequency modes. The higher the crystal symmetry, the smaller the number of elastic, dielectric, and piezoelectric constants. The definitions of the piezoelectric constants follow directly from these equations, that is,

$$e_{33} = \left(\frac{-\partial T_3}{\partial E_3} \right)_S = \left(\frac{\partial D_3}{\partial S_3} \right)_E \tag{14.3}$$

The most important characteristic of a piezoelectric material as a transducer is the coupling factor, which is the square root of the ratio of electrical (mechanical) work that can be done under ideal conditions to the total energy stored from the mechanical (electric) source. Coupling factors exist for many different configurations of electric field and mechanical stress. With the piezoelectric ceramics, by far the most important transducer materials, the important coupling factors are as follows:

1. k_p, planar mode of a thin disk with electric field and polarization parallel to the thickness and stress uniform in the plane ($T_1 = T_2$; all other stress zero).
2. k_{31}^l, longitudinal mode of a bar with electric field and polarization parallel to thickness and stress parallel to length (T_1; all other stress zero). LE_t mode.

TABLE 14.1. CHARACTERISTICS OF PIEZOELECTRIC MATERIALS

Ceramics					10^{-12}m^2/N		10^{-12}C/N		
$Pb(Zr, Ti)O_3$	k_{33}^l	k_{33}^t	k_p	$\epsilon_{33}^l/\epsilon_0$	s_{33}^E	s_{11}^E	d_{33}	d_{31}	Q_M
Navy Type 1	0.70	0.51	−0.58	1300	15.5	12.3	289	−123	500
Navy Type 2	0.71	0.49	−0.60	1700	18.8	16.4	374	−171	75
Navy Type 3	0.64	0.48	−0.51	1000	13.5	11.5	225	−97	1200
Moderate-Q filter material	0.58	0.44	−0.44	1000	12.5	10.5	195	−78	650
High-Q filter material	0.38	0.30	−0.25	425	9.3	9.0	71	−27	1400
$PbNb_2O_6$ ceramic	0.38	0.37	−0.07	225	25	—	85	−9	11

Crystals

Crystal	Cut	Mode[b]	k	d (10^{-12}C/N)	e (C/m^2)	ϵ/ϵ_0
Quartz	X	TE	0.10	—	0.17	4.6
	AT	TS	−0.09	—	−0.095	4.6
Rochelle salt	X-45	LE_t	0.65	275	—	350
Zinc oxide	Z	TE	0.28	—	1.1	8.8
$LiNbO_3$	Z	TE	0.16	—	1.3	29
	X[a]	"TS"	0.61	—	13.6	44

[a] X-cut $LiNbO_3$ does not have a pure TS mode.

[b] e and ϵ^s given for high-frequency modes (TE, TS), d and ϵ^T given for low-frequency modes (LE_t).

3. k_{33}^l, longitudinal mode of a rod with electric field, polarization, and stress parallel to length (T_3; all other stress zero). LE_p mode.
4. k_{33}^t, thickness extensional mode of a thin plate with polarization and strain parallel to thickness (S_3; all other strain zero). TE mode.

A number of piezoelectric ceramics based on lead titanate zirconate have been developed with characteristics optimized for various applications. Donor additives (Bi^{3+} for Pb^{2+} or Nb^{5+} for Zr, Ti^{4+}) give enhanced permittivity and lower aging with poorer linearity. These are the so-called soft ceramics, identified as Navy Type 2. They are generally used as generator devices. Acceptor additives (Fe^{3+} for Ti, Zr^{4+}) give lower permittivity but enhanced linearity. These are the so-called hard ceramics, identified as Navy Type 3. They are generally used as motor devices. Navy Type 1 material has higher permittivity based on a lowering of the Curie point for an isovalent substitution (Sr^{2+} for Pb^{2+}). Variable valence additives such as uranium and chromium are used to produce the filter compositions.

Table 14.1 lists some parameters of a few important piezoelectric materials.

14.2-3 Motor Devices—Conversion of Electrical to Mechanical Power

Motor applications include such devices as loudspeakers, audible alarms, sonar radiating transducers, ultrasonic transducers (cleaners, nebulizers, bonders, etc.), actuators, and other applications in which mechanical work is done. Desirable characteristics of the piezoelectric material are as follows:

1. High electromechanical coupling.
2. High permittivity for low-frequency applications, low permittivity for high-frequency applications.
3. Good linearity (low mechanical and dielectric losses).
4. High mechanical strength.
5. Reasonably good time and temperature stability.

Efficiency does not depend on high electromechanical coupling, but other factors being equal, efficiency is higher when electromechanical coupling is high. Lack of exact linearity between electric field and charge and between mechanical stress and strain leads directly to dielectric and mechanical losses, respectively, and consequent decrease in efficiency and generation of power-limiting heat. High permittivity is important at low frequencies because this provides high strain per electric field input. This means a high piezoelectric d constant.

$$d = k\sqrt{\epsilon^T s^E} = \left(\frac{\partial S}{\partial E}\right)_T = \left(\frac{\partial D}{\partial T}\right)_E \tag{14.4}$$

In general, with piezoelectric materials, k and ϵ vary much more than s and therefore k and ϵ should both be high for motor applications.

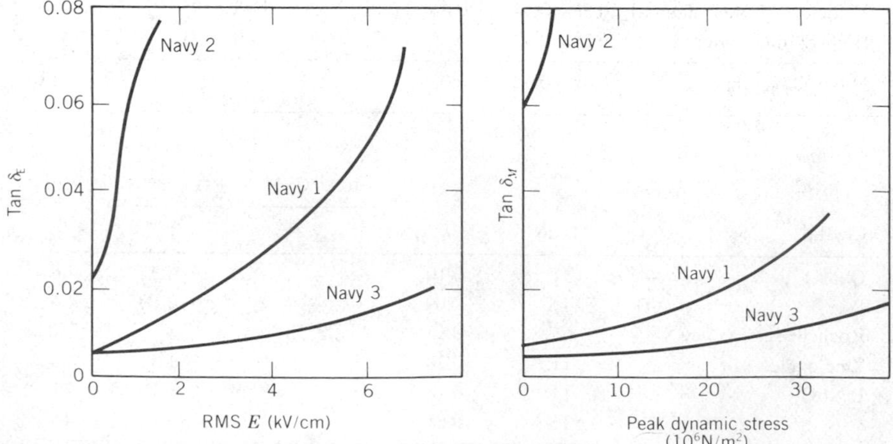

Fig. 14.7 Tan δ_E vs. electric field and tan δ_M vs. mechanical stress for several lead zirconate titanate compositions.

At high frequency, nevertheless, the element impedance may become very low, creating problems of electrical matching. In this circumstance it is desirable that ϵ be relatively low.

Dielectric and mechanical loss factors are amplitude dependent with the piezoelectric ceramics. Figure 14.7 shows amplitude dependence for three commonly used ceramic lead titanate zirconate compositions. Navy Type 3 material has superior characteristics and is used in the highest power ultrasonic and sonar applications. Navy Type 1 material is used in less demanding motor applications. Navy Type 2 material cannot be used in these applications but with related materials is used in sonic alarms and loudspeakers where high permittivity and coupling are especially important and driving power is relatively much lower.

A radiating sonar transducer is shown schematically with its equivalent circuit in Fig. 14.8. Typically the bias bolt provides a static mechanical bias sufficient to prevent tensile stress in the piezoelectric elements at the highest driving amplitude. Its effect on permittivity especially must be provided for in the design. The front mass usually has cross-sectional area five to ten times that of the ceramic element to reduce acoustic mismatch to the water load. Actual transducers are fluid loaded on only one side and end masses are adjusted independently to change the transducer Q. In the figure there is considerable simplification, as the stiffness and mass of the bias bolt, compliance of the end masses, mass of the ceramic elements, and mass loading by the fluid load are all neglected. The devices are operated at or near resonance. The resonant frequency is given by

$$f = \frac{1}{2\pi\sqrt{MC/2}} \tag{14.5}$$

A special value of Q_M is that for perfect matching. In this case the acoustic load and electric driver impedance are set so that they correspond to the image impedance of the transducer considered as a bandpass filter, with an inductor equal to $1/w_R^2 C_0$ connected across the transducer. Under these conditions

$$\frac{1}{Q_L} = \frac{k}{\sqrt{1-k^2}} = \frac{\mathrm{BW}}{\sqrt{f_2 f_1}} = \frac{f_2 - f_1}{\sqrt{f_2 f_1}} \tag{14.6}$$

where f_2 and f_1 are the upper and lower cutoff frequencies and BW is the bandwidth.

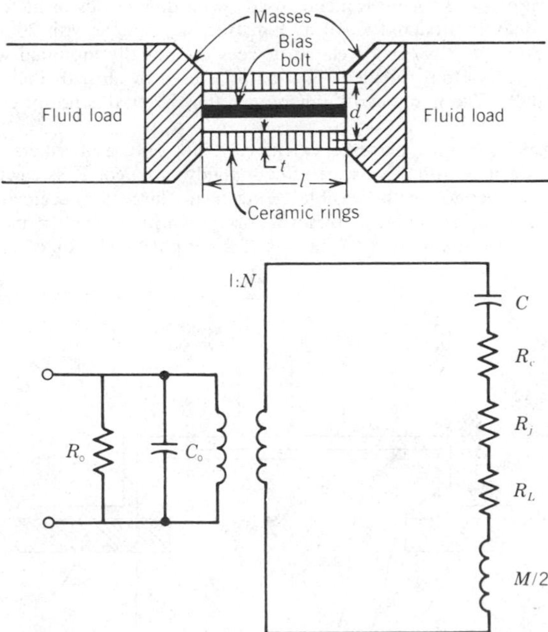

Fig. 14.8 Equivalent circuit and mechanical configuration of a symmetric sonar-type transducer. M represents the mass at each end, C the compliance of the ceramic, C_0 the axially clamped capacitance, R_0 the dielectric loss, R_c and R_j the mechanical loss in the ceramic and elsewhere respectively, and R_L the acoustic load.

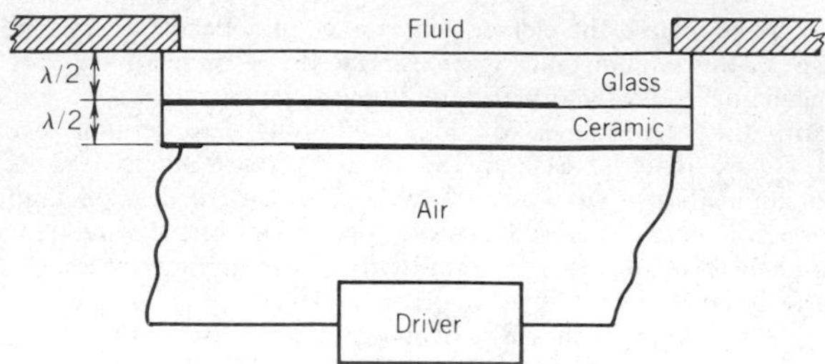

Fig. 14.9 Nebulizing transducer.

Radiating power ranges up to about 10 W/cm^3 of piezoelectric material, but higher values can be obtained at low duty cycle with sufficient mechanical bias. Limiting factors are usually mechanical strength and heat.

In general, ultrasonic transducers operate at higher values of loaded Q (range around 40) than do sonar radiators (range below 10). With the latter, dielectric and mechanical losses are of about equal importance, whereas with the former, mechanical losses are more important. Figure 14.9 shows a typical nebulizing transducer for operation at about 1.3 MHz. In this case a half-wave glass plate is bonded to the fluid-loaded side; it prevents electrode erosion and improves matching. The electrode is only in the central portion to trap the desired acoustic energy at the center away from the mounting at the edge. This trapping is discussed briefly in Section 14.2-5. A device of this type can nebulize about 10 cm^3 of water per minute with 25 W power input.

Figure 14.10 shows one type of audible alarm transducer. The device shown resonates in the umbrella mode. Its fundamental frequency is related to the diameter and vane thickness as follows

$$f \sim 1.9\left(\frac{t}{D^2}\right)\sqrt{Y/[\rho(1-\sigma^2)]} \tag{14.7}$$

where Y = Young's modulus, N/m^2, of vane,
ρ = density, kg/m^3, of vane,
σ = Poisson's ratio ($\sim$0.3 for most metals used), with dimensions in meters.

Sound pressure levels may approach 100 dB at 1 m from the source with 20 V drive at resonance (reference zero dB = 20×10^{-6} N/m^2). Related devices are nodally mounted with the out-of-phase component of the acoustic output shielded. The material used in most of these devices has relative permittivity of about 3000. The nonlinearity inherent in this material is not of concern because input power is low.

Piezoelectric loudspeakers are now used extensively, but only as tweeters above about 2 kHz. These are broadband devices with resonance effects highly damped. Piezoelectric units working at lower frequencies are not economically feasible owing to the large displacements and size required. Piezoelectric loudspeakers utilize either a bimorph or unimorph driver. In most cases the cone is attached to the center of the driver and to a frame at the periphery. Motion of the piezoelectric driver

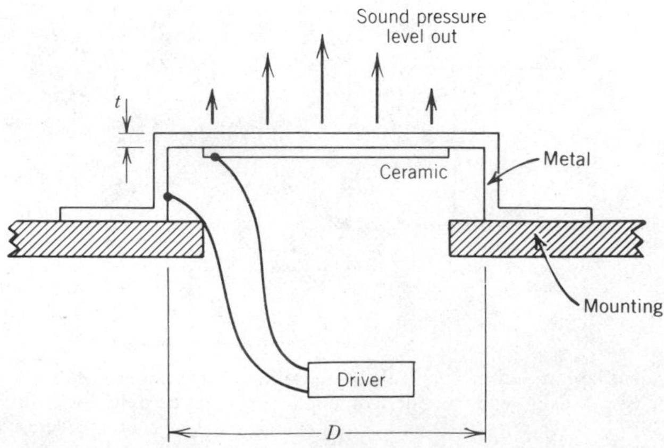

Fig. 14.10 Piezoelectric acoustic generator.

moves the cone by inertia. Acoustic pressure over 95 dB at 1 m and 20 V drive can be obtained using a horn coupled to a very small cone.

There have been few actuator applications of piezoelectrics but activity with such devices as pumps, valves, and relays has been considerable. Generally, devices requiring high displacement utilize bending elements and those requiring high force use stacks of rings or disks. With the latter, static free displacement of about 4.2×10^{-4} cm and a blocked force of 1230 N (275 lb) can be obtained with a stack 2.5 cm long and 1 cm^2 in area. If the stack contains 20 elements in parallel, the voltage required is 1000 V ($E = 8$ kV/cm). Such devices have been used to tune lasers in servo systems. A typical one-element cantilever bender 2.5-cm long, 1.27-cm wide, with element and backing each 0.375-mm thick, can give 180×10^{-4} cm free displacement and 1.4 N blocked force with 1100 V (30 kV/cm) drive. In both cases Navy Type 3 material is used and the driving field is in the same direction as poling.

14.2-4 Generator Devices—Conversion of Mechanical to Electric Power

Generator applications include such devices as gas ignitors, accelerometers, phonograph pickups, microphones, shell fuses, sonar hydrophones, and other applications where electrical work is done. Desirable characteristics of the piezoelectric material are as follows:

1. High electromechanical coupling.
2. Reasonably good time and temperature stability.
3. High permittivity for low-frequency applications.

For gas ignitor applications, the material must also have high mechanical strength and it must resist depolarization at high stress levels. For explosive fuse applications, high dielectric and mechanical strength are required for maximum energy extraction.

The first consumer-oriented application of piezoelectric transducers was in phonograph pickups and related headphones and microphones. Since about 1950, piezoelectric ceramics have replaced Rochelle salt in these applications. Piezoelectric phonograph pickups translate a groove modulation on the order of 1/40 mm into about 0.2 V. The voltage generated in a cantilever-mounted series ceramic bimorph with length l and total thickness t is related to the displacement δ as follows.

$$V = \tfrac{3}{8} g_{31} Y(t^2/l^2)\delta \tag{14.8}$$

where $Y = 1/s_{11}^E$. For $l = 1$ cm, $t = 0.05$ cm, and $\delta = 0.025$ mm, this is about 10 V. High permittivity and a highly damped mounting structure with a judicious blending of several damped resonances give a fairly broad and smooth frequency response. Note that the displacement of the piezoelectric bender element is considerably less than that of the needle of a phonograph cartridge because of substantial system compliance to allow tracking at low force.

Sonar hydrophones are usually also broadband transducers. High capacitance is important for depth applications to minimize effects of cable loading, unless the hydrophone has an internal amplifier. Linearity is not important because signal amplitudes are low, but in some cases operation at great ocean depth is required and increasing the sensitivity means that static stress is amplified. It is important that the material not undergo depolarization at high stress. Navy Type 2 is the material composition usually used. For optimum signal-to-noise ratio the usual figure of merit is the product of the appropriate piezoelectric constants d and g. Recently PVF_2 has been used in special hydrophones.

Piezoelectric ignition devices are of two types. In one type the piezoelectric element is squeezed and in the other it is struck. The latter is less expensive and is therefore used much more widely. In Europe and Japan these devices are very broadly used in homes for ignition of gas appliances; pilot lights are rare. In the United States many of these devices are also used, but generally outside the home, as pilot lights are still in general use there. The voltage is generated by a stress level of about 5×10^7 N/m^2. Navy Type 2 material ($g_{33} \sim 25 \times 10^{-3}$ Vm/N) is used. Elements are about 1.5 cm long.

$$V = g_{33} T_3 l \sim 19\ kV \tag{14.9}$$

With squeeze ignitors, stress levels are similar but Navy Type 1 material is used because Navy Type 2 material would quickly depolarize.

A related application for flash cameras consists of a Navy Type 2 element about 6 mm^2 in area by 2.5 mm thick sandwiched between metal caps. When struck by the shutter mechanism this unit develops 3 to 4 kV, causing ignition of a special high-impedance lamp.

Piezoelectric ceramics are also used in certain applications related to fusing. A stress wave passes through the transducer on impact, destroying the internal polarization. Depending on the electrical termination, very high voltage or current is developed. Under optimum conditions energy as high as about 1.5 J/cm^3 can be obtained.

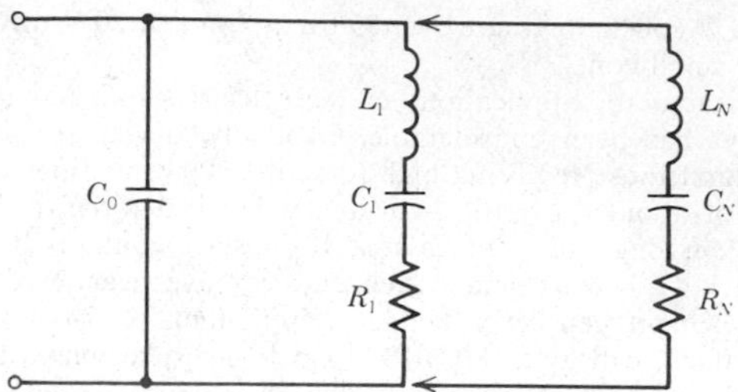

Fig. 14.11 Equivalent circuit of a piezoelectric two-terminal resonator.

14.2-5 Piezoelectric Circuit Elements

Circuit applications of piezoelectric elements include primarily filters and clock resonators. Microprocessor applications of resonators have proliferated in the last several years. For these applications the primary requirements are:

1. Temperature stability of resonant frequency.
2. Time stability of resonant frequency (low aging).
3. High mechanical Q.
4. Electromechanical coupling matched to the application.

Quartz is by far the best available piezoelectric material for narrow-band filters and high-quality clock oscillators. Both natural and synthetic quartz are readily available at low cost. Nevertheless there are lead titanate zirconate ceramic compositions that are satisfactory for a great many oscillator applications at lower cost. The ceramic materials also provide very high-quality wide-band filtering in the 200 to 600 kHz range and very inexpensive filtering for consumer frequency-modulated (FM) radio at 10.7 MHz.

The equivalent circuit of a piezoelectric two-terminal resonator is shown in Fig. 14.11. The series resonance f_s occurs at the frequency at which $wL_1 = 1/wC_1$. Parallel resonance occurs at a higher frequency f_p, where $1/wC_0 = wL_1 - (1/wC_1)$. The two frequencies are related as follows:

$$k_{\text{eff}}^2 = \frac{f_p^2 - f_s^2}{f_p^2} = \frac{C_1}{C_1 + C_0} = \frac{1}{1 + r} \tag{14.10}$$

where $r = C_0/C_1$.

The effective coupling factor k_{eff} is related to the static or material coupling factors in a relatively complex manner, dependent on stress distribution at resonance. When stress is uniform, as is nearly the case with the heavily mass-loaded sonar transducer of Fig. 14.8, then k_{eff} of that transducer is equal to the material coupling factor. This is k_{33}^l for the configuration in Fig. 14.8. Usually $k_{\text{eff}} \sim 0.9 k_{\text{mat}}$. In all cases where $k_{\text{eff}} \neq k_{\text{mat}}$ there are overtones that can be represented by additional "mechanical" branches (shown schematically in Fig. 14.11).

Quartz and piezoelectric ceramics have several major differences with respect to filter and resonator applications:

1. Quartz has a mechanical Q about 10^3 higher.
2. The stability of the resonant frequency is about 10^3 better for quartz. This includes effects of temperature, aging, drive level, shock, and vibration.
3. Ceramics have over 10^2 higher permittivity.
4. Ceramics have k^2 up to 10^2 as high.

Resonators

Precision oscillators use quartz resonators. Although the accuracy is inferior to that for atomic frequency standards, all radio frequency standards specify quartz slave oscillators. Piezoelectric resonators operate in a low impedance mode near f_s or in a high impedance mode closer to f_p. In the latter case, load or stray capacitance of the oscillator circuit shifts the frequency of operation. This is an important consideration with quartz (much less so with ceramic) resonators. Quartz resonators have been used from a few kHz to over 200 MHz. Up to about 50 kHz flexure modes are used. From 50 kHz to about 500 kHz, face-shear modes are used. At higher frequencies, thickness shear (usually

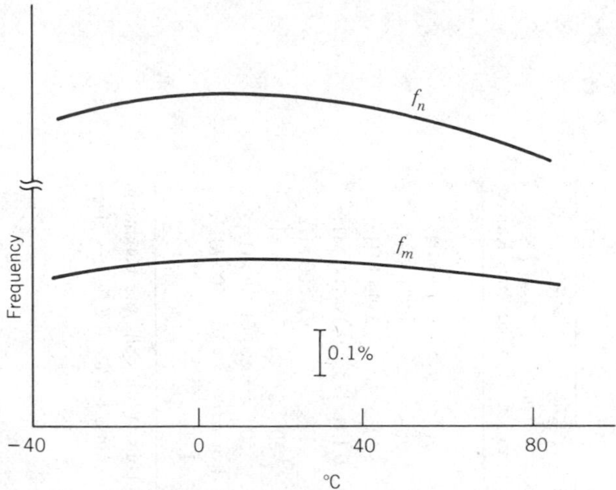

Fig. 14.12 Frequencies of minimum f_m and maximum f_n impedance vs. temperature for a lead zirconate titanate filter composition (disk, planar mode).

AT-cut)† is used, overtones typically above about 25 MHz. Typically for AT-cut quartz, $r \sim 250$ at the fundamental and $\sim 250n^2$ $(n = 3, 5, 7 \ldots)$ at overtones. With low-frequency modes, f_s forms an exact harmonic sequence, but with the high-frequency thickness modes, f_p forms the exact harmonic sequence. Different cuts are used in the various frequency ranges. At AT-cut (thickness shear) mode has a third-order frequency-temperature characteristic, and orientation can be optimized for specific temperature ranges. Total variation can easily be held to less than 10^{-5} over a 75°C range. With the NT-cut‡ (length-width flexure) used in tuning forks in watches frequency-temperature variation is better than 10^{-6}/°C. Most quartz crystal units have $Q_M = 1/2\pi f_s R_1 C_1$ in the range of 10^4 to 10^5, but values greater than 10^6 have been obtained with AT-cut units near 1 MHz. Most quartz resonators are mounted in hermetically sealed metal cans, minimizing environmental effects. With low-frequency units, aging is ascribed to the mounting and is in the range of 10^{-5} per year. The aging of higher frequency units is ascribed to mass transfer. This is of the order of 10^{-5} per year for general purpose units to 10^{-7} per year for the best evacuated glass enclosures. Specifications on resonator frequency are commensurate with the stability. In most cases, circuit adjustments are made for precise frequency.

Where frequency need not be more precise than about 0.5% and economy is of major concern, ceramic resonators can be used. In the range from a few kHz to about 100 kHz, length-width flexure (tuning forks) are used. From about 100 kHz to about 1 MHz the extensional mode of bars or more commonly the planar mode of disks or contour mode of square plates is used. From 1 MHz to about 15 MHz, the thickness extensional mode is used. Total frequency variation with temperature is about 0.2% (Fig. 14.12) for the low-frequency devices and 0.5% for thickness mode units. Aging totals below 0.2% for five years. Many applications require less than 1% accuracy. Values of mechanical Q range from about 500 to 2000 with low values of Q accompanying high coupling (about 0.50 for k_p and k_{33}^t) and high values of Q occurring with low coupling (about 0.25 for k_p and k_{33}^t).

Filters

Piezoelectric filters compete more directly with mechanical filters than with inductance-capacitance (LC) filters because of the limited Q of inductors. Quartz offers the highest Q. Metal resonators used in mechanical filters have a Q at least of an order of magnitude less, and ceramic resonators have a lower Q but one still substantially higher than that of inductors. Filters of course use resonators, and in the case of coupled mode and SAW (surface acoustic wave) devices more than one pole is obtained on a single wafer. General data for filter circuits are listed in Table 14.2.

Use of coupled-mode devices has come about through an understanding of energy trapping, which has also resulted in much cleaner resonant responses in single as well as multiple resonators. With certain resonant modes, including thickness shear and thickness extension of plates and width extension of bars, energy can be trapped under and near the electrode due to mechanical and

† See ref. 7, pp. 83, 99.
‡ See ref. 7, pp. 83, 97.

TABLE 14.2. GENERAL DATA FOR FILTER CIRCUITS

	k	% Bandwidth[a]	Q	Temperature Coefficient (ppm/°C)	Frequency Range
Ceramic $Pb(Zr, Ti)O_3$					
Flexural	0.2–0.3	1–5	500–2000	20	5–100 kHz
Longitudinal	0.2–0.5	1–15	500–2000	20	0.1–1 MHz
Thickness extension	0.2–0.4	1–10	300–1500	40	3–15 MHz
Quartz (flexural, longitudinal, thickness shear)	0.1	0.01–0.4	10^5–10^6	0.1–1	0.03–200 MHz
Mechanical	—	0.3–3[b]	$1–2 \times 10^4$	1	0.3–500 kHz
LC	—	2–50 or higher	20–200	50	1 kHz–100 MHz
Active	—	2–50 or higher	100–500	50	1–50 kHz
Surface wave (quartz or $LiTaO_3$)	Complex	2–40 (with coils)	—	1 quartz 25 $LiTaO_3$	10–1500 MHz

[a] For piezoelectric filters the maximum relative bandwidth = $k^2/2(1 - k^2)$ without tuning coils. With coils the maximum relative bandwidth = $k/\sqrt{1 - k^2}$. The maximum bandwidth without coils can be increased to almost twice the figures given with the ladder configuration.

[b] Without tuning coils on transducers.

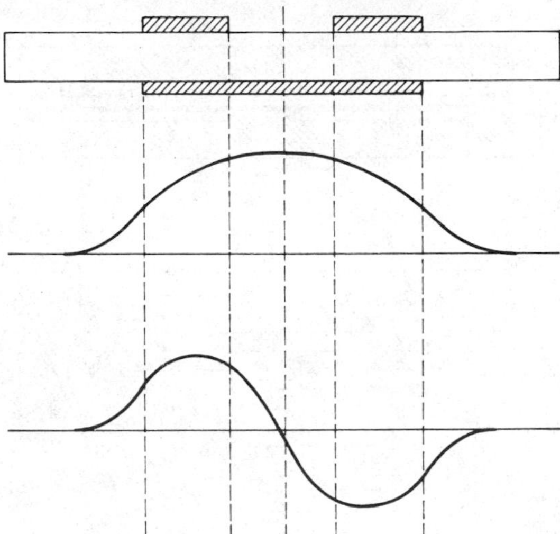

Fig. 14.13 Lateral distribution of thickness shear or thickness extensional stress for a split resonator at the first symmetric and first antisymmetric modes.

electrical loading. The mechanical and electrical loading effects are approximately equal with quartz; the electrical effect is much larger in high-k piezoelectric ceramics. By proper choice of electrode size and mass, only a single mode can be trapped, resulting in a very clean frequency response. The modes that can be trapped are inharmonic overtones occurring between f_s and f_p. When an electrode is split into two parts, the first symmetric and first antisymmetric modes can be trapped (Fig. 14.13). The three-terminal device then constitutes a two-pole filter with input and output electrodes. Spacing between the two coupled resonators adjusts coupling and thus bandwidth.

There are several major types of piezoelectric bandpass filters, including ladder filters, cascaded single-pole three-terminal resonators, lattice and cascade-lattice filters, coupled-mode and cascaded coupled-mode filters, and surface-wave filters. The three major design techniques are image parameter, computer optimization, and synthesis. The basis for piezoelectric filtering lies in the inductive reactance of a single-pole resonator between f_r (close to f_s) and f_a (close to f_p). In general the image parameter method is adequate for ladder filters, but optimization must be applied to image parameter designs of lattice filters. Network synthesis is used in design of coupled-mode and cascaded single-pole resonators.

The classic filter functions that have attenuation poles only at zero and infinite frequency are Butterworth, Chebychev, Gaussian, and Bessel. A Butterworth filter has optimally flat passband.

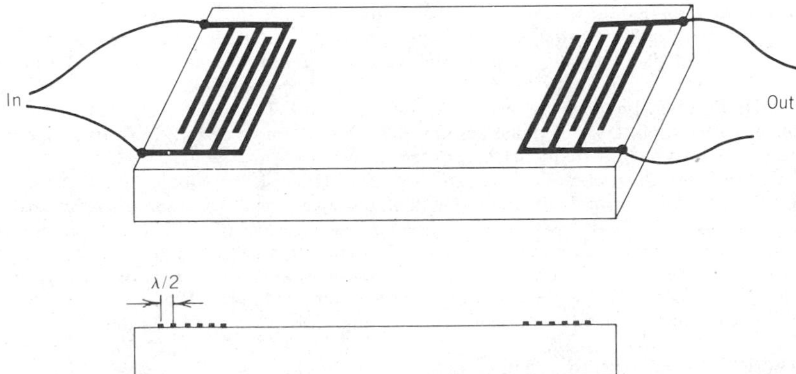

Fig. 14.14 Surface wave device on a piezoelectric substrate. Input transducer launches surface acoustic wave, which propagates to output transducer. The acoustic wave is confined within a few wavelengths of the surface.

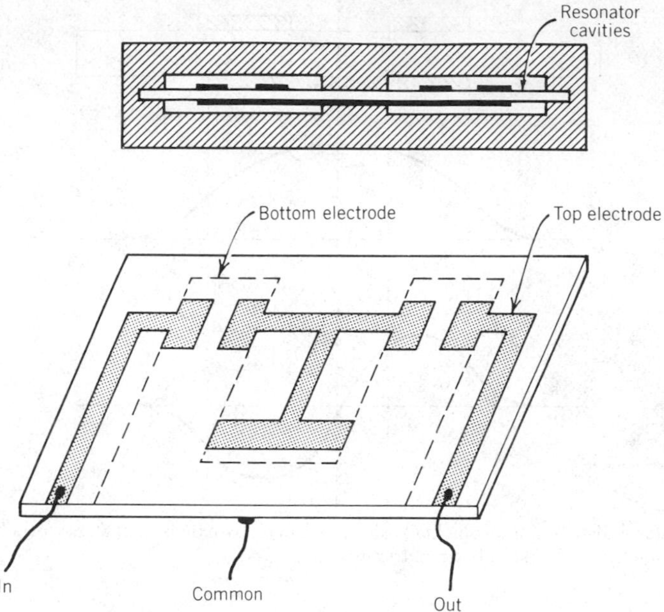

Fig. 14.15 Construction of a ceramic frequency-modulation (10.7 MHz) coupled-mode filter.

Chebychev filters have optimally steep skirts for a maximum allowable ripple with equal ripple in the passband; as the allowable ripple approaches zero the Chebychev filter becomes a Butterworth filter. For the gaussian filter the derivative with respect to frequency of the phase function is a polynomial having all zeros at the bandpass center. The Bessel filter has nearly linear phase response in an equal ripple sense.

Quartz filters are usually made using combinations of single-mode resonators (lattice) or by cascading two-resonator coupled-mode devices. Western Electric now makes a telephone filter, however, that consists of eight coupled resonators on a single wafer. The eight-pole response has a 3.2 kHz bandwidth at 8 MHz.

Quartz discrete and coupled-mode filters have bandwidths in the range 0.01 to 0.4% without inductors and to several percent with inductors. Below about 5 MHz, discrete resonators are used; both coupled-mode and discrete devices are used in the approximate range of 5 to 200 MHz.

More recently, surface acoustic-wave devices have come into general use, primarily for filtering, but they are also used as signal processors, acoustooptic modulators, and so on. Quartz is the predominant substrate, but other materials are lithium niobate and lithium tantalate crystals, zinc oxide films, and even modified lead titanate ceramic. A SAW device has an input interdigital transducer and an output transducer defined by precise electrode fingers (Fig. 14.14). Fourier-transform pair and digital filter design procedures are well understood and precise. The placement and length of the interdigital electrodes determine the filter response. Center frequency ranges from about 10 MHz to 1500 MHz and maximum bandwidth to 40%. Typical insertion loss is higher than with other piezoelectric filters, but phase response can be extremely linear. SAW filter applications include television intermediate frequency stages, television games, and radar.

Ceramic filters use flexural-mode devices below 100 kHz, disks or square plates (planar mode) from about 200 to 1000 kHz, and coupled-mode thickness extensional devices in the megahertz range. Bandwidths are listed in Table 14.2. The most widely used piezoelectric filter is a cascade of two coupled-mode pairs on a single ceramic wafer at 10.7 MHz with a bandwidth of 2 to 3% (Fig. 14.15). With a coupling capacitor on an unpoled section of the same wafer, a four-pole response is achieved. Almost all FM radios now utilize these devices. From about 300 to 600 kHz, as many as 17 discrete resonators are cascaded in a ladder configuration, offering very high-quality filtering.

Bibliography

Berlincourt, D., *Ferroelectrics* **10**:111–119 (1976).

Cady, W. G., *Piezoelectricity*, Vols. 1 and 2, Dover, New York, 1964.

"Guide to Dynamic Measurements of Piezoelectric Ceramics with High Electromechanical Coupling," IEC-STD, pub. 483 (1976).

Jaffe, H. and D. Berlincourt, *Proc. IEEE* **53**:1372–1386 (1965).

Landolt-Börnstein, K. H. and A. M. Hellwege, eds., New Series Group III, Vol. 11, Springer-Verlag, Berlin, 1979. (See Ch. 3 by W. R. Cook and H. Jaffe.)

Mason, W. P., *Electromechanical Transducers and Wave Filters*, Van Nostrand, New York, 1948.

Mason, W. P., *Piezoelectric Crystals and Their Application to Ultrasonics*, Van Nostrand, New York, 1950.

Mattiat, O. E., ed., *Ultrasonic Transducer Materials*, Plenum, New York–London, 1971.

"Measurements of Piezoelectric Ceramics" (ANSI C 83.24-1962), IEEE-Std 171 (1961).

Onoe, M. and H. Jumonji, *Elec. Comm. Eng.* (Japan) **48**:84–93 (1965).

"Piezoelectric Ceramic for Sonar Transducers," MIL-STD-1376 (Ships) (1977).

Piezoelectric Ceramics, distributed by Electronic Components and Materials Div., N. V. Philips, Eindhoven, Netherlands (1968).

Proc. IEEE **65**, special issue on surface wave devices and applications (May 1976).

Proc. IEEE **67**, special issue on miniaturized filters (Jan. 1979).

Reference Data for Radio Engineers, 5th ed., Howard W. Sams, Indianapolis (1972). (See Chs. 7–9.)

Shockley, W., D. R. Curran, and D. J. Koneval, *J. Acoust. Soc. Am.* **41**:981–993 (1967).

"Standard on Piezoelectricity," IEEE-Std 176 (1977).

CHAPTER 15
OPTICAL DEVICES AND DISPLAYS

L. S. WATKINS

AT&T Technologies
Princeton, New Jersey

PETER PLESHKO

IBM Corporation
Kingston, New York

MARTIN WOLF

University of Pennsylvania
Philadelphia, Pennsylvania

15.1 OPTICAL SOURCES

L. S. Watkins

15.1-1 Nature of Light

Light is electromagnetic radiation in the part of the spectrum to which the eye is sensitive. Figure 15.1 shows the whole spectrum; the visible portion is the small section in the wavelength range of 400 to 700 nm. Some of the information presented in Sections 15.1, 15.2, and 15.4 is also applicable to the ultraviolet (UV) and infrared (IR) portions of the spectrum.

Since light is electromagnetic radiation, it can be completely characterized by Maxwell's[1] equations. We will not go through the rather complicated analysis but just describe some of the more important properties of light.

Phase Velocity

The phase velocity of light (the velocity of planes of constant phase, i.e., wavefronts) is

$$v = c/\sqrt{\epsilon\mu} \tag{15.1}$$

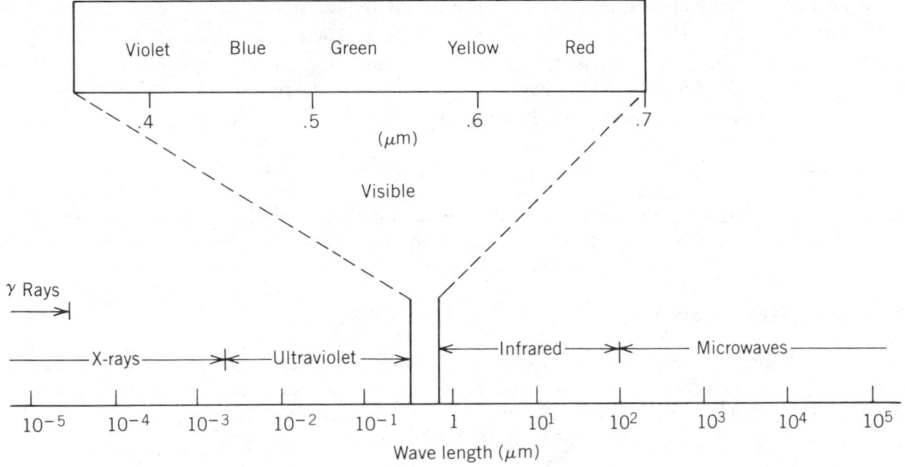

Fig. 15.1 Range of the visible portion of the electromagnetic spectrum.

where c is the velocity of light in vacuum and is a constant of value 299,796 km/s. The denominator is a term in optics referred to as the refractive index of a medium

$$n = \sqrt{\epsilon\mu} \tag{15.2}$$

where ϵ is the dielectric constant (permittivity) and μ is the magnetic permeability. The wavelength of the light is thus $\lambda = v/\nu$, where λ is the wavelength in the medium and ν is the frequency. The refractive index in general varies with wavelength; this is the dispersion property of a medium.

Absorption

Media can also absorb light, a process that can be described in two ways. The light flux transmitted through a medium is

$$I = I_0 e^{-\alpha x} \tag{15.3}$$

where x is the distance traveled in the medium by a wave of incident energy I_0 and α is the absorption coefficient, usually in reciprocal centimeters.

Absorption can also be represented in the refractive index as an imaginary term

$$\bar{n} = n(1 + ik) \tag{15.4}$$

where k is the attenuation index, $i = \sqrt{-1}$, and $\bar{n}$ is the complex refractive index. The two equations are related in the following manner:

$$\alpha = \frac{4\pi}{\lambda_0} nk \tag{15.5}$$

λ_0 is the light wavelength in vacuum.

Group Velocity

The velocity of energy transmission by a wave is the group velocity and is given by

$$u = v - \lambda \frac{dv}{d\lambda} \tag{15.6}$$

λ is the wavelength of the light in medium and equals λ_0/n. In vacuum the group and phase velocities are the same. In all cases $u \leqslant c$, the velocity of light in vacuum.

Polarization

Light propagation in isotropic nonguiding structures is by TEM (transverse electromagnetic) waves, where the electric and magnetic field vectors are perpendicular to the propagation direction and orthogonal to each other. Light can have a number of different polarization (direction of the electric field vector) states:

Unpolarized. The polarization of the light is random and constantly changing.

Linearly Polarized. The electric vector is confined to one direction.

Elliptically Polarized. The electric vector rotates either left-handedly or right-handedly at the frequency of the light; the magnitude of the electric field vector traces out a stationary ellipse.

Circularly Polarized. The special case when the magnitude of the field vector is constant.

15.1-2 Geometric Optics

The wavelength of light is quite small, so that for many situations it can be approximated to zero and the propagation of light energy described in terms of light rays. This branch of optics is called geometric optics, since ray behavior is described in geometric terms.

Properties of Light Rays

Refraction. When light strikes a boundary and goes from one medium to another, it experiences a change in propagation velocity, as described by the refractive index. This results in a change in propagation direction, given by Snell's law

$$n_1 \sin \theta_1 = n_2 \sin \theta_2 \tag{15.7}$$

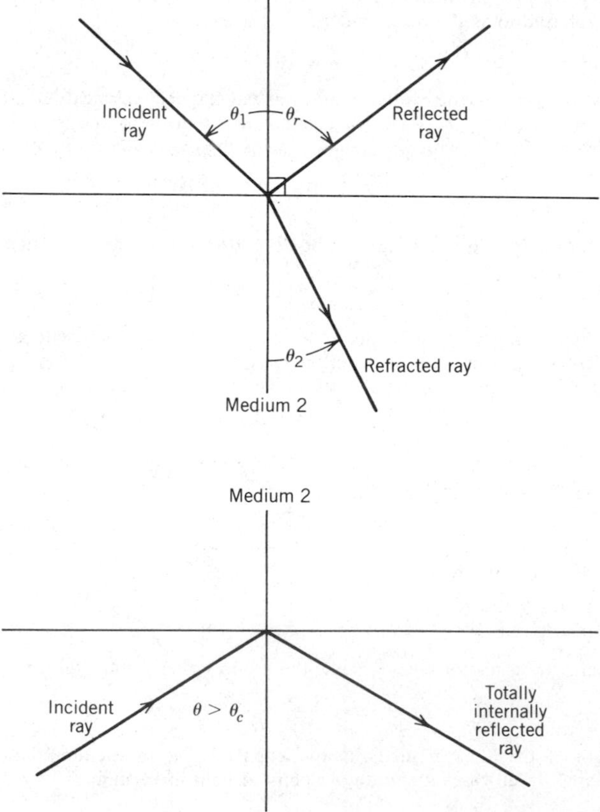

Fig. 15.2 Diagrams of light rays: (a) Trajectory when incident at an interface between two media of different refractive indices. Light is both reflected and refracted. θ_1 is incidence angle, θ_2 is refraction angle, and θ_r is reflection angle. (b) Condition when light is totally internally reflected. Medium 1 has a larger refractive index than does medium 2. θ_c is critical angle.

Figure 15.2 is a light ray diagram. θ_1 is the incidence angle and θ_2 is the refraction angle. The incident ray, the refracted ray, and the surface or boundary normal all lie in the plane of incidence. n_1 and n_2 are the respective refractive indices of the media.

Critical Angle. When light travels from a dense to a less dense medium there is an incidence angle θ_1 for which $\sin\theta_2 = 1$. This incidence angle is called the critical angle (θ_c). For angles greater than $\theta_1 = \theta_c$, the light is totally internally reflected at the interface with no light being transmitted into the second medium. The critical angle is given by $\theta_c = \sin^{-1}(n_2/n_1)$.

Reflection. In addition to light being refracted a portion of it is reflected at an interface. The angle θ_r at which it is reflected is equal to the incidence angle. The amount of light that is reflected and transmitted at the interface can be calculated from wave optics; the results are given by the Fresnel formulae

$$R_p = \frac{\tan^2(\theta_1 - \theta_2)}{\tan^2(\theta_1 + \theta_2)} \tag{15.8a}$$

$$R_s = \frac{\sin^2(\theta_1 - \theta_2)}{\sin^2(\theta_1 + \theta_2)} \tag{15.8b}$$

$$T_p = \frac{\sin 2\theta_1 \sin 2\theta_2}{\sin^2(\theta_1 + \theta_2)\cos^2(\theta_1 - \theta_2)} \tag{15.8c}$$

$$T_s = \frac{\sin 2\theta_1 \sin 2\theta_2}{\sin^2(\theta_1 + \theta_2)} \tag{15.8d}$$

R_p and T_p are the reflection and transmission coefficients for light linearly polarized parallel to the plane of incidence. R_s and T_s are the same coefficients respectively for the orthogonal polarization.

Normal Incidence. For this situation there is no polarization distinction, giving

$$R = \left(\frac{n-1}{n+1}\right)^2 \tag{15.9a}$$

$$T = \frac{4n}{(n+1)^2} \tag{15.9b}$$

Brewster Angle. When $\theta_1 + \theta_2 = \pi/2$, the denominator in the equation for R_p goes to ∞. The incidence angle for this is given by

$$\theta_1 = \tan^{-1}\left(\frac{n_2}{n_1}\right) \tag{15.10}$$

and is called Brewster's angle. At this angle the reflection for the parallel polarization $R_p = 0$.

Metallic Mirror Reflection. The conduction property of the metal results in the refractive index having a large imaginary component. At normal incidence the reflection is given by the same Fresnel equations.

$$R = \frac{|\bar{n} - 1|^2}{|\bar{n} + 1|^2} \tag{15.11}$$

except that the refractive index is a complex quantity of the form given in Eq. (15.4).

At nonnormal incidence, the reflection is similarly given by the Fresnel Eq. (15.8). Here the refraction angles θ_2 are complex (as given by Snell's law). The effect is to introduce a phase change in the S and P polarization on reflection so that incident linearly polarized light will in general become elliptically polarized.

Prism Deflection. When light passes through a prism it is deflected as shown in Fig. 15.3. The deflection angle is given by

$$\gamma = \theta_1 + \psi_2 - \alpha \tag{15.12}$$

where α is the angle between the two surfaces. ψ_2 can be determined from Snell's law and the geometric construction shown in Fig. 15.3. The minimum deviation γ_m occurs when $\theta_1 = \psi_2$. The measurement of this deviation angle is an accurate method of measuring the refractive index of the

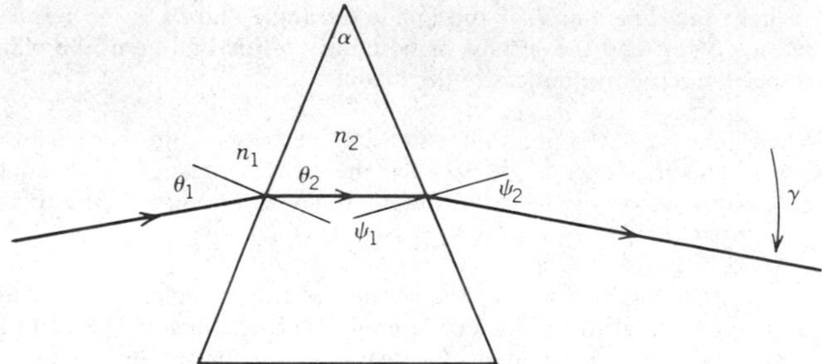

Fig. 15.3 Deflection of a light ray by a prism of refractive index n_2 in a surrounding medium of refractive index n_1.

prism material as

$$\frac{n_2}{n_1} = \frac{\sin\frac{1}{2}(\alpha + \gamma_m)}{\sin\frac{1}{2}\alpha} \tag{15.13}$$

Thin Prism. If α is small, some approximations can be made and the prism deflection given as

$$\gamma = \left(\frac{n_2}{n_1} - 1\right)\alpha \tag{15.14}$$

This assumes that the prism is used near its minimum deviation angle.

Thin Lens. This is a lens whose thickness is small compared with its diameter and the distances associated with its optical properties. Figure 15.4 shows image formation of P_2 from object P_1 by a convex lens. The gaussian lens formula relates object (P_1) and image (P_2) positions to the lens focal length f.

$$\frac{1}{f} = \frac{1}{s_1} + \frac{1}{s_2} \tag{15.15}$$

If the distance to the object from the lens is smaller than the focal length, the image is no longer real, as shown by the dashed lines. For a concave lens the image is not real and the focal length f is negative.

The power of a lens is given by

$$P = \frac{1}{f} \qquad \left(\frac{1}{f} = \text{diopter in units of m}^{-1}\right) \tag{15.16}$$

The lensmaker's formula gives the focal length of a thin lens in terms of the refractive index and the surface curvatures

$$\frac{n_1}{f} = (n_2 - n_1)\left(\frac{1}{r_1} - \frac{1}{r_2}\right) \tag{15.17}$$

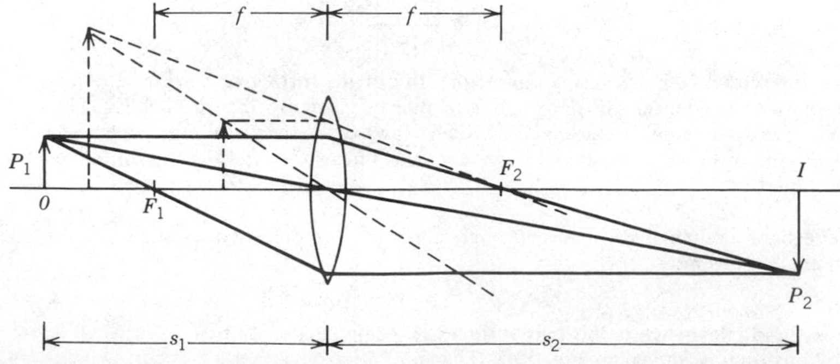

Fig. 15.4 Formation of an image P_2 of an object P_1 with use of a thin lens. Dotted lines show formation of a virtual image when distance to the object from the lens is less than the focal length.

where n_1 = refractive index of surrounding medium
n_2 = lens' medium index
r_1 and r_2 = surface radii
Positive radii have center of curvature to the right of the surface.

The lens f number describes the diameter of the lens in relation to its focal length

$$f_{\text{number}} = \frac{f}{d}$$

There is a similar term called the working f number and it relates the lens diameter to the object distance s_1

$$f_{\text{number}} = \frac{s_1}{d}$$

In microscopy another term is used to describe the lens aperture. This is the numerical aperture (NA)

$$\text{NA} = \frac{nd}{2s_1} = \frac{n}{2 \times f_{\text{number}}}$$

n is the refractive index of the medium between the lens and the object.

15.1-3 Incoherent Light

Light can be coherent or incoherent, depending on the source. Common sources of light provide incoherent light, because they consist of many independent radiators. For example, in a fluorescent tube each radiation atom is essentially independent. Thus there is no fixed phase relation between the waves from these atoms, and the light is incoherent. In a laser the light is generated in a resonant cavity and the resulting beam can have well-defined phase fronts; this light is coherent.

Brightness and Illumination

The flux density of light radiating from a point source falls off with the square of the distance from the source.

Figure 15.5 shows a surface being illuminated by an extended source. The flux dE falling on an incremental area dA from an incremental area dS on the source P is

$$dE = \frac{BdA \cos\theta \, ds \cos\Psi}{r^2} \tag{15.18}$$

The constant B is characteristic of the source and is called the luminance, or photometric brightness. The units are candles per square centimeter and dE is the luminous flux or power, in lumens. The illuminance of the surface or flux density is

$$dI = \frac{dE}{dA} \quad (\text{lm/cm}^2) \tag{15.19}$$

Two systems for quantifying light energy are commonly used. The radiometric unit, which is the one more familiar to electrical engineering, is watts. However, white light is also defined in photometric units using the candela and relating it to a black-body source.

The comparison is done using the "relative visibility curve," which relates the sensitivity of the average eye to the wavelength of light. This sensitivity is maximum near a wavelength of 0.55 μm; the constant relating the luminous flux to the radiant flux at this wavelength is 685 lm/W. It must be

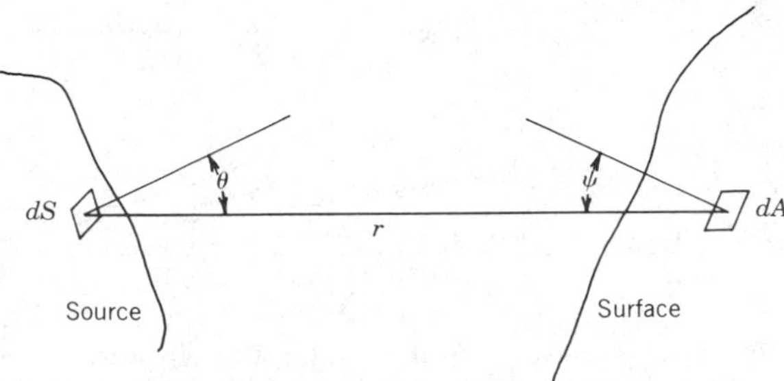

Fig. 15.5 Diagram of an arbitrary surface being illuminated by a source and the elemental source element dS to calculate flux dE falling on surface element dA.

remembered that the lumen has a spectral distribution given by black-body radiation. Radiometric units are absolute and refer to the energy of electric fields.

Source radiance is given either as luminance, which can be directly related to temperature and emissivity, or as watts per square centimeter per steradian per nanometer. This is the power emitted per unit surface area over a unit of solid angle for an increment of spectral range.

Thermal Sources

When objects are heated, they emit radiation. As the temperature is increased, the amount of radiation emitted increases and the spectral distribution changes. A black body is defined as a surface that absorbs all radiation on it. Kirchhoff's law of radiation states

$$\frac{W}{a} = \text{const} = WB \tag{15.20}$$

that the ratio of emitted radiation W to absorbed radiation a is constant at a given temperature. Thus a black body is a standard emission surface with which other surfaces can be compared.

The energy distribution of a black body for different temperatures is given in Fig. 15.6. The 2000K temperature is typical for a tungsten filament lamp. (The newer tungsten halide lamps are hotter, closer to 3000K.) The sun's temperature is 6000K.

The shape of the spectral emission curve is given by Planck's[2] law

$$W = \frac{c_1}{\lambda^5}[\exp(c_2/\lambda T) - 1]^{-1} \tag{15.21}$$

$$c_1 = 3.7413 \times 10^{-12}\text{W} \cdot \text{cm}^2$$

$$c_2 = 1.4380 \text{ cm K}$$

$$W = \text{W} \cdot \text{cm}^{-2} \cdot \text{increment of wavelength (cm}^{-1})$$

The radiation from a black-body source is inherently incoherent, with atoms or molecules emitting radiation independently.

There are very few materials that are true black bodies. Carbon lampblack is one. The emissivity of a surface is defined as the ratio of that actual radiation emitted to black-body radiation. Table 15.1 is a list of common materials and their emissivities.

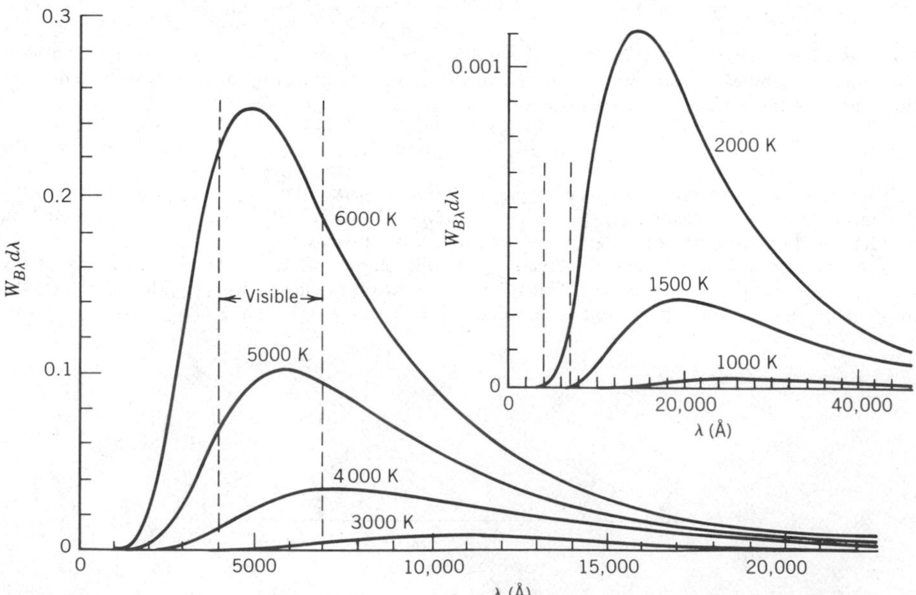

Fig. 15.6 Black-body radiation curves plotted to scale. The abscissas give the wavelengths in angstroms and the ordinates give the energy in calories per square centimeter per second in a wavelength interval $d\lambda$ of 1 Å. For numerical values see *Smithsonian Physical Tables*, 8th ed., p. 314. From Jenkins and White,[2] with permission.

TABLE 15.1. EMISSIVITIES OF COMMON MATERIALS

Material	Temperature (°C)	Emissivity
Tungsten	2000	0.28
Nickel-chromium (80:20)	600	0.87
Lampblack	20–400	0.96
Polished silver	200	0.02
Glass	1000	0.72
Platinum	600	0.1
Graphite	3600	0.8
Aluminum (oxidized)	600	0.16
Carbon filament	1400	0.53

Tungsten Filament Lamp. In this source a tungsten filament is heated to produce light. The filament is protected from oxidation and vaporization by an inert gas. More recently the use of a quartz envelope and a halide atmosphere has permitted the filament to run at higher temperatures. As can be seen from Eq. (15.21) and Fig. 15.6, this increases the luminance of the source and gives a more even distribution of light over the visible spectrum.

Standard Light Source (Black-Body Equivalent). The emissivity of materials suitable for incandescence is less than 1, so that to produce a black-body standard a black-body equivalent is used.

This is achieved by having an enclosed space with a small opening into it. If the area of the opening is much smaller than the enclosed area, the radiation from the opening is very close to that of black body for the temperature of the cavity, provided the interior surface has reasonable emissivity, for example, 0.5. The units of luminance are candelas per square centimeter. Black-body radiation at the melting point of platinum is defined as $1/60\ \text{cd/cm}^2$.

Arc Lamp. Higher temperatures can be obtained by generating an arc between two electrodes. This heats the gas to a temperature of 6000K or more, close to the temperature of white light from the sun. The temperature depends on the current flowing through the arc, gas pressure and composition, and other factors. Figure 15.7 shows the output spectra of two types of arc lamp compared with that of a

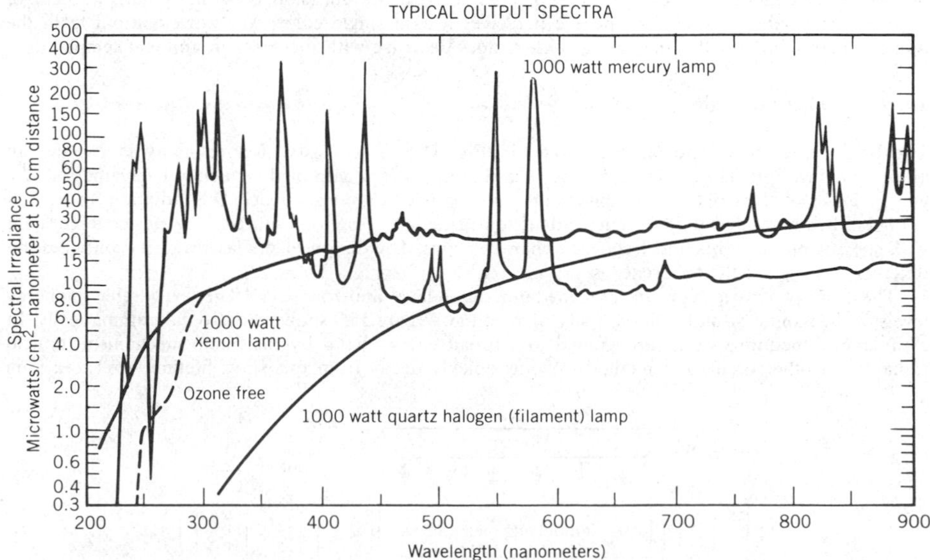

Fig. 15.7 Comparison of output from two arc lamps containing mercury or xenon gas with output from a tungsten filament lamp. The higher temperature arc lamps give more output at the shorter wavelengths. Reproduced from the Oriel Corporation Catalog, 1979, with permission.

quartz halogen tungsten filament lamp. The inert xenon gas gives essentially a white light source. The use of mercury vapor in the arc gives more light in the UV range because of excitation and fluorescence of the mercury atoms.

Fluorescent Source

When a gas is excited, either by a DC discharge or RF excitation, the electrons in it move to higher orbits and the atoms thus are raised to excited states. When the atoms relax back to the ground state, they give off the energy they have absorbed from the excitation. Energy emission can be partially in the form of light; the wavelength of the light is characteristically related to the excited state and the gas involved. Since many excited states are involved, a number of different wavelengths are associated with a particular gas.

Low-pressure fluorescent lamps are used to provide light of specific wavelengths. If one wavelength only is required, optical filters or a spectrometer can be used to isolate it. Typically, the luminance of a low-pressure lamp is small, and its primary purpose is to provide light of narrow spectral width at a specific wavelength.

The fluorescent lamp is very efficient, since in it a high proportion of the input energy is converted to light. White light is obtained from a fluoresent light by coating of the inside of the gas envelope with various types of phosphor. The gas in the light is a mercury-argon mixture, which when excited produces ultraviolet and violet radiation that excites the phosphor.

Since the radiation is generated by fluorescence and phosphorescence, the white-light spectral distribution is characteristic of the elements involved and does not follow Planck's radiation law.

Light-Emitting Diodes

Light can be emitted from a semiconductor material when an electron and hole recombine. The wavelength of the light is related to the amount of energy released on recombination when the electron goes from the conduction band down to the valence band. The frequency of the light is given by

$$h\nu \doteqdot E_g \tag{15.22}$$

where E_g is the bandgap of the semiconductor and h is Planck's constant (6.626×10^{-34} J/s).

Only semiconductors with "direct" bandgaps emit light. For light emission to occur, the conduction band must be populated with many electrons (and the valence band with holes). This is achieved by forward biasing a *pn* junction and injecting electrons and holes into the junction region.

The light from the junction is incoherent and is emitted over a large solid angle. The light has a dominant wavelength, as expressed in Eq. (15.22), with the emission covering a narrow band of wavelengths (~100 nm) so that the eye perceives it as a single color. The power output from the diode is approximately linear with current; it does decrease with increases in ambient temperature.

15.1-4 Coherent Light

The basic structure of the laser is shown in Fig. 15.8. It consists of a resonant cavity and an amplifying medium. The system is arranged so that the energy gain of the medium is greater than the energy losses of the cavity, and light oscillation occurs. As a consequence, the radiation, instead of being emitted by a multitude of independent sources, comes from one source. It is thus coherent light with specific phase fronts and frequency characteristics. The design of the laser and its configuration determine how good the coherence is.

There are two basic types of laser medium: three level and four level. The levels refer to energy levels of the atomic or molecular amplifying medium. Figure 15.9 shows the two diagrammatically. In a three-level medium, atoms are excited to a broad energy band by the pump mechanism (optical radiation or other oscillation method). Atoms quickly decay from this short lifetime level to a long

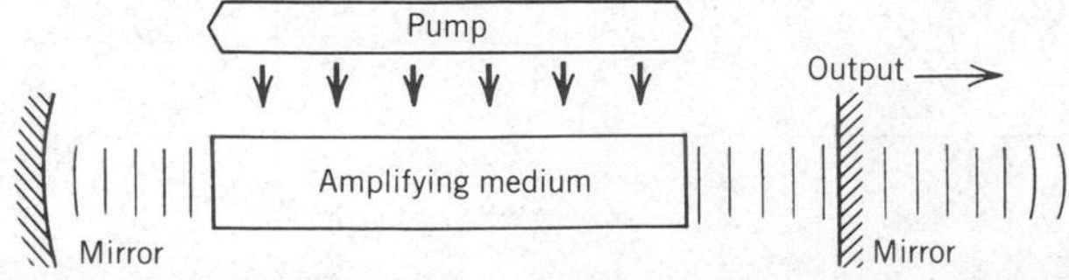

Fig. 15.8 Schematic of the basic laser structure, which consists of a resonator and an amplifying medium. The pump is used to excite atoms or molecules in the amplifying medium.

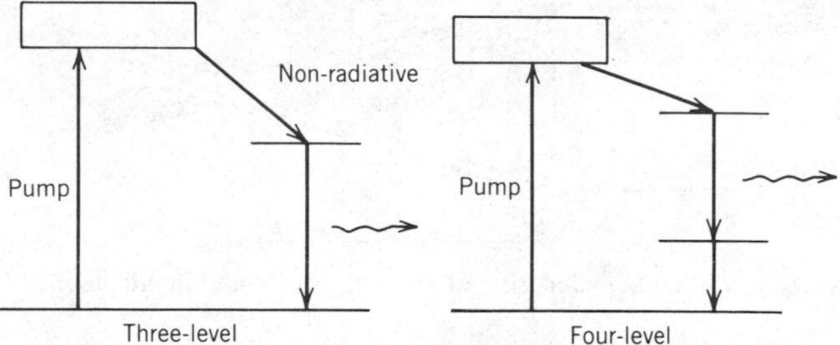

Fig. 15.9 Simplified energy diagram for three-level and four-level laser media.

lifetime level, the upper laser level. Atoms return to the ground state by spontaneous emission (incoherent light) of a photon, or can be stimulated to emit a photon by an existing photon. The latter mechanism results in two photons in phase (coherent amplification). If there are atoms in the lower laser level, absorption of a photon can raise them to the upper laser level. The wavelength of the photon is given by Eq. (15.22), where E_g is now the energy difference between the upper and lower laser levels.

In the four-level system, the lower laser level (short lifetime) is above the ground state. Atoms then decay from this level to the ground state rapidly. The laser medium exhibits gain when there are more atoms in the upper laser level than in the lower level, since the probability of absorption and stimulated emission are equal. Gain is in general more difficult to achieve in a three-level system, since over 50% of the atoms have to be excited out of the ground state into the upper laser level.

Resonator

Figure 15.10 shows a typical two-mirror laser resonator. Curved and flat mirrors are used to produce the desired beam properties. The fundamental longitudinal resonant frequencies (modes) of the cavity are given by

$$\nu = \frac{nc}{2l} \tag{15.23}$$

Since the cavity is much longer than the light wavelength, it is a multimode cavity with many resonances $\Delta\nu = c/2l$ apart. Unless specially designed not to, most lasers will oscillate on a number of these longitudinal modes.

The phase front properties of a laser are normally specified as single TEM_{00} or multimode (transverse). For a single-mode laser beam there is a uniphase front with a near gaussian distribution in energy given by

$$I(r) = I_0 \exp\left[2\left(\frac{-r^2}{w_0^2}\right)\right] \tag{15.24}$$

The divergence of the beam is given by

$$\theta_{\frac{1}{2}} = \frac{\lambda}{\pi w_0} \tag{15.25}$$

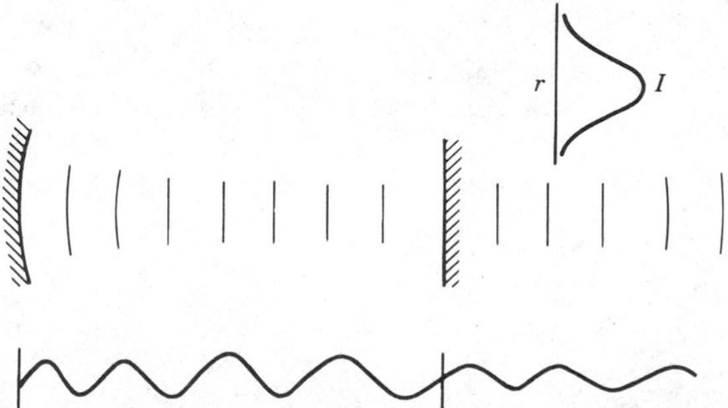

Fig. 15.10 Schematic of a laser resonator formed by a curved and a flat mirror. One mirror is partially transmitting to provide an output beam.

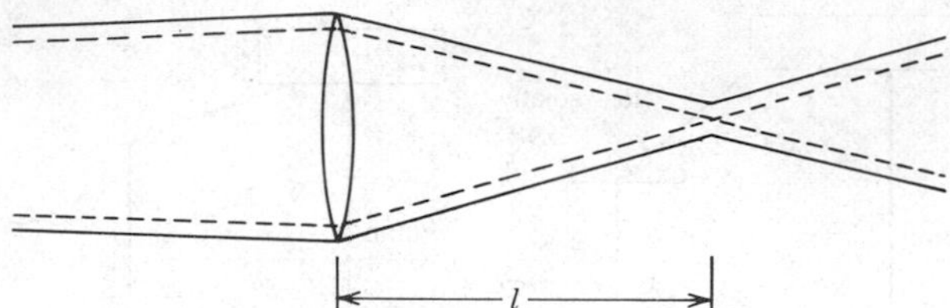

Fig. 15.11 Lens focusing of a beam from a multimode laser.

If the beam is focused by a lens, the focused spot size is given by

$$w_f = \frac{\lambda l}{\pi w_d} \tag{15.26}$$

where l is the distance from the lens to the position of the focused spot and w_d is the beam radius entering the lens.

In a multimode laser the beam has a more complicated phase distribution, with accompanying intensity fluctuations. Multimode beams are normally specified in terms of a beam diameter and a beam divergence σ. When focused by a lens, as shown in Fig. 15.11, the spot size is approximately given by

$$d = l\sigma \tag{15.27}$$

where l is defined as before.

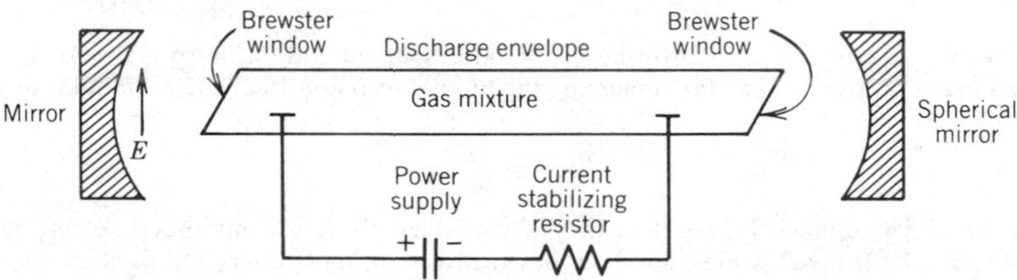

Fig. 15.12 Schematic representation of a gas laser using DC discharge excitation and a discharge envelope with Brewster windows. The Brewster windows theoretically have 100% transmission for laser modes whose E vector, as shown, is in the plane of the paper. From A. Yariv, *Quantum Electronics*, Wiley, New York, 1967, with permission.

TABLE 15.2. TYPES OF LASERS, THEIR ASSOCIATED WAVELENGTHS, AND THEIR TYPICAL POWER OUTPUT

Gas	Wavelength	Power, Continuous Wave (CW)
He-Ne	0.633 μm	1–20 mW
Argon	UV, visible	1 W
Krypton	Visible	1 W
He-Cd	0.325 μm, 0.442 μm	2 mW, 40 mW
CO_2	10.6 μm	10 W–1 kW
N_2	0.337 μm	100 kW (pulsed)
Solid State		
Ruby	0.694 μm	1–100 W
Nd-YAG (or glass)	1.06 μm	1–100 W
Semiconductor		
Ga-As	0.85 μm	10 mW
In-Ga-As-P	1.3 μm	5 mW

Gas Lasers

In gas lasers the amplifying medium is a gas or gas mixture. The most common is helium-neon; one design is shown in Fig. 15.12. A DC discharge excites the helium electrons to various upper states. Resonant transfer excites the neon atoms to provide an inversion condition with an upper state more populated than lower states. This results in energy gain at a wavelength corresponding to the energy difference between these states.

The two mirrors provide a resonant cavity with oscillation at a number of longitudinal modes. One of the mirrors is partially transmitting, giving the output laser beam. The energy gain of the laser medium is quite low and so it is important that losses be kept small. The mirrors are multilayer dielectric and the tube windows are at the Brewster angle. (In some cases the mirrors are inside the tube.) Table 15.2 gives the more popular gases available and their associated wavelengths.

Solid State Lasers

The solid state laser was the first to be operated back in 1959 by Theodore Mainman.[3] The basic principles are the same as those for gas lasers except that the excitation of the laser medium is optical by means of either flash lamp (only pulsed operation) or high-intensity incandescent lamps. The laser is in the form of a rod being illuminated from the side. The ends are polished and either mirrors are put directly on these surfaces or antireflection coatings are used with external mirrors. Energy gain in solid state lasers tends to be higher than in gas lasers; however, to optimize power and minimize component damage inside the cavity, low-loss coatings and components are still used.

Semiconductor Lasers

In a light-emitting diode light is generated by the spontaneous recombination of hole-electron pairs. The device is normally configured so that the light escapes before significant amounts can be reabsorbed and generate an electron-hole pair. If the current through the junction is increased sufficiently, inversion can be obtained where there are more electrons in the conduction band than in the valence band. In this case light in the junction will stimulate carriers to recombine in phase, giving light amplification.

A semiconductor laser is shown in Fig. 15.13. The resonant cavity is formed by cleaving (or polishing) the crystal perpendicular to the junction. The end faces usually have enough reflection because of the high crystal refractive index to provide cavity mirrors. Additional coatings can be used to increase the reflection.

The light beam is normally elliptical in cross section and so the divergence is also different for the two axes. Depending on the design the beam can be multimode or single-mode pulsed or continuous wave.

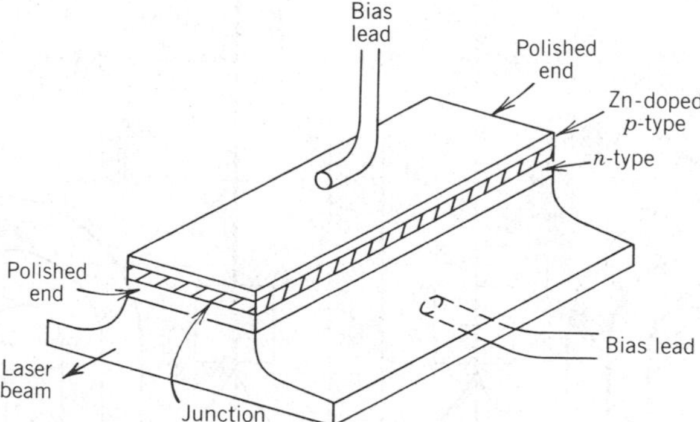

Fig. 15.13 A typical *pn* junction laser in Ga-As. Two parallel faces are polished and serve as the resonator reflectors. From A. Yariv, *Quantum Electronics*, Wiley, New York, 1967, with permission.

15.1-5 Applications of Sources

Both tungsten filament lamps and fluorescent lamps are used for general illumination purposes, the fluorescent being more efficient. In situations where bright highly collimated light is required, such as in searchlights and projectors, the arc source is best, the tungsten filament being the next preference where convenience is a significant factor.

Arc and fluorescent lamps with mercury vapor are also used as UV radiation sources for curing photoresists and other polymers.

Light-emitting diodes produce light of a single color and have found major application as indicator lamps. The infrared diodes are used more in optical communications and for optical isolators, encoders, and security devices such as burglar alarms.

Lasers have found many applications: High-power lasers are used for industrial processes such as welding, cutting, heat treating, annealing, and in situations where conventional approaches will not work. Lower power lasers are applied in measurement and inspection systems when their special properties of spectral purity and collimation with useful intensity can be used. In all uses of lasers, there must be understanding of the safety precautions required. This is especially true for higher power lasers.

References

1 M. Born and E. Wolf, *Principles of Optics*, 3rd ed., Pergamon, New York, 1965.

2 F. A. Jenkins and H. E. White, *Fundamentals of Optics*, 3rd ed., McGraw-Hill, New York, 1957.

3 T. H. Mainman, *Nature* **187**:493 (1960).

Bibliography

Mauro, J. A., *Optical Engineering Handbook*, General Electric Co., 1966.

RCA, *Electro-Optics Handbook*, EOH-11, 1974.

Smith, W. J., *Modern Optical Engineering*, McGraw-Hill, New York, 1966.

Yariv, A., *Quantum Electronics*, Wiley, New York, 1967.

Yariv, A., *Introduction to Optical Electronics*, Holt, Rinehart and Winston, New York, 1971.

15.2 OPTICAL FIBERS AND CONNECTORS

L. S. Watkins

15.2-1 Fibers

Recent developments in fabrication of low absorption loss, high silica content glass have permitted the optical fiber (or light pipe) to be considered for telecommunication purposes. Losses that were once hundreds of decibels per kilometer have been reduced to 1 dB/km and below.

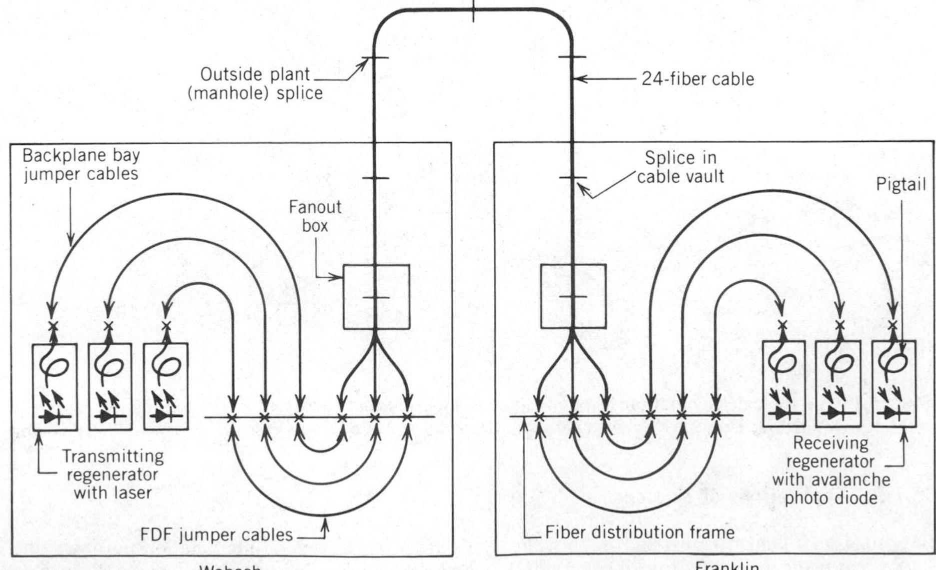

Fig. 15.14 Basic components of a fiber-optic communication link. From Wolaner, in *Fiber Optics*, Bendow and Mitra, eds., Plenum, New York, 1979, with permission.

Figure 15.14 shows the basic components of an optical fiber link. The output from a light source, currently a light-emitting diode (LED) or a laser, is amplitude modulated by a digital or analog signal, achieved by direct modulation of the source drive current. The light is coupled into a fiber and through various connectors into a fiber cable. The cable is normally a series of sections spliced together to form the link. The fiber at the other end is then routed to a detector and the modulated light signal is converted into a modulated detector current to provide the signal output. If the link is long, repeaters consisting of a receiver and laser regenerator for each fiber are used.

Types

There are three basic types of optical fiber: step-index multimode, graded-index multimode, and single mode. Figure 15.15 shows the optical configuration of these. The step-index multimode fiber has a core glass with a higher refractive index than the cladding. Light rays traveling in the core with incidence angles greater than the critical angle are totally internally reflected at the core/cladding interface and are trapped in the fiber. The situation is similar for the single-mode fiber except that because of the smaller index difference, fewer light rays are trapped. In the graded-index fiber, light rays are also trapped because of the index gradient which bends them toward the center. The refractive index in a graded fiber is normally described by

$$n = n_1\left[1 - \Delta\left(\frac{r}{a}\right)^{\alpha}\right] \tag{15.28}$$

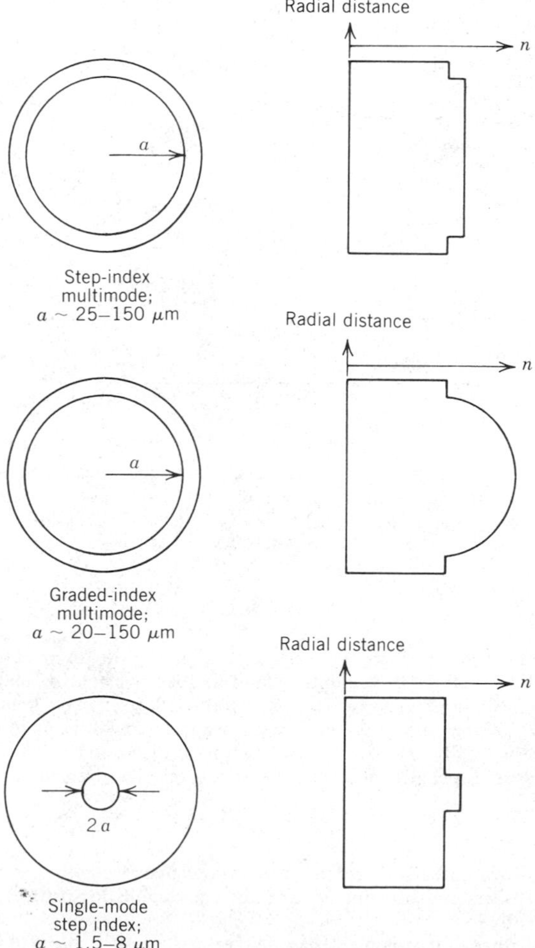

Fig. 15.15 Cross section and refractive index profile of the three basic fiber types. From Botez and Herskowitz, *Proc. IEEE* **68**:689 (1980), with permission.

where n_1 = refractive index at center
r = radius
a = fiber core radius
α = profile exponent

The rays in the graded-index fiber all travel at approximately the same velocity, whereas for the step-index multimode fiber the rays with the smaller incidence angle travel slower. Thus a light pulse traveling down this fiber is more spread out, resulting in a lower allowable pulse transmission rate (or lower signal bandwidth). Wave optics analysis of fibers shows that the light is transmitted down the fiber in modes, hence the term "multimode fiber." Single-mode fiber permits only one mode to propagate and as a consequence it has very high bandwidth characteristics.

Loss

The attenuation in fibers is specified in decibels per kilometer for an operating wavelength. Figure 15.16 shows a typical fiber spectral loss curve.

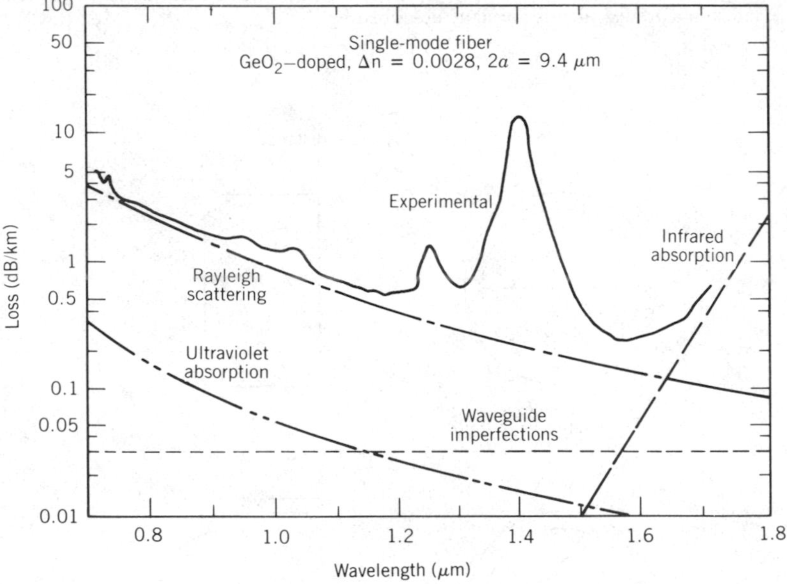

Fig. 15.16 Loss spectra for a typical low-loss graded-index high-silica fiber. Also indicated are the fundamental limiting phenomena and their spectral characteristics. The peaks of 1.25 μm and 1.4 μm are due to hydroxide ions in the glass. From Li, *Proc. IEEE* **68**:1175 (1980), with permission.

Bandwidth

This is specified for multimode fibers as megahertz kilometer at an operating wavelength. The bandwidth is limited by modal delay fluctuations and to a lesser extent by material dispersion. As a consequence bandwidth is not affected by source spectral width for narrow linewidths (10 nm) typical for a laser. For LEDs (300 nm) there may be a reduction in bandwidth due to source spectral width. Single-mode fibers are normally specified in terms of pulse dispersion in picoseconds per kilometer nanometer. Here the source spectral width directly affects the fiber transmission bandwidth.

Cables

Immediately on being drawn, fibers are coated with a protective plastic to preserve their intrinsic high strength. A number of fibers are subsequently arranged in a cable. Figure 15.17 shows cross sections of various cables currently in use.

The mechanism of fiber failure is for a flaw (either existing or induced) to grow and propagate under fiber tension. The tension for this growth process depends inversely on the original flaw size.

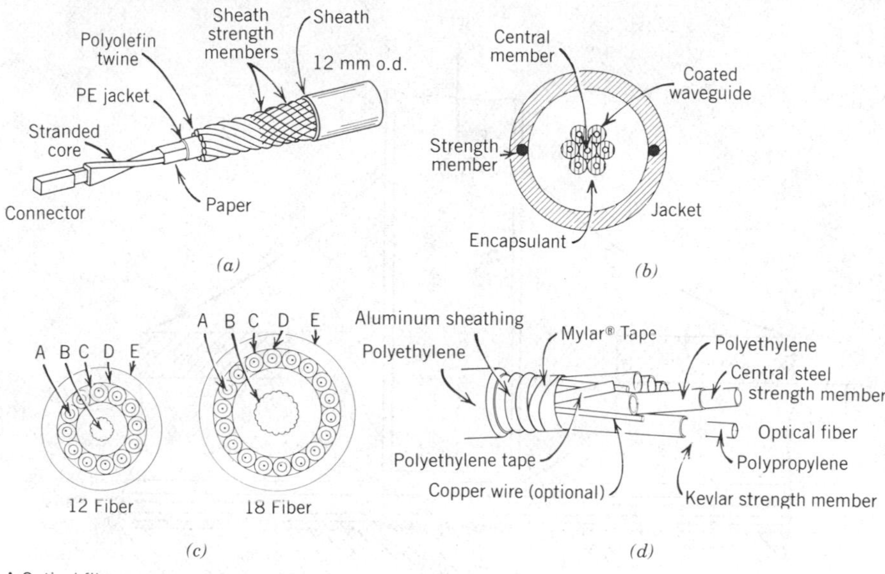

Fig. 15.17 Various types of cable structures. All of them provide strength members to reduce the tensile load on the fiber. (*a*) From American Telephone and Telegraph Company, 1978, with permission. (*b*) From Schwartz, Gloge, and Kempf, in *Optical Fiber Telecommunications*, Miller and Chynoweth, eds., Academic, New York, 1979, with permission. (*c*) From Belden Corp., *Fiber Optics*, 1978, with permission. (*d*) From Valtec, a Philips-M/A-Com Venture, with permission.

The fiber cable design and fabrication process therefore produces a fiber with a minimum allowable flaw size and puts it in a cable, which restricts the tension applied to the fiber and prevents subsequent flaw-inducing damage to it. Cables therefore have a specified tensile strength, above which the fiber is liable to break. The cable design also has to minimize microbending loss, a mechanism in which light is coupled out of the fiber by small radius bends or kinks impressed on it by the cable structure.

Coupling Light From Source to Fiber

The amount of light coupled from a source into a fiber depends on the beam spot size and divergence of the laser and on the core diameter and numerical aperture of the fiber.

The numerical aperture of a fiber for small angles is given by

$$\mathrm{NA} \doteqdot n_1 2\Delta = n_0 \sin\theta \tag{15.29}$$

where θ is the largest incidence angle of the rays that are trapped in the core of the fiber and n_0 is the refractive index where θ is measured (in air $n_0 = 1$). An LED source is incoherent and radiates into a half-space solid angle; the light can be coupled by either imaging the emitting area onto the fiber with a lens or putting the LED up against the fiber, as shown in Fig. 15.18.

For optimum coupling, the source area imaged onto the core must be smaller than the core diameter. The maximum light coupled is given by the coupling efficiency

$$\eta = \frac{(\mathrm{NA})^2}{1 + 2/\alpha} \tag{15.30}$$

and indicates that, as would be expected, the higher the numerical aperture the more light that is coupled into the fiber.

The semiconductor laser output is more directional and can essentially be completely coupled into the fiber. The beam divergence from the laser is normally too large to be directly coupled and a lens or lenses are required to match the source to the fiber. In addition the laser, especially if multimode,

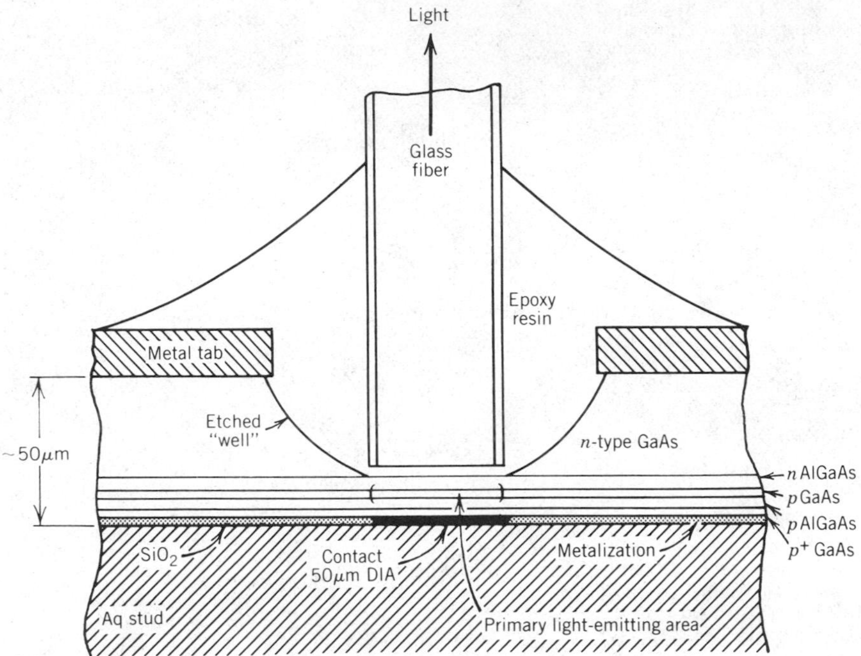

Fig. 15.18 Arrangement for coupling light from light-emitting diode into a fiber by butt coupling to emitting surface. From Burrus, Casey Jr., and Li, in *Optical Fiber Communications*, Miller and Chynoweth, eds., Academic, New York, 1979, with permission.

(spatially) has elliptical output beams with differing divergences in the two axes. This requires astigmatic optics for optimum coupling. As shown in Fig. 15.19 for the laser diode, the coupling optics must form a beam which focuses to a spot of diameter less than the core diameter and have a convergence angle θ less than that derived from the numerical aperture, Eq. (15.29). For a single-mode fiber, optimum coupling occurs when the spot diameter equals the beam diameter of the propagating mode in the fiber.

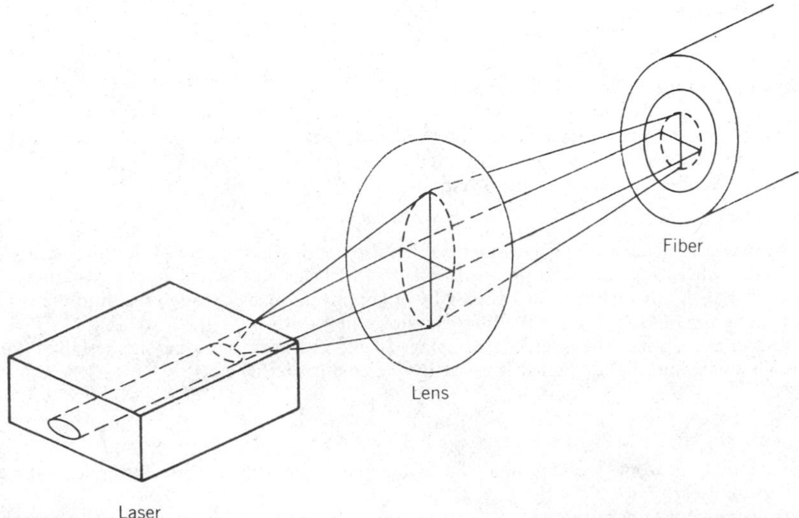

Fig. 15.19 Coupling of ellipitical beam from laser into a fiber. Lens is a cylindrical lens which focuses in the vertical direction only to produce a circular spot onto the fiber core end face.

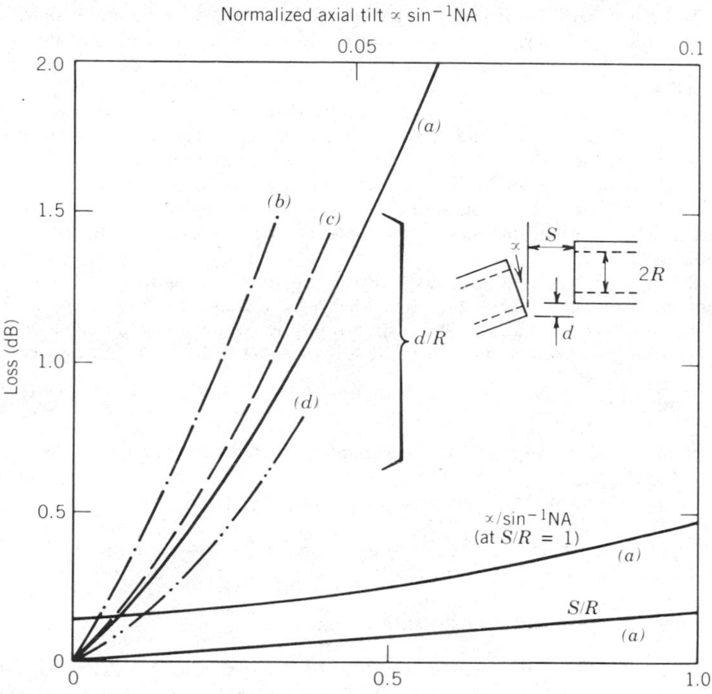

Fig. 15.20 Losses generated by fiber misalignment: (*a*) Results of measurements over short fiber lengths with a light-emitting diode source 4; (*b*) curves 5 assuming uniform power distribution; (*c*) curve assuming gaussian distribution with long length of fiber out of joint; and (*d*) curve assuming short length of fiber out of joint. From Dalgleish, *Proc. IEEE* **68**:1226 (1980), with permission.

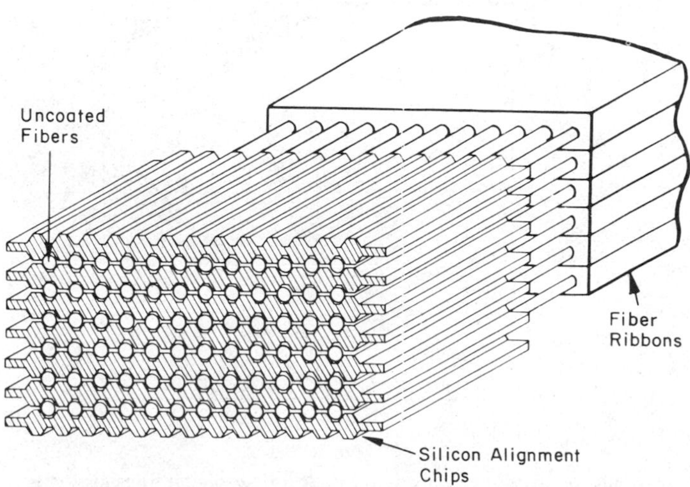

Fig. 15.21 Array splice for ribbon-type cables. From Dalgleish, *Proc. IEEE* **68**:1226 (1980), with permission.

Detection of Transmitted Light. The output beam from a fiber diverges from the core with a half angle given by Eq. (15.29). This can be detected by a suitable sensor being butt coupled close to or in contact with the fiber, ensuring that the beam is totally within the detector's sensitive area.

Permanent Splicing. Light coupling between fibers is achieved by butt coupling to align the cores of the fibers so that the light transmission path is continuous. Differences in fiber parameters, such as core diameter, numerical aperture, and profile, will cause splice loss. Also misalignments of the fibers specifically offset, axial tilt, and separation will cause scattering out of light.

Figure 15.20 shows both measured and computed results for misalignment loss for LED or laser-excited fibers. These results assume there is no reflection loss at the fiber end faces. Such loss adds approximately 0.3 dB to the loss; it can be eliminated by using a refractive index matching medium between the fibers.

Two basic techniques are used to permanently join fibers: fusion splicing and adhesive bonding in an alignment chip. In fusion splicing the fibers are butted together and heat is applied (arc or torch) to fuse the ends together. Figure 15.21 shows an array splice where grooved silicon chips are used for alignment. Epoxy is used to clamp the fibers and also provides the index matching medium.

15.2-2 Connectors

There are a number of types of connectors for coupling single fibers. In all cases the fiber is mounted coaxially in a larger cylindrical or conical plug and the fiber end faces are polished flat. The two ends are then fitted and clamped into a mating sleeve to provide the required alignment and to butt the fiber end faces together.

Losses for permanent splices can be as low as 0.1 to 0.5 dB, whereas losses for connectors tend to be somewhat higher, in the 0.3 to 1.0 dB range.

15.2-3 Communication Link Considerations

Fiber-optic communication links are best suited for digital communications. Special care is required to obtain the linear modulation from laser or LED sources that is required for acceptable cross-talk

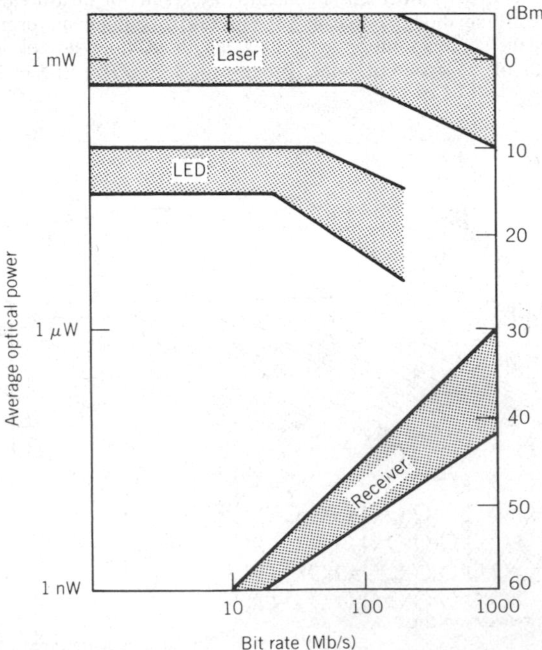

Fig. 15.22 Transmission margin vs. bit rate for optical-fiber digital transmission systems. Upper two bands represent available (launched) transmitted power from lasers and light-emitting diodes; lower band represents required received power for 10^{-9} error probability. From Gloge and Li, *Proc. IEEE* **68**:1269 (1980).

TABLE 15.3. FIBER-OPTICS SYSTEMS CONFIGURATIONS AND PERFORMANCES

Requirements/Technology	Digital Telecommunications	Analog Transmission	Local Data Communications: Point-to-Point Data Link, Low Speed	Point-to-Point Data Link, High Speed	Passive Data Bus, Linear	Passive Data Bus, Star
Requirements						
Distance between repeaters	> 4 km	0–2 km	0.1–2 km	0.1–0.5 km	0.2–2 km	0.1–0.5 km
Bandwidth (typical)	1.5–300 MBd	0–10 MHz	DC–1 MBd	1–100 MBd	DC–1 MBd	1–20 MBd
Signal-to-noise ratio at bandwidth or BER[a]	> 20–30 dB; $< 10^{-9}$	> 40–70 dB	10^{-7}	10^{-9}	10^{-9}	10^{-9}
Acquisition time	Slow OK	Slow OK	Slow OK	Slow OK	Within a few bits	Within a few bits
Available-power-supply voltage constraints	Minor	Minor	+ 5 V	+ 5 or − 5.2 V	+ 5 V	+ 5 or − 5.2 V
Cost/end (optical-electrical interface) to compete economically	< 0.5–$2000 (2-way)	TBD[b]	< $10	< $50	< $50	< $50
Performance monitor	Desirable-essential	Desirable	Desirable	Desirable	Desirable	Desirable
Received optical level dynamic range capability	20–50 dB	10–20 dB	> 20 dB	> 20 dB	> 40 dB	> 20 dB
Applicable Technologies to Meet Requirements						
Emitter	Laser or LED	LED (or laser)	LED	LED	LED/laser	LED/laser
Detector	APD[c] or *pin*	*pin* (or APD)	*pn* or *pin*	*pin*	*pin*	*pin*
Fiber						
Core size/cladding size (μm)	50/125	50/125 or larger	200/-	100/140	100/140, or 200/-	100/140
Bandwidth	—	—	> 10 MHz-km	> 100 MHz-km	> 10 MHz-km	> 20 MHz-km
Attenuation	< 10–< 4 dB/km	Various, < 10 dB between *T* and *R*	< 30 dB/km	< 10 dB/km	< 10 dB/km	< 10 dB/km

Source: Prsonick et al., *Proc. IEEE* **68**:1254 (1980), with permission.
[a] BER, bit error rate.
[b] To be determined.
[c] ADP, avalanche photodiode.

performance. A simple comparison between fiber and coaxial signal-to-noise ratios can be made as follows:

$$\text{SNR}_c = \frac{v^2 e^{-\alpha}}{Z_0 4KTB} = 1.6 \times 10^{19} \frac{e^{-\alpha}}{B}$$

$$\text{SNR}_f = \frac{P_0 K^2 e^{-\alpha}}{2h\nu B} = 6.25 \times 10^{14} \frac{e^{-\alpha}}{B}$$

Thus for equal transmission attenuation, the coax link has higher inherent signal-to-noise ratio.

The fiber transmission loss is independent of modulation bandwidth and is much lower than coaxial loss at higher frequencies. Hence the fiber has clear economic advantages for higher transmission rate digital communications.

Figure 15.22 shows the typical range of transmission margins vs. bit rate for lasers and LEDs over typical multimode graded-index fibers (50 μm core diameter, 0.2 NA). This shows that for low bit rate short-link systems, LEDs are quite suitable, whereas for optimum high performance lasers should be used.

15.2-4 Applications

Table 15.3 lists various systems configurations and performances. The typical fiber loss curve of Figure 15.16 indicates that the optimum operational wavelengths are 1 to 3 μm and 1.55 μm. Early available sources were GaAs based and operated at 850 nm, indicating typical transmission losses of 3 to 5 dB/km. Sources and detectors tailored to the 1 to 3 μm- and subsequently 1.55-μm wavelength bands are starting to become available now.

Table 15.4 shows a typical loss budget calculation for a fiber link. The transmitter output power includes the loss incurred in coupling into the fiber. A separate bandwidth calculation has to be done using the source, detector, and fiber bandwidth (combined with link length) to determine overall transmission bandwidth.

TABLE 15.4. LOSS BUDGET FOR FIBER LINK

Loss	Worst Case	Best Case
Transmitter average output power	− 17 dBm	− 14 dBm
Five optical connectors	− 9 dB	− 2.5 dB
Graded-index cable, 2 km	− 16 dB	− 8 dB
Receiver coupling loss	− 1 dB	− 1 dB
Time degradation	− 3 dB	0 dB
Temperature degradation (0–50°C)	− 2.5 dB	0 dB
Average power at detector	− 48.5 dBm	− 25.5 dBm
∴ Minimum required receiver sensitivity	− 48.5 dBm	
Minimum optical dynamic range (assuming 2-dB receiver variation)	25 dB	
Minimum receiver saturation level	− 25.5 dBm	

Bibliography

Bendow, B. and S. S. Mitra, *Fiber Optics*, Plenum, New York, 1979.

Botez, D. and G. J. Herskowitz, "Components for Optical Communications Systems: A Review," *Proc. IEEE* **68**:689 (1980).

Elion, G. R. and H. A. Elion, *Fiber Optics in Communications Systems*, Marcel Dekker, New York, 1978.

Gloge, D., *Optical Fiber Technology*, IEEE, 1976.

Midwinter, J. E., *Optical Fibers for Transmission*, Wiley, New York, 1979.

Miller, S. E. and A. G. Chynoweth, *Optical Fiber Telecommunications*, Academic, New York, 1979.
Proc. IEEE, special issue on optical-fiber communications (Oct. 1980).

15.3 OPTICAL DISPLAYS

Peter Pleshko

Visual displays employ many diverse physical phenomena. In this section emphasis is put on the physics and engineering aspects of a few of the many available technologies. Primary emphasis is on the cathode-ray tube (CRT), the main display technology currently in use, and on some other technologies currently in high-volume production: liquid crystals, light-emitting diodes, plasma and vacuum fluorescent displays. Brief mention is also made of electroluminescent and electrochromics as some of the emerging display technologies.

15.3-1 Optical Display Concepts and Parameters

Visual displays are output devices that function at the "man-machine" interface as electrooptical transducers. The section that follows discusses the psychophysical characteristics of the human eye and the photometric parameters (Table 15.5) used to quantify these characteristics.

Psychophysical Characteristics of Human Eye and Photometric Parameters Used for Quantification

Spectral Response. The spectral response of the average human eye is shown in Fig. 15.23. The light-adapted (photopic) response is the one most used. It is a function of the cones in the retina and occurs after the eye has adapted to a background illumination of at least 3 cd/m^2 (nits, nt). The dark-adapted response (scotopic) occurs when the eye has adapted to at most 3×10^{-5} nt.

Color Representation. Any color can be produced by a combination of the three monochromatic components red, green, and blue. The Commission Internationale de l'Eclairage (CIE) has adopted a standard of colorimetry which represents the attributes of color by a three-dimensional diagram. The three idealized primarics are shown in Fig. 15.24. The tristimulus values X, Y, and Z are read at the wavelengths shown in the figure. The normalization of chromatic coordinates is calculated by

$$x = X/(X+Y+Z)$$
$$y = Y/(X+Y+Z)$$
$$z = Z/(X+Y+Z)$$

so that their sum is unity

$$x + y + z = 1$$

A chromaticity diagram developed in 1931 is shown in Fig. 15.25. A later diagram (1960) with uniform chromaticity scales is also in wide use.[1]

Symbol Size and Resolution. Symbol size S as a function of viewing distance D is expressed as a function of the angle α (in minutes of arc) subtended at the eye, in which $\tan\alpha$ is approximated by the angle itself

$$\frac{S}{D} = \frac{2\pi\alpha}{360 \times 60}$$

Some studies recommend a symbol size of 15 min under good viewing conditions[2] and 21 to 25 min under poor viewing conditions.[3]

Illuminance. Ambient illumination, or illuminance, is expressed in lumens per square meter (lux), which is the international unit, or lumens per square foot (footcandles), which is the unit commonly used in the United States. A typical office illumination level is 590 lux or 55 fc.

Luminance (Brightness). Luminance is a measure of the radiated power per unit area emitted by a display device in the visible wavelength region. The international unit of luminance is candelas per square meter (nits), although the unit commonly used in the United States is the footlambert. Luminance is averaged over space and time.

Contrast and Gray Scale. Contrast is created by spatial modulation of luminance. For a luminance of the symbol or pixel L_p and a background or nonemitting area luminance L_b, the contrast ratio CR

TABLE 15.5. PHOTOMETRIC QUANTITIES AND UNITS

Quantity	Symbol	Defining Equation[a]	SI Unit	Symbol
Luminous energy	Q, Q_v	$Q_v = \int K(\lambda) Q_{e\lambda}\, d\lambda$	Lumen second	lm s
Luminous density	w, w_v	$w = \partial Q / \partial V$	Lumen second per cubic meter	lm s/m^3
Luminous flux	Φ, Φ_v	$\Phi = \partial Q / \partial t$	Lumen	lm
Luminous flux density at a surface				
Luminous exitance	M, M_v	$M = \partial \Phi / \partial A$	Lumen per square meter	lm/m^2
Illuminance	E, E_v	$E = \partial \Phi / \partial A$	Lux	lx
Luminous intensity	I, I_v	$I = \partial \Phi / \partial \omega$ (ω = solid angle through which flux from point source is radiated)	Candela	cd
Luminance	L, L_v	$L = \partial^2 \Phi / \partial \omega\, (\partial A \cos \theta) = \partial I / (\partial A \cos \theta)$ (θ = angle between line of sight and normal to emitting surface considered)	Nit	nt
Luminous efficacy	K	$K = \Phi_v / \Phi_e$	Lumen per watt	lm/W
Spectral luminous efficacy	$K(\lambda)$	$K(\lambda) = \Phi_{v\lambda} / \Phi_{e\lambda}$	Lumen per watt	lm/W
Luminous efficiency	$V(\lambda)$	$V = K(\lambda) / K(\lambda_{max})$ $K(\lambda_{max})$ = maximum value of $K(\lambda)$ function	(Numeric)	—

Conversion Factors for Luminous Exitance Quantities

Luminous Exitance Quantity	lm/m²	lm/ft²	lm/cm²
1 lm/m² =	1	0.0929	1×10^{-4}
1 lm/ft² =	10.764	1	0.001076
1 lm/cm² =	1×10^{4}	929	1

Conversion Factors for Illuminance Quantities

Illuminance Quantity	lux (lx)	footcandle (fc)	phot (ph)
1 lux (lm/m²) =	1	0.0929	1×10^{-4}
1 footcandle (lm/ft²) =	10.764	1	0.001076
1 phot (lm/cm²) =	1×10^{4}	929	1

Conversion Factors for Luminance Quantities

Luminance Quantity	nit (nt)	stilb (sb)	apostilb (asb)	lambert (L)	milli-lambert (mL)	foot-lambert (fL)	candela per square foot (cd/ft²)
1 nit (cd/m², or lm/sr/m²) =	1	1×10^{-4}	3.1416	3.1416×10^{-4}	0.31416	0.2919	0.0929
1 stilb (cd/cm²) =	1×10^{4}	1	3.1416×10^{4}	3.1416	3141.6	2919	929
1 apostilb (π^{-1} cd/m²) =	0.3183	3.183×10^{-5}	1	1×10^{-4}	0.1	0.0929	0.02957
1 lambert (π^{-1} cd/cm²) =	3183	0.3183	1×10^{4}	1	1000	929	295.7
1 millilambert =	3.183	3.183×10^{-4}	10	0.001	1	0.929	0.2957
1 footlambert (π^{-1} cd/ft²) =	3.426	3.426×10^{-4}	10.764	0.0010764	1.0764	1	0.3183
1 candela per square foot =	10.764	0.0010764	33.82	0.003382	3.382	3.1416	1

[a] A = area, m²; ω = solid angle, sr; V = volume, m³; t = time, s.

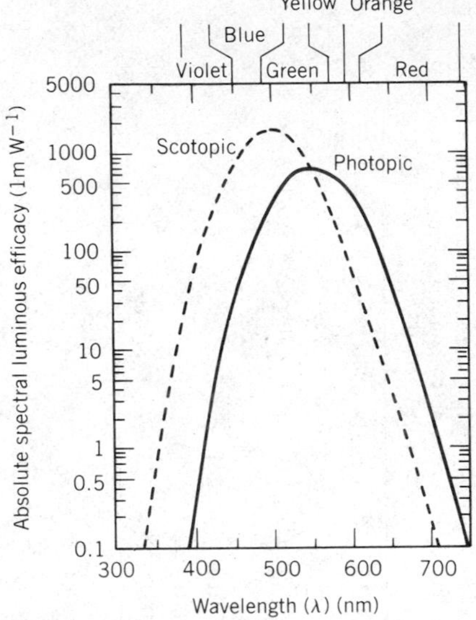

Fig. 15.23 Luminous efficacy of the human eye.

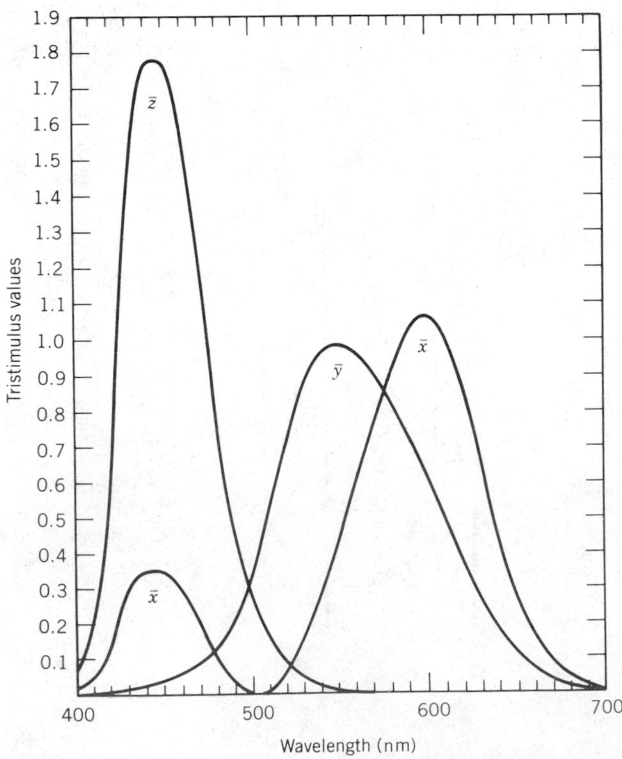

Fig. 15.24 Standard primary colors (Commission Internationale de l'Eclairage).

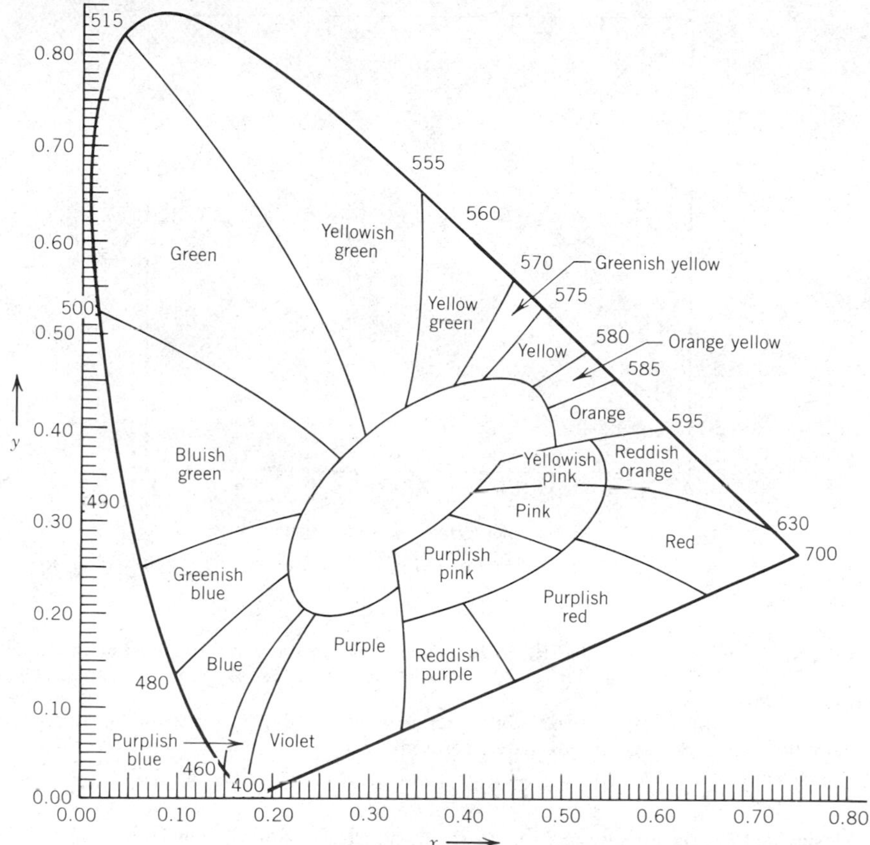

Fig. 15.25 Chromaticity diagram (Commission Internationale de l'Eclairage, 1931).

is given as

$$\mathrm{CR} = \frac{L_p}{L_b}$$

Gray scale is a measure of the range of luminance values available from a display device. The number of gray shades N_g is related to the contrast ratio CR by

$$N_g = 1 + 6.644 \log(\mathrm{CR})$$

where a shade of gray is defined as a luminance ratio of $\sqrt{2}$ between two successive levels of luminance.

Flicker. Flicker is the visual perception of a change in luminance occurring at a frequency below the critical frequency corresponding to the integration time constant of the human eye.

Typically, refresh displays without interlacing operate at frequencies of 40 to 60 Hz.

Display System Characteristics

Modulation Transfer Function (MTF). This cathode-ray-tube (CRT) characteristic is obtained by measuring the sine wave frequency response of the CRT system. The transfer function characterizes the visual output as a function of the electrical input, in the X direction only. The MTF is given by[4]

$$\mathrm{MTF} = \frac{\text{output peak value}}{\text{input peak value}} \cdot \frac{\text{input signal average value}}{\text{output signal average value}}$$

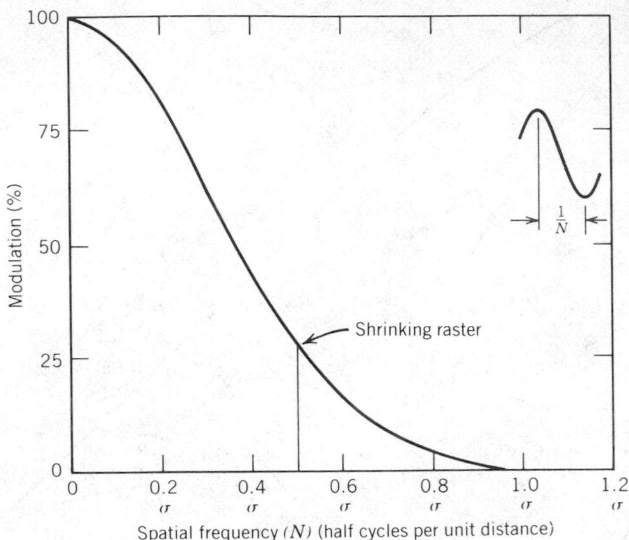

Fig. 15.26 Modulation transfer function (MTF) plot for a cathode-ray tube (CRT).

A typical MTF curve is shown in Fig. 15.26, where σ is the standard deviation of an assumed gaussian luminance distribution for a CRT spot. Other methods of measuring resolution are described in Ref. 5.

The MTF area (MTFA) is defined as the area between the display system MTF and a visual sine-wave threshold curve[4] and approximately corresponds to the amount of information that can be displayed and seen.

Multiplexing and Matrix Addressing. These two terms refer to two different aspects of the same topic. Multiplexing refers to the time-sharing of the data bus (Y drivers) in a matrix display, shown in Fig. 15.27, among the various X lines. Thus it refers to the temporal, or duty cycle, aspect of the drive waveforms.

Matrix addressing refers more to the amplitude requirements that are imposed on the electrical signals on both X and Y lines due to the existence of sneak paths and the electrooptic behavior of a particular display device. However, in common usage, "multiplexing" and "matrix addressing" are used interchangeably to refer to both the temporal and the amplitude characteristics of matrix addressing waveforms.

Sneak path considerations lead to the restriction that for maximum multiplexing capability, the X and Y lines must be driven by voltage (low-impedance) sources.[6] To achieve uniform on and off luminance levels on the devices in an $X - Y$ matrix whose output luminance responds to the peak voltage value, either 2 : 1 or 3 : 1 selection waveforms can be applied (see Fig. 15.27).

15.3-2 Cathode-Ray Tubes (CRTs)

Basic Principles and Optical Effect

The basic electrostatic focus cathode-ray tube (CRT) is shown in Fig. 15.28. The electron gun provides the electron-beam source and the beam focusing. Deflection of the beam is provided by either magnetic deflection yokes or electrostatic deflection plates. Between the screen and the deflection region is a drift region through which the electron beam traverses a path determined by the deflection forces.

The electrooptical effect utilized in the CRT is known as cathodoluminescence. Electrons generated by the electron gun are accelerated by a high field and impinge on the phosphor-coated screen. The energy of the incident electrons excites the electrons in the phosphor crystalline material, which produce radiant energy as they return to their unexcited state. The phosphor electrons are displaced to discrete levels, depending on the crystalline structure of the material. Displacement produces radiation in the visible range, characteristic of the particular phosphor material employed.

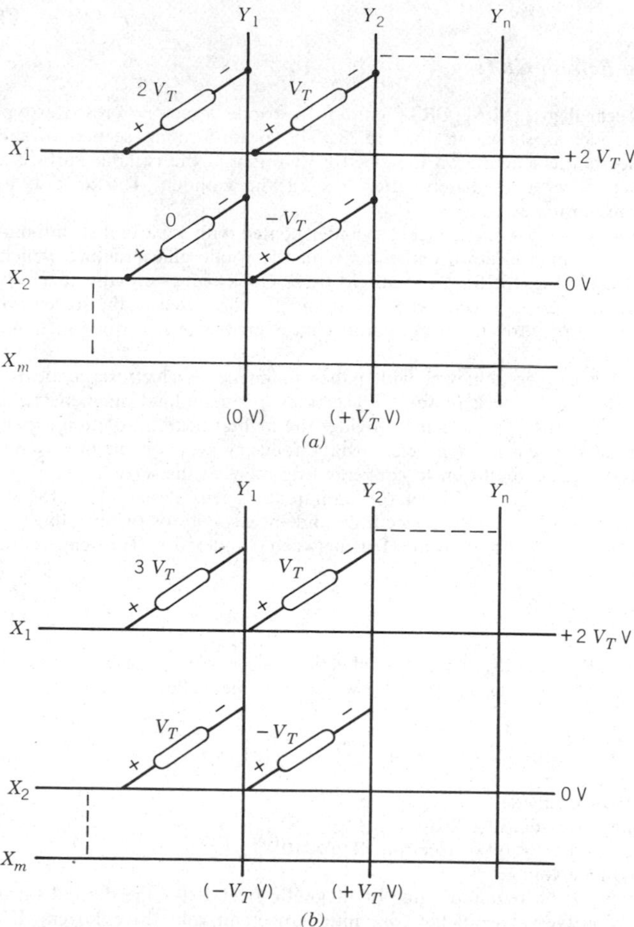

Fig. 15.27 Selection amplitudes for multiplexing waveforms. (*a*) 2 : 1 Peak selection. On, $2V_T$ V; off, $|\pm V_T|$, 0 V. (*b*) 3 : 1 Peak selection. On, $3V_T$ V; off, $|\pm V_T|$ V.

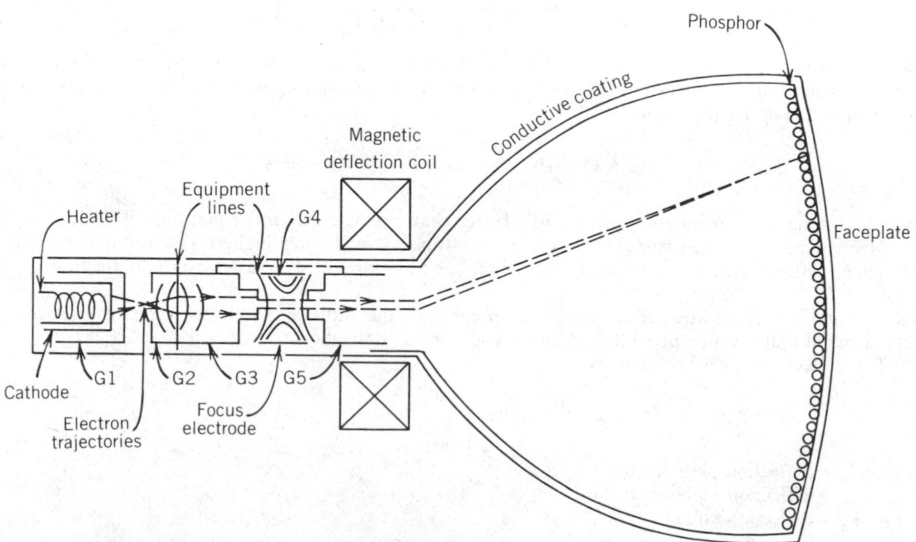

Fig. 15.28 Electrostatic focus, magnetic deflection cathode-ray tube (CRT).

Monochromatic Refresh CRTs [7]

Electron Gun Technology. Most CRTs employ a triode structure crossover-type gun with an oxide-coated cathode, as shown in Fig. 15.28. The beam-forming section of the crossover gun constitutes an electron lens acting on the electrons emitted at the cathode surface, causing them to converge to a crossover immediately after leaving the cathode. This is also an area of high space-charge repulsion forces.

The cathode is typically a small nickel cylinder coated with a barium-strontium-calcium oxide a few (2) mils thick for high electron emission. With the triode gun structure, which consists of the cathode (G1 and G2 in Fig. 15.28), the focus electrode G4 is kept essentially at zero bias while the G3 and G5 electrodes are kept at several kilovolts (typically connected to the screen potential). The G2 voltage may be used to control the beam cutoff characteristics of the triode section.

Focusing. Focusing can be achieved with either magnetic or electrostatic fields. With magnetic focusing, a coil around the neck of the CRT creates a longitudinal magnetic field which exerts a focusing force on the beam. This scheme provides the highest beam resolution capability but is more costly than electrostatic focusing. With electrostatic focusing, which is the most widely used, voltages applied to electrodes cause the beam to converge to a focus at the screen.

Most commonly used is the Einzel (or unipotential) lens design, Fig. 15.28, because of the negligible current drawn by the focus electrode and insensitivity of focus voltage to screen voltage. The electrode G4 is designed to focus at a bias between 0 and 550 V. This lens has severe aberrations which degrade the resolution.

Electron-Beam Deflection

Magnetic. Most CRTs employ magnetic deflection yokes to provide both horizontal and vertical deflection of the electron beam because of low cost. The deflection angle λ provided by the yoke is given by[8]

$$\lambda = \sin^{-1}\left[L_f B \sqrt{e/(2V_a m)} \right]$$

where L_f = length of magnetic field region, m
B = magnetic flux density, Wb/m^2
e/m = charge-to-mass ratio of electron, 1.76×10^{11} C/kg
V_a = final anode voltage

The production of the required value of magnetic field B for the desired maximum deflection angle λ_m depends on the values of the yoke inductance and yoke drive current. If the yoke driving voltage is V_y and the sweep time to deflect the beam by the angle λ is $T_s/2$ in seconds, then the yoke inductance L in henrys can be stated as[9]

$$L \leqslant \frac{1.76 \times 10^4 V_y^2 T_s^2 D_t}{V_a D_c^2 \tan(\lambda_m/2) \sin^2 \lambda_m}$$

where D_t is the inside diameter of the tube neck in inches and D_c is the inside diameter of the yoke coil in inches. The yoke current required to drive the inductance L in henrys, given above, as a function of the deflection angle is

$$I \geqslant 5.33 \times 10^{-3} D_c \left[\frac{V_a \tan(\lambda_m/2)}{D_t L} \right]^{1/2} \sin \lambda$$

In actual designs, currents are typically 30% larger than predicted by this equation.[9]

Magnetically deflected, magnetically focused tubes provide the highest resolution capability. Magnetic deflection cannot provide speeds as fast as those provided by electrostatic deflection.

Electrostatic. Electrostatic deflection is implemented by the addition, within the tube, of orthogonal sets of metal plates which provide horizontal and vertical deflection.

The deflection angle λ is given by[7]

$$\lambda = \tan^{-1}\left[\frac{V_d}{2V_a} \cdot \frac{L_p}{d} \right]$$

where L_p = deflection plate length
d = deflection plate separation
V_d = deflection voltage
V_a = final anode voltage

TABLE 15.6. SELECTED CATHODE-RAY TUBE PHOSPHOR CHARACTERISTICS

Phosphor Type	Color	Color Coordinates x	Color Coordinates y	Chemical Composition	Flicker Frequency (Hz)	Life to Half Brightness (C/cm^2)	Efficiency (lm/W)	Decay Time to 10% (ms)
P1	Green	0.200	0.715	Zn_2SiO_4:Mn	42	> 100	31	25
P2	Green	0.260	0.528	ZnSCdS:Cu	50	12	28	7
P4	White	0.280	0.270	ZnS:Ag + ZnSCdS:Ag	50	5	43	0.06
P20	Yellow	0.405	0.553	ZnSCdS:Ag	50	25	65	6.5
P22	Green	0.229	0.599	ZnSCdS:Ag	50	25	50	6
P22	Blue	0.150	0.049	ZnS:Ag	50		5	4.8
P22	Red	0.658	0.342	$Y_2O_3:Eu^{+3}$	50		12	1.5
P22	Red	0.667	0.332	$YVO_4:Eu^{+3}$	50		7	1.5
P31	Green	0.248	0.583	ZnS:Cu	50	25	45	7
P39	Green	0.200	0.710	Zn_2SiO_4:MnAs	31	> 25	15	400
P43	Green	0.281	0.591	$Gd_2O_2S:Tb^{+3}$	50	> 50	41.5	1.2
P44	Green	0.300	0.596	$La_2O_2S:Tb^{+3}$	50	> 50	20.5	1.7
P45	White	0.230	0.305	$Y_2O_2S:Tb^{+3}$	50	> 50	21	1.8
RP20	White	0.211	0.300	$Tb:Sr_3(PO_4)_2:0.5\%TbF_3$	42		7.7	70
	Yellow-green	0.340	0.550	$CaS:Ce^{+3}$	47		42	4
	Green	0.267	0.692	$SrGa_2S_4$:Eu	50		26.7	12.5

The units used only have to be consistent, since ratios of voltages and distances are used in the expression for λ.

Electrostatic deflection provides the highest deflection speeds and is normally used in conjunction with electrostatic focus, for the achievement of moderate resolution.

Phosphor Screen. Monochromatic CRTs have a uniform phosphor applied to the inside of the CRT faceplate. Aluminization of the phosphor coating side that faces the electron gun increases the efficiency and contrast of the screen and makes the screen a stable equipotential surface. At voltages below 7 kV, aluminization is not used, since the electron-beam energy is not sufficient to penetrate through the aluminum layer. Commonly used phosphors and their characteristics are listed in Table 15.6.

Performance Characteristics

Luminance. The relationship between the time-average luminance L_{av}, the duty cycle per spot position D, the peak current density of the spot J_p (in A/cm^2),[10] and the potential of the screen with respect to the cathode V_s, is given by

$$L_{av} = KDJ_pV_s^n$$

where K is a constant and n can vary from 1 to 2, typically being 1.5.

Cathode Current Density. The limit on peak cathode current density J_p is given by[8,11]

$$J_p = \frac{3.1\,K_m \times 10^{-6}V_d^{1.5}}{d_c^2}\ \text{A/cm}^2$$

where K_m = modulation constant typically equal to 3.5
d_c = grid aperture diameter, in.
V_d = grid drive voltage

Peak cathode loading is plotted in Fig. 15.29.

Maximum values of oxide cathode current densities are limited by the effects of temperature of operation on life. For a pulse average cathode loading of 0.3 to 0.5 A/cm^2, the cathode has a 10,000-hr mean time between failures (MTBF). This yields a cathode current of 600 μA for a 0.020-in. diameter cathode.

Spot Size. The spot diameter d_s on the phosphor screen is limited by four principal effects. The first effect on the spot diameter, d_o, is due to fundamental electron imaging optics considerations, that is,

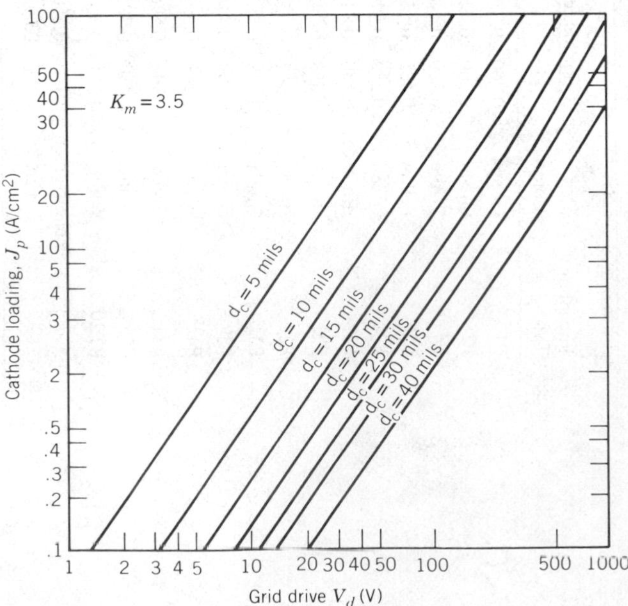

Fig. 15.29 Peak cathode loading.

magnification. For a thin aberrationless electron lens, the undeflected electron-beam diameter at the electron-gun crossover d_c is magnified by the magnification factor M as

$$d_o = d_c(M) = d_c\left(\frac{0.814}{\theta\sqrt{V_s}}\right)$$

where θ is the half angle of convergence in radians and V_s is the potential of the screen with respect to the cathode.[11]

The second effect is due to spherical aberrations d_{sa}, due to the inability of the triode and/or the focus lens to focus an image at the same point. The third effect d_{sc} is due to space charge repulsion forces, which essentially vary linearly with beam current. The fourth effect d_{os} is due to optical scattering within the phosphor screen, which for a typical phosphor particle size is about 3 mil[7] of equivalent diameter.

These aberrations are independent and can be combined in quadrature, yielding the CRT spot diameter

$$d_s = \left(d_0^2 + d_{\text{sa}}^2 + d_{\text{sc}}^2 + d_{\text{os}}^2\right)^{1/2}$$

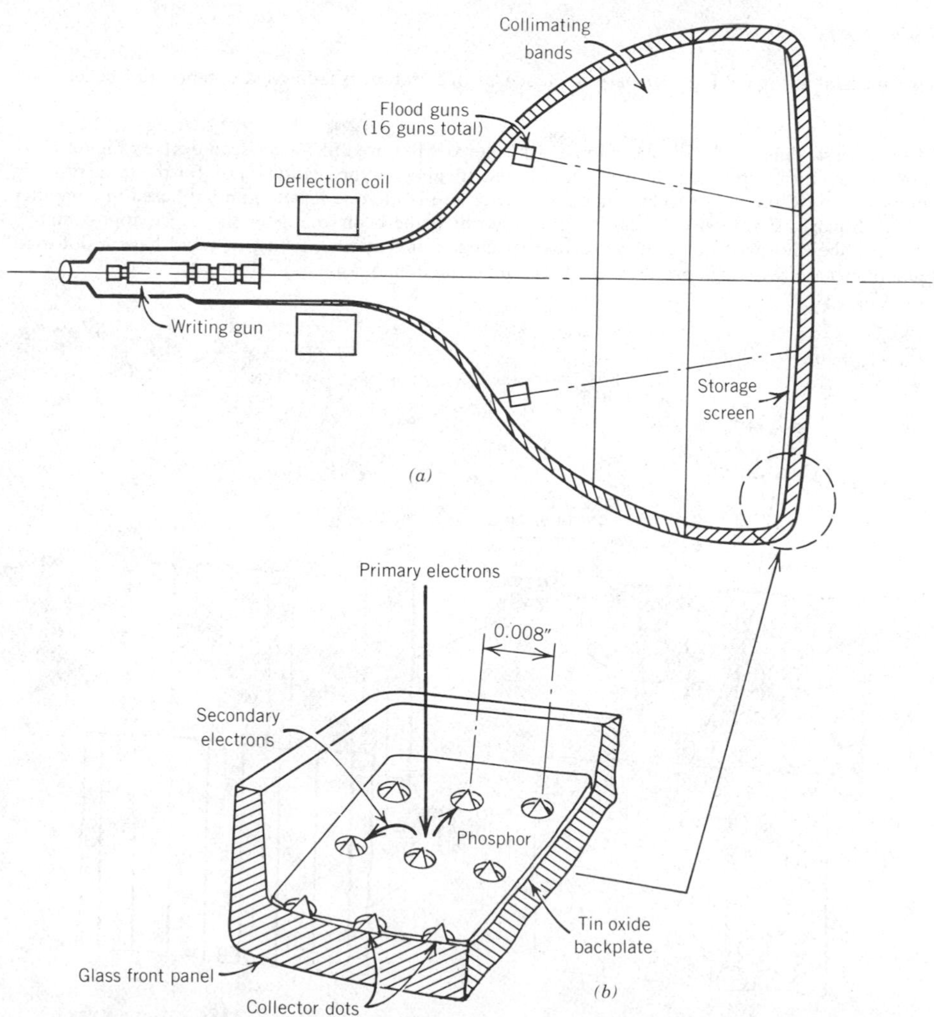

Fig. 15.30 Tektronix bistable storage cathode-ray tube (CRT). (*a*) Direct view storage CRT. (*b*) Screen structure.

Phosphor Life. Pfahnl's law[12] of phosphor aging is given by

$$L_a = \frac{L_0}{1 + CN}$$

where L_a = aged luminance
L_0 = initial luminance
N = number of electrons deposited per square centimeter
C = "burn parameter" (cm^{-2}) corresponding to number of electrons per square centimeter needed to reduce intensity by factor of two

Table 15.6 gives values of C for different phosphors. Browning of the glass faceplate is another life-limiting factor to be considered.

Bistable Storage CRT

This type of display device requires a more complex screen than other CRTs to incorporate a storage mechanism. Fig. 15.30 shows the basic structure of one kind of device, a Tektronix bistable storage tube.[13] In this device the flood gun provides the sustaining electron source needed to continue the excitation of the phosphor after the writing beam has moved on to another location. For a 19-in. diagonal CRT, the number of addressable points is 4096 × 3120.

Color CRTs

Two types of color CRT devices are discussed in this section: shadow-mask tubes and penetration tubes.

Shadow-Mask Tubes.[14–17] The shadow-mask color CRT is used for raster-scan devices. Figure 15.31 shows the relationship of the electron beam to the openings in the shadow mask for the three primary colors in two different gun technologies. The delta gun is the one most commonly used in computer output displays. For good color purity and uniformity, the beam diameter should be approximately 2.5 times the phosphor dot center-to-center spacing. Conventional delta-gun tubes have a dot-triad pitch of 660 μm, with high performance tubes in production at a pitch of 310 μm.

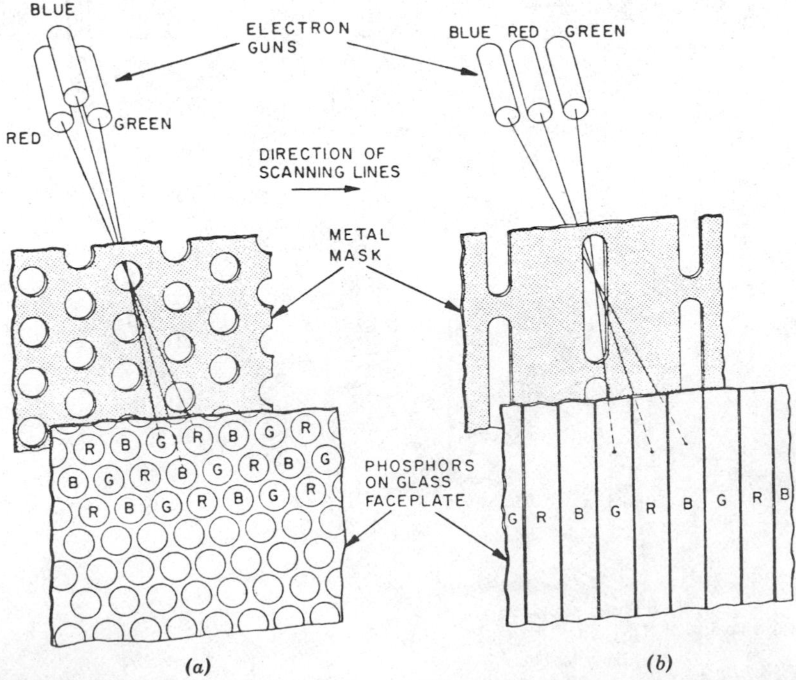

Fig. 15.31 Two shadow-mask technologies. (*a*) Round holes and delta guns. (*b*) Slit openings and in-line guns.

The in-line gun currently dominates in receiver designs. This is due to its advantage over the delta gun design in achieving self-convergence via specially shaped deflection fields which achieve convergence of the three beams over the entire screen area. Conventional in-line tubes have a triad pitch of 840 μm.

Beam Penetration Tubes.[18,19] In this tube technology, multiple (usually two) phosphor types with different characteristic color emissions are selectively excited depending on the value of anode voltage. With a two-phosphor-layer screen tube, the anode voltage is typically switched from 10 to 12 kV to 16 to 18 kV to go from red to green with intermediate values of voltage for orange and yellow (Fig. 15.32).

The amount of information that can be displayed on this kind of tube is limited by the low brightness of the red phosphor. Thus this technology is used in stroke or vector generation systems that have higher duty cycles per picture element compared with raster scan systems.

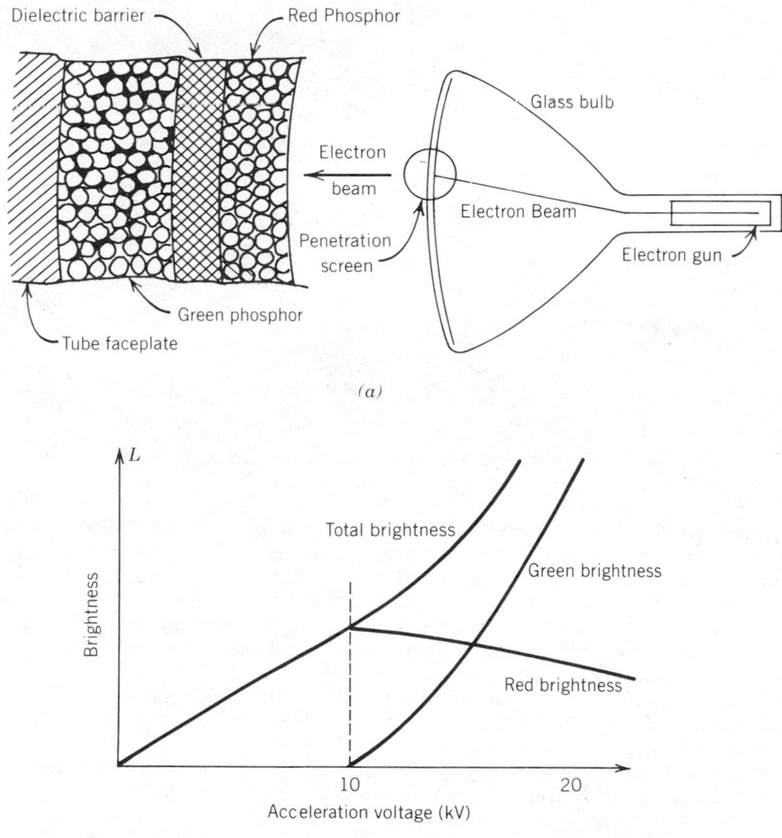

Fig. 15.32 Beam penetration tube and characteristics of phosphor. (*a*) Color penetration tube (simplified). (*b*) Phosphor brightness vs. accelerator voltage, P49 screen. Courtesy of DuMont Division of Thompson-CSF Components Corp.

15.3-3 Liquid Crystal Displays (LCDs)

Materials and Optical Effect[20–23]

Liquid crystal materials are organic compounds with a temperature range within which they exhibit crystalline properties that can be used to produce an optical effect. Characteristic temperature ranges are shown in Fig. 15.33.

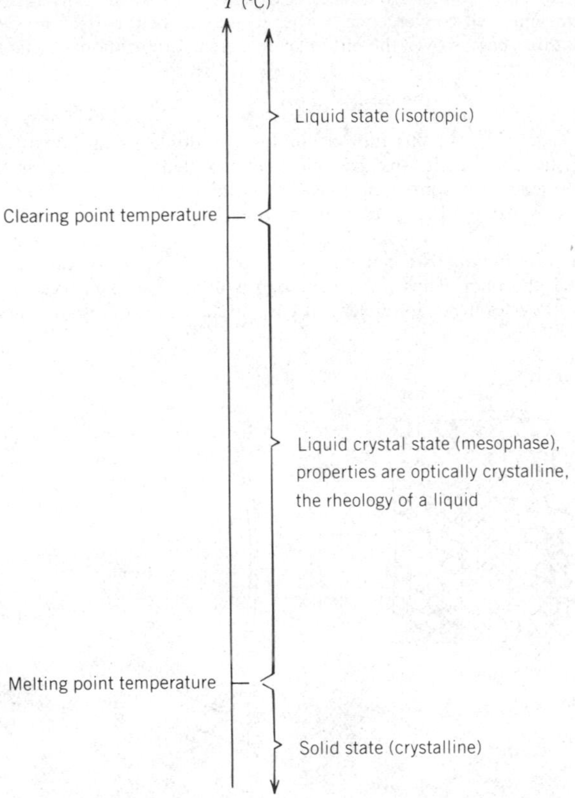

Fig. 15.33 Liquid crystal temperature-phase characteristics.

A device is formed by sandwiching a thin layer (12 μm) of a liquid crystal material between two glass plates, as shown in Fig. 15.34. For surface alignment, either a surfactant is used or the surface of the plates is coated with a ridged dielectric layer such that the molecules at the surface align themselves uniformly and predictably either parallel to the surface (homogeneous alignment) or perpendicular to the surface (homeotropic alignment). The molecules in the rest of the liquid crystal layer then conform to the surface alignment condition and exhibit a smooth progressive change in alignment from one surface to the other if they are aligned differently.

Liquid crystal materials can be synthesized with either positive or negative dielectric anisotropy materials, defined in Fig. 15.35. The alignment of the molecules with an applied field for the two different types of dielectric anisotropy is shown in this figure.

Several modes of operation for producing electrooptical effects can be obtained with a nematic liquid crystal material whose molecular order is shown in Fig. 15.36.

Dynamic Scattering Mode (DSM)[24]

Optical Effect. Applying an electric field across the liquid crystal layer causes the molecules (with negative dielectric anisotropy) to orient parallel to the surface. At the same time their orientation is upset by the current that flows due to the intentional presence of a conductive dopant in the molecular solution. This current flow causes random molecular orientation and thus scattering of incident light, Fig. 15.37.

Device Configuration. The device (Fig. 15.34) consists of two glass substrates, typically 1/8 to 1/16-in. thick with indium-tin oxide transparent conductors on both substrates. For alignment, a surfactant, a rubbing of the substrate, or each substrate is coated with a thin layer of SiO_2 deposited at an oblique angle to produce the desired alignment orientation. Devices are usually aligned homeotropically at the surface with a small angle of departure from the perpendicular. The amount

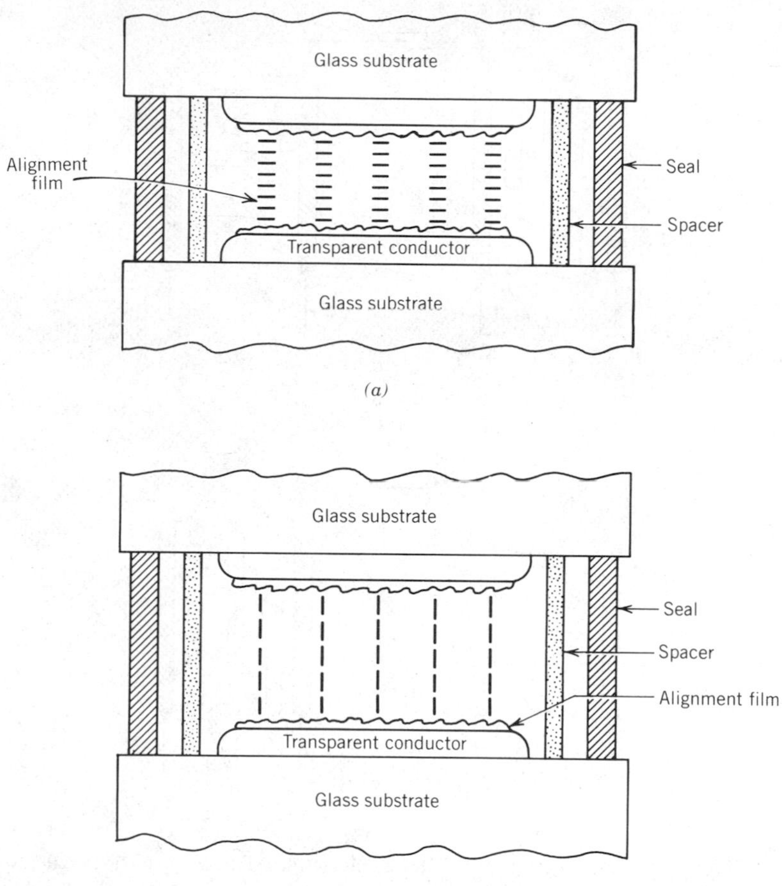

Fig. 15.34 Basic liquid crystal cell with two types of alignment. (*a*) Homogeneous (||) alignment. (*b*) Homeotrophic (⊥) alignment.

of dopant is inversely proportional to the cutoff frequency of the device and is limited by solubility characteristics.

Twisted Nematic Mode (TNM)

Optical Effect. Due to the optical birefringent property of liquid crystals, polarized light interacts with the molecular orientation of the material and polarizers and causes the incident polarized light to be either transmitted, reflected, or blocked. This optical effect does not require the flow of electrical current.

Device Configuration. Figure 15.38 shows relative orientations for reflective and transmissive cells for a light-on-dark display or a dark-on-light display.

Guest-Host Mode (GHM)

Optical Effect. The anisotropic absorption property of the pleochroic dye (guest) molecules mixed with a nematic (host) liquid crystal material, Fig. 15.39, produces the different optical states of the cell. The dye molecules align themselves parallel to the host molecules, whose orientation changes with the application of an electric field.

The addition of about 5% of a cholesteric compound[25] to the mixture, together with homeotropic surface alignment, eliminates the need for a polarizer.

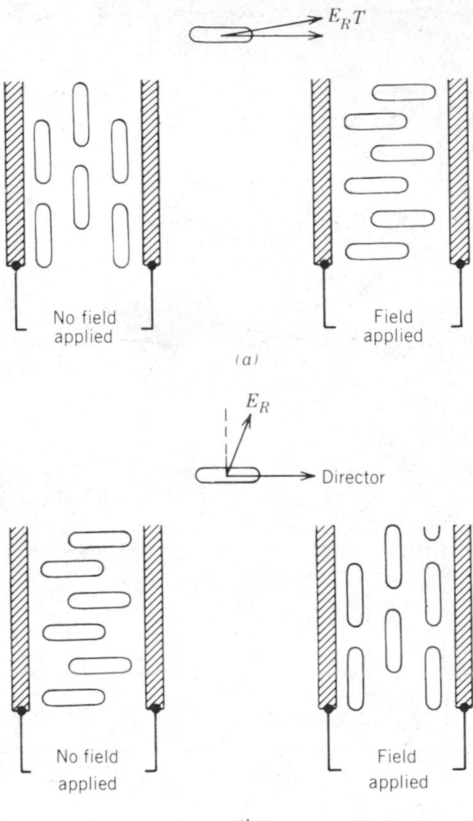

Fig. 15.35 Positive (*a*) and negative (*b*) dielectric anisotropy liquid crystal material response to applied voltage.

Device Configuration. The display cell is shown in simplified form in Fig. 15.39, with transparent electrodes on two glass substrates, spaced about 10 μm apart. The nematic host material has a positive dielectric anisotropy, which causes the molecules to orient parallel to an applied electric field and effects minimum absorption by the dye molecules which align parallel to the host molecules. With no applied field, the cholesteric additive gives the LC layer a twisted structure which produces maximum absorption and appears dark. The molecular surface orientation is perpendicular.

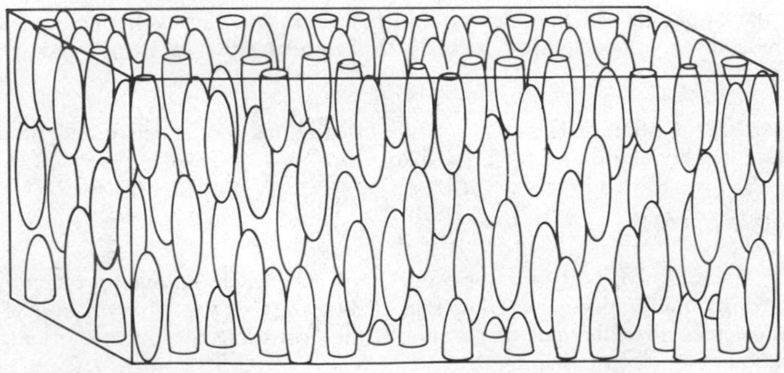

Fig. 15.36 Molecular order of nematic liquid crystal material.

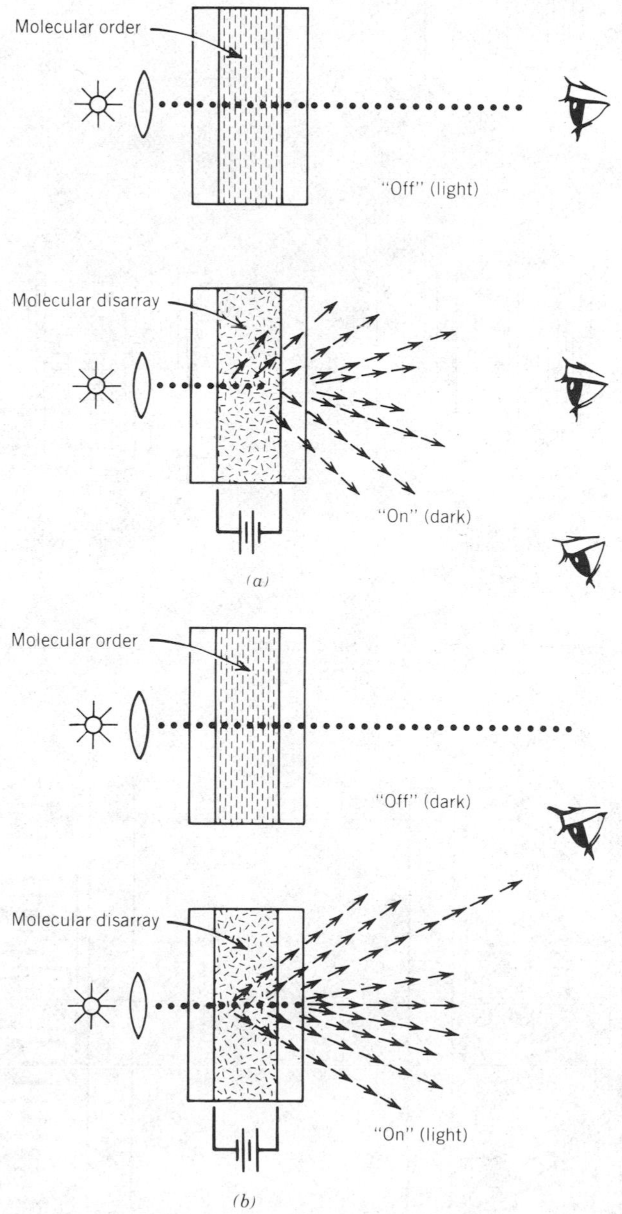

Fig. 15.37 Dynamic scattering transmissive modes of operation of liquid crystal displays. (*a*) specular mode. (*b*) Diffusing mode.

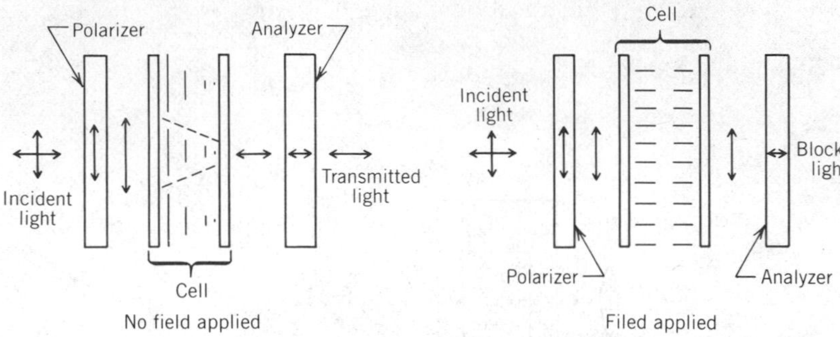

Polarizer
Analyzer
Incident
light
Transmitted
light
Cell
No field applied
Cell
Incident
light
Blocked
light
Polarizer
Analyzer
Filed applied

(a)

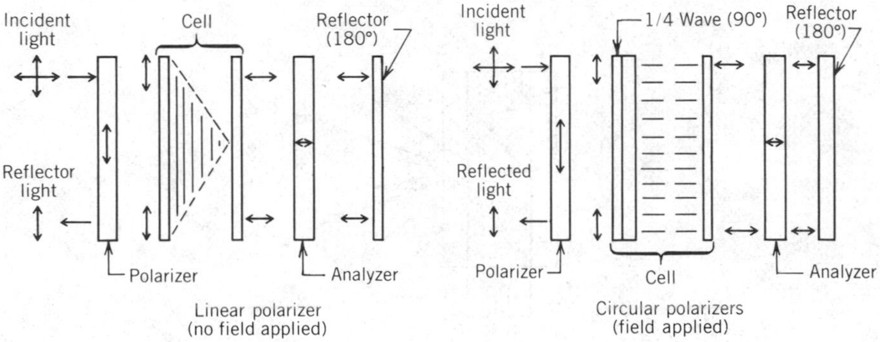

Incident
light
Cell
Reflector
(180°)
Reflector
light
Polarizer
Analyzer
Linear polarizer
(no field applied)
Incident
light
1/4 Wave (90°)
Reflector
(180°)
Reflected
light
Polarizer
Cell
Analyzer
Circular polarizers
(field applied)

(b)

Fig. 15.38 Reflective and transmissive twisted nematic mode (TNM) cells: (*a*) Transmissive. (*b*) Reflective.

Homeotropic surface alignment
Cholesteric phase

(a) (b)

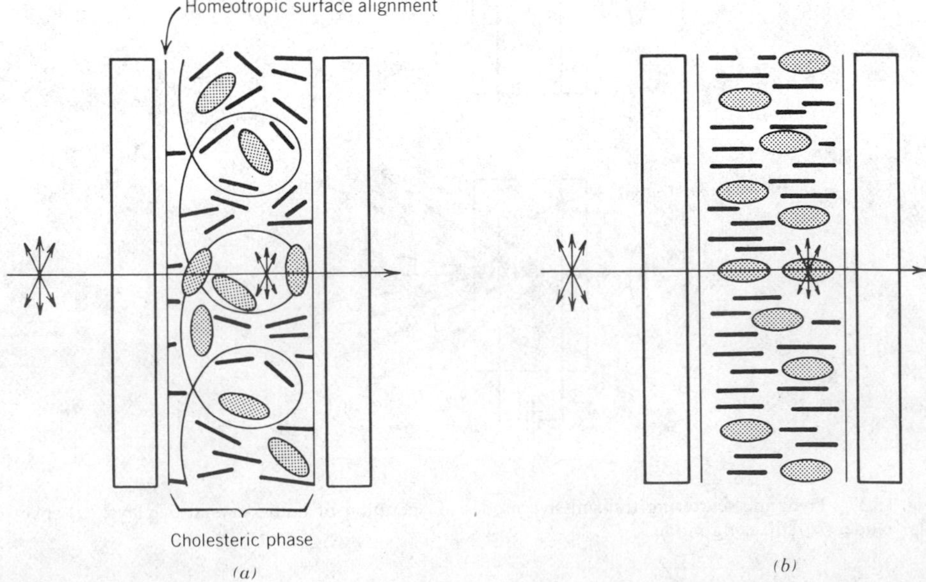

Fig. 15.39 Optical effect of guest-host mode. (*a*) Unenergized state. All polarizations of incoming light are strongly attenuated. (*b*) Energized state. All polarizations of incoming light are transmitted with little attenuation.

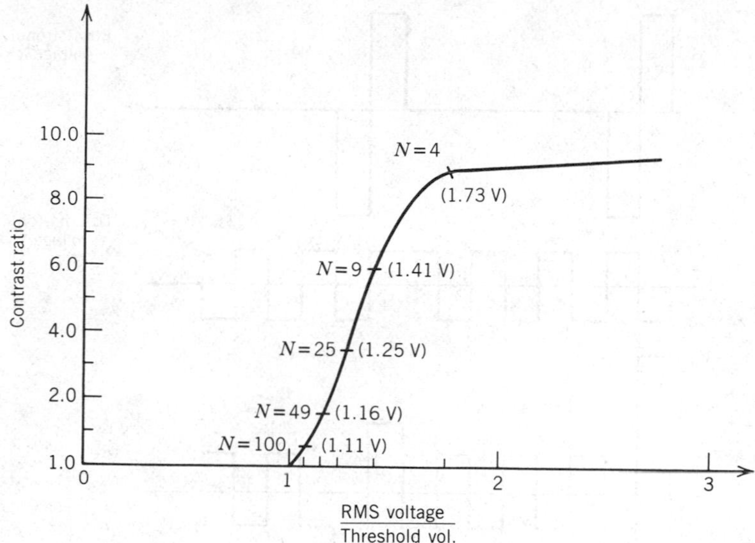

Fig. 15.40 Liquid crystal display contrast ratio vs. normalized root-mean-square (RMS) applied-voltage characteristic.

Performance Characteristics

Resolution. Resolutions of 30 to 50 lines per inch have been reported, with up to 200 scanned lines.

Viewing Angle. The electrooptical effects of all three modes of operation of LCDs can be characterized by the contrast voltage curve shown in Fig. 15.40. It should be noted that the voltage axis represents the root-mean-square (RMS) applied voltage. This characteristic varies with cell spacing, temperature, and angle of light measurement. The angular distribution of light for an LCD cell is also a function of drive voltage.

Power. The typical power requirement for a DSM LCD is 10 $\mu A/cm^2$ (DC) at 1×10^6 V/m. The TNM field-effect type of LCD typically requires 1 $\mu A/cm^2$ at 0.5×10^6 V/m. The GHM-type requirements are between those of the other two types. DC current flow through an LCD cell causes severe degradation due to electrolysis.

Frequency and Time Response. The critical (cutoff) frequency f_c, in hertz, for a dynamic scattering liquid crystal material is related to the resistivity ρ (ohm-cm), as[24]

$$f_c = \frac{1.05 \times 10^{12}}{\rho}$$

Typical response times and their dependence on both applied voltage and cell thickness are given by

$$t_{\text{rise}} \propto \frac{\eta d^2}{(V_a - V_t)^2}$$

where η = viscosity
d = cell thickness
V_a = applied voltage
V_t = threshold voltage of LC

$$t_{\text{fall}} \propto \eta d^2$$

Typical response times are in the 100 to 200 ms range at room temperature (25°C).

Multiplexing.[26] Referring to Fig. 15.41, RMS scanning waveforms are shown where, to maximize the duty cycle, the strobe waveforms should be applied to the lines on the smaller dimension of the matrix of display elements to be multiplexed.

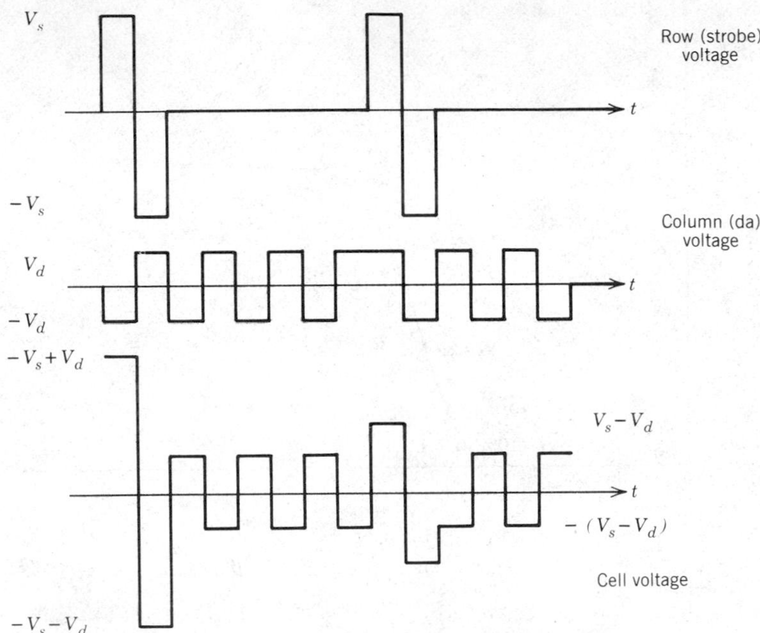

Fig. 15.41 Liquid crystal scanning waveforms based on RMS voltage criteria.

The effect of duty cycle, $1/N$, on contrast is shown in Fig. 15.40. The contrast ratio vs. RMS applied-voltage curve for a given device determines what voltage V_{on} is required to achieve the minimum acceptable contrast. From this same characteristic the threshold voltage V_{off} (V_t used previously) is determined. These two voltages fix the maximum number of scanned lined that can be driven to meet the contrast requirement. The maximum number of scanned lines is given by

$$N = \left[\frac{Q^2 + 1}{Q^2 - 1} \right]^2$$

where $Q = V_{on}/V_{off}$. The voltage levels to achieve this number of scanned lines are given by

$$V_d = V_{off} \left[\frac{Q^2 + 1}{2} \right]^{1/2}$$

$$V_s = \sqrt{N}\, V_d$$

Practical limitations due to tolerances seem to limit the number of scanned lines N to between 60 and 120. With use of a double scanned matrix,[27] the number of lines in the scan dimension can be doubled at the expense of additional drive circuits and a more complex conductor pattern.

The rise-time response of a multiplexed display increases by approximately $\sqrt{N}$, producing a slowly developing panel image.

Temperature Effects. The TNM cell threshold voltage decreases by 0.5% per degree centigrade. (Temperature has no effect on the DSM cell threshold voltage.) The rise and fall times are strong functions of the temperature due to the temperature-viscosity characteristics of the LC material. Typical effects of temperature on response times would be such that a fall time value of 140 ms at 25°C would become 500 ms at 0°C.

Panel Life. The life of LC devices is limited by the surface alignment life and by penetration of moisture through the cell seal material. With evaporated alignment layers and hermetic glass frit seals, life is quoted as 30 to 50 khr.

15.3-4 Light-Emitting Diode Displays (LEDs)[22,28]

Materials and Optical Effect

The light-emitting diode (LED) is a semiconductor *pn*-junction device capable of light emission in the visible spectrum under forward-bias operation. The basic operating mechanism is injection luminescence, which produces visible radiation by a two-step process. The injection of minority carriers across the junction, Fig. 15.42, is followed by radiative recombination.

The wavelength λ of the emitted radiation is related to the bandgap E (in electron volts, eV) of the

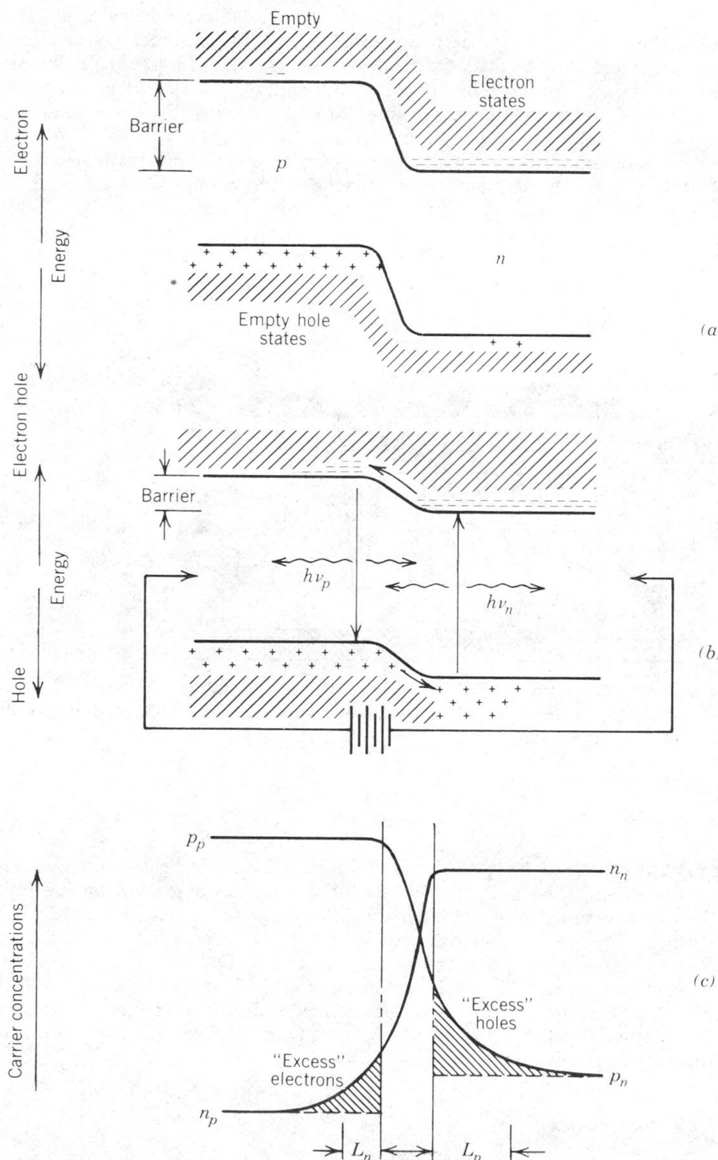

Fig. 15.42 Light-emitting diode *pn*-junction operation: (*a*) zero bias; (*b*) forward bias reduces potential barrier height; (*c*) forward bias with majority and minority carrier concentrations on *n* and *p* sides. From Craford,[29] with permission.

semiconductor via

$$\lambda = \frac{12{,}600}{E}\ \text{Å}$$

For radiation in the visible range, that is, from 4000 to 7200 Å, the bandgap has to lie in the range of 1.75 to 3.15 eV. Because many semiconductors in the III-V group have bandgaps in the visible range, they predominate in LED manufacture.

Device Configuration

LED fabrication technology is similar to integrated circuit technology, utilizing epitaxial deposition, selective diffusion of impurities, and photolithographic delineation of metallurgy to form individual or arrays of devices. Figure 15.43 shows a typical device configuration for a ternary alloy $GaAs_{1-x}P_x$ device. This device is formed from two binary compound materials, one of which has a low bandgap (GaAs) and the other having a large bandgap (GaP). By varying the proportion of the alloy constituents, the bandgap can be monotonically varied from the value of 1.44 eV for GaAs to the value of 2.26 eV corresponding to that of GaP. This alloy behaves as a direct bandgap semiconductor for values of $X \leqslant 0.46$ ($E = 1.99$ eV), yielding an efficient emitter in the visible range.

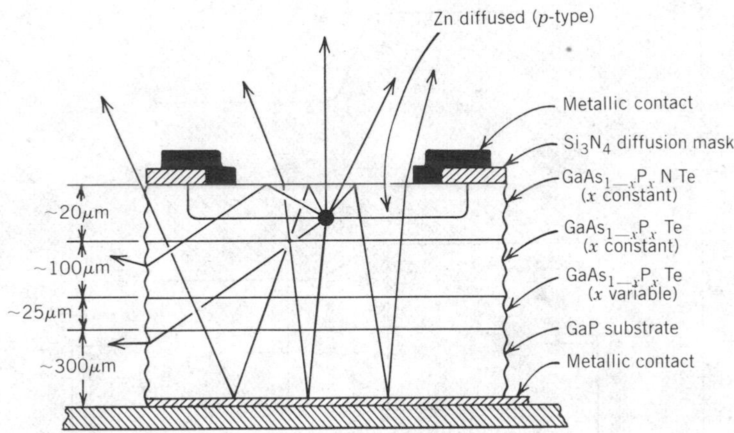

Fig. 15.43 Schematic cross section of $GaAs_{1-x}P_x$ device structure. Nitrogen doping produces orange, yellow, or green emission. Light emitted downward is reflected from the metallic contact. From Craford,[29] with permission.

Device Characteristics

LEDs have a current-voltage (I-V) relationship similar to conventional semiconductor diodes.

Typical I-V curves for $GaAs_{1-x}P_x$ diodes with different values of x are shown in Fig. 15.44. As can be seen, the forward voltage drop across the diode increases with increasing bandgap. Table 15.7 lists the electrical and optical characteristics of several LEDs. Operating these diodes above the reverse breakdown voltage generally results in severe degradation of the reverse characteristic.

The luminance-current density (L-J) characteristic of an LED is shown in Fig. 15.45. Over most of the range, this characteristic is fairly linear with a slope β in nits per amps per square centimeter with a practically negligible offset at low currents. At high current levels the efficiency of recombination starts to drop off due to thermal effects.

Performance Characteristics

Luminance. The typical luminance-current density characteristic with luminance expressed in nits shown in Fig. 15.45 can be approximated over its linear range by

$$L = \beta \frac{I}{A_j} \qquad \frac{(\text{ma})}{(\text{cm}^2)}$$

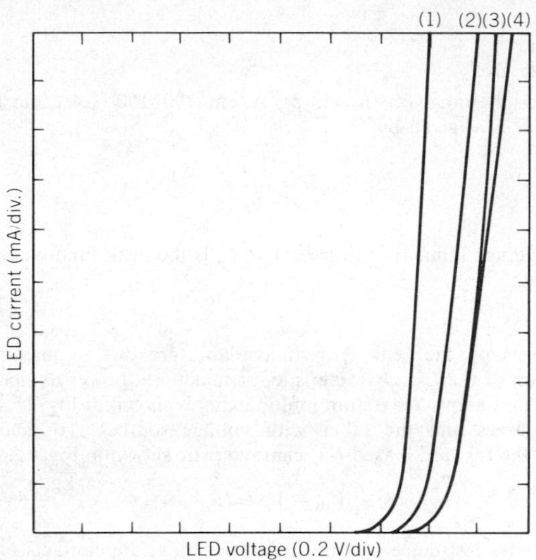

Fig. 15.44 Current-voltage characteristics for various color emitters. (1) Red, $x = 0.4$. (2) Orange, $x = 0.65$. (3) Yellow, $x = 0.86$. (4) Green, $x = 1.0$. From Craford,[29] with permission.

TABLE 15.7. $GaAs_{1-x}P_x$ LIGHT-EMITTING DIODE CHARACTERISTICS

x	Color	Peak Wavelength, λ (Å)	Threshold, V_T (V)
0.40	Red	6490	1.4
0.65	Orange	6320	1.5
0.86	Yellow	5890	1.6
1.00	Green	5650	1.6

Source. M. G. Craford, *Proc. SID* **18**:2 (1977), with permission.

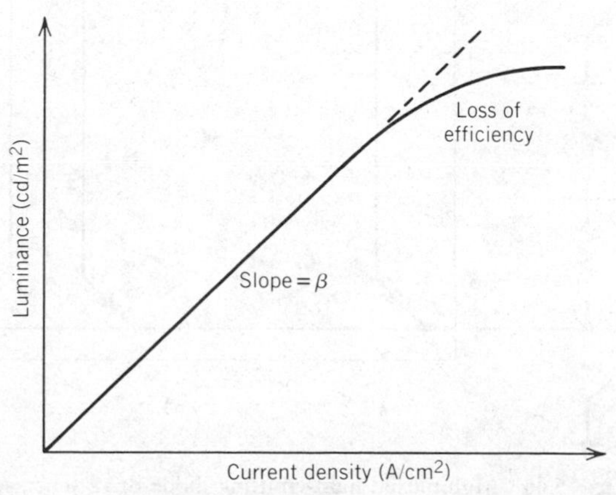

Fig. 15.45 Light-emitting diode luminance-current density characteristic.

where β = slope
I = diode current
A_j = diode junction area

Typical values of β lie in the range 308 to 513 nt/A/cm^2 (90–150 fL/A/cm^2).[29]

The diode duty factor D is given by

$$D = \frac{L_{av}}{L_p}$$

where L_{av} is the time-average luminance desired and L_p is the peak luminance corresponding to the peak diode drive current.

Multiplexing. Limitations on the peak current available are due to maximum allowable power supply voltage, saturation of the L-I characteristic, or maximum power dissipation.

The drive condition that allows maximum multiplexing is shown in Fig. 15.46 in which one axis is driven with current sources and the other with voltage sources. To determine the maximum multiplexing capability, the forward biased I-V characteristic shown in Fig. 15.44 is approximated as

$$V_{fb} = V_t + IR_d$$

where R_d is the diode series resistance (in ohms) that occurs at the contacts to the diode and across the bulk semiconductor layers of the diode and I is the forward current through the diode. This resistance is in the 1 to 10 Ω range.[28]

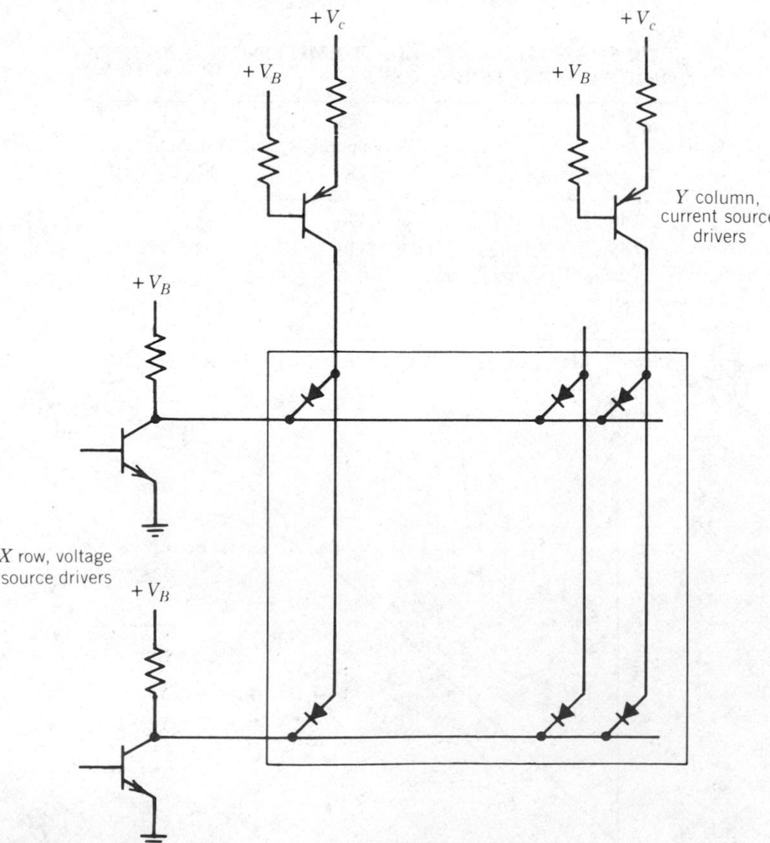

Fig. 15.46 Multiplexed light-emitting diode drive scheme.

The maximum forward-bias voltage across the diode is limited by the reverse-breakdown voltage of the diode V_B, and results in

$$(I_p)_{\max} \cong \frac{V_B}{R_d}$$

The number of lines N that can be scanned in a multiplexed array is related to the refresh period T_R and the time required to load a line of data T_L

$$N = \frac{T_R}{DT_R + T_L} = \frac{1}{L_{av}/L_p + T_L/T_R}$$

Combining all previous results for L_p and I_p

$$N_{\max} = \frac{1}{(L_{av}R_dA_j)/(\beta V_B) + T_L/T_R}$$

This constraint does not apply to directly driven LEDs, that is, LEDs driven individually.

Another limitation on the maximum size of an LED matrix array (m rows and n columns) that can be driven is basically due to the maximum current each one of the m horizontal drivers has to absorb from n diodes driven in parallel. This current is limited by driver-device current capability and by ohmic drops along the lines interconnecting the diodes.

Temperature Effects. Both the diode luminous efficacy and the forward voltage vary with temperature.

Typical temperature coefficients of luminance are $-1.0\%/°C$. The temperature coefficient of the forward voltage drop is typically -2.0 mV/°C; the wavelength of the radiation also changes with temperature.

Response Times. The response times of an LED are in the 10- to 10,000-ns range. The turn-off time T_{off} of an LED is the slower of the two response times and is dependent on the charge storage time constant τ_f and the ratio of forward I_f to reverse I_r current, as in a conventional diode

$$T_{off} = \frac{I_f}{I_R}\tau_f$$

Life. The life of LED devices is generally limited by bond failures caused by vibration, corrosion, and flexing due to temperature excursions. Depending on the packaging, life is quoted from 50 to 250 khr. With good bond and package technology, life is limited by surface leakage, diffusion of contaminants, and the formation of intrinsic nonradiative recombination centers.

15.3-5 Plasma Displays (PDPs)

Materials and Optical Effect

When a voltage exceeding the gas breakdown voltage is applied across a gas-discharge cell, ionization takes place and a cathode glow (fall region) is formed, across which the major part of the applied voltage appears. To minimize the value of breakdown voltage, Penning mixtures are used, that is, mixtures with a trace amount of an additive gas whose ionization potential is approximately equal to (but less than) the metastable potential of the host gas.

Neon is used as the host gas for light emission in the visible range (5850 Å), with argon and xenon as Penning additives for high efficacy. When the gas breaks down, argon and neon ions generated in the discharge bombard the cathode surface, causing secondary electrons to be emitted, which sustains the discharge. During this process, neon atoms are excited, and as they decay to their unexcited state they produce the characteristic neon-orange emission.

For neon–argon, the breakdown voltage is fairly constant in the range of 0.1 to 1.0% argon, with the highest luminous efficacy at 0.1%. The breakdown voltage of a DC plasma cell is given by[30]

$$V_B = \frac{(V_B)_{\min}}{\ln\left[\dfrac{2.718\,pd}{pd_n}\right]}\left(\frac{pd}{pd_n}\right)$$

where p is the gas pressure in torrs (or millimeters of mercury) and d is the gap in centimeters. The quantity pd_n is a characteristic cathode fall constant. For neon plus 0.5% argon it is equal to 1.44

TABLE 15.8. LN (1/Γ) VALUES FOR VARIOUS CATHODE MATERIALS WITH ARGON ION IMPINGING ON THE CATHODE

Cathode Material	ln (1/γ)
Mo (sputtered)	2.6
Ni (sputtered)	3.6
Mg	2.45
MgO	0.562
CeO_2	0.89
La_2O_3	0.59

torr-cm. The minimum breakdown voltage $(V_B)_{min}$ is given by[31]

$$(V_B)_{\min} = V_i + \frac{1}{\eta_m}\ln(1/\gamma)$$

where V_i = gas ionization voltage, 16.6 V
η_m = gas ionization efficiency coefficient, 0.029
γ = secondary emission coefficient of cathode material for neon plus 0.5% argon

Published values of $\ln(1/\gamma)$ for several electrode materials and neon plus 0.5% argon are listed in Table 15.8.

DC Plasma (Gas Discharge) Display[32]

Device Configuration and Characteristics. In this display device the electrodes are directly in contact (DC coupled) with the gas. A simple device configuration is shown in Fig. 15.47.

Performance Characteristics

Multiplexing. Most DC plasma display panels are operated in a refresh mode. For a refresh frame time of T_R seconds and a device whose ionization time T_i is short compared with the pulsewidth DT_R, that is

$$DT_R > T_i$$

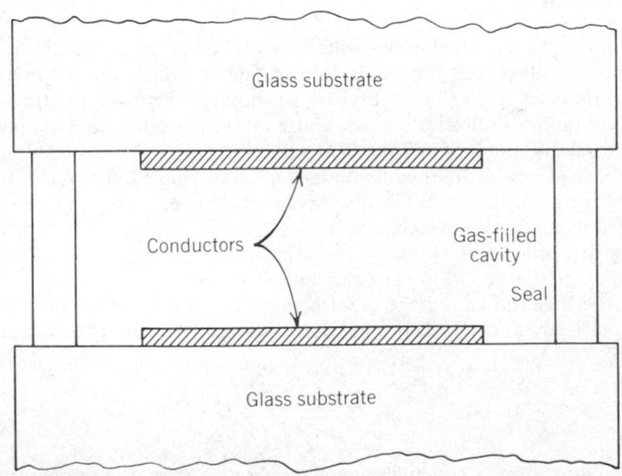

Fig. 15.47 DC plasma display cell cross section.

the average luminance is given by

$$L_{av} = L_p D$$

where L_p is the peak luminance and D is the duty factor. The peak luminance is proportional to the peak current, that is,

$$L_p = KI_p$$

Therefore the duty factor is given by

$$D = \frac{L_{av}}{KI_p}$$

Taking into account the duty factor and the data load time T_L with a refresh period T_R, the value of the number of scanned lines N is given by

$$N = \frac{1}{D + T_L/T_R}$$

where T_L is calculated by multiplying the number of Y lines by the input data shift rate. Typical values of N are 200 to 300 lines, with a time-averaged brightness of 86 to 172 nt (25–50 fL).

Resolution. The resolution limit of a DC plasma display is limited fundamentally by the requirement that the physical gap allow for a cathode fall region and practically by manufacturing tolerances. Panels currently in production have a resolution of 2 lines/mm (50 line/in.).

Panel Life. In DC plasma displays, panel life is limited by cathode sputtering. The sputtering rate at a given pressure increases superlinearly with peak current. Mercury vapor is added to extend cathode life, and panel life is specified at 30 to 150 khr.

AC Plasma Display[33]

Device Configuration and Characteristics. This device is AC coupled to the discharge, as shown in Fig. 15.48. As a consequence, the applied-bias waveform has to be an AC-"sustain" waveform, as shown in Fig. 15.49. With every alternation of the sustain waveform, the sustain voltage is of such polarity as to add algebraically to the wall voltage developed on the cell walls by the discharge produced on the previous cycle, if the cell was once previously written. The sum of the wall voltage and applied sustain voltage is then sufficient to cause the cell to fire again. The gas breakdown then causes current to flow, which charges up the wall capacitance in opposition to the applied voltage and extinguishes the discharge. However, on the next alternation, the applied voltage is again algebra-

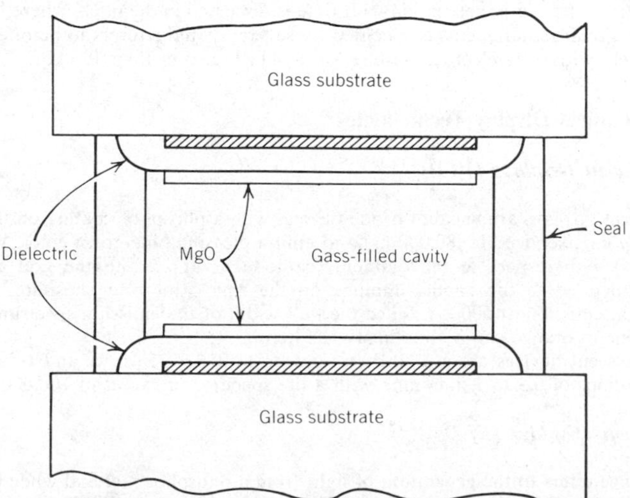

Fig. 15.48 AC plasma display cell cross section.

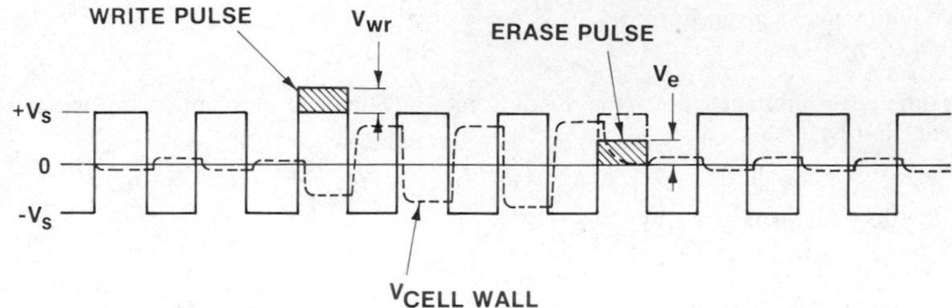

Fig. 15.49 Sustain, write, erase, and wall voltage waveforms across an AC plasma display cell.

ically additive to the wall voltage and causes the cell to fire. This is the memory mechanism of the AC plasma display cell.

The typical cell parameters for a 2 line/mm (50 line/in.) display are a 25-μm dielectric thickness with 2.32 pF/mm^2, a magnesium oxide overcoat material that has a secondary emission coefficient in the range 0.5 to 0.6, and a gap of 100 to 125 μm. The gas mixture is typically neon plus 0.1% argon at 350 torr pressure.

The characteristic voltages of this cell are the minimum sustain voltage $(V_s)_{min}$ and the maximum sustain voltage $(V_s)_{max}$. The minimum sustain voltage is that value of voltage at which a cell, once written, will remain on. The maximum sustain voltage is that value of voltage at which a cell, once erased, will remain off.

Performance Characteristics

Luminance. The average luminance L_{av} is proportional to the time-integral of the current and therefore proportional to the sustain frequency f_{sus}, in hertz, the wall capacitance C_{wall}, and the sustain voltage operating level $(V_s)_{op}$, in the operating range between $(V_s)_{min}$ and $(V_s)_{max}$, that is

$$L_{av} \propto f_{sus} C_{wall}[(V_s)_{op} - (V_s)_{min}]$$

Typical values of time-averaged luminance are 100 nt (30 fL) at a 40-kHz sustain frequency.

Resolution. Resolution of up to 4.8 lines/mm (120 lines/in.) have been demonstrated and 720 × 432 line panels with 2.8 lines/mm (70 lines/in.) are in manufacture.[17]

Temperature and Altitude. Changes in ambient temperature and altitude affect the gas pressure and panel gap. With a temperature rise from 20 to 60°C, operating voltages typically change +0.43%.

Panel Life. The reduction of panel margin with time is discussed in detail elsewhere.[34] Panel lifetime as determined by circuit requirements is specified by several manufacturers to be in excess of 30,000 power-on hr (POH), with 0.1% electrical failure, or a MTBF greater than 100 khr.

15.3-6 Other Optical Display Technologies

Vacuum Fluorescent Displays (VFDs)[35,36]

Vacuum fluorescent displays are vacuum triode devices with a phosphor coating on the anode. They have an extremely long service life (80,000 hr) and emit a pleasing blue-green color. When a positive voltage (10–40 V) with respect to the filament (cathode) is applied to the grid and anode, the electrons are accelerated to the anode, impinge on the zinc oxide-zinc phosphor, and excite the phosphor into light emission at 5000 Å. Since the half width of the emission spectrum is about 1000 Å, colors from blue to orange can be obtained via filtering.

Vacuum fluorescent devices are available in sizes up to 256 × 256 rows and columns of picture elements at resolutions of 0.5 to 1 line/mm, with a life specified in excess of 100,000 hr.

Electroluminescent Displays (ELDs)[17,28]

Electroluminescence refers to the generation of light from a phosphor material when an electric field is placed across it. The material most used in AC electroluminescence is thin-film zinc sulfide. AC electroluminescence development has made more progress than DC electroluminescence. For AC EL devices, thin films of zinc sulfide with a manganese activator (< 1.0%) are used. For DC EL devices,

TABLE 15.9. DISPLAY TECHNOLOGY ATTRIBUTES AND APPLICATIONS

	Cathode-Ray Tube (CRT) Refresh	Light-Emitting Diode (LED)	Plasma Display		Liquid Crystal Display (LCD) Twisted Nematic Mode	Vacuum Fluorescent Display (VFD)	Electroluminescent Display (ELD)
			AC	DC			
Attributes							
No. of lines	525–800	100 × 100	1024 × 1024	320 × 240	175 × 175	200 × 42	180 × 240
Resolution (line/mm)	1–4	1–2	2.4–3.32	0.64–2	4	0.5–1	2–4
Luminance (nt)	170–500	35–350	35	35	Reflective	35	100
Contrast	50	10	30	30	12	10	20
Color	Full	R, O, Y, G	Orange	O, G, R, B		B–G	O–Y
Efficacy (lm/W)	2–7	$0.06^{R,G}$	0.1–0.3	0.07 (0.2)	NA	5.2	1.0
Life (MTBF[a])	> 10 khr	50–250 khr	> 100 khr	30–150 khr	30–50 khr	> 100 khr	> 100 khr
Applications							
Single character			X	X	X	X	X
One line	X	X	X	X	X	X	X
Multiline	X	X	X	X	X	X	X
Full page	X		X				
Multipage	X		X				

[a] MTBF, mean time between failures.

zinc sulfide powder phosphors with a copper activator are used. These devices are emissive (active) and require transparent electrodes. Resolutions of 4 lines/mm (100 line/in.) are currently being developed. The emission is yellow. The voltages required for this device are in the 200 to 300-V range.

Electrochromic Displays (ECDs)[21]

Electrochromism refers to materials that exhibit a change in the absorption spectrum of molecules as a function of electrolysis, oxidization, and reduction. Various inorganic and organic materials exhibit solid-solid (WO_3), liquid-solid (viologens), or liquid-liquid phase transitions. These are low-voltage (1.5 V) and relatively high-current nonemissive (passive) display devices.

15.3-7 Display Technology Attributes and Applications

Some display technology attributes and applications are shown in Table 15.9.

References

1 D. Judd, B. Judd, and Gunther Wyszecki, *Color in Business, Science, and Industry*, 3rd ed., Wiley, New York, 1975, p. 298.

2 S. L. Smith, "Letter Size and Legibility," *Human Factors* **21**(6):661–670 (1979).

3 A. T. Buckler, *A Review of the Literature on the Legibility of Alphanumerics on Electronic Displays*, U.S. Govt. Report No. ADA 040625, May 1977.

4 H. L. Snyder, in L. M. Biberman, ed., *Perception of Displayed Information*, Plenum, New York, 1973, p. 97.

5 S. Sherr, *Electronic Displays*, Wiley, New York, 1979.

6 A. Sobel, "Some Constraints on the Operation of Matrix Displays," *IEEE Trans. Elect. Dev.*, **ED-18**(9) (1971).

7 N. H. Lehrer, "CRT's: An Overview of Their Performance," Conference Record 1980 Biennial Display Research Conference, pp. 114–119.

8 H. Moss, *Narrow Angle Electron Guns and Cathode Ray Tubes*, Academic, New York, 1968.

9 H. C. Masterman, "Know Deflection Yoke Specs," *Elect. Des.* (Feb. 15, 1979).

10 F. Christiansen and H. Patil, "Optimizing CRT's for Display Applications," *Proc. SID* **20**(2):89–93 (1979).

11 F. G. Oess, "CRT Considerations for Raster Dot Alphanumeric Presentations," *Proc. SID* **20**(2):81–88 (1979).

12 A. Pfahnl, "Aging of Electronic Phosphors in Cathode Ray Tubes," *Adv. Elect. Tube Tech.*, pp. 204–208 (Sep. 1960).

13 C. Curtin et al., "Large Screen Display for Bistable Storage of Up to 17,000 Characters," *Int. Symp. Digest*, pp. 98–99 (1973).

14 A. M. Morrell, "Color Picture Tube Design Trends," *Proc. SID* **22**(1):3–9 (1981).

15 A. M. Morrell et al., *Color Television Picture Tubes*, Academic, New York, 1974.

16 M. Takata and R. Hirai, "Color Computer Display Tubes," *Int. Symp. Dig.*, pp. 164–165 (1980).

17 W. F. Goede, "A Review of Display Technology in Japan," *Information Display*, pp. 8–10 (Nov. 1980).

18 J. Bun, "Comparative Evaluation of High Resolution Color CRT's," *Int. Symp. Dig.*, pp. 78–79 (1980).

19 J. P. Galves, "Multicolor and Multipersistence Penetration Screens," *Proc. SID* **29**(2):95–104 (1979).

20 G. J. Sprokel, ed., *The Physics and Chemistry of Liquid Crystal Devices*, Plenum, New York, 1979.

21 A. R. Kmetz and F. K. von Willisen, eds., *Nonemissive Electrooptic Displays*, Plenum, New York, 1976.

22 J. I. Pankove, ed., *Topics in Applied Physics: Display Devices*, Vol. 40, Springer-Verlag, New York, 1980.

23 E. P. Raynes, "Recent Advances in Liquid Crystal Materials and Display Devices," Conference Record of 1978 Biennial Display Research Conference, pp. 8–11.

24 Z. Blank et al., "Design of a Dynamic Scattering Liquid Crystal Material System for Multiplex Operation," Conference Record of 1978 Biennial Display Research Conference, pp. 44–51.

25 D. L. White and G. N. Taylor, "A New Absorptive Mode Reflective Liquid Crystal Display Device," *J. Appl. Physics* **45**:4718–4723 (1974).

26 P. M. Alt and P. Pleshko, "Scanning Limitations of Liquid Crystal Displays," *IEEE Trans. Elect. Dev.* **ED-2**(2):146–155 (Feb. 1974).

27 E. Kaneko et al., "Liquid Crystal Television Display," 1978 *Int. Symp. Dig.*, pp. 92–93 (1978).

28 J. I. Pankove, ed., *Topics in Applied Physics: Electroluminescence*, Vol. 17, Springer-Verlag, New York, 1977.

29 M. G. Craford, "Recent Developments in Light Emitting Diode Technology," Conference Record of 1976 Biennial Display Research Conference, p. 66.

30 J. D. Cobine, *Gaseous Conductors—Theory and Engineering Applications*, Dover, New York, 1958.

31 J. R. Acton and J. D. Swift, *Cold Cathode Discharge Tubes*, Academic, New York, 1963.

32 R. Cola et al., in B. Kazan, ed., *Advances in Image Pickup and Display*, Vol. 3, Academic, New York, 1977.

33 T. N. Criscimagna and P. Pleshko, in J. I. Pankove, ed., *Topics in Applied Physics: Display Devices*, Vol. 40, Springer-Verlag, New York, 1980.

34 P. Pleshko, "AC Plasma Display Panel Aging Model and Lifetime Calculations," *IEEE Trans. Elect. Dev.* **ED-28**(6) (June 1981).

35 K. Kasano et al., "A 240-Character Vacuum Fluorescent Display and Its Drive Circuitry," *Proc. SID* **21**(2):107–112 (1980).

36 T. Nakamura, "Itron VFD's Become Word Processing Displays," *J. Electr. Eng.*, pp. 57–60 (May 1980).

15.4 OPTICAL SENSORS

L. S. Watkins

Light is an electromagnetic wave and exhibits wavelike properties such as interference, refraction, and reflection at interfaces with difference density media, as discussed in Section 15.1. When interacting electronically with a medium, the light must be quantized according to Planck's theory. The light quantum is called a photon and has energy

$$E = h\nu \tag{15.31}$$

ν is the light frequency and h is Planck's constant, 6.56×10^{-34} J-s. The energy of each photon is very small and it gets smaller with larger wavelengths and vice versa.

15.4-1 Vacuum Photodevices

Vacuum photodevices make use of the photoemissive effect, where light falling on a surface causes emission of electrons. The electrons are collected at a positively biased anode. Figure 15.50 shows the energy diagram for a metal at 0K. The electron in the level closest to the Fermi level will be ejected from the metal surface if it receives a minimum amount of energy, equal to the work function ϕ for the material. Thus

$$E > \phi \tag{15.32}$$

for the particular photoemissive surface, and the work function determines the longest wavelength at which a photoemissive detector will operate. Figure 15.51 shows some typical photocathode spectral response curves.

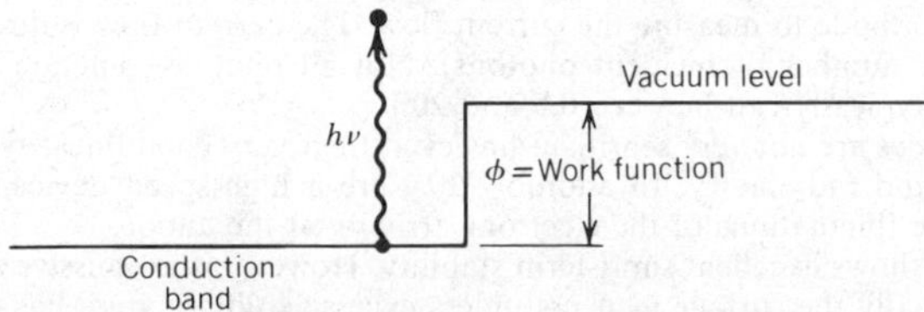

Fig. 15.50 Vacuum photocathode. The vacuum level corresponds to the energy of an electron at rest an infinite distance from the cathode. The work function is the minimum energy required to lift an electron from the metal to the vacuum. Only photons with energy can be detected.

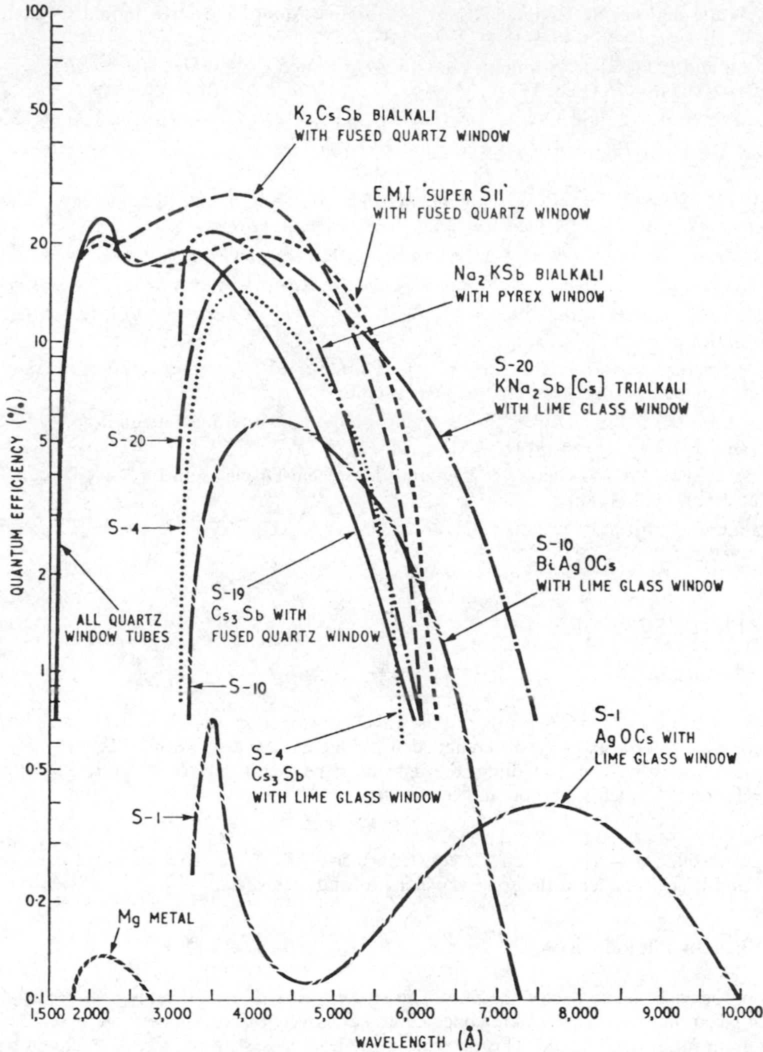

Fig. 15.51 Typical spectral sensitivity curves of some commonly used photocathodes. Individual cathodes show quite wide variations and so these plots give only approximate indication of performance. From Dance, 1969, with permission.

Vacuum Photodiode

The vacuum photodiode is the simplest device, consisting of a negatively biased photocathode and a positive anode. Light is shined onto the photocathode surface and a circuit is completed between the anode and the photocathode to measure the current flow. The current flow is directly proportional to the light intensity (or number of incident photons). Not all photons generate photoelectrons, and quantum efficiencies typically run between 0.5 and 20%.

Vacuum photodiodes are not very sensitive; however, their very good linearity make them suitable for light calibration and radiometry. In addition they are a high-speed device, the rise time being limited by transit-time fluctuations of the electrons arriving at the anode.

The vacuum tube shows excellent short-term stability. However, the emissive surface fatigues with exposure to light. Usually the surface recovers unless excessive illumination has occurred.

The output current from a phototube is

$$i = \frac{Pe\eta}{h\nu} \tag{15.33}$$

where P = incident optical power
η = quantum efficiency
e = electron charge

15.4-2 Gas-Filled Photodiodes

To increase light sensitivity, the vacuum phototube can be filled with about 0.1 mm pressure of argon. In this case the photoelectrons, accelerating under the influence of the anode voltage, ionize the argon molecules and create more electrons. Typically, multiplication factors of 5 to 10 can be achieved, depending on the applied anode voltage.

The best applications for gas-filled photodiodes are as simple light sensors. The ionization process results in an increasingly nonlinear response with light level and also a low-frequency response ≈ 10 kHz.

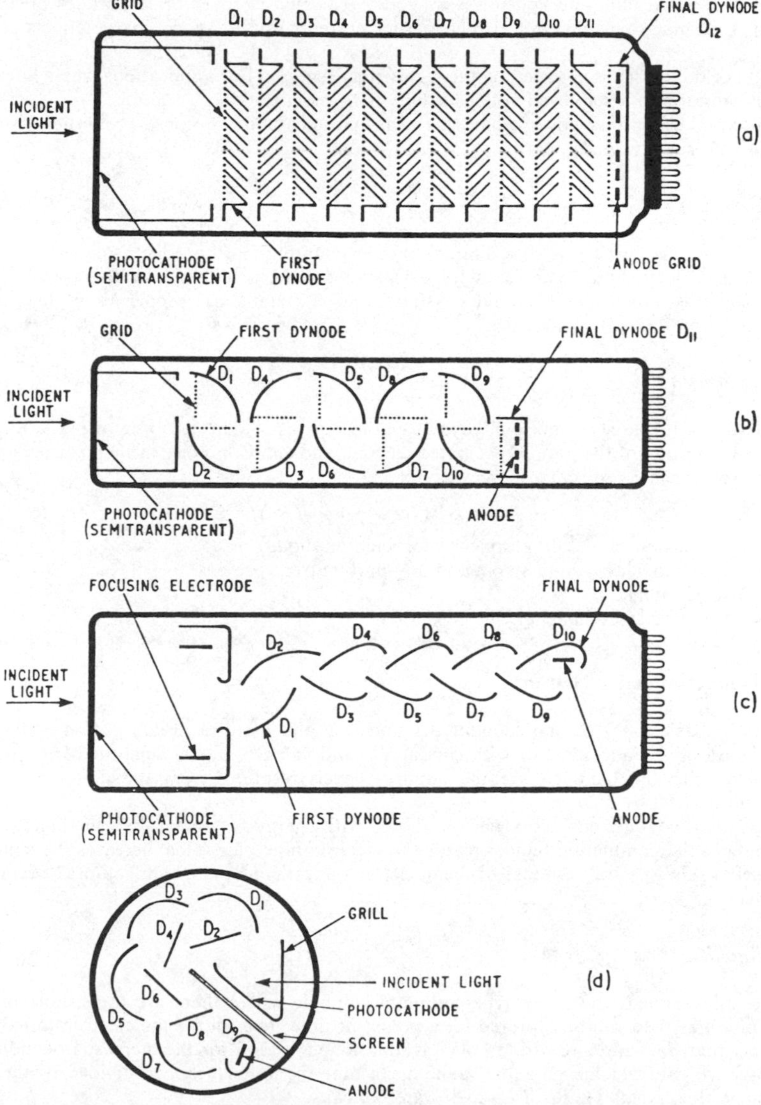

Fig. 15.52 Types of dynodes used in photomultiplier tubes: (*a*) venetian-blind structure, (*b*) bosc and grid system, (*c*) focused structure, and (*d*) circular cage-focused dynode system. From Dance, 1969, with permission.

15.4-3 Photomultipliers

The photomultiplier is the most sensitive detector for visible radiation. Series of dynodes are used to generate secondary emission, and electron gains of up to 10^8 can be achieved with only modest degradation of the linearity and high-frequency response of vacuum photodiodes. There are a number of different designs of photomultipliers with dynode structures that optimize one or the other of these requirements.

Figure 15.52 shows four different types of structure. The first two are unfocused dynodes; these are more efficient at collecting electrons. The latter two are focusing types and tend to result in less transit time variation and so higher frequency response.

Figure 15.53 shows the normal bias circuit. A potentiometer chain provides the voltages for the dynodes; typically the high-voltage supply ranges from 700 V to 3 kV, depending on the design. The output voltage is taken across R_L. The capacitors provide decoupling where high-frequency response is required and prevent saturation from the potentiometer resistors.

The linearity of a photomultiplier is very good, typically 3% over three decades of light level. Saturation is normally encountered at high output currents due to space charge effects at the last dynode.

The gain of the device is adjusted by the high voltage supply. For applications where linearity and stability are important, a stabilized voltage supply is necessary.

The spectral response is governed by the same photocathode response properties discussed in connection with Fig. 15.24.

Photon Counting

For detection of very low levels of radiation, photon counting can be done using photomultipliers. Since up to 10^8 electrons are generated for each photoelectron, individual photon arrivals can be detected. There is a considerable field of study into the statistical properties of light fields as measured by photon counting statistics.[1]

Noise Mechanism

There are two main sources of noise in vacuum photo devices: quantum noise, owing to the discrete and fluctuating nature of the photon-generated current, and Johnson noise in the load resistor. Noise power is given by these respective sources as

$$\delta \bar{v}^2 = GR(2ei\,\Delta f) + 4kTR\,\Delta f \qquad (15.34)$$

where G = current gain in photomultiplier (1 for vacuum diode)
i = current, including both signal and any dark current present
R = load resistance
Δf = signal frequency band

15.4-4 Photoconductive Sensors

There are two basic types of photoconductors: intrinsic and extrinsic. Figure 15.54 is a simplified energy diagram of a semiconductor with conduction and valence bands. Separated from them by a small amount into the bandgap are the impurity levels resulting from the donor and acceptor impurities introduced by doping.

Intrinsic photoconduction occurs when a photon with energy $h\nu$ greater than the bandgap excites an electron into the conduction band, creating an electron-hole pair. This increases the conductivity of the material. The spectral response of the detector is governed by the bandgap of the semiconductor material.

Charge Amplification

In a semiconductor material like n-type cadmium sulphide where there are traps, holes under the influence of a bias field will be captured for a period of time. This allows the electron to move to the anode rather than recombining with a hole, giving a longer period for the increase in conduction. A photoconductive gain is achieved equal to the mean time the hole is trapped divided by the electron transit time in the crystal. Gains of up to 10^4 are typical.

The response time of these sensors is consequently slow, ~10 ms, and the output is quite nonlinear.

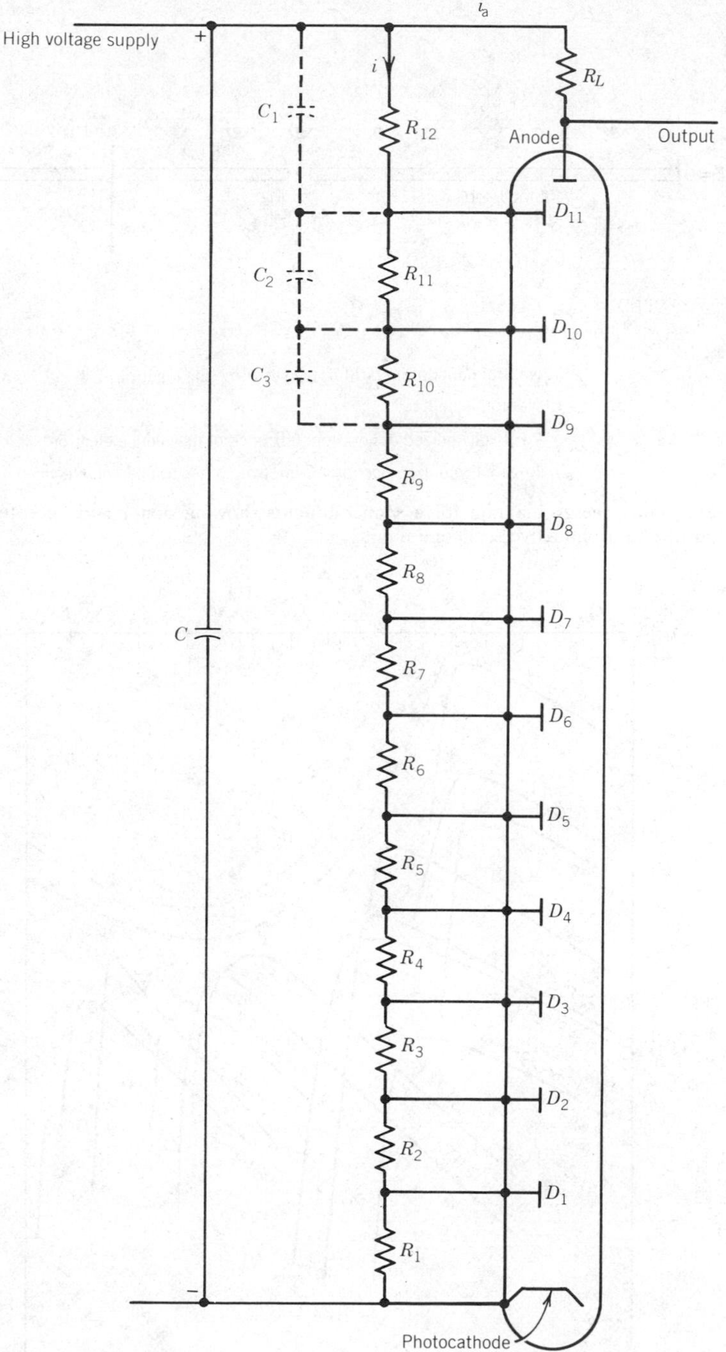

Fig. 15.53 The basic circuit to provide voltages required on the photomultiplier tube and to obtain output voltage. From Dance, 1969, with permission.

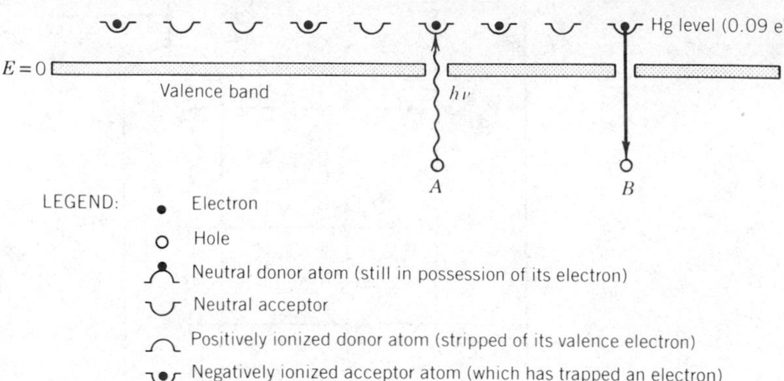

Fig. 15.54 Simplified energy diagram for a semiconductor showing donor and acceptor impurity levels involved in photoconductive semiconductors.

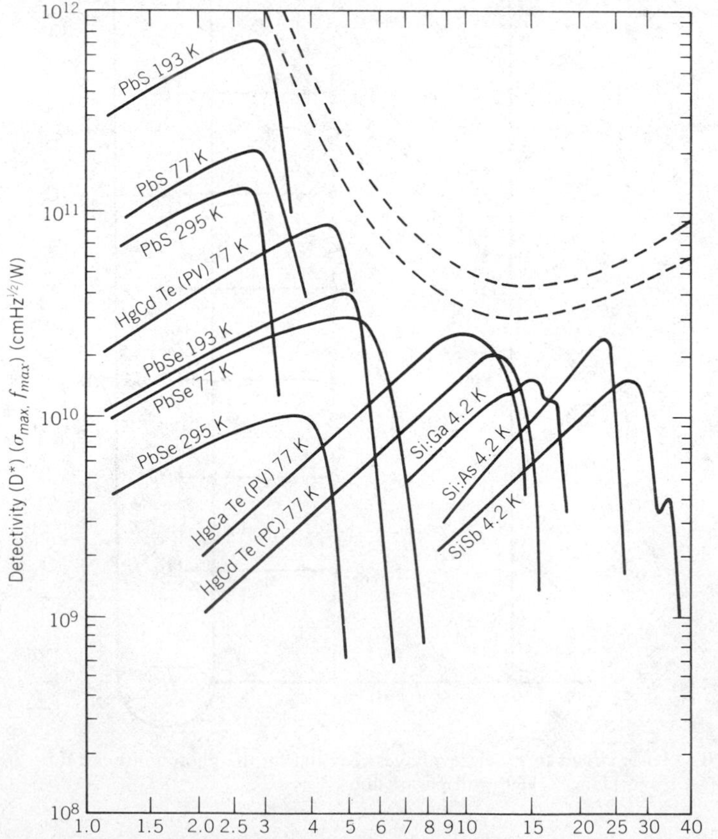

Fig. 15.55 Relative sensitivity of a number of commercial photoconductors. Data adapted from publications of the Santa Barbara Research Center, with permission.

Extrinsic Effect

The operation of an extrinsic photoconductor is similar to that of an intrinsic detector except that the photon excites an electron from the valence band into the acceptor level corresponding to the hole of the acceptor atom. As a result the energy required is much smaller, the reason why these detectors have applications for longer wavelengths. Figure 15.55 shows typical responses of a number of these devices.

Since the doping material determines the acceptor energy level, both the host material and the dopant are identified. The energy level is usually small, especially when longer wavelengths are being detected and the level is populated heavily by thermal excitation at room temperature. Thus for useful detection the devices have to be operated at liquid N_2 temperatures or lower.

The response of a photoconductor can be written as the current flowing

$$i = \frac{P\eta\tau_0 ev}{h\nu d} \tag{15.35}$$

where P = optical power at frequency ν
h = Planck's constant
v = drift velocity $= \mu E$ where μ = mobility and E = electric field
η = quantum efficiency (at freq ν)
τ_0 = lifetime of carriers
e = charge on electron

The charge amplification can be written as

$$\frac{\tau_0}{\tau_d} \tag{15.36}$$

where $\tau_d = d/v$, the drift time for a carrier to go across the crystal.

Noise

The main source of noise is generation recombination. The noise is analogous to shot noise in vacuum devices and is due to the random nature of photon arrival. The noise current is given by

$$\bar{i}^2 = \frac{4ei(\tau_0/\tau_d)\,\Delta f}{1 + 4\pi^2\nu^2\tau_0^2} \tag{15.37}$$

and includes the fact that carriers are both being created and recombined.

When cooled detectors are not being used, Johnson noise must be added in as well [Eq. (15.34)].

15.4-5 Junction Photodetectors

Types of Junction Photodetectors

***Pn* Photodiodes.** Photodiodes are now the most prevalent type of photodetectors. In them a *pn* junction is formed in the semiconductor material. Figure 15.56 shows the energy diagram of a detector with a reverse bias. Light with energy greater than the bandgap creates electrons in the *p* region and holes in the *n* region. If these are within the diffusion length of the junction they move toward it and are then swept across by the field. Light falling in the junction area creates hole pairs which are separated by the field. In each case an electron charge is contributed to the external circuit. If there is no external bias, the movement of the carriers creates an external voltage, the *p* material becoming positive. The maximum voltage obtainable is equal to the difference between the fermi levels in the *p* and *n* material and approaches the bandgap E.

***Pin* Photodiodes.** The carriers generated in the junction region experience the highest field and get separated most rapidly and so give the fastest response. The *pin* diode has an intermediate thick intrinsic layer and is designed to absorb the light here where the highest field exists. This minimizes slower carrier diffusion times in the *p* and *n* regions.

The signal current generated by incident light power P is

$$i = \frac{Pe\eta}{h\nu} \tag{15.38}$$

similar to that of a vacuum device. Similarly, the output current is linear with incident light level. In addition there is dark current, which is due to the thermal generation of carriers.

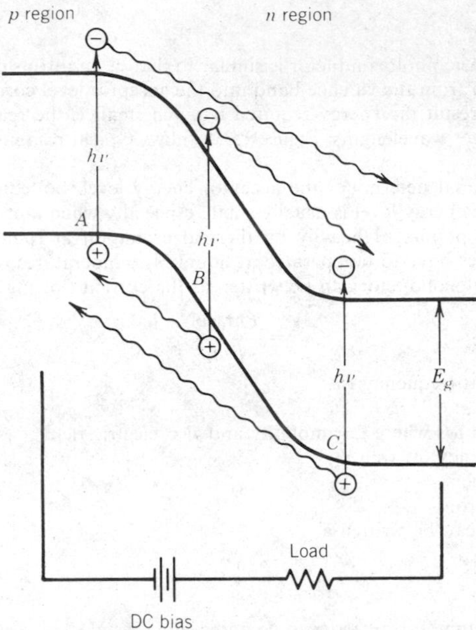

Fig. 15.56 Energy diagram showing the three types of electron-hole pair creation by absorbed photons that contribute to current flow in a *pn*-junction photodiode.

Avalanche Photodiode. If the reverse bias of a photodiode is increased close to the breakdown voltage, carriers will be accelerated in the depletion region and have sufficient energy to excite other electrons into the conduction band, resulting in a multiplication effect (avalanche gain). Gains of 50 are typical, although values in excess of 2500 have been reported.

Avalanche diodes are specially designed to have uniform junction regions to handle the high applied fields.

Solar Cell. This is basically a large-area *pn* junction in silicon optimized to detect solar radiation. In the solar cell, antireflection coatings are used to optimize light retention.

Phototransistors. This is a normal transistor in which light is shined onto the collector base (reverse-biased) junction. The light acts in a manner similar to the base, injecting carriers into the junction. Thus the collector current from the *pn* junction is amplified over the normal photo current.

Photo Field Effect Transistor (FET). This is an FET in which light is used to generate carriers that change the gate voltage. FETs can have very high sensitivities.

Characteristics of Junction Photodetectors

Frequency Response. The frequency response of junction photodiodes is limited by three factors: (1) The diffusion time for carriers generated in the *p* and *n* regions to reach the junction, (2) the junction capacitance and its combination with the equivalence circuit resistance, and (3) the transit time of carriers across the junction.

Noise. There are two main sources of noise, Johnson noise due to thermal effects in the load resistance and shot noise from the carrier generation process. The noise current is

$$\bar{i}^2 = 2ei\,\Delta f + \frac{4kT\,\Delta f}{R} \tag{15.39}$$

i is the average current, including dark current. In photodiodes Johnson noise usually dominates. When there is avalanche gain, the shot noise term is much larger and can be comparable. The signal-to-noise ratio increases with increasing avalanche gain until the two noise terms are comparable. After this both signal and shot noise increase in the same proportion (in practice, shot noise increases more). There are also additional contributions from contact noise, thermal generation recombination noise, and so on, in the semiconductor material.

Detectivity. The performance of a detector is often stated as the detectivity D^*. This is given as

$$D^* = \frac{\sqrt{A \, \Delta f}}{\text{NEP}} \tag{15.40}$$

where NEP = noise equivalent power (radiation incident to give signal-to-noise ratio of 1)
A = detector area
Δf = signal bandwidth

The radiation is also specified in terms of wavelength or equivalent black-body conditions. This provides a figure of merit with which to compare various detectors.

Spectral Sensitivity. Table 15.10 shows the spectral sensitivity range for common semiconductor photodiode detectors. The long wavelength limit is governed by the energy gap of the material. The short wavelength is limited by the absorption of the material and any windows, since the radiation has to penetrate into the junction region to create carriers.

TABLE 15.10. SPECTRAL SENSITIVITY RANGE FOR COMMON SEMICONDUCTORS

Semiconductor	Wavelength Range (μm)
Ge	0.4–1.9
Si	0.1–1.2
GaP	2–4.5
InSb	1.0–5.5
InAs	1.0–3.8
PfSnTe	2.0–18
HgCdTe	1.0–12

15.4-6 Imaging Photodetectors

Types of Imaging Photodetectors

Image Intensifier Tubes. The combination of a photoemissive surface, an electron accelerator, and a phosphor material permit visible images to be formed from an incident light image on the photoemissive surface. This can have two applications, either brightening of a very weak image for night vision purposes or conversion of a nonvisible image, such as one in infrared light, into a visible image.

Figure 15.57 shows two types of device. In both, electrons are emitted from the surface in proportion to the incident light image. The electrons are accelerated and focused on the phosphor screen, where an image is formed. Luminance gains are typically 50 to 100; however, a number of tubes can be used in series to magnify the gain even more. Spectral sensitivities are those of available photoemissive materials, as shown in Fig. 15.51.

Image Orthicon Camera Tube. There are two basic types of television camera tube. The image orthicon tube uses the photoemissive effect. Figure 15.58 shows the layout of the tube. Light falling on the photocathode causes electrons to be emitted, which are attracted toward the positively biased target. The target is a mesh and the electrons pass through, landing on the glass electron target screen. This causes secondary electrons to be emitted which are also collected by the target. The result is positively charged glass on which a charge image is formed that is a direct replica of the light image on the photocathode.

A low-velocity electron beam is raster scanned across the target to neutralize the charge. Surplus electrons return through the multiplier stages and generate a current into the signal-output terminal. The current strength is thus inversely related to the light level at each point in the scan. The tube is very sensitive because of the charge storage and the electron multiplication effects.

Vidicon Camera Tube. The second type of TV sensor is a simpler image tube that uses the photoconductive effect (Fig. 15.59). The scanning electron beam maintains the rear side of the

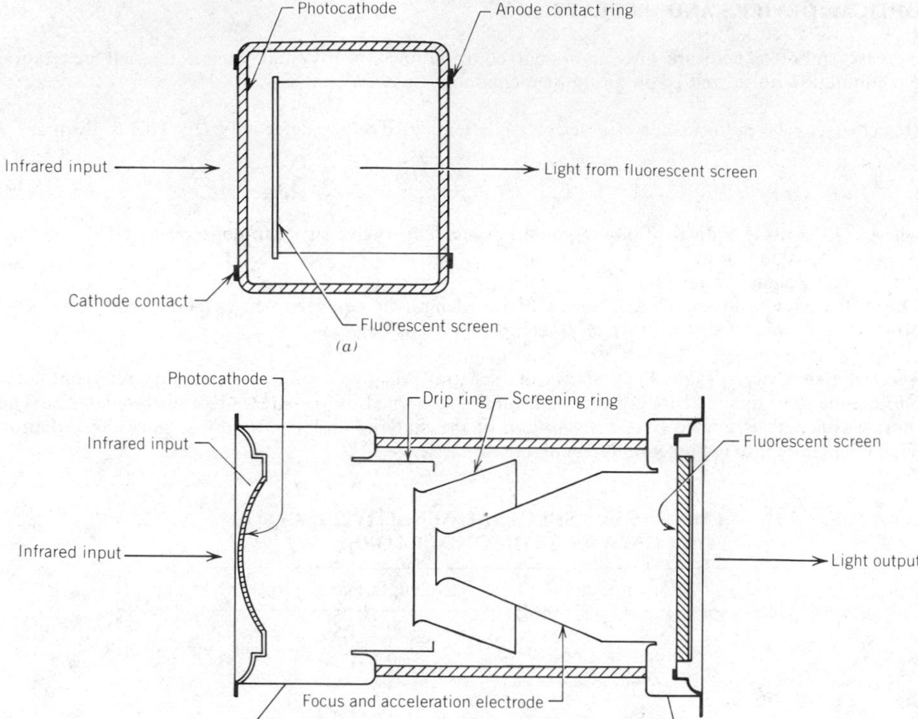

Fig. 15.57 (*a*) and (*b*). Two types of image intensifier tubes.

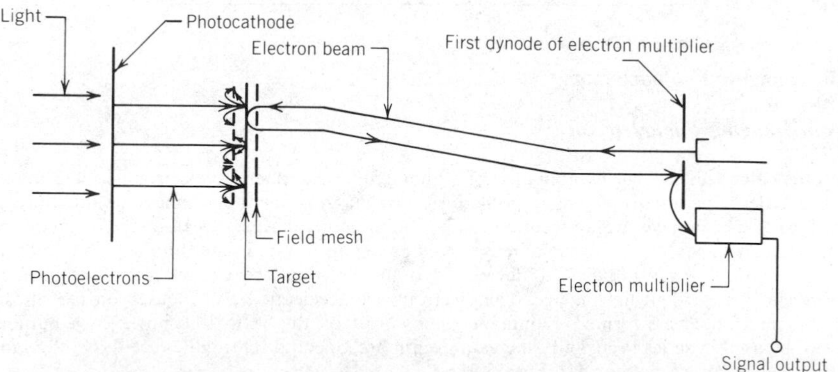

Fig. 15.58 Schematic diagram of an image orthicon camera tube. From RCA, *Electro-Optics Handbook*, RCA, Solid State Division, Lancaster, PA, 1974, with permission.

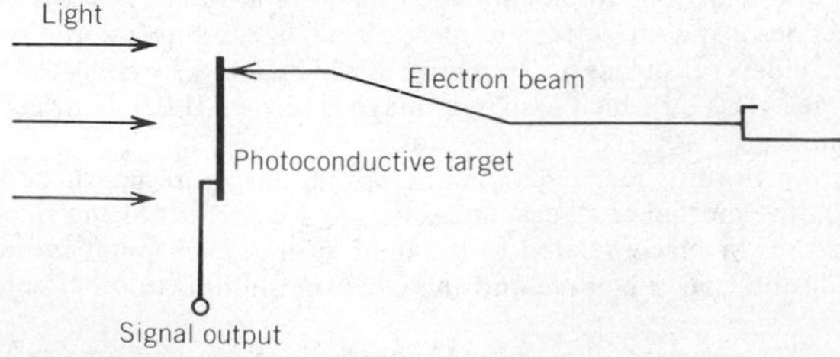

Fig. 15.59 Schematic diagram of a vidicon camera tube. From RCA, *Electro-Optics Handbook*, RCA, Solid State Division, Lancaster, PA, 1974, with permission.

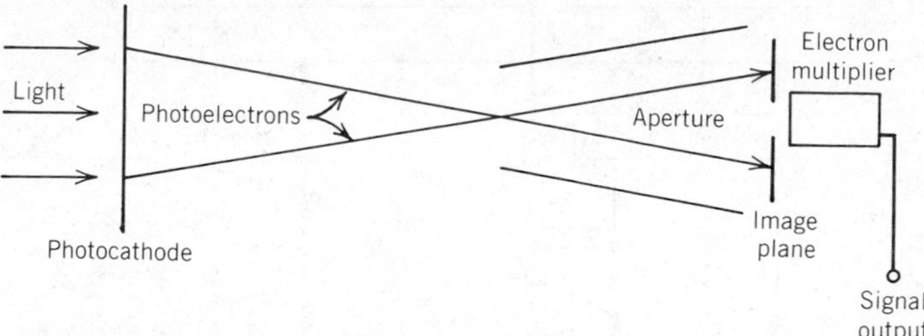

Fig. 15.60 Schematic diagram of an image dissector camera tube. Magnetic field causes deflection scanning. From RCA, *Electro-Optics Handbook*, RCA, Solid State Division, Lancaster, PA, 1974, with permission.

photoconductor at 0 V. Between scans the light image increases the conductivity, causing the rear side to charge at varying rates toward the 30-V bias voltage, forming a charge image. The raster-scanned electron beam, in charging the surface back to 0 V, provides a current flow into the output in proportion to the replaced charge. The output is thus a signal proportional to the light level at each point in the scan.

The main disadvantages of the vidicon tube are its longer response time and smaller dynamic range. With the development recently of longer wavelength photoconductive films, infrared wavelength vidicons with sensitivities below 1 μm have become available. The advantages are its simplicity and ease of use.

Image Dissector Camera Tube. The image dissector tube, shown in Fig. 15.60, uses an electrostatic imaging lens to focus the photoelectrons from the photocathode onto the image plane. A pinhole with an electron multiplier detects the electrons coming from a particular point on the photocathode. This device is relatively insensitive compared with other devices; however, it does not have to be raster scanned and thus is useful for tracking applications or where random access to an image is required.

Diode Arrays. A recent development has been the solid-state array detectors. These can be in either linear array or two-dimensional array form.

Figure 15.61 shows the schematic diagram for a linear array. The device consists of an array of *pn*-junction photodiodes. The diode has its own capacitance and/or added capacitance connected in parallel. Light falling on the diode charges up the capacitance, the charge being the integral product of the intensity of the light and the time the diode is exposed.

These diodes are sequentially accessed and the charge is removed by local integrated circuitry built on the device. The output is thus a sequence of charge pulses. The charge is proportional to the light intensity and the display time, which is normally the time between scans when the diode is accessed.

Since diode arrays are solid state, they are very rugged detectors and the location of the diodes is accurately known and fixed. Thus these devices are very suitable for accurate linear or two-dimensional optical measurements. The devices can be very sensitive and can have variable sensitivity, since the diode integrates the photoelectron signal until accessed. The spectral sensitivity is that of the photodiode material; the majority of devices at present are silicon based because of the advanced state of silicon integrated-circuit technology. In addition to normal visible and near-infrared spectral sensitivities, applications to detection of x-rays have been achieved.[2]

Modulation Transfer Function (MTF)

This term is used to describe the resolution of imaging. It is analogous to "frequency response" for an electronic circuit. Its use permits prediction of the imaging performance of a series of optical and electrooptic devices. Figure 15.62 shows a typical MTF curve. The transfer function operates on the light intensity or flux and is measured as a function of linear spatial frequency, for example, cycles per millimeter. Thus if the response of a TV camera is needed, the following equation is appropriate

$$M_{\mathrm{TOT}} = M_L \times M_v$$

where M_L and M_v are the lens and the vidicon tube MTFs, respectively.

If the camera were imaging a black and white striped image in which the density variation from black to white was sinusoidal, the amplitude of the sinusoidal video signal would be M_{TOT} times the DC output from a solid white image.

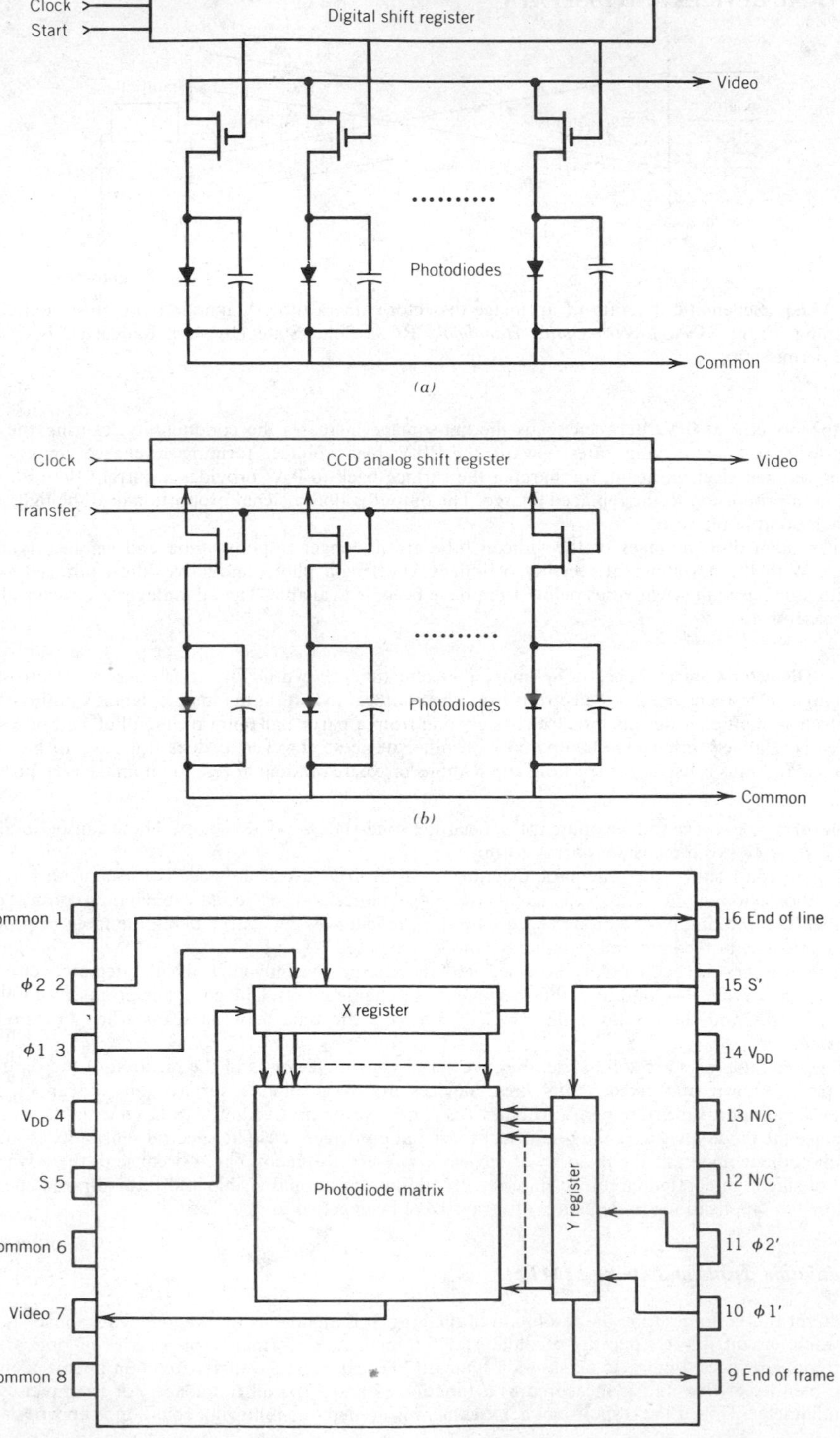

Fig. 15.61 Schematic diagram of diode array sensors, (*a*), (*b*) Linear sensors showing different circuits for clocking out individual charges on the photodiode capacitors. (*c*) Arrangement for a two-dimensional sensor. From Reticon Catalog, 1977, Reticon Corp., subsidiary of EG&G, Inc., CA, with permission.

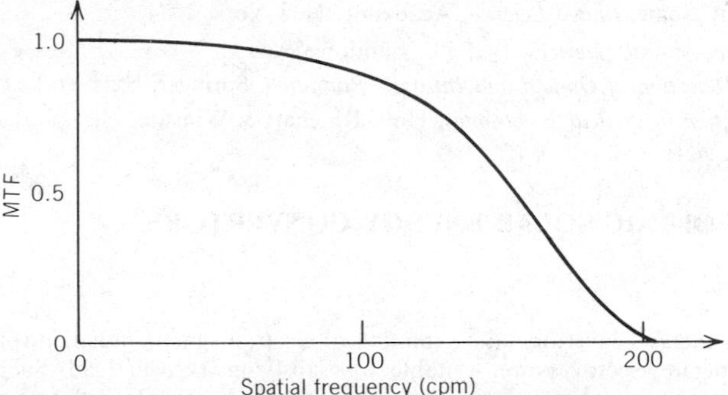

Fig. 15.62 Example of a modulation transfer function (MTF) curve for an optical imaging system. Image components at 130 cpm would only be half as bright, while components at 200 cpm would not be resolved at all.

15.4-7 Applications of Detectors

The following list summarizes applications of optical sensors.

1. **Vacuum photodiodes.** Radiometry and high-speed detectors for wavelengths less than 1.1 μm.
2. **Gas-filled photodiodes.** Simple on-off light sensors for wavelengths less than 1.1 μm, optical encoders, tachometers, alarms, proximity sensors, etc.
3. **Photomultipliers.** Medium-speed detectors and where very high sensitivity is required for wavelengths less than 1.1 μm. Used in nuclear scintillation counters, light scattering, and other light measurement applications.
4. **Intrinsic photoconductors with charge amplification.** Simple on-off light sensors for visible and near-infrared wavelengths; same applications as no. 2 plus light meters.
5. **Extrinsic photoconductors.** Infrared detectors, normally special applications because of the need for liquid N_2 cooling.
6. **Junction photodiodes.** General purpose visible and near-infrared detector. Same as no. 2 plus optoisolators, light meters, optical communications, solar cells, etc.
7. ***Pin* diodes.** For higher speed applications, mainly in optical communications and pulsed spectroscopy.
8. **Avalanche photodiodes.** For more sensitive applications to obtain the ultimate in signal-to-noise ratio, primarily in optical communications.
9. **Phototransistors and photo FETs.** Simple light detector where more sensitivity is required; same as no. 2.
10. **Image intensifier tubes.** Used to brighten images or for infrared viewing. Snooper scopes for night viewing, astronomy, plasma physics, x-ray fluoroscopy, bioluminescence, nuclear particle track viewing in a scintillator, etc.
11. **Image orthicon cameras.** Primary TV camera sensor, mainly studio applications.
12. **Vidicon cameras.** General purpose TV camera tube for CCTV and portable camera, also infrared viewing applications.
13. **Image dissectors.** Tracking applications, guidance and target acquisition.
14. **Diode arrays.** Image measurement applications, optical gauging and control, optical readers.

References

1 B. Crosignani, P. DiPorto, and Bartolotti, *Statistical Properties of Scattered Light*, Academic, New York, 1975.

2 Reticon Corp., subsidiary of EG&G, Inc., Application notes #101.

Bibliography

Amos, S. W. and D. C. Birkinshaw, *Television Engineering: Principles and Practice*, ILIFFE, London, 1963.

Barrows, W. E., *Light, Photometry and Illuminating Engineering*, McGraw-Hill, New York, 1951.

Daintz, J. C. and R. Shaw, *Image Science*, Academic, New York, 1974.

Dance, J. B., *Photoelectronic Devices*, ILIFFE, London, 1969.

Kingston, R. H., *Detection of Optical and Infrared Radiation*, Springer, New York, 1978.

Yariv, A., *Introduction to Optical Electronics*, Holt, Rinehart & Winston, New York, 1971.

15.5 PHOTOVOLTAIC SOLAR ENERGY CONVERTORS

Martin Wolf

Photovoltaic solar energy convertors utilize the flux of electromagnetic radiation (photons) available in sunlight to generate electric power suitable for satisfying a given load. Such loads may, as terrestrial loads, range from a watch or pocket calculator (DC, uncontrolled power in the milliwatt range) to a feeder line into a utility grid (AC, highly frequency and voltage controlled, many megawatts), or they may be the power needs of a spacecraft.

The convertors form a part of a system that contains, in the sequence of the energy flow, the following subsystems (Fig. 15.63):

Optical concentrator (optional)
Photovoltaic array or array field (the converter itself)
Energy storage (optional)
Power conditioning and control

Three types of photovoltaic solar power systems can be discerned (Fig. 15.64). These have specific technical problems which still need to be resolved (Table 15.11), as well as the common problem of further reduction of their capital cost to permit the production of electric energy at prices that are competitive with those of the utilities. In a few special applications a photovoltaic array is used by itself; it may be as small as a few interconnected and encapsulated solar cells (occasionally just one cell), as used to operate a pocket calculator or certain toys.

Commercially traded systems are now principally either complete installed systems or their components. The photovoltaic component primarily traded is the module. Several types are shown in Fig. 15.65. Some modules have circular cells, which result in a low packing factor, while others have

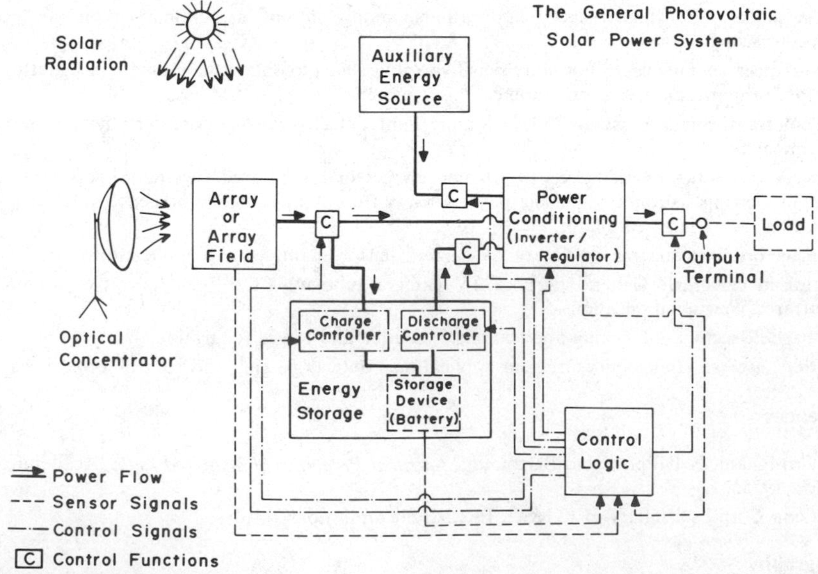

Fig. 15.63 Block diagram of the general photovoltaic solar power system. The system contains, as a minimum, the array or array field. The other blocks are optional. The load may be a local electric power consumer or a utility grid.

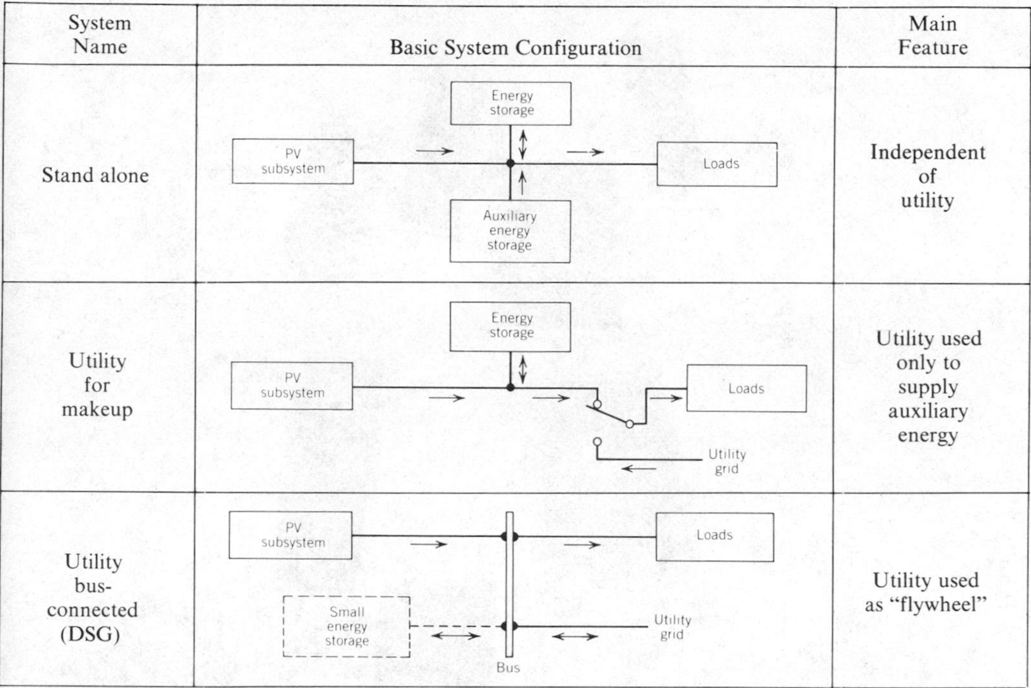

Fig. 15.64 Block diagrams for the three major types of photovoltaic solar power systems, classified according to the system's relationship to the utility grid. "DSG" stands for the currently used term "distributed storage or generation."[1]

TABLE 15.11. MAJOR PROBLEMS WITH PHOTOVOLTAIC SOLAR POWER SYSTEMS

System Type	Problem Area
Stand alone	Energy storage
Utility for makeup with major energy storage	Energy storage, principally battery: life, efficiency, cost
Utility bus connected	Grid interface: stability, fault protection, personnel safety Long range: capacity limit for DSG (distributed storage or generation)

rectangular or square cells to obtain a high packing factor. One module has square cells with cropped corners which permit better utilization of round semiconductor wafers while still resulting in a rather high packing factor.

15.5-1 (Photovoltaic) Array

The array is the basic energy-converting unit of the photovoltaic system. It is composed of "modules" that may be electrically connected in series and in parallel to attain the desired array voltage and current, respectively (Fig. 15.66).

The array contains a mechanical structure that facilitates mounting of the modules and of their electrical interconnections. If located on the ground, the array also includes adequate footings to

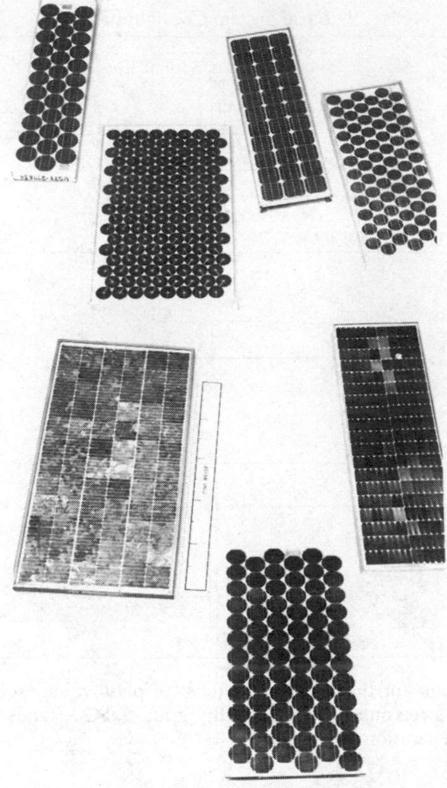

Fig. 15.65 Commercial photovoltaic modules available from various manufacturers. Some modules have circular cells, which results in a low packing factor, while others have rectangular or square cells to obtain a high packing factor. From DOE/JPL brochure on the low-cost solar array project, Jet Propulsion Laboratory Report No. 5101-178 Rev. C. (Jul. 1981), with permission.

assure the integrity and stability of the array under wind and snow loads. Alternately, if the array is mounted on a building, attachment elements to the building structure are provided.

The array usually contains means for ready connection or disconnection of individual modules, for example, for replacement or repair. (A module maintains its output voltage and its capability to deliver current up to the value of its short-circuit current as long as light falls on it, even if it is disconnected from the array.)

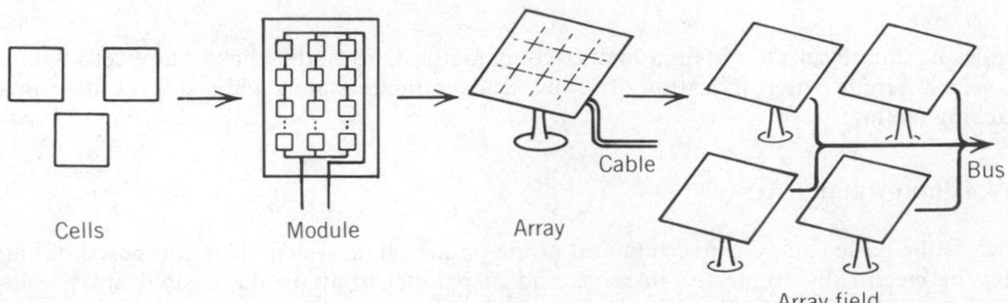

Fig. 15.66 Illustration of the nomenclature commonly used to describe elements in the hierarchy from solar cell to array field.

Fig. 15.67 Residence in Potomac, Maryland, that has a 2.3-kW stand-alone photovoltaic system. The photovoltaic array of 36 modules covers the lower approximately 40% of the south-facing surface of the roof. From D. Best, *Solar Age* 7(7):25 (Jul. 1982), with permission.

The design of an array or of an array field is site and load specific, and the construction involves substantial field work.

Flat Arrays

If an optical concentrator is not used, then the first surface of a solar collector to intercept solar radiation is generally planar. Photovoltaic arrays without devices for optical concentration are therefore designated flat arrays. When mounted for fixed orientation they are especially well suited for incorporation into a building structure (Fig. 15.67).

Concentrator Arrays

Any type of optical concentrator that has been found useful for solar-thermal energy conversion has also been considered for photovoltaic conversion. This includes one-dimensionally and two-dimensionally concentrating systems of the reflective and refractive types (Fig. 15.68). For the latter, Fresnel lenses are preferred because of the reduced mass required. Concentration ratios from barely above unity to several thousand have been used with photovoltaic arrays. The trend seems to be toward high-ratio concentrators (concentration ratio 50 or higher) (Fig. 15.69).

For photovoltaic concentrator systems, the convertor array generally contains one or more series-conected strings of solar cells to step up the output voltage from the 0.5- to 1-V level of the individual cell, rather than to parallel-connect the cells which would multiply the already high current levels. However, any irradiance nonuniformity has a much greater influence on the performance of the cells in a series string than it would have on a parallel-connected group of cells. Thus, irradiance uniformity in the focal surface is generally an important criterion in the selection of a concentrator for a photovoltaic system, together with the arrangement of the solar cells in this focal surface.

As concentrator arrays increasingly exclude indirect solar radiation with increasing concentration ratio, their application is advantageous primarily in arid climates (Fig. 15.70). Also, because of their higher maintenance requirements in comparison to those of flat arrays, they are used more often in attended stations, which means large systems.

Tracking Arrays

The power output of a flat array varies with the incidence angle i of the radiation, approximately proportional to cos i. For $i \leqslant 60°$, the approximation is usually very good. This cosine relationship

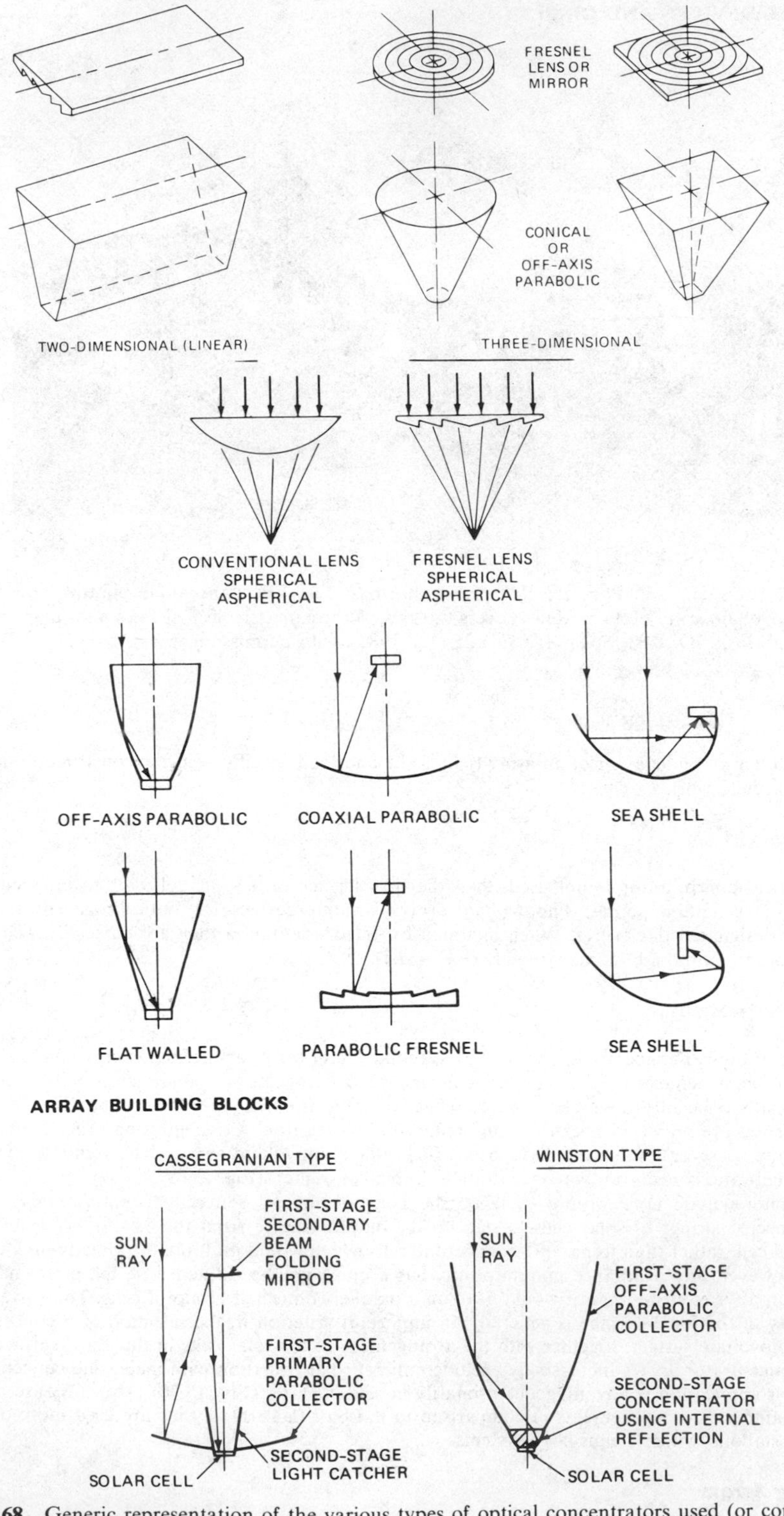

FRESNEL
LENS OR
MIRROR
CONICAL
OR
OFF-AXIS
PARABOLIC
TWO-DIMENSIONAL (LINEAR)
THREE-DIMENSIONAL
CONVENTIONAL LENS
SPHERICAL
ASPHERICAL
FRESNEL LENS
SPHERICAL
ASPHERICAL
OFF-AXIS PARABOLIC
COAXIAL PARABOLIC
SEA SHELL
FLAT WALLED
PARABOLIC FRESNEL
SEA SHELL
ARRAY BUILDING BLOCKS
CASSEGRANIAN TYPE
WINSTON TYPE
SUN
RAY
FIRST-STAGE
SECONDARY
BEAM
FOLDING
MIRROR
FIRST-STAGE
PRIMARY
PARABOLIC
COLLECTOR
SECOND-STAGE
LIGHT CATCHER
SOLAR CELL
SUN
RAY
FIRST-STAGE
OFF-AXIS
PARABOLIC
COLLECTOR
SECOND-STAGE
CONCENTRATOR
USING INTERNAL
REFLECTION
SOLAR CELL

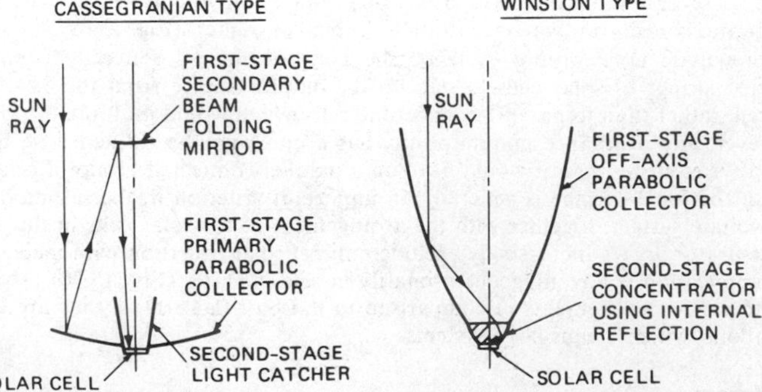

Fig. 15.68 Generic representation of the various types of optical concentrators used (or considered for use) with photovoltaic solar energy convertors. The Winston-type concentrator is often called a compound parabolic concentrator. From Rauschenbach,[2] pp. 264–266, with permission.

assumes that the array temperature remains constant with the change in incidence angle, and that the indirect radiation from the sky forms an insignificant contribution to the total solar irradiance H. "Tracking" implies keeping the incidence angle i at all times as close to zero as practical and economical. This means changing the orientation of the array relative to the earth surface continuously or at practical intervals. Tracking is done either manually or, more frequently, automatically. When an optical concentration ratio above 5 is applied, tracking of the concentrator array is a must. Even for flat arrays, the present price relationship between modules and the mechanical and electronic tracking components, including their installation, indicates the use of tracking to be economically advantageous in many climates. Two-dimensional tracking (e.g., azimuth and elevation or declination and hour angle) is predominant (Fig. 15.71), although automatic tracking of the hour angle in an equatorially mounted system, with periodic manual adjustment of the declination angle, is occasionally used.

In temperate climates where higher turbidity of the atmosphere and partial cloud cover occur during much of the year, resulting in a substantial indirect component of incident solar radiation, the performance advantage of tracking systems is reduced. Consequently, fixed mounted flat arrays are often preferred because of their greater simplicity and application versatility, which result in lower initial cost and reduced maintenance requirements.

Array Field

The term "array field" is used in large systems. The field is composed of a number of arrays, of the bus system that connects them, and often of a load control and switching facility (Fig. 15.72).

15.5-2 Balance of System (BOS)

All parts of the system beyond the solar modules and the optical concentrators, if the latter are used, are commonly lumped together under the name "balance of system" (BOS).

Area-Related Balance-of-System (ARBOS)

For a free-standing array or array field, a structure has to be provided to which the modules can be bolted. This structure needs to be appropriately anchored in the ground and to be of sufficient mechanical strength to maintain the integrity of the array under a variety of environmental influences, particularly wind and snow loads. In addition, a cable system is needed to connect the array(s) to the power conditioning and control subsystem. As this part of the system is, in several aspects, proportional to the area of the array, it is generally known under the name "area-related balance-of-system" (ARBOS).

Fig. 15.69 (*a*) The 2.5-kW concentrator array designed and built by Martin-Marietta as part of the photovoltaic power system for two small villages in Saudi Arabia. The array uses 30 × 30-cm Fresnel lenses and single-crystal silicon cells designed for application in concentrator systems. The individual array, consisting of 34 modules, is 12.8 m long and 2.75 m high, has an effective concentration ratio of 40, and has an output of 2.275 kW under normalized operating conditions. (*a*) Photo of a complete array. From D. D'Alessandro, *Solar Age* **4**(12):14 (Dec. 1979). with permission.

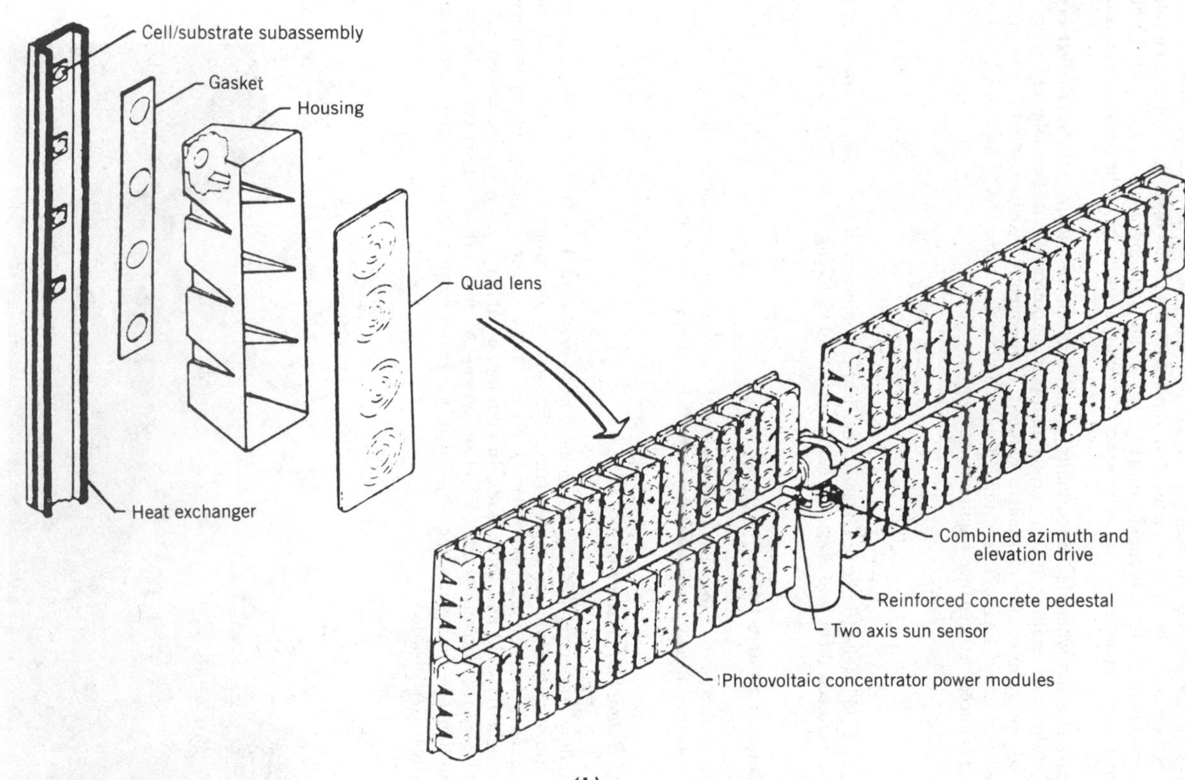

(b)

Fig. 15.69 (*b*) Exploded view, showing detail of the Fresnel lenses, their mounting, and the two axis tracking mechanism. From Sandia laboratories,[3] with permission.

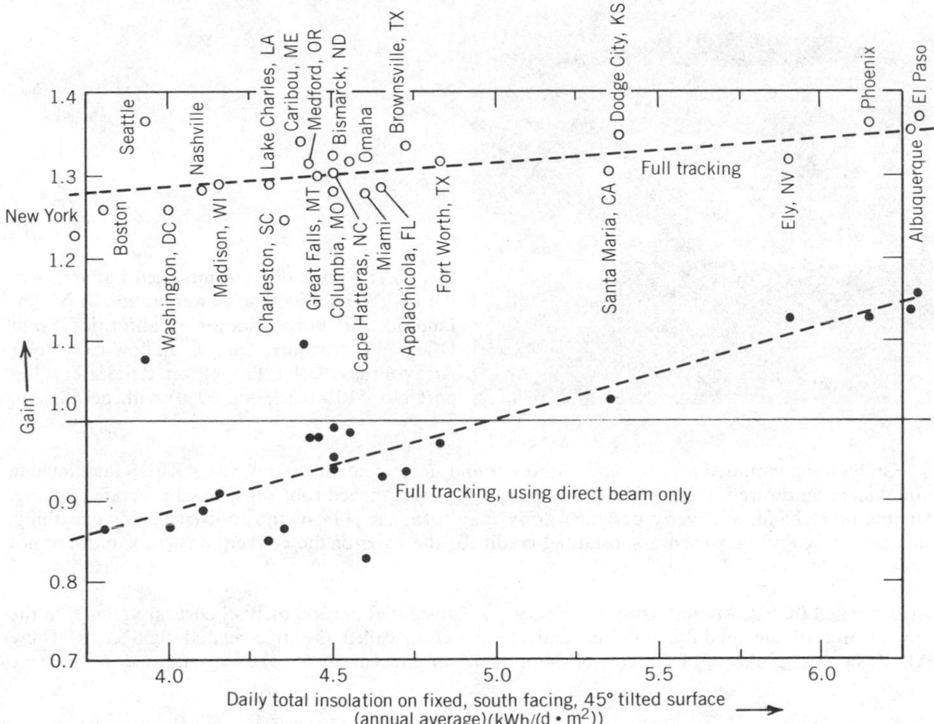

Fig. 15.70 Gain in energy flux input attainable by using a fully tracking flat array or a fully tracking high-optical-concentration ratio system that can utilize only the direct solar beam. The gain is expressed relative to the daily insolation (annual average) available on a fixed, south-facing, 45° tilted surface as it has been observed at 25 stations in various climate zones. A flat array that tracks fully provides a significant gain over a fixed mounted array in all climates, while tracking high-ratio concentrator systems provide gain only in arid climates (with the exception of Seattle and Caribou, Maine).[4]

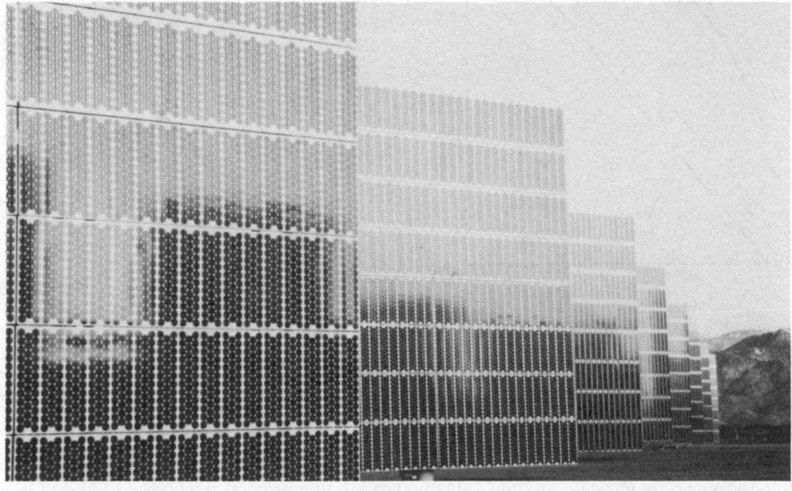

Fig. 15.71 A 10 × 10-m, flat, two-dimensional tracking array that is part of a 1-MW photovoltaic station at Hesperia, California, supplying power into the Southern California Edison Co. grid. The station was built by ARCO Solar in about a year. Courtesy of E. Berman, ARCO Solar.

Fig. 15.72 Field of fixed mounted flat arrays of the 60-kW photovoltaic power station at Mount Laguna Air Force Station, California. From DOE/JPL brochure, "Status of Low-Cost Solar Array Project", Jet Propulsion Laboratory Report No. 5101-143. (Jan. 1980) with permission.

On building-mounted arrays, such as those found in residential systems, the ARBOS may include attachment hardware for the modules and possibly a strengthened roof substructure, besides cabling. On the other hand, a cleverly designed array may form the skin of the roof (replacing sheathing, shingles, etc.), and may earn a substantial credit for the parts of the conventional roof that are not needed in this case.

Area-Related BOS Costs and Array Efficiency. A substantial portion of BOS costs are related to the frontal area of the modules installed and are therefore called the area-related BOS costs. These ARBOS costs—exclusive of the cost of the modules or concentrators—are concentrated in the array

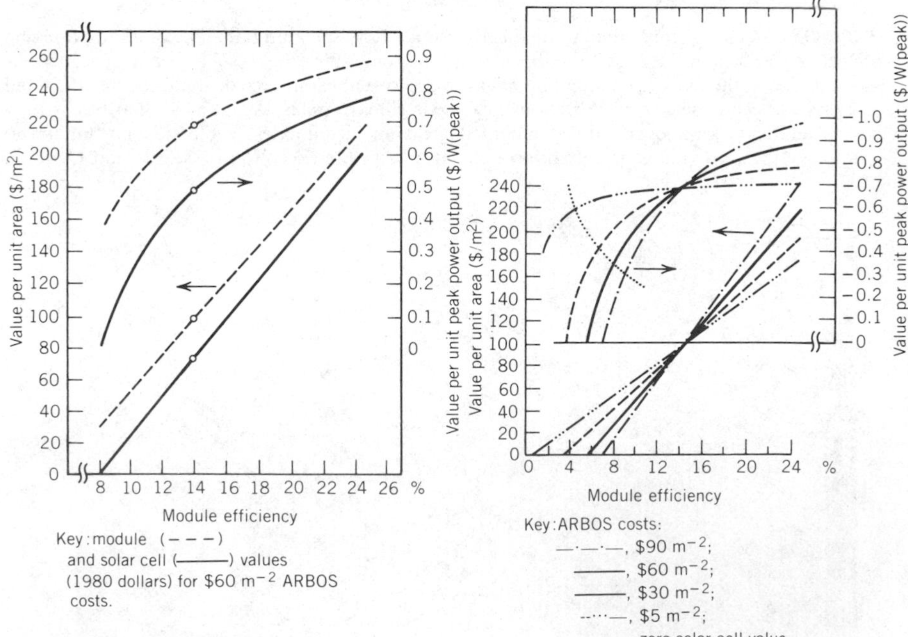

Fig. 15.73 Module value in dependence on module efficiency, with area-related balance-of-system (ARBOS) cost as parameter. The module value is that module price at which the total cost of a given system equals that of a reference system of equal output. The value is obtained in a system with the same ARBOS cost per unit area but a different module efficiency, which results in a differing area needed for the array or array field. The total system cost determines the price of the energy delivered, which is the ultimate measure for competitiveness. From Wolf,[5] with permission.

and array field, much of which costs are actually proportional to the module area. A part of the array costs (usually minor) is independent of the module area. As ARBOS costs make in many cases a significant contribution to the total system cost, they form a strong economic driver to high module efficiency. To satisfy a given load and in light of the limited available energy flux in sunlight, higher module efficiency translates into reduced array area and consequently lower total ARBOS costs for the entire system (Fig. 15.73). It may be noted that land costs usually form an insignificant part of the ARBOS costs.

Energy Storage Subsystem

To satisfy the loads of most solar energy utilization systems, electrical power is needed during periods of insufficient solar radiation (nights, cloudy days, etc.). In many systems provision is therefore made for storage of excess electrical energy available during times of high "insolation" by including an "energy storage subsystem" in the photovoltaic system design. In larger systems it is economic to choose a storage capacity that can supply energy for about 24 hours. Provision is then usually made for supply of auxiliary electrical energy for periods of inadequate insolation that exceed the capacity of the storage subsystem. In other systems, particularly stand-alone systems of smaller power rating, energy storage capacity equivalent to the energy needs of the load for a longer period, as long as a month or more, is provided. In this case no source for auxiliary energy is needed.

The availability of commercial devices for cost-effective electric power storage is limited. The best compromise between the desired attributes is still the lead-acid battery. Because of the many deep discharge cycles that the storage device usually experiences as part of a photovoltaic power system, the more reliable, longer life, industrial types of batteries are often preferred, although they command a higher price than, for example, automobile batteries. Additional devices that have been considered for electrical energy storage in photovoltaic systems include other electrolytic batteries, such as NiFe, NiCd, and AgZn batteries; reversible fuel cells; high-speed flywheels in vacuum enclosure; compressed gases, particularly air, stored in underground caverns; and superconducting inductive devices. However, none of these devices has been developed sufficiently to permit incorporation into commercial systems.

Auxiliary Energy Subsystem

Since the energy storage subsystem of a photovoltaic power system may have limited capacity, another source of electrical energy is needed for extended periods of inadequate insolation. Such auxiliary energy may be drawn from the utility grid by temporarily connecting only the loads to the grid, without intertying the photovoltaic system itself. Where this is not economical, or where a grid is not available, a motor-generator set or a fuel cell, with local storage of a suitable fuel, may be incorporated into the photovoltaic system as an auxiliary energy subsystem.

Power Conditioning Subsystem

The electrical output of a photovoltaic array is considerably affected by environmental influences, albeit reversibly. Therefore at least some voltage regulation is required. The output current is essentially proportional to the solar irradiance, while the output voltage decreases with increasing array temperature by approximately 0.4 to 0.5%/K. In addition the voltage increases approximately logarithmically with the solar irradiance.

Voltage regulation is most simply achieved by connecting an electrolytic storage battery of adequate capacity in parallel to the array (including a series-connected diode to prevent battery discharge through the array). This type of voltage regulation, however, entails operation at the lowest expected array operating voltage rather than near the maximum power output point.

At the other end of the spectrum in regulation are the so-called maximum power point trackers, which might suitably be called variable DC-to-DC transformers. These regulators provide a match between the impedance of the array at its operating point for maximum power output and that of the load. They generally apply variable pulse modulation to the input (DC) current and utilize reactive elements for energy storage and either integration of the output pulse train to DC or filtering to achieve an AC output of acceptable harmonics content. Such regulators or regulator/invertors can be designed to operate with only a few percent power loss, albeit at a higher price than lower efficiency power conditioners, of which a wide variety are available.

If the photovoltaic power system is to be intertied with the utility grid, the power-conditioning subsystem has to operate under much more stringent conditions than in other systems. Besides the need for narrower voltage tolerances, tight frequency and phase control have to be maintained, and the acceptable harmonic content is usually smaller. Consequently, a regulator/invertor for a utility-grid-connected photovoltaic power system will be more sophisticated and thus more costly than a

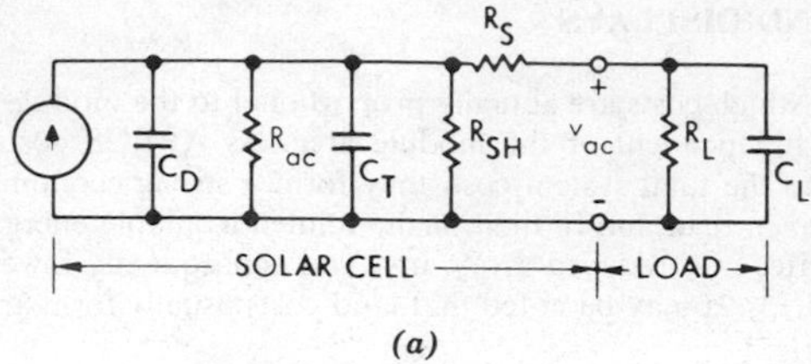

(a)

V_{dc} (mV)	R_{ac} (Ω)	R_s (Ω)	R_{SH} (kΩ)	C_D (μF)	C_T (μF)
550	0.2–2	0.1–0.5	5–50	2	0.06
350	1–10	0.1–0.5	5–50	0.2	0.06

(b)

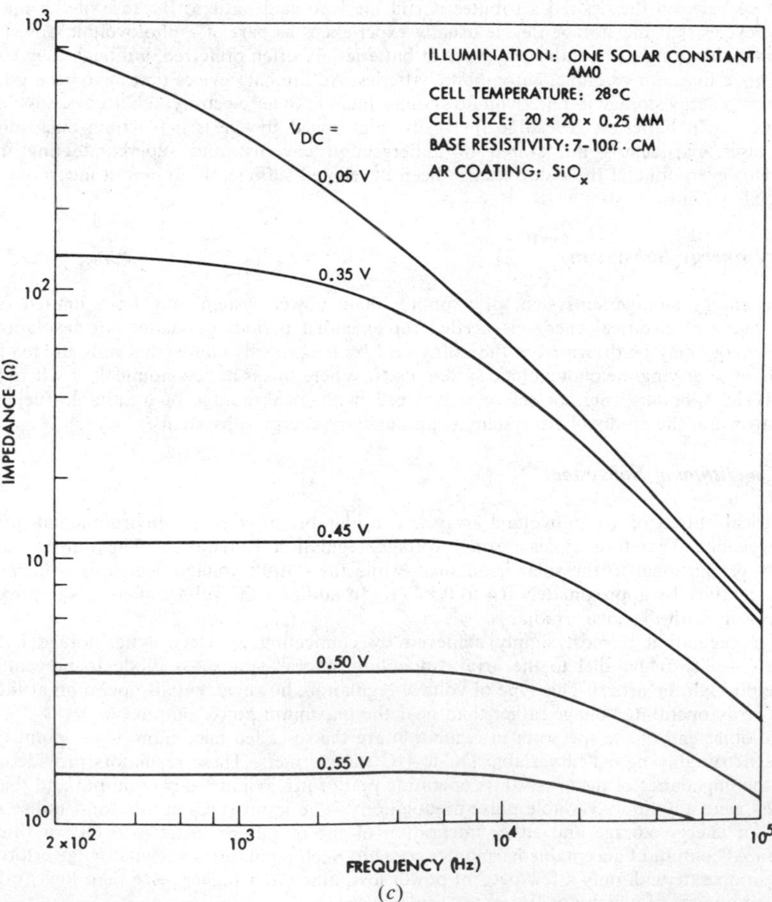

(c)

Fig. 15.74 AC properties of solar cells that are of importance for interfacing with the power conditioning unit: (a) AC small-signal equivalent circuit of the solar cell, (b) low-frequency AC parameters at two cell operating points, and (c) solar cell impedance as a function of frequency. From Rauschenbach,[2] p. 61, with permission.

power conditioner for a stand-alone system. For the design of the interface between the array and a pulse-modulating regulator, the AC characteristics of the array and their frequency dependence are of importance (Fig. 15.74).

15.5-3 Module

The photovoltaic module is the unit that is somewhat standardized, semi-mass-produced in factories (current annual worldwide production about 10 MW peak rating), and sold from catalogs. The module sizes, which range from a fraction of a square meter to a few square meters, have been chosen so that the modules can be readily shipped and handled and so that they simultaneously form a logical unit with respect to voltage and current rating for the most commonly used photovoltaic systems.

Module Electrical Ratings

The voltage and current rating of the module is determined by the number of individual solar cells used in the module and by their series and parallel connections (Fig. 15.75). Voltage ratings in the 6- to 30-V range are common. Some modules have provision for several output voltages by permitting external series or parallel connection of internally connected groups of solar cells.

Maximum Power Output. The maximum power output $P_{M,\text{mp}}$ of a module is described by

$$P_{M,\text{mp}} = H \cdot A_M \cdot \eta_M \quad \text{W} \tag{15.41}$$

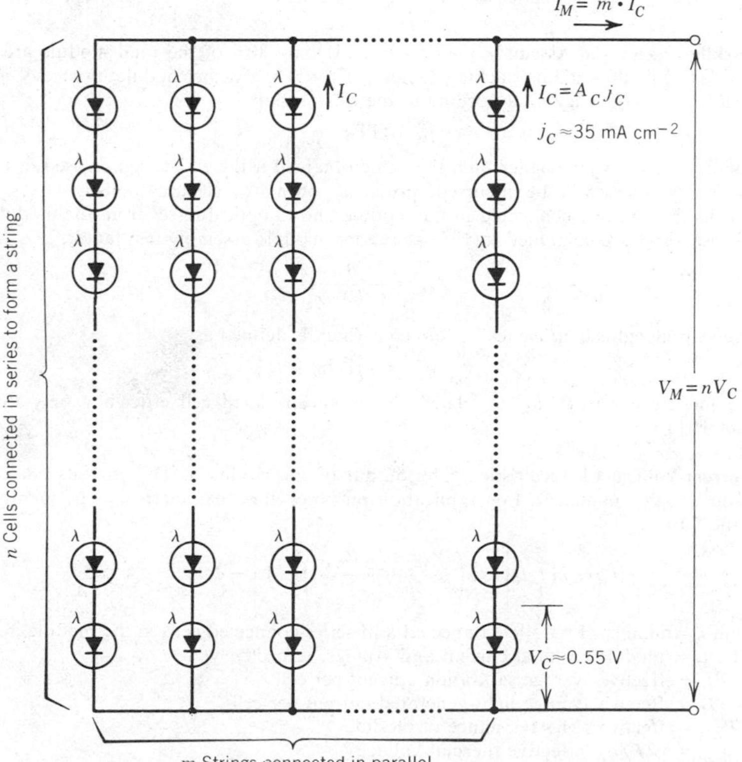

Fig. 15.75 Schematic of the series-parallel arrangement of solar cells within a module. An array may similarly be composed of series-parallel-connected modules, each module consisting of series-parallel-connected cells, as shown. Subscripts C and M refer to cell and module, respectively; A_c is the area of the individual cells.

where A_M = total area of module that intercepts flux from solar radiation
H = solar irradiance
η_M = module conversion efficiency

Thus a module of 1 m² area and a rated efficiency η_M of 12% has a rated peak power output $P_{M,\mathrm{mp,r}}$ of 120 W and, at 12-V rated output voltage $V_{M,\mathrm{mp,r}}$, delivers a current $I_{M,\mathrm{mp}}$ of 10 A.

Conversion Efficiency. Conversion efficiency is generally defined by inversion of Eq. (15.41):

$$\eta = \frac{P_{\mathrm{mp}}}{A \cdot H} \tag{15.42}$$

where P_{mp} and A relate to the appropriate values of photovoltaic systems, arrays, modules, or cells.

The efficiency or power output ratings for terrestrial applications are generally based on a "rated" solar irradiance H_r of 1.0 kW/m², which approximates the peak solar flux available at sea level under airmass 1 (AM1) conditions. Some organizations prefer to base efficiency ratings on a spectral distribution of sunlight corresponding to airmass 1.5 or 2, as more closely resembling the spectral distributions available during much of the year, (see ASTM standards E891-82 and E892-82). The ratings also apply at a temperature of 25 or 28°C, or at a set of "normal operating conditions."

Modules or cells for space application are rated at airmass 0 with irradiance of 1.353 kW/m² at 28°C. Because of the difference in spectral distribution, silicon solar cells show an airmass 1 efficiency approximately 1.1 times as high as their airmass 0 efficiency.

All conversion efficiency ratings are based on the total area A of the device considered and on the total radiant solar energy flux, that is, the entire solar spectral distribution

$$H = \int_0^{\infty} H(\lambda)\, d\lambda \tag{15.43}$$

where $H(\lambda)$ is the spectral irradiance and λ is the wavelength.

Module Packing Factor and Assembly Loss Factor. If only 80% of the total module area actually comprises solar cells, the module packing factor (PF) is 0.8 and the module efficiency η_M will be smaller than the cell efficiency η_C according to the relationship

$$\eta_M \leqslant (\mathrm{PF}) \cdot \eta_C \tag{15.44}$$

η_M will usually be somewhat smaller than the value obtained if the equals sign is used in Eq. (15.44) because assembly losses will be incurred, primarily owing to electrical mismatch between the individual cells that are interconnected in the module and to optical losses from the module window material. These losses are combined to yield the term "module assembly loss factor," ℓ_M

$$\ell_M = \frac{\eta_M}{(\mathrm{PF}) \cdot \eta_C} \tag{15.45}$$

Often an "encapsulated cell efficiency" is quoted, which is defined as

$$\eta_{\mathrm{EC}} = \ell_M \cdot \eta_C \tag{15.46}$$

so that the module efficiency η_M is related to the encapsulated cell efficiency only through the packing factor (PF).

Module Current-Voltage Characteristic. The output of the module is DC, and its current-voltage characteristic is highly nonlinear. For application purposes, this characteristic is frequently approximated by the form

$$I_M \approx m\left\{ I_0\left[\exp\left(\frac{V_M - I_M R_{s,M}}{n V_{\mathrm{th,eff}}}\right) - 1\right] - I_L \right\} \quad \mathrm{A} \tag{15.47}$$

where m and n = number of parallel-connected and series-connected cells in the module, respectively
I_M and V_M = module terminal current and voltage, respectively
I_0 = effective average saturation current per cell
I_L = effective average light-generated current per cell
$R_{s,M}$ = effective series resistance of module
$V_{\mathrm{th,eff}} = NkT/q$, "effective thermal voltage"
kT/q = 26 mV is the "thermal voltage" of the ideal diode at 300K.

The factor N facilitates approximation of the actual, more complicated I-V characteristic by a single exponential. The value of N is usually between 1 and 2, often closer to 2, but a value as high as 5 or 6 is sometimes seen. In the representation of Eq. (15.47), I_0 is not related to the saturation current of the "ideal" *pn*-junction diode but is orders of magnitude larger. Its value depends on the magnitude of N. It has to be observed that the approximation (15.47) is valid only when all cells in the module

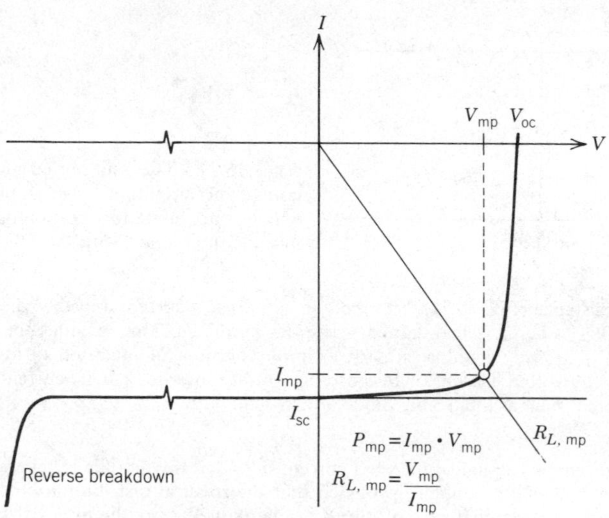

Fig. 15.76 General current-voltage characteristic of a solar cell or photovoltaic module. The part of the characteristic of interest is shown in the fourth quadrant of the current-voltage plane, which corresponds to the cell's (or the module's) performance as a generator.

have approximately equal *I-V* characteristics, when all cells experience the same irradiance, and when all cell interconnections are intact and of approximately the same resistance.

Output Power. The power output of the module is given by

$$P_M = I_M \cdot V_M \qquad \text{W} \tag{15.48}$$

Introducing I_M from Eq. (15.47) into Eq. (15.48) and finding the maximum of the product for a given value of I_L yields the maximum power point voltage $V_{M,\text{mp}}$ and, using this value in Eq. (15.47), the maximum power point current $I_{M,\text{mp}}$ and as their product the maximum power output $P_{M,\text{mp}}$. The maximum power output is often expressed by

$$P_{M,\text{mp}} = I_{M,\text{sc}} \cdot V_{M,\text{oc}} \cdot (\text{FF}) \qquad \text{W} \tag{15.49}$$

FF is called the fill-factor; it is determined by a large number of device attributes (Fig. 15.76).

Open Circuit Voltage. The open circuit voltage $V_{M,\text{oc}}$ is obtained from Eq. (15.47) by setting $T_M = 0$

$$V_{M,\text{oc}} = nV_{\text{th,eff}} \ln\left(\frac{I_L + I_0}{I_0}\right) \qquad \text{V} \tag{15.50}$$

Short-Circuit Current. The current I_M for $V_M = 0$ according to Eq. (15.47) is defined as the short-circuit current $I_{M,\text{sc}}$. In many practical cases it is

$$I_{M,\text{sc}} \approx -mI_L \qquad \text{A} \tag{15.51}$$

The approximation (15.51) is valid when the voltage drop across the series resistance is

$$I_{M,\text{sc}} \cdot R_{s,M} \ll nV_{\text{th,eff}} \ln \frac{I_L}{I_0} \qquad \text{A} \tag{15.52}$$

Simplified Equivalent Circuit. Equation (15.47) suggests the commonly used simplified equivalent circuit of a constant-current generator, producing $m \cdot I_L$, shunted by a forward-biased diode through which the current

$$I_{D,M} = m \cdot I_0 \left[\exp\left(\frac{V_M - I_M R_{s,M}}{nV_{\text{th,eff}}}\right) - 1\right] \qquad \text{A} \tag{15.53}$$

flows as a result of the diode bias voltage $V_{D,M} = V_M - I_M R_{s,M}$, with the series resistance $R_{s,M}$ which, to permit an expression of the form of Eq. (15.47) or (15.53), is assumed to be a single

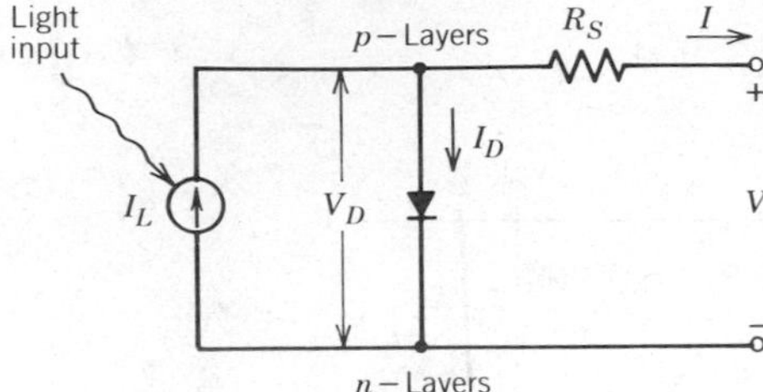

Fig. 15.77 The simplified lumped-constant equivalent circuit of a solar module (or cell), which is adequate for most considerations of the module interaction with the DC load.

lumped-constant resistance (Fig. 15.77). The module will draw a current and thus dissipate energy as long as a voltage $V_M > V_{M,oc}$ is maintained across its terminals. This includes the nonilluminated case. Thus when $V_M > V_{M,oc}$ could occur, for example, because of inclusion of an energy storage subsystem in the photovoltaic system or because of a utility intertie, a back-current limiting device has to be incorporated, such as a blocking diode or a switch.

Light-Generated Current. The light-generated current $m \cdot I_L$ of the module is generally proportional to the total irradiance H of the sunlight, provided that the spectral distribution of the incident light remains constant and that irradiance uniformity is maintained over the area of the module. This proportionality usually holds over a substantial number of decades in irradiance, up to moderate concentration ratios ($\sim 100\times$), in most currently produced modules composed of crystalline silicon or gallium-arsenide solar cells.

Partial Shading or Localized Interconnect Failures. A potential failure mechanism to be avoided is the development of "hot spots" because of nonuniform illumination, particularly partial shading, or as a result of localized interconnect failure. In these cases individual cells can be driven into reverse-bias condition, where they dissipate substantial amounts of power generated by the remaining

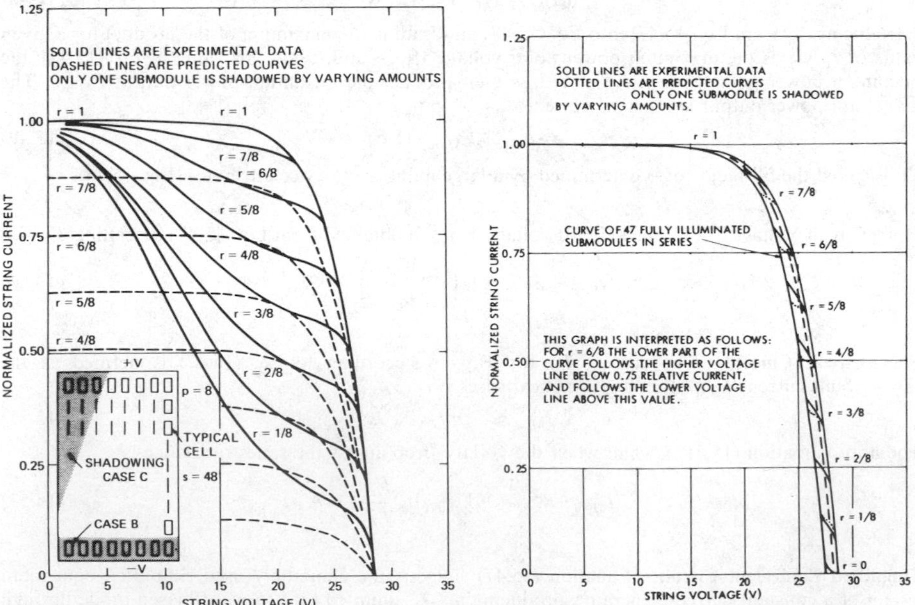

Fig. 15.78 Current-voltage (I-V) curves of a string of 48 series-connected submodules which each have eight parallel-connected cells. Shading is according to case C in the insert and the parameter r expresses the fraction of cells that are fully illuminated in the most heavily shaded submodule. (a) I-V curves for a module without shunt diodes. (b) I-V curves for the same module with shunt diodes. From Rauschenbach,[2] pp. 77–78, with permission.

(*a*)

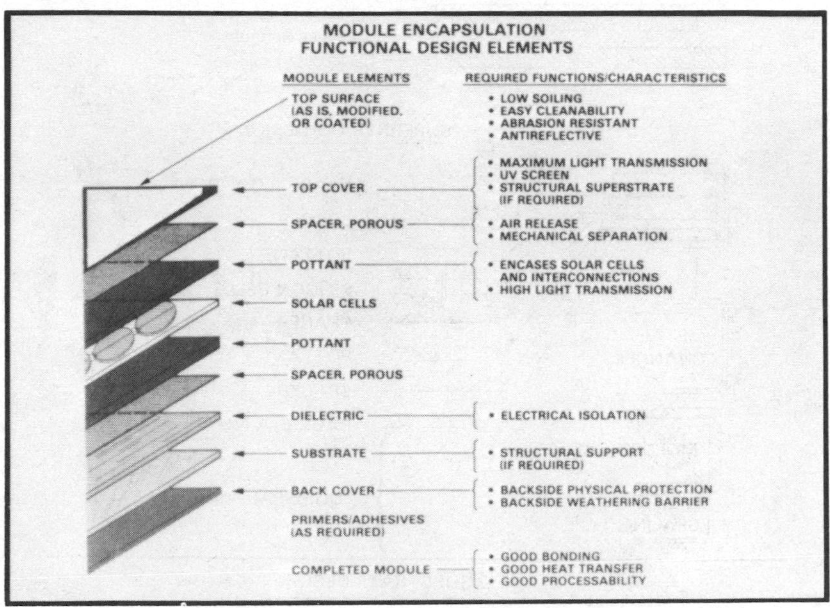

(*b*)

Fig. 15.79 (*a*) Illustration of major hazards to which photovoltaic solar modules are exposed and against which the encapsulation is designed to protect the solar cells. (*b*) Elements of the solar cell encapsulation that are part of the module assembly and the purpose and major characteristics of each. From DOE/JPL brochure on the low-cost solar array project, Jet Propulsion Laboratory Report No. 5101-178, Rev. C (Jul. 1981), with permission.

cells of the series string. This can lead to the formation of hot spots with sometimes catastrophic consequences to the module encapsulation or the cells themselves. Appropriate series/parallel arrangement of the cells and/or provision of bypass diodes can eliminate the hot-spot problem. In addition, the provision of bypass diodes can considerably increase the available power output under partial shading or localized interconnect failure conditions (Fig. 15.78).

Module Encapsulation

An important function of module encapsulation is to provide mechanical strength and reduction of environmental influences (Fig. 15.79). The typical operating design life for a terrestrial module is 20 years. The module not only has to be mechanically self-supporting through all handling but has to be able to withstand, without damage, all reasonable wind, hail, and snow loads when it is attached to the supporting structure of the array. A 3-mm-thick sheet of glass is usually adequate for providing this strength (structural superstrate) (Fig. 15.80).

The second function of the encapsulation is, as a minimum, to prevent electrolytic corrosion from altering the solar cells and their interconnections, which constitute a multimaterial system. This

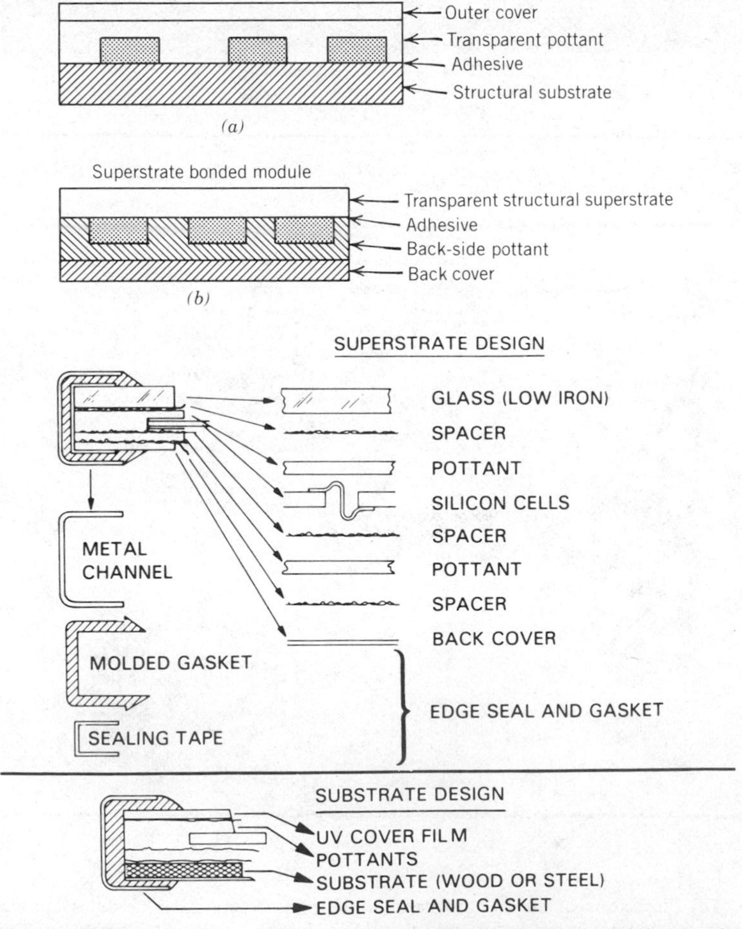

Fig. 15.80 Design of superstrate-bonded (*a*) and substrate-bonded (*b*) modules. In the first, the window forms the principal structural element to assure the mechanical integrity of the module. In the second, the structural substrate provides the mechanical strength, while the window serves only for protection from environmental influences. From "Photovoltaic Module Encapsulation Design and Materials Selection: Vol. I," pp. 3-2 to 3-3, Jet Propulsion Laboratory Report No. 5101-177 (Jun. 1982) with permission.

means excluding moisture penetration. As practically all plastics are permeable by moisture, an aluminum foil is often used together with a plastic film, such as Mylar, for the rear encapsulation. EVA (ethyl vinyl acetate) has evolved as the preferred, low-cost bonding material. Bonding of the edges of the assembly requires special attention to avoid delamination.

An important aspect of design and material selection is the limitation of stress during the severe temperature cycling which the modules experience in their operaton. Such stresses can lead to fatigue failure of electrical connections or of the moisture barrier.

Module Terminals

The module normally has a terminal box at which the connection of the module to the array bus system is made. The terminal box needs to meet applicable safety standards, as reflected in electrical codes. Attention needs to be paid to encapsulation requirements against degrading environmental influences and to protection against accidental electrical shock. The solar cells, and consequently the module, carry a voltage at their terminals whenever they are illuminated.

15.5-4 Solar Cells

A large number of semiconducting materials exist which have properties potentially suitable for photovoltaic solar energy conversion. However, only a few materials have so far been successfully applied to this purpose, and the only cells commercially available are prepared from single-crystal or large polycrystalline silicon. The latter material is often designated semicrystalline silicon.

Basic Mechanisms of Solar Cell Operation

All approaches to solar cell design are based on the utilization of two principal mechanisms. The first is the generation, by absorption of photons from the solar spectrum, of charge carriers that can move freely in the semiconducting material. The second is the separation of these mobile carriers, which contain equal numbers of positive and negative charges, by means of an electrostatic potential step provided within the semiconducting material (Fig. 15.81) or at its surface. This potential step is usually called potential barrier. As a result of this separation, the charge carriers can flow through an electrical circuit external to the device and can create a voltage across the terminals of the device.

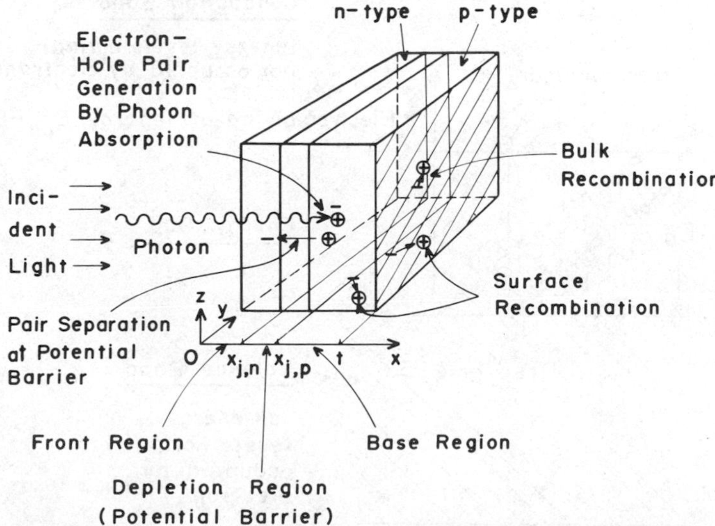

Fig. 15.81 The principal operating mechanisms of photovoltaic solar cells and the nomenclature used in the description of these mechanisms. Depicted is a cell with a *pn* junction as the potential barrier and with distinct *n*-type and *p*-type semiconducting layers as the front and the base regions.

To achieve high efficiency in this process that converts photons from the free flux of energy from the sun into electrical power that can be dissipated in an arbitrary load, a number of detailed physical mechanisms must be optimized. However, some fundamental physical performance limitations are connected with the two principal operating mechanisms which cannot be overcome by improved technological approaches.

Fundamental Performance Limitations. The first limitation is connected with the process of charge carrier generation by photon absorption. Only photons of energy equal to or greater than the bandgap energy E_G of the chosen semiconductor can interact with the valence band electrons and thus generate free electron-hole pairs (Fig. 15.82). This is reflected in the absorption spectra of the semiconductors (Fig. 15.83). These absorption characteristics leave out a substantial photon flux in the long wavelength tail of the solar spectrum, which is not absorbed (Fig. 15.84).

Any energy of the photons that exceeds the energy gap of the semiconductor as present in photons of light wavelengths smaller than those corresponding to the energy gap will normally be dissipated as heat in the semiconductor, since the electron-hole pairs tend to keep only the minimum energy required for their existence, that is, the energy of the bandgap. Only when the energy of the photon is greater than three times the energy gap can there be a generation of two electron-hole pairs per photon. Thus once a semiconducting material has been selected for photovoltaic conversion, the maximum fraction of the energy content of the solar flux that can be utilized in the generation of free charge carriers (electron-hole pairs) is fixed (Fig. 15.85).

The fraction of the total photon flux that can thus lead to electron-hole pair generation results in the ideal light-generated current density $j_{L,\mathrm{id}}$

$$j_{L,\mathrm{id}} = q\int_0^{\lambda = hc/E_G} N_{\mathrm{Ph}}(\lambda)\,d\lambda \qquad \mathrm{A\ cm^{-2}} \tag{15.54}$$

The ideality lies in the assumption that all photons of $E_{\mathrm{Ph}} \geqslant E_G$ in the solar radiation actually enter the semiconductor, that they are absorbed therein, that each photon creates an electron-hole pair, and that each electron-hole pair created will actually be separated to contribute to the light-generated current (Fig. 15.86).

The second group of (two) fundamental performance limitations is connected with the current-voltage characteristic that results from the solar cell being a semiconductor device with a potential barrier. While the electron-hole pairs generated by photon absorption each have the bandgap energy E_G, which corresponds to a voltage $V_G = E_G/q$ (q = electronic charge = 6.10^{-19} A s), the maximum voltage that can be obtained from the solar cell is the open circuit voltage V_{oc}. This gives rise to the

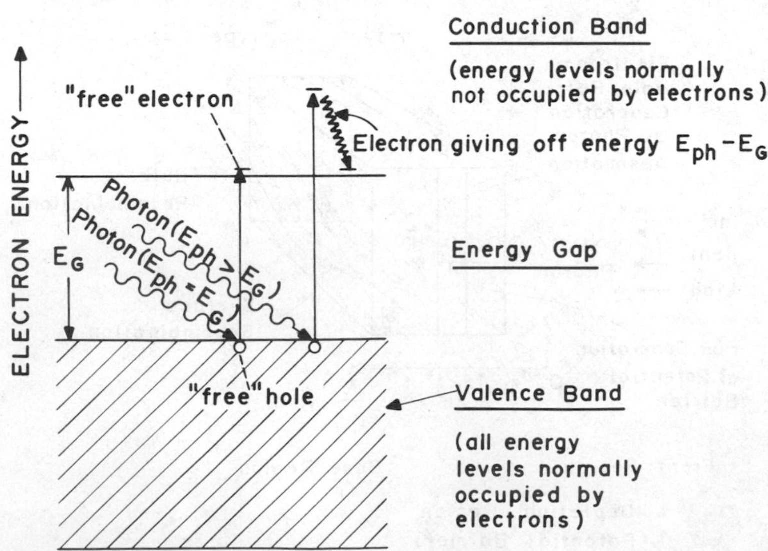

Fig. 15.82 Energy band diagram of a semiconductor, showing the energy gap between the valence band and the conduction band and the process of photon absorption with generation of a free electron-hole pair.

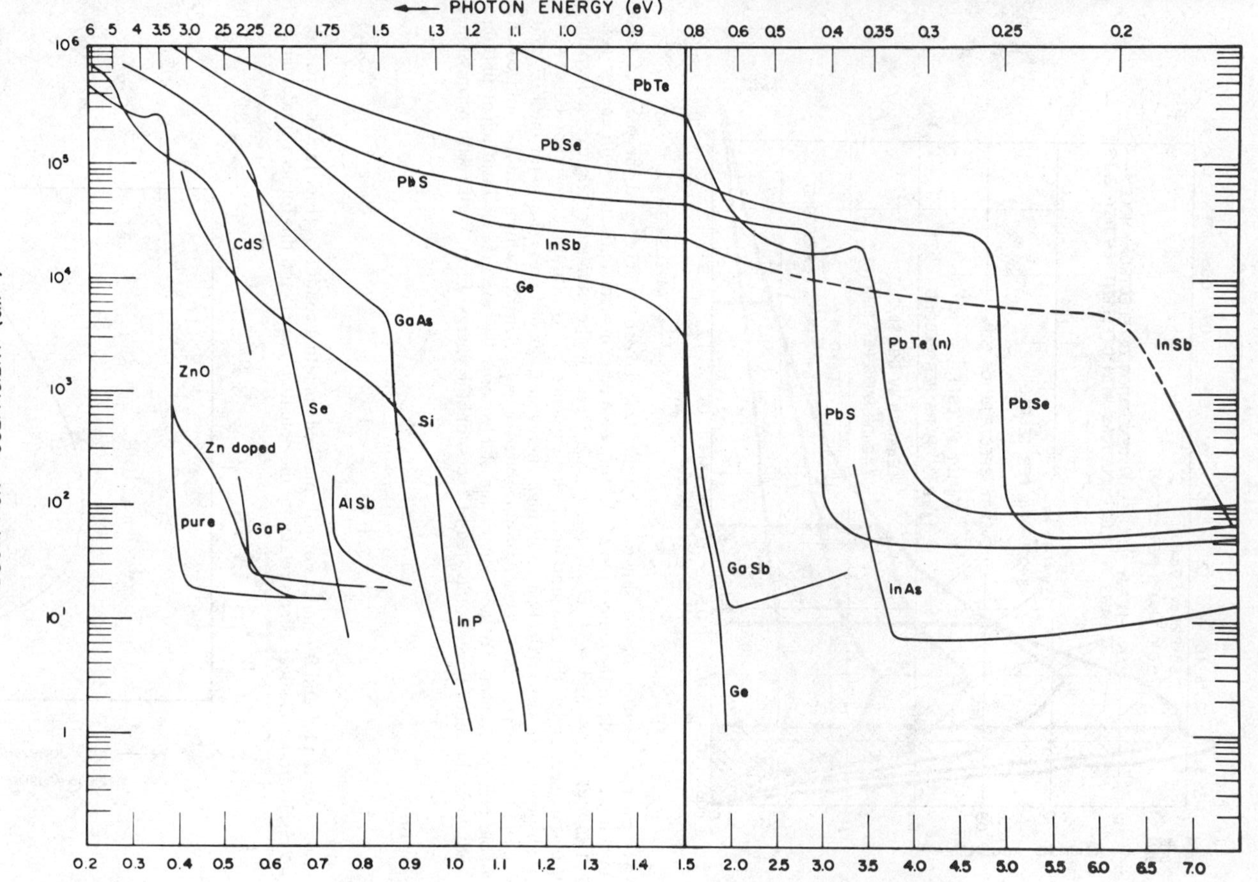

Fig. 15.83 Absorption spectra of various semiconductors. Each semiconductor exhibits fall off of the absorption ("absorption edge") at a typical wavelength, which corresponds to the energy gap E_G.

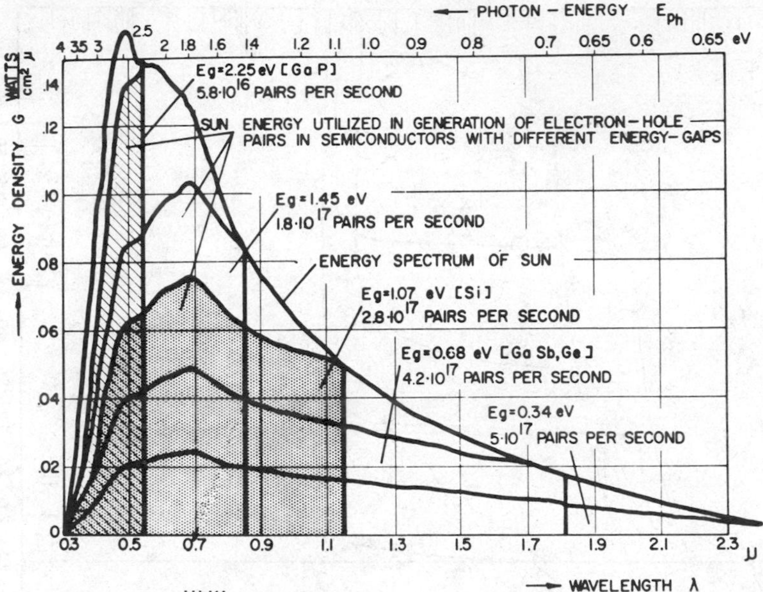

Fig. 15.84 Energy spectrum of the sun on a bright clear day at sea level, and the parts of this spectrum that are useable in the generation of electron-hole pairs in semiconductors with energy gaps of 2.25, 1.45, 1.07, 0.68, and 0.34 eV. Listed for each case is the number of electron-hole pairs generated, obtained under the assumption of the existence of an abrupt absorption edge with complete absorption and zero reflection on its high energy side. Shaded areas shown for the 1.07- and 2.25-eV energy gap depict the wavelength-integrated energy density, which can be utilized in electron-hole pair generation. To the right of the shaded area lies the area of insufficient photon energy and above the shaded area is the area of excess photon energy. From Wolf,[6] with permission.

voltage factor

$$VF = \frac{V_{oc}}{V_G} < 1 \tag{15.55}$$

VF is usually between 0.5 and 0.7 under normal solar cell operating conditions (Fig. 15.87). A part of this factor results from statistical effects related to the potential barrier, and this forms a fundamental limitation (Fig. 15.88).

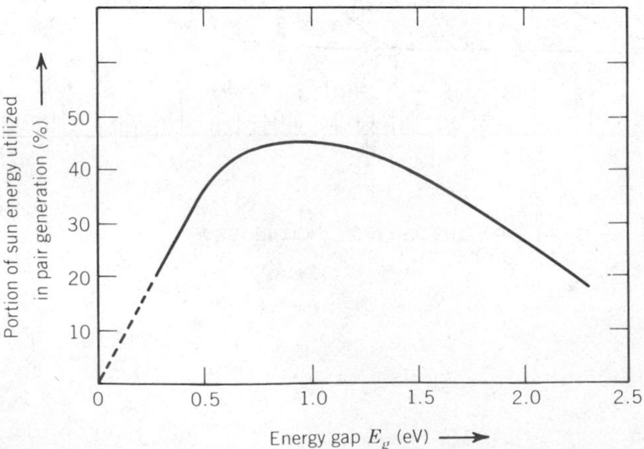

Fig. 15.85 Fraction of the energy contained in the solar radiation flux that can be used in the generation of electron-hole pairs, shown as a function of the energy gap. From Wolf,[6] with permission.

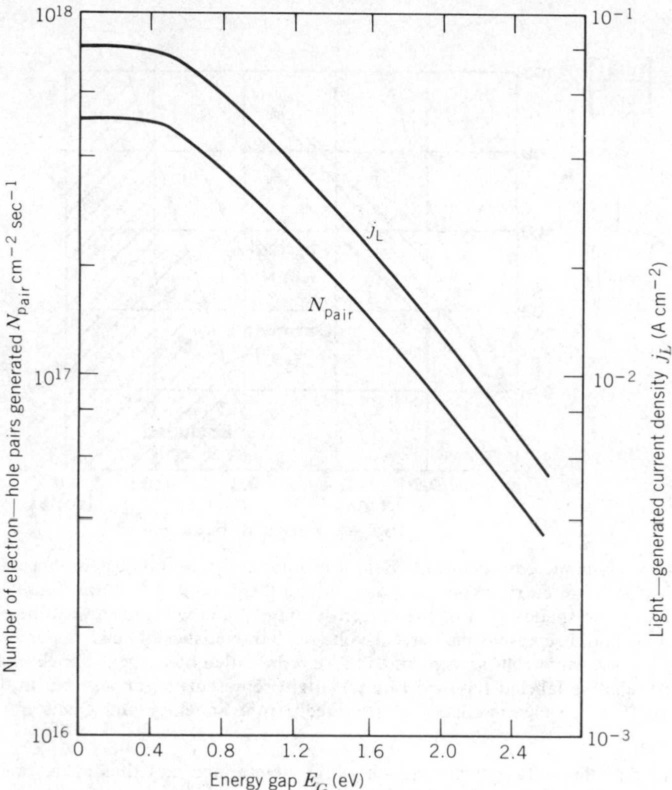

Fig. 15.86 Number of electron-hole pairs generated (per unit area and unit time) plotted as a function of the energy gap corresponding to data in Figs. 15.84 and 15.85. This number gives the maximum theoretically possible light-generated current density j_L, which is also shown as a function of the energy gap. From Wolf,[6] with permission.

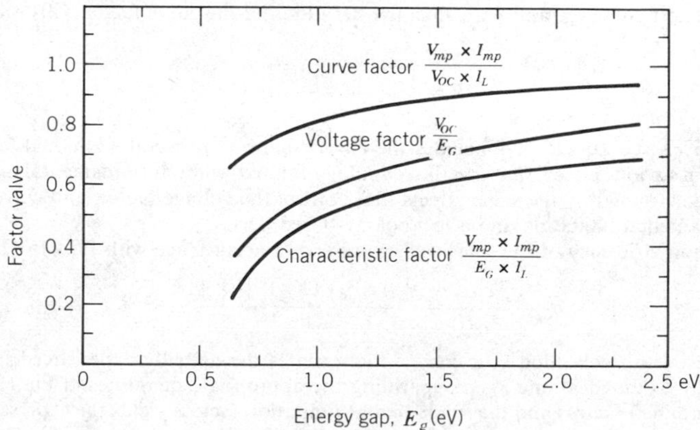

Fig. 15.87 Voltage factor and curve factor as functions of the energy gap. Both factors relate to the ideal diode current-voltage relationship and thus provide upper limits to the actually achievable values. Both contain material- and technology-dependent parameters, such as the minority carrier lifetime and mobility. The values obtained for these parameters in semiconductors of differing bandgaps are often far from the values expected on the basis of fundamental physical limitations. In an attempt to eliminate this technology-based influence, the best values achieved on the furthest developed materials, silicon and gallium-arsenide, are usually applied for all materials of differing energy gaps, as was done in the preparation of these curves. From Wolf,[6] with permission.

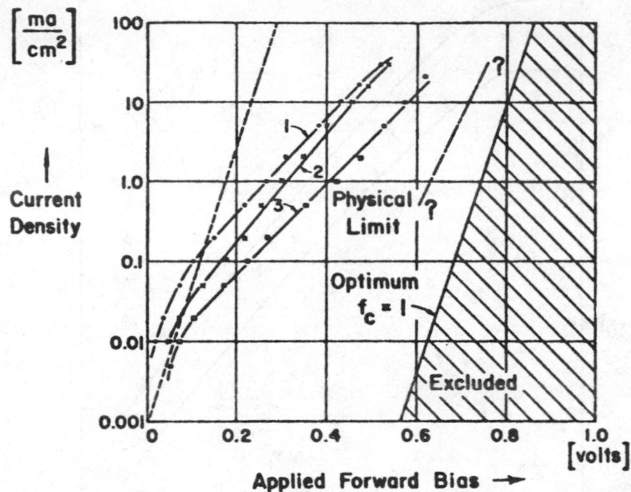

Fig. 15.88 Thermodynamic considerations yield a minimum saturation current density j_0 for silicon and consequently an ideal current-voltage characteristic ("Optimum $f_c = 1$" in figure). The slope of this curve is q/kT. The shaded part of the current-voltage plane will thus always be excluded. The curves labeled 1, 2, and 3 represent the current-voltage characteristics of cells prepared around 1960. The characteristics of current cells are similar to those represented by curve 3. Shockley and Queisser[7] speculated that the curve labeled Physical Limit?? might represent the upper limit to the practically achievable current-voltage characteristics. Reproduced from Shockley and Queisser,[7] with permission.

The remainder of the voltage factor is technology determined and thus open for improvement. The major influence on this technology-determined part of the voltage factor is the recombination of electron-hole pairs in excess of the fundamentally required one.

The power maximally extractable from a unit area of solar cell is

$$P_{\mathrm{mp}} = j_{\mathrm{mp}} \cdot V_{\mathrm{mp}} \qquad \mathrm{W\ cm^{-2}} \tag{15.56}$$

Here $j_{\mathrm{mp}} < j_L$ and $V_{\mathrm{mp}} < V_{\mathrm{oc}}$. If only the idealized current-voltage characteristic of a device with a potential barrier, such as that given by Shockley's diffusion theory for *pn*-junction devices, is considered for determining j_{mp} and V_{mp}, then the definition of the curve factor (CF) is obtained (Fig. 15.87).

$$\mathrm{CF} = \frac{(j_{\mathrm{mp}} \cdot V_{\mathrm{mp}})_{\mathrm{id}}}{j_L \cdot V_{\mathrm{oc}}} < 1 \tag{15.57}$$

The fill-factor (FF) [see Eq. (15.71)] includes the curve factor and several additional factors, all < 1, which result from various effects that are all technology limited, without fundamental limitations. The curve factor is determined by the same effects that control the voltage factor, and correspondingly it includes a fundamental limitation and a technology-based part.

The conversion efficiency of the solar cell, expressed in accordance with Eq. (15.42), is then

$$\eta_C = \frac{P_{C,\mathrm{mp}}}{A_C H} = \frac{\gamma_C \cdot j_{L,\mathrm{id}} \cdot (\mathrm{VF}) \cdot (\mathrm{FF}) \cdot E_G}{q \cdot H} \tag{15.58}$$

where γ_C is the overall collection efficiency, which relates the actually experienced light-generated current density j_L to the ideal one, $j_{L,\mathrm{id}}$. Separating the appropriate quantities in Eq. (15.58) into the technology-determined factors and the fundamental-limitation factors yields the expression

$$\eta_C = \underbrace{\left[\gamma_C (\mathrm{VF})_{\mathrm{technol}} \cdot \frac{(\mathrm{FF})}{(\mathrm{CF})} \cdot (\mathrm{CF})_{\mathrm{technol}} \right]}_{\text{technologically determined factors}} \cdot \eta_{\mathrm{fund}} \tag{15.59}$$

where the fundamentally based efficiency η_{fund} contains 3 factors

$$\eta_{\mathrm{fund}} = (\mathrm{VF})_{\mathrm{fund}} \cdot (\mathrm{CF})_{\mathrm{fund}} \cdot \frac{j_{L,\mathrm{id}} \cdot E_G}{qH} = (\mathrm{VF})_{\mathrm{fund}} \cdot (\mathrm{CF})_{\mathrm{fund}} \cdot \frac{E_G \int_0^{\lambda = hc/E_G} N_{\mathrm{Ph}}(\lambda)\, d\lambda}{\int_0^{\infty} N_{\mathrm{Ph}}(\lambda) \frac{hc}{\lambda}\, d\lambda} \tag{15.60}$$

The last of these three factors represents the losses due to insufficient and excess photon energy, which were presented in Fig. 15.85 as functions of the energy gap for devices that are prepared with one dominating bandgap and a single potential barrier. The basic voltage factor $(VF)_{fund}$ is about 0.76 (for silicon at 300K) and the basic curve factor $(CF)_{fund}$ is about 0.96.

The fundamental losses, in combination, for silicon solar cells thus amount to ~68% of the available solar energy flux. Because of their energy gap dependence, fundamental losses have a strong influence on the selection of semiconductors to be used for solar cells. However, the technology-determined losses today present an additional factor near 0.5, and these command more interest because of the possibility of considerably reducing them through expenditure of engineering efforts, in some cases to near-negligible levels. Principal causes of technology-based losses connected with light-generated current, open-circuit voltage, and fill factor are as follows:

1. Light-generated current

(a) Optical surface properties (reflection)
(b) Contact coverage
(c) Incomplete absorption (thickness)
(d) Recombination outside depletion region (bulk and surface, including contacts)
(e) "Dead layers"

2. Open-circuit voltage

(a) Recombination outside depletion region (bulk and surface, including contacts)
(b) Bandgap narrowing
(c) "Current leakage"

3. Fill factor

(a) Same as (a) for open-circuit voltage
(b) Same as (b) for open-circuit voltage
(c) Same as (c) for open-circuit voltage
(d) Recombination in depletion region
(e) Series resistance

Multibandgap Systems. The possibility exists for avoidance of part of the fundamental losses by use of two or more semiconductors of different bandgaps, each including a potential barrier. In this way a better match between the solar spectrum and that part of the spectrum that can be utilized for electron-hole pair generation can be achieved. Use of a very large number of semiconductors with different bandgaps should ideally reduce the losses due to insufficient and excess photon energy to very small values. As this would greatly reduce the light-generated current in each cell of the system, the voltage and curve factors would be decreased, which would somewhat counteract the benefits of using a large number of semiconductors. An intermediate number of semiconductors thus should lead to an optimum multibandgap system.

There are two basic approaches to the realization of multibandgap systems. One is use of spectrally selective optical filters to split up the incident beam and to direct the individual parts of the solar spectrum to the cells with the appropriate bandgaps (Fig. 15.89). The other approach utilizes the

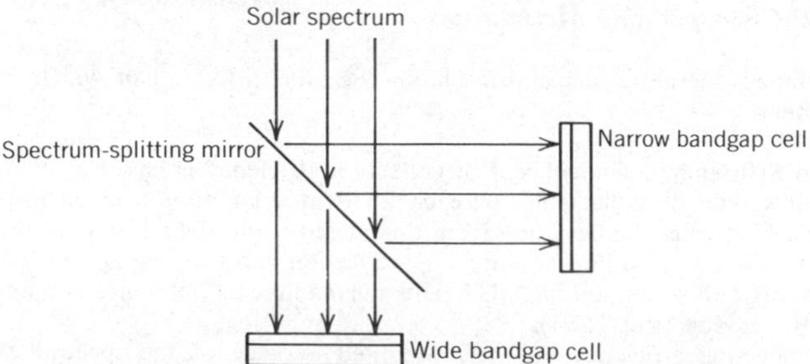

Fig. 15.89 Schematic representation of a multibandgap solar cell system using optical beam splitting, for example, via a dichroic mirror. The approach requires that the two cells be geometrically separated so as to intercept two separate optical beams. While this necessitates an awkward arrangement, it carries the advantage that the cells can be easily cooled and easily matched to separate electrical circuits. From Fonash,[8] p. 175, with permission.

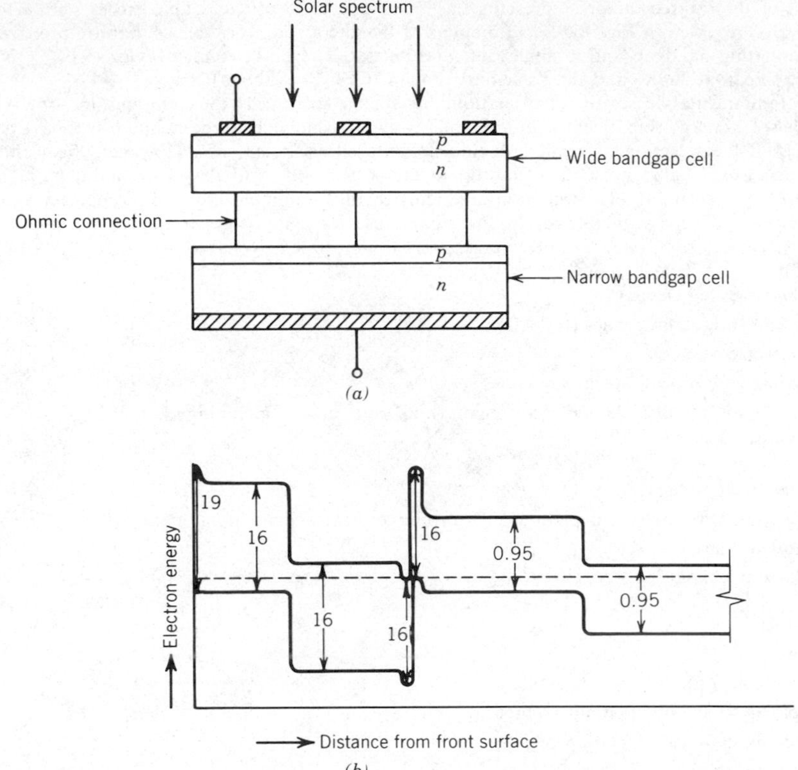

Fig. 15.90 (*a*) Physical arrangement of a cascaded multibandgap solar cell system, using the example of the two-cell stack. (*b*) Energy band diagram of a two-cell stack. The two cells in the stack have to be electrically connected through ohmic contacts or their equivalents (e.g., tunnel junctions), while at the same time their interfaces have to be designed for maximum optical transmission. From Fonash,[8] pp. 175–177, with permission.

fact that photons of energy less than bandgap energy are not absorbed in a semiconductor. The semiconductor can thus act as an optical filter, transmitting the longer wavelength part of the solar spectrum (Fig. 15.90). In this scheme the cells are mounted behind each other, the cell with the largest bandgap being exposed to the incident solar radiation and the remaining cells being arranged with consecutively decreasing bandgaps [Fig. 15.90(*b*)]. This approach is frequently called the stacked cell or cascaded cell system.

Additional Solar Cell Operating Mechanisms

A number of detailed operating mechanisms have been identified, all of which have some losses associated with them.

Overall Collection Efficiency. The concept of collection efficiency is based on the fact that ideally one charge carrier should be collectable for each incident solar photon of energy $E_{\text{Ph}} \geqslant E_G$. The collection efficiency expresses the deviation from this ideality, and the addition of the word "overall" designates that all effects of influence, such as reflection, absorption in the window or in the antireflection coating, and so on, are included. This means that the reference is the total photon flux at the outside of the module (Fig. 15.91).

The overall collection efficiency γ is the weighted average of the spectral overall collection efficiency $\gamma(\lambda)$, which is the more fundamental quantity

$$\gamma = \frac{\int_0^{\lambda = hc/E_G} N_{\text{Ph}}(\lambda)\gamma(\lambda)\,d\lambda}{\int_0^{\lambda = hc/E_G} N_{\text{Ph}}(\lambda)\,d\lambda} = \frac{\int_0^{\lambda = hc/E_G} H(\lambda)\gamma(\lambda)\lambda\,d\lambda}{\int_0^{\lambda = hc/E_G} H(\lambda)\lambda\,d\lambda} \tag{15.61}$$

1 One electron-hole pair per photon with $h\nu \geqslant E_G$ (loss due to insufficient photon energy).
2 The electron-hole pair keeps only energy E_G (loss due to excess photon energy). (**1** and **2** together cause ~50% loss of energy content of solar spectrum, if E_G is chosen.)
3 Separation of electrons and holes by built-in potential barrier (electrical generator; leads to ideal light-generated current).
4 Electron-hole pairs are generated in front, space-charge, and base regions. Some pairs recombine before reaching the potential barrier; leads to collection efficiency ~90%.
5 Some photons are lost due to reflection, absorption without electron-hole pair generation (in antireflection coating, ITO film, etc.) or transmission through wafer/film; leads to overall collection efficiency.

Fig. 15.91 Physical mechanisms leading to light-generated current $j_L(\approx j_{sc})$.

$\gamma(\lambda)$ depends on the device design and the electronic material parameters as they result from the fabrication processes applied.

In fact, only a fraction of the incident photons with $E_{\text{Ph}} \geqslant E_G$ is actually absorbed in the semiconducting material. Some photons are reflected; some are absorbed in certain layers of the device in which carrier generation does not occur, such as in antireflection coatings; some do not get to the semiconductor because part of its surface is shaded by metal contacts or otherwise made a "not active surface"; and finally, some that get into the semiconductor are transmitted through it and not absorbed. (Fig. 15.92). Thus the overall collection efficiency actually comprises the following factors

$$\gamma = (1 - S)(1 - R)(1 - A)(1 - T) \cdot \eta_{\text{coll}} \tag{15.62}$$

where S = fraction of projected surface of device that is shaded or otherwise made nonactive
R = reflectance of active area
A = any absorptance ahead of semiconductor
T = transmittance or fraction of photons that entered semiconductor but are not absorbed in it

S could be distributed over R and A but for practical reasons is usually considered separately. In addition, S is generally independent of spectral effects, while R, A, and T are averages of spectrally dependent quantities, weighted by the spectral distribution of the available light, so that

$$\gamma = (1 - S)\frac{\int_0^{\lambda = hc/E_G}[1 - r(\lambda)][1 - a(\lambda)][1 - t(\lambda)]\eta_{\text{coll}}(\lambda)H(\lambda)\lambda\, d\lambda}{\int_0^{\lambda = hc/E_G} H(\lambda)\lambda\, d\lambda} \tag{15.63}$$

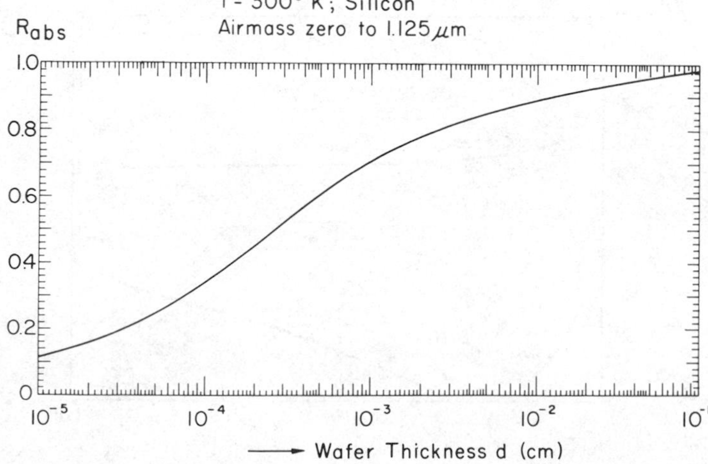

Fig. 15.92 Fraction R_{abs} of the total number of photons contained in the solar flux that is absorbed in a single pass through a silicon layer of thickness d. Data are based on the spectral distribution of airmass 0 sunlight up to the wavelength of 1.125 μm. The relatively large thickness needed to absorb a high fraction of the available photon flux results from the "indirect" bandgap of silicon.

where η_{coll} is the internal collection efficiency within the semiconductor itself, usually just called collection efficiency.

From Eq. (15.63) are derived the definitions

$$R = 1 - \frac{\int_0^{\lambda = hc/E_G} [1 - r(\lambda)] H(\lambda) \lambda \, d\lambda}{\int_0^{\lambda = hc/E_G} H(\lambda) \lambda \, d\lambda} \tag{15.64}$$

$$A = 1 - \frac{\int_0^{\lambda = hc/E_G} [1 - r(\lambda)][1 - a(\lambda)] H(\lambda) \lambda \, d\lambda}{(1 - R) \int_0^{\lambda = hc/E_G} H(\lambda) \lambda \, d\lambda} \tag{15.65}$$

$$T = 1 - \frac{\int_0^{\lambda = hc/E_G} [1 - r(\lambda)][1 - a(\lambda)][1 - t(\lambda)] H(\lambda) \lambda \, d\lambda}{(1 - R)(1 - A) \int_0^{\lambda = hc/E_G} H(\lambda) \lambda \, d\lambda} \tag{15.66}$$

and

$$\eta_{coll} = \frac{\int_0^{\lambda = hc/E_G} [1 - r(\lambda)][1 - a(\lambda)][1 - t(\lambda)] \eta_{coll}(\lambda) H(\lambda) \lambda \, d\lambda}{(1 - R)(1 - A)(1 - T) \int_0^{\lambda = hc/E_G} H(\lambda) \lambda \, d\lambda} \tag{15.67}$$

Antireflection Treatments. Most semiconductors have a high dielectric constant and consequently a high refractive index, which in turn results in a high optical reflectance. For silicon, this reflectance is above 30%. To reduce the impact of the reflectance on overall collection efficiency, the solar cell front surface must be given an antireflection treatment.

Single-layer antireflection interference coatings are commonly used with a quarter wavelength thickness and a refractive index n_c

$$n_c \approx \sqrt{n_1 \cdot n_2} \tag{15.68}$$

where n_1 and n_2 are the refractive indices of the two materials bordering on the coating. Such coatings provide, on silicon solar cells, a reflectance of 7 to 10%. Multilayer antireflection coatings, as used on very high-performance devices, reduce the reflectance to 2 to 3% (Fig. 15.93).

Another approach to reduction of reflection is structuring of the front surface into a multitude of pyramids, each about 10 to 30 μm high, rather than use of the more common planar front surface. Such "texturing" of the front surface reduces its reflectance to the 10 to 12% range; applying a single-layer antireflection coating in addition to texturing reduces reflectance to 2 to 3%. Texturing is

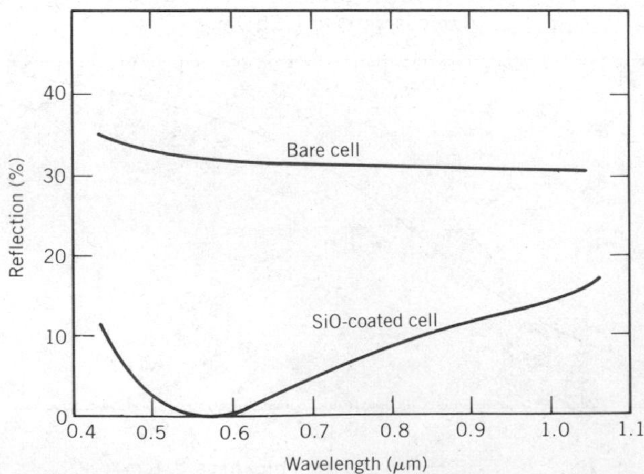

Fig. 15.93 Reflectance of a solar cell with an untreated silicon surface ("bare") boundaring to air, shown as a function of wavelength, and for comparison, reflectance of a cell equipped with an SiO antireflection coating. From Ralph and Wolf,[9] with permission.

easily accomplished by applying a chemical etch to single-crystal silicon of (100) crystalline orientation of the front surface. In addition to causing reduced reflection, the texturing has the advantage of causing oblique penetration of the photons into the solar cell and thus increasing their chance for absorption.

As far as the module is concerned, there usually is additional reflection at the surface of the transparent cover material. In some cases, antireflection treatments are applied there also, particularly texturing of the front surface of glass covers.

Absorption Without Electron-Hole Pair Generation. The primary seats of absorption without electron-hole pair generation are in the antireflection coating and, with regard to the module, in the transparent cover material, which usually is glass, and possibly in the adhesive used to bond the cover glass to the cells. The amount of cover absorption often is reduced by the use of glass of low iron content rather than the usual window glass. Use of higher quality antireflection coatings on the cells usually reduces absorption there to negligible values.

Other locations of unproductive absorption can be passivation layers on solar cells, which are sometimes used to improve the electronic properties, and transparent conductive coatings, which are often applied to thin-film solar cells. For the latter coatings, a tradeoff is made between electrical conductance and optical transmission. In addition, some solar cells that include very high impurity concentrations near the front surface have exhibited near-zero collection efficiency from this very highly doped layer, which therefore has obtained the name "dead layer."

Electron-Hole Pair Generation. In the discussion of the fundamental limitations, photon absorption was treated as if all photons with energy equal to or greater than the bandgap energy could be absorbed in the solar cell. In reality, the onset of absorption at the energy gap is not that abrupt. The absorption coefficient is a function of wavelength (Fig. 15.83) so that the average photon penetration depth ranges from a tenth of a micrometer or less to centimeters in the spectral range of interest. In some semiconductors, called indirect bandgap semiconductors, such as silicon, this change of $\alpha(\lambda)$ is very gradual and extends throughout the solar spectrum, while in direct bandgap semiconductors, such as gallium-arsenide, most of this change of $\alpha(\lambda)$ occurs in a narrow wavelength band near the bandgap wavelength. Consequently, in all semiconductors substantial generation of electron-hole pairs occurs right below the front surface, while in some a sizeable fraction of the photons penetrates deep into the device and a smaller fraction is even transmitted through it, depending on the thickness of the semiconductor used (Fig. 15.92). To reduce the size of this transmitted fraction, some silicon solar cells have been fabricated with optical back surface reflectors to provide the photons a second path through the device and thus increase their chance for absorption.

Charge Carrier Collection. The purpose of the collection process is to turn the photon absorption process into a useful current-generating mechanism and to prevent recombination of the electron-hole pairs. The means for accomplishing collection is an electrostatic potential step built into the semiconducting device (Fig. 15.82). Such a potential step, generally called potential barrier, occurs at a *pn* junction, which is a transition between *p*-type and *n*-type semiconducting materials; at a Schottky barrier, which is a metal-to-semiconductor transition; at a metal–insulator–semiconductor (MIS) or an electrolyte–semiconductor interface; or at any corresponding structure. The potential step occurs over a thin layer in which the concentration of mobile charges is normally decreased in comparison to the majority carrier concentrations in the adjacent, electrically neutral regions of the semiconductor. Consequently, such a layer is generally called a depletion region or less specifically a space charge region.

A potential step across a layer of finite thickness creates an electric field within this layer. The direction of the electric field is such that minority carriers present in a region adjacent to the potential barrier will be accelerated by the field away from this region, if they reach the edge of the layer with the field. These minority carriers thus end up on the other side of the depletion region, where they are majority carriers, that is, charge carriers of the same type as those resulting from the dopant impurity imbedded in that region.

The existence of such a minority carrier sink at the edge of a region where electron-hole pairs are generated results in a density gradient of the minority carriers in that region and consequently diffusion of these carriers toward the sink. This flow of charge carriers constitutes a current. The charge carrier transport by diffusion can be assisted by a properly directed electrostatic field built into the neutral regions of the solar cell. Such carrier transport due to an electric field is called drift. In total, four currents result from this mechanism, one each originating in the front region and the base region and two originating in the depletion region [Fig. 15.94(*a*)]. The sum of these four currents is the light-generated current. It is important that the solar cell be designed and constructed so that the vast majority of the minority carriers generated by photon absorption will reach the depletion

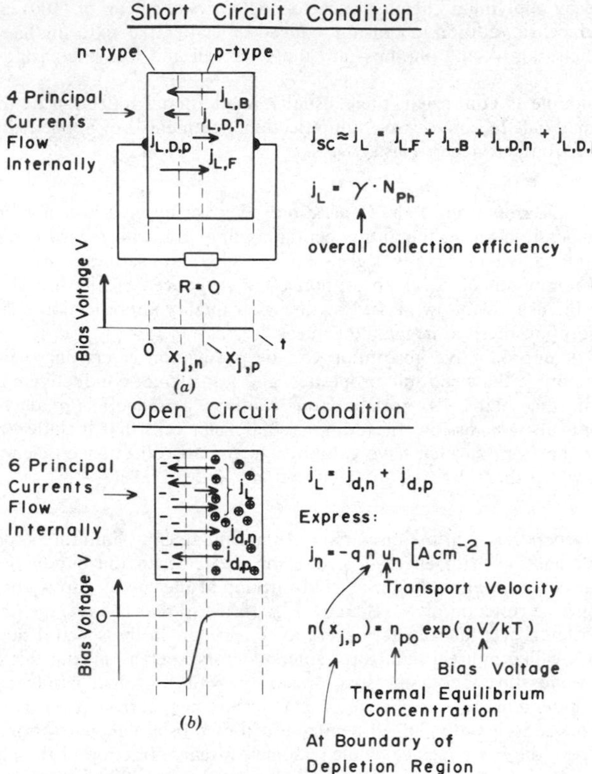

Fig. 15.94 Current components flowing in a *pn*-junction solar cell with distinct front, base, and depletion layers, for the short-circuit condition (*a*) and the open-circuit condition (*b*). The short-circuit condition has only the four principal light-generated current components, while in the open-circuit condition, two additional injection current components flow, labeled j_d.

region before recombining with majority carriers. The fraction of the generated minority carriers thus collected represents the collection efficiency η_{coll}.

The collection process leads to an accumulation of "excess" majority carriers on both sides of the potential barrier, denoting an excess over the concentration present in the thermal equilibrium condition. If an external connection of zero resistance is provided between the semiconducting materials on the two sides of the potential barrier, the excess majority carriers flow through this external circuit back to the other side. There they enter again as minority carriers and recombine with the plentiful majority carriers. This flow of charge carriers through the external circuit constitutes a measurable current. As this current represents the entire flow across the potential barrier of carriers resulting from photon absorption, it is called the light-generated current, which when expressed for a unit device area, is the current density j_L. This physical picture validates the concept of a constant-current generator in the equivalent circuit (Fig. 15.77).

The Open-Circuit Condition. In the absence of any external connection, the accumulation of the excess majority carriers on both sides of the potential barrier has to lead to a steady-state situation in which the light-generated charge carrier flow j_L across the potential barrier is counteracted by a carrier flow of equal magnitude but opposite direction, so that there is zero net current flow. In reality, this current has two components, one each originating in the front region and the base region [Fig. 15.94(*b*)]. There can be additional sources of such opposite current flow. Since these sources are undesired, the resulting currents are called excess currents. These will be discussed later.

The principal current flow opposite to the light-generated current comes about as follows. As long as the concentration of the majority carriers is adequately small so that the carriers do not interact with each other and can thus be treated like an ideal gas, their energy distribution is described by the Boltzmann distribution. Thus in the thermal equilibrium condition, the fraction $\exp(-qV_0/kT)$ of the majority carriers in a region has enough energy to pass over the potential step of magnitude V_0. In the thermal equilibrium condition, which includes no change of the height V_0 of the potential barrier by an applied voltage nor any generation of carriers by photon absorption, equal numbers of charge carriers flow across the barrier in both directions, so that there is no net current flow. If the potential barrier is a *pn* junction with the majority carrier density $n_n(x_{j,n})$ existing at the *n*-type boundary of the depletion region, which is located at distance $x_{j,n}$ from the front surface of the device (see Fig. 15.94*b*), then the electron concentration at the *p*-type boundary of the potential barrier, located at $x_{j,p}$, will be

$$n_{p0}(x_{j,p}) = n_n(x_{j,n}) \cdot \exp(-qV_0/kT) \qquad \text{cm}^{-3} \tag{15.69}$$

The electrons are minority carriers on this *p* side, and their concentration, given by Eq. (15.69), represents their thermal equilibrium concentration n_{p0} throughout the *p* region, if the latter is uniformly doped. A corresponding relationship exists for the holes on the *n* side of the potential barrier. For simplicity of discussion, however, the existence of these holes and the resultant current will be neglected in the following.

In the open-circuit condition, the excess majority carriers accumulating on both sides of the potential barrier lower the height of this barrier by the potential difference V, which is measurable as a bias voltage at the device's terminals. In this case the voltage is called the open-circuit voltage V_{oc}. Thus the fraction of majority carriers that can surmount the potential barrier and appear as minority carriers on the other side of the depletion region will be increased by the factor $\exp(qV_{oc}/kT)$. This increase causes excess minority carriers to exist at the boundary of the depletion region in the concentration

$$n_p(x_{j,p}) - n_{p0} = n_{p0}[\exp(qV_{oc}/kT) - 1] \tag{15.70}$$

These excess minority carriers are subject to recombination in the *p* region.

When the thickness of the *p* region equals at least several diffusion lengths L_n, then the concentration of the excess minority carriers injected across the potential barrier varies with the distance inside the *p* region as the result of recombination according to

$$n_p(x) - n_{p0} = [n_p(x_{j,p}) - n_{p0}]\exp[-(x - x_{j,p})/L_n] \tag{15.71}$$

It is

$$L_n = (D_n \cdot \tau_n)^{1/2} \tag{15.72}$$

with D_n being the diffusion constant and τ_n the minority carrier lifetime for electrons (both for *p*-type material). This distribution causes diffusion of the minority carriers away from the depletion region boundary into the *p* region

$$-D_n \frac{dn}{dx} = \frac{D_n}{L_n}[n_p(x_{j,p}) - n_{p0}]\exp[-(x - x_{j,p})/L_n] \tag{15.73}$$

assuming no other transport mechanism to be active. The minority carrier flux of Eq. (15.73), multiplied by $-q$, represents a current density j_d, generally called the diffusion current. As the distribution of the injected minority carriers according to Eq. (15.71) is based on recombination in the bulk of the *p* region, and this injection current is in reality sustained by the disappearance of the injected carriers via recombination, this current is equally appropriately called a recombination current. It is to be remembered that a similar injection current, although possibly of quite different magnitude, flows from the *p* region to the *n* region.

With these injection currents, a current flow is established that is of opposite direction to the light-generated current. The lowering of the potential barrier proceeds until the recombination current equals the light-generated current, and a steady-state but nonthermal equilibrium is established. As the bias voltage is connected with a lowering of the potential barrier, it corresponds to a forward bias voltage in a diode or rectifier (direction of easy current flow).

As the recombination rate in a given region is proportional to the excess minority carrier concentration within the region, and inversely proportional to the minority carrier lifetime, a longer minority carrier lifetime [smaller transport velocity, $u_n(x_{j,p})$ in Fig. 15.95] requires a higher minority carrier concentration for an equal recombination rate which, in turn, causes an equal recombination current. A higher minority carrier concentration within the respective region also means a proportionally higher minority carrier concentration at the interface of that region with the depletion region, and consequently a greater reduction of the barrier height, that is, a larger open-circuit voltage. Consequently, a reduction of recombination not only leads to improved collection efficiency but also to increased output voltages. In fact, while the collection efficiency saturates with increasing minority carrier lifetimes, no such effect exists for the technologically determined part of the voltage factor.

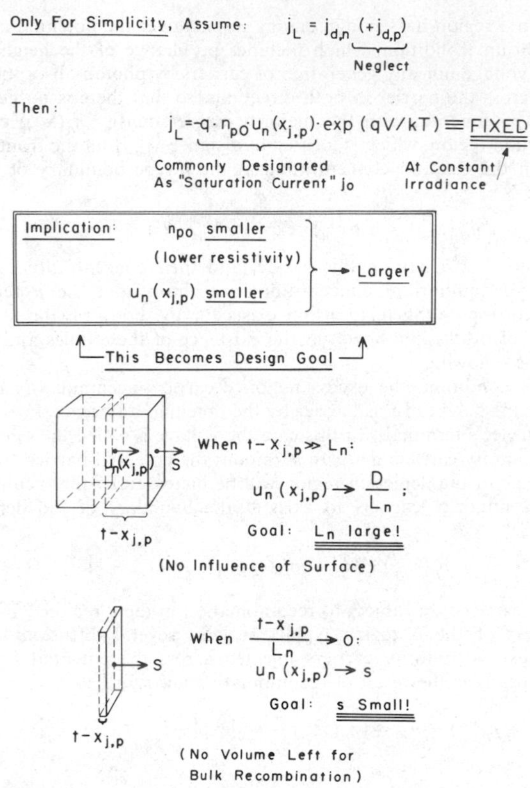

Only For Simplicity, Assume: $j_L = j_{d,n}$ $(+j_{d,p})$
Neglect
Then:
$j_L = -q \cdot n_{po} \cdot u_n(x_{j,p}) \cdot \exp(qV/kT) \equiv$ FIXED
Commonly Designated As "Saturation Current" j_0
At Constant Irradiance
Implication: n_{po} smaller
(lower resistivity)
$u_n(x_{j,p})$ smaller
→ Larger V
This Becomes Design Goal
When $t - x_{j,p} \gg L_n$:
$u_n(x_{j,p}) \rightarrow \frac{D}{L_n}$;
Goal: L_n large!
(No Influence of Surface)
$t - x_{j,p}$
When $\frac{t - x_{j,p}}{L_n} \rightarrow 0$:
$u_n(x_{j,p}) \rightarrow s$
Goal: s Small!
(No Volume Left for Bulk Recombination)

Fig. 15.95 Principal considerations in the design of high-efficiency solar cells.

Ideal Current-Voltage Characteristic. In neither the short-circuit nor the open-circuit voltage condition does the solar cell deliver any electrical power. However, with any finite, nonzero load resistance, electrical power can be delivered from the solar cell and, with a particular resistance value, the output reaches the maximum power point. Thus the current-voltage characteristic between short circuit and open circuit is of importance. Because of the energy distribution of the charge carriers according to Boltzmann's law, as discussed previously, the dependence of the injection current across the potential barrier becomes an exponential function of the voltage by which the barrier height is reduced [Eq. (15.70)]. This is the ideal current-voltage characteristic of the solar cell [Eq. (15.47), but with N in $V_{\text{th,eff}}$ reduced to 1, so that $V_{\text{th,eff}}$ becomes V_{th}]. This leads directly to the curve factor (CF) for the maximum power point, with the technologically determined part again dominated by excess recombination resulting from crystal defects.

Real Current-Voltage Characteristics. A variety of additional effects influence the I-V characteristic and reduce the fill factor below the curve factor. Consequently, for cell analysis purposes, the equivalent circuit of a solar cell is better represented by Fig. 15.96 than by Fig. 15.77, which is adequate for system design purposes.

A most important contribution to the current-voltage characteristic comes from the recombination within the depletion region, which is not considered in the diffusion current discussed previously. This current is represented by the second diode in Fig. 15.96, and its I-V characteristic differs from that of the diffusion current by containing factor $1 \leqslant N \leqslant 2$ in the exponent. With $j_{0.2} \gg j_{0.1}$, it dominates the solar cell I-V characteristic at low voltages but usually does not affect the open-circuit voltage. In very good solar cells, $j_{0.2}$ is small enough so that this second diode has only a negligible effect on the maximum power point (Fig. 15.97).

Other mechanisms of current conduction past the potential barrier can also lead to reduced voltages of the solar cell. When they result in a linear current-voltage relationship, they are attributed to a shunt resistance R_{sh}. They are, in fact, often caused by a resistive path, such as one created by unintentional metal deposition across the edge of the device.

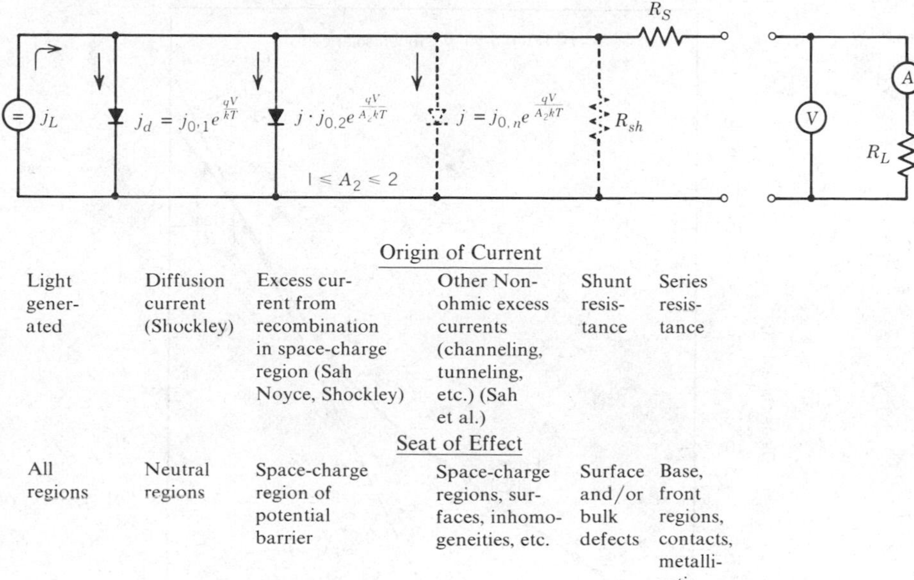

[See W. Shockley and W. T. Read, *Phys. Rev.* **87**, 835 (1952); R. N. Hall, *Phys. Rev.* **87**, 387 (1952); and C.-T. Sah, R. N. Noyce, and W. Shockley, *Proc. IRE*, **45** 1228 (1957); C.-T. Sah, *IEEE Trans. Electron Devices*, **ED9**, 94 (1962).]

Fig. 15.96 Generalized equivalent circuit diagram for the performance analysis of "real" solar cells. The circuit includes a number of physical mechanisms that cause "excess current" flow.

Other mechanisms lead to nonlinear, some to exponential, current-voltage relationships. They can be caused by inhomogeneities in the device, such as "spikes" at the potential barrier, by surface channels, or by tunneling through thin space-charge regions.

By good device design and careful fabrication, all these excess-current-producing mechanisms can be reduced to levels where their influence on solar cell performance is essentially negligible.

Minority Carrier Recombination. Some recombination of electron-hole pairs takes place by direct transition across the bandgap and results in reemission of a photon. This process is particularly effective in the direct bandgap semiconductors. It is a fundamental process which determines the basic limitation of the voltage and curve factors.

However, orders of magnitude more recombination events occur through recombination centers within the bandgap, and they result in the dissipation of the electron-hole pair energy as heat rather than as photon energy. These centers are generally connected with defects in the crystal structure, including certain impurities. This excess recombination influences the collection efficiency and determines the technology-dependent parts of the voltage and curve factors through the diffusion current. Consequently, much of the design and fabrication effort for solar cells is concerned with limiting the excess recombination.

The general parameter used to quantify recombination in the volume of the semiconductor is the recombination rate U. It describes the loss of excess minority carriers per unit volume and time. U, in its most general form, may be expressed as a power series in the excess minority carrier concentration $n' = n - n_0$

$$U = -\frac{dn'}{dt} = \beta_1 n' + \beta_2 n'^2 + \beta_3 n'^3 + \dots \qquad \text{cm}^{-3}\,\text{s}^{-1} \tag{15.74}$$

At low-level injection, which means that the excess minority carrier concentration is small compared with the majority carrier concentration, the first term predominates. At high-level injection, that is, the excess minority carrier concentration is large compared with the thermal equilibrium majority carrier concentration, the second term expresses the radiative recombination prevalent in the direct bandgap semiconductors, and the third term depicts Auger recombination. The linear (first) term in Eq. (15.74) expresses both recombination via recombination centers and Auger recombination at high majority carrier concentrations. Auger recombination involves two majority carriers and

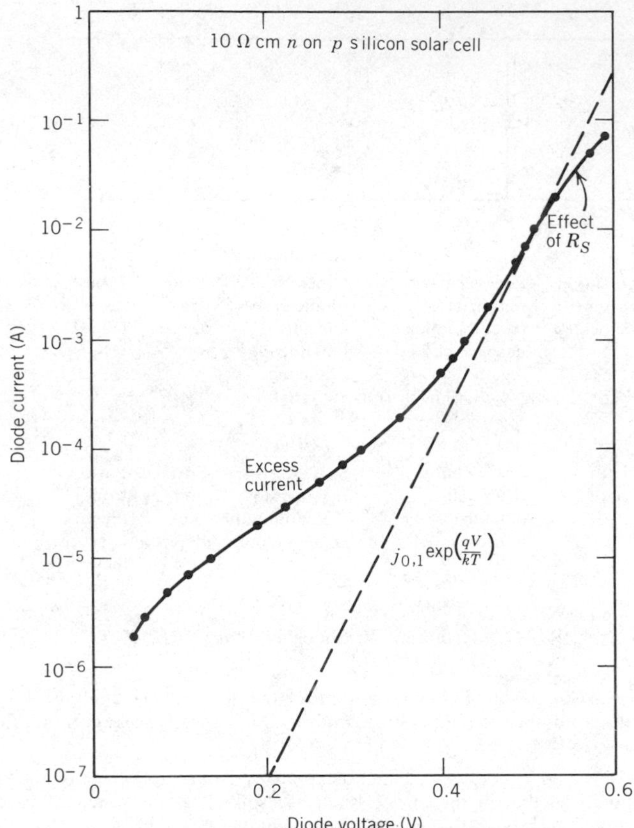

Fig. 15.97 Current-voltage characteristic of a silicon solar cell taken without illumination, shown as a solid line with measurement points as dots. The ideal current-voltage relationship is represented by the straight dashed line. The difference in current between that measured on a real device and that of the ideal characteristic is called the excess current. It is seen that excess current dominates at low voltages, while the real characteristic is, at high voltages, primarily influenced by the effects of series resistance.

occurs with or without recombination centers, but it always generates heat rather than photons. The linear term is of primary interest in solar cell analyses.

The inverse of the proportionality constant β_1 is called the minority carrier lifetime τ, which expresses the average duration of the time periods between the generation (or injection) of minority carriers and their recombination with the majority carriers. In general, τ ranges between nanoseconds and milliseconds, depending on the type of semiconductor and its crystalline perfection. For recombination through a single type of recombination center that can be represented by a single localized energy level E_r in the bandgap, U is given by

$$U = -\frac{np - n_i^2}{\tau_{p0}(n + n_1) + \tau_{n0}(p + p_1)} \qquad \text{cm}^{-3}\,\text{s}^{-1} \tag{15.75}$$

where $n_i = p_i$ is the intrinsic carrier concentration, referring to undoped material, and n_1 (or p_1) the electron (or hole) concentration that would exist if the Fermi level would be at energy level E_r (Fig. 15.98).

The saturation lifetimes for holes and electrons are, respectively,

$$\tau_{p0} = \frac{1}{\sigma_p v_{\text{th}} N_r} \qquad \text{s} \tag{15.76a}$$

$$\tau_{n0} = \frac{1}{\sigma_n v_{\text{th}} N_r} \qquad \text{s} \tag{15.76b}$$

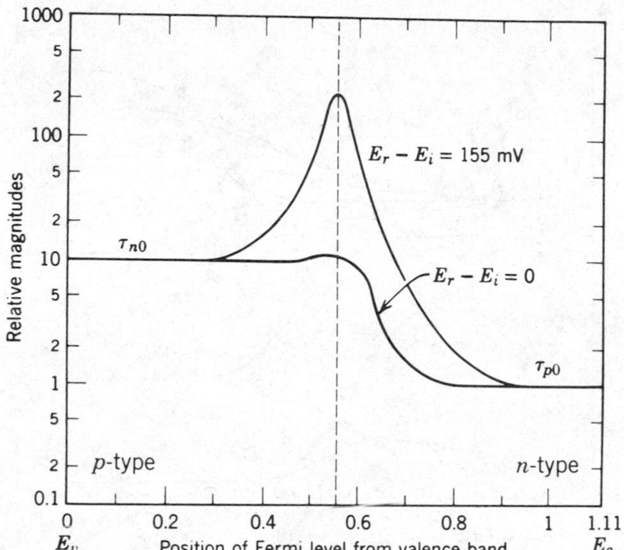

Fig. 15.98 Variation of the minority carrier lifetime with the position of the Fermi level in the energy gap, according to Shockley-Read-Hall recombination theory. For deep recombination levels, that is, levels whose energy is near the center of the energy gap ($E_r - E_i \approx 0$), the minority carrier lifetime is relatively independent of the Fermi level position, while for shallow centers that is, centers with energy closer to the edges of the allowed energy bands, the minority carrier lifetimes can increase considerably at higher resistivities. To relate the graph to resistivity, it may be observed that the resistivity increases approximately exponentially with increasing distance of the Fermi level from the respective band edge. From Ghandhi, *Semiconductor Power Devices*, Wiley, New York, 1977, p. 7, with permission.

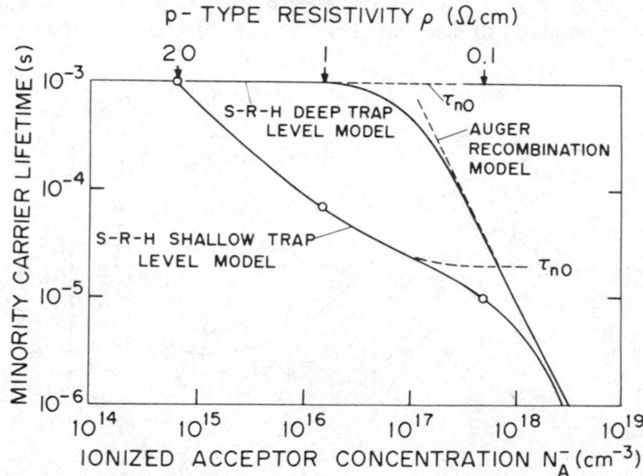

Fig. 15.99 Two models for dependence of minority carrier (electron) lifetime on the impurity concentration in *p*-type silicon, according to the Shockley-Read-Hall theory, with the addition of Auger recombination at the higher impurity concentrations. The curves have been selected to yield the same minority carrier lifetimes at 20 Ω cm. The deep trap level model gives a saturation lifetime τ_{n0} of 10^{-3} s, while the shallow trap level model yields one of $2 \cdot 10^{-5}$ s. The differences in minority carrier lifetime occur primarily in the intermediate impurity concentrations, which are of interest in solar cell design. As deep levels generally form more effective recombination centers, the deep trap level model represents a silicon in which a much lower recombination center concentration has been achieved than in the material represented by the shallow trap level model. Adapted from Bowler and Wolf,[10] with permission.

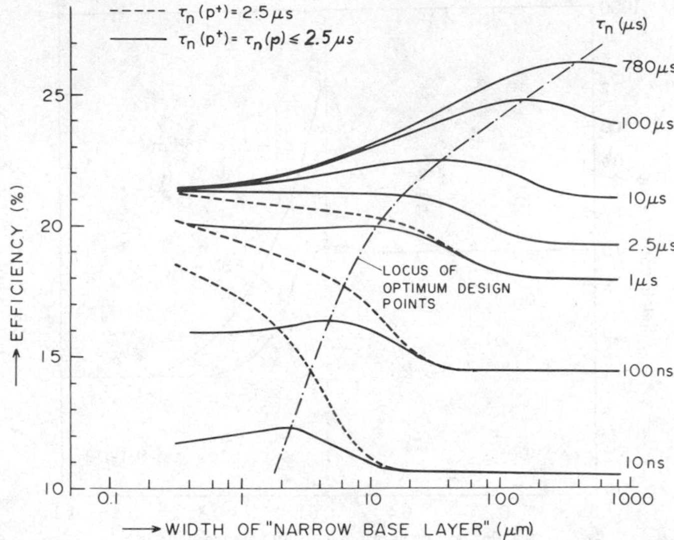

Fig. 15.100 Dependence of conversion efficiency η on minority carrier lifetime in the narrow, less heavily doped base layer and on the width of this layer. For the upper four curves and the dotted portions of the lower three curves, the minority carrier lifetime in the more heavily doped base layer has been held constant at 2.5 μs, the value dictated by Auger recombination. The three lower solid curves have been computed with identical lifetimes for both layers, assuming that a high recombination center density dominates the recombination mechanisms in both layers. The design of the front region has been held fixed in these calculations. From Bowler and Wolf,[10] with permission.

where σ = effective capture cross sections, cm^2

v_{th} = thermal velocity, cm s^{-1}

N_r = density of recombination centers, cm^{-3}

For shallow centers ($E_r \ll E_G$), τ is a function of the resistivity, with $\tau \geqslant \tau_0$ at the higher resistivities, while for deep centers ($E_r \approx E_G/2$), $\tau \approx \tau_0$ for all resistivities, down to the low values of resistivity where Auger recombination starts to dominate (Fig. 15.99). τ_0 refers to either τ_{n0} or τ_{p0}. The strong

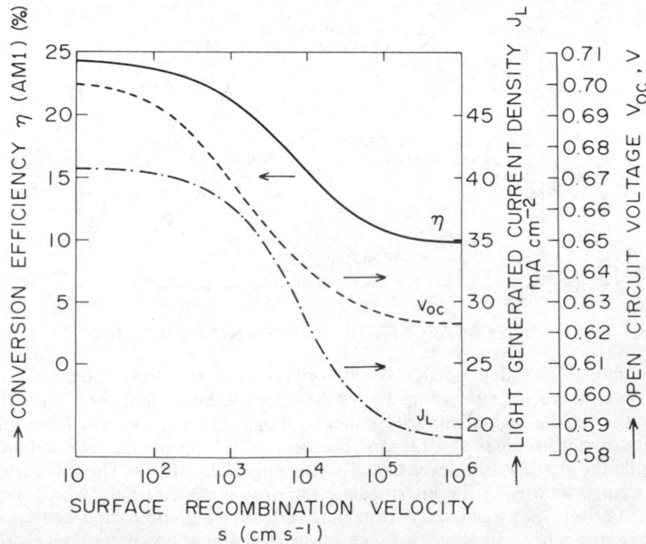

Fig. 15.101 Dependence of light-generated current density j_L, open-circuit voltage V_{oc}, and conversion efficiency η on the surface recombination velocity s, which was here assumed to be equal at both the front and the back surfaces of the cell. From Bowler and Wolf,[10] with permission.

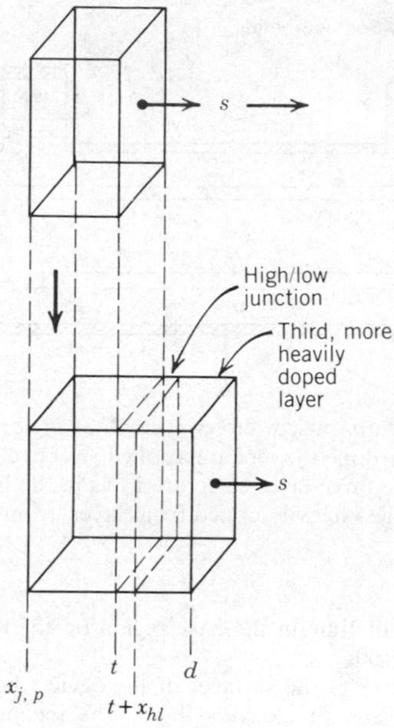

Contact needed at $x = t$ or at $x = d$
Remedy: Add two more layers
Effect: $s \approx 10^6$ cm/s
Wanted: $s = u(t) \leqslant 20$ cm/s
Effect of high/low junction:

$$\frac{u(t)}{u(t+x_{hl})} \geqslant \frac{p_p(t)}{p_p(t+x_{hl})}$$

High/low junction alone:

$$\text{With} \quad p_p(t) = 2.10^{16}\ \text{cm}^{-3}$$

$$p_p(t+x_{hl}) = 2.10^{18}\ \text{cm}^{-3}$$

(limited by Auger recombination)

$$\frac{u(t)}{u(t+x_{hl})} = 10^{-2}$$

Thus needed:

$$u(t+x_{hl}) \leqslant 2000\ \text{cm/s}$$

But $s = u(d) = 10^6$ cm/s
Add "thick" third layer

$$d-(t+x_{hl}) > L_n$$

so that

$$u(t+x_{hl}) = \frac{D_n}{L_n} \leqslant 2000\ \text{cm/s}$$

Achievable with

$$L_n \geqslant 3.3 \cdot 10^{-3}\ \text{cm}$$

(at $D_n = 6.5\ \text{cm}^2\text{V}^{-1}\ \text{s}^{-1}$)
Requires $\rightarrow \tau_0 \geqslant 2\ \mu\text{s}$

Fig. 15.102 Effects of a high/low junction layer and a third, more heavily doped base layer in achieving desired low transport velocities at the back side of the less heavily doped base layer.

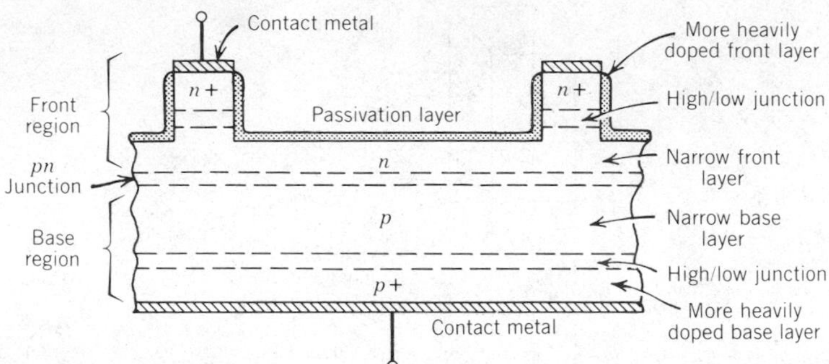

Fig. 15.103 Solar cell design with raised three-layer front-region structure where the high/low junctions and third, more heavily doped layers are applied selectively only under the ohmic contacts and grid lines. The purpose of the three-layer structure is, as in the base region, to shield the primary active region that is, the narrow, less heavily doped front layer, from the influence of the high surface recombination.

influence of the minority carrier lifetime in the base region on the maximally achievable conversion efficiency is illustrated in Fig. 15.100.

Other recombination takes place at the surfaces of the device through centers that are associated with the discontinuity of the crystal that occurs there. The recombination rate is expressed by a minority carrier (electron) current density

$$j_n = qn_p u_n \qquad \text{A cm}^{-2} \tag{15.77}$$

which flows into the respective surface with the transport velocity u_n (cm s^{-1}). In this case the transport velocity is called a surface recombination velocity, generally expressed by s (cm s^{-1}). The impact of the surface recombination velocity on the key performance parameters of a solar cell is shown in Fig. 15.101. The surface recombination velocity on open surfaces can usually be reduced by special surface treatments, including application of passivation layers which often consist of oxides. On metal semiconductor interfaces that are to serve as low resistance ohmic contacts, high surface recombination velocities are generally needed. There a shielding of the contact surface from the volume that contains excess minority carriers may be required. Shielding is possible by use of a thin insulating layer through which only the majority carriers can tunnel or of a high/low junction, usually with a more heavily doped layer interspersed between the junction and the contact surface (Figs. 15.102 and 15.103).

Generally four paths are available for the reduction of recombination:

1. Increase minority carrier lifetime (fewer defects).
   (a) Avoid heavy doping (Auger recombination).
2. Reduce volume for recombination.
3. Reduce surface recombination velocities.
4. Shield volumes containing excess minority carriers from surfaces (or regions) with high recombination rates (such as ohmic contacts).

All of these paths are pursued simultaneously in modern high efficiency solar cell design.

Series Resistance. A significant loss mechanism that influences the *I-V* characteristic at the higher current levels (Fig. 15.97), is due to series resistance (R_s) in the device. A major contribution to series resistance results from the lateral flow of the current in the front region, which is the layer between the light-exposed surface and the depletion region [*n* layer or *p* layer, depending on type of solar cell (*n* on *p* or *p* on *n*)]. This layer generally has a relatively high sheet resistance, from about 20 to a few hundred ohms. This resistance contribution is substantially reduced by an overlay of a metallic grid contact structure, which consists of a multitude of narrow, usually parallel grid fingers (Fig. 15.104).

Overlay of a metal structure over the front of the device causes a loss of light-sensitive area without reducing the frontal surface of the device, resulting in a loss of potential light-generated current. This loss is usually expressed through the "shading factor." The front metal structure can be designed so that the shading loss is kept in the 2 to 3% range, for a combined front-layer series resistance and shading loss of 5 to 6% of the ideal power output.

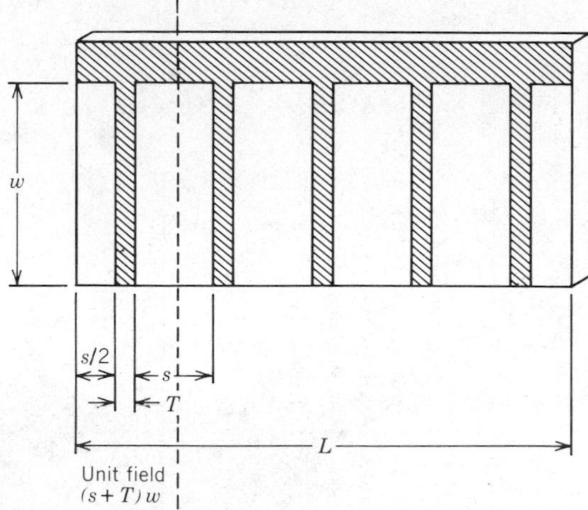

Fig. 15.104 Configuration of a front contact structure, including parallel-arranged grid lines. From Wolf,[6] with permission.

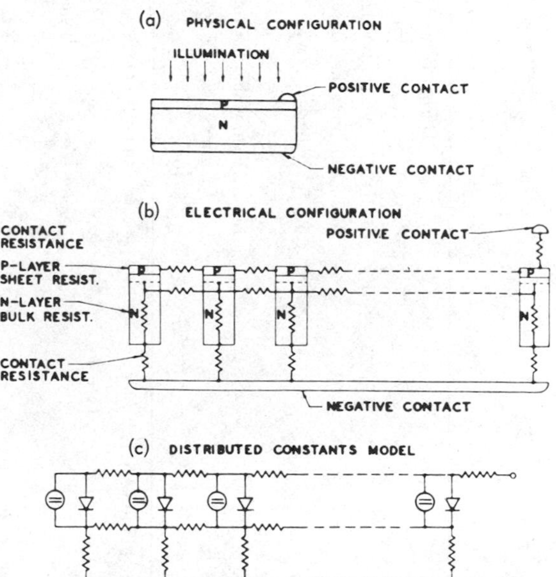

Fig. 15.105 "Distributed constant" version of solar cell equivalent circuit, which provides a more accurate representation than does the single-lumped-constant series resistance model. (From Wolf and Rauschenbach,[11] with permission.

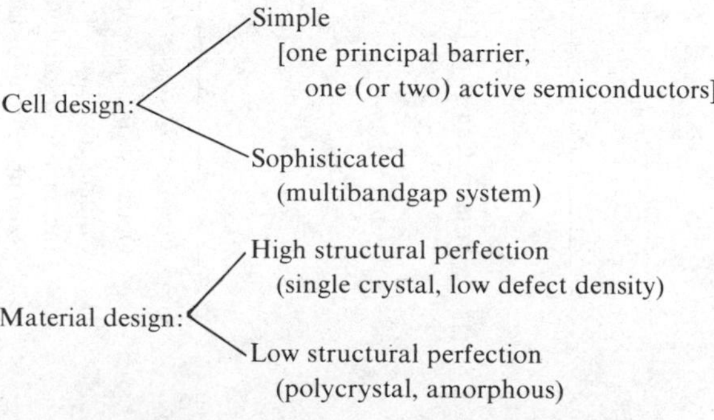

Fig. 15.106. Basic solar cell design approaches.

Primary Causes of Losses	Symbol	Design Parameters: Present Goal	Design Parameters: First Milestone	1982 Best Verified Cell (Westinghouse, 1 cm^2)
		Base: 200 μm wide $\tau_{n,p} = 950\ \mu s$ $\tau_{n,p}^{+} = 2.6\ \mu s$ Front: 2 μm wide $s = 10^3$ cm s^{-1} $\tau_{p,n} = 10\ \mu s$	Base: 200 μm wide $\tau_{n,p} = 95\ \mu s$ $\tau_{n,p}^{+} = 0.26\ \mu s$ Front: 2 μm wide $s = 10^3$ cm s^{-1} $\tau_{p,n} = 0.1\ \mu s$	(Averages of test results from Westinghouse, SERI, and JPL)
Light-generated current: Fundamental limit (AM1)	$j_{L,id}$	44 mA cm^{-2}		
Optical surface properties (reflection)	$(1 - R)$	0.97	0.97	0.95 (±0.02)
Contact coverage	S	0.966	0.966	0.97
Incomplete absorption (thickness); Recombination outside depletion region (bulk and surface, including contacts)	η_{coll}	0.95	0.92	0.88 (Distribution assumed)
"Dead layers"	$(1 - A)$	1.0	1.0	1.0 (Distribution assumed)
Overall collection efficiency	γ	0.89	0.86	0.81
Light-generated current (AM1)	$j_L = \gamma j_{L,id}$	39.2 mA cm^{-2}	37.9 mA cm^{-2}	35.5 mA cm^{-2}

Open-circuit voltage: Fundamental limit	$(VF)_{fund} = 0.76$		0.836 V		
Recombination outside depletion region (bulk and surface, including contacts); Bandgap narrowing	$(VF) = (VF)_{techn}(VF)_{fund}$	0.65		0.60	0.54
"Current leakage"	(R_{sh})	1.0		1.0	1.0
Open-circuit voltage	$V_{oc} = (VF) \cdot E_G$	0.715 V		0.661 V	0.594 V
Fill factor: Fundamental limit	$(CF)_{fund}$		0.96		
Recombination outside depletion region (bulk and surface, including contacts)	$(CF) = (CF)_{techn}(CF)_{fund}$	0.85		0.84	0.84
Bandgap narrowing; "Current leakage"	(R_{sh})	1.0		1.0	1.0
Recombination in depletion region	$(CF)_{add'l}$	0.97		0.97	0.975
Series resistance	(R_s)	0.98		0.98	0.975
Fill factor	(FF)	0.81		0.80	0.80
Resulting conversion efficiency	η	0.226		0.200	0.172

Fig. 15.107. Analysis of the influence of various loss mechanisms on performance of the best recently prepared silicon solar cells and reduction of losses planned in cells presently under development.

The combined front-layer loss can be limited to 5 to 6% on larger area cells (width > ~1 cm) only when a front-layer metallization of high conductivity is applied, such as silver, copper, or aluminum of "bulk conductivity." Such bulk conductivity is achieved by electroplating or vacuum deposition processes. Substantially lower conductivity is attained by, for example, powder metallurgy processes (thick film). Also, for solar cells in which the current path through the metal along the front layer exceeds approximately 2 cm in length, the use of bus lines is required, in which the thickness is built up by overlay of a bulk conductor, such as a wire or ribbon. The thickness achieved is greater than that achievable by a deposition process.

In some approaches, the sheet resistance of the front layer is reduced by application of a transparent conductive coating composed of a material such as indium tin oxide (ITO). This approach, however, cannot reduce the sheet resistance below 3 to 10 Ω without encountering substantial non-carrier-generating absorption. It should be noted that the single-lumped-constant model for the series resistance (Figs. 15.77 and 15.96) is only an approximation to the actual situation, which is better represented by Fig. 15.105. Only when the total series resistance is held to a very low value, such as to result in Joule losses of only a few percent, then the representation by a single lumped R_s is an adequate approach.

Developments in Progress

The preceding discussions of operating mechanisms and associated losses connected with solar cells suggest essentially two basic design targets with two principal choices for each, allowing four permutations (Fig. 15.106). All four permutations are currently pursued.

The first permutation utilizes a simple device combined with high material perfection. After selection of the appropriate semiconductor according to its bandgap to minimize the basic losses (or two semiconductors for heterojunction designs), the high material perfection is pursued to reduce recombination. The remainder of the effort is aimed at minimizing all other technologically determined losses.

The classical example of this approach is the single-crystal silicon solar cell. This cell is commercial with present conversion efficiencies for terrestrial cells in the 12 to 15% range, while 18 to 19% has been achieved in laboratory cells. The primary aims of current development efforts are the reduction of production costs and further improvement in efficiency. The collection efficiency already has been brought to about 90%. Consequently, efforts now are mainly directed toward reducing the influence of various recombination effects, so as to bring the open-circuit voltage from near 0.6 to near 0.7 V, with consequent attainment of efficiencies above 20% (Fig. 15.107).

Other semiconducting materials, such as gallium arsenide and cadmium telluride, have been used in the same approach. In the first, conversion efficiencies approaching 20% (AM1) have been achieved by use of a wider bandgap semiconductor for a "window layer," which is instrumental in reducing surface recombination effects but which does not really form a part of the active semiconductor device.

The second permutation involves the sophisticated-device approach (multibandgap system) combined with high material perfection. Here, efficiencies above 25%, possibly up to 40%, are expected without optical concentration. This is clearly, at least in the earlier development stages, the highest cost approach. Nevertheless, it is certain to be the most economical approach for systems using high-ratio optical concentrators. In the long run this approach may also become economical for application in flat-array systems. Using the optical beam splitting method, efficiencies in the high 20% range have been achieved. In the stacked-cell approach, however, efficiencies of developmental systems have not yet exceeded those of silicon solar cells. The best material candidates for these systems are the III-V and II-VI compound semiconductors, in part used as ternary or even quaternary compounds. The major unresolved problem in the stacked-cell approach is connected with the interface between the individual cells of the stack. Another question is connected with potential mismatches of the cells in the stack under varying operating conditions. This may necessitate extracting the power separately from each cell, rather than from the series-connected stack as a whole.

On the other end of the spectrum is the simple device with material of low structural perfection, primarily thin films of polycrystalline or amorphous semiconductors. The driving force for this approach is the potential of lower processing costs. Because of the reduced structural perfection, higher recombination rates than achievable in the high structural perfection approaches will have to be accepted, with consequent efficiencies in the 6 to 12% range. The materials of primary current interest for this category are copper-indium-diselenide ($CuInSe_2$) and an amorphous silicon-hydrogen alloy. In both, 10% efficiency (air mass 1) has been achieved in laboratory cells. The copper sulfide/cadmium sulfide (Cu_2S/CdS) approach has been essentially disbanded because of persistent stability problems.

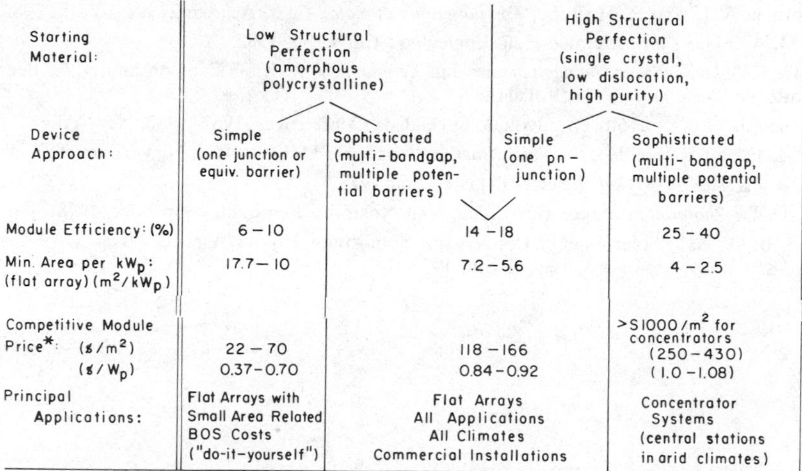

Starting Material:
Low Structural Perfection (amorphous polycrystalline)
High Structural Perfection (single crystal, low dislocation, high purity)
Device Approach:
Simple (one junction or equiv. barrier)
Sophisticated (multi-bandgap, multiple potential barriers)
Simple (one pn-junction)
Sophisticated (multi-bandgap, multiple potential barriers)
Module Efficiency: (%)
6 – 10
14 – 18
25 – 40
Min. Area per kWp: (flat array) (m²/kWp)
17.7 – 10
7.2 – 5.6
4 – 2.5
Competitive Module Price*: ($/m²) ($/Wp)
22 – 70
0.37 – 0.70
118 – 166
0.84 – 0.92
>$1000/m² for concentrators
(250 – 430)
(1.0 – 1.08)
Principal Applications:
Flat Arrays with Small Area Related BOS Costs ("do-it-yourself")
Flat Arrays All Applications All Climates Commercial Installations
Concentrator Systems (central stations in arid climates)
* FOR $ 1200 / kWp INSTALLED FLAT ARRAY AT $ 50/m² AREA-RELATED BOS COSTS.

Fig. 15.108 Future directions for photovoltaics.

Combining the low material perfection approach with the sophisticated device approach should lead to efficiencies in the range of those attainable by the simple-device/high-material-perfection approach. Thus, competition in this middle-efficiency range may ultimately develop, and it is too early to predict which of the approaches can lead to lower manufacturing costs. Various amorphous semiconductors are currently thought to be suitable for such stacked-cell multibandgap systems, and a developmental amorphous SiC/SiH cell stack has already yielded above 8% efficiency (Fig. 15.108).

References

1 F. U. Wetzler, "Connecting Renewable Power Sources into the System," *IEEE Spect.* **19**, (11): 42–45 (Nov. 1982).

2 H. S. Rauschenbach, *Solar Cell Array Design Handbook*, Van Nostrand Reinhold, New York, 1980.

3 Sandia Laboratories, Report No. SAND 79-0557 (Apr. 1979).

4 E. C. Boes, in J. F. Kreider and F. Kreith, eds., *Solar Energy Handbook*, McGraw-Hill, New York, 1981, pp. 2-46, 2-52, 2-53.

5 M. Wolf, "Module and Solar Cell Values as Function of Efficiency," *Solar Cells* **3**:327–396 (1981).

6 M. Wolf, "Limitations and Possibilities for Improvement of Photovoltaic Solar Energy Converters. Considerations for Earth's Surface Operation," *Proc. IRE* **48**:1246–1263 (Jul. 1960).

7 W. Shockley and H. Queisser, "Detailed Balance Limit of Efficiency of *pn* Junction Solar Cells," *J. Appl. Phys.* **32**:510 (1961).

8 S. J. Fonash, *Solar Cell Device Physics*, Academic, New York, 1981.

9 E. L. Ralph and M. Wolf, in Record of 4th Photovoltaic Specialists Conf., Cleveland, OH, June 1964, Interagency Advanced Power Group PIC-SOL 20915, pp. B-7-1-15.

10 D. L. Bowler and M. Wolf, "Interactions of Efficiency and Material Requirements for Terrestrial Silicon Solar Cells," *IEEE Trans. Comp. Hyb. and Man. Technol.* **CHMT-3**:464–472 (Dec. 1980).

11 M. Wolf and H. Rauschenbach, "Series Resistance Effects on Solar Cell Measurements," *Adv. Ener. Conv.* **3**:455–479 (1963).

Bibliography

Backus, C. E., ed., *Solar Cells*, IEEE Press, New York, 1976.

Dickinson, W. C. and P. N. Cheremisinoff, eds., *Solar Energy Technology Handbook*, Marcel Dekker, New York, 1980.

Fahrenbruch, A. L. and R. H. Bube, *Fundamentals of Solar Cells*, Academic, New York, 1983.

Green, M. A., *Solar Cells*, Prentice-Hall, Englewood Cliffs, NJ, 1982.

Hovel, H. J., *Solar Cells*, in *Semiconductors and Semimetals*, R. K. Willardson and A. C. Beer, eds., Academic, New York, 1975, Vol. 11.

IEEE, Records of 6th to 16th Photovoltaic Specialists Conferences, 1967–1982, New York.

Kreider, J. F. and F. Kreith, eds., *Solar Energy Handbook*, McGraw-Hill, New York, 1981.

Neville, R. C., *The Solar Cell*, Elsevier, Amsterdam, 1978.

Pulfrey, D. L., *Photovoltaic Power Generation*, Van Nostrand Reinhold, New York, 1978.

Seraphin, B. O., ed., "Solar Energy Conversion; Solid-State Physics Aspects," Vol. 31 of *Topics in Applied Physics*, Springer-Verlag, Berlin, 1979.

CHAPTER 16
ELECTRICAL TRANSDUCERS

GERALD WEISS

Polytechnic Institute of New York, Brooklyn, N.Y.

16.1 PRINCIPLES, CLASSIFICATION, TERMINOLOGY

16.1-1 Nomenclature

The term *transducer* has been applied to devices or combinations of devices that convert signals or energy from one physical form into another form. More specifically, in measuring systems a transducer is defined as a device that provides a usable output in response to a specified *measurand*. The measurand is "a physical quantity, property, or condition which is measured" and the output is "the electrical quantity, produced by a transducer, which is a function of the specified measurand." [These and other definitions employed in this handbook are taken from *Electrical Transducer Nomenclature and Terminology*, ANSI MC6.1-1975 (ISA-S37.1-1969).]

16.1-2 Transducer Elements

In some transducers, generation of electrical output from the physical measurand takes place in two stages. There is a *sensing element*, which responds directly to the measurand, and a *transduction element*, in which the electrical output originates. As an example, many types of pressure transducers consist of a sensing element that converts pressure into mechanical displacement and couples to a transduction element that generates an electrical output in response to the displacement. A third element of the transducer system, electrical circuitry for signal conditioning and signal processing, may be packaged integrally with the transducer or located remotely. In solid-state integrated circuit transducers the electronic and transducer elements may be fabricated during the same processing steps, employing conventional (e.g., silicon) materials technology.

16.1-3 Passive and Active Transducers

In a passive transducer the action of the measurand produces a change in a passive electrical circuit element, resistance, inductance, or capacitance. Such transducers require an external electric source for excitation. A few transducer types are active and generate an ouput voltage of their own, for example, transducers based on the thermoelectric or piezoelectric effect. The self-generated output is usually at a low level and requires amplification.

16.1-4 Analog and Digital Transducers

By far the majority of transducers have an analog output, defined as an "output which is a continuous function of the measurand, except as modified by the resolution of the transducer" (ISA-S37.1-1969). In passive transducers the analog output is often ratiometric, that is, the information is contained in the ratio of the electrical transducer output voltage to some reference voltage such as the transducer excitation. Voltage analog signals may be DC or AC (amplitude modulation). Current analog output finds application in process control. Some transducers provide a pulse analog output, that is, a pulse rate proportional to the measurand. Inductive and capacitive transducers are sometimes employed in a circuit that generates a variable frequency signal, with frequency proportional to the measurand (frequency modulation). These pulse analog and frequency analog signals are easily converted to digital signals by counting pulses or cycles and storing the count. Transducers may be employed as on-off devices with only two values of output; a liquid level switch is an example. The shaft encoder is an example of a transducer that has a digital output in the form of a parallel binary signal.

16.1-5 Electrical Circuits

Electrical and electronic circuits in the transducer system provide a variety of functions:

Generation of excitation or reference voltage and frequency

Generation of output signal, typically by a bridge circuit or potentiometric circuit

Signal conditioning, that is, amplification of low level outputs and adjustment of voltage (or current) output values to a standard range

Noise suppression, filtering, and ground isolation

Signal conversion, such as AC/DC, or analog/digital

Signal processing, such as linearization of inherently nonlinear outputs

Bridge and potentiometer circuits may be either deflection or null type. In a deflection circuit the bridge or potentiometer output becomes the transducer output, after appropriate signal conditioning and processing. In a null circuit the bridge (or potentiometer) output is employed to adjust another circuit element so as to null the bridge (or potentiometer). The output of such a self-balancing or servo-type transducer is some suitable variable associated with the self-balancing action.

The electronics may be located entirely inside the transducer package, may be entirely separate, or may be divided in the two locations. Where the transducer system consists of several packages, the interconnections provided by the user are part of the total measurement system; correct wiring, shielding, and grounding are essential to achieve the specified performance.

16.1-6 Function Characteristic

Every transducer type has an ideal output-measurand relationship, described by a theoretical equation or by a tabular (numerical) or graphical representation. This ideal transfer characteristic is in many instances a linear one, that is, it is represented by a straight line. The slope of the straight line is the transfer ratio or "transfer function" of this transducer. In the case of an inherently nonlinear ideal characteristic, the transfer ratio is sometimes used to describe the transducer behavior over a small range of the measurand.

The behavior of a real transducer deviates from that of the ideal one, and in consequence the real transducer will indicate a value of the measurand which is in error. There are static and dynamic error components.

Static Error

The static error is the difference between the actual calibration curve determined under steady state conditions and the theoretical function. The word *conformity* is sometimes used for this difference. When the theoretical function is a straight line, the term *linearity* applies. The transducer linearity describes "the closeness of the calibration curve to a specified straight line" (ISA-S37.1-1969).

Definitions of linearity vary, depending on which straight line is used in the comparison:

Independent Linearity. The best straight-line fit
Zero-Based Linearity. The best straight line passing through the point of minimum output
End-Point Linearity. The line passing through the points of minimum and maximum output
Absolute Linearity. The theoretically defined straight line

Additional errors due to changes in environmental conditions, aging, or other factors must be added to the linearity error exhibited by the static characteristics. The allowable total error is usually expressed in the form of an error band about the theoretical function. The width of the error band is expressed as a percentage. Commonly it is a percent of maximum output, which means that the width of the error band is constant at all output levels. Occasionally the error band is expressed as a percent of actual output; in some cases a combination is used or different error bands are specified over various output ranges.

Dynamic Error

The dynamic characteristics of transducers are described in the same terms as those employed in circuit and system analysis, using time response, frequency response, or transfer function (Laplace transform) notation and concepts.

16.2 TRANSDUCERS FOR SOLID-MECHANICAL MOTION

16.2-1 Scope and Definitions

The following subsections deal with transducers for mechanical displacement and its derivatives and with force and related quantities. *Displacement* is defined as "the change in position of a body or point with respect to a reference point" (ISA-S37.1-1969). A distinction is made between translational and rotational quantities, that is, the transducers to be considered here are for linear (translational) displacement, velocity, and acceleration; for force; for angular (rotational) displacement, velocity, and acceleration; and for torque. The term displacement used alone usually refers to the translational variable. Motion measurement may be *relative*, with respect to another body, or *absolute*, with respect to inertial space.

Pressure is force acting on a surface, measured as force per unit area. In solid mechanics, the term *stress* is employed for applied force per unit area. Tensile stress is defined to be positive and compressive stress negative. *Strain* is defined as the elongation of the material subjected to stress, divided by the original length.

The term *vibration* refers to a motion that varies in magnitude with time, with frequent reversals in direction; in instrumentation technology, mechanical vibration measurement usually refers to acceleration, sometimes to velocity. The term *shock* refers to a sudden aperiodic motion, and shock measurement specifically refers to acceleration.

16.2-2 Relative Displacement Transducers, General

Displacement transducers, translational and rotational, play a crucial role in instrumentation because mechanical displacement serves as a secondary, derived variable for many measurands, such as acceleration, force, stress, pressure, temperature, and fluid flow. The amount of displacement in these derived measurements is usually very small. As in other measuring devices, most displacement transducers respond to a range of inputs from zero to full scale. *Angular* displacement is a unique physical quantity which need not necessarily have any limit, and there are some rotational displacement transducers that have "infinite" range (continuous rotation).

The translational transducers to be described are used for motion in the range of a few centimeters or smaller. The ranging methods used for very-long-distance measurements—radar, sonar, and optical—are described elsewhere in this handbook.

16.2-3 Resistance Potentiometers

The resistance potentiometer consists of a resistance element with moving contact (Fig. 16.1). With a fixed voltage excitation, the output voltage is a specified function of the contact (wiper) position. If the resistance is uniformly distributed, the output is proportional to the input within a specified accuracy tolerance (*linear* potentiometer, where the term linear now means proportional, not translational). Special, nonlinear function potentiometers are also available, either by a specially

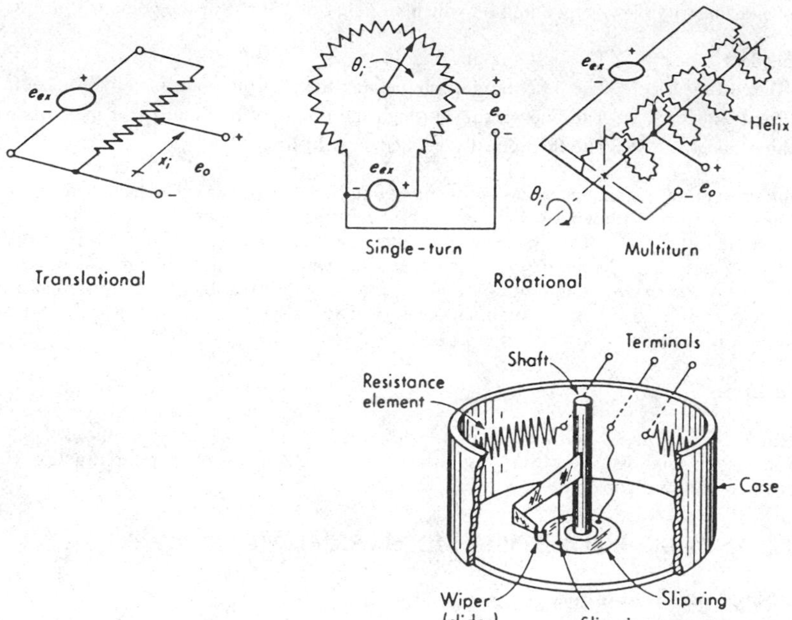

Fig. 16.1 Potentiometer displacement transducer. From McGraw-Hill, New York, with permission.

shaped resistance element or through external resistors connected to specially provided taps. With a variable input voltage, the potentiometer is an analog multiplier.

The standard commercially available precision potentiometer is a rotational device, with bearings and contacts designed to provide long life for the given application. The single-turn potentiometer experiences a complete range of resistance variation in something less than one full turn, typically 350°. Such a single-turn device may be *continuous rotation* in a mechanical sense or may have internal stops that limit the mechanical travel. Multiturn potentiometers have the resistance element arranged in a helical spiral, typically five or ten turns. Translational (linear motion) potentiometers are also available.

There are two basic types of resistance element. In the wirewound potentiometer, the element is made of turns of wire wound on a form. The wiper contacts only a small portion of each turn. The resistance between wiper and end and the corresponding output voltage therefore change in discrete steps (Fig. 16.2). The output resolution is a function of the number of wire turns that can be

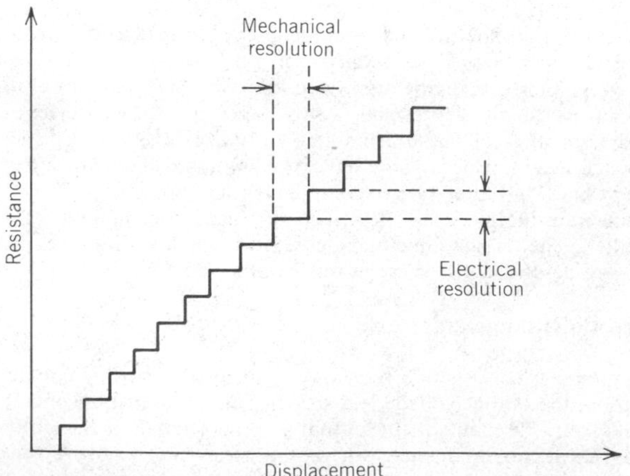

Fig. 16.2 Resolution of wire-wound potentiometers.

provided; it is affected by the length of the element and the resistance value. The nonwirewound potentiometer has a resistance element made of conductive plastic or deposited film conductor. Such potentiometers have infinite resolution, but they do have imperfections such as micro nonlinearities and contact resistance variations, which are described by an output smoothness specification.

16.2-4 Inductive Displacement Transducer Principles

Variable inductance displacement transducers can be classified as moving coil or moving iron. In the moving-coil transducer, the motion to be measured is made to change the relative position of several coils, leading to a change in mutual inductance that can be used to produce an output signal. In the moving-iron or *reluctive* transducer, the motion is employed to change the reluctance of a magnetic flux path. This in turn causes a change in self-inductance or mutual inductance that can be translated into a change in electrical output.

The eddy-current-type inductive displacement transducer responds to the relative location of a conductive metal object (target) in an AC magnetic field. It is frequently employed for proximity sensors, particularly in on-off measurements (proximity switches).

A back torque is generated whenever the relative motion to be measured is associated with a change of energy storage in the magnetic field. That is, reluctive transduction is reversible, and a device can be built that produces output motion in response to an electrical input. This principle is employed in solenoids, vibrators, and torquers.

16.2-5 Differential Transformers

The differential transformer is an iron-core transformer with movable core. The motion of the core changes the inductive coupling between the primary winding and one or more secondary windings, thus producing an output voltage related to the core displacement. The basic circuit of Fig. 16.3 is employed in most of the available configurations. For a particular position of the core called the *null* or *balance*, the two secondary voltages are equal and opposite and the net transformer output is zero. When the core is displaced from the null, the two secondary voltages are no longer identical and the transformer produces an output voltage. Motion of the core in the opposite direction produces a similar effect with 180° phase reversal of the output. The differential transformer has the advantage

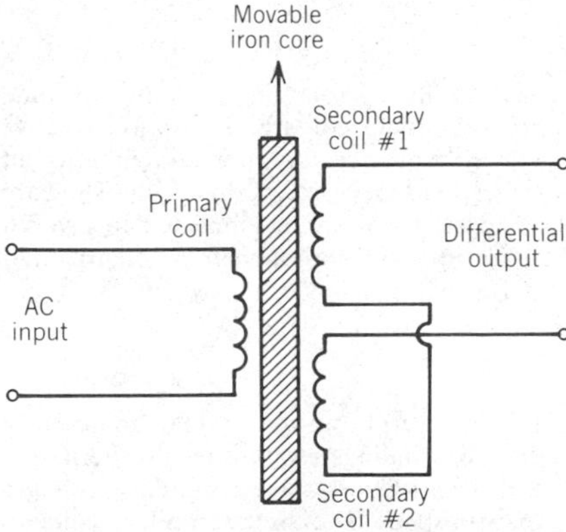

Fig. 16.3 Differential transformer, basic circuit.

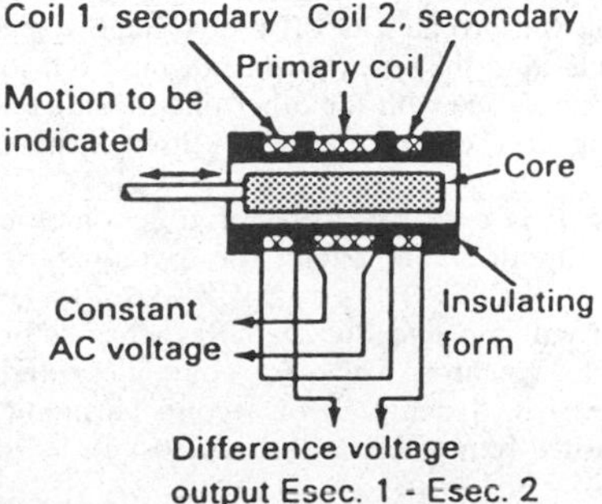

Fig. 16.4 Linear variable differential transformer (LVDT). From Schaevitz Engineering, with permission.

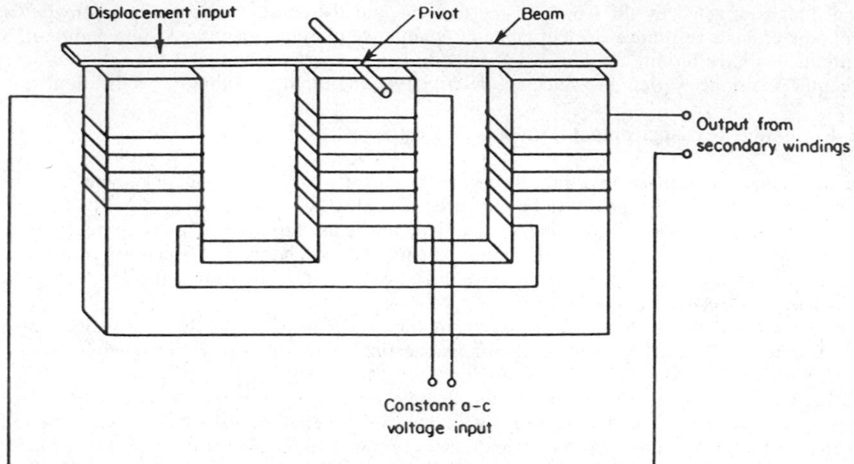

Fig. 16.5 *E*-pickoff differential transfomer. From McGraw-Hill, New York, with permission.

of lack of contacts and infinite resolution. The standard commercial form, the linear variable differential transformer (LVDT), is a translational device whose output varies linearly with core position over a specified range of motion (Fig. 16.4). An older device, the *E*-pickoff transformer (Fig. 16.5), does not have as good a linearity as the LVDT and is used primarily as a null sensor. Both the LVDT and the E-pickoff are also available in rotational form. The differential transformer is an AC device. It can be used in standard DC instrumentation by providing an AC oscillator (carrier generator) to drive the input winding and a demodulator to convert the output into DC. Some commercial designs provide the entire circuitry as one package built into the transformer housing (so-called DC-LVDT).

16.2-6 Synchros

The synchro is a rotational position transducer with true continuous rotation capability (infinite angle) and inherent high angular accuracy, of the order of minutes of arc. It is an inductive transducer, with construction similar to that of a three-phase wound rotor induction motor, but single-phase AC is used. Synchros are known under various trade names (Selsyn, Autosyn). They are normally employed in pairs, attached to separate shafts, and electrically interconnected to give an indication of the difference of the two shaft angles. There are two categories, *control* synchros and *torque* synchros.

Control Synchros

In control synchros, the standard pair consists of a generator or control transmitter (CX) and control transformer (CT), Fig. 16.6. The voltage impressed on the CX windings establishes an alternating magnetic field fixed in direction with respect to the rotor. This induces three secondary voltages which are in time phase but whose magnitudes are 120° apart in space. The three secondary voltages represent a code that is uniquely related to the angular position of the rotor. When these three secondary voltages are impressed on the corresponding windings of a CT, the magnetic field direction of the CX is repeated inside the CT. The CT output winding will produce an off-null voltage when the CT rotor is not exactly at right angles to the magnetic field, and this voltage can be employed to energize a motor that will drive one of the two shafts into coincidence with the other one (positional servomechanism). A control *differential transmitter* (CDX), Fig. 16.7, can be used when three or more shafts are involved.

A synchro *resolver* is a device with two-phase windings. It is used primarily for trigonometric function generation and as an electromechanical analog computing component, for trigonometric computations and coordinate conversion (Fig. 16.8).

Slip rings and brushes are required to bring electric current into and out of the synchro rotor windings. Various other schemes are functionally similar to synchros. They are contactless and frictionless, employing a variable reluctance principle (Microsyn, Synchrotel) or second harmonic voltage generation (Magnesyn). Other schemes employ deposited-film techniques (Inductosyn).

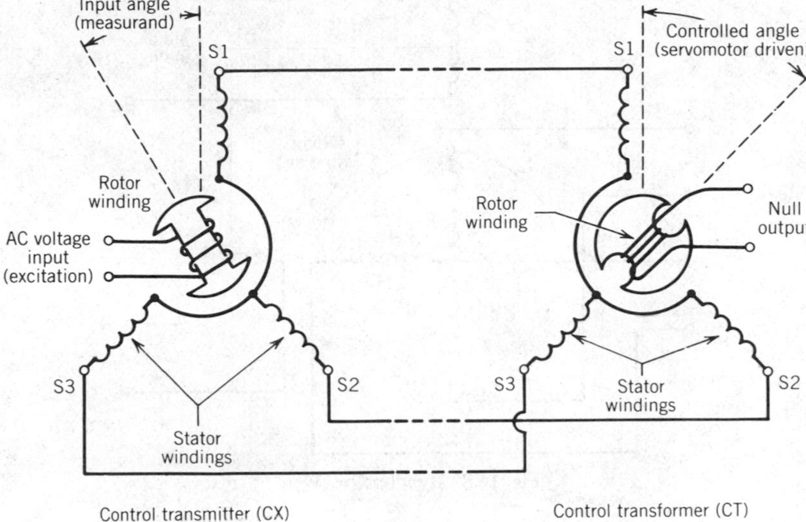

Fig. 16.6 Control synchro pair (CX-CT).

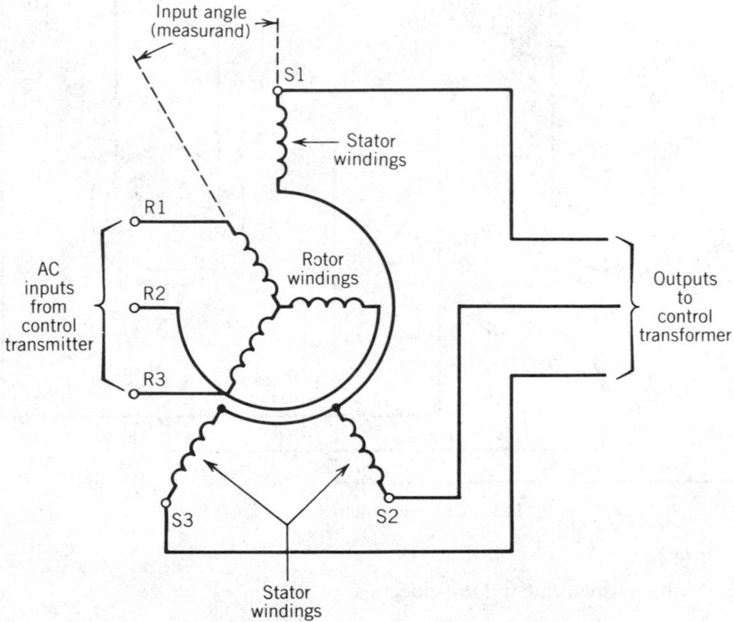

Fig. 16.7 Synchro control differential transmitter (CDX).

Torque Synchros

Torque synchros are an older type of synchro than control synchros. A standard torque synchro pair consists of a synchro generator or torque transmitter (TX) and a synchro motor or torque receiver (TR), Fig. 16.9. Both components are energized from the line. Currents flow in the stators if the two shafts are not aligned, and this results in torque generation in both machines. If one shaft is suitably restrained, the other shaft will be driven into coincidence with it. Torque differential transmitters (TDX) and differential receivers (TDR) are also made.

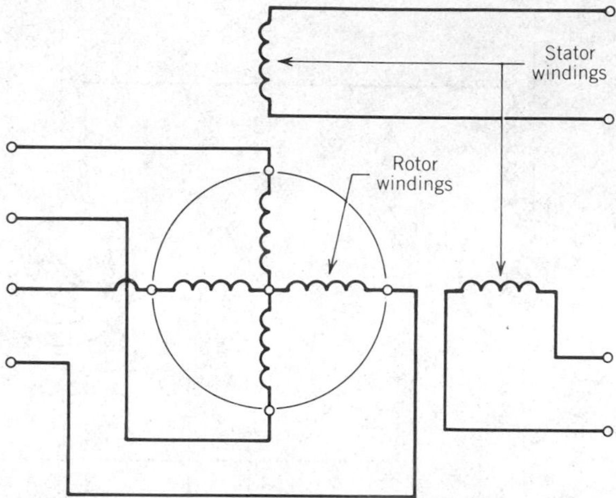

Fig. 16.8 Synchro resolver.

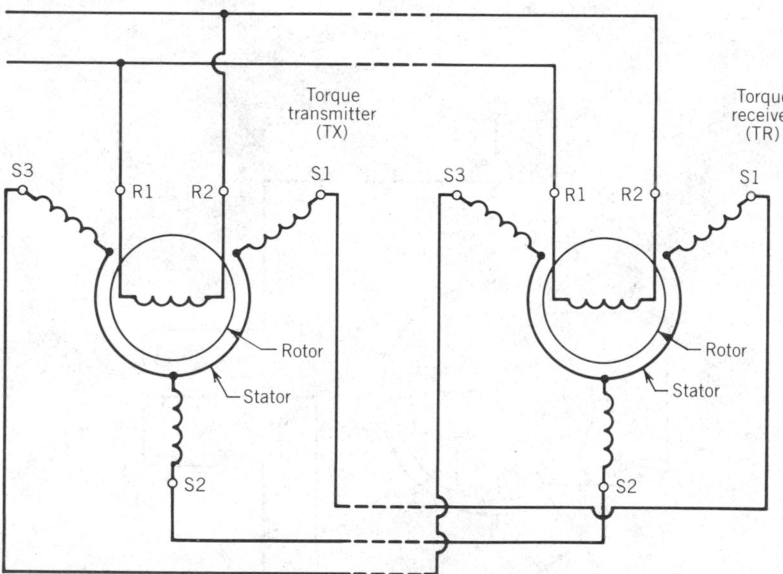

Fig. 16.9 Torque synchro pair (TX-TR).

16.2-7 Capacitive Displacement Transducers

Relative motion can be employed to change capacitance. A common implementation is a proximity switch, which detects the motion of an object that has a permittivity (dielectric constant) different from the surrounding medium. Capacitive transduction is also used for continuous displacement sensing. The measurand can be used to displace an electrode with respect to a second, stationary electrode; this scheme is used in some pressure transducers. Another method is to have the measurand move a piece of dielectric material between fixed electrodes.

The variable capacitance can be measured by a fixed-frequency AC bridge. The capacitance can be employed as the frequency-determining element in an electronic oscillator circuit. In the latter scheme, a change in measured displacement causes a change in output frequency (frequency modulation). This "frequency analog" is easily converted to digital form by pulse-counting methods.

To the extent that the measured motion produces a change in electric field energy storage, capacitive transduction is reversible. Practical capacitive torquers require high voltages.

16.2-8 Digital Shaft Encoders

Digital encoders are divided into two classes, *absolute* encoders and *incremental* encoders. The absolute encoder is a displacement transducer with output in the form of a parallel digital word; it is an electromechanical analog-to-digital converter. The incremental encoder produces a single electrical output pulse for a specified motion, and it can be employed to sense either velocity or displacement.

Absolute Encoders

The heart of the absolute encoder is a patterned disk (for rotation) or strip (for translation) with a number of tracks, one track for each binary digit (bit) in the output, Fig. 16.10. The pattern can be read by a number of different sensing schemes: optically, by electrical contact (via brushes), magnetically, or capacitively. An optical encoder disk is built from alternating opaque and transparent segments which are sensed by an array of light sources, focusing system, and photosensors. A brush encoder has alternating conductive and nonconductive segments read by an array of metal brushes in contact with the disk. A magnetic encoder has alternating magnetized and unmagnetized segments sensed by tiny toroidal coil reading heads. Resolution of commercially available rotational encoders is from 6 to 20 bits in 1- to 10-in. disk sizes. Because of the difficulty of fabricating small disks with accurately spaced segments, the higher resolution encoders are sometimes multiturn units, with two disks geared together. The rotational encoder (also called a shaft encoder) is a true "infinite angle" continuous rotation transducer. Translational encoders are also available.

Absolute encoders must be provided with a way of overcoming the spurious transients (switching ambiguities) that arise when transition from one number to the next number requires a change in more than one digit, for example, a transition from natural binary 0111 to natural binary 1000. These switching ambiguities are inevitable in disks employing a polystrophic digital code, because it is not possible to fabricate encoders with such mechanical perfection that all sensors consistently operate in synchronism. One solution is to use a monostrophic code (a code in which only one digit changes per transition), for example, the Gray code. Another method employed in older brush-type encoders is to employ two brushes per track arranged in various configurations known as U-scan and V-scan. The output of commercially available encoders may be in any of the digital codes employed in computers, such as natural binary, binary code, binary coded decimal (BCD), excess-3, and so forth, with electronic logic packaged within the transducer housing.

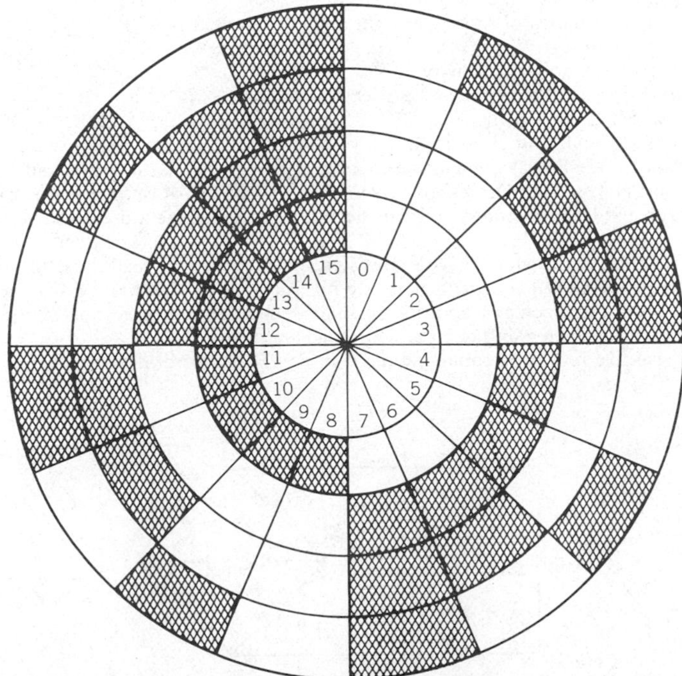

Fig. 16.10 Encoder disk.

Incremental Encoders

The incremental encoder in its simplest form has a single on-off track, and one output pulse is generated for each segment. Basically this transducer is a *pulse tachometer*, that is, it has an electrical output whose pulse rate (i.e., frequency) is proportional to the mechanical speed. Bidirectional sensing can be obtained by providing a second track offset by one half of the segment length. This pulse tachometer becomes a relative displacement transducer if the output pulses are stored and counted. For absolute motion measurement the counter must be reset to zero at the appropriate angle. Some incremental encoders have a third track with a zero reference marker or index that can be employed for this reset. The incremental counter is vulnerable to loss of pulses or spurious pulses due to noise, power interruption, or other causes. These count errors, when they occur, will persist until the reference marker is sensed and the electronic counter is reset.

16.2-9 Strain Gages

Resistance strain gages are resistors subjected to a change in physical size. They are employed to measure very small displacements, either in direct measurement of motion or in secondary measurements where such motion is the result of an applied force, pressure, or acceleration. The phenomena on which strain gage action are based can be summarized in the following relations

$$R = \frac{\rho l}{A} \tag{16.1}$$

$$\frac{\Delta R}{R} = \frac{\Delta l}{l} - \frac{\Delta A}{A} + \frac{\Delta \rho}{\rho} \tag{16.2}$$

$$= \frac{\Delta l}{l}(1 + 2\upsilon) + \frac{\Delta \rho}{\rho} = k\frac{\Delta l}{l} \tag{16.3}$$

$$k = \frac{\Delta R/R}{\Delta l/l} = 1 + 2\upsilon + m \tag{16.4}$$

$$m = \frac{\Delta \rho/\rho}{\Delta l/l} = \Pi_1 E \tag{16.5}$$

where R = resistance
ρ = resistivity
l = length of resistor parallel to current flow
A = cross-sectional area of resistor at right angles to current flow
υ = Poisson's ratio
k = gage factor or strain sensitivity
m = term displaying the piezoresistive effect
Π_1 = piezoresistive coefficient
E = Young's modulus (of elasticity)

Poisson's ratio υ of material remains very constant within the elastic range and has a value of about 0.3 for most metals. The piezoresistive coefficient m can be positive or negative; can be uneven, that is, vary with the level of strain; and can be more or less temperature sensitive depending on material.

In wire resistance strain gages the piezoresistive phenomenon is small. Useful resistance wire alloys with low temperature coefficient of resistivity, such as the Advance or Constantan copper-nickel alloys, have a gage factor of about 2. In semiconductor strain gages the piezoresistive effect predominates, and gage factors in the 50 to 200 range are obtained.

The strain sensitive resistor is commonly used in a Wheatstone bridge circuit, Fig. 16.11. In an

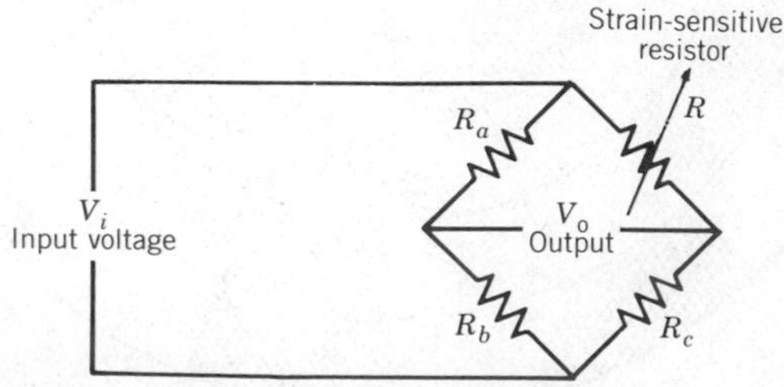

Fig. 16.11 Strain gage bridge.

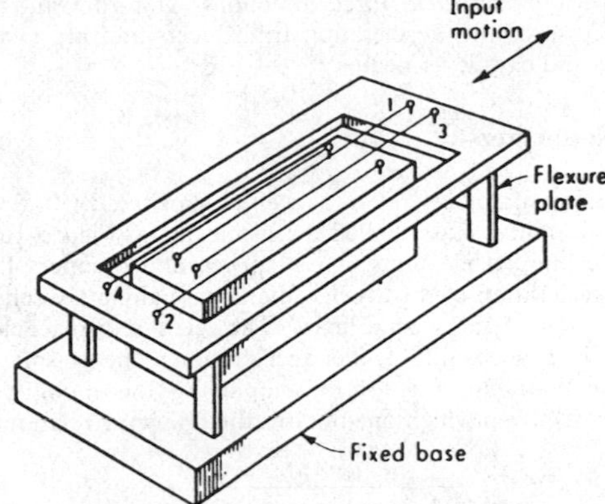

Fig. 16.12 Unbonded strain gage. From McGraw-Hill, New York, with permission.

ideal equal-arm bridge with zero source resistance and infinite detector resistance the output voltage resulting from the change in a single resistor arm is

$$V_0/V_i = (1/4)\,\Delta R/R \tag{16.6}$$

$$V_0/V_i = (1/4)k(\Delta l/l) \tag{16.7}$$

where V_i equals the bridge excitation voltage and V_0 equals the bridge output voltage. The circuit therefore measures fractional elongation or strain, rather than absolute displacement or motion. Achievable output voltage values are limited by the strain gage factor value and by the limitation imposed on the bridge excitation voltage V_i to avoid undue self-heating. For wire resistance strain gages, typical full-scale output of 2 to 3 mV is obtained over the elastic range. Such a small output requires careful signal conditioning. The output can be doubled by employing two gages on opposite sides of the bridge, or gages in "push-pull," one in tension and the other in equal compression. The output of semiconductor strain gage transducers is of the order of 100 mV full scale.

Wire resistance strain gages have inherently excellent linearity and low temperature sensitivity. In semiconductor strain gages, both the linearity and the temperature sensitivity can be controlled with proper semiconductor doping. The temperature stability of strain gage transducers is further improved by employing a dummy gage as one of the Wheatstone bridge arms.

A variety of physical embodiments and configurations are commercially available. Figure 16.12 shows an unbonded strain gage in which four preloaded strain wires are connected in a Wheatstone bridge. Figure 16.13 gives examples of bonded wire and metal-foil strain gages. A strain rosette is an

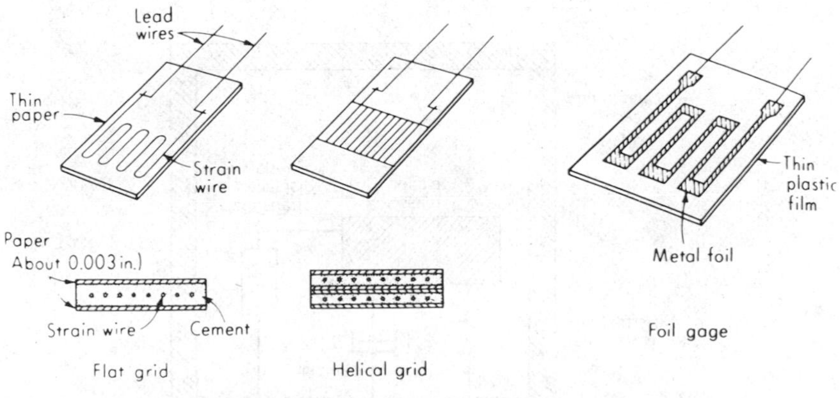

Fig. 16.13 Bonded strain gages. From Doebelin, *Measurement Systems Applications and Design*, McGraw-Hill, New York, 1971, p. 228, with permission.

assembly of gages that measures strain in three directions. Thin-film and integrated-circuit strain gages are used in force, pressure, and acceleration transducers and are available in such forms as diffused silicon diaphragms and cantilever beams.

16.2-10 Absolute Accelerometers

Accelerometers are used to measure acceleration as well as shock and vibration. The sensing element consists of an elastically restrained mass, called a seismic mass. Figure 16.14 illustrates the basic operation in terms of three lumped elements, mass, spring, and damper. The input variable to be measured is the absolute acceleration of the accelerometer case along the sensitive axis, that is, in the direction of freedom of motion of the seismic mass. The relative displacement of the seismic mass with respect to the case is a measure of the acceleration. The seismic accelerometer can be represented by an electrical analog, Fig. 16.15, employing the mobility method (i.e., velocity $\rightarrow$ voltage and force $\rightarrow$ current), from which one obtains the following relation:

$$\frac{Z}{A} = \frac{M}{Ms^2 + Bs + K} \qquad (16.8)$$

$$= \frac{1}{s^2 + 2\xi\omega_n s + \omega_n^2} \qquad (16.9)$$

where $Z = X - Y$ = displacement of seismic mass relative to accelerometer case, where X = displacement of accelerometer case along sensitive axis and Y = displacement of seismic mass in inertial space
A = acceleration of accelerometer case along sensitive axis
M = seismic mass
M_c = mass of accelerometer case
K = spring stiffness, force/displacement
B = viscous damping coefficient, force/velocity
$\omega_n = \sqrt{K/M}$ = undamped natural frequency
$\zeta = B/2\sqrt{KM}$ = damping ratio
s = Laplace transform variable

The steady-state response of this basic accelerometer mechanism is inversely proportional to ω_n^2. Its dynamic response is uniform for frequencies extending from zero to some high limiting value determined by the natural frequency ω_n. Hence there is a tradeoff between sensitivity and bandwidth. The upper frequency limit is commonly defined as the value at which the response has changed by 5%, which is approximately 0.2 ω_n for an undamped accelerometer and 0.5 ω_n for a 0.7 damping ratio. Some accelerometer mechanisms are oil filled to provide maximum flatness; others are inherently undamped, with damping ratio of 0.05 or less.

The practical implementation of this measurement principle requires sensing of the mechanical displacement. This can be accomplished by use of previously described displacement transducers or by a piezoelectric transducer. In some cases the displacement transducer elements do double duty: a differential transformer core serves as the seismic mass or unbonded strain wires serve as springs.

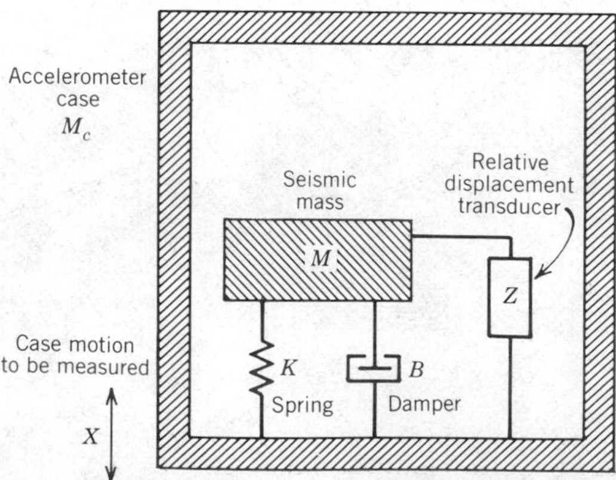

Fig. 16.14 Seismic accelerometer.

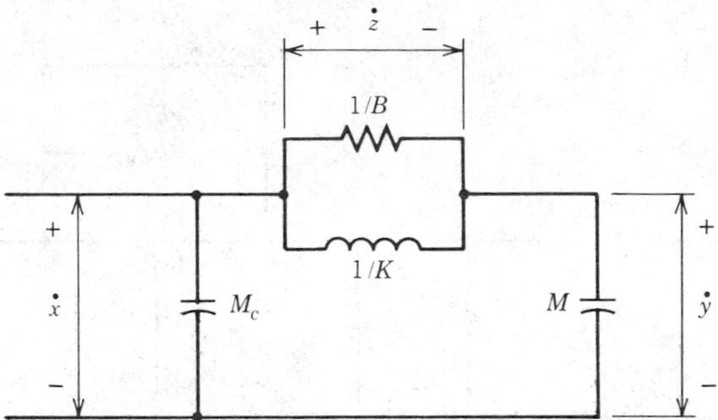

Fig. 16.15 Electrical analog of a seismic accelerometer.

16.2-11 Cantilever Beam Accelerometer

A common implementation of the seismic accelerometer principle is the cantilever beam, which serves a dual function of mass and spring. The beam deflection, related to acceleration as in Eq. (16.9), is usually sensed by semiconductor strain gages directly bonded to the beam.

The cantilever beam configuration is also the basis of the more recent integrated-circuit accelerometers in which the transducer and electronics are fabricated on the same silicon crystal. Both bipolar and metal oxide semiconductor (MOS) technologies are used. A deflectable cantilever beam (microbeam) is formed on the crystal by conventional silicon technology. The movement of the beam is then detected by piezoresistive, piezoelectric, or capacitive techniques. In the piezoresistive version, piezoresistive sensors are integrated into the beam at appropriate locations. In the piezoelectric version, the beam is overcoated with a piezoelectric film (zinc oxide). In the capacitive version, the beam is plated with metal and forms one side of a capacitor, the other side being the substrate.

16.2-12 Piezoelectric Transducers

Piezoelectricity is the phenomenon of coupling between elasticity and electric field in certain types of solid crystals. When a piezoelectric crystal is deformed, it generates an electric charge, and conversely, when an electric charge is applied the crystal is deformed. Materials used are natural crystals such as quartz, synthetic crystals such as ammonium dihydrogen phosphate, polarized ferroelectric ceramics such as barium titanate, and semiconductors such as zinc oxide. Piezoelectricity is exploited both for displacement transducers (mechanical-to-electrical conversion) and for acoustic generators (electrical-to-mechanical conversion).

As displacement transducer, the equivalent circuit of Fig. 16.16 is appropriate, and the response is

$$v = \frac{kRs}{RCs + 1} x \tag{16.10}$$

where k = electromechanical coupling factor
s = Laplace transform variable
C, R = respectively, combined capacitance and leakage resistance of transducer, cable, and load

The transducer therefore does not respond to steady-state (i.e., constant) displacements and senses dynamic displacements with uniform sensitivity over a frequency range above a cutoff value $1/RC$.

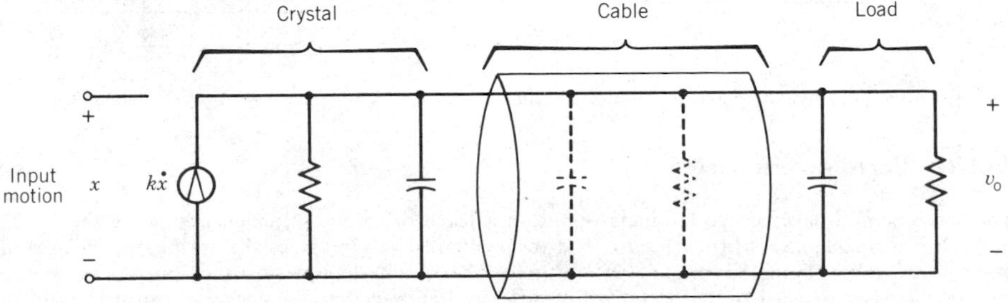

Fig. 16.16 Piezoelectric transducers, equivalent circuit.

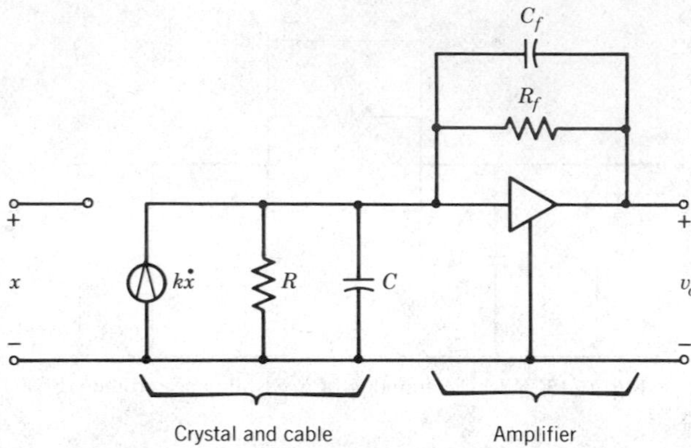

Fig. 16.17 Charge amplifier.

A charge amplifier (Fig. 16.17) is often employed to provide a definite cutoff frequency $1/R_fC_f$ independent of the electrical parameters of the cable

$$v_0 = -\frac{kR_fs}{R_fC_fs + 1}x \tag{16.11}$$

16.2-13 Piezoelectric Accelerometers

The piezoelectric crystal serves both as the displacement transducer and the elastic restraint member of the seismic mechanism. Figure 16.18 shows two typical configurations, compression mode and shear mode. These devices are used as vibration and shock transducers and in other applications where the absence of response to constant accelerations is immaterial. The useful frequency range extends from a lower limit determined by the parasitic circuit elements (or by the charge amplifier parameters) to an upper limit determined by the seismic mechanism. The high natural frequency ω_n is somewhat counterbalanced by the extremely low damping of this inherently undamped device. Typical frequency response range is 2 to 5000 Hz.

Integrated circuit accelerometers have been built with piezoelectric films (see Section 16.2-11); silicon itself is not piezoelectric.

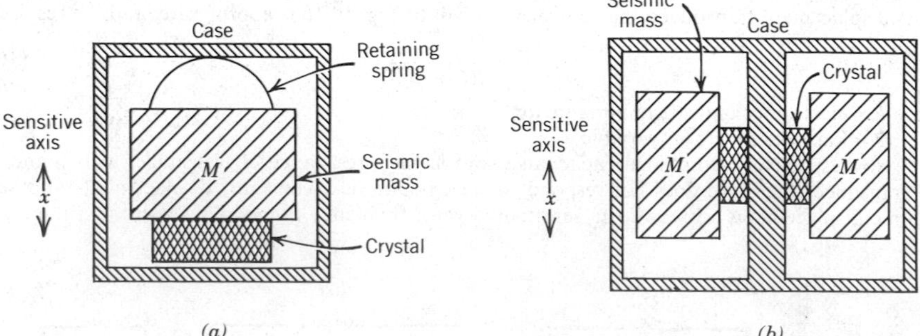

Fig. 16.18 Piezoelectric accelerometer configurations: (*a*) compression mode, (*b*) shear mode.

16.2-14 Servo Accelerometers

The servo or null-balance accelerometer employs a force or torque transducer to bring the seismic mass displacement back to (near) zero. The torquer effectively replaces the spring, which in some cases is entirely eliminated. Figure 16.19 shows the block diagram of such an accelerometer. Acceleration is measured by the torquer current. The static accuracy is dependent primarily on the torquer. Natural frequency and damping are set by electrical parameters of the feedback loop.

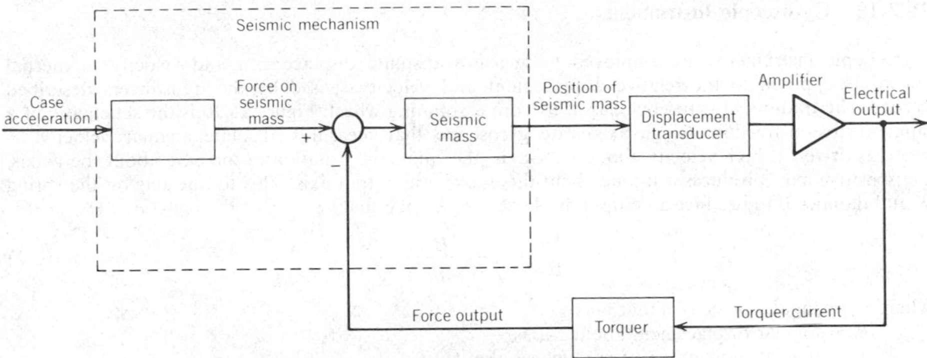

Fig. 16.19 Servo accelerometer block diagram.

16.2-15 Velocity Transducers

Velocity transducers can be classified as analog and pulse. They can be rotational (tachometer) or translational. Analog velocity sensors are predominantly of the electromagnetic type: A voltage is induced in a wire moving in a constant magnetic field. The translational devices are known as moving-coil or moving-magnet pickups. The rotational version is a small DC generator with permanent magnet rotor. Other electromagnetic versions are the AC permanent-magnet generator and the AC induction generator.

Pulse tachometers are sometimes called digital tachometers, but in fact their output is "pulse rate analog," that is, the output pulse rate is proportional to the velocity. Pulse tachometers are available in many physical configurations, some of which were described in Section 16.2-8. The output of pulse tachometers can be converted to a parallel digital word by counting the number of high-frequency clock pulses between successive tachometer output pulses or by counting the number of tachometer output pulses during a specified base period. The pulse tachometer output can also be converted to an analog voltage by use of a frequency-to-voltage converter.

The flyball velocity pickup and the eddy-current drag-cup tachometer are examples of sensors that produce an output displacement proportional to the measured velocity. They are used in some hand-held tachometers.

16.2-16 Force, Torque, and Weight Transducers

Force, torque, and weight can be measured by a variety of methods based on comparison of the unknown with a known force. Such methods are used primarily for steady forces. Sensors for dynamic measurements are predominantly based on application of the unknown to an elastic member and measurement of the resulting deflection or strain. The primary element that converts a force or weight into an elastic deflection is called a load cell. Such a cell can have various physical forms, such as the proving ring and cantilever beam. Dynamically, these devices can be approximated by a lightly damped second-order system, but in many devices the "mass" and "spring" are distributed and cannot be separately identified. Strain gages are the predominant displacement sensors employed in load cells. In a similar fashion, shaft torques are measured by sensing of the deflection along a fixed length of shaft. The sensors can be attached to the shaft itself or a special instrumented shaft section (torquemeter) can be installed. Slip rings are provided if the measurement is made on a rotating shaft.

16.2-17 Torquers

Force and torque generators (torquers, force motors) are electrical-to-mechanical transducers. As employed in measurements, they are basically electric actuators or motors with limited displacement. Torquers are used in force-balance-type nulling schemes, for example, in the servo accelerometer and in gyroscopic instruments. In these applications the output torque-to-input current (or voltage) relation of the torquer must be held to a precise value. The most common torquer designs are based on the D'Arsonval (moving-coil) principle or on the variable reluctance (moving-iron) effect.

16.2-18 Gyroscopic Instruments

Gyroscopic instruments are employed to measure absolute displacement and velocity in inertial space, as opposed to the relative displacement and velocity sensed by the transducers described earlier. The traditional gyroscope has at its core a spinning wheel. Figure 16.20 is the schematic of a single-degree-of-freedom (single-axis) rate gyroscope that measures absolute angular velocity. A wheel is driven at high velocity about the x-axis, the spin axis. Rotation of the base about the y-axis, the sensitive axis, produces a torque about the z-axis, the output axis. This torque acts on the spring K and damper B to produce an output displacement θ_0 such that

$$\frac{\theta_0}{\Omega_y} = \frac{H}{Js^2 + Bs + K} \tag{16.12}$$

where Ω_y = angular velocity about y-axis
θ_0 = angular displacement about z-axis
H = angular momentum of gyro wheel about x-axis
J = moment of inertia of gyroscope about z-axis
K = torsional spring constant for z-axis motion
B = rotational viscous friction for z-axis motion

Accurate velocity measurement requires precise wheel speed, precision spring, and small displacement in order to avoid cross-coupling effects. The output displacement can be sensed by standard transducers, for example, a differential transformer or E-pickoff.

The single-degree-of-freedom rate-integrating gyro measures absolute angular displacement. Its configuration is identical to that of the rate gyro except that it has no spring restraint. Its output-input relation in terms of the input axis rotation θ_y is

$$\frac{\theta_0}{\theta_y} = \frac{H}{Js + B} \tag{16.13}$$

Precision displacement measurement requires that the damping constant B be precise. A standard design uses a precisely temperature-controlled fluid as a flotation and damping medium.

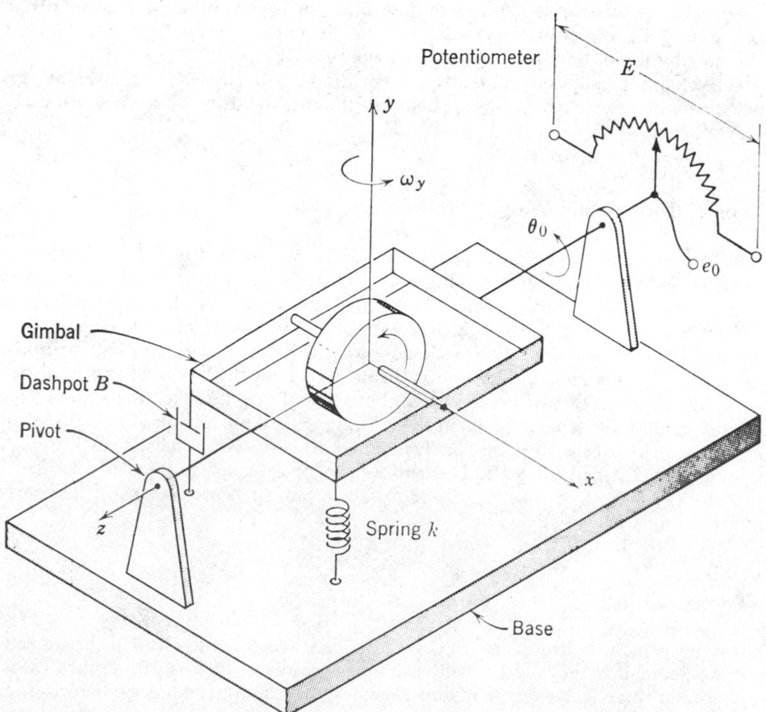

Fig. 16.20 Single-axis rate gyroscope. From Clark, *Introduction to Automatic Control Systems*, Wiley, New York, 1962, p. 126, with permission.

In addition to the displacement pickoff, most gyros also have a torquer mounted on the output axis. This permits introduction of correction torques or employment of the gyro in a null-balance or servo mode.

Three such single-degree-of-freedom gyros mounted in mutually perpendicular directions are required to measure the rotational displacement of a rigid body in space. An alternative method in traditional gyroscopics is to employ a two-degree-of-freedom gyro (free gyro). A vertical gyro is a two-degree-of-freedom gyro with its spin axis maintained parallel to the local vertical; a gravity-sensing device such as a pendulum or electrolytic switch is usually packaged with this instrument. A directional gyro has its spin axis lined up with the horizontal, that is, perpendicular to the local vertical.

The laser gyroscope is based on two laser beams which travel along a closed optical path in opposite directions. Changes in gyro rate alter the apparent path length of the two beams. This difference is detected by an optical interferometer.

16.3 TRANSDUCERS FOR FLUID-MECHANICAL QUANTITIES

16.3-1 Scope and Definitions

The following subsections deal with instruments for the sensing and measurement of fluid flow, pressure, level, density, and viscosity. This includes liquids and gases and mixtures thereof, and other forms of matter that have fluid-like properties, such as powdered solids and slurries.

Instruments of this type find application in industry and laboratories, in oceanographic and meteorological work, and in medicine, biology, and bioengineering. The same principles are applicable in all situations. The practical examples discussed in this section are usually drawn from industry. Examples from other areas can be found in applicable sections of the handbook.

Until fairly recently, industrial measurements of fluid-mechanical quantities were largely carried out mechanically. Signal processing, signal transmission, and power actuation were predominantly pneumatic. In the past decade, electronic methods have come to the forefront, and transducers that exploit the recent advances in solid-state electronics have reached the market. This handbook section focuses primarily on these electrical and electronic transducers.

16.3-2 Pressure Measurement Principles

Pressure is defined as the force exerted on a unit area (newton per square meter, or pascal). The pressure of a liquid or gas may be defined as *absolute* or as *relative* to some reference pressure. A frequently employed pressure reference is atmospheric pressure, either the local ambient pressure or a standard atmosphere ($\sim$100 kPa); the term "gage pressure" is employed for such a measurand. The measurand "differential pressure" represents the relative pressure between two variables, both of which may be varying or unknown. A further distinction is made between pressure switches, which are on-off devices actuated when a predetermined pressure value is sensed, and continuous pressure sensors.

Vacuum is negative gage pressure, below ambient. "High vacuum" refers to the pressure region near absolute zero, typically less than 10^{-3} torr (0.13 Pa).

Differential pressure is a derived measurand in some flow, level, density, and viscosity measurement schemes. It is also used as a means of determining altitude above sea level and water depth. In industrial control usage, a pressure transducer is often referred to as a pressure transmitter.

16.3-3 Pressure-Sensing Elements

The most common pressure measurement principle is to have the pressure deform or displace a solid elastic membrane and sense the resulting deformation or displacement. An elastic diaphragm is a simple example of a sensing element for relative pressure measurements. For gage pressure measurements, one side of the diaphragm is exposed to ambient pressure and the other side to the process pressure; for differential measurements, both sides are exposed to process pressures. The diaphragm must be able to withstand exposure to the measured fluid.

For absolute pressure measurements, the diaphragm is part of a sealed capsule that has been evacuated (aneroid capsule).

Originally, pressure-measuring devices were purely mechanical. Mechanical amplification was required to obtain a displacement large enough to provide a pointer reading or to drive a conventional electromechanical displacement transducer. Pressure bellows and pressure tubes (such as the Bourdon tube) are elastic devices especially designed to provide an amplified output motion.

16.3-4 Conventional Pressure Transducers

A wide range of combinations of sensing and transduction elements is employed in available pressure transducers. Resistance potentiometers, differential transformers, and reluctive and capacitive transducers are available. Piezoelectric crystals are used for dynamic pressure measurements. Strain wires have been employed in a scheme where the strain causes a change in the resonant frequency of the wire; this change is made to produce an electrical output in the form of a frequency analog (FM). Semiconductor strain gages are employed either in the form of a separate strain gage assembly or integral with the deflecting membrane, as described later.

The *force-balance* pressure transducer is a feedback (closed-loop) device. The amplified displacement of the sensing membrane is employed to energize a force motor or torquer which drives the membrane back to its null position. The input to the force motor is therefore related to the measurand and serves as the transducer output. The advantage of the feedback scheme is that it minimizes the effect of inaccuracies in the sensing and transducing elements, such as elastic nonlinearity, hysteresis, and temperature sensitivity. The transducer accuracy depends on the calibration of the force motor. Pneumatic and electromechanical force-balance pressure transducers are available.

16.3-5 Integrated-Circuit Pressure Transducers

Various pressure transducer designs are based on conventional integrated-circuit batch-processing technology. Silicon is etched to form a pressure diaphragm. Various strain-sensing regions are provided on the diaphragm, and electronic amplification can be integrated on the same crystal. The most prevalent forms are piezoresistive and capacitive.

The most widely used device is the piezoresistive pressure diaphragm where strain-sensitive resistance patterns are precisely located on a silicon wafer; see Fig. 16.21. The patterns are usually organized as a full-wave bridge, which cancels the effect of temperature gradients. This transducer has low output impedance and is reasonably stable.

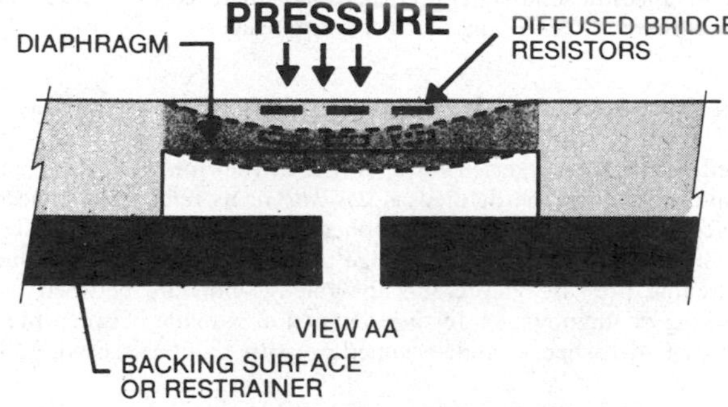

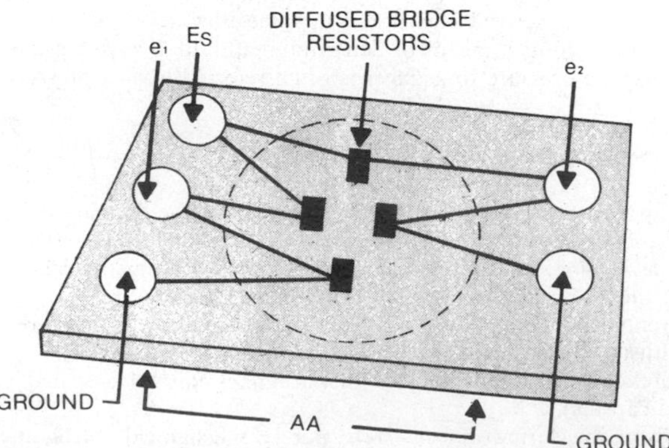

Fig. 16.21 Piezoresistive pressure diagram. From Hall, *Instr. Contr. Sys.*, p. 60 (Apr. 1981), with permission.

The diaphragm can also be fabricated so as to be one side of a variable capacitor. This transducer is extremely stable but requires on-chip amplification. Other schemes that have been used are based on piezojunction sensing, that is, pressure-induced changes in the current gain or reverse current of a *pn* junction.

In many applications the semiconductor diaphragm requires protection against contamination by the process fluid and is separated from it by some kind of barrier, such as a stainless steel diaphragm.

16.3-6 Vacuum Transducers

The unit commonly used in vacuum work is the torr: 1 torr = 1 mm Hg = 1000 μm $\approx$ 0.02 lb/in.^{2}(psi) $\approx$ 130 Pa. The standard deflection-type pressure transducers described earlier can cover negative pressure ranges down to 1 torr. Diaphragm sensors with a capacitance transduction element are available that cover a useful measurement range of 10^{-5} to 1 torr; this represents essentially a refinement of standard technology. Entirely different principles are involved in several classes of vacuum pressure transducers that basically "count gas molecules."

Thermal conductivity vacuum transducers are based on the theoretical linear relationship between gas pressure and thermal conductivity existing at pressures low enough so that the mean free path of a gas molecule is large compared with the pertinent dimension of the measuring apparatus. The transducer consists of a glass tube open to the gas to be measured that contains a resistance heating element supplied with a constant electrical current. The temperature of the heating element depends on the total heat loss from it; the conduction component of this heat loss is a measure of gas pressure, for gas of a given composition. The temperature is measured by a thermocouple or thermopile, by a thermistor, or by combining the functions of heating and measurement in a resistance thermometer (Pirani gage). Thermal conductivity gages have a useful range down to 10^{-5} torr.

In the ionization vacuum transducer a stream of electrons is used to ionize a sample of the gas. The number of positive gas ions so formed is directly proportional to the electron current and to the gas pressure

$$i_i = S i_e p \tag{16.14}$$

where i_i = ion current (transducer output)
i_e = electron current
p = gas pressure
S = ionization gage sensitivity

The sensitivity S is constant for a given gas over a wide range of pressures. The ionization current is developed by one of two methods. In the hot-cathode (thermionic) ionization gage, electrons are emitted by a filamentary cathode and the ion current is collected at the anode. In the cold-cathode ionization gage, the electrons are pulled out of the cathode by a high electrostatic field. (Alpha particles are employed as the electron source in one design.) Typical range of measurement is 10^{-9} to 10^{-3} torr; various refinements have brought the detectable pressure down to 10^{-14} torr.

The output current of ionization transducers is very small, and considerable signal-processing circuitry is required to achieve a useful measuring instrument.

16.3-7 Flow Measurement Principles

The measurand called flow rate may actually be one of three different physical quantities: (1) the linear velocity (meters per second) of the fluid at a particular point (a vector quantity with magnitude and direction, measured with respect to a reference that may be stationary or moving), (2) the volumetric flow rate (cubic meters per second) through a cross-sectional area, which is the surface integral of the linear flow rate over the area, and (3) the mass flow rate (kilograms per second) through a cross-sectional area, which is the surface integral of velocity multiplied by density.

In a few instances the flow to be measured is unconfined, for example, wind velocity. In most cases the flow is confined, either in open channels (rivers and partly filled pipes with one free liquid boundary) or closed channels (full pipes), and the rate to be measured is unidirectional, along the axis of the channel.

With some exceptions (such as in meteorology and oceanography), the user is interested in the volume or mass flow rate. While many applications require a mass flow measurement, mass flow is a difficult quantity to sense and most flowmeters are volumetric. Mass flow must then be determined via a simultaneous density measurement or computation; a few flowmeter types measure mass directly.

Virtually all volumetric flowmeters are based on sensing of flow velocity and are then calibrated in terms of volume. In some types the flow velocity is sensed or sampled at a point (or rather over a small area), and the volumetric rate determination requires a knowledge of the velocity profile, that is, the distribution of fluid velocity over the channel cross section. Such a flowmeter reading would have to be corrected if the velocity profile is variable or if it is not identical to the one on which the calibration is based. In other flow sensors the output represents an average of velocities over the profile. Some flowmeter types measure volume directly.

Velocity profile in closed channels is determined by the Reynolds number, a dimensionless ratio defined by

$$\mathrm{Re} = VD\rho/\mu \tag{16.15}$$

where V = flow velocity, m/s
D = pipe diameter, m
ρ = density, kg/m^3
μ = dynamic viscosity, Pa · s

At Reynolds numbers below 2000, flow is laminar and the velocity profile is parabolic. At Reynolds numbers above 10,000, flow is turbulent and the velocity profile is substantially uniform over the diameter. The transition from laminar to turbulent flow is gradual, and the flow pattern in the Reynolds number range from 2000 to 10,000 is not clearly defined. In most flowmeter installations, it is desirable that the flow sensor location be preceded and followed by several diameters of straight pipe of uniform cross section, in order that the velocity profile for which the flowmeter is calibrated can become fully developed.

Since both the density and the viscosity are characteristics of a particular material, a flowmeter calibrated for one liquid may be in error when used with a different liquid. Furthermore, changes in temperature and pressure will also cause changes in the Reynolds number and thereby affect the flowmeter calibration. Flow measurement accuracy is also predicated on an a priori knowledge of the physical composition of the fluid. For example, the volumetric output reading of a flowmeter based on a flow velocity sensor assumes that the entire pipe is full of liquid; the reading will be in error if gas bubbles are entrained with the liquid.

16.3-8 Pressure Head and Velocity Head-Flow Sensing Elements

Head-type flowmeters consist of a flow-sensing element that interacts directly with the stream to induce velocity and pressure changes and a secondary element that senses and transduces the pressure changes.

Pressure head elements are restrictions inserted in the flow which cause the velocity to increase as flow passes through the narrow throat. Pressure decreases as the velocity rises. The pressure drop across the restriction is related to flow:

For turbulent flow

$$Q = k\sqrt{\Delta P/\rho} \tag{16.16}$$

For laminar flow

$$Q = k\Delta P/\mu \tag{16.17}$$

where Q = volumetric flow rate
ΔP = differential pressure
ρ = density
μ = viscosity
k = flowmeter constant

Various configurations of flow restrictions are employed: orifice plates, Venturi tubes, flow nozzles, flow tubes, and elbows.

Velocity head elements create stagnation points where the velocity drops to zero and pressure rises. The pressure vs. flow relationship obeys the same equations as in the pressure head sensor. The pitot tube is the simplest velocity head element; it is usually configured in the form of a probe which measures the component of fluid velocity parallel to it. It can be employed to measure the velocity of an unconfined stream, for example, wind velocity, or airspeed of an aircraft.

Weirs and flumes are pressure-head sensing elements employed in open-channel flow measurements. Weirs are damlike structures placed across the flowing stream, with a notch cut out of the upper edge to create a flow path. The level of liquid in the notch is related to the flow rate; the relationship is a 2.5 or 1.5 power law, depending on the shape of the notch. Flumes are primary elements that restrict the width of the channel, producing a change in level dependent on flow. Parshall flumes are most widely used. The secondary sensor for these open-channel flow measurements can be a level transducer or a pressure transducer.

16.3-9 Rotameters

The variable-area constant-pressure flowmeter or rotameter consists of a vertical tube of tapered diameter in which a float takes on a vertical position corresponding to the (upward) flow rate through the tube. Some rotameters have glass tubes for direct observation of the float position. In other cases the float position can be sensed by a suitable displacement sensor, for example, a linear differential transformer. The rotameter is popular because it is often the lowest in cost for small pipe sizes (up to 5 cm diameter). Accuracy is of the order of 2% of full scale.

16.3-10 Flowmeters with Moving Parts

Several flow sensor types operate with moving parts exposed to the fluid. They measure flow by obtaining an angular displacement or velocity dependent on flow rate.

The *positive displacement flowmeter* has a rotor consisting of chambers of fixed volume. The angle of rotation is directly proportional to the volume of fluid. These meters are normally employed to measure total volume rather than rate, for example, in the metering of water and fuel. Accuracy is high, particularly in the large sizes, where 0.1% accuracy is obtained in liquids and 0.5% in gas. The positive-displacement flowmeter is actually a positive-displacement fluid motor designed for low leakage, friction, and inertia. A related scheme is the metering pump, a specially designed positive-displacement pump that measures flow while operating as a pump; its accuracy is of the order of 1%.

The *turbine flowmeter* has a rotor in the pipe which is driven by the fluid. The angular velocity is proportional to flow rate. A magnetic pickup is normally used to sense speed, so that the electrical output is in the form of a frequency (FM) or a pulse rate. This meter is used with clean fluids. Accuracy is of the order of 0.5% of actual rate.

The *angular momentum flowmeter* is a true mass flowmeter. A representative example of the type is the impeller-turbine flowmeter. An impeller driven at constant speed by a synchronous motor imparts an angular momentum to the fluid proportional to the mass flow rate. A spring-restrained turbine located downstream from the impeller converts this momentum into a displacement proportional to mass flow rate, which is sensed by a displacement transducer.

The *target flowmeter* has a cantilevered vane in the flow path which is deflected by the moving fluid. The displacement is measured by strain gages.

16.3-11 Magnetic Flowmeter

The magnetic flowmeter is based on Faraday's law. Its basic elements, shown in Fig. 16.22, are a fluid with electrical conductivity, a length of insulated straight pipe, a set of electromagnet coils to produce a magnetic field perpendicular to the direction of fluid flow, and a pair of electrodes located on the pipe with their axis perpendicular to both the magnetic field and the moving fluid. The output voltage is

$$E = BDv \times 10^{-4} \quad \text{V} \tag{16.18}$$

where B = flux density, T
D = pipe diameter, m
v = fluid velocity, m/s

Since the flux density depends on the current flowing through the electromagnet, the output is

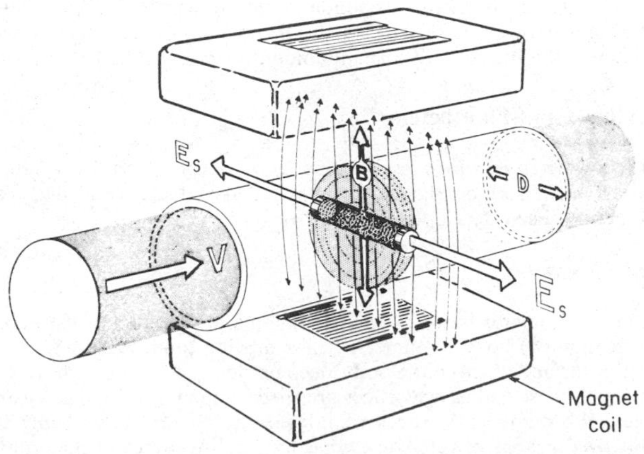

Fig. 16.22 Magnetic flowmeter principle. From Fath, "Fischer & Porter Tech. Information" 10D-14 (Apr. 1977), with permission.

ratiometric, that is, the flow rate information must be determined from the ratio of the flowmeter output voltage E to a reference voltage E_r derived from the magnet coil current. The basic measurement is that of an average fluid velocity, independent of pressure, density, and viscosity variations. The calibration in terms of volumetric flow can be made accurate over wide ranges of velocity profile by suitable shaping of the magnetic field. The flow sensor output voltage is quite small, of the order of millivolts, and careful signal conditioning has been the key to successful magnetic flowmeter designs. Accuracy is of the order of 0.5% of full scale.

Fluid conductivity has an effect on the overall performance because the conductivity enters into the flow sensor behavior as the source resistance for the output voltage. The lower the conductivity, the higher the internal resistance and the more difficult the signal conditioning problem. Present designs permit measurement of fluids down to 10 μS/m (0.1 μS/cm), which includes a wide variety of liquids including organic solvents.

The magnetic field excitation can be DC, AC, or pulsed. DC excitation in most liquids produces polarization, and this results in a large constant DC output upon which the small velocity-dependent output is superimposed. Therefore DC excitation is not practical except in the metering of liquid metal flow. AC magnetic field excitation eliminates the polarization problem but results in significant zero-flow (null) output due to stray electric currents in the fluid itself and to pickup voltages in the external circuitry. In pulsed operation the magnet coils are energized by a low-frequency square wave. A particular design employs a unidirectional square wave, that is, an on-off cycle of DC voltage. During the on period of magnet coil excitation the flowmeter output represents signal-plus-noise; during the off period the output represents the noise alone. By subtracting the off-period output from the on-period output one obtains an output that consists of the flow-dependent signal with a minimum of noise contamination. At zero flow, this subtraction completely eliminates the residual pickup and provides an exact zero output.

The magnetic flowmeter is "obstructionless" and capable of handling dirty, corrosive, and otherwise difficult fluids and slurries. The only items in direct contact with the fluid are the pipe wall and the electrodes. Built-in ultrasonic electrode cleaning is provided in some designs. The pipe section itself must be made of nonmagnetic material, except in some designs where the magnet coils are inside the pipe, which then forms a magnetic return path. An insulating pipe liner is required except where fiberglass or other nonconducting pipe material is employed.

16.3-12 Thermal Flowmeters

There are two types of flow-sensing principles that employ heat. The anemometer flowmeter senses the heat loss of an electrically self-heated resistance thermometer due to the fluid flow. In the constant-current anemometer, a resistor carrying current immersed in a fluid reaches a temperature determined by the balance of i^2R heat generation and convective heat loss. The resistance of the temperature-sensitive resistor is then a measure of the flow velocity. Resistances used are thin wires (hot-wire anemometer), thin films, and thermistors. Circuits using thermocouples have also been used.

The other device, to which is given the name "thermal flowmeter," consists of an electrical heater immersed in the fluid with two equidistant temperature sensors, one upstream and one downstream. The differential temperature is a function of the mass flow rate. The heater and temperature sensors can also be mounted external to the pipe (boundary-layer flowmeter); such a flowmeter approaches the ideal case of noninvasive measurement. In the thermal flowmeter, heat is actually a tracer; an analogous application is the injection of a small amount of radioactive tracer (nucleonic flowmeter).

16.3-13 Oscillating-Fluid Flowmeters

This is a class of flowmeter in which oscillations, vortices, or swirls are created that are linearly related to flow velocity. These oscillations are then detected electronically. Various methods of creating oscillations have been developed.

Vortex-Shedding Flowmeter

The vortex-shedding flowmeter is based on the formation and detection of Karman vortex trails when a bluff (i.e., nonstreamlined) body is immersed in a moving fluid. At a sufficiently high flow rate, eddies formed at the surface of the bluff body become detached in an orderly fashion (Fig. 16.23). The frequency at which these eddies or vortices are shed by the bluff body is linear with flow velocity over a range of Reynolds numbers. Vortices are not shed at low flow rates, hence the vortex-shedding flowmeter is usable over a range of velocities, typically 15:1. Accuracy of the order of 1% of full scale is achieved. The response is independent of Reynolds number above a minimum limit.

When a vortex sheds from one side of the bluff body, the velocity at that point increases and the pressure decreases; on the other side of the bluff body the velocity decreases and the pressure

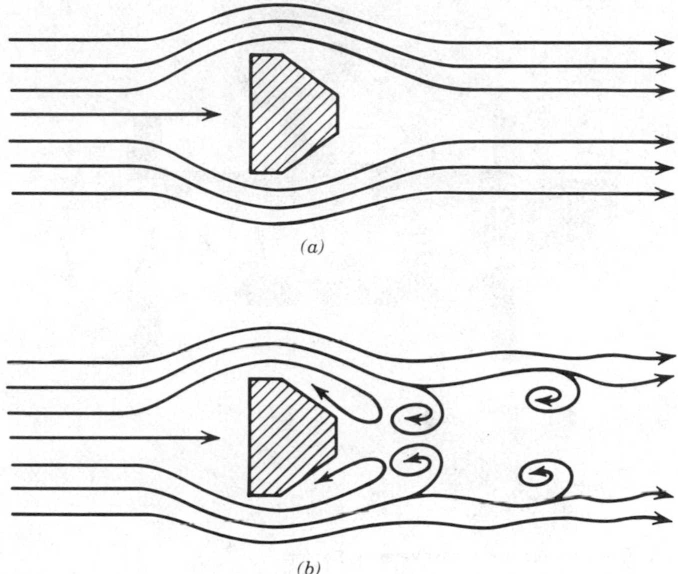

Fig. 16.23 Von Kármán vortex trails: (*a*) low-velocity flow past nonstreamlined body, (*b*) vortex formation at higher flow velocity.

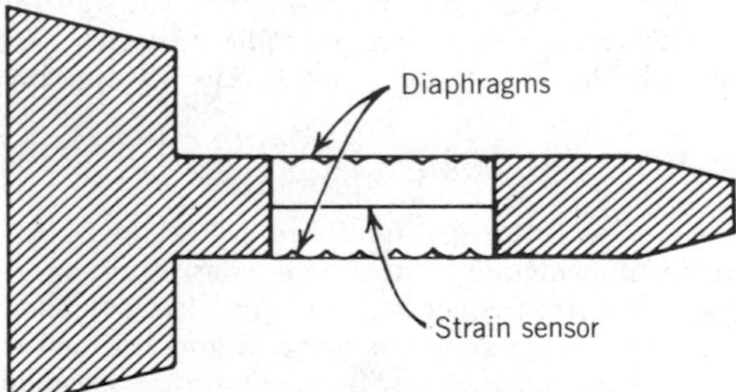

Fig. 16.24 Vortex generation flow transducer.

increases. The alternating flow changes or pressure changes can be sensed by a suitable transducer built into the vortex generator body, such as a pair of thermistors or a strain sensor (Fig. 16.24). Another scheme employs an ultrasonic beam which is transmitted across the flow; each vortex passing through the beam impresses on it one cycle of amplitude modulation.

Vortex Precession Flowmeter

The vortex precession flowmeter ("swirlmeter"), Fig. 16.25, is based on the behavior of a rotating fluid in a piping enlargement. The fluid to be measured enters the metering section through a swirler, a fixed set of curved blades which impart to the fluid a rotation or swirl proportional to the flow velocity. The center of rotation of the vortex moves away from the center line and precesses about it at a frequency related to the swirl rate. The frequency of the alternating flow is sensed by a self-heating thermistor. A flow straightener (deswirler) at the exit of the metering section isolates the flowmeter from downstream conditions. The swirlmeter is used primarily in gas measurement. Accuracy is better than 1% of actual rate.

Wall Attachment Flowmeter

The wall attachment flowmeter is based on the creation of eddies when a fluid emerges from a blunt opening. It the blunt opening is on one side of the pipe only, the pressure difference due to the

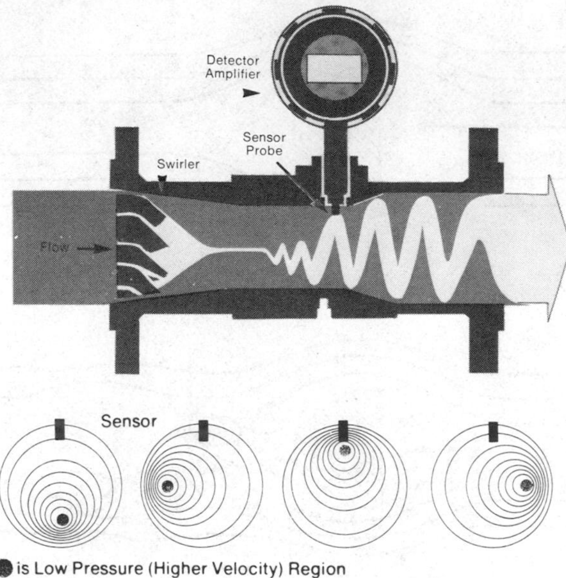

Fig. 16.25 Vortex precession flowmeter. From Fischer & Porter Catalog IOS (1974), with permission.

different flow patterns will cause the stream to be pushed against the wall and the eddies to adhere to it. In one design, a pair of feedback passages causes the eddies to attach themselves alternately to the top wall and to the bottom wall. The rate of alternation is sensed with a pressure sensor.

16.3-14 Ultrasonic Flowmeters

The interaction of sound waves with moving fluid can be employed to measure flow rate. An ultrasonic flowmeter is an instrumentation system that consists of one or more electroacoustic transducers operating in the megahertz frequency range and electronic circuitry to extract the flow rate information. The attraction of such a system is that it is nonobstructive and that there exists the potential for a true noninvasive measurement, with the flow sensors completely outside the pipe (clamp-on flowmeter). These advantages have motivated considerable development effort, and presently several types of ultrasonic flowmeters are on the market that provide accuracies of the order of 0.5% of full scale for flow velocities of 1 m/s full range or higher.

The *transit time difference meter* employs two transducers located relatively upstream and downstream, at opposite sides of the pipe (Fig. 16.26). Acoustic waves are launched upstream and downstream in an alternating sequence and the transit time is measured

$$t_d = \frac{L}{c + V\cos\theta} \tag{16.19}$$

$$t_u = \frac{L}{c - V\cos\theta} \tag{16.20}$$

where t_d = transit time of downstream sound wave
t_u = transit time of upstream sound wave
c = acoustic velocity (speed of sound in fluid)
V = average flow velocity of fluid
L = distance between transducers
θ = angle between sonic path and fluid flow direction

The difference of the reciprocals of the transit times determines a frequency difference Δf

$$\Delta f = \frac{1}{t_d} - \frac{1}{t_u} = \frac{2V\cos\theta}{L} \tag{16.21}$$

proportional to fluid velocity and independent of the speed of sound in the fluid, and therefore also unaffected by parameters such as temperature and composition, which affect the speed of sound. This ideal performance is approximated in practice.

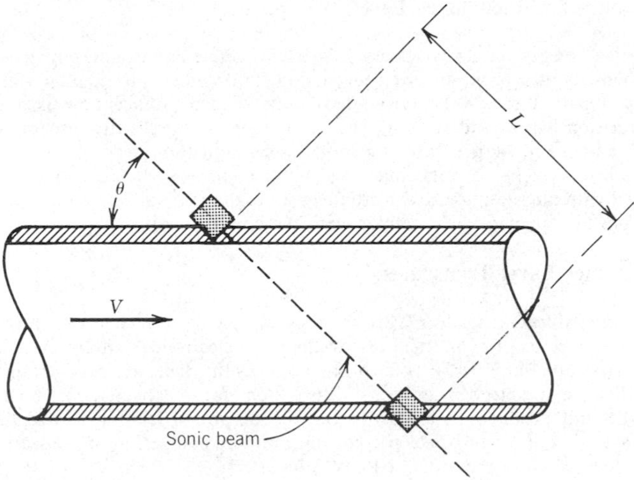

Fig. 16.26 Ultrasonic flowmeter, transit time difference scheme.

The *ultrasonic doppler flowmeter* is based on the principle that the frequency of ultrasonic waves reflected by scattering particles in the moving fluid is shifted in proportion to the velocity of the scatterers. These scatterers may be solid particles suspended in the fluid or small entrained gas bubbles; they may be an inherent part of the fluid to be metered or introduced deliberately for the purpose of ultrasonic doppler metering. It is assumed that the scatterers travel at the same velocity as the fluid itself. Doppler flowmeters require only a single transducer which may be wettable, that is, internal to the metering section, or clamp on, that is, external. The clamp-on version can be permanent, with an epoxy adhesive, or portable, with the transducer attached with acoustic putty or silicone grease.

A third version of currently manufactured ultrasonic flowmeters is the ultrasonically scanned vortex-shedding flowmeter discussed earlier. Other schemes of employing ultrasonics in flow measurements have been studied in the laboratory.

16.3-15 Level Measurement Principles

Level measurement deals with the sensing of material and product levels in containers, pipes, and vessels. The materials can have the form of liquids, slurries, powders, and granular solids. In many cases, level is an indirect measurand for volume or mass. The simplest measurement is the sensing of liquid level in a stationary vessel under static (no-flow) conditions. In the most general sense, level measurement is the sensing of an interface, such as liquid/gas, liquid/liquid, or liquid/solid. The sensing problem becomes difficult where the interface is not sharply defined, such as in the case of a foaming, boiling, or carbonated liquid or a stream with suspended particles or with a layer of scum. This range of application problems has led to the development and use of a wide variety of different level transducers.

One distinguishes between contact methods, where the transducer is entirely or partially immersed in the process fluid, and noncontact measurements, where the entire transducer is outside the vessel whose level is being measured. The latter schemes are used in the case of difficult fluids, such as sticky fluids that would adhere to an immersed transducer.

Level switches (point-level sensing) are used when it is desired to obtain an indication when a predetermined level has been reached. The term *continuous-level sensing* refers to the transducer whose output is a continuous function of the measurand.

16.3-16 Mechanical Type Level-Sensing Elements

A wide variety of mechanical devices exist. Floats, displacers, and sight gages translate the level measurand into a mechanical displacement that can be displayed on a dial or sensed electrically. Liquid level can also be sensed by measurement of the pressure difference between liquid near the bottom of the vessel and the gas above the liquid. A weight-and-cable device is used to measure the level of bulk material in tanks and silos. Rotating or vibratory paddle-type level switches detect the presence or absence of granular solids.

16.3-17 Resistance and Inductance Level Transducers

Conductivity level-sensing elements can be used with conductive liquids and powders. The conductivity level switch consists of a probe or electrode; rising fluid comes into contact with the probe and closes an electrical path through the (grounded) tank. The resistance tape is a continuous-level sensor; it is a precision wirewound resistor. The submerged portion of the resistor is shorted by the conducting fluid, and the resistance variation provides an indication of level.

The inductive level switch is a noncontact sensor for conductive liquids, solids, and slurries. The inductor energized with alternating current produces a magnetic field. The circuit detects the change in inductance when the magnetic field couples with the conductive fluid.

16.3-18 Capacitance Level Transducers

Capacitance level transducers (Fig. 16.27) are the most widely used in electronic level measurements. The sensing element consists of cylindrical or parallel-plate electrodes that are immersed in the fluid. As the fluid level rises, the fluid displaces the ambient air as the dielectric between the electrodes, and in view of the difference in dielectric constants, causes a change in capacitance. The measuring circuit senses this change and produces a usable output signal proportional to level. Temperature and composition changes that affect the dielectric constant require compensation. Linearity is 0.1% of full scale and overall accuracy is of the order of 1% of full scale.

For conductive fluids, the fluid itself serves as one side of the capacitor; the second capacitor plate is located above the fluid and parallel to the surface. This is a noncontact sensing method that is not affected by fluid coating of the electrode.

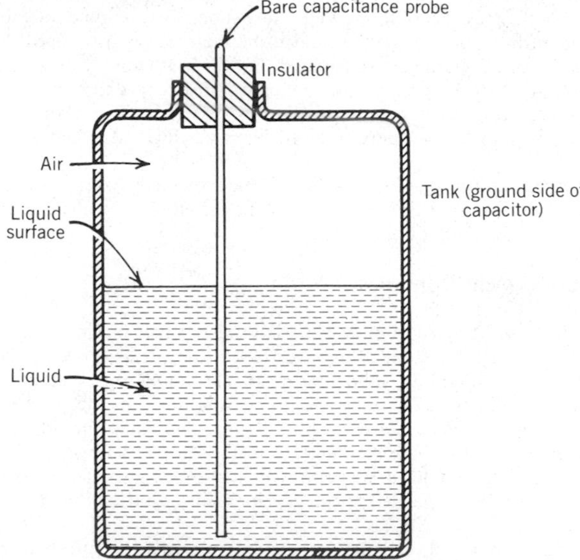

Fig. 16.27 Capacitance level transducer.

16.3-19 Microwave, Optical, and Nuclear Radiation Level Gauges

Level measurements can also be based on the transmission, reflection, and absorption of microwave, optical, or nuclear beams by the process fluid.

Microwave level switches represent a noncontact method but require nonmetallic tanks or a nonmetallic window in the tank. The measuring system consists of a transmitter that directs microwave energy into the tank and a receiver that responds to the transmitted energy. Air transmits microwaves with virtually no loss; water and water solutions absorb almost all the energy. The suitability of other process materials depends on dielectric constant and electrical conductivity.

Various optical schemes are used as level switches. In the refractive sensor, a probe terminates in a crystal that provides total reflection of light. A light beam is generated by a light-emitting diode and is detected by a phototransistor. When the probe is immersed in a fluid having a refractive index different from that of air, the beam is no longer reflected but instead is refracted into the fluid. The phototransistor senses the absence of the reflected beam. Infrared light is used and the transducer is not affected by the opacity of the fluid.

Nuclear radiation level measurement is noncontact and the equipment is entirely external to the process vessel. The system consists of a source of radiation, radioisotope or radium, and a nuclear radiation detector. The operation depends on the differences in nuclear absorption of the material above and below the interface to be sensed. Both point sensing and continuous sensing are available.

16.3-20 Ultrasonic Level Transducers

Ultrasonic level transducers are made in a variety of configurations to sense the level of liquids and of granular solids. Both point sensing and continuous sensing are available. Several basic measurement methods are employed.

The echo-ranging (sonar) level measuring system (Fig. 16.28) is based on the reflection of sound from an interface where there is a large difference in acoustic impedance. Sound pulses generated by an acoustic transducer are reflected from the interface and received by the transducer. The electronics measures the time it takes the sound wave to travel to and from the interface and provides correction for temperature variations and other factors that influence the velocity of sound. With the transducer mounted in the air or vapor space above a tank of liquid or granular solid, the measurement is entirely unaffected by the properties of the process fluid. The transducer can also be mounted under the liquid; such placement often results in greater accuracy for difficult interface measurements.

Sonic ranging may also be used for point level measurement. In addition, there are ultrasonic level switches based on the changes in sonic transmission characteristics of the fluid, similar to the wave schemes described in the preceding subsection. Both contact and noncontact types are available.

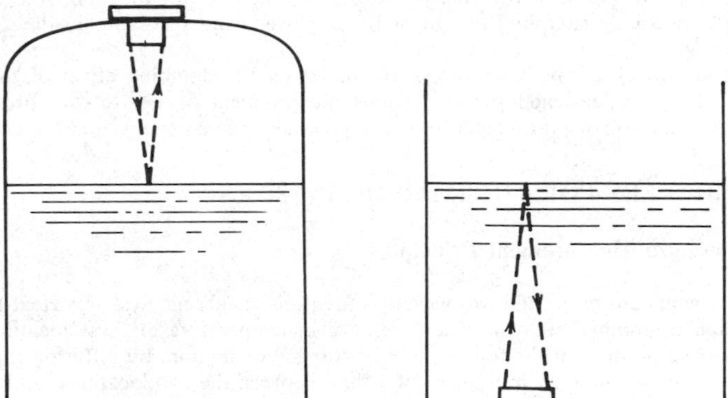

Fig. 16.28 Ultrasonic level sensing by echo ranging.

16.3-21 Density Transducers

Density is defined as the quantity of matter per unit volume (kilograms per cubic meter). Specific gravity is defined as the density relative to that of a reference material, usually water under standard conditions (1000 kg/m^3).

The density of a material of known composition is often inferred from measurements of other physical parameters, such as temperature or pressure. In cases where the composition of the material is not well known, a direct measurement of density is required. In the case of liquids, such direct density measurements employ principles similar to those described for level transducers.

Mechanical transducers for liquid density comprise floats (hydrometers), displacers (also used for gases), and hydrostatic head sensors (differential pressure gauges). The output of these transducers is a mechanical displacement that can be viewed directly or transmitted remotely by the agency of an electrical transducer.

The *vibration-element density transducer* consists of a plate or reed immersed in the fluid. The density of the fluid determines the natural frequency and damping of the immersed object. This can be detected by one of two measurement modes, steady-state vibration at the natural frequency or step excitation with observation of the damping.

Ultrasonic density measurement requires an electroacoustic transducer (piezoelectric or magnetostrictive) immersed in the fluid, plus considerable electronic signal processing. When the transducer is energized with steady state (AC), its current-voltage relationship depends on the acoustic impedance of the fluid, which in turn is related to the product of density and acoustic velocity.

Nuclear radiation density sensing is similar to the nuclear level sensing application, with the difference that the nuclear beam is entirely confined to the fluid. This method is noncontacting and all the equipment is outside the process vessel.

16.3-22 Viscosity Transducers

Viscosity is the property of a fluid that characterizes its resistance to flow. The absolute (dynamic) viscosity (pascal-seconds; 1 mPa · s = 1 centipoise) is the constant of proportionality between applied stress and resulting shear velocity. Kinematic viscosity (square meters per second; $1\ \text{mm}^2/\text{s} = 1$ centistoke) is the ratio of dynamic viscosity over density. The proportionality between shear velocity and applied stress holds for Newtonian fluids; for nonNewtonian fluids the concept of apparent viscosity is used. Viscosity measuring devices are known as viscometers or rheometers.

The principal methods of detecting viscosity are mechanical. The capillary tube viscometer is based on the relationship between pressure, flow, and kinematic viscosity in a capillary (Hagen-Poiselle law). In the laboratory version of the instrument, a sample of fixed volume is allowed to fall through a capillary tube and the time of fall is measured. In the inline version of this scheme, a constant flow pump draws a sample from the process stream, forces it through a capillary tube, and returns it to the process stream; the pressure drop through the capillary tube indicates the viscosity; a differential pressure gauge is employed to convert this pressure drop into a usable output. In both types of measurement the sample must be held at a constant temperature.

In the rotating viscometer, a cylinder or disk is rotated in the fluid under test through a spring. The deflection of the spring measures the dynamic viscosity. Measurements can be made at different speeds, which permits determination of the properties of a nonNewtonian fluid over a wide range. For inline measurements, a sampling line must be employed, and the sampled flow rate must be constant and laminar.

Vibrating reed and ultrasonic viscometers are based on the damping effect of the fluid on a vibrating body. They are designed for continuous measurement of viscosity in process streams, subject to maintenance of constant temperature and pressure or to compensation for such variations.

16.4 TRANSDUCERS FOR THERMAL QUANTITIES

16.4-1 Temperature Measurement Principles

Heat energy and temperature are the two variables involved in thermal systems. Heat is transferred from one location to another because of a difference in temperature at these locations. The heat transfer can take one or more of the following three forms: conduction, by diffusion through a solid or a stagnant fluid; convection, by movement of a fluid between the two locations; and radiation, by electromagnetic waves.

Temperature is an "intensive" quantity: The combination of two bodies of the same temperature results in exactly the same temperature. The idea of a standard unit that can be indefinitely multiplied or subdivided to generate any arbitrary magnitude, as employed in the concepts of length, mass, and time, cannot be applied to the concept of temperature. The temperature of a body is a measure of the mean kinetic energy of its molecules; therefore in theory temperature is a derived quantity based on mass, length, and time. But since these kinetic energies cannot be measured, an independent standard for temperature must be defined.

The absolute thermodynamic temperature scale is defined in terms of an ideal Carnot cycle, or by a constant-volume or constant-pressure gas thermometer using an ideal gas; the unit is the kelvin. The International Practical Temperature Scale is an approximation based on a set of fixed points, with interpolation equations based on specific standard transducers. The equation

$$T(\text{K}) = t(°\text{C}) + 273.15 \tag{16.22}$$

can be considered an engineering approximation of the absolute thermodynamic kelvin temperature T(K) by the practical, empirical Celsius (centigrade) temperature scale t(°C).

Temperature measurements are either contacting or noncontacting. In the contacting measurement, the temperature-sensing element is at the same temperature as that of the material to be measured, and heat transfer between them takes place by conduction. Contacting temperature transducers are made in two forms: surface temperature transducers, which are attached to the surface of the material whose temperature is to be measured, and immersion probes, which are inserted into fluids whose temperature is to be measured. In the noncontacting measurement, heat transfer is by radiation, and the sensing-element temperature is correlated to the unknown temperature by calibration.

The accuracy and speed of response of a contacting temperature transducer depend on its thermal capacity (product of mass and specific heat) and the thermal conductance between it and the material to be measured (product of heat transfer coefficient and heat transfer area, watts per degree Celsius). The ratio of thermal capacity and thermal conductance gives the thermal time constant.

It is sometimes necessary to protect a temperature transducer from direct contact with corrosive material by embedding it in a sheath or installing it in a pressure-sealed chamber (thermowell). This introduces additional thermal capacities and thermal resistances which affect the accuracy and slow down the response time. Other effects that must be considered are sensor self-heating (in the case of passive sensors), heat losses through the sensor support (in immersion probes), velocity effects (in the temperature measurement of moving fluids), and the size of the sensor (that is, its thermal capacity) relative to the total thermal capacity of the material to which it is attached.

16.4-2 Thermal Expansion Temperature Transducers (Thermometers)

A number of temperature transducers are based on the phenomenon of thermal expansion of solids, liquids, and gases. The resulting displacement can be employed as a direct visual display or can be sensed with an electrical transducer. The principal categories are:

1. **Liquid-in-glass and liquid-in-metal thermometers.**

2. **Pressure thermometers.** These consist of a temperature bulb, an interconnecting capillary tube, and an elastic pressure-sensing element such as a Bourdon tube, bellows, or capsule. The entire system is filled with liquid or gas. A change in temperature sensed by the bulb is translated into a pressure change in the elastic element and results in a deflection

3. **Bimetallic thermometers.** These consist of strips of metal of different thermal expansion coefficients bonded together. A change in temperature causes a differential expansion of the two sides of the bonded strip, which causes the strip to bend. The resulting mechanical displacement is a function of the temperature. Bimetallic strips are frequently employed for point sensing (thermoswitch, thermostat)

16.4-3 Thermoelectric Transduction Principles

The thermoelectric transducer, or thermocouple, is based on the Seebeck effect: A current flows in a closed circuit composed of dissimilar metal conductors exposed to different temperatures. The observed Seebeck current is actually the result of two reversible thermoelectric phenomena, the (converse of the) Peltier and Thomson effects. The Peltier electromotive force (emf) is the thermoelectric voltage generated at the junction of dissimilar metal conductors. The Thomson emf is the thermoelectric voltage generated in a single homogeneous conductor exposed to a temperature gradient along its length. The magnitude and sign of the Peltier and Thomson voltages are functions of the materials and the temperatures. The irreversible phenomena of heat conduction and i^2R loss also enter into the overall thermoelectric conversion. It is not possible to predict the performance of a thermocouple entirely from theoretical considerations. Practical thermoelectric instrumentation is based on calibration of specific materials and on experimentally established thermocouple laws.

Law of Homogeneous Material

A thermoelectric current cannot be sustained in a circuit of a single homogeneous material by the application of heat alone, regardless of variations in conductor cross section. As a result, two different materials are required, and the current depends only on the temperatures at the junctions and is independent of temperatures elsewhere as long as the two materials are homogeneous (Fig. 16.29*a*).

Law of Intermediate Materials

The net thermoelectric voltage in a circuit composed of several dissimilar materials is zero as long as the entire circuit is at a uniform temperature. One consequence is that a third material C can be inserted into a thermocouple circuit consisting of two materials A and B without introducing new thermal emfs, as long as the two newly introduced junctions are at identical temperatures; this applies if C is introduced by breaking A (Fig. 16.29*b*) or between A and B (Fig. 16.29*c*). Another

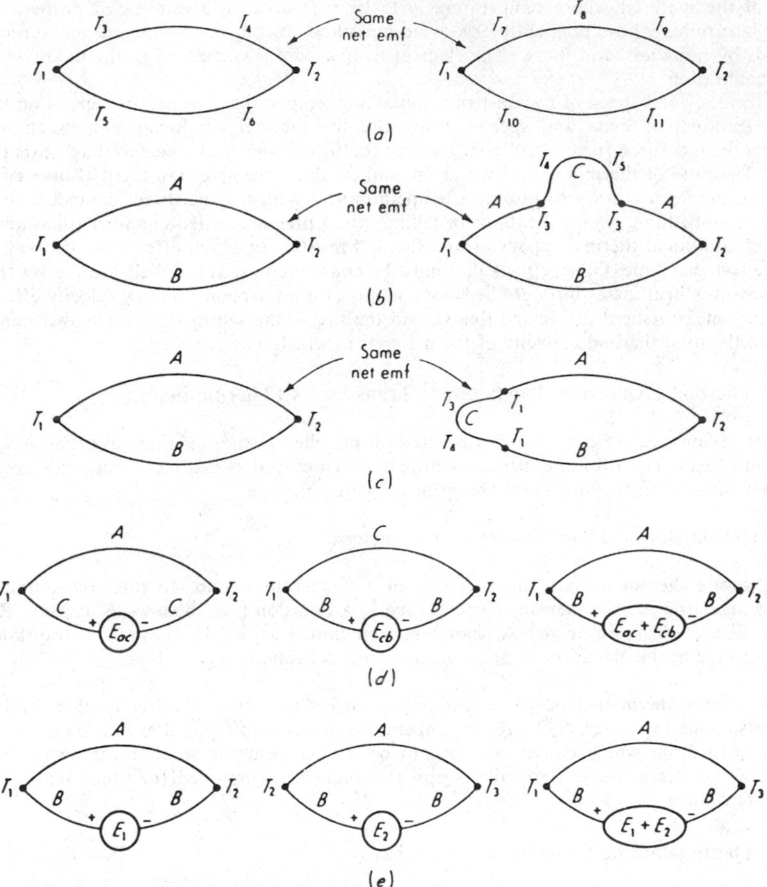

Fig. 16.29 Thermocouple laws, (*a*)–(*e*). From McGraw-Hill, New York, with permission.

consequence is that if the thermal emf of two materials A and C is E_{AC}, and that of materials B and C is E_{BC}, then the thermal emf of materials A and B is the algebraic sum $E_{AC} + E_{BC}$ (Fig. 16.29*d*).

Law of Intermediate Temperatures

If a thermocouple produces emf E_1 when its junctions are at temperatures T_1 and T_2, and E_2 when at T_2 and T_3, then it will produce the algebraic sum $E_1 + E_2$ when the junctions are at T_1 and T_3 (Fig. 16.29*e*). As a result, a thermocouple calibrated for one reference temperature can be employed with a different reference temperature by use of an appropriate voltage correction.

The electrical energy exhibited by the product of thermoelectric voltage and resulting current flow originates in heat absorption. Since heat is absorbed at the junction of dissimilar conductors (Peltier effect), the junction must be somewhat cooler than the surrounding medium whose temperature is being measured. The heat transfer, which is very small in thermocouples because of the minute current flow, can be enhanced by forcing current through the junction from an external source. This is the basis of thermoelectric refrigeration. The Peltier thermal effect is reversible, with the direction of heat flow depending on the direction of current flow across the junction.

16.4-4 Thermocouples

A practical thermocouple circuit (Fig. 16.30) consists of two wires of special alloys joined at one end, the sensing junction. The alloy circuit is terminated and connected to ordinary copper wiring at the reference junction. Assuming that the alloy wiring is homogeneous, particularly in the critical

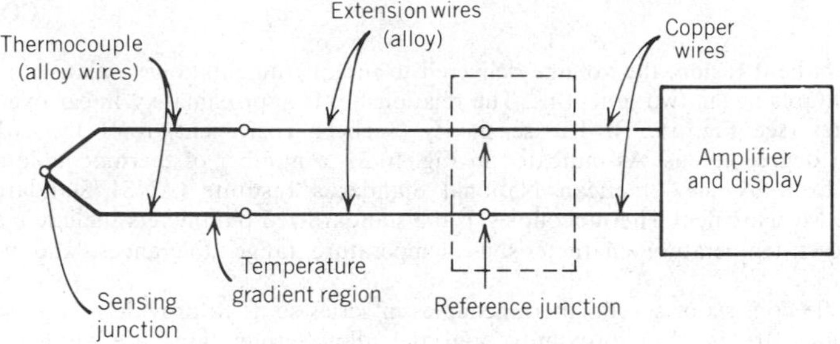

Fig. 16.30 Thermocouple circuit.

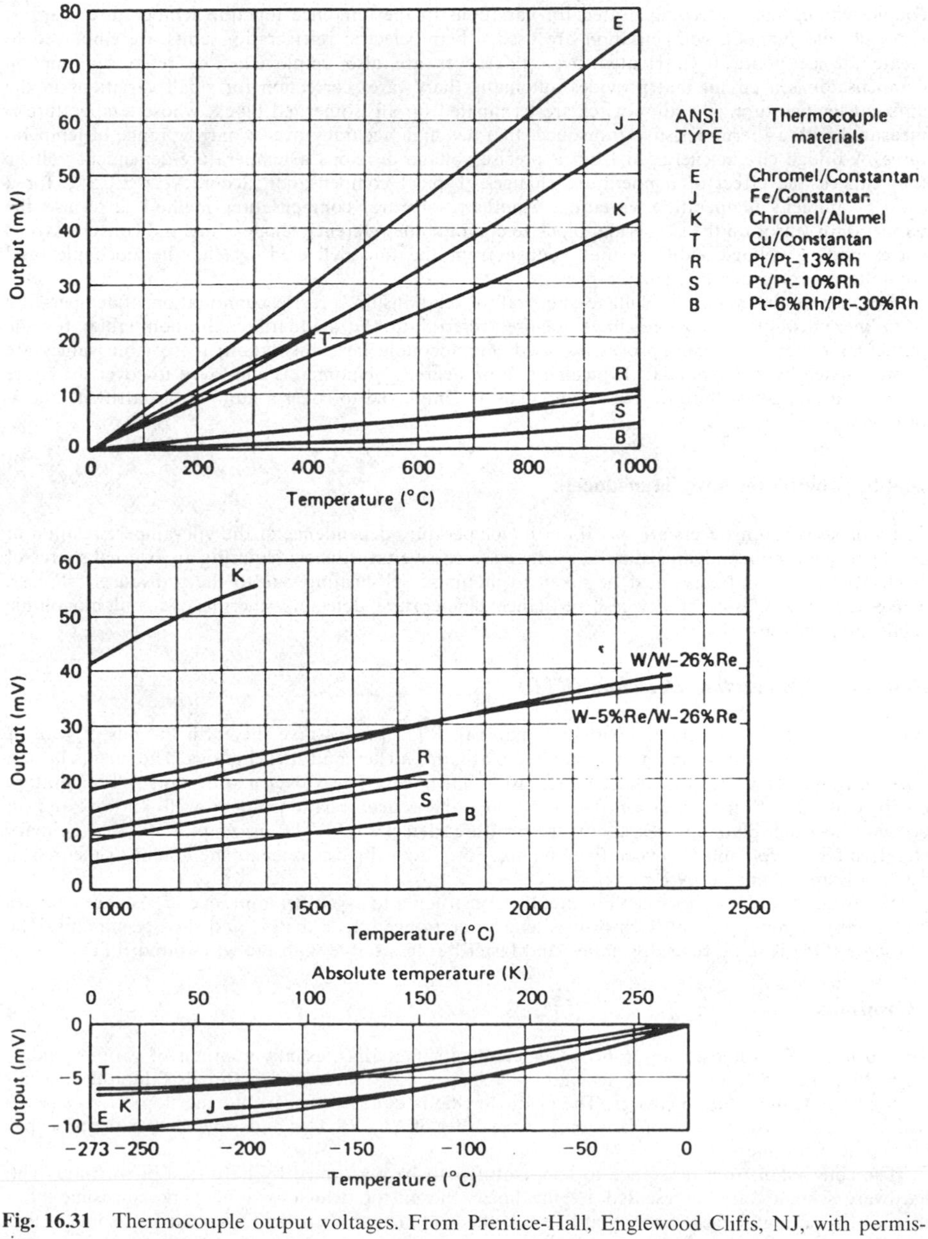

ANSI TYPE	Thermocouple materials
E	Chromel/Constantan
J	Fe/Constantan
K	Chromel/Alumel
T	Cu/Constantan
R	Pt/Pt-13%Rh
S	Pt/Pt-10%Rh
B	Pt-6%Rh/Pt-30%Rh

Fig. 16.31 Thermocouple output voltages. From Prentice-Hall, Englewood Cliffs, NJ, with permission.

temperature gradient region, the voltage delivered to the instrumentation is related to the difference in the temperatures at the two junctions. The relationship is approximately linear over a wide range of temperatures (see Fig. 16.31). The sensitivity (Seebeck coefficient) is of the order of tens of microvolts per degree celsius. As indicated in Fig. 16.31, a number of thermocouple materials have been standardized by the American National Standards Institute (ANSI Standard C96.1-1975, "Temperature Measurement Thermocouples"); the standardized parameters include material composition, voltage-vs.-temperature characteristics, temperature range, tolerances, and wire insulation color code.

A thermopile consists of several thermocouples in series so as to provide a large sensitivity. The sensing junctions are in close proximity, and the alloy terminations are all held at the same temperature.

The reference junction must be held at a known fixed temperature, or alternatively the thermocouple reading must be compensated for variations in the reference junction temperature. Various types of cold junction compensation are used. Thermoelectric refrigeration units are employed to create an ice point, 0°C. Heated oven references are also employed. The reference junction compensator is a circuit that provides automatic (hardware) correction for small variations in the ambient temperature. The alloy wires are terminated on an isothermal block, whose temperature is measured with a thermoresistive transducer that has high accuracy over a narrow range of temperatures. A bridge circuit energized from a precise voltage develops a temperature-dependent voltage that cancels the effect of temperature changes. Typical compensation accuracy is ±0.5°C for a ±10°C ambient temperature variation. Another, software, compensation method is to use the temperature sensor on the isothermal block to compute the reference temperature and convert it to its corresponding voltage, subtract that voltage from the one delivered by the thermocouple, and compute the corresponding temperature.

Because of the nonlinear voltage-temperature relationship, precise computation of temperature from thermocouple voltage readings requires referral to the standard calibration tables for the particular material. In microprocessor-based thermocouple instrumentation, these table values are approximated by a polynomial. Typically, a ninth-degree polynomial is employed to cover the entire temperature range, or third-degree polynomials are employed to cover a number of narrow temperature bands.

16.4-5 Thermoresistive Transducers

Thermoresistive transducers are based on the temperature dependence of the volt-ampere relation of conductor and semiconductor materials. These are passive transducers requiring an external source of electrical energy, and care must be taken to minimize self-heating due to the transducer i^2R loss. There are three classes of devices: resistance temperature detectors, thermistors, and monolithic semiconductor sensors.

Resistance Temperature Detectors (RTD)

RTDs are made from metallic conductor material. RTD probes take the form of coils of wire or metal films. The most common materials are platinum, nickel, and nickel alloys. The metals have a positive resistance-vs.-temperature characteristic that is very linear over a wide range: Temperature coefficients are of the order of 0.4%/°C. The theoretical relationship is well known, and in consequence the platinum resistance temperature detector (PRTD) is employed as a primary standard for interpolation between fixed points. For practical measurements the nonlinear theoretical relation is approximated by a power series.

Wirewound RTD probes must be carefully constructed to avoid strain-induced resistance changes and electromagnetic pickup. Such probes tend to be rather fragile, bulky, and slow responding. The film-type RTD is more rugged, smaller, and faster but less stable than the wirewound RTD.

Thermistors

Thermistors are made from amorphous semiconductor materials, usually mixtures of various oxides, formed into flakes, beads, disks, rods, and other shapes. Most thermistor materials exhibit decreasing resistance as temperature increases. The characteristic is quite nonlinear, but the slope is very large, with temperature coefficients of resistance of the order of 4%/°C. This high sensitivity is the principal attraction of the thermistor.

The conversion from resistance to temperature can be performed by hardware or software. The hardware schemes are represented by the linear thermistor, which is a network consisting of a thermistor and matching resistors that linearize the overall characteristics. These circuits come in two forms, linear thermistor voltage dividers, which are three or four terminal networks, and linear thermistor resistance networks (two terminal). Linearity error is of the order of 0.25°C over a 100°C temperature range. Software schemes are based on various empirical relations, such as

$$1/T = a_0 + a_1 \ln R + a_3(\ln R)^3 \tag{16.23}$$

$$T = \frac{b_1}{\ln R - b_0} - b_2 \tag{16.24}$$

where R = resistance of thermistor, Ω
T = temperature, K
a, b = curve-fitting constants for particular thermistor

Equation (16.23) can give a fit within ±0.02°C over a 100°C span; Eq. (16.24) is less accurate but faster to execute.

Monolithic Semiconductor Sensors

Monolithic semiconductor temperature transducers are based on the temperature dependence of the volt-ampere relation of a forward-biased *pn* junction

$$i = I_s[\exp(qv/kT) - 1] \tag{16.25}$$

or

$$v = \frac{kT}{q}\ln(i/I_s) \qquad \text{provided } i/I_s \gg 1 \tag{16.26}$$

where i = current through junction
v = voltage across junction
T = temperature, K
I_s = saturation current
k = Boltzmann's constant = 1.38×10^{-23} J/K
q = charge of electron = 1.6×10^{-19} C

When the current i is held constant by use of a current source and the value I_s is controlled by appropriate semiconductor geometry, such as the use of matched transistor pairs, the voltage is inherently proportional to absolute temperature. In practice, accuracy of the order of ±0.5°C over a 100°C temperature range is obtained. The semiconductor temperature sensor is employed where linearity over a small temperature range is desired, for example, in the hardware thermocouple reference junction compensator. Temperature-sensitive Zener diodes have also found use as temperature-sensing elements.

16.4-6 Radiation Thermometers (Pyrometers)

Radiation pyrometers are noncontacting temperature transducers that respond to radiative heat transfer. Two basic principles underlie the measurement: (1) the Stefan-Boltzmann law, which relates the total radiant energy emission to the temperature, and (2) Plank's equation, which relates the spectral distribution of the radiated energy to the temperature. The radiation emitted by the object whose temperature is to be measured (called the target) is focused by means of lenses, mirrors, or optical fibers and possibly filters on some sort of radiation detector. The wavelengths over which the focusing system and detector operate may cover the visible (0.3–0.72 μm) and the near infrared spectrum (0.7 to about 40 μm).

Radiation detectors are either thermal or photon devices. Thermal detectors consist of a collector body that is blackened to absorb a maximum of radiation of all wavelengths, with an attached temperature transducer that measures the collector temperature. Thermocouples, thermopiles, resistance temperature detectors, and thermistors configured for this application are all used; the specialized RTD and thermistor forms are also called bolometers. Photon detectors are photovoltaic or photoconductive transducers that respond to visible or infrared radiation; since they operate at an atomic level they are much faster than thermal detectors.

Pyrometers are calibrated against a standard that approximates a perfect radiating body (black body). The actual measurement must be corrected for the reduced emissivity of the target (emissivity is 1 for a blackbody). Emissivity depends on the shape and texture of the target and is a function of wavelength. Uncertainty about target emissivity is a principal source of radiation pyrometer error. Target reflectance and transmittance are also factors in the calibration; if the target material is not solid, or is thin, the pyrometer will "see" through the surface. The reflectance and transmittance of the intervening medium between target and detector (air, vapor) is another source of variability.

Optical pyrometers compare the brightness of the target with that of a reference source contained in the instrument. The measurement may be made over a narrow band in the optical region, or broadband in the infrared region. There are two basic techniques, null balance, where the reference source is continually adjusted until it is matched to the target, and fixed reference. Spectroscopic methods are employed in measuring the temperature of hot gases and of stars.

16.4-7 Heat-Flux Transducers

Heat flux is the heat transfer rate per unit area (watts per square meter). Direct heat transfer rate measurements can be made for convective and radiant heat transfer. The basic measuring scheme in both cases is to allow the heat flux to warm up a sensing disk and to measure a temperature difference by thermocouple, thermopile, or thin-film temperature sensitive resistor. The measured temperature is correlated to the heat transfer rate by theory and calibration. Calorimeters are used to measure total heat transfer, convection plus radiation if both are present. Radiometers are used to measure radiant heat transfer only. A typical radiometer is essentially a calorimeter with an optical window covering the sensor to shield it from convection.

The slug-type heat-flux sensor is a thick metallic disk of known mass, thermally insulated from its housing. The front surface is exposed to the heat flux and a thermocouple is attached to the back surface. The heat transfer rate is related to the rate of change of temperature.

The foil calorimeter (Gardon gage) has a sensor consisting of a thin membrane bonded at its periphery to a copper heat sink. The thermal energy impinging on the membrane is transferred radially to the heat sink, producing a temperature difference between the center and the rim of the foil directly proportional to heat transfer rate. This difference is measured thermoelectrically. In practice, the foil is made of a thermoelectric alloy such as Constantan. A thin copper wire attached to the center of the foil constitutes one thermojunction, and the copper-Constantan contact around the rim constitutes the other junction.

Thermopile heat-flux sensors consist of series-connected thermocouples, with the sensing junctions exposed to the heat flux and the cold junctions attached to a heat sink. In the wafer type a thermal-insulating barrier is placed between the two sets of junctions. Heat flow produces a temperature drop across the wafer which is sensed by the thermopile. In the suspension type radiometer the hot junctions are attached to a receiving disk suspended by the thermopile wires; heat is conducted along the wires to the heat sink, producing a temperature difference.

16.5 TRANSDUCERS FOR ACOUSTIC QUANTITIES

16.5-1 Acoustic Transducer Principles

Acoustics deals with the propagation of pressure fluctuations in solid and fluid media. The properties of an acoustic wave at a specific point in space and time can be described in terms of particle displacement, velocity, and pressure. Receiving transducers (such as microphones and hydrophones) convert acoustic variations into an electrical quantity. Transmitting transducers (such as loudspeakers and earphones) convert electrical signals into acoustic variations. Some transducers are designed to serve both as transmitter and receiver (transceivers). Acoustic transducers can be viewed as antennas, and their design considerations parallel those for antennas. Directionality is one of the principal parameters. The coupling of the transducer to the medium involves considerations of acoustic impedance and radiation resistance.

The basic transduction methods employed in acoustic transducers are the same as the ones used for displacement, velocity, and pressure measurements: variable resistance, variable reluctance (moving iron), moving coil, variable capacitance, and piezoelectric. Magnetostrictive transducers are used, particularly at high power levels.

16.5-2 Sound-Pressure Transducers

Sound-pressure level (SPL) is defined in decibels (dB)

$$\mathrm{SPL} = 20\log(p/p_{\mathrm{ref}}) \qquad \mathrm{dB} \tag{16.27}$$

where p is the rms (root-mean-square) pressure to be measured and p_{ref} is some reference rms pressure. An accepted standard reference value for noise measurements is $p_{\mathrm{ref}} = 2 \times 10^{-5}$ Pa (2×10^{-4} μbar or dyne/cm^2 or 3×10^{-9} psi). This 0-dB level approximates the normal human hearing threshold at 1000 Hz. The average human pain threshold is 144 dB. Extreme noise, such as that produced by large rocket engines, may be of the order of 170 dB. Atmospheric pressure is 194 dB. Sound-pressure transducers are thus expected to detect pressure fluctuations representing 10^{-10} of the ambient pressure, and to be capable of a measurement range of 10^9:1.

Microphones used for sound-pressure transduction are basically special-purpose gage-pressure transducers. The usual sensing element is a thin diaphragm that converts pressure into displacement. A capillary tube or hole connects both sides of the diagram so that static (ambient or atmospheric) pressure on both sides is equalized and only the dynamic (i.e., fluctuating, sound) pressure is sensed. The transduction element is usually of the capacitive-, piezoelectric-, or moving coil-type.

The sound-level meter is a self-contained measurement system, usually portable and battery operated. It consists of a microphone, an electronic amplifier, electrical networks, and an indicating meter.

16.5-3 Magnetostrictive Transducers

Magnetostriction is the phenomenon of coupling between elasticity and magnetism in certain materials. A sample of such material exposed to a magnetic field exhibits a dimensional change, and vice versa. The effect is most pronounced in nickel and other ferromagnetic alloys. Magnetostrictive transducers are employed in high-power-level acoustic devices, ultrasonic equipment, and sonar. For details, refer to Chapter 14 of this handbook.

16.6 TRANSDUCERS FOR OPTICAL AND INFRARED QUANTITIES

Visible light is electromagnetic radiation in the wavelength range between 400 nm (7.5×10^{14} Hz) for violet and 760 nm (3.9×10^{14} Hz) for red. The ultraviolet radiation bands cover the range from 400 nm to approximately 10 nm. Near infrared designates the wavelengths from 760 nm to approximately 3 μm (10^{14} Hz).

The intensity of radiation is described by three units: luminous flux, luminous (flux) intensity, and illuminance. The basic quantity is luminous intensity; the SI unit, the candela (cd), is defined in terms of blackbody radiation at the temperature of freezing platinum. A 100-W inside-frosted tungsten filament light bulb has an intensity of approximately 120 cd. Luminous flux describes the radiant power; the SI unit, the lumen (lm), is the luminous flux emitted from 1 cd in a solid angle of 1 (steradian) sr (1 cd = 1 lm · sr). Illuminance is the luminous flux per unit area; the SI unit, the lux (lx), is defined as lumens per square meter.

Transducers are either light sources that convert electrical or other energy forms into light energy or light-sensing elements that convert light energy into electrical energy. The principal light sources are filament lamps, discharge and arc lamps, light-emitting diodes, lasers, and phosphor screens. The principal light-sensing elements are photoemissive tubes and photoconductive devices (photodiodes and phototransistors). For details refer to Chapter 15 of this handbook.

16.7 TRANSDUCERS FOR NUCLEAR RADIATION

16.7-1 Scope and Definitions

Nuclear radiation is the emission of particles and radiation from atomic nuclei. The most common forms of nuclear emanations are alpha particles (helium nuclei), beta particles (electrons), neutrons, x-rays, and gamma rays (photons, electromagnetic radiation of very short wavelength). Various units are employed to describe nuclear phenomena.

1. **Activity.** The bequerel (Bq) is the activity of a radionuclide having one spontaneous nuclear transition per second. The older non-SI unit curie (Ci) is defined as the amount of radioactive material that exhibits the same number of disintegrations per second as 1 g of radium (1 Ci = 3.7×10^{10} Bq).
2. **Exposure.** Exposure describes the amount of ionization produced in air at standard conditions; the SI unit is the coulomb per kilogram (C/kg). The older non-SI unit roentgen (R) is defined as the radiation that produces 1 esu (electrostatic unit) of ions or ion pairs in 1 cm^3 of air; 1 R = 2.58×10^4 C/kg.
3. **Absorbed dose.** The gray (Gy) is the energy imparted by ionizing radiation to a mass of matter corresponding to 1 J/kg. The older non-SI unit rad (rd) is a dose of 100 ergs/g (1 rd = 0.01 Gy).
4. **Dose equivalent.** Dose equivalent is measured by the rem (*r*oentgen *e*quivalent *m*an). One rem is that amount of radiation of any type that produces the same biological effect (which needs to be specified) as is obtained from 1 rd of 200 kV x-rays (peak voltage applied to the x-ray tube). The rem can also be defined as the product rem = (rad)(RBE), where RBE is the relative biological effectiveness, which must be stated in terms of a specific biological phenomenon.

Detection and measurement of nuclear emission requires that the emanation be allowed to interact with the nuclear radiation transducer. The interaction with matter of the various types of nuclear radiation is quite different. Therefore the response of a transducer can be tailored to the desired measurand by suitable choice of materials, such as the material in the window that admits the radiation through the transducer case and the material in the transducer itself. Neutron detection requires the use of a conversion material that emits charged particles or photons when a neutron impinges.

The two principal electrical transduction methods are charge separation, that is, ionization in gases and semiconductors, and scintillation, that is, conversion of nuclear radiation into light which is then sensed photoelectrically.

16.7-2 Gas-Filled Ionization Detector Principles

The basic detector is a gas-filled tube, usually in the form of a cylindrical chamber that acts as the cathode and a central wire that acts as anode. Radiation, admitted through a window of appropriate transparency, causes the gas to become ionized. The voltage applied to the electrodes causes separation of the ionized charges and produces an ionization current which can be measured. The output current or voltage is a measure of the incoming radiation.

The gas tube exhibits six regions of operation when the voltage is increased with a given amount of ionization (ionization pulse input), shown for two different input pulse levels A and B in Fig. 16.32. In region I (recombination region), the amount of charge collected is proportional to the voltage; that is, the charge collected is only a fraction of that produced by ionization, and the remaining charges have undergone ion recombination. The output current therefore does not represent a measurement of the incoming radiation.

In region II (ion chamber region), saturation has set in. All the charges formed by ionization are collected, and the output current is a measure of the incoming radiation, independent of the detector voltage.

In region III (proportional region), gas amplification occurs: Additional ions are produced due to secondary and tertiary ionization (avalanche ionization). The amount of this amplification is a function of voltage alone, and the current output is proportional to the incoming radiation. This region can be used for nuclear measurements with a regulated voltage supply.

In region IV (limited proportionality region), the proportionality between output current and input radiation gradually disappears, that is, the value of the gas amplification factor is a function of the size of the incoming radiation pulse. As the voltage is increased further to the Geiger threshold, the gas ionization avalanche covers the entire electrode and the output curves for various size inputs merge.

In the following region V (Geiger-Mueller region), a single ionizing event produces a cascade ionization effect, regardless of the size of the event, and each new event produces a new current output pulse. Therefore this region can be used to count discrete nuclear emission events but cannot

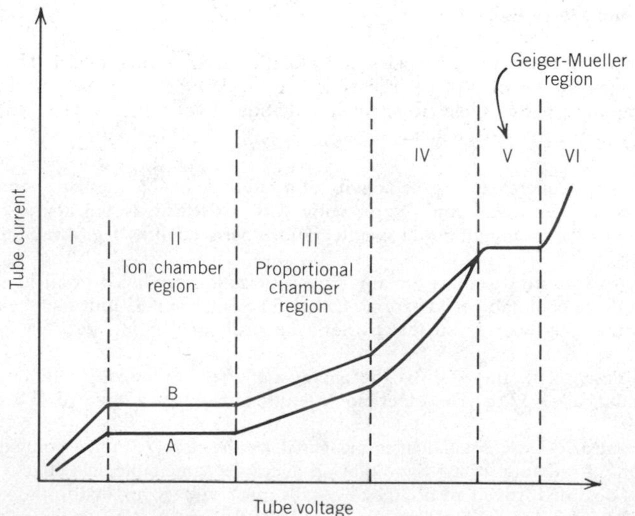

Fig. 16.32 Gas-filled ionization detector operating regions. A, low incident radiation level; B, high incident radiation level.

be used to discriminate between type and size of events. The final region VI (continuous discharge region), is not useful for measurement because a single ionization event sets off a continuous flow of current.

16.7-3 Gas-Filled Ionization Detectors

The ionization chamber operates in region II (Fig. 16.32). It has a very low output current which requires amplification but does not need a highly stable voltage supply. The chamber can be used as a current detector, with a load resistance, or as a charge detector, with a load capacitance. In the latter case the load capacitor is charged by the ion current and the total charge, measured with an electrometer circuit, indicates the accumulated dosage. Depending on variations in chamber and circuit design, the ion chamber can be used to measure direct radiation intensity (continuous current flow) or discrete ionization events (current pulses).

Proportional counters are ionization chambers and circuits specially designed to operate in region III. The output current pulses are much larger. A stabilized voltage source is required.

The Geiger counter (Geiger-Mueller tube) operates in region V and requires 500 to 2000 V for its operation; the voltage need not be precisely regulated. The tube is filled with a special gas or vapor which acts as a quenching agent, to ensure that the electrical discharge caused by a single nuclear input pulse is of finite duration. Typical output pulses have a 1 μs rise time and a duration of several microseconds. After the quenching, the counter requires a period of about 100 μs to recover before it can detect another event. The tube has a useful life of 10^8 to 10^{10} counts.

16.7-4 Semiconductor Ionization Detectors

A reverse-biased *pn* junction can be employed as nuclear radiation detector. The nuclear particles cause ionization by creation and separation of electron-hole pairs in the depletion region. The response is proportional to the impinging energy, regardless of the type of radiation. Principal advantages of the semiconductor detectors are high energy resolution and fast response. The small detector size has advantages and disadvantages. A large detection volume is required in some applications. In the solid-state detectors the sensitive volume depends on the sensitive area, that is, the crystal size, and on the depletion depth. Currently available detectors have sensitive areas over the range of 1 mm^2 to 20 cm^2 and depletion depths of 15 μm to 1 cm. The most common type of semiconductor radiation sensors are:

Surface-Barrier Detectors. These are *pn* silicon or germanium detectors where the *pn* junction is right at the surface of the detector.

Lithium-Drifted Detectors. A greater depletion depth is obtained by drifting lithium ions into silicon or germanium *pn*-junction detectors.

Cadmium-Telluride Detectors. These have high sensitivity because of the high atomic number of the constituent elements.

16.7-5 Scintillation Detectors

Scintillation detectors consist of a sensing element that converts ionizing radiation into visible or ultraviolet light and a transduction element that converts the light into an electrical output. The sensing element in the scintillation detector is a body of fluorescent material (or fluor) which emits ultraviolet or visible light on exposure to ionizing radiation. The flash of light following an ionizing event is called a scintillation. The fluor can be an inorganic or organic crystal, a solid plastic, or a plastic in solution. Different fluors are suitable for different types of radiation. Among the more common materials are zinc sulfide, anthracene, naphthalene, and sodium iodide with a trace of thallium, called NaI(Tl).

The transducing element is normally a photomultiplier tube. Some scintillation detectors incorporate a pulse-height analyzer circuit that can be adjusted so that the output accepts only pulses falling into a specified amplitude range. Such a detector will then count only ionizing events in a certain energy range (energy window).

16.8 TRANSDUCERS FOR MAGNETIC FIELDS

16.8-1 Scope and Definitions

The properties of a magnetic field can be described by the parameters of magnetizing force, magnetic flux, and magnetic flux density. Transducers are available to measure the magnetic flux density in air or gases. The unit of magnetic flux is the weber (Wb); the older cgs unit is the maxwell or line; 1 maxwell = 10^{-8} Wb. The unit of flux density is the tesla (T) equivalent to Wb/m^2. The cgs unit is the gauss (Γ) (maxwell/cm^2), $\Gamma = 10^4$ T. In geomagnetic measurements the gamma (γ) is frequently employed, $1\ \gamma = 10^{-9}$ T. A magnetic flux density-measuring device is called a gaussmeter even though the unit gauss is obsolete.

16.8-2 Hall Effect Transducer

When a magnetic field is applied across a current-carrying conductor, an electric voltage will be developed orthogonally to both current and magnetic field, as shown schematically in Fig. 16.33. The relationship between the variables is

$$E_z = R_h \frac{I_x B_y}{d} \tag{16.28}$$

where I_x = current in x direction
B_y = magnetic flux density in y direction
E_z = voltage in z direction
d = thickness of Hall element in z direction
R_h = Hall coefficient

The phenomenon is independent of frequency, down to DC. The Hall effect can be utilized in the design of analog multipliers where both I_x and B_y are variable; such multipliers are the basis of the watt transducer used for electric power measurements. When a fixed current I_x is used, the Hall effect device becomes a magnetic field transducer.

The magnitude of the Hall effect is most pronounced in certain semiconductor elements and compounds. Bismuth and germanium have large Hall coefficients but suffer from the drawback of large temperature sensitivity. Indium arsenide is a preferred material. Practical Hall probes have the element mounted on a substrate and protected by encapsulation, with an overall thickness of the order of 1 mm.

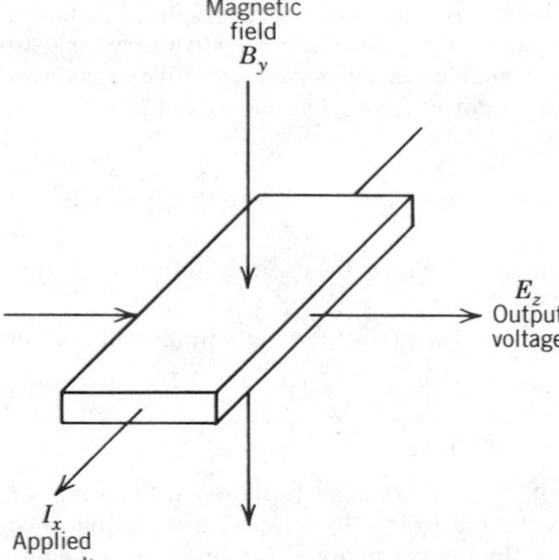

Fig. 16.33 The Hall effect.

16.8-3 Magnetoresistive Transducer

Electrical conductivity is dependent on the magnetic field in which the conductor is immersed. This phenomenon, called magnetoresistance, is most pronounced in bismuth and in such semiconductors as indium arsenide and indium antimonide, typically 20%/T in Bi and several orders larger in the semiconductors. The resistance change is proportional to the square of the flux density change except at very high densities.

16.8-4 Magnetic Field Sensitive Transistor

The currents flowing in a *pn*-junction device can be altered by magnetic induction. This effect has been exploited in the two-collector *npn* transistor which is arranged so that the magnetic field perpendicular to the plane of the collectors causes an unbalance of the two collector currents (magnetotransistor or magnistor).

16.8-5 Fluxgate Transducer

The fluxgate transducer is a saturable core device that can detect a constant (DC) magnetic field. A pair of identical AC-excited iron-core transformers is connected with secondaries in phase opposition. The DC magnetic field to be measured is additive to the flux in one transformer and subtractive in the other, and causes a second harmonic differential secondary output voltage roughly proportional to the magnitude of the DC field. This arrangement is the basis of the fluxgate magnetometer used for airborne magnetic surveying and detection.

16.9 TRANSDUCERS FOR ANALYSIS INSTRUMENTATION

16.9-1 Scope and Definitions

Analysis instrumentation deals with the detection, identification, and measurement of the physical and chemical constituents of mixtures. The methods employed range from Archimedes' specific gravity measurement to use of the mass spectrometer. Many of the methods are indirect, that is, what is being measured are bulk properties such as density or conductivity from which the composition can be inferred from an a priori knowledge of the constituents. A number of the measuring schemes require sampling. The original, off-line laboratory sampling methods have been developed in many cases into on-line industrial equipment.

The following paragraphs describe electrical transducers employed in a number of indirect bulk property measurement schemes. Electrochemical transducers are described in Section 16.10.

16.9-2 Electrical Conductivity Transducers

Solution conductivity systems are employed to analyze the composition of ionic solutions, particularly aqueous solutions. The conductivity of water ranges from 0.055 μS/cm for ultrapure water to 0.85 S/cm for a 20% HCl solution; the equivalent resistivity range is 18.3 MΩ-cm to 1.18 Ω-cm. There is also a substantial temperature effect on conductivity which is quite nonlinear and depends on the material and the concentration. Two types of transducers are employed: the electrode or contacting type and the electrodeless or induction type. Contacting electrodes find their use in the low conductivity range, up to 20 mS/cm. Electrodeless transducers are employed in the higher conductivity range, down to 1 mS/cm. They are designed principally for liquids containing oily, fibrous, and abrasive solids that would cause fouling or abrading of electrodes, and for fluids that are very conductive or highly corrosive and would cause electrical polarization problems or electrochemical action at electrode surfaces.

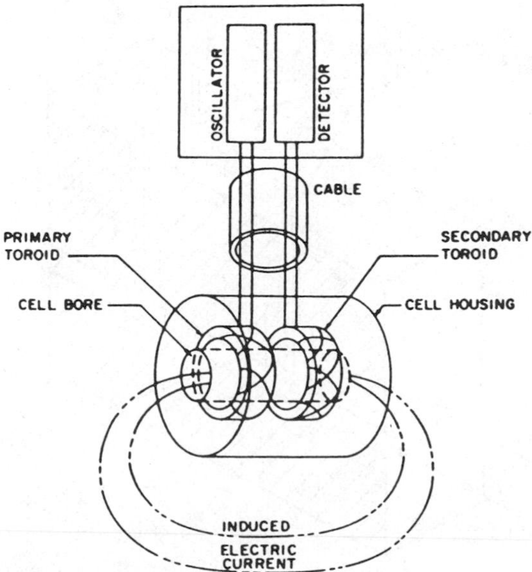

Fig. 16.34 Electrodeless conductivity sensor. Courtesy of the Foxboro Company.

The contacting conductivity transducer is a parallel plate "liquid resistor" in the form of a chamber or cell configured so that it can be inserted in piping or vessel walls. The resistor usually forms one arm of a bridge circuit which is energized with low-frequency AC or square wave. Different cell sizes are used to avoid very large or very small resistance values, which are difficult to measure; these cell sizes are represented by the parameter called cell factor or cell constant. A cell factor of 1 designates a liquid resistor in the form effectively of a cube with 1-cm sides; such a cell would make the bridge "direct reading" in that the measured conductance in siemens equals the desired conductivity in siemens per centimeter. For low-conductivity liquids a cell with larger electrodes and smaller electrode spacing is used, and a cell factor less than unity has to be applied to convert the measured conductance value into conductivity. Cell factors of 0.01, 0.1, 1, 10, and 100 are standard. Platinum, nickel, titanium, and graphite have been used for electrode material.

The electrodeless transducer, illustrated schematically in Fig. 16.34, consists of two toroidally wound coils which are inductively coupled by a "loop" of solution current. An alternating voltage applied to one toroid induces in the fluid an electric current whose magnitude depends on the solution conductivity. This solution loop current in turn induces a voltage in the second toroid which is a function of conductivity.

Both types of conductivity cells are built integrally with temperature sensors, either thermistors or thin-film RTDs. These thermoresistive transducers are used as part of a temperature compensation network (hardware compensation) or as the input to a software compensation scheme based on interpolation between discrete points of the conductivity-vs.-temperature characteristic stored in memory.

16.9-3 Thermal Conductivity Transducer

A thermal conductivity detector (TCD) is employed as the detecting device in some gas chromatographs. The measurement principle is hot-wire anemometry. The thermal conductivity cell, illustrated in Fig. 16.35, consists of a large metal block (heat sink) in which flow passages and cavities are drilled for the flow of air over temperature-sensitive resistors. A sample of the gas to be analyzed is passed over the pair of sensing resistors, while a reference gas of known properties is passed over a second pair of reference resistors. The four resistors, either metal filaments or thermistors, are connected in a Wheatstone bridge. The gas flow rates and temperature must be closely controlled. Under these circumstances, the bridge unbalance voltage is a measure of the difference in thermal conductivity of the sample and reference gases.

Except for a few anomalous compounds, the thermal conductivity of gases does not show a great variation; with respect to air, it lies in the range of 0.3 to 1.5. Thus the result of the measurement is nonspecific unless the gas to be analyzed is known to be a binary mixture. The anomalous gases are hydrogen and helium, which have a thermal conductivity six or seven times larger than that of air. As

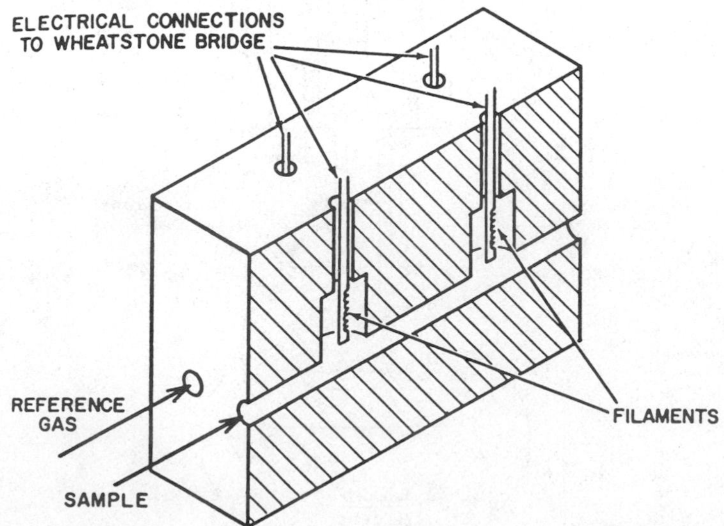

Fig. 16.35 Thermal conductivity cell. From Chilton, Radnor, PA, with permission.

a result it is possible to use the TCD to measure the amount of hydrogen or helium in a gas mixture against a background of hydrocarbons or air. Accuracies of the order of 5% are achieved.

16.9-4 Magnetic Susceptibility Transducer

Measurement of magnetic susceptibility is the basis of some gas analyzers, particularly for molecular oxygen. The measure susceptibility is defined by

$$\mu_r = 1 + X/\mu_0 \tag{16.29}$$

where X = susceptibility, H/m
μ_r = relative permeability with respect to free space
μ_0 = permeability of free space = $4\pi \times 10^{-7}$ H/m

The value of X is positive for paramagnetic materials and negative for diamagnetic ones. For most gases, the numerical value of X is of the order of 10^{-5} or smaller. Molecular oxygen is anomalous in that it is strongly paramagnetic with a susceptibility of the order of 10^{-3}. The value is also strongly temperature dependent.

The deflection-type paramagnetic O_2 detector consists of a nonmagnetic test body suspended in a constant magnetic field and free to rotate about a single axis. A temperature-controlled sample of the gas to be analyzed is admitted into the test chamber. A force is developed on the test body proportional to the difference in magnetic susceptibilities of the test body and the gas in which it is immersed. The force causes the test body to deflect.

The thermal-type paramagnetic O_2 detector depends on the fact that the susceptibility of O_2 decreases with increasing temperature. A sample of the gas is admitted into a ring chamber or dead-ended cavity where it is exposed to a magnetic field and heated. A convection current is set up due to the difference in diffusion rate of heated and unheated O_2, and this flow is detected by an anemometer-type thermal flowmeter.

16.9-5 Sound-Velocity Transducer

Sound-velocity transducers are employed in analysis of the composition of liquids or of liquids with solid suspensions. A typical probe consists of two piezoelectric transducers and a reflector in a single housing which can be inserted into the liquid. Ultrasonic pulses emitted from the sending transducer are reflected by the reflector and sensed by the receiving transducer. The time of flight of the pulses is measured electronically. An integral temperature transducer is sometimes provided.

16.9-6 Moisture Transducers

Moisture in Solids

The moisture content of powdered solids can be determined by measurement of electrical conductivity or dielectric constant. The conductivity cell can be used as a moisture transducer for nonconducting solid powders. The dielectric constant is obtained through a capacitance measurement; the dielectric constant of water, 80, is much larger than that of most other materials. The transducers are made in the form of small cells that sample the material to be analyzed, or as flow-through gages. Microwave and nuclear gages are also employed in moisture measurements.

Moisture in Fluids

The moisture content of gases and vapors is expressed by the concept of humidity. Relative humidity (% RH) is the ratio of the partial pressure of the water vapor actually present to the water vapor pressure required for saturation at the given temperature. Dew point is the temperature to which the mixture must be cooled to produce saturation. Absolute water content is often expressed in parts per million.

Psychrometers

Psychrometers use two temperature transducers to measure dry bulb and wet bulb temperatures. From these two measurements the relative humidity can be computed through use of psychrometric charts. The electronic psychrometer takes the form of a hollow tube through which gas is driven by a fan. The dry bulb temperature is measured with a thermistor or platinum RTD located near the inlet of the tube. A second, matched temperature detector is mounted downstream with a reservoir and wick assembly to measure wet bulb temperature.

Hygrometers

Mechanical hygrometers employ as a transducer a fiber that changes mechanical dimensions with exposure to moisture. Human or animal hair was first used for this purpose. Electronic hygrometers use the change of electrical conductivity of hygroscopic material as a sensing mechanism. Lithium chloride is a commonly used material (Dunmore cell). An ion-exchange sensor made of sulfonated polystyrene is another type of hygrometric sensing element (Pope cell). A temperature-sensing element is usually packaged integrally into the humidity transducer probe.

The aluminum oxide transducer is a moisture-sensitive capacitor. A layer of porous aluminum oxide forms the dielectric between two metal electrodes. As the porous layer absorbs water, the dielectric constant of the transducer changes. This transducer can be employed for moisture measurements in liquids, such as in hydrocarbons.

The *electrolytic hygrometer* is an electrolytic cell coated with a thin film of hygroscopic material, such as phosphorous pentoxide. A voltage applied to the cell dissociates the absorbed water into hydrogen and oxygen. The current flowing in the cell is determined by the number of water molecules dissociated. This transducer is used primarily for dry gas measurements in the low parts-per-million region.

The *saturated salt* type of dew point detector takes advantage of the special properties of lithium chloride, which shows a sharp decrease in electrical resistance when its own relative humidity increases above 11%. A self-regulating scheme is employed. Electrical voltage is applied to the lithium chloride element. As the humidity of the surrounding gas increases, the lithium chloride absorbs moisture, which lowers its resistance, increases the current flowing through it, and generates heat. As the lithium chloride temperature increases, moisture is driven off, increasing the resistance, reducing the current, and reducing the self-heating. This self-regulation or feedback maintains the lithium chloride near 11% RH regardless of the humidity of the surrounding atmosphere. The final temperature of the lithium chloride element is related to the desired dew point. A temperature-sensing element is built into the transducer for this purpose.

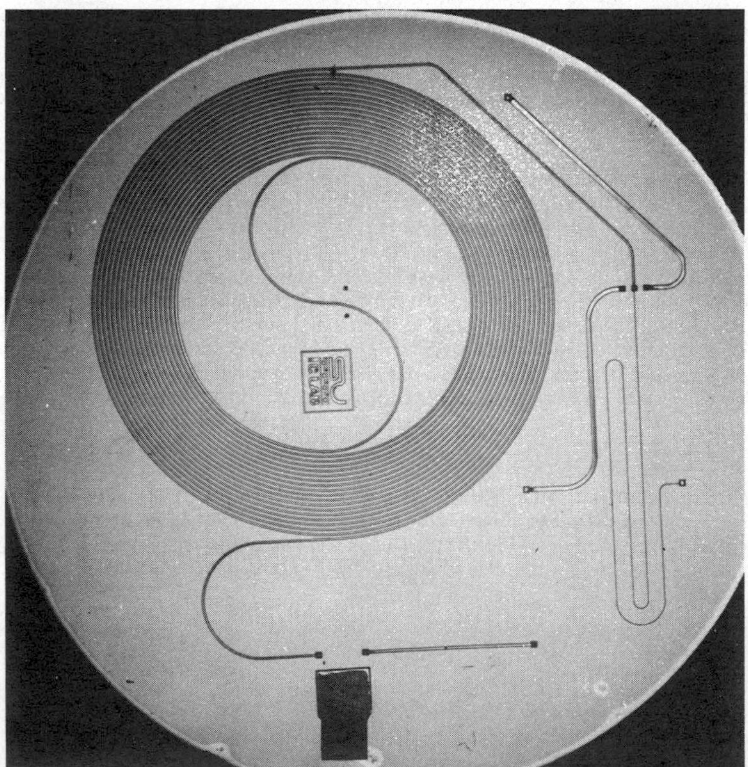

Fig. 16.36 Silicon wafer-type gas chromatography column. From Barth, *IEEE Spect.*, p. 33 (Sep. 1981), with permission.

The *optical condensation hygrometer* measures dew point by the detection of water condensation on a mirror whose temperature is being lowered. A complete instrument includes a thermoelectric heat pump for cooling the mirror, photoelectric detection of condensation, and a platinum RTD thermometer.

The integrated-circuit surface impedance measurement device allows dew point monitoring by the measurement of sheet resistance.

16.9-7 Spectroscopic Methods

Analytical procedures based on the microscopic properties of particular atoms or molecules include mass spectroscopy, flame ionization detection (FID), chemiluminescence, optical emission, ultraviolet and infrared spectroscopy, electron spin resonance (ESR), nuclear magnetic resonance (NMR), and x-ray analysis. Many of the original laboratory methods have been elaborated into industrial on-line instrumentation.

16.9-8 Chromatography

Chromatography is a process of separating the constituents of a mixture so that they can be identified and measured sequentially. Conventional chromatographs are large, complex, and costly instrument systems. Figure 16.36 shows an experimental miniature gas chromatography column etched on a 5-cm-diameter silicon wafer. The front side shows the shallow grooves that form the column. Also etched on the wafer on the front or reverse side are valve seats, capillaries for injection of test gas samples, and connections to a detachable thermal conductivity detector.

16.10 ELECTROCHEMICAL TRANSDUCERS

16.10-1 Scope and Definitions

Any material that changes its electronic properties with chemical composition can be used as the basis for an electrochemical sensor (ECS). On a historical basis, one distinguishes between conventional electrochemical electrodes (electrometric transducers) and the new chemically sensitive electronic devices (CSED). The conventional electrode serves as a transducer for ions. The CSED family comprises devices that sense ionic and nonionic species.

16.10-2 Ionic Transducer Principles

An electrical potential is developed across a solid membrane separating two liquids of different ionic content. The basic phenomenon can be expressed by the Nernst equation

$$E = \frac{RT}{nF} \ln \frac{a_1}{a_2} \tag{16.30}$$

where E = voltage across membrane
R = gas constant = 8.32 Joules per mol degree
T = temperature, K
F = Faraday constant = 9.65×10^4 Coulombs per mol
n = ionic charge ($n = 1$ for monovalent ion, etc.)
a_1 = activity of ion to be measured in solution on one side of membrane
a_2 = activity of same ion in solution of other side of membrane

Ionic activity is related to ionic concentration C by

$$a = \gamma C \tag{16.31}$$

where γ is the activity coefficient which is unity for highly dilute electrolytes. If the ionic activity of the electrolyte on one side of the sensitive membrane is known (reference solution), then the activity of the same ion in the electrolyte on the other side of the membrane (the measurand) can be determined from the voltage, provided that no other ions are present that would cause additional voltages to be generated and would thus create an interference with the measurand.

To employ such a membrane as a transducer, the fundamental problem is to make electrical contact with the electrolytes. If one were to insert metal electrodes into the two solutions for this purpose, additional voltages would be produced in the circuit as a result of the chemical reactions at the electrode surfaces. A stable and reproducible ionic transducer requires a rather elaborate connection scheme consisting of so-called half cells coupled to the electrolytes by salt bridges or buffer solutions. A half cell consists of an electrode immersed in an electrolyte with which it is in chemical equilibrium. Only cells with electrochemically reversible half-cell reactions give reproducible

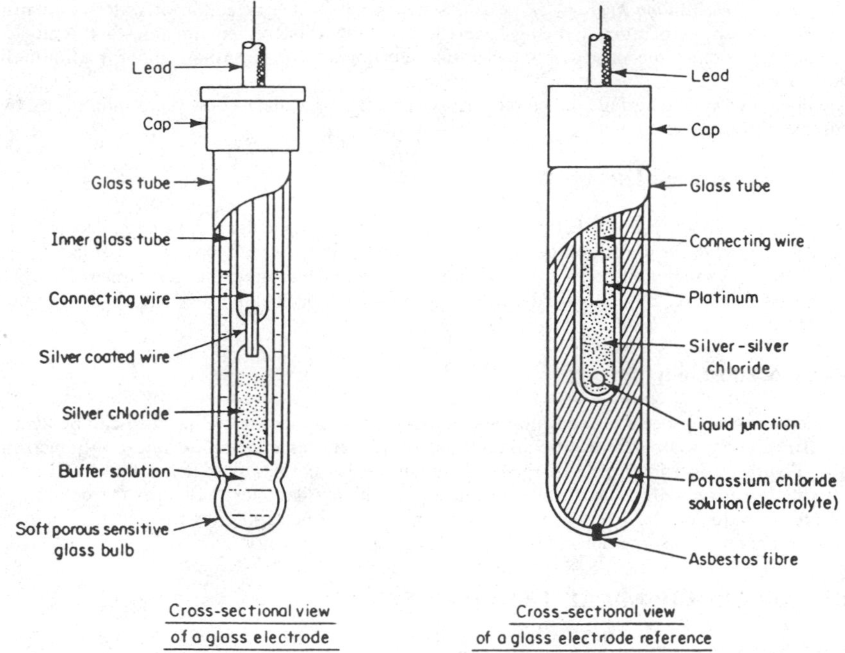

Fig. 16.37 Glass electrodes for ionic transducer. From McGraw-Hill, New York, with permission.

results and are usable. A pair of such half cells, chosen for stability and reproducibility, serves as the contact-making mechanism with the two sides of the sensitive membrane. The basic scheme, shown in Fig. 16.37, consists of two electrodes, a sensing electrode that contains an internal reference electrode as a contact-making device to the reference solution and a reference electrode. The most common reference cells are the calomel cell (mercury in contact with mercurous chloride/potassium chloride paste) and the silver/silver chloride cell.

The output voltage at 25°C, using common logarithms, is

$$E = E^\circ + \frac{0.059}{n} \log \gamma C \qquad \text{V} \tag{16.32}$$

where E° is a constant for the particular configuration of materials. The source impedance of the sensing membrane is of the order of 10^8 Ω, and a high-input impedance voltage measurement is required. Traditionally a potentiometer-type null-balance circuit was used, and analytical chemists still refer to such ionic measurements by the name *potentiometry*.

16.10-3 pH Transducers

Hydrogen ion concentration is customarily defined by the logarithmic measure pH

$$\text{pH} = \log \frac{1}{C_{\text{H}^+}} \tag{16.33}$$

In pure water some of the molecules disassociate. At 25°C the concentration of both H^+ and OH^- ions is 10^{-7} mol/l. That is, the pH of chemically neutral water is 7. pH < 7 is defined as acidic and pH > 7 as alkaline. These numbers are temperature dependent, because the amount of H_2O dissociation increases with temperature. At 100°C pure water has a hydrogen ion concentration of 10^{-6}, that is, the pH is 6, but the water is still neutral. In an aqueous solution the product $(C_{H+})(C_{OH-})$ remains constant when the concentration of either ion species is increased by addition of a solute.

At 25°C the electrode output voltage is 0.059(pH). Thus a pH change of 0.1 (a 25% change in C_{H+}) produces an output of approximately 6 mV. Careful signal conditioning is required to obtain a reliable and reproducible output.

The sensing membrane of the conventional pH electrode is made of glass. It has been found that the glass membrane is almost a perfect H^+-ion-specific transducer. It responds to pH virtually without interference by any other ions present in the solution, except at very high values, above pH = 12, when sodium (and other alkali metal) ions cause interference. Antimony metal electrodes are available for abrasive solutions or for chemicals that attack glass. In industrial practice, the sensing and reference electrodes are available separately or mounted in combination inside one assembly. A built-in temperature sensor (platinum RTD) is sometimes provided. Various types of self-cleaning methods, such as ultrasonic or flow turbulence, can be provided.

16.10-4 Ion-Selective (pION) Transducers

The same principle that governs pH electrodes can be applied to the measurement of other ions, by proper choice of sensitive membrane material. pION transducers are much less ion specific than is the pH electrode, that is, the presence of ions of species other than the measurand may cause substantial interference. The membranes employed are special glasses, crystalline solids, and polymers. There are some ions for which no suitable selective membrane has been found and where special electrodes incorporating an ion exchanger have been built.

16.10-5 Oxidation-Reduction Potential Transducers

The oxidation-reduction potential (ORP) or redox potential transducer uses a noble metal (platinum or gold) electrode as the sensing element. An oxidizing solution provides a positive output and a reducing solution provides a negative output that follows the Nernst equation

$$E = E^\circ + \frac{RT}{nF} \ln \frac{(\text{OX})}{(\text{Red})} \tag{16.34}$$

The result is pH and temperature dependent. The E° term depends on the reference electrode, not on the ORP electrode, which is inert.

16.10-6 Ionized Gas Transducers

The concentration of oxygen in a gas mixture can be detected by an ionic cell maintained at a sufficiently high temperature to ionize the oxygen. Porous platinum conductors are fired on the inner and outer surfaces of a ceramic, such as yttrium-stabilized zirconium dioxide. One surface is exposed to the measurand and the other to ambient air. The voltage across the electrodes indicates the relative concentration according to the Nernst equation. The cell temperature must be held constant (550–1300°C) and the complete transducer system includes the heaters and a thermocouple temperature-sensing element.

16.10-7 Other Electrometric Methods

Coulometry is based on the electrolysis of ionic compounds. The quantity of electricity is proportional to the amount of liberated ions, according to Faraday's law. Coulometry can be carried out with constant current or constant voltage. Constant current has the advantage that only time has to be measured. Constant voltage permits a choice of setting of the half-cell potential and therefore good separation of ion species.

Polarography can be used to measure the ionic concentrations in a solution containing many ion species. The sensing electrode must be very small (microelectrode) in order to create a large depletion region. Ions reach the electrode only by diffusion, the rate of diffusion being specific for each species. Voltage is applied in the form of a ramp waveform, and the resulting current will show a number of plateaus. The voltage at which a new current rise begins identifies the type of ion species, while the height of the current plateau measures the concentration.

16.10-8 Chemically Sensitive Electronic Devices (CSEDs)

Research is in progress on semiconductor ionic transducers such as the ion-controlled diode (ICD) and the ion-selective field transistor (ISFET). These are MOS structure devices where the gate is replaced by an ionic solution that is coupled to the semiconductor by a coating of a chemically sensitive membrane.

Among the experimental nonionic CSEDs is a family of gas-sensitive MOSFETs. A catalytic metal such as palladium or platinum is used as the gate metal; it adsorbs molecules of the gas to be measured, hydrogen, methane, or carbon monoxide, for example. The adsorbed gas causes a change in the work function at the metal-insulator interface which affects the transistor current.

Bibliography

Ahrendt, W. R. and C. J. Savant, Jr., *Servomechanism Practice*, 2nd ed., McGraw-Hill, New York, 1960.

Allocca, J. A. and A. Stuart, *Transducers, Theory and Applications*, Reston, Reston, VA, 1984.

Anderson, N. A., *Instrumentation for Process Measurement and Control*, 3rd ed., Chilton, Radnor, PA, 1980.

Beckwith, T. G., N. Lewis Buck, and Roy D. Marangoni, *Mechanical Measurements*, 3rd ed., Addison-Wesley, Reading, MA, 1982.

Benedict, R. P., *Fundamentals of Temperature, Pressure, and Flow Measurement*, 2nd ed., Wiley, New York, 1977.

Boyes, G., ed., *Synchro and Resolver Conversion*, Analog Devices, Norwood, MA, 1980.

Canfield, E. B., *Electromechanical Control Systems and Devices*, Wiley, New York, 1965.

Cheremisinoff, P. and H. J. Perlis, *Analytical Measurements and Instrumentation for Process and Pollution Control*, Ann Arbor Science, Ann Arbor, MI, 1981.

Cheung, P. W., D. G. Fleming, W. H. Ko, and M. R. Neuman, eds., *Theory, Design, and Biomedical Applications of Solid State Chemical Sensors*, CRC, Boca Raton, FL, 1978.

Chien, C. L. and C. R. Westgate, eds., *The Hall Effect and Its Applications*, Plenum, New York, 1980.

Cho, C. H., *Measurement and Control of Liquid Level*, Instrument Society of America, Research Triangle Park, NC, 1982.

Cobbold, R. S., *Transducers for Biomedical Measurements: Principles and Applications*, Wiley, New York, 1974.

Considine, D. M. and S. D. Ross, eds., *Process Instruments and Controls Handbook*, 2nd ed., McGraw-Hill, New York, 1974.

Covington, A. K., ed., *Ion Selective Electrode Methodology*, Vols. I and II, CRC, Boca Raton, FL, 1979.

Cracknell, A. P., ed., *Remote Sensing Applications in Marine Science and Technology*, Reidel, Dordrecht, 1983.

Davis, S. A. and B. K. Ledgerwood, *Electromechanical Components for Servomechanisms*, McGraw-Hill, New York, 1961.

DeCarlo, J. P., *Fundamentals of Flow Measurement*, ISA, Research Triangle Park, NC, 1983.

DeMarre, D. A. and D. Michaels, *Bioelectronic Measurements*, Prentice-Hall, Englewood Cliffs, NJ, 1983.

Doebelin, E. O., *Measurement Systems: Application and Design*, 3rd ed., McGraw-Hill, New York, 1983.

Fleming, D. G. and B. N. Feinberg, *Handbook of Engineering in Medicine and Biology*, Vol. II, Instruments and Measurements, CRC, Boca Raton, FL, 1978.

Fleming, D. G., W. H. Ko, and M. R. Neuman, eds., *Indwelling and Implantable Pressure Transducers*, CRC, Boca Raton, FL, 1977.

Fritschen, L. J. and L. W. Gay, *Environmental Instrumentation*, Springer-Verlag, New York, 1979.

Gagnepain, J. J. and T. Meeker, eds., *Piezoelectricity*, Gordon & Breach, New York, 1982.

Gillum, D. R., *Industrial Pressure Measurement*, ISA, Research Triangle Park, NC, 1982.

Gillum, D. R., *Industrial Level Measurement*, ISA, Research Triangle Park, NC, 1984.

Harvey, G. F., ed., *ISA Transducer Compendium*, UMI, Charlotte, NC, 1980.

Herbert, J. M., *Ferroelectric Transducers and Sensors*, Gordon & Breach, New York, 1982.

Herceg, E. E., *Handbook of Measurement and Control*, Schaevitz Engineering, Camden, NJ, 1976.

Jones, E. B., *Instrument Technology*, Vol. 1, Measurement of Pressure, Level, Flow, Temperature, 3rd ed., Newnes-Butterworth, Woburn, MA, 1974.

Kerlin, T. W. and R. L. Shepard, *Industrial Temperature Measurement*, ISA, Research Triangle Park, NC, 1982.

Lion, K. S., *Elements of Electrical and Electronic Instrumentation*, McGraw-Hill, New York, 1975.

Lipták, B. G., ed., *Instrument Engineers Handbook*, Vols. I, II, and Supplement, Chilton, Radnor, PA, 1969–1972.

Lynnworth, L. C., *Physical Acoustics*, in Vol. XIV, W. P. Mason and R. N. Thurston, eds., Academic, New York, 1979.

Lyons, J. L., *The Designer's Handbook of Pressure-Sensing Devices*, Van Nostrand Reinhold, New York, 1980.

Miller, R. W., *Flow Measurement Engineering Handbook*, McGraw-Hill, New York, 1983.

Moore, R. L., ed., *Basic Instrumentation Lecture Notes and Study Guide*, Vol. 1, *Measurement Fundamentals*, Vol. 2, *Process Analyzers and Recorders*, 3rd ed., Prentice-Hall, Englewood Cliffs, NJ, 1983.

Neubert, H. K. P., *Strain Gauges; Kinds and Uses*, St. Martin's, London, 1967.

Neubert, H. K. P., *Instrument Transducers: An Introduction to Their Performance and Design*, Oxford University, Oxford, England, 1975.

Neuman, M. R., D. G. Fleming, W. H. Ko, and P. W. Cheung, eds., *Physical Sensors for Biomedical Applications*, CRC, Boca Raton, FL, 1980.

Norton, H. N., *Sensor and Analyzer Handbook*, Prentice-Hall, Englewood Cliffs, NJ, 1982.

O'Higgins, P. J., *Basic Instrumentation: Industrial Measurements*, McGraw-Hill, New York, 1966.

Peggs, G. N., *High Pressure Measurement Techniques*, Elsevier, Amsterdam, 1983.

Perry, A. E., *Hot Wire Anemometry*, Oxford University, Oxford, England, 1982.

Perry, C. C. and H. R. Lissner, *The Strain Gage Primer*, 2nd ed., McGraw-Hill, New York, 1962.

Quinn, T. J., ed., *Temperature*, Academic, New York, 1984.

Rollo, F. D., ed., *Nuclear Medicine Physics, Instrumentation, and Agents*, Mosby, St. Louis, 1977.

Savet, P. H., ed., *Gyroscopes, Theory and Design*, McGraw-Hill, New York, 1961.

Scott, R. W., ed., *Developments in Flow Measurements*, Vol. 1, Elsevier, Amsterdam, 1982.

Sheingold, D. H., ed., *Transducer Interfacing Handbook,* Analog Devices, Norwood, MA, 1980.

Smith, E., *Principles of Industrial Measurement for Control Applications*, ISA, Research Triangle Park, NC, 1984.

Soisson, H. E., *Instrumentation in Industry*, Wiley, New York, 1975.

Spitzer, D., *Industrial Flow Measurement*, ISA, Research Triangle Park, NC, 1984.

Spitzer, F. and B. Howarth, *Principles of Modern Instrumentation*, Holt, Rinehart & Winston, New York, 1972.

Spooner, R. B., ed., *Hospital Instrumentation: Care and Servicing for Critical Care Units*, ISA, Research Triangle Park, NC, 1977.

Sydenham, P. H., *Transducers in Measurement and Control*, ISA, Research Triangle Park, NC, 1980.

Thomas, J. D. R., ed., *Ion-Selective Electrode Reviews*, Vols. 1–4, Pergamon, Elmsford, NY, 1980–1983.

Tompkins, W. J., and J. G. Webster, eds., *Design of Microcomputer-Based Medical Instrumentation*, Prentice-Hall, Englewood Cliffs, NJ, 1981.

Traister, R., *Principles of Biomedical Instrumentation and Monitoring*, Reston, Reston, VA, 1981.

Vesely, J., D. Weiss, and K. Stulik, *Analysis with Ion-Selective Electrodes*, Wiley, New York, 1978.

Webster, J. G., ed., *Medical Instrumentation: Application and Design*, Houghton Mifflin, Boston, 1978.

Weiss, M. D., *Biomedical Instrumentation*, Chilton, Radnor, PA, 1973.

Welkowitz, W. and S. Deutsch, *Biomedical Instruments, Theory and Design*, Academic, New York, 1976.

Wexler, A., ed., *Humidity and Moisture Measurement and Control in Science and Industry*, Vol. 4, Krieger, Melbourne, FL, 1982.

Window, A. L., and G. S. Holister, *Strain Gauge Technology*, Elsevier, Amsterdam, 1982.

Woolvet, G. A., *Transducers in Digital Systems*, IEE, London, 1979.

Periodicals and Conference Proceedings

Association for the Advancement of Medical Instrumentation
Medical Instrumentation (bimonthly).
Proceedings of the Annual Meeting.

BHRA Fluid Eng'g Co., *Advances in Flow Measurement Techniques*, Proceedings of the 1981 International Conference, Cranfield, UK.

Institute of Electrical and Electronics Engineers (IEEE)
Transactions on Biomedical Engineering.
Transactions on Electron Devices.
Special issues on Solid State Sensors, Actuators, Interface Devices:
Vol. ED-16 No. 10, October 1969.
Vol. ED-26 No. 12, December 1979.
Vol. ED-29 No. 1, January 1982.

Transactions on Industrial Electronics.
Transactions on Instrumentation and Measurement.

Institute of Measurement and Control (London)
Transactions (quarterly).

Institution of Electrical Engineers (IEE), London
Physical Measurement and Instrumentation (monthly).
Electrical Measurement and Instrumentation (monthly).
International Conference on Automotive Electronics (1979).

Instrument Society of America (ISA): Research Triangle Park, NC 27709
Instrumentation Technology (monthly).
ISA Transactions (quarterly).
Advances in Instrumentation (annual).
Aerospace and Test Instrumentation Proceedings.
Analysis Instrumentation.
Biomedical Sciences Instrumentation.
Engineering in Medicine and Biology.
Instrumentation in the Aerospace Industry.
Instrumentation in the Chemical and Petroleum Industries.
Instrumentation in the Food Industry.
Instrumentation in the Mining and Metallurgy Industry.
Instrumentation in the Power Industry.
Instrumentation in the Pulp and Paper Industry.
Instrumentation in the Textile Industry.
Instrumentation Symposium for the Process Industries.
Proceedings of the Institute of Environmental Sciences.
Flow—Its Measurement and Control in Science and Industry, Vol. 2, St. Louis, MO, March 1981.
International Symposium on Temperature.

International Measurement Confederation
IMEKO Congress Proceedings (triennial).
Flow Measurement of Fluids, Proceedings of the FLOMENKO Conference, Groningen, Netherlands, Sept. 1978, North Holland.

North Atlantic Treaty Organization
Proceedings NATO Advanced Study Institute on Chemically Sensitive Electronic Devices, Hightstown, NJ, 1980, Elsevier, Amsterdam.

Society of Automotive Engineers
Proceedings of the Annual Congress:
1976 (SP404): Sensors for Electronic Systems.
1977 (SP418): The Automotive Application of Sensors.
1978 (SP427): Automotive Application of Sensors.
1979 (SP441): Automotive Sensors.
1980 (SP458): Sensors for Automotive Systems.
1981 (SP486): Sensors.
1982 (SP511): Sensors and Actuators.
1983 (SP536): Sensors and Actuators.
1984 (SP567): Sensors and Actuators: New Approaches.
Proceedings of the International Congress on Transportation Electronics (P-111) 1982.

Technicon Corporation, Advances in Automated Analysis
Proceedings of the 7th Technicon International Congress, December 1976, New York, Vol. 2, Industrial Symposia, 1977, Mediad, NY.

Sensors, The Journal of Machine Perception (monthly), North American Technology, Peterborough, NH.

Sensors and Actuators (quarterly), Elsevier Sequoia, Lausanne, Switzerland.

Standards and Miscellaneous Publications

American Society for Testing & Materials (ASTM), *Manual on the Use of Thermocouples in Temperature Measurement*, 2nd ed., 1974.

General Eastern Instruments Corporation, *Selecting Humidity Sensors for Industrial Processes*, 1982, Watertown, MA 27709.

Hewlett-Packard Company, *Practical Temperature Measurements*, Application Note 290, August 1980.

Instrument Society of America
Standards and Practices for Instrumentation, 7th ed., 1983.
Directory of Instrumentation, Vol. 6, 1984–85.

Microswitch Division of Honeywell, Inc., *Hall Effect Transducers—How to Apply Them as Sensors*, 1982, Freeport, IL.

Technical Database Corp., *Industrial Sensor Directory*, 1983, Conroe, TX 77305.

US Department of the Navy, *Encoders, Shaft Angle to Digital*, Military Standardization Handbook MIL-HDBK-231(AS), 1 Jul. 1970.

Variable Resistive Components Institute, *Terms and Definitions*, Precision Potentiometer Standard vrci-p-100A, 1974, Evanston, IL 60203.

CHAPTER 17
CONSTRUCTION TECHNIQUES

GUSTAVE R. GASCHNIG

Raytheon Company, Bedford, Massachusetts

17.1 PRINTED CIRCUITS

17.1-1 Single- and Double-Sided Boards

A printed circuit board can be either single or double sided. This means that either one or both sides of the dielectric base material are copper clad. Often a number of single-sided sheets are laminated together to obtain a multilayer circuit. In such cases a glass cloth prepreg is used and molded under heat and pressure for multilayer printed circuitry.

The main advantages of printed circuits over conventional component interconnections by discrete wires are reduction in assembly cost and achievement of greater uniformity. Other benefits are better packaging densities, reduced assembly weight, and lower rejection rates owing to the elimination of human error.

Basically, the manufacture of printed circuits starts with the preparation of a master artwork of the circuit pattern. The artwork can be done either on glass, for more stability, or on film, for easier handling and storage. Either way, it must be converted into a printing master (mask or screen). The very simplest type of etched printed circuit is a copper-clad laminate printed with a resist in the pattern of the desired circuit. To accomplish this, artwork is converted into a screen for screen printing of the resist pattern or into a photomask that is used to expose the laminate precoated with the photoresist to ultraviolet light. After exposure the unwanted resist is dissolved, leaving the desired printed pattern (Fig. 17.1).

The circuit board is then further processed. The laminate is treated with a copper etchant such as ammonium persulfate, chromic acid and sulfuric acid, cupric chloride, or ferric chloride, the remaining resist acts as an etch stop, and only the desired conductor pattern remains. Etching of the conductor pattern offers good pattern definition. In total, many steps are involved in the production of high-quality, reliable circuit boards, including cleaning, inspection, drying, baking, hole drilling, exposing, developing, stripping, and sanding.

17.1-2 Multilayer Board

Multilayer printed circuits (Fig. 17.2) are composed of a series of separate conducting circuit planes separated by insulating materials such as prepreg and bonded together into relatively thin homogeneous constructions with internal and external connections to each level of the circuitry as required by the specific design. In many cases these boards offer significant space savings and may even permit replacement or simplification of complex wiring or connection systems. A distinct advantage

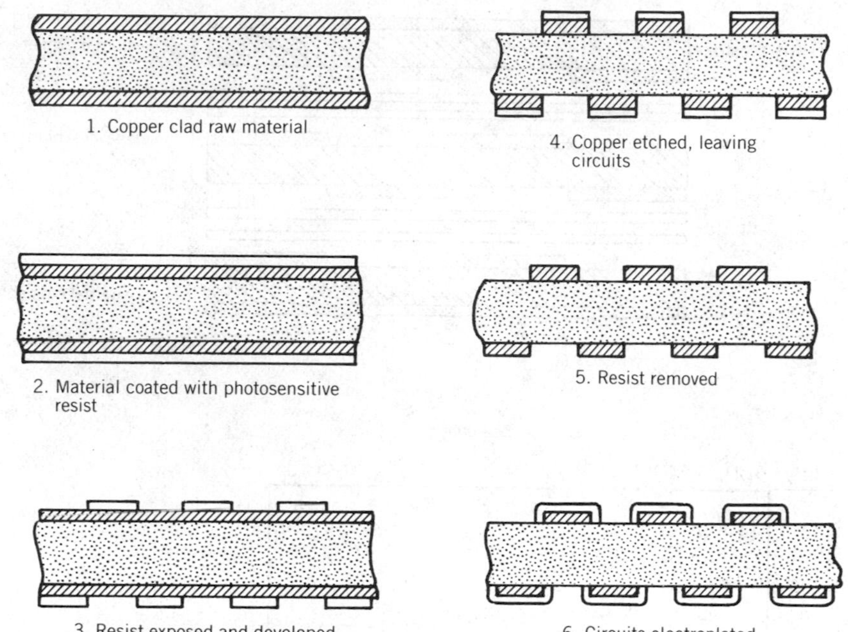
1. Copper clad raw material
2. Material coated with photosensitive resist
3. Resist exposed and developed
4. Copper etched, leaving circuits
5. Resist removed
6. Circuits electroplated

Fig. 17.1 Negative photomask printed circuit process.

of multilayer circuits is that internal and external shielding or ground layers can be incorporated directly into the circuit board to provide electrical decoupling and to minimize interference. In addition, such circuits can provide heat sinking as well as controlled impedance conductors or striplines.

Rigid multilayer boards are made of the same basic materials as the single- and double-sided boards. After the boards are etched and processed they are bonded together under heat and pressure with use of multilayer laminating presses. Prepreg materials serve as the bonding and insulating layers. Epoxy-glass and polyimide-glass are presently used for this purpose.

The plated-through hole-type of multilayer board allows interconnection to and between layers by means of a hole that is plated with a conductive material. The clearance hole-type of multilayer provides access to terminal pads on each layer above it. The built-up type of multilayer achieves interlayer connections by sequential metal deposition of conductive patterns with or without the need for holes going through the entire thickness of the board.

The use of multilayer boards as interconnection devices for integrated circuits results in great reductions in the overall size and weight of an electronic system. It also aids in achieving even heat distribution and the reliable removal of heat, which in a system using integrated circuits can pose a serious design problem. With multilayer boards, all interconnections can be placed on internal layers and a heat sink of thick, solid copper or other material can be provided on the outer surfaces.

17.1-3 Flexible Circuits

Flexible printed circuitry is similar to the rigid-type circuitry found in single-layer, double-layer, and multilayer boards except that the support dielectric material is flexible. Conductors are often bonded or laminated between two dielectric protective layers (Fig. 17.3). Flexible circuits can be fabricated in single-sided, double-sided, multilayer, and shielded types.

Plastic film, such as polyester and polyimide types, is commonly used for the production of flexible circuits, as are many other dielectrics. These usually include such other types of films as polytetrafluoroethylene (PTFF), fluorinated ethylene propylene (FEP), polyvinyl chloride, polyethylene, polypropylene, polyparabanic, and thin fabrics such as PTFF-coated glass cloth, epoxy-glass cloth, and FEP-glass cloth.

Flexible circuitry offers design features and advantages such as space savings, weight savings, and elimination of connections that might be required if rigid boards were used. Production processes for flexible circuitry may be similar to those for rigid boards. Normally some consideration must be given

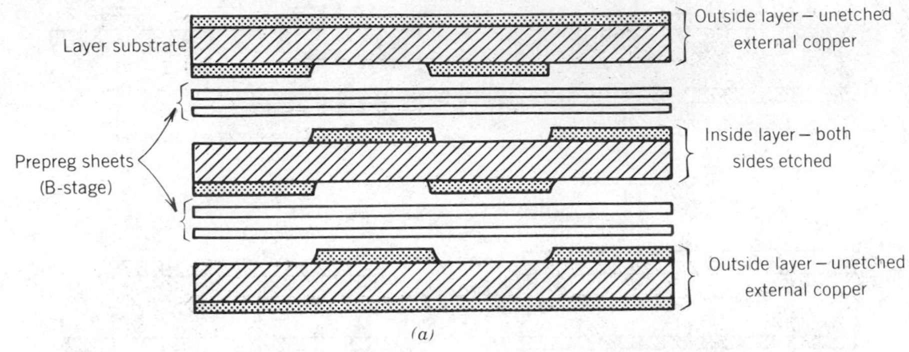

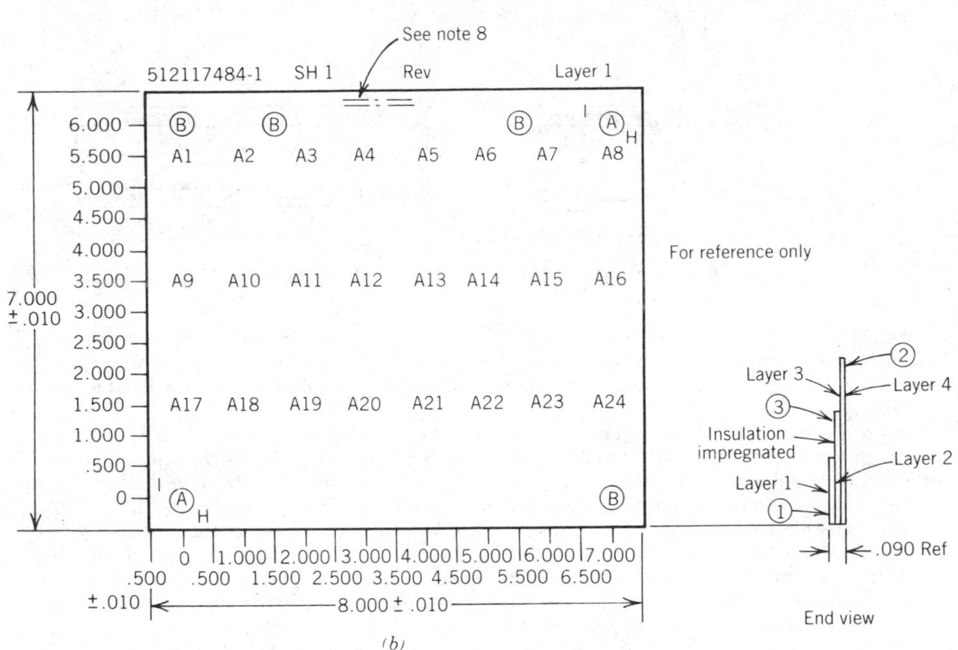

Fig. 17.2 (*a*) Exploded view of multilayer board prior to lamination step (six copper layers); (*b*) Plan view, after lamination.

to the cleaning solutions, etchants, and processing temperatures involved, since some of the substrates and adhesives used in flexible construction may be more susceptible to damage than those used in rigid products.

New materials for flexible printed circuits are being introduced continuously because of the extreme interest being shown in these products by the printed circuit industry. Furthermore, technology also is changing rapidly.

A typical cable system for data communications is an undercarpet wiring system. Equivalent to an 85-Ω coaxial cable, it is only 0.090 in. high and contains AWG (American Wire Gauge) 30, silver-plated copper conductors; aluminum-polyester laminated tape shielding; and an AWG 30, silver-plated copper, solid drain wire. In addition, 50-conductor (25-pair) telephone undercarpet cable is available. It is designed for interconnecting telephone sets via a distribution box or baseboard connector to a private automatic branch exchange (PABX), call director, or distribution closet. The cable is 0.040 in. high and is available in lengths up to 35 ft with standard 50-pin telephone connectors on each end. Costs can be reduced by as much as 40% compared with the cost of conventional systems.

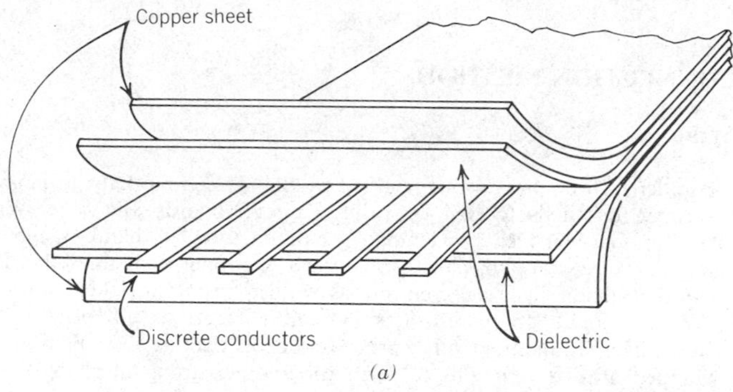

(a)

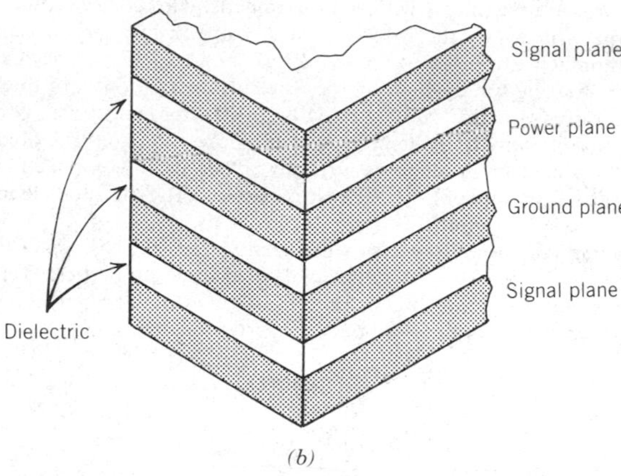

(b)

Fig. 17.3 Flexible printed wiring structures: (*a*) simple multilayer, (*b*) four-layer.

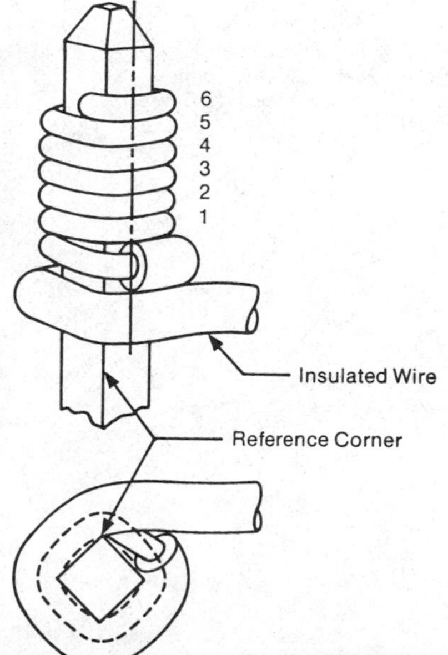

Fig. 17.4 Solderless wrapped connection on a square terminal post.

17.2 INTERCONNECTION METHODS

17.2-1 Wire Wrap

Wire wrapping, or solderless wrapping, was originated by Bell Telephone Labs to improve mechanical connections between terminals located on printed circuit boards and wire. Wire-wrap is a registered trademark of Gardner-Denver Company, manufacturer of hand, semiautomatic, and automatic machinery for wire-wrapping panels. Solderless wrapping dominates the discrete wiring field. Claims of over 300 billion electrical connections with no reported field failures are made by Gardner-Denver. What accounts for wire-wrap's incredible record is an airtight, gastight, contamination-free wire contact termination with a pressure lock as tight as 130,000 lb/in.^{2}. The wire is actually stretch wrapped around terminals to form these super-strong mechanical and electrical connections without the need of soldering. A large contact area helps with reliability and further strength.

One of the requirements for the wire-wrap pins is sharp corners so that proper contact is achieved. Typical applications consist of 0.025 in. square pins on a multiple of a 0.025 in. grid. Pins are normally phosphor-bronze and plated with various metals. The configuration consists of a square post with sharp corners on which the wire is tightly wrapped in a helix around the post, making high-pressure metal contacts at the sharp corners (Figs. 17.4 and 17.5). The pressure contact in turn produces a very high shearing force which scrapes the surface film on both the wire and the post. In addition, solid-state diffusion adds to the initially satisfactory electrical connection by actually increasing the mechanical strength at the point of contact as time goes by. Solderless wire wrapping has been rated as the single most reliable, currently available interconnection technique. The US Naval Avionics Facility at Indianapolis (NAFI) Report No. TR-1242 states there is no known failure rate established for solderless wrapping.

Electronic packaging commonly involves wire sizes 26, 28, and 30 AWG. Kynar insulation is a general-purpose type with good cut-through and cold-flow properties. Teflon has good high-temperature properties but is inferior in cut-through and cold-flow properties. Tefzel is a material in the Kynar/Teflon group that combines the best properties of both.

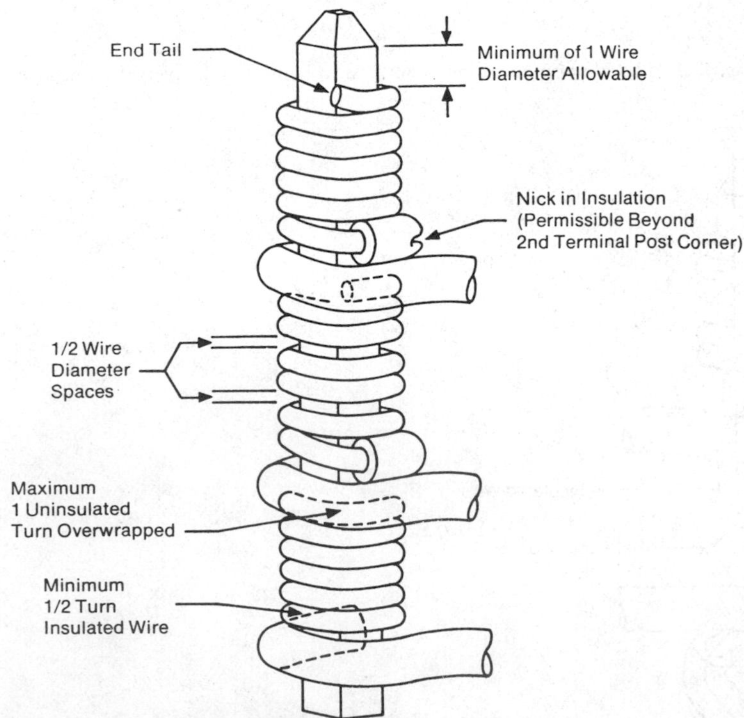

Fig. 17.5 Acceptable multiple wrap connections.

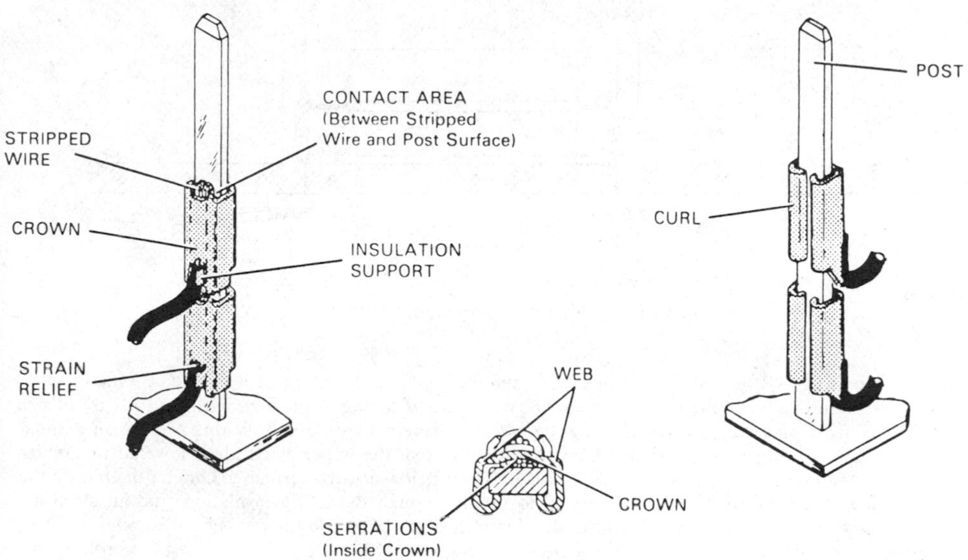

Fig. 17.6 Features of clip termination.

17.2-2 Clip Termination

Clip terminations have become acceptable in the electronics industry owing to their flexibility, reliability, serviceability, and density in point-to-point wiring. One of the main reasons for the popularity of this wiring technique is the capability for use of either solid or stranded wire. Wherever vibration is encountered, stranded wire is especially desirable. Where flexibility is necessary, for example, in routing and formation of wire runs to match chassis configurations, stranded wire is also preferred. Unusual wire types, including twisted-pair wires, can be clip terminated.

The termination consists of a stranded or solid wire connected to a rectangular post by means of a phosphor-bronzed clip and retained under sustained pressure (Fig. 17.6). When the clip and wire are applied to the post, mechanical energy is stored in the clip in the form of elastic stresses. Therefore the mechanical connection is a compressed spring which maintains a continual force between the wire and the post, even though environmental conditions may be extreme. This is why many companies use the technique, particularly in airborne equipment.

The clip and wire are applied simultaneously to the post by means of a tool that may be hand held or mounted on a numerically controlled machine (similar to the tool used for wire wrapping). In actual operation of the tool, the insulation is stripped from the wire, preformed, and applied to the post under the clip. As this happens, the wire and post surfaces are mechanically cleaned to obtain large contact surfaces. This technique combines the features of wire stripping and clip termination with the maintainability and serviceability of a separable connection. The wiring head then moves on a programmed path to the next selected terminal position, routing the wire as it travels. In the next position the wire is cut off, stripped, and clipped to the terminal. Clips can be positioned at any of three levels.

Wire size ranging from 22 to 28 AWG may be used with a 0.031×0.062-in. post and 28 to 32 AWG with a 0.022×0.036-in. post. Mil-W-16878 type E or ET wire is recommended. Clip termination may be used on a 0.100-in. grid and does not require that post centers be on a multiple of any grid. Random post location such as in posts used for input-output (I/O) or discrete components may be automatically wired.

17.2-3 Stitch Welding

As an alternate to printing of circuit boards and wire wrapping, stitch welding was originally developed for aerospace applications. The technique employs a cold-flowable plastic-coated nickel wire as an interconnection medium. Wiring is accomplished by feeding of insulated wire through a hollow electrode. This enables any number of points to be interconnected before the wire is cut (Fig. 17.7). The wire is pulse welded to the pin after the insulation is forced away from the joint by means of electrode pressure.

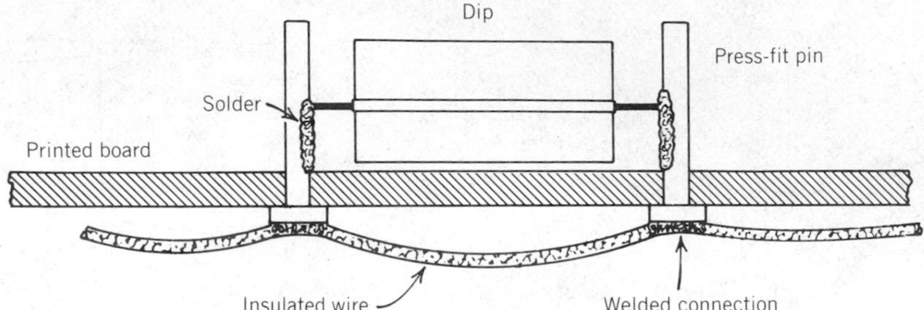

Fig. 17.7 Welded pin connections (stitch welding).

Nickel wire is used in conjunction with stainless steel pads or pins. The choice of these two materials allows a wide process margin so that welds are as strong as the wire itself. In the case of size 30 AWG wire, pull strengths on the order of 5 to 6 lb have been obtained. Wiring is done on manual or semiautomatic machines. When a connection is desired, the upper electrode is lowered pneumatically to the desired pin. A lower electode rises up from below a work surface to contact the pin on the other side of the board. Pressure is applied to break through the teflon insulation, and an electrical pulse causes a diffusion bond to occur at the junction of the wire and the pin. The electrodes are then positioned consecutively to all of the other common points in a given signal string. The continuously fed wire eliminates all need for stripping or cutting wires to prescribed lengths. All work is performed on one Z level, eliminating the need for programmed welding by levels. After the wiring step, components are attached to the other end of the pin in a variety of ways depending on the type of pin used.

The mechanics of making a wiring board are relatively simple. A universal printed wiring artwork is made which typically has only ground and power conductors on opposite sides of the board. The boards are manufactured and holes are drilled in a predetermined pattern. Pins are selectively press fit into these holes. The power and ground planes are etched away so that the pins are electrically isolated. At this stage, the boards would be ready for the signal, or stitch-weld, wiring.

Rewelding to the same pin can be done several times without any degradation of the weld. The direct point-to-point wiring possible with this process assures minimum wire lengths and very few parallel runs. Direct interconnects are possible on 0.050-in. centers.

17.2-4 Multiwire System

The multiwire system has established itself as a proven discrete wiring technique. The process is an automated interconnection system in which polyimide-insulated magnet wire is laid down on an adhesive-coated substrate with or without etched power and ground planes. The terminations are formed by drilling through the wire and board and plating the sides of the hole. The resulting plated-through holes are then used for component insertion and subsequent soldering operations. In

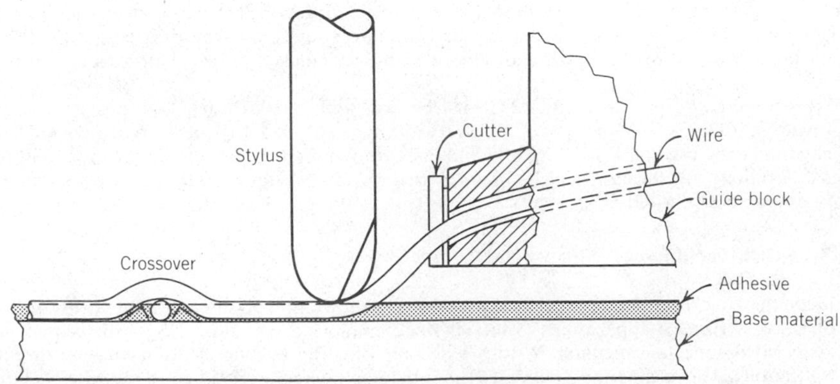

Fig. 17.8 Multiwire interconnection system.

essence, this technique combines conventional plated-through hole-printed wiring technology with point-to-point wiring (Fig. 17.8).

The wire used is size 34 AWG polyimide with an insulation thickness of 1.2 mil. This wire, which has a diameter of 6.3 mil, is equivalent electrically in current-carrying capability to a 2.4-mil width of 1-oz copper (1.4 mil thick). Systems may be wired on 100-mil grid patterns and wiring is performed at 250 in./s. A numerically controlled foot, which employs ultrasonic energy, embeds the wire in a thermoplastic adhesive on the surface of the board. After the wire is in place, a thin (0.4-mil) glass prepreg is laminated to the board surface to provide protection.

The use of insulated wires permits unlimited crossover on each wiring plane, thereby simplifying the design and providing high interconnection density. Because wires are separated by 2.4 mil of polyimide insulation, breakdown voltage at a crossover is in excess of 1200 V DC.

One of the advantages of the multiwire system is that controlled transmission line characteristics can be built into the board, holding the impedance to within ±10% of the design value. This specification is comparable to transmission line construction in multilayer boards and is superior to it in wire-wrapped boards, where the lack of constraint on the wire dress precludes a tightly controlled transmission line characteristic. With this process, the entire logic interconnection of the system is made by computer. The computer designs the routing of conductors and directs their physical placement by a numerically controlled wiring machine. This process provides ease of design changes and field repairability not normally inherent in other interconnection techniques.

17.2-5 Infobond System

The basic system consists of the substitution of interconnection solder lands for solderless wrap pins. This process can achieve high wiring densities on a printed board.

The Infobond system is a double-sided board with the wiring on one side and components soldered to the other. Circuit design and layout are controlled by punched paper tape. Polyamide-polyurethane-insulated size 38 AWG wire is fed through a thermal compression bonding tip and is automatically wired from bonding land to bonding land until a wiring net is complete. There the wire is machine cut and a new net started.

The wiring machine has six board wiring stations. All are controlled by a numerically controlled unit and punched paper tape. Up to six boards are pin registered on the machine table. The insulated copper wire is fed through the compression bonding tip of each of the six stationary heads. The heads are servo controlled and brought into contact with the prefluxed bonding lands on the boards and all six wires are soldered simultaneously, through the insulation. As the heads press the wires on the bonding lands, a controlled burst of heat is applied to thermally strip the wire insulation in the contact region and also solder the wires to the pretinned lands. Air blasts cool the bonds quickly and also prevent the insulation from stripping beyond the bond region.

Next the bonds are mechanically tested by applying a preset pulling force through the heads. Electrical continuity between bonds and wires is then checked. If it is OK, the table moves on to the next location. At the end of the wiring net, the wires are cut at each board. An automatic check is made to verify that the cuts were successful. Components are then added and each board again tested.

Advantages of the Infobond system lie in the areas of repairability, density, design change, and effective cost. Standard tools are available for engineering changes in the field. The elimination of wrapping terminals allows an increase in circuit density, and the use of a small-size wire allows placement of the wire closer to voltage and ground planes, which in turn affords significant noise reduction.

17.2-6 Solder Wrapping

In the solder-wrap system, a special computer-controlled wiring stylus strings a fine insulated wire to the solder tails or leads of sockets or pins in a regularly patterned printed circuit board having wiring guides between the rows of leads. The wires are soldered to the tails by a probe that thermally strips away the insulation at the soldering point. Solder is actually fed to the wrapped joint. After soldering is completed, the loop of connections is cut at the proper places.

Solder wrapping produces boards with an extremely low profile and high wiring and packaging densities. It is one of the first systems to make an in-progress check of wiring accuracy. It is employed for wiring circuit cards and backplanes for computers, minicomputers, and peripherals. Another advantage of this technique is that it has a manual version with special hand tools for doing prototype work. The hand-wiring tool designed to speed breadboarding consists of a stainless steel housing, a replaceable spool of wire, and an extended tip for guiding the wire. For breadboarding, the component leads are inserted into the plated-through holes of prototype boards and are wired with the hand tool by making a connection at each appropriate point. Wire tension is adjustable and replacement wire spools are easily changeable. The connections are then soldered.

Both a single-station and a four-station machine are available. They are controlled by either paper tape or floppy disk. Each station is capable of automatically performing the three operations of stringing, soldering, and cutting and can achieve 300 connections per hour. In operation, specifically patterned printed boards are loaded onto an indexing table and the stringing, or interconnection cycle, is performed. The connections are formed by a single, continuous wire being wrapped once around each protruding component lead, as per point-to-point wiring instructions; hence the wired connections are electrically common until the final cycle.

The machine can sense a missed wrap during stringing by monitoring the wire tension. The soldering probe serves a dual function by performing a wire-to-lead continuity check as it solders each lead. The machine also monitors the wire feed and checks for damaged or missing insulation.

Repair is accomplished in the following manner. A typical finished board of the process consists of an array of dual inline packages (DIPs), which are solder-wrapped connections. The outline procedure for DIP replacement is as follows: Clip the component leads flush with the board on the component side, form the new DIP leads for planar soldering, and solder the new component to the plated-through-hole land pattern on the component side such as with a flatpack.

17.3 CIRCUIT PACKAGES

17.3-1 Flat Pack

A flat pack consists of a subassembly composed of two or more stages made up of integrated circuits and thin-film components mounted on a ceramic substrate (Fig. 17.9). This semiconductor network is enclosed in a shallow rectangular package with the connecting leads projecting from edges of the package. Therefore leads may be spot welded to terminals on a substrate or soldered to a printed circuit board. Flat packs have thin rectangular shapes and usually are less than 0.1 in. thick. Leads may be brought out on all sides for greater freedom of system design and more compact system packaging. Flat packs typically have metal top and bottom plates with the chip mounted between. Leads pass through glass-to-metal or ceramic-to-metal seals.

Flat packs are harder to handle than DIPs because the delicate leads are easily bent accidentally. Flat packs thus are usually hand mounted. Most standard logic circuits available in DIPs are also available in flat packs. More complex circuits in flat packs cost more than those in DIPs. The main advantage of flat packs is low profile.

17.3-2 Dual-Inline Packages

DIPs are by far the most widely used integrated-circuit (IC) package configuration. DIPs can have from 4 to 64 pins arranged in two parallel rows (Fig. 17.10). The device chip can be encapsulated in a

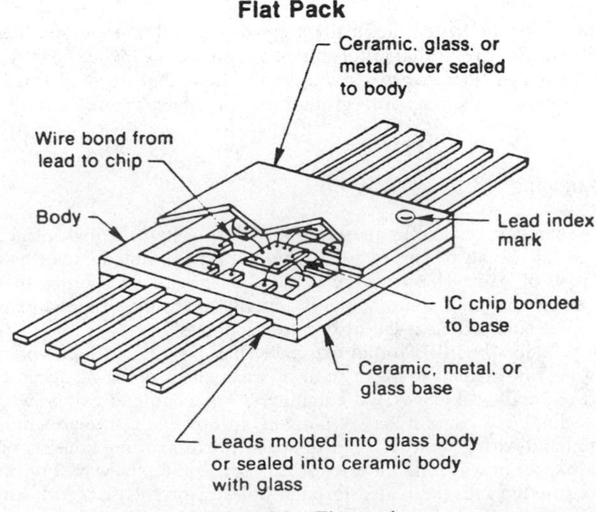

Fig. 17.9 Flat pack.

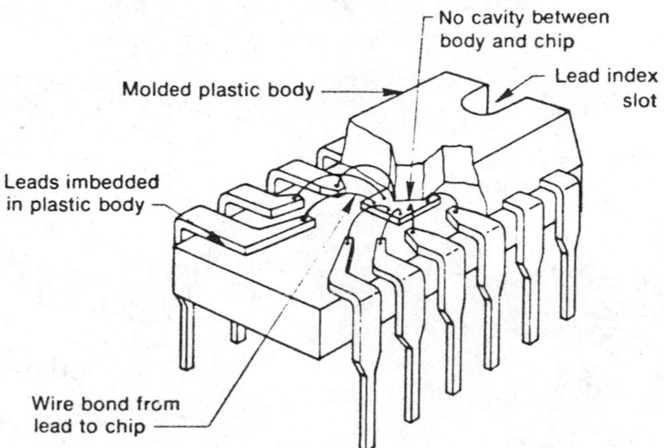

Fig. 17.10 Dual inline package (DIP).

plastic body or hermetically sealed in a ceramic body. Plastic bodies are sufficient for most commercial and industrial applications. More costly ceramic housings provide greater reliability.

DIPs can be either soldered to a printed circuit board or plugged into soldered sockets. Although body dimensions vary among manufacturers, standard DIP leads are spaced on 0.100-in. centers. Plastic DIPs are for use in temperatures of 0 to 70°C; ceramic DIPs can be used in temperatures ranging from −55 to +125°C. DIPs are easy to handle and to insert in printed circuit boards. Also their wide spacing and exposed rugged leads make them easy to troubleshoot with minimum danger of shorting or lead damage. Virtually all types of digital circuits are available in DIPs, from simple gate arrays to complete microprocessors.

17.3-3 Hermetic Chip Carriers

As electronic circuits increase in density, speed, and complexity and larger chips and packages are required, interconnection methods that use printed circuit real estate most efficiently are sought. One

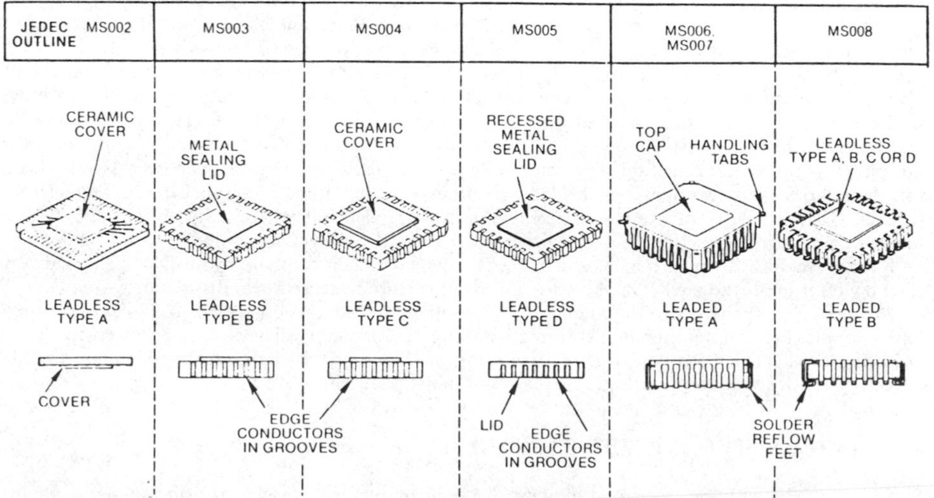

Fig. 17.11 The 50-mil-center terminal spacing JEDEC (Joint Electron Devices Engineering Council) outlines for chip carriers.

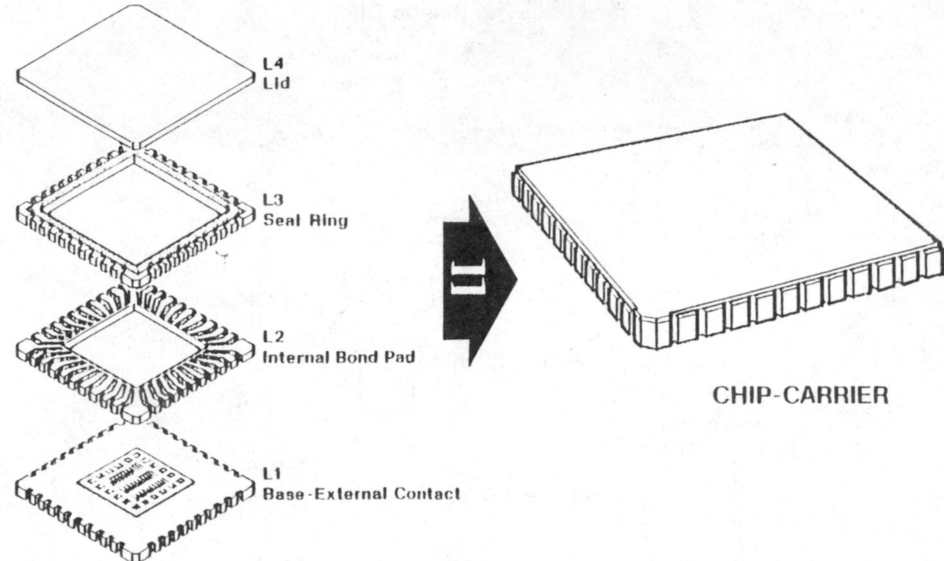

Fig. 17.12 Typical leadless hermetic chip carrier [JEDEC (Joint Electron Devices Engineering Council) type C 40-mil center]. L = layer.

solution is the use of chip carriers, including the leadless chip carrier (LCC), that can be directly mounted to printed circuit boards without sockets. However, to obtain this more efficient use of printed wiring board real estate, more heat must be removed than previously was the case. In addition, a metal material with a low thermal coefficient of expansion is required to maintain solder joint reliability with ceramic chip carriers. At present, multilayer ceramic material is commonly used as a substrate. It has the advantage of matching the thermal expansion of ceramic chip carriers but is, at present, available in limited sizes. New and future electronic systems that use densely packed chip carriers will need boards of 50 in.^{2} or more. At present, ceramic substrates cannot be mass produced that large.

A look at the basic concept of ceramic chip carriers is in order. The original chip carriers are square multilayer ceramic packages with a pattern of gold metallization pads on the bottom and an internal cavity into which the semiconductor or integrated circuit can be bonded. On the top edge of the cavity is a metal sealing surface; a conventional Kovar seal lid is bonded to the top.

The packages are fabricated in sheet or wafer form with the connections from internal planes to the external pads made by metallized via holes. These prepunched holes also provide a row of perforations that can be used as separation lines when the ceramic sheet is separated into individual pieces. One method of attaching the chip carriers to a larger interconnection circuit is by reflow soldering to a set of pre-solder-coated lands on a ceramic mother board. The packages can also be assembled to the ceramic substrate by passage through a hot air oven or by use of hot air guns or a hot plate. A more recent method is by vapor-phase soldering. Chip carriers have been standardized into four types (and two subtypes, leaded or leadless) by the Joint Electron Devices Engineering Council (JEDEC) with lead terminal spacings of 0.040 and 0.050 in. (See Fig. 17.11.) Efforts are under way to decrease the terminal spacing to 0.020 and 0.025 in. which would increase I/O density. Still, the present-size chip carriers have a 5 : 1 size reduction when compared with DIPs. Chip carriers not only offer improved packaging but also improved performance and heat dissipation.

Leadless chip carriers have handling advantages over leaded carriers owing to the fact that they have no pins that can bend or break during handling and assembly. They are also less costly. They are being considered for many large-scale integration (LSI) and very-large-scale integration (VLSI) circuits. They are available in plastic, ceramic, and metal. (See Fig. 17.12.)

17.4 INTEGRATED CIRCUIT PACKAGING

In the case of monolithic integrated circuits, the semiconductor devices [transistor, diode, medium-scale integration (MSI), or LSI chips] are usually assembled and sealed within special packages referred to as the TO-5 type, the flat pack, or the DIP. The processes used in the assembly operations within these packages generally include gold-silicon eutectic die bonding and thermocompression or ultrasonic wire bonding to assure the required mechanical and electrical connections between the semiconductor chips and appropriate internal bonding pads. After assembly these packages often are hermetically sealed for maximum protection of the semiconductor chips.

The use of unpackaged active device chips (beam lead, flip chip, chip and wire) assembled onto a thick-film substrate can provide an increase in packaging density and module complexity at a lower cost by allowing sealing at the module level instead of individual sealing of each component. The most efficient thick-film integrated-circuit packages in terms of heat transfer and weight utilize the ceramic substrate as the package base.

LSI is a constant objective of semiconductor manufacturers. A practical system for achieving some of the advantages of LSI is the interconnection of monolithic circuits on specially designed MSI chips. Hybrid circuit technology can be used to fabricate the required multilayer interconnection patterns and packages at a minimum tooling cost.

VLSI devices are now becoming a reality, but packages to house these high-performance chips are either still in the research-and-development phase or the prototype fabrication state. Packages with pin counts in the 100 to 300 range are projected with power on the order of 15 W and rise times approaching 0.5 ns. Larger chip sizes (1 cm^2) combined with a larger number of terminals counteract the package size reduction. VLSI packaging and assembly design involve many interrelated factors. These include package fabrication, environment, electrical performance, materials, and compatible assembly.

High-technology packages must effectively transfer chip performance to the system. Some of the environmental factors include cooling mechanisms, ambient temperature levels, second-level interconnect spacings, and package-mounting techniques. Packages can either be inserted or surface mounted to boards. Surface attachment allows mounting of components on both sides of the board. Advanced boards include advantages such as finer lines and spacings with reduction in the number of layers. Combining the advantages of fine lines and spaces and smaller via holes is necessary to take full advantage of the high-density chip carrier.

Today the 100-mil pin array package is adaptable to standard printed circuit boards that include a connector option. However, the future will bring still higher density possibilities with surface-mounted packages, such as chip carriers, or direct tape-bonded devices. The closer the device packaging becomes, the shorter the signal lengths required for higher system speed.

Standardization will lead to lower or the same cost. Volume packaging will both drive cost down and provide a stabilized base for improved reliability.

Chip, package, and board must all be considered in a systems approach to packaging design of integrated circuits. Such factors as temperature, environment, chip designs, and electrical characteristics play an important part in the package design for achievement of increased functional density and high performance.

17.5 ELECTROMECHANICAL DEVICES

17.5-1 Switches

Switches are devices that make or break connections in an electrical or electronic circuit. In computing systems they are also used to make selections; the toggle switch, for example, completes a conditional jump. Switches are usually manually operated but can also work by mechanical, thermal, electromechanical, barometric, hydraulic, or gravitational means.

The varieties, configurations, and electrical and mechanical possibilities of switches are almost infinite. They include open or skeleton type, enclosed general purpose, feather touch, modular, magnetic reed, cam follower, miniaturized, reset, one-way pulse, toggle, subminiature, sealed, rotary, and printed circuit board switches.

Switches are frequently custom made to meet the needs of a particular application. Each application usually requires some modification in switch movement, terminal type, size, electrical/mechanical life, or other characteristic. Switch movements and forces must be coordinated with the movements and forces of a mating mechanical member. Since normal manufacturing tolerances cause unit-to-unit variations in the mating member, the switch must somehow accommodate these variations. In addition, switches are frequently the last components designed in a unit; consequently, they must fit in the limited space remaining.

The switch manufacturer must know what movements and forces are available to operate a switch, what space is available for the switch, and what actual electrical load is involved. Several other considerations allow the switch manufacturer to propose the best quality-price combination from his line. These include (1) the switch pole and throw arrangement, for example, single-pole single-throw (SPST), double-pole single-throw, (2) the electrical and mechanical life required (they are not always the same), (3) the general application description and environmental conditions, for example, moisture, atmospheric contamination, (4) the approvals required, for example, Underwriters Laboratories (UL), military, and (5) the required switch quantity, very important for analysis of the tooling required.

Selection criteria for buying or using switches include:

Electrical Load. Low-voltage, low-current applications where no arc is likely to occur are best met with gold contacts. For driving of inductive loads, diode suppression across the load will prevent energy discharge across the switch contacts. High-voltage high-energy loads generally require silver-cadmium oxide contacts.

Contact Bounce. In many applications, contact bounce in a switch creates no problem. When interfacing to logic circuits, however, a debounce circuit should not be overloaded.

Operating Force. Operating forces for the switch should be specified low enough so that it can be operated easily, yet high enough so that it can withstand environmental vibration.

Mounting Means. Several industry standards have evolved for mounting of switches. Some switches are available with "ears" capable of locking into panes or doors, which is a handy feature for safety interlock switches.

Operating Means. Is the switch operated? Manually? By a lever mechanism system? A spring system? A rotating cam? Another mechanism?

Levers and Actuators. Various lever types with or without rollers are available. Buttons should be of a material that can sustain many operations with minimal wear; acetol and nylon are suitable.

Pretravel and Overtravel. These refer to the distance that the actuator moves before operation and the remaining distance that the actuator may continue to move. A third term, "differential," describes the hysteresis, to prevent teasing of the switch. Specifications for these should be reviewed to determine the switch's suitability for the specific application.

Certification. To supply major markets of today's world, a manufacturer must have the approval of the UL, the Canadian Standards Association (CSA), the British Standards Institute (BSI), and the Deutscher Elecktrotechniker Verband (VDE). The CEE (International Commission for the Approval of Electrical Equipment) certification scheme is also available.

Terminals. Standard terminals are screw, solder, quick connect, and printed circuit board. Wire leads are available primarily for sealed switches.

Environment. Switches used in areas where they will be unprotected from weather, silicones, extreme dust, grease, or spray may need to be sealed.

Advances in technology continue to give birth to new technologies and products. One such new product is the membrane switch. Most designs using front panel switching can utilize membrane switch technology. Many uses can be found in business machine industries, point of sale (POS) terminals, electronic toys, and appliances.

The basic switch consists of interconnections and contacts deposited on two layers of material separated by a spacer layer with openings in it to create a contact gap (Fig. 17.13). The spacer layer seals the contact system via adhesives on both sides. The switch assembly can be mounted to a panel with mounting holes, studs, or an adhesive mounting sheet, depending on the type of switch and the design application.

A decorative graphic overlay is specified and secured to the top of the switch with adhesive. Touching a defined switch position depresses the membrane, causing the conductive material on its lower surface to touch the contact points below and act as a shorting bar to close the switch (Fig. 17.14). When the switch is released, the resilient flexible membrane returns to its original position and the contact is broken.

Membranes by nature are resistant to contamination, humidity, and corrosive environments. The installed cost of membrane switches is often less than that of conventional mechanical or electronic switches. Membrane switches are available only as normally open, momentary-contact devices. Although versatile, they cannot provide latching (push on, pull off), alternate-action (push on, push off), or key-interlock mechanisms.

Manufacturers offer a large selection of standard colors, sizes, and styles with legends and symbols. Any custom artwork may be silk screened onto the panel subsurface. Membrane switches are available in either flexible or rigid construction. Flexible panels use a thin polymeric substrate and need to be mounted on a flat rigid surface (Fig. 17.13*a*). Rigid construction normally involves a printed circuit board substrate on which the contacts and leads have been etched (Fig. 17.13*b*). The latter affords high switch and interconnect densities. In addition, components may be mounted to the printed circuit board to make an entire component package. Many types of terminations are available including solder pins, edgecards, solder tabs, and wire-wrap terminations.

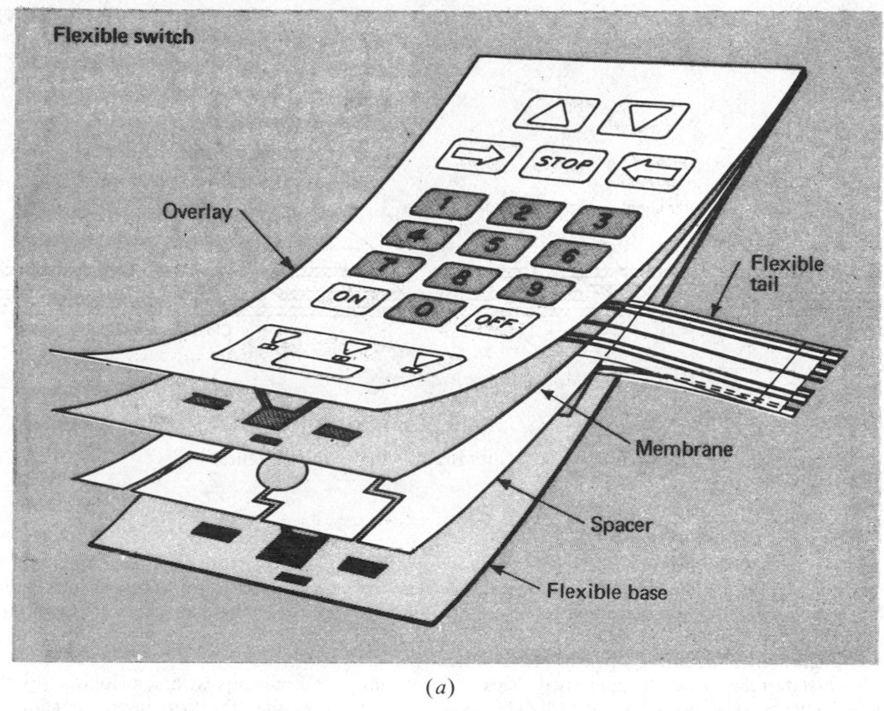

(a)

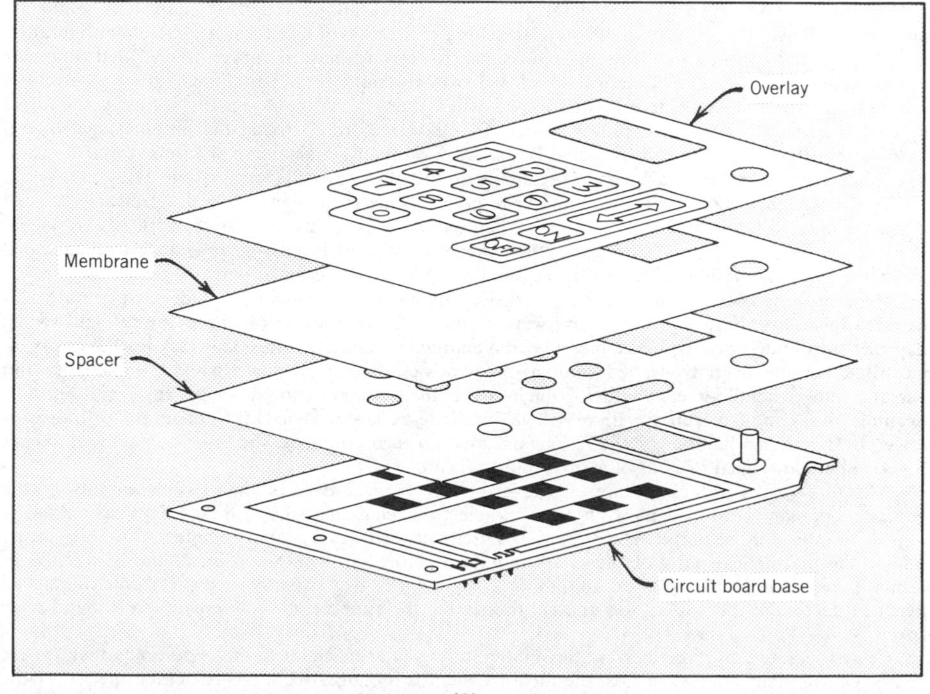

(b)

Fig. 17.13 Membrane switches. (a) Flexible switch has a thin polymeric substrate. (b) Rigid switch has a printed circuit board base.

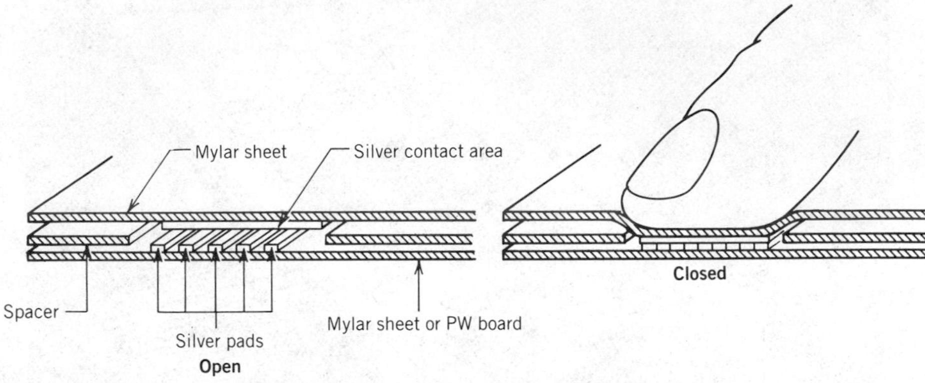

Fig. 17.14 Basic membrane touch switch configuration.

17.5-2 Relays

A relay is an electrically controlled device that opens and closes electrical contacts to effect the operation of other devices in the same or another electrical circuit. Control of an electromechanical relay is most commonly achieved with the application of a specified voltage or current to two input terminals. A coil within the relay translates control-signal electrical energy to mechanical energy, causing mechanical action to open or close contacts.

Solid-state relays operate external devices by controlling load currents with semiconductors that act as switches. The use of semiconductors as switches requires that the load characteristics and power being switched be defined in much more detail than for electromechanical relays. The controlled power must be specified as AC or DC; in the case of DC, the polarity or current direction through the switch must be specified.

The most commonly encountered relay contact configurations and their equivalent solid-state circuits are shown in Fig. 17.15. Solid-state switching of a DC load in a form A contact configuration is normally accomplished by using the input signal to forward bias an appropriately rated transistor into a saturated condition. For normally closed-switch action (form B), a small driving current is required to maintain the output switch closed when there is no input signal. Normally this drive current is taken from the load power source. It is diverted from driving the output stage to stop load-current flow when the control signal is applied. Combining the circuits of forms A and B results in load-current control equivalent to a single-pole double-throw switch (forms C and D).

Solid-state relays have faster switching in response to the input signal than do electromechanical relays. "Contact" life is practically limitless when necessary precautions against overloads are taken. With no moving parts, most solid-state relays are encapsulated to provide resistance to high shock and vibration. High reliability is thus achieved and maintenance costs are greatly reduced.

Mercury-wetted contact relays are fast, sensitive high-capacity switching devices that operate with extreme reliability. Basically, a mercury-wetted contact relay consists of one or more glass switch capsules surrounded by a coil. The mercury film eliminates contact bounce that may lead to errors in electronic circuits. Mercury-wetted contact relays possess both high power-gain characteristics and versatile power-handling capabilities, which make them suitable for a wide range of uses. For example, they can be driven at a 25-mW level by diode-transistor logic (DTL) or transistor-transistor logic (TTL) and can handle a 250-VA solenoid load on their contacts. The same contacts can pass a microvolt analog signal or control a small motor.

Armature relays are essentially simple electromechanical devices that switch circuits at the command of small control signals. Physical and electrical characteristics of these relays can be varied almost infinitely in response to different requirements. However, interdependence is very strong among the mechanical parts of the relay and the electrical system. Selection of the best relays in terms of required performance at minimum cost involves two primary steps: (1) determining all pertinent factors of the application and (2) translating the application requirements to a sound relay specification.

Relays operated on direct current have inherently greater mechanical life expectancy than do arc relays. Of the many sources of DC, the most frequent is probably rectified AC. Often the AC ripple influences relay operation. Some DC relays can tolerate ripple; others need filtering.

Most manufacturers build standard relays with a particular mounting but are able to supply a range of other mountings. Many open and most enclosed relays can be mounted on plugs for use

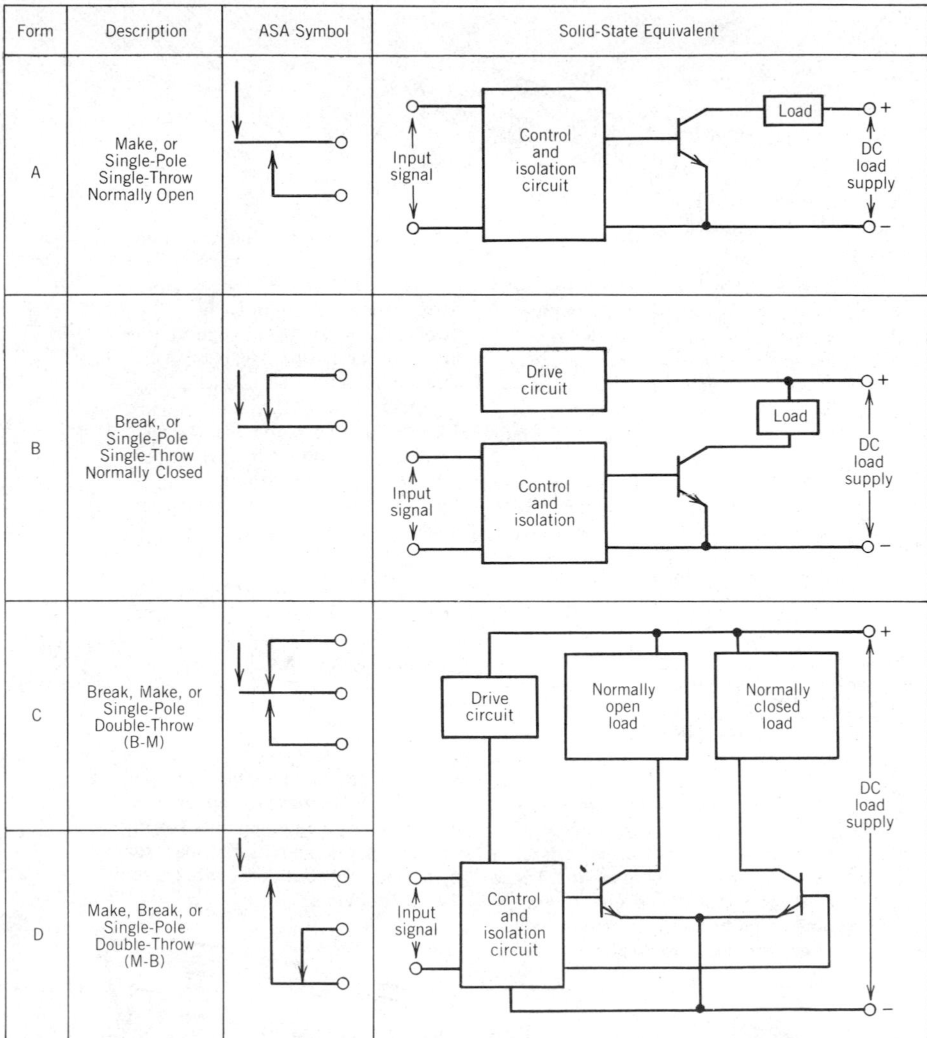

Fig. 17.15 Equivalent solid-state switching circuits for commonly used relay-contact forms.

with sockets. The most common terminals supplied on relays are screw, solder, plug in, printed circuit, and quick connect. Power relays are usually supplied with screw terminals; however, quick-connect terminals are also common. Most relays other than power type are supplied with solder-type connections, but can usually be supplied with other types.

Enclosures protect the relays from severe dust and atmospheric conditions as well as from mechanical injury. Relay suppliers provide finishes for the component parts of the relay that protect the parts from most ordinarily encountered environmental conditions.

Most open-type relays have small areas that are large enough for the manufacturer to print its company name and the part number, but have little room for anything else. Enclosed relays usually provide enough flat surfaces so that more information as well as wiring schematics can be supplied. Silk screening can be economically applied on small areas.

17.5-3 Connectors

The purpose of a connector is to make a secure yet easily removable connection between mating conductors. It is important that these connections offer low resistance and be electrically "invisible."

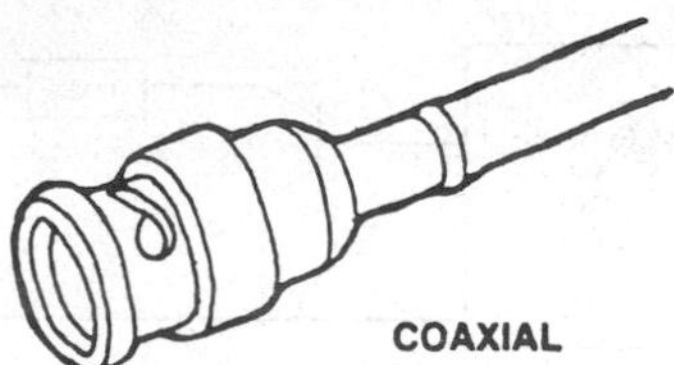

COAXIAL

Coaxial cable is a transmission line consisting of either a solid or stranded center conductor surrounded by a dielectric covered with a conductive braid and a layer of insulation. Coaxial cable connectors terminate both center conductor and the braid. They are used mainly in radio-frequency circuits and in computer transmission lines. Coaxial connectors are specified for either nonconstant impedance or are impedance-matched to the cables they connect. The type of cable used often determines the type of connector selected.

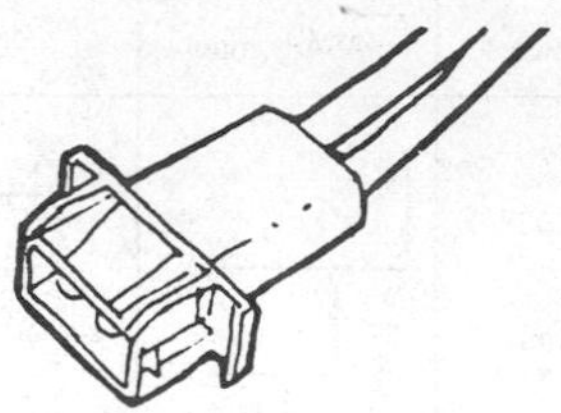

A **soft-shell** connector has no metal (or hard) shell. The term is normally applied to low-cost connectors used for consumer or commercial applications. Usually molded from an inexpensive thermoplastic like nylon, these connectors are relatively large for the number of contacts they hold. This design allows the use of large contacts for high current capacities and permits loose manufacturing tolerances which reduce production costs.

HIGH VOLTAGE

High-voltage connectors are used primarily to protect personnel and equipment from contact with or arcing from high-voltage terminations. These connectors usually have a protective outer shell or boot, and are typically rated from 6,000 to 60,000 V.

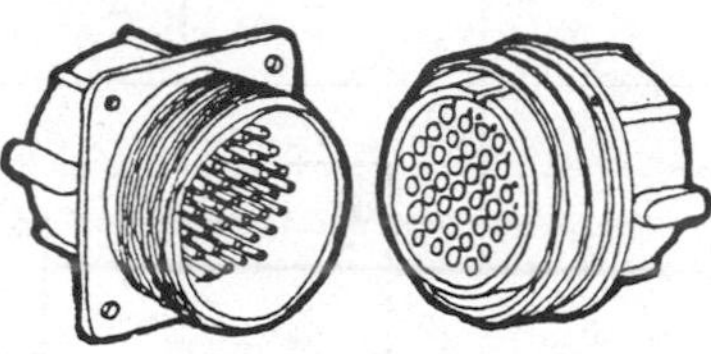

CIRCULAR

A **circular** connector consists of a plug and receptacle that are connected to each other by a coupling ring. The ring can be a screw-on type, bayonet-type, or a push-pull configuration with from 2 to 85 contacts.

FLEXIBLE CABLE

Flexible cable connectors are available for both flat-conductor and round-conductor flat cable. In addition to connectors that terminate the wires by soldering, welding, or crimping, sandwich-like, two-piece connectors are available that pierce the insulation to make connections. This type connector allows rapid connection of round wire or flat cable, flat to flat, and flat to round conductor cable.

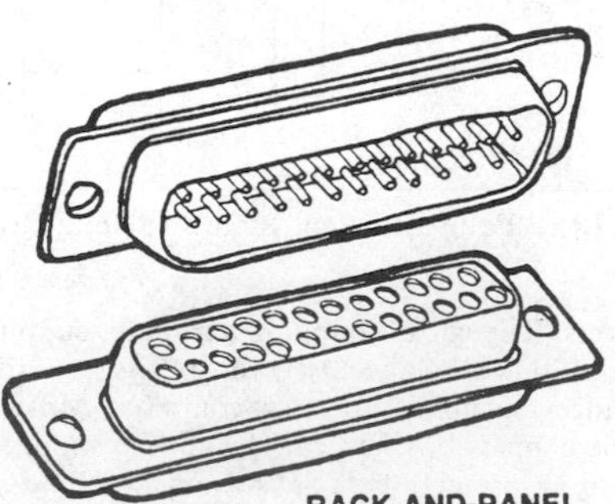

RACK-AND-PANEL

Rack-and-panel connectors are usually square or rectangular in shape with a large number of contacts for connecting racks or drawers of electrical or electronic equipment to a back panel. Several locking mechanisms are available for these blind-mating connectors.

Fig. 17.16 Common connectors.

The basic parts of any connector, regardless of its classification, are mating plugs and receptacles, each containing pins or sockets in some insulating medium. Both plugs and receptacles may have protective metal or plastic shells.

The largest categories of connectors are circular (or cylindrical), rack-and-panel (rectangular), printed circuit (both edge and two piece) and coaxial (primarily radio frequency) (Fig. 17.16). The classical connectors are the metal-shelled cylindricals and rack-and-panel units, both of which have been available for more than 40 years. Now offered in miniature form, primarily for military applications, they are high-contact density precision products.

Radio-frequency coaxial connectors have also been available for more than 40 years. Dimensioning and materials selection are set by the strict requirements set up for RF transmission. With the conversion of circuitry from metal chassis to printed circuit cards, the demand for printed circuit connectors was created. With the acceptance of the printed circuit board came the need for many different kinds of adapters, sockets, and backplanes, giving rise to a whole new interconnection subsection in the connector field.

The demand for high-density, more flexible wiring and faster, more efficient means of terminating low-voltage and signal-level cables has made ribbon cable, more commonly called flat flexible cable, increasingly attractive. Various rack-and-panel PC edge-type connectors, adapters, and sockets have been adapted to mass-termination insulation displacement construction.

Coaxial connectors are most frequently used in the transmission of audio, video, and radio-frequency transmission up to the microwave frequencies. Available in standard or miniature sizes, these connectors are discrete products. They are designed to accommodate the central or coaxial wire and metallic shielding. The shielding minimizes radio-frequency interference, either from the cable (transmission) or into the cable (reception). Coaxial connectors are also made as inserts for multiwire cylindrical connectors.

Circular connectors are most often used to make external connections between the various units or components of an electronic system, whether it be radio, radar, sonar, weapons, data transmission, or an industrial machine. The external cylindrical shell offers the best protection from shock, vibration, and environmental influence and is easily coupled or uncoupled by various methods.

Rack-and-panel connectors were initially designed for use within protective cabinetry for the interconnection of various chassis and modules in a complete system. Military standard rack-and-panel units may have the same pins and sockets as comparable quality cylindrical units. The rectangular shells permit the highest possible wiring density per unit of space available, but pose certain fastening problems. Plug and receptacle are mechanically coupled with internal screws or external clamps. There are numerous plastic-shelled connectors with high-density contacts that are more properly known as rectangular connectors.

Contacts must permit high electrical conductivity and yet be easily engaged or disengaged. This is accomplished with spring retention action in high-quality high-density connectors. Pins are machined from brass or nickel-silver and contacts are made of nonferrous metals with spring properties, such as beryllium-copper, phosphor-bronze, or nickel-silver. The socket is designed so that a flexible contact surface is able to grip the pin with sufficient force for high conductivity following many engagements and disengagements.

Gold plating is the preferred surface of mating contacts because it permits a sliding action and resists corrosion, oxidation, and other forms of contamination. A plating thickness of 50 μin. is still a requirement for military specification connectors.

The pins and sockets are assembled in shells by inserting them in spaces, which set the center-to-center dimensions. Some degree of flexibility is desirable to prevent misalignment or damage to pins and sockets. The voltage levels and frequencies of the signals being transmitted determine minimum contact spacing.

Most high-grade connectors have removable contacts that are attached to the wires of the cable by crimping. Once crimped, the wires with their contacts are inserted into the bodies of the connector with special tools. Poke-home contacts are inserted with a hand tool that compresses the springs. The springs lock the contacts into position within the shell. Because the pins and sockets can be removed easily, inspection is facilitated and field changes or repairs are readily performed.

There are hermetically sealed connectors that call for the individual attachment of wires to contacts by welding or soldering. The contacts are not flexible and special care must be taken in handling these products with glass-to-metal seals.

Printed circuit board (PCB) connectors are designed to permit multiple contacts between a PCB or card and a cable. The simplest of these is the card-edge connector, a slotted plastic block fitted with rows of electrically isolated, metal spring contacts. When the edge of the PC card is inserted, contact is made between the conducting fingers on the card and the wires or cable attached to the tail ends of the contacts (Fig. 17.17).

PCB connectors are available with as few as 6 and as many as 50 or more contacts. Connectors for double-sided boards are said to have dual readout. Most edge connectors are designed to accept 1/16 in. (0.062 in.)-thick PC boards or cards.

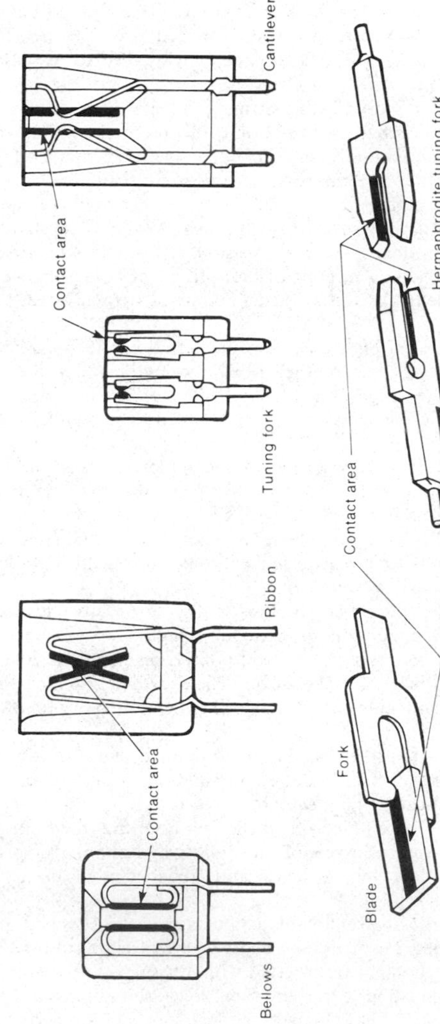

Fig. 17.17 Common circuit board contact forms.

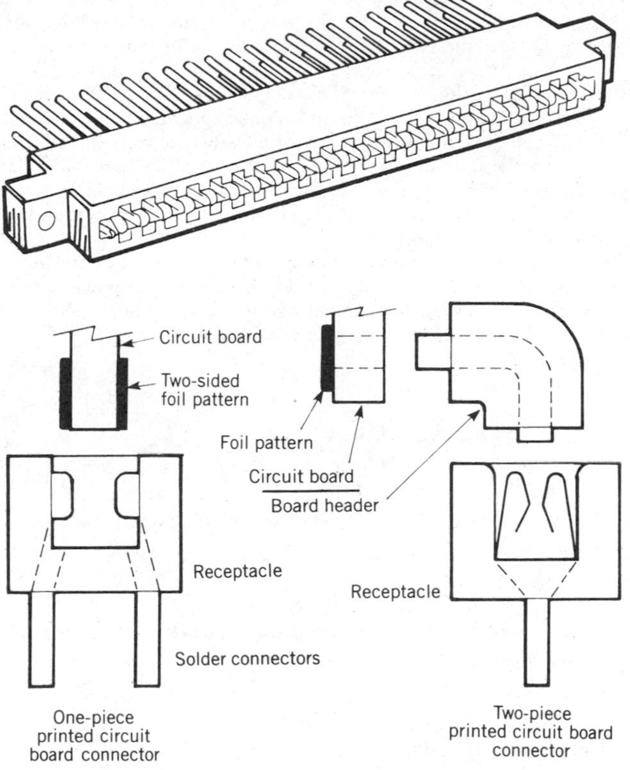
Circuit board
Two-sided
foil pattern
Foil pattern
Circuit board
Board header
Receptacle
Receptacle
Solder connectors
One-piece
printed circuit
board connector
Two-piece
printed circuit board
connector

Fig. 17.18 Circuit board connector contacts.

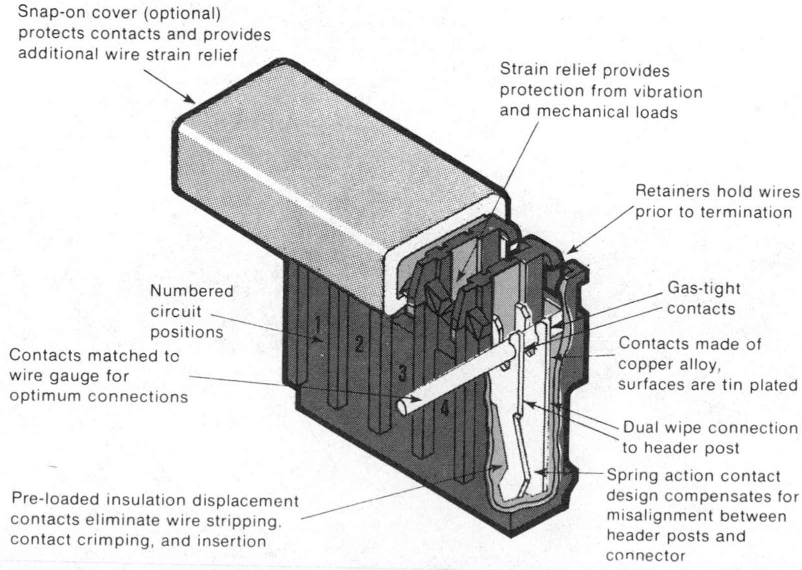
Snap-on cover (optional)
protects contacts and provides
additional wire strain relief
Strain relief provides
protection from vibration
and mechanical loads
Retainers hold wires
prior to termination
Numbered
circuit
positions
Gas-tight
contacts
Contacts matched to
wire gauge for
optimum connections
Contacts made of
copper alloy,
surfaces are tin plated
Dual wipe connection
to header post
Spring action contact
design compensates for
misalignment between
header posts and
connector
Pre-loaded insulation displacement
contacts eliminate wire stripping,
contact crimping, and insertion
1
2
3
4

Fig. 17.19 Typical features of mass-termination insulation displacement connectors.

Two-piece PCB connectors are used where there is some concern that the contacts may be separated by vibration or shock (Fig. 17.18). Various materials are used for the bodies of PCB connectors. These include the thermosetting resins diallyl phthalate, melamine, and phenolics, thermoplastics, and nylon and polycarbonate.

Mass-termination insulation displacement (IDC) connectors (Fig. 17.19) include PCB connectors, transitions, headers, sockets, adapters, input/output connectors, and integrated-circuit DIP sockets and plugs. These products all have in common a set of movable contacts embedded in the connector, permitting them to be rapidly and simply attached to standardized flat-ribbon cable with a single compression action. The rows of contacts, which may or may not have gold plating, can be caused to move in union by means of external pressure through the insulating layer around the wires, wedging them securely on the wires.

The most generally overlooked contribution to connector reliability or unreliability is the handling of the connectors during installation and maintenance. Mishandling; physical damage, particularly to contact pins; dirt; improper or careless attachment of wire or cable; improper mounting; and so on, can result in reliability problems with an otherwise reliable connector.

Bibliography

Assembly Engineering, Hitchcock, Wheaton, IL.

Circuits Manufacturing, Benwill, Boston, MA.

Electronic Packaging & Production, Cahners, Denver, CO.

Harper, C. A., *Handbook of Electronic Packaging*, McGraw-Hill, New York, 1969.

Harper, C. A., *Handbook of Wiring, Cabling and Interconnecting for Electronics*, McGraw-Hill, New York, 1972.

Industry Week, Penton, Cleveland, OH.

Institute for Interconnecting and Packaging Electronic Circuits (ICP), Evanston, IL.

Insulation/Circuits, Lake, Libertyville, IL.

Machine Design, Penton, Cleveland, OH.

Microelectronic Manufacturing & Testing, Lake, Libertyville, IL.

Production, Production, Bloomfield Hills, MI.

Solid State Technology, Cowan, Port Washington, NY.

CHAPTER 18
MICROWAVE DEVICES

HING-LOI A. HUNG

COMSAT Laboratories, Clarksburg, Maryland
George Washington University, Washington, DC

Microwave techniques have been used increasingly in such diverse applications as terrestrial communications, radio astronomy, satellite communications, and radar systems as well as in industrial applications and consumer products such as microwave ovens. Components used in these applications operate with wavelengths in the range of centimeters, 30 to around 0.1 cm (frequencies 1–300 GHz), and their circuit dimensions are generally of the order of magnitude of or less than these wavelengths. The microwave portion of the frequency spectrum is shown in Fig. 18.1. As the growth of microwave technology is accelerating, transmission lines of distributed elements in waveguides, coaxial cables, and striplines are being partly replaced by components with even smaller dimensions. These components include lumped-element circuits, microstripline circuits, and active devices, using the more recent silicon and gallium-arsenide thin- and thick-film semiconductor technologies in the microwave integrated-circuit (MIC) and the monolithic microwave integrated-circuit (MMIC) approaches. Present state-of-the-art developments of active components have resulted in microwave amplifier and oscillator tubes and their solid-state counterparts for frequencies over 100 GHz.

18.1 PASSIVE COMPONENTS

Microwave components that guide electromagnetic waves through a dielectric medium or perform various functions, such as combining, dividing, or phase shifting EM waves, are listed under transmission lines, reciprocal devices, and nonreciprocal devices. They are, in general, components in which dissipation is caused by factors including conduction loss, dielectric loss, radiation, and material power absorption.

18.1-1 Transmission Lines

Transmission lines used for propagating microwave energy either in the "conventional" form of two conductors or as metallic waveguides can be analyzed either by solving Maxwell's equations with appropriate boundary conditions or by using the distributed-circuit theory, in terms of voltage, current, and impedance only. Table 18.1 lists some of the useful results relating series resistance R, shunt conductance G, series inductance L, and shunt capacitance C with the characteristic impedance of the transmission line Z_0 and of the load Z_L. These distributed parameters are expressed per unit length and the wave is assumed to propagate in the positive z direction. Wave transmission and reflection are also related to parameters such as voltage standing-wave ratio and transmission line characteristic impedance.

The *Smith chart*[1] shown in Fig. 18.2 graphically relates impedance to reflection coefficient or to standing-wave ratio (given in the horizontal scale) and position of a voltage minimum. It consists of

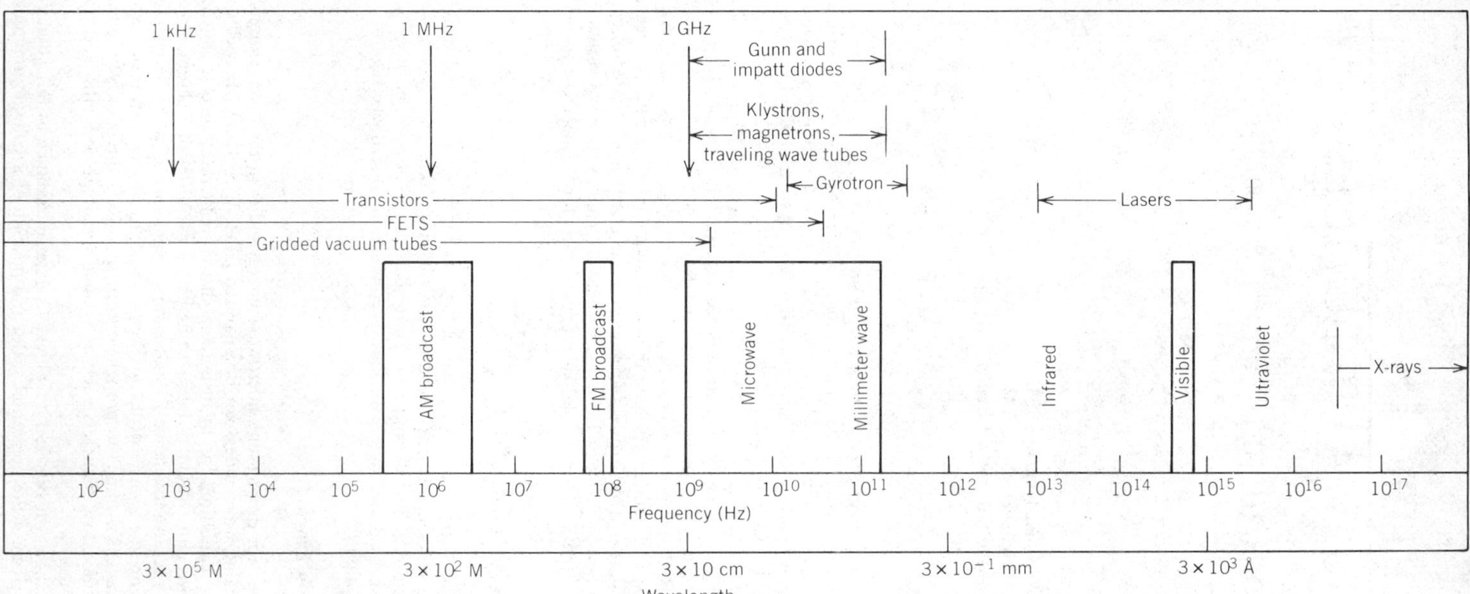

Fig. 18.1 Microwave region of the electromagnetic spectrum.

loci of constant resistance and reactance plotted on a polar diagram in which the radius corresponds to the magnitude of the reflection coefficient and the angle corresponds to the phase of the reflection coefficient, referred to a general point along the line. By combinations of operations on the chart, the user can devise complicated impedance-matching techniques. Currently, complex computer analysis and synthesis programs[2,3] allow further analysis and optimization of the design of the matching circuits[27] to specific requirements.

Hollow conducting *rectangular waveguides* that can support transverse electric (TE) and transverse magnetic (TM) waves are most commonly used for low-loss and high-power applications. They are dimensioned (nominally with the height half or slightly less than half the width) to allow the lowest order (dominant) mode, such as TE_{10}, to propagate with the least amount of attenuation. In this mode only the electric field component whose half-sinusoidal field pattern spans the guide with maximum magnitude at the center is transverse to the direction of energy flow. Table 18.2 gives the

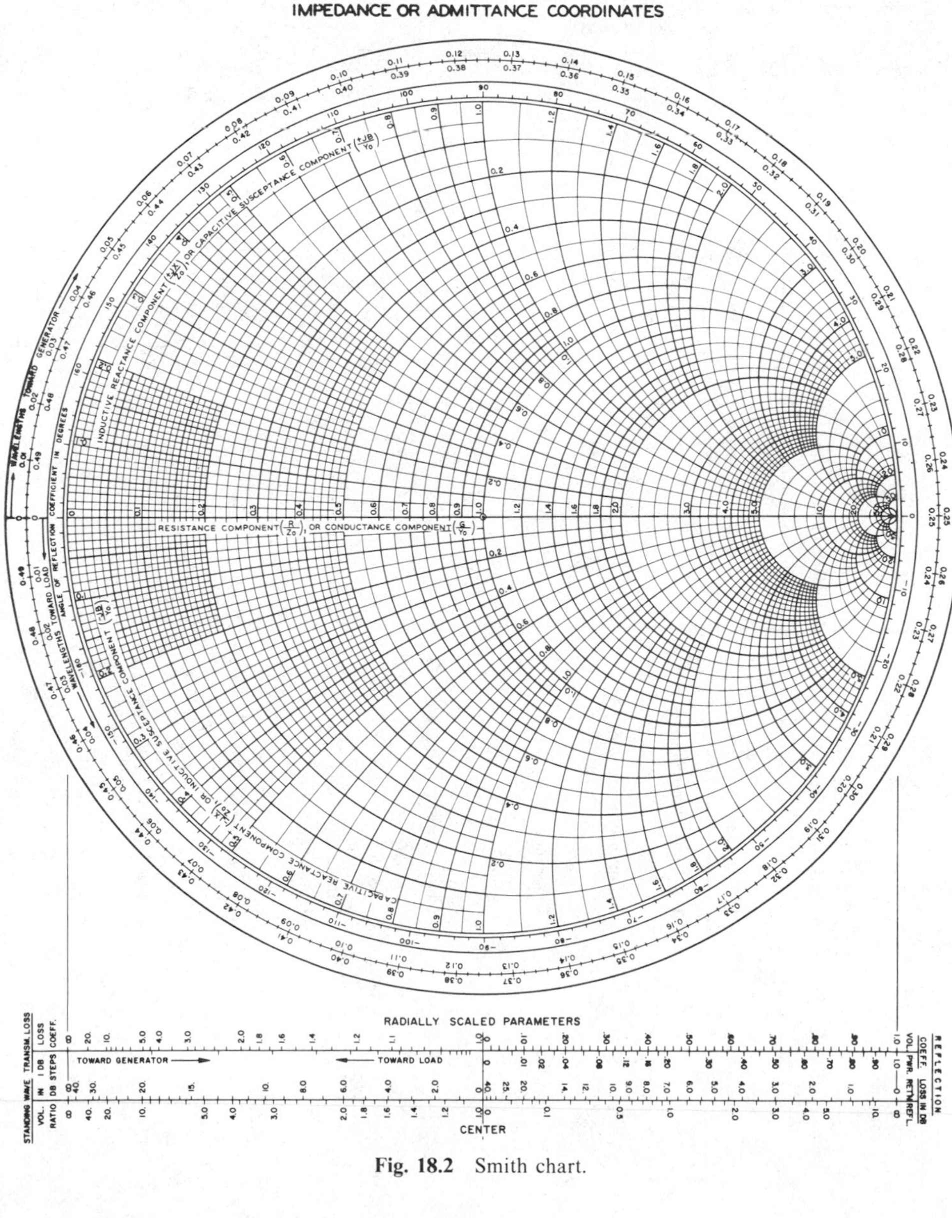

IMPEDANCE OR ADMITTANCE COORDINATES

RADIALLY SCALED PARAMETERS

Fig. 18.2 Smith chart.

TABLE 18.1. TRANSMISSION LINE CHARACTERISTICS

	Parameters	Lossless Case	Low-Loss Case (Approximate Results, $a - i$)	Lossy Case
a	Propagation constant $\gamma = \alpha + j\beta$	$j\omega\sqrt{LC}$	$\alpha + j\beta$ (below)	$\sqrt{(R + j\omega L)(G + j\omega C)}$
b	Attenuation constant α	0	$R/2Z_0 + GZ_0/2$	Real part of γ
c	Phase constant β	$\omega\sqrt{LC} = \omega/v = 2\pi/\lambda$	$\omega\sqrt{LC}\left[1 - \frac{RG}{4\omega^2 LC} + \frac{G^2}{8\omega^2 C^2} + \frac{R^2}{8\omega^2 L^2}\right]$	Imaginary part of γ
d	Characteristic impedance Z_0	$\sqrt{\frac{L}{C}}$	$\sqrt{\frac{L}{C}}\left[1 + j\left(\frac{G}{2\omega C} - \frac{R}{2\omega L}\right)\right]$	$\sqrt{\frac{R + j\omega L}{G + j\omega C}}$
e	Input impedance Z_i at l from load with Z_L	$Z_0\left[\frac{Z_L + jZ_0 \tan \beta l}{Z_0 + jZ_L \tan \beta l}\right]$		$Z_0\left[\frac{Z_L + Z_0 \tanh \gamma l}{Z_0 + Z_L \tanh \gamma l}\right]$
f	Impedance of shorted line	$jZ_0 \tan \beta l$	$Z_0\left[\frac{\alpha l + j \tan \beta l}{1 + j\alpha l \tan \beta l}\right]$	$Z_0 \tanh \gamma l$
g	Impedance of open line	$-jZ_0 \cot \beta l$	$Z_0\left[\frac{1 + j\alpha l \tan \beta l}{\alpha l + j \tan \beta l}\right]$	$Z_0 \coth \gamma l$
h	Impedance of quarter-wave line	$\frac{Z_0^2}{Z_L}$	$Z_0\left[\frac{Z_0 + Z_L \alpha l}{Z_L + Z_0 \alpha l}\right]$	$Z_0\left[\frac{Z_L \tanh \alpha l + Z_0}{Z_0 \tanh \alpha l + Z_L}\right]$

i	Impedance of line, integral number of half wavelengths	Z_L	$Z_0\left[\frac{Z_L + Z_0\alpha l}{Z_0 + Z_L\alpha l}\right]$	$Z_0\left[\frac{Z_L + Z_0 \tanh \alpha l}{Z_0 + Z_L \tanh \alpha l}\right]$
j	Voltage along line $V(-l)$	$V_i \cos \beta l + jI_i \sin \beta l$		$V_i \cosh \gamma l + I_i Z_0 \sinh \gamma l$
k	Current along line $I(-l)$	$I_i \cos \beta l + j\frac{V_i}{Z_0} \sin \beta l$		$I_i \cosh \gamma l + \frac{V_i}{Z_0} \sinh \gamma l$
l	Reflection coefficient Γ	$(Z_L - Z_0)/(Z_L + Z_0)$		$(Z_L - Z_0)/(Z_L + Z_0)$
m	Standard-wave ratio ρ	$(1 + \lvert\Gamma\rvert)/(1 - \lvert\Gamma\rvert)$	$(1 + \lvert\Gamma\rvert)/(1 - \lvert\Gamma\rvert)$	$(1 + \lvert\Gamma\rvert)/(1 - \lvert\Gamma\rvert)$
n	Γ at a point, impedance real (R)		$\Gamma = \frac{R - Z_0}{R + Z_0}$	
o	$R > Z_0$ (at voltage maximum)		$\rho = R/Z_0$	
p	$R < Z_0$ (at voltage minimum)		$\rho = Z_0/R$	
q	Reflected power, P_r Incident power, P_i		$P_r/P_i = \lvert\Gamma\rvert^2 = \left(\frac{\rho - 1}{\rho + 1}\right)^2$	
r	Transmitted power, P_t Incident power, P_i		$P_t/P_i = 1 - \lvert\Gamma\rvert^2 = \frac{4r}{(1 + r)^2}$	

TABLE 18.2. RIGID RECTANGULAR WAVEGUIDE DATA (DOMINANT MODE: TE_{10})

Electronic Industries Association Equivalent Waveguide Number (WR No.)	Frequency Range (GHz)	Wavelength Range (CM)	Cutoff Frequency (GHz)	Cutoff Wavelength (CM)	$2\lambda/\lambda_c$ Range	λ_g/λ Range	Theor. CW Power Rating (MW)
975	0.75–1.12	39.95–26.76	0.605	49.53	1.61–1.08	1.70–1.19	27.0–38.5
770	0.96–1.45	31.23–20.67	0.766	39.12	1.60–1.06	1.66–1.18	17.2–24.1
650	1.12–1.70	26.76–17.63	0.908	33.02	1.62–1.07	1.70–1.18	11.9–17.2
510	1.45–2.20	20.67–13.62	1.157	25.91	1.60–1.05	1.67–1.18	7.5–10.7
430	1.70–2.60	17.63–11.53	1.372	21.84	1.61–1.06	1.70–1.18	5.2–7.5
340	2.20–3.30	13.63–9.08	1.736	17.27	1.58–1.05	1.78–1.22	3.1–4.5
284	2.60–3.95	11.53–7.59	2.078	14.43	1.60–1.05	1.67–1.17	2.2–3.2
229	3.30–4.90	9.08–6.12	2.577	11.63	1.56–1.05	1.62–1.17	1.6–2.2
187	3.95–5.85	7.59–5.12	3.152	9.510	1.60–1.08	1.67–1.19	1.4–2.0
159	4.90–7.05	6.12–4.25	3.711	8.078	1.51–1.05	1.52–1.19	0.79–1.0
137	5.85–8.20	5.12–3.66	4.301	6.970	1.47–1.05	1.48–1.17	0.56–0.71
112	7.05–10.00	4.25–2.99	5.259	5.700	1.49–1.05	1.51–1.17	0.35–0.46
90	8.20–12.40	3.66–2.42	6.557	4.572	1.60–1.06	1.68–1.18	0.20–0.29
75	10.00–15.00	2.99–2.00	7.868	3.810	1.57–1.05	1.64–1.17	0.17–0.23

Theor. Attenuation (dB/100 ft)	Material	MIL-W-85G Type Number RG( )/U	Flange Choke UG( )/U	Flange Cover UG( )/U	Inside Dimension (in.)	Tolerance ±	Outside Dimension (in.)	Tolerance ±	Wall Thickness Nominal
.137–.095	Alum.	204			9.750–4.875	.010	10.000–5.125	.010	0.125
.201–.136	Alum.	205			7.700–3.850	.005	7.950–4.100	.005	0.125
.317–.212	Brass	69		417A	6.500–3.250	.005	6.660–3.410	.005	0.080
.269–.178	Alum.	103		418A					
					5.100–2.550	.005	5.260–2.710	.005	0.080
.588–.385	Brass	104		435A	4.300–2.150	.005	4.460–2.310	.005	0.080
.501–.330	Alum.	105		437A					
.877–.572	Brass	112		553	3.400–1.700	.005	3.560–1.860	.005	0.080
.751–.492	Alum.	113		554					
1.102–.752	Brass	48	54A	53	2.840–1.340	.005	3.000–1.500	.005	0.080
.940–.641	Alum.	75	585	584					
					2.290–1.145	.005	2.418–1.273	.005	0.064
2.08–1.44	Brass	49	148B	149A	1.872–0.872	.005	2.000–1.000	.005	0.064
1.77–1.12	Alum.	95	406A	407					
					1.590–0.795	.004	1.718–0.923	.004	0.064
2.87–2.30	Brass	50	343A	344	1.372–0.622	.004	1.500–0.750	.004	0.064
2.45–1.94	Alum.	106	440A	441					
4.12–3.21	Brass	51	52A	51	1.122–0.497	.004	1.250–0.625	.004	0.064
3.50–2.74	Alum.	68	137A	138					
6.45–4.48	Brass	52	40A	39	0.900–0.400	.003	1.000–0.500	.003	0.050
5.49–3.83	Alum.	67	136A	135					
					0.750–0.375	.003	0.850–0.475	.003	0.050

TABLE 18.2. (*Continued*)

Electronic Industries Association Equivalent Waveguide Number (WR No.)	Frequency Range (GHz)	Wavelength Range (CM)	Cutoff Frequency (GHz)	Cutoff Wavelength (CM)	$2\lambda/\lambda_c$ Range	λ_g/λ Range	Theor. CW Power Rating (MW)
62	12.4–18.00	2.42–1.66	9.486	3.160	1.53–1.05	1.55–1.18	0.12–0.16
51	15.00–22.00	2.00–1.36	11.574	2.590	1.54–1.05	1.58–1.18	0.080–0.107
42	18.00–26.50	1.66–1.13	14.047	2.134	1.56–1.06	1.60–1.18	0.043–0.058
34	22.00–33.00	1.36–0.91	17.328	1.730	1.57–1.05	1.62–1.18	0.034–0.048
28	26.50–40.00	1.13–0.75	21.081	1.422	1.59–1.05	1.65–1.17	0.022–0.031
22	33.00–50.00	0.91–0.60	26.342	1.138	1.60–1.05	1.67–1.17	0.014–0.020
19	40.00–60.00	0.75–0.50	31.357	0.956	1.57–1.05	1.63–1.16	0.011–0.015
15	50.00–75.00	0.60–0.40	39.863	0.752	1.60–1.06	1.67–1.17	0.0063–0.0090
12	60.00–90.00	0.50–0.33	48.350	0.620	1.61–1.06	1.68–1.18	0.0042–0.0060
10	75.00–110.00	0.40–0.27	59.010	0.508	1.57–1.06	1.61–1.18	0.0030–0.0041
8	90.00–140.00	0.333–0.214	73.840	.406	.64–1.05	1.75–1.17	0.0018–0.0026
7	110.00–170.00	0.272–0.176	90.840	.330	1.64–1.06	1.77–1.18	0.0012–0.0017
5	140.00–220.00	0.214–0.136	115.750	.259	1.65–1.05	1.78–1.17	0.00071–0.00107
4	170.00–260.00	0.176–0.115	137.520	.218	1.61–1.05	1.69–1.17	0.00052–0.00075
3	220.00–325.00	0.136–0.092	173.280	.173	1.57–1.06	1.62–1.18	0.00035–0.00047

Theor. Attenuation (dB/100 ft)	Material	MIL-W-85G Type Number RG( )/U	Flange Choke UG( )/U	Flange Cover UG( )/U	Inside Dimension (in.)	Tolerance ±	Outside Dimension (in.)	Tolerance ±	Wall Thickness Nominal
9.51–8.31	Brass	91	541	419	0.622–0.311	.0025	0.702–0.391	.003	0.040
— —	Alum.	—	—	—					
6.14–5.36	Silver	107	—	—					
					0.510–0.255	.0025	0.590–0.335	.003	0.040
20.7–14.8	Brass	53	596	595	0.420–0.170	.0020	0.500–0.250	.003	0.040
17.6–12.6	Alum.	121	598	597					
13.3–9.5	Silver	66	—	—					
					0.340–0.170	.0020	0.420–0.250	.003	0.040
— —	Brass	—	600	599	0.280–0.140	.0015	0.360–0.220	.002	0.040
— —	Alum.	—	—	—					
21.9–15.0	Silver	96	—	—					
— —	Brass	—		383	0.224–0.112	.0010	0.304–0.192	.002	0.040
31.0–20.9	Silver	97		—					
					0.188–0.094	.0010	0.268–0.174	.002	0.040
— —	Brass	—		385	0.148–0.074	.0010	0.228–0.154	.002	0.040
52.9–39.1	Silver	98		—					
— —	Brass	—		387	0.122–0.061	.0005	0.202–0.141	.002	0.040
93.3–52.2	Silver	99		—					
					0.100–0.050	.0005	0.180–0.130	.002	0.040
152–99	Silver	138	—	—	0.080–0.040	0.0003	0.156 Dia	.001	—
163–137	Silver	136	—	—	0.065–0.0325	0.00025	0.156 Dia	.001	—
308–193	Silver	135	—	—	0.051–0.0255	0.00025	0.156 Dia	.001	—
384–254	Silver	137	—	—	0.043–0.0215	0.00020	0.156 Dia	.001	—
512–348	Silver	139	—	—	0.034–0.0170	0.00020	0.156 Dia	.001	—

TABLE 18.3. CHARACTERISTICS OF PARALLEL CONDUCTOR TRANSMISSION LINES

	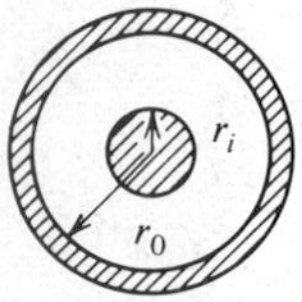 r_i, r_0	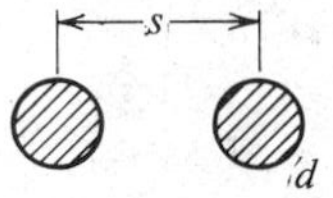 s, d	$p = s/d$, $q = s/D$; 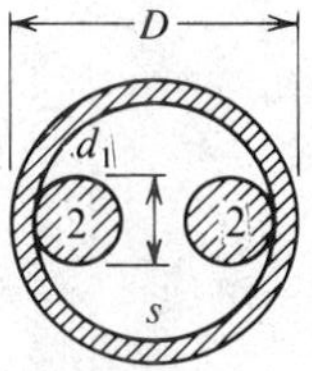 D, d_1, s	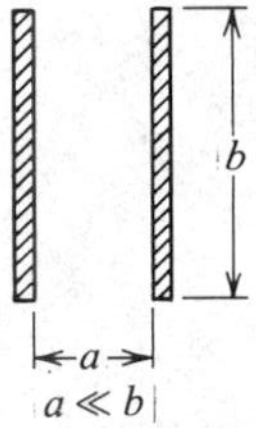 a, b; $a \ll b$
Capacitance, C (F/m)	$\dfrac{2\pi\epsilon^{[a]}}{\ln(r_0/r_i)}$	$\dfrac{\pi\epsilon}{\cosh^{-1}(s/d)}$	—	$\dfrac{\epsilon b}{a}$
Inductance, L (H/m)	$\dfrac{\mu^{[b]}}{2\pi}\ln(r_0/r_i)$	$\dfrac{\mu}{\pi}\cosh^{-1}(s/d)$	—	$\mu\dfrac{a}{b}$
Conductance, G (mho/m)	$\dfrac{2\pi\sigma}{\ln(r_0/r_i)} = \dfrac{2\pi\omega\epsilon''^{[c]}}{\ln(r_0/r_i)}$	$\dfrac{\pi\sigma}{\cosh^{-1}(s/d)} = \dfrac{\pi\omega\epsilon''}{\cosh^{-1}(s/d)}$	—	$\dfrac{\sigma b}{a} = \dfrac{\omega\epsilon'' b}{a}$
Resistance, R (ohm/m)	$\dfrac{R_s^{[d]}}{2\pi}\left(\dfrac{1}{r_0} + \dfrac{1}{r_i}\right)$	$\dfrac{2R_s}{\pi d}\left[\dfrac{s/d}{\sqrt{(s/d)^2 - 1}}\right]$	$\dfrac{2R_{s2}}{\pi d}\left[1 + \dfrac{1 + 2p^2}{4p^4}(1 - 4q^2)\right] + \dfrac{8R_{s3}}{\pi D}q^2\left(1 + q^2 - \dfrac{1 + 4p^2}{8p^4}\right)$	$2\dfrac{R_s}{b}$

Characteristic impedance Z_0 (ohm)	$\frac{\eta^{[e]}}{2\pi}\ln(r_0/r_i)$	$\frac{\eta}{\pi}\cosh^{-1}(s/d)$	$\frac{\eta}{\pi}\left\{\ln\left[2p\left(\frac{1-q^2}{1+q^2}\right)\right]-\frac{1+4p^2}{16p^4}(1-4q^2)\right\}$	$\eta\frac{a}{b}$
Z_0 for air dielectric	$60\ln(r_0/r_i)$	120	120 {same as above}	$120\pi\frac{a}{b}$
Attenuation due to conductor α_c	$R/2Z_0$			
Attenuation due to dielectric α_d	$GZ_0/2 = \sigma\eta/2 = \pi/\lambda^{[f]}(\epsilon''/\epsilon')$			
Total attenuation (dB/m)	$8.686(\alpha_c + \alpha_d)$			
Phase constant for low-loss line β	$\omega\sqrt{\mu\epsilon'} = 2\pi/\lambda$			

Source. Adapted from Ramo et al.,[8] pp. 444–445, with permission.

[a] $\epsilon = \epsilon' - j\epsilon''$ permittivity, F/m.

[b] μ = permeability, H/m.

[c] $\epsilon'' = \sigma/\omega$ loss factor of dielectric.

[d] R_s = skin effect conductor resistivity, Ω.

[e] $\eta = \sqrt{\mu/\epsilon}$, Ω.

[f] λ = wavelength in dielectric.

characteristics of different waveguides optimized for specific range of frequency. Their recommended frequency of operation is nominally 30% above the cutoff frequency ($f_c = c/2a$, where c is the velocity of light and a is the width of the waveguide) at the low frequency end, and 30% below the next higher order mode ($f_c' = c/a$, nominally) at the high end of the frequency range. In the table, attenuation and power handling capability of the waveguides are also presented. To increase the peak power rating of the waveguides above the maximum values recommended for normal use, the waveguides can be pressurized with dry air or with sulfur-hexafluoride to achieve even higher rating. Additional data on rectangular waveguide are given in Refs. 4 and 5.

Cylindrical waveguides[6] have unique applications such as in rotating joints for antenna feeds. Their lowest cutoff frequency modes are TE_{11} and TM_{01}. However, the TE_{0k}, for any number k, yields decreasing attenuation for a waveguide of fixed size as frequency is increased. Oversized guides in the TE_{01} circular electric mode were experimented with for use in long-distance millimeter-wave telephone carrier transmission, despite the extra precision needed to prevent occurrence of extraneous modes. These modes can occur if the waveguide becomes slightly elliptical or its axis is curved. Properly installed waveguides can offer attenuation of only a few decibels per mile at millimeter waves above 40 GHz. However, this particular application of cylindrical waveguides was not implemented; instead fiber optic techniques have been used.

The *ridge waveguide*[4,7] has a central ridge added to the top or bottom (or both) of a rectangular waveguide section. Because of the capacitive effect at the center, the cutoff frequency and the effective impedance are lowered. This structure offers impedance matching and broader bandwidth in a single mode, at the expense of a slight increase in transmission loss.

The *radial transmission line*[8] guides electromagnetic energy radially between two parallel, circularly conducting plates. This type of transmission has found applications in solid-state oscillators in which the diode is embedded between the plates inside a waveguide cavity. Other possible applications are in power combiner circuits and sectoral horns.

Parallel conductor transmission line characteristics of a number of different configurations are listed in Table 18.3. The most common one is the coaxial line, largely because of ease of construction and good shielding between electromagnetic fields inside and outside the line. Only impedance values from 30 to 100 Ω or higher are attainable. The shielded parallel line provides even lower impedance. The coaxial lines offer physical flexibility and exhibit higher losses, especially at high frequencies, compared with rectangular waveguide structures. Detailed characteristics and data are given in Refs. 4 and 9.

The *strip transmission line*[10,11] consists of a center conductor strip in a dielectric material between two ground planes. A cross-sectional configuration of this line, together with other types, is shown in Fig. 18.3. Propagation fields in stripline are in the transverse electromagnetic (TEM) mode. Because of the nonplanar configuration of the stripline, its major applications have been limited to discrete components.

The *microstripline*[12–14] consists of a strip conductor separated from the ground plane by a polished dielectric layer, such as alumina, fused silica, or duroid. All circuit definition is performed in

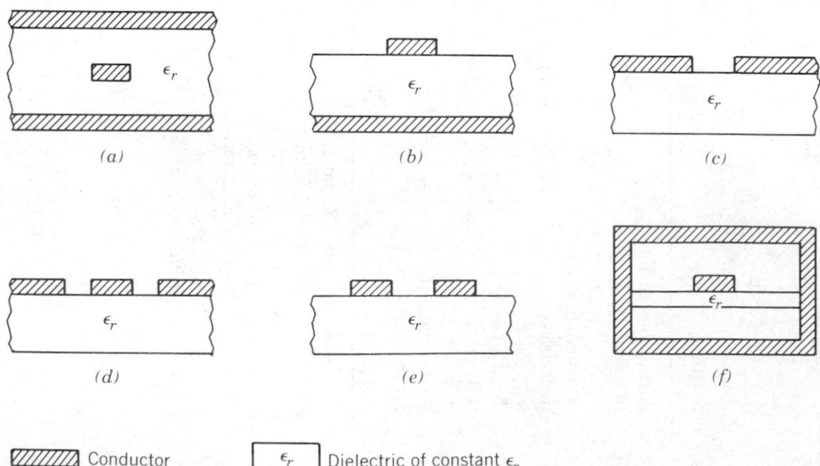

Fig. 18.3 Cross-sectional configurations of various dielectric transmission lines: (*a*) stripline, (*b*) microstrip, (*c*) slotline, (*d*) coplanar waveguide (CPW), (*e*) coplanar strip, (*f*) suspended substrate.

the plane of the strip conductor. The impedance and length of the lines determine the circuit properties. Because most of the field is confined to the dielectric below the conductor strip, the normal mode of propagation is quasi TEM. Slightly higher electrical loss and less power-handling capabilities are offered by this configuration, compared with those of the stripline.

The *suspended substrate transmission line*[4] consists of a substrate with a strip of conductor on the upper face suspended in a metal enclosure. The transmission characteristics are determined by the substrate thickness, dielectric constant, air space height, and width of the strip conductor.

Slotline and *coplanar waveguides*,[12] the other forms of distributed circuits, are useful in specific applications. Both types of transmission lines have conductors on only one side of the substrate, allowing shunt mounting of devices without requiring through holes, as in the case of microstriplines. They also have longitudinal as well as transverse radio-frequency magnetic fields and polarization properties that are useful for nonreciprocal ferrite devices. These transmission lines are gaining acceptance in microwave integrated circuits (MICs).

Figures 18.4 through 18.7 show the characteristics of stripline, microstripline, slotline, coplanar waveguide, and coplanar strips.

Circuits of the *lumped-element form*[16] use components that behave as capacitors, inductors, or resistors. Because they are a small fraction of a wavelength in size, they offer size advantage over microstriplines at the lower microwave frequency range, as in some MIC applications.

Microwave networks are characterized by scattering parameters (or S-parameters).[17,18] Incident and reflected-wave variables rather than voltages and currents, whose definitions are not unique, are used. To obtain the parameter, reflection and transmission measurements are made at ports terminated with matched characteristic impedances. This approach avoids the uncertainty of open and short circuits used in low-frequency characterizations. For a two-port network, the S-parameter relationship is given as follows

$$\begin{aligned} b_1 &= S_{11}a_1 + S_{12}a_2 \\ b_2 &= S_{21}a_1 + S_{22}a_2 \end{aligned} \tag{18.1}$$

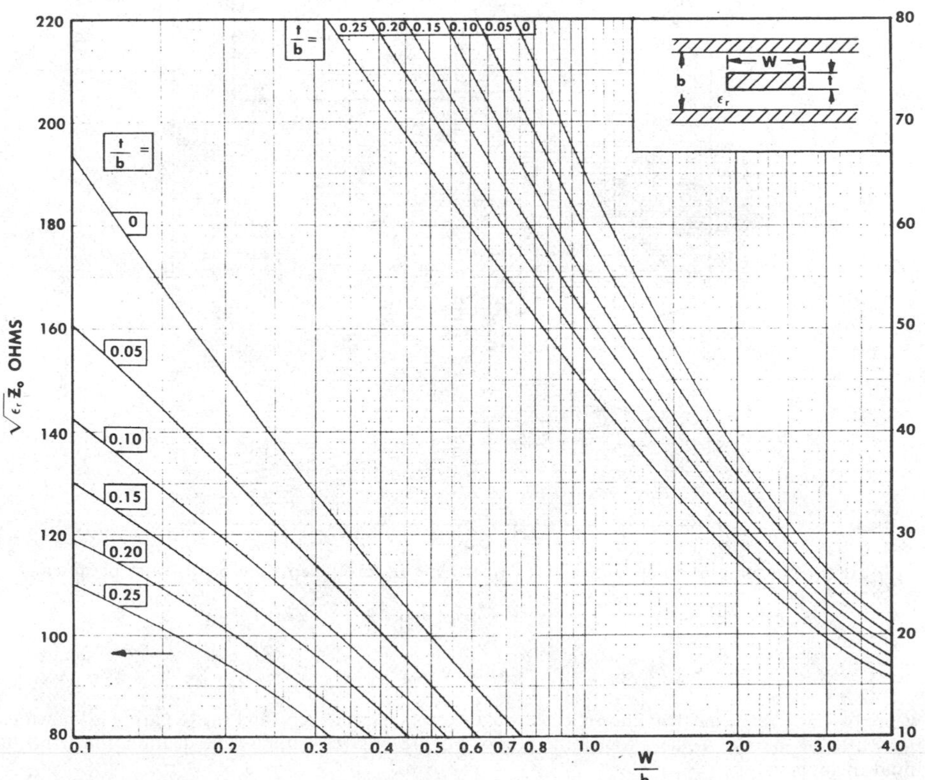

Fig. 18.4 Stripline characteristic impedance vs. w/b reproduced from Saad,[4] with permission.

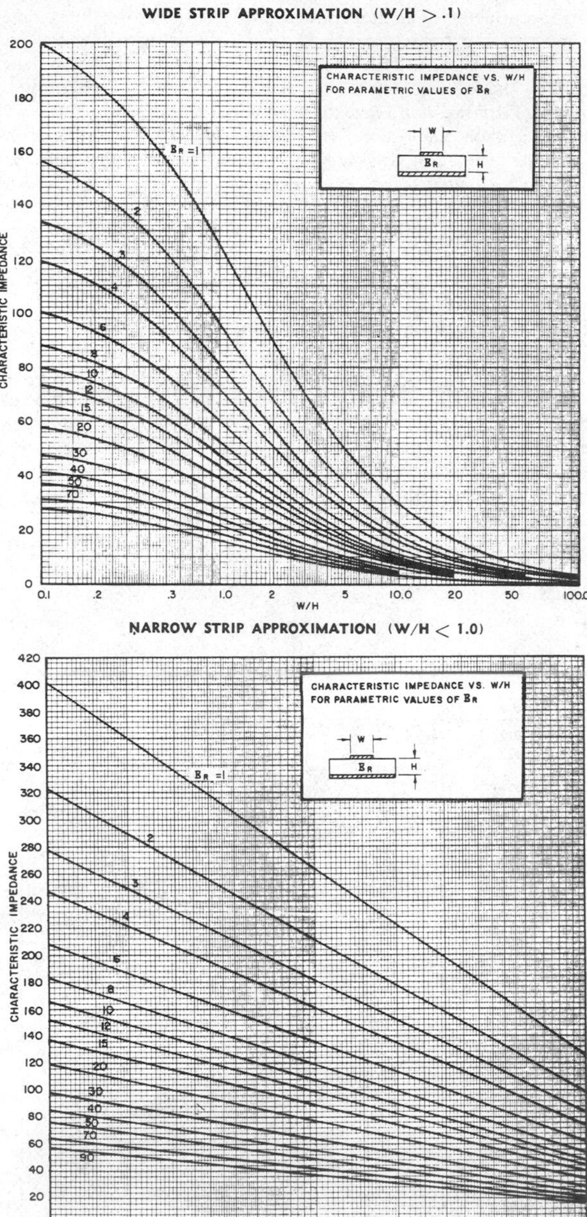

Fig. 18.5 Characteristic impedance of microstripline. Reproduced from Saad,[4] with permission.

or

$$[b] = [S][a]$$

where $[a]$ and $[b]$ are incident and reflected wave vectors, respectively, and the $[S]$ array is known as the scattering matrix. A linear transformation of Eq. (18.1) gives output wave quantities in terms of input quantities

$$\begin{bmatrix} b_2 \\ a_2 \end{bmatrix} = [T] \begin{bmatrix} a_1 \\ b_1 \end{bmatrix} \tag{18.2}$$

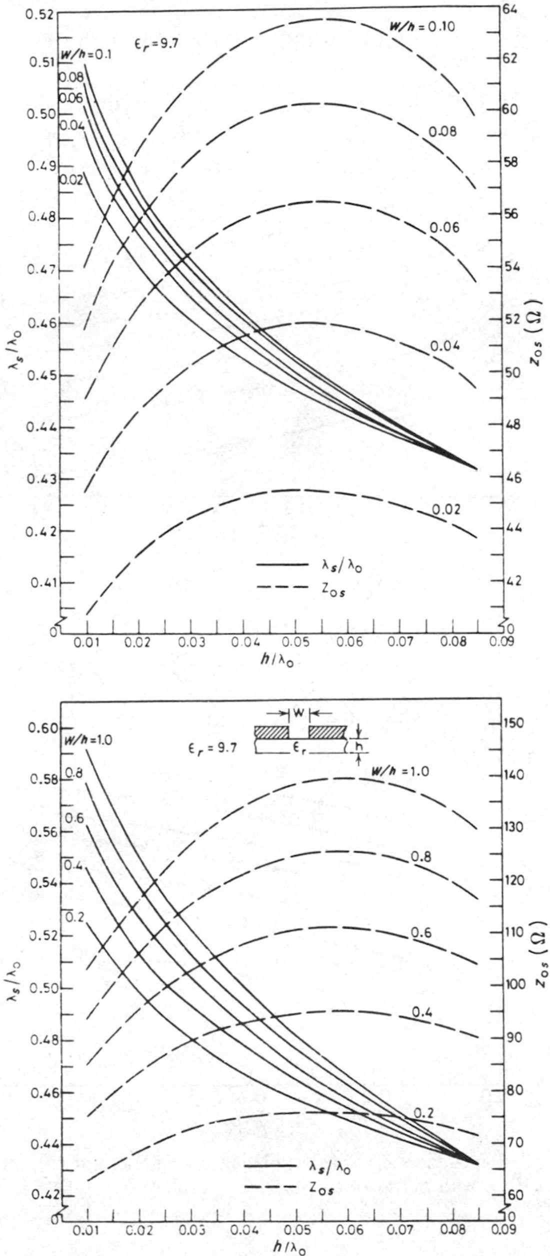

Fig. 18.6 Relative wavelength and characteristic impedance vs. ratio of dielectric thickness and wavelength of slotline. Reproduced from Gupta et al.,[12] with permission.

The $[T]$ array, or transmission matrix, is related to the $[S]$ matrix. Thus a complete microwave network system can be analyzed at the terminal ports, if each component is characterized by the S- or T-matrix and then cascaded.

18.1-2 Reciprocal Devices

Reciprocal devices have electrical characteristics such that transmission properties through the device are the same for different propagation directions.

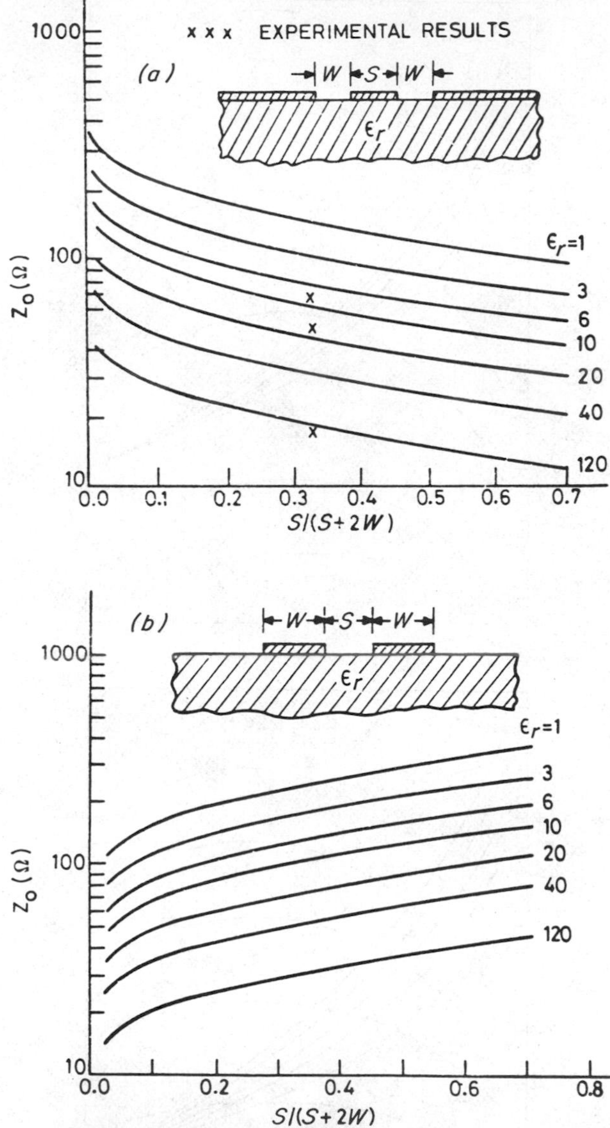

Fig. 18.7 Characteristic impedance Z_0 of (*a*) coplanar waveguide and (*b*) coplanar strips. Reproduced from Wen,[15] p. 1089, with permission.

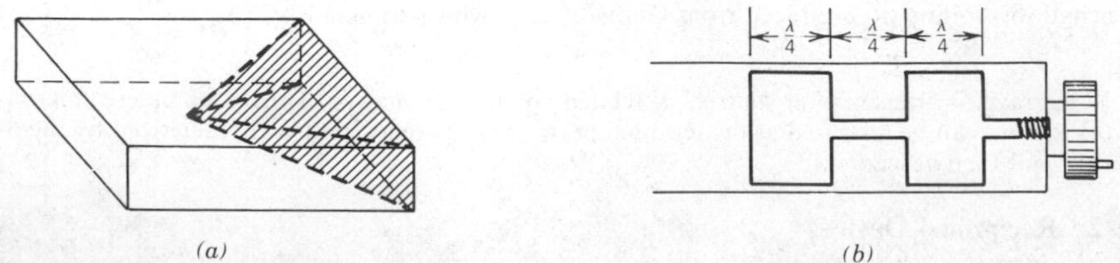

Fig. 18.8 Terminations: (*a*) matched load with tapered wedge and (*b*) sliding RF short circuit.

Terminations[19] such as matched load and variable short circuit, as shown in Fig. 18.8, are used in impedance matching and microwave measurements. The matched load, a tapered wedge of lossy material inserted into the waveguide, absorbs the incident power with minimal reflection. An input standing-wave ratio of 1.01 or smaller can usually be achieved with an overall length of one or more wavelengths.

The variable short termination reflects all the incident power. The phase of the reflected wave is varied by changing the position of the short circuit, thereby causing a variation in the reactance of the termination.

Attenuators[19] may be of the fixed or the variable type and are used to adjust the power level in microwave systems. One type consists of a thin, tapered resistive card inserted with variable depth into the guide in a longitudinal slot cut in the center of the broad wall of a rectangular guide. A higher precision type utilizes an adjustable length of waveguide operating below its cutoff frequency. The most satisfactory attenuator is the rotary type, consisting of two rectangular-to-circular waveguide tapered transitions, together with an intermediate section of circular waveguide that is free to rotate. By combining the effects of polarization of field propagation and the absorption due to the resistive cards in the three sections, precision attenuation can be obtained as a function of angle of rotation. Attenuators in coaxial systems, generally, rely on properly matched resistor circuits.

The *phase shifter*[20] produces an adjustable change in the phase angle of the wave transmitted through it. The linear phase shifter generates phase variation through the movement of dielectric slabs inside a waveguide. The precision rotary phase changer has a construction similar to that of the rotary attenuator, with the resistive cards replaced by a half-wave plate and two quarter-wave outer plates.

Directional couplers[8,21] are four-port components for sampling the forward and reverse power flows in a guide in a separable scheme, as illustrated in Fig. 18.9. Waves that are coupled to another parallel guide through a series of holes nominally spaced a quarter wave apart are in phase for the forward-traveling waves and out of phase for the reverse waves. Varying the dimensions of the holes and their spacing allows achievement of different coupling factors (ratio of power levels in the primary and secondary lines) (3 to 40 dB) and frequency bandwidths. Other configurations such as the cross-guide, stripline, and microstrip couplers are used in microwave measurements.

The *hybrid junction*[8,21] may be used as a four-port bridge, transmit-receive, or balanced-mixer network. The magic T is a combination of an E-plane T and an H-plane T. An E-plane (H-plane) T is a waveguide T in which the axis of its side arm is parallel to the E (H) field of the main guide. The coaxial line hybrid ring "rat race" structure has four ports a quarter wave apart. The basic principle for both networks is that the division of power and the phase changes resulting from the two paths to each output port cause the additions and cancellations designed. Schematics of microwave hybrids are shown in Fig. 18.10.

The *resonator*[5,8] is generally formed from a metallic waveguide enclosure or from shorted- or open-circuit lengths of transmission lines. Electromagnetic energy is stored in the resonator structure. The finite conductivity of the metallic walls and the medium give rise to power and thus result in some equivalent resistance. Transmission-line resonators have a lower value of Q (more lossy) than does the waveguide. Resonant circuits are used extensively in tuned amplifiers, oscillators, filter networks, wavemeters, and other devices. The commonly used resonators are the rectangular cavity, cylindrical cavity, and reentrant cavity. Recently the dielectric resonator has found more application in microwave systems.[22] Because of the advances in dielectric materials, reduction in resonator size and improvements in frequency stabilities over life and temperature have been achieved.

Filters[23] are employed in all frequency ranges to provide very-low-loss transmission within the desired passband and to reject out-of-band spurious signals. Three common types—low-pass, high-

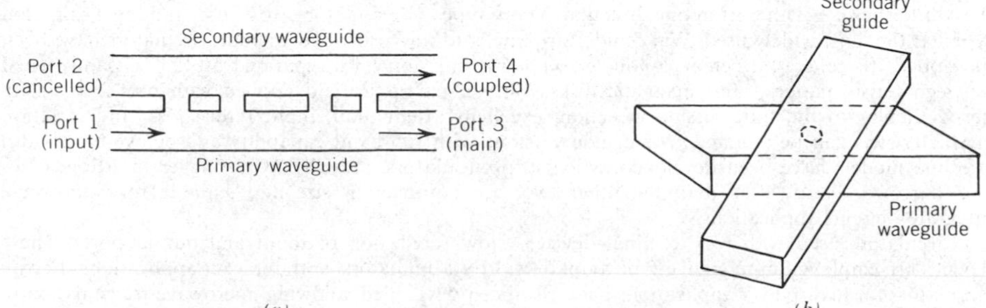

Fig. 18.9 Directional couplers: (*a*) Four holes, (*b*) Bethe hole.

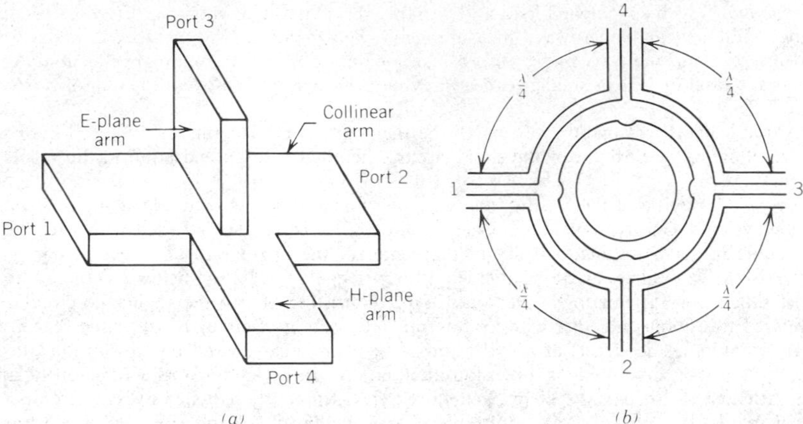

Fig. 18.10 Microwave hybrids: (*a*) magic tee, (*b*) hybrid ring.

pass, and bandpass—have been realized in waveguide, coaxial, stripline, and microstrip transmission lines. Waveguide structures offer low-loss high-power capabilities, whereas the transmission line types provide compact size advantage. The number of sections (or poles) used in the design depends on the requirement of band edge steepness and stop-band attenuations. Recently filter designs using both active elements and dielectric resonators have been developed.

Other components such as stub tuners, multiple quarter-wave sections, capacitive irises, and inductive-centered vanes are used for impedance matching, such that power flow in the desired frequency range from one type of transmission medium or structure to another is maximized.

18.1-3 Nonreciprocal Components

The development of ferrite materials[24] has led to their application in a number of nonreciprocal microwave devices.[25] These have electrical properties such as variation of the transmission coefficient through the device with propagation direction. The principle of operation is mostly based on Faraday rotation. When a ferrite medium, properly biased with a static magnetic field, is placed in a circularly polarized electromagnetic field, a differential phase shift for the two directions of propagation can be obtained. This phase shift, together with 3-dB power divider circuits, can also provide nonreciprocal attenuation in the designed frequency band.

Gyrators, useful in phase arrays, are two-port devices that have a 180°-phase shift difference for transmission from one direction compared with that in the reverse direction. This is achieved with an axially directed magnetic field to a ferrite rod inside a waveguide, as illustrated in Fig. 18.11*a*.

Isolators permit unattenuated transmission in the forward direction but provide very high attenuation in the reverse direction. The isolator is often used to couple a signal-from-source generator to a load of either passive or active networks. All the available power can be delivered to the load while effects such as power output variation and frequency pulling (changing) with variations in load impedance are avoided. Typical performance for an isolator are insertion loss of less than 1 dB, isolation of 20 to 30 dB, and 10% to octave bandwidths.

Circulators, shown in Fig. 18.11*b* and *c*, are multiport junctions in which the wave is coupled from the nth to the $(n + 1)$th port in one direction. Many types of circulators are in use, and the Y-junction type[26] is the most widely used. Waveguide, stripline, and microstripline junction circulators have been developed. In general, a center conductor with transmission lines separated at 120° is sandwiched between ferrite material and magnetic disks. With a magnetic field applied with proper intensity perpendicular to the transmission direction, excellent attenuation, input return loss, and isolation characteristics can be attained. Waveguide versions with full-band capability and octave-bandwidth stripline models have been developed. Microstrip circulators, though having higher insertion losses and narrower bandwidths than the other two, are compact in size and suitable for microwave integrated-circuit applications.

Circulators used with two terminal devices allow separation of input and output ports. These devices are employed in reflective-type amplifiers, PIN attenuators, and duplexer applications. Ferrite circulators for high-power applications have also been developed, allowing microwave transmission in the kilowatts continuous-wave (CW) range.

Limiters use the nonlinear properties of ferrites to limit the high level (kilowatts) of microwave power transmissions. Generally they are used in conjunction with PIN diodes.

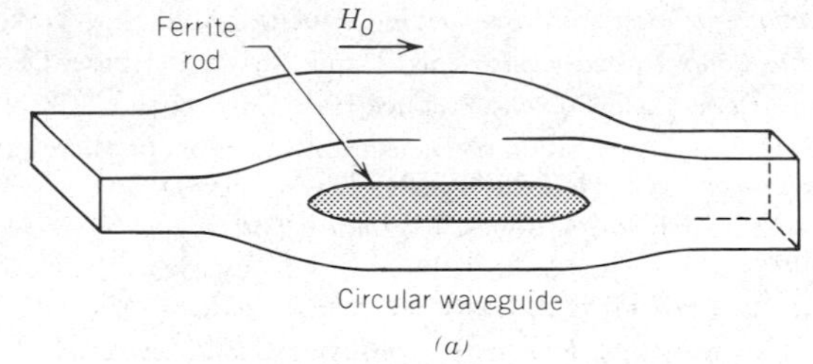

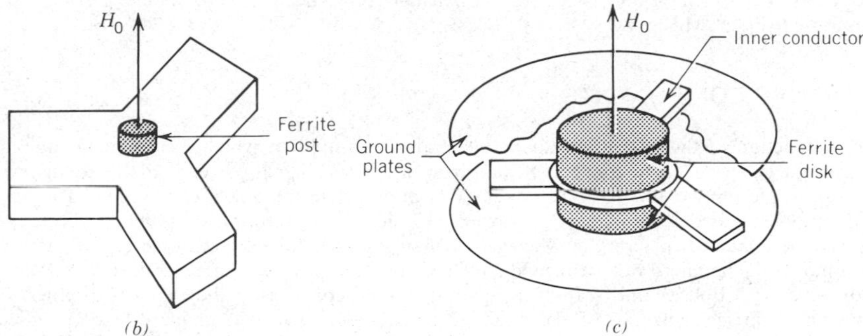

Fig. 18.11 Nonreciprocal devices: (*a*) gyrator, (*b*) waveguide three-port junction circulator, (*c*) stripline three-port junction circulator.

References

1 P. H. Smith, "Transmission-Line Calculator," *Electronics* **12**:29–31 (Jan. 1939); "An Improved Transmission-Line Calculator," *Electronics* **17**:130 (Jan. 1944).

2 Super-Compact, FILSYN and AMPSYN, Compact Software Inc., A COMSAT Technology Product Company, Palo Alto, CA.

3 L. W. Nagel, "SPICE 2-A, Computer Program to Simulâte Semiconductor Circuits," memo ERL-M520 Selectronic Research Laboratory, University of California, Berkeley, May 1975.

4 T. Saad, ed., *Microwave Engineer's Handbook*, Vols. 1 and 2, McGraw-Hill, New York, 1971.

5 N. Marcuritz, ed., *Waveguide Handbook*, McGraw-Hill, New York, 1951.

6 C. G. Montgomery et al., *Principles of Microwave Circuits*, McGraw-Hill, New York, 1947.

7 S. Hopfer, "The Design of Ridged Waveguide," *IRE Trans.* **MTT-3**:20–29, (Oct. 1955).

8 S. Ramo, J. R. Whinnery, and T. Van Duzer, *Fields and Waves in Communication Electronics*, Wiley, New York, 1965.

9 T. Moreno, *Microwave Transmission Design Data*, McGraw-Hill, New York, 1948.

10 H. Howe, *Stripline Circuit Design*, Artech House, Dedham, MA, 1974.

11 I. J. Bahl and R. Garg, "A Designer's Guide to Stripline Circuits," *Microwaves* **17**:90–96 (1978).

12 K. C. Gupta, R. Garg, and I. J. Bahl, *Microstrip Lines and Slotlines*, Artech House, Dedham, MA, 1979.

13 T. G. Bryant and J. A. Weiss, "Parameters of Microstrip Transmission Lines and of Coupled Pairs of Microstrip Lines," *IEEE Trans.* **MTT-16**:1021 (Dec. 1968).

14 H. A. Wheeler, "Transmission-Line Properties of Parallel Strips Separated by a Dielectric Sheet," *IEEE Trans. Micro. Theory Tech.*, **MTT-3**:172–185 (Mar. 1965).

15 C. P. Wen, "Coplanar Waveguide: A Surface Strip Transmission Line Suitable for Non-Reciprocal Gyromagnetic Device Application," *IEEE Trans. Micro. Theory Tech.* **MTT-17**: 1087–1090 (Dec. 1969).

16 M. Caulton, "Lumped Elements in Microwave Integrated Circuits" in L. Young and H. Sobol, eds., *Advances in Microwaves*, Vol. 8, Academic, New York, 1974.

17 H. J. Carlin and A. B. Giordano, *Network Theory*, Prentice-Hall, Englewood Cliffs, NJ, 1964.

18 Hewlett Packard, "S-Parameter Design," Application Notes No. 155, Hewlett Packard, Palo Alto, CA, 1972.

19 R. E. Collin, *Foundations for Microwave Engineering*, McGraw-Hill, New York, 1966.

20 L. R. Whicker, *Ferrite Control Components*, Vols. 1 and 2, Artech House, Dedham, MA, 1974.

21 S. Y. Liao, *Microwave Devices and Circuits*, Prentice-Hall, Englewood Cliffs, NJ, 1980.

22 J. K. Plourde and C. L. Ren, "Application of Dielectric Resonators in Microwave Components," *IEEE Trans. Micro. Theory Tech.* **MTT-29**(8):754–770 (Aug. 1981).

23 G. Matthaei, L. Young, and E. M. T. Jones, *Microwave Filters, Impedance-Matching Networks, and Coupling Structures*, Artech House, Dedham, MA, 1980.

24 W. H. Von Anlock, ed., *Handbook of Microwave Ferrite Materials*, Academic, New York, 1965.

25 B. Lax and K. J. Button, *Microwave Ferrites and Ferromagnetics*, McGraw-Hill, New York, 1962.

26 H. Bosma, "Junction Circulators," *Adv. Micro.* **6**:125–257 (1971).

27 K. C. Gupta, R. Garg, and R. Chadha, *Computer Aided Design of Microwave Circuits*, Artech House, Dedham, MA, 1981.

18.2 ACTIVE COMPONENTS

The basic function of the microwave source generator or amplifier, whether continuous, pulsed, or swept as is the generator, is to convert DC power to microwave signal power available to an external load. In principle, only a few basic equations with appropriate boundary conditions are needed to solve the problems associated with microwave electronics. An electron-beam (microwave tube)-type device can be analyzed in terms of Maxwell's equations and the Lorentz force law (dynamics of electron motion). The microwave semiconductor-type problem involves, in general, solving Poisson's, Schrödinger's, and Boltzmann's transport equations while considering the impurity profiles, band structure, and carrier dynamics of the material, respectively. The two major categories of active components are microwave tubes and semiconductor devices. While microwave tubes still dominate very-high-power applications in microwave systems, solid-state microwave devices are gradually replacing tubes, offering lower noise, better linearity performance, smaller size, and higher reliability.

18.2-1 Microwave Tubes

Common types of microwave tubes are shown in Table 18.4. They can be divided into four groups: planar tubes, linear-beam (O-type) tubes, cross-field (M-type) tubes, and most recently developed, gyrotrons. Details of the development and performance of some of these microwave tubes are given in Refs. 1 and 2.

Planar Tubes

Vacuum tubes such as *tetrodes* and *triodes*[3,4] are limited in their operation to lower microwave frequencies because of the effects of lead inductance, interelectrode capacitance, electron transit time, and gain-bandwidth product limitations. In general, triodes are preferred for applications above 1 GHz and tetrodes for those below 1 GHz. Figure 18.12 shows a schematic of a microwave triode.

Planar triode tubes consist of alumina or beryllia ceramic envelopes with penetrating metal members which support the electrodes. These disk-shaped members are usually made of kovar and are plated with copper or silver to reduce conduction loss. Various oxide-coated cathodes or dispenser cathodes (a metal sponge impregnated with a barium compound) are used to increase emission density and electrical conductivity and thereby to improve output power. Grids consist of wires of a gold-coated refractory metal, such as molybdenum or tungsten, for high thermal conductivity and tensile strength. The heat-sunk anodes are usually made of copper.

Resonant cavity circuits for oscillator and amplifier applications have been developed in waveguides, coaxial lines, and striplines. RF average powers in the tenths of watts and peak pulse powers of the order of kilowatts have been obtained up to 4 GHz. The DC plate to RF conversion efficiency ranges from 60% pulsed at 1 GHz to a few percent CW at 5 GHz. With an annular circuit for combining a number of tubes, a substantial increase in output power has been achieved at 1 GHz.[5]

Linear Beam Tubes

Linear-beam tubes include klystrons and traveling-wave tube (TWT) structures.

The *reflex klystron* oscillator is schematically shown in Fig. 18.13. The thermionic cathode emits electrons, and the anode voltage accelerates them. The resulting electron beam passes through the anode and the two grids of the cavity. The beam returns through the cavity after being reflected by

TABLE 18.4. COMMON TYPES OF MICROWAVE TUBES

	Linear-Beam Tubes (O-Tubes)		Cross-Field Tubes (M-Tubes)		Gyrotrons	
Planar Tubes	Cavity	Slow-Wave Structure	Resonant Structure	Nonresonant Structure	Cavity	Slow-Wave Structure
Triode amplifier and oscillator	Two-cavity klystron amplifier	Coupled-cavity TWT[a] amplifier (TWTA)		Forward-wave cross-field amplifier (FWCFA)	Gyro-klystron amplifier	Gyro TWT amplifier
Tetrode	Multicavity klystron amplifier	Backward-wave amplifier (BWA)				
		Low-noise helix TWTA				
	Reflex klystron oscillator	Coupled-cavity TWT oscillator	Magnetron oscillator	M-carcinotron oscillator	Gyrotron oscillator	
	Two-cavity klystron oscillator	Backward-wave oscillator (BWO)		M-backward wave oscillator (M-BWO)		
	Hybrid twystron amplifier					

[a] TWT, traveling wave tube.

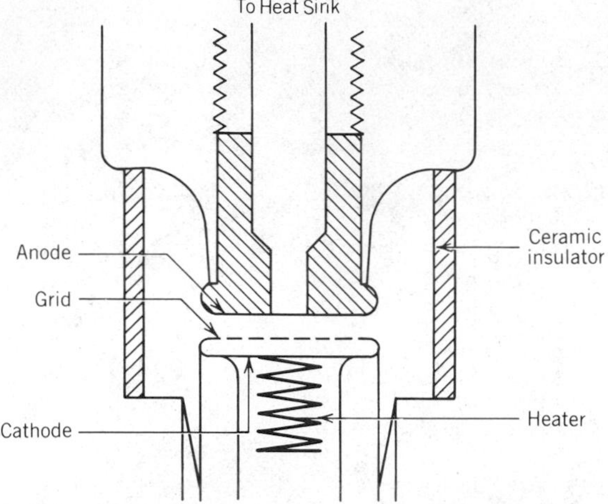

Fig. 18.12 Simplified schematic of a microwave triode.

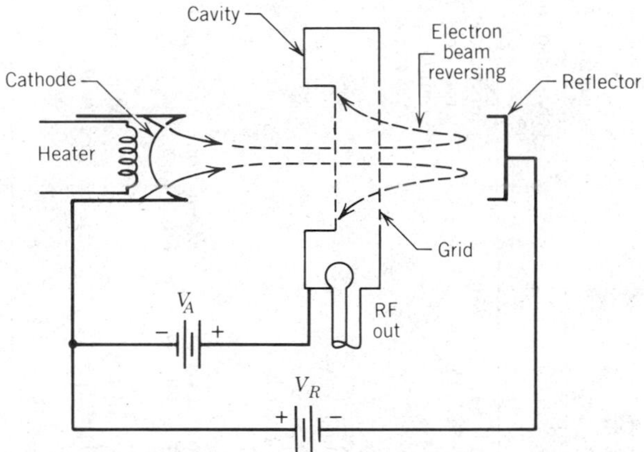

Fig. 18.13 Schematic of a reflex klystron oscillator.

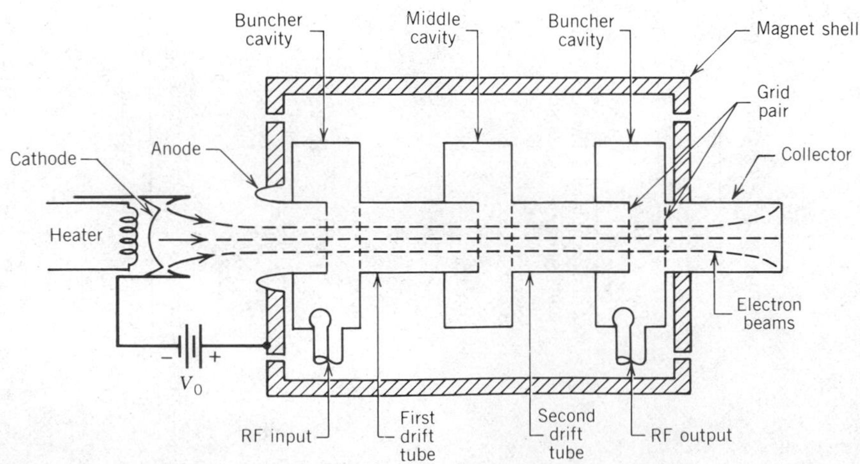

Fig. 18.14 Schematic of a three-cavity klystron amplifier.

the electric field owing to the potential on the reflector, which is negative with respect to the cathode. With a transit time in the reflecting field of $n + 3/4$ cycles (where n is an integer) at the period corresponding to the resonant frequency of the cavity and with proper loading, microwave power will be generated.

Cavity tuning can be achieved by an internal tuner, which requires that the vacuum wall of the tube be deformed to result in changes of capacitive loading. An external tuner through a resonant cavity outside the vacuum envelope offers an alternative. Electronic tuning of the power and frequency modes can also be achieved by varying the reflector voltage.

Reflex klystrons are used as signal sources, local oscillators for receivers, and low-power transmitters. A typical performance is less than 500 mW output at frequencies from 1 to 25 GHz with efficiencies from 20 to 30%.

Two-cavity klystron oscillators can function more efficiently and achieve higher output than reflex klystrons. The feedback required for oscillation is attained through an iris structure that couples the cavities. Since the resonant frequency of each cavity and the phase shift in feedback path must be adjusted when oscillation frequency is changed, this type of oscillator is usually fixed frequency with a narrow tuning range.

The *multicavity klystron amplifier* in one of its forms consists of *three-cavity* resonators connected by drift tubes, as shown in Fig. 18.14. An accelerated electron beam travels through the first cavity, where it is modulated in velocity by the field from the input RF signal. In the second drift tube, bunching of electrons of different velocities occurs. At the third cavity, the electron beam is modulated in density, and an RF current is impressed across the third cavity. When the cavity is tuned to the signal frequency, power may be removed from it, and the amplified output signal results. The addition of more cavities, some detuned, allows an accumulative signal gain of as much as 60 dB and broad banding effects of up to 35% tuning to be attained.

Performance of a two-cavity amplifier of an average power up to 500 kW, a pulsed power of 30 MW, an efficiency of 40%, and a power gain of 30 dB at 10 GHz has been reported.[5]

The commonly used periodic slow-wave structure for the *traveling wave tube amplifier* (*TWTA*)[6] is helical, as shown in Fig. 18.15. The electromagnetic wave travels at nearly the velocity of light along the helix wire, while the electron beam, focused by a constant magnetic field, travels down the axis. Because the electrons and the RF wave travel at the same average velocity, continuous interactions are possible that can transfer energy from the electron beam to the RF traveling wave, resulting in RF amplification.

The low-frequency amplification limit is set by nonsynchronization, when the wave travels faster than the beam. The high-frequency limit is set by the reduction in coupling between the helix and the beam, which occurs because the wave amplitude varies appreciably over the beamwidth as the wavelength approaches the helix diameter. The main advantage of the TWT over the klystron is its relatively broad frequency band of operation. Other slow-wave structures such as disk-loaded or disk-on-rod-type are dispersive, that is, the phase velocities vary with frequency.

Helix TWTAs, like klystrons, are designed for both CW and pulsed operation. Units can provide gains from 30 to 60 dB over an octave in frequency. The average power-handling capability ranges from milliwatts to kilowatts.

Low-noise TWTAs are being replaced by solid-state amplifiers such as bipolar transistor or MESFET amplifiers, which offer lower noise, simpler circuitry, smaller size, and higher reliability.

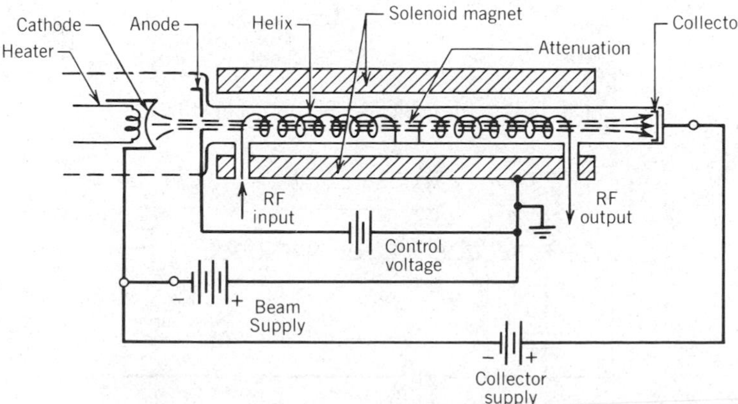

Fig. 18.15 Schematic of O-type helix traveling wave tube.

Coupled-cavity TWTs[7,8] are devices capable of generating very high microwave power, up to 10 MW. Two types of coupled-cavity circuits in TWTs have been demonstrated. The cloverleaf and centipede forward-wave circuits exhibit negative mutual inductive coupling between the cavities and operate with the fundamental space harmonic. The circuits are normally employed for megawatt peak power applications. The second type, in which the first space harmonic circuit involves positive mutual coupling between cavities, is for both pulsed and CW applications. The coupled-cavity TWTA is bandwidth limited, because gain drops off fairly rapidly at the band edges. An example of the cavity-coupled structure is given in Fig. 18.16.

The *twystron amplifier*,[9] a hybrid device illustrated in Fig. 18.17, consists of a multicavity stagger-tuned klystron input driver section and a traveling wave output section. The combination of the high-broadband-gain klystron and high-efficiency traveling wave tube characteristics, resulting in flat gain performance, is used in high-power radar transmitter applications. Peak power from 1 to 10 MW and average power from 1 to 30 kW at S and C bands are achievable.

The linear beam (O-type) *backward-wave oscillator* (BWO) will result if a TWTA is provided with a properly phased continuous feedback along the circuit. A schematic of a BWO with the helical

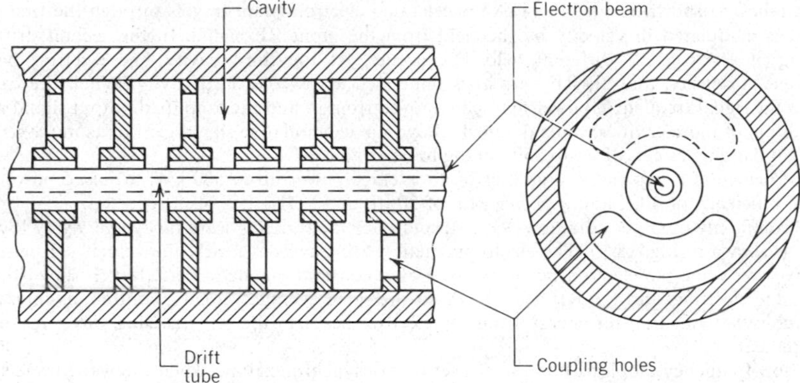

Fig. 18.16 Cavity-coupled structure in a traveling-wave tube.

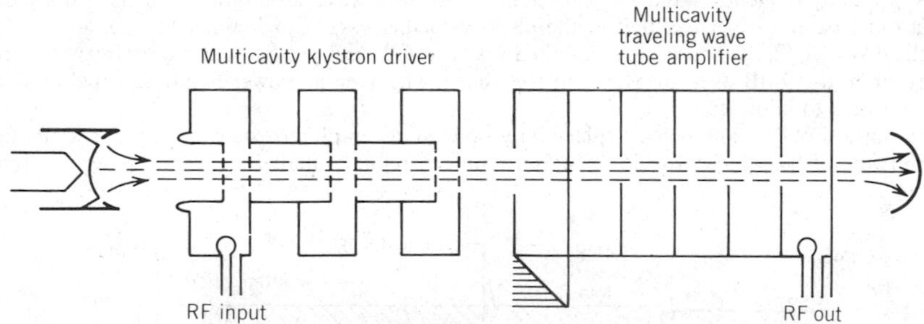

Fig. 18.17 Schematic of a twystron amplifier.

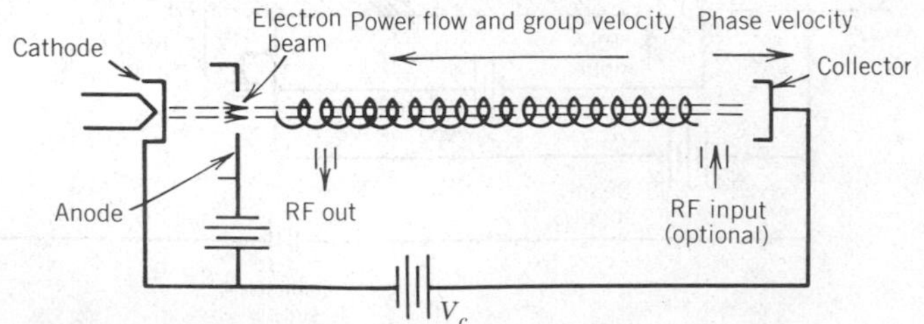

Fig. 18.18 Simplified representation of a helix backward-wave oscillator or amplifier.

slow-wave structure is shown in Fig. 18.18. The BWO has its output coupling at the cathode end. The tube is capable of oscillating at any beam voltage and thus can be tuned over a large frequency range by adjusting the voltage.

When used as an amplifier, the device can be adjusted to provide a narrow bandwidth, the center frequency of which can be altered by changing the beam voltage.

The BWO has relatively low AM, FM, and spurious noise. Among its many applications is use as a wideband electronically tunable generator for swept-frequency instruments. Power BWOs have also served as signal sources from the K-band to over 100 GHz. In this power application, inductively coupled types of cavities, for example, are used instead of helical circuits.

Cross-Field Tubes

The basic structure of a conventional *magnetron*[10] consists of a number of identical resonators arranged in a cylindrical pattern, as shown in Fig. 18.19. As in other microwave tubes, part of the energy in a magnetron gained by an electron accelerated by a DC electrical potential is converted into RF energy. The DC and RF potentials act on the electrons simultaneously in the interactive space of a magnetron, whereas in the O-type TWT and the BWO, electrons are preaccelerated before they enter the interactive structure; thus they are kinetic energy converters. Furthermore, a magnetic field perpendicular to the electric field is essential for the operation of a magnetron, whereas a magnetic field is applied to confine the beam in the linear beam devices. The magnetron can be tuned by introducing additional metallic parts that interfere with the electric or magnetic field of the cavities. The AC power may be coupled out from one of the cavities by a coaxial line or loop or by means of a waveguide.

Magnetron oscillators are capable of generating high peak RF pulses, such as 40 MW at 10 GHz with 50 kV DC voltage. The average power outputs are as much as 300 W at X-band, and the DC-to-RF conversion efficiency ranges from 40 to 70%. Magnetrons are used as beacons and transmitters; in Doppler, airborne, and missile systems; and in the largest commercial applications, microwave ovens.

M-carcinotron and M-backward-wave oscillators[11] are nonresonant injection-beam types. A linear version is shown in Fig. 18.20. The electrons emitted from the cathode are directed into the interaction region between the slow wave structure and the negative electrode (also known as sole) by a DC magnetic field. The electron beam interacts with a backward-wave space harmonic of the circuit. As is the case with linear-beam BWOs, frequency tuning can be obtained by varying the anode-to-sole voltage. These oscillators have been used as power signal sources with average power over 100 W at X-band. Since the M-carcinotron is a cross-field device, efficiencies ranging from 30 to 60% are achievable.

The *cross-field amplifiers* (CFAs)[12] can be grouped by their electron stream source as emitting-sole or injected-beam types. Their mode of operation can be either forward- or backward-wave type. The CFAs have low to moderate gain and bandwidth and high efficiency and are mostly used in radar applications.

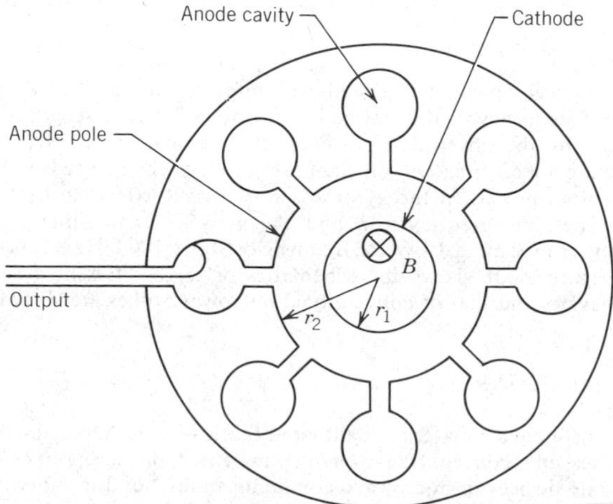

Fig. 18.19 Schematic of a cylindrical magnetron.

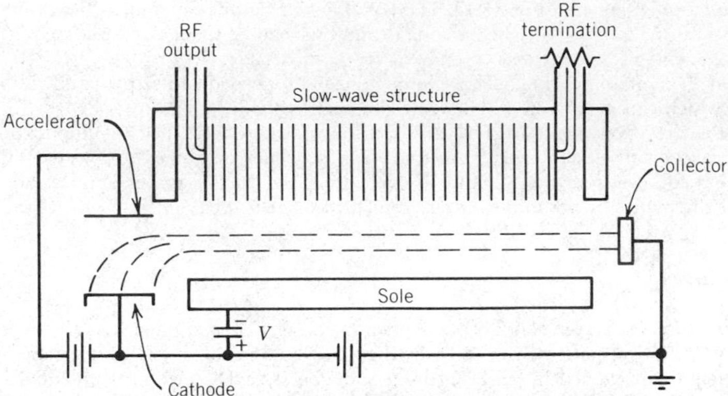

Fig. 18.20 Schematic of a linear M-carcinotron oscillator.

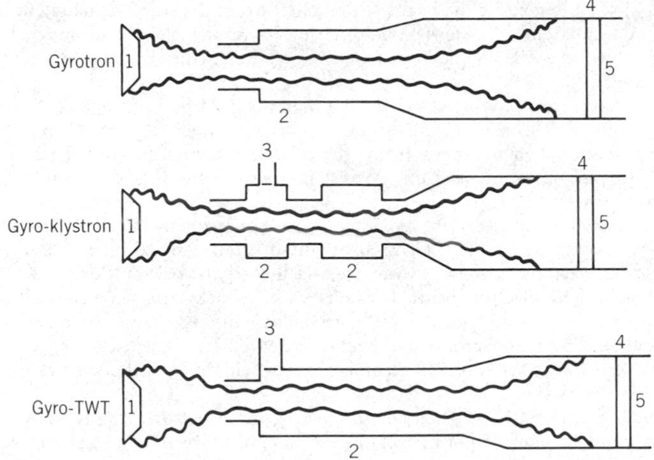

Fig. 18.21 Schematic of gyro devices. 1, cathode; 2, interaction circuits; 3, input waveguide; 4, output window; 5, output waveguide. Reprinted from Jory,[13] with permission.

Gyrotrons

The *gyrotron*[13] in recent developments has a predicted output of hundreds of kilowatts at millimeter wavelengths. The output frequency of this device is determined by the strength of the magnetic field in the tube rather than by the physical dimensions of a resonant structure. Also the beam and microwave circuits require a large area and an elaborate cooling scheme needs to be implemented for achievement of high output power. In the gyrotron, a relativistic effect causes the bunching of the electron beam. Two types, the gyro-klystron and the gyro-TWT amplifier structures, are being investigated. Output power exceeding 1 kW for frequencies above 100 GHz has been demonstrated in gyrotron oscillators. Figure 18.21 shows the schematics of three different gyro device structures. Performance of gyro devices and that of conventional microwave tubes are shown in Figs. 18.22 and 18.23.

18.2-2 Semiconductor Devices

Recent advances in semiconductor material quality and fabrication technologies and the introduction of new device structures and concepts have greatly increased the importance and expanded the applications of solid-state devices in microwave communications[14] and in other systems.[15] Semiconductor devices and their associated microwave circuits can be designed for generating, amplifying, detecting, modulating, and switching microwave signals. A microwave field effect transistor or diode

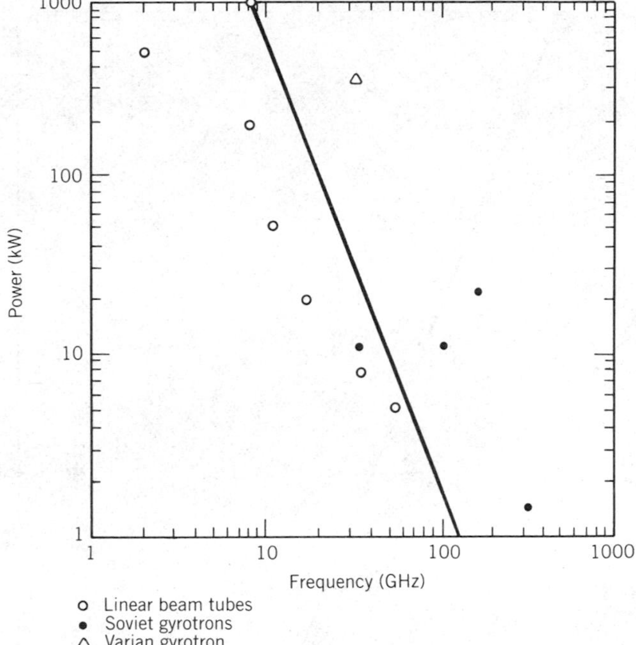

Fig. 18.22 Continuous wave performance of conventional linear beam tubes and gyrotrons. Reprinted from Jory,[13] with permission.

can be used for mixing two signals to produce signals with the sum and difference frequencies in a heterodyne scheme; the frequency of the modulated incoming signal (carrying information) can be down- or up-converted. Finally, harmonic signals can be generated to attain power at a frequency that is a multiple of the fundamental. Table 18.5 lists the various microwave solid-state devices used today.

The development of semiconductor devices[16,17] has, in general, been in two categories. The first approach is the steady extension of the upper frequency limits of existing devices such as bipolar transistors and field effect transistors by virtue of the increasingly fine control of electrode sizes and improvement of material qualities. The second approach is the more erratic progress of new devices based on different structures and new or modified principles. Examples include transfer electron devices and ballistic and permeable-base transistors.

Nonlinear and Plasma Diodes

Varactors. Varactors (*var*iable re*actors*)[18] derive their usefulness from the nonlinearity of capacitance, that is, the nonlinearity of the charge-voltage characteristics. This variable reactance exists as a result of the variation of the width of the depletion region with external bias voltage. The fast response of the reactive nonlinearity and the low RF loss of these diodes allow their use in a number of microwave applications.

The two widely used semiconductors for varactors are silicon (Si) and gallium arsenide (GaAs), the latter being preferred. They are chosen because of their relatively low dielectric constant for minimizing capacitance, their high mobility (for at least one carrier) for achieving low electrical resistance, and their large bandgaps for low saturation currents and reasonable thermal conductivities.

Varactor diodes can be fabricated with diffusion process into n-type epitaxial layer on n^+ substrates. The types of GaAs varactors currently in use are the diffused and the Schottky barrier structures incorporating either epitaxially grown material or nonepitaxial substrates. A typical diode and its equivalent circuit are shown in Fig. 18.24.

The Manley–Rowe power relations[19] are a useful analytical tool for predicting and understanding the principles of varactor applications such as parametric amplifiers, harmonic generators, and frequency up- and down-converters. These general equations relate power flowing into and out of

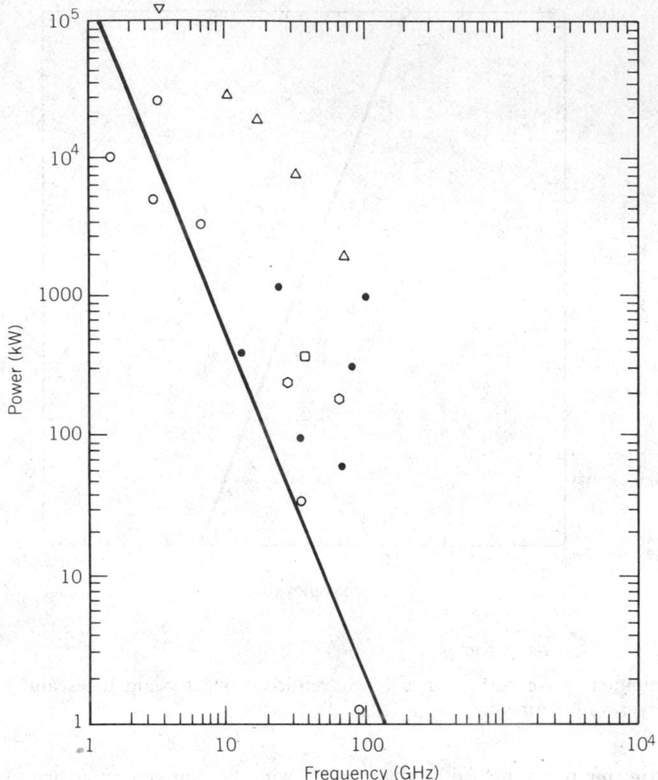

○ Linear beam tubes
● Soviet gyrotrons
□ Naval Research Lab gyrotrons
△ Naval Research Lab ultra relativistic gyrotrons
○ Varian gyrotrons
▽ MIT Relativistic nagnetron

Fig. 18.23 Pulsed performance of conventional linear beam tubes and gyrotrons. Reprinted from Jory,[13] with permission.

nonlinear reactance elements and predict power gain and conversion efficiency. For any lossless nonlinear reactance excited by two signals resulting in current and voltage with frequency components, the relations are

$$\sum_{m=0}^{\infty} \sum_{n=-\infty}^{\infty} \frac{mP_{mn}}{m\omega_1 + n\omega_2} = 0$$

$$\sum_{n=0}^{\infty} \sum_{m=-\infty}^{\infty} \frac{nP_{mn}}{m\omega_1 + n\omega_2} = 0$$

where m and n are integers and P_{mn} is the average power flowing into the single-valued nonlinear reactance at frequencies $\pm|m\omega_1 + n\omega_2|$.

The frequency capability of a varactor is limited by two factors. The first is the dielectric relaxation frequency of the semiconductor, which is inversely proportional to the dielectric constant. The second is the finite value of the saturated velocity of the electrons.

Varactor diodes are used in various functions. Varying the reactance through an externally applied bias allows a microwave signal to be *switched* or *modulated*. When the diode is placed in an oscillator cavity circuit, the frequency of oscillation can be tuned by the variable capacitance of the

TABLE 18.5. COMMON MICROWAVE SOLID-STATE DEVICES

Nonlinear and Plasma Diodes	Tunnel Devices	Transfer Electron Devices (TEDs)	Avalanche, Barrier-Injected Transit Time	Transistors	Quantum Electronic Devices
Varactors	Tunnel diodes	Gunn effect	IMPATT (impact ionization avalanche transit time)	BJT (bipolar junction transistor)	MASER (microwave amplification by stimulated emission of radar)
Schottky					
Epitaxy		LSA (Limited-space-charge-accumulation)	TRAPATT (trapped plasma avalanche-triggered transit)	FET (field effect transistor)	
				JFET (junction FET)	
Varistors				MESFET (metal semiconductor FET)	
Point contact			BARITT (barrier injection and transit time)		
Schottky					
PIN					

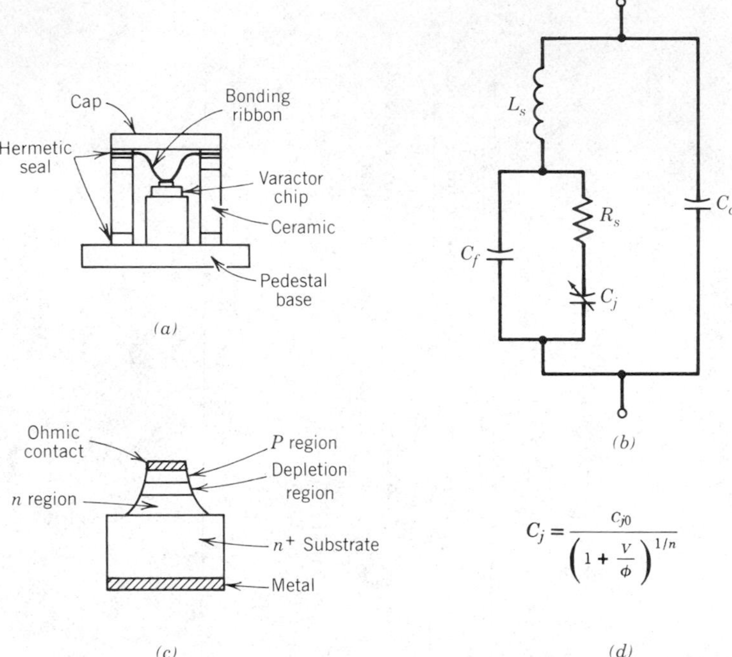

Fig. 18.24 Varactor diode: (*a*) packaged device, (*b*) simplified equivalent circuit, (*c*) device chip, (*d*) junction capacitance relationship: C_{j0}, capacitance at zero bias; V, bias voltage; ϕ, contact potential; n, factor depending on doping profile (2 for abrupt junction, 3 for linearly graded junction).

diode by changing the bias voltage. These applications are usually limited to low power levels (< 0 dBm). Besides the burnout problem, the capacitance characteristics of some diodes may change appreciably as the power level of the RF signal is increased.

Extensive use has been made of varactor *harmonic generators*[20,21] in solid-state microwave sources at high frequencies, where direct generation is difficult or inconvenient. In general, a crystal-controlled transistor oscillator provides a stable reference frequency for the multiplier chain. The oscillator can be either oven or electronically temperature stabilized over environmental changes or life.

Harmonic generators work on the principle that whenever a sinusoidal source drives a nonlinear reactance, the capacitance varies instantaneously with current or voltage. This device can generate harmonics with efficiencies limited mainly by circuit loss and approaching 100%. In comparison, devices that use nonlinear resistance effects, such as varistors, have efficiencies proportional to $1/n^2$ (where n is the order of the harmonic) because of inherent resistance loss.

Other applications of varactor diodes are in the parametric amplification and up-conversion of one of the two microwave signals of different frequencies applied to the diodes.

The *parametric amplifier*[18,22] makes use of a nonlinear element for mixing the pump signal, the input signal, and the idler frequency. The most common application of this device is in a nondegenerate configuration in low-noise amplifiers. The frequency of the pumped source is usually several times the idler frequency. Since the output and input signal are of the same frequency at the same port, a nonreciprocal device such as a circulator generally separates the output from the input. Another technique for separating the output and input power uses hybrid circuits. An equivalent circuit of a parametric amplifier is given in Fig. 18.25.

In the low-microwave-frequency range, the bandwidth of a nondegenerate amplifier with a 20-dB gain can be over 20%. Limited by the low self-resonant frequency of packaged varactors, the maximum bandwidth of higher frequency amplifiers is about 10%. Improvements are being made to extend this limit.

A *degenerate parametric amplifier* permits input termination to be common to both the signal and the idler bands when they partly or entirely overlap because the pump frequency is nearly twice the signal frequency.

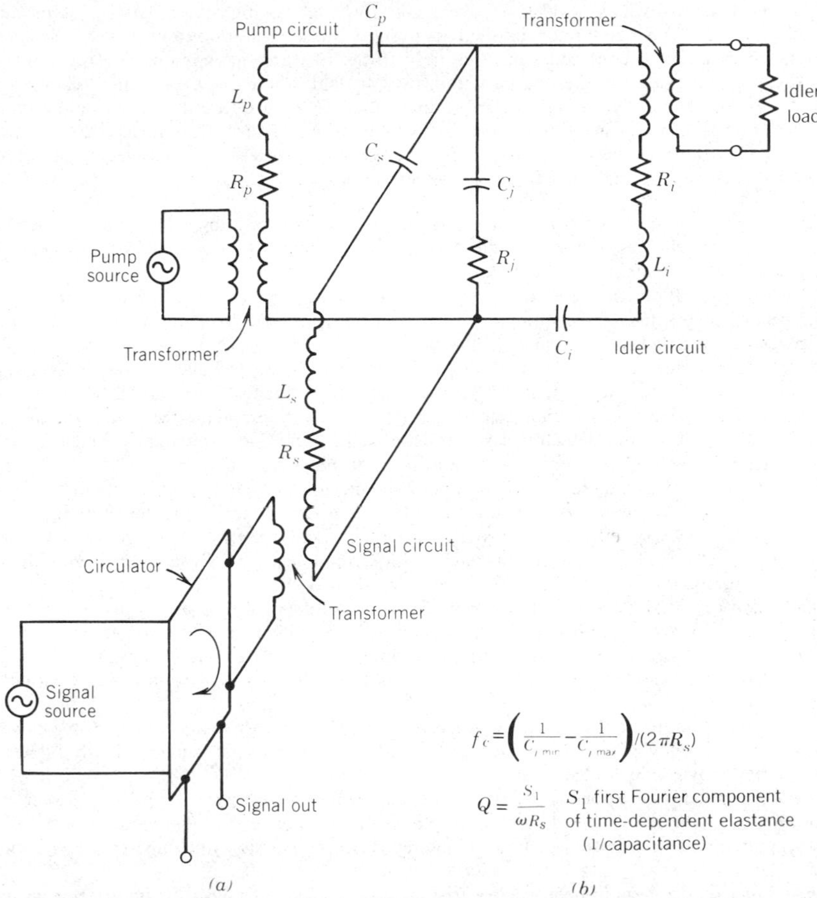

Fig. 18.25 Parametric amplifier: (*a*) general equivalent circuit, (*b*) dynamic figures of merit.

Amplifier design depends on system requirements such as very low noise, large bandwidth, high reliability, and good degree of linearity. The criteria for choosing a varactor include capacitive impedance of the same order as the circulator, a high dynamic quality factor at the signal frequency, a self-resonant frequency as high as attainable, and a constant varactor characteristic for the range of temperature of operation, particularly for low-temperature applications.

The effective input noise temperature of varactor parametric amplifiers can be substantially lowered by cooling the amplifier to liquid-nitrogen temperature and even further to liquid-helium temperature.

Parametric amplifiers have been used satisfactorily in radio-observatory radiometric systems, satellite ground stations, ground-communication systems, and long-range radars. However, the gain and transmission characteristics are susceptible to pump source and circuit impedance variations and to environmental temperature changes. For some low-noise applications, parametric amplifiers are gradually being replaced by the fast advancing microwave field effect transistor amplifiers (FETA) operating in an uncooled or cooled configuration. The microwave FETA offers simpler circuitry, higher gain stability over temperature, and lower cost than the parametric amplifiers.

A basic up-converter circuit produces an output signal, f_0, at a frequency higher than one of the two lower frequency inputs, f_l, after mixing, the other frequency being f_u. The upper sideband up-converter (USUC) generates a signal at $f_0 = f_u + f_l$, and the lower sideband up-converter (LSUC) generates a signal at $f_0 = f_u - f_l$.

Small-signal up-converters, wherein the pump signal is the only large signal, are generally used in low-noise applications. Large-signal up-converters, wherein all three frequencies correspond to large amplitudes of the varactor voltage and current, are designed for high efficiency and high power output.

Step-recovery or *snap-back* diodes[23,24] are generally used in high-order frequency multipliers (commonly times eight), when circuit simplicity is desired and maximum output power is not needed. The reactance variation is confined almost entirely to the forward-bias region. The transition time, that is, the time required for the charge withdrawal to end when the bias voltage changes from forward to reverse, characterizes the quality of these diodes. With optimum design of the impurity profile, this transition time can be reduced from typical values of several hundred seconds to a few picoseconds. The transition period, a harmonic-rich switching transient, allows frequency multiplying without idler circuits, although idler frequencies may exist.

Varistors. *Varistors* (*var*iable res*istors*) are characterized by their nonlinear current-voltage characteristics. The nonlinear resistance can be derived from metal-to-semiconductor contacts, as in point-contact diodes or Schottky barrier diodes.

Two major applications of these diodes are in detector and mixer circuits, as shown in Figs. 18.26 and 18.27, since they satisfy the requirements for good performance: low-minority-carrier injection, low series resistance, and low junction capacitance.

The detector diode[25] can be used in the front stage of a microwave receiver to convert an amplitude-modulated microwave signal directly to the low-frequency information-bearing modulation envelope. The sensitivity of such a scheme is limited by the noise generated within the diode and the amplifier that follows the detector. Series resistance and junction capacitance limit the detector sensitivity. Another factor is the microwave input/output impedance match.

Varistor diodes used as a mixer in a heterodyne scheme can improve the sensitivity of the receiver. The modulated signal (frequency or amplitude modulated) is first mixed with a local oscillator signal. The resulting difference signal (or intermediate-frequency signal) can be set sufficiently high above the $1/f$ noise spectrum of the diode. The intermediate frequency (IF) signal is then amplified and detected accordingly.

Single-ended or unbalanced mixers use a single diode, whereas double-ended or balanced mixers use two diodes. Balanced mixers have three advantages: achieving cancellation of reflected local oscillator power from the mixer toward the input terminals, obtaining cancellation of local-oscillator amplitude-modulated noise at the IF output port, and attaining very high values of intermodulation rejection.

Upper (lower) single-sideband mixers receive a band of frequencies above (or below) the local-oscillator frequency. The conversion loss for a double-sideband is usually slightly higher than that for the best single-sideband mixers.

In contrast to varactor upper and lower single-sideband up-converters, upper and lower single-sideband down-converters using varistors usually have similar conversion-loss values.

Other applications of varistors include frequency down conversion and high-speed limiting and sensing.

Point-contact diodes[26] (millimeter-wave varistor diodes) are mostly used for millimeter waves. As a result of the small point-contact area, the capacitance may be made very small; consequently, the cutoff frequency can be very high. Parasitic inductance and capacitance arising from the leads and the support for the active region can be minimized.

Point-contact varistors are also used in a clipper circuit for limiting comparatively low-power signals in FM receivers. The advantage of varistors over varactors for this purpose is that the varistor does not exhibit amplitude-to-phase modulation conversion.

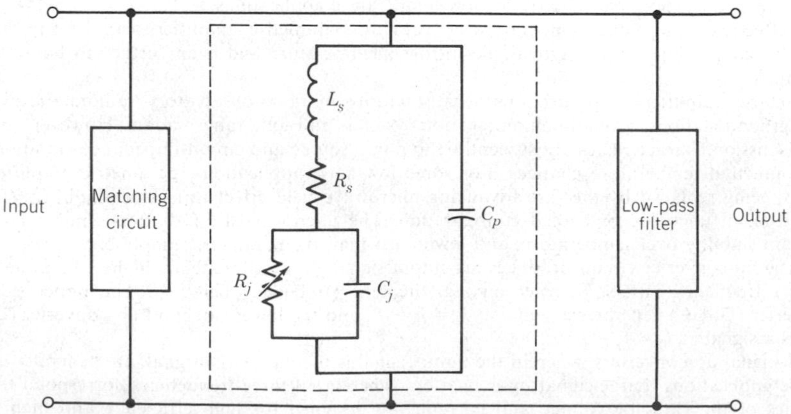

Fig. 18.26 A varistor detector showing equivalent circuit of the diode.

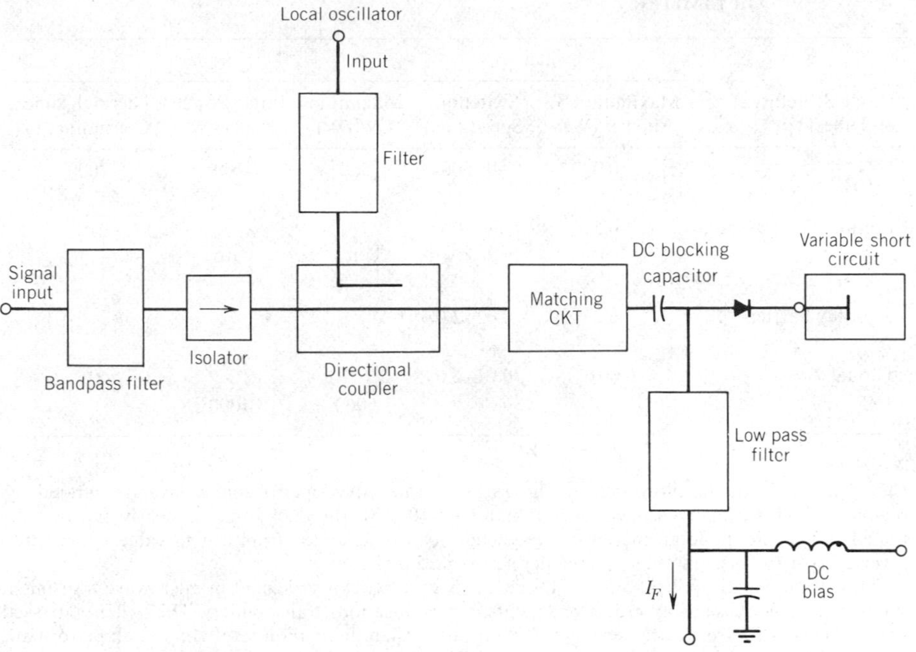

Fig. 18.27 Unbalanced (single-ended) diode mixer circuit.

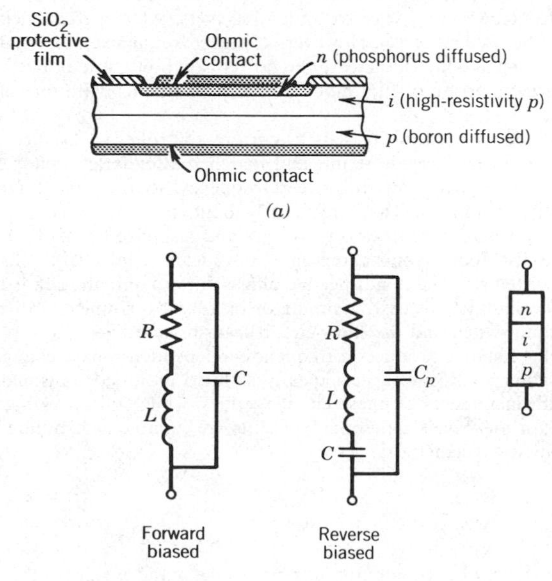

Fig. 18.28 PIN diode: (*a*) typical wafer structure, (*b*) equivalent circuit.

TABLE 18.6. COMPARISON OF PERFORMANCE OF MICROWAVE DIODES AS SWITCH OR LIMITER

	Switch		Limiter		
Diode Structures or Tube Type	Maximum CW Power (W)	Switching Speed (μs)	Maximum CW (W)	Input Power Peak (kW)	Thermal Time Constant (μs)
PIN	50.0	50–500	500	100	2000
Epi-junction varactor	0.5	1–100	} 10	} 10	} 20
Schottky barrier	0.2	0.02–2.0			
Tunnel	0.001	0.01–2.0			
T/R tube			5000	1000	

PIN. The *PIN* diode, illustrated in Fig. 18.28, has heavily doped p and n layers separated by a region of high-resistivity semiconductor material (10–200 μm thick) that is nearly intrinsic. The impedance of the diode at microwave frequencies can be changed from a high value under zero or reverse bias to a very low value at a moderate forward bias.

One of the major applications of PIN diodes is as a switching element in microwave transmission systems such as phase-array radars or satellite communications transponders. The switches, basically reactive networks, use a difference in reflection rather than dissipation to obtain switch performance. Thus little power is dissipated by the small-size diode to control relatively large amounts of power. In general, at power levels higher than about 1 W only PIN diodes, which typically have a larger wafer area than other types and a high breakdown voltage, can withstand the large RF voltages and currents. Typical operating power levels and switching speeds of PIN diodes and those of other diode types are given in Table 18.6.

PIN diodes are also used as switched limiters[27] in applications such as protectors to prevent high transmitter power from damaging sensitive radar receivers. At low frequencies (< 500 MHz), the shunt PIN diode can be used passively. However, at higher frequencies, external bias must be used to provide adequate limiting action. To increase the power capability, it may be necessary to use several diodes in parallel. A comparison of PIN diode limiter performance with that of other devices is also shown in Table 18.6.

When a PIN diode is forward biased, a current-controlled resistance in parallel with the capacitive effect in the intrinsic region results and matched attenuator circuits can be formed.[28] PIN attenuators are used in leveler loops for sweep-frequency sources, pulse modulators, and pump sources for parametric amplifiers. They are usually built in compact configuration with dynamic ranges up to 80 dB with bias currents of 0 to 10 mA and response times of 30 to 100 nsec.

PIN diodes have also found applications in electronically controllable phase-shifter circuits for phased-array radar systems.[29] The nondispersive phase-shifter configuration is basically a switchable length of transmission line. In this case, a circulator or a hybrid coupler can be used to direct signal in and out of the diode switch and the short-circuit transmission lines.

The dispersive phase shifter produces a frequency-independent phase change within the designed band. One configuration consists of a periodically shunted mounted transmission line arranged for binary operation with increments of phase change values, while still providing low voltage standing wave ratio (VSWR) for any phase states. Such circuits are capable of handling kilowatts of power in 120-μsec pulses in the low gigahertz range.

Tunnel Devices

Tunnel diodes[30,31] have very heavy impurity doping on the p and n sides and an abrupt transition at the junction. When the diode is forward biased to certain voltages, the electrons can tunnel (by virtue of quantum mechanics) through the potential barrier at the junction to energy states on the opposite side. The tunneling phenomenon is a majority carrier-effect which, together with the energy-band structure, gives rise to a negative resistance over a portion of the forward current-voltage characteris-

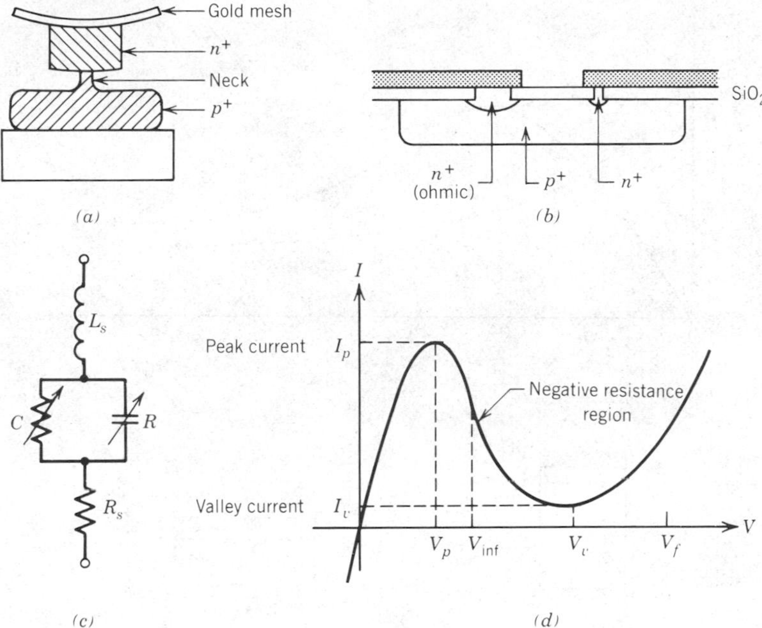

Fig. 18.29 Tunnel diode: (*a*) typical structure, (*b*) planar structure, (*c*) equivalent circuit, (*d*) *I-V* characteristics.

tic. Even at small reverse voltages, a large current conduction occurs. The most commonly used material is germanium. Other materials such as gallium-arsenide, gallium-antimonide, and silicon have been studied. Figure 18.29 illustrates a common microwave tunnel diode structure, its equivalent circuit, and the forward *I-V* characteristic.

Because of their fast operation mechanism, tunnel diodes have been used in different applications up to the millimeter-wave region. Diodes constructed with low peak currents have nonlinear *I-V* characteristics with high curvature around zero bias and a large reverse current which allow the devices to function passively in detector, mixer, and low-noise frequency conversion applications. These devices are usually called *backward diodes*.

Tunnel diodes actively used in the negative resistance range of the forward-biased characteristic have found applications in microwave medium-noise amplifiers and low-cost oscillators. They are generally referred to as *Esaki diodes*, after the person who pioneered the analytical study of these structures.

As diode characteristics are very sensitive to doping profiles and energy levels in semiconductors, extreme attention is required in the mechanical and electrical design of a tunnel-diode circuit to prevent mechanical stress, high-level radiation, or static discharges on the device. Special consideration is needed to ensure circuit stability and gain variation (as with an amplifier) over temperature changes. Lastly, tunnel diodes have low power-handling capability. As a result, they are gradually being replaced by microwave transistors in amplifiers and oscillator applications and by Schottky-barrier diodes in mixer and detector uses.

Bulk Effect Devices

Transfer Electron Devices (TEDs). TEDs[32–34] are bulk material structures with ohmic contacts. They have no *pn* junctions, as illustrated in Fig. 18.30. Their operation is based on the differential negative conductance effect that occurs in a number of semiconductors. *n*-Type gallium arsenide (GaAs) and *n*-type idium phosphide (InP)[35] materials are the most intensively investigated. The effect has also been observed in germanium (Ge) binary, ternary, and quaternary compounds. TEDs operate with "hot" electrons, those with energies much higher than the thermal energy.

The Gunn effect devices, as TEDs are sometimes called, were discovered experimentally by J. B. Gunn in 1963. Coherent microwave fluctuations of current through the *n*-type GaAs and InP

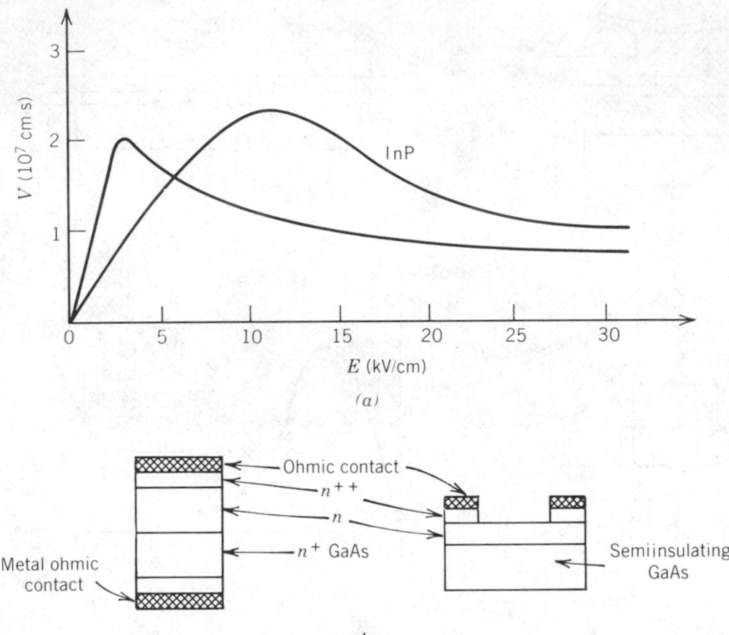

Fig. 18.30 Bulk effect devices: (*a*) Velocity vs. electric field of GaAs and InP devices showing the negative differential mobility. (*b*) Schematic geometries of samples.

materials were observed when the applied voltage exceeded a certain critical value of several thousand volts per centimeter. The frequency of oscillation was approximately equal to the reciprocal of the transit time across the length of the sample. Early theoretical predictions by Ridly and Watkins and by Hilsum had stated that a negative differential resistance occurs because of the reduction of electron drift velocity with increased electric field owing to the gradual transfer of a fraction of the conduction-band electrons from a low-energy, high-mobility state to a higher energy, low-mobility state. Kroemer pointed out that the theoretical transferred-electron effect was consistent with the experimental Gunn effect. Later, Copeland's computer and experimental work demonstrated oscillations that were not transit-time limited. Thus the bulk property was extended to other modes of operation.

Several different modes of operation have been studied depending on impurity doping density and uniformity, active region dimension, cathode contact property, bias condition, and circuit configuration. The uniform-field mode implies that no buildup of internal space change occurs, and the current-voltage characteristics of the device are directly related to the velocity-field relationship. The ideal maximum efficiencies are 30% (with GaAs) and 45% (with InP) and are independent of the operating frequency when the frequency is lower than the reciprocals of the energy relaxation and the internal scattering times.

A TED of lightly doped or short sample with an n_0L product less than 10^{12} cm^{-2} (where n_0 is the doping concentration and L is the length of the sample) exhibits a stable field distribution and positive DC resistance. When it is connected to a parallel resonant circuit with appropriate load resistance, the TED can oscillate in the transit-time accumulation mode with almost 10% efficiency.

The transit-time dipole-layer mode occurs when the n_0L product is larger than 10^{12} cm^{-2}. Mature dipole layers are formed near the cathode as a result of space-change perturbations in the material increasing exponentially in space and time, propagate to the anode. The device is generally placed in shunt with a high-Q resonant circuit.

The quenched dipole-layer mode results with a TED in a resonant circuit if the high-field dipole-layer was quenched before it arrived at the anode. Oscillation can occur at frequencies f of the order of the dielectric relaxation frequency and higher than the transit time frequency, that is, when $fL > 2 \times 10^7$ cm/s. The theoretical efficiency can be 13%.

Another important mode of operation, especially for high peak-power pulse application, is the limited-space-charge-accumulation (LSA) mode. The electric field across the device rises from below the threshold and falls back again quickly enough that high-field dipole layers do not have time to be formed. Only the primary accumulation layer occurs near the cathode and is quenched during one RF cycle. For GaAs and InP, the limiting ratios are $10^4 < n_0/f < 10^5$ s/cm^3. Thus the mode of operation is highly circuit dependent. The highest operating frequency of LSA TEDs is much less than that of transit-time devices, with an upper limit of about 20 GHz for GaAs and a higher frequency for InP.

TEDs, with their reasonably low-noise (AM and FM) and broadband characteristics, have been extensively used in local oscillators for microwave instruments, intrusion alarms, and pump sources for parametric amplifiers. Power TED amplifiers and oscillators have found applications in radar systems. A summary of their performance for both CW and pulsed operation as a function of frequency is given in Fig. 18.31. In general, these TEDs have demonstrated an operating frequency

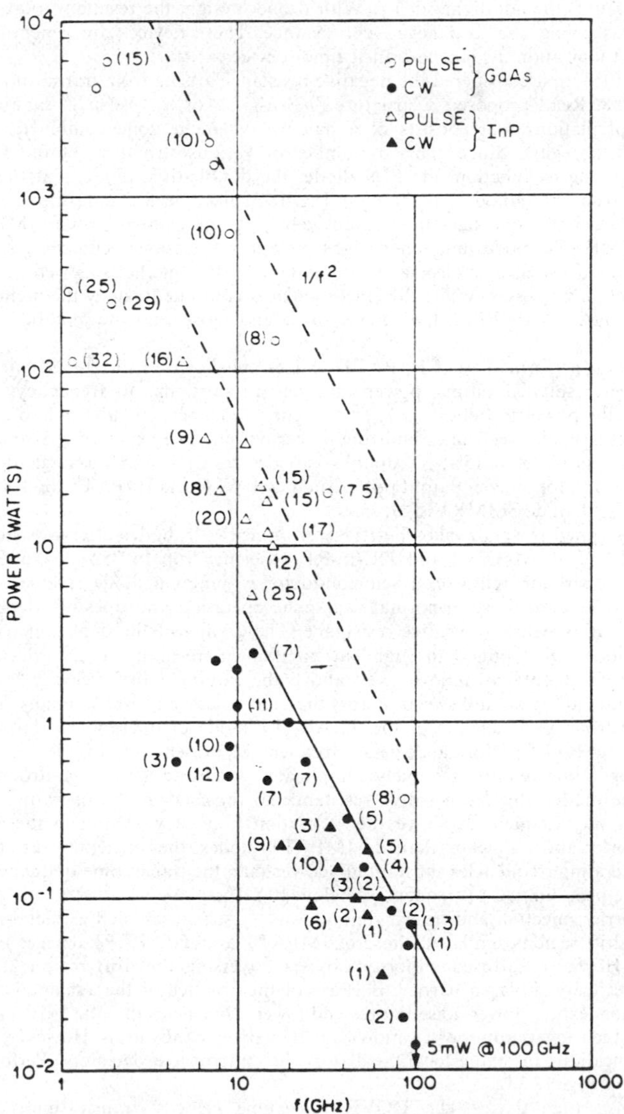

Fig. 18.31 Summary of transfer electron device (TED) performance. The numbers in parentheses indicate the DC-to-RF conversion efficiencies in percent. Reprinted from Sze,[16] p. 34, with permission.

range from below 1 GHz to over 100 GHz. Continuous power from a few milliwatts to over 2 W with efficiencies up to 15% and pulsed power from 1 W to 6 kW with efficiencies from a few percent to 30% have been attained. TEDs have an estimated upper-frequency limit around 150 GHz.

TEDs have also found applications in high-speed digital and analog operations.[36,37] One configuration of these functional devices is with nonuniform cross-sectional area. The second consists of one or more additional electrodes along the length of the device to control the current waveforms.

Avalanche and Barrier-Injected Transit-Time Devices

The operation of avalanche transit-time devices[38,39] depends on two mechanisms to produce negative resistance at microwave frequency. The avalanche breakdown due to impact ionization generates an avalanche current delayed with respect to terminal voltage. The transit time of these charges through the drift region causes the additional delay. With proper design, the resultant total phase difference can be 180°, thus giving rise to a negative resistance. These devices are generally referred to as IMPATT (*imp*act ionization *a*valanche *t*ransit *t*ime) devices.

Shockley, in 1954, first considered the negative resistance arising from transit time in semiconductor diode. In 1958, Read proposed a structure (p^+-n-ν-n^+ or n^+-p-π-p^+) capable of generating high-frequency oscillations that consists of a narrow avalanche zone combined with a relatively high-resistance drift region. Since 1965, a number of various structures of the IMPATT family, including the one-side *pn* junction, the PIN diode, the double-drift diode, and the modified Read structures (lo-hi-lo and hi-lo diodes) as shown in Fig. 18.32 have been demonstrated.

Both Si and GaAs materials have been extensively used and gemanium (Ge) IMPATTs have also been reported. IMPATT diode junction designs include fabrication techniques, such as diffusion, double epitaxy by diffusion, or molecular beam epitaxy (MBE) methods as well as Schottky-barrier and ion-implantation processes. While the silicon devices continue to show the highest output-power capabilities, the GaAs IMPATTs exhibit lower noise and those with the modified Read structures, higher efficiency.

Some of the best performances of IMPATT devices are shown in Fig. 18.33. For CW operations, thermal limitation results in output power inversely proportional to frequency, while electronic limitation causes the power to fall off as $1/f^2$ at higher frequencies. Under pulsed conditions, where the thermal effect is small, electronics limit the power output. The expected maximum efficiency for single-drift Si devices is about 15%, for double-drift devices 21%, and for single-drift GaAs diodes about 38%. In general, for a given output power such as 1 W, GaAs IMPATT diodes exhibit about 10 dB lower FM noise than do Si IMPATT devices.

TRAPATT (*tra*pped *p*lasma *a*valanche-*t*riggered *t*ransit)[40,41] diodes offer very high pulsed power (as much as 1.2 kW at 1.1 GHz) and DC-to-RF efficiency (up to 75% at 0.6 GHz) microwave generation. Their operation relies on a semiconductor *pn*-junction diode reverse biased to current densities much larger than those in normal avalanche operation and does not depend on transport and injection delays to cause a negative resistance. The doping of the depletion region is generally such that the diodes will "punch through" at breakdown, resulting in DC electric fields in the depletion region just before breakdown well above the saturated drift-velocity level. Avalanching occurs at the high-field region and sweeps across the diodes, leaving it substantially filled with plasma of holes and electrons. A schematic of the TRAPATT mode of operation is shown in Fig. 18.34. State-of-the-art TRAPATT performance versus frequency is presented in Fig. 18.35.

TRAPATT operation requires relatively complicated circuit design and control of device properties to match the diode effective negative resistance to the load at the operating frequency while reactively terminating frequencies above the oscillation frequency. Although these devices exhibit higher pulsed power and efficiency than do IMPATT diodes, they generally result in higher noise measures and operating frequencies substantially lower than the transit time frequency, and thus their applications have been limited to frequencies below 10 GHz.

BARITT (*bar*rier *i*njection and *t*ransit *t*ime) diode[42,43] structures such as metal-n-metal, p-n-ν-p, and p-n-p have drift regions similar to those of IMPATT and TRAPATT devices but with a transit angle of $3\pi/2$. However, diffused minority carriers traversing the drift region are thermionically injected from the forward-biased barrier. Because of the absence of the avalanche phenomena, the BARITT oscillators exhibit lower noise figures and lower efficiencies than do IMPATT devices. Their use is limited by their low output power and relatively narrow bandwidths. However, BARITT diodes have found applications in stable local oscillators and microwave detectors. Performance of these diodes is shown in Fig. 18.33.

Another transit time device, the DOVETT (*do*uble *v*elocity *t*ransit *t*ime) diode, has been proposed. In this device, the carrier velocity near the injection terminal is much lower than that close to the collection terminal. As a result, the transit time delay in the low-velocity injection region increases. DOVETT diodes are expected to yield higher negative resistance and thus better efficiencies than BARITT diodes.[44,45]

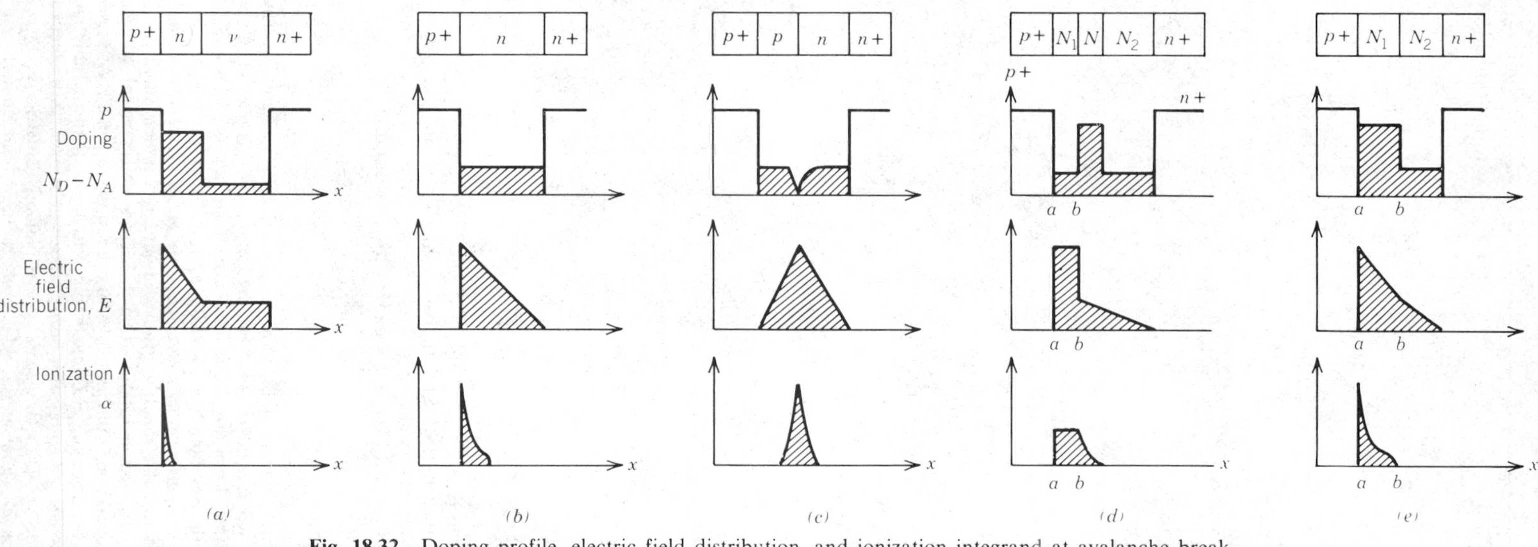

Fig. 18.32 Doping profile, electric field distribution, and ionization integrand at avalanche breakdown of various IMPATT (impact ionization avalanche transit time) diode structures: (*a*) Read, (*b*) abrupt *pn*, (*c*) double drift, (*d*) modified Read lo-hi-lo, (*e*) modified Read hi-lo.

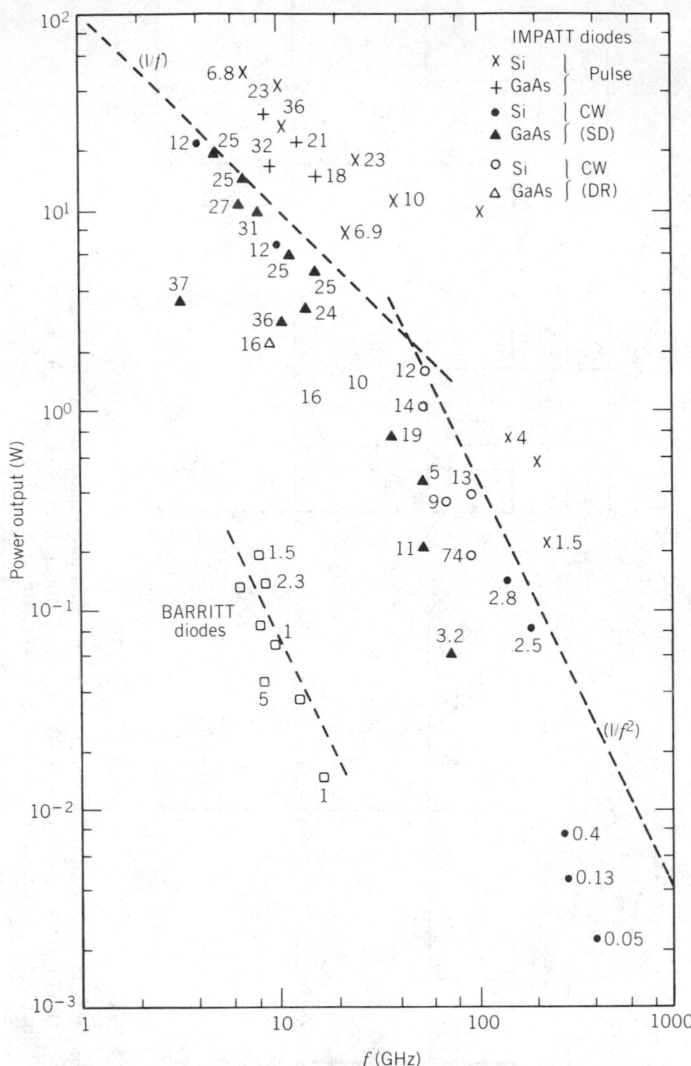

Fig. 18.33 Performance of IMPATT (impact ionization avalanche transit time) and BARITT (barrier injection and transit time) diodes. Efficiency in percent is given by the number against each experimental point. Reprinted from Sze,[16] p. 51, with permission.

Microwave Transistors

Microwave amplification utilizing the previously presented two-terminal devices requires a circulator or hybrid circuits to separate the output from the input signal. Transistors (*trans*fer re*sistors*), bipolar or unipolar (field effect), are three-terminal devices capable of delinearating the input from the output. In bipolar devices both the majority and minority carriers are involved in the current conduction process, whereas in unipolar devices only majority carriers predominate.

Microwave BJTs (*b*ipolar *j*unction *t*ransistors) require much smaller dimensions (micrometer range) and have more stringent fabrication processes and thermal considerations than do the low-frequency devices, although the principles of operation are similar. At present almost all BJTs are of the silicon *npn* type, since the electron mobility is higher than the hole mobility. To reduce the collector series resistance, an epitaxial n on n^+ substrate is used. A protective SiO_2 layer is usually deposited on the surface. These transistors have planar geometries of three general shapes[46,47]—interdigitated, overlayed, and mesh—which characterize the emitter and base configurations

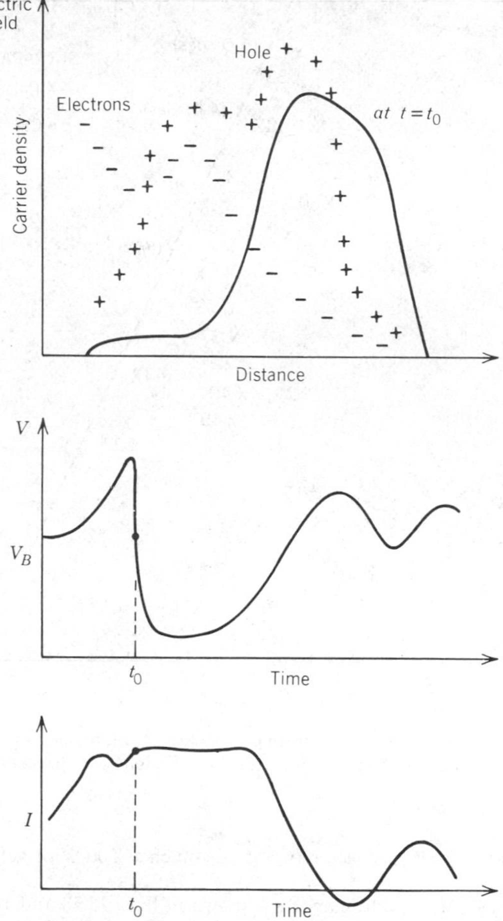

Fig. 18.34 Numerical calculation of electric field carrier density vs. distance, voltage and current vs. time of a TRAPATT (trapped plasma avalanche-triggered transit) diode. Reprinted from Evans,[41] p. 1061, with permission.

(Fig. 18.36). These formats allow reduction of emitter-strip width and base-layer thickness for high-frequency operation while maintaining current-power capability.

An important figure of merit is the maximum oscillation frequency, $f_{\max}$, derived by extrapolating the unilateral power gain versus frequency curve to unity. Considering that the emitter and base electrode pattern of a transistor is defined in terms of electrode spacing S, length L, and width W and the per unit area base resistance r_0 and collector capacitance C_0, $f_{\max}$ are given by

$$f_{\max} \simeq \frac{1}{2S}\left(\frac{f_T}{2\pi r_0 C_0}\right)^{1/2}$$

where f_T, the cutoff frequency at which the common emitter short-circuit current gain becomes unity, is inversely proportional to the emitter-to-collector delay time.[48]

The past decade has seen the development of numerous approaches for increasing the performance of bipolar transistors by reducing the intrinsic collector-base capacitance or shrinking the lateral geometry by implementing various self-aligning techniques. Also different structures to reduce the effective emitter-strip width and the base resistance have been implemented, resulting in high cutoff frequencies and low noise figures.

Figure 18.37 shows the output power performance of bipolar transistors. As noted, the power falls off as $1/f^2$ as a result of carrier saturation velocity and the limitations of avalanche breakdown. For

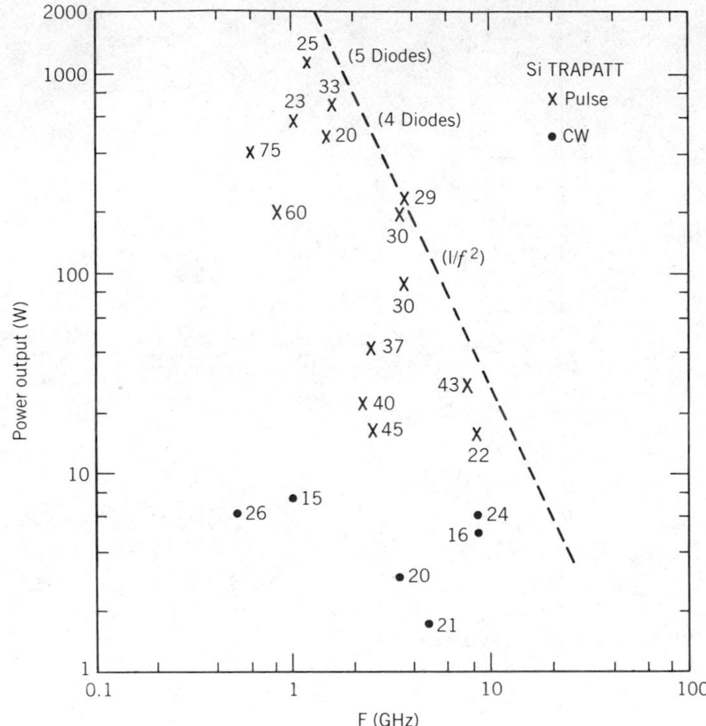

Fig. 18.35 Performance of TRAPATT (trapped plasma avalanche-triggered transit) diode. The number against each experimental point indicates the efficiency in percent. Reprinted from Sze,[16] p. 30, with permission.

CW operation, 1.5 W at 10 GHz has been reported. As much as 500 W of pulsed power at 1 GHz can be achieved.

The low noise and oscillator performance are shown in Figs. 18.38 and 18.39. The silicon bipolar transistor remains the best discrete device for oscillator applications below the x-band because of its low $1/f$ noise properties. However, for low noise and power applications above the lower microwave range, the microwave field effect transistor has gradually become the dominant device.

Other bipolar transistor-like three-terminal structures have been explored, including the heterojunction transistor[49] using an $Al_xGa_{1-x}As$ emitter, a p-type GaAs base, and an n-type GaAs collector; hot-electron transistors such as a metal/insulator (M-I-M-I-M) structure;[50] and a permeable-base transistor[51] using a four-layer GaAs/metal process. All of these structures are aimed at improving performance and increasing frequency of operation.

The FET (*f*ield-*e*ffect *t*ransistor)[52,53] is essentially a voltage-controlled resistor. Currently, planar gallium-arsenide (GaAs) MESFETs (*me*tal *s*emiconductor *f*ield *e*ffect *t*ransistors) dominate most microwave applications. A metal-semiconductor (Schottky-barrier) rectifying contact instead of a *pn* junction as in the case of JFET (*j*unction *f*ield *e*ffect *t*ransistor) is used for the gate electrode. In the normal common source amplifier configuration, the gate is reverse biased and the input impedance is high. When a microwave signal is applied at the gate, the depletion layer width is modulated. This modulation causes the conduction channel to vary and this in turn causes the large output drain current variation across the load. The majority of the devices use n-type GaAs material because the mobility of electrons is higher in this material than that of holes and carriers in silicon. Oscillations in MESFET devices close to 60 GHz have been observed.

MESFETs offer simpler fabrication processes, lower device resistance for low-noise application, and good heat dissipation for high-power operation compared with JFETs. JFETs, on the other hand, can be fabricated with a heterojunction or a buffered-layer gate configuration, offering the potential of improved high-frequency performance.

A cross-sectional schematic of a MESFET is shown in Fig. 18.40. An equivalent circuit model for small signal operation showing the parasitic elements is also given. MESFETs with recessed gate

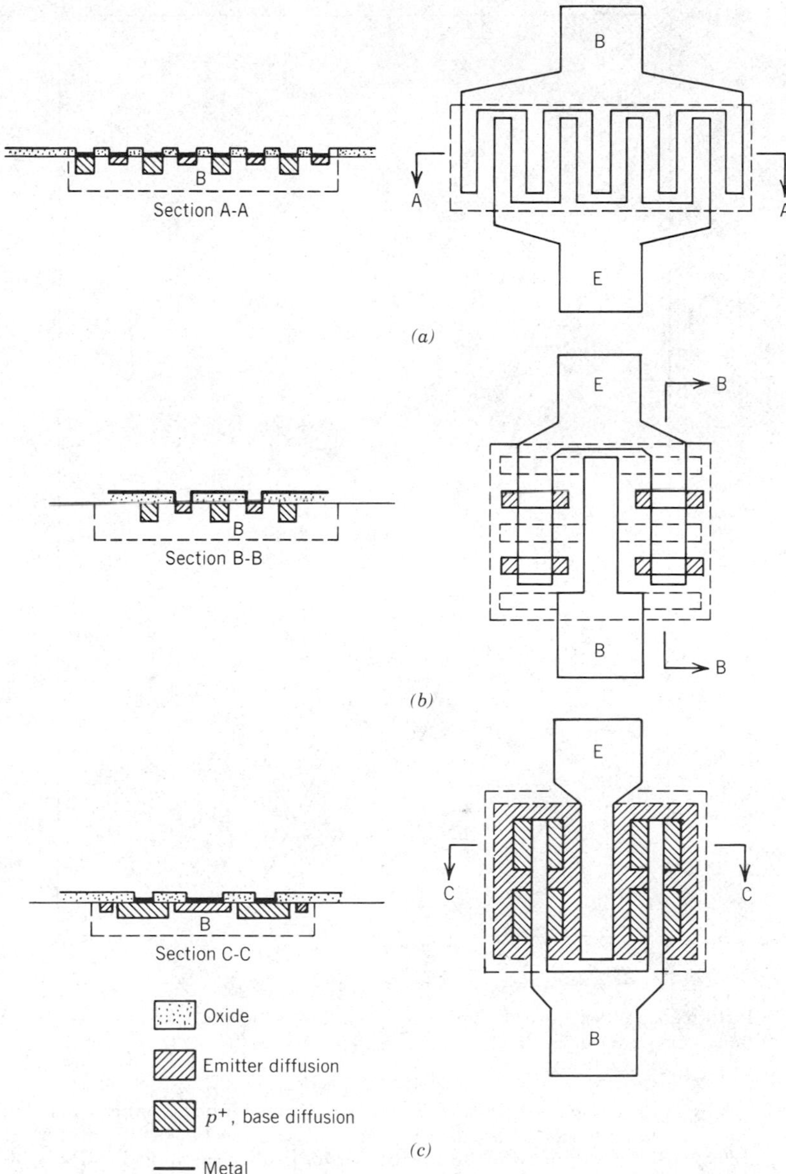

Fig. 18.36 General types of microwave transistor geometry: (*a*) interdigitated, (*b*) overlay, (*c*) mesh. From Cooke,[46] with permission.

length of less than a quarter of a micron have been fabricated, and this development has improved the low-noise performance and has extended the high-frequency applications of these devices. New fabrication techniques using electron-beam lithography and ion implantation as well as material growth methods such as molecular beam epitaxy (MBE), metal-oxide chemical vapor deposition (MOCVD), and liquid-encapsulated Czochralski (LEC) techniques are being used to achieve the small dimensions and manufacturing capability for the devices. In power MESFET technology, devices with gate widths of several millimeters and drain breakdown voltages of over 40 V have been demonstrated, through proper device design including recessed gate structure, via hole-plated heat sink, and flip-chip mounting techniques. An on-chip matching method that allows impedance matching very close to the device has resulted in multiple-cell chips achieving broadband operation.

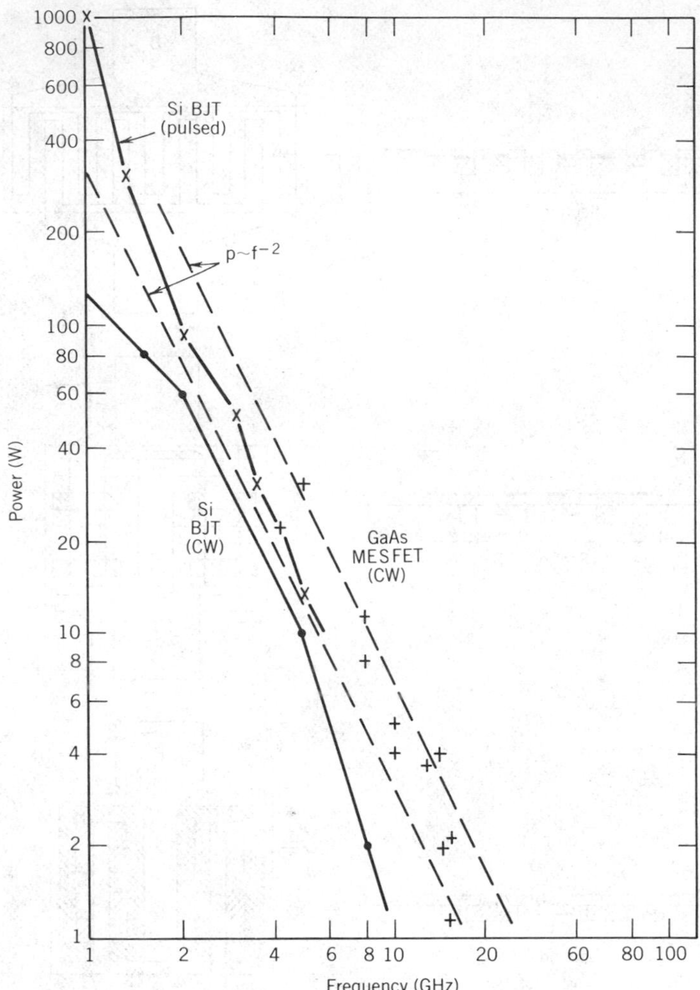

Fig. 18.37 Performance of silicon BJT (bipolar junction transistor) and gallium-arsenide MESFET (field effect transistor) power devices.

MESFET chips and packaged devices have mostly been used with planar transmission lines such as microstripline and lumped-element circuits. Advances in device technology have resulted in the dominance of this type of device in low-noise applications for frequencies up to 30 GHz. With use of cooling techniques using thermoelectric or cryogenic devices, noise figures comparable to those of cooled parametric amplifiers have been demonstrated. Figure 18.38 shows the performance of MESFETs as a function of frequency.

MESFET power amplifiers with devices of multiple interdigitated finger configuration are gradually replacing traveling-wave tube amplifiers (TWTAs) in the 4-GHz communications band. Amplifiers of power output as high as 40 W are available at C-band. Power outputs up to about 2 W in the 20-band have been demonstrated. Solid-state power amplifiers (SSPAs), as these MESFET amplifiers are sometimes referred to, offer better nonlinearity characteristics and reliability than the TWTAs. However, DC-to-RF conversion efficiencies are still lower, even though SSPAs require much lower DC voltage. The present status of these devices is given in Fig. 18.37.

Besides low-noise and high-power amplification uses, MESFETs have also found applications in microwave switches, mixers, power dividers, and combiners and oscillators. Studies are still being made to improve the $1/f$ noise problem associated with GaAs MESFET oscillators. A dual-gate configuration (two parallel gates between source and drain terminals) has been used for both mixer

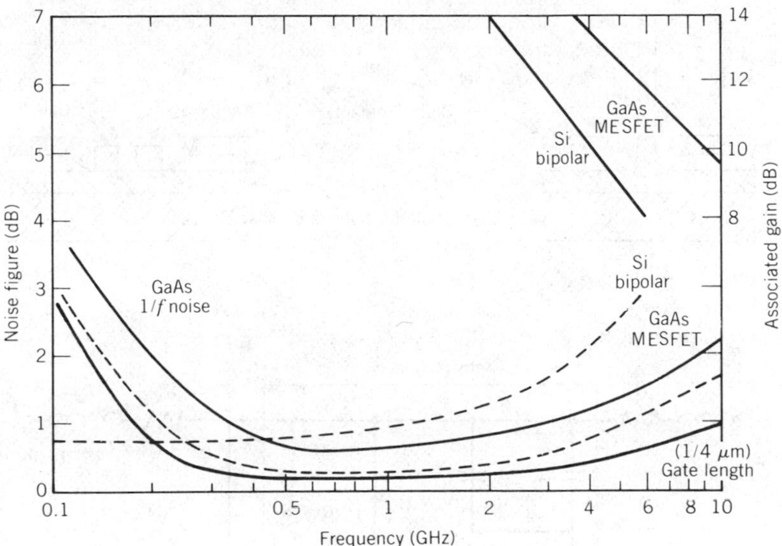

Fig. 18.38 Low noise performance of silicon BJT (bipolar junction transistor) and gallium-arsenide MESFET (field effect transistor).

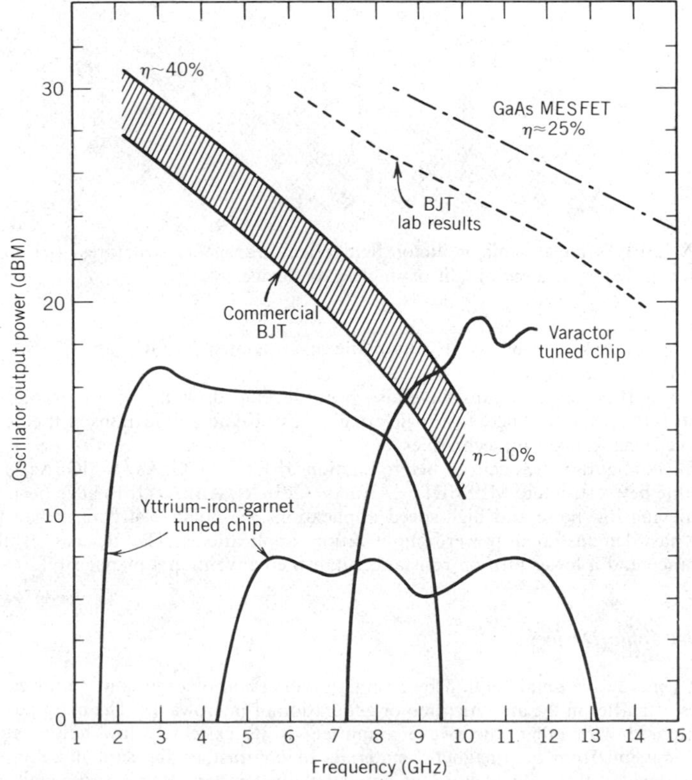

Fig. 18.39 Oscillator performance of BJT (bipolar junction transistor) and gallium-arsenide MESFET (field effect transistor).

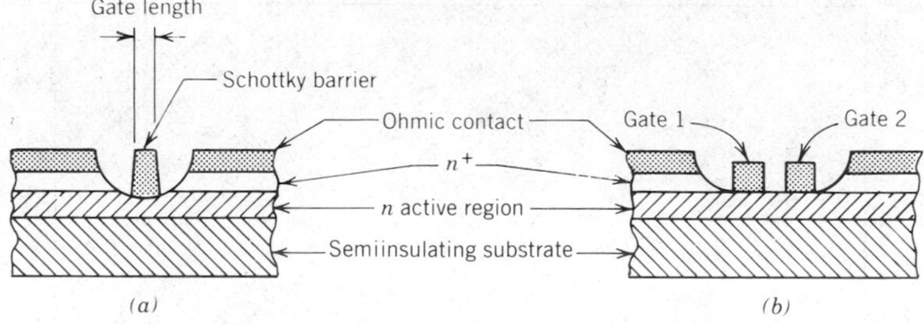

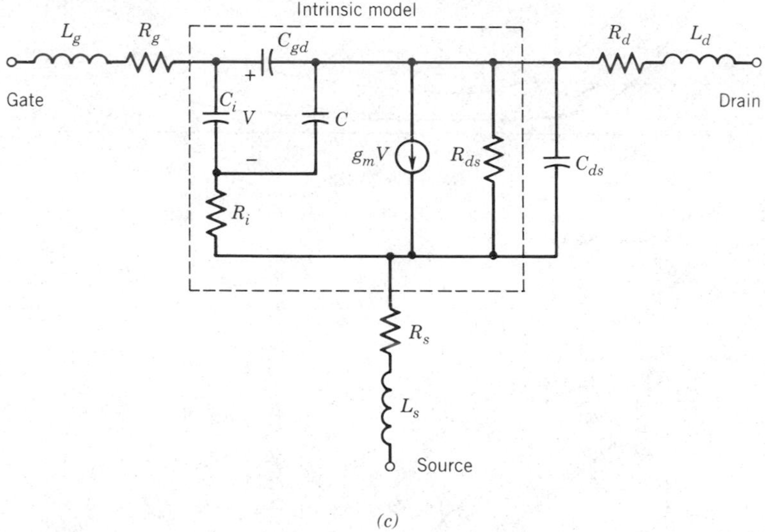

Fig. 18.40. MESFET (metal semiconductor field effect transistor) structures: (*a*) single recessed gate, (*b*) dual gate. (*c*) Equivalent circuit of single-gate device.

and switch applications as well as in automatic gain control (AGC) amplifiers, providing an additional control of the device.

MESFET and JFET logic circuits have also been developed, with the more rapid growth using GaAs material rather than Si. These fast logic circuits have found applications in the development of new microwave signal-processing techniques.

Other FET configurations, such as heterojunction JFET[54] (*p*-GaAs/*p*-AlGaAs/*n*-GaAs structure) and double heterostructure MESFET[55] (AlInAs/GaInAs/AlInAs/InP) have been reported and aimed at achieving low-noise and high-speed applications. Nonplanar structures like the V-groove silicon FET has demonstrated power amplification applications. The devices exhibit a higher transconductance and a lower turn-on resistance than the conventional planar FET.

Quantum Electronic Devices

The MASER (*m*icrowave *a*mplification by *s*timulated *e*mission of *r*adiation)[56] utilizes the quantum characteristics of matter in the amplification or generation of microwaves. The main advantage of the MASER compared with other microwave amplifiers is its extremely low noise caused only by spontaneous emission from paramagnetic material. In contrast to the shot noise in conventional electron-beam devices, this kind of noise at liquid helium temperatures is very small at microwave frequencies. MASERS have been used as low-noise receivers for amplifying weak signals in radio astronomy and satellite earth station applications. They also operate as an oscillator, providing a very

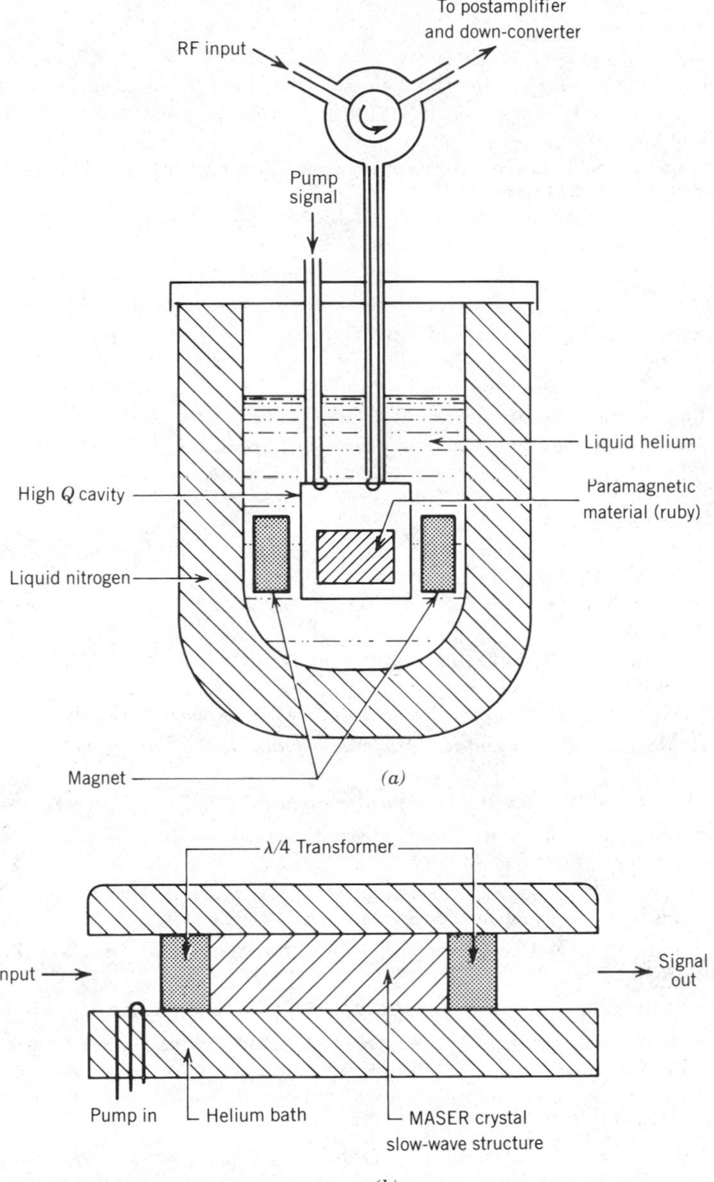

Fig. 18.41 Schematic of MASER (microwave amplification by stimulated emission of radiation) device: (*a*) cavity MASER, (*b*) simple traveling-wave MASER.

accurate time or frequency standard because of their extreme frequency stability and narrow bandwidth characteristics.

A schematic diagram of a ruby three-level cavity-MASER for low-noise application is shown in Fig. 18.41. The resonant cavity, limited by its usable bandwidth, can be replaced by a slow-wave structure. The electromagnetic field in this traveling-wave MASER[57] is weakly coupled to the structure and interaction occurs for an extended distance and time resulting in high signal gain. The ammonia two-level MASER operating at a sufficiently high gas-beam density functions as an oscillator and provides an accurate frequency standard.[58]

References

1 *Proc. IEEE*, special issue on microwave tubes, **61**(3) (Mar. 1973).

2 E. Okress, ed., *Crossed-Field Microwave Devices*, Vols. 1 and 2, Academic, New York, 1961.

3 D. R. Hamilton, J. K. Knipp, and J. B. H. Kuper, *Klystrons and Microwave Triodes*, McGraw-Hill, New York, 1948.

4 J. E. Beggs and N. T. Lavoo, "High Performance Experimental Power Triodes," *IEEE Trans. Elect. Dev.* **ED-13**:502 (May 1966).

5 D. H. Priest, "Annular Circuits for High Power Multiple Tube RF Generators at VHF and UHF," *Proc. IRE* **38**:515–520 (May 1950).

6 S. Y. Liao, *Microwave Devices and Circuits*, Prentice-Hall, Englewood Cliffs, NJ, 1980.

7 J. R. Pierce, *Traveling Wave Tubes*, D. Van Nostrand, Princeton, NJ, 1950.

8 J. T. Mendel, "Helix and Coupled-Cavity Traveling-Wave Tubes," *Proc. IEEE* **61**(3):280–289 (Mar. 1973).

9 A. Staprans et al., "High-Power Linear-Beam Tubes," *Proc. IEEE* **61**(3):299–330 (Mar. 1973).

10 G. B. Collins, *Microwave Magnetrons*, McGraw-Hill, New York, 1948.

11 J. V. Gewartowsky and H. A. Watson, *Principles of Electron Tubes*, D. Van Nostrand, Princeton, NJ, 1965.

12 J. F. Skowran, "The Continuous-Cathode (Emitting-Sole) Cross-Field Amplifier," *Proc. IEEE* **61**:330–356 (Mar. 1973).

13 H. Jory, "Gyro-Device Developments and Applications," Intl. Electron Devices Meet., *Tech. Digest*, pp. 182–185 (1981).

14 H. A. Watson, *Microwave Semiconductor Devices and Their Circuit Applications*, McGraw-Hill, New York, 1969.

15 R. Bowers and J. Frey, "The Impact of Solid State Devices," *Adv. Electron. Electr. Phys.* **38**:147–194 (1975).

16 S. M. Sze, *Physics of Semiconductor Devices*, 2nd ed., Wiley, New York, 1981.

17 M. J. Howes and D. V. Morgan, eds., *Microwave Devices Device Circuit Interactions*, Wiley, New York, 1976.

18 P. Penfield Jr. and R. P. Rafuse, *Varactor Applications*, MIT, Cambridge, MA, 1962.

19 J. M. Manley and H. E. Rowe, "Some General Properties of Nonlinear Elements, I: General Energy Relations," *Proc. IRE* **44**:904 (1956).

20 D. Leenov and A. Uhlin Jr., "Generation of Harmonics and Subharmonics at Microwave Frequencies with *P-N* Junction Diodes," *Proc. IRE* **47**:1724 (1959).

21 R. P. Rafuse, "Recent Developments in Parametric Multipliers," *Proc. Nat. Electron. Conf.* **19**:461 (1963).

22 K. K. N. Chang, *Parametric and Tunnel Diodes*, Prentice-Hall, Englewood Cliffs, NJ, 1964.

23 J. L. Moll, S. Krakaner, and R. Shen, "*P-N* Junction Charge Storage Diodes," *Proc. IRE* **50**:45–53 (1962).

24 H. T. Friis, "Analysis of Harmonic Generator Circuits for Step-Recovery Diodes," *Proc. IEEE* **55**:1192–1194 (1967).

25 A. M. Cowley and H. O. Sorensen, "Quantitative Comparison of Solid-State Microwave Detectors," *IEEE Trans. Micro. Theory Tech.* **MTT-14**:588–602 (Dec. 1966).

26 C. A. Burrus, "Millimeter-Wave Point Contact and Junction Diodes," *Proc. IEEE* **54**:575–587 (1966).

27 D. Leenov, "The Silicon PIN Diode as a Microwave Radar Protector at Megawatt Levels," *IEEE Trans. Elect. Dev.* **ED-11**:53 (1964).

28 J. K. Hunton and A. G. Ryals, "Microwave Variable Attenuators and Modulators Using P-i-n Diodes," *IEEE Trans. Micro. Theory Tech.* **MTT-10**:262 (1962).

29 R. W. Burns and L. Stark, "P-i-n Diodes Advance High-Power Phase Shifting," *Microwaves* **4**:38 (1965).

30 W. F. Chow, *Principles of Tunnel Diode Circuits*, Wiley, New York, 1964.

31 S. P. Gentile, *Basic Theory and Application of Tunnel Diodes*, D. Van Nostrand, Princeton, NJ, 1962.

32 L. F. Eastman, *Gallium Arsenide Microwave Bulk and Transit-Time Devices*, Artech House, Dedham, MA, 1973.

33 B. G. Bosch and R. W. H. Engelmann, *Gunn-Effect Electronics*, Wiley, New York, 1975.

34 H. W. Thim, "Solid State Microwave Sources," in C. Hilsun, ed., *Handbook on Semiconductors*, Vol. 4, *Device Physics*, North-Holland, Amsterdam, 1980.

35 H. D. Rees and K. W. Gray, "Indium Phosphide: A Semiconductor for Microwave Devices," *Solid State Elect. Dev.* **1** (1976).

36 M. Shoji, "Functional Bulk Semiconductor Oscillator," *IEEE Trans. Elect. Dev.* **ED-14**:535 (1967).

37 H. Hartnagel, *Gunn-Effect Logic Devices*, American Elsevier, New York, 1973.

38 G. Haddad, *Avalanche Transit-Time Devices*, Artech House, Dedham, MA, 1973.

39 H. A. Haus, H. Statz, and R. A. Pucel, "Optimum Noise Measure of IMPATT Diodes," *IEEE Trans. Micro. Theory Tech.* **MTT-19** (Oct. 1971).

40 B. C. DeLoach Jr. and D. L. Scharfetter, "Device Physics of TRAPATT Oscillators," *IEEE Trans. Elect. Dev.* **ED-17**:9 (1970).

41 W. J. Evans, "Circuits for High-Efficiency Avalanche Diode Oscillators," *IEEE Trans. Micro. Theory Tech.* **MTT-17**:1060 (1969).

42 H. W. Ruegg, "A Proposed Punch-Through Microwave Negative Resistance Diode," *IEEE Trans. Elect. Dev.* **ED-15**:577 (1968).

43 D. J. Coleman Jr. and S. M. Sze, "The Baritt Diode—A New Low Noise Microwave Oscillator," IEEE Device Res. Conf., Ann Arbor, MI, June 28, 1971; "A Low-Noise Metal-Semiconductor-Metal (MSM) Microwave Oscillator," *Bell Syst. Tech. J.* **50**:1695 (1971).

44 J. E. Sitch, A. Majerfeld, P. N. Robson, and F. Hasegawa, "Transit-Time-Induced Microwave Negative Resistance in GaAlAs-GaAs Heterostructure Diodes," *Electron. Lett.* **11**:457 (1975).

45 J. E. Sitch and P. N. Robson, "Efficiency of BARITT and DOVETT Oscillators," *Sol. State Electron Dev.* **1**:31 (1976).

46 H. F. Cooke, "Microwave Transistor: Theory and Design," *Proc. IEEE* **59**:1163 (1971).

47 C. P. Snapp, "Bipolars Quietly Dominate," *Micro. Sys. News*, Nov. 1979, pp. 45–67.

48 R. L. Pritchard, J. B. Angell, R. B. Adler, J. M. Early, and W. M. Webster, "Transistor Internal Parameters for Small-Signal Representation," *Proc. IRE* **49**:725–738 (Apr. 1961).

49 W. Shockley, U. S. Patent 2569347, 1951.

50 C. A. Mead, "Tunnel-Emission Amplifier," *Proc. IRE* **48**:359 (1960).

51 C. O. Bozler, G. D. Alley, R. A. Murphy, D. C. Flanders, and W. T. Lindley, "Fabrication and Microwave Performance of the Permeable Base Transistor," *IEEE Tech. Dig.*, Int. Electron Dev. Mtg., 1979, p. 384.

52 C. A. Liechti, "Microwave Field-Effect Transistor," *IEEE Trans. Micro. Theory Tech.* **MTT-24** (June 1976) (special issue on microwave field effect transistors).

53 J. V. DiLorenzo and D. Khandelwal, eds., *GaAs FET Principles and Technology*, Artech House, Dedham, MA, 1982.

54 S. Umebachi, K. Ashahi, M. Inoue, and G. Kano, "A New Heterojunction Gate GaAs FET," *IEEE Trans. Electron Dev.* **ED-12**:613 (1975).

55 J. Barnard, C. E. C. Wood, and L. F. Eastman, "Double Heterojunction $Ga_{0.47}In_{0.53}As$ MESFETs with Submicron Gates," *IEEE Electron Dev. Lett.* **EDL-1**:174 (1980).

56 A. E. Siegman, *Microwave Solid-State Masers*, McGraw-Hill, New York, 1964.

57 R. W. Degrasse, E. O. Schulz-DuBois, and H. E. D. Scovil, "The Three-Level Solid State Traveling-Wave Maser," *Bell Syst. Tech. J.* **38**:105 (1959).

58 J. P. Gordon, H. J. Zeiger, and C. H. Townes, "The Maser—New Type of Microwave Amplifier Frequency Standard and Spectrometer," *Phys. Rev.* **99**:1264–1274 (1955).

18.3 MICROWAVE INTEGRATED CIRCUITS (MICs)

As many of the solid-state low-noise power and control devices were gradually replacing vacuum tubes in microwave systems, there was still a need to improve the compactness of the overall size, especially in the sophisticated airborne phase-array radars. Hybrid microwave integrated circuits were first introduced to satisfy this requirement. More recently, research and development of monolithic microwave integrated circuits have been expanded to include other applications, such as analog and digital microwave signal processing.

18.3-1 Hybrid Microwave Integrated Circuits[1,2]

Two classes of hybrid MICs include distributed transmission line and lumped-element circuits. A combination of these two approaches is also employed. In these MICs, the active elements are fabricated independently from the passive components and impedance-matching circuits at the input and output of the device with generally different metallic and semiconductor materials.

The distributed circuits mostly employ microstripline configurations for the matching circuits on substrates such as fused silica, alumina, and duroid. However, coplanar waveguide, slotline, and suspended substrate approaches have been gradually gaining acceptance in MIC applications. Passive components, resistors, and capacitors for DC bias or blocking purposes and others can be incorporated into MICs by direct deposition or bonding of chip components.

The lumped-element form of circuits achieves impedance matching with components that are a small fraction of wavelength in dimension. These components perform as resistor, inductor, and capacitor and their values are independent of frequency over the designed bandwidth. This approach has been made possible by new fabrication technologies, such as ion-milling for metallization and advanced lithography allowing fine-line definitions. Interconnections among passive lumped-element chips can be achieved with wire-bonding; they can also be deposited simultaneously with some of the printed components such as inductors. As in the distributed-circuit case, active devices in chip form or in hermetically sealed packages, which are fabricated separately from the circuits, are normally wire-bonded and/or soldered in place.

The hybrid MICs offer flexibility in circuit manufacturing, as changes requiring tuning of circuits and addition of components are possible after the microwave component is fabricated.[3] Complete subsystems: receivers, up- and down-converters with individual components such as preamplifiers, mixers, modulators, attenuators, and local oscillators have been fabricated on hybrid MICs.

18.3-2 Monolithic Microwave Integrated Circuits (MMICs)[4–6]

Analog microwave technology has progressed rapidly from hybrid MICs to *m*onolithic *m*icrowave *i*ntegrated *c*ircuits (MMICs). The monolithic approach requires higher integration of all active devices and passive matching circuits and interconnections formed on the surface of an active epitaxial layer or a semiinsulating substrate by some fabrication technique, such as ion-implantation, epitaxy, evaporation, sputtering, diffusion, or other. Most of the MMICs are fabricated employing silicon (Si), silicon-on-sapphire (SOS) and gallium-arsenide (GaAs) semiconductor materials.

The commonly used active elements are bipolar transistors and MESFETs. Because of the lower mobility properties of silicon, Si MMICs have been limited in use to the lower microwave frequency range, where GaAs circuits have found applications up to millimeter waves.

The monolithic approach offers the following advantages: small size and weight, low cost and improved reliability, reproducibility of flexible circuit design, and broadband and multifunction performance on a single chip. Many circuit functions available with GaAs MMICs are not probable with use of conventional hybrid MIC technology. Examples are the need for many small gate-width FETs on a circuit and situations where the design of the discrete active devices is not suitable for the uses to which they are being put.

Extensive work has been performed on monolithic amplifiers for low-noise and high-power applications. Efforts are also being directed toward a complete monolithic transmitter and receiver with T/R switches for radar systems. With use of lumped-elements and/or active matching approaches, broadband amplifiers have been fabricated with frequency from DC to 12 GHz and a few decibels of power gain in stages.[7] The concept of active matching with use of FETs to replace passive lumped elements or distributed transmission lines for circuit impedance matching and filtering is gradually gaining acceptance in MMICs, as device technologies and modeling improve and as demand increases for reduction of circuit real-estate to achieve better cost-effectiveness and reliability. The monolithic circuits are especially applicable to millimeter-wave applications, since most of the parasitics such as bonding-wire inductance and stray capacitance, which critically degrade high-frequency performance, can be minimized. A complete front-end receiver on a single GaAs chip is now possible.[8]

References

1 L. Young and H. Sobol, eds., *Advances in Microwaves*, Vol. 8, Academic, New York, 1974.

2 H. Sobol, "Applications of Integrated Circuit Technology to Microwave Frequencies," *Proc. IEEE* **59**:1200–1211 (Aug. 1971).

3 R. S. Pengelley, "Hybrid vs. Monolithic Microwave Circuits," *Micro. Sys. News*, Jan. 1983, pp. 77–114.

4 R. A. Pucel, "Design Consideration for Monolithic Microwave Circuits," *IEEE Trans. Micro. Theory Tech.* **MTT-29**:513–534 (June 1981).

5 J. E. Davey and J. G. Oakes, eds., joint special issue on GaAs ICs, *IEEE Trans. Micro. Theory Tech.* **MTT-30** (July 1982).

6 R. A. Murphy, ed., *IEEE 1982 Microwave and Millimeter-Wave and Monolithic Circuits Symposium*, 1982.

7 E. Strid and K. R. Gleason, "A DC-12 GHz Monolithic GaAs FET Distributed Amplifier," *IEEE Trans. Micro. Theory Tech.* **MTT-30**:969–975 (July 1982).

8 A. Chu et al., "GaAs Monolithic Circuit for Millimeter Wave Receiver Applications," *ISSCC Dig. Tech. Papers*, 1980, pp. 144–145.

PART 4

PASSIVE CIRCUITS

CHAPTER 19
LINEAR ELECTRIC CIRCUITS

JEFFREY HANTGAN

State University of New York, Stony Brook, New York

YACOV SHAMASH

MARTIN GAZOURIAN

Florida Atlantic University, Boca Raton, Florida

19.1 FUNDAMENTALS OF ELECTRIC CIRCUITS

Jeffrey Hantgan

19.1-1 Circuit Elements, Electrical Parameters, and Their Voltage-Current Relationships

Classification of Circuits and Electrical Parameters

Electrical networks or circuits, which are formed by the interconnection of electrical circuit components, can be classified into various categories (i.e., linear or nonlinear, lumped or distributed, passive or active, time invariant or time varying) by examining the mathematical models used in the description of their behavior. For example, a linear network is a circuit composed of only linear components and described by a set of linear integro-differential equations.

To determine whether a network with no initial energy storage is portwise linear, two arbitrary excitations $e_1(t)$ and $e_2(t)$ can be used. If the responses to the excitations $e_1(t)$ and $e_2(t)$ are $r_1(t)$ and $r_2(t)$, then the response to an excitation $K_1e_1(t) + K_2e_2(t)$ in a portwise linear network will be $K_1r_1(t) + K_2r_2(t)$. A network that is not linear is called nonlinear.

Usually the excitations and responses of lumped electrical networks are described using voltages and currents, although other quantities such as electric charge, magnetic flux, or power flow might be used. These quantities are a subset of the International System of Units (SI). The SI system contains seven basic quantities or units from which all other quantities are derived (Table 19.1).

The set of electrical circuit components can also be described using the terms "lumped" and "distributed." For lumped components the electrical response is immediate, since their physical size is much smaller than the wavelength of the highest operating frequency. Typical lumped elements are resistors, inductors, capacitors, and transformers. Distributed components, however, experience a delay in the electrical response since their spatial dimensions must be explicitly considered. Typical distributed elements are transmission lines and waveguides. Mathematically, lumped circuit elements are described using ordinary differential equations, while distributed circuit elements are described using partial differential equations.

TABLE 19.1. INTERNATIONAL SYSTEM OF UNITS (SI)

Quantity	Unit	Symbol/Abbreviation	Composition	Derivation
		Base Units		
Length (l)	meter (m)	m/l		
Mass	kilogram	kg		
Time	second (s)	s/t		
Electric current	ampere	A/i		$i = dq/dt$
Temperature (degrees)	kelvin	K		
		Important Derived Quantities		
Charge	coulomb	C/q	A − s	$\int i(t)\,dt$
Energy	joule	J/W	$\text{kg} - \text{m}^2/\text{s}^2$	$\int p\,dt = \int v \cdot i dt$
Power	watt	W/p	$\text{kg} - \text{m}^2/\text{s}^3$	$p = dw/dt = v \cdot i$
Voltage	volt	V/v	$\text{kg} - \text{m}^2/\text{m}^2 - \text{A}$	$v = dw/dq$

A passive network denotes a system that is unable to deliver more energy to an external circuit than it initially stores. To determine if a circuit is passive, the total energy

$$w(t) = \int_{-\infty}^{t} v(t)i(t)\,dt \qquad \text{J}$$

absorbed by the network must be examined for all possible excitations. If the total energy absorbed by the circuit is greater than or equal to zero for all time (i.e., $w(t) \geqslant 0$), the network is passive. A circuit that is not passive is called active.

Whenever an excitation $e(t)$ produces a unique response $r(t)$ in a circuit independent of the excitation's time of application, the network is called time invariant. If a network is not time invariant it is called time varying. The most commonly used circuits are time invariant.

A linear time-invariant circuit can be analyzed in the frequency domain using such techniques as (1) generalized phasor, (2) Fourier transforms, and (3) Laplace transforms. (See Chapter 4.) These techniques simplify the mathematical computation by transforming a real integro-differential equation in the time domain into an algebraic equation in the complex frequency domain. The generalized phasor technique is simple to use but applicable only when the forcing function can be represented by an exponential function (DC, exponentials, sinusoids, and damped sinusoids). The Fourier transform method extends the generalized phasor technique to include certain nonsinusoidal and nonperiodic forcing functions, but the method is limited to initially relaxed circuits. This Fourier transform technique, however, can be used to yield the steady-state response to a sinusoidal excitation by assuming the driving function to exist for all time. The Laplace transform method allows a wider variety of excitations than do either of the previous methods while automatically incorporating the initial conditions into the solution for the complete response.

The circuit symbol for a two-terminal element is shown in Fig. 19.1. The direction of the assumed conventional current (flow of positive charge) is indicated by an arrow, while the polarity markings (+, −) indicate that terminal A is assumed to be at a potential of $v(t)$ volts greater than terminal B. Clearly the reference directions of the voltage and the current in a two-terminal device are arbitrary. For convenience, the current can be assigned to flow through the device from the terminal of higher potential to the terminal of lower potential. This convention, called the asserted reference direction or load reference convention, provides for a power flow into the device whenever the actual $v \cdot i$ product is positive.

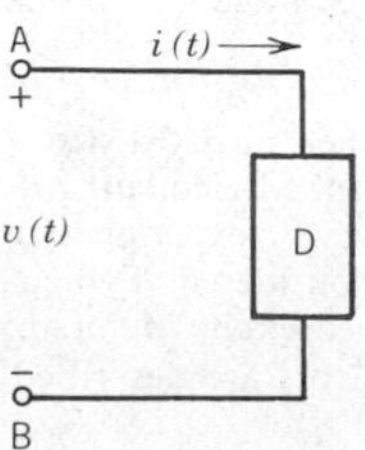

Fig. 19.1 Circuit symbol for a two-terminal element.

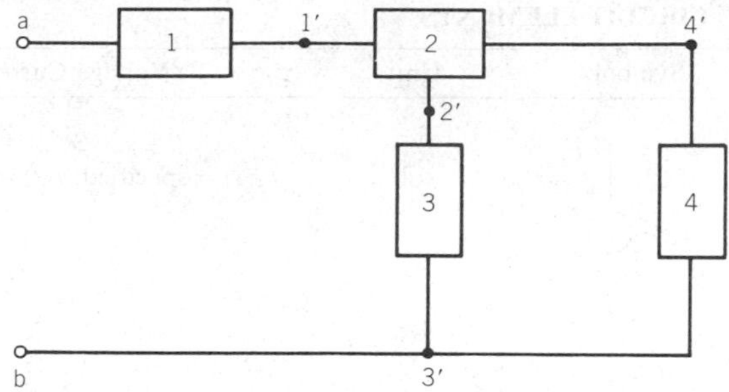

Fig. 19.2 A possible electrical network.

Electrical networks are formed by interconnecting electrical devices together at their terminals. Figure 19.2 displays a possible network. A junction where two or more devices are joined is referred to as a node. The electrical components numbered 1, 3, and 4 are two-terminal devices, while the component numbered 2 is a three-terminal device. A two-terminal device or a group of devices that is accessed through two terminals only (i.e., the network of Fig. 19.2 between points a and b) is referred to as a branch. Since an electrical signal can be applied or measured between any two terminals, a two-terminal pair of an electrical device or network is often called a port (i.e., points a, b in Fig. 19.1). When a sequence of branches or devices is connected in succession a path is formed. By connecting devices 2 and 3 or 2 and 4 of Fig. 19.2 a path is formed from node 1′ to 3′. A closed path from node 2′ back to 2′ can be formed in Fig. 19.2 by tracing a path through 2, 3, and 4. Closed paths are referred to as loops or circuits.

Basic Ideal Circuit Elements

In describing lumped electrical networks, several ideal basic circuit elements will be needed. Some of these basic ideal elements are listed in Table 19.2 and are accompanied by their respective mathematical voltage-current (v-i) description.

Different types of ideal sources are displayed in Table 19.2. These ideal sources, which are active devices, can be divided into two sets: independent and dependent. The independent sources can be further divided into two subgroups: voltage and current. In the case of an ideal voltage source, a prescribed voltage will be maintained between the terminals independent of the load or current supplied. For the ideal current source, the current flow through the device is prescribed and independent of the load. Therefore, any voltage can be furnished by an ideal current source.

The dependent sources are divided into four subgroups: voltage-controlled voltage source (VCVS), current-controlled voltage source (CCVS), voltage-controlled current source (VCCS), and current-controlled current source (CCCS). In each of these cases the terminal conditions (a voltage across the source or the current through the source) are controlled by an existing voltage or current located elsewhere within the electrical network. For example, the CCVS maintains a voltage across the dependent source independent of the current through this dependent source, but functionally dependent on an existing current located within the circuit elsewhere.

The voltage-current characteristic for the linear resistor is shown in Fig. 19.3 as a straight line. This straight line relation can be expressed by Ohm's law as

$$v(t) = Ri(t) \quad \text{or} \quad i(t) = Gv(t) \tag{19.1}$$

where $R = 1/G$. R is the resistance in ohms (Ω) and G is the conductance in mhos (℧) of the resistor. Both R and G are constants. The circuit symbol for a resistor is shown in Table 19.2.

Two special types of resistors are the short circuit and the open circuit. By definition, a short circuit can support any current, has no voltage drop between its terminals, and is described as having zero resistance or infinite conductance. Conversely, an open circuit can support any voltage between its terminals, has no current flowing through its terminals, and is described as having zero conductance or infinite resistance. The characteristics for an open-circuit element and a short-circuit element are displayed in Fig. 19.4.

Using these definitions of open circuit and short circuit, the independent voltage source and the independent current source are reexamined. Consider an independent voltage source that supports any current for a prescribed voltage $v(t)$. For the special case $v(t) = 0$, the independent voltage source appears to be a short circuit. Therefore an independent voltage source can be regarded as a generalized short circuit. Employing a similar argument, consider an independent current source that

TABLE 19.2. IDEAL CIRCUIT ELEMENTS

Device	Symbol	Unit	Voltage-Current Description
Active Devices			
Voltage source	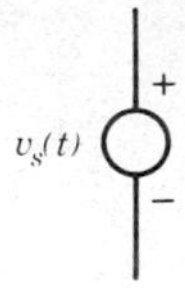 $v_s(t)$, +, −	volt	$v_s(t)$ = specified; $i(t)$ = anything
Current source	 $i_s(t)$	ampere	$i_s(t)$ = specified; $v(t)$ = anything
Dependent voltage source	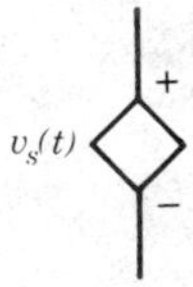 $v_s(t)$, +, −	volt, μ (dimensionless) r-ohm	Voltage-controlled voltage source (VCVS): $v_s(t) = \mu v_c(t)$; $i(t)$ = anything; $v_c(t)$ = a controlling voltage Current-controlled voltage source (CCVS): $v_s(t) = ri_c(t)$; $i(t)$ = anything; $i_c(t)$ = a controlling current
Dependent current source	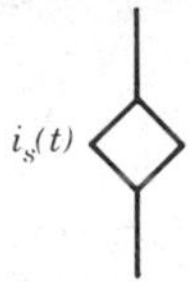 $i_s(t)$	ampere, β (dimensionless) siemen	Voltage-controlled current source (VCCS): $i_s(t) = gv_c(t)$; $v(t)$ = anything; $v_c(t)$ = a controlling voltage Current-controlled voltage source (CCVS): $i_s(t) = \beta i_c(t)$; $v(t)$ = anything; $i_c(t)$ = a controlling current
Passive Devices			
Resistor	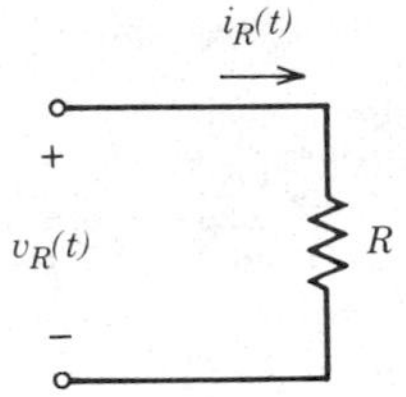 $i_R(t)$, +, $v_R(t)$, −, R	ohm	$v(t) = i(t)R$
Inductor	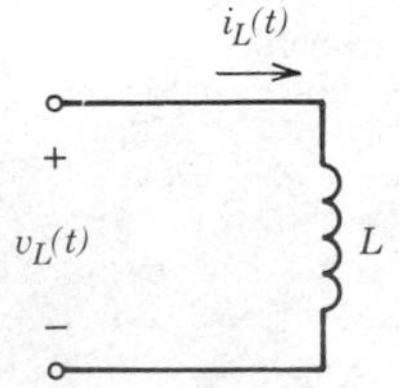 $i_L(t)$, +, $v_L(t)$, −, L	Henry	$v_L(t) = L_1 \dfrac{di_L(t)}{dt}$ $i(t) = \dfrac{1}{L}\int_0^t v(\tau)\,d\tau + i_0$
Capacitor	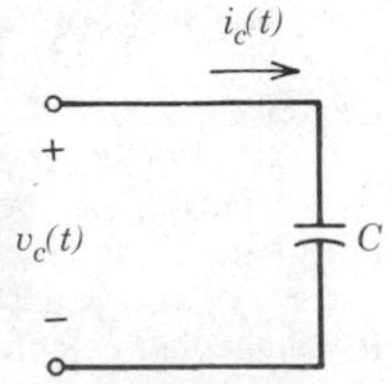 $i_c(t)$, +, $v_c(t)$, −, C	farad	$i_C(t) = C \dfrac{dV_C(t)}{dt}$ $v(t) = \dfrac{1}{C}\int_0^t i(\tau)\,d\tau + V_0$

TABLE 19.2. (*Continued*)

Device	Symbol	Unit	Voltage-Current Description
Passive Devices			
Coupled coil	$i_1(t)$, $i_2(t)$, M, $+$ $v_1(t)$ $-$, L_1, L_2, $+$ $v_2(t)$ $-$		$v_i(t) = L_1 \dfrac{di_1(t)}{dt} + M \dfrac{di_2(t)}{dt}$ $v_2(t) = M \dfrac{di_1(t)}{dt} + L_2 \dfrac{di_2(t)}{dt}$
Ideal transformer	$i_1(t)$, $n:1$, $i_2(t)$, $+$ $v_1(t)$ $-$, $+$ $v_2(t)$ $-$		$\dfrac{v_2(t)}{v_1(t)} = n \qquad \dfrac{i_1(t)}{i_2(t)} = -n$

supports any voltage for a prescribed current $i(t)$. For the special case $i(t) = 0$, the independent current source appears to be an open circuit. Therefore an independent current source is regarded as a generalized open circuit.

From Eq. (19.1) it is observed that the voltage and current waveforms of the resistor are identical, except for their amplitude. The resistor given by Eq. (19.1) is a linear time-invariant electrical device. An examination of the total energy absorbed by the resistor

$$w_R(t) = \int_{-\infty}^{t} v(\tau)i(\tau)\,d\tau = \int_{-\infty}^{t} Ri^2(\tau)\,d\tau > 0$$

reveals that this device is passive. The resistor always dissipates energy, since the area under the integral between any two times is positive $[i^2\tau \geqslant 0]$.

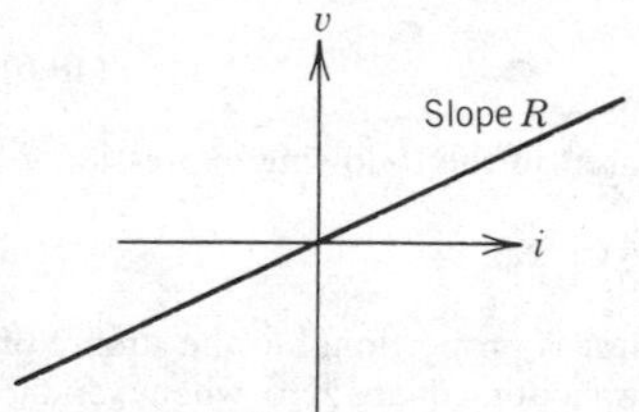

Fig. 19.3 Voltage-current characteristic of a linear time-invariant resistor.

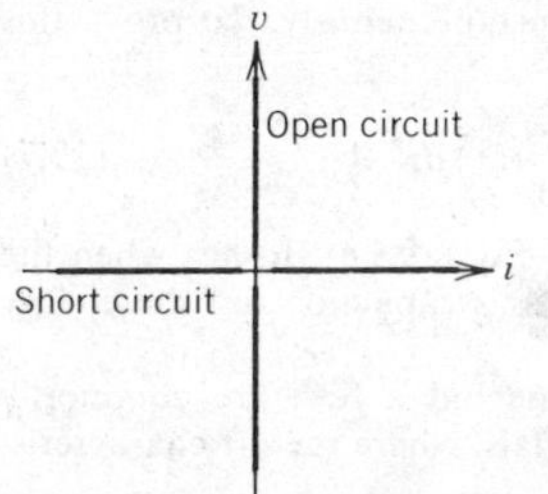

Fig. 19.4 Voltage-current characteristics of a short circuit and an open circuit.

An ideal inductor, sometimes referred to as a coil, stores energy in its magnetic field. Faraday showed that changing the flux linkages in a coil will induce a voltage at the coil's terminals. Faraday's law can be stated as

$$v(t) = \frac{d\lambda}{dt} \tag{19.2}$$

where the flux linkage λ is the product of the flux and the number of turns of the coil. In a linear inductor, the flux is linearly proportional to the current. Therefore the relation between the terminal voltage and the current through an inductor can be obtained by rewriting Faraday's law as

$$v(t) = L\frac{di}{dt} \tag{19.3}$$

where L is the inductance, a constant measured in Henrys. The circuit symbol for the inductor is shown in Table 19.2. An inductor described by Eq. (19.3) is a linear time-invariant electrical device.

To obtain the total energy absorbed by an inductor, Eq. (19.3) is used in the following expression

$$w_L(t) = \int_{-\infty}^{t} v(\tau)i(\tau)\,d\tau = \int_{-\infty}^{t} (Li)\,di = \tfrac{1}{2}Li^2(t)$$

This equation demonstrates that the total energy absorbed is proportional to the square of the final current. Since the total energy $w_L(t)$ absorbed by an inductor equals zero whenever the current $i(t)$ through the inductor equals zero, the inductor can be used as an energy storage device. An electrical device capable of storing energy but not capable of dissipating energy is called lossless. The inductance L is a measure of the inductor's energy storage ability.

Since the energy stored in an inductor cannot change instantaneously, the current $i(t)$, which is a direct measure of the energy stored in the inductor, cannot change instantaneously. This result can easily be proven by multiplying both sides of Eq. (19.3) by dt and integrating

$$i(t) = \frac{1}{L}\int_{-\infty}^{0} v(\tau)\,d\tau + \frac{1}{L}\int_{0}^{t} v(\tau)\,d\tau = I_0 + \frac{1}{L}\int_{0}^{t} v(\tau)\,d\tau \tag{19.4}$$

For a nonsingular voltage function $v(t)$, the right-hand integral of Eq. (19.4) vanishes when the limits of integration are from zero to zero plus. Therefore the current through an inductor and the energy stored in an inductor cannot change instantaneously.

An ideal capacitor stores energy in its electric field. It can be formed using two electrodes separated by a dielectric. In a linear time-invariant capacitor, the charge produced on the electrodes is proportional to the voltage applied

$$q(t) = Cv(t) \tag{19.5}$$

where the capacitance C is a constant measured in farads. The circuit symbol for a capacitor is shown in Table 19.2.

By definition, the current is the time derivative of the charge. Therefore the voltage-current relation for a capacitor can be derived by differentiating Eq. (19.5)

$$i(t) = \frac{dq(t)}{dt} = C\frac{dv(t)}{dt} \tag{19.6}$$

To obtain the total energy absorbed by a capacitor, Eq. (19.6) is used in the following expression

$$w_c(t) = \int_{-\infty}^{t} v(\tau)i(\tau)\,d\tau = \int_{-\infty}^{v} Cv\,dv = \tfrac{1}{2}Cv^2(t)$$

This equation illustrates that the total energy absorbed by a capacitor is proportional to the square of the final voltage. Since the total energy $w_c(t)$ absorbed by the capacitor equals zero whenever the voltage $v(t)$ across the capacitor equals zero, the capacitor is lossless and can be used as an energy storage element. The capacitance C is a measure of the capacitor's energy storage ability.

Since the energy stored in a capacitor cannot change instantaneously, the voltage $v(t)$, which is a direct measure of the energy stored in the capacitor, cannot change instantaneously. To prove this result, multiply both sides of Eq. (19.6) by dt and integrate

$$v(t) = \frac{1}{C}\int_{-\infty}^{0} i(\tau)\,d\tau + \frac{1}{C}\int_{0}^{t} i(\tau)\,d\tau = V_0 + \frac{1}{C}\int_{0}^{t} i(\tau)\,d\tau \tag{19.7}$$

For a nonsingular current function $i(t)$, the right-hand side integral of Eq. (19.7) vanishes when the limits of integration are from zero to zero plus. Hence the voltage across a capacitor and the energy stored in a capacitor cannot change instantaneously.

Coupled coils, ideal transformers, and dependent sources, to name but a few, are commonly employed circuit devices. These devices consist of more than two terminals, where the v-i characteristics for one set of terminals are related to or coupled to the v-i characteristics at another set of terminals.

Consider the transformer pictured in Table 19.2. Here two inductors, L_1 and L_2, share a common magnetic field. The induced voltage across one coil is a superposition of two individually produced voltages resulting from the time-dependent currents i_1 and i_2. The mathematical equations expressing the relationship between port voltages and currents for the linear time-invariant transformer are

$$\begin{aligned} v_1(t) &= L_1 \frac{di_1(t)}{dt} + M \frac{di_2(t)}{dt} \\ v_2(t) &= M \frac{di_1(t)}{dt} + L_2 \frac{di_2(t)}{dt} \end{aligned} \tag{19.8}$$

L_1 and L_2 are the self-inductance of coils 1 and 2 in Henrys and M is the mutual inductance in Henrys. Mutual inductance is related to self-inductance through the expression

$$M = k\sqrt{L_1 L_2} \tag{19.9}$$

where k is the coefficient of coupling and has a minimum value of -1 and a maximum value of $+1$. A transformer possessing a coupling coefficient equal to unity is called perfect.

The signs associated with the terms in Eq. (19.8) are for the transformer given in Table 19.2 and are controlled by the sense of the windings. The sense of the windings is indicated by reference dots placed on the transformer or shown in an accompanying figure. When both currents enter or both currents leave the windings at the dotted terminals, the sign affixed to the mutual inductance will be positive.

The total energy delivered to a transformer is

$$\begin{aligned} w(t) &= \int_{-\infty}^{t} [v_1(\tau)i_1(\tau) + v_2(\tau)i_2(\tau)]\, d\tau \\ &= \tfrac{1}{2}\left[L_i i_1^2(t) + 2\, M i_1(t) i_2(t) + L_2 i_2^2(t)\right] \end{aligned} \tag{19.10}$$

Since the coupling coefficient squared must be less than or equal to 1, the transformer is a lossless passive device capable of energy storage.

An ideal transformer, which is different from a perfect transformer, is characterized by a turns ratio n. For the reference directions given, the ideal transformer of Table 19.2 possesses the following v-i characteristics

$$v_1(t) = n v_2(t) \quad \text{and} \quad i_2(t) = -n i_1(t) \tag{19.11}$$

The total energy delivered to the ideal transformer is zero for all time. Hence the ideal transformer is passive, lossless, and incapable of energy storage.

Impedance

As mentioned earlier, a linear time-invariant circuit can be analyzed in the frequency domain using one of several transform methods. Each of these transform methods employs the concept of impedance. This concept is of utmost importance in circuit analysis and synthesis.

The impedance $Z(s)$ is defined as the ratio of the transform voltage to the transform current at a terminal pair when the network is initially relaxed, or

$$Z(s) = \frac{1}{Y(s)} = \frac{V(s)}{I(s)} \tag{19.12}$$

The reciprocal of impedance is admittance. When impedance or admittance is referred to generically, the term "immittance" is used. For lumped linear time-invariant networks the impedance function $Z(s)$ is a real rational function.

In the analysis of a circuit, the frequency of excitation determines one of the values assumed by the complex frequency variable s. For a

$$\left.\begin{array}{rl} \textit{DC or constant excitation} & s = 0 \\ \textit{Exponential excitation } (e^{-\alpha t}) & s = -\alpha \\ \textit{Sinusoid excitation } (\cos \omega t) & s = j\omega \\ \textit{Damped sinusoid excitation } (e^{-\alpha t}\cos \omega t) & s = -\alpha + j\omega \end{array}\right\} \tag{19.13}$$

The frequency response of a network is obtained by examining the plots $|Z(s)|$ (magnitude of $Z(s)$) and $\angle Z(s)$ (angle of $Z(s)$) for varying values of the complex frequency variable s.

TABLE 19.3. IMPEDANCE FUNCTIONS OF VARIOUS IDEAL CIRCUIT ELEMENTS

Element	Diagram	Formulas	
Resistor	$I_R(s)$ →, $V_R(s)$, R	$Z_R(s) = R$	$V_R(s) = RI_R(s)$
Inductor	$I_L(s)$ →, $V_L(s)$, sL	$Z_L(s) = sL$	$V_L(s) = sLI_L(s)$
Capacitor	$I_C(s)$ →, $V_C(s)$, $\frac{1}{sC}$	$Z_C(s) = \frac{1}{sC}$	$V_C(s) = I_C(s)/sC$
Coupled coils	$I_1(s)$ →, sM, ← $I_2(s)$, $V_1(s)$, sL_1, sL_2, $V_2(s)$	$V_1(s) = sL_1I_1(s) + sMI_2(s)$	$V_2(s) = sMI_1(s) + sL_2I_2(s)$

Table 19.3 summarizes the impedance functions for some previously mentioned ideal circuit elements.

19.1-2 Fundamental Circuit Theorems

Kirchhoff's Laws

In the previous section, emphasis was placed on the voltage-current relationship for individual circuit elements. If these and other circuit elements are interconnected, various network theorems and methods of analysis will be needed to determine the actual voltages and currents within a network.

Kirchhoff formulated two fundamental laws that are the cornerstone of circuit analysis. They deal solely with the topology of a network and are completely independent of any voltage-current relationship for an electrically neutral device.

Kirchhoff's current law (KCL) is a rewording of the conservation of charge hypothesis. Consider a node (Fig. 19.5) that connects two branches. One branch directs a current into the node (I_{in}), while the other branch guides a current out of the node (I_{out}). Since the node must be electrically neutral at all times, conservation of charges requires

$$I_{in} = I_{out} \quad \text{or} \quad I_{in} - I_{out} = 0$$

Now let the current I_{out} be represented by the current I'_{out} ($I_{out} = -I'_{out}$), then

$$I_{in} + I'_{out} = 0$$

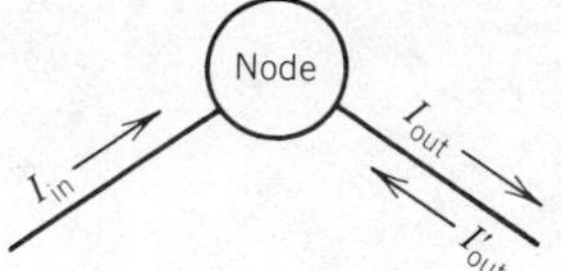

Fig. 19.5 Circuit node.

Kirchhoff's current law states that the algebraic sum of all branch currents entering (leaving) a node must equal zero at any given instant

$$\sum i(t) = 0 \tag{19.14}$$

In the algebraic current sum at a particular node, a plus sign is affixed to the magnitude of those branch currents entering (leaving) a node while a minus sign is affixed to the magnitude of those branch currents leaving (entering) a node. Apply KCL to each of the four nodes for the circuit shown in Fig. 19.6 results in

$$\left.\begin{array}{ll} \textit{Node A} & -i_1(t) - i_2(t) - i_3(t) = 0 \\ \textit{Node B} & i_1(t) + i_4(t) + i_5(t) = 0 \\ \textit{Node C} & i_2(t) + i_6(t) - i_4(t) = 0 \\ \textit{Node D} & i_3(t) - i_5(t) - i_6(t) = 0 \end{array}\right\} \tag{19.15}$$

The voltage across branch 2 in Fig. 19.7, called the branch voltage, is labeled v_{12} when node 1 is assumed to be at a higher potential than node 2, and v_{21} when node 2 is assumed to be at a higher potential than node 1. Clearly $v_{12} = -v_{21}$. Kirchhoff's voltage law (KVL) states that the algebraic sum of the branch voltages around any closed loop is zero at any given instant

$$\sum v(t) = 0 \tag{19.16}$$

Since voltage is a measure of the energy delivered to or received from an electric charge moving through an electric field, KVL is a rewording of the law of conservation of energy.

To apply KVL, the closed loop must be traversed in a single direction. When the branch is traversed from a node of higher potential to a node of lower potential, a plus (minus) sign is affixed to the magnitude of the branch voltage; a minus (plus) sign is affixed to the magnitude of the branch voltage when the branch is traversed from a node of lower potential to a node of higher potential.

Applying KVL to Fig. 19.7

$$v_{12}(t) + v_{23}(t) + v_{31}(t) = 0 \tag{19.17}$$

Let node 3 be arbitrarily chosen as a datum or reference node. The voltages $v_2(t)$ ($v_2 = v_{23}$) and $v_1(t)$ ($v_1 = v_{13}$) are called node voltages, and the branch voltage $v_{12}(t)$ specified as the difference of two node voltages will then be

$$v_{12}(t) = v_1(t) - v_2(t) \tag{19.18}$$

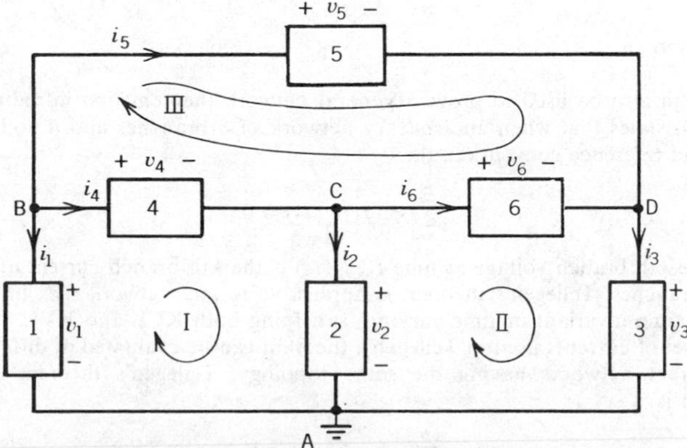

Fig. 19.6 A circuit with four nodes and six branches.

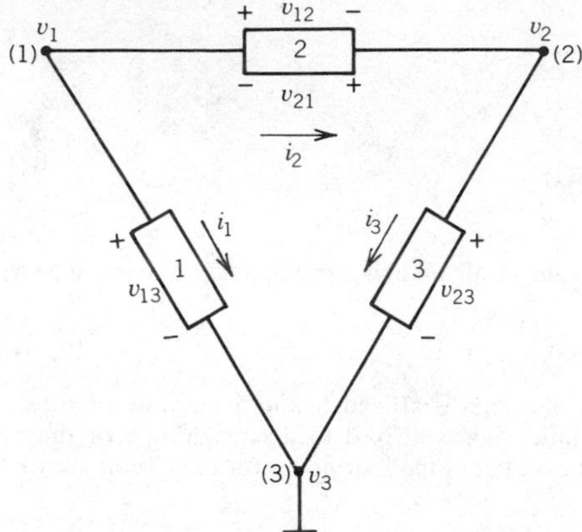

Fig. 19.7 Three-branch circuit.

Application of KVL to the loops specified in Fig. 19.6 yields

$$\left.\begin{array}{ll} \textit{Loop I} & -v_1(t) + v_4(t) + v_2(t) = 0 \\ \textit{Loop II} & -v_2(t) + v_6(t) + v_3(t) = 0 \\ \textit{Loop III} & v_5(t) - v_6(t) - v_4(t) = 0 \end{array}\right\} \tag{19.19}$$

Equations (19.14) and (19.16) state Kirchhoff's laws for time-dependent voltage and time-dependent current variables. Kirchhoff's laws hold for transform quantities as well. Assume for the circuit shown in Fig. 19.6 that the transforms of the time-domain variables $v_k(t)$ and $i_k(t)$ are $V_k(s)$ and $I_k(s)$. Then Eq. (19.15) and (19.19) can be rewritten as

$$\left.\begin{array}{ll} \textit{Loop I} & -V_1(s) + V_4(s) + V_2(s) = 0 \\ \textit{Loop II} & -V_2(s) + V_6(s) + V_3(s) = 0 \\ \textit{Loop III} & V_5(s) - V_6(s) - V_4(s) = 0 \end{array}\right\} \tag{19.20}$$

and

$$\left.\begin{array}{ll} \textit{Node A} & -I_1(s) - I_2(s) - I_3(s) = 0 \\ \textit{Node B} & I_1(s) + I_4(s) + I_5(s) = 0 \\ \textit{Node C} & I_2(s) + I_6(s) - I_4(s) = 0 \\ \textit{Node D} & I_3(s) - I_5(s) - I_6(s) = 0 \end{array}\right\} \tag{19.21}$$

Kirchhoff's laws are fundamental in circuit analysis. When Kirchhoff's laws and the element v-i characteristics are used in conjunction, the resulting system of equations completely describes the network.

Tellegen's Theorem

KCL and KVL can also be used to prove a general network theorem first introduced by Tellegen. Tellegen's theorem states that when an arbitrary network of b branches and n nodes satisfies KCL, KVL, and the load reference convention, then

$$\sum_{k=1}^{b} v_k(t_1) i_k(t_2) = 0 \tag{19.22}$$

where $v_k(t_1)$ is the kth branch voltage at time t_1, $i_k(t_2)$ is the kth branch current at time t_2, and the sum is over all branches. Tellegen's theorem is applicable to any network (i.e., linear or nonlinear, passive or active, time invariant or time varying) satisfying both KCL and KVL. In fact, the set of voltages and the set of currents used in Tellegen's theorem can be evaluated at different times or can come from different networks having the same topology. Tellegen's theorem expressed in the frequency domain is

$$\sum_{k=1}^{b} V_k(s) I_k(s) = 0 \tag{19.23}$$

As an example of Tellegen's theorem, the network of Fig. 19.7 is used resulting in

$$v_{13}i_1 + v_{12}i_2 + v_{23}i_3 = 0 \tag{19.24}$$

Rewriting the branch voltages of Eq. (19.24) in terms of the node voltages and then regrouping yields

$$v_1(i_1 + i_2) + v_2(i_3 - i_2) = 0 \tag{19.25}$$

Tellegen's theorem has thus been demonstrated for Fig. 19.7, since the multiplying sums are equal to zero by KCL.

Duals

KCL and KVL bear a striking resemblance to each other. If the words "currents" and "node" in KCL are replaced respectively by the words "voltages" and "loop," KVL is obtained. This is a familiar occurrence in network analysis, where a simple interchange of words or symbols in one expression yields a valid second expression. Expressions or equations obtained by such a procedure are called duals. Duals are important in network analysis because, given the appropriate set of transforms, a solution obtained for one network automatically yields the solution for the dual network.

Equivalent Circuits

Often the amount of computation needed in the analysis of electrical circuits is efficiently reduced by the use of equivalent circuits. Circuits are called equivalent when their terminals exhibit the same v-i characteristics. Two network theorems employing equivalent circuits to reduce the amount of computational effort in the analysis of electrical networks are Thévenin and Norton. These are dual theorems that automatically yield source transformations.

Thévenin's and Norton's Theorems

Consider network N of Fig. 19.8, which has been broken up into two parts, N_a and N_b. Assume analytical or graphical equations can be obtained for the port v-i characteristics of network N_a and network N_b separately. Provided no coupling exists internally between networks N_a and N_b, the two equations can be solved simultaneously. This procedure yields the actual port voltage and port current when networks N_a and N_b are connected.

The conditions of the previous paragraph are the most general for application of Thévenin's or Norton's theorems. However, computation efficiency usually requires network N_a to be linear time invariant. If network N_a is linear time invariant, then it can be replaced by the circuit of Fig. 19.9a for Thévenin's theorem or that of Fig. 19.9*b* for Norton's theorem.

To determine the Thévenin equivalent circuit (Fig. 19.9*a*), a transform voltage and a transform current are sought at the terminals of network N_a under special load conditions. These special load conditions are used to determine the open-circuit voltage $V_{oc}(s)$ and the short-circuit current $I_{sc}(s)$. Once these values are obtained, the Thévenin voltage and Thévenin impedance are

$$\begin{aligned} V_{Th}(s) &= V_{oc}(s) \\ Z_{Th}(s) &= V_{oc}(s)/I_{sc}(s) \end{aligned} \tag{19.26}$$

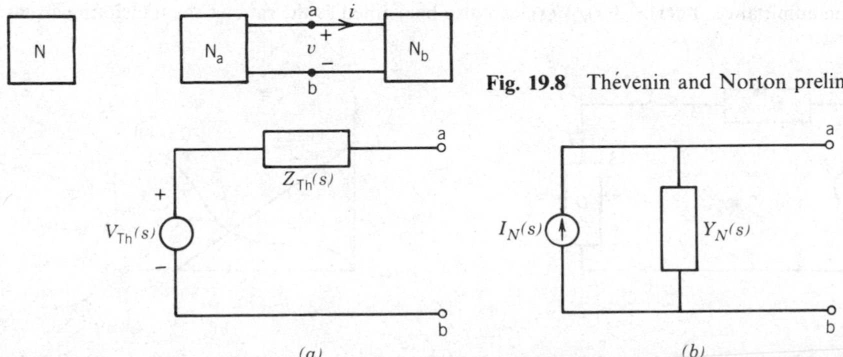

Fig. 19.8 Thévenin and Norton preliminaries.

Fig. 19.9 Thévenin (*a*) and Norton (*b*) equivalents.

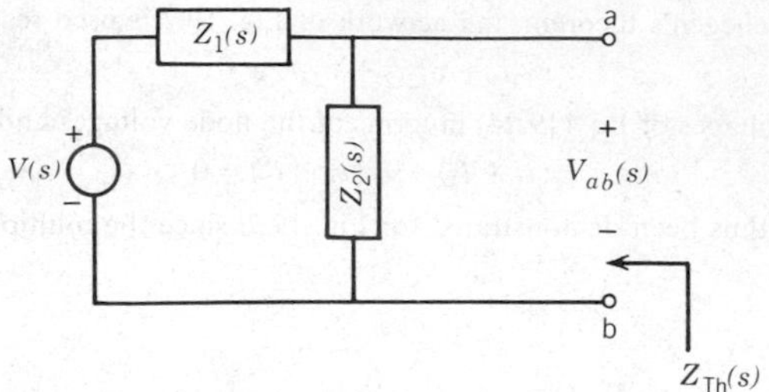

Fig. 19.10 Example circuit for Thévenin's theorem.

The Thévenin impedance, $Z_{Th}(s) = V(s)/I(s)$, can also be defined as the ratio of the transform voltage to the transform current at the terminals of network N_a when all independent voltage sources within N_a are shorted and all independent current sources within N_a are opened.

As an example, the Thévenin equivalent of Fig. 19.10 is determined. The open-circuit voltage is

$$V_{oc}(s) = \frac{Z_2}{Z_1 + Z_2} V(s)$$

and the short-circuit current is

$$I_{sc}(s) = V(s)/Z_1$$

The Thévenin impedance is then

$$Z_{Th}(s) = V_{oc}(s)/I_{sc}(s) = Z_1 Z_2/(Z_1 + Z_2)$$

which can also be obtained by shorting the independent source $V(s)$ and examining the impedance seen looking to the left of terminals a, b.

Figure 19.11a shows the Thévenin equivalent circuit of Fig. 19.10 connected to a load device D under the following conditions: (1) The voltage source is a DC source and (2) Z_1 and Z_2 are resistors. The node voltage V across the load and and the current I through the load are related by the following expression

$$V = V_{Th} - R_{Th} \cdot I$$

The graph of this linear equation, called a load line, is shown in Fig. 19.11b, as is the v-i characteristic of the load. The intersection of these two curves yields the voltage across and the current through the load device when the load device is connected to the terminals a, b in Fig. 19.10. This point of intersection is referred to as the quiescent point (Q point).

To determine the Norton equivalent circuit (Fig. 19.9b), a transform voltage and a transform current are sought at the terminals of network N_a under special load conditions. These special load conditions are used to determine the open-circuit voltage $V_{oc}(s)$ and the short-circuit current $I_{sc}(s)$. Once these values are obtained, the Norton current and Norton admittance are

$$\begin{aligned} I_N(s) &= V_{oc}(S)/Z_{Th}(s) = I_{sc}(s) \\ Y_N(s) &= 1/Z_{Th}(s) = I_{sc}(s)/V_{oc}(s) \end{aligned} \tag{19.27}$$

The Norton admittance, $Y_N(s) = I(s)/V(s)$, can also be defined as the ratio of the transform current

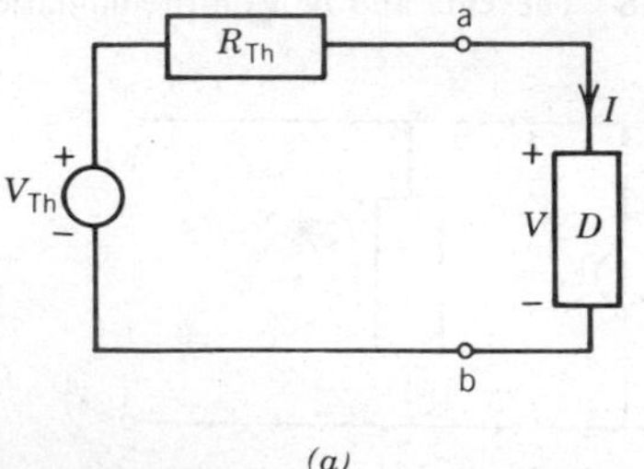

(a)

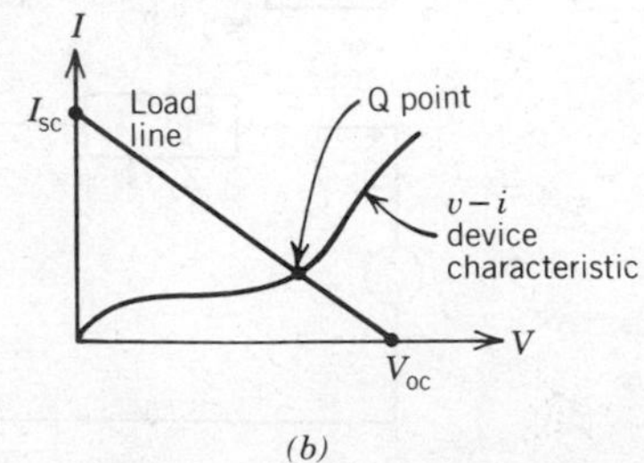

(b)

Fig. 19.11 Thévenin equivalent of circuit in Fig. 19.10 (a) and graph of load-line and voltage-current (v–i) characteristic of circuit (b). Point of intersection is called quiescent (Q) point.

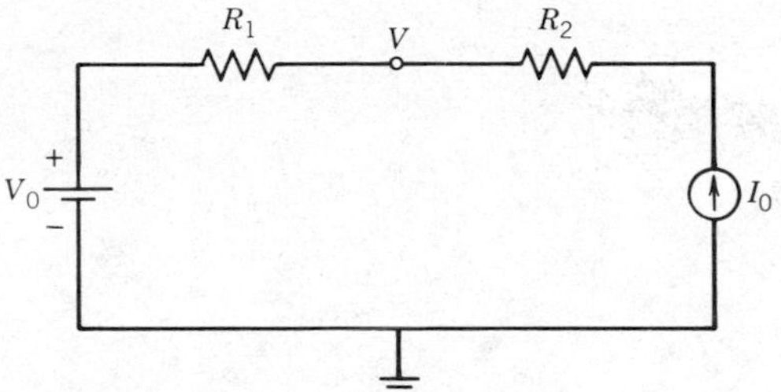

Fig. 19.12 Circuit for superposition illustration.

to the transform voltage at the terminals of network N_a when all independent voltage sources within N_a are shorted and all independent current sources within N_a are opened.

An immediate consequence of Thévenin's and Norton's theorems are source transformations. Figure 19.9*a*, Fig. 19.9*b*, Eq. (19.26), and Eq. (19.27), when viewed together, show how a voltage (current) source can be transformed into a current (voltage) source automatically.

Superposition Theorem

Consider the linear circuit of Fig. 19.12. The node voltage V determined by KCL

$$V = R_1 I_0 + V_0$$

has two components. Each results from an independent source acting alone. The component $R_1 I_0$ is found by shorting the independent voltage source, and the component V_0 is found by open circuiting the independent current source. This result is an application of the superposition principle for linear networks. The superposition theorem states that a response in a linear network resulting from a number of independent sources and initial conditions acting simultaneously can be obtained by adding together the separate responses resulting from each independent source or initial condition acting alone. An independent source or initial condition acts alone when all other independent sources and initial conditions are removed. Independent voltage sources are removed by replacing them with a short circuit and independent current sources are removed by replacing them with an open circuit.

Reciprocity

Multiport networks can be classified as reciprocal or nonreciprocal. A reciprocal network has the special property that when a response r_i at the ith port is initiated by an excitation e_j at the jth port, then

$$r_i e_j = r_j e_i \tag{19.28}$$

In this definition the network is assumed to have no internal independent sources and is initially relaxed. In addition, the topology of the network must remain unaltered by the interchange of response and excitation. Furthermore, if a voltage (current) source that is considered a generalized short (open), is used as the excitation, then the response must be a measured current (voltage); otherwise the topology of the network will be altered upon interchange of response and excitation.

19.1-3 Applications of Fundamental Circuit Theorems

Series Circuits

The material in previous sections included basic definitions or fundamental theorems. In this section utility concepts are presented. These concepts are extremely useful tools for the analysis of networks; when applied with some forethought they can greatly reduce the amount of computation required for a solution. Throughout the ensuing discussion the concepts of impedance, $Z(s)$, admittance, $Y(s)$, and phasors will be used to generalize the results.

Figure 19.13 illustrates a series circuit. A series circuit comprises two or more circuit elements carrying the same current. If KCL is applied at node A, the currents through device 1 and device 2 are observed to be equal. Applying KVL to the circuit of Fig. 19.13

$$V = V_1 + V_2 = IZ_1 + IZ_2 = I(Z_1 + Z_2) = IZ_T \tag{19.29}$$

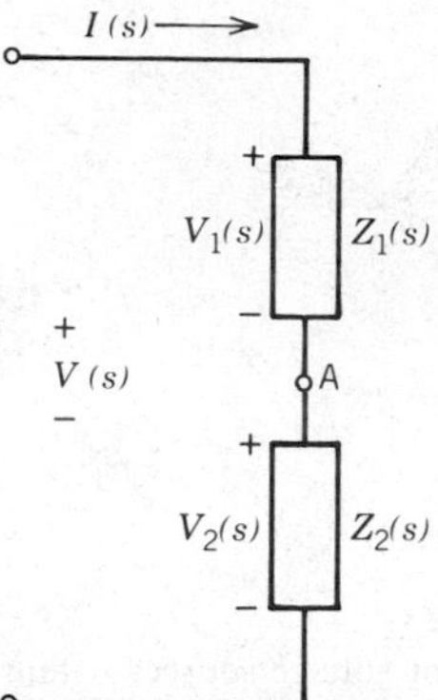

Fig. 19.13 Series circuit.

In general, m impedances in series can be replaced by an equivalent impedance

$$Z_T = Z_1 + Z_2 + \cdots + Z_m = \sum_i^m Z_i \tag{19.30}$$

The v-i characteristics observed at the terminals of a series circuit and at the terminals of the equivalent impedance are identical.

A common use for the series circuit is as a voltage divider. For Fig. 19.13, the voltage across the ith branch is

$$V_i = \frac{Z_i}{Z_1 + Z_2} V = \frac{Z_i}{Z_T} V \tag{19.31}$$

The voltage divider has many useful applications as an actual device, or it can be used to simplify the number of computations in the analysis of a circuit.

Parallel Circuits

Figure 19.14 illustrates a parallel circuit. The parallel circuit is the dual of the series circuit. A parallel circuit comprises two or more circuit elements constrained by the same voltage. With KVL applied to the loop shown, the voltages across the devices are the same. Applying KCL to the circuit of Fig. 19.14

$$I = I_1 + I_2 = VY_1 + VY_2 = V(Y_1 + Y_2) = VY_T \tag{19.32}$$

In general, m admittances in parallel can be replaced by an equivalent admittance

$$Y_T = Y_1 + Y_2 + \cdots + Y_m = \sum_i^m Y_i \tag{19.33}$$

The v-i characteristics observed at the terminals of a parallel circuit and at the terminals of the equivalent admittance are identical.

A common use for the parallel circuit is as a current divider. Referring to Fig. 19.14, the current through the ith branch is

$$I_i = \frac{Y_i}{Y_T} I \tag{19.34}$$

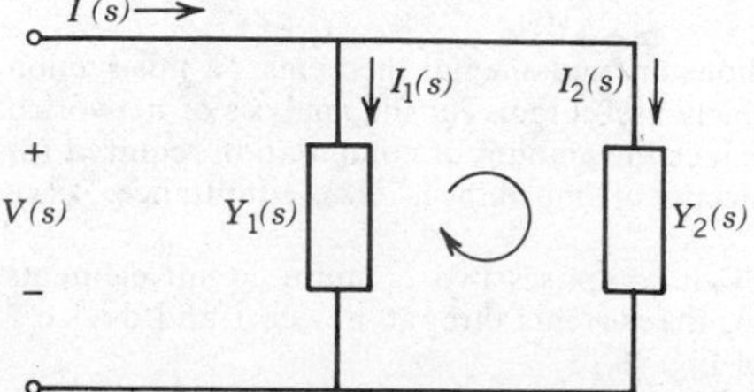

Fig. 19.14 Parallel circuit.

The current divider has many useful applications and can simplify the number of computations needed in the analysis of a circuit.

Complex Power

A common problem in engineering is power transfer. Many times the power is transmitted in sinusoidal form. Often it is desirable to maximize the power transferred from a source to a load. The aim of maximizing power transfer is accomplished by impedance matching. These ideas are illustrated here using one-port devices, but they can be generalized for use with multiport devices.

If a linear system is excited by a periodic function, the response will be a periodic function with the same period. Assume that the excitation and response are respectively given by the sinusoids

$$\begin{aligned} v(t) &= V_m \cos(\omega t + \theta) \\ i(t) &= I_m \cos(\omega t + \psi) \end{aligned} \tag{19.35}$$

where V_m and I_m are the maximum values of the sinusoidal time functions of frequency ω having phase angles θ and ψ. The average value of a periodic function is defined

$$f_{\text{ave}} = \frac{1}{T} \int_0^T f(\tau)\, d\tau \tag{19.36}$$

where T is the time of one period. The average value summarizes a periodic function over one period using a DC constant, for example,

$$f_{\text{ave}} \cdot T = \int_0^T f(\tau)\, d\tau$$

For a sinusoid, the average value over one period is zero.

The effective value or root mean square (RMS) value of a periodic function $f(t)$ is

$$f_{\text{rms}} = \sqrt{\frac{1}{T} \int_0^T f^2(\tau)\, d\tau} \tag{19.37}$$

In engineering, power measurements are facilitated using RMS values. The RMS value of a sinusoid is

$$\text{RMS}_{\text{sinusoid}} = A_m / \sqrt{2} = A_{\text{eff}} \tag{19.38}$$

where A_m is the maximum value of the sinusoid. In general, phasors will be specified as effective value phasors and not maximum value phasors.

For the voltage and current waveform specified by Eq. (19.35), the instantaneous power delivered to a network is

$$p(t) = V_m I_m \cos(\omega t + \theta) \cos(\omega t + \psi) = \tfrac{1}{2} V_m I_m \cos(\theta - \psi) + \tfrac{1}{2} V_m I_m \cos(2\omega t + \theta + \psi) \quad \text{W} \tag{19.39}$$

The energy absorbed by the network over one period is then

$$w = \int_0^T p(\tau)\, d\tau = \frac{T}{2} V_m I_m \cos(\theta - \psi) \quad \text{J}$$

The average value of power dissipated within the network for sinusoidal excitation is

$$\begin{aligned} P_{\text{ave}} &= w/T \\ &= \tfrac{1}{2} V_m I_m \cos(\theta - \psi) \qquad \text{W} \\ &= V_{\text{eff}} I_{\text{eff}} \cos(\theta_z) \end{aligned} \tag{19.40}$$

Clearly the average value of power in terms of effective value phasors is the product of their magnitudes multiplied by the power factor ($\cos\theta_z$). The angle θ_z is simply the impedance angle or the difference in phase between the voltage and the circuit. Conventionally, the power factor is said to be lagging if the current lags the voltage and leading if the current leads the voltage. If the impedance Z is expressed as

$$Z = |Z| e^{j\theta_z} = R + jX$$

where $R = |Z| \cos\theta_z$ and $X = |Z| \sin\theta_z$, the average power absorbed by the network will be

$$P_{\text{ave}} = I_{\text{eff}}^2 R \qquad \text{W} \tag{19.41}$$

The concept of complex power can easily be defined at this point. The peak power borrowed from

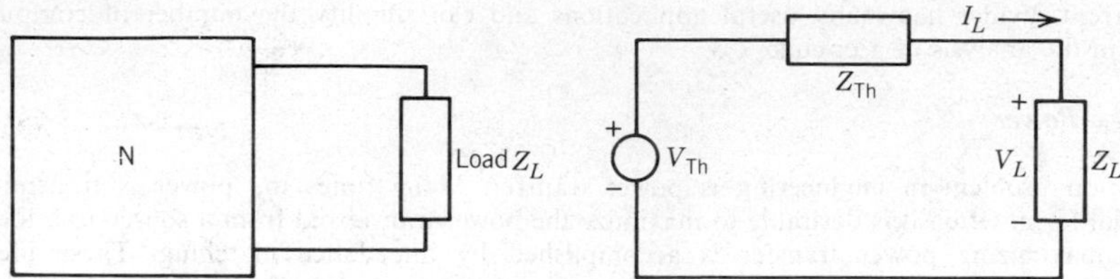

Fig. 19.15 Network and its Thévenin equivalent.

a circuit, called reactive power, is

$$\begin{aligned} P_X &= V_{eff} I_{eff} \sin\theta_z \qquad \text{VARS (volts-amps reactive)} \\ &= I_{eff}^2 X \end{aligned} \tag{19.42}$$

Reactive power will be positive for an inductive network and negative for a capacitive network. Both average and reactive powers can be related by a single expression

$$\begin{aligned} P_A &= V_{eff} I_{eff} \cos\theta_z + jV_{eff} I_{eff} \sin\theta_z \\ &= V_{eff} I_{eff} e^{j\theta_z} \\ &= \overline{V}_{eff} \overline{I}^*_{eff} \qquad \text{VA} \end{aligned} \tag{19.43}$$

The power P_A is called the complex power or the apparent power. If the rules of complex algebra are followed, complex power in different portions of a network can be added directly.

Maximum Power Transfer and Impedance Matching

Consider the network of Fig. 19.15 and its Thévenin equivalent. It is desired to maximize the power delivered by the network N to the adjustable load Z_L at a specified frequency. From Eq. (19.41), the power dissipated in the load is

$$P_L = |I_L|^2 R_L = \frac{|V_{Th}|^2 R_L}{(R_{Th} + R_L)^2 + (X_{Th} + X_L)^2} \tag{19.44}$$

First P_L is maximized by taking the partial derivative with respect to X_L ($\partial P_L / \partial X_L = 0$), resulting in

$$X_L = -X_{Th} \tag{19.45}$$

Substituting this value for X_L into Eq. (19.44) and maximizing P_L ($\partial P_L / \partial R_L = 0$) again requires

$$R_L = R_{Th} \tag{19.46}$$

Hence when the load is conjugately matched to the source

$$Z_L = Z^*_{Th} = R_{Th} - jX_{Th} \tag{19.47}$$

the maximum power

$$P_{max} = \frac{1}{4} \frac{|V_{Th}|^2}{R_{Th}} \tag{19.48}$$

is transferred to the load.

Bibliography

Aatre, V., *Network Theory and Filter Design*, Wiley Eastern Limited, New Delhi, 1981.

Balabanian, N. and T. Bickart, *Electrical Network Theory*, Wiley, New York, 1969.

Balabanian, N. and T. Bickart, *Linear Network Theory: Analysis, Properties, Design and Synthesis*, Matrix, Beaverton, OR, 1981.

Bode, H., *Network Analysis and Feedback Amplifier Design*, Krieger, Huntington, NY, 1975.

Brenner, E. and M. Javid, *Analysis of Electric Circuits*, McGraw-Hill, New York, 1967.

Carlin, H. and A. Giordano, *Network Theory: An Introduction to Recoprical and Nonreciprocal Circuits*, Prentice-Hall, Englewood Cliffs, NJ, 1964.

Chan, S., et al., *Analysis of Linear Networks and Systems*, Addison-Wesley, Reading, MA, 1972.
Chen, W., *Active Network and Feedback Amplifier Theory*, McGraw-Hill, New York, 1980.
Chirlian, P., *Basic Network Theory*, McGraw-Hill, New York, 1969.
Desoer, C. and E. Kuh, *Basic Circuit Theory*, McGraw-Hill, New York, 1969.
Guillemin, E., *Introductory Circuit Theory*, Wiley, New York, 1953.
Guillemin, E., *Synthesis of Passive Networks*, Krieger, Huntington, NY, 1977.
Kuh, E. and R. Rohrer, *Theory of Linear Active Networks*, Holden-Day, San Francisco, 1967.
Smith, R., *Circuits, Devices, and Systems*, Wiley, New York, 1973.
Van Valkenburg, M., *Network Analysis*, Prentice-Hall, Englewood Cliffs, NJ, 1974.
Van Valkenburg, M., *Linear Circuits*, Prentice-Hall, Englewood Cliffs, NJ, 1982.
Weinberg, L., *Network Analysis and Snythesis*, Krieger, Huntington, NY, 1975.

19.2 ANALYSIS OF DC AND AC CIRCUITS

Yacov Shamash and Martin Gazourian

In this section it is assumed that the reader is well acquainted with the circuit theorems described in Section 19.1. The purpose of this section is to demonstrate with examples how these theorems are used in analyzing DC and AC circuits. We will begin by applying the theorems to DC circuits. A brief outline of complex numbers is then given, followed by analysis of the steady-state behavior of AC circuits.

19.2-1 DC Circuits

DC circuits contain resistors, capacitors, inductors, dependent sources, and independent DC sources. These circuits will be analyzed using nodal analysis, mesh analysis, Thévenin's theorem, and the superposition theorem. Note that for DC circuits the capacitors can be replaced by open circuits and the inductors by short circuits. This follows from the current-voltage relationship of the capacitor and inductor, Eqs. (19.3) and (19.6).

Example 1. For the circuit shown in Fig. 19.16*a*, find V_c and I_R.

Solution. Replacing the capacitor by an open circuit and the inductor by a short circuit results in the equivalent circuit of Fig. 19.6*b*. Using KVL around the loop we obtain

$$2 = (1 + 3)I_R \Rightarrow I_R = 0.5 \text{ A}$$

Since V_c is equal to the voltage across the 3-Ω resistor, it follows that

$$V_c = 3 \times I_R = 1.5 \text{ V}$$

Nodal Analysis

It is well known that if we know the voltage at every node in the circuit with respect to a common node (usually at 0 V), we can find all the element currents and voltages in the circuit.

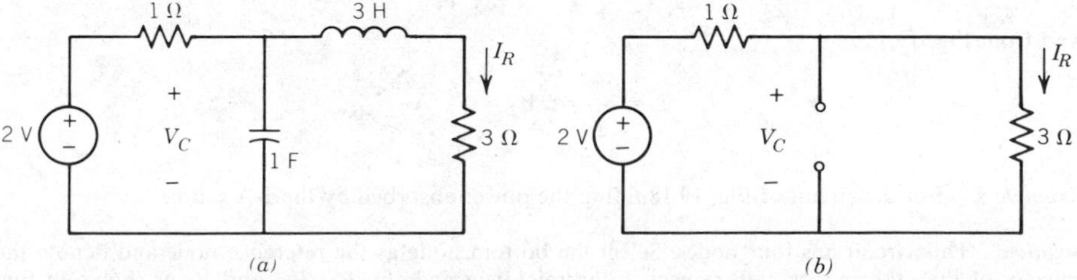

Fig. 19.16 Circuits for Example 1: (*a*) original circuit, (*b*) equivalent circuit.

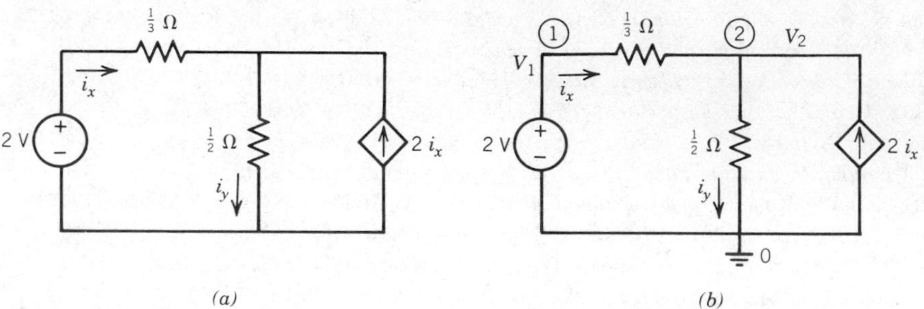

Fig. 19.17 Circuit for Example 2: (*a*) original circuit, (*b*) nodes and node voltages of the circuit.

The nodal analysis technique allows us to find these voltages in a systematic way. We will demonstrate the implementation of the technique with two examples.

Example 2. Find the current i_y in the circuit of Fig. 19.17*a*.

Solution. This circuit has three nodes. We select the bottom node as the reference node and define the node voltages at nodes 1 and 2 as V_1 and V_2, respectively. (See Fig. 19.17*b*.)

In nodal analysis our objective is to find the node voltages V_1 and V_2. Since we have two unknowns (V_1 and V_2), we need two independent equations involving V_1 and V_2. From the circuit of Fig. 19.17*b* we see that

$$V_1 = 2\text{ V} \tag{19.49}$$

This is one equation (a very simple one!).

The second equation is obtained by applying KCL at node 2. Thus

$$\frac{V_2 - V_1}{\frac{1}{3}} + \frac{V_2}{\frac{1}{2}} = 2i_x \tag{19.50}$$

But

$$i_x = \frac{V_1 - V_2}{\frac{1}{3}}$$

Hence Eq. (19.49) becomes

$$\frac{V_2 - V_1}{\frac{1}{3}} + \frac{V_2}{\frac{1}{2}} = 2\frac{(V_1 - V_2)}{\frac{1}{3}}$$

or

$$-9\,V_1 + 11\,V_2 = 0 \tag{19.51}$$

Equations (19.49) and (19.51) are two equations in terms of V_1 and V_2. Solving these equations gives

$$V_1 = 2\text{ V}, \qquad V_2 = \frac{18}{11}\text{ V}$$

And from Fig. 19.17*a*

$$i_y = \frac{V_2}{\frac{1}{2}} = 2V_2 = \frac{36}{11}\text{ A}$$

Example 3. For the circuit of Fig. 19.18*a*, find the power absorbed by the 1-A source.

Solution. This circuit has four nodes. Select the bottom node as the reference node and denote the voltages of the other nodes with respect to the reference node by V_1, V_2, and V_3 as shown in Fig. 19.18*b*.

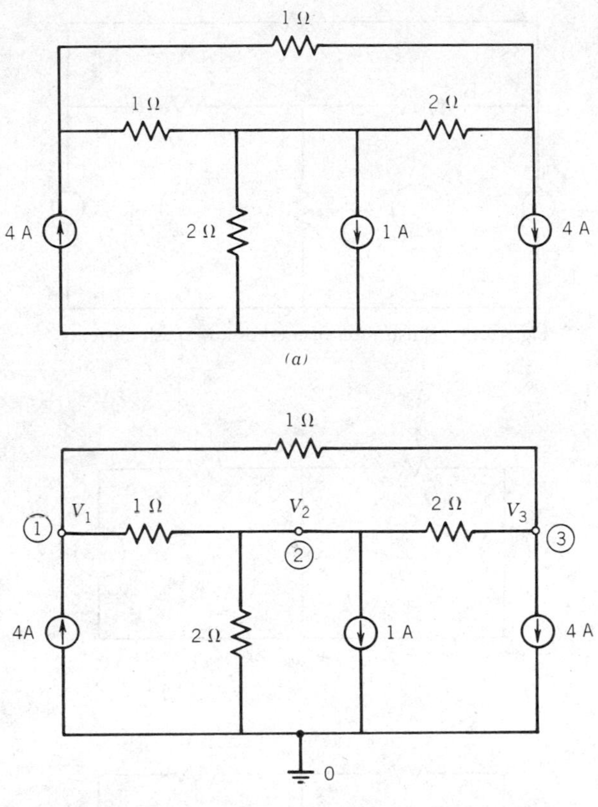

Fig. 19.18 Circuit for Example 3 (nodal analysis): (*a*) original circuit, (*b*) defining nodes and node voltages of the circuit.

We need three independent equations involving V_1, V_2, and V_3. These are obtained by applying KCL at nodes 1, 2, and 3 as follows

Node 1 $$\frac{V_1 - V_2}{1} + \frac{V_1 - V_3}{1} = +4 \quad \text{or} \quad 2V_1 - V_2 - V_3 = 4 \tag{19.52}$$

Node 2 $$\frac{V_2}{2} + \frac{V_2 - V_1}{1} + \frac{V_2 - V_3}{2} = -1 \quad \text{or} \quad -2V_1 + 4V_2 - V_3 = -2 \tag{19.53}$$

Node 3 $$4 + \frac{V_3 - V_2}{2} + \frac{V_3 - V_1}{1} = 0 \quad \text{or} \quad -2V_1 - V_2 + 3V_3 = -8 \tag{19.54}$$

Solving the three equations (19.52), (19.53), and (19.54) yields $V_1 = \frac{4}{3}$ V, $V_2 = -\frac{2}{15}$ V, and $V_3 = -\frac{6}{5}$ V. The voltage across the 1-A source is V_2 and hence the power absorbed by the source is

$$P = vi = -\tfrac{2}{15} \times = -\tfrac{2}{15}\ \text{W}$$

Mesh Analysis

We can solve planar network problems by the use of the mesh analysis technique.[1] A mesh in a circuit is defined as a closed loop that does not contain any other closed loops. A mesh current is defined as a circulatory current that flows around the mesh. Figure 19.19 shows a circuit with three meshes and associated mesh currents i_1, i_2, and i_3.

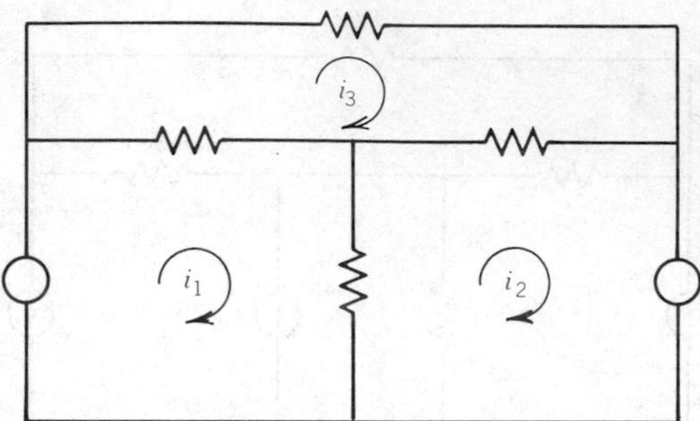

Fig. 19.19 Illustration of meshes and mesh currents.

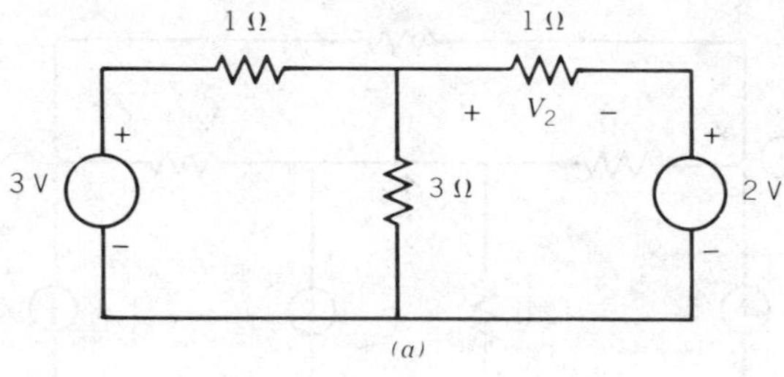

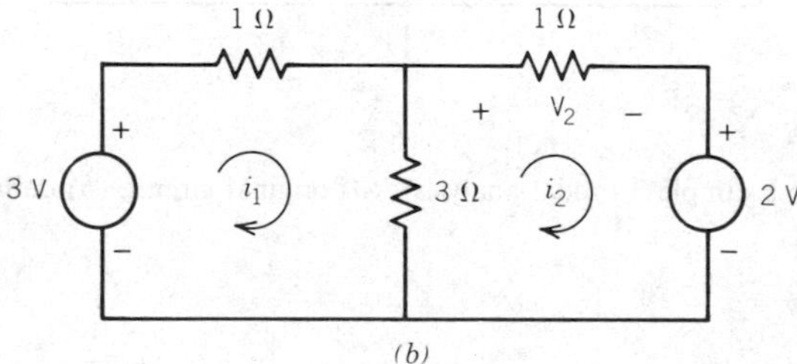

Fig. 19.20 Circuit for Example 4: (*a*) original circuit, (*b*) defining meshes and mesh currents.

For a circuit with L meshes, we need L independent equations in terms of the L mesh currents. We then solve for the L mesh currents. The branch currents and voltages in the circuit can be obtained from these mesh currents.

Example 4. For the circuit shown in Fig. 19.20*a* find the voltage V_2.

Solution. The mesh currents i_1 and i_2 are chosen in the clockwise direction. (See Fig. 19.20*b*.) The two equations necessary to find i_1 and i_2 are obtained by applying KVL around the two meshes.

$$\textit{Mesh 1} \qquad 1(i_1) + 3(i_1 - i_2) = 3 \qquad \text{or} \qquad 4i_1 - 3i_2 = 3 \tag{19.55}$$

$$\textit{Mesh 2} \qquad 1(i_2) + 3(i_2 - i_1) = -2 \qquad \text{or} \qquad -3i_1 + 4i_2 = -2 \tag{19.56}$$

Solving Eqs. (19.55) and (19.56) for i_1 and i_2 gives $i_1 = \frac{6}{7}$ A, $i_2 = \frac{1}{7}$ A. From Fig. 19.20*b*

$$V_2 = 1(i_2) = \frac{1}{7} \text{ V}$$

Example 5. Find V_1 and i_x in the circuit shown in Fig. 19.21, where the mesh currents i_1, i_2, and i_3 are as shown.

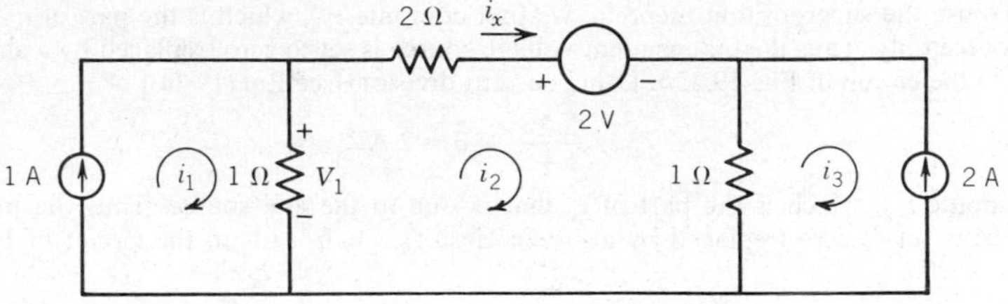

Fig. 19.21 Circuit for Example 5.

Solution. The only mesh currents flowing through the 1-A and 2-A sources are i_1 and i_3, respectively. Hence

$$i_1 = 1 \text{ A} \qquad \text{and} \qquad i_3 = -2 \text{ A}$$

Applying KVL around mesh 2 yields

$$1(i_2 - i_1) + 2(i_2) + 1(i_2 - i_3) = -2 \qquad \text{or} \qquad -i_1 + 4i_2 - i_3 = -2 \tag{19.57}$$

Substituting $i_1 = 1$ and $i_3 = -2$ into Eq. (19.57) yields

$$i_2 = -\frac{3}{4} \text{ A}$$

We can find V_1 and i_x from the mesh currents. Thus

$$V_1 = 1(i_1 - i_2) = \frac{1}{4} \text{ V} \qquad \text{and} \qquad i_x = i_2 = -\frac{3}{4} \text{ A}$$

Superposition Theorem

The superposition theorem is stated in Section 19.1-2. The next two examples demonstrate how it is used in calculating a response in a linear network resulting from two or more independent sources.

Example 6. In the circuit of Fig. 19.22*a* find the current i_x.

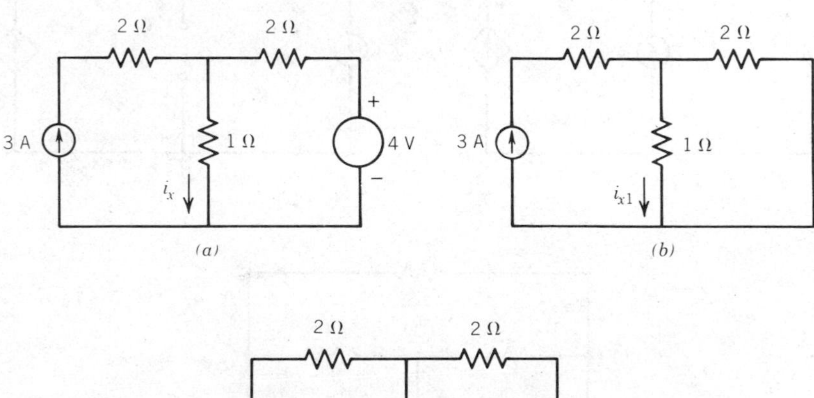

Fig. 19.22 Circuit for Example 6 illustrating the superposition theorem: (*a*) complete circuit response, (*b*) response due to 3-A source only, (*c*) response due to 4-V source only.

Solution. To use the superposition theorem, we first compute i_{x1}, which is the part of i_x that is due to the 3-A source only. Thus the independent voltage source is set to zero (replaced by a short circuit) which leads to the circuit of Fig. 19.22*b*. Using current division [see Eq. (19.34)]

$$i_{x1} = \frac{2}{1+2} \times 3 = 2 \text{ A}$$

Next we compute i_{x2}, which is the part of i_x that is due to the 4-V source. Thus the independent current source is set to zero (replaced by an open circuit) which leads to the circuit of Fig. 19.22*c*. Hence

$$i_{x2} = \frac{4}{2+1} = \tfrac{4}{3} \text{ A}$$

It follows from the superposition theorem that

$$i_x = i_{x1} + i_{x2} = 2 + \tfrac{4}{3} = \tfrac{10}{3} \text{ A}$$

It should be noted that any dependent sources must be included in the circuit when applying the superposition theorem in all the phases.

Example 7. In the circuit of Fig. 19.23*a* find the current i_x.

Solution. Using the superposition theorem, we can compute the response i_{x1} due to the 2-A source alone and then i_{x2} due to the 4-A source alone. Thus replacing the 4-A source by an open circuit, as shown in Fig. 19.23*b*, and applying KCL at node 2

$$i_1 = i_{x1} + 2i_{x1} = 3i_{x1}$$

Then applying KCL at node 1 we get

$$2 + i_1 = i_{x1}$$

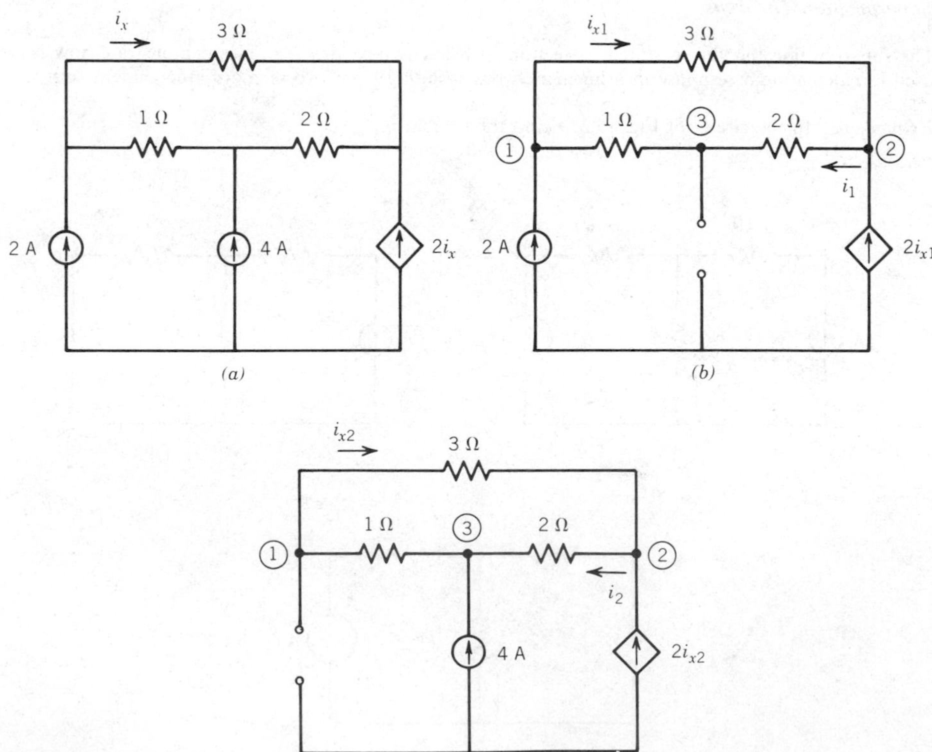

Fig. 19.23 Circuit for Example 7 illustrating the superposition theorem: (*a*) complete circuit response, (*b*) response due to 2-A source only, (*c*) response due to 4-V source only.

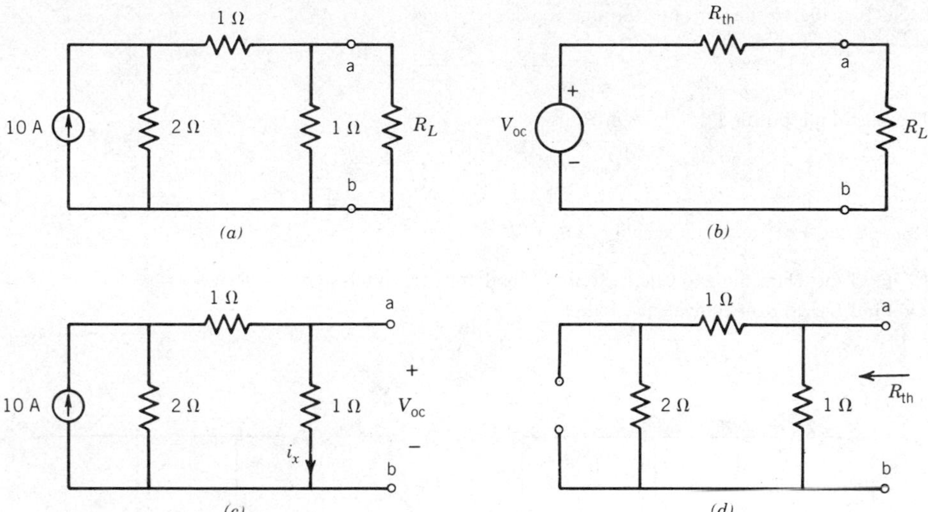

Fig. 19.24 Circuit for Example 8 illustrating Thévenin's theorem: (*a*) complete circuit, (*b*) Thévenin equivalent circuit, (*c*) finding V_{oc}, (*d*) finding T_{Th}.

or

$$i_{x1} - 3i_{x1} = 2, \qquad i_{x1} = -1 \text{ A}$$

Replacing the 2-A source by an open circuit, we get the circuit of Fig. 19.23*c*. Applying KCL at node 2 in Fig. 19.23*c* we get

$$i_2 = i_{x2} + 2i_{x2} = 3i_{x2}$$

and applying KCL at node 3 we get

$$i_2 = i_{x2} - 4 \qquad i_{x2} = -2 \text{ A}$$

Hence by the superposition theorem

$$i_x = i_{x1} + i_{x2} = -1 - 2 = -3 \text{ A}$$

Thévenin's Theorem

This theorem together with Norton's theorem is stated in Section 19.2-1. In this section we will show how the Thévenin equivalent circuit can be obtained for a number of different circuits.

Example 8. For the circuit of Fig. 19.24*a* find R_L such that the maximum power is delivered to R_L and calculate the maximum power delivered to the load.

Solution. To apply the maximum power theorem [see Eq. (19.46), which states that $R_L = R_{Th}$], we must find the Thévenin equivalent resistance of the circuit. Replacing the circuit to the left of terminals a, b by its Thévenin equivalent circuit, we obtain the circuit of Fig. 19.24*b* as follows.

To find V_{oc}, open circuit terminals a, b as shown in Fig. 19.24*c*. By current division

$$i_x = \frac{2}{2+2} \times 10 = 5 \text{ A}$$

$$V_{oc} = 1(i_x) = 5 \text{ V}$$

To find R_{Th} we set all the independent sources to the left of terminals a, b to zero, that is, current sources become open circuits and voltage sources become short circuits. For the circuit in this example this yields the circuit of Fig. 19.24*d*. Using a series-parallel combination of resistors we get

$$R_{Th} = 1 \| (1 + 2) = \frac{3}{4} \ \Omega$$

Hence for maximum power transfer

$$R_L = R_{Th} = \frac{3}{4} \ \Omega$$

From Fig. 19.24*b* the current i through R_L is

$$i = \frac{V_{oc}}{R_{Th} + R_L} = \frac{5}{\frac{3}{4} + \frac{3}{4}} = \frac{10}{3} \text{ A}$$

The maximum power P_{max} through R_L is

$$P_{max} = i^2 R_L = \frac{100}{9} \times \frac{3}{4} = \frac{25}{3} \text{ W}$$

Example 9. For the circuit shown in Fig. 19.25*a*:

1. Find the Thévenin equivalent circuit to the left of terminals a, b.
2. Find the Norton equivalent circuit.

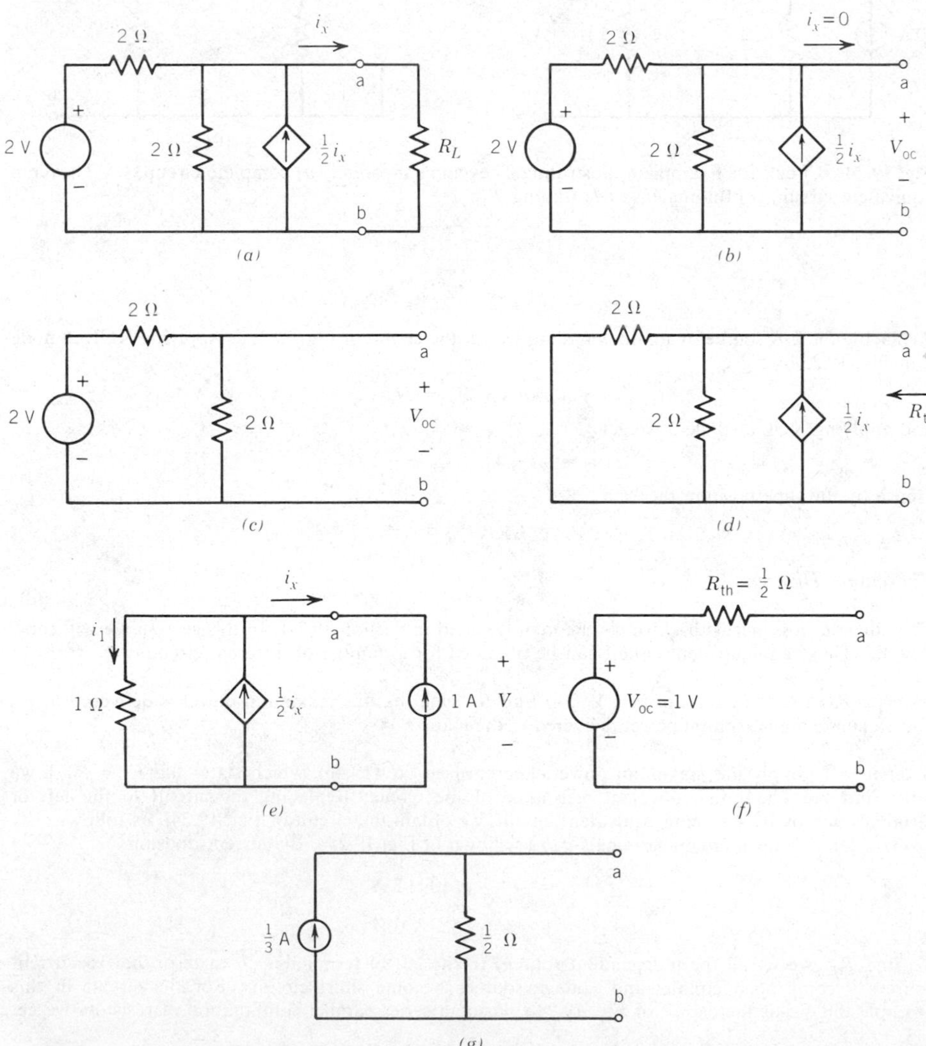

Fig. 19.25 Circuit for Example 9 illustrating the Thévenin and the Norton theorem: (*a*) complete circuit for Example 9, (*b*) computing V_{oc}, (*c*) computing V_{oc}, (*d*) circuit for determining R_{Th}, (*e*) computing R_{Th} using a 1-A source, (*f*) Thévenin equivalent circuit, (*g*) Norton equivalent circuit.

Solution.

1. To find V_{oc} we open circuit terminals a, b as shown in Fig. 19.25*b*. Clearly $i_x = 0$, hence the dependent current source has a value of zero and may be replaced by an open circuit. Thus the circuit reduces to that of Fig. 19.25*c*. By voltage division

$$V_{oc} = \frac{2}{2+2} \times 2 = 1 \text{ V}$$

To find R_{th} we set the independent voltage source to zero (replaced by a short circuit) which results in the circuit of Fig. 19.25*d*. We cannot use series-parallel combinations of resistors to find R_{th} because of the presence of the dependent source. However, if we apply a 1-A source at terminals a, b, then the voltage V across the source will be equal to R_{th}.[1] Thus combining the two resistors in Fig. 19.25*d* and applying a 1-A source at terminals a, b yields the circuit of Fig. 19.25*e*. Clearly

$$i_x = -1 \text{ A}$$

and the current i_1 through the 1-Ω resistor is

$$i_1 = \frac{1}{2} i_x - i_x = -\frac{1}{2} i_x = \frac{1}{2} \text{ A}$$

$$V = 1\left(\frac{1}{2}\right) = R_{th} = \frac{1}{2} \ \Omega$$

Thus the Thévenin equivalent circuit of Fig. 19.25*a* is as shown in Fig. 19.25*f*.

2. By use of source transformation on the Thévenin equivalent circuit of Fig. 19.25*f* we obtain the circuit shown in Fig. 19.25*g*, the Norton equivalent circuit.

19.2-2 Complex Numbers

Analysis of AC circuits requires knowledge of complex numbers. In this section we will review complex numbers to the extent needed to solve AC circuits.

Forms of Complex Numbers

We designate complex numbers by the use of capital letters $A, B, C, \ldots$. Two common forms are used to represent complex numbers. A complex number in rectangular form is described as

$$A = a + jb \tag{19.58}$$

where a and b are real numbers and represent the real and imaginary parts of A, respectively. The imaginary operator, denoted as j, is defined by

$$j^2 = -1 \quad \text{or} \quad j = \sqrt{-1} \tag{19.59}$$

An equivalent representation of complex numbers is the exponential form, given by

$$A = |A| e^{j\theta} \tag{19.60}$$

where

$$|A| = \sqrt{a^2 + b^2} \tag{19.61}$$

$$\theta = \tan^{-1} \frac{b}{a} \tag{19.62}$$

and

$$a = |A| \cos \theta \tag{19.63}$$

$$b = |A| \sin \theta \tag{19.64}$$

To illustrate this, the equivalent exponential representation of the number $A = 3 + j3$ is

$$A = \left(\sqrt{3^2 + 3^2}\right) \exp j\left(\tan^{-1} \frac{3}{3}\right)$$

$$A = \sqrt{18}\, e^{j45°}$$

By definition the conjugate A^* of a complex number A is obtained by replacing j with $-j$ in A. Therefore in the rectangular form

$$A^* = a - jb \tag{19.65}$$

and in exponential form

$$A^* = |A|e^{-j\theta} \tag{19.66}$$

Mathematical Operations Using Complex Numbers

To add or subtract complex numbers we use the rectangular form and add or subtract the respective real and imaginary parts of the complex numbers. Thus if $A = a + jb$ and $B = c + jd$, then

$$A + B = (a + c) + j(b + d) \tag{19.67}$$

and

$$A - B = (a - c) + j(b - d) \tag{19.68}$$

To multiply or divide complex numbers we have the option of using either the rectangular or the exponential form. Let

$$A = |A|e^{j\theta}$$
$$B = |B|e^{j\phi}$$

Then the product of the two complex numbers is given by

$$AB = (|A|e^{j\theta})(|B|e^{j\phi})$$

or

$$AB = |A||B|e^{j(\theta+\phi)} \tag{19.69}$$

The ratio A/B is given by

$$\frac{A}{B} = \frac{|A|e^{j\theta}}{|B|e^{j\phi}}$$

or

$$\frac{A}{B} = \frac{|A|}{|B|}e^{j(\theta-\phi)} \tag{19.70}$$

If the complex numbers A and B are given in rectangular form, that is,

$$A = a + jb$$
$$B = c + jd$$

then the product is given by

$$AB = (a + jb)(c + jd) = ac + jbc + (j^2)bd + jbc$$

and noting that $j^2 = -1$, we find

$$AB = (ac - bd) + j(ad + bc) \tag{19.71}$$

while the ratio is given by

$$\frac{A}{B} = \frac{a + jb}{c + jd}$$

Multiplying both numerator and denominator by the conjugate of B leads to

$$\frac{A}{B} = \frac{a + jb}{c + jd} \cdot \frac{c - jd}{c - jd}$$

or

$$\frac{A}{B} = \frac{(ac + db) + j(bc - ad)}{c^2 + d^2} \tag{19.72}$$

Example 10. Find the product of a complex number A and its conjugate A^*.

Solution. If $A = |A|e^{j\theta}$, then $A^* = |A|e^{-j\theta}$. Hence

$$AA^* = |A|e^{j\theta}|A|e^{-j\theta} = |A|^2e^{j(\theta-\theta)} = |A|^2e^{j0} = |A|^2$$

Thus the product of a complex number with its conjugate is a real number.

Example 11. If $A = 1 + j2$ and $B = 2 - j2$, find the ratio A/B.

Solution.

$$\frac{A}{B} = \frac{1 + j2}{2 - j2}$$

Multiplying the numerator and denominator by the conjugate of B allows us to reduce the denominator to a purely real number [see Eq. (19.72)]. Thus

$$\frac{A}{B} = \frac{(1 + j2)(2 + j2)}{(2 - j2)(2 + j2)} = \frac{-2 + j6}{8} = -\frac{1}{4} + j\frac{3}{4}$$

19.2-3 AC Analysis

In this section we consider the response of circuits to one or more independent sinusoidal sources. We will assume that all sources have been "on" for a long time and that the circuit is in the steady state (i.e., all transients have died out). We will solve these AC circuits using phasors. This procedure transforms a time-domain circuit into the phasor domain (often called the frequency domain) when the analysis is simplified. The procedure is detailed below.

Step 1. Transform all independent voltage and current sinusoidal sources from the time domain to the phasor domain, using

$$\underset{\text{(time domain)}}{A\cos(\omega t + \theta)} \Leftrightarrow \underset{\text{(phasor domain)}}{Ae^{j\theta}} \tag{19.73}$$

Notationally the phasor form of a voltage $v(t)$ is denoted by V and a current $i(t)$ by I.

Step 2. Replace R, L, and C by their respective impedances (see Table 19.3 with $s = j\omega$)

$$R \Leftrightarrow R$$
$$L \Leftrightarrow j\omega L$$
$$C \Leftrightarrow \frac{-j}{\omega C}$$

where ω is the frequency of the source. Note that if there are independent sources of different frequencies the superposition theorem must be used in analyzing the circuit.

Step 3. Solve the phasor domain circuit using mesh analysis, nodal analysis, Thévenin's theorem, or any other circuit analysis technique.

Step 4. Put the answer in exponential form and then transform it back to the time domain using Eq. (19.73).

Example 12. For the circuit of Fig. 19.26a find $v(t)$.

Solution. The source frequency is $\omega = 2$ rad/s. Thus the inductor impedance is

$$j\omega L = j2\ \Omega$$

The equivalent phasor domain circuit of Fig. 19.26a is shown in Fig. 19.26b.

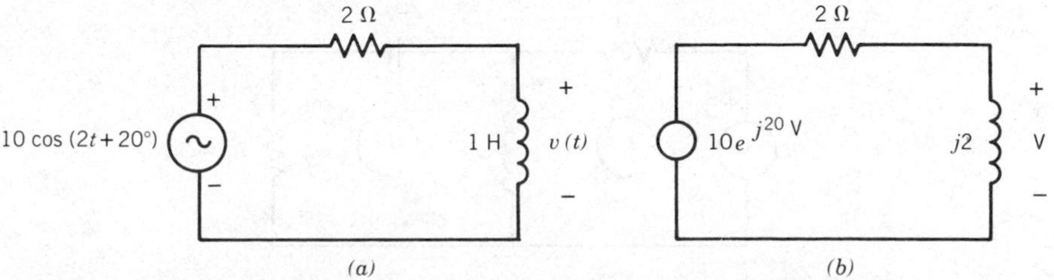

Fig. 19.26 Circuits for Example 12—AC analysis: (a) circuit in time domain, (b) circuit in phasor domain.

Using voltage division, Eq. (19.31),

$$V = \left(\frac{j2}{2+j2}\right)(10e^{j20}) = \left(\frac{j}{1+j}\right)(10e^{j20})$$

To convert V back to the time domain it must be expressed in exponential form. Converting the term $j/(1+j)$ to exponential form yields

$$V = \left(\frac{e^{j90°}}{\sqrt{2}\,e^{j45°}}\right)10e^{j20°} \qquad \text{or} \qquad V = \frac{10}{\sqrt{2}}\,e^{j65°} = 7.07e^{j65°}$$

Converting to the time domain using Eq. (19.73) leads to

$$v(t) = 7.07\cos(2t + 65°)$$

Example 13. In the circuit of Fig. 19.27*a*:

1. Find the current $i(t)$.
2. Find the average power delivered to R_L.

Solution. The equivalent phasor domain circuit is shown in Fig. 19.27*b*. Mesh analysis will be used to solve this circuit.

1. Define the phasor mesh currents I_1 and I_2 as shown in Fig. 19.27*c*. Applying KVL around each mesh yields

$$\textit{Mesh 1} \qquad 2(I_1) + j3(I_1 - I_2) = 2 \tag{19.74}$$

$$\textit{Mesh 2} \qquad -j(I_2) + 2(I_2) + j3(I_2 - I_1) = 0 \tag{19.75}$$

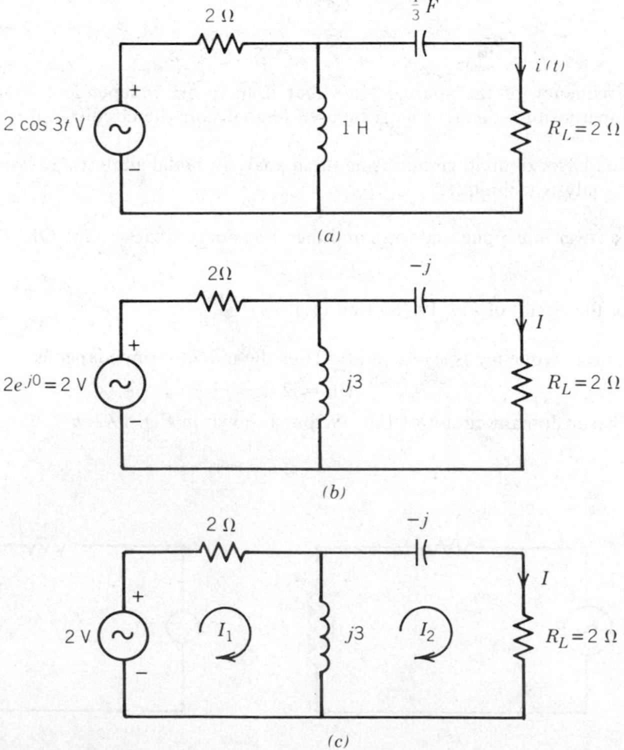

Fig. 19.27 Circuits for Example 13—AC analysis: (*a*) circuit in time domain, (*b*) circuit in phasor domain, (*c*) assigning mesh currents.

Simplifying Eqs. (19.74) and (19.75) yields

$$(2+j3)I_1-(j3)I_2=2 \tag{19.76}$$

and

$$(-j3)I_1+(2+j2)I_2=0 \tag{19.77}$$

Solving these two simultaneous equations for I_1 and I_2 gives

$$I_1=\frac{2}{7+10j} \qquad \text{and} \qquad I_2=\frac{6j}{(2+j2)(7+j10)}$$

or in exponential form

$$I_1=0.164e^{j55^\circ}\,\text{A} \qquad \text{and} \qquad I_2=0.174e^{-j10^\circ}\,\text{A}$$

Recognizing from Fig. 19.27c that branch current $I=I_2$ and converting I to the time domain we obtain

$$i(t)=0.174\cos(3t-10^\circ)$$

2. To find the average power delivered to R_L we use Eq. (19.40)

$$P_{\text{avg}}=\tfrac{1}{2}V_mI_m\cos(\theta)$$

where V_m and I_m are the magnitudes of the phasor voltage and current, respectively, and θ is the phase difference between the phasor voltage and phasor current. For a resistive load, the angle θ is zero (hence $\cos\theta=1$) and $V=IR$. Hence

$$P_{\text{avg}}=\tfrac{1}{2}I_m^2R$$

Substituting $I_m=0.174$ and $R=R_L=2\ \Omega$ gives

$$P_{\text{avg}}=\tfrac{1}{2}(0.174)^2\times 2=0.03\ \text{W}$$

Example 14. For the circuit in Fig. 19.28a:

1. Find the Thévenin equivalent circuit to the left of terminals a, b.
2. Find the effective (rms) value of the voltage and current in the load for $Z_L=(1+j)$ and also find the average power delivered to Z_L.
3. Find the impedance Z_L such that Z_L absorbs maximum power, and then compute this power.

Solution.

1. Converting to the phasor domain we obtain the circuit of Fig. 19.28b. To find V_{oc}, open circuit terminals a, b as shown in Fig. 19.28c. Since $I=0$, there is no current through the capacitor. The 2-A source current flows through the 1-Ω resistor. Hence

$$I_X=2\ \text{A}$$

and applying KVL we obtain

$$-1(I_X)-2I_X+V_{\text{oc}}=0 \quad \text{or} \quad V_{\text{oc}}=3I_X=6\ \text{V}$$

To find Z_{Th} we open circuit the current source as shown in Fig. 19.28d. Since the network contains a dependent source we cannot compute Z_{Th} by using series-parallel combinations of impedances. Thus, place a 1-A source across the a, b terminals as shown in Fig. 19.28e. Then

$$Z_{\text{Th}}=V$$

From the figure we can see that

$$I_X=1$$

and applying KVL around the loop we get

$$-(1-j)I_X-2I_X+V=0$$

or

$$V=(3-j)I_X=(3-j), \qquad Z_{\text{Th}}=(3-j)\ \Omega$$

Hence the Thévenin equivalent circuit to the left of the a, b terminals is shown in Fig. 19.28f.

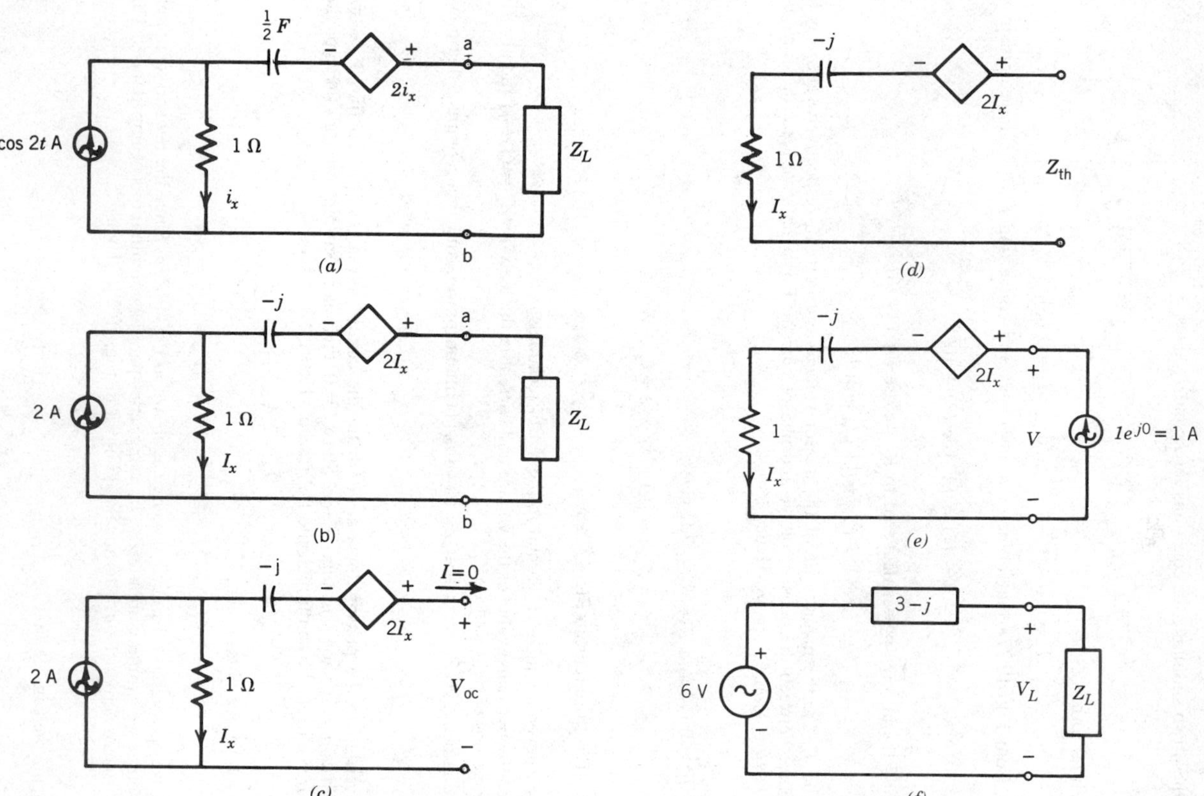

Fig. 19.28 Circuit for Example 14—AC analysis: (*a*) circuit in time domain, (*b*) circuit in phasor domain, (*c*) computing V_{oc}, (*d*) to find Z_{Th}, (*e*) computing Z_{Th}, (*f*) Thévenin equivalent circuit.

2. From the circuit of Fig. 19.28*f*

$$V_L = \frac{Z_L}{Z_L + Z_{\text{Th}}} \times 6 = \frac{(1+j)}{(1+j)+(3-j)} 6 = \frac{1+j}{4} 6 = \frac{3}{\sqrt{2}} e^{j45°} = 2.12 e^{j45°} \text{ V}$$

From Eq. (19.38)

$$V_{L\text{eff}} = \frac{2.12}{\sqrt{2}} = 1.5 \text{ V (rms)}$$

Also

$$I_L = \frac{V_L}{Z_L} = \frac{3(1+j)}{2(1+j)} = 1.5 \text{ A}$$

and

$$I_{L\text{eff}} = \frac{1.5}{\sqrt{2}} = 1.06 \text{ A (rms)}$$

The average power delivered to Z_L is given by Eq. (19.40)

$$P_{\text{avg}} = V_{\text{eff}} I_{\text{eff}} \cos \theta$$

where θ is the phase difference between the voltage and current phasors. In this case $\theta = 45°$. Hence

$$P_{\text{avg}} = 1.5 \times 1.06 \times \cos 45° = 1.12 \text{ W}$$

3. For maximum power transfer, Eq. (19.47)

$$Z_L = Z^*_{\text{th}} = (3+j)\ \Omega$$

In this case using voltage division in Fig. 19.28*f*

$$V_L = \frac{Z_L}{Z_L + Z_{\text{th}}} \times 6 = \frac{3+j}{(3+j)+(3-j)} \times 6 = (3+j) = \sqrt{10}\, e^{j(\tan^{-1} 1/3)} = 3.16 e^{j18.4°}$$

and

$$I_L = \frac{V_L}{Z_L} = 1 \text{ A}$$

Hence

$$P_{\text{avg}}^{(\text{max})} = \frac{1}{2} \times 1 \times 3.16 \cos 18.4° = 1.5 \text{ W}$$

Example 15. For the circuit of Fig. 19.29*a* compute:

1. The power factor of the load.
2. The average power delivered to the load.
3. The reactive power delivered to the load.
4. The complex power P_A delivered to the load.

Solution. The first step in the solution is to obtain the equivalent phasor circuit as shown in Fig. 19.29*b*, where $\omega = 1$. The load impedance Z_L is given by

$$Z_L = -j + [(1-j) \| (1+j)] = (1-j) \quad \Omega = \sqrt{2} \angle -45°$$

1. The power factor of the load is given by

$$\cos(-45°) = \frac{1}{\sqrt{2}} = 0.707$$

The power factor is a leading one because the current leads the voltage by 45°.

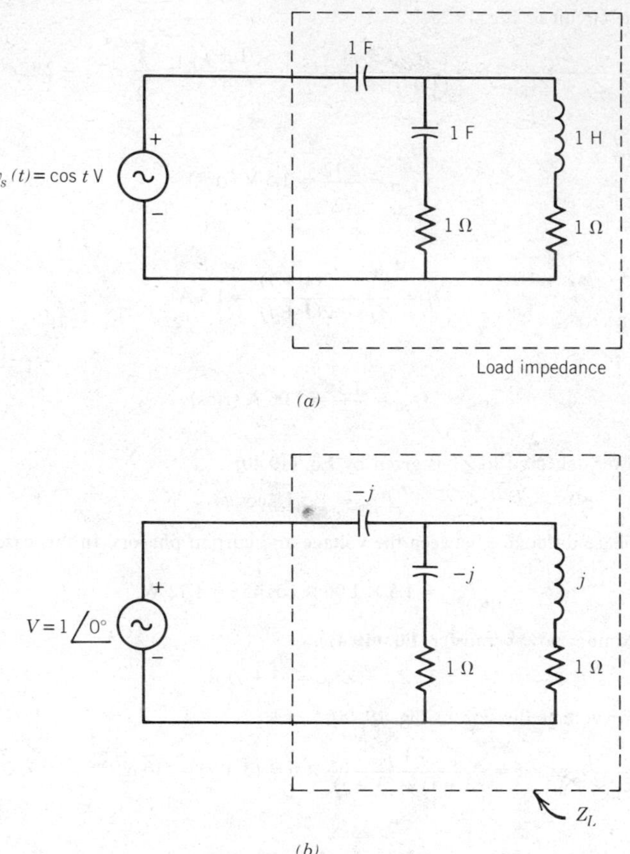

Fig. 19.29 Circuits for Example 15: (*a*) time-domain circuit, (*b*) phasor-domain circuit.

2. The load current I is given by

$$I = \frac{V_s}{Z_L} = \frac{1\angle 0°}{\frac{1}{\sqrt{2}}\angle -45°} = \sqrt{2}\angle 45° \text{ A}$$

$$P_L = \tfrac{1}{2}V_m I_m \cos\theta = \tfrac{1}{2}\times 1 \times \sqrt{2} \times \cos(-45°) = \tfrac{1}{2} \text{ W}$$

3. $Q_L = \frac{1}{2}V_m I_m \sin(-45°) = -\frac{1}{2}$ VAR
4. The complex power P_A is given by

$$P_A = P_L + jQ_L = \tfrac{1}{2}(1-j) \text{ VA}$$

Example 16. For the circuit shown in Fig. 19.30*a*:

1. Find the transfer function V_2/V_1.
2. If $v_1(t) = 10\cos 2t$, find the steady state response $V_2(t)$.

Solution. The circuit has two inductors that are mutually coupled. The coupling relationship is defined in Table 19.2. The first step is to derive the equivalent phasor circuit as shown in Fig. 19.30*b*. Note that all the variables in the circuit are functions of the input frequency ω.

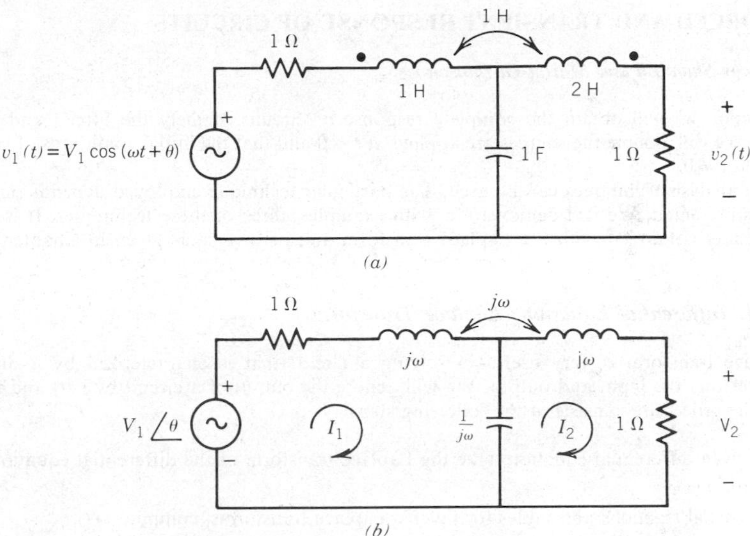

Fig. 19.30 Circuits for Example 15 to illustrate mutual coupling: (*a*) circuit in time domain, (*b*) circuit in phasor domain.

1. The circuit contains two meshes. Hence, we define the mesh currents I_1 and I_2 as shown. Then applying KVL around the two meshes we get

$$\textit{Mesh 1} \qquad V_1(j\omega) = \left(1 + j\omega + \frac{1}{j\omega}\right)I_1 - \frac{1}{j\omega}I_2 - \underbrace{j\omega I_2}_{\text{due to coupling}}$$

$$\textit{Mesh 2} \qquad 0 = \underbrace{-j\omega I_1}_{\text{due to coupling}} = \frac{1}{j\omega}I_1 + \left(2j\omega + \frac{1}{j\omega} + 1\right)I_2$$

Solving these two equations we get

$$I_2 = \frac{(j\omega)^2 + 1}{(j\omega)^3 + 3(j\omega)^2 + 2(j\omega) + 2}V_1$$

But

$$V_2 = \bar{I}_2 \times 1 = \frac{(j\omega)^2 + 1}{(j\omega)^3 + 3(j\omega)^2 + 2(j\omega) + 2}V_1$$

$$\frac{V_2}{V_1} = \frac{1 - \omega^2}{-j\omega^3 - 3\omega^2 + 2j\omega + 2} = \frac{1 - \omega^2}{(2 - 3\omega^2) + j(2\omega - \omega^3)}$$

2. In this case, $\omega = 2$ rad/s and $V_1 = 10\angle 0°$.
Hence

$$V_2 = (10\angle 0°)\left[\frac{1 - 4}{(2 - 3 \times 4) + j(4 - 8)}\right] = \frac{-30}{-10 - j4} = \frac{30}{10 + j4} = 2.79\angle 21.8°$$

$$v_2(t) = 2.79\cos(2t - 21.8°) \text{ V}$$

References

1 W. Hayt Jr. and J. E. Kemmerly, *Engineering Circuit Analysis*, McGraw-Hill, New York, 1982.
2 L. S. Bobrow, *Elementary Linear Circuit Analysis*, Holt, Rinehart & Winston, New York, 1983.

19.3 FORCED AND TRANSIENT RESPONSE OF CIRCUITS

Yacov Shamash and Martin Gazourian

In this section we will obtain the complete response of circuits, namely the forced and transient responses. We will assume the sources are applied at $t = 0$ and that the initial conditions of the circuit are given at $t = 0$.

Several analysis techniques can be used. The particular technique employed depends on how the problem is presented. We will demonstrate, with examples, three of these techniques. It is assumed that the reader is familiar with the Laplace transform and its inverse as given in Chapter 4 of this handbook.

Method 1: Differential Equation / Laplace Transform

The Laplace transform is very useful in solving a circuit that is characterized by a differential equation relating the input and output. We will denote the output of a circuit by $y(t)$ and the input by $x(t)$. The procedure consists of the following steps:

1. For a given differential equation, take the Laplace transform of the differential equation.
2. Solve for $Y(s)$.
3. Using partial fractions and tables for inverse Laplace transforms, compute $y(t)$.

The next two examples illustrate the specific details of the method.

Example 17. For the differential equation

$$\frac{dy(t)}{dt} + 2y(t) = x(t) \tag{19.78}$$

find $y(t)$ given the initial condition $y(0) = 0$ and the input $x(t) = u(t)$ (the unit step).

Solution. Taking the Laplace transform of both sides of Eq. (19.78) yields

$$[sY(s) - y(0)] + 2Y(s) = \frac{1}{s}$$

where $1/s$ is the Laplace transform of the unit step (see Table 4.4). Substituting $y(0) = 0$ and solving for $Y(s)$ gives

$$Y(s) = \frac{1}{s(s+2)} \tag{19.79}$$

Expanding the right-hand side of Eq. (19.79) into a partial fraction leads to

$$Y(s) = \frac{0.5}{s} - \frac{0.5}{s+2} \tag{19.80}$$

Taking the inverse Laplace transform of both sides of Eq. (19.80) (see Table 4.4) gives

$$y(t) = (0.5 - 0.5e^{-2t})u(t)$$

Example 18. For the circuit of Fig. 19.31:

1. Find the differential equation relating the input $v_s(t)$ and the voltage $v_c(t)$.
2. Find $v_c(t)$ for $t \geq 0$, given that $v_c(0) = -\frac{1}{8}$ V and $v_s(t) = (\sin 2t)u(t)$.

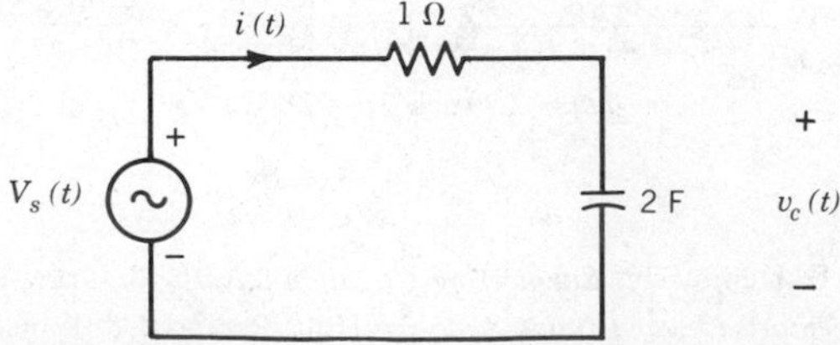

Fig. 19.31 Circuit for Example 18—to determine the complete response of a circuit.

Solution.

1. Applying KVL around the loop we obtain

$$v_s(t) = 1 \times i(t) + v_c(t) \tag{19.81}$$

But

$$i(t) = C\frac{dv_c(t)}{dt} = 2\frac{dv_c(t)}{dt}$$

Hence Eq. (19.81) becomes

$$v_s(t) = 2\frac{dv_c(t)}{dt} + v_c(t) \tag{19.82}$$

Equation (19.82) is a differential equation that relates the input $v_s(t)$ and the output $v_c(t)$.

2. For $v_s(t) = (\sin 2t)u(t)$, Eq. (19.82) becomes

$$(\sin 2t)u(t) = 2\frac{dv_c(t)}{dt} + v_c(t) \tag{19.83}$$

Taking the Laplace transform of both sides of Eq. (19.83) and using Table 4.4 we get

$$\frac{1}{s^2+4} = 2sV_c(s) - 2v_{c0} + V_c(s)$$

and substituting $v_{c0} = -\frac{1}{8}$ V and solving for $V_c(s)$ we get

$$V_c(s) = \frac{-s^2}{8(s^2+4)(s+0.5)}$$

Expanding $V_c(s)$ into partial fractions gives

$$V_c(s) = \frac{-1/128}{5+0.5} + \frac{8/128}{s^2+4} - \frac{172/128s}{s^2+4}$$

Using Table 4.4 for inverse Laplace transforms yields

$$v_c(t) = \left(-\frac{1}{128}e^{-0.5t} + \frac{8}{128}\sin 2t - \frac{127}{128}\cos 2t\right)u(t) \text{ V}$$

Inspection of $v_c(t)$ reveals the transient and forced responses of $v_c(t)$. The term $(-(1/128)e^{-0.5t})u(t)$ decays to zero as $t \to \infty$, and hence defines the transient solution of $v_c(t)$. Consequently the other terms $[(8/128)\sin 2t - (127/128)\cos 2t]u(t)$ constitute the forced part of the voltage $v_c(t)$, that is, $v_c(t)$ is of the same form as the input forcing function $v_s(t) = \sin 2tu(t)$.

Method 2: s-Domain Modeling

Often the circuit is given with a defined input, a defined output, and a set of initial conditions. In this situation the differential equation relating the input and output is not available or is difficult to obtain. However, we can solve this circuit by transforming it to the s-domain and solving it using standard techniques such as mesh analysis, nodal analysis, Thévenin's theorem, or another circuit analysis technique.

Before we present some examples to illustrate s-domain modeling, we must show how inductors and capacitors with initial conditions are transformed in the s-domain.

Consider the inductor shown in Fig. 19.32*a*. For the inductor

$$v(t) = L\frac{di(t)}{dt} \tag{19.84}$$

Taking the Laplace transform of both sides of Eq. (19.84) yields the s-domain equation for the inductor

$$V(s) = sLI(s) - Li(0) \tag{19.85}$$

Equation (19.85) can be rewritten

$$I(s) = \frac{1}{sL}V(s) + \frac{1}{s}i(0) \tag{19.86}$$

Equations (19.85) and (19.86) are modeled with the two circuits shown in Fig. 19.32*b* and *c*, respectively.

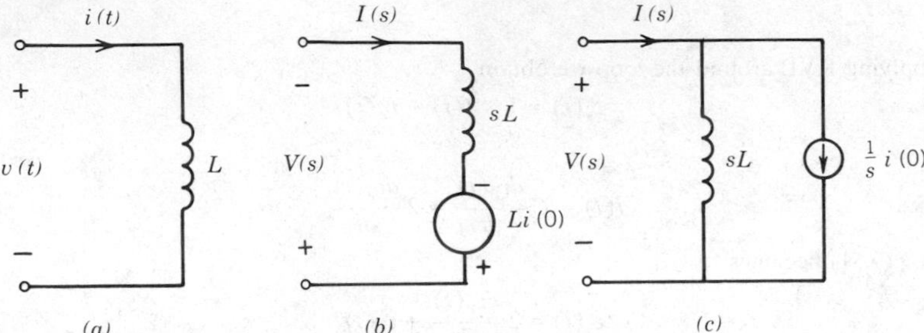

Fig. 19.32 s-Domain modeling of an inductor: (*a*) time-domain model, (*b*) s-domain model using a voltage source, (*c*) s-domain model using a current source.

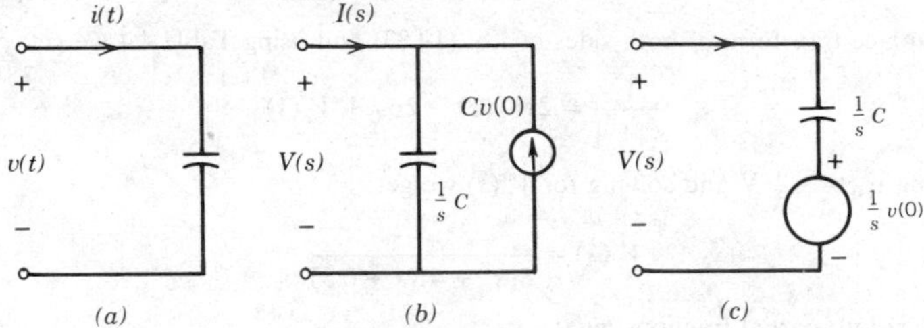

Fig. 19.33 s-Domain modeling of a capacitor: (*a*) time-domain model, (*b*) s-domain model using a current source, (*c*) s-domain model using a voltage source.

For the capacitor shown in Fig. 19.33*a*

$$i(t) = C\frac{dv(t)}{dt} \tag{19.87}$$

Taking the Laplace transform of both sides of Eq. (19.87) yields the *s*-domain equation for the capacitor

$$I(s) = sCV(s) - Cv(0) \tag{19.88}$$

which can be rewritten in the form

$$V(s) = \frac{1}{sC}I(s) + \frac{1}{s}v(0) \tag{19.89}$$

Equations (19.88) and (19.89) are modeled with the two circuits shown in Fig. 19.33*b* and *c*, respectively.

When we transform a circuit to the *s* domain we can use any of the models shown in Fig. 19.32*b* and *c* and Fig. 19.33*b* and *c*. Typically we select the models that will simplify the analysis.

Example 19. For the circuit of Fig. 19.34*a* find $i(t)$, given that $v_s(t) = u(t)$ and $v_c(0) = 1$ V.

Solution. We will use the model for the capacitor shown in Fig. 19.33*c*. Since the Laplace transform of $u(t)$ is $1/s$, the *s*-domain equivalent circuit is given in Fig. 19.34*b* with the mesh currents $I_1(s)$ and $I_2(s)$. Applying KVL around mesh 1 and mesh 2 we have

$$\textit{Mesh 1} \qquad I_1(s) + \frac{1}{s} + \frac{1}{s}[I_1(s) - I_2(s)] = \frac{1}{s}$$

or

$$\left(1 + \frac{1}{s}\right)I_1(s) - \frac{1}{s}I_2(s) = 0 \tag{19.90}$$

$$\textit{Mesh 2} \qquad 2I_2(s) + \frac{1}{s}[I_2(s) - I_1(s)] = \frac{1}{s}$$

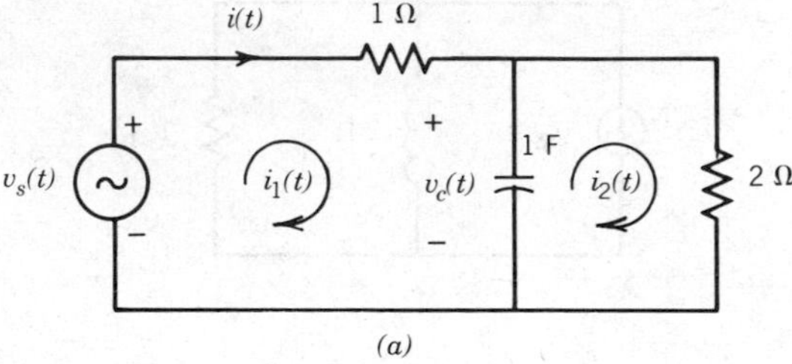

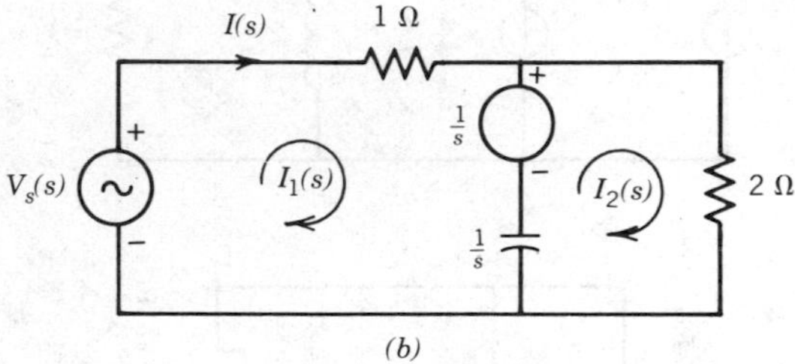

Fig. 19.34 Circuit for Example 19: (*a*) time-domain circuit, (*b*) s-domain equivalent circuit.

or

$$-\frac{1}{s}I_1(s) + \left(2 + \frac{1}{s}\right)I_2(s) = \frac{1}{s} \tag{19.91}$$

Solving the two simultaneous equations (19.90) and (19.91) for $I_1(s)$ yields

$$I_1(s) = \frac{1}{2s(s + \frac{3}{2})}$$

Expanding into partial fractions and noting that

$$I(s) = I_1(s)$$

we obtain

$$I(s) = \frac{\frac{1}{3}}{s} - \frac{\frac{1}{3}}{(s + \frac{3}{2})} \tag{19.92}$$

Taking the inverse Laplace transform of Eq. (19.92) we obtain

$$i(t) = \frac{1}{3}\left(1 - e\frac{-3t}{2}\right)u(t)\text{ A}$$

Example 20. For the circuit shown in Fig. 19.35*a* find the voltage $v(t)$ given that $i_s(t) = u(t)$ A and $i_L(0) = -1$ A.

Solution. Using the model for the inductor of Fig. 19.32*c*, we show the *s*-domain equivalent of the circuit of Fig. 19.35*a* in Fig. 19.35*b*. Combining the two sources into one source and combining the two impedances into one impedance gives us the circuit of Fig. 19.35*c*. Hence

$$V(s) = I(s)Z(s) = \frac{2}{s} \times \frac{2s}{2s+1} = \frac{4}{2s+1} = \frac{2}{s + \frac{1}{2}}$$

Taking the inverse Laplace transform of $V(s)$ we obtain

$$v(t) = 2e^{-\frac{1}{2}t}u(t)\text{ V}$$

Example 21. Find the transfer function of the circuit of Fig. 19.36*a* where the input and output are defined as $v_i(t)$ and $v_0(t)$, respectively.

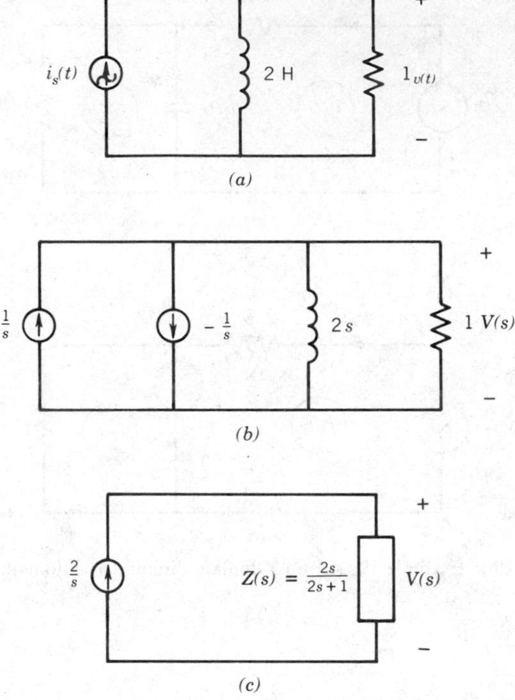

Fig. 19.35 Circuit for Example 20: (*a*) circuit in time domain, (*b*) equivalent circuit in s domain, (*c*) reduced equivalent circuit.

Solution. The transfer function of a system is defined as the ratio of the Laplace transform of the output to the Laplace transform of the input, with all the initial conditions equal to zero. Thus for this example the transfer function $H(s)$ is given by

$$H(s) = \frac{V_0(s)}{V_i(s)} \tag{19.93}$$

with the initial conditions set to zero. The *s*-domain equivalent circuit of Fig. 19.36*a* is shown in Fig. 19.36*b*.

The parallel combination of the capacitor and the 2-Ω resistor is $2/(1 + 2s)$. Hence using voltage division we obtain

$$V_0(s) = V_i(s)\left[\frac{2/(1+2s)}{(s/2) + 2/(1+2s)}\right]$$

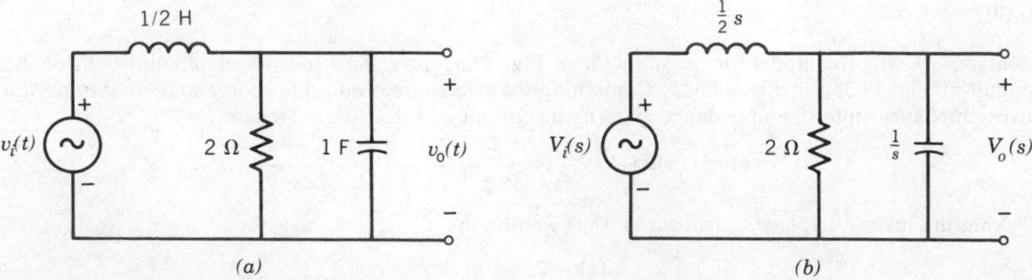

Fig. 19.36 Circuit for Example 21 to derive the transfer function of a circuit: (*a*) time domain circuit, (*b*) s-domain circuit.

Simplifying and dividing by $V_i(s)$ yields the transfer function

$$H(s) = \frac{V_i(s)}{V_0(s)} = \frac{2}{s^2 + 0.5s + 2}$$

Method 3: State-Variable Analysis

In this section the reader is expected to be familiar with matrix algebra as presented in Chapter 2. The state-variable representation is another time-domain characterization of circuits. It uses a set of internal variables $(q_1, q_2, \ldots, q_n)$ in addition to the input-output variables to describe the circuit. Formally *the state of a system is defined as a set of variables, called state variables, such that the knowledge of these variables at* $t = t_0$, *together with the input for* $t > t_0$, *completely determines the behavior of the system for any time* $t > t_0$, that is, if we know $q_1(t_0), q_2(t_0), \ldots, q_n(t_0)$ and the inputs for $t \geqslant t_0$ we can determine any set of defined outputs. The following example illustrates this concept of state.

Example 22. Consider the series *RLC* circuit shown in Fig. 19.37. From basic circuit theory we know that the behavior of this circuit for $t \geqslant t_0$ is completely determined by the input voltage $v_s(t)$ for $t \geqslant t_0$ and the inductor current $i_L(t)$ and capacitor voltage $v_c(t)$ at $t = t_0$. Hence it follows from the definition of the state of a circuit that $i_L(t)$ and $v_c(t)$ are a set of state variables for this circuit, that is

$$q_1(t) = i_L(t), \qquad q_2(t) = v_c(t)$$

In general we can represent a circuit in terms of its state variables, say $q_1(t), q_2(t), \ldots, q_n(t)$ by a set of first-order differential equations which in matrix form are

$$\mathbf{q}(t) = A\mathbf{q}(t) + B\mathbf{x}(t) \tag{19.94}$$

$$\mathbf{y}(t) = C\mathbf{q}(t) + D\mathbf{x}(t) \tag{19.95}$$

where $\mathbf{q}(t)$ is the state vector, $\mathbf{x}(t)$ is the input vector, and $\mathbf{y}(t)$ is the output vector. The matrices A, B, C, and D are of compatible dimensions. Equation (19.94) is called the state equation of the circuit and Eq. (19.95) is called the output equation of the circuit. The $(n \times n)$ matrix A is known as the state matrix of the circuit.

Example 23. For the *RLC* circuit of Fig. 19.37 the circuit equations are

$$L\frac{di_L(t)}{dt} + Ri_L(t) + v_c(t) = v_s(t) \tag{19.96}$$

$$C\frac{dv_c(t)}{dt} = i_L(t) \tag{19.97}$$

Define the state variables $q_1(t)$ and $q_2(t)$ by

$$q_1(t) = i_L(t)$$
$$q_2(t) = v_c(t)$$

Using Eqs. (19.96) and (19.97) the state equations become

$$\dot{q}_1(t) = -\frac{R}{L}q_1(t) - \frac{1}{L}q_2(t) + \frac{1}{L}v_s(t)$$

$$\dot{q}_2(t) = \frac{1}{C}q_1(t)$$

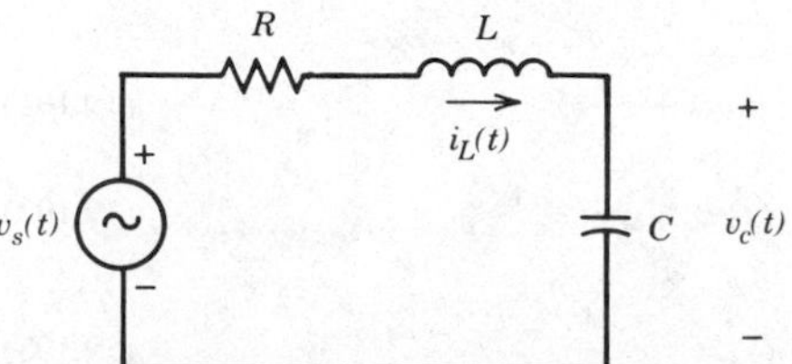

Fig. 19.37 Circuit for Example 22.

which can be written in matrix form as

$$\begin{bmatrix} \dot{q}_1(t) \\ \dot{q}_2(t) \end{bmatrix} = \begin{bmatrix} -\dfrac{R}{L} & -\dfrac{1}{L} \\ \dfrac{1}{C} & 0 \end{bmatrix} \begin{bmatrix} q_1(t) \\ q_2(t) \end{bmatrix} + \begin{bmatrix} \dfrac{1}{L} \\ 0 \end{bmatrix} v_s(t) \tag{19.98}$$

Comparing Eqs. (19.94) and (19.98) we get

$$A = \begin{bmatrix} -\dfrac{R}{L} & -\dfrac{1}{L} \\ \dfrac{1}{C} & 0 \end{bmatrix}, \qquad B = \begin{bmatrix} \dfrac{1}{L} \\ 0 \end{bmatrix}$$

If we consider the output $y(t)$ of this circuit to be the voltage across the capacitor, then

$$y(t) = v_c(t) = q_2(t)$$

Hence the output matrices C and D of Eq. (19.95) are

$$[C] = [0\ 1], \qquad [D] = [0]$$

In electrical circuits consisting of resistors, capacitors, and inductors it is natural to identify the state variables with the inductor currents and capacitor voltages. It should be noted that the choice of state variables for a given circuit is not unique.[1] Before we proceed to solve the state equations, we present a cookbook approach to finding the state equations for linear circuits:

1. Define the state variables as the inductor currents $i_{L1}(t), i_{L2}(t), \ldots, i_{Lm}(t)$ and capacitor voltages $v_{c1}(t), v_{c2}(t), \ldots, v_{cp}(t)$. You can arbitrarily select current direction and voltage polarities.

2. Replace each inductor by a current source. That is, replace L_1 with a current source labeled i_{L1}, L_2 with current source i_{L2}. and so on. Similarly replace each capacitor $C_1, C_2, \ldots, C_p$ by voltage sources $v_{c1}, v_{c2}, \ldots, v_{cp}$.

3. Using standard analysis techniques (mesh, nodal, etc.), solve for the voltages v_{L1}, $v_{L2}, \ldots, v_{Lm}$ across the current sources $i_{L1}, i_{L2}, \ldots, i_{Lm}$, and solve for the currents $i_{c1}, i_{c2}, \ldots, i_{cp}$ through the voltage sources $v_{c1}, v_{c2}, \ldots, v_{cp}$.

4. Replace v_{L1} with $L_1[di_{L1}(t)/dt]$, v_{L2} with $i_2[di_{L2}/dt]$, and so on. Replace i_{c1} with $C_1[dv_{c1}(t)/dt]$, i_{c2} with $C_2[dv_{c2}(t)/dt]$, and so on.

Example 24. Derive a state-variable representation for the circuit of Fig. 19.38*a*.

Solution. Define the state variables $i_L(t)$ and $v_c(t)$. Replacing L and C with a current source $i_L(t)$ and voltage source $v_c(t)$ reduces the circuit of Fig. 19.38*a* to that of Fig. 19.38*b*. Solving for $i_C(t)$ and $v_L(t)$ gives

$$i_C(t) = -i_L(t) + \frac{v_s(t) - v_c(t)}{2} \tag{19.99}$$

$$v_L(t) = v_c(t) - 3i_L(t) \tag{19.100}$$

Substituting

$$i_C(t) = C\frac{dv_c(t)}{dt} = \tfrac{1}{2}\,\frac{dv_c(t)}{dt}$$

$$v_L(t) = L\frac{di_L}{dt} = \frac{di_L(t)}{dt}$$

Equations (19.99) and (19.100) become

$$\frac{dv_c(t)}{dt} = -2i_L(t) - v_c(t) + v_s(t) \tag{19.101}$$

$$\frac{di_L(t)}{dt} = v_c(t) - 3i_L(t) \tag{19.102}$$

Also

$$v_0(t) = v_c(t) \tag{19.103}$$

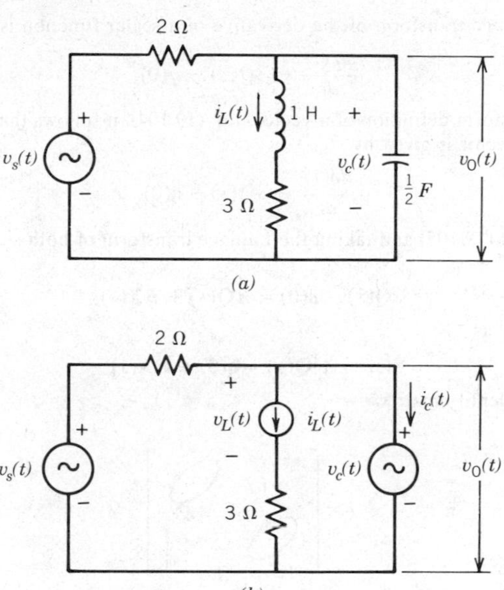

Fig. 19.38 Circuit for Example 23: (*a*) RLS circuit with one input and one output, (*b*) equivalent circuit with inductor and capacitor replaced by a current source and a voltage source, respectively.

Letting $v_c(t) = q_1(t)$ and $i_L(t) = q_2(t)$, the state vector equations become

$$\mathbf{q}(t) = A\mathbf{q}(t) + Bx(t)$$
$$y(t) = C\mathbf{q}(t) + Dx(t)$$

where

$$A = \begin{bmatrix} -1 & -2 \\ 1 & -3 \end{bmatrix}, \qquad B = \begin{bmatrix} 1 \\ 0 \end{bmatrix}, \qquad C = [1\ 0], \qquad D = [0]$$

$$\mathbf{q}(t) = \begin{bmatrix} q_1(t) \\ q_2(t) \end{bmatrix}, \qquad x(t) = v_s(t) \qquad \text{and} \qquad y(t) = v_O(t)$$

Solution of the State-Variable Equations

A linear time-invariant circuit may be represented by its state equations and output equations, Eqs. (19.94) and (19.95), respectively. Equation (19.94) represents a set of n first-order linear differential equations. If we solve Eq. (19.94) for $\mathbf{q}(t)$, then we can easily solve for the output $\mathbf{y}(t)$ by substituting $\mathbf{q}(t)$ into the algebraic equation (19.95).

The solution of the set of first-order differential equations, Eq. (19.94), requires that we are given the input functions $\mathbf{x}(t)$ and a set of initial conditions $\mathbf{q}(t_0)$. For notational simplicity we will assume the initial conditions are given at $t_0 = 0$, and the input $\mathbf{x}(t) = \mathbf{0}$ for $t < 0$. Consequently the solution of Eq. (19.94) for the state vector $\mathbf{q}(t)$ will be zero for $t < 0$.

There are several approaches for solving Eq. (19.94). A simple approach is to utilize the Laplace transform. The Laplace transform of a vector is determined by finding the Laplace transform of each component of the vector. For example,

$$\mathscr{L}[\mathbf{q}(t)] = \mathscr{L}\begin{bmatrix} q_1(t) \\ q_2(t) \\ \vdots \\ q_n(t) \end{bmatrix} = \begin{bmatrix} Q_1(s) \\ Q_2(s) \\ \vdots \\ Q_n(s) \end{bmatrix} = \mathbf{Q}(s) \tag{19.104}$$

We know that the Laplace transform of the derivative of a scalar function is

$$\frac{dq(t)}{dt} \leftrightarrow sQ(s) - q(0)$$

Using the Laplace transform definition of a vector, Eq. (19.104), it follows that the Laplace transform of the derivative of a vector is given by

$$\frac{d\mathbf{q}(t)}{dt} \leftrightarrow s\mathbf{Q}(s) - \mathbf{q}(0) \tag{19.105}$$

Using Eqs. (19.104) and (19.105) and taking the Laplace transform of both sides of Eq. (19.94) results in

$$s\mathbf{Q}(s) - \mathbf{q}(0) = A\mathbf{Q}(s) + B\mathbf{X}(s) \tag{19.106}$$

which may be rewritten as

$$[sI - A]\mathbf{Q}(s) = \mathbf{q}(0) + B\mathbf{X}(s) \tag{19.107}$$

where I is the $(n \times n)$ identity matrix

$$I = \begin{bmatrix} 1 & & & 0 \\ & 1 & & \\ & & \ddots & \\ 0 & & & 1 \end{bmatrix} \tag{19.108}$$

Postmultiplying Eq. (19.107) by $(sI - A)^{-1}$ yields

$$\mathbf{Q}(s) = (sI - A)^{-1}\mathbf{q}(0) + (sI - A)^{-1}B\mathbf{X}(s) \tag{19.109}$$

Define the fundamental matrix as

$$\phi(s) = (sI - A)^{-1} \tag{19.110}$$

Equation (19.109) then becomes

$$\mathbf{Q}(s) = \phi(s)\mathbf{q}(0) + \phi(s)B\mathbf{X}(s) \tag{19.111}$$

To find $\mathbf{q}(t)$ we must find the inverse Laplace transform of $\mathbf{Q}(s)$ as given in Eq. (19.111). Note that

$$\phi(t) = \mathcal{L}^{-1}[\Phi(s)] \triangleq \text{transition matrix} \tag{19.112}$$

Example 25. The state equations of a system are given by

$$\dot{\mathbf{q}}(t) = \begin{bmatrix} -1 & -2 \\ 1 & -4 \end{bmatrix}\mathbf{q}(t) + \begin{bmatrix} 0 \\ 0 \end{bmatrix}x(t)$$

1. Find $\phi(t)$.
2. Determine the state vector $\mathbf{q}(t)$ for $t \geqslant 0$, given that

$$\mathbf{q}(0) = \begin{bmatrix} 1 \\ 0 \end{bmatrix}.$$

Solution.

1. From Eq. (19.110) the fundamental matrix is given by

$$\Phi(s) = (sI - A)^{-1}$$

$$\Phi(s) = \left[s\begin{bmatrix} 1 & 0 \\ 0 & 1 \end{bmatrix} - \begin{bmatrix} -1 & -2 \\ 1 & -4 \end{bmatrix}\right]^{-1}$$

$$\Phi(s) = \begin{bmatrix} s+1 & 2 \\ -1 & s+4 \end{bmatrix}^{-1}$$

$$\Phi(s) = \frac{\begin{bmatrix} s+4 & -2 \\ 1 & s+1 \end{bmatrix}}{(s+1)(s+4)+2} = \frac{\begin{bmatrix} s+4 & -2 \\ 1 & s+1 \end{bmatrix}}{s^2+5s+6}$$

recognizing that $s^2 + 5s + 6 = (s+2)(s+3)$

$$\Phi(s) = \begin{bmatrix} \dfrac{(s+4)}{(s+2)(s+3)} & \dfrac{-2}{(s+2)(s+3)} \\ \dfrac{1}{(s+2)(s+3)} & \dfrac{(s+1)}{(s+2)(s+3)} \end{bmatrix}$$

Expanding each element of the matrix into a partial fraction expansion we get

$$\Phi(s) = \begin{bmatrix} \dfrac{2}{s+2} + \dfrac{-1}{s+3} & \dfrac{-2}{s+2} + \dfrac{2}{s+3} \\ \dfrac{1}{s+2} - \dfrac{1}{s+3} & \dfrac{-1}{s+2} + \dfrac{2}{s+3} \end{bmatrix}$$

Taking the inverse Laplace transform of each element of the matrix we obtain

$$\Phi(t) = \begin{bmatrix} (2e^{-2t} - e^{-3t})u(t) & (-2e^{-2t} + 2e^{-3t})u(t) \\ (e^{-2t} - e^{-3t})u(t) & (-e^{-2t} + 2e^{-3t})u(t) \end{bmatrix}$$

and check

$$\Phi(0) = \begin{bmatrix} 1 & 0 \\ 0 & 1 \end{bmatrix}$$

2. To solve for $\mathbf{q}(t)$ we make use of Eq. (19.111), where in this example $B = [0]$. It follows that

$$Q(s) = \Phi(s)\mathbf{q}(0)$$

Hence taking inverse Laplace transforms

$$\mathbf{q}(t) = \Phi(t)\mathbf{q}(0) = \begin{bmatrix} (2e^{-2t} - e^{-3t})u(t) & (-2e^{-2t} + 2e^{-3t})u(t) \\ (e^{-2t} - e^{-3t})u(t) & (-e^{-2t} + 2e^{-3t})u(t) \end{bmatrix}\begin{bmatrix} 1 \\ 0 \end{bmatrix}$$

$$= \begin{bmatrix} (2e^{-2t} - e^{-3t})u(t) \\ (e^{-2t} - e^{-3t})u(t) \end{bmatrix}$$

Example 26. For the system described by

$$\dot{\mathbf{q}}(t) = \begin{bmatrix} -1 & -2 \\ 0 & -3 \end{bmatrix}\mathbf{q}(t) + \begin{bmatrix} 0 \\ 1 \end{bmatrix}x(t), \qquad \mathbf{q}(0) = \begin{bmatrix} 2 \\ 0 \end{bmatrix}$$

1. Determine the fundamental matrix $\Phi(s)$.
2. Using Eq. (19.111) obtain the state vector $\mathbf{Q}(s)$, given $x(t) = u(t)$.
3. Find $\mathbf{q}(t)$ by taking the inverse Laplace transform of $\mathbf{Q}(s)$.

Solution.

1. From Eq. (19.110)

$$\Phi(s) = (sI - A)^{-1}$$

$$\Phi(s) = \left[s\begin{bmatrix} 1 & 0 \\ 0 & 1 \end{bmatrix} - \begin{bmatrix} -1 & -2 \\ 0 & -3 \end{bmatrix}\right]^{-1}$$

$$\Phi(s) = \begin{bmatrix} (s+1) & 2 \\ 0 & s+3 \end{bmatrix}$$

$$\Phi(s) = \frac{\begin{bmatrix} (s+3) & -2 \\ 0 & (s+1) \end{bmatrix}}{(s+1)(s+3)}$$

$$\Phi(s) = \begin{bmatrix} \dfrac{1}{(s+1)} & \dfrac{-2}{(s+1)(s+3)} \\ 0 & \dfrac{1}{(s+3)} \end{bmatrix}$$

2. From Eq. (19.111)

$$\mathbf{Q}(s) = \Phi(s)\mathbf{q}(0) + \Phi(s)BX(s)$$

In this example $X(s) = 1/s$, hence

$$\mathbf{Q}(s) = \Phi(s)\mathbf{q}(0) + \Phi(s)B \cdot 1/s$$

$$\mathbf{Q}(s) = \begin{bmatrix} \dfrac{1}{(s+1)} & \dfrac{-2}{(s+1)(s+3)} \\ 0 & \dfrac{1}{(s+3)} \end{bmatrix} \begin{bmatrix} 2 \\ 0 \end{bmatrix} + \begin{bmatrix} \dfrac{1}{(s+1)} & \dfrac{-2}{(s+1)(s+3)} \\ 0 & \dfrac{1}{(s+3)} \end{bmatrix} \begin{bmatrix} 0 \\ \dfrac{1}{s} \end{bmatrix}$$

$$\mathbf{Q}(s) = \begin{bmatrix} \dfrac{2}{s+1} + \dfrac{-2}{s(s+1)(s+3)} \\ 0 \qquad \dfrac{1}{s(s+3)} \end{bmatrix}$$

$$\mathbf{Q}(s) = \begin{bmatrix} \dfrac{2s^2+6s-2}{s(s+1)(s+3)} \\ \dfrac{1}{s(s+3)} \end{bmatrix}$$

3.

$$\mathbf{q}(t) = \begin{bmatrix} \mathscr{L}^{-1}\left[\dfrac{2s^2+6s-2}{s(s+1)(s+3)}\right] \\ L^{-1}\left[\dfrac{1}{s(s+3)}\right] \end{bmatrix}$$

$$\mathbf{q}(t) = \begin{bmatrix} \mathscr{L}^{-1}\left[\dfrac{-\frac{2}{3}}{s} + \dfrac{3}{s+1} + \dfrac{-\frac{1}{3}}{s+3}\right] \\ \mathscr{L}^{-1}\left[\dfrac{\frac{1}{3}}{s} + \dfrac{-\frac{1}{3}}{s+3}\right] \end{bmatrix}$$

$$\mathbf{q}(t) = \begin{bmatrix} (-\frac{2}{3} + 3e^{-t} - \frac{1}{3}e^{-3t})u(t) \\ (\frac{1}{3} - \frac{1}{3}e^{-3t})u(t) \end{bmatrix}$$

Transfer Functions of Systems Described in State-Variable Form

The transfer function $H(s)$ of a single-input/single-output system with input $X(s)$ and output $Y(s)$ was defined in Example 21 by

$$H(s) = \frac{Y(s)}{X(s)} \tag{19.113}$$

The transfer function of a system is defined with the system at rest (i.e., zero initial conditions). Consequently we set $\mathbf{q}(0) = \mathbf{0}$. Taking the Laplace transform of the state equations (19.94) and (19.95) with $\mathbf{q}(0) = \mathbf{0}$ we obtain

$$s\mathbf{Q}(s) = A\mathbf{Q}(s) + B\mathbf{X}(s) \tag{19.114}$$

$$\mathbf{Y}(s) = C\mathbf{Q}(s) + D\mathbf{X}(s) \tag{19.115}$$

Solving for $\mathbf{Q}(s)$ in Eq. (19.114) we obtain

$$\mathbf{Q}(s) = (sI - A)^{-1}B\mathbf{X}(s)$$

and substituting into Eq. (19.115) leads to

$$\mathbf{Y}(s) = [C(sI - A)^{-1}B + D]\mathbf{X}(s) \tag{19.116}$$

Comparing Eqs. (19.113) and (19.116) it follows that the matrix transfer function of the system is

$$[H(s)] = C(sI - A)^{-1}B + D \tag{19.117}$$

Equation (19.117) may be written as

$$[H(s)] = C\frac{\text{adj}[sI - A]B}{\det[sI - A]} + D \tag{19.118}$$

where

$$(sI - A)^{-1} = \frac{\text{adjoint of } (sI - A)}{\text{determinant of } (sI - A)}$$

It is clear from Eq. (19.118) that the poles of $H(s)$ correspond to those values of s for which

$$\det[sI - A] = 0 \tag{19.119}$$

If A is an $(n \times n)$ matrix, then Eq. (19.119) will be an nth-order polynomial equation commonly called the characteristic equation of the system. The roots of the characteristic equation are referred to as the eigenvalues or characteristic roots of the system.

Assuming that the numerator and denominator polynomials of Eq. (19.118) do not have any common factors, then the poles of the transfer function $H(s)$ are the roots of Eq. (19.119), that is, the eigenvalues of the system. If there are cancellations between the numerator and denominator polynomials of Eq. (19.118), then the poles of $H(s)$ will consist of those eigenvalues that remain after the cancellation.

Since a system is defined as being stable if all the poles of $H(s)$ are in the right half s-plane,[2] it follows that the system is stable if the eigenvalues of the system have negative real parts.

Example 27. A system is described by the following state matrices:

$$A = \begin{bmatrix} 3 & 2 \\ -3 & -4 \end{bmatrix}, \quad B = \begin{bmatrix} 1 \\ 0 \end{bmatrix}, \quad C = [0 \quad 1], \quad D = [0]$$

1. Find the characteristic equation of the system.
2. Find the eigenvalues of the system.
3. Find $H(s)$.
4. Identify the poles of $H(s)$.
5. Determine if the system is stable.

Solution.

1. The characteristic equation is given by Eq. (19.119). Thus

$$\det[sI - A] = \left| \begin{bmatrix} s & 0 \\ 0 & s \end{bmatrix} - \begin{bmatrix} 3 & 2 \\ -3 & -4 \end{bmatrix} \right|$$

$$\det[sI - A] = \begin{vmatrix} s-3 & -2 \\ 3 & s+4 \end{vmatrix} = (s-3)(s+4) + 6$$

$$\det[sI - A] = s^2 + s - 6$$

2. The eigenvalues of the system are the roots of the characteristic equation, that is, the roots of

$$s^2 + s - 6 = 0$$
$$(s-2)(s+3) = 0$$

Eigenvalues are $s_1 = 2$ and $s_2 = -3$.

3. The transfer function is given by Eq. (19.118)

$$H(s) = \frac{[0 \quad 1]\text{adj}[sI - A]\begin{bmatrix} 1 \\ 0 \end{bmatrix}}{(s^2 + s - 6)} + [0]$$

$$H(s) = \frac{[0 \quad 1]\begin{bmatrix} s+4 & 2 \\ -3 & s-3 \end{bmatrix}\begin{bmatrix} 1 \\ 0 \end{bmatrix}}{(s^2 + s - 6)}$$

$$H(s) = \frac{[0 \quad 1]\begin{bmatrix} s+4 \\ -3 \end{bmatrix}}{(s^2 + s - 6)}$$

$$H(s) = \frac{-3}{(s^2 + s - 6)}$$

4. Clearly the poles of the system are the root of the denominator of $H(s)$, that is, $s_1 = 2$ and $s_2 = -3$. Hence it is seen that the poles of the system correspond to its eigenvalues.
5. The system is unstable since one of its poles has a negative real part.

References

1 R. E. Ziemer, W. H. Tranter, and D. F. Fannin, *Signals and Systems: Continuous and Discrete*, Macmillan, New York, 1983.

2 J. C. Reid, *Linear System Fundamental: Continuous and Discrete, Classic and Modern*, McGraw-Hill, New York, 1983.

CHAPTER 20
PASSIVE FILTERS

ICARIUS E. PYROS

New York Institute of Technology, Old Westbury, New York

20.1 INTRODUCTION

A network that is inserted in a circuit for the purpose of frequency discrimination is called a filter. Filters select a narrow band of frequencies without reduction in magnitude and totally suppress all other frequencies.

Figure 20.1 and the list that follows define the terms often used in the development of filter circuits.

Driving point or input impedance Z_i. The impedance seen by the generator across terminals a, b. This impedance will be equal to the ratio of the voltage across the terminals E_{ab} to the current supplied to the terminals I_{in} with all other voltage sources removed.

$$Z_i = \frac{E_{ab}}{I_{\text{in}}} \tag{20.1}$$

The driving point or input admittance is the reciprocal of the driving point impedance.

$$Y_i = \frac{1}{Z_i} = \frac{I_{\text{in}}}{E_{ab}} \tag{20.2}$$

Transfer impedance Z_t. The ratio of the voltage applied to one loop of a circuit E_{cd} to the current flowing in a second loop I_{in}.

$$Z_t = \frac{E_{cd}}{I_{\text{in}}} \tag{20.3}$$

The transfer admittance is the reciprocal of the transfer impedance.

$$Y_t = \frac{1}{Z_t} = \frac{I_{\text{in}}}{E_{cd}} \tag{20.4}$$

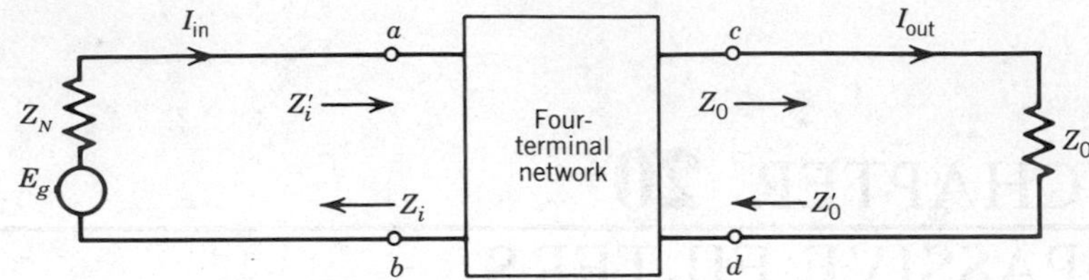

Fig. 20.1 Filter terms.

Short-circuit impedance. The impedance looking into one end of a circuit with the other end short circuited.

Open-circuit impedance. The impedance looking into one end of a circuit with the other end open circuited.

Image impedances. Z_i and Z_0 are said to be image impedances when the two following conditions are fulfilled:

1. $Z_i = Z_i'$ when the output terminals are closed through Z_0.
2. $Z_0 = Z_0'$ when the input terminals are closed through Z_i.

Characteristic impedance. The one particular case of image impedance where the input and output impedances are the same. In addition to the conditions described for image impedance:

$$Z_0 = Z_i \tag{20.5}$$

20.2 ATTENUATION AND PHASE FACTORS

In the circuit of Fig. 20.2, the network is adjusted so that the impedance looking into the network is equal to the impedance of the load:

$$Z_{01} = \frac{V_2}{I_2}$$

$$Z_{01} = \frac{V_1}{I_1}$$

Therefore

$$\frac{V_1}{I_1} = \frac{V_2}{I_2}$$

and

$$\frac{V_1}{V_2} = \frac{I_1}{I_2} \tag{20.6}$$

Since current and voltage are phasor quantities

$$I_1 = |I_1|e^{j\theta_{I_1}}$$

$$I_2 = |I_2|e^{j\theta_{I_2}}$$

$$V_1 = |V_1|e^{j\theta_{V_1}}$$

$$V_2 = |V_2|e^{j\theta_{V_1}}$$

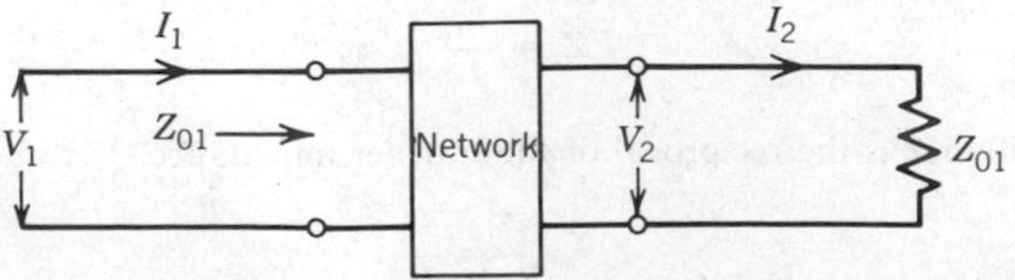

Fig. 20.2 Matched filter network.

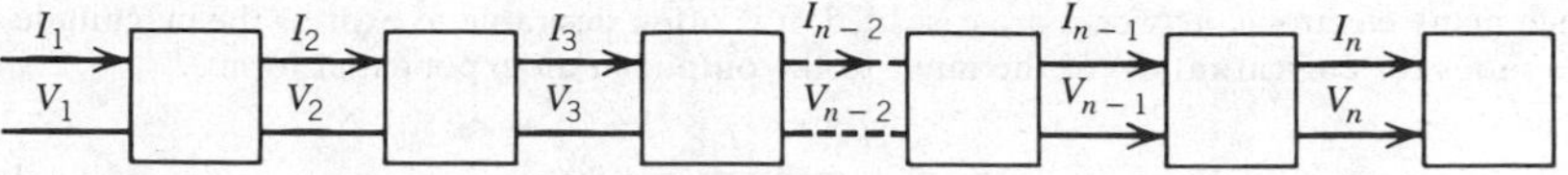

Fig. 20.3 Cascaded networks.

Substituting:

$$\frac{I_1}{I_2} = \frac{|I_1|e^{j\theta_{I_1}}}{|I_2|e^{j\theta_{I_2}}} = \frac{|I_1|}{|I_2|}e^{j(\theta_{I_1}-\theta_{I_2})} = \frac{|I_1|}{|I_2|}e^{j\beta} \tag{20.7}$$

where $\beta = (\theta_{I_1} - \theta_{I_2})$ (phase shift).

Similarly:

$$\frac{V_1}{V_2} = \frac{|V_1|e^{j\theta_{V_1}}}{|V_2|e^{j\theta_{V_2}}} = \frac{|V_1|}{|V_2|}e^{j(\theta_{V_1}-\theta_{V_2})}$$

For the same Z the angle θ_{I_1} and θ_{I_2} associated with the current I_1 and I_2 and the θ_{V_1} and θ_{V_2} associated with V_1 and V_2 will be related as follows:

$$\theta_{I_1} - \theta_{I_2} = \theta_{V_1} - \theta_{V_2} = \beta$$

and

$$\frac{V_1}{V_2} = \frac{|V_1|}{|V_2|}e^{j\beta} \tag{20.8}$$

In a circuit in which networks are cascaded (Fig. 20.3), if

$$Z_{01} = Z_{02} = Z_{03} = \cdots = Z_{0(n-1)} = Z_{0n}$$

then

$$\frac{V_1}{I_1} = \frac{V_2}{I_2} = \frac{V_3}{I_3} = \cdots = \frac{V_{n-1}}{I_{n-1}} = \frac{V_n}{I_n} \qquad \frac{V_1}{V_2} = \frac{I_1}{I_2} \qquad \frac{V_2}{V_3} = \frac{I_2}{I_3} \qquad \frac{V_{n-1}}{V_n} = \frac{I_{n-1}}{I_n}$$

$$\frac{|V_1|}{|V_n|} = \left(\frac{|V_1|}{|V_2|}\right)\left(\frac{|V_2|}{|V_3|}\right)\left(\frac{|V_3|}{|V_4|}\right)\cdots\left(\frac{|V_{n-1}|}{|V_n|}\right) \tag{20.9}$$

If

$$\frac{|V_1|}{|V_2|} = e^{\alpha_{12}} \qquad \text{and} \qquad \frac{|V_2|}{|V_3|} = e^{\alpha_{23}}$$

then

$$\frac{|V_{n-1}|}{|V_n|} = e^{\alpha_{(n-1)n}}$$

and

$$\frac{|V_1|}{|V_n|} = (e^{\alpha_{12}})(e^{\alpha_{23}})(e^{\alpha_{34}})\cdots(e^{\alpha_{(n-1)n}}) = e^{\alpha_{12}+\alpha_{23}+\cdots+\alpha_{(n-1)n}}$$

Taking the natural log of both sides:

$$\ln\frac{|V_1|}{|V_n|} = \alpha_{12} + \alpha_{23} + \cdots + \alpha_{(n-1)n}$$

If we substitute I_1/I_2 for V_1/V_2, and so on, in Eq. 20.9 we have

$$\frac{|I_1|}{|I_n|} = \left(\frac{|I_1|}{|I_2|}\right)\left(\frac{|I_2|}{|I_3|}\right)\cdots\left(\frac{|I_{n-1}|}{|I_n|}\right)$$

and

$$\ln\frac{|I_1|}{|I_2|} = \alpha_{12} + \alpha_{23} + \alpha_{34} + \cdots + \alpha_{(n-1)n}$$

$$\ln\frac{|V_1|}{|V_2|} = \ln\frac{|I_1|}{|I_2|} = \alpha_{12}$$

Since many electronic networks are cascaded, it is often desirable to express the magnitude of the voltage ratios (or current ratios) of the input to the output in an exponential form:

$$e^{\alpha} = \frac{|V_1|}{|V_2|} = \frac{|I_1|}{|I_2|} = \alpha_{12} \tag{20.10}$$

It has been shown that

$$\frac{V_1}{V_2} = \frac{|V_1|}{|V_2|} e^{j\beta}$$

Substituting $|V_1|/|V_2|$ for e^{α}:

$$\frac{V_1}{V_2} = e^{\alpha} e^{j\beta}$$

Similarly

$$\frac{I_1}{I_2} = \frac{|I_1|}{|I_2|} e^{j\beta} = e^{\alpha} e^{j\beta}$$

If

$$\frac{V_1}{V_2} = \frac{I_1}{I_2} = e^{\gamma}$$

then

$$e^{\gamma} = e^{\alpha} e^{j\beta} = e^{(\alpha + j\beta)}$$

$$\gamma = \alpha + j\beta \tag{20.11}$$

α is referred to as the attenuation factor and expresses the reduction in magnitude of the voltages, currents, and power in the circuit. β is referred to as the phase factor and expresses the change in phase of the output to input (wavelength factor). γ is referred to as the propagation factor and represents the change in magnitude and phase.

Units of attenuation

$$\frac{|V_1|}{|V_2|} = \frac{|I_1|}{|I_2|} = e^{\alpha}$$

$$\ln \frac{|V_1|}{|V_2|} = \ln \frac{|I_1|}{|I_2|} = \alpha$$

The ratio of the natural logarithm of the voltage or current is measured in nepers. Let N = number of nepers:

$$\ln \frac{|V_1|}{|V_2|} = \ln \frac{|I_1|}{|I_2|} = N$$

It is also possible to express the power relationship for the cascade circuit.

$$\frac{P_1}{P_n} = \frac{I_1^2 R_{01}}{I_n^2 R_{0n}}$$

For characteristic impedance

$$R_{01} = R_{0n} \tag{20.12}$$

$$\frac{P_1}{P_n} = \frac{I_1^2}{I_n^2}$$

$$\ln\left[\frac{P_1}{P_n}\right] = 2 \ln\left[\frac{I_1}{I_n}\right] = 2N \quad \text{and} \quad \frac{P_1}{P_n} = e^{2N}$$

The ratio of the logarithm to the base 10 of the power ratio is known as the "bel."

$$\log\left[\frac{P_1}{P_n}\right] = B$$
$$\frac{P_1}{P_n} = 10^B \tag{20.13}$$

where B = number of bels. Let the number of decibels $D = 10B$:

$$\frac{P_1}{P_n} = 10^{D/10} \tag{20.14}$$

Equating Eqs. (20.13) and (20.14) and taking the log we have

$$e^{2N} = 10^{D/10} \qquad \log e^{2N} = \log 10^{D/10}$$
$$2N \log e = \frac{D}{10} \log 10 \qquad \therefore \qquad D = 20N \log e = 8.686N$$

Since the number of decibels equals 8.686 times the number of nepers, each decibel must be 1/8.686 of one neper, or

$$\text{dB} = \frac{\text{nepers}}{8.686}$$

The voltage and current in terms of decibels are:

$$D = 20 \log\left[\frac{E_1}{E_n}\right] \text{dB}$$

It should be noted that the half-power point is often called the 3 dB point. Since $P_0 = 2P_1$,

$$D = 10 \log 2$$
$$D = 10(0.3010) = 3.01 \text{ dB}$$

20.3 *T* AND π SECTIONS

It has been stated that the purpose of the filter is to discriminate frequencies. The filter is not intended to change the overall impedance of the circuit. The basic type filters are the T and the π filter (Fig. 20.4).

$$Z_{\text{in "}T\text{"}} = \frac{Z_1}{2} + \frac{[Z_1/2 + Z_{0T}][Z_2]}{Z_1/2 + Z_{0T} + Z_2} \qquad Z_{\text{in "}\pi\text{"}} = \frac{(2Z_2)[Z_1 + (2Z_2)(Z_{0\pi})/2Z_2 + Z_{0\pi}]}{2Z_2 + Z_1 + [(2Z_2)(Z_{0\pi})/2Z_2 + Z_{0\pi}]}$$

Since it is not intended for the filter to change the impedance across a, b,

$$Z_{\text{in}} = Z_{0T} \qquad Z_{\text{in}} = Z_{0\pi}$$

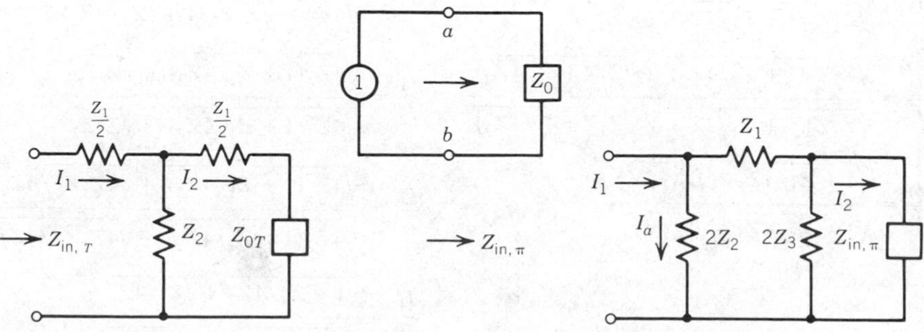

Fig. 20.4 *T* and π filters.

Substituting and solving for Z_0:

$$Z_{0T} = \sqrt{Z_1 Z_2\left(1 + \frac{Z_1}{4Z_2}\right)} \tag{20.15}$$

$$Z_{0\pi} = \sqrt{\frac{Z_1 Z_2}{(1 + Z_1/4Z_2)}} \tag{20.16}$$

Short-circuit impedance:

$$Z_{\text{sc},T} = \frac{Z_1}{2} + \frac{Z_1/2(Z_2)}{Z_1/2 + Z_2} \qquad Z_{\text{sc},\pi} = \frac{(2Z_2)(Z_1)}{2Z_2 + Z_1}$$

Open-circuit impedance:

$$Z_{0c,T} = \frac{Z_1}{2} + Z_2$$

$$Z_{0T} = \sqrt{Z_{0c,T} Z_{\text{sc},T}}$$

$$E_1 = I_1 Z_{0T}$$

$$E_1 = \frac{I_1 Z_1}{2} + \frac{I_2 Z_1}{2} + I_2 Z_{0T}$$

$$E_1 = \frac{I_1 Z_1}{2} + I_2\left[\frac{Z_1}{2} + Z_{0T}\right]$$

$$I_1 Z_{0T} = \frac{I_1 Z_1}{2} + I_2\left[\frac{Z_1}{2} + Z_{0T}\right]$$

$$I_1\left[Z_{0T} - \frac{Z_1}{2}\right] = I_2\left[\frac{Z_1}{2} + Z_{0T}\right]$$

$$\frac{I_1}{I_2} = \frac{[Z_1/2 + Z_{0T}]}{[Z_{0T} - Z_1/2]}$$

$$Z_{0c,\pi} = \frac{(2Z_2)(Z_1 + 2Z_2)}{Z_1 + 4Z_2}$$

$$Z_{0\pi} = \sqrt{Z_{0c,\pi} Z_{\text{sc},\pi}}$$

$$E_1 = I_1 Z_{0\pi}$$

$$E_1 = (I_1 - I_a)Z_1 + I_2 Z_{0\pi}$$

$$(I_1 - I_a)Z_1 = E_1 - I_z Z_{0\pi}$$

$$E_1 - (I_1 - I_a - I_2)(2Z_2) = E_1 - I_2 Z_{0\pi}$$

$$I_2(Z_{0\pi} + 2Z_2) = (I_1 - I_a)2Z_2$$

But $I_a = \dfrac{E_1}{2Z_2} = \dfrac{I_1 Z_{0\pi}}{2Z_2}$

Substituting this value, we have

$$I_2(Z_{0\pi} + 2Z_2) = I_1 2Z_2 - I_1 \frac{Z_{0\pi} 2Z_2}{2Z_2}$$

$$I_2(2Z_2 + Z_{0\pi}) = I_1(2Z_2 - Z_{0\pi})$$

$$\frac{I_1}{I_2} = \frac{2Z_2 + Z_{0\pi}}{2Z_2 - Z_{0\pi}}$$

Now substituting the value of Z_{0T} and $Z_{0\pi}$ of Eqs. (20.15) and (20.16) we have

$$\frac{I_1}{I_2} = \frac{\sqrt{Z_1 Z_2(1 + Z_1/4Z_2)} + Z_1/2}{\sqrt{Z_1 Z_2(1 + Z_1/4Z_2)} - Z_1/2}$$

$$\frac{I_1}{I_2} = \frac{\sqrt{Z_1 Z_2(1 + Z_1/4Z_2)} + \sqrt{Z_1^2/4}}{\sqrt{Z_1 Z_2(1 + Z_1/4Z_2)} - \sqrt{Z_1^2/4}}$$

$$\frac{I_1}{I_2} = \frac{\sqrt{Z_1 Z_2(1 + Z_1/4Z_2)} + \sqrt{Z_1 Z_2(Z_1/4Z_2)}}{\sqrt{Z_1 Z_2(1 + Z_1/4Z_2)} - \sqrt{Z_1 Z_2(Z_1/4Z_2)}}$$

$$\frac{I_1}{I_2} = \frac{\sqrt{1 + Z_1/4Z_2} + \sqrt{Z_1/4Z_2}}{\sqrt{1 + Z_1/4Z_2} - \sqrt{Z_1/4Z_2}}$$

$$\frac{I_1}{I_2} = \frac{2Z_2 + \sqrt{Z_1 Z_2/1 + (Z_1/4Z_2)}}{2Z_2 - \sqrt{Z_1 Z_2/1 + (Z_1/4Z_2)}}$$

$$\frac{I_1}{I_2} = \frac{\sqrt{4Z_2^2} + \sqrt{Z_1 Z_2/1 + (Z_1/4Z_2)}}{\sqrt{4Z_2^2} - \sqrt{Z_1 Z_2/1 + (Z_1/4Z_2)}}$$

$$\frac{I_1}{I_2} = \frac{\sqrt{4Z_2^2(1 + Z_1/4Z_2)} + \sqrt{Z_1 Z_2}}{\sqrt{4Z_2^2(1 + Z_1/4Z_2)} - \sqrt{Z_1 Z_2}}$$

$$\frac{I_1}{I_2} = \frac{\sqrt{4Z_2^2(1 + Z_1/4Z_2)} + \sqrt{4Z_1 Z_2^2/4Z_2}}{\sqrt{4Z_2^2(1 + Z_1/4Z_2)} - \sqrt{4Z_1 Z_2^2/4Z_2}}$$

$$\frac{I_1}{I_2} = \frac{\sqrt{1 + Z_1/4Z_2} + \sqrt{Z_1/4Z_2}}{\sqrt{1 + Z_1/4Z_2} - \sqrt{Z_1/4Z_2}}$$

Therefore I_1/I_2 for "T" is equal to I_1/I_2 for "π":

$$\frac{I_1}{I_2} = \left[\frac{\sqrt{1 + Z_1/4Z_2} + \sqrt{Z_1/4Z_2}}{\sqrt{1 + Z_1/4Z_2} - \sqrt{Z_1/4Z_2}}\right]\left[\frac{\sqrt{1 + Z_1/4Z_2} + \sqrt{Z_1/4Z_2}}{\sqrt{1 + Z_1/4Z_2} + \sqrt{Z_1/4Z_2}}\right]$$

$$\frac{I_1}{I_2} = \left[\frac{\left(\sqrt{1 + Z_1/4Z_2} + \sqrt{Z_1/4Z_2}\right)^2}{\left(\sqrt{1 + Z_1/4Z_2}\right)^2 - \left(\sqrt{Z_1/4Z_2}\right)^2}\right] = \left[\frac{\left(\sqrt{1 + Z_1/4Z_2} + \sqrt{Z_1/4Z_2}\right)^2}{(1 + Z_1/4Z_2 - Z_1/4Z_2)}\right]$$

$$\frac{I_1}{I_2} = \left[\sqrt{1 + \frac{Z_1}{4Z_2} + \frac{Z_1}{4Z_2}}\right]^2$$

$$\ln\frac{I_1}{I_2} = 2\ln\left[\sqrt{1 + \frac{Z_1}{4Z_2}} + \sqrt{\frac{Z_1}{4Z_2}}\right]$$

$$10\log\frac{I_1}{I_2} = 20\log\left[\sqrt{1 + \frac{Z_1}{4Z_2}} + \sqrt{\frac{Z_1}{4Z_2}}\right]$$

It has been shown that $I_1/I_2 = e^{\alpha + j\beta}$ and $\alpha + j\beta = \ln I_1/I_2$. Substituting for $\ln I_1/I_2$:

$$\alpha + j\beta = 2\ln\left(\sqrt{1 + \frac{Z_1}{4Z_2}} + \sqrt{\frac{Z_1}{4Z_2}}\right) \tag{20.17}$$

The frequencies that are allowed to reach the load (Z_0) with no attenuation are located in the passband. Before discussing the limits of $Z_1/4Z_2$ that would result in no attenuation, it is important to review some of the basic concepts associated with logarithms. Assume a phasor quantity of $Ae^{j\theta}$ having a magnitude A and an angle θ. Let

$$\alpha + j\beta = \ln Ae^{j\theta}$$

$$\alpha + j\beta = \ln A + \ln e^{j\theta}$$

$$\alpha + j\beta = \ln A + j\theta \ln e$$

$$\alpha + j\beta = \ln A + j\theta$$

$$\alpha = \ln A \quad \text{and} \quad \beta = \theta$$

In these equations, A represents a positive number and is measured in radians. An analysis of this equation will lead to the following conclusions:

1. The natural logarithm of a phasor quantity is a complex quantity.
2. When $A = 1$, $\alpha = \ln 1$ and thus $\alpha = 0$.
3. When $A < 1$, $\alpha =$ negative.
4. When $A > 1$, $\alpha =$ positive.
5. When the phasor quantity is a positive quantity ($\theta = n\pi$ where n is an even integer), the imaginary component is equal to zero.
6. Even when the phasor quantity is an imaginary number [$\theta = n(\pi/2)$ where n is an integer], the natural logarithm of the phasor quantity contains a real component.

For attenuation to occur, α must be a real positive number; for no attenuation to occur, either α must be equal to zero or must be a negative number (physically impossible). It may be seen from the previous analysis that no attenuation will occur when $A \leqq 1$.

Since A represents a positive quantity, the given limits may be expressed as follows:

$$0 \leqq A \leqq 1$$

Let

$$\left[\sqrt{1+\frac{Z_1}{4Z_2}}+\sqrt{\frac{Z_1}{4Z_2}}\right]=Ae^{j\theta} \qquad \left|\sqrt{1+\frac{Z_1}{4Z_2}}+\sqrt{\frac{Z_1}{4Z_2}}\right|=A$$

For $A \leqq 1$:

$$0 \leqq \left|1+\frac{Z_1}{4Z_2}\right| \leqq 1$$

Subtracting 1 from each side of the inequality:

$$-1 \leqq \frac{Z_1}{4Z_2} \leqq 0$$

The boundary frequency at which the attenuation becomes real is called the cutoff frequency. At this point

$$\frac{Z_1}{4Z_2}=0 \qquad \frac{Z_1}{4Z_2}=-1$$

All frequencies that are attenuated are located in the stopband:

$$a \neq 0$$

Since no attenuation is desired in the passband region, filters are designed with a minimum of resistance and

$$Z_1 \approx \alpha_1 \qquad \text{and} \qquad Z_2 \approx Z_2$$

For low-pass filters, the horizontal components are inductors, since ωL is small at low frequencies. The vertical component is a capacitor, since $1/\omega C$ is large at low frequencies (Fig. 20.5*a*).

For high-pass filters, the horizontal components are capacitive, since $1/\omega C$ is small at higher frequencies. The vertical component is an inductor, since ωL is large at high frequencies (Fig. 20.5*b*).

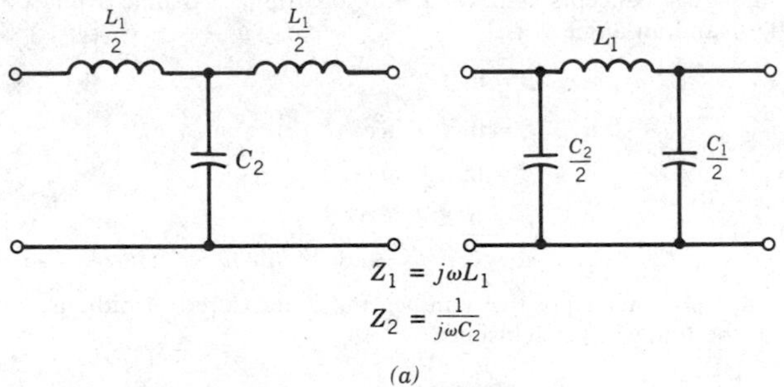

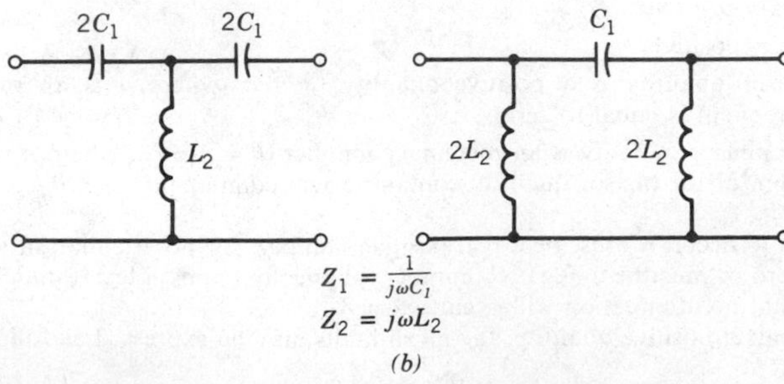

Fig. 20.5 Low-pass (*a*) and high-pass (*b*) filters.

Substituting in equations for cutoff $\omega = \omega_0$ and $f = f_0$, for low-pass filters:

$$\frac{Z_1}{4Z_2} = -1 \qquad \frac{Z_1}{4Z_2} = 0$$

$$\frac{j\omega_c L_1}{4/j\omega_c C_2} = -1 \qquad \frac{j\omega_c L_1}{4/j\omega_c C_2} = 0$$

$$\frac{-\omega_c^2 L_1 C_2}{4} = -1 \qquad \frac{\omega_c^2 L_1 C_2}{4} = 0$$

$$\omega_c = \frac{2}{\sqrt{L_1 C_2}} \qquad \omega_c = 0$$

$$f_c = \frac{1}{\pi\sqrt{L_1 C_2}} \qquad f_c = 0$$

$$\text{Lower limit} \quad f_c = 0 \tag{20.18}$$

$$\text{Upper limit} \quad f_c = \frac{1}{\pi\sqrt{L_1 C_2}} \tag{20.19}$$

and for high-pass filters:

$$\frac{Z_1}{4Z_2} = -1 \qquad \frac{Z_1}{4Z_2} = 0$$

$$\frac{1/j\omega_c C_1}{4j\omega_c L_2} = -1 \qquad \frac{1/j\omega_c C_1}{4j\omega_c L_2} = 0$$

$$-\frac{1}{4\omega_c^2 C_1 L_2} = -1 \qquad \frac{1}{4\omega_c^2 C_1 L_2} = 0$$

$$\omega_c = \frac{1}{2\sqrt{C_1 L_2}} \qquad \omega_c = \infty$$

$$f_c = \frac{1}{4\pi\sqrt{C_1 L_2}} \qquad f_c = \infty$$

$$\text{Lower limit} \quad f_c = \frac{1}{4\pi\sqrt{C_1 L_2}} \tag{20.20}$$

$$\text{Upper limit} \quad f_c = \infty \tag{20.21}$$

20.4 CONSTANT *K* FILTERS

Filters that work into fixed resistive loads are called constant K filters (Fig. 20.6). Let $Z_{1K} = Z_1$ and $Z_{2K} = Z_2$, $(Z_{1K})(Z_{2K}) = R_K^2$, where R_K is a constant independent of frequency.

For low-pass K filters:

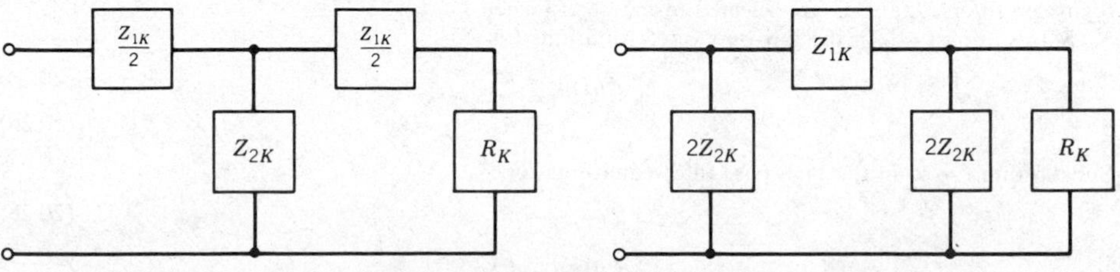

Fig. 20.6 Constant *K* filters.

For low-pass K filters:

$$Z_{1K} = j\omega L_{1K}$$

$$Z_{2K} = \frac{1}{j\omega C_{2K}}$$

$$[Z_{1K}][Z_{2K}] = [j\omega L_{1K}]\left[\frac{1}{j\omega C_{2K}}\right]$$

$$R_K^2 = \frac{L_{1K}}{C_{2K}}$$

For "T" / For "π"

$$f_c = \frac{1}{\pi\sqrt{L_{1K}C_{2K}}} \qquad f_c = \frac{1}{\pi\sqrt{L_{1K}C_{2K}}}$$

$$\omega_c = 2\pi f_c \qquad \omega_c = 2\pi f_c$$

$$\omega_c = \frac{2}{\sqrt{L_{1K}C_{2K}}} \qquad \omega_c = \frac{2}{\sqrt{L_{1K}C_{2K}}}$$

$$Z_{0TKL} = R_K\sqrt{1 - \frac{f^2}{f_0^2}} \qquad Z_{0\pi KL} = \frac{R_K}{\sqrt{1 - f^2/f_c^2}} \tag{20.22}$$

For high-pass K filters:

$$Z_{1K} = \frac{1}{j\omega C_{1K}}$$

$$Z_{2K} = j\omega L_{2K}$$

$$[Z_{1K}][Z_{2K}] = \left[\frac{1}{j\omega C_{1K}}\right][j\omega L_{2K}]$$

$$R_K^2 = \frac{L_{2K}}{C_{1K}}$$

For "T" / For "π"

$$f_c = \frac{1}{4\pi\sqrt{C_{1K}L_{2K}}} \qquad f_c = \frac{1}{4\pi\sqrt{C_{1K}L_{2K}}}$$

$$\omega_c = 2\pi f_c \qquad \omega_c = 2\pi f_c$$

$$\omega_c = \frac{1/2}{\sqrt{C_{1K}L_{2K}}} \qquad \omega_c = \frac{1/2}{\sqrt{C_{1K}L_{2K}}}$$

$$Z_{0TKH} = R_K\sqrt{1 - \frac{f_c^2}{f^2}} \qquad Z_{0\pi KH} = \frac{R_K}{\sqrt{1 - f_c^2/f^2}} \tag{20.23}$$

For a constant K filter, Z_0 is a pure resistance component. From the equations given it can be seen that the value of Z_0 varies with frequency. In practice the values of L and C are selected to match the load Z_0 at only one frequency. For low-pass filters, L and C are selected to match Z_0 when $f = 0$. For high-pass filters, L and C are selected to match Z_0 when $f = \infty$.

Substituting $f = 0$ in the low-pass filter equations for Z_0:

$$Z_{0TKL} = R_K \tag{20.24}$$

$$Z_{0\pi KL} = R_K \tag{20.25}$$

Substituting $f = \infty$ in the high-pass filter equations for Z_0:

$$Z_{0TKH} = R_K \tag{20.26}$$

$$Z_{0\pi KH} = R_K \tag{20.27}$$

Solution of L and C for a constant K filter, first for low pass:

$$\frac{L_{1K}}{C_{2K}} = R_K^2 \qquad \frac{L_{1K}^2}{L_{1K}C_{2K}} = R_K^2$$

$$\frac{L_{1K}}{\sqrt{L_{1K}C_{2K}}} = R_K \qquad \frac{\pi L_{1K}}{\pi\sqrt{L_{1K}C_{2K}}} = R_K$$

$$\pi L_{1K} f_c = R_K \qquad L_{1K} = \frac{R_K}{\pi f_c}$$

$$L_{1K} = \frac{2R_K}{2\pi f_c}$$

$$L_{1K} = \frac{2R_K}{\omega_c} \tag{20.28}$$

Similarly:

$$\frac{C_{2K}L_{1K}}{C_{2K}^2} = R_K^2 \qquad \frac{\sqrt{C_{2K}L_{1K}}}{C_{2K}} = R_K$$

$$\frac{\pi\sqrt{C_{2K}L_{1K}}}{\pi C_{2K}} = R_K \qquad \frac{1}{\pi f_c C_{2K}} = R_K$$

$$C_{2K} = \frac{1}{\pi f_c R_K} \qquad C_{2K} = \frac{2}{2\pi f_c R_K}$$

$$C_2 = \frac{2}{\omega_c R_K} \tag{20.29}$$

and for high pass:

$$\frac{L_{2K}}{C_{1K}} = R_K^2 \qquad \frac{L_{2K}^2}{L_{2K}C_{1K}} = R_K^2$$

$$\frac{L_{2K}}{\sqrt{L_{2K}C_{1K}}} = R_K \qquad \frac{4\pi L_{2K}}{4\pi\sqrt{L_{2K}C_{1K}}} = R_K$$

$$4\pi L_{2K} f_c = R_K \qquad L_{2K} = \frac{R_K}{4\pi f_c}$$

$$L_{2K} = \frac{R_K/2}{2\pi f_c}$$

$$L_{2K} = \frac{R_K/2}{\omega_c} \tag{20.30}$$

Similarly:

$$\frac{C_{1K}L_{2K}}{C_{1K}^2} = R_K^2 \qquad \frac{\sqrt{C_{1K}L_{2K}}}{C_{1K}} = R_K$$

$$\frac{4\pi\sqrt{C_{1K}L_{2K}}}{4\pi C_{1K}} = R_K \qquad \frac{1}{4\pi f_c C_{1K}} = R_K$$

$$C_{1K} = \frac{1}{4\pi f_c R_K} \qquad C_{1K} = \frac{1/2}{2\pi f_c R_K}$$

$$C_{1K} = \frac{1/2}{\omega_c R_K} \tag{20.31}$$

The constant K filter has the following two limitations:

1. Z_0 is not sufficiently constant over passband.
2. Attenuation is gradual in the stopband immediately after cutoff.

20.5 *M*-DERIVED FILTERS

In order to correct the limitations of the constant K filter, the "M"-derived filter (Fig. 20.7) was developed. The M-derived filter provides for a more constant value of Z_0 over the passband range and for a sharper cutoff. It should be kept in mind that the characteristic impedance of the M-derived filter is the same as that for the constant K filter.

$$Z_{0TM} = Z_{0TK} \qquad Z_{0\pi M} = Z_{0\pi K}$$

In addition, the cutoff frequency of the M-derived filter is the same as that for the constant K filter. Let

$$Z_{1M} = MZ_{1K}$$

$$Z_{2M} = \frac{(1 - M^2)}{4M} Z_{1K} + \frac{Z_{2K}}{M}$$

Rearranging terms:

$$\frac{Z_{1M}}{2} = M\frac{Z_{1K}}{2}$$

$$2Z_{2M} = 2\frac{(1 - M^2)}{4M} Z_{1K} + 2\frac{Z_{2K}}{M} \tag{20.32}$$

$$2Z_{2M} = \frac{(1 - M^2)}{2M} Z_{1K} + 2\frac{Z_{2K}}{M} \tag{20.33}$$

For infinite attenuation the vertical leg is in series resonance. This is illustrated in Fig. 20.8.

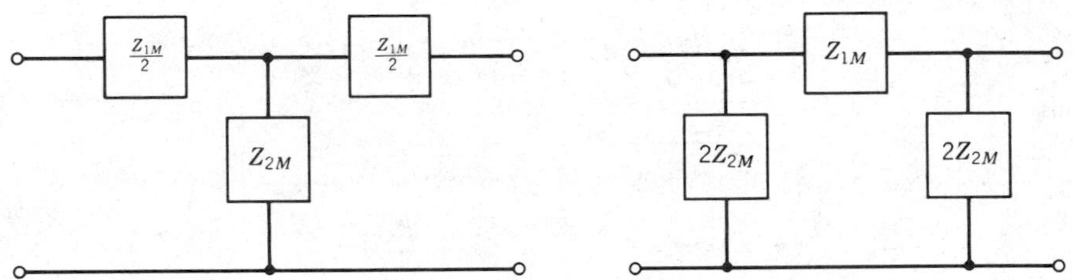

Fig. 20.7 M-derived filters.

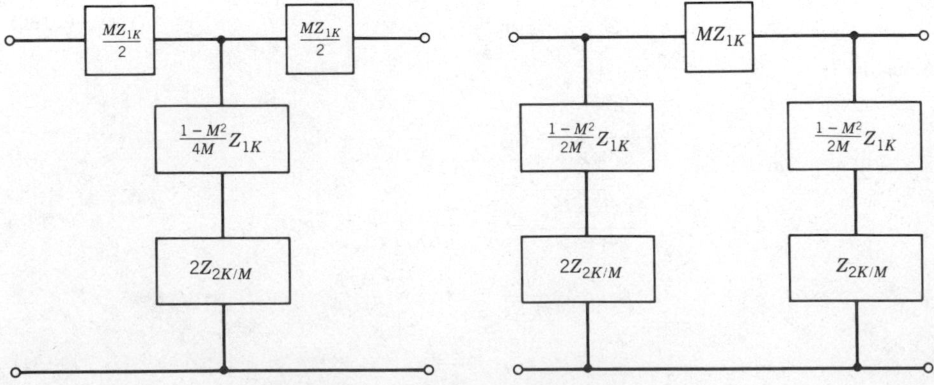

Fig. 20.8 Conditions for infinite attenuation.

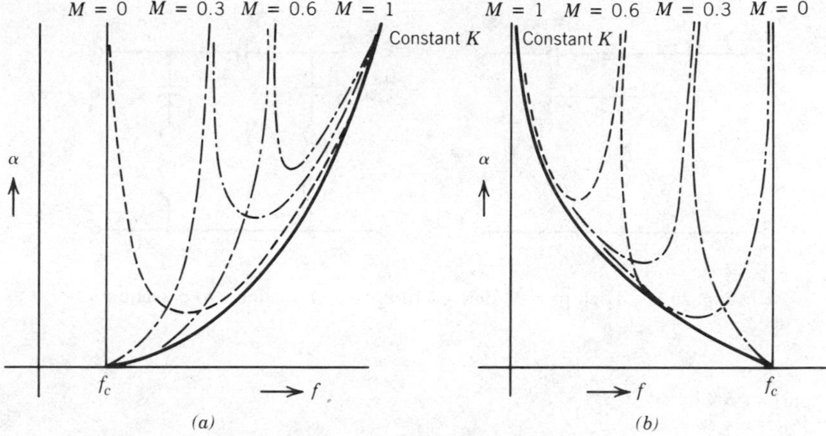

Fig. 20.9 Attenuation curves for a high-pass (*a*) and a low-pass (*b*) *M*-derived filter.

The attenuation curves for a high-pass and a low-pass *M*-derived filter are shown in Fig. 20.9. From these curves and the equations already developed, it can be seen that the constant *K* filter may be considered one particular *M*-derived filter, where $M = 1$. *T*- and π-section *M*-derived low-pass filters are shown in Fig. 20.10.

For low-pass filters:

π section

$$\omega_\infty \frac{(1 - M^2)L_{1K}}{4M} = \frac{1}{\omega_\infty M C_{2K}} \qquad \omega_\infty \frac{(1 - M^2)L_{1K}}{2M} = \frac{2}{\omega_\infty M C_{2K}}$$

$$\omega_\infty^2 = \frac{1}{\dfrac{(1 - M^2)}{4M} M L_{1K} C_{2K}} \qquad \omega_\infty^2 = \frac{1}{\dfrac{(1 - M^2)}{4M} M L_{1K} C_{2K}}$$

Thus:

$$\omega_{\infty T} = \omega_{\infty\pi} = \omega_\infty = \frac{1}{\sqrt{\dfrac{1 - M^2}{4} L_{1K} C_{2K}}}$$

$$f_\infty = \frac{\omega_\infty}{2\pi} = \frac{1}{\dfrac{2\pi}{2}\sqrt{(1 - M^2)L_{1K}C_{2K}}} \tag{20.34}$$

$$f_\infty = \frac{1}{\pi\sqrt{(1 - M^2)L_{1K}C_{2K}}} \tag{20.35}$$

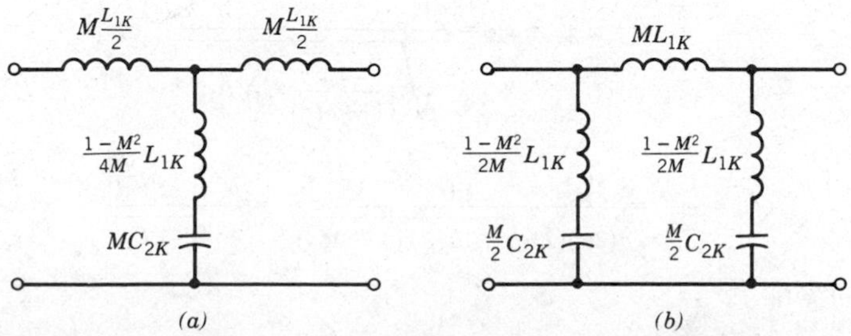

Fig. 20.10 *M*-derived low-pass filters: (a) *T*-section, (b) π section.

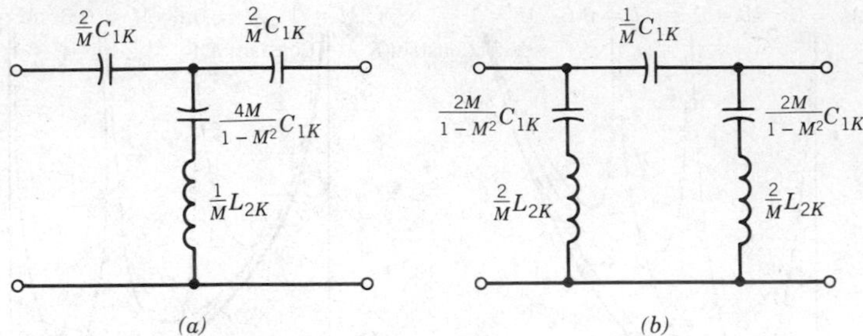

Fig. 20.11 High pass M-derived filters: (*a*) T section, (*b*) π section.

For a constant K filter:

$$f_{cT} = \frac{1}{\pi\sqrt{L_{1K}C_{2K}}} \quad \Bigg| \quad f_{c\pi} = \frac{1}{\pi\sqrt{L_{1K}C_{2K}}}$$

Assuming the same cutoff frequency in an M-derived section:

$$f_{\infty\,\text{low}} = \frac{1}{\sqrt{(1-M^2)}}\,\frac{1}{\pi\sqrt{L_{1K}C_{2K}}}$$

$$f_{\infty\,\text{low}} = \frac{f_c}{\sqrt{(1-M^2)}}$$

Solving for M:

$$M = \sqrt{1 - \frac{f_c^2}{f_\infty^2}} = \frac{\sqrt{f_\infty^2 - f_c^2}}{f_\infty} \tag{20.36}$$

For high-pass filters (Fig. 20.11):

$$\textit{T}\text{-section} \quad \Bigg| \quad \pi \text{ section}$$

$$\frac{1}{\omega_\infty\left[\dfrac{4M}{(1-M^2)}\right]C_{1K}} = \omega_\infty\frac{L_{2K}}{M} \quad \Bigg| \quad \frac{(1-M^2)}{(2M)}\,\frac{1}{\omega_\infty C_{1K}} = \frac{2\omega_\infty L_{2K}}{M}$$

$$\omega_\infty^2 = \frac{1}{\left[\dfrac{4M}{(1-M^2)}\right]C_{1K}\dfrac{L_{2K}}{M}} \quad \Bigg| \quad \omega_\infty^2 = \frac{1}{\left[\dfrac{4M}{(1-M^2)}\right]C_{1K}\dfrac{L_{2K}}{M}}$$

Thus:

$$\omega_{\infty T} = \omega_{\infty\pi} = \omega_\infty = \frac{1}{\sqrt{\dfrac{4}{(1-M^2)}C_{1K}L_{2K}}}$$

$$f_\infty = \frac{\omega_\infty}{2\pi} = \frac{1}{(2\pi)(2)\sqrt{C_{1K}L_{2K}/(1-M^2)}} = \frac{\sqrt{(1-M^2)}}{4\pi\sqrt{C_{1K}L_{2K}}}$$

For a constant K filter:

$$f_{cT} = \frac{1}{4\pi\sqrt{C_{1K}L_{2K}}} \quad \Bigg| \quad f_c\pi = \frac{1}{4\pi\sqrt{C_{1K}L_{2K}}}$$

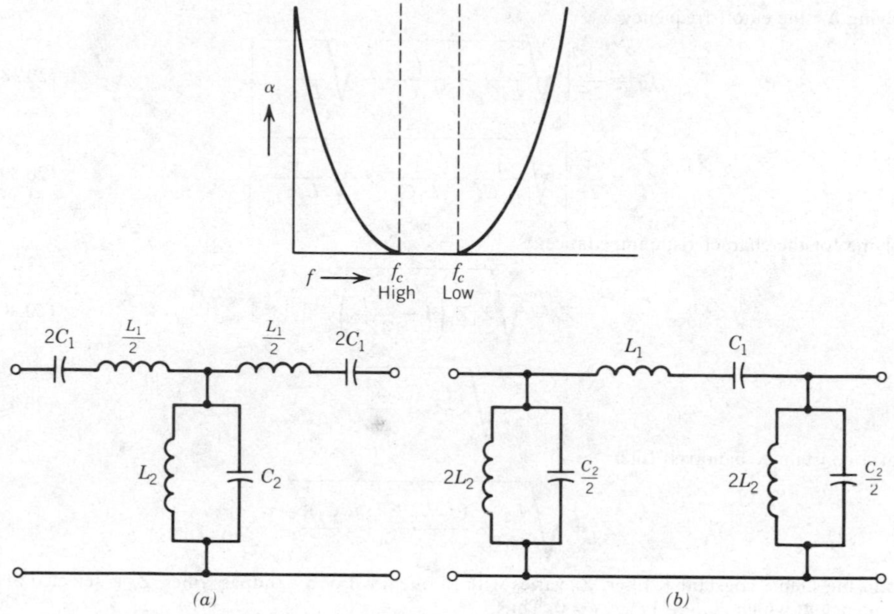

Fig. 20.12 *T (a)* and *π (b)* section bandpass filters.

Assuming the same cutoff frequency in an *M*-derived section:

$$f_{\infty \text{ high}} = \left[\sqrt{(1 - M^2)}\right]\left[\frac{1}{4\pi\sqrt{C_{1K}L_{2K}}}\right]$$

$$f_{\infty \text{ high}} = \left[\sqrt{(1 - M^2)}\right][f_c]$$

Solving for *M*:

$$M = \sqrt{1 - \frac{f_\infty^2}{f_c^2}} = \frac{\sqrt{f_c^2 - f_\infty^2}}{f_c} \tag{20.37}$$

Very often it is desired to select one band of frequencies. This band selection may be obtained by connecting a low- and a high-pass filter section together.

A bandpass filter (Fig. 20.12) may be obtained by placing a high-pass filter in series with a low-pass filter. The cutoff frequency of the high-pass filter is lower than the cutoff frequency of the low-pass filter.

$$Z_1 = j\omega L_1 + \frac{1}{j\omega C_1} \qquad \frac{1}{Z_2} = \frac{1}{j\omega L_2} + j\omega C_2$$

For a constant *K* bandpass filter:

$$Z_1 Z_2 = \frac{L_2(\omega^2 L_1 C_1 - 1)}{C_1(\omega^2 L_2 C_2 - 1)}$$

for $Z_1 Z_2 = R_K^2$, where R_K is independent of frequency.

$$\omega^2 L_1 C_1 - 1 = \omega^2 L_2 C_2 - 1$$

$$L_1 C_1 = L_2 C_2$$

$$Z_1 Z_2 = \frac{L_2}{C_1} = \frac{L_1}{C_2}$$

$$\alpha = 0 \qquad \text{when} \quad -1 \leq \frac{Z_1}{4Z_2} \leq 0$$

Solving for the cutoff frequency:

$$f_{cL} = \frac{1}{2\pi}\left[\sqrt{\frac{1}{L_1C_1} + \frac{1}{L_1C_2}} - \sqrt{\frac{1}{L_1C_2}}\right] \tag{20.38}$$

$$f_{cH} = \frac{1}{2\pi}\left[\sqrt{\frac{1}{L_1C_1} + \frac{1}{L_1C_2}} + \sqrt{\frac{1}{L_1C_2}}\right] \tag{20.39}$$

Solving for the characteristic impedance:

$$Z_0 = \sqrt{Z_1Z_2\left(1 + \frac{Z_1}{4Z_2}\right)} \tag{20.40}$$

$$Z_0 = \sqrt{Z_1Z_2 + \frac{Z_1^2}{4}} \tag{20.41}$$

For a constant K bandpass filter:

$$Z_0 = \sqrt{\frac{L_2}{C_1} + \frac{(j\omega L_1 + 1/j\omega C_1)^2}{4}}$$

As in the simple constant K filter, Z_0 varies with frequency. For a bandpass filter, Z_0 is selected at a frequency at which $j\omega L_1 + 1/j\omega C_1 = 0$. Thus:

$$Z_0 = \sqrt{\frac{L_2}{C_1}} = \sqrt{\frac{L_1}{C_2}} = R_K \tag{20.42}$$

The relationship between the constant K-type filter and the M-derived filter is the same as that for low-pass and high-pass filters.

A band-elimination filter (Fig. 20.13) may be obtained by placing a high- and a low-pass filter in

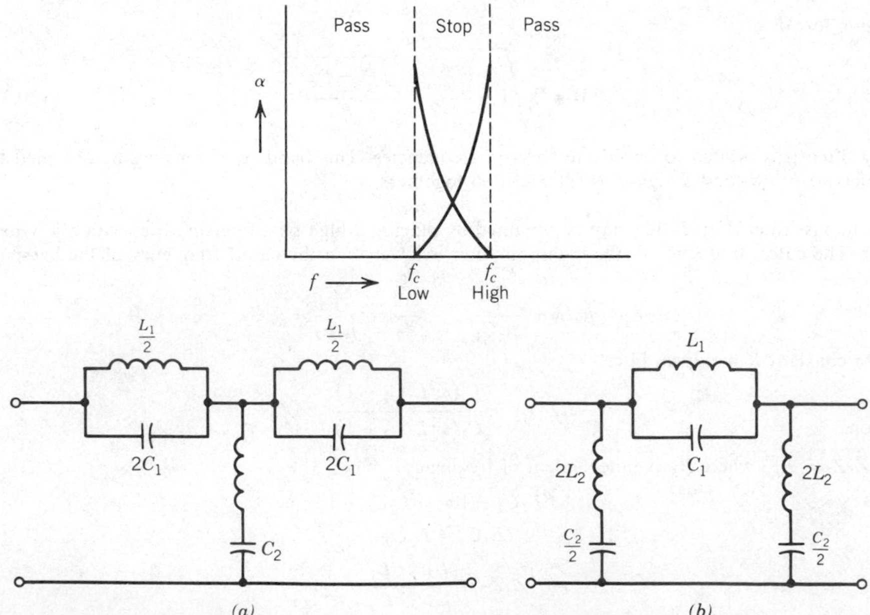

Fig. 20.13 T (a) and π (b) section band-elimination filters.

parallel. The stopbands of two filters overlap in the range of the undesired frequencies.

$$\frac{1}{Z_1} = \frac{1}{j\omega L_1} + j\omega C_1 \qquad Z_2 = j\omega L_2 + \frac{1}{j\omega C_2}$$

For a constant K band-elimination filter:

$$Z_1 Z_2 = \frac{L_1(\omega^2 L_2 C_2 - 1)}{C_2(\omega^2 L_1 C_1 - 1)}$$

for $Z_1 Z_2 = R_K^2$.

Similarly:

$$L_1 C_1 = L_2 C_2$$

$$Z_1 Z_2 = \frac{L_1}{C_2} = \frac{L_2}{C_1}$$

$$\alpha = 0 \qquad \text{when} \quad -1 \leqq \frac{Z_1}{4Z_2} \leqq 0$$

Solving for the cutoff frequency:

$$f_{cL} = \frac{1}{2\pi}\left[\sqrt{\frac{1}{L_2 C_2} + \frac{1}{L_2 C_1}} - \sqrt{\frac{1}{L_2 C_1}}\right] \tag{20.43}$$

$$f_{cH} = \frac{1}{2\pi}\left[\sqrt{\frac{1}{L_2 C_2} + \frac{1}{L_2 C_1}} + \sqrt{\frac{1}{L_2 C_1}}\right] \tag{20.44}$$

Solving for the characteristic impedance:

$$Z_0 = \sqrt{Z_1 Z_2\left(1 + \frac{Z_1}{4Z_2}\right)} \tag{20.45}$$

$$Z_0 = \sqrt{Z_1 Z_2 + \frac{Z_1^2}{4}} \tag{20.46}$$

For a constant K band-elimination filter:

$$Z_0 = \sqrt{\frac{L_1}{C_2} + \frac{(j\omega L_2 + 1/j\omega C_2)^2}{4}}$$

For a band-elimination filter, Z_0 is selected at a frequency at which $j\omega L_2 + 1/j\omega C_2 = 0$. Thus:

$$Z_0 = \sqrt{\frac{L_1}{C_2}} = \sqrt{\frac{L_2}{C_1}} = R_K \tag{20.47}$$

The relationship between the constant K-type filter and the M-derived filter is the same as that for low-pass and high-pass filters.

20.6 IDEAL LOW-PASS FILTER

For most filter applications, the loss response is of prime importance. An ideal low-pass filter loss response (α vs. ω) is shown in Fig. 20.14. Note that the loss in the passband $(\alpha_P) = 0$ up to the cutoff frequency (ω_C) and is infinite for frequencies higher than ω_C in the stopband (α_S).

This ideal response cannot be realized. In practice, α_P has some upper limit in the range $0 \leqslant \omega \leqslant \omega_P$ and α_S has some lower limit for $\omega_P \leqslant \omega \leqslant \infty$ (Fig. 20.15). Loss (α) in decibels is given by

$$\alpha = 10 \log|P_{\max}/P_d| = 20 \log|H(j\omega)| \tag{20.48}$$

where $P_{\max}$ is available generator power and P_d is output power of the filter.

If the source resistance and load resistance are unequal, zero loss cannot be achieved for zero frequency. Therefore a constant loss is unavoidable (Fig. 20.15). This changes Eq. (20.48) by adding the multiplier $10^{\alpha/20}$ to $H(j\omega)$ (Fig. 20.16).

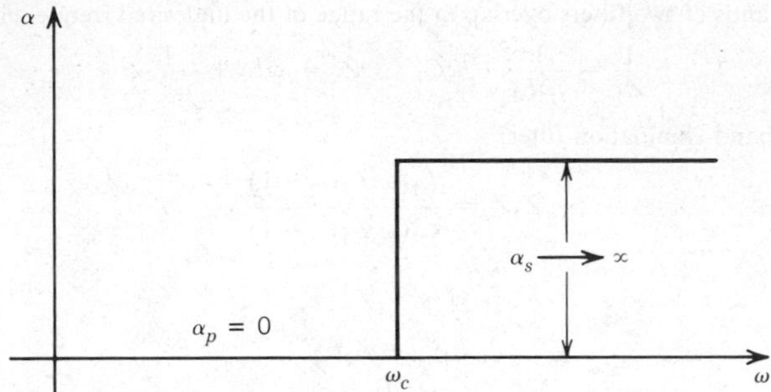

Fig. 20.14 Ideal low-pass filter.

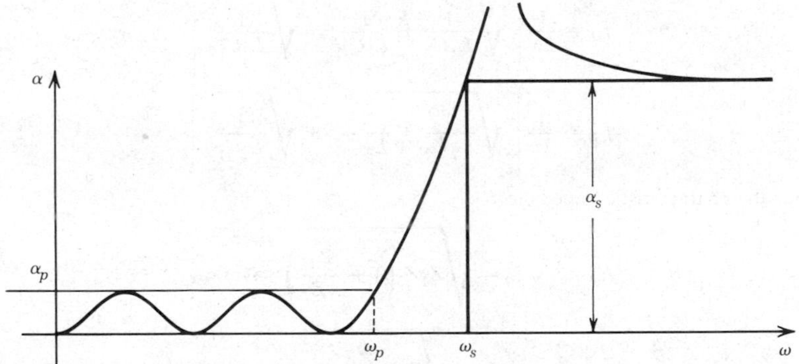

Fig. 20.15 Practical low-pass filter.

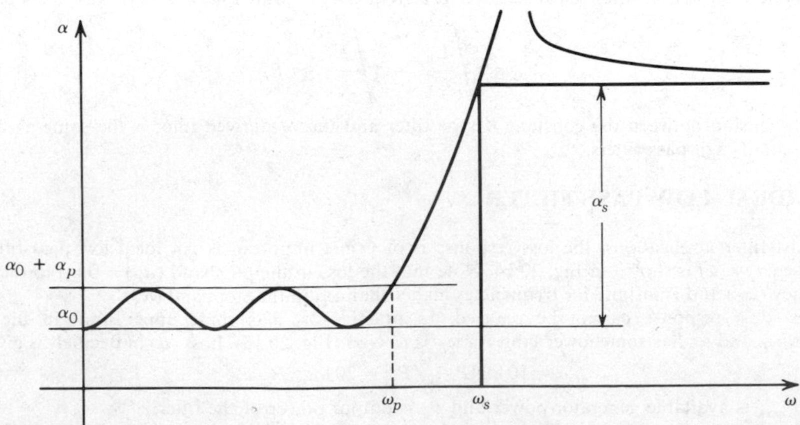

Fig. 20.16 Flat loss filter response.

20.7 USE OF CHARACTERISTIC FUNCTION

While the previous specifications relate to the loss response $\alpha(\omega)$, the solution to the approximation problem is carried out using the characteristic function $K(\omega)$ rather than $\alpha(\omega)$ for the following reasons.

1. $K(\omega)$ is a polynomial or a rational function of ω and $\alpha(\omega)$ is not. Approximations with polynomials or rational functions are relatively congenial.

2. $\alpha(\omega) = 10\log[1 + |K(j\omega)|^2]$. If $\alpha = 0$, $|K|^2 = 0$. If $\alpha = \infty$, $|K|^2 = \infty$. If $dx(\omega)/d\omega <> 0$, $d|K|^2/d\omega <> 0$. If $\alpha' = \alpha'' = \ldots \alpha^n$, $(|K|^2)' = (|K|^2)'' = (|K|)n = 0$. Thus the frequency-dependent properties of $|K^2|$ duplicate the curves of $\alpha(\omega)$, differing only in their vertical scales.

3. Filters have low loss and high loss in specified frequency bands. This can be designed in a one-dimensional problem if we use $K(j\omega)$, since the low-loss passband can be obtained by placing all the zeros of $K(s)$ on the $j\omega$ axis. The high-loss stopband is generated by placing all the poles of $K(s)$ on the real axis.

Therefore $|K|^2$ is convenient when loss characteristics are of interest.

20.8 BUTTERWORTH FILTERS

If in the characteristic equation $K(s) = \pm Cs^n$ we have a Butterworth filter, C is some constant and n = the number of natural modes or the total number of loss poles (including those at 0 or infinity). The proof follows.

For a minimally flat approximation, $|H(j\omega)|^2$ is an even function of ω.

$$|H(j\omega)|^2 = H(j\omega)H(-j\omega) = H^2_{\text{even}}H^2_{\text{odd}} \tag{20.49}$$

Therefore $\alpha(\omega)$ is an even function of ω and ω^2 is the independent variable.

For a maximally flat passband:

$$\alpha(\omega^2) = d\alpha(\omega^2)/d(\omega^2) = \ldots d^{n-1}\alpha(\omega^2)/d(\omega^2)^{n-1} = 0 \tag{20.50}$$

In Eq. (20.50), all relations are valid for $\omega^2 = 0$. Therefore the equation may be rewritten in terms of $|K|^2$:

$$|K|^2 = (|K|^2)' = (|K|^2)'' = \ldots (|K|^2)^{n-1} = 0 \tag{20.51}$$

where $\omega^2 = 0$ and the indicated differentiations are with respect to ω^2.

Equation (20.51) imposes m conditions on $|K|^2$, which must have n free parameters to satisfy same. If we let $|K|^2$ be a rational function of ω^2, then

$$|K|^2 = \frac{C^2\omega^{2n} + a_{n-1}\omega^{2(n-1)} + \ldots a_1\omega^2 + a_0}{Q_m(\omega^2)} \tag{20.52}$$

where Q_m is an mth degree polynomial in ω^2 with a nonzero constant term. Then by choosing the n coefficients zero,

$$a_0 = a_1 = \ldots a^{n-1} = 0$$

we get

$$|K|^2 = \frac{C^2\omega^{2n}}{Q_m(\omega^2)}$$

Since $C^2 > 0$, the condition $|K|^2 \geqslant 0$ requires that $Q_m(\omega^2) = 1$. Then $|K|^2 = C^2\omega^{2n}$. Therefore $K(s) = \pm Cs^n$.

To realize this filter, one must have the following information:

1. The degree n of the filter.
2. The characteristic function $K(s)$.
3. The filter function $H(s)$.
4. The loss poles contained in H and K.

Therefore to design a Butterworth filter given by

$$\alpha(w) = 10\log(1 + C^2\omega^{2n}) \tag{20.53}$$

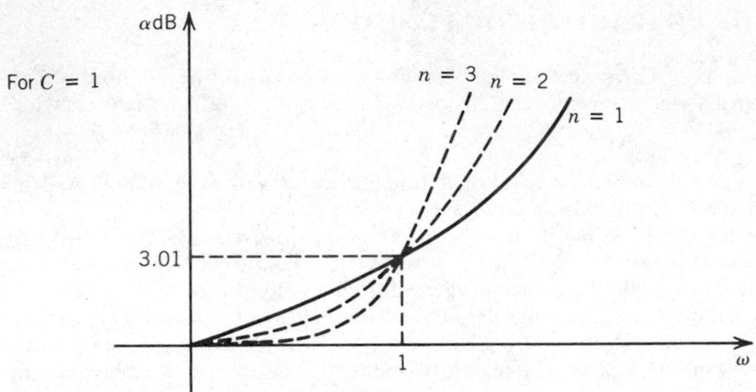

Fig. 20.17 Loss response for Butterworth filter.

C and n must be chosen to satisfy the following conditions:

$$\alpha(\omega) \leqslant \alpha_P \qquad \text{for} \quad |\omega| \leqslant \omega_P \tag{20.54}$$

$$\alpha(\omega) \geqslant \alpha_S\alpha_S \qquad \text{for} \quad |\omega| > \omega_S \tag{20.55}$$

The number of elements needed for a realizable filter increases with n. Therefore n should be an integer and as small as possible. Increasing the value of n gives lower power loss in the passband and higher loss in the stopband.

Figure 20.17 shows the effect of n on the response of a Butterworth filter where

$$\alpha(\omega) = 10\log(1 + C^2\omega^{2n})$$

where C and n are chosen to satisfy the following conditions:

$$\alpha(\omega) \leqslant \alpha_P \qquad \text{for} \quad |\omega| \leqslant \omega_P$$

$$\alpha(\omega) \geqslant \alpha_S \qquad \text{for} \quad |\omega| \geqslant \omega_S \tag{20.56}$$

If the requirements of Eq. (20.55) are met with equalities, then

$$\alpha_P = 10\log(1 + C^2\omega_P^{2n}) \tag{20.57}$$

$$\alpha_S = 10\log(1 + C^2\omega_S^{2n}) \tag{20.58}$$

$$C^2\omega_P^{2n} = 10^{\alpha_P/10} - 1$$

$$\alpha^2\omega_S^{2n} = 10^{\alpha_S/10} - 1$$

Dividing Eq. (20.57) by (20.58) and solving for n we obtain

$$n \geqslant \frac{\log k_1}{\log k} \tag{20.59}$$

where

$$k_1 = \left[\frac{10^{\alpha_P/10} - 1}{10^{\alpha_S/10} - 1}\right]^{1/2}, \qquad k = \frac{\omega_P}{\omega_S}$$

and

$$C = \sqrt{\frac{10\alpha_S/10 - 1}{\omega_S}} \tag{20.60}$$

To compute $H(s)$ with $F(s) = Cs^n$ and $p(s) = 1$ using the Field Keller equation

$$e(s)e(s-1) = 1 + C^2(-1)^n s^{2n}$$

Therefore the roots of $e(s)e(-s)$ appear as roots of the equation

$$s^{2n} = C^{-2}(-1)^{n-1} = \frac{e^{j\pi(n-1+2k)}}{C^2}, \qquad k = 1, 2, \ldots$$

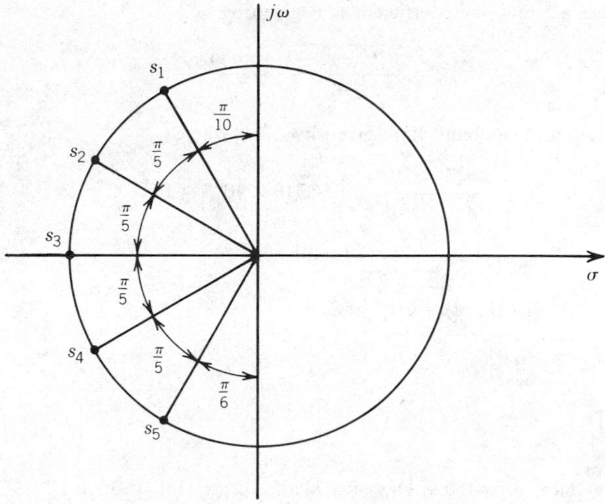

Fig. 20.18 Butterworth filter, $n = 5$: Natural modcs.

Therefore

$$s_k = C^{-1/n}\exp\left[\pi/2 + \frac{\pi(2k-1)}{2n}\right], \qquad k = 1, 2, \ldots, 2n \tag{20.61}$$

But since $e(s)$ is strictly a Hurwitz polynomial, only the left half plane (LHP) roots $s_1, s_2, \ldots, s_n$ qualify as zeros. These roots lie on equal angles on a half circle of radius $C^{1/n}$ in the left half of the s plane. This result is illustrated in Fig. 20.18 for $n = 5$. Thus

$$H(s) = e(s) = \pm C \prod_{k=1}^{n} (s - s_k) \tag{20.62}$$

$$K(s) = F(s) = \pm Cs^n \tag{20.63}$$

Example 20.1. Find n, $H(s)$, and $K(s)$ for a Butterworth filter with the specifications $\alpha_P \leqslant 0.1$ dB for $f \leqslant 3$ mHz, $\alpha_S \geqslant 60$ dB for $f \geqslant 24$ mHz.

Solution.

$$k_1 = \left[\frac{10^{\alpha_p/10} - 1}{10^{\alpha_s/10} - 1}\right]^{1/2}, \qquad k = \frac{\omega_P}{\omega_S}$$

$$k_1 = \sqrt{\frac{10^{0.1} - 1}{10^6 - 1}} = 0.15262 \times 10^{-3}$$

$$k = 3/12 = 0.25$$

$$n \geqslant \frac{\log k_1}{\log k} \geqslant \frac{\log(0.15262 \times 10^{-3})}{\log(0.25)} \geqslant 6.3388$$

Let $n = 7$; if passband specs are met:

$$C = \frac{\sqrt{10^{\alpha_P/10} - 1}}{\Omega_P^n} = \frac{0.1526}{\Omega_P^7}$$

where Ω_p^n is the normalized passband limit normalized by dividing ω by ω_0, where ω_0 is given by

$$\omega_0 = \frac{2\pi(3 \times 10^6)}{(0.15262)^{1/2}} = 24.656 \times 10^6$$

If the stopband specs are met, the coefficient C is given by

$$C = \frac{\sqrt{10^{\alpha_S/10} - 1}}{\Omega_S^n} = \frac{999.995}{\Omega_S^7}$$

where Ω_s^n is the normalized stopband limit given by ω/ω_0 where

$$\omega_0 = \frac{2\pi(24 \times 10^6)}{(999.9995)^{1/2}} = 56.310 \times 10^6 \qquad \text{for} \quad C = 1$$

The normalized values of $s_k(k = 1 \ldots 7)$ are found by

$$s_k(\text{normalized}) = \exp\left[j\pi/2 + \frac{\pi(2k-1)}{14} \right] \qquad k = 1 \ldots 7$$

$$K(s) = s^7$$

$$H(s) = \prod_{k=1}^{7} (s - s_k)$$

$$= (s = 1)(s^2 + 0.4450 + 1)(s^2 + 1.24705 + 1)(s^2 + 1.80915 + 1)$$

$$= s^2 + 4.4940s^6 + 10.0978s^5 + 14.590s^4 + 14.5920s^3 + 10.0975s^2 + 4.4940s + 1$$

where s and s_k are normalized. To denormalize, replace s by s/ω_0.

20.9 CHEBYCHEV FILTERS

The characteristics of a Chebychev filter are shown in Fig. 20.19. Note that the passband exhibits equal ripples. For this case

$$|K|^2 = k^2 \cos^2 n\mu(\omega)$$

where $\mu(\omega) = \cos^{-1}(\omega/\omega_P)$. Here it is seen that $|K|^2$ is a polynomial in $(\omega/\omega_P)^2 = \cos^2\mu s$. Since

$$\cos n\mu = \text{Re}\left[(e^{j\mu})^n \right]$$

$$= \text{Re}[\{(\cos \mu + j \sin \mu)^n\}]$$

$$= \cos^n\mu - (n/2)(\cos^{n-2}\mu)(1 - \cos^2\mu) + (n/4)(\cos^{n-4}\mu)(1 - \cos^2\mu)^2 + \ldots$$

Also for $|\omega| < \omega_P$, μ is real and hence $|K|^2 \leqslant k_P^2$. As ω goes from 0 to ω_P and μ goes from $\pi/2$ to 0, $|K|^2$ oscillates n times between 0 and R_P^2 for $\omega > \omega_P$ is monotonic increasing function of ω^2.

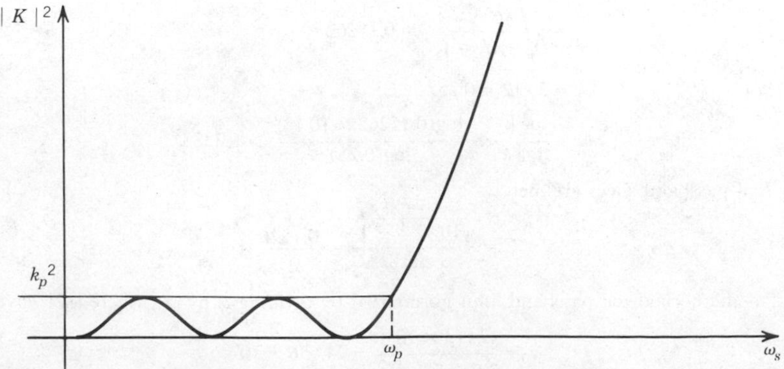

Fig. 20.19 Equal-ripple filter response.

For optimality, $P_n(\omega^2) = n$th order polynomial in ω^2 if it is restricted to the values $0 \leqslant P_n(\omega^2) \leqslant k_P^2$ for $\omega^2 \leqslant \omega_P^2$. Therefore

$$T_n(\alpha) = \cos(n \cos^{-1}\alpha) \tag{20.64}$$

which is the entering expression of $|K|^2$, the expression for an nth order Chebychev polynomial. Hence the name.

For $\omega^2 \geqslant \omega_P$

$$\omega = \omega_P \cos\,\mu = \omega_P \cosh(j\mu) \tag{20.65}$$

Therefore μ is imaginary and is given by

$$\mu = -j \cosh^{-1}(\omega/\omega_P) \tag{20.66}$$

Hence

$$|K|^2 = k_P^2 \cosh^2(nj\mu) = k_P^2[n \cosh^{-1}(\omega/\omega_P)]$$

Therefore $|K|^2$ is real and greater than k_P^2. Therefore

$$\begin{aligned} |K|^2\omega = \omega_P &= 10^{\alpha_P/10} - 1 \leqslant k_P^2 \\ |K|^2\omega = \omega_s &= 10^{\alpha_P/10} - 1 \geqslant k_P^2 \cosh[h^2(z)] \\ z &= (n \cosh^{-1})\omega_S/\omega_P \end{aligned} \tag{20.67}$$

Therefore

$$n \geqslant = \frac{\cosh^{-1}(1/K_1)}{\cosh^{-1}(1/K)}$$

where K and K_1 are as given for the Butterworth filter.

If we replace ω by s/j and let $\mu = v + j\omega$ and $s = \sigma + j\omega$, then

$$\begin{aligned} |K(s)|^2 &= k_P^2 \cos^2 n\mu(s) = k_P^2 T_n^2(s/j\omega_P) \\ \mu(s) &= \cos^{-1}(s/j\omega_P) \end{aligned} \tag{20.68}$$

Then from the Field and Keller equation

$$k_P^2 \cos^2[n(v_k + j\omega_k) + 1] = 0, \qquad k = 1, 2, \ldots, n \tag{20.69}$$

where $\mu_k = v_k + j\omega_k$ is the kth natural mode in the complex μ-plane.

Using $\cos(x + jy) = \cos x \cosh y - \sinh y$ we obtain

$$s_k = 0_k + j\omega_k = j\omega_P \cos(v_k + j\omega_k) \tag{20.70}$$

$$\sigma_k = \omega_P \sin v_k \sinh \omega_k \tag{20.71}$$

$$\omega_k = \omega_P \cos v_k \cosh \omega_k \tag{20.72}$$

The locus on which the s_k's lie is given by

$$\frac{\sigma_k^2}{\omega_P^2 \sinh^2 \omega_k} + \frac{\omega_k^2}{\omega_P^2 \cosh^2 \omega_k} = 1 \qquad \text{(ellipse)}$$

The zeros of $H(s) = e(s)$ we know. Its constant factor may be found by

$$\lim_{\omega \to \infty} |H|^2 = \lim_{W \to \infty} |K|^2 = \lim_{\omega \to \infty} k_P^2(\omega/\omega_P)^{2n}$$

Hence

$$H(s) = e(s) = \pm k_P \prod_{k=1}^{n} \left(\frac{s}{\omega_P} - \sin v_k \sinh \omega_k - j \cos v_k \cosh \omega_k \right) \tag{20.73}$$

with reflection zeros at $\omega_P \cos\{[(2k-1)/n]/2)\}$ for $k = 1, 2, \ldots, n$.

The Chebychev polynomials needed in $k(s)$ are easily generated if we let $\omega_P = 1$.

$$T_{m+1}(\omega) = \cos(m\mu + \mu) = \cos m\mu \cos\,\mu - \sin m\mu \sin\,\mu$$

$$T_{m-1}(\omega) = \cos m\mu \cos\,\mu + \sin m\mu \sin\,\mu$$

$$T_{m+1}(\omega) + T_{m-1}(\omega) = 2\omega T_m$$

All $T_n(\omega)$ can be readily found.

Example 20.2. Find n, $k(s)$, and $H(s)$ for a Chebychev filter that satisfies the specifications of Example 20.1 (Butterworth filter).

Solution.

$$n \geqslant \frac{\cosh^{-1} 6552.2212}{\cosh^{-1}} = \frac{9.48071}{2.06344} = 4.59462$$

$n = 5$ is sufficient.

Ω_P^n normalized = 1. By choosing $\omega_0 = 2\pi(3 \times 10^6)$ and solving for T_S we get

$$T_S = 16\omega^5 - 20\omega^3 + 5\omega$$

Hence

$$K_P = (10^{1/10} - 1)^{1/2} = 0.152262$$

$$K(s) = (0.15262)(16s^2 + 20s^3 + 5s)$$

$$= 2.4419s^5 + 3.052s^3 + 0.763s$$

The normalized modes may be obtained by letting

$$s_k = \sigma_k + j\omega_k$$

where $\sigma_k = \sin v_k \sinh \omega_k$

$$\omega_k = \cos v_k \cosh \omega_k$$

$$v_k = \pm \frac{(2k-1)}{n}$$

$$\omega_k = \pm \frac{1 \sinh(1/k_2)}{n}$$

For $n = 5$, $K = 0.1562$.

Finally, $H(s)$ is given by

$$H(s) = 0.1562(s + 0.538)(s + 0.1665 - j1.080)(s + 0.1665 + j1.08)(s + 0.435 - j0.667)(2s + 0.435 + j0.667)$$

Bibliography

Balbanian, N., *Network Synthesis*, Prentice-Hall, Englewood Cliffs, NJ, 1958.

Budak, A., *Passive and Active Network Analysis and Synthesis*, Houghton Mifflin, Boston, 1974.

Cauer, W., "New Theory and Design of Wave Filters," *Physics* **2**:242–268 (1932).

Craig, J. W., *Design of Lossy Filters*, MIT Press, Cambridge, MA, 1970.

Hansell, G. E., *Filter Design and Evaluation*, Van Nostrand Reinhold, New York, 1969.

Huelsman, L. P., *Active Filters: Lumped, Distributed, Digital and Parametric*, McGraw-Hill, New York, 1970.

Humphrey, D. S., *The Analysis, Design and Synthesis of Electric Filters*, Prentice-Hall, Englewood Cliffs, NJ, 1970.

Javid, M. and E. Brenner, *Analysis, Transmission and Filtering of Signals*, McGraw-Hill, New York, 1963.

Orchard, H. J., "The Roots of the Maximally Flat-Delay Polynomials," *IEEE Trans. Cir. Theory* **CT12**:452–454 (1965).

Temes, C. C. and S. K. Mitra, *Modern Filter Theory and Design*, Wiley, New York, 1973.

Thompson, W. E., "Delay Networks Having Maximally Flat Frequency Characteristics," *Proc. IEEE*, Vol. 96, Pt. 3, pp. 487–490 (Nov. 1949).

Zverev, A. I., *Handbook of Filter Synthesis*, Wiley, New York, 1967.

CHAPTER 21
ANALYSIS OF MAGNETIC SYSTEMS†

GORDON R. SLEMON

Department of Electrical Engineering
University of Toronto

Ferromagnetic materials are used in essentially all electrical machines and transformers. There are several interrelated reasons for this widespread use. In a closed magnetic path, a large magnetic flux can be established and controlled by the application of a very small magnetomotive force from an encircling coil. In a magnetic path with an air gap, the effect of the magnetomotive force of the coil is concentrated at the gap, allowing the production of an intense magnetic field in a restricted volume of space. In addition, the forces between sections of magnetic material in a magnetic path with an air gap are several orders of magnitude greater than could be obtained with a similar volume of conductor material only.

The saturation and hysteresis properties of ferromagnetic materials can be exploited to produce a variety of useful devices. The relationship between B and H is nonlinear, multivalued, history dependent, and often time dependent. Formidable difficulties may therefore be expected in analysis. This chapter begins with a consideration of various approximate models for the B-H characteristic which may lead to useful prediction of performance with a minimum of analytical effort.

In addition to the complexity in the B-H relationship of the material, there is also complexity in the shapes of magnetic structures in electric apparatus. The latter part of the chapter considers means by which the complex magnetic field of the structure may be reduced, for analytical purposes, to a simple magnetic or electric equivalent circuit.

21.1 APPROXIMATE MODELS FOR *B-H* CHARACTERISTICS

Consider the uniformly wound torus of magnetic material shown in Fig. 21.1. The potential difference v across the coil terminals is

$$v = Ri + \frac{d\lambda}{dt} \qquad \text{V} \tag{21.1}$$

where R is the coil resistance and λ is the flux linkage of the coil. If the voltage v is to be expressed as a function of current and time only, the relation between the flux linkage λ and the current i must be known.

† Reprinted with permission from Gordon R. Slemon, *Magnetoelectric Devices*, Wiley, New York, 1966.

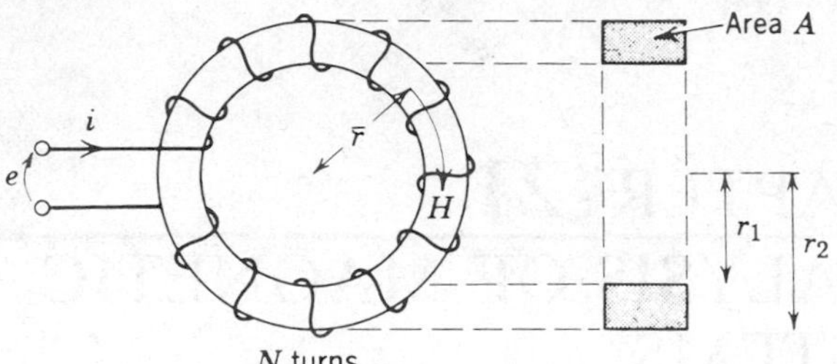

Fig. 21.1 Uniformly wound torus.

The magnetic field intensity H in the torus may be related to the current i by use of the circuital law. If the ratio of the inner radius r_1 to the outer radius r_2 of the torus is near unity, the magnetic field within the torus may be assumed to be uniform in intensity, its average magnitude being related to the coil current by

$$i = \frac{\bar{l}H}{N} \qquad \text{A} \tag{21.2}$$

where $\bar{l}$, the mean length of the flux path, is equal to $2\pi\bar{r}$. Using the same assumption of uniformity of the magnetic field over the cross-sectional area A, the flux linkage λ is related to the magnetic flux density B by

$$\lambda = NAB \qquad \text{Wb} \tag{21.3}$$

Combining Eqs. (21.2) and (21.3) gives

$$\frac{\lambda}{i} = \left(\frac{N^2A}{\bar{l}}\right)\left(\frac{B}{H}\right) \qquad \text{Wb/A} \tag{21.4}$$

This shows that the flux linkage-current relationship for the coil has the same shape as the B-H characteristic of the material. If the ordinate of the B-H characteristic is multiplied by NA and the abscissa by $\bar{l}/N$, the λ-i characteristic is obtained. It must be stressed that this applies only in situations where uniformity of the magnetic field can be assumed.

Since the magnetic flux ϕ in the core is equal to the product BA, and the magnetomotive force $\mathscr{F}$ of the winding is equal to $H\bar{l}$, the B-H relationship may also be rescaled to produce the relation between magnetic flux and magnetomotive force for the magnetic element.

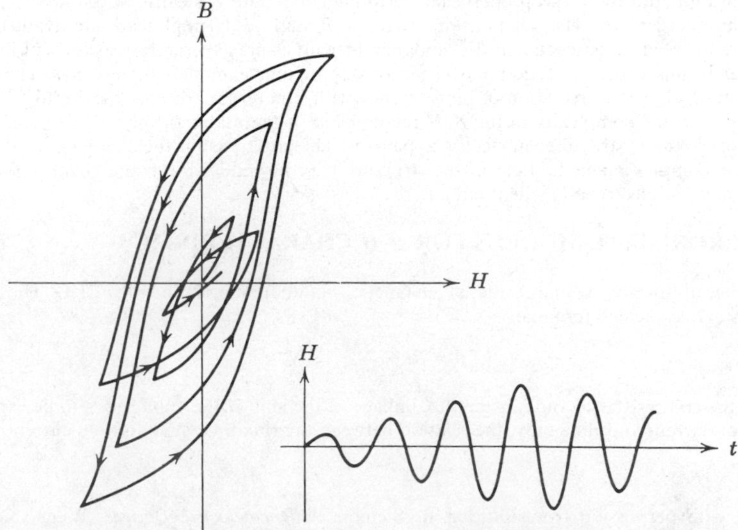

Fig. 21.2 Locus of the B-H characteristic for an alternating intensity of H of variable magnitude.

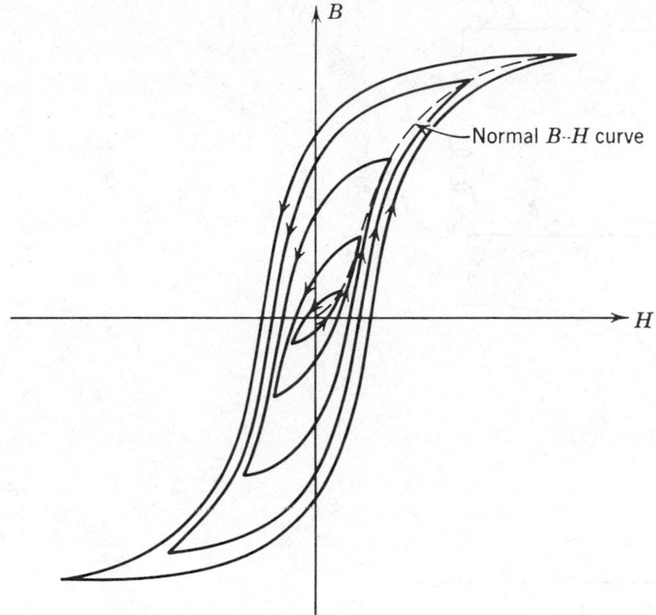

Fig. 21.3 A set of hysteresis loops and the normal magnetization curve for a ferromagnetic material.

The complexity of attempting an accurate representation of the B-H relationship is demonstrated in Fig. 21.2. This figure shows the B-H locus that may be followed as the magnetic field intensity H alternates through a number of cycles with variable peak magnitude. Note that at a given value of magnetic field intensity, the flux density may have any value within a wide range. To determine the appropriate value, it is necessary to know the history of the B-H locus. It is obviously impossible to record the loci for all possible histories; thus actual B-H loci are seldom used in analysis. Simplified approximations generally give results of adequate accuracy.

For alternating current of constant peak magnitude, the closed, symmetrical B-H loops shown in Fig. 21.3 represent the behavior of the material. The analytical difficulty in using such loops is that it is necessary to know, in advance, the peak amplitude of either B or H to decide which loop is applicable.

Normal Magnetization Curve

Several simple and useful approximations can be obtained if the hysteresis effect in the material is neglected. Without its memory, the B-H relationship becomes single valued. The most commonly used approximation, known as the normal magnetization curve, is shown in Fig. 21.3. This curve is the locus of the tips of a set of symmetrical hysteresis loops. This is the curve that is most readily available in the descriptive literature on most types of soft magnetic material. It is obtained by use of a flux meter, which measures the total change in the flux of a sample when its exciting current is reversed.

Numerical techniques are applicable to analyses that involve the normal magnetization curve. As a simple example, consider the circuit of Fig. 21.4*a* in which the constant voltage V is applied to a coil at time $t = 0$ by closing the switch. The coil is represented by its resistance R and by the flux linkage λ vs. current i curve of Fig. 21.4*b*. This curve is derived from the normal magnetization curve of Fig. 21.3 by rescaling the B-H curve using Eqs. (21.2) and (21.3). At $t = 0$, the current i is zero. We want to find the current i as a function of time t. From Eq. (21.1), we have

$$\left(\frac{d\lambda}{dt}\right)_{t=0} = V \tag{21.5}$$

Assuming that this rate of change of flux linkage remains approximately constant for a short interval of time Δt, the flux linkage λ_1 at time Δt can be approximated by

$$\lambda_1 = V\Delta t \tag{21.6}$$

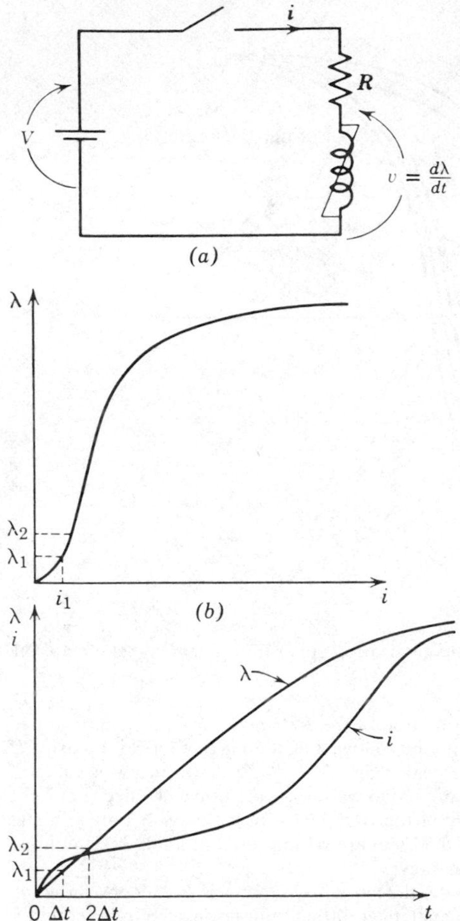

Fig. 21.4 (*a*) Circuit. (*b*) Flux linkage λ-*i* characteristic. (*c*) Flux linkage and current as functions of time *t*.

Referring to the λ-*i* curve, the corresponding current is i_1. The slope of the λ-*t* curve at $t = \Delta t$ can now be determined as

$$\left(\frac{d\lambda}{dt}\right)_{t=\Delta t} = V - Ri_1 \tag{21.7}$$

and the flux linkage at time $2\Delta t$ can be approximated by

$$\lambda_2 = \lambda_1 + \Delta t\left(\frac{d\lambda}{dt}\right)_{t=\Delta t} \tag{21.8}$$

Continued repetition of this calculation results in the data for the curves of flux linkage and current as functions of time, plotted in Fig. 21.4*c*.

This simple numerical method of solution is adequate for many nonlinear systems. Reference should be made to books on numerical analysis for more elaborate techniques that provide either a greater accuracy, fewer steps in the computation, or both.

Piecewise Linearization of the Normal Magnetization Curve. In many analyses the normal magnetization curve (Fig. 21.3) may be adequately represented by the *B-H* characteristic of Fig. 21.5. This consists of a linear portion in the range $-B_k < B < B_k$ having an unsaturated relative permeability μ_n and two linear portions for the ranges $|B| > B_k$, each having a slope $\mu_s\mu_0$ where μ_s is termed the saturated relative permeability. Within the unsaturated range, the flux linkage-current relationship of

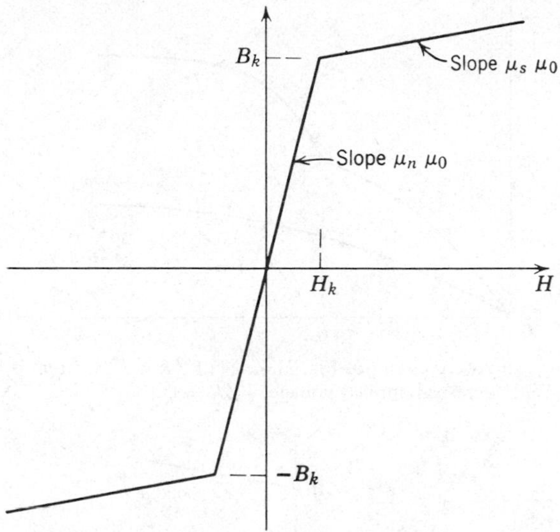

Fig. 21.5 Piecewise linearization of *B-H* curve.

a coil (such as that of Fig. 21.1) may be expressed as the unsaturated value of inductance L_n where, from Eq. (21.4)

$$L_n = \frac{\lambda}{i} = \frac{N^2 A}{\bar{l}} \mu_n \mu_0 \quad \text{H} \tag{21.9}$$

In this expression, A is the core area, $\bar{l}$ the mean length of the flux path, and N the number of turns. If the magnitude of the coil current i is less than $i_k = H_k \bar{l}/N$, the voltage-current relationship of the coil may be expressed by the linear differential equation

$$v = Ri + L_n \frac{di}{dt} \quad \text{V} \tag{21.10}$$

If the magnitude of the current exceeds i_k, the voltage-current relationship becomes

$$v = Ri + L_s \frac{di}{dt} \quad \text{V} \tag{21.11}$$

where

$$L_s = \frac{N^2 A}{\bar{l}} \mu_s \mu_0 \quad \text{H} \tag{21.12}$$

Figure 21.6 shows the current vs. time curve for the circuit of Fig. 21.4*a* for two values of applied voltage V. In the lower curve the current does not reach the value i_k; in the upper curve this value is exceeded. Since Eqs. (21.10) and (21.11) are of the first order and have constant coefficients, all terms of the solutions are simple exponentials. To arrive at the solution for case 2 in Fig. 21.6, Eq. (21.10) is used until $i = i_k$. The final conditions (i_k, t_k) of Eq. (21.10) are then used as the initial conditions in the solution of Eq. (21.11), which applies for $i > i_k$.

The process of simplifying the *B-H* relation may be carried further for those situations where the magnetic field intensity H is negligible as long as the flux density B does not approach its saturation value. The unsaturated relative permeability μ_n of Fig. 21.5 may then be set at infinity. This approximation, shown in Fig. 21.7*a*, proves to be very useful in the analysis of devices, such as saturable reactors, which operate far into the saturated region of the *B-H* curve.

Sometimes the further simplification of making the saturated relative permeability μ_s equal to zero is justified. Physically, we know that even a perfectly grain-oriented material cannot have a value of μ_s less than unity. This approximation, shown in Fig. 21.7*b*, is applicable in those situations where the saturated inductance L_s of Eq. (21.12) is negligible in relation to the other parameters of the system under analysis.

To demonstrate the use of these simple linearized models, consider the circuit of Fig. 21.8*a* in which the voltage $v = \hat{V} \cos \omega t$ is applied to a coil at $t = 0$. Suppose the λ-i characteristic of the coil is

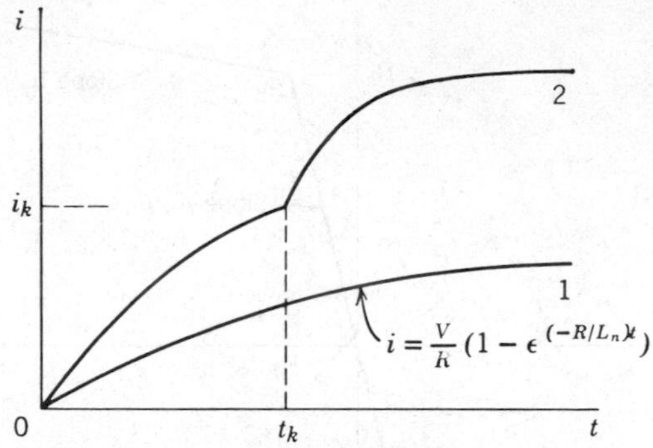

Fig. 21.6 Current-time curves for circuit of Fig. 21.4*a*. (1) $V/R < i_k$. Current is insufficient to reach saturated region. (2) With increased applied voltage, $V/R > i_k$.

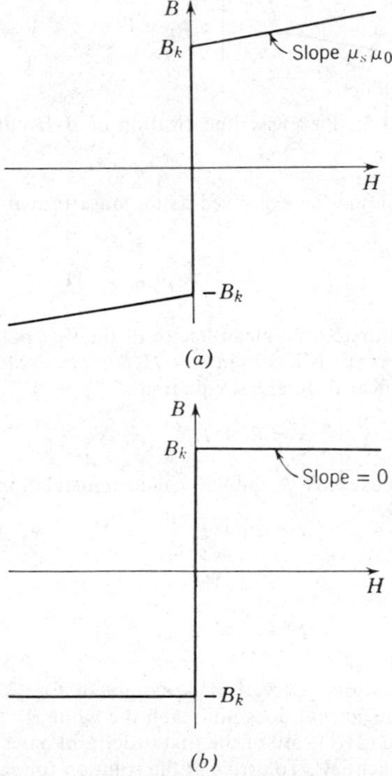

Fig. 21.7 (*a*) Linearization of *B-H* curves, similar to Fig. 21.6, with $\mu_n = \infty$. (*b*) $\mu_n = \infty$ and $\mu_n = 0$.

represented by the idealized model of Fig. 21.8*b*. The voltage-current relationship is given by the equation

$$Ri + \frac{d\lambda}{dt} = v$$

$$= \hat{V} \cos \omega t \qquad \text{V} \tag{21.13}$$

For $\lambda_k > \lambda > -\lambda_k$, the current i is zero and all the applied voltage in Eq. (21.13) is absorbed as a rate of change of flux linkage. If $\lambda = 0$ at $t = 0$, the flux linkage can be expressed initially as

$$\lambda = \int_0^t \hat{V} \cos \omega t \, dt$$

$$= \frac{\hat{V}}{\omega} \sin \omega t \qquad \text{Wb} \tag{21.14}$$

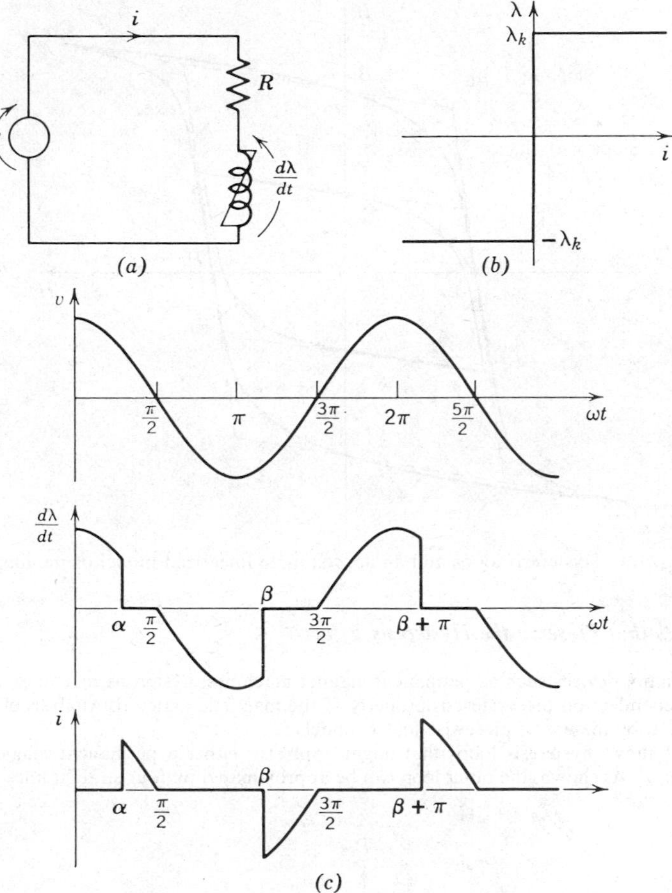

Fig. 21.8 (*a*) Circuit with $v = \hat{V} \cos \omega t$. (*b*) Idealized relation between flux linkage λ and current i. (*c*) Waveforms of v, $d\lambda/dt$, and i.

Equation (21.14) describes the flux linkage λ until $\omega t = \alpha$ where λ reaches its critical value λ_k. Thus

$$\alpha = \sin^{-1} \frac{\omega \lambda_k}{V} \tag{21.15}$$

Since no further increase in flux linkage can occur, $d\lambda/dt$ is zero for the rest of the interval during which v is positive. Thus for $\alpha < \omega t < \pi/2$

$$i = \frac{\hat{V}}{R} \cos \omega t \tag{21.16}$$

After $\omega t = \pi/2$, the flux linkage is given by

$$\lambda = \lambda_k + \int_{\pi/2\omega}^{t} v \, dt \tag{21.17}$$

until λ reaches its negative value of $-\lambda_k$ at $\omega t = \beta$. From this point until the end of the period of negative applied voltage, the current is again given by Eq. (21.16). The waveforms of the applied voltage V, the emf $d\lambda/dt$, and the current i are shown in Fig. 21.8*b*.

This example demonstrates one of the useful properties of a saturable magnetic core. If the core can be represented by the idealized *B-H* relationship of Fig. 21.7*b*, and the coil resistance negligible, all the applied voltage is absorbed by the coil until the core reaches saturation. The coil then becomes a short circuit, switching all the applied voltage to the element connected in series. This property is exploited in saturable reactors, magnetic amplifiers, and magnetic frequency multipliers.

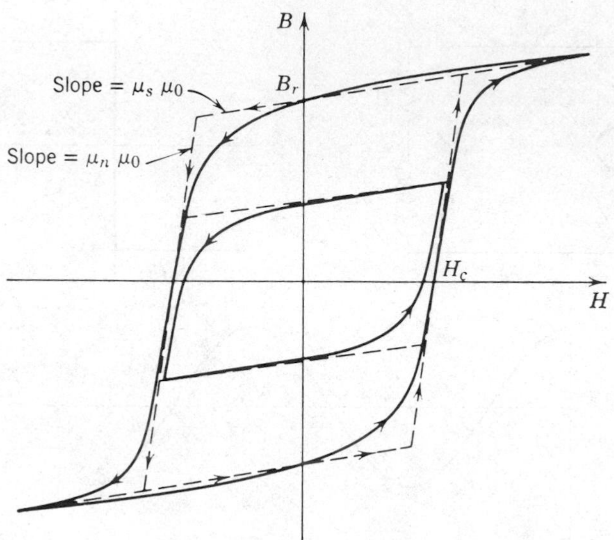

Fig. 21.9 Hysteresis loops and an approximate linearized model of the loops.

Linear Models that Preserve the Hysteresis Effect

Operation of many devices such as permanent-magnet machines, hysteresis machines, and magnetic amplifiers is dependent on the hysteresis property of the magnetic material. Analysis of these devices can be expedited by the use of piecewise linear models.

Figure 21.9 shows hysteresis loops that might apply for either a permanent magnetic or a soft magnetic material. As shown, the outer loop can be approximated by four straight lines, two of which

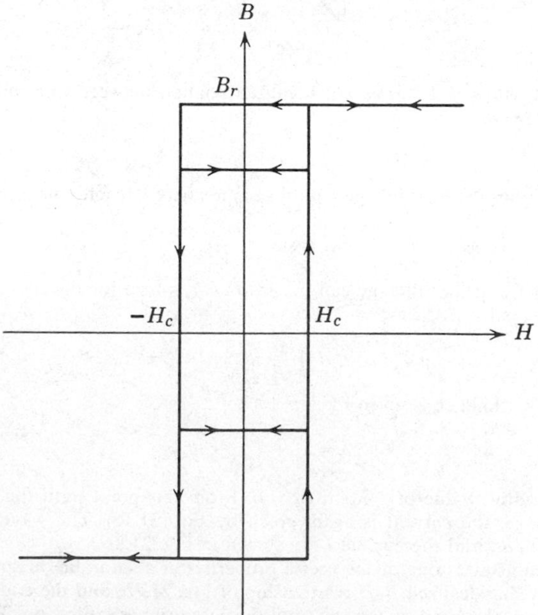

Fig. 21.10 Idealized model for the *B-H* loops of a grain-oriented material, including the hysteresis effect.

have a slope $\mu_n\mu_0$, the other two of which have a slope $\mu_s\mu_0$. It may be seen that the approximation can be reasonably accurate, except that the corners of the loop. The enclosed areas of the actual loop and its linearized model can be made approximately equal by a judicious choice of the model, resulting in a model having the same hysteresis loss as the actual loop.

As the maximum flux density is decreased, the width of the hysteresis loop often decreases only slightly. In such circumstances a linearized model having lines of the same slopes as for the major loop may be used to represent the smaller loops. One of these smaller loops with its linearized model is shown in Fig. 21.9.

With good grain-oriented magnetic materials, the sides of the hysteresis loop are essentially vertical, and the saturated portions approach an incremental slope of μ_0. It is often possible to represent such a loop with acceptable accuracy by the simple linearized model of Fig. 21.10. This model is similar to that of Fig. 21.7*b*, except that the hysteresis property has been retained. Any excursion in magnetic field intensity in the range $-H_c < H < H_c$ is assumed to occur along a horizontal locus of constant flux density, while a change of flux density is assumed to occur only at $H = \pm H_c$.

To demonstrate the use of this model, let us consider the system of Fig. 21.11*a*, which acts as a pulse counter. Suppose a number of irregularly spaced rectangular pulses of amplitude V and time

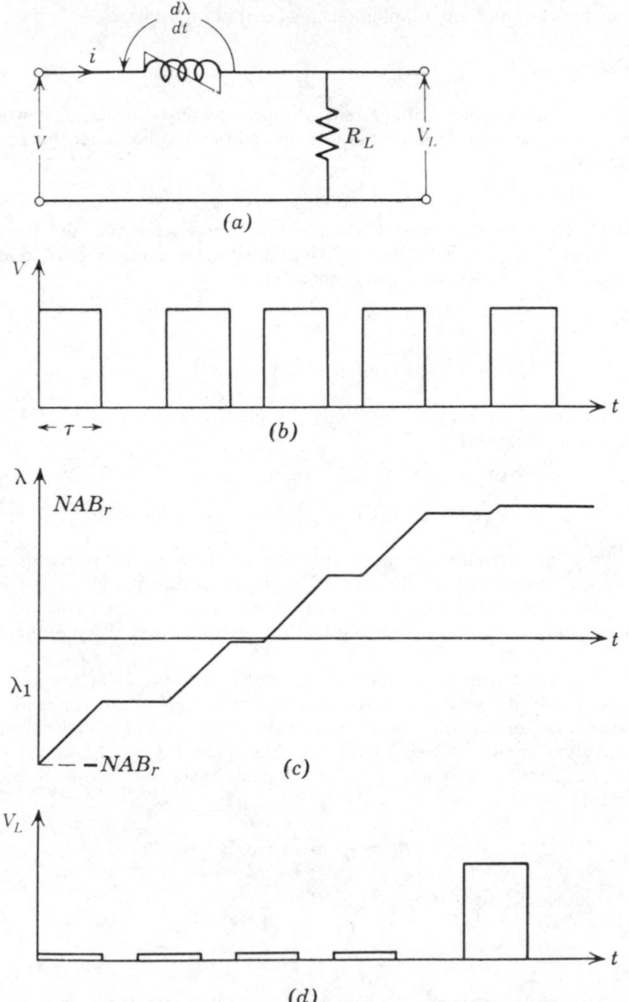

Fig. 21.11 (*a*) Pulse counter circuit. (*b*) Pulses of applied voltage. (*c*) Flux linkage. (*d*) Output voltage.

duration τ are applied to a load resistance R_L in series with a coil having a rectangular-loop core. The coil resistance is considered to be negligible in comparison with R_L. Suppose the core is initially set to have negative residual flux density B_r by use of a negative coil current. When a pulse of voltage is applied, the current i rises to the value

$$i_c = \frac{H_c \bar{l}}{N} \qquad \text{A} \tag{21.18}$$

where $\bar{l}$ is the mean path length of the core and N the number of coil turns. The rate of change of coil flux linkage λ is given by

$$\frac{d\lambda}{dt} = V - R_L i_c \tag{21.19}$$

Integrating Eq. (21.19) yields the flux linkage λ_1 at the end of the first pulse as

$$\lambda_1 = -NAB_r + (V - R_L i_c)\tau \qquad \text{Wb} \tag{21.20}$$

At the cessation of the pulse, the current and the magnetic field intensity return to zero. Subsequent pulses cause the flux linkage λ to increase by an amount $(V - R_L i_c)\tau$ for each pulse. When λ reaches its positive saturation value NAB_r, all the applied voltage appears across the load resistance R_L, as shown in Fig. 21.11*d*. By judicious choice of core dimensions and turns, this system gives an output pulse following the application of any number (such as ten) of input pulses.

Impedance Model

Most AC apparatus operates with a voltage that is approximately sinusoidal. It would therefore be useful to have simple representation for a nonlinear inductor operating under this condition. Suppose a sinusoidal voltage of

$$v = \hat{V} \sin \omega t \qquad \text{V} \tag{21.21}$$

is applied to a coil whose resistance is negligible and whose core has the flux linkage-current characteristic of Fig. 21.12*a*. This characteristic is obtained by rescaling a *B-H* loop of the material using Eqs. (21.2) and (21.3). The flux linkage of the coil is

$$\begin{aligned} \lambda &= \int e \, dt \\ &= \frac{\hat{V}}{\omega} \sin\left(\omega t - \frac{\pi}{2}\right) \qquad \text{Wb} \end{aligned} \tag{21.22}$$

The current i in the coil has the periodic but nonsinusoidal form shown in Fig. 21.12*b*. This current may be expressed as a Fourier series.

$$\begin{aligned} i &= \hat{I}_1 \sin(\omega t - \theta) + \text{odd harmonic terms} \\ &= \hat{I}_r \sin \omega t + \hat{I}_x \sin\left(\omega t - \frac{\pi}{2}\right) + \text{odd harmonic terms} \end{aligned} \tag{21.23}$$

In many analyses the harmonic components of the current can be ignored and only the fundamental-frequency component is preserved. One reason is that if the voltage is sinusoidal, the harmonic currents deliver no net power. If necessary, the behavior of the harmonics may be studied separately after a first approximation to the solution has been obtained using fundamental-frequency quantities only.

In Eq. (21.23) the fundamental component i_1 of the current i has been separated into a component i_r in phase with the voltage v and a component i_x lagging the voltage v by $\pi/2$ radians. The relation between the fundamental-frequency components of voltage and current can be represented by the equivalent circuit of Fig. 21.12*c*. In this figure V_1, I_1, I_r, and I_x are the phasors corresponding to v, i_1, i_r, and i_x, respectively. The hysteresis losses in the core are equal to the loss in the resistance R_h where

$$R_h = \frac{V_1}{I_r} = \frac{\hat{V}}{\hat{I}_r} \qquad \Omega \tag{21.24}$$

The magnetizing reactance X_0 is given by

$$X_0 = \frac{V_1}{I_x} = \frac{\hat{V}}{\hat{I}_x} \qquad \Omega \tag{21.25}$$

The hysteresis loop and waveforms of Fig. 21.12*a* and *b* apply for only one value of applied voltage. As the voltage is varied in magnitude, the equivalent circuit parameters R_h and X_0 also change. The reactance X_0 drops rapidly in value as the core enters its saturated region. The resistance

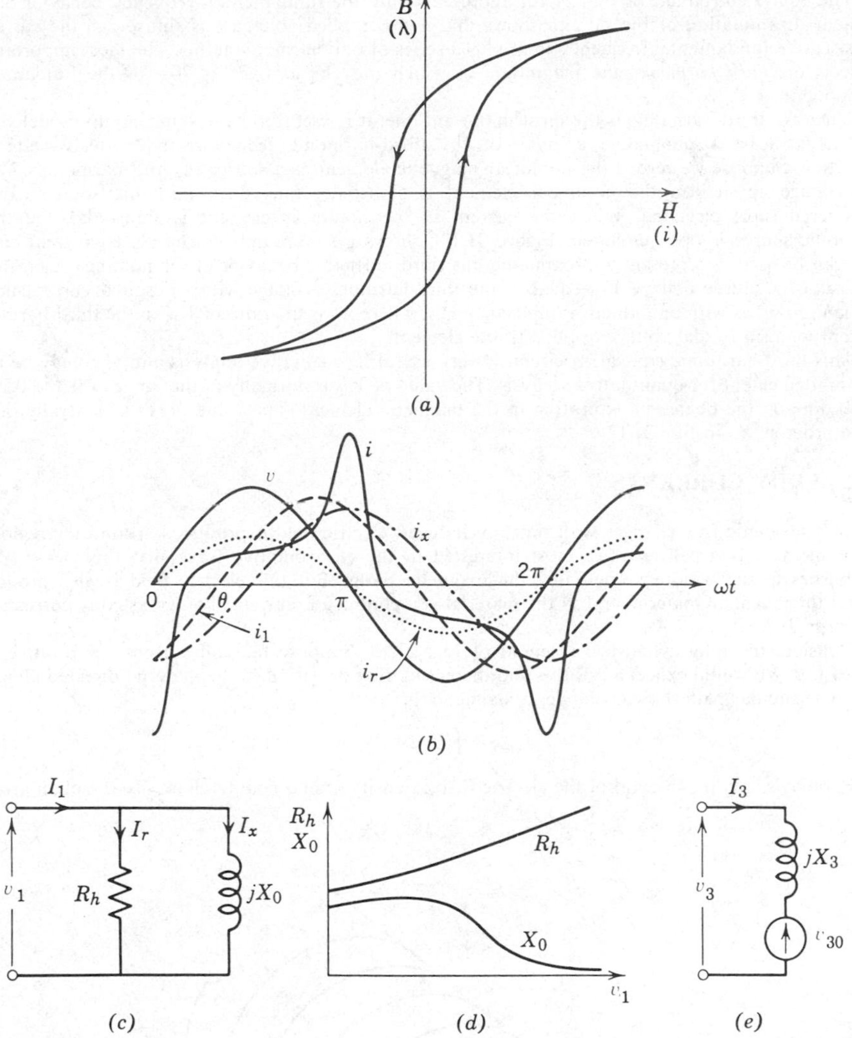

Fig. 21.12 (*a*) *B-H* or λ-*i* loop. (*b*) Waveforms of applied voltage v, current i, fundamental component i_1, in-phase component i_r, and quadrature component i_x. (*c*) Fundamental-frequency equivalent circuit. (*d*) Variation of equivalent circuit parameters with applied voltage V_1. (*e*) Third harmonic equivalent circuit.

R_h generally tends to rise with increasing applied voltage, indicating that the hysteresis losses are proportional to a power of the applied voltage which is less than 2. It may be shown that in a core having the idealized *B-H* loop of Fig. 21.10, the hysteresis losses at a given frequency are directly proportional to the applied voltage until saturation is reached. For this condition, R_h is directly proportional to V_1.

The values of the parameters R_h and X_0 could be derived from the *B-H* loops of the material as described in relation to Fig. 21.12. This process is very tedious. The parameters can usually be derived directly from published data of the power loss and reactive volt-amperes per unit of volume of the material when tested at constant frequency and variable sinusoidal flux density.

When using the equivalent circuit of Fig. 21.12*c* to represent a nonlinear magnetic element, the parameters R_h and X_0 should be adjusted to the values appropriate for the applied voltage V_1. If this voltage V_1 varies over only a small range of magnitude, it is often possible to assume R_h and X_0 to be constant at appropriate average values.

The equivalent circuit of Fig. 21.12*c* represents only the fundamental-frequency behavior of the element. Examination of Fig. 21.12*b* shows that if the applied voltage v is sinusoidal, the current i consists of a fundamental frequency term plus a series of odd harmonic terms. The most important of these is the *third harmonic*, the magnitude of which may be as high as 70% of the fundamental component.

Since the third harmonic is the most important one, it is useful to have some circuit model which permits at least a qualitative analysis of the third-harmonic behavior under steady-state AC operation. Suppose we regard the nonlinear magnetic element as a source of third harmonics. When the voltage applied to the nonlinear element is sinusoidal, this third-harmonic source can be considered short circuited. When the current in the nonlinear element is sinusoidal, the third-harmonic source is open circuited. Figure 21.12*e* shows a simple third-harmonic equivalent circuit that can be used to represent approximately the third-harmonic behavior of the nonlinear element. It consists of a source voltage V_{30} equal to the third-harmonic voltage with sinusoidal current in the element, in series with an inductive impedance jX_3, where X_3 is the ratio of V_{30} to the third-harmonic current with sinusoidal voltage applied to the element.

This third-harmonic equivalent circuit is very useful in qualitative analysis and also may be used to a limited extent for quantitative analysis. The value of V_{30} is normally in the range of 0.3 to 0.7 V_1, depending on the degree of saturation in the magnetic element. The value of X_3 is generally of the same order as X_0 in Fig. 21.12*c*.

21.2 EDDY CURRENTS

When a magnetic flux changes with time, an induced electric field is produced around the region of changing flux. Normally we are most interested in the electromotive force that this electric field establishes in the windings encircling the magnetic paths. But this electric field is also produced within the magnetic material and, if the material is a conductor, currents known as eddy currents are established.

Consider the long cylindrical solenoid of Fig. 21.13. Suppose the coil current i is positive and increasing. We would expect a positive and increasing flux density B in the material directed along its axis. The circular path shown encircles a magnetic flux ϕ of

$$\phi_{\text{encl.}} = \int \vec{B} \cdot d\vec{A} \qquad \text{Wb} \tag{21.26}$$

By Faraday's law, the integral of the electric field intensity $\vec{\mathscr{E}}$ in a counterclockwise direction around

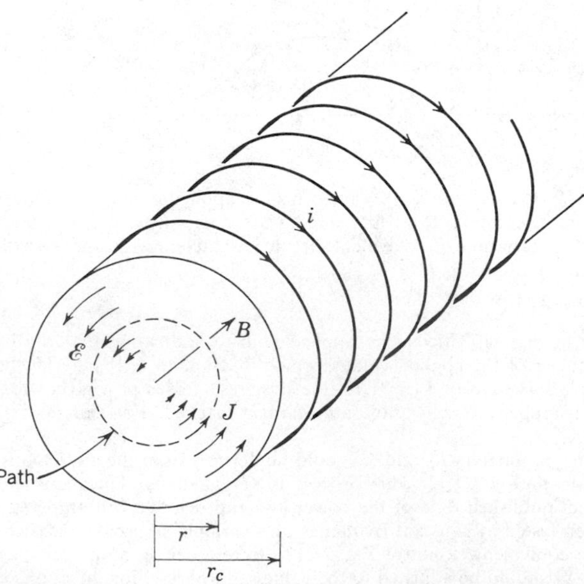

Fig. 21.13 Eddy currents in a magnetic core. (The current i is increasing with time.)

this path is equal to the rate of change of this flux. Because of the circular symmetry,

$$\mathscr{E} = \frac{1}{2\pi r}\frac{d\phi_{\text{encl.}}}{dt} \qquad \text{V/m} \tag{21.27}$$

If the material has a resistivity ρ, this electric field intensity sets up a colinear current density

$$J = \frac{\mathscr{E}}{\rho} \qquad \text{A/m}^2 \tag{21.28}$$

This eddy current density is in a direction that causes it to oppose the change in the enclosed magnetic flux. The opposition to the change in flux is greatest along the axis of the solenoid ($r = 0$), since all the eddy current encircles this axis. The effect becomes zero at the periphery of the solenoid where $r = r_c$.

Therefore one effect of eddy currents is to cause the time-varying magnetic flux density to be nonuniform within the material. An alternating magnetic flux tends to be concentrated toward the outside surface of the material, since the effect of eddy currents in preventing variation of magnetic flux is greatest near the central axis. This is known as the *magnetic skin effect*. If a magnetic material is to be used to best advantage, the magnetic flux density should be reasonably uniform over its cross-sectional area. Thus there is a practical limit to the thickness of a solid conducting magnetic material that should be used at any given frequency of operation.

A second effect of the eddy currents is to produce power loss in the material. The eddy current loss per unit of volume of material is

$$p = \rho J^2 \qquad \text{W/m}^3 \tag{21.29}$$

One means of controlling eddy current effects is the use of high-resistivity materials. Pure iron has a resistivity of about 10^{-7} Ω-m. The addition of about 4% silicon to the iron increases its resistivity to about 6×10^{-7} Ω-m. The ferrite materials are oxides rather than metals and have very high resistivities. For example, nickel-zinc ferrite has a resistivity of about 10^{-4} Ω-m.

With metallic magnetic materials, the principal means of controlling eddy currents is the use of thin sheets of laminations. Figure 21.14 shows how a toroidal core can be made from a long, thin strip of magnetic material. The surfaces of the material are covered with a thin insulating coating. When the magnetic flux in the core changes, an electric field is set up in the material, as in Fig. 21.13. But the resultant eddy currents cannot flow from layer to layer and are restricted to paths within the cross section of the strip.

We now derive an approximate expression for the eddy current losses within a laminated magnetic material. Figure 21.15 shows an enlarged cross section of the strip used in the toroidal core of Fig. 21.14. We assume that the eddy currents are not large enough to influence significantly the magnetic field within the lamination. The flux density is considered uniform. Consider the closed path shown within the lamination in Fig. 21.15. The sides of this path are at distance x from the centerline of the lamination. This path encloses a magnetic flux of

$$\phi_x = 2xyB \qquad \text{Wb} \tag{21.30}$$

Since $y \gg x$, the change of this magnetic flux may be assumed to produce an electric field of constant magnitude down one side of the path and up the other. By Faraday's law,

$$\mathscr{E}_x 2y = \frac{d\phi_x}{dt} \qquad \text{V} \tag{21.31}$$

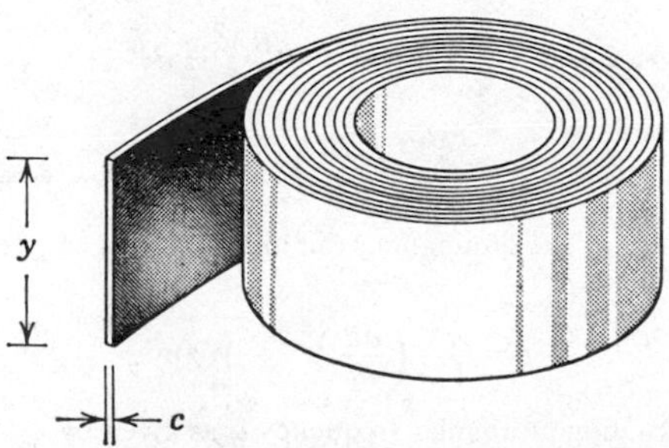

Fig. 21.14 Toroidal core made from a thin strip of material.

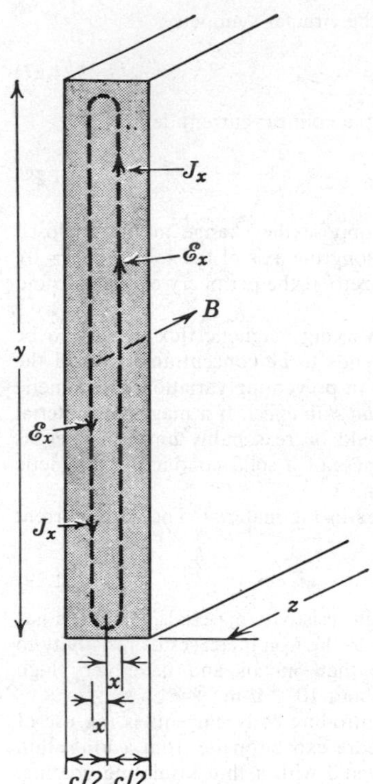

Fig. 21.15 Determination of eddy-current loss in a lamination.

Combining Eqs. (21.30) and (21.31) gives

$$\mathscr{E}_x = x\frac{dB}{dt} \qquad \text{V/m} \tag{21.32}$$

The current density at a distance x from the center plane of the lamination is therefore

$$J_x = \frac{\mathscr{E}_x}{\rho}$$

$$= \frac{x}{\rho}\frac{dB}{dt} \qquad \text{A/m}^2 \tag{21.33}$$

The total power loss in the lamination of thickness c, height y, and length z is found by integrating the loss density of Eq. (21.29) over the volume V of the lamination.

$$P = \int \rho J^2\, dV$$

$$= \int_{-c/2}^{c/2} \rho\left(\frac{x}{\rho}\frac{dB}{dt}\right)^2 yz\, dx$$

$$= \frac{c^3 yz}{12\rho}\left(\frac{dB}{dt}\right)^2 \qquad \text{W} \tag{21.34}$$

Averaged over the volume cyz of the lamination, the instantaneous eddy current power loss per unit volume is

$$p = \frac{c^2}{12\rho}\left(\frac{dB}{dt}\right)^2 \qquad \text{W/m}^3 \tag{21.35}$$

If the flux density B is alternating at angular frequency ω as given by

$$B = B \sin \omega t \tag{21.36}$$

the average eddy current power loss per unit volume is

$$\bar{p} = \frac{1}{2\pi}\int_0^{2\pi}\frac{c^2\omega^2B^2}{12\rho}\cos^2\omega t\, d\omega t$$

$$= \frac{c^2\omega^2B^2}{24\rho} \qquad \text{W/m}^3 \tag{21.37}$$

It is noted that this loss is proportional to the square of the lamination thickness, the square of the frequency, and the square of the maximum flux density. At power frequencies, laminations having a thickness of about 0.3 to 0.6 mm are often used. As the operating frequency is increased, the lamination thickness is generally reduced. For audio frequency apparatus, a lamination thickness of about 0.02 to 0.05 mm is used. For still higher frequencies, cores are often made of powdered iron and nickel molded to shape with an insulating adhesive.

In Section 21.1 an impedance model was developed to represent the AC properties of a coil with a ferromagnetic core. It is convenient to include the eddy current properties within this model. For a core of cross-sectional area A, mean length of flux path $\bar{l}$, lamination thickness c, and turns N, the total eddy current power loss is, from Eq. (21.35),

$$P = \frac{c^2}{12\rho}\left(\frac{dB}{dt}\right)^2 A\bar{l} \qquad \text{W} \tag{21.38}$$

The induced emf in the coil is

$$e = \frac{d\lambda}{dt}$$

$$= NA\frac{dB}{dt} \qquad \text{V} \tag{21.39}$$

By substituting from Eq. (21.39) into (21.38), the power loss can be expressed in terms of the induced voltage as

$$P = \frac{c^2}{12\rho}\,\frac{\bar{l}}{N^2A}\,e^2$$

$$= \frac{e^2}{R_e} \qquad \text{W} \tag{21.40}$$

Thus as shown in Fig. 21.16, the eddy current effect can be included in the model as a resistance

$$R_e = \frac{N^2A}{\bar{l}}\,\frac{12\rho}{c^2} \qquad \Omega \tag{21.41}$$

This resistance applies for any value of frequency for which the assumption of uniform flux density in the material is valid.

The resistances R_e and R_h in Fig. 21.16 representing eddy and hysteresis losses are normally combined into one loss element R_0. This resistance is normally determined from published values of total power loss per unit of volume or weight of material at a given frequency and flux density.

The foregoing analysis of eddy currents is based on an assumption of uniform flux density over the cross section of the material. This assumption may be reasonably valid from a macroscopic point of view. But a change in the average value of flux density actually results from a complex displacement of microscopic domain walls. The predictions of Eq. (21.37)—that eddy current losses vary as the square of the lamination thickness, the square of the frequency, and the square of the flux density—are often found to be inaccurate when compared with measured values of loss.

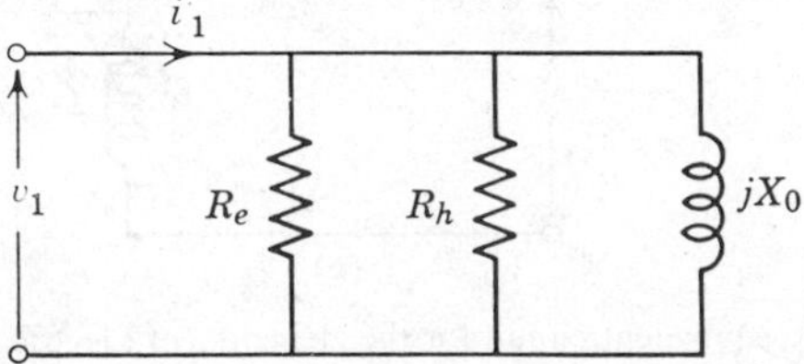

Fig. 21.16 Equivalent circuit for a magnetic element similar to that of Fig. 21.12c, but with resistance R_e to represent eddy current effect.

21.3 EQUIVALENT CIRCUITS FOR COMPLEX MAGNETIC SYSTEMS

When we encounter a complex system of electrical elements and wish to analyze its performance, our normal approach is to develop an electric equivalent circuit for the system. We put into the equivalent circuit only those parameters that are considered significant in influencing the performance. The equivalent circuit is thus a simplified mathematical model of the real system. Having developed an adequate equivalent circuit, we employ the well-known techniques of electric circuit analysis to determine its behavior. The accuracy with which the solution of the electric circuit behavior represents the performance of the real system is limited only by the adequacy of the equivalent circuit model.

In electrical machines, transformers and other electrical devices, ferromagnetic material is used in a wide variety of shapes. Various parts of multilimbed magnetic structures are encircled by coils. In this section we show how such complex magnetic systems may be represented approximately by equivalent circuits. The system is first represented by a magnetic equivalent circuit, which is then transformed into an equivalent electric circuit. The methods of electric circuit analysis may then be used.

Derivation of Magnetic Equivalent Circuits

Let us first derive a magnetic equivalent circuit for a simple magnetic system. Figure 21.17*a* shows a torus of magnetic material with a winding of N turns. The magnetomotive force around the magnetic path is

$$\mathscr{F} = Ni \qquad \text{A} \tag{21.42}$$

This magnetomotive force establishes a magnetic field intensity H, which in turn produces a magnetic flux density B in the material. Integration of this flux density over the cross-sectional area A of the core gives the magnetic flux ϕ. The cause-effect relationship between the magnetomotive force $\mathscr{F}$ and the magnetic flux ϕ may be expressed symbolically by the equivalent magnetic circuit of Fig. 21.17*b*. The magnetic properties of the material and the dimensions of the core determine its reluctance $\mathscr{R}$. Under the idealized condition where the relative permeability can be regarded as constant, that is,

$$B = \mu_r \mu_0 H \tag{21.43}$$

the reluctance can be expressed as

$$\mathscr{R} = \frac{\mathscr{F}}{\phi}$$

$$= \frac{\bar{l}}{\mu_r \mu_0 A} \qquad \text{A/Wb} \tag{21.44}$$

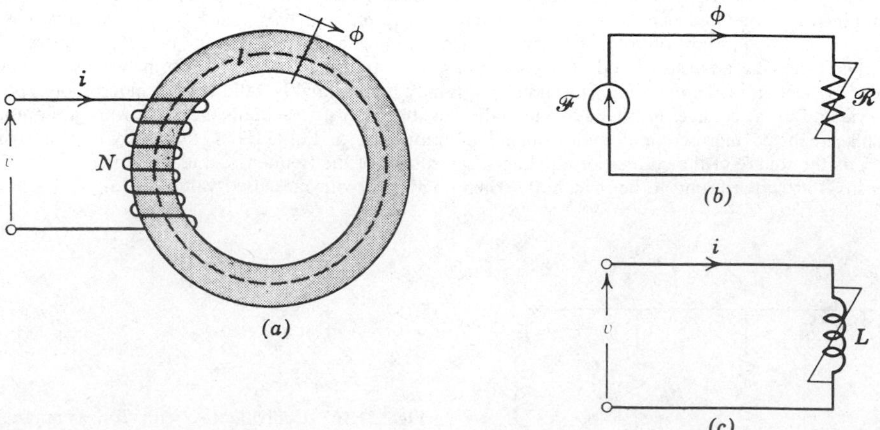

Fig. 21.17 (*a*) Simple magnetic element. (*b*) Magnetic equivalent circuit for the element. (*c*) Electric equivalent circuit for the element.

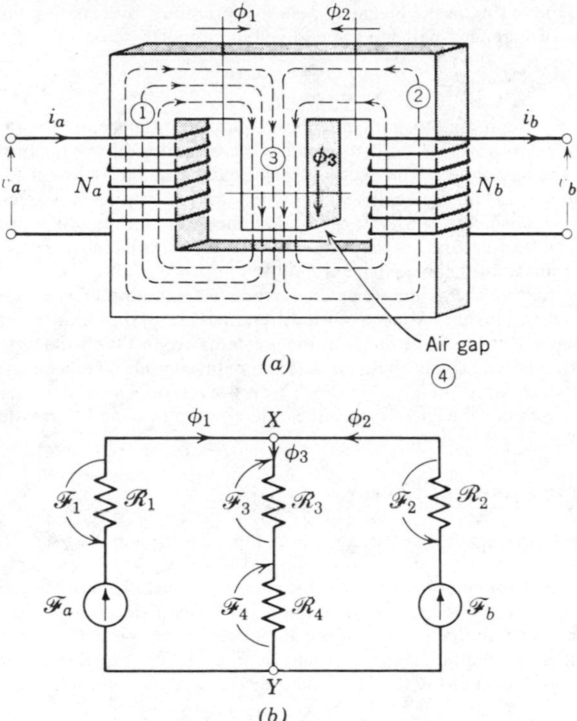

Fig. 21.18 (*a*) Magnetic system. (*b*) Magnetic equivalent circuit for the system.

In general the reluctance of a ferromagnetic core is nonlinear. The reluctance symbol is then used in a magnetic equivalent circuit merely to denote a magnetic element for which an $\mathscr{F}$-ϕ relation exists. This relation may be obtained by rescaling the *B-H* characteristic for the material using the expressions

$$\phi = BA \qquad \text{Wb} \tag{21.45}$$

and

$$\mathscr{F} = H\bar{l} \qquad \text{A} \tag{21.46}$$

For purposes of analysis, any of the approximate models for the *B-H* characteristic developed in Section 21.1 may be used.

Let us now consider, as an example, the magnetic system of Fig. 21.18*a*. This system consists of a three-legged magnetic core, two of the legs having windings and the third leg having an air gap. Basically, this is a complex three-dimensional magnetic field problem. But by the use of simplifying assumptions, the magnetic field can be reduced to a magnetic circuit of lumped reluctances. Let us assume that, except in the air gap, all magnetic flux is confined to the magnetic material. The leakage flux in the air paths around the windings is considered negligible.

The magnetic system may now be divided into four sections, each of which has a uniform flux over its length. Three of these sections represent magnetic paths in the material, and the fourth represents the air-gap path. Each section may be represented by a reluctance that relates the flux to the magnetomotive force required to establish that flux along the length of the section.

Figure 21.18*b* shows the magnetic equivalent circuit that results from the foregoing assumptions. Reluctances $\mathscr{R}_1$, $\mathscr{R}_2$, and $\mathscr{R}_3$ represent the three paths in the magnetic material carrying magnetic fluxes ϕ_1, ϕ_2, and ϕ_3, respectively. The air gap is represented by the linear reluctance $\mathscr{R}_4$, and its magnetic flux is ϕ_3. The circuital law applies to any closed path in magnetic field system. In a magnetic equivalent circuit this law is represented by the following relation: Around any closed path, the total magnetomotive force of the windings is equal to the sum of the products of reluctance and flux.

$$\sum \mathscr{F} = \sum \mathscr{R}\phi \qquad \text{around closed path} \tag{21.47}$$

The continuity of magnetic flux in the magnetic field is represented by equating the sum of the fluxes entering any junction of magnetic paths in the equivalent circuit to zero,

$$\sum_{\text{junction}} \phi_{\text{into}} = 0 \tag{21.48}$$

Normal methods of circuit analysis may now be employed to determine the fluxes in Fig. 21.18*b* for a given set of magnetomotive forces. If the relative permeability of the magnetic material can be considered constant, the reluctance of the sections may be determined by use of Eq. (21.44), in which $\bar{l}$ is the mean length of the flux path in each section and A is the cross-sectional area. If the permeability cannot be considered constant, each reluctance element may be represented by a graph of the relation between its flux and its magnetomotive force. Graphical or trial-and-error methods may then be used for analysis of the equivalent circuit.

It should be noted that all the assumptions are introduced in the process of deriving the magnetic equivalent circuit from the magnetic system. With different assumptions, a different equivalent circuit is obtained. For example, if the leakage fluxes in the air paths around the windings in Fig. 21.18*a* had not been neglected, the reluctances of these air leakage paths would have been connected across the respective magnetomotive forces in Fig. 21.18*b*. There is therefore no unique equivalent magnetic circuit for a magnetic system. The chosen circuit should contain just the information required for the problem to be solved.

Derivation of Electric Equivalent Circuits

A magnetic equivalent circuit, such as that shown in Fig. 21.18*b*, is most useful in the analysis and design of a device. However, if the device is connected to other electric elements, it is desirable to have an equivalent circuit for the device from which the relationships between the terminal voltages and currents can be obtained directly. In this section it is shown that the electric equivalent circuit can be derived directly and uniquely from the magnetic equivalent circuit.

First, let us consider the simple magnetic circuit of Fig. 21.17*b*, which relates two variables—the coil magnetomotive force $\mathscr{F}$ and the flux ϕ by the reluctance parameter $\mathscr{R}$.

$$\mathscr{F} = \mathscr{R}\phi \qquad \text{A} \tag{21.49}$$

In the equivalent electric circuit, the variables are the voltage v between the coil terminals and the current i in the coil. Let us assume that the core reluctance is constant and that the coil resistance is negligible. The electric circuit variables are related to the magnetic circuit variables by the two relations

$$i = \frac{\mathscr{F}}{N} \qquad \text{A} \tag{21.50}$$

and

$$v = N\frac{d\phi}{dt} \qquad \text{V} \tag{21.51}$$

By substituting from Eqs. (21.49) and (21.50) into Eq. (21.51), the relation between the two electric circuit variables is

$$\begin{aligned} v &= N\frac{d}{dt}\left(\frac{\mathscr{F}}{\mathscr{R}}\right) \\ &= \frac{N}{\mathscr{R}}\frac{d\mathscr{F}}{dt} = \frac{N^2}{\mathscr{R}}\frac{di}{dt} \\ &= L\frac{di}{dt} \qquad \text{V} \end{aligned} \tag{21.52}$$

Thus the reluctance parameter $\mathscr{R}$ in the magnetic circuit is replaced by an inductance parameter L in the electric equivalent circuit of Fig. 21.17*c*. The value of the inductance is inversely proportional to the value of the reluctance.

If the reluctance parameter is not constant, it can be represented by a curve relating magnetomotive force and flux. The corresponding nonlinear inductance can be represented by a curve relating the core flux linkage $\lambda = N\phi$ to the current i. Approximations such as those shown in Figs. 21.4*b*, 21.7, 21.8*b*, 21.9, and 21.10 may be used for this relation where appropriate.

Let us now consider the more complex magnetic circuit of Fig. 21.18*b*. Suppose, for the present, that each of the fluxes ϕ_1, ϕ_2, and ϕ_3 links an N-turn coil. The corresponding voltages v_1, v_2, and v_3 produced in these coils are given by

$$v_1 = N\frac{d\phi_1}{dt}, \qquad v_2 = N\frac{d\phi_2}{dt}, \qquad v_3 = \frac{d\phi_3}{dt} \tag{21.53}$$

At node X in Fig. 21.18b, the flux variables are related, from Eq. (21.48), by

$$\sum_{\substack{\text{into}\\ \text{node}}} \phi = \phi_1 + \phi_2 - \phi_3 = 0 \tag{21.54}$$

The flux variables at node Y are related by the same equation. From Eq. (21.53), the corresponding voltage variables must be related by the expression

$$v_1 + v_2 - v_3 = 0 \tag{21.55}$$

Now consider the left-hand mesh of the magnetic circuit of Fig. 21.18b. From Eq. (21.47), the magnetomotive force variables are related by

$$\mathscr{F}_a = \mathscr{F}_1 + \mathscr{F}_3 + \mathscr{F}_4 \tag{21.56}$$

Let us consider each of these magnetomotive force components to be produced by corresponding components of current in N-turn coils. These current components are then related by the expression

$$i'_a = i_1 + i_3 + i_4 \tag{21.57}$$

Around the right-hand mesh of Fig. 21.18b, the magnetomotive force relation is

$$\mathscr{F}_b = \mathscr{F}_2 + \mathscr{F}_3 + \mathscr{F}_4 \tag{21.58}$$

The relation between the corresponding current variables is

$$i'_b = i_2 + i_3 + i_4 \tag{21.59}$$

The prime symbols are added to i'_a and i'_b to distinguish them from i_a and i_b in Fig. 21.18a.

Each reluctance in the magnetic circuit relates a magnetic flux variable ϕ and a magnetomotive force variable $\mathscr{F}$. From Eqs. (21.50), (21.51), and (21.52), the corresponding voltage variable v and current variable i can be related by an inductance parameter. For example, the relation $\mathscr{F}_1 = \mathscr{R}_1\phi_1$ in the magnetic circuit corresponds to the relation

$$v_1 = L_1 \frac{di_1}{dt} \tag{21.60}$$

Equations (21.55), (21.57), (21.59), and (21.60) describe the electric equivalent circuit shown in Fig. 21.19a. For each of the two independent meshes in the magnetic circuit, there is an independent node in the electric circuit. The branch currents entering these two nodes (designated as A and B) are related by Eqs. (21.57) and (21.59). For each independent node in the magnetic circuit there is a corresponding mesh in the electric circuit. The branch voltages around the one independent mesh in the electric circuit are related by Eq. (21.55). For each reluctance branch in the magnetic circuit, there is a corresponding inductance branch in the electric circuit. For each magnetomotive force source in the magnetic circuit, there is a corresponding coil current in the electric circuit.

The form of the electric circuit of Fig. 21.19a may be derived directly from the magnetic circuit of Fig. 21.18b by use of the topological principle of duality. This topological technique is demonstrated in Fig. 21.19b. A node is marked within each mesh of the magnetic circuit, and a reference node is marked outside the circuit. These nodes are then joined by branches, one of which passes through each element of the magnetic circuit. It is observed that the form of the resulting network of branches is identical to the form of the electric circuit of Fig. 21.19a. For each reluctance in a mesh of the magnetic circuit, there is an inductance connected to the corresponding node of the electric circuit. Where a reluctance is common to two meshes in the magnetic circuit, the corresponding inductance interconnects the corresponding nodes of the electric circuit. For each magnetomotive force, there is a corresponding driving current; for each flux, there is a corresponding voltage between nodes.

The reluctances corresponding to ferromagnetic parts of the magnetic system may represent nonlinear relationships between their fluxes and magnetomotive forces. Each reluctance element in the magnetic circuit has a corresponding inductance element in the electric circuit. Thus each inductance element may represent a similar nonlinear relationship between the flux linkage λ in a coil of N turns encircling the particular branch of the magnetic system and the current i in an N-turn coil that produces the magnetomotive force for the branch. The rate of change of the flux linkage produces the voltage variable in the electric circuit. The nonlinearities in the magnetic circuit are therefore preserved in the electric equivalent circuit.

Any of the nonlinear models suggested by Figs. 21.4b, 21.7, 21.8b, 21.9, and 21.10 may be used to represent these nonlinear inductance elements. The choice depends on the problem under study. For use with alternating voltages, each nonlinear inductance element can be represented by a circuit model of the form shown in Fig. 21.16. This consists of an inductive reactance in parallel with a resistance representing the hysteresis and eddy current losses of the element. Both the reactance and the resistance are generally nonlinear functions of the voltage across the branch.

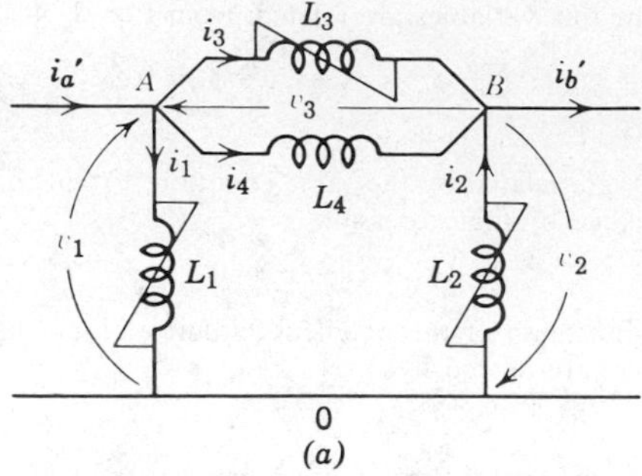

(a)

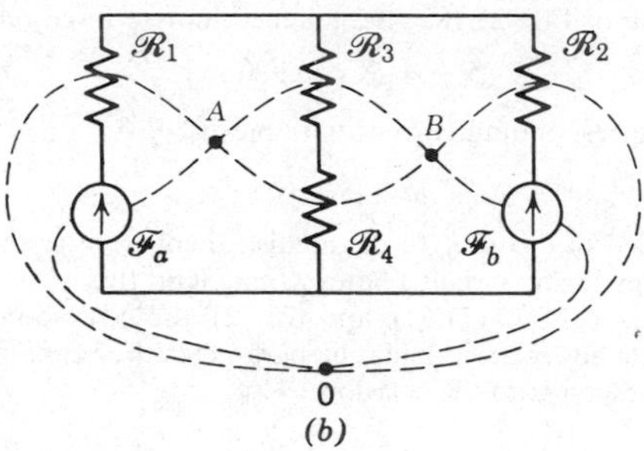

(b)

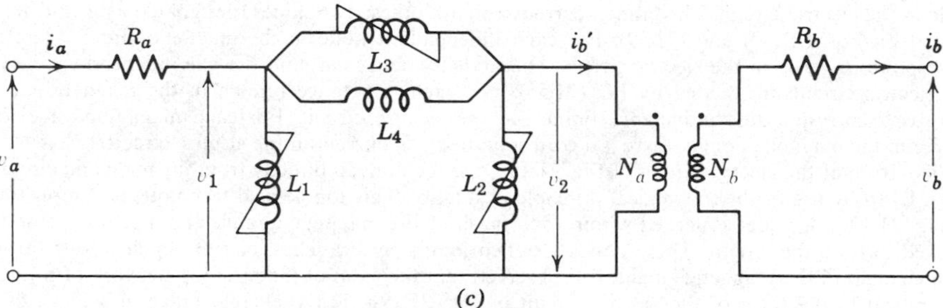

(c)

Fig. 21.19 Equivalent electric circuit for the magnetic circuit of Fig. 21.18: (*a*) Elementary form of circuit, (*b*) topological technique of derivation, (*c*) circuit including ideal transformer and winding resistances.

The electric equivalent circuit of Fig. 21.19*a* was developed with the assumption that all windings had N turns. Since the number of turns is generally different in the various windings, it is necessary to add ideal transformers at the terminals of the equivalent electric circuit to obtain the actual induced voltages and the actual currents in the windings. The reference number of turns N is normally made equal to the number of turns in one of the windings; no ideal transformer is then required for this winding. In Fig. 21.19*c*, N has been made equal to N_a.

The resistances R_a and R_b of the two windings have also been added to the equivalent circuit in Fig. 21.19*c*. The terminal voltages of the two windings are v_a and v_b; the induced voltages in the windings are v_1 and $v_2(N_b/N_a)$.

When an appropriate equivalent circuit has been developed for a magnetic device, the performance of the device may be predicted by use of the techniques normally employed in circuit analysis. When convenient, elements may be transformed across the ideal transformers.

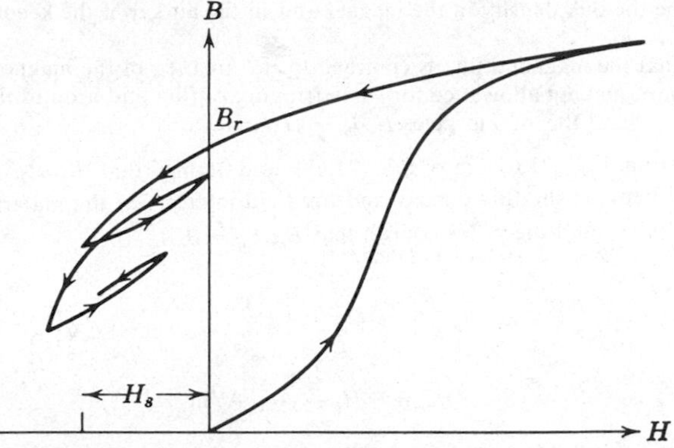

Fig. 21.20 B-H locus of permanent-magnet material during initial magnetization and subsequent application of demagnetization field intensity.

21.4 ANALYSIS OF PERMANENT-MAGNET SYSTEMS

A permanent-magnet material is one that can maintain a constant magnetic orientation of its domains in spite of a substantial externally applied field. Figure 21.20 shows the B-H locus that might be followed by a closed path of permanent-magnet material that is magnetized from an initially unmagnetized state. A large pulse of magnetic field intensity H is applied to the path, and on its removal the flux density remains at the residual value B_r. Suppose a reversed magnetic field intensity of magnitude H_s is now applied to the path. On its removal and reapplication, the B-H locus follows a minor loop, as shown. In most analyses this minor loop may be considered a single straight line. Application of a reversed magnetic field intensity of magnitude less than H_s causes an excursion along this line. If the reversed magnetic field intensity has a magnitude greater than H_s, the locus of operation moves down to a lower and more or less parallel line. The incremental slope of these lines representing minor loops is known as the *recoil permeability*. For Alnico magnets it is in the range of 3 to 5 μ_0, whereas for ferrite magnets it may be as low as 1.1 μ_0.

After a magnet has been initially magnetized, usual practice is to stabilize it by subjecting it to a demagnetizing magnetic field intensity H_s that is somewhat larger than the magnet is expected to encounter in service. As long as this demagnetizing intensity is not subsequently exceeded, the magnet should operate along an essentially straight-line minor loop.

Permanent-Magnet Systems with Air Gaps

Figure 21.21*a* shows a permanent magnet with an air gap. Suppose this magnet has been magnetized with a soft iron keeper in its air gap leaving it in the residually magnetized state denoted as *a* in Fig.

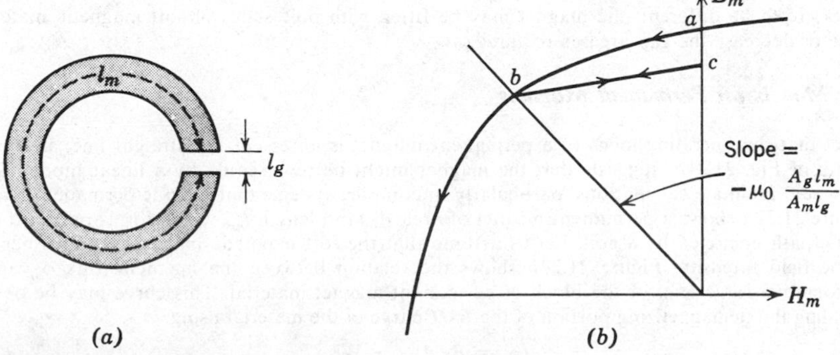

Fig. 21.21 (*a*) Permanent magnet with an air gap. (*b*) Graphical analysis of air-gap magnet.

21.21*b*. What will be the flux density in the magnet and in the air gap if the keeper is now withdrawn from the air gap?

Let us assume that the magnetic flux is confined to the area A_m of the magnet and to an effective area A_g of the air gap (making allowance for some fringing of flux and around the gap). Application of the circuital law around the system gives $H_m l_m + H_g l_g = 0$, or

By substituting from Eq. (21.61) into Eq. (21.62) and noting that $B_g = \mu_0 H_g$, we arrive at the following relation between the flux density and the field intensity in the material:

The continuity of flux around the path requires that $B_m A_m = B_g A_g$, or

$$B_m = B_g \frac{A_g}{A_m} \qquad \text{T} \tag{21.62}$$

$$H_m = -H_g \frac{l_g}{l_m} \qquad \text{A/m} \tag{21.61}$$

$$B_m = -\mu_0 \frac{A_g}{A_m} \frac{l_m}{l_g} H_m \qquad \text{T} \tag{21.63}$$

The second relation between B_m and H_m is the *B-H* curve of Fig. 21.21*b*. As shown, the operating point of the material is at the intersection of the *B-H* curve and the straight line representing Eq. (21.63). As the magnet keeper is removed, the operating point of the material moves along the locus *a-b*. If the keeper is reinserted, the operating point of the material moves along the essentially straight-line recoil locus *b-c*.

The foregoing analysis shows that the operating point of an air-gap magnet is determined by the demagnetization portion of its *B-H* loop and by the dimensions of the magnet. The corresponding design problem is to choose the operating point and dimensions of the material so that a given air-gap field may be produced with a minimum of magnet material. Suppose we wish to obtain a flux density B_g in an air gap of length l_g and cross-sectional area A_g. By using Eqs. (21.61) and (21.62), the required volume of magnet material is

$$\begin{aligned} V_m &= A_m l_m \\ &= \left(\frac{B_g A_g}{B_m}\right)\left(\frac{-H_g l_g}{H_m}\right) \\ &= \frac{B_g^2 V_g}{\mu_0 |B_m H_m|} \qquad \text{m}^3 \end{aligned} \tag{21.64}$$

Thus to produce a flux density B_g in an air gap of volume V_g, a minimum volume of magnet material is required if the material is opreated where the magnitude of the product $B_m H_m$ is greatest. This product is a measure of the energy that can be supplied per unit volume of material to an air-gap field. It is known as the energy product of the material and its value is in the range of 5000 to 50,000 J/m^3 in the better permanent-magnet materials.

When an operating point such as *b* in Fig. 21.21*b* has been chosen, the length l_m and area A_m of the magnet material may be chosen to make the intersection of lines occur at that point. In the simple magnet of Fig. 21.21*a*, the areas A_g of the air gap and A_m of the material are nearly equal. The air-gap flux density must then be essentially the same as the flux density in the material. If these flux densities are to be different, the magnet may be fitted with pole shoes of soft magnetic material to increase or decrease the gap area as required.

Linear Models for Permanent Magnets

The fact that the operating locus of a permanent magnet is an essentially straight line, as shown in locus *b-c* of Fig. 21.21*b*, suggests that the magnet might be represented by a linear model. Such a model would facilitate calculations, particularly in complex systems that include permanent magnets.

Figure 21.22*a* shows a permanent magnet of area A_m and length l_m, which forms part of a closed magnetic path encircled by a coil. Let us assume that the soft magnetic material requires negligible magnetic field intensity. Figure 21.22*b* shows the relation between the magnetic flux ϕ_m and the magnetomotive force $\mathcal{F}_m$ of the block of permanent-magnet material. This curve may be obtained by rescaling the demagnetizing portion of the *B-H* curve of the material using

$$\phi_m = B_m A_m \qquad \text{Wb} \tag{21.65}$$

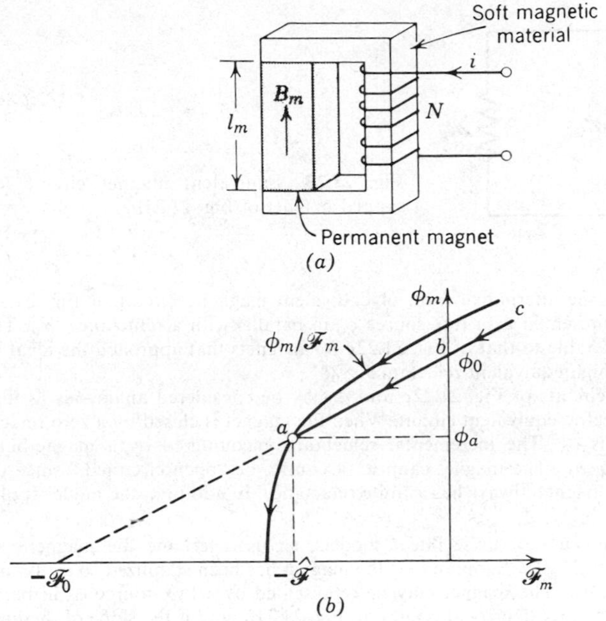

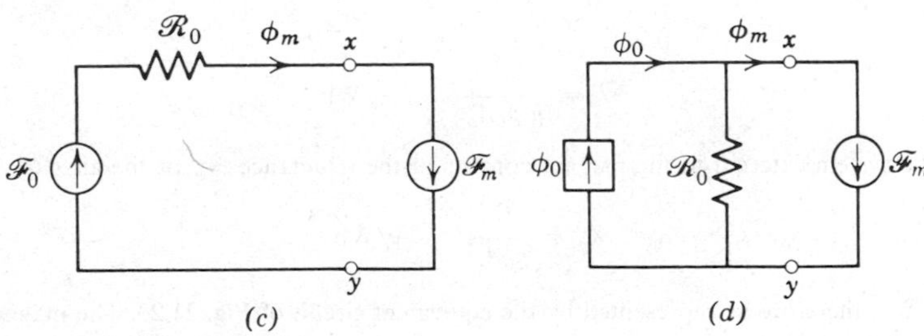

Fig. 21.22 (*a*) Closed system containing a permanent magnet and a coil. (*b*) Graphical analysis. (*c*) Equivalent circuit with magnetomotive force source. (*d*) Equivalent circuit with magnetic flux source.

and

$$\mathcal{F}_m = H_m l_m \qquad \text{A} \tag{21.66}$$

Suppose that the magnet is initially magnetized using a positive current i and then stabilized by application of a negative current sufficient to make $\mathcal{F}_m = Ni = -\hat{\mathcal{F}}$. This brings the operating point on the $\phi_m - \mathcal{F}_m$ locus to point a, where the magnet flux is ϕ_a. If, in the subsequent operation of this magnet, the magnitude of the magnetomotive force applied in the negative direction does not exceed $\hat{\mathcal{F}}$, the magnet operates along the locus a-b-c. This locus may be closely approximated by a straight line of slope $1/\mathcal{R}_0$ denoted by the expression

$$\mathcal{F}_m = -\mathcal{F}_0 + \mathcal{R}_0\phi_m \qquad \text{A (for } \mathcal{F}_m > -\hat{\mathcal{F}}) \tag{21.67}$$

This equation describes the equivalent magnetic circuit of Fig. 21.22*c*. The magnet is represented as a source of magnetomotive force $\mathcal{F}_0$ in series with a reluctance $\mathcal{R}_0$. The part of the system external to the magnet is simply a magnetomotive force $\mathcal{F}_m = Ni$ in this case.

Equation (21.67) may be divided through by $\mathcal{R}_0$ and rewritten in the form

$$\phi_m = \phi_0 + \frac{\mathcal{F}_m}{\mathcal{R}_0} \qquad \text{Wb (for } \phi_m > \phi_a) \tag{21.68}$$

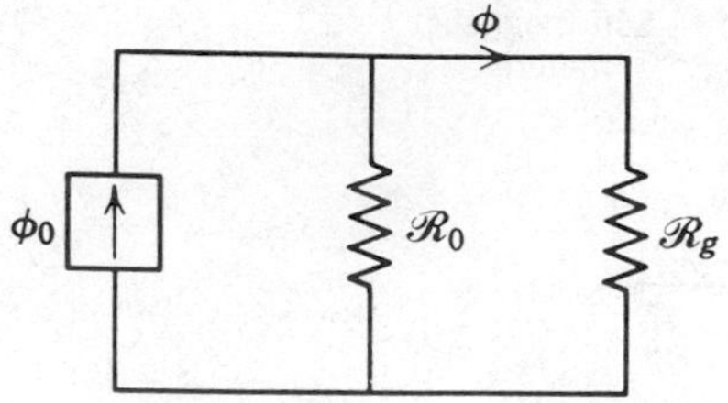

Fig. 21.23 Equivalent magnet circuit for the air-gapped magnet of Fig. 21.21*a*.

This equation describes the alternative form of equivalent magnetic circuit of Fig. 21.22*d*. In this circuit the magnet is represented as a flux source ϕ_0 in parallel with a reluctance $\mathscr{R}_0$. This form of equivalent circuit is preferable to that of Fig. 21.22*c* for magnets that approach the ideal behavior of constant flux and very high equivalent reluctance $\mathscr{R}_0$.

The two equivalent circuits of Fig. 21.22*c* and *d* may be considered analogous to the Thévenin and Norton forms of electric equivalent circuit. When the magnet is closed by a zero reluctance path, its "short-circuit" flux is ϕ_0. The incremental reluctance encountered by a magnetomotive force applied to the magnet is $\mathscr{R}_0$. The magnet cannot, of course, be "open circuited," since the air path between the ends of the magnet always has a finite reluctance. In addition, the model applies only for $\mathscr{F}_m > -\hat{\mathscr{F}}$.

As an example of the use of these linear models, let us determine the magnetic flux in the air-gapped magnet of Fig. 21.21*a*. Suppose that the magnet has been stabilized so as to operate along the locus *b-c* in Fig. 21.21*b*. The magnet may be represented by a flux source ϕ_0 in parallel with a reluctance $\mathscr{R}_0$. If B_0 is the flux density at point *c* in Fig. 21.21*b*, and if the slope of the line *b-c* is the recoil permeability $\mu_r\mu_0$,

$$\phi_0 = B_0 A_m \qquad \text{Wb} \tag{21.69}$$

and

$$\mathscr{R}_0 = \frac{l_m}{\mu_r \mu_0 A_m} \qquad \text{A/Wb} \tag{21.70}$$

The magnetic system external to the magnet consists of the reluctance $\mathscr{R}_g$ of the air gap, where

$$\mathscr{R}_g = \frac{l_g}{\mu_0 A_g} \qquad \text{A/Wb} \tag{21.71}$$

The system may therefore be represented by the equivalent circuit of Fig. 21.23. The magnetic flux ϕ in the air gap is given by

$$\phi = \frac{\mathscr{R}_0}{\mathscr{R}_0 + \mathscr{R}_g}\phi_0 \qquad \text{Wb} \tag{21.72}$$

Bibliography

Boast, W. B., *Principles of Electric and Magnetic Fields*, Harper and Row, New York, 1948. (Includes chapters on flux plotting and estimation of reluctances.)

Cunningham, W. J., *Introduction to Nonlinear Analysis*, McGraw-Hill, New York, 1958.

Katz, H. W., *Solid State Magnetic and Dielectric Devices*, Wiley, New York, 1959.

Parker, R. J. and R. J. Studders, *Permanent Magnets and Their Applications*, Wiley, New York, 1962.

Slemon, G. R., "A Method of Approximate Steady-State Analysis for Nonlinear Networks," *Proc. Inst. Elec. Eng.* **100(1)**:275 (1953).

Stanton, R. G., *Numerical Methods for Science and Engineering*, Prentice-Hall, Englewood Cliffs NJ. 1961.

Further Readings

Chang, S. S. L., *Energy Conversion*, Prentice-Hall, Englewood Cliffs NJ, 1963.

Dekker, A. J., *Electrical Engineering Materials*, Prentice-Hall, Englewood Cliffs, NJ, 1959.

Hadfield, D., *Permanent Magnets and Magnetism*, Swift Levick, London, and Wiley, New York, 1962.

Halliday, D. and R. Resnick, *Physics for Students of Science and Engineering*, Part II, Wiley, New York, 1962.

Ham, J. M. and G. R. Slemon, *Scientific Basis of Electrical Engineering*, Wiley, New York, 1961.

Jackson, W., *The Insulation of Electrical Equipment*, Chapman and Hall, London, 1954.

Messerle, H. K., "Dynamic Circuit Theory," *Trans. Am. Inst. Elec. Eng.* **59**:567 (1960).

Morrish, A. H., *The Physical Principles of Magnetism*, Wiley, New York, 1965.

Newhouse, V. L., *Applied Superconductivity*, Wiley, New York, 1964.

Roters, H. C., *Electromagnetic Devices*, Wiley, New York, 1941.

Skilling, H. H., *Electromechanics: A First Course in Electromechanical Energy Conversion*, Wiley, New York, 1962.

Sproull, R. L., *Modern Physics*, 2nd ed., Wiley, New York, 1963.

Whitehead, S., *Dielectric Breakdown of Solids*, Oxford University Press, London, 1953.

CHAPTER 22
COMPUTER-AIDED CIRCUIT ANALYSIS

MILTON ROSENSTEIN

New York Institute of Technology
Old Westbury, New York

22.1 INTRODUCTION

The use of a computer dramatically extends the engineer's ability to analyze (and design) complex networks. Though theoretically all networks can be solved by hand, drudgery and the probability of error make large-scale analytic solutions impractical. The computer removes these limitations and can effect a qualitative as well as quantitative change.

In virtually all modern engineering environments, engineers have access to computer facilities, either a main frame, a microcomputer, or both. Increasingly sophisticated network-analysis software is becoming available, as is improved hardware. Engineers have no choice but to integrate these tools into their own work. The reward is a significant increase in productivity, accuracy, and creativity.

The major problems with integrating computer technology into the workplace are intelligent use of available software and efficient use of the computer. Though some sophisticated network-analysis programs can be used by naive users, such use can be as dangerous as use of incompletely understood formulas. Also, use of large programs for small tasks can be inefficient. For most effective use of the computer, engineers must fully understand the programs they use, must use large commercial network-analysis programs when these are appropriate, and must compose their own programs when required. Consequently this section emphasizes the concepts and principles that underlie computer analysis rather than presenting a detailed description of available software.

22.2 COMPUTER ANALYSIS: DC NETWORKS

Computer analysis of multinode networks (DC or AC) is based on matrix algebra. Typically, in commercial software information is entered into the computer that permits it to form an incidence matrix describing the topology of the network, a branch conductance matrix describing the elements in each branch, and branch source vectors describing the current and voltage sources in each branch. These results are then used to establish the nodal conductance and excitation matrices. These matrices in turn are solved by matrix algebra, giving the desired circuit solutions.

Often the data-entry procedures of commercial software are tedious and slow, paced apparently for the least knowledgeable user. The engineer can usually meet the computer at a higher level, entering circuit matrices in toto instead of piecemeal. This is one of the advantages of a deeper understanding of computer network analysis.

Incidence Matrix

The incidence matrix describes the network topology. It is a matrix representation of the branches between nodes and gives no information about the makeup of each branch. In effect the matrix represents the circuit as an oriented graph, specifying the orientation of each branch and the nodes to which the branch is incident.

For a graph having n nodes and b branches, the complete incidence matrix A is an $n \times b$ rectangular matrix defined as follows:

$$a_{ij} = \begin{matrix} 1 & \text{if branch } j \text{ incident at node } i \text{ and oriented away} \\ -1 & \text{if branch } j \text{ incident at node } i \text{ and oriented toward} \\ 0 & \text{if branch } j \text{ not incident at node } i \end{matrix}$$

In circuit terms, orientation may be interpreted as the assumed direction of current flow. Figure 22.1 shows a circuit, its topology, and two matrices: a complete incidence matrix and an incidence matrix. Since each column of the complete incidence matrix sums to zero, all rows are not independent. The last row of Fig. 22.1c can be omitted without loss of information. The complete incidence matrix less one row (usually the ground node row) is called the incidence matrix (Fig. 22.1d).

Some important circuit relations (Kirchhoff's laws) may be verified in terms of the incidence matrix. Kirchhoff's current law (currents at a node sum to zero) is expressed by

$$\mathbf{A}i = 0 \tag{22.1}$$

and is verified as follows using Fig. 22.1.

$$\begin{bmatrix} -1 & 1 & 1 & 0 & 0 & 0 \\ 0 & -1 & 0 & 1 & 1 & 0 \\ 0 & 0 & -1 & -1 & 0 & 1 \end{bmatrix} \begin{bmatrix} i_1 \\ i_2 \\ i_3 \\ i_4 \\ i_5 \\ i_6 \end{bmatrix} = \begin{bmatrix} 0 \\ 0 \\ 0 \\ 0 \\ 0 \\ 0 \\ 0 \end{bmatrix} \tag{22.2}$$

We can also convert the node voltages v_n to branch voltages v_b, measured in respect to the ground or

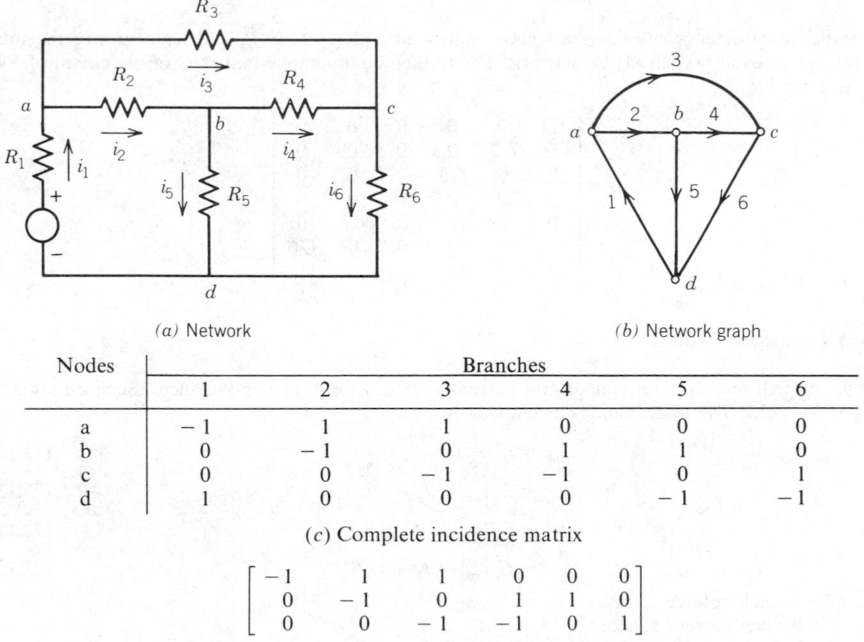

(a) Network

(b) Network graph

Nodes	Branches					
	1	2	3	4	5	6
a	−1	1	1	0	0	0
b	0	−1	0	1	1	0
c	0	0	−1	−1	0	1
d	1	0	0	0	−1	−1

(c) Complete incidence matrix

$$\begin{bmatrix} -1 & 1 & 1 & 0 & 0 & 0 \\ 0 & -1 & 0 & 1 & 1 & 0 \\ 0 & 0 & -1 & -1 & 0 & 1 \end{bmatrix}$$

(d) Incidence matrix

Fig. 22.1 Derivation of incidence matrix: (a) network, (b) network graph, (c) complete incidence matrix, (d) incidence matrix.

datum node by

$$\mathbf{V}_b = \mathbf{A}^T\mathbf{V}_n \tag{22.3}$$

and verified as follows

$$\mathbf{A}^T = \begin{bmatrix} -1 & 0 & 0 \\ 1 & -1 & 0 \\ 1 & 0 & -1 \\ 0 & 1 & -1 \\ 0 & 1 & 0 \\ 0 & 0 & 1 \end{bmatrix} \tag{22.4a}$$

Then applying Eq. (22.3) we get

$$\mathbf{V}_b = \begin{bmatrix} -1 & 0 & 0 \\ 1 & -1 & 0 \\ 1 & 0 & -1 \\ 0 & 1 & -1 \\ 0 & 1 & 0 \\ 0 & 0 & 1 \end{bmatrix} \begin{bmatrix} v_{n1} \\ v_{n2} \\ v_{n3} \end{bmatrix} \tag{22.4b}$$

which equals

$$v_{b1} = -v_{n1} \tag{22.5}$$

$$v_{b2} = v_{n1} - v_{n2} \tag{22.6}$$

$$v_{b3} = v_{n1} - v_{n3} \tag{22.7}$$

$$v_{b4} = v_{n2} - v_{n3} \tag{22.8}$$

$$v_{b5} = v_{n2} \tag{22.9}$$

$$v_{b6} = v_{n3} \tag{22.10}$$

Branch Conductance Matrix

The branch conductance matrix is a square matrix of order $b \times b$. Its elements specify the total conductance in each branch of the network. The branch conductance matrix G of the circuit of Fig. 22.1a is given by

$$\mathbf{G} = \begin{bmatrix} G1 & 0 & 0 & 0 & 0 & 0 \\ 0 & G2 & 0 & 0 & 0 & 0 \\ 0 & 0 & G3 & 0 & 0 & 0 \\ 0 & 0 & 0 & G4 & 0 & 0 \\ 0 & 0 & 0 & 0 & G5 & 0 \\ 0 & 0 & 0 & 0 & 0 & G6 \end{bmatrix} \tag{22.11}$$

where $G = 1/R$.

Branch Excitation Matrix

A simple branch may contain voltage and current sources as well as conductances. Such a branch is shown in Fig. 22.2. The total branch current (i_{bk}) is given by

$$i_{bk} = i_{ek} - J_k \tag{22.12}$$

$$= G_k v_{ek} - J_k \tag{22.13}$$

$$= G_k(v_{bk} + E_k) - J_k \tag{22.14}$$

$$= G_k v_{bk} - J_k + G_k E_k \tag{22.15}$$

where E_k = branch voltage source
J_k = branch current source
i_{bk} = branch current
v_{bk} = branch voltage
as shown in Fig. 22.2.

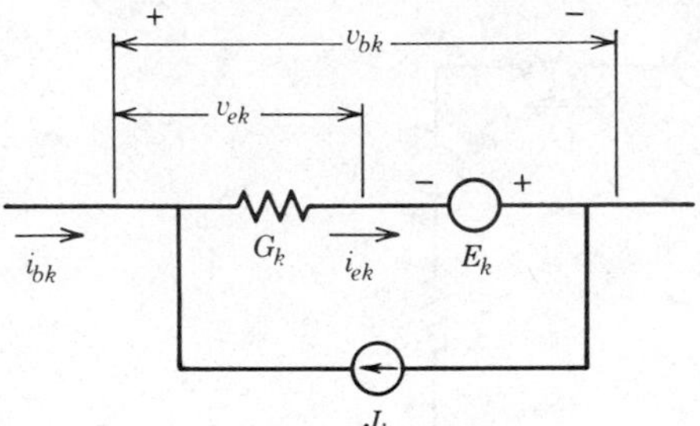

Fig. 22.2 A simple branch.

In matrix notation Eq. (22.15) becomes

$$\mathbf{I}_b = \mathbf{GV}_b - \mathbf{J} + \mathbf{GE} \tag{22.16}$$

where $\mathbf{I}_b$ = column matrix of branch currents
$\mathbf{E}$ = column matrix of branch voltage sources
$\mathbf{G}$ = branch conductance matrix
$\mathbf{J}$ = column matrix of branch current sources

Solving for Node Voltages

Equations 22.1, 22.3, and 22.16 can be combined to yield the solution in terms of node voltages (Jensen and Watkins, 1974). This result is given by

$$\mathbf{V}_n = (\mathbf{AGA}^T)^{-1}\mathbf{A}(\mathbf{J} - \mathbf{GE}) \tag{22.17}$$

Recognizing $\mathbf{AGA}^T$ as the nodal conductance matrix $\mathbf{G}_n$ and $\mathbf{A}(\mathbf{J} - \mathbf{GE})$ as the equivalent current vector $\mathbf{J}_n$, Eq. (22.17) becomes

$$\mathbf{V}_n = \mathbf{G}_n\mathbf{J}_n \tag{22.18}$$

Hand solutions of nodal networks usually begin with Eq. (22.18). It is evident that if $\mathbf{G}_n$ is easily derived, a considerable amount of computer manipulation can be saved by starting with $\mathbf{G}_n$ and $\mathbf{J}_n$ and using the computer to do the tedious matrix inversion and multiplication.

Example 22.1 Complete Computer Solution. Given the circuit of Fig. 22.3*a*, find the node voltages V_1 and V_2. (This example is done by hand.) From the oriented graph (Fig. 22.3*b*), the incidence matrix **A** is determined.

$$\mathbf{A} = \begin{bmatrix} -1 & 1 & 1 & 0 \\ 0 & 0 & -1 & 1 \end{bmatrix}$$

Next the branch conductance matrix **G** is found

$$\mathbf{G} = \begin{bmatrix} 1 & 0 & 0 & 0 \\ 0 & 3 & 0 & 0 \\ 0 & 0 & 1 & 0 \\ 0 & 0 & 0 & 2 \end{bmatrix}$$

The branch source vectors are given by

$$\mathbf{J} = \begin{bmatrix} 0 \\ 1 \\ 0 \\ 0 \end{bmatrix} \qquad \mathbf{E} = \begin{bmatrix} 0 \\ 0 \\ 5 \\ 10 \end{bmatrix}$$

The nodal conductance matrix **G** is found

$$\mathbf{G} = \mathbf{AGA}^T = \begin{bmatrix} -1 & 1 & 1 & 0 \\ 0 & 0 & -1 & 1 \end{bmatrix} \begin{bmatrix} 1 & 0 & 0 & 0 \\ 0 & 3 & 0 & 0 \\ 0 & 0 & 1 & 0 \\ 0 & 0 & 0 & 2 \end{bmatrix} \begin{bmatrix} -1 & 0 \\ 1 & 0 \\ 1 & -1 \\ 0 & 1 \end{bmatrix}$$

$$\mathbf{G} = \begin{bmatrix} 5 & -1 \\ -1 & 3 \end{bmatrix}$$

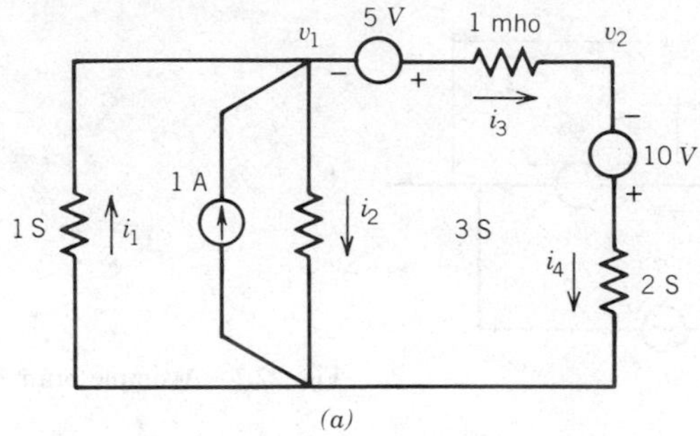

(a)

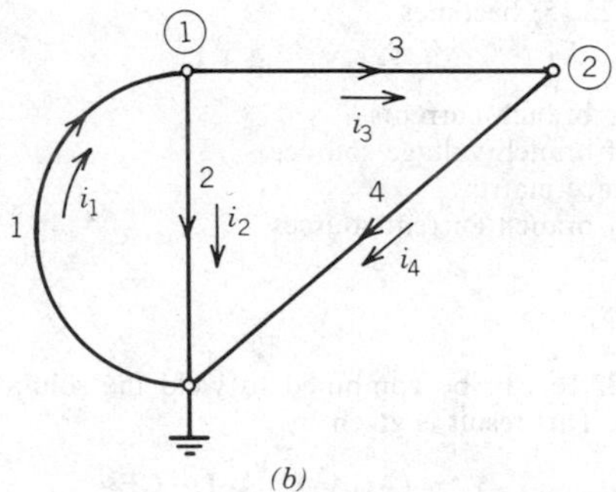

(b)

Fig. 22.3 Circuit for Example 22.1: (a) circuit, (b) oriented graph.

The equivalent current vector $\mathbf{J}_n$ is computed from $\mathbf{J}_n = \mathbf{A}(\mathbf{J} - \mathbf{GE})$.

$$\mathbf{J}_n = \begin{bmatrix} -1 & 1 & 1 & 0 \\ 0 & 0 & -1 & 1 \end{bmatrix} \left[\begin{bmatrix} 0 \\ 1 \\ 0 \\ 0 \end{bmatrix} - \begin{bmatrix} 1 & 0 & 0 & 0 \\ 0 & 3 & 0 & 0 \\ 0 & 0 & 1 & 0 \\ 0 & 0 & 0 & 2 \end{bmatrix} \begin{bmatrix} 0 \\ 0 \\ 5 \\ 10 \end{bmatrix} \right]$$

$$\mathbf{J}_n = \begin{bmatrix} -4 \\ -15 \end{bmatrix}$$

Finally the node voltages are computed from

$$\mathbf{V}_n = \mathbf{G}_n \mathbf{J}_n$$

$$1/14 \times \begin{bmatrix} 3 & 1 \\ 1 & 5 \end{bmatrix} \begin{bmatrix} -4 \\ -15 \end{bmatrix} = \begin{bmatrix} -1.92 \\ -5.64 \end{bmatrix}$$

General Branch

The general branch (Fig. 22.4) includes a dependent current source J_{dk} and a dependent voltage source E_{dk}. J_{dk} is a current introduced into branch k by a voltage in branch j and is given by

$$J_{dk} = gm_{kj} V_{ej} \tag{22.19}$$

The effect of branch j on branch k is represented by an off-diagonal transconductance gm_{kj} introduced into the branch conductance matrix at position k, j. Similarly E_{dk} is a voltage in branch k due to a voltage in branch j and is given by

$$E_{dk} = \mu_{kj} V_{ej} \tag{22.20}$$

Performing a Norton's transformation on the branch containing G_k and E_{dk} gives

$$Je_{dk} = \mu_{kj} G_k V_{ej}$$

This effect can be represented by introducing an off-diagonal $\mu_{kj} G_k$ into the branch conductance matrix at position k, j. Subsequently the procedure is as shown in Example 22.1.

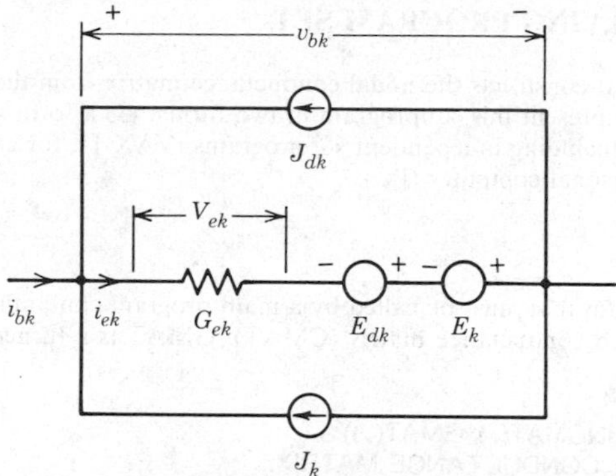

Fig. 22.4 A general branch.

Example 22.2. For the circuit of Fig. 22.5, determine the branch conductance matrix.

The branch conductance matrix is composed of the branch conductance matrix of Fig. 22.3 plus the two off-diagonal elements introduced by the two dependent sources $gm_{2,1}V_1$ and $\mu_{4,3}G_2V_3$ as shown in the following.

$$\mathbf{G}_1 = \begin{bmatrix} 1 & 0 & 0 & 0 \\ 2 & 3 & 0 & 0 \\ 0 & 0 & 1 & 0 \\ 0 & 0 & 1.5 & 2 \end{bmatrix}$$

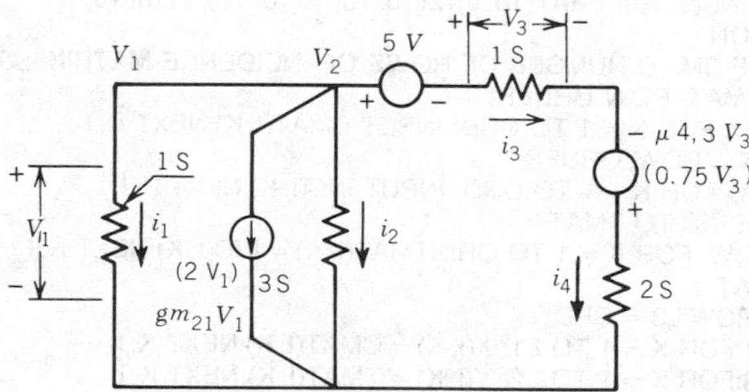

Fig. 22.5 Circuit with dependent sources.

Computer Solutions

The solution of the general DC network is given by

$$\mathbf{V}_n = (\mathbf{A}\mathbf{G}_1\mathbf{A}^T)^{-1}\mathbf{A}(\mathbf{J} - \mathbf{G}_1\mathbf{E}) \tag{22.21}$$

This solution involves several matrix operations: matrix multiplication, transposition, subtraction, and inversion. Of the four, matrix inversion is the most time consuming, with matrix multiplication next. Matrix inversion and multiplication can be simplified by using special numeric techniques. For the computer to operate efficiently these special techniques are essential.

Available numeric procedures for solving matrix systems include the Gauss elimination procedure, the Gauss–Jordan procedure, LU factorization, and others. Analysis of these techniques is beyond the scope of this chapter (see Jensen and Watkins, 1974 Chapter 7; Calahan, 1972 Chapter 2). However, as an example of an automated network solver, we present a set of subprograms that can be used either separately or in combination to solve networks.

22.3 NETWORK-SOLVING PROGRAM SET

BNET is a subprogram that constructs the nodal conductance matrix from the incidence and branch conductance matrices. We present this subprogram in two forms: (1) a form suitable for a computer supporting BASIC PLUS including independent subprograms (VAX 11, for example) and (2) a form suitable for the typical personal computer (PC).

Form 1

This version is a subprogram that must be called by a main program containing the incidence matrix (INCMAT) and the branch conductance matrix (CMAT). GMAT is returned as the nodal conductance matrix.

```
10 SUB BNET(CMAT(, ), INCMAT(, ), GMAT(, ))
20 REM BUILDS NODAL CONDUCTANCE MATRIX.
30 REM TRANSPOSES INCIDENCE MATRIX; RESULT PLACED IN INCT
40   MAT INCT = TRN(INCMAT)
50 REM CMAT × INCT (G × A^-1) RESULT PLACED IN TEMP
60   MAT TEMP = CMAT * INCT
70 REM INCMAT × TEMP (A × G × TRN(A))
80 MAT GMAT = INCMAT * TEMP
90   SUBEND
```

Form 2

This version is a full program that prints out the nodal conductance matrix. It must be chained or combined with other programs to produce a complete network solver.

```
10 REM BNET: BUILDS NODAL CONDUCTANCE MATRIX
20 REM CMAT; BRANCH CONDUCTANCE MATRIX. INC; INCIDENCE MATRIX.
30 REM TMAT; TRANSPOSED INCIDENCE MATRIX. GMAT; NODAL CONDUCTANCE MATRIX
35 DIM CMAT(10, 10), INC(9, 10), TMAT(10, 9), Z(10, 10), X(10, 10), Y(10, 10)
40 REM INPUT SECTION
50   INPUT"ORDER OF CMAT, NUMBER OF ROWS OF INCIDENCE MATRIX"; ORD, ROW
60   PRINT "ENTER CMAT; ROW ORDER"
70   FOR I = 1 TO ORD: FOR K = 1 TO ORD: INPUT CMAT(I, K):NEXT K, I
80   PRINT "ENTER INC; ROW, ORDER"
90   FOR I = 1 TO ROW:FOR K = 1 TO ORD: INPUT INC(I, K):NEXT K, I
100 REM TRANSPOSE INC TO TMAT
110   FOR I = 1 TO ROW: FOR K = 1 TO ORD:TMAT(K, I) = INC(I, K):NEXT K, I
120 REM CMAT × TMAT
130   L1 = ORD:L2 = ROW:L3 = ORD
140   FOR I = 1 TO L1:FOR K = 1 TO L1: X(I, K) = CMAT(I, K):NEXT K, I
150   FOR I = 1 TO L3:FOR K = 1 TO L2: Y(I, K) = TMAT(I, K):NEXT K, I
160   GOSUB 1000
170 REM INC × (CMAT × TMAT)
180   FOR I = 1 TO L2:FOR K = 1 TO ORD:X(I, K) = INC(I, K):NEXT K, I
190   FOR I = 1 TO L2:FOR K = 1 TO L2:Y(I, K) = Z(I, K):NEXT K, I
200 GOSUB 1000
210 REM ON LEAVING SUB 1000, Z CONTAINS GMAT
220   FOR I = 1 TO L2:FOR K = 1 TO L2: PRINT Z(I, K); : NEXT K, I
230 END


1000 REM SUBROUTINE: INNER MULTIPLICATION
1010 REM L1; COLUMNS OF MAT1. L2; ROWS OF MAT2. L3 = L1
1020 REM X; MAT1. Y; MAT2. Z; RESULT
1030   FOR I = 1 TO L1: FOR K = 1 TO L2
1040   SUM = 0
1050     FOR J = 1 TO L3: SUM = SUM + X(I, J) * Y(J, K): NEXT J
1060   Z(I, K) = SUM
1070   NEXT K, I
1080   RETURN
```

XMAT is a subprogram that computes the total excitation matrix $\mathbf{J}_n$ (current vector) from the branch excitations $\mathbf{J}_k$ and $\mathbf{E}_k$. BASIC PLUS and PC versions are presented. For the BASIC PLUS version the main program must supply the incidence, branch conduction, and excitation matrices. The PC program is complete in itself.

BASIC PLUS Version.

```
10 SUB XMAT(JMAT(, ), EMAT(, ), INCMAT(, ), JNMAT(, ), CMAT(, ))
20 REM SOLVES Jn = A(J - GE)
30   MAT TEMP1 = CMAT * EMAT
40   MAT TEMP2 = JMAT - TEMP1
50   MAT JNMAT = INCMAT * TEMP2
60   SUBEND
```

PC Program. The PC program is shown in Fig. 22.6. Input consists of the branch conductance, incidence, branch current, and branch voltage matrices. Output is the equivalent current vector.

Once the nodal conductance matrix $\mathbf{G}_n$ and the equivalent current vector are known, the circuit can be solved for node voltages $\mathbf{V}_n$. The following BASIC PLUS subprogram finds the node voltages by inverting the $\mathbf{G}_n$ matrix and multiplying $\mathbf{J}_n$ by it.

```
10 SUB REALMAT(GN(, ), JN(, ), VN(, ))
20 REM GN:BRANCH COND.MAT, JN:CURRENT VECTOR:VN:NODE VOLTAGES
30 REM INVERT GN
40 MAT TEMP = INV(GN)
50 REM SOLVE
60 MAT VN = TEMP * JN
70 SUBEND
```

The PC version (Fig. 22.7) uses the Gauss–Jordan method to solve the matrix set. $\mathbf{J}_n$ is added to $\mathbf{G}_n$, producing an augmented matrix of order n by $n + 1$. The $\mathbf{G}_n$ section of the augmented matrix is reduced to an identity matrix (a square matrix with all main diagonal elements = 1 and all

```
10 REM COMPUTES EQUIVALENT CURRENT VECTOR:
20 DIM INC(9,10),GMAT(10,10),JMAT(10,1), EMAT(10,1), Z(10,1)
25 REM INC:INCIDENCE MAT. G:BRANCH COND. MAT. JMAT:BRANCH CURRENT MAT
26 REM JN:EQUIV. CURRENT VECTOR
30 REM INPUT SECTION
40 INPUT"ORDER,NO.-NODES,NO.-BRANCHES";XOD,NN,NB
50 PRINT"INCIDENCE MAT":FORI=1 TO NN:FOR K=1 TO NB:INPUT INC(I,K):NEXTK,I
60 PRINT"COND. MAT":FORI=1 TO XOD:FOR K=1TO XOD:INPUT G(I,K):NEXT K,I
70 PRINT"CURRENT EXCIT. MAT":FORI=1 TO XOD:INPUT JMAT(I,1):NEXT I
75 PRINT"VOLTAGE EXCIT.":FOR I=1 TO XOD:INPUT EMAT(I,1):NEXT I
80 REM COMPUTE JN=INC(JMAT-G*EMAT)
85 REM G*EMAT
90 FOR I=1 TO XOD:FOR K=1 TO XOD:X(I,K)=G(I,K):NEXT K,I
100 FOR I=1TO XOD:Y(I,1)=EMAT(I,1):NEXT I:L1=XOD:L2=1:L3=XOD
110 GOSUB 1000
120 REM JMAT-G*EMAT
130 FORI=1 TO XOD:JMAT(I,1)=JMAT(I,1)-Z(I,1):NEXT I
140 REM INC*REST
150 FOR I=1 TO NN:FOR K=1 TO NB:X(I,K)=INC(I,K):NEXT K,I
160 FOR I=1 TO XOD:Y(I,1)=JMAT(I,1):NEXT I:L1=NN :L2=1:L3=XOD
170 GOSUB 1000
180 PRINTOUT
190 FOR I=1 TO NN:PRINT Z(I,1):NEXT I
200 END
1000 REM SUBROUTINE MAT MULTIPLICATION
1010 REM X:MAT 1,Y:MAT 2, Z:RESULT
1020 FOR I=1 TO L1: FOR K=1 TO L2:SUM=0
1030 FOR J=1 TO L3:SUM=SUM+X(I,J)*Y(J,K):NEXT J
1040 Z(I,K)=SUM:NEXT K,I
1050 RETURN
```

Fig. 22.6 Equivalent current vector program.

```
10 REM REALMAT:SOLVES REAL MATRIX SETS
20 DIM REAL(4,5)
30 INPUT"ORDER";XOD
40 PRINT"ENTER AUGMENTED MATRIX"
50 FOR I=1 TO XOD:FOR K=1 TO XOD+1:INPUT REAL(I,K):NEXTK,I
60 GOSUB 1000
70 PRINT"NODE VOLTAGE MATRIX"
80 FOR I=1 TO XOD:PRINTREAL(I,XOD+1):NEXT I
90 END
1000 REM SUBROUTINE: COMPUTES NODE VOLTAGES
1010 REM CHECK-MAIN DIAGONAL ZERO
1020 FOR I=1 TO XOD:AR=REAL(I,I):IF AR=0 THEN PRINT"MAIN DIAG-0":RETURN
1040 REM GAUSS-JORDAN ALGORITHM
1050 FOR J=1 TO XOD+1:REAL(I,J)=REAL(I,J)/AR:NEXT J
1060 REM SKIPS ZERO ELEMENTS
1070 FOR K=1 TO XOD:IF K-I=0 GOTO 1100
1080 BR=REAL(K,I)
1090 FOR J=1 TO XOD+1:REAL(K,J)=REAL(K,J)-BR*REAL(I,J):NEXT J
1100 NEXT K
1110 NEXT I
1120 REM RESULT IN LAST COLUMN OF MATRIX-REAL
1130 RETURN
```

Fig. 22.7 Personal computer (PC) program REALMAT.

```
5 REM SOLVES COMPLEX MATRICES--COMPMAT DISCA
10 DIM  R(4,5), M(4,5)
20 INPUT "ORDER";LIM
30 PRINT"REAL AND IMAG IMMIT.LAST COL.EXCIT."
40 FORI=1TOLIM:FORK=1TOLIM+1: INPUT R(I,K):NEXTK,I
70 FORI=1TOLIM:FORK=1TOLIM+1: INPUT M(I,K):NEXTK,I
90 GOSUB 1000
95 PRINT"      REAL MAT RESULT"
100 FORI=1TOLIM:PRINT: FORK=1TOLIM+1:PRINTR(I,K);:NEXTK,I
105 PRINT
110 PRINT"   IMAG MAT RESULT"
120 FORI=1TOLIM:PRINT:FORK=1TOLIM+1:PRINTM(I,K);:NEXTK,I
130 END
1000 REM COMPMAT SOLVES COMPLEX MATRICES
1010 REM CHECKS FOR MAIN DIAGONAL 0
1020 FLAG=1
1030 FORI=1TOLIM:AR=R(I,I):AI=M(I,I):AA=(AR^2+AI^2)
1040 IFAA=0 THENPRINT"0 MAIN DIAG":FLAG=0:RETURN
1050 REM GAUS-JORDAN ALGORITHM
1060 FORJ=1TOLIM+1
1070 R(I,J)=(R(I,J)*AR+M(I,J)*AI)/AA
1080 M(I,J)=M(I,J)*AR/AA-R(I,J)*AI/AR+M(I,J)*(AI^2)/(AA*AR)
1090 NEXT J
1100 FORK=1TOLIM
1110 IF(K-I)=0THEN1165
1120 BR=R(K,I):BI=M(K,I)
1130 FORJ=1TOLIM+1
1140 R(K,J)=R(K,J)-(BR*R(I,J)-BI*M(I,J))
1150 M(K,J)=M(K,J)-(BR*M(I,J)+BI*R(I,J))
1160 NEXTJ
1165 NEXTK
1167 NEXTI
1170 REM REAL -LAST COLUM OF MAT R,IMAG-LAST COL MAT M
1180 RETURN
```

Fig. 22.8 Personal computer (PC) program COMPMAT.

off-diagonal elements = 0). Since

$$\mathbf{J}_n = \mathbf{G}_n \mathbf{V}_n$$

then

$$\mathbf{X} = \mathbf{I}\mathbf{V}_n = \mathbf{V}_n \tag{22.22}$$

where $\mathbf{I}$ is the identity matrix and $\mathbf{X}$ is the $\mathbf{J}_n$ matrix transformed by operations converting $\mathbf{G}_n$ to $\mathbf{I}$ (Calahan, 1972; Jacquez, 1970).

The procedures described thus far for the analysis of DC networks hold for steady-state AC networks as well with a few additional considerations. The branch conductance matrix becomes a complex branch admittance matrix $\mathbf{Y}_n(j\omega)$. This results in a possibly complex and frequency-sensitive nodal admittance matrix given by

$$\mathbf{Y}_n(j\omega) = [\mathbf{A}\mathbf{Y}_1(j\omega)\mathbf{A}^T] \tag{22.23}$$

The equivalent current vector $\mathbf{J}_n(j\omega)$ may also be complex and frequency sensitive. It is given by

$$\mathbf{J}_n(j\omega) = \mathbf{A}[J - \mathbf{Y}_1(j\omega) \times \mathbf{E}] \tag{22.24}$$

where the source vectors $\mathbf{J}$ and $\mathbf{E}$ may be complex and frequency sensitive. The incidence matrix $\mathbf{A}$ is unchanged over the DC case. In addition, mutual inductive coupling may exist, resulting in off-diagonal elements in $\mathbf{Y}_1$ without the presence of dependent sources. These complicating factors mean that networks can only be solved (by computer) for one $j\omega$ at a time and that a language supporting complex arithmetic is desirable. See Jensen and Watkins (1974) for an AC analysis program using complex arithmetic.

However, it is possible to handle complex matrices in a language that does not directly support it. Figure 22.8 shows a PC program (in BASIC) that solves AC networks. The nodal admittance matrix and the equivalent current vectors are entered as two augmented matrices, one real and one containing the imaginary elements. A modification of the Gauss–Jordan method is then applied.

22.4 SENSITIVITY CHECKING

The sensitivity of an output parameter (usually voltage) to perturbations in the values of internal components is an important circuit measure. This is defined by the sensitivity coefficient S given by

$$S = \partial v_0 / \partial P_j \tag{22.25}$$

where v_0 is the output voltage and P_j is any circuit component.

At first blush, it appears that computer usage makes feasible a straightforward procedure merely requiring computation of v_0 for small internal changes of component values by the methods already discussed. This method, known as the perturbation method, is practical for small networks but becomes impractical for larger networks. Two, more efficient methods, are available. One, involving matrix differentiation based on the equations developed previously (Jensen and Watkins, 1974), is mathematically messy but conceptually simple. The second, which we find more congenial, is based on Tellegen's theorem (Calahan, 1972). Here we discuss only the second method.

Tellegen's Theorem

Tellegen's theorem (Tellegen, 1952) is defined by

$$\sum_{k=1}^{n_b} i_k \hat{v}_k = 0 \qquad \sum_{k}^{n_b} \hat{i}_k v_k = 0 \tag{22.26}$$

where n_b = number of branches in each of two circuits (a, b)

i_k, v_k = branch currents and voltages of circuit "a"

$\hat{i}_k, \hat{v}_k$ = branch currents and voltages of circuit "b"

The theorem applies to any two circuits with identical topologies. A special case occurs when the theorem is applied to a single circuit ($i_k = \hat{i}_k$ and $v = \hat{v}_k$). Then

$$\sum_{k=1}^{n_b} i_k v_k = 0 \tag{22.27}$$

which is the expression of conservation of instantaneous power in a network.

Application to Sensitivity Analysis. Consider two networks (a, b) with identical graphs, i_k, v_k being the variables of "a"; $\hat{i}_k$, $\hat{v}_k$ of "b." If we perturb the is and the vs in a, Tellegen's theorem holds, giving

$$(v_k \textit{ perturbed}) \qquad \sum_k [v_k(t) + dv_k(t)]\hat{i}_k(t) = 0 \tag{22.28}$$

and

$$(i_k \text{ perturbed}) \qquad \sum_k [\hat{v}_k(t)(i_k(t) + di_k(t)] = 0 \tag{22.29}$$

Equations (22.28) and (22.29) require

$$\sum_k dv_k(t)\hat{i}_k(t) = 0 \tag{22.30}$$

$$\sum_k \hat{v}_k(t)di_k(t) = 0 \tag{22.31}$$

Applying Tellegen's theorem to resistive networks with only input voltage sources yields

$$(dv_S\hat{i}_S - di_S\hat{v}_S) + (dv_{R1}\hat{i}_{R1} - di_{R1}\hat{v}_{R1}) + (dv_{R2}\hat{i}_{R2} - di_{R2}\hat{v}_{R2}) + \ldots (dv_0\hat{i}_0 - di_0\hat{v}_0) = 0 \tag{22.32}$$

where v_0 is the output voltage. When v_S is set equal to zero, i_0 is set equal to -1, and all corresponding branch resistances in "b" are set equal to those in "a," Eq. (22.32) reduces to

$$dv_0/dR_p = -i_{R_p}\hat{i}_{R_p} \tag{22.33}$$

(Calahan, 1972).

An efficient sensitivity analysis can be carried out using Eq. (22.33). An adjoint circuit is first established by replacing v_S by a short circuit and adding an ideal current source, $i = -1$ A, across the output resistânce. Using the methods discussed above, both circuits are solved for branch currents. The negative products of the currents through corresponding branches of the primary and adjoint networks give the sensitivity coefficients of the primary network.

Example 22.3 Sensitivity Check. Solving network A and adjoint network B in Fig. 22.9 for branch currents gives:

Network A	Network B	Coefficient
I_1 1.78 A	0.285 A	− 0.507
I_2 1.43 A	− 0.815 A	1.17
I_3 0.357 A	0.285 A	− 0.102
I_4 0.357 A	− 0.570 A	0.203

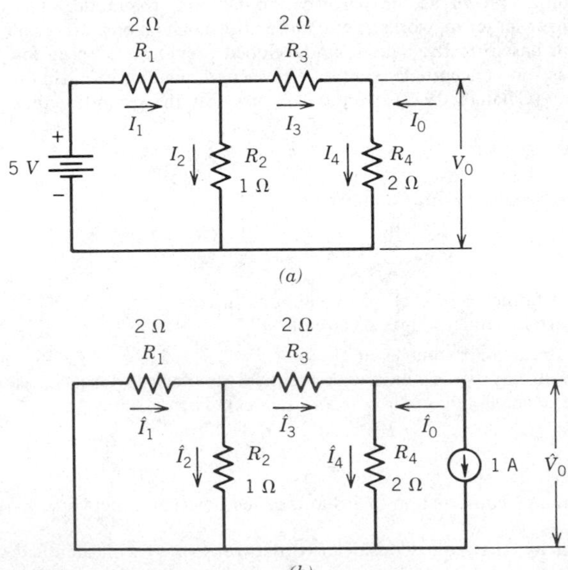

Fig. 22.9 Circuit for Example 22.3: (*a*) primary network, A, (*b*) adjoint network, B.

```
10 REM SENSCHECK-DC SENSITIVITY CHECK
20 REM INPUT AUGMENTED NODAL CONDUCTANCE,GN(I) AND INCIDENCE MATRICES,IN(I)
30 REM BRANCH CONDUCTANCE VECTROR-G(I)
40 REM OUTPUT SENSITIVIY VECTOR
50 REM INPUT SECTION
60 INPUT"ORDER,NODES+VS,BRANCHES-VS ";XOD,NN,NB
70 PRINT"AUGMENTED CONDUCTANCE MATRIX"
80 FORI=1 TO XOD:FOR K=1 TO XOD+1: INPUT GN(I,K):NEXTK,I
90 PRINT"BRANCH CONDUCTANCE VECTOR"
100 FOR I=1 TO NB:INPUT G(I):NEXT
110 PRINT"PRIMARY INCIDENCE MATRIX"
120 FOR I= 1TO NN:FOR K=1 TO NB:INPUT IN(I,K):NEXTK,I
130 REM COMPUTE NODE VOLTAGES FOR PRIMARY CIRCUIT
140 FORI=1TOXOD:FORK=1TOXOD+1:GS(I,K)=GN(I,K):NEXTK,I:GOSUB1000
150 REM TRANSFORM NODE VOTAGES TO BRANCH VOLTAGES
160 GOSUB 2000
170 FOR I=1 TO NB:VP(I)=VB(I):NEXT
180 REM COMPUTE NODE VOLTAGES FOR ADJOINT CIRCUIT
190 REM CHANGE J COLUMN IN AUG. COND.MATRIX
200 FORI=1TO XOD-1:GN(I,XOD+1)=0:NEXT
210 GN(XOD,XOD+1)=-1
220 FORI=1TOXOD:FORK=1TOXOD+1:GS(I,K)=GN(I,K):NEXTK,I:GOSUB 1000
230 REM TRANSFORM NODE VOTAGES TO BRANCH VOLTAGES
240 GOSUB 2000
250 REM COMPUTE SENSITIVITY VECTOR
255 PRINT:PRINT"SENSITIVITIES IN BRANCH ORDER"
260 FOR I=1 TO NB:S(I)=-G(I)^2*VP(I)*VB(I):PRINTS(I):NEXT
270 END
1000 REM SUBROUTINE REALMAT
1010 REM CHECK MAIN DIAGONAL ZERO
1020 FOR I=1 TO XOD:AR=GS(I,I):IF AR =0 THEN PRINT"MAIN DIAG-0":RETURN
1040 REM GAUSS-JORDAN ALGORITHM
1050 FOR J=1 TO XOD+1:GS(I,J)=GS(I,J)/AR:NEXT J
1060 REM SKIPS ZERO ELEMENTS
1070 FOR K=1 TO XOD:IF K-I=0 GOTO 1100
1080 BR=GS(K,I)
1090 FOR J=1 TO XOD+1:GS(K,J)=GS(K,J)-BR*GS(I,J):NEXT J
1100 NEXT K
1110 NEXT I
1120 REM RESULT IN LAST COLUMN OF GS
1130 FOR I=1 TO XOD:VN(I,1)=GS(I,XOD+1):NEXT I
1140 REM RESULT IN VN
1150 RETURN
2000 REM SUBROUTINE TRANSFORMS NODE VOLTAGES TO BRANCH VOLTAGES
2010 REM TRANSPOSES INCIDENCE MATRIX
2020 FOR I=1 TO NB:FORK=1 TO NN:IT(I,K)=IN(K,I):NEXT K,I
2030 REM MULTIPLIES TRANSPOSED IN AND VN
2032 FORJ=1TOXOD:TEMP(J,1)=VN(J,1):NEXT:FORJ=1TOXOD :VN(J+1,1)=TEMP(J,1):NEXT
2035 VN(1,1)=GN(1,XOD+1)/G(1)
2040 FOR I=1 TO NB:SUM=0:FOR K=1 TO NN:SUM=SUM+IT(I,K)*VN(K,1):NEXT K
2050 VB(I)=SUM:NEXT I
2060 RETURN
```

Fig. 22.10 Personal computer (PC) program SENSCHECK.

A PC program that prints out a vector of branch sensitivities for a DC circuit with one input voltage is shown as Fig. 22.10. The program exploits the fact that the nodal conductance matrix is the same for primary and adjoint networks of this type. It is therefore only necessary to input the augmented primary matrix, solve for the node voltages, convert these to branch voltages, and repeat the procedure for the adjoint network with the current vector changed. After a correction is made for branch 1 by reinserting V_s, branch sensitivities are computed from

$$dv_0/dR_p = -V_p\hat{V}_p/R_p^2 \tag{22.34}$$

22.5 OTHER TOPICS

In this limited introduction to computer-aided network analysis, a number of topics have been necessarily ignored. Some additional topics of interest to the network designer are described here.

Transient Analysis

Analytic solutions of dynamic networks generate sets of differential equations. Computers are generally unable to find analytic expressions easily but are limited to numerical results. However, a numerical solution or a graph output vs. time is usually sufficient. A number of algorithms exist for finding numerical solutions. These include the Euler, Milne, Runge–Kutta, and other methods. Stark (1970), Jacquez (1970), and Jensen and Watkins (1974) provide easily followed expositions. For advanced numerical techniques see Calahan (1972).

Tolerance Analysis

Sensitivity analysis can be used to determine the maximum tolerances on circuit components for an allowed variation in output (Calahan, 1972; Dantzig, 1963; Butler, 1971).

Nonlinear Network Analysis

Analysis of nonlinear networks is based on iterative procedures. By nature the computer is suited for these tasks. Efficient algorithms exist for the analysis of nonlinear networks (Calahan, 1972; Carnahan, Luther, and Wilkes, 1969; Katzenelson, 1965).

Automatic Design

Design goals for networks can be reached by iterative methods that are made to converge on the goals by repeated parameter adjustment. The designer must choose the network configuration, but the tedious computation after each adjustment is efficiently handled by computer. Optimization can be carried out in the s plane or the frequency domain (Calahan, 1964; Calahan, 1972; Temes and Calahan, 1967).

Mainframe Programs (ECAP)

Network-solving programs are available for most mainframe computers (ECAP, IBM; CIRC, Xerox; SCEPTRE, VAX). These programs use the basic principles and procedures described in this chapter but offer additional flexibility and options. As a typical example of these programs, we summarize the features of ECAP.

ECAP is divided into four sections: a language section used to describe the network to be analyzed and DC, AC, and transient analysis sections. The analysis is always linear. Nonlinear devices must be transformed into piecewise linear circuits before entry. ECAP performs sensitivity and tolerance analysis, computes and plots frequency response, and permits the modification of circuit elements.

Circuit description is entered as follows. The circuit nodes are labeled starting with the ground node as N0. Branches are also labeled. The user must supply the following information to the language section of the program:

1. Kind of analysis desired.
2. Circuit description.
3. Parameters to be solved.

Circuit description contains the branch number, the nodes between which the component is connected, and the value of the component and its tolerance. For DC and AC analysis, a single value specifies each excitation. For transient analysis, wave forms can be specified by entering the time increment and the amplitudes at succeeding increments.

Equation Formulation

Though some techniques for equation formulation have been described, many advanced and more general techniques exist. These are based on more advanced applications of graph theory and utilize state variable analysis (Calahan, 1972; Pottle, 1966; Seshu and Reed, 1961; Branin and Wang, 1970).

Bibliography

Branin, F. H. and K. U. Wang, "A New Hybrid Formulation of the Network Equations," IBM Report TR21.409, Kingston, NY (Dec. 1970).

Butler, E. M., "Large Change Sensitivities for Statistical Design," *Bell Syst. Tech. J.* **50(4)**:1209–1224 (Apr. 1971).

Calahan, D. A., "Computer Solution of the Network Realization Problem," Proc. 2nd Allerton Conference, University of Illinois, pp. 175–200, 1964.

Calahan, D. A., *Computer Aided Network Design*, McGraw-Hill, New York, 1972.

Carnahan, B., H. A. Luther, and J. O. Wilkes, *Applied Numerical Methods*, 1969.

Dantzig, G. B., *Linear Programming and Extensions*, Princeton University Press, Princeton, NJ, 1963.

Jacquez, J. A., *A First Course in Computing and Numerical Methods*, Addison-Wesley, Reading, MA, 1970.

Jensen, R. W. and B. O. Watkins, *Network Analysis Theory and Computer Methods*, Prentice-Hall, Englewood Cliffs, NJ, 1974.

Katzenelson, J., "An Algorithm for Solving Non-Linear Resistive Networks," *Bell Syst. Tech. J.* **44**:1605–1620 (Nov. 1965).

Levin, H., *Introduction to Computer Analysis*, Prentice-Hall, Englewood Cliffs, NJ, 1970.

Pottle, C., "State Space Techniques for General Active Network Analysis," Ch. 3 in Kuo and Kaiser, eds., *System Analysis by Digital Computer*, Wiley, New York, 1966.

Seshu, S. and M. B. Reed, *Linear Graphs and Electrical Networks*, Addison-Wesley, Reading, MA, 1961.

Stark, P. A., *Introduction to Numerical Methods*, Macmillan, New York, 1970.

Tellegen, B. D. H., "A General Network Theorem with Applications," *Philips Res. Rpt.*, No. 7, pp. 259–269.

Temes, G. C. and D. A. Calahan, "Computer Aided Network Optimization—The State of the Art," *Proc. IEEE* (Nov. 1967).

PART 5

ACTIVE CIRCUITS

CHAPTER 23
OPERATIONAL AMPLIFIERS

YUSUF Z. EFE

The Cooper Union for Advancement of Science and Art
New York, New York

23.1 IDEAL OPERATIONAL AMPLIFIERS

Operational amplifiers were introduced in the early 1940s. They were designed with vacuum tubes and were used to accomplish addition, subtraction, and other mathematical operations; hence the name "operational amplifier," or "op-amp."

The op-amp is a direct-coupled high-gain amplifier that also has provision for external feedback. Through the external feedback, the response of the op-amp can be controlled virtually independently of its internal parameters. Figure 23.1*a* represents a conventional op-amp.

Input terminals labeled (−) and (+) are called *inverting* and *noninverting terminals*, respectively. These terminals are also called differential input terminals because the output voltage v_0 depends on the difference in voltage between them. That is,

$$v_0 = A_0(v_2 - v_1) \tag{23.1}$$

where A_0 is the open-loop voltage gain of the op-amp. The equivalent circuit model of the op-amp is shown in Fig. 23.1*b* where r_i represents the input resistance and r_0 represents the output resistance of the op-amp. Typical values are $A_0 > 100{,}000$, $r_i > 100\ \text{k}\Omega$, $r_0 < 100\ \Omega$.

The ideal op-amp possesses the following characteristics:

1. The open-loop gain $A_0 = \infty$.
2. The input resistance $r_i = \infty$.
3. The output resistance $r_0 = 0$.
4. The bandwidth $BW = 0$.
5. The output voltage is zero when the input voltage is zero, that is, $v_0 = 0$ when $v_i = 0$.

From these ideal characteristics two very important additional properties can be deduced.

6. Since the input resistance is infinite, no current flows into the amplifier input terminals.
7. When feedback is employed, the differential input voltage reduces to zero.

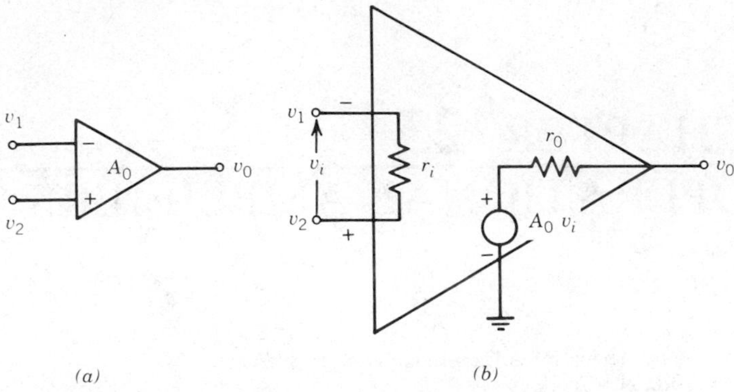

Fig. 23.1 Operational amplifier: (*a*) symbol, (*b*) equivalent circuit.

23.2 BASIC OPERATIONAL AMPLIFIER CONFIGURATIONS

Three basic configurations for op-amps are:

1. The inverting configuration, the inverter
2. The noninverting configuration, the noninverter
3. The differential configuration

Practically all other op-amp circuits are based in some way on these configurations.[1-4]

Inverter

The input signal is applied to the inverting terminal through a resistance R_1 and the output is fed back to this terminal (feedback) through R_2, as shown in Fig. 23.2. Applying Kirchhoff's current law (KCL) at the node v_x

$$\frac{v_i - v_x}{R_1} + \frac{v_0 - v_x}{R_2} = i_b \tag{23.2}$$

For an ideal op-amp (or for an op-amp having characteristics close to those of an ideal op-amp), $i_b = 0$. This means that the voltage across the input resistance r_i is zero or can be approximately equal to zero. That is, the $(-)$ input terminal is considered internally connected to ground, virtual ground. With these approximations, Eq. (23.2) becomes

$$\frac{v_i}{R_1} + \frac{v_0}{R_2} = 0 \tag{23.3}$$

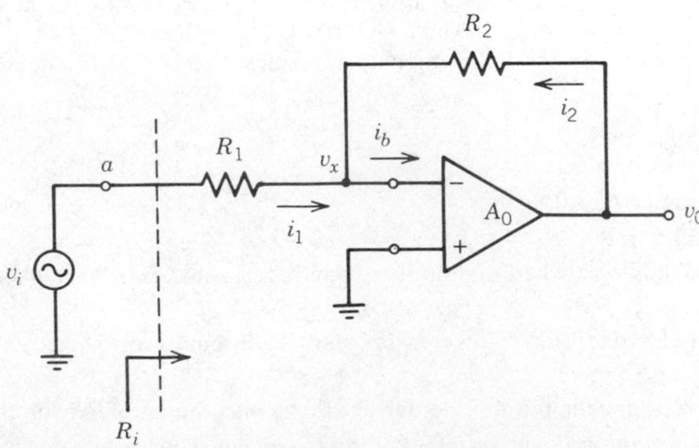

Fig. 23.2 Inverting operational amplifier.

Thus the voltage gain with feedback A_{CL}, *called closed-loop* gain, is found to be

$$A_{CL} = \frac{v_0}{v_i} = -\frac{R_2}{R_1} \tag{23.4}$$

where the (−) sign indicates the inverting property of the circuit.

The input resistance R_i is then found to be (recall that $v_x = 0$)

$$R_i = \frac{v_i}{i_1} = R_1 \tag{23.5}$$

Practical Inverting Op-Amp. Equations (23.4) and (23.5) are valid only if $A_0 = \infty$. In a practical op-amp, however, $A_0 \neq \infty$, $r_i \neq \infty$, and $r_0 \neq 0$. In this case the op-amp in Fig. 23.2 is replaced by its equivalent circuit, shown in Fig. 23.1*b*.

With use of the fact that $i_b = v_x/r_i$, $A_0 = v_0/v_x$, and $A_{CL} = v_0/v_i$ and manipulating Eq. (23.2), the overall closed-loop gain is found to be

$$A_{CL} = \frac{G_1 A_0}{G_i + G_2(1 - A_0) + g_i} \tag{23.6}$$

where $G_1 = 1/R_i$
$G_2 = 1/R_2$
$g_i = 1/r_i$

Input and Output Resistances. The input resistance between terminal *a* and the ground is found to be

$$R_i = R_1 + r_i \| R_t \tag{23.7}$$

where

$$R_t = \frac{R_2 + r_0}{1 + A_0}$$

The output resistance R_0 between the output terminal and ground is

$$R_0 = \frac{v_0}{i_0} = 1 \Bigg/ \left[\frac{1 + R_1 A_0/(R_1 + R_2)}{r_0} + \frac{1}{R_1 + R_2} \right] \tag{23.8}$$

Note that when $A_0 = \infty$ or $r_0 = 0$, the output resistance reduces to zero (as expected).

Noninverter

The input signal is applied to the noninverting terminal (Fig. 23.3). Since no input current flows into either input terminals, that is, since $r_i = \infty$ and $v_i - v_x = 0$, then

$$v_i = v_x = \frac{R_1}{R_1 + R_2} v_0$$

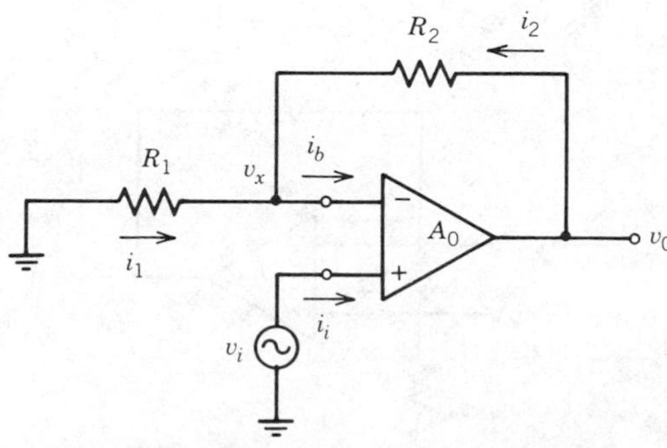

Fig. 23.3 Noninverting operational amplifier.

and

$$A_{CL} = \frac{v_O}{v_i} = 1 + \frac{R_2}{R_1} \tag{23.9}$$

Hence $1 \leqslant A_{CL} \leqslant \infty$ for a noninverting amplifier. $A_{CL} = 1$ when $R_2 = 0$. Under this condition R_1 may be deleted, and the circuit is then called a *voltage follower*.

The input resistance of the noninverting amplifier is $R_i = v_i/i_i = \infty$, since $i_i = 0$.

Practical Noninverting Op-Amp. Equation (23.9) is valid only if $A_0 = \infty$. In the case of a practical op-amp, the overall closed loop gain is

$$A_{CL} = \frac{A_0}{1 + A_0 R_1/(R_1 + R_2)} \tag{23.10}$$

which gives A_{CL} in Eq. (23.9) as $A_0 \to \infty$.

Input and Output Resistances. The input resistance defined as $R_i = v_i/i_i$ is

$$R_i = \frac{A_0 r_i R_1}{R_1 + R_2} \tag{23.11}$$

and the output resistance is

$$R_0 = \frac{v_0}{i_0} = 1 \Bigg/ \left[\frac{R_1 + R_2 + r_i A_0}{r_0(R_1 + R_2)} + \frac{1}{R_1 + R_2} \right] \tag{23.12}$$

Note that when A_0 or r_i approaches infinity, $R_i \to \infty$ and $R_0 \to 0$ (as expected).

Differential Amplifier

The differential op-amp is a combination of the two previous configurations, as shown in Fig. 23.4. Using superposition, one can show that

$$v_0 = \frac{R_{22}}{R_{21}} v_2 - \frac{R_{12}}{R_{11}} v_1 \tag{23.13}$$

By adjusting R_{21} and R_{22} such that $R_{22} = R_{12} = R_2$ and $R_{21} = R_{11} = R_1$, Eq. (23.13) gives

$$A_{CL} = \frac{v_0}{v_2 - v_1} = \frac{R_2}{R_1} \tag{23.14}$$

This is the gain of the amplifier for differential mode signals, that is, for $v_1 \neq v_2$.

Ideally the output of an ideal differential amplifier should be given by

$$v_0 = A_{CL}(v_2 - v_1) \tag{23.15}$$

That is, any signal common to both inputs should have no effect on v_0. The quantity used to measure how much the common-mode signal is suppressed relative to the input-difference voltage is called the *common-mode rejection ratio* (CMRR). Let $A_1 = -(R_{12})/(R_{11})$ and $A_2 = (R_{22})/(R_{21})$ and let

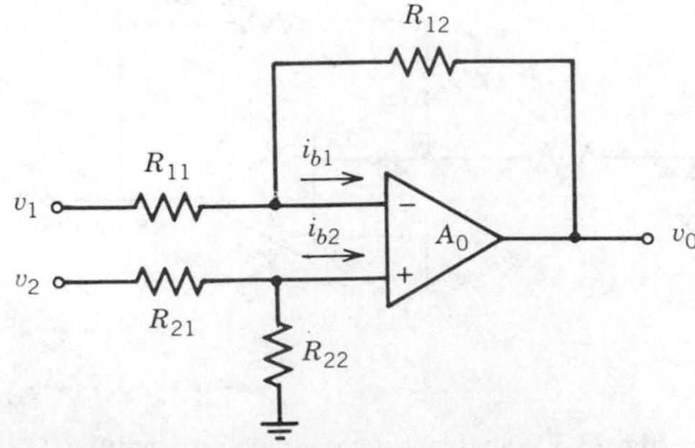

Fig. 23.4 Differential operational amplifier.

$A_d = (A_2 - A_1)/2$ and $A_c = A_1 + A_2$. Then the CMRR is defined as

$$\text{CMRR} = \left|\frac{A_d}{A_c}\right| \quad \text{or} \quad \text{CMRR}_{dB} = 20\log\left|\frac{A_d}{A_c}\right| \tag{23.16}$$

and it can be shown that

$$v_0 = A_d v_d\left(1 + \frac{1}{\text{CMRR}}\cdot\frac{v_c}{v_d}\right) \tag{23.17}$$

where $v_d = v_2 - v_1$ and $v_c = (v_1 + v_2)/2$. It can be observed from Eq. (23.17) that the amplifier should be designed such that $\text{CMRR} \gg v_c/v_d$ in order that the common-mode signal can be rejected effectively.

Input Resistance. The input resistances for inverting and noninverting terminals are not the same. They are given by

$$R_i[\text{for } (-) \text{ input}] = R_{11} \tag{23.18}$$

$$R_i[\text{for } (+) \text{ input}] = R_{21} + R_{22} \tag{23.19}$$

23.3 APPLICATIONS OF BASIC CONFIGURATIONS

Applications of the basic configurations arc countless. Only several commonly used circuits will be discussed here.

Inverting Summing Amplifier

The output of an inverting summing amplifier (Fig. 23.5) is proportional to the linear sum of the input voltages.

Since $v_x = 0$, application of the KCL at node x yields

$$\frac{v_1}{R_{11}} + \frac{v_2}{R_{12}} + \frac{v_3}{R_{13}} = -\frac{v_0}{R_2}$$

or

$$v_0 = -\left(\frac{R_2}{R_{11}}v_1 + \frac{R_2}{R_{12}}v_2 + \frac{R_2}{R_{13}}v_3\right) \tag{23.20}$$

Observe that each input is scaled independently by R_2, R_{11}, R_{12}, R_{13}.

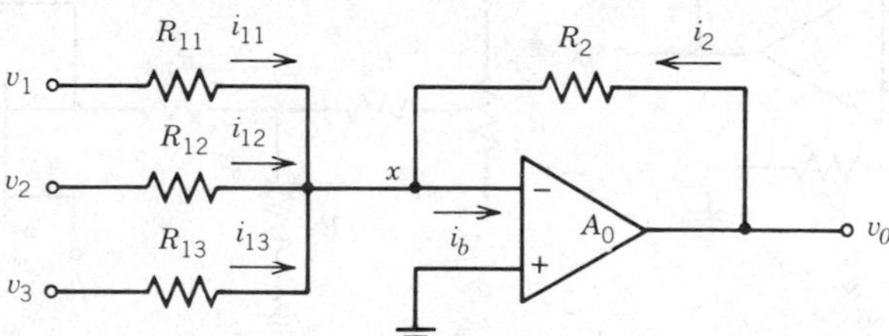

Fig. 23.5 Inverting summing amplifier.

Audio Mixer

Isolation is an important characteristic of the configuration shown in Fig. 23.5. Isolation is the result of each signal source seeking virtual ground potential at the summing node x. Thus the input signals v_1, v_2, v_3 do not interact. This is a very desirable feature, particularly for an audio mixer. Hence the intensity of each signal can be controlled independently from others by adjusting the resistance in the signal path.

Inverting Averaging Amplifier

The output of an averaging amplifier is proportional to the average of all input signals. This averaging process is achieved by adjusting the input resistances R_{11}, R_{12}, R_{13}, and R_2, as in Fig. 23.5.

For example, if $R_{11} = R_{12} = R_{13} = R$ and $R_2 = R/3$, then Eq. (23.20) yields

$$v_O = -(v_1 + v_2 + v_3)/3 \tag{23.21}$$

which is the desired result.

23.4 CHARACTERISTICS OF OPERATIONAL AMPLIFIERS

Manufacturers specify pertinent electrical characteristics of their op-amps. They usually provide this information on data sheets as maximum ratings, typical values, or minimum ratings. These characteristics can be divided into DC and AC characteristics. The two types are reviewed in this section. A typical data sheet is given in Appendix 23.1 at the end of this chapter.

DC Characteristics

Knowledge of the following DC characteristics is important in understanding the behavior of an op-amp in a circuit. The errors caused by these characteristics can then be eliminated or at least reduced.

Offset Voltage. The ideal op-amp produces a zero output voltage for a zero differential input. However, unavoidable imperfections in circuit components within real amplifiers cause a voltage at the output when the input voltage is zero. This voltage is called the output offset voltage. The differential input voltage required between input terminals to obtain zero output is called input offset voltage, V_{os}. The effect of this voltage in an inverter can be observed from the following equation

$$v_O = \left(1 + \frac{R_2}{R_1}\right)V_{os} - \frac{R_2}{R_1}v_i \tag{23.22}$$

where $(1 + R_2/R_1)$ is called the *noise gain*. The effect of the input offset voltage can be reduced by adding a small DC voltage at the input and adjusting its magnitude and polarity to give a zero output voltage when $v_i = 0$. This process is called nulling the output offset voltage. In some op-amps, special terminals are provided to null the output offset voltage. If such terminals are not available, then the nulling techniques, known as universal nulling techniques, shown in Fig. 23.6 for both inverting and noninverting amplifiers, can be used.

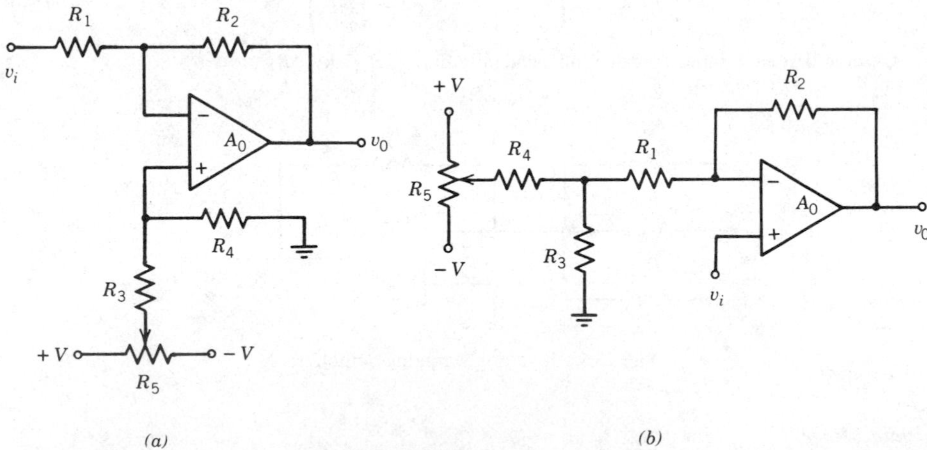

Fig. 23.6 Universal offset voltage nulling circuits: (*a*) inverting operational amplifier, (*b*) noninverting op-amp.

Input Bias Current. In an ideal op-amp, input currents are zero. In reality, the input terminals conduct small amounts of DC currents I_{b_1} and I_{b_2} to bias the internal transistors. The input bias current of an op-amp is defined as the average of two input currents when the output is nulled to zero, that is,

$$I_b = (I_{b_1} + I_{b_2})/2 \tag{23.23}$$

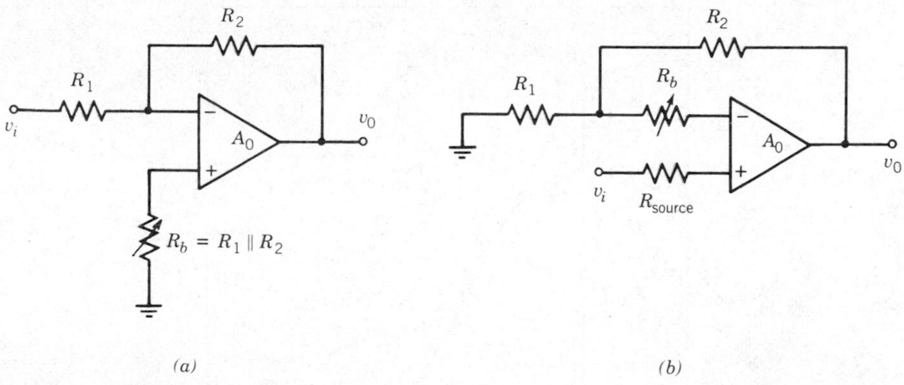

Fig. 23.7 Cancellation of effect of input bias current on output for (*a*) inverting operational amplifier, (*b*) noninverting op-amp.

For a typical op-amp this current is in the range of 10 to 100 nA. The cancellation of the effect of the input bias current on output is shown in Fig. 23.7.

As with the input offset voltage, the input bias current varies with temperature. However, if the resistances seen by both inputs are made identical, input bias current changes with temperature may be neglected.

Input Offset Current. The difference in magnitudes between I_{b_1} and I_{b_2} is called the input offset current, I_{os}.

$$I_{os} = |I_{b_1} - I_{b_2}| \tag{23.24}$$

To minimize error in the output voltage due to offset current, a resistor R_b, shown in Fig. 23.7, is connected. The typical I_{os} is less than 25% of the input bias current I_b, defined in Eq. (23.23).

AC Characteristics

The AC operation of an op-amp is not free from error. Errors depend on whether AC output voltage is small signal or large signal. If only small AC output signals are present, the performance of the circuit is limited by noise and frequency response of the op-amp. If large AC output signals are present, then the op-amp characteristic called slew rate affects the performance of the op-amp.

Frequency Response. The open-loop gain of an op-amp varies with frequency, which makes the device useful only over a finite range of frequencies. Typical open-loop gain vs. frequency and phase response of an op-amp is given in Fig. 23.8. Observe that the open-loop gain is practically constant at low frequencies. This constant gain is called DC gain, A_{DC}. As the frequency increases, the gain of the op-amp decreases, which makes the device bandwidth limited. This limitation is due to the inevitable stray capacitances present inherently internal to the op-amp.

Note that the gain response has two corner points (B and C) at 10 and 10^6 Hz. These points are said to correspond to poles of the transfer function of the op-amp and to be determined by approximating the gain plot by straight asymptotic lines (Bode approximation). For the cases in which the open-loop gain frequency response $A_0(f)$ contains one pole or two poles, the gain can be expressed, respectively, as

$$A_0(f) = \frac{A_{DC}}{1 + j\dfrac{f}{f_p}} \tag{23.25}$$

and

$$A_0(f) = \frac{A_{DC}}{\left(1 + j\dfrac{f}{f_{p_1}}\right)\left(1 + j\dfrac{f}{f_{p_2}}\right)} \tag{23.26}$$

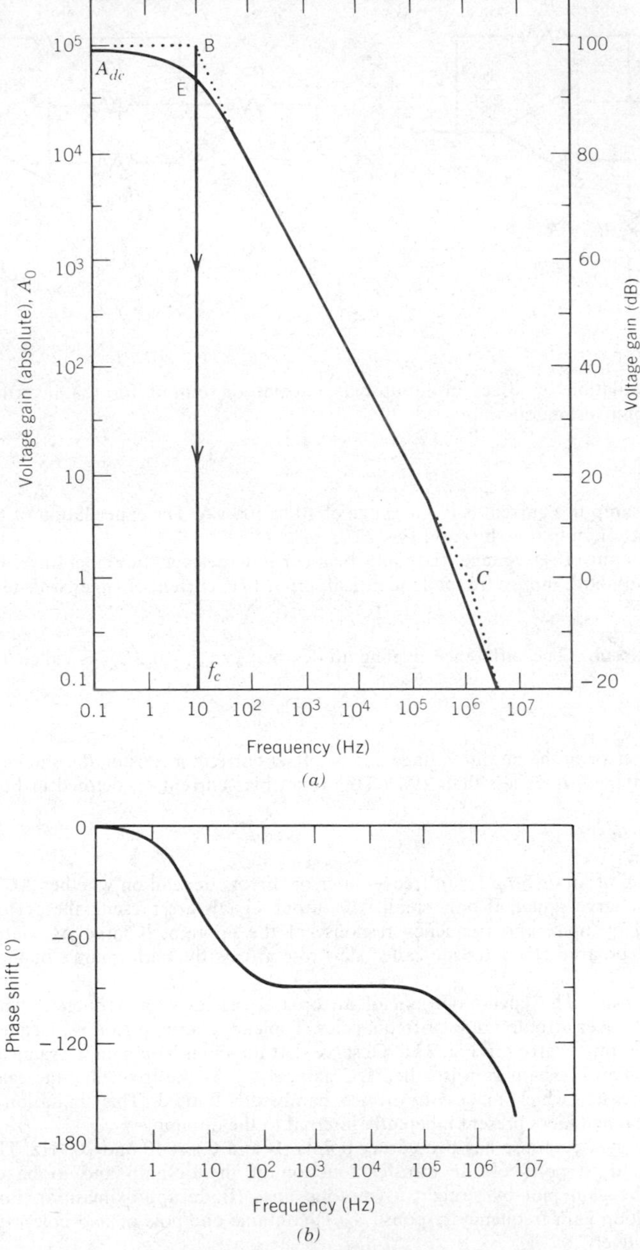

Fig. 23.8 Frequency response of an op-amp: (*a*) open-loop gain plot, (*b*) phase shift plot.

Bandwidth, BW. The low-frequency point (E in Fig. 23.8) where the gain is down 3 dB from the DC value is called open-loop −3-dB bandwidth, or cutoff frequency, f_c. The gain at f_c is given by

$$A_c = \frac{A_{\mathrm{DC}}}{\sqrt{2}} \tag{23.27}$$

$f_c = 10$ Hz in Fig. 23.8*a*.

Unity Gain-Bandwidth, B_u. The frequency at which the gain is unity is called unity gain bandwidth. $B_u = 10^6$ Hz = 1 MHz in Fig. 23.8*a*. If the unity gain bandwidth is not given, it can be computed by

$$B_u = \frac{0.35}{t_r} \tag{23.28}$$

where t_r is the transient rise time in seconds and B_u is in hertz.

Slew Rate. In the large-signal dynamic behavior of the amplifier, it is important to know how fast the output can change when a large signal is applied to its inputs. The maximum rate of change of the op-amp output voltage is called the slew rate. The slew rate of an op-amp represents the maximum slope that the step response can have. Typical slew rates range from 0.1 V/μs to over 1000 V/μsec. A typical data sheet for an op-amp, that is, for the μA741, is shown in Appendix 23.1 at the end of this chapter.

23.5 INSTRUMENTATION AMPLIFIER

The instrumentation amplifier is a high-accuracy differential op-amp that can faithfully amplify low-level signals in the presence of high common-mode noise. Other features are high input impedance, low offset and drift, low nonlinearity, stable gain, and low output impedance. The instrumentation amplifier shown in Fig. 23.9 has these properties. High input impedance at both inputs is guaranteed by use of op-amps A_1 and A_2 in noninverting configurations. When the differential amplifier A_3 is balanced, high CMRR is obtained. Using superposition, the voltages at a and b due to inputs v_1 and v_2 are found to be

$$v_a = \left(1 + \frac{R_1}{R_G}\right)v_1 - \frac{R_1}{R_G}v_2 \tag{23.29}$$

$$v_b = \left(1 + \frac{R_3}{R_G}\right)v_2 - \frac{R_3}{R_G}v_1 \tag{23.30}$$

If op-amp A_3 is balanced, that is, if $R_5R_6 = R_4R_7$, and if $R_1 = R_3$, then the output v_O is

$$v_O = \frac{R_5}{R_6}\left(1 + \frac{2R_1}{R_G}\right)(v_2 - v_1) \tag{23.31}$$

Hence the gain of the complete amplifier can be varied by a single resistor R_G, as indicated by the dashed line in Fig. 23.9. Note, however, that the variation of the gain is not linear with respect to R_G.

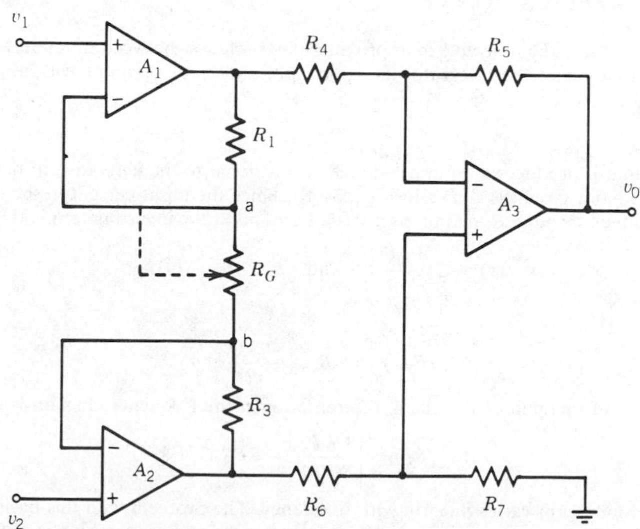

Fig. 23.9 Variable-gain differential input instrumentation amplifier.

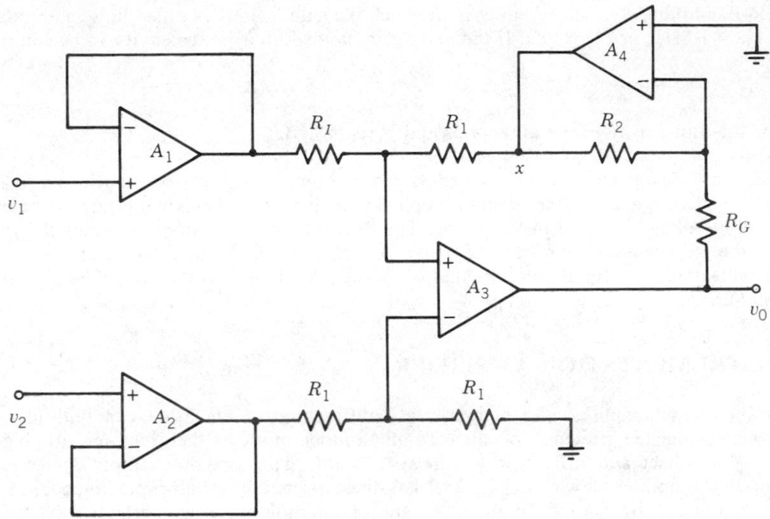

Fig. 23.10 Linear-gain controlled instrumentation amplifier.

The gain of the instrumentation amplifier shown in Fig. 23.10 can be varied linearly with respect to R_G. In this case, the input of op-amps A_1 and A_2 are at unity gain.

The gain of the amplifier is controlled by the additional op-amp A_4, and the output can be shown to be

$$v_0 = -\frac{R_G}{R_2}(v_1 - v_2) \tag{23.32}$$

Hence the gain of the circuit is adjusted linearly by the potentiometer R_G. However, the output resistance of A_4 causes a mismatching in the resistors, which degrades the CMRR. The effect of this output resistance is minimized by feedback at low frequencies, but it has significant effect at high frequencies.

23.6 LINEAR APPLICATIONS

In many op-amp circuits the objective is to provide linear relation between input and output signals. Only the differentiator and the integrator as linear applications of op-amps will be presented here.

Differentiator

An ideal differentiator produces an output voltage proportional to the derivative of the input voltage. The basic differentiator circuit is shown in Fig. 23.11*a*. Since the input current to the ideal op-amp is zero and the feedback through R_f maintains a virtual ground at the inverting terminal, then

$$i(t) = C_1 \frac{dv_i(t)}{dt} \quad \text{and} \quad v_0(t) = -i(t)R_f \tag{23.33}$$

Hence

$$v_0(t) = -R_f C_1 \frac{dv_i(t)}{dt} \tag{23.34}$$

The magnitude of the gain of this ideal differentiator in the frequency domain is given by

$$|A_{CL}(f)| = \left| \frac{v_0(f)}{v_i(f)} \right| = 2\pi R_f C_1 f \tag{23.35}$$

Hence the closed-loop gain varies linearly with frequency. The problem with this basic differentiator, however, is the input impedance of the circuit. It is a pure capacitance, which is not acceptable for most signal sources. Also high-frequency noise would obscure the differentiated signal, and the circuit

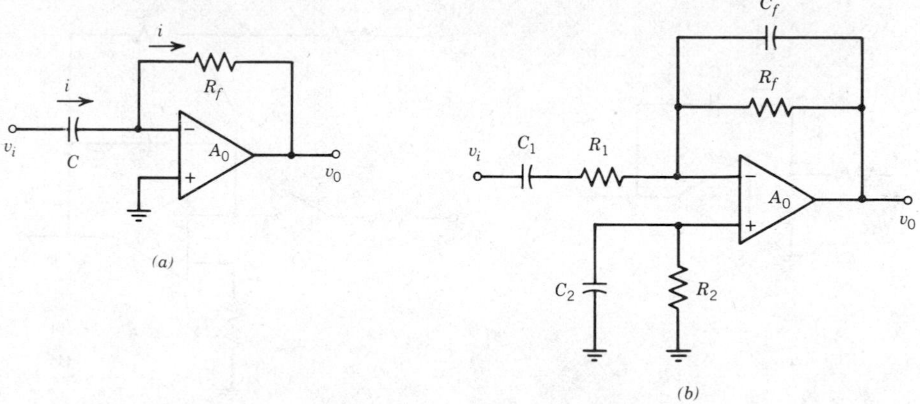

Fig. 23.11 Differentiator circuit: (*a*) basic, (*b*) modified.

has a tendency toward instability. The modified circuit shown in Fig. 23.11*b* is the preferred means of eliminating these problems. R_2 is included to prevent the input bias current from producing a DC offset at the op-amp output. Capacitor C_2 is needed to bypass the thermal noise of R_2 to ground. If C_f is not included, the closed-loop gain is found to be

$$A_{CL}(f) = \frac{-j(f/f_d)}{1 + j(f/f_1)} \tag{23.36}$$

where

$$f_d = 1/(2\pi R_f C_1), \qquad f_1 = 1/(2\pi R_1 C_1) \tag{23.37}$$

The true differentiation is achieved for the frequencies $f_d \leqslant f \leqslant f_1$. Beyond this range the circuit acts as a voltage amplifier.

When C_f is included, the closed-loop gain frequency response is

$$A_{CL}(f) = \frac{-i(f/f_d)}{[1 + i(f/f_1)][1 + i(f/f_2)]} \tag{23.38}$$

where $f_2 = 1/2\pi R_f C_f$. Note that the inclusion of R_1 and C_f introduced two poles, one due to R_1 and the other due to C_f. This provides a stable system and reduces the high frequency noise.

Integrator

An ideal integrator produces an output voltage proportional to the integral of the input voltage signal. The basic integrator circuit is shown in Fig. 23.12*a*. The feedback around the op-amp is provided by the capacitor C_f which maintains a virtual ground at the inverting input of the op-amp. Thus

$$v_0(t) = -\frac{1}{C_f}\int_0^t i(\tau)\,d\tau + v_0(0) \tag{23.39}$$

where

$$i(t) = \frac{v_i(t)}{R_1}$$

and hence

$$v_0(t) = -\frac{1}{R_1 C_f}\int_0^t v_i(\tau)\,d\tau + v_0(0) \tag{23.40}$$

The magnitude of the gain in the frequency domain is then given by

$$|A_{CL}(f)| = \left|\frac{V_0(f)}{V_i(f)}\right| = \frac{1}{2\pi R_1 C_f f} \tag{23.41}$$

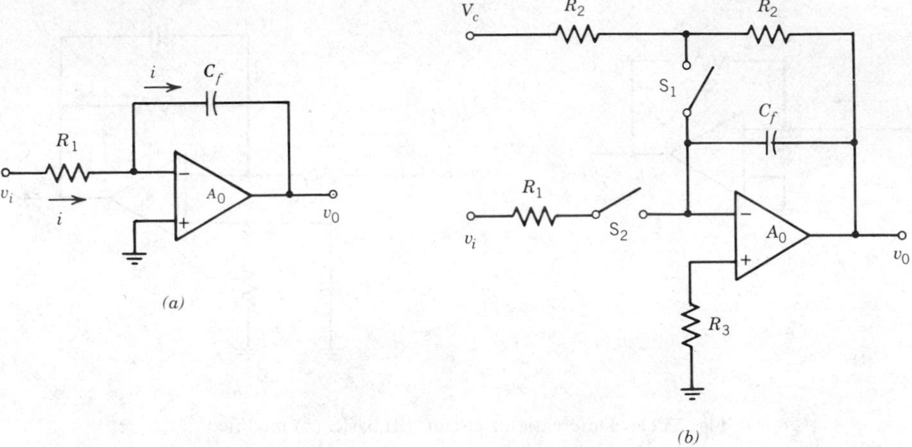

Fig. 23.12 Integrator circuit: (*a*) basic, (*b*) with manual reset.

The finite gain and bandwidth affect the response of the integrator. The sources of error in an integrator are input offset voltage V_{os} and input offset current I_{os}. Because of these DC errors, the output of the integrator consists of two components, as indicated in Eq. (23.42).

$$v_0(t) = \underbrace{-\frac{1}{R_1 C_f}\int v_i(t)\,dt}_{\text{signal term}} + \underbrace{\frac{1}{R_1 C_f}\int V_{os}\,dt + \frac{1}{C_f}\int I_{os}\,dt + V_{os}}_{\text{error terms}} \tag{23.42}$$

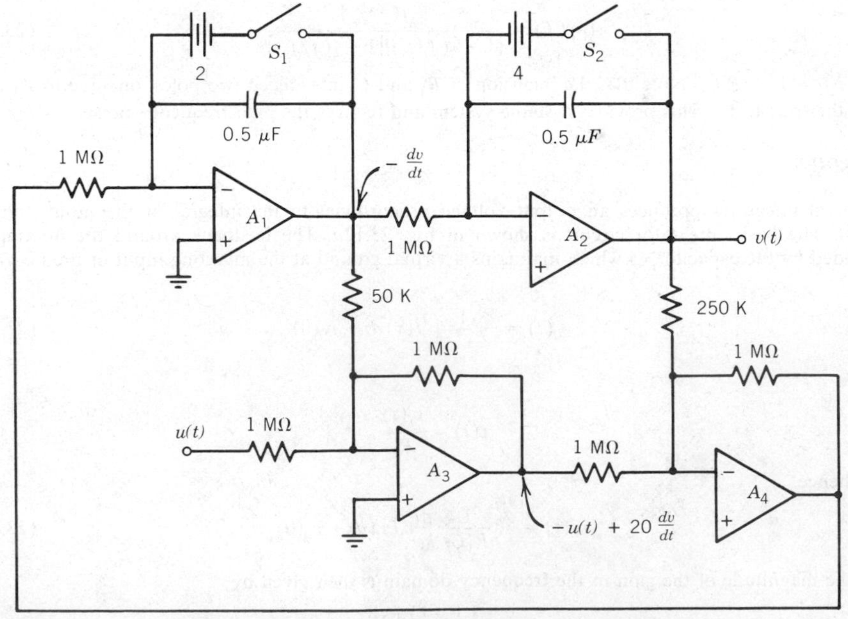

Fig. 23.13 Analog computer realizing Eq. (23.43).

The ramp voltage errors caused by V_{os} and I_{os} in the second and third terms of this equation will continue to increase until the amplifier reaches its saturation voltage (or the limit set by the external circuitry). If the resistance R_3 in Fig. 23.12*b* is not included, the error caused by I_{os} will be replaced by a larger error due to the input bias current I_b.

The integration process in an integrator can be initiated and terminated by a simple switching circuit, manually, as shown in Fig. 23.12*b*. When the switch S_1 is closed, the capacitor C_f is charged and the output voltage rises to the negative of V_c (reset mode). If S_1 is opened and S_2 closed, the circuit begins integration of the signal $v_i(t)$ beginning at the value of $-V_c$ (integrate mode). If both switches are held open, the output voltage will hold its latest value (hold mode).

For a good integrator, the feedback capacitor must be chosen with a dielectric leakage current less than the bias current of the op-amp. Also, for long-term integration chopper-stabilized op-amps are used.

As an example, let us consider the implementation of a second-order differential equation in an analog computer.

$$\frac{d^2v(t)}{dt^2} + 20\frac{dv(t)}{dt} + 4v(t) = u(t) \tag{23.43}$$

Let the initial conditions be

$$\frac{dv(0)}{dt} = 2, \qquad v(0) = 4 \tag{23.44}$$

It can be shown that the circuit in Fig. 23.13 will realize this differential equation.

Observe that op-amps A_1 and A_2 operate as integrators and op-amps A_3 and A_4 function as adders.

23.7 NONLINEAR APPLICATIONS

Although for many analog applications linear relationships are sought, most analog computations require nonlinear characteristics. They are considered linear over certain limited ranges (to be useful). Nonlinear signal-processing applications include comparators, logarithmic and antilogarithmic amplifiers, and others. If nonlinear elements are used in the feed-forward and/or feed-back paths, entire families of nonlinear circuits can be constructed. Basic building blocks of mathematical functions are then used for myriad signal-processing applications.

Comparator

Comparators, as the name implies, are used to compare a signal with respect to a reference signal. An ideal comparator and its transfer characteristic are shown in Fig. 23.14. The signal is applied to the

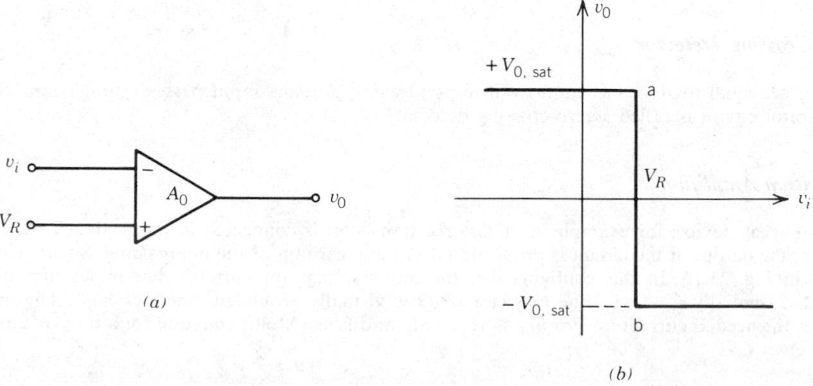

Fig. 23.14 Basic comparator: (*a*) symbol, (*b*) transfer characteristic.

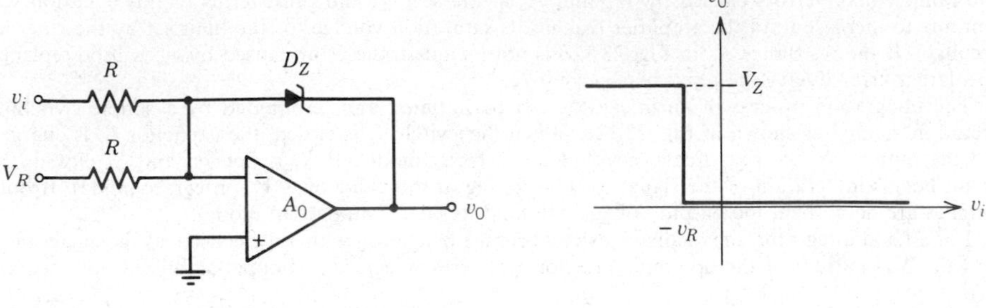

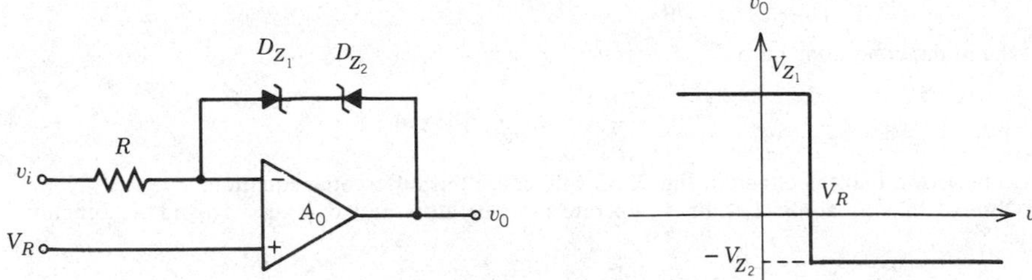

Fig. 23.15 Comparators with feedback limiters.

inverting terminal (inverting comparator). Hence

$$v_0 = \begin{cases} V_{0,\text{sat}} & \text{for} \quad v_i < V_R \\ 0 & \text{for} \quad v_i = V_R \\ -V_{0,\text{sat}} & \text{for} \quad v_i > V_R \end{cases} \tag{23.45}$$

If v_i and V_R are interchanged in Fig. 23.14*a*, a noninverting comparator results. The op-amp is operated in an open-loop condition. Therefore the slope of the line *ab* is infinite for an ideal op-amp and A_0 for a practical op-amp. The output is limited by the saturation levels $+V_{0,\text{sat}}$ and $-V_{0,\text{sat}}$. These saturation voltages can be varied by the power supply voltages applied to the op-amp.

It is usually desired to control the output swing independently of the power supply voltages. Two such circuits and their transfer characteristics are shown in Fig. 23.15.

Zero-Crossing Detector

If V_R is set equal to zero, the output will respond every time the input passes through zero. Such a comparator circuit is called a zero-crossing detector.

Logarithm Amplifier

If a nonlinear device, for example, a diode or a transistor, is connected in the feedback path of an op-amp, the output of the circuit is proportional to the logarithm of the input signal. Such a circuit is shown in Fig. 23.16. In this configuration, the input voltage (or current) determines the collector current. Essentially $i_1 = -i_C$, the collector of Q is virtually grounded, and $v_0 = v_{\text{BE}}$. The op-amp supplies the needed current i_E. For $h_{\text{FE}} \gg 1$, $i_C \cong i_E$ and Eber–Moll's equation for a bipolar transistor model gives

$$i_C = \alpha_N I_{\text{ES}} \left[\exp\left(-\frac{qV_{\text{EB}}}{kT} \right) - 1 \right] \tag{23.46}$$

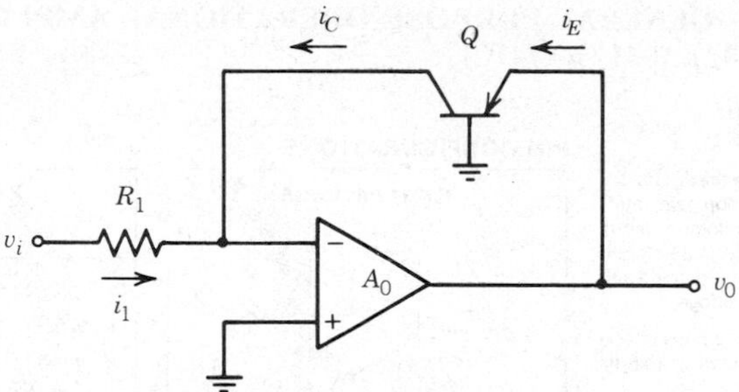

Fig. 23.16 Logarithm amplifier circuit.

where $v_{EB} = v_0$
α_N = normal (forward) current gain
I_{ES} = reverse saturation current
$k = 1.380 \times 10^{-23}$ J/°K (Boltzmann constant)
T = absolute temperature, K

For $v_{BE} < 100$ mV, we have

$$i_C \cong \alpha_N I_{ES} \exp\left(-\frac{qV_{EB}}{kT}\right) \tag{23.47}$$

The op-amp holds $i_C = -i_1 = -(v_i)/(R_1)$, and

$$v_0 = -\frac{kT}{q} \ln\left(-\frac{v_i}{R_1 \alpha_N I_{ES}}\right) \tag{23.48}$$

Thus the output voltage is proportional to the logarithm of the input voltage. A dynamic range of six to eight decades can be achieved with this circuit. It should be noted that the output voltage v_0 is temperature dependent due to the scale factor kT/q and the reverse saturation current I_{ES}. Both of these temperature effects can be reduced by using temperature compensating circuits.

Antilogarithm Circuits

If the positions of the input resistor and transistor in Fig. 23.16 are exchanged, the circuit shown in Fig. 23.17 can be used as an inverse logarithm (antilogarithm) amplifier. Again, this circuit suffers from temperature sensitivity. Also we want an output v_0 such that

$$v_0 = K_1 e^{K_2 v_i} \tag{23.49}$$

where K_1 and K_2 are constants. However, the circuit shown will not give $v_0 = K_1$ for $v_i = 0$. If the op-amp is carefully offset trimmed, $v_0 = 0$ for $v_i = 0$. Hence if the op-amp is properly biased, v_0 can be made equal to K_1 for $v_i = 0$. It must then be ensured that the operation will not be affected by kT/q of Q.

Other nonlinear applications include multipliers, dividers, square-law circuits, and so on. Because of limited space these applications will not be covered here. The interested reader may refer to the references listed at the end of this chapter.

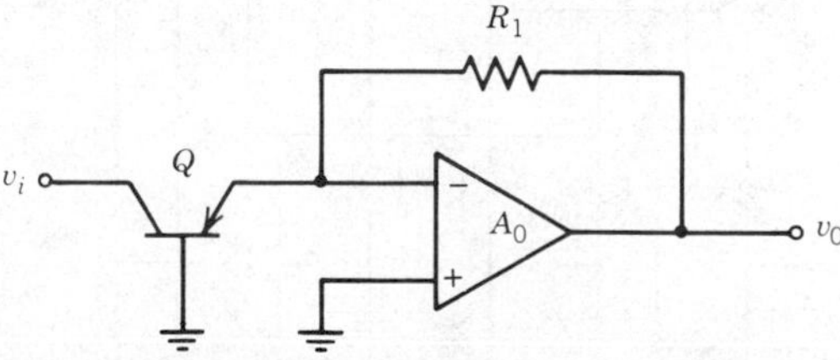

Fig. 23.17 Antilogarithm amplifier circuit.

APPENDIX 23.1. GENERAL PURPOSE OPERATIONAL AMPLIFIER (MC1458/MC1558/μA741/μA741C)

μA741/741C,
MC1458/1558-F,N,H

DESCRIPTION

The μA741 is a high performance operational amplifier with high open loop gain, internal compensation, high common mode range and exceptional temperature stability. The μA741 is short-circuit protected and allows for nulling of offset voltage.

The MC1558/MC1458/SA1458 consist of a pair of 741 operational amplifiers on a single chip.

FEATURES

- Internal frequency compensation
- Short circuit protection
- Excellent temperature stability
- High input voltage range
- No latch-up
- 1558/1458 are 2 "op amps" in space of one 741 package
- MC1558 Mil std 883A,B,C available
- μA741 Mil std 883A,B,C available

PIN CONFIGURATIONS

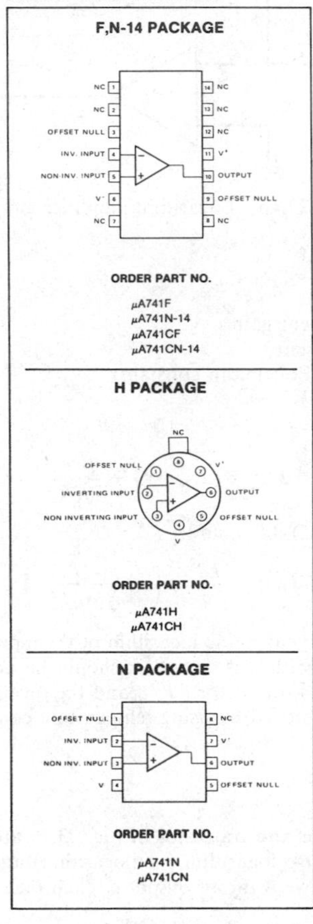

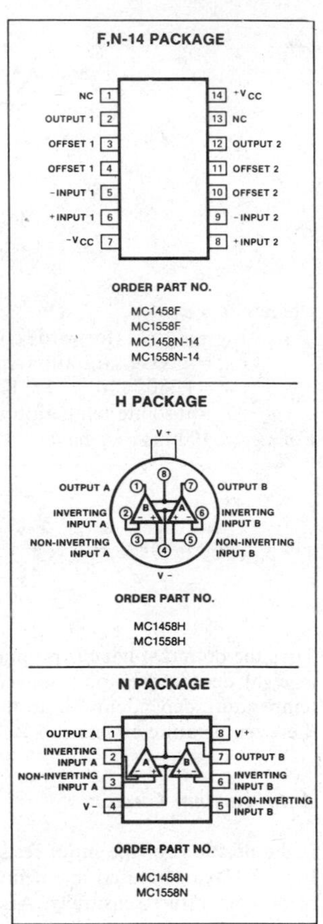

EQUIVALENT SCHEMATIC

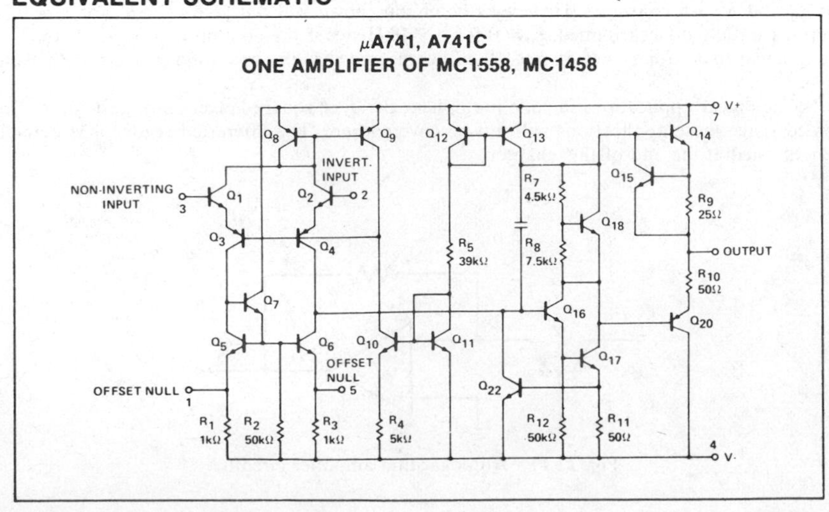

ABSOLUTE MAXIMUM RATINGS

PARAMETER	RATING	UNIT
Supply voltage		
µA741C		
MC1458	±18	V
µA741, MC1558	±22	V
Internal power dissipation, N-14	600	mW
N package	500	mW
H package[1]	800	mW
F package	1000	mW
Differential input voltage	±30	V
Input voltage[2]	±15	V
Output short-circuit duration	Continuous	
Operating temperature range		
µA741C, MC1458	0 to +70	°C
	-40 to +85	°C
µA741, MC1558	-55 to +125	°C
Storage temperature range	-65 to +150	°C
Lead temperature (soldering 60sec)	300	°C

NOTES

1. Ratings based on thermal resistances, junction to ambient, of 208°C/W, 240°C/W, 150°C/W, 110°C/W for N-14, N, H and F packages respectively, and a maximum junction temperature of 150°C.
2. For supply voltages less than ±15V, the absolute maximum input voltage is equal to the supply voltage.

µA741/741C,
MC1458/1558-F,N,H

DC ELECTRICAL CHARACTERISTICS $T_A = 25°C$, $V_S = ±15V$, unless otherwise specified.

	PARAMETER	TEST CONDITIONS	µA741			µA741C			UNIT
			Min	Typ	Max	Min	Typ	Max	
V_{OS}	Offset voltage	$R_S = 10kΩ$		1.0	5.0		2.0	6.0	mV
		$R_S = 10kΩ$, over temp.		1.0	6.0			7.5	mV
I_{OS}	Offset current			20	200		20	200	nA
		Over temp.						300	nA
		$T_A = +125°C$		7.0	200				nA
		$T_A = -55°C$		20	500				nA
I_{BIAS}	Input bias current			80	500		80	500	nA
		Over temp.						800	nA
		$T_A = +125°C$		30	500				nA
		$T_A = -55°C$		300	1500				nA
V_{OUT}	Output voltage swing	$R_L = 10kΩ$	±12	±14		±12	±14		V
		$R_L = 2kΩ$, over temp.	±10	±13		±10	±13		V
A_{VOL}	Large signal voltage gain	$R_L = 2kΩ$, $V_O = ±10V$	50	200		20	200		V/mV
		$R_L = 2kΩ$, $V_O = ±10V$, over temp.	25			15			V/mV
	Offset voltage adjustment range			±30			±30		mV
P_{SRR}	Supply voltage rejection ratio	$R_S \leq 10kΩ$					10	150	µV/V
		$R_S \leq 10kΩ$, over temp.		10	150				µV/V
CMRR	Common mode rejection ratio								dB
		Over temp.	70	90					dB
I_{CC}	Supply current			1.4	2.8		1.4	2.8	mA
		$T_A = +125°C$		1.5	2.5				mA
		$T_A = -55°C$		2.0	3.3				mA
V_{IN}	Input voltage range	(µA741, over temp.)	±12	±13		±12	±13		V
R_{IN}	Input resistance		0.3	2.0		0.3	2.0		MΩ
P_d	Power consumption			50	85		50	85	mW
		$T_A = +125°C$		45	75				mW
		$T_A = -55°C$		45	100				mW
R_{OUT}	Output resistance			75			75		Ω
I_{SC}	Output short-circuit current			25			25		mA

µA741/741C,
MC1458/1558-F,N,H

DC ELECTRICAL CHARACTERISTICS (Cont'd) T_A = 25°C, V_S = ±15V, unless otherwise specified.

	PARAMETER	TEST CONDITIONS	MC1558			UNIT
			Min	Typ	Max	
V_{OS}	Offset voltage	R_S = 10kΩ R_S = 10kΩ, over temp.		1.0	5.0 6.0	mV mV
I_{OS}	Offset current	 Over temp.		20	200 500	nA nA
I_{BIAS}	Input bias current	 Over temp.		80	500 1500	nA nA
V_{OUT}	Output voltage swing	R_L = 10kΩ R_L = 2kΩ, over temp.	±12 ±10	±14 ±13		V V
A_{VOL}	Large signal voltage gain	R_L = 2kΩ, V_O = ±10V R_L = 2kΩ, V_O = ±10V, over temp.	50 25	100		V/mV V/mV
	Offset voltage adjustment range			±30		mV
P_{SRR}	Supply voltage rejection ratio	R_S ≤ 10kΩ		30	150	µV/V
CMRR	Common mode rejection ratio			70	90	dB
I_{CC}	Supply current			2.3	5.6	mA
V_{IN} R_{IN}	Input voltage range Input resistance	(µA741, over temp.)	±12	±13		V MΩ
P_d	Power consumption			70	150	mW
 R_{OUT} I_{SC}	Channel separation Output resistance Output short-circuit current			120 25		dB Ω mA

DC ELECTRICAL CHARACTERISTICS (Cont'd) T_A = 25°C, V_S = ±15V, unless otherwise specified.

	PARAMETER	TEST CONDITIONS	MC1458			UNIT
			Min	Typ	Max	
V_{OS}	Offset voltage	R_S = 10kΩ R_S = 10kΩ, over temp.		2.0	6.0 7.5	mV mV
I_{OS}	Offset current	 Over temp.		20	200 300	nA nA
I_{BIAS}	Input bias current	 Over temp.		80	500 800	nA nA
V_{OUT}	Output voltage swing	R_L = 10kΩ R_L = 2kΩ, over temp.	±12 ±10	±14 ±13		V V
A_{VOL}	Large signal voltage gain	R_L = 2kΩ, V_O = ±10V R_L = 2kΩ, V_O = ±10V, over temp.	25 15	200		V/mV V/mV
	Offset voltage adjustment range			±30		mV
P_{SRR}	Supply voltage rejection ratio	R_S ≤ 10kΩ		30	170	µV/V
CMRR	Common mode rejection ratio			70	90	dB
I_{CC}	Supply current			2.3	5.0	mA
V_{IN} R_{IN}	Input voltage range Input resistance	(µA741, over temp.)	±12	±13		V MΩ
P_d	Power consumption			70	170	mW
 I_{SC}	Channel separation Output short-circuit current			120 25		dB mA

µA741/741C,
MC1458/1558-F,N,H

AC ELECTRICAL CHARACTERISTICS T_A = 25°C, V_S = ±15V, unless otherwise specified.

PARAMETER	TEST CONDITIONS	µA741, µA741C			MC1558, MC1458			UNIT
		Min	Typ	Max	Min	Typ	Max	
Parallel input resistance	Open loop, f = 20Hz				0.3			MΩ
Parallel input capacitance	Open loop, f = 20Hz		1.4					pF
Common mode input impedance	f = 20Hz					200		MΩ
Equivalent input noise voltage	A_V = 100, R_S = 10kΩ, B_W = 1.0kHz f = 1.0kHz					45		nV√Hz
Power bandwidth	A_V = 1, R_L = 2.0kΩ, THD ≤ 5% V_{OUT} = 20Vp-p					14		kHz
Phase margin						65		degrees
Gain margin						11		dB
Unity gain crossover frequency	Open loop		1.0			1.0		MHz
Transient response unity gain	V_{IN} = 20mV, R_L = 2kΩ, C_L ≤ 100pf							
Rise time			0.3			0.3		µs
Overshoot			5.0			5.0		%
Slew rate	C ≤ 100pf, R_L ≥ 2k, V_{IN} = ±10V		0.5			0.8		V/µs

TYPICAL PERFORMANCE CHARACTERISTICS

OUTPUT VOLTAGE SWING AS A FUNCTION OF SUPPLY VOLTAGE

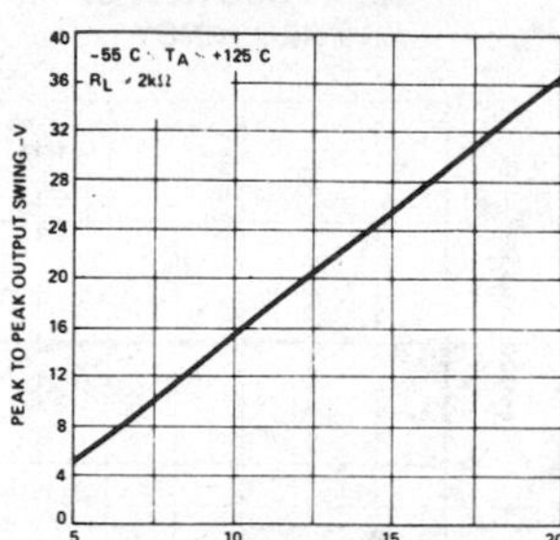

INPUT COMMON MODE VOLTAGE RANGE AS A FUNCTION OF SUPPLY VOLTAGE

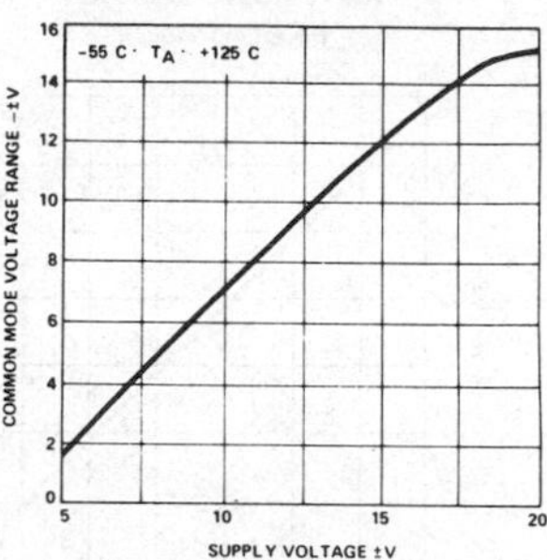

POWER CONSUMPTION AS A FUNCTION OF SUPPLY VOLTAGE

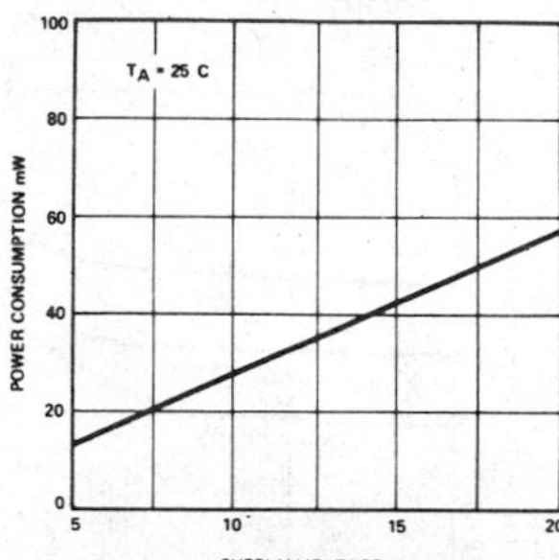

INPUT BIAS CURRENT AS A FUNCTION OF AMBIENT TEMPERATURE

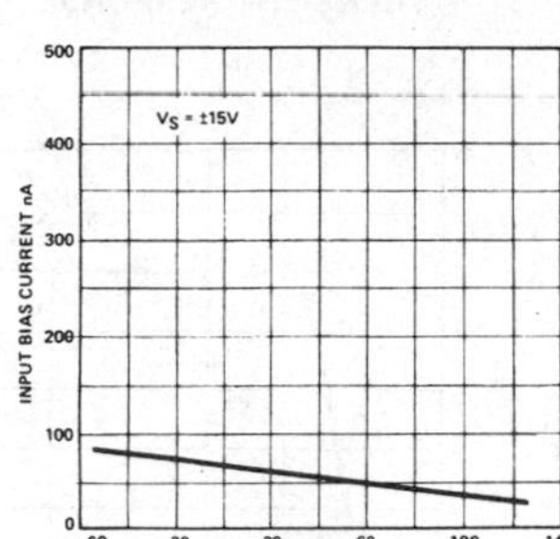

INPUT RESISTANCE AS A FUNCTION OF AMBIENT TEMPERATURE

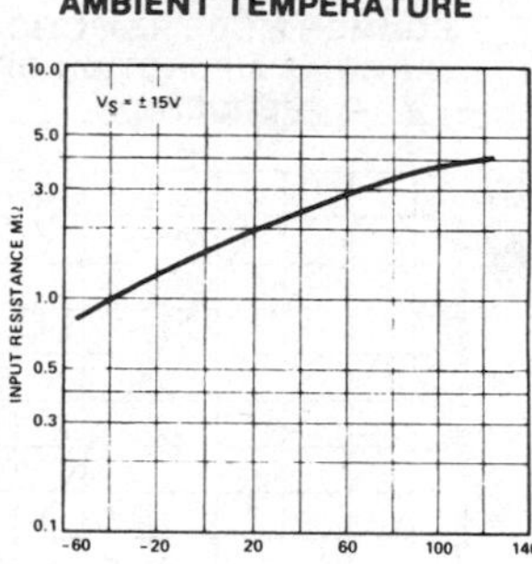

INPUT OFFSET CURRENT AS A FUNCTION OF SUPPLY VOLTAGE

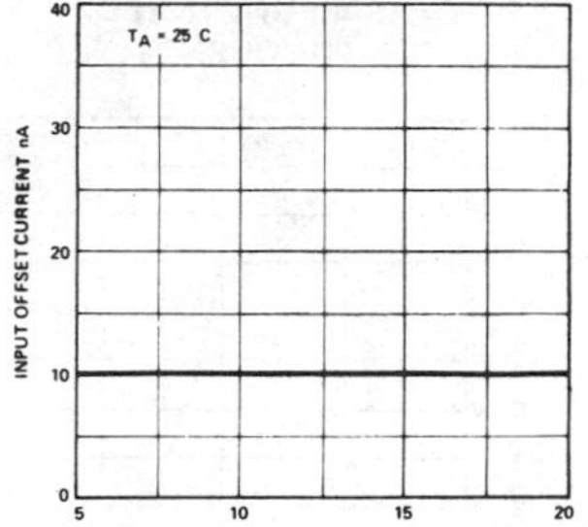

INPUT OFFSET CURRENT AS A FUNCTION OF AMBIENT TEMPERATURE

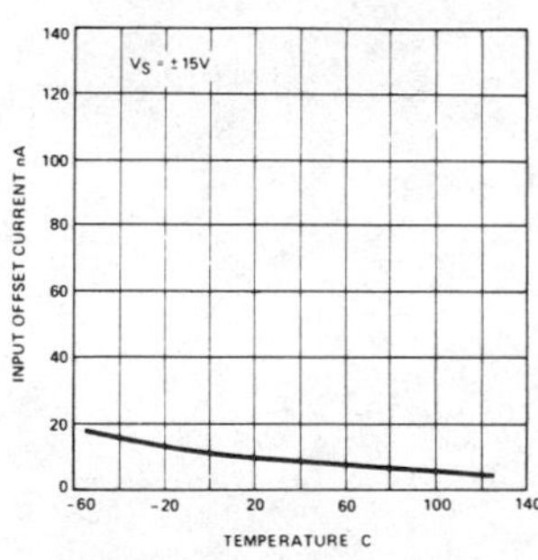

POWER CONSUMPTION AS A FUNCTION OF AMBIENT TEMPERATURE

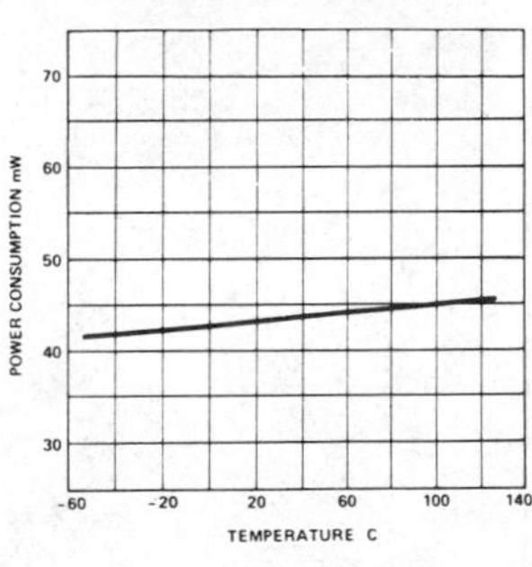

OUTPUT VOLTAGE SWING AS A FUNCTION OF LOAD RESISTANCE

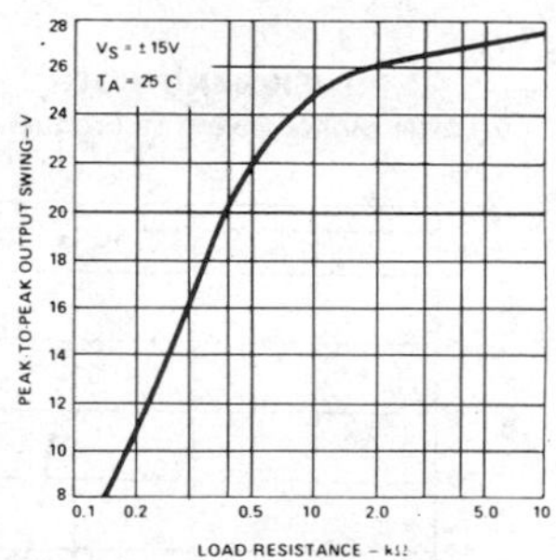

OUTPUT SHORT-CIRCUIT CURRENT AS A FUNCTION OF AMBIENT TEMPERATURE

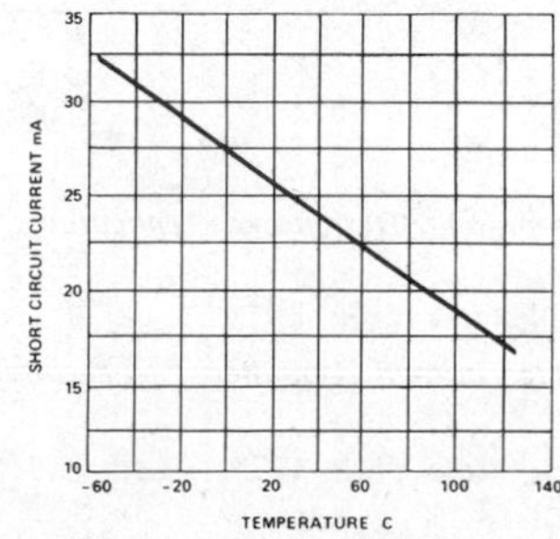

INPUT NOISE VOLTAGE AS A FUNCTION OF FREQUENCY

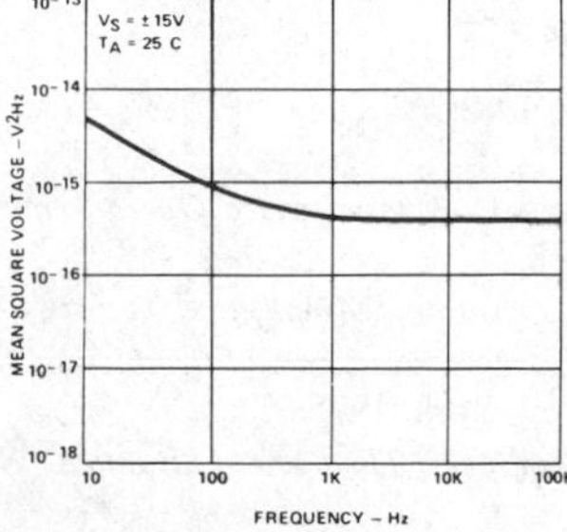

INPUT NOISE CURRENT AS A FUNCTION OF FREQUENCY

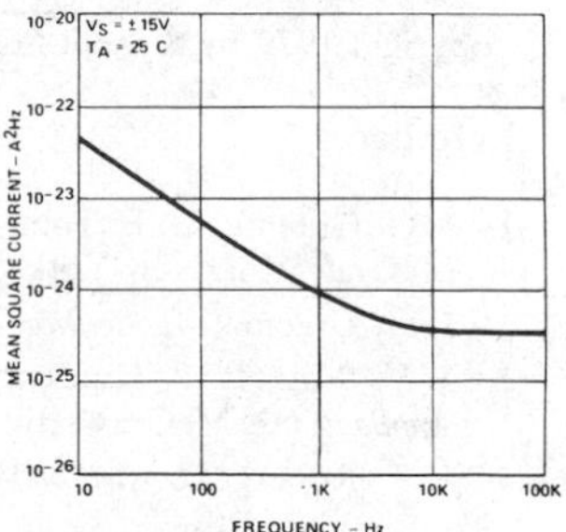

TYPICAL PERFORMANCE CHARACTERISTICS (Cont'd)

BROADBAND NOISE FOR VARIOUS BANDWIDTHS

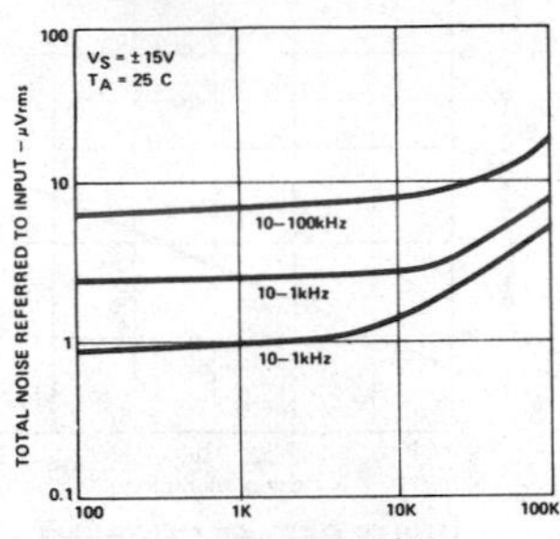

OPEN LOOP VOLTAGE GAIN AS A FUNCTION OF FREQUENCY

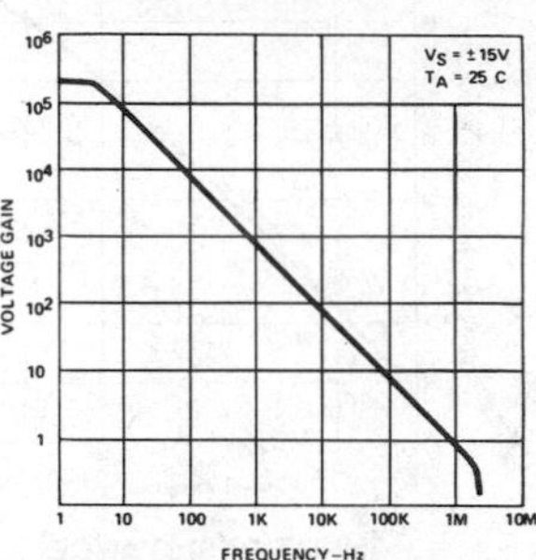

OPEN LOOP PHASE RESPONSE AS A FUNCTION OF FREQUENCY

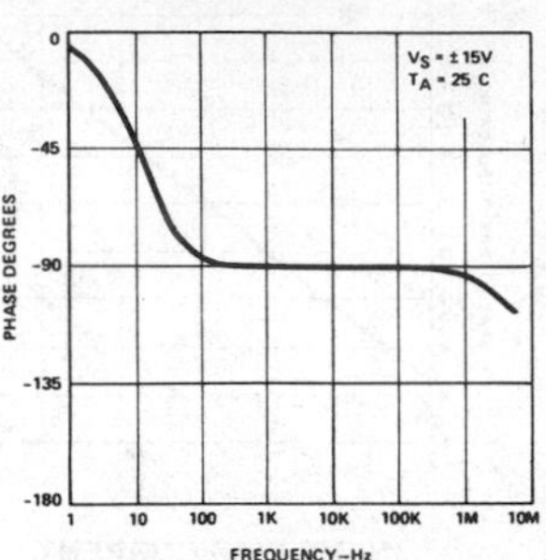

OUTPUT VOLTAGE SWING AS A FUNCTION OF FREQUENCY

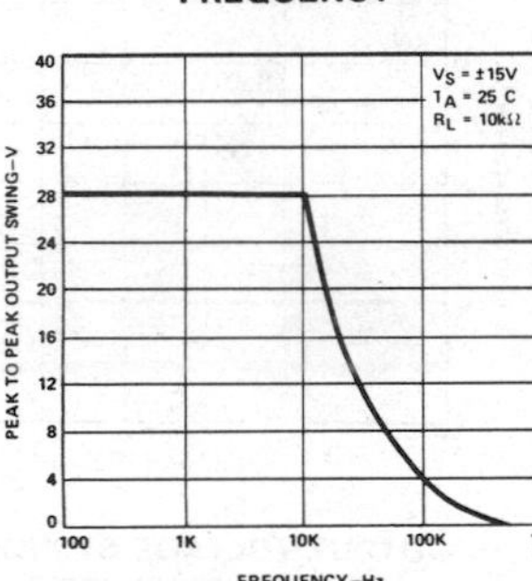

COMMON MODE REJECTION RATIO AS A FUNCTION OF FREQUENCY

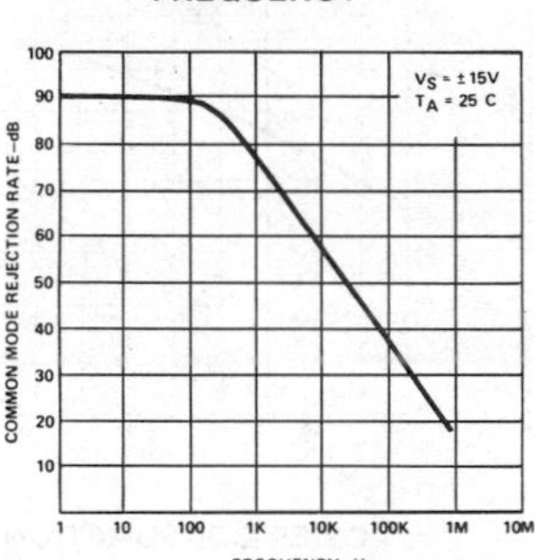

TRANSIENT RESPONSE

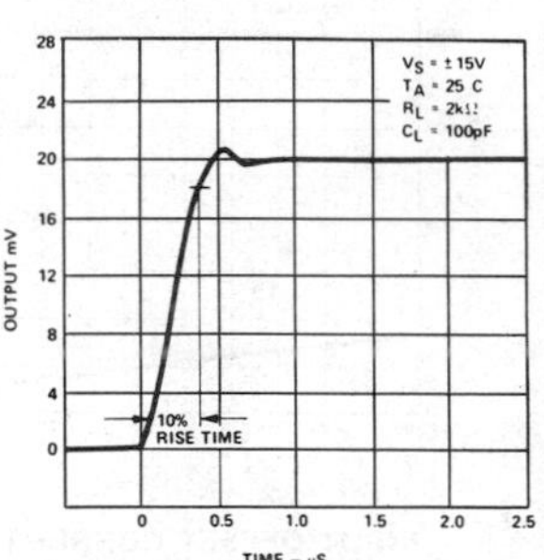

POWER BANDWIDTH (Large Signal Swing vs Frequency)

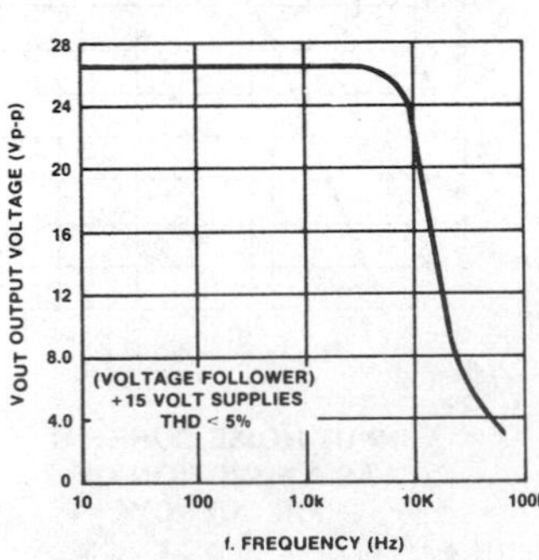

References

1 J. G. Graeme, G. E. Tobey, and L. P. Huelsman, *Operational Amplifiers, Design and Applications*, McGraw-Hill, New York, 1971.

2 J. G. Graeme, *Application of Operational Amplifiers*, McGraw-Hill, New York, 1973.

3 J. V. Wait, L. P. Huelsman, and G. A. Korn, *Introduction to Operational Amplifier Theory and Applications*, McGraw-Hill, New York, 1975.

4 J. K. Roberge, *Operational Amplifiers, Theory and Practice*, Wiley, New York, 1975.

CHAPTER 24

WAVEFORM GENERATORS AND RELATED CIRCUITS

YUSUF Z. EFE

The Cooper Union for Advancement of Science and Art
New York, New York

24.1 OSCILLATORS

Oscillators are circuits whose output is a periodic signal. The output of an oscillator can be a sinusoidal signal or a nonsinusoidal signal, for example, square wave, triangular wave. Various kinds of sinusoidal and nonsinusoidal oscillators are discussed here.

24.1-1 Sinusoidal Oscillators

A number of circuit configurations provide sinusoidal outputs even without an input signal excitation. Consider the feedback system shown in Fig. 24.1. This is a positive feedback circuit (see Chapter 33). The gain of this circuit is given by

$$G = \frac{A}{1 - \beta A} \tag{24.1}$$

As indicated in Chapter 33, the open-loop gain βA approaches unity, the closed loop gain G approaches infinity, and a finite output voltage can result at the absence of the input signal. Thus the condition for sinusoidal oscillation can be stated as

$$\beta A = |\beta A| \underline{/360^\circ} \tag{24.2}$$

at a single frequency ω_0, where $k = 0, 1, 2, 3, \ldots$. This means that the feedback amplifier will be unstable at a frequency ω_0, providing a sinusoidal output with no input present. Then a signal of frequency ω_0 can be transmitted without change in its magnitude or phase.

Oscillations may also occur in a negative feedback system. When several amplifier stages are connected forming a negative feedback, reactive effects around the loop may cause an additional 180° phase shift, changing negative feedback to positive feedback. This then may cause oscillation. Compensating circuits are used to prevent such oscillations.

In practice, the magnitude of the open-loop gain $|\beta A|$ is made slightly larger than unity. In this case the amplitude of the output oscillation will grow at first. The increase in amplitude is limited by the nonlinearity of the active device associated with the amplifier A. The oscillation may be initiated by a transient voltage generated by turning the power on or by the presence of noise. Several sinusoidal oscillator circuits are described here.

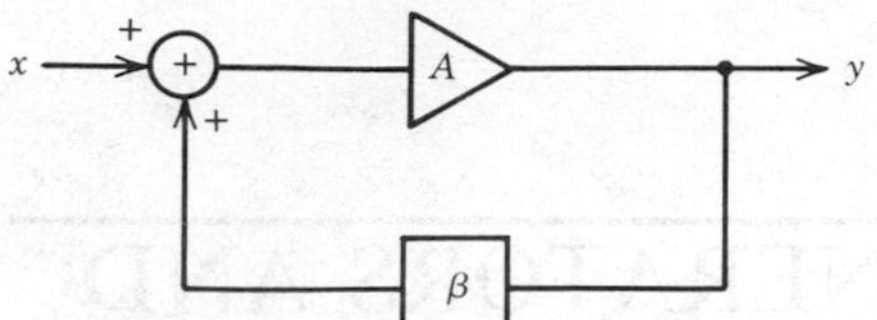

Fig. 24.1 Positive feedback circuit.

Phase-Shift Oscillator

An oscillator, in general, requires positive feedback in which the output signal is fed back in phase to sustain the input. The common-emitter stage in Fig. 24.2 provides a 180° phase reversal between the input signal at its base and the output signal at its collector. The three-stage RC phase-shifting network provides an additional 180° phase shift, satisfying the phase-angle condition for oscillation.

Since the input impedance of the transistor $R_i = h_{ie} \| R_b$, where $R_b = R_1 \| R_2$ and is normally less than R, the series resistance R_1 is added so that $R = R_i + R_1$. The condition for oscillation is then $I_3/I_b \geqslant 1\angle 0$. The loop equations are found to be

$$\begin{bmatrix} R_C + R - jX_C & -R & 0 \\ -R & 2R - jX_C & -R \\ 0 & -R & 2R - jX_C \end{bmatrix} \cdot \begin{bmatrix} I_1 \\ I_2 \\ I_C \end{bmatrix} = \begin{bmatrix} -h_{ie} I_b R_C \\ 0 \\ 0 \end{bmatrix} \tag{24.3}$$

where $X_C = 1/\omega_0 C$, thus

$$\frac{I_3}{I_b} = -\frac{R_C R^2 h_{fe}}{\Delta} \geqslant 1\angle 0° \tag{24.4}$$

where Δ is the determinant of the coefficient matrix and is given by

$$\Delta = (R^3 + 3R_C R^2 - R_C X^2 - 5RX^2) - jX_C(6R^2 + 4R_C R - X_C^2) \tag{24.5}$$

The frequency of oscillation ω_0 is determined by setting the imaginary part of Δ equal to zero

$$\omega_0^2 = \frac{1}{C^2(6R^2 + 4R_C R)} \tag{24.6}$$

Substitution of Eq. (24.6) into Eq. (24.4) yields

$$h_{fe} \geqslant 23 + \frac{29R}{R_C} + \frac{4R_C}{R} \tag{24.7}$$

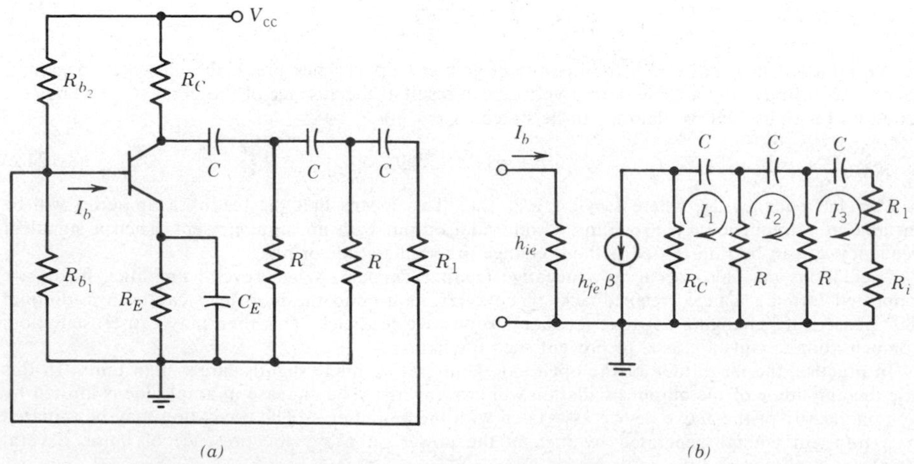

Fig. 24.2 (*a*) Transistor phase-shift oscillator. (*b*) Equivalent circuit.

Letting $\alpha = R_C/R$, the minimum value for h_{fe} is 44.5. Therefore, a transistor with $h_{fe} < 44.5$ cannot be used to design this phase-shift oscillator.

Phase-shift oscillators are useful for generating audio frequencies. For higher frequencies, other types of oscillators should be used.

Wien–Bridge Oscillator

The Wien–Bridge oscillator, shown in Fig. 24.3, is another example of an RC oscillator. The resistors R_1 and R_2 are used for amplitude stabilization. From this figure we have

$$v_O = A(v_y - v_x) \tag{24.8}$$

$$v_x = \frac{R_2}{R_1 + R_2} v_O \tag{24.9}$$

$$v_y = \frac{Z_2}{Z_1 + Z_2} v_O = \frac{j\omega CR}{1 - \omega^2 C^2 R^2 + j\omega^3 CR} v_O \tag{24.10}$$

and the open-loop gain

$$\beta A = \frac{v_O}{v_O'} \tag{24.11}$$

or

$$\beta A = A\left[\frac{j\omega CR}{1 - \omega^2 C^2 R^2 + j\omega CR} - \frac{R_2}{R_1 + R_2}\right] \tag{24.12}$$

To determine the frequency of oscillation, that is, to satisfy the phase angle condition for oscillation, the imaginary part of Eq. (24.12) is made equal to zero. Therefore

$$\omega_0 = \frac{1}{CR} \tag{24.13}$$

Substitution of Eq. (24.13) in Eq. (24.12) to determine the magnitude condition $|\beta A| \geqslant 1$ gives

$$A \geqslant \frac{3(R_1 + R_2)}{R_1 - 2R_2} \tag{24.14}$$

Tuning can be accomplished by varying the capacitors C, the resistors R, or both. Observe, however, that as R_1 approaches $2R_2$, the required gain approaches infinity. For a low distortion, the

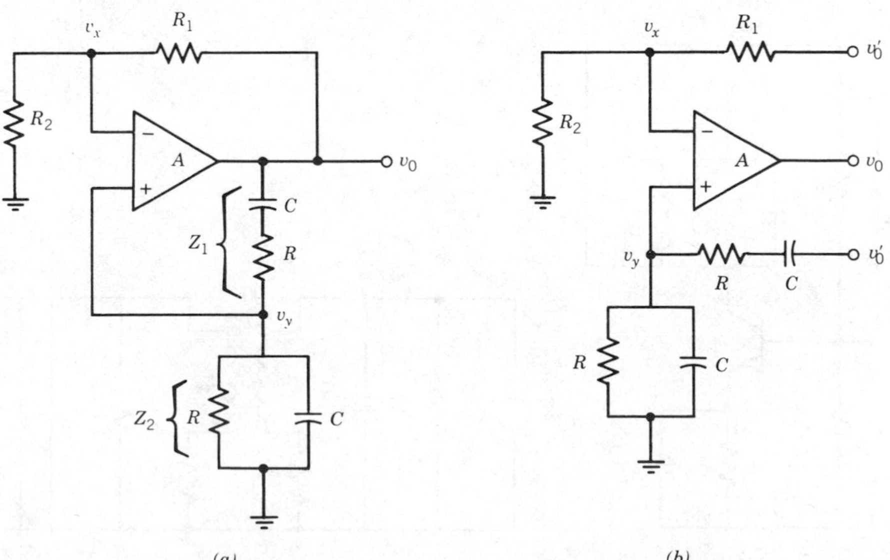

Fig. 24.3 (*a*) Wien-Bridge oscillator circuit. (*b*) Circuit redrawn to determine the open-loop gain βA.

amplitude of oscillation must be limited. This can be done by using a nonlinear resistor for R_1 such that as the amplitude of oscillation increases, R_1 decreases to force $R_1 \cong 2R_2$. This process limits the size of the oscillation, since the circuit will stop oscillation if $R_1 = 2R_2$.

Tuned Oscillators

A tuned-collector oscillator is shown in Fig. 24.4. A field effect transistor (FET) can also be used as the active device. Tuned oscillators can operate in the class A or class C modes.

The equivalent circuit of the tuned-collector oscillator is shown in Fig. 24.4*b*. The resistors R_1 and R_2 are the effective resistances of L_1 and L_2, respectively, R_i is the input impedance, and C_i is the effective input capacitance of the load. Assume that C_b and C_E are short circuits at the signal frequency, R_{b_2} is large so that it can be considered to be open circuit, and $1/\omega C_i \gg R_i$ so that C_i can be ignored. Then

$$I_1 = -\frac{j\omega M}{R_i + R_2 + j\omega L_2} I_3 \cong \frac{j\omega M}{R_i + R_2} \cdot I_3 \tag{24.15}$$

Writing the loop equations we have

$$\begin{bmatrix} h_0 + \dfrac{1}{j\omega C} & -\left(\dfrac{1}{j\omega C} + \dfrac{j\omega M g_m R_i}{h_o(R_i + R_2)}\right) \\ -\dfrac{1}{j\omega C} & R_1 + \dfrac{\omega^2 M^2}{R_i + R_2} + j\left(\omega L_1 - \dfrac{1}{\omega C}\right) \end{bmatrix} \cdot \begin{bmatrix} I_2 \\ I_3 \end{bmatrix} = \begin{bmatrix} 0 \\ 0 \end{bmatrix} \tag{24.16}$$

The frequency of oscillation is obtained by setting the imaginary part of the determinant of the coefficient matrix

$$\omega_0^2 = \frac{(1 + R_1 h_0)(R_i + R_2)}{L_1 C(R_i + R_2) - M^2 h_0} \tag{24.17}$$

and from the real part

$$g_m \geqslant \frac{C(R_i + R_2)}{R_i M}\left[\frac{\omega_0^2 M^2 C + g_0 L_1(R_i + R_2)}{C(R_i + R_2)} + R_1\right] \tag{24.18}$$

The transistor in Fig. 24.4 is connected in common-emitter configuration. However, other configurations may be used to design a tuned oscillator. Also, a tuning circuit may be placed at the input of the transistor.

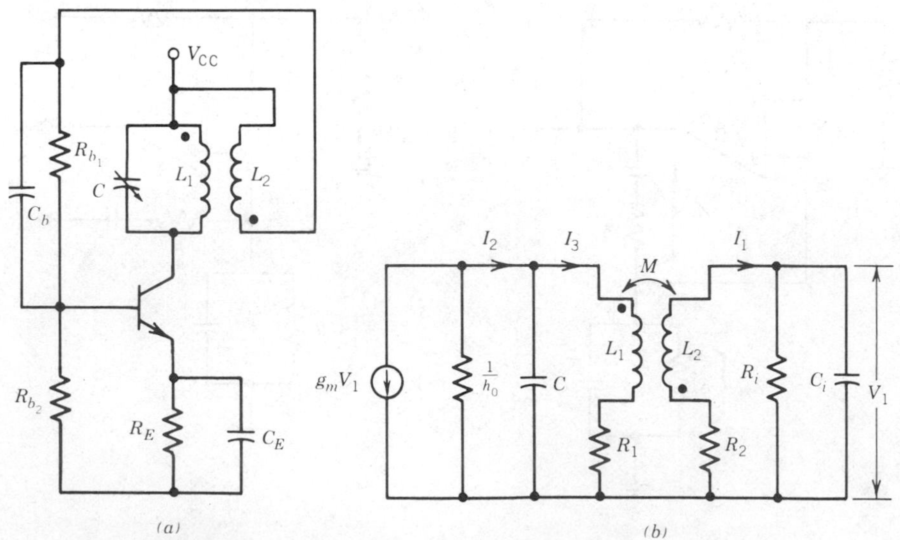

Fig. 24.4 (*a*) Tuned-collector oscillator. (*b*) Equivalent circuit.

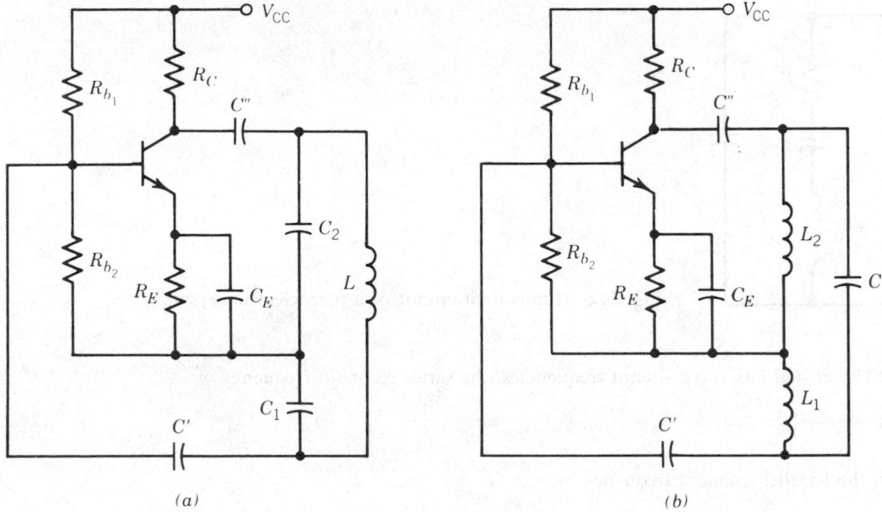

Fig. 24.5 (*a*) Colpits oscillator. (*b*) Hartley oscillator.

Colpits and Hartley Oscillators

The Colpits oscillator, shown in Fig. 24.5*a*, is one of the widely used oscillators. The feedback circuit consists of L, C_1, and C_2. The Colpits oscillator is used in very-high-frequency circuits. Analysis of this circuit yields

$$\text{Frequency of oscillation} \quad \omega_0^2 = \frac{1}{h_{ie} R_0 C_1 C_2} + \frac{1}{L C_{eff}} \tag{24.19}$$

where $C_{eff} = (C_1 C_2)/(C_1 + C_2)$ and the magnitude condition yields

$$\beta \geqslant \frac{C_2}{C_1} + \frac{h_{ie}}{R_0} \cdot \frac{C_1}{C_2} \tag{24.20}$$

The Hartley oscillator, shown in Fig. 24.5*b*, is practically identical to the Colpits oscillator except that the capacitances C_1 and C_2 are replaced with the inductors L_1 and L_2 and the inductance L is replaced with a capacitance C in the tuned circuits. The analysis of the Hartley oscillator circuit yields

$$\text{Frequency of oscillation} \quad \omega_0^2 = \frac{R_0 h_{ie}}{h_{ie} R_0 C (L_1 + L_2) + L_1 L_2} \tag{24.21}$$

and the minimum requirement for h_{fe} of the transistor is

$$h_{fe} \geqslant \frac{h_{ie} L_2}{R_0 L_1} + \frac{L_1}{L_2} \tag{24.22}$$

where R_0 is the output impedance. These formulas assume that the amplifiers in both Colpits and Hartley oscillators are operated in class-A mode. For a good frequency stability, high-Q circuits must be used.

Crystal-Controlled Oscillators

A number of crystals exhibit the piezoelectric effect. That is, if a piezoelectric crystal is strained mechanically, a voltage is produced between its parallel faces. Conversely, when a voltage is applied across the faces of such a crystal, the crystal will expand or contract depending on the polarity of the voltage applied. A piezoelectric crystal represents an element that behaves as a high-Q resonant circuit. Thus piezoelectric oscillators have good frequency stability. The equivalent circuit of a piezoelectric crystal is shown in Fig. 24.6. The capacitance C_s represents the capacitance of the supporting and wiring system.

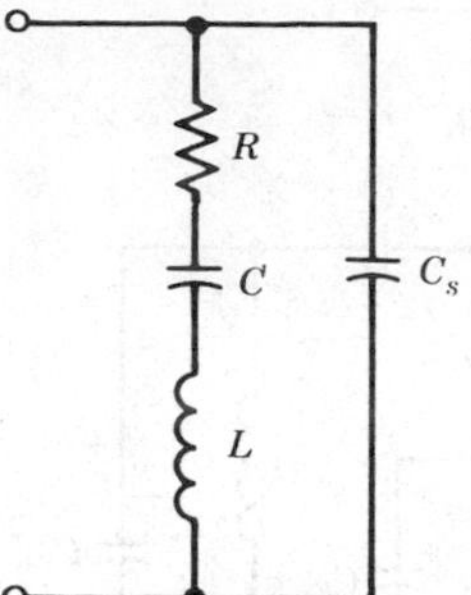

Fig. 24.6 Equivalent circuit of a piezoelectric crystal.

The crystal has two resonant frequencies, the series resonant frequency

$$\omega_s = \frac{1}{\sqrt{LC}} \tag{24.23}$$

and the parallel resonant frequency

$$\omega_p = \frac{1}{\sqrt{LC_T}} \tag{24.24}$$

where C_T is total capacitance around the loop. The crystal oscillator circuits used in communication are designed to maintain a frequency tolerance of $\pm 0.0005\%$ or better, and the Q of the crystal can be as high as 10^5, while the Q of a coil typically is in the range from 50 to 100.

Stability of Oscillators

An oscillator is considered stable if its amplitude and its frequency of oscillation both remain constant during operation.

Amplitude Stability. Recall that the condition for oscillation is that $\beta A = 1 \angle 0°$. If the magnitude of the open-loop gain $|\beta A|$ is less than unity, the oscillation will stop. Such a decrease in magnitude may be caused by aging, operating point changes of the active device, temperature, and other factors. For this reason oscillator circuits are designed such that $|\beta A|$ is slightly greater than unity at the frequency of oscillation. When the amplitude of the output signal increases, the active device lowers the gain to the required value. For good stability, the change in gain with output voltage amplitude should be large and an increase in amplitude must cause the gain to decrease. That is, $\Delta A / \Delta v_0$ must be a large negative number for an oscillator to be stable.

Frequency Stability. The oscillation frequency of an oscillator may also drift. In some applications, 1 to 2% of drift may be tolerable. However, in others the frequency must be constant at all times. The oscillation frequency depends not only on circuit elements of the tuned circuit but also on the parameters of the active device. For example, the parameters of the active device vary with bias voltage, temperature, and age. Another cause of frequency drift is power-supply voltage variations. Hence, for good frequency stability the effects of all these parameters must be minimized. Frequency stabilization of an oscillator circuit thus is a complex process.

If, however, a group of elements are identified as being the cause of the most frequency instability in the oscillator, that is, the phase angle $\theta(\omega)$ changes rapidly with the change of values of these parameters, then attention is concentrated on these elements. In this case, $d\theta(\omega)/d\omega$ serves as a measure of the independence of the frequency of all other elements of the circuit. The frequency of stability is improved as $d\theta(a)/d\omega$ increases. As $d\theta(\omega)/d\omega \rightarrow \infty$, the frequency of oscillation depends only on this group of elements.

It can be shown that $d\theta(\omega)/d\omega$ at $\omega = \omega_0$ is, in general, proportional to the quality factor of the circuit, Q. Hence a tuned-circuit oscillator with a high Q will have excellent frequency stability. It is for this reason that crystal oscillators have excellent frequency stabilization.

24.1-2 Nonsinusoidal Oscillators

The output of a nonsinusoidal oscillator can be a square, pulse, triangular, or sawtooth waveform. Such a waveform can be generated by operational amplifiers, comparators, integrators, differentiators, and associated circuitry. The upper limit of usable speed is determined by the response time of the active devices used in the circuit. Several nonsinusoidal oscillators are presented in this section.

Square-Wave Generator

The square-wave generator shown in Fig. 24.7*a* produces a waveform shown in Fig. 24.7*b*. This circuit is also known as the astable multivibrator as it has two quasistable states. That is, the output v_O remains in one state for a time T_1 and then changes to the second state abruptly for a time T_2. Hence the period of the square wave is $T = T_1 + T_2$.

Observe that a fraction of the voltage at point x is fed back to the noninverting input of the operational amplifier A_1. The fraction is determined by R_2 and R_3 and is

$$\gamma = \frac{R_3}{R_2 + R_2} \tag{24.25}$$

Therefore the differential input voltage v_i is

$$v_i = v_c - \gamma v_x \tag{24.26}$$

When $v_i > 0$, $v_x = -V_z$; when $v_i < 0$, $v_x = +V_z$. Hence the capacitor C will be charged exponentially toward V_z through the integrator formed by R_4, C. The voltage at x will remain constant at $v_x = V_z$ until $v_c = \gamma v_x = \gamma V_z$. When $v_c > \gamma v_x$, the output reverses itself abruptly such that $v_x = -V_z$. The capacitor now discharges exponentially toward $-V_z$. Since the op-amp A_2 is merely a voltage follower used as a buffer, $v_O = v_x$.

For $0 < t < T_1$ it can be shown that

$$v_c(t) = V_z[1 - (1 + \gamma)e^{-t/(R_4C)}] \tag{24.27}$$

At the point of transition (positive to negative) $t = T_1 = T/2$, $v_c = \gamma v_x = \gamma V_z$, then it is found that

$$T = 2R_4C \ln \frac{1 + \gamma}{1 - \gamma} \tag{24.28}$$

The square-wave generator shown is excellent for fixed-frequency applications in the audio frequency range. The frequency may be trimmed by varying R_4. The frequency stability depends primarily on C and zener diodes. To expand the range of frequency, the op-amp A_1 must be selected carefully. If the output of the op-amp (when saturated) is constant and symmetrical, then R_1 and the zener diodes may be omitted.

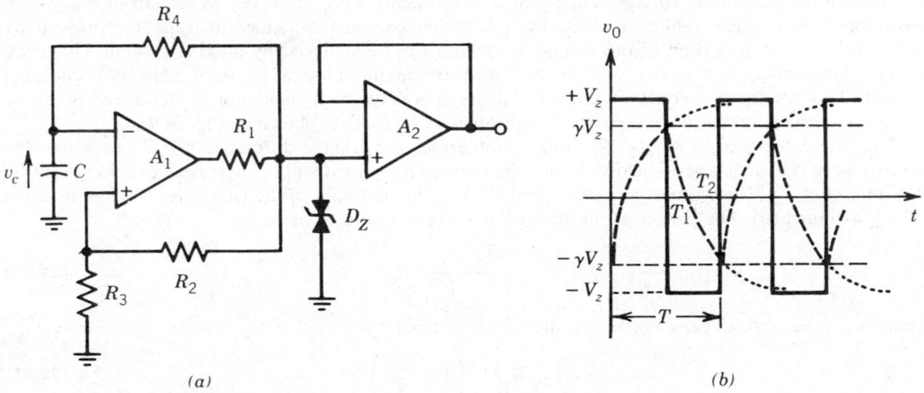

Fig. 24.7 Square-wave generator: (*a*) circuit diagram, (*b*) output waveform.

Pulse Generators

Pulse waveforms are commonly used for timing and sampling. A pulse generator circuit similar to the square-wave circuit is shown in Fig. 24.8. The resistor R_4 of Fig. 24.7*a* in the negative feedback loop is replaced by a resistance-diode network.

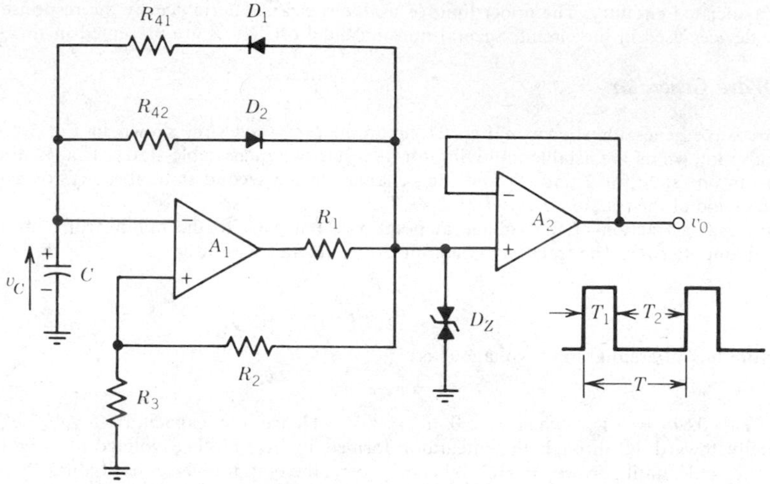

Fig. 24.8 Pulse generator.

When the output is positive, D_1 conducts and the capacitor C is charged through R_{41}. When the output is negative, D_2 conducts and the capacitor C is charged through R_{42}. If $R_{41} < R_{42}$, then $T_1 < T_2$. Hence positive-going pulses are obtained. If the diodes are reversed, or if $R_{42} < R_{41}$, then negative-going pulses are obtained. The width of the pulses is determined by

$$T_1 = R_{41}C \ln \frac{1+\gamma}{1-\gamma}, \qquad T_2 = R_{42}C \ln \frac{1+\gamma}{1-\gamma} \tag{24.29}$$

The period of the pulse train is $T = T_1 + T_2$.

Triangular-Wave Generator

A triangular-wave generator is obtained when a square wave is integrated. In Fig. 24.7, this is achieved by R_4 and C. When the capacitor voltage v_c integrates up to γV_z, the comparator reverses the slope of the integration voltage. The result is a triangular wave. However, as seen from Fig. 24.7*b*, the slope of v_c is quite nonlinear. Actually it is rather exponential. Linearity can be improved by using only the initial portion of the voltage v_c, which can be achieved by making γ small. However, better triangular-wave linearity can be achieved by maintaining a constant capacitor charging current. This results in a constant rate of change of voltage with time. For achievement of better control and greater precision, a separate integrator may be used, as shown in Fig. 24.9.

The integrator formed by A_2, R_f, and C_f integrates the voltage difference $v_s - V_s$ in which the polarity of v_s is changed periodically. Hence the voltage integrated will be increased and decreased by the amount of V_s. The symmetry is controlled by V_s. The midpoint of the triangular wave is adjusted by V_{os} as indicated. The period of the triangular wave is found to be

$$T = \frac{2V_{pp}V_z}{V_z^2 - V_s^2} \cdot R_f C_f \tag{24.30}$$

where V_{pp}, the peak-to-peak voltage of the wave, is given by

$$V_{pp} = 2V_z\left(1 + \frac{R_3}{R_2}\right) \tag{24.31}$$

Hence the amplitude of the triangular wave is adjusted by the ratio R_3/R_2 and V_z. When V_{pp} is fixed, the frequency of the oscillation is determined by R_f and/or C_f.

Sawtooth Generator

Sawtooth generators are similar to triangular-wave generators in that a linear ramp is generated, as shown in Fig. 24.10*b*. A sawtooth waveform is used in sweep testing and display applications. To obtain such a waveform, a pulse train is integrated. In this case, T_1 and T_2 are given by Eq. (24.29).

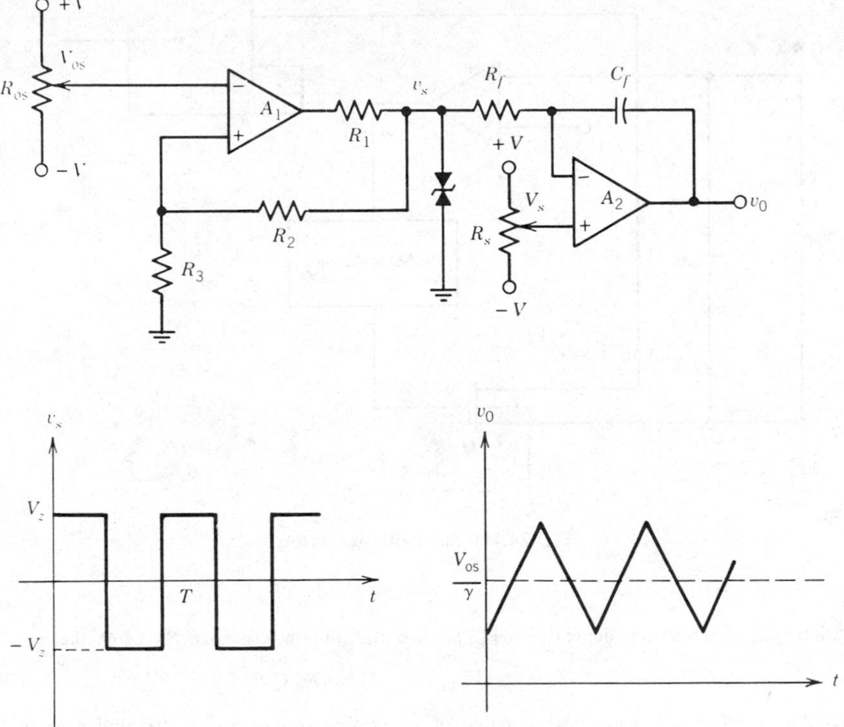

Fig. 24.9 Triangular-wave generator.

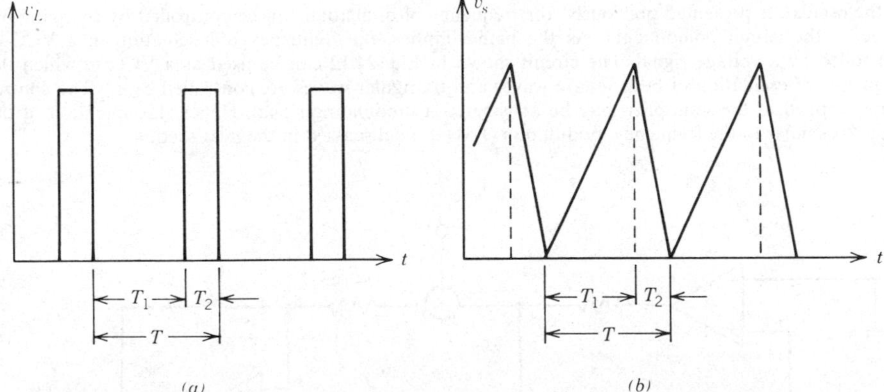

Fig. 24.10 (*a*) Pulse train. (*b*) Sawtooth waveform obtained from the pulse train in (*a*).

Another approach to obtaining a sawtooth waveform is shown in Fig. 24.11. If I is a constant current source, then

$$v_c(t) = \frac{I}{C}t \tag{24.32}$$

which is a linear ramp function. Voltage $v_c(t)$ is then applied to comparator A_1. When v_c reaches a predetermined amplitude, V_{REF}, the comparator triggers the monostable which acts as a driver for Q. This, in turn, discharges capacitor C. The cycle then repeats itself. Amplifier A_2 acts as a buffer. The

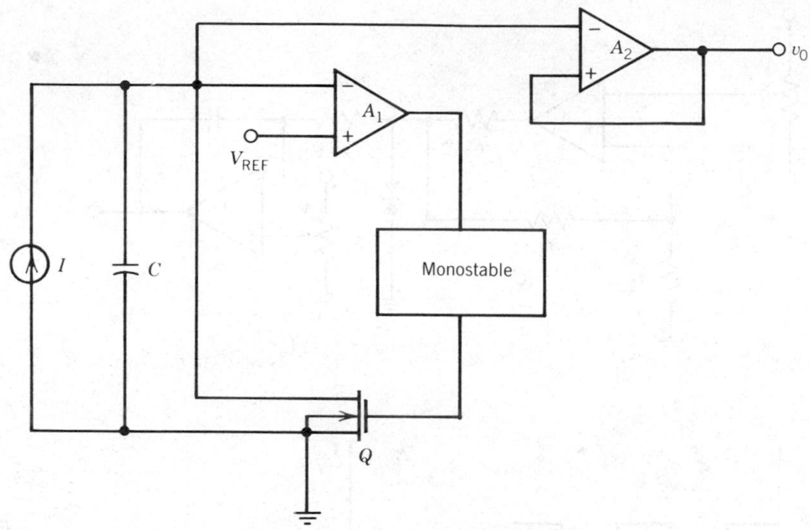

Fig. 24.11 Sawtooth generator.

repetition rate, or frequency, depends on V_{REF}, current I, the on-resistance R_{ds} of Q, that is,

$$T_1 \cong \frac{C}{I} \cdot V_{REF}, \qquad T_2 \cong 4R_{ds}C \tag{24.33}$$

Several multifunction generators capable of generating square waves, triangular waves, sine waves, and so on are available (Intersil 8038, Signetics 566, Exar 2206, etc.).

Voltage-Controlled Oscillator (VCO)

In the oscillators presented previously, the frequency of oscillation can be controlled by changing the value of the circuit components. As the name implies, the frequency of oscillation in a VCO is controlled by a voltage signal. The circuit shown in Fig. 24.12 can be used as a VCO in which the frequency of oscillation of both square waves and triangular waves are controlled by v_c. The control signal applied to the multiplier may be considered a modulating signal. Hence the operation of the circuit is similar to the frequency modulation (FM) to be discussed in the next section.

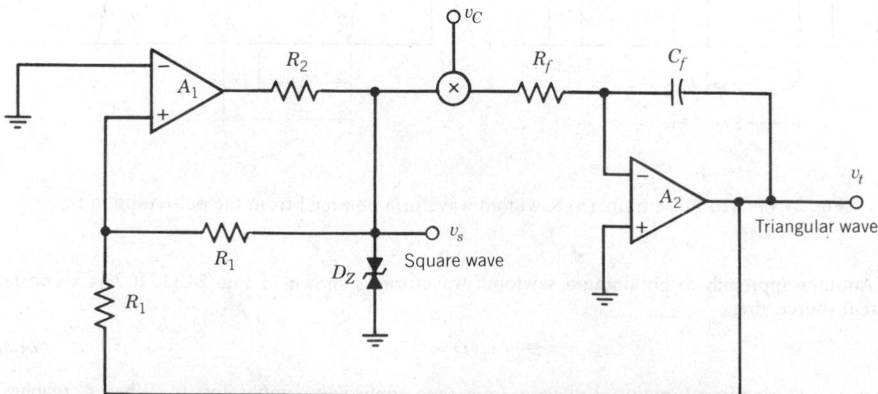

Fig. 24.12 Frequency controlled square-wave and triangular-wave oscillator.

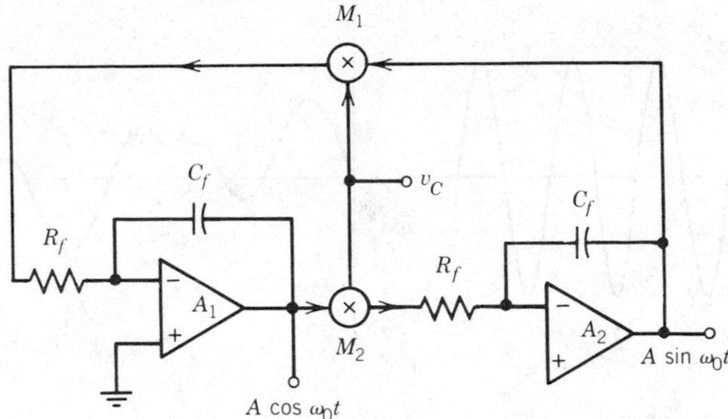

Fig. 24.13 Sine-wave generator whose frequency is controlled by v_c.

The oscillation of frequency is given by

$$f_0 = \frac{v_c}{20\pi R_f C_f} \tag{24.34}$$

Note that the oscillation frequency is a linear function of the control voltage v_c.

The frequency of oscillation of a sine-wave oscillator can be controlled similarly. The circuit shown in Fig. 24.13 is a sinusoidal VCO whose frequency is controlled by the voltage v_c. The frequency of oscillation is given by Eq. (24.34). The devices M_1 and M_2 are analog multipliers.

24.2 MODULATION

Communication systems transmit information from one place to another. However, the frequency of transmission is of great importance, since it affects the size of the antenna significantly. For example, if frequencies to be transmitted are below 100 kHz, the size of the antennas at the transmitting and receiving ends becomes very large. For frequencies above 30 MHz, the transmission is essentially line of sight. In a radio transmitter, an oscillator generates the basic radio frequency signal, known as the *carrier*. This signal has a frequency much greater than the highest frequency to be transmitted. The carrier itself has no "intelligence." The intelligence, that is, the information to be transmitted, is added to the carrier before the transmission. The process of charging the carrier with intelligence is called *modulation*. The intelligence is recovered at the receiving end by demodulating (detecting) the received signal.

The carrier may be modulated in any of several ways such as in *amplitude*, in *frequency*, or in *phase*. The corresponding modulation is referred to as amplitude modulation (AM), frequency modulation (FM), or phase modulation (PM). We also classify modulation as being either continuous modulation or pulse modulation. Continuous modulation is denoted when the carrier is a continuous sinusoidal wave and pulse modulation is denoted when the carrier is a pulse (sampled data).

24.2-1 Amplitude Modulation

Amplitude modulation (AM) is the process of varying the magnitude of the carrier in accordance with that of another signal. When the carrier is modulated by the signal to be transmitted, its amplitude has the same shape as that of the modulating waveform. This is illustrated in Fig. 24.14.

Let the carrier waveform be

$$x_c(t) = X_c \cos \omega_c t \tag{24.35}$$

where $\omega_c = 2\pi f_c$ is the carrier frequency. Suppose that $x_m(t)$ is the modulating signal. Then amplitude modulation is accomplished in the following manner

$$x(t) = [X_c + kx_m(t)]\cos \omega_c t \tag{24.36}$$

where k is a constant. To study amplitude modulation, assume that the modulating signal is a sinusoid and is expressed as

$$x_m(t) = X_m \cos \omega_m t \tag{24.37}$$

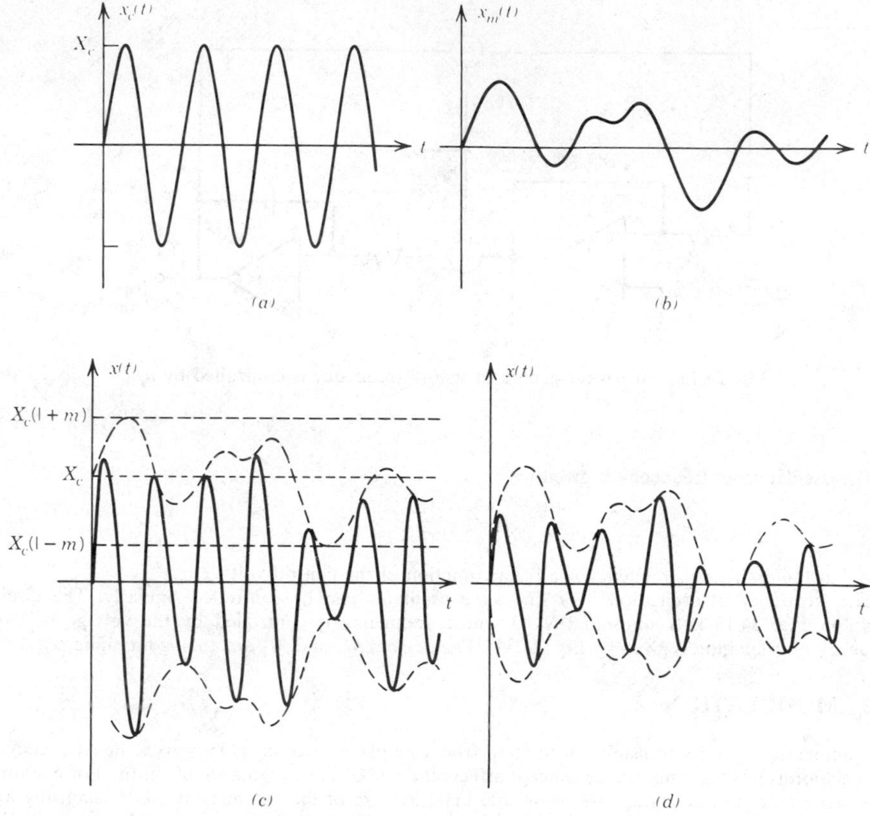

Fig. 24.14 Amplitude modulation: (*a*) carrier, (*b*) modulating signal, (*c*) amplitude-modulated carrier, (*d*) overmodulated carrier.

Thus

$$x(t) = X_c[1 + m\cos\omega_m t]\cos\omega_c t \tag{24.38}$$

where $m = (kX_m)/X_c$ is called the *index of modulation*. Observe that the amplitude of the modulated signal, $x(t)$, varies in time, which underscores the meaning of amplitude modulation. In practice, $0 \leqslant m \leqslant 1$. When $m = 1$, the amplitude of the modulated signal varies between $2X_c$ and 0. In this case the carrier is said to be 100% modulated. If $m > 1$, the carrier is completely interrupted for a time, that is, the envelope of the carrier no longer has the same form as the modulating signal, and the carrier is said to be overmodulated (Fig. 24.14*d*). It should be noted that the carrier frequency ω_c must be much greater than the rate of variation of $x(t)$ for an envelope to be observed.

Expansion of Eq. (24.38) and application of the trigonometric relations yields

$$x(t) = X_c\cos\omega_c t + \tfrac{1}{2}mX_c\cos(\omega_c + \omega_m)t + \tfrac{1}{2}mX_c\cos(\omega_0 - \omega_m)t \tag{24.39}$$

Observe that the carrier is unchanged and that two additional frequencies, $(\omega_c + \omega_m)$ and $(\omega_c - \omega_m)$, called sidebands, have been produced during modulation. Hence the bandwidth of the signal $x(t)$ is given by

$$B = \frac{2\omega_m}{2\pi} = 2f_m \tag{24.40}$$

$$\begin{aligned} &\textit{Carrier} && X_c\cos\omega_c t \\ &\textit{Upper sideband} && \tfrac{1}{2}mX_c\cos(\omega_c + \omega_m)t \\ &\textit{Lower sideband} && \tfrac{1}{2}mX_c\cos(\omega_c - \omega_m)t \end{aligned} \tag{24.41}$$

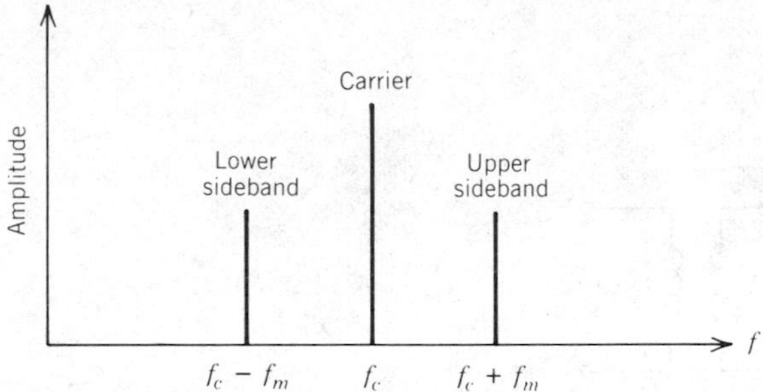

Fig. 24.15 Frequency spectrum of an amplitude-modulated signal.

The frequency spectrum of an amplitude-modulated waveform is shown in Fig. 24.15.

Therefore a tuned amplifier used to amplify a modulated carrier should have sufficient bandwidth to include both sideband frequencies. Observe that the modulating frequency is not a frequency component in the modulated wave. If the tuned circuits do not have sufficient bandwidth, the highest modulating frequencies will not be reproduced by the receiving device.

The root-mean-square (rms) power contained in the carrier and sidebands is determined as follows:

$$\textit{Carrier power} \quad P_c = M\frac{V_c^2}{2} \tag{24.42}$$

$$\textit{Sideband power} \quad P_{sb} = M\frac{m^2V_c^2}{4} \tag{24.43}$$

where M is a constant measured in siemens. Note that

$$P_{sb} = \frac{m^2}{2}P_c \tag{24.44}$$

For $m = 1$ (i.e., in the case of 100% modulation), the sideband power is only one half that of the carrier power for sinusoidal modulating waveforms. It differs from other waveforms. For example, if the carrier is 100% modulated by a square wave, the amplitude of the signal will be $2V_c$ for half of the period and 0 for the other half. The total power is then

$$P_c + P_{sb} = \frac{M}{2}\left(\frac{4V_c^2}{2}\right) = MV_c^2 \tag{24.45}$$

where P_c is given by Eq. (24.42). Hence for 100% modulation by a square wave the sideband power is

$$P_{sb} = P_c \tag{24.46}$$

Amplitude modulation is inefficient, since much of the power is contained in the carrier. To increase efficiency, the carrier is suppressed. At the receiver end, an oscillator generates a signal that replaces the carrier. However, if the frequency of the oscillator does not match the frequency of the carrier, excessive distortion results. A procedure used to transmit video in TV broadcasting is to partially suppress the carrier and one sideband. This, too, reduces power requirements and bandwidth. This is called *vestigial sideband transmission*. The effect of distortion on the TV picture in this case is not critical. *Single sideband transmission* is also used if the above problems are not critical. The modulation is said to be linear when the envelope of the modulated wave has the same waveform as that of the modulating signal.

Modulating Circuits

A large variety of circuits provide amplitude modulation. To illustrate the basic principles, several typical transistor circuits are presented. The most commonly used transistor AM circuits can be divided into two groups: collector modulation and base modulation. In either case, the amplifiers are often operated class C (or class B if the power level is not too high).

Collector Modulation. A typical collector-modulated transistor circuit using a class-C RF amplifier is shown in Fig. 24.16. The base drive must be large enough to saturate the transistor at the peak of

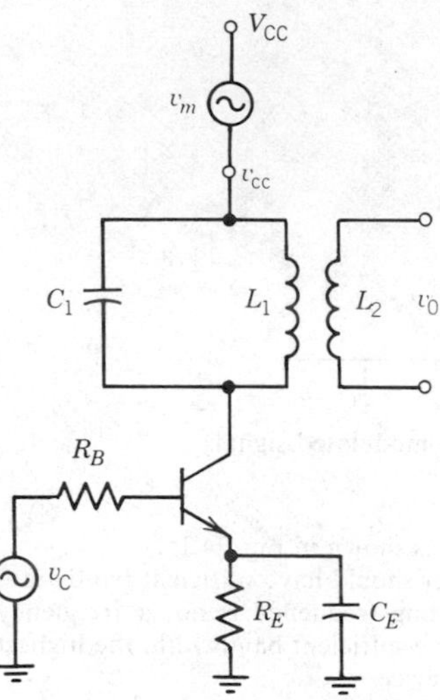

Fig. 24.16 Collector modulator. Modulating signal: $v_m = V_m \cos \omega_m t$; carrier signal: $v_c = V_c \cos \omega_c t$.

the modulation cycle to provide linear modulation. The harmonic distortion produced by the amplifier is removed by the tuned circuit associated with the amplifier. The amplitude of the output signal, when the transistor is saturated, is equal to the power supply voltage. Thus a change in the power supply voltage changes the output signal proportionally

$$v_{CC} = V_{CC} + v_m = V_{CC} + V_m \cos \omega_m t$$

and the AM waveform is given by

$$v_0 = V_{CC}\left(1 + \frac{V_m}{V_{CC}} \cos \omega m t\right) \cos \omega_c t \tag{24.47}$$

For a 100% modulation, $V_m = V_{CC}$.

In collector modulation, since the transistor is driven into saturation excess charge is stored in the base. Thus collector current cannot decrease until the excess charge is removed; in fact, this current may even increase when it would normally decrease. This out-of-phase current would decrease the efficiency of the amplifier.

Base Modulation. Amplitude modulation can also be obtained by varying the base current. A typical circuit is shown in Fig. 24.17. The capacitors C and C_0 are chosen so that they present low impedance to the carrier signal and high impedance to the modulating signal. Then the output signal v_0 is found to be

$$v_0 = K v_C [V_{BB} - V_{BE,ON} + v_m]$$

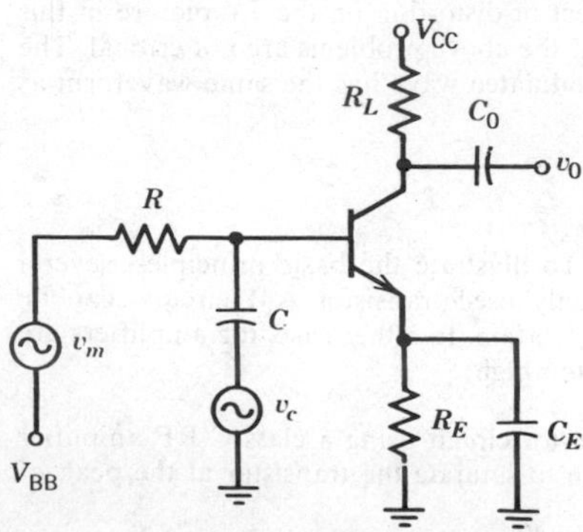

Fig. 24.17 Class-A base modulator. Modulating signal: $v_m = V_m \cos \omega_m t$; carrier signal: $v_c = V_c \cos \omega_c t$.

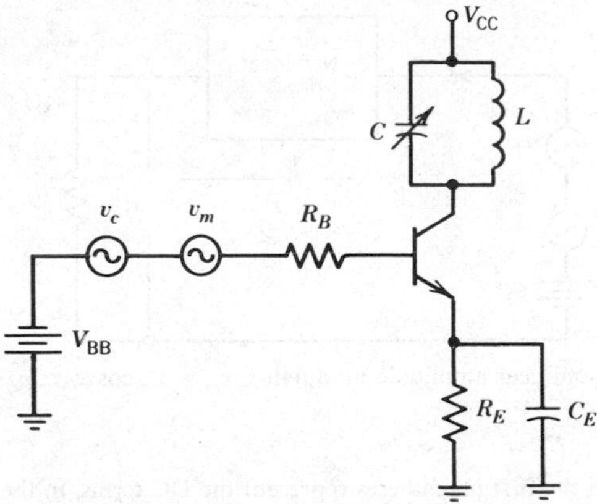

Fig. 24.18 Class-C (or class-B) base modulator.

where

$$K = \frac{\beta R_L}{0.026R}$$

Letting $V = V_{BB} - V_{BE,ON}$ yields $v_0 = KVv_C(1 + v_m/V)$ or

$$v_0 = E_C(1 + m\cos\omega_m t)\cos\omega_c t \tag{24.48}$$

where $E_C = -KV \cdot V_C$ and $m = V_m/V$.

Base modulation has the advantage of significantly reducing power requirements on the modulator. The main disadvantages are that more distortion is produced and that the modulator's efficiency is lower than that of collector modulators. Hence when the power level is high, collector modulators are often used.

To improve efficiency and obtain a higher index of modulation, a class-C (or class-B) amplifier may be used (Fig. 24.18).

The amplifier operates as a class-C amplifier when $V_{BB} = 0$. The collector resonator circuit, formed by L and C, resonates at the frequency ω_c. The output voltage v_0 can be shown to be

$$v_0 = KV_C\left[\phi(t) - \tfrac{1}{2}\sin 2\phi(t)\right] \tag{24.49}$$

where

$$K = \frac{\beta R}{\pi R_B} \quad \text{and} \quad \phi(t) = \cos^{-1}\frac{(V_{BB} - V_m\cos\omega_m t)}{C}$$

R = resonant collector impedance.

It must be noted that adjustment of the base-modulated amplifier is more critical than that of the collector-modulated amplifier and that a high degree of linearity is more difficult to accomplish.

Nonlinear Modulation

The nonlinear modulator shown in Fig. 24.19 is often used in low-power systems to obtain an amplitude modulation.

It will be assumed that the nonlinear device has an input/output characteristic given by

$$i_D = Kv_D^2 \tag{24.50}$$

This device is then called square-low modulator. If the load resistance R_L is chosen so that $v_L = i_D R_L$ has no significant effect on v_D, then

$$v_D \cong V_0 + V_m\cos\omega_m t + V_C\cos\omega_c t \tag{24.51}$$

Using Eq. (24.51) in Eq. (24.50) and manipulating yields

$$i_D = K\left(V_0^2 + \tfrac{1}{2}V_C^2 + \tfrac{1}{2}V_m^2\right) + \frac{K}{2}\left(V_C^2\cos 2\omega_c t + V_m^2\cos 2\omega_m t\right) + 2KV_0(V_C\cos\omega_c t + V_m\cos\omega_m t)$$
$$+ KV_C V_m[\cos(\omega_c - \omega_m)t + \cos(\omega_c + \omega_m)t] \tag{24.52}$$

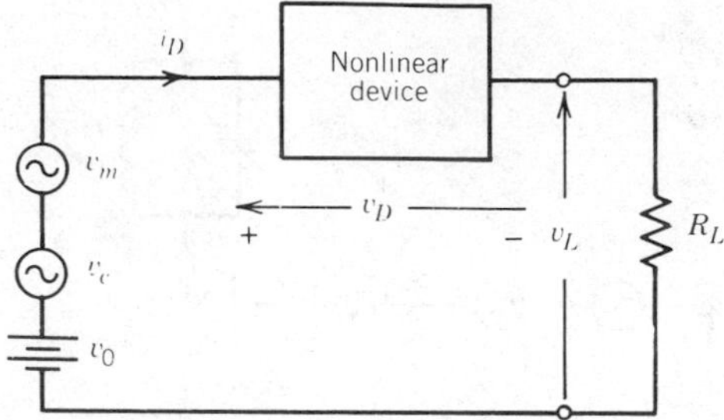

Fig. 24.19 Nonlinear amplitude modulator. $v_m = V_m \cos \omega_m t$; $v_c = V_c \cos \omega_c t$.

Note that the terms in the first parentheses represent the DC terms, in the second parentheses the second harmonic terms, in the third parentheses the fundamental terms, and finally in the last parentheses the sideband frequencies. If $\omega_c \gg \omega_m$ and if the signal is passed through a bandpass filter that passes only those signals close to ω_c, the resulting output voltage would be

$$v_L = R_L K \{2V_0(V_C \cos \omega_c t + V_m \cos \omega_m t) + V_C V_m [\cos(\omega_c - \omega_m)t + \cos(\omega_c + \omega_m)t]\} \tag{24.53}$$

where the amplitude modulation index m is given by

$$m = \frac{E_m}{E_0}$$

It is interesting to observe that when the DC bias $V_0 = 0$, the carrier frequency term in Eq. (24.53) disappears and only the upper and lower sideband frequencies remain. The resulting modulation is called double sideband suppressed carrier (DSBSC).

Square-low amplitude modulation can be achieved by operating a class-A amplifier in the nonlinear region of its characteristic. However, the circuit is very inefficient and the distortion should be very high. Hence, the use of square-low modulation is limited.

Single Sideband Transmission. The sidebands carry the intelligence in a modulated wave, but sideband power is, in general, less than carrier power. Moreover, both upper and lower sidebands carry the same information. Therefore, for increased efficiency the carrier and one sideband may be eliminated. This type of transmission is called *single sideband transmission*; the power and bandwidth requirements are reduced for a specific transmission efficiency. In single sideband transmission, the carrier is inserted in the receiver end to recover the intelligence. However, the local oscillator providing the carrier must have good frequency stability in order for the carrier to recover the modulating signal. A balanced modulating circuit, shown in Fig. 24.20, can be used to suppress the

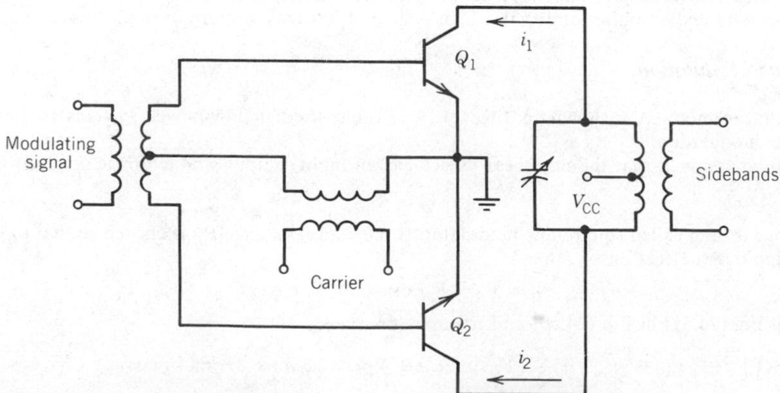

Fig. 24.20 Balanced modulating circuit.

carrier and to select one of the two sidebands. Observe that the carrier is applied in phase to the inputs of the two transistors. These transistors are operated in class B, which causes the carrier to be suppressed. The modulating signal, on the other hand, is applied in opposite phase to the two inputs, and the output is tuned to the desired sideband frequency to eliminate essentially the modulating signal. Additional filtering may be required to eliminate the undesired sideband.

The effective output voltage can be shown to be

$$v_0 = A\{\cos[(\omega_c + \omega_m)t] + \cos[(\omega_c - \omega_m)t]\} \tag{24.54}$$

Observe that the output contains only sideband frequencies, as expected. Only one of these sidebands is transmitted.

24.2-2 Frequency and Phase Modulation

In frequency modulation (FM), the frequency of a carrier is varied in accordance with that of a modulating signal. Consider the sinusoidal signal

$$v(t) = V(t) \cdot \cos[\omega_c t + \phi(t)] \tag{24.55}$$

In amplitude modulation, $\phi(t)$ is kept constant and $V(t)$ is varied proportionately to the modulating signal $v_m(t)$. In this as in preceding sections, we investigate the case in which $V(t) = V$ constant and the phase angle $\phi(t)$ is varied proportionally to $v_m(t)$. Let

$$\theta(t) = \omega_c t + \phi(t) \tag{24.56}$$

where $\phi(t)$ is called the instantaneous phase angle. The instantaneous angular frequency is defined as

$$\omega_i(t) = \frac{d\theta(t)}{dt} \tag{24.57}$$

If the instantaneous angular frequency $\omega_i(t)$ varies in accordance with the modulating signal, the resulting modulation is called frequency modulation, that is,

$$\omega_i(t) = \omega_c + K_f v_m(t) \tag{24.58}$$

where K_f is a constant. Using Eq. (24.58) in Eq. (24.57) it is found that

$$\theta(t) = \omega_c t + K_f \int v_m(t)\,dt \tag{24.59}$$

Thus the frequency-modulated signal is

$$v(t) = V\cos\left[\omega_c t + K_f \int v_m(t)\,dt\right] \tag{24.60}$$

Let the modulating signal $v_m(t)$ be given by

$$v_m(t) = V_m \cos \omega_m t \tag{24.61}$$

Substitution of Eq. (24.61) in Eq. (24.59) yields

$$\theta(t) = \omega_c t + \frac{K_f V_m}{\omega_m} \sin \omega_m t \tag{24.62}$$

where

$$m_f = \frac{K_f V_m}{\omega_m} \tag{24.63}$$

is called the index of frequency modulation.

$$v(t) = V\cos(\omega_c t + m_f \sin \omega_m t) \tag{24.64}$$

In the case of phase modulation, PM, the phase angle $\theta(t)$ is varied in accordance with a modulating signal. For example, let

$$\phi(t) = K_2 v_m(t) \tag{24.65}$$

Then the phase-modulated signal will be of the form [with $v_m(t) = V_m \sin \omega_m t$]

$$v(t) = V\cos[\omega_c t + K_2 v_m(t)] \tag{24.66}$$

or

$$v(t) = V\cos(\omega_c t + K_2 V_m \sin\omega_m t) \tag{24.67}$$

where

$$m_p = K_2 V_m \tag{24.68}$$

is called the *index of phase modulation* and thus

$$v(t) + V\cos(\omega_c t + m_p \sin \omega_m t) \tag{24.69}$$

Observe that Eqs. (24.64) and (24.69) are essentially the same. Let m_A denote the m_f or m_p. Then, expanding Eq. (24.69), we have

$$v(t) = V[\cos \omega_c t \cos(m_A \sin \omega_m t) - \sin \omega_c t \sin(m_A \sin \omega_m t)] \tag{24.70}$$

The expressions $\cos(m_A \sin \omega_m t)$ and $\sin(m_A \sin \omega_m t)$ can be expressed in terms of Bessel functions as follows

$$\cos(m_A \sin \omega_m t) = J_0(m_A) + 2J_2(m_A)\cos 2\omega_m t + 2J_4(m_A)\cos 4\omega_m t + \dots \tag{24.71}$$

$$\sin(m_A \sin \omega_m t) = 2J_1(m_A)\sin \omega_m t + 2J_3(m_A)\sin 3\omega_m t + \dots \tag{24.72}$$

Substitution of Eqs. (24.71) and (24.72) in Eq. (24.70) and manipulation yields

$$\begin{aligned} v(t) = V\{&J_0(m_A)\cos \omega_c t + J_1(m_A)[\cos(\omega_c + \omega_m)t - \cos(\omega_c - \omega_m)t] \\ &+ J_2(m_A)[\cos(\omega_c + 2\omega_m)t + \cos(\omega_c - 2\omega_m)t] + \dots\} \end{aligned} \tag{24.73}$$

where the $J_i(m_A)$s are Bessel functions of the first kind.

Observe that $v(t)$ contains the carrier and an infinite number of side frequencies spaced at intervals of ω_m on either side of the carrier frequency ω_c. This implies that FM and PM require infinite bandwidth. However, the coefficients $J_i(m_A)$ rapidly decrease for values of $i > m_A$, causing the amplitudes of the side frequencies for the higher values of i to become negligibly small. The practical bandwidth is found to be

$$\text{BW} = 2(m_A + 1) f_{m,\max} \tag{24.74}$$

where $f_{m,\max}$ is the highest modulating frequency. It should be noted that the bandwidth requirements of a frequency- or phase-modulating system are, in general, greater than the bandwidth requirements of the corresponding amplitude-modulating system. However, in the FM and PM systems (together also called angle modulation), m_A can be increased without producing nonlinear distortion so that the detected signal can be increased without increasing the power. Also, the FM system is less susceptible than the AM system to adjacent channel and on-frequency interference from other radio stations. Furthermore, the FM system is less sensitive to impulse type of noise.

It is interesting to observe from Eqs. (24.60) and (24.66) that FM and PM are closely related. For example, in a PM system, application of the modulating signal to an integrator and of its output to the input of the PM system causes the output to be FM modulated. Similarly, an FM system can be converted to a PM system by differentiating the modulating signal before it is applied to the FM modulator.

FM and PM Modulator Circuits

In sinusoidal oscillators, where an LC circuit determines the frequency of oscillation, one can generate an FM signal by varying either L or C. That is, if the value of L or C is made a function of the modulating signal $v_m(t)$, an FM system is obtained. These circuits are commonly called reactance modulator. Figure 24.21 shows typical transistor reactance modulator and FET circuits.

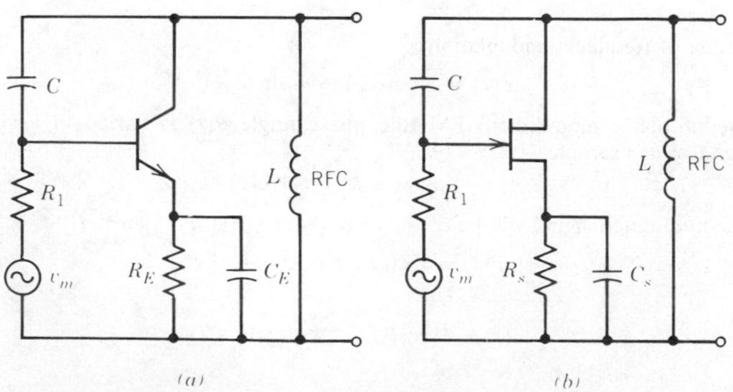

Fig. 24.21 (*a*) Transistor reactance modulator. (*b*) Field effect transistor reactance modulator.

Assuming that the bypass capacitors C_S and C_E are short circuits and that the radio frequency choke, RFC, is an open circuit at the carrier frequency, $h_{re} = 0$. Then the output admittance Y_0 is found to be

$$Y_0 = \frac{KR + 1}{R + 1/i\omega C} + \frac{1}{r_0} \tag{24.75}$$

where

$$\textit{For transistor} \quad K = \frac{h_{fe}}{h_{ie}}, \qquad r_0 = \frac{1}{h_{oe}}, \qquad r_i = h_{ie}$$

$$\textit{For FET} \quad K = g_m, \qquad r_0 = r_d, \qquad r_i = \infty$$

and $R = R_1 \| r_i$. If $1/(\omega C) \gg R$ and $KR \gg 1$, then Eq. (24.75) reduces to

$$Y_0 = \frac{1}{r_0} + jKR\omega C = \frac{1}{r_0} + j\omega C_{eff} \tag{24.76}$$

Hence the output admittance is a resistance r_0 in shunt with a capacitance C_{eff}. If the g_m of the FET or the h_{fe} of the transistor varied, the effective capacitance C_{eff} would also vary. The frequency $\omega_m \ll \omega_C$, hence $1/(\omega_m C)$ is very large and $\omega_m L$ is small at ω_m. Therefore $v_m(t)$ essentially affects only the input bias of the transistor or FET, which changes g_m or h_{fe} and thus C_{eff} varies, causing the output admittance to vary. It is for this reason that these circuits are called *reactance modulators*.

Diode Reactance Modulator Circuit. A reverse-biased semiconductor diode (varactor) can also be used to provide phase modulation. Such a circuit and its linear model is shown in Fig. 24.22. The capacitance of a varactor is given by

$$C = Kv^{-n} \tag{24.77}$$

where K and n are constants for a given varactor and $0 < n < 1$. R represents the combination of R_1 and the reverse resistance of the diode D. Note that V_{DD} should be such that the diode D is always reverse biased and C_1 acts as a short circuit at ω_m. The resonant frequency of the circuit is then given by

$$\omega_0 = \frac{1}{\sqrt{LC}} \tag{24.78}$$

Using Eqs. (24.77) and (24.78) with $v = V_{DD} + v_m$ we have

$$\omega_0 = \frac{1}{\sqrt{KL}} V_{DD}^{n/2} \left(1 + \frac{v_m}{V_{DD}}\right)^{n/2} \tag{24.79}$$

Assume that $|v_m/V_{DD}| < 1$, and expanding into a Taylor's series we have

$$\omega_0 = \frac{V_{DD}^{n/2}}{\sqrt{KL}} \left[1 + \frac{n}{2} \cdot \left(\frac{v_m}{V_{DD}}\right) + \frac{n(n-2)}{8} \cdot \left(\frac{v_m}{V_{DD}}\right)^2 + \cdots \right] \tag{24.80}$$

If $|v_m/V_{DD}| \ll 1$, then Eq. (24.80) can be approximated by

$$\omega_0 \cong \frac{V_{DD}^{n/2}}{\sqrt{KL}} \left(1 + \frac{n}{2} \cdot \frac{v_m}{V_{DD}}\right) \tag{24.81}$$

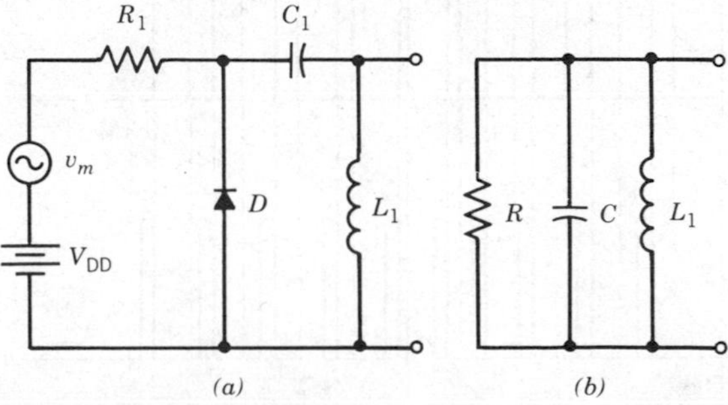

Fig. 24.22 (*a*) Diode reactance modulator. (*b*) Its linear model.

If $v_m = V_m \sin \omega_m t$, the instantaneous frequency ω_i can be written as

$$\omega_i = \frac{V_{DD}^{n/2}}{\sqrt{KL}}\left(1 + \frac{nV_m}{2V_{DD}} \cdot \sin \omega_m t\right) \tag{24.82}$$

Thus phase modulation is obtained with the index of phase modulation given by

$$m_p = \frac{nV_m}{2V_{DD}} \tag{24.83}$$

To obtain frequency modulation, the modulating signal is first integrated before it is applied to this circuit.

24.2-3 Pulse Modulation

The previous sections have dealt with continuous-wave modulations, that is, the carrier was a continuous sinusoidal wave whose parameters were varied proportional to the parameters of the modulating signal. This section discusses pulse modulation, in which sampled data are used instead of continuous signals.

The sampling theorem states "if a bandlimited signal is sampled at a rate at least as high as twice the highest frequency in the spectrum, the original signal can be recovered." Therefore the information can be transmitted in sampled data form and then converted into analog form. In pulse modulation, the sampled values directly modulate a periodic pulse train. For example, the amplitude, width, or portion of the pulse waveform can be varied proportionally to the sampled signal. This is referred to as pulse-amplitude modulation (PAM), pulse-width modulation (PWM), or pulse-position modulation (PPM), respectively. See Fig. 24.23.

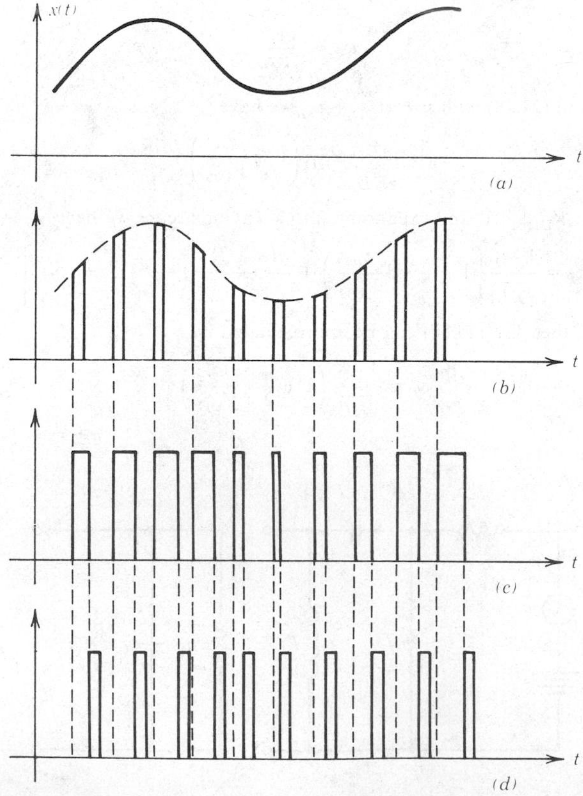

Fig. 24.23 Types of pulse modulation: (*a*) modulating signal, (*b*) pulse-amplitude modulation, (*c*) pulse-width modulation, (*d*) pulse-position modulation.

Observe that in PAM, the period and duty cycle remain constant as the amplitude of the pulses varies corresponding to the amplitude of the modulating signal. In PWM, the amplitude and period remain constant while the width of the pulses varies in accordance with the width of the modulating signal. Finally, in PPM both the amplitude and width are kept constant while the position of the pulses varies in accordance with that of the modulating signal.

Pulse-Amplitude Modulation (PAM)

A circuit performing PAM is shown in Fig. 24.24*a*. A pulse train v_p is applied to the base of the transistor Q. This pulse train switches the transistor on and off. The modulating voltage v_m varies from 0 to $-V_m$. When the amplitude of the pulse reaches $v_p = V_p$, the transistor is off and the output $v_O = -(v_m + V_{REF})$. When $v_p = 0$, the transistor is saturated and $v_O = -(V_{REF} + 2V_{CE,SAT})$. Hence the amplitude of the output pulses will be varied in accordance with that of the modulating signal, as illustrated in Fig. 24.24*b*. (Recall that $V_{CE,SAT} \leqslant 0.2$ V for a switching transistor.)

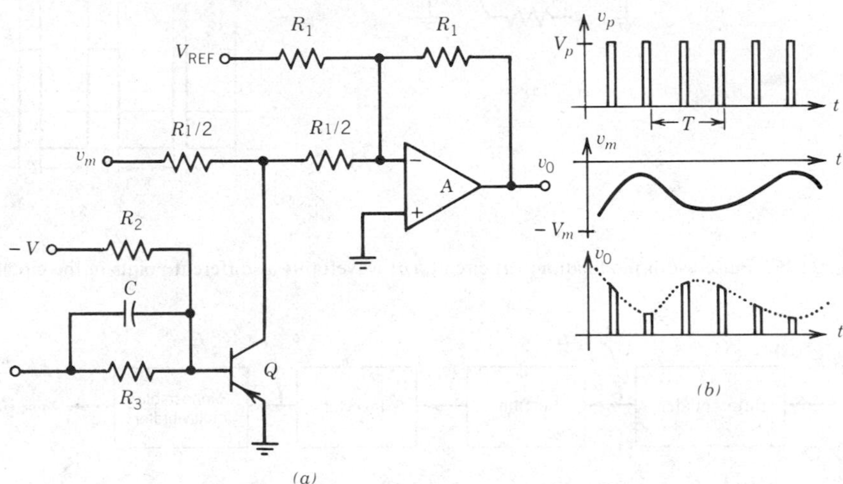

Fig. 24.24 Pulse-amplitude modulation: (*a*) circuit, (*b*) waveforms.

Pulse-Width Modulation (PWM)

PWM is also known as pulse-duration modulation (PDM). A circuit performing PWM is shown in Fig. 24.25*a*. The slowly varying modulating signal v_m controls the width of pulses, and the carrier input pulses are usually square wave.

Hence the output of the integrator v_1 is a triangular wave. A negative modulating signal is applied to the comparator A_2 whose output swings may be limited by two zener diodes V_{z_1} and V_{z_2}. The output PWM signal is obtained, as illustrated in Fig. 24.25*b*. Note that the widths of pulses $T_1, T_2, \ldots$ are different. The width of the pulse can be determined by

$$T_i = T \cdot \frac{V_p - v_m}{2V_p} \tag{24.84}$$

Hence the duration of the output pulse is proportional to the modulating signal. PWM is often used where the remote proportional control of a position or a position rate is desired.

Pulse-Position Modulation (PPM)

PPM is closely related to PWM. In fact, it can be generated from PWM, as illustrated in Fig. 24.26. A pulse-width modulated signal is first applied to a differentiator whose output is alternating impulses $x_1(t)$, as shown in Fig. 24.26*c*. These pulses are then rectified by a half-wave rectifier and inverted, providing the impulses $x_2(t)$ corresponding to negative edges of the PWM signal, as seen in Fig. 24.26*d*. Hence the impulses are positioned according to the modulating signal. This impulse train is then used to trigger a monostable multivibrator, giving the PPM signal $x_{PPM}(t)$.

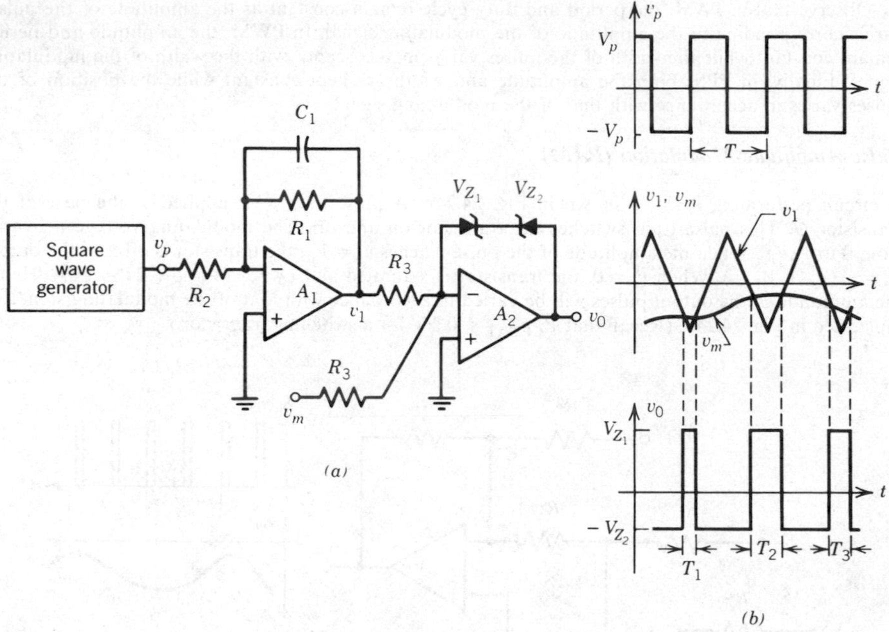

Fig. 24.25 Pulse-width modulation: (*a*) circuit, (*b*) waveforms at different points in the circuit.

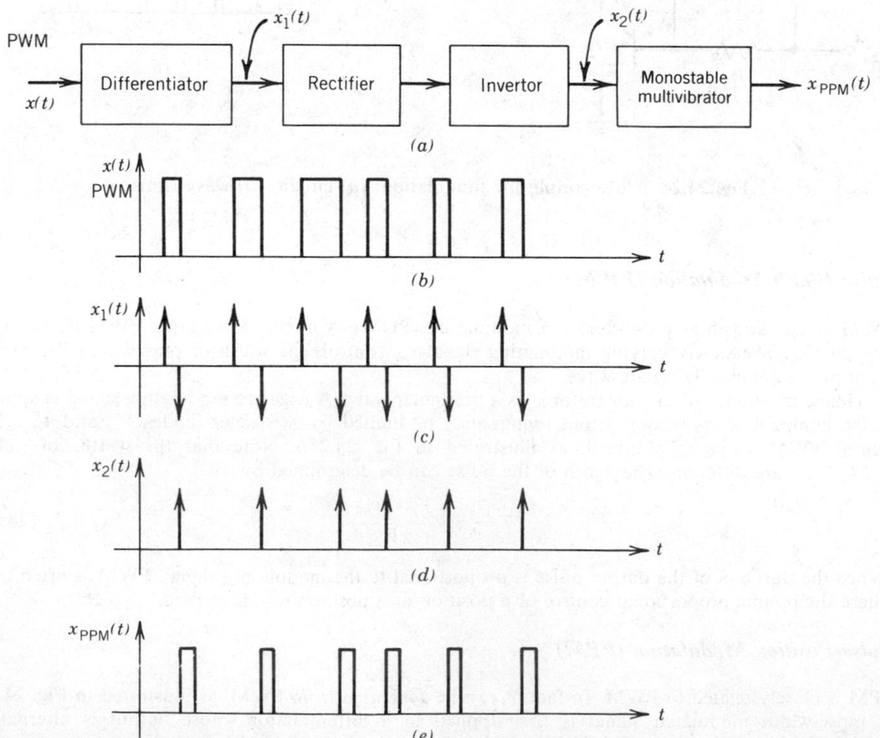

Fig. 24.26 (*a*) Generation of pulse-position modulation (PPM) from pulse-width modulation (PWM). (*b, c, d, e*) Pulse waveforms at different points.

The modulation techniques discussed up to now are categorized as analog modulation. There are a number of digital communication techniques, for example, pulse-code modulation (PCM), amplitude-shift keying (ASK), frequency-shift keying (FSK), phase-shift keying (PSK). These are discussed in Chapter 48.

24.3 DEMODULATORS

In Section 24.2 several modulation processes (in which electrical information is superimposed on a carrier for transmitting intelligence information) have been discussed. At the receiving end, the modulated carrier is received and amplified to a level at which the original signal can be recovered. The recovery process is known as demodulation or detection. In this section demodulation circuits (for each modulation method) are discussed.

24.3-1 AM Demodulators

Linear Diode Demodulator

The linear diode AM demodulator is simply a half-wave rectifier containing an RC filter, as shown in Fig. 24.27*a*. The demodulator circuit is designed to filter out the variations at the carrier frequency ω_c and retain the variations at the modulating signal ω_m. At the positive peak of the input signal, the capacitor charges to this peak voltage. If

$$RC \gg \frac{2\pi}{\omega_c} \tag{24.85}$$

where ω_c is the carrier frequency, as the input signal slightly falls off (i.e., $v_O > v_i$), the diode will be cut off and the capacitor will discharge through R. This will continue until $v_O = v_i$, at which time the

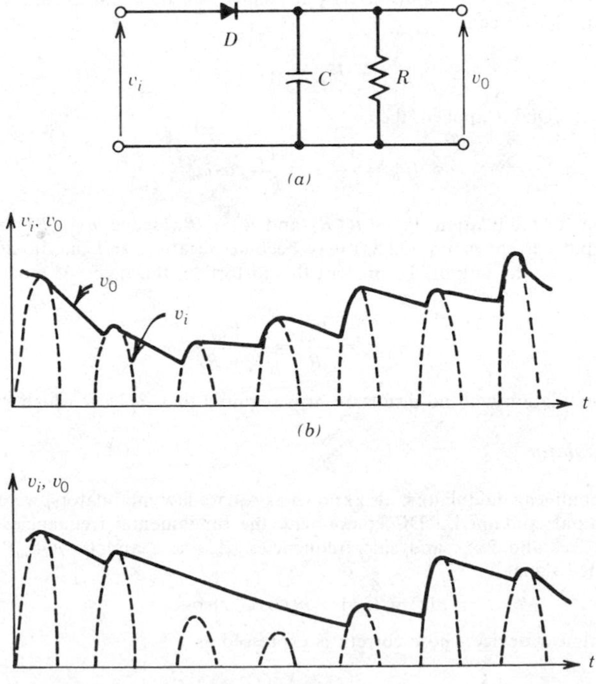

Fig. 24.27 Linear amplitude-modulation (AM) diode demodulator: (*a*) circuit, (*b*) input and output waveforms, (*c*) input and output waveforms conditions when Eqs. (24.86) and (24.87) are not satisfied.

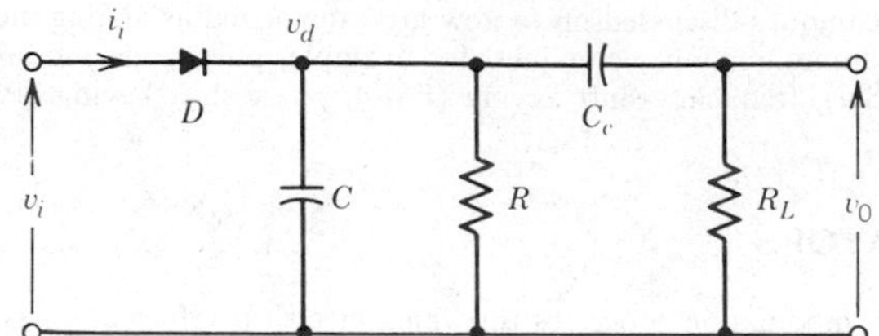

Fig. 24.28 Linear amplitude-modulation modulator with RC-coupled load.

diode will start conducting and the cycle will repeat itself. The output voltage will therefore have a "rippled" waveform, as shown in Fig. 24.27*b*. If the RC time constant is too large, the output will not be able to fall off quickly enough and hence the waveform in Fig. 24.27*c* will be obtained. Observe that, in this case, the output voltage will not follow the variations in amplitude of the input signal v_i. To avoid this problem, the following condition must be satisfied

$$RC \ll \frac{2\pi}{\omega_{m,\max}} \tag{24.86}$$

where $\omega_{m,\max}$ is the highest frequency in the modulating signal. The rippled output voltage is filtered out by subsequent amplification stages, and the final output signal becomes a smooth signal corresponding to the original waveform used as the modulating signal at the transmitting end.

The output signal of the demodulator will, generally, have a DC component that may be undesirable. To block this component, a coupling capacitor C_c is included, as shown in Fig. 24.28. The value of C_c is chosen such that it will act effectively as a short circuit at all modulating signal frequencies.

If an unmodulated carrier is applied, the voltage v_d will be a DC signal whose magnitude is equal to the amplitude of the carrier V_c, and the output voltage will be equal to zero. Hence the average current drawn from the source is

$$I_{i,\text{av}} = \frac{V_c}{R} \tag{24.87}$$

If the modulated signal is applied, then

$$I_{i,\text{av}} = \frac{V_c}{R} + m\frac{V_c}{R_{\text{ac}}}\cos\omega_m t \tag{24.88}$$

where m is the index of modulation, $R_{\text{ac}} = R \parallel R_L$ and $R_{\text{ac}} < R$. Hence mV_c/R_{ac} may become larger than V_c/R, the input current in Eq. (24.88) may become negative, and the diode will not conduct. This causes distortion at the output. To prevent this distortion, the index of modulation m must be limited such that

$$m_{\max} = \frac{R_{\text{ac}}}{R} = \frac{R_L}{R + R_L} \tag{24.89}$$

To obtain 100% modulation with no distortion, it is required that $R_L \gg R$ which yields $m \cong 1$.

Nonlinear Demodulator

In Section 24.2-1 nonlinear modulators, also known as square-law modulators, were presented. It was shown that the output contained a DC component, the fundamental frequencies ω_c and ω_m, their second harmonics $2\omega_c$ and $2\omega_m$, and side frequencies $(\omega_c - \omega_m)$ and $(\omega_c + \omega_m)$. Suppose that an amplitude-modulated signal

$$v(t) = V_c(1 + m\cos\omega_m t)\cos\omega_c t \tag{24.90}$$

is applied to a nonlinear device whose current is expressed as

$$i(t) = Kv(t)^2 \tag{24.91}$$

or

$$i(t) = K[V_c(1 + m\cos\omega_m t)\cos\omega_c t]^2 \tag{24.92}$$

Manipulation of this equation yields the following:

$$\textit{DC term} \quad KV_c^2/2 \tag{24.93}$$

$$\textit{Fundamental term} \quad KV_c^2 m \cos \omega_m t \tag{24.94}$$

$$\textit{Second-harmonic term} \quad \tfrac{1}{4} KV_c^2 m^2 \cos 2\omega_m t \tag{24.95}$$

The output also contains the following high-frequency terms: ω_c, $2\omega_c$, $2\omega_c - \omega_m$, $2\omega_c + \omega_m$, $2(\omega_c - \omega_m)$, $2(\omega_c + \omega_m)$. These frequencies are all above the audio frequency range and are eliminated by the demodulator filter and amplification stages. The DC component is eliminated by the coupling capacitor. Hence the output contains only the desired signal $KV_c^2 m \cos \omega_m t$ (the fundamental component) and the second-harmonic term (which causes an amplitude distortion).

Nonlinear demodulators are used in mixers that are operated in the nonlinear region.

24.3-2 FM Demodulators

As has been discussed in Section 24.2, frequency and phase modulation can be obtained from one another. For example, a frequency-modulating circuit can be converted to a phase-modulating circuit by differentiating the modulating signal before applying it to the frequency modulator. Hence an FM demodulator can be converted to a PM demodulator by integrating the output.

Slope Demodulator

A resonant circuit tuned to a frequency slightly different from the carrier frequency of the FM signal can be used as an FM demodulator. Consider the frequency response of a resonant circuit, as shown in Fig. 24.29. Observe that the center frequency ω_0 of the resonant circuit is set less than the carrier frequency ω_c of the FM signal. As the frequency varies about ω_c, the output voltage varies. The variation of the output with frequency along the slope AB can be considered linear if $\Delta\omega_c$ is small. Thus an amplitude-modulated signal is produced from a frequency-modulated signal. Then this signal is detected by an amplitude-modulation demodulator, as discussed earlier. Therefore the final output will be demodulated FM signal.

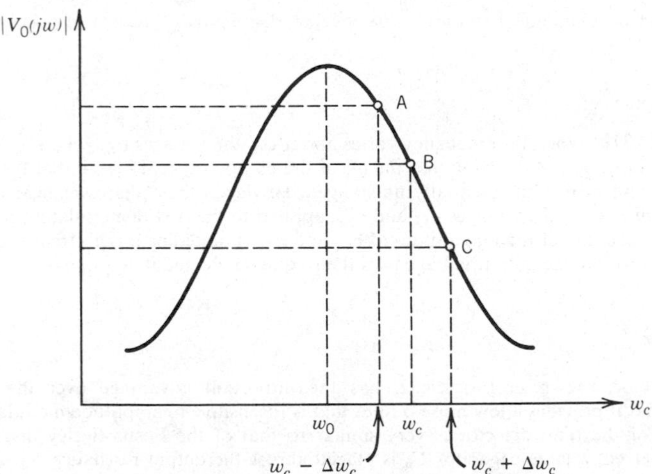

Fig. 24.29 Response of a tuned amplifier.

Phase-Shift Demodulator

For improved linearity, other types of demodulators must be used. The Foster-Seeley phase-shift discriminator is a quite linear FM demodulator. Its circuit is shown in Fig. 24.30. When the secondary circuit is tuned to resonance at ω_0, the voltage v_1 leads v_i by 90° and v_2 lags v_i by 90°, as

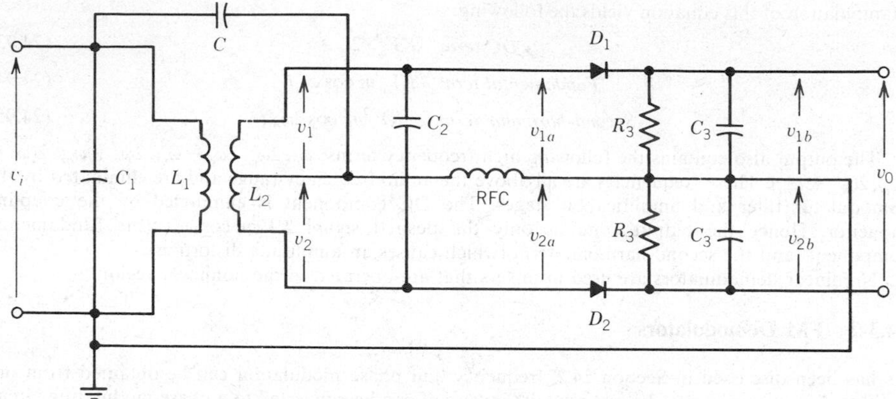

Fig. 24.30 Foster–Seeley phase-shift discriminator.

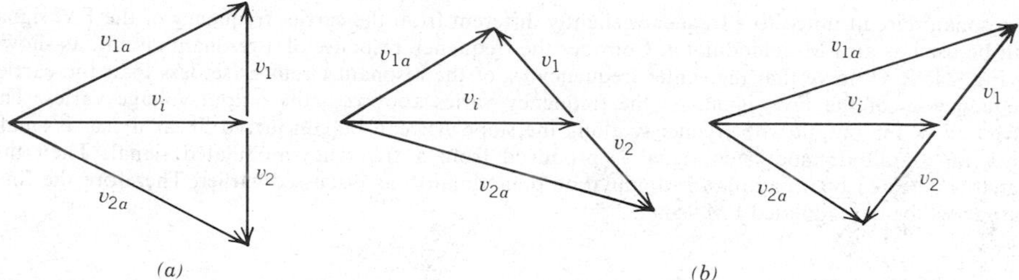

Fig. 24.31 Phasor diagrams for the Foster–Seeley discriminator (*a*) at resonance and (*b*) off resonance.

shown in Fig. 24.31*a*. When the frequency varies about ω_0, the voltages $v_{1a} = v_1 + v_i$ and $v_{2a} = v_2 + v_i$ vary as shown in Fig. 24.31*b*. Note that the *Q*s of the circuit are kept low so that the magnitudes of the voltages v_1 and v_2 remain essentially unchanged. However, their phases change with frequency. Hence the magnitudes of the voltages v_{1a} and v_{2a}, applied to the AM demodulator section, vary with frequency. The variation of the magnitudes of v_{1a} and v_{2a} is quite linear with frequency over a range of frequency. Therefore the output voltage v_0 is the frequency demodulated signal.

Ratio Detector

The ratio detector, shown in Fig. 24.32, has the important advantage over the Foster–Seeley discriminator that it provides a low noise output that is insensitive to amplitude modulation.

The circuit of the ratio detector is very similar to that of the Foster-Seeley discriminator. The diode D_2 is reversed, a large capacitor C_4 is placed across the output resistance R_4, and the output voltage is taken from the common points of C_3s and R_4s, as indicated. Because of the large capacitance across the series combination of the load, the total load voltage v_3 remains unchanged as the amplitude of the input signal varies. However, the ratio of the two load voltages changes with frequency in such a way that their sum is forced to remain practically constant. Hence the name "*ratio detector*." Since the RF circuitry is the same as that for the Foster–Seeley discriminator, the phasor diagrams for v_{1a}, v_{2a}, and v_i are the same as shown in Fig. 24.31. The voltage v_3 is maintained practically constant by making the circuit time constant RC_4 large (i.e., larger than the period of the lowest modulating frequency).

In contrast to the Foster–Seeley discriminator, the ratio detector has a DC component in the output. This DC voltage may be used for automatic gain control (AGC) which may be desirable

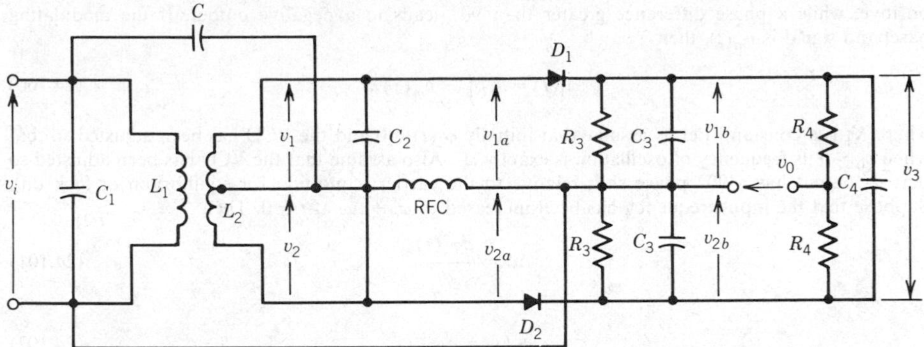

Fig. 24.32 Ratio detector.

when limiters are not used. The Foster–Seeley discriminator is slightly more linear; hence it produces less nonlinear distortion than the ratio detector. Also it can be shown that the output of the discriminator is twice as large.

Phase-Locked Loop

The phase-locked loop, or PLL, can be used as an FM or PM demodulator. The PLL is a feedback system comprising a phase comparator, a lowpass filter, an error amplifier in the forward signal path, and a voltage-controlled oscillator (VCO) in the feedback path, as shown in Fig. 24.33.

The phase comparator is, in actuality, a multiplier circuit that mixes the input signal with the VCO signal. The basic principle of PLL operation can be briefly explained as follows: Let

$$v_1 = v_1 \cos[\omega_c t + \phi_1(t)] \tag{24.96}$$

$$v_2 = v_2 \cos[\omega_c t + \phi_2(t)] \tag{24.97}$$

where $\phi_1(t)$ and $\phi_2(t)$ are used to represent change in frequencies. The output of the phase comparator is

$$v_e = v_1 \cdot v_2 = \tfrac{1}{2} V_1 V_2[\cos(2\omega_c t + \phi_1 + \phi_2) + \cos(\phi_1 - \phi_2)] \tag{24.98}$$

The lowpass filter removes the sum frequency component but passes the DC component, which is then amplified and fed back to the VCO. Hence

$$v_0 = \tfrac{1}{2} A V_1 V_2 \cos(\phi_1 - \phi_2) \tag{24.99}$$

When the phase difference $\theta = \phi_1 - \phi_2$ is 90°, the output of the phase comparator $v_e = 0$. This is called a state of equilibrium. As this phase difference becomes less than 90° the output becomes

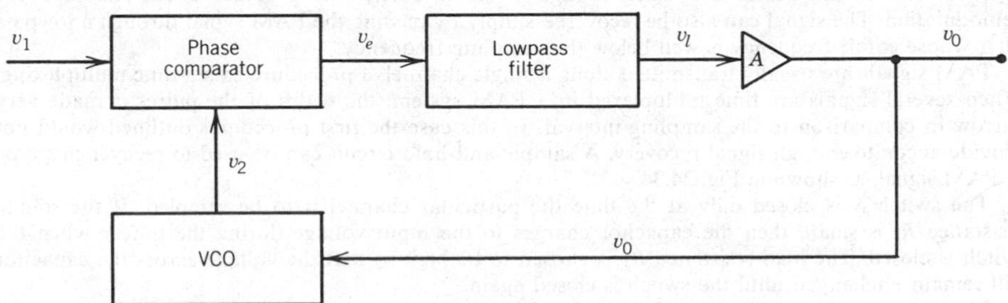

Fig. 24.33 Block diagram of a phase-locked loop.

positive, while a phase difference greater than 90° leads to a negative output. If the modulating baseband signal is $v_m(t)$, then

$$\phi_1(t) = K_1 \int_{-\infty}^{t} v_m(\tau)\, d\tau \tag{24.100}$$

where K_1 is a constant. Let us assume that initially $\phi_1(t) = 0$ and the VCO has been adjusted so that when $v_0 = 0$ its frequency of oscillation is exactly ω_c. Also assume that the VCO has been adjusted so that its output has a 90° phase shift relative to the carrier (condition for equilibrium or lock on). Suppose that the input frequency has been increased to $\omega_c + \Delta\omega$ at $t = 0$. Then

$$\Delta\omega = \frac{d\phi_1(t)}{dt} \tag{24.101}$$

or

$$\phi_1(t) = \Delta\omega \cdot t \tag{24.102}$$

That is, the change in frequency causes the phase $\phi_1(t)$ to change linearly with time. The phase difference at the phase comparator input will produce a positive output voltage v_e, which causes the frequency of the VCO to increase. A new equilibrium (lock-on) point will then be established, as the frequency of the VCO has been increased to equal the frequency of the input signal. The output of the VCO is given by

$$v_2 = V_2 \cos\left[\omega_c t + K_0 \int_{-\infty}^{t} v_0(\tau)\, d\tau\right] \tag{24.103}$$

where K_0 is a constant.

If the input and VCO frequencies are to be the same at lock-on condition, it is required that

$$\frac{d\phi(t)}{dt} = \frac{d}{dt} K_0 \int_{-\infty}^{t} v_0(\tau)\, d\tau \tag{24.104}$$

Using Eq. (24.101) in Eq. (12.104) we have

$$v_0(t) = \frac{\Delta\omega}{K_0}$$

Observe that the output voltage is proportional to the frequency change, $\Delta\omega$, as required in our FM demodulator.

It is assumed that the input signal and the VCO are at the same frequency ω_c. Initially this may not be the case. If the frequency difference between the input signal and the VCO is less than the closed-loop bandwidth of the PLL, the PLL will lock quickly. If the range of frequency differences is larger than the closed-loop bandwidth of the PLL, the VCO frequency is swept at an approximate rate to search for the signal.

The PLL can also be used as a demodulator for phase-modulated signals by integrating the input of the VCO. This is possible because the input of the VCO is proportional to the frequency change of the PLL input, and the integral of this signal is also proportional to the change in phase of the PLL input.

24.3-3 Pulse Amplitude Demodulators

The pulse amplitude-modulated (PAM) signal can be recovered by applying the signal to an AM demodulator. If the pulses are close enough to each other, this provides a reasonable PAM demodulation. The signal can also be recovered simply by passing the PAM signal through a lowpass filter whose cutoff frequency is well below the sampling frequency.

PAM signals are usually transmitted along a single channel, a procedure called time multiplexing. When several signals are time multiplexed in a PAM system, the width of the pulses is made very narrow in comparison to the sampling interval. In this case the first procedures outlined would not provide accurate enough signal recovery. A sample-and-hold circuit can be used to recover this type of PAM signal, as shown in Fig. 24.34.

The switch S is closed only at the time the particular channel is to be sampled. If the source resistance R_s is small, then the capacitor charges to the input voltage during the time τ when the switch is closed. The load resistance R_L is chosen to be high so that the voltage across the capacitor will remain unchanged until the switch is closed again.

The stepwise output signal can be smoothed by passing the signal through a lowpass filter. The sample-and-hold circuit provides an efficient, reliable, and noise-free PAM demodulator.

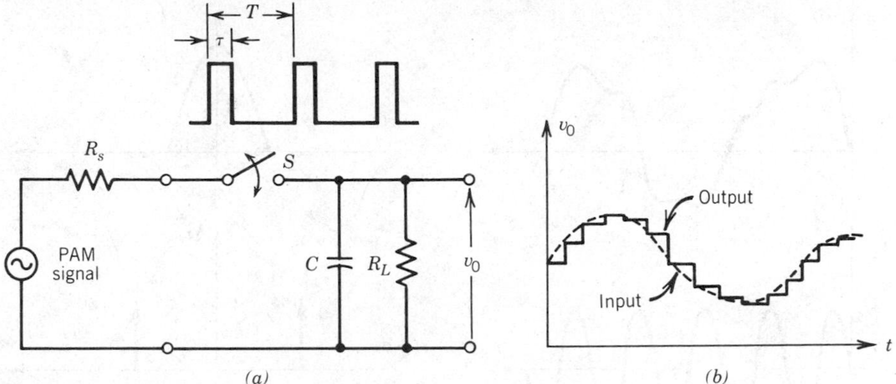

Fig. 24.34 (*a*) Sample-and-hold circuit as a pulse amplitude-modulated (PAM) signal demodulator. (*b*) Input and output signals.

24.4 FREQUENCY CONVERTERS

24.4-1 General Principles

In processing communication signals, it is often desirable or convenient to translate the frequency range from one region to another. For example, suppose that several different signals in the same spectral range are to be transmitted through a single communication channel. At the receiving end they are to be recovered separately. This multiple transmission (multiplexing) can be accomplished by translating each of the signals to a different frequency range. If these frequency ranges do not overlap, each signal at the receiving end can be separated with use of bandpass filters.

When signals are transmitted through free space instead of wires, antennas are used. For efficient transmission, the length of the antenna should be of the order of magnitude of the wavelength of the signal being transmitted. Hence a signal having a frequency of 1000 Hz requires an antenna length of the order of 300,000 m, which obviously is impractical. The length can be reduced to a practical length if the frequency of the signal is translated to a higher frequency range.

Other reasons necessitate frequency translation for practical or efficient signal processing in communication systems. The reader can find these in most communication books.[1-5] In this section, the general principles of frequency translation, also called frequency conversion or frequency mixing, will be discussed. The systems performing frequency translations are called frequency converters or frequency mixers.

A signal can be translated to another spectral range by being multiplied with suitable sinusoidal signal. The process of frequency translation is illustrated in Fig. 24.35. Consider a bandlimited signal $x(t)$ whose spectrum is $X(\omega)$ in Figs. 24.35*a* and 24.35*b*, respectively. When the signal $x(t)$ is multiplied by the sinusoidal signal $y(t) = \cos \omega_c t$ (see Figs. 24.35*b*, *c*), it can be shown that (as illustrated in Figs. 24.35*e*, *f*) the spectra of the resulting signal is the shifted version of the original signal by the amount of $\pm \omega_c$.

Consider, for example, a modulated signal $x(t)\cos \omega t$. Multiplication of this signal by the auxiliary sinusoidal signal $\cos \omega_c t$ yields

$$\begin{aligned} f(t) &= x(t) \cdot \cos \omega t \cos \omega_c t \\ f(t) &= \tfrac{1}{2}x(t)\cos(\omega + \omega_c)t + \tfrac{1}{2}x(t)\cos(\omega - \omega_c)t \end{aligned} \tag{24.105}$$

Observe that the product consists of the sum and difference frequencies, $\omega + \omega_c$ and $\omega - \omega_c$, each multiplied by $x(t)$. Assuming that $\omega \neq \omega_c$, multiplication has translated the signal spectra into two new frequencies. The frequency range occupied by the original signal is called the baseband. The process of multiplying a signal with an auxiliary sinusoidal signal is called mixing or heterodyning. The part of the translated signal within the range from ω_c to $\omega_c + \omega_M$ is called the upper-sideband signal, and the part in the range from $\omega_c - \omega_M$ to ω_c is called the lower-sideband signal. A basic generalized frequency converter is shown in Fig. 24.36. The local oscillator provides the auxiliary sinusoidal signal $\cos \omega_c t$, which is also called the mixing signal, the heterodyning signal, or the carrier signal.

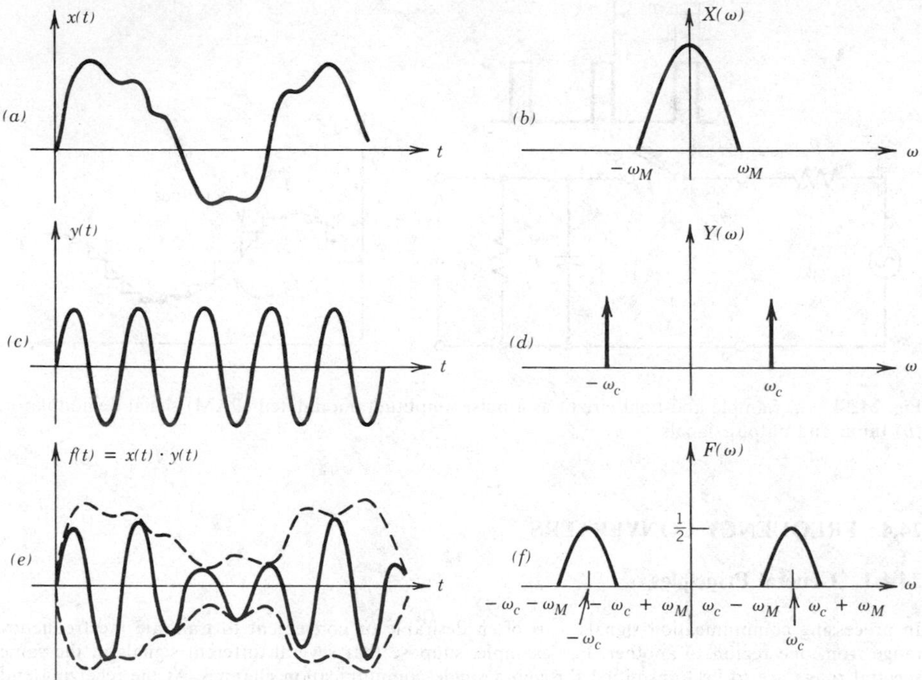

Fig. 24.35 Illustration of frequency translation: (*a*) bandlimited signal, (*b*) frequency spectra of $x(t)$, (*c*) $y(t) = \cos\omega_c t$, (*d*) frequency spectra of $y(t)$, (*e*) modulated signal $f(t) = x(t) \cdot y(t)$, (*f*) frequency spectra of $f(t)$, $F(\omega)$.

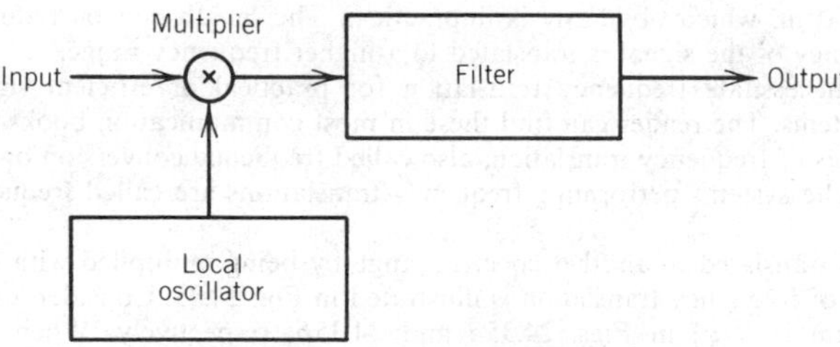

Fig. 24.36 Frequency converter circuit.

Recovery of Baseband Signal

Consider the signal $x(t)$ whose frequency spectrum has been translated by multiplying $x(t)$ by $\cos\omega_c t$. The recovery of $x(t)$ may be accomplished simply by multiplying the translated signal with $\cos\omega_c t$. This can be shown as follows

$$[x(t) \cdot \cos\omega_c t] \cdot \cos\omega_c t = x(t)\cos^2\omega_c t = \tfrac{1}{2}x(t) + \tfrac{1}{2}x(t)\cos 2\omega_c t \tag{24.106}$$

Hence the baseband signal $x(t)$ is recovered. Note, however, that during this process another signal whose frequency spectrum extends from $2\omega_c - \omega_M$ to $2\omega_c + \omega_M$ has been introduced. Since $\omega_c \gg \omega_M$ for most applications, the frequency spectrum of this double-frequency signal is widely separated from the baseband signal. Therefore the double-frequency signal can be removed by a lowpass filter. It should be noted that in this method of recovery, the baseband signal (i.e., using a second multiplication of the receiving end), a signal exactly synchronous with the auxiliary signal at the transmitting end, should be available at the receiving end. Such systems are called synchronous or

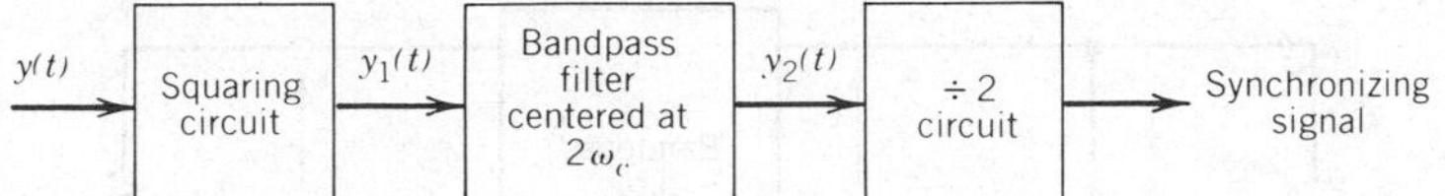

Fig. 24.37 Generation of synchronizing signal.

coherent systems. Generation of such a synchronous signal at the receiving end is not always feasible. A commonly used method is shown in Fig. 24.37. To demonstrate generation of the synchronizing signal, let the baseband signal be $\cos \omega t$ and the received signal be

$$y(t) = A \cos \omega t \cdot \cos \omega_c t \tag{24.107}$$

Then

$$\begin{aligned} y_1(t) = y^2(t) &= A^2 \cos^2 \omega t \cdot \cos \omega_c t \\ &= \frac{A^2}{4}\left[1 + \tfrac{1}{2}\cos 2(\omega_c + \omega)t + \tfrac{1}{2}\cos 2(\omega_c - \omega)t + \cos 2\omega t + \cos 2\omega_c t\right] \end{aligned} \tag{24.108}$$

The bandpass filter selects the frequency component $(A^2/4) \cdot \cos 2\omega_c t$ and rejects all the other components in Eq. (24.108). The "divide-by-two" circuit provides the necessary synchronizing signal. Hence the output of the divide-by-two circuit is used to multiply the received signal to recover the baseband signal, $\cos \omega t$.

We have seen that the spectrum of any signal may be translated $\pm\omega_c$ rad/s by multiplying it with any periodic signal whose fundamental frequency is ω_c rad/s.

Let the periodic signal $p_T(t)$ be represented by the Fourier series

$$P_T(t) = \sum_{k=-\infty}^{\infty} A_k e^{jk\omega_c t} \tag{24.109}$$

where $\omega_c = 2\pi / T$. Multiplying $p_T(t)$ by $x(t)$ yields

$$x(t) p_T(t) = \sum_{k=-\infty}^{\infty} A_k x(t) e^{jk\omega_c t} \tag{24.110}$$

Taking the Fourier transform of both sides of Eq. (24.110) gives

$$F\{x(t) \cdot p_T(t)\} = \sum_{k=-\infty}^{\infty} A_k X(\omega - k\omega_c) \tag{24.111}$$

Observe that the spectrum of $x(t)p_T(t)$ contains $F(\omega)$, $F(\omega - \omega_C)$, $F(\omega - 2\omega_C)$, . . . , and so on; that is, the resulting spectrum contains $F(\omega)$ and $F(\omega)$ translated by $\pm\omega_C, \pm 2\omega_C, \ldots, \pm k\omega_C, \ldots$. The spectrum centered around ω_c can be selected by using a bandpass filter whose center frequency is ω_c.

24.4-2 Mixers

The operation of multiplying a signal with an auxiliary sinusoidal (or, in general, periodic) signal is called mixing or heterodyning, and the circuits performing this operation are called mixers. Therefore a mixer can be viewed as an amplitude modulator. A basic mixer using an electromechanical switch is shown in Fig. 24.38*a* and the signals at different points are shown in Fig. 24.38*b*.

This type of mixer is also known as a chopper-type modulator whose operation can be explained as follows. The switch driven by the carrier signal $A \cos \omega_c t$ alternates between terminals at the rate of ω_c rad/s. For one half of the period the switch is open and allows the slowly varying input signal to be applied to the bandpass filter. During the next half period the switch grounds the input terminals of the filter. Thus the result is the chopped form of the input signal at a frequency of ω_c. This chopping process is equivalent to multiplying the input signal $v(t)$ by a unity magnitude pulse train with a frequency of ω_c. The desire amplitude-modulated signal $v_0(t)$ is obtained by passing the chopped signal through a bandpass filter whose center frequency is ω_c. The electromechanical switch in Fig. 24.38*a* can be replaced by an electronic switch using diodes, transistors, or FETs. At high frequencies, diodes are usually used as switches. Two examples are the shunt and series mixers illustrated in Fig. 24.39. The diode bridge, in both cases, acts as a switch. Nodes a and b are alternately short circuited and open circuited by the diode bridge. If the diode characteristics are identical, the bridge will be balanced, eliminating the carrier signal passing through when the diodes are conducting.

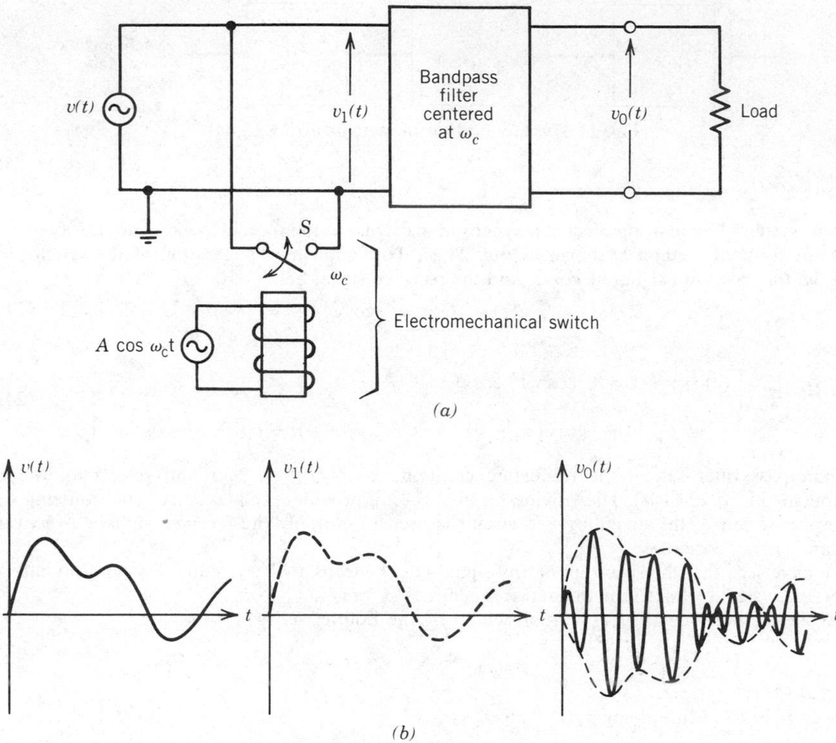

Fig. 24.38 (*a*) Mixer with electromechanical switch. (*b*) Signals at different points.

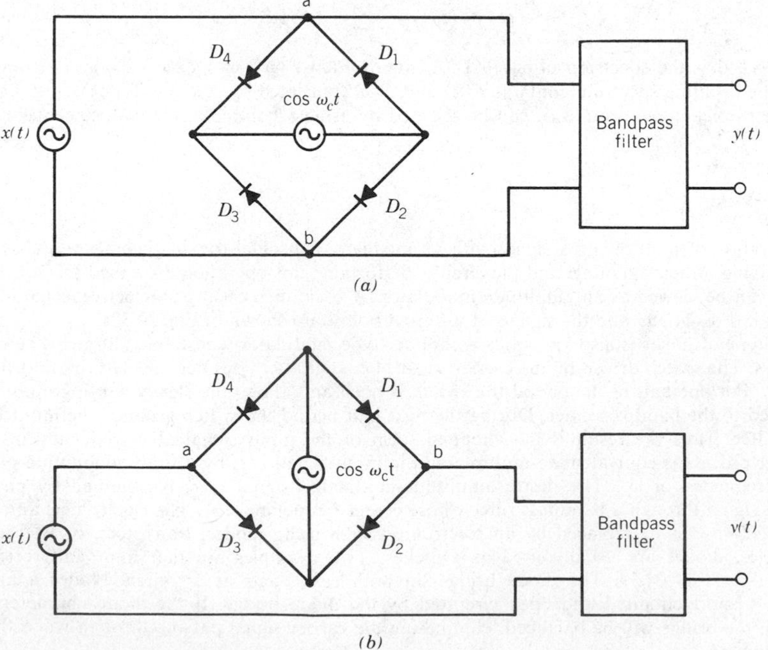

Fig. 24.39 (*a*) Shunt mixer. (*b*) Series modulator.

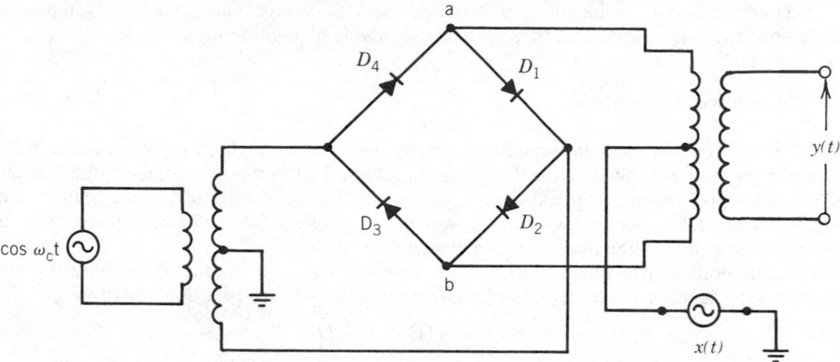

Fig. 24.40 Doubly balanced mixer.

Doubly Balanced Mixers

A popular configuration known as a doubly balanced mixer (or ring modulator) is shown in Fig. 24.40. The advantage of this circuit over the previous two circuits is that it does not require ideal components. The doubly balanced mixers are used up to, and beyond, 1 GHz. This circuit requires matched-diode characteristics and accurately center-tapped transformers. Inaccuracies in center taps and mismatch in diode characteristics result in nonideal performance, and some carrier may appear in the output. Analysis of this circuit shows that the doubly balanced mixer effectively multiplies to $x(t)$ by a rectangular wave carrier switching between $+1$ and -1. This eliminates the DC component in the carrier. As a result, the input spectrum to the filter will not contain the modulating-signal spectrum. Therefore the desired signal can be recovered by using a lowpass filter.

Chopper Amplifier

An excellent application of the principle of frequency conversion is the construction of amplifiers operating at DC or very low frequencies. It is very difficult to design amplifiers to amplify very low frequencies, since the size of the required coupling capacitors becomes very large. Consequently, to amplify DC signals or very low frequency signals, direct coupling is used, which introduces drift in the quiescent operating point of the amplifier. The chopper amplifier, which essentially shifts the spectrum of the input signal from a low-frequency range to a suitable high-frequency range where it can be easily amplified, can be used to overcome this problem. A chopper amplifier circuit is shown in Fig. 24.41. A mechanical switch (chopper) is used for modulation and demodulation. Since the processes of modulation and demodulation both need the carrier to be of the same frequency, the same chopper must be used for modulation and demodulation, as indicated in Fig. 24.41. The DC component in the input signal is removed by the capacitor C. Therefore the spectrum of the input signal is the same as that of the input signal but shifted by $\pm\omega_c$, $\pm 2\omega_c$, $\pm 3\omega_c$, . . . , and so on. This signal is amplified, and again multiplied, by a periodic square wave generated by the chopper at the

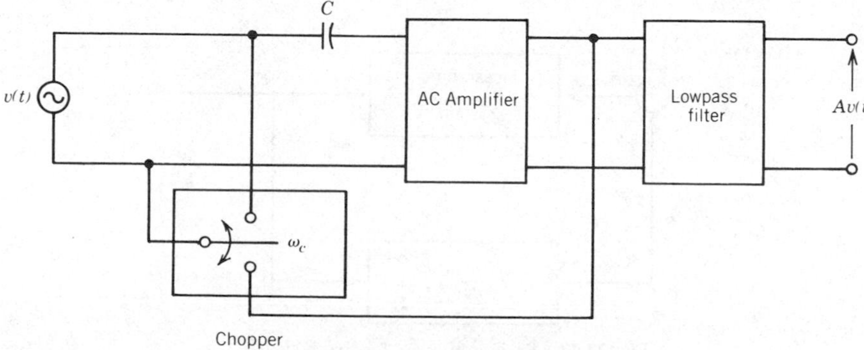

Fig. 24.41 Chopper amplifier.

output. Consequently, the spectrum is retranslated back to $\omega = 0$. The primary advantage of the chopper amplifier is that it allows construction of very stable high-gain amplifiers.

24.4-3 Parametric Converters

Parametric amplifiers are used in amplifying circuits where the amplifying power comes from the variation of a parameter in the circuits. In linear amplifiers the basic power source is the bias battery or DC power supply, whereas in parametric amplifiers the basic electrical energy comes from a steady-state sinusoidal power source. When parametric amplifiers are used for frequency conversion they are referred to in the literature as parametric converters.

Consider the nonlinear feedback system shown in Fig. 24.42. Assume that the nonlinear system in the feedback loop functions as a second-order polynomial (the square-root nonlinearity)

$$r(t) = a_1 y(t) + a_2 y^2(t) \tag{24.112}$$

Let the input be a sinusoid

$$x(t) = A \cos \omega t \tag{24.113}$$

and the transfer function of the linear time-invariant system be $H(s)$. When the feedback loop is broken, the output is given by

$$y(t) = B \cos(\omega t + \theta) \tag{24.114}$$

where

$$B = A \cdot |H(j\omega)| \quad \text{and} \quad \theta = \angle H(j\omega) \tag{24.115}$$

Hence in the open-loop system there is no frequency conversion. The output of the nonlinear system (when the feedback is broken) is

$$r(t) = a_1 B \cos(\omega t + \theta) + a_2 B^2 \cos^2(\omega t + \theta) \tag{24.116}$$

or

$$r(t) = \frac{a_2 B^2}{2} + a_1 B \cos(\omega t + \theta) + \frac{a_2 B^2}{2} \cos(2\omega t + 2\theta) \tag{24.117}$$

Since $e(t) = x(t) - y(t)$ is applied to the linear time-invariant system (feedback is closed), the output is given by

$$y(t) = \frac{a_2 B^2 |H(0)|}{2} + B \cos(\omega t + \theta) - a_1 B |H(j\omega)| \cos \cdot (\omega t + 2\theta) - \frac{a_2 B^2 |H(j2\omega)|}{4} \cos(2\omega t + \phi) \tag{24.118}$$

where

$$\theta = 2\theta + \angle H(j2\omega) \tag{24.119}$$

Observe that the output of the system contains a DC term, a term containing the input frequency, and a term with the second harmonic. This output has been obtained when the nonfeedback output is fed back through the nonlinear system. If the output in Eq. (24.118) is fed back, it can be shown that the output will contain a DC term and terms witn the input frequency, second, third, and fourth harmonics of the input frequency. As this feedback process is repeated, it can be shown that the output will have all harmonics of the input frequency up to some maximum. For example, in the case

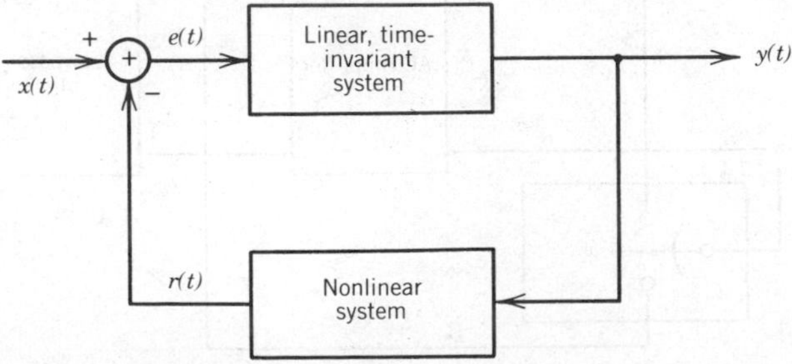

Fig. 24.42 Basic nonlinear feedback system.

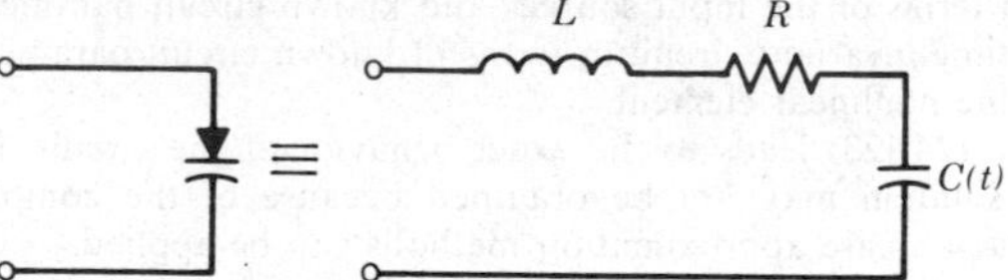

Fig. 24.43 Circuit model of a varactor diode.

of a square-law nonlinearity, this maximum is 2^{n-1}, where n is the number of feedback iterations processed. The energy, in this process, comes from the input signal, which is at a different frequency.

When the system in Fig. 24.42 is driven by two sinusoidal inputs having two different frequencies, then the output will contain all harmonics of both frequencies and all sums and differences of all of these harmonics. For example, if the input is

$$x(t) = A_1 \cos \omega_1 t + A_2 \cos(\omega_2 t + \theta) \tag{24.120}$$

then the output will be of the form

$$y(t) = \sum_{j=0}^{\infty} \sum_{i=0}^{\infty} B_{ij} \cos[|j\omega_1 + i\omega_2|t + \theta] \tag{24.121}$$

The above procedure can be extended to η distinct input signals.

Most widely used parametric converters in microwave systems make use of the varactor diode as the nonlinear element. A varactor is a diode that has a voltage-dependent junction capacitance. A circuit model for the varactor diode is shown in Fig. 24.43. It consists of a linear resistor, an inductor, and a nonlinear capacitance, all connected in series. The time variation of the varactor capacitance is accomplished by a local oscillator called a pump. A basic parametric converter having two inputs at two different frequencies is shown in Fig. 24.44. The input source signals to be amplified are the basic steady-state power source; the nonlinear capacitance of the varactor diode provides necessary energy conversion.

The heuristic approach to analysis of nonlinear parametric converters does not always yield the response of a particular circuit. However, it provides reasonable approximations for most practical applications. Depending on the nonlinear function of the nonlinear element, the exact analysis may become complex. In this case the previously discussed heuristic approach may be performed, initially, to simplify the analysis.

The Thévenin equivalent circuit below the nonlinear element, the varactor, is shown in Fig. 24.44*b*. Kirchhoff's voltage law gives

$$v_T(t) = \int_{t0}^{t} z_T(t-\tau) i(\tau)\, d\tau + v(t) \tag{24.122}$$

where $v_T(t)$ is the open-circuit Thévenin voltage and $z(t)$ is the impulse response of the linear time-invariant circuit as seen from the terminals of the varactor diode Manipulation of Eq. (24.122) yields the following nonlinear integral equation

$$v_0(t) = s(t) + \int_{t0}^{t} h(t-\tau) F[v_0(\tau)]\, d\tau \tag{24.123}$$

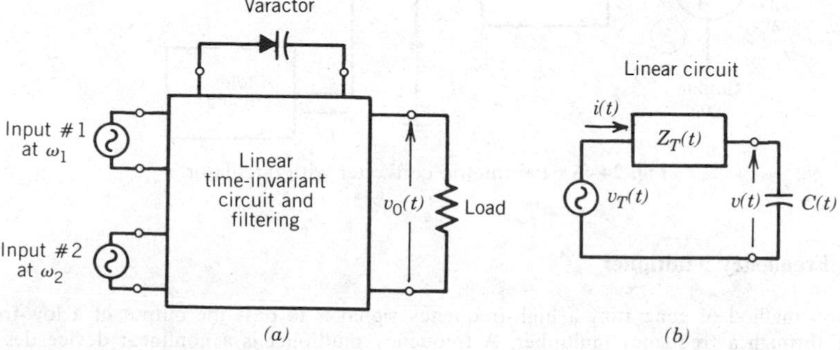

Fig. 24.44 (*a*) Basic parameteric converter circuit. (*b*) Its Thévenin equivalent circuit.

where $s(t)$ is known in terms of the input sources and known circuit parameters, $h(t)$ is the impulse response of the linear time-invariant circuit in terms of known circuit parameters, and $F[v_0(t)]$ is the nonlinear function of the nonlinear element.

The solution of Eq. (24.123) leads to the exact behavior of the circuit. However, as mentioned previously, the exact solution may not be obtained because of the complexity of the nonlinear function $F[v_0(t)]$. In such a case approximation methods can be applied.

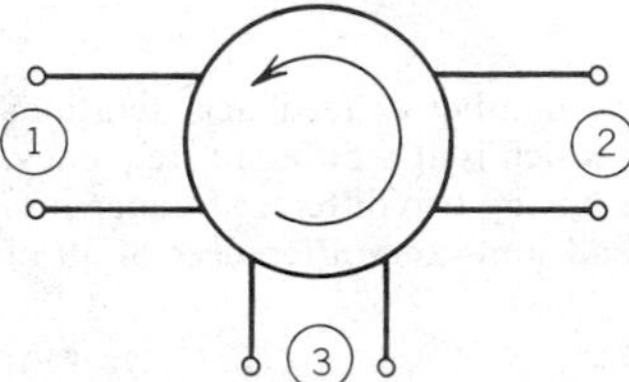

Fig. 24.45 Circulator.

Parametric Converter with Circulator

A circulator is a three-part circuit having a cyclic power-transmission property. Power entering one part is transmitted to an adjacent part in cyclic order, as shown by the circular arrow in Fig. 24.45. For the circulator shown, the cyclic order is 132. When a signal is incident at port 1, with all ports match terminated (no reflections), no signal is transmitted to port 3. Similarly, a signal incident at port 2 is completely transmitted to port 1, and a signal incident at port 3 is completely transmitted to port 2. This cyclic property can be seen best by using scattering-parameter analysis.

At ultra high frequency (UHF) and microwave frequencies, the circulator-type parametric converters are widely used. A parametric converter with circulator is shown in Fig. 24.46. The input signal and the amplified output signal are separated by the circulator. The pump and idler sources are coupled to the varactor to drive the nonlinear varactor capacitance. The signal and pump currents mix in the nonlinear varactor to produce signals at many frequencies. The idler tuning circuit provides a signal that may be required for specific circuit consideration to achieve the desired device performance. For example, the idler current mixes with the pump current to produce the signal frequency.

The parametric converter with circulator is essentially a negative-resistance device at the signal frequency. For example, the action of the pump and idler currents in the varactor presents a negative resistance at the signal port. The system remains stable as long as the magnitude of the negative resistance is less than that of the source resistance.

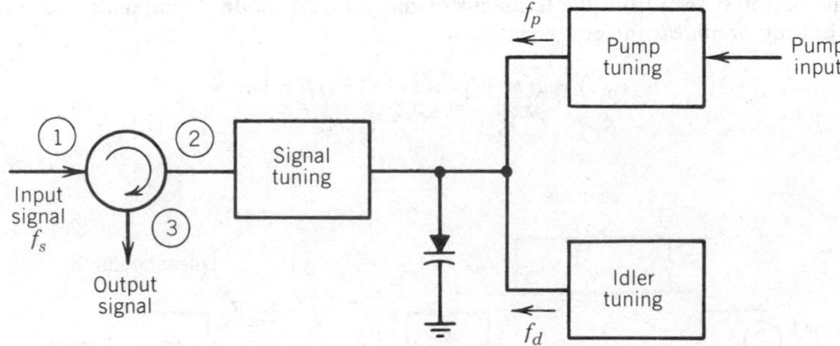

Fig. 24.46 Parametric converter with circulator.

24.4-4 Frequency Multiplier

A common method of generating a high-frequency signal is to pass the output of a low-frequency oscillator through a frequency multiplier. A frequency multiplier is a nonlinear device designed to multiply the frequencies of the signal applied by a given factor.

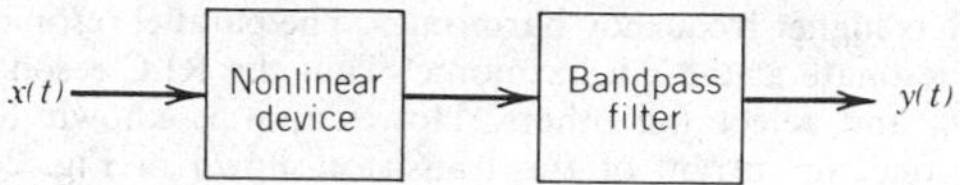

Fig. 24.47 Basic frequency multiplier.

As an example, consider a nonlinear device having a square-law characteristic given by

$$y(t) = ax^2(t) \tag{24.124}$$

Let the input signal be the FM signal given by

$$x(t) = A\cos(\omega_c t + k\sin\omega_m t) \tag{24.125}$$

The output is found to be

$$y(t) = \frac{aA^2}{2} + \frac{aA^2}{2}\cos(2\omega_c t + 2k\sin\omega_m t) \tag{24.126}$$

The DC term can be removed by a lowpass filter. Observe from the second term that both the carrier frequency and the modulation index have been multiplied by 2. Using an nth-law nonlinear device, the frequency can be multiplied by a factor of η. If the purpose of the frequency multiplication is to increase the modulation index, the result may be a very high carrier frequency that is undesirable. To avoid this problem, a frequency converter may be used to control the value of the carrier frequency.

A block diagram of a frequency multiplier is shown in Fig. 24.47. The frequency multiplication is also used to obtain a wideband FM waveform from a narrowband one. This process is known as Armstrong indirect FM transmission.[1]

A simple transistor-frequency multiplier is shown in Fig. 24.48. The transistor operates in the class-C mode, which means that the transistor is in the cutoff region for more than half of the period of the input signal. When the input signal nears the positive peak, the transistor is driven into the active region or even into saturation. The collector current will be in the form of pulses, as indicated in Fig. 24.48b. Although the collector current waveform has the same fundamental frequency as has

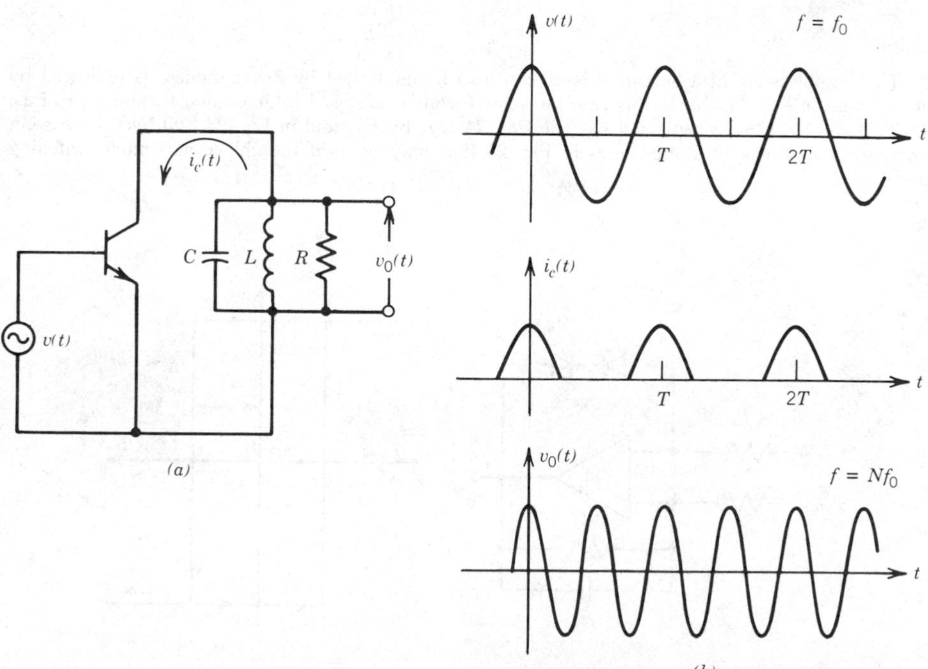

Fig. 24.48 Transistor-frequency multiplier: (a) circuit schematic, (b) waveforms of various signals.

the driving signal, it also has higher frequency harmonics. The parallel resonant circuit formed by the RLC portion is tuned to resonate at the nth harmonic. Thus the RLC resonant circuit picks out the signal with frequency nf_0 and reject the others. However, it is known that with most periodic waveforms, such as the collector-current of the transistor shown in Fig. 24.48, the amplitudes of harmonics progressively decrease with increasing harmonic numbers. Consequently, as the order of multiplication increases the output signal becomes progressively smaller. Thus, for higher order multiplication, several multipliers may be cascaded. The transistor-frequency multiplier of Fig. 24.48 can provide multiplication by 2 to 5.

24.5 DETECTORS

24.5-1 Level Detector

A level detector, also known as a Schmitt trigger, determines if an input voltage is greater or less than a reference voltage. For slowly varying input signals the output will also change slowly. This becomes a serious disadvantage when the output of a detector is used to trigger a logic circuit requiring fast trigger pulses. As an example, a noninverting level detector is shown in Fig. 24.49*a*.

When $v_y < v_x$ so that $v_0 = V_{0,\text{sat}}$, R_2 will feed back a signal and the new reference signal at x will be

$$v_x = V_1 = \frac{R_2 V_R}{R_1 + R_2} + \frac{R_1 V_{0,\text{sat}}}{R_1 + R_2} \tag{24.127}$$

If v_i is increased further, v_0 remains constant at $V_{0,\text{sat}}$. When $v_i = V_1$, the output regeneratively switches to $v_0 = -V_{0,\text{sat}}$ and remains at this value as long as $v_i > V_1$. This portion of the transfer characteristic corresponds to the portion abcde in Fig. 24.49*b*.

If $v_i > V_1$ so that $v_0 = -V_{0,\text{sat}}$, and v_i is decreased, then

$$v_x = V_2 = \frac{R_2 V_R}{R_1 + R_2} + \frac{R_1 \cdot V_{0,\text{sat}}}{R_1 + R_2} \tag{24.128}$$

At $v_i = V_2$, the output will regeneratively switch to $v_0 = +V_{0,\text{sat}}$, as indicated in Fig. 24.49*b*, by the portion edfba. Note that $V_2 < V_1$, and the difference between V_1 and V_2 is called the *hysteresis*, V_H

$$V_H = V_1 - V_2 = \frac{2 R_2 V_{0,\text{sat}}}{R_1 + R_2} \tag{24.129}$$

The output swing of a Schmitt trigger can also be controlled by Zever diodes, as indicated by dotted lines in Fig. 24.49*a*. In this case the same formulas are used to determine V_1 and V_2 in Eqs. (24.127) and (24.128) by replacing $V_{0,\text{sat}}$ in Eq. (24.127) by V_{Z1} and in Eq. (24.128) by V_{Z2}. Also, a capacitor C_2, shown by dotted lines in Fig. 24.49*a*, may be used to achieve maximum switching speed.

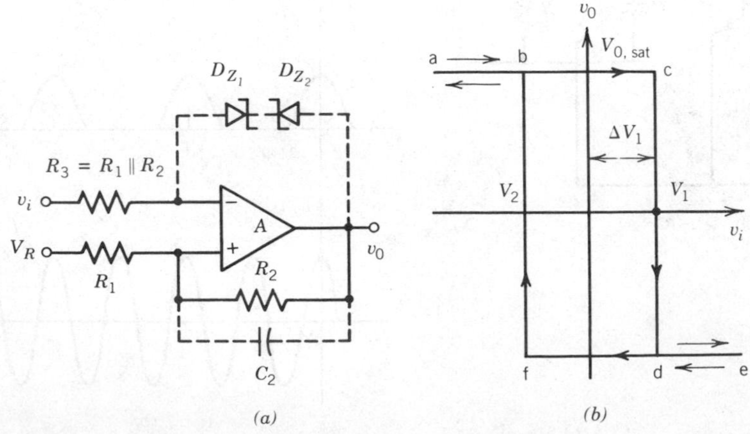

Fig. 24.49 Noninverting level detector: (*a*) circuit schematic, (*b*) transfer characteristic.

Zero-Crossing Detector

If V_R is set equal to zero, the output will respond every time the input passes through zero. Such a circuit is called a zero-crossing detector.

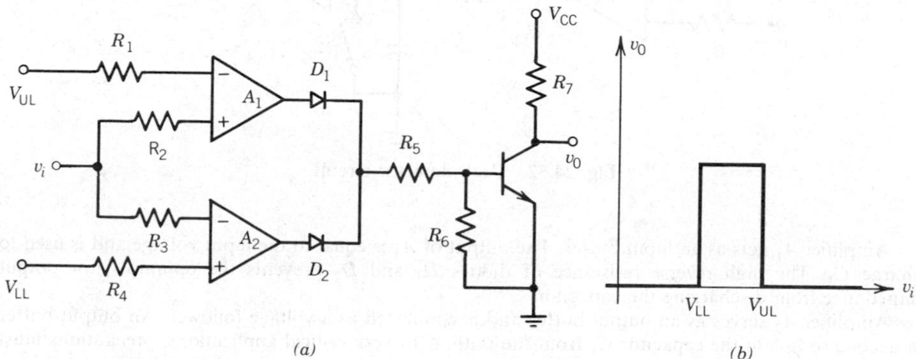

Fig. 24.50 Double-ended limit detector: (*a*) circuit schematic, (*b*) transfer characteristic.

Double-Ended Limit Detector

This type of detector is also known as a window comparator and is used to detect whether a given voltage is within a prescribed voltage limit. This can be accomplished by simply combining the outputs of two comparators, one indicating greater than a lower limit, V_{LL}, the other indicating less than an upper limit, V_{UL}. If the applied voltage v_i is within the range, that is, if $V_{LL} \leqq v_i \leqq V_{UL}$, the output will be true, or high. If the applied voltage is not within the range, the output of one of the comparators will not be true, that is, it will be low. A window comparator and its transfer characteristic are shown in Fig. 24.50.

24.5-2 Peak Detector

On occasion an instrument design requires a circuit that will find and hold the maximum value (the peak) of an input waveform. Figure 24.51 illustrates the requirement of such a circuit. The solid line of the graph represents a continuously varying signal voltage as a function of time. The dotted line represents the output of a peak detection circuit.

The circuit in Fig. 24.52 is an example of a peak detection circuit. This circuit produces an output voltage v_O which increases only when the slope of the input voltage v_i is both increasing and greater than any previous v_i. When the slope of v_i is either zero or decreasing, the value of v_O remains at the highest previous level of v_i. If perfect operational amplifiers, diodes, and capacitors are used, the output v_O will remain at the highest previous level of v_i until the circuit is reset by the switch S_1.

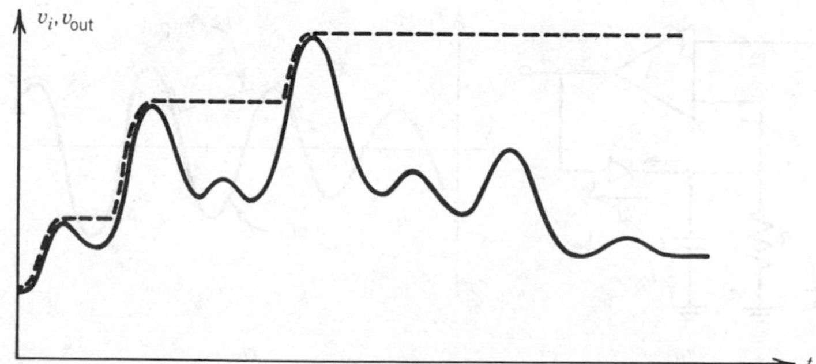

Fig. 24.51 Input and output signals of a peak detector.

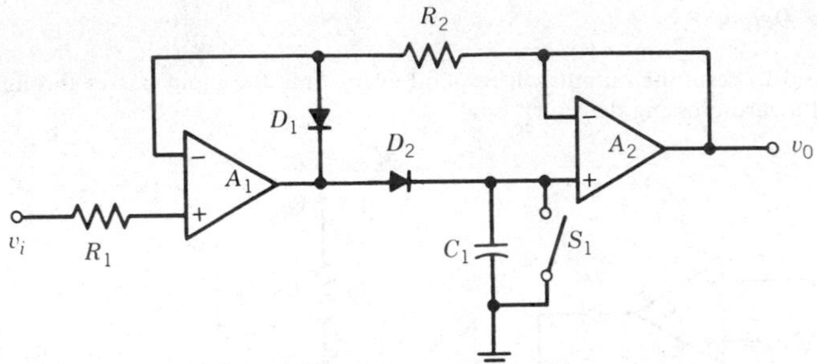

Fig. 24.52 Peak detector circuit.

Amplifier A_1 acts as an input buffer. The output of A_1 is equal to the input voltage and is used to charge C_1. The high reverse resistance of diodes D_1 and D_2 prevents the op-amp's low output impedance from discharging the capacitor.

Amplifier A_2 serves as an output buffer and is connected as a voltage follower. An output buffer is needed to isolate the capacitor C_1 from the output. In very critical applications, precautions must be taken to maintain the charge on C_1. First, a high-quality capacitor should be used, such as polystyrene or another low-leakage type. Second, amplifier A_2 should be of the FET input type which has very high input impedance, and more important, low input current. Since the voltage discharge rate is

$$\frac{\Delta v_C}{\Delta t} = \frac{I}{C_1} \tag{24.130}$$

the lower the input current, the lower the drift in voltage across C_1.

Resistor R_2 is included to allow A_1 to be clamped in the off state by diode D_1. This results in faster recovery. Switch S_1 presents the only low impedance path to ground and is used to discharge C_1.

While this circuit represents single-quadrant operation, the polarity of operation can be reversed if D_1 and D_2 are reversed.

24.5-3 Envelope Detector

An envelope detector circuit provides an output that follows the peaks of an AM signal. Therefore the circuits discussed in Section 24.3-1 can be used as envelope detectors. In this section, another envelope detector that uses an operational amplifier is presented. Consider the circuit shown in Fig. 24.53. As the input voltage v_i increases and becomes sufficiently positive, the diode D starts conducting, and the capacitor C charges toward the peak of v_i. The charging current is

$$i_c = i_d - \frac{v_x}{R} \tag{24.131}$$

where $v_x \cong v_i$ for good-quality high-gain operational amplifiers. When v_i reaches its peak, it starts

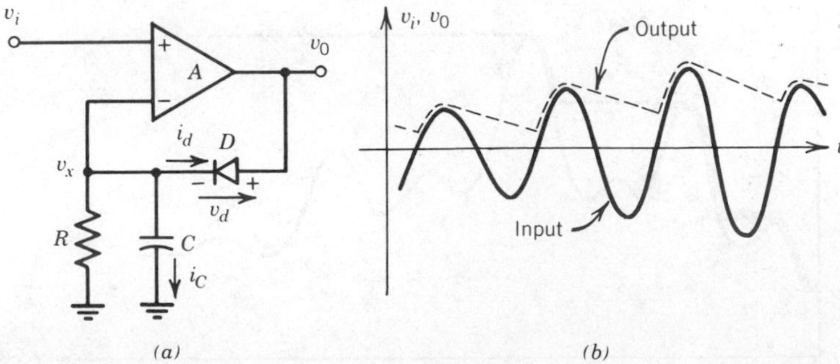

Fig. 24.53 Envelope detector: (*a*) circuit schematic, (*b*) input and output waveforms.

decreasing. Since the voltage across the capacitor cannot change suddenly, the diode D is reverse biased. Hence D cuts off. The capacitor C discharges through the resistance R. If the time constant $\tau = RC$ is selected appropriately, the discharge of the capacitor can be made slow, resulting in the waveform at the output in Fig. 24.53*b*. Hence the output is the envelope of the AM modulated signal.

24.5-4 Phase Detector

The phase demodulators discussed in Section 24.3-2 can also be used as phase detectors. The circuits discussed here are widely used for phase detection. Consider the circuit shown in Fig. 24.54. The switch can be a FET or a transistor switch. Assume that the output of the VCO is a square wave and is used to close and open the switch (i.e., the switch is closed when the VCO output voltage is positive and open when the VCO output voltage is negative). Further, assume that the frequencies of the input signal and the VCO are the same (i.e., $f_1 = f_2 = f$, where f_2 is the frequency of the VCO).

Let the input voltage be given by

$$v_i = V\sin(\omega t + \theta_i) \tag{24.132}$$

where θ_i is the phase of v_i. Assume that the output of the VCO is positive, that is, $v_2 > 0$, when

$$-\frac{\pi}{2} \leqq \omega t + \theta_2 \leqq \frac{\pi}{2} \tag{24.133}$$

or equivalently when

$$-\frac{1}{\omega}\left(\frac{\pi}{2} + \theta_2\right) \leqq t \leqq \frac{1}{\omega}\left(\frac{\pi}{2} - \theta_2\right) \tag{24.134}$$

where θ_2 is the phase of the output of the VCO. Then the output of the circuit is determined by

$$v_O = \frac{\omega}{2\pi} \cdot \int_{-(1/\omega)(\pi/2+\theta_2)}^{(1/\omega)(\pi/2-\theta_2)} V\sin(\omega t + \theta_i)\,dt \tag{24.135}$$

Integration of this equation yields

$$v_O = \frac{V}{\pi}\sin(\theta_i - \theta_2) \tag{24.136}$$

Note that the output is proportional to $\sin(\theta_i - \theta_2)$, and when $\theta_i = \theta_2$ the output will be zero. Hence the circuit in Fig. 24.54 is used to detect the phase of the input signal compared with the phase of the VCO circuit. This circuit is also known in the literature as a phase comparator.

The circuit shown in Fig. 24.55 makes use of zero-crossing detectors with hysteresis, differentiators, and a bistable op-amp.

The circuits R_{k1}, R_{k2}, R_{k3}, and A_k (with $k = 1, 2$) act as a zero detector (with A_2 being the inverting type). Thus v_{21} lags v_i by 180°, and the outputs v_{11} and v_{21} are square waves. These square waves are differentiated by C_{k1} and R_{k4}. Diodes D_{k1} and D_{k2} are used to pass only the positive pulses resulting from the differentiation. These positive pulses are used to trigger the bistable circuit formed by the elements R_{31}, R_{32}, R_{33}, R_{34}, A_3, D_Z, and D_{Z2}. For example, pulses applied through v_{12} cause the bistable op-amp to go to a low state (i.e., $v_{31} = -V_{Z1}$) and pulses through v_{22} cause the bistable op-amp to go to a high state (i.e., $v_{31} = +V_{Z2}$). The output of the bistable op-amp, v_{31}, is averaged by the lowpass filter formed by R, C. If the Zener diodes D_{Z1} and D_{Z2} are identical, that is, if $V_{Z1} = V_{Z2} = V_Z$, it can be shown that the output voltage is

$$v_O = (\theta_i - \theta_R)V_Z \tag{24.137}$$

where θ_i is the phase of the input signal v_i and θ_R is the phase of the reference signal v_R. Observe that

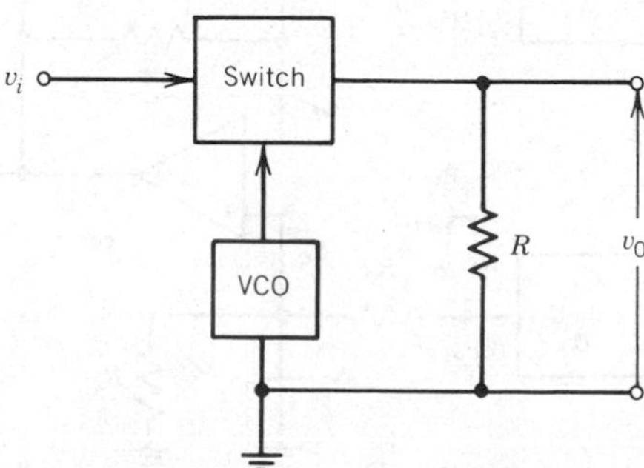

Fig. 24.54 Phase detector.

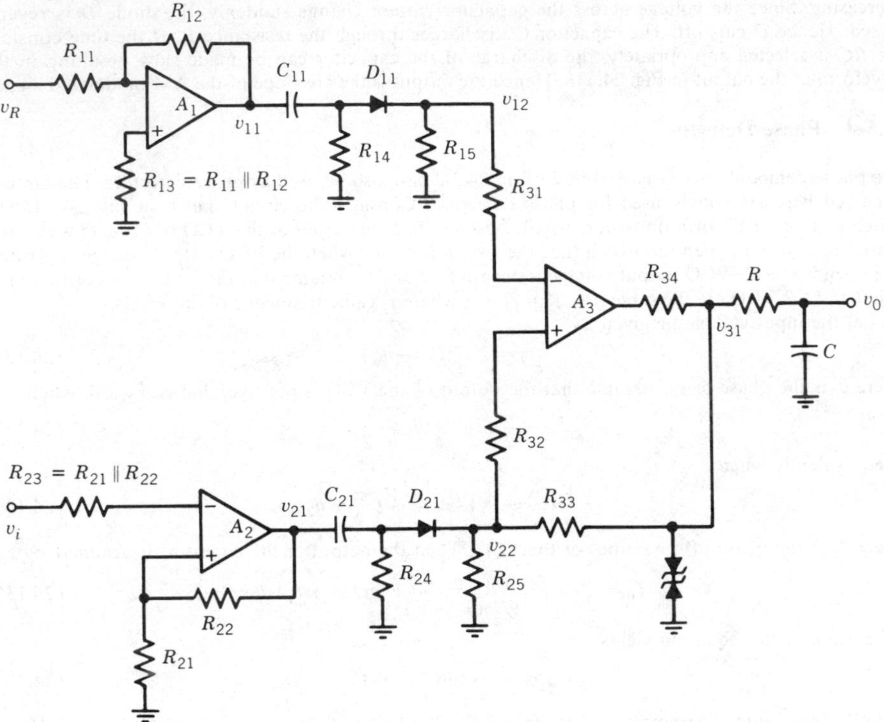

Fig. 24.55 Phase detector using zero-crossing detectors, differentiators, and a bistable operational amplifier.

the output voltage is directly proportional to $(\theta_i - \theta_R)$. Therefore $v_0 = 0$ indicates that $\theta_i = \theta_R$; $v_0 > 0$ indicates that $\theta_i > \theta_R$.

24.5-5 Frequency Detector

The FM demodulators discussed in Section 24.3-2 are considered to be frequency detectors as well. Here a frequency detector that can be used as a frequency comparator or frequency difference detector is presented. Consider the circuit shown in Fig. 24.56. Let the signal v_R be the reference

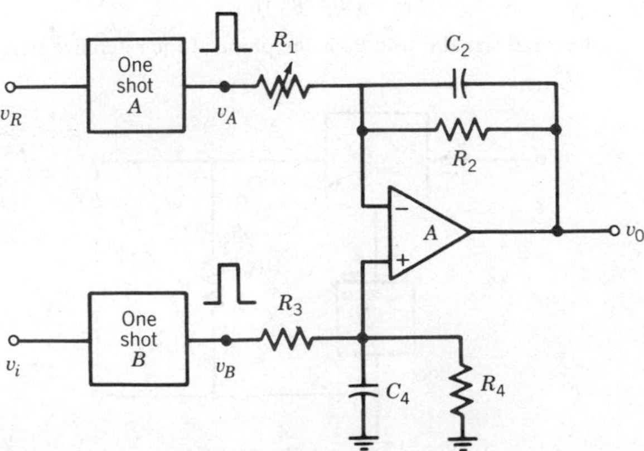

Fig. 24.56 Frequency detector.

signal with a frequency of f_R and v_i be the input signal with unknown frequency of f_i. Each of these signals triggers a one shot, once every cycle. These one shots are adjusted so that the width and amplitude of the pulses at v_A and v_B are identical (for a balanced operation). The components R_1 through R_4, C_2, C_4, and the op-amp form a lowpass filter. The pulse train at v_A is applied to the inverting input of the lowpass filter through R_1, and the pulse train at v_B is applied to the noninverting input of the lowpass filter through R_3. Let the amplitudes of v_A and v_B be V, that is, the pulses swing between 0 and V, and let their widths be T. Then the average values of v_A and v_B are given by

$$v_{A,\text{Avg}} = VTf_R \tag{24.138}$$

$$v_{B,\text{Avg}} = VTf_i \tag{24.139}$$

If R_1 through R_4 are selected such that $R_1 = R_2$ and $R_3 = R_4$, then it can be shown that the output of the lowpass filter is given by

$$v_0 = VT(f_i - f_R) \tag{24.140}$$

Thus if $f_i = f_R$ for a balanced circuit, the output $v_0 = 0$; if $v_0 > 0$, then $f_i > f_R$; if $v_0 < 0$, then $f_i < f_R$. It must be noted that for a precise operation, the input offset voltage and the input offset current of the op-amp must be nulled—otherwise the offset voltage at the output can cause an error.

References

1 F. G. Stremler, *Introduction to Communication Systems*, Addison-Wesley, Reading, MA, 1977.

2 A. B. Carlson, *Communication Systems*, McGraw-Hill, New York, 1975.

3 B. P. Lathi, *Signals, Systems, and Communication*, Wiley, New York, 1965.

4 W. D. Gregg, *Analog and Digital Communications*, Wiley, New York, 1977.

5 J. M. Wozencraft and I. M. Jacobs, *Principles of Communication Engineering*, Wiley, New York, 1967.

CHAPTER 25
WAVEFORM SHAPING AND TIMING CIRCUITS

YUSUF Z. EFE

The Cooper Union for Advancement of Science and Art
New York, New York

25.1 WAVEFORM SHAPING

Certain complex waveforms are often needed in communication or instrumentation applications. To produce these, usually a simple waveshape is generated and then some operation is performed on this shape to obtain the desired waveform. Generation of the desired waveform usually involves the use of active as well as passive elements. The process in which the shape of a nonsinusoidal signal is modified to a desired waveform through the use of a linear network is called linear wave shaping. Shaping of a nonsinusoidal signal by means of a nonlinear network is called nonlinear wave shaping. In this section both types of waveshaping are discussed.

25.1-1 Linear Waveshaping

In this subsection the time responses to various input waveforms of linear passive circuits containing one or more energy storage elements are examined. Although exact analysis is possible, this is usually very tedious, and in practice reasonable approximations are utilized which yield results acceptable for most practical applications.

RC Differentiator

A series *RC* circuit, shown in Fig. 25.1, provides an output approximately proportional to the derivative of the input signal. This circuit is a basic high-pass filter. As the frequency of the input signal increases, the reactance of the capacitor decreases. Hence the higher frequency components in the input signal appear at the output with lower attenuation than that of the lower frequency components. At very high frequencies the capacitor acts virtually as a short circuit, and therefore the input signal appears at the output with almost no attenuation.

The behavior of the *RC* circuit can be best described by the following differential equation

$$v_0(t) = Ri(t) = v_i(t) - \frac{1}{C}\int i(t)\,dt \tag{25.1}$$

If the time constant $\tau = RC$ is very small compared with the time required for the input signal to change appreciably, the voltage across R will be very small compared with the voltage across the

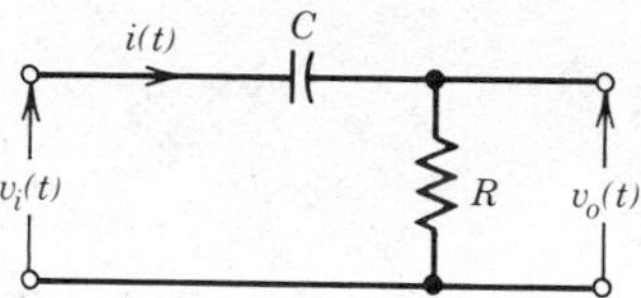

Fig. 25.1 *RC* differentiator circuit.

capacitor C. Then Eq. (25.1) can be approximated by

$$v_i(t) \cong \frac{1}{C}\int i(t)\,dt \tag{25.2}$$

Thus

$$i(t) \cong C\frac{dv_i(t)}{dt} \tag{25.3}$$

or

$$v_0(t) = RC\frac{dv_i(t)}{dt} \tag{25.4}$$

Hence the output is proportional to the derivative of the input voltage. At discontinuities, the variations of the input signal would appear at the output, and the circuit would only roughly approximate a differentiator. Hence if the input signal does not contain instantaneous jumps and the time constant is sufficiently small, the circuit of Fig. 25.1 can be used as a reasonably good differentiator. The time response of this circuit to various input signals is presented next.

Sinusoidal Input. Consider the sinusoidal input signal

$$v_i(t) = V_m \sin \omega t \tag{25.5}$$

The output is found to be

$$v_0(t) = \frac{V_m R}{\sqrt{R^2 + 1/(\omega^2 C^2)}}\sin(\omega t + \theta) \tag{25.6}$$

where

$$\theta = \tan^{-1}(1/\omega RC) \tag{25.7}$$

If $\omega RC \ll 1$, then $\theta \cong 90°$, and the output can be approximated as

$$v_0(t) \cong V_m \omega RC \cos \omega t \tag{25.8}$$

which agrees with the expected value in Eq. (25.4). Since the output will be a small fraction of the input for a satisfactory differentiation, the output frequently has to be amplified by a high-gain amplifier.

Step-Voltage Input. If the input is a step voltage, that is,

$$v_i(t) = \begin{cases} 0 & \text{for} \quad t < 0 \\ V & \text{for} \quad t \geqslant 0 \end{cases} \tag{25.9}$$

then the solution of the differential equation yields

$$v_0(t) = Ve^{-t/\tau} \tag{25.10}$$

where $\tau = RC$. Note that the voltage across the capacitor cannot change instantaneously. Hence if the capacitor initially carries no charge, then $v_C(0^-) = v_C(0^+) = 0$ and the output voltage at $t = 0^+$ will be $v_0(0^+) = V$. As $t \to \infty$, the capacitor will be charged to V and thus $v_0(t)$ will approach zero. This agrees with the fact that the derivative of a constant is zero. However, it must be noted that in the vicinity of $t = 0$, this is not true and the waveform shown in Fig. 25.2 is obtained.

If $t > 5\tau$, the output falls off by 99% of its value at $t = 0^+$. Hence even though the steady state is approached as $t \to \infty$, for most practical applications it can be assumed that the final value will be reached after $t > 5\tau$.

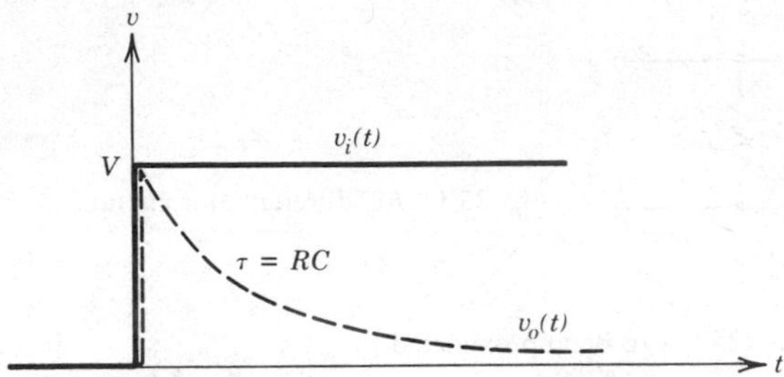

Fig. 25.2 Step-voltage response of a differentiator.

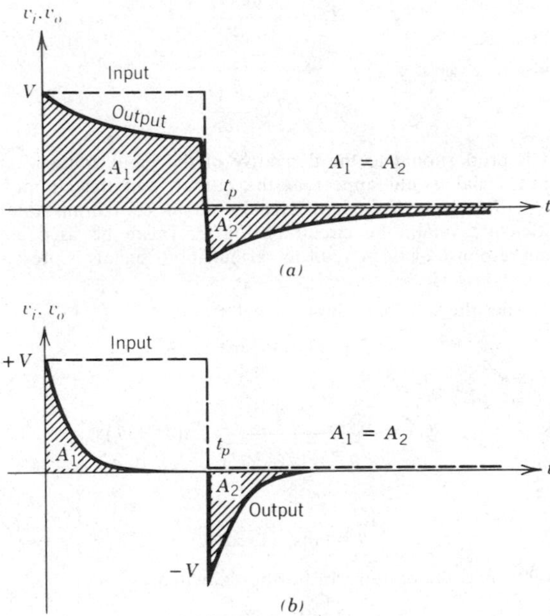

Fig. 25.3 Response of an RC differentiator (a) for $RC \gg t_p$, (b) for $RC \ll t_p$.

Pulse Input. If a pulse of width t_p and amplitude V is applied, the response for $t < t_p$ is the same as that for the step input, that is,

$$v_0(t) = Ve^{-t/RC} \qquad \text{for} \quad t < t_p \tag{25.11}$$

At the end of the pulse, the input falls instantaneously from V to 0. The output must also drop by V (since the voltage across the capacitor cannot change instantaneously). It can be shown that

$$v_0(t) = V(e^{-t_p/RC} - 1) \cdot e^{-(t-t_p)/RC} \qquad \text{for} \quad t > t_p \tag{25.12}$$

Hence the shape of the pulse will be distorted. The amount of distortion depends on the time constant $\tau = RC$. The output waveforms for two different RCs are shown in Fig. 25.3. It can be shown that the area below the x axis is always equal to the area above the x axis (e.g., $A_1 = A_2$). Observe that if $RC \ll t_p$, the input pulse has been distorted completely and, at the output, positive going and negative going spikes are observed.

Square-Wave Input. The square wave is considered to be periodically repeated pulses. Therefore the resulting output responses, with different time constants, can be obtained from the results using pulse input (presented earlier) very easily.

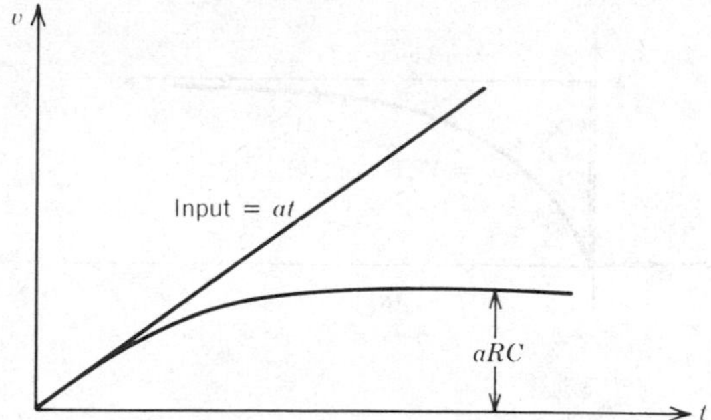

Fig. 25.4 Response of an RC differentiator to a ramp input.

Ramp Input. Consider the ramp input waveform

$$v_i(t) = \begin{cases} at & \text{for } t > 0 \\ 0 & \text{for } t \leqslant 0 \end{cases} \tag{25.13}$$

Then Eq. (25.1) gives

$$a = \frac{v_0(t)}{RC} + \frac{dv_0(t)}{dt} \tag{25.14}$$

The solution of this equation is

$$v_0(t) = aRC(1 - e^{-t/RC}) \tag{25.15}$$

The response of an RC differentiator to a ramp input is shown in Fig. 25.4. Note that for $t \ll RC$, the output follows linearly the input ramp. However, as t gets larger, output departure from linearity is observed. As $t \to \infty$, $v_0(t) = aRC$ will be reached and the output will be leveled, as indicated in the figure.

RC Integrator

The RC circuit in Fig. 25.5 provides an output approximately proportional to the integral of the input signal. This circuit is a basic low-pass filter. The high-frequency components in the input signal are attenuated highly, since the reactance of the capacitor decreases with increasing frequency. At very high frequency the capacitor acts as a virtual short circuit, and the output falls off to zero.

The behavior of this circuit can be described by the following differential equation

$$v_0(t) = v_i(t) - Ri(t) = \frac{1}{C}\int i(t)\, dt \tag{25.16}$$

When the output voltage is small compared with the input, the following approximations yield sufficiently close results to the exact values

$$v_i(t) \cong Ri(t) \tag{25.17}$$

$$v_0(t) \cong \frac{1}{RC}\int v_i(t)\, dt \tag{25.18}$$

Hence the output voltage is proportional to the integral of the input voltage. These approximations can be achieved by selecting the time constant $\tau = RC$ very large compared with the time required for the input signal to change appreciably.

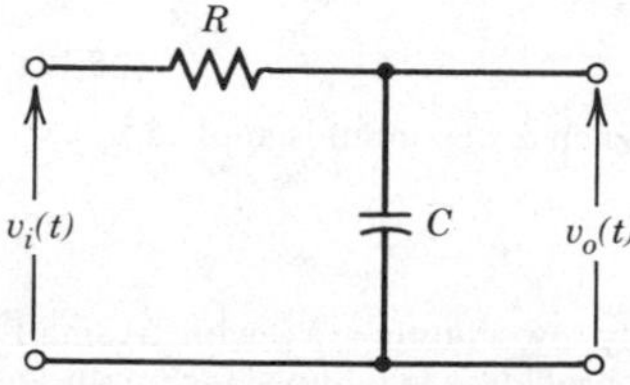

Fig. 25.5 RC integrator circuit.

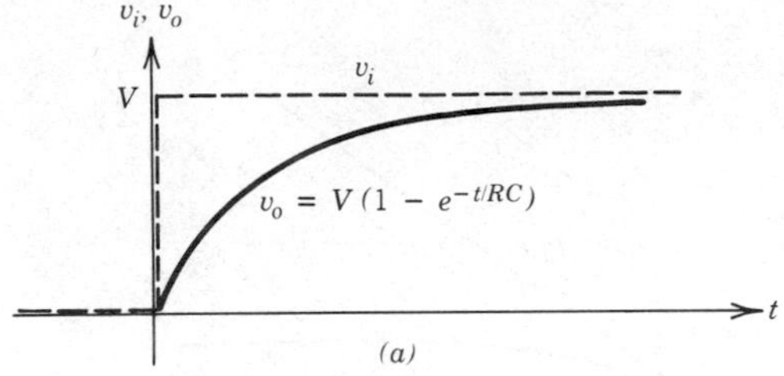

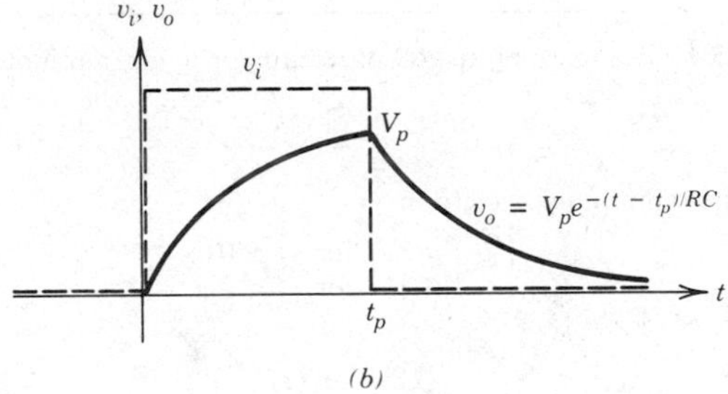

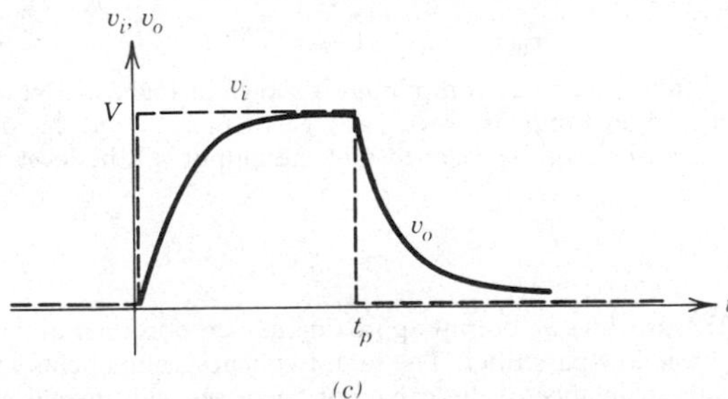

Fig. 25.6 Responses of an *RC* integrator to (*a*) a step-input, (*b*) and (*c*) pulse inputs.

The response of an *RC* integrator to different types of input signals can be examined in the same manner as has been done for the *RC* differentiator. Because of space limitations, these are only summarized in Fig. 25.6 (without proof). The reader may follow the same procedures presented for the *RC* differentiator.

Differentiators and integrators can also be designed by using series *RL* circuits. Derivations and results are similar to those presented for the *RC* cases.

Attenuator

Attenuators are used to reduce the amplitude of a signal waveform. A simple resistive attenuator is shown in Fig. 25.7*a*, in which C_0 represents the stray capacitance loading of the output. Because of this capacitance loading, the resistive attenuator cannot respond to a step input instantaneously. The time constant of the response is given by

$$\tau = C_0[R_2 \| (R_1 + R_2)] \tag{25.19}$$

and the output will asymptotically approach the value that follows when a step input is applied

$$v_0(\infty) = \frac{R_2}{R_1 + R_2} \cdot V \tag{25.20}$$

The response of the attenuator can be improved significantly by shunting R_1 with a small capacitor C_1 as shown in Fig. 25.7*b*. Because the capacitor C_1 will provide a low-impedance path at

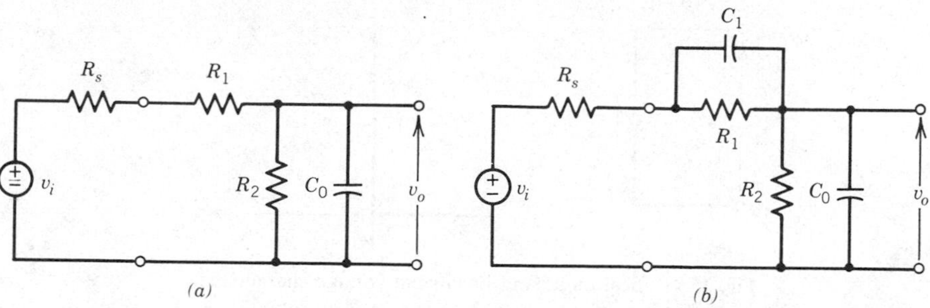

Fig. 25.7 Attenuator circuits: (*a*) resistive, (*b*) compensated.

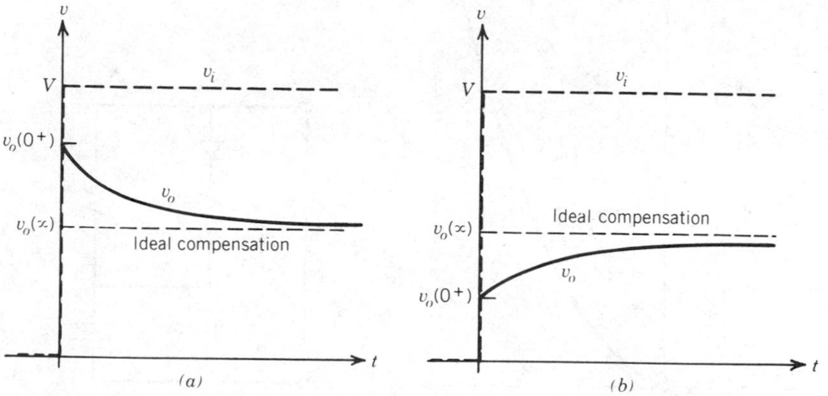

Fig. 25.8 Response of an attenuator to a step input (*a*) $R_1C_1 > R_2C_2$, overcompensated, (*b*) $R_1C_1 < R_2C_2$, undercompensated.

$t = 0^+$, the output will reach steady state much faster. The value of the output voltage at $t = 0^+$ is given by

$$v_0(0^+) = \frac{C_1}{C_1 + C_0} V \tag{25.21}$$

For perfect compensation, $v_0(0^+) = V_0(\infty)$. Hence from Eqs. (25.20) and (25.21) we obtain

$$R_1C_1 = R_2C_0 \tag{25.22}$$

The responses of a compensated attenuator to a step input for $R_1C_1 > R_2C_0$ and $R_1C_1 < R_2C_0$ are given in Fig. 25.8. It is assumed in these responses that the source resistance of the signal source driving the attenuator is zero. If not, the output signal will be distorted at $t = 0^+$. Namely the output will not reach $v_0(\infty)$ instantaneously. Instead it will reach $v_0(\infty)$ with a time constant τ_1, where

$$\tau_1 = R_s \frac{C_1C_0}{C_1 + C_0} \tag{25.23}$$

The time constant for the output reaching its steady-state value is

$$\tau_2 = (R_1 \| R_2)(C_1 + C_0) \tag{25.24}$$

If $R_s \ll (R_1 + R_2)$, the τ_1 can be made much smaller than τ_2 so that the rise at $t = 0^+$ would be considered abrupt.

25.1-2 Nonlinear Waveshaping

Waveform shapings performed by means of nonlinear passive circuits are relatively limited. Nonlinear properties of diodes and transistors can be used very effectively to accomplish more complex and flexible waveshaping.

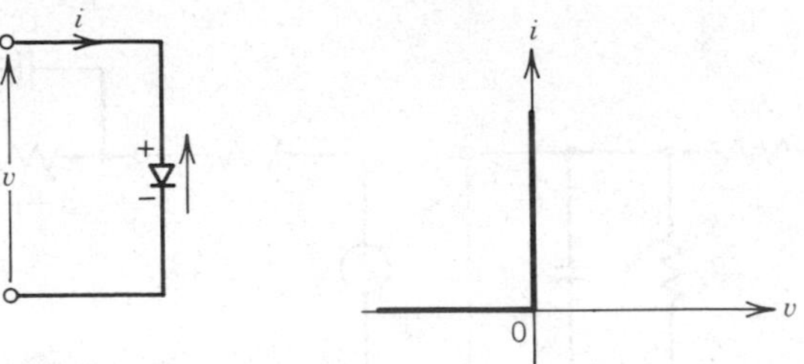

Fig. 25.9 Ideal diode and its current-voltage characteristic.

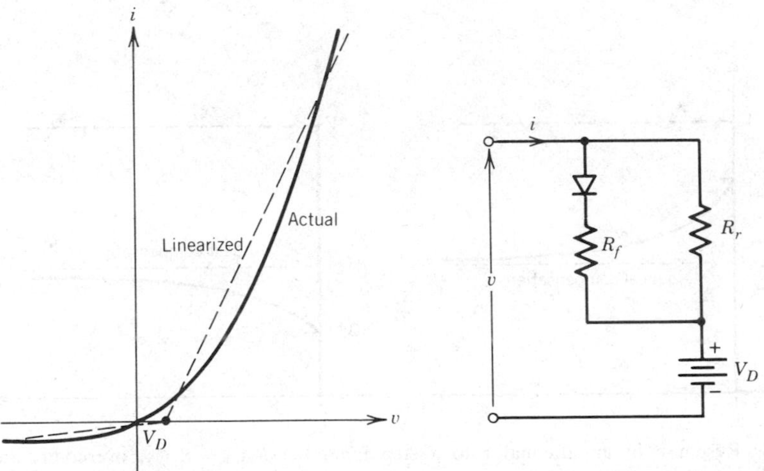

Fig. 25.10 The current-voltage characteristic of a physical diode and its piecewise linear model.

Piecewise Linear Models for Diode

An ideal diode is similar to a voltage-controlled switch. For example, when an ideal diode is forward biased it acts as a short circuit; when it is reverse biased it behaves as an open circuit. The symbol and the current-voltage (i-v) characteristic are shown in Fig. 25.9.

The physical diode has the i-v characteristic shown in Fig. 25.10. This curve can be approximated by using the piecewise-linearization technique, which is also illustrated in Fig. 25.10. This model is not unique, and represents only one possibility among many others. The model indicates that when the diode is forward biased, $v > V_D$, the diode conducts and is equivalent to a resistor R_f (called forward resistance) series battery V_D. When the diode is reverse biased, the diode is equivalent to a very large resistance R_r (called reverse resistance). For a good diode, $R_r \gg R_f$. The cut-in voltage, V_D, is approximately 0.7 V for silicon and 0.3 V for germanium.

Various possible piecewise linear models of a physical diode are given in Table 25.1. The reader may generate many others. Note that the symbol (—▷⊢—) represents an ideal diode and (—▶⊢—) represents a physical diode.

Clipping Circuits

Circuits used to select part of an arbitrary waveform that lies below or above some reference voltage level are called *clipping* circuits. Clipping circuits are also known as limiters, slicers, and amplitude selectors.

Consider the diode circuit in Fig. 25.11*a*. When the diode is reverse biased, the shunt branch is opened and the input signal is transmitted through R_1 to the output. In this case the output will not

TABLE 25.1. PIECEWISE LINEAR MODELS OF A PHYSICAL DIODE

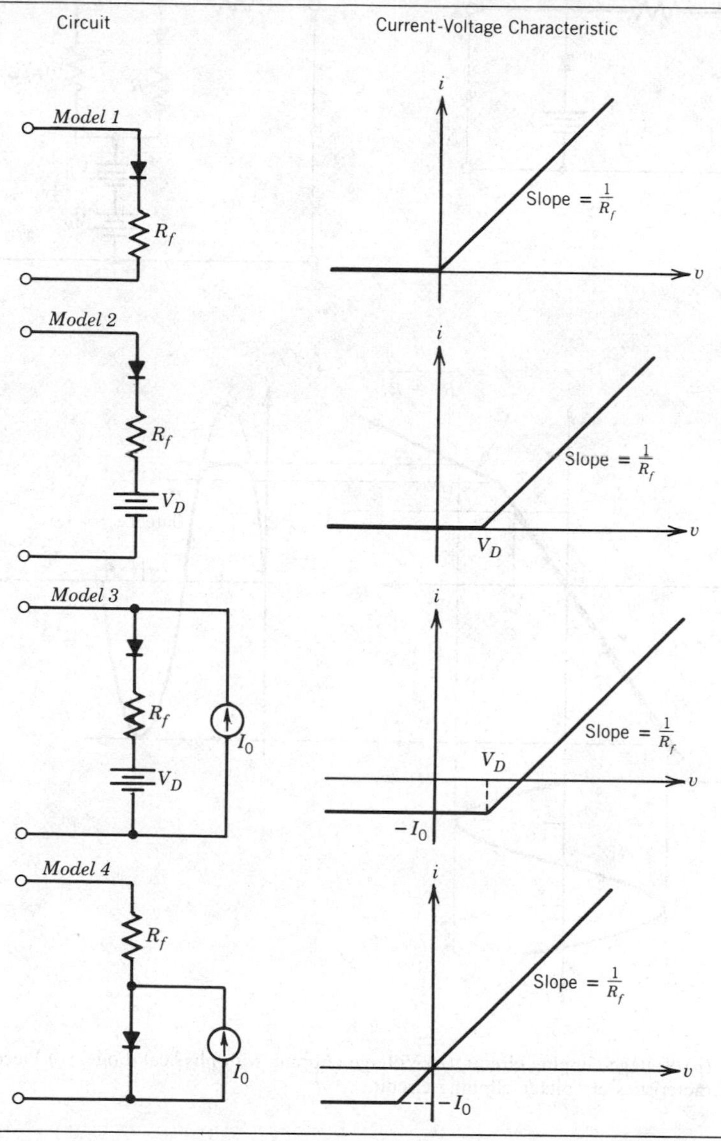

be distorted, and $v_0 = v_i$. When v_i is increased beyond the reference voltage V_R, the diode is forward biased and acts as a short circuit. Therefore $v_0 = V_R$. The transfer characteristic of this circuit, with an ideal diode, is shown in Fig. 25.11c. The circuit, with the actual physical diode model, is shown in Fig. 25.11b and results in the following:

$$v_0 = \frac{R_r}{R_1 + R_r} v_i + \frac{R_1}{R_1 + R_r} V_R \tag{25.25}$$

If R_f is very large, then $v_0 \cong v_i$. When the diode is reverse biased ($v_i < V_R$), then

$$v_0 = \frac{R_f}{R_1 + R_f} v_i + \frac{R_1}{R_1 + R_f} V_R \tag{25.26}$$

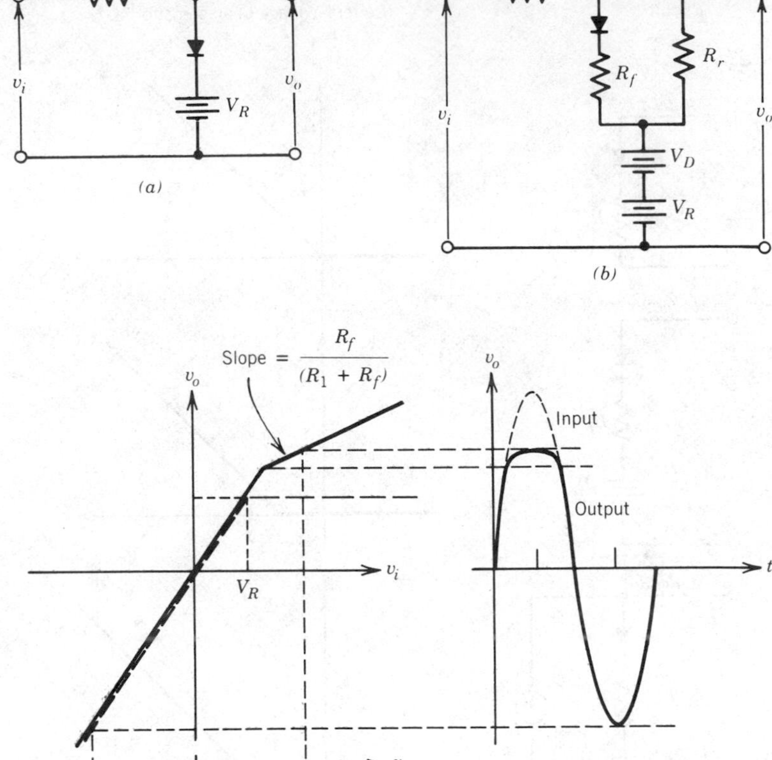

Fig. 25.11 (*a*) Voltage-clipping circuit. (*b*) Voltage clipping with physical diode. (*c*) Piecewise linear transfer characteristics of voltage-clipping circuit.

When the diode is forward biased ($v_i > V_R$), then

$$V_R = V_D + V_R \tag{25.27}$$

Suppose a sinusoidal signal with an amplitude larger than $(V_D + V_R)$ is applied to this circuit. Observe from Fig. 25.11*c* that the output exhibits a suppression of the positive peak of the signal. If $R_f \ll R$, then the slope of the straight line after the breakpoint will be very small. This results in even more suppression of the positive peaks of the outputs, and the peaks will appear to be clipped off.

If the diode in Fig. 25.11*a* is reversed, the piecewise linear transfer characteristic of the circuit in Fig. 25.12 is obtained. In this case the portion of the waveforms more positive than $V_R - V_D$ is transmitted with no distortion while the other portions are severely suppressed. Some diode clipping circuits and their corresponding output waveforms are illustrated in Fig. 25.13.

Clipping at two different levels can also be achieved by using two diodes. Such a circuit and its transfer characteristic are shown in Fig. 25.14. The states of the diodes under certain conditions are

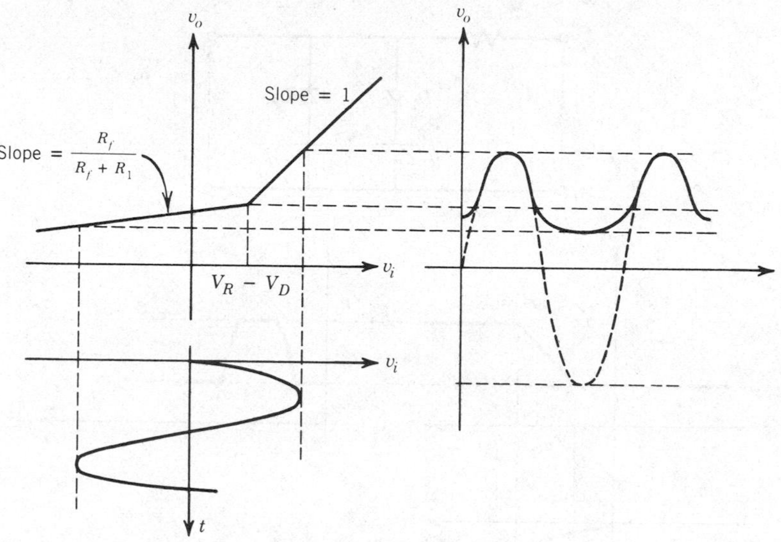

Fig. 25.12 Piecewise linear transfer characteristic of the circuit in Fig. 25.11*a* with the diode reversed.

summarized as follows. (Note that it is assumed that $V_{R_1} > V_{R_2}$.)

Input v_i	Output v_O	States of Diodes
$v_i \leqslant V_{R_1}$	$v_O = V_{R_2}$	D1 off, D2 on
$V_{R_2} < v_i < V_{R_1}$	$v_O = v_i$	D1 off, D2 off
$v_i \geqslant V_{R_2}$	$v_O = V_{R_1}$	D1 on, D2 off

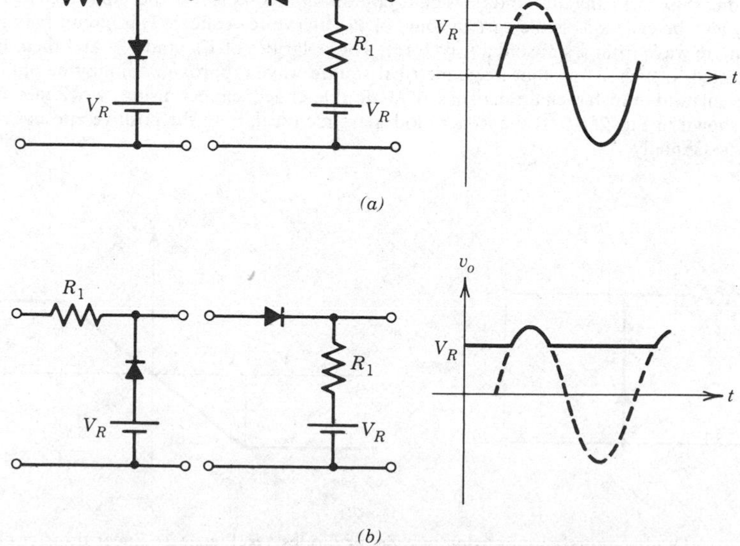

Fig. 25.13 (*a* and *b*) Various diode clipping circuits.

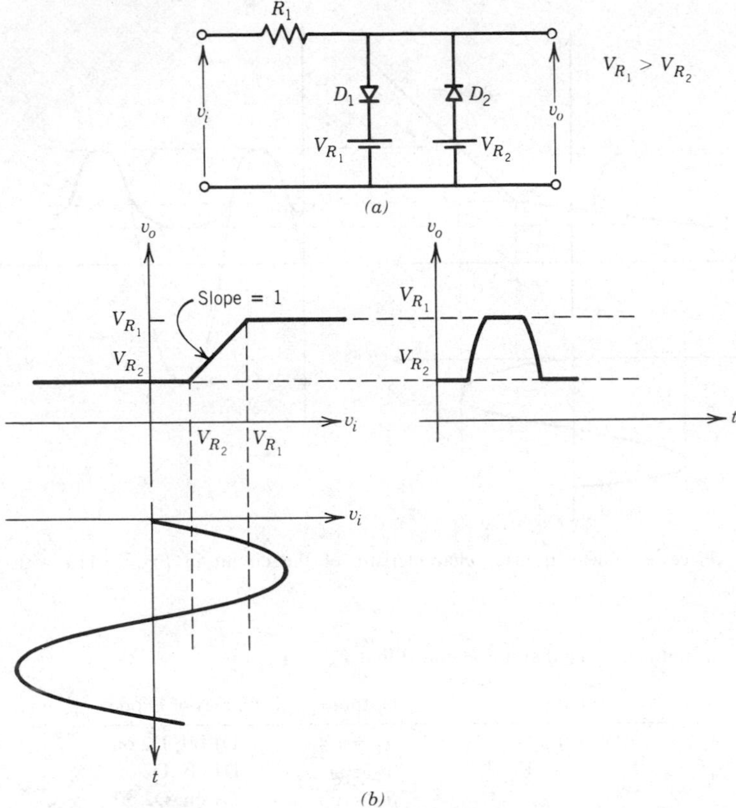

Fig. 25.14 (*a*) Double-diode clipping circuit. (*b*) Piecewise linear transfer characteristics of the circuit.

Observe from Fig. 25.14 that the output signal contains only a slice of the input signal. Hence the circuit may also be called, as is the case in some of the literature, a slicer. This circuit may be used to obtain a square wave from a sinusoidal waveform. The polarities of V_{R_1} and V_{R_2} and their amplitudes can be adjusted in such a way that a symmetrical square wave (approximate) may be obtained.

The circuit and transfer characteristics of a double-ended clipper using two Zener diodes (in series) are shown in Fig. 25.15. If the Zener diodes are identical, both the positive and negative peaks will be clipped equally.

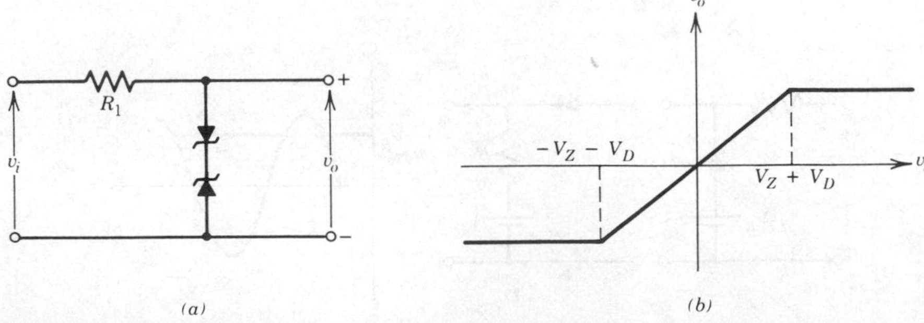

Fig. 25.15 (*a*) Double-ended circuit using two Zener diodes. (*b*) Piecewise linear transfer characteristics of circuit.

Observe that there is no energy-storing element involved in clipping circuits. However, if the frequency of the input signal is high, the output may also be distorted due to stray capacitances.

Transistor Clipping Circuit. The transistor circuit in Fig. 25.16*a* can also be used as a double-ended clipper. When v_i is sufficiently negative, T1 is OFF and T2 is ON in its active region. As v_i increases, both T1 and T2 will be ON in their active region. As v_i continues to increase, the emitter of T1 will follow its base. However, since the base of T2 is fixed with the voltage V_{BB2} at a certain level of v_i, T2 will be OFF. The transfer characteristics of this circuit are shown in Fig. 25.16*b*. The governing equations are given as follows. The total emitter current I is given by

$$I = I_1 + I_2 = \frac{V_{\text{BB2}} + V_{\text{EE}} - V_{\text{BE2}}}{R_e} \tag{25.28}$$

Since in general, $V_{\text{BB2}} + V_{\text{EE}} \gg \text{V}_{\text{BE2}} \simeq 0.2$ V, Eq. (25.28) reduces to

$$I \cong \frac{V_{\text{BB2}} + V_{\text{EE}}}{R_e} \tag{25.29}$$

Therefore the total emitted current I is essentially constant.

When T2 is OFF (i.e., $I_2 = 0$), $I_1 = I$ and the output swings to its upper limit of $V_{\text{0,U}} = V_{\text{CC}}$. When T1 is OFF (i.e., $I_1 = 0$), $I = I_2$ and the output swings to its lower limit of

$$V_{\text{0,L}} = V_{\text{CC}} - R_C I_{C2} \tag{25.30}$$

and

$$I_2 = \frac{\beta + 1}{\beta} \cdot I_{C2} \tag{25.31}$$

From Fig. 25.16*a* we have

$$v_i = V_{\text{BB2}} + V_{E1} - V_{E2}$$

Assume that the input reaches to its upper limit $V_{i,\text{U}}$ when $I_1 = 0.9I$ and $I_2 = 0.1I$, and to its lower limit when $I_1 = 0.1I$ and $I_2 = 0.9I$. Then it can be shown that

$$v_i = V_{\text{BB2}} + \eta V_T \ln \frac{I_1}{I_2} \tag{25.32}$$

where $\eta = 1$ for germanium and $\eta = 2$ for silicon, $V_T = 26$ mV at 25°C. Hence

$$V_{i,\text{U}} = V_{\text{BB2}} + \eta V_T \ln \rho \tag{25.33}$$

$$V_{i,\text{L}} = V_{\text{BB2}} + \eta V_T \ln \rho \tag{25.34}$$

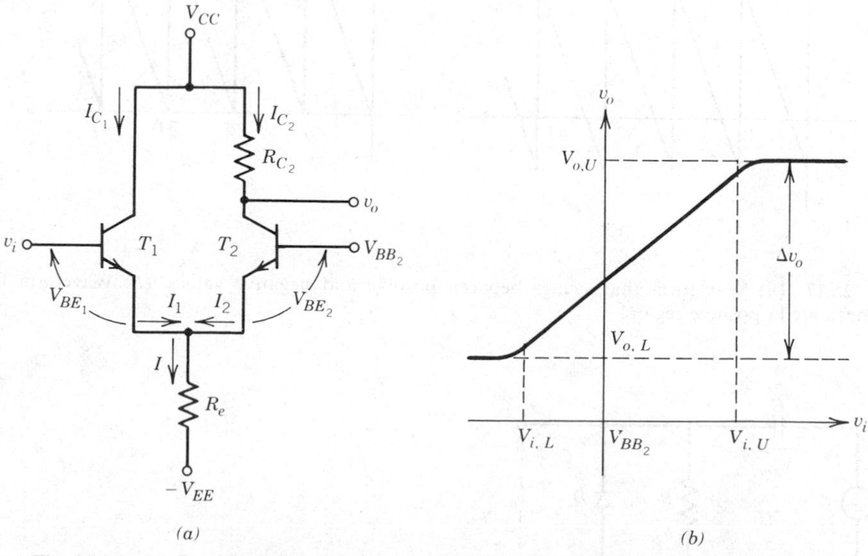

Fig. 25.16 (*a*) Double-ended transistor clipping circuit. (*b*) Its transfer characteristics.

The total input swing is then given by

$$\Delta v_i = 2\eta V_T \ln \rho \tag{25.35}$$

and the total output swing is

$$\Delta v_O = R_C I_{C2} \tag{25.36}$$

Because the current switches from T1 to T2 while the total emitted current I remains essentially constant, the circuit in Fig. 25.16*a* is also known as a *current-mode switch*.

Clamping Circuits

It is often desired that a waveform never becomes negative (or positive) even though the signal swings positive and negative. The circuit that accomplishes this is called a *clamping circuit*. For example, suppose that a waveform, such as that of Fig. 25.17*a* has been produced by a circuit and it is desired that the waveform should never swing negative (as in Fig. 25.17*b*). This may be accomplished by adding a DC component to it. However, in general, V_{1m} varies in time. Therefore a fixed DC power supply cannot be used. The clamping circuit shown in Fig. 25.18 can be used to obtain the waveform of Fig. 25.17*b*.

When v_i is negative, the diode will conduct and the capacitor C will be charged to V_{1m} with the polarity shown (it is assumed that the diode D is an ideal diode). Assume also that $RC \gg T$ where T is the period of the input signal. As v_i rises above $-V_{1m}$, the voltage $V_{1m} + v_i$ reverse biases the diode and cuts it off. Hence the voltage across the capacitor remains essentially constant at V_{1m} since $RC \gg T$, and

$$v_O = V_{1m} + v_i \tag{25.37}$$

This means that the input waveform has been shifted by the desired amount, V_1, and the output waveform shown in Fig. 25.17*b* is obtained at the output. It is important to note that the capacitor charges at a very fast rate, since the diode is ideal (i.e., forward resistance of an ideal diode is zero). If the voltage V_{1m} varies, the capacitor voltage gradually follows the new value of V_1, and the output voltage will not go negative.

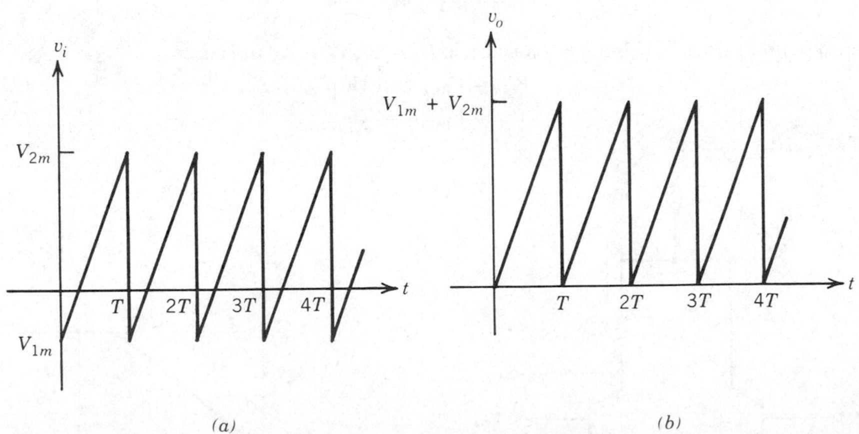

Fig. 25.17 (*a*) Waveform that swings between positive and negative values. (*b*) Waveform that swings only in positive region.

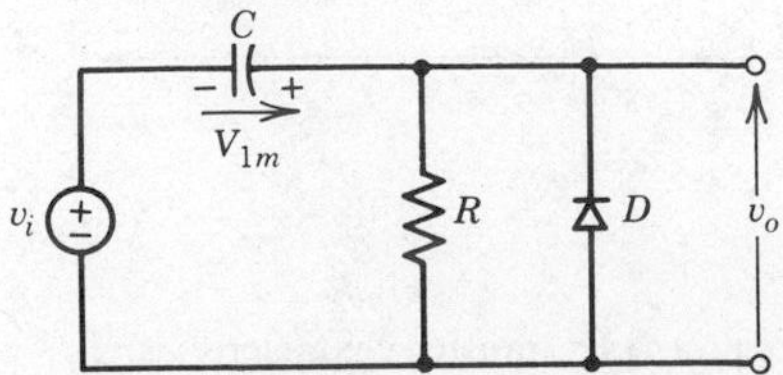

Fig. 25.18 Diode clamping circuit.

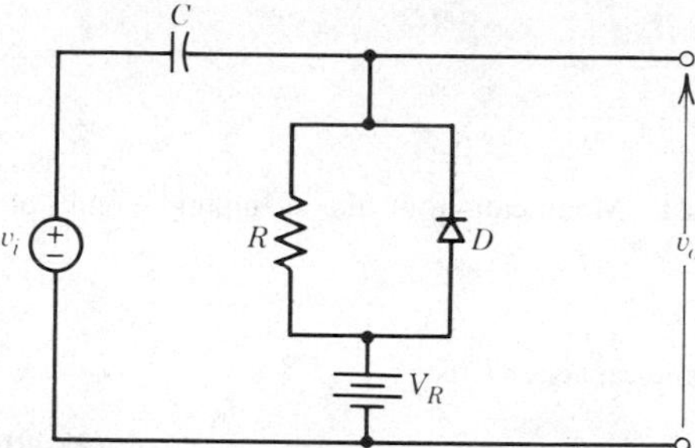

Fig. 25.19 Clamping circuit that shifts the level to V_R.

If the diode is not ideal, the output waveform may be slightly distorted and some small negative voltage would be observed.

Frequently it is desired that the signal level be shifted so that its minimum value is not zero but some predetermined value V_R (reference voltage). The circuit shown in Fig. 25.19 can be used for this purpose. In this case the output voltage can be shown to be

$$v_0 = V_R + V_{1m} + v_i \tag{25.38}$$

Clamping Circuit Theorem. Consider the square wave shown in Fig. 25.20*a* applied to a clamping circuit. The general form of the output waveform is shown in Fig. 25.20*b*. In the steady state, let A_f be the area of the output waveform when the diode is forward biased and A_r be the area of the output waveform when the diode is reverse biased. For a periodic waveform such as is shown in Fig. 25.20, the net change in the charge stored in the capacitor C over any cycle must be zero. The charge flowing into the capacitor is

$$Q_f = \int_0^{T_1} \frac{v_{of}}{R_f}\, dt \tag{25.39}$$

where v_{of} is the time-varying output voltage when the diode is conducting and R_f represents the

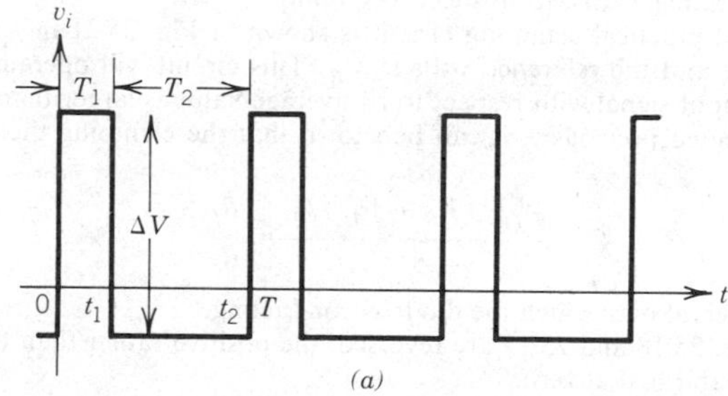

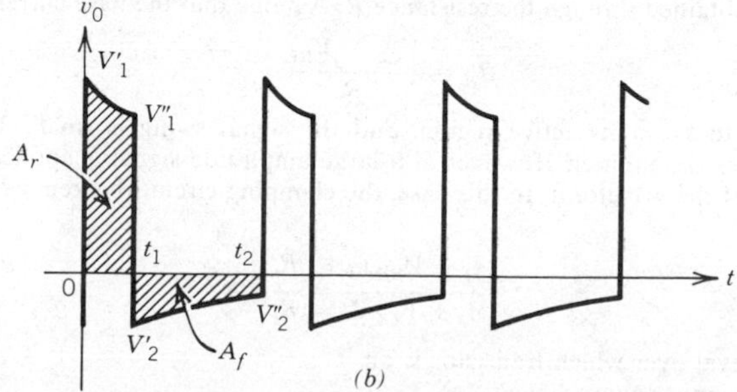

Fig. 25.20 (*a*) Square wave input applied to a clamping circuit. (*b*) Clamped output waveshape.

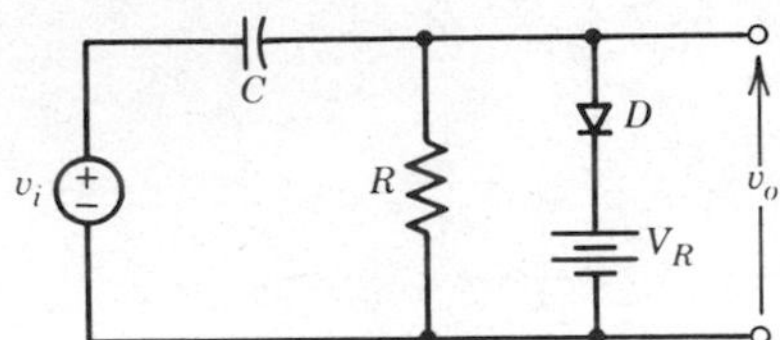

Fig. 25.21 Modification of the clamping circuit of Fig. 25.19.

forward diode resistance. The charge flowing out of the capacitor C is

$$Q_r = \int_0^{T_2} \frac{v_{or}}{R_r}\, dt \tag{25.40}$$

where v_{or} is the voltage across the output when the diode is cut off. Since the net change in the charge stored in C over the cycle is zero

$$Q_f = Q_r \tag{25.41}$$

Hence using Eqs. (25.39) and (25.40) it can be shown that

$$\frac{R_f}{R_r} = \frac{\int_0^{T_1} v_{of}\, dt}{\int_0^{T_2} v_{or}\, dt} \tag{25.42}$$

or

$$\frac{R_f}{R_r} = \frac{A_f}{A_r} \tag{25.43}$$

Hence it can be concluded that for any input waveform, the ratio of the area under the output waveform in the forward direction to that in the reverse direction is constant. This is known as the clamping circuit theorem.

For an ideal diode, $R_r = \infty$; hence Eq. (25.43) yields $A_f = 0$, which is expected as indicated in Fig. 25.17*b*.

If the circuit is modified by including a fixed reference voltage V_R, as in Fig. 25.19, the clamping theorem in Eq. (25.43) remains valid. In this case, however, the areas A_f and A_r are measured with respect to the level V_R rather than with respect to ground.

Another widely used practical clamping circuit is shown in Fig. 25.21 in which the resistor R is placed across the diode and the reference voltage V_R. This circuit will operate perfectly well if the negative swing of the input signal with respect to its average value is larger than V_R. With use of the similar procedure presented previously, it can be shown that the clamping theorem will take on the form

$$\frac{A_f - (V_R + V_D)T_1}{A_r} = \frac{R_f}{R_r} \tag{25.44}$$

where T_1 is the time interval over which the diode is conducting.

If the diodes in Figs. 25.18 and 25.19 are reversed, the positive rather than the negative extremity of the signal will be established at zero.

Transistor Clamping Circuit. Consider the transistor amplifier, shown in Fig. 25.22, in which fixed-current bias is obtained through the resistance R. Assume that the base current

$$I_\beta = \frac{V_{CC} - V_{BE}}{R} \tag{25.45}$$

forces the transistor to be in its active region, and the signal swing is small. Then the transistor behaves as a small-signal amplifier. However, if a large amplitude signal is applied, clamping occurs at the positive poles of the waveform. In this case, the clamping circuit theorem can be shown to take on the form

$$\frac{A_f - V_{BE}T_1}{A_r + V_{CC}T} = \frac{R_f}{R_r} \tag{25.46}$$

where T_1 = time interval over which transistor is on
T = period of input square wave
V_{BE} = base-emitter voltage when transistor is on

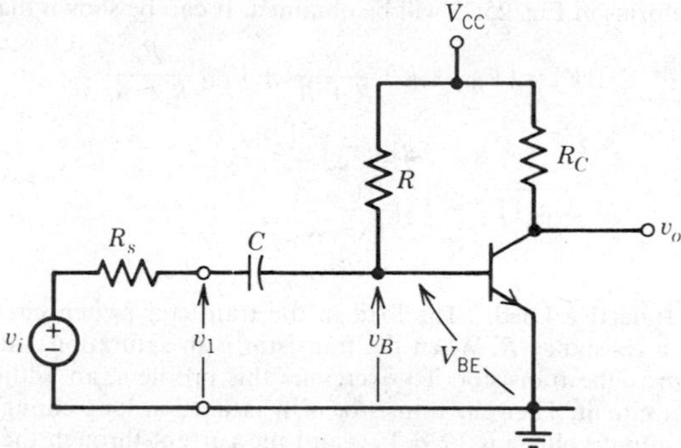

Fig. 25.22 Transistor clamping circuit.

Switching Circuits

Transistors are widely used in applications where they operate as switches. The transistor clamping circuit (Fig. 25.22) can be used as a switch. This is accomplished by driving the transistor between cutoff and saturation (clamp). For example, the transistor can be held in saturation in the absence of an input signal by the current supplied through the resistance R. Waveforms at various points are shown in Fig. 25.23. The base-to-emitter junction of the transistor can be viewed as a diode. Hence the base circuit is precisely the clamping circuit. The transistor in this circuit can be viewed as a switch that opens (cutoff) and closes (saturation) and is controlled by the square-wave input signal. If the time constant $\tau = C(R + R_s)$ with which the capacitor C charges at the positive peak of the input signal is small compared with the time interval T_2 over which the base voltage returns to a practically

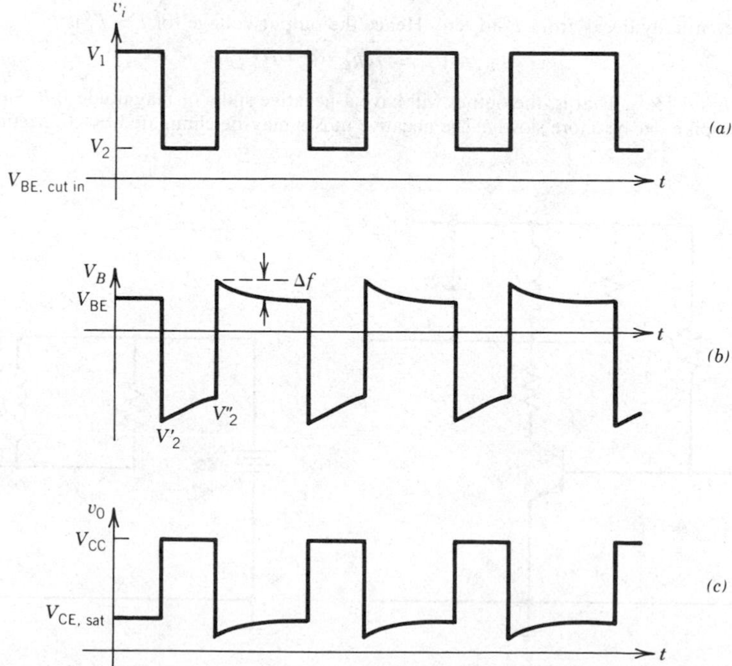

Fig. 25.23 (*a*, *b*, and *c*) Waveforms produced at various points of the transistor switch circuit of Fig. 25.22.

constant level, the waveforms in Fig. 25.23 will be obtained. It can be shown that

$$V_2' = (V_{BE} - V)\frac{R}{R + R_s} + V_{CC}\frac{R_s}{R + R_s} \tag{25.47}$$

$$V_2'' = V_{CC} - (V_{CC} - V_2')e^{-T_1/\tau} \tag{25.48}$$

$$\Delta_f \cong (V_2'' - V_2')\frac{r_{bb'}}{R_s} \tag{25.49}$$

Transistor Switch with Inductive Load. The load in the transistor switch circuit in Fig. 25.24 is an inductor L shunted by a resistance R. When the transistor is in saturation, the collector current can be large enough to destroy the transistor. To overcome this problem, an additional resistance R_C is included in the collector circuit. Then the transistor is in saturation long enough so that all transients have disappeared, the output voltage is $V_0 = V_{CC}$ and the current through the inductor is

$$I_0 = \frac{V_{CC} - V_{CE,sat}}{R_C} \tag{25.50}$$

at the instant ($t = 0^+$) the transistor is driven to cutoff. The current through the inductor at $t = 0^+$ is I_0 (since the current in an inductor cannot change instantaneously). Hence from Fig. 25.24*b*, the output voltage is given by

$$v_0 = V_{CC} + I_0 R e^{-t/\tau} \tag{25.51}$$

where $\tau = L/R$. That is, driving the transistor from saturation to cutoff produces a spike of amplitude $I_0 R$ at its collector. The magnitude of this spike may become very large, exceeding the maximum collector breakdown voltage. The output waveform for a square-wave input (Fig. 25.25*a*) of a transistor switch with inductive load is shown in Fig. 25.25*b*.

When the transistor is returned to saturation at $t = T_1^+$, the inductor acts (initially) as an open circuit, and the current I_0' through R is given by

$$I_0' = \frac{V_{CC} - V_{CE,sat}}{R + R_C} \tag{25.52}$$

which will eventually decay from I_0' to zero. Hence the output voltage for $t > T_1$ is

$$v_0 = V_{CC} - I_0' R e^{-(t-T_1)/\tau'} \tag{25.53}$$

where $\tau' = L/(R \| R_C)$. That is, the output will have a negative spike of magnitude $I_0' R$. Since $\tau' > \tau$, the negative spike decays more slowly. The negative pulses may be eliminated by connecting a diode

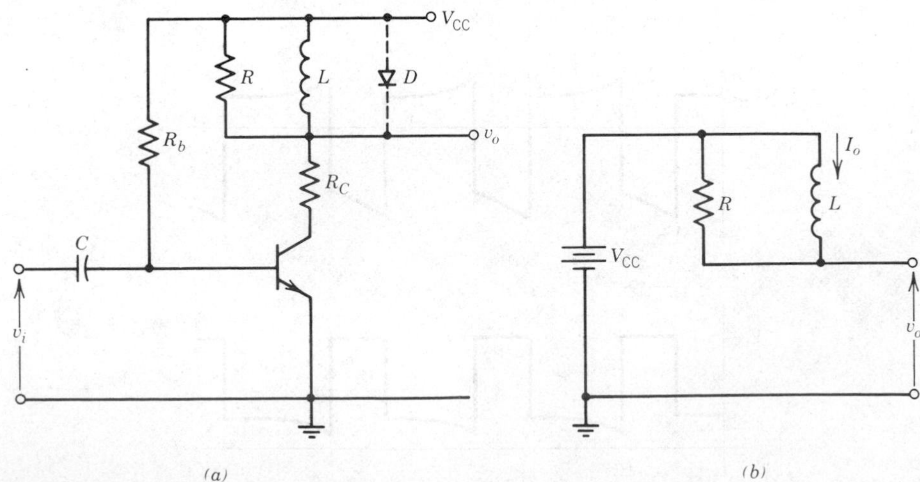

Fig. 25.24 (*a*) Transistor switch with inductive load. (*b*) Equivalent circuit when transistor is cut off.

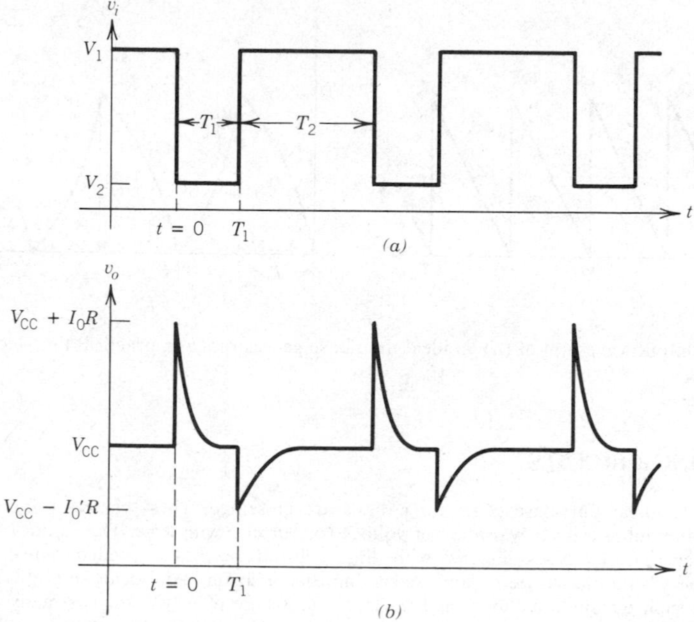

Fig. 25.25 (*a*) Square-wave input. (*b*) Output waveform of a transistor switch with inductive load.

across the inductor (as indicated by dashed lines in Fig. 25.24*a*). In this case the resistance R may be omitted.

Nonsaturating Switches. In the transistor switching circuits discussed previously, the transistor is driven from saturation to cutoff or from cutoff to saturation. However, when a transistor is saturated, excess charge is stored in the base-emitter and base-collector junctions. Therefore if the transistor in saturation is suddenly driven to cutoff, these excess charges must be removed before the switch will open again. This, of course, slows down the switching operation. One way of increasing the switching speed is to prevent the transistor from going into saturation. The principle of obtaining improved switching speed is shown in Fig. 25.26. The collector voltage now may drop while the diode D conducts. In this case

$$v_{CB} = -V_D + V_B \tag{25.54}$$

to prevent saturation, $V_{CB} > 0$. That is, if $V_B > V_D$, the transistor will not be saturated. If the diode D is a germanium diode, then the voltage source V_B can be realized by the use of a forward-biased silicon diode. If D is a silicon diode, then two silicon diodes in series may be used to realize V_B.

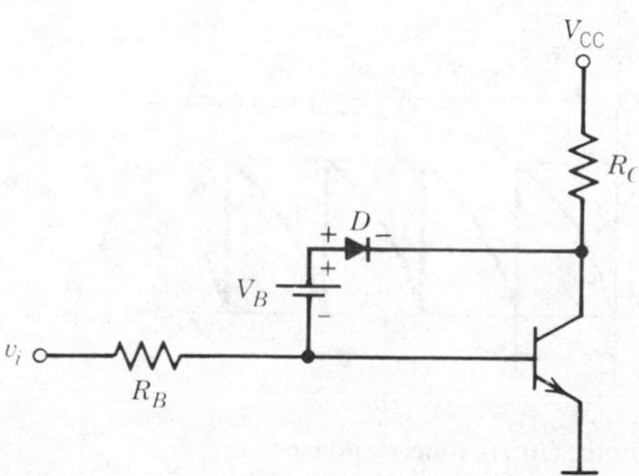

Fig. 25.26 Nonsaturating transistor circuit.

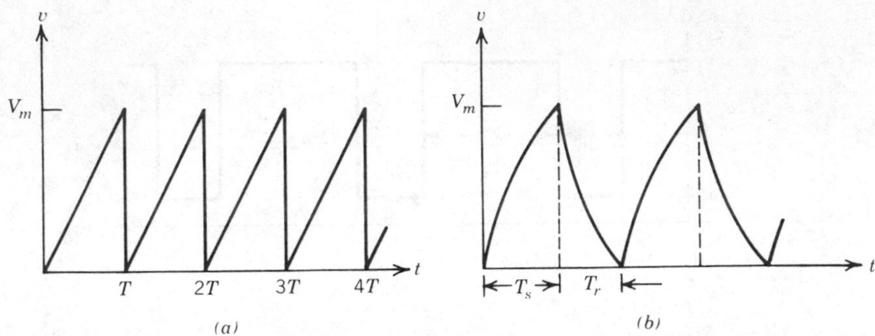

Fig. 25.27 Output waveform of (*a*) an ideal time-base generator, (*b*) a practical time-base generator.

25.2 TIMING CIRCUITS

In this section, linear time-base generator circuits are presented. These circuits provide an output waveform that exhibits a linear variation of voltage (or current) with time. That is, ideally the output voltage (or current) increases linearly with time until it reaches a predetermined final value, "instantaneously" returns to zero, and starts increasing again. An ideal and a typical output waveform of such circuits are shown in Fig. 25.27. This type of waveform has many applications, such as in the display of time-varying signals on oscilloscope screens, radar, television receivers and transmitters, and control systems. These circuits are also known as sweep generators or time-base generators.

25.2-1 Linear Voltage Sweep

Common to all linear voltage sweep circuits are (1) production of some form of exponential charging of a capacitor and (2) discharge of the capacitor at the proper time. Differences among sweep circuits involve techniques used to linearize the sweep, active elements used, and techniques used to discharge the capacitor.

A simple voltage sweep circuit and its time response are shown in Fig. 25.28. At $t = 0$, the switch S is opened and the capacitor is charged toward the voltage

$$V = IR \tag{25.55}$$

exponentially. That is, the output voltage is

$$v_s = V(1 - e^{-t/RC}) \tag{25.56}$$

At the time $t = T_s$, denoted as the sweep time, the sweep amplitude reaches the value V_s and the switch is closed. Note that the sweep is not linear, but it is exponential. If, however, the amplitude of

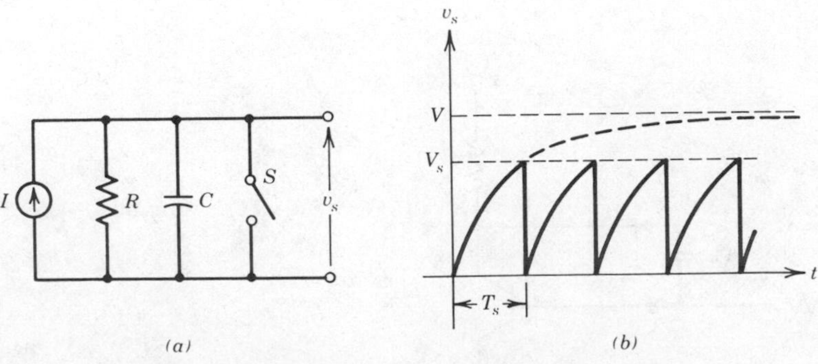

Fig. 25.28 (*a*) Voltage sweep circuit. (*b*) Its time response.

the sweep voltage is restricted to a small portion of its steady-state value V, the sweep voltage v_s can better approximate a linear rise. Better approximation can also be obtained by increasing the time constant $\tau = RC$. To see this, let us expand the exponential of Eq. (25.56) in series form so that

$$v_s = \frac{Vt}{RC}\left(1 - \frac{1}{2!}\cdot\frac{t}{RC} + \frac{1}{3!}\cdot\frac{t^2}{R^2C^2} - \cdots\right) \tag{25.57}$$

If $t/RD \ll 1$, then

$$v_s \cong \frac{V}{RC}t \tag{25.58}$$

Hence the sweep voltage is linear with time (as is desired), and the amplitude of the sweep voltage is given by

$$V_s = \frac{V}{RC}T_s \tag{25.59}$$

Therefore it can be concluded that if a sweep is to be reasonably linear, the time constant $\tau = RC$ must be large compared with the sweep time T_s.

Sweep Linearity

The quality of sweep is measured by its departure from the ideal sweep. Two possible choices of the measurement of sweep linearity are illustrated in Fig. 25.29. The sweep linearity error in Fig. 25.29*a* is called transmission error and is defined by

$$e_t = \frac{V_s' - V_s}{V_s'}\cdot 100\% \tag{25.60}$$

The sweep linearity error in Fig. 25.29*b* is called displacement error and is defined by

$$e_d = \frac{(v_s - v_s')\text{max}}{V_s}\cdot 100\% \tag{25.61}$$

With use of Eqs. (25.56) through (25.59) it can be shown that the sweep linearity errors for the circuit in Fig. 25.28*a* are

$$e_t = \frac{1}{2}\cdot\frac{T_s}{RC}\cdot 100\% \tag{25.62}$$

and

$$e_d = \frac{1}{8}\cdot\frac{T_s}{RC}\cdot 100\% \tag{25.63}$$

If there were no source or load resistance in Fig. 25.28*a*, the capacitor C would be charged by a

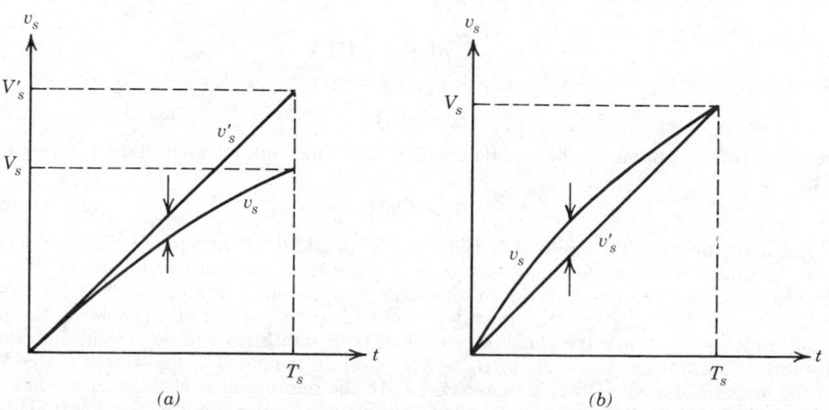

Fig. 25.29 Two methods of defining sweep linearity: (*a*) transmission error, (*b*) displacement error.

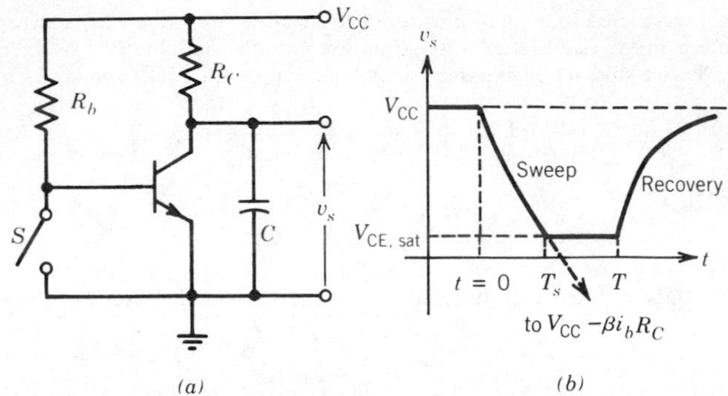

Fig. 25.30 (*a*) Simple transistor voltage sweep circuit. (*b*) Output waveform produced.

constant current I and the voltage across the capacitor would be

$$v_s = \frac{1}{C}\int I\,dt = \frac{1}{C}t \tag{25.64}$$

Observe that this is the ideal linear sweep. [See Eq. (25.58).]

Transistors can be used as a current source. A simple transistor voltage sweep circuit is shown in Fig. 25.30*a*. When the switch S is closed, the transistor is cut off and the sweeping capacitor C is charged to V_{CC} through R_C with a time constant $\tau_s = R_C C$. When the switch is opened at $t = 0$, the transistor is driven into saturation and the charged capacitor keeps it in saturation. When the switch is closed at $t = T$, the transistor is again cut off and the capacitor charges back (recovers) to V_{CC} with the time constant $\tau_r = R_C C$.

The circuit can be designed so that the magnitude of the steady-state value of v_s, after the switch is closed, is much larger than that of the magnitude of the sweep voltage $V_{CC} - V_{CE,sat}$. That is,

$$|V_{CC} - \beta I_b R_C| \gg |V_{CC} - V_{CE,sat}| \tag{25.65}$$

where

$$I_b = \frac{V_{CC} - V_{BE}}{R_b} \tag{25.66}$$

In this case the sweep voltage for $0 < t < T_s$ can be assumed to be linear for most practical applications, and

$$T_s = \frac{V_{CE,sat} - V_{CC}}{-\beta I_b R_C} \cdot \tau_s \tag{25.67}$$

As a numerical example, let $R_b = R_C = 10$ K, $V_{CC} = 10$ V, $V_{CE,sat} = 0.2$ V, $V_{BE} = 0.7$ V, and $\beta = 50$. Then

$$|V_{CC} - \beta I_b R_C| = 455 \text{ V}$$
$$|V_{CC} - V_{CE,sat}| = 9.8 \text{ V}$$
$$T_s = 0.021\tau_s$$

The recovery can be assumed to be complete within five time constants; hence the recovery time is given by

$$T_r = 5R_C C \tag{25.68}$$

The recovery time can be improved considerably if an additional recharging current is supplied. This can be done by placing a second transistor in the circuit, as shown in Fig. 25.31. When the switch S is open, the transistor T_2 is driven into saturation and the sweeping capacitor is charged to the voltage $V_{CC} - V_{CE,sat}$. At $t = 0$, the switch is closed and T_2 is cut off. The sweep starts, and continues until $t = T_s$. When the switch is reopened, both transistors will be on and the capacitor charges toward the voltage $V_{CC} + R_C\,(\beta_2 I_{b_2} - \beta_1 I_{b_1})$, which is limited to the voltage $V_{CC} - V_{CE,sat}$ within the recovery time T_r. Observe from Fig. 25.31*b* the improvement of the recovery time. Also note that both sweeping and recovering sections of the output waveform are quite linear. Therefore

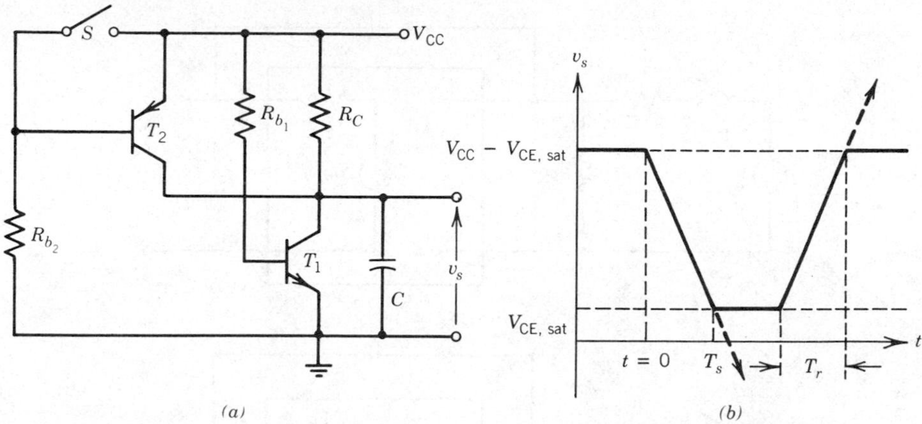

Fig. 25.31 (*a*) Two-transistor sweeping circuit to improve recovery time. (*b*) Waveform produced at the output.

either part can be used as the sweep, the time constants for both parts are the same, and

$$\tau_s = \tau_r = R_C C \tag{25.69}$$

Miller Sweep

The basic sweep circuit in Fig. 25.28 produces an exponential sweep output. A small portion of this output can be considered linear, which is acceptable for many practical applications. Miller sweep circuits make use of negative feedback for improving the linearity considerably. A Miller sweep circuit is shown in Fig. 25.32*a*, where R_i is the input resistance, R_0 is the output resistance, and A is the gain of the amplifier. The amplifier can be an operational amplifier or a transistor. If the input circuit is replaced by its Thévenin equivalent, the circuit in Fig. 25.32*b* is obtained where

$$V_1 = \frac{R_i}{R_s + R_i} \cdot V \quad \text{and} \quad R_1 = R_i \| R_s \tag{25.70}$$

It can be shown that

$$v_i = \frac{1}{1 + (1 - A)(R_1/R_0)} \cdot V_1 + \frac{1}{[R_1 + R_0/(1 - A)](1 - A)C} \int_0^t V_1\, dt \tag{25.71}$$

and the sweep voltage at the output is

$$v_s = Av_i = \frac{AV_1}{1 + (1 - A)(R_1/R_0)} + \frac{AV_1}{[R_1 + R_0/(1 - A)](1 - A)C}\, t \tag{25.72}$$

In general, the gain of the amplifier is a very large negative number and R_0 (the output resistance) is very small. Therefore we can assume that $|1 - A| \cong |A|$, that $R_1 \gg R_0/(1 - A)$, and that Eq. (25.72) reduces to

$$v_s \cong -\frac{R_0}{R_1} V_1 - \frac{V_1}{R_1 C}\, t \tag{25.73}$$

Thus the output is a linear negative-going voltage sweep with a sweep speed of

$$\text{Sweep speed} = -\frac{V_1}{R_1 C} \tag{25.74}$$

Nonlinearity of better than 0.1% can easily be achieved with this circuit. A positive-going sweep can be obtained by simply reversing the polarity of the source voltage V in Fig. 25.33*a*.

Transistor Miller Sweep Circuit. A simple transistor Miller circuit is shown in Fig. 25.33*a*. The capacitor C is connected in a feedback loop around the amplifier formed by T_1. The transistor acts as a switch. When T_1 is on, all of the current through R_1 goes to ground, the transistor T_1 is cut off,

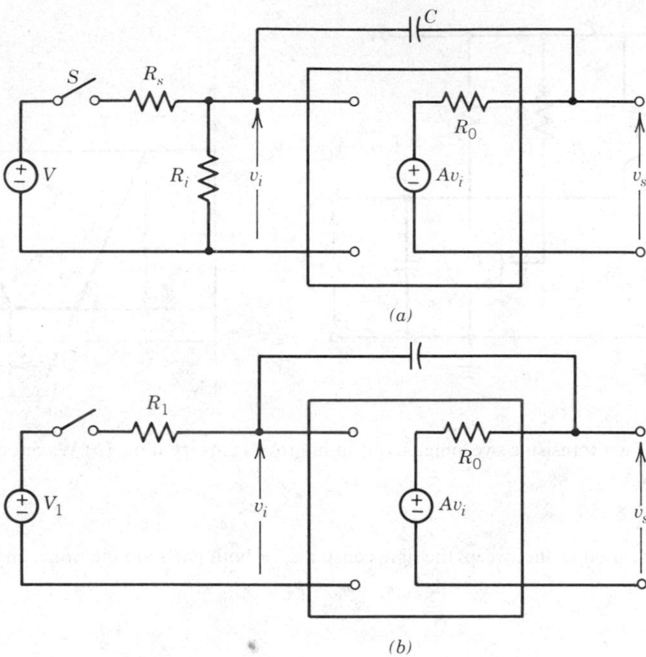

Fig. 25.32 (*a*) Miller sweep circuit. (*b*) Input circuit is replaced by its Thévenin equivalent.

and the output of T_1 is clamped at V_{CC}. When T_2 is off, T_1 is turned on and operates in its active region and a collector current flows through R_C. The negative feedback loop is closed through the capacitor C to the base of T_1, which tends to minimize the base current of T_1. Hence the collector voltage (sweep voltage) v_s produces a negative-going linear voltage sweep, as indicated in Fig. 25.33*b*. The sweep voltage can be shown to be

$$v_s \cong \frac{V_{CC}}{R_b C} t \tag{25.75}$$

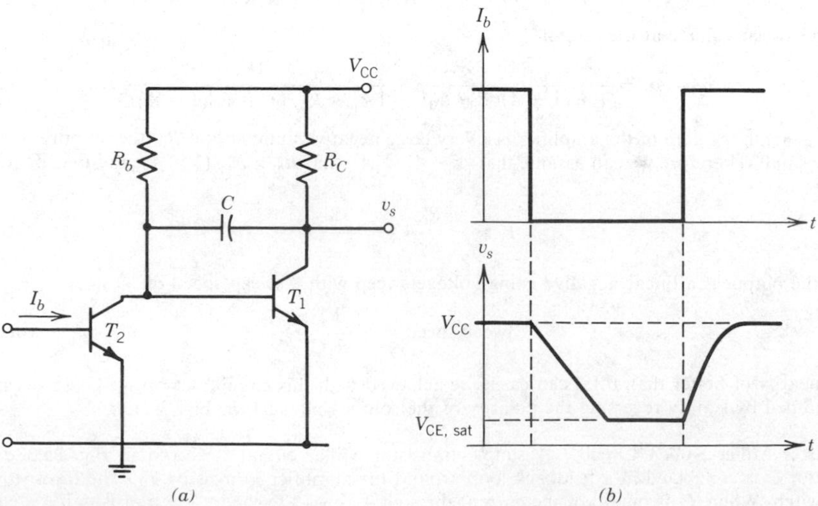

Fig. 25.33 (*a*) Transistor Miller sweep circuit. (*b*) Output sweep voltage waveform.

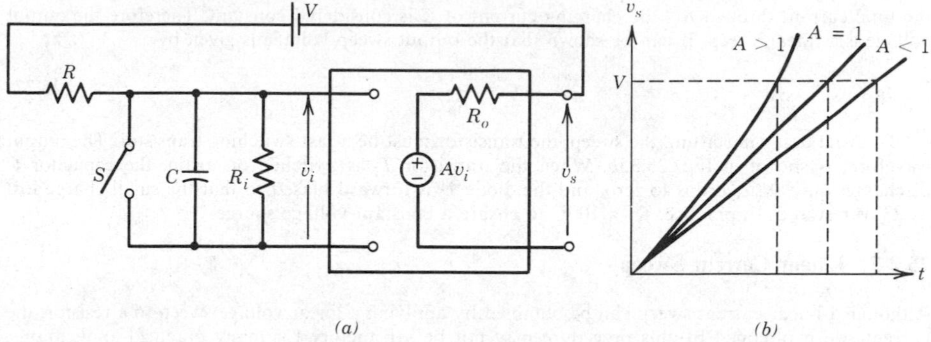

Fig. 25.34 (*a*) Bootstrap sweep circuit. (*b*) Output sweep waveforms for $A = 1$ and $A \neq 1$.

Bootstrap Sweep

A typical bootstrap sweep circuit is shown in Fig. 25.34. This circuit differs from the Miller sweep circuit in that the capacitor C is not a part of the feedback loop. The amplifier in the circuit is a common emitter follower; therefore $A \cong 1$. Neglecting the small output resistance R_0, and proceeding as in the case of the Miller sweep, it can be shown that

$$v_s = \frac{V}{RC} t \tag{25.76}$$

Note that the sweep is independent of the input resistance R_i ($R \ll R_i$).

It is important to keep the gain of the amplifier as close to unity as possible to achieve good linearity in the sweep. The output sweep for $A = 1$, $A > 1$, and $A < 1$ are shown in Fig. 25.34*b*.

Transistor Bootstrap Sweep Circuit. A transistor bootstrap circuit is shown in Fig. 25.35*a*. The transistor T_2 again behaves like a switch. The capacitor C is charged through the resistance R_b from the voltage across C_1.

Assume that the switch T_2 is on (in saturation). Then the transistor T_1, connected in emitter-follower configuration, is cut off. The capacitor C_1 is charged to the voltage V_{CC} through the diode D and the emitter resistance R_e.

When the transistor T_2 is cut off, the capacitor C is charged by the current through the resistance R_b, causing the transistor T_1 to turn on. Since T_1 is an emitter-follower configuration, the output voltage v_s follows the voltage across C. Because the base current of T_1 is very small compared with

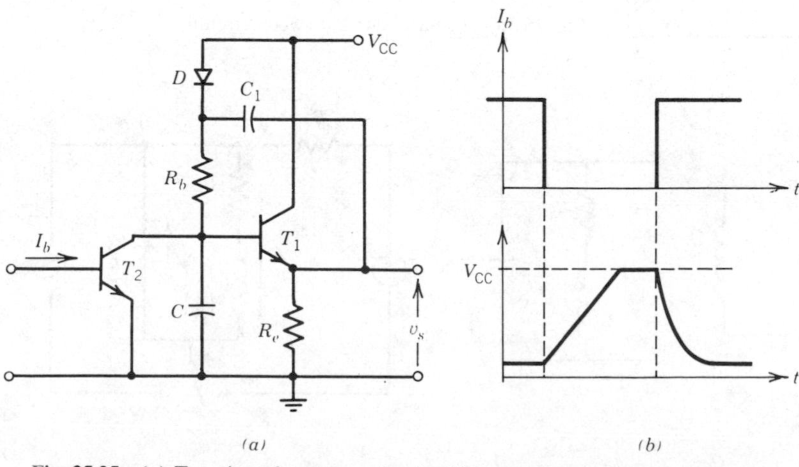

Fig. 25.35 (*a*) Transistor bootstrap circuit. (*b*) Output sweep voltage waveform.

the total current through R_b, the charging current of C is considered constant. Therefore the output voltage is a linear sweep. It can be shown that the output sweep voltage is given by

$$v_s = \frac{V_{CC}}{R_b C} t \tag{25.77}$$

To avoid delay in starting the sweep, the transistor must be a fast-switching transistor. The output waveform is shown in Fig. 25.35*b*. When the transistor T_2 is switched on again, the capacitor C discharges quickly, v_s drops to zero, and the diode D is forward biased so that the small charge lost by C_1 is restored. In practice, $C_1 \geqslant 100C$ to ensure a constant voltage source.

25.2-2 Linear Current Sweep

Although a linear current sweep can be obtained by applying a linear voltage sweep to a resistor, the current sweep obtained by this procedure may not be satisfactory for many practical applications. For example, deflection of a high-energy electron beam in television circuits requires a very large sweep amplitude. Hence magnetic deflection, which is produced by a linear current sweep in television receivers and radars, is preferred.

Linear current sweep circuits are similar to those of linear voltage sweep circuits except that the sweep capacitor is replaced with an inductor as the basic sweep element. A basic current sweep circuit is shown in Fig. 25.36. The inductor is assumed to be ideal and the voltage source provides a constant voltage. When the switch is closed momentarily, we have

$$V = L\frac{di}{dt} \tag{25.78}$$

Hence

$$i = \frac{V}{L} t \tag{25.79}$$

However, all practical coils have winding resistance and interwinding stray capacitance. Therefore a practical coil can be modeled, as shown in Fig. 25.37*a*, and the actual current sweep circuit will have the form shown in Fig. 25.37*b*.

Assume that the capacitor C_1 is initially discharged, and the current through the inductor is zero [i.e., just before the switch is closed $v_C(0^-) = 0$ and $i_L(0^-) = 0$]. Immediately after the switch is closed, the complete circuit current V/R_s will flow into the capacitor C_1. The time constant for this

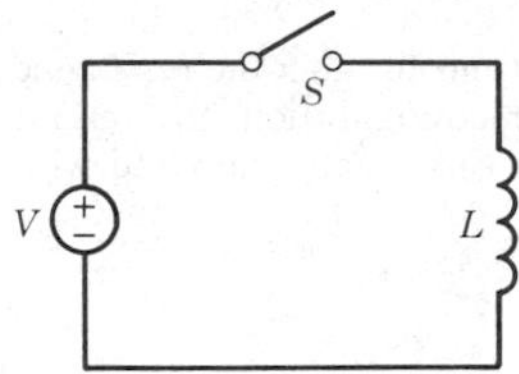

Fig. 25.36 (*a*) Basic current sweep circuit.

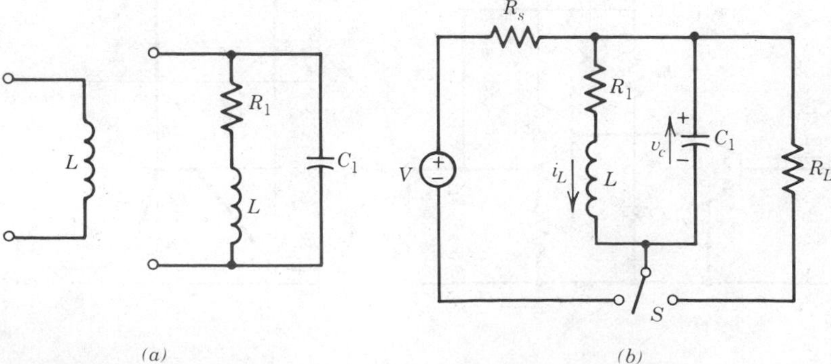

Fig. 25.37 (*a*) Practical inductor and its equivalent model. (*b*) Current sweep circuit.

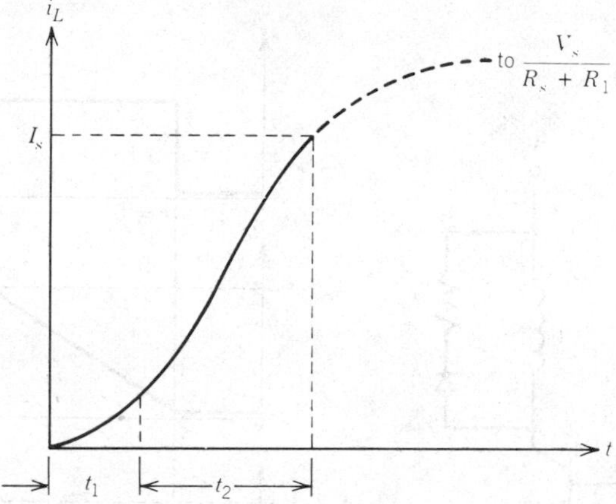

Fig. 25.38 Current sweep waveform indicating the delay and the linear sweep.

charging is $\tau_1 = R_s C_1$. Because of the initial charge time caused by the stray capacitance, the current sweep is delayed. At the end of the delay, which is assumed to be $t_1 = 4\tau_1$, the circuit enters into its sweeping region. The time response of this circuit (neglecting the effect of the stray capacitance) is expressed in the exponential form

$$i_L = \frac{V_s}{R_s + R_1}(1 - e^{-t/\tau_s}) \tag{25.80}$$

where $\tau_s = L/(R_s + R_1)$. If $t/\tau s \ll 1$, then Eq. (25.80) can be approximated by

$$i_L \cong \frac{V_s}{L} t \tag{25.81}$$

That is, for good sweep linearity the sweep interval t_2 must be limited to a small value. This imposes a restriction on the coil design. Namely, τ_1 must be much smaller than τ_s. This is easily accomplished, since the stray capacitance is, in general, very small, assuring that $\tau_1 \ll \tau_s$. The wave shape, indicating the delay and the sweep sections, is shown in Fig. 25.38.

Transistor Current Sweep

A simple transistor current sweep circuit is shown in Fig. 25.39*a*. The transistor in this circuit operates as a switch. When the transistor is on (saturated) (assuming that the saturation resistance of the transistor is small), the inductor current increases linearly (as discussed previously). During this interval, the diode D is cut off. At the end of the sweep (i.e., at $t = T_s$), the transistor is cut off and the inductor current flows through the diode's resistance R_d until this current falls to zero. This decay is given by

$$i_L = I_L e^{-R_D(t-\tau_s)/L} \tag{25.82}$$

where

$$I_L = \frac{V_{CC}}{L} t \tag{25.83}$$

is the sweep amplitude and R_D is the sum of the damping and the diode forward resistance. The input voltage, inductor current, and collector voltage waveforms are shown in Fig. 25.39*b*. Observe that when the transistor is cut off at the instant $t = T_s$, a spike of amplitude $I_L R_d$ is produced. This spike must be limited to the value below the collector-base breakdown voltage, which can be done by limiting the size of the resistance R_d.

The nonlinearity in the current sweep circuit is caused primarily by the series resistance with the inductor. The reason is the fact that as the inductor current increases, the current in the series resistance also increases. Hence the voltage across the coil and the rate of change of current decrease. The sweep circuit in Fig. 25.40*a* is used to compensate for the voltage developed across the coil's resistance.

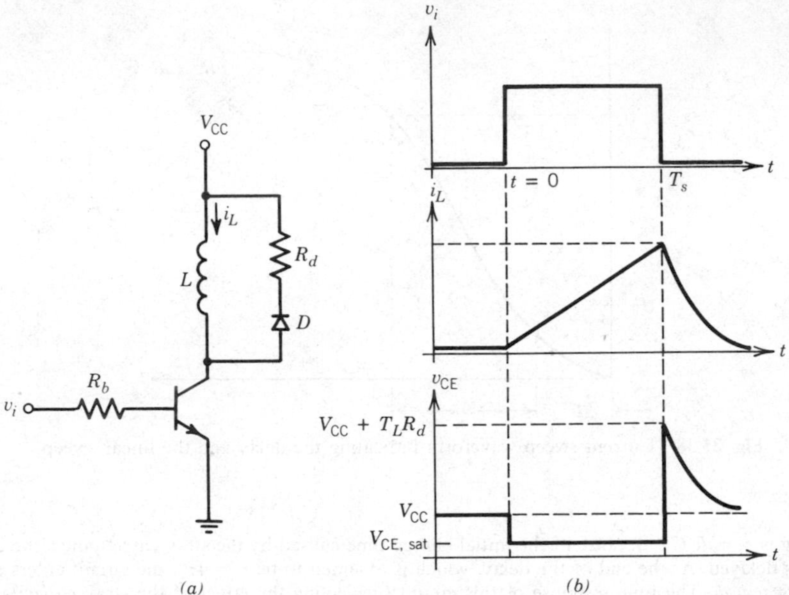

Fig. 25.39 (*a*) Transistor current sweep circuit. (*b*) Input voltage inductor current and collector voltage waveforms.

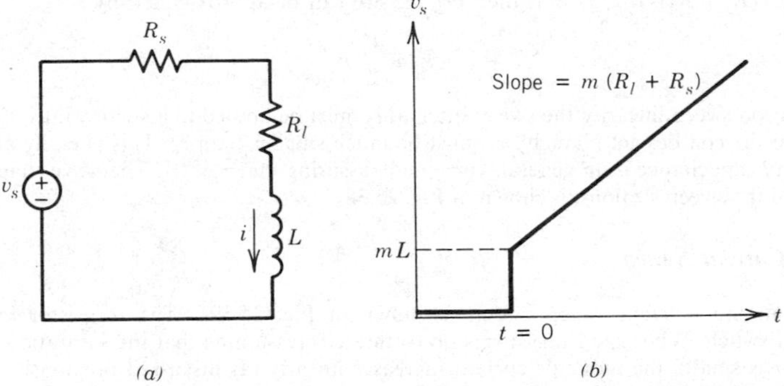

Fig. 25.40 (*a*) Current sweep circuit. (*b*) Trapezoidal waveform.

The loop equation gives

$$v_s = (R_s + R_l)I + L\frac{di}{dt} \tag{25.84}$$

If the current is to be linear (i.e., if $i = mt$), then v_s will have the form

$$v_s = Lm + m(R_s + R_l)t \tag{25.85}$$

This implies that the source voltage will have a step at $t = 0$, and a ramp with a slope of $m(R_s + R_l)$, as shown in Fig. 25.40*b*. This waveform is called a *trapezoidal waveform*. The trapezoidal waveform can be generated in a voltage sweep circuit by simply adding a resistance R, as shown in Fig. 25.41.

To illustrate this, assume that the switch S is opened at $t = 0$. Then the output voltage v_s is given by

$$v_s = V - \frac{R_s}{R_s + R} V \cdot e^{-t/(R_s + R)C} \tag{25.86}$$

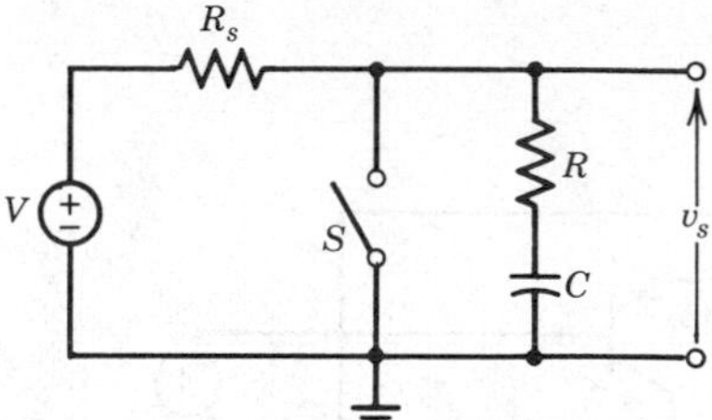

Fig. 25.41 Trapezoidal waveform generator.

If $t/(R_s + R)C \ll 1$, then Eq. (25.86) can be approximated by

$$v_s \cong V - \frac{R_s}{R_s + R} V\left[1 - \frac{t}{(R_s + R)C}\right] \tag{25.87}$$

Furthermore, in general $R_s \gg R_1$ and v_s becomes

$$v_s \cong \frac{R}{R_s} V + \frac{1}{R_s C} Vt \tag{25.88}$$

Hence the output voltage waveform is a trapezoidal waveform (as desired). This trapezoidal waveform should not be applied directly to an inductor—otherwise the voltage v_s will not be given by Eq. (25.86). It should be applied through the active device, such as a transistor.

25.3 555 TIMER INTEGRATED CIRCUIT

The 555 timer integrated circuit (IC) is a monolithic timing circuit that has a wide variety of applications. Some of these are listed in Appendix 25.1, the Fairchild μA555 data sheet. This section provides information on the interval architecture and two basic operating modes (i.e., monostable and astable) of the 555 timer IC.

25.3-1 Internal Architecture of 555

The 555 timer IC package consists of two voltage comparators (the threshold and trigger comparators), a control flip-flop, a discharge transistor Q_D, a resistor voltage divider network, and an output inverter buffer (Fig. 25.42). The resistive network, consisting of three equal resistors (5 K each), acts as a voltage divider providing reference voltages for both comparators, as indicated. The outputs of the comparators are applied to the RS flip-flop. When the trigger voltage falls below $1/3\ V_{CC}$, the trigger comparator sets the RS flip-flop, which drives the output to a "high" state. In normal operations the threshold pin monitors the capacitor voltage of the *RC* timing circuit. At the time the capacitor voltage exceeds $2/3\ V_{CC}$, the threshold comparator resets the flip-flop, driving the output to a "low" state. At this time, the transistor Q_D turns on and discharges the external timing capacitor. The timing cycle is now completed. The next timing cycle starts when another negative pulse arrives at the trigger input.

25.3-2 Monostable (One-Shot) Operation

The circuit configuration of the 555 timer IC is shown in Fig. 25.43. Initially, before the sequence of events starts, the control flip-flop holds the transistor Q_D "on," causing the external capacitor C_1 to be shorted to ground. When a voltage below $1/3\ V_{CC}$ is sensed by the trigger comparator (at the negative edge), the control flip-flop is set, releasing the short circuit across C_1 by turning Q_D "off." At this time, the output goes "high." The voltage across C_1 starts rising exponentially toward V_{CC} with a time constant of R_1C_1. When this voltage reaches $2/3\ V_{CC}$, the threshold comparator resets the flip-flop. This, in turn, sets Q_1 "on" and discharges C_1, and the output goes "low." Hence the timer goes back to its initial "standby" state awaiting another negative-going trigger input pulse.

Ignoring the capacitor leakage, the time it takes the capacitor voltage to reach the $2/3$-V_{CC} level can be shown to be

$$T = 1.1\ R_1C_1 \tag{25.89}$$

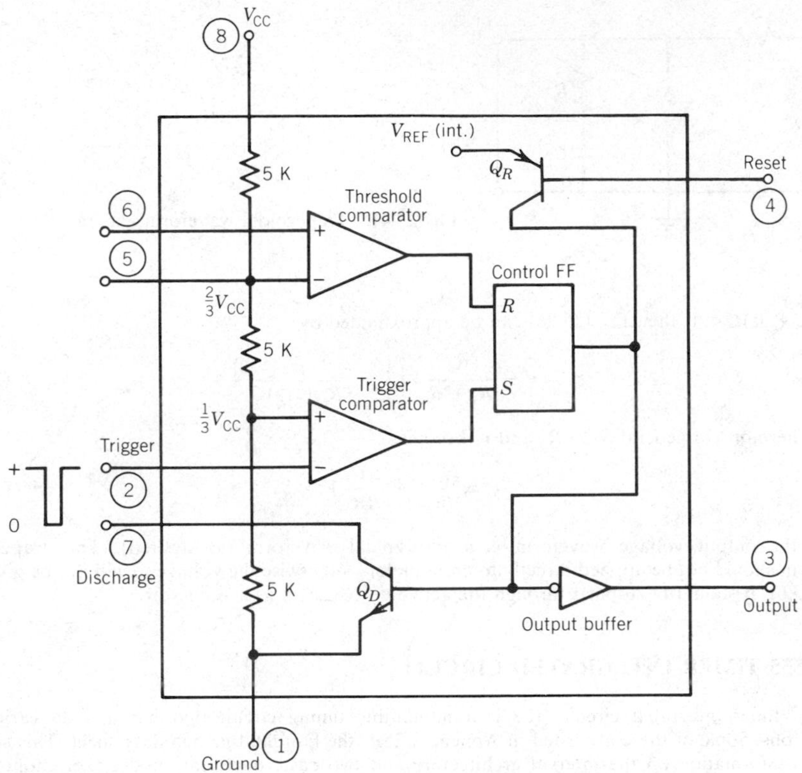

Fig. 25.42 Internal architecture of the 555 timer integrated circuit. Numbers in circles represent the pin assignments.

where T is in seconds, R_1 is in ohms, and C_1 is in farads. Thus the output stays "high" for the time 1.1 R_1C_1 s. Manufacturers of the 555 timer provide curves to select proper R_1 and C_1 values to accomplish desired pulse width at the output (see Appendix 25.1).

When the timer is to be used in noisy environments, it is recommended that a 0.01-μF capacitor be placed between pin 5 and the ground (Fig. 25.43). Otherwise the timing indicated in Eq. (25.89) may not hold.

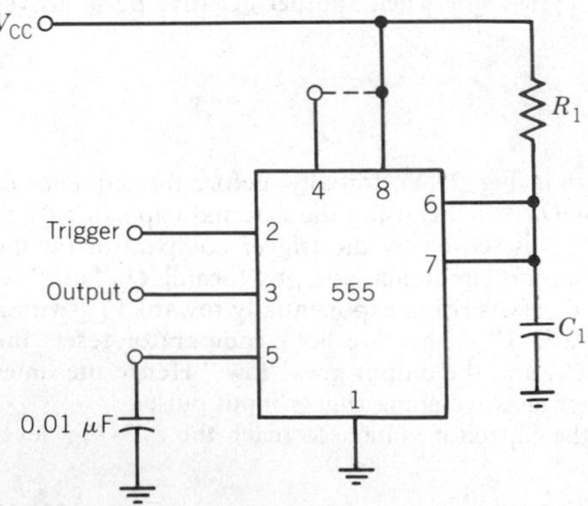

Fig. 25.43 Monostable operation of the 555 timer integrated circuit.

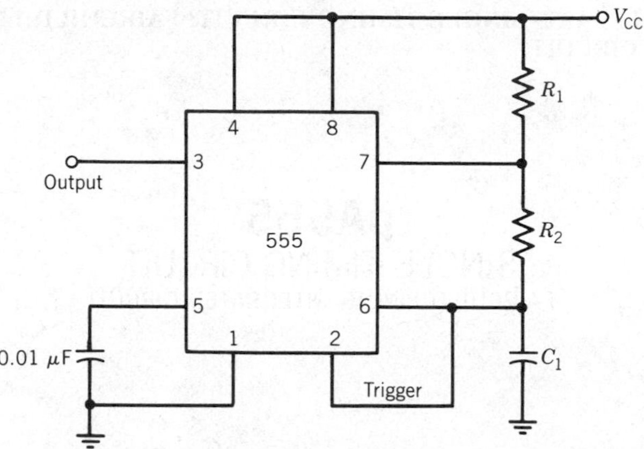

Fig. 25.44 Astable operation of the 555 timer integrated circuit.

25.3-3 Astable (Free-Running or Oscillatory) Operation

The circuit diagram for an astable operation is shown in Fig. 25.44. Note that the trigger input is now tied to the treshold pin and a resistance, R_2, is added.

At the time the power is applied to the circuit, the capacitor C_1 is discharged, causing the trigger to be "low." This automatically triggers the timer, which charges the capacitor through R_1 and R_2. When the capacitor voltage reaches the threshold level of 2/3 V_{CC}, the output goes "low" and Q_D turns on. The timing capacitor now discharges through R_2. As soon as the voltage across the capacitor falls to the level of 1/3 V_{CC}, the trigger comparator sets the flip-flop and automatically retriggers the timer. The frequency of the oscillation is given by

$$f = \frac{1.443}{(R_1 + 2R_2)C_1} \tag{25.90}$$

By proper selection of the values for R_1 and R_2, the desired duty cycle can be obtained. A duty cycle of less than 50% can be obtained by inserting diodes to the circuit, as shown in Fig. 25.45. Note that without these diodes, C_1 charges through $R_1 + R_2$ and discharges through R_2. With the diodes included, C_1 charges through R_1 and discharges through R_2; hence a duty cycle of less than 50% can be obtained.

Proper values for R_1 and R_2 can be obtained from the curves that manufacturers provide in the data sheets (see Appendix 25.1).

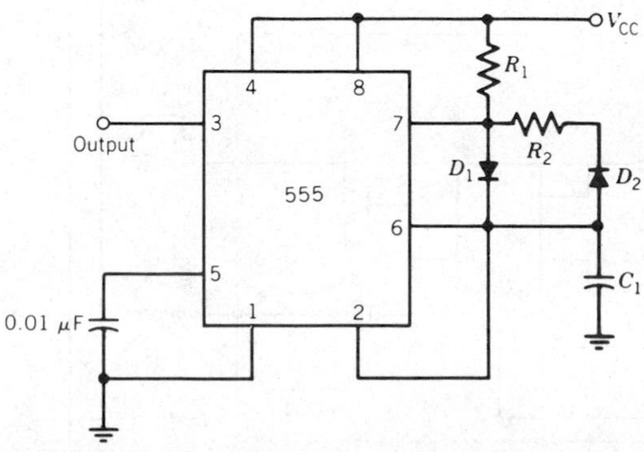

Fig. 25.45 Circuit obtaining less than 50% duty cycle.

APPENDIX 25.1 μA555 SINGLE TIMING CIRCUIT: FAIRCHILD LINEAR INTEGRATED CIRCUIT

μA555
SINGLE TIMING CIRCUIT
FAIRCHILD LINEAR INTEGRATED CIRCUIT

GENERAL DESCRIPTION – The μA555 Timing Circuit is a very stable controller for producing accurate time delays or oscillations. In the time delay mode, the delay time is precisely controlled by one external resistor and one capacitor; in the oscillator mode, the frequency and duty cycle are both accurately controlled with two external resistors and one capacitor. By applying a trigger signal, the timing cycle is started and an internal flip-flop is set, immunizing the circuit from any further trigger signals. To interrupt the timing cycle a reset signal is applied ending the time-out.

The output, which is capable of sinking or sourcing 200 mA, is compatible with TTL circuits and can drive relays or indicator lamps.

- MICROSECONDS THROUGH HOURS TIMING CONTROL
- ASTABLE OR MONOSTABLE OPERATING MODES
- ADJUSTABLE DUTY CYCLE
- 200 mA SINK OR SOURCE OUTPUT CURRENT CAPABILITY
- TTL OUTPUT DRIVE CAPABILITY
- TEMPERATURE STABILITY OF 0.005% PER °C
- NORMALLY ON OR NORMALLY OFF OUTPUT
- DIRECT REPLACEMENT FOR SE555/NE555

ABSOLUTE MAXIMUM RATINGS

Supply Voltage	+18 V
Power Dissipation (Note 1)	600 mW
Operating Temperature Ranges	
μA555TC/HC	0°C to +70°C
μA555HM	–55°C to +125°C
Storage Temperature Range	–65°C to +150°C
Lead Temperature	
Plastic Mini DIP(9T) (Soldering, 10 s)	260°C
Metal Can (5T) (Soldering, 60 s)	300°C

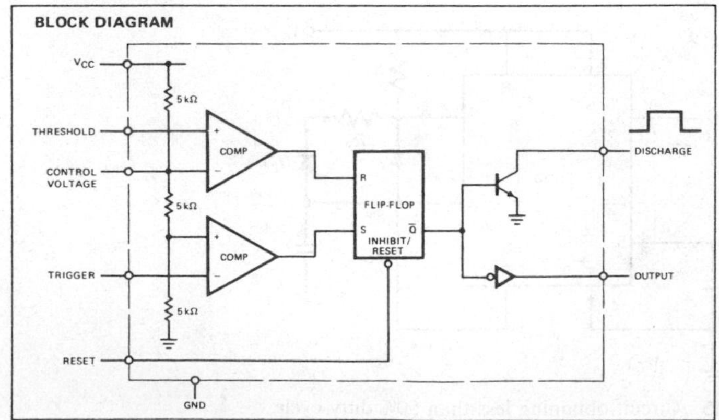

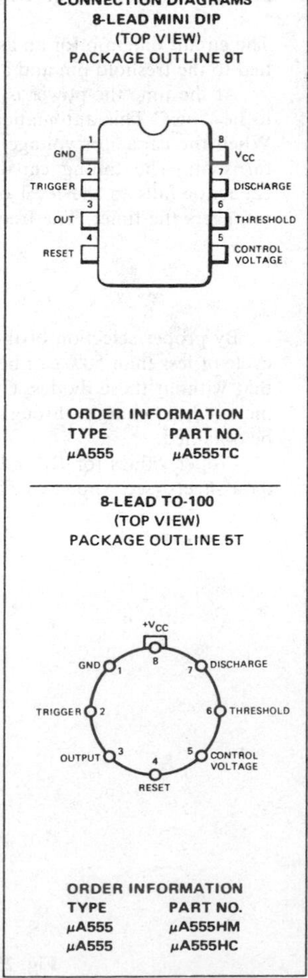

FAIRCHILD LINEAR INTEGRATED CIRCUITS • μA555

ELECTRICAL CHARACTERISTICS (T_A = 25°C, V_{CC} = +5.0 V to +15 V, unless otherwise specified)

PARAMETER	TEST CONDITIONS	μA555HM			μA555TC/HC			UNITS
		MIN	TYP	MAX	MIN	TYP	MAX	
Supply Voltage		4.5		18	4.5		16	V
Supply Current	V_{CC} = 5.0 V, R_L = ∞		3.0	5.0		3.0	6.0	mA
	V_{CC} = 15 V, R_L = ∞ LOW State (Note 1)		10	12		10	15	mA
Timing Error								
Initial Accuracy	R_A, R_B = 1 kΩ to 100 kΩ		0.5	2.0		1.0		%
Drift with Temperature	C = 0.1 μF (Note 2)		30	100		50		ppm/°C
Drift with Supply Voltage			0.05	0.2		0.1		%V
Threshold Voltage			2/3			2/3		X V_{CC}
Trigger Voltage	V_{CC} = 15 V	4.8	5.0	5.2		5.0		V
	V_{CC} = 5.0 V	1.45	1.67	1.9		1.67		V
Trigger Current			0.5			0.5		μA
Reset Voltage		0.4	0.7	1.0	0.4	0.7	1.0	V
Reset Current			0.1			0.1		mA
Threshold Current	Note 3		0.1	0.25		0.1	0.25	μA
Control Voltage Level	V_{CC} = 15 V	9.6	10	10.4	9.0	10	11	V
	V_{CC} = 5.0 V	2.9	3.33	3.8	2.6	3.33	4.0	V
Output Voltage Drop (LOW)	V_{CC} = 15 V							
	I_{SINK} = 10 mA		0.1	0.15		0.1	0.25	V
	I_{SINK} = 50 mA		0.4	0.5		0.4	0.75	V
	I_{SINK} = 100 mA		2.0	2.2		2.0	2.5	V
	I_{SINK} = 200 mA		2.5			2.5		V
	V_{CC} = 5.0 V							
	I_{SINK} = 8.0 mA		0.1	0.25				V
	I_{SINK} = 5.0 mA					0.25	0.35	V
Output Voltage Drop (HIGH)	I_{SOURCE} = 200 mA							
	V_{CC} = 15 V		12.5			12.5		V
	I_{SOURCE} = 100 mA							
	V_{CC} = 15 V	13	13.3		12.75	13.3		V
	V_{CC} = 5.0 V	3.0	3.3		2.75	3.3		V
Rise Time of Output			100			100		ns
Fall Time of Output			100			100		ns

NOTES:
1. Supply Current is typically 1.0 mA less when output is HIGH.
2. Tested at V_{CC} = 5.0 V and V_{CC} = 15 V.
3. This will determine the maximum value of $R_A + R_B$. For 15 V operation, the max total R = 20 MΩ.
4. For operating at elevated temperatures the device must be derated based on a +125°C maximum junction temperature and a thermal resistance of +45°C/W junction to case for TO-5 and +150°C/W junction to ambient for both packages.

TYPICAL PERFORMANCE CURVES

MINIMUM PULSE WIDTH REQUIRED FOR TRIGGERING

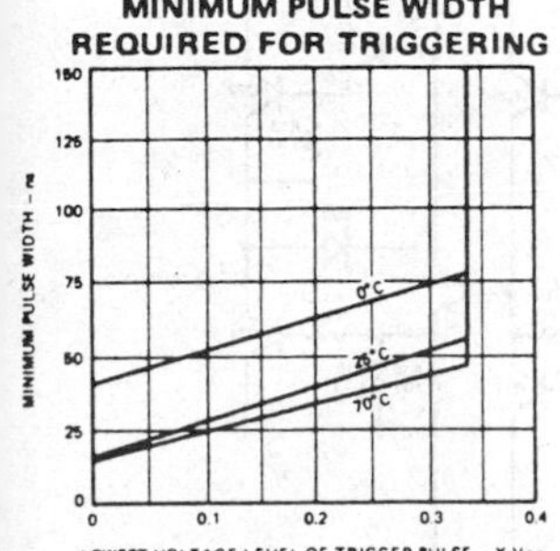

TOTAL SUPPLY CURRENT AS A FUNCTION OF SUPPLY VOLTAGE

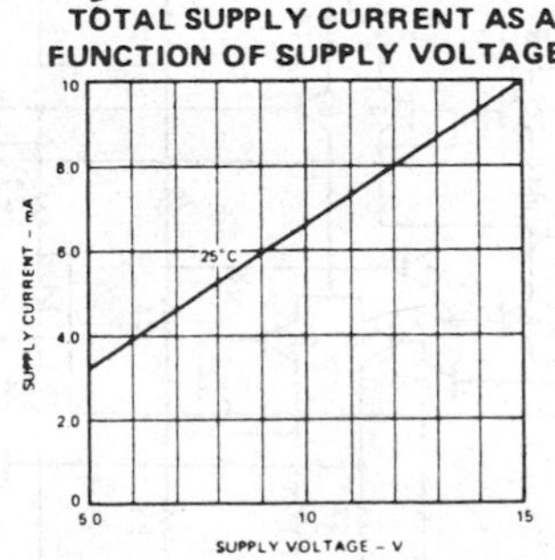

HIGH OUTPUT VOLTAGE AS A FUNCTION OF OUTPUT SOURCE CURRENT

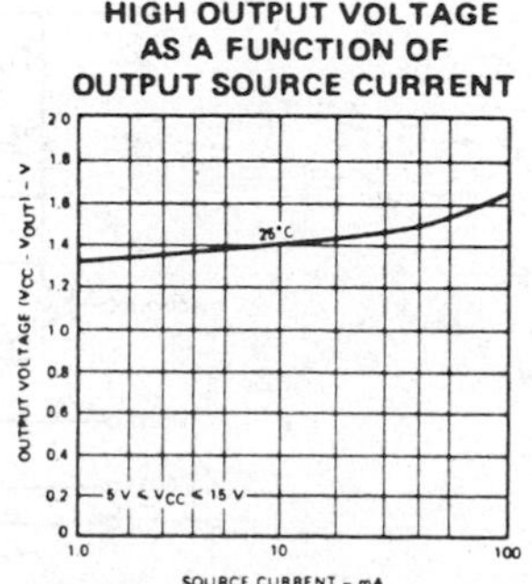

FAIRCHILD LINEAR INTEGRATED CIRCUITS • μA555

TYPICAL PERFORMANCE CURVES (Continued)

LOW OUTPUT VOLTAGE AS A FUNCTION OF OUTPUT SINK CURRENT

V_{CC} = 5 V

25°C

OUTPUT VOLTAGE – V

SINK CURRENT – mA

LOW OUTPUT VOLTAGE AS A FUNCTION OF OUTPUT SINK CURRENT

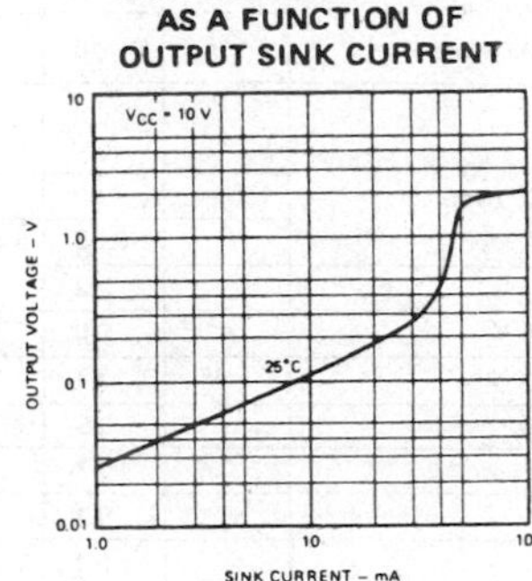

LOW OUTPUT VOLTAGE AS A FUNCTION OF OUTPUT SINK CURRENT

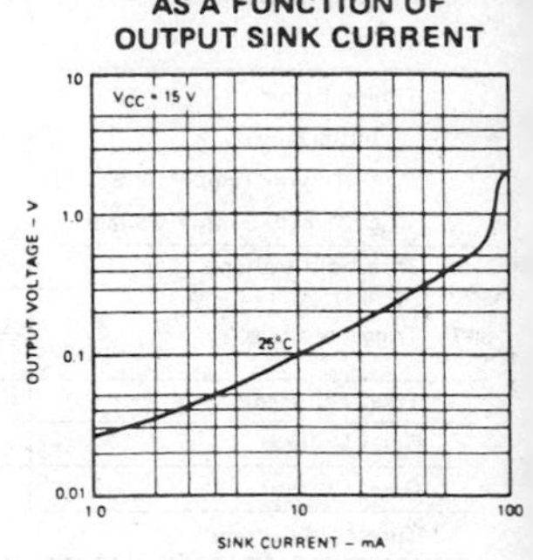

DELAY TIME AS A FUNCTION OF SUPPLY VOLTAGE

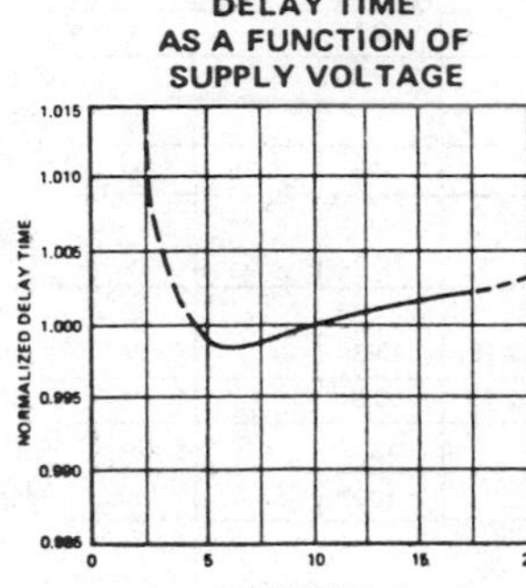

DELAY TIME AS A FUNCTION OF AMBIENT TEMPERATURE

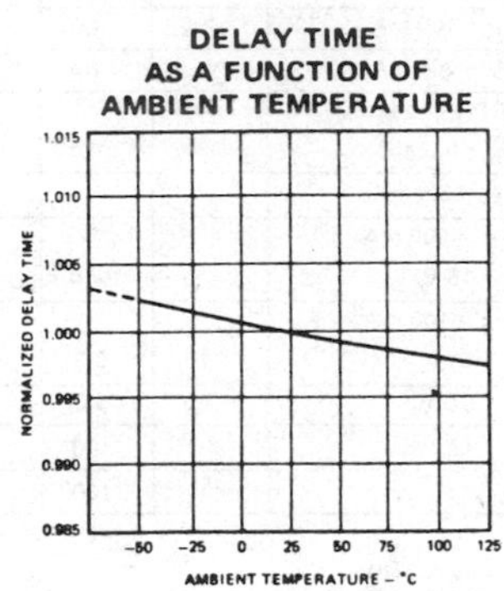

PROPAGATION DELAY AS A FUNCTION OF VOLTAGE LEVEL OF TRIGGER PULSE

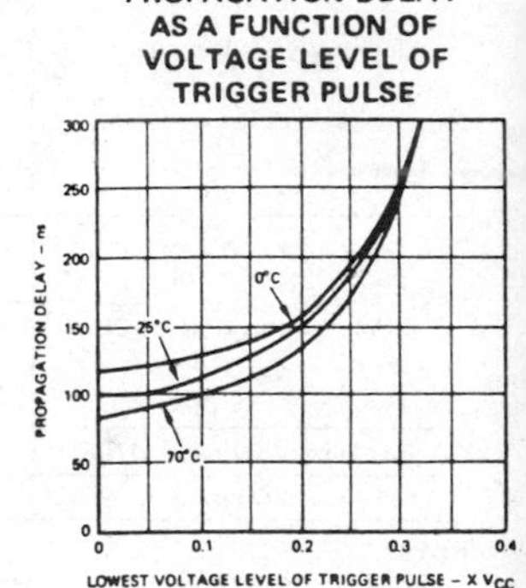

EQUIVALENT CIRCUIT

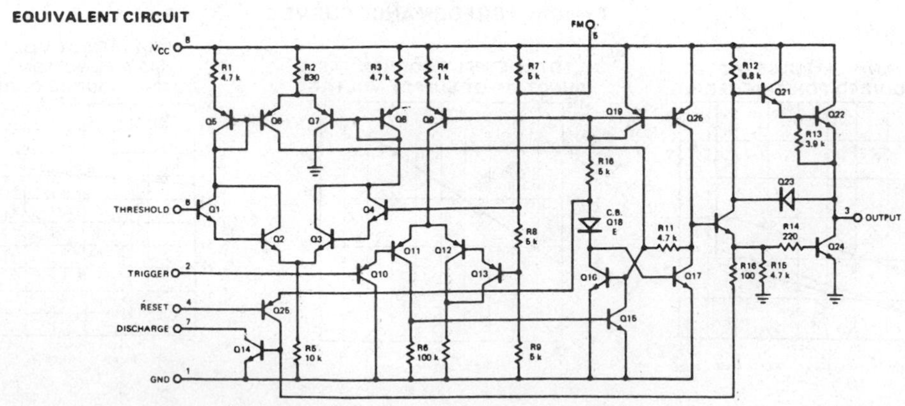

FAIRCHILD LINEAR INTEGRATED CIRCUITS • µA555

TYPICAL APPLICATIONS

MONOSTABLE OPERATION

In the monostable mode, the timer functions as a one-shot. Referring to Figure 1 the external capacitor is initially held discharged by a transistor inside the timer.

When a negative trigger pulse is applied to lead 2, the flip-flop is set, releasing the short circuit across the external capacitor and drives the output HIGH. The voltage across the capacitor, increases exponentially with the time constant τ = R1C1. When the voltage across the capacitor equals 2/3 V_{CC}, the comparator resets the flip-flop which then discharges the capacitor rapidly and drives the output to its LOW state. Figure 2 shows the actual waveforms generated in this mode of operation.

The circuit triggers on a negative-going input signal when the level reaches 1/3 V_{CC}. Once triggered, the circuit remains in this state until the set time has elapsed, even if it is triggered again during this interval. The duration of the output HIGH state is given by t = 1.1 R1C1 and is easily determined by Figure 3. Notice that since the charge rate and the threshold level of the comparator are both directly proportional to supply voltage, the timing interval is independent of supply. Applying a negative pulse simultaneously to the Reset terminal (lead 4) and the Trigger terminal (lead 2) during the timing cycle discharges the external capacitor and causes the cycle to start over. The timing cycle now starts on the positive edge of the reset pulse. During the time the reset pulse is applied, the output is driven to its LOW state.

When Reset is not used, it should be tied high to avoid any possibility of false triggering.

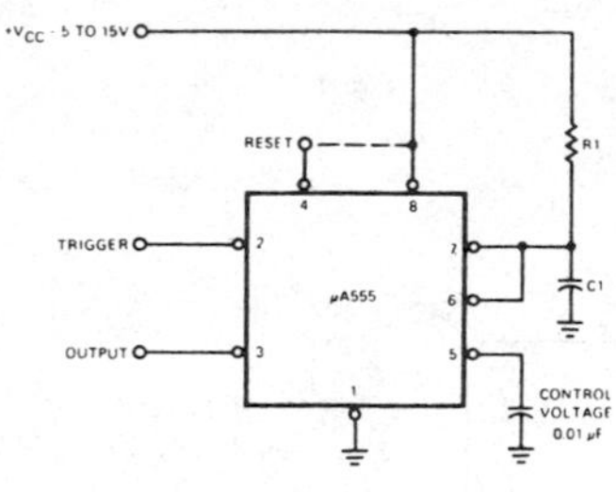

Fig. 1

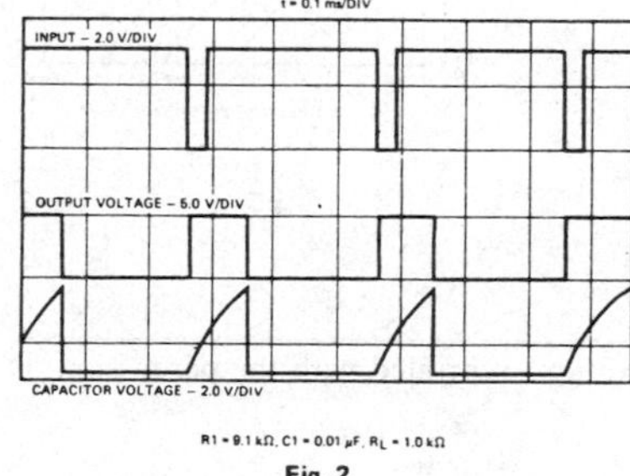

Fig. 2

TIME DELAY AS A FUNCTION OF R1 AND C1

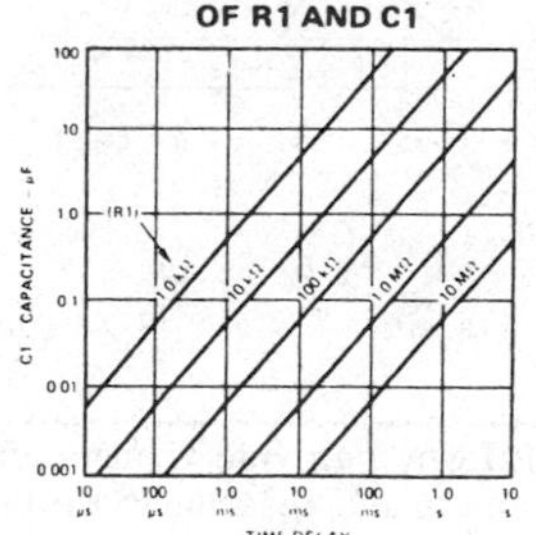

Fig. 3

ASTABLE OPERATION

When the circuit is connected as shown in Figure 4 (leads 2 and 6 connected) it triggers itself and free runs as a multivibrator. The external capacitor charges through R1 and R2 and discharges through R2 only. Thus the duty cycle may be precisely set by the ratio of these two resistors.

In the astable mode of operation, C1 charges and discharges between 1/3 V_{CC} and 2/3 V_{CC}. As in the triggered mode, the charge and discharge times and therefore frequency are independent of the supply voltage.

Figure 5 shows actual waveforms generated in this mode of operation.

The charge time (output HIGH) is given by:

$$t_1 = 0.693\ (R1 + R2)\ C1$$

and the discharge time (output LOW) by:

$$t_2 = 0.693\ (R2)\ C1$$

Thus the total period T is given by:

$$T = t_1 + t_2 = 0.693\ (R1 + 2R2)\ C1$$

The frequency of oscillation is then:

$$f = \frac{1}{T} = \frac{1.44}{(R1 + 2R2)\ C1}$$

and may be easily found by Figure 6.

The duty cycle is given by:

$$D = \frac{R2}{R1 + 2R2}$$

FAIRCHILD LINEAR INTEGRATED CIRCUITS • μA555

TYPICAL APPLICATIONS (Cont'd)

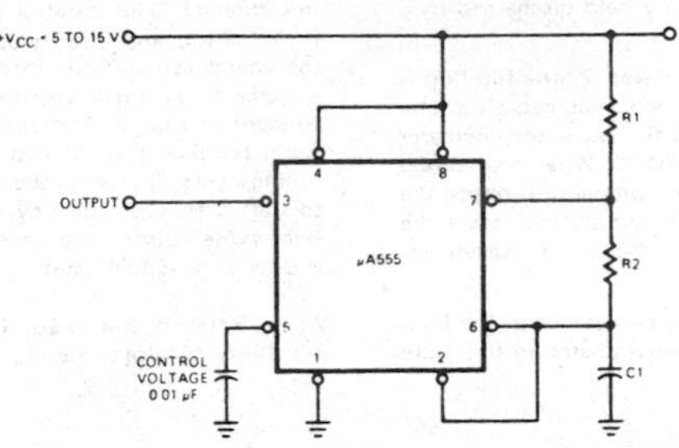

Fig. 4

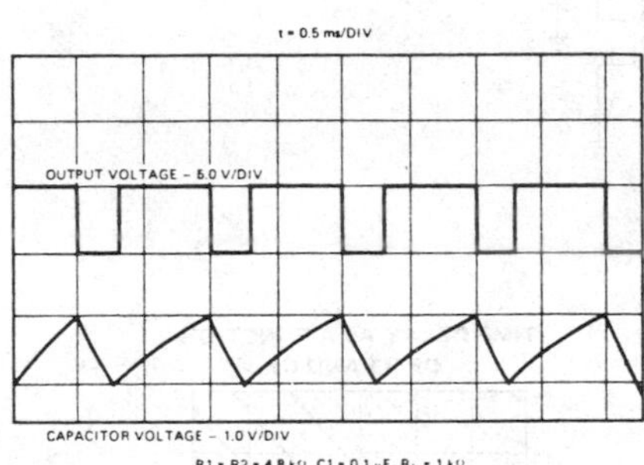

Fig. 5

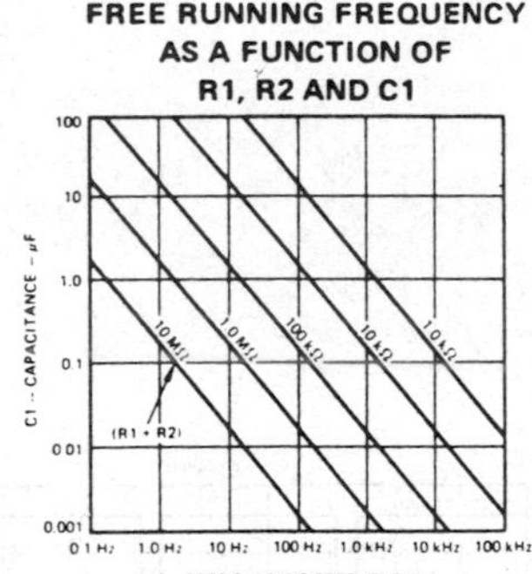

Fig. 6

Bibliography

Angleo, E. J., Jr., *Electronics: BJTs, FETs, and Microcircuits*, McGraw-Hill, New York, 1969.

Chance, B. et al., *Waveforms*, MIT Radiation Laboratory Series, Vol. 19, McGraw-Hill, New York, 1949.

Clarke, K. K. and M. V. Joyce, *Transistor Circuit Analysis*, Addison-Wesley, Reading, MA, 1961.

Gregory, R. O. and J. C. Bowers, "Simple Square-Wave Generator," *Electronics* **35**(51):47 (Dec. 21, 1962).

Hamilton, D. J., "A Transistor Pulse Generator for Digital Systems," *IRE Trans. Electron. Comp.* **EC-7**(3):244–249 (Sept. 1958).

Harris, J. N., *Digital Transistor Circuits*, SEEC, Vol. 6, Wiley, New York, 1966.

Millman, J. and C. C. Halkias, *Electronic Devices and Circuits*, McGraw-Hill, New York, 1967.

Millman, J. and H. Taub, *Pulse, Digital and Switching Waveforms*, McGraw-Hill, New York, 1965.

Strauss, L., *Wave Generation and Shaping*, McGraw-Hill, New York, 1970.

Wallmark, J. T. and H. Johnson, *Field Effect Transistors*, Prentice-Hall, Englewood Cliffs, NJ, 1966.

CHAPTER 26
ACTIVE FILTERS

YUSUF Z. EFE

The Cooper Union for Advancement of Science and Art
New York, New York

Passive filters are composed of passive elements (e.g., resistors, capacitors, and inductors) and are generally denoted as *RLC* filters. The major problem in passive filters is inductor size: Inductors become very bulky at low frequencies. In active filters, which are designed with resistors, capacitors, and active devices [i.e., operational amplifiers (op-amps)], inductors are eliminated. Hence *active filters* also are known as *inductorless filters*.

The advantages of active filters are as follows:

1. They are small in size and weight and have low cost.
2. They usually have very high input impedance and very low output impedance and hence provide excellent isolation capability, that is, their responses are practically independent of load and source impedances.
3. Because of their excellent isolation characteristics, they can be cascaded easily to obtain desired filter responses.
4. The quality factor, Q, of active filters can extend up to about 500.
5. They can provide both voltage and current gain (or loss).
6. All types of filter characteristics can be realized by them.
7. Their useful range of frequencies is much wider than those of passive filters. For example, frequencies from 0.001 Hz to 1 MHz are possible. (Beyond 1 MHz, passive filters become practical and economical.)

However, active filters have certain disadvantages compared with passive filters:

1. They need a power supply that can generate noise.
2. The input signal is usually limited to the range of -10 to $+10$ V, and the output current is limited to a few milliamps.
3. They are sensitive to temperature changes and component aging.

26.1 FILTER TRANSFER FUNCTIONS

The general form of a network transfer function is

$$H(s) = \frac{a_n s^n + a_{n-1} s^{n-1} + \cdots + a_1 s + a_0}{b_m s^m + b_{m-1} s^{m-1} + \cdots + b_1 s + b_0} \tag{26.1}$$

where all coefficients a_i and b_i are real. This function can be factored into biquadratics as

$$H(s) = \prod_{i=1}^{N} K_i \frac{a_i s^2 + b_i s + c_i}{d_i s^2 + e_i s + f_i} \tag{26.2}$$

where K_i = constant
$a_i = 1$ or 0
$d_i = 1$ or 0

In the case of a real pole, $d_i = 0$ and $e_i = 1$, and for a real zero, $a_i = 0$, $b_i = 1$. Thus for a complex pole-zero pair, the biquadratic (also called the second-order filter function) can be represented by

$$H(s) = K \frac{s^2 + as + b}{s^2 + cs + d} \tag{26.3}$$

Many types of second-order filter characteristics can be obtained from Eq. (26.3). They are shown in Table 26.1.

The general design procedure is initial realization of a biquadratic corresponding to the filter type and then cascading of several biquads to obtain the desired filter characteristc. Therefore we shall concentrate on the design of second-order active filters.

The biquadratic in Eq. (26.3) can be written as

$$H(s) = K \cdot \frac{s^2 + (\omega_z / Q_z) s + \omega_z^2}{s^2 + (\omega_p / Q_p) s + \omega_p^2} \tag{26.4}$$

where K = constant
ω_z, ω_p = zero and pole undamped frequencies, respectively
Q_z, Q_p = Qs of complex zeros and poles, respectively

The 3-dB bandwidth (BW) is given by

$$\text{BW} = \frac{\omega_p}{Q_p} \tag{26.5}$$

The transfer function $H(s)$ realized by active RC networks must satisfy the following properties:

1. $H(s)$ is a rational function in s with real coefficients.
2. All poles of $H(s)$ must be in the left half s plane.
3. The zeros of $H(s)$ can be anywhere in the s plane.
4. The poles on the $j\omega$ axis must be simple.
5. Complex poles and zeros occur as conjugate pairs.

26.2 SENSITIVITY

The designer usually has a choice of many circuits that can realize a given transfer function. If exact values were to be used for the various components, there would be almost no difference among

TABLE 26.1. SECOND-ORDER FILTER FUNCTIONS

Filter Type	Transfer Function
Low pass	$\dfrac{K}{s^2 + cs + d} = \dfrac{K}{s^2 + (\omega_p/Q_p)s + \omega_p^2}$
High pass	$\dfrac{Ks^2}{s^2 + cs + d} = \dfrac{Ks^2}{s^2 + (\omega_p/Q_p)s + \omega_p^2}$
Bandpass	$\dfrac{Ks}{s^2 + cs + d} = \dfrac{Ks}{s^2 + (\omega_p/Q_p)s + \omega_p^2}$
Band reject	$\dfrac{K(s^2 + b)}{s^2 + cs + d} = \dfrac{K(s^2 + b)}{s^2 + (\omega_p/Q_p)s + \omega_p^2}$

realizations. However, actual values of components differ from nominal values owing to tolerances, temperatures, aging, and so on. As a result, the response of a designed filter may deviate from that of its nominal form. To minimize the deviation, components are selected that have small tolerances and very low temperature and aging effects. This, of course, increases the cost of the filter significantly. An alternative approach is to select the filter circuit such that its response is less sensitive to changes.

The sensitivity S of a quantity y with respect to the change in a quantity x is defined by

$$S_x^y = \lim_{\Delta x \to 0} \frac{\Delta y / y}{\Delta x / x} = \frac{x}{y} \cdot \frac{\partial y}{\partial x} = \frac{\partial(\ln y)}{\partial(\ln x)} \tag{26.6}$$

in which y can be identified with ω_p, ω_z, Q, and gain of the filter and x can be identified with the element values, the gain(s) of the operational amplifier(s), and the parameters of the active element(s). For example,

$$S_R^{\omega_p} = \frac{R}{\omega_p} \cdot \frac{\partial \omega_p}{\partial R} \tag{26.7}$$

$$S_C^{H(s)} = \frac{C}{H(s)} \cdot \frac{\partial H(s)}{\partial C} \tag{26.8}$$

$$S_x^{A(\omega)} = \frac{x}{A(\omega)} \cdot \frac{\partial A(\omega)}{\partial x} \tag{26.9}$$

where $A(\omega)$ is the gain in decibels and is defined as

$$A(\omega) = 20 \log|H(j\omega)| \tag{26.10}$$

and x represents any elements or parameters of the active elements in the filter. For small changes in x, it can be shown that the change in gain, $\Delta A(\omega)$, is given by

$$\Delta A(\omega) = \frac{\partial A(\omega)}{\partial(\ln x)} \cdot \frac{\Delta x}{x} \tag{26.11}$$

and the total change in gain (in decibels) due to the simultaneous variation in all the elements is given by

$$\Delta A(\omega) = \sum_{i=1}^{n} \frac{\partial A(\omega)}{\partial(\ln x_i)} \cdot \frac{\Delta x_i}{x_i} \tag{26.12}$$

For performance of a sensitivity analysis using a computer, the topological description of the elements and expected changes of elements are entered. Then the quantity $\partial A(\omega)/\partial(\ln x_i)$ in Eq. (26.12) is computed by perturbing the ith element, keeping the others unchanged. This is repeated for every element in the filter circuit and then Eq. (26.12) is used to determine the total change in gain. This approach provides reasonably good results for the gain sensitivity of the circuit, if the changes in components are small.

Another approach that does not require small changes in the components is the computer algorithm called the Monte Carlo method. The computer program is included in some general-purpose circuit analysis computer programs, for example, in SCEPTRE and ASTAP, which are widely used.

Some relationships useful in sensitivity computations follow:

$$S_x^{y_1 y_2} = S_x^{y_1} + S_x^{y_2} \tag{26.13}$$

$$S_x^{y_1/y_2} = S_x^{y_1} - S_x^{y_2} \tag{26.14}$$

$$S_x^{y_1+y_2} = \frac{y_1 S_x^{y_1} + y_2 S_x^{y_2}}{y_1 + y_2} \tag{26.15}$$

$$S_x^y = -S_x^{1/y} = -S_{1/x}^y \tag{26.16}$$

$$S_x^{cy} = S_{cx}^y = S_x^y \tag{26.17}$$

$$S_x^{y^n} = n S_x^y \tag{26.18}$$

$$S_x^y = S_x^{|y|} + j \arg y S_x^{\arg y} \tag{26.19}$$

$$S_x^{\arg y} = \frac{1}{\arg y} \operatorname{Im} S_x^y \tag{26.20}$$

$$S_x^{|y|} = \operatorname{Re} S_x^y \tag{26.21}$$

Various sensitivities of a given filter can be computed using these relationships.

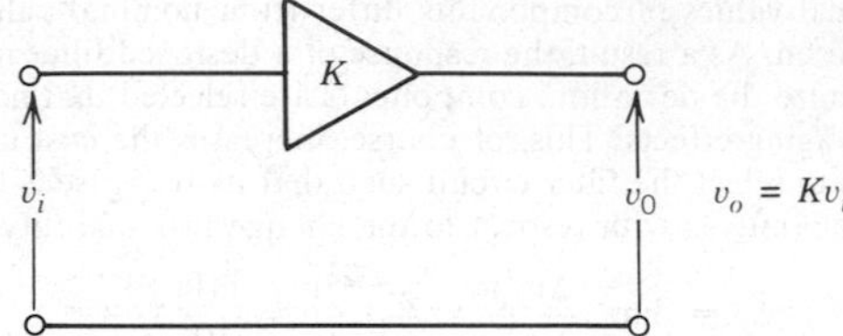

Fig. 26.1 Symbol for a voltage-controlled voltage source (VCVS).

26.3 ACTIVE DEVICES IN ACTIVE FILTERS

Voltage-Controlled Voltage Source

An active device in an *RC* active filter can be considered a controlled source. However, a voltage-controlled voltage source (VCVS) is commonly used. An ideal VCVS possesses the following characteristics:

1. It has infinite input impedance.
2. It has zero output impedance.
3. Its output voltage is a constant multiplier of its input voltage (i.e., $v_o = Kv_i$, where K is referred to as the gain). If $K > 0$, the VCVS is said to be noninverting; if $K < 0$, the VCVS is said to be inverting.

The commonly used symbol for a VCVS is shown in Fig. 26.1. Inverting and noninverting VCVSs can be realized with use of an operational amplifier, as illustrated in Fig. 26.2.

A general single-amplifier *RC* active filter is shown in Fig. 26.3. Depending on the *RC* network, the filter response can be low-pass, bandpass, high-pass, band-reject, and so on. These types of filters are discussed in the following subsections.

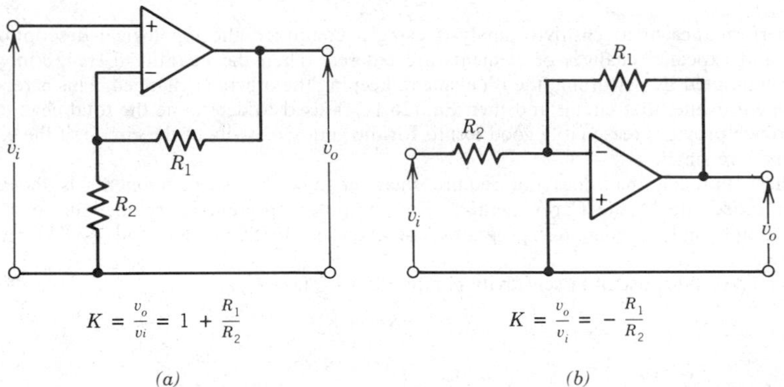

Fig. 26.2 Voltage-controlled voltage source (VCVS): (*a*) noninverting, (*b*) inverting.

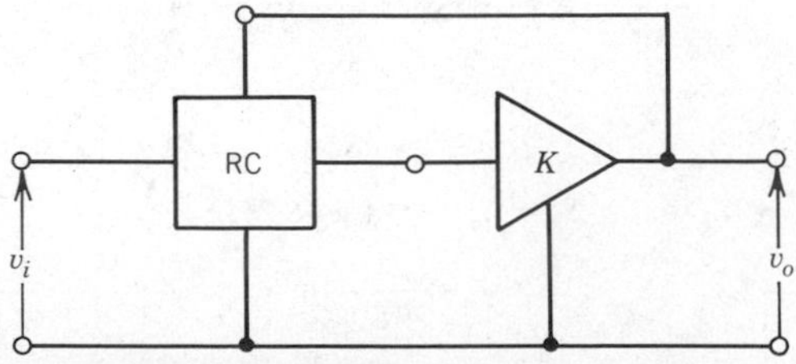

Fig. 26.3 Single-amplifier *RC* active filter.

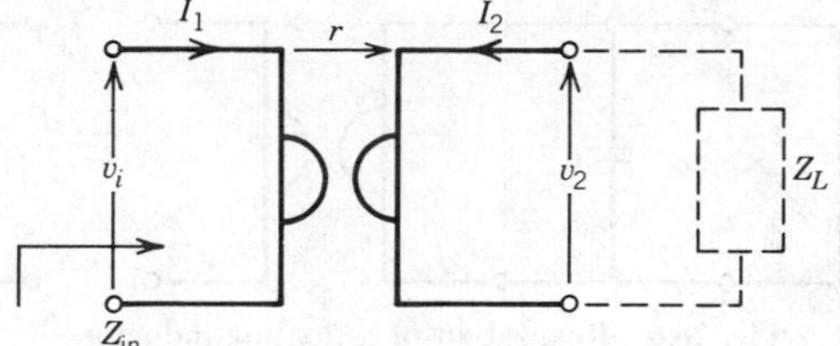

Fig. 26.4 Symbolic representation of an active gyrator.

Gyrator

Another device commonly used in the design of active networks is an active gyrator, a two-port device shown symbolically in Fig. 26.4. The active gyrator is defined by (using the y-parameter definition)

$$\begin{bmatrix} I_1 \\ I_2 \end{bmatrix} = \begin{bmatrix} 0 & -\dfrac{1}{r} \\ \dfrac{1}{r} & 0 \end{bmatrix} \cdot \begin{bmatrix} V_1 \\ V_2 \end{bmatrix} \tag{26.22 }$$

where r is referred to as the gyration resistance. The most useful property of the gyrator is that the input impedance at either port is proportional to the reciprocal of the impedance connected at the other port. That is, it can be shown that

$$Z_{\text{in}} = \frac{V_i}{I_i} = \frac{r^2}{Z_2} \tag{26.23}$$

As an example, if the load impedance is a capacitance C, then the input impedance is

$$Z_{\text{in}} = sr^2C \tag{26.24}$$

which corresponds to an inductor of value $L = r^2C$. Hence the gyrator terminated with a capacitor can be used to realize an inductor. This process is used to eliminate inductors in the passive RLC filters, as will be demonstrated later. The physical realization of a gyrator using operational amplifiers is shown in Fig. 26.5.

It can be shown that the y-parameter matrix for this circuit is

$$\begin{bmatrix} \dfrac{1/A_1}{(1+1/A_1)R_1} & \dfrac{-1}{(1+1/A_1)R_1} \\ \dfrac{R_3 + (R_3 + R_4)[1/(A_1) - 1/(A_2)]}{(1+1/A_1)[R_4 + (R_3+R_4)/A_2]R_2} & \dfrac{1/A_1}{(1+1/A_1)R_2} \end{bmatrix} \tag{26.25}$$

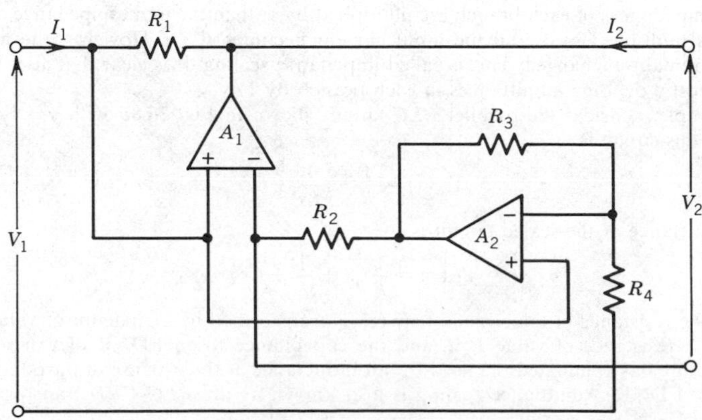

Fig. 26.5 Realization of a gyrator.

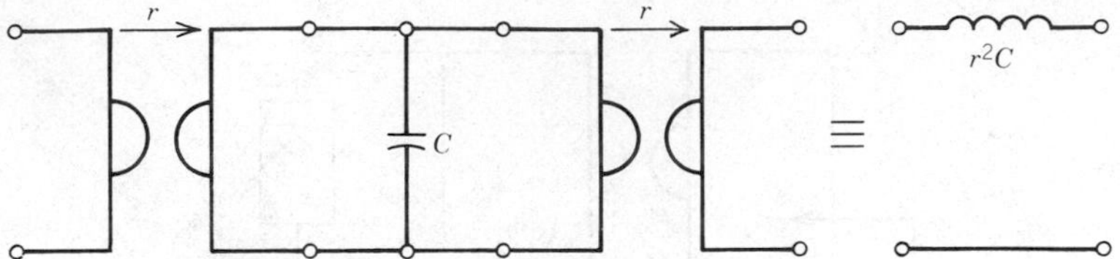

Fig. 26.6 Realization of a floating inductor.

If identical operational amplifiers are used (i.e., the open-loop gains of the operational amplifiers are equal), then $A_1 = A_2 = A_3 = A$, and also $R_1 = R_2 = R_3 = R_4 = R$. Then Eq. (26.25) reduces to

$$\frac{1}{(1+1/A)R} \cdot \begin{bmatrix} \frac{1}{A} & -1 \\ \frac{1}{1+2/A} & \frac{1}{A} \end{bmatrix} \tag{26.26}$$

Hence as A becomes large, Eq. (26.26) reduces to Eq. (26.22). The gyrator realized in Fig. 26.5 has a limitation on its use, since one of its input terminals is grounded. This implies that when an inductor is realized by this gyrator, one of the terminals of the inductor will be grounded. This may not be desirable for a low-pass filter or for some other *RLC* filters. To realize an inductor whose terminals are not grounded (a "floating inductor"), two gyrators with a capacitor C are cascaded, as shown in Fig. 26.6.

Frequency-Dependent, Negative-Resistance Device

The frequency-dependent, negative-resistance (FDNR) device is a one-port active element whose admittance is defined by

$$Y(s) = \frac{I(s)}{V(s)} = s^2 D \tag{26.27}$$

where D is a positive-real constant with units of farad-seconds. Note that with $s = j\omega$, Eq. (26.27) becomes

$$Y(j\omega) = -\omega^2 D \tag{26.28}$$

Thus the admittance is real, frequency dependent, and negative. The symbol for the FDNR device and its realization are shown in Fig. 26.7.

The input admittance of the FDNR device shown in Fig. 26.7 is found to be

$$Y(S) = RC^2 s^2 = Ds^2$$

where $D = RC^2$.

It is known from network theory that if the impedance of each branch in a network is scaled by γ, that is, if the impedances of each branch are multiplied by γ, then the input impedance of the original network Z_{in} is multiplied by γ and the input current becomes I_{in}/γ. However, the node voltage's signal sources remain unchanged. This is called impedance scaling. It is clear that admittance scaling is accomplished by dividing admittances in each branch by $1/\gamma$.

As an example, consider the parallel *RLC* circuit shown in Fig. 26.8*a* with $\gamma = 1/s$. The input admittance of this circuit is

$$Y_{in}(s) = \frac{1}{R} + \frac{1}{Ls} + Cs$$

The input admittance of the scaled circuit is

$$Y'_{in}(s) = \frac{1}{R} \cdot s + \frac{1}{L} + Cs^2$$

Observe that the resistance in the original network is transformed to a capacitor of value $1/R$ F, the inductance to a resistance of value L Ω, and the capacitance to an FDNR of value C. Note that admittance scaling has eliminated the need for an inductance at the expense of introduction of a new active element, FDNR. Admittance scaling is also known as the *RLC-CRD* transformation in the literature.

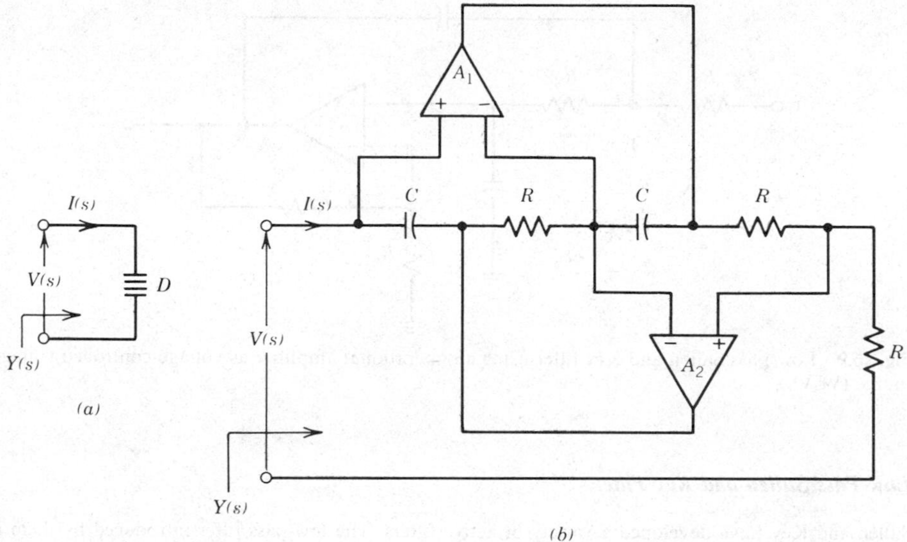

Fig. 26.7 Frequency-dependent, negative-resistance (FDNR) device: (*a*) symbol, (*b*) realization using operational amplifiers.

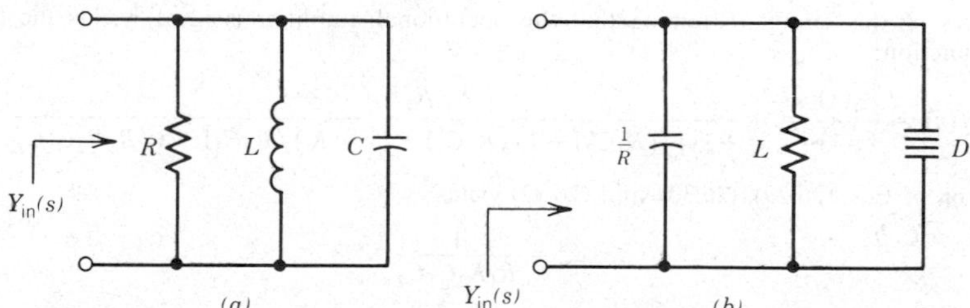

Fig. 26.8 (*a*) Parallel *RLC* circuit. (*b*) Same circuit after admittance scaling.

26.4 REALIZATION OF LOW-PASS ACTIVE FILTERS

In this subsection a variety of low-pass active filter designs are presented. These are only suggested designs. Other circuits realizing the same low-pass filters are also described in the literature. The filters presented here have proven to have satisfactory filter responses and sensitivities.

The general form of the low-pass filter transfer function is

$$H(s) = \frac{H_0\omega_n^2}{s^2 + 2\zeta\omega_n s + \omega_n^2} \tag{26.29}$$

where ζ = damping factor
ω_n = undamped natural frequency
H_0 = DC gain
and

$$Q = \frac{1}{2\zeta} \tag{26.30}$$

It will be assumed that the reader knows how to obtain the necessary transfer function, in the form of Eq. (26.29), starting from filter specifications. This topic is covered in Chapters 19 and 20.

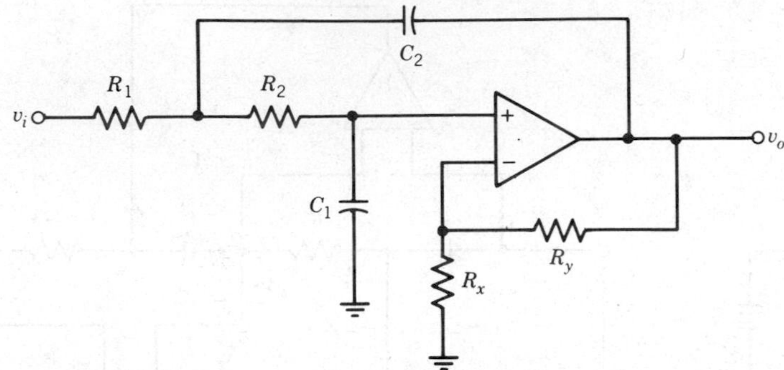

Fig. 26.9 Low-pass Sallen and Key filter using an operational amplifier as voltage-controlled voltage source (VCVS).

Low-Pass Sallen and Key Filter

Sallen and Key have developed a variety of active filters. The low-pass filter introduced by them is shown in Fig. 26.9. An operational amplifier is used together with the resistances R_x and R_y to form the VCVS where the gain is given by

$$K = 1 + \frac{R_y}{R_x} \tag{26.31}$$

Analysis of this circuit (assuming that the operational amplifier is ideal) yields the following transfer function:

$$H(s) = \frac{V_o(s)}{V_i(s)} = \frac{K/R_1R_2C_1C_2}{s^2 + s[1/(R_1C_2) + 1/(R_2C_2) + (1-K)/R_2C_1] + 1/R_1R_2C_1C_2} \tag{26.32}$$

Comparison of Eqs. (26.29), (26.30), and (26.32) yields

$$\omega_n^2 = \frac{1}{R_1R_2C_1C_2} \tag{26.33}$$

$$Q = \frac{1}{2\zeta} = \left[\left(\frac{R_2C_1}{R_1C_2}\right)^{1/2} + \left(\frac{R_1C_1}{R_2C_2}\right)^{1/2} + (1-K)\left(\frac{R_1C_2}{R_2C_1}\right)^{1/2}\right]^{-1} \tag{26.34}$$

$$H_0 = K \tag{26.35}$$

Using the sensitivity definition and relations provided in Section 26.2 it can be shown that

$$S_{R_1}^{\omega_n} = S_{R_2}^{\omega_n} = S_{C_1}^{\omega_n} = S_{C_2}^{\omega_n} = -\frac{1}{2} \tag{26.36}$$

$$S_{R_1}^{K} = S_{R_2}^{K} = S_{C_1}^{K} = S_{C_2}^{K} = -1 \tag{26.37}$$

$$S_{K}^{\omega_n} = S_{R_x}^{\omega_n} = S_{R_y}^{\omega_n} = 0 \tag{26.38}$$

$$S_{R_1}^{Q} = -S_{R_2}^{Q} = -\frac{1}{2} + Q\left(\frac{R_2C_1}{R_1C_2}\right)^{1/2} \tag{26.39}$$

$$S_{C_1}^{Q} = -S_{C_2}^{Q} = -\frac{1}{2} + Q\left[\left(\frac{R_1C_1}{R_2C_2}\right)^{1/2} + \left(\frac{R_2C_1}{R_1C_2}\right)^{1/2}\right] \tag{26.40}$$

$$S_{K}^{Q} = QK\left(\frac{R_1C_2}{R_2C_1}\right)^{1/2} \tag{26.41}$$

$$S_{K}^{H_0} = 1 \tag{26.42}$$

Design Procedure

The general procedure in the design of an active filter is selection of component values that meet the desired specifications for the filter. Such specifications include bandwidth ω_n [or Q, see Eq. (26.5)], and the VCVS' gain K.

Next the values of components R_1, R_2, C_1, C_2, and K are determined by means of Eqs. (26.33) and (26.34). However, since here only three equations must satisfy five quantities, either two of the quantities should be fixed or additional arbitrary relations among them should be introduced. These arbitrary relations or fixed values must be selected in such a way that the sensitivities computed by means of Eqs. (26.36) through (26.42) are as low as possible. Three design methods are widely used.

Design Method 1

Given: H_0, ω_n, Q
Choose: $R_1 = R_2 = R$ (where R = a convenient value) and K
Calculate: C_1, C_2

The VCVS is commonly chosen to be the voltage follower, that is, $K = 1$. Hence with use of Eqs. (26.33) and (26.34) it can be shown that

$$C_1 = \frac{1}{2Q\omega_n R} \tag{26.43}$$

and

$$C_2 = \frac{2Q}{\omega_n R} \tag{26.44}$$

The sensitivities corresponding to this design are given in Table 26.2. The primary disadvantage of this method is that widely different capacitor values are required, as can be seen from

$$\frac{C_2}{C_1} = 4Q^2 \tag{26.45}$$

Hence this method would provide practically acceptable values for C_1 and C_2 if the filter is a low-Q ($Q < 5$) filter.

Design Method 2

Given: H_0, ω_n, Q
Choose: $C_1 = C_2 = C$ (where C = a convenient value) and $K = 2$
Calculate: R_1, R_2

The value of $K = 2$ can be accomplished by selecting $R_x = R_y$, as can be seen from Eq. (26.31). With use of Eqs. (26.33) and (26.34) it can be shown that

$$R_1 = \frac{Q}{C\omega_n} \tag{26.46}$$

TABLE 26.2. SENSITIVITIES FOR LOW-PASS SALLEN AND KEY FILTERS

Sensitivity	Design 1	Design 2	Design 3
$S^{\omega_n}_{R_1, R_2, C_1, C_2}$	$-\frac{1}{2}$	$-\frac{1}{2}$	$-\frac{1}{2}$
$S^Q_{R_1} = -S^Q_{R_2}$	0	$\frac{1}{2}$	$-\frac{1}{2} + Q\sqrt{\frac{n}{m}}$
$S^Q_{C_1} = -S^Q_{C_2}$	$\frac{1}{2}$	$\frac{1}{2} + Q^2$	$-\frac{1}{2} + Q\left(\frac{1}{\sqrt{mn}} + \frac{m}{n}\right)$
S^Q_K	$2Q^2$	$2Q^2$	$Q\sqrt{\frac{n}{m}}$

and

$$R_2 = \frac{1}{QC\omega_n} \tag{26.47}$$

The sensitivities corresponding to this design are given in Table 26.2. In this case the spread in resistance values is observed, since

$$\frac{R_1}{R_2} = Q^2 \tag{26.48}$$

However, this spread is not as critical as in the case of capacitors, because the manufacture of resistances in integrated-circuit form is much easier than in capacitor form.

Design Method 3

Given: H_0, ω_n, Q
Choose: $m = R_2/R_1, n = C_2/C_1, K = 1$
Calculate: R_1, R_2, C_1, C_2

With use of Eqs. (26.33) and (26.34) it can be shown that

$$\omega_n^2 = \frac{1}{mnR_1^2C_1^2} \tag{26.49}$$

and

$$Q = \frac{\sqrt{mn}}{m+1} \tag{26.50}$$

With n fixed and m varying, it can be seen that Q is maximum when $m = 1$. This yields

$$Q_{\max} = \tfrac{1}{2} \cdot \sqrt{n}$$

or

$$\frac{C_2}{C_1} \equiv 4Q_{\max}^2 \tag{26.51}$$

Observe that for $m = 1$, the ratio of C_2 to C_1 can be very large. To overcome this problem and keep the capacitor values within the range of standard available values, m is selected such that

$$m \leqslant 1/(4Q^2) \tag{26.52}$$

Then the value of n is calculated by using Eq. (26.50). To determine element values, select $C_1 = C$ (a convenient value). The values of the other elements are then computed as follows:

$$C_2 = nC_1 \tag{26.53}$$

$$R_1 = \frac{1}{\omega_n C_1 \sqrt{mn}} \tag{26.54}$$

$$R_2 = mR_1 \tag{26.55}$$

The sensitivities corresponding to this design method are given in Table 26.2. As indicated earlier, there are only three equations relating the specifications of ω_n, Q, and H_0, while there are five quantities to be determined. Therefore many more design procedures can be devised than those provided here.

26.5 REALIZATION OF HIGH-PASS ACTIVE FILTERS

The transfer function of a high-pass filter is given by

$$H(s) = \frac{V_o(s)}{V_i(s)} = \frac{H_0 s^2}{s^2 + (\omega_n/Q)s + \omega_n^2} = \frac{H_0 s^2}{s^2 + 2\zeta\omega_n s + \omega_n^2} \tag{26.56}$$

where H_0 is the gain at high frequencies and ω_n, Q are as defined previously. Recall that the high-pass transfer function in Eq. (26.56) is closely related to the low-pass transfer function with use of the low-pass to high-pass transformation. Namely, every s in the low-pass transfer function is replaced by $1/s$. This transformation corresponds to replacement of every resistance R_1 in the low-pass circuit by a capacitor of value $C_i = 1/R_i$ and every capacitor C_i by a resistance of value

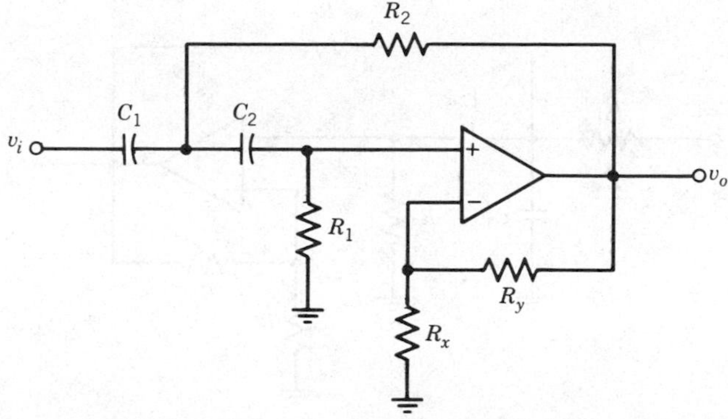

Fig. 26.10 High-pass Sallen and Key filter using an operational amplifier as voltage-controlled voltage source (VCVS).

$R_j = 1/C_j$. When these replacements are applied to the low-pass filter in Fig. 26.9, the high-pass Sallen and Key filter shown in Fig. 26.10 is obtained. Note that the low-pass to high-pass transformation does not affect the VCVS gain K. Therefore the resistances R_x and R_y are not transformed to capacitances.

With use of the transformations just described and manipulation, the following transfer function is obtained:

$$H(s) = \frac{V_o(s)}{V_i(s)} = \frac{Ks^2}{s^2 + s[1/R_1C_2 + 1/R_1C_1 + (1-K)/R_2C_1] + 1/R_1R_2C_1C_2} \tag{26.57}$$

Comparison of this equation with Eq. (26.56) yields

$$\omega_n^2 = \frac{1}{R_1R_2C_1C_2} \tag{26.58}$$

$$\zeta = \frac{1}{2}\left[\left(\frac{R_2C_1}{R_1C_2}\right)^{1/2} + \left(\frac{R_2C_2}{R_1C_1}\right)^{1/2} + (1-K)\left(\frac{R_1C_2}{R_2C_1}\right)^{1/2}\right] \tag{26.59}$$

$$H_0 = K = 1 + \frac{R_y}{R_x} \tag{26.60}$$

Some of the parameter sensitivities for this network are as follows:

$$S_{R_1}^{\omega_n} = S_{R_2}^{\omega_n} = S_{C_1}^{\omega_n} = S_{C_2}^{\omega_n} = -\frac{1}{2} \tag{26.61}$$

$$S_{R_1}^{\zeta} = \frac{1}{2} - Q \cdot \frac{C_1 + C_2}{\omega_n R_1 C_1 C_2} \tag{26.62}$$

$$S_{R_2}^{\zeta} = \frac{1}{2} - Q \cdot \frac{1-K}{\omega_n R_2 C_1} \tag{26.63}$$

$$S_{C_1}^{\zeta} = \frac{1}{2} - Q \cdot \frac{(1-K)R_1 + R_2}{\omega_n R_1 R_2 C_1} \tag{26.64}$$

$$S_{C_1}^{\zeta} = \frac{1}{2} - Q \cdot \frac{1}{\omega_n R_1 C_2} \tag{26.65}$$

$$S_K^{\zeta} = -Q \cdot \frac{K}{\omega_n R_2 C_1} \tag{26.66}$$

$$S_K^{H_O} = 1 \tag{26.67}$$

Design Procedure

The design procedures for high-pass filters are similar to the procedures used for low-pass filters, which have already been discussed.

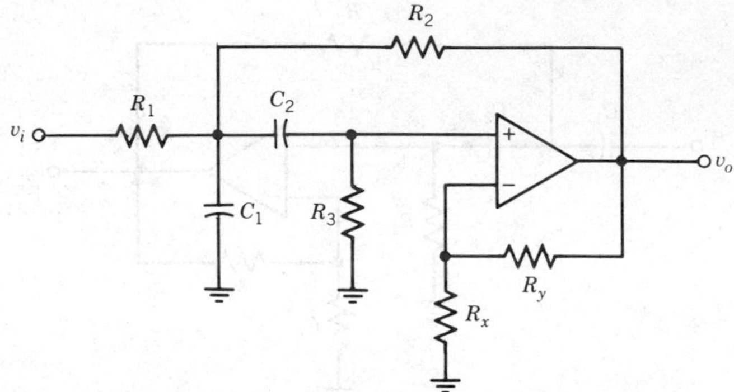

Fig. 26.11 Bandpass Sallen and Key filter using an operational amplifier as a voltage-controlled voltage source (VCVS).

26.6 REALIZATION OF BANDPASS ACTIVE FILTERS

The transfer function of a bandpass filter is given by

$$H(s) = \frac{H_0(\omega_n/Q)s}{s^2 + (\omega_n/Q)s + \omega_n^2} \tag{26.68}$$

where H_0 is the gain at the resonance and ω_n is the resonant frequency. The bandpass Sallen and Key filter realizing the transfer function in Eq. (26.68) is shown in Fig. 26.11.

An analysis of this circuit gives the following transfer function:

$$H(s) = \frac{V_o(s)}{V_i(s)} = \frac{Ks/R_1C_1}{s^2 + s[1/R_1C_1 + 1/R_3C_2 + 1/R_3C_1 + (1-K)/R_2C_1] + (R_1 + R_2)/R_1R_2R_3C_1C_2} \tag{26.69}$$

Comparison of this equation with Eq. (26.68) yields the following:

$$\omega_n^2 = \frac{R_1 + R_2}{R_1R_2R_3C_1C_2} \tag{26.70}$$

$$Q = \left(\frac{1}{R_1C_1} + \frac{1}{R_3C_2} + \frac{1}{R_3C_1} + \frac{1-K}{R_2C_1}\right)^{-1} \sqrt{\frac{R_1 + R_2}{R_1R_2R_3C_1C_2}} \tag{26.71}$$

$$H_0 = \frac{K}{R_iC_i} \cdot \frac{Q}{\omega_n} \tag{26.72}$$

$$K = 1 + \frac{R_y}{R_x} \tag{26.73}$$

Design Procedure

The design procedures discussed for the low-pass filter apply also to the design of the bandpass filter.

General Comments

Some comments about the filter designs discussed are in order. Sallen and Key filters are clearly a positive-gain filter. The advantages of such a filter are:

1. Design procedures and relations are simple.
2. Element values and their spread can be controlled quite early.
3. Low VCVS gain can be used which can be readily stabilized.

The disadvantages of the Sallen and Key structure are:

1. It has high sensitivities for high-Q network functions.
2. If the poles are very close to the $j\omega$ axis (high-Q poles), a small increase in the gain may move these poles to the right of the s-plane, causing the circuit to become unstable.

Sensitivities can be reduced by employing negative-gain filter structures. That is, an inverting-type VCVS can be used. The negative-gain filter is always stable. However, its element value spread is usually large; design procedures are complex; and the filter requires high VCVS gains, which are difficult to stabilize compared with low VCVS gains. Since the disadvantages of the negative-gain filter are more serious than disadvantages of the positive-gain filter, positive-gain filters are commonly used.

Another type of commonly used active filter is the multiple-feedback filter structure. Typical multiple feedback low-pass, high-pass, and bandpass filters are shown in Fig. 26.12. These filters are also known as infinite-gain filters, since the VCVS has an infinite gain (ideally). It must be pointed out that the overall gain of the filter is finite. We shall discuss here only the design of the low-pass filter shown in Fig. 26.12a. Analysis of this circuit yields the following transfer function:

$$H(s) = \frac{V_o(s)}{V_i(s)} = \frac{-1/R_1R_3C_1C_2}{s^2 + s(1/R_1 + 1/R_2 + 1/R_3)(1/C_1) + 1/R_2R_3C_1C_2} \tag{26.74}$$

Comparison of this equation with Eq. (26.29) yields

$$\omega_n^2 = \frac{1}{R_2R_3C_1C_2} \tag{26.75}$$

$$Q = \sqrt{\frac{C_1}{C_2}}\left[\left(\frac{R_3}{R_2}\right)^{1/2} + \left(\frac{R_2}{R_3}\right)^{1/2} + \frac{(R_2R_3)^{1/2}}{R_1}\right]^{-1} \tag{26.76}$$

$$|H_0| = \frac{R_2}{R_1} \tag{26.77}$$

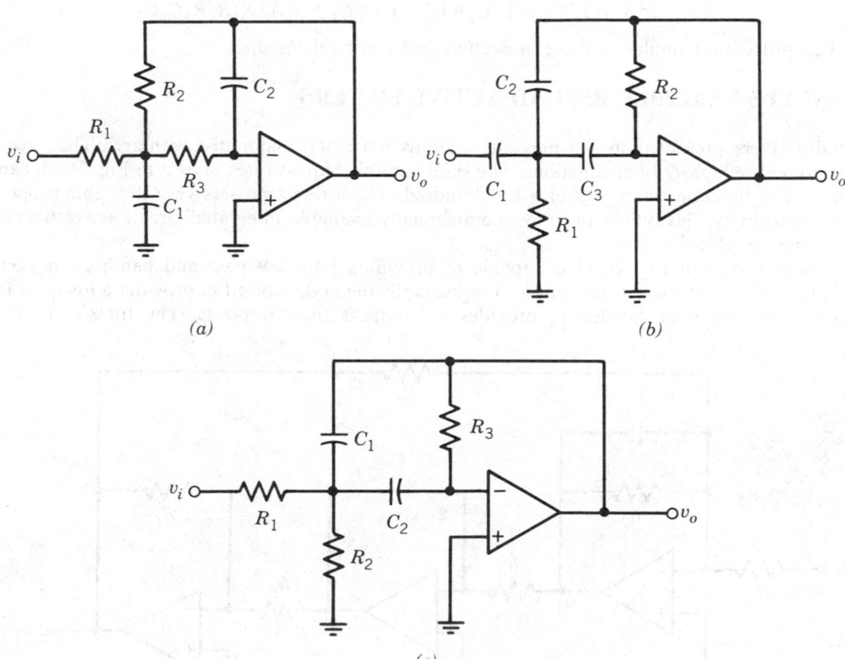

Fig. 26.12 Infinite-gain (multiple-feedback) filter structures: (a) low-pass, (b) high-pass, (c) bandpass.

Sensitivity functions can be shown to be given by

$$S_{R_2}^{\omega_n} = S_{R_3}^{\omega_n} = S_{C_1}^{\omega_n} = S_{C_2}^{\omega_n} = -\frac{1}{2} \tag{26.78}$$

$$S_{R_1}^{\omega_n} = 0 \tag{26.79}$$

$$S_{C_1}^{Q} = -S_{C_2}^{Q} = \frac{1}{2} \tag{26.80}$$

$$S_{R_1}^{Q} = Q\left[\frac{1}{R_1}\cdot\left(\frac{R_2R_3C_2}{C_1}\right)^{1/2}\right] \tag{26.81}$$

$$S_{R_2}^{Q} = -\frac{1}{2}Q\left[\left(\frac{R_2C_2}{R_3C_1}\right)^{1/2} + \frac{1}{R_1}\cdot\left(\frac{R_2R_3C_2}{C_1}\right)^{1/2} - \left(\frac{R_3C_2}{R_2C_1}\right)^{1/2}\right] \tag{26.82}$$

$$S_{R_3}^{Q} = -\frac{1}{2}Q\left[\left(\frac{R_3C_2}{R_2C_1}\right)^{1/2} + \frac{1}{R_1}\left(\frac{R_2R_3C_2}{C_1}\right)^{1/2} - \left(\frac{R_2C_2}{R_3C_1}\right)^{1/2}\right] \tag{26.83}$$

It can be observed that the Q sensitivities of this circuit are significantly lower than the sensitivity of Sallen and Key filters. The major disadvantages of these circuits is that the number of external elements is larger than that of Sallen and Key filters. Also the passive elements must be of high quality and low tolerance to accomplish proper tuning.

The transfer functions of the high-pass and bandpass filters shown in Fig. 26.12*b* and 26.12*c*, respectively, are as follows.

High-pass filter

$$H(s) = \frac{-s^2C_1/C_2}{s^2 + s(C_1/C_2C_3 + 1/C_2 + 1/C_3)1/R_2 + 1/R_1R_2C_2C_3} \tag{26.84}$$

Bandpass filter

$$H(s) = \frac{-s/R_1C_1}{s^2 + s(1/C_1 + 1/C_2)(1/R_3) + (R_1 + R_2)/R_1R_2R_3C_1C_2} \tag{26.85}$$

Design procedures similar to those in Section 26.4 can be devised.

26.7 STATE-VARIABLE BIQUAD ACTIVE FILTERS

The active filters presented in the previous sections have one major disadvantage: They are not suitable to realize high-Q filter functions. The state-variable biquad filter, shown in Fig. 26.13, can be used to realize functions with a Q of a few hundred. These filters possess very desirable properties (e.g., low sensitivity, flexibility) and most commercially available integrated-circuit active filters are based on this structure.

The filter shown in Fig. 26.13 is capable of providing both low-pass and bandpass responses, depending on the node used as the output. For example, the node labeled v_3 provides a low-pass filter response, while the node labeled v_1 provides a bandpass filter response. The transfer functions

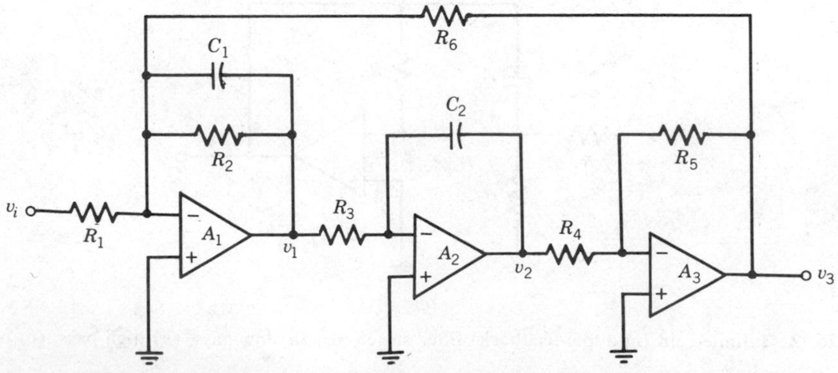

Fig. 26.13 State-variable biquad active filter realizing both low-pass and bandpass filters.

corresponding to the low-pass and bandpass filter responses are given as follows:

$$H_{LP}(s) = \frac{V_3(s)}{V_i(s)} = \frac{-1/R_1R_3C_1C_2}{s^2 + (1/R_2C_1)s + 1/R_3R_6C_1C_2} \tag{26.86}$$

$$H_{BP}(s) = \frac{V_1(s)}{V_i(s)} = \frac{-s/R_1C_1}{s^2 + (1/R_2C_1)s + 1/R_3R_6C_1C_2} \tag{26.87}$$

The disadvantage of this circuit is that three operational amplifiers are required.

Universal Active Filter

The state-variable filter structure shown in Fig. 26.14 uses three operational amplifiers and simultaneously provides low-pass, high-pass, and bandpass filter outputs. Such filters are useful as general purpose building blocks and are commonly called universal active filters. The structure is relatively insensitive to Q variations owing to change in its elements' values and is capable of realizing Qs of a few hundred.

The transfer functions for each filter function follow. Universal filters with an internal structure similar to that shown in Fig. 26.14 are commercially available from several manufacturers (e.g., National Semiconductor, model AF100; Integrated Micro-System, model μAR-2000; Baldwin Electronics, model FS-50).

The transfer functions are

Low-pass filter section

$$H_{LP}(s) = \frac{K_{LP}}{s^2 + b_1 s + b_2} \tag{26.88}$$

where

$$K_{LP} = \frac{R_4(R_2 + R_3)}{R_2R_5R_6C_1C_2(R_1 + R_4)} \tag{26.89}$$

$$b_1 = \frac{R_1(R_2 + R_3)}{R_2R_6C_1(R_1 + R_4)} \tag{26.90}$$

$$b_2 = \frac{R_3}{R_2R_5R_6C_1C_2} \tag{26.91}$$

High-pass filter section

$$H_{HP}(s) = \frac{K_{HP}s^2}{s^2 + b_1 s + b_2} \tag{26.92}$$

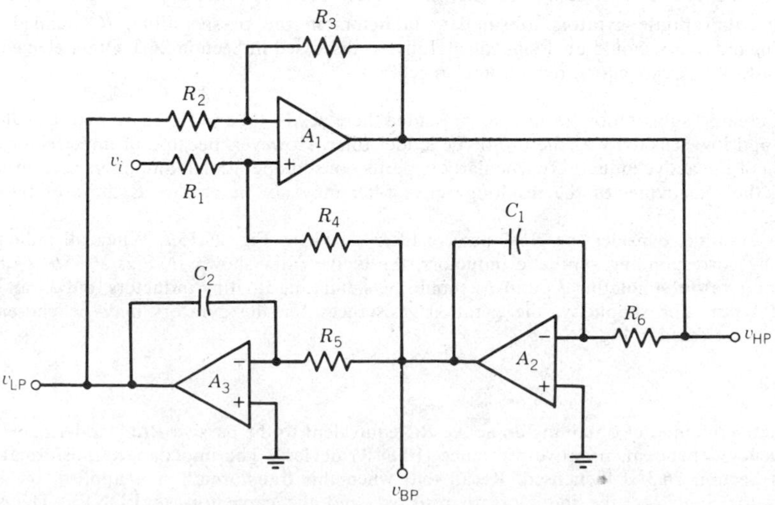

Fig. 26.14 Second-order universal active filter.

where

$$K_{HP} = \frac{R_4(R_2 + R_3)}{R_2(R_1 + R_4)} \tag{26.93}$$

and b_1 and b_2 are given by Eqs. (26.90) and (26.91), respectively.

Bandpass filter section

$$H_{BP}(s) = \frac{K_{BP}s}{s^2 + b_1 s + b_2} \tag{26.94}$$

where

$$K_{BP} = -\frac{R_4(R_2 + R_3)}{R_2 R_6 C_1(R_1 + R_4)} \tag{26.95}$$

and b_1 and b_2 are given by Eqs. (26.90) and (26.91), respectively.

All of these filters can be tuned by varying R_5 and R_6 and/or C_1 and C_2. The Q can be adjusted by varying R_4.

26.8 ACTIVE FILTER DESIGNS BY INDUCTOR SIMULATION

In network theory it is known that double-terminated passive LC ladder networks can be designed to possess very low sensitivity due to variations of element values. Because of the difficulty of realizing inductors in integrated-circuit form, passive filters can be realized only with use of discrete components. Invention of the gyrator and the frequency-dependent, negative-resistance (FDNR) device has created tremendous interest in the design of active filters based on their passive counterparts. Such design procedures are called passive network simulation methods. The filters created have the advantage of possessing low sensitivities (i.e., they have the same sensitivity properties of the passive prototype network). Since tables of element values of passive filters are readily available, the design procedures for passive network simulation is quite straightforward.

Two recently used methods of designing active filters using passive filters as prototypes are described here.

Method 1

This method is based on the direct replacement of inductors in the passive filter by a gyrator terminated with an appropriate capacitor. The inductance, simulated by using a gyrator, is denoted as the *synthetic inductance*. The steps in applying this method are as follows.

1. Using the specifications, design the RLC normalized prototype ladder passive filter.
2. Design appropriate gyrators to simulate inductors in the passive filter. (Grounded and/or floating inductors should be designed carefully, as discussed in Section 26.3. Other elements, such as resistors and capacitors, remain unchanged.)

The inductor substitution technique presented here leads to a realization that has the same topology and low sensitivity as the prototype ladder filter. However, because of imperfections in the realization of the active inductor (particularly imperfections in operational amplifiers used in realizing gyrators), the sensitivities of the resulting active filter may not be as low as those of the passive prototype.

As an example, consider the RLC passive filter shown in Fig. 26.15*a*. When all inductors are replaced by corresponding synthetic inductors, the active filter shown in Fig. 26.15*b* is obtained. Recall that gyrators simulating L_1 and L_3 should be simulating floating inductors, but L_2 and L_4 are grounded types. For simplicity, the gyration resistances for the gyrators may be chosen to be identical, as indicated in Fig. 26.15*b*.

Method 2

An alternative method of obtaining an active RC equivalent to the passive RLC ladder network is to use frequency-dependent, negative-resistance (FDNR) devices. The impedance transformation discussed in Section 26.3 is then used. Recall that when this transformation is applied, the resistors transform to capacitors, the inductors to resistors, and the capacitors to FDNRs. The steps in applying this method are as follows.

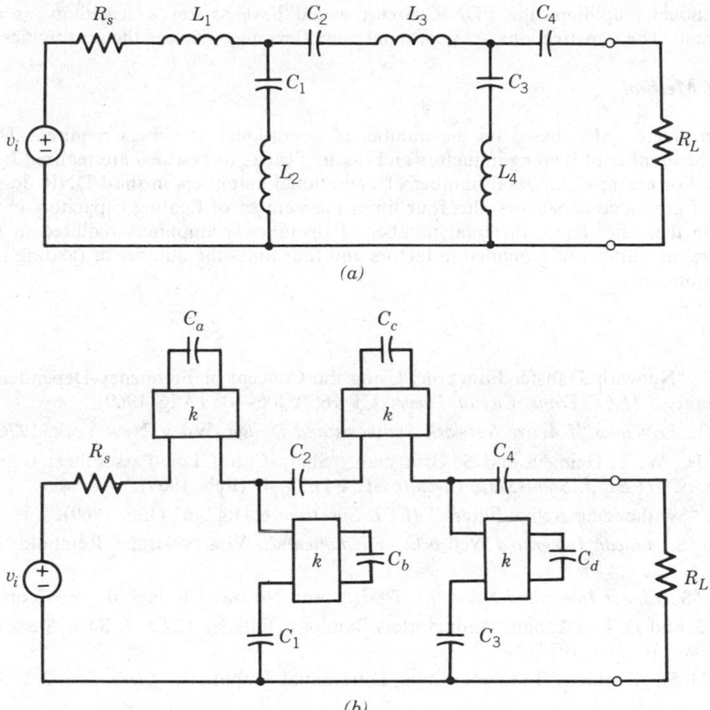

Fig. 26.15 (*a*) *RLC* passive ladder network. (*b*) Equivalent active filter. $C_a = L_1/k^2$, $C_b = L_2/k^2$, $C_c = L_3/k^2$, $C_d = L_4/k^2$.

1. Using the specifications, design an *RLC* normalized prototype ladder passive filter.
2. Use *RLC-CRD* transformation to obtain a network containing *R*s, *C*s, and FDNRs.
3. Design the FDNRs as discussed in Section 26.3, and tune them properly.

As an example, consider the low-pass filter shown in Fig. 26.16*a*. When the indicated replacements are made, the FDNR realization of Fig. 26.16*b* is obtained. Again, except for the imperfections

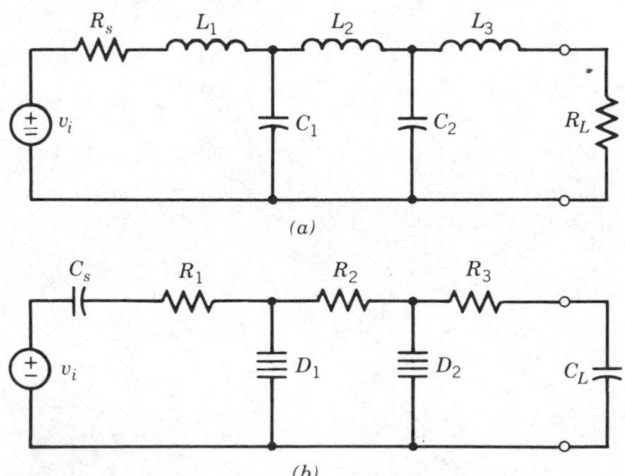

Fig. 26.16 (*a*) Passive low-pass ladder filter. (*b*) Equivalent frequency-dependent, negative-resistance (FDNR) realization. $C_s = 1/R_s$, $R_1 = L_1$, $R_2 = L_2$, $R_3 = L_3$; $C_L = 1/R_L$, $D_1 = C_1$, $D_2 = C_2$.

of the operational amplifiers, the FDNR circuit would have as low a sensitivity as the passive prototype circuit. The imperfections in operational amplifiers may increase the sensitivities slightly.

Selection of Method

The selection of method is based on the number of operational amplifiers required. This number depends on the number of floating inductors and floating capacitors (which are realized by cascading two FDNRs. For example, the total number of operational amplifiers in the FDNR design is twice the number of grounded capacitors plus four times the number of floating capacitors in the passive prototype. On the other hand, the total number of operational amplifiers required in the gyrator design is twice the number of grounded inductors and four times the number of floating inductors in the passive prototype.

Bibliography

Bruton, L. T., "Network Transfer Functions Using the Concept of Frequency-Dependent Negative Resistance," *IEEE Trans. Circuit Theory* **CT 16**(3):406–408 (Aug. 1969).

Daryanani, G., *Principles of Active Network Synthesis and Design*, Wiley, New York, 1976.

Holmes, W. H., W. E. Heinlein and S. Grützman, "Sharp-Cutoff Low-Pass Filters Using Floating Gyrators," *IEEE J. Solid-State Circuits* **SC 4**(1):38–49 (Feb. 1969).

Mitra, S. K., "Synthesizing Active Filters," *IEEE Spectrum* **6**(1):47–63 (Jan. 1969).

Moschytz, G. S., *Linear Integrated Networks, Fundamentals*, Von Nostrand Reinhold, New York, 1974.

Moschytz, G. S., *Linear Integrated Networks, Design*, Von Nostrand Reinhold, New York, 1975.

Orchard, H. J. and D. F. Sheahan, "Inductorless Bandpass Filters," *IEEE J. Solid-State Circuits* **SC 5**(3):108–118 (Jun. 1970).

Riordan, R. H. S., "Simulated Inductors Using Differential Amplifiers," *Elect. Lett.* **3**(2):50–51 (Feb. 1967).

Sallen, R. P. and E. L. Key, "A Practical Method of Designing RC Active Filters," *IRE Trans. Circuit Theory* **CT 2**(1):74–75 (Mar. 1955).

Sedra, A. S. and P. O. Brackett, *Filter Theory and Design: Active and Passive*, Matrix, Champaign, IL, 1978.

Strauss, L., *Wave Generation and Shaping*, McGraw-Hill, New York, 1970.

Thomas, L. C., "The Biquad: Part I—Some Practical Design Considerations," *IEEE Trans. Circuit Theory* **CT 18**(3):350–357 (May 1971).

Thomas, L. C., "The Biquad: Part II—A Multipurpose Active Filtering System," *IEEE Trans. Circuit Theory* **CT 18**(3):358–361 (May 1971).

CHAPTER 27
POWER SUPPLIES

YUSUF Z. EFE

The Cooper Union for Advancement of Science and Art
New York, New York

27.1 INTRODUCTION

Essentially all of the electronic devices discussed in this section require a DC power source. In some cases one or more batteries may supply adequate DC power. However, in most instances the battery is neither the most economical nor the most convenient means of obtaining DC power. Its relatively short lifetime, relatively large volume requirements (in many instances), and frequent recharging requirement preclude the use of the battery. The major source of power to electronic circuits is the DC power supply that converts an AC input voltage to a DC output voltage.

The primary characteristics that must be considered in the design of a power supply are

1. The *DC output voltage* (or voltages) required by the circuit.
2. The *maximum current* requirement of the load.
3. The variation of the DC output voltage allowed. This is known as the *voltage regulation* of a power supply and is expressed by Eq. (27.1):

$$\%\ \text{Regulation} = \frac{V_{DC}(\text{no load}) - V_{DC}(\text{full load})}{V_{DC}(\text{no load})} \times 100 \tag{27.1}$$

   In this equation, V_{DC} (no load) is the load voltage when the load current is zero and V_{DC} (full load) is the load voltage when the rated maximum DC current is drawn from the power supply.
4. The output voltage of a DC power supply contains a residual AC component (called ripple) which is superimposed on the DC output of a regulated power supply. The figure of merit that is used to describe the amount of ripple at the output is called the ripple factor, r, and is defined by Eq. (27.2):

$$r = \frac{\text{rms value of AC ripple voltage}}{\text{DC load voltage}} \tag{27.2}$$

Several unregulated and regulated DC power supplies will be discussed here.

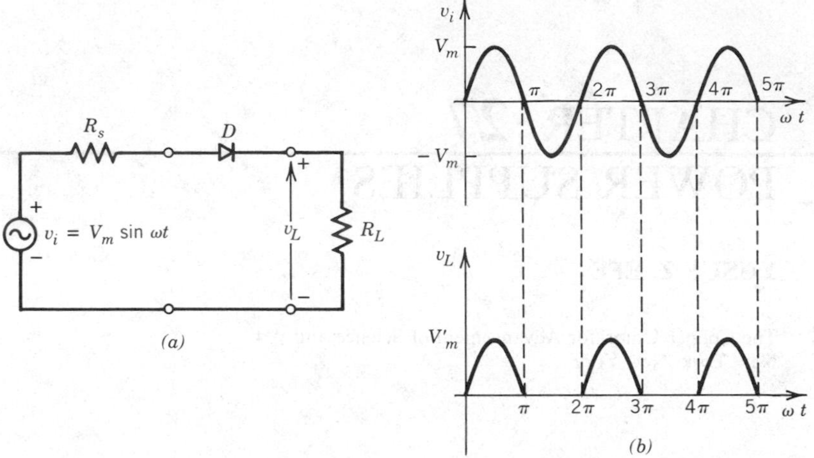

Fig. 27.1 (*a*) Half-wave rectifier circuit. (*b*) Input and output waveform.

27.2 UNREGULATED DC POWER SUPPLIES

Half-Wave Rectifier

Diodes may be used to convert AC currents and voltages into DC currents and voltages. This process is called *rectification*, and such a circuit is called a *rectifier*. The diode in Fig. 27.1*a* allows current in one direction so that the output in Fig. 25.1*b* is obtained. This circuit is called a *half-wave rectifier*, since half of the AC waveform has been blocked.

Let the forward diode resistance be R_f and the reverse-biased resistance be very large so that it can be ignored. Then the average, or DC, output voltage (neglecting diode voltage drop) can be obtained as follows:

$$V_{\mathrm{DC}} = \frac{1}{2\pi} \cdot \int_0^{2\pi} \frac{V_m R_L}{R_L + R_s + R_f} \sin \omega t \cdot d(\omega t) \tag{27.3}$$

or

$$V_{\mathrm{DC}} = \frac{R_L}{R_L + R_s + R_f} \cdot \frac{V_m}{\pi} = \frac{V'_m}{\pi} \tag{27.4}$$

If $R_L \gg R_s + R_f$, then

$$V_{\mathrm{DC}} = \frac{V_m}{\pi} \tag{27.5}$$

The ripple factor of a half-wave rectifier is given by

$$r = \sqrt{\frac{\pi^2}{4} - 1} = 1.21 \tag{27.6}$$

This is a very large ripple factor. For example, if an audio amplifier were powered by such a half-wave rectifier, a very high hum would result. It will be shown that use of filters at the output will reduce the hum caused by the presence of large AC components in the output.

Full-Wave Rectifier

As can be seen in the half-wave rectifier, only alternate half cycles have been utilized. The performance of a rectifier can be improved by rectification of every half cycle. The resulting rectifier is called a *full-wave rectifier* and is shown in Fig. 27.2.

Observe that a full-wave rectifier basically consists of two half-wave rectifiers connected to a single load, R_L. A center-tapped transformer is needed to provide AC voltages for each half-wave rectifier section. During the positive portion of the secondary waveform, the point *a* is positive with respect to the point *b*. Hence D_1 will be forward biased and D_2 reverse biased. During the next half-cycle the point *b* will be positive with respect to the point *a*, causing D_2 to be forward biased and D_1 reverse biased. Thus the output waveform shown in Fig. 27.2*b* is obtained.

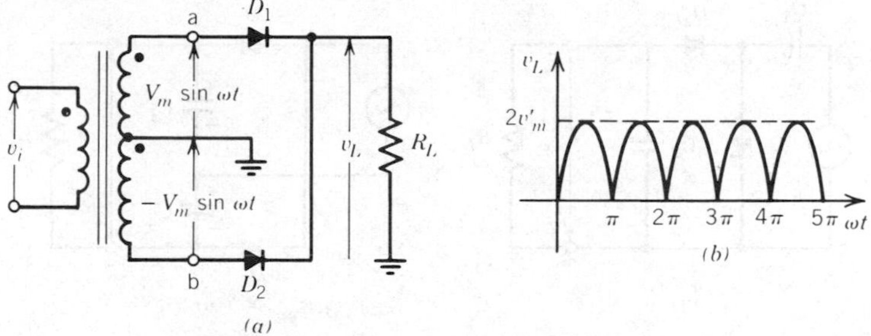

Fig. 27.2 (*a*) Full-wave rectifier. (*b*) Output waveform.

The DC value of the output voltage can be shown to be

$$V_{DC} = 2 \cdot \frac{R_L}{R_L + R_t + R_f} \cdot \frac{V_m}{\pi} = 2 \cdot \frac{V'_m}{\pi} \tag{27.7}$$

If $R_L \gg R_t + R_f$, then

$$V_{DC} = 2 \cdot \frac{V_m}{\pi} \tag{27.8}$$

Thus the full-wave rectifier produces double the direct voltage of the half-wave rectifier [see Eq. (27.5)]. The ripple-factor can be shown to be

$$r = \sqrt{\frac{\pi^2}{8} - 1} = 0.483 \tag{27.9}$$

Note that the ripple factor has been significantly improved compared with that of the half-wave rectifier [see Eq. (27.6)]. However, it is still high for many applications. The major disadvantage of the full-wave rectifier is that a center-tapped transformer is required.

Bridge Rectifier

A bridge rectifier, shown in Fig. 27.3, can be used to obtain a full-wave rectification without use of a transformer. The bridge rectifier possesses two disadvantages: (1) it requires four diodes and (2) since two of the diodes are always in series with the load, the voltage drop due to diodes is doubled and may not be neglected. These disadvantages, however, are not very serious, since semiconductor diodes are inexpensive compared with a transformer, and the voltage drop is relatively low and can be neglected for most practical applications.

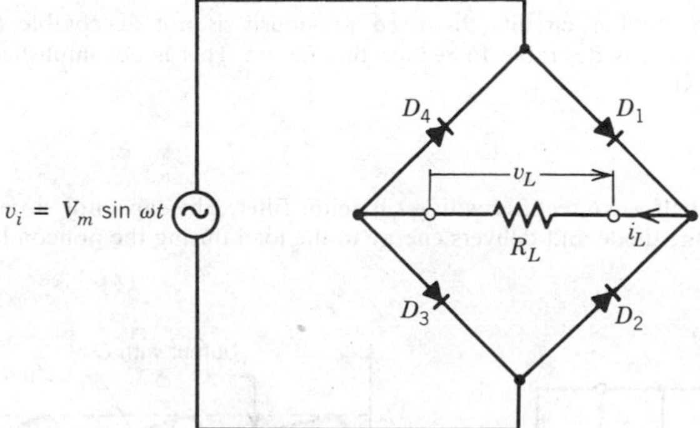

Fig. 27.3 Bridge rectifier circuit.

Voltage Multipliers

The DC output voltage V_{DC} in rectifier circuits is limited by the peak value of the AC voltage applied, V_m. Note that $V_{DC} < V_m$. It is possible to develop peak output voltage that is a multiple of the peak value of the input signal. The circuits in Fig. 27.4 can be used for voltage doubling.

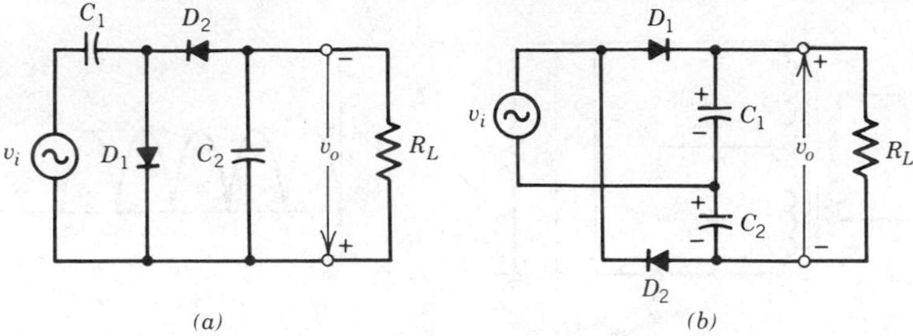

Fig. 27.4 Voltage doubler circuits. $v_i = V_m \sin \omega t$.

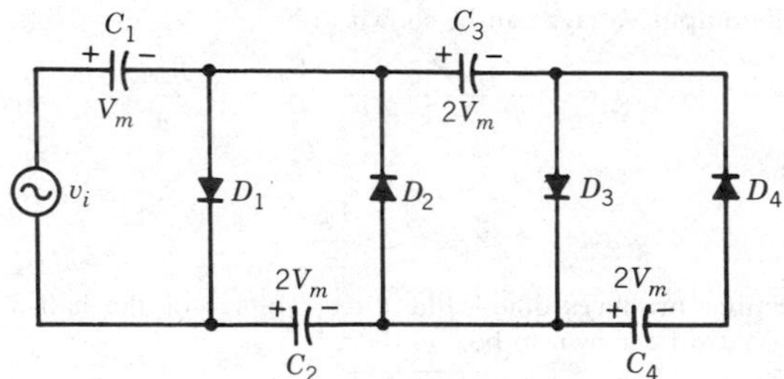

Fig. 27.5 Voltage quadrupler.

In Fig. 27.4*a*, during the positive half cycle D_1 conducts and C_1 charges to V_m. Assume that the voltage drop across each diode is negligible when it is conducting. During the next (negative) half cycle, the voltage across D_1 is $2V_m$, D_2 conducts, and C_2 is charged to the peak value of $2V_m$. Hence this circuit is a voltage doubler.

The circuit in Fig. 27.4*b* consists of two half-wave rectifiers connected in series. On the positive half cycle, D_1 conducts and C_1 is charged to V_m. The ripple frequency in this circuit is twice the AC supply frequency. The regulation of this circuit can be shown to be better than that of the circuit in Fig. 27.4*a*.

By connecting additional diode-capacitor circuits it is possible to obtain rectifier circuits that triple and quadruple the input voltage. A voltage quadrupler is shown in Fig. 27.5.

27.3 POWER SUPPLY FILTERS

The ripple factor in rectifier circuits discussed previously is not acceptable for many practical applications. Therefore it is desirable to reduce this factor. This is accomplished with use of filter circuits, discussed next.

Capacitor Filter

Figure 27.6 shows a half-wave rectifier with a capacitor filter. The capacitor stores energy during the conduction cycle of the diode and delivers energy to the load during the nonconducting cycle. Hence

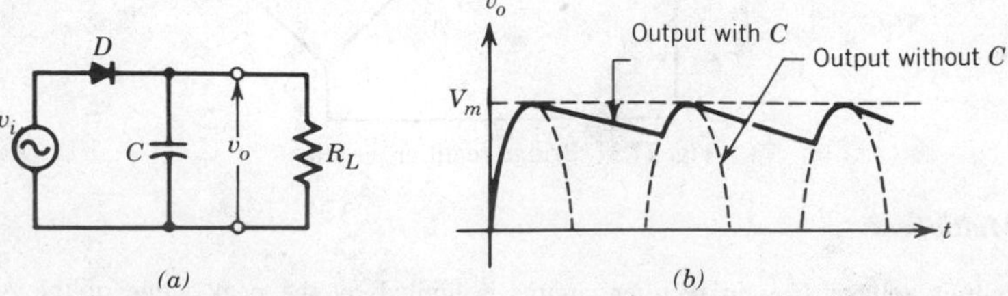

Fig. 27.6 (*a*) Half-wave rectifier with a capacitor filter. (*b*) Output waveform with and without *C*.

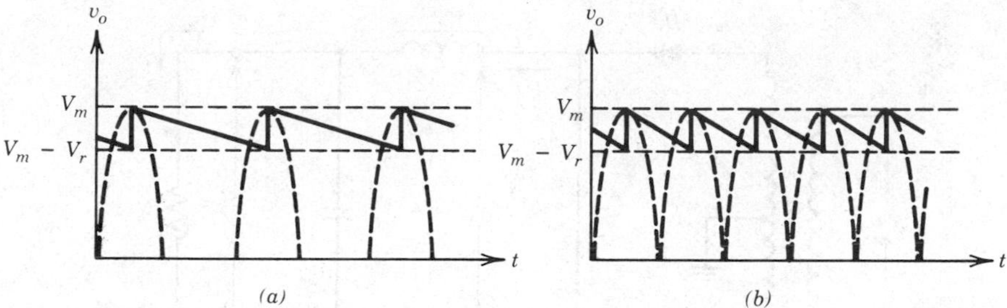

Fig. 27.7 Approximate output voltage of capacitor filter: (*a*) for a half-wave rectifier, (*b*) for a full-wave rectifier.

there is always a current passing through the load, which considerably reduces the ripple. The output waveform of a half-wave rectifier with a capacitor filter is shown in Fig. 27.6*b*. If $R_L C \gg T$, where T is the period of the input AC signal, then the load voltage and current will have a low ripple in their waveform.

Exact analysis of this half-wave rectifier circuit is quite complicated. An approximate analysis, indicated in Fig. 27.7*a*, provides reasonably accurate results when the ripple factor is small. This analysis yields

$$V_{DC} = \frac{V_m}{1 + 1/(2fR_L C)} \tag{27.10}$$

Ripple factor

$$r = \frac{1}{2\sqrt{3}\, fR_L C} \tag{27.11}$$

Ripple voltage

$$V_r = \frac{V_m}{fR_L C} \tag{27.12}$$

where f is the frequency of the input AC signal. If a capacitor filter is used with a full-wave rectifier, it can be shown that (see Fig. 27.7*b*)

$$V_{DC} = \frac{V_m}{1 + 1/(4fR_L C)} \tag{27.13}$$

$$\frac{1}{4\sqrt{3}\, fR_L C} \tag{27.14}$$

$$V_r = \frac{V_m}{2fR_L C} \tag{27.15}$$

Observe from the above equations that the full-wave rectifier requires one half as large a capacitance as does the half-wave rectifier for the same V_{DC}, r, and V_r.

In addition, note that the magnitude of the ripple voltage V_r varies inversely with the time constant $R_L C$, and any changes in the load current cause changes in the ripple voltage. As the load current increases, the DC output voltage decreases considerably. Therefore the voltage regulation of the capacitor filter is very poor.

L-Section Filter

A full-wave rectifier with an *L*-section filter is shown in Fig. 27.8. This type of filter is used to improve the ripple factor and regulation. In Fig. 27.8, the voltage v_s can be expressed in terms of its Fourier series as

$$v_s = V_m \left[\frac{2}{\pi} - \frac{4}{\pi} \sum_{\substack{k \text{ even} \\ k \neq 0}} \frac{\cos k\omega t}{(k-1)(k+1)} \right] \tag{27.16}$$

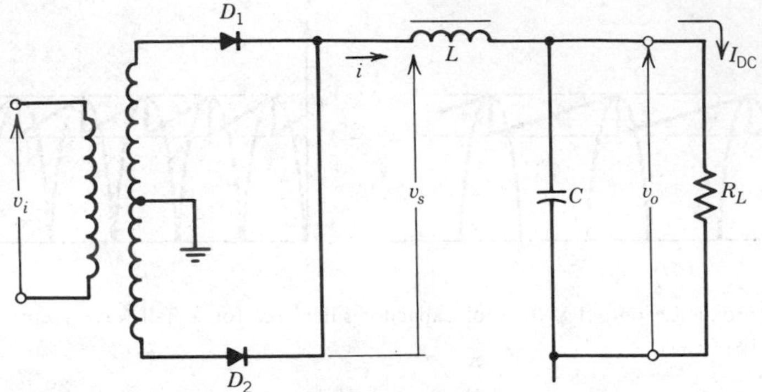

Fig. 27.8 Full wave rectifier with L-section filter.

In most of the practical applications, $1/\omega C \ll R_L$ and $\omega L \gg 1/\omega C$ at the frequency ω and its harmonics. Hence the approximate circuit analysis for the first two harmonics yields

$$V_{o2,\mathrm{eff}} \approx \frac{V_m}{3\pi\sqrt{2}\,\omega^2 LC} \tag{27.17}$$

$$V_{o4,\mathrm{eff}} \approx \frac{V_m}{60\pi\sqrt{2}\,\omega^2 LC} \tag{27.18}$$

where $V_{o2,\mathrm{eff}}$ and $V_{o4,\mathrm{eff}}$ are the effective values of the second and fourth harmonic components, respectively, in the output voltage. Note that $V_{o4,\mathrm{eff}}$ is 1/20 of $V_{o2,\mathrm{eff}}$. The higher order harmonic components are even smaller. Therefore we can approximate the effective value of the AC component by $V_{o2,\mathrm{eff}}$ and thus

$$V_{\mathrm{DC}} = \frac{V_m}{\pi} - RI_{\mathrm{DC}} \tag{27.19}$$

$$R = \frac{1}{6\sqrt{2}\,\omega^2 LC} \tag{27.20}$$

where R is the sum of the diode, the inductor L, and the transformer resistance. Observe that the ripple is independent of load—a very desirable feature.

A bleeder resistance is normally included in a rectifier circuit shunted across the DC output terminals. The functions of this resistance are threefold: (1) It bleeds the charge from the filter capacitor when the rectifier is turned off, (2) it improves the regulation, and (3) it acts as a minimum load, preventing the development of excessive voltage in the filter inductance.

The preceding discussion of the full-wave rectifier with L-section filter assumes that a current flows through the inductance continuously. If the current were to be cut out for a part of each cycle, this analysis would no longer be valid. The current through the inductance is continuous if the following condition is satisfied

$$L_c \geqslant \frac{R_L}{3\omega} \tag{27.21}$$

where L_c is called the *critical inductance*.

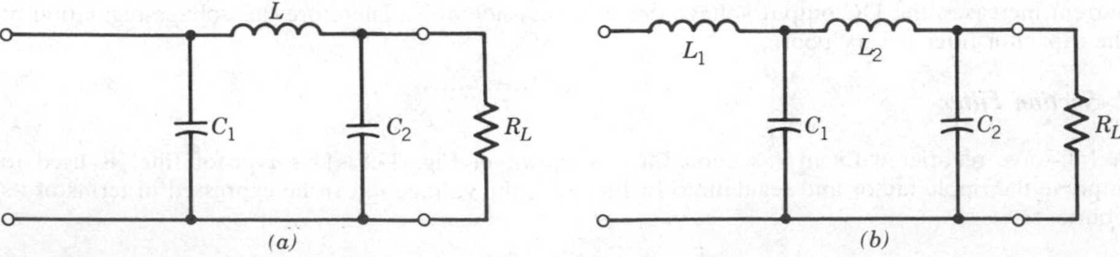

Fig. 27.9 (*a*) Π-section filter. $r = \sqrt{2}/\omega^3 LC_1C_2R_L$. (*b*) Multiple L-section filter. $r = \sqrt{2}/3 \cdot 1/\omega^4 L_1L_2C_1C_2R_L$.

Filtering can be improved by increasing the number of energy-storing elements. Two such circuits and their corresponding ripple factors are shown in Fig. 27.9. The first is known as the capacitance input filter or the Π-section filter, and the second is known as the inductance input filter or the multiple L-section filter.

27.4 REGULATED DC POWER SUPPLIES

In many electronic systems the voltage regulation of the power supplies previously discussed is not acceptable, since DC output voltage varies directly with AC input voltage. The first and foremost function of any voltage regulator is to maintain the specified voltage at its output. Such maintenance should be expected over a specified range of load currents. There are, however, factors such as temperature and input voltage changes that tend to adversely affect voltage regulation. To compound the situation, there is some interaction between the previously mentioned performance degrading factors. This section describes several regulated DC power supply circuits which improve voltage regulation.

Zener Diode Regulators

The Zener diode can be used as a voltage regulator in rectifier circuits. Such a regulator circuit and the current-voltage characteristic of the Zener diode are shown in Fig. 27.10. The Zener diode in this circuit is operated in the breakdown region. Note that the voltage across the diode is practically independent of the current through it. As the unregulated DC voltage V_{DC} varies, the output voltage across the load is kept at $V_{L,DC} = V_z$. Observe that $V_{L,DC}$, the DC voltage across the load, hardly changes at all even though V_{DC} may change significantly. Therefore the regulation in the Zener diode regulator is excellent. However, this regulator is inefficient because the voltage drop across R_0 varies to offset changes in v_L. If v_L tends to increase, then i_z increases significantly which, in turn, increases i so that the voltage drop across R_0 compensates for the increase in voltage. Therefore large amounts of power are dissipated in the Zener diode itself and in R_0. The efficiency of such regulation circuits is well below 50%. For this reason Zener diode regulators are useful only in low-power circuits.

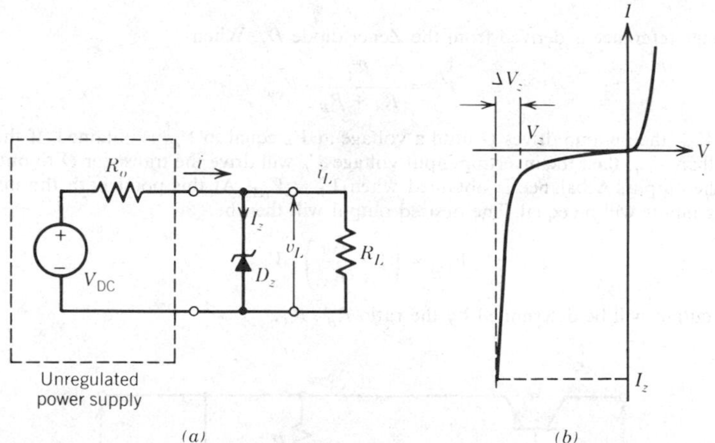

Fig. 27.10 (*a*) Zener diode regulator. (*b*) Zener diode current voltage characteristics.

Emitter Follower for Zener Diode Regulator

Usually, poor regulation of a power supply is caused by its high internal impedance. An emitter follower can be used to convert this high impedance to low impedance, as shown in Fig. 27.11*a*.

Zener diode regulators are also restricted to relatively small output currents. Emitter follower regulators provide current amplification, hence removing this restriction.

In higher current applications, more than one emitter follower can be used, as shown in Fig. 27.11*b*. This circuit has the advantage that its regulation may be poor. For relatively high-load

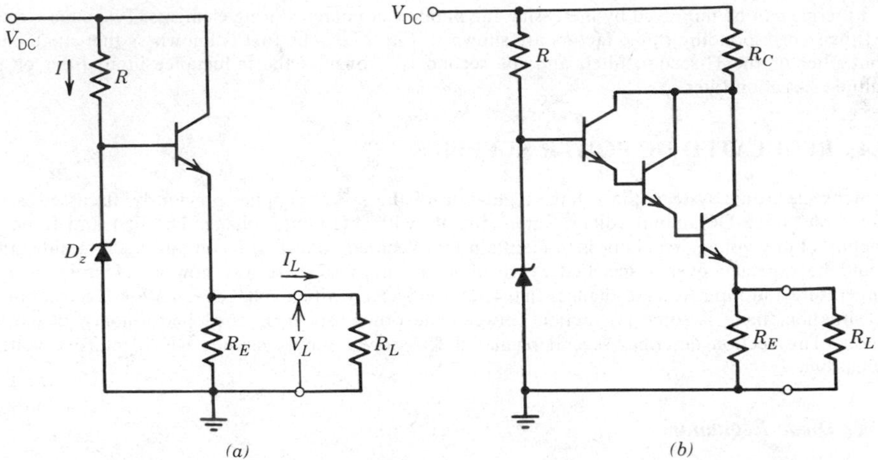

Fig. 27.11 (*a*) Emitter follower with Zener diode regulator. (*b*) Multiple emitter followers.

current, the transistor dissipation can become excessive, and the voltages $V_{BE(on)}$ and V_z, which both affect the output voltage, are temperature dependent, which can lead to poor regulation.

Series Voltage Regulators

A voltage regulator, in general, consists of an op-amp, voltage references (e.g., Zener diodes), and a series-pass element. The circuit in Fig. 27.12 is a typical series voltage regulator in which the series-pass element (the transistor Q) acts as a variable resistor. In this case the series-pass element dissipates the excess voltage ($V_{in} - V_{out}$). In many cases of high power requirements (high input-output voltage differential and/or large load currents), an external series-pass transistor may be utilized.

The voltage reference is derived from the Zener diode D_z. When

$$V_F = \frac{R_A}{R_A + R_B} \cdot V_{out} \tag{27.22}$$

is less than V_{ref}, the op-amp drives Q until a voltage at V_F equal to V_{ref} is obtained. If the voltage at V_F is more than V_{ref}, then the inverting input voltage V_F will drive the transistor Q to obtain a lower voltage at the output. A balance is obtained when $V_F = V_{ref}$. At this point both the inverting and noninverting inputs will be equal. The desired output will then be

$$V_{out} = \left(1 + \frac{R_B}{R_A}\right) \cdot V_{ref} \tag{27.23}$$

Clearly the output will be determined by the ratio $R_B : R_A$.

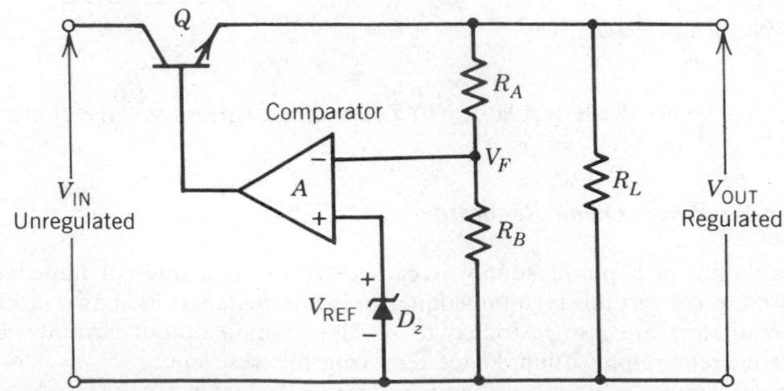

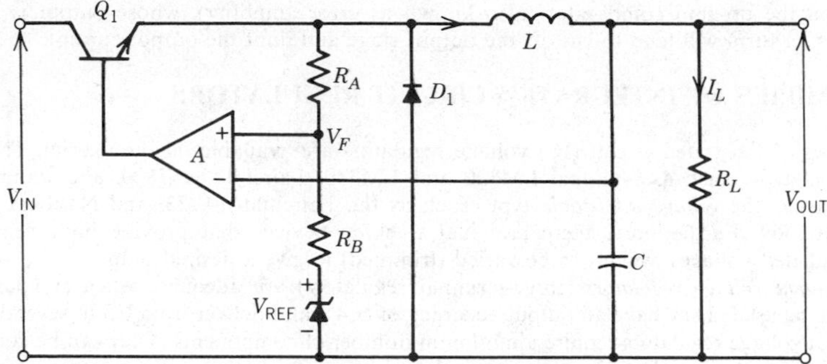

Fig. 27.13 Basic switching voltage regulator.

Series-pass regulators are inherently low in efficiency. This is due to the power dissipated (wasted) by the series-pass transistor Q.

Switching Voltage Regulators

Switching regulators use a high-frequency switch to turn the series-pass transistor on and off. Figure 27.13 shows a typical switching regulator circuit. The higher the frequency, the smaller the components for a specified output power capability. The main drawbacks of the switching regulator are the use of an unusually large number of external components and of an inductor. The inductor, however, can be made small if high frequencies are used in the design.

In this circuit configuration, regulation is accomplished by controlling the duty cycle of the transistor Q_1. The transistor is switched on or off so that its operation never lies in its linear region. Hence power dissipation is at a minimum. Diode D_1 is inserted so that it will conduct when Q_1 is off. When Q_1 is on, D_1 is reverse biased and therefore does not conduct. Load current I_L that flows through L charges the capacitor C. When V_{out} reaches V_F, the comparator turns Q_1 off. The inductor current begins to decrease. When V_{out} reaches a value of slightly less than V_{ref}, the comparator switches Q_1 back to on. The voltage output is a function of the duty cycle of the switching circuit

$$V_{out} = \frac{t_{on}}{t_{on} + t_{off}} \cdot V_{in} \tag{27.24}$$

where t_{on} and t_{off} are turn-on and turn-off times, respectively.

Short-Circuit Protection

In many voltage regulators, a second transistor is added for current limiting purposes. Figure 27.14 is a simplified schematic of such an arrangement.

The base-emitter of Q_2 will become forward biased at a particular level of I_L, due to the external current sensing resistor R_{sc}. When this occurs, the collector of Q_2 will sink most of the available

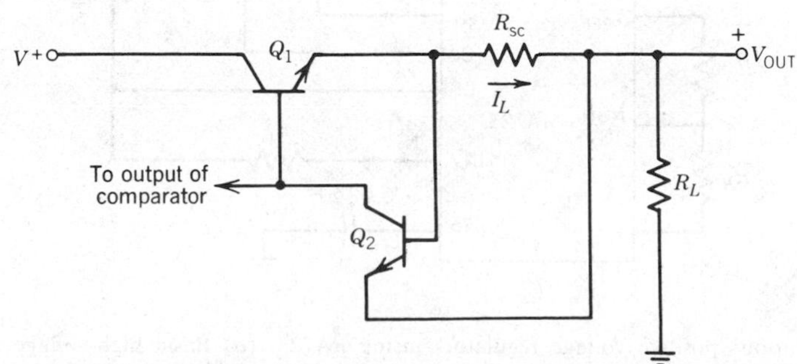

current from the op-amp comparator (also known as error amplifier), whose output is a current source. This, in turn, will tend to cut off the output stage and limit the output current.

27.5 FAMILIES OF INTEGRATED CIRCUIT REGULATORS

Many classes of integrated circuit (IC) voltage regulators are available on the market. There is the *single voltage* type, such as National LM320 and LM340, Fairchild μA7800, and Lambda 1400. Another class is the *voltage adjustable* type, such as the Fairchild μA723, and National Motorola LM105, MC1569. Furthermore, there are dual tracking devices that provide both negative and positive regulated voltages which can be varied (trimmed) to give a desired output.

Fixed output voltage regulators (three-terminal regulators) are adequate when standard output voltages can be used. They have an output accuracy of ±4% and deliver from 0.5 to several amperes. Fixed output voltage regulators require a minimum number of components. They can be converted to give variable outputs at higher power levels by the addition of external components.

Several IC voltage regulator circuits are discussed in this section, and the μA723 is used in the illustrative examples. The data sheet of the μA723 is provided in Appendix 27.1 at the end of this chapter.

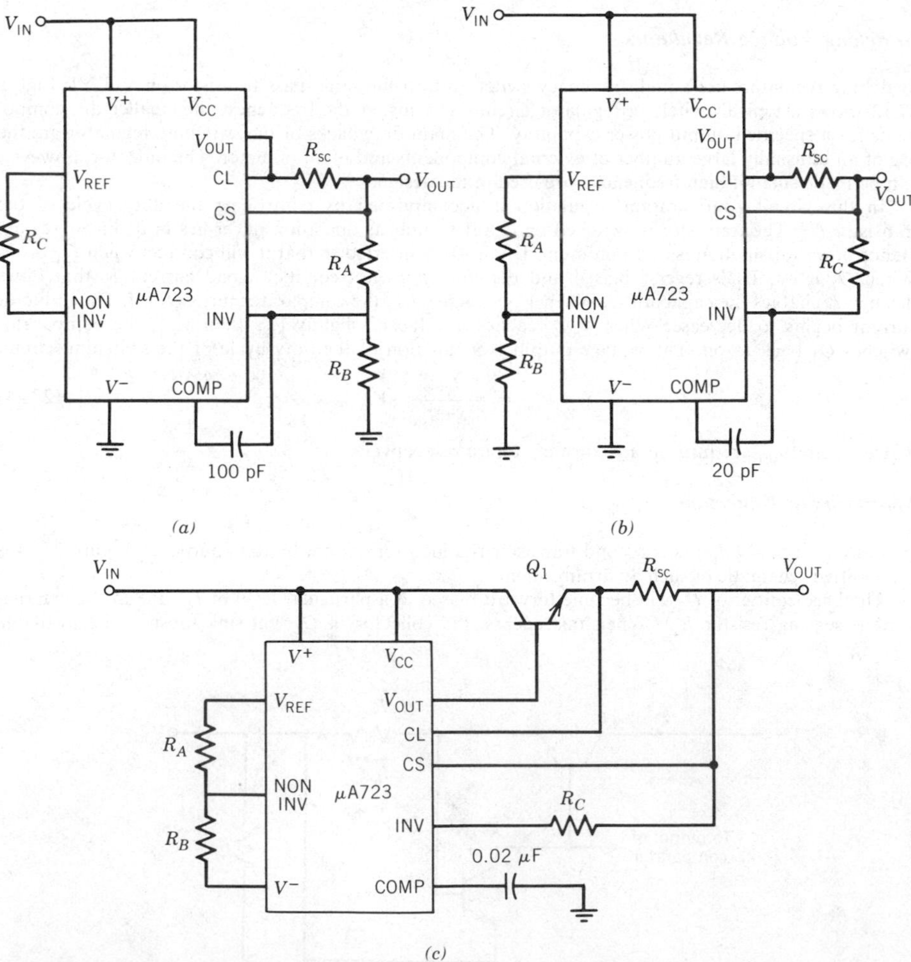

Fig. 27.15 Various positive voltage regulators, using μA723: (*a*) Basic high-voltage regulator, 7 V < V_{out} < 37 V, (*b*) basic low-voltage regulator, 2 V < V_{out} < 7 V, (*c*) basic +5-V-, 5-A-voltage regulator.

Positive Voltage Regulators

A basic positive voltage regulator gives a regulated output that is positive with respect to ground. Such a circuit is shown in Fig. 27.15. Generally the values of R_A and R_B set a division ratio of the output with respect to the reference voltage. The basic equations that govern the values of R_A and R_B and other parameters follow.

1. For a minimum temperature drift choose

$$R_C = R_A \| R_B \tag{27.25}$$

R_C may be eliminated by shorting V_{ref} and NON INV terminals for minimum component count.

2. Current limiting

$$I_{sc} = \frac{V_{sense}}{R_{sc}} \cong \frac{0.65}{R_{sc}} \tag{27.26}$$

3. Output voltage

$$V_{out} = \left(1 + \frac{R_A}{R_B}\right) V_{ref} \quad \text{for Fig. 27.15}a \tag{27.27}$$

$$V_{out} = \left(1 + \frac{R_A}{R_B}\right)^{-1} V_{ref} \quad \text{for Fig. 27.15}b \tag{27.28}$$

where $V_{ref} \cong 7.15$ V (typical).

Consider that a voltage regulator with an output voltage $V_{out} = 5$ V will be designed. Current is to be limited to $I_{sc} = 100$ mA. The unregulated input voltage is $V_{in} = 30$ V (V_{in} can be as high as 40 V for the μA723).

The values of R_A and R_B should be such that R_A (and R_B) $\leqslant 15$ kΩ so that V_{ref} will not be loaded appreciably. While a current of at least 5 mA can be drawn from V_{ref}, excessive current drain will cause poor regulation at low V_{in} and high dissipation at high V_{in}, which are undesirable. Selecting $R_B = 10$ kΩ, Eq. (27.28) yields a value of $R_A \cong 4$ kΩ ($V_{ref} = 7.15$ V, $V_{out} = 5$ V). Using Eq. (27.25), we obtain $R = 4\text{ k}\Omega \| 10\text{ k}\Omega = 2.9$ kΩ. Using Eq. (27.26), we obtain $R_{sc} = 0.65/0.1 = 6.5$ Ω.

Negative Voltage Regulators

A negative voltage regulator gives a regulated output that is negative with respect to ground. Such a negative voltage regulator is shown in Fig. 27.16. This configuration makes use not only of its V_{ref} but

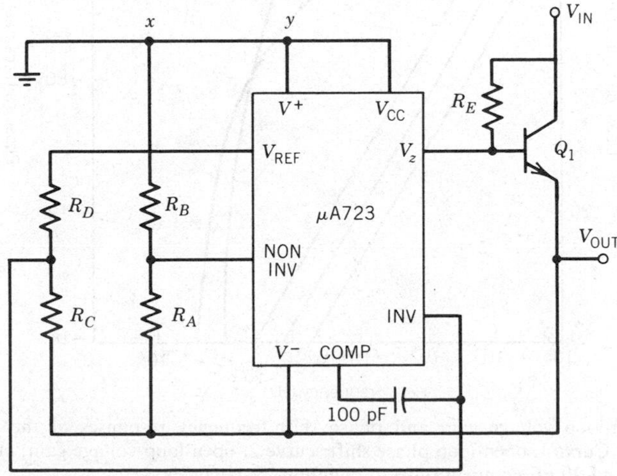

Fig. 27.16 Negative voltage regulator.

of an internal 6.2-V Zener diode as well. Contrary to positive regulators, in negative regulators the inverting and noninverting inputs are reversed and the external pass transistor Q_1 acts as a level-shifted emitter follower from V_{out}. Good regulation is achieved, since the regulator is driven from its own output. Regulation is controlled by the h_{FE} of Q_1 and the load regulation of the IC.

The resistor R_E should be of sufficient value to drive the maximum load current required, through Q_1 at V_{in}(min). However, the value of R_E should limit the current through the zener [at a minimum load and V_{in}(max)] to 10 mA. This places a constraint on the lower limit of h_{FE} of Q_1. For large current requirements, Q_1 may be replaced with a Darlington device. A limitation of this circuit is its inability to provide an output voltage of less than 9 V.

The output voltage for the regulator in Fig. 27.16 is given by

$$V_{out} = -V_{ref} \cdot \frac{R_C(R_A + R_B)}{R_A(R_C + R_D)} \tag{27.29}$$

For negative output voltages less than 9 V, the $V+$ and V_{CC} terminals both must be connected to a positive supply such that the voltage between $V+$ and $V-$ is greater than 9 V. In addition, the connection between x and y should be removed. The inputs to the internal amplifier should never be more positive than V_{out}. As V_{ref} is approximately 7.2 V, it should be divided to below 5 V. If we let $R_C = R_D$, the input to the inverting terminal will be 3.6 V and will hence be satisfactory.

Frequency Compensation. The stability of a voltage regulator can be assured in two steps. As a first step, the adequate AC and DC performance of both the internal gain stages (usually very high gain) and external active devices (if used) must be established. A second step is provision of the necessary compensation using standard operational amplifier techniques.

A case in point is the μA723, although the analysis can be used for other devices as well. Figure 27.17 illustrates open-loop phase shift and open-loop voltage gain vs. frequency of the μA723 voltage gain stage. An increased rate of phase shift is shown in curve 1 starting from approximately 10 kHz. This is mostly due to the beta fall off (at high frequencies) in the amplifier's output stage. Curve 2 is the accompanying open-loop voltage gain. These demonstrate that external frequency compensation is needed for stability regardless of whether external active devices are used.

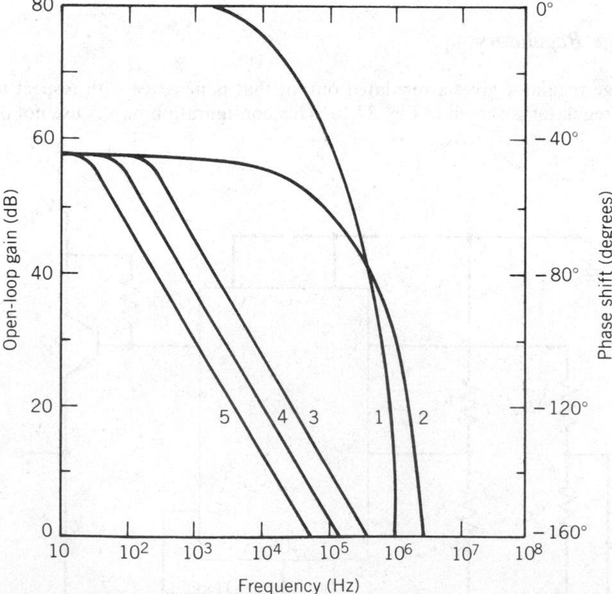

Fig. 27.17 Open-loop voltage gain and phase shift frequency responses of the μA723 precision voltage regulator. Curve 1, open-loop phase shift; curve 2, open-loop voltage gain; curves 3, 4, and 5, gain with 5, 10, and 20 nF compensation to common, respectively.

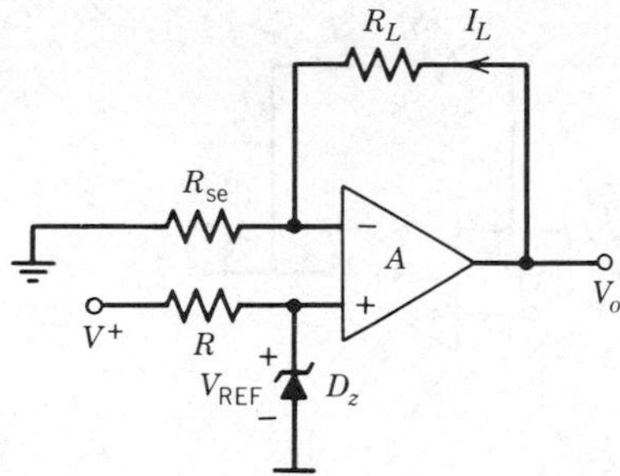

Fig. 27.18 Basic noninverting current regulator.

Recommended for frequency compensation at unity gain is either a 5-nF capacitor from the compensation terminal to the V − terminal or a 20-pF capacitor from the compensation terminal to the inverting input. If compensation is achieved by using the 20-pF capacitor, it is important that the inverting input be isolated by some independence from the rest of the circuit.

When external series-pass devices are to be used, their 3-dB bandwidth must be considered, since these devices usually have a lower bandwidth than the μA723.

Current Regulators

A current regulator circuit using an op-amp as the control element is shown in Fig. 27.18. The circuit is similar to the basic voltage regulator circuit. Actually, with a fixed value of R_L there is no difference. The primary difference in a current regulator is that the current rather than the voltage is maintained constant.

For example, if V_{ref} and R_{se} are fixed, the current flowing in R_L (load resistance) and in R_{se} (sensing resistance) will remain constant and is determined by

$$I_L = \frac{V_{ref}}{R_{se}} \tag{27.30}$$

The expression is true regardless of the value of R_L, since the op-amp is ideal. Hence regulation of the current in the load is accomplished.

Voltage-Controlled Current Source (VCCS)

Sometimes the ability to convert a voltage signal to a proportional output current is useful. This can be achieved by replacing V_{ref} in Fig. 27.18 with a variable voltage. As seen in Eq. (27.30), the sensitivity of the voltage-to-current conversion is inversely proportional to the sensing resistance R_{se}. Hence when high accuracy of load current is required, this resistance should be a precision type.

Basic voltage-to-current converter circuits with floating load (i.e., neither terminal of the load is grounded) are shown in Fig. 27.19. The current flowing through the load in each circuit is indicated alongside the circuits.

DC-to-DC Converters

Many electronic circuits are designed using operational amplifiers that require +15 V and −15 V and other integrated circuits requiring 5 V. In such situations, ±15 V can be obtained by using a DC-to-DC converter. The circuit shown in Fig. 27.20 can be used to convert 5 V DC to ±15 V. The pulse train from the astable multivibrator turns the transistor on and off. At the instant the transistor is switched off, a voltage spike of about four to five times $V_{CC} = +5$ V develops at the collector of Q. This voltage is rectified by the diode D_2, filtered by the RC filter formed by R_2 and C_2, and regulated by the Zener diode with $V_z = V$. Simultaneously a negative voltage spike appears in L_2. This voltage is also rectified by D_1, filtered by R_1 and C_1, and regulated to obtain −V volts. Although this circuit (as it stands) provides only about 10 mA current, it can be boosted to the desired level by using current amplification. This circuit provides voltages higher than the DC voltage available. Voltages lower than the available DC voltage can also be obtained by using series-voltage regulators (discussed earlier). A variety of circuits available in the literature and manufacturer's data provide circuits to be used for both purposes.

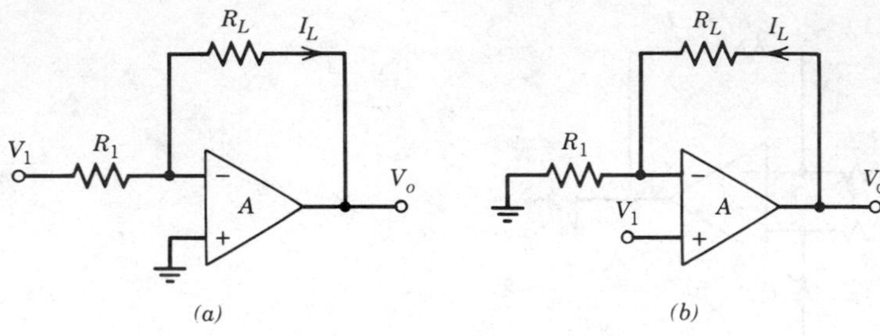

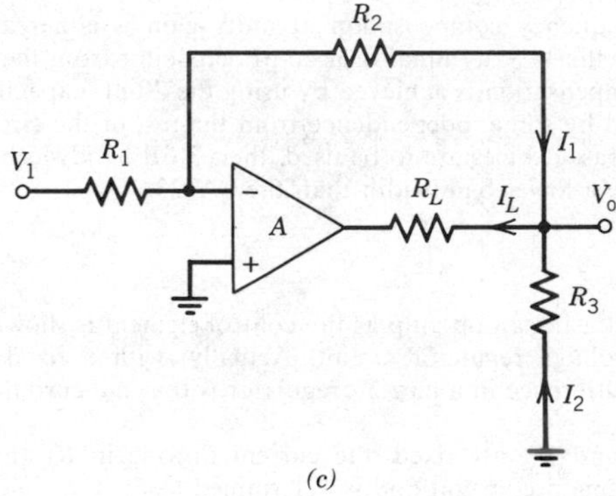

Fig. 27.19 Voltage-to-current converter circuits: (*a*) Inverting configuration, $I_L = V_1/R_1$; (*b*) noninverting configuration, $I_L = V_1/R_1$; (*c*) inverting current amplifier, $I_L = (1 + R_2/R_3)(V_1/R_1)$.

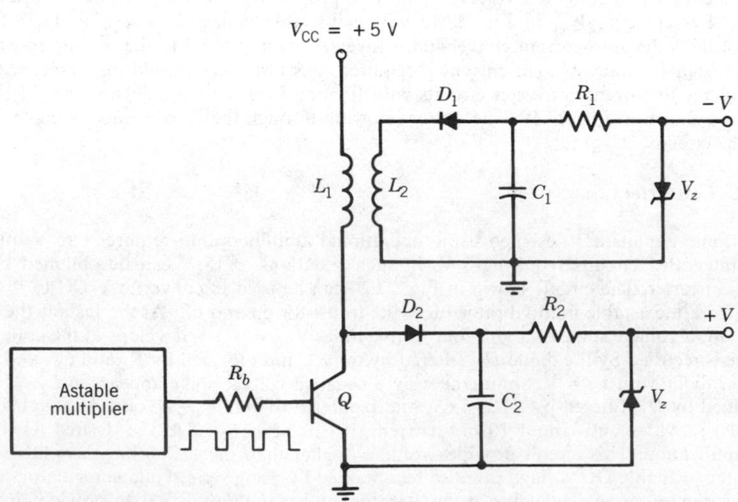

Fig. 27.20 Dual voltage DC-to-DC converter.

APPENDIX 27.1. μA723 PRECISION VOLTAGE REGULATOR

A Schlumberger Company

μA723 Precision Voltage Regulator

Linear Products

Description

The μA723 is a Monolithic Voltage Regulator constructed using the Fairchild Planar epitaxial process. The device consists of a temperature-compensated reference amplifier, error amplifier, power-series pass transistor and current-limit circuitry. Additional NPN or PNP pass elements may be used when output currents exceeding 150 mA are required. Provisions are made for adjustable current limiting and remote shutdown. In addition to the above, the device features low standby current drain, low temperature drift and high ripple rejection. The μA723 is intended for use with positive or negative supplies as a series, shunt, switching or floating regulator. Applications include laboratory power supplies, isolation regulators for low level data amplifiers, logic card regulators, small instrument power supplies, airborne systems and other power supplies for digital and linear circuits.

- **POSITIVE OR NEGATIVE SUPPLY OPERATION**
- **SERIES, SHUNT, SWITCHING OR FLOATING OPERATION**
- **0.01% LINE AND LOAD REGULATION**
- **OUTPUT VOLTAGE ADJUSTABLE FROM 2 TO 37 V**
- **OUTPUT CURRENT TO 150 mA WITHOUT EXTERNAL PASS TRANSISTOR**

Absolute Maximum Ratings

Pulse Voltage from V+ to V−, (50 ms) (μA723)	50 V
Continuous Voltage from V+ to V−	40 V
Input/Output Voltage Differential	40 V
Differential Input Voltage	±5 V
Voltage Between Non-Inverting Input and V−	+8 V
Current from V_Z	25 mA
Current from V_{REF}	15 mA
Internal Power Dissipation (Note)	
Metal	800 mW
DIP	1000 mW
Storage Temperature Range	−65°C to +150°C
Operating Temperature Range	
Military (μA723)	−55°C to +125°C
Commercial (μA723C)	0°C to +70°C
Pin Temperature (Soldering)	
Metal, Ceramic DIP (60 s)	300°C
Molded DIP (10 s)	260°C

Notes on following pages.

Connection Diagram
10-Pin Metal

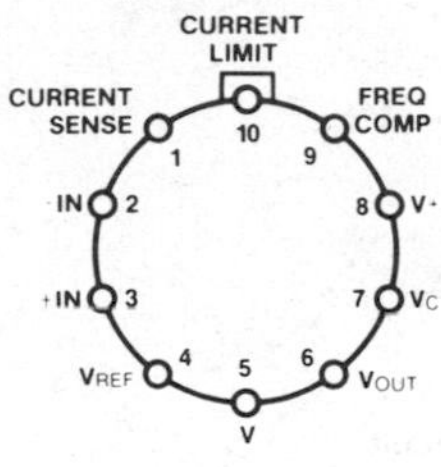

(Top View)

Pin 5 connected to case.

Order Information

Type	Package	Code	Part No.
μA723	Metal	5X	μA723HM
μA723C	Metal	5X	μA723HC

Connection Diagram
14-Pin DIP

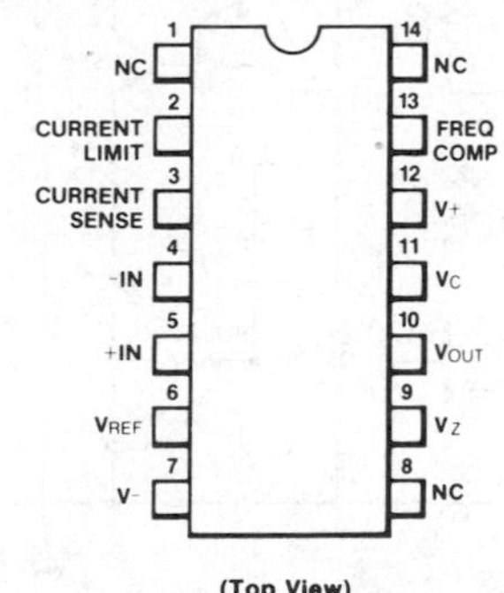

(Top View)

Order Information

Type	Package	Code	Part No.
μA723	Ceramic DIP	6B	μA723DM
μA723C	Ceramic DIP	6B	μA723DC
μA723C	Molded DIP	9B	μA723PC

Block Diagram

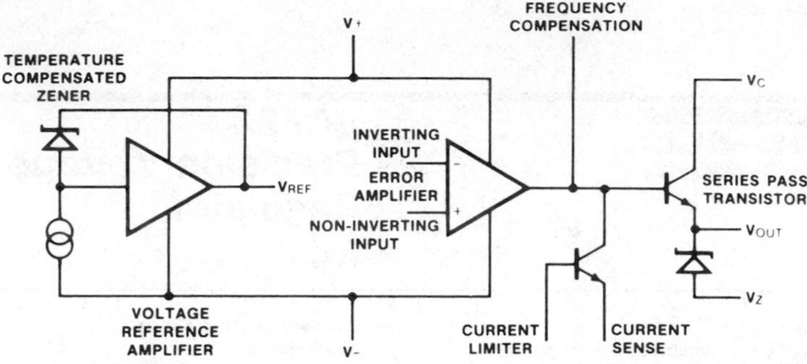

Equivalent Circuit

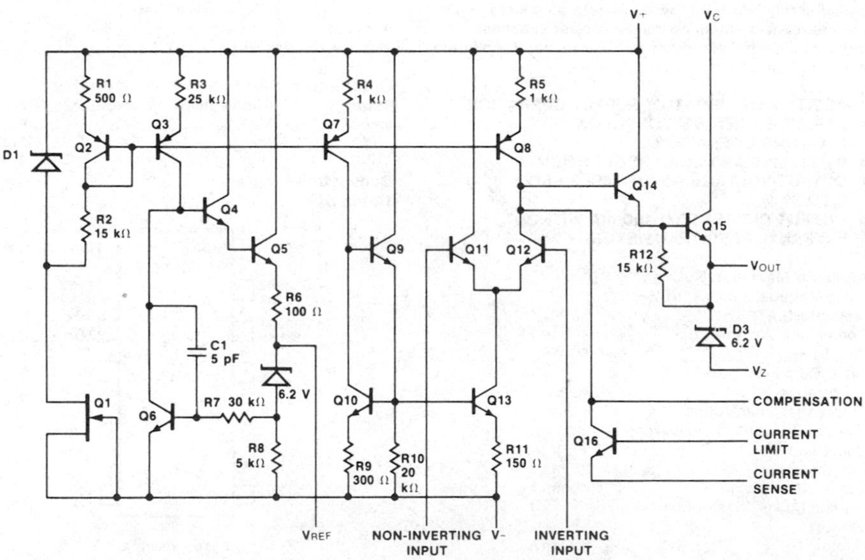

Note

1. Rating applies to ambient temperatures up to 25°C. Above 25°C ambient derate based on the following thermal resistance values:

	θJA	
	*Typ	Max
TO-5	150	190
Molded DIP	80	90
Ceramic DIP	95	105

μA723

Electrical Characteristic $T_A = 25°C$, $V_{IN} = V+ = V_C = 12$ V, $V- = 0$, $V_{OUT} = 5$ V, $I_L = 1$ mA, $R_{SC} = 0$, C1 = 100 pF, $C_{REF} = 0$, unless otherwise specified. Divider impedance as seen by error amplifier ≤ 10 kΩ connected shown in *Figure 1*. Line and load regulation specifications are given for the condition of constant chip temperature. Temperature drifts must be taken into account separately for high dissipation conditions.

Characteristic	Condition	Min	Typ	Max	Unit
Line Regulation	$V_{IN} = 12$ V to $V_{IN} = 15$ V		0.01	0.1	$\%V_O$
	$V_{IN} = 12$ V to $V_{IN} = 40$ V		0.02	0.2	$\%V_O$
	$-55°C \leq T_A \leq +125°C$, $V_{IN} = 12$ V to $V_{IN} = 15$ V			0.3	$\%V_O$
Load Regulation	$I_L = 1$ mA to $I_L = 50$ mA		0.03	0.15	$\%V_O$
	$-55°C \leq T_A \leq +125°C$, $I_L = 1$ mA to $I_L = 50$ mA			0.6	$\%V_O$
Ripple Rejection	f = 50 Hz to 10 kHz		74		dB
	f = 50 Hz to 10 kHz, $C_{REF} = 5\ \mu F$		86		dB
Average Temperature Coefficient of Output Voltage	$-55°C \leq T_A \leq +125°C$		0.002	0.015	%/°C
Short Circuit Current Limit	$R_{SC} = 10\ \Omega$, $V_O = 0$		65		mA
Reference Voltage	$I_{REF} = 0.1$ mA	6.95	7.15	7.35	V
Reference Voltage Change With Load	$I_{REF} = 0.1$ mA to 5 mA			20	mV
Output Noise Voltage	BW = 100 Hz to 10 kHz, $C_{REF} = 0$		20		μV_{rms}
	BW = 100 Hz to 10 kHz, $C_{REF} = 5\ \mu F$		2.5		μV_{rms}
Long Term Stability			0.1		%/1000 hrs
Standby Current Drain	$I_L = 0$, $V_{IN} = 30$ V		2.3	3.5	mA
Input Voltage Range		9.5		40	V
Output Voltage Range		2.0		37	V
Input/Output Voltage Differential		3.0		38	V

μA723C

Electrical Characteristic $T_A = 25°C$, $V_{IN} = V+ = V_C = 12$ V, $V- = 0$, $V_{OUT} = 5$ V, $I_L = 1$ mA, $R_{SC} = 0$, C1 = 100 pF, $C_{REF} = 0$, unless otherwise specified. Divider impedance as seen by error amplifier ≤ 10 kΩ connected as shown in *Figure 1*. Line and load regulation specifications are given for the condition of constant chip temperature. Temperature drifts must be taken into account separately for high dissipation conditions.

Characteristic	Condition	Min	Typ	Max	Unit
Line Regulation	$V_{IN} = 12$ V to $V_{IN} = 15$ V		0.01	0.1	$\%V_O$
	$V_{IN} = 12$ V to $V_{IN} = 40$ V		0.1	0.5	$\%V_O$
	$0°C \leq T_A \leq 70°C$, $V_{IN} = 12$ V to $V_{IN} = 15$ V			0.3	$\%V_O$
Load Regulation	$I_L = 1$ mA to $I_L = 50$ mA		0.03	0.2	$\%V_O$
	$0°C \leq T_A \leq 70°C$, $I_L = 1$ mA to $I_L = 50$ mA			0.6	$\%V_O$
Ripple Rejection	f = 50 Hz to 10 kHz		74		dB
	f = 50 Hz to 10 kHz, $C_{REF} = 5\ \mu F$		86		dB
Average Temperature Coefficient of Output Voltage	$0°C \leq T_A \leq 70°C$		0.003	0.015	%/°C
Short Circuit Current Limit	$R_{SC} = 10\ \Omega$, $V_O = 0$		65		mA
Reference Voltage	$I_{REF} = 0.1$ mA	6.80	7.15	7.50	V
Reference Voltage Change With Load	$I_{REF} = 0.1$ mA to 5 mA			20	mV
Output Noise Voltage	BW = 100 Hz to 10 kHz, $C_{REF} = 0$		20		μV_{rms}
	BW = 100 Hz to 10 kHz, $C_{REF} = 5\ \mu F$		2.5		μV_{rms}
Long Term Stability			0.1		%/1000 hrs
Standby Current Drain	$I_L = 0$, $V_{IN} = 30$ V		2.3	4.0	mA
Input Voltage Range		9.5		40	V
Output Voltage Range		2.0		37	V
Input/Output Voltage Differential		3.0		38	V

Typical Performance Curves for μA723 and μA723C

Line Regulation as a Function of Input/Output Voltage Differential

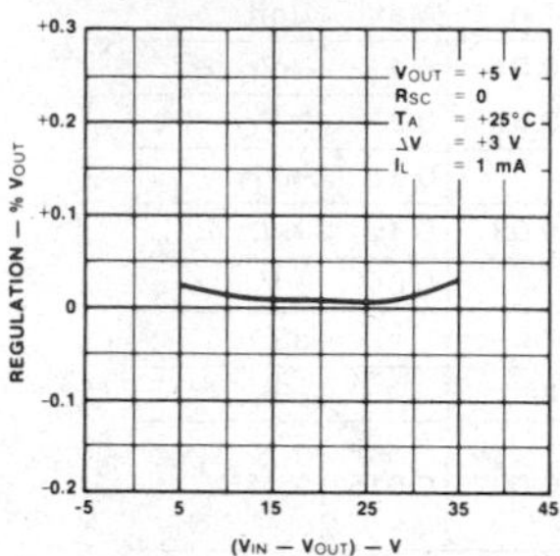

Current Limiting Characteristics as a Function of Junction Temperature

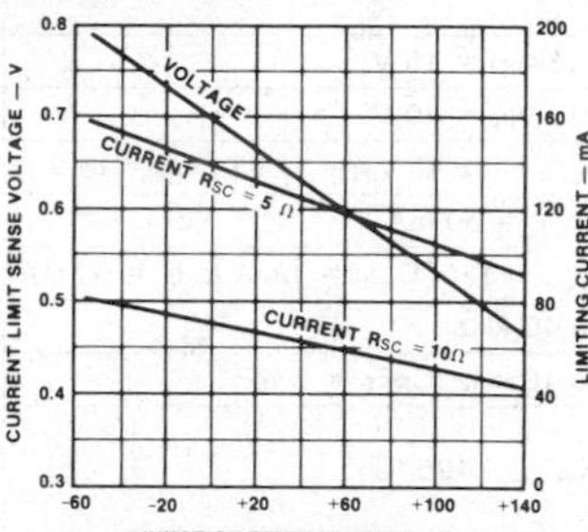

Load Transient Response

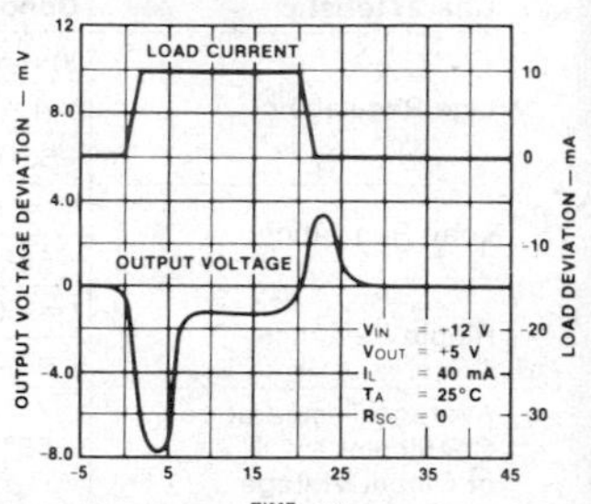

Load Regulation as a Function of Input/Output Voltage Differential

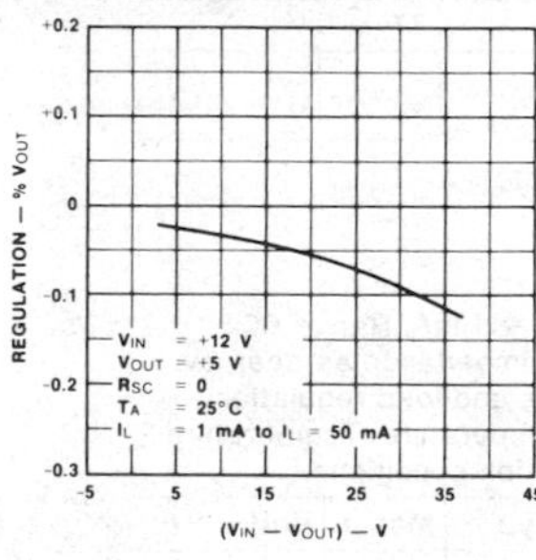

Line Transient Response

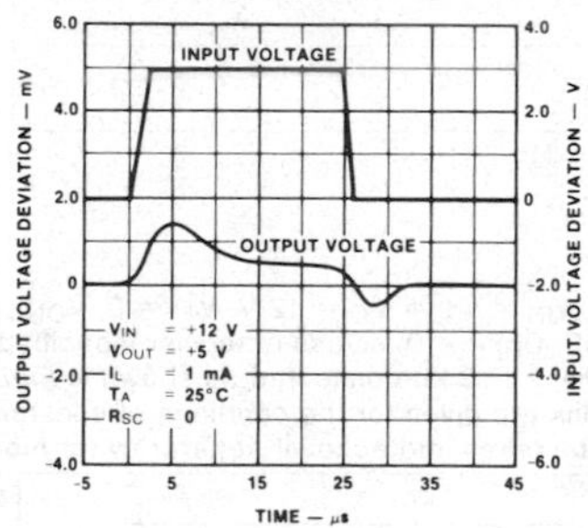

Output Impedance as a Function of Frequency

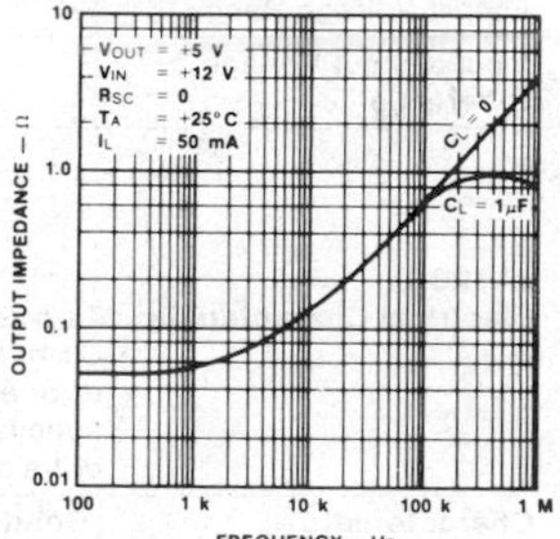

Typical Performance Curves for μA723

Maximum Load Current as a Function of Input-Output Voltage Differential

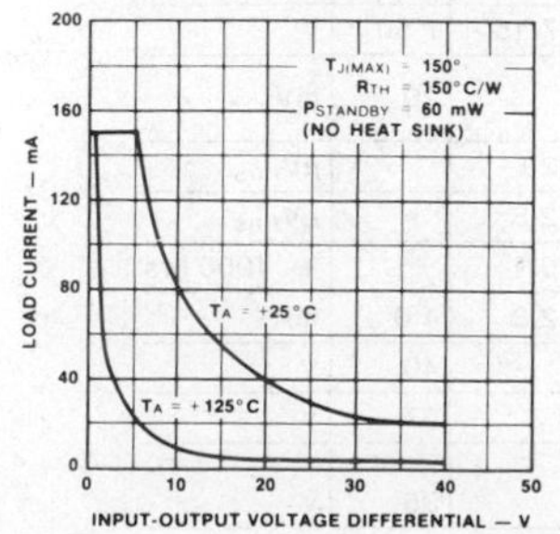

Load Regulation Characteristics Without Current Limiting

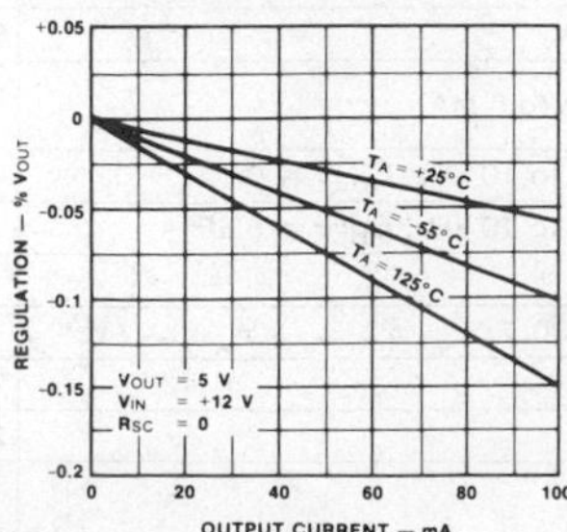

Load Regulation Characteristics With Current Limiting

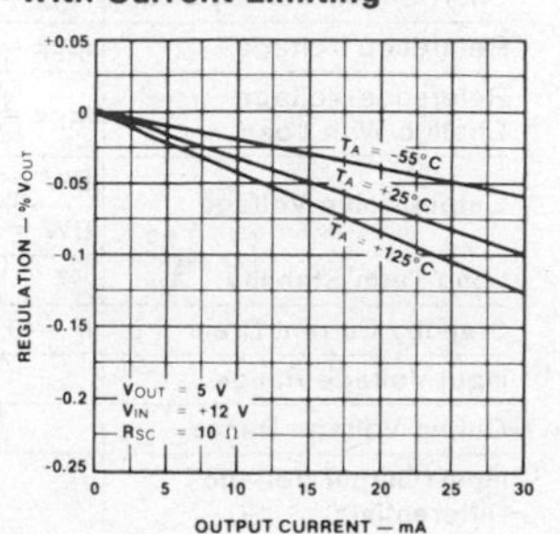

Typical Performance Curves (Cont.)

Load Regulation Characteristics With Current Limiting

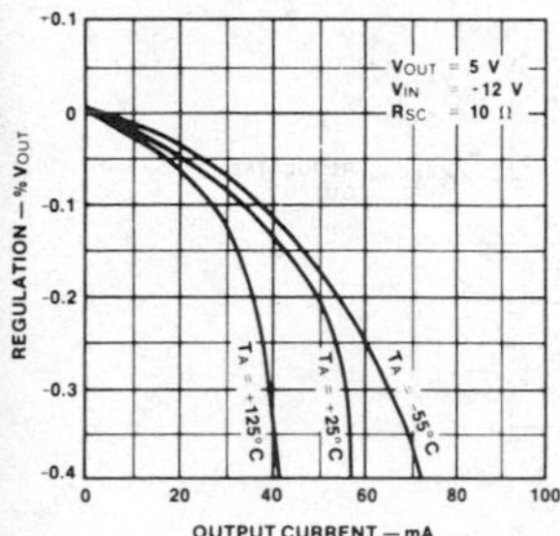

Current Limiting Characteristics

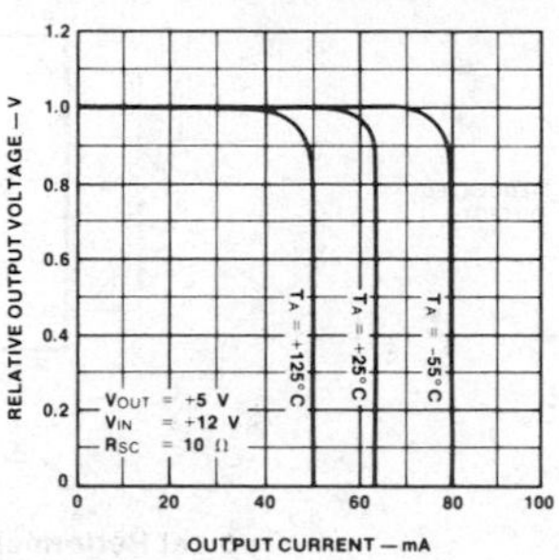

Standby Current Drain as a Function of Input Voltage

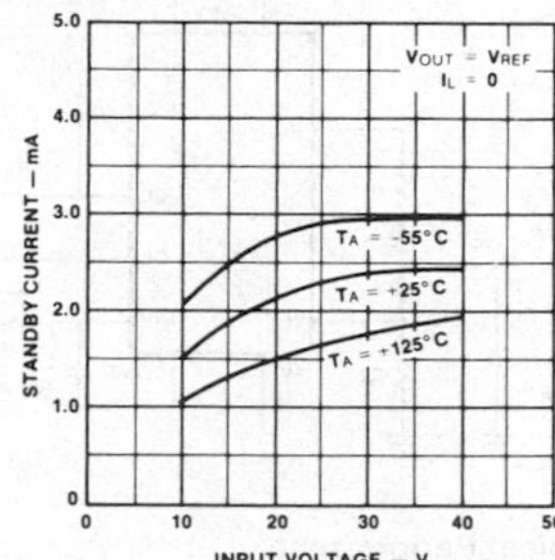

Typical Performance Curves for μA723C

Maximum Load Current as a Function of Input/Output Voltage Differential

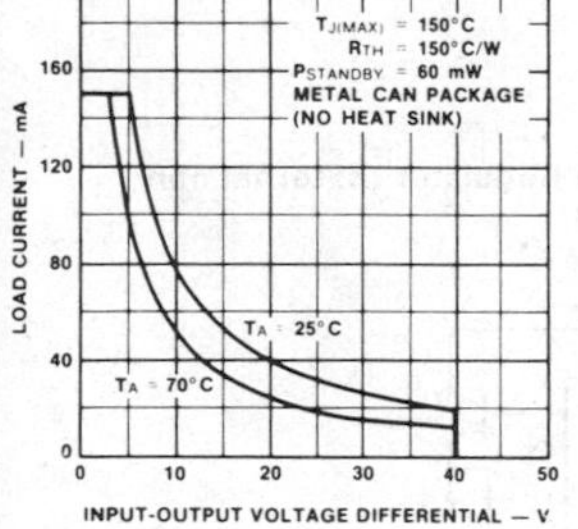

Load Regulation Characteristics Without Current Limiting

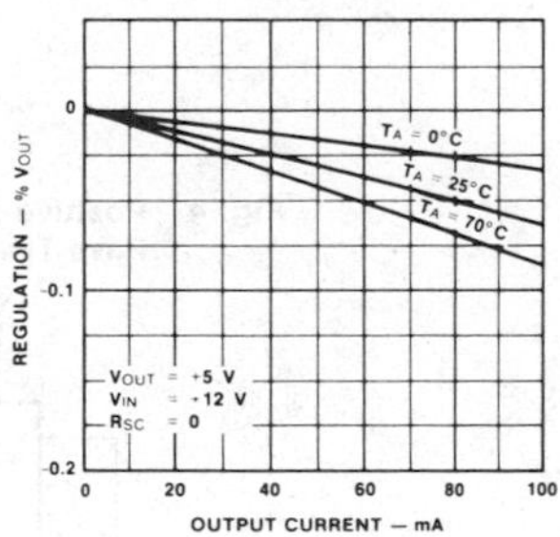

Current Limiting Characteristics

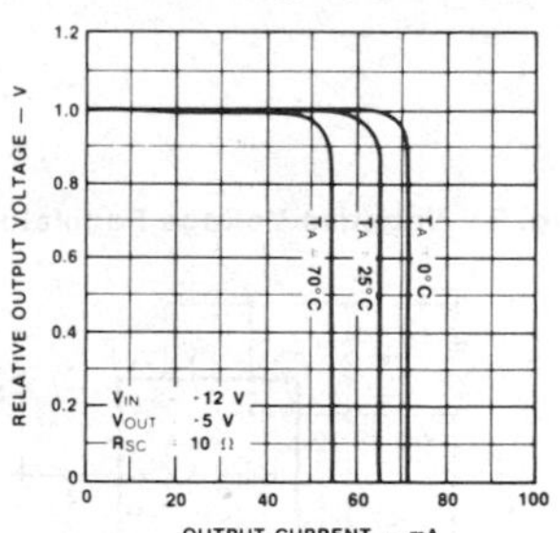

Maximum Load Current as a Function of Input/Output Voltage Differential

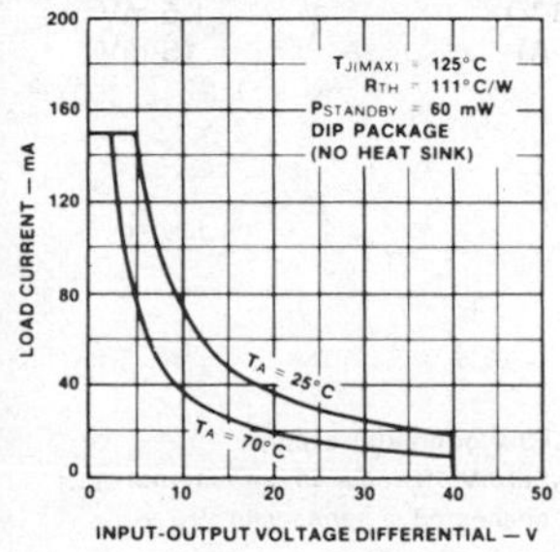

Load Regulation Characteristics With Current Limiting

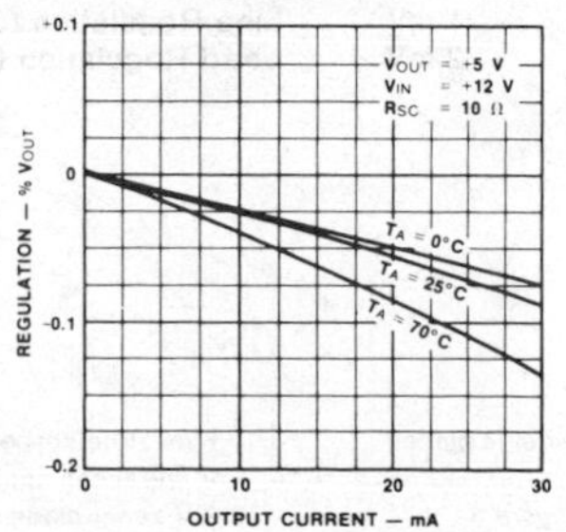

Standby Current Drain as a Function of Input Voltage

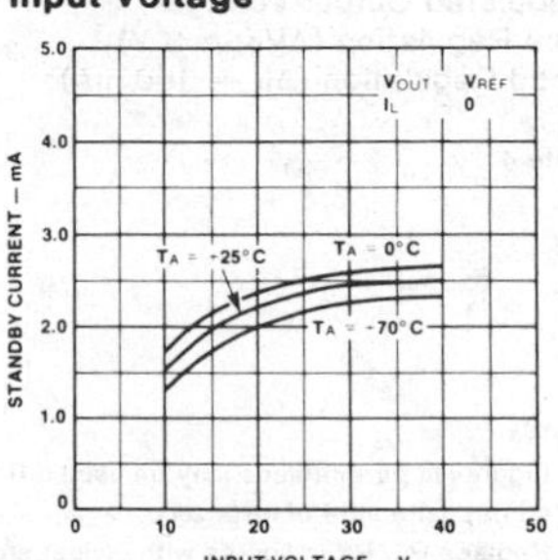

Typical Applications

Fig. 1 Basic Low Voltage Regulator (V_{OUT} = 2 to 7 V)

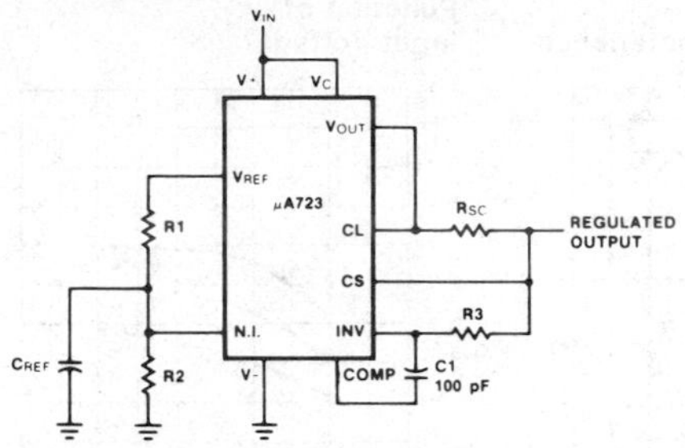

Typical Performance

Regulated Output Voltage	5 V
Line Regulation (ΔV_{IN} = 3 V)	0.5 mV
Load Regulation (ΔI_L = 50 mA)	1.5 mV

Note

$R_3 = \frac{R_1 R_2}{R_1 + R_2}$ for minimum temperature drift.

Fig. 2 Basic High Voltage Regulator (V_{OUT} = 7 to 37 V)

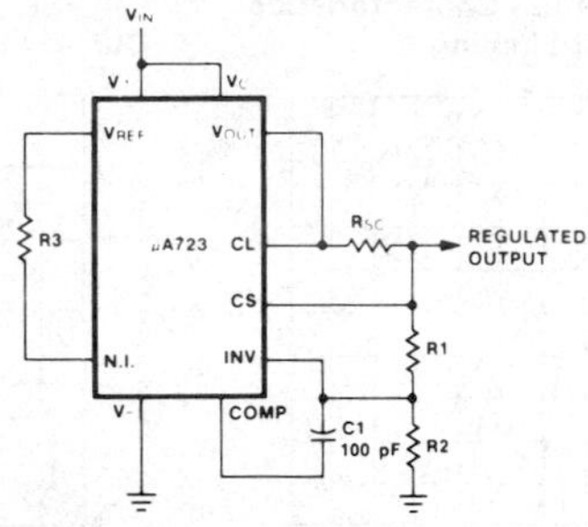

Typical Performance

Regulated Output Voltage	15 V
Line Regulation (ΔV_{IN} = 3 V)	1.5 mV
Load Regulation (ΔI_L = 50 mA)	4.5 mV

Note

$R_3 = \frac{R_1 R_2}{R_1 + R_2}$ for minimum temperature drift.

R_3 may be eliminated for minimum component count.

Fig. 3 Negative Voltage Regulator

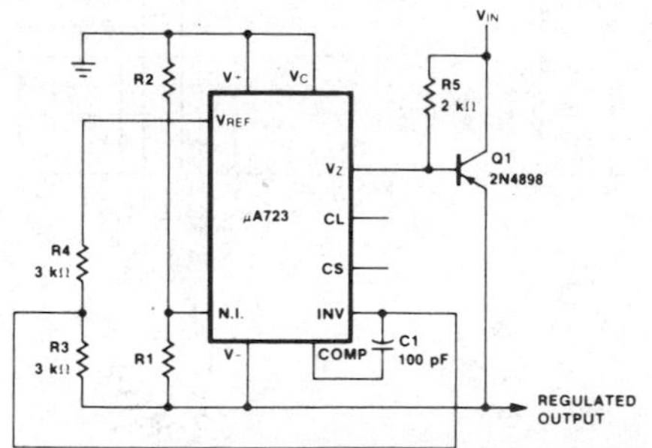

Typical Performance

Regulated Output Voltage	−15 V
Line Regulation (ΔV_{IN} = 3 V)	1 mV
Load Regulation (ΔI_L = 100 mA)	2 mV

Note 4

Fig. 4 Positive Voltage Regulator (External npn Pass Transistor)

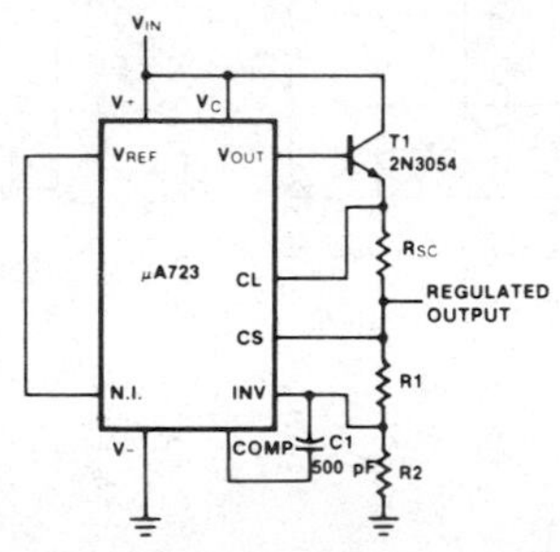

Typical Performance

Regulated Output Voltage	+15 V
Line Regulation (ΔV_{IN} = 3 V)	1.5 mV
Load Regulation (ΔI_L = 1 A)	15 mV

Notes

1. Figures in parentheses may be used if R1/R2 divider is placed on opposite side of error amp.
2. Replace R1/R2 in figures with divider shown in *Figure 8*.
3. V+ must be connected to a +3 V or greater supply.
4. For metal can applications where V_Z is required, an external 6.2 V zener diode should be connected in series with V_{OUT}.

Typical Applications (Cont.)

Fig. 5 Positive Voltage Regulator (External pnp Pass Transistor)

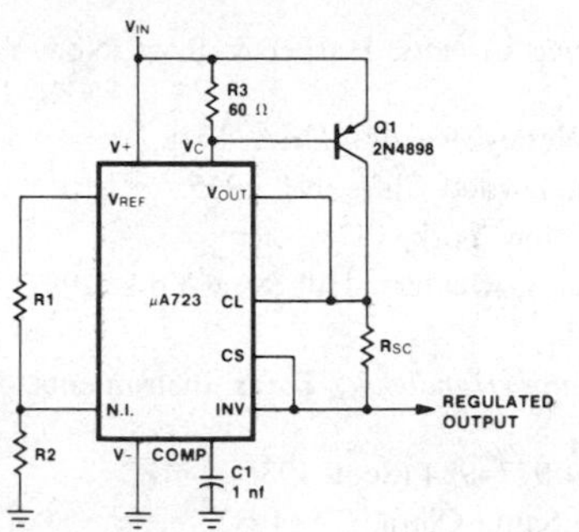

Typical Performance

Regulated Output Voltage	+5 V
Line Regulation (ΔV_{IN} = 3 V)	0.5 mV
Load Regulation (ΔI_L = 1 A)	5 mV

Fig. 6 Foldback Current Limiting

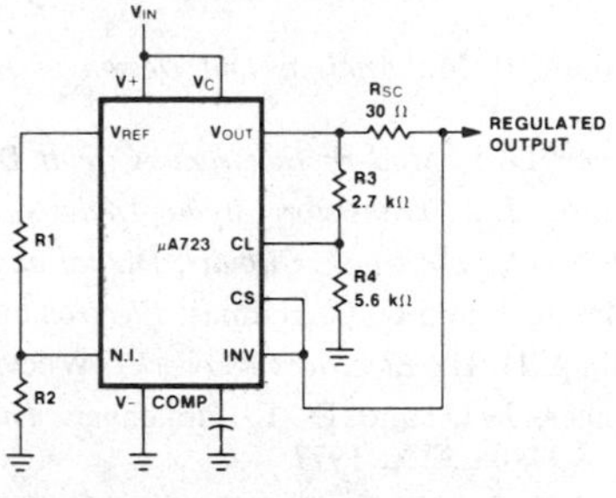

Typical Performance

Regulated Output Voltage	+5 V
Line Regulation (ΔV_{IN} = 3 V)	0.5 mV
Load Regulation (ΔI_L = 10 mA)	1 mV
Short-Circuit Current	20 mA

Fig. 7 Remote Shutdown Regulator with Current Limiting

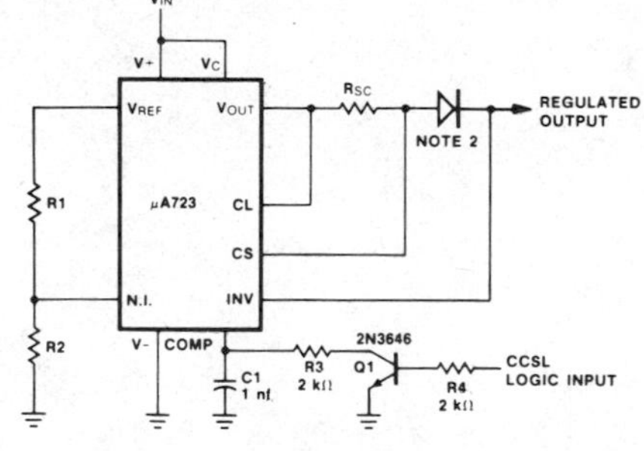

Fig. 8 Output Voltage Adjust

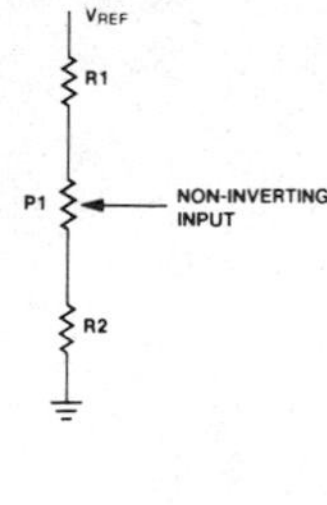

Notes

Current limit transistor may be used for shutdown if current limiting is not required. Add if $V_{OUT} > 10$ V

Typical Performance

Regulated Output Voltage	+5 V
Line Regulation (ΔV_{IN} = 3 V)	0.5 mV
Load Regulation (ΔI_L = 50 mA)	1.5 mV

Bibliography

Chirlian, P. M., *Electronic Circuits, Physical Principles, Analysis and Design*, McGraw-Hill, New York, 1971.

Chirlian, P. M., *Analysis and Design of Integrated Electronic Circuits*, Harper & Row, New York, 1981.

Corner, D. J., *Modern Electronic Circuit Design*, Addison-Wesley, Reading, MA, 1976.

Corning, J. J., *Transistor Circuit Analysis*, Prentice-Hall, Englewood Cliffs, NJ, 1965.

Holt, C. A., *Electronic Circuits, Digital and Analog*, Wiley, New York, 1978.

Millman, J. and C. C. Halhias, *Electronic Devices and Circuits*, McGraw-Hill, New York, 1967.

Shilling, H. H., *Electrical Networks*, Wiley, New York, 1978.

Spencer, J. D. and D. E. Pippenger, *The Voltage Regulator Handbook*, Texas Instruments, Inc., Dallas, TX, 1977.

Stout, M. B., "Analysis of Rectifier Circuits," *Elec. Eng*. **54**:977–984 (Sept. 1935).

Voltage Regulator Handbook, National Semiconductor Co., Santa Clara, CA, 1982.

Wilson, E. C. and R. T. Windecker, "DC Regulated Power Supply Design," *Sol. State J*. **2**(11):37–46 (Nov. 1961).

PART 6

DIGITAL CIRCUITS

CHAPTER 28
GATES

PETER GRAHAM

Florida Atlantic University, Boca Raton, Florida

28.1 INTRODUCTION

Chapter 6 introduced Boolean algebra and its application to logic design, including the formulation and simplification of switching functions. In this chapter attention is directed toward the implementation of switching functions using electronic gates. Truth tables and symbols for the commonly available gates are presented along with descriptive introductions of variations such as "three-state," "Schmitt trigger," and "open collector." Definitions of gate specification parameters are included to facilitate the comparison of gates manufactured using the various technologies like transistor-transistor logic (TTL), emitter-coupled logic (ECL), and complementary metal oxide semiconductor (CMOS) logic. The basic circuits of these technologies are analyzed and some recommendations are given regarding their choice and application. The chapter closes with a brief introduction to integrated injection logic (IIL) and charge-coupled devices (CCD).

28.2 GATE SYMBOLS

The basic building blocks of digital computers are electronic circuits whose input and output voltage levels can have only two possible values, one corresponding to a logical 0 and the other to a logical 1. The simplest of these circuits, whose output voltage levels are related to the input levels by the logical functions AND, OR, and INVERT, are called gates. It can be said that even the most complex digital circuits are, for the most part, interconnections of gate circuits.

The three basic functions (AND, OR, INVERT) are represented pictorially by the symbols given in Fig. 28.1. The truth tables of the operations are included for easy reference. (See Chapter 6 for details on Boolean algebra.) While the equations of Fig. 28.1 comprise the three basic logical functions, electronic gates more readily perform the combined operations NAND and NOR shown in Fig. 28.2.

It is common practice to replace the inverter symbol by a little circle (referred to as the bubble), as shown on the NAND and NOR gate symbols. Alternative symbols for NAND and NOR arise from the deMorgan theorem identities $\overline{A \cdot B} = \overline{A} + \overline{B}$ and $\overline{A + B} = \overline{A} \cdot \overline{B}$. These are illustrated in Fig. 28.3, which demonstrates a simple rule concerned with pushing bubbles through gates.[1] A bubble pushed from a gate output to the gate inputs appears at all gate inputs and changes the gate symbol

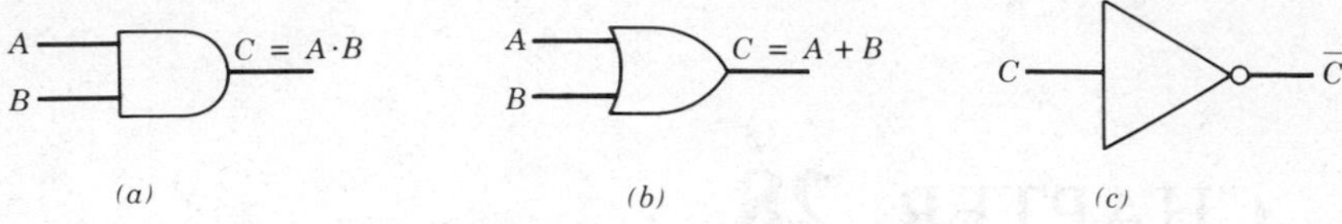

Fig. 28.1 Basic logical functions: (*a*) AND, (*b*) OR, (*c*) INVERT.

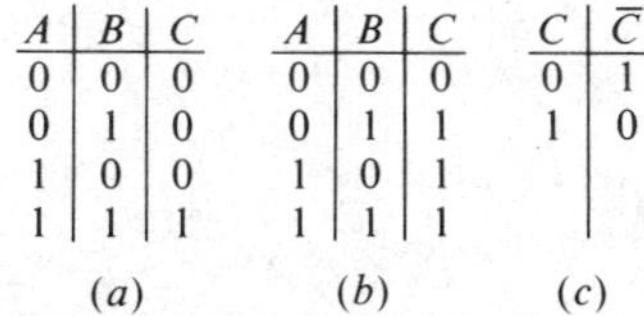

A	B	C
0	0	0
0	1	0
1	0	0
1	1	1

(*a*)

A	B	C
0	0	0
0	1	1
1	0	1
1	1	1

(*b*)

C	$\overline{C}$
0	1
1	0

(*c*)

A, B → C; $C = \overline{AB} = \overline{A} + \overline{B}$

(*a*)

A, B → C; $C = \overline{A + B} = \overline{A}\,\overline{B}$

(*b*)

Fig. 28.2 Combined operations: (*a*) NAND, (*b*) NOR.

A	B	C
0	0	1
0	1	1
1	0	1
1	1	0

(*a*)

A	B	C
0	0	1
0	1	0
1	0	0
1	1	0

(*b*)

$AB = C$ → $\overline{\overline{A} + \overline{B}} = AB = C$

(*a*)

$A + B = C$ → $\overline{\overline{A}\,\overline{B}} = A + B = C$

(*b*)

Fig. 28.3 Implementation of AND, OR gates in terms of NAND, NOR gates.

A	B	C
0	0	0
0	1	0
1	0	0
1	1	1

A	B	$\overline{A}$	$\overline{B}$	$\overline{A} + \overline{B}$	$\overline{\overline{A} + \overline{B}}$
0	0	1	1	1	0
0	1	1	0	1	0
1	0	0	1	1	0
1	1	0	0	0	1

(*a*)

A	B	C
0	0	0
0	1	1
1	0	1
1	1	1

A	B	$\overline{A}$	$\overline{B}$	$\overline{A}\,\overline{B}$	$\overline{\overline{A}\,\overline{B}}$
0	0	1	1	1	0
0	1	1	0	0	1
1	0	0	1	0	1
1	1	0	0	0	1

(*b*)

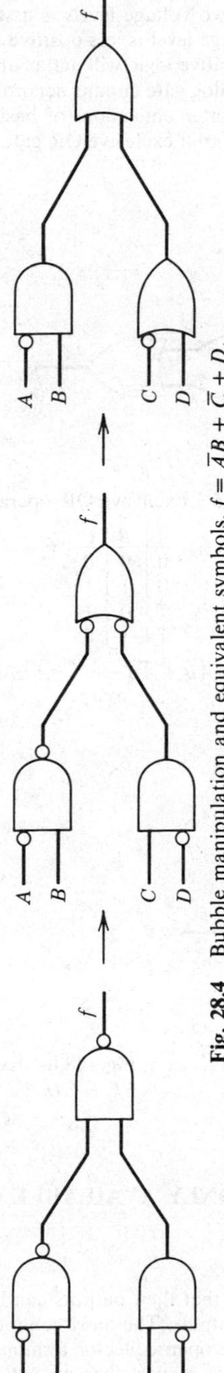

Fig. 28.4 Bubble manipulation and equivalent symbols. $f = \bar{A}B + \bar{C} + D$.

from OR to AND or from AND to OR. If an input in the original symbol has a bubble, this bubble is canceled by a bubble pushed through the gate from the output. An example of this technique is given in Fig. 28.4.

The bubble symbol also is used to differentiate between positive and negative logic. In positive logic systems the more positive of the two voltage levels is associated with the logical 1, whereas in negative logic systems the logical 1 voltage level is less positive than the logical 0 level. A gate circuit that performs the AND operation in positive logic will act as an OR gate for negative logic, and vice versa. It is a universal convention to catalog gate circuits according to their functions in positive logic.

Complex circuits, which in fact are interconnections of basic gates, are also classified as gates in data books.[2,3] Examples are the quad 2-input exclusive OR gate shown in Fig. 28.5 and the AND-OR gate of Fig. 28.6.

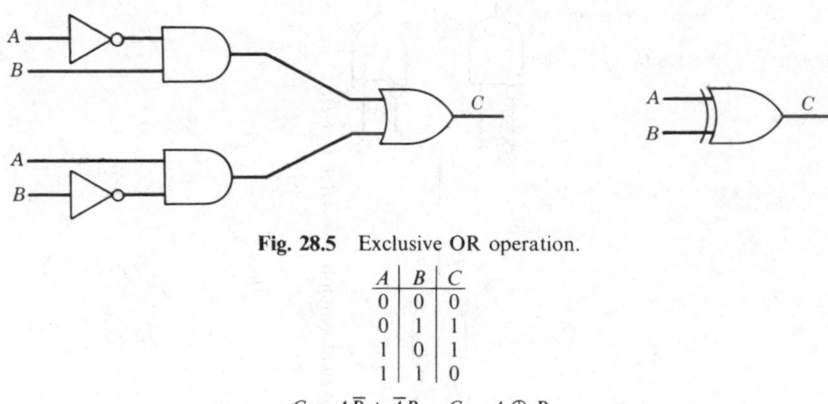

Fig. 28.5 Exclusive OR operation.

A	B	C
0	0	0
0	1	1
1	0	1
1	1	0

$$C = A\overline{B} + \overline{A}B \Rightarrow C = A \oplus B$$

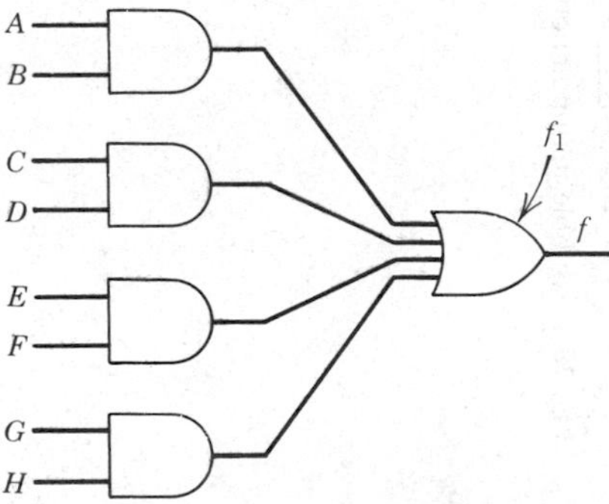

Fig. 28.6 Expandable gate. $f = f_1 + AB + CD + EF + GH$.

28.3 VARIATIONS OF COMMONLY AVAILABLE GATES

Wired Logic Gates

Some electronic gates are designed such that their outputs can be directly interconnected to generate a logic function of the individual gate outputs. The most common examples of this type of capability occur in gates whose output is either the open collector terminal or the open emitter terminal of the output transistor. Open collector outputs require that an external "pull-up" resistor be connected from the positive supply voltage to the collector of the output *npn* transistor. Several such outputs can share the same pull-up resistor, which results in the ANDing together of the individual output functions. Similarly, *npn* open emitter outputs can share a common "pull-down" resistor connected to the negative supply voltage, the individual outputs then becoming ORed together. In spite of the fact

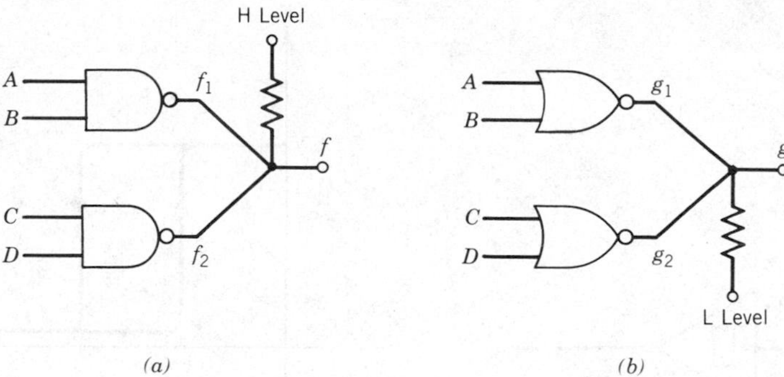

Fig. 28.7 Wired logic gates: (*a*) AND, $f = f_1 f_2 = (\overline{A} + \overline{B})(\overline{C} + \overline{D})$, (*b*) OR, $g = g_1 + g_2 = \overline{AB} + \overline{CD}$.

that the open collector connection implements the AND operation, any direct interconnection of outputs is often referred to as wired-OR logic.[1,4]

Wired logic must be used cautiously. The designer must make certain that the outputs are appropriate for direct interconnection and that the wired logic generates the desired function. As an additional precaution in asynchronous applications, only outputs of gates on the same chip should be wired, otherwise there is the possibility of timing problems arising from unequal gate delays. Examples of wired logic are illustrated in Fig. 28.7.

Three-State Logic Gates

In addition to having logical 0 and logical 1 outputs, some gates have a third condition wherein the output terminal presents a very high impedance to its load. In essence the third state, which is initiated by a separate input, disconnects the gate output from the load. Three-state devices are mandatory in systems where two or more outputs time share the same data line.

Any output can be converted to a three-state output by insertion of a three-state buffer between the output and its load. The buffer is equivalent to a switch that opens and closes in response to the logic level on its enable input. Three-state inverting buffers that are equivalent to an inverter cascaded with a three-state buffer also are available. Symbols for three-state buffers are shown in Fig. 28.8.

Fig. 28.8 Three-state logic symbols: (*a*) Three-state inverting buffer, (*b*) Three-state non-inverting buffer. E = enable input.

Schmitt Trigger Gates

Most gates have a single valued voltage-out versus voltage-in characteristic; that is, for any given set of voltage levels on the gate inputs there is a unique voltage level at the gate output. This is not true, however, for Schmitt trigger gates (Fig. 28.9). A Schmitt trigger inverter, for example, has an input-output transfer characteristic of the form given in Fig. 28.9*b*.[3] Observe that an output shift from high to low requires a higher level input voltage than that required to shift the output from low to high. This so-called hysteresis property of the transfer characteristic is accomplished by using positive feedback from the output to the input, which results in a very fast transition of the output

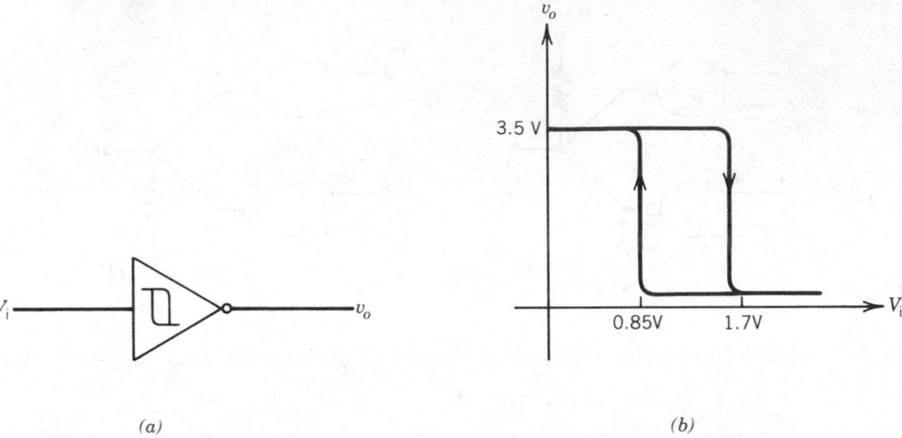

Fig. 28.9 Schmitt trigger inverter: (*a*) Logic symbol, (*b*) voltage transfer characteristic.

level in response to a relatively slow variation of the input level through a critical value. Schmitt trigger gates are used to exploit either the fast transition time or the improved immunity to noise. A multi-input Schmitt trigger gate consists of a normal implementation of the gate cascaded with a Schmitt trigger buffer at the output.

Expandable Gates

The AND-OR gate shown in Fig. 28.6 is an example of an expandable gate. Expandable gates have at least one input that is closer to the output than are the other inputs. The output OR gate of this example has a fifth input which is brought out to a pin of the integrated circuit package. This allows any user-designated logic function to be ORed together with the AND gate outputs.

28.4 GATE SPECIFICATION PARAMETERS

There are a variety of technologies for fabricating electronic gates in integrated circuits. The effect is that several different logic circuit families are available to the user. Which family is used is usually determined by the application, with speed, power consumption, and the number of circuits per unit chip area being the most important considerations. A number of parameters are defined and used in circuit specifications which assist the designer in making a choice.[2,3] In a majority of cases, logic circuits of any complexity are interconnections of basic gate circuits. The interconnections could be external to the gate chips, but for the most part, gates are interconnected as part of the fabricating process. Circuit complexity of a chip is specified using four rather broad classifications:

Small-Scale Integration (SSI). The inputs and outputs of every gate are available for external connection at the chip pins (with the exception that exclusive OR and AND-OR gates are considered SSI).

Medium-Scale Integration (MSI). Several gates are interconnected to perform somewhat more elaborate logic functions such as flip-flops, counters, multiplexers, and so on.

Large-Scale Integration (LSI). Several of the more elaborate circuits associated with MSI are interconnected within the integrated circuit to form a logic system on a single chip. Chips such as calculators, digital clocks, and small microprocessors are examples of LSI.

Very-Large-Scale Integration (VLSI). This designation is usually reserved for chips having a very high density, 1000 or more gates per chip.[5] These include the large single chip memories, gate arrays, and microcomputers.

Specifications of logic speed require definitions of switching times. These definitions can be found in the introductory pages of most data manuals. Four of them pertain directly to gate circuits. These

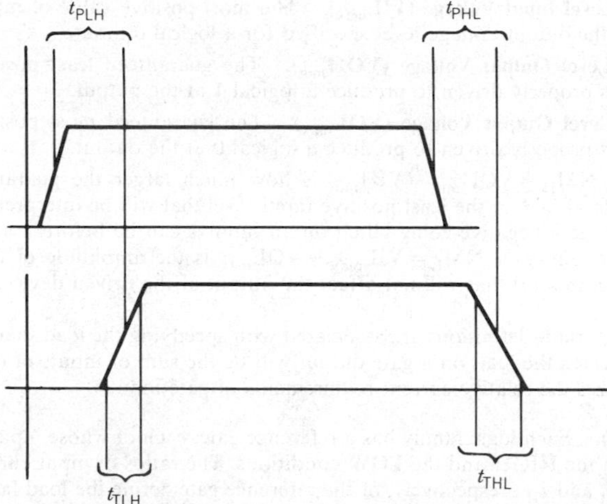

Fig. 28.10 Definitions of switching times.

are (see also Fig. 28.10):

LOW-to-HIGH Propagation Delay Time (t_{PLH}). The time between specified reference points on the input and output voltage waveforms when the output is changing from low to high.

HIGH-to-LOW Propagation Delay Time (t_{PHL}). The time between specified reference points on the input and output voltage waveforms when the output is changing from high to low.

Propagation Delay Time (t_{PD}). The average of the two propagation delay times: $t_{PD} = t_{PDH} + t_{PDL}$.

LOW-to-HIGH Transition Time (t_{TLH}). The rise time between specified reference points on the LOW to HIGH shift of the output waveform.

HIGH-to-LOW Transition Time (t_{THL}). The fall time between specified reference points on the HIGH to LOW shift of the output waveform. The reference points usually are 10% and 90% of the voltage level difference in each case.

Power consumption, driving capability, and effective loading of gates are defined in terms of currents.

Supply Current, Outputs High (I_{xxH}). The current delivered to the chip by the power supply when all outputs are open and at the logical 1 level. The xx subscript depends on the technology.

Supply Current, Outputs Low (I_{xxL}). The current delivered to the chip by the supply when all outputs are open and at the logical 0 level.

Supply Current, Worst Case (I_{xx}). When the output level is unspecified, the input conditions are assumed to correspond to maximum supply current.

Input HIGH Current (I_{IH}). The current flowing into an input when the specified HIGH voltage is applied.

Input LOW Current (I_{IL}). The current flowing into an input when the specified LOW voltage is applied.

Output HIGH Current (I_{OH}). The current flowing into the output when it is in the HIGH state.

Output LOW Current (I_{OL}). The current flowing into the output when it is in the LOW state.

The most important voltage definitions are concerned with establishing ranges on the logical 1 (HIGH) and logical 0 (LOW) voltage levels.

Minimum High-Level Input Voltage (VIH_{min}). The least positive value of input voltage guaranteed to result in the output voltage level specified for a logical 1 input.

Maximum Low-Level Input Voltage (VIL_{max}). The most positive value of input voltage guaranteed to result in the output voltage level specified for a logical 0 input.

Minimum High-Level Output Voltage (VOH_{min}). The guaranteed least positive output voltage when the input is properly driven to produce a logical 1 at the output.

Maximum Low-Level Output Voltage (VOL_{max}). The guaranteed most positive output voltage when the input is properly driven to produce a logical 0 at the output.

Noise Margins. $NM_H = VOH_{min} - VIH_{min}$ is how much larger the guaranteed least positive output logical 1 level is than the least positive input level that will be interpreted as a logical 1. It represents how large a negative-going glitch on an input 1 can be before it affects the output of the driven device. Similarly $NM_L = VIL_{max} - VOL_{max}$ is the amplitude of the largest positive-going glitch on an input 0 that will not affect the output of the driven device.

Finally, three important definitions are associated with specifying the load that can be driven by a gate. Since in most cases the load on a gate output will be the sum of inputs of other gates, the first definition characterizes the relative current requirements of gate inputs.

Load Factor (LF). Each logic family has a reference gate, each of whose inputs is defined to be a unit load in both the HIGH and the LOW conditions. The ratios of input currents I_{IH} and I_{IL} of other gates to I_{IH} and I_{IL}, respectively, of the reference gate define the load factors of those gates. The second definition characterizes the relative driving abilities of gate outputs.

Drive Factor (DF). A gate output has drive factors for both the HIGH and the LOW output conditions. These factors are defined as the respective ratios of I_{OH} and I_{OL} of the gate to I_{IH} and I_{IL} of the reference gate.

Fan-Out. For a given gate the fan-out is defined as the maximum number of inputs of the same type of gate that can be properly driven by that gate output. When gates of different load factors are interconnected, fan-out must be adjusted accordingly. For example, a gate with HIGH/LOW drive factors of 25/12.5 driving gates with load factors of 1.0/1.0 will allow a fan-out of 12 ($12.5/1 \rightarrow 12$), corresponding to the smaller of the HIGH and LOW drive factor:load factor ratios. If the gate with the drive factors 25/12.5 has load factors of 1.25/1.25, the defined fan-out for that particular gate family will be 10 ($12.5/1.25 \rightarrow 10$).

28.5 BIPOLAR TRANSISTOR GATES

A logic circuit using bipolar junction transistors (BJTs) can be classified either as saturated logic or as nonsaturated logic. A saturated logic circuit contains at least one BJT that is saturated in one of the stable modes of the circuit. In nonsaturated logic circuits none of the transistors is allowed to saturate. Since bringing a BJT out of saturation requires a few additional nanoseconds (called the storage time),[4,6] nonsaturated logic is faster. The fastest circuits available at this time are emitter-coupled logic (ECL), with transistor-transistor logic (TTL) having Schottky diodes connected to prevent the transistors from saturating (Schottky TTL) being a fairly close second. Both of these families are nonsaturated logic. All TTL families other than Schottky are saturated logic.

Transistor-Transistor Logic (TTL)

TTL evolved from resistor-transistor logic (RTL) through the intermediate step of diode-transistor logic (DTL).[4,7] All three families are cataloged in data books published in 1968,[8] but only TTL is still available.

The basic circuit of the standard TTL family is typified by the 2-input NAND gate shown in Fig. 28.11*a*. To estimate the operating levels of voltage and current in this circuit, assume that any transistor in saturation has $V_{CE} = 0.2$ V and $V_{BE} = 0.75$ V. Let drops across conducting diodes also be 0.75 V and transistor current gains (when nonsaturated) be about 50.

As a starting point, let the voltage levels at both inputs A and B be high enough that T_1 operates in the reversed mode. In this case the emitter currents of T_1 are negligible,[4] and the current into the base of T_1 goes out the collector to become the base current of T_2. This current is readily calculated by observing that the base of T_1 is at $3 \times 0.75 = 2.25$ V so there is a 2.75-V drop across the 4 kΩ resistor. Thus $I_{B1} = I_{B2} = 0.7$ mA, and it follows that T_2 is saturated. With T_2 saturated, the base of T_3 is at $V_{CE2} + V_{BE4} = 0.95$ V. If T_4 is also saturated, the emitter of T_3 will be at $V_{D3} + V_{CE4} = 0.95$ V, and T_3 will be cut off. The voltage across the 1.6-kΩ resistor is $5 - 0.95 = 4.05$ V, so the collector current of T_2 is about 2.5 mA. This means the emitter current of T_2 is 3.2 mA. Of this, 0.75 mA goes through the 1-kΩ resistor, leaving 2.45 mA as the base current of T_4. Since the current gain of T_4 is about 50, it will be well into saturation for any collector current less than 100 mA, and the output at

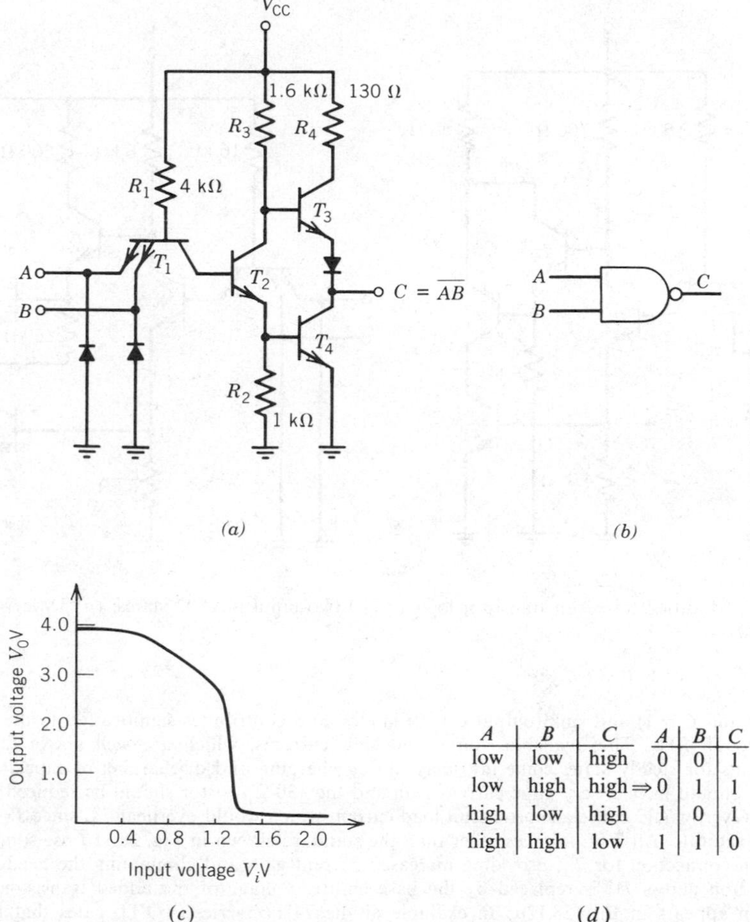

A	B	C	A	B	C
low	low	high	0	0	1
low	high	high ⇒	0	1	1
high	low	high	1	0	1
high	high	low	1	1	0

Fig. 28.11 Two-input transistor-transistor logic (TTL) NAND gate type 7400: (*a*) Circuit, (*b*) symbol, (*c*) voltage transfer characteristic (V_i to both inputs), (*d*) truth table.

C is a logic 0. The corresponding levels required at the inputs are estimated from $V_{BE4} + V_{BE2} + V_{EC1}$, or about 1.7 V.

Now let either or both of the inputs be dropped to 0.2 V. T_1 is then biased to saturation in the normal mode, so the collector current of T_1 extracts the charge from the base region of T_2. With T_2 cut off, the base of T_4 is at 0 V and T_4 is cut off. T_3 will be biased on by the current through the 1.6-kΩ resistor (R_3) to a degree regulated by the current demand at the output C. The drop across R_3 is quite small for light loads, so the output level at C will be $V_{CC} - V_{BE3} - V_{D3}$, which will be about 3.5 V corresponding to the logical 1.

The operation is summarized in the truth table of Fig. 28.11*c*, identifying the circuit as a 2-input NAND gate. The input-output voltage transfer characteristic (Fig. 28.11*d*), where V_i is applied to inputs A and B simultaneously, is derived in detail elsewhere.[4] The sloping portion of the characteristic between $V_i = 0.55$ and 1.2 V corresponds to T_2 passing through the active region in going from cutoff to saturation.

Diodes D_1 and D_2 are present to damp out "ringing" that can occur, for example, when fast voltage level shifts are propagated down an appreciable length (20 cm or more) of microstripline formed by printed circuit board interconnections. Negative overshoots are clamped to the 0.7 V across the forward-biased diode.

The series combination of the 130-Ω resistor T_3, D_3, and T_4 in the circuit of Fig. 28.11*a*, forming what is referred to as the totem-pole output circuit, provides a low impedance drive in both the

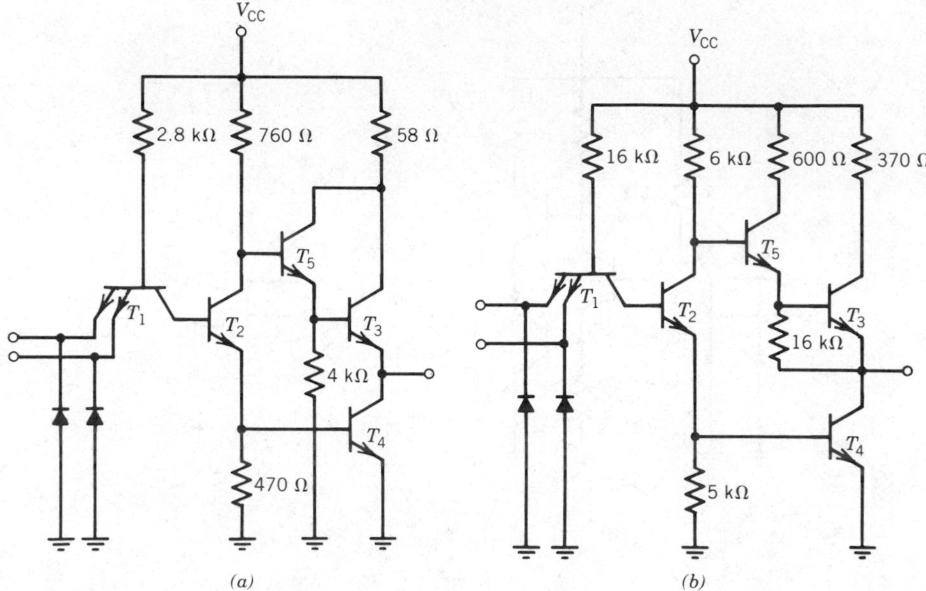

Fig. 28.12 Modified transistor transistor logic (TTL) two-input NAND states: (*a*) Type 74H00, (*b*) type 74L00.

source (output $C = 1$) and sink (output $C = 0$) modes, and contributes significantly to the relatively high speed of TTL. The available source and sink currents, which are well above the normal requirements for steady state, come into play during charging and discharging of capacitive loads. Ideally T_3 should have a very large current gain and the 130-Ω resistor should be reduced to 0. The latter, however, would cause a short-circuit load current which would overheat T_3, since T_3 would be unable to saturate. All TTL families other than the standard shown in Fig. 28.11*a* use some form of Darlington connection for T_3, providing increased current gain and eliminating the need for diode D_3. The drop across D_3 is replaced by the base emitter voltage of the added transistor T_5. This connection appears in Fig. 28.12*a*, an example of the 74Hxx series of TTL gates that trades off increased power consumption for increased speed, and in Fig. 28.12*b*, a gate from the 74Lxx series that sacrifices speed to lower power dissipation.[2]

All TTL families have examples of open collector gates, referred to in the section on wired logic gates. The 7403 2-input open collector NAND gate circuit appears in Fig. 28.13. The appropriate

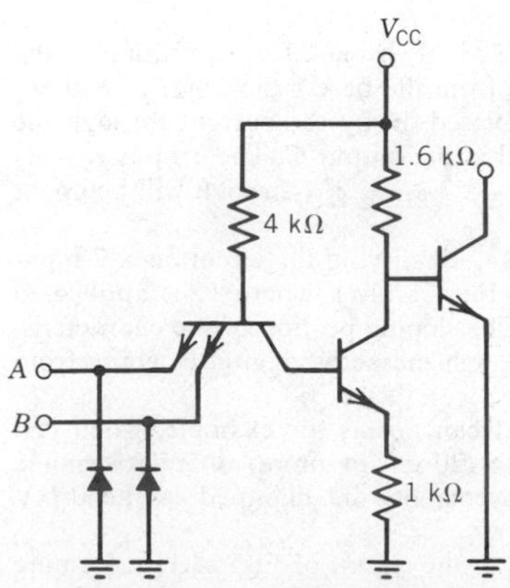

Fig. 28.13 Open collector two-input NAND gate.

value for the external pull-up resistor depends on the gate and the application. The maximum dissipation allowed for the open collector transistor sets a lower limit on the resistance value. For wired-AND applications the resistance range depends on how many outputs are being wired and on the load being driven by the wired outputs. Formulas are given in the data books.[2,9] Since the open collector configuration does not have the speed enhancement associated with an active pull-up, the low to high propagation delay (t_{PLH}) is about double that of the totem-pole output. It should be observed that totem-pole outputs should not be wired, since excessive currents in the active pull-up circuit could result.

Nonsaturated TTL. Two TTL families, the Schottky (74Sxx) and the low-power Schottky (74LSxx), can be classified as nonsaturating logic. The transistors in these circuits are kept out of saturation by the connection of Schottky diodes with the anode to the base and the cathode to the collector.[4] Schottky diodes are formed from junctions of metal and an n-type semiconductor, the metal fulfilling the role of the p-region. Since there are thus no minority carriers in the region of the forward-biased junction, the storage time required to bring a pn junction out of saturation is eliminated. The forward-biased drop across a Schottky diode is around 0.3 V. This clamps the collector at 0.3 V less than the base, maintaining V_{CE} above the saturation threshold of 0.3 V. Circuits for the 2-input NAND gates 74LS00 and 74S00 are given in Fig. 28.14.[2,3,9] The special transistor symbol is a short-form notation indicating the presence of the Schottky diode.

Note that both of these circuits have an active pull-down transistor T_6 replacing the pull-down resistance connected to T_2 in Fig. 28.12. The addition of T_6 decreases the turn-on and turn-off times of T_4. In addition, the transfer characteristic for these devices is improved by the squaring off of the sloping region between $V_i = 0.55$ and 1.2 V (see Fig. 28.11*d*). This happens because T_2 cannot become active until T_6 turns on, which requires at least 1.2 V at the input.[4]

The diode AND circuit of the 74LS00 in place of the multiemitter transistor will permit maximum input levels substantially higher than the 5.5 V specified for all other TTL families.[2,9] Input leakage currents for 74LSxx are specified at $V_i = 10$ V, and input voltage levels up to 15 V are allowed. The 74LSxx has the additional feature of the Schottky diode D_1 in series with the 100-Ω output resistor. This allows the output to be pulled up to 10 V without causing a reverse breakdown of T_5. The relative characteristics of the several versions of the TTL 2-input NAND gate are compared in Table 28.1. The 74F00 represents one of the new technologies that have introduced improved Schottky TTL in recent years.[10,11]

TTL Design Considerations. Before undertaking construction of a logic system, the wise designer consults the information and recommendations provided in the data books of most manufacturers.[2,3,9] Some of the more significant tips are provided here for easy reference.

1. Power supply, decoupling, and grounding. The power supply voltage should be 5 V with less than 5% ripple factor and better than 5% regulation. When packages on the same printed circuit board are supplied by a bus there should be a 0.05-μF decoupling capacitor for every five to ten packages between the bus and the ground. If a ground bus is used, it should be as wide as possible and should surround all the packages on the board. Use of a ground plane, if possible, is very desirable. If a long ground bus is used, both ends must be tied to the common system ground point.

2. Unused gates and inputs. If a gate on a package is not used, its inputs should be tied either high or low, whichever results in the least supply current. For example, the 7400 draws three times the current with the output low as with the output high, so the inputs of an unused 7400 gate should be grounded. An unused input of a gate, however, must be connected so as not to affect the function of the remaining inputs. For a 7400 NAND gate, such an input must either be tied high or paralleled with a used input. It must be recognized that paralleled inputs count as two when determining the fan-out. Inputs that are tied high can be connected either to V_{CC} through a 1-kΩ or more resistance (for protection from supply voltage surges) or to the output of an unused gate whose input will establish a permanent output high. Several inputs can share a common protective resistance. Low-power Schottky TTL requires no resistance, since 74LSxx inputs tolerate up to 15 V without breakdown. If inputs of low-power Schottky are connected in parallel and driven as a single input, the switching speed is decreased, in contrast to the situation with other TTL families.[11]

3. Interconnection. Use of line lengths of up to 10 in. (5 in. for 74S) requires no particular precautions, except that in some critical situations lines cannot run side by side for an appreciable distance without causing cross talk due to capacitive coupling between them. For transmission line connections, a gate should drive only one line, and a line should be terminated in only one gate input. If overshoots are a problem, a 25- to 50-Ω resistor should be used in series with the driving gate input

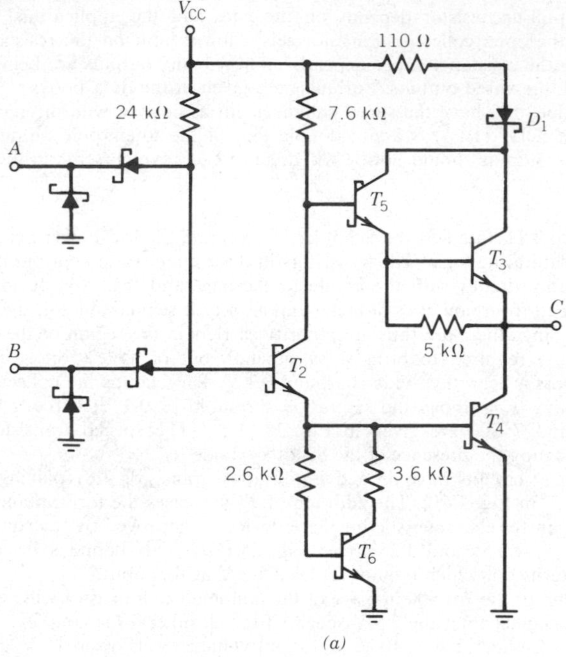

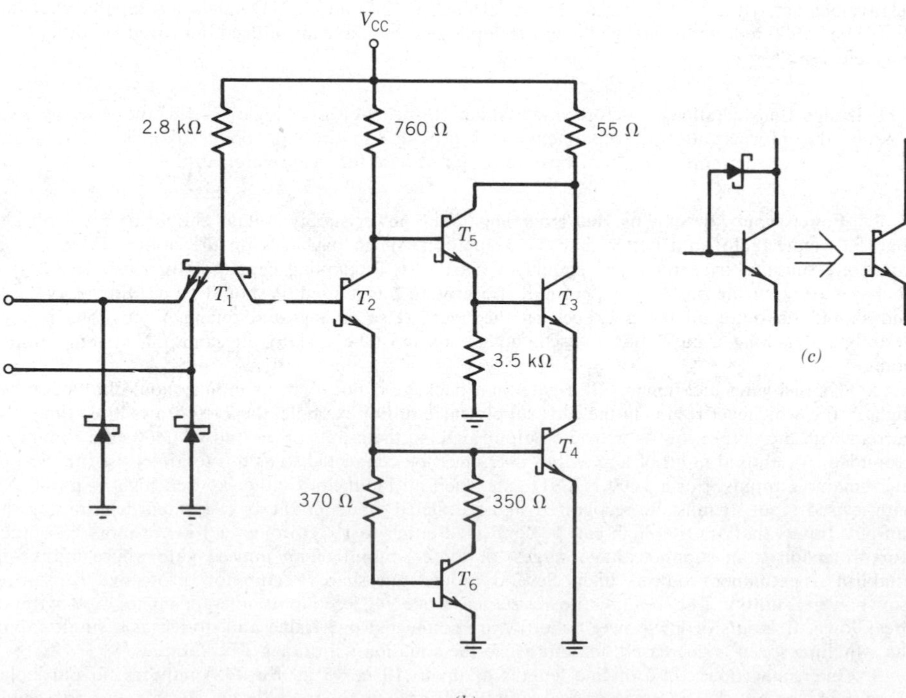

Fig. 28.14 Transistor-transistor logic (TTL) nonsaturated logic: (*a*) Type 74LS00 two-input NAND gate, (*b*) type 74S00 two-input NAND gate, (*c*) significance of the Schottky transistor symbol.

TABLE 28.1. COMPARISON OF TRANSISTOR-TRANSISTOR LOGIC (TTL) 2-INPUT NAND GATES

	Supply Current		Propagation Delay Time		Noise Margins				
TTL Type	I_{CCH}[a] (mA)	I_{CCL} (mA)	t_{PLH} (ns)	t_{PHL} (ns)	NM_H (V)	NM_L (V)	Load Factor, H/L	Drive Factor, H/L	Fan-Out
74F00	2.8	10.2	2.9	2.6	0.7	0.3	0.5/0.375	25/12.5	33
74S00	10	20	3	3	0.7	0.3	1.25/1.25	25/12.5	10
74H00	10	26	5.9	6.2	0.4	0.4	1.25/1.25	12.5/12.5	10
74LS00	0.8	2.4	9	10	0.7	0.3	0.5/0.25	10/5	20
7400	4	12	11	7	0.4	0.4	1/1	20/10	10
74L00	0.44	1.16	35	31	0.4	0.5	0.25/0.1125	5/2.25	20

[a] See text for explanation of abbreviations.

and the receiving gate input pulled up to 5 V through a 1-kΩ resistor. Driving and receiving gates should have their own decoupling capacitors between the V_{CC} and ground pins. Parallel lines should have a grounded line separating them to avoid cross talk.

4. Mixing TTL subfamilies. Even synchronous sequential systems often have asynchronous features such as reset, preset, load, and so on. Mixing high-speed 74S TTL with lower speed TTL (74LS for example) in some applications can cause timing problems resulting in anomalous behavior; such mixing is to be avoided with rare exceptions. The exceptions must be carefully analyzed.

Emitter Coupled Logic (ECL)

ECL is a nonsaturated logic family where saturation is avoided by operating the transistors in the common collector configuration. This feature, in combination with a smaller difference between the HIGH and LOW voltage levels (less than 1 V) than other logic families, makes ECL the fastest logic available at this time. The circuit diagram of a widely used version of the basic 2-input ECL gate is given in Fig. 28.15.[12] The power supply terminals V_{CC1}, V_{CC2}, V_{EE}, and V_{TT} are available for flexibility in biasing. In normal operation V_{CC1} and V_{CC2} are connected to a common ground, V_{EE} is biased to -5.2 V, and V_{TT} is biased to -2 V. With these values the nominal voltage for the logical 0 and 1 are, respectively, -1.75 and -0.9 V.[12] Operation with the V_{CC} terminals grounded maximizes the immunity from noise interference.[4]

A brief description of the operation of the circuit will verify that none of the transistors saturates. For the following discussion, V_{CC1} and V_{CC2} are grounded, V_{EE} is -5.2 V, and V_{TT} is -2 V. Diode drops and base emitter voltages of active transistors are 0.8 V.

First observe that the resistor-diode (D_1 and D_2) voltage divider establishes a reference voltage of -0.55 V at the base of T_3, which translates to -1.35 V at the base of T_2. When either or both of the inputs A and B are at the logical 1 level of -0.9 V, the emitters of T_{1A}, T_{1B}, and T_2 will be 0.8 V lower, at -1.7 V. This establishes the base emitter voltage of T_2 at $-1.35-(-1.7)=0.35$ V, so T_2 is cut off. With T_2 off, T_4 is biased into the active region, and its emitter will be at about -0.9 V, corresponding to a logical 1 at the $(A+B)$ output. Most of the current through the 365-Ω emitter resistor, which is $[-1.7-(-5.2)]/0.365=9.6$ mA, flows through the 100-Ω collector resistor, dropping the base voltage of T_5 to -0.96 V. Thus the voltage level at the output terminal designated $\overline{(A+B)}$ is -1.76 V corresponding to a logical 0.

When both A and B inputs are at the LOW level of -1.75 V, T_2 will be active, with its emitter voltage at $-1.35-0.8=-2.15$ V. The current through the 365-Ω resistor becomes $[-2.15-(-5.2)]/0.365=8.2$ mA. This current flows through the 112-Ω resistor pulling the base of T_4 down to -0.94 V, so that the $(A+B)$ output will be at the LOW level of -1.75 V. With T_{1A} and T_{1B} cut off, the base of T_5 is close to 0 V, and the $\overline{(A+B)}$ output will therefore be at the nominal HIGH level of -0.9 V.

Observe that the output transistors T_4 and T_5 are always active and function as emitter followers, providing the low-output impedances required for driving capacitive loads. As T_{1A} and/or T_{1B} turn on, and T_2 turns off as a consequence, the transition is accomplished with very little current change in the 365-Ω emitter resistor. It follows that the supply current from V_{EE} does not undergo the sudden

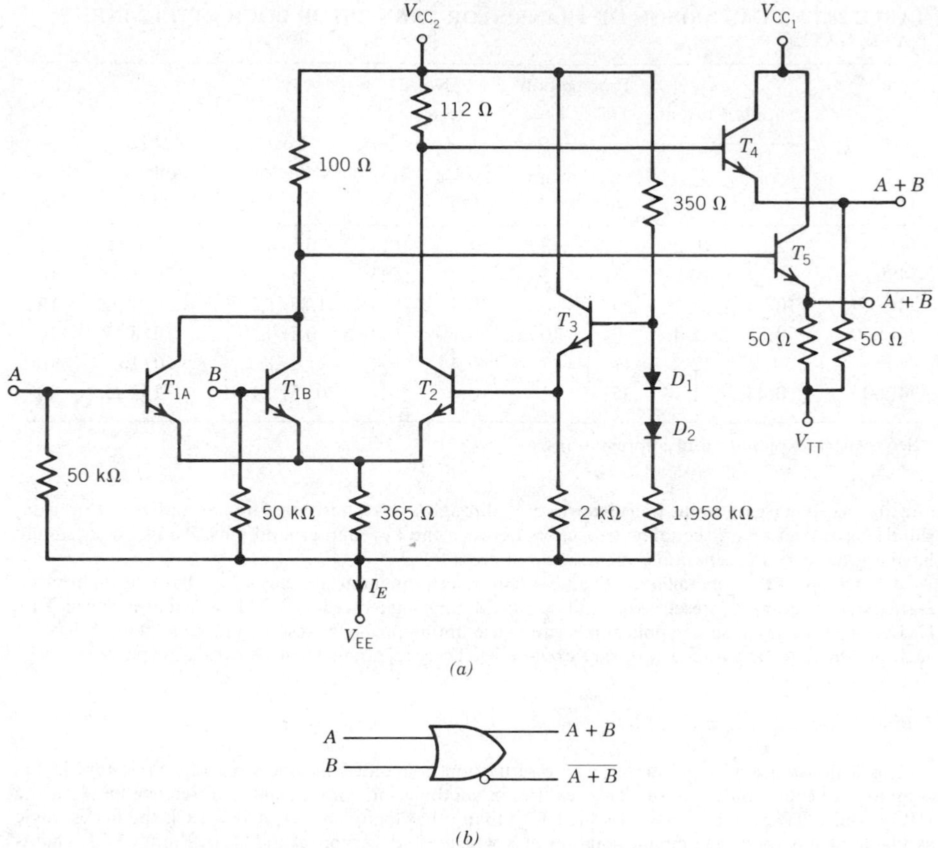

Fig. 28.15 Emitter-coupled logic basic gate (ECL 10102): (*a*) Circuit, (*b*) symbol.

increases and decreases prevalent in TTL, thus eliminating the need for decoupling capacitors. This is a major reason why ECL can be operated successfully with the low noise margins which are inherent in logic having a relatively small difference between the HIGH and LOW voltage levels (see Table 28.2). The relatively small level shifts between LOW and HIGH also permit low propagation times without excessively fast rise and fall times. This reduces the effects of residual capacitive coupling between gates, thereby lessening the required noise margin. For this reason the faster ECL (100xxx) should not be used where the speed of the 10xxx series is sufficient. A comparison of three ECL series is given in Table 28.2. The propagation times t_{PLH} and t_{PHL} and transition times t_{TLH} and t_{THL} are defined in Fig. 28.10. Transition times are from 20 to 80% levels.

The 50-Ω pull-down resistors shown in Fig. 28.15 are connected externally. The outputs of several gates can therefore share a common pull-down resistor to form a wired-OR connection. The open emitter outputs also provide flexibility for driving the transmission lines, the use of which in most cases is mandatory for interconnecting this high-speed logic. A twisted pair interconnection can be driven using the complementary outputs $(A + B)$ and $(\overline{A + B})$ as a differential output. Such a line should be terminated in an ECL line receiver (10114).[13]

Since ECL is used in high-speed applications, special techniques must be applied in the layout and interconnection of chips on circuit boards. Users should consult design handbooks published by the suppliers before undertaking the construction of an ECL logic system.[14]

While ECL is not compatible with any other logic family, interfacing buffers, called translators, are available. In particular, the 10124 converts TTL output levels to ECL complementary levels, and the 10125 converts either single-ended or differential ECL outputs to TTL levels. Among other applications of these translators, they allow the use of ECL for the highest speed requirements of a system while the rest of the system uses the more rugged TTL. Another translator is the 10177, which

TABLE 28.2. COMPARISON OF EMITTER-COUPLED LOGIC (ECL) QUAD 2-INPUT NOR GATES ($V_{TT} = V_{EE} = 5.2$ V, $V_{CC1} = V_{CC2} = 0$ V)

ECL Type	Power Supply Terminal V_{EE} (V)	Power Supply Current I_E (mA)	Propagation Delay Time t_{PLH}[a] (ns)	t_{PHL} (ns)	Transition Time t_{TLH}[b] (ns)	t_{THL}[b] (ns)	Noise Margins NM_H (V)	NM_L (V)	Test Load
ECL II									
1012	− 5.2	18[c]	5	4.5	4	6	0.175	0.175	Fan-out of 3
95102	− 5.2	11	2	2	2	2	0.14	0.145	50 Ω
10102	− 5.2	20	2	2	2.2	2.2	0.135	0.175	50 Ω
ECL III									
1662	− 5.2	56[c]	1	1.1	1.4	1.2	0.125	0.125	50 Ω
100102[d]	− 4.5	55	0.75	0.75	0.7	0.7	0.14	0.145	50 Ω
11001[e]	− 5.2	24	0.7	0.7	0.7	0.7	0.145	0.175	50 Ω

[a] See text for explanation of abbreviations.

[b] 20% to 80% levels.

[c] Maximum value (all other values typical).

[d] Quint 2-input NOR/OR gate.

[e] Dual 5/4-input NOR/OR gate.

converts the ECL output levels to *n*-channel metal oxide semiconductor (NMOS) levels. This is designed for interfacing ECL with *n*-channel memory systems.

28.6 COMPLEMENTARY METAL OXIDE SEMICONDUCTOR (CMOS) LOGIC

Metal oxide semiconductor (MOS) technology is prevalent in LSI systems due to the high circuit densities possible with these devices. *p*-Channel MOS was used in the first LSI systems, and it still is the cheapest to produce because of the higher yields achieved due to the longer experience with PMOS technology. PMOS, however, is largely being replaced by NMOS (*n*-channel MOS), which has the advantages of being faster (since electrons have greater mobility than holes) and having TTL compatability.[5] In addition, NMOS has a higher function/chip area density than PMOS; the highest density in fact of any of the current technologies. Use of NMOS and PMOS, however, is limited to LSI and VLSI fabrications. The only MOS logic available as SSI and MSI is CMOS (complementary MOS).

CMOS is faster than NMOS and PMOS, and it uses less power per function than any other logic. While it is suitable for LSI, it is more expensive and requires somewhat more chip area than NMOS or PMOS. In many respects it is unsurpassed for SSI and MSI applications. It is as fast as standard TTL, and has the largest noise margin of any logic type. Being compatible with low-power Schottky TTL, its inability to drive high current demand loads ($I_{OL_{max}} = 0.36$ mA) is not a serious limitation.

Either the NAND gate or the NOR gate qualifies as the basic 2-input CMOS gate, since the circuits of both of these are equally simple (see Fig. 28.16). All the field effect transistors are enhancement types which are cut off when the gate to source voltage is 0 V. Referring to Fig. 28.16*a*, and assuming that V_{SS} is at ground potential, observe that with A and B both at a positive level (logical 1), the *n*-channel FETs T_3 and T_4 will both be on and the *p*-channel FETs T_1 and T_2 will both be off. This causes the output C to be 0 V (logical 0). If A and/or B are low, T_4 and/or T_3 will be off, and T_1 and/or T_2 will be on. This disconnects the output C from ground and connects it to V_{DD} (logical 1). Thus the circuit corresponds to a 2-input NAND gate for positive logic. Similar reasoning establishes the circuit of Fig. 28.16*b* as a 2-input positive logic NOR gate.

A unique advantage of CMOS is that for all input combinations the steady state current from V_{DD} to V_{SS} is almost zero because at least one of the series FETs is open. Since CMOS circuits of any complexity are interconnections of the basic gates, the quiescent currents for these circuits are extremely small, an obvious advantage which becomes a necessity for the practicality of digital watches, for example, and one which alleviates heat dissipation problems in high-density chips. Also

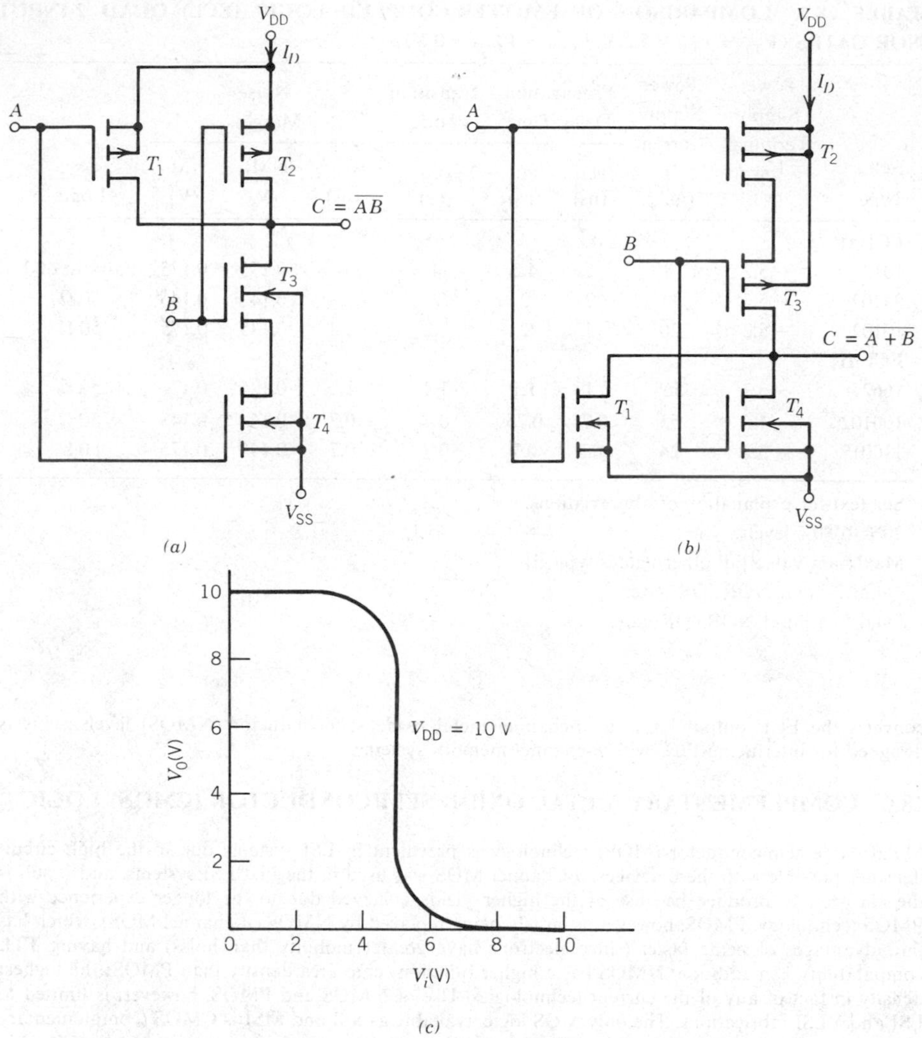

Fig. 28.16 Complementary metal oxide semiconductor (CMOS) NAND gate (*a*), NOR gate (*b*), and inverter transfer characteristic (*c*).

a noteworthy feature of CMOS digital circuits is the absence of components other than FETs. This attribute, which is shared by PMOS and NMOS, accounts for the much higher function/chip area density than is possible with TTL or ECL. During the time the output of a CMOS gate is switching there will be current flow from V_{DD} to V_{SS}, partly due to the charging of junction capacitances and partly because the path between V_{DD} and V_{SS} closes momentarily as the FETs turn on and off.[4] This causes the supply current to increase in proportion to the switching frequency in a CMOS circuit. Manufacturers specify that the supply voltage for standard CMOS can range over 3 V $\leqslant V_{DD} - V_{SS} \leqslant$ 18 V, but switching speeds are slower at the lower voltages, mainly due to the increased resistances of the "on" transistors. The output switches between low and high when the input is midway between V_{DD} and V_{SS}, and the output logical 1 level will be V_{DD} and the logical 0 level V_{SS} (Fig. 28.16*c*). If CMOS is operated with $V_{DD} = 5$ V and $V_{SS} = 0$ V, the V_{DD} and V_{SS} levels will be almost compatible with TTL except that the TTL totem pole output high of 3.4 V is marginal as a logical 1 for CMOS. To alleviate this, when CMOS is driven with TTL a 3.3-kΩ pull-up resistor between the TTL output and the common V_{CC}, V_{DD} supply terminal should be used. This raises V_{OH} of the TTL output to 5 V.[13]

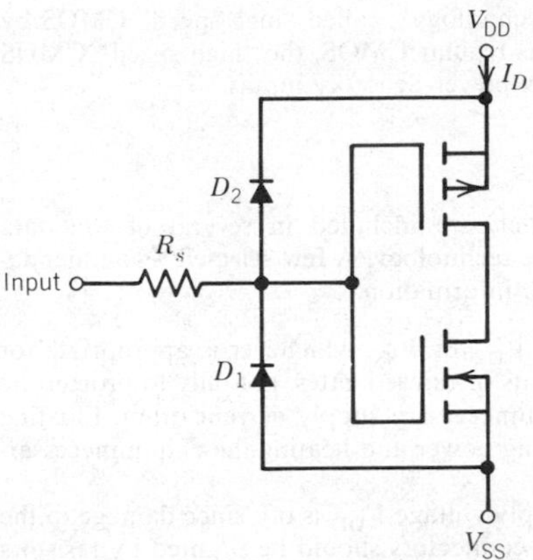

Fig. 28.17 Diode protection of input transistor gates. $200\Omega \leqslant R_s \leqslant 1.5\ k\Omega$.[15,16]

All CMOS inputs are diode protected to prevent static charge from accumulating on the FET gates and causing punch-through of the oxide insulating layer. A typical configuration is illustrated in Fig. 28.17. Diodes D_1 and D_2 clamp the transistor gates between V_{DD} and V_{SS}. Care must be taken to avoid input voltages that would cause excessive diode currents. For this reason manufacturers specify an input voltage constraint from $V_{SS} - 0.5$ V to $V_{DD} + 0.5$ V. The resistance R_s helps protect the diodes from excessive currents but is introduced at the expense of switching speed, which is deteriorated by the time constant of this resistance and the junction capacitances. Propagation times for the 2-input NAND gate are typically 125 ns at $V_{DD} - V_{SS} = 5$ V, 50 ns at $V_{DD} - V_{SS} = 10$ V, and 40 ns at $V_{DD} - V_{SS} = 15$ V. Owing to the relatively high output resistance of CMOS in both the HIGH and LOW input states, these propagation times are quite sensitive to capacitive loading.

Characteristics at 25°C, including typical propagation and transition times (defined for 10 to 90% points in Fig. 28.10), are given in Table 28.3. The fan-out figure given is conservative, being based upon the standard test load capacitance for time measurements and the guaranteed maximum capacitance at any CMOS gate input of 7.5 pF. The fan-out figure can be exceeded without deleterious effect other than an increase in the switching times. Based on the worst-case ratio of available drive current to input current, the fan-out would be of the order of 400.

It should be noted that Table 28.3 gives data for both a B-series gate (designated xxxxB) and a UB-series gate (xxxxUB). UB-series gates that do not have the additional buffer stages which are included between the input and output of the B-series gates have slower transition times (due to the lower gain) and lower noise margins (also due to the lower gain).

TABLE 28.3. COMPARISON OF COMPLEMENTARY METAL OXIDE SEMICONDUCTOR (CMOS) LOGIC QUAD 2-INPUT NAND GATES ($V_{SS} = 0$ V)

CMOS Type	Power Supply Terminal V_{DD} (V)	Power Supply Current $I_{D\max}$ (μA)	Transition Time t_{TLH}[a] (ns)	Transition Time t_{THL} (ns)	Propagation Delay Time t_{PLH} t_{PHL} (ns)	Noise Margins NM_H (V)	Noise Margins NM_L (V)	Test Load, C_L/R_L (pF/kΩ)	Fan-out
	5	1	100	100	125	1.45	1.45	50/0.2	6
4011B	10	2	50	50	50	2.95	2.95	50/0.2	6
	15	4	40	40	40	3.95	3.95	50/0.2	6
	5	1	180	100	90	0.95	0.95	50/0.2	6
4011UB	10	2	90	50	50	1.95	1.95	50/0.2	6
	15	4	65	40	40	2.45	2.45	50/0.2	6

[a] See text for explanation of abbreviations.

The reader should be aware of a recently introduced technology[17] called "high speed" CMOS by several manufacturers. For the same power consumption as regular CMOS, the "high speed" CMOS is as fast as low-power Schottky and can fan out to 10 low-power Schottky inputs.

CMOS Design Considerations

Design and handling recommendations for CMOS, which are included in several of the data books,[15,16,18] should be consulted by the designer using this technology. A few selected recommendations are included here to illustrate the importance of such information.

1. All unused CMOS inputs should be tied either to V_{DD} or V_{SS}, whichever is appropriate for proper operation of the gate. This rule applies even to inputs of unused gates, not only to protect the inputs from possible static charge buildup, but to avoid unnecessary supply current drain. Floating gate inputs will cause all the FETs to be conducting, wasting power and heating the chip unnecessarily.[13]

2. CMOS inputs should never be driven when the supply voltage V_{DD} is off, since damage to the input-protecting diodes could result. Inputs wired to edge connectors should be shunted by resistors to V_{DD} or V_{SS} to guard against this possibility.[18]

3. Slowly changing inputs should be conditioned using Schmitt trigger buffers to avoid oscillations that can arise when a gate passes slowly through the transition region.[16]

4. Wired-AND configurations cannot be used with CMOS gates, since wiring an output HIGH to an output LOW would place two series FETs in the "on" condition directly across the chip supply.[18]

5. Capacitive loads greater than 5000 pF across CMOS gate outputs act as short circuits and can overheat the output FETs at higher frequencies.[18]

6. Designs should be used that avoid the possibility of having low impedances (such as generator outputs) connected to CMOS inputs prior to power-up of the CMOS chip. The resulting current surge when V_{DD} is turned on can damage the input.[15]

While this list of recommendations is incomplete, it should alert the CMOS designer to the value of the information supplied by the manufacturers.

28.7 CHOOSING A LOGIC FAMILY

A logic designer planning a system using SSI and MSI chips will find that an extensive variety of circuits is available in all three technologies: TTL, ECL, and CMOS. The choice of which technology will dominate the system is governed by what are often conflicting needs, namely speed, power consumption, noise immunity, and the ease of interfacing with system inputs and outputs. Tables 28.1, 28.2, and 28.3 give a good indication of how the three technologies relate to at least the first three of these considerations. Sometimes a decision is clear-cut. If the need for a low static power drain is paramount, CMOS is the only choice. When the system clock frequency exceeds 100 MHz, ECL must be used. At frequencies below 80 MHz, it is difficult to justify the use of ECL rather than Schottky TTL unless ECL is compatible with the system inputs and outputs. Until the introduction of "high speed CMOS," low-power Schottky in many cases was the best compromise between speed and power consumption. However, the "high speed" CMOS, probably, is superior in every respect.[17]

At frequencies below 10 MHz, CMOS is usually the best choice. In addition to having the obvious advantages of low power consumption and good noise immunity, CMOS has an extensive catalog of MSI and LSI devices, which adds flexibility to the design process. These devices include building blocks such as counters and shift registers of many stages, static memories, and even single chip microcomputers. Integration on this scale is not practical in TTL or ECL due to the lower chip densities and much higher power dissipation of these technologies. Even in systems requiring the drive capability of TTL, CMOS can be used throughout most of the design, with CMOS buffers (e.g., 4050B) being used for TTL interfacing at the outputs.

28.8 INTEGRATED INJECTION LOGIC (IIL)

IIL is a bipolar gate configuration which in basic gate form is similar to direct coupled transistor logic (DCTL).[4] In essence this circuit consists of *npn* transistor inverters that share common collect or load resistors to form wired-AND logic (Fig. 28.18*a*) implementing a positive NOR gate. These gates are not marketed in SSI, since the major attribute of IIL is that it is a high-density form of bipolar logic. The high density is accomplished by replacing the collector resistors (R_{ca}, R_{cb}, and R_{cc} in Fig. 28.18.*a*) by active loads in the form of *pnp* lateral transistors (T_a, T_b, and T_c in Fig. 28.18*b*). A single

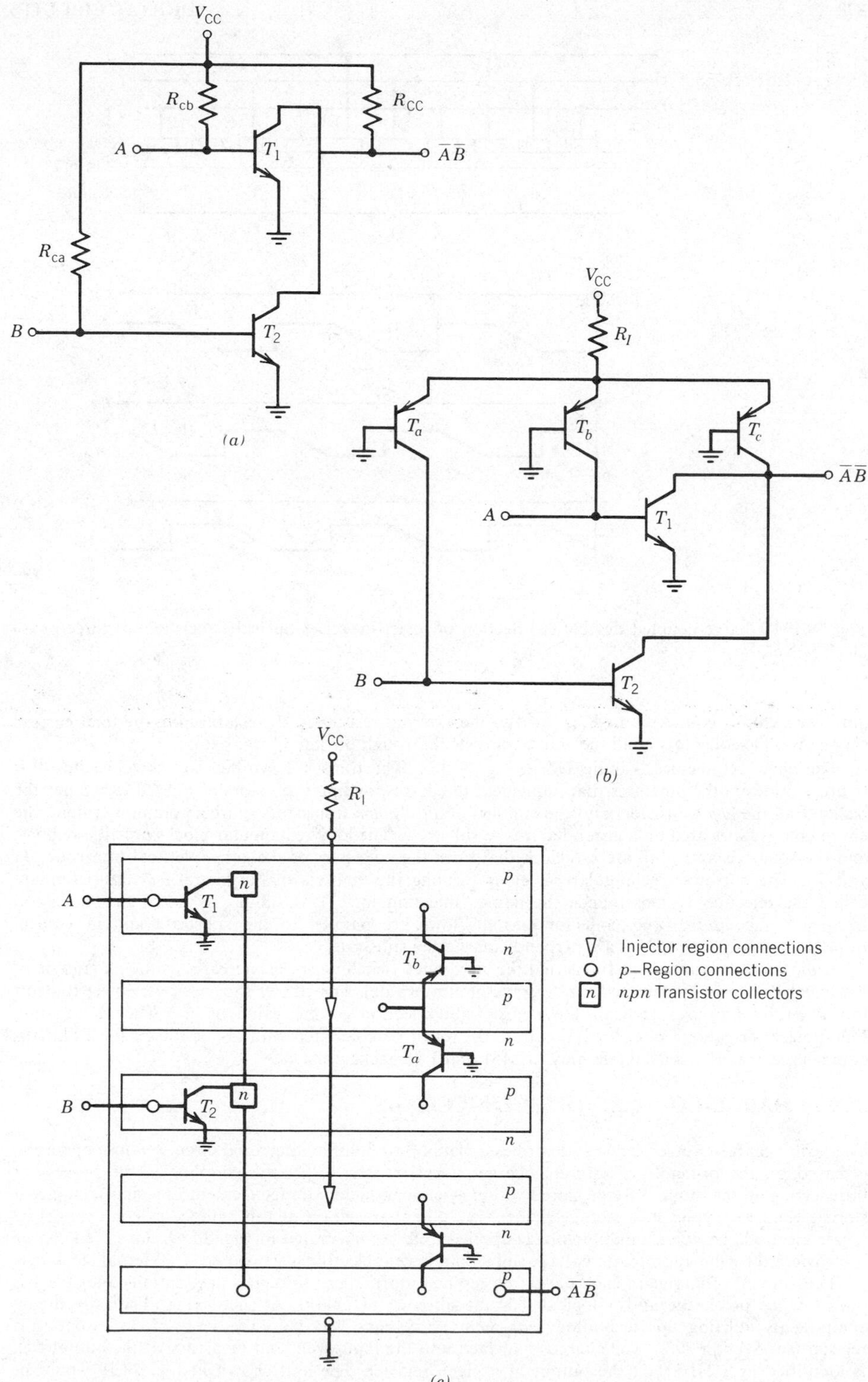

Fig. 28.18 Evaluation of integrated injection logic (IIL): (*a*) Direct coupled transistor logic (OCTL) NOR gate, (*b*) DCTL NOR gate with injection *pnps*, (*c*) physical structure of IIL (transistors are symbolic).

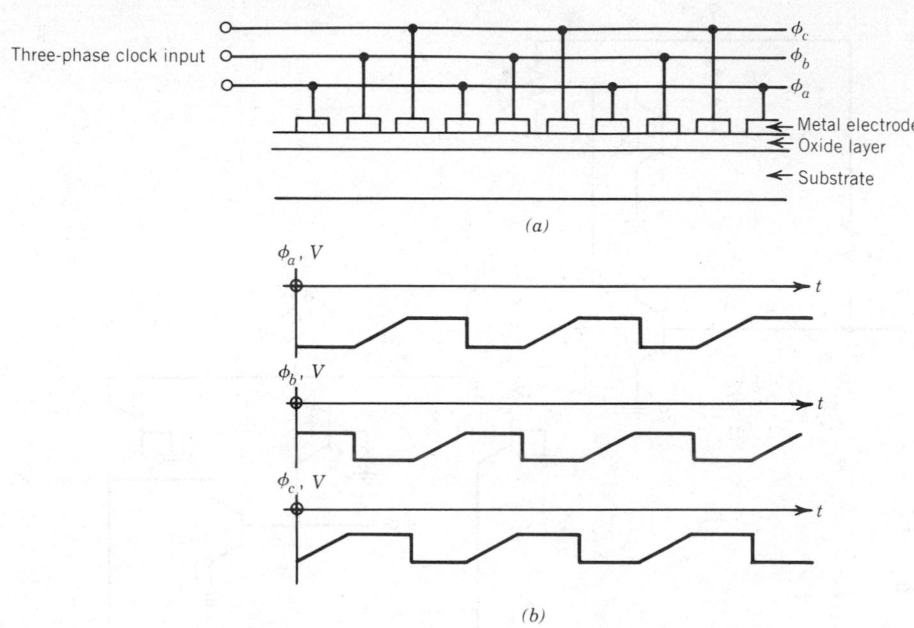

Fig. 28.19 Charge-coupled device: (*a*) Section of series-in, series-out shift register, (*b*) three-phase clock waveform.

pin of the chip is connected to V_{CC} through the external resistance R_I, establishing the total current supplied to the collectors of all the active transistors through the lateral transistors.

The chip architecture is indicated in Fig. 28.18*c*. The transistor symbols are sketched into this figure to indicate the junctions that implement the corresponding transistors of Fig. 28.18*b*. Since the bases of all the *pnp* transistors and the emitters of all the *npn* transistors share a common ground, the entire chip is fabricated on a grounded *n*-type substrate. The long rectangular blocks are all *p*-regions, and the square islands ([*n*]) are *n*-regions that form the collectors of the active vertical transistors T_1 and T_2. The narrower rectangular *p*-regions forming the emitters of the lateral *pnp* transistors are called the injection regions (hence the name "injection logic"). It should be noted that in lateral transistors the emitter and collector current flows are parallel to the substrate and in vertical transistors these current flows are perpendicular to the substrate.

A particular advantage of IIL is that the maximum switching speed increases as the magnitude of the injected current is increased (up to some limiting value). This allows the speed vs. power tradeoff to be set by the user through the appropriate selection of the value of R_I. The propagation time-power per gate product for IIL is from 0.1 to 0.7 pJ compared with about 100 pJ for TTL. It is emphasized that IIL is available only in MSI and LSI packages.

28.9 CHARGE-COUPLED DEVICES (CCD)

The term "charge-coupled device" describes a class of monolithic integrated circuits whose operation is based on the principle of discrete charge-packet transfer.[19] Charge coupling is the process of transferring all the mobile charge stored within a semiconductor storage element to a similar adjacent storage element. Typically, a storage element is a potential well created in a doped silicon channel by a gate electrode having a suitable voltage applied to it. As illustrated in Fig. 28.19, the stored charge is transferred by the appropriate voltage shifts on adjacent electrodes generated by external clocking.

The amount of charge in each packet represents information. In digital applications a logical 1 is the presence of charge and a logical 0 is the absence of charge, or vice versa. The only digital components utilizing this technology are memory systems based on the series-in, series-out shift register (see Section 29.2). The charge is injected into the input well, and its arrival at the output well is identified as a HIGH at the output of a given register. The unit shown in Fig. 28.19 would be driven by a 3-phase clock. Every third electrode is interconnected and brought out to one of the clocks. Two other "every-third-electrode" sets are similarly interconnected and driven by the other two clock phases. Negative clock voltages are used to transfer positive charge packets. Clocking rates are up to 10 MHz.

A CCD can operate as an analog shift register, since the amount of charge in a given packet can be directly proportional to an analog voltage level. Thus a stream of packets of magnitudes corresponding to a sampled analog signal at the input can be shifted through the register, simulating a delay line. The CCD 311 is 130-bit register that can be clocked over the frequency range of 10 kHz to 15 MHz providing a delay time of 13 μs to 13 ms.[19]

References

1 W. I. Fletcher, *An Engineering Approach to Digital Design*, Prentice-Hall, Englewood Cliffs, NJ, 1980.

2 *TTL Data Book*, Fairchild Camera and Instrument Corp., California, 1978.

3 *The TTL Data Book for Design Engineers*, Texas Instruments, Inc., Texas, 1980.

4 H. Taub and D. Schilling, *Digital Integrated Electronics*, McGraw-Hill, New York, 1977.

5 L. A. Levanthal, *Introduction to Microprocessors: Software, Hardware, Programming*, Prentice-Hall, Englewood Cliffs, NJ, 1978.

6 C. Belove and D. Schilling, *Electronic Circuits, Discrete and Integrated*, 2nd Ed., McGraw-Hill, New York, 1979.

7 L. Strauss, *Wave Generation and Shaping*, McGraw-Hill, New York, 1970.

8 *The Integrated Circuit Data Book*, Motorola Semiconductor Products, Inc., Arizona, 1968.

9 *Low Power Schottky TTL*, Vol. 9, Ser. A, Motorola Semiconductor Products, Inc., Arizona, 1977.

10 *Fairchild Advanced Schottky TTL*, Fairchild Camera and Instrument Corp., California, 1980.

11 *Supplement to the TTL Data Book for Design Engineers*, 2nd Ed., Texas Instruments, Inc., Texas, 1981.

12 *ECL Data Book*, Fairchild Camera and Instrument Corp., New York, 1977.

13 P. Horowitz and W. Hill, *The Art of Electronics*, Cambridge University Press, Cambridge, England, 1980.

14 W. R. Blood, *MECL System Design Handbook*, 3rd Ed., Motorola Semiconductor Products, Inc., Arizona, 1982.

15 *CMOS Integrated Circuits*, Motorola, Inc., Arizona, 1978.

16 *CMOS Data Book*, Fairchild Camera and Instrument Corp., New York, 1977.

17 *QMOS High-Speed CMOS Logic*, RCA Corp., New Jersey, 1982.

18 *RCA Integrated Circuits*, RCA Corp., New Jersey, 1976.

19 *MOS/CCD Data Book*, Fairchild Camera and Instrument Corp., California, 1975.

CHAPTER 29
FLIP-FLOPS AND REGISTERS

THOMAS W. PARSONS

Hofstra University, Hempstead, New York

29.1 FLIP-FLOPS

A flip-flop, or bistable multivibrator, is a logic circuit that has two stable states. It can be set to one or the other of these states by application of a suitable excitation signal to its inputs, and it remains in the chosen state after the excitation has been removed. A flip-flop is thus capable of "remembering" the applied excitation.

Basic Cross-Coupled Circuit

All flip-flops are based on the cross-coupled circuit, also called a latch, (Fig. 29.1).[1,2,3] The operation of one version of a standard *S-R* flip-flop (Fig. 29.1*a*) can be seen with the aid of the timing diagram in Fig. 29.2*a*. Inputs *R* and *S* are normally 0. If a logic 1 is applied to *S*, output $\overline{Q}$ will go to 0. This 0 is immediately applied to the input of the upper NOR gate. Both inputs of this gate are now 0, so *Q*, the output, will be a 1. The flip-flop is now said to be set.

The 1 from *Q* is also applied to the input of the lower NOR. If the *S* input is now returned to 0, the presence of the 1 from *Q* keeps $\overline{Q}$ set to 0. Hence the flip-flop retains its setting after the inputs have returned to 0.

If a 1 is applied to input *R*, output *Q* will go to 0 and output $\overline{Q}$ will go to 1. The flip-flop is now said to be reset. If *R* returns to 0, the outputs will retain these new values.

If both inputs were to be set to 1, both outputs would be 0. Since *Q* and $\overline{Q}$ are always intended to be complements (as the designation $\overline{Q}$ implies), *R* and *S* are normally never permitted to be set to 1 simultaneously.

The standard logic symbol for an *S-R* flip-flop is shown in Fig. 29.1*b*. The $\overline{Q}$ output is designated by a small right triangle.

Another version of this basic circuit is shown in Fig. 29.1*c*, with its corresponding timing diagram in Fig. 29.2*b*. In this version the inputs are normally 1 and the flip-flop is set or reset by application of a 0 to the appropriate input. We can distinguish between this flip-flop and that of Fig. 29.1*a* in two ways. We can label the inputs $\overline{S}$ and $\overline{R}$, as in Fig. 29.1*c*, or we can indicate an inverting input, as shown in the logic symbol for the *S-R* flip-flop in Fig. 29.1*d*. In this version, in order to ensure that *Q* and $\overline{Q}$ are complements, inputs $\overline{S}$ and $\overline{R}$ are not permitted to be both 0.

In any flip-flop, the inputs have to be present for some minimum time for the flip-flop to respond. We can see this by considering the timing involved in the setting operation just described (see Fig. 29.2). The applied signals will take a finite amount of time to propagate through the NOR gates; let this time be t_{pd}. The 1 applied to the *S* input must be present long enough for this input to propagate back to the lower NOR and latch the flip-flop into its set state. To do this the input must first pass the lower NOR, making $\overline{Q}$ go to 0 after a time t_{pd}. As $\overline{Q}$ goes to 0, the upper NOR starts toward 1, so

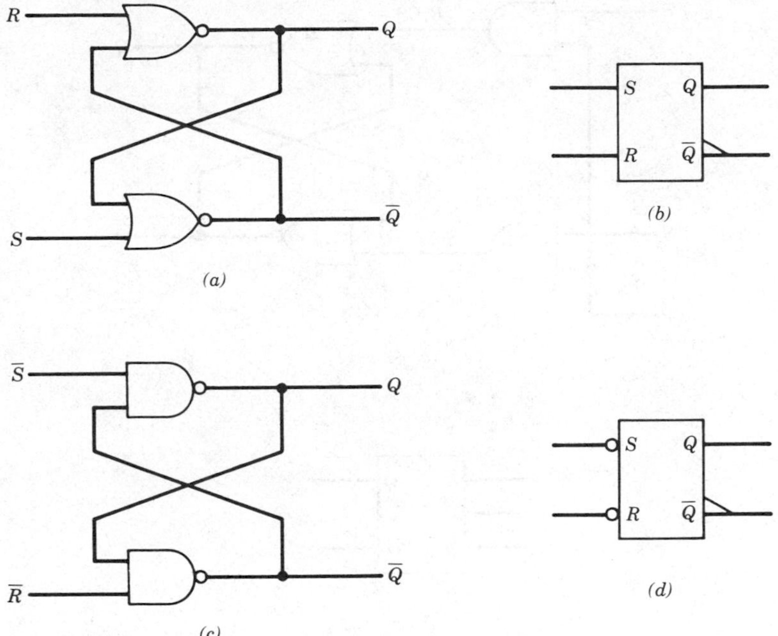

Fig. 29.1 Basic cross-coupled (latch) circuit: (*a*) With true inputs, using NOR gates; (*b*) standard symbol; (*c*) with complemented inputs, using NAND gates; (*d*) standard symbol.

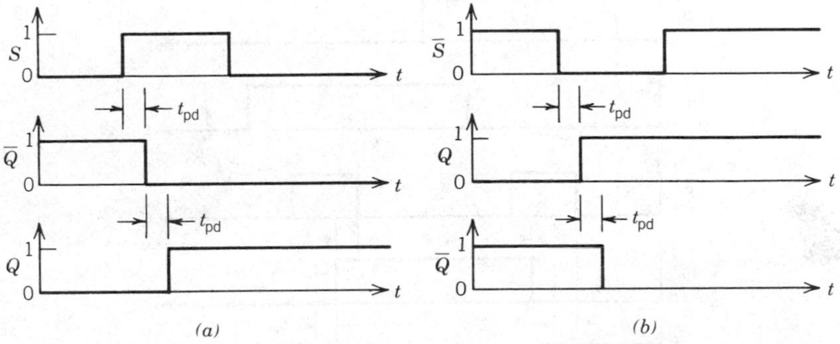

Fig. 29.2 Timing of cross-coupled latch: (*a*) with NOR gates, (*b*) with NAND gates.

that the change in Q is delayed t_{pd} seconds from the change in $\overline{Q}$. Hence the total duration of the S input must be at least two delays, or $2t_{pd}$. If the signal lasts less than this time, $\overline{Q}$ may change state briefly or may begin to do so, but Q will not, and when S returns to 0 the flip-flop will return to its initial state. A similar argument can be made for the circuit version in Fig. 29.1*c*, as illustrated by the timing diagram in Fig. 29.2*b*.

Clocked Flip-Flop

In practice it is usually desirable to synchronize the operation of a flip-flop with a clock pulse. An elementary clocked *S-R* flip-flop is shown in Fig. 29.3*a*. Note that the flip-flop is permitted to observe and respond to its control inputs only while the clock pulse is 1. While the clock pulse is 0, the flip-flop remains in the state to which it was set during the clock pulse. The use of clocked

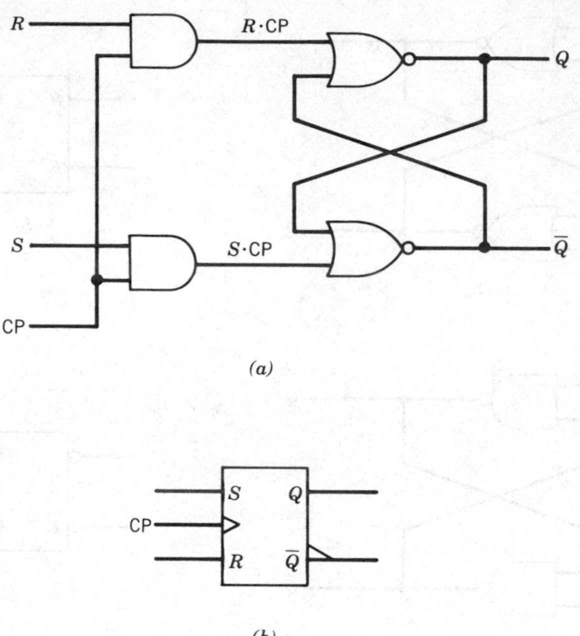

Fig. 29.3 Clocked *S-R* flip-flop (*a*) and its standard symbol (*b*).

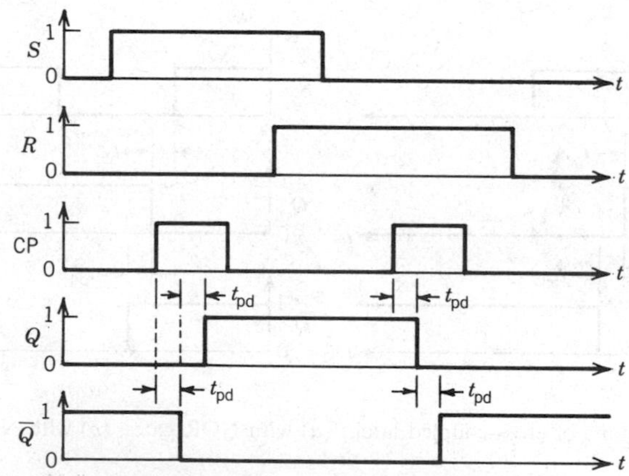

Fig. 29.4 Timing diagram for clocked *S-R* flip-flop.

flip-flops is the rule rather than the exception, and for the balance of this discussion, all flip-flops will be assumed to be clocked unless specifically stated to be otherwise.

The operation of the clocked *S-R* flip-flop can be seen in the timing diagram in Fig. 29.4. Note that the flip-flop changes state at the beginning of the clock pulse, that is, on its rising edge. The clock pulse has to last long enough for the effect of the new input value to propagate through the flip-flop. The minimum width is determined by the same analysis as that for the simple latch, discussed previously, and again turns out to be $2t_{pd}$. In all clocked flip-flops (with the exception of the edge-triggered flip-flop, to be described), the inputs must remain stable (i.e., must not change) over the duration of the clock pulse.

TABLE 29.1. STATE TABLE FOR *R-S* FLIP-FLOP

R	S	$Q(t+1)$
0	0	$Q(t)$
0	1	1
1	0	0
1	1	Not permitted

The operation of a flip-flop can be characterized by a state table giving the value of Q following a clock pulse as a function of the excitation applied to its control inputs. The state table for an S-R flip-flop is shown in Table 29.1. In this table, $Q(t)$ is the value of Q before the clock pulse and $Q(t+1)$ is the value after the clock pulse.

The graphic symbol for an S-R flip-flop is shown in Fig. 29.3*b*. The clock pulse input is represented by a small triangle within the box and the $\overline{Q}$ output is distinguished by a small right triangle outside the box.

D and *J-K* Flip-Flops

Besides the S-R flip-flop there are two other commonly used types: the D and J-K flip-flops.

The D flip-flop is shown in Fig. 29.5 and its timing diagram is shown in Fig. 29.6. This flip-flop is essentially a NAND-gate S-R flip-flop with its inputs connected by an inverter, as shown in Fig. 29.5*a*. When a clock pulse is applied, the logic level present at the D input is transferred to the Q output; between clock pulses the D input is ignored. A state table for the D flip-flop is shown in Table 29.2.

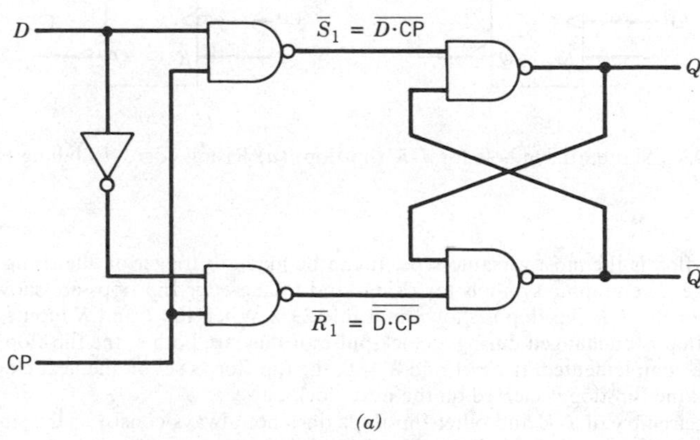

(*a*)

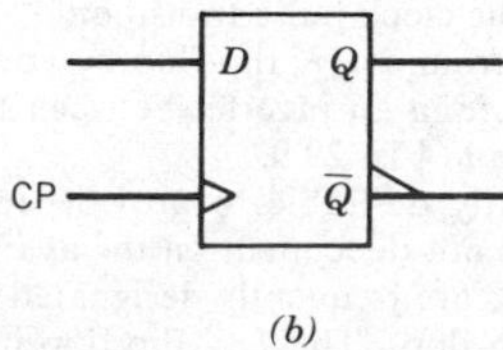

(*b*)

Fig. 29.5 Clocked D flip-flop (*a*) and its standard symbol (*b*).

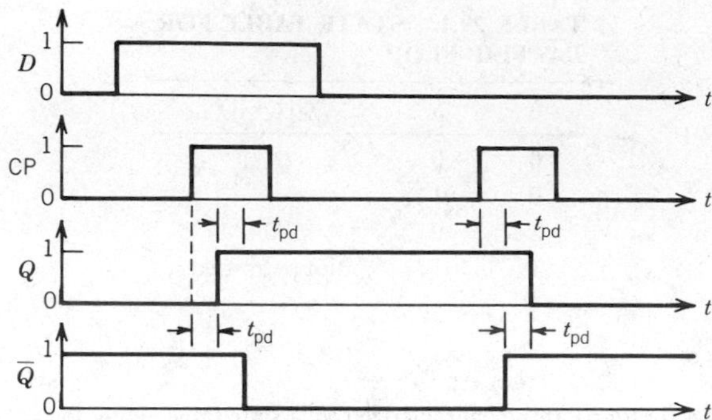

Fig. 29.6 Timing diagram for clocked D flip-flop.

TABLE 29.2. STATE TABLE FOR *D* FLIP-FLOP

D	$Q(t+1)$
0	0
1	1

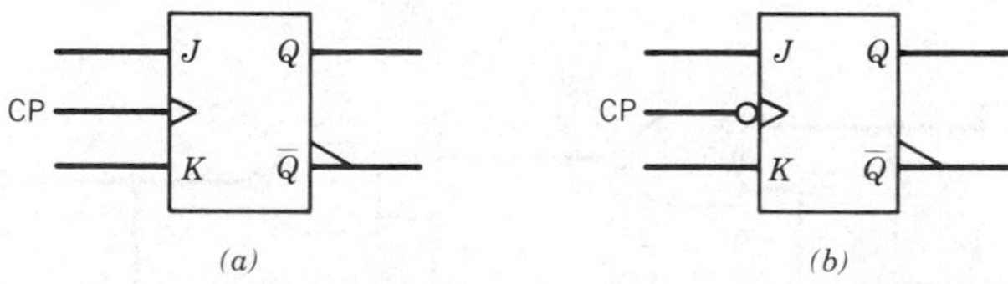

Fig. 29.7 Standard symbols for J-K flip-flop: (*a*) Rising edge, (*b*) falling edge.

The J-K flip-flop is the most versatile type. It can be made to trigger on the rising or falling edge of the clock pulse. The graphic symbols for rising- and falling-edge flip-flops are shown in Fig. 29.7. The state table for the J-K flip-flop is shown in Table 29.3. When the J and K inputs are both 0, the state of the flip-flop is unchanged during a clock pulse; if they are both 1, the flip-flop is toggled, that is, its outputs are complemented. If $J = 1$ and $K = 0$, the flip-flop is set on the next clock pulse, and if $J = 0$ and $K = 1$, the flip-flop is cleared on the next clock pulse.

The internal circuitry of J-K and other flip-flops does not always consist of the simple NOR-gate and NAND-gate forms shown in Fig. 29.1. The delay from an input to an output can involve several gate delays so that the overall time needed for the flip-flop outputs to stabilize is more than $2t_{pd}$. The timing diagrams in Fig. 29.8 assume an effective time delay t_{eff} for the outputs to stabilize after the appropriate edge of the clock pulse. For simplicity, the output transitions are shown as occurring simultaneously, delayed by t_{eff} from the clock pulse transition.

A toggling flip-flop can be made from a J-K flip-flop by connecting its J and K inputs together, and a D flip-flop can be made by putting an inverter between the J and K inputs, as with the S-R flip-flop. These connections are shown in Fig. 29.9.

In clocked flip-flops it is frequently desirable to provide an auxiliary set or reset input that is asynchronous, that is, an input that is not dependent on the availability of a clock pulse and that can override the other inputs. These inputs are frequently designated preset and clear (to distinguish them from the set and reset of the S-R flip-flop). The J-K flip-flop in Fig. 29.10*a* shows preset and clear inputs as they are usually drawn. In many types of flip-flop these inputs are complemented, as shown in Fig. 29.10*b*.

TABLE 29.3. STATE TABLE FOR *J-K* FLIP-FLOP

J	K	$Q(t+1)$	$\overline{Q}(t+1)$
0	0	$Q(t)$	$\overline{Q}(t)$
0	1	1	0
1	0	0	1
1	1	$\overline{Q}(t)$	$Q(t)$

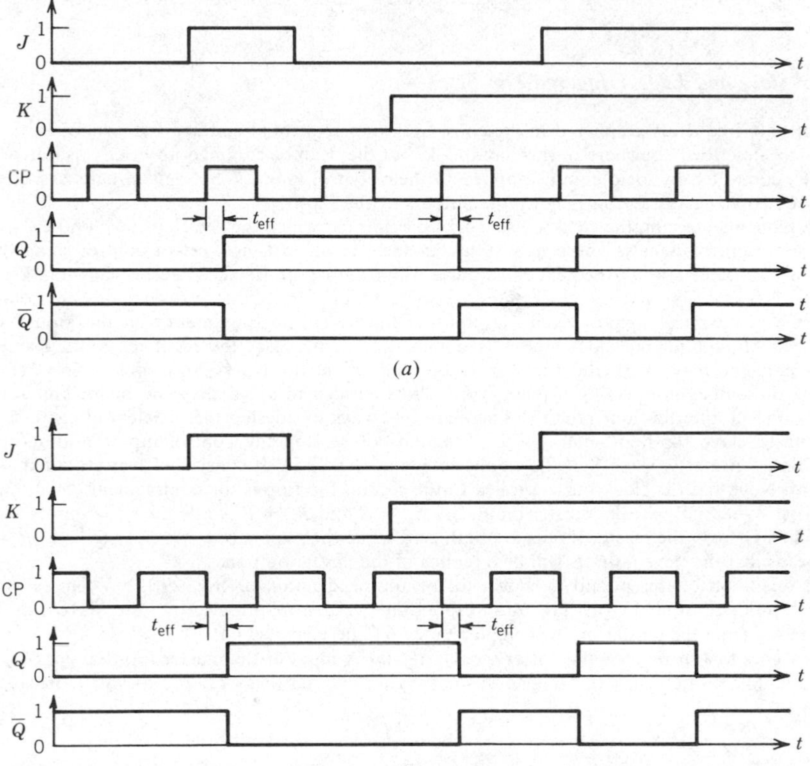

Fig. 29.8 Timing diagrams for *J-K* flip-flop: (*a*) triggering on rising edge of clock pulse, (*b*) triggering on falling edge.

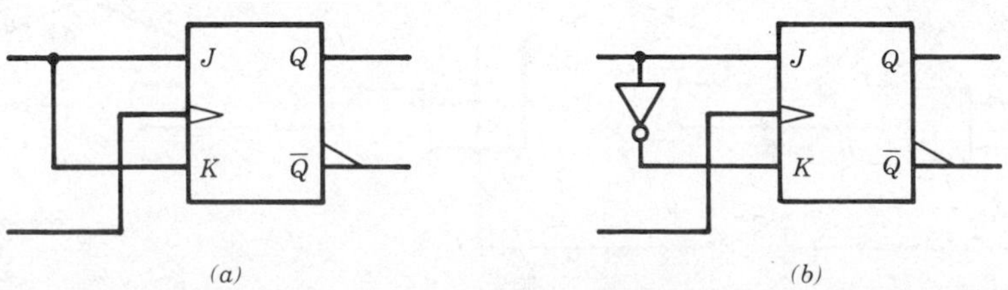

Fig. 29.9 Connections of *J-K* flip-flop to implement a toggling flip-flop (*a*) and a *D* flip-flop (*b*).

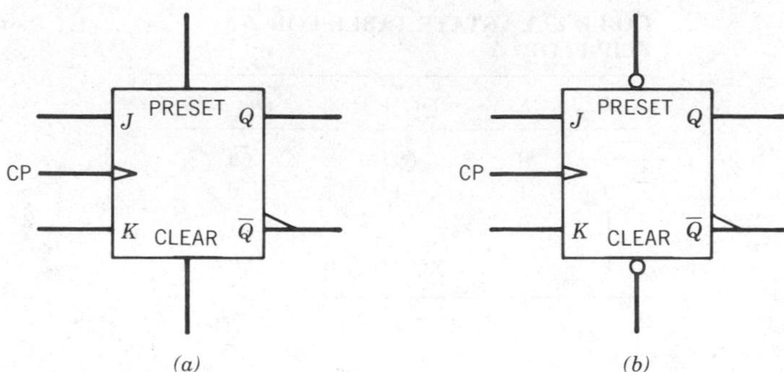

Fig. 29.10 *J-K* flip-flop with present and clear inputs: (*a*) noninverting, (*b*) inverting.

Master-Slave and Edge-Triggered Flip-Flops

Flip-flops are frequently employed in the construction of sequential circuits, such as counters. These circuits are described elsewhere in this handbook, but they can be summed up briefly as one or more flip-flops controlled by logic circuits applied to their control inputs. The logic circuits are driven in part by external signals and in part by the outputs of the flip-flops.

In a sequential circuit, the clock pulse marks the point at which the flip-flops take on new states in response to control signals. These new states, in turn, result in a new set of control signals which determine the transitions on the next clock pulse. With simple flip-flops of the sort shown in Fig. 29.2, however, it is essential that the clock pulse be short enough that the state transitions are completed before the new control signals reach the inputs. Otherwise the requirement that the inputs remain unchanged throughout the clock pulses is not met. In some cases the feedback path through the control circuitry may cause the flip-flop to oscillate while the clock pulse is 1. There are many practical difficulties in selecting a pulse width short enough to avoid these problems and still long enough to clock the flip-flops reliably. There are two ways to sidestep the problem altogether.

A master-slave flip-flop separates the functions of sensing the control inputs and setting the output states. A master-slave *S-R* flip-flop is shown in Fig. 29.11. It consists of two cascaded clocked flip-flops. Note that the clock pulse applied to the second flip-flop is the complement of that applied to the first. When CP = 1, the master inputs are $S_m = S$ and $R_m = R$, while the slave inputs are both zero. When CP = 0, the master inputs are both zero while the slave inputs are $S_s = Q_m$ and $R_s = \overline{Q}_m$. This means that the slave outputs will be a replica of the master outputs.

The operation of this circuit is shown in the timing diagram of Fig. 29.12. When CP = 1, the master flip-flop is able to observe and respond to its control inputs, but the slave flip-flop is disconnected from the master by the complemented CP, here labeled CP_s.

If CP goes to 1 at time t_1, then after a delay t_{pd} the *S* input of the master flip-flop goes to 1, and after an additional delay t_{pd} $\overline{Q}_m$ goes to 0; finally after another delay t_{pd} Q_m goes to 1. Because CP_s

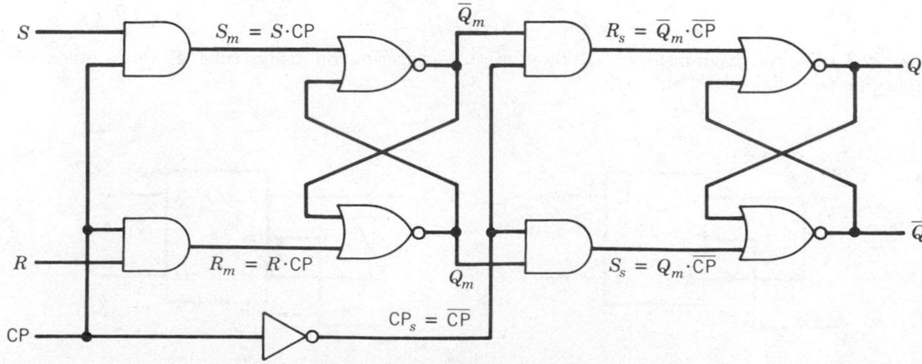

Fig. 29.11 Master-slave *S-R* flip-flop.

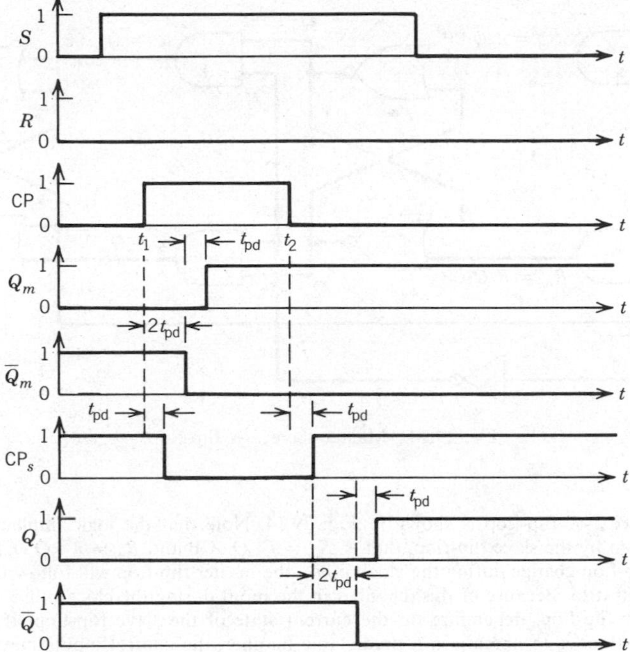

Fig. 29.12 Timing diagram for master-slave flip-flop.

has been at 0 ever since $t_1 + t_{pd}$, however, the slave flip-flop does not respond to the new states of Q_m and $\overline{Q}_m$. When CP returns to 0 at some later time t_2, the master flip-flop is disconnected from the S and R inputs and can no longer respond to them. At time $t_2 + t_{pd}$, however, CP_s goes to 1 and the slave flip-flop changes state in response to the values of Q_m and $\overline{Q}_m$. Outputs Q and $\overline{Q}$ now reflect the new state resulting from the inputs that were at S and R while CP was 1. If in response to this new state S or R takes on a new value, Q and $\overline{Q}$ will not be affected by it until a new clock pulse has occurred, because until then the master flip-flop is disconnected from S and R.

Note that in this master-slave flip-flop, the control inputs are passed to the master flip-flop on the rising edge of the clock pulse, but the slave flip-flop changes state on the falling edge of the clock pulse. Master-slave flip-flops can be made in which the master reads the inputs on the falling edge of the clock pulse and the slave changes state on the rising edge; one such is shown in Fig. 29.13.

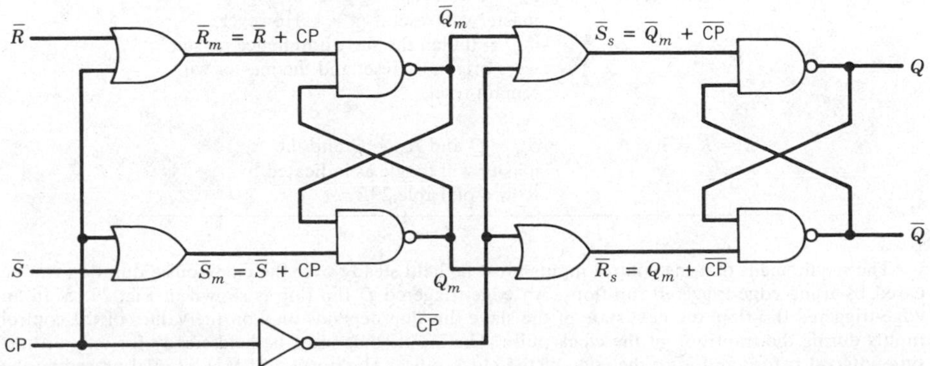

Fig. 29.13 Master-slave *S-R* flip-flop reading data on falling edge of clock pulse and changing state on rising edge.

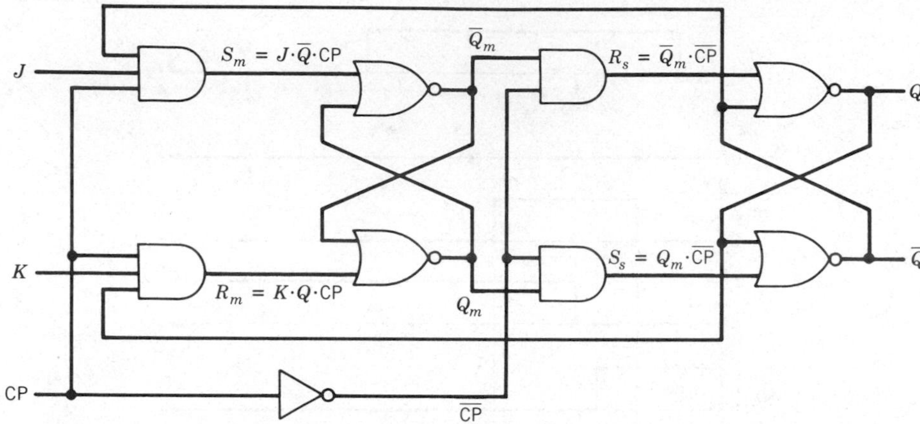

Fig. 29.14 Master-slave *J-K* flip-flop.

A master-slave *J-K* flip-flop is shown in Fig. 29.14. Note that the logic implementing the *J-K* operation is driven by the slave flip-flop, that is, $S_m = J \cdot \overline{Q} \cdot \text{CP}$ and $R_m = K \cdot Q \cdot \text{CP}$. If the control inputs of this flip-flop change during the clock pulse, the master flip-flop will follow the change if the *J-K* inputs permit it to. Because of this, changes in the input during the clock pulse may or may not affect the master flip-flop, depending on the current state of the slave flip-flop. If a system using master-slave flip-flops is to operate in a predictable manner, the control inputs may be held steady over the duration of the clock pulse.

The state table in Table 29.3 can be verified from the circuit in Fig. 29.14 by considering the logic equations for the inputs to the master flip-flop when CP = 1:

If	Then
$J = K = 0$	$S_m = R_m = 0$ and the master will not change
$J = 1$ and $K = 0$	$S_m = \overline{Q}$ and $R_m = 0$ and the master will set if $\overline{Q} = 1$. However, if $\overline{Q} = 0$ then the slave and hence master were originally set and the master will remain set
$J = 0$ and $K = 1$	$S_m = 0$ and $R_m = Q$ and the master will reset if $Q = 1$. However, if $Q = 0$ then the slave and hence master were originally reset and the master will remain reset
$J = K = 1$	$S_m = \overline{Q}$ and $R_m = Q$ and the master will toggle as indicated by Row 4 of Table 29.3

The requirement that the control inputs must be held steady over the clock-pulse duration can be eased by using edge-triggered flip-flops. An edge-triggered *D* flip-flop is shown in Fig. 29.15. In an edge-triggered flip-flop, the next state of the slave flip-flop depends only on the values of the control inputs during the transition of the clock pulse. The inputs thus must be held steady for only a short time interval before and after the edge of the clock pulse. The portion of this interval preceding the clock pulse transition is known as the setup time (see Fig. 29.16), and the portion following the transition is called the hold time.

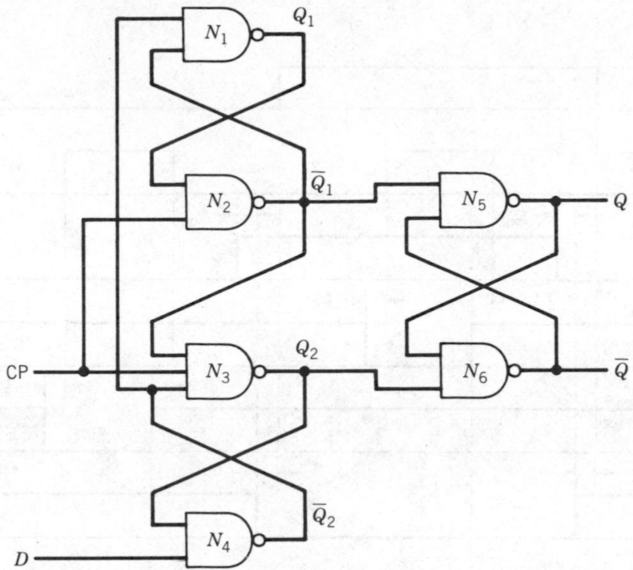

Fig. 29.15 Edge-triggered *D* flip-flop triggering on rising of clock pulse.

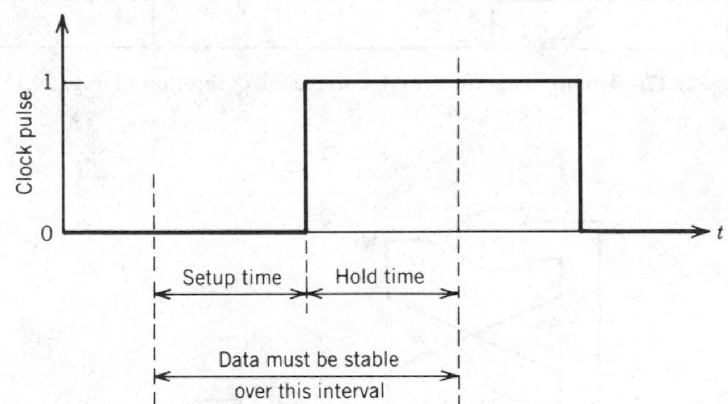

Fig. 29.16 Setup and hold times for edge-triggered flip-flop of Fig. 29.15.

The operation of the edge-triggered flip-flop can be described with the aid of the timing diagram in Fig. 29.17. In this flip-flop, Q_1 and $\overline{Q}_2$ work together to hold the state of the *D* input. When *D* goes to 1, $\overline{Q}_2$ goes to 0 and Q_1 goes to 1. Later, when *D* goes back to 0, $\overline{Q}_2$ returns to 1 and Q_1 to 0. Note that these transitions do not require a clock-pulse transition to enable them. Signals Q_1 and $\overline{Q}_2$ serve to arm N_2 and N_3 for the clock pulse.

When the clock pulse is 0, outputs $\overline{Q}_1$ and Q_2 are held at 1 and the slave flip-flop (N_5 and N_6) remains unchanged. When the clock pulse makes its transition to 1, then either Q_2 and $\overline{Q}_1$ goes to 0, depending on the states of Q_1 and $\overline{Q}_2$. If $D = 1$, then $\overline{Q}_2 = 0$ and $Q_1 = 1$, and the clock-pulse transition causes $\overline{Q}_1$ to go to 0, setting output Q to 1 and $\overline{Q}$ to 1. If $D = 0$, then $\overline{Q}_2 = 1$ and $Q_1 = 0$, so the clock pulse causes Q_2 to go to 0 instead, setting the outputs Q to 0 and $\overline{Q}$ to 1.

The setup time is clearly the time required for the *D* input to propagate to Q_1; this is $2t_{pd}$. The hold time is then the time required for N_2, N_3, N_5, and N_6 to respond to the clock pulse transition; this is $3t_{pd}$.

The flip-flop of Fig. 29.15 triggers on the rising edge of the clock pulse. Edge-triggered flip-flops can also be made to trigger on the falling edge by substituting NOR for NAND gates, as shown in Fig. 29.18.

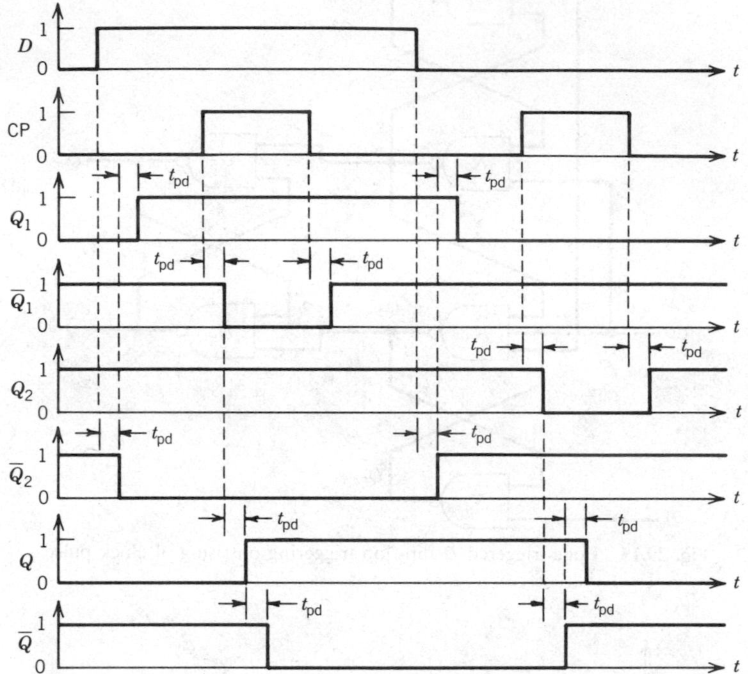

Fig. 29.17 Timing diagram for edge-triggered *D* flip-flop of Fig. 29.15.

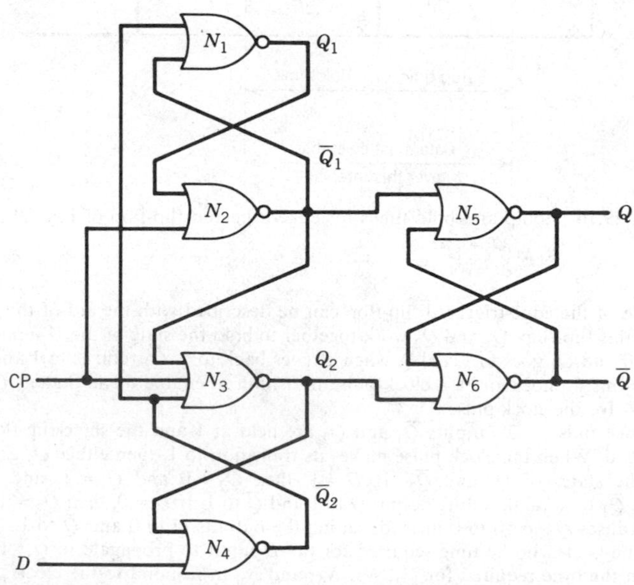

Fig. 29.18 Edge-triggered *D* flip-flop triggering on falling edge.

29.2 REGISTERS

A register is a set of flip-flops used to hold a string of binary digits. Registers are used for many purposes. The register may be holding its contents for subsequent transmittal elsewhere or in order to perform specific operations on them. A buffer register is one that holds data in order to ease timing constraints between other components of a larger system. An accumulator is a register with associated hardware to perform arithmetic or logical operations on its contents.

Data Registers[2]

A simple register is shown in Fig. 29.19. It consists of a set of D flip-flops which may be triggerable on either the rising or falling edge of the clock pulse, depending on the application. Data are applied to the D inputs. When data are to be read into the register, a clock pulse is applied to the CP input by enabling the LOAD input. The Q outputs of the register are normally made available to the user; in some applications it may be desirable to make the $\overline{Q}$ outputs available as well. A CLEAR input is usually provided, connected to the asynchronous clear inputs of the individual flip-flops, as shown in Fig. 29.19.

Shift Registers[1,2,3]

A shift register is a register that can shift its contents to the right or to the left. Uses of shift registers include implementation of shift operations in computer central processing units and conversion between serial and parallel data.

A simple shift register is shown in Fig. 29.20. Its operation can be seen with the aid of the timing diagram in Fig. 29.21. (This diagram assumes that flip-flops are triggered on the rising edge of the clock pulse and that each flip-flop requires an effective time t_{eff} for its outputs to stabilize after being clocked.) The register's contents are shifted one position to the right with every clock pulse. (When no shift is desired, the clock pulses are blocked by disabling the shift enable input.) The vacated bit position at the left end of the register is set to the level applied to the serial input. The Q outputs of the flip-flops are available as parallel outputs. Since data can be applied only by way of the serial input, this register is termed a serial-in, parallel-out register.

Notice that the output of each flip-flop is identical to that of the flip-flop immediately to its left but is delayed by one clock pulse. Because of this, shift registers can also be used as digital delay elements, provided the required delay is an integer number of clock pulses in length.

The right shift in Fig. 29.20 is obtained by connecting the input of each flip-flop to the output of the flip-flop on its left. To shift left, the inputs would have to be connected to the flip-flops to the right. By multiplexing the inputs, either operation can be obtained. An example of this configuration is shown in Fig. 29.22. This register has three possible functions: shift right, shift left, and parallel

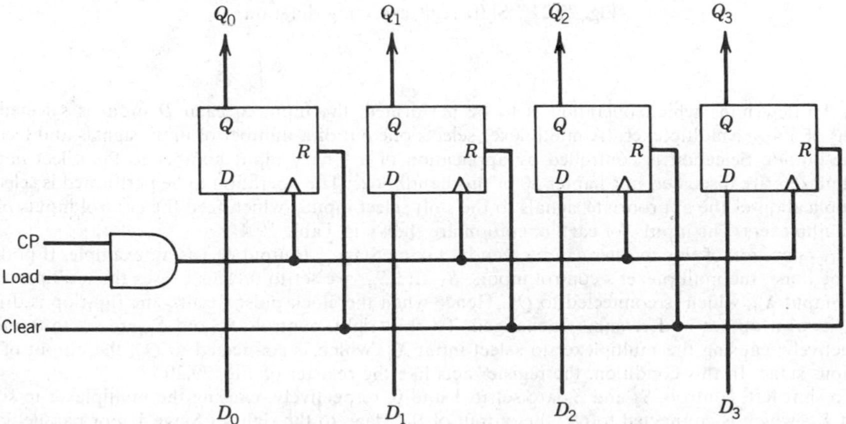

Fig. 29.19 Basic data register. Inputs at $D_0 - D_4$ are transferred to outputs when clock pulse is allowed.

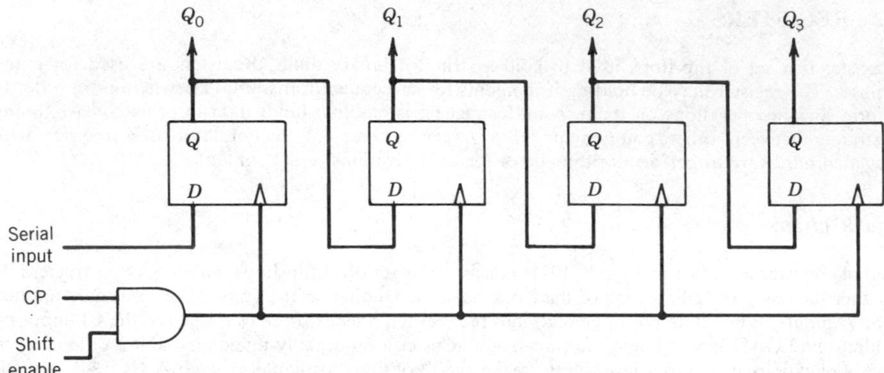

Fig. 29.20 Simple shift-right shift register.

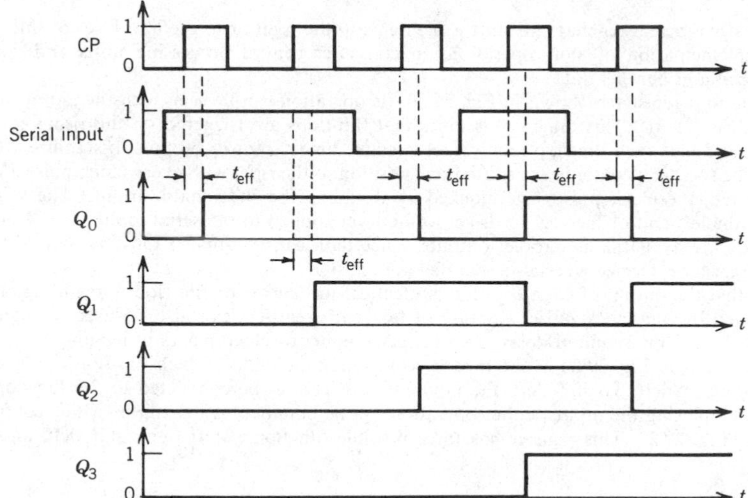

Fig. 29.21 Shift register timing diagram.

load. To determine which operation is to be performed, the input to each D input is selected by means of a 4×1 multiplexer. (A multiplexer selects one out of a number of input signals and feeds it to the output. Selection is controlled by application of a binary input number to the select inputs. Multiplexers are discussed in Chapter 30 of this handbook.) The operation to be performed is selected by application of the appropriate signals to the shift select inputs, which feed the control inputs of all the multiplexers. The inputs for each operation are shown in Table 29.4.

The operation of this register can be seen by taking Stage 1 (output Q_1) as an example. If nothing is to be done, the multiplexer's control inputs, S_1 and S_0, are set to 0. This causes the multiplexer to select input X_0, which is connected to Q_1. Hence when the clock pulse occurs, the flip-flop is driven from its own output and remains unchanged. To shift right, controls S_1 and S_0 are set to 0 and 1, respectively, causing the multiplexer to select input X_1, which is connected to Q_0, the output of the previous stage. In this condition, the register acts like the register of Fig. 29.20.

To shift left, controls S_1 and S_0 are set to 1 and 0, respectively, causing the multiplexer to select input X_2, which is connected to Q_2, the output of the stage to the right of Stage 1. For parallel load, S_1 and S_0 are both set to 1, causing the multiplexer to select X_3, which connects the D input to the corresponding parallel input line, I_1.

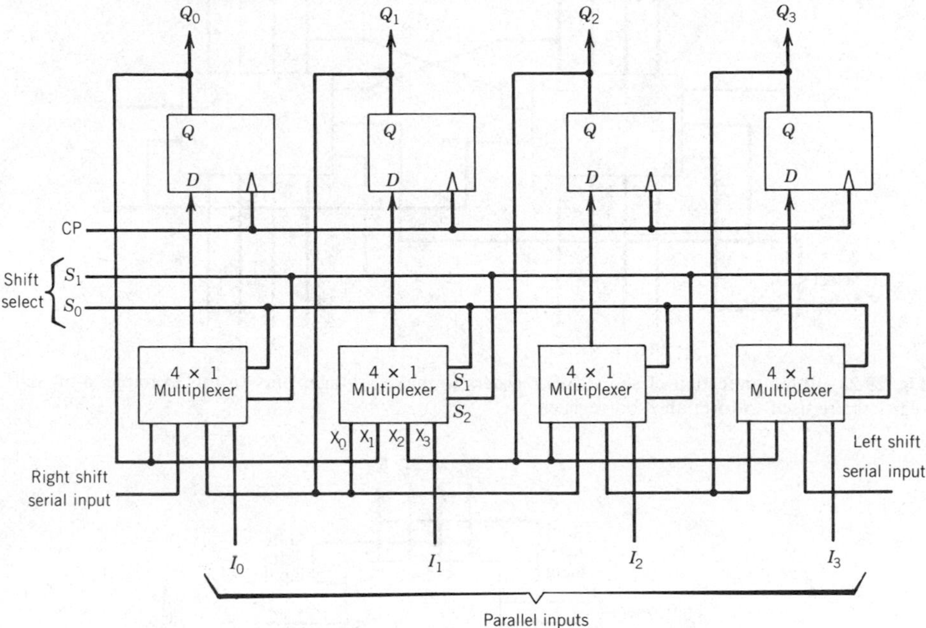

Fig. 29.22 Universal shift-left, shift-right shift register with parallel load.

TABLE 29.4. CONTROL INPUTS FOR SHIFT REGISTER OF FIG. 29.22

S_1	S_0	Function
0	0	No change
0	1	Shift right
1	0	Shift left
1	1	Parallel load

Shift Register Applications

1. **Interconnection of shift registers.** Shift registers are commonly available in 4-bit and 8-bit sizes. If a larger register is required, it can be formed from smaller registers by interconnecting them as shown in Fig. 29.23. The select and clock-pulse inputs are paralleled, and the serial input of each register is connected to the end bit of the adjacent register.
2. **Circular shift registers.** In some applications it is desirable to rotate the contents of the register. In this case, the bit shifted out of the right end of the register is applied to the serial input at the left. This connection is shown in Fig. 29.24.
3. **Serial/parallel interconversion.** Serial data can be converted to parallel data by applying them to the serial input of a shift register, as shown in Fig. 29.25. In this example the serial data are assumed to take the form of 4-bit words. The register is a shift-right register clocked at the bit rate of the serial input, and a parallel 4-bit word is available at the output terminals of the register every four clock pulses.

   To convert parallel data to serial, a shift register can be loaded using parallel inputs, as shown in Fig. 29.26. The register is set to the shift-right mode and clocked at the desired bit rate. The

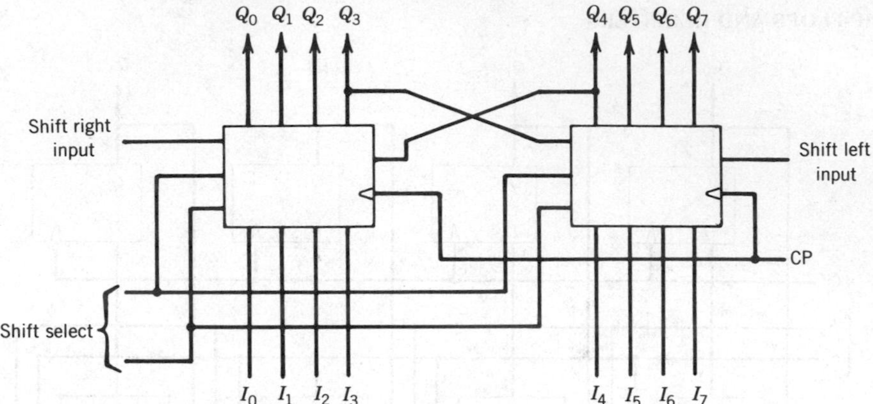

Fig. 29.23 Interconnection of short shift registers to make a longer one. In this example, 4-bit shift registers are used to form an 8-bit register.

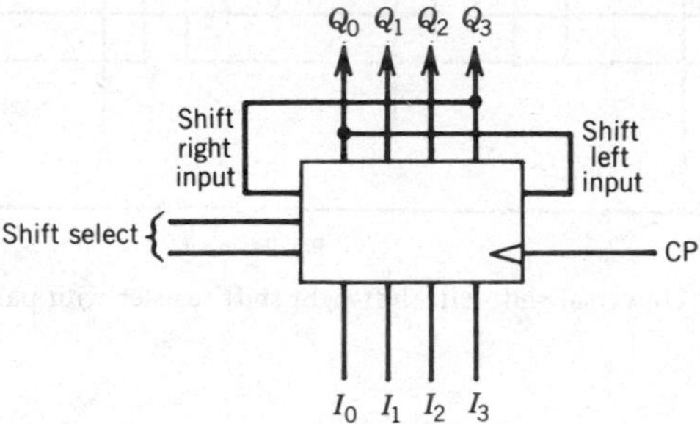

Fig. 29.24 Connection of bidirectional shift register to permit rotation of contents.

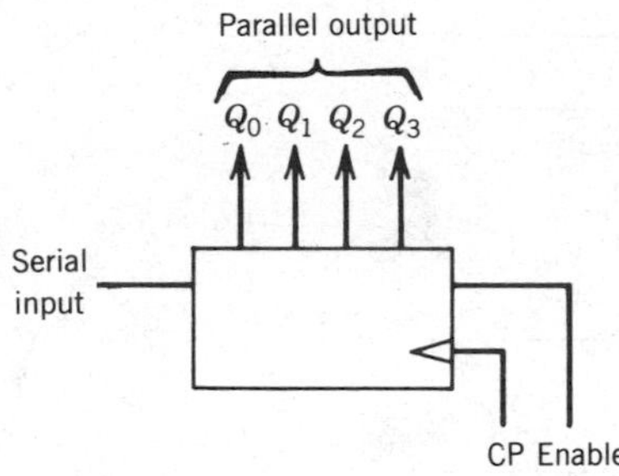

Fig. 29.25 Use of right-shift shift register for serial-to-parallel conversion.

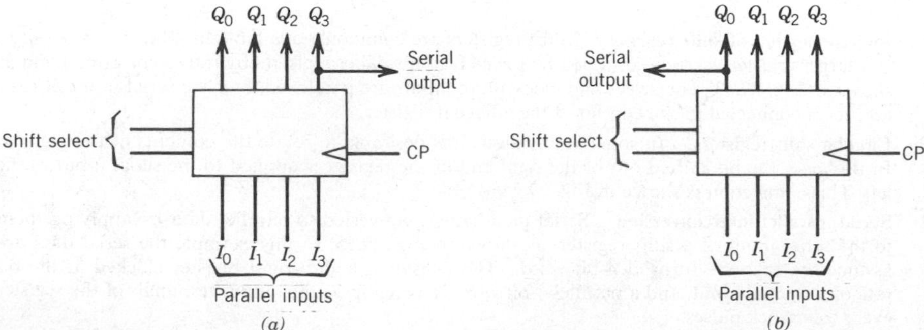

Fig. 29.26 Use of shift register for parallel-to-serial conversion: (*a*) High-order bit first, (*b*) low-order bit first. The serial inputs are not used and not shown.

serial data are available at the Q_3 output of Fig. 29.26*a*. In this configuration the least significant bit (LSB) of the input data appears first and the most significant bit (MSB) appears last. If the opposite sequence is required, the register is set to the shift-left mode and the output taken from Q_0, as shown in Fig. 29.26*b*.

References

1 F. J. Hill and G. R. Peterson, *Introduction to Switching Theory and Logical Design*, 3rd ed., Wiley, New York, 1981.

2 M. M. Mano, *Digital Logic and Computer Design*, Prentice-Hall, Englewood Cliffs, NJ, 1979.

3 H. Taub and D. Schilling, *Digital Integrated Electronics*, McGraw-Hill, New York, 1977.

CHAPTER 30
DIGITAL AND ARITHMETIC FUNCTION DEVICES

PETER GRAHAM

Florida Atlantic University, Boca Raton, Florida

30.1 INTRODUCTION

Chapters 28 and 29 dealt with logic devices such as gates and flip-flops, which are generally described as small-scale integrated (SSI) circuits. This chapter introduces circuits consisting of interconnections of several gates and/or flip-flops, which constitute medium-scale integration (MSI) devices commonly used as building blocks for digital systems. These include counters, timers, decoders, encoders, and the arithmetic logic unit. Examples are provided to indicate the variety of devices available, and a few applications are given to illustrate how the use of such devices can reduce the total package count of a system.

30.2 COUNTERS

A simple modulo N counter is a sequential circuit that changes state with each successive pulse at its single input, generating a prescribed sequence of N different states within the circuit, and repeating this sequence for every set of N input pulses.[1] The N memory states are implemented by M flip-flops ($M \geqslant \log_2 N$), which are clocked in response to the counter input pulses. (See Chapter 29, Section 29.1, for a description of clocked flip-flops.) The flip-flop control inputs are all internally connected and are appropriate logical functions of the flip-flop outputs, as shown in Fig. 30.1. The counter outputs likewise are functions of the flip-flop states.

Most counters have, in addition to the pulse input line, one or more inputs which in some manner control the operation of the counter. Many but not all counters also have a sufficient number ($\geqslant \log_2 N$) of output lines to identify all N states of the sequence. Counters are available in a variety of forms on medium-scale integrated (MSI) circuit chips in all three of the widely used MSI technologies, transistor-transistor logic (TTL), emitter-coupled logic (ECL), and complementary metal oxide semiconductor (CMOS) logic (see Chapter 28). Classifying these various counters requires a number of definitions.

Definitions Associated With Outputs

Counters with enough output lines to encode all N states generate a modulo-N sequence of outputs. The standard method of characterizing an output sequence is the ordering of output lines to represent

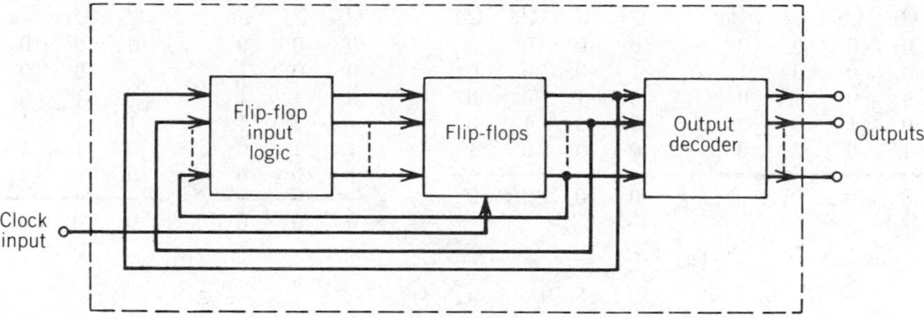

Fig. 30.1 Block diagram of a single-mode counter.

bits of a binary number, from the least significant bit (LSB) to the most significant bit (MSB). The output sequence can then be identified with a sequence of binary numbers that defines the "output code" of that counter. The two most widely used output codes lead to the following definitions of counters:

M-Stage (or M-Bit) Binary Counter. A modulo-N ($N = 2^M$) counter whose M outputs generate the natural binary number sequence.

Binary Coded Decimal (BCD) Decade Counter. A 4-stage modulo-10 counter whose four outputs generate the binary equivalent of the decimal number sequence 0 through 9.

Each of the above counters count in the natural binary sequence. The decade counter is actually a special-case binary counter that is modulo 10. Associated with each are the following simple definitions:

Up Counter. A counter that outputs the natural binary sequence of numbers in increasing order.

Down Counter. A counter that sequences through the natural binary numbers in decreasing order.

If a binary or BCD counter is not identified as either of the above, it usually is an up-counter.

The ring counter (or Johnson counter) is another type of counter that is identified with a particular form of output.[1] Two types of ring counters can be defined:

Standard Ring Modulo-N Counter. A counter that has N output lines, ordered Q_0 through Q_{N-1}, which are sequentially asserted HIGH, each for one input pulse period.

Twisted Ring Modulo-N Counter. A counter that has N output lines, ordered Q_0 through Q_{N-1}, each of which is asserted HIGH for $N/2$ consecutive input pulse periods. The level shifts of the outputs occur sequentially, with Q_i shifting one clock pulse before Q_{i+1}.

The last two definitions are clarified in Fig. 30.2, using modulo-5 and modulo-6 counters, respectively.

Definitions Associated With Pulse Input

Data books giving counter specifications usually refer to the counting input line as the "clock pulse input" or sometimes more simply as the "clock input." The latter term will be used here.

The count of some counters advances when the clock input shifts from LOW to HIGH, while other counters advance on the HIGH-to-LOW shift. These are called, respectively, rising edge and falling edge initiated inputs. Some counters are claimed to have both options (the 4017B, for example), but this is accomplished by the provision of a separate input that simply inverts the incoming pulse before applying it to the counter. A designer must recognize this when checking for possible timing problems in a system utilizing the falling edge input.

Two important distinctions must be made regarding the internal operation of counters. These lead to the following two definitions:

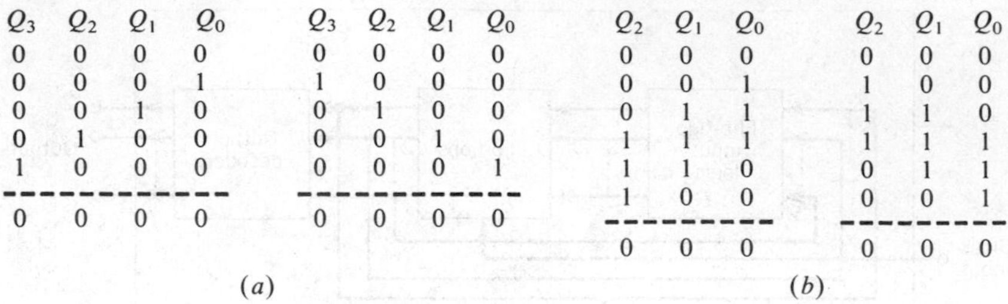

Q_3	Q_2	Q_1	Q_0
0	0	0	0
0	0	0	1
0	0	1	0
0	1	0	0
1	0	0	0
0	0	0	0

Q_3	Q_2	Q_1	Q_0
0	0	0	0
1	0	0	0
0	1	0	0
0	0	1	0
0	0	0	1
0	0	0	0

(*a*)

Q_2	Q_1	Q_0
0	0	0
0	0	1
0	1	1
1	1	1
1	1	0
1	0	0
0	0	0

Q_2	Q_1	Q_0
0	0	0
1	0	0
1	1	0
1	1	1
0	1	1
0	0	1
0	0	0

(*b*)

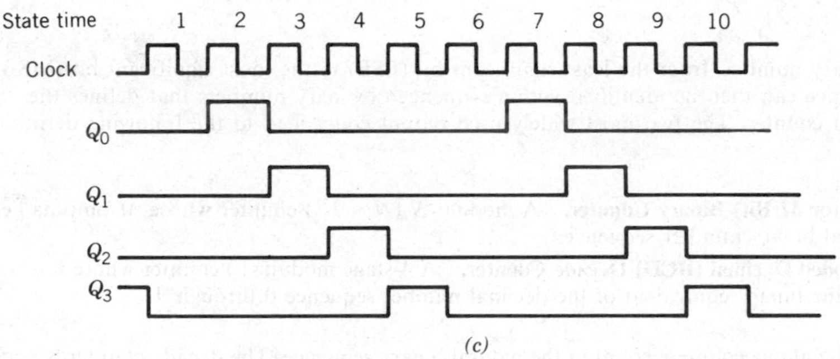

(*c*)

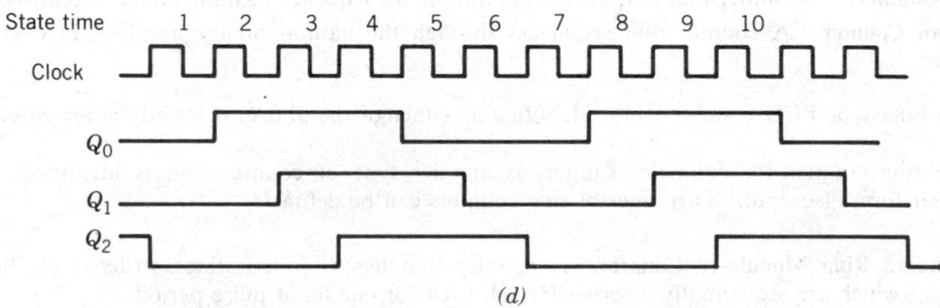

(*d*)

Fig. 30.2 (*a*) Two versions of standard ring counting. (*b*) Two versions of twisted ring counting. (*c*) Timing diagram for standard ring. (*d*) Timing diagram for twisted ring.

Synchronous Counter. A counter whose clock input is applied to all the memory state flip-flop clock inputs, so that all the flip-flops destined to change state at a particular input pulse do so simultaneously. An example is shown in Fig. 30.3.

Ripple Counter. A counter whose clock input is applied only to the least significant bit storage flip-flop, with the output of the LSB flip-flop driving the clock input of the next-to-LSB flip-flop, and so on, as shown in Fig. 30.4.

MSI packaged ripple counters invariably use master-slave flip-flops which change state on the falling edge of the flip-flop clock input. Consequently these counters advance on the falling edge of the counter clock input. It is important to note that since the flip-flops of ripple counters do not change state simultaneously, neither do the counter outputs. Combinational logic functions of such outputs are therefore subject to unplanned glitches and must not be used for asynchronous applications such as reset or load functions. As an example, consider the output function $Z = Q_1 \overline{Q}_0$

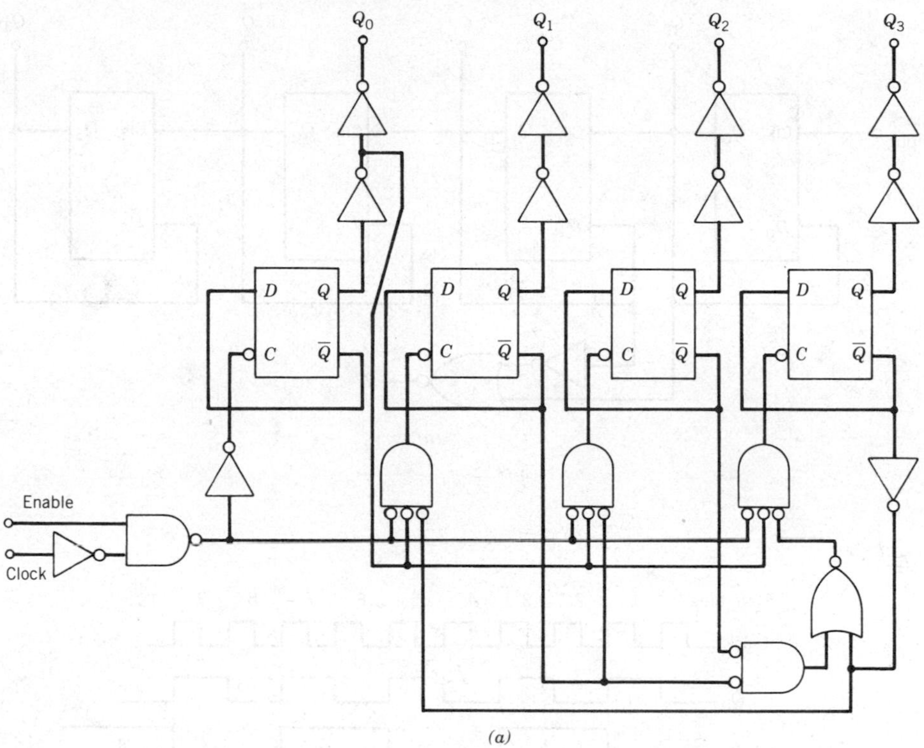

(a)

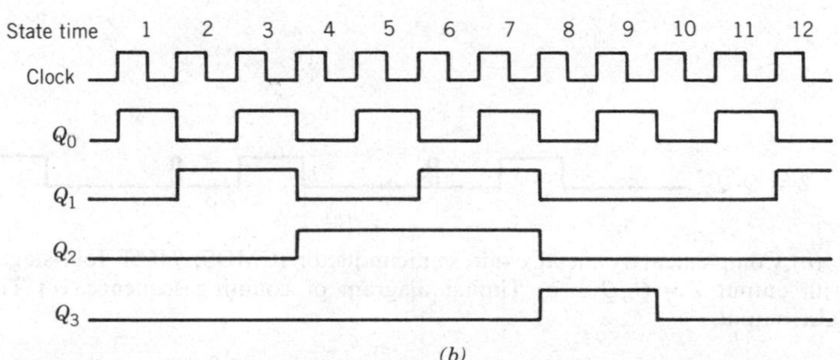

(b)

Fig. 30.3 (*a*) Complementary metal oxide semiconductor (CMOS) 4518B synchronous decade counter (reset not shown). (*b*) Timing diagram of counting sequence (enable input HIGH).

of the ripple counter of Fig. 30.4. It is legitimately HIGH during clock state time 2. However, at the end of clock state time 3, Q_0 shifting from HIGH to LOW clocks Q_1 from HIGH to LOW. Typical propagation delay for a CMOS D flip-flop operating at 5 V is 220 ns, so that the delay in Q_1 down shifting causes Z to be illegally HIGH for that duration. Since the typical asynchronous reset minimum pulse width for the same flip-flop is only 65 ns,[2,3] the 220-ns glitch could have serious consequences.

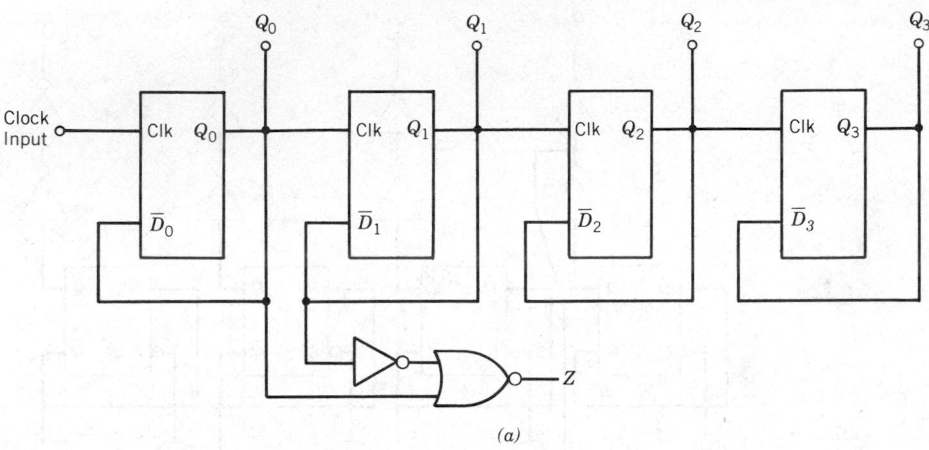

(a)

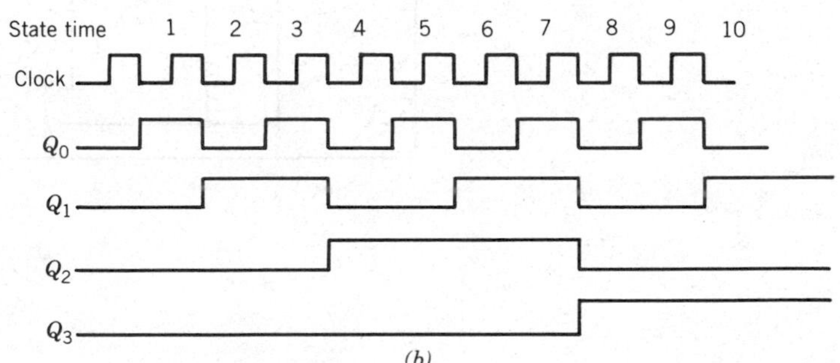

(b)

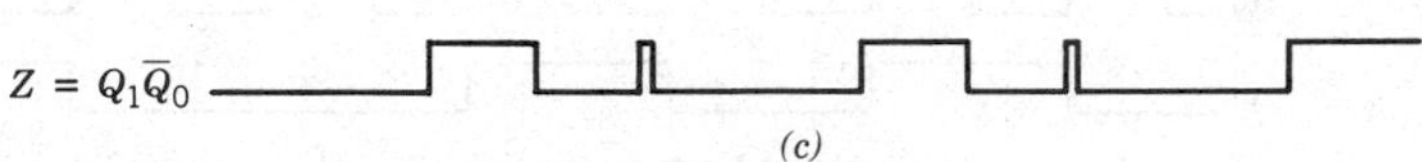

(c)

Fig. 30.4 (*a*) Complementary metal oxide semiconductor (CMOS) 74C93 four-stage binary ripple counter with output $Z = \overline{Q}_0 Q_1$. (*b*) Timing diagram of counting sequence. (*c*) Timing diagram showing false output.

Most if not all synchronous counters use rising-edge triggered flip-flops as storage elements and thus advance on the rising edge of the clock input, as shown in the timing diagram of Fig. 30.3*b*.

Control Inputs and Outputs

As an illustration of the variety of control inputs and outputs available in MSI technology, refer to the CMOS 4029B counter (Fig. 30.5). This versatile counter can be programmed by control inputs to be an up or a down counter (the U/D input), and to count modulo 10 in BCD or modulo 16 in natural binary (B/D input). There are four data input lines (P_0-P_3 in Fig. 30.5) through which any 4-bit binary number can be loaded into the counter. The loading occurs when the preset-enable (PE)

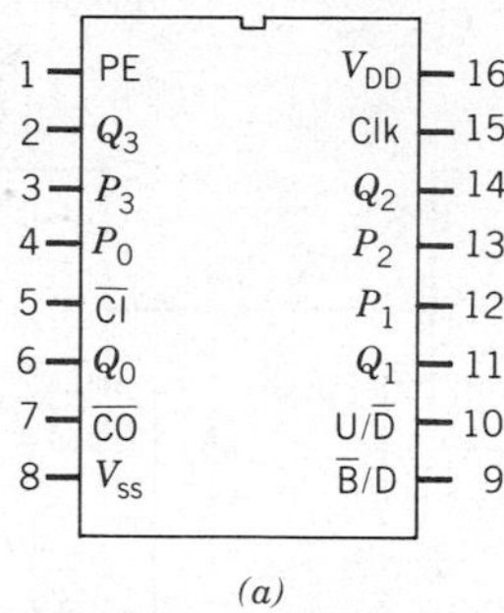

(a)

$\overline{CI}$	$U/\overline{D}$	$\overline{B}/D$	PE	Action
1	×	×	0	No count
0	1	0	0	Count-up binary
0	1	1	0	Count-up decade
0	0	0	0	Count-down binary
0	0	1	0	Count-down decade
×	×	×	1	Parallel load

(b)

Fig. 30.5 (a) *Pin* assignments for complementary metal oxide semiconductor (CMOS) counter 4029B. (b) Truth table for control input actions. × = don't care.

input is HIGH, independently of the clock input status. This is called asynchronous loading (in contrast to synchronous loading that occurs on the next rising edge of the clock input after the load command is enabled). When the PE input is HIGH the clock is inhibited so that counting does not advance until the next rising clock input after the PE input goes LOW. The input designated carry input ($\overline{CI}$), when HIGH, also inhibits the clock. The main use of this input is for cascading counters in applications such as that shown in Fig. 30.6. The carry output ($\overline{CO}$) goes LOW only when the counter reaches its maximum count (1111 for binary, 1001 for BCD) in the up-count mode. In the down-count mode ($U/\overline{D} = 0$), $\overline{CO}$ becomes a borrow, going LOW only when the counter is in the 0000 state. Thus, counter 2 of Fig. 30.6 has its clock input enabled only when counter 1 contains 1000, so that counter 2 is decremented only once every 10 input pulses. The complete descriptive name for the type 4029B counter is "a presettable, cascadable, up/down, binary/decade 4-stage synchronous counter with asynchronous load."

Many counters are available that possess some or all of these features with variations. Some counters have separate clocks for up count and down count. As mentioned previously, some have synchronous load; these may, in addition, have synchronous or asynchronous clear (tantamount to loading all zeros). Most multimode counters are either binary or BCD; some can be programmed to be either (the CMOS 4569B, for example).

There are large-scale integration chips consisting of several counters. An example is the CMOS 4534B,[2] a 24-pin chip that has five cascaded BCD counters allowing it to count through the decimal equivalent range of 0 to 99,999. The chip has only four output lines, but on-chip multiplexers (see Section 30.3), complete with the necessary scanning oscillator, sequentially present the outputs of the counters to these lines in order from the LSB to the MSB. Carry-in and carry-out lines permit the cascading of 4534B chips to realize 10-digit or more decade counters.

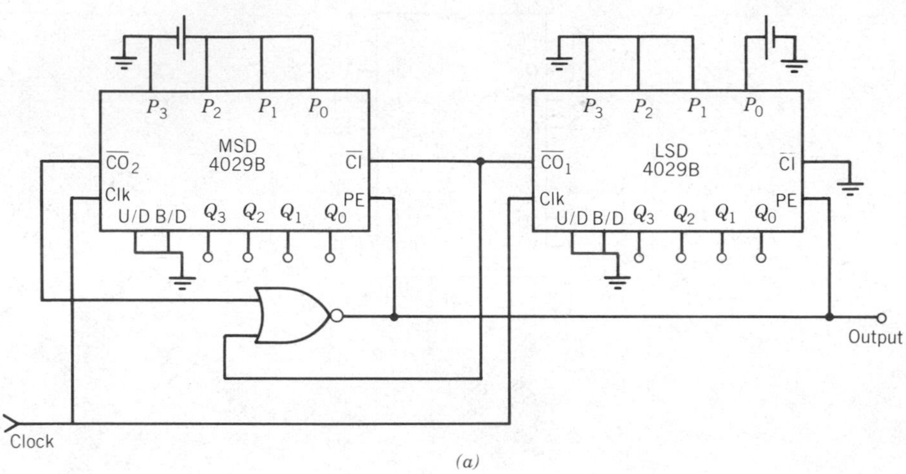

(a)

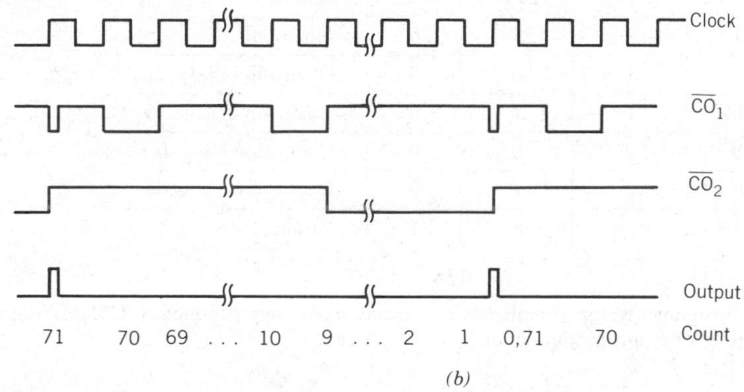

(b)

Fig. 30.6 Cascaded down counters set for divide-by-71: (*a*) Interconnection diagram, (*b*) timing diagram. MSD = most significant digit; LSD = least significant digit.

Divide-by-N Counters

Many applications require that the frequency of a clock pulse generator be divided by some integral number. It is conceptually simple to recognize that a modulo-N counter can be made to divide by N, because the counter will go through any given state only once in every N clock pulses. Thus designing a divide-by-N system is equivalent to forcing a counter into the desired modulo-N mode. Any counter having an asynchronous reset and an output from every stage can be forced back to the zero state from any other state by activation of the reset with the appropriate logic. For example, a modulo-16 binary counter can be converted to modulo 10 by activation of the reset input as soon as the MSB and the next to LSB are simultaneously HIGH. The effect is that the counter at the tenth clock pulse momentarily enters the 1010 state, activating its reset and switching to 0000 in the few nanoseconds necessary for the asynchronous reset to be effective. Care must be taken in applying this method to ensure that the momentary glitch into the 1010 state (see Fig. 30.7) does not cause a problem, and one must avoid cascading counters of different speeds, for example, low-power Schottky TTL and Schottky TTL, since the different reset times can lead to unexpected results.

Presettable counters can be connected to achieve a desired divide-by-N system by connection of appropriate levels to the data inputs and use of the carry outputs to assert the preset enable (PE) inputs. Fig. 30.6 shows two CMOS 4029 binary/decade, up/down counters connected to divide by 71. The mode control inputs are connected for decade (B/D = 0) countdown ($\overline{\text{U/D}}$ = 0) operation, so that each $\overline{\text{CO}}$ output is HIGH except when that counter is storing 0000. Since the $\overline{\text{CI}}$ input inhibits

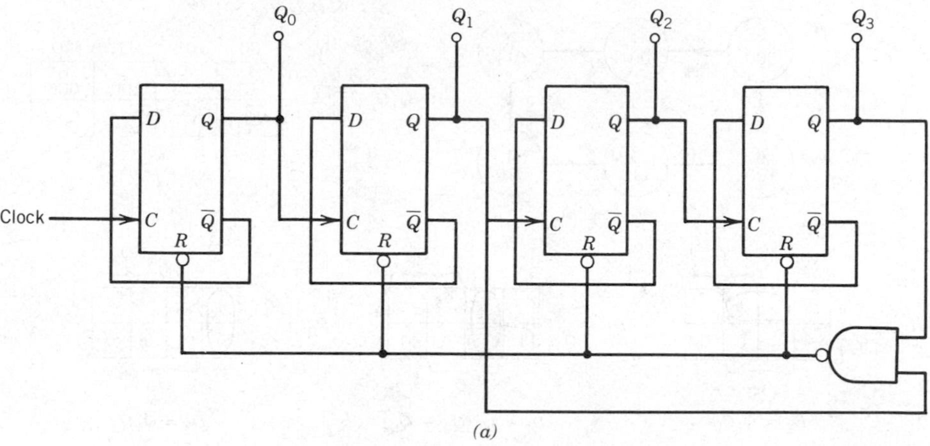

(a)

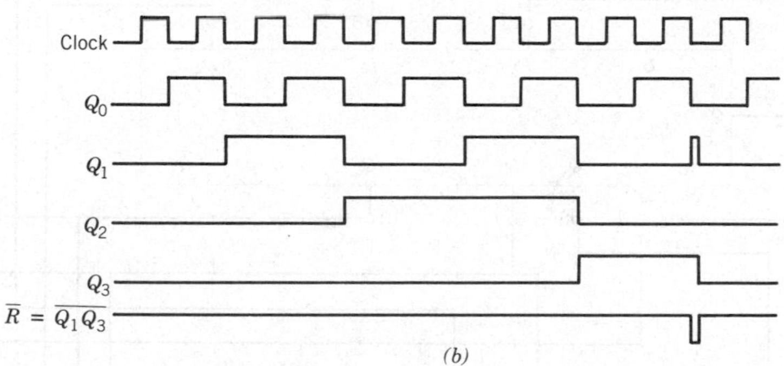

(b)

Fig. 30.7 (*a*) Ripple counter with reset programmed for modulo 10. (*b*) Timing diagram showing momentary existence of the 1010 state.

counting when HIGH, the most significant digit (MSD) counter down counts only on the rising clock edges when the least significant digit (LSD) counter holds 0000. The NOR gate is connected so that PE is HIGH only when both counters hold 0000. Since the PE input is asynchronous, the counters are both 0 only momentarily, so the sequence synchronized to the clock is 2, 1, 71, 70, . . . , and there is one output pulse for every 71 clock pulses.

Design of Synchronous Counters

The need may arise for a special counter not available in an MSI package. If a particular output sequence is required, one solution is to design a decoder that converts the standard binary or BCD outputs of an available counter into the desired output sequence. An alternative that provides greater flexibility and often is cheaper is to custom design the required counter around a quad-D flip-flop chip such as the TTL 74175,[4,5] which contains four rising-edge triggered flip-flops with a common clock input. This is easily done with the aid of Karnaugh maps, as illustrated in Fig. 30.8. The example is a modulo-6 synchronous counter whose counting sequence is defined in the state diagram of Fig. 30.8*a*. The digits 1 through 6 on the Karnaugh map (Fig. 30.8*b*) correspond to the state numbers in Fig. 30.8*a*. Entered into these numbered cells are the binary combinations corresponding to the next state in the sequence. Since these are the flip-flop states required after the next rising edge of the clock pulse, the functions corresponding to the next state maps define the logic required for the *D* inputs of the flip-flops. With this design the output sequence is obtained directly from the flip-flop outputs.[1]

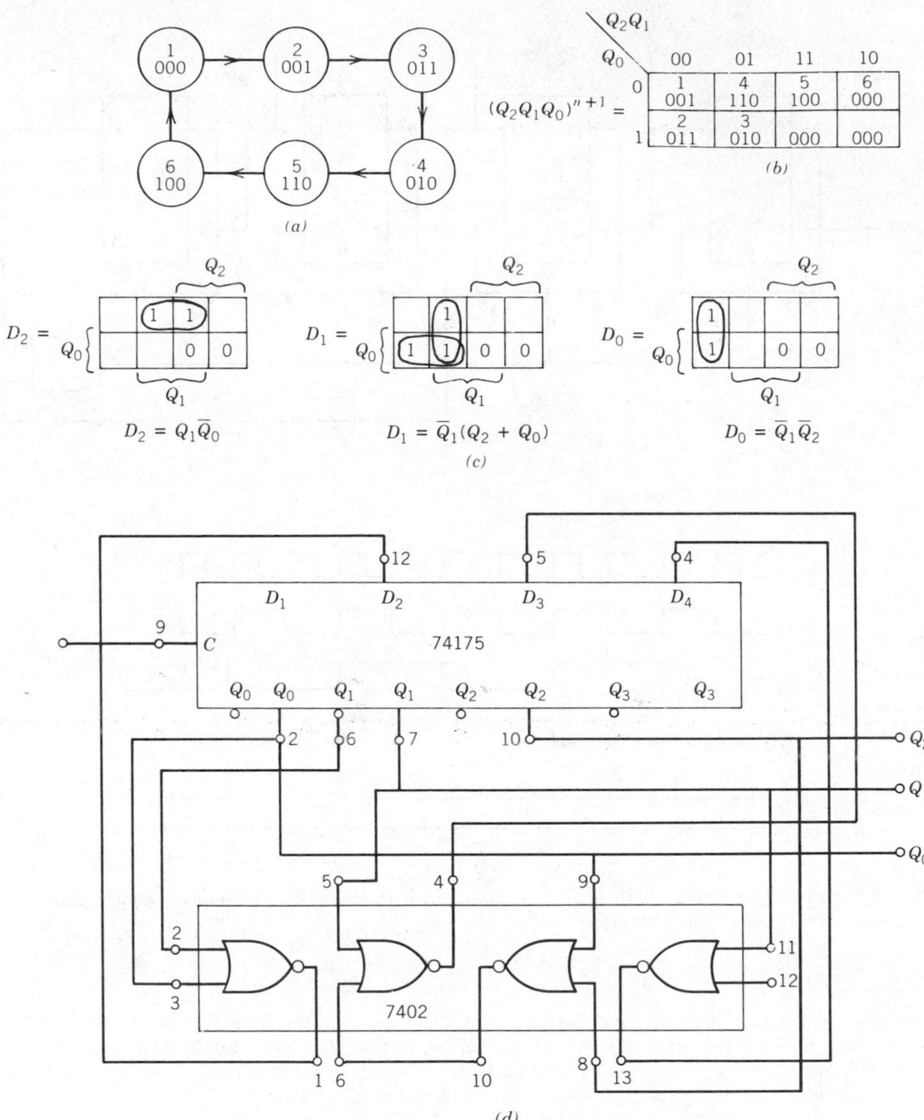

Fig. 30.8 (*a*) State diagram of a modulo-6 counter. (*b*) Karnaugh mapping of the state diagram. (*c*) Karnaugh maps for the *D* flip-flop control inputs. (*d*) Implementation of the modulo-6 counter.

30.3 TIMERS

A timer is a device designed primarily to generate adjustable time delays.[6,7] The most common timing circuit is the monostable multivibrator. In its usual form, the monostable (or one-shot, as it is often called), in response to a voltage level shift at its input, will output a voltage pulse of adjustable duration. The pulse may be positive-going or negative-going depending on the circuit, and for a given set of conditions will have a constant amplitude. The pulse duration is typically established by a suitable choice of values of a resistor and a capacitor. Several types of timers are available in integrated circuit form, some of which are discussed in the following paragraphs.

A timing circuit (Fig. 30.9) that is relatively simple in concept is marketed by several manufacturers under the classification "555 timer." Specifications of the 555 usually are to be found in linear

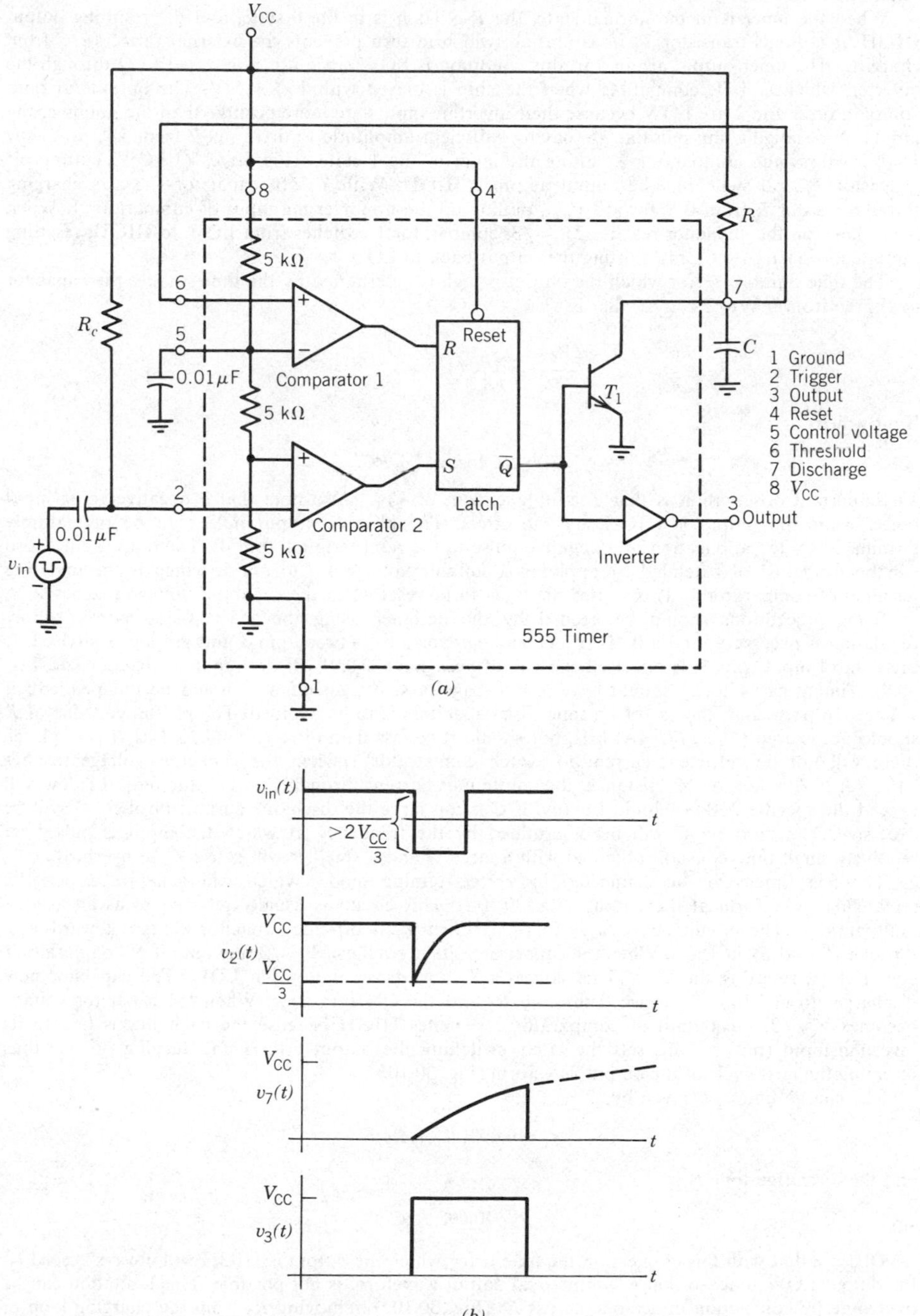

Fig. 30.9 555 Timer monostable multivibrator connection: (*a*) Logic diagram, (*b*) timing diagram.

integrated circuit data books.[8,10] In Fig. 30.9*a* the timer chip circuit is shown inside the dashed line, with the external connections being those required for monostable operation. In this mode the circuit functions as follows.[6,7]

When the timer is in the normal state, the *R-S* latch is in the 0 state, and the resulting output HIGH at $\overline{Q}$ holds transistor T_1 in saturation, which in turn prevents the external capacitor C from charging. The timer output at pin 3 in this condition is LOW since it is connected to $\overline{Q}$ through the inverter, which is TTL compatible when the chip is biased with $V_{CC} = 5$ V. The outputs at both comparators 1 and 2 are LOW because their inverting inputs are more positive than the noninverting inputs. A negative-going pulse at V_{in} having sufficient amplitude to drive pin 2 from V_{CC} to below $V_{CC}/3$ will switch comparator 2, setting the latch to the 1 state. This drives $\overline{Q}$ LOW, cutting off transistor T_1 and switching the output at pin 3 HIGH. With T_1 off, capacitor C starts charging through resistor R from 0 V toward V_{CC}, pulling up the noninverting input of comparator 1. When the voltage on the capacitor reaches $2V_{CC}/3$, comparator 1 switches from LOW to HIGH, resetting the latch, discharging C, and shifting the output back to LOW.

The time duration T for which the output is high is determined by the time it takes the capacitor to charge from 0 V to $2V_{CC}/3$, that is

$$\frac{2V_{CC}}{3} = V_{CC}(1 - e^{-T/RC})$$

from which

$$T = RC \ln 3 \cong 1.1\ RC$$

An important observation is that T is independent of V_{CC}. Also note that a negative-going input pulse, while the output is HIGH, has no effect. The positive output pulse can be prematurely terminated by the application of a negative pulse to the reset terminal (pin 4). The output pulse also can be shortened or extended by applying a suitable voltage V_m to pin 5, which is the inverting terminal of comparator 1. This causes the latch to be reset when the capacitor voltage reaches V_m.

Some precautions should be heeded by the designer using the 555.[11] Good power supply regulation is necessary, and a 0.01-μF decoupling capacitor between pin 8 and ground is advised. If the control input (pin 5) is not used the circuit *must* have a 0.01-μF decoupling capacitor (see Fig. 30.9). Timing capacitor C should have low leakage, and its capacitance should be independent of voltage. In particular, the use of ceramic disk capacitors is to be avoided. The resistance value of R should not exceed ($V_{CC}/175\ \mu$A) MΩ, nor should it be less than ($V_{CC}/30$ mA) kΩ. If R is too large there will not be sufficient current to switch comparator 1 when the capacitor voltage reaches $2V_{CC}/3$. If R is too low a resistance, the continuous current through T_1 when the output is low will exceed the specified 30-mA limit. Finally, if C is too large the discharge current through T_1 will be excessive. The limit on C will be determined by the frequency at which the input is pulsed. A relatively small time constant obtained with a large C and a small R will cause T_1 to overheat.

The 555 timer can be connected in a free-running mode, which will generate a periodic rectangular waveform at the output (Fig. 30.10*a*). This circuit is usually referred to as an astable multivibrator. The circuit operates as follows. When T_1 is off, the capacitor charges toward V_{CC} through R_1 and R_2 in series. When the capacitor voltage reaches $2V_{CC}/3$, the output of comparator 1 goes HIGH, resetting the latch. This saturates T_1 and drives the output LOW. The capacitor now discharges from $2V_{CC}/3$ toward 0 through R_2 and the ON transistor. When the capacitor voltage becomes $V_{CC}/3$, the output of comparator 2 switches HIGH because the capacitor is tied to its inverting input (pin 2). This sets the latch, switching the output HIGH and turning off T_1, thus initiating the next cycle of the output waveform (Fig. 30.10*b*).

The charge time t_1 is given by

$$t_1 = 0.693(R_1 + R_2)C$$

and the discharge time t_2 is

$$t_2 = 0.693\ R_2C$$

Observe that with this connection the time t_1 for which the output is HIGH will always exceed t_2, the output LOW time, so that a symmetrical output waveform is not possible. This limitation can be overcome by connection of a diode across R_2 (Fig. 30.10*c*), removing R_2 from the charging loop of the capacitor. The additional diode in series with R_2 places a diode in both the charge and the discharge paths, causing the ratio $t_1 : t_2$ to equal the resistance ratio $R_1 : R_2$. This circuit permits a wide range of duty cycles [where duty cycle is defined as (pulse duration)/(period)]. To ensure that oscillations will start in any of these free-running circuits, R_2 should be greater than 3.0 kΩ.

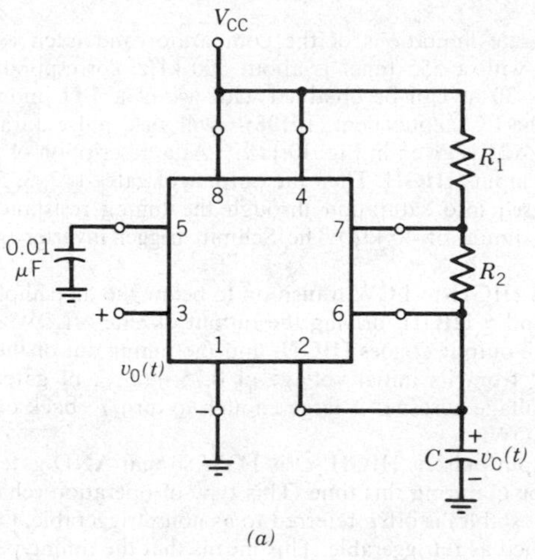

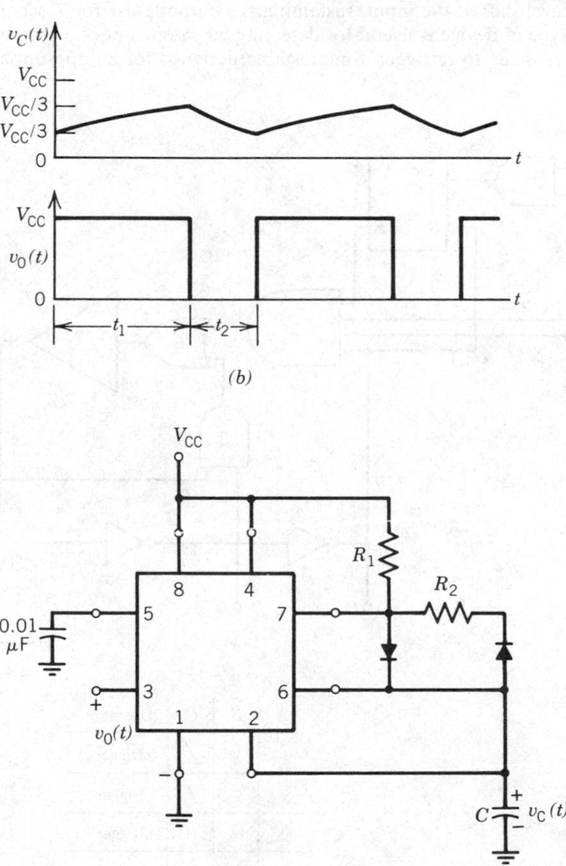

Fig. 30.10 Astable connection of 555 timer: (*a*) Typical circuit, (*b*) voltage waveforms, (*c*) modification to increase duty cycle range.

Short-Duration Timers

Owing primarily to the slew-rate limitations of the comparator and latch circuits, the maximum frequency reliably obtainable with a 555 timer is about 500 kHz, corresponding to pulsewidths of about 1 μs. Pulses as short as 30 ns can be obtained with use of a TTL monostable multivibrator (74121, for example),[4] while the ECL equivalent (10198)[12] will yield pulse durations of 10 ns or less.

The logic diagram of the 74121 is given in Fig. 30.11*a*.[13] As a description of the operation, assume that the initial condition is all inputs HIGH. Then the outputs of gates 1, 2, 3, 5, and 6 will be LOW, and transistor T_1 will be biased into saturation through the timing resistance R. (The saturation requirement limits R to a maximum of 40 kΩ.) The Schmitt trigger inverter (gate 4) output will be HIGH.

Now let input A undergo a HIGH-to-LOW transition to below the threshhold of 0.8 V. This shifts the outputs of gates 1, 2, 3, and 6 HIGH, driving the output of gate 4 LOW and pulling down the base of T_1, turning T_1 off. The output Q goes HIGH, and the timing out of the RC circuit begins as C starts to charge through R from its initial voltage of $0.75 - (V_{OH}$ of gate 4) toward V_{CC}. The output pulse ends when the voltage across C is large enough to turn T_1 back on, shifting the outputs of gates 3, 5, and 6 back to LOW.

Observe that while the output pulse is HIGH, $\overline{Q}$ is LOW so that AND gate 3 will not respond to any changes at inputs A, B, or C during this time. This type of operation (characteristic also of the 555 timer connected as a monostable) is often referred to as nonretriggerable. Other monostables (the 74122, for example) are classified as retriggerable. This means that the timing period (T) is reinitiated by the appropriate level shift at the input, sustaining the output pulse for T seconds beyond the final input change. This type of device is useful for detecting when one out of a sequence of events fails to occur. The events are made to retrigger a monostable designed for an appropriate period T, so that

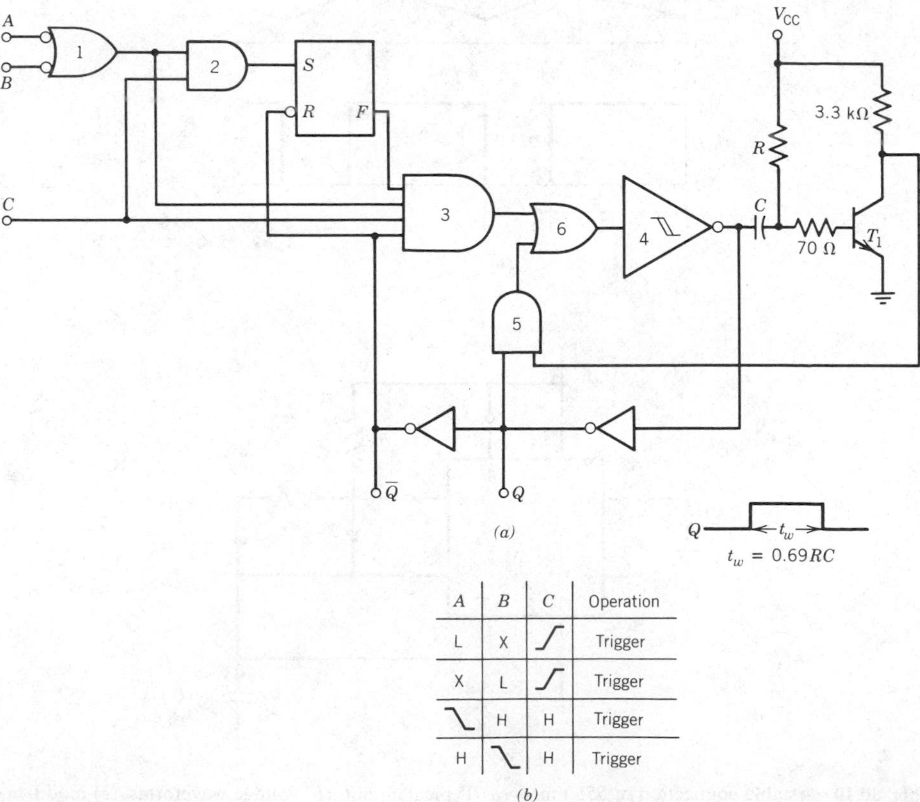

A	B	C	Operation
L	X	╱	Trigger
X	L	╱	Trigger
╲	H	H	Trigger
H	╲	H	Trigger

Fig. 30.11 Transistor-transistor logic (TTL)-type 74121 monostable multivibrator: (*a*) Logic diagram, (*b*) truth table for control inputs. × = Don't care; ╲ = high-to-low transition, ╱ = low-to-high transition.

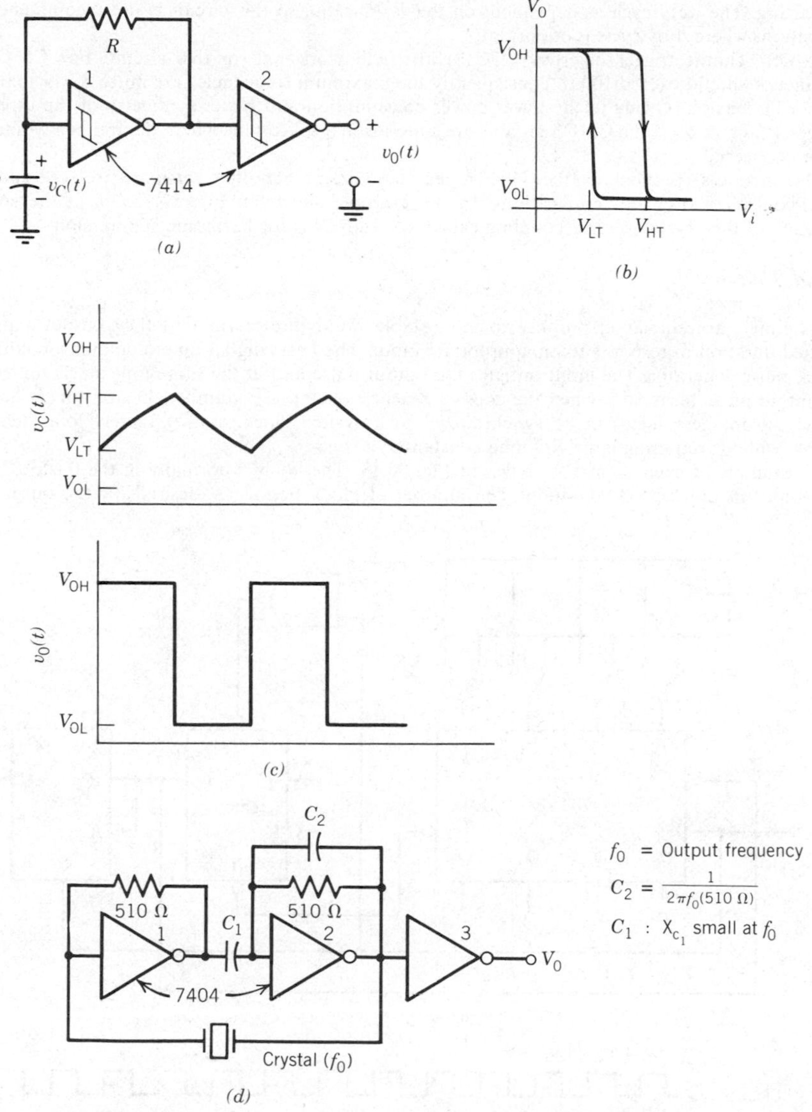

Fig. 30.12 Schmitt trigger inverter: (*a*) Used in astable multivibrator, (*b*) transfer characteristic, (*c*) capacitor and output waveforms, (*d*) used in crystal-controlled astable multivibrator.

the output remains HIGH until an event is missing. It should be noted that if the retriggerable operation is not required, the wise designer will use a nonretriggerable circuit.

High-frequency astable multivibrators can be constructed using Schmitt trigger inverters (Fig. 30.12*a*).[1] The operation is readily explained by assuming that the output of inverter 1 has just switched from LOW to HIGH. The voltage across capacitor C at this instant is the LOW threshhold V_{LT} (Fig. 30.12*b*) of the inverter transfer function. The capacitor starts charging through R from this initial value toward V_{OH} until its voltage reaches the HIGH threshhold V_{HT}. At this instant the inverter output downshifts to V_{OL}, and the capacitor commences discharging toward V_{OL}. The next cycle begins when the capacitor voltages again become V_{LT}. The frequency depends on the transfer function and is particularly sensitive to V_{HT}-V_{LT}, so it varies somewhat from chip to chip. A good estimate, however, for the TTL inverter 7414 is about $1/3RC$ Hz. Frequencies from < 1 kHz to 30 MHz are obtainable with this chip. Values for resistance R should be less than 1.4 kΩ to ensure self-starting. The duty cycle is dependent on the R/C ratio, so this circuit is not recommended for applications where duty cycle is important.

CMOS Schmitt trigger inverters (47C14) also will work well in this circuit. For CMOS, the resistance R should exceed 10 kΩ, consequently the maximum frequencies are much lower than those of the TTL version. Owing to the lower power consumption and better symmetry of the threshhold voltages V_{HT} and V_{LT}, the CMOS circuits are superior in applications where the frequency limitation is of no concern.

TTL inverters (particularly the 7404 in the standard or Schottky versions) are well suited for crystal-controlled oscillator applications. In the example shown in Fig. 30.12*d*, f_0 is the specified frequency of the crystal, C_1 is a coupling capacitor, and C_2 is for harmonic suppression.

Digital Timers

Digital timers are similar in purpose to monostable multivibrators in that they output a pulse of specified duration in response to an appropriate input. The basic digital timer contains a counter and a clock pulse generator. The input initiates the output pulse and at the same time starts the counter. The output pulse terminates when the counter reaches some preset number. Digital timers have two distinct advantages; they can be synchronized to a system clock, and extremely long delays are possible without requiring large RC time constants.

An example of such a timer is shown in Fig. 30.13. The latch is normally in the 0 state, holding the counter in the LOAD condition. The number of clock periods N desired for the output pulse

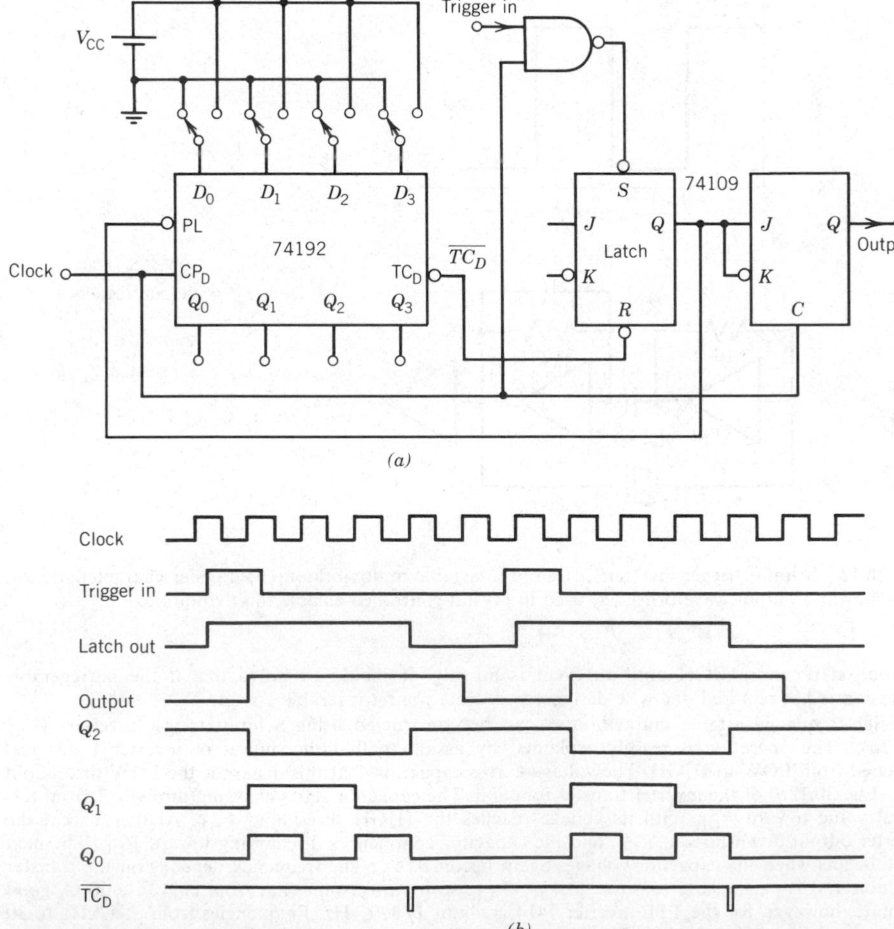

Fig. 30.13 Digital timer: (*a*) Circuit, (*b*) timing diagram, timer set for four clock periods.

duration is selected using suitable levels on the data inputs D_0-D_3. When the trigger input is enabled with a HIGH, the next rising edge of the clock pulse sets the latch, converting the counter from LOAD to COUNT, and setting up the *JK* flip-flop. The next rising edge of the clock pulse decrements the counter to $N - 1$ and changes the state of the *JK* flip-flop, shifting the output HIGH. $N - 1$ clock periods later the rising edge of the clock pulse decrements the counter to 0, causing the borrow output, $\overline{TC_D}$, to go LOW, resetting the latch. The rising edge of the next clock pulse resets the *JK* flip-flop to 0; returning the output to LOW and completing the output pulse whose duration is exactly *N* clock pulse periods. This sequence is illustrated in Fig. 30.13*b* for $N = 4$.

Digital timers are available in MSI circuit packages. An example is the CMOS 4154B,[2] which contains a 16-stage binary counter and an internally generated clock whose frequency is controllable over a range of 0 to 100 kHz by an externally connected resistance and capacitance. The counter is modulo-2^N, where *N* can be selected to be 8, 10, 13, or 16. This establishes the timing pulse range to be from 2.5 ms to many hours. The CMOS 4154B can also be operated with an external clock at frequencies up to 6 MHz.

30.4 MULTIPLEXERS

An *n*-input digital multiplexer is a combinatorial gate circuit with *n* data inputs, each of which can be selected to appear at the single output.[1] An 8-input multiplexer, for example, will have 8 data lines, one output line, and three data-select lines which provide the 8 different addresses required. To keep the diagram relatively simple, the multiplexer shown in Fig. 30.14 has only four inputs. The extension to any number of data lines is obvious.

The data lines are labeled D_0-D_3 in Fig. 30.14, and S_0 and S_1 are the data-line-select inputs. For example, when S_0 and S_1 are both LOW and the enable input is LOW, data on line D_0 appears at the output *Y*. The enable input permits multiplexer outputs to be ORed together to expand the number of data lines.

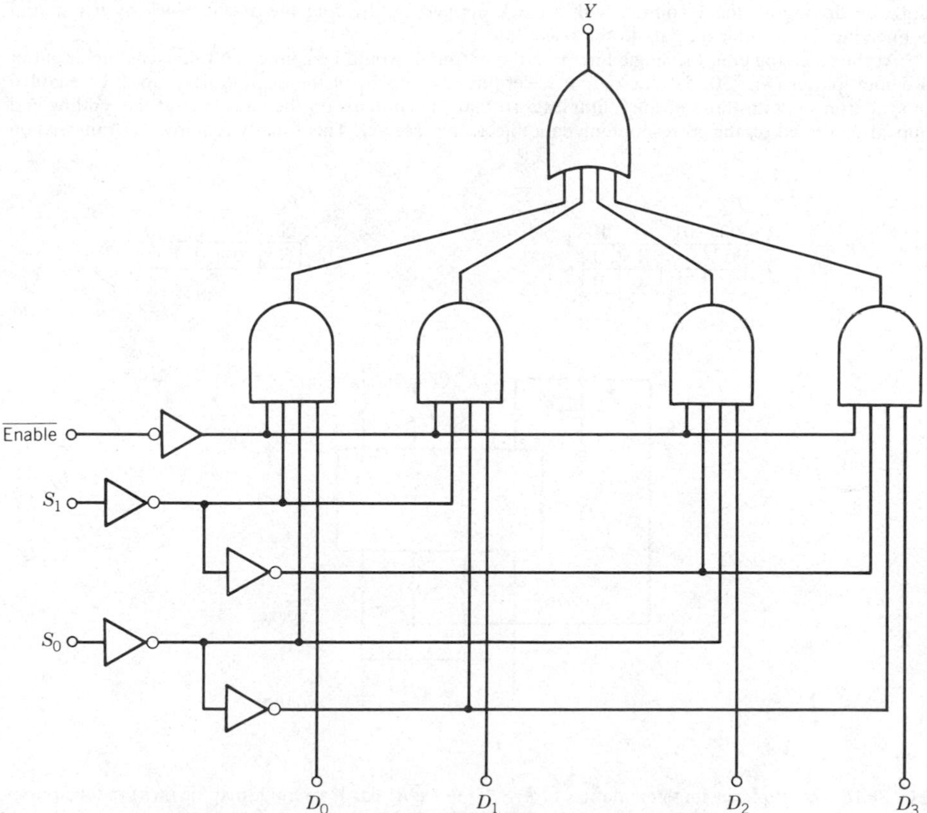

Fig. 30.14 A four-input multiplexer (one half of a 74153).

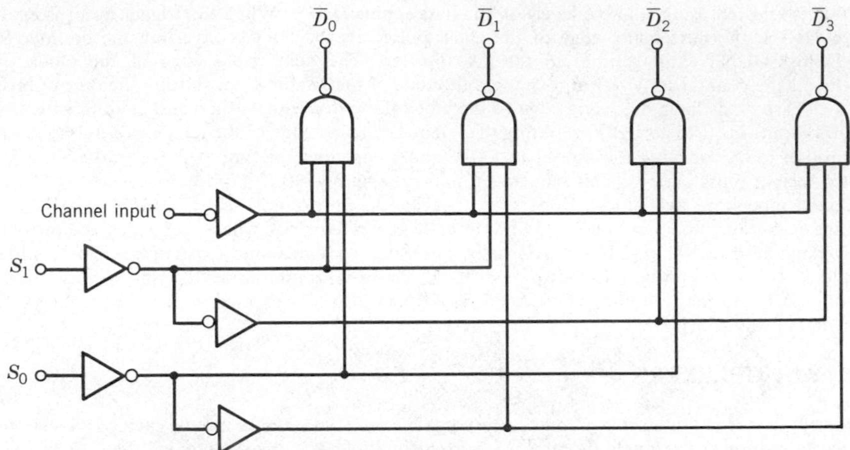

Fig. 30.15 Four-channel demultiplexer (one half of a 74139).

When used in a typical time-division multiplexing application, the device of Fig. 30.14 would have four different data streams at its inputs. The data on any one line might be binary words of b bits each presented as a sequence of 1's and 0's synchronized with a master clock. The time between words on a given line must be long enough for the other three lines to be sampled and for generation of a separating code between words on a given line. If the separating code word were c bits, each address would be applied to the data select inputs S_0 and S_1 for $b + c$ clock pulses. This is readily done by driving a binary counter with a clock derived by dividing the master clock by $b + c$, and connecting the counter outputs to the select lines.

At the receiving end, the single line from the output Y would be connected to the channel input of a demultiplexer (Fig. 30.15). The output select lines S_0 and S_1 of the demultiplexer must be enabled in synchronism with those of the multiplexer so that the contents on the data lines at the sending end appear unaltered on the corresponding data line at the receiver. This usually requires the transmission

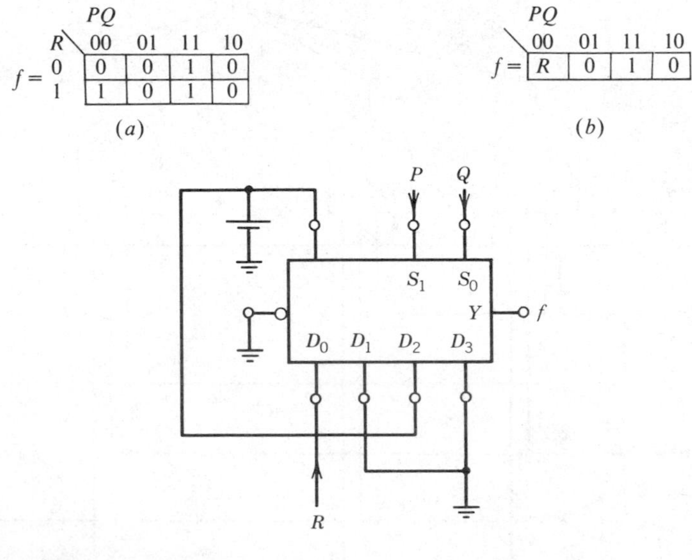

Fig. 30.16 Multiplexer implementation of $f = PQ + \overline{P}\overline{Q}R$: ($a$) Karnaugh map of f, (b) reduced map of (a), (c) implementation using one half of a 74153.

of a synchronizing pulse. The particular connection shown in Fig. 30.15 is half of a 74139.[4] Note that the channel input is inverted and the outputs are inverted, the two inversions cancelling one another.

Multiplexers in Combinatorial Logic Design

Four AND gates driving an OR gate, as shown in Fig. 30.14, indicates that these MSI circuits can be applied to realizations of combinatorial logic. The 4-bit multiplexer, for example, lends itself directly to implementing three variable Boolean functions. To illustrate, consider the function $f = PQ + \overline{P}\overline{Q}R$ shown on map (*a*) of Fig. 30.16. The simplest approach in general is to reduce the map size to the dimension corresponding to the number of select bits by variable entry mapping.

Variable entry mapping is a means of reducing the dimensions of a Karnaugh map by entering functions into the map cells instead of simply 1's and 0's.[1] Generally speaking, an n-variable Boolean function can be mapped onto an m-variable Karnaugh map ($m < n$) by associating m of the variables with the map coordinates, and entering functions of the remaining $n - m$ variables onto the map cells. The reduction process is as follows (see Fig. 30.17). Having selected the m-variables that are to be its coordinates (A, B, and C in Fig. 30.17*a* and A and C in Fig. 30.17*b*), draw and label the reduced map. Each cell of the reduced map will be associated with a group of 2^{n-m} cells which form a submap of the original. The submap for a given cell is identified on the original map as the

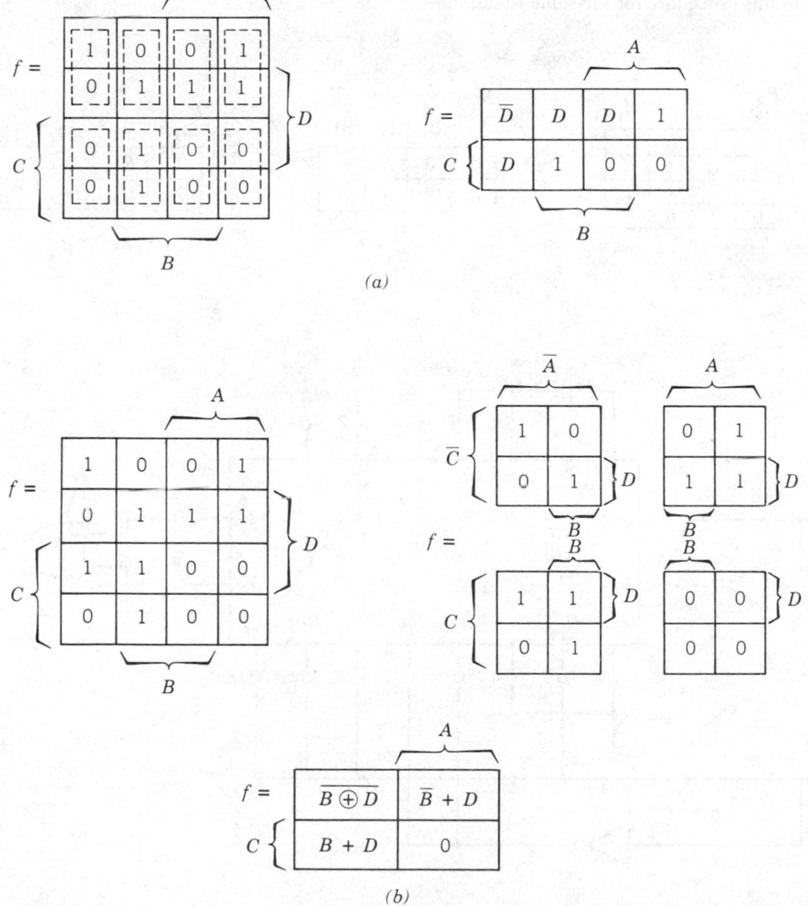

Fig. 30.17 Variable entry maps of $f = \overline{B}\overline{C}\overline{D} + \overline{A}BC + \overline{A}CD + A\overline{C}D + B\overline{C}D$: (*a*) Reduction from four to three map variables, (*b*) reduction from four to two map variables.

intersection of the coordinates of that cell. The function to be entered into the cell is found by simplifying its corresponding submap. To help clarify this concept, the original map is exploded in Fig. 30.17*b* into the four submaps (which when combined yield the original map). The submap at the top left expands to $\overline{B \oplus D}$. Similarly, the remaining submaps expand to $\overline{B} + D$, $B + D$, and 0. These are all shown in the squares of the reduced map.

Referring to the example of Fig. 30.16, it is seen that R has been removed as a map variable and has become an entry variable. The variables P and Q are connected to the data select inputs S_1 and S_0, Q going to S_0 because it is the least significant map variable. This allows each of the cells of the reduced map to be associated with a data line input, as shown in Fig. 30.16*c*. To complete the circuit, the function entered in a given cell of the reduced map is connected to the corresponding data line, and the implemented function is taken from the output Y of the multiplexer.

Multiplexers with enable inputs can be combined to implement logic functions of more variables. If the data select lines of two multiplexers are paralleled, then the combination of one enable line and the complement of the other form a third data select line. Since the enable lines are asserted LOW, the four data lines associated with the uncomplemented enable are the least significant bits. In Fig. 30.18 the four-variable function $f(SPQR) = S\overline{P}\overline{Q}\overline{R} + SPR + \overline{P}QR + \overline{S}PQ\overline{R}$ is implemented using the two halves of a 74153 multiplexer. It should be noted that the two sections have common select lines S_1 and S_0 and separate enable lines $\overline{E}_a$ and $\overline{E}_b$.

Note also that the number of multiplexer inputs must equal the number of cells on the reduced map of the desired function. Thus a five-variable function generates a four-variable reduced map that has 16 cells, and therefore 16 multiplexer inputs are required. An alternative that results in a tradeoff between multiplexers and external logic is to reduce the map by one more variable. Figure 30.19 illustrates this procedure for the same function f.

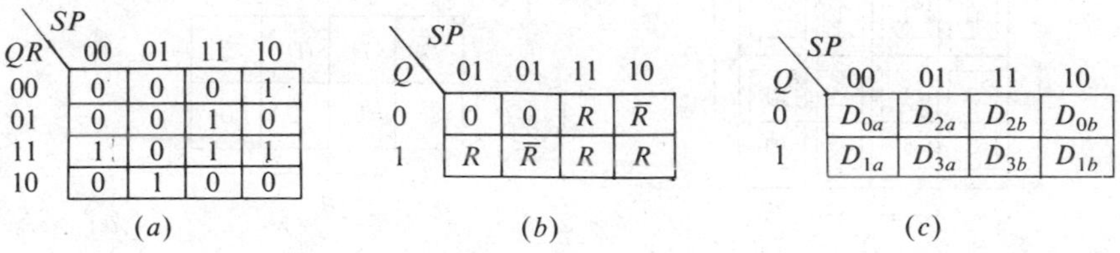

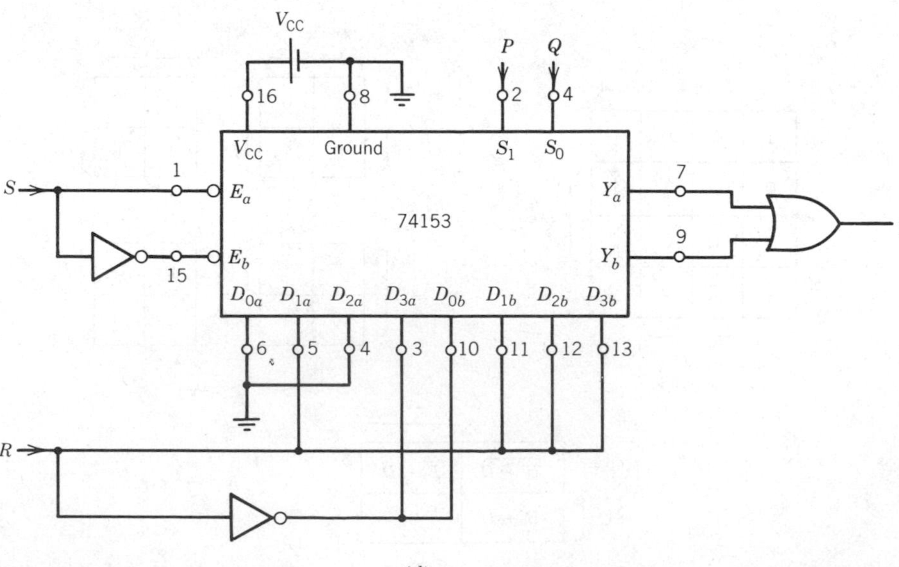

Fig. 30.18 Multiplexer implementation of $f = S\overline{P}\overline{Q}\overline{R} + SPR + \overline{P}QR + SPQ\overline{R}$: (*a*) Karnaugh map, (*b*) reduced map, (*c*) association of data input lines with map cells, (*d*) implementation with 74153.

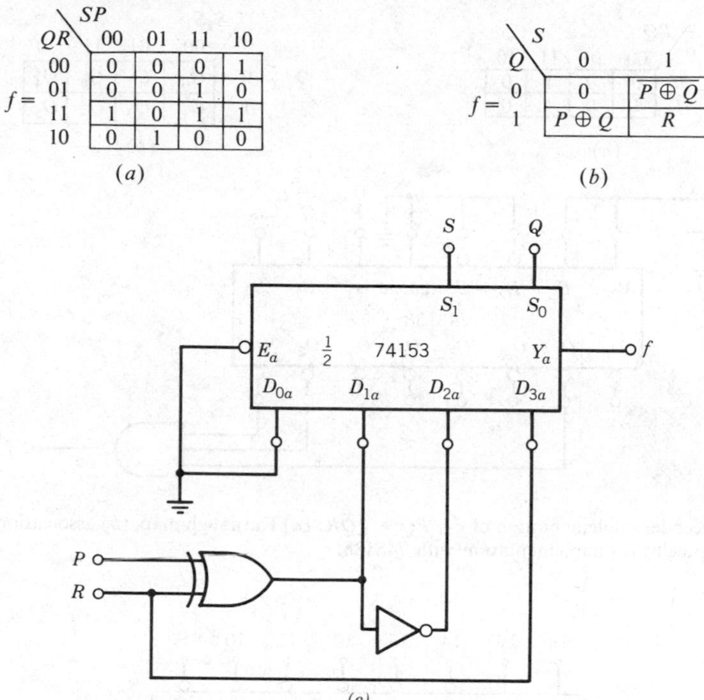

Fig. 30.19 Alternative multiplexer implementation of $f = S\overline{P}\overline{Q}\overline{R} + SPR + \overline{P}QR + SPQ\overline{R}$: (*a*) Karnaugh map, (*b*) reduced map, (*c*) implementation with one half of a 74153.

30.5 DECODERS, ENCODERS, AND CODE CONVERTERS

Decoders, encoders, and code converters are devices that fall into similar classifications by virtue of the functions they perform.[1] They can be defined as follows:

Decoder. A combinational circuit with m inputs and n outputs, where $n \leqslant 2^m$, which at any given time has only one output line active. The particular line that is active is determined by which member of the input code set is presented at the input lines.

Encoder. A combinational circuit with m inputs and n outputs, where $m \leqslant 2^n$, which at any given time will have only one input line active. Which member of the code set appears at the output depends on the particular input line that is active.

Code Converter. A combinational circuit with m inputs and n outputs which has a different set of input codes and output codes. For each member of the input code set there is a corresponding member of the output code set on a one-to-one basis.

Decoder Applications

The demultiplexer shown in Fig. 30.15 is also classified as a 1 of 4 decoder (or as a 2-line to 4-line decoder), and it follows that any decoder with an enable input to serve as a data line can be used as a demultiplexer.

A decoder for which $n = 2^m$ can be used for implementing combinational logic circuits. The method becomes apparent upon identifying such a decoder as a "minterm recognizer" in the sense that the particular output line that is asserted identifies that its corresponding minterm is applied to the output-select lines. This is illustrated with the example shown in Fig. 30.20, where a 1 of 8 decoder (74S138) is used to implement a three-variable function. The three variables, P, Q, and R, are connected respectively to select lines S_2, S_1, and S_0. Thus each cell of the Karnaugh map (Fig. 30.20*b*) is associated with one of the output lines. Since the function is to be logical 1 when the output 0_1, 0_6, or 0_7 is selected, the function is implemented by ORing these outputs together. Note that a NAND gate is used because the outputs are active LOW. While this scheme appears to be quite

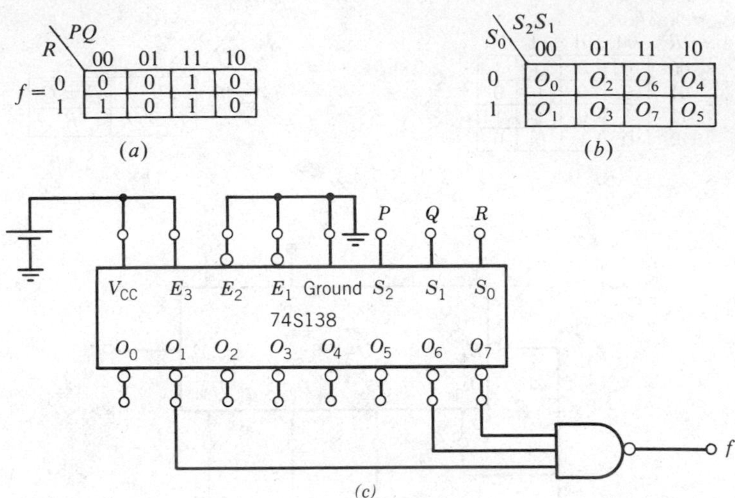

Fig. 30.20 Decoder implementation of $f = PQ + \overline{P}\overline{Q}R$: (*a*) Karnaugh map, (*b*) association of output lines with map cells, (*c*) implementation with 74S138.

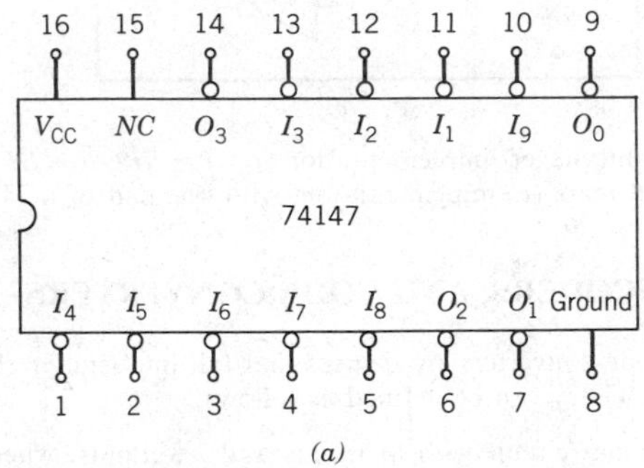

(*a*)

Inputs									Outputs			
I_1	I_2	I_3	I_4	I_5	I_6	I_7	I_8	I_9	O_3	O_2	O_1	O_0
H	H	H	H	H	H	H	H	H	H	H	H	H
×	×	×	×	×	×	×	×	L	L	H	H	L
×	×	×	×	×	×	×	L	H	L	H	H	H
×	×	×	×	×	×	L	H	H	H	L	L	L
×	×	×	×	×	L	H	H	H	H	L	L	H
×	×	×	×	L	H	H	H	H	H	L	H	L
×	×	×	L	H	H	H	H	H	H	L	H	H
×	×	L	H	H	H	H	H	H	H	H	L	L
×	L	H	H	H	H	H	H	H	H	H	L	H
L	H	H	H	H	H	H	H	H	H	H	H	L

(*b*)

Fig. 30.21 74147 Ten-line to four-line priority encoder: (*a*) Pinout diagram, (*b*) truth table (× = don't care).

wasteful of the capabilities of the decoder, one must recognize that, within fan-out limits, any number of different functions of the three select variables can be implemented with the addition of one NAND gate per function. The enable inputs on the 74S138 permit these units to be interconnected to increase the number of function variables. For example, four decoders can be used to generate a set of five variable functions.[1] It should be noted that each five-variable function could require a NAND gate of up to 16 inputs.

Encoder Applications

Encoders are primarily used to interface keypads to the logic that is to be keypad driven. An example of such an encoder is the 74147, which converts ten inputs to a four-line BCD output. There are in fact only nine data input lines, each of which is active LOW. The BCD zero output is activated when all nine input lines are HIGH (corresponding to none active). The 74147 is designated as a priority encoder. This means that if more than one input line is active (LOW), the output code will correspond to the highest order active input line. If line 5 and line 8 are both LOW, for example, the output code (which is also active LOW) will be LHHH, (1000 → 8). The priority feature allows several actions to be programmed simultaneously and executed in the sequence established by the line orders. The completion of each action must deactivate the highest order line before the next output code will be asserted. The truth table is shown in Fig. 30.21 (× corresponds to "don't care").

Code Converter Applications

A simple example of a code converter is the so-called BCD-to-seven-segment display decoder. This has four inputs that are activated simultaneously. The seven output lines are the proper HIGH-LOW combination to drive the light-emitting diode segments which replicate the coded decimal number. An example is shown in Fig. 30.22. Other code converter examples are BCD-to-excess-three-coded

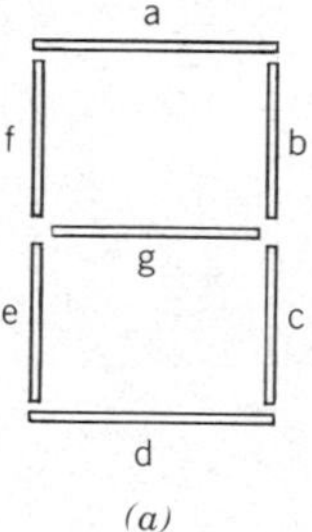

(*a*)

Inputs				Outputs (active low)							Display
I_3	I_2	I_1	I_0	$\bar{a}$	$\bar{b}$	$\bar{c}$	$\bar{d}$	$\bar{e}$	$\bar{f}$	$\bar{g}$	
L	L	L	L	L	L	L	L	L	L	H	0
L	L	L	H	H	L	L	H	H	H	H	1
L	L	H	L	L	L	H	L	L	H	L	2
L	L	H	H	L	L	L	L	H	H	L	3
L	H	L	L	H	L	L	H	H	L	L	4
L	H	L	H	L	H	L	L	H	L	L	5
L	H	H	L	H	H	L	L	L	L	L	6
L	H	H	H	L	L	L	H	H	H	H	7
H	L	L	L	L	L	L	L	L	L	L	8
H	L	L	H	L	L	L	H	H	L	L	9

(*b*)

Fig. 30.22 Binary coded decimal (BCD)-to-seven-segment display decoder: (*a*) Segment identification, (*b*) truth table.

decimal, BCD-to-excess-three-unit-distance-coded-decimal and/or BCD-with-parity-bit. The latter is a four-input five-output code converter.

30.6 BINARY ARITHMETIC DEVICES

Adders

A full adder (FA) is a three-input two-output combinational circuit that adds two 1-bit binary numbers. The three inputs include the two bits to be added and the carry input, provided to allow the cascading of adders to sum *n*-bit binary numbers. One output is the sum of the two bits and the carry input, and the other output is the carry generated by this sum. A half adder (HA) is an adder with no carry input. Truth tables and logical diagrams of full and half adders are presented in Fig. 30.23.[1]

The parallel addition of two *n*-bit binary numbers requires $n - 1$ full adders and one half adder, as shown in Fig. 30.24*a* (for $n = 8$). The numbers to be added are normally in storage registers and are simultaneously gated to the adder inputs at the start of the add operation. Since the adders are combinational circuits, the addition is performed in a few gate delays (typically 25 ns for standard TTL to as fast as 2 ns for ECL). The time required for a complete add operation, however, is considerably longer due to the carry-ripple effect. The worst case carry-ripple delay occurs when the binary equivalent of 1 is added to a binary number consisting of all 1's. In this case the carry generated by the half adder must propagate stage by stage all the way through to the MSB adder. With a typical carry propagation delay of 18 ns/stage, this would amount to 144 ns for an 8-bit adder.

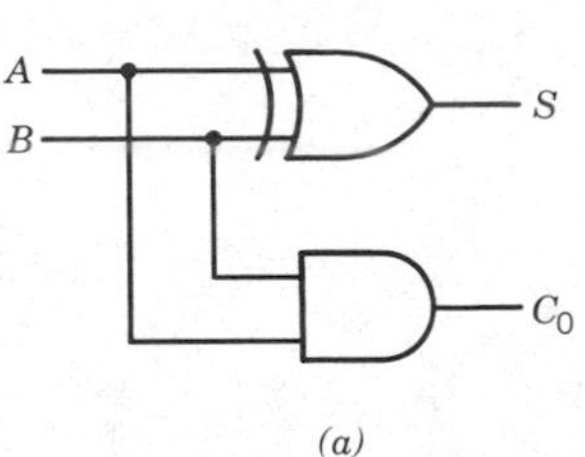

(*a*)

A	B	S	C_0
0	0	0	0
0	1	1	0
1	0	1	0
1	1	0	1

$$S = A \oplus B$$
$$C_0 = AB$$

(*a*)

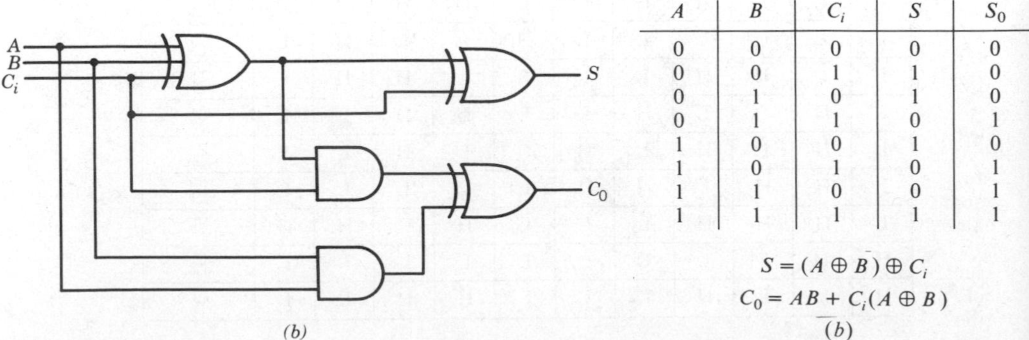

(*b*)

A	B	C_i	S	S_0
0	0	0	0	0
0	0	1	1	0
0	1	0	1	0
0	1	1	0	1
1	0	0	1	0
1	0	1	0	1
1	1	0	0	1
1	1	1	1	1

$$S = (A \oplus B) \oplus C_i$$
$$C_0 = AB + C_i(A \oplus B)$$

(*b*)

Fig. 30.23 Adders: (*a*) Circuit and truth table for half adder, (*b*) circuit and truth table for full adder.

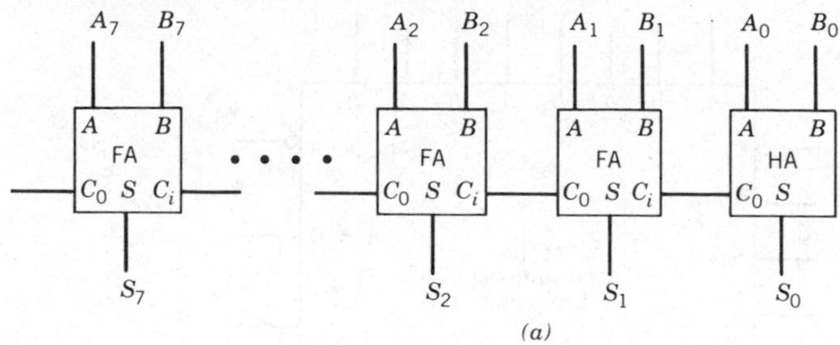

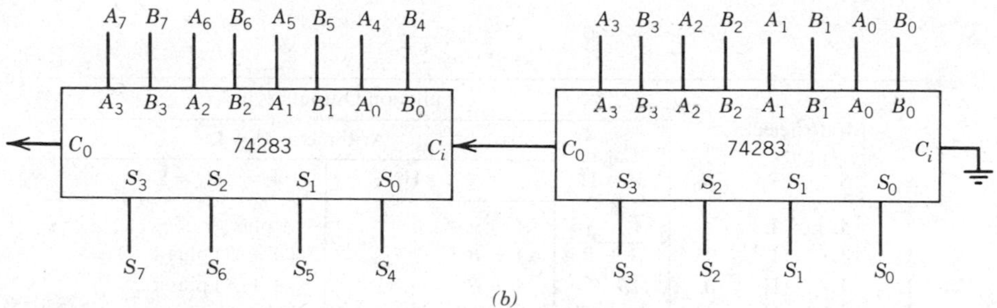

Fig. 30.24 8-Bit adders: (*a*) 8-Bit adder using seven full adders and one half adder; worst case propagation delay = 72 *ns*, (*b*) 8-Bit adder using two 4-bit carry lookahead adders; worst case propagation delay = 18 *ns*.

This delay time can be reduced by using 4-bit binary adders with carry lookahead. These adders are combinational circuits with nine inputs (one carry input and 8 inputs for two 4-bit numbers) and five outputs (one carry output in addition to the 4-bit sum). Since the carry output is a combinational function of all the inputs, the ripple effect through the four bits is eliminated so the maximum propagation delay from C_{in} to C_{out} is 18 ns. Thus implementing an 8-bit adder using two 4-bit carry-lookahead adders will reduce the propagation delay from 144 to 36 ns. The two 8-bit adders are compared in Fig. 30.24.

An alternative to including all the lookahead carry logic on the adder integrated circuit is to provide carry-generate and carry-propagate outputs to be used as inputs to a lookahead carry generator.[14] This makes it possible to have lookahead carry around more than 4 bits. Extremely fast additions using this method can be performed by emitter-coupled logic. For example, eight 4-bit arithmetic logic units (ECL10181) in the add mode, in conjunction with two lookahead carry blocks (ECL10179), will complete the add operation on two 32-bit numbers in the typical time of 19 ns.[15]

Subtractors

In principle, subtractors (both full and half) can be designed as combinational circuits using the same approach as for adders. The truth table is set up for subtraction and the carries become borrows. Since in most computing situations subtraction is actually accomplished through the addition of the 1's or 2's complement (depending on the system) of the subtrahend and the minuend, subtractors as such are not available in MSI packages. The subtraction operation is included among the capabilities of the 4-bit arithmetic logic unit (ALU) packaged as the TTL74181 in a 24-pin DIP. Even in this unit the subtract is performed by 1's complementing (generating $A - B - 1$) and requires an end-around carry into the LSB position to complete the $A - B$ operation. As indicated in Fig. 30.25, this ALU will perform 14 arithmetic operations in addition to add and subtract ($M = L$), and 16 logic functions ($M = H$).

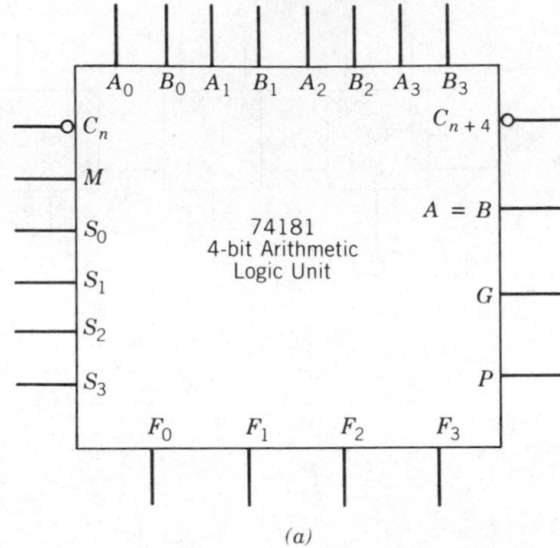

(*a*)

Mode Select Inputs				Function Outputs F_i		
				Logic,	Arithmetic $M = \text{L}$	
S_3	S_2	S_1	S_0	$M = \text{H}$	$C_n = \text{H}$	$C_n = \text{L}$
L	L	L	L	$\overline{A}$	A	A plus 1
L	L	L	H	$\overline{A + B}$	$A + B$	$(A + B)$ plus 1
L	L	H	L	$\overline{A}B$	$A + \overline{B}$	$(A + \overline{B})$ plus 1
L	L	H	H	Logic 0	minus 1	0
L	H	L	L	$\overline{A}B$	$A + A\overline{B}$	A plus $A\overline{B}$ plus 1
L	H	L	H	$\overline{B}$	$(A + B)$ plus $A\overline{B}$	$(A + B)$ plus $A\overline{B}$ plus 1
L	H	H	L	$A \oplus B$	A minus B minus 1	A minus B
L	H	H	H	$A\overline{B}$	AB minus 1	$A\overline{B}$
H	L	L	L	$\overline{A} + B$	A plus AB	A plus AB plus 1
H	L	L	H	$\overline{A \oplus B}$	A plus B	A plus B plus 1
H	L	H	L	B	$(A + \overline{B})$ plus AB	$(A + \overline{B})$ plus AB plus 1
H	L	H	H	AB	AB minus 1	AB
H	H	L	L	Logic 1	A plus A	A plus A plus 1
H	H	L	H	$A + \overline{B}$	$(A + B)$ plus A	$(A + B)$ plus A plus 1
H	H	H	L	$A + B$	$(A + \overline{B})$ plus A	$(A + \overline{B})$ plus A plus 1
H	H	H	H	A	A minus 1	A

(*b*)

Fig. 30.25 Transistor-transistor logic (TTL) 74181 arithmetic logic unit: (*a*) Input and output designations, (*b*) output functions. *Key*: A_i, B_i = Operand inputs; C_n = Carry input; C_{n+4} = Carry output; M = Logic/arithmetic input; S_i = Mode select inputs; F_i = Function outputs; G = Carry generate output; P = Carry propagate output.

Multipliers

The simplest scheme for the multiplication of binary numbers is best illustrated using an example. Figure 30.26*a* shows the multiplication of the binary equivalent of 6 (the multiplicand) by the binary equivalent of 5 (the multiplier). Observe that the operation includes the forming of the product of each bit of the multiplier and the complete multiplicand. These so-called partial products are shifted left *m* stages, where *m* depends on the significance of the multiplier bits. The shifted partial products are then summed to form the final product.

Implementing this algorithm requires shift registers that will hold the multiplicand and the multiplier; an accumulator register, which is of the combined lengths of the multiplier and multipli-

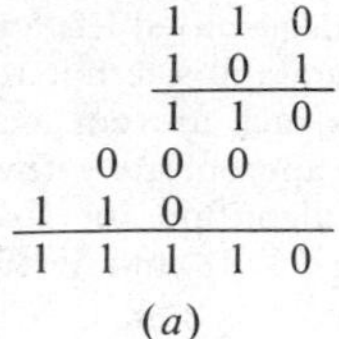

(*a*)

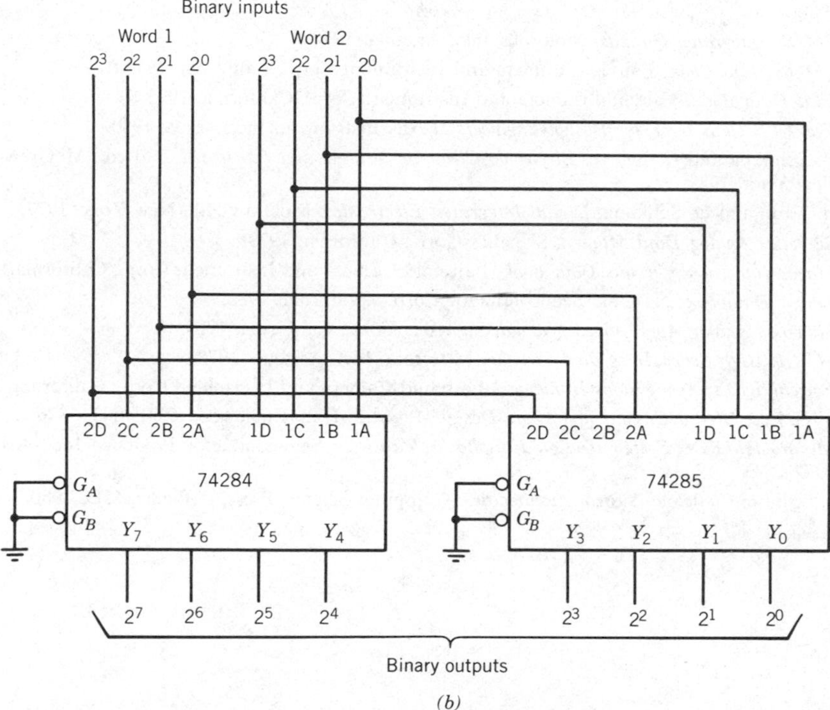

(*b*)

Fig. 30.26 Binary multiplication: (*a*) Example of multiplication operation, (*b*) 4 × 4 parallel binary multiplier.

cand registers; and an adder with as many bits as the accumulator. The only modification of the procedure illustrated in Fig. 30.26*a* is that the addition is performed by adding each new partial product to the total of the previously formed partial products in the accumulator. Observe that each 1 in the multiplier requires a shift and add operation, while each 0 requires just a shift operation. In a synchronous system, this means that the time required to complete the multiplication of two numbers will be N times the duration of each shift and add operation, where N is the number of bits in the multiplier.

This process can be accelerated by use of a more complex combinational logic circuit to form partial products involving more than just one bit of the multiplier at a time. For example, TTL 4-bit parallel multipliers 74284 and 74285, connected as shown in Fig. 30.26*b*, form the 8-bit product of two 4-bit binary numbers in about 20 ns. Four of these 8-bit product units can be interconnected with eight adders and four arithmetic logic units (ALUs) in a configuration called the Wallace tree[16] to produce an 8-bit × 8-bit multiplier having 16 outputs.

Division

Division in binary is accomplished by subtracting the divisor from the dividend until the remainder is less than the divisor and counting the number of subtractions required (N, say). The MSB of the quotient is then $N - 1$. The remainder is then shifted left and the process repeated to obtain the next significant figure of the quotient. The operation is terminated when either the remainder is 0 or the desired precision is obtained. Dividers as such are not available as integrated circuits. Division in calculators and computers is done by the appropriate software that utilizes circuits that perform the other arithmetic operations. A division algorithm for fixed-point binary integers is discussed in Chapter 59 of this handbook (refer to Figs. 59.18 and 59.19).

References

1 W. I. Fletcher, *An Engineering Approach to Digital Design*, Prentice-Hall, Englewood Cliffs, NJ, 1980.

2 *CMOS Integrated Circuits*, Motorola, Inc., Arizona, 1978.

3 *CMOS Data Book*, Fairchild Camera and Instrument Corp., California, 1975.

4 *TTL Data Book*, Fairchild Camera and Instrument Corp., California, 1975.

5 *The TTL Data Book for Design Engineers*, Texas Instrument, Inc., Texas, 1980.

6 D. Schilling and C. Belove, *Electronic Circuits, Discrete and Integrated*, 2nd ed., McGraw-Hill, New York, 1979.

7 H. Taub and D. Schilling, *Digital Integrated Electronics*, McGraw-Hill, New York, 1977.

8 *Signetics Analog Data Manual*, Signetics Corp., California, 1979.

9 *Linear Integrated Circuits Data Book*, Fairchild Camera and Instrument Corp., California, 1976.

10 *Linear Databook*, National Semiconductor Corp., California, 1982.

11 *Signetics Analog Applications Manual*, Signetics Corp., California, 1979.

12 *MECL High Speed Integrated Circuits*, Motorola, Inc., Arizona, 1978.

13 *Fairchild TTL Applications Handbook*, Fairchild Camera and Instrument Corp., California, 1973.

14 S. C. Lee, *Digital Circuits and Logic Design*, Prentice-Hall, Englewood Cliffs, NJ, 1976.

15 *Motorola MECL System Design Handbook*, Motorola Semiconductor Products, Inc., Arizona, 1972.

16 J. S. Base, *Computer System Architecture*, Computer Science Press, Potomac, MD, 1980.

CHAPTER 31

DIGITAL-TO-ANALOG (D/A) AND ANALOG-TO-DIGITAL (A/D) CONVERTERS

MICHAEL BALTRUSH

New Jersey Institute of Technology, Newark, New Jersey

The digital computer uses 1's and 0's to represent internal values and manipulates these with digital techniques. However, the "world" outside the computer is analog in nature, that is, the values produced range over a wide continuum. To transform data from one realm to another, discrete to continuous or continuous to discrete, digital-to-analog (D/A) converters or analog-to-digital (A/D) converters are used. The manufacturers listed in Table 31.1 produce various conversion equipment. This equipment ranges from individual integrated circuits, which are complete converters in themselves, to complete subsystems that can be used for data acquisition.

31.1 WEIGHTED RESISTOR AND WEIGHTED LADDER D/A CONVERTERS

The D/A converter is a device that accepts a digital word and a reference analog voltage or current (see Fig. 31.1) and produces an analog voltage or current. The digital word can be binary coded, offset binary coded, binary coded decimal (BCD) or signed magnitude or an inverted representation of one of these. The code represents a fraction with the radix point to the left of the most significant bit (MSB), called A_0. The least significant bit (LSB) is called A_{n-1}, where n is the number of bits in the digital word. The full-scale value of the reference analog value is multiplied by the fraction to produce the analog value output.

One of the simplest D/A converter circuits is the weighted resistor D/A converter[1,2] shown in Fig. 31.2. The digital word controls the switches where a logical 1 indicates a switch to the V_{ref} and a logical 0 a switch to ground. Successive resistors are weighted by a factor of 2 producing binary weighted contributions to the analog output, shown in Eq. (31.1)

$$V_{out} = V_{ref} * (A_{n-1} * 2^{-(n-1)} + A_{n-2} * 2^{-(n-2)} + \cdots + A_1 * 2^{-1} + A_0) \tag{31.1}$$

However, maintaining accurate resistor ratios over a large range, 4096 to 1 for a 12-bit converter, makes the D/A converter difficult to manufacture.

A way to reduce the number of resistors is to use a limited number of values with suitable attenuation, Fig. 31.3.[2] Within each bank of four resistors the weights are factors of 2. Between the banks are the attenuation resistors. These can be configured to produce a binary weighting (if $r = 8R$) or a BCD weighting (if $r = 4.8R$) for different input codes.

TABLE 31.1. A SAMPLE OF CONVERTER VENDORS

Analog Devices, Inc.
PO Box 280
Norwood, MA 02062

Burr-Brown Research Corp.
6730 S. Tucson Blvd.
Tucson, AZ 85706

Datel Intersil, Inc.
11 Cabot Blvd.
Mansfield, MA 02048

Fairchild Semiconductor
464 Ellis
Mountain View, CA 94042

Motorola Semiconductor Products, Inc.
PO Box 20912
Phoenix, AZ 85036

RCA
Route 202
Somerville, NJ 08876

Signetics
811 E. Arques Ave.
Sunnyvale, CA 94086

Texas Instruments, Inc.
PO Box 20219
Dallas, TX 75220

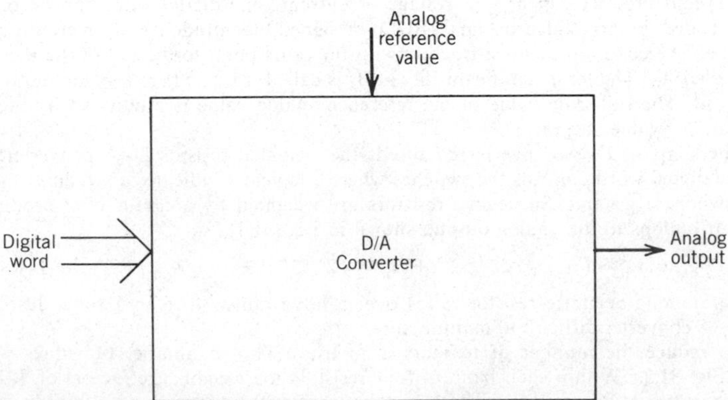

Fig. 31.1 Block diagram of a digital-to-analog (D/A) converter.

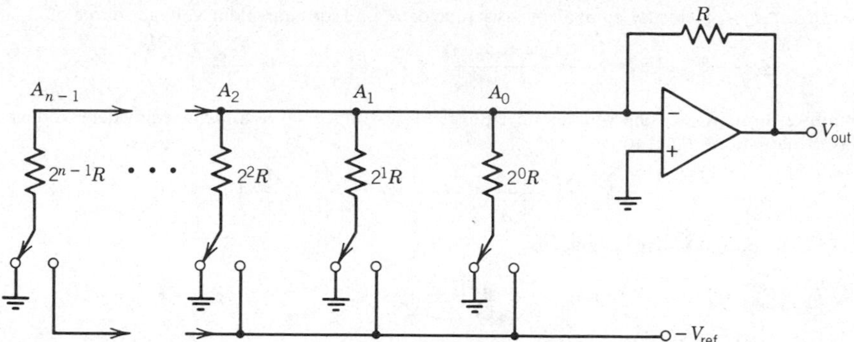

Fig. 31.2 Weighted resistor digital-to-analog (D/A) converter consisting of a summing amplifier and a weighted resistor network.

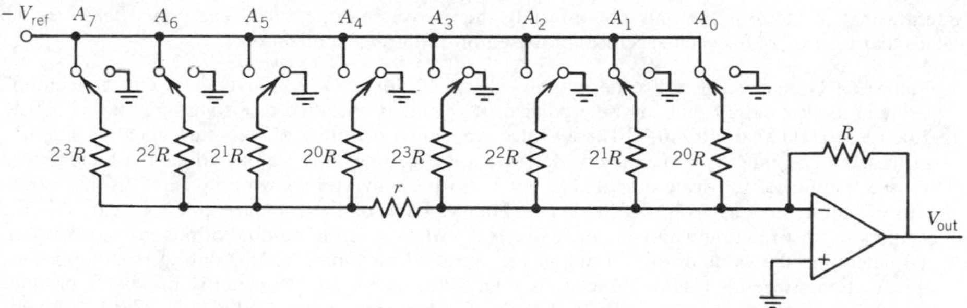

Fig. 31.3 8-Bit weighted resistor digital-to-analog (D/A) converter. $r = 8R$ for straight 8-bit binary converter; $r = 4.8R$ for two-decimal-digit binary coded decimal (BCD) converter.

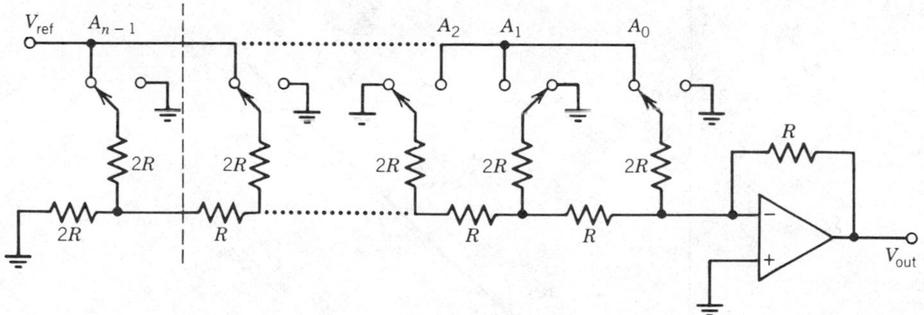

Fig. 31.4 $R - 2R$ latter digital-to-analog (D/A) converter.

Carrying the reduction in number of resistors as far as possible, the resistor values can be reduced to the pair of resistor values R and $2R$, Fig. 31.4. This converter is a popular design well suited to integrated circuits. Each bit of the digital word, representing a switch closure, contributes to the output in the following way. The Thévenin equivalent resistance at the dashed line shown in Fig. 31.4 is R and the equivalent voltage source is

$$V_{ref} * \frac{A_{n-1}}{2}$$

Repeating at A_{n-2} yields an equivalent resistance of R and an equivalent voltage source of

$$V_{\text{ref}} * \frac{(A_{n-1}/2 + A_{n-2})}{2} = V_{\text{ref}} * \left(\frac{A_{n-1}}{2^2} + \frac{A_{n-2}}{2} \right)$$

Continuing through A_0, one obtains an equivalent resistance of R and an equivalent voltage (by means of induction) equal to

$$V_{\text{ref}} * \left(\frac{A_{n-1}}{2^n} + \frac{A_{n-2}}{2^{n-1}} + \cdots \frac{A_1}{2^2} + \frac{A_0}{2^1} \right)$$

Therefore the output voltage is given by

$$V_{\text{out}} = \left(\frac{-V_{\text{ref}}}{2} \right) * (A_{n-1} * 2^{-(n-1)} + A_{n-2} * 2^{-(n-2)} + \cdots A_1 * 2^{-1} + A_0) \qquad (31.2)$$

31.2 SPECIFICATIONS FOR D/A CONVERTERS

In the choice of a D/A converter, some understanding of the specifications supplied by the manufacturer is necessary. The following are terms used by manufacturers in specifying their converters. Each of the terms is defined or explained. The discussion associated with each term assumes that the term is the only one affecting the converter's operation. The user determines the values and tolerances for each specification based on the user's application.

Full-Scale Output. The full-scale output voltage V_0 for a D/A converter is the "maximum" value of analog output that can be produced. It is usually specified as a round number: 2.500 V, 5.000 V, 10.000 V, or 10.240 V. These values are standard in that off-the-shelf units are supplied in these ranges, but any value may be designed for. Figure 31.5 is a plot of the input binary word vs. the fraction of full-scale output, V_0. For the 3-bit converter shown only eight distinct values can be input, ranging from 000 to 111 in binary. Each of these input values is treated as the numerator of a fraction whose denominator is 2^n, where n is the number of bits in the converter. To determine the value of analog output the specified maximum analog output is multiplied by this fraction. Since the resulting fraction is always less than 1, the converter is not able to produce the maximum value of analog output. A 12-bit D/A converter specified with a 5.000-V full-scale output produces 4.9987 V $[(2^{12} - 1)/2^{12}) * V_0$ or $(4095/4096) * V_0]$.

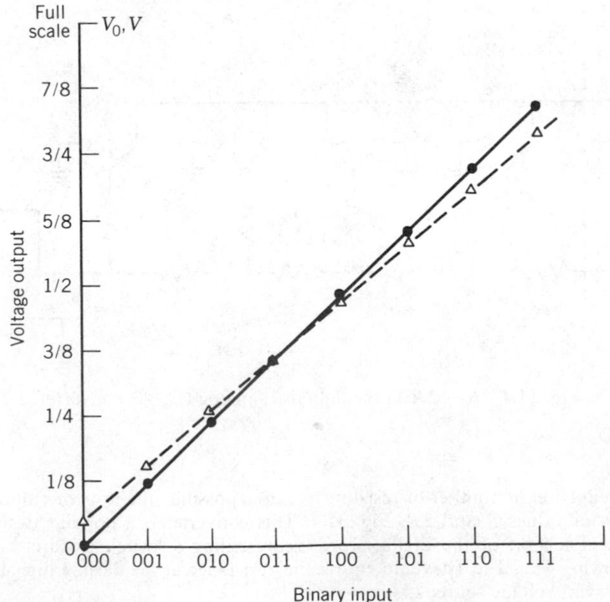

Fig. 31.5 Ideal transfer function for a 3-bit digital-to-analog (D/A) converter (solid line) and a transfer function with offset and gain errors (dashed line).

Resolution. The resolution of a D/A converter is the smallest analog change that is produced at the output. The D/A converter resolution is determined by the number of bits in the digital input word. For a binary-coded D/A voltage converter the resolution in volts is $V_0/2^n$. For the 3-bit converter in Fig. 31.5, the smallest change that could be produced in the output is 1/8 of V_0 ($V_0/2^3$). For a 12-bit D/A converter with a 5.000-V full-scale output, the resolution is 1.221 mV ($5/2^{12}$) or one part in 4096.

Offset Error. Offset error is the measured analog output of the D/A converter when its digital input code is zero. In Fig. 31.5, the dotted line connecting the triangles intercepts the voltage output axis at a point not equal to 0. This difference is the offset error. Owing to component mismatches, the digital value input to produce an analog 0 output is seldom exact; therefore adjustments are usually available to eliminate or minimize this error.

Gain. The gain of a D/A converter is usually set to 1. For the 3-bit converter, Fig. 31.5, a gain of 1 would produce a 7/8 full-scale analog value at a digital input value of 111. This is shown by the solid line connecting the dots in Fig. 31.5, which has a gain of 1. Each produced voltage is equal to its predicted value. If the gain is other than 1, the input digital values will not produce the desired analog output. In Fig. 31.5 the gain is less than 1 for the line connecting the triangles. For an input value of 111, the maximum input value, the converter produces a value less than the one specified, 7/8 of V_0. Adjustments are usually available to minimize this error.

Linearity Error. The linearity error of a D/A converter measures the deviation of the converter's actual output from a line fitted to the measured end points. In Fig. 31.6 the dots that do not fall on the line connecting the measured end points (those associated with the 000 and 111 inputs) have individual linearity errors. The linearity error magnitudes are the distance of the dots from the line. This error is specified by a fraction of LSB or percentage of full-scale output. The largest value of linearity error is the one specified.

Differential Linearity Error. This is the maximum difference between each analog output step and the ideal step size of one LSB.

Monotonicity. A D/A converter is monotonic if increasing digital codes produce increasing analog outputs. If this does not happen the converter is said to be nonmonotonic.

Absolute Accuracy Error. The absolute accuracy error of a D/A converter is the difference between the full-scale set point and an absolute voltage standard usually traceable to the National Bureau of Standards.

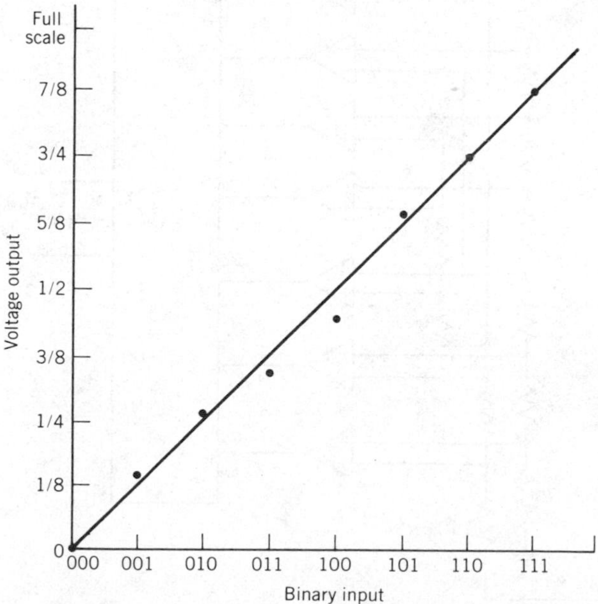

Fig. 31.6 Nonlinear transfer function with a line fitted to the end points.

Relative Accuracy. Relative accuracy is the difference between the specified value and actual value measured for a given digital input word. This assumes that offset and gain errors have been minimized.

Stability. The performance of a D/A converter varies with time, temperature, and power supply voltage variations. The converter's resistance to these factors is its stability. For example, a typical coefficient for gain from a manufacturer's data sheet is 20 ppm/°C. Many of the other specifications have associated temperature coefficients and aging constants.

Settling Time. In an ideal D/A converter, a digital input code change from one value to another produces an instantaneous change in analog output. In a real converter, time delays are introduced by various components, and the output requires a certain amount of time to reach the new value. This time is the converter's settling time.

In D/A converters, a large analog output transient, glitch, may occur for a small digital input change. This is most pronounced at major bit transitions. For example, a transition from binary 1000 to binary 0111 may pass through binary 0000 if the digital switching devices turn off faster than they turn on. This transient binary value of 0000 produces the output glitch. Glitches can be minimized by proper design or additional circuitry in the D/A converter.

31.3 PARALLEL AND SUCCESSIVE APPROXIMATION A/D CONVERTERS

An A/D converter takes an unknown analog signal and converts it to a binary number consisting of n bits that can be manipulated by a digital computer.

The parallel A/D converter, shown in Fig. 31.7, is the fastest of the A/D converters.[1,3] It uses hardware to perform the conversion in parallel. It is also called a flash converter because of its high speed. In a 3-bit converter, the unknown analog value is simultaneously compared with seven different values. The comparator outputs are fed to combinational logic, which produces the 3-bit digital value. The parallel n-bit A/D converter is used when high speed is necessary but low resolution is acceptable, since the number of comparators required is $2n - 1$.

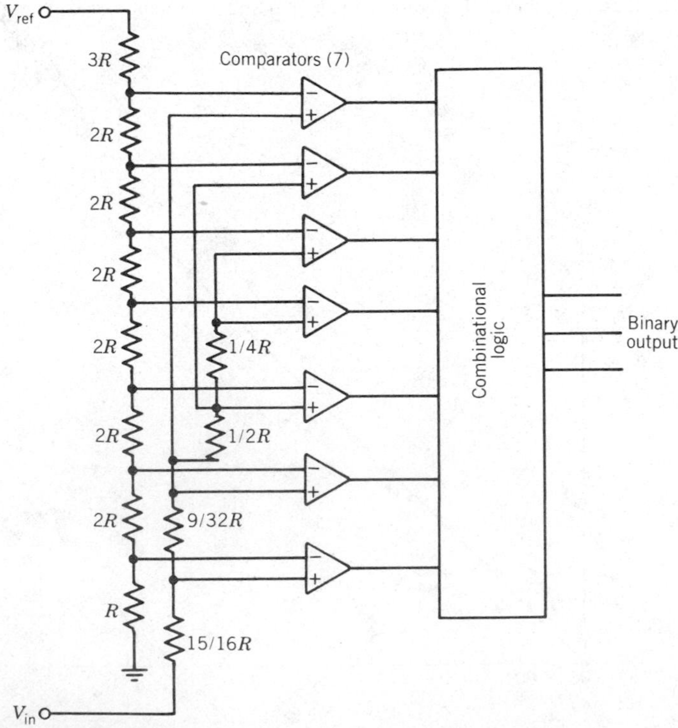

Fig. 31.7 Block diagram of a 3-bit parallel analog-to-digital (A/D) converter.

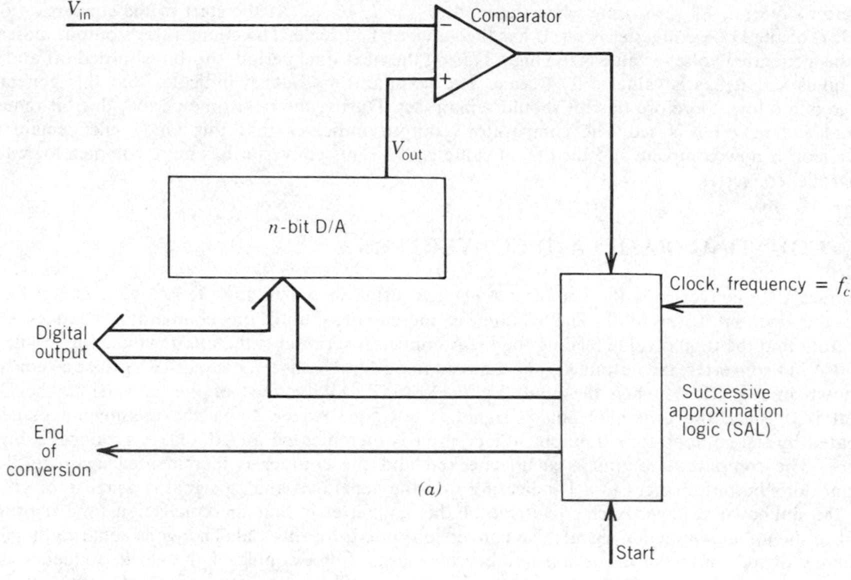

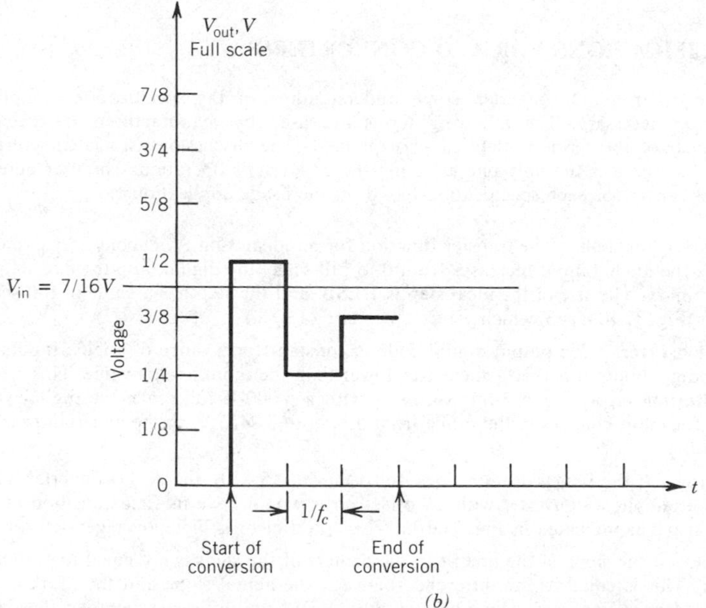

Fig. 31.8 (*a*) Block diagram of successive approximation analog-to-digital (A/D) converter. (*b*) Output waveform V_{out} with 3-bit digital-to-analog (D/A) converter for $V_{in} = 7/16$ V.

A fast A/D converter without the parallel A/D converter's hardware overhead is the successive approximation A/D converter of Fig. 31.8.[1,3,4] This converter uses a binary search procedure to perform the conversion in n clock periods, where n is the number of bits in the digital code. The successive approximation logic (SAL) sets each bit in turn and compares the resulting D/A converter output with the unknown voltage. If the D/A converter output with the bit on is higher than the unknown voltage, the bit is turned off; otherwise the bit remains on. The next bit is treated the same until all bits have been tried. For a 3-bit A/D converter, Fig. 31.8*b* is a plot of the contained D/A converter's output, V_{out}, vs. time when the input $V_{in} = 7/16$ V.[1] At the start of the conversion only the MSB of the D/A converter is set. It has the value of 1/2 scale. The comparator's output indicates that this generated voltage value is too high. During the next time period, this bit is turned off and the next bit is set; it has a value of 1/4 scale. The comparator's output indicates that this generated voltage is too low, therefore this bit should remain set. During the next time period, this bit remains set and the next bit is set. The comparator's output indicates that this bit should remain set. Conversion is now complete and the digital value is 011. This converter has more complex logic than the parallel converter.

31.4 COUNTING (RAMP) A/D CONVERTERS

A simple A/D converter is the counting A/D converter shown in Fig. 31.9.[1,3] Conversion begins when the counter is set to 0. The counter is incremented until the comparator changes state, indicating that the digital value feeding the D/A converter represents the unknown analog input. For a 3-bit A/D converter, the output V_{out} of the contained D/A converter is plotted against a time axis, as shown in Fig. 31.9*b*, when the input $V_{in} = 11/16$ V. At the start of the conversion, the D/A output is 0. Since the generated analog signal is not equal to or above the unknown voltage, as indicated by the comparator's 0 output, the counter is incremented and the D/A produces a higher voltage. The comparator output is again checked and the counter is incremented again until the comparator's output changes to a 1 indicating that the generated analog signal is equal to or greater than the unknown voltage. A characteristic of this converter is that the conversion time is proportional to the unknown analog signal. The conversion time for a full-scale change is equal to the clock frequency divided into the maximum number of counts. For example, if the clock frequency is 10 MHz, the maximum throughput rate for a 10-bit resolution converter is somewhat less than 10 KHz [100 μs per conversion (10 MHz/2^{10})].

31.5 SPECIFICATIONS FOR A/D CONVERTERS

In the choice of an A/D converter, some understanding of the specifications supplied by the manufacturer is necessary. The following terms are used by manufacturers in specifying their converters. Each of the terms is defined or explained. The discussion associated with each term assumes that that term is the only one affecting the converter's operation. The user determines the values and tolerances for each specification based on the user's application.

Ideal Transfer Function. The transfer function for an ideal 3-bit A/D converter is shown in Fig. 31.10*a*. As the analog input increases from 0 to full scale, the digital output value steps from 000 to 111 in binary. The size of the ideal step is 1 LSB, and the step has a value of full scale divided by 2^n. (For Fig. 31.10 $n = 3$, which yields a step size of 1/8.)

Quantization Error. The output digital code is constant for a range of analog inputs equal to 1 LSB. The digital output is first higher, then lower than the input analog value. This error is called the quantization error. For a 3-bit converter with a 5.000-V full scale, the digital value 001 is produced for input voltages in the range from 0.3175 to 0.9425 V, or the uncertainty is 0.625 V or 1/8 full scale.

Offset Error. If the first transition does not occur at 0.5 LSB, the A/D converter has an offset error. For example, a converter with an offset error would have its first transition at other than 1/16 full scale, as indicated in Fig. 31.10*a*. This offset error is highly exaggerated for clarity.

Gain Error. If the slope of the line fit to the centers of the steps is not equal to 1, then there is a gain error. This is equal to the difference between the actual slope and the ideal slope. In Fig. 31.10*b* the dashed line has a slope greater than 1 and the difference between the dashed line's slope and the ideal line's slope is the gain error.

Differential Linearity Error. A/D converters have linearity errors similar to those for D/A converters. Differential linearity error is the difference between the actual code step and the ideal step size of 1 LSB.

Missing Code Error. An error specific to A/D converters is the missing code error. The output code jumps from value a to value $a + 2$, bypassing the $a + 1$ code value.

Monotonicity. An A/D converter may not be monotonic. If the output code decreases as the analog input value increases, the converter exhibits nonmonotonic behavior. If the input analog voltage changes from 1/2 full scale to 5/8 full scale, the digital output should change from 100 to 101. However, a nonmonotonic converter's behavior could produce a digital output of 100 to 011.

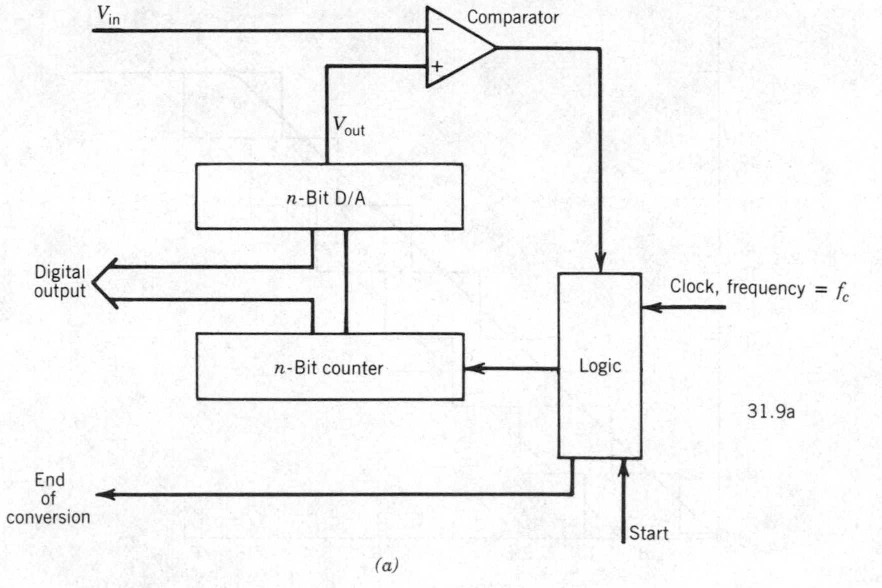

(a)

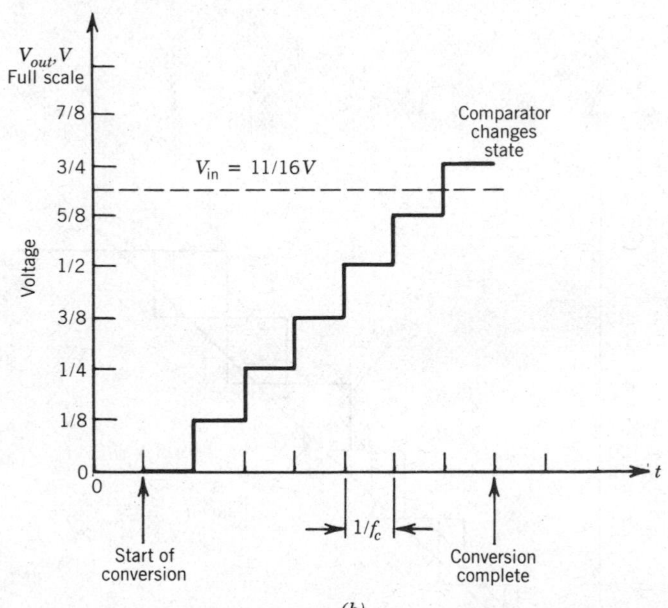

(b)

Fig. 31.9 (*a*) Block diagram of a counting analog-to-digital (A/D) converter. (*b*) Output waveform V_{out} with 3-bit digital-to-analog (D/A) converter for $V_{in} = 11/16$ V.

The digital output's value has decreased (from 1/2 to 3/8 full scale), while the analog input has increased (from 1/2 to 5/8 full scale).

Stability. The stability of an A/D converter is affected by time, temperature, and power supply voltage variations. Converter specifications include temperature coefficients and aging parameters for those specifications affected. A manufacturer's typical value for the gain temperature coefficient is 15 ppm/°C.

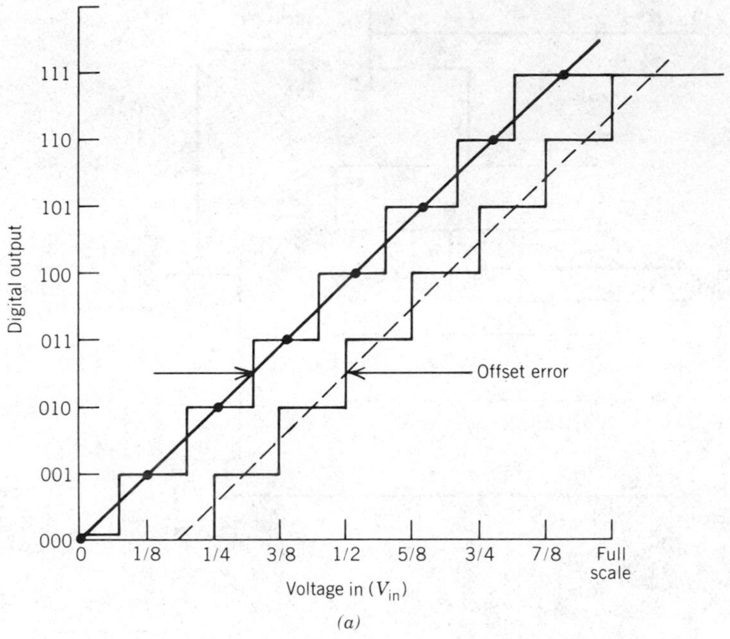

(a)

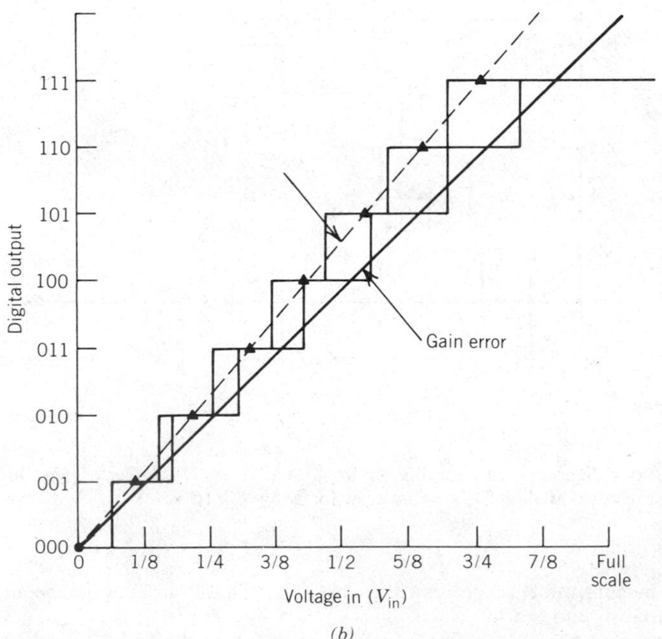

(b)

Fig. 31.10 Transfer functions for a 3-bit A/D converter. (*a*) Ideal transfer function (solid line) and transfer function with an offset error (dashed line), (*b*) transfer function with a gain error (dashed line).

References

1 D. H. Sheingold, ed., *Analog-Digital Conversion Handbook*, Analog Devices, Inc., Norwood, MA, 1972.

2 R. C. Jaeger, "Tutorial: Analog Data Acquisition Technology—Part I—Digital-to-Analog Conversion," *IEEE Micro*. **2**(2):20–37 (May 1982).

3 R. C. Jaeger, "Tutorial: Analog Data Acquisition Technology—Part II—Analog-to-Digital Conversion," *IEEE Micro*. **2**(3):46–57 (Aug. 1982).

4 H. Taub and D. Schilling, *Digital Integrated Electronics*, McGraw-Hill, New York, 1977.

Bibliography

Bruck, D. B., *Data Conversion Handbook*, Hybrid Systems Corp., 1974.

Dooley, D. J., ed., *Data Conversion Integrated Circuits*, IEEE Press, New York, 1980.

Hoeschele, D. F., *Analog-Digital and Digital-Analog Conversion Techniques*, Wiley, New York, 1968.

Schmid, H., *Electronic Analog/Digital Conversion*, Van Nostrand, New York, 1970.

CHAPTER 32
DIGITAL FILTERS

THOMAS W. PARSONS

Hofstra University, Hempstead, New York

32.1 DESCRIPTION

A digital filter is a system for filtering sampled signals whose samples are represented in numerical form (typically as binary numbers). The filtering operation is performed by direct computations on the signal samples. Where equations are used to analyze or characterize the operation of analog filters, the equations used in digital filtering *are* the filter—that is, the filter carries out the very arithmetic operations specified by the equations.

Digital filters offer several advantages over their analog counterparts:

1. **Dynamic range.** The upper limit is set by the size of the number the digital hardware is able to represent, the lower limit by quantization noise and roundoff errors. The range between these limits depends only on the word lengths used (i.e., the number of bits used in the binary representations of the signal). If the word lengths can be made large enough, in principle the dynamic range has no limit. Limits are set in practice by tradeoffs among speed, performance, and cost requirements.
2. **Freedom from component problems.** Filter parameters are represented by binary numbers and do not drift with time. Increasing the word length allows any desired degree of accuracy to be obtained. Within the limits of that accuracy, the filter performs exactly as designed. There are no problems of component tolerances or drift and none associated with nonideal behavior of resistors, capacitors, inductors, or amplifiers. There are also no problems of input or output impedance or loading effects between stages.
3. **Switchability.** If the filter parameters are kept in registers (as opposed to being hard wired), the contents of these registers can be changed at will and instantaneously (i.e., between consecutive sample times). Hence filters can be made perfectly switchable. A single switchable filter can also be time multiplexed to process multiple inputs.
4. **Adaptability.** A digital filter can be implemented either in hardware or as a computer program. Hardware digital filters are also easily adapted to computer control of their parameters.

Digital filters are members of a class of systems known as sampled-data systems. A system receives one or more input signals and produces one or more outputs. (Only systems with one input and one output will be considered here.) We cope with the complexity of real systems by finding simple systems whose behavior suitably approximates that of the system under consideration. The two principal simplifications of interest here are the assumption of linearity and the use of sampling.

Linear Systems

If we use $H[\ ]$ to represent what the system does (the system function), then we may define linearity as follows:

Given $x_1(t)$ and $x_2(t)$ with a and b as constant factors, we can write

$$y(t) = H[ax_1(t) + bx_2(t)] \tag{32.1a}$$

$$= aH[x_1(t)] + bH[x_2(t)] \tag{32.1b}$$

$$= ay_1(t) + by_2(t) \tag{32.1c}$$

The test of linearity is whether one can go from Eq. (32.1b) to Eq. (32.1c).

Two other simplifications commonly made with linear systems are the following:

1. Time invariance. $H[\ ]$ does not change with time. If $y(t) = H[x(t)]$, then $H[x(t-\tau)] = y(t-\tau)$, where τ is a positive or negative time delay.

2. Causality. Output cannot anticipate input. More precisely, for linear time-invariant systems, given two possible inputs to the system $x_1(t)$ and $x_2(t)$, then if $x_1(t) = x_2(t)$ for $t \leqslant t_0$, then $H[x_1(t)] = H[x_2(t)]$ for $t \leqslant t_0$.

All of these assumptions are normally applicable to digital filters. The only exceptions to this statement are the following:

1. Noncausal systems can be modeled on the computer to a limited extent.
2. The fact that digital filter parameters can be switched arbitrarily at will means that filters can be time varying if this is desired. Switching usually involves selection of parameter values from a limited repertoire; in this case each set of parameters can be analyzed separately as a time-invariant system, although transient behavior after parameter switching may require special study.
3. Limitations in precision and dynamic range inherent in digital representation of numbers can limit the applicability of the linearity assumption.

In the following material, we will not consider noncausal and time-varying systems. We will assume linearity and confront its limits in the section on quantization effects.

Sampled-Data Systems

A sampled-data system can be viewed as a continuous-time system that is observed only at discrete times t_i and must therefore be treated as a discrete-time system. The times t_i are typically equally spaced. In that case $t_i = iT$, where T is the interval between samples. Instead of writing x as $x(t_i)$, we write either $x(i)$ or x_i. The sampling rate, f_s, is equal to $1/T$.

Choice of the sampling rate is a fundamental problem in the design of sampled-data systems. It is desirable to sample at as low a rate as possible, but if the rate is too low, we may be unable to reconstruct the corresponding continuous function from its samples. The sampling theorem states that any continuous-time signal can be completely characterized by, and perfectly reconstructed from, equally spaced samples, provided $f_s \geqslant 2f_m$, where f_m is the highest frequency component present in the original signal. This minimum rate is commonly known as the Nyquist rate.[1-4]

A continuous function sampled at less than the Nyquist rate is said to be undersampled. In an undersampled signal, any component at a frequency $f > f_s/2$ will appear to be at a lower frequency $f' = f_s - f$. This phenomenon is known as aliasing.[1-3] When data are sampled for analysis in the computer or for digital processing, it is standard practice to provide protection by means of an *antialiasing filter*. This is an analog filter, placed before the sampling hardware, that rejects any frequency components higher than $f_s/2$. The systems engineer must take care that the sampling frequency is high enough that no essential information is wiped out by the antialiasing filter.

Time-Domain Analysis

Sampled-data systems can be characterized in the time domain by a linear constant-coefficient difference equation of the form

$$\begin{aligned} y(n) + a_1 y(n-1) + a_2 y(n-2) + \cdots + a_p y(n-p) \\ = b_0 x(n) + b_1 x(n-1) + b_2 x(n-2) + \cdots + b_q x(n-q) \end{aligned} \tag{32.2}$$

If we have access only to the input and output of the system, we can characterize the system by its response to a standard input. Define the "unit impulse function" $\delta(n)$ as follows:

$$\delta(n) = \begin{cases} 1 & n = 0 \\ 0 & \text{otherwise} \end{cases} \tag{32.3}$$

Then the "impulse response" $h(n)$ is the response of the system to an input $\delta(n)$.[1-3] If $x(n)$ is any other input signal, then the response to $x(n)$ can be found with the aid of $h(n)$

$$y(n) = \sum_{k=-\infty}^{\infty} x(k)h(n-k) = x(n) * h(n) \tag{32.4}$$

The operation $x(n) * h(n)$ is known as the convolution of $x(n)$ and $h(n)$. The result is that if we know the impulse response of a linear system, we can compute its response to any other input whatsoever. The system's input/output behavior is thus completely characterized by its impulse response.

The concept of impulse response also provides a useful extension to the definition of a causal system: A linear time-invariant system is causal if $h(n)$ is 0 for $n < 0$.

Frequency-Domain Analysis

For analysis of sampled-data systems in the frequency domain, it is convenient to use the z-transform representation. Let the z-transforms of the input and output signals be defined in the usual way. Then using the shifting and linearity properties of the z-transform (see Chapter 4 in this handbook), we can write thc z-transform of Eq. (32.2) as follows:

$$Y(z) \sum_{i=0}^{p} a_i z^{-i} = X(z) \sum_{j=0}^{q} b_j z^{-j} \tag{32.5}$$

Defining the transfer function of the system as

$$H(z) = Y(z)/X(z) \tag{32.6}$$

then $H(z)$ is a ratio of polynomials in z

$$H(z) = N(z)/D(z) \tag{32.7a}$$

where

$$N(z) = \sum_{j=0}^{q} b_j z^{-j} \tag{32.7b}$$

and

$$D(z) = \sum_{i=0}^{p} a_i z^{-i} \tag{32.7c}$$

$N(z)$ has q roots, for each of which $H(z) = 0$; these roots are called the zeros of H. Similarly $D(z)$ has p roots for which $H(z) = \infty$; they are called the poles of H. Since any polynomial is defined (to within a multiplicative constant) by its roots, H is completely characterized (except for a gain factor) by its poles and zeros.

Rewriting Eq. (32.6) as

$$Y(z) = X(z)H(z) \tag{32.8}$$

one obtains Eq. (32.4) by virtue of the convolution property of the z-transform. We recognize $h(n)$ as the impulse response of the system; hence $H(z)$ is the z-transform of the impulse response.

The object of filter design is normally the realization of some specified frequency response. The frequency response of a system can be found by evaluating $H(z)$ on the unit circle

$$H(e^{j\omega}) = H(z)|_{z=e^{j\omega}} \tag{32.9}$$

where

$$\omega = 2\pi f/f_s \tag{32.10}$$

Note that angular distances around the unit circle map linearly onto frequency. The DC term corresponds to the point $(1 + j0)$ and that the point $(-1 + j0)$ corresponds to $f_s/2$. Note also that as a result of this definition the frequency response is periodic with period 2π, corresponding to a periodicity in f of f_s. In this chapter, ω will be considered as limited to the range $[-\pi, \pi]$ unless otherwise stated.

Equation (32.9) can be rewritten

$$H(e^{j\omega}) = \sum_{n=-\infty}^{\infty} h(n)e^{-j\omega n} \tag{32.11}$$

$h(n)$ can then be considered the Fourier series expansion of the periodic function $H(e^{j\omega})$. Then

$$h(n) = \frac{1}{2\pi}\int_{-\pi}^{\pi} H(e^{j\omega})e^{j\omega n}d\omega \tag{32.12}$$

Equation (32.12) could conceivably be used as the starting point for the design of a digital filter; this possibility is pursued in Section 32.5.

$H(e^{j\omega})$ is a continuous function of frequency. In many practical cases a discrete approximation to $H(e^{j\omega})$ is sufficient. The discrete Fourier transform (DFT) of $h(n)$ provides such an approximation. The DFT of a sequence of N points is defined as follows:

$$H(k) = \sum_{n=0}^{N-1} x(n)W^{-nk} \tag{32.13a}$$

where $W = e^{j2\pi/N}$. The inverse DFT is given by

$$h(n) = \frac{1}{N}\sum_{k=0}^{N-1} H(k)W^{nk} \tag{32.13b}$$

See Chapter 4 of this handbook for a discussion of the DFT.

Linear-Phase Systems

There are many signal-processing applications in which phase relations are important and must not be disturbed by filtering. For such purposes a filter with zero phase shift would be ideal; in practice one must settle for a filter whose phase shift is proportional to frequency. Such a filter is called linear phase. It is easy to show what such a filter must be like.

If the phase of the frequency response is proportional to the frequency, then it must be possible to factor $H(e^{j\omega})$ into a pure real function $R(\omega)$ and a linear-phase factor $e^{ja\omega}$

$$H(e^{j\omega}) = R(\omega)e^{ja\omega} \tag{32.14}$$

From elementary Fourier transform theory, if a filter's frequency response is $R(\omega)$, then its impulse response $r(n)$ must be an *even* function of time, as shown in Fig. 32.1*a*. Such a filter is not causal. If $r(n)$ is of finite duration, however, the filter can be made causal by shifting $r(n)$ to the right, as shown in Fig. 32.1*b*. But the linear-phase factor corresponds to shifting $r(n)$ in time; hence the linear-phase filter can be viewed as our ideal zero-phase filter shifted to make it causal. $h(n)$, the shifted $r(n)$, is no longer an even function, but it remains symmetrical in the following sense: If the impulse response is

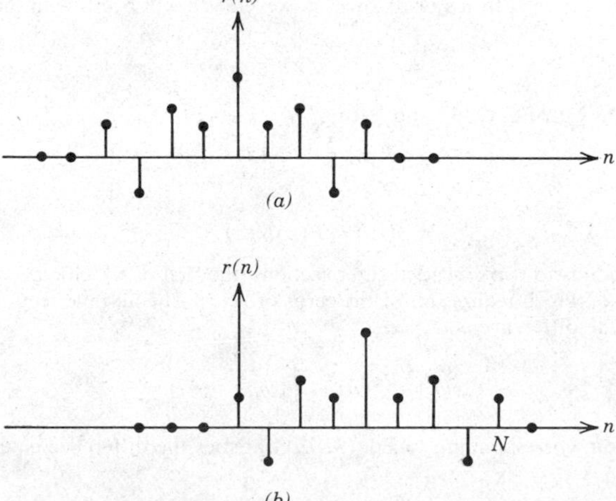

Fig. 32.1 Even impulse response: (*a*) Corresponding to real, even frequency response, (*b*) shifted to make filter causal.

N samples long, then

$$h(n) = h(N - 1 - n) \tag{32.15}$$

This symmetry is the necessary and sufficient condition for phase linearity.

Stability

A system is stable if a bounded (i.e., finite-amplitude) input always produces a bounded output. A discrete-time system is stable if

$$\sum_{k=-\infty}^{\infty} |h(k)| < \infty \tag{32.16}$$

or if all poles of $H(z)$ are inside the unit circle in the (complex) z plane.

Finite-Impulse- and Infinite-Impulse-Response

A fundamental distinction in sampled-data systems, and particularly in digital filters, is the duration of the impulse response. We refer to finite-impulse-response (FIR) systems and infinite-impulse-response (IIR) systems.

If, in the difference equation, Eq. (32.2), the coefficients a_i are all 0 for $i > 0$, then the equation reduces to

$$y(n) = b_0 x(n) + b_1 x(n-1) + \cdots + b_q x(n-q) \tag{32.17}$$

Such a system cannot possibly have an impulse response more than q samples long, because if the impulse occurred more than q samples ago, the entire right-hand side of Eq. (32.17) is zero. Hence such a system is an FIR system. If Eq. (32.17) is rewritten using z-transforms, it will be immediately apparent that an FIR filter is also an all-zero system. That is, $H(z)$ has no poles except at $z = 0$.

If any of the higher a-coefficients in Eq. (32.2) is not zero, however, then the impulse response can last forever, because even if the impulse occurred more than q samples ago, it is "remembered" in the output samples $y(n-1)$, $y(n-2)$, ... Hence the system will be an IIR system and $H(z)$ will have finite-frequency poles.

32.2 STRUCTURES

It is customary to represent the flow of data within a digital filter by a flow diagram. These diagrams apply whether the filter is to be realized in hardware or software. For a hardware filter, a flow diagram represents the actual interconnection of the filter components; for a software filter, the diagram provides a conceptual model of the filter's operation and can be used as a point of departure for programming.

Since the FIR filter is a special case of the IIR filter, the two types have some structures in common. Each also has special forms of its own, however. We will begin with the IIR case.

IIR Structures[1,2]

Direct Form. Starting with Eq. (32.7) and letting

$$P(z) = 1/D(z) \tag{32.18}$$

one can write

$$Y(z) = P(z)N(z) \tag{32.19}$$

This equation corresponds to two cascaded filter sections, the first of which realizes all of the poles of $H(z)$ and the second of which realizes all of the zeros of $H(z)$. The all-pole section, corresponding to Eq. (32.18), satisfies the difference equation

$$p(n) = x(n) - \prod_{i=1}^{N} b_i p(n-i) \tag{32.20}$$

and the all-zero section, corresponding to Eq. (32.19), satisfies the difference equation

$$y(n) = \sum_{j=1}^{M} a_j p(n-j) \tag{32.21}$$

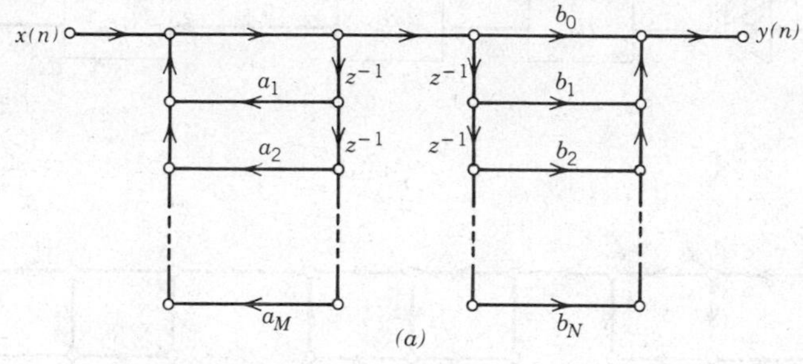

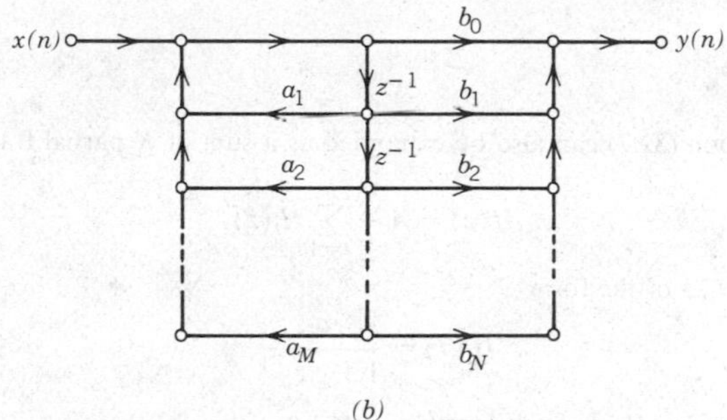

Fig. 32.2 Direct-form infinite-impulse-response (IIR) filter structure: (*a*) Part on the left realizes the poles, the part on the right the zeroes, (*b*) structure resulting from consolidation of the two delay paths in (*a*).

In these equations M is the number of zeros in $H(z)$ and N is the number of poles. These equations lead to the "direct form" structure (also known as the canonical form) shown in Fig. 32.2*a*, where the structure on the left corresponds to Eq. (32.20) and the structure on the right to Eq. (32.21). A simpler version of this structure is shown in Fig. 32.2*b*.

Cascade Form. If the numerator and denominator polynomials of Eq. (32.7) are factored, we can write

$$H(z) = g \prod_{i=1}^{P} H_i(z) \tag{32.22}$$

Here g is a gain factor and P is a number that depends on how the factorization is done. Since we must work with real coefficients, complex roots are factored as conjugate pairs. If there are M_1 real zeros, M_2 conjugate zero pairs, N_1 real poles, and N_2 conjugate pole pairs, then P is the maximum of $(M_1 + M_2)$ and $(N_1 + N_2)$. Each $H_i(z)$ is formed from one factor of $N(z)$ and one factor of $D(z)$ and is, in general, a biquadratic:

$$H_i(z) = \frac{1 + a_i z^{-1} + b_i z^{-2}}{1 + c_i z^{-1} + d_i z^{-2}} \tag{32.23}$$

For real zeros $b_i = 0$, and for real poles $d_i = 0$, and if M and N are not equal, some numerators (or denominators) will be unity. Equation (32.22) leads to the cascade form shown in Fig. 32.3(*a*). In this figure each box contains one second-order section of the form of Eq. (32.23). The individual sections themselves are realized in direct form; hence the overall structure is as shown in Fig. 32.3(*b*).

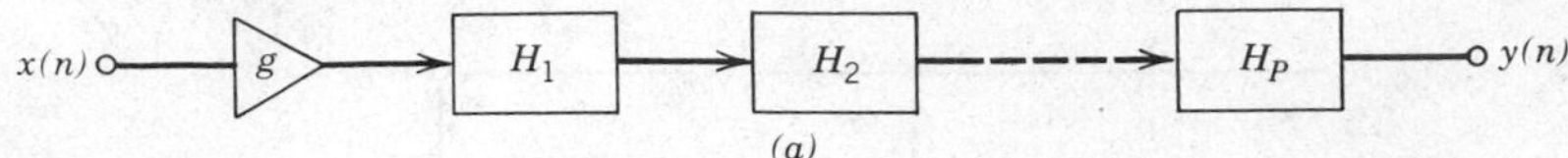

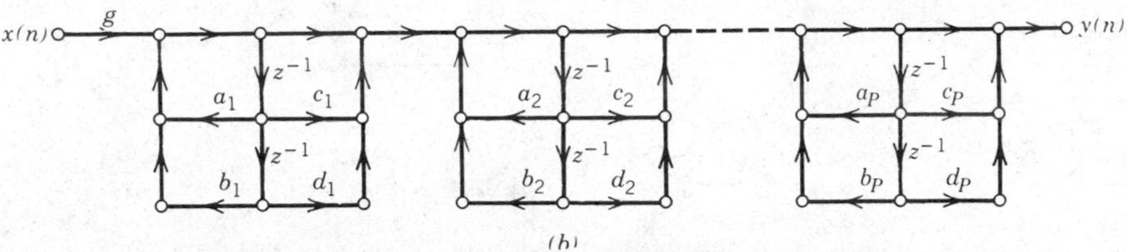

Fig. 32.3 Cascade-form infinite-impulse-response (IIR) filter structure: (*a*) Gain term is followed by cascaded order-2 or order-1 direct-form sections, (*b*) cascade form in detail showing makeup of individual sections.

Parallel Form. Equation (32.7) can also be expanded as a sum of K partial fractions

$$H(z) = A + \sum_{i=1}^{K} H_i(z) \tag{32.24}$$

where for real poles, H_i is of the form

$$H_i(z) = \frac{a_i}{1 + c_i z^{-1}} \tag{32.25a}$$

and for complex pole pairs, H_i is of the form

$$H_i(z) = \frac{a_i + b_i z^{-1}}{1 + c_i z^{-1} + d_i z^{-2}} \tag{32.25b}$$

The number of sections K is the sum of $N_1 + N_2$. The corresponding structure is the parallel form shown in Fig. 32.4. As with the cascade form, each section is realized in direct form.

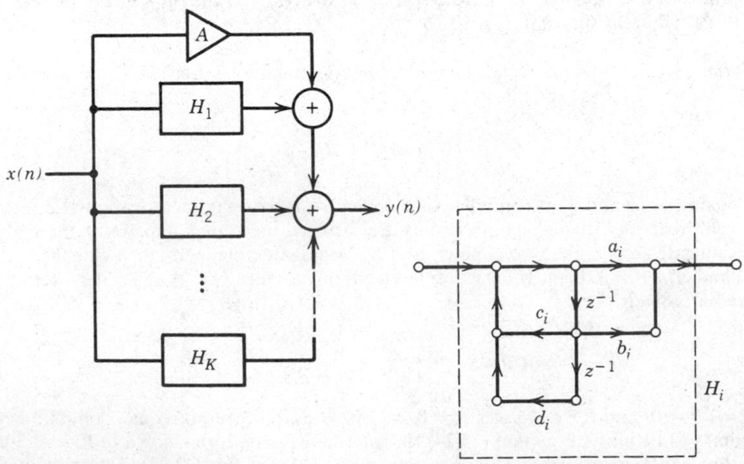

Fig. 32.4 Parallel-form infinite-impulse-response (IIR) structure derived from partial-fraction expansion of $H(z)$. A typical $H_i(z)$ section is shown on the right.

Comparison of Structures. Although the direct-form structure is the most appealing by reason of its simplicity and its direct relation to the original difference equation, it is not recommended for filters of order higher than 2. This is because the direct form is extremely sensitive to coefficient errors. Since the coefficients are always represented with finite precision, some error in their value is inevitable. These errors will affect the performance, and occasionally the stability, of the filter. Hence the use of forms that are minimally sensitive to these errors is important. The direct form must be used for order 2 or less, but in all other cases direct-form structure performance must be analyzed carefully, preferably by computer simulation.

The cascade form has the advantage that pole and zero locations are directly related to the coefficients of each section, as given in Eq. (32.24). The complex roots of $1 + az^{-1} + bz^{-2}$ can be located by inspection; the radial distance from the origin is equal to $\sqrt{b}$ and the real part of the root is $-a/2$. In the case of filters whose parameters are to be computed and changed in real time, this implementation may therefore be particularly convenient. On the other hand, the gain of each section depends on its coefficients, and the signal level in each stage must be controlled by additional gain elements to avoid overflow or underflow.

The parallel form is free of most of the problems mentioned. Pole locations can be found as in the cascade form, although the locations of the zeros cannot.

FIR Structures[1,2]

Direct Form. The direct and cascade forms previously described are readily adapted to FIR filters. Since the pole-producing and zero-producing portions of these structures are distinct, FIR forms are derived by simply leaving out the pole-producing portions of the IIR forms. The FIR direct form is shown in Fig. 32.5*a*.

Cascade Form. In the cascade form, each section is likewise an all-zero quadratic (or linear) section, and the structure is therefore simplified in the same way that the direct form is simplified. A FIR cascade-form structure is shown in Fig. 32.5*b*.

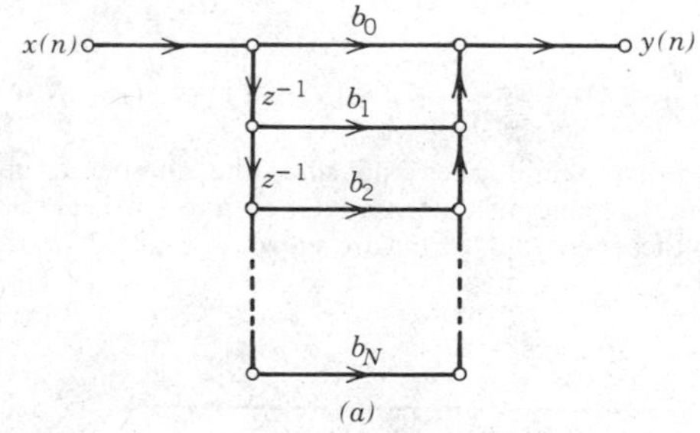

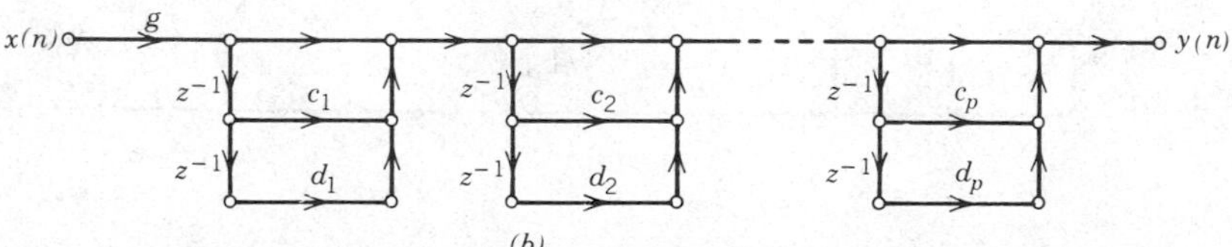

Fig. 32.5 Finite-impulse-response (FIR) filter structures: (*a*) Direct-form, (*b*) cascade form.

Frequency-Sampling Form. The parallel IIR form was derived from a partial-fraction expansion of $H(z)$. Since partial-fraction expansions are based on poles, this approach is clearly out of the question for FIR filters. A parallel form is possible, however, but it is derived in an entirely different way. A detailed explanation is given in Section 32.5; here we only observe that this form is based on one particular design procedure which takes as its point of departure the Fourier transform of the impulse response. The frequency-sampling form is shown in Fig. 32.6.

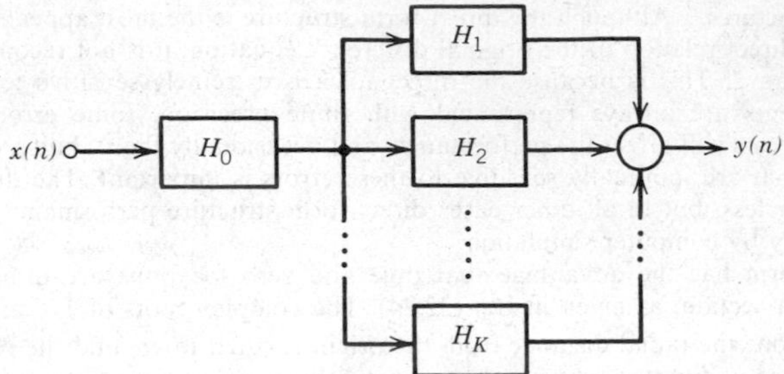

Fig. 32.6 Frequency-sampling finite-impulse-response (FIR) filter structure. (For detailed structure, see Fig. 32.23.)

Linear-Phase Form. Many applications require filters whose phase response is linear with frequency. We showed in Section 32.1 that linear-phase FIR filters have a symmetrical impulse response. That is, for an order-N filter

$$h(n) = h(N - 1 - n) \tag{32.26}$$

This symmetry requirement leads to certain economies in implementation. In an FIR filter, the h's correspond one-to-one with the b's of Eq. (32.17). Because of the symmetry in $h(n)$, Eq. (32.17) becomes

$$y(n) = b_0 x(n) + b_1 x(n-1) + \cdots + b_1 x(n - N + 1) + b_0 x(n - N) \tag{32.27a}$$

This can be written for N even, as

$$y(n) = \sum_{i=0}^{N/2-1} b_n[x(n-1) + x(n - N + i)] \tag{32.27b}$$

or, for N odd, as

$$y(n) = b_{N/2} x(n - N/2) + \sum_{i=0}^{(N-1)/2-1} b_n[x(n-1) + x(n - N + i)] \tag{32.27c}$$

Since each term in the sum requires a multiplier and since the number of terms is reduced by approximately one half, considerable economies in computation and hardware are achieved. Direct-form structures corresponding to these rewritten forms are shown in Figs. 32.7*a* and 32.7*b*.

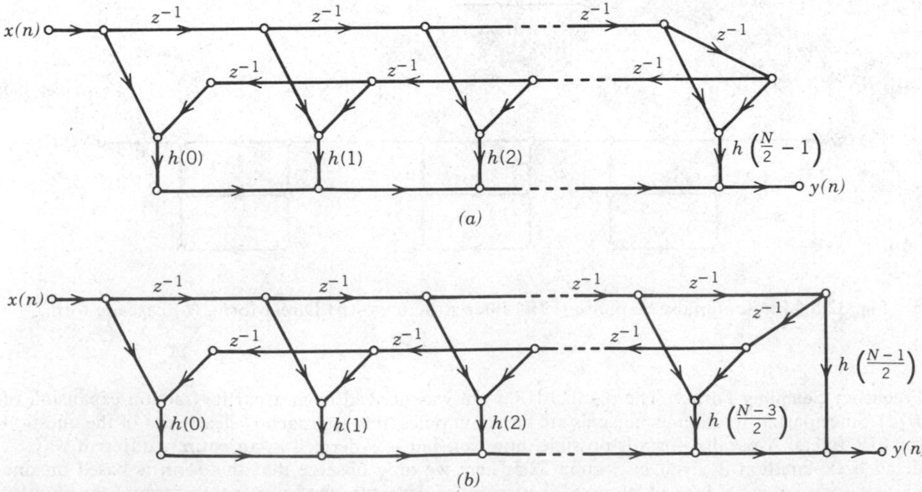

Fig. 32.7 Linear-phase direct-form finite-impulse-response (FIR) structures: (*a*) Even order, (*b*) odd order.

Comparison of Structures. The advantages and disadvantages of the direct and cascade forms are substantially the same for FIR filters as for IIR filters. The direct form obviously has no stability problems, but coefficient-quantization errors may still degrade performance. The cascade structure still offers the most direct control over locations of zeros.

Linear-phase filters are a special case. Because of the symmetry of the filter, errors in quantizing coefficients do not disturb phase linearity, but may still degrade performance otherwise. If linear-phase filters are realized in cascade form, the sensitivity to coefficient quantization will be less, but the errors may destroy the linear-phase characteristic. These issues will be taken up at greater length in Section 32.5.

Notice that in all filter structures shown, the only arithmetic operations required are multiplication and addition.

32.3 NUMBER REPRESENTATION AND QUANTIZATION EFFECTS

All signals that are passed through digital filters are in the form of samples represented as binary numbers. The digital representation of these samples, and of intermediate values inside the filter, must necessarily use a finite word length. The multiplications, and possibly the additions, performed during filtering in general yield intermediate results that are longer than the available word lengths and must thus be rounded off. Similarly, the filter coefficients are also represented by binary numbers of finite length. Because of these limitations, the sample values passed through the filter contain small inaccuracies which show up as noise, and the filter coefficients as stored in the filter will differ from the design values. The effects of these errors depend in part on the number system used in the implementation of the filter.

Overflow and Roundoff Errors

Arithmetic operations on n-bit binary numbers can give results longer than n bits in two ways. When numbers are multiplied, the product will, in general, be $2n$ bits long. If n-bit numbers are added, the sum may be $n + 1$ bits long. In either case, the result normally has to be shortened to n bits again before computation can proceed. To shorten a number is to throw away information and to introduce an inaccuracy in the computation.

For example, consider the one-zero filter in Fig. 32.8, which uses 6-bit binary numbers. The delayed input sample, $x(n-1)$, is multiplied by the coefficient a and added to the incoming sample, $x(n)$. Suppose the coefficient and samples have the binary values shown in the figure. Then multiplying 0.101001 by 0.110111 yields a product of 0.100011001111. This product must be shortened to 6 bits to be used in the filter; hence there is a resultant roundoff error of -0.000000001111 (about -0.65%). Similarly, the shortened value, 0.100011, added to 0.100110 (the incoming sample), results in a sum of 1.001001. This is again in excess of the 6-bit limit and represents an overflow, which cannot be treated by simply throwing away the low-order bit.

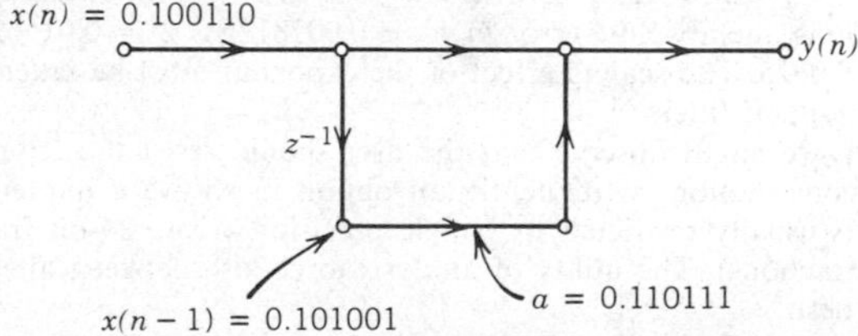

Fig. 32.8 Single-zero filter with data and coefficient represented by 6-bit fractions.

Truncation and Rounding. If the extra bits of a product are simply discarded, the result is said to be truncated. If the result is set to the n-bit number nearest the original value, the result is said to be rounded. Truncation is computationally simpler, but (1) the cost difference is trivial; (2) truncation errors may be biased, while roundoff errors are not; and (3) the errors introduced by truncation tend to be greater than those introduced by rounding off. We will consider only rounding off here.

Rounding is done as follows: When excess bits are discarded, test the most significant bit of the part being thrown away. If this bit is a 1, increment the n-bit result; if the bit is a 0, leave the result unchanged. For example, the binary fraction 0.011001101011, rounded off to 6 bits, is 0.011010, because the discarded part, 101011, begins with a 1. This rule introduces a small bias, since if the number is exactly between the two nearest values, it is always rounded up: 0.011000100000 rounds to

0.011001, with an error of 0.0000001. Such values are normally so infrequent that the bias is negligible.

If a binary fraction is rounded to n bits, the resulting error is at most $2^{-(n+1)}$. That is

$$-2^{-(n+1)} \leqslant e \leqslant 2^{-(n+1)} \tag{32.28}$$

The effect of roundoff error is analyzed statistically. We assume that all errors within this range are equally likely, hence that roundoff error is a zero-mean random variable with variance $\sigma_n^2 = 2^{-2n}/12$.[4]

Overflow. If adding two n-bit fractions yields a sum of 1 or greater, an overflow occurs. In floating-point arithmetic the result can be renormalized by shifting right and incrementing the exponent. In shifting, the LSB is lost and a roundoff error results. In fixed-point arithmetic, overflow raises three problems: (1) The number has to be divided by 2 to bring its magnitude down to less than 1, (2) the loss of the LSB leads to roundoff error, and (3) some scaling is necessary elsewhere in the filter to compensate for the fact that the number is half its true value. The side effects of these problems are troublesome enough that the common practice is to scale signal levels in advance to avoid the possibility of overflow.

Roundoff Errors in Digital Filters

Fixed-Point Arithmetic. If fixed-point overflow is avoided by scaling the signal levels, then roundoff errors will occur only with multiplication. On each new multiplication, the product will in general be different and a different roundoff error will be introduced. The errors thus appear as a sequence of small random variables added to the system; the end result is an increase in the noise level in the system. This "roundoff noise" can be reduced to any desired level by increasing the word length. The purpose of roundoff-noise analysis is to find the shortest word length compatible with given dynamic-range requirements. Details are given in Refs. 1 and 2.

These analyses depend on the assumption that noise sources are uncorrelated with one another and with the signal. This is essentially true if the signal itself is wideband noise but is not true for simple deterministic signals like constants or sinusoids. In such cases it is necessary to back up the analysis by a computer simulation. Fortunately many common signals such as speech or music can be treated as approximately random.

It should be borne in mind that absolute accuracy is not required in these analyses. The goal of the analysis is to decide on a word length; this length must be an integer number of bits; each additional bit increases the dynamic range by a factor of 2; hence accuracies of 50% are often adequate.

Floating-Point Arithmetic. Error analysis for floating-point arithmetic is complicated by two considerations. First, the need to normalize after additions means that roundoff errors may result from addition as well as multiplication. Second, the structure of the floating-point number means that a roundoff error is not an absolute quantity but the product of the error in the fraction and 2^e. For example, if two numbers $V_1 = 0.578125 \times 2^1$ and $V_2 = 0.578125 \times 2^3$ both have an error of $1/128 = 0.0078125$ in their fractions, the absolute error in V_1 is $0.0078125 \times 2^1 = 0.015625$, while the error in V_2 is $0.0078125 \times 2^3 = 0.0625$. The scaling effect of the exponent must be taken into account in any noise analysis of floating-point filters.[5,6]

As a practical point, we might observe that the user should resist the temptation to analyze to excess. Floating-point computation is frequently an option in software implementations. Here the choice of word length is usually restricted to "single precision" (e.g., 24-bit fractions) and "double precision" (e.g., 56-bit fractions). The utility of analysis to choose between alternatives that are this far apart is dubious at best.

Limit Cycles and Oscillations

A further consequence of finite word length is the possibility of oscillation in a theoretically stable filter. This is best illustrated by an example. Suppose a one-pole filter with the transfer function

$$y(n) = -0.6y(n-1) + x(n)$$

is implemented using 4-bit fractions for the data. If the input is an impulse [Eq. (32.3)] scaled to an amplitude of 0.5 at $n = 0$, then y takes on the values shown in Table 32.1. (The values are given as signed-magnitude binary fractions to show the effects of rounding.) After $n = 5$, the filter is in a limit cycle and the resulting oscillations in y are termed limit-cycle oscillations.

Limit-cycle oscillations have been studied for first- and second-order systems by Jackson.[7] For first-order systems of the form

$$y(n) = x(n) - ay(n-1) \tag{32.29}$$

TABLE 32.1. LIMIT-CYCLE OSCILLATIONS IN THE FILTER $y(n) = -0.6y(n-1) + x(n)$ USING 4-BIT SAMPLES AND NO GUARD BITS

n	$y(n)$	$y(n)$ (Rounded)
0	0.10000000	0.1000
1	− 0.01001101	− 0.0101
2	0.00110000	0.0011
3	− 0.00011101	− 0.0010
4	0.00010011	0.0001
5	− 0.00001001	− 0.0001
6	0.00001001	0.0001
7	− 0.00001001	− 0.0001
8	0.00001001	0.0001

with n-bit fractions used with rounding, limit-cycle oscillations can occur for y in the region $(-k, k)$, where k is the largest b-bit fraction satisfying

$$k \leqslant \frac{0.5 \times 2^{-b}}{1 - |a|} \tag{32.30}$$

For $|a| < 0.5$, no oscillations can occur.

For second-order systems of the form

$$y(n) = x(n) - py(n-1) - qy(n-2) \tag{32.31}$$

again with b-bit fractions used with rounding, limit cycle oscillations can occur in the region $(-k, k)$, where k is the largest b-bit fraction satisfying

$$k \leqslant \frac{0.5 \times 2^{-b}}{1 - |q|} \tag{32.32}$$

Again, limit-cycle oscillation can be prevented by making $|q| < 0.5$.

In practice, however, the coefficients are determined by design requirements and cannot easily be altered to avoid limit-cycle oscillation. Instead one uses Eqs. (32.30) and (32.32) to find the maximum amplitude of these oscillations and append enough guard bits (i.e., extra-low-order bits beyond the design word-length requirement) so that the oscillations take place in the guard bits only. For example, if in the example above, 6-bit fractions were used for the internal variables of the filter, the oscillation would affect only the sixth bit and would never appear in the 4-bit output, as shown in Table 32.2. Here the filter enters a limit cycle at $n = 7$, but because only bit 6 is affected, the 4-bit output rounds off to 0.

TABLE 32.2. LIMIT-CYCLE OSCILLATIONS IN THE FILTER $y(n) = -0.6y(n-1) + x(n)$ WITH TWO GUARD BITS

	$y(n)$		
n	Product	6 Bits	Output
0	0.100000000	0.100000	0.1000
1	− 0.010011010	− 0.010011	− 0.0101
2	0.001011011	0.001011	0.0011
3	− 0.000110100	− 0.000111	− 0.0010
4	0.000100010	0.000100	0.0001
5	− 0.000010011	− 0.000010	− 0.0001
6	0.000001010	0.000001	0.0000
7	− 0.000000100	− 0.000001	0.0000
8	0.000000100	0.000001	0.0000
9	− 0.000000100	− 0.000001	0.0000

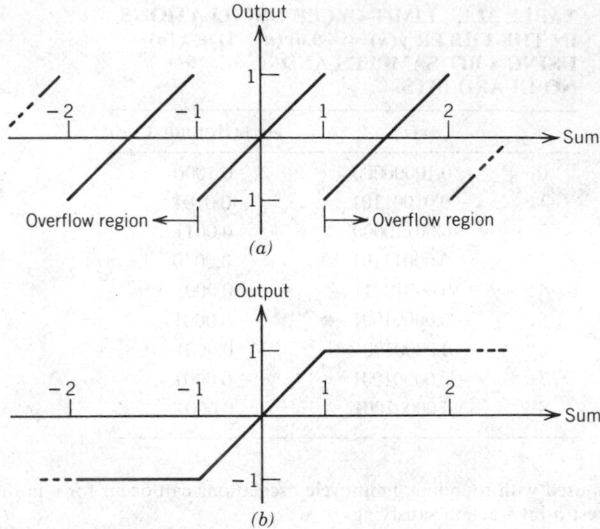

Fig. 32.9 Transfer characteristic of (*a*) conventional 2's-complement binary adder, (*b*) saturating adder.

A different oscillation problem can arise from overflows in addition. The filter has supposedly been designed to prevent such overflows, but in practice any system can be overloaded on occasion. If 2's-complement arithmetic is used, overflow of a positive number may "wrap around" to produce a spurious negative value, and similarly for overflow of a negative number. For example, the 6-bit binary fractions 0.111011 and 0.101110 add up to 1.101001. The sum exceeds 1 and the carry out of the high-order bit sets the sign bit to 1, indicating a negative number. (See Chapter 55 of this handbook for further information on binary addition.) Such an adder's transfer characteristic thus looks like Fig. 32.9*a*. These unexpected changes of sign can result in instability and large-amplitude oscillations. The solution is to use a saturating adder. In a saturating adder, an overflow in either direction results in an output value equal to the largest possible magnitude, as shown in Fig. 32.9*b*.

Coefficient Quantization Effects

Filter coefficient values, like signal samples, are represented by finite-precision numbers and thus also have quantization errors.[1,2] These errors cause errors in the positions of the filter's poles and zeros, and the filter does not perform exactly as intended. The degree of displacement of the poles and zeros from their design values depend on (1) the word-length used for the coefficients and (2) the filter structure. We will analyze coefficient quantization in second-order filters and will show that in this case it has the effect of limiting the choice of pole and zero locations.

IIR Filters. A direct-form realization of a pair of complex poles is shown in Fig. 32.10. The poles of this structure are given by the roots of the denominator polynomial. This polynomial can be rewritten in terms of the polar coordinates of the roots

$$1 + 2\rho \cos \theta z^{-1} + \rho^2 z^{-2} \qquad (32.33)$$

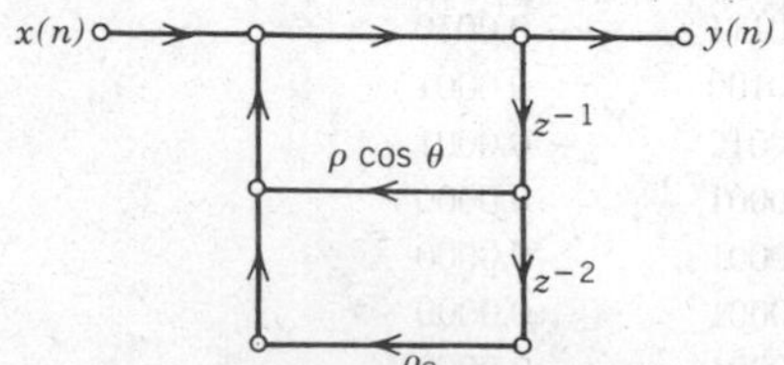

Fig. 32.10 Direct-form realization of single complex pole pair. Coefficients are shown in terms of polar coordinates of poles.

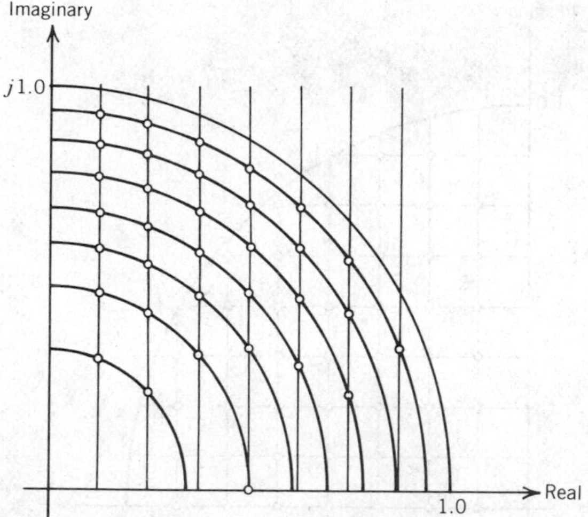

Fig. 32.11 First quadrant of unit circle, showing possible pole locations using structure of Fig. 32.10 with three-bit quantization.

If the coefficients are represented by (e.g.) 4-bit, 2's-complement numbers, then ρ can take on only seven values, ranging from $\sqrt{0.001_2}$ through $\sqrt{0.111_2}$. With the same quantization, $\rho \cos \theta$ can take on only 15 values, ranging from -0.111_2 to $+0.111_2$.

With this quantization, therefore, the only possible root locations are at the intersections between a family of 7 circles with a family of 15 straight lines. These intersections are shown for the first quadrant of the unit circle in Fig. 32.11. This quantization is much coarser, of course, than would be encountered in practice, but it highlights two important characteristics: (1) The spacing of the circles is not uniform—the circles are more closely spaced as they approach the unit circle. This means that the density of possible pole locations increases as we approach the unit circle, as we would like it to do. In spite of this, however, (2) there are relatively few intersections in the important low-frequency region around $(1 + j0)$.

An alternate way of realizing second-order poles, due to Gold and Rader,[4] offers a different choice of pole locations. This structure uses cross-coupled first-order sections, as shown in Fig. 32.12. Its transfer function is

$$H(z) = \frac{\rho z^{-1}(\sin \theta + \rho \sin 2\theta)}{1 + 2\rho \cos \theta z^{-1} + \rho^2 z^{-2}} \tag{32.34}$$

This has the same poles as the structure of Fig. 32.10. Here, however, the filter coefficients are $\rho \cos \theta$ and $\rho \sin \theta$, the real and imaginary coordinates of the pole position; hence the resulting set of possible pole locations is the uniform lattice shown in Fig. 32.13. Note that this lattice is denser (i.e., the possible pole locations are closer together) near $(1 + j0)$ than they are in Fig. 32.11 and less dense near $(0 + j1)$. Other things being equal, the designer should choose the structure that provides the greater density in the region of interest.

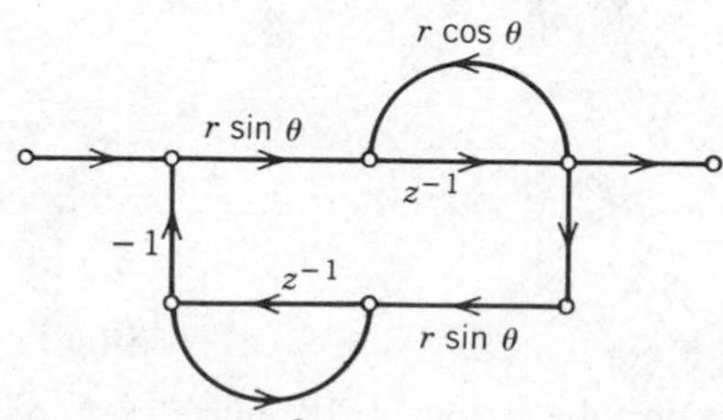

Fig. 32.12 Alternate realization of single complex pole pair based on cross-coupled first-order sections.

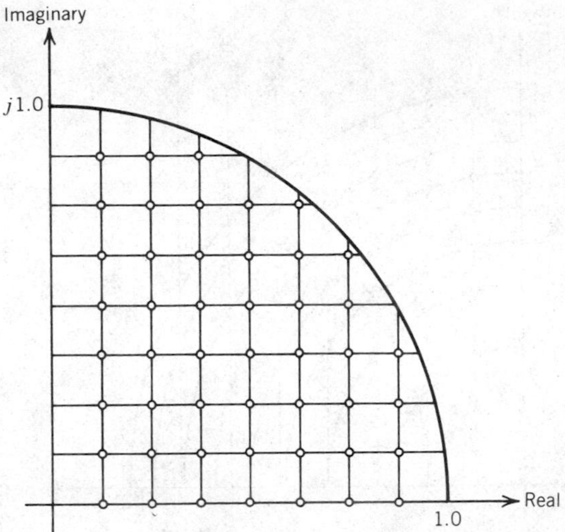

Fig. 32.13 First quadrant of unit circle, showing possible pole locations using structure of Fig. 32.12 with 3-bit quantization.

This limited selection of pole locations is most serious in high-order narrow-band filters where many poles have to be placed in a small region very close to the unit circle; in such a case even small errors in position can degrade performance seriously and very long words may be required regardless of the configuration used.

In higher-order direct-form filters, the analysis is extremely involved and cannot be summed up in terms of possible pole positions, because there is interaction among the poles. Specifically, Kaiser[8] has shown that displacement of the poles from their desired value is greatest when large numbers of poles are located close together. This situation is particularly apt to arise in high-order narrow-band filters. For this reason the direct-form structure is not normally recommended for filter orders greater than 2. Since pole interactions are not found in second-order structures, they also do not occur in cascade or parallel structures.

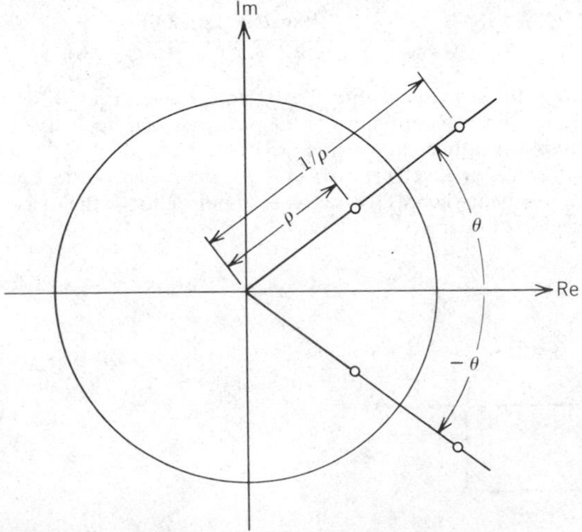

Fig. 32.14 Locations of zeroes in linear-phase finite-impulse-response (FIR) filters.

FIR Filters. Most of the foregoing analysis carries over to FIR filters. We discuss only the question of linear phase here. Quantizing a set of symmetrical coefficients does not destroy their symmetry; hence quantization errors in the direct form do not destroy phase linearity. They do displace the zeros from their design values, and in high-order filters may move unit-circle zeros off the unit circle.

If a cascade realization is used, quadratic realizations of unit-circle zeros will always lie on the unit circle. This is because these zeros result from a factor $(z^{-2} + az^{-1} + 1)$ and a coefficient of 1 has no quantization error. In linear-phase filters, however, zeros off the unit circle come in groups of four, with polar coordinates $(\rho, \pm\theta)$, $(1/\rho, \pm\theta)$ (Fig. 32.14). If these pairs are realized by quadratic sections, the limited choice of root locations usually prevents, for a given $(\rho, \pm\theta)$, placement of a corresponding pair exactly at the reciprocal points $(1/\rho, \mp\theta)$. Hence if phase linearity is a critical requirement, it may be better to implement fourth-order terms in direct form.

32.4 INFINITE-IMPULSE-RESPONSE (IIR) FILTER DESIGN TECHNIQUES

Digital filter designs generally fall into two categories. There are those done by hand, with only the aid of a pocket calculator, and there are those that require a computer. The first category of designs is simpler; the design equations are short and well motivated. The first category is also older and tends (with some exceptions) to yield an inferior product. Computer-aided designs are difficult to describe concisely and generally require reference to the programs used, which are too lengthy to be given in their entirety here. Therefore the designs that can be done by hand are described at greater length here than are the computer-aided designs, since any description of the latter eventually breaks off with a reference to the program to be used.

There are two general approaches to designing IIR digital filters. The first is to start with a known analog filter that satisfies the design requirements and to find a corresponding digital filter. The second is to approximate the desired filter performance directly in the digital domain or, equivalently, in the z-transform domain. These approximation methods require solution of sets of nonlinear equations and are best adapted for computer-aided design. We will begin by discussing the methods that start with an equivalent analog design.

Impulse Invariance

Since a linear system is completely specified by its impulse response, one possible design strategy is to derive a digital filter whose impulse response is a sampled version of the impulse response of a given analog filter.[1,3] That is, if $g(t)$ is the analog impulse response and $h(n)$ is the digital impulse response, then

$$h(n) = g(nT) \tag{32.35}$$

where T is the sampling interval.

As in any sampling operation, however, one must give careful consideration to the relation between sampling rate and bandwidth. The operation of sampling $g(t)$ can be thought of as a multiplication by an impulse train with spacing $T = 1/f_s$. But by elementary Fourier transform theory, this operation convolves the frequency response $G(f)$ with another impulse train with spacing $f_s = 1/T$.[9] The result is a replication of $G(f)$ at intervals of f_s, as shown in Fig. 32.15. Unless $G(f)$ is suitably band limited, aliasing will result from these replications. (The impulse response will be undersampled.) Hence the utility of this design approach is limited to filters whose frequency response has a sharp cutoff well below $f_s/2$. Its main appeal is its intuitive and computational simplicity.

The procedure for impulse-invariant design is as follows:

1. Start with the transfer function of the corresponding analog filter, $G(s)$.
2. Find the partial-fraction expansion of $G(s) = \sum_i G_i(s)$.
3. For each $G_i(s)$, find the corresponding $H_i(z)$.
4. Use a parallel-form filter made up of quadratic blocks corresponding to $H_i(z)$.

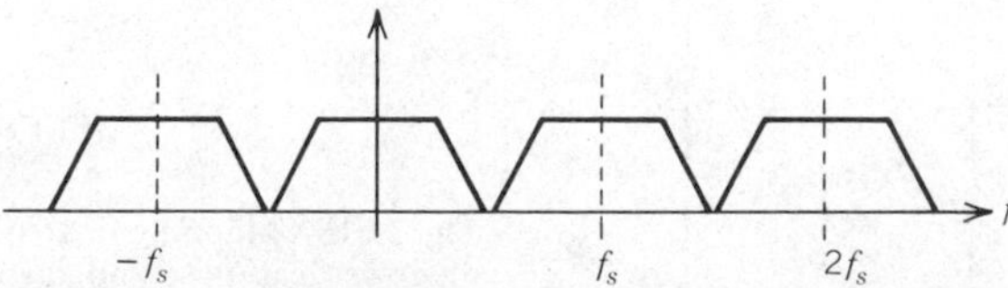

Fig. 32.15 Replications of impulse response at intervals of f_s as a result of sampling.

Bilinear Transformation

In the z-transform domain, the unit circle corresponds to the $j\omega$ axis in the Laplace transform domain. Deriving a digital filter from an equivalent analog filter can be viewed as finding a mapping from the one domain to the other that will wrap the $j\omega$ axis around the unit circle. Impulse invariance does just this, in fact, with the qualification that a single stretch of the $j\omega$ axis, from $-\pi f_s$ to πf_s, goes around the unit circle once. The aliasing problem, in this interpretation, arises because all additional stretches of the $j\omega$ axis, from $(2n-1)\pi f_s$ to $(2n+1)\pi f_s$, $n = \ldots, -2, -1, 0, 1, 2, \ldots,$ also wrap around the unit circle. What is needed is a function that will map the entire $j\omega$ axis onto the unit circle just once. The function that does this is the bilinear transformation

$$s = 2f_s \frac{1 - z^{-1}}{1 + z^{-1}} \tag{32.36}$$

If this function is substituted for s in $G(s)$, the result will be a function which, when evaluated around the unit circle, will take on precisely the values $G(s)$ does when evaluated on the $j\omega$ axis.

Use of the transformation of Eq. (32.36) maps an infinite range of frequencies onto the finite unit circle. Naturally this mapping distorts the frequency scale. Call the radian frequency in the z-transform domain ω_d and call the corresponding radian frequency in the analog domain ω_a. In the z-transform domain, ω_d corresponds to angular distance around the unit circle. This angular distance will be $\omega_d T$ radians, where T is the sampling period. Evaluating Eq. (32.36) on the unit circle, $z = e^{j\omega_d T}$

$$\omega_a = 2/T \tan(\omega_d T/2) \tag{32.37}$$

The practical consequence of this is that, when deriving a digital filter from an analog design using the bilinear transformation, we must *predistort* the frequencies of interest in the analog domain so that, after they have been transformed, they will come out in the right places in the z-transform domain.

This is shown in Fig. 32.16. The digital-domain frequency is plotted on the horizontal axis and the analog-domain frequency on the vertical axis. The solid curve is a plot of Eq. (32.37). Below the digital-domain axis a hypothetical frequency response is shown, and the necessary predistortion in the time domain can be seen to the left of the analog-domain axis.

The procedure for designing an IIR filter using the bilinear transformation is thus as follows:

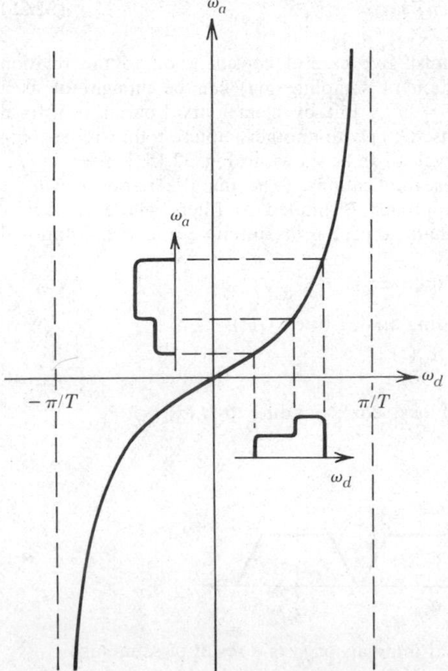

Fig. 32.16 Mapping of frequencies between analog (vertical axis) and digital (horizontal axis) domains. A possible frequency response and its predistorted analog equivalent are shown.

1. Predistort all critical frequencies using Eq. (32.37).
2. Design (or find in a table) the analog filter transfer function $G(s)$, using the predistorted frequencies.
3. Find the corresponding digital-filter transfer function $H(z)$ by substituting Eq. (32.36) into $G(s)$.
4. Derive the appropriate digital section, either by factorization or by partial-fraction expansion of $H(z)$.

Example. Using $f_s = 10$ kHz, design a two-pole Butterworth filter with a corner frequency of 1 kHz.

1. Frequency predistortion. $T = 0.0001$ s, $\omega_d = 6283.2$ rad/s. Hence $\omega_a = 2 \times 10^4 \tan(6283.2 \times 10^{-4}/2) = 6498.4$ rad/s(1034.25 Hz).
2. Analog design. A two-pole low-pass Butterworth filter with $\omega_a = 6498.4$ has the transfer function

$$G(s) = \frac{4.2229 \times 10^7}{s^2 + 9.19011 \times 10^3 s + 4.2229 \times 10^7}$$

3. Bilinear transformation.

$$H(z) = \frac{4.2229 \times 10^7}{4 \times 10^8 \left(\dfrac{1 - z^{-1}}{1 + z^{-1}}\right)^2 + 1.83802 \times 10^8 \left(\dfrac{1 - z^{-1}}{1 + z^{-1}}\right) + 4.2229 \times 10^7}$$

$$= 0.067455 \frac{1 + 2z^{-1} + z^{-2}}{1 - 1.14298z^{-1} + 0.412802z^{-2}}$$

4. Implementation. Since the polynomials are quadratics, any implementation will reduce to a single order-2 direct form, as shown in Fig. 32.17.

In this example, $G(s)$ has two complex poles at $s = 4595.05 + j4595.05$ and a double zero at $s = \infty$. In the digital filter, the poles are at $z = 0.57149 + j0.29360$ and the zeros are at $z = -1$. The frequency response of this filter is shown in Fig. 32.18.

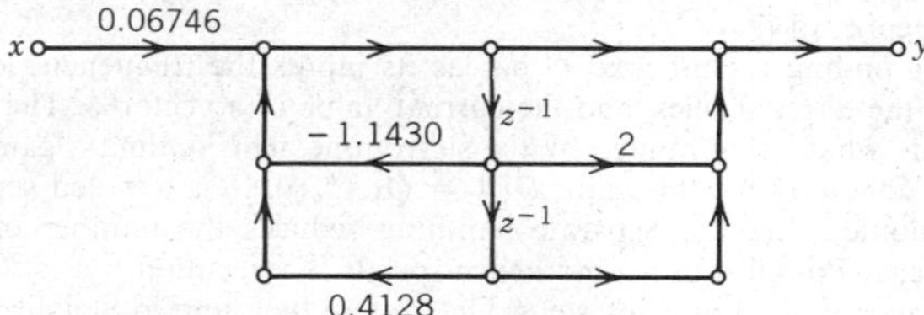

Fig. 32.17 Realization of a two-pole Butterworth filter by means of bilinear transformation.

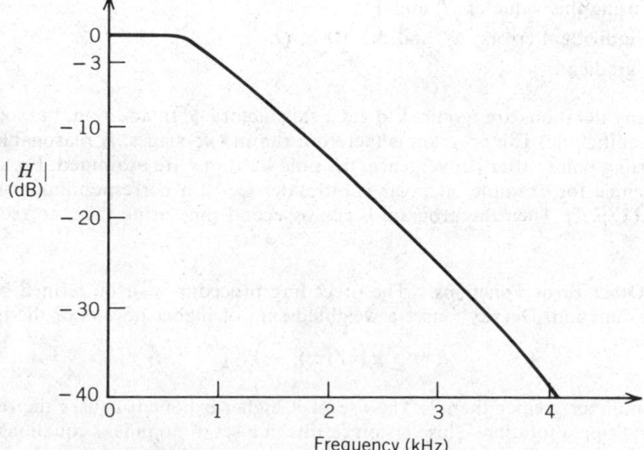

Fig. 32.18 Frequency response of filter of Fig. 32.17.

Because it warps the frequency scale, the bilinear transformation is best adapted to filters whose frequency responses are essentially piecewise constant, as opposed to filters whose response must be some straight-line function of frequency (e.g., differentiators).

Direct Design Techniques

We discuss three techniques for designing IIR filters directly from a specified frequency response. These design procedures are computationally very laborious and in general are not practical without the aid of a computer.

Minimization of Mean Squared Error (MSE). In the first technique, due to Steiglitz,[10] the filter gain is specified at M discrete frequencies, $\{z_i\}$, and the coefficients are determined that minimize the errors at these points. For minimum coefficient sensitivity and for analytic tractability, assume a cascade structure of N sections

$$Y(z) = AH(z) \tag{32.38}$$

A is a gain constant where H is a product of biquadratic forms given by

$$H(z) = \prod_{k=1}^{N} \frac{1 + a_k z^{-1} + b_k z^{-2}}{1 + c_k z^{-1} + d_k z^{-2}} \tag{32.39}$$

Let Y_i^d be the specified magnitude of the frequency response at a frequency z_i and let $Y(z_i)$ be the digital-filter approximation. Then the MSE is defined as

$$Q(\boldsymbol{\theta}) = \sum_{i=1}^{M} [|Y(z_i)| - Y_i^d]^2 \tag{32.40}$$

where $\boldsymbol{\theta}$ is the vector of unknown parameters

$$\boldsymbol{\theta} = \{a_1, b_1, c_1, d_1, \ldots, a_N, b_N, c_N, d_N, A\} \tag{32.41}$$

and the sum is taken over the M points at which Y has been specified.

Finding this minimum requires solving a set of simultaneous nonlinear equations. Steiglitz uses the Fletcher-Powell algorithm, which can be found in the IBM scientific subroutine package.[11] This program requires a user-supplied subroutine that computes Q and the partial derivatives of Q with respect to the parameters grad Q.

The subroutine for finding Q and grad Q has as its inputs the frequencies of interest $\{z_i\}$, the specified responses at these frequencies, and the current value of a vector $\boldsymbol{\phi}$. The vector $\boldsymbol{\phi}$ comprises all of $\boldsymbol{\theta}$ except the gain, which is computed by the subroutine. This optimum gain is written A^*, and Steiglitz uses $\hat{Q}(\boldsymbol{\phi})$ to denote Q for this gain: $\hat{Q}(\boldsymbol{\theta}) = Q(A^*, \boldsymbol{\phi})$. A is handled separately because its optimum can be computed directly. Separate handling reduces the number of parameters to be optimized by the Fletcher-Powell routine and hence speeds its execution.

The subroutine comprises the following steps. Details can be found in Steiglitz[10] and in Peled and Liu.[2]

1. Calculate the transfer function, exclusive of the gain, for all specified frequencies.
2. Compute A^* using this value of H_i and Y_i^d.
3. Calculate the individual errors, E_i, and the MSE, $\hat{Q}$.
4. Calculate the gradient.

In general, many iterations are required to get a satisfactory $\boldsymbol{\phi}$. In addition, the solution may have poles outside the unit circle. The program is therefore run in two stages. A reasonable initial guess is provided as a starting point; after convergence, the pole locations are examined. For any poles found outside the unit circle for example, at polar coordinates (ρ, θ), a corresponding pole is substituted inside, that is, at $(1/\rho, \theta)$. Then the program is run a second time using these corrected locations as the initial guess.

Minimization of Other Error Functions. The preceding procedure can be refined by using a more complicated error function. Deczky[12] uses a weighted sum of higher powers of the individual errors

$$E = \sum_i a_i[|Y(z_i)| - Y_i^d]^{2p} \tag{32.42}$$

where p is some number greater than 1. The use of a higher exponent causes the resultant error to tend toward an equiripple solution. This version results in a set of nonlinear equations, as before, that are solved in the same way. Deczky's method can also be used to obtain a specified group delay.

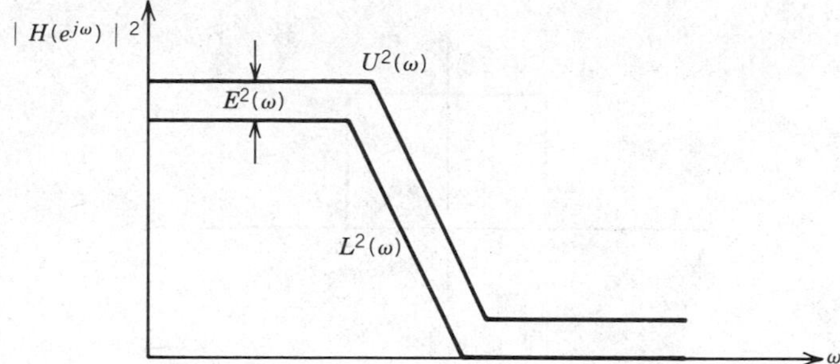

Fig. 32.19 Definition of gain limits for linear-programming filter design.

Linear Programming. Linear programming is a technique for finding values $\{x_1, x_2, \ldots, x_n\}$ that maximize the linear combination

$$R = \sum_{j=1}^{n} c_j x_j \tag{32.43}$$

subject to the linear constraints

$$\sum_{j=1}^{n} a_{ij} x_j \leqslant b_i \qquad i = 1, 2, \ldots, m \tag{32.44}$$

Thajchayapong and Rayner[13] have shown a way to apply this technique to digital filter design. The magnitude-squared transfer function $|H(z)|^2$, evaluated on the unit circle, can be written

$$|H(e^{j\omega})|^2 = P(\omega, k_p)/Q(\omega, l_q) \tag{32.45}$$

where

$$P(\omega, k_p) = \sum_{p=0}^{M} K_p \cos(p\omega T) \tag{32.46}$$

$$Q(\omega, l_q) = 1 + \sum_{q=1}^{N} l_q \cos(q\omega T) \tag{32.47}$$

These expressions are linear in k_p and l_q.

The filter gain is specified by upper and lower limits $U^2(\omega)$ and $L^2(\omega)$, their difference is $E^2(\omega)$. (See Fig. 32.19.) The constraints are based on the requirement that P/Q lie between these limits. The function R is chosen in such a way that maximizing it forces the actual response toward $U^2(\omega)$ in the passband and toward $L^2(\omega)$ in the stopband.

These constraints are evaluated at each frequency of interest and the function R is maximized using the simplex technique. If the resulting R is negative, then either the order of the filter must be increased or the error specifications must be relaxed.

After the optimum has been obtained, the specified design is given in terms of $\cos(n\omega T)$. The design is completed by substituting $0.5(z^n + z^{-n})$ for $\cos(n\omega T)$. The resulting polynomials in z refer to $|H(z)|^2$ and will have roots outside as well as inside the unit circle; in going from $|H(z)|^2$ to $H(z)$ we discard the roots outside the unit circle for stability and minimum phase. This method has been generalized by Rabiner et al.[14]

32.5 FINITE-IMPULSE-RESPONSE (FIR) FILTER DESIGN TECHNIQUES

One of the chief attractions of FIR filters is the fact that they can easily be made to have an exactly linear phase response. In many applications, linear phase is a necessity; hence the bulk of FIR design techniques have been developed with linear-phase filters in view.

Fourier-Transform Techniques: Windowing

A natural starting point for FIR filter design is the specified frequency response, taking advantage of the relation between this response and the impulse response of the filter.

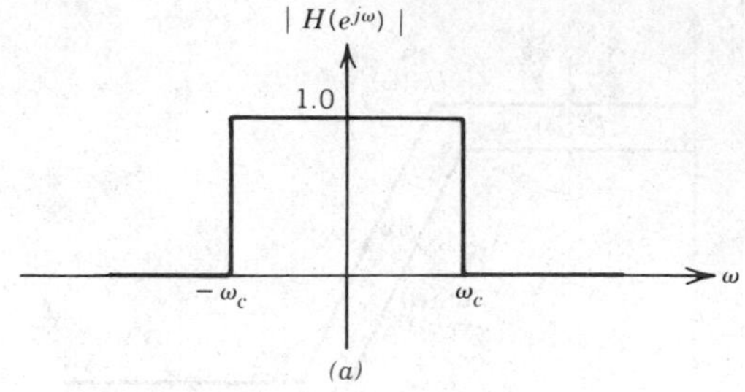

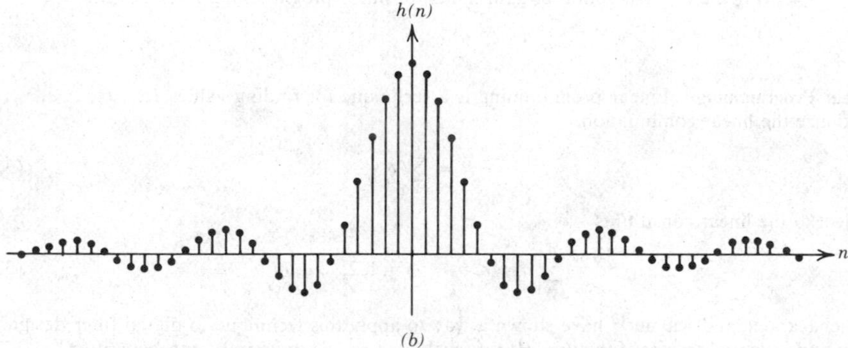

Fig. 32.20 (*a*) Ideal low-pass filter characteristic with cutoff at ω_c. (*b*) Impulse response corresponding to ideal low-pass filter.

If a digital filter has an impulse response $h(n)$, then its frequency response $H(e^{j\omega})$ is given by Eq. (32.11). If ω is scaled to the sampling frequency, then $H(e^{j\omega})$ is periodic with period 2π and $h(n)$ can be obtained from Eq. (32.12).

Consider the case of an ideal low-pass filter, as shown in Fig. 32.20*a*. This response is

$$H(e^{j\omega}) = \begin{cases} 1 & |\omega| < \omega_c \\ 0 & \text{otherwise} \end{cases} \tag{32.48}$$

where ω_c is the normalized cutoff frequency $2\pi f_c/f_s$. Applying Eq. (32.12), one obtains

$$h(n) = \frac{1}{2\pi}\int_{-\omega_c}^{\omega_c} e^{j\omega n} d\omega = \frac{\sin \omega_c n}{\pi n} \tag{32.49}$$

whose general appearance will resemble Fig. 32.20*b*.

This solution is not quite ready to use as it stands: (1) The impulse response is not finite and (2) the system is noncausal (because the impulse response is nonzero for negative time). A finite, causal approximation can be found, however, by windowing the impulse response, that is, truncating it for $|n|$ greater than some cutoff time t_c, and by shifting the response in time until the system is causal. These two steps are shown in Fig. 32.21*a* and 32.21*b*.

Windowing the impulse response causes a departure from the specified frequency response, as shown in Fig. 32.21*c*: (1) The passband response is no longer flat, but shows ripples that increase steadily in amplitude until the cutoff frequency, (2) the stopband response is no longer zero, and (3) the transition between passband and stopband is no longer abrupt.

To minimize these ripples, we multiply the original infinite-impulse-response by a windowing function other than a pure rectangular pulse. There is no finite window function whose transform has no sidelobes, but functions can be found whose transforms have very small sidelobes. If one of these windowing functions is used, the ripples in the frequency response will be correspondingly reduced.

Many windowing functions have been explored.[15] A few of the best-known ones are listed in Table 32.3 and shown in Fig. 32.22, along with their transforms. In all of these windowing functions the side lobes are all much smaller than those resulting from the rectangular window and the main

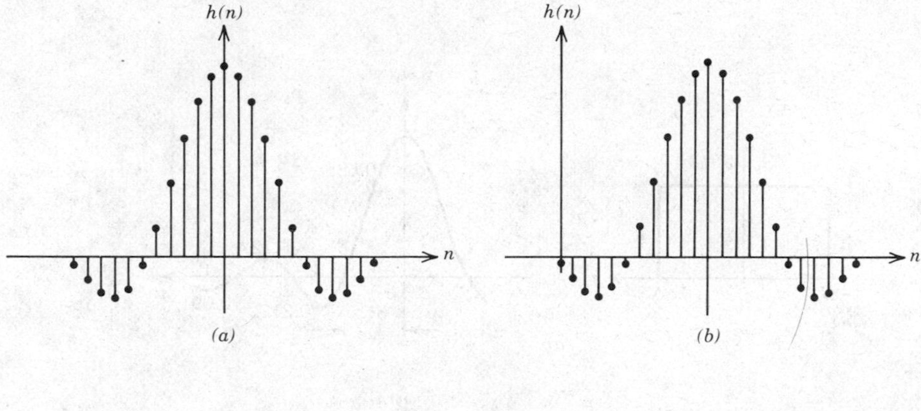

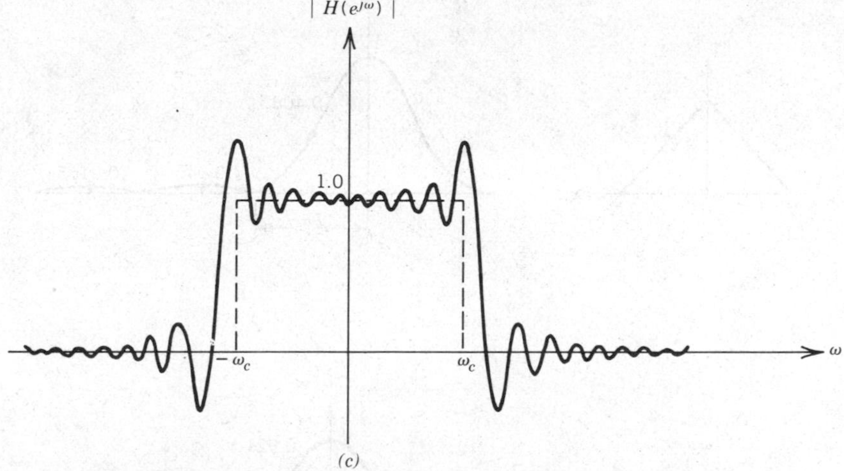

Fig. 32.21 (*a*) Truncated version of impulse response in Fig. 32.20*b*. (*b*) Truncated response shifted so as to make system causal. (*c*) Filter frequency response resulting from truncation of impulse response.

TABLE 32.3. COMMON WINDOWING FUNCTIONS

Name	Description[a]
Rectangular	$w(k) = 1$
Fejer-Bartlett	$w(k) = 1 - \|2k/N\|$
Hanning	$w(k) = (1 + \cos \pi k/N)/2$
Hamming	$w(k) = 0.54 + 0.46 \cos \pi k/N$
Kaiser	$w(k) = \dfrac{I_0\left[N\theta_a\sqrt{1-(k/N)^2}\right]}{I_0[N\theta_a]}$

[a] For all windows, $w(k) = 0$ for $|k| > N$.

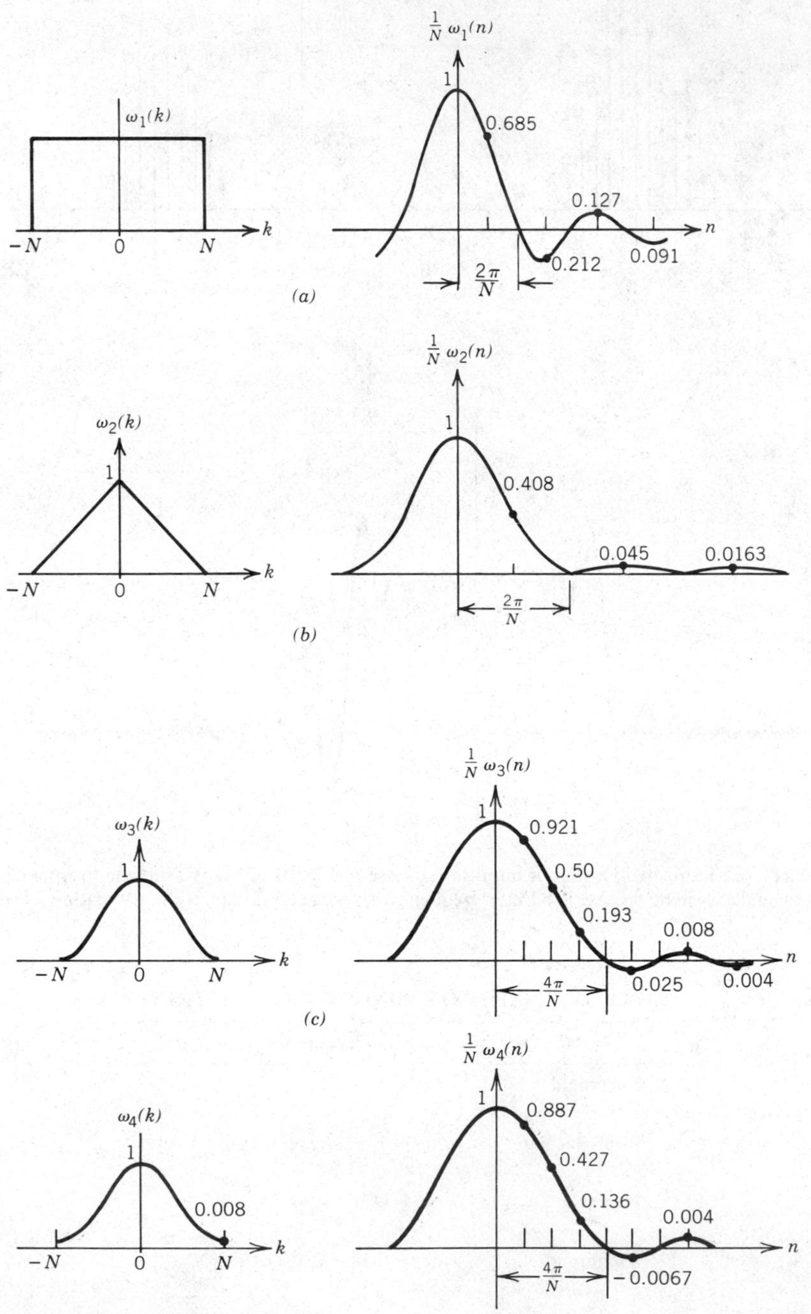

Fig. 32.22 Common window functions and their transforms: (*a*) Rectangular, (*b*) triangular (Fejér-Bartlett), (*c*) Hanning, (*d*) Hamming, (*e*) Kaiser.

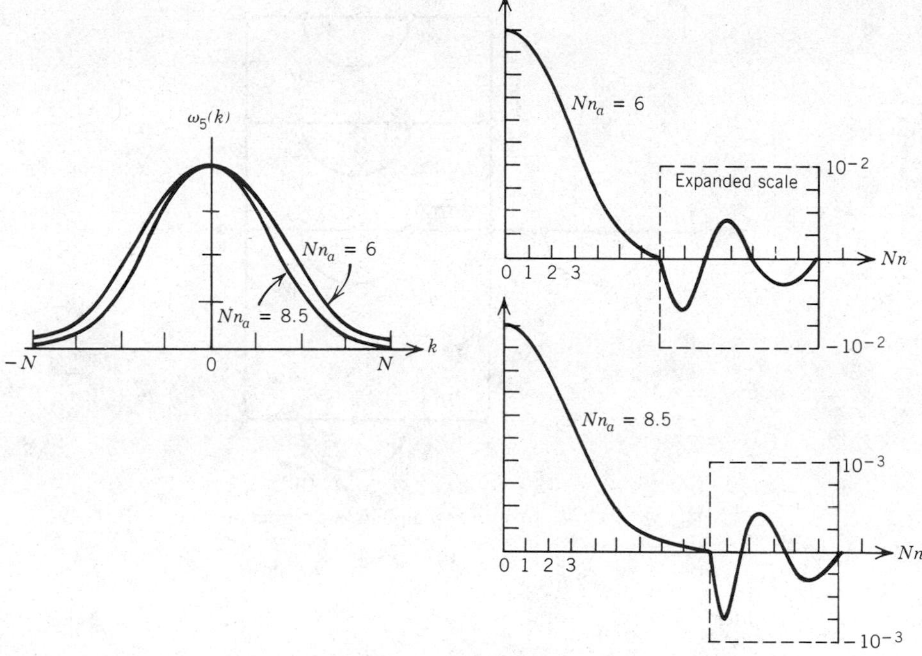

(e)

Fig. 32.22 (Continued)

lobes are all wider than that resulting from the rectangular window. These points are true of all windows; the search for the ideal windowing function is a search for the best tradeoff between sidelobe amplitude and main lobe width. No FIR filter designed by Fourier transformation and windowing is optimal. The appeal of the technique lies entirely in its simplicity and economy.

The Fourier transform design of FIR filters can be summarized as follows:

1. Write the equation for the desired frequency response.
2. Find the impulse response by means of Eq. (32.49).
3. Select a windowing function and a window width to meet the required ripple and transition-width specifications. Window the impulse response accordingly.
4. Shift the impulse response to make it causal.

Frequency Sampling Design

The transfer function of a digital filter can be found from its impulse response form the relation [refer to Eq. (32.13a)]

$$H(z) = \sum_{n=0}^{N-1} h(n)z^{-n} \tag{32.50}$$

The impulse response can be found from the frequency response by means of the inverse DFT [Eq. (32.13b)]

$$h(n) = \frac{1}{N} \sum_{k=0}^{N-1} H(k)W^{nk} \tag{32.51}$$

Combining these two relations

$$H(z) = \frac{1}{N} \sum_{n=0}^{N-1} z^{-n} \sum_{k=0}^{N-1} H(k)W^{nk} \tag{32.52}$$

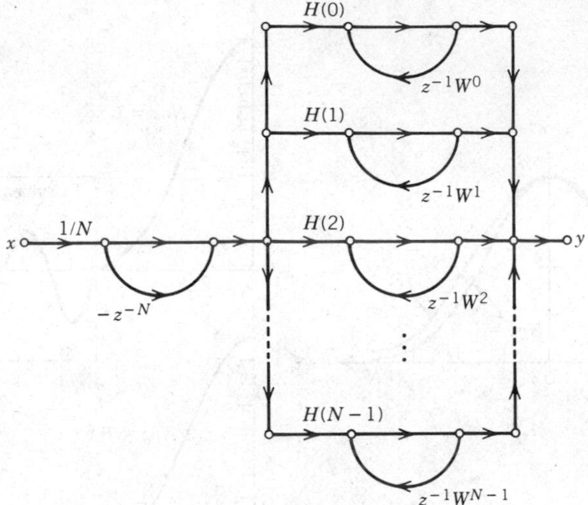

Fig. 32.23 Basic frequency-sampling filter structure.

which can be rewritten

$$H(z) = \frac{1 - z^{-N}}{N} \sum_{k=0}^{N-1} \frac{H(k)}{1 - z^{-1}W^k} \tag{32.53}$$

This form can be realized by the parallel structure shown in Fig. 32.23. Notice that the multipliers $H(k)$ are simply equally spaced samples of the desired frequency response. The design procedure for this filter consists simply of substituting samples of the desired frequency response into Eq. (32.53).

As usual, this design requires some additional work to make it practical. First, the $(1 - z^{-1}W^k)$ denominator places a pole on the unit circle and leads to a marginally stable system. In practice, the poles are placed just inside the circle by making the denominator $(1 - az^{-1}W^k)$, where a is a constant just slightly less than unity (e.g., $1 - 2^{-12}$). Second, the samples $H(k)$ are complex numbers. In practice, conjugate $H(k)$'s are combined to yield real multipliers, with some increase in complexity. Third, good accuracy requires many closely spaced samples and hence a large number of parallel branches. The design is attractive, however, for narrow-band filters in which only a few samples are nonzero, since branches having zero multipliers can be omitted. Finally, the ripple response of this filter is poor. It can be greatly improved by leaving one or more samples in the transition bands unconstrained and setting these to optimize the ripple response. This is best done with the aid of a computer, using a gradient-search program.[16]

Optimum Filter Design

The concept of designing by fitting the frequency response to a predetermined set of points lies behind modern computer-aided optimum filter design. This design yields equiripple performance in both passbands and stopbands. It is an iterative process where each iteration yields an approximation to the desired filter. The ripple performance of the approximation is then compared with the specification and a new approximation developed as a result.

Modern optimum design procedures take the form of computer programs. We will describe the general basis of their operation and refer the reader to published listings. These programs may also be available on line through time-sharing services.

For simplicity, assume the frequency response $H(e^{j\omega})$ is pure real and the impulse response $h(n)$ is symmetric about $n = 0$. (This noncausal filter is readily made causal by a simple shift when the design is finished.) Then $h(n)$ is defined over a region $-M < n < M$, and $H(e^{j\omega})$ can be described in terms of cosine functions

$$H(e^{j\omega}) = h(0) + \sum_{n=1}^{M} 2h(n)\cos(n\omega) \tag{32.54}$$

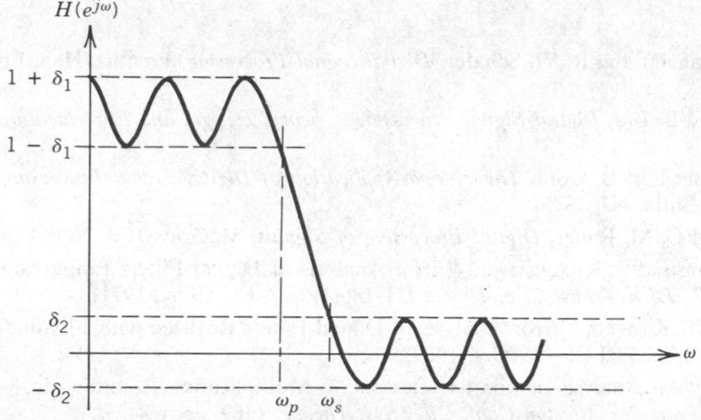

Fig. 32.24 Optimum finite-impulse-response (FIR) filter design specifications: δ_1, passband ripple; δ_2, stopband ripple; ω_p, end of passband; ω_s, start of stopband.

Let the desired equiripple filter be as shown in Fig. 32.24. δ_1 is the passband tolerance, δ_2 the stopband tolerance, ω_p the end of the passband, and ω_s the start of the stopband.

The optimum design procedure can now be viewed as the problem of finding a set of multipliers of $\cos(n\omega)$ that will cause the sum in Eq. (32.54) to fit within the tolerances of Fig. 32.24. Solutions to this problem are Herrmann,[17] Herrmann and Schuessler,[18] Hofstetter et al.,[19–21] Parks and McClellan,[22,23] and Rabiner.[24,25] These solutions all use the same basic approach.

The points of maximum error are initially assumed to be equally spaced in frequency, as shown in Fig. 32.25. A frequency response is then computed that passes through these points, as shown by the curve in Fig. 32.25. The actual maximum errors do not, in general, occur at the expected points. The optimum procedure is iterative, however, and on each new iteration the positions of maximum error are taken to be the positions found in the previous iteration. Thus, for example, on pass 1 the actual extrema were found at the locations indicated by the arrows; thus on pass 2 the maximum error points are shifted to the arrows, as indicated by the hollow circles in the figure. Pass 2 then computes a new frequency response that passes through the hollow circles. Then the actual maximum errors are found for this new response, and a new iteration is made. The process repeats until there is no shift between the expected errors and the actual ones. From the final response curve, the multipliers of Eq. (32.54) can be obtained; these give the required impulse response.

Copies of a computer program written by Parks and McClellan appear in Refs. 22, 23, and 26.

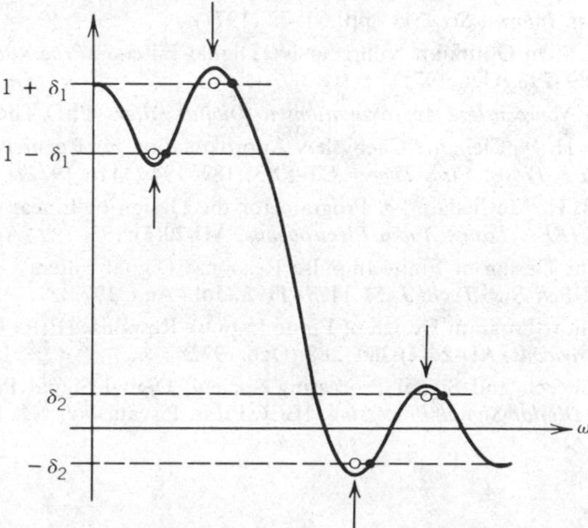

Fig. 32.25 Assumed locations of maximum frequency-response errors in first and second iterations of optimum filter design.

References

1 A. V. Oppenheim and R. W. Schafer, *Digital Signal Processing*, Prentice-Hall, Englewood Cliffs, NJ, 1975.

2 A. Peled and B. Liu, *Digital Signal Processing: Theory, Design, and Implementation*, Wiley, New York, 1976.

3 L. R. Rabiner and B. Gold, *Theory and Application of Digital Signal Processing*, Prentice-Hall, Englewood Cliffs, NJ, 1975.

4 B. Gold and C. M. Rader, *Digital Processing of Signals*, McGraw-Hill, New York, 1969.

5 E. P. F. Kan and J. K. Aggarwal, "Error Analysis of Digital Filters Employing Floating-Point Arithmetic," *IEEE Trans. Circ. Theory* **CT-18**(11):678–681 (Nov. 1971).

6 B. Liu and T. Kaneko, "Error Analysis of Digital Filters Realized with Floating-Point Arithmetic," *Proc. IEEE* **57**:1735–47 (Oct. 1969).

7 L. Jackson, "An Analysis of Limit Cycles due to Multiplication Rounding in Recursive Digital Filters," in *Proc. 7th Allerton Conf. Circ. Sys. Theory*, 1969, pp. 69–78.

8 J. F. Kaiser, "Some Practical Considerations in the Realization of Linear Digital Filters," *Proc. 3rd Allerton Conf. Circ. Sys. Theory*, pp. 621–623 (Oct. 20–22, 1965).

9 A. Papoulis, *The Fourier Integral and its Applications*, McGraw-Hill, New York, 1962.

10 K. Steiglitz, "Computer-Aided Design of Recursive Digital Filters," *IEEE Trans. Audio Electroacous.* **AU-18**(2):123–129 (Jun. 1970).

11 "System/360 Scientific Subroutine Package, Version III, Programmer's Manual," IBM Data Processing Div., White Plains, NY, Document H20-0205-3, pp. 221–225 (1968).

12 A. G. Deczky, "Synthesis of Recursive Digital Filters Using the Minimum *p*-Error Criterion," *IEEE Trans. Audio Electroacous.* **AU-20**(4):257–263 (Oct. 1972).

13 P. Thajchayapong and P. J. W. Rayner, "Recursive Digital Filter Design by Linear Programming," *IEEE Trans. Audio Electroacous.* **AU-21**(2):107–112 (Apr. 1973).

14 L. R. Rabiner et al., "Linear Programming Design of IIR Digital Filters with Arbitrary Magnitude Function," *IEEE Trans. Acous., Speech, Sig. Proc.* **ASSP-22**(2):117–123 (Apr. 1974).

15 F. J. Harris, "On the Use of Windows for Harmonic Analysis with the Discrete Fourier Transform," *Proc. IEEE* **66**(1):51–83 (Jan. 1978).

16 L. R. Rabiner et al., "An Approach to the Approximation Problem for Nonrecursive Digital Filters," *IEEE Trans. Audio Electroacous.* **AU-18**(2):83–106 (Jun. 1970).

17 O. Herrmann, "On the Design of Nonrecursive Digital Filters with Linear Phase," *Elec. Lett.* **6**(11):328–329 (1970).

18 O. Herrmann and H. W. Schuessler, "Design of Nonrecursive Digital Filters with Minimum Phase," *Elec. Lett.* **6**(11):329–330 (1970).

19 E. Hofstetter et al., "A New Technique for the Design of Nonrecursive Digital Filters," *Proc. 5th Ann. Princeton Conf. Inform. Sci. Sys.*, pp. 64–72 (1971).

20 E. Hofstetter et al., "On Optimum Nonrecursive Digital Filters," *Proc. 9th Allerton Conf. Circ. Sys. Theory*, pp. 789–798 (Oct. 1971).

21 J. Siegel, *Design of Nonrecursive Approximations to Digital Filters*, PhD Thesis, MIT, June, 1972.

22 T. W. Parks and J. H. McClellan, "Chebyshev Approximation for Recursive Digital Filters with Linear Phase," *IEEE Trans. Circ. Theory* **CT-19**(3):189–194 (Mar. 1972).

23 T. W. Parks and J. H. McClellan, "A Program for the Design of Linear Phase Finite Impulse Response Filters," *IEEE Trans. Audio Electroacous.* **AU-20**(3):195–199 (Aug. 1972).

24 L. R. Rabiner, "The Design of Finite Impulse Response Digital Filters Using Linear Programming Techniques," *Bell Sys. Tech. J.* **51**:1177–1198 (Jul.–Aug. 1972).

25 L. R. Rabiner, "Linear Program Design of Finite Impulse Response (FIR) Digital Filters," *IEEE Trans. Audio Electroacous.* **AU-20(4)**:280–288 (Oct. 1972).

26 IEEE Acoustics, Speech and Signal Processing Society, Digital Signal Processing Committee, eds., *Programs for Digital Signal Processing*, IEEE Press, Piscataway, NJ, 1981.

PART 7

SYSTEMS ENGINEERING, AUTOMATIC CONTROL, AND MEASUREMENTS

CHAPTER 33
SYSTEMS ENGINEERING

RICHARD G. COSTELLO

The Cooper Union for Advancement of Science and Art
New York, New York

33.1 DEFINITION OF SYSTEM

Defining the term "system" is not a simple task, because the word pervades our present technological society and appears in many different areas with just as many different meanings. A dictionary is a good starting point in the search for meaning. Turning to page 2562 of *Webster's New International Dictionary* (2nd Edition Unabridged), one finds 15 definitions for the word "system." Skipping over legal systems, the Dewey decimal system, the AMTRACK system of trains, and the alimentary and nervous systems leaves a large set of definitions for system that have a direct engineering application. Three of the most pertinent are:

An aggregation or assemblage of objects united by some form of regular interaction or interdependence; a group of diverse units so combined by nature or art as to form an integral whole, and to function, operate, or move in unison and, often, in obedience to some form of control . . . .

An organized or methodically arranged set of ideas; a complete exhibition of essential principles or fact, arranged in a rational dependence or connection . . . .

A formal scheme or method governing organization, arrangement, etc., of objects or material, or a mode of procedure; a definite or set plan of ordering, operating, or proceeding . . . .

In a general sense these are fine definitions, but they lack certain specifications that are usually understood to be applied to all engineering, scientific, or technical systems. Some of these important specifications are:

1. A system must be capable of description. If it can't be described, it isn't a system (at least in the engineering, scientific, or technical sense). The description may contain approximations with probabilistic or stochastic data, so that parameters or values may change or be random variables or even be unobservable.

2. A system must produce an output, perform an action, or reach an objective. Most systems also require an input. Control systems, computer systems, economic systems, and management systems all have inputs and outputs.

3. A system must be governed by rules of operation whose goal is to minimize or maximize some function of the inputs and outputs. Control systems are generally designed to minimize errors, management systems to maximize profit, and so on. Optimization theory deals with finding the best rule or control to produce the desired minimization or maximization for the particular system being optimized.

With this background, the definitions of systems engineering found in Shinners[1] will be quoted and used in this chapter as the generally accepted definition for the systems engineer:

A large complex system consisting of control, computer, and communications systems, functioning in a highly integrated and interdependent manner to achieve overall sound performance, reliability, schedule, cost, maintainability, power consumption, weight, and life expectancy, is referred to as the systems engineering problem.

For further definitions of a system with both a mathematical and an intuitive slant, refer to Klir,[2] especially Sections 1.3 and 1.2. For a wide selection of practical, technical, and mathematical system background material, refer to Glorioso and Osorio.[3]

33.2 SYSTEM CATEGORIZATION BY FUNCTION

The following system types are designed, altered, used, or applied by the electrical engineer.

1. **Control systems.** Servomechanisms; robots; regulators of pressure, temperature, velocity, and so on; process controllers, numerical machine tool controllers, manufacturing plant controllers.
2. **Computer systems.** Direct digital control systems; system architecture and logical design; system programs; multiprocessing systems; time-sharing systems; system networks; dedicated microcomputer instrument or control systems.
3. **Communication systems.** Analog radio, TV, and phone systems; satellite relay systems; navigation systems; network and packet switching systems; digital communication systems; coding and error-correcting systems; enciphering and deciphering systems; modulating and demodulating systems; signal processing and conditioning systems.
4. **Management systems.** CPM (critical path method) or PERT (program evaluation and review technique); linear and dynamic programming methods of maximization; simplex algorithm for maximization subject to constraints; economic modeling systems.
5. **Human operator systems.** Man-machine interface systems, vehicular control systems; bioengineering systems; medical diagnostic systems; artificial limbs and organs; electromyographic control systems; life-support systems.
6. **Radar systems.** Pulsed doppler; police radar; air traffic control systems; radar altimeter systems; side-looking radar mapping systems; ship and harbor radar navigation systems; air-search radar systems.
7. **Navigation and guidance systems.**† LORAN (long range aid to navigation); DME (distance-measuring equipment) for aircraft guidance; SONAR (sound navigation and ranging) for underwater navigation; VOR (visual omni range) for aircraft guidance; glideslope indicators for aircraft landing; satellite navigation systems for submarines, aircraft, missiles, and aerospace vehicles; sun sensor and canopus (star) sensor for satellites and planetary exploration craft; microwave guidance systems.

33.3 CONTROL SYSTEMS

Control systems are discussed in detail in Chapter 34, dealing with automatic control. Definitions of some of the terms used for control systems are given here.

Plant. The plant is the physical device that is controlled by a control system. It may be a single device, such as a motor, or it may be a collection of devices, such as the motor, gears, cables, and weights that make up an elevator plant. In its largest sense a plant may be a chemical process, a power generator, or a spacecraft. Systems containing various plant types are discussed in Chapter 34, Section 34.4.

† Navigation and guidance are discussed in Chapters 50 through 54.

Feedback Control System. This system contains a feedback loop that feeds the output signal back to the reference input and compares these two signals to find the difference or error between the input and the output. The feedback control system minimizes this error, which results in the system output being related to or controlled by the system reference input. Feedback control systems are discussed in Chapter 34, Sections 34.3 through 34.9.

Servomechanism or Servo. A servo is a feedback control system with a mechanical output, such as position, angle, velocity, or acceleration. The power-steering mechanism of a vehicle is a servomechanism, and a robot welder contains many servomechanisms—one for each degree of freedom of motion. Servo theory is discussed in Chapter 34, Sections 34.3 through 34.9, and servo hardware is discussed in Chapter 36.

Regulator or Automatic Regulator. A regulator is a feedback control system with a constant or nearly constant reference input. The task of the regulator is to keep the output constant, in response to unknown system disturbances. A home-heating system with a thermostatic controller is an example of a temperature-regulating system. Regulator theory is discussed in Chapter 34, Sections 34.3 and 34.4.

Process Control System. A process control system is an automatic regulating system in which the regulated output is a process variable such as flow rate, pH, density, temperature, or pressure. Process control is widely used in industry, particularly in the chemical and food processing industries. One of the most widely used control algorithms in the process industry is the PID (position, integral, derivative) controller. Industrial process control is discussed in Chapter 35.

33.4 COMPUTER SYSTEMS

Computer systems are discussed in Chapters 55 through 65. Digital measurements of control system variables are discussed in Chapter 37. Computer simulations are discussed in Chapter 38.

Computers and control systems are becoming increasingly interrelated, so that it is difficult to judge the separation between the two. Chapter 35 discusses automotive control and the applications of computers therein. Computers are also becoming the heart of many communication systems. Message-switching computers are both computer and communications devices, and cannot be separated into one or the other.

33.5 COMMUNICATION SYSTEMS

As mentioned in Section 33.4, communication systems are becoming increasingly intertwined with computer systems. This is especially true in such areas as packet switching, message switching, time domain and frequency domain multiplexing (TDM and FDM), and various coding areas meant to either correct errors or protect the security of a communication link by various enciphering and deciphering algorithms. Communication systems are discussed in Chapters 47 through 49.

Control systems and servomechanisms are also intermeshed with satellite communication systems (Chapter 49), because the attitude or orientation of satellites in space must be accurately controlled so that the antennas can point at the proper receiving areas on earth. For deep-space communications, the receiving antenna on earth (e.g., the Goldstone dish antenna in California) must point at and follow the spacecraft. Control systems are discussed in Chapter 34.

33.6 MANAGEMENT SYSTEMS

The primary purpose of a management system is provision of information and prediction of possible difficulties along with provision of possible solutions to appropriate management levels.[4] A management system consists of the following: a systematic plan; a method of gathering data; a method of making reasonable estimates of costs, time needed to complete subtasks, and partitioning of resources or work effort; and finally a method of processing and presenting all the information obtained to the appropriate management level.

A management system may be purely software, with the necessary data processing being carried out and the reports being written by hand. Although hand-written reports provide valuable information to management, they are not what is generally meant by a management system. Typically, a modern management system such as an MIS (management information system) provides information that varies weekly, daily, or even hourly. To report such rapid changes, some systematic data-processing algorithm is applied to potentially large amounts of variable data using a digital computer or computer system.

Some management systems merely condense data obtained from many remote sources (e.g., sales offices or warehouses) and produce up-to-the-minute reports including graphs or trend-line presentations of sales, profits, losses, inventories, accounts receivable, and other pertinent business indicators. A larger system breaks this information up into segments or divisions naturally related to the business being managed. An even larger system seeks to cross correlate one management parameter with another by performing a multiple regression analysis on the data sets available for analysis. Thus a manager might request to see the relationship or correlation coefficients between profits for the last quarter and the number of units manufactured, or overhead expenses. An MIS of this nature consists of a digital computer, various data input stations and output display stations, and a custom tailored set of computer programs and statistical packages.

We will consider here four important system management tools: PERT, CPM, economic models, and linear programming.

PERT (Program Evaluation and Review Technique)

An entirely different class of management systems provides more than up-to-date digested information. This class helps management to schedule inordinately long or complicated sequential or interrelated tasks. The most well-known method of this type is PERT, or the essentially similar CPM (critical path method).

PERT is an acronym for program evaluation and review technique, developed in 1958 by the US Navy to manage the Polaris submarine missile system (fleet ballistic missile system). PERT finds the critical path, or bottleneck, that limits the early completion of a complex project. CPM is an essentially similar process. The difference between PERT and CPM will be discussed later.

The concept of critical path is best understood by referring to a PERT diagram, such as the one shown in Fig. 33.1. Figure 33.2 illustrates the nomenclature of Fig. 33.1.

The PERT diagram is similar to a signal flow graph (discussed in Section 34.4). The branches or arrows indicate the amount of work time or activity needed to proceed from one circled event to the next. All events must be completed, and all paths must be followed. Each connecting arrow or activity is labeled with four time values in the form shown in Fig. 33.2. The times must all be measured in the same units, such as days, weeks, or months. These times must be estimated in performing a PERT analysis. The accuracy of the time estimates directly influences the accuracy of the PERT chart and the conclusions drawn from the chart. The accuracy is further influenced by the

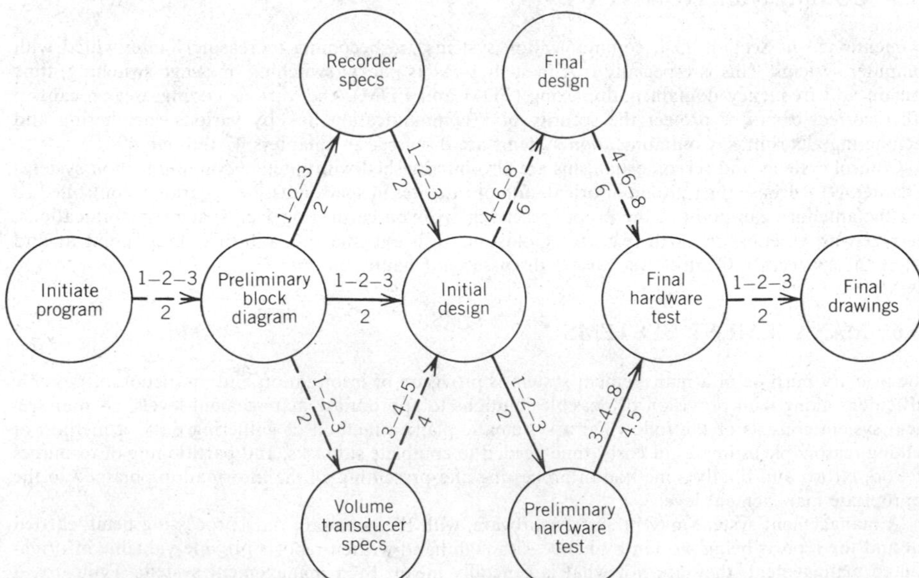

Fig. 33.1 PERT (program evaluation and review technique) diagram for the design of an automatic recording Warburg respirometer. Dotted line = critical path = longest time to complete all necessary tasks; solid line = noncritical path. Notation: $(1-2-3)/2$, $(4-6-8)/6$, etc. $= (t_1 - t_2 - t_3)/t_e$; $t_e = (t_1 + (4)t_2 + t_3)/6$. See Fig. 33.2 for definitions of the t factors.

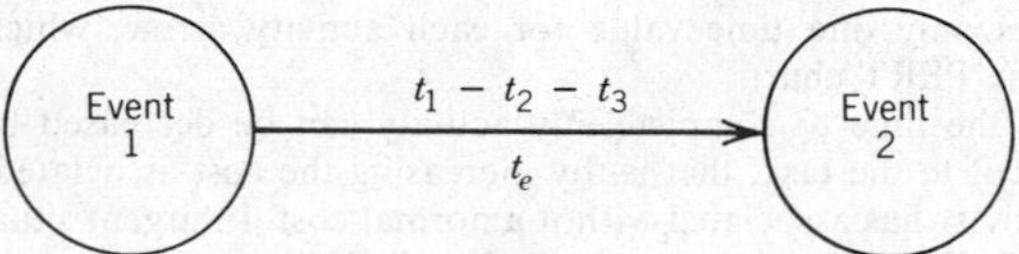

Fig. 33.2 Activity arrow, labeled with work times for PERT (program evaluation and review technique) diagrams. t_1 = minimum time to complete activity; t_2 = most likely time to complete activity; t_3 = maximum time to complete activity; t_e = statistically expected time, based on beta probability distribution with standard deviation of $1/6(t_3 - t_1)$. Shinners[5] shows $t_e \cong (t_1 + 4t_2 + t_3)/6$.

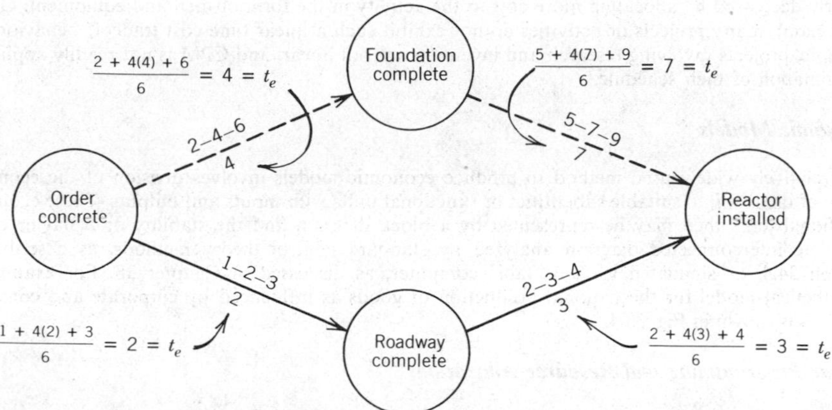

Fig. 33.3 PERT (program evaluation and review technique) chart "slack" time example. Allocation of men and resources from path B (solid line) to path A (dotted line) may save six weeks. Path A (critical path) = 4 + 7 = 11 weeks statistically expected time t_e; path B = 2 + 3 = 5 weeks (t_e); "slack" time = 11 − 5 = 6 weeks.

degree to which the assumption of a beta distribution with standard deviation $\sigma = (t_3 - t_1)/6$ applies to the actual time estimates.[1]

Once the PERT chart is constructed, each possible path connecting the beginning to the end of the project is examined, and the total time for each path is summed up. The path with the longest time is known as the critical path. Since all paths must be covered for accomplishment of all events, the longest time path is the one to be improved on. The method for improvement is simple. Workers or resources from an activity that does not form part of the critical path and is completed before the critical path are in excess or "slack," and these may be shifted to some portion of the critical path to speed up its completion.

Consider the highly simplified segment of a PERT chart shown in Fig. 33.3. The critical path requires a statistically expected time of 4 + 7 = 11 weeks, whereas the parallel noncritical path requires only 2 + 3 = 5 weeks. The slack time in the noncritical path is therefore 6 weeks (11 − 5 weeks). Allocation of men and resources from the noncritical path to the critical path makes possible a saving of up to six weeks. In reality, much less time than six weeks may be saved since different special skills and specialized resources may be required for the two different paths. However, if the times are measured in months or years rather than weeks, new men can be hired or trained and specialized equipment can be acquired.

CPM (Critical Path Method)

Although PERT and CPM both have the goal of finding the critical path for the completion of a project, CPM was originally differentiated from PERT. Both methods find and analyze the critical path from an activity event flow graph, as shown in Fig. 33.1. In this respect they are identical. The main difference between CPM and PERT is in time estimation.

PERT assigns three probabilistic activity times and calculates an expected time (denoted t_e) on the basis of a beta probability distribution. CPM assigns a single deterministic time, which is assumed to be accurately known, on the basis of experience with similar projects (e.g., building a house or a road). Thus CPM uses only one time value for each activity arrow, which is equivalent to the expected time (t_e) of the PERT chart.

CPM assumes that the time to complete any activity can be decreased by devoting more men, materials, and equipment to the task, that is, by increasing the cost associated with the activity. The normal time for the activity has associated with it a normal cost. In urgent situations, the time may be decreased by a "crash" effort. The shortened time is called the crash time, and the associated larger cost is called the crash cost. Thus once the critical path is found from the PERT/CPM chart, CPM assigns a time-cost tradeoff relationship to each activity, which in general is different for each. Often this relationship is linear, and the process of linear programming may be applied to the entire set of activities along the critical path. In this manner optimum time-cost tradeoff for each activity branch may be found in order to decrease the project completion time by a specified number of days. CPM is often used in managing large programs (i.e., construction projects) where activity times can be linearly decreased by allocating more cost to the activity in the form of men and equipment. On the other hand, many projects or activities do not exhibit such a linear time-cost tradeoff behavior. For example, projects involving research and invention are not linear, and CPM is not readily applicable to estimation of their schedule.

Economic Models

One relatively widely used method to produce economic models involves division of the economic entity of interest into suitable subentities or functional units with inputs and outputs. These economic functional units then may be represented by a block diagram and the stability or behavior of the complete interconnected diagram analyzed by standard control theory methods, as described in Section 34.4, or simulated via a suitable computer, as discussed in Chapter 38. For example, a hypothetical model for the national production of goods as influenced by corporate and consumer tax rates is shown in Fig. 33.4.

Linear Programming and Resource Allocation

Linear programming is a mathematical approach to system management that uses linear equations to express the relationships of profit and the apportionment of resources via constraint equations. It was developed following World War II and was applied to the Berlin airlift of 1948.

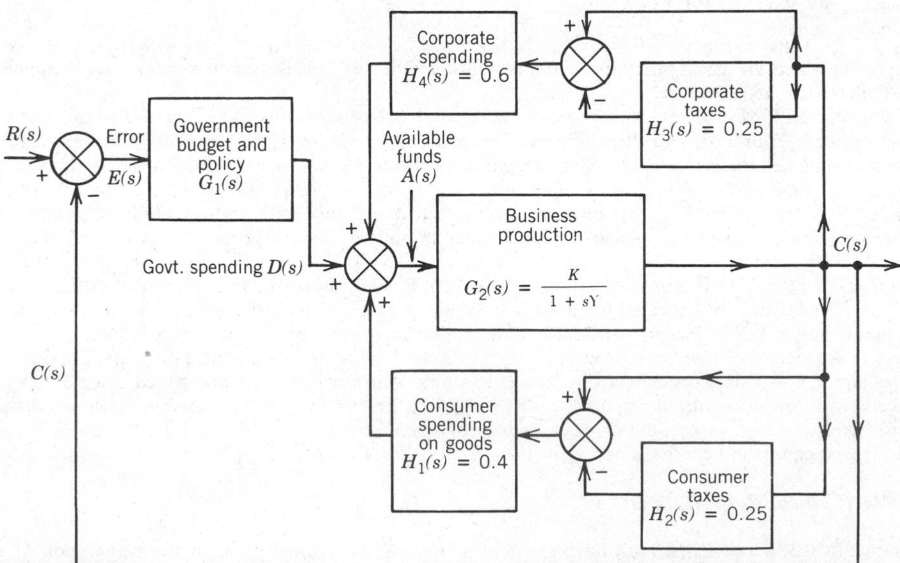

Fig. 33.4 Economic management system model: Block diagram for the national production of goods. $R(s)$ = desired production of goods, in dollars; $C(s)$ = national production of goods, in Dollars. The feedback gains $H_1(s)$, $H_2(s)$, $H_3(s)$, and $H_4(s)$, along with the forward gain K and time constant Y, must be adjusted in accordance with government policy $G_1(s)$ and actual economic conditions.

For example, it can be applied to determine how many shirts and how many pairs of pants a clothing manufacturer should produce to maximize profits from the limited amount of cloth, thread, and buttons on hand. Consider the following hypothetical requirements for producing shirts or pants, where S is the number of shirts produced and P is the number of pairs of pants. S and P are called the decision variables.

Resource	Amount of Resource	Needed for 1 Shirt	Needed for 1 Pair of Pants	Constraint Equation
Cloth	1000 ft^2	5 ft^2	7 ft^2	$5S + 7P \leqslant 1000$
Thread	2000 yd	7 yd	15 yd	$7S + 15P \leqslant 2000$
Buttons	1000	10	2	$10S + 2P \leqslant 1000$

Assume the manufacturer knows that the profit on each shirt will be \$2.00 and on each pair of pants \$4.00. The profit, or objective function, is then

$$\text{Profit} = \text{objective function} = 2S + 4P \qquad \$$$

The constraint equations are written as "equal or less than" because not all of a resource, such as cloth, may be used if the total production is limited by a small amount of another resource, such as buttons.

These constraint equations for cloth, thread, and buttons may be plotted on a set of orthogonal axes assigned to the various final products which are, in this case, the number of shirts S and the number of pairs of pants P that are to be produced. The three plots are shown in Fig. 33.5, along with a fourth plot that superimposes all three of the constraint plots along with the further constraint that the number of shirts or pants manufactured cannot be a negative number. Note that the possible solutions to the allocation problem all fall within the shaded area in the figure. This shaded area is called the region of feasibility, or the region of feasible solutions.

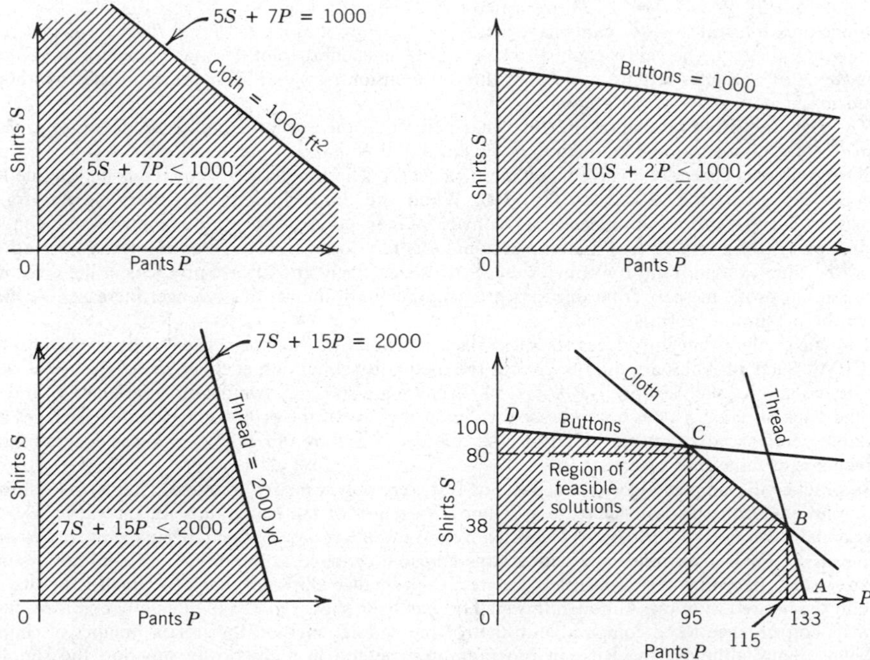

Fig. 33.5 Linear programming example of resource allocation. Profit (objective function) = $2S + 4P$. S and P are decision variables. The profit will always be a maximum (or a minimum) along the boundary and at one of the vertices of the region of feasible solutions, as a consequence of the linear constraints and the resulting convex region of possible or feasible solutions. Thus the possible solutions occur at the vertices A, B, C, or D. Superimposing all three constraints, with $S \geqslant 0$ and $P \geqslant 0$, yields a convex region of feasible solutions.

When all the constraints are combined in the fourth plot, a closed region results that is convex outward and has the vertices $0, A, B, C, D$. The fundamental theorem of linear programming is that the profit, or objective function, is maximized at one of the constraint vertices A, B, C, or D. Obviously nothing is maximized at 0, which corresponds to 0 shirts, 0 pants, and thus 0 profit. Since the profit function is given by $2S + 4P$, profit will increase if either S or P is increased. Thus any number in the interior region of the constraint polygon cannot maximize the profit, since S and P can be increased until the limiting constraint line is reached. Along any constraint line either S or P increases for some direction of movement along the line. Therefore the profit increases (in very special circumstances it might stay constant) for some direction of movement along the constraint line. The movement and profit increase terminates at the same vertex. Thus the maximum (and minimum) profit occurs at a vertex of the linear programming constraint polygon.

In this case there are four vertices to be tested, A, B, C, and D, for the maximum profit. Tabulating the number of shirts S, the number of pants P, and the profit at each vertex yields

Vertex	S	P	Profit $(2S + 4P)$($)
A	0	133 (133.33)	532
B	38 (38.46)	115 (115.38)	536
C	80 (80.88)	95 (95.59)	540
D	50	0	100
Arbitrary mid point	50	50	300

The values of S and P in the table are found by solving the pairs of equations that bind each constraint. For example, point B occurs when the cloth constraint $(5S + 7P \leqslant 1000)$ intersects the thread constraint $(7S + 15P \leqslant 2000)$. S and P at point B are found by solving the simultaneous equations

$$5S + 7P = 1000$$
$$7S + 15P = 2000$$

for which $S = 38.46$ and $P = 115.38$ (truncated values 38 and 115).

Since only integral shirts or pants have market value, the solutions for S and P are truncated (not rounded) to integers. The values could also be read off an accurate plot. However plots for more than three items or variables require more than three dimensions, so that such problems are invariably solved mathematically and not graphically.

The resulting profit tabulation indicates that point C is the optimum choice to maximize profits. Point C corresponds to the manufacture of 80 shirts and 95 pairs of pants.

Real problems generally involve many more variables. The simultaneous equations that define the vertices may grow from 2×2 to 100×100. When the dimensionality increases, the constraint boundary shifts from a two-dimensional convex planar polygon to a three-dimensional convex gemlike polyhedron to a convex multidimensional abstract form. In 1947, Dantzig devised a method called the simplex algorithm. It examines the vertices selectively and always proceeds in the direction of maximum profit increase from one vertex to another until the profit no longer increases. At that vertex the maximum profit is found.

Dantzig's[6] algorithm introduces so-called slack variables (not connected with the slack time for the CPM). Such variables are used to change the inequality equations, such as $5S + 7P \leqslant 1000$, into equality equations, such as $5S + 7P + X = 1000$. There are still only two decision variables, S and P, and the slack variable X does not increase the dimensionality of the problem. Computer packages are available commercially that are based on the simplex algorithm and can solve linear programming problems with thousands of constraints.

In practice the simplex algorithm behaves as if it were polynomially bounded. That is, the number of simplex iterations tends to increase as a linear function of the constraints. In the early 1980s a polynomially bounded algorithm was reported by the three Russian mathematicians Shor, Iudin, and Nemirouski.[7] Their algorithm starts with a large ellipsoid centered at the origin and does not assume linearity. The algorithm then constructs progressively smaller ellipsoids, each of which is known to contain the desired solution. Although this method has been shown to be polynomially bounded, tests show its convergence to be comparable to that of the simplex method for certain families of simple problems. Thus although the Russian programming method is theoretically superior, the simplex algorithm remains the most heavily used linear programming method in the early 1980s.

33.7 HUMAN OPERATOR CONTROL SYSTEMS

Human operator control systems consist of a controlled plant, a display, and a human controller performing some operation with the display. The most common examples concern vehicles guided by human operators. The controlled plant in this case is the ship, plane, automobile, or spacecraft being guided. The display is a collection of instruments and CRT (cathode ray tube) displays. The human operator is the helms person, pilot, or driver of the vehicle. Other examples of human operator control systems include tracking of a target on a CRT or PPI (plan-position-indicator) display, operation of a manually controlled elevator, aiming of a tank gun, and manual control of part or all of any large complex system in times of emergency.

The formal analysis of human operator control systems started with Tustin's 1947 paper[8] dealing with the aiming of a tank gun. Tustin proposed various linear models for the human operator, one of which is shown in Fig. 33.6. He introduced the term "remnant" to describe that part of the output of the human operator that was not linearly related to the operator's input and hence was not produced by his or her linear model. The remnant or noise signal is shown in Figs. 33.7 and 33.8.

The biggest area of interest in human operator modeling is manual control of piloted aircraft or aerospace vehicles. Attention has focused on these manual control systems because the vehicles involved are the most difficult to control, having the shortest time constants and the fastest response time of any human guided vehicles. At the very onset of pilot modeling, Tustin and others recognized that the human operator is nonlinear or, at best, quasilinear; the development of human operator models has followed the developments in nonlinear control theory. Selected human operator models are given in Table 33.1, and a brief description follows. Although the majority of these models were proposed for aircraft control, the models are equally applicable to other human control tasks, such as driving a car or a ship or manually controlling a nuclear power plant.

Linear Transfer Function

In Tustin's[8] linear transfer function, the input signal is three sinusoids, relatively prime in frequency, summed up to produce a random-appearing input to circumvent the predictive ability of a human operator. A linear transfer function is fit to the set of computed input-output data at various frequencies. Tustin's model is shown in Fig. 33.6.

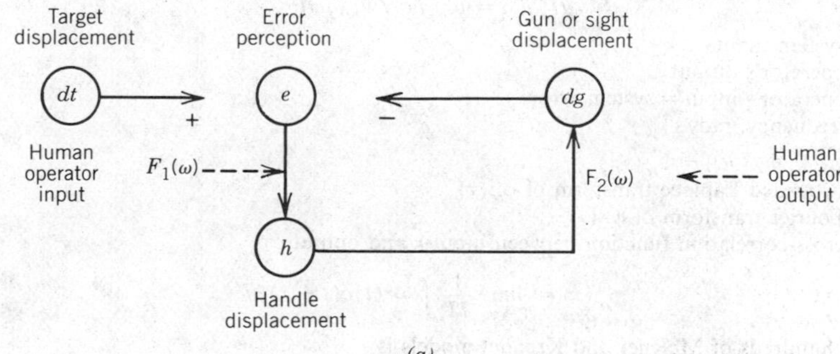

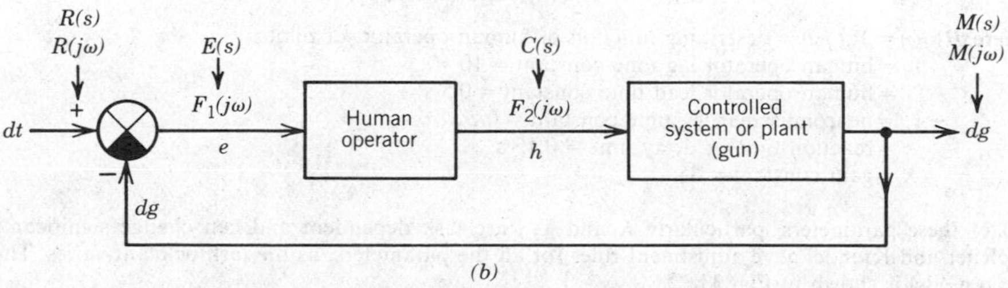

Fig. 33.6 Compensatory tracking model of Tustin[8]: (*a*) original form, (*b*) equivalent block diagram, $dt = r(t)$ = system input in time domain; $e = e(t)$ = system error in time domain; $h = c(t)$ = human operator's output in time domain; $dg = m(t)$ = system output in time domain; $F_1(j(\omega) = E(s)|_{s=j\omega}$ = human operator's input in frequency domain; $F_2(j\omega) = C(s)|_{s=j\omega}$ = human operator's output in frequency domain; $M(j\omega) = M(s)|_{s=j\omega}$ = system output in frequency domain; $R(j\omega) = R(s)|_{s=j\omega}$ = system input in frequency domain.

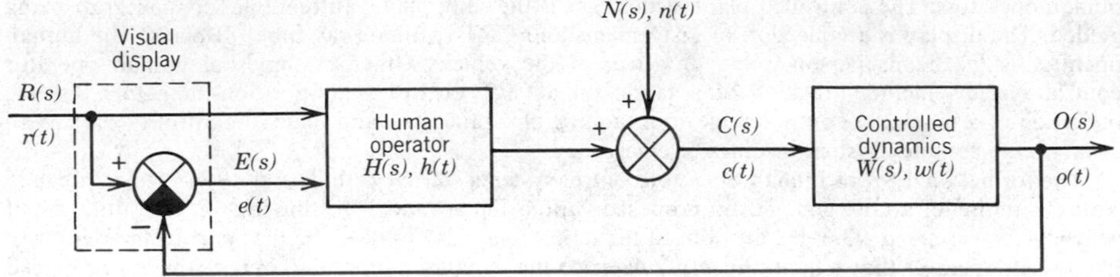

Fig. 33.7 Block diagram for the visual pursuit system. $H(s) =$ linear approximation (describing function of human operator); $N(s) =$ that part of human operator's output not linearly related to input (remnant or noise); $W(s) =$ controlled dynamics, or plant; $R(s) =$ system input; $E(s) =$ system error; $C(s) =$ human operator's output; and $O(s) =$ system output. To convert any Laplace domain function, such as $H(s)$, to a frequency domain function, simply formally substitute $j\omega$ for s. Example:

$$H(s) = \frac{25\epsilon^{-0.1s}}{(s+10)}$$

$$H(j\omega) = H(s)|_{s=j\omega} = \frac{25\epsilon^{-0.1j\omega}}{(j\omega+10)}$$

Loop gain $= H(s)W(s)|_{s=j\omega} = H(j\omega)W(j\omega)$

Quasilinear Describing Function

The quasilinear describing function of McRuer et al.[9] is similar in concept to Tustin's model but is vastly advanced in computational approach. The transfer function of the human operator, expressed as a Fourier transform $H(j\omega)$, is obtained as the ratio of the Fourier transforms of two cross correlations, or

$$H(j\omega) = \Phi_{rc}(j\omega)/\Phi_{re}(j\omega)$$

where $r =$ system input
$c =$ operator's output
$e =$ operator's input = system error
$\omega =$ frequency, rad/s
$j = \sqrt{-1}$, $\Phi_{rc}(j\omega) = \Phi_{rc}(s)|_{s=j\omega}$
$\Phi_{rc}(s) =$ two-sided Laplace transform of $\phi_{rc}(\tau)$
$\Phi_{rc}(j\omega) =$ Fourier transform of $\phi_{rc}(\tau)$
$\phi_{rc}(\tau) =$ cross-correlation function between input r and output c[26]

$$\phi_{rc}(\tau) = \lim_{\Upsilon\to\infty} \frac{1}{2\Upsilon}\int_{-\Upsilon}^{\Upsilon} r(t)c(t+\tau)\,dt$$

One of hundreds of McRuer and Krendel models is

$$H(j\omega) = \frac{K\epsilon^{-j\omega\tau}(1+j\omega\Upsilon_L)}{(1+j\omega\Upsilon_n)(1+j\omega\Upsilon_I)}$$

where $H(j\omega) = Y_p(j\omega) =$ describing function of human operator, or pilot
$\Upsilon_I =$ human operator lag time constant = 10 s
$\Upsilon_L =$ human operator lead time constant = 0.5 s
$\Upsilon_n =$ neuromuscular lag time constant = 0.2 s
$\tau =$ reaction time or delay time = 0.15 s
$K =$ gain constant = 35

All of these parameters, particularly K and Υ_L, are task dependent and can change significantly. McRuer and Krendel gave adjustment rules for all the parameters, as the task or plant varies. Their basic model is shown in Fig. 33.8.

Important Simplification. If $r(t)$ is sinusoidal, then the transfer function of the human operator $H(j\omega)$ is given by

$$\frac{\Phi_{rc}(j\omega)}{\Phi_{re}(j\omega)} = \frac{\mathscr{F}[c(t)]}{\mathscr{F}[e(t)]} = H(j\omega)$$

where $\mathscr{F}[\;] =$ Fourier transform operator. The Fourier transform of a time function is much easier

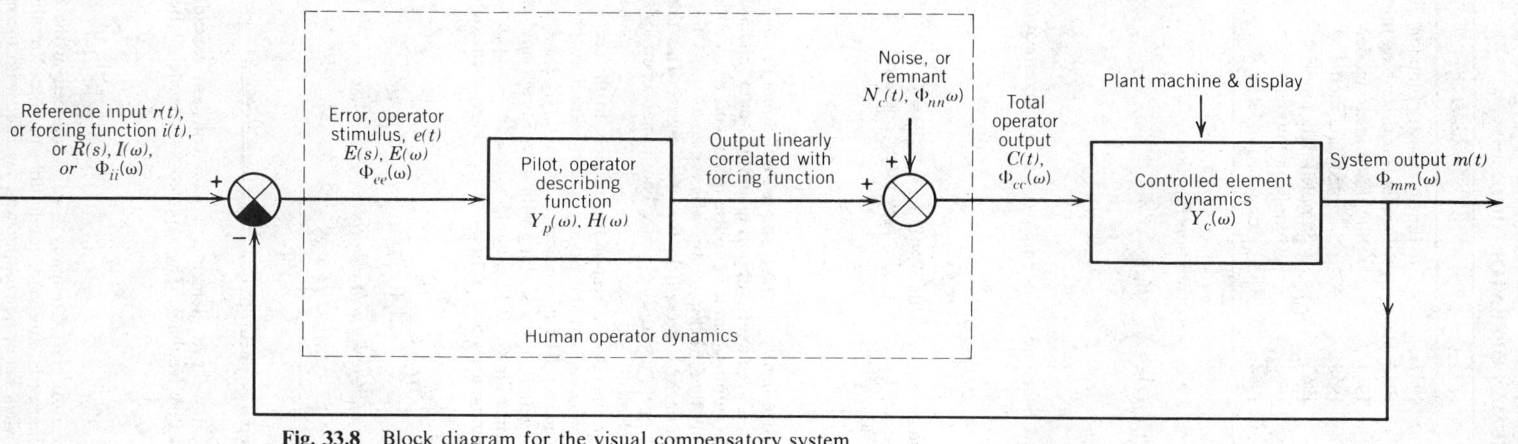

Fig. 33.8 Block diagram for the visual compensatory system.

$$Y_p(j\omega) = H(j\omega) = (\Phi rc(s))/(\Phi re(s))|_{s=j\omega}$$

where $\Phi_{rc}(s)$ = two-sided Laplace transform of $\phi_{rc}(\tau)$, the cross-correlation function between the system input r and the operator's output c, mathematically: $\phi_{rc}(\tau) = \lim_{T\to\infty} 1/T\int_0^T r(t)c(t+\tau)\,dt$ and $\Phi_{rc}(s) = \mathscr{L}\{\phi_{re}(\tau)\}$ = cross-power spectral density between r and c. $\Phi_{ii}(\omega) = \Phi_{ii}(s)|_{s\to j\omega}$ = power spectral density; $\Phi_{ii}(s) = \mathscr{L}\{\phi_{ii}(\tau)\} = \int_{-\infty}^{\infty}\phi_{ii}(\tau)\epsilon^{-s\tau}dt$ = two-sided Laplace transform of $\phi_{ii}(\tau)$; $\phi_{ii}(\tau) = \lim_{T\to\infty} 1/T\int_0^T i(t)i(t+\tau)\,dt$ = autocorrelation function of $i(t)$. Loop gain = $Y_p(j\omega)Y_c(j\omega)$.[9]

TABLE 33.1. SELECTED HUMAN OPERATOR MODELS

Investigator	Date	Model Type
Tustin[8]	1947	Linear transfer function
McRuer and Krendel[9]	1957–1967	Quasilinear describing function
Soliday and Schohan[40]	1964	Simplified linear model
Ward[39]	1958	Sampled-data model
Bekey[10]	1962	Sampled-data model
Goodyear Aircraft Co.[11]	1962	Nonlinear model
Knoop and Fu[12]	1964	Adaptive model

to evaluate than the Fourier transform of a cross-correlation function.

$$\mathscr{F}[f(t)] = \int_{-\infty}^{\infty} f(t)e^{-j\omega t}\,dt$$

Simplified Linear Model

The simplified model of Soliday and Schohan[40] was employed in an investigation of high-speed low-level aircraft flight.

$$H(s) = \frac{\epsilon^{-s\tau}}{(1 + s\Upsilon_I)}$$

This is identical to McRuer and Krendel's model with $K = 1$, $\Upsilon_L = 0$, $\Upsilon_n = 0$. In 1963 K.0KI used an identical simplified model for the transfer function of the bow plane operator of a submarine, in a simulation of the depth control of a submarine.

Sampled-Data Model

Experimental psychologists, such as Hick in 1947,[41] have pointed out discontinuous behavior on the part of the human operator, and various arguments support a sampled-data model. It is known that human sense organs quantize their outputs, emitting pulse trains of a repetition rate related nonlinearly to stimulus intensity, and that tracking data often display a 1- to 2-Hz sinusoidal component when no 1- to 2-Hz component is present in the input signal. This unexplained signal could be accounted for by a sampler operating at 2 to 4 Hz. The human operator's delay could be accounted for by a sample-and-hold element.

Finally, the inability of the human operator to respond separately to the second of two closely spaced individual signals, a phenomenon termed the psychological refractory period, can also be readily explained by a sampled-data model. The psychological refractory period is simply the time between samples. Ward[39] used a sample time of 0.5 s, along with a zero-order hold element, for most of his work, while Bekey[10] used a sample time of 0.33 s, along with a first-order hold network.

Bekey's model consisted of a sampling switch, a first-order hold element, and the following human operator transfer function:

$$H(s) = \frac{K}{1 + s\Upsilon_N}$$

First-order hold element

$$\left[\frac{1 - e^{-s\Upsilon_s}}{s}\right]^2 \frac{(1 + s\Upsilon_s)}{\Upsilon_s}$$

where Υ_s = sample period = 0.33 s and Υ_N = neuromuscular lag time constant.

Bekey's model consisted of the above two elements, a gain constant cascaded with an impulse sampling switch, and unity negative feedback. The sampled-data model of Bekey is shown in Fig. 33.9.

Nonlinear Model

The nonlinear model of the Goodyear Aircraft Company[11] was produced in the era of the giant analog computers. It was an evolutionary trial-and-error model that produced results so good that a test pilot could not tell immediately when his or her output was switched with the model's output.[11]

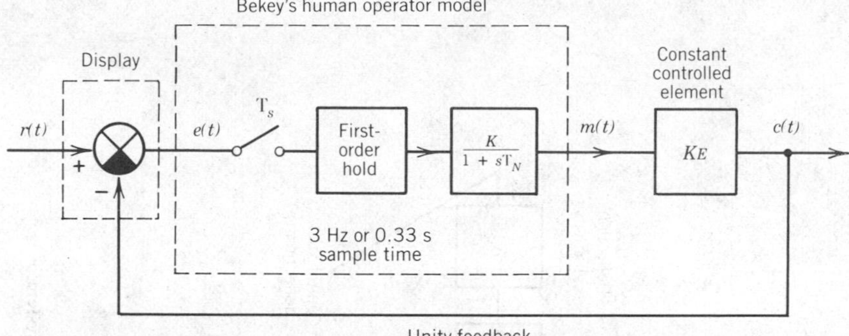

Fig. 33.9 Sampled-data model of Bekey[10] for compensatory human operator tracking, with constant dynamics. Υ_s = sampling period, impulse sampling at 3 Hz, or $\Upsilon_s = 0.33$ s; $r(t)$ = stationary random-appearing input to system; $e(t)$ = system error; K = human operator gain constant; Υ_N = neuromuscular lag time constant; $m(t)$ = human operator's output; $c(t)$ = system output; K_E = controlled element transfer function; First-order hold = $[(1 - e^{-s\Upsilon_s})/s]^2(1 + s\Upsilon_s)/\Upsilon_s$.

The Goodyear model is shown in Fig. 33.10. It includes well-known physiological factors, for example, perception threshold, reaction time, acceleration and rate estimation, and anticipation.

Adaptive Model

One of the most useful abilities of the human operator is the ability to alter the form of his or her response in accordance with changes in the system being controlled. The linear models of McRuer and Krendel, for example, change structure for different types of controlled elements, as shown in the following table. (The typical numerical values given can easily change by a factor of two or more for different operators.)

Controlled Element Transfer Function, or Plant	Constant Characteristics, Delay $e^{-s\tau}$ and neuromuscular lag $(1 + s\Upsilon_n)$	Adaptive Human Operator Characteristic	Verbal Description of Adaptive Characteristic
Constant	$\frac{Ke^{-0.15s}}{(1+0.2s)}$	$\times \frac{(1+s)}{(1+10s)}$	Lag-lead plus variable gain K
First order C_1/s	$\frac{Ke^{-0.15s}}{(1+0.2s)}$	$\times \frac{1}{1}$	Pure variable gain K
Second order C_1/s^2	$\frac{Ke^{-0.15s}}{(1+0.2s)}$	$\times \frac{(1+10s)}{(1+s)}$	Lead-lag plus variable gain K

The adaptive model of Knoop and Fu[12] is of the model-matching type, containing a plant model, as shown in Fig. 33.11. The plant model is adjusted to correspond to the controlled element, as the parameters of the controlled element vary.

Dual-Mode Model

All of the previous models are intended primarily to handle input signals that are filtered random noise, or summed sinusoids of different frequencies. They do not accurately predict an operator's response to discontinuous steplike signals (such as a sudden gust of wind). Although some of the previous models can be adjusted to produce reasonable step responses, they do not produce reasonable continuous tracking responses. The dual-mode model of Costello[13] contains a linearized

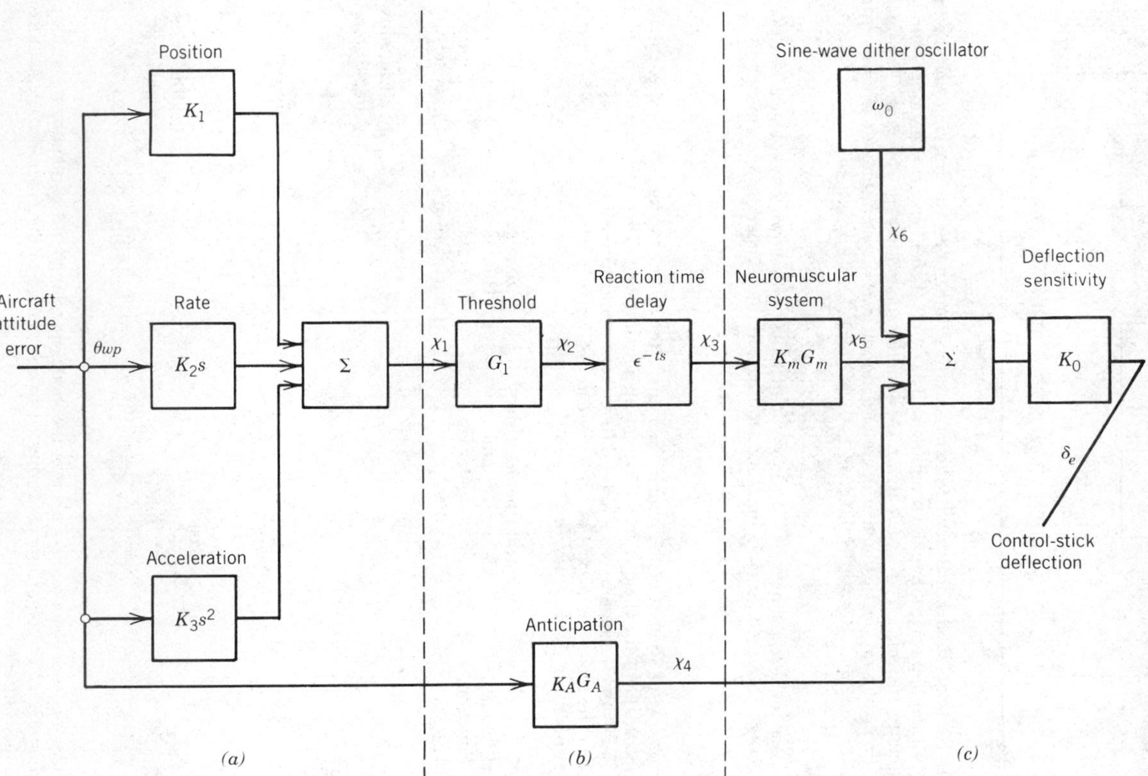

Fig. 33.10 Nonlinear model of the Goodyear Aircraft Co:[11] (*a*) Linear computation system, (*b*) nonlinear and nonminimum phase operations, (*c*) motor system.

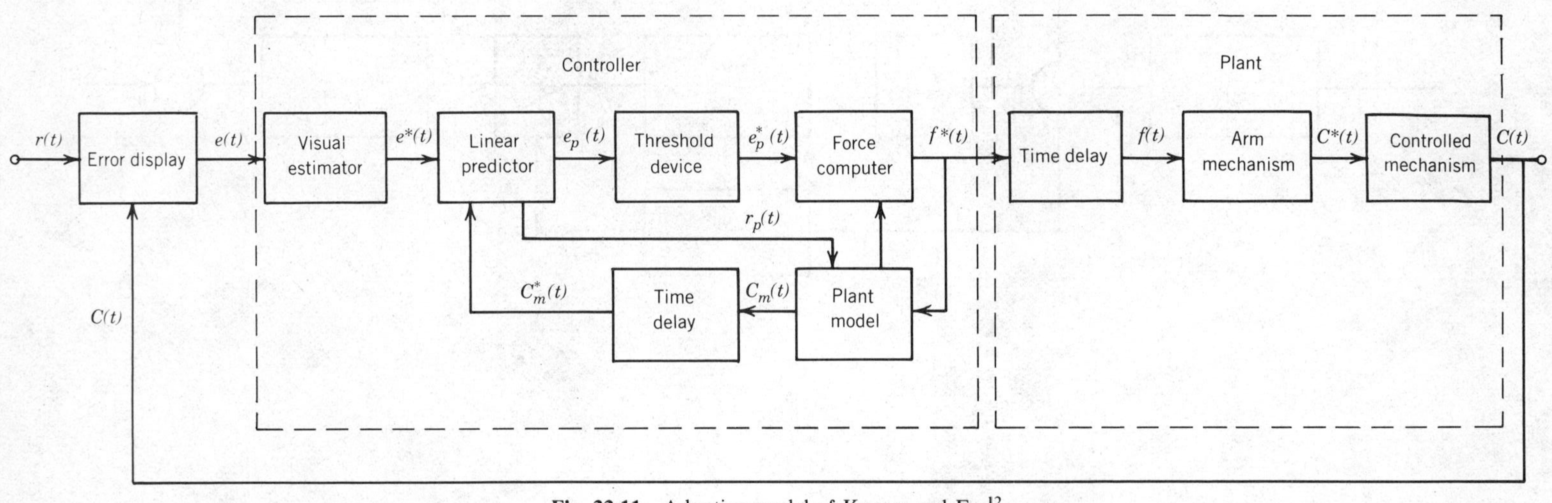

Fig. 33.11 Adaptive model of Knoop and Fu.[12]

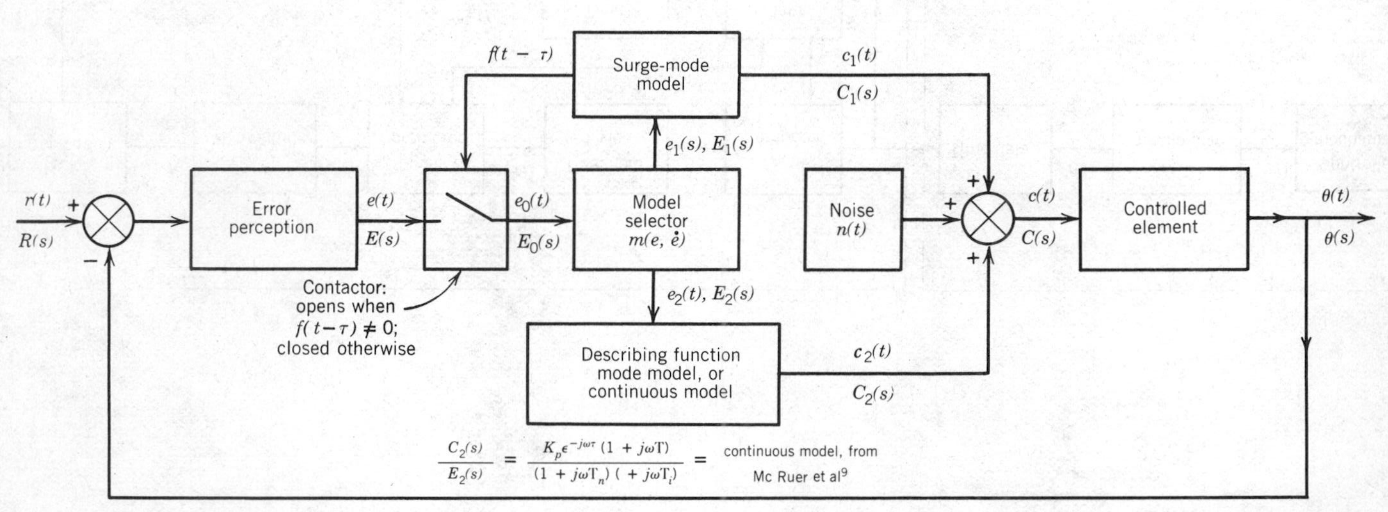
r(t)
R(s)
+
−
Error perception
e(t)
E(s)
Contactor: opens when f(t−τ) ≠ 0; closed otherwise
f(t − τ)
e0(t)
E0(s)
Surge-mode model
Model selector m(e, ė)
e1(s), E1(s)
e2(t), E2(s)
Describing function mode model, or continuous model
c1(t)
C1(s)
c2(t)
C2(s)
Noise n(t)
+
+
+
c(t)
C(s)
Controlled element
θ(t)
θ(s)
C2(s)/E2(s) = Kp ε^(−jωτ) (1 + jωT) / (1 + jωTn) (+ jωTi) = continuous model, from Mc Ruer et al9

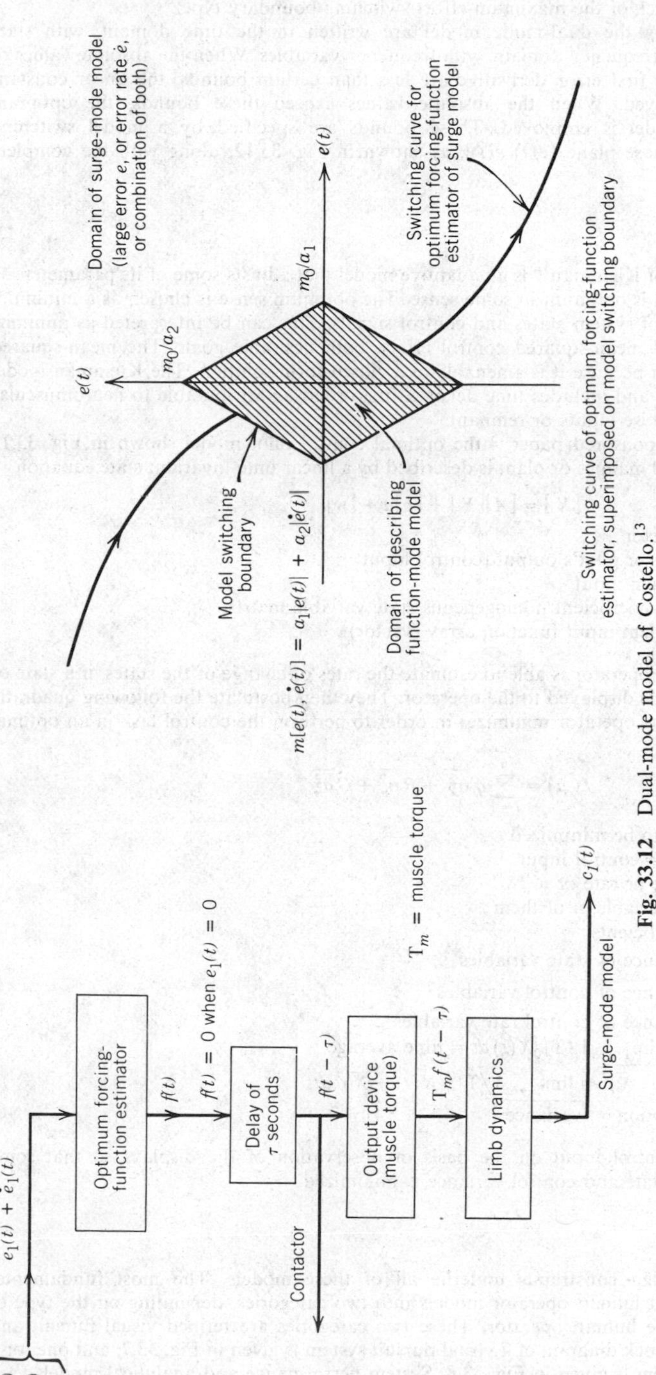

$e_1(t)$
$\dot{e}_1(t)$
$e_1(t) + \dot{e}_1(t)$
Optimum forcing-function estimator
$f(t)$
$f(t) = 0$ when $e_1(t) = 0$
Delay of τ seconds
Contactor
$f(t - \tau)$
Output device (muscle torque)
$T_m f(t - \tau)$
T_m = muscle torque
Limb dynamics
Surge-mode model
$c_1(t)$
$m[e(t), \dot{e}(t)] = a_1|e(t)| + a_2|\dot{e}(t)|$
Model switching boundary
Domain of describing function-mode model
Domain of surge-mode model (large error e, or error rate $\dot{e}$, or combination of both)
$\dot{e}(t)$
m_0/a_2
m_0/a_1
$e(t)$
Switching curve for optimum forcing-function estimator of surge model
Switching curve of optimum forcing-function estimator, superimposed on model switching boundary

Fig. 33.12 Dual-mode model of Costello[13]

constant coefficient model of the type popularized by McRuer and Krendel and an optimum second-order controller model, of the maximum-effort switching-boundary type.

The equations describing the dual-mode model are written in the time domain, with state variables, rather than in the frequency domain, with frequency variables. When the absolute values of the system error and of the first error derivative are less than certain bounds, the linear constant coefficient model is employed. When the absolute values exceed these bounds, the optimum second-order controller model is employed. These bounds are specified by a model switching boundary constructed in phase plane $[e(t), \dot{e}(t)]$, as shown in Fig. 33.12, along with the complete dual-mode model.

Optimal Control Model

The optimal control model of Kleinman[14] is an adaptive model that adjusts some of its parameters so that the output of the model is optimum in some sense. The optimum sense is chosen as a minimum mean-square-weighted sum of system states and control signals. This can be interpreted as minimizing mean-squared error and mean-squared control effort, both desirable goals. The mean-squared mathematical form is chosen because it is amenable to a closed-form solution. The Kleinman model is continuous and stochastic and includes time delay, a first-order lag attributable to neuromuscular dynamics; it also includes noise inputs or remnant.

In a Wright–Patterson sponsored paper,[15] the optimal control pilot model shown in Fig. 33.13 was developed. The vehicle dynamics or plant is described by a linear time-invariant state equation

$$[\dot{X}] = [A][X] + [b]u + [w]$$

where $[X]$ = vehicle state vector
u = human operator or pilot's output (control input)
$[w]$ = white noise system input
$[A]$ = square constant coefficient homogeneous state variable matrix
$[b]$ = constant coefficient input function array (vector)

The authors assume that the operator is able to estimate the rates of change of the states, if a state or linear combination of states is displayed to the operator. They then postulate the following quadratic performance index, which the operator minimizes in order to perform the control task in an optimal manner:

$$J(\mu) = \sum_{i=1}^{n} q_i \overline{\sigma_{X_i}^2} + R\overline{\sigma_u^2} + G\overline{\sigma_{\dot{u}}^2}$$

where $J(\mu)$ = cost function to be minimized
μ = pilot's output = control input
$\dot{\mu}$ = time derivative or rate of μ
X_i = vehicle state variable, n of them
q_i, R, G = weighting coefficients
$\overline{\sigma_{X_i}^2}$ = averaged variance of state variables
$\overline{\sigma_u^2}$ = averaged variance of control variables
$\overline{\sigma_{\dot{u}}^2}$ = averaged variance of control rate variables
Here $\overline{X}$ = mean of $X = \lim_{T\to\infty} 1/T\int_0^T X(t)\,dt$ = time average
σ^2 = variance $= \overline{(X - \overline{X})^2} = \lim_{T\to\infty} 1/T\int_0^T (X(t) - \overline{X})^2\,dt$
σ = standard deviation $= \sqrt{\text{variance}}$

The operator chooses a control input on the basis of observation of the displays so that some weighted sum of averaged state and control variance is minimized.

Design Constraints

Certain parameters or design constraints underlie all of these models. The most fundamental constraint divides all current human operator models into two categories, depending on the type of information displayed to the human operator. These two categories are termed visual pursuit and compensatory tracking. A block diagram of a visual pursuit system is given in Fig. 33.7, and one for a compensatory tracking system is given in Fig. 33.8. System performance and analytical models vary markedly between the two configurations.

Visual Pursuit. This results when both input and output are displayed to the human operator and the operator attempts to match the system's input and output. Aiming a shotgun at a clay bird is a visual pursuit task.

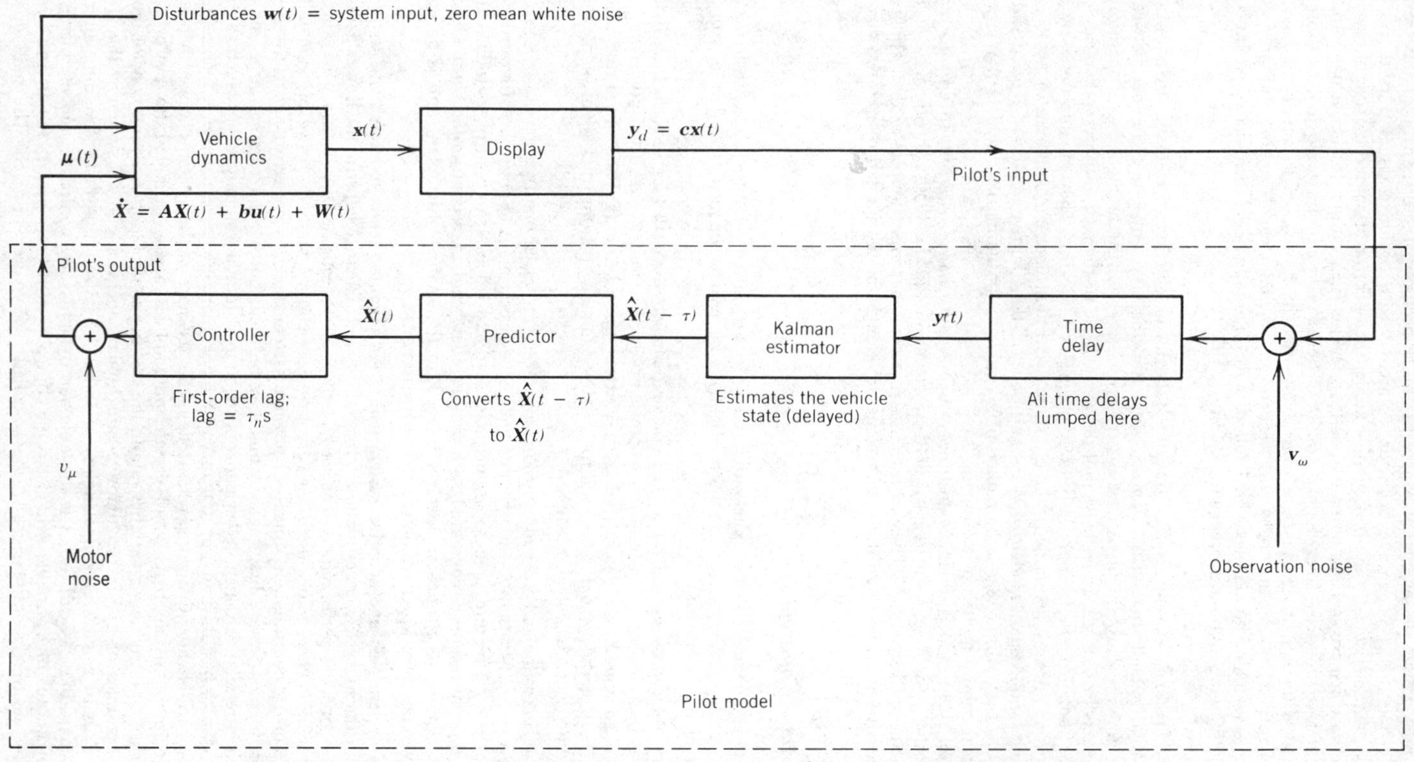

Fig. 33.13 Optimal control model of Baron and colleagues.[15] Boldface symbols represent vectors or matrices. A circumflex over a symbol represents an estimated quantity.

A special form of visual pursuit occurs when future values of the input are visible or known. The term "precognitive" is employed to describe this case. Driving a car or reaching for a doorknob are examples of precognitive visual pursuit, since the future position of the road or of the knob are known to the operator. Precognitive behavior, compared with ordinary visual pursuit, possesses several unique characteristics. In the precognitive mode, the operator often executes rapid previously learned motions, which are carried out in an open-loop fashion (i.e., without further visual correction, once the motion is initiated). Such highly nonlinear behavior is not tractable to straightforward control analysis. Consequently it has not been investigated in as detailed a manner as has visual pursuit or compensatory tracking behavior. Models that include visual pursuit and precognitive behavior require switching boundaries which delineate the predominately linear visual pursuit control from the predominately nonlinear precognitive control.[13,16]

Compensatory Tracking. Compensatory tracking does not involve actual tracking or following per se, as does visual pursuit. In a visual compensatory tracking task, only the difference between the system input and the output, that is, the system error, is presented to the human operator. The operator then acts in such a manner as to null the error. Most of the past and presently used control models of the human operator apply only to compensatory tracking. They include the early linear models of Tustin[8] (already mentioned), the quasilinear constant coefficient models of McRuer and colleagues,[9] the sampled-data models of Ward and Bekey,[10] the adaptive models of Knoop and Fu[12] and of Wasaff,[17] the multiple regression models of Elkind,[18] the optimum filter models of Wierwille,[19] and the optimal control models of Kleinman.[14]

From all of these models, certain constraints on the performance of a human operator can be deduced. North[20] and Fogel[21] stated most of these constraints in the 1950s, and their basic nature remains unchanged in the 1980s. These constraints, and some current examples of their application, are enumerated in the following (along with some advantages of the human operator system over purely automatic control systems).

Human Operator Constraints

1. **Delay or reaction time.** Typically this time is 0.5 s for an inattentive automobile driver, 0.25 s for an attentive automobile driver, 0.15 s for a fighter pilot, athlete, or racing car driver, with 0.10 s being the limit of champions.
2. **Maximum frequency response for a visual manual compensatory tracking task such as flying an airplane by instruments.** Typically this response is 10 rad/s for a −3 dB (half-power) fall off in gain, although some tracking performance may extend up to 20 rad/s. The general rule of thumb is that 1 Hz (6.28 rad/s) marks the limit for sustained human operator tracking in simple manual control (unity gain controlled dynamics).
3. **Phase-lead generation and order of controlled plant.** A zero-order plant (dynamics = K) requires zero-phase lead generation by the operator and is the easiest plant to control. A first-order plant (dynamics = K/s) requires some operator phase lead and is harder to control, as indicated by a lower frequency response, compared with a zero-order plant. A second-order plant (dynamics = K/s^2) requires much greater operator phase lead than a first-order plant and is difficult to control, with an even lower maximum frequency response.
4. **Range of system gain and adaptive system gain.** An operator tends to prefer and to perform better with the highest gain compatible with stability and linearity constraints. Lower gains lead to sluggish or slow performance. The early tank guns investigated by Tustin were aimed by cranking a handwheel many times—a very low gain control method. However, gain must be limited to insure controllability under all possible system states. When errors are relatively small and input frequencies are low, higher gains may be used, but when errors are large and input frequencies are relatively high, lower gains must be used to avoid nonlinearities (such as driving the error display past its limits) or to avoid unstable gain and phase relationships. One solution is to employ variable adaptive gains, which change as the state of the system changes. Modern jet aircraft employ this approach, and certain automobiles use a variable-ratio power-steering mechanism that introduces a variable gain in the steering control loop. Currently one of the most extreme examples of a variable adaptive-gain human operator control system is the General Dynamics' F-16 single-seat single-engine jet fighter. By itself this fast-responding aircraft is statically unstable, or nearly so. The unstable aerodynamics are controlled and the plane made flyable by an adaptive variable-gain electronic control system that operates the aircraft's control surface by electrical control signals—the so-called fly-by-wire approach. The pilot does not directly control the F-16 but rather sends commands to the electronic control system, which directly controls the plane.

Probably the most famous adaptive-gain human operator-controlled but computer-stabilized vehicle is the United States' space shuttle. The space shuttle is fly-by-wire, controlled with four computers "voting" on the appropriate action.

5. **Crossover model gain adjustment.** The crossover model of McRuer and colleagues[9] combines the Bode stability criterion (discussed in Section 34.4) with the observed human operator preference for systems with controlled dynamics of the form K/s. "Crossover" refers to the open-loop gain magnitude crossing over the 0-dB level, which corresponds to the critical unity gain level. The crossover model states that in the region of stable control, the operator adjusts the gain so that in the region of 0 dB, the open-loop gain magnitude has the form K/s. Symbolically

$$Y_p(j\omega)Y_c(j\omega) = H(s)W(s)|_{s=j\omega} \equiv \frac{K}{s}e^{-s\tau}|_{s=j\omega_c}$$

where $Y_p(j\omega) = H(s)$ = operator or pilot's transfer function
$Y_c(j\omega) = W(s)$ = controlled plant or aircraft transfer function
ω_c = crossover frequency, rad/s (At $\omega = \omega_c$, $|Y_p(j\omega_c)Y_c(j\omega_c)| = 0$ dB)
τ = operator's reaction time, s
For useful tracking, ω_c must be greater than any input frequency.

Human Operator Advantages[21]

1. **Pattern recognition.** No machine or computer has yet come near the ability of the human operator to recognize patterns and to extract signals from noise in a real-time operating environment.
2. **Multimode sensors.** Ordinary control systems cannot compare either in the number of sensory modes or the degree of resolution or sensitivity with the human operator. It is possible to build a controller with greater sensitivity or resolution in one sensory mode (such as temperature, color, or sound perception), but only at relatively great cost and only for one mode at a time. In addition, the human operator contains sensor and information channels capable of recognizing danger or anomalous system behavior (often before catastrophe strikes).
3. **Cost, size, weight, maintainability, and availability.** Compared with any control system that can duplicate all the control actions typically undertaken by a human operator, the human operator is:
   (a) Lighter in weight
   (b) Lower in cost
   (c) Relatively easier to maintain
   (d) Relatively easier to program or train
   (e) Inexpensive to produce

Only in limited specialized situations have automatic pilots or robot welders taken over from the human operator. The automatic pilots or robot welders cannot handle emergency or unusual situations. In fact, their principal advantage is that they handle repetitive, standard, step-by-step operations that bore the human operator (and consequently lead to poor human operator performance).

33.8 SYSTEM CATEGORIZATION BY MATHEMATICAL MODEL

Systems can be categorized by the form of the mathematical model describing their characteristics, such as

Linear differential equations
Laplace transform equations
State equations
Difference equations
z-Transform equations
Discrete-time equations
Nonlinear differential equations

Table 33.2 summarizes several of the mathematical models used to describe the characteristics of systems.

TABLE 33.2. SUMMARY OF COMMON MATHEMATICAL APPROACHES EMPLOYED IN SYSTEM ANALYSIS

Equation Type	Example	Solution Methods
Linear differential equation	$\frac{d^2}{dt^2}C(t) + 2\frac{d}{dt}C(t) + 3C(t) = r(t)$ Initial condition: $C(t)\|_{t=0} = c(0)$ and $\frac{d}{dt}C(t)\|_{t=0} = \dot{c}(0)$	Trial and error, infinite power series, numerical integration approximations, Laplace Transforms
Laplace transform equation	$s^2C(s) - sc(0) - \dot{c}(0) + 2sC(s) - 2c(0) + 3C(s) = R(s)$ For zero initial conditions: $s^2C(s) + 2sC(s) + 3C(s) = R(s)$	Algebraic manipulations, tables of transforms, partial fraction expansions, inversion integrals
State equation	$\begin{bmatrix} \dot{X}_1 \\ \dot{X}_2 \end{bmatrix} = \begin{bmatrix} 0 & 1 \\ -3 & -2 \end{bmatrix} \begin{bmatrix} X_1 \\ X_2 \end{bmatrix} + \begin{bmatrix} 0 \\ r(t) \end{bmatrix}$ $\quad X_1 = C(t)$, $X_2 = \frac{d}{dt}C(t)$	Matrix methods, transition matrix exponential, Laplace transforms, linear systems theory
Difference equation	$X(K+2) + 2X(K+1) + 3X(K) = r(K)$ Initial condition: $X(K)\|_{K=0} = X(0)$ and $X(K)\|_{K=1} = X(1)$	Infinite series summation, convolution summation, numerical solution, z-transforms
z-Transform equation	$z^2X(z) - z^2X(0) - zX(1) + 2zX(z) - 2zX(0) + 3X(z) = R(z)$ For zero initial conditions: $z^2X(z) + 2zX(z) + 3X(z) = R(z)$	Algebraic manipulation, partial fraction expansions, tables of transforms, inversion integrals
Discrete time equation	$X(K\Upsilon + 2\Upsilon) + 2X(K\Upsilon + \Upsilon) + 3X(K\Upsilon) = r(K\Upsilon)$ (Υ = constant sample time) Initial conditions: $X(K\Upsilon)\|_{K=0} = X(0)$ and $X(K\Upsilon)\|_{K=1} = X(\Upsilon)$	Same methods as for difference equations. Note that K is an integer, whereas $K\Upsilon$ is not, except for the special case $\Upsilon = 1$ s.
Nonlinear differential equation	$\frac{d^2}{dt^2}C(t) - [1 - C(t)^2]\frac{dC(t)}{dt} + C(t) = 0$ (Vanderpol equation)	No solution, in general; phase plane analysis, describing functions, linearization approximations, Lyapunov method, Popov's method, numerical integration

On the basis of the form of the system's mathematical model and the structure of the system, systems are generally categorized and referred to as follows:

Linear systems
Nonlinear systems
Continuous systems
Discrete systems
Single-input single-output systems
Multiple-input multiple-output systems
Deterministic systems
Stochastic systems
Optimal systems
Adaptive systems

Each of these will be discussed briefly from a systems viewpoint.

Linear Systems

A linear system is described or modeled by a linear equation. The principal identifying characteristic of a linear system is that it satisfies the law of superposition. This law states that the system output for the sum of two or more inputs is equal to the sum of the outputs for each input applied separately. It follows that if a system input $r(t)$ produces an output $C(t)$, then an input $Kr(t)$ produces an output $KC(t)$. Thus if the input to a linear system is doubled, the output is doubled.

Linear systems are analyzed by all of the mathematical methods mentioned in Section 33.8 and Table 33.2, except those for nonlinear differential equations.

Another implication of the superposition theorem is that the output of a linear system is given by the convolution of its input with the system's impulse response. When Laplace transforms are taken, convolution becomes multiplication. Thus in the Laplace domain, the transform of a linear system's output is given by the product of its input transform with the transform of the system's impulse response. This is the basis of linear system analysis using transfer functions, which are the Laplace transforms of system impulse responses. Table 34.6 presents the convolution integral in conjunction with superposition and Laplace transforms.

Linear systems may be continuous-time systems or discrete-time systems. Discrete-time systems are discussed later in this chapter.

An excellent introduction to linear systems is given by Kuo.[22,23] An equally excellent introduction to superposition and convolution is given by Dertouzos and co-workers.[24]

Nonlinear Systems

A nonlinear system is described by a nonlinear equation. The principal identifying characteristics of a nonlinear equation are:

Trigonometric, exponential, or other nonlinear functions

Trigonometric

$$C(t) = \sin r(t)$$

Functional exponential

$$C(t) = e^{r(t)}$$

Integer exponential

$$C(t) = r(t)^2$$

Discontinuous equations

$$C(t) = 2r(t) \qquad r(t) > 1$$
$$C(t) = 0 \qquad r(t) \leqslant 1$$

Differential equations with variable coefficients, or products of derivatives

Linear

$$\frac{d}{dt}C(t) + C(t) = 0$$

Nonlinear, variable coefficient C(t)

$$C(t) \cdot \frac{d}{dt}C(t) + C(t) = 0$$

Linear second-order differential equation

$$\frac{d^2}{dt^2}C(t) + \frac{d}{dt}C(t) + C(t) = 0$$

Nonlinear differential equation due to trigonometric term, $A \sin C(t)$

$$\frac{d^2}{dt^2}C(t) + \frac{d}{dt}C(t) + A \sin C(t) = 0$$

Linear differential equation with nonlinear driving function $A \sin \omega t$

$$\frac{d^2}{dt^2}C(t) + \frac{d}{dt}C(t) + C(t) = A \sin \omega t$$

Nonlinear product or square of first-order derivative

$$\frac{d^2}{dt^2}C(t) + \left[\frac{d}{dt}C(t)\right]^2 + C(t) = 0$$

Nonlinear systems do not behave in accordance with the principle of superposition. Doubling the input does not double the output, as it would for a linear system. All real systems exhibit some form of nonlinearity such as saturation, dead zone, backlash, hysteresis, stiction, and/or coulomb friction. Nonlinear systems are discussed in detail in Sections 34.8 and 34.9.

A thorough introduction to nonlinear systems is given by Gibson.[25] Graham and McRuer[26] cover the subject and emphasize the description of correlation functions and power spectral density, used in their later work on nonlinear as well as linear human pilot models.

Continuous Systems

Continuous systems are described by equations or models in which time varies in a continuous fashion. Continuous systems are often denoted as analog systems and are usually described by differential equations, state equations, or Laplace transform equations. A true continuous system does not contain discontinuous elements, such as relays, switches, or digital devices. However, a piecewise continuous system can contain these elements. Between any discontinuous changes, such as turn on or turn off of a rocket engine, the piecewise continuous system is analyzed on a continuous basis. Continuous systems can be either linear or nonlinear.

D'Azzo and Houpis[27] give a wide-ranging introduction to continuous systems.

Discrete Systems

A discrete system is described by difference equations rather than by differential equations. Within a discrete system, some of the variables vary in a discrete, steplike fashion. Filters often are used to smooth this steplike behavior before the system output is produced, so that the output of a discrete system is often a smooth continuous function of time. Typically a discrete system samples a time function to produce sampled data which are then processed. Discrete systems usually contain one or more of the following elements:

Sampling switch (impulse sampler, or sample and hold)
Hold element (zero-order hold or "boxcar" generator; first- or second-order hold)
A-to-D and D-to-A converters (A = analog or continuous signal, D = digital code, held constant for one sampling time)
Relays or switches
Digital signal processor (either a filter, or a specialized controller, or a full computer system)

The growth of discrete systems parallels that of digital computers, especially microcomputers. Before 1970 the vast majority of hardware systems were continuous in nature. By the early 1980s discrete systems were taking over many military and aerospace applications, especially in the guidance and control areas. In the commercial arena, direct digital control, numerical machine tool control, and industrial robots are all examples of discrete systems.

A comprehensive introduction to discrete systems is given by Cadzow.[28]

Single-Input Single-Output Systems

Single-input single-output systems can be reduced to one expression, relating the system input to the system output. Examples are regulator systems, such as a temperature, pressure, or velocity control system. The input is the desired temperature, pressure, or velocity and the output is the regulated temperature, pressure, or velocity.

A single-input single-output system may be a component of a much larger system containing other signals. In reality there can be no single-input system since, in addition to the intended input, there are always additional extraneous disturbance inputs generated by wind gusts, sea waves, temperature variations, or any other external fluctuation that affects the system (e.g., noise).

A text well stocked with practical control system examples, such as that by Shinner,[5] gives many examples of single-input single-output systems.

Multiple-Input Multiple-Output Systems

In a multiple-input multiple-output system, each output is affected by one or more inputs. One output may also be affected by other outputs. Most large complex systems are of the multiple-input multiple-output variety. For example, a satellite attitude control system typically controls three angular coordinates and their rates. Therefore the system has six outputs.

In a noninteracting multiple-input multiple-output system, changes in one input affect only one output. Such a system can be conceptually separated into a set of single-input single-output systems, all contained within one large framework. For example, a process control system might have several inputs, such as raw materials, temperatures, flow rates, and pressures, and might produce several product outputs. It is desirable that each of the outputs be separately controlled by a single input for ease of varying production. Thus a noninteracting multiple-input multiple-output system is desired. Usually this is difficult if not impossible to accomplish. Interaction between the inputs occurs, and several or all inputs must be adjusted to vary one of the outputs in a desired fashion. In his truly outstanding text, Ogata[29] discusses this problem, using state space transfer matrices. In his slim but fact-filled text, Owens[30] also covers multiple-input multiple-output systems.

Deterministic Systems

The output of a deterministic system is uniquely determined for all time from its initial conditions and input signals. It is desirable that a system be deterministic, so that there will be no uncertainty in the system output. A system described by a differential or difference equation with constant coefficients and containing no disturbance or noise inputs is a deterministic system. In reality there are always extraneous disturbance inputs, which produce random fluctuations in the system output. Hence to have a deterministic system in practice, these random external disturbances must be negligible.

The majority of control theory texts, particularly introductory texts such as that of D'Azzo and Houpis,[27] deal exclusively with deterministic systems.

Stochastic Systems

Stochastic systems are subjected to or employ random variables or processes, and they are analyzed by statistical methods. The output of a stochastic system cannot be uniquely determined at any instant of time, but the expected response of the output can be determined within a calculated variance. Typically a stochastic system has a noisy input signal. The desired goal in system design is a system that responds to the underlying input signal and not to the noise that is corrupting this signal. The noise effect can be minimized by filtering. The filtering process may employ signal averaging, signal correlation, analog or digital filters, or weighted feedback or Kalman filters. The parameters of the required filter depend on the system-controlled plant and the characteristics of the input and noise signals. If any of these change significantly, the required filter parameters also change. Thus to be truly effective, a stochastic system should be able to change its filtering parameters in accordance with changes in the statistical properties of the noise-corrupted signals. Although such adaptive stochastic systems have been discussed,[31] few have been built. The overwhelming majority of stochastic systems assume constant stationary† noise statistics, particularly white (all frequency) noise with a gaussian bell-shaped probability distribution. Occasionally colored (frequency-band-limited) noise is assumed.

Two reasons for assuming a stationary gaussian probability distribution are:

1. Many random disturbances and noise signals are closely approximated by a gaussian probability distribution, which is stationary or does not change if the time origin is shifted.

2. The stationary gaussian probability distribution is mathematically attractive. Its autocorrelation function[26]

$$\phi_{nn}(\tau) = \lim_{\Upsilon \to \infty} \frac{1}{2\Upsilon} \int_{-\Upsilon}^{\Upsilon} n(t+\tau) n(t)\, dt \tag{33.1}$$

is an impulse function $K\delta(\tau)$ that has a Laplace transform of simply K, where K is the weight or area under the Dirac delta impulse $\delta(\tau)$. K is given by $\phi_{nn}(0)$, which is called the mean-square value of $n(t)$.

Stochastic systems with random inputs are often analyzed by utilizing correlation functions and power density spectrum functions. The latter are also called the power spectrum or the power spectral density. The correlation functions are time functions (cross correlation for two signals, autocorrelations for one signal with itself). The power spectral density functions are Fourier or Laplace transforms of the autocorrelation functions, and hence are frequency functions. The transforms of cross correlation time functions are called cross-power spectral density or cross-spectral density functions; these are also frequency functions.[32] Some of the symbols used for these quantities are:

Time signals. $r(t)$ and $c(t)$

Expressions for the correlation and power spectrum density terms are:

Autocorrelation. $\phi_{rr}(\tau)$ and $\phi_{cc}(\tau)$ [see $\phi_{nn}(\tau)$ given previously for integral definition]

Cross correlation

$$\phi_{rc}(\tau) = \lim_{\Upsilon \to \infty} \frac{1}{2\Upsilon} \int_{-\Upsilon}^{\Upsilon} r(t) c(t+\tau)\, dt \tag{33.2}$$

Power spectral density

$$\Phi_{rr}(\omega) \equiv s_{rr}(\omega) \equiv s_r(\omega) = s_r(f) \equiv Y_r(f) \equiv Y_{rr}(\omega) \equiv Y_{rr}(j\omega) = \Phi_{rr}(s)$$

Two-sided Laplace transform

$$\Phi_{rr}(s) = \int_{-\infty}^{\infty} \phi_{rr}(\tau) e^{-s\tau}\, d\tau \qquad \text{where} \quad s = \sigma + j\omega \tag{33.3}$$

Fourier transform

$$\Phi_{rr}(\omega) = \Phi_{rr}(s)|_{s=j\omega} = \int_{-\infty}^{\infty} \Phi_{rr}(\tau) e^{-j\omega\tau}\, d\tau \tag{33.4}$$

Cross power spectral density

$$\Phi_{rc}(s) \equiv \Phi_{rc}(j\omega) \equiv \Phi_{rc}(\omega) = s_{rc}(\omega) \equiv Y_{rc}(\omega)$$

$$\Phi_{rc}(s) = \int_{-\infty}^{\infty} \phi_{rc}(\tau) e^{-s\tau}\, d\tau \tag{33.5}$$

$$\Phi_{rc}(\omega) = \int_{-\infty}^{\infty} \phi_{rc}(\tau) e^{-j\omega\tau}\, d\tau = \Phi_{rc}(s)|_{s=j\omega} \tag{33.6}$$

† A stationary function's statistical properties do not change if the time origin is shifted.

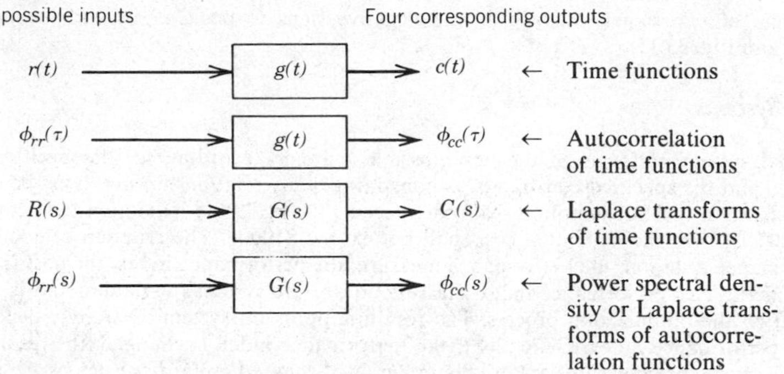

Fig. 33.14 Transfer functions.[33]

For Fig. 33.14, the deterministic and stochastic relationships are as follows:

$$C(s) = G(s)R(s) = R(s)G(s) \tag{33.7}$$

$$\Phi_{cc}(s) = G(s)G(-s)\Phi_{rr}(s) = G(-s)G(s)\Phi_{rr}(s) \tag{33.8}$$

$$c(t) = \int_0^t g(\tau)r(t-\tau)\,d\tau = \int_0^t r(\tau)g(t-\tau)\,d\tau \qquad \text{Convolution integral } r(t)=0,\ t<0 \tag{33.9}$$

$$\phi_{cc}(\tau) = \frac{1}{2\pi j}\int_{c_1-j\infty}^{c_1+j\infty} G(s)G(-s)\Phi_{rr}(s)e^{s\tau}\,ds \qquad \text{Laplace inversion integral; } c_1 = \text{constant chosen for convergence—generally } c_1 = 0 \tag{33.10}$$

Negative frequency

$$\Phi_{rr}(s) = \Phi_{rr}(-s) \tag{33.11}$$

$$\Phi_{rc}(s) = \Phi_{cr}(-s) \tag{33.12}$$

Sum or difference of signals[34] *(see Fig. 33.15)*

$$e(t) = r(t) - c(t) \tag{33.13}$$

$$E(s) = R(s) - C(s) \tag{33.14}$$

$$\phi_{ee}(\tau) = \phi_{rr}(\tau) + \phi_{cc}(\tau) - \phi_{rc}(\tau) - \phi_{cr}(\tau) \tag{33.15}$$

$$\phi_{ee}(s) = \phi_{rr}(s) + \phi_{cc}(s) - \phi_{rc}(s) - \phi_{cr}(s) \tag{33.16}$$

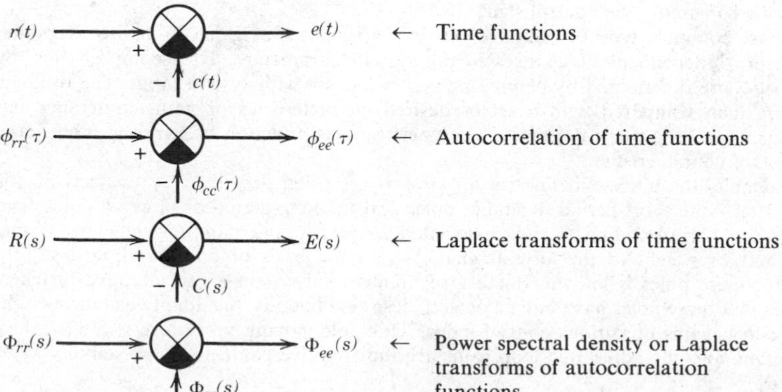

Fig. 33.15 Sum of difference of signals.[33] Only one of the four possible sets of signals is used at one time.

For the sum of two signals, change all the negative signs to positive signs in the preceding four equations and Fig. 33.15.

Optimal Systems

An optimal system performs in the best possible manner according to the specified criterion of "goodness" and the specified constraints. A constraint is any relationship that must be satisfied, such as "the maximum fuel consumption shall not exceed 1000 kg," "the maximum temperature shall not exceed 1000°C," or "the maximum cost shall not exceed \$1000." The criterion of goodness is called the performance criterion, the performance measure, the performance index, the cost function, or the figure of merit. This performance index is a function of the system's variables, and is maximized or minimized by the optimization process. The resulting optimum system is strongly dependent on the choice of performance index. Generally if the performance index is changed, the resulting system is changed. Performance measures are usually integrals of squared quantities, such as the integral of the squared error, which should be minimized. When state variable matrix equations are employed, the performance measure is generally taken as an integral of a positive definite function (which is usually a quadratic form). In the scalar case, a quadratic form reduces to a squared variable.

A two-dimensional quadratic form for the vector $X = [X_1, X_2]^T$ is

$$F(X) = X^T A X = [X_1, X_2]\begin{bmatrix} a_{11} & a_{12} \\ a_{21} & a_{22} \end{bmatrix}\begin{bmatrix} X_1 \\ X_2 \end{bmatrix}$$

Multiplying out

$$F(X) = [X_1, X_2]\begin{bmatrix} a_{11}X_1 + a_{12}X_2 \\ a_{21}X_1 + a_{22}X_2 \end{bmatrix} = a_{11}X_1^2 + a_{21}X_2X_1 + a_{12}X_1X_2 + a_{22}X_2^2$$

Now $F(X) = F(X_1, X_2)$ is positive definite if $F(X) > 0$ for $X \neq 0$, and $F(X) = 0$ only for $X = 0$.

For the scalar case, $F(X) = F(X_1) = a_{11}X_1^2$.

The integral of quadratic forms or squared variables is employed for a cost function or performance index, because such an integral is monotonic and independent of the sign of the variables. Thus if the variable were error related, either a positive or a negative error would increase the cost function. The only way to minimize such a cost function is to keep the error near zero. A negative error cannot cancel out a later positive error. The procedures for optimizing a system are discussed in detail in Sections 34.11 through 34.14. Ogata[29] gives a succinct but extremely clear introduction to optimal systems (Sections 7.3–7.6).

Adaptive Systems

Adaptive systems alter their internal parameters or structure in accordance with changes in the control task. These changes may be due to changing system dynamics brought about by either internal or external variations.[34] Examples of such variations are the change in vehicle's mass owing to fuel consumption or the change in aerodynamic control forces owing to changes in vehicle altitude and/or velocity. These changes may also be due to variations in the type of control signals and the type of noise corrupting the control signal.

The most common type of adaptive system varies internal gains to try to keep the system's characteristics independent of changes in the system parameters. To accomplish this, the system characteristics are determined by comparing system output with system input. The resultant characteristics are then compared with a set of desired characteristics or with a reference model. The adaptive gains are adjusted to optimize some performance function or figure of merit based on this comparison of characteristics.

For example, the pitch dynamics of an aircraft are often described by two sets of poles in the complex plane—the short-period dynamics poles and the long-period or phugoid poles. The position of these poles, corresponding to their associated frequency and damping, changes as the altitude, velocity, and/or weight of the aircraft changes. On the basis of pilot-handling tests, the desired position for these poles is known. If the measurement subsystem of an adaptive system controller determines that these poles have moved outside desirable bounds, the adaptive controller changes the adjustable loop gains of various control loops. This pole-moving procedure is often called stability augmentation, and the adjustable loop gains are the adaptive controls of the stability augmentation system.

To measure or determine system characteristics, a special test input is usually applied. Two methods can be used. In the first method a deterministic input, such as a pulse that approximates a Dirac delta input can be applied when the system is near equilibrium. The resultant output will then

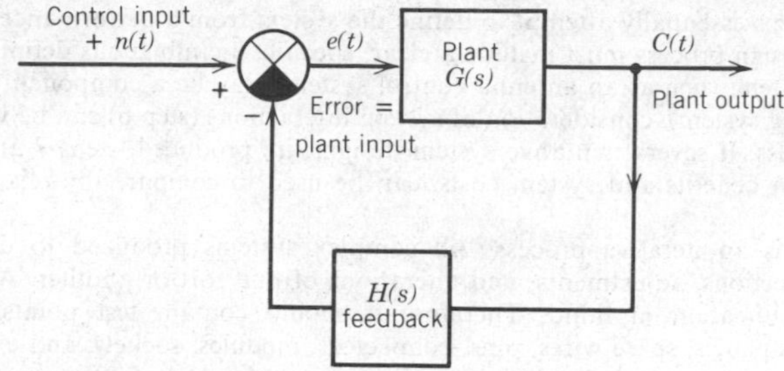

Fig. 33.16 Determination of $G(s)$ from power spectral density functions, with random noise input $n(t) = n$. $\Phi_{nc}(s) = \int_{-\infty}^{\infty}\phi_{nc}(\tau)e^{-s\tau}d\tau$ (two-sided Laplace transform); $G(s) = \Phi_{nc}(s)/\Phi_{ne}(s)$; $\phi_{nc}(\tau) = \lim_{T\to\infty}(1/2T)\int_{-T}^{T}n(t)c(t+\tau)dt$ (Cross-correlation).

approximate the system's weighting function. If such disturbances cannot be tolerated, a purposely generated stationary random noise signal of zero mean can be added to the actual control input, and the system transfer function can then be determined from power spectral density methods, namely[26]

$$G(s) = \frac{\Phi_{nc}(s)}{\Phi_{ne}(s)}$$

where $G(s)$ = transfer function of plant to be determined
n = noise input to system
e = input of plant
c = output of plant
$\Phi_{nc}(s)$ = cross-power spectrum, noise-to-plant output
$\Phi_{ne}(s)$ = cross-power spectrum, noise-to-plant input
as defined in this chapter for stochastic systems.

The corresponding block diagram is shown in Fig. 33.16.

Again, Ogata[29] gives a brief, lucid, and relatively complete introduction to adaptive systems. If one were to seek out a single book that covers all of the control theory topics discussed in Section 33.8, it would be difficult, if not impossible, to find a text superior to Ogata's, even though this book is now over a decade old.

33.9 OVERVIEW OF SYSTEM DESIGN PROCESS

The system design process starts with a theoretical formulation of the problem, which includes definition of some or all of the following 14 steps:

1. What is the system supposed to do?
2. What are the system inputs?
3. What are the system outputs?
4. What are the system constraints?
5. What are the alternatives for each of the first four steps?
6. What effects will the system have on devices that interact with it, and vice versa?
7. What are the anticipated benefits of the system?
8. What are the anticipated system costs?
9. What provisions should or can be made for future modifications and unanticipated contingencies?
10. What are the principal components or subsystems of the system?
11. What specifications must the subsystems or components meet for the system to perform satisfactorily?
12. What is the anticipated time for the design, construction, testing, and procurement of each system component or subsystem?
13. What resources (men, machines, facilities, capital) must be assigned to the system?
14. What is the relative importance of each of the previous steps to the particular system? Can any steps be omitted?

The first five steps essentially attempt to define the system from a performance standpoint. To be successful, every design process must include a clear, specific, unambiguous definition of the system. Since any given system, such as an antenna control system, may be a component of a larger system (e.g., a radar landing system), consideration of system interactions (step 6) can be vitally important in the definition process. If several tentative system designs are produced, steps 7 and 8, evaluation of the expected system benefits and system costs, can be used to compare the relative merits of each system design.

System design is an iterative process. All complex systems produced to date have required modifications, corrections, adjustments, and alterations of one sort or another. A system should be designed with modification in mind. Therefore it should contain test points, modular plug-in components, access panels, spare wires, pins, connectors, modules, sockets, and extra memory. Only with this philosophy can a system be designed that is successful and of minimum cost.

Once a rough theoretical system design has been settled on, it is typically divided into subsystems, along functional lines. The entire step process is then reapplied to the subsystems, in an iterative manner, and details are added to the design. If the system is large, complex, and costly, each subsystem may be designed in a sequential order determined by a management program, described in Section 33.6. As the theoretical design proceeds, each component should be periodically examined for the following three key factors:[35]

1. Suitability. Will it do the job?
2. Feasibility. Can it be manufactured?
3. Acceptability. Is its performance worth the expected costs in terms of time, materials, men, and money?

Various designs can be evaluated or judged by applying the following techniques:

1. **Figures of merit.** Desirable performance factors, such as strength, accuracy, power.
2. **Optimization methods.** Discussed in Sections 34.11 through 34.14; used to maximize a figure of merit.
3. **Model building and testing.** With use of physical or mathematical models.
4. **Computer simulation.** Used in conjuction with either mathematical or physical models.
5. **Comparison testing.** Used when several different subsystem designs are carried through to completion.

Once a theoretical design has been produced in the form of block diagrams, computer programs, blueprints, models, specifications, and reports, the actual physical system design process starts. Each component or subsystem must be physically designed or purchased so that all system specifications are met. This is also an iterative process, where the physical components are tested and then redesigned if necessary to meet the required specifications.

In practice, the theoretical design process may turn out to be faulty at some point, and a subsystem that was originally judged theoretically suitable, feasible, and acceptable proves to be difficult if not impossible to physically design and produce. In this case a completely new theoretical design is required. In either case a physical system design is eventually completed, and prototype subsystems are built and interconnected.

The complete physical system must then be tested to determine if specifications are met, to determine if problems develop that require redesign, and to check for system reliability, maintainability, and availability. If all appears satisfactory, the job is still not over. Efforts should be made to lower the system's initial cost and life-cycle cost (which includes its cost to operate) without lowering performance, or to improve performance without increasing costs. In addition, long-term reliability, maintainability, and availability testing should continue, especially as design changes are introduced. Hence the system design process for large complex systems is iterative and never really ends. The objective of improving the system's performance; improving reliability, maintainability, and availability; and reducing the cost of operation continues throughout the life of the system.

For further details of managing the system design process, refer to Hillier and Lieberman,[36] Cook and Russell,[37] or Huysmans.[38]

References

1 S. M. Shinners, *A Guide to Systems Engineering and Management*, Lexington Books, Lexington, MA, 1976.

2 G. J. Klir, *An Approach to General Systems Theory*, Van Nostrand Rienhold, New York, 1969.

3 R. M. Glorioso and F. C. Osorio, *Engineering Intelligent Systems*; *Concepts, Theory & Applications*, Digital Press, Bedford, MA, 1980.

4 K. N. Dervitsiotis, *Operations Management*, McGraw-Hill, New York, 1981.

5 S. M. Shinners, *Modern Control System Theory and Application*, 2nd ed., Addison-Wesley, Reading, MA, 1978.

6 G. B. Dantzig, *Linear Programming and Extensions*, Princeton University Press, Princeton, NJ, 1963.

7 R. G. Eland, "The Allocation of Resources by Linear Programming," *Scient. Am.*, pp. 124–126 (Jul. 1981).

8 A. Tustin, "The Nature of the Operator's Response in Manual Control and Its Implications for Controller Design," *J. IEEE* **94** (part IIA):190–202 (1947).

9 D. T. McRuer, D. Graham, and E. Krendel, "Human Pilot Dynamics in Compensatory Systems," Wright Patterson Air Force Flight Dynamics Laboratory, Report AFFDL-TR-65-15 (Jul. 1965).

10 G. A. Bekey, *Sampled Data Models of the Human Operator in a Control System*, PhD Dissertation, University of California, Los Angeles, 1962.

11 Goodyear Aircraft Co., *Final Report, Human Dynamics Study*, Goodyear Aircraft Co., Akron, OH, Report No. GER-4750, 1962.

12 D. Knoop and K. Fu, *An Adaptive Model of the Human Operator in a Control System*, PhD Dissertation, Purdue University, West Lafayette, IN, 1964.

13 R. G. Costello, "The Surge Model of the Well-Trained Human Operator in Simple Manual Control," *IEEE Trans. Man-Machine Sys.* **MMS-9(1)**:2–9 (Mar. 1968).

14 D. L. Kleinman et al., "An Optimal Control Model of Human Response, Part I: Theory and Validation," *Automatica* **6**, (3), 357–369, (May 1970).

15 S. Baron, D. L. Kleinman, D. C. Miller, W. H. Levison, J. I. Elkind, "Application of Optimal Control Theory to the Prediction of Human Performance in a Complex Task," Wright Patterson Air Force Flight Dynamics Lab, Report AFFDL-TR-69-81 (Mar. 1970).

16 L. R. Young and J. L. Meiry, "Bang-Bang Aspects of Manual Control in High Order Systems," *IEEE Trans. Auto. Cont.*, **AC-10(3)**:336–341 (Jul. 1965).

17 C. R. Wasaff, "A Human Adaptive Control Scheme," presented at 7th IEEE Symp. on Human Factors in Electronics, Minneapolis, MN, May 5–6, 1966.

18 J. I. Elkind, "Further Studies of Multiple Regression Analysis of Human Pilot Dynamic Response: A Comparison of Analysis Techniques and Evaluation of Time Varying Measurements," Bolt Beranek and Newman, Inc., Cambridge, MA, tech. documentary.

19 W. W. Wierwille, "A Theory of the Optimal Deterministic Characterization of the Time-Varying Dynamics of the Human Operator," NASA, Washington DC, Contractor Rept. CR-170, 1965.

20 J. D. North, "The Design and Interpretation of Human Control Experiments," Part I, *Ergonomics* **1**(4):314–327 (Aug. 1958).

21 L. J. Fogel, "The Human Computer in Flight Control," *Memorandum*, Convair, Div. General Dynamics Corp., San Diego, CA, pp. 1–19 (May 1957).

22 B. C. Kuo, *Linear Networks and Systems*, McGraw-Hill, New York, 1967.

23 B. C. Kuo, *Automatic Control Systems*, 4th ed., Prentice-Hall, Englewood Cliffs, NJ, 1982.

24 M. L. Dertouzos, M. Athans, R. N. Spann, and S. J. Mason, *Systems, Networks, and Computations, Basic Concepts*, McGraw-Hill, New York, 1972.

25 J. E. Gibson, *Nonlinear Automatic Control*, McGraw-Hill, New York, 1963.

26 D. Graham and D. McRuer, *Analysis of Nonlinear Control Systems*, Wiley, New York, 1961.

27 J. J. D'Azzo and C. H. Houpis, *Linear Control System Analysis and Design*, 2nd ed., McGraw-Hill, New York, 1981.

28 J. A. Cadzow, *Discrete Time Systems*, Prentice-Hall, Englewood Cliffs, NJ, 1973.

29 K. Ogata, *Modern Control Engineering*, Prentice-Hall, Englewood Cliffs, NJ, 1970.

30 D. H. Owens, *Multivariable and Optimal Systems*, Academic Press, New York, 1981.

31 D. Pierre, *Optimization Theory with Applications*, Wiley, New York, 1969.

32 Y. Takahashi, M. Rabins, and D. Auslander, *Control and Dynamic Systems*, Addison-Wesley, Reading, MA, 1970.

33 S. S. L. Chang, *Synthesis of Optimum Control Systems*, McGraw-Hill, New York, 1961 (see Section 3-5).

34 J. M. Mendel and K. S. Fu, *Adaptive Learning and Pattern Recognition Systems, Theory and Applications*, Academic Press, New York, 1970.

35 C. R. Mischke, *Mathematical Model Building*, Iowa State University Press, Ames, 1980.

36 F. S. Hillier and G. J. Lieberman, *Introduction to Operations Research*, 3rd ed., Holden-Day, San Francisco, 1980.

37 T. M. Cook and R. A. Russell, *Introduction to Management Science*, Prentice-Hall, Englewood Cliffs, NJ, 1977.

38 J. Huysmans, *The Implementation of Operations Research*, Wiley, New York, 1970.

39 J. R. Ward, *The Dynamics of a Human Operator in a Control System: A Study Based on the Hypothesis of Intermittency*, PhD Thesis, Department of Aeronautical Engineering, University of Sydney, Australia, 1958.

40 S. M. Soliday and B. Schohan, "Task Loading of Pilots in Simulated Low Altitude, High Speed Flight," **7**(1):45–53 (Feb. 1965).

41 N. Glickman, T. Inouye, S. E. Telser, R. W. Keston, F. K. Hick, and M. K. Fahnestack, "Physiological Adjustments of Human Beings to Sudden Changes in Environment," **19**(8) (August 1947).

Bibliography

Baker, K. R., *Introduction to Sequencing and Scheduling*, Wiley, New York, 1974.

Buffa, E. S., *Modern Production Management*, 5th ed., Wiley, New York, 1977.

Costello, R. G. and T. J. Higgins, *IEEE Trans. Human Fact. Electr.* **HFE-7**:174–181 (Dec. 1966).

Drucker, P., *Management: Tasks, Responsibilities, Practices*, Harper & Row, New York, 1974.

Duncan, A. J., *Quality Control and Industrial Statistics*, 4th ed., Irwin, Homewood, IL, 1974.

Hillier, F. and G. J. Lieberman, *Introduction to Operations Research*, 2nd ed., Holden-Day, San Francisco, 1974.

Jury, E. I. and T. Pavlidis, *IEEE Trans. Autom. Contr.* **AC-8**:210–217 (1963).

Keeney, R. L. and H. Raiffa, *Decision Analysis With Multiple Objectives*, Wiley, New York, 1976.

Levin, R. I. and C. A. Kirkpatrick, *Planning and Control with PERT/CPM*, McGraw-Hill, New York, 1966.

Marshall, P. W., et al., *Operations Management: Text and Cases*, Irwin, Homewood, IL, 1975.

Meadows, D. H., et al., *The Limits to Growth*, New American Library, New York, 1972.

Moder, J. J. and C. R. Phillips, *Project Management with CPM and PERT*, 2nd ed., Reinhold, New York, 1970.

Peterson, R. and E. A. Silver, *Decisions Systems for Inventory Management and Production Planning*, Wiley, New York, 1977.

Riggs, J. L., *Production Systems: Analysis Planning and Control*, 2nd ed., Wiley, New York, 1976.

Riggs, J. L., *Engineering Economics*, McGraw-Hill, New York, 1977.

Riggs, J. L. and M. S. Inoue, *Introduction to Operation Research and Management Science*, McGraw-Hill, New York, 1975.

Spencer, M. H., et al., *Managerial Economics*, 4th ed., Irwin, Homewood, IL, 1975.

Summers, L. G. and K. Ziedman, "A Study of Manual Control Methodology, with Annotated Bibliography," *NASA Contractor Report CR-125*, Washington, DC, 1964.

Tersine, R. J., *Materials Management and Inventory Systems*, Elsevier, New York, 1976.

Vaughn, R. C., *Quality Control*, Iowa State University Press, Ames, 1974.

Wagner, H. M., *Principles of Operations Research*, Prentice-Hall, Englewood Cliffs, NJ, 1974.

Young, L. R. and L. Stark, "Biological Control Systems—A Critical Review & Evaluation," *NASA Contractor Report CR-190*, Washington, DC, 1965.

CHAPTER 34
AUTOMATIC CONTROL

RICHARD G. COSTELLO

The Cooper Union for Advancement of Science and Art
New York, New York

34.1 INTRODUCTION

Automatic control is a branch of system theory concerned with the control of physical entities or devices. The configurations of the devices vary with their purpose and function. Examples include control of the flow of water through a turbine, regulation of the temperature in a diffusion furnace, and stabilization of the orientation of a communications satellite.

The early Greeks designed automatic control systems of the regulation type, intended to keep the fluid level in a container constant even though some fluid was periodically removed. A similar design is used in the present-day tank-type flush toilet, where the fluid level in the tank is kept constant even though most of the fluid is removed each time the toilet is flushed. This is one of the earliest examples of a closed-loop feedback control system of the regulator type. In such a system the fluid level in the container is sensed by a float that feeds back the information via a mechanical linkage to a valve and admits more fluid from a reservoir or pipe when the fluid level falls.

The prolific author Hero of Alexandria described a "wine dispenser" he designed in the year A.D. 50, which kept a wine cup full by means of a float regulator.[1]

According to Mayr,[1] the earliest automatic control system was incorporated in a water clock designed and built by the Greek mechanician Ktesibios, working in the service of the Egyptian King Ptolemy II in Alexandria about 300 BC. The automatic control was a float valve that kept the level in a small elevated tank constant. The small tank, in turn, contained a small orifice through which the water trickled down into a much larger tank, which slowly filled up. The water level in the larger tank raised a float attached to a pointer that indicated the hours. Ktesibios' regulator float was remarkably similar to the present-day float valve used in automobile carburetors, over 2000 years later.

Since these early beginnings, automatic control was developed to the point where it is found in almost every aspect of our daily lives. Table 34.1 lists significant developments in the area of automatic control, in chronologic order, from antiquity to World War I.

TABLE 34.1. CHRONOLOGICAL DEVELOPMENT OF AUTOMATIC CONTROL FROM ANTIQUITY TO WORLD WAR I

Year	Device	Inventor or Developer; Place	Description
Ca 250 BC (Hellenistic period)	Feedback float regulator	Ktesibios; Alexandria, Greece	Water clock employing feedback float regulator. Credited as first example of feedback control system.[a] Feedback device, sensor, and control device all embodied in the same physical element, the float. Basically identical devices are found in float valve compartment of modern automobile carburetor; specifically, float-operated needle valve
~1st century AD	First modern control system	Heron; Alexandria, Greece	Heron's "Pneumatica"[b] contains drawings and explanations for about 100 devices. His float regulator (Pneumatica II.31) contains innovation of separating sensing device (float) from control device (valve) so that system formally fits definition of feedback system in modern sense[c]
9th century AD	On-off control device	The Banū Mūsā (Three sons of Mūsā); Baghdad	Brothers Mūsā wrote at least 21 papers. One, "Kitāb al-Hiyal" (Book of Ingenious Mechanisms) contains about 100 drawings and explanations of devices. Numbers 83 and 84 describe nonlinear on-off control device used to maintain constant liquid level; it can be considered one of first examples of on-off control system. Banū Mūsā were familiar with Heron of Alexandria's work. They added significant development to Heron's float regulators with throttling valve, which required no constant force to be kept closed. (Heron's regulators worked only at low pressures, since control force had to

			directly oppose fluid pressure via flat plate held against end of supply pipe.) Example of modern throttling valve is standard water-supply gate valve used as main shut-off valve for most domestic water supply systems. Mūsā valve was topologically similar to laboratory stopcock valve or hydraulic control spool valve, since it did not contain threaded screw drive of gate valve
Ca 1650 AD	First thermostatic control system	Cornelius Drebbel; Alkmaar, Holland	According to all available evidence,[d] first feedback system to be invented in modern Europe and independent of ancient models is temperature regulator invented by Drebbel. He designed and built thermostatically controlled laboratory furnaces called Athanors and thermostatically controlled incubators. Temperature was controlled by damper valve that regulated supply of air to fire. Damper was mechanically linked to control rod floating on mercury contained in U-tube. Closed end of U-tube was connected to glass cylinder filled with alcohol which was inserted in furnace or incubator. As temperature increased, alcohol expanded and pushed mercury up U-tube, which in turn lifted floating control rod which shut damper. This restricted airflow to fire and lowered furnace or incubator temperature
1680 AD	Steam safety valve, or pressure regulator	Denis Papin; Blois, France	Papin, one of inventors of steam engine, made first pressure cooker in 1680 by placing heavy weight on lid of pot. It was to be used for digesting or softening bones so that

TABLE 34.1. (*Continued*)

Year	Device	Inventor or Developer; Place	Description
			housewife could "extract nourishing juices from bones which would otherwise have been abandoned as but poor prey by ye hungry dogs." He thus originated steam safety valve, or pressure regulator, which is, in modified form, one of most common regulators in use today. Most modern steam or pressure relief valves utilize spring pressure, rather than force of weight, to regulate controlled pressure. However, unmodified weight-controlled pressure cooker is still around after 300 years
1783 AD	Thermostat (rod type)	Bonnemain; Paris, France	Evolution from Drebbel's rather delicate laboratory temperature regulator to ruggedly engineered practical temperature controller was accomplishment of French engineer, Bonnemain. Rather than employing fluid-filled glass tube to sense temperature changes, he applied idea[e] long known to clockmakers, namely, different rates of thermal expansion of dissimilar metals. Bonnemain's design employed long solid steel rod surrounded by somewhat shorter cylinder of metal with high coefficient of thermal expansion. At far end steel rod and outer cylinder were hermetically joined together for perhaps 0.5 in. (1 cm). This far end and most of length of cylinder were immersed in boiler or furnace. Seal was provided between outer cylinder and boiler, several inches from outside end. At outside end, cylinder would climb up steel rod as temperature increased,

			due to greater lengthening or expansion of outer cylinder compared with steel rod. Distance was amplified by mechanical linkage and resultant motion was employed to control furnace flue damper. Exact same configuration, fast-expanding metal cylinder over steel bar core, is employed in whole class of present-day temperature controllers, as typified by thermostatic controller for domestic gas-fired hot water heater
1788	First well-known control system, flyball centrifugal speed governor	James Watt; Scotland	Watt did not invent the flyball governor, or patent it. He perfected and applied flyball governor (or centrifugal pendulum) invented by English millwrights 20 or 30 years earlier. Millwrights built grist mills, for grinding grain between two large stone disks rotating in horizontal plane. Disks had to be kept separated a small amount to allow grain to fit between them. When mill was stopped or stones were turning slowly (they were powered by waterwheels or windmills), stones had to separate a larger distance. Top stone was raised by device called lift-tenter. Lift-tenter utilized force of two lead weights (balls) which spun around vertical axis and rose horizontally when in motion and fell when stopped. Falling balls, via linkage, raised millstone minute amount. Millwrights also used flyball or centrifugal pendulum to regulate speed of windmills. Thomas Mead patented flyball governor[f] for windmills in 1787, one year before Watt built his governor for steam engines. Watt's governor is remembered because

TABLE 34.1. (*Continued*)

Year	Device	Inventor or Developer; Place	Description
			many were built, and they operated effectively. Watt engine was sensation, and spinning flyballs were its hallmark. It quickly spread over all of Europe, replacing older, slower, less efficient Newcomen engine. At first governor was concealed to protect device, since firm of Boulton & Watt did not attempt to patent it. (Boulton provided financing and was majority owner. Watt invented engines.) Within two decades, entire engineering world knew of Watt's governor, and name of Watt is now associated worldwide with invention of speed control. Six feedback devices listed here before Watt's are far more original but have remained extremely obscure. (primarily because they were not produced in large quantities). Watt can be regarded as father of feedback control, because his rapidly multiplying flyball governors spread concept of feedback around world. Working flyball-governed steam engine can be seen at Smithsonian Institute in Washington, DC
1830 AD	Bimetallic strip thermostat	Andrew Ure; Glasgow, Scotland	Ure coined the word "thermostat" to describe "an apparatus for regulating temperature in vaporization, distillation and other processes" which he patented in 1830. Ure's flat-strip thermostat is used in present-day home-heating thermostats, although strip is usually coiled up like large watch spring to enable longer and hence more sensitive bimetallic strip to fit in small enclosure

1858	Power steering, steam assisted	Isambard Kingdom Brunel; Portsmouth, England	Brunel was engineer who designed suspension bridges for railroads and later designed ships to extend reach of Great Western railway, his employer. His first ship, "Great Western" (1838), was first steamship built to make regular transatlantic crossings. His second ship (1845), "Great Britain," was largest ship afloat at time and was first large iron steamship as well as first large ship using screw propeller. His third ship, named "Great Eastern," was so large it required power steering. Great Eastern had displacement of 18,914 tons gross, six sailmasts, paddle engines built by Scott-Russell & Co., and screw engines built by James Watt & Co. Her famous accomplishment was laying of transatlantic cable. Very few know of her other significant accomplishment, application of feedback-controlled power steering mechanism. This position control system used control valve that can be considered, in basic concept, forerunner of present-day hydraulic spool valve and present-day vehicular power-steering system. Control valve of Great Eastern's power steering is said to have originated in 1713 with Humphrey Potter, whose job was to open and then close at exactly right moments valves that admitted and exhausted steam of manually controlled Newcomen steam pumping engine. Potter is said to have noticed that admission of steam was always required when piston was in one particular position and that exhaust was needed at another, and this prompted him to conceive of linking piston's motion to control valves.

TABLE 34.1. *(Continued)*

Year	Device	Inventor or Developer; Place	Description
			If story is true, Potter may have been first individual to lose his job to automation.[g]
1869	Air brake (first major pneumatic control system)	George Westinghouse; Central Bridge, New York	Westinghouse air brake is not a closed-loop automatic control system. It is manual control system in which human operator closes control loop by sensing error and performing control function. Airbrake system is open-loop pneumatic amplifier, which provides large braking forces at many wheels simultaneously, in response to relatively insignificant control force. Operator's control force opens air valve, which allows pressurized air to flow to remote air cylinders which then move in response to air pressure, and apply braking force. Trains still use air brakes, as do many large trucks. Since working fluid is air, small leaks do not pose very serious problem, as they do in hydraulic systems. Automobile can run out of hydraulic brake fluid because of small leak, but train does not run out of air because of small leak. In addition, cars of train may be readily disconnected, as may trailer trucks, without hydraulic fluid being spilled. Little escaping air causes no bother or mess. Also pneumatic controls can be used in combustible atmospheres, since they can be designed to be spark-free. For these reasons, pneumatic controls have proliferated from air brake to modern pneumatic process control valves, which are used in large numbers by industry, and may be seen in large numbers under hood of most automobiles built during 1970s and 1980s

1872	Servo, servomechanism	Joseph Farcot; France	Farcot invented another steam-powered steering mechanism. Describing it, he stated, "We thought it necessary to give a new and characteristic name to this novel mechanism, and have called it a SERVO or enslaved motor." Thus Farcot coined terms "servo" and "servo-motor" in 1872, which are still commonly applied to describe powered position controller
1914–1919	First gyro control systems	Elmer Ambrose Sperry; Cortlandt, New York	Although gyroscopic top had been known since antiquity, Sperry developed it as control device, and his name is wedded to it, viz., Sperry gyrocompass. In 1911 first Sperry gyrocompass was installed (on US battleship Delaware). From this application developed entire new area of control, aiming and guidance of projectiles, ships, torpedoes, aircraft, and eventually, although not within Sperry's lifetime, missiles and space vehicles. Modern cruise ships sail smoothly, thanks to Sperry's gyroscopic ship stabilizer, which is automatic roll control system, and they are kept on course by Sperry's gyropilot, an automatic yaw or steering control system

[a] Ref. 2, pp. 11–16.
[b] Over 100 manuscripts are preserved, all prepared after the 12th century, according to Mayr.[2]
[c] "A Feedback Control System is a control system which tends to maintain a prescribed relationship of one system variable to another by comparing functions of these variables and using the difference as a means of control." 1951 definition, American Institute of Electrical Engineers, "Proposed Symbols and Terms of Feedback Control Systems."
[d] Ref. 2, pp. 56–64.
[e] Differential expansion was used by John Harrison in a temperature-compensated clock pendulum designed in 1726.[3]
[f] Ref. 4.
[g] Ref. 5.

TABLE 34.2. CONTROL ORIGINS

Era of Earliest Appearance	Control Type	Examples
300–200 BC	Mechanical	Pressure regulators, mechanical speed governors, float valves, pneumatic and hydraulic control, temperature regulators
19th century	Electrical	Relays, motors, slidewires, switches, selsyns, thermostats, pressurestats
20th century	Electronic	Servo amplifiers (tube or transistor), photocells, electronic transducers, auto pilots, guidance servos, robotics
Late 20th century	Digital	Stepper motors, A/D and D/A converters, digital computers, digital filters, direct digital control, numeric control of machine tools, digital robotics, visual input control systems

Table 34.2 summarizes control origin by chronologic era, while Table 34.3 lists five existing areas of control and gives multiple examples and Table 34.4 enumerates seven evolving areas of control and gives numerous examples.

34.2 CLASSICAL AUTOMATIC CONTROL

Classical automatic control deals with the analysis and design of linear open-loop control systems using block diagrams, Laplace transform transfer functions, and frequency domain methods.

An open-loop control system is shown in Fig. 34.1. A relationship exists between the output and the input, so that changing the input causes the output to change in a known fashion. The input-output relationship may be called system gain, system transfer function, system weighting function, or system plant function. In the example shown in Fig. 34.1, a water valve controls the flow of water in a shower. The input to the system is the angular position of the valve handle, and the output is the water flow rate. As the handle is turned, the input angle θ_i increases, as does the water flow rate Q_w, over a certain limited range. Over this range Q_w is proportional to θ_i and may be described by the following linear equation:

$$Q_w = K \cdot \theta_i$$

where Q_w = system output = water flow rate, gal/min
θ_i = system input = water valve angle, rad
K = constant of proportionality, gain, or transfer function

This is the simplest possible open-loop system, with a gain or transfer function that is simply a constant over the range for which the system is linear.

The linear range is quite small for a standard domestic shower valve, and often does not exceed 1 rad of angle, while the valve handle can often be turned dozens of radians. However, a linear model of the valve is usually used (as shown in Fig. 34.1) even though the valve is distinctly nonlinear, because the linear model is amenable to a linear mathematical analysis and linear differential equations are readily solved in closed form. The Laplace transform transfer function is the most common tool used by the control engineer to solve linear differential equations. It cannot be used in general to solve nonlinear differential equations. Nonlinear differential equations usually cannot be solved at all in closed form. Thus before the age of the computer, very heavy emphasis was placed on obtaining linearized models of systems, even if the system were nonlinear. In fact, all physical systems become nonlinear as their respective limits of torque, acceleration, temperature, voltage, or other

TABLE 34.3. EXISTING AREAS OF CONTROL

Type of Control	Examples
Regulators,[a] automatic regulators, set-point controllers	Temperature, pressure, humidity, liquid level, volume flow rate, specific gravity, velocity,[b] acceleration,[b] position,[b] voltage, current, frequency, phase, light level, sound level, magnetic flux density, torque, force
Servos,[c] servomechanisms, pursuit control	Generalized control system operates on same variables as does regulator system but both input and output are in general time-varying functions. Typical control variables are same as those listed for regulators
Numerical control,[d] machine tool control	Turret lathe; jig borer; multispindle drill press; vertical lathe; three-axis milling machine; spot welder; two-axis metal flame cutter; two-axis laser cloth cutter, metal cutter, wood carver; shaper; tool or bit changers; transfer lines, pallet control
Direct digital control[e]	Automated assembly line, chemical process control, electric power plant control, traffic control systems
Human operator control[f]	Aircraft pilot flying "on instruments" or "flying blind"; operator guiding any vehicle with respect to a visual reference of command signal, e.g., automobile driver keeping car centered in one lane of twisting country road, or submarine helmsman keeping heading constant by watching heading display

[a] The input to a regulator-type feedback control system is a constant, the set point. The purpose of a regulator is to keep the controlled variable constant.
[b] Linear or angular units.
[c] The input to a servomechanism or generalized control system is a variable. The controlled variable should linearly follow the input variable, except for a change in coordinates or units. Thus the input may be an angle or a voltage while the output may be a velocity or a position.
[d] Numerical control refers to automatically controlled machines that are self-sufficient with their own source of control signals, either via a punch-paper tape, internal electronic memory, or other stored information.
[e] Direct digital control refers to a distributed system containing many machines or components that respond to and communicate with a central digital controller, usually a mainframe computer.
[f] Human operator control refers to a closed-loop control system in which one or more elements consist of a human operator.

TABLE 34.4. EVOLVING AREAS OF CONTROL

Area	Examples
Economic control	International monetary control, regulated markets, input/output matrices, world dynamics, federal monetary control
Physiological control	Pacemakers, programmable hormone injectors, insulin injectors, myoelectric artificial limb control, direct or induced neural stimulation or reception/computer mind interface
Biologic control	Predator/pest control, ecological cyclic control, land and wildlife management
Multidimensional or large-scale system control	Country-wide power distribution networks, worldwide communication networks, space system control, air traffic control
Human operator	Helicopter pilot eyesight aiming system, fingertip space vehicle control, myoelectric no-movement control, voice-operated controllers
Fuzzy, uncertain probabilistic control	Radar guidance, weapon guidance and control, weather control, stock market control, population control, noise or perturbation minimization, war games, sonar guidance, infrared guidance, navigation and correlation control
Robotics	Automated manufacturing and material-loading device using servomechanisms, microprocessors, and sensor devices

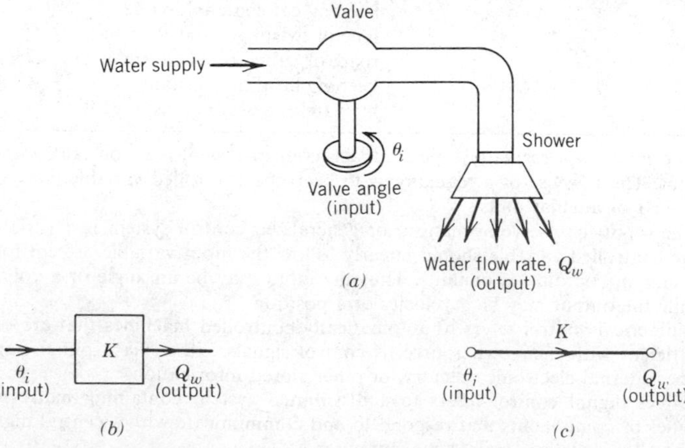

Fig. 34.1 Open-loop control system; water faucet flow-rate: (*a*) Sketch of the physical system, (*b*) block diagram, (*c*) flow graph. Approximate linear model equation: $Q_w = K \cdot \Theta_i$, where Q_w = system output = water flow rate, gal/min; Θ_i = system input = water valve angle, rad; and K = constant of proportionality, gain, or transfer function.

physical parameter are approached. This important fact should be kept in mind. The real world is nonlinear.

All linear models of physical systems are approximations. Any model analyzed with the Laplace transform transfer function approach is an approximation. As limits are approached, these models fail.

The distinguishing characteristic of a closed-loop control system is a comparison between the desired value of some system variable and the actual physical value of that variable. The difference between these two values is termed the error. The error signal is then applied as the input to certain system components configured so as to drive the error signal toward zero. As the error is reduced, the desired value of the chosen system variable is approached. Most often the variable whose value is desired is the system output. Figure 34.2 illustrates a closed-loop system used to heat a home.

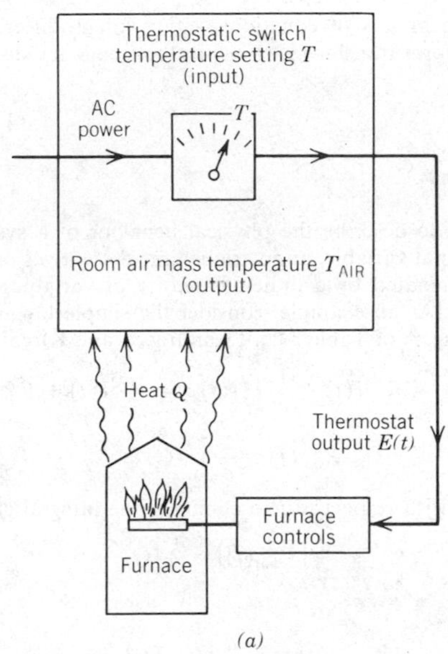

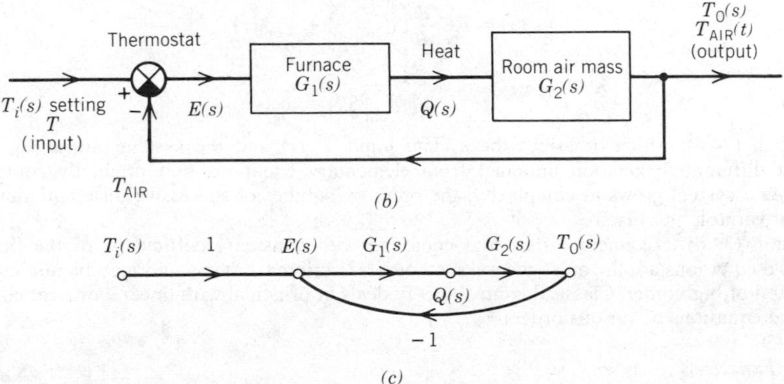

Fig. 34.2 Closed-loop control system; home heating system: (*a*) Sketch of physical system, (*b*) block diagram, (*c*) signal flow graph. $E(t)$ = switching signal related to temperature error, where $E(t)$ = "On" if $T_{AIR} < T - \delta$ and $E(t)$ = "Off" if $T_{AIR} > T - \alpha$; δ = differential temperature of thermostat turn-on, °F, typically 1 to 5°F; α = anticipator temperature of thermostat turn-off, °F, typically 0 to 5°F; T_{AIR} = room air temperature, °F $\Rightarrow T_0(s)$; T = thermostat temperature setting, °F $\Rightarrow Ti(s)$; Q = heat, BTU/hr; $T_0(s)$ = output temperature, Laplace transform of $T_{AIR}(t)$; $T_i(s)$ = input temperature, Laplace transform of $T(t)$.

Closed-loop systems can be designed to be very accurate, so that the effects of internal component variations and of random disturbances are reduced. This is the principal advantage of a closed-loop control system. The principal disadvantage of such a system is that it may be unstable, even though all its components are individually stable.

The systems just discussed have single inputs and single outputs. Examples of such systems are the thermostatic control of a home furnace, which is a closed-loop system, or control of a water faucet flow rate, which is an open-loop system.

Multiple input/output systems are discussed in Section 33.8.

34.3 CLASSICAL CONTROL SYSTEM REPRESENTATIONS†

Three methods are commonly used to describe a linear time-invariant automatic control system:

1. A mathematical equation or sets of equations; either integrodifferential equations in the time domain or Laplace transform transfer functions in the frequency domain.
2. Block diagrams.
3. Signal flow graphs.

Integrodifferential Equations

These equations can be used to describe the physical behavior of a system in terms of the system input variable, the system output variable, and various time derivatives or integrals of these variables. Most often, integrals are eliminated by a judicious choice of variables or by differentiation of an equation containing integrals. As an example, consider the simple low-pass or phase lag or quasiintegrating network shown in Fig. A of Table 34.5. From Fig. A and Kirchhoff's voltage law:

$$V_{in}(t) = R \cdot I(t) + \frac{1}{c}\int_0^t I(t)\,dt = R \cdot I(t) + V_{out}(t) \tag{34.1}$$

$$V_{out}(t) = \frac{1}{c}\int_0^t I(t)\,dt \tag{34.2}$$

Differentiating Eq. (34.2) with respect to t to eliminate the integral gives

$$\frac{d}{dt}[V_{out}(t)] = \frac{1}{c} I(t) \tag{34.3}$$

Rearranging Eq. (34.3)

$$I(t) = C\frac{d}{dt}[V_{out}(t)] \tag{34.4}$$

Inserting Eq. (34.4) in Eq. (34.1)

$$V_{in}(t) = RC\frac{d}{dt}[V_{out}(t)] + V_{out}(t) \tag{34.5}$$

In dot notation

$$V_{in} = RC\dot{V}_{out} + V_{out} \tag{34.6}$$

Equation (34.6), which describes the system input $V_{in}(t)$ and the system output $V_{out}(t)$, is a first-order differential equation obtained from elementary equations that originally contained an integral. As a system grows in complexity, the order or number of successive differentiations of the dependent variable increases.

Equation (34.6) is a linear differential equation, with constant coefficients, of the first order. Differential equations are the most general method of describing systems and may be nonlinear, time varying, and of high order. Classical control theory deals in principal with linear, constant coefficient, differential equations of various orders.

Transfer Functions

Transfer functions are simply the Laplace transforms of the integrodifferential equations relating the input and output variables of a system. Transfer functions are formulated as the ratio of the output

† Ogata,[6] Kuo,[7] and Shinners[8] thoroughly cover classical control system representations. Ogata's coverage of signal flow graphs is particularly lucid.

TABLE 34.5. DIFFERENTIAL EQUATIONS AND TRANSFER FUNCTIONS

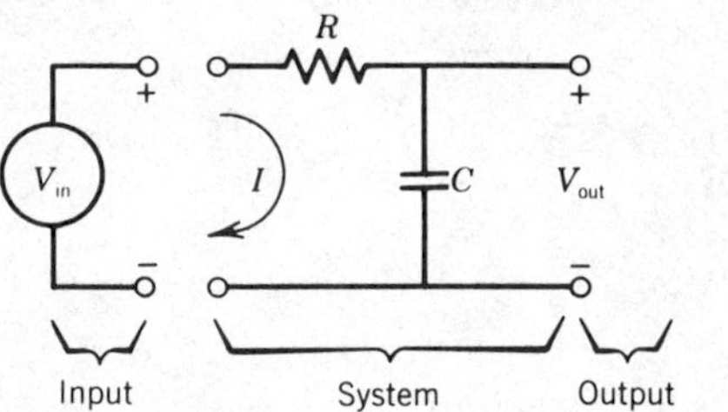

Fig. A Low-pass, phase-lag, or quasiintegrating network

R = resistance, a constant, Ω
C = capacitance, a constant, F
V_{in} = input voltage source with zero internal impedance, V
V_{out} = output voltage, delivered to infinite impedance load, V
s = Laplace transform variable, s^{-1}
t = time, s
I = input current, A

Type	Equation
Differential equation	$V_{in}(t) = RC\frac{d}{dt}[V_{out}(t)] + V_{out}(t)$
Laplace transform of differential equation	$V_{in}(s) = RC[s \cdot V_{out}(s)] + V_{out}(s)$ $V_{in}(s) = V_{out}(s) * (RCs + 1)$
Transfer function or gain $G(s)$	$\frac{V_{out}(s)}{V_{in}(s)} = \frac{1}{RCs+1} = G(s)$
Weighting function, $W(t)$, or inverse Laplace transform of transfer function (L^{-1} = inverse Laplace transform operator)	$W(t) = L^{-1}\left\{\frac{V_{out}(s)}{V_{in}(s)}\right\} = L^{-1}\{G(s)\}$ $W(t) = \frac{1}{RC}e^{-t/RC}$ for $t \geqslant 0$
Two systems in series, cascaded transfer functions	$\xrightarrow[V_{in}(s)]{} \boxed{G_1(s)} \rightarrow \boxed{G_2(s)} \xrightarrow[V_{out}(s)]{} = \xrightarrow[V_{in}(s)]{} \boxed{G_3(s)} \xrightarrow[V_{out}(s)]{}$ $G_1(s)G_2(s) = G_3(s)$
Two systems in series, cascaded weighting functions	$\xrightarrow[V_{in}(t)]{} \boxed{W_1(t)} \rightarrow \boxed{W_2(t)} \xrightarrow[V_{out}(t)]{} = \xrightarrow[V_{in}(t)]{} \boxed{W_3(t)} \xrightarrow[V_{out}(t)]{}$ where $W_3(t) = \int_0^t W_1(\tau)W_2(t-\tau)\,d\tau$ = convolution integral, τ = dummy variable or $W_3(t) = \int_0^t W_1(t-\tau)W_2(\tau)\,d\tau$ = equivalent convolution integral

TABLE 34.6. SHORT TABLE OF LAPLACE TRANSFORMS[a]

Laplace Transform $F(s)$	Time Function $f(t)$	Comment
$F(s+a)$	$e^{-at}f(t)$	Multiplicative exponential decay
$e^{-as}F(s)$	$f(t-a)U_s(t-a)$	$U_s(t-a)$ = delayed unit step; time delay
$aF(as)$	$f(t/a)$	Time scale
$sF(s) - \lim_{t\to 0+} f(t)$	$(d/dt)f(t) = f'(t) = \dot{f}(t)$	Time derivative
$s[sF(s) - \lim_{t\to 0+} f(t)] - \lim_{t\to 0+} f'(t)$	$(d^2/dt^2)f(t) = f''(t) = \ddot{f}(t)$	Second derivative
$(1/s)F(s)$	$\int_0^t f(T)\,dT$	Integration in time domain, $f(t) = 0$, for $t < 0$[b]
$F_1(s)F_2(s)$	$\int_0^t f_1(t-T)f_2(T)\,dT = \int_0^t f_2(t-T)f_1(T)\,dT$	Convolution integral
$\lim_{s\to\infty} sF(s)$	$\lim_{t\to 0+} f(t)$	Initial value theorem
$\lim_{s\to 0} sF(s)$ [$F(s)$ has no poles on $J\omega$ axis or in right half plane]	$\lim_{t\to\infty} f(t)$	Final value theorem [if no limit exists, $sF(s)$ may give finite result, which is wrong]
1	$\delta(t)$	Dirac delta function, $\delta(0)\to\infty$ $\delta(t) = 0,\ t \neq 0$
$1/s$	$U_s(t)$	Unit step function
e^{-as}	$\delta(t-a)$	Delayed Dirac delta function
K/s	K	Constant
$1/(s+a)$	e^{-at}	Exponential decay
$1/s^2$	t	Unit ramp

$1/s^3$	$t^2/2$	Unit parabola (note the 1/2)
$1/s^n$	$t^{n-1}/(n-1)!$	Powers of t
$\omega/(s^2+\omega^2)$	$\sin(\omega t)$	Pure sinusoid
$s/(s^2+\omega^2)$	$\cos(\omega t)$	Pure cosinusoid
$(\omega_n^2)/(s^2+2\zeta\omega_n s+\omega_n^2)$ simple normalized quadratic	$\left(\omega_n/\sqrt{1-\zeta^2}\right)\epsilon^{-\zeta\omega_n t}\sin\left(\omega_n\sqrt{1-\zeta^2}\,t\right)$	Damped sinusoid
Factored quadratic, general form of preceding quadratic $(A+jB)^c/(s+\alpha+j\omega)+(A-jB)/(s+\alpha-j\omega)$ or $(M\angle\theta)/(s+\alpha+j\omega)$ + conjugate (θ = angle of $+j\omega$ term) 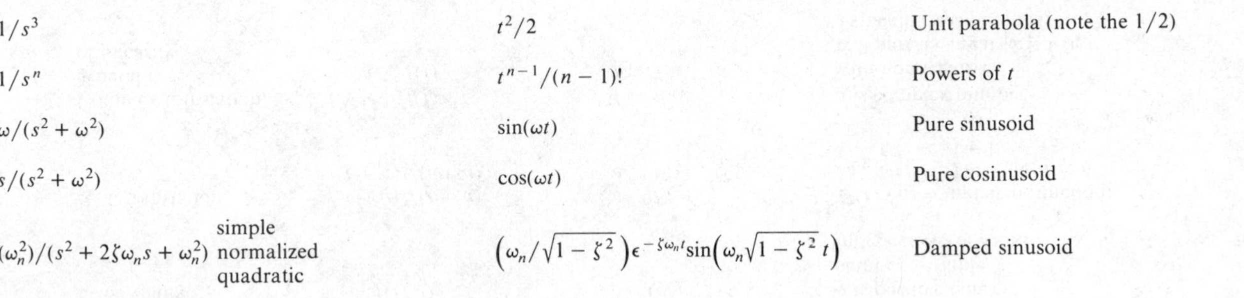 $M=\sqrt{A^2+B^2}$ = magnitude $\theta° = \tan^{-1}(B/A)$	$2Me^{-\alpha t}\sin(\omega t+90°-\theta°)$ Note: $\omega=\omega_n\sqrt{1-\zeta^2}$ $\alpha=\zeta\omega_n$ $\omega_n^2=\alpha^2+\omega^2$ $\zeta=\cos\phi \doteq$ damping factor; this follows from equating $s^2+2\zeta\omega_n s+\omega_n^2=(s+\alpha)^2+\omega^2$	Phase shifted damped sinusoid $j\omega$ axis, ω, s plane, ω_n, ϕ, σ axis, $-\alpha$, $-\zeta\omega_n$
Last two entries are related, with partial fraction expansion yielding $A=0,\ B=\omega_n^2/2\omega,\ M=\omega_n^2/2\omega$ $\theta=90°,\ 2M=\omega_n^2/\omega=\omega_n^2/\omega_n\sqrt{1-\zeta^2}$ or $2M=\omega_n/\sqrt{1-\zeta^2}$, $\alpha=\zeta\omega_n$, $\omega=\omega_n\sqrt{1-\zeta^2}$ (The simple normalized quadratic is a special case of the factored quadratic)	$2Me^{-\alpha t}\cos(\omega t-\theta°)$ $-2Me^{-\alpha t}\cos(\omega t\pm 180°-\theta°)$ $-2Me^{-\alpha t}\sin(\omega t-90°-\theta°)$	Equivalent forms of phase-shifted damped sinusoid Be sure to convert degrees to radians when evaluating sinusoids

[a] After 42 years, the Laplace transform table of choice is still that of Gardner and Barnes.[9]
[b] See footnote of Table 34.10, when $f(t)$ exists for negative time.
[c] Table 34.14 gives a step-by-step evaluation of complex residues $A+jB$ in order to find the negatively damped phase-shifted sinusoid $C_4(t)$ of Table 34.13.

TABLE 34.7. SELECTED PROPERTIES OF LAPLACE TRANSFORMS, ANNOTATED

Title	$f(t)$	$F(s)$	Comments
Linearity	$af(t)$ $af_1(t) \pm bf_2(t)$	$aF(s)$ $aF_1(s) \pm bF_2(s)$	a and b are constants or variables independent of t and s
Scale change	$f(t/a)$	$aF(as)$	a is positive constant or variable independent of t and s
Real translation	$f(t-a)U(t-a)$ $f(t+a)U(t+a)$	$e^{-as}F(s)$ $e^{+as}F(s)$	$U(t-a)$ = unit step function; e.g., $U(t-a)=0, \quad t<a$ $U(t-a)=1 \quad t>a$
Complex translation, general case of linearity	$e^{-at}f(t)$ $e^{+at}f(t)$	$F(s+a)$ $F(s-a)$	a is complex number with nonnegative real part, i.e., real $\{a\} > 0$; Typically, a is pure real

Real nth-order differentiation	$(d^n/dt^n)f(t)$		$s^nF(s) - \sum_{k=1}^{n} s^{n-k} \frac{d^{k-1}}{dt^{k-1}} f(t)\|_{t=0}$	
Complex differentiation	$-tf(t)$		$(d/ds)F(s)$	
Differentiation with respect to second independent variable	$(\partial/\partial a)f(t,a)$		$(\partial/\partial a)F(s,a)$	
Complex integration	$(1/t)f(t)$		$\int_s^\infty F(s)\,ds$	
Integration with respect to second independent variable	$\int_{a_1}^{a_2} f(t,a)\,da$		$\int_{a_1}^{a_2} F(s,a)\,da$	a is second variable independent of t and s
Commutativity of Laplace transforms over real and imaginary parts of $f(t)$ or $F(s)$	Real $\{f(t)\}$ Imag $\{f(t)\}$	$\Leftrightarrow$ $\Leftrightarrow$	Real $\{F(s)\}$ Imag $\{F(s)\}$	

divided by the input, so that they represent a gain function. In the calculation of a transfer function, all initial conditions are set to zero. This ensures that the system output is equal to zero when the initial value of the input is zero. The principal reason for using transfer functions in the Laplace transform frequency domain is that this permits the algebraic combination of system elements (e.g., two systems connected in series can be replaced by a new single system whose transfer function is simply the product of the two component transfer functions). This is a spectacular simplification compared with the manipulations required to produce the composite differential equation resulting from two cascaded systems described by differential equations. The principal drawback to transfer functions is that they are limited to linear differential equations (or linear systems) by the mathematical constraints of the Laplace transform.

The inverse Laplace transform of a transfer function is a time function, often called a weighting function, and is given by the convolution integral of the component weighting functions. This is summarized in Table 34.5, which also lists the various differential equation, Laplace transform, and transfer function methods for the representation of systems. Laplace transforms of common time functions are presented in Table 34.6, and certain properties of Laplace transforms are given in Table 34.7.

Block Diagrams and Signal Flow Graphs

These are graphic representations of interconnected transfer functions. Each block of a block diagram contains a transfer function. Information or signals travel in one direction only through a block. The direction of signal travel is indicated by arrows. It is assumed that blocks do not interact, and that cascaded or paralleled blocks do not load down or affect the transfer function of other blocks. If blocks do interact or load down each other, they cannot be combined by the laws of block diagram simplification.

In the case of interaction, a new transfer function must be derived from the basic differential equations describing the interacting systems. An example of two interacting transfer functions is given in Fig. 34.3.

Basic block diagram relationships are shown in Table 34.8, and block diagram simplification is shown in Table 34.9. All of the block diagrams shown have a single output. They contain a transfer function that is the ratio of the block output divided by the block input (in the Laplace transform domain). In practice each block corresponds to a physically identifiable component of a system. The signals that enter and leave the block are sometimes not physically available (e.g., observable), but are internal to a physical component that has been modeled by two or more blocks.

Table 34.10 defines these signals along with other common block diagram terms and symbols. The Laplace transforms given in Tables 34.6 and 34.7 are used in conjunction with block diagram analysis. Table 34.11 presents six common rules used for block diagram or Laplace transform arithmetic, in both scalar and matrix forms. Finally, Table 34.12 presents the rules for signal flow graph simplification and gives an example of evaluation of Mason's gain formula for the transfer function of a signal flow graph. Block diagrams and signal flow graphs are essentially interchangeable representations, with a block corresponding to an arrow or directed branch of a signal flow graph, as illustrated in Table 34.8.

34.4 ANALYSIS AND COMPENSATION OF CLASSICAL CONTROL SYSTEMS

The responses of a control system vary in accordance with the mathematical form of the system model and the particular input signal applied. With use of the Laplace transform approach, the step, ramp, and acceleration responses are calculated and graphically presented here for various mathematical control system models described as type 0, type 1, and type 2 systems.

Control system errors are then described in terms of the classical error coefficients, derived via the Laplace transform final value theorem, as are the classical methods employed for determining control system stability, including the Routh–Hurwitz test, the Nyquist plot, the Bode plot, and the root locus plot. Finally, system compensation is covered.

Order and Types of Control Systems, and Their Responses

Consider the control system shown in Fig. 34.4. Its closed-loop system transfer function $C(s)/R(s)$ is given by the following:

$$\frac{C(s)}{R(s)} = \frac{G(s)}{1 + G(s)H(s)} \tag{34.7}$$

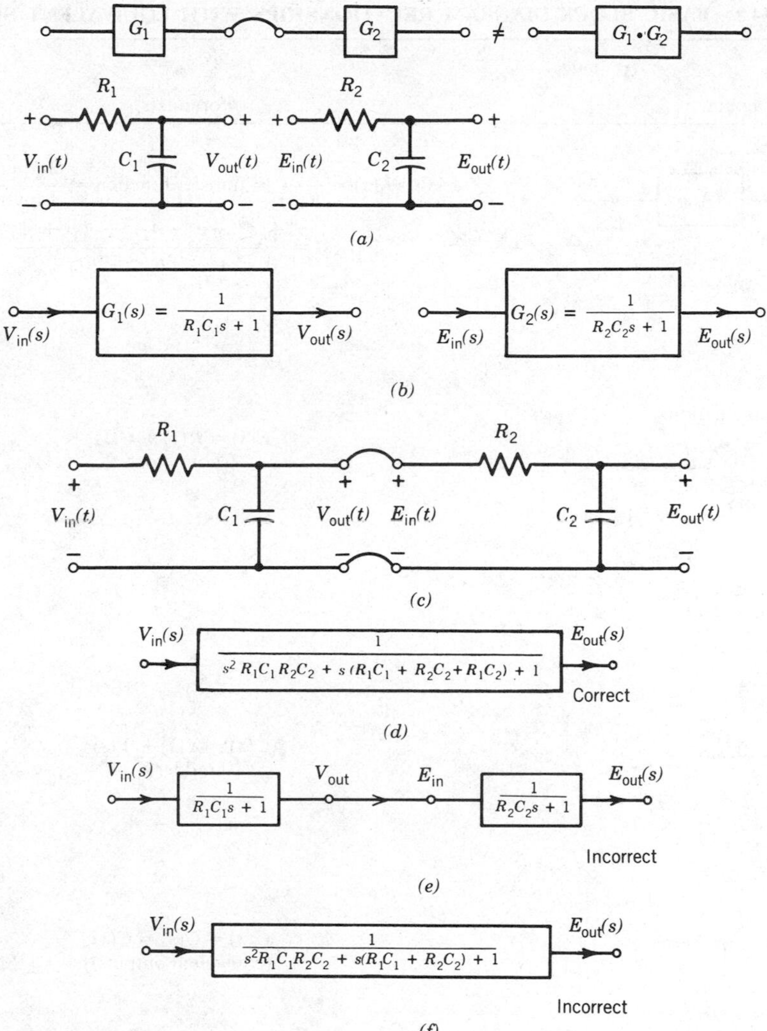

Fig. 34.3 Example of two interacting transfer functions that cannot be simply multiplied out:

(*a*) Two phase-lag networks. $V_{in}(t)$ = input of network 1, $E_{in}(t)$ = input of network 2, $V_{out}(t)$ = output of network 1, $E_{out}(t)$ = output of network 2.

(*b*) Transfer functions of two phase-lag networks, derived for zero source impedance and infinite load impedance.

(*c*) Cascaded system: $V_{out}(t)$ is connected to $E_{in}(t)$. Network 2 loads down network 1 with a noninfinite load impedance; hence the transfer functions are changed and connecting the blocks is not allowed. A new transfer function must be derived.

(*d*) Correct transfer function of the network shown in (*c*), taking network interaction or loading into account.

Compare the result here with the incorrect cascade obtained if the two original transfer functions of *b* were multiplied together.

(*e*) Incorrect cascade of interacting transfer functions [multiplied out in (*f*)]. Loading effect invalidates the individual transfer functions.

(*f*) Incorrect cascade of interacting block diagrams, multiplied out. Note the missing term R_1C_2, which corresponds to the interaction of the two networks.

TABLE 34.8. BASIC BLOCK DIAGRAM RELATIONSHIPS WITH EQUIVALENT SIGNAL

Block Diagram	Formula(s)

Basic

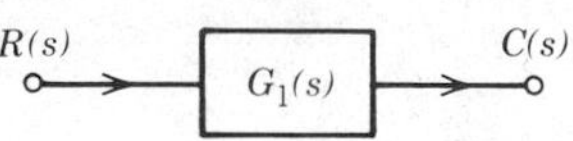

$$G_1(s) = \frac{C(s)}{R(s)} = \text{transfer function} = \frac{\text{Output(s)}}{\text{Input(s)}}$$

$$G_1(s) = \frac{A_m s^m + \cdots A_2 s^2 + A_1 s + A_0}{B_n s^n + \cdots B_2 s^2 + B_1 s + B_0}$$

Signal combinations

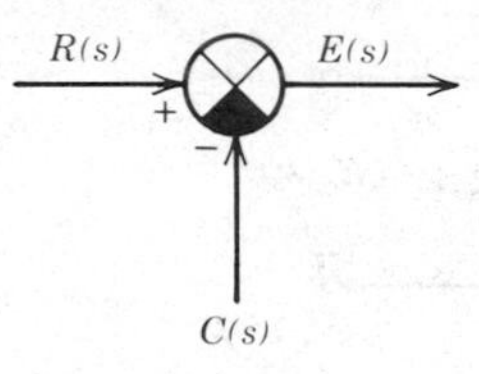

$$E(s) = R(s) - C(s)$$

(subtractor)

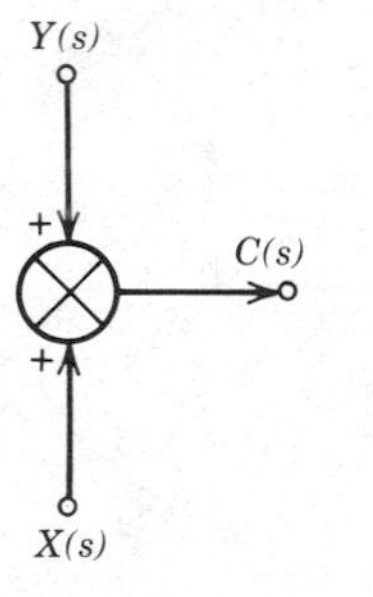

$$C(s) = X(s) + Y(s)$$

(adder)

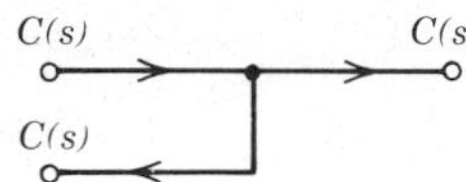

$$C(s) = C(s) = C(s)$$

(equivalent outputs)

Basic negative feedback loop, reduced to $G_1(s)$

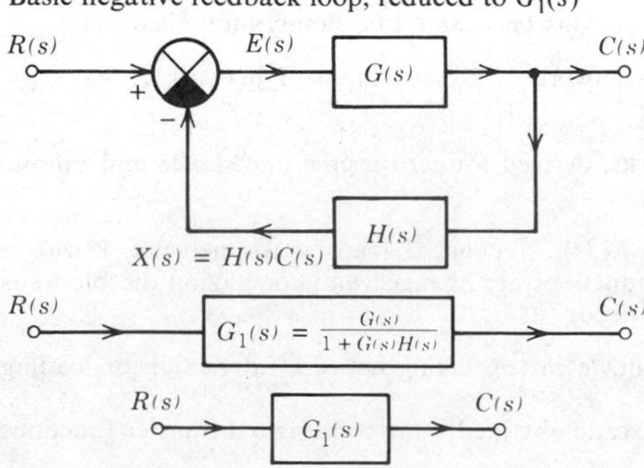

$$E(s) = R(s) - X(s)$$

$$X(s) = H(s)C(s)$$

$$E(s) = R(s) - H(s)C(s)$$

$$C(s) = G(s)E(s) \text{ or } E(s) = \frac{C(s)}{G(s)}$$

$$\frac{C(s)}{G(s)} = R(s) - H(s)C(s)$$

$$C(s) = G(s)R(s) - G(s)H(s)C(s)$$

$$C(s) + G(s)H(s)C(s) = G(s)R(s)$$

$$C(s)[1 + G(s)H(s)] = G(s)R(s)$$

$$\frac{C(s)}{R(s)} = \frac{G(s)}{1 + G(s)H(s)} = G_1(s)$$

FLOW GRAPHS (SCALAR QUANTITIES)

Equivalent Signal Flow Graph	Values
$R(s) \xrightarrow{G_1(s)} C(s)$	$A, B =$ constants $s =$ Laplace transform variable, $1/s$ $m =$ order of numerator polynomial $n =$ order of denominator polynomial For bounded response without infinite impulses: $n > m$

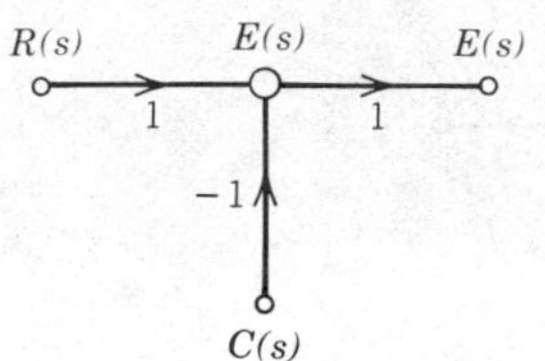

$E(s) = R(s) - C(s)$
(subtractor)

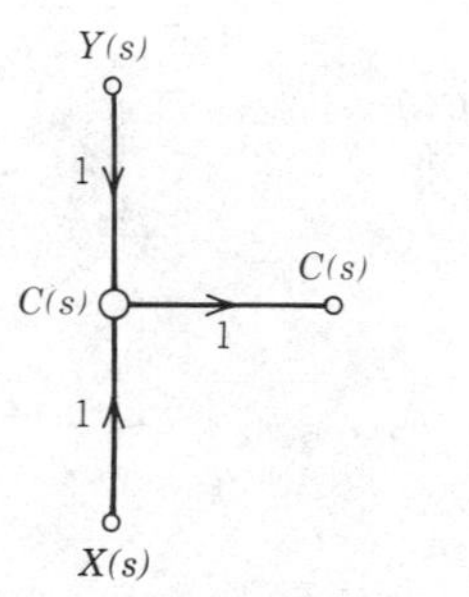

$C(s) = X(s) + Y(s)$
(adder)

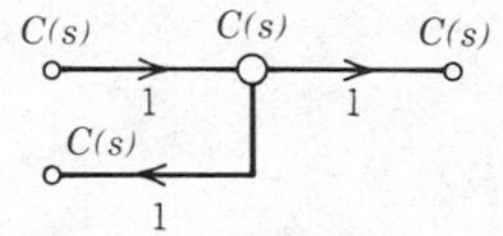

$C(s) = C(s) = C(s)$
(equivalent outputs)

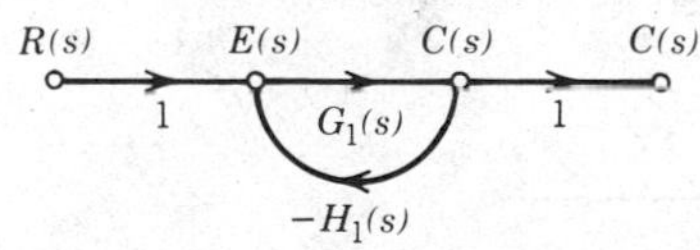

$R(s) =$ reference input signal
$C(s) =$ controlled output signal
$E(s) =$ error signal
$X(s) =$ feedback signal (only in block diagram)
$G(s) =$ forward gain or plant transfer function, open loop

$$R(s) \xrightarrow{G_1(s) = \dfrac{G(s)}{1 + G(s)H(s)}} C(s)$$

$$R(s) \xrightarrow{G_1(s)} C(s)$$

$H(s) =$ feedback gain or transfer function
$s =$ Laplace transform variable
$G(s)H(s) =$ loop gain
$1 + G(s)H(s) =$ characteristic equation
$G_1(s) =$ closed-loop gain or transfer function

TABLE 34.9. BLOCK DIAGRAM SIMPLIFICATION (SCALAR QUANTITIES)

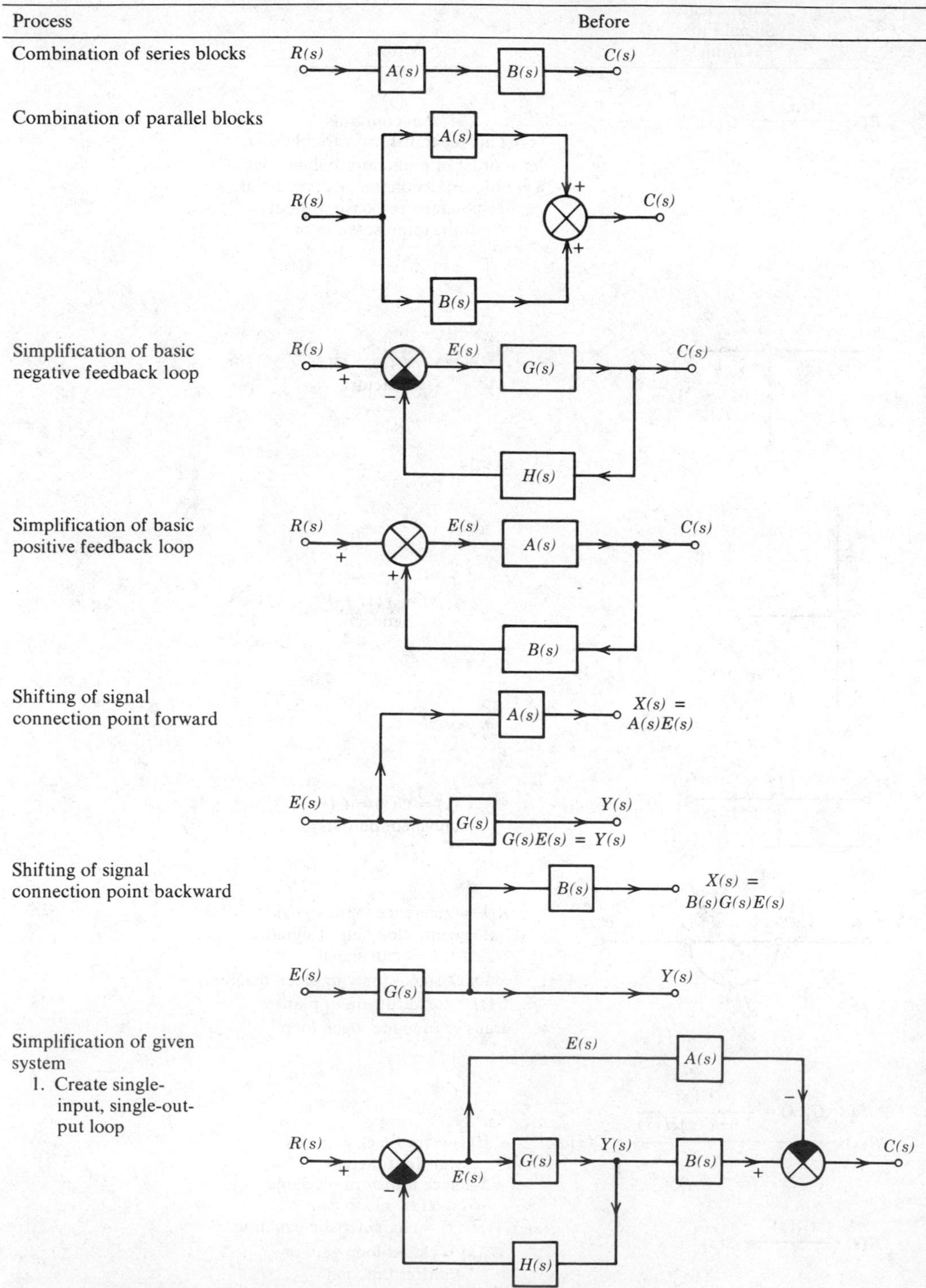

After	Relationship

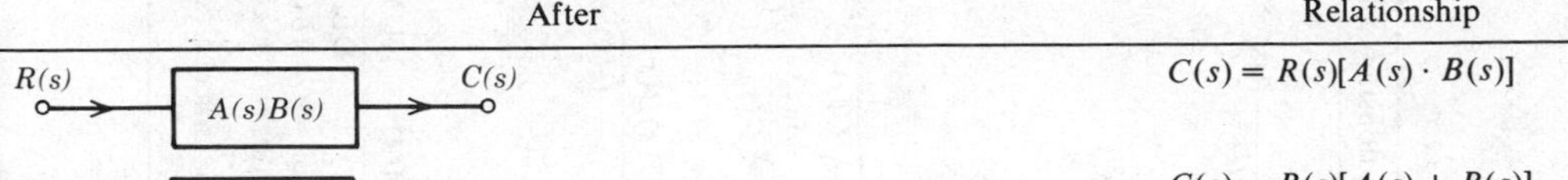

$C(s) = R(s)[A(s) \cdot B(s)]$

$C(s) = R(s)[A(s) + B(s)]$

R(s) $\frac{G(s)}{1+G(s)H(s)}$ C(s)

$$C(s) = R(s)\left[\frac{G(s)}{1 + G(s)H(s)}\right]$$

R(s) $\frac{A(s)}{1-A(s)B(s)}$ C(s)

$$C(s) = R(s)\left[\frac{A(s)}{1 - A(s)B(s)}\right]$$

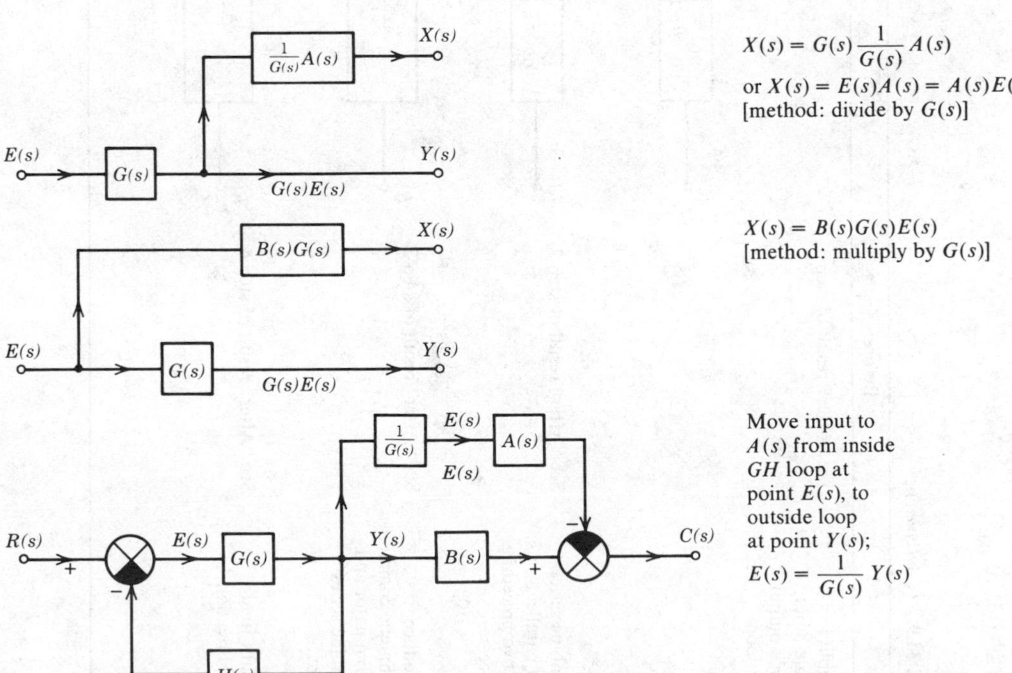

$X(s) = G(s)\frac{1}{G(s)}A(s)$
or $X(s) = E(s)A(s) = A(s)E(s)$
[method: divide by $G(s)$]

$X(s) = B(s)G(s)E(s)$
[method: multiply by $G(s)$]

Move input to $A(s)$ from inside GH loop at point $E(s)$, to outside loop at point $Y(s)$;
$E(s) = \frac{1}{G(s)}Y(s)$

TABLE 34.9. (*Continued*)

Process	Before	After	Relationship
2. Simplify single-input, single-output GH loop	See "After" result of step 1	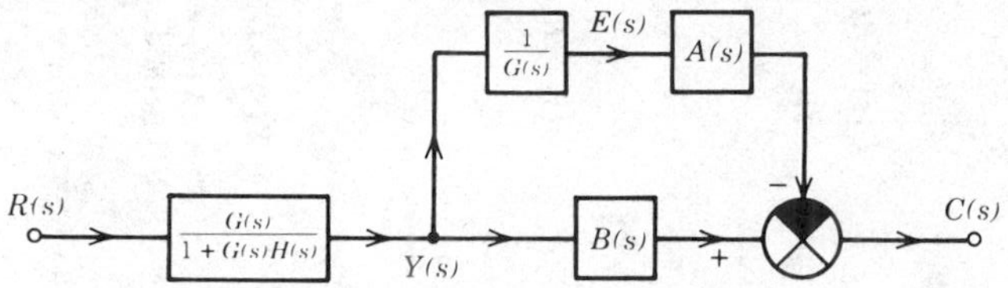 $R(s)$ → $\frac{G(s)}{1+G(s)H(s)}$ → $Y(s)$; $Y(s)$ → $\frac{1}{G(s)}$ → $E(s)$ → $A(s)$ (−); $Y(s)$ → $B(s)$ (+); → $C(s)$	$Y(s) = \frac{G(s)}{1 + G(s)H(s)} R(s)$
3. Combine series and parallel terms; see two preceding processes	See "After" result of step 2	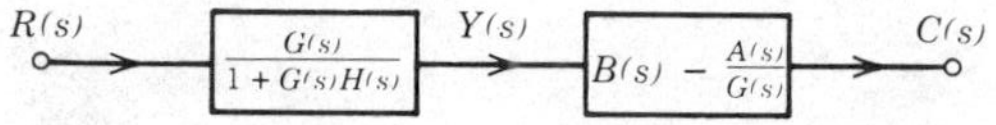 $R(s)$ → $\frac{G(s)}{1+G(s)H(s)}$ → $Y(s)$ → $B(s) - \frac{A(s)}{G(s)}$ → $C(s)$	$C(s) = B(s)Y(s) - \frac{1}{G(s)} A(s)Y(s)$ or $C(s) = \left[B(s) - \frac{A(s)}{G(s)} \right] Y(s)$
4. Introduce arithmetic common denominator and combine series terms	See "After" result of step 3	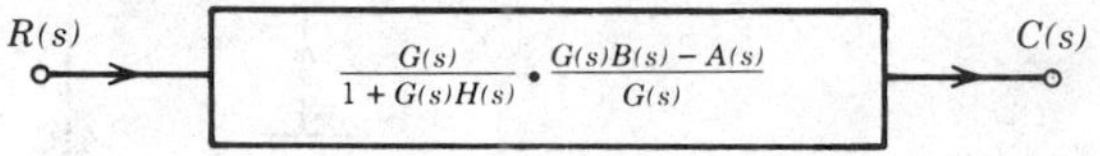 $R(s)$ → $\frac{G(s)}{1+G(s)H(s)} \cdot \frac{G(s)B(s)-A(s)}{G(s)}$ → $C(s)$	$B(s) - \frac{A(s)}{G(s)} = \frac{G(s)B(s) - A(s)}{G(s)}$
5. Result: final simplification	See "After" result of step 4	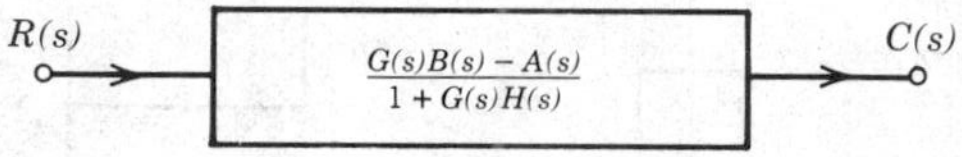 $R(s)$ → $\frac{G(s)B(s)-A(s)}{1+G(s)H(s)}$ → $C(s)$	$\frac{G(s)}{\text{DEN}(s)} \cdot \frac{\text{NUM}(s)}{G(s)} = \frac{\text{NUM}(s)}{\text{DEN}(s)}$ NUM = numerator = $G(s)B(s) - A(s)$ DEN = denominator = $1 + G(s)H(s)$ or $\frac{G(s)}{G(s)} = \frac{1}{1} = 1$, cancel $G(s)$

TABLE 34.10. BLOCK DIAGRAM SIMPLIFICATION AND DEFINITION OF TERMS

Symbol	Arbitrary Example	Definition
$R(s)$	$R_a(s) = \dfrac{\omega_0}{s^2 + \omega_0^2}$	Input to control system in Laplace transform notation. Often called reference input
$C(s)$	$C_a(s) = \dfrac{(s + \omega_0)}{s^2(s^2 + \omega_1^2)(s + \omega_2)}$	Output from control system in Laplace transform notation. Often called controlled output
$G(s)$	$G_a(s) = \dfrac{1}{s(s + \omega_0)}$	Gain of system element or block, generally in forward direction from input to output. Often called plant gain or plant transfer function, or forward gain; expressed in Laplace domain. ω_0, ω_1, ω_2 are angular frequency constants (rad/s)
$H(s)$	$H_a(s) = \dfrac{1}{s}$ [for unity feedback $H(s) = 1$]	Gain of system element or block, generally in reverse or feedback direction, from output to input. Often called feedback gain or feedback transfer function. Expressed in Laplace domain
$E(s)$	$E_a(s) = R_a(s) - C_a(s)$	Error of unity feedback control system, in Laplace domain. Difference between desired reference input $R(s)$ and actual controlled output $C(s)$. When $H_a(s)$ is not unity, input to $G(s)$ is still often termed $E(s)$, but $E_a(s)$ is now $R_a(s) - H_a(s)C_a(s)$, which is no longer difference (or error) between R_a and C_a. For nonunity feedback, $E(s)$ is related to error but is not identical to it
$A(s)$	$A_a(s) = s^2 + \omega_0^2$	Gain of system element, or forward amplification, in Laplace domain. Exact analog of $G(s)$ using different symbol, common to electronics analysis. ω_0 = frequency (rad/s)
$B(s)$	$B_a(s) = \dfrac{s + 1/T_1}{s + 1/T_2}$	Gain of transfer function of system element or block, in Laplace domain. T_1, T_2 = time constants (s)
$X(s)$ $Y(s)$		Variable, or signals, or intermediate outputs, or internal inputs existing within system; expressed in Laplace domain
$r(t)$		Inverse Laplace transform of $R(s)$. Input to control system in time domain.
$c(t)$		Inverse Laplace transform of $C(s)$. Output from control system in time domain.

TABLE 34.10. (*Continued*)

Symbol	Arbitrary Example	Definition
$f(t)$	$\frac{1}{2\pi j}\int_{C_0-j\infty}^{C_0+j\infty}F(s)e^{st}\,ds$ $=\sum$ residues $F(s)e^{st}\vert_{\text{poles of } F(s)}$	Inverse Laplace transform of $F(s)$ in region for which convergence exists. C_0 = constant, often equal to zero
$F(s)$	$F(s)=\int_0^{\infty}f(t)e^{-st}\,dt$	One-sided Laplace transform of $f(t)$ in region for which integral converges
s		Laplace operator. s = Laplace transform variable s as operator corresponds to differentiation in time domain, since $sF(s)-f(t)\vert_{t=0}$ is Laplace transform of derivative of $f(t)$. $1/s$ as operator corresponds to integration in time domain, since $F(s)/s$ is Laplace transform of $\int_0^t f(t)\,dt$[a]

[a] When $f(t)$ has a value for t less than zero, the expression for the Laplace transform of $\int f(t)$ becomes

$$\frac{F(s)}{s}+\frac{\int_{-\infty}^{0}f(t)\,dt}{s}=\text{Laplace transform of }\int f(t)\,dt$$

If $f(t)$ contains an impulse at $t=0$, there are two different results, one for $t=0-$ and one for $t=0+$.

TABLE 34.11. BASIC RULES OF BLOCK DIAGRAM OR LAPLACE TRANSFORM ARITHMETIC

Rules for Scalar Quantities	Rules for Matrix Quantities
$A(s)B(s)=B(s)A(s)$	$A(s)B(s)\neq B(s)A(s)$
$A(s)+B(s)=B(s)+A(s)$	$A(s)+B(s)=B(s)+A(s)$
$A(s)[X(s)+Y(s)]=A(s)X(s)+A(s)Y(s)$	$A(s)[X(s)+Y(s)]=A(s)X(s)+A(s)Y(s)$
$A(s)[X(s)+Y(s)]=X(s)A(s)+Y(s)A(s)$	$A(s)[X(s)+Y(s)]\neq X(s)A(s)+Y(s)A(s)$
$[X(s)+Y(s)]A(s)=X(s)A(s)+Y(s)A(s)$	$[X(s)+Y(s)]A(s)=X(s)A(s)+Y(s)A(s)$
$C(s)=R(s)\frac{G(s)}{1+G(s)H(s)}=\frac{G(s)}{1+G(s)H(s)}R(s)$[a]	$C(s)=[I+G(s)H(s)]^{-1}G(s)R(s)$ with terms in this order and this order only. I = identity matrix; $[\text{Matrix}]^{-1}$ = Inverse of matrix. Matrix must be square and have nonzero determinant

[a] Diagram for basic negative feedback loop in Table 34.8 illustrates this rule.

TABLE 34.12. SIGNAL FLOW GRAPH SIMPLIFICATION (SCALAR QUANTITIES)

Process	Before	After	Relationship
Combination of series branches	$R(s)$ $A(s)$ $B(s)$ $C(s)$	$R(s)$ $C(s)$ $A(s)\cdot B(s)$	$C(s) = [A(s)B(s)]R(s)$
Combination of parallel branches	$A(s)$ $R(s)$ $C(s)$ $B(s)$	$R(s)$ $C(s)$ $A(s) + B(s)$	$C(s) = [A(s) + B(s)]R(s)$
Simplification of basic negative feedback loop	$R(s)$ $E(s)$ $C(s)$ $C(s)$ 1 $G_1(s)$ 1 $-H_1(s)$	$R(s)$ $\frac{G_1(s)}{1+G_1(s)H_1(s)} = G(s)$ $C(s)$ 1 $E(s)$ $C(s)$ 1 $R(s)$ $G(s)$ $C(s)$	$C(s) = \{G_1(s)/[1 + G_1(s)H_1(s)]\}R(s)$ $= G(s)R(s)^a$
Mason's gain formula	$A(s)$ $R(s)$ $E(s)$ $Y(s)$ $C(s)$ $C(s)$ 1 $G_1(s)$ $G_2(s)$ 1 $-H_1(s)$ $-H_2(s)$	$R(s)$ $G(s)$ $C(s)$	$G(s) = \sum \frac{P_k\Delta_k}{\Delta} = \frac{P_1\Delta_1 + P_2\Delta_2}{\Delta}$ All k forward paths; For two forward paths
	Forward path P_1 A $R(s)$ $C(s)$ 1 1	$P_1 = A$	$G(s) = \frac{A + G_1G_2}{1 + G_1H_1 + G_1G_2H_2 + AH_2}$ (with s notation dropped)
	$R(s)$ Forward path P_2 $C(s)$ 1 G_1 G_2 1	$P_2 = G_1G_2$	

TABLE 34.12. *(Continued)*

Derivation of $G(s)$ immediately above, using Mason's gain formula:

Definition of Loops

L_i = all single loops = L_1, L_2, L_3

L_iL_j = all combinations of two loops having *no* common nodes, i.e., nontouching loops. Zero here.

$L_iL_jL_k$ = all combinations of three nontouching loops. Zero here.

Loop 1 = $\underline{\underline{L_1 = -G_1H_1}}$

Loop 2 = $\underline{\underline{L_2 = -G_1H_2H_2}}$

Loop 3 = $\underline{\underline{L_3 = -AH_2}}$

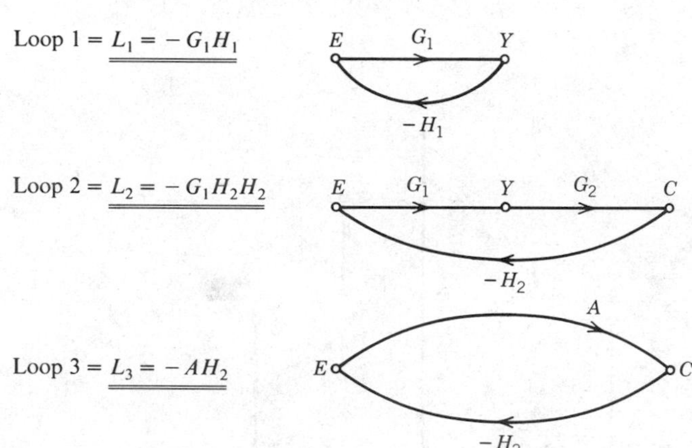

$L_1L_2 = L_2L_3 = L_1L_3 = 0$ since all these loops touch each other at node E.

$L_1L_2L_3 = 0$ since all these loops touch each other at node E.

Definition of Δ (= graph determinant), Δ_K (subdeterminant)

$$\Delta = 1 - \sum_i L_i + \sum_{ij} L_iL_j - \sum_{ijk} L_iL_jL_k \ldots$$

$$\Delta = 1 - \sum_i (L_1 + L_2 + L_3) + \sum_{ij}(0) - \sum_{ijk}(0) \ldots$$

$$\Delta = 1 - (-G_1H_1 - G_1G_2H_2 - AH_2) + 0 - 0 + \ldots$$

$$\underline{\underline{\Delta = 1 + G_1H_1 + G_1G_2H_2 + AH_2}} = \text{graph determinant}$$

$\Delta_k = \Delta$ with all terms containing any element or branch that touches path P_k completely removed. That is, loop gain term $G_1G_2H_2$ is completely removed if one or two of the three elements touch path P_k. For the example shown, for $K = 1$, $P_k = P_1 = A$, and

$$\Delta_1 = 1 + \cancel{G_1H_1} + \cancel{G_1G_2H_2} + \cancel{AH_2} \quad \text{or} \quad \underline{\underline{\Delta_1 = 1}}$$

Three terms removed, since all three touch path $P_1 = A$ at node E. Note that loop G_1H_1 is removed even though it does not contain any branch of $P_1 = 1 \cdot A \cdot 1$. One single touch at node E is sufficient. Similarly $\underline{\underline{\Delta_2 = 1}}$, $P_2 = G_1G_2$, for same reason, i.e., all loops touch path P_2 at node E.

$$P_1\Delta_1 = A \cdot 1 = \underline{\underline{A}}$$

$$P_2\Delta_2 = G_1G_21 = G_1G_2$$

---

[a]This simplification is generally not used, but rather is incorporated in Mason's gain formula.
[b]The directed gain of unity between $R(s)$ and $E(s)$ is required. Omitting it is a common source of error. The omission changes nontouching loops into loops that touch. Reference 6, K. Ogata, *Modern Control Engineering*, pp. 124–128.

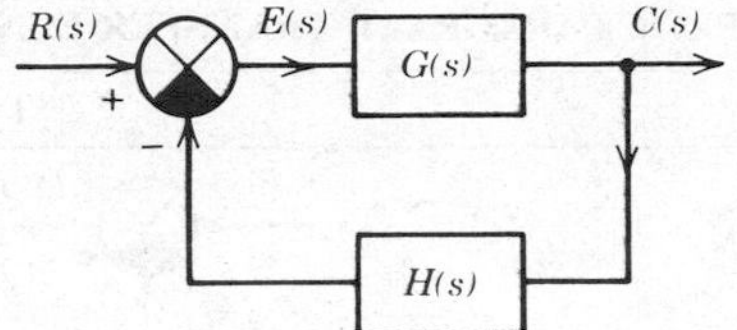

Fig. 34.4 General feedback control system.

The characteristic equation of the general feedback control system is obtained by setting the denominator equal to zero

$$1 + G(s)H(s) = 0 \tag{34.8}$$

The order of a control system refers to the order of the system's characteristic equation, given by Eq. (34.8). If the characteristic equation is multiplied out and expressed in polynomial form, the highest power of s appearing in the polynomial is the order of the control system. (First-, second-, third-, and fourth-order systems are illustrated later in Table 34.19.) The error is defined as $E(s) = R(s) - H(s)C(s)$ (see Fig. 34.4). In the case of unity feedback, $H(s) = 1$, and $E(s) = R(s) - C(s)$.

The *type* of a control system refers to the highest power of s that can be factored out of the denominator of $G(s)H(s)$. The general form of $G(s)H(s)$ is

$$G(s)H(s) = \frac{(s + Z_1)(s + Z_2)\cdots}{s^n(s + P_1)(s + P_2)\cdots} \tag{34.9}$$

where Z_i and P_i are constants that may be complex numbers. Z refers to zeros and P refers to the poles of $G(s)H(s)$.

The type of the control system described by Eq. (34.9) is numerically equal to n. Type 0 and type 1 systems are generally stable, and are often encountered in practice. Type 2 systems are difficult to stabilize, in general, especially for overall gains exceeding unity over a broad frequency range. However, type 2 systems are occasionally employed because they can track an acceleration input (with a finite error), and can produce a zero steady-state error for a velocity input (with a finite error) and a zero steady-state error for a position input. Type 0 systems can track a position input (with a finite error). Type 3 and higher systems are usually not encountered in practice, as they are extremely difficult to stabilize.

The type of a system has an intuitive or mathematical interpretation. A type n system integrates the error signal n times. For example, a type 1 system cannot have a finite steady-state position error since, in the steady state, the output is constant. If there were a position error, however small, the one integration of that error would produce a growing output ramp, which contradicts the requirement for a constant steady-state output. The only signal whose first integral is a constant is a zero signal. Hence the error of a one integration or a type 1 system must be zero in the steady state for a constant input. The responses of type 0, type 1, and type 2 systems to step, ramp, and acceleration (parabolic) inputs are shown in Table 34.13. Examination of this table reveals several important facts. First, the unit step response of a type 0 system always displays a steady-state error. In the table, the error is twice as large as the actual output. That is, the output signal is 1/3 unit and the error signal is 2/3 unit. (Calculation of control system error is discussed next.) A practical type 0 system would have a much higher forward gain than unity, which would reduce the error a corresponding amount.

The unit step response $C_3(t)$ of the type 1 system of Table 34.13 shows zero steady-state error, while the unit ramp response $C_5(t)$ of the type 1 system shows a constant steady-state error of 2 units. (These results are calculated later in Table 34.15 and in Table 34.20.) Type 1 systems are used very often in practice when either a zero steady-state error is necessary (as in a mechanical positioning system) or the ability to track a ramp (or velocity) function is necessary (as in a weapon-aiming system). The constant steady-state ramp error is compensated for by adding sufficient lead into the aiming system (just as a wild-fowl hunter leads the shotgun aiming point). In both of these aiming systems, the ramp function is a ramp of azimuth or angle, which continually increases as the target is tracked.

The type 2 system of Table 34.13 appears benign, with $G(s) = 1/s^2(s + 2)$. Yet its closed-loop response is unstable for all inputs. As previously stated, type 2 systems are difficult to stabilize, and the type 2 system illustrated is unstable. However, this is not to imply that all type 2 systems are unstable. One of the shortcuts to predicting system stability can be applied in this case. The characteristic equation of $G(s) = 1/s^2(s + 2)$, with $H(s) = 1$, is $s^3 + 2s^2 + 0s + 1$. The zero coefficient, or missing term in the characteristic equation, immediately indicates instability. This is corroborated by factoring $s^3 + 2s^2 + 1 = 0$, using an HP-41c calculator and the Mathematics Appli-

TABLE 34.13. BLOCK DIAGRAM, TRANSFER FUNCTION, AND RESPONSES

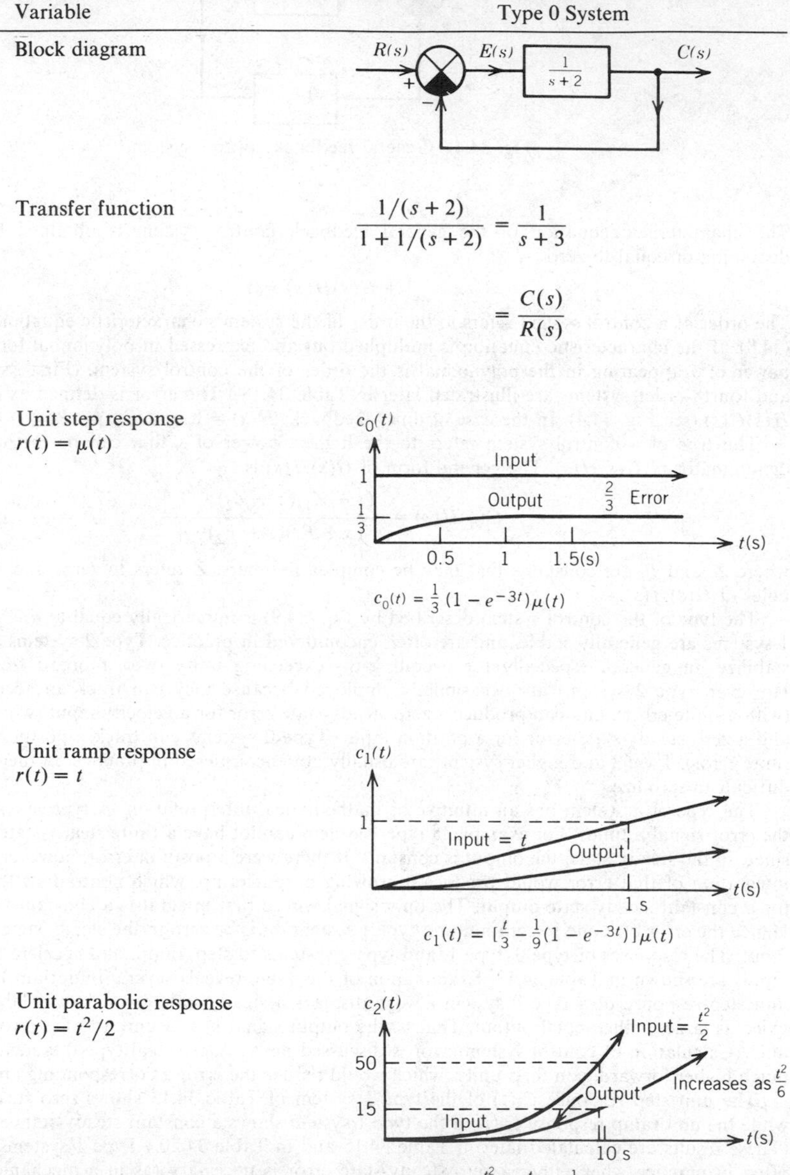

[a]Missing s^1 term, therefore unstable.

cation Pac's Polynomial Solutions Program. The factors are

$$(s + 2.205569)(s - 0.102785 + j0.665457)(s - 0.102785 - j0.665457) = 0$$

The two complex poles have positive real parts, namely $+0.102785$. This positive real part produces an exponential response envelope, which grows up toward infinity at the rate $e^{+0.102785t}$. Remember that if $(s - A)$ is a factor of a polynomial, then $+A$ is a root of the polynomial.

OF TYPE 0, 1, AND 2 SYSTEMS TO UNIT STEP, RAMP, AND PARABOLIC INPUTS

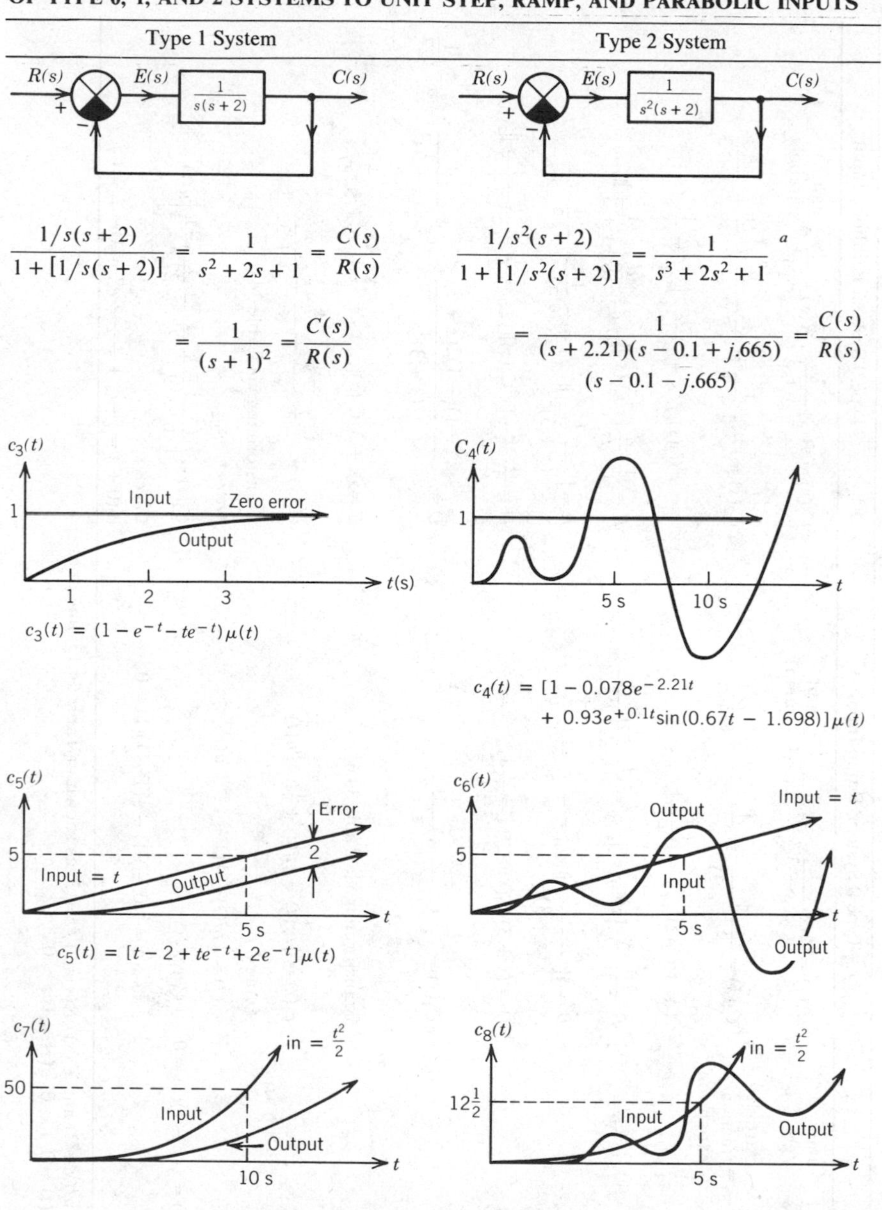

Table 34.14 illustrates the technique used to solve for the response of a type 0, type 1, and type 2 system using the Laplace transform partial fraction method. The table includes the output $c(t)$ solutions for the step, ramp, and parabolic responses of a type 0 system, $c_0(t)$, $c_1(t)$ and $c_2(t)$, respectively. Table 34.15 presents output solutions for a step and ramp response of the type 1 system, $c_3(t)$ and $c_4(t)$, respectively, and the unstable step response of the type 2 system, $c_5(t)$. All of these responses are plotted in Table 34.13.

TABLE 34.14. CALCULATION OF OUTPUT $c(t)$ SOLUTION TO UNIT STEP, RAMP, AND PARABOLIC INPUT OF TYPE 0 SYSTEM[a]

Type 0 System, Step Input	Type 0 System, Ramp Input	Type 0 System, Parabolic Input
$C_0(s) = R(s)\frac{1}{s+3}$, $R(s) = \frac{1}{s} =$ unit step	$C_1(s) = R(s)\frac{1}{s+3}$, $R(s) = \frac{1}{s^2} =$ unit ramp	$C_2(s) = R(s)\frac{1}{s+3}$, $R(s) = \frac{1}{s^3} =$ unit parabola $= \frac{t^2}{2}$
$C_0(s) = \frac{1}{s}\frac{1}{s+3} = \frac{A}{s} + \frac{B}{s+3} =$ partial fraction expansion	$C_1(s) = \frac{1}{s^2(s+3)} = \frac{A}{s^2} + \frac{B}{s} + \frac{D}{s+3}$	$C_2(s) = \frac{1}{s^3}\frac{1}{s+3} = \frac{A}{s^3} + \frac{B}{s^2} + \frac{D}{s} + \frac{E}{s+3}$
$A = sC_0(s)\vert_{s=0} = \frac{1}{s+3}\vert_{s=0} = \frac{1}{3}$	$A = s^2C_1(s)\vert_{s=0} = \frac{1}{s+3}\vert_{s=0} = \frac{1}{3}$	$A = s^3C_2(s)\vert_{s=0} = \frac{1}{s+3}\vert_{s=0} = \frac{1}{3}$
$B = (s+3)C_0(s)\vert_{s=-3} = \frac{1}{s}\vert_{s=-3} = -\frac{1}{3}$	$B = \frac{1}{1!}\frac{d}{ds}[s^2C_1(s)]\vert_{s=0} = \frac{1}{1!}\frac{d}{ds}[\frac{1}{s+3}]\vert_{s=0}$	$B = \frac{1}{1!}\frac{d}{ds}[s^3C_2(s)]\vert_{s=0} = \frac{d}{ds}[\frac{1}{s+3}]\vert_{s=0}$
$C_0(s) = \frac{\frac{1}{3}}{s} + \frac{-\frac{1}{3}}{s+3}$	$B = 1[\frac{-1}{(s+3)^2}]\vert_{s=0} = -\frac{1}{3^2} = -\frac{1}{9}$	$B = [\frac{-1}{(s+3)^2}]\vert_{s=0} = -\frac{1}{3^2} = -\frac{1}{9}$
Transforming via Table 34.6	$D = (s+3)C_1(s)\vert_{s=-3} = \frac{1}{s^2}\vert_{s=-3} = \frac{1}{9}$	$D = \frac{1}{2!}\frac{d^2}{ds^2}[s^3C_2(s)]\vert_{s=0} = \frac{1}{2}\frac{d^2}{ds^2}[\frac{1}{s+3}]\vert_{s=0}$
$c_0(t) = \frac{1}{3}u(t) - \frac{1}{3}e^{-3t}u(t)$[b]	$C_1(s) = \frac{\frac{1}{3}}{s^2} + \frac{-\frac{1}{9}}{s} + \frac{\frac{1}{9}}{s+3}$	$D = \frac{1}{2}\frac{(-1)(-1)2(s+3)}{(s+3)^4}\vert_{s=0} = \frac{1}{(s+3)^3}\vert_{s=0} = \frac{1}{27}$
Answer		
$c_0(t) = \frac{1}{3}(1 - e^{-3t})u(t)$	Transforming via Table 34.6	$E = (s+3)C_2(s)\vert_{s=-3} = \frac{1}{s^3}\vert_{s=-3} = \frac{-1}{27}$
For $t \gg 0$, $c_0(t) = \frac{1}{3}$	$c_1(t) = \frac{1}{3}tu(t) - \frac{1}{9}u(t) + \frac{1}{9}e^{-3t}u(t)$	$C_2(s) = \frac{\frac{1}{3}}{s^3} + \frac{-\frac{1}{9}}{s^2} + \frac{\frac{1}{27}}{s} + \frac{-\frac{1}{27}}{s+3}$
Check	*Answer*	Transforming via Table 34.6
$c_0(t)\vert_{t=0} = 0?$	$c_1(t) = \frac{1}{3}tu(t) - \frac{1}{9}(1 - e^{-3t})u(t)$	*Answer*
$c_0(0+) = \frac{1}{3}(1 - e^{0+})u(0+) = \frac{1}{3}(1-1)(1) = 0$	$\surd$ *Check*	$C_2(t) = \left[\frac{t^2}{6} - \frac{t}{9} + \frac{1}{27} - \frac{1}{27}e^{-3t}\right]u(t)$
	$c_1(0+) = \frac{1}{3}(0+)(1) - \frac{1}{9}(1-1)(1) = 0\surd$	*Check*
		$c_2(0+) = \frac{0}{6} - \frac{0}{9} + \frac{1}{27} - \frac{1}{27}e^0 = \frac{1}{27} - \frac{1}{27} = 0\surd$

[a] For all three cases, $G(s) = 1/(s+2)$ and $C(s)/R(s) = 1/s+3$, as shown in Table 34.13, line one.
[b] $u(t) =$ unit step function. $u(t) = 0$, $t < 0$. $u(t) = 1$, $t > 0$.

TABLE 34.15. CALCULATION OF OUTPUT $c(t)$ SOLUTION TO STEP AND RAMP INPUT OF TYPE 1 SYSTEM $G(s) = 1/s(s+2)$ AND $C(s)/R(s) = 1/(s+1)^2$ AND TO STEP INPUT OF TYPE 2 SYSTEM $G(s) = 1/s^2(s+2)$, $H(s) = 1$

Type 1 System, Step Input, $G(s) = \dfrac{1}{s(s+2)}$, $H(s) = 1$

$$\frac{C_3(s)}{R(s)} = \frac{G(s)}{1 + G(s)H(s)}$$

$$\frac{C_3(s)}{R(s)} = \frac{1}{s^2 + 2s + 1} = \frac{1}{(s+1)^2}$$

$$C_3(s) = R(s)\frac{1}{(s+1)^2},\ R(s) = \frac{1}{s}$$

$$C_3(s) = \frac{1}{s}\,\frac{1}{(s+1)^2} = \frac{A}{(s+1)^2} + \frac{B}{s+1} + \frac{D}{s}$$

$$A = (s+1)^2 C_3(s)|_{s=-1} = \frac{1}{s}|_{s=-1} = -1$$

$$B = \frac{1}{1!}\frac{d}{ds}[(s+1)^2 C_3(s)]|_{s=-1} = \frac{d}{ds}[\frac{1}{s}]|_{s=-1}$$

$$B = \frac{-1}{s^2}|_{s=-1} = -1$$

$$D = sC_3(s)|_{s=0} = \frac{1}{(s+1)^2}|_{s=0} = 1$$

$$C_3(s) = \frac{-1}{(s+1)^2} + \frac{-1}{s+1} + \frac{1}{s}$$

Transforming via Table 34.6

Answer

$c_3(t) = [-te^{-t} - e^{-t} + 1]u(t)$

Check

$c_3(t)|_{t=0+} = 0?$

$c_3(0t) = [-(0)e^0 - e^0 + 1] = 0 - 1 + 1 = 0\surd$

For $t \gg 0$, $c_3(t) = 1$

Type 1 System, Ramp Input, $G(s) = \dfrac{1}{s(s+2)}$, $H(s) = 1$

$$C_5(s) = R(s)\frac{1}{(s+1)^2},\ R(s) = \frac{1}{s^2} \qquad \text{unit ramp} \Rightarrow \frac{1}{s^2} \Rightarrow t$$

$$C_5(s) = \frac{1}{s^2}\,\frac{1}{(s+1)^2} = \frac{A}{s^2} + \frac{B}{s} + \frac{D}{(s+1)^2} + \frac{E}{s+1} \quad \text{(partial fraction expansion)}$$

$$A = s^2 C_5(s)|_{s=0} = \frac{1}{(s+1)^2}|_{s=0} = \frac{1}{1} = 1$$

$$B = \frac{1}{1!}\frac{d}{ds}[s^2 C_5(s)]|_{s=0} = \frac{d}{ds}\left[\frac{1}{(s+1)^2}\right]|_{s=0} = \frac{-2}{(s+1)^3}|_{s=0} = \frac{-2}{1} = -2$$

$$D = (s+1)^2 C_5(s)|_{s=-1} = \frac{1}{s^2}|_{s=-1} = \frac{1}{1} = 1$$

$$E = \frac{1}{1!}\frac{d}{ds}[(s+1)^2 C_5(s)]|_{s=-1} = \frac{d}{ds}\left[\frac{1}{s^2}\right]|_{s=-1} = \frac{-2}{s^3}|_{s=-1} = \frac{-2}{-1} = +2$$

$$C_5(s) = \frac{1}{s^2} + \frac{-2}{s} + \frac{1}{(s+1)^2} + \frac{2}{s+1}$$

Inverting via Table 34.6

Answer

$C_5(t) = (t - 2 + te^{-t} + 2e^{-t})u(t)$

Check

$C_5(+0) = 0?$

$C_5(0+) = (0 - 2 + 0 + 2) = 0\surd$

For $t \gg 0$,

$C_5(t) = t - 2$

TABLE 34.15. (*Continued*)

Type 2 System. Step Input. $G(s) = \dfrac{1}{s^2(s+2)}$, $H(s) = 1$, $R(s) = \dfrac{1}{s}$

$$C_4(s) = R(s)\frac{1}{s^3 + 2s^2 + 1} = R(s)\frac{1}{(s+2.21)(s-0.1+j0.665)(s-0.1-j0.665)} \quad \text{(approximate roots used)}$$

$$\frac{C_4(s)}{R(s)} = \frac{G(s)}{1 + G(s)H(s)}$$

$$\frac{C_4(s)}{R(s)} = \frac{1}{s^3 + 2s^2 + 1}$$

$$C_4(s) = \frac{1}{s}\,\frac{1}{(s+2.21)(s-0.1+j0.67)(s-0.1-j0.67)} = \frac{A}{s} + \frac{B}{s+2.21} + \frac{M\angle\theta}{s-0.1+j0.67} + \frac{M\angle-\theta}{s-0.1-j0.67} \quad \text{(conjugate term)}$$

$$A = sC_4(s)|_{s=0} = \left.\frac{1}{s^3 + 2s^2 + 1}\right|_{s=0} = \frac{1}{1} = 1 \; [\text{Note: } (s-0.1+j.067)(s-0.1-j0.67) \cong s^2 - 0.206s + 0.453]$$

$$B = (s+2.21)C_4(s)|_{s=-2.21} = \left.\frac{1}{s(s^2 - 0.206s + 0.453)}\right|_{s=-2.21} = \frac{1}{-12.8} = -0.078$$

$M\angle\theta$ = magnitude at angle θ of complex residue of pole $s = 0.1 - j0.67$

$$M\angle\theta = (s-0.1+j0.67)C_4(s) = \left.\frac{1}{s(s+2.21)(s-0.1-j0.67)}\right|_{s=+0.1-j0.67}$$

$$M\angle\theta = \frac{1}{(0.1-j0.67)(2.31-j0.67)(-j1.34)} = \frac{1}{0.677\angle-81.5^\circ \; 2.41\angle-16.2^\circ \; 1.34\angle-90^\circ}$$

Sketches of polar conversions

$M\angle\theta = 0.465\angle 187.3^\circ$

Inserting A, B, and $M\angle\theta$ into $C_4(s)$ yields:

$$C_4(s) = \frac{1}{s} + \frac{-0.078}{s+2.21} + \frac{0.465\angle 187.3^\circ}{s-0.1+j0.67} + \text{conjugate}$$

(Note that $M\angle\theta$ is numerator of $+j\omega$ term, with denominator of $s - 0.1 + j0.67$, which has pole at $s = +0.1 - j0.67$.)
Conjugate term is not used. Transforming via relationship

$$\frac{M\angle\theta}{s+\alpha+j\omega} + \text{conj.} \Rightarrow [2Me^{-\alpha t}\sin(\omega t + 90° - \theta°)]\,\mu_s(t)$$

where: $\alpha = -0.1$

$\omega = 0.67$

$2M = 2(0.465) = 0.93$

$\theta° = 187.3°$

$90° - \theta° = -97.3°$

$-97.3° = -1.698$ rad

$$\frac{1}{s} \Rightarrow 1\mu_s(t)$$

$$\frac{K}{s+\beta} \Rightarrow Ke^{-\beta t}\mu_s(t)$$

Answer

$C_4(t) = [1 - 0.078e^{-2.21t} + 0.93e^{0.1t}\sin(0.67t - 1.698\text{ rad})]\,\mu_s(t)^b$

Check

$C_4(0+) = 0?$

$C_4(0+) = 1 - 0.078e^{0} + 0.93e^{0}\sin(-97.3°) = 1 - 0.078 + 0.93(-0.99) \cong 0.00\surd$

Tabulated Values of $C_4(t)$, plotted in Table 34.13

t	0	1	2	3	4	5	6	7	8	9	10	11	12	13	14	15
$C_4(t)$	0.0	0.11	0.60	0.02	2.15	2.5	2.2	1.3	−0.03	−1.1	−1.4	−0.6	1.2	3.3	4.7	4.7

[a] Since s^1 term is missing in the characteristic equation of $C_4(s)$, $C_4(t)$ will be unstable, as indicated by the negative real part of the complex roots (-0.1), which will produce a growing exponential response $e^{+0.1t}$

[b] Multiplicative function $\mu_s(t)$ indicates that $C_4(t) = 0$ for $t < 0$.

It is important to emphasize that all type 2 systems are not unstable, and a stable type 2 system does have the virtue of having a zero steady-state error to step and ramp inputs, and a finite following error to an acceleration (parabolic) input. This is discussed further in the next section.

Control System Errors

The purpose of a control system is to accurately control certain variables, such as temperature, position, or velocity. One way of judging the performance of a control system is to determine (e.g., measure, simulate, or calculate) the error of the system in response to various standard types of input signals. If the standard test inputs (e.g., step, ramp, acceleration) are identical to actual operational inputs, such tests of system output error are directly applicable to judging actual system performance errors. However, if the actual operational inputs are not similar to the standard test inputs, the results of such tests are not directly applicable to judging actual system performance errors. In many practical cases, the errors for standard test inputs do serve as an excellent guide for estimating actual system performance errors.

Usually the errors of actual systems are larger than the errors produced by the simplest standard test input, the unit step input, because actual inputs contain additional components which add to the step, including noise. Hence if a system's deterministic unit step error response is unsatisfactory, its error response to actual system inputs containing deterministic and stochastic inputs is usually worse, and the system is unacceptable.

In general, therefore, the standard test inputs should be viewed as providing a minimum error performance standard. One glaring exception to this rule occurs when the actual system inputs are radically different in frequency spectrum from the standard test inputs. In cases where the actual frequency spectrum is band limited, it is possible for a system to produce acceptably small operational errors even though the errors to standard test inputs containing the entire frequency spectrum (e.g., a step input) are unacceptably large. An example of this behavior occurs in high-performance aircraft control systems, where the normal pilot input can be viewed as a random band-limited signal having a maximum frequency content in the neighborhood of 1 Hz. A step input of perhaps one third of the maximum control stick excursion from its midpoint would, in general, produce an unacceptable transient response in addition to unacceptably large errors because the unit step input signal has a frequency spectrum that contains all frequencies. Most high-performance aircraft filter out any high-frequency signals introduced into their control loop. This filtering is produced electronically in a fly-by-wire or electrically controlled aircraft or mechanically, hydraulically, or pneumatically by suitable low-pass filters. Springy control cables, torsionally compliant control surfaces, and flexible structural members all tend to round off the sharp corner of the step response, which lowers the high-frequency content of the signal. In addition, the pilot tends to act as a low-pass filter with a frequency response in the neighborhood of 1 Hz.

The definition of system error in unity and nonunity feedback systems is different. Referring to Table 34.16 for the nonunity feedback case, the error is taken as the output from the comparitor (or subtractor), which compares the reference input $R(s)$ with the feedback signal $X(s)$. Therefore the error $E(s)$ is

$$E(s) = R(s) - X(s) \tag{34.10}$$

or

$$E(s) = R(s) - H(s)C(s) \tag{34.11}$$

The error defined by Eq. (34.11) is not the difference between the system's input and the output, which is the usual definition for the unity-feedback case as shown in Table 34.16. [The error of Eq. (34.11) is the difference between the input and the feedback signal.]

It is possible for two systems to have identical closed-loop transfer functions and different errors for identical inputs and outputs. This will occur if one system has different forward and feedback gains from the other system. Consider the standard unity feedback system pictured in Table 34.16. For a unity-feedback system, the error $E_1(s)$ is

$$E_1(s) = R(s) - X_1(s) \tag{34.12}$$

or

$$E_1(s) = R(s) - C(s) \tag{34.13}$$

Observe that Eq. (34.13) differs from Eq. (34.11). If $R(s)$ and $C(s)$ are identical in both cases, and $H(s)$ is not unity, then $E(s)$ will differ from $E_1(s)$. This is clearly shown in Table 34.17, which illustrates the input, output, feedback, and error responses vs. time for two systems with the same transfer functions but with different forward gains and feedback transfer functions. Referring to this table, observe that the inputs and outputs are identical but that the errors are different.

TABLE 34.16. ERROR OF NONUNITY AND UNITY FEEDBACK SYSTEMS

System	Block Diagram	Calculation of Error
Nonunity feedback system	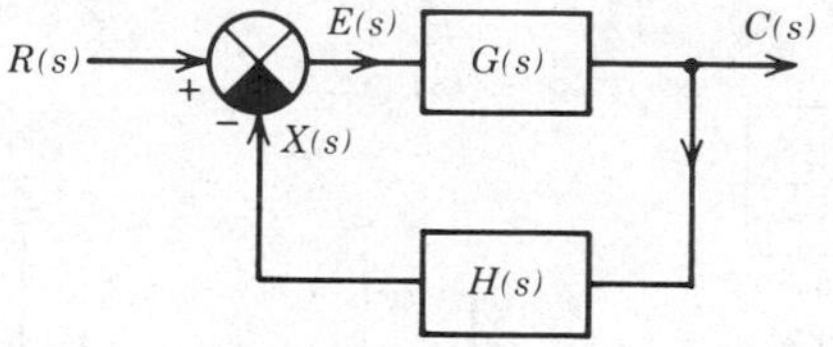 $R(s)$ → (+) summing junction (−) → $E(s)$ → $G(s)$ → $C(s)$; feedback through $H(s)$ gives $X(s)$. $X(s) = H(s)C(s)$	$C(s) = E(s)G(s)$ $E(s) = \text{Error} = C(s)\dfrac{1}{G(s)}$ $E(s) = R(s) - X(s)$ $E(s) = R(s) - H(s)C(s)$ Also $\dfrac{E(s)}{R(s)} = \dfrac{C(s)}{R(s)} \cdot \dfrac{1}{G(s)} = \dfrac{G(s)}{1 + G(s)H(s)} \cdot \dfrac{1}{G(s)}$ $\dfrac{C(s)}{R(s)} = \dfrac{G(s)}{1 + G(s)H(s)}$ $\dfrac{E(s)}{R(s)} = \dfrac{1}{1 + G(s)H(s)}$ $E(s) = \dfrac{R(s)}{1 + G(s)H(s)}$
Unity feedback system	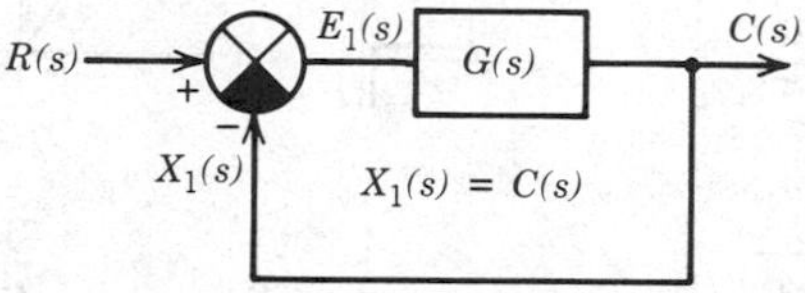 $R(s)$ → (+) summing junction (−) → $E_1(s)$ → $G(s)$ → $C(s)$; unity feedback gives $X_1(s)$. $X_1(s) = C(s)$	$C(s) = E_1(s)G(s)$ $E_1(s) = R(s) - X_1(s)$ $E_1(s) = R(s) - C(s)$ $E_1(s) = C(s) \cdot \dfrac{1}{G(s)}$ Also $\dfrac{E_1(s)}{R(s)} = \dfrac{C(s)}{R(s)} \cdot \dfrac{1}{G(s)} = \dfrac{G(s)}{1 + G(s)} \cdot \dfrac{1}{G(s)}$ $= \dfrac{1}{1 + G(s)}$ $\dfrac{C(s)}{R(s)} = \dfrac{G(s)}{1 + G(s)}$ $E_1(s) = \dfrac{R(s)}{1 + G(s)}$

The standard test inputs are the unit step, the unit ramp, and the unit parabolic input. When these inputs are applied to a type 0, a type 1, and a type 2 system, the steady-state errors can be readily found by applying the Laplace transform final-value theorem $[\lim_{t\to\infty} e(t) = \lim_{s\to 0} sE(s)]$. From Table 34.16

$$E(s) = R(s)\frac{1}{1 + G(s)H(s)} \tag{34.14}$$

Thus for a unit step input (position input) $r_p(t) = u(t)$ (or $R_p(s) = 1/s$), and $E(s)$ becomes

$$E_p(s) = \frac{1}{s}\,\frac{1}{1 + G(s)H(s)} \tag{34.15}$$

The final value of $e(t)$, as t goes to infinity, is then given by

$$\lim_{t\to\infty} e(t) = e_{ss}(t) = \lim_{s\to 0} s \cdot E_p(s) = \lim_{s\to 0} \frac{s}{s} \cdot \frac{1}{1 + G(s)H(s)}$$

or

$$\lim_{s\to 0} \frac{1}{1 + G(s)H(s)} = \frac{1}{1 + \lim_{s\to 0} G(s)H(s)} \tag{34.16}$$

where $e_{ss}(t)$ = steady-state error as $t \to \infty$.

The position error constant K_p is defined directly from Eq. (34.16) as

$$K_p = \lim_{s\to 0} G(s)H(s) \tag{34.17}$$

TABLE 34.17. DIFFERENT ERRORS PRODUCED BY TWO SYSTEMS WITH IDENTICAL TRANSFER FUNCTIONS

Variable	Nonunity Feedback Error, Unit Step Input	Unity Feedback Error, Unit Step Input
Block diagram	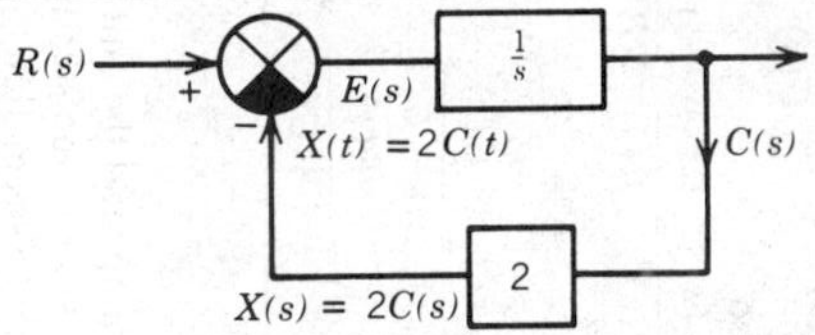 $R(s)$, +, −, $E(s)$, $\frac{1}{s}$, $C(s)$, $X(t) = 2C(t)$, 2, $X(s) = 2C(s)$	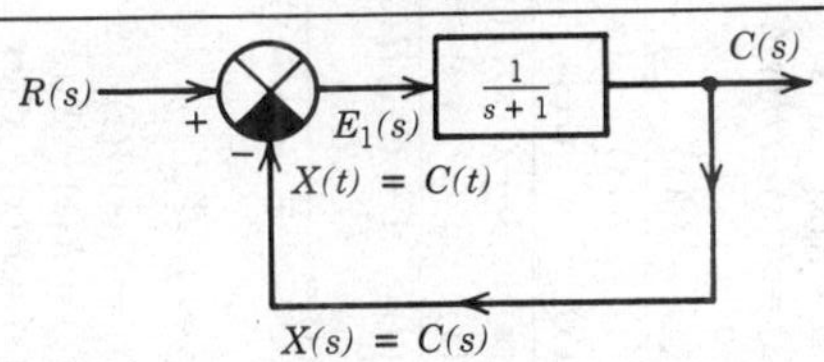 $R(s)$, +, −, $E_1(s)$, $\frac{1}{s+1}$, $C(s)$, $X(t) = C(t)$, $X(s) = C(s)$
Closed-loop transfer function	$\frac{C(s)}{R(s)} = \frac{G}{1+GH} = \frac{1/s}{1+2/s} = \boxed{\frac{1}{s+2}}$ $G = 1/s,\ H = 2$	$\frac{C(s)}{R(s)} = \frac{G}{1+GH}\ \frac{1/(s+1)}{1+[1/(s+1)]} = \boxed{\frac{1}{s+2}}$ $G = \frac{1}{s+1},\ H = 1$
Input $r(t)$	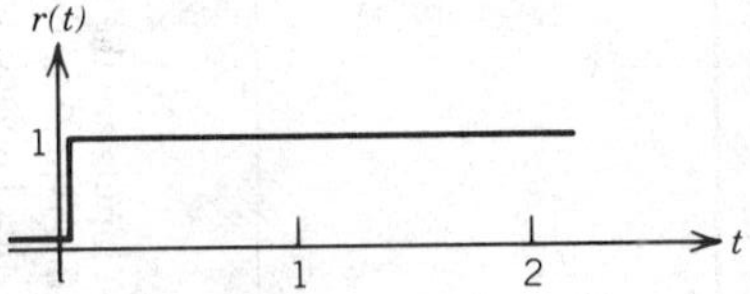 $r(t)$, 1, t, 1, 2	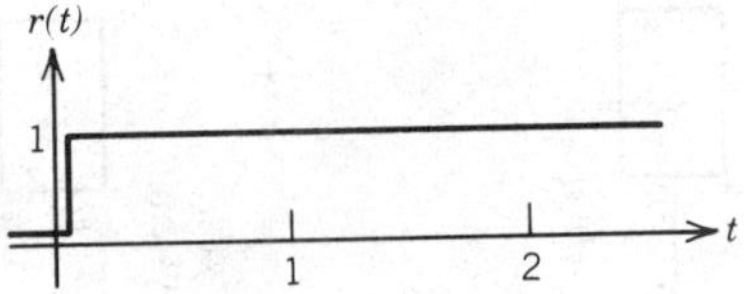 $r(t)$, 1, t, 1, 2
Output $c(t)$	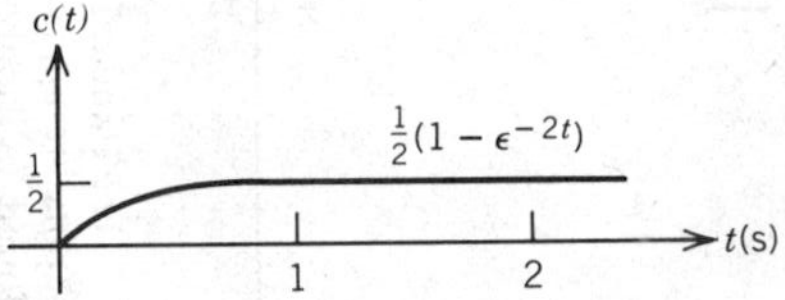 $c(t)$, $\frac{1}{2}$, $\frac{1}{2}(1-\epsilon^{-2t})$, t(s), 1, 2	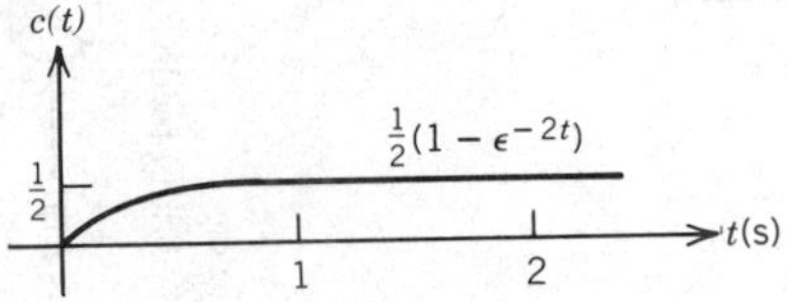 $c(t)$, $\frac{1}{2}$, $\frac{1}{2}(1-\epsilon^{-2t})$, t(s), 1, 2

Feedbacks $x(t)$

Error $e(t) = r(t) - x(t)$

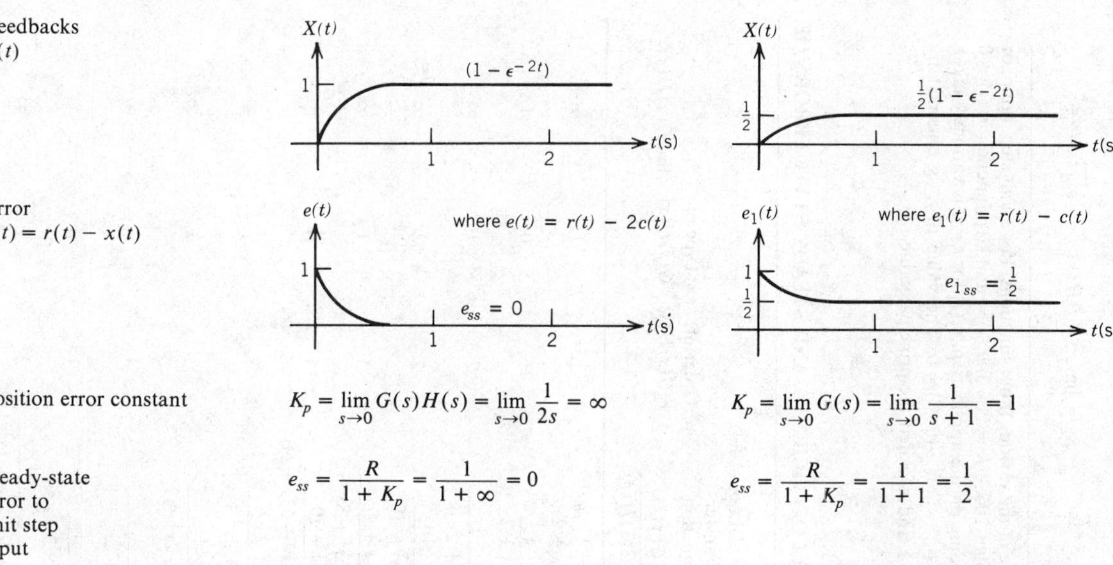

Position error constant	$K_p = \lim_{s\to 0} G(s)H(s) = \lim_{s\to 0} \frac{1}{2s} = \infty$	$K_p = \lim_{s\to 0} G(s) = \lim_{s\to 0} \frac{1}{s+1} = 1$
Steady-state error to unit step input	$e_{ss} = \frac{R}{1+K_p} = \frac{1}{1+\infty} = 0$	$e_{ss} = \frac{R}{1+K_p} = \frac{1}{1+1} = \frac{1}{2}$

TABLE 34.18. ERROR CONSTANTS FOR STEP, RAMP, AND PARABOLIC INPUTS

Input	Input Time Function	Input Laplace Transform	Error Constant	Steady-State Errors in Time Domain
Step or position	$Ru_s(t)^a$	R/s	$K_p = \lim_{s\to 0} G(s)H(s)$	$e_{ss}(t) = \dfrac{R}{1+K_p}$
Ramp or velocity	$Rtu_s(t)$	R/s^2	$K_v = \lim_{s\to 0} s \cdot G(s)H(s)$	$e_{ss}(t) = \dfrac{R}{K_v}$
Parabola or acceleration	$\dfrac{Rt^2}{2} u_s(t)$	$\dfrac{R}{s^3}$	$K_a = \lim_{s\to 0} s^2 G(s)H(s)$	$e_{ss}(t) = \dfrac{R}{K_a}$

[a] R is a constant indicating the amplitude of the various input functions. For a unit step, ramp, or parabolic input, $R = 1$. For a step of height 206.3, $R = 206.3$. $u_s(t)$ = unit step function. $u_s(t) = 0$ for $t < 0$ and $u_s(t) = 1$ for $t > 0$. Depending on the application, $u_s(t)$ for $t = 0$ is variously taken as 0, $\frac{1}{2}$, or 1. One will often see $u(0+) = 1$ or $u(0-) = 0$. The value of $\frac{1}{2}$ is obtained by averaging the limits $u_s(0+)$ and $u_s(0-)$, and is called the principal value.

TABLE 34.19. TYPE, ORDER, ERROR COEFFICIENTS, AND STEADY-STATE ERRORS OF

Forward Gain $G(s)$	Feedback Gain $H(s)$	Closed-Loop Transfer Function $\dfrac{G(s)}{1+G(s)H(s)}$	Order of $1+G(s)H(s)$	Type of $G(s)H(s)$	K_p $\lim_{s\to 0} G(s)H(s)$
$\dfrac{1}{s}$	2	$\dfrac{1}{s+2}$ [d]	1	1	∞
$\dfrac{1}{s+1}$	1	$\dfrac{1}{s+2}$ [d]	1	0	1
$\dfrac{1}{s+2}$	$\dfrac{1}{s}$	$\dfrac{s}{s^2+2s+1}$	2	1	∞
$\dfrac{10}{s^2}$	1	$\dfrac{10}{s^2+10}$	2	2	∞
$\dfrac{1}{s(s+2)}$	$\dfrac{3}{s+4}$	$\dfrac{s+4}{s^3+6s^2+8s+3}$	3	1	∞
$\dfrac{1}{(s+1)(s+2)}$	$\dfrac{s+10}{s+20}$	$\dfrac{s+20}{s^3+23s^2+63s+50}$	3	0	$\frac{1}{4}$
$\dfrac{1}{s^2(s+2)}$	s	$\dfrac{1}{s^3+2s^2+s}$	3	1	∞
$\dfrac{1}{s^2(s+2)}$	1	$\dfrac{1}{s^3+2s^2+1}$ [e]	3	2	∞
$\dfrac{1}{s^2(s+2)}$	$\dfrac{1}{s}$	$\dfrac{s}{s^4+2s^3+1}$ [e]	4	3	∞
$\dfrac{1}{s+2}$	1	$\dfrac{1}{s+3}$	1	0	$\frac{1}{2}$
$\dfrac{1}{s(s+2)}$	1	$\dfrac{1}{s^2+2s+1} = \dfrac{1}{(s+1)^2}$	2	1	∞

[a] R = constant (input amplitude), often 1.
[b] $\mu(t)$ = unit step function.
[c] $[t^2/2]\mu(t)$ transforms to $1/s^2$.
[d] For explanation of two different errors produced by identical transfer functions, see Table 34.17.
[e,f] Unstable system; results are meaningless. Errors are ∞. Errors are *not* zero, as predicted by error coefficients,[e] or $2R$.[f] All errors are ∞.
[g] $C_0(t)$, $C_3(t)$, and $C_5(t)$ are plotted in Table 34.13. $R/1.5$, 0, and $2R$ are final values of the error.

Therefore Eq. (34.16) can be written as

$$e_{ss}(t) = \frac{1}{1 + K_p} \qquad \text{(for unit step input)} \tag{34.18}$$

In a similar fashion, the error for a unit ramp or velocity input, $r_v(t) = tu(t)$ (or $R_v(s) = 1/s^2$) and a unit parabolic or acceleration input $r_a(t) = (t^2/2)u(t)$ [or $R_a(s) = 1/s^3$] can be found.† For these errors, a velocity error constant K_v and an acceleration error constant K_a are defined. These definitions and the resultant errors are summarized in Table 34.18.

The most common source of error in using the error constants is omitting or forgetting that a unit parabolic input is $(1/2)t^2$. Forgetting the $1/2$ leads to an extra factor of 2 in the resultant error. The equations in Table 34.18 are applied to 11 systems of type 0 to 3 and of orders 1 to 4 in Table 34.19. The error constants and the resultant steady-state errors are tabulated, along with the order, type, and closed-loop transfer function for each system.

To illustrate the procedures used, the errors for $C_0(t)$, $C_3(t)$, and $C_5(T)$ of Table 34.13 are calculated in Table 34.20.

It is important to emphasize that a system must be stable for these error constants to have meaning. If a system has roots either in the right half of the s plane or on the $j\omega$ axis (e.g., if the roots

† For details see, for example, Kuo.[7]

A CONTROL SYSTEM

K_v $\lim_{s\to 0} sG(s)H(s)$	K_a $\lim_{s\to 0} s^2G(s)H(s)$	Step Error[a] $r(t) = R\mu(t)$ $e_{ss} = R/(1+K_p)$	Ramp Error[b] $r(t) = Rt\mu(t)$ $e_{ss} = R/K_v$	Parabolic Error[c] $r(t) = [Rt^2/2]\mu(t)$ $e_{ss} = R/K_A$
2	0	0[d]	$\frac{R}{2}$	∞
0	0	$\frac{R}{2}$[d]	∞	∞
$\frac{1}{2}$	0	0	$2R$	∞
∞	10	0	0	$\frac{R}{10}$
$\frac{3}{8}$	0	0	$\frac{8}{3}R$	∞
0	0	$\frac{4}{5}R$	∞	∞
$\frac{1}{2}$	0	0	$2R$	∞
∞	$\frac{1}{2}$	[e]	[e]	[f]
∞	∞	[e]	[e]	[e]
0	0	$\frac{R}{1.5} = 1 - C_0(t)\|^{g}_{t\to\infty}$	∞	∞
$\frac{1}{2}$	0	$0 = 1 - C_3(t)\|^{g}_{t\to\infty}$	$2R = t - C_5(t)\|^{g}_{t\to\infty}$	∞

TABLE 34.20. ERROR CONSTANT CALCULATIONS FOR CERTAIN STEP AND RAMP RESPONSES SHOWN IN TABLE 34.13[a]

Response	Error Constant Calculation
$C_0(t)$, for $G(s) = \dfrac{1}{s+2}$, $H(s) = 1$ Unit step input $R = 1$	$K_p = \lim_{s\to 0} G(s)H(s) = \lim_{s\to 0} \dfrac{1}{s+2} = \dfrac{1}{2}$ $e_{ss}(t) = \dfrac{1}{1+K_p} = \dfrac{1}{1+\frac{1}{2}} = \dfrac{2}{3} = \dfrac{1}{1.5} \to \left(\text{or } \dfrac{R}{1.5}\right)$
$C_3(t)$, for $G(s) = \dfrac{1}{s(s+2)}$, $H(s) = 1$ Unit step input $R = 1$	$K_p = \lim_{s\to 0} G(s)H(s) = \lim_{s\to 0} \dfrac{1}{s(s+2)} = \infty$ $e_{ss}(t) = \dfrac{1}{1+K_p} = \dfrac{1}{1+\infty} = 0 \to (\text{or } 0 \cdot R)$
$C_5(t)$, for $G(s) = \dfrac{1}{s(s+2)}$, $H(s) = 1$ Unit ramp input $R = 1$	$K_v = \lim_{s\to 0} sG(s)H(s) = \lim_{s\to 0} \dfrac{1}{s+2} = \dfrac{1}{2}$ $e_{ss}(t) = \dfrac{1}{K_v} = \dfrac{1}{\frac{1}{2}} = 2 \to (\text{or } 2 \cdot R)$

[a] $C_0(t)$, $C_3(t)$, and $C_5(t)$ are plotted in Table 34.13 and the final values of the errors are shown. These values are calculated here directly from the error coefficients. These errors are also tabulated in the last two rows of Table 34.19.

of the system's characteristic equation have positive real parts), then the error constants are meaningless (even though they have perfectly reasonable values). For example, Table 34.19 shows two systems where $G(s) = 1/s^2(s+2)$ whose errors are infinite because the systems are unstable; yet the predicted steady-state errors are finite or zero.

Control System Stability

Many methods exist for analyzing or predicting the stability of feedback control systems, on the basis of examination of certain characteristics of a mathematical model or equation describing the system. Some of the most commonly used methods are intuitively satisfactory; they can be understood by examination of the very common model of a single-loop feedback control system shown in Fig. 34.5. Both figures are equivalent and produce the same mathematical expression for the gain, or transfer function, of the closed-loop system.

Transfer function

$$\frac{C(s)}{R(s)} = \frac{G(s)}{1 + G(s)H(s)} \tag{34.19}$$

where s = Laplace transform variable
$s = \sigma + j\omega$
E = error = $R(s) - H(s)C(s)$

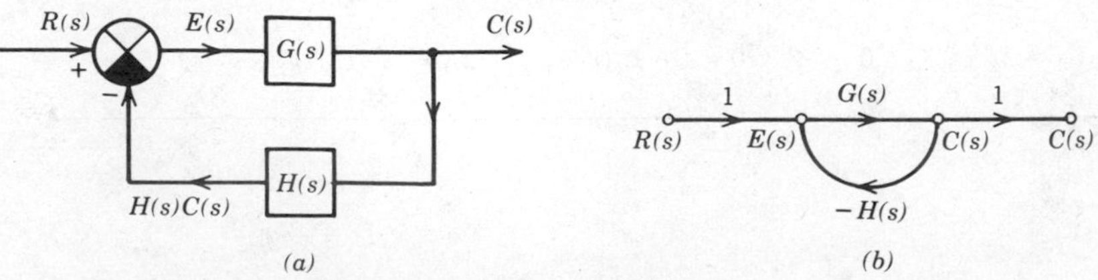

Fig. 34.5 Closed-loop system; *(a)* Block diagram, *(b)* signal flow graph.

Equation (34.19) is derived in Table 34.8 dealing with block diagram reduction and in Table 34.12 dealing with signal flow graphs.

Basic Concept of Stability Predictions. The basic concept of stability prediction involves the examination of $G(s)H(s)$ in the denominator of Eq. (34.19). $G(s)H(s)$ is called the loop gain, or the open-loop gain. If for some value of s, say $s = s_0$, the numerical value of $G(s)H(s)$ becomes the real number minus one, then Eq. (34.19) becomes

$$\frac{C(s)}{R(s)} = \frac{G(s)}{1 + G(s)H(s)}\bigg|_{s=s_0} = \frac{G(s_0)}{1-1} = \frac{G(s_0)}{0} \quad \begin{matrix}\text{undefined}\\ \text{or}\\ \text{unstable}\\ \text{or}\\ \text{infinite}\end{matrix} \tag{34.20}$$

Thus if the quantity $G(s)H(s)$ ever takes on the value -1, the feedback control systems pictured in Fig. 34.5 exhibit an unbounded response. Since stability of a linear system is defined as a bounded output for a bounded input, the systems become unstable. Three techniques for analyzing the stability of a feedback control system use the equation $G(s)H(s) = -1$. They are the Bode plot method, the Nyquist technique, and the root locus method. The Bode and Nyquist techniques are both frequency domain methods, wherein the general Laplace variable $s = \sigma + j\omega$ is replaced by the pure imaginary variable $j\omega$ (where ω is the real frequency measured in radians per second and j is the square root of -1). The root locus technique uses both the real (σ) and imaginary ($j\omega$) parts of the Laplace variable s.

The Bode, Nyquist, and root locus methods all determine the stability of closed-loop systems by examining the open-loop gain $G(s)H(s)$. This is a great simplification, because the open-loop system equations are often in factored form whereas the closed-loop system equations are often in unfactored form. The factored equations are much easier to deal with, which partially explains the popularity of these three methods.

Bode Technique. The Bode technique employs two logarithmic plots of $G(s)H(s)$. The first one is a plot of $20\log|G(j\omega)H(j\omega)|$ versus $\log\omega$. Usually a logarithmic scale is used for the ω axis, so that no actual logarithms need be evaluated. From these two plots, it is relatively easy to determine whether the condition $G(j\omega)H(j\omega) = -1$ (0 dB at an angle $\pm 180°$) ever occurs. The actual application of the Bode plot technique involves a host of contingencies that are discussed in Ogata,[6] Kuo,[7] Shinners,[8] D'Azzo and Houpis,[10] and in other standard texts dealing with classical control theory. The procedures for constructing a Bode plot follow.

1. Reduce a complex system to an equivalent single-loop feedback system with forward gain $G(s)$ and feedback gain $H(s)$.

2. Form the loop gain $G(s)H(s)$ and factor the numerator and denominator into a first-order real term and a quadratic second-order term, plus a gain constant and an s^n term. For example:

$$G(s)H(s) = \frac{740(s+2)}{s(s+15)(s^2+6s+25)}$$

3. Find the poles and zeros of $G(s)H(s)$ by inspection for first-order terms. For second-order terms use

$$(s^2 + 2\zeta\omega_n s + \omega_n^2)$$

where ω_n = natural frequency of second-order pole (or zero) and ζ = damping factor.

Thus for the given $G(s)H(s)$:

First-Order Zeros	First-Order Poles	Second-Order Poles
$\omega = 2$ rad/s	$\omega = 0$	$\omega_n = \sqrt{25} = 5$ rad/s
	$\omega = 15$ rad/s	$\zeta = 6/2\omega_n = 6/(2)(5) = 0.6$

Note that s is replaced by $j\omega$ and then the j is dropped when referring to angular frequency ω, in radians per second (not hertz or cycles per second). If hertz or cycles per second units are desired later on, use $\omega = 2\pi f$, where f = frequency in hertz. For the example given there are no second-order zeros, merely for the sake of simplicity.

4. Calculate the magnitude of $G(j\omega_0)H(j\omega_0)$ at a frequency ω_0 that is equal to or smaller than any of the finite pole or zero frequencies. This is accomplished by merely eliminating all terms except the constants and the power of s, or s^n, term. For example, choose $\omega_0 = 1$ rad/s, which is smaller than $\omega = 2$, $\omega = 5$, $\omega = 25$ for the given $G(s)H(s)$. Thus

$$\text{Mag} = |G(s)H(s)|_{s=j\omega_0} = \left|\frac{740(2)}{s(15)(25)}\right|_{s=j\omega_0=j1} = \frac{740(2)}{(1)(15)(25)} = 3.95$$

This procedure works because the Bode straight-line magnitude plot approximates a term such as $(j1+2)$ as being exactly (2). The value 3.95 just calculated is an *exact* value on the straight-line Bode plot. The straight-line Bode plot approximates the exact or actual values of $G(s)H(s)$, where $s = j\omega$.

5. Convert this magnitude to decibels (dB) using

$$\text{dB} = 20\log[|G(j\omega)H(j\omega)|] \qquad \text{at } \omega = \omega_0$$

Thus for the given example

$$\text{Mag} = 20\log(3.95) = 20(0.5962) = 11.9\text{ dB}$$

$$\text{or Mag} \cong 12\text{ dB} \qquad \text{at } \omega_0 = 1\text{ rad/s}$$

This will be the starting point for the Bode magnitude plot of $G(s)H(s)$ and is the only calculation required. For comparison, the exact computer-calculated value of $G(s)H(s)$ evaluated at $s = j1$ is 12.9657 dB at an angle of $-81.2853°$. The exact value of 12.9657 is larger than the straight-line plot of 11.9 dB, because of the effect of the zero term $(s+2)$. Pole terms, such as $(s+15)$, have the opposite effect, causing the exact value to be less than the straight-line Bode plot value. The maximum errors occur exactly at corner frequencies, and are +3 dB for single zeros and -3 dB for single poles. At an octave (factor of 1/2 or 2) away from the corner, the error drops to ± 1 dB, which corresponds to the difference between 12.96 and 11.9 dB just discussed. Second-order terms also deviate from the straight-line plot near the corner frequency. The errors vary with the damping factor ζ and can generally be neglected for $0.5 < \zeta < 0.7$. For $\zeta \cong 1.0$, simply use double the error for a single-pole term, or -6 dB at the corner, -2 dB at the octave points. At $\zeta = 0.7$, the error is -3 dB. At $\zeta = 0.5$, the error is 0 dB, while at $\zeta = 0.3$, the error is approximately 4 dB and at $\zeta = 0.2$, approximately 8 dB. Analytically, the error at the corner for a quadratic Bode magnitude term is given by $20\log(1/2\zeta)$. Observe that if ζ approaches zero, the error approaches infinity—the term resonates at the corner frequency. Fig. 34.6 illustrates the behavior of a quadratic pole factor, near the corner frequency, while Fig. 34.7 gives the Bode plot with a quadratic pole factor for this example.

6. Lay out a logarithmic frequency scale, using semilog paper. Choose the decade frequencies so as to include all the pole and zero frequencies, plus a decade above and below these frequencies. For the given example, $\omega = 2$, 15, and 25 rad/s, so the chosen ω axis would be 0.1 to 1000 rad/s, or four decades. In a pinch, two decades could be used, namely 1 to 100 rad/s, but the complete phase plot, which extends a complete decade above and below each corner frequency, could not be drawn. The magnitude plot does not need these extra decade extensions.

7. Lay out a vertical decibel magnitude scale on the linear semilog axis, generally with major divisions of 10 or 20 dB. At a minimum, extend from at least +20 to 0 to -20 dB. A typical decibel scale might run from +40 to -60 dB. The region about 0 dB is of most importance.

8. On the logarithmic frequency scale, draw small "up arrows" at each zero frequency and small "down arrows" at each pole frequency. For a quadratic factor, draw a double-headed arrow. Each arrow signifies a change in straight-line magnitude slope of 20 dB per decade, in the direction of the arrow. Thus for the given system

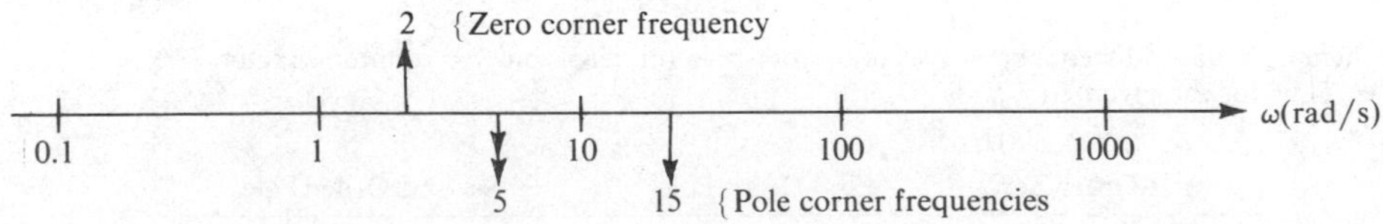

9. Using the calculated starting point of $G(j1)H(j1) = 11.9 \cong 12$ dB, the straight-line magnitude plot may be immediately sketched by drawing a line of slope $-20(n)$ dB/decade through the starting point, up to the first arrow, where the slope is changed as indicated.

The n factor corresponds to the exponent in the pole term s^n, and is $n = 1$ for the given example. Thus the plot for the example starts at a slope of -20 dB/decade, goes to $\omega = 2$ and then changes to $-20 + 20 = 0$ dB/decade. The slope of 0 dB/decade goes to $\omega = 5$ and changes to $-20 \cdot 2 = -40$ dB/decade, which goes to $\omega = 15$. At $\omega = 15$ the slope changes to $-40 + -20 = -60$ dB/decade and goes unchanged toward infinity. The completed magnitude plot is shown in Fig. 34.7.

10. To construct the phase plot, draw two opposite arrows, two decades apart, centered on the corner frequency. The first arrow goes up for zeros and down for poles. Each arrow indicates a

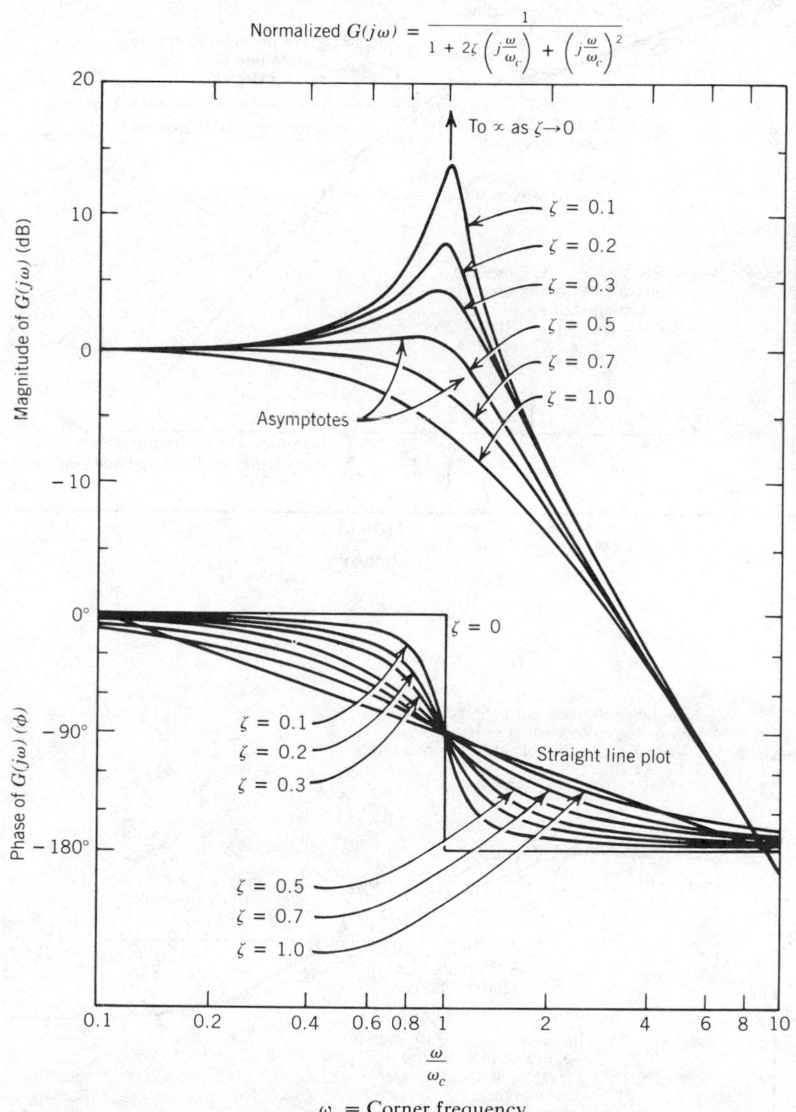

Fig. 34.6 Normalized Bode magnitude and phase plots for error correction near the corner of a quadratic second-order pole. $s = j\omega$ (i.e., s replaces $j\omega$), ω = radian frequency variable, ξ = damping ratio ($0 < \xi < 1$) for complex poles ω_n = natural or resonant frequency or corner frequency in radians per second, a constant. Note that if $\xi > 1$, the quadratic term can be factored into two real-pole terms. The quadratic is called overdamped. If ξ is negative, the quadratic term exhibits unstable behavior, with a step response envelope that grows exponentially toward infinity. $G(s) = \omega_n^2/(s^2 + 2\zeta\omega_n s + (\omega_n)^2)$; $G(j\omega) = (\omega_n)^2/((j\omega)^2 + 2\xi\omega_n(j\omega) + (\omega_n)^2)$. From Ogata,[6] reproduced with permission.

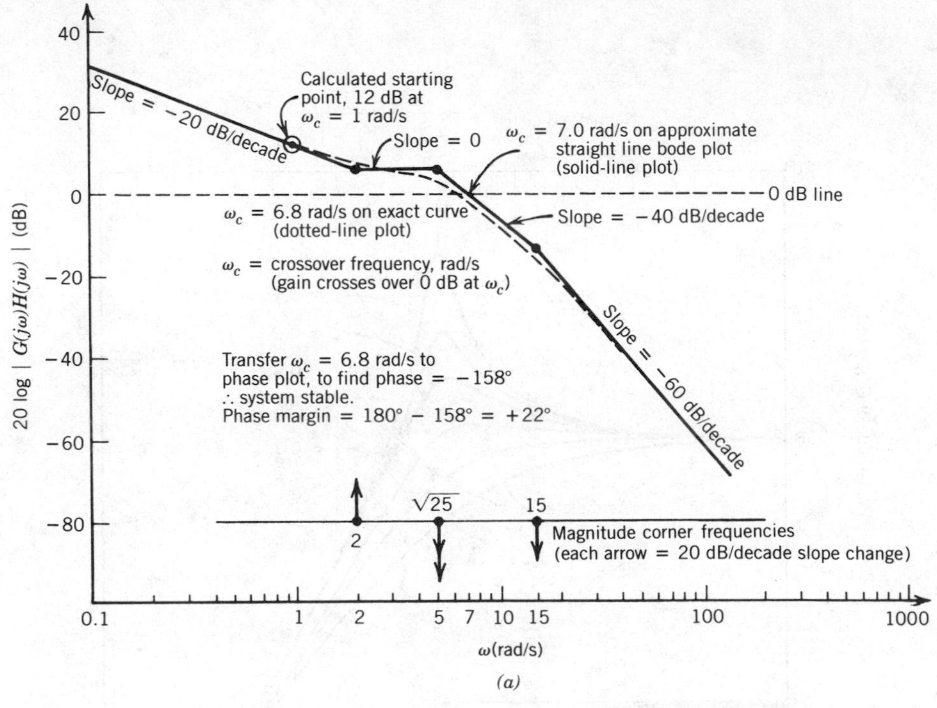

(a)

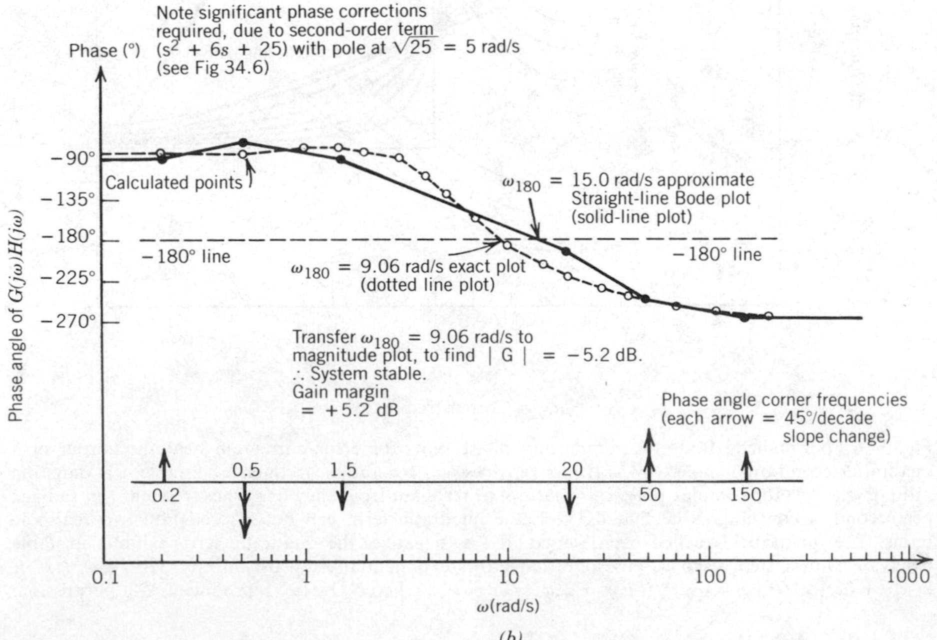

(b)

Fig. 34.7 Bode magnitude plot (*a*) and phase plot (*b*) of $G(s)H(s) = 740(s + 2)/[s(s + 15)(s^2 + 6s + 25)]$.

change in phase plot slope of 45°/decade, in the direction of the arrow. Thus for the given system

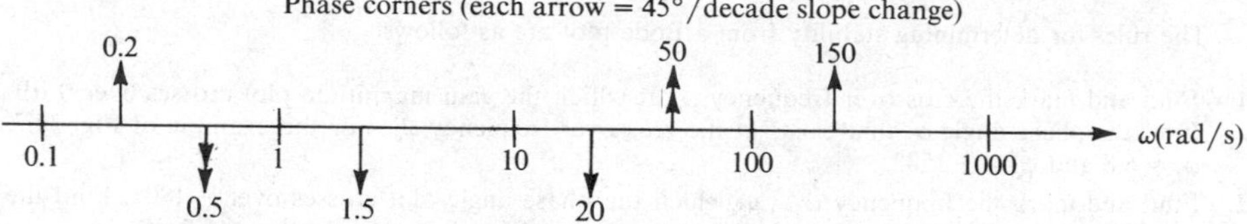

11. Calculate the phase angle of $G(s)H(s)|_{s=j\omega}$ as ω approaches 0 and then infinity. Thus for the given $G(s)H(s)$ (where the symbol $j0$ denotes an arbitrarily small imaginary quantity, while $j\infty$ denotes an arbitrarily large imaginary quantity)

$$\text{Angle } G(j0)H(j0) = \text{angle}\frac{740(j0+2)}{(j0)(j0+15)[(j0)^2+6j0+25]} = \text{angle}\frac{740(2)}{(j0)(15)(25)}$$

$$= \text{angle}\frac{0°+0°}{90°+0°+0°} = \frac{0°}{90°} = -90°$$

$$\text{Angle } G(j\infty)H(j\infty) = \text{angle}\frac{740(j\infty+2)}{j\infty(j\infty+15)[(j\infty)^2+6j\infty+25]} = \text{angle}\frac{740(j\infty)}{(j\infty)(j\infty)(j\infty)^2}$$

$$= \text{angle}\frac{0°+90°}{90°+90°+180°} = \frac{90°}{360°} = -270°$$

12. Start the phase plot at a constant slope of 0°/decade with a value of the angle of $G(j0)H(j0)$, which is $-90°$ for the given example. Sketch the phase plot using the arrow method discussed for the magnitude plot. The completed phase plot is also shown in Fig. 34.7.

13. As a check, see if the final 0°/decade constant slope of the phase plot has a phase value equal to $G(j\infty)H(j\infty)$, which is $-270°$ for the given example. A correct phase plot "closes" to an angle of $\pm 90 \cdot K°$, $K = 0, 1, 2, 3, \ldots$

14. If a phase corner frequency occurs within an octave of the frequency at which the magnitude plot crosses over 0 dB (the "crossover frequency"), calculate the exact value of the phase curve at this frequency, since significant phase errors are likely. Bode magnitude straight-line plots are usually accurate within a few decibels, whereas Bode phase straight-line plots are usually inaccurate and require significant (10 to 30°) correction when two or three poles and zero terms fall within a decade (or even a two-decade) range, as shown by Fig. 34.7. Quadratic factors, with small ζ (ζ less than 0.3) complicate the phase plot. All of these complications are illustrated by the Bode phase plot for the given example, shown in Fig. 34.7. For $\zeta = 1$, the quadratic factor phase plot changes phase by 180° over two decades, or it has a linear slope of $2 \cdot 45°$/decade, or 90°/decade. As ζ approaches zero, this 180°-phase change occurs over zero decades; that is, the phase tends to change by 180° over a very small frequency range centered on the corner frequency ω_n, for small values of ζ. However, for $\zeta > 0.3$, the straight-line phase plot is adequate, as long as the 0-dB-gain crossover does not occur within one octave of ω_n. This is illustrated in Fig. 34.6, which shows normalized phase plots for various values of ζ.

15. Determine the phase margin and gain margin. From the Bode plot (Fig. 34.7) corrected or exact curves one obtains

Gain Crossover Point	180° Phase Angle Point
$\lvert\text{Gain}\rvert = 0$ dB	Phase angle = $-180°$
Frequency = ω_c, say 6.8 rad/s	Frequency = ω_{180}, say 9.06 rad/s
Phase angle = ϕ_c, say $-158°$	$\lvert\text{Gain}\rvert = G(\omega_{180})$, say -5.2 dB
Phase margin = 180° − 158° = 22°	Gain margin = +5.2 dB

Then:

(a) Gain margin is that increment in gain that may be added, so that the resultant gain is 0 dB, at a phase angle of $-180°$. Thus if 5.2 dB is added to -5.2 dB, the result is 0 dB; and the gain margin is simply $-G(\omega_{180}) = -(5.2) = +5.2$ dB. For a stable system the gain margin must be positive.

(b) Phase margin is that decrement or negative increment in phase angle that may be added, so that the resultant phase angle is $-180°$ at 0 dB gain. Thus if $-158°$ is decremental by 22°, a phase angle of $-180°$ results. Thus the phase margin is $+22°$. Be careful of the multiple minus signs involved. Phase margin = 180° + (phase angle at 0 dB gain) = 180° + ϕ_c = 180° + (−158°) = 22°. For a stable system the phase margin must be positive.

The rules for determining stability from a Bode plot are as follows:

1. Find and mark the crossover frequency ω_c at which the gain magnitude plot crosses over 0 dB. Find the phase angle ϕ_c that exists at the crossover frequency ω_c. For the example of Fig. 34.7, $\omega_c = 6.8$ and $\phi_c = -158°$.
2. Find and mark the frequency ω_{180} at which the phase angle plot crosses over −180°. Find the magnitude $|G(\omega_{180})|$ at ω_{180}. For the given example, $\omega_{180} = 9.06$, $|G(\omega_{180})| = -5.2$ dB.
3. If there is only one value for ω_c and one value for ω_{180}, the Bode plot stability determination is unambiguous. If there are two or more values, it is best to construct a Nyquist plot from the Bode plot values of magnitude and phase vs. frequency. The Nyquist plot can be used to determine conditionally stable systems in an unambiguous fashion by simply checking to see whether the point $-1 + j0$ is enclosed by the Nyquist plot.
4. The system is stable if
   (a) At 0 dB, or $\omega = \omega_c$, the phase angle is less negative than −180° (say −158°), *and*
   (b) At −180° phase angle, or $\omega = \omega_{180}$, the gain magnitude is less than 0 dB (say −6 dB).
5. The system is unstable if
   (a) At 0 dB, or $\omega = \omega_c$, the slope of the magnitude plot is equal to or more negative than −60 dB/decade. For this test, only the magnitude curve is needed. This test is actually another form of test (a), since a slope of −60 dB/decade implies a phase shift more negative than −180°. For the given system, shown in Fig. 34.7, the slope of the magnitude plot near 0 dB gain is approximately −40 dB/decade. Thus instability has not been shown. Note that this test does not prove stability; it only proves instability. A system may be unstable with a slope of −40 dB/decade or of −20 dB/decade near 0 dB gain.

Note. Another method of drawing the Bode plots involves using a programmable calculator to calculate the value of $G(j\omega)$, both magnitude and phase, as a function of individual values of ω, such as $\omega = 0.1, 0.3, 1, 3, 10, 30, 100$, and so on. These data may then be plotted. Some calculators have plotters; the HP-41c, for example, can directly draw Bode plots. Of course a large computer system can also draw Bode plots on a display screen or on a digital plotter.

Nyquist Technique. The Nyquist technique is quite similar to the Bode technique except that a polar plot of the phase angle and magnitude of $G(j\omega)H(j\omega)$ is made as a function of radian frequency ω. Usually a linear radial scale is used for the Nyquist polar plot, as opposed to the logarithmic or decibel magnitude scale used for Bode plots. The Nyquist plot is readily examined to determine whether the condition $G(j\omega)H(j\omega) = -1$ ever occurs. The application of the Nyquist technique also involves a host of contingencies, which follow from Cauchy's principle of the argument and the conformal mapping techniques that form the mathematical basis for the Nyquist method. (These contingencies are also discussed in Ogata,[6] Kuo,[7] Shinners,[8] and D'Azzo and Houpis.[10]) A Nyquist plot example with detailed explanations is given in Fig. 34.8.

Under typical circumstances, $G(s)H(s)$ has no poles or zeros in the right half of the s plane. This implies that the open-loop system $G(s)H(s)$ alone is stable. However, the closed-loop system $G(s)/[1 + G(s)H(s)]$ may be unstable. The principal criterion employed for determining closed-loop system stability via either the Bode or Nyquist technique is not just the test for $G(j\omega)H(j\omega) = -1$, but rather a test for $|G(j\omega)H(j\omega)| \geqslant 1$, when the phase angle of $G(j\omega)H(j\omega)$ is exactly −180° [or the equivalent angles $-180° \pm N(360°)$]. For example, consider the case $G(j\omega_0) = -3$, $H(j\omega_0) = +1$. Both the Nyquist and Bode methods correctly predict an unstable closed-loop system, yet a formal substitution in Eq. (34.19) produces

$$\frac{G(j\omega_0)}{1 + G(j\omega_0)H(j\omega_0)} = \frac{-3}{1 + (-3)(1)} = \frac{-3}{-2} = 1.5$$

This result is incorrect. The system does not exhibit a gain of 1.5 but rather an unbounded gain. An intuitive explanation for this conflicting result follows from an examination of the transient conditions that result when the system is first turned on.

Before turn-on, $G(s)$ is zero. After turn-on, the magnitude of $G(s)$ grows from 0 toward 3 at a phase angle of 180°. Between these two points the magnitude of $G(s)$ must pass through unity. At that instant the closed-loop gain $G(s)/[1 + G(s)H(s)]$ becomes unbounded.

In actual practice the system becomes nonlinear, and the nonlinearities limit the system response to some large but bounded level. The open-loop gain never reaches −3. At a gain of −1, linear operation ceases. The minus sign or 180° phase shift of $G(s)H(s) = -1$ corresponds to an inversion of the intended negative feedback into positive feedback. The positive feedback produces the observed instability. A common example of this occurs with public address systems, in which some of the audio output is fed back to the audio input. In this case the audio amplifier and loudspeaker act as the open-loop gain $G(s)$, and the audio path between loudspeaker and audio input device, which is usually a microphone, acts as the feedback gain $H(s)$. The magnitude of $H(s)$ is generally much less than 1, while the magnitude of $G(s)$ is greater than 1. For some particular length of audio feedback path, the phase angle of $G(s)H(s)$ will be $-180°$. If the magnitude of $G(s)H(s)$ over this path reaches unity, the public address system will "howl." The howl is a limit-cycle response (nonlinear oscillation) of the amplifier which becomes nonlinear as its output tries to grow without bound. The common cure is to lower the gain, so that total loop gain is less than 1 at the point at which the phase angle is $-180°$.

In its most general sense, the Nyquist plot is a conformal mapping from a closed-loop path in the s plane to a closed-loop path in the $G(s)H(s)$ plane. Both the s plane and the $G(s)H(s)$ plane are complex planes, with real and imaginary axes. "Conformal" means that angles are preserved but shapes are not. The path chosen in the s plane encloses the entire right half of the s plane in order to test for any possible closed-loop poles occurring for positive real values of s. Numeric values of s along this closed path are inserted into the expression for $G(s)H(s)$, and $G(s)H(s)$ is evaluated. The resulting real and imaginary values for $G(s)H(s)$ are plotted. These points, one for each value of s, describe a locus in the $G(s)H(s)$ plane. An example is given in Fig. 34.8.

Once the locus of the open-loop gain $G(s)H(s)$ is obtained, the stability of the closed-loop system containing $G(s)H(s)$ is determined by applying the Nyquist stability rules, given in Table 34.21.

The basic principle behind the Nyquist rules stems from Cauchy's principle of the argument, which effectively leads to the following results. For every pole of $G(s)/[1 + G(s)H(s)]$ encircled by the locus in the s plane, there will be an encirclement N of the point $-1 + j0$ in the $G(s)H(s)$ plane Both encirclements will be in the same direction. The positive direction for encirclements is usually defined as clockwise. If this definition is changed to counterclockwise, the Nyquist stability criterion remains unchanged, while the direction of a positive encirclement is reversed. Either definition works. In most cases $G(s)H(s)$ has no open-loop poles in the right half of the s plane, and the Nyquist rule reduces to a simple test to see if the point $-1 + j0$ is encircled in the $G(s)H(s)$ plane, as s ranges

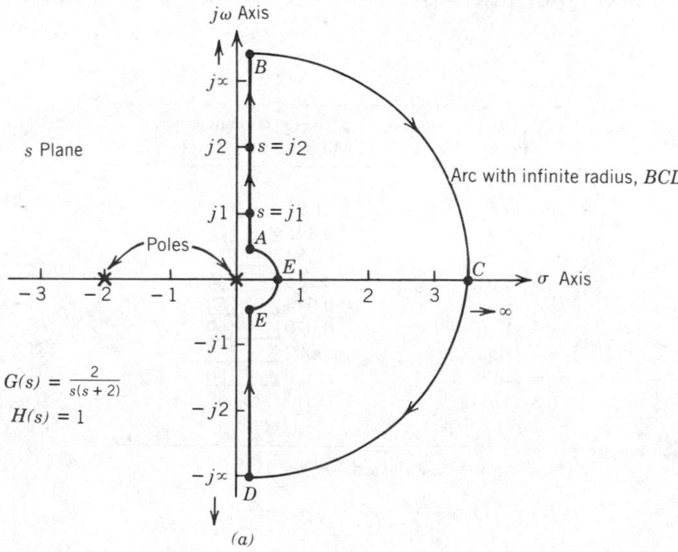

Fig. 34.8 Nyquist plot of $G(s)H(s) = 2/[s(s + 2)]$. (*a*) s-Plane locus. Path AB = principal path, for which $s = j\omega, 0 < \omega < \infty$; path EFA = arc with small radius, detouring around pole(s) at origin or on $j\omega$ axis, so that there are no poles within s-plane contour. If $G(s)H(s)$ has no poles or zeros in the right half of the s plane, only the principal path $AB(s = j0 \rightarrow j\infty)$ need be plotted in the $G(s)H(s)$ plane.

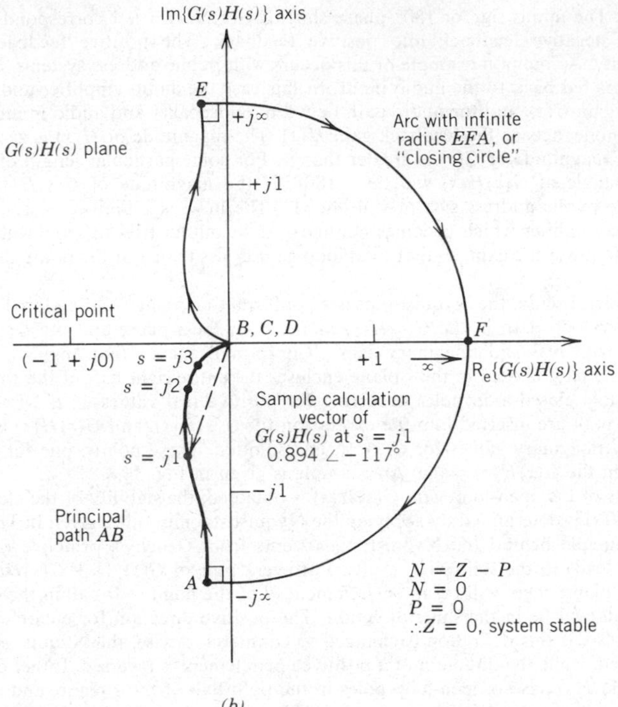

Fig. 34.8 (*b*) Stability determination, using Case I of Table 34.21. P = poles of $G(s)H(s)$ in right half plane = 0 [s plane in part (*a*) of figure]. N = number of clockwise encirclements of critical point $(-1 + j0)$ in $G(s)H(s)$ plane ($N = 0$). $N = Z - P$, Nyquist criterion. Hence $Z = 0$ and the system is stable. Only the principal path need be plotted, to see if it encircles or encloses the critical point $(-1 + j0)$. Z = zeros of $1 + G(s)H(s)$ = poles of $G(s)/[1 + G(s)H(s)]$. For stability, $Z = 0$.

Point	$G(s)H(s)$ in Magnitude $\underline{/\text{ Angle}}$
$s = j0 = A$	$\infty \underline{/-90^\circ} = A$
$s = j0.5$	$1.94 \underline{/-104^\circ}$
$s = j1.0$	$0.894 \underline{/-117^\circ} = (*)^a$
$s = j2.0$	$0.354 \underline{/-135^\circ}$
$s = j3.0$	$0.185 \underline{/-146^\circ}$
$s = j5.0$	$0.074 \underline{/-158^\circ}$
$s = j10.0$	$0.020 \underline{/-169^\circ}$
$s = j100.0$	$0.0002 \underline{/-179^\circ}$
$s = j\infty = B$ [part (*a*) of figure]	$0.0 \underline{/-180^\circ} = B$ [part (*b*) of figure]

[a](*) = sample calculation for $s = j1.0$

$$G(j1)H(j1) = \frac{2}{s(s+2)}\Big|_{s=j1} = \frac{2}{(j1)(j1+2)}$$

$$G(j1)H(j1) = \frac{2}{1\underline{/90^\circ}\,2.236\underline{/26.6^\circ}} = \frac{2}{2.236\underline{/116.6^\circ}}$$

$$= 0.894\underline{/-116.6^\circ}$$

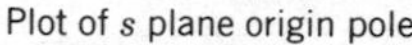

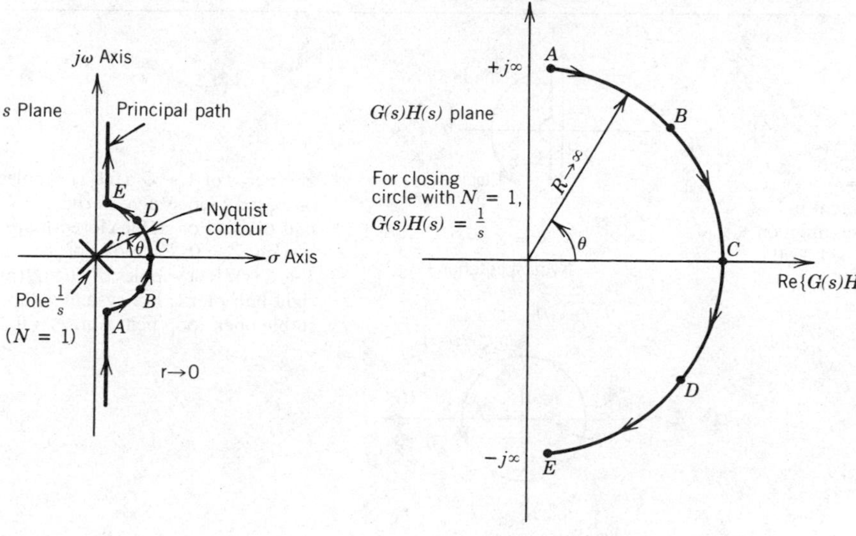

(c)

Arc($ABCDE$) = Semicircular detour about pole $1/s$ with infinitesimal radius r
Let $s = re^{j\theta} = r\angle\theta$ where $r \to O$

Arc($ABCDE$) = one half-circle of infinite radius R; $R = 1/r$ For pole $1/s$ $G(s)H(s) = 1/s = 1/re^{j\theta} = R^{-j\theta}$ $= R^{j\phi}$ $R = 1/r \to \infty$ $\phi = -\theta$ = one-half circle, reverse direction for general nth-order pole $1/s^N$ $R = 1/r^N \to \infty$; $\phi = -N\theta = N$ half circles

Fig. 34.8 (c) Plot characteristics—closing circle—for $G(s)H(s) = K/s^N(s + P_1)(s + P_2)\ldots(s + P_M)(s^2 + 2\zeta\omega_n s + \omega_n^2)$. The pole at the origin, $1/s^N$, will produce N half circles of infinite radius in the $G(s)H(s)$ plane, circling in opposite direction to the vanishingly small detour about pole in s plane. These so-called closing circles in $G(s)H(s)$ plane meet the principal part of the $G(s)H(s)$ locus at a right angle.

TABLE 34.21. NYQUIST LOCUS RULES FOR STABILITY

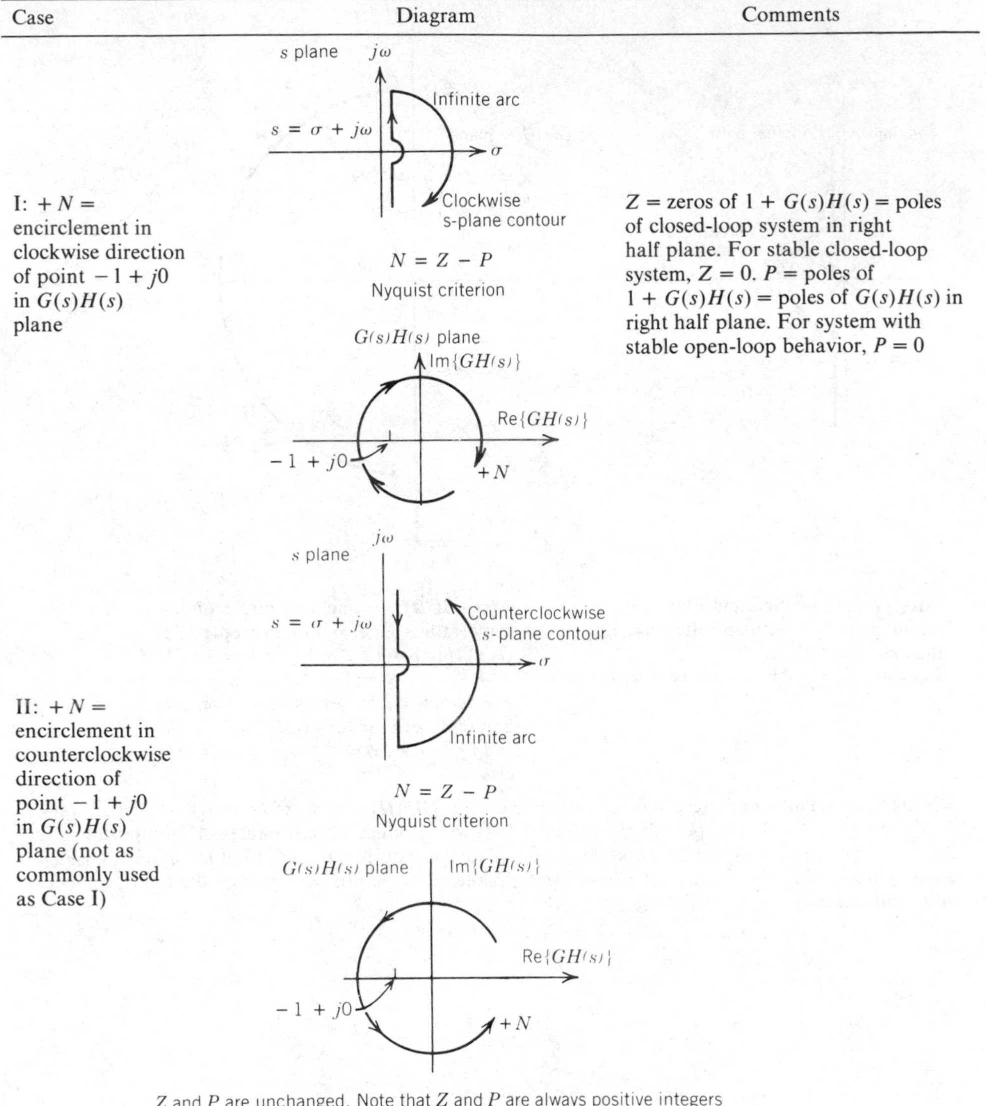

Case	Diagram	Comments
I: $+N =$ encirclement in clockwise direction of point $-1 + j0$ in $G(s)H(s)$ plane	$N = Z - P$ Nyquist criterion	$Z =$ zeros of $1 + G(s)H(s) =$ poles of closed-loop system in right half plane. For stable closed-loop system, $Z = 0$. $P =$ poles of $1 + G(s)H(s) =$ poles of $G(s)H(s)$ in right half plane. For system with stable open-loop behavior, $P = 0$
II: $+N =$ encirclement in counterclockwise direction of point $-1 + j0$ in $G(s)H(s)$ plane (not as commonly used as Case I)	$N = Z - P$ Nyquist criterion	

Z and P are unchanged. Note that Z and P are always positive integers (or zeroes), whereas N may be a positive, zero, or negative integer.

around its closed contour. If the -1 point is encircled in the $G(s)H(s)$ plot, then $G(s)/[1 + G(s)H(s)]$ has a pole for some value of s within the s plane locus, which includes all values of s with positive real parts. This positive real factor will produce an unstable closed-loop system response.

In reality the Nyquist rules are expressed in terms of the zeros Z of $1 + G(s)H(s)$. These zeros are the poles of $G(s)/[1 + G(s)H(s)]$, as defined in Table 34.21, which sets forth the Nyquist rule or criteria. In both cases shown in Table 34.21, the Nyquist rule or criterion is the same, namely

$$N = Z - P \tag{34.21}$$

In all cases, for stability Z must equal zero, indicating that the characteristic equation $1 + G(s)H(s)$ has no zeros and hence that the closed-loop transfer function $G(s)/[1 + G(s)H(s)]$ has no poles.

Note that an unstable open-loop system $G(s)H(s)$ with $P \neq 0$ may become a stable closed-loop system if $Z = 0$ or if $N = P$. Shinners[8] gives such an example, along with many others.

When constructing a Nyquist plot without the use of computer or calculator, it is often simplest to first construct Bode plots of $|G(j\omega)H(j\omega)|$ vs. ω, and $\phi(\omega)$ = phase angle of $G(j\omega)H(j\omega)$ vs. ω. Then by choosing values of ω, one can readily obtain pairs of points $|G(j\omega)H(j\omega)|$ vs. $\phi(\omega)$ for the Nyquist plot.

Root Locus Method. The root locus method, developed by Evans,[11] is a plot of a system's closed-loop pole locations in the s plane as the forward gain constant is varied, usually from $K = 0$ to $K = \infty$. The forward plant gain is taken as $KG(s)$, where K is the gain constant to be selected. The corresponding closed-loop transfer function, from Table 34.8, the basic negative feedback loop, with $G(s)$ replaced by $KG(s)$, is

$$G_1(s) = \frac{KG(s)}{1 + KG(s)H(s)} \tag{34.22}$$

where K = gain constant of $G(s)$
$G(s)$ = forward plant gain or transfer function, open loop
$H(s)$ = feedback gain or transfer function
$G_1(s)$ = closed-loop gain

Poles of $G_1(s)$ occur at the zeros or roots of

$$1 + KG(s)H(s) = 0 \tag{34.23}$$

or

$$KG(s)H(s) = -1 = 1\angle \pm 180° \pm K360° \qquad K = 0, 1, 2, \ldots \tag{34.24}$$

The equation for the roots involves a complex or vector quantity $KG(s)H(s)$ equated to a real constant -1 or equivalently 1 at an angle $\pm 180°$. There are thus two requirements for a particular value of s, say s_0, to satisfy the root locus equation, namely

Magnitude condition

$$|KG(s_0)H(s_0)| = |-1| = 1 \tag{34.25}$$

Angle condition

$$\text{Angle}\,[KG(s_0)H(s_0)] = \text{angle}\,[G(s)H(s)] = \pm 180°(2k + 1) \qquad k = 0, 1, 2, \ldots \tag{34.26}$$

In almost all cases the angle condition uses the angle $-180°$, rather than the equivalent angles of, say, $+180°$ or $-540°$. This occurs because the phase shift of most real systems ranges between 0° and $-270°$, although larger ranges are met in practice. Examples and very thorough discussions of the techniques of constructing a root locus plot are given by Ogata,[6] Kuo,[7] and Shinners.[8] The rules for constructing a root locus plot follow.

1. Real axis locii. Plot the n poles of $KG(s)H(s)$ and the m finite zeros of $KG(s)H(s)$. Zeros at infinity are not plotted. Root locii will lie on the real axis, starting at a pole and ending at a zero, when the total number of real poles and real zeros of $G(s)H(s)$ to the *right* of the locii section is odd, for $K > 0$. (For the complementary root locii, for which $K < 0$, change right to left.) If there are locii on the real axis between two poles, then a break-away point exists between the poles; and if there are locii between two zeros, then a break-in point exists between the zeros. If $G(s) = K(s + 1)/[(s + 2)(s + 3)]$, the zeros are $Z_1 = -1$, $Z_2 = \infty$ and the poles are $P_1 = -2$, $P_2 = -3$. Locii lie on the real axis between $-\infty$ and -3 and between -2 and -1.

2. Starting and ending points. Put arrows on the locii as they are sketched in. The arrows indicate increasing gain K. Locii start at poles, with arrows pointing away, and $K = 0$ at the pole. Locii end at zeros, with $K = \pm\infty$, with arrows pointing toward the zero. There are usually zeros at infinity, which are approached asymptotically. If $G(s) = K(s + 1)/[(s + 2)(s + 3)]$

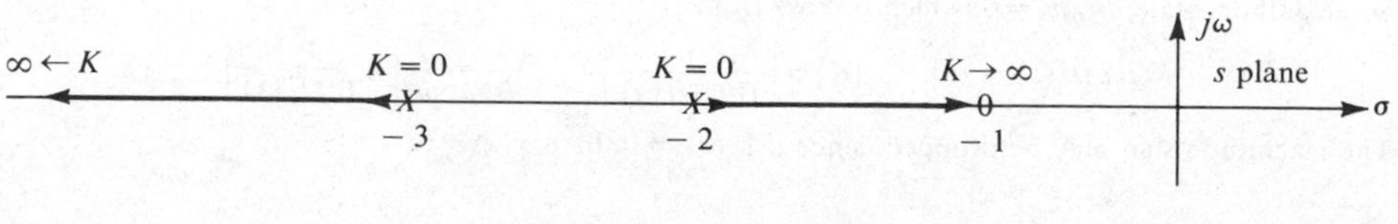

$\infty \leftarrow K$ | $K = 0$ | $K = 0$ | $K \to \infty$ | $j\omega$ | s plane | σ

-3 | -2 | -1

3. Number of locii. One locus starts at each pole. There are thus n locii, where n is the number of poles or the order of $G(s)H(s)$. If $G(s)H(s) = K(s+1)/[s^2(s+2)(s+3)]$ there are four locii.

4. Angle of asymptotes. For $K > 0$, the first angle = 180°/number of asymptotes = $180°/(n-m)$. For $K < 0$, the first angle = 360°/number of asymptotes = $360°/(n-m)$, where n = number of finite poles and m = number of finite zeros: The remaining asymptotes are always equally spaced in angle, for example, for three asymptotes, the asymptotes are 120° apart (360°/3). See below:

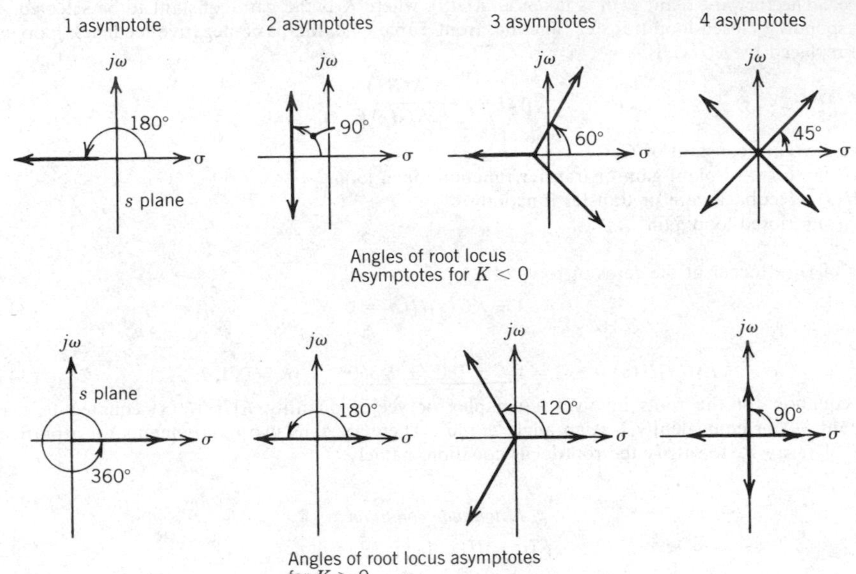

5. Intersection of asymptotes. The asymptotes always intersect on the real axis, at a point $\sigma = \sigma_A$

$$\sigma_A = \frac{\sum P - \sum Z}{n - m}$$

where P = value of s at pole of $G(s)H(s)$
Z = value of s at zero of $G(s)H(s)$
n = number of poles (finite)
m = number of zeros (finite)

$$\text{If } G(s) = \frac{K(s+1)}{s^2(s+2)(s+3)}, \qquad \sigma_A = \frac{\sum(0+0-2-3) - \sum(-1)}{4-1}$$

$$\sigma_A = \frac{-5+1}{3} = \frac{-4}{3}$$

6. Number of asymptotes. The number of asymptotes is equal to $n - m$, the pole-zero excess, which is also the number of zeros at infinity.

7. Break-away or break-in points. First, estimate the location of any break-away or break-in points, since the following equation gives a necessary but not a sufficient condition. That is, several possible values are generally found, some or none of which are actually break-away points for the locus being plotted. All break-away points for $K > 0$ and $K < 0$, are found by finding the roots of

$$\frac{d}{ds}\left[\frac{1}{G(s)H(s)}\right] = 0 \qquad \text{or} \qquad \frac{d}{ds}[G(s)H(s)] = 0$$

Some authors state $dk/ds = 0$, which follows from

$$KG(s)H(s) = -1 \qquad |K| = \left|\frac{1}{G(s)H(s)}\right| \qquad \frac{dk}{ds} = \frac{d}{ds}\left|\frac{1}{G(s)H(s)}\right|$$

The magnitude sign may be dropped, since if $|x(s)| = 0$, then $x(s) = 0$.

8. $j\omega$ axis intersection. Intersection with the $j\omega$ axis may be found from the auxiliary equation above a row of zeros in the Routh–Hurwitz array, as discussed in Example 34.2, Fig. 34.12, p. 1060.

9. Calculation of K on the locii. The magnitude of K at any point $s = s_K$ on the root locii may be found from

$$|K| = \frac{1}{|G(s)H(s)|}\bigg|_{s=s_K}$$

10. Angles of departure and arrival. These angles may be found by picking a test point very close to the pole or zero and applying the angle condition, which results in

$$\text{Test point } s = s_T \qquad \angle G(s_T)H(s_T) = -180° \text{ or } (2K+1)\pi \qquad K = 0 \pm 1 \pm 2 \cdots$$

and

$$\angle G(s_T)H(s_T) = \angle\text{zeros} - \angle\text{poles} = \sum_{i=1}^{m} \angle s + Zi - \sum_{j=1}^{n} \angle s + Pj = -180°$$

All the angles of the zeros and poles to the test point s_T can be measured with a protractor or found analytically, except for the angle from the pole (or zero) that is very close to the test point. This angle is called θ and is the only unknown angle in the angle condition. It is thus readily found. For the complementary root locus, the $-180°$ angle condition changes to 0° or 360° or $2K\pi$, $K = 0 \pm 1 \pm 2 \cdots$.

Nichols Chart. The Nichols chart is used to convert open-loop gain and phase data into closed-loop gain and phase data for unity feedback control systems. The chart graphically converts $G(j\omega)$ into $G(j\omega)/[1 + G(j\omega)]$. The open-loop gain (in decibels) and phase (in degrees) are plotted via a rectangular external coordinate system. The closed-loop gain and phase corresponding to this point are read off an internal curvilinear coordinate system. The external coordinates 0 dB at $-180°$ phase correspond to infinite closed-loop decibels and the center of the curvilinear internal coordinate system.

Table 34.22 presents Bode plots, Nyquist plots, Nichols charts, and root locus plots for 12 common transfer functions.

Other Methods. Two other methods are commonly employed to determine system stability by manipulating Eq. (34.19). The Bode, Nyquist, and root locus methods use only the term $G(s)H(s)$ of Eq. (34.19) and test for $G(s)H(s) = -1$. The remaining two methods require that $1 + G(s)H(s)$ be multiplied out and put in the form of one numerator polynomial $N(s)$ over one denominator polynomial $D(s)$. The roots of $N(s)$ are then the poles of the closed-loop system transfer function

$$\frac{G(s)}{1 + G(s)H(s)} = \frac{G(s)}{N(s)/D(s)} = \frac{G(s)D(s)}{N(s)} \tag{34.27}$$

Thus the numerator of the characteristic equation $1 + G(s)H(s)$ becomes the denominator of the closed-loop system transfer function $G(s)/[1 + G(s)H(s)]$. The roots of $N(s)$ are called the poles or eigenvalues of the system. If any of these roots are positive real numbers or complex numbers with a positive real part, the system will exhibit an unbounded step response, produced by a term or factor of the form $\exp(A \cdot t)$, where A is the positive real quality.

The conceptually simplest method employed to determine system stability is to factor the numerator $N(s)$ of the system's characteristic equation, or equivalently, find the roots of $N(s) = 0$. If any of these roots have positive real parts, the system is unstable. This approach is sometimes called the direct method of Liapunov.

An indirect method exists that tests for the existence of any roots with positive real parts, without actually finding the roots. This method is called the Routh–Hurwitz test. Both Routh and Hurwitz were mathematicians who competed for the Adams Prize, offered for the solution of a stability problem in celestial mechanics. In 1877 Routh won the prize with his paper, "Essay on the stability of a given state of motion," in which he developed an array computation to test for the existence of any roots of a polynomial with positive real parts. The Routh array test is shown in Fig. 34.9.

At the same time, Hurwitz developed an equivalent test utilizing a matrix approach. For an nth order polynomial, the Hurwitz matrix is $N \times N$. The first row of the Hurwitz matrix is the second row of the Routh array, and vice versa. Every following pair of Hurwitz rows is constructed from the previous pair by shift of the original pair one column to the right and insertion of a leading zero in each row. Any terms that would fall outside the $N \times N$ matrix boundaries are omitted. The leading principal minors 1×1, 2×2, $3 \times 3, \ldots, N \times N$ are extracted from the $N \times N$ matrix and their determinants evaluated, as shown in Fig. 34.10. If all these determinants are positive, the system is stable.

Example 34.1 First Routh–Hurwitz Array Example, Stable System. Given

$$1 + G_1(s)H_1(s) = \frac{N_1(s)}{D_1(s)} = \frac{(s+1)(s+2)(s+3)(s+4)}{D(s)} = \text{a stable characteristic equation}$$

or

$$N_1(s) = s^4 + 10s^3 + 35s^2 + 50s + 24$$

or

$$N_1(s) = A_4s^4 + A_3s^3 + A_2s^2 + A_1s + A_0$$

To test $N_1(s)$ for any possible roots with a positive real part, form the Routh array, as shown in Fig. 34.9. The number of roots with positive real parts is equal to the number of sign changes between the coefficients of the first column of the Routh array. Since these coefficients are all positive (1, 10, 30, 42, 24), the system is stable.
The complexities and special cases for the Routh array test are illustrated by examples for (1) an unstable system (Example 34.2), (2) a system that produces a row of zeros in the Routh array (Example 34.3), and (3) a system that produces a zero first column coefficient (Example 34.4).

Example 34.2 First Method of Liapunov and Routh–Hurwitz Methods Second Example, Unstable System.
Given

$$G(s) = \frac{s+1}{s(s^2+s+4)} \quad \text{and} \quad H(s) = \frac{s+2}{s+5} \tag{34.28}$$

as shown in Fig. 34.11. The Bode, Nyquist, and root locus methods use the open-loop gain, $G(s)H(s)$, to predict stability.

$$G(s)H(s) = \frac{s+1}{s(s^2+s+4)} \cdot \frac{s+2}{s+5} = \frac{(s+1)(s+2)}{s(s+5)(s^2+s+4)} \tag{34.29}$$

The Routh–Hurwitz method and the first method of Liapunov use the numerator of $1 + G(s)H(s)$, $N(s)$.

$$1 + G(s)H(s) = 1 + \frac{(s+1)(s+2)}{s(s+5)(s^2+s+4)} = \frac{N_2(s)}{D_2(s)} \tag{34.30}$$

Note. $N_2(s)$ is the particular value of $N(s)$ used in this example. Combining terms:

$$1 + G(s)H(s) = \frac{s(s+5)(s^2+s+4) + (s+1)(s+2)}{s(s+5)(s^2+s+4)} = \frac{N_2(s)}{D_2(s)} \tag{34.31}$$

Compare Eqs. (34.29) with (34.31). Note how the addition of a single term, 1, produces a great number of terms in the numerator of $1 + G(s)H(s)$, compared with the relatively simple numerator of $G(s)H(s)$.

$$\begin{aligned} N_2(s) &= s(s+5)(s^2+s+4) + (s+1)(s+2) \\ N_2(s) &= (s^4 + 2s^3 + 9s^2 + 20s) + (s^2 + 3s + 2) \\ N_2(s) &= s^4 + 2s^3 + 10s^2 + 23s + 2 \end{aligned} \tag{34.32}$$

Factoring $N_2(s)$, as required by the first method of Liapunov, is often difficult, especially as the order of $N(s)$ increases. Using an HP-41c calculator and a program from the HP mathematics package (HP-41c Mathematics Pac, Polynomial Solutions)

$$N_2(s) = (s + 0.09045)(s + 2.1416)(s - 0.1160 + j3.211)(s - 0.1160 - j3.211) \tag{34.33}$$

Thus the roots of $N_2(s)$, the numerator of the closed-loop system's characteristic equation $1 + G(s)H(s)$, are

$$\lambda_1 = -0.09045 \qquad \lambda_3 = 0.1160 - j3.211 \qquad \lambda_2 = -2.1416 \qquad \lambda_4 = 0.1160 + j3.211 \tag{34.34}$$

It is important to remember that the numerator $N_2(s)$ of $1 + G(s)H(s)$ is the denominator of the closed-loop system transfer function, as shown in Fig. 34.11. Thus the roots of $N(s)$ are the *zeros* of $1 + G(s)H(s)$, which in turn are the *poles* of the closed-loop transfer functions $G(s)/[1 + G(s)H(s)]$.

If any of these roots have a positive real part, the system will exhibit an unbounded response for a bounded input and will behave in an unstable fashion, driven by a term or factor of the form $e^{\lambda t}$, where λ is the positive real part.

Referring to the four roots of $N_2(s)$ given by Eq. (34.34), two roots are observed to have positive real parts, namely

$$\lambda_3 = 0.1160 - j3.211 \quad \text{and} \quad \lambda_4 = 0.1160 + j3.211$$

Thus in accordance with the first method of Liapunov, the closed-loop control system described by Eq. (34.28) is unstable, with two unstable or right-half plane poles.

In the Routh–Hurwitz method, the same $N(s)$, or numerator of the characteristic equation $1 + G(s)H(s)$, is used that is used by the direct method of Liapunov. Obtaining $N_2(s)$ from Eq. (34.32), the first two rows of the Routh tabulation are written down, using the coefficients of $N_2(s)$. The remaining rows are then calculated sequentially, resulting in Fig. 34.12.

There are two sign changes, from $+2$ to $-3/2$ and then from $-3/2$ to 25 2/3. Hence according to the Routh–Hurwitz tabulation, $N(s)$ has two roots with positive real parts, which will produce an unstable system. This agrees with the results of the first method of Liapunov, wherein the two roots with positive real parts were actually found. One method serves as a check on the other method.

If a computer or calculator capable of factoring a fourth-order (or higher) polynomial is not available, the Routh–Hurwitz method may be used to determine whether $N(s)$ has any roots with positive real parts. Note that the Routh–Hurwitz method does not, in general, find the roots of $N(s)$, it merely checks for the existence of any roots with positive real parts, without actually finding the roots. In the special case of purely imaginary roots, or roots with no real part, the imaginary roots (but not all the roots) can be found by an additional procedure, illustrated in the third Routh–Hurwitz example, shown in Fig. 34.13.

	s^4	$+10s^3$	$+35s^2$	$+50s$	$+24$
s^4	1		35		24
s^3	10		50		0
s^2	$\dfrac{(10)(35)-(1)(50)}{10} = 30$		$\dfrac{(10)(24)-(1)(0)}{10} = 24$		0
s^1	$\dfrac{(30)(50)-(10)(24)}{30} = 42$		$\dfrac{(30)(0)-(10)(0)}{30} = 0$		0
s^0	$\dfrac{(42)(24)-(30)(0)}{42} = 24$		0		0

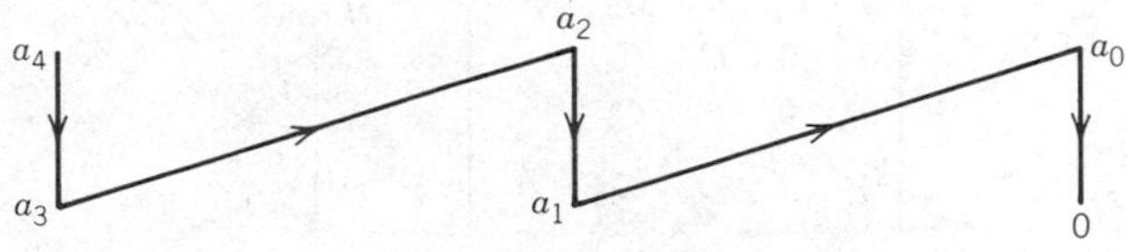

Fig. 34.9 First Routh–Hurwitz example (Example 34.1 in text) for $N_1(s) = s^4 + 10s^4 + 35s^2 + 50s + 24 = 0 = A_4s^4 + a_2s^3 + A_2s^2 + A_1s + A_0$. The coefficients of the first two rows are obtained directly from $N(s)$ as shown in the figure. The coefficient of each succeeding row is found from the previous two rows by multiplying out a 2×2 determinant formed of the first and second columns, then the first and nth columns—with the normal determinant signs *reversed*. The values thus obtained are divided by the previous coefficient of the first column. Thus to find the s^2 row coefficients

$$\text{column 1 coeff.} = \frac{\begin{vmatrix} 1 & 35 \\ 10 & 50 \end{vmatrix}}{10} = \frac{(A_3)(A_2) - (A_4)(A_1)}{A_2} = 30$$

$$\text{column 2 coeff.} = \frac{\begin{vmatrix} 1 & 24 \\ 10 & 0 \end{vmatrix}}{10} = \frac{(A_3)(A_0) - (A_4)(0)}{A_2} = 24$$

Note the sign convention $\begin{vmatrix} - & + \end{vmatrix}$ which is opposite that for a 2×2 determinant.

$$N(s) = s^4 + 10s^3 + 35s^2 + 50s + 24 = 0$$

Fourth-order Hurwitz determinant

$$D_4 = \begin{vmatrix} 10 & 50 & 0 & 0 \\ 1 & 35 & 24 & 0 \\ 0 & 10 & 50 & 0 \\ 0 & 1 & 35 & 24 \end{vmatrix} = 4 \times 4 \text{ Hurwitz determinant}$$

$$D_1 = |10| = 10 > 0$$

$$D_2 = \begin{vmatrix} 10 & 50 \\ 1 & 35 \end{vmatrix} = 350 - 50 = 300 > 0$$

$$D_3 = \begin{vmatrix} 10 & 50 & 0 \\ 1 & 35 & 24 \\ 0 & 10 & 50 \end{vmatrix} = \begin{array}{l} 10 \cdot 35 \cdot 50 + 1 \cdot 10 \cdot 0 + 0 \cdot 50 \cdot 24 \\ -(0 \cdot 35 \cdot 0 + 24 \cdot 10 \cdot 10 + 50 \cdot 50 \cdot 1) \end{array}$$

$$D_3 = 17{,}500 - 2{,}400 - 2{,}500 = 12{,}600 > 0$$

To evaluate D_4, expand by minors along column 4

$$D_4 = -0\begin{vmatrix} 1 & 35 & 24 \\ 0 & 10 & 50 \\ 0 & 1 & 35 \end{vmatrix} + 0\begin{vmatrix} 10 & 50 & 0 \\ 0 & 10 & 50 \\ 0 & 1 & 35 \end{vmatrix} - 0\begin{vmatrix} 10 & 50 & 0 \\ 1 & 35 & 24 \\ 0 & 1 & 35 \end{vmatrix} + 24\begin{vmatrix} 10 & 50 & 0 \\ 1 & 35 & 24 \\ 0 & 10 & 50 \end{vmatrix}$$

or $D_4 = 24\ D_3$, $D_3 = 302{,}400 > 0$.

In summary:

$D_1 =$	10	All $D_n > 0$
$D_2 =$	300	∴ all roots of $N(s)$ have negative
$D_3 =$	12,600	real part
$D_4 =$	302,400	∴ stable characteristic equation

For a general nth-order characteristic equation with real time-invariant constant coefficients

$$N(s) = C_0 s^n + C_1 s^{n-1} + C_2 s^{n-2} + \cdots + C_{n-2}s^2 + C_{n-1}s + C_n = 0$$

the form of the $N \times N$ Hurwitz determination D_n is

$$\begin{vmatrix} C_1 & C_3 & C_5 & C_7 & \cdots \\ C_0 & C_2 & C_4 & C_6 & \cdots \\ 0 & C_1 & C_3 & C_5 & \cdots \\ 0 & C_0 & C_2 & C_4 & \cdots \\ 0 & 0 & C_1 & C_3 & \cdots \\ 0 & 0 & C_0 & C_2 & \cdots \\ \vdots & \vdots & \vdots & \vdots & \ddots\ C_n \end{vmatrix} \quad N \text{ rows}$$

$\longleftarrow N \text{ columns} \longrightarrow$

Fig. 34.10 Hurwitz matrix test. The same fourth-order stable characteristic equation as used for the Routh array is used here.

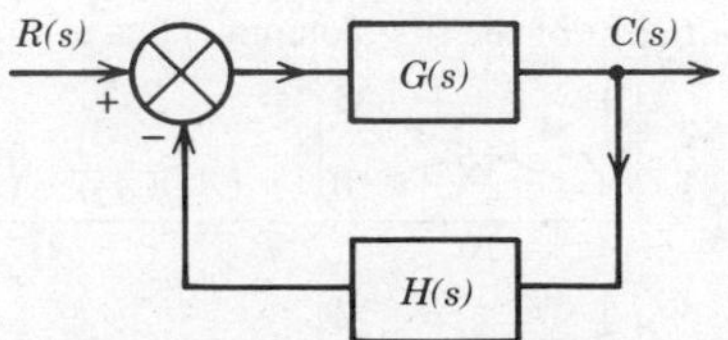

Fig. 34.11 Numerator of $1 + G(s)H(s)$ is denominator of the closed-loop transfer function. Open-loop gain = $G(s)H(s)$. Characteristic equation = $1 + G(s)H(s) = N(s)/D(s)$ = numerator(s)/denominator(s). Closed-loop transfer function = $G(s)/[1 + G(s)H(s)] = G(s)/[N(s)/D(s)] = [G(s)D(s)]/N(s)$.

TABLE 34.22.[a]

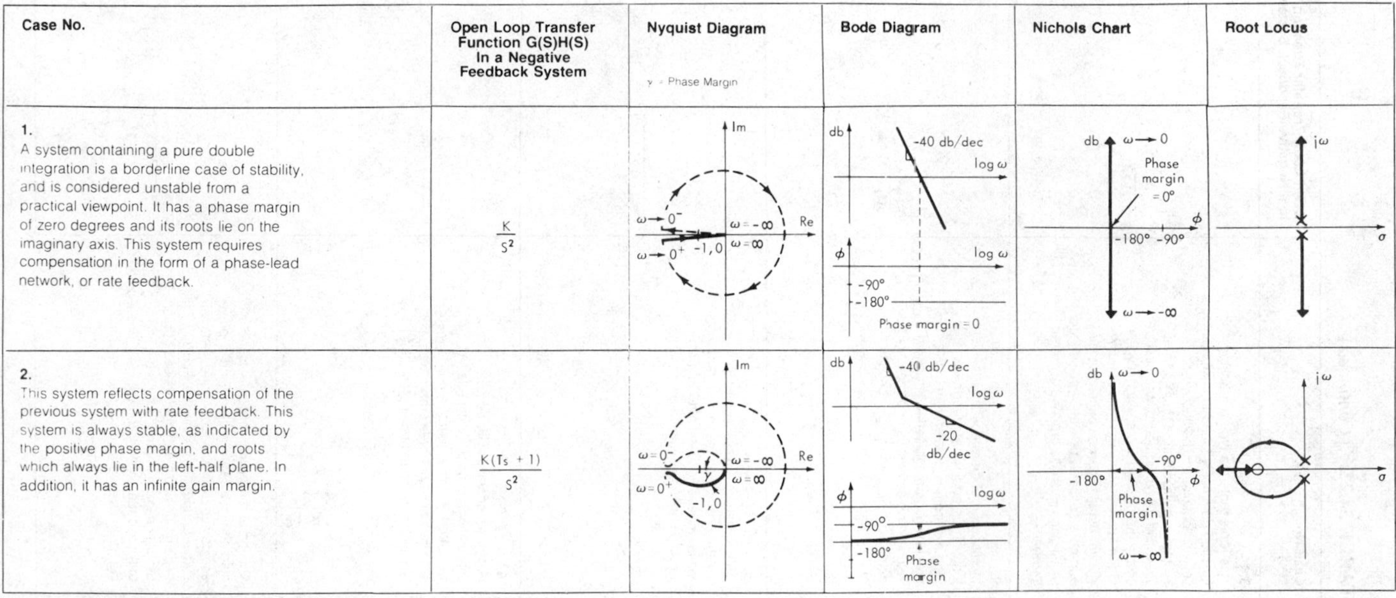

Case No.	Open Loop Transfer Function G(S)H(S) In a Negative Feedback System	Nyquist Diagram	Bode Diagram	Nichols Chart	Root Locus
1. A system containing a pure double integration is a borderline case of stability, and is considered unstable from a practical viewpoint. It has a phase margin of zero degrees and its roots lie on the imaginary axis. This system requires compensation in the form of a phase-lead network, or rate feedback.	$\frac{K}{S^2}$				
2. This system reflects compensation of the previous system with rate feedback. This system is always stable, as indicated by the positive phase margin, and roots which always lie in the left-half plane. In addition, it has an infinite gain margin.	$\frac{K(Ts+1)}{S^2}$				

[a] Ref. 12. With the permission of *Control Engineering*.

TABLE 34.22. (*Continued*)

Case No.	Open Loop Transfer Function G(S)H(S) In a Negative Feedback System
3. This represents the type of transfer function commonly found in instrument servomechanisms. The second-order system is always stable (from a linear viewpoint), as indicated by the positive phase margin and roots which always lie in the left-half plane. Its gain margin is infinite. Stability can be increased by adding a cascade network (phase-lead, phase lag, or phase-lag-lead) or rate feedback.	$\frac{K}{S(T_1S+1)}$
4. Modification of the preceding transfer function by adding rate feedback. The phase margin is greater than in the preceding case, indicating improved stability characteristics and greater damping. The gain margin remains infinite.	$\frac{K(T_1S+1)}{S(T_2S+1)}, \quad T_1>T_2$
5. This system is always stable, as indicated by the positive phase margin and roots which always lie in the left-half plane. In addition, it has an infinite gain margin.	$\frac{K}{(T_1S+1)(T_2S+1)}, \quad T_1>T_2$
6. A system containing three pure integrations is always unstable. As indicated, it has a negative phase margin of minus ninety degrees, and two roots always lie in the right-half plane (for K>o.)	$\frac{K}{S^3}$
7. This type of transfer function is found in field-controlled servomotors, and in power servomechanisms. Although the system is stable for the gains shown, it would become unstable at higher values of gain. Stability can be improved by the addition of a cascade network (phase lead, phase lag, or phase lag-lead) or rate feedback.	$\frac{K}{S(T_1S+1)(T_2S+1)}, \quad T_1>T_2$

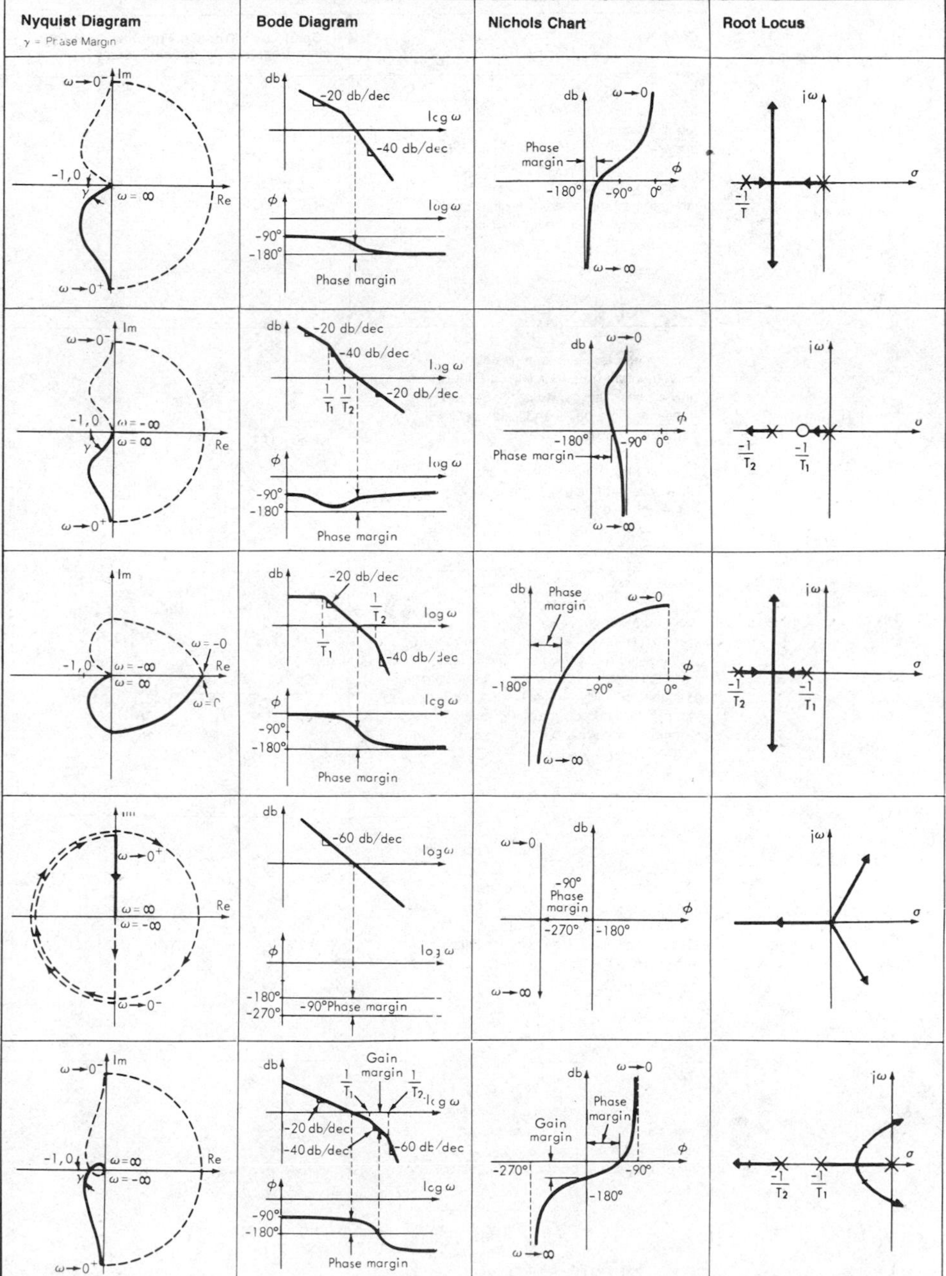
Nyquist Diagram
γ = Phase Margin
Bode Diagram
Nichols Chart
Root Locus
ω→0⁻
Im
Re
-1,0
ω = ∞
ω = -∞
ω→0⁺
ω = -0
ω = 0
db
-20 db/dec
-40 db/dec
-60 db/dec
log ω
ϕ
-90°
-180°
-270°
Phase margin
Gain margin
-90°Phase margin
ω→∞
0°
jω
σ
-1/T
-1/T₁
-1/T₂
1/T₁
1/T₂

TABLE 34.22. (*Continued*)

Case No.	Open Loop Transfer Function G(S)H(S) In a Negative Feedback System
8. Modification of the preceding transfer function, with higher gain for greater accuracy and the addition of rate feedback, results in this transfer function. As shown, the system is stable, as indicated by the phase margin and roots which always lie in the left-half plane.	$\frac{K(T_1S+1)}{S(T_2S+1)(T_3S+1)}, \quad T_2>T_1>T_3$
9. This system represents an instrument servomechanism containing rate feedback and a phase-lag network. It is capable of achieving high gains and very good accuracy. As indicated in Table 1, the system is stable as reflected by the positive phase margin, and roots which always lie in the left-half plane for all values of gain.	$\frac{K(T_1S+1)(T_2S+1)}{S(T_3S+1)(T_4S+1)}, \quad T_3>T_4>T_1>T_2$
10. Stability of this third-order system is dependent on the value of the gain. For the gain selected in the illustration of Table 1, the system is shown as being stable. For higher values of gain, it would become unstable. Stability of this system (for the gain shown) can be improved by the addition of rate feedback or a phase-lead network.	$\frac{K}{(T_1S+1)(T_2S+1)(T_3S+1)}, \quad T_1>T_2>T_3$
11. Modification of the transfer function, with higher gain (for greater accuracy) and the preceding addition of rate feedback results in this transfer function. As indicated, the system is stable as reflected by the positive phase margin and roots which always lie in the left-half plane for all values of gain.	$\frac{K(T_1S+1)}{(T_2S+1)(T_3S+1)(T_4S+1)}, \quad T_2>T_3>T_1>T_4$
12. This fourth-order system, containing three pure integrations, is always unstable even with the presence of two zeros. An additional phase-lead network, or rate feedback, would be required to stabilize it.	$\frac{K(T_1S+1)(T_2S+1)}{S^3(T_3S+1)}, \quad T_3>T_1>T_2$

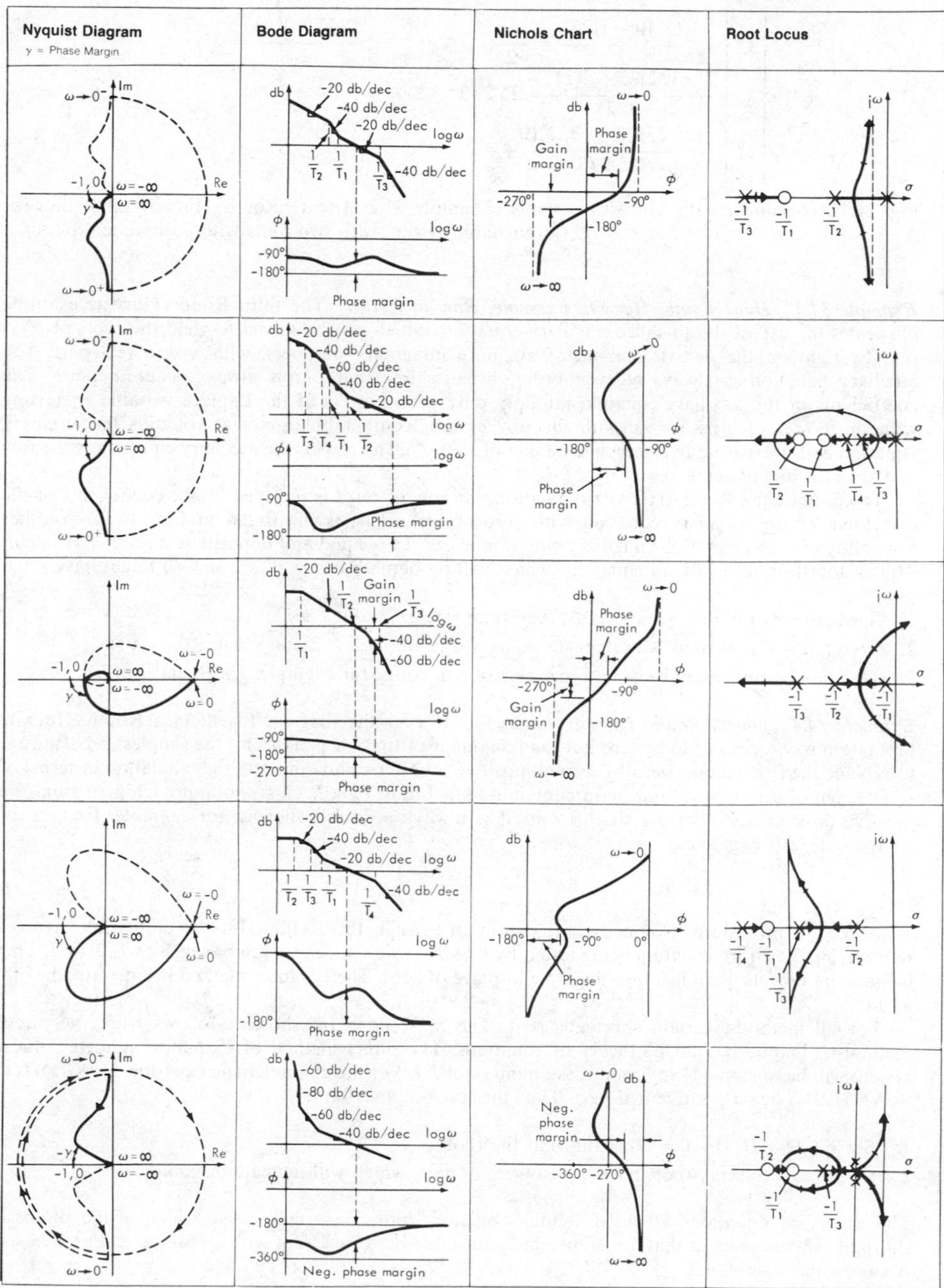

Nyquist Diagram
γ = Phase Margin
Bode Diagram
Nichols Chart
Root Locus
Phase margin
Gain margin
Neg. phase margin

Sign changes		s^4 $+2s^3$	$+10s^2$ $+23s$	$+2$
+	s^4	1	10	2
+	s^3	2	23	0
−	s^2	$\frac{(2)10-1(23)}{2}=\frac{-3}{2}$	$\frac{2(2)-1(0)}{2}=2$	
+	s^1	$\frac{(-3/2)23-2(2)}{(-3/2)}=25\ 2/3$	0	0
+	s^0	$\frac{(25\ 2/3)2-(3/2)(0)}{(25\ 2/3)}=2$	0	0

Fig. 34.12 Second Routh–Hurwitz example (Example 34.2 in text). Routh–Hurwitz tabulation for $N_2(s)=s^4+2s^3+10s^2+23s+2=0$ (an unstable system with two roots with positive real parts).

Example 34.3 Third Routh–Hurwitz Example, Row of Zeros. The third Routh–Hurwitz example illustrates the use of the so-called auxiliary equation, which can be solved to yield the roots of $N(s)$ that lie right on the $j\omega$ axis and hence are pure imaginary numbers, with a zero real part. The auxiliary equation is always of even order, because imaginary roots always occur in pairs. The coefficients of the auxiliary equation multiply only even powers of the Laplace variable s, starting with the power of s associated with the row of the Routh tabulation that contains the auxiliary equation and decreasing by successive factors of s^2 to s^0. The row of the auxiliary equation is the row just above a row of zeros.

To continue the Routh–Hurwitz tabulation, the row of zeros is replaced by the coefficients of the derivative of the auxiliary equation with respect to s. This is illustrated in Fig. 34.13. Another possibility of root pairs of the auxiliary equation exists. Two equal and opposite real roots may occur. Thus a fourth-order auxiliary equation (which will be biquadratic in s^4, s^2, and s^0) might have

1. Two pairs of pure imaginary roots, for example, $s\pm j1$ and $s\pm j3$.
2. Two pairs of real roots with opposite signs, for example, $s\pm 2$ and $s\pm 4$.
3. One pair of imaginary roots and one pair of real roots, for example, $s\pm j6$ and $s\pm 7$.

Example 34.4 Fourth Routh–Hurwitz Example. If ever the first coefficient in a Routh–Hurwitz tabulation row turns out to be zero but the remainder of the row is nonzero, the simplest procedure is to replace the first column zero by a small positive number ϵ and continue the tabulation in terms of ϵ. The sign of any first-column term containing ϵ is found by allowing ϵ to approach zero from the positive direction, so that the sign of ϵ itself is always positive, whereas, for example, the sign of $(+10\epsilon-1)/\epsilon$ is negative

$$\lim_{\epsilon\to+0}\frac{10\epsilon-1}{\epsilon}\to\frac{-1}{\epsilon}\to-\infty$$

A second procedure to remedy a first-column zero in the Routh–Hurwitz tabulation involves multiplying the entire equation being tested by $(s+k)$, where k is any number such as 1, 2, or 3. This is much more work than just inserting an ϵ in place of zero. The epsilon ϵ method is illustrated in Fig. 34.14.

For all methods, certain shortcuts exist. The most powerful shortcut follows from the direct method of Liapunov and the theory of equations. The direct method of Liapunov indicates that a system will be unstable if any root of the numerator $N(s)$ of the characteristic equation $1+G(s)H(s)=N(s)/D(s)$ has a positive real part. Thus the test becomes:

1. Set $1+G(s)H(s)=0=N(s)$ and find the roots.
2. Check to see if any roots have a positive real part, which will indicate instability.

The theory of equations allows a solution by inspection in certain cases, which is the ultimate shortcut. These are cases that are immediately unstable because a root with a positive real part exists. A case is unstable if:

1. One or more coefficients of the polynomial $N(s)$ are zero, for example, $s^4+3s^2+1=0$. (No s term, the coefficient of s is zero.)
2. The coefficients of $N(s)$ do not all have the same sign, for example, $s^4+3s^2-2s+1=0$. (The coefficient of s has a minus sign.)

A good discussion of the Routh–Hurwitz technique is given by Ogata,[6] Kuo,[7], Shinners,[8] and D'Azzo and Houpis.[10]

Control System Compensation

Without proper conpensation a control system will most likely operate in a marginal or unstable manner. A marginal response might exhibit a bit too much overshoot or a bit too large position error. These performance inadequacies can usually be rectified by addition of a suitable compensation network to the errant system.

In more extreme cases the control system may be totally inoperative—it is unstable. Again it is usually possible to correct instability by adding a suitable compensation network to the unstable system (see Table 34.23).

It is generally desirable to add the compensation network to a part of the system where energy levels are low. That is, if a complex system contains many elements, such as a radar, a signal processor, an electrical amplifier, an electromechanical actuator, and/or a gimbaled rocket engine, it is wisest to add the control system compensation to the signal processor or the electrical amplifier, where signals are at the microwatt to milliwatt power level. It is unwise and perhaps impossible to add system compensation to the rocket motor, where signals are at the kilowatt to the megawatt level.

The effect of adding a system compensation network is to change the system loop gain(s). Thus a system compensation network has the same transfer function anywhere within a given control loop, as shown in Table 34.23. However, the physical construction of the compensator is vastly different, depending on the type of signal (electrical, hydraulic, pneumatic, mechanical) and the power level of the transmitted signal. Compare the forward path compensation with the feedback compensation in Table 34.23. In both cases the feedback loop gain is $A(s)G(s)H(s)$, yet the input to the compensation network is $E(s)$ in the case of forward path compensation, and $H(s)C(s)$ in the case of feedback compensation. Different inputs require different physical networks.

Complex compensating networks are usually electrical in nature and operate on low-level control signals, with amplitudes of perhaps 10^{-3} to 10^{+2} V and power levels of perhaps 10^{-6} to 10^{+2} W. Hydraulic or pneumatic compensation is used primarily when the system control signals are all hydraulic or pneumatic. It is important not to confuse a hydraulic or pneumatic damper or shock absorber with a hydraulic or pneumatic compensation network. The hydraulic or pneumatic damper alters the transfer function of the system element to which it is attached. It does not add another transfer function in series with that element. On the other hand, a hydraulic or pneumatic compensator, such as a hydraulic low-pass or phase-lag filter, adds an additional transfer function in series with the system element to which it is attached. As previously mentioned, hydraulic or pneumatic compensating networks are used principally in hydraulically or pneumatically operated systems, whereas hydraulic or pneumatic dampers are used in all types of systems, including electrically operated systems.

Phase Lead and Phase Lag Compensators. As explained in the discussion on system stability, a feedback control system becomes unstable (from the viewpoint of classic control theory) when the system gain exceeds unity at a frequency for which the feedback is positive. Normally the feedback is negative. The feedback becomes positive when a total of $-180°$ of phase lag is produced by the elements within the feedback loop.

Consequently there are two approaches to improving or producing system stability. The first approach is to add positive phase shift or phase lead, so that $-180°$ of phase lag is no longer produced when the gain is unity. The second method involves reducing the gain magnitude beneath unity (or 0 dB) when the phase lag is 180° or more. (A discussion and review of decibels is given at the end of this section.)

The second method, altering or decreasing the system phase lag, can be accomplished by inserting a phase lead filter somewhere in the feedback loop, as illustrated in Table 34.23.† The second method, lowering the system gain, can be accomplished at all frequencies with a simple attenuator, or the system gain can be lowered only in a selected frequency region by utilizing a phase-lag filter. Note that a phase-lag filter is used to lower the gain. It is not used to increase the phase lag, which is undesirable (and is indeed the major problem with using a phase-lag filter as an attenuator).

Using a simple attenuator to lower the gain is often not desirable, since lowering the gain at zero frequency [zero frequency gain $= \lim_{s\to 0} G(s)H(s)$] lowers all the steady-state error coefficients, which increases the steady-state errors. Recall that the steady-state errors are inversely related to the limit: $\lim_{s\to 0} G(s)H(s)$ (see the discussion of control system errors). A phase-lag filter does not lower

† Alternatively, rate feedback can be added in parallel with a unity feedback loop to produce an ideal lead network without a pole factor. This is discussed later.

Signs		s^5	$+3s^4$	$-3s^3$	$-9s^2$	$-4s$	-12
+	s^5	1		-3		-4	
+	s^4	3		-9		-12	
Skip row of zeros	s^3	$\frac{3(-3)-1(-9)}{3}=0$	$\frac{3(-4)-1(-12)}{3}=0$	0	← Row of zeros		
+	s^3	Aux. eq. yields 4	Aux. eq. yields -6	(Replace zeros via auxiliary equation)			
−	s^2	$\frac{4(-9)-3(-6)}{4}=\frac{-18}{4}$	$\frac{4\cdot(-12)-3\cdot 0}{4}=-12$				
−	s^1	$\frac{-18/4(-6)-4(-12)}{-18/4}=\frac{-50}{3}$	0				
−	s^0	$\frac{-50/3(-12)}{-50/3}=-12$	0				

Fig. 34.13 Third Routh–Hurwitz example (Example 34.3 in text). Auxiliary equation, row of zeros. $N_3(s)=s^5+3s^4-3s^3-9s^2-4s-12=0$. Since one or more coefficients exhibit a sign change, there is at least one root with a positive real part. The Routh–Hurwitz tabulation can be used to find the exact number of roots with a positive real part. Solving the auxiliary equation

$$3s^4-9s^2-12=0$$

yields (divide by 3, solve the biquadratic for two roots, s_1 and s_2, each squared)

$$(s_1, s_2)^2=\frac{3\pm\sqrt{9-4(-4)}}{2}=\frac{3\pm 5}{2}=\frac{8}{2},\frac{-2}{2}$$

$$(s_1, s_2)^2=+4,-1 \quad \text{or} \quad \text{roots}=+2,-2,+j,-j$$

One root has a positive real part, namely $+2$. This is also indicated by the single change of sign from $+4$ in the s^3 row to $-18/4$ in the s^2 row of the Routh tabulation. Each change of sign corresponds to one root of $N(s)$ with a positive real, or unstable, part, namely $+2$.

As a final check, $N_3(s)$ is factored, utilizing a root-finding program (HP Mathematics Pac, Polynomial Solutions).

$$N_3(s) = (s-2)(s+2)(s-j)(s+j)(s+3)$$

The five roots of $N_3(s)$ then are

$$\lambda_1 = +2 \quad \lambda_3 = +j \quad \lambda_5 = -3$$
$$\lambda_2 = -2 \quad \lambda_4 = -j$$

Note that the four roots λ_1 to λ_4 agree, as they should, with the four roots found from the auxiliary equation. The final root $\lambda_5 = -3$ has a negative real part, namely -3. Thus only one root, $\lambda_1 = +2$, has a positive real part, which agrees with the single first-column sign change of the Routh–Hurwitz tabulation. Thus the single Routh array sign change, the auxiliary equation solution, and the root-finding program all agree, as they should—there is one root of $N_3(s)$ with a positive real part. Auxiliary equation, s^4 row: Note: The auxiliary equation is obtained from the coefficients above the row of zeros, starting with an s^4 term, which leads the row, and then decreasing exponents by two; so that the terms are

$$(3)s^4 + (-9)s^2 + (-12)s^0 = 3s^4 - 9s^2 - 12 = 0$$

or

$$s^4 - 3s^2 - 4 = 0, \text{ then}$$

$d/ds(s^4 - 3s^2 - 4)$ is $4s^3 - 6s$, which replaces the row of zeros. See the second s^3 row. Note that any row may be divided by a positive number without affecting the validity of the test. Similarly, any auxiliary equation may be divided by a positive number without affecting the validity of the test.

Signs		s^4 $+2s^3$	$+3s^2$ $+6s$	$+4 = N_4(s)$
+	s^4	1	3	4
+	s^3	2	6	0
+	s^2	$\dfrac{(2)3-1(6)}{2} = 0 \rightarrow \epsilon$ (0 replaced by ϵ)	$\dfrac{(2)4-1(0)}{2} = 4$	0
−	s^1	$\dfrac{(\epsilon)6-2(4)}{\epsilon} = \dfrac{-8}{\epsilon}$	0	0
+	s^0	4	0	0

Fig. 34.14 Fourth Routh–Hurwitz example (Example 34.4 in text), First element zero is replaced by epsilon ϵ and there are two unstable complex roots, $N_4(s) = s^4 + 2s^3 + 3s^2 + 6s + 4 = 0$. ϵ is a small positive number, hence ϵ is positive, and $-8/\epsilon$ is negative. The signs of the first columns are $+, +, +, -, +$, reading from s^4 to s^0. There are thus two sign changes and two roots with a positive real part. As a check, a root-finding program yields the following factors:

$$N(s) = (s + 1.000)(s + 1.4779)(s - 0.239 + j1.628)(s - 0.239 + j1.628)$$

The two roots with a positive real part are thus

$$\lambda_1 = +0.239 - j1.628$$
$$\lambda_2 = +0.239 + j1.628$$

A common error is to forget the sign change between a factor, such as $(s + 1.000)$, and the corresponding root, $s = -1.000$.

---

the gain at zero frequency but only at some higher frequency (above the so-called corner frequency of the filter). Hence a phase-lag filter does not affect a system's steady-state errors as a simple gain attenuator does. For this reason, phase-lag filters are used. To reemphasize, phase-lag filters are not used to produce phase lag. They are used to lower the gain, but only at frequencies above the corner frequency of the filter. See Fig. 34.15 for the frequency response of a phase-lag filter.

Phase-lead filters, unlike phase-lag filters, are used for their effect on system phase angle. As mentioned previously, a phase-lead filter is used to add positive phase lead to the negative phase lag generated by various control system components so that the total phase angle becomes more positive than $-180°$ at the gain crossover point. For example, if the system phase angle at gain crossover (where the gain crosses over the 0-dB or unity gain level) happens to be $-200°$, a phase-lead filter producing $+60°$ of phase lead would improve the system phase angle to $-200° + 60° = -140°$. This would change an unstable system ($-200°$ at 0 dB) to a stable system ($-140°$ at 0 dB).

Note that a simple passive resistor-capacitor phase-lead filter, as shown in Fig. 34.16, has a constant attenuation, $a = (R_1 + R_2)/R_2$, which must be canceled out by either adding an additional amplifier of gain a or by increasing the system gain by a. (See also Fig. 34.15.)

Phase-Lead Compensation Design. Phase-lead filters have problems associated with their design, since the filter affects the system gain as well as the system phase shift. Therefore, in general, a real filter must be designed for a significantly larger phase lead than the minimum phase lead required. In some complex cases, a simple phase-lead filter will not improve system stability at all, because the added phase lead is negated by the added frequency-dependent gain. The reason a phase-lead filter works as well as it does depends on one simple fact: A phase-lead filter starts producing significant phase lead at a frequency one decade below the frequency at which the phase-lead filter starts to produce significantly increasing gain. This is illustrated in Fig. 34.16.

The phase-lead filter must be designed so that the decade of frequency for which the filter produces significant phase lead occurs in the region where the uncompensated system has a phase lag exceeding (or close to) $-180°$, along with a gain magnitude exceeding unity, or 0 dB.

Phase Lag Compensation Design. The problem associated with a phase-lag filter as a system compensation element is that it produces phase lag. Phase lag is bad for a feedback control system because it makes the system less stable. The phase-lag filter is used to produce a gain attenuation over a desired frequency range, which increases system stability. The attendant phase lag must be positioned in frequency where it will do no harm. Generally this means that the phase lag must occur at frequencies for which the system phase shift can accept the additional phase lag (of the phase-lag filter) and still not become more negative than $-180°$ at a gain of 0 dB. Sometimes this can be accomplished by placing the lag filter's corner frequencies far below all other corner frequencies. If this is impossible, a lead-filter compensator should be tried. If this does not work, a lead-lag compensator, which combines the phase lead of the lead filter with the attenuation of the lag-filter,

TABLE 34.23. SYSTEM COMPENSATION

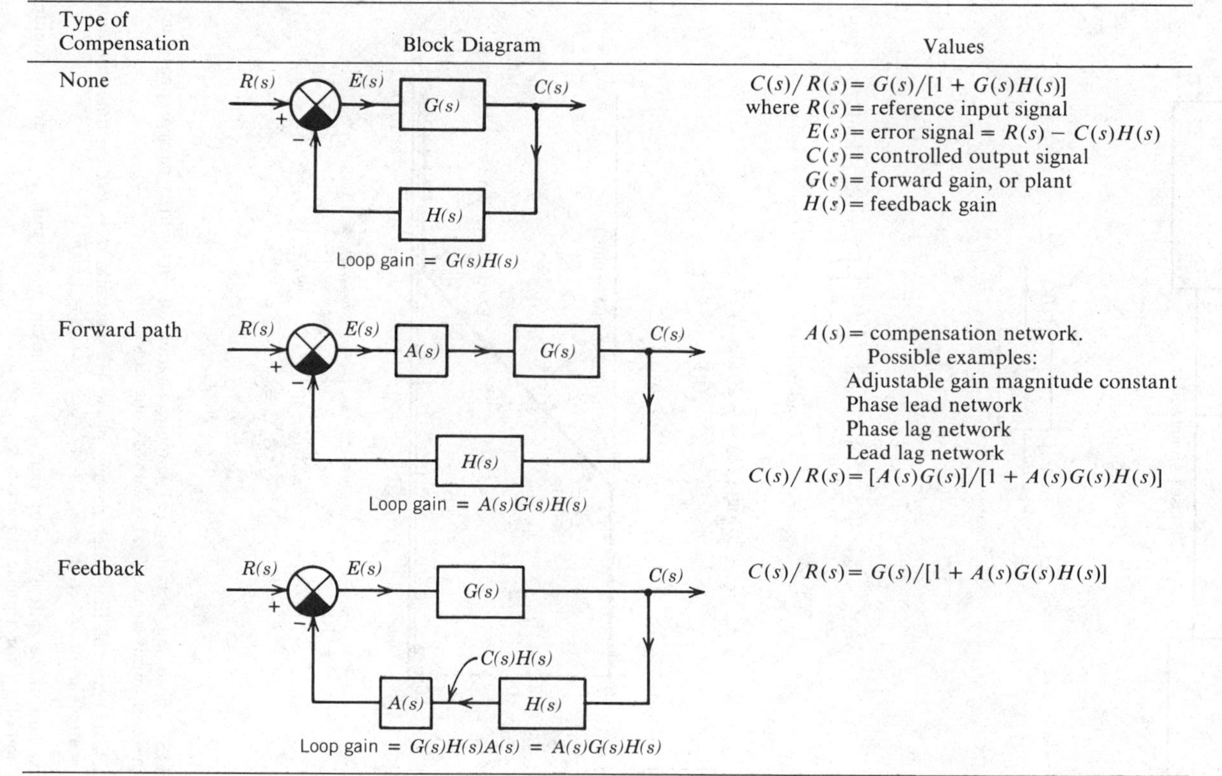

Type of Compensation	Block Diagram	Values
None		$C(s)/R(s) = G(s)/[1 + G(s)H(s)]$ where $R(s)$ = reference input signal $E(s)$ = error signal = $R(s) - C(s)H(s)$ $C(s)$ = controlled output signal $G(s)$ = forward gain, or plant $H(s)$ = feedback gain
Forward path		$A(s)$ = compensation network. Possible examples: Adjustable gain magnitude constant Phase lead network Phase lag network Lead lag network $C(s)/R(s) = [A(s)G(s)]/[1 + A(s)G(s)H(s)]$
Feedback		$C(s)/R(s) = G(s)/[1 + A(s)G(s)H(s)]$

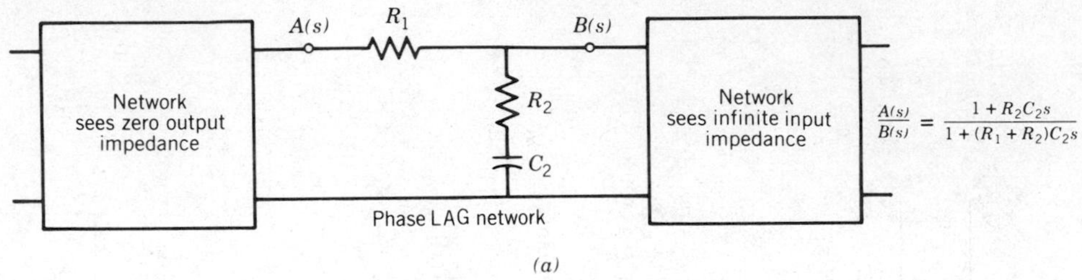

(a)

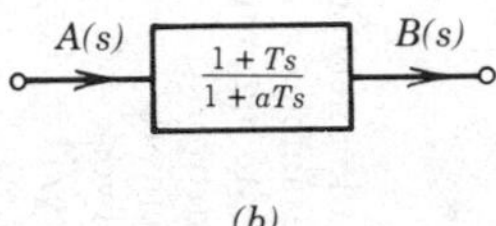

(b)

$$\underset{\text{(lag)}}{\phi\max} = \sin^{-1}\left(\frac{1-a}{1+a}\right) = -\sin^{-1}\left(\frac{a-1}{a+1}\right) = -\phi_{\max}\ (\text{lead})$$

$$\omega_{\max} = \sqrt{\omega_z \omega_p} = \frac{1}{T\sqrt{a}}\quad (\text{lead or lag})$$

$$\text{dB}_{\max} = +20\log(1/a) = -20\log(a)$$

(c)

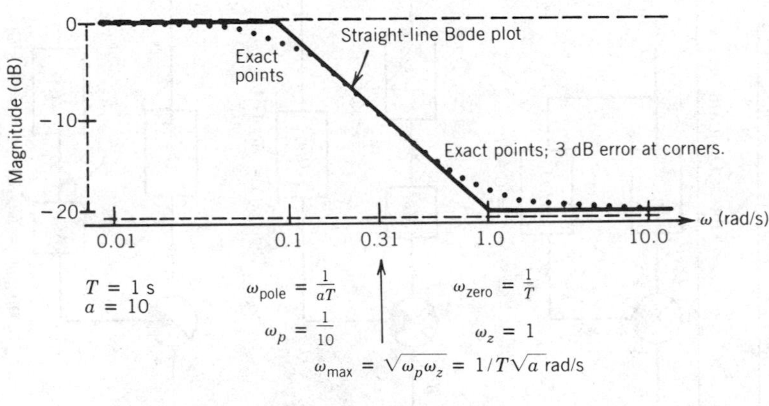

(d)

Fig. 34.15 Phase lag network. (*a*) Circuit diagram of first-order passive phase lag network. $A(s)$ = network input signal, $B(s)$ = network output signal.

(*b*) Block diagram of first-order passive phase lag network. $T = R_2C_2$ = time constant, $a = 1 + R_1/R_2$ = attenuation, $a \geqslant 1, \omega_z = \omega_{\text{zero}} = 1/T, \omega_p = \omega_{\text{pole}} = 1/aT$.

(*c*) Maximum phase shift $\phi_{\max}$, frequency of maximum phase shift $\omega_{\max}$, and maximum network attenuation $\text{dB}_{\max}$ for a phase lag network.

(*d*) Magnitude response for a phase lag network

$$\text{Lag}(s) = \frac{1+s}{1+10s} = \frac{1+Ts}{1+aTs}\qquad T = 1, a = 10$$

$$\text{Mag} = 20\log\left|\frac{1+s}{1+10s}\right|_{s=j\omega}$$

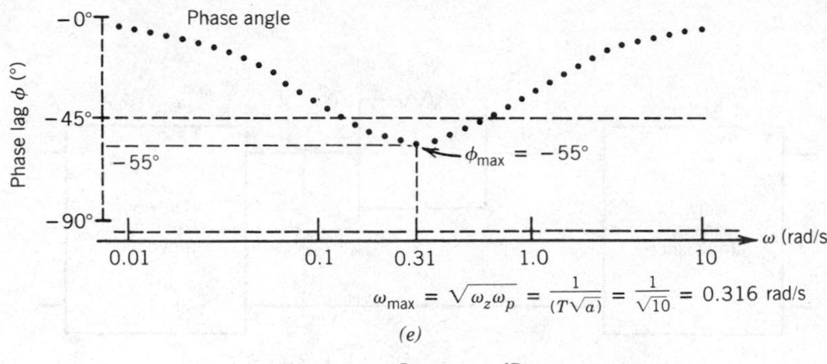

(e)

a	$\pm \Phi_{max}$	dB_{max}
1.0	0.0	0.0
2.0	19.5	− 6.0
3.0	30.0	− 9.5
4.0	36.9	− 12.0
5.0	41.8	− 14.0
6.0	45.6	− 15.6
7.0	48.6	− 16.9
8.0	51.1	− 18.1
9.0	53.1	− 19.1
10.0	54.9	− 20.0
11.0	56.4	− 20.8
12.0	57.8	− 21.6
13.0	59.0	− 22.3
14.0	60.1	− 22.9
15.0	61.0	− 23.5
16.0	61.9	− 24.1
17.0	62.7	− 24.6
18.0	63.5	− 25.1
19.0	64.2	− 25.6
20.0	64.8	− 26.0

$$\text{Lead network} = \frac{1}{a}\,\frac{1 + aTs}{1 + Ts}$$

$$\phi_{max} = \sin^{-1}\left(\frac{a-1}{a+1}\right) \qquad a > 1$$

Lead (positive angle) max attenuation = 20 log (a) as $\omega \to 0$

$$\text{Lag network} = \frac{1 + Ts}{1 + aTs}$$

$$\phi_{max} = \sin^{-1}\left(\frac{1-a}{1+a}\right) \qquad a > 1$$

Lag (negative angle) max attenuation = 20 log(a) as $\omega \to \infty$

(f)

(*e*) Phase response for a phase lag network

$$\text{Lag}(s) = \frac{1 + s}{1 + 10s} = \frac{1Ts}{1 + aTs} \qquad T = 1, a = 10$$

a = attenuation, $\phi_{max} = -55°$ at $\omega_{max} = 0.316$ rad/s.

(*f*) Magnitude of maximum phase lead or lag vs. attenuation a (use negative angles for a lag network). Maximum attenuation decibels are also tabulated. The equation for ϕ_{max} indicates that for a given value of a (the pole: zero frequency ratio), a lead and a lag network have the same phase shift magnitude but opposite signs. For example, if $a = 10$, then a phase lead network has a phase shift of $+55°$ while a phase lag network has a shift of $-55°$. ω_{max} is identical for either a phase lead or a phase lag network. ω_{max} is midway between the network's pole and zero, on a log ω axis, or ω_{max} is the geometric mean of ω_{zero} and ω_{pole}; $\omega_{max} = \sqrt{\omega_z \omega_p}$. dB_{max}, the network attenuation, is the same for both networks. It occurs at low frequencies for a lead network and at high frequencies for a lag network, where "low" and "high" are relative to ω_{max} (rad/s), the network's center frequency.

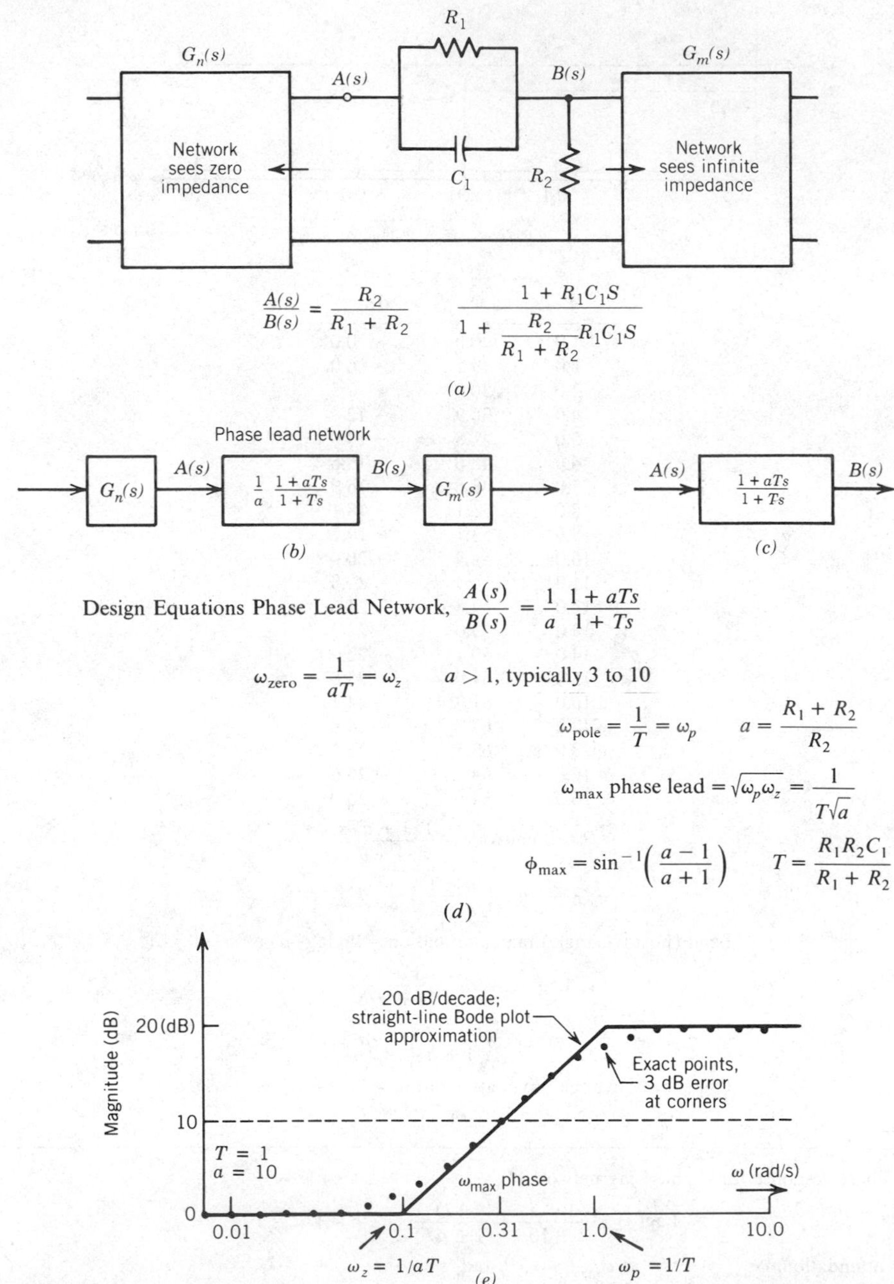

Fig. 34.16 First-order passive phase-lead network: (*a*) Circuit diagram of network containing R_1, C_1, and R_2. A gain of a (to cancel out the network attenuation of a) is usually incorporated into $G_n(s)$ or $G_m(s)$ either as a buffer amplifier (which also will produce the required impedance) or as an adjustment of the gain constants of $G_n(s)$ or $G_m(s)$. (*b*) Block diagram. a = attenuation = $(R_1 + R_2)/R_2$, $a > 1$; T = time constant = $(R_1R_2C_1/(R_1 + R_2)$; $A(s)$ = network input signal; $B(s)$ = network output signal; $G_n(s), Gm(s)$ = original uncompensated gains or transfer functions. (*c*) Simplified network. Gain of a added externally, which cancels out the $1/a$ attenuation factor shown in (*b*). (*d*) Design equations. (*e*) Bode magnitude plot of $(1 + 10s)/(1 + s)$ for a simplified network of the form $(1 + aTs)/(1 + Ts)$. Magnitude = $20 \log \left|\frac{1 + 10s}{1 + s}\right|_{s=j\omega} \cdot \omega_z = 0.1$, $\omega_p = 1.0$, $\omega_{max} = 0.31$.

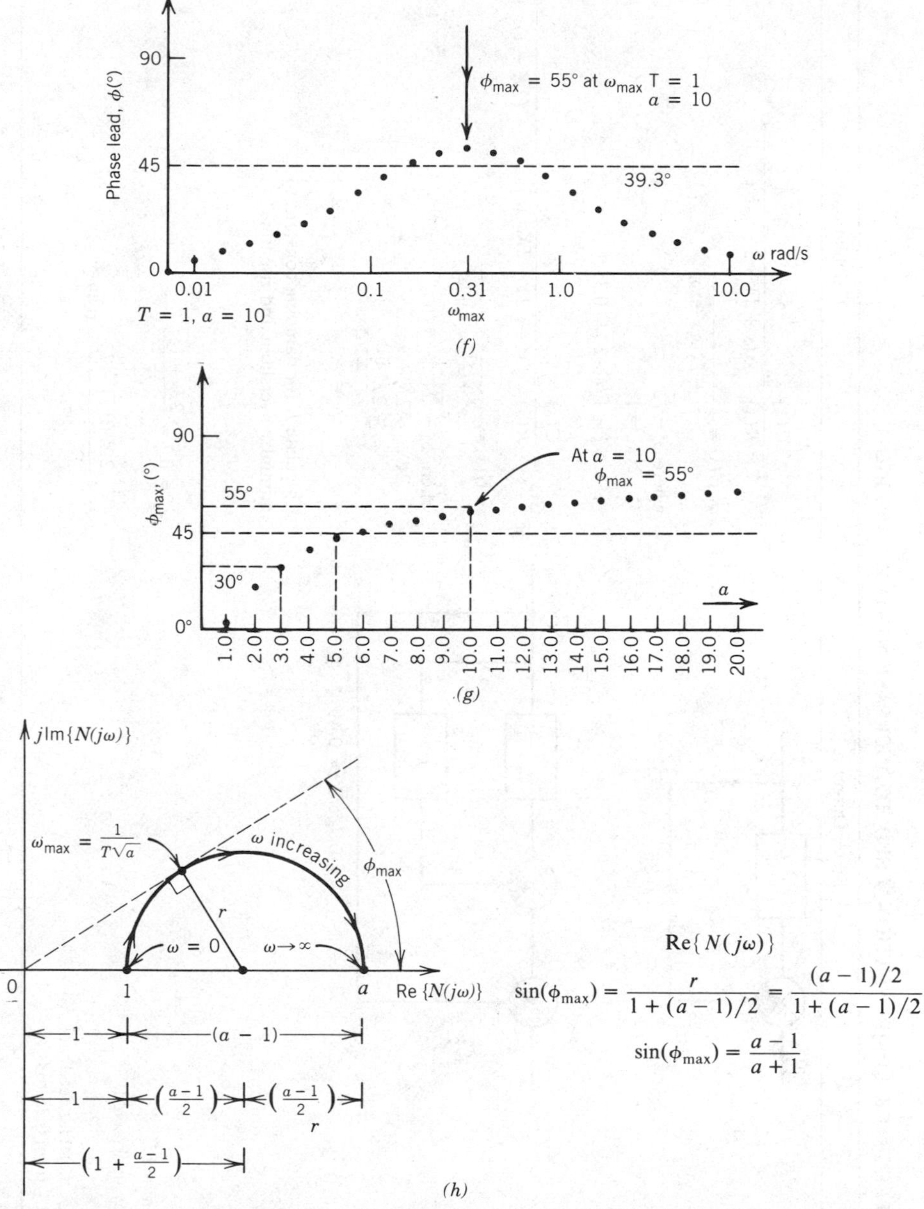

$$\sin(\phi_{max}) = \frac{\text{Re}\{N(j\omega)\}}{\dfrac{r}{1+(a-1)/2}} = \frac{(a-1)/2}{1+(a-1)/2}$$

$$\sin(\phi_{max}) = \frac{a-1}{a+1}$$

(*f*) Bode phase plot of $(1 + 10s)/(1 + s)$. $\omega_m = \sqrt{\omega_z \omega_p} = \omega_{max}$, $\omega_m = 0.31 = \sqrt{0.1 - 1.0}$. (*g*) Plot of ω_{max} vs. a, $\phi_{max} = \sin^{-1}[(a-1)/(a+1)]$. At $a = 3$, $\phi_{max} = 30.0°$. At $a = 6$, $\phi_{max} = 45.6°$. Numerically tabulated values of phase lead ϕ_{max} vs. network attenuation a are given in Fig. 34.15*f*. Note that a is equal to the pole:zero ratio, ω_p/ω_z. At $a = 1$, the network reduces to a unity gain, with zero phase lead and 0-dB attenuation (20 log 1 = 0 dB). The typical range for a is 3 to 10. Values less than 3 give too little phase lead to be useful, while values greater than 10 produce relatively small increases in phase lead for a given increment of attenuation. (*h*) Polar plot of simplified network $N(s)$. $N(s) = (1 + aTS)/(1 + TS)$, $s = j\omega$. Geometric construction of $\sin(\phi_{max}) = (a-1)/(a+1)$. $a > 1$, $\omega_{zero} = 1/aT$, $\omega_{pole} = 1/T$, ω_{max} = midpoint on log ω phase plot = geometric mean. $\omega_{max} = \sqrt{\omega_z \omega_p}$, $\omega_{max} = \sqrt{(1/aT)(1/T)}$, $\omega_{max} = 1/(T\sqrt{a})$.

TABLE 34.24. MAJOR LOOP FEEDBACK COMPENSATION, USED TO INCREASE SYSTEM DAMPING

Type of System	Block Diagram	Calculations
Uncompensated Damping constant = 0.35 = ζ, natural frequency = $r/s = \sqrt{2}$	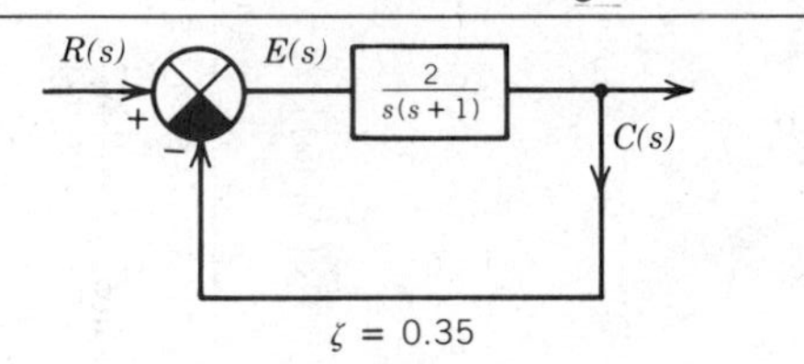 $\zeta = 0.35$	$G(s) = \frac{C(s)}{R(s)} = \frac{2}{s(s+1)+2} = \frac{2}{s^2+s+2}$ or $G(s) = \frac{2}{s^2 + 2\zeta\omega_n s + \omega_n^2}$ where $\omega_n = \sqrt{2}$ $2\zeta\omega_n = 1$ $\zeta = 1/(2\sqrt{2}) = 0.35$
Compensated Add tachometer or derivative minor-loop feedback to increase damping to ζ' (an arbitrary desirable value if ζ' is chosen, namely $\zeta' = 0.7$)	R(s), E(s), 2 / s(s + 1), Ks, C(s) $\zeta = 0.7$ for $K = 0.49$	$G(s) = \frac{C(s)}{R(s)} = \frac{2}{s(s+1)+2Ks+2} = \frac{2}{s^2+(2K+1)s+2}$ or $G(s) = \frac{2}{s^2 + 2\zeta'\omega_n s + \omega_n^2}$ where $\omega_n = \sqrt{2}$ $2\zeta'\omega_n = 2K+1$ $\zeta' = \frac{2K+1}{2\sqrt{2}}$ Calculate K for damping constant to minimize overshoot and rise time
Results 1. Compensated damping constant = 0.7 2. Compensated natural frequency unaffected, $\omega_n = \sqrt{2}$ 3. Added compensation of $0.49s = Ks$. This feedback changes the damping constant ζ from 0.35 to 0.7		$\zeta' = 0.7 = \frac{2K+1}{2\sqrt{2}}$ $K = \frac{2\sqrt{2}\,(0.7) - 1}{2} = 0.49$

may be the solution. See Fig. 34.15 for the schematics and characteristics of the phase-lag network.

For the lead, the lag, and the lead-lag filters, a good rule of thumb is that the filter's poles and zeros should be approximately one decade apart in frequency. If none of these approaches works, a high-order filter is required or the system components must be changed.

Minor-Loop Feedback Compensation. Changing the system transfer functions is commonly achieved by introducing damping. Damping can be physically introduced by hydraulic, pneumatic, or electromechanical dampers. Damping can also be introduced by adding another feedback loop incorporating a differentiator element, such as a tachometer, or a rate gyroscope. For a tachometer, $V_{out} = K \cdot \text{RPM} = K\dot{\theta} = Ks\theta(s)$. This is called minor-loop feedback. (See Table 34.24.)

Pole-Zero Cancellation. Another method of changing and thus stabilizing a system's transfer function involves the conceptually simple approach of canceling out a pole by adding on a multiplicative term that contains a zero factor identical to the pole factor. Mathematically this is a perfect solution. In reality it is not so perfect, because the zero may not exactly cancel out the pole, due to component tolerances and variations as well as the effect of changing system state and age. An example of pole-zero cancellation is shown in Fig. 34.17.

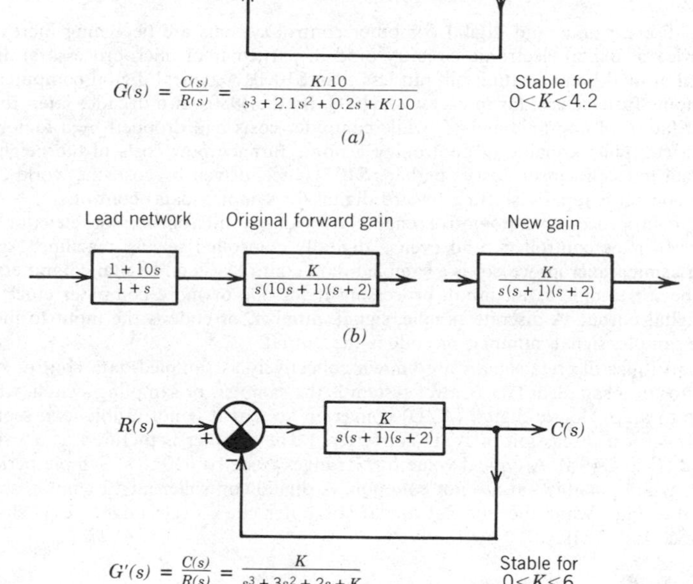

Before Pole Cancellation

	$s^3 + 2.1s^2 + 0.2s + K/10$	
s^3	1	0.2
s^2	2.1	$K/10$
s^1	$\dfrac{0.42 - K/10}{2.1}$	$(K < 4.2)$
s^0	$K/10$	$(K > 0)$

After Pole Cancellation

	$s^3 + 3s^2 + 2s + K$	
s^3	1	2
s^2	3	K
s^1	$\dfrac{6 - K}{3}$	$(K < 6)$
s^0	K	$(K > 0)$

(*d*)

Fig. 34.17 Pole-zero cancellation: (*a*) Uncompensated system. Maximum gain $K = 4.2$. (*b*) Lead network compensation, cancels pole at (10s + 1). (*c*) Compensated system. Maximum gain $K = 6.0$. Therefore stability is improved. (*d*) Routh–Hurwitz arrays.

Note Concerning Decibels. Decibels (dB) are evaluated as $20\log|(G)j\omega|$. $|G(j\omega)|$ is a magnitude, not a power function $[G(j\omega) = G(s)|_{s=j\omega}$ evaluated at $s = j\omega]$. For power functions, where gains are the ratio of power in: power out, such as the dBm, decibels are evaluated as $10\log|G(j\omega)|$, and dBm = decibels relative to 1 mW, or

$$\text{dBm} = 10\log\left|\frac{P_{\text{out}}}{10^{-3}}\right|$$

The prefix "deci" means one tenth. Thus a decibel is one tenth of a bel, which is a unit of power, and the "10 log" form is easy to remember. The "20 log" form is also easy to remember if one recalls that, for voltage across a resistance, power $= V^2/R$, or power varies with voltage squared. V is a voltage or magnitude function, and $10\log|V^2| = 20\log|V|$. The unit of power gain, the bel, was named after Alexander Graham Bell, the inventor of the telephone, about 1876. The bel is not used, because it represents a very large power gain, namely a factor of ten for each bel. A gain of 5 bel $= 10^5 =$ 10,000.

Note that a gain of 2 corresponds to a gain of 6 dB in magnitude $[20\log 2 = 20(0.3) = 6]$ and a gain of 3 dB in power $[10\log 2 = 10(0.3) = 3]$. It is important not to confuse these two factors, since both 6 dB and 3 dB correctly correspond to a gain of 2. However, the gains are defined differently, one being a magnitude or voltage gain (6 dB) and the other being a power gain (3 dB). For the Bode magnitude plots of this section, a gain of 2 corresponds to 6 dB, and a gain of 1/2 corresponds to − 6 dB.

34.5 SAMPLED-DATA CONTROL

Sampled-data, discrete time, and digital computer control systems are becoming increasingly common, as the price of digital electronic circuitry (and in particular of microprocessors) drops. In the 1960s, a typical annual home heating bill ran less than $1000. A typical digital computer capable of controlling a home furnace cost far in excess of $1000. By the 1980s, two decades later, fuel costs had increased by a factor of approximately 5, while computer costs had dropped by a factor of perhaps 1000, so that a computer capable of controlling a home furnace now costs in the neighborhood of $100. The actual microprocessor cost is perhaps $10. Hence, driven by cost, the world of consumer and industrial control is rapidly shifting toward digital (or sampled-data) control.

There are microprocessor automotive engine controllers, microprocessor elevator controllers, microprocessor furnace controllers, and even a digitally controlled sewing machine. Any feedback control system using a microprocessor is a sampled-data control system. Digital computers operate by taking a number or sample as an input, processing it for one or more computer clock cycles, and providing a digital output. A discrete sample, signal, number, or code is the input to the computer, and a discrete sample, signal, number, or code is the output.

Systems that utilize discrete signals are known collectively as sampled-data control systems. The identifying factor of a sampled-data control system is the sampler, or sampling switch, which is often incorporated into an analog-to-digital (A/D) converter, so that it is not visible as a separate entity. (See Section 37.4 for a discussion of A/D converters.) The sampler is pictured as a switch, closing every T seconds (Fig. 34.18). A typical value for T ranges from 1 to 10^{-6} s. Sample periods as short as 10^{-9} s or 1 ns are possible but are not common. A digital controller might employ a sample time of a few microseconds, while the internal digital computer clock cycle might be as short as a few nanoseconds. See Fig. 34.19.

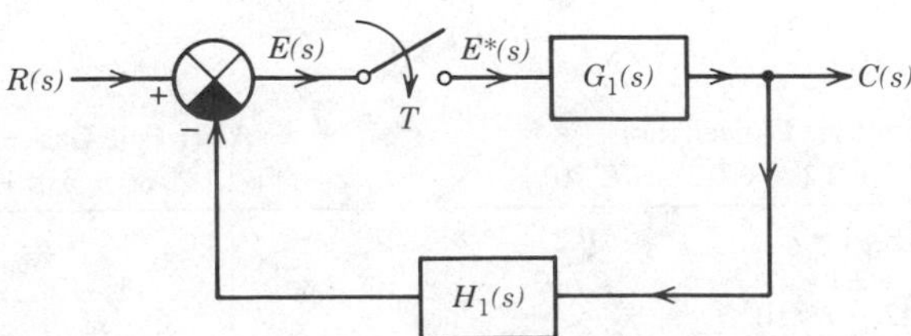

Fig. 34.18 Sampled-data control system.
$E^*(s)$ = Laplace transform of $E^*(t)$; $E^*(t)$ = sampled values of $E(t)$ at times $t = nT$; $E^*(t) = E(0)\,\delta(0) + E(T)\delta(t-T) + E(2T)\delta(t-2T) + \cdots + E(nT)\delta(t-nT)$; $G_1(s)$ = forward gain or plant transfer function; $H_1(s)$ = feedback gain or transfer function; $R(s)$ = reference input signal; $C(s)$ = controlled output signal; $\delta(nT)$ = Dirac delta function, where $\int_{-\infty}^{\infty} f(t)\delta(nT)\,dt = f(nT)$, $\delta(nT) = 0\ (nT \neq 0)$, $\delta(0) = \infty$, $\delta(t-nT)|_{t=nT} \to \infty$, $\int_{-\infty}^{\infty}\delta(0)\,dt = 1$ = weight of impulse.

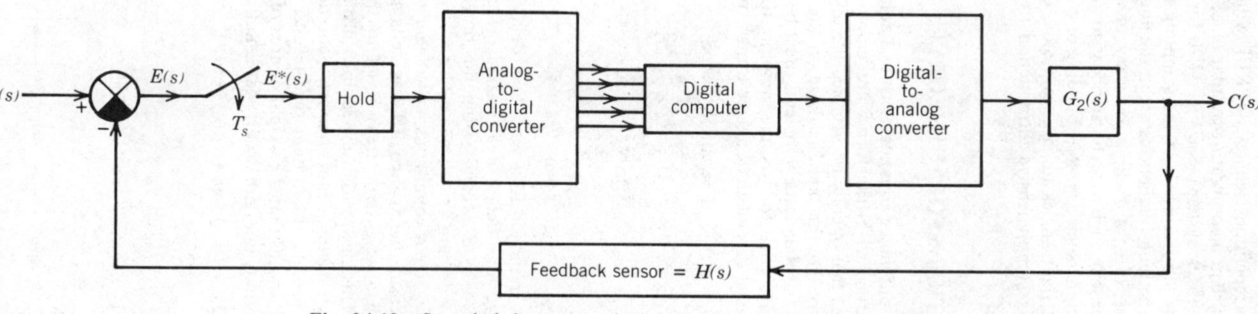

Fig. 34.19 Sampled-data control system using a digital controller.

The basic concepts of sampled-data control theory were developed during World War II, in an effort to analyze and design radar (*rad*io *a*utomatic *r*anging) systems. Radars operate by transmitting a brief pulse and then receiving a weakened return pulse reflected from any target within the radar beam's path. The delay between transmitted pulse and received return signal is proportional to the distance between the radar antenna and the target.

An operating radar uses a pulse repetition frequency (PRF), which becomes the sample switch frequency for any tracking control system incorporating a radar (e.g., the radar tracks the target and represents the comparator of the feedback system). See Fig. 34.20 for a block diagram of a tracking radar. The sampling switch period is the reciprocal of the sampling frequency, or $T = 1/\text{PRF}$. (Of interest is the fact that radar pulse repetition frequencies are in the audio frequency range and can be directly sensed as a mysterious noise, often a clicking, by a person directly exposed to a beam of sufficient intensity. The proper response in such a circumstance is to get out of the beam immediately either by entering a conducting structure (such as a steel building or a car) or by running at top speed away from the beam, preferably at right angles to the beam axis. Chapter 51 presents a detailed discussion of radar principles. For a comprehensive introduction to sampled-data control, Tou's[13] early text is one of the best.

34.6 CONTROL SYSTEM z-TRANSFORM THEORY

z-Transforms are employed to solve difference equations, just as Laplace transforms are employed to solve differential equations. In particular, z-transforms are employed to generate sampled-data transfer functions for control systems that operate on sampled (or discrete) signals.

The constraints on the z-transform are the same as those on the Laplace transform, namely (1) the equations must be linear and (2) the equation coefficients must be constant and not time varying. In addition, for the z-transform analysis that follows, the sampling time T must be a constant.

The z-transform is, in fact, a special case of the Laplace transform, with z being defined as

$$z = e^{sT} \qquad T = \text{sample period} \tag{34.35}$$

or equivalently

$$s = \frac{1}{T}\ln z \tag{34.36}$$

Inverting Eq. (34.35) yields

$$z^{-1} = e^{-sT} \tag{34.37}$$

In the Laplace domain, e^{-sT} corresponds to a pure time delay of T seconds. Hence in the z domain, z^{-1} corresponds to a pure time delay of T seconds. This is the intuitive basis of the z-transform. Each sample is delayed by T seconds from the previous sample (e.g., $0, T, 2T, 3T, 4T, \ldots, nT$, which becomes in the z domain $1, z^{-1}, z^{-2}, z^{-3}, z^{-4}, \ldots, z^{-n}$). This is best illustrated by an example. Refer to Fig. 34.21, where a continuous, curvilinear function $f(t)$ is sampled every T seconds. The z-transform is derived using the artifice of Dirac delta sampling, where the samples are of infinitesimal time duration, with a weight (or area) equal to the value of the function being sampled at the time of the sample. This ensures that the samples are constant, since the function $f(t)$ cannot change in zero time. Although the delta functions are theoretically of infinite amplitude, only the areas of the delta functions are used to obtain numerical results, and the area of the Dirac delta function is defined to be unity. $2\delta(t)$ means that the weight or area of the delta function is multiplied by two. The problem of two times infinity does not occur. Mathematically

$$\delta(t) = 0 \qquad t \neq 0 \quad \text{and} \quad \delta(t) \to \infty \qquad t = 0$$

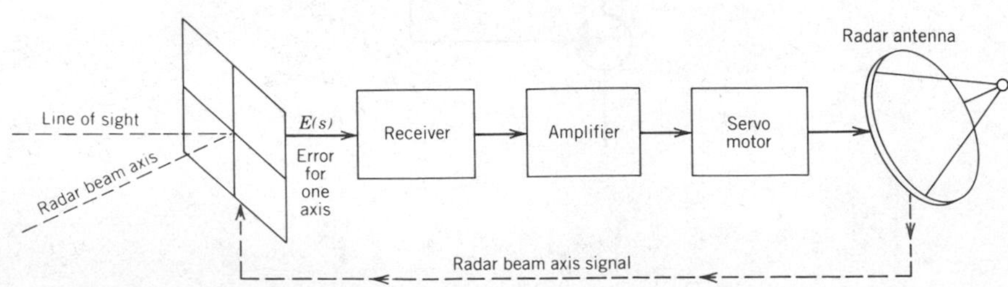

Fig. 34.20 Basic tracking radar configuration, one axis.

also

$$\int_{-\infty}^{+\infty}\delta(t)\,dt=1 \qquad \text{and} \qquad \int_{0^-}^{0^+}\delta(t)\,dt=1 \tag{34.38}$$

The limits $0-$ to $0+$ may be used, since $\delta(t)$ has a value only at $t=0$. That value is infinite, but the integral is finite and equal to unity. For an impulse weight of two

$$\int_{-\infty}^{\infty}2\delta(t)=2 \qquad \text{and} \qquad \int_{-\infty}^{\infty}f(t)\delta(t)\,dt=f(0) \tag{34.39}$$

and

$$\int_{-\infty}^{\infty}f(t)\delta(t-T)\,dt=f(T)$$

Figure 34.21 shows an infinite string or comb (the impulses look like the teeth on a comb ↑↑↑↑↑) of delta functions. Each delta function selects or samples the value of the continuous function $f(t)$ at a time where the delta function "turns on," namely, when the argument of the delta function is zero. That is, $\delta(t-nT)$ "turns on" at $t=nT$, to produce $\delta(0)$ which, in turn, produces a sample at a time of nT seconds.

It is important not to confuse the Dirac delta function with the Kronecker delta function. Both are depicted as $\delta(t)$. The Dirac delta function has an amplitude of infinity, whereas the Kronecker delta function has an amplitude of unity. Unless otherwise specified, throughout this section the term "delta function" and the symbol $\delta(t)$ refer to the Dirac delta function.

The sampled function $\delta^*(t)$ of Fig. 34.21 is composed of delta functions of weights 0.1, 0.5, 0.7, 1.0, and so on. The delta functions are shown with a height proportional to their area or weight. Mathematically, the heights are all infinite. In reality, the heights are exactly as shown. No Dirac delta samplers with infinite amplitudes exist in the real world. As often occurs, theory and reality diverge. The theory is retained because it yields correct results and simplifies the mathematics.

Figure 34.21 then illustrates the Laplace transform of $f^*(t)$, $F^*(s)$; obtained by utilizing the transform pair[13]

$$k\delta(t-nT)\Rightarrow ke^{-nTs} \qquad T=\text{sample time} \tag{34.40}$$

which results in

$$F^*(s)=\sum_{n=0}^{\infty}f(nT)e^{-nTs} \tag{34.41}$$

Equation (34.41) is simply a sum of sampled values $f(nT)$, each value being delayed nT seconds by the factor e^{-nTs}. $f(nT)$ is a number such as 0.1 or 0.5 or -100.86.

Finally, the substitution for z is made, as defined by Eq. (34.37) and repeated here:

$$z^{-1}=e^{-sT} \qquad \text{or} \qquad z=e^{sT} \tag{34.37}$$

This substitution results in the z-transform for $f(t)$

$$F(z)=\sum_{n=0}^{\infty}f(nT)z^{-n} \tag{34.42}$$

If this series converges, $F(z)$ can be expressed in a closed form. If the series does not converge, Eq. (34.42) still yields the z-transform, which is unbounded as $n\to\infty$.

As a simple example, consider the z-transform of a unit step $u(t)$, where $u(t)$ equals 1 for $t\geqslant 0$. Inserting $u(t)$ into Eq. (34.42)

$$z[u(t)]=F(z)=\sum_{n=0}^{\infty}u(t)z^{-n}=\sum_{n=0}^{\infty}1z^{-n} \tag{34.43}$$

Writing out the series

$$F(z)=\sum_{n=0}^{\infty}z^{-n}=1+z^{-1}+z^{-2}+z^{-3}+\cdots+z^{-\infty}$$

Factoring out z^{-1}

$$F(z)=1+z^{-1}(1+z^{-1}+z^{-2}+\cdots+z^{-\infty})$$

Observe that the series in parentheses is the series representation of $F(z)$ for a unit step, shown two lines previously. Thus

$$F(z)=1+z^{-1}[F(z)]$$

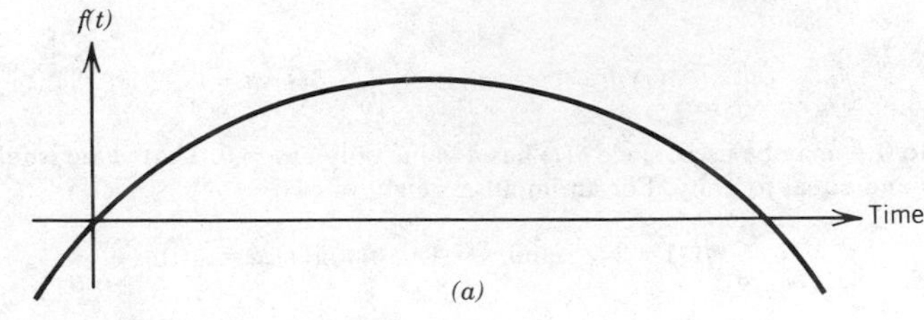

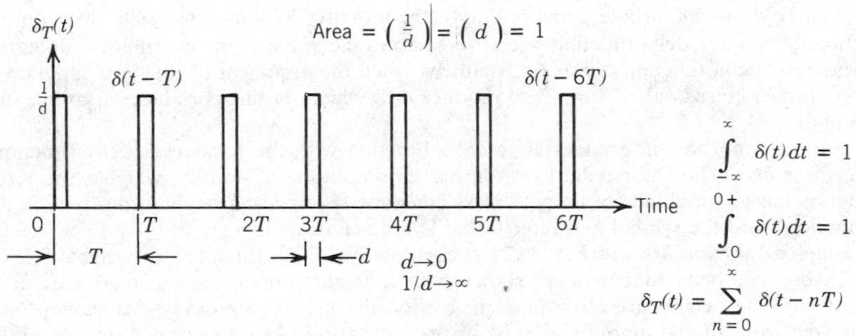

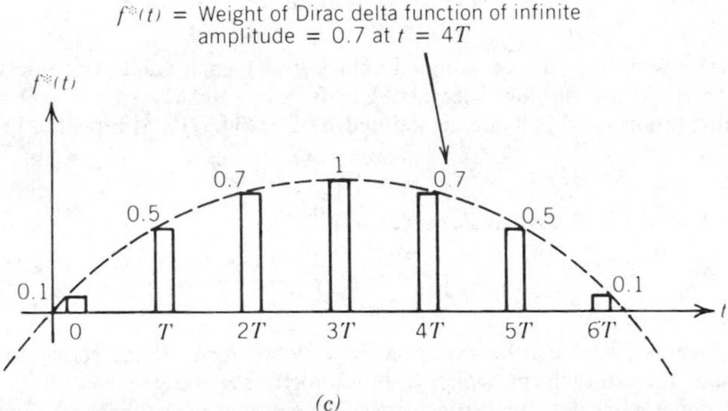

Fig. 34.21 Sampling and the z-transform. (a) Continuous time function $f(t)$, which is to be sampled. (b) Dirac delta sampler $\delta_T(t)$. δ = Dirac delta functions, T = sample period. (c) Sampled function $f^*(t)$. $f^*(t) = f(t) \cdot \delta_T(t)$, where $f(t)$ is interpreted as the weight of the impulse at time nT. The weight is, in turn, interpreted as the area of the impulse.

$$f^*(t) = \sum_{n=0}^{\infty} f(nT)\delta(t - nT)$$

$$f^*(t) = 0.1\delta(t) + 0.5\delta(t - T) + 0.7\delta(t - 2T) + 1\delta(t - 3T) + \cdots$$

$$\text{Laplace transform } f^*(t) \Rightarrow F^*(s) = 0.1e^{0s} + 0.5e^{-Ts} + 0.7e^{-2Ts} + 1e^{3Ts} + \cdots$$

Since $L[k \cdot \delta(t - nT)] = ke^{-nTs}$, L = Laplace transform operator and $L[\delta(t)] = 1$; $L[\delta(t - nT)] = 1e^{-nTs}$.

$F^*(s)$	or $F^*(s) = \sum_{n=0}^{\infty} f(nT)e^{-nTs}$
z-transform $z = e^{st}$ $F(z)$	$F(z) = 0.1z^0 + 0.5z^{-1} + 0.7z^{-2} + 1z^{-3} + \cdots$ or $F(z) = \sum_{n=0}^{\infty} f(nT)z^{-n}$
Summary, in series form	$f^*(t) = 0.1\delta(t) + 0.5\delta(t-T) + 0.7\delta(t-2T) + 1\delta(t-3T) + \cdots$ $F^*(s) = 0.1e^{0s} + 0.5e^{-Ts} + 0.7e^{-2Ts} + 1e^{-3Ts} + \cdots$ $F(z) = 0.1z^0 + 0.5z^{-1} + 0.7z^{-2} + 1z^{-3} + \cdots$
Summary, in summation form	$f^*(t) = \sum_{n=0}^{\infty} f(nT)\delta(t-nT)$ $F^*(s) = \sum_{0}^{\infty} f(nT)e^{-nTs}$ $F(z) = \sum_{0}^{\infty} f(nT)z^{-n}$

(*d*)

(*d*) Summary of *Z*-transform of a sampled function.

Rearranging terms

$$F(z)(1 - z^{-1}) = 1$$

Solving for $F(z)$ we obtain the closed form representation for $F(z)$ of a unit step:

$$F(z) = \frac{1}{1 - z^{-1}} = \frac{z}{z - 1}$$

The series converges for $|z^{-1}| < 1$. In a similar manner the z-transforms of various time functions can be derived. A short list of z-transforms is compiled in Table 34.25.

Note that the z-transform can be readily inverted[6] to the infinite-series form by dividing the denominator of $F(z)$ into the numerator of $F(z)$ in such a manner as to produce negative powers of z. For the case of the unit step, where $F(z) = z/(z-1)$

$$\begin{array}{r l} & 1 + z^{-1} + z^{-2} + z^{-3} + \cdots \\ z - 1 \overline{)\, z} & \\ \underline{z - 1} & \\ 1 & \\ \underline{1 - z^{-1}} & \\ z^{-1} & \\ \underline{z^{-1} - z^{-2}} & \\ z^{-3} & \end{array} \qquad (34.44)$$

The coefficients of the infinite series expansion of $F(z)$ in powers of z^{-n} are the values of the time function $f(t)$ at the times $t = nT$, or $f(nT)$. Thus to convert a z-transform infinite series into a time-domain result, simply equate the z^{-n} coefficients with the values of $f(nT)$. $F(z) = 1 + z^{-1} + z^{-2} + z^{-3} + \cdots$ becomes $f(0) = 1$, $f(T) = 1$, $f(2T) = 1$, $f(3T) = 1, \ldots$. All the coefficients are 1 for the unit step function $u(t) \Rightarrow z/(z-1)$. Note that the values of $f(t)$ at times between the sample instants $T, 2T, 3T, \ldots$ are not known and might in fact oscillate widely. Such behavior can be discovered by using the modified z-transform to find the value of $f(t)$ at times between the sample instants. For a discussion of the modified z-transform, see Section 34.7.

z-Transforms can be directly applied to generate digital computer algorithms or to solve difference equations that describe control processes. Example 34.5 illustrates application of a partial-fraction expansion method to a z-transform, to determine the time solution for a given

TABLE 34.25. TABLE OF z-TRANSFORMS

Laplace Transform $F(s)$	Time Function $f(t), t > 0$	z-Transform $F(z)$
1	$\delta(t)$ [a]	1
e^{-kTs} [c]	$\delta(t - kT)$	z^{-k}
$\frac{1}{s}$	$u_s(t)$ [b]	$\frac{z}{z-1}$
$\frac{1}{s^2}$	t	$\frac{Tz}{(z-1)^2}$
$\frac{2}{s^3}$	t^2	$\frac{T^2 z(z+1)}{(z-1)^3}$
$\frac{1}{s+a}$	e^{-at}	$\frac{z}{z-e^{-aT}}$
$\frac{1}{(s+a)^2}$	te^{-at}	$\frac{Tze^{-aT}}{(z-e^{-aT})^2}$
$\frac{a}{s(s+a)}$	$1 - e^{-at}$	$\frac{z(1-e^{-aT})}{(z-1)(z-e^{-aT})}$
$\frac{1}{(s+a)(s+b)}$	$\frac{1}{(b-a)}(e^{-at} - e^{-bt})$	$\frac{1}{(b-a)}\left[\frac{z}{z-e^{-aT}} - \frac{z}{z-e^{-bT}}\right]$
$\frac{a}{s^2(s+a)}$	$t - \frac{1}{a}(1 - e^{-at})$	$\frac{Tz}{(z-1)^2} - \frac{(1-e^{-aT})z}{a(z-1)(z-e^{-aT})}$
$\frac{a}{s^3(s+a)}$	$\frac{1}{2}\left(t^2 - \frac{2}{a}t + \frac{2}{a^2}u_s(t) - \frac{2}{a^2}e^{-at}\right)$	$\frac{T^2 z}{(z-1)^3} + \frac{(aT-2)Tz}{2a(z-1)^2} + \frac{z}{a^2(z-1)} - \frac{z}{a^2(z-e^{-aT})}$
$\frac{a^2}{s(s+a)^2}$	$u_s(t) - (1+at)e^{-at}$	$\frac{z}{z-1} - \frac{z}{z-e^{-aT}} - \frac{aTe^{-aT}z}{(z-e^{-aT})^2}$

$\frac{a^2}{s^2(s+a)^2}$	$t - \frac{2}{a}u_s(t) + \left(t + \frac{2}{a}\right)e^{-at}$	$\frac{1}{a}\left[\frac{(aT+2)z - 2z^2}{(z-1)^2} + \frac{2z}{z - e^{-aT}} + \frac{aTe^{-aT}z}{(z-e^{-aT})^2}\right]$
$\frac{\omega}{s^2+\omega^2}$	$\sin \omega t$	$\frac{z \sin \omega T}{z^2 - 2z\cos \omega T + 1}$
$\frac{s}{s^2+\omega^2}$	$\cos \omega t$	$\frac{z(z - \cos \omega T)}{z^2 - 2z\cos \omega T + 1}$
$\frac{\omega}{s^2-\omega^2}$	$\sinh \omega t$	$\frac{z \sinh \omega T}{z^2 - 2z\cosh \omega T + 1}$
$\frac{s}{s^2-\omega^2}$	$\cosh \omega t$	$\frac{z(z - \cosh \omega T)}{z^2 - 2z\cosh \omega T + 1}$
$\frac{\omega}{(s+a)^2+\omega^2}$	$e^{-at}\sin \omega t$	$\frac{ze^{-aT}\sin \omega T}{z^2 - 2ze^{-aT}\cos \omega T + e^{-2aT}}$
$\frac{a^2+\omega^2}{s[(s+a)^2+\omega^2]}$	$1 - e^{-at}\sec\phi\cos(\omega t + \phi)$ $\quad \phi = \tan^{-1}(-a/\omega)$	$\frac{z}{z-1} - \frac{z^2 - ze^{-aT}\sec\phi\cos(\omega T - \phi)}{z^2 - 2ze^{-aT}\cos \omega T + e^{-2aT}}$
$\frac{s+a}{(s+a)^2+\omega^2}$	$e^{-at}\cos \omega t$	$\frac{z^2 - ze^{-aT}\cos \omega T}{z^2 - 2ze^{-aT}\cos \omega T + e^{-2aT}}$

Source. Adapted from Tou[13] and Ogata.[6]

[a] $\delta(t)$ = Dirac delta function. $\delta(t) = 0$ for $t \neq 0$, $\delta(t) \to \infty$ for $t = 0$, $3\delta(t)$ = Dirac delta function with a weight of 3.

$$\int_{0-}^{0+} \delta(t)\,dt = 1 \qquad \int_{-\infty}^{+\infty} \delta(t)\,dt = 1$$

$$\int_{-\infty}^{+\infty} \delta(t)f(t)\,dt = f(0) \qquad \int_{-\infty}^{+\infty} \delta(t-T)f(t)\,dt = f(T)$$

[b] $u_s(t)$ = unit step function. $u_s(t) = 0$ for $t < 0$, $u_s(t) = 1$ for $t > 0$.

[c] $z = e^{sT}$ where T = sampling period of z-transform.

Note: $s = (1/T)\ln z$ where s = Laplace transform variable. ω = Bilinear transform variable, used to map the inside of the unit circle in the z domain, into the right half plane of the ω domain, via $z = (1+\omega)/(1-\omega)$. To check for stability of $f(z)$, replace z by $(1+\omega)/(1-\omega)$, solve for $f(\omega)$, and apply the Routh–Hurwitz test to $f(\omega) = 0$.

difference equation. Although this example is relatively simple, it illustrates the procedure used and the various notations that appear in the literature.

Note that the partial-fraction expansion is performed on $F(z)/z$ and not on $F(z)$ alone. After the expansion of $F(z)/z$ is completed, both sides of the resulting equation are multiplied by z to give a partial-fraction expansion of $F(z)$ that contains a z in the numerator of each fraction. This allows the z-transforms of Table 34.25 to be used directly.

Example 34.5 z-Transform Solution of a Difference Equation.[14] Consider the first order difference equation:

$$Y(k) = Bu(k) + AY(k-1)$$

where $u(k) = 0,\ k = -1, -2, -3, \ldots$
$u(k) = 1,\ k = 0, 1, 2, 3, \ldots$
$B, A =$ constants

This can be converted to a sampled-time function as follows

$$Y(nT) = Bu(nT) + AY[(n-1)T] \qquad (\text{let } T = 1 \text{ s})$$

The equivalent time function is

$$Y(t) = Bu(t) + AY(t-T)$$

The equivalency holds only at the sample instants $t = nT = nk$.

With use of Table 34.25, the z-transform is readily found as

$$Y(t) \Rightarrow Y(z)$$

$$u(t) \Rightarrow \frac{z}{z-1}$$

$$Y(t-T) \Rightarrow z^{-1}Y(z)$$

Inserting these

$$Y(z) = B\frac{z}{z-1} + Az^{-1}Y(z)$$

Solving for $Y(z)$ by collecting terms:

$$Y(z)(1 - Az^{-1}) = B\frac{z}{z-1}$$

Dividing out and multiplying by z/z:

$$Y(z) = B\frac{z}{z-1}\,\frac{1}{1-Az^{-1}} = B\frac{z}{z-1}\,\frac{z}{z-A}$$

Since a z in the numerator is needed for z-transforms (see Table 34.25), expand $Y(z)/z$ as a partial fraction group and then multiply both sides by z to get $Y(z)$:

$$\frac{Y(z)}{z} = \frac{Bz}{(z-1)(z-A)}$$

Expanding by partial fractions:

$$\frac{Y(z)}{z} = \frac{C}{z-1} + \frac{D}{z-A} = \frac{Bz}{(z-1)(z-A)}$$

Solving for C and D via the residue method:

$$C = \frac{Y(z)}{z}(z-1)\bigg|_{z=1} = \frac{Bz}{z-A}\bigg|_{z=1} = \frac{B}{1-A}$$

$$D = \frac{Y(z)}{z}(z-A)\bigg|_{z=A} = \frac{Bz}{z-1}\bigg|_{z=A} = \frac{BA}{A-1} = \frac{-BA}{1-A}$$

Inserting C and D back into $Y(z)/z$:

$$\frac{Y(z)}{z} = \frac{B}{1-A}\,\frac{1}{z-1} + \frac{-BA}{1-A}\,\frac{1}{z-A}$$

Factoring out $B/(1-A)$

$$\frac{Y(z)}{z} = \frac{B}{1-A}\left[\frac{1}{z-1} + \frac{-A}{z-A}\right]$$

Multiplying both sides by z:

$$Y(z)=\frac{B}{1-A}\left[\frac{z}{z-1}+\frac{-Az}{z-A}\right]$$

$Y(nT)$ is found using the inverse z-transforms (from Table 34.25):

$$\frac{z}{z-1}\Rightarrow 1u(t)$$

$$\frac{z}{z-A}\Rightarrow A^{n}u(t)^{\dagger} \qquad \begin{aligned}&n=\text{sample number;}\\&n=0,1,2,\ldots\end{aligned}$$

Thus $Y(nT)=B/(1-A)[1-A\cdot A^{n}]u(nT)$.

Simplifying

$$Y(nT)=\frac{B}{1-A}[1-A^{n+1}]=B\frac{1-A^{n+1}}{1-A} \qquad n\geqslant 0$$

In terms of k, the sample number

$$Y(K)=B\frac{1-A^{k+1}}{1-A} \qquad \text{where} \quad T=1\text{ sec}, k\geqslant 0$$

This closed-form expression can be recognized as the formula for the sum of a finite geometric series with a first term B and a ratio of A, containing $K+1$ terms, that is,

$$Y(k)=B+AB+A^2B+A^3B+\ldots A^KB$$

or

$$Y(k)=B(1+A+A^2+A^3+\ldots A^K)$$

Example 34.6 z-Transform Solution of Closed-Loop Sampled Data Control System. Determine the output $C^*(t)$ at the sampling instants of the sampled data control system shown in Fig. 34.22. Note that $C(t)$ is equal to $C^*(t)$ only at the sampling instants. Consider the input function $r(t)$ to be a unit step.

The given system is first transformed to the z domain. Referring to Fig. 34.22, the z-transform is derived with the use of the z-transform pair:

$$\frac{a}{s(s+a)}\Rightarrow\frac{(1-e^{-at})z}{(z-1)(z-e^{-at})}$$

(Given $T=1/10$ sec, $a=10$, then $e^{-at}=e^{-1}=0.368$)

$$F(s)=\frac{K}{s(s+10)}$$

which can therefore be written as

$$F(s)=(0.1K)\frac{10}{s(s+10)}\Rightarrow(0.1K)\frac{(1-0.368)z}{(z-1)(z-0.368)}=F(z)$$

Simplifying:

$$F(z)=\frac{(0.1K)(0.632)z}{(z-1)(z-0.368)}=\frac{0.0632Kz}{z^2-1.368z+0.368}$$

The open-loop gain $F(z)$ is then converted to the closed-loop gain $G(z)$ by utilizing the standard block diagram reduction technique

$$G(z)=\frac{F(z)}{1+F(z)}=\frac{N(z)/D(z)}{1+N(z)/D(z)}=\frac{N(z)}{D(z)+N(z)}$$

where $N(z)=$ numerator of $F(z)$
$D(z)=$ denominator of $F(z)$
$G(z)=$ closed-loop gain
$F(z)=$ open-loop gain

†This transform is obtained from $z/(z-e^{-aT})\rightarrow e^{-at}$ by noting that $t=nT$ and letting $e^{-aT}=A$. Then $z/(z-A)\rightarrow e^{-at}=e^{-anT}=(e^{-aT})^n=A^n$.

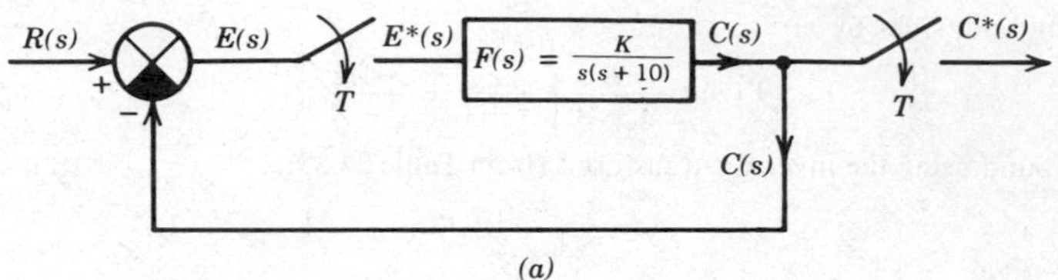

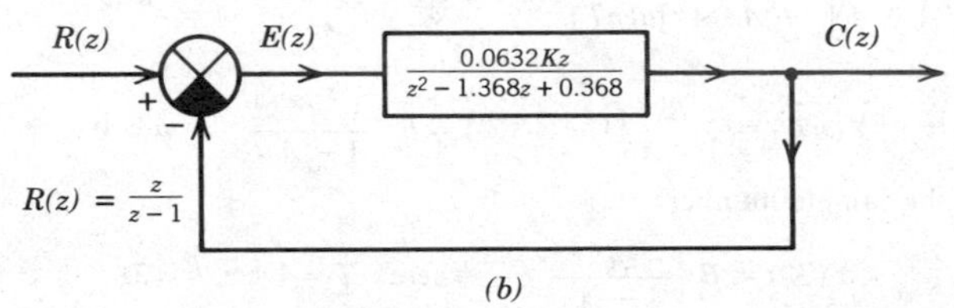

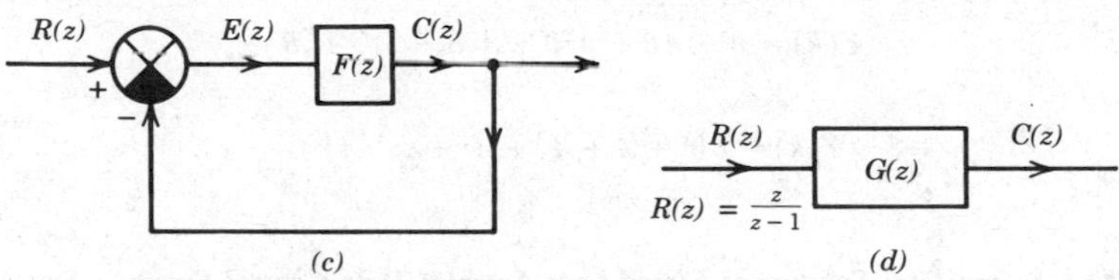

$$C(z) = R(z)G(z) = \frac{z}{z-1}\,\frac{0.0632(20)z}{z^2 + [0.0632(20) - 1.368]z + 0.368} \qquad (\text{for } K = 20)$$

$$C(z) = \frac{z}{z-1}\,\frac{1.264z}{z^2 - 0.104z + 0.368}$$

$$C(z) = \frac{1.264z^2}{(z-1)(z^2 - 0.104z + 0.368)}$$

(e)

Fig. 34.22 Block diagram of the sampled-data control system of Example 34.6. (*a*) s-Domain representation. $R(s) = 1/s$, $T = 1/10$ sec. (*b*) z-Domain representation. $R(z) = z/z - 1$. Obtained by the *z*-transforms (see Table 34.25)

$$\frac{a}{s(s+a)} \Rightarrow \frac{(1 - e^{-at})}{(z-1)(z - e^{-at})},$$

$$\frac{1}{s} \Rightarrow \frac{z}{z-1}$$

$$e^{-at} = e^{-10\frac{1}{10}} = e^{-1} = 0.368$$

$$(1 - e^{-aT})/a = (1 - 0.368)/10 = 0.632/10 = 0.0632$$

$$(z-1)(z - 0.368) = z^2 - 1.368z + 0.368$$

(*c*) Redrawing of (*b*), where $F(z) = \dfrac{0.0632Kz}{z^2 - 1.368z + 0.368} = \dfrac{N(z)}{D(z)}$ (*d*) Simplification of (*c*), where $G(z) = F(z)/[1 + F(z)]$.

$$G(z) = \frac{N(z)/D(z)}{1 + N(z)/D(z)} = \frac{N(z)}{D(z) + N(z)} = \frac{0.0632K z}{z^2 + (0.0632K - 1.368)z + 0.368}$$

For stability, $0 < K < 43.2$ (from Routh–Hurwitz test using bilinear transformation; explained in detail in Example 34.6). Let $K = 20$. (*e*) Finding $C(z)$.

or

$$G(z) = \frac{0.0632Kz}{z^2 + (0.0632K - 1.368)z + 0.368}$$

This system, like any feedback control system, should be examined for stability. The standard Routh–Hurwitz array test cannot be directly applied to z-transform equations. However, the standard Routh–Hurwitz test can be employed if the *bilinear transformation* is first applied to the characteristic equation (denominator set equal to zero) of the closed-loop transfer function of $G(z)$. The bilinear transformation, $z = (1 + w)/(1 - w)$, converts the transcendental equation in z ($z = e^{sT}$) into a polynomial equation in w, which can be tested by the Routh–Hurwitz array.

The characteristic equation in the z domain is

$$z^2 + (0.0632K - 1.368)z + 0.368 = 0$$

Substituting the bilinear transform, $(z = 1 + w)/(1 - w)$

$$\frac{(1 + w)^2}{(1 - w)^2} + (0.0632K - 1.368)\frac{1 + w}{1 - w} + 0.368 = 0$$

Simplifying:

$$(1 + w)^2 + (0.0632K - 1.368)(1 + w)(1 - w) + 0.368(1 - w)^2 = 0$$

$$1 + 2w + w^2 + (0.0632K - 1.368)(1 - w^2) + 0.368(1 - 2w + w^2) = 0$$

$$w^2(1 - 0.0632K + 1.368 + 0.368) + w(2 - 0.736) + (1 + 0.0632K - 1.368 + 0.368) = 0$$

$$w^2(2.736 - 0.0632K) + w(1.264) + 0.0632K = 0$$

Applying the Routh–Hurwitz array test

	$w^2(2.736 - 0.0632K) + w(1.264) + 0.0632K$	
w^2	$(2.736 - 0.0632K)$	$0.0632K$
w^1	1.264	0
w^0	$0.0632K$	0

For stability there must be no sign changes in the first column of the Routh–Hurwitz array. Therefore

$$0.0632K > 0 \qquad \text{or} \qquad K > 0$$

and

$$(2.736 - 0.0632K) > 0 \qquad \text{or} \qquad K < \frac{2.736}{0.0632} = 43.29$$

Thus K must be greater than zero and less than 43.29 for stable behavior. Arbitrarily choosing $K = 20$ allows the desired unit step response to be numerically evaluated. Inserting $K = 20$ into the expression for $G(z)$ yields

$$G(z) = \frac{0.0632(20)z}{z^2 + [0.0632(20) - 1.368]z + 0.368} = \frac{1.264z}{z^2 - 0.104z + 0.368}$$

Finding $C(z)$ from $G(z)$ and $R(z)$:

$$C(z) = R(z)G(z) = \frac{z}{z - 1}\,\frac{1.264z}{z^2 - 0.104z + 0.368}$$

$$C(z) = \frac{1.264z^2}{(z - 1)(z^2 - 0.104z + 0.368)}$$

Factoring the quadratic:

$$C(z) = \frac{1.264z^2}{(z - 1)(z - 0.052 + j0.604)(z - 0.052 - j0.604)}$$

Performing a partial fraction expansion of $C(z)/z$ (so that the z can later be multiplied out to give a z in the partial fraction numerators, which is required for obtaining the inverse z transform):

$$\frac{C(z)}{z} = \frac{A}{z - 1} + \frac{B}{z - 0.052 + j0.604} + \frac{B\text{ conjugate}}{z - 0.052 - j0.604}$$

Evaluating A by the residue technique:

$$A = \frac{C(z)}{z}(z-1)\Big|_{z=1} = \frac{1.264z}{z^2 - 0.104z + 0.368}\Big|_{z=1}$$

$$A = \frac{1.264}{1 - 0.104 + 0.368} = \frac{1.264}{1.264} = 1.000$$

Evaluating B by the residue technique:

$$B = \frac{C(z)}{z}(z - 0.052 + j0.604)\Big|_{z=0.052-j0.604}$$

$$B = \frac{1.264z}{(z-1)(z-0.052-j0.604)}\Big|_{z=0.052-j0.604} = \frac{1.264(0.052-j0.604)}{(-0.948-j0.604)(-j1.208)}$$

$$B = -0.500 + j0.261 = 0.564\measuredangle 152.4°$$

Inserting A and B back into the partial fraction expansion for $C(z)/z$ and multiplying out z:

$$C(z) = \frac{z}{z-1} + \frac{z(-0.5+j0.261)}{z-0.052+j0.604} + \frac{z(-0.5-j0.261)}{z-0.052-j0.604}$$

Inverting $C(z)$ using the z-transforms (from Table 34.25):

$$\frac{z}{z-1} \Rightarrow u(t)$$

$$\frac{z}{z-e^{-aT}} \Rightarrow e^{-at} \qquad T = 1/10 \text{ sec}$$

The conjugate poles $(0.052 \pm j0.604)$ must be put in the form

$$e^{-aT} = e^{-a/10} = e^{-(X+jY)/10}$$

where $a = X + jY$ and $T = 1/10$ s.

Equating terms:

$$-e^{-a/10} = -0.052 + j0.604$$

Expanding:

$$e^{-a/10} = e^{-(X+jY)/10} = e^{-X/10}e^{-jY/10} = 0.052 - j0.604 = 0.606\measuredangle -1.485 \text{ rad}$$

Equating magnitudes and angles:

$$e^{-X/10} = \text{magnitude} = 0.606 \qquad e^{-jY/10} = \text{angle} = -1.485 \text{ rad}$$

Solving for X and Y:

$$-\frac{X}{10} = \ln 0.6062 = -0.500 \qquad -\frac{Y}{10} = -1.485 \text{ rad}$$

where $X = +5.00$ and $Y = +14.85$.

Thus

$$z - 0.052 + j0.604 = z - e^{-a/10} = z - e^{-(X+jY)T}$$

and

$$z - 0.052 - j0.604 = \text{conjugate} = z - e^{-(X-jY)T}$$

where $T = 1/10$ s
$X = 5$
$Y = 14.85$

Inserting these values into $C(z)$ yields

$$C(z) = \frac{z}{z-1} + \frac{z(-0.5+j0.261)}{z-e^{-(5+j14.85)T}} + \frac{z(-0.5-j0.261)}{z-e^{-(5-j14.85)T}}$$

The inverse z-transforms previously given for $z/(z-1)$ and $z/(z-e^{-aT})$ may now be directly applied:

$$c^*(t) = 1 + (-0.5+j0.261)e^{-(5+j14.85)t} + (-0.5-j0.261)e^{-(5-j14.85)t}$$

Performing rectangular-to-polar conversions:

$$c^*(t) = 1 + (0.564e^{j152.4°})e^{-(5+j14.85)t} + (0.564e^{-j152.4°})e^{-(5-j14.85)t}$$

Rearranging terms and factoring the complex exponentials:

$$c^*(t) = 1 + 1.128e^{-5t}\,\frac{e^{-j(14.85t-152.4°)} + e^{+j(14.85t-152.4°)}}{2}$$

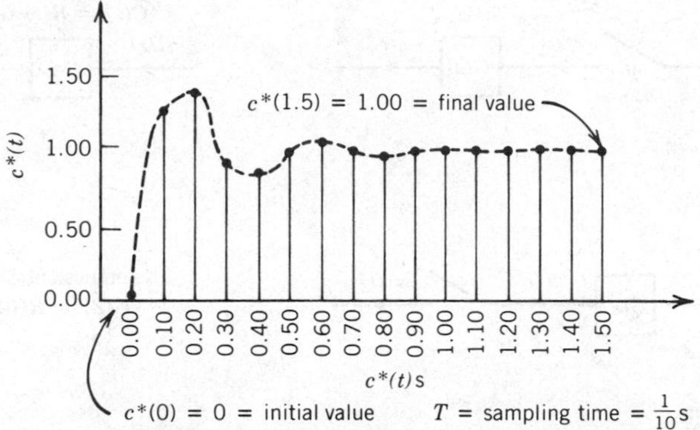

Fig. 34.23 Plot of the unit step response $c^*(t) = 1 + 1.128e^{-5t}\cos(14.85t - 152.4°)$ at the sampling time $T = 0.1$ s, where $c^*(t)$ = sampled values of $c(t)$. Valid *only* at the sampling instants $t = 0, 0.1, 0.2, 0.3, \ldots$ s. This is the sample time response of $C(z)$ shown in Fig. 34.22*e*. Note that $c^*(t)$ is derived from $C(z) = 1.264z^2/(z-1)(z^2 - 0.104z + 0.368)$ in Example 34.6.

Expressing $c^*(t)$ as a sinusoid:

$$\frac{e^{-jX} + e^{+jX}}{2} \Rightarrow \cos X$$

where the units are $14.85t$(rad) and $-152.4°(= -2.66 \text{ rad})$:

$$c^*(t) = 1 + 1.128e^{-5t}\cos(14.85t - 152.4°)$$

or

$$c^*(t) = 1 + 1.128e^{-5t}\cos(14.85t - 2.66)$$

Check:

$$c(0) = 1 + 1.128(1)(-0.8862) = 0.00036\surd \text{ (approximately zero)}$$

The values at the sampling instants, from $t = 0$ to $t = 1.5$ s, with 1/10-s sampling, are plotted in Fig. 34.23.

It is very tempting to assume that $c^*(t)$ is valid for all time, not just at the sampling instants. The mind quickly joins the dotted line of the calculated sampled-data response into a continuous time function. For this particular example, $c^*(t)$ is a very good approximation of the actual system response between the sample instants, because the sampling time ($t = 1/10$ s) is much shorter than one half the system's period of oscillation. If the sampling time were increased to 10 s, this would not be true, and "connecting the dots" would *not* produce a good approximation of the actual system response between the sampling instants. An example of misleading sampled-data calculations, produced by a sampling time equal to one half a system's natural periods of oscillation, is shown in Fig. 34.24.

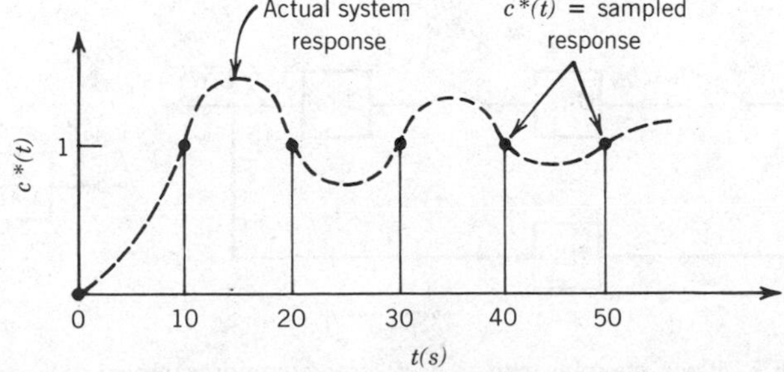

Fig. 34.24 Sketch of a possible misleading sampled-data calculated response, produced by a sampling time T of 10 s, which is equal to one half of the system's natural periods of oscillation. The sampled response values for $t > 0$ are always unity, whereas the actual system response oscillates. To correctly predict the oscillatory response, T should be decreased to perhaps one tenth of the system's oscillatory period or to approximately 2 s for the response shown.

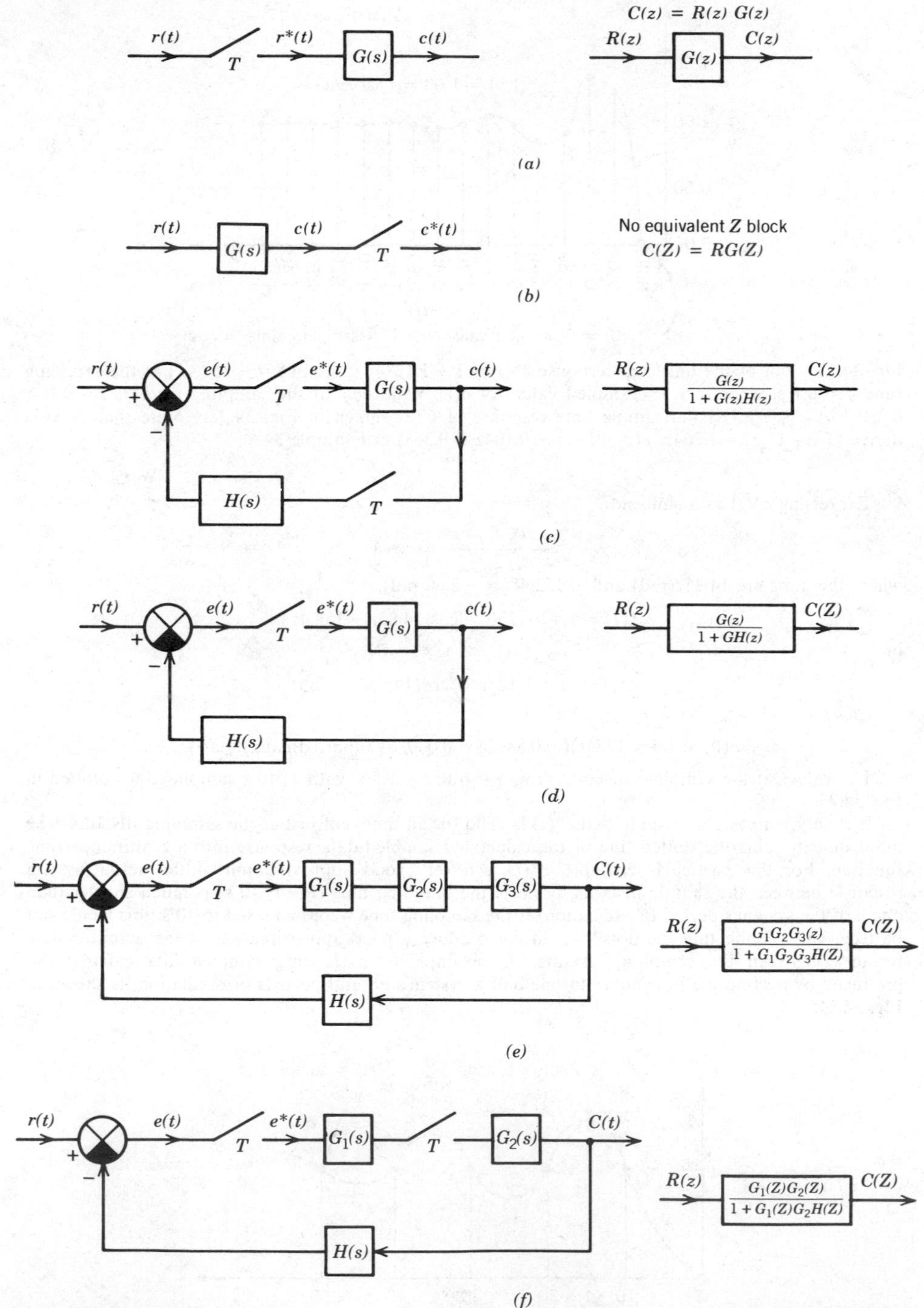

Fig. 34.25*a* through *j* Block diagram reduction of some common sampled-data control systems. In (d) note that $GH(z)$ = one z-transform, *not* given by $G(z)$ times $H(z)$. It is the single z-transform of the quantity $G(s)H(s)$. In (g), when the error is not sampled, $R(z)$ does not exist as a separate z-transform component and hence no simplified transfer function relating $C(z)$ to $R(z)$ exists. The z-transform for $C(z)$ for two such systems is presented in (h).

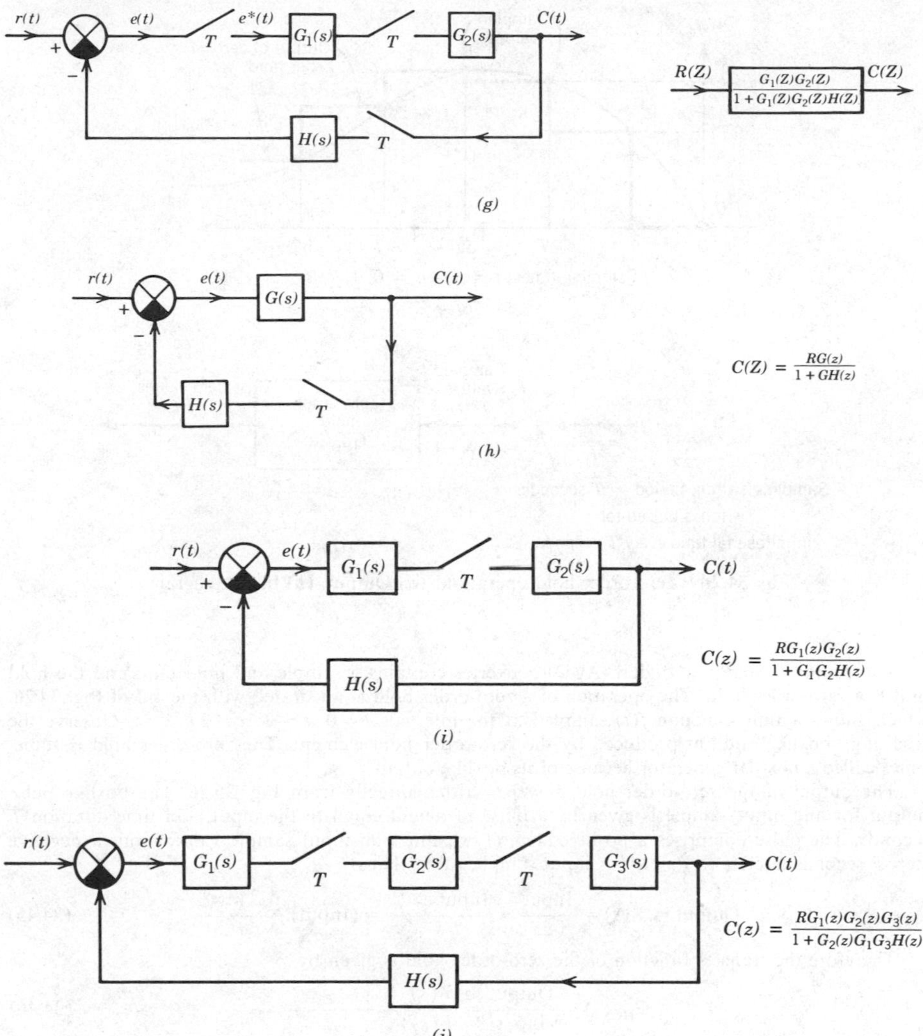

34.7 z-TRANSFORM BLOCK DIAGRAMS INCLUDING HOLD ELEMENTS

z-Transform block diagrams can be combined (e.g., reduced) exactly as Laplace transform block diagrams are combined as long as the sampling time is constant and in synchronism throughout a control system. See Tables 34.8 and 34.9 for a summary of the rules in the s domain and Fig. 34.25 for specific examples of z-transform block diagram reductions.

Certain elements are used only in sampled-data systems and not in continuous systems. The most common unique z-transform block (element) is the "hold element." The two most popular hold elements are the "zero-order hold" and the "first-order hold."

The purpose of the zero-order hold device is to convert the almost instantaneous sample, of theoretically zero-time duration, into a constant level value over the duration of the intersampling period T. Essentially it operates as a clamp. The resultant constant-level value equals the sampled value and has a slope of zero (as do all constants). The slope of zero is responsible for the name zero-order hold.

A first-order hold would hold the first derivative (the slope) as well as the initially sampled magnitude of the sampled function. First-order holds are far less common than are zero-order holds.

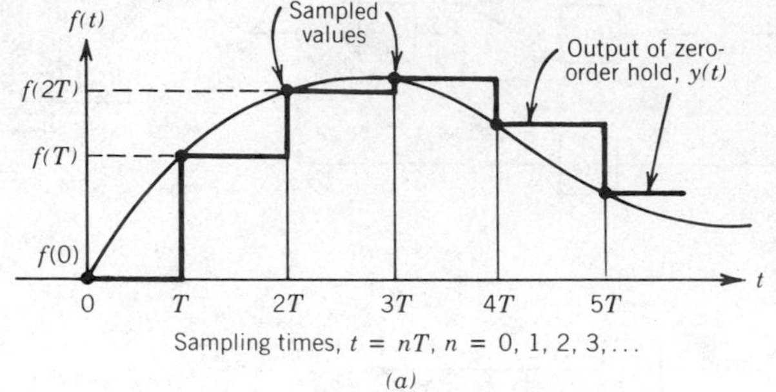

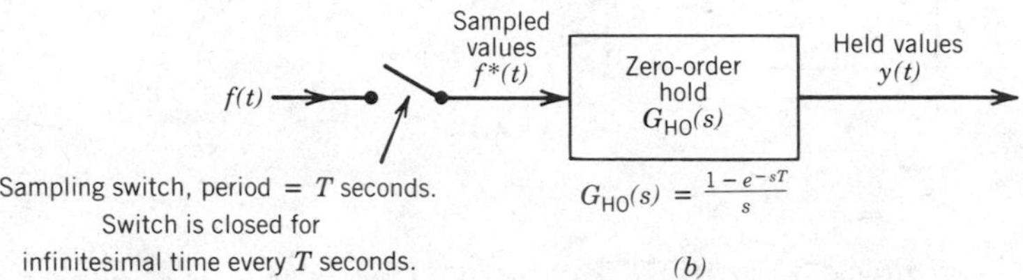

Fig. 34.26 Zero-order hold operation: (*a*) Output, (*b*) block diagram.

The standard analog-to-digital (A/D) converter contains a sample and hold unit, and the hold unit is a zero-order hold. The operation of a zero-order hold is illustrated with the aid of Fig. 34.26, which shows a time function $f(t)$ sampled at the intervals $t = 0$, $t = T$, $t = 2T, \ldots$. Observe the resulting "boxlike" output produced by the zero-order hold element. The zero-order hold is sometimes called a box-car generator because of its boxlike output.

The output of the zero-order hold may be written directly from Fig. 35.26. The boxlike pulse output for any input sample is given by a pulse of height equal to the input, and time duration T seconds. The pulse comprises a positive step, $(1/s)$, times an input sample value minus a negative step T seconds later: $(-1/s)e^{-sT} \times$ input sample value. Thus:

$$\text{Output} = Y(s) = \frac{\text{Input}}{s} - \frac{\text{Input}\, e^{-sT}}{s} = (\text{Input})\,\frac{1-e^{-sT}}{s} \tag{34.45}$$

Therefore the transfer function of the zero-order hold is given by

$$G_{\text{HO}}(s) = \frac{\text{Output}}{\text{Input}} = \frac{Y(s)}{F(s)} = \frac{1-e^{-sT}}{s} \tag{34.46}$$

It is common practice to cascade the zero-order hold $G_{\text{HO}}(s)$ with the system plant or forward gain transfer function $G_P(s)$, without a sampling switch between $G_{\text{HO}}(s)$ and $G_P(s)$. In such a situation the transform of $G_{\text{HO}}(s)G_P(s)$ must be found as a single z-transform and not as a product of z-transforms. That is, $G_{\text{HO}}(s)$ and $G_P(s)$ must be first multiplied out in the s domain and a single (lengthy) z-transform found for the resulting function of s. The process is straightforward, as shown:

$$\begin{aligned} Z[G_{\text{HO}}(s)G_P(s)] &= Z\left[\frac{1-e^{-sT}}{s}\,G_P(s)\right] \\ &= Z\left[(1-e^{-sT})\,\frac{G_P(s)}{s}\right] \\ &= (1-z^{-1})Z\left[\frac{G_P(s)}{s}\right] \end{aligned} \tag{34.47}$$

The simplification results from the fact that the delay term $(1-e^{-sT})$ and its z-transform $(1-z^{-1})$ can be factored out from their respective functions. The net result is that a z-transform of $G_P(s)$ divided by s is required, rather than a z-transform of $G_P(s)$ alone.

The influence of the presence of sampler switches upon the z-transform transfer function of a block diagram is illustrated in Fig. 34.25. Whenever synchronized sampler switches isolate elements in a loop, the z-transform of these isolated elements is taken, one at a time, on an individualized basis. Whenever there is no sampler between two or more elements, these elements must be treated as a combined single entity with one z-transform taken for the resulting composite transfer function in s. Thus if 2, 3, 4 or more transfer function blocks are cascaded in the s domain, with no sampler switch between these blocks, only one z-transform block results. Whenever separate z-transform blocks appear, it is understood that the corresponding Laplace transform blocks are separated by synchronized sampler switches. To illustrate the effect of a zero-order hold and also the effect of varying the sampling period, the following example is provided.

Example 34.7. Given a unity feedback error sampled system containing a zero-order hold and having a plant transfer function of $G_P(s) = 1/(s+1)$

Find:
1. The equivalent z-transform simplified block diagram, relating $C(z)$ to $R(z)$.
2. For a unit step input $[R(z) = z/(z-1)]$, find $C_a(z)$ for $T = 1$ sec, $C_b(z)$ for $T = 1/2$ sec, and $C_c(z)$ for $T = 1/10$ sec. Also find the corresponding sampled-output functions in the time domain, $C_a^*(t)$ for $T = 1$ sec, $C_b^*(t)$, for $T = 1/2$ sec, and $C_c^*(t)$ for $T = 1/10$ sec, and plot the respective sampled responses vs. time.
3. Repeat 1. and 2. for $T = 1$ sec, with the zero-order hold removed.
4. Calculate the response of a continuous-time unsampled system, plot the time response, and compare results.

Step 1. The block diagram for the first step is shown in Fig. 34.27a. Referring to this figure

$$G(s) = G_{\mathrm{HO}}(s)G_P(s) = \frac{1-e^{-sT}}{s}\,\frac{1}{s+1} = (1-e^{-sT})\frac{1}{s(s+1)} \tag{34.48}$$

Referring to the table of z-transforms, Table 34.25

$$(1-e^{-sT})F(s) \Rightarrow (1-z^{-1})F(z)$$

$$\frac{a}{s(s+a)} \Rightarrow \frac{z(1-e^{-aT})}{(z-1)(z-e^{-aT})} \qquad \text{let } a = 1$$

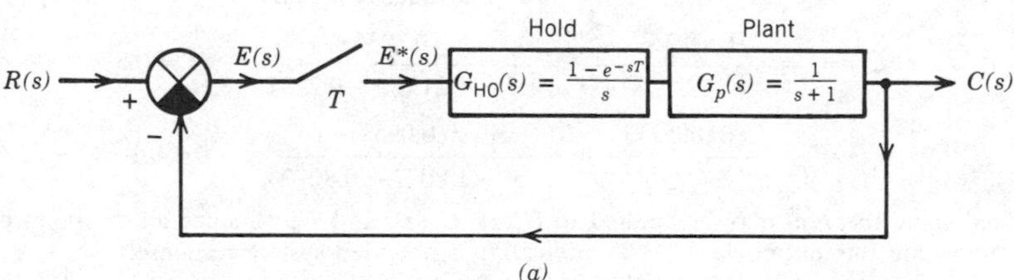

(a)

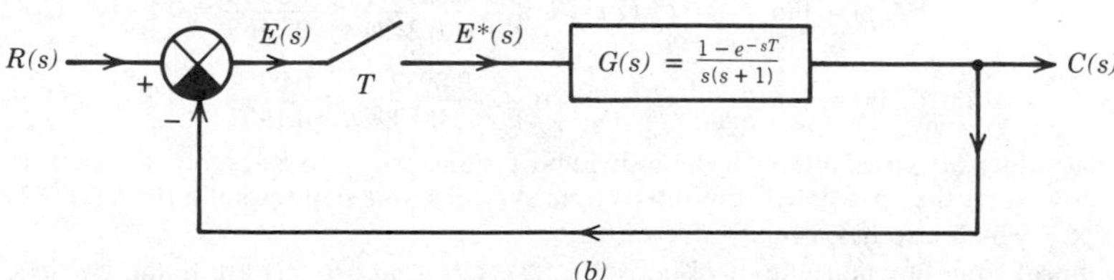

(b)

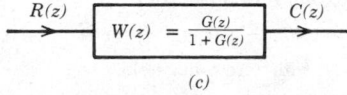

R(z) $W(z) = \frac{G(z)}{1+G(z)}$ C(z)

(c)

Fig. 34.27 (a) Block diagram of an error-sampled unity feedback system containing a zero-order hold, for Example 34.7. (b) Combining blocks. (c) In z domain, simplified as per Fig. 34.25d. $G(z) = Z\{G(s)\} = (1-z^{-1})[z(1-e^{-T})/(z-1)(z-e^{-T})] = 1-e^{-T}/z-e^{-T}$.

Applying these two transform pairs to $G(s)$ with $a = 1$ yields $G(z)$:

$$G(z) = (1 - z^{-1})\frac{z(1 - e^{-T})}{(z-1)(z - e^{-T})} = \frac{(z-1)}{z}\,\frac{z(1 - e^{-T})}{(z-1)(z - e^{-T})}$$

Canceling the z terms:

$$G(z) = \frac{(z-1)(1 - e^{-T})}{(z-1)(z - e^{-T})} = \frac{1 - e^{-T}}{z - e^{-T}}$$

Now from Fig. 34.25d, the equivalent reduced z-transform function is

$$W(z) = \frac{G(z)}{1 + GH(z)} = \frac{G(z)}{1 + G(z)} \quad \text{(since } H = 1\text{)}$$

Substituting the value for $G(z)$

$$W(z) = \frac{(1 - e^{-T})/(z - e^{-T})}{1 + (1 - e^{-T})/(z - e^{-T})} = \frac{1 - e^{-T}}{z + (1 - 2e^{-T})}$$

This completes step 1, since $C(z) = R(z)W(z)$.

The reduced z-domain block diagram is a single block containing $W(z)$, with input $R(z)$ and output $C(z)$.

The specified input function is a unit step, for which $R(s) = 1/s$ and $R(z) = z/(z-1)$. Multiplying this value of $R(z)$ by $W(z)$ yields the output $C(z)$.

$$C(z) = \frac{z}{z-1}\,\frac{1 - e^{-T}}{z + (1 - 2e^{-T})}$$

Step 2. In step 2, three specific sampling times are to be used, which will produce three different outputs, $C_a(z)$, $C_b(z)$, and $C_c(z)$ for $T = 1$ sec, $T = 1/2$ sec, and $T = 1/10$ sec. Evaluating numerically yields

$$C_a(z) = \frac{z(0.6321)}{(z-1)(z+0.2642)} = \frac{z(0.6321)}{z^2 - 0.7358z - 0.2642} \qquad T = 1$$

$$C_b(z) = \frac{z(0.3935)}{(z-1)(z-0.2131)} = \frac{z(0.3935)}{z^2 - 1.2131z + 0.2131} \qquad T = 0.5$$

$$C_c(z) = \frac{z(0.0952)}{(z-1)(z-0.8097)} = \frac{z(0.0952)}{z^2 - 1.8097 + 0.8097} \qquad T = 0.1$$

The final-value theorem may be applied to $C_a(z)$, $C_b(z)$, and $C_c(z)$, since all the poles of these functions lie within the unit circle ($|z| \leqslant 1$), indicating a bounded system response.

$$\lim_{t\to\infty} C_a^*(t) = \lim_{z\to 1}\frac{z-1}{z}C_a(z) = \lim_{z\to 1}\frac{0.6321}{z + 0.2642} = \frac{0.6321}{1.2642} = 0.5$$

$$\lim_{t\to\infty} C_b^*(t) = \lim_{z\to 1}\frac{z-1}{z}C_b(z) = \lim_{z\to 1}\frac{0.3935}{z - 0.2131} = \frac{0.3935}{0.7869} = 0.5$$

$$\lim_{t\to\infty} C_c^*(t) = \lim_{z\to 1}\frac{z-1}{z}C_c(z) = \lim_{z\to 1}\frac{0.0952}{z - 0.8097} = \frac{0.0952}{0.1903} = 0.5$$

The final values all agree, although the individual z-transforms are markedly different. In step 4. it will be shown that the unsampled continuous time system's unit step response for $G(s) = 1/(s+1)$ is $\frac{1}{2}(1 - e^{-2t})$, which also has a final value of 0.5.

The sampled time function output responses $C_a^*(t)$, $C_b^*(t)$, and $C_c^*(t)$ are found by dividing the denominator of $C(z)$ into the numerator of $C(z)$, to obtain a series of powers of z^{-n}. The results are

$$C_a(z) = 0.63z^{-1} + 0.47z^{-2} + 0.51z^{-3} + 0.50z^{-4} + 0.50z^{-5} + \cdots$$

for $T = 1.0$ sec, $z^{-1} =$ 1-sec delay, with zero-order hold.

$$C_b(z) = 0.39z^{-1} + 0.47z^{-2} + 0.49z^{-3} + 0.49z^{-4} + 0.50z^{-5} + \cdots$$

for $T = 0.5$ sec, $z^{-1} =$ 0.5-sec delay, with zero-order hold.

$$C_c(z) = 0.095z^{-1} + 0.17z^{-2} + 0.23z^{-3} + 0.27z^{-4} + 0.30z^{-5} + 0.32z^{-6} + 0.34z^{-7} + \cdots$$

for $T = 0.1$ sec, $z^{-1} =$ 0.1-sec delay, with zero-order hold.

The corresponding time responses are (t measured in secs):

$C_a^*(t) =$	0.63	0.47	0.51	0.50	0.50
$t =$	1	2	3	4	5
$C_b^*(t) =$	0.39	0.47	0.49	0.49	0.50
$t =$	0.5	1	1.5	2	2.5
$C_c^*(t) =$	0.095	0.17	0.23	0.27	0.30
$t =$	0.1	0.2	0.3	0.4	0.5

This completes step 2.

Step 3. The response to $C_0(z)$, with no zero-order hold, and $T = 1$ s, will be found next. Referring to Fig. 34.28:

$$G_0(s) = \frac{1}{s+1}$$

Using the z-transform

$$\frac{1}{s+a} \Rightarrow \frac{z}{z - e^{-aT}}$$

$G_0(z)$ is obtained by substituting $a = 1$:

$$G_0(z) = \frac{z}{z - e^{-T}}$$

The reduced block diagram in the z domain is

$$W_0(z) = \frac{G_0(z)}{1 + G_0(z)} = \frac{z}{2z - e^{-T}} = \tfrac{1}{2}\frac{z}{(z - 0.5e^{-T})}$$

The output $C_0(z)$ is given by

$$C_0(z) = R(z)W_0(z) = \frac{z}{(z-1)}\left[\tfrac{1}{2}\frac{z}{(z - 0.5e^{-T})}\right]$$

Substituting in the sampling time, $T = 1$ sec:

$$C_0(z) = \frac{0.5z^2}{(z-1)(z-0.1839)} = \frac{0.5z^2}{z^2 - 1.1839z + 0.1839}$$

Applying the final-value theorem to $C_0(z)$:

$$\lim_{t\to\infty} C_0^*(t) = \lim_{z\to 1} \frac{z-1}{z} C_0(z) = \lim_{z\to 1} \frac{0.5z}{z - 0.1839} = \frac{0.5}{0.8951} = 0.6127$$

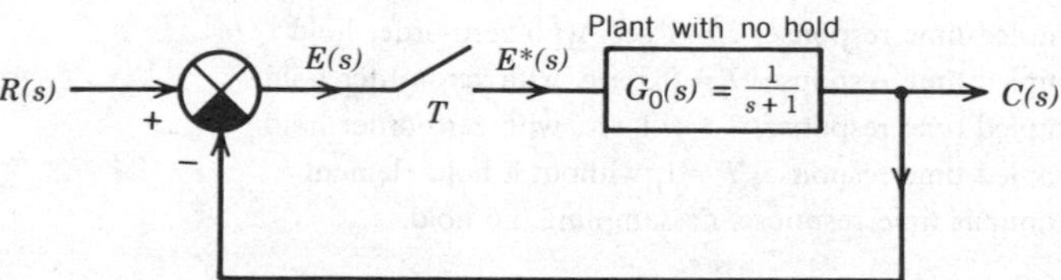

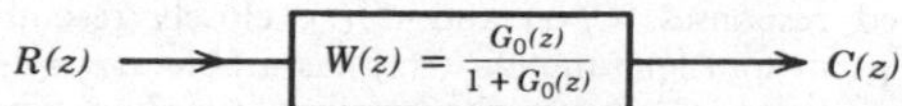

Fig. 34.28 Block diagram of an error-sampled unity feedback system, with no zero-order hold, for Example 34.7, step 3. $G_0(z) = z[G(s)] = z/(z - e^{-T})$.

Observe that the final value of the sampled system with no zero-order hold is 0.6127, which does not agree with the final value of the continuous system or the three other identical sampled systems having a zero-order hold (but different sampling times). These last four related systems all have a final value of 0.5 units, for a unit step input. Note that a holding device is an essential part of almost all digitally controlled sampled-data control systems in order to interface the sampler (or digital computer) with an analog portion of the control system. Therefore this case (no-hold circuit) is not a practical one.

The sampled time-response output function, $C_0^*(t)$, is found by dividing the denominator of $C_0(z)$ into the numerator of $C_0(z)$. The result is

$$C_0(z) = 0.5 + 0.59z^{-1} + 0.62z^{-2} + 0.61z^{-3} + 0.61z^{-4} + \cdots$$

for $T = 1$ sec, $z^{-1} = 1$-sec. delay, no zero-order hold.

The corresponding sampled-time response $C_0^*(t)$ is (with t measured in seconds):

$C_0^*(t) = 0.5$	0.59	0.61	0.61	0.61
$t = 0+$	1	2	3	4

This completes step 3. Note that $C_0^*(t)$ has a value of 0.5 units at $t = 0+$. That is, $C_0^*(t)$ is zero for $t < 0$, and 0.5 units for $t > 0$ ($t = 0+$). This jump of 0.5 units in zero time is a consequence of the artifice of impulse sampling and the lack of a hold element. Real systems cannot jump a finite amount in zero time, although such behavior is often predicted by z-transform analysis.

Step 4. The unit step response of a continuous-time unity feedback system with $G(s) = 1/(s+1)$ is found in step 4, for purposes of comparison. The output $C(s)$ is given by

$$C(s) = R(s)\frac{G(s)}{1+G(s)} = \frac{1}{s}\,\frac{1/(s+1)}{1+1/(s+1)} = \frac{1}{s}\,\frac{1}{(s+2)}$$

Performing a partial fraction expansion

$$C(s) = \frac{1/2}{s} + \frac{-1/2}{s+2}$$

Using the Laplace transform

$$\frac{K}{s} \to K\mu(t), \qquad \frac{K}{s+a} \to Ke^{-at}$$

yields the continuous-time output function $C(t)$:

$$C(t) = \tfrac{1}{2} - \tfrac{1}{2}e^{-2t} = \tfrac{1}{2}(1 - e^{-2t})$$

and

$$\lim_{t\to\infty} C(t) = \tfrac{1}{2}(1-0) = \tfrac{1}{2}$$

The five output functions are plotted vs. time in Fig. 34.29. The functions plotted are:

$C_a^*(t)$ = sampled-time response, $T = 1$ sec, with zero-order hold
$C_b^*(t)$ = sampled time response, $T = 0.5$ sec, with zero-order hold
$C_c^*(t)$ = sampled time response, $T = 0.1$ sec, with zero-order hold
$C_0^*(t)$ = sampled time response, $T = 1$, without a hold element
$C(t)$ = continuous time response, no sampling, no hold

Note that the input for all plots is a unit step, and the plant transfer function for all plots is $1/(s+1)$.

Observations, drawn from a comparison of these plots, are:

1. Two of the sampled responses, $C_b^*(t)$ and $C_c^*(t)$, closely resemble the exact continuous response $C(t)$. The system plant transfer function $G(s)$ has a pole at $s = -1$. $C_b^*(t)$ and $C_c^*(t)$ have sampling times of less than 1 sec. $C_c^*(t)$ has the smallest sampling time, one tenth that of the numerical value of the real part of the pole magnitude, $s = -1$; that is, $1/10$ s. $C_c^*(t)$, with the smallest sampling time, most closely resembles the continuous system time response. However, $C_c^*(t)$ requires many more calculations to cover a given time span.

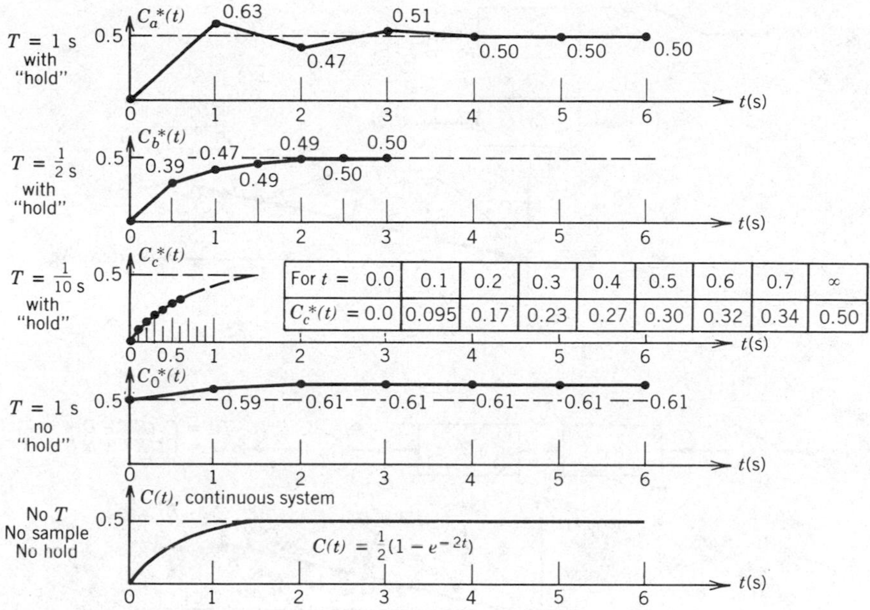

Fig. 34.29 Results of Example 34.7, step 4. *Five plots of the unit step response of a system with* $G(s) = 1/(s+1)$, $H(s) = 1$, $R(s) = 1/s$. The first three responses, $c_a^*(t)$, $c_b^*(t)$, and $c_c^*(t)$, are for an error-sampled system with a zero-order hold, with sampling times of $T = 1$, 1/2, and 1/10 s, respectively. The fourth plot, $c_0^*(t)$, is for an error-sampled system, with $T = 1$ s, but without a zero-order hold. For the first four plots, only the seven dots, "·", are valid values for $c_a^*(t)$, $c_b^*(t)$, $c_c^*(t)$, and $c_0^*(t)$. The fifth plot, $c(t)$, is the step response of a continuous system with no sampling or hold elements.

2. $C_a^*(t)$ has a sampling time that is not less than the real part of the transfer function pole at $s = -1$. The response $C_a^*(t)$ exhibits an overshoot, whereas the continuous-time response has no overshoot.

3. $C_0^*(t)$, without a hold device, exhibits a response that differs markedly from the continuous system time response at both small values and large values of time. $C_0^*(t)$ jumps up to 0.5 units at $t = 0+$ secs, whereas the continuous response is zero at $t = 0+$ secs. Also the final value of $C_0^*(t)$ is 0.61 units, whereas the continuous system's final value is 0.50 units, for a unit step input. As mentioned previously, all common digitally controlled sampled-data systems employ a hold device and, therefore, this case is not a practical one.

34.8 MODIFIED z-TRANSFORM, $F(z,m)$

The ordinary z-transform $f(z)$, when inverted to the time domain, yields the sample sequence $f^*(t) = f^*(kT)$ which is valid only at the sampling instants $t = 0, T, 2T, 3T, \ldots, kT, \ldots$. Assuming for illustrative purposes that the sample period T is 1 s, the ordinary z-transform tells us nothing about the system response at 1/3 or 1/2 or $\pi/2$ or 6 7/8 s. Time response data will be obtained only at $0, 1, 2, 3, \ldots$ s. The modified (or delayed) z-transform provides information on the response between sampled periods.

The modification consists of conceptually adding a delay of D seconds to the function being evaluated, where D varies from zero to one sample period T. This is shown in Fig. 34.30. To obtain a sample at 1/3 T seconds, a delay of 2/3 T seconds is needed, as shown in Fig. 34.31c. In general, to obtain a sample at mT seconds where m is between 0 and 1, the delay D must be $D = T(1-m)$ seconds, as shown in Fig. 34.32. The delay D does not appear anywhere in the modified z-transform; only the related parameter m appears. In actual applications, m is the sample period shift. A modified z-transform $F(z,m)$ with $m = \frac{1}{3}$ produces sample results $f^*(t,m)$ at time 1/3 $T, (1+\frac{1}{3})T, (2+\frac{1}{3})T, (3+\frac{1}{3})T, \ldots, (K+m)T$ seconds, as shown in Fig. 34.31. This figure also shows the response of the modified z-transform for $m = 0$ and $m = 1$. Note that the sampled values for $m = 0$ and $m = 1$ are

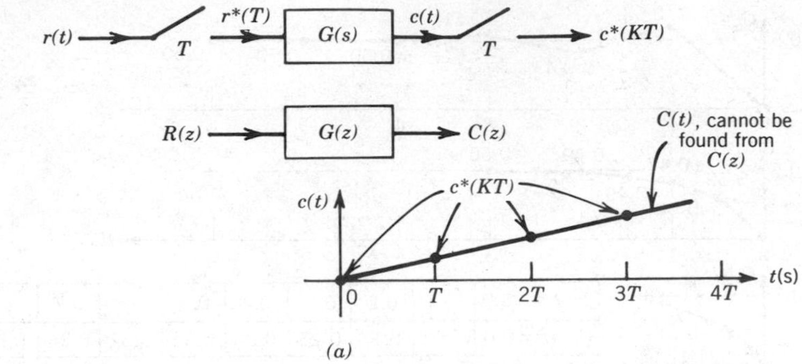

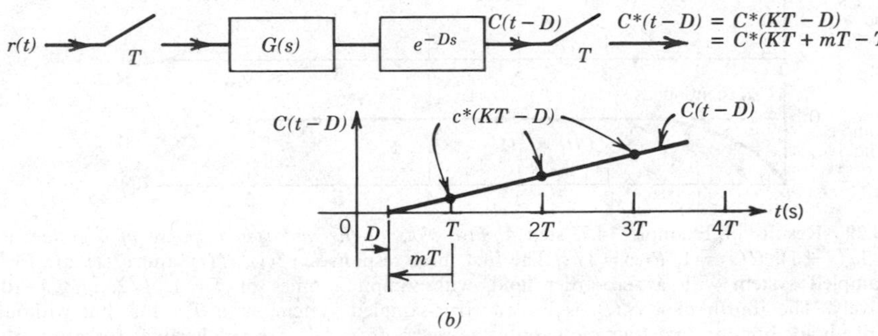

Fig. 34.30 Modified z-transform concept. (a) Ordinary z-transform, $f(t) \Rightarrow F(z)$, $F(z) \Rightarrow f^*(KT)$ $r(t) \Rightarrow R(z)$, $G(s) \Rightarrow G(z)$, $C(z) \Rightarrow C^*(KT)$ $K =$ sample numbers. The inverse z-transform of $C(z)$ yields $C^*(T)$ which is valid only at the instants of time $t = 0, T, 2T, 3T, \ldots$. The superscript star (*) denotes a sampled time function, which is a sequence of numerical values. (b) Modified z-transform, $f(t) \Rightarrow F(z, m)$, $F(z, m) \Rightarrow f^*(KT + mT)$. $r(t) \Rightarrow R(z)$, $G(s) \Rightarrow G(z, m)$, $C(z, m) \Rightarrow C^*(KT + mT)$, $0 \leqslant m \leqslant 1$, $D =$ delay, used conceptionally. It does not appear in transforms. $D = T(1 - m)$ seconds.

identical except for a delay of T seconds when $m = 0$. A brief glossary of the terminology used for the modified z-transform is shown in Table 34.26.

The impulse-sampled function $f^*(t)$ is given by (see Table 34.26):

$$f^*(t) = \sum_{K=0}^{\infty} f(t)\delta(t - KT) \tag{34.49}$$

The modified or delayed impulse-sampled function $f^*(t, m)$ is given by (Table 34.26):

$$f^*(t, m) = f^*(t - D) = \sum_{K=0}^{\infty} f(t - D)\delta(t - KT) = \sum_{K=0}^{\infty} f(KT - D)\delta(t - KT) \tag{34.50}$$

From Fig. 34.32:

$$D + mT = T \quad \text{or} \quad D = T(1 - m) = T - mT \tag{34.51}$$

Inserting Eq. (34.51) into Eq. (34.50):

$$f^*(t, m) = \sum_{K=0}^{\infty} f(KT - T + mT)\delta(t - KT)$$

or

$$f^*(t, m) = \sum_{K=0}^{\infty} f(KT + mT - T)\delta(t - KT) \tag{34.52}$$

The delay of one period, $-T$, produces a z^{-1} factor in $F(z, m)$:

$$f^*(t, m) \Rightarrow F(z, m) = z^{-1} \sum_{K=0}^{\infty} f(KT + mT)z^{-K} \tag{34.53}$$

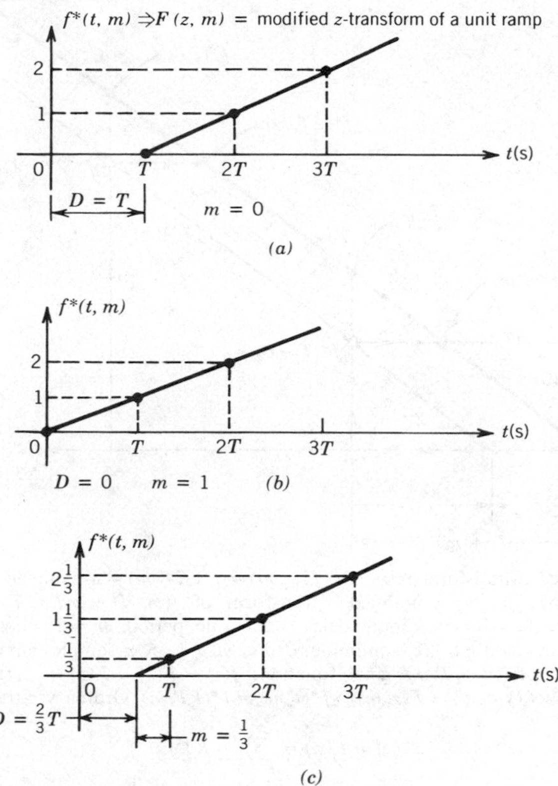

Fig. 34.31 Unit ramp modified z-transform response for $m = 0, 1/3, 1$. (*a*) $m = 0$, $D = T =$ one sample delay (assume $T = 1$ s), $\bullet = f^*(t, m) = f^*(t, 0) \Rightarrow F(z, 0)$. (*b*) $m = 1$, $D = T(1 - m) = 0 =$ no delay (assume $T = 1$ s), $\bullet = f^*(t, m) = f^*(t, 1) = f^*(t) \Rightarrow F(z) = F(z, 1)$, ordinary z-transform for functions continuous at $t = 0$. (*c*) $m = 1/3$, $D = T(1 - m) = 2/3\ T$ seconds (assume $T = 1$ s), $\bullet = f^*(t, m) = f^*(t, \frac{1}{3}) \Rightarrow F(z, \frac{1}{3})$. $f^*(t, m) =$ a sampled value of $f(t)$ at the instant $t = KT + mT$. When $m = 1/3$, $f^*(t, m)$ gives the sampled value at $1/3\ T$ later than $f^*(t) \Rightarrow F(z)$, the ordinary z-transform. Note that the delay of $2/3\ T$ produces a result at $1/3\ T$, relative to the undelayed response.

The ordinary z-transform is

$$f^*(t) \Rightarrow F(z) = \sum_{K=0}^{\infty} f(KT)z^{-K} \tag{34.54}$$

An important consequence of Eq. (34.53) should be noted. If $F(z, m)$ is inverted to $f^*(t, m)$ by dividing the denominator of $F(z, m)$ into the numerator of $F(z, m)$ the resulting series consists of terms $z^{-1}f(KT + mT)z^{-K}$ or $f(KT + mT)z^{-(K+1)}$. Thus a factor of z^{-4} corresponds to the *third* sample, shifted toward increasing time by mT seconds $[z^{-4} \Rightarrow -(K+1) = -4$ or $K + 1 = 4$ or $K = 3]$. In general, when inverting by long division, a factor of z^{-n} corresponds to the $(n-1)$th sample interval for modified z-transforms, whereas a factor of z^{-n} corresponds to the nth sample for ordinary z-transforms. This is illustrated by Example 34.8, where the inverse z-transform of a unit ramp is taken, first by long division of the ordinary z-transform and then by long division of the modified z-transform. The results are plotted in Fig. 34.33, where is it pointed out that a factor of z^{-n} corresponds to a time of nT seconds for the ordinary z-transform $[z^{-2} \Rightarrow 2T$ seconds]; whereas a factor of z^{-n} corresponds to a time of $(n - 1 + m)T$ seconds for the modified z-transform $[z^{-2} \Rightarrow (2 - 1 + \frac{1}{3})T = \frac{4}{3}T$ seconds, for $m = \frac{1}{3}]$.

Example 34.8 Modified z-Transforms of a Ramp. First consider the ordinary z-transform of a unit ramp $f(t) = t\mu(t)$. From the z-transform Table 34.25:

$$F(z) = \frac{Tz}{(z-1)^2} = \frac{Tz}{z^2 - 2z + 1} \qquad \text{for} \quad f(t), t \geqslant 0$$

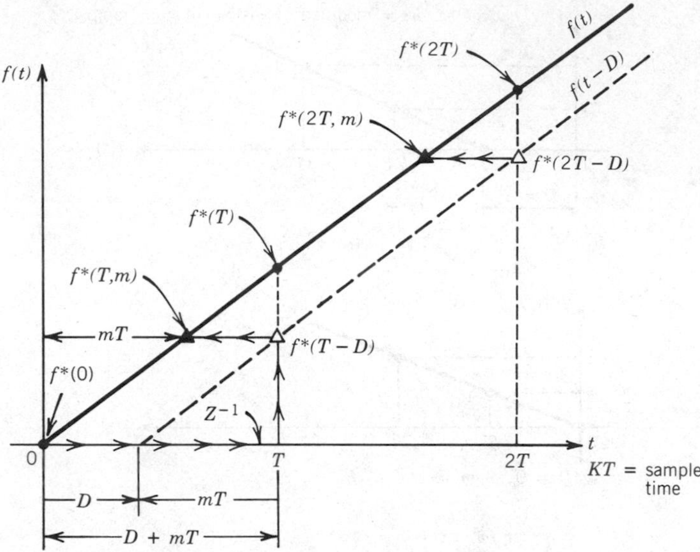

Fig. 34.32 Modified z-transform response. $f^*(t,m) = f^*(KT,m) =$ a sampled value of $f(t)$ at the instant $t = KT + mT$, $f(z,m) =$ modified z-transform of $f(t)$, $D + mT = T =$ sample period, $D = T(1-m) =$ delay, $m = 0 =$ maximum delay $= T =$ one period, $m = 1 =$ no delay $=$ ordinary z-transform (for functions that are continuous), $0 \leqslant m \leqslant 1$, $K =$ sample number $= 0, 1, 2, 3, \ldots$, $z^{-1} =$ delay of one period $= T$. Original function: $f(t) \Rightarrow F(s) \Rightarrow F(z) \Rightarrow f^*(t) = f^*(KT)$. Delayed function: $f(t-D) \Rightarrow F(s)e^{-Ds} \Rightarrow F(z,m) \Rightarrow f^*(t,m) = f^*(kT,m)$. Ordinary z-transform:

$$f^*(t) \Rightarrow F(z) = \sum_{K=0}^{\infty} f(KT)z^{-K}$$

Modified z-transform:

$$f^*(t,m) \Rightarrow F(z,m) = z^{-1} \sum_{K=0}^{\infty} f(KT + mT)z^{-K}$$

Key:
● = Undelayed sampled values, $f^*(t) = f^*(KT)$
▲ = Delayed sampled values, $f^*(t,m) = f^*(KT,m)$
△ = Undelayed sampled values on delayed function, for illustrative purposes only

---

Inverting via long division:

$$\begin{array}{r l}
 & Tz^{-1} + 2Tz^{-2} + 3Tz^{-3} + 4Tz^{-4} + \ldots \\
f^*(KT) \Rightarrow z^2 - 2z + 1 \big/ & Tz \\
 & \underline{Tz - 2T + Tz^{-1}} \\
 & 2T - Tz^{-1} \\
 & \underline{2T - 4Tz^{-1} + 2Tz^{-2}} \\
 & 3Tz^{-1} - 2Tz^{-2} \\
 & \underline{3Tz^{-1} - 6Tz^{-2} + 3Tz^{-3}} \\
 & 4Tz^{-2} - 3Tz^{-3}
\end{array}$$

Each z^{-1}, z^{-2}, z^{-3} corresponds to a time delay of $T, 2T, 3T$ seconds and $f^*(T) = T$, $f^*(2T) = 2T$,

TABLE 34.26. GLOSSARY OF MODIFIED z-TRANSFORM TERMINOLOGY

Term	Definition
$f(t)$	A continuous time function
$f^*(t)$	Sampled time function, a sequence of numbers that are the weights of a train of impulse functions of period T
$f^*(t,m)\,[=f^*(t-D)]$	Delayed sample time function; a sampled value of $f(t)$ at the instant $t = KT - D$ or $t = KT + mT - T = (K-1)T + mT$
D	Time function delay $= T(1-m)$ seconds
T	Sample period
$F(z)$	z − Transform of $f(t)$
$F(z,m)$	Modified z-transform of $f(t)$
$\delta(t)$	Dirac delta function, impulse function; $\delta(t) = 0$ for $t \neq 0$, $\delta(t) \to \infty$ for $t = 0$, $\int_{0-}^{0+} \delta(t)\,dt = 1$
Impulse-sample function $f^*(t)$	$f^*(t) = \sum_{K=0}^{\infty} f(t)\delta(t-KT)$
Modified or delayed impulse-sample function $f^*(t,m)$	$f^*(t,m) \equiv f^*(t-D) = \sum_{K=0}^{\infty} f(t-D)\delta(t-KT) = \sum_{K=0}^{\infty} f(KT-D)\delta(t-KT)$

TABLE 34.27. SHORT TABLE OF MODIFIED z-TRANSFORMS

$f(t)$	$f(s)$	$f(z,m)^a$	Term
$\mu_s(t)$	$\frac{1}{s}$	$\frac{1}{z-1}$	Step
$\delta(t)$	1	0	Impulse
$\delta(t-kT)$	e^{-kTs}	z^{m-1-k}	Delayed impulse
t	$\frac{1}{s^2}$	$\frac{mT}{z-1}+\frac{T}{(z-1)^2}$	Ramp
$\frac{t^2}{2!}$	$\frac{1}{s^3}$	$\frac{T^2}{2}\left[\frac{m^2}{z-1}+\frac{2m+1}{(z-1)^2}+\frac{2}{(z-1)^3}\right]$	Parabola
e^{-at}	$\frac{1}{s+a}$	$\frac{e^{-amT}}{z-e^{-aT}}$	Exponential decay
$1-e^{-at}$	$\frac{a}{s(s+a)}$	$\frac{1}{z-1}-\frac{e^{-amT}}{z-e^{-aT}}$	Time constant of $1/a$
$\sin \omega t$	$\frac{\omega}{s^2+\omega^2}$	$\frac{z\sin(m\omega T)+\sin(1-m)\omega T}{z^2-2z\cos(\omega T)+1}$	Sine
$\cos \omega t$	$\frac{s}{s^2+\omega^2}$	$\frac{z\cos(m\omega T)-\cos(1-m)\omega T}{z^2-2z\cos(\omega T)+1}$	Cosine
$e^{-at}\sin \omega t$	$\frac{\omega}{(s+a)^2+\omega^2}$	$\frac{[z\sin(m\omega T)+e^{-aT}\sin(1-m)\omega T]e^{-amT}}{z^2-2ze^{-aT}\cos(\omega T)+e^{-2aT}}$	Damped sine
$e^{-at}\cos \omega t$	$\frac{s}{(s+a)^2+\omega^2}$	$\frac{[z\cos(m\omega T)-e^{-aT}\cos(1-m)\omega T]e^{-amT}}{z^2-2ze^{-aT}\cos(\omega T)+e^{-2aT}}$	Damped cosine

Source. Adapted from Tou.[13]
[a] Delay $=(1-m)T$: $m=1=$ no delay; $m=0=$ one period delay; $T=$ sample time.

$f^*(3T)=3T$, and so on. Thus the inverse of the ordinary z-transform yields a sequence of sample values $T, 2T, 3T, \ldots$ which lie on the unit ramp at the sample times $T, T2, T3, \ldots$. To find the response at, say, $\frac{1}{3}T, (1+\frac{1}{3})T, (2+\frac{1}{3})T$, the modified z-transform is required. From the modified z-transform Table 34.27, based on Tou's† work:

$$F(z,m)=\frac{mT}{z-1}+\frac{T}{(z-1)^2} \qquad \text{for} \quad f(t)=t,\ t\geqslant 0$$

Let $m=1/3$. Then

$$F(z,\tfrac{1}{3})=\frac{\frac{1}{3}T}{z-1}+\frac{T}{(z-1)^2}=\frac{\frac{1}{3}T(z-1)+T}{(z-1)^2}=\frac{\frac{1}{3}Tz+\frac{2}{3}T}{z^2-2z+1}$$

Inverting via long division:

$$\begin{array}{r} \tfrac{1}{3}Tz^{-1}+\tfrac{4}{3}Tz^{-2}+\tfrac{7}{3}Tz^{-3}+\ldots \\ f^*(KT+\tfrac{1}{3}T)\Rightarrow z^2-2z+1\overline{\big)\ \tfrac{1}{3}Tz+\tfrac{2}{3}T\qquad\qquad} \\ \tfrac{1}{3}Tz-\tfrac{2}{3}T+\tfrac{1}{3}Tz^{-1}\qquad \\ \hline \tfrac{4}{3}T-\tfrac{1}{3}Tz^{-1}\qquad \\ \tfrac{4}{3}T-\tfrac{8}{3}Tz^{-1}+\tfrac{4}{3}Tz^{-2} \\ \hline \tfrac{7}{3}Tz^{-1}-\tfrac{4}{3}Tz^{-2} \end{array}$$

Recalling the discussion following Eqs. (34.53) and (34.54), the factor $\frac{1}{3}Tz^{-1}$ corresponds to $K=0$ or $f^*(\frac{1}{3}T)$, the factor $\frac{4}{3}Tz^{-2}$ corresponds to $K=1$ or $f^*(T+\frac{1}{3}T)$, and the factor $\frac{7}{3}Tz^{-3}$

†Tou's 1959 text[13] contains one of the most comprehensive modified z-transform tables available anywhere.

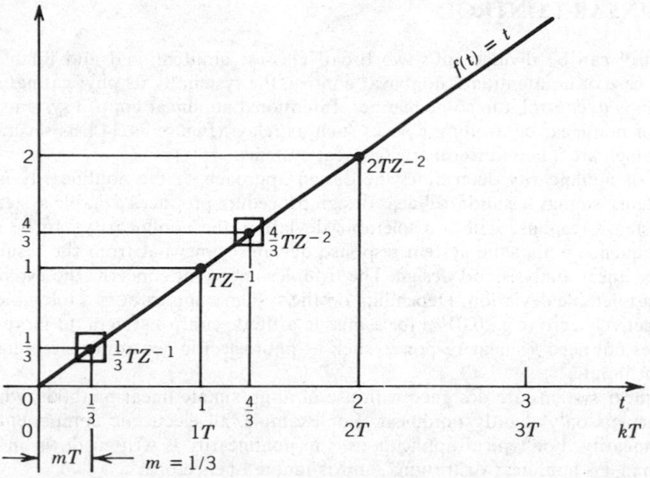

Fig. 34.33 Results of Example 34.8, unit ramp response, calculated by the ordinary z-transform and by the modified z-transform with $m = 1/3$ and $T = 1$ s. $f^*(t, m) =$ ▣ $= f^*(t, \frac{1}{3})$, $f^*(t) = \bullet$. Sampled time functions valid only at sample points • or ▣.

Key:

• = ordinary z-transform $f(z)$

$z^{-2} \Rightarrow 2T = 2$ s

▣ = modified z-transform $F(z, m)$

$z^{-2} \Rightarrow (2 - 1 + \frac{1}{3})T = \frac{4}{3}$ s

$z^{-n} \Rightarrow (n - 1 + m)T$ s

---

corresponds to $K = 2$ or $f^*(2T + \frac{1}{3}T)$. There is an extra z^{-1} term present in the modified z-transform. The resulting time samples are thus, for an ordinary z-transform of $f(t) = t$, and inversion via long division (T = sample time, chosen as 1 s)

Terms of $F(z)$	Sample No. K	Time of Sample KT	Magnitude of Time Response $f^*(KT)$
Tz^{-1}	1	$T = 1$	1
$2Tz^{-2}$	2	$2T = 2$	2
$3Tz^{-3}$	3	$3T = 3$	3
$4Tz^{-4}$	4	$4T = 4$	4
Az^{-h}	n	$nT = n$	A

and for a modified z-transform of $f(t) = t$, $m = 1/3$, inversion via long division

Terms of $F(z, m)$	Sample No. K	Time of Sample $(K - 1 + m)T$	Magnitude of Time Response $f^*(KT + \frac{1}{3}T)$
$\frac{1}{3}Tz^{-1}$	0	$(0 + \frac{1}{3})T = 1/3$	$\frac{1}{3}$
$\frac{4}{3}Tz^{-2}$	1	$(1 + \frac{1}{3})T = 4/3$	$\frac{4}{3}$
$\frac{7}{3}Tz^{-3}$	2	$(2 + \frac{1}{3})T = 7/3$	$\frac{7}{3}$
Az^{-n}	$n - 1$	$(n - 1 + \frac{1}{3})T$	A

These results are plotted in Fig. 34.33.

34.9 NONLINEAR CONTROL

Nonlinear control can be divided into two broad classes: unintentional and intentional nonlinear systems. In the case of unintentional nonlinear control, the system by its physical nature is nonlinear, and the problem is to control it in some manner. Intentional nonlinear control systems result from the designer's use of nonlinear controlling devices such as relays. Figure 34.34 lists several unintentional nonlinearities which are inherent to otherwise linear systems.

The degree of nonlinearity determines the design approach. If the nonlinearity is small in these quasilinear systems, so that a standard linear design procedure produces a stable system and correctly predicts the system's response within a tolerable deviation, the nonlinearity can be simply ignored. The only consequence is that the system response deviates somewhat from the results predicted by the approximate linear analysis and design. The troublesome point concerns the explicit specification of just what is a tolerable deviation. Depending on the system's application, a tolerable deviation may range from effectively zero (e.g., 0.01%) for a missile attitude control system, to more than 50% for a system that does not need a linear response, such as photoelectric control of street lights in response to ambient light input.

Many nonlinear systems are designed with use of approximate linear methods, when the designer knows the system is only slightly nonlinear. For example, all electronic components display some degree of nonlinearity. For typical applications, the nonlinearity is written down in a specification, such as "less than 1% nonlinear distortion," and is ignored henceforth.

However, in many other cases the unintentional nonlinearity cannot be ignored. When a linear design procedure cannot successfully predict system stability, the nonlinearity obviously cannot be ignored. In such a situation, specific nonlinear analysis and design procedures must be employed. The phase-plane procedure, the describing function method, Liapunov's stability criterion, and Popov's circle criterion are presented here as examples of how to analyze such highly nonlinear systems.

The second broad class of nonlinear control concerns the analysis and design of systems wherein a nonlinearity is intentionally introduced. A nonlinearity may be introduced because it improves performance in some fashion, or because it lowers the overall cost of the system. One of the clearest examples of this is a minimum-time "bang-bang" controller. A bang-bang controller is effectively a controlled switch (or relay) that produces maximum system effort at all times, until the system error is zero. Thus a vehicle traveling from point A to point B controlled by a bang-bang controller would accelerate at maximum rate until, on the basis of measurements or calculations, the bang-bang controller would switch the vehicle to maximum deceleration. In a correctly designed system the vehicle would reach zero velocity at point B, at which point the controller would shut off. In an automobile, this would correspond to flooring the accelerator or gas pedal at point A and holding it to the floor until a point somewhere between points A and B, when the brakes would be slammed on full force to bring the vehicle to a stop exactly at point B. In normal linear control, the gas and brake would be applied gradually. In bang-bang control, either the gas is full on or the brake is full on. Hence the name, bang-bang control. Bang-bang control is time optimal for a plant with the general transfer function

$$\frac{C(s)}{U(s)} = \frac{1}{a_n s^n + a_{n-1} s^{n-1} + \cdots + a_1 s + a_0} \tag{34.55}$$

It is the quickest way to get from point A to point B. It can be proven that, at most, $n - 1$ switches in sign are required, where n is the order of the system's transfer function, given by Eq. (34.55).[15,16]

For an automobile of mass M with plant transfer function $1/Ms^2$, $n = 2$, so that one switching action is required, from gas to brake. However, this is not the method generally employed to control an automobile (with the exception of the cars used in drag racing), because time-optimal behavior is not the most important criterion for controlling an automobile. Longevity of the vehicle engine; probability of engine and driver survival; reasonable fuel, tire, and repair costs; and compliance with speed limits are all considered more important than minimization of travel time.

However, if minimum travel time or minimum machine cycle time is required, then the highly nonlinear bang-bang controller would be the appropriate design choice.[15,16] For example, minimum response time is required when slewing a deck-mounted missile or gun-positioning system aboard a ship. In such a case, where a nonlinear control scheme is intentionally chosen, the various methods of analyzing and designing nonlinear systems previously mentioned would have to be employed. For example, bang-bang controllers can be designed using the phase-plane method, when the system is second order (which is the case for acceleration/deceleration control of an automobile or a spacecraft along a single axis).

Unique Characteristics of Nonlinear Systems

The most obvious characteristic of a nonlinear system follows from the fact that since the system is nonlinear, the principle of superposition does not apply. Thus if the input to a nonlinear system is doubled, the output is not doubled. In fact, the output may jump to zero, or it may jump to a high-frequency oscillation. The first instance might occur if a system were equipped with an automatic shut-down device. Such a switch, or any switch for that matter, is a nonlinear device. The second instance can occur in a system that exhibits limit cycles or jump resonances. A limit cycle is a nonlinear, often nonsinusoidal, cyclical, bounded oscillation in the output of a nonlinear system, which occurs only for certain amplitude ranges of the system's output. A jump resonance is a discontinuous jump in the output amplitude of a nonlinear system, as the frequency of the output signal changes. At the jump frequency, the system output is multivalued depending on whether the jump has occurred. Both these phenomena, jump resonance and limit cycle oscillation, are amplitude dependent.

In a linear system, the system's frequency response is independent of amplitude. For example, for a linear audio amplifier, the amplifier's frequency response is the same at low volume as at medium or high volume. For a nonlinear system this is not true. This is one of the most significant characteristics of a nonlinear system: The frequency response of such a system varies with amplitude and, in general, is not constant, since amplitude varies. Thus a nonlinear amplifier may sound good at medium volume but terrible at high volume, because its frequency response changes.

Another unusual nonlinear phenomenon is that of subharmonic oscillation. If the input to a nonlinear system has a frequency f_0 and the output has a lower frequency f_0/n, where n is an integer greater than one, the system exhibits subharmonic oscillation. A subharmonic oscillation depends on initial conditions as well as on the amplitude and frequency of the input function. In general, it is necessary to give the system a metaphorical kick, say by suddenly changing the amplitude or the frequency of the input function, to initiate subharmonic oscillations.[17]

Limit cycles can be affected by two other nonlinear phenomena. The limit cycle can be inhibited or stopped by frequency quenching (or asynchronous quenching). This consists of simply applying a large amplitude input of a frequency not integrally related to the limit cycle frequency, and then lowering the input amplitude. Thus a limit cycle of 1000 Hz could be quenched by applying an input of 314 or 1414 Hz. The other limit cycle effect is just the opposite. The limit cycle continues, but at a different frequency.[6] "Frequency entrainment" refers to a limit cycle locking into step with an applied input signal that is slightly different in frequency from the free-running limit cycle. Nonlinear oscillators, such as the Van der Pol oscillator, can lock to an input signal in this manner. The Van der Pol equation is second order, with a damping term that is negative for small amplitudes and positive for large amplitudes. The amplitude of the limit cycle of the Van der Pol system depends only on the constants K_1 and K_2 of Eq. (34.56), the Van der Pol equation:

$$\ddot{X} - \underbrace{K_1(1 - X^2)\dot{X}} + K_2X = 0 \qquad K_1 > 0, K_2 > 0$$

$$-(1 - X^2) = \text{Damping term: negative for } |X| < 1,\ \text{positive for } |X| > 1 \tag{34.56}$$

The Van der Pol equation produces limit cycle oscillations that are relatively easy to simulate on either an analog or a digital computer.

To summarize, nonlinear systems do not follow the rule of superposition and they exhibit certain unique behavior modes that are not displayed by linear systems (and that cannot be explained by a linear analysis procedure). The unique characteristics of nonlinear systems that have been briefly discussed are:

1. Limit cycles.
2. Jump resonance.
3. Subharmonic oscillation.
4. Amplitude-dependent frequency response.
5. Frequency entrainment.
6. Asyncronous quenching.

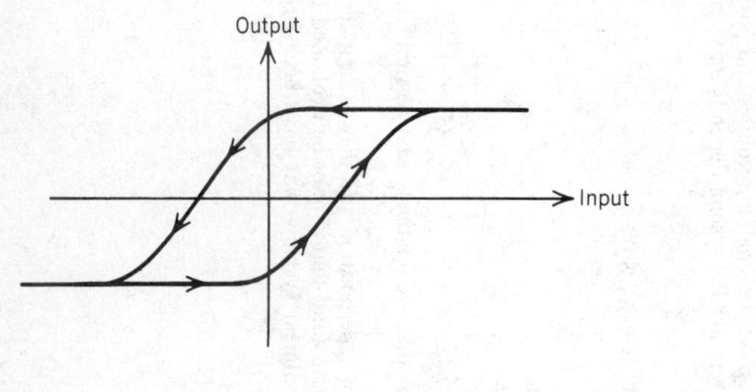
Output
Input

(a)

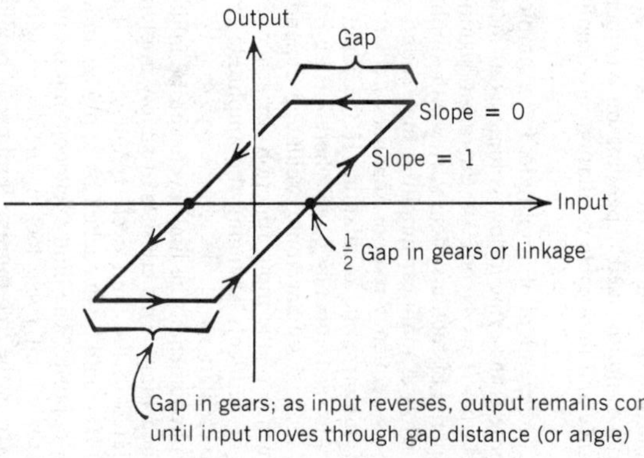
Output
Gap
Slope = 0
Slope = 1
Input
½ Gap in gears or linkage

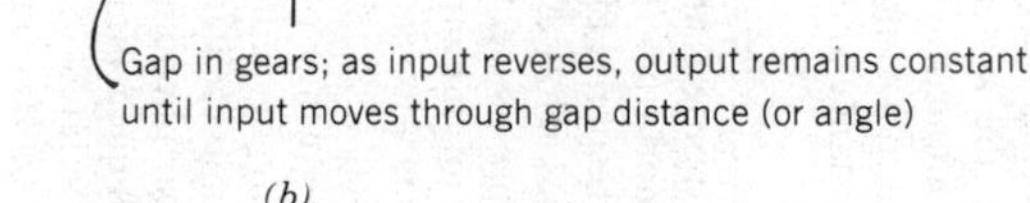
Gap in gears; as input reverses, output remains constant
until input moves through gap distance (or angle)

(b)

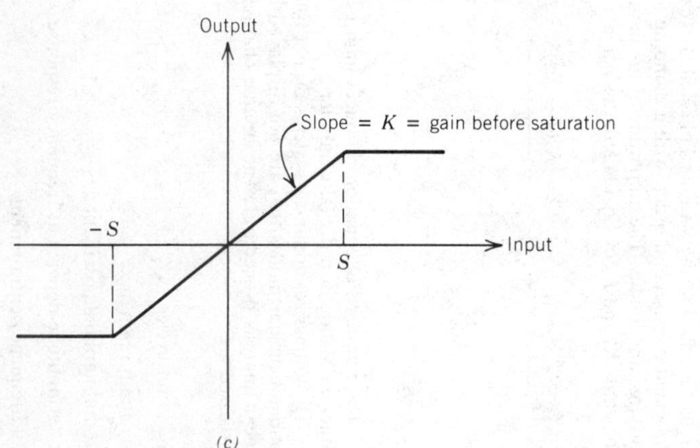
Output
Slope = K = gain before saturation
−S
S
Input

(c)

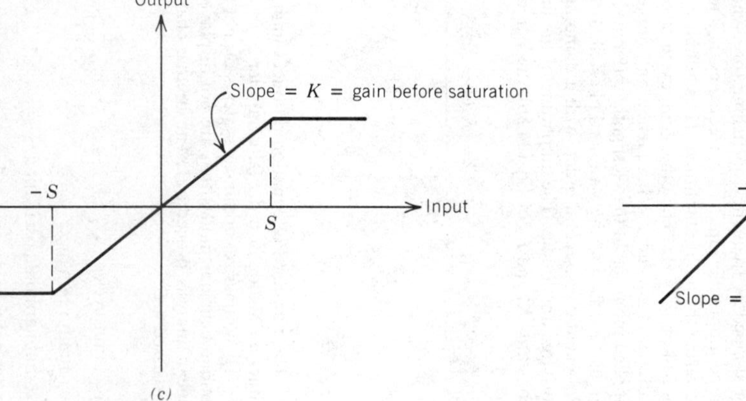
Output
Slope = K = gain after dead zone exceeded
−D
D
Input
Slope = K
D = dead zone level for input signal

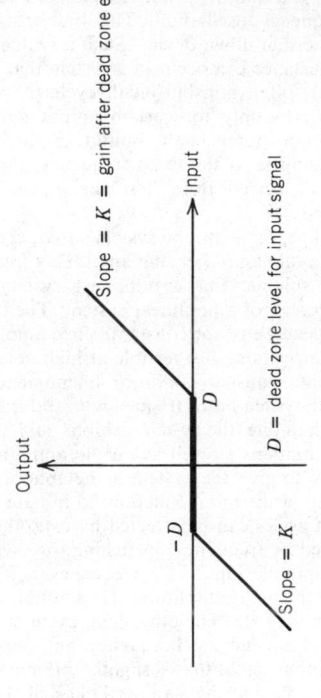

(d)

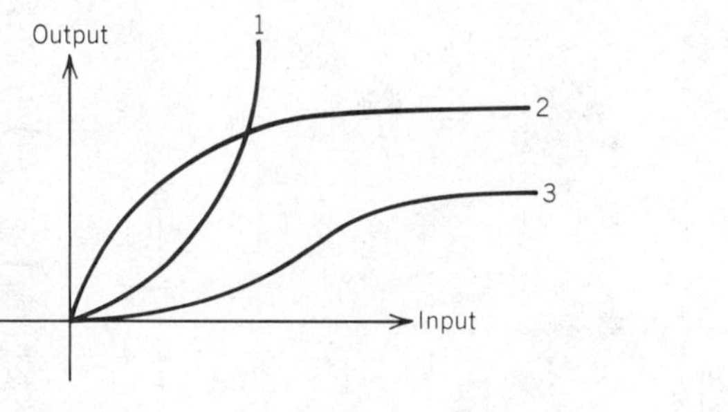

(e)

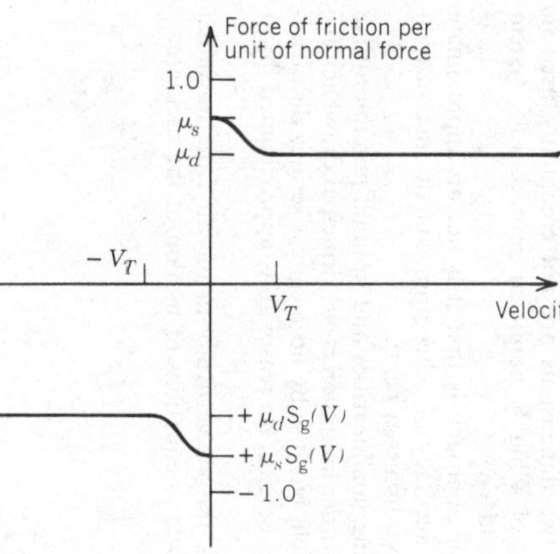

(f)

Fig. 34.34 Unintentional nonlinearities inherent to certain control elements or sometimes occurring in others: (*a*) Hysteresis; can occur in magnetic and other devices in which the output depends on the direction of the input change as well as on the magnitude of the input. (*b*) Backlash; can occur whenever there is loose play in a moving connected set of elements, such as in gears, clutches, couplings, linkages. (*c*) Saturation; can occur in any physical device, but commonly in amplifiers and torquers, that cannot deliver an output above a certain level. s = saturation level for the input signal (for output signal, saturation level is Ks). (*d*) Dead zone; can occur in any linear sensor that requires a certain minimum signal before that signal can be sensed, or in any actuator, such as a motor, torquer, or syncrocontrol transmitter/motor, that requires a minimum input before any change in output occurs. (*e*) Nonlinear input-output relationship; can occur in a device usually regarded as linear, for example, an amplifier, spring, damper, resistor. 1, 2, and 3 may not be symmetric; only first quadrant is shown for illustration. (*f*) Nonlinear friction, static friction. Friction force = $\mu \times$ normal force. μs = coefficient of static friction (that of tire on road typically is 0.9). μd = coefficient of dynamic or sliding friction (that of skidding tire on road typically is 0.8); also called coefficient of kinetic friction. Sg = signum function, Sg(V) = +1 for positive velocity and −1 for negative velocity. V_T = transition velocity, a value typically much smaller than 1/100 mph (effectively zero velocity). When velocity reverses or is negative, the frictional force also reverses or is negative. The coefficient of friction is still called positive, the sign inversion being provided by the sign of the velocity V.

Phase-Plane Analysis

The phase-plane analysis method is a useful method for analyzing the transient response and stability of nonlinear second-order systems. The phase plane is a cartesian plane with an abcissa chosen to be one state variable and an ordinate chosen to be another state variable of a second-order system. The abcissa is usually taken as x and the ordinate as $\dot{x}$. Although both state variables are functions of time, time does not appear explicitly (but can be determined implicitly). Theoretically, the concepts of the phase-plane technique can be extended from two dimensions to three (or more) and hence to three (or more) state variables. However, more than two dimensions are not commonly used in practice. The usual case of application is the analysis of a highly nonlinear second-order system having two state variables x and $\dot{x}$, or equivalently x_1 and x_2.

The principal advantage of the phase-plane analysis approach is that it is not an approximate linearization method and hence is not limited by the accuracy of other approximate linearization methods (e.g., the often-used describing-function method discussed next).

Phase-plane plots are particularly useful for switching nonlinearities and optimal minimum-time controller design for second-order systems. The second-order limitation is the principal drawback for the conventional, two-dimensional, phase-plane approach, but it is by no means a serious disadvantage, since many practical systems are second order or can be reasonably approximated by a second-order system.

As an illustration of the technique, consider a rock thrown vertically in the air from an initial height $x(0) = x_0$ with an initial velocity $\dot{x}(0) = V_0$. Newton's equations of motion of the rock are

$$F = MA$$

or

$$-mg = m\frac{d^2x}{dt^2}$$

or

$$-g = \frac{d^2x}{dt^2}$$

where A = upward acceleration = $\frac{d^2x}{dt^2}$
x = upward position
$\dot{x}$ = upward velocity = $\frac{dx}{dt}$
F = upward force = $-mg$ (gravity pulls down)
m = mass of rock
mg = weight of rock = downward force
$x(0) = x_0$ = initial position
$\dot{x}(0) = V_0$ = initial velocity
g = acceleration of gravity, constant
K_1, K_2 = constants of integration

Integrating once

$$\int -g\,dt = -gt = \frac{dx}{dt} + K_1 = \int \frac{d^2x}{dt^2}\,dt \tag{34.57}$$

Integrating again

$$\int -gt\,dt = -g\frac{t^2}{2} = x + K_1t + K_2 = \int \left(\frac{dx}{dt} + K_1\right) dt \tag{34.58}$$

Rewriting Eq. (34.57) and replacing dx/dt by $\dot{x}$

$$\dot{x} = -K_1 - gt$$

Obviously, inserting $t = 0$ yields

$$\dot{x}(0) = -K_1 \qquad \text{or} \qquad \dot{x}(0) = V_0 = -K_1$$

Thus the expression for $\dot{x}$ becomes

$$\dot{x} = V_0 - gt \tag{34.59}$$

From Eq. (34.58), with $t = 0$

$$0 = x(0) + 0 + K_2 \qquad \text{or} \qquad K_2 = -x_0 \equiv -x(0) \tag{34.58}$$

Inserting the values for K_1 and K_2 into Eq. (34.58) yields the equation for x

$$x = V_0 t + x_0 - g\frac{t^2}{2} \tag{34.60}$$

The standard analysis procedure would be then to plot the velocity $\dot{x}$ vs. time [Eq. (34.59)] and the position x vs. time [Eq. (34.60)]. The phase-plane approach, however, involves a plot of $\dot{x}$ vs. x. Hence the time parameter t must be eliminated from Eqs. (34.59) and (34.60). One method is to pick a value of time and then calculate $\dot{x}$ and x. This yields one point on the trajectory of the rock, as plotted in the $\dot{x}, x$ plane. The process is repeated to create a complete trajectory via a "connect-the-dots" approach for a digital computer algorithm.

Since this problem is relatively simple, $\dot{x}$ may be related to x analytically by eliminating t. This is not the usual method for constructing a phase-plane plot, because the problem is not usually this simple. However, the analytical method clearly points out the structure of various phase-plane trajectories. Practical graphical methods of constructing the plots for complex systems will be given later.

Equation (34.59) can be solved for t:

$$t = \frac{\dot{x} - V_0}{-g} = \frac{V_0 - \dot{x}}{g}$$

This value may be substituted into Eq. (34.60) to eliminate t:

$$x = V_0\left[\frac{V_0 - \dot{x}}{g}\right] + x_0 - \frac{g}{2}\left[\frac{V_0 - \dot{x}}{g}\right]^2$$

After considerable manipulation, the result for $\dot{x}$ is

$$\dot{x} = \pm\sqrt{V_0^2 + 2g(x_0 - x)} \tag{34.61}$$

This result allows families of trajectories to be drawn in the $\dot{x}, x$ plane [for various initial conditions $\dot{x}(0), x(0) = V_0, x_0$]. Several such trajectories are shown in Fig. 34.35. Since the motion of the rock will stop when the rock hits the ground, no trajectories are shown for negative height x. However, if the rock were thrown upward from the edge of a cliff, a level taken as the origin of coordinates, then a negative height corresponding to the rock falling below the cliff would be perfectly reasonable.

A generalized phase-plot, allowing a rock to be thrown from below as well as above the reference level for positive or negative velocity, is shown in Fig. 34.36. This second figure is a typical

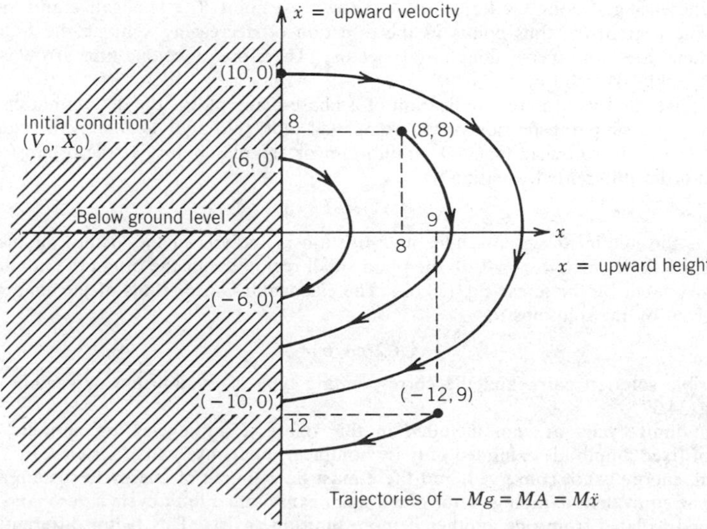

Fig. 34.35 Phase-plane trajectories of a rock thrown vertically upward from an initial height X_0, with an initial velocity V_0. Arrows indicate increasing time. Negative X is below ground, i.e., $X \geqslant 0$ is a hard constraint.

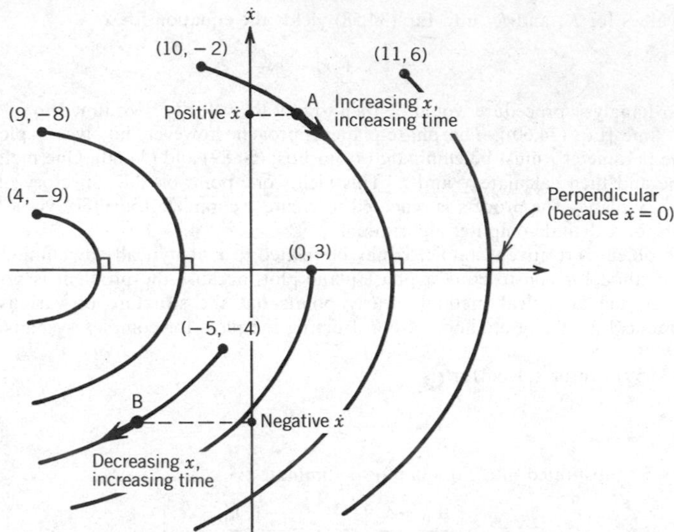

Fig. 34.36 Complete phase-plane trajectories of a rock thrown vertically upward. Initial condition: $(v_0, x_0) = [\dot{x}(0), x(0)]$. The rock may fall below the reference level $x = 0$ (e.g., the rock may be thrown from the edge of a cliff). Also, determination of the direction of increasing time along the trajectories. At time arrow A, $\dot{x}$ is positive and therefore X is increasing (as time increases). At time arrow B, $\dot{x}$ is negative and decreasing (as time increases).

phase-plane plot, with the entire phase plane filled with trajectories. Note that arrows are indicated on the trajectories to show the direction of increasing time. Once an initial condition is specified, a particular trajectory is specified. The system dynamics, described by the state variable pair $x, \dot{x}$, continually move along this phase-plane trajectory in the direction of the time arrows.

The procedure for assigning the direction of the time arrows is as follows. Pick a point where the trajectory has a relatively small slope and a relatively large $\dot{x}$ magnitude, such as the point A of Fig. 34.36. At point A, $\dot{x}$ is positive and therefore x is increasing. The time arrow thus points in the direction of increasing x along the trajectory, as shown. At point B, $\dot{x}$ is negative and therefore x is decreasing. The time arrow thus points in the direction of decreasing x along the trajectory. The direction of time does not reverse along any trajectory. Therefore, once one time arrow is found, the rest may be quickly sketched in.

One immediate and very useful application of a phase-plane plot is the determination of stability or limit cycles by a simple inspection of the completed plot. (No analytical work is required, just a quick look.) There are six basic types of equilibrium or singular points at the origin for the basic linear second-order differential equation

$$\ddot{x} + a\dot{x} + bx = 0 \tag{34.62}$$

As long as the nonlinear system under investigation is analytic in the vicinity of the origin (no jumps or infinite slopes near $\dot{x}, x = 0, 0$), then in a small region about the origin the nonlinear system can be approximated by the linear Eq. (34.62). The eigenvalues, λ, or roots of the characteristic Eq. (34.62) are given by the solutions to

$$\lambda^2 + \lambda a + b = 0 \tag{34.63}$$

The six possible solution pairs, and the corresponding type of equilibrium or singular points, are shown in Fig. 34.37.

Note that limit cycles are not included in this figure. Limit cycles are periodic motions or oscillations of fixed amplitude exhibited only by nonlinear, nonconservative systems. In a nonconservative system, energy is not conserved, and there must be a source of energy to sustain a limit-cycle oscillation. The equivalent damping factor of a system exhibiting a limit cycle is zero, and limit cycles are unique and isolated from one another if more than one exists. This factor differentiates a limit cycle from an undamped linear system oscillation. The phase-plane plot of $\lambda_1, \lambda_2 = \pm j\omega$ is a set of ovals (or circles), where the singular point is called a center, as shown in Fig. 34.37.

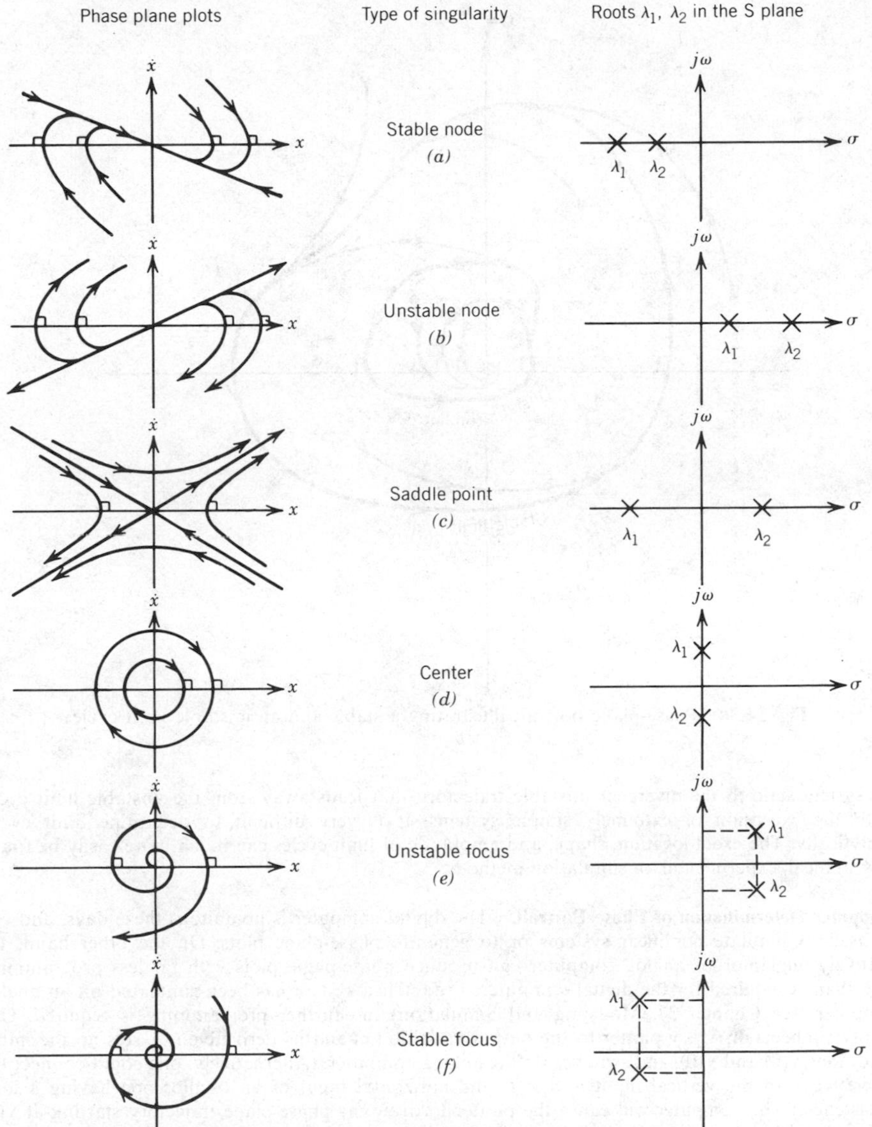

Fig. 34.37 Six possible equilibrium or singular points of the phase plane. The roots are eigenvalues of $\ddot{x} + a\dot{x} + bx = 0$, or $(S - \lambda_1)(S - \lambda_2) = 0$.

There are an infinite number of such closed oscillatory trajectories, depending on the particular initial conditions. Thus the linear system can exhibit oscillations of any amplitude, all with frequency ω. An example of limit cycles existing on the phase plane is illustrated in Fig. 34.38. Notice that the limit cycles have fixed amplitudes (they do not exhibit an infinite number of closed trajectories of varying amplitude). The pendulum of a grandfather clock exhibits such limit-cycle behavior. The pendulum may be started up with a large or a small amplitude swing. The pendulum's swing eventually stabilizes to one fixed constant amplitude, which is independent of the initial conditions at the start of swing. The amplitude of such a limit-cycle oscillation depends on the internal workings of the system, not on the initial conditions. Limit cycles may be stable or unstable, as shown in Fig. 34.38. Unstable limit cycles cannot be observed because system perturbations or noise quickly shifts

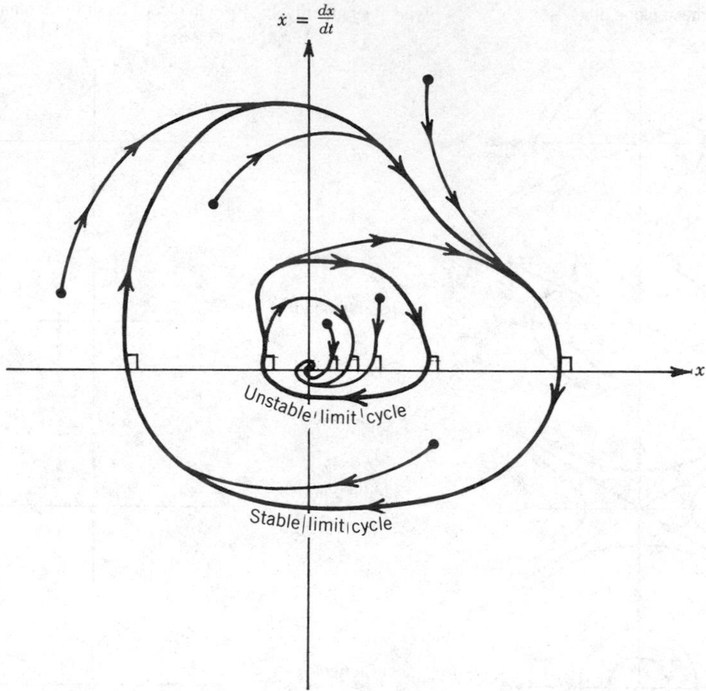

Fig. 34.38 Phase-plane portrait illustrating a stable and an unstable limit cycle.

the system state to the divergent unstable trajectory that leads away from the unstable limit cycle. With the exception of extremely simple systems, it is very difficult to determine limit cycles analytically. The exact location, shape, and amplitude of limit cycles can much more easily be found by graphical, experimental, or simulation methods.[18]

Computer Determination of Phase Portrait. The digital computer is ubiquitous these days, and can be used to simulate nonlinear systems or to generate phase-plane plots. On the other hand, the relatively unglamorous analog computer can produce phase-plane plots with far less programming time than is required for the digital computer. In fact, if a system has been simulated on an analog computer (see Chapter 38, Modeling and Simulation), no further programming is required. One simply connects an $x - y$ plotter to the output variable (x) and its derivative ($\dot{x}$), sets up the initial conditions $x(0)$ and $\dot{x}(0)$, and switches on the analog computer. (Alternatively, one could connect the derivative $\dot{x}$ to the vertical input and x to the horizontal input of an oscilloscope having a long persistence.) The computer will cause the plotter to draw the phase-plane trajectory starting at $x(0)$ and $\dot{x}(0)$ and thus to generate one trajectory of the complete phase portrait. The operator then repeats the procedure with different initial conditions to complete the phase-plane portrait. Unless specialized simulation programs are available for the digital computer, this type of computer will require vastly more programming effort than an analog computer to produce a phase plot. Even if such programs are available, the time necessary to master their operation will in all probability greatly exceed the time necessary to master use of the analog computer. This is just one reason for the continued existence of analog computers in the age of digital computers.

Construction of Phase-Portrait Using Method of Isoclines. The method of isoclines is the standard procedure used to hand calculate a phase portrait. The construction depends on finding an isocline, which is a line or locus having constant trajectory slope in the $(x, \dot{x})$ plane. Consider the general form of a nonlinear-second order state equation:

$$\begin{aligned} \dot{x}_1 &= x_2 \\ \dot{x}_2 &= f(x_1, x_2) \end{aligned} \tag{34.64}$$

The velocity $\dot{x}_1$ (or x_2) is plotted on the ordinate or y axis, and the position x_1 is plotted on the abcissa or x axis. The slope of a trajectory at any point is then given by m:

$$m = \text{trajectory slope} = \frac{dy}{dx} = \frac{d\dot{x}_1}{dx_1} = \frac{dx_2}{dx_1} = \frac{dx_2/dt}{dx_1/dt} = \frac{\dot{x}_2}{\dot{x}_1} \tag{34.65}$$

Different constant values of the trajectory slope m may be chosen, such as $m = 0, 1, 2, 3, -1, -2, \infty$, and inserted into Eq. (34.65), which can then be used to plot the locus of all points that have the slope m. In many cases the locus of constant trajectory slope m is a straight line, so that the isocline is a straight line with slope a.

It is important not to confuse the slope of the isocline a with the slope of the trajectories m, which cross the isocline. The best method is to draw short hash marks or dashes of slope m all along the isocline of slope a. The dashes then represent the direction of the trajectories. If the dashes are extended until they connect, a rough polygonal phase portrait will result. The procedure is illustrated in Example 34.9.

Example 34.9. Construct the phase-plane plot of

$$\ddot{x} + 2\dot{x} + 3x = 0$$

Defining the state variables as

$$x_1 = x$$
$$x_2 = \dot{x}$$

then

$$\begin{aligned} \dot{x}_1 &= x_2 \\ \dot{x}_2 &= \ddot{x} = -2\dot{x} - 3x = -2x_2 - 3x_1 \end{aligned} \tag{34.66}$$

The isocline equation, Eq. (34.65), yields

$$m = \text{trajectory slope} = \frac{\dot{x}_2}{\dot{x}_1} = \frac{-2x_2 - 3x_1}{x_2} = -2 - 3 \cdot \frac{x_1}{x_2} = -2 - 3\left(\frac{x}{\dot{x}}\right)$$

Now $x_1 = x$ will be the x axis of the plot and $x_2 = \dot{x}_1 = \dot{x}$ will be the y axis of the plot. Hence constant ratios of x_1/x_2 or x/y correspond to the straight lines through the origin of the phase plane with slope $a = dy/dx = y/x = \dot{x}/x$. The simplest procedure is to pick integral ratios of $\dot{x}/x$ = isocline slopes a, and to tabulate the corresponding trajectory slopes given by evaluating Eq. (34.65).

$\dot{x}/x$ = Isocline Slope a,	$x/\dot{x}$	$-3(x/\dot{x})$	$-2-3(x/\dot{x})$ = Trajectory Slope = Hash Mark Slope = m
0	∞	$-\infty$	$-\infty$
1	1	-3	-5
3	$\frac{1}{3}$	-1	-3
∞	0	0	-2
-3	$-\frac{1}{3}$	$+1$	-1
$-\frac{3}{2}$	$-\frac{2}{3}$	$+2$	0
-1	-1	$+3$	$+1$
$-\frac{1}{3}$	-3	$+9$	$+7$

Using the isocline slopes of a from this table and showing trajectory hash marks of slope m crossing the isoclines, the phase trajectory in Fig. 34.39 can be constructed. Once the isoclines and the trajectory hash marks have been drawn, construction of the trajectories is a relatively simple procedure. From a specified initial condition, such as point A, the hash mark of slope -5 is extended about half way to the next isocline, point B. Then the trajectory slope is shifted to the next value, $m = -\infty$, and continues with a constant slope of $-\infty$ from B to C. At C the trajectory slope is shifted to $m = +7$, and it continues to D. This process quickly generates a complete trajectory, with initial condition A.

It is important to emphasize that the accuracy of the resulting trajectory depends on using as many a isoclines as possible. Time arrows are put on, with use of the method discussed in conjunction with Eq. (34.61). A complete set of trajectories is readily constructed by repeating the process for different initial conditions.

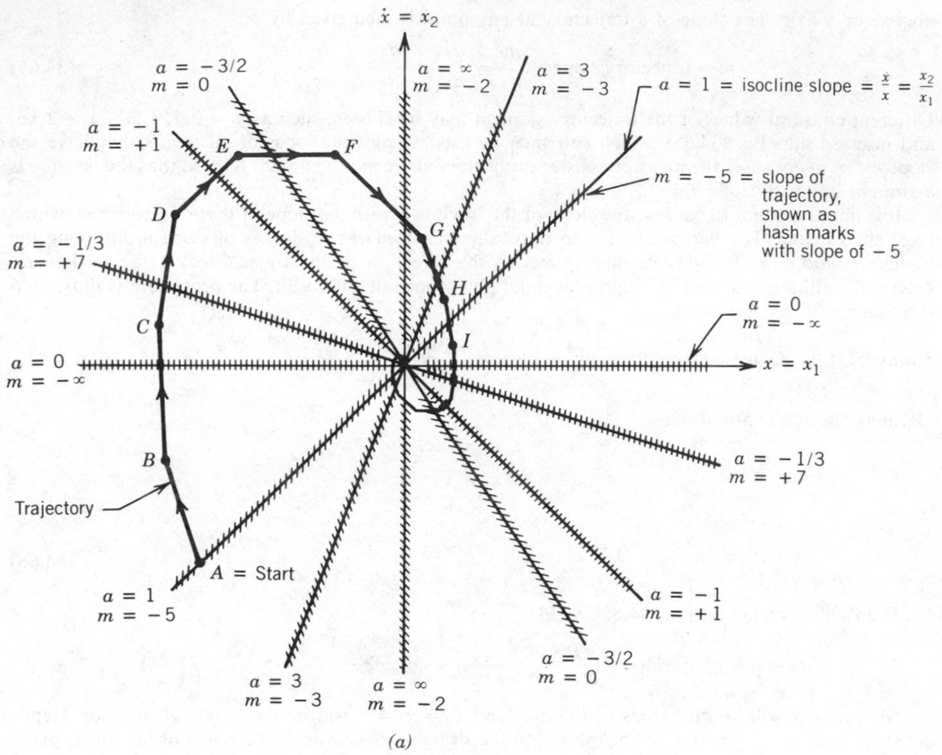

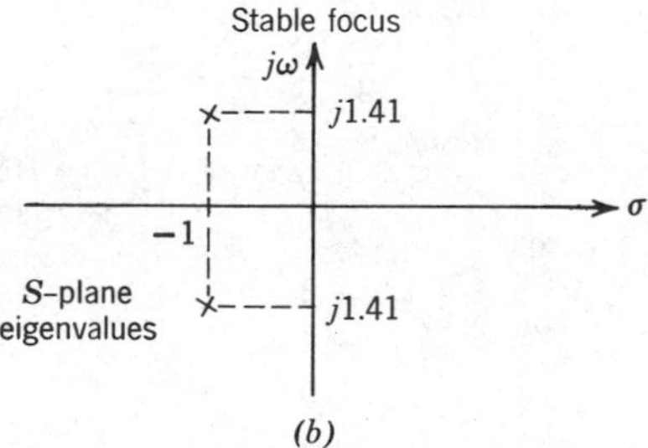

Fig. 34.39 Isocline locus for example 34.9: (*a*) Phase-plane plot of $\ddot{x} + 2\dot{x} + 3x = 0$. The isoclines are the radial lines, with a slope of *a*. The short hash marks, with a slope of *M*, indicate the trajectory slope at a particular isocline. Arrows indicate increasing time, from *A* to *B* to *C*, and so on. (*b*) S-plane eigenvalues, stable focus. $\lambda^2 + 2\lambda + 3 = 0$; $\lambda_1 = -1.0 + j1.41$; $\lambda_2 = -1.0 - j1.41$.

For a complex nonlinear system, the isoclines are often curves, such as parabolas, or are discontinuous, as for a system that switches to one value for $x > 0$ and to a different value for $x < 0$. For example, consider a rocket in space. Assume that it uses forward and reverse ion thrusters that consume very little mass, so that the mass of the rocket remains essentially constant. The equation of motion of the rocket along its straight line flight path is

$$\pm F = MA \tag{34.67}$$

where M = mass of rocket
$A = \ddot{x}$ = acceleration of rocket
$+F$ = forward thrust
$-F$ = reverse thrust

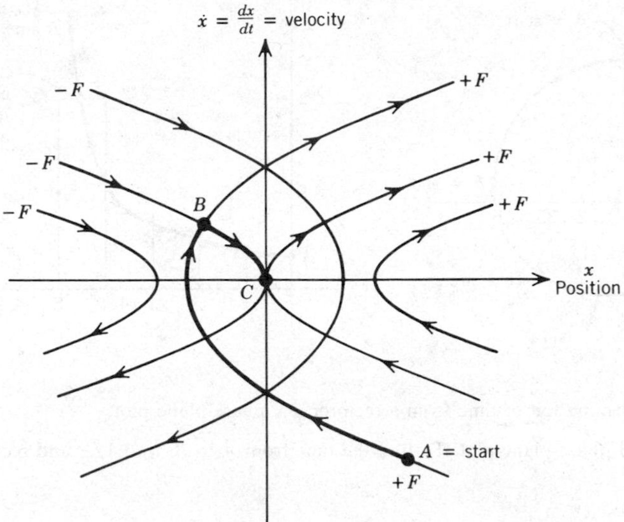

Fig. 34.40 Phase-plane plot of a rocket with forward $(+F)$ and reverse $(-F)$ thrusters. The minimum time path to the origin $= A - B - C$. B = switching point from $+F$ to $-F$.

The trajectory of the rocket in the phase plane $(x, \dot{x})$ is derived for $-F$ in Eqs. (34.57) through (34.67) in conjunction with the discussion of the trajectory of a thrown rock that obeys the identical equation: $-mg = -F = MA$. The trajectories for $+F$ are simply reverse reflections about the $\dot{x}$ axis, so that $+x$ becomes $-x$. This is shown in Fig. 34.40, which is obtained from Fig. 34.36 by adding on a reversed plot for $+F$. It is possible to get from any point in the phase plane to any other point by switching from a $+F$ (or $-F$) trajectory to a $-F$ (or $+F$) trajectory at the proper point. A minimum time trajectory is shown in Fig. 34.40 as a heavy line from point A (large position x, large negative velocity $-\dot{x}$) to rest (0 velocity, 0 position) at point C. Switching takes place from forward $(+F)$ thrust (trajectory AB) to reverse $(-F)$ thrust (trajectory BC) at switching point B. To remain at rest at C, the thrust must be shut off, $F = 0$. It can be proven that for any second-order nonlinear system, at most one switching is required for a minimum time trajectory.[17]

Determining Trajectory Time. It is possible to determine the time necessary to traverse any segment of a phase-plane trajectory, such as the time necessary to go from point B to point C in Fig. 34.40. The procedure is based on the following manipulation of the phase-plane variables:

$$y \text{ axis} = x_2 = \dot{x}_1 = \dot{x} = \frac{dx}{dt}$$

$$x \text{ axis} = x_1 = x \tag{34.68}$$

$$t_{AB} = \int_A^B dt = \int_A^B \frac{dt}{dx}\,dx = \int_A^B = \frac{1}{x_2}\,dx_1 = \int_A^B \frac{1}{\dot{x}}\,dx$$

To evaluate the integral of Eq. (34.68) graphically, a new plot must be constructed. A new y axis is required equal to the reciprocal of the phase-plane axis, namely, $1/x_2$ or $1/\dot{x}$. The x axis remains the same. The area under the curve plotted on the $(1/\dot{x}, x)$ axis is equal to the time difference between the end points A and B. A problem occurs when $\dot{x}$ or x_2 approaches zero, so that the reciprocal $1/\dot{x}$ or $1/x_2$ approaches infinity. The solution is to limit the closeness of approach to zero to some formally finite value (e.g., between 1 and 10% of the total range). If a graphical area is broken up into 10 to 20 regions with an equal x-axis increment, this problem is readily avoided. An example is illustrated in Fig. 34.41.

Describing Function Method

Perhaps the most direct method for dealing with nonlinear systems involves approximation of the nonlinear systems by a linear system. This approach is of value in direct proportion to that amount of

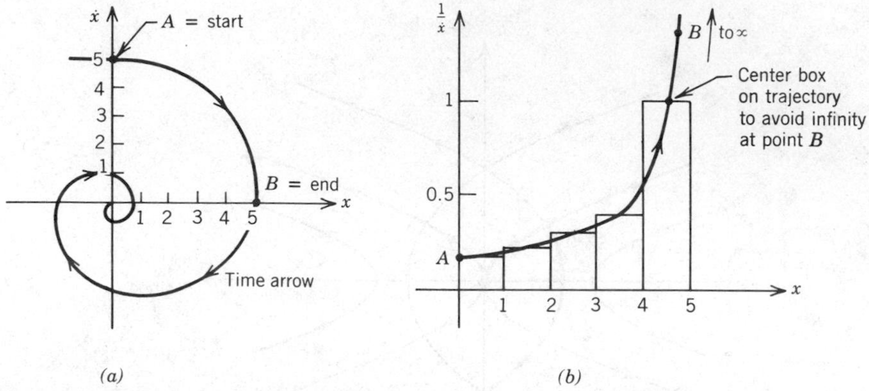

Fig. 34.41 Determination of time from a reciprocal $\dot{x}$ phase-plane plot.

(*a*) Conventional phase-plane plot. To find the time from A to B, find $1/\dot{x}$ and plot vs. x.

x	$\dot{x}$	$1/\dot{x}$
0	5	0.20
1	4.8	0.21
2	4.2	0.24
3	3.4	0.29
4	2	0.50
5	0	∞
4.5	1	1

(*b*) Reciprocal phase-plane plot. Time = area under $1/\dot{x}$ vs. x plot. $\int(1/\dot{x})dx = \int(1/(dx/dt))\,dx = \int(dt/dx)\,dx = \int dt = t$.

Box	Height	Area
1	1/4.9	0.204
2	1/4.5	0.222
3	1/3.8	0.263
4	1/2.7	0.370
5	1/1	1.00

Note that $x = 5$, $\dot{x} = 0$, $1/\dot{x} = \infty$ is excluded. The midpoint $x = 4.5$, $\dot{x} = 1$, $1/\dot{x} = 1$ is used for the height of the box between $x = 4$ and $x = 5$. The total area from A to $B = 0.204 + 0.222 + 0.263 + 0.370 + 1.00 = 2.06$ s. The units of x and $\dot{x}$ must be consistent, that is, if x is in meters, then $\dot{x}$ must be in meters per second.

the system performance that can be accounted for by the linear portion. This approach works well for systems that are only slightly nonlinear, with perhaps 75% or more of the system's output signal being linearly related to the system input. When less than one half of the system's output is so related, a linearization approximation is very poor and generally not worthwhile.

The describing function method is one possible linear approximation of a nonlinear system. The approach is straightforward. A sinusoid of arbitrary amplitude, frequency, and phase is applied as an input to the nonlinearity. The arbitrary phase is usually chosen as zero degrees, so that the input function is

$$R(t) = R\sin(\omega_0 t + \theta_r) \tag{34.69}$$

where θ_r = phase of $R(t)$ (usually 0°)
R = magnitude of $R(t)$
ω_0 = frequency of $R(t)$

The steady-state system output $C(t)$ is then obtained. $C(t)$ usually will be a nonsinusoidal nonlinear function of $R(t)$, with a steady-state periodic behavior. This nonsinusoidal output is then broken down into an infinite sum of sinusoids, with use of the Fourier series expansion. All the terms are disregarded, except for the fundamental term of frequency ω_0, thus

$$C(t) = C_0 + C_1\sin(\omega_0 t + \phi_1) + C_2\sin(2\omega_0 t + \phi_2) + \ldots$$

$$C(t) \cong C_1\sin(\omega_0 t + \phi_1)$$

The describing function is defined as the ratio of the fundamental frequency component of the output signal to the input signal (which has only a fundamental component). Analytically, the describing function is defined by the following relation:

$$\text{Des Fn} = N = \frac{C_1 e^{j\phi_1}}{Re^{j\theta_r}} = \frac{C_1}{R} e^{j(\phi_1 - \theta_r)} \tag{34.70}$$

If θ_r is chosen to be zero, then

$$\text{Des Fn} = N = \frac{C_1}{R} e^{j\phi_1} \tag{34.71}$$

The describing function is a complex number of magnitude C_1/R and with a phase of ϕ_1 degrees. In general C_1 is a function of R and of various system parameters. Several common nonlinearities and the corresponding describing functions are shown in Fig. 34.42. To use a describing function, the nonlinearity is replaced by the complex gain and phase given by the describing function.

As an illustration of procedure for obtaining a describing function, the describing function for a switching or relay nonlinearity (see Fig. 34.43*a*) is developed in Example 34.10.

Example 34.10 Evaluation of the Describing Function of a Switching Nonlinearity. Recalling Eq. (34.71), and referring to Fig. 34.43:

$$\text{Des Fn} = \frac{C_1}{R} e^{j\phi_1} \tag{34.71}$$

From the rules for combining sinusoids:

$$C_1 = \sqrt{A_1^2 + B_1^2} \tag{34.72}$$

$$\phi_1 = \tan^{-1}\frac{A_1}{B_1} \tag{34.73}$$

From the equations for the coefficient of a Fourier series expansion:

$$B_n = \widetilde{2c(t)\sin(n\omega_0 t)} \qquad \widetilde{f(t)} = \text{average value of } f(t) \tag{34.74}$$

$$A_n = \widetilde{2c(t)\cos(n\omega_0 t)} \qquad n = 1, 2, 3, \ldots \tag{34.75}$$

The 2's are needed, since the average value of $\sin^2 x$ or $\cos^2 x$ is $1/2$, as is obvious from $\sin^2 x + \cos^2 x = 1$. The DC or zero frequency term is an anomaly. It is given by $A_0/2$. Since only A_1 and B_1 are needed, with all higher harmonic frequency terms being disregarded, set $n = 1$:

$$B_1 = \widetilde{2c(t)\sin(\omega_0 t)} = 2\frac{1}{2\pi}\int_0^{2\pi} c(t)\sin(\omega_0 t)d(\omega_0 t) \tag{34.76}$$

$$A_1 = \widetilde{2c(t)\cos(\omega_0 t)} = 2\frac{1}{2\pi}\int_0^{2\pi} c(t)\cos(\omega_0 t)d(\omega_0 t) \tag{34.77}$$

Since the output $c(t)$ is in time phase with the input sinusoid and displays all the symmetries of a sine wave, $c(t)$ will have only sine wave components in its Fourier series expansion. This occurs because the average value of a sine wave and of a cosine wave of identical or different frequencies and zero phase shift is always zero. Analytically $c(t)$ is an odd function, that is, $c(-t) = -c(t)$. For any odd function, the Fourier series expansion will contain only sine terms, which are also odd functions.

Thus $A_1 = 0$, since there are no cosine terms in the expansion of the odd function $c(t)$. Evaluating B_1 by formally substituting the graphical values of $c(t)$ from Fig. 34.43*c* into Eq. (34.76):

$$B_1 = \frac{1}{\pi}\int_0^{2\pi} c(t)\sin(\omega_0 t)d(\omega_0 t) = \frac{1}{\pi}\int_0^{\pi} M\sin(\omega_0 t)d(\omega_0 t) + \frac{1}{\pi}\int_{\pi}^{2\pi} -M\sin(\omega_0 t)d(\omega_0 t) \tag{34.78}$$

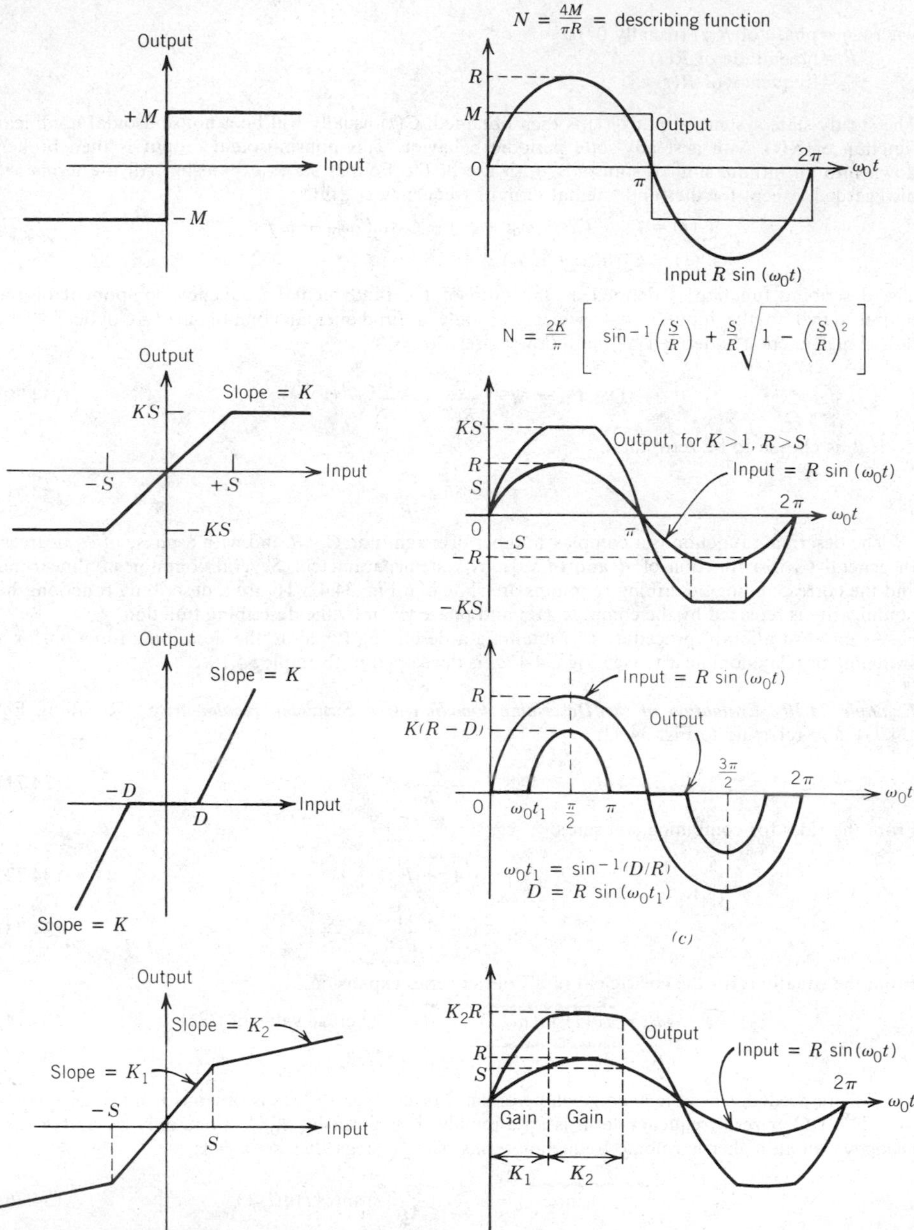

Fig. 34.42 Describing functions N for some common nonlinearities with a sinusoidal input driving function of $R\sin(\omega t)$. In general, N is a function of the input magnitude R and the input frequency ω_0; $N = N(R, \omega_0)$. However, for the four nonlinearities tabulated, N is only a function of R; $N = N(R)$. (*a*) Switching function. M = output magnitude, R = input magnitude, $R\sin(\omega_0 t)$ = input function. (*b*) Saturation. S = input saturation level, R = input magnitude, KS = maximum output level. Note that if $R > S$ or $S/R > 1$, there is no saturation and $N = K$. (*c*) Dead zone. D = input dead zone magnitude, R = input magnitude. $N = K - (2K/\pi)[\sin^{-1}(D/R) + (D/R)\sqrt{1-(D/R)^2}]$. Note that if $R < D$ or $D/R > 1$, the output is zero and $N = 0$. (*d*) High-low gain. S = input gain switching magnitude, R = input magnitude, K_1 = high gain, K_2 = low gain. For $R > S$, $N = K_2 + [2(K_1 - K_2)\pi][\sin^{-1}(S/R) + (S/R)\sqrt{1-(S/R)^2}]$. Note that if $R < S$ or $S/R > 1$, the gain does not switch and $N = K_1$. Drawn on the basis of information from Ogata.[6]

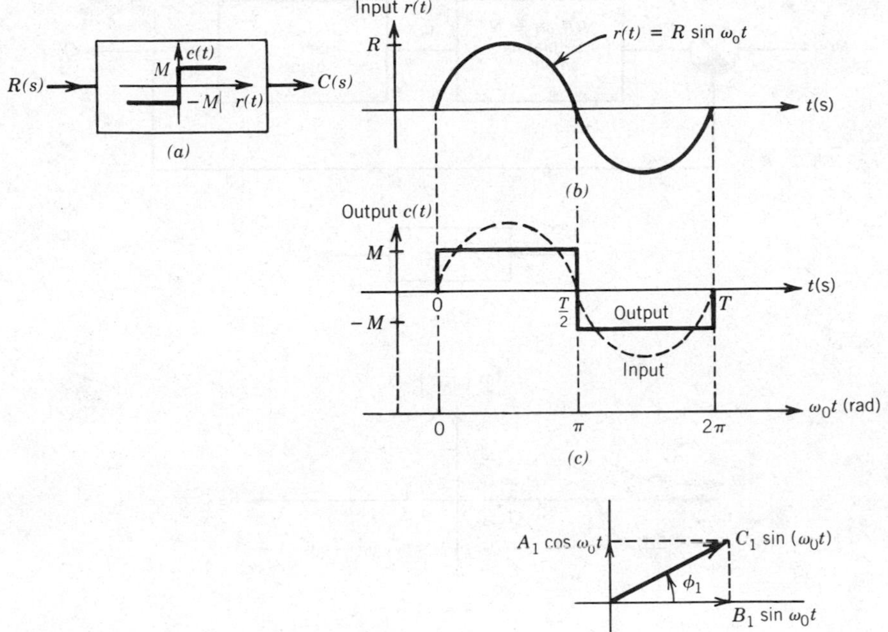

Fig. 34.43 Evaluation of the describing function of switching nonlinearity. See Example 34.10. (*a*) Block containing a switching nonlinearity with an output magnitude of M. (*b*) Input to nonlinearity. (*c*) Output from nonlinearity. (*d*) Combination of vectors.

Performing the integrations [using $\int \sin(ax)d(ax) = -\cos(ax)$]:

$$B_1 = \frac{1}{\pi}[-M\cos(\omega_0 t)]_0^{\pi} + -\frac{1}{\pi}[-M\cos(\omega_0 t)]_{\pi}^{2\pi} \tag{34.79}$$

Evaluating the limits of the variable ($\omega_0 t$):

$$\begin{aligned} B_1 &= \frac{-M}{\pi}[\cos(\pi) - \cos(0)] + \frac{M}{\pi}[\cos(2\pi) - \cos(\pi)] \\ B_1 &= \frac{-M}{\pi}[-1 - 1] + \frac{M}{\pi}[1 - -1] = \frac{4M}{\pi} \end{aligned} \tag{34.80}$$

In this rather special case, B_1 turns out to be independent of the magnitude of the input driving function R. Usually this is not true. In addition, since there is no A_1 term, the phase shift angle ϕ_1 is zero, since $\tan^{-1}(A_1/B_1) = \tan^{-1}(0) = 0$ [in Eq. (34.73)].

Thus inserting Eq. (34.80) into Eq. (34.72):

$$C_1 = \sqrt{A_1^2 + B_1^2} = \sqrt{0 + B_1^2} = B_1 = \frac{4M}{\pi} \tag{34.81}$$

Finally, the describing function N is evaluated from Eq. (34.71):

$$N = \frac{C_1}{R}e^{j\phi} = \frac{4M/\pi}{R}e^{j0} = \frac{4M}{\pi R} \qquad \text{since} \quad e^{j0} = e^0 = 1 \tag{34.82}$$

In summary, the describing function just determined, and all of the describing functions listed in Fig. 34.42, are complex values having a magnitude and phase. They relate the linear (fundamental) component of a nonlinear system's output to a linear sinusoidal input driving function. In general they are functions of the magnitude and frequency of the sinusoidal input.

A linear system produces a sinusoidal output of identical frequency but differing magnitude and phase, when driven by a sinusoidal input. The linear system's magnitude and phase are given by formally substituting $j\omega_0$ for the variable S in the system's closed-loop transfer function $G(s)$, where ω_0 is the frequency of the applied input. On the other hand, a nonlinear system can produce a zero frequency (or a DC output) as well as a multitude of harmonic frequency outputs, in addition to the fundamental component (or linear frequency) output. The describing function is simply the ratio of the linearly related fundamental frequency component of the output to the sinusoidal input driving

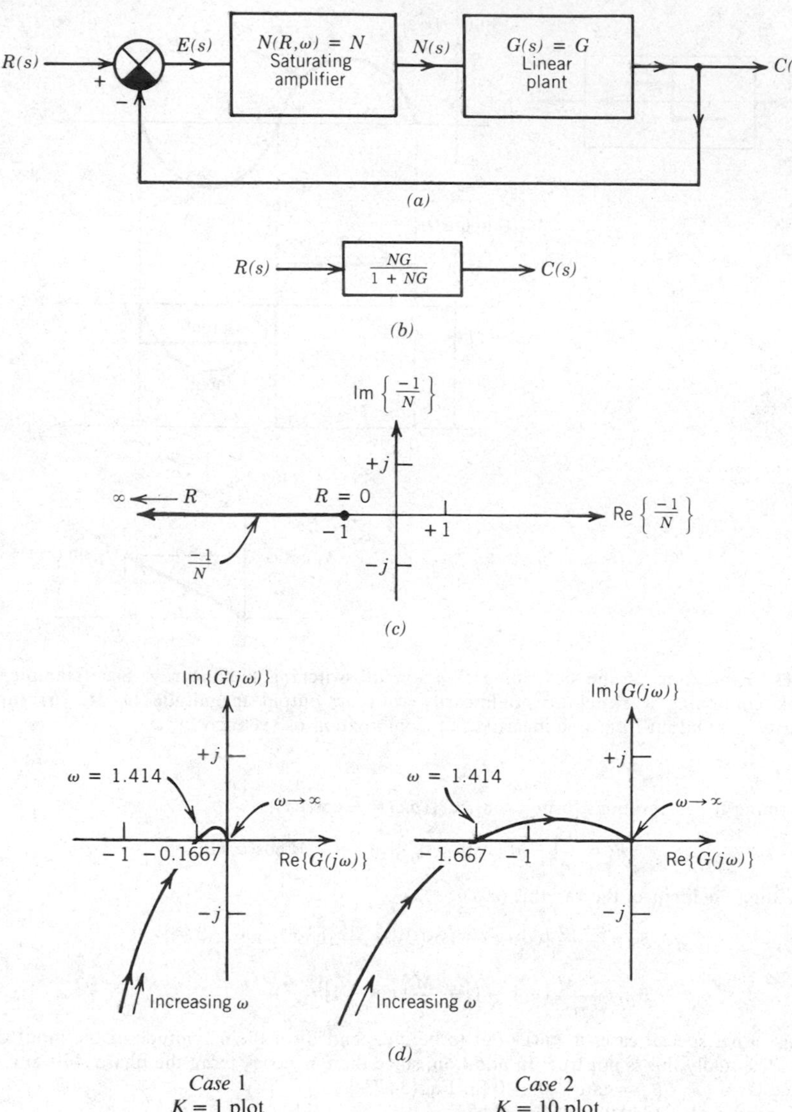

Fig. 34.44 Stability of a nonlinear control system, with examples of the existence of a stable limit cycle and of no limit cycle. $N = N(R, \omega)$ = describing function of the nonlinear element; $G = G(s)$ = linear forward gain, or plant; $s = j\omega$ for plots; $R(s)$ = input to system; $C(s)$ = system output; $E(s)$ = error signal = input to nonlinearity; $N(s)$ = nonlinearity output; S = saturation level of nonlinearity, a constant. (*a*) Block diagram. (*b*) Equivalent closed-loop transfer function. (*c*) Plot of saturation nonlinearity, expressed as $-1/N$, in the complex plane. In this case $N(R, \omega) = N(R)$; that is, N is not a function of ω. See Fig. 34.42*b*. For saturation, $N = N(R) = 2K/\pi\left[\sin^{-1}(s/R) + s/R\sqrt{1-(s/R)^2}\right]$. (*d*) Plot of $G(j\omega)$ in the complex plane. Let $G(s) = K/[s(s+1)(s+2)]$ = linear plant.
(*e*) Combined plot of $-1/N$ and $G(j\omega)$ for $K = 1$ and $K = 10$. $N = N(R, \omega) = N(R)$ = the describing function of a saturation nonlinearity. R = magnitude of the system input, $R\sin(\omega t)$. Point A ($K = 10$) denotes the intersection of the $-1/N$ locus and the $G(j\omega)$ locus at $\omega = 1.414$, indicating the existence of a limit cycle. The topology of the intersection indicates that the limit cycle is stable. (See Fig. 34.45.) Point B ($K = 1$) denotes the absence of a limit cycle; no intersection, no limit cycle. The system is stable for $K = 1$, since the $-1 + j0$ point is not encircled by the $G(j\omega)$ locus.

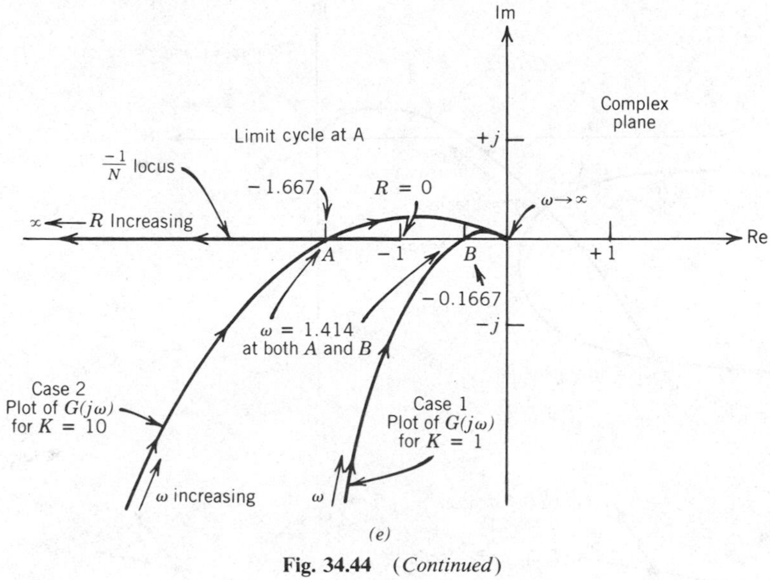

(e)

Fig. 34.44 (*Continued*)

function. All the higher frequency and lower frequency components of the nonlinear output are neglected.

Nonlinear System Stability. One of the principal applications of the describing function method is prediction of system stability. Consider the system of Fig. 34.44, which contains a nonlinear element N, an amplifier that saturates for input voltages that exceed $\pm S$ volts, in series with a linear plant $G(s)$. The characteristic equation of the system is

$$1 + NG = 0 \tag{34.83}$$

or equivalently

$$G = \frac{-1}{N} \tag{34.84}$$

or, specifying the arguments,

$$G(j\omega) = \frac{-1}{N(R, \omega)} \qquad \text{with} \quad s \to j\omega \tag{34.85}$$

Equation (34.85) can be directly employed to check for the stability of a nonlinear system of the form shown in Fig. 34.44, and to find any possible frequencies of oscillation. The method is straightforward. A plot of $G(j\omega)$ is made on rectangular (real and imaginary) coordinates, as a function of ω (see Fig. 34.44*d*). This is identical to the Nyquist plot procedure. However, instead of checking to see whether the point $(-1 + j0)$ is encircled, the test becomes a check to see whether the locus or plot of $1/N$, made on the same axes, is intersected by the $G(j\omega)$ plot. If the two plots intersect, the system is unstable with a limit cycle of frequency approximately equal to the frequency of $G(j\omega)$ at the intersection point. Such an intersection (at point A) is shown in Fig. 34.44*e* for $K = 10$. Note that when K is decreased to 1, the intersection no longer occurs and the system becomes stable.

The limit cycle for $K = 10$ may exhibit sustained oscillations or may diverge, depending on the direction of the $-1/N(R, \omega)$ locus as it crosses the $G(j\omega)$ locus. This is illustrated in Fig. 34.45, which indicates that the limit cycle of Fig. 34.44 is a stable one.

If the $G(j\omega)$ locus and the $-1/N(R, \omega)$ locus intersect at a very shallow angle or are tangent at some point, the accuracy of the test is probably poor. The accuracy is not poor, however, if it is known that the high-frequency components in the nonlinear output are rapidly attenuated. In either case, when the $G(j\omega)$ locus and the $-1/N(R, \omega)$ locus intersect at a shallow angle, the results of the test should be confirmed by another method of nonlinear system stability analysis, or by simulation of the nonlinear system.

Ogata,[6] Shinners,[8] Gibson,[17] and Graham and McRuer[18] discuss describing functions in much greater detail.

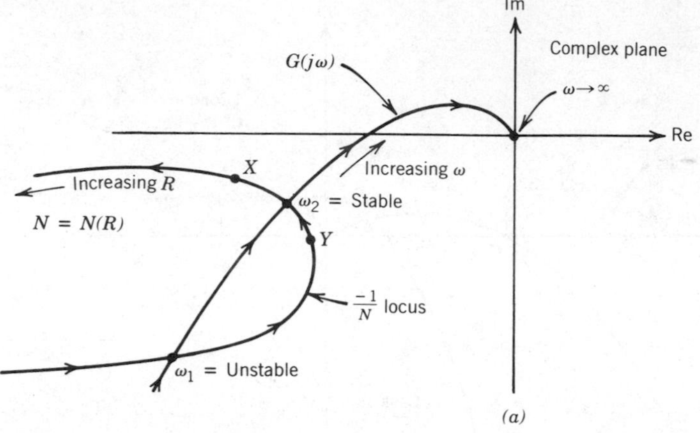

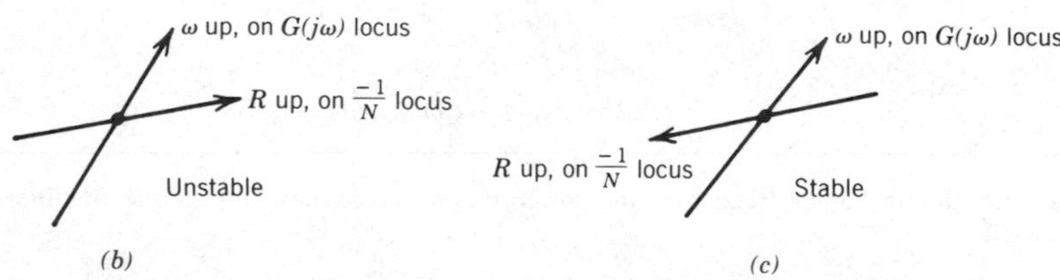

Fig. 34.45 Determination of stable and unstable limit cycles from intersections of $G(jw)$ and $-1/N$ locii. (*a*) Complex plane plot. (*b*) Unstable intersection. (*c*) Stable intersection.

Explanation: The $-1/N$ locus plays the part of the $(-1 + j0)$ point in the Nyquist stability test. Consider point Y. It is enclosed by the $G(j\omega)$ locus and hence is "unstable," and signal amplitudes (R) will grow. As R grows, operation on the $-1/N$ locus moves to point ω_2. Now consider point X. It is not enclosed by the $G(j\omega)$ locus and hence is "stable," so that signal amplitudes will decrease. As R decreases, operation on the $-1/N$ locus moves to point ω_2. Thus ω_2 corresponds to a stable limit cycle, since perturbations away from ω_2 are driven back to ω_2. Conversely, ω_1 corresponds to an unstable limit cycle, by an analogous argument. Note that ω_1 and ω_2 are only approximate limit-cycle frequencies, depending on the adequacy of the describing function approximation to the actual nonlinearity.

Liapunov's Stability Criterion

Liapunov's (or Lyapunov's) paper, "On the general problem of stability of motion," was published in Russian, by Kharkov, in 1892. It took almost 50 years for Liapunov's method to gain worldwide acceptance, probably because of the language problem, which is still evident in the various spellings of his name.

Liapunov's stability criterion differs in fundamental approach from all other stability tests in that stability is determined from a measurement of the rate of change of energy. It can be applied to linear as well as nonlinear systems, but since the standard linear system stability tests yield significant additional information, such as the degree of stability, Liapunov's stability criterion has become identified primarily with the determination of nonlinear system stability.

The fundamental basis of the Liapunov stability criterion can be explained in the following intuitive manner. Consider a mechanical system that somehow drives a mass in response to a control signal. The moving mass has a kinetic energy $\frac{1}{2}mV^2$. If this energy steadily increases toward infinity, the velocity of the mass will also increase up toward infinity, and the system will be unstable. The derivative of the kinetic energy with respect to time of this mechanical system is positive, since the kinetic energy is increasing with time. On the other hand, if the derivative of the kinetic energy with respect to time is always negative, the kinetic energy will always decrease toward zero and the system will be stable because the velocity of the mass will also decrease toward zero.

The kinetic energy cannot become negative, because it is a function of a squared variable, namely

V^2. A function such as $\frac{1}{2}mV^2$, which is always positive and can be zero only when its argument V is itself zero, is called a positive definite function.

The basis of the Liapunov stability test is to find a so called Liapunov function $V(x)$ that is a positive definite function of a system's state variable (x), and then to take the time derivative of that function. If the time derivative is always negative, the system is stable. This is the criterion for the Liapunov stability test.

The actual procedure involves some specialized terminology and considers types of stability. The terminology, which can be readily mastered, is presented in Table 34.28. However, a fundamental factor complicates the application of the Liapunov test, and it cannot be circumvented. The complication resides in the choice of the positive definite Liapunov function. Any arbitrary positive definite function can be chosen, with arbitrary coefficients. In the intuitive example just discussed, where $\frac{1}{2}mV^2$ was the Liapunov function, an infinite number of other functions, such as $4mV^6$ or $\frac{1}{16}V^8$ (where $V = x_1$ = a state variable) could have been chosen. The usual procedure is to pick one function after the other, or to somehow generate a Liapunov function, by working backward from a generalized $dV(x)/dt$ that is negative definite.

One very popular choice for a positive definite Liapunov function has a quadratic form. A positive definite quadratic form contains every state variable squared and none, some, or all of the possible cross product terms, such as

Quadratic form Liapunov functions $V(x) = X^TAX,\ X = x_1, x_2, x_3, \ldots$

$$V_1(x_1, x_2) = 2x_1^2 + x_2^2 = [x_1x_2]\begin{bmatrix} 2 & 0 \\ 0 & 1 \end{bmatrix}\begin{bmatrix} x_1 \\ x_2 \end{bmatrix} \tag{34.86}$$

$$V_2(x_1, x_2) = x_1^2 + Kx_1x_2 + 4x_2^2 = [x_1x_2]\begin{bmatrix} 1 & \dfrac{K}{2} \\ \dfrac{K}{2} & 4 \end{bmatrix}\begin{bmatrix} x_1 \\ x_2 \end{bmatrix} \tag{34.87}$$

$$V_3(x_1, x_2) = 2x_1^2 + 10x_1x_2 + 3x_2^2 = [x_1x_2]\begin{bmatrix} 2 & 5 \\ 5 & 3 \end{bmatrix}\begin{bmatrix} x_1 \\ x_2 \end{bmatrix} \tag{34.88}$$

$$V_4(x_1, x_2, x_3) = \{x_1^2 + 4x_1x_2 + 3x_2^2 + 12x_1x_3 + 5x_3^2 + 6x_2x_3\} = [x_1x_2x_3]\begin{bmatrix} 1 & 2 & 6 \\ 2 & 3 & 3 \\ 6 & 3 & 5 \end{bmatrix}\begin{bmatrix} x_1 \\ x_2 \\ x_3 \end{bmatrix} \tag{34.89}$$

$$V_5(x_1, x_2) = 7x_1^2 = [x_1x_2]\begin{bmatrix} 7 & 0 \\ 0 & 0 \end{bmatrix}\begin{bmatrix} x_1 \\ x_2 \end{bmatrix} \tag{34.90}$$

A quadratic form may be positive definite, positive semidefinite, or sign indefinite if the coefficients of the squared terms are positive. See Table 34.28 for the definition of terms. If the coefficients of the squared terms are negative, the quadratic form may be negative definite, negative semidefinite, or sign indefinite.

A straightforward test (Sylvester's criterion) determines the sign definiteness of a quadratic form in terms of the symmetric matrix $(a_{ij} = a_{ji})$ representation of the quadratic form, as shown in Eqs. (34.86) to (34.90). The test consists of evaluating the principal minors $(\Delta_1, \Delta_2, \Delta_3, \ldots)$ of the coefficient matrix of the quadratic form. The principal minors are $1 \times 1, 2 \times 2, 3 \times 3, \ldots,$ starting at the a_{11} term and extending down along the major diagonal to add the a_{22} term and then the a_{33} term, until the determinant of the entire matrix becomes the last principal minor. If the principal minors are greater than zero, the quadratic form is positive definite. An illustration follows.

Quadratic Form

$$x^TAx \tag{34.91}$$

$$A_a = \begin{bmatrix} 2 & 0 \\ 0 & 1 \end{bmatrix} \quad \begin{matrix} \Delta_1 = |2| = 2 & \Delta_2 = \begin{vmatrix} 2 & 0 \\ 0 & 1 \end{vmatrix} = 2(1) - 0(0) = 2 \\ \Delta_1 > 0 & \Delta_2 > 0 \quad \therefore \text{positive definite (all } \Delta > 0) \end{matrix} \tag{34.92}$$

$$A_b = \begin{bmatrix} 1 & \dfrac{K}{2} \\ \dfrac{K}{2} & 4 \end{bmatrix} \quad \Delta_1 = |1| = 1 \quad \Delta_2 = \begin{vmatrix} 1 & \dfrac{K}{2} \\ \dfrac{K}{2} & 4 \end{vmatrix} = 1(4) - \frac{K}{2}\left(\frac{K}{2}\right) = 4 - \frac{K^2}{4} \tag{34.93}$$

(*Continued on page 1122*)

TABLE 34.28. LIAPUNOV STABILITY CRITERION, DEFINITION OF TERMS

Term	Definition
System equation	$\dot{x}_n = f_n(x_n t)$ (where $\dot{x} \equiv dx/dt$, t = time) $\dot{x}_n = \begin{bmatrix} \dot{x}_1 \\ \dot{x}_2 \\ \vdots \\ \dot{x}_n \end{bmatrix}$ $f_n = \begin{bmatrix} f_1 \\ f_2 \\ \vdots \\ f_n \end{bmatrix}$ $x = \begin{bmatrix} x_1 \\ x_2 \\ \vdots \\ x_n \end{bmatrix}$
State vector	X_n, n-dimensional vector
System equation function	f_n, n-dimensional vector whose elements are functions of $x_1, x_2, \ldots, x_n$ and t
Equilibrium point	$f_n(x_n, t) = 0$;[a] all derivatives are equal to zero, and all state variables are constant at equilibrium point
Linear time-invariant system	$\dot{x}_n = A \cdot x_n$ or $f_n(x_n, t) = A \cdot x_n$; that is, system function consists of array of constants
Liapunov function	$V(x_n, t)$, or $V(x_n)$ if time is not explicit variable, which is usual case. Liapunov function must be (a) Positive definite (b) Scalar [there is only one equation $V(x_n, t)$ which when evaluated yields single scalar number] (c) Differentiable, to the extent that it has continuous first partial derivatives
Global	Liapunov function may be global, or extend over all state space, if $V(x_n, t) \rightarrow \infty$ as $\lVert X_n \rVert \rightarrow \infty$
Norm	$\lVert X_n \rVert = (x_1^2 + x_2^2 + \cdots + x_n^2)^{\frac{1}{2}}$ $= (x_n^T x_n)^{\frac{1}{2}} (x_n^T$ = transpose of $x_n)$
Time derivative of Liapunov function	$\dot{V}(x_n, t) = \dot{V}(x_n) = \dfrac{dV(x_n)}{dt}$ $\dfrac{dV}{dt} = \dfrac{\partial V}{\partial x_1}\dfrac{dx_1}{dt} + \dfrac{\partial V}{\partial x_2}\dfrac{dx_2}{dt} + \cdots + \dfrac{\partial V}{\partial x_n}\dfrac{dx_n}{dt}$ or $\dfrac{dV}{dt} = \dfrac{\partial V}{\partial x_1}\dot{x}_1 + \dfrac{\partial V}{\partial x_2}\dot{x}_2 + \cdots + \dfrac{\partial V}{\partial x_n}\dot{x}_n$ but $\dot{x}_1, \dot{x}_2, \ldots, \dot{x}_n$ are simply rows $f_1, f_2, \ldots, f_n$ of system equation. Thus $\dfrac{dV}{dt} = \dfrac{\partial V}{\partial x_1} f_1(x_n) + \dfrac{\partial V}{\partial x_2} f_2(x_n) + \cdots + \dfrac{\partial V}{\partial x_n} f_n(x_n)$
Positive definite	Scalar function $V(x_n)$ is positive definite in region R of state space including origin if (a) $V(x_n) > 0$ $\quad x_n \neq 0$ (b) $V(0_n) = 0$ $\quad$ i.e., $V(x_n)$ is zero *only* when all x_n equal zero Example: $V(x_n) = x_1^2 + x_2^2$ $\quad$ positive definite $V(x_n) = (x_1 + x_2)^2$ $\quad$ not positive definite, since $V(x_n) = 0$ when $x_1 = -x_2$; say $x_1 = 3$ and $x_2 = -3$. This is example of positive semidefinite function
Positive semidefinite	Scalar function $V(x_n)$ is positive semidefinite in region R

TABLE 34.28. (*Continued*)

Term	Definition
	of state space including origin, if (a) $V(x_n) \geqslant 0$ (b) $V(0_n) = 0$ Note that $V(x_n)$ may equal zero anywhere in state space, not just at origin
Sign indefinite	Scalar function is sign indefinite in region R of state space including origin if (a) $V(x_{n1}) > 0$ (b) $V(x_{n2}) < 0$ change in sign may occur in any part of region R. Simply put, sign-indefinite function is sometimes positive and sometimes negative
Negative definite	$V(x_n)$ is negative definite if $-V(x_n)$ is positive definite
Negative semidefinite	Function $V(x_n)$ is negative semidefinite if $-V(x_n)$ is positive semidefinite
Asymptotic stability, theorem I[b]	System is asymptotically stable if, when perturbed away from zero, state variables shrink back to zero (original equilibrium point) as time increases. Equilibrium point is given by $f_n(0, t) = 0$, as specified in definition for equilibrium point. System will be asymptotically stable if Liapunov function, $V(x_n, t)$ or $V(x_n)$, can be found such that (a) $V(x_n, t)$ or $V(x_n)$ is positive definite (b) $\dot{v}(x_n, t)$ or $\dot{V}(x_n)$ is negative definite
Global asymptotic stability, theorem II (or asymptotic stability in large)	Follows from asymptotic stability, theorem I, with added requirement (c): (a) $V(x_n)$ is positive definite (b) $\dot{V}(x_n)$ is negative definite (c) $V(x_n) \to \infty$ as $\lVert x_n \rVert \to \infty$
Stability in sense of Liapunov, theorem III[c]	Weaker form of stability, wherein system can oscillate forever in limit cycle and, as long as state variables do not increase toward infinity, system is called stable in sense of Liapunov. Oscillator is stable in sense of Liapunov. Conditions are same as those for asymptotic stability, theorem I, with condition (b) being relaxed from negative definite to negative semidefinite: (a) $V(x_n)$ is positive definite (b) $\dot{V}(x_n)$ is negative semidefinite or zero
Asymptotic stability, theorem IV	(a) $V(x_n)$ is positive definite (b) $\dot{V}(x_n)$ is negative semidefinite (c) $\dot{V}(x_n)$ does not vanish identically along any trajectory in state space x_n except at origin, $x_n = 0$, as $t \to \infty$ Requirements (b) and (c) of theorem IV allow $\dot{V}(x_n)$ to omit some of state variables, as long as $\dot{V}(x_n)$ never becomes identically zero as t approaches infinity (see Example 34.12)

TABLE 34.28. (*Continued*)

Term	Definition
Instability, theorem V	This follows from asymptotic stability, theorem I, with sign of derivative of Liapunov function changed from negative to positive. Positive derivative corresponds to increase without end, which is instability. System will be unstable and origin will be unstable equilibrium point if (a) $V(x_n, t)$ or $V(x_n)$ is positive definite (b) $\dot{V}(x_n, t)$ or $\dot{V}(x_n)$ is positive definite

[a] By a transformation of coordinates ($x_n^* = x_n \pm K_n$), the equilibrium point can always be translated to the origin, $x_n = 0$, or $f_n(0_n, t) = 0$. The equilibrium point or singular point can be stable or unstable.

[b] This is Liapunov's main stability theorem, also known as Liapunov's second method or as the direct method of Liapunov. The significant point is that the differential equations describing the nonlinear system (the system equation, first definition) do not have to be solved to determine stability via Liapunov's second method. Liapunov's first method is to solve the system equations or a linearized Taylor series approximation of the system equations. If the roots of the linearized characteristic equation are non-zero and have negative real parts, that is, if the roots lie in the left half of the s plane, the nonlinear system is asymptotically stable in the sense of Liapunov. There is nothing startlingly new about Liapunov's first method. It is Liapunov's second method that is unique, and extremely powerful if the right $V(x_n, t)$ can be found. Finding the right $V(x_n, t)$ is the problem.

[c] It is possible to prove asymptotic stability, rather than just stability in the sense of Liapunov, utilizing the less restrictive requirement that $\dot{V}(x_n)$ be negative semidefinite, by adding in the further requirement (c) $V(x_n) \to \infty$ as $\|x_n\| \to \infty$.

If $K < 4$,

$$\Delta_1 > 0 \qquad \Delta_2 > 0 \qquad \therefore \text{positive definite if } K < 4$$

$$A_c = \begin{bmatrix} 2 & 5 \\ 5 & 3 \end{bmatrix} \qquad \begin{array}{ll} \Delta_1 = |2| = 2 & \Delta_2 = \begin{vmatrix} 2 & 5 \\ 5 & 3 \end{vmatrix} = 2(3) - 5(5) = -19 \\ \Delta_1 > 0 & \Delta_2 < 0 \quad \therefore \text{sign indefinite (some } \Delta \text{ are } > 0, \text{ some } \Delta \text{ are } < 0) \end{array} \tag{34.94}$$

$$A_d = \begin{bmatrix} 1 & 2 & 6 \\ 2 & 3 & 3 \\ 6 & 3 & 5 \end{bmatrix} \qquad \begin{array}{l} \Delta_1 = |1| = 1 > 0 \\ \Delta_2 = \begin{vmatrix} 1 & 2 \\ 2 & 3 \end{vmatrix} = 1(3) - 2(2) = -1 < 0 \\ \Delta_3 = \begin{vmatrix} 1 & 2 & 6 \\ 2 & 3 & 3 \\ 6 & 3 & 5 \end{vmatrix} = 1(3)5 + 2(3)6 + 6(2)3 \\ \qquad -6(3)6 - 3(3)1 - 5(2)2 = -50 < 0 \\ \therefore \text{sign indefinite} \end{array} \tag{34.95}$$

$$A_e = \begin{bmatrix} 7 & 0 \\ 0 & 0 \end{bmatrix} \qquad \begin{array}{l} \Delta_1 = |7| = 7 > 0 \\ \Delta_2 = \begin{vmatrix} 7 & 0 \\ 0 & 0 \end{vmatrix} = 7(0) - 0(0) = 0 \\ \therefore \text{positive semidefinite (all } \Delta > 0 \text{ or } = 0) \end{array} \tag{34.96}$$

$$A_f = \begin{bmatrix} -7 & 0 \\ 0 & 0 \end{bmatrix} \qquad \begin{array}{l} \Delta_1 = |-7| = -7 < 0 \\ \Delta_2 = \begin{vmatrix} -7 & 0 \\ 0 & 0 \end{vmatrix} = -7(0) - 0(0) = 0 \\ \therefore \text{negative semidefinite (all } \Delta < 0 \text{ or } = 0) \end{array} \tag{34.97}$$

It is possible to generate a matrix A that will always yield a positive definite quadratic form by picking any square matrix D with positive coefficients and evaluating

$$A = D^T D \tag{34.98}$$

For example, choose

$$D = \begin{bmatrix} 1 & 3 \\ 2 & 4 \end{bmatrix}$$

then

$$A = D^T D = \begin{bmatrix} 1 & 2 \\ 3 & 4 \end{bmatrix} \begin{bmatrix} 1 & 3 \\ 2 & 4 \end{bmatrix} = \begin{bmatrix} 5 & 11 \\ 11 & 25 \end{bmatrix} \tag{34.99}$$

The quadratic form associated with the matrix x of Eq. (34.99) is positive definite, as may be demonstrated by applying Sylvester's criterion:

$$\Delta_1 = |5| = 5 > 0$$

$$A = \begin{bmatrix} 5 & 11 \\ 11 & 25 \end{bmatrix} \qquad \Delta_2 = \begin{vmatrix} 5 & 11 \\ 11 & 25 \end{vmatrix} = 5(25) - 11(11) = 4 > 0$$

$$\therefore \text{positive definite (all } \Delta_i > 0)$$

Example 34.11. Determine whether the following quadratic form is negative definite.

$$V_1(x_n) = -x_1^2 - 5x_2^2 - 10x_3^2 + 2x_1x_2 - 4x_2x_3 - 6x_1x_3$$

Procedure. Show that $-V(x_n)$ is positive definite.

$$-V_1(x_n) = x_1^2 + 5x_2^2 + 10x_3^2 - 2x_1x_2 + 4x_2x_3 + 6x_1x_3$$

The quadratic form can be written as a symmetric matrix:

$$-V_1(x_n) = [x_1 x_2 x_3] \begin{bmatrix} 1 & -1 & 3 \\ -1 & 5 & 2 \\ 3 & 2 & 10 \end{bmatrix} \begin{bmatrix} x_1 \\ x_2 \\ x_3 \end{bmatrix} = x^T A x$$

Applying Sylvester's criterion to A:

$$\Delta_1 = |1| = 1 > 0$$

$$\Delta_2 = \begin{vmatrix} 1 & -1 \\ -1 & 5 \end{vmatrix} = 1(5) - (-1)(-1) = 4 > 0$$

$$\Delta_3 = \begin{vmatrix} 1 & -1 & 3 \\ -1 & 5 & 2 \\ 3 & 2 & 10 \end{vmatrix} = \left\{ \begin{array}{c} 1(5)10 + (-1)(2)3 + 3(-1)(2) \\ -3(5)3 - 2(2)1 - 10(-1)(-1) \end{array} \right\} = -21 < 0$$

Since Δ_1 and $\Delta_2 > 0$ but $\Delta_3 < 0$, the quadratic form is sign indefinite. Repeat for

$$V_2(x_n) = -7x_1^2 - 5x_2^2 - 10x_3^2 + 2x_1x_2 - 4x_2x_3 - 6x_1x_3$$

$$-V_2(x_n) = 7x_1^2 + 5x_2^2 + 10x_3^2 - 2x_1x_2 + 4x_2x_3 + 6x_1x_3$$

$$-V_2(x_n) = [x_1 x_2 x_3] \begin{bmatrix} 7 & -1 & 3 \\ -1 & 5 & 2 \\ 3 & 2 & 10 \end{bmatrix} \begin{bmatrix} x_1 \\ x_2 \\ x_3 \end{bmatrix}$$

$$\Delta_1 = |7| = 7 > 0$$

$$\Delta_2 = \begin{vmatrix} 7 & -1 \\ -1 & 5 \end{vmatrix} = 7(5) - (-1)(-1) = 34 > 0$$

$$\Delta_3 = \begin{vmatrix} 7 & -1 & 3 \\ -1 & 5 & 2 \\ 3 & 2 & 10 \end{vmatrix} = \left\{ \begin{array}{c} 7(5)10 + (-1)(2)3 + 3(-1)2 \\ -3(5)3 - 2(2)7 - 10(-1)(-1) \end{array} \right\} = 255 > 0$$

Since all the Δ_i are greater than zero, $-V_2(x_n)$ is positive definite, and $V_2(x_n)$ is then negative definite.

Example 34.12. Determine the stability at the origin for the following linear system:

$$\begin{bmatrix} \dot{x}_1 \\ \dot{x}_2 \end{bmatrix} = \begin{bmatrix} 0 & 1 \\ -2 & -3 \end{bmatrix} \begin{bmatrix} x_1 \\ x_2 \end{bmatrix}$$

Multiplying out:

$$\dot{x}_1 = x_2$$

$$\dot{x}_2 = -2x_1 - 3x_2$$

Since the system is linear, the following Liapunov energy function is the appropriate choice:

$$V(x_n) = a_1x_1^2 + a_2x_2^2 \qquad a_1, a_2 > 0$$

Then

$$\dot{V}(x_n) = \frac{\partial V}{\partial x_1}\dot{x}_1 + \frac{\partial V}{\partial x_2}\dot{x}_2 \qquad \text{(Time derivative of Liapunov function, Table 34.28)}$$

Substituting for $\partial V/\partial x_1$ and $\partial V/\partial x_2$:

$$\dot{V}(x_n) = (2a_1x_1)\dot{x}_1 + (2a_2x_2)\dot{x}_2$$

Substituting for $\dot{x}_1$ and $\dot{x}_2$:

$$\dot{V}(x_n) = (2a_1x_1)x_2 + (2a_2x_2)(-2x_1 - 3x_2)$$

Multiplying out:

$$\dot{V}(x_n) = 2a_1x_1x_2 - 4a_2x_1x_2 - 6a_2x_2^2$$

Require that $a_2 > 0$ and $2a_1x_1x_2 - 4a_2x_1x_2 = 0$. Therefore $a_2 > 0$ and $a_1 = 2a_2$.

Choose $a_2 = 1$, then $a_1 = 2$ and the result is $\dot{V}(x_n) = -6(1)x_2^2 = -6x_2^2$ = negative semidefinite, since x_1 is not present.

However, it can be shown that $\dot{V}(x_n)$ can vanish only at the origin $x_1 = x_2 = 0$, and hence, by theorem IV of Table 34.28, the system is asymptotically stable. Furthermore, the system is asymptotically stable in the large since $V(x_n) \to \infty$ as $\|x_n\| \to \infty$ (Theorem II).

As an illustration that $\dot{V}(x_n)$ can vanish only at the origin, examine $\dot{V}(x_n) = -6x_2^2$. If $\dot{V}(x_n)$ were to be identically zero for all time, then x_2 would have to be zero and $\dot{x}_2$ would also have to be zero; otherwise x_2 would move away from zero. The state equations are

$$\dot{x}_1 = x_2$$
$$\dot{x}_2 = -2x_1 - 3x_2$$

Substituting $x_2 = 0$ and $\dot{x}_2 = 0$:

$$\dot{x}_1 = 0$$
$$0 = -2x_1 - 3(0)$$

By inspection, $x_1 = 0$ and $\dot{x}_1 = 0$.

Thus $\dot{V}(x_n)$ can be identically zero for all time only at the origin, which satisfies the asymptotic stability requirement of theorem IV of Table 34.28.

Example 34.13. Determine the stability at the origin of the following nonlinear system:

$$\dot{x}_1 = x_2$$
$$\dot{x}_2 = -2x_1^3 - 3x_2$$

Procedure. Select positive definite Liapunov function such that its derivative contains an x_1^3 term, so that cancellation can be arranged. Choose

$$V(x_n) = a_1x_1^4 + a_2x_2^2$$

Then, differentiating,

$$\dot{V}(x_n) = (4a_1x_1^3)\dot{x}_1 + (2a_2x_2)\dot{x}_2$$

Substituting in $\dot{x}_1$ and $\dot{x}_2$

$$\dot{V}(x_n) = (4a_1x_1^3)x_2 + (2a_2x_2)(-2x_1^3 - 3x_2)$$
$$\dot{V}(x_n) = 4a_1x_1^3x_2 - 4a_2x_1^3x_2 - 6a_2x_2^2$$

Assuming that $a_2 = 1$ and $a_1 = 1$ (the choice is arbitrary), then:

$$\dot{V}(x_n) = -6x_2^2 = \text{positive semidefinite}$$

With use of the same arguments as those used in Example 34.12, the system is asymptotically stable in the large, through theorem IV of Table 34.28.

Summary. The important points concerning Liapunov's direct or second method are:

1. A Liapunov function is a generalized energy function ($mV^2/2, CV^2/2, LI^2/2, KX^2/2$) and has the following properties:
   (a) It is not a unique function for any given system.
   (b) It is a scalar function, which evaluates to a simple number.

(c) It is positive definite. This means

$$V(x_n, t) \quad \text{or} \quad V(x_n) > 0$$
$$V(0, t) \quad \text{or} \quad V(0) = 0$$

(d) For stable systems, the time derivative of the Liapunov function is negative.

2. There is a big difference between stable and asymptotically stable. A stable system can have a bounded limit-cycle oscillation, which is stable in the sense that the oscillation does not grow. Asymptotic stability is the conventional form of stability, with all motion ending as t goes to infinity.
3. If the origin of a given system is stable, then Liapunov functions exist that can prove stability, although finding even one may be difficult. Inability to find a Liapunov function that proves stability does not prove instability. The stability conditions obtained for a particular V function are usually sufficient, but not necessary.
4. Liapunov's second method can be applied to linear as well as nonlinear systems with either constant or time-varying coefficients. However, it is generally used only for nonlinear systems analysis, since many methods are available for analyzing linear systems that provide more information.
5. The real power of Liapunov's method can be found in the ingenious methods developed to find the requisite Liapunov function, or in applying the method in a novel way. See the following.

(a) Linear system $Q - P$ method†

$$\dot{x}_n = Ax_n$$
$$V(x_n) = x_n^* P x_n$$
$$\dot{V}(x_n) = x_n^*(A^*P + PA)x_n$$
$$\dot{V}(x_n) = -x_n^* Q x_n \text{ for stability}$$
$$Q = -(A^*P + PA) \text{ for stability}$$

where x_n^* = conjugate transpose and Q, P = positive definite hermitian matrix.

(b) Krasovskii's theorem‡

$$\dot{x}_n = f_n(x_n)$$
$$V(x_n) = f_n^*(x_n) f_n(x_n) = \text{norm squared}$$
$$F_n(x_n) = \text{jacobian of } f_n(x_n)$$
$$\hat{F}_n(x_n) = F_n^*(x_n) + F_n(x_n) = \text{definition}$$
$$\dot{V}(x_n) = f_n^*(x_n)\hat{F}_n(x_n) f_n(x)$$

which is asymptotically stable if $F(x_n)$ is negative definite. $f_n^*(x_n)$ = conjugate transpose of $f_n(x_n)$, jacobian = partial derivative of each row with respect to each variable.

$$\begin{bmatrix} \frac{\partial}{\partial x_1} f_1 & \frac{\partial}{\partial x_2} f_1 \\ \frac{\partial}{\partial x_1} f_2 & \frac{\partial}{\partial x_2} f_2 \end{bmatrix} = F_n(x_n) = \text{jacobian of } f_n(x_n)$$

for a 2×2 matrix.

(c) Variable gradient method§

$$x_n = f_n(x_n)$$
$$\dot{V} = (\nabla V)' \dot{x}$$

or

$$\dot{V} = \begin{bmatrix} \frac{\partial V}{\partial x_1} & \cdots & \frac{\partial V}{\partial x_n} \end{bmatrix} \begin{bmatrix} \dot{x}_1 \\ \vdots \\ \dot{x}_n \end{bmatrix}$$

† See Ogata,[6] p. 726.
‡ See Ogata,[6] p. 735.
§ See Lasalle and Lefschetz,[19] p. 59, or Ogata,[6] p. 737.

where ∇V = gradient of V

$$\nabla V = \begin{bmatrix} \dfrac{\partial V}{\partial x_1} \\ \vdots \\ \dfrac{\partial V}{\partial x_n} \end{bmatrix}$$

and $(\nabla V)'$ = transpose of ∇V; asymptotically stable if the nth dimensional curl of ∇V equals zero. For $n = 3$

$$\frac{\partial \nabla V_2}{\partial x_1} = \frac{\partial \nabla V_1}{\partial x_2}, \quad \frac{\partial \nabla V_3}{\partial x_1} = \frac{\partial \nabla V_1}{\partial x_3}, \quad \frac{\partial \nabla V_3}{\partial x_2} = \frac{\partial \nabla V_2}{\partial x_3}, \quad \text{or} \quad \frac{\partial \nabla V_i}{\partial X_j} = \frac{\partial \nabla V_j}{\partial X_i} \quad (i, j = 1, 2, \ldots, n)$$

(d) The circle criterion of Popov is developed from Liapunov's direct method and is presented next.

Circle Criterion of Popov

One interpretation of a stability theorem for nonlinear systems developed by Popov[20] is particularly easy to employ. The following interpretation is conservative and lumps together various types of nonlinearities that Popov treated as individual cases (single-valued nonlinearities, passive hysteresis, and active hysteresis). The lumping together increases the conservative effect but greatly simplifies the application of the criterion. The net result is that the following Popov circle criterion is sufficient but not necessary. If the $G(j\omega)$ locus does not cut or enclose the Popov circle, the nonlinear system is stable, just as if the Popov circle were the $(-1 + j0)$ point of the Nyquist stability criterion for linear systems. On the other hand, if the $G(j\omega)$ locus does cut or enclose the Popov circle, the system is not

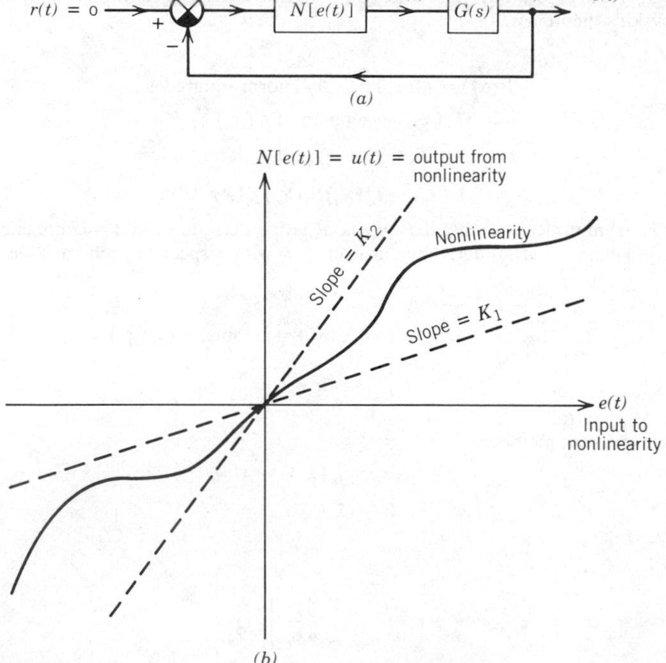

Fig. 34.46 Nonlinear closed-loop control system for Popov stability analysis. Conditions: (1) $G(s)$ is linear and is output stable (has more poles than zeros, has no pole zero cancellations, and all the poles have negative real parts) and (2) the nonlinearity $N[e(t)]$ is bounded by two lines of slope K_1, and K_2, which pass through zero, as shown in (*b*). $\infty \geqslant K_2 > K_1 \geqslant 0$.

necessarily unstable. This Popov criterion can be used to prove that a nonlinear system of the form shown in Fig. 34.46 is stable. It cannot be used to show that the system is unstable.

Note that the figure places several constraints on the system to be analyzed. First, the system must contain a linear transfer function that is output stable. This restriction is satisfied if all the poles of $G(s)$ lie in the left half of the s plane, and if $G(s)$ has more poles than zeros. Second, the nonlinearity is bounded by two straight lines of slopes K_1 and K_2, as shown in Fig. 34.46. If K_2 equals ∞ and K_1 equals 0, the bounds open up to include the entire first and third quadrants, which fits just about any conceivable nonlinearity. Unfortunately, in such a case the Popov criterion becomes very stringent, and the Popov circle degenerates into the entire left half of the $G(j\omega)$ plane. This is the case for an ideal relay, as shown in Fig. 34.47, which shows K_1 and K_2 slopes for several other nonlinearities, including saturation, and a relay with a dead zone. Once the K_1 and K_2 slopes are numerically evaluated for the given nonlinearity, the procedure is as follows:

1. Draw a plot of the imaginary vs. the real part of $G(j\omega)$ as ω varies from zero to infinity. This is identical to the Nyquist plot for linear systems.
2. Plot the points $-1/K_1$ and $-1/K_2$.
3. Pass a circle through these two points, centered between them. The circle's center can be found with a compass or analytically.

$$\text{Center} = \frac{-1/K_1 - 1/K_2}{2} = \frac{-(K_2 + K_1)/K_1K_2}{2} = \frac{-(K_1 + K_2)}{2K_1K_2} \tag{34.100}$$

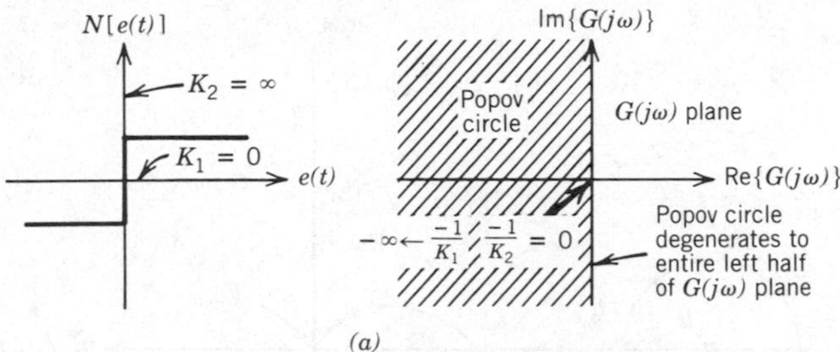

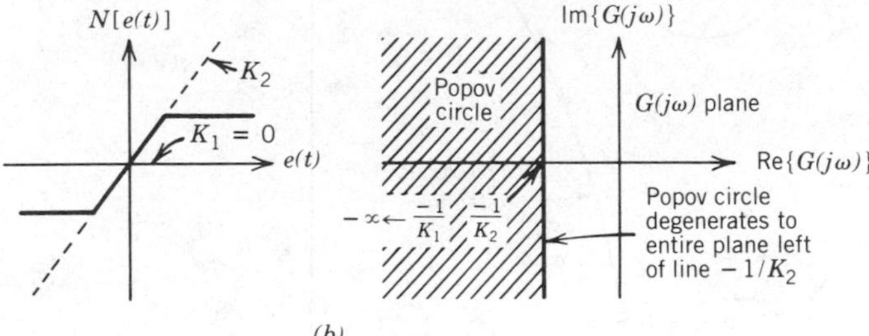

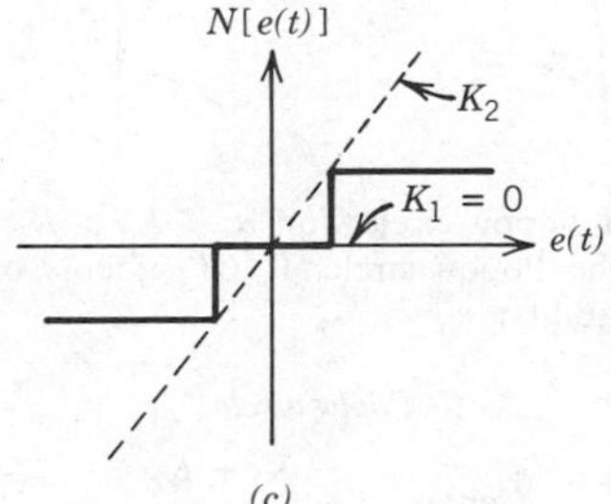

Fig. 34.47 Several common nonlinearities and their limiting Popov slopes, K_1 and K_2, and Popov circles. (*a*) Ideal relay. (*b*) Saturation. (*c*) Relay with dead zone. The Popov circle is identical to that in (*b*). Note that in the case depicted here, the Popov circle has infinite radius and degenerates to a vertical line at $-1/K_2$.

4. Determine system stability by inspection. If the circle is not cut or enclosed by the $G(j\omega)$ plot, the system is stable.
5. If the circle is cut or enclosed by the $G(j\omega)$ locus, the test is inconclusive—the system may be stable or unstable.

Note that if the system is linear so that $N[e(t)]$ becomes a linear gain of unity, then $K_2 = K_1 = 1$. Inserting $K_1 = K_2 = 1$ into Eq. (34.100) yields

$$\text{Center} = \frac{-(K_1 + K_2)}{2K_1K_2} = -\frac{1+1}{2(1)(1)} = -\frac{2}{2} = -1 \tag{34.101}$$

The Popov circle thus degenerates to the point $(-1) = (-1 + j0)$ for a linear system and becomes the Nyquist stability criterion.

An example of a Popov plot is shown in Fig. 34.48. Note that if $K_1 = 0$, which is the case for all the nonlinearities shown in this figure, then the Popov circle degenerates to a vertical line through the point

$$\text{Re}\{G(j\omega)\} = -\frac{1}{K_2} \tag{34.102}$$

Example 34.14. Given a linear actuator with transfer function $G_A(s)$, driven by a DC amplifier which saturates for input signals exceeding 10 V, find the maximum allowable gain of the nonlinear

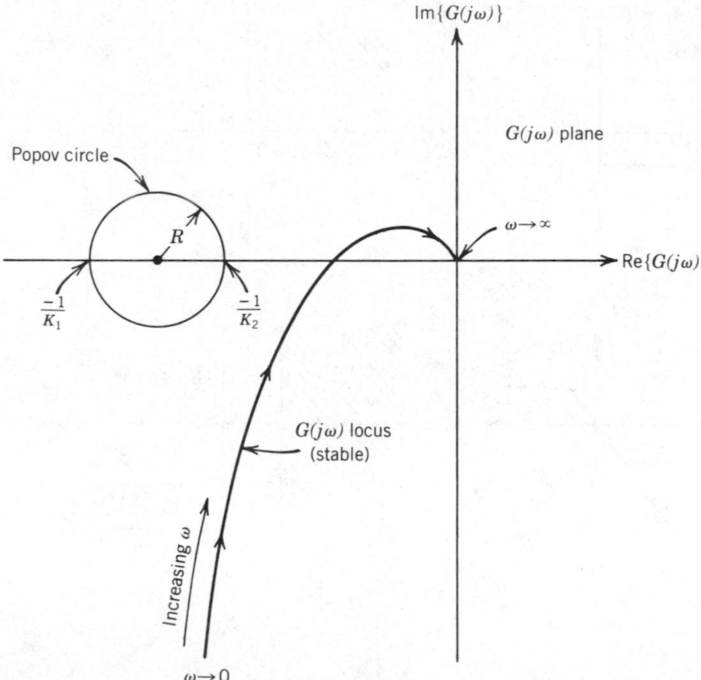

Fig. 34.48 Plot of $G(j\omega)$ with a Popov circle, for $\infty > K_2 > K_1 > 0$. The system is stable if the $G(j\omega)$ locus does not intersect the Popov circle. If $G(j\omega)$ cuts or encloses the circle, the test is inclusive and does not prove instability.

Popov circle

$$\text{Center:} \quad -\frac{K_1 + K_2}{2K_1K_2}$$

$$\text{Radius:} \quad R = \frac{K_2 - K_1}{2K_1K_2}$$

DC amplifier. Compare this gain with the maximum allowable gain if the amplifier did not saturate and was linear.

$$G_A(s) = \frac{1}{s(s+1)(s+10)} \quad \text{(given)} \tag{34.103}$$

$$G_A(j\omega) = \frac{1}{j\omega(1+j\omega)(10+j\omega)} \tag{34.104}$$

Finding the real and the imaginary parts of $G(j\omega)$:

$$G_A(j\omega) = \frac{-11}{(10-\omega^2)^2 + 121\omega^2} + j\frac{-(10-\omega^2)}{\omega\left[(10-\omega^2)^2 + 121\omega^2\right]} \tag{34.105}$$

Evaluating points for the Popov plot:

ω	$G(j\omega)$	$\text{Re}\{G(j\omega)\}$	$\text{Im}\{G(j\omega)\}$
0.001	40 dB $\angle -90°$	-0.11	-100.0
0.01	20 dB $\angle -91°$	-0.11	-10.0
0.1	0.04 dB $\angle -96°$	-0.11	-1.0
1.0	-23 dB $\angle -141°$	-0.05	-0.04
10.0	-63 dB $\angle -219°$	-0.0005	$+0.0004$
$\sqrt{10} = 3.16$	-41 dB $\angle -180°$	-0.009091	0.0

The nonlinearity (see Fig. 34.49) is bounded by $K_1 = 0$, $K_2 = K$, where K is the gain of the DC amplifier before saturation, thus

$$\frac{-1}{K_1} \to -\infty \quad \text{and} \quad \frac{-1}{K_2} = \frac{-1}{K} \tag{34.106}$$

Consequently the Popov circle degenerates into a veritical line at $\text{Re}\{G(j\omega)] = -1/K$. From Fig. 34.49, the maximum magnitude of the negative value of $\text{Re}\{G(j\omega)\}$ is -0.11.

From Eq. (34.102), to avoid an intersection of the $G(j\omega)$ locus and the Popov circle:

$$\text{Re}\{G(j\omega)\} = -0.11 > \frac{-1}{K_2} = \frac{-1}{K} \tag{34.107}$$

or

$$0.11 < \frac{1}{K} \quad \text{(minus signs reverse an inequality)}$$

or

$$K < \frac{1}{0.11} = 9.1 \tag{34.108}$$

Thus the DC amplifier gain must be less than 9.1 for the nonlinear system to be stable in accordance with the Popov criterion. Note that the actual level of saturation of the DC amplifier was not used at all. This is one reason the Popov method is conservative. With a gain of 9.1, the system is stable whether the DC amplifier saturates at 1, 10, or 100 V. This means that in reality a larger gain than 9.1 is allowable, as the saturation level increases. The Popov method indicates that if the gain is 9.1 or less, the system is stable. It does not indicate that if the gain exceeds 9.1, the system becomes unstable.

The maximum gain for a nonsaturating amplifier is found using the Routh–Hurwitz array, with the gain K lumped in with $G_A(s)$.

$$1 + G_A(s)H(s) = 1 + \frac{K}{s(s+1)(s+10)} \quad \text{since} \quad H(s) = 1 \tag{34.109}$$

From this equation, the characteristic equation $D(s)$ of the linear system is:

$$s(s+1)(s+10) + K = s^3 + 11s^2 + 10s + K \tag{34.110}$$

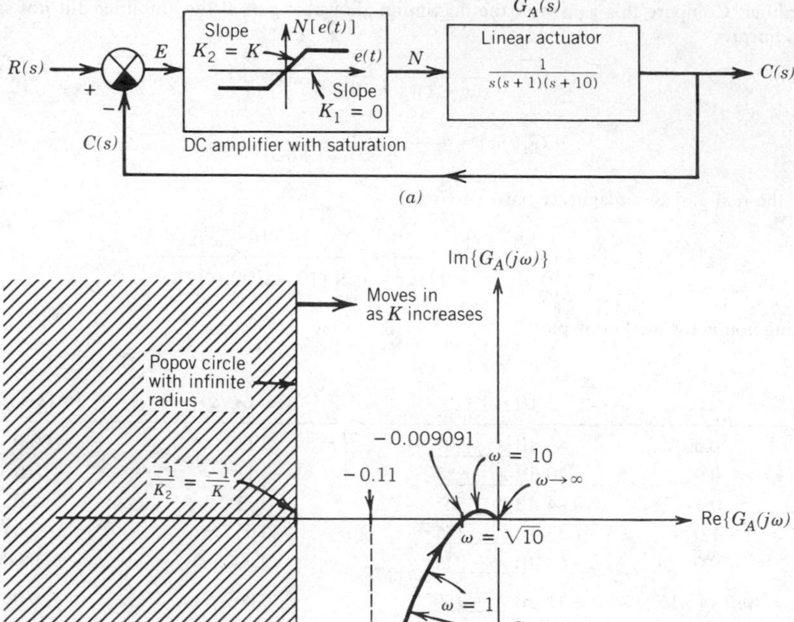

Fig. 34.49 Example 34.14, solution via Popov plot. (*a*) Block diagram of saturating DC amplifier of gain K and linear actuator $G_A(s)$. (*b*) Popov plot. With saturation the gain K can be increased until the degenerate straight line Popov "circle" becomes tangent to the $G_A(j\omega)$ locus. The result, $K = 1/0.11 = 9.1$, is conservative. K = amplifier gain. Maximum K with saturation $= 1/0.11 = 9.1$, maximum K with no saturation $= 1/0.009091 = 110$.

Carrying out the Routh–Hurwitz array test:

	$s^3 + 11s^2$	$+ 10s + K$
s^3	1	10
s^2	11	K
s^1	$\frac{110 - K}{11}$	$(K < 110)$
s^0	K	$(K > 0)$

For stability, $(110 - K)/11 > 0$ and $K > 0$. By inspection of the s^1 row of the Routh–Hurwitz array, the maximum value of the DC amplifier gain is 110 if no saturation occurs. This is 12 times higher than the gain of 9.1 allowed in the case of saturation.

The result of the Routh–Hurwitz test may be confirmed by referring to the Nyquist plot segment of the Popov plot, namely the plot of $G(j\omega)$. From the tabulated data following Eq. (34.104), the plot of $G(j\omega)$ for unity gain crosses the real axis at

$$G(j\omega)|_{\omega_0=\sqrt{10}} = -41 \text{ dB} \angle -180 = -0.009091 + j0.0 \tag{34.111}$$

The linear system will become unstable if sufficient gain is added so that $G(j\omega_0)$ becomes $(-1 + j0)$.

Thus the maximum stable gain K for a purely linear system is:

$$K(-0.009091 + j0) = -1 + j0$$

or

$$K = \frac{-1}{-0.009091} = 110$$

This Nyquist plot result agrees exactly with the Routh–Hurwitz array result, which serves as a check on the mathematics.

34.10 MODERN CONTROL

The unqualified term "modern control" refers to a method of analysis and design that returns to the basic differential equations that describe a system. Generally in modern control design, and nth-order system is described by a set of n first-order differential equations, which are expressed in matrix form and called state equations. This is opposed to classical control, where the system is described by one nth-order differential equation and analyzed using frequency domain (or Laplace transform) techniques.

Modern control has risen to prominence as the availability of digital computers capable of solving a set of nth-order coupled differential equations has risen. A principal strength of the modern control state variable approach is that the matrix notation used is compact, and one single matrix equation holds for any order of system.

Modern control employs another fundamental concept different from the methods of classical control. This is the concept of "optimization," which is achieved by the minimization of a "Cost Function" C. The design specifies certain positive functions of system performance, such as the square of the error or the square of the control effort, to formulate a cost function. The cost function, which is always positive, is then integrated to yield the system "Performance Index" J. Since J is the integral of a positive (or zero) quantity, such as the system error squared, J always increases with time (or remains constant). If J is minimized, the integral of the error is minimized, which is one possible optimal design procedure. Note that this procedure bears certain resemblances to Liapunov's direct method for determining system stability. To be specific, a function of the system's variables is formulated which is always positive, and then the behavior of that function is examined.

If modern control design has any fault, it lies in the designer's choice of cost function. If the designer chooses simply to minimize error by using a cost function C_1

$$C_1 = e_n(t)^2 \tag{34.112}$$

where $e_n(t)$ = system error vector with n components, and associated performance index J_1

$$J_1 = \int_0^\infty C_1\, dt = \int_0^\infty e_n(t)^2\, dt \tag{34.113}$$

then it is possible (and indeed likely) that an unsatisfactory or unrealizable design may result which requires control efforts that cannot be provided (e.g., a very large, impulsive, or infinite control effort may be required). This problem can be overcome by insertion of a term constraining the control effort, $U(t)$, into the performance index. Thus a better performance index J_2 might be

$$J_2 = \int_0^\infty \left[e_n(t)^2 + U_m(t)^2 \right] dt \tag{34.114}$$

where $U_m(t)$ = control vector with m components.

However, the designer might decide that one factor of the cost function is more important to satisfactory system operation than the other, so that the cost function components $e_n(t)^2$ and $U_m(t)^2$ must be given an appropriate weighing. This can be accomplished by introduction of weighting constants K_1 and K_2, so that the improved performance index J_3 becomes

$$J_3 = \int_0^\infty \left[K_1 e_n(t)^2 + K_2 U_m(t)^2 \right] dt \tag{34.115}$$

This is a workable performance index, which was used in 1952 by Bryson[21] in conjunction with the control of a homing missile. The control effort term $U_m(t)$ had special significance in Bryson's case, since the missile's rudder control effort was supplied by a hydraulic accumulator. A hydraulic accumulator is a sealed container containing hydraulic fluid under pressure. Once the fluid is used up, the missile becomes uncontrollable. In such a case, the designer's dilemma is evident: A missile that misses the target is unsatisfactory. Therefore the error must be more heavily weighted than the control effort. However, a missile that is uncontrollable cannot hit the target. Therefore the control effort must be more heavily weighted than the error! These contradictory requirements cannot both be met simultaneously, and hence optimal control is seen to require engineering tradeoffs (just as traditional design procedures require such tradeoffs).

The designer must clearly specify exactly what he or she wishes to accomplish and at what cost. Since a clear statement of a problem does indeed seem to produce a good portion of the solution, the optimal control approach is logically appealing because it requires a clear statement of the system design goals. These then become the system's cost function and performance index.

In the design of large-scale systems (where optimal control finds its largest application), the performance index/cost function approach enjoys another significant advantage: The designer deals with a single, albeit complicated, cost function in order to specify system performance. In the classical control theory approach, the designer would have to deal with a multitude of performance criteria, such as stability, rise time, settling time, steady-state errors, percent overshoot, damping, response to disturbances, and others. All of these performance criteria are replaced by the single cost function. The optimization logic, based on the calculus of variations, is typically embedded in a computer program that selects the best combination(s) of the control parameters and control gains to minimize the chosen cost function. However, few real systems are designed by optimal control alone. Optimal control serves as an excellent starting point for the design of a system, but since the designer may have overlooked some important performance factor when specifying the system performance index, it is almost mandatory to check or alter the supposedly optimal design by applying various classical control methods (e.g., steady-state error analysis, frequency response, or root locus analysis or, most likely, a complete system simulation study). Simulation, the testing of a physical or, more probably, a computer model of the newly designed system has almost become a requisite of design by modern control methods (just as it was almost always a requisite of design by classical control methods).

It is important to emphasize that the performance index of Eq. (34.115) does not require stability, and it may very well turn out that the optimal control system designed is unstable! Therefore stability should always be checked, as well as the other classical control functions just mentioned.

State Space Equations

The state-space equations employed by modern control theory are a set of n-coupled first-order differential equations describing an nth-order system. The state variables need not be physically measurable or observable quantities, but realistically it is best if most or all of the state variables are measurable, since the optimal control law may require feedback of all the state variables (with appropriate weighting). For a purely mechanical system, the choice of state variables would almost always be the system's position and velocity for a second-order system, or the position, velocity, and acceleration for a third-order system. For a passive electrical network, the choice of the state variables would almost always be the currents through each inductor and the voltage across each capacitor.

For a general nth-order system with output variable $c(t)$ and input variable $r(t)$, described by the differential equation

$$\frac{d^n}{dt^n}c(t) + a_1\frac{d^{n-1}}{dt^{n-1}}c(t) + \cdots + a_{n-1}\frac{d}{dt}c(t) + a_n c(t) = r(t) \tag{34.116}$$

the state variables are almost always chosen as

$$\left.\begin{aligned} x_1 &= c(t) \\ x_2 &= \dot{c}(t) = \frac{d}{dt}c(t) = \frac{d}{dt}x_1 = \dot{x}_1 \\ x_3 &= \ddot{c}(t) = \frac{d^2}{dt^2}c(t) = \frac{d}{dt}\frac{d}{dt}c(t) = \frac{d}{dt}x_2 = \dot{x}_2 \\ &\vdots \\ x_n &= \frac{d^{n-1}}{dt^{n-1}}c(t) = \dot{x}_{n-1} \end{aligned}\right\} \tag{34.117}$$

The set of equations in Eq. (34.117) gives $(n-1)$ of the required n first-order differential state equations, in very simple form, namely,

$$\begin{aligned} \dot{x}_1 &= x_2 \\ \dot{x}_2 &= x_3 \\ \dot{x}_3 &= x_4 \\ &\vdots \\ \dot{x}_{n-1} &= x_n \end{aligned}$$

The last equation for $\dot{x}_n$ is found by solving for the highest order (nth) derivative of the basic differential equation, Eq. (34.116), and then substituting in $x_1 = c(t)$, $x_2 = \dot{c}(t)$, $x_3 = \ddot{c}(t)$, and so on, resulting in

$$
\begin{aligned}
\dot{x}_1 &= x_2 \\
\dot{x}_2 &= x_3 \\
\dot{x}_3 &= x_4 \\
&\vdots \\
\dot{x}_{n-1} &= x_n \\
\dot{x}_n &= -a_n x_1 - a_{n-1} x_2 - \cdots - a_1 x_n + r(t)
\end{aligned}
\tag{34.118}
$$

Equation (34.118) is usually written in matrix form

$$
\begin{bmatrix} \dot{x}_1 \\ \dot{x}_2 \\ \dot{x}_3 \\ \vdots \\ \dot{x}_{n-1} \\ \dot{x}_n \end{bmatrix} = \begin{bmatrix} 0 & 1 & 0 & 0 & \cdots & 0 \\ 0 & 0 & 1 & 0 & \cdots & 0 \\ 0 & 0 & 0 & 1 & \cdots & 0 \\ \vdots & \vdots & \vdots & \vdots & \cdots & \vdots \\ 0 & 0 & 0 & 0 & \cdots & 1 \\ -a_n & -a_{n-1} & -a_{n-2} & -a_{n-3} & \cdots & -a_1 \end{bmatrix} \begin{bmatrix} x_1 \\ x_2 \\ x_3 \\ \vdots \\ x_{n-1} \\ x_n \end{bmatrix} + \begin{bmatrix} 0 \\ 0 \\ 0 \\ \vdots \\ 0 \\ 1 \end{bmatrix} U(t) \tag{34.119}
$$

where $r(t) = U(t) =$ the input function.

In compact matrix notation, Eq. (34.119) is written as

State equation

$$
\dot{\mathbf{x}}_n = \mathbf{A}\mathbf{x}_n + \mathbf{B}\mathbf{U} \tag{34.120}
$$

where the output $c(t) = x_1$ is given by

$$
c(t) = [1 \quad 0 \quad \cdots \quad 0] \begin{bmatrix} x_1 \\ x_2 \\ \vdots \\ x_n \end{bmatrix} \tag{34.121}
$$

or the matrix notation

Output equation

$$
c(t) = \mathbf{C}\mathbf{x}_n \tag{34.122}
$$

The problem is best clarified by an example.

Example 34.15. Given a system described by the differential equation

$$
\dddot{y} + 9\ddot{y} + 26\dot{y} + 24y = 3U \tag{34.123}
$$

where $U(t)$ is the system input $[U(t) = U]$ and $y(t)$ is the system output $[y(t) = y]$, find the state-variable description of the system.

Step 1. Define the n state variables by

$$
x_1 = y, \qquad x_2 = \dot{y}, \qquad x_3 = \ddot{y} \tag{34.124}
$$

Note that no state variable is defined for the highest order derivative $\dddot{y}$, since $\dddot{y}$ is equated with $\dot{x}_3$.

Step 2. Obtain the first $(n-1)$ state variable equations directly from the definitions of step 1.

$$
\begin{aligned}
\dot{x}_1 &= \dot{y} = x_2 \\
\dot{x}_2 &= \ddot{y} = x_3
\end{aligned}
$$

Step 3. Solve for the highest-order derivative $\dddot{y}$, and substitute state variables for the differential equation variables $\ddot{y}$, $\dot{y}$ and y.

$$
\begin{aligned}
\dddot{y} &= -24y - 26\dot{y} - 9\ddot{y} + 3U \\
\dot{x}_3 &= -24x_1 - 26x_2 - 9x_3 + 3U
\end{aligned}
$$

Step 4. Write the state equation in matrix form:

$$\begin{bmatrix} \dot{x}_1 \\ \dot{x}_3 \\ \dot{x}_3 \end{bmatrix} = \begin{bmatrix} 0 & 1 & 0 \\ 0 & 0 & 1 \\ -24 & -26 & -9 \end{bmatrix} \begin{bmatrix} x_1 \\ x_2 \\ x_3 \end{bmatrix} + \begin{bmatrix} 0 \\ 0 \\ 3 \end{bmatrix} U \tag{34.125}$$

$$[\dot{\mathbf{x}}_n] = [\mathbf{A}][\mathbf{x}_n] + [\mathbf{B}]U$$

and

$$y = [1 \quad 0 \quad 0] \begin{bmatrix} x_1 \\ x_2 \\ x_3 \end{bmatrix} = [\mathbf{C}][\mathbf{x}_n]$$

It is extremely important to note that state variables are not unique, and that if the definition of the state variables were changed [Eq. (34.124)], the final state variable equation [Eq. (34.125)] would be different. Several valid and different state variable assignments are:

Set 1	Set 2	Set 3
$x_1 = y$	$x_1 = \dot{y}$	$x_1 = y$
$x_2 = \dot{y}$	$x_2 = y$	$x_2 = \dot{x}_1 + K_2U = \dot{y} + K_2U$
$x_3 = \ddot{y}$	$x_3 = \ddot{y}$	$x_3 = \dot{x}_2 + K_1U = \ddot{y} + K_2\dot{U} + K_1U$

Set 1 is the original choice, which produces Eq. (34.125). Set 2 merely shifts around the elements of Eq. (34.125) and serves no useful purpose. Set 3 is very useful because if the input driving function contains derivative terms, set 3 can be used to eliminate these derivatives. Then, the same A matrix of Eq. (34.125) results, with a different B matrix obtained from the added (KU) terms in set 3. (K_1U) would eliminate any first derivative term and (K_2U) any second derivative term. This is illustrated in Example 34.16.

Example 34.16. Repeat Example 34.15 with a term $5\dot{U}$ added to the input driving function. Therefore the system's differential equation is

$$\dddot{y} + 9\ddot{y} + 26\dot{y} + 24y = 3U + 5\dot{U} \tag{34.126}$$

Step 1.

$$x_1 = y$$
$$x_2 = \dot{x}_1 = \dot{y}$$
$$x_3 = \dot{x}_2 + K_1U = \ddot{y} + K_1U \qquad \text{or} \qquad \ddot{y} = x_3 - K_1U$$

Step 2. Obtain relations for $\dot{x}_n$:

$$\dot{x}_1 = \dot{y} = x_2$$
$$\dot{x}_2 = \ddot{y} = x_3 - K_1U$$
$$\dot{x}_3 = \dddot{y} + K_1\dot{U} \qquad \text{or} \qquad \dddot{y} = \dot{x}_3 - K_1\dot{U}$$

Step 3. Solve for $\dddot{y}$ and substitute in x_1, x_2, x_3

$$\dddot{y} = -24y - 26\dot{y} - 9\ddot{y} + 3U + 5\dot{U}$$
$$\dot{x}_3 - K_1\dot{U} = -24x_1 - 26x_2 - 9(x_3 - K_1U) + 3U + 5\dot{U}$$

Step 4. Eliminate the $\dot{U}$ term by a proper choice of K_1. Here, choose $K_1 = -5$.

$$\dot{x}_3 + 5\dot{U} = -24x_1 - 26x_2 - 9(x_3 + 5U) + 3U + 5\dot{U}$$

eliminated by $K_1 = -5$

$$\dot{x}_3 = -24x_1 - 26x_2 - 9x_3 - 45U + 3U$$
$$\dot{x}_3 = -24x_1 - 26x_2 - 9x_3 - 42U$$

Step 5. Write the complete state equation set by inserting the K_1 value(s) back into step 2, which yields $\dot{x}_2 = x_3 + 5U$ and $\dot{x}_1 = x_2$, and then use the $\dot{x}_3$ equivalent of step 4.

$$\begin{bmatrix} \dot{x}_1 \\ \dot{x}_2 \\ \dot{x}_3 \end{bmatrix} = \begin{bmatrix} 0 & 1 & 0 \\ 0 & 0 & 1 \\ -24 & -26 & -9 \end{bmatrix} \begin{bmatrix} x_1 \\ x_2 \\ x_3 \end{bmatrix} + \begin{bmatrix} 0 \\ 5 \\ -42 \end{bmatrix} [U] \tag{34.127}$$

$$[\dot{\mathbf{x}}_n] = [\mathbf{A}][\mathbf{x}_n] + [\mathbf{B}_1]U$$

and

$$y = [1 \quad 0 \quad 0] \begin{bmatrix} x_1 \\ x_2 \\ x_3 \end{bmatrix} = [\mathbf{C}][\mathbf{x}_n]$$

If two derivative terms of the input driving function were to be eliminated, the highest order derivative (K_2U) would be eliminated first, and the resulting value of K_2 would then be inserted back into step 4 to eliminate the first-order derivative (K_1U). This process is a straightforward algebraic procedure that can be solved in general terms. However, the general terms are difficult to memorize and offer relatively little computational advantage over the $K_1, K_2, \ldots, K_n$ method just described, for any problem that is simple enough to be solved without the aid of a computer. The general form is derived by Kuo.[7]

Solutions of the State Equations

The general state equation (where $\mathbf{x}_n$ = state vector, $\mathbf{U}$ = input vector, and $\mathbf{A}, \mathbf{B}$ = constant matrices)

$$\dot{\mathbf{x}}_n = \mathbf{A}\mathbf{x}_n + \mathbf{B}\mathbf{U} \tag{34.128}$$

can be solved by straightforward Laplace transform techniques, as long as the rules of matrix algebra are followed. Taking the Laplace transform:

$$s\mathbf{x}_n(s) - \mathbf{x}_n(t)|_{t=0} = \mathbf{A}\mathbf{x}_n(s) + \mathbf{B}\mathbf{U}(s)$$

Shifting terms:

$$s\mathbf{x}_n(s) - \mathbf{A}\mathbf{x}_n(s) = \mathbf{x}_n(t)|_{t=0} + \mathbf{B}\mathbf{U}(s)$$

Factoring:

$$[s\mathbf{I} - \mathbf{A}]\mathbf{x}_n(s) = \mathbf{x}_n(0) + \mathbf{B}\mathbf{U}(s)$$

where $\mathbf{I}$ = identity matrix,

$$\mathbf{I} = \begin{bmatrix} 1 & 0 & 0 \\ 0 & 1 & 0 \\ 0 & 0 & 1 \end{bmatrix}$$

for a 3×3 matrix. Premultiplying by $[s\mathbf{I} - \mathbf{A}]^{-1}$, the matrix inverse of $[s\mathbf{I} - \mathbf{A}]$:

$$x_{\mathbf{n}}(s) = [s\mathbf{I} - \mathbf{A}]^{-1} x_{\mathbf{n}}(0) + [s\mathbf{I} - \mathbf{A}]^{-1}\mathbf{B}\mathbf{U}(s)$$

Take the inverse Laplace transform and note that $x_n(0)$, the initial condition of the state variable, is a vector of constants, whereas $U(s)$ is a function of (s). Therefore the second term of $x_n(s)$ contains a product of Laplace transforms:

$$\mathbf{x}_n(t) = L^{-1}\{[s\mathbf{I} - \mathbf{A}]^{-1}\} x_{\mathbf{n}}(0) + L^{-1}\{[s\mathbf{I} - \mathbf{A}]^{-1}\mathbf{B}\mathbf{U}(s)\} \tag{34.129}$$

where L^{-1} = inverse Laplace transform operator.

The state transition matrix, $\mathbf{\Phi}(t)$, is denoted as the inverse Laplace transform of the inverse of the matrix $[s\mathbf{I} - \mathbf{A}]$:

$$\mathbf{\Phi}(t) = L^{-1}\{[s\mathbf{I} - \mathbf{A}]^{-1}\} \tag{34.130}$$

Inserting Eq. (34.130) into Eq. (34.129) [and recalling that the inverse Laplace transform of $F_1(s)F_2(s)$ is given by the convolution integral $\int_0^t f_1(\Upsilon) f_2(t - \Upsilon)\, d\Upsilon$] yields

$$\mathbf{x}_n(t) = \mathbf{\Phi}(t)\mathbf{x}_n(0) + \int_0^t \mathbf{\Phi}(\Upsilon)\mathbf{B}\mathbf{U}(t - \Upsilon)\, d\Upsilon \tag{34.131}$$

Equation (34.131) is the general solution for any system described by state variables in the form of Eq. (34.128). Except in simple cases, the evaluation of $\mathbf{x}_n(t)$ is a formidable task. A simple case is presented in the following example.

Example 34.17. Given the system

$$\begin{bmatrix} \dot{x}_1 \\ \dot{x}_2 \end{bmatrix} = \begin{bmatrix} 0 & 1 \\ -2 & -3 \end{bmatrix} \begin{bmatrix} x_1 \\ x_2 \end{bmatrix} + \begin{bmatrix} 0 \\ 1 \end{bmatrix} U(t) \qquad U(t) = \text{unit step function} \tag{34.132}$$

First, the state transition matrix $\mathbf{\Phi}(t)$ is found. From Eq. (34.130)

$$\mathbf{\Phi}(t) = L^{-1}\{[s\mathbf{I} - \mathbf{A}]^{-1}\}$$

Evaluating $[s\mathbf{I} - \mathbf{A}]$ by using the $[A]$ matrix of Eq. (34.132):

$$[s\mathbf{I} - \mathbf{A}] = \begin{bmatrix} s & 0 \\ 0 & s \end{bmatrix} - \begin{bmatrix} 0 & 1 \\ -2 & -3 \end{bmatrix} = \begin{bmatrix} s & -1 \\ +2 & s+3 \end{bmatrix}$$

Taking the inverse of $[s\mathbf{I} - \mathbf{A}]$:

$$[s\mathbf{I} - \mathbf{A}]^{-1} = \frac{\begin{bmatrix} s+3 & 1 \\ -2 & s \end{bmatrix}}{\begin{vmatrix} s & -1 \\ +2 & s+3 \end{vmatrix}} = \frac{\begin{bmatrix} s+3 & 1 \\ -2 & s \end{bmatrix}}{s(s+3)+2}$$

$$[s\mathbf{I} - \mathbf{A}]^{-1} = \frac{\begin{bmatrix} s+3 & 1 \\ -2 & s \end{bmatrix}}{s^2+3s+2} = \begin{bmatrix} \dfrac{s+3}{(s+1)(s+2)} & \dfrac{1}{(s+1)(s+2)} \\ \dfrac{-2}{(s+1)(s+2)} & \dfrac{s}{(s+1)(s+2)} \end{bmatrix} \tag{34.133}$$

Expanding each term of the 2×2 matrix of Eq. (34.133) into partial fractions:

$$[s\mathbf{I} - \mathbf{A}]^{-1} = \begin{bmatrix} \dfrac{2}{s+1} + \dfrac{-1}{s+2} & \dfrac{1}{s+1} + \dfrac{-1}{s+2} \\ \dfrac{-2}{s+1} + \dfrac{2}{s+2} & \dfrac{-1}{s+1} + \dfrac{2}{s+2} \end{bmatrix}$$

Taking the inverse Laplace transform of $[s\mathbf{I} - \mathbf{A}]^{-1}$ by utilizing the transform pair $K/(s+a) \rightarrow Ke^{-at}$:

$$\mathbf{\Phi}(t) = L^{-1}\{[s\mathbf{I} - \mathbf{A}] = \begin{bmatrix} 2e^{-t} - e^{-2t} & e^{-t} - e^{-2t} \\ -2e^{-t} + 2e^{-2t} & -e^{-t} + 2e^{-2t} \end{bmatrix} \tag{34.134}$$

This completes the evaluation of the first term of Eq. (34.129) and Eq. (34.131) for the system response. The first term of Eq. (34.129) and Eq. (34.131), which is driven by the initial conditions $x_n(0)$, is often called the homogeneous response of the system. It is the system response when no input function is present (only initial conditions are present). The response to the input term is found next.

The second term of Eq. (34.131) is

$$\int_0^t \mathbf{\Phi}(\Upsilon)\mathbf{B}\mathbf{U}(t - \Upsilon)\, dt$$

where $\mathbf{\Phi}(\Upsilon)$ is given by Eq. (34.134) (with t replaced by Υ), $\mathbf{B}$ is a given quantity in Eq. (34.132), namely $\mathbf{B} = \left[\begin{smallmatrix} 0 \\ 1 \end{smallmatrix}\right]$, and $\mathbf{U}(t)$ is the unit step function $\mathbf{U}(t) = 1$ for $t > 0$, so that $\mathbf{U}(t - \Upsilon)$ also equals one, over the range of integration $\Upsilon =$ "0" to "t." Inserting all of these values yields

$$\int_0^t \begin{bmatrix} 2e^{-\Upsilon} - e^{-2\Upsilon} & e^{-\Upsilon} - e^{-2\Upsilon} \\ -2e^{-\Upsilon} + 2e^{-2\Upsilon} & -e^{-\Upsilon} + 2e^{-2\Upsilon} \end{bmatrix} \begin{bmatrix} 0 \\ 1 \end{bmatrix} [1]\, d\Upsilon$$

Carrying out the matrix multiplications first, and then integrating, yields

$$\int_0^t \begin{bmatrix} e^{-\Upsilon} - e^{-2\Upsilon} \\ -e^{-\Upsilon} + 2e^{-2\Upsilon} \end{bmatrix} d\Upsilon = \begin{bmatrix} \frac{1}{-1}e^{-\Upsilon} - \frac{1}{-2}e^{-2t} \\ \frac{-1}{-1}e^{-\Upsilon} + \frac{2}{-2}e^{-2\Upsilon} \end{bmatrix} \begin{matrix} \Upsilon = t \\ \Upsilon = 0 \end{matrix} \tag{34.135}$$

Evaluating the limits:

$$\begin{bmatrix} -e^{-\Upsilon} + \frac{1}{2}e^{-2\Upsilon} \\ e^{-\Upsilon} - e^{-2\Upsilon} \end{bmatrix} \begin{matrix} \Upsilon = t \\ \Upsilon = 0 \end{matrix} = \begin{bmatrix} -e^{-t} + \frac{1}{2}e^{-2t} - \left(-e^0 + \frac{1}{2}e^0\right) \\ e^{-t} - e^{-2t} - (e^0 - e^0) \end{bmatrix}$$

Noting the $e^0 = 1$, and simplifying:

$$\begin{bmatrix} -e^{-t} + \frac{1}{2}e^{-2t} + \frac{1}{2} \\ e^{-t} - e^{-2t} \end{bmatrix} = \text{response to unit step}$$

Combining the first term of Eq. (34.131) [given by Eq. (34.134)], with the second term [given by Eq. (34.135)] yields the total response $\mathbf{x}_n(t)$ for the second-order system described by Eq. (34.132):

$$\mathbf{x}_n(t) = \begin{bmatrix} x_1(t) \\ x_2(t) \end{bmatrix} = \begin{bmatrix} 2e^{-t} - e^{-2t} & e^{-t} - e^{-2t} \\ -2e^{-t} + 2e^{-2t} & -e^{-t} + 2e^{-2t} \end{bmatrix} \begin{bmatrix} x_1(0) \\ x_2(0) \end{bmatrix} + \begin{bmatrix} \frac{1}{2} - e^{-t} + \frac{1}{2}e^{-2t} \\ e^{-t} - e^{-2t} \end{bmatrix} \tag{34.136}$$

In many typical control situations, the initial conditions $x_1(0)$ and $x_2(0)$ are zero, prior to the application of the unit-step test input. In such a situation, the response of the system described by Eq. (34.132) becomes

$$\mathbf{x}_n(t) = \begin{bmatrix} x_1(t) \\ x_2(t) \end{bmatrix} = \begin{bmatrix} \frac{1}{2} - e^{-t} + \frac{1}{2}e^{-2t} \\ e^{-t} - e^{-2t} \end{bmatrix} \tag{34.137}$$

If the system were a mechanical one, x_1 would be the output position and x_2 would be the output velocity in accordance with the first row of the state equation, $\dot{x}_1 = x_2$, of Eq. (34.132). A plot of $x_1(t)$ = system position and $x_2(t)$ = system velocity is given in Fig. 34.50.

Matrix Exponential. If the system being investigated were of first order, then the matrix **A** of Eq. (34.129) would be 1×1, that is, it would consist of a single constant (scalar), for example, a.

For the scalar case, where the matrix **A** reduces to the single constant (scalar) a, the expression for $\mathbf{\Phi}(t)$, Eq. (34.129), becomes

$$\mathbf{\Phi}(t) = L^{-1}\{[sI - A]^{-1}\} = L^{-1}\left\{\frac{1}{s-a}\right\}$$

which results in

$$\mathbf{\Phi}(t) = e^{+at}$$

In the nonscalar case, the transition matrix $\mathbf{\Phi}(t)$ can also be expressed as an exponential in series form:

$$\mathbf{\Phi}(t) = e^{+[\mathbf{A}]t} \tag{34.138}$$

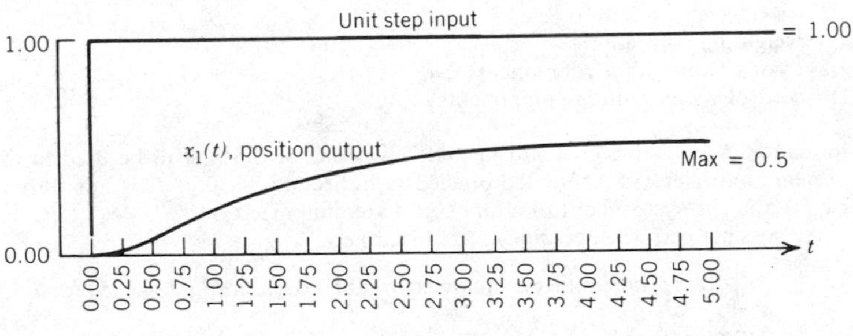

(a)

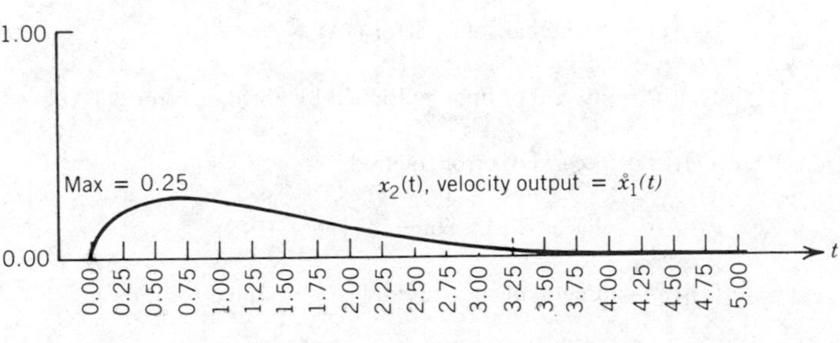

(b)

Fig. 34.50 Plots of $x_1(t)$ (a) and $x_2(t)$ (b) for the system of Example 34.17. $\dot{\mathbf{X}} = \mathbf{AX} + \mathbf{BU}$, where $\mathbf{A} = \begin{bmatrix} 0 & 1 \\ -2 & -3 \end{bmatrix}$, $\mathbf{B} = \begin{bmatrix} 0 \\ 1 \end{bmatrix}$, U = unit step. $x_1(t) = \frac{1}{2} - e^{-t} + \frac{1}{2}e^{-2t}$ = position reponse to unit step input and $x_2(t) = e^{-t} - e^{-2t}$ = velocity response to unit step input. Initial conditions: $x_1(0) = 0$, $x_2(0) = 0$.

In this case the matrix exponential can be evaluated in closed form from Eq. (34.129) [as illustrated in Example 34.17 via Eq. (34.134) for $\Phi(t)$], or it can be evaluated from an infinite series for e^{+x}, expressed in matrix form. Thus

$$e^{+x} = 1 + x + \frac{1}{2!}x^2 + \frac{1}{3!}x^3 + \cdots$$

and

$$e^{[\mathbf{A}]t} = I + [\mathbf{A}]t + \tfrac{1}{2!}[\mathbf{A}][\mathbf{A}]t^2 + \tfrac{1}{3!}[\mathbf{A}][\mathbf{A}][\mathbf{A}]t^3 + \cdots \tag{34.139}$$

It is possible to evaluate the state transition matrix $\Phi(t)$ from $e^{[\mathbf{A}]t}$ given by Eq. (34.139), particularly if a computer is programmed to numerically sum the series. If the series does not converge, a finite result cannot be obtained in this manner.

Since the state transition matrix $\Phi(t)$ obeys the properties of an exponential, the following relations apply to the state transition matrix:

1. $\Phi(t) = e^{[\mathbf{A}]t}$ = matrix exponential $= I + [\mathbf{A}]t + \frac{1}{2!}[\mathbf{A}]^2t^2 + \ldots$.
2. $\Phi(0) = e^{[\mathbf{A}]0} = [I]$ = identity matrix.
3. $\Phi^{-1}(t) = e^{-[\mathbf{A}]t} = e^{[\mathbf{A}](-t)} = \Phi(-t)$ = matrix inverse of $\Phi(t)$.
4. $\Phi(t_1 + t_2) = e^{[\mathbf{A}](t_1+t_2)} = e^{[\mathbf{A}]t_1}e^{[\mathbf{A}]t_2} = \Phi(t_1)\Phi(t_2)$.
5. $[\Phi(t)]^2 = e^{[\mathbf{A}]t}e^{[\mathbf{A}]t} = e^{[\mathbf{A}](2t)} = \Phi(2t)$.
6. $[\Phi(t)]^n = \Phi(nt)$.

34.11 OPTIMAL CONTROL TECHNIQUES

Optimal control, and the concept of the cost function and performance index, were introduced in Section 34.10 on modern control [Eqs. (34.112) through (34.115)] in connection with modern control theory. "Modern control" and "optimal control" are effectively synonymous to some writers, with "modern control" being the more encompassing term.

The performance index J_3, given by Eq. (34.115) and repeated here, is only one possible choice.

$$J_3 = \int_0^\infty \left[K_1 e_n(t)^2 + K_2 U_m(t)^2 \right] dt \tag{34.115}$$

where K_1, K_2 = weighting constants
$e_n(t)$ = error vector with n components
$U_m(t)$ = control vector with m components

The performance index J_3 is significant in that it was one of the first to be used in the optimal design of a system controller. As mentioned previously in Section 34.10, if J_3 is minimized, then the system error $e_n(t)$ and the system control effort $U_m(t)$ are minimized.

Several other possible performance indices are listed here:[8]

$$\int_0^\infty e_n^2(t)\,dt = \text{integral square error (ISE) [used in Eq. (34.115) for } J_3] \tag{34.140}$$

$$\int_0^\infty t e_n^2(t)\,dt = \text{integral time-multiplied square error (ITSE)} \tag{34.141}$$

$$\int_0^\infty |e_n(t)|\,dt = \text{integral absolute error (IAE)} \tag{34.142a}$$

$$\int_0^\infty t|e_n(t)|\,dt = \text{integral of time multiplied by absolute error (ITAE)} \tag{34.142b}$$

$$\int_0^\infty [e_n^2(t) + \dot{e}_n^2(t)]\,dt = \text{integral of error squared plus error rate squared (IES + ERS)} \tag{34.143}$$

$$\int_0^\infty t[|e_n(t)| + |\dot{e}_n(t)|]\,dt = \text{integral of time-multiplied absolute error rate plus absolute error rate (ITAE + AER)} \tag{34.144}$$

$$\int_0^\infty \mathbf{x}_n'(t)\mathbf{Q}\mathbf{x}_n(t)\,dt = \text{quadratic performance index} \tag{34.145}$$

$\mathbf{x}_n'(t)$ = transpose of $\mathbf{x}_n(t)$. $\mathbf{x}_n(t)$ = system state vector. If the desired final value is the origin, then $\mathbf{x}(t)$

plays the role of the system error. $\mathbf{Q}$ = positive definite (or semidefinite) matrix

$$\int_0^\infty [\mathbf{x}_n'(t)\mathbf{Q}\mathbf{x}_n(t) + \mathbf{U}_m'(t)\mathbf{R}\mathbf{U}_m(t)]\,dt$$

$$= \text{quadratic performance index weighting system state vector } \mathbf{x}_n(t), \tag{34.146}$$

by positive definite (or semidefinite) matrix $\mathbf{Q}$, and weighting system control vector $\mathbf{U}_m(t)$ by positive definite (or semidefinite) matrix R.

For example, if $\mathbf{Q}$ and $\mathbf{R}$ chosen as diagonal matrices:

$$\mathbf{Q} = \begin{bmatrix} q_1 & 0 & 0 \\ 0 & q_2 & 0 \\ 0 & 0 & q_3 \end{bmatrix} \qquad \mathbf{R} = \begin{bmatrix} r_1 & 0 \\ 0 & r_2 \end{bmatrix} \qquad U_m'(t) = \text{transpose of control vector } \mathbf{U}_m(t)$$

then Eq. (34.146) becomes

$$\int_0^\infty \left[q_1 x_1^2(t) + q_2 x_2^2(t) + q_3 x_3^2(t) + r_1 U_1^2(t) + r_2 U_2^2(t)\right] dt \tag{34.147}$$

If the system error is defined as

$$e_n(t) = d_n(t) - x_n(t) \tag{34.148}$$

where $d(t)$ is the desired final state, and if the desired final state is the origin of state space $d_n(t) = 0$, then Eq. (34.148) becomes

$$e_n(t) = -x_n(t) \quad \text{and} \quad e_n^2(t) = x_n^2(t) \tag{34.149}$$

Thus in these circumstances the $x(t)$ in Eq. (34.148) becomes $e(t)$ and the quadratic performance index of Eq. (34.147) becomes identical in form with the original performance index J_3 given by Eq. (34.115), which was used in 1952 by Bryson.[21] The difference is one of generality. The quadratic forms $\mathbf{x}_n'\mathbf{Q}\mathbf{x}_n$ and $\mathbf{U}_m'\mathbf{R}\mathbf{U}_m$ are more general than x_n^2 and U_m^2. The quadratic forms are advantageous in that they allow the second method of Liapunov (see Section 34.9) to be applied to the design of an optimal controller. This controller will be stable in the sense of Liapunov. It is discussed here.

Optimization Using Second Method of Liapunov

Consider the system

$$\dot{\mathbf{x}}_n(t) = \mathbf{A}\mathbf{x}_n(t) \tag{34.150}$$

where A = stable matrix, in which all eigenvalues of A have negative real parts, and $x_n(\infty) = 0$ (i.e., the system is asymptotically stable).

The performance index to be minimized is the quadratic performance index of Eq. (34.145), namely

$$J = \int_0^\infty \mathbf{x}_n'(t)\mathbf{Q}\mathbf{x}_n(t)\,dt = \int_0^\infty \mathbf{x}'\mathbf{Q}\mathbf{x}\,dt \tag{34.151}$$

where $\mathbf{x}_n(t)$ has been replaced by $\mathbf{x}$ to simplify the following exposition.

A positive definite Liapunov function is selected:

$$V(x) = x'Px \tag{34.152}$$

Since it is postulated that the system $\dot{\mathbf{x}} = \mathbf{A}\mathbf{x}$ is asymptotically stable, $\dot{\mathbf{V}}(x)$ must be negative definite. Let us assume that a $\mathbf{Q}$ can be found, so that the negative definite Liapunov function $\dot{\mathbf{V}}(x)$ is

$$\dot{\mathbf{V}}(x) = -\mathbf{x}'\mathbf{Q}\mathbf{x} \tag{34.153}$$

where $\mathbf{x}'$ = transpose of $\mathbf{x}$.

The required relationship between $\mathbf{P}$ and $\mathbf{Q}$ is found next from Eqs. (34.152) and (34.153):

$$\dot{\mathbf{V}}(\mathbf{x}) = -\mathbf{x}'\mathbf{Q}\mathbf{x} = \frac{d}{dt}\mathbf{V}(\mathbf{x}) = \frac{d}{dt}(\mathbf{x}'\mathbf{P}\mathbf{x}) \tag{34.154}$$

Using the chain rule of differentiation, and noting that $\mathbf{P}$ is a matrix of constants

$$\left[\frac{d}{dt}(f_1 f_2) = \dot{f}_1 f_2 + f_1 \dot{f}_2\right], \qquad \left[\frac{d}{dt} f_1 = \dot{f}_1\right]$$

$$\frac{d}{dt}(\mathbf{x}'\mathbf{P}\mathbf{x}) = \dot{\mathbf{x}}'\mathbf{P}\mathbf{x} + \mathbf{x}'\mathbf{P}\dot{\mathbf{x}} \tag{34.155}$$

Combining Eqs. (34.154) and (34.155):

$$-\mathbf{x}'\mathbf{Q}\mathbf{x} = \dot{\mathbf{x}}'\mathbf{P}\mathbf{x} + \mathbf{x}'\mathbf{P}\dot{\mathbf{x}} \tag{34.156}$$

But, from Eq. (34.150), $\dot{\mathbf{x}} = \mathbf{A}\mathbf{x}$, and it is also recalled that the transpose of a matrix product is given by $(BC)' = C'B'$, or $\dot{\mathbf{x}}' = \mathbf{x}'\mathbf{A}'$.

Inserting these values into Eq. (34.156):

$$-\mathbf{x}'\mathbf{Q}\mathbf{x} = \mathbf{x}'\mathbf{A}'\mathbf{P}\mathbf{x} + \mathbf{x}'\mathbf{P}\mathbf{A}\mathbf{x} = \mathbf{x}'(\mathbf{A}'\mathbf{P} + \mathbf{P}\mathbf{A})\mathbf{x} \tag{34.157}$$

Thus equating the quadratic form matrices on the extreme ends of Eq. (34.157):

$$-\mathbf{Q} = \mathbf{A}'\mathbf{P} + \mathbf{P}\mathbf{A} \qquad \text{or} \qquad \mathbf{Q} = -[\mathbf{A}'\mathbf{P} + \mathbf{P}\mathbf{A}] \tag{34.158}$$

Thus the assumption that the system $\dot{\mathbf{x}} = \mathbf{A}\mathbf{x}$ is asymptotically stable enabled $\mathbf{Q}$ to be found in terms of $\mathbf{P}$ and $\mathbf{A}$ by utilizing Liapunov's second method [as given by Eq. (34.158)]. Furthermore, the performance index J of Eq. (34.151) is readily evaluated as follows. Utilizing Eq. (34.154), $-\mathbf{x}'\mathbf{Q}\mathbf{x} = d/dt(\mathbf{x}'\mathbf{P}\mathbf{x})$:

$$J = \int_0^\infty \mathbf{x}'\mathbf{Q}\mathbf{x}\,dt = \int_0^\infty -\frac{d}{dt}(\mathbf{x}'\mathbf{P}\mathbf{x})\,dt = -\mathbf{x}'\mathbf{P}\mathbf{x}|_0^\infty$$

or

$$J = -\mathbf{x}'(\infty)\mathbf{P}\mathbf{x}(\infty) - -\mathbf{x}'(0)\mathbf{P}\mathbf{x}(0)$$

Since the system is stable and $\mathbf{x}(\infty) \to 0$ as postulated, the term $-\mathbf{x}'(\infty)\mathbf{P}\mathbf{x}(\infty) \to 0$, and

$$J = +\mathbf{x}'(0)\mathbf{P}\mathbf{x}(0) \tag{34.159}$$

Thus the performance index J is given by $\mathbf{x}'(0)\mathbf{P}\mathbf{x}(0)$. If it is desired to minimize J with respect to some variable parameter contained in the state variable matrix $\mathbf{A}$, then one minimizes $\mathbf{x}'(0)\mathbf{P}\mathbf{x}(0)$ with respect to the variable parameter instead of minimizing an elaborate integral for J. $\mathbf{P}$ is found from $\mathbf{A}$ and $\mathbf{Q}$ (which are given quantities) by evaluating Eq. (34.158), $\mathbf{Q} = -[\mathbf{A}'\mathbf{P} + \mathbf{P}\mathbf{A}]$.

In this homogeneous problem, there is no input driving function. The system responds in accordance with its initial conditions, $\mathbf{x}(0)$, which can be considered the input, and where the final value is $\mathbf{x}(\infty) \to 0$. If a system has an input, this approach can be readily applied by defining $\mathbf{x}$ as the system error, which must go to zero for this technique to be applicable.

The following example illustrates the procedure.

Example 34.18[6] Optimizing Damping Coefficient for Second-Order System.

The second-order system is shown in Fig. 34.51. It is a type 1 system with a single factor of s in the denominator of $G(s)$, and hence the system has zero position error. Thus the error may be identified with the state variables and satisfy the condition $x_n(\infty) \to 0$ utilized in the derivation of the performance index J in Eq. (34.159). From Fig. 34.51:

$$\frac{C(s)}{R(s)} = \frac{G}{1+GH} = \frac{G}{1+G} = \frac{1}{s(s+2d)+1} = \frac{1}{s^2+2ds+1} \tag{34.160}$$

where d is the damping constant of the normalized second-order characteristic equation $s^2 + 2d\omega_n s + \omega_n^2$, where $\omega_n = 1$ and $d > 0$.

The problem is to find the optimal value of d that minimizes the performance index J, when the system is subject to a unit step input $r(t) = 1$, $t > 0$.

$$J = \int_0^\infty \mathbf{x}'\mathbf{Q}\mathbf{x}\,dt \qquad \text{[given in Eq. (34.159)]} \tag{34.161}$$

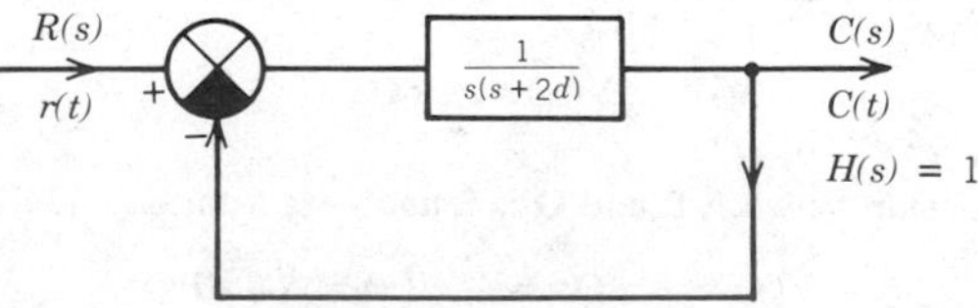

Fig. 34.51 Second-order system with variable damping d.

$$\frac{C(s)}{R(s)} = \frac{G}{1+GH} = \frac{G}{1+G} = \frac{1}{s(s+2d)+1} = \frac{1}{s^2+2ds+1}$$

$$C(s)[s^2 + 2sd + 1] = R(s)$$

Taking the inverse Laplace transform:

$$\ddot{c}(t) + 2d\dot{c}(t) + \dot{c}(t) = r(t)$$

where

$$x = x_n(t) = \begin{bmatrix} x_1 \\ x_2 \end{bmatrix} = \begin{bmatrix} x \\ \dot{x} \end{bmatrix} = \begin{bmatrix} e \\ \dot{e} \end{bmatrix}$$

and

$$Q = \begin{bmatrix} 1 & 0 \\ 0 & 1 \end{bmatrix} \qquad \text{(given)}$$

It is now necessary to find the state-variable equation $\dot{\mathbf{x}} = \mathbf{Ax}$. From Eq. (34.160) (by cross multiplying):

$$C(s)[s^2 + 2ds + 1] = R(s) \tag{34.162}$$

Taking the inverse Laplace transform:

$$\ddot{c}(t) + 2d\dot{c}(t) + c(t) = r(t) \tag{34.163}$$

From the initial discussion, the state variable $x(t)$ is to be chosen as the system error, $e(t) = x(t) = r(t) - c(t)$. Thus

$$\begin{aligned} x(t) &= r(t) - c(t) \\ \dot{x}(t) &= 0 - \dot{c}(t) \qquad \text{since } r(t) = 1 \text{ for } t > 0 \\ \ddot{x}(t) &= 0 - \ddot{c}(t) \end{aligned} \tag{34.164}$$

Inserting these state variables into Eq. (34.163):

$$-\ddot{x}(t) - 2d\dot{x}(t) - x(t) = 0$$

or

$$\ddot{x}(t) + 2d\dot{x}(t) + x(t) = 0 \tag{34.165}$$

Note that Eq. (34.165) is in the required homogeneous form, with no input function $r(t)$ appearing. The input has been incorporated into the initial conditions of $x_n(t)$, which are

$$\begin{aligned} x(t)|_{t=0+} &= r(0+) - c(0+) = 1 - 0 = 1 \\ \dot{x}(t)|_{t=0+} &= -\dot{c}(0+) = 0 \end{aligned} \tag{34.166}$$

It is assumed in this example that the system starts at rest, so that $c(0+) = \dot{c}(0+) = 0$. Equations (34.165) and (34.166) may now be put in standard state-variable form by defining

$$\begin{aligned} x_1 &= x(t) \\ x_2 &= \dot{x}(t) \end{aligned} \tag{34.167}$$

Then, from Eqs. (34.167) and (34.165):

$$\begin{aligned} \dot{x}_1 &= \dot{x}(t) = x_2 \\ \dot{x}_2 &= \ddot{x}(t) = -2d\dot{x}(t) - x(t) = -2dx_2 - x_1 \end{aligned} \tag{34.168}$$

or in matrix form:

$$\begin{bmatrix} \dot{x}_1 \\ \dot{x}_2 \end{bmatrix} = \begin{bmatrix} 0 & 1 \\ -1 & -2d \end{bmatrix} \begin{bmatrix} x_1 \\ x_2 \end{bmatrix} \quad \text{and} \quad \begin{bmatrix} x_1(0) \\ x_2(0) \end{bmatrix} = \begin{bmatrix} 1 \\ 0 \end{bmatrix} \tag{34.169}$$

Equivalently

$$\dot{\mathbf{x}} = \mathbf{Ax} \qquad \text{where } \mathbf{A} = \begin{bmatrix} 0 & 1 \\ -1 & -2d \end{bmatrix} \tag{34.170}$$

Since Eq. (34.170) is homogeneous and $\mathbf{A}$ is a stable matrix (e.g., $x_n(\infty) \to 0$), the value of J is given by Eq. (34.159), repeated below,

$$J = \mathbf{x}'(0)\mathbf{P}\mathbf{x}(0) \tag{34.159}$$

where $\mathbf{P}$ is given by (34.158), repeated below:

$$\mathbf{Q} = -[\mathbf{A}'\mathbf{P} + \mathbf{PA}] \qquad \text{or} \qquad \mathbf{A}'\mathbf{P} + \mathbf{PA} = -\mathbf{Q} \tag{34.158}$$

Inserting the known values of $\mathbf{A}$ and $\mathbf{Q}$ into Eq. (34.158):

$$\begin{bmatrix} 0 & -1 \\ 1 & -2d \end{bmatrix} \begin{bmatrix} P_{11} & P_{12} \\ P_{12} & P_{22} \end{bmatrix} + \begin{bmatrix} P_{11} & P_{12} \\ P_{12} & P_{22} \end{bmatrix} \begin{bmatrix} 0 & 1 \\ -1 & -2d \end{bmatrix} = \begin{bmatrix} -1 & 0 \\ 0 & -1 \end{bmatrix} \tag{34.171}$$

Since $\mathbf{P}$ is the matrix of a quadratic form, $\mathbf{P}$ is chosen to be symmetric, with $P_{12} = P_{21}$.

Multiplying Eq. (34.171) out yields

$$\left[\begin{array}{c|c} -P_{12} & -P_{22} \\ \hline P_{11}-2dP_{12} & P_{12}-2dP_{22} \end{array}\right] + \left[\begin{array}{c|c} -P_{12} & P_{11}-2dP_{12} \\ \hline -P_{22} & P_{12}-2dP_{22} \end{array}\right] = \left[\begin{array}{c|c} -1 & 0 \\ \hline 0 & -1 \end{array}\right] \tag{34.172}$$

Equating the "corners" or the elements of Eq. (34.172):

$$-P_{12}+-P_{12}=-1 \tag{34.173a}$$

$$-P_{22}+P_{11}-2dP_{12}=0 \tag{34.173b}$$

$$P_{11}-2dP_{12}-P_{22}=0 \tag{34.173c}$$

$$P_{12}-2dP_{22}+P_{12}-2dP_{22}=-1 \tag{34.173d}$$

Equations (34.173b and c) are identical. Hence only three independent equations result, which is a consequence of setting $P_{12}=P_{21}$. Solving Eq. (34.173a) yields

$$P_{12}=\tfrac{1}{2} \tag{34.174a}$$

Inserting $P_{12}=\frac{1}{2}$ into Eq. (34.173d) yields

$$\tfrac{1}{2}-2dP_{22}+\tfrac{1}{2}-2dP_{22}=-1 \quad \text{or} \quad -4dP_{22}=-2 \quad \text{or} \quad P_{22}=\frac{1}{2d} \tag{34.174b}$$

Inserting $P_{12}=\frac{1}{2}$ and $P_{22}=\frac{1}{2}d$ into Eq. (34.173b) yields

$$-\frac{1}{2d}+P_{11}-2d\frac{1}{2}=0 \quad \text{or} \quad P_{11}=d+\frac{1}{2d} \quad \text{or} \quad P_{11}=\frac{2d^2+1}{2d} \tag{34.174c}$$

Therefore the result for the P matrix is

$$\mathbf{P}=\begin{bmatrix} \dfrac{2d^2+1}{2d} & \dfrac{1}{2} \\ \dfrac{1}{2} & \dfrac{1}{2d} \end{bmatrix} \tag{34.175}$$

The performance index $J=\mathbf{x}'(0)\mathbf{P}\mathbf{x}(0)$ is

$$J=[1 \quad 0]\begin{bmatrix} \dfrac{2d^2+1}{2d} & \dfrac{1}{2} \\ \dfrac{1}{2} & \dfrac{1}{2d} \end{bmatrix}\begin{bmatrix} 1 \\ 0 \end{bmatrix} \tag{34.176}$$

or

$$J=[1 \quad 0]\begin{bmatrix} \dfrac{2d^2+1}{2d} \\ \dfrac{1}{2} \end{bmatrix}=\frac{2d^2+1}{2d}$$

To minimize the performance index J with respect to the damping ratio d, set $\partial J/\partial d=0$

$$\frac{\partial J}{\partial d}=\frac{2d(4d)-(2d^2+1)2}{(2d)^2}=0 \tag{34.177}$$

(The derivative was evaluated using $(U/V)'=(VU'-UV')/V^2$, where $U'=\partial U/\partial d$, $V'=\partial V/\partial d$.)

Solving for d:

$$8d^2-4d^2-2=0 \quad \text{or} \quad 4d^2=2 \quad \text{or} \quad d=\frac{1}{\sqrt{2}}=0.707 \tag{34.178}$$

Thus the optimal value of the damping coefficient d for the given normalized second-order system and given performance index $\mathbf{Q}$ matrix is $d=0.707$, a value used well before the advent of optimal control. The value of the damping ratio depends on the weighting of x_1 and x_2, the error and error rate, as specified by the $\mathbf{Q}$ matrix. If only the error is weighted, with the error rate totally ignored, so that $\mathbf{Q}$ is chosen as

$$\mathbf{Q}_1=\begin{bmatrix} 1 & 0 \\ 0 & 0 \end{bmatrix} \tag{34.179}$$

then the optimal solution for the damping ratio d becomes

$$d_1=\tfrac{1}{2} \tag{34.180}$$

TABLE 34.29. OPTIMAL DAMPING RATIO d VS. WEIGHTING OF ERROR AND ERROR RATE (K_1 = ERROR WEIGHT, K_2 = ERROR RATE WEIGHT) FOR SECOND-ORDER SYSTEM OF FIG. 34.51

K_2/K_1	d	Significance of d
0	0.5	Minimum value of d that can be optimal
1	0.707	Classical optimal damping
3	1	Critical damping
>3	>1	Overdamped, slow response

It is possible to repeat the solution for the damping ratio d when arbitrary weights K_1 and K_2 are applied to the error and error rate, so that Q is chosen as Q_{12}, where

$$Q_{12} = \begin{bmatrix} K_1 & 0 \\ 0 & K_2 \end{bmatrix} \tag{34.181}$$

The result for d is d_{12}, where

$$d_{12} = \sqrt{\frac{K_1 + K_2}{4K_1}} = \frac{1}{2}\sqrt{1 + K_2/K_1} \tag{34.182}$$

where K_1 = error weight and K_2 = error rate weight.

The following observations concerning the damping ratio d of the normalized second-order system shown in Fig. 34.51 may be drawn from Eq. (34.182) which gives the optimal value of d that minimizes the quadratic performance index:

$$J = \int_0^\infty [e \quad \dot{e}] \begin{bmatrix} K_1 & 0 \\ 0 & K_2 \end{bmatrix} \begin{bmatrix} e \\ \dot{e} \end{bmatrix} dt \tag{34.183}$$

where e = error = X_1 and $\dot{e}$ = error rate = x_2.

Observations From Eq. (34.182)

1. The minimum value of the damping ratio d is 0.5, which occurs for $K_2/K_1 = 0$ or for no weight ($K_2 = 0$) applied to the error rate.

2. Optimal damping of $d = 0.707$ occurs for $K_2/K_1 = 1$, or for equal weighting of the error and error weight.

3. Critical damping of $d = 1$ occurs for $K_2/K_1 = 3$, or for three times as much weight applied to the error rate as to the error.

4. If error rate is weighted by more than three times the error weight, the optimal system becomes overdamped, and the damping ratio d exceeds unity.

These results are summarized in Table 34.29.

34.12 CALCULUS OF VARIATIONS

The calculus of variations is as old as calculus, dating back to Newton and 17th century Western Europe. Newton's famous calculus-of-variations problem was the determination of the shape of a solid of revolution having the least resistance to motion when moved through a fluid. This problem is of great interest to the designers of submarines and aircraft and is still being studied three centuries after Newton.

The calculus of variations deals with the determination of the form, shape, or value of an unknown quantity f, so that some integral may assume a maximum or minimum value J.[22]

Note that this is exactly the optimal control problem, where the integral to be minimized is the performance index J (also called a functional), and the quantity to be determined as a consequence of minimizing J is the optimal system control input U (which is the unknown quantity f). The names of the principal contributors to the calculus of variations reads like a "Who's Who" of mathematicians and includes Newton, Bernoulli, Euler, Lagrange, Legendre, Jacobi, and Weirstrass. The method of Euler will be discussed since it represents the original backbone of the calculus of variations. Only the first variation, which utilizes first derivatives, will be dealt with (in an example). The second variation, which utilizes second derivatives, will be developed but not discussed at length.

The term "variation" refers to the varying value of the integral J, as the function being integrated, f, is changed. The simplest general problem of the calculus of variations is to minimize (or maximize) the integral J_1:

$$J_1 = \int f(x, y, y')\, dx \tag{34.184}$$

where $y' = dy/dx$.

J is varied by letting y increase by an infinitesimal amount of KV, a procedure used by Lagrange, resulting in

$$y + KV \tag{34.185}$$

where K = infinitesimal arbitrary constant and V = any regular uniform function of x, say $V = h(x)$, with regular derivatives within the range of integration.

The varied integral J_2 is thus

$$J_2 = \int f(x, y + KV, y' + KV')\, dx \tag{34.186}$$

where $y' + KV' = dy/dx + K(dV/dx)$ and K is a constant and $V = h(x)$.

The change, or difference, or variation between J_2 and J_1 is

$$\Delta J = J_2 - J_1 \tag{34.187}$$

Substituting Eqs. (34.184) and (34.186) into Eq. (34.187) we get

$$\Delta J = \int [f(x, y + KV, y' + KV') - f(x, y, y')]\, dx \tag{34.188}$$

Expanding $f(x, y + KV, y' + KV')$ in a Taylor series about the point x, y, y', in powers of K, the result for a single variable expansion about x_0 is

$$F(x_0 + h) = F(x_0) + hf'(x_0) + \frac{h^2}{2!} f''(x_0) + \ldots \tag{34.189}$$

where

$$f'(x_0) = \frac{d}{dx} f(x_0) \qquad \text{and} \qquad f''(x_0) = \frac{d^2}{dx^2} f(x_0) \ldots$$

Equation (34.188) has two terms containing the increments KV and KV' of the two variables y and y'. Thus a Taylor series in two variables is necessary, utilizing partial derivatives with respect to the two variables y and y'. The principal difference, other than doubling the number of terms, is that when two variables are used cross partial derivative terms (of the form $\partial^2/\partial y\, \partial y'$) appear.

Performing the Taylor series expansion:

$$\begin{aligned} f(x, y + KV, y' + KV') &= f(x, y, y') + KV \frac{\partial}{\partial y} f(x, y, y') + KV' \frac{\partial}{\partial y'} f(x, y, y') \\ &\quad + \frac{K^2V^2}{2!} \frac{\partial^2}{\partial y^2} f(x, y, y') + \frac{K^2(V')^2}{2!} \frac{\partial}{\partial y'} f(x, y, y') \\ &\quad + (KV)(KV') \frac{\partial^2}{\partial y\, \partial y'} f(x, y, y') + \text{higher order terms} \end{aligned} \tag{34.190}$$

When Eq. (34.190) is inserted back into ΔJ [Eq. (34.188)], the terms $f(x, y, y')$ cancel out and ΔJ may be written as

$$\Delta J = KI_1 + \tfrac{1}{2}K^2 I_2 + R_3 \tag{34.191}$$

where $I_1 = \displaystyle\int \left(V \frac{\partial f}{\partial y} + V' \frac{\partial f}{\partial y'} \right) dx$ = first variation

$I_2 = \displaystyle\int \left[V^2 \frac{\partial^2 f}{\partial y^2} + 2VV' \frac{\partial^2 f}{yy'} + (V')^2 \frac{\partial^2 f}{\partial y'^2} \right] dx$ = second variation

R_3 = remainder terms of third and higher orders

$f = f(x, y, y')$

Equation (34.191) is obtained directly from the two variable Taylor expansions of Eq. (34.190). The first term is canceled out. Next, by simply factoring out the K and K^2 terms, the first-order derivatives are collected into one integral (the first-variation integral), and the remaining second-order derivatives are collected into another integral (the second variation).

The first-variation integral is now modified by integrating by parts. The basic rule for integrating by parts is

$$\int_a^b U\,dV = UV|_a^b - \int_a^b V\,dU \tag{34.192a}$$

The second term of the first-variation integral, Eq. (34.191), is

$$\int V' \frac{\partial f}{\partial y'}\,dx \tag{34.192b}$$

Let

$$U = \frac{\partial f}{\partial y'}$$

$$dV = V'\,dx = \frac{dV}{dx}\,dx = dV$$

Then

$$dU = \frac{d}{dx}\,U\,dx = \frac{d}{dx}\left(\frac{\partial f}{\partial y'}\right)dx$$

$$V = V$$

Substituting these values into the integral by parts results in

$$\int_a^b V' \frac{\partial f}{\partial y'}\,dx = \frac{\partial f}{\partial y'}\,V\bigg|_a^b - \int_a^b V \frac{d}{dx}\left(\frac{\partial f}{\partial y'}\right)dx \tag{34.193}$$

V represents the variation in the variable y, where y can vary everywhere except at its fixed end points or limits a and b. At these end points, y is fixed and cannot vary; hence $y + KV = y + 0$ and $V = 0$ at the end points. Thus

$$\frac{\partial f}{\partial y'}\,V\bigg|_a^b = 0 \cdot \frac{\partial f}{\partial y'}\bigg|_b - 0 \cdot \frac{\partial f}{\partial y'}\bigg|_a = 0 \tag{34.194}$$

Then Eq. (34.193) becomes

$$\int V' \frac{\partial f}{\partial y'}\,dx = -\int V \frac{d}{dx}\left(\frac{\partial f}{\partial y'}\right)dx \tag{34.195}$$

When this new value for the second term of the first-variation integral I_1 is inserted back into Eq. (34.191) for I_1, the result after factoring out V is

$$I_1 = \int V\left\{\frac{\partial f}{\partial y} - \frac{d}{dx}\left(\frac{\partial f}{\partial y'}\right)\right\}dx \tag{34.196}$$

Returning to Eq. (34.191) for the total variation $\Delta J = KI_1 + \frac{1}{2}K^2 I_2 + R_3$, note that since K is an infinitesimal, K^2 is negligible compared with K. The first variation I_1 dominates ΔJ, so that $\Delta J \simeq KI_1$. If K changes sign, ΔJ changes sign (e.g., one variation would increase ΔJ and another would decrease ΔJ). But a maximum of ΔJ is characterized by a decrease for all possible variations, and a minimum of ΔJ is characterized by an increase for all possible variations. Consequently I_1 must be zero for ΔJ to be a minimum or a maximum. This result, $I_1 = 0$, is

$$I_1 = 0 = \int V\left\{\frac{\partial f}{\partial y} - \frac{d}{dx}\left(\frac{\partial f}{\partial y'}\right)\right\}dx \tag{34.197}$$

V may be any arbitrary regular function of x and may be positive whenever $[\partial f/\partial y - (d/dx)(\partial f/\partial y')]$ is positive and negative whenever that term is negative. Thus the product of V and $\{\partial f/\partial y - (d/dx)(\partial f/\partial y')\}$ could always be positive, and hence the integral could not vanish (unless the bracketed term was everywhere identically zero). The final result is that the original integral is minimized, or maximized, if the first variation I_1 is zero, which occurs if

$$\frac{\partial f}{\partial y} - \frac{d}{dx}\left(\frac{\partial f}{\partial y'}\right) = 0 \tag{34.198}$$

where $f = f(x, y, y')$; $y' = dy/dx$.

This equation was first derived by Euler in 1744 and is often called the characteristic equation, or sometimes Euler's equation (along with various other equations he derived). It represents the first of several essential tests to be satisfied if an integral of the form of Eq. (34.184) is to have a maximum or a minimum. Euler's calculus-of-variations equation (the characteristic equation) can be simplified if the integral $f(x, y, y')$ does not contain all three possible variables x, y, and y'.

Case I No y Variable. $f(x, y, y')$ becomes $f(x, y')$. Then $\partial f/\partial y = 0$, and inserting this into Eq. (34.198) yields

$$\frac{d}{dx}\left(\frac{\partial f}{\partial y'}\right) = 0 \tag{34.199}$$

Since the derivative of a constant is zero, $(d/dx)(K_1) = 0$, the final result is

$$\frac{\partial f}{\partial y'} = K_1 \tag{34.200}$$

where K_1 = arbitrary constant
$f(x, y, y') = f(x, y')$ (no y variable)
$y' = \dfrac{dy}{dx}$

Case II No x Variable. $f(x, y, y')$ becomes $f(y, y')$. Then, taking the total partial derivative of $f(y, y')$ with respect to x:

$$\frac{\partial f}{\partial x} = \frac{\partial}{\partial y} f(y, y') \frac{dy}{dx} + \frac{\partial}{\partial y'} f(y, y') \frac{dy'}{dx}$$

or

$$\frac{\partial f}{\partial x} = \frac{\partial f}{\partial y} y' + \frac{\partial f}{\partial y'} y''$$

where $dy'/dx = (d/dx)(dy/dx) = d^2y/dx^2 = y''$.

From Eq. (34.198):

$$\frac{\partial f}{\partial y} = \frac{d}{dx}\left(\frac{\partial f}{\partial y'}\right)$$

Substituting the expression for $\partial f/\partial y$ into the preceding for $\partial f/\partial x$:

$$\frac{\partial f}{\partial x} = \frac{d}{dx}\left(\frac{\partial f}{\partial y'}\right) y' + \frac{\partial f}{\partial y'} y''$$

Integrating with respect to x, and noting that all terms are constant with respect to x except $d/dx(\partial f/\partial y')$, results in

$$f = \frac{\partial f}{\partial y'} y' + K_2 \tag{34.201}$$

where K_2 = arbitrary constant, the result of $\int(\partial f/\partial y')y''\,dx = K_2$ between constant x limits
$f(x, y, y') = f(y, y')$ (no x variable)
$y' = \dfrac{dy}{dx}$

These results of the calculus of variations will now be employed in an illustrative example. An arbitrary curve, lying in a plane, connects points a and b. The calculus of variations will be employed to find the equation of the curve having the shortest distance between a and b. Although the problem is comparatively simple and the straight-line result is well known, the procedure illustrates the application of the calculus of variations in a comprehensible manner. Other examples are given by Forsyth[23] and Bliss.[24]

Example 34.19 Calculus of Variations. Find the equation of the curve, lying in one plane, that represents the shortest distance between two points.

The two points, a and b, and the arbitrary curve joining them are shown on an $x - y$ coordinate system in Fig. 34.52.

From the figure, the increment ds in arc length along the arbitrary curve connecting a and b is

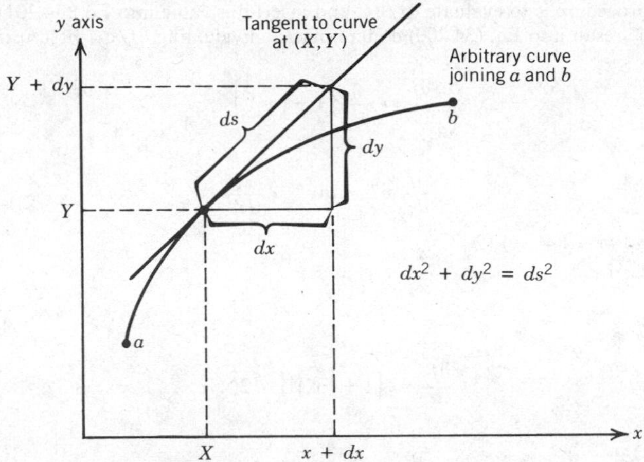

Fig. 34.52 Calculus of variations, Example 34.19, to find the shortest distance curve joining points a and b. $ds = \sqrt{dx^2 + dy^2}$ = approximate arc length increment, for an increment of x equaling dx.

given by

$$ds = [dx^2 + dy^2]^{\frac{1}{2}} \tag{34.202}$$

Factoring out dx^2:

$$ds = \left[\left(1 + \frac{dy^2}{dx^2}\right)\right]^{\frac{1}{2}} dx \tag{34.203}$$

Denoting dy/dx by y':

$$ds = [1 + (y')^2]^{\frac{1}{2}} dx \tag{34.204}$$

Integrating ds to get the total distance between points a and b:

$$\int_a^b ds = \text{distance} = \int_a^b [1 + (y')^2]^{\frac{1}{2}} dx = J = \int f\, dx \tag{34.205}$$

To find the shortest distance curve between points a and b, the integral J must be minimized. This minimization can be achieved by utilizing the calculus of variations, and in particular Euler's equation as given by Eq. (34.198):

$$\frac{\partial f}{\partial y} - \frac{d}{dx}\left(\frac{\partial f}{\partial y'}\right) = 0 \tag{34.198}$$

where $f = f(x, y, y') = [1 + (y')^2]^{\frac{1}{2}}$ = integrand and $y' = dy/dx$. Note that in this particular case the integrand f is only a function of dy/dx, or y'.

The integrand f is not a function of x, nor is it a function of y. Thus, as discussed in the development of Euler's equation, Euler's equation simplifies to the two special cases given by Eqs. (34.200) and (34.201):

$$\frac{\partial f}{\partial y'} = K_1 \qquad (\text{no } y \text{ appears in } f) \tag{34.200}$$

or

$$f(x, y, y') \to f(x, y') \qquad \text{where} \quad y' = dy/dx$$

$$f = y' \frac{\partial f}{\partial y'} + K_2 \qquad (\text{no } x \text{ appears in } f) \tag{34.201}$$

or

$$f(x, y, y') \to f(y, y') \qquad \text{where} \quad y' = dy/dx$$

The solution procedure is to evaluate $\partial f/\partial y'$ and insert this value into Eq. (34.201) containing K_2, and then insert the result into Eq. (34.200) containing K_1. Evaluating $\partial f/\partial y'$ first, using f as specified in Eq. (34.205)

$$\frac{\partial f}{\partial y'} = \frac{\partial}{\partial y'}[1 + (y')^2]^{\frac{1}{2}}$$

Recalling

$$\frac{d}{dy'}x^m = mx^{m-1}\frac{dx}{dy'}$$

where $m = 1/2$ and $m - 1 = -1/2$

$$x = [1 + (y')^2]$$

$$\frac{dx}{dy'} = 2(y')$$

or

$$\frac{\partial f}{\partial y'} = \tfrac{1}{2}[1 + (y')^2]^{-\frac{1}{2}}2(y')$$

or

$$\frac{\partial f}{\partial y'} = y'[1 + (y')^2]^{-\frac{1}{2}} \tag{34.206}$$

Inserting this result into Eq. (34.201), along with the value of $f = [1 + (y')^2]^{\frac{1}{2}}$

$$f = y'\frac{\partial f}{\partial y'} + K_2$$

$$f = [1 + (y')^2]^{\frac{1}{2}} = y'y'[1 + (y')^2]^{-\frac{1}{2}} + K_2$$

Multiplying through by $[1 + (y')^2]^{\frac{1}{2}}$:

$$[1 + (y')^2] = (y')^2 + K_2[1 + (y')^2]^{\frac{1}{2}}$$

Canceling the $(y')^2$ terms:

$$1 = K_2[1 + (y')^2]^{\frac{1}{2}}$$

Solving for K_2:

$$K_2 = [1 + (y')^2]^{-\frac{1}{2}} = \text{constant} \tag{34.207}$$

Now if $[1 + (y')^2]^{-\frac{1}{2}}$ is constant, then $[1 + (y')^2]$ must be constant and y' must be constant. Let us represent the last constant by m. Then

$$y' = m$$

or

$$\frac{dy}{dx} = m \tag{34.208}$$

or

$$dy = m\,dx$$

or

$$\int dy = \int m\,dx$$

or

$$y = mx + C \tag{34.209}$$

where C = constant of integration and $m = y' = dy/dx$.

Equation (34.209), the equation of a straight line of slope m and y intercept C, is the desired result. It can also be obtained by substituting Eq. (34.206) for $\partial f/\partial y'$ into Eq. (34.200), namely $\partial f/\partial y' = K_1$. The result is

$$\frac{\partial f}{\partial y'} = y'[1 + (y')^2]^{-\frac{1}{2}} = K_1 \tag{34.210}$$

From Eq. (34.207), $[1 + (y')^2]^{-\frac{1}{2}} = K_2$. Substituting this result into Eq. (34.210):

$$y'K_2 = K_1 \tag{34.211}$$

Therefore y' is a constant, as was shown previously, and the same final result is obtained: The equation of the minimum distance curve connecting two points in a plane is the straight line

$$y = mx + C \tag{34.212}$$

where $m = dy/dx$ = slope and C = constant = y intercept.

34.13 DYNAMIC PROGRAMMING

Dynamic programming, one of the most useful optimization techniques available, was principally developed by Bellman in the 1950s.[25] The central concept of dynamic programming involves a "multistage decision process"[26] whereby a complex problem is broken up into n individual stages. Each stage involves an independent decision process, whose results are used in the next stage. Thus the solution progresses by a succession of decision processes, each one building on its predecessors. Each decision process for a given problem is related to or "embedded" in a class of similar problems, which grow in complexity as n increases.

An important constraint on the type of linear or nonlinear problems that can be solved by dynamic programming relates to the separability or independence of the individual decision processes, a requirement described by the term "Markovian process." A Markovian process has the property that after any number of decisions, each of which affects the state of the system, the effects of the remaining decisions depend solely on the state of the system at the time those decisions are made. Thus past decisions have no effect on future decisions aside from determining the present state of the system. Note that many dynamic programming problems are solved in a backward manner, so that the meaning of past and future with respect to real time can be reversed without the independence of the decisions constituting the Markovian process being changed.

Bellman described this independent decision process as his "principle of optimality," which states that an optimal policy has the property that whatever the initial state and decision are, the remaining decisions must constitute an optimal policy with regard to the state resulting from the first decision.

Using this step-by-step decision process, or optimal policy, dynamic programming produces "recurrence relations" which relate one optimal solution to the next, and so on.

Figure 34.53 illustrates how an optimal trajectory or decision process is made up of segments that are also optimal trajectories. Thus if the path $ABCDE$ shown in the figure is optimal, then the optimal path between AB is simply a truncated segment of $ABCDE$, and not some alternate nonoptimal path, such as path AXB (shown in the figure).

An example of a four-stage discrete dynamic programming problem is given by the four steps of Fig. 34.54, where the optimal path is found by a backward solution process. For this problem, with unique initial and final states, a forward solution may also be used, as shown in Fig. 34.55. Note that in step 3 of Fig. 34.54, the paths 2-5-6 and 3-4-6 were eliminated, so that future cost calculations involving these two paths were also eliminated. This is the advantage of dynamic programming—not

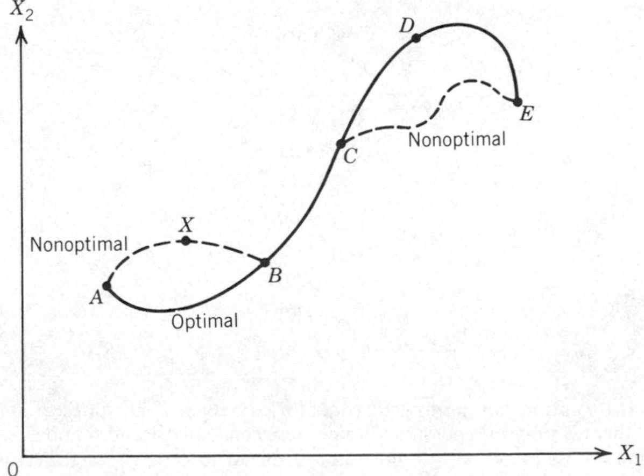

Fig. 34.53 Optimal trajectories in state space. The optimal trajectories here (assuming $ABCDE$ has been shown to be optimal) are (1) AB, (2) BC, (3) CD, (4) DE, (5) ABC, (6) BCD, (7) CDE, (8) $ABCD$, (9) $BCDE$. The nonoptimal trajectories are (1) AXB in place of AB, (2) CE in place of CDE, (3) $AXBCE$ in place of $ABCDE$.

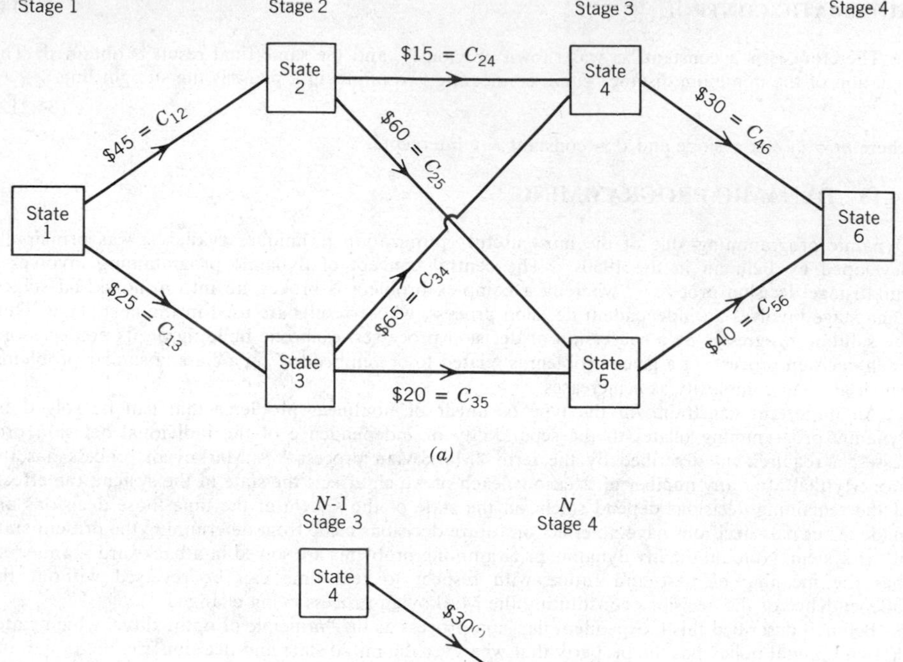

Fig. 34.54 Dynamic programming example. (*a*) Step 1. State transition costs. Costs might be air fares between geographic states, and mathematical states might be geographic states. The problem is to find the lowest cost route or trajectory from state 1 (stage 1) to state 6 (stage 4).

Cost Table

C_{ij}	Cost
C_{12}	$\rightarrow 45$
C_{13}	$\rightarrow 25$
C_{24}	$\rightarrow 15$
C_{25}	$\rightarrow 60$
C_{34}	$\rightarrow 65$
C_{35}	$\rightarrow 20$
C_{46}	$\rightarrow 30$
C_{56}	$\rightarrow 40$

(*b*) Step 2. Backward solution minimum cost trajectories between the states of stage $N - 1 = 3$ and stage $N = 4$. Since there is only one possible choice between states 4 and 6 and 5 and 6, that choice is taken as the minimum cost route $^{(*)}$(= total minimum cost to state 6), or trajectory.

Cost Table

Trajectory	Cost
4-6	30^{(*)}$
5-6	40^{(*)}$

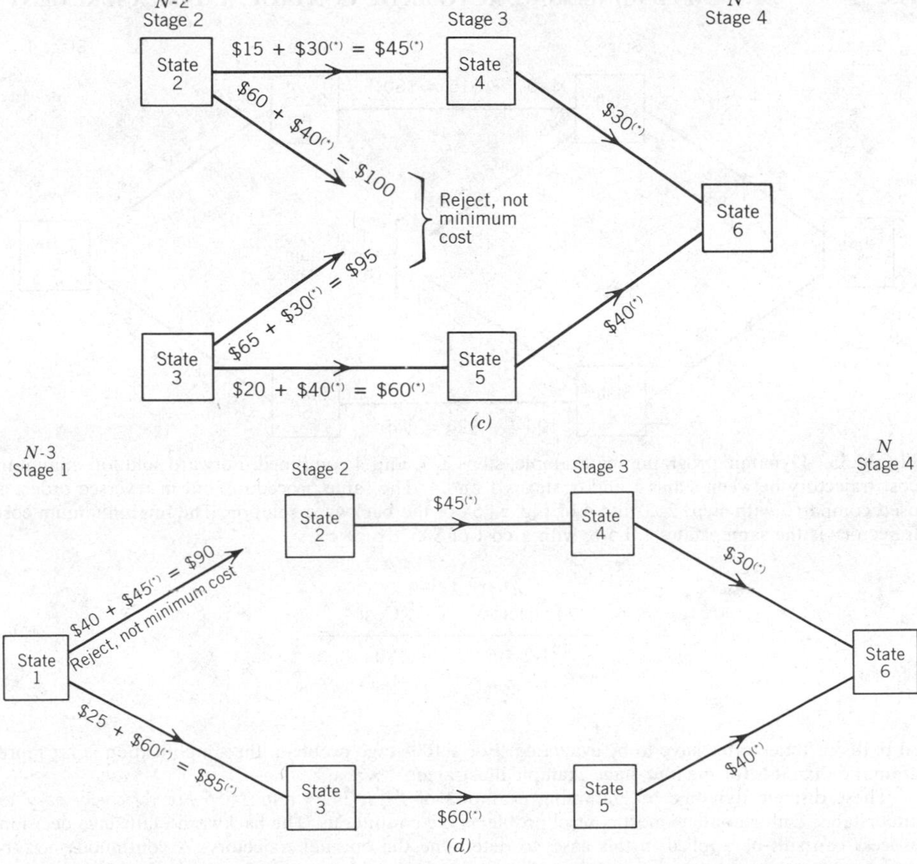

(c)

(d)

(*c*) Step 3. Backward solution minimum total cost trajectories between the state of stage $N - 2 = 2$ and stage $N = 4$. The minimum cost values [$30^{(*)}$ and $40^{(*)}$] [$^{(*)}$ = total minimum cost to state 6] of the previous decision stage are used to calculate the total costs from states 2 to 6 and 3 to 6. From the four results, the two minimum cost paths are selected.

Cost Table

Trajectory	Cost
2-4-6	\$ 45$^{(*)}$
2-5-6	\$100
3-4-5	\$ 95
3-5-6	\$60$^{(*)}$

(*d*) Step 4. Backward solution minimum total cost trajectory between stage $1 = N - 3$ and stage $4 = N$. Note that the trajectories 1-2-5-6 and 1-3-4-6 never had to be evaluated, since paths 2-5-6 and 3-4-6 had been eliminated in step 3. The minimum cost values [$45^{(*)}$ and $60^{(*)}$] of the previous decision stage are used to calculate the total cost from stages 1 to 6. From the two results, the minimum is chosen. The final minimum cost result is trajectory 1-3-5-6.

Cost Table

Trajectory	Cost
1-2-4-6	\$90
1-3-5-6	\$85$^{(*)}$

$^{(*)}$ = total minimum cost to state 6.

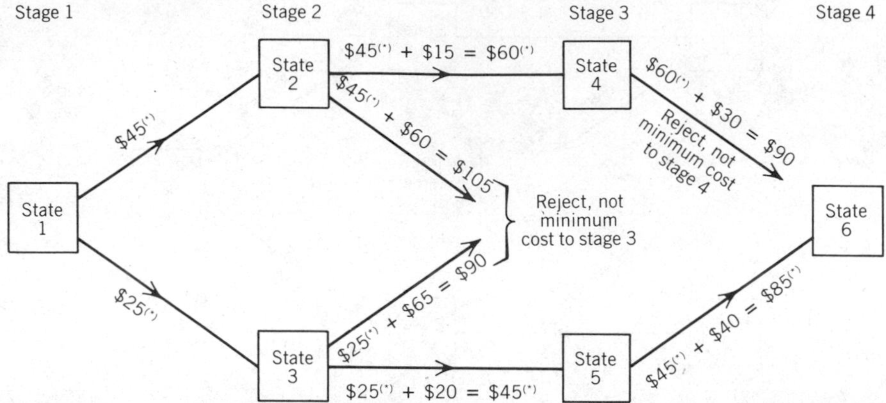

Fig. 34.55 Dynamic programming example, steps 2, 3, and 4 combined. Forward solution minimum cost trajectory between states 1 and 6, stages 1 and 4. The same procedure, but in reversed order, is used compared with steps 2, 3, and 4 of Fig. 34.54 of the backward solution. The final minimum cost trajectory is the same, states 1-3-5-6, with a cost of $85.

Cost Table

Trajectory	Cost
1-2-4-6	$90
1-3-5-6	$85(*)

all paths or trajectories have to be evaluated. For a 100-stage problem, the simplification is far more dramatic than it is for the four-stage example illustrated.

These discrete dynamic programming examples of Figs. 34.54 and 34.55 are relatively easy to understand. Unfortunately, most control problems are continuous. The backward multistage decision process can still be applied in this case, to determine the optimal trajectory. A continuous performance measure is employed which is a time integral of some error or loss function of the system variables. The goal is then to minimize this integral, and hence minimize the error or loss. For example, a performance measure P might be chosen as the integral of the sum of the squares of the system error and control input u:

$$P = \min \int_{t_0}^{t_f} [(c - r)^2 + u^2]\, dt \tag{34.213}$$

where P = performance measure to be minimized
c = system output
t_0 = initial time
t_f = final time
r = system input
u = control effort (such as fuel consumption)
$c - r$ = system error

Note that all of these variables are functions of time; thus $c = c(t)$, $r = r(t)$, and so on.

The continuous integral of Eq. (34.213), denoted briefly by the integral sign $\int$ can be broken up into n parts each of duration Δt seconds. Each of these parts can then be handled as part of an n-stage dynamic programming problem.

For the first increment, assuming $t_0 = 0$:

$$P = \int_0^{t_f} = \int_0^{\Delta t} + \int_{\Delta t}^{t_f} \tag{34.214}$$

An n-stage approximation could then be numerically evaluated as follows:

$$P = \int_0^{t_f} = \int_0^{\Delta t} + \int_{\Delta t}^{2\Delta t} + \int_{2\Delta t}^{3\Delta t} + \int_{(n-1)\Delta t}^{n\Delta t = t_f} \tag{34.215}$$

Since each of the integrals of Eq. (34.215) is evaluated over a very short interval Δt, each integral can be approximated by the product of the integrand and Δt, evaluated at the given limits. For example, evaluating the integrand given by Eq. (34.213) at the limits $2\Delta t$ to $3\Delta t$ yields

$$\int_{2\Delta t}^{3\Delta t} \cong \{[c(t) - r(t)]^2 + u(t)^2\}\Big|_{t=2\Delta t} \cdot (\Delta t) \tag{34.216}$$

or

$$\int_{2\Delta t}^{3\Delta t} \cong \{[c(2\Delta t) - r(2\Delta t)]^2 + u(2\Delta t)^2\}\, \Delta t \tag{34.217}$$

Invoking the principle of optimality, each integral of Eq. (34.215) must in turn be minimized, in order to minimize the performance measure P. But each integral can be arithmetically evaluated, as given by Eq. (34.217). Thus the problem reduces to minimizing n arithmetic expressions of the form of Eq. (34.217). The control effort $u(t)$ is generally constrained, and $u(t)$, $c(t)$, and $r(t)$ are assigned discrete values for the purpose of numerical evaluation. Detailed examples of this type are given by Mickle and Sze,[27] and by Elgerd.[28] If a continuous rather than discrete solution is desired, it can be shown that the performance measure P is minimized by the solution to the Hamilton–Jacobi equation (which yields the optimal control policy).

In problems of this type, the system is usually described by state variables $x(t)$ (see Section 34.10). The performance measure P to be minimized is written in terms of these state variables $x(t)$ and the control input(s) $u(t)$.

$$P = \int_{t_0}^{t_f} g(x, u, t)\, dt \tag{34.218}$$

where t = time
$u = u(t)$ = control vector
$x = x(t)$ = state variable vector
g = integrand function
P = performance measure to be minimized

Now let us denote the minimum value of the performance measure P by the special symbol J. J will then be a function of the initial time t_0 and the initial state $x(t_0) = x_0$.

$$J = J(x_0, t_0) = \int_{t_0}^{t_f} g(x^*, u^*, t)\, dt \tag{34.219}$$

where x^* = optimal state variable trajectory = $x^*(t)$ and u^* = optimal control = $u^*(t)$. The state variable equations describing the system are

$$\dot{x} = f[x, u, t] \tag{34.220}$$

where f = state variable function. The resulting Hamilton–Jacobi[29] equation is

$$\frac{\partial J(x_0, t_0)}{\partial t_0} + g(x_0^*, u_0^*, t_0) + \left[\frac{\partial J(x_0, t_0)}{\partial x_0}\right]^T f(x_0^*, u_0^*, t_0) = 0 \tag{34.221}$$

Several examples demonstrating the solution of the Hamilton–Jacobi equation are given by Kirk,[30] Moore,[31] and Sage.[9]† These are all relatively simple problems, because the Hamilton–Jacobi equation, like most partial differential equations, cannot be easily solved in general and may be impossible to solve for some particular cases. For this reason the Hamilton–Jacobi equation is quite often used as a check on the optimality of a particular solution rather than as the means employed to obtain the solution.[32] A good background of the various programming methods is given by Hadley,[33] McMillan,[34] and Owens.[35]

34.14 LAGRANGE MULTIPLIER METHOD

The Lagrange multiplier method is used to maximize or minimize a given function $f(x)$, subject to various constraints $C_i(x)$ that restrict the range of the components of the vector x.[36]

Function to maximize or minimize

$$f(\mathbf{x}) = f(x_1, x_2, x_3, \ldots) \tag{34.222}$$

ith Constraint equation

$$C_i(\mathbf{x}) = C_i(x_1, x_2, x_3, \ldots) \tag{34.223}$$

When the constraint is satisfied, $C_i(x) = 0$.

†Example 4.4-1, p. 77; Example 4.4-2, p. 79; Example 4.4-3, p. 81.

λ_i is the Lagrange multiplier. Each $C_i(\mathbf{x})$ is multiplied by λ_i, that is, by $\lambda_1 C_1(x) + \lambda_2 C_2(x)$, and so on.

Lagrange function

$$L = f(x) + \sum_i \lambda_i C_i(x) \tag{34.224}$$

which is differentiated. The maximization or minimization of $f(x)$, subject to the constraints $C_i(x)$, is given by the solution to the following equations:

$$\begin{aligned} &\frac{\partial}{\partial_{x_1}} L = 0 \qquad \frac{\partial}{\partial_{\lambda_1}} L = 0 \\ &\frac{\partial}{\partial_{x_2}} L = 0 \qquad \frac{\partial}{\partial_{\lambda_2}} L = 0 \\ &\frac{\partial}{\partial_{x_3}} L = 0 \qquad \frac{\partial}{\partial_{\lambda_3}} L = 0 \\ &\quad\vdots \qquad\qquad\quad \vdots \end{aligned} \tag{34.225}$$

In many problems there may be only one constraint and hence only one λ, that is, λ_1, while there may be many components of x, that is, x_1, x_2, and x_3.

The equations $(\partial/\partial_{\lambda})L = 0$ simply give back the corresponding ith constraint equations. Note that when the constraints are all satisfied, the term $\sum_i \lambda_i C_i$ becomes zero, and the maximization of L of Eq. (34.215) becomes the ordinary maximization of $f(x)$.

The procedure is best illustrated with two simple examples. Note that all derivatives are partial derivatives with all the variables treated as constants, except the one variable involved in the differentiation.

Example 34.20 Lagrange Multiplier Example 1.

Problem. Maximize the volume of an open-top cylindrical tank, subject to the constraint that the amount of building material available has an area of A square meters.

Step 1. Find the function to maximize, $f(x_1, x_2, x_3, \ldots)$. There are two variables, H and R (see Fig. 34.56*a*). Let $x_1 = R$, $x_2 = H$.

$$f(x) = f(x_1, x_2) = \text{tank volume}$$
$$f(x) = \pi R^2 H = \pi x_1^2 x_2$$

Step 2. Find the constraint equation and set it equal to zero. There is only one constraint (tank surface area equals A), and consequently one λ.

$$C_1 = \text{tank area} - A = 0$$
$$C_1 = \text{side area plus bottom area} - A = 0$$
$$C_1 = 2\pi RH + \pi R^2 - A = 0$$
$$C_1(x_1, x_2) = 2\pi x_1 x_2 + \pi x_1^2 - A = 0$$

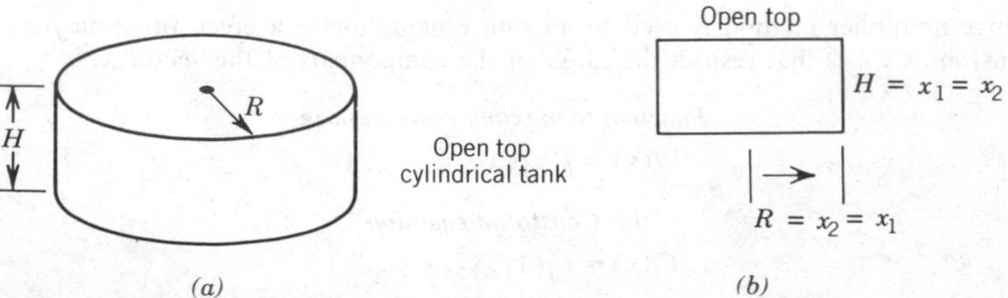

Fig. 34.56 Diagram for Example 34.20, Lagrange Multiplier Example 1. (*a*) Open-top cylindrical tank. R = tank radius, H = tank height. (*b*) Side view of tank. $x_1 = x_2$ or $H = R$ for maximum volume of an open-top cylindrical tank.

Step 3. Set up the Lagrange function.

$$L = f(x) + \lambda_1 C_1(x_1, x_2)$$

$$L = \pi x_1^2 x_2 + \lambda_1(2\pi x_1 x_2 + \pi x_1^2 - A)$$

Step 4. Take the partial derivatives of L with respect to x_1, x_2, and λ_1 and set them equal to zero.

$$\frac{\partial L}{\partial x_1} = 2\pi x_1 x_2 + \lambda_1(2\pi x_2 + 2\pi x_1) = 0$$

$$\frac{\partial L}{\partial x_2} = \pi x_1^2 + \lambda_1(2\pi x_1) = 0$$

$$\frac{\partial L}{\partial_{\lambda_1}} = 2\pi x_1 x_2 + \pi x_1^2 - A = 0$$

Step 5. Solve for x_1, x_2, and λ. From $\partial L/\partial x_2$, divide by πx_1, to obtain

$$x_1 + 2\lambda_1 = 0 \qquad \text{or} \qquad x_1 = -2\lambda_1$$

Insert $x_1 = -2\lambda_1$ into $\partial L/\partial x_1$, to obtain

$$2\pi(-2\lambda_1)x_2 + \lambda_1[(2\pi x_2 + 2\pi(-2\lambda_1)] = 0$$

Multiply out:

$$-4\pi\lambda_1 x_2 + 2\pi\lambda_1 x_2 - 4\pi\lambda_1^2 = 0$$

Collect terms:

$$-2\pi\lambda_1 x_2 - 4\pi\lambda_1^2 = 0$$

Cancel out $(-2\pi\lambda_1)$:

$$x_2 + 2\lambda_1 = 0 \qquad \text{or} \qquad x_2 = -2\lambda_1$$

But it was previously found that $x_1 = -2\lambda_1$. Therefore $x_1 = x_2$, or the tank radius R equals the tank height H. This gives the general solution for the shape of a maximum-volume open-top tank. Note that the tank is rather squat, only one half as high as it is wide. The actual dimensions, subject to the constraint tank area $= A$, are now readily found from

$$\partial L/\partial\lambda_1 = 2\pi x_1 x_2 + \pi x_1^2 - A = 0$$

Insert $x_2 = x_1$ into $\partial L/\partial\lambda_1$:

$$2\pi x_1 x_1 + \pi x_1^2 = A$$

Collect terms:

$$3\pi x_1^2 = A$$

Solve for x_1:

$$x_1 = \sqrt{A/3\pi} = x_2$$

Note that the actual value of λ_1 was not needed to solve for x_1 and x_2. λ_1 is readily found from $x_1 = x_2 = -2\lambda_1$ or from $\lambda_1 = -2x_1 = -2\sqrt{A/3\pi}$. A side view of the tank is shown in Fig. 34.56*b*.

Example 34.21 Lagrange Multiplier Example 2.

If the problem is repeated for a closed-top tank, the area of the top must be added to the constraint equation (see Fig. 34.57). This adds a second πR^2 term in step 2. Consequently, the Lagrange function of step 3 becomes:

$$L = \pi R^2 H + \lambda(\pi R^2 + \pi R^2 + 2\pi RH - A)$$

Solving, as just outlined, yields

$$R = -2\lambda, \qquad H = -4\lambda, \qquad \text{or} \qquad H = 2R$$

Note that the closed-top tank appears square in a side view (Fig. 34.57), compared with the open-top tank (Fig. 34.56*b*), which is one-half as high as it is wide. Both of these tanks have the maximum volume possible, subject to the constraint of a fixed surface area A. The volumes are different, however. The different solutions arise because two different problems are formulated, an open-top tank and a closed-top tank.

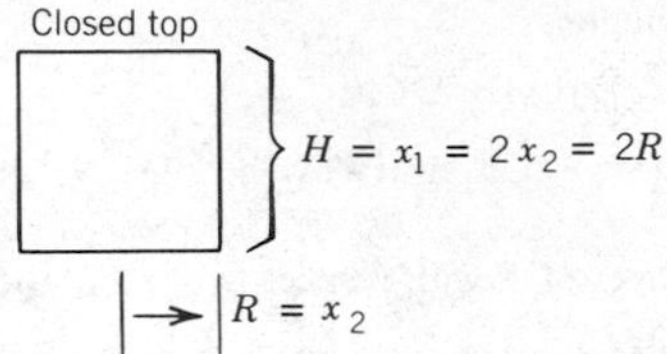

Fig. 34.57 Diagram for Example 34.21, Lagrange Multiplier Example 2. Side view of closed-top cylindrical tank. $H = -4\lambda$, $R = -2\lambda$, $H = 2R$, or $x_1 = 2x_2$. $H = x_1 = 2x_2 = 2R$ for maximum volume of a closed top cylindrical tank.

The Lagrange technique can be used to find the optimum solution to a specified design problem subject to the constraints. It is up to the designer to formulate the optimum design, that is, to choose an open-top cylindrical tank, or a rectangular tank, or an ellipsoidal tank, or another type. The Lagrange maximization or minimization technique optimizes a chosen design but it does not produce the design.

References

1 O. Mayr, "The Origins of Feedback Control," *Scient. Am.*, pp. 110–118 (Oct. 1970).

2 O. Mayr, *The Origins of Feedback Control*, MIT Press, Cambridge, 1970.

3 A. R. Ramsey, "The Thermostat or Heat Governor," *Trans. Newcomen Soc.* **25**:53–54 (1946), cited in Ref. 2.

4 T. Mead, "Regulation for Wind and Other Mills," British Patent (Old Series) No. 1484, 1787, cited in Ref. 2.

5 R. H. MacMillan, *Automation, Friend or Foe?*, Cambridge University Press, Cambridge, 1956.

6 K. Ogata, *Modern Control Engineering*, Prentice-Hall, Englewood Cliffs, NJ, 1970 (very comprehensive).

7 B. C. Kuo, *Automatic Control Systems*, 4th ed., Prentice-Hall, Englewood Cliffs, NJ, 1982.

8 S. M. Shinners, *Modern Control System Theory and Application*, 2nd ed., Addison-Wesley, Reading, MA, 1978.

9 M. F. Gardner and J. L. Barnes, *Transients in Linear Systems*, Wiley, New York, 1942 (14th printing 1958; went out of print in 1977. Look for a used copy for extensive table of 140 Laplace transform pairs).

10 J. J. D'Azzo and C. H. Houpis, *Linear Control System Analysis and Design; Conventional and Modern*, McGraw-Hill, New York, 1975.

11 W. R. Evans, "Graphical Analysis of Control Systems," *AIEE Trans.* **67** (part II):547–551 (1948).

12 S. M. Shinners, "How to Approach the Stability Analysis and Compensation of Control Systems," *Control Eng.* (May 1978).

13 J. T. Tou, *Digital and Sampled-Data Control Systems*, McGraw-Hill, New York, 1959.

14 J. A. Cadzow, *Discrete Time Systems*, Prentice-Hall, Englewood Cliffs, NJ, 1973.

15 R. Bellman, I. Glicksberg, and O. Gross, "On the Bang-Bang Control Problem," *Q. Appl. Math.* **14**:11–18 (1956).

16 J. P. LaSalle, in *Contributions to Differential Equations*, Vol. 5, Princeton University Press, Princeton, NJ, 1960, pp. 1–24.

17 J. E. Gibson, *Nonlinear Automatic Control*, McGraw-Hill, New York, 1963.

18 D. Graham and D. McRuer, *Analysis of Nonlinear Control Systems*, Wiley, New York, 1961.

19 J. LaSalle and S. Lefschetz, *Stability by Liapunov's Direct Method, with Applications*, Academic, New York, 1961.

20 V. M. Popov, "Absolute Stability of Nonlinear Systems of Automatic Control," *Automat. Rem. Contr.* **22**:857–875 (Aug. 1961).

21 A. E. Bryson Jr. and Y. C. Ho, *Applied Optimal Control*, Blaisdell, Waltham, MA, 1969.

22 H. Sagan, *Introduction to the Calculus of Variations*, McGraw-Hill, New York, 1969.

23 A. R. Forsyth, *Calculus of Variations*, Dover, New York, 1960.

24 G. A. Bliss, *Calculus of Variations*, published for the Mathematical Association of America by Open Court, Chicago, 1935.

25 R. Bellman, *Dynamic Programming*, Princeton University Press, Princeton, NJ, 1957.

26 R. Bellman and R. Kalaba, *Dynamic Programming and Modern Control Theory*, Academic, New York, 1965.

27 M. H. Mickle and T. W. Sze, *Optimization in Systems Engineering*, Intext, Scranton, PA, 1972.

28 O. I. Elgerd, *Control Systems Theory*, McGraw-Hill, New York, 1967.

29 A. P. Sage, *Optimum Systems Control*, Prentice-Hall, Englewood Cliffs, NJ, 1968.

30 D. E. Kirk, *Optimal Control Theory, An Introduction*, Prentice-Hall, Englewood Cliffs, NJ, 1970.

31 J. B. Moore, *Linear Optimal Control*, Prentice-Hall, Englewood Cliffs, NJ (Example 2.2, p. 21), 1971.

32 M. Athans and P. L. Falb, *Optimal Control, An Introduction to the Theory and Its Applications*, McGraw-Hill, New York, 1966. (Contains an extensive bibliography of early optimal control papers and texts.)

33 G. F. Hadley, *Nonlinear and Dynamic Programming*, Addison-Wesley, Reading, MA, 1964.

34 C. McMillan Jr., *Mathematical Programming*, Wiley, New York, 1970.

35 D. H. Owens, *Multivariable and Optimal Systems*, Academic, New York, 1981.

36 H. Hancock, *Theory of Maxima and Minima*, Dover, New York, 1960.

Bibliography

Anderson, B. D. O. and J. B. Moore, *Linear Optimal Control*, Prentice-Hall, Englewood Cliffs, NJ, 1971.

Atherton, D. P. A., *Stability of Nonlinear Systems*, Wiley, New York, 1981.

Auslander, D. M., Y. Takahashi, and M. J. Rabins, *Introducing Systems and Control*, McGraw-Hill, New York, 1974.

Ayres, F., *Theory and Problems of Matrices*, Schaum, New York, 1962.

Bennett, A. A., W. E. Milne, and H. Bateman, *Numerical Integration of Differential Equations*, Dover, New York, 1956.

Blackman, R. B. and J. W. Tukey, *The Measurement of Power Spectra*, Dover, New York, 1959.

Brewer, J. W., *Control Systems: Analysis, Design, and Simulation*, Prentice-Hall, Englewood Cliffs, NJ, 1974.

Brigham, E. O., *The Fast Fourier Transform*, Prentice-Hall, Englewood Cliffs, NJ, 1974.

Brogan, W. L., *Modern Control Theory*, Quantum, New York, 1974. (Good inexpensive text.)

Chang, S. S. L., *Synthesis of Optimum Control Systems*, McGraw-Hill, New York, 1961.

Chen, C. F. and I. F. Haas, *Elements of Control System Analysis*, Prentice-Hall, Englewood Cliffs, NJ, 1968. (Excellent visual presentation.)

Chen, C. T., *Introduction to Linear System Theory*, Holt, Rinehart and Winston, New York, 1970.

Chirlian, P. M., *Signals, Systems, and the Computer*, Intext, New York, 1973.

Cochin, I., *Analysis and Design of Dynamic Systems*, Harper & Row, New York, 1980. (Excellent diagrams of components, valves, loudspeakers, gyroscopes, brake systems, auto and aircraft suspension, and control systems.)

Cooper, L. and D. Steinberg, *Introduction to Methods of Optimization*, W. B. Saunders, Philadelphia, 1970.

D'Angelo, H., *Linear Time-Varying Systems: Analysis and Synthesis*, Allyn & Bacon, Boston, 1970.

Del Toro, V., *Electromechanical Devices for Energy Conversion and Control Systems*, Prentice-Hall, Englewood Cliffs, NJ, 1968.

Desoer, C. A., *Notes for a Second Course on Linear Systems*, Van Nostrand Reinhold, New York, 1970.

DiStefano III, J. J., A. R. Stubberud, and I. J. Williams, *Theory and Problems of Feedback and Control Systems*, Schaum, New York, 1967. (Good low-cost introductory book.)

Dorf, R. C., *Time-Domain Analysis and Design of Control Systems*, Addison-Wesley, Reading, MA, 1965.

Dorf, R. C., *Modern Control Systems*, Addison-Wesley, Reading, MA, 1967. (Very good coverage of root locus plots and an excellent table of transfer function Bode, Nyquist, Nichols, and root locus plots.)

Eveleigh, V. W., *Introduction to Control Systems Design*, McGraw-Hill, New York, 1972. (Table of 56 Laplace transforms.)

Frederick, D. K. and A. B. Carlson, *Linear Systems in Communication and Control*, Wiley, New York, 1971. (Clear diagrams of waveforms, Fourier spectra, convolution, sampling.)

Grabbe, E. M., S. Ramo, and D. E. Wooldridge, *Handbook of Automation, Computation, and Control*, Vol. 1, *Control Fundamentals*, 1958, Vol. 3, *Systems and Components*, Wiley, New York, 1961.

Gupta, S. C. and L. Hasdorff, *Fundamentals of Automatic Control*, Wiley, New York, 1970.

Hermes, H. and J. P. LaSalle, *Functional Analysis and Time Optional Control*, Academic, New York, 1969.

Horowitz, I. M., *Synthesis of Feedback Systems*, Academic, New York, 1963.

Huggins, W. H. and D. R. Entwisle, *Introductory Systems and Design*, Blaisdell, Waltham, MA, 1968. (Very thorough treatment of signal flow graphs.)

Jolley, L. B. W., *Summation of Series*, 2nd revised ed., Dover, New York, 1961.

Kailath, T., *Linear Systems*, Prentice-Hall, Englewood Cliffs, NJ, 1980.

Katz, P., *Digital Control Using Microprocessors*, Prentice-Hall, Englewood Cliffs, NJ, 1981.

Kuo, B. C., *Discrete-Data Control Systems*, Prentice-Hall, Englewood Cliffs, NJ, 1970.

Kuo, B. C., *Digital Control Systems*, Holt, Rinehart and Winston, New York, 1980.

Lathi, B. P., *Signals, Systems, and Controls*, Intext, New York, 1974. (Very good appendix treating partial fraction expansion with complex and multiple roots, Bode plots, and matrix algebra.)

Merriam III, C. W., *Optimization Theory and the Design of Feedback Control Systems*, McGraw-Hill, New York, 1964.

Murphy, G. J., *Control Engineering*, Boston Technical, Cambridge, MA, 1965. (Relatively extensive table of components and their transfer functions, good section on carrier control systems.)

Ogata, K., *State Space Analysis of Control Systems*, Prentice-Hall, Englewood Cliffs, NJ, 1967.

O'Higgins, P. J., *Basic Instrumentation, Industrial Measurements*, McGraw-Hill, New York, 1966.

Owens, D. H., *Multivariable and Optimal Systems*, Academic, London, 1981.

Papoulis, A., *Signal Analysis*, McGraw-Hill, New York, 1977.

Perkins, W. R. and J. B. Cruz Jr., *Engineering of Dynamic Systems*, Wiley, New York, 1969.

Polak, E. and E. Wong, *Notes For a First Course on Linear Systems*, Van Nostrand Reinhold, New York, 1970.

Power, H. M. and R. J. Simpson, *Introduction to Dynamics and Control*, McGraw-Hill, London, 1978.

Rubinstein, M. F., *Patterns of Problem Solving*, Prentice-Hall, Englewood Cliffs, NJ, 1975. (Excellent wide spectrum, thought-provoking text. Control theory is only one part of it. Covers game theory, biocontrol, decision theory, probability, and modeling.)

Rubio, J. E., *The Theory of Linear Systems*, Academic, New York, 1971.

Saucedo, R. and E. E. Schiring, *Introduction to Continuous and Digital Control Systems*, Macmillan, New York, 1968.

Slemon, G. R., *Transducers, Transformers, and Machines*, Wiley, New York, 1966.

Solodovnikov, V. V., *Introduction to the Statistical Dynamics of Automatic Control Systems*, Dover, New York, 1960.

Stephenson, R. E., *Computer Simulation for Engineers*, Harcourt Brace Jovanovich, New York, 1971.

Takahashi, Y., M. J. Rabins, and D. M. Auslander, *Control and Dynamic Systems*, Addison-Wesley, Reading, MA, 1970.

Tou, J. T., *Modern Control Theory*, McGraw-Hill, New York, 1964.

Turnball, H. W., *The Theory of Determinants, Matrices and Invariants*, Dover, New York (soft cover), 1960.

CHAPTER 35
CONTROL SYSTEM APPLICATIONS

ROGER D. FRUECHTE

General Motors Research Laboratories
Warren, Michigan

THOMAS E. MARLIN

Exxon Research and Engineering Company
Florham Park, New Jersey

35.1 AUTOMOTIVE CONTROL SYSTEMS

Roger D. Fruechte

35.1-1 Background

Today's automobiles employ many control systems not thought possible only a few years ago. The primary motivation for development of these control systems is the need to meet exhaust emission standards. However, because of the recent availability of low-cost electronics, primarily the microcomputer, a number of other control systems have evolved that improve fuel economy or enhance driveability and driver convenience. This section gives an overview of automotive control applicable to conventional spark-ignited gasoline engines.

The Environmental Protection Agency of the US government has established standards for three constituents in the exhaust stream: (1) oxides of nitrogen (NO_X), (2) carbon monoxide (CO), and (3) unburned hydrocarbons (HC). The current method for establishing exhaust standards is the Federal Test Procedure,[1,2] which uses driving patterns that are typical of a trip around downtown Los Angeles. This test is conducted on a chassis dynamometer and involves a vehicle following a specified velocity trajectory. The test is categorized into three phases. Phase 1, the cold-transient phase, starts with the engine cold after an overnight "soak," and is 3.6 miles long. Phase 2, the cold-stabilized phase, is 3.9 miles long and has more stop-and-go driving than the first, with speeds rarely above 30 mph and some 13 idle periods. Phase 3, the hot-transient phase, is a repeat of phase 1 except that the engine starts warm.

The exhaust sampling system consists of a positive displacement pump which draws a constant volume of exhaust gases and dilution air into collection bags through a mixing chamber and heat exchanger. The bags are collected at the end of each phase. Emission calculations, in grams per mile, are based on the weighted formula $(0.43C_T + C_S + 0.57H_T)/7.5$, where C_T, C_S, and H_T are mass measurements of the given pollutants collected during cold-transient, cold-stabilized, and hot-transient phases, respectively. The weighting factors 0.43 and 0.57 on the cold-transient and

		1970 1971	1972	1973 1974	1975 1976	1977	1978	1979	1980	1981	1982	1983	1984	1985
EXHAUST EMISSIONS	HC (G/MI)	4.1	3.0		1.5				0.41					
	CO (G/MI)	34	28		15				7.0	3.4				
	NO_X (G/MI)	N. R.	N. R.	3.1		2.0				1.0				
FUEL (MPG) ECONOMY		N. R.					18	19	20	22	24	26	27	27.5

Fig. 35.1 US emission control and fuel economy requirements.

hot-transient phases are based on data reflecting typical conditions accompanying the average engine start (some would be hot, some would be cold).

The exhaust emission standards have become progressively more stringent since 1970, as shown in Fig. 35.1. In addition, federal regulations on fuel economy, called CAFE for *c*orporate *a*verage *f*uel *e*conomy, have also been established beginning in 1978. Although the most significant approach to meeting fuel economy regulations has been weight reduction, additional gains are achieved if the engine is operated in its most efficient manner. This may not be the most desirable thing to do, however, because of the direct interaction between emission production and fuel economy. A typical tradeoff between hydrocarbons, oxides of nitrogen, and fuel economy is shown in Fig. 35.2. For a given engine vehicle combination without an emission aftertreatment device, any attempt to improve the fuel economy (while keeping the NO_X level constant) by operating the engine more efficiently will result in increased HC. In view of this tradeoff, control systems are needed to appropriately control the engine such that emissions standards are met without excessive degradation of either fuel economy or driveability, that is, the ability of the engine/vehicle to follow driver commands.

In conventional spark-ignited engines, the driver controls the position of a throttle valve through a direct mechanical linkage, and this controls the flow rate of air into the intake manifold. Fuel proportional to the air flow is mixed with the air stream via either a venturi carburetor or fuel injectors. The piston motion ingests the combustible mixture into the cylinders and torque is produced when a spark at the spark plug electrode causes the mixture to burn. The gases resulting from this combustion pass through the exhaust manifold and out the tail pipe.

Three primary control variables are used for exhaust emission control: air-fuel ratio (A/F), spark advance (SA), and exhaust gas recirculation (EGR). Each of these is described here with regard to its effect on both emissions and fuel economy.

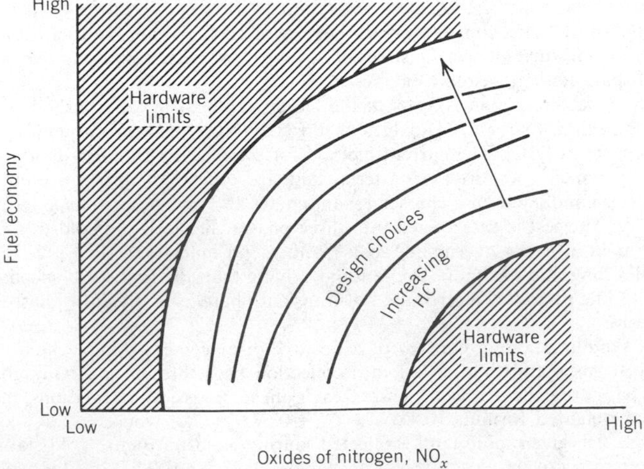

Fig. 35.2 Fundamental fuel-exhaust emissions tradeoff. HC, unburned hydrocarbons.

Air-Fuel Ratio

Air-fuel ratio is defined as the mass ratio of air to fuel inducted into the engine, that is, A/F = (mass flow rate of air)/(mass flow rate of fuel). The value of A/F for which there is theoretically sufficient oxygen to oxidize all of the fuel is called the stoichiometric ratio. For spark ignition engines, this is about 14.7/1. Generally, as A/F is increased (leaning of the mixture), CO production decreases because increased oxygen tends to guarantee more complete combustion. HC also decreases for increasing A/F up to about 18/1. For larger values, mixture homogeneity and flame propagation speeds are affected, so that HC emissions begin to increase. NO_X is formed when the combustion temperature is greater than 1950°C. It tends to have its maximum production at an A/F slightly greater than the stoichiometric ratio. Fuel economy generally increases with increasing A/F.

Spark Advance

Initiation of the spark pulse by the ignition system relative to top dead center (TDC) position of the cylinders is termed spark advance (SA). Historically, SA has been varied as a function of speed and load (manifold vacuum) in the distributor to maintain engine efficiency. Spark timing generally must be advanced for increased speed to compensate for the shorter time to initiate combustion, and retarded for increased load to compensate for increased flame speed due to a reduction in the amount of residual exhaust gas in the combustion chamber. For emission control in its simplest form, the distributor SA calibration curve is altered so that the spark initiates combustion later in the cycle than is done normally. Although the peak combustion temperature is lowered by spark retard, which results in a reduction of NO_X, the temperature of the exhaust gas leaving the cylinder is actually increased, allowing additional oxidation of unburned HC in the exhaust manifold. Spark retard can cause a significant reduction in engine-cycle efficiency and fuel economy.

Exhaust Gas Recirculation

Exhaust gas recirculation (EGR) is the introduction into the intake manifold of a portion of the exhaust gas. The dilution of the combustion mixture reduces flame speed and peak combustion temperature and this results in lower oxides of nitrogen. However, because of the reduced exhaust temperatures, EGR increases HC emissions slightly. EGR in a small amount improves efficiency; however, additional quantities reduce efficiency, increase cyclic variability, and ultimately lead to misfire.

35.1-2 Engine Modeling for Control Algorithm Development

In the past, coordination of engine control variables over the entire operating regime of the engine has been done via on-engine testing. For the most part, the controls are open-loop schedules as functions of engine speed, load, and temperature. More recently, mathematical modeling has played an important role in the calibration process as well as in control system design.

Static Models

Static models, often referred to as engine maps, relate input and output variables of an engine at a constant operating condition. They are useful in the design of control strategy to meet emission standards while maximizing fuel economy and they also provide a base from which a transient control strategy can be designed for the enhancement of driveability.

One method[3,4] for obtaining static models is to collect data on an engine connected to a dynamometer at a limited number of fixed speed and load (torque) conditions. These conditions are selected such that they are representative of those encountered in the course of a particular driving schedule. Tests are first run on a vehicle that translates the demands of a vehicle driving schedule into an engine speed/torque vs. time density matrix. Given a specific vehicle package, this matrix remains nearly invariant (depending on changes in the transmission shift schedule) regardless of engine calibrations of SA, A/F, and EGR. At each of the appropriately chosen speed/load points, the input variables are slowly varied and measurements of fuel consumption and emissions are made. Each of the control variables can alter torque production from the engine, so that as the input variable is changed, throttle position must also be changed to maintain constant torque.

Linear regression analysis[5] is then used to fit smooth surfaces to experimental data for brake-specific values of fuel consumption (FC), NO_X, CO, and HC at each of the speed/load points. The obtained analytical expressions are quadratic functions in SA, A/F, and EGR. In general terms, the equation for the response of BSFC† is

$$\mathrm{BSFC} = \alpha_0 + \alpha_1(\mathrm{SA}) + \alpha_2(A/F) + \alpha_3(\mathrm{EGR}) + \alpha_{11}(\mathrm{SA})^2 + \alpha_{22}(A/F)^2 + \alpha_{33}(\mathrm{EGR})^2 + \alpha_{12}(\mathrm{SA} \times A/F) + \alpha_{23}(A/F \times \mathrm{EGR}) + \alpha_{31}(\mathrm{EGR} \times \mathrm{SA}) \tag{35.1}$$

Linear Dynamic Models

Static models may be perfectly adequate for determining initial steady-state schedules of EGR, SA, and A/F but their usefulness is somewhat limited for determining appropriate control during a transient maneuver or if a closed-loop control system is used that must account for the dynamics of the engine sensors and actuators. For these conditions dynamic models must be developed. This section describes the development of linear dynamic models for which an extensive body of literature is available on control system design.

One approach is to treat the engine as a "black box" whereby engine outputs are related to engine inputs by gain and phase plots as a function of frequency. This approach relies heavily on experimental measurements and does not consider the details of the physical process. This makes the modeling process much easier; however, interrelationships between measured variables are not easily determined and some variables of interest are not easily measured dynamically.

A second approach relies on a state-variable formulation of the intake process, combustion process, and load dynamics, linearization of the equations, and then use of frequency response and static tests to fit the constants in the linearized equations. This approach is particularly useful if several variables are used in the feedback control law or if any control optimization schemes are to be employed.

As an example, an engine model at the idle condition is derived with throttle angle α and load torque T_L as the only input variables. At idle, air flow across the throttle blade is essentially sonic; hence, air flow into the intake manifold $\dot{M}_I$ is dependent only on the throttle angle, or

$$\dot{M}_I = f_1(\alpha) \tag{35.2}$$

The mass of air in the intake manifold M_M is the integral of the difference between the flow in and the flow out $\dot{M}_O$, or

$$M_M = \int \left(\dot{M}_I - \dot{M}_O\right) dt \tag{35.3}$$

where $\dot{M}_O$ is a function of the manifold air mass and engine speed N, that is,

$$\dot{M}_O = f_2(M_M, N) \tag{35.4}$$

The manifold pressure P_M is primarily a function of manifold air mass, or

$$P_M = f_3(M_M) \tag{35.5}$$

For constant A/F and SA, the indicated torque T_I is proportional to the quantity of air ingested into the cylinders but delayed in time by τ_D to account for the piston going through the intake and compression stroke. Thus

$$T_I = f_4\left[\frac{\dot{M}_O(t-\tau_D)}{N(t-\tau_D)}\right] \tag{35.6}$$

Finally, the engine speed is related to the net accelerating torque by

$$N = \frac{1}{J}\int (T_I - T_F - T_L)\, dt \tag{35.7}$$

where J is the inertia of engine and load, T_L is an applied engine load, and T_F is the friction torque given by

$$T_F = f_5(N) \tag{35.8}$$

Linearizing Eqs. (35.2) to (35.8) at a specific operating condition for small signal perturbations in α and T_L gives a set of linear equations that is represented by the block diagram in Fig. 35.3.

Nonlinear Dynamic Models

Linear perturbation models have the advantage of simplicity and are particularly suited to stability studies or for use of the wealth of theory dealing with optimal control designs. Nonlinearities that

† BSFC is brake-specific fuel consumption. It is the fuel consumed in grams per hour divided by the measured power output of the engine. Thus BSFC has units of grams per kilowatt-hour.

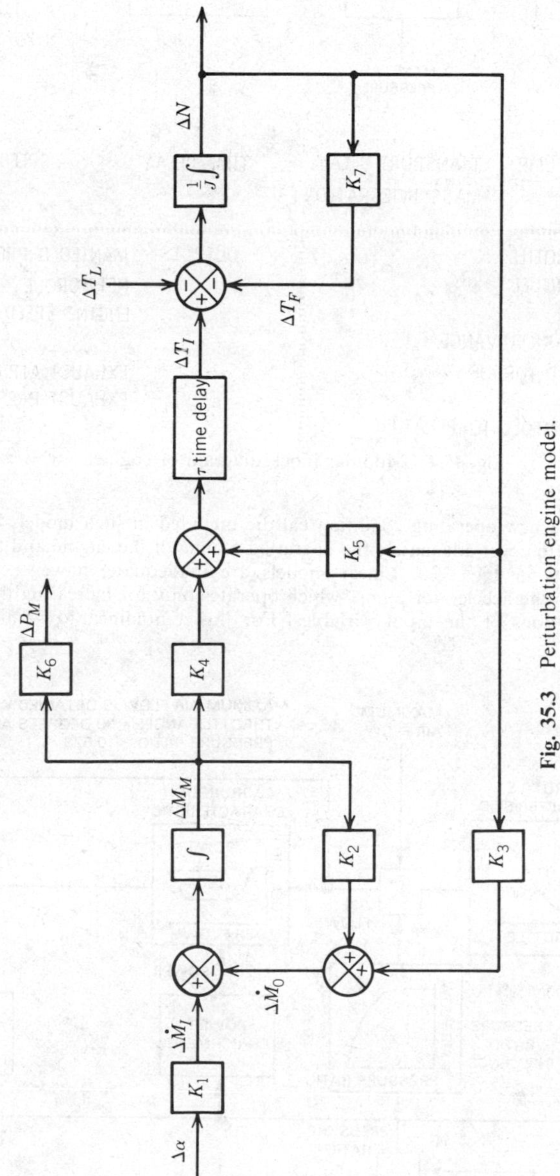

Fig. 35.3 Perturbation engine model.

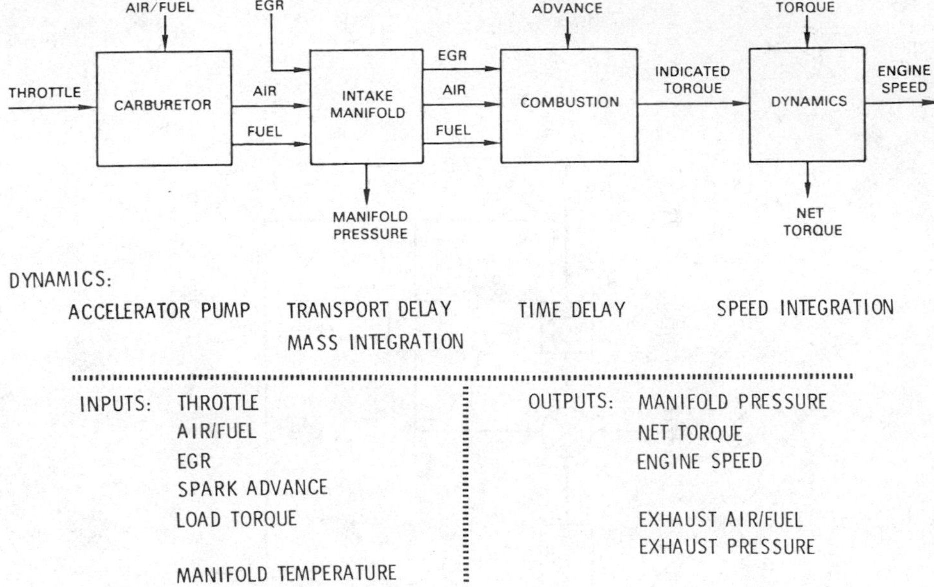

Fig. 35.4 Modular block diagram of engine.

result from going to a new operating condition can be included in such models by development of several linear models to cover adequately the operating regime of the engine and then use of smooth functions for the K's in Fig. 35.3. Linear models are inadequate, however, to address severe nonlinearities such as the accelerator pump, which operates only for increasing throttle rates, or the large signal perturbations of the input variables. For this a nonlinear dynamic engine model is necessary.

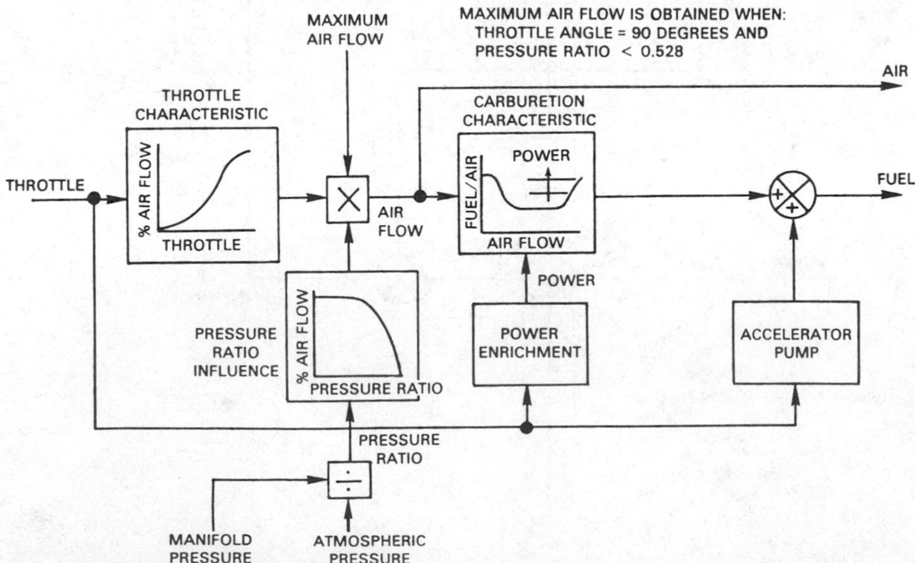

Fig. 35.5 Carburetor model. From "A Mathematical Engine Model for Development of Dynamic Engine Control," Paper 800054, presented at the 1980 SAE International Congress and Exhibition, Detroit, MI, Feb. 1980 by D. J. Dobner, © 1980 Society of Automotive Engineers, Inc., reprinted with permission.

One approach to building a nonlinear dynamic engine model is to consider the engine's principal processes in modular form.[6] A model of this form, shown in the block diagram of Fig. 35.4, allows changes to be made in components of the engine and in the resulting new models of these components without drastically affecting the remainder of the model. Each of the modules are modeled to a sufficient degree of complexity such that the model characterizes the inputs and outputs of the module over the expected operating range of each of the variables and for all frequencies less than 10 Hz. As an example, Fig. 35.5 shows the general details of the carburetor model. Its independent input is throttle position (α) and its dependent outputs are rates of air ($\dot{M}_I$) and fuel ($\dot{M}_F$) that flow into the intake manifold. Note that with the exception of the accelerator pump, no dynamics are associated with the carburetor.

Although a great deal of testing would seem to be required to obtain all the model parameters, generally for control system design it is not necessary to have *precise* models. For example, the throttle characteristic in Fig. 35.5 can be obtained by testing a carburetor on a flow bench or engine. Once obtained, it should apply for all conventional carburetors having circular throttle bores.

35.1-3 Closed-Loop Control Systems in Automobiles

Three closed-loop control systems currently are used on many production automobiles. The first, stoichiometric A/F control, addresses the emissions problem. The second, idle speed control, improves fuel economy. And the third, closed-loop knock limiting, enhances driveability. The three are described here.

Stoichiometric Air-Fuel Control

Probably the most significant contribution to emission control on the automobile is the catalytic converter, an exhaust treatment device. Before 1981 only the oxidizing catalyst was needed. It used small amounts of platinum to oxidize HC and CO in the exhaust. Beginning in 1981, when the emission standard for NO_X was halved from 1980 (see Fig. 35.1), the three-way catalyst was required for many cars.[7] The three-way catalyst performs the oxidation of HC and CO as before and reduces NO_X to N_2 and O_2, using small amounts of rhodium as the catalytic element.

The efficiency characteristics of the three-way catalyst as a function of air-fuel ratio (A/F) is shown in Fig. 35.6. When the A/F ratio is lean (excess of oxygen), control of HC and CO entering the converter is very good but control of NO_X is poor. When the A/F ratio is rich (deficiency of oxygen), control of NO_X is very good but control of HC and CO is poor. At the stoichiometric value, a narrow window exists where the control of all three pollutants is quite good.

The purpose of closed-loop air-fuel control is to maintain the A/F of the engine exhaust gas at the stoichiometric mixture, since no open-loop fuel metering system is capable of achieving regulation of A/F to the required ± 0.1 of a ratio. A simplified block diagram is shown in Fig. 35.7. The

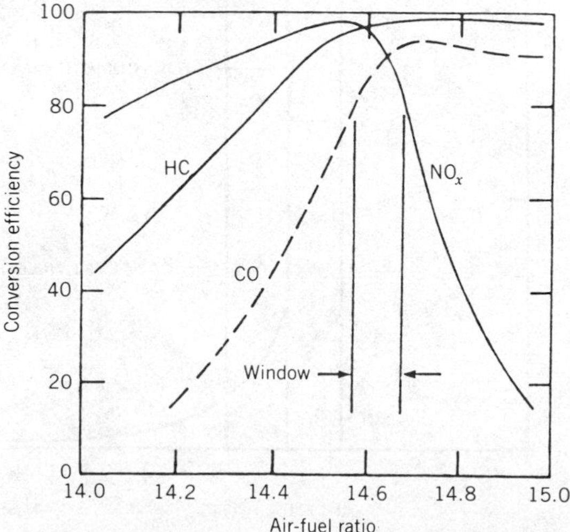

Fig. 35.6 Efficiency of three-way catalytic converter.

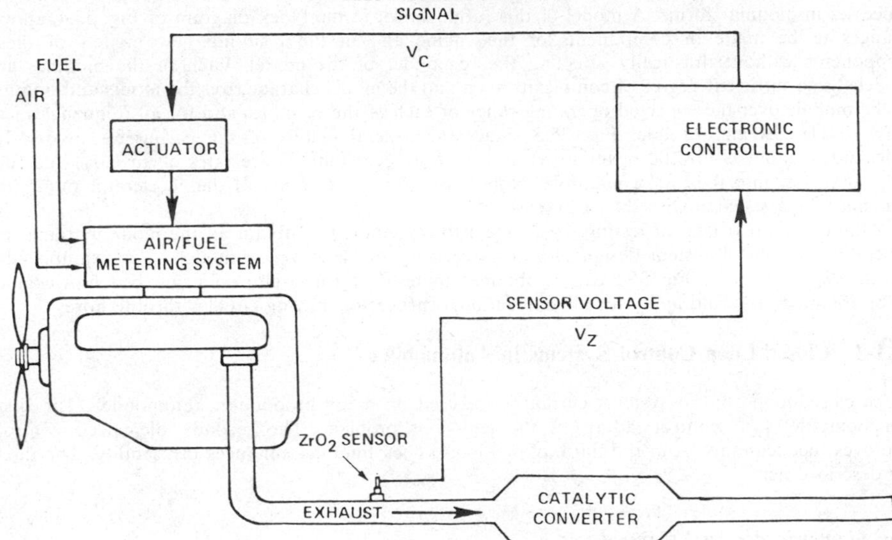

Fig. 35.7 Stoichiometric air-fuel control system.

principal components of the system are an exhaust gas sensor, an electronic controller, and a means for adjusting the fuel via a carburetor or electronic fuel injection system.

The typical exhaust gas sensor, which is placed directly in the engine exhaust-gas stream, consists of a zirconium element coated with platinum. The zirconium and platinum act as an electrochemical cell when heated by the hot exhaust gases. The sensor's internal impedance, output voltage, and time of response are all functions of temperature. As shown in Fig. 35.8, a very low voltage is generated when the air-fuel mixture is leaner than the stoichiometric mixture. As the air-fuel mixture becomes richer and passes through the stoichiometric mixture, a voltage up to 900 mV is generated.

The actual adjustment of fuel to the engine is generally done by adjusting the position of metering

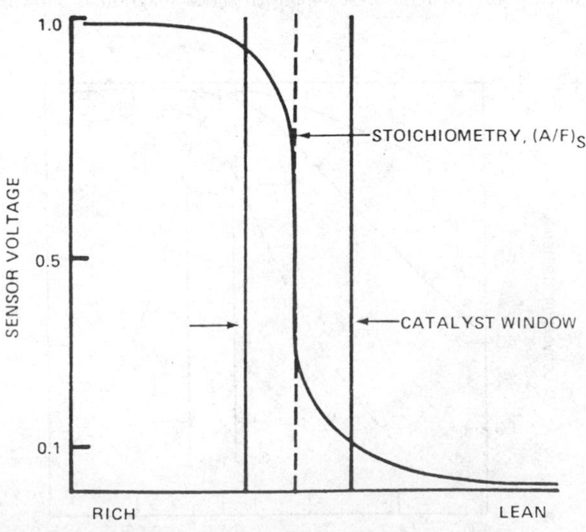

Fig. 35.8 Zirconium-oxide sensor characteristic.

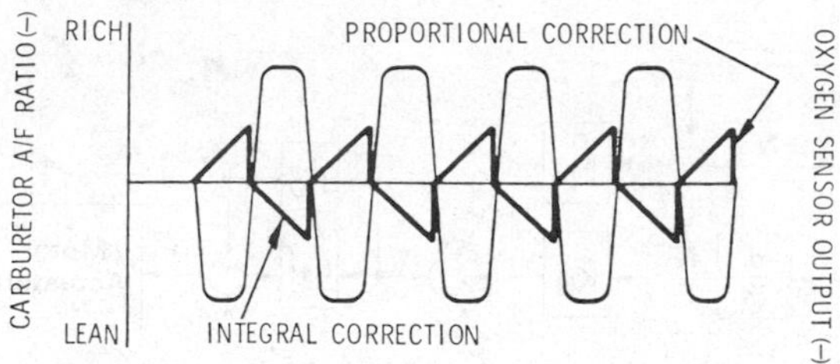

Fig. 35.9 Typical traces of oxygen sensor and air-fuel ratio commands. From G. W. Niepoth and S. P. Stonestreet, "Closed-Loop Engine Control," *IEEE Spectrum*, Nov. 1977, p. 55, reprinted with permission.

rods with an electromechanical actuator in the case of the carburetor or by adjusting the pulse width to the fuel injectors in the case of electronic fuel injection. The engine receives the change in air-fuel mixture through the intake manifold, burns the mixture, and passes it down the exhaust manifold past the exhaust gas sensor and through the catalytic converter. A dominant characteristic of the engine is the time delay in the induction, combustion, and exhaust process. This must be compensated for in the control strategy so that the air-fuel mixture stays as close as possible to the stoichiometric mixture under all conditions.

The controller must take advantage of both open-loop and closed-loop strategies. A typical closed-loop strategy is use of a form of proportional plus integral control. As long as a rich or lean mixture is indicated, the control signal ramps in the corrective direction at a fixed rate, which may be a function of speed and load. When the exhaust oxygen sensor switches, the control signal steps in the opposite direction of the previous signal by a fixed amount and begins to ramp in the corrective direction. This results in a limit cycle, as indicated by the traces in Fig. 35.9. Generally a single set of ramp and step calibrations is insufficient for the entire operating regime of the engine, and gain scheduling is required. One of the reasons for this is that the engine time delay changes significantly with respect to flow rates through the engine. The allowance of gain variations for different engine speed and load conditions allows the limit cycle amplitude to be minimized. An open-loop control strategy is required when (1) the oxygen sensor is warming up and is not yet providing an output to the controller and (2) enrichment is desired for full throttle accelerations. The open-loop signals are generally scheduled as a function of load, speed, and coolant temperature.

Idle Speed Control

Nearly 19% of the 1975 Federal Test Procedure driving schedule is with the engine at idle. Numerous tests have shown that as much as 0.25 km/1 (0.6 miles/gal) fuel economy improvement is possible when the idle speed is lowered approximately 100 r/min from normal curb idle settings. However, lowering the idle speed by 100 r/min can result in frequent engine stalls when accessory loads such as air conditioning or power steering lockup are suddenly applied. One method for circumventing this problem is an idle speed control (ISC) system.[7]

A simplified block diagram of an ISC system is shown in Fig. 35.10. With the application of an

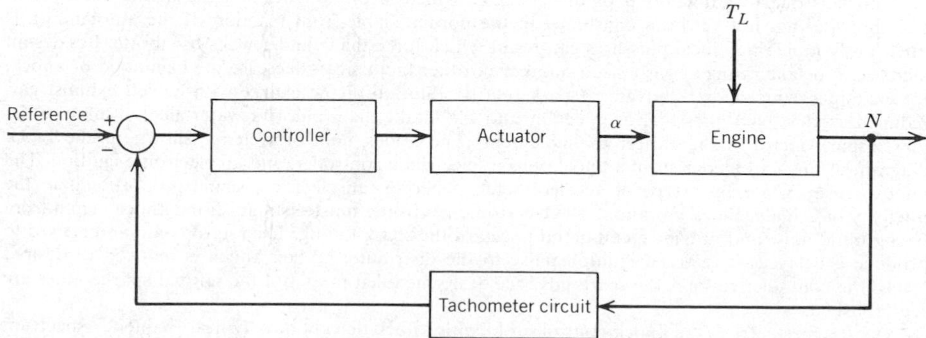

Fig. 35.10 Idle speed control system.

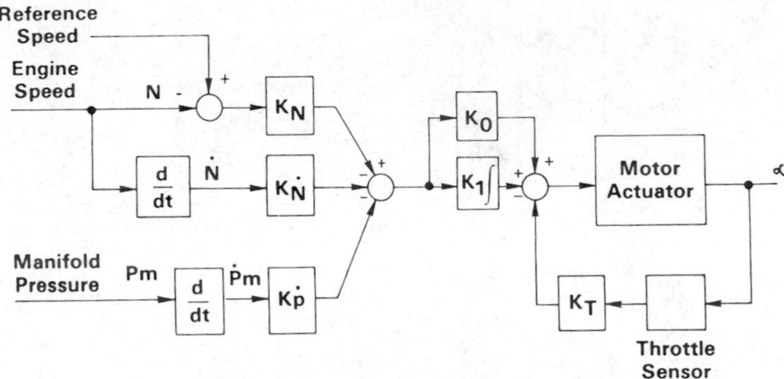

Fig. 35.11 Idle speed control strategy.

engine load T_L, the engine speed N begins to drop. This change in speed is sensed by a tachometer circuit which generates a voltage that is proportional to the change in engine speed. A controller operates on the difference between the desired idle speed and the measured speed and provides a voltage to an actuator. The actuator may be of the type that positions the throttle in response to controller commands for carbureted engines or, as in some fuel-injected engines, may control the orifice size of an air bypass path around the throttle blade. In either event the engine speed returns to the desired value.

An approach to designing the controller is to use the engine model just described and models of the tachometer and actuator and then use analytical control techniques. Linearized models are applicable, since idle speed control operates in a limited speed range and the effects of speed-dependent nonlinearities are minimal. The simplest controller would be just a proportional term, K_N. The problem is that K_N must be so large for it to have any effect on engine stalls that loop instability is likely. In addition, it may be desirable to reduce the loop sensitivity to parameter variations with inner feedback loops. Since both throttle position a and manifold pressure P_M are measurable quantities, they also can be used as feedback variables.

A control algorithm that makes use of these variables is shown in Fig. 35.11. The integrator is included to eliminate steady-state errors. The effect of adding speed rate $\dot{N}$ and manifold pressure rate $\dot{P}_M$ to the control law is illustrated by actual traces of engine speed for an application of the power-steering load in Fig. 35.12. Increasing speed feedback alone nearly causes sustained oscillations, but as $\dot{N}$ and $\dot{P}_M$ are added to the control law, the loop becomes progressively more stable.

Knock Limiting

Spark-ignited engines with high compression ratios generally have higher efficiency (hence better fuel economy) than those engines with lower compression ratios. There previously has been no incentive to raise compression ratios, however, since the increased efficiency gains of the vehicle are offset by energy losses at the refinery in producing the higher octane unleaded fuel required by high compression ratio engines. Use of the lower octane gasoline, such as 91 Research Octane Number (RON), with a high compression ratio engine produces a phenomenon called detonation or knock.

Knock is due to autoignition of the end gas, which is the air-fuel mixture in the combustion chamber that has not yet been consumed in the normal flame-front reaction. If the autoignition is sufficiently rapid, high local pressures can result which force the cylinder walls to vibrate. Besides an increase in octane rating of gasoline, a number of other factors can decrease the likelihood of knock, for example, reduced spark advance (spark retard), dilution of the charge with cooled exhaust gas, reduced exhaust back pressure, or injection into the intake manifold of a water-alcohol mixture. Of these, spark retard is the easiest to implement. The knock control system,[8] shown in the block diagram of Fig. 35.13, consists of a knock sensor, electronic controller, and an electronic ignition. The knock sensor, generally a type of accelerometer, generates an electrical signal proportional to the intensity of cylinder block vibration. The electronic controller must distinguish the knock signal from background noise and provide a command to retard the spark timing. The retard command is used to produce a delayed or retarded ignition pulse to the distributor. When knock is reduced to desired levels, the controller restores the spark advance at a scheduled rate until the normal spark values are reestablished.

The toughest part of the knock control problem is knock detection. A typical frequency spectrum of the knock sensor output is shown in Fig. 35.14. The solid line shows the output without knock and

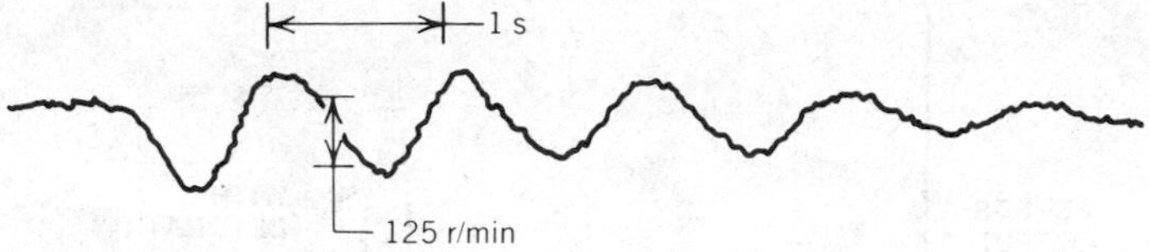

Feedback: Speed

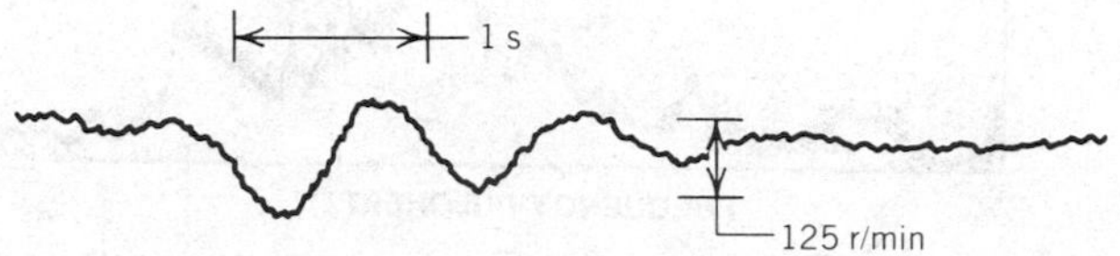

Feedback: Speed + speed rate

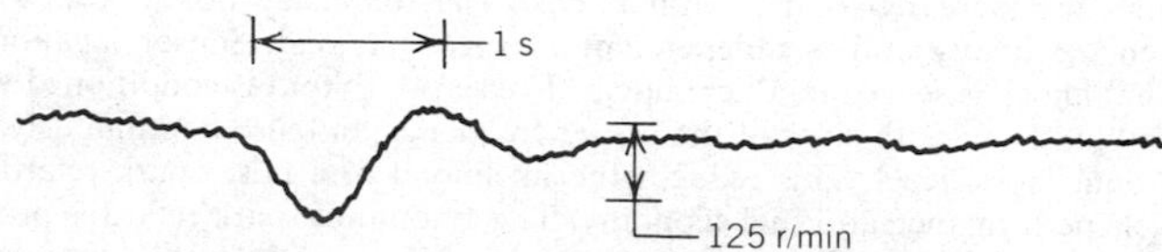

Feedback: Speed + speed rate + manifold pressure rate

Fig. 35.12 Engine speed response to load application.

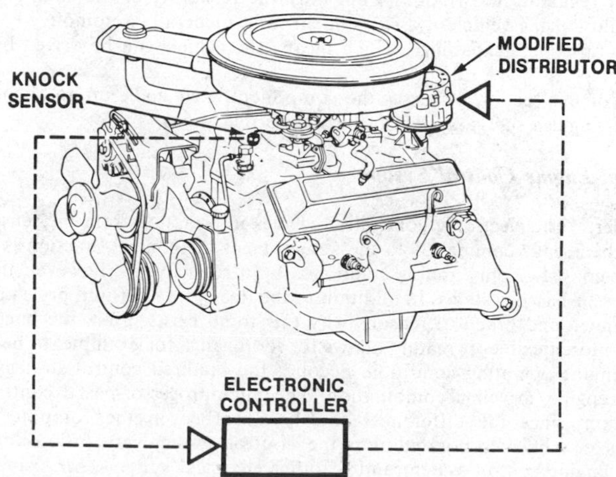

Fig. 35.13 Knock-limiting control system. From "Energy Conservation with Increased Compression Ratio and Electronic Knock Control," Paper 790173 presented at the 1979 SAE International Congress and Exposition, Detroit, MI, Feb. 1979 by J. H. Currie, D. S. Grossman, and J. J. Gumbleton, © 1979 Society of Automotive Engineers, Inc., reprinted by permission.

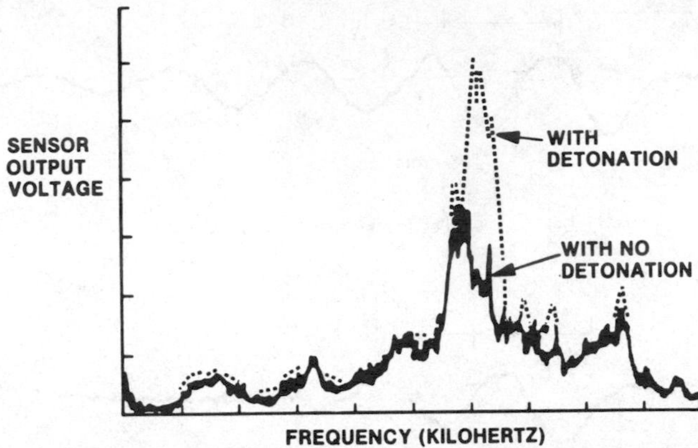

Fig. 35.14 Knock sensor output vs. frequency. From "Energy Conservation with Increased Compression Ratio and Electronic Knock Control," Paper 790173 presented at the 1979 SAE International Congress and Exposition, Detroit, MI, Feb. 1979 by J. H. Currie, D. S. Grossman, and J. J. Gumbleton, © 1979 Society of Automotive Engineers, Inc., reprinted by permission.

the dashed line shows the increased output with knock. The frequency of the knock vibration is unique to a given engine family and is independent of speed or load. Sensor location is critical, however, to assure that knock is sensed in all cylinders. The sensor output is conditioned with a bandpass filter with sharp cutoffs on both sides of the center frequency and also continuously averaged to establish the background noise level. This reduces the likelihood of a false spark retard, which can result in loss of vehicle performance and fuel economy. The amount of spark retard is proportional to the intensity of the detected knock. To maintain vehicle operation, the spark timing is retarded rapidly when knock is detected and then readvanced slowly. More rapid changes can cause uneven engine power surges that can be felt by the driver.

35.1-4 CONTROL SYSTEM IMPLEMENTATION

The question of how to implement a control system on an automobile involves many tradeoffs relative to cost, environmental conditions, reliability, and availability of hardware. As an example, typical environmental conditions for which the control system must operate are shown in Table 35.1.

The question of reliability must address not only the reliability of the components but also the safety and driveability of the vehicle if a component fails. Generally automotive control systems are designed with a "limp-home" capability, which means the vehicle can be driven but with degraded performance.

The remainder of this section discusses the components that make up an engine control system and factors that distinguish such a system from other applications.

Components of an Engine Control System

Electronic Controller. The electronic controller receives information from the sensors, processes the information, and then sends commands to the actuators. If the control function is relatively simple, the controller may consist of only analog components. In recent years, however, the trend has been toward microprocessor-based systems. In addition to the decreasing per-unit price of the microprocessor, there are technical and practical reasons why this trend exists. First, the microprocessor-based controller is much more flexible. It readily allows for loop gains, for example, to be nonlinear as well as a function of engine operating condition. Second, the required control strategy may be slightly different for each engine/driveline combination. The microprocessor-based controller reduces part number proliferation, since the differences can be absorbed in the computer memory. Third, additional features can be incorporated into the control system with little additional cost. Self-diagnostics of the engine control system and of other electrical systems, for example, are relatively easy to add without additional hardware.

In a typical application, read-only-memory (ROM) is used for storing look-up tables that allow adjustments to either the gains or control commands as functions of other sensed variables. The central processing unit (CPU) executes the engine control program stored in ROM and, using a

TABLE 35.1. ENVIRONMENTAL EXTREMES FOR AUTOMOTIVE CONTROL SYSTEMS

Environmental Variable	Extreme Operating Conditions
Temperature	− 40 to + 85°C when located away from engine compartment; − 40 to + 120°C when located within engine compartment but not on engine itself
Humidity	0.1 to 200 g/kg, or equivalently, − 40 to + 65°C dew point temperature
Corrosive and contaminating chemicals	Salt spray, fine dust, gravel bombardment, engine and transmission oils, brake fluid, axle grease, washer solvent, gasoline, degreasers, soap, steam, battery acid
Vibration and shock	Up to ± 10 g (rms) vibration, dependent on location and mounting orientation; higher shock levels up to ± 50 g (peak) are encountered during shipping, handling, installation, and road bumps
Electromagnetic impulses	Load dump voltage spikes up to 125 V of 1-s duration, alternator field decays of − 100 V of 200-ms duration, and ignition impulses of 3 V (peak) of 10-μs duration

Source. Ronald K. Jurgen, "The Automobile for Better or Worse," *IEEE Spectrum*, Nov. 1977, p. 32.

scratch-pad memory (RAM), provides commands to the input/output (I/O) controller. The I/O controller processes the commands to match the requirements of a particular actuator, for example, adjust pulse width, adjust voltage level, adjust frequency, and so on.

Sensors. Besides the oxygen sensor and knock sensor described previously, numerous other sensors are required by an engine control system, as indicated in Table 35.2. Nearly all of the sensors in Table 35.2 provide analog signals that must be processed by an analog-to-digital (A/D) converter prior to being used by the computer. The exception is the engine speed and crankshaft position sensor. A magnetic pickup, either mounted in the distributor or near a specially designed toothed wheel on the crankshaft, provides a train of pulses to the electronic controller. To determine engine speed from a distributor pickup, for example, a pulse is sent to the controller every 90° of crankshaft

TABLE 35.2. ENGINE CONTROL SENSORS

Variable to Sense	Range	Common Types
Throttle angle position	0 to 90°	Potentiometer
Crank angle position	0 to 360°	Magnetic pickup
Manifold absolute pressure	30 to 100 kPa	Silicon strain gauge
Barometric pressure	80 to 100 kPa	Silicon strain gauge
Intake air flow	5 to 180 g/s	Hot wire anemometer
Fuel flow	0.3 to 12 g/s	Vane
Air temperature	− 40 to + 85°C	Thermistor
Coolant temperature	− 40 to + 120°C	Thermistor
Engine speed	400 to 5000 r/min	Magnetic pickup

rotation (for an eight-cylinder engine). With use of a reference clock in the controller, the number of reference clock pulses that occur for every distributor pulse is inversely related to engine speed. Generally some form of averaging is required on the speed computation to prevent large deviations in control commands that depend heavily on engine speed.

Actuators. Actuators receive the control commands from the controller and adjust the engine input variable. Table 35.3 summarizes the actuators that would be used in typical engine control systems.

Historically, manifold vacuum has provided the power for automotive accessories and controls, for example, the vacuum windshield wiper, vacuum advance for spark timing, power assisted brakes, and so on. This also is the case for many engine control actuators. In recent years, however, cars have become smaller because they have better fuel economy. The smaller engines operate closer to their maximum torque capability, resulting in lower intake manifold vacuum. Thus vacuum as a power source has been significantly reduced. For this reason the trend is toward electromechanical actuators.

Filter Requirements. Engine torque is produced by a train of pulses caused by the firing of each cylinder every two revolutions of the crankshaft. The result is that engine speed is not a smooth function, even at a so-called steady-state operating condition. Likewise, manifold pressure and air flow consist of an average value plus a train of pulses that arise because of piston motion.

The frequency range of these undesirable pulsations is as low as 16.6 Hz for a four-cylinder engine operating at 500 r/min to as high as 333.3 Hz for a eight-cylinder engine operating at 5000 r/min. Unfortunately it is the lower frequencies that give the most trouble because (1) they are close to the frequency bandwidth of most engine control systems (generally 1–2 Hz) and (2) the peak-to-peak excursions of the measured engine variables are inversely related to the frequency of pulsations for constant load.

Wherever possible, analog filters are used prior to any digital processing by a microprocessor-based controller to prevent aliasing. Generally first-order filters are used because of reduced complexity with a cutoff frequency in the neighborhood of 10 Hz. Likewise, first-order digital filters are generally used to keep the required number of RAM words to a minimum.

Factors That Affect Closed-Loop Performance

As with any control system design, many aspects of the hardware can contribute to poor dynamic or static performance or even loop instability if not properly taken into account. To name just a few,

TABLE 35.3. ENGINE CONTROL ACTUATORS

Actuator	Types in Use
Exhaust gas recirculation valve	(a) Exhaust back pressure and manifold pressure determine pintle position; no electronics (b) Electropneumatic controller determines amount of vacuum to actuator; pintle position feedback required
Carburetor	Air-fuel ratio determined by metering rod position; controlled by pulse width modulation of solenoid or by step motor
Fuel injector	Quantity of fuel determined by pulse width to electromagnetic injector; injectors may be mounted above throttle plate or at each intake valve
Spark pulse	Timing of pulse determined by controller; electronic ignition required
Throttle actuator	(a) Throttle position determined by manifold pressure bleed to spring-loaded actuator; no electronics (b) DC motor with gearbox
Air bypass valve	Pintle position determined by pulse width modulation of solenoid or by step motor

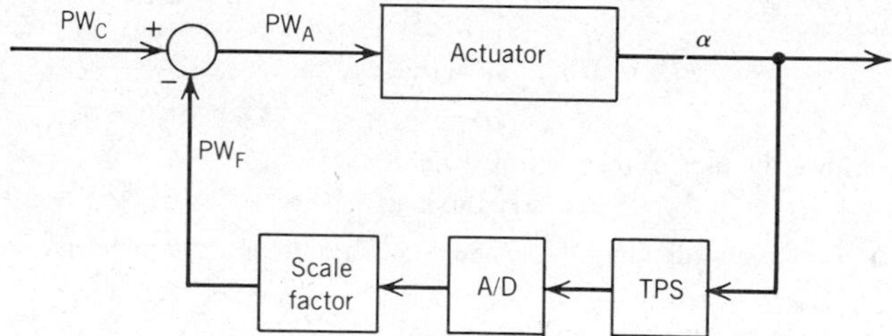

Fig. 35.15 Actuator with throttle position sensor (TPS) feedback.

such things as loop time of the digital control algorithm; inherent update rate of measured signals; time constants of sensors, actuators, or filters; and underflow or overflow of digital filters all can reduce the stability margin of a closed-loop system. Quantization errors of A/D converters not only can contribute to loop instability but also can cause excessive steady-state errors.

As an example, suppose it is desired to use throttle position feedback around an idle speed actuator to reduce ISC sensitivity, to changes in actuator parameters (see Fig. 35.15). The throttle position sensor (TPS) is needed for A/F control over the entire range of throttle motion (0–90°), but for idle speed, control is needed over 0 to 20°. The idle speed control algorithm computes a pulse width PW_C based on the difference between desired and actual engine speed. This pulse width is modified by PW_F from the TPS feedback path such that $PW_A = PW_C - PW_F$ is the actual pulse width issued to the actuator every 50 ms. The desired feedback gain is $PW_F/\Delta\alpha = 34$ ms/°. Since the D/A converter-TPS combination for an 8-bit A/D converter has a scale factor of 255 bits/90°, an internal scale factor of 12 ms/bit is required. Thus PW_F can have values that are only multiples of 12 ms (up to 50 ms).

Better resolution can be achieved by amplifying the output from the TPS signal and then using an additional A/D channel, although the additional hardware means additional cost. Suppose the amplifier gain is chosen to be 3. Then the scale factor of the D/A converter-amplifier-TPS combination has a scale factor of 255 bits/30°, and to obtain a feedback gain of 34 ms/° now requires an internal scale factor of 4 ms/bit. Since PW_F can now have values that are multiples of 4 ms, the improved resolution reduces the likelihood of a limit cycle.

35.1-5 ADVANCED CONCEPTS

Lagrange Method for Optimal Engine Calibrations

Engine calibrations of SA, A/F, and EGR in the past have generally been accomplished through a trial-and-error procedure of on-vehicle testing. This is an extremely time-consuming process. In addition, a set of calibrations for one engine-vehicle combination did not necessarily carry over to another engine-vehicle combination even if it differed only slightly in displacement, torque, and so on. Such an inefficient use of manpower and capital has provided the motivation for systematic application of advanced techniques for the determination of "optimal" calibrations.

The optimal engine control problem[9–12] is to find the set of controls that minimizes the total fuel consumed over a driving schedule such as the FTP, subject to emission constraints. One approach[12] to this problem is to use the engine models that relate emissions production and fuel consumption as a function of the input variables at selected operating points that represent the FTP driving schedule (see Section 35.2-1). Let M be the number of time points in the driving schedule, $\mathbf{x}_i$ the engine operating points defined by speed and load at time t_i, $\mathbf{u}_i$ the engine controls at time t_i, $\mathrm{FC}(\mathbf{x}_i, \mathbf{u}_i)$ the fuel consumption, and $\mathrm{HC}(\mathbf{x}_i, \mathbf{u}_i)$, $\mathrm{NO}_X(\mathbf{x}_i, \mathbf{u}_i)$, and $\mathrm{CO}(\mathbf{x}_i, \mathbf{u}_i)$ the rate of production of the emission variables. Then the optimal control problem is to find $\mathbf{u}_i$ that minimizes $\sum_{i=1}^{M} \mathrm{FC}(\mathbf{x}_i, \mathbf{u}_i)$ subject to the emission constraints

$$\begin{aligned} &\sum_{i=1}^{M} \mathrm{HC}(\mathbf{x}_i, \mathbf{u}_i) \leqslant \mathrm{HC}_{\max} \\ &\sum_{i=1}^{M} \mathrm{NO}_X(\mathbf{x}_i, \mathbf{u}_i) \leqslant \mathrm{NO}_{X\max} \\ &\sum_{i=1}^{M} \mathrm{CO}(\mathbf{x}_i, \mathbf{u}_i) \leqslant \mathrm{CO}_{\max} \end{aligned} \tag{35.9}$$

Additional constraints that must be satisfied are the equations that relate the engine variables $\mathbf{x}$ to the control variables **u**

$$h(\mathbf{x}, \mathbf{u}) = 0, \qquad \mathbf{u} = (\mathbf{u}_1, \mathbf{u}_2, \ldots, \mathbf{u}_M) \qquad \mathbf{x} = (\mathbf{x}_1 \mathbf{x}_2, \ldots, \mathbf{x}_M) \tag{35.10}$$

and the upper and lower bounds of the control variables

$$\mathbf{u}_L \leqslant \mathbf{u}_i \leqslant \mathbf{u}_U \tag{35.11}$$

The Lagrangian is formed by adjoining the exhaust constraints onto the objective function, that is,

$$L = \sum_{i=1}^{M} \mathrm{FC}(\mathbf{x}_i, \mathbf{u}_i) + \lambda_1 \left[\sum_{i=1}^{M} \mathrm{HC}(\mathbf{x}_i, \mathbf{u}_i) - \mathrm{HC}_{\max} \right] + \lambda_2 \left[\sum_{i=1}^{M} \mathrm{NO}_X(\mathbf{x}_i, \mathbf{u}_i) - \mathrm{NO}_{X_{\max}} \right] + \lambda_3 \left[\sum_{i=1}^{M} \mathrm{CO}(\mathbf{x}_i, \mathbf{u}_i) - \mathrm{CO}_{\max} \right] \tag{35.12}$$

The Lagrangian problem, then, is to find **u** that minimizes L subject to the constraints on the engine relations and the control bounds. The solution of the Lagrangian problem is also the solution to the engine calibration problem and, in fact, decomposes into M smaller problems; that is, the controls that minimize L also solve the minimization problem at each time point t_i. Details of solution methods can be found in Chapter 34.

The optimal set of controls is only valid at the M selected operating points (characterized by speed and load). Although these points are representative of the FTP driving schedule, polynomial approximations to the optimal control laws are required to provide a control strategy for operating conditions not included in the M points and to provide a smooth transition of the controls that otherwise might cause erratic behavior of the engine.

Linear-Quadratic Optimization Method

The Lagrange method for optimal engine calibrations does not account for dynamics inherent in the engine, actuators, or sensors. The transients in the FTP driving schedule are relatively slow (compared with the component dynamics); thus for emission calibrations this technique seems to be adequate. The problem is that achieving these calibrations in practice may be difficult because of a deterioration in driveability. That is, because of sensor and actuator dynamics, engine stumble, sag, or hesitation may occur. These are undesirable phenomena that relate to the response of the vehicle to the driver's accelerator pedal commands.

Numerous researchers[13–16] have attempted to compensate for component dynamics. A technique that has received much attention in the aerospace industry over the last decade or so and may some day find application in the automotive industry is *l*inear *q*uadratic (LQ) optimization. This "drive-by-wire" concept (compared with "fly-by-wire" in the aerospace industry) makes use of a microcomputer system to control engine inputs (including the throttle) in response to the driver's requested throttle command. This is fundamentally different from conventional control in that here the driver has no direct link to the throttle position. To provide fail-safe operation in practice, the LQ control would likely add or subtract control perturbations around nominally controlled mechanical connections (at least for the throttle control).

The most compelling reason for use of LQ optimization is the uncomplicated structure of the controller. The controller design requires that linearized engine models in state-variable form be created. To limit the number of models in practice, models are created at points that are representative of engine operation and then smooth functions are used to allow for changes in the model parameters. The controller provides commands for throttle position, EGR, A/F, and SA.

The LQ problem is formulated as an infinite time regulator problem where $\delta\dot{\mathbf{u}}(t)$ is chosen to minimize the quadratic functional

$$J = \int [\delta\mathbf{x}^T \mathbf{S} \delta\mathbf{x} + \delta\mathbf{u}^T \mathbf{Q} \delta\mathbf{u} + \delta\dot{\mathbf{u}}^T \mathbf{R} \delta\dot{\mathbf{u}}]\, dt \tag{35.13}$$

subject to the linearized dynamic equation

$$\delta\dot{\mathbf{x}} = \mathbf{A}\delta\mathbf{x} + \mathbf{B}\mathbf{u} \tag{35.14}$$

The state vector $\delta\mathbf{x}$ includes all the engine, actuator, and sensor dynamics; however, since we are only interested in the driveability question, the penalty matrix **S** need only penalize the states that represent engine torque and speed. The solution to this optimization problem is the familiar state feedback control law

$$\delta\dot{\mathbf{u}} = -\mathbf{K}_1 \delta\mathbf{x} - \mathbf{K}_2 \delta\mathbf{u} \tag{35.15}$$

where

$$\mathbf{K}_1 = \mathbf{R}^{-1}\mathbf{P}_2 \quad \text{and} \quad \mathbf{K}_2 = \mathbf{R}^{-1}\mathbf{P}_3 \tag{35.16}$$

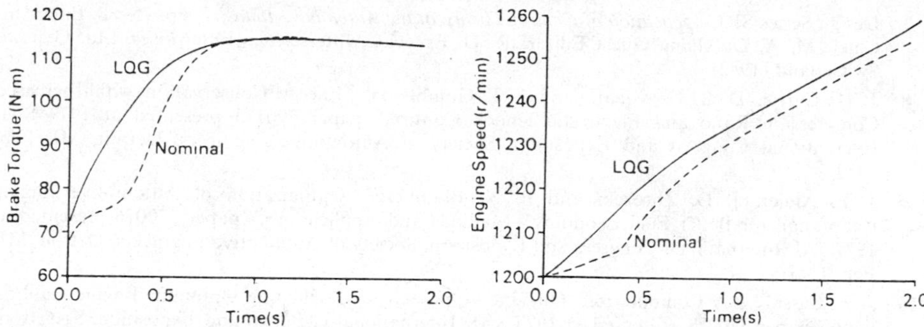

Fig. 35.16 Simulated (LQG) linear quadratic gaussian problem and nominal response to step throttle input.

The matrices $\mathbf{P}_2$ and $\mathbf{P}_3$ are determined from the solution to the algebraic matrix Riccati equation

$$\begin{aligned} \mathbf{P}_1\mathbf{A} + \mathbf{A}\mathbf{P}_1 + \mathbf{S} - \mathbf{P}_2^T\mathbf{R}^{-1}\mathbf{P}_2 &= \mathbf{0} \\ \mathbf{B}^T\mathbf{P}_1 + \mathbf{P}_2\mathbf{A} - \mathbf{P}_3\mathbf{R}^{-1}\mathbf{P}_2 &= \mathbf{0} \\ \mathbf{P}_2\mathbf{B} + \mathbf{B}^T\mathbf{P}_2^T + \mathbf{Q} - \mathbf{P}_3\mathbf{R}^{-1}\mathbf{P}_3 &= \mathbf{0} \end{aligned} \tag{35.17}$$

For most practical applications the full state, $\delta\mathbf{x}$, is not available to the controller, either because some of the states are not measurable or because the control designer chooses not to measure certain states owing to sensor costs. In any event a state reconstructor must be used, such as the Kalman filter.[17] The Kalman filter finds the minimum mean square error state estimate for a linear system corrupted by white gaussian plant and measurement noise. It uses the measured states and the known system inputs and provides the optimum state estimate $\delta\hat{\mathbf{x}}$. Thus in the control law of Eq. (35.15), $\delta\mathbf{x}$ is replaced by $\delta\hat{\mathbf{x}}$. Because of the assumed statistical characterization of the noise, the LQ problem has now become an LQG (*l*inear *q*uadratic *g*aussian) problem.

To illustrate the advantage of LQ control over conventional control in a transient maneuver, consider a 20% step increase in throttle command corresponding to a 48 to 56 km/h acceleration maneuver. The LQ controller increases SA while coordinating the EGR command and actual throttle opening. Preventing an abrupt opening of the throttle while sharply increasing the initial fuel rate reduces the deviation in A/F at the cylinders. A comparison of the torque and speed response with conventional control is given in Fig. 35.16. For conventional control, the throttle snap causes the air flow to change rapidly. Because the fuel flow changes more slowly there is a "leaning" of the A/F at the cylinders. This results in a slight sag in torque, which the driver may find unacceptable.

References

1 "EPA Dynamometer Driving Schedule," *Title 40 United States Code of Federal Regulations*, Part 86, Appendix 1, Rev. of July 1, 1978.

2 "Test Procedures," *Title 40 United States Code of Federal Regulations*, Part 86.177-5 through 86.177-15, Rev. of July 1, 1978.

3 L. S. Vora, "Computerized Five Parameter Engine Mapping," paper 770079 presented at 1977 SAE International Congress and Exposition, Society of Automotive Engineers, Detroit, MI, Feb. 1977.

4 J. A. Tennant, R. A. Giacomazzi, J. D. Powell, and H. S. Rao, "Development of Engine Models via Automated Dynamometer Tests," paper 790179 presented at 1979 SAE International Congress and Exposition, Society of Automotive Engineers, Detroit, MI, Feb. 1979.

5 D. M. Young and R. T. Gregory, *A Survey of Numerical Mathematics*, Vol. I, Addison-Wesley, Reading, MA, 1972.

6 D. J. Dobner, "A Mathematical Engine Model for Development of Dynamic Engine Control," paper 800054 presented at 1980 SAE International Congress and Exposition, Society of Automotive Engineers, Detroit, MI, Feb. 1980.

7 F. E. Coats, Jr. and R. D. Fruechte, "Dynamic Engine Models for Control Development. Part II: Application to Idle Speed Control," *Int. J. Vehicle Design*, Technological Advances in Vehicle

Design Series SP3, *Application of Control Theory in the Automotive Industry*, pp. 75–88, Editor-in-Chief: M. A. Dorgham, Guest Editor: R. D. Fruechte, Interscience Enterprises, Ltd. Geneva, Switzerland (1982).

8 J. H. Currie, D. S. Grossman, and J. J. Gumbleton, "Energy Conservation with Increased Compression Ratio and Electronic Knock Control," paper 790173 presented at 1979 SAE International Congress and Exposition, Society of Automotive Engineers, Detroit, MI, Feb. 1979.

9 J. E. Aiuler, J. D. Zbrozek, and P. N. Blumberg, "Optimization of Automotive Engine Calibration for Better Fuel Economy—Methods and Applications," paper 770076 presented at 1977 SAE International Congress and Exposition, Society of Automotive Engineers, Detroit, MI, Feb. 1977.

10 J. F. Cassidy, "A Computerized On-Line Approach to Calculating Optimum Engine Calibrations," paper 770078 presented at 1977 SAE International Congress and Exposition, Society of Automotive Engineers, Detroit, MI, Feb. 1977.

11 A. R. Dohner, "Transient System Optimization of an Experimental Engine Control System Over the Federal Emissions Driving Schedule," paper 780286 presented at 1978 SAE International Congress and Exposition, Society of Automotive Engineers, Detroit, MI, Feb. 1978.

12 H. S. Rao, A. I. Cohen, J. A. Tennant, and K. L. Van Voorhies, "Engine Control Optimization Via Nonlinear Programming," paper 790177 presented at 1979 SAE International Congress and Exposition, Society of Automotive Engineers, Detroit, MI, Feb. 1979.

13 J. F. Cassidy Jr., M. Athans, and W. H. Lee, "On the Design of Electronic Automotive Engine Controls Using Linear Quadratic Control Theory," *IEEE Trans. Auto. Contr.* **AC-25**(5):901–912 (Oct. 1980).

14 J. B. Lewis, "Automotive Engine Control: A Linear Quadratic Approach," MS Thesis, Massachusetts Institute of Technology, Cambridge, MA, 1980.

15 D. L. Stivender, "Engine Air Control—Basis of a Vehicular Systems Control Hierarchy," paper 780346 presented at 1978 SAE International Congress and Exposition, Society of Automotive Engineers, Detroit, MI, Feb. 1978.

16 R. L. Woods, "An Air Modulated Fluidic Fuel-Injection System," *Trans. ASME J. Dynamic Sys. Measure. Contr.* **101**:71–76 (Mar. 1979).

17 R. E. Kalman, "A New Approach to Linear Filtering and Prediction Problems," *Trans. ASME J. Basic Eng.* **82**:34–35 (Mar. 1980).

35.2 INDUSTRIAL PROCESS CONTROL SYSTEMS

Thomas E. Marlin

35.2-1 Importance of Process Control

This section presents the basic concepts and sample implementations of automatic control as applied in the process industries. The process industries are characterized by large, complex plants requiring excellent automatic control (process control) for safe, profitable operation. Most plants process fluid, slurries, or solids through an integrated set of unit operations such as furnaces, chemical reactors, physical separation units, and heat exchangers. Large amounts of material, up to hundreds of tons per hour, are processed at extreme pressures, over 100 bar, and temperatures ranging from −200 to over 1000°C. Adding to the control challenges are the stringent product specifications, which can require impurity compositions below ten parts per million. Major business areas typically included in the process industries are petroleum refining, chemical plants, paper mills, pharmaceuticals, food processing, steel plants, and utility plants.

A basic goal common to these industries is achievement of desired production rates of material within specified quality restrictions in a safe and environmentally acceptable manner. For purposes of process control, this goal can be expanded into the control objectives given in Table 35.4. This table is a valuable checklist that can be referred to when developing a preliminary control system design for regulating temperatures, pressures, flow rates, and so on. The relationship between general objective and specific control design is often not obvious. Therefore the process control designer must have a good understanding of the plant's economic driving forces, of steady-state and dynamic operation, and of automatic control theory.

Naturally a key control objective is to increase the plant's profitability by, for example, increasing yields of higher value products and reducing consumption of fuel. An increase in profitability is often achieved through control that results in plant operation close to optimum conditions. Figure 35.17

TABLE 35.4. PROCESS CONTROL OBJECTIVES

Operate process in safe, stable manner
Design control systems that operator can monitor, comprehend, and when necessary, intervene selectivly
Prevent large deviations from product specifications during upsets
Allow operator to change one set point (reference value) without unduly upsetting other controlled variables
Prevent large, rapid changes in manipulated variables that might violate operating constraints or upset integrated or downstream units
Operate process consistent with product quality objectives, e.g., deviations in quality of one product might be much more costly than in another product
Control product qualities at values that maximize operating profit when considering product rates and values and energy consumption

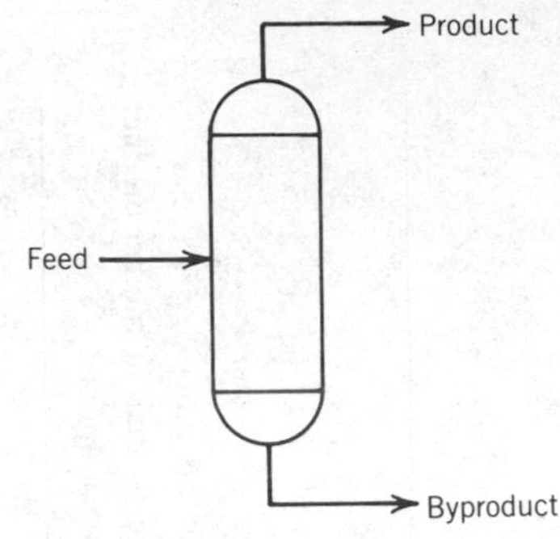

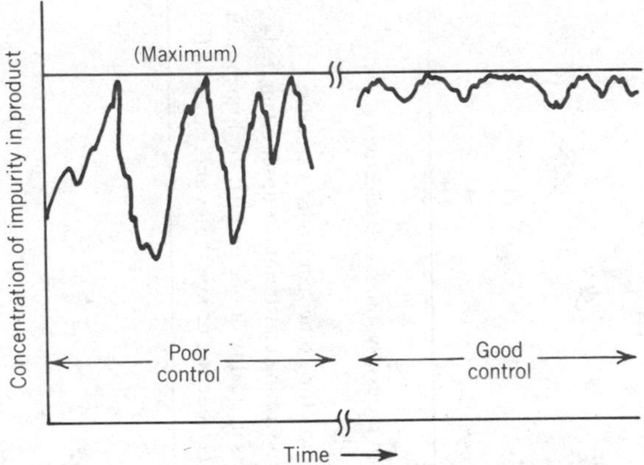

Fig. 35.17 Distillation column separating a feed into a valuable product stream with a maximum impunity level and a lower-level by-product stream.

shows a distillation column that separates a feed into a valuable product stream with a maximum impurity level and a lower value byproduct stream. Improved control through methods discussed in this section results in upgrading of a larger percentage of the feed to the higher value product while observing the maximum impurity limitation. Experience has demonstrated that improved process control can capture large economic credits in a wide range of plants such as refineries,[1] chemical plants,[2,3] utility plants,[4] paper mills,[5] and steel plants.[6]

To capture these credits, process control is typically applied in all process units of an entire plant. Most measured variables are transmitted from sensors throughout the plant to the centralized control house. The measurements are displayed in a specially designed control room for one or more plant operators who are responsible for the plant's operation. Analog or digital calculations for up to several hundred closed-loop controllers are performed by computing equipment in the control room. The resulting signals are transmitted to the final control elements, mostly valves, to effect changes in the process variables. Naturally the plant operator has the option to intervene in any closed-loop system and directly adjust the manipulated variable (usually referred to as manual or open-loop operation). Automatic safety and alarm systems also exist to prevent injury to personnel or damage to equipment.

The design of a plant-wide control system is clearly a complex task beyond the scope of this discussion. However, the basic concepts required to analyze smaller plant sections and design control systems are explained here, and these concepts, tempered with considerable experience, are sufficient to implement process control successfully in most process industry plants. The next section presents how to characterize the typical dynamic response (transfer function) of a control loop. Subsequent sections present use of this characterization in the design of single-variable and multivariable process control systems and in determination of adjustable parameters (called tuning). Then extensions to improve regulatory and optimizing control are presented. Finally, some general considerations in process control design are summarized and demonstrated with examples.

35.2-2 Process Dynamics

A thorough understanding of dynamic response without control, termed open-loop response, is required to design a control system for a process. For most industrial processes, dynamic response is the result of complex physiochemical processes that would take up to thousands of differential equations to model in detail.[7] Fortunately, the control designer is usually interested in the relationships between manipulated and controlled variables over a limited range of operation, and these relationships can often be characterized by linear equations with only a few coefficients. These equations are used directly in the design and tuning of the control systems described hereafter.

Self-Regulation

Key factors in the dynamic response of an open-loop process are stability and self-regulation. Nearly all industrial processes are open-loop asymptotically stable (defined in Chapter 34) except for the non-self-regulatory processes that will be described; therefore no other open-loop unstable process is considered in this section.

Self-regulation is a property inherent in most processes that causes them to attain a new steady state after a change has been made in a manipulated variable. This property is best explained with reference to the simple level processes in Fig. 35.18. For the self-regulating process, the level increases as a result of an increase in the flow into the tank. The increased hydrostatic head causes an increase in the flow out. Eventually the balance between the flow in and out is reestablished, and the level attains a new, altered steady state. For the non-self-regulatory process, the flow in and flow out are not coupled through the level, and the level changes (until it overflows the tank) at the constant rate proportional to the difference between the flow in and out. As a result, level processes are often referred to as pure integrators. Level change is the most frequently occurring non-self-regulatory industrial process, and their control is addressed with a separate tuning method. Therefore the remainder of this section emphasizes self-regulatory processes.

Process Models

The most frequently encountered self-regulatory process is overdamping (defined in Chapter 34), that is, the dynamic response is not oscillatory and never exceeds its new steady state when responding to a step change. For a limited region around a normal operating condition, this process can be modeled

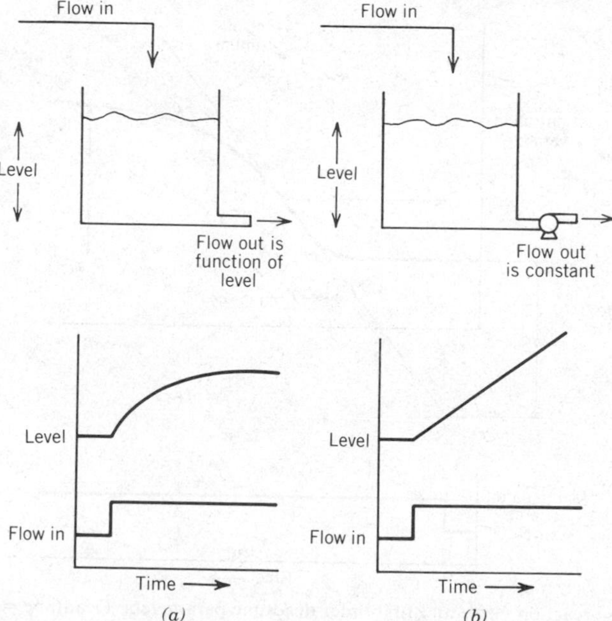

Fig. 35.18 Examples of self-regulatory (*a*) and non-self-regulatory (*b*) tank levels.

by simple transfer function models such as the following:

First-order lag and deadtime

$$\frac{C(s)}{M(s)} = \frac{Kpe^{-\theta s}}{(1+\tau s)}$$

Second-order lag and deadtime

$$\frac{C(s)}{M(s)} = \frac{Kpe^{-\theta s}}{(1+\tau_1 s)(1+\tau_2 s)}$$

where $C(s)$ = change in controlled variable
Kp = steady-state process gain
$M(s)$ = change in manipulated variable
θ = deadtime
τ, τ_1, τ_2 = time constants
s = Laplace variable

The deadtime is the time between when the manipulated variable is changed and when the controlled variable begins to respond. For the design and tuning of controllers, a first-order lag and deadtime process provides an adequate approximation.

Process Reaction Curve

The parameters in a first-order deadtime model—gain, time constant, and deadtime—can be determined from the empirical response of the controlled variable to a step change in the manipulated variable. This response, termed the process reaction curve, is shown in Fig. 35.19 with the method for calculating model parameters.[8] The process reaction curve can be obtained by manually adjusting the manipulated variable; several process reaction curves with manipulated variable changes of different sizes and directions should be obtained to ensure that the data are valid and the process is nearly linear. Naturally the change in manipulated variable must be large enough to produce a change in controlled variable substantially greater than the signal's noise without being so large as to constitute a significant process upset.

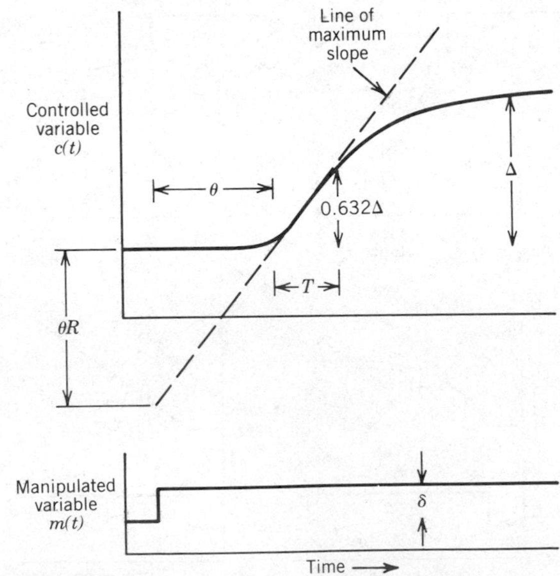

Fig. 35.19 Process reaction curve and first-order deadtime parameters. Deadtime = θ, time constant = t or Kp/R, process gain = Δ/δ.

The process reaction curve method is the simplest and most widely applied method for estimating model parameters. However, it requires the intentionally introduced upset to last until the process has reached a new steady state. For the few cases in which such a long upset is not acceptable, the parameters can be estimated from the response of the process to a modified impulse (square wave); the estimation procedure requires a small computer program to evaluate the Fourier integral transform.[9] Alternate methods[10] also exist for oscillatory and higher order processes not addressed by the method in Fig. 35.19.

Both the step and impulse response methods require that the process be operated in open loop. Statistical methods have been proposed which periodically perturb the closed-loop control system and estimate the model parameters. These methods have not been widely employed because of the required periodic process upsets and the difficulty in estimating parameters with correlated noise in the closed-loop system.[11]

If plant tests are not possible or if the plant is being designed, use of the empirical methods for determining model coefficients is not feasible. For some processes, rough estimates of the parameters can be easily obtained. For example, a time constant might be related to the holdup in the drum (volume/flow), or the deadtime might be the transportation lag in a pipe (length/velocity). However, the process dynamics of many industrial processes are the results of complex interactions among chemical reactions, phase equilibria, and multiple holdups. Therefore for complex processes like those in chemical reactors and distillation columns requiring detailed analysis, the usual recourse is to perform a detailed simulation of the process dynamics.

Finally, the emphasis here on linear models with constant coefficients should not be taken to imply that such process models are very accurate, especially over a wide range of operating conditions. Rather the success of these models in process control indicates that the control methods explained in the following sections are not overly sensitive to model errors. However, large errors due to process nonlinearities, unmeasured load upsets, and measurement noise can degrade control performance significantly.

35.2-3 Single-Variable Feedback Control

A typical process plant has several hundred controlled variables, many of which can be regulated by single-input, single-output feedback controllers. In addition, nearly all advanced control designs in the process industries (see Sections 35.2-5 and 35.2-6) retain single-variable controllers at the lowest level of a control hierarchy. Fortunately, decades of plant experience have demonstrated that good control performance can be obtained for a wide range of processes with one linear feedback control algorithm, the proportional integral derivative, or PID.

The performance of this controller can be altered to conform to many application needs by adjusting (tuning) a few parameters. Due to its great success, the algorithm presented in this section is available in nearly all commercial analog and digital instrumentation and computer control systems; therefore the explanations and tuning guidelines presented are applicable with little or no alteration on these systems. In this section the basic concepts of the PID controller are presented and tuning guidelines are explained. Then moderate alteration in the algorithm and alternate tuning methods commonly employed in response to special process circumstances are explained. Finally, filtering and validity checking of the measured input signal are discussed.

PID Controller

The relationship of the PID controller to other components in the feedback system is shown in Fig. 35.20. The controller operates on an error signal which is the difference between the measured variable and the set point (reference). The general goal of the controller is to maintain the error within reasonable limits and usually to attain a zero error. The controller achieves this goal by calculating the manipulated variable; this calculation is most easily understood as the sum of three modes. The first mode is proportional control, which is calculated according to the following equation:

$$m_p(t) = K_c e(t) + \text{constant} \tag{35.18}$$

where K_c = proportional gain and $100/K_c$ = proportional band. The proportional mode provides an instantaneous correction in the manipulated variable for an error signal. However, a purely proportional controller cannot reduce the error to zero; this sustained error is called proportional offset. Attempting to reduce the error to a small value by increasing the gain leads to closed-loop instability for all but simple first-order processes without deadtime.

Proportional offset can be eliminated by the integral mode (also called reset), which is calculated

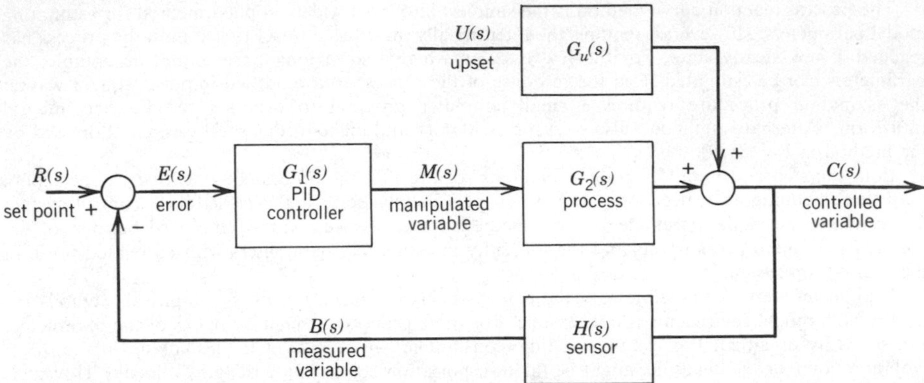

Fig. 35.20 Block diagram of single-variable control system using a proportional-integral-derivative (PID) controller.

according to the following equation:

$$m_I(t) = \frac{K_c}{K_I}\int e(t)\,dt + \text{constant} \tag{35.19}$$

where K_I = integral time. The integral mode continuously changes the manipulated variable until the error is zero, but it does not respond quickly to the first appearance of a non-zero error. Additional improvement results from the derivative or rate mode, which is calculated as follows:

$$m_d(t) = K_c K_D \frac{d}{dt} b(t) + \text{constant} \tag{35.20}$$

where K_D = derivative time. The derivative mode recognizes the rate of change of the measured variable and corrects the manipulated variable appropriately.

The PID controller combines the three modes into one algorithm, given here in continuous and discrete forms:

$$m(t) = K_c\left(e(t) + \frac{1}{K_I}\int e(t)\,dt + K_D\frac{db(t)}{dt}\right) + \text{constant} \tag{35.21}$$

$$m_n = K_c\left(e_n + \frac{\Delta t}{K_I}\sum_i e_i + \frac{K_D}{\Delta t}(b_n - b_{n-1})\right) + \text{constant} \tag{35.22}$$

where n = execution number and Δt = execution period. This controller is most commonly employed as a PID or as a PI, which requires K_D to be equal to zero. Naturally, negative feedback requires proportional gains of different signs for different processes. The sign is usually entered separately through a switch that indicates the "sense" of the controller, and the controller gain is always positive.

When the controller is initially placed in automatic from manual, the controller should initialize the control calculation of $m(t)$ to the current value of the manipulated variable to prevent a process upset. This initialization is usually termed bumpless transfer. It is during this initialization that the "constant" term is evaluated. Other initialization options, such as setting the set point equal to the current measured variable, are available in some digital systems, but their use is very application- and digital-system specific.

Controller Tuning

The ability of the controller in automatic operation to achieve its control objectives is highly dependent on the values of the tuning parameters K_c, K_I, and K_D. For most processes the objective is to maintain the controlled value near the set point as upsets and set point changes occur, while also limiting the variation in the manipulated variable. A sample response of a single-variable system to a set point change is given in Fig. 35.21, with useful terms explained. Ascertaining the correct tuning parameters for each process is a stepwise procedure involving (1) determination of initial control parameters, (2) their implementation and observation of the system performance, and (3) adjustment of the parameters to obtain desired performance. Fortunately the controller performance is not highly

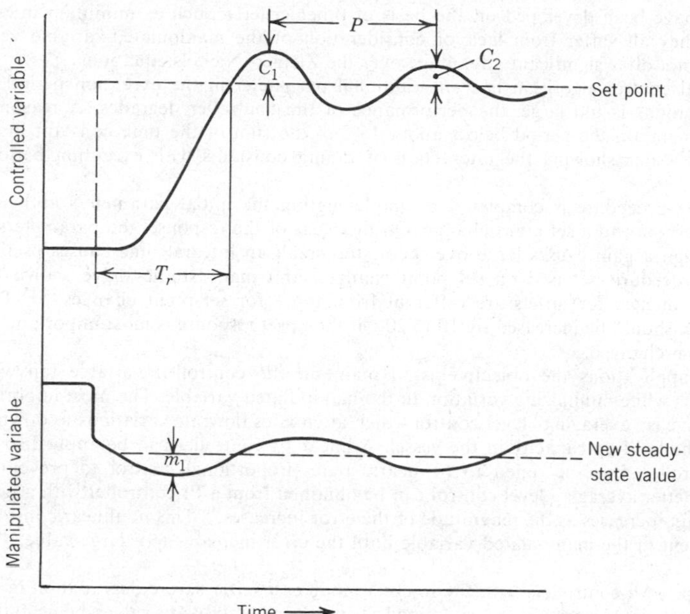

Fig. 35.21 Closed-loop response to a set point change. C_1 = controlled-variable overshoot, C_1/C_2 = decay ratio, m_1 = manipulated variable overshoot, p = period, T_r = rise time.

sensitive to controller tuning; therefore small changes in the process (e.g., nonlinearities) and tuning errors do not significantly degrade performance.

The initial controller parameters for overdamped processes can be calculated directly from the open-loop process reaction curve described in Fig. 35.19. Reasonable initial parameters can be calculated using the well-known Ziegler-Nichols equations that follow.[12]

Proportional

$$K_c = \frac{\delta}{\theta R} \tag{35.23}$$

PI

$$K_c = \frac{0.90\delta}{\theta R} \qquad K_I = 3.3\theta \tag{35.24}$$

PID

$$K_c = \frac{1.20\delta}{\theta R} \qquad K_I = 2.0\theta \qquad K_D = 0.5\theta \tag{35.25}$$

Some care must be taken in using these equations. First, the 4 : 1 decay ratio resulting from these constants is too oscillatory for some process applications. For these processes, it would be advisable to increase the integral time for the initial estimate by about 25%. Second, the derivative time is rather large, especially since most industrial measurements have significant noise; therefore the initial derivative time should be decreased by 50%. Finally, the Ziegler–Nichols equations yield very large gains for processes with small deadtimes. While the resulting tuning parameters theoretically provide good controlled-variable response, the large variations in the manipulated variable, approaching bang-bang control, are not usually acceptable. Where this limitation applies, the product of the initial controller gain K_c and the process gain K_p should not be greater than about 1.2, and the integral time should not be less than three quarters of the sum of the deadtime and first-order lag. Other initial tuning rules have been developed on the basis of other criteria, such as minimum integral squared error,[13] but they all suffer from lack of consideration of the manipulated-variable variation and, therefore, do not offer significant advantage over the Ziegler–Nichols equations.

The digital control algorithm has an additional parameter in the execution period. If the time between executions is too large, the performance of the controller degrades. A reasonable rule of thumb is to maintain the period below about 15% of the sum of the time constant plus deadtime; detailed correlations showing the interaction of tuning constants with execution period are available.[13,14]

The tuning procedure is completed by implementing the initial parameters and observing the closed-loop response to a set point change. On the basis of the response, the parameters can be fine tuned. Too high a gain causes large overshoot; too small an integral time causes oscillations. This fine-tuning procedure is based on set point changes, and many studies have shown that optimal controller parameters for upsets are different from those for set point changes.[14,15] Basically, the controller gain should be increased by 10 to 20% if the upset response is most important[14] and the set point is seldom changed.

In some applications the objective is to maintain the controlled variable (anywhere) within specified limits while minimizing variation in the manipulated variable. The most important example of this objective is "averaging" level control which attenuates flowrate variations to downstream units by utilizing the hold-up capacity in the vessel. A linear PI controller can be applied to this process, but the controller must be tuned to be nearly pure proportional control to prevent closed-loop oscillations. Better averaging level control can be obtained from a PI controller with a nonlinear gain so that the gain increases as the magnitude of the error increases.[16] This nonlinear controller provides little adjustment in the manipulated variable until the error increases to a large value.

Reset Windup. All controllers with the integral mode can suffer severe degradation in performance when the manipulated variable is constrained. Typical constraints are valves being fully opened or closed and speed limitation on the valve or other final control element. When the manipulated variable is at an upper or lower limit, the error signal remains non-zero for an extended period, and the integral of the error becomes very large. This large integral value results in the manipulated variable remaining at its constraint until the integral control term has been reduced by a large overshoot in the controlled variable. This problem is termed reset windup, and various approaches are available to reduce its effect in digital and analog systems. Basically, these approaches ensure that the calculated controller output and the actual manipulated value do not deviate by too large an amount.[17,18]

Adaptive Tuning. All of the preceding material assumed a linear process, that is, that the parameters in the process reaction curve are constant. Naturally, all physical processes are nonlinear, and some important industrial processes are so nonlinear that the controller tuning constants must be frequently adjusted to compensate for changes in operating conditions. When this adjustment is performed automatically in the closed-loop system, it is termed adaptive tuning. (Adaptive control is discussed further in Chapters 33 and 34.) One method of adaptive tuning involves choosing a nonlinear final control element, usually a valve, to compensate for a predictable process nonlinearity. A typical example is the use of an equal percentage valve in a flow control loop to compensate for the change of pump head with flow.[19] A similar adaptive tuning method is alteration of controller parameters according to deterministic rules as process conditions change. This method, termed gain scheduling, can be applied when the process gain changes with process throughput or product purity.[20] The most sophisticated and potentially most powerful approach for adaptive tuning involves on-line estimation of the process model parameters and recalculation of the tuning parameters. This method has not been widely applied because of the need to frequently force (upset) the process and because of the difficulties in statistically estimating parameters in a closed-loop system, but successful industrial applications have been reported, for example, temperature control of a process furnace.[21]

Input Processing. The measured signal to the controller is corrupted by measurement and transmission noise and by high-frequency process variation. If too large, this noise causes undesirable variation in the manipulated variable, especially when the controller gain or derivative time is large. Therefore it is common to filter the input signal to attenuate noise. The simplest and most widely used filter is the exponential filter, which is simply a first-order lag:

$$b'(s) = \frac{b(s)}{1 + \tau_f s} \tag{35.26}$$

where τ_f = filter time constant. This constant is adjusted to reduce the noise. Too small a value does not significantly attenuate the noise; too large a value introduces a significant lag into the feedback control system. Simpler tuning and better noise rejection have been attributed to the addition of a nonlinear gain to either the filter[22] or the controller.[14] The nonlinear feature allows heavy damping for noise with small amplitude and fast response for large input signal changes.

Another important method of input signal conditioning is validity checking. A common goal of such checking is to ensure that the sensor is sending a signal by comparing the measurement with its expected range, which has a live zero (normally 4 to 20 mA). Other checks involve identifying questionable measurements which, for example, do not change or change at too high a rate. Clearly the control engineer must have a good understanding of the process dynamics and noise characteristics to design and tune these checks for which no standard algorithms yet exist.

35.2-4 Multivariable Feedback Control

Each process unit operation like that of a furnace, distillation column, or chemical reactor has several manipulated and controlled variables. In the most general sense the control system regulates all of the controlled variables by adjusting all of the manipulated variables. Since a change in one manipulated variable can and often does influence more than one controlled variable, the design of multivariable feedback control schemes is considerably more complex than that of single-variable control. The simplest multivariable feedback design, a combination of single-variable controllers, functions well in many processes if the controlled and manipulated variables are paired correctly. In some cases process interactions are so strong that compensation must be included in the control system.

Interaction Analysis

When sufficient empirical experience with similar processes is not available, a method of analyzing the process interactions is needed for the design of a multivariable control system. The method should indicate how controlled and manipulated variables are paired and whether single-loop control performance is expected to suffer from strong process interactions. A method termed relative gain analysis[23–25] is the most widely used in the process industries for multivariable control. Steady-state relative gain analysis is explained with reference to the simple blending process shown in Fig. 35.22.

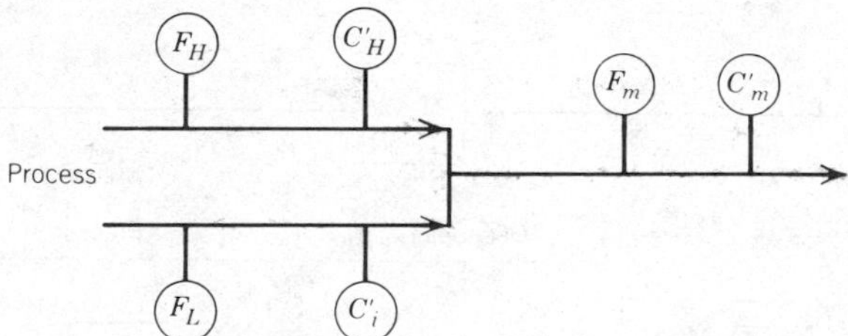

Fig. 35.22 Relative gain for blending process. F_H = flow of stream H; C'_H = property of stream H; F_L = flow of stream L; C'_L = property of stream L; F_m = flow of mixed stream; C'_m = property of mixed stream.

Model

$$F_M = F_H + F_L \qquad C_m = \frac{F_H C_H}{F_H + F_L} \qquad \text{with} \qquad C_H = C'_H - C'_L, \quad C_m = C'_m - C'_L$$

Relative gain

	F_H	F_L
F_m	C_m/C_H	$1 - C_m/C_H$
C_m	$1 - C_m/C_H$	C_m/C_H

Sample values

$C_H = 1000 \qquad F_H = 100$

$C_m = 100 \qquad F_H = 1000$

	F_H	F_L
F	0.10	0.90
C_m	0.90	0.10

The process mixes two streams of different compositions to obtain a blended stream of desired composition and flowrate. The relative gain relating manipulated variable j and controlled variable i is defined as

$$\lambda_{ij} = \frac{\left(\dfrac{\partial c_i}{\partial m_j}\right)_{m_{K \neq j}}}{\left(\dfrac{\partial c_i}{\partial m_j}\right)_{c_{K \neq i}}} \tag{35.27}$$

The relative gain values can be calculated from models, as is done in Fig. 35.22, or from changing the inputs one at a time in the plant or a complex simulator. A matrix M is formed from the resulting relationships, which are $\Delta c_i/\Delta m_j$, with all other m_K's constant. The transpose of the inverse of M is then calculated and called C. The relative gain is the product of each element of M with its corresponding element (same row and column) of C.[24]

A relative gain value of zero indicates that the manipulated variable has no influence on the controlled variable; a value of one indicates that the manipulated variable influences the controlled variable and no other. For easy analysis, the relative gain values are usually presented in an array as shown in Fig. 35.22. The example indicates that relatively weak interaction will occur and good control should be obtained if two single-loop controllers are paired as follows: C_m to F_H and F_m to F_L. The other possible pairing indicates very strong interaction and poor performance. Generally, relative gains below 0.70 or above about 1.5 indicate questionable single-loop performance, perhaps requiring decoupling (discussed in this section). Note that the relative gain does not consider the response of various pairings to upsets, and this important factor might be overriding in selecting the pairing.[26] Additional details on this method and its application to various processes like distillation[27] and extensions to consider linear dynamics[28] are available in the literature.

One High-Priority Variable. In many cases the control objectives indicate that one of the interacting controlled variables is much more important than the others. In these cases single-variable control is

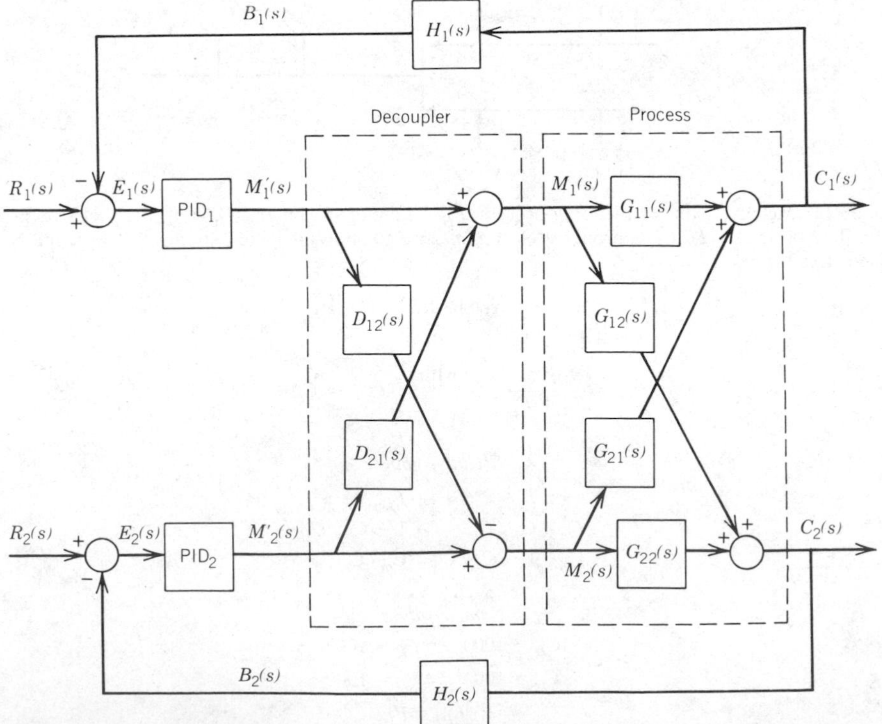

Fig. 35.23 Block diagram of multivariable control system with decoupler. PID = proportional-integral-derivative algorithm.

adequate even if the relative gain array indicates strong interaction. The pairing should be chosen to provide a high relative gain and fast response for the key controlled variable, and its controller should be tightly tuned as described in Section 35.2-3. The process interactions are attenuated by tuning the less important controllers loosely with low controller gains and large integral times.

Decoupling

The approach of using several single-variable controllers is not appropriate for processes with strong interaction requiring good control of several variables. The standard method for process control system design is to retain the basic PID algorithm but to alter the algorithms' outputs in a manner that compensates for process interactions. The method is explained for a two-variable system; extensions to higher order systems follow easily.[9] As shown in Fig. 35.23, the calculation of the error

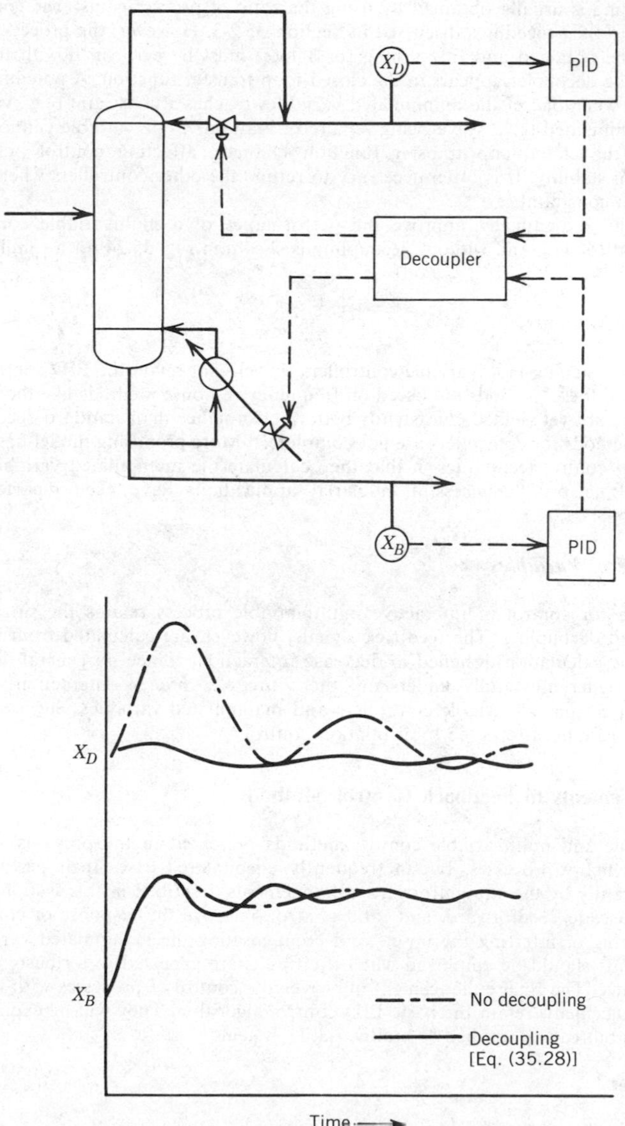

Fig. 35.24 Response of distillation control with and without decoupling for a set point change in X_B. PID = proportional-integral-derivative algorithm. From W. Luyber, *AIChE Journal* **16**:198–203 (Mar. 1970).

for each controller is the same as that for single-variable control. The outputs of the controllers (m_i'), however, are not implemented as the process manipulated variables. They are inputs to decoupling calculations which compensate for the process interaction terms $G_{12}(s)$ and $G_{21}(s)$. The decoupling terms given in Eq. (35.28) allow each controller output to affect only one controlled variable.[29]

$$D_{12}(s) = \frac{G_{12}(s)}{G_{22}(s)}$$
$$D_{21}(s) = \frac{G_{21}(s)}{G_{11}(s)} \tag{35.28}$$

All transfer functions can be determined empirically from process reaction curves. Studies have shown that if the numerator and denominator of the decoupling terms have similar dynamics, adequate decoupling is usually obtained by using the ratio of process gains. The controllers can be tuned according to the procedures discussed in Section 35.2-3. However, the process reaction curves (using m_i') must be obtained and fine tuning for a loop must be performed with the decoupler in operation, since the decoupler appears in the closed-loop transfer function. A potential problem with decoupling arises when one of the manipulated variables reaches a constraint (e.g., valve fully open) or cannot be manipulated (e.g., valve being repaired). Naturally, one variable cannot be controlled. More seriously, the closed-loop transfer function changes, affecting control performance and, sometimes, system stability. It is often necessary to retune the other controllers when a manipulated variable reaches a constraint.

Decoupling can dramatically improve the performance of a multivariable control system. A comparison of control with and without decoupling is shown in Fig. 35.24 for a simulated distillation column.[29]

Modern Control Methods

New methods for designing multivariable controllers that do not retain the PID controllers are being developed. Some of these methods are based on frequency response methods like the inverse Nyquist array.[30] They have not yet yielded consistently better performance than standard decouplers, perhaps because design methods for controllers are not complete.[31] More promising model-based methods are similar to optimal control techniques in that they calculate the manipulated variables to achieve a desired control trajectory.[32] Successful industrial applications have been reported on catalytic crackers[33] and boilers.[34]

Calculated Control Variables

A final approach for control of interactive, multivariable process retains the single-variable PID controllers without decoupling. The feedback signals, however, are calculated from many measured variables, with the calculation designed to decrease interactions. Since no general design method is available, the designer must fully understand the sources of process interaction and the correct coupling between calculated-variable controllers and manipulated variables. Successful applications of this approach have been reported in distillation control.[35,36]

35.2-5 Enhancements to Feedback Control Methods

The single-variable and multivariable control methods presented in the previous sections provide good control for many processes, but in frequently encountered cases their performance can be improved significantly by the straightforward enhancements described in this section. The first three enhancements, cascade, feedforward, and ratio control, improve the response of control systems to upsets by measuring or inferring the upsets and compensating the manipulated variable. Naturally these enhancements should be employed when the upsets are expected to seriously degrade control system performance. The final enhancement improves the control of processes with large deadtimes. All of these enhancements retain the basic PID control algorithm. They will be explained for single-variable systems but can be applied to multivariable systems.

Cascade Control

Cascade control systems have the potential for identifying and correcting upsets in the manipulated variable before such upsets significantly affect the variable to be controlled. A standard single-variable system is shown in Fig. 35.25*a* with the physical process separated into two components. Upsets occurring in either process $G_2'(s)$ or $G_2(s)$ are not sensed and corrected until they affect the

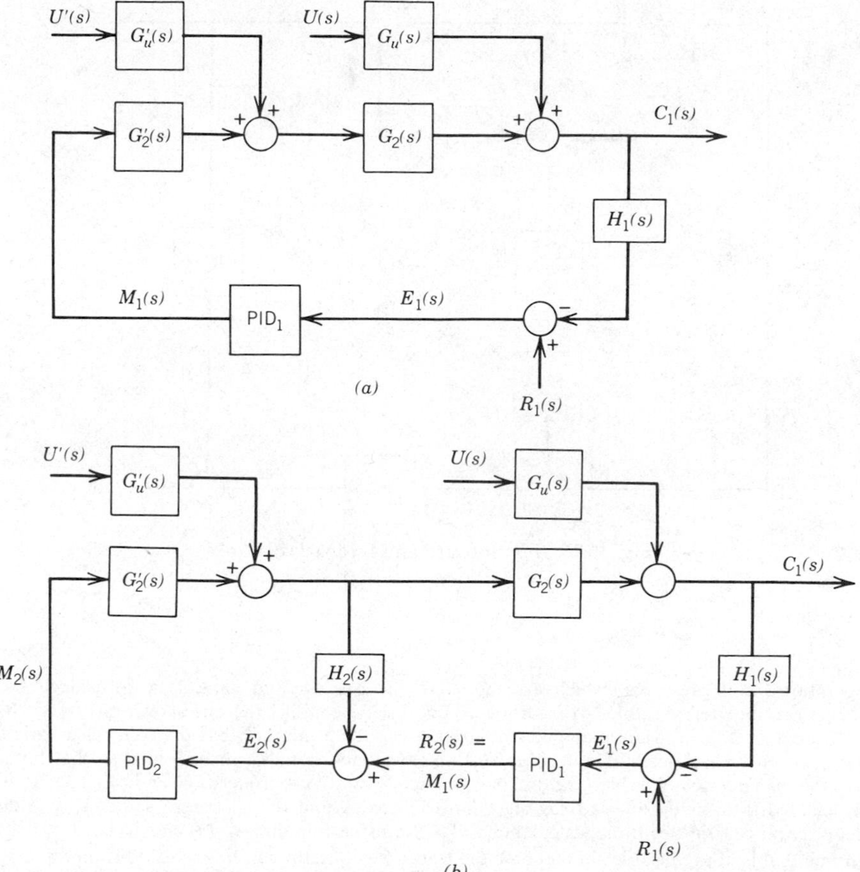

Fig. 35.25 Comparison of single-variable (*a*) and cascade (*b*) control systems.

controlled variable. The cascade control system in Fig. 35.25*b* ascertains that the upset $U'(s)$ has occurred by measuring the output of the secondary process, $G_2'(s)$. The controller can take action quickly to correct for the upset. The major controlled variable, called the primary variable, is regulated by the primary controller, which adjusts the set point of the secondary controller. Clearly the process must have a measurable secondary variable that responds to the adjustment of the manipulated variable for cascade control to be feasible. The advantage of the cascade system is realized only when the secondary process responds more rapidly than the primary. A rough rule of thumb is that the secondary process should have a dynamic response (deadtime plus time constant) not more than one third that of the primary process.[37]

The two cascade controllers use the control algorithms described in Section 35.2-3. Naturally the secondary controller must have the ability to receive a set point from the operator or remotely from the primary controller; usually a cascade switch is provided for the set point source. In a cascade control system, the secondary controller must be tuned first with its set point adjusted by the operator (cascade open). After the secondary is satisfactorily tuned, the cascade system can be placed in operation and the primary loop tuned. Since the primary controller's integral mode will reduce its error signal to zero, the secondary control needs only proportional control; however, it usually has an integral mode and is tuned conventionally for operation when the cascade is open. A cascade system has more controller outputs and therefore is very susceptible to reset windup, discussed in Section 35.2-6.

Feedforward Control

Feedforward methods can be used to compensate for the effects of measurable upsets that cannot be corrected by a fast inner-loop cascade. Typical upsets are feed flowrate, composition, utility

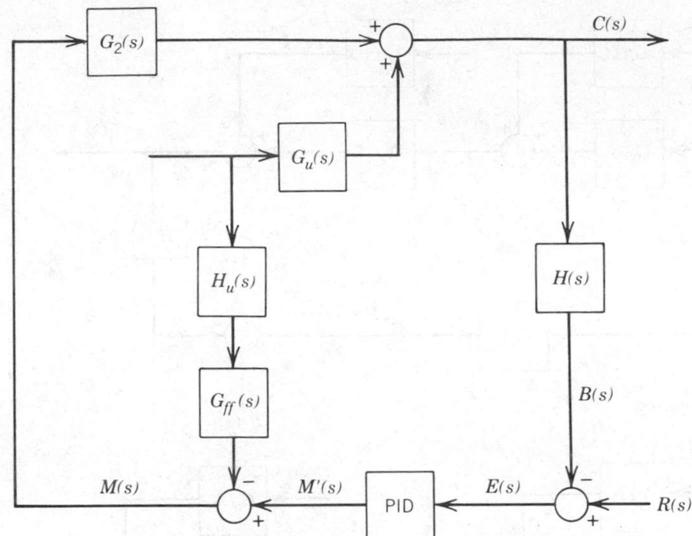

Fig. 35.26 Feedforward and feedback control.

$$G_{ff}(s) = \frac{G_u(s)}{G_2(s)} = \frac{K_u(1 + t_2 s)}{K_2(1 + t_u s)} e^{+(\theta_2 - \theta_u)s}.$$

temperatures, and pressures, which are inputs to the process and cannot be influenced by the process's manipulated variables. A feedforward control system, like the one shown in Fig. 35.26, has the objective of adjusting the manipulated variable in such a manner that the controlled variable is not influenced by the upset. The feedforward and feedback signals can be added so that both can adjust the manipulated variable. The design of the feedforward controller is given in Fig. 35.26; it can be calculated from gain and lead/lag algorithms in analog and digital control systems, with digital systems required for deadtime calculations. The transfer function $G_{ff}(s)$ might have a negative deadtime that is not realizable; in that case the lead time constant can be increased somewhat. When the numerator and denominator in $G_{ff}(s)$ have nearly the same deadtimes and time constants, good compensation is usually obtained by steady-state feedforward compensation calculated from the process gains alone.

Ratio Control

A simple strategy similar to feedforward control is employed when control objectives require that two variables be maintained at a desired ratio. A ratio control algorithm measures the two variables and controls their ratio by adjusting a manipulated variable that affects only the dependent variable. As shown in Fig. 35.27, the dependent variable is maintained in the desired ratio to the independent or "wild" variable. Ratio control algorithms are available in both analog and digital commercial control systems. Usually the ratio is calculated from the independent variable. Therefore it may be necessary to filter the independent variable so that measurement noise is not propagated to the controller input. Ratio control is often used in quality control of blending processes and in maintaining product and utility (e.g., steam) flows proportional to the feedrate of many processes.

Deadtime Compensation

The previous enhancements dealt with improving the response to measured upsets; control performance can also be degraded by particular process dynamic characteristics. Especially important is the frequently encountered existence of large process deadtimes caused by transportation delays or a combination of many first-order processes. When a large deadtime exists in a closed-loop system, the controller must be tuned with a small gain and long integral time to ensure stability, and such tuning yields a slow response to upsets and set point changes.[8,14] A feedback design called deadtime compensation is able to speed the response of closed-loop systems.[38] This scheme involves a model-based calculation of the controller input signal and a feedback signal to correct the model. As

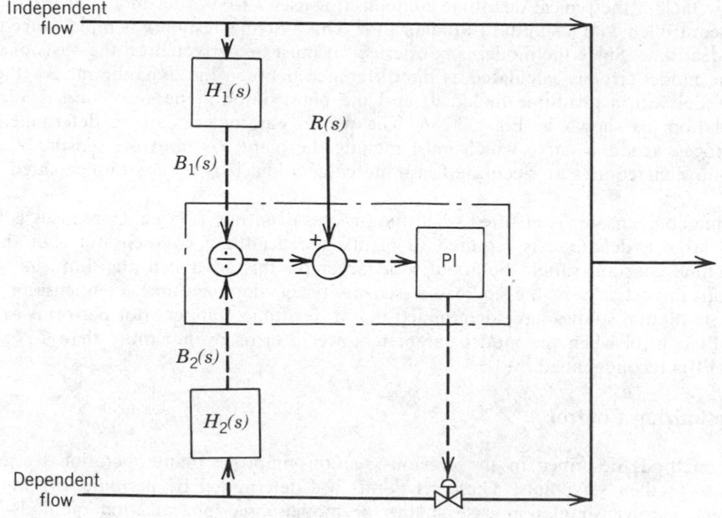

Fig. 35.27 Blending process with ratio controller (in box).

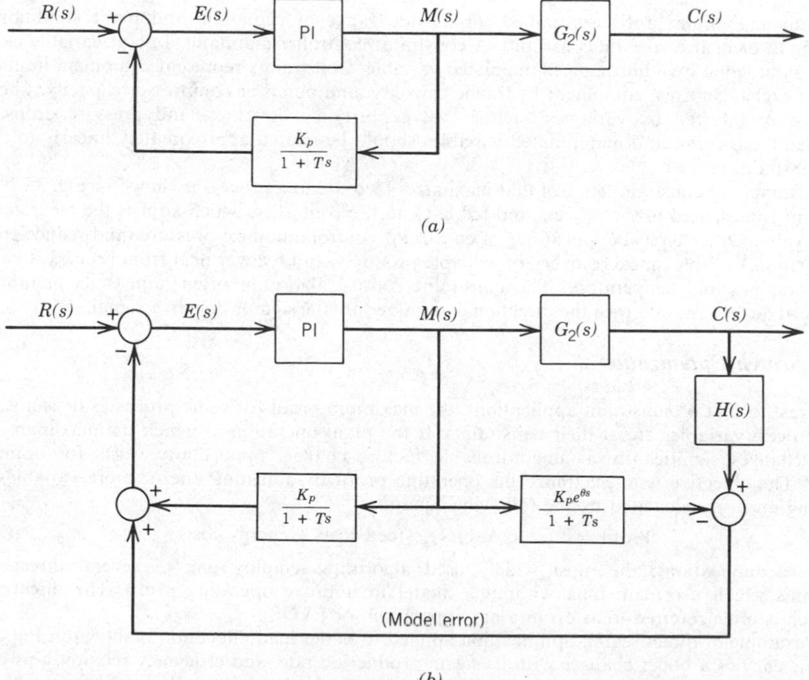

Fig. 35.28 Deadtime compensation: (*a*) Model-based part, (*b*) complete.

shown in Fig. 35.28*a*, the typical deadtime compensator uses a first-order model of the process, and the model is controlled with a standard PI algorithm. This system's response is rapid, since the system contains no deadtime. Since the model is not perfect, it must be corrected on the basis of the plant's response. The model error is calculated as the difference between the dynamic model (the previous first-order model with a deadtime included) and the plant output. The correction is added to the model calculation, as shown in Fig. 35.28*b*. The model parameters can be determined from an open-loop process reaction curve which must include the plant, G_2, and the sensor, H, dynamics. Since the controller requires an accurate deadtime calculation, it is always implemented via digital calculations.

A deadtime compensator is justified when the process deadtime is large. Consensus is lacking on exactly how large a deadtime is required to justify the deadtime compensator, but the ratio of deadtime to time constant should be about 1 or larger for this advanced algorithm to be justified. Also of great importance is the increased sensitivity of the deadtime compensator to model inaccuracy. Simulation studies have determined that a deadtime compensator performs as well as or better than PI control when the model parameters are in error by not more than 25%; for larger uncertainty, PI is recommended.[39]

35.2-6 Optimizing Control

The control methods described in the previous sections improve plant operation by maintaining variables close to their set points. These set points are determined by periodic, off-line economic analyses which involve simulation case studies or model-based optimization methods like linear programming. Typically these analyses are performed infrequently, every few days or weeks, because they are so time consuming. When important factors in plant profitability such as feed quality, equipment constraints, or energy efficiency change frequently, many set points can be far from their true optimum values. Therefore the opportunity exists to capitalize on existing process flexibility by frequently reoptimizing selected variables on-line. Two of the simplest, most widely employed methods of on-line optimization are described in this section. More complex (and less frequently used) approaches employing sophisticated mathematical techniques are beyond the scope of this section.[40]

Constraint Control

Frequently maximum profit is attained for a wide range of economic and plant conditions by operating at or near a specific constraint. A constraint controller maintains a plant variable close to its constraint value by adjusting a manipulated variable. Constraints represent equipment limitations like heat exchanger area, equipment hydraulic capacity, and pump or compressor capacity. They are measured or inferred by variables such as valve positions, flowrates, and pressure drops. The controlled (constraint) and manipulated variables should be related approximately linearly to ensure good constraint control.

An example of constraint control that maximizes feedrate to a process is shown in Fig. 35.29. The constraint is measured in the process and fed back to the controller, which adjusts the feedrate at its desired value. Other typical applications of constraint control minimize pressure (and reduce energy) in distillation,[41–43] maximize recovery of valuable products, and recover heat from process streams at the highest possible temperature. The constraint control algorithm often adjusts its manipulated variable at a constant rate with the direction determined by the sign of the error signal.

Direct Search Optimization

In contrast to that of constraint applications, the maximum profit for some processes does not occur when process variables are at their constraints. If the plant operation at which the maximum profit occurs changes significantly as uncontrollable factors change, opportunity exists for optimizing control. The objective is to maximize the operating profit by adjusting one or more variables. The operating profit is calculated by the following equation:

$$\text{Profit} = \text{Product values} - (\text{feed costs} + \text{energy costs}) \tag{35.29}$$

For these applications, the most widely used algorithms employ one of several direct-search algorithms which use plant data without a model to improve operating profit. The direct-search approach is often referred to as evolutionary operation, or EVOP.

An example of direct-search optimization applied to boiler load allocation is shown in Fig. 35.30. The efficiency of a boiler changes with its steam production rate, and efficiency relationships can be different for two boilers.[44] Therefore correct allocation of total steam demand between the two boilers can result in reduced fuel consumption. The direct-search optimizer would minimize the unit

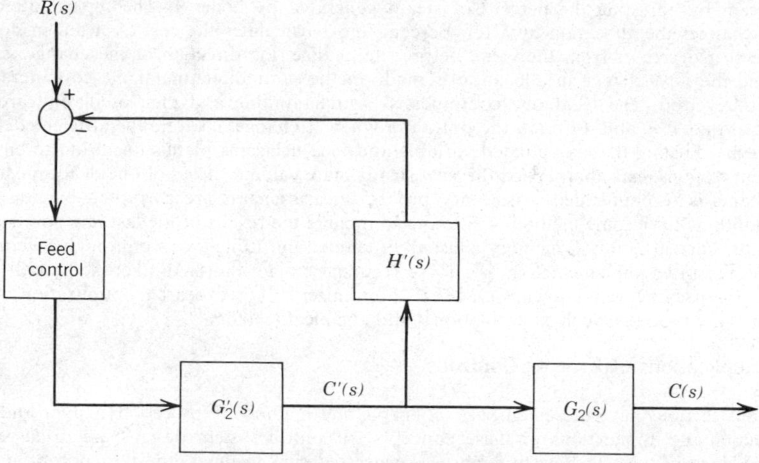

Fig. 35.29 Feed maximization constraint control.

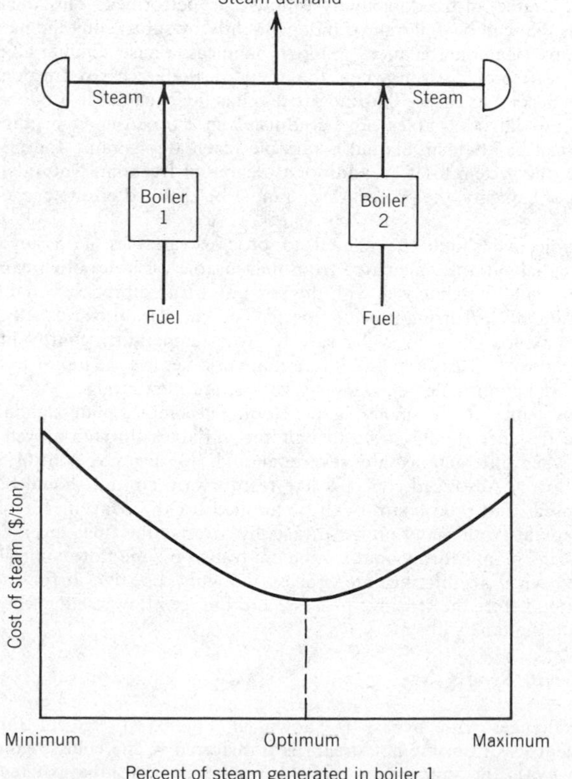

Fig. 35.30 Direct-search optimization applied to boiler load allocation.

cost of steam by adjusting the percent of steam generated by boiler 1. The typical direct-search algorithm changes the allocation by a few percent and recalculates the cost of steam in dollars per ton. If the cost decreases from the value before the change, the direction of the change is deemed correct, and the next change in allocation is made in the same direction. If the cost increases, the direction is reversed. The technique continues to search, finding and then oscillating around the optimum, so that it is able to track the optimum when it changes. Naturally sufficient delay must exist between adjusting the manipulated variable and measuring the plant's operation to ensure that the transient response has approached the new steady-state value; a delay of one deadtime and three time constants is recommended. Also since process measurements are corrupted by noise, direct-search algorithms have some method of filtering to include the results of the last few points searched as well as the current point. A memory is usually included by fitting a curve through several points; this procedure can be implemented in batch[45] or recursive[46] algorithms. Another method of including memory is to use the well-known COMPLEX optimizer.[47] Direct-search optimization has been studied on many processes such as distillation[47] and chemical reactors.[46]

35.2-7 Applications of Process Control

The previous sections have presented basic concepts and methods in process dynamics and control. In this section the applications of these concepts to control system designs are discussed. First, general guidance is given in how to analyze a process design to determine the appropriate control system. Then a few frequently used control designs are demonstrated by application on important process unit operations. All feedback controllers shown in this section use the standard PI algorithm.

Considerations in Process Control Design

As mentioned several times in this chapter, the most important task in process control design is establishment of control objectives. The general set of objectives given in Table 35.4 can be used as a checklist for establishing process needs. To determine whether the control objectives are possible in the steady state, a degrees-of-freedom analysis must be performed. This analysis determines the number of variables that can be determined independently by subtracting the number of constraining physical relationships (material balances, energy balances, phase equilibria, etc.) from the total number of system variables.[48] Even when the theoretical degrees of freedom exist, steady-state simulations may be needed to ensure that the process has the capacity to attain all control objectives, that is, that the manipulated variables can be adjusted in a broad enough range so that all desired process operations can be attained. Should a variable reach a constraint, for example, a flow attains its maximum value, the system loses an additional degree of freedom. Naturally if the process lacks degrees of freedom or capacity, the process design must be changed or one or more control objectives must be deleted.

Each control objective should be related to process variables. If a key variable cannot be measured, an inferential variable calculated from measurable variables often can be formulated. The inferential variable should correlate well with the key unmeasured process variable and should not be too sensitive to commonly occurring process upsets. The correlation can usually be determined by a few steady-state simulation cases, as is the case for tray temperature control in distillation product quality inferential control.[49] Measured and inferential variables are the inputs to feedback controllers, and signals to the final control elements, mostly valves, are the outputs.

Applying the principle of parsimony in developing a simple, understandable, well-performing control system, the designer should strive to pair the variables through a system of single-variable controllers. Only when interactions are severe enough to degrade control performance should decoupling be employed. Advanced concepts like feedforward control should be applied to mitigate the influence of upsets, and ratio control can be applied for the convenience of the plant operator. Finally, integrated plants with many processes usually arrange the flow, level, and pressure loops in such a manner that the plant's throughput can be controlled by one flow controller. Occasionally this rule is not followed when an intermediate process is highly sensitive to flowrate changes. In these cases a tank is located before the sensitive process, and the feed flow is adjusted infrequently and at a slow rate to regulate the tank's inventory.

Heat Exchanger with Bypass

A simple heat exchanger with bypass is shown in Fig. 35.31. Except for minimal losses to surroundings, all heat loss from the hot stream is transferred to the cold stream; therefore only one degree of freedom exists for temperature control. In the example, the exit temperature of the hot stream is controlled by adjustment of the bypass of the hot stream around the exchanger. The

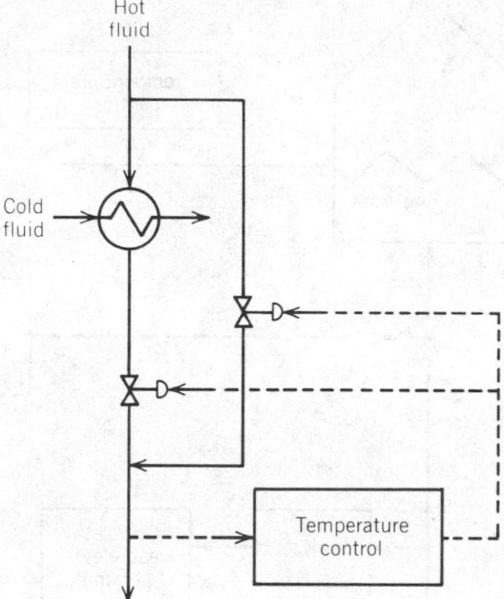

Fig. 35.31 Heat exchanger with bypass.

controller sends the same signal to both valves, and one valve opens with an increasing signal while the other closes with an increasing signal. Thus the bypass can be adjusted from zero to the total hot stream flowrate. The hot stream is bypassed so that the temperature control system can react rapidly to upsets. If the bypass were on the cold stream, the response of the hot exit temperature would be slower because the dynamic response of the heat exchanger would be included in the closed-loop system.

Furnace with Fuel-Fired Cascade

A control design that achieves good regulation of the furnace stream's outlet temperature as fuel gas composition changes is shown in Fig. 35.32. A change in fuel gas composition affects the energy released in combustion and, therefore, the controlled temperature. A "calorimetric" analyzer can determine the heating value of the fuel per unit flow. Therefore control of the product of heat released per flowrate times the flowrate provides rapid correction for upsets in the fuel gas composition. The temperature controller adjusts the secondary controller's set point to regulate the outlet temperature.

Pressure Control with Split Range

Steam systems are used throughout the process industries to power large machines (mostly compressors and pumps) and to provide heat transfer. Steam is generated in boilers and distributed at several pressure levels. Controllers at each pressure level (called a header) sense the pressure and manipulate flows in or out of the header. Steam systems are usually quite complex; the simplified section of a steam system in Fig. 35.33 demonstrates split-range control of two manipulated variables. The control objectives are to control the header pressure and to minimize the letdown flow through valve V2, thereby maximizing the flow through the turbine and maximizing energy to the process. With the controller output expressed as 0 to 100%, valve V1 opens proportionally to the output signal from 0 to 50% and remains fully open from 50 to 100%. Valve V2 remains closed from 0 to 50% and opens proportionally to the output from 50 to 100%; at 100% V2 is fully open. The resulting control achieves the stated objectives. For example, if the header pressure is above its set point, the signal from the controller decreases, causing (1) V2 to close slightly if the output is in the range of 50 to 100% (V1 is fully open) or (2) V1 to close slightly if the output is in the range of 0 to 50% (V2 is fully closed). Note that the two valves should have about the same maximum flowrates so that the process gains for both closed-loop systems are nearly equal and constant controller tuning is satisfactory. The objectives for split-range control could be achieved by two feedback controllers with different set points, but this approach costs more and involves the potential for controller interaction.

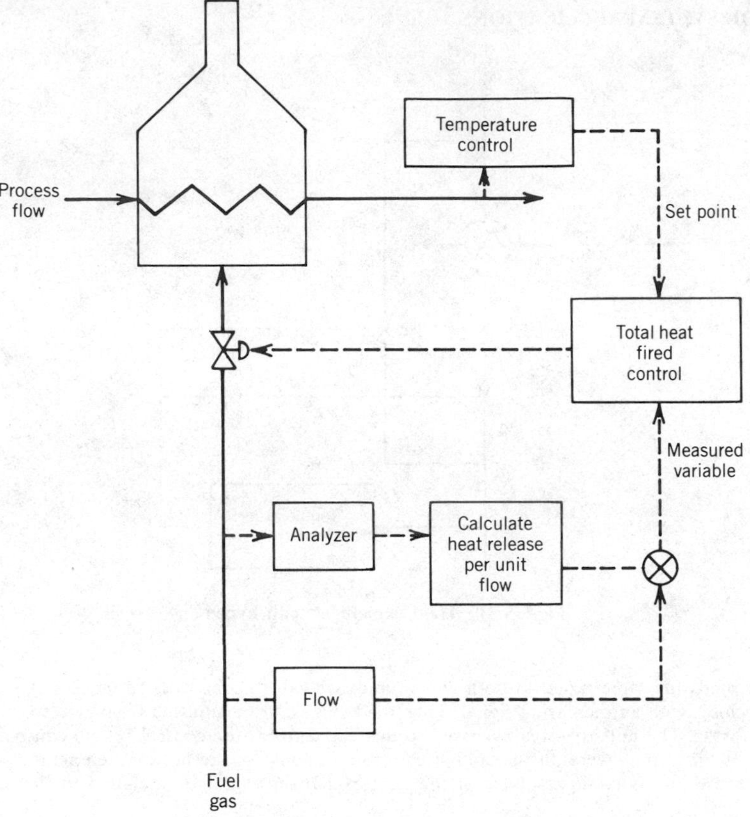

Fig. 35.32 Furnace with head-fired cascade.

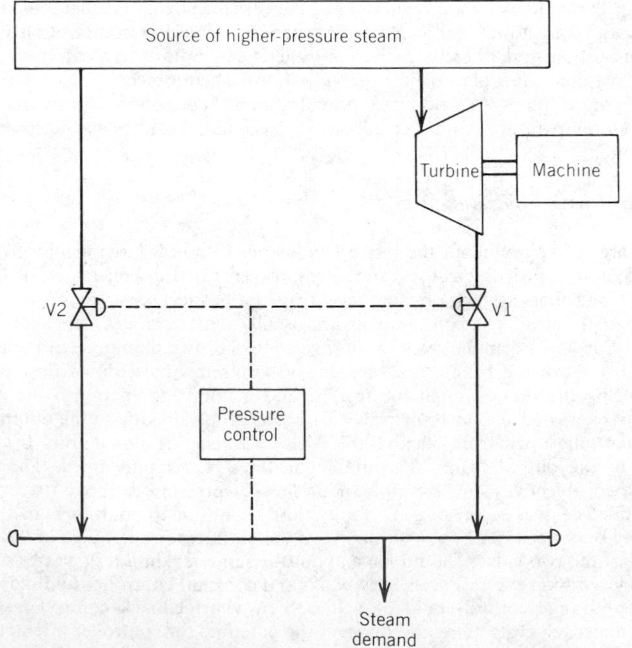

Fig. 35.33 Steam header pressure control with split-range output.

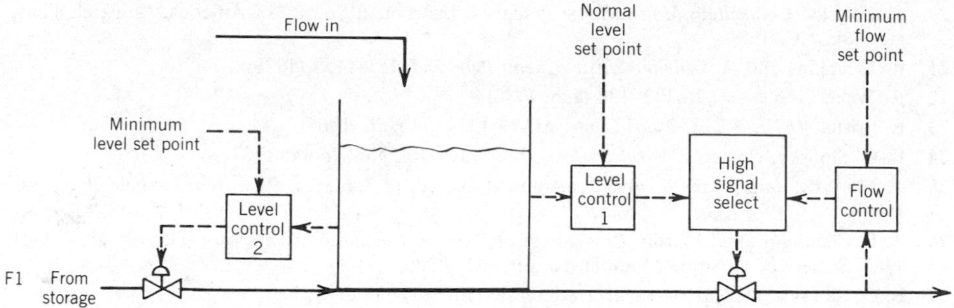

Fig. 35.34 Level-flow process with signal select.

Level Control with Signal Select

The simple level control system shown in Fig. 35.34 demonstrates the frequently used signal select approach for a simple control system with one manipulated variable. The control objectives are (1) to control the level by adjusting the flow out (with F1 zero, if possible), (2) to never violate the minimum flow out (to another sensitive process), and (3) to never allow the level to fall below its minimum value. Level controller 1 and the flow controller send signals to a high signal select, and only the higher of the two controller outputs reaches the valve. (Naturally, low signal selects are also used for other applications.) If the flow in decreases, the flow controller prevents the level controller from decreasing the flow below its minimum. The level decreases until level controller 2 begins to manipulate F1 to prevent the level from falling below its minimum value. Since one of the two controllers sending its output to the signal select is always prevented from adjusting the manipulated variable, control systems with signal selects are particularly prone to suffer from windup, and anti-reset windup must be designed into the system.

References

1 R. Rinne et al., *Hydrocar. Proc.* **61**:141–148 (Mar. 1982).

2 R. Saporita and R. McCue, *Oil Gas J.* **76**:69–71 (Sep. 4, 1978).

3 J. Bayer and R. Clay, paper cc-79-102 presented at Nat. Petro. Refiners Assoc. 1979 Computer Conference, Oct. 28–31, 1979.

4 J. D. Robnett, paper 81-IPC-PWR-2 presented at Industrial Power Conference (ASME), Oct. 18–21, 1981.

5 H. Tari, in H. R. Van Navta Lemke and H. B. Verbruggen, eds., *Digital Computer Applications to Process Control*, IFAC and North-Holland, The Hague, 1977, pp. 65–77.

6 H. Broekhuis et al., in H. R. Van Navta Lemke and H. B. Verbruggen, eds., *Digital Computer Applications to Process Control*, IFAC and North-Holland, The Hague, 1977, pp. 43–58.

7 D. M. Himmelblau and K. B. Bischoff, *Process Analysis and Simulation: Deterministic Systems*, Wiley, New York, 1968.

8 A. Lopez et al., *Instru. Technol.* **14**:57–62 (Nov. 1967).

9 W. L. Luyben, *Process Modeling, Simulation, and Control for Chemical Engineers*, McGraw-Hill, New York, 1973.

10 D. Graup, *Identification of Systems*, Van Nostrand Reinhold, New York, 1972.

11 K. Astrom et al., *Automatica* **13**:457–476 (Sep. 1977).

12 J. Ziegler and N. Nichols, *Trans. ASME* **64**:759 (1942).

13 A. Lopez et al., *Instru. Contr. Sys.* **42**:89–95 (Feb. 1969).

14 H. Fertik, paper 74-503 presented at ISA Conference, Oct. 28–31, 1974.

15 P. Harriott, *Process Control*, McGraw-Hill, New York, 1964.

16 J. Shunta and W. Fehervari, *Instr. Technol.* **23**:43–48 (Jan. 1976).

17 H. Fertik and C. Ross, paper 10-1-ACOS-67 presented at 22nd ISA Conference, Chicago, Sept. 11–14, 1967.

18 G. Shinsky, *Instr. Contr. Sys.* **44**:9 (Aug. 1971).

19 D. Wolter, *Instr. Technol.* **24**:55–63 (Oct. 1977).

20 G. Shinsky, *Controlling Multivariable Processes*, Instrument Society of America, Research Triangle Park, NC, 1981.

21 P. Tompkins and A. Corripio, *Instr. Chem. Petr. Ind.* **15**:51–59 (1979).

22 R. Weber, *AIChE J.* **26**:132–134 (Jan. 1980).

23 E. Bristol, *IEEE Trans. Auto. Contr.* **AC-11**:133–134 (Jan. 1966).

24 F. G. Shinsky, *Process Control Systems*, McGraw-Hill, New York, 1967.

25 T. McAvoy, *Interaction Analysis*, Instrument Society of America, Research Triangle Park, NC, 1983.

26 T. F. Cheung and T. Marlin, *Proceedings of Summer Simulation Conference*, Denver, Jul. 19–21, 1982, Society of Computer Simulation, pp. 462–472.

27 F. G. Shinsky, *Distillation Control*, McGraw-Hill, New York, 1977.

28 L. Tung and T. Edgar, *AIChE J.* **27**:690–693 (Jul. 1981).

29 W. Luyben, *AIChE J.* **16**:198–203 (Mar. 1970).

30 H. Rosenbrock, in T. J. Williams and E. J. Kompass, eds., *Multivariable Control Systems*, Dun-Donnelley, New York, 1975, pp. 1–24.

31 C. Schwanke et al., *ISA Trans.* **16**:69–81, 1977.

32 C. Garcia et al., *IEC PPD* **21**:308–323 (Apr. 1982).

33 D. Prett, paper 51b presented at AIChE 86th National Meeting, Apr. 1979.

34 R. Mehra et al., in T. F. Edgan and T. E. Seborg, eds., *Process Control 2*, United Engineering Trustees, 1982, pp. 287–310.

35 R. Weber et al., *Proceedings of American Control Conference*, June 1982, Washington, D. C., Vol. I, pp. 87–90.

36 A. Waltz, *InTech* **27**:41–44. (Aug. 1980).

37 R. Perry and C. Chilton, *Chemical Engineer's Handbook*, Wiley, New York, 1973.

38 O. Smith, *ISA J.* **6**:28 (Feb. 1959).

39 C. Hang et al, paper CI79-611 presented at ISA National Conference, Oct. 22–25, 1979.

40 F. Larman et al., in H. R. Van Navta Lemke and H. B. Verbruggen, eds., *Digital Computer Applications to Process Control*, IFAC and North-Holland, The Hague, 1977.

41 A. Maarleveld et al., *Automatica* **6**:51–58 (Nov. 1970).

42 F. G. Shinsky, *Distillation Control*, McGraw-Hill, New York, 1977.

43 H. Kister et al., *Trans. Inst. Chem. Engr.* **57**:43–48 (Jan. 1979).

44 C. Cho, *Instrumentat. Technol.* **25**:55–58 (Oct. 1978).

45 A. Morshedi et al., *IEC PDD* **16**:473–478 (1977).

46 C. Garcia, *AIChE J.* **27**:960–968 (Nov. 1981).

47 Y. Sawaragi et al., *Automatica* **7**:509–516 (1971).

48 E. Gilliland et al., *Ind. Engr. Chem.* **34**:551–557 (May 1942).

49 T. Tolliver et al., *InTech* **27**:75–80 (Sep. 1980).

CHAPTER 36
ROBOTICS

J. RONALD BAILEY

IBM Corporation
Boca Raton, Florida

36.1 INTRODUCTION

Background

A robot is defined by the Robot Institute of America as "a reprogrammable, multi-functional manipulator, designed to move material, parts, tools, or special devices, through variable programmed motions for the performance of a variety of tasks."

From a historical viewpoint, the origins of the industrial robot can be linked to the continued development of automated machinery, which began with the Industrial Revolution in the 1760s. In the first wave of mechanization, the focus was on parts manufacturing with human intelligence and control. The machines listed here are representative of this first phase of industrialization:

Time	Machine
1775	Cylinder borer (Wilkinson)
1779	Metal working lathe (Maudslay)
1800	Gun stock lathe (Blanchard)
1804	Jacquard loom (Punch cards)
1818	Milling machine (Whitney)
1820s	Planer
1840s	Turret lathe
1860s	Universal milling machine
1870s	Automatic screw machine
1880s	Hobbing machine

As industrial development continued into the twentieth century, the focus of automation gradually shifted to manufacturing technology with human intelligence and numerical control.

Time	Event
1913	Moving assembly line (Ford)
1924	Transfer machine (Morris Motors)
1952	NC (numerically controlled) milling machine (Parsons with MIT)
1955	APT (application programming of tools)—NC programming language
1957	Component inserter (IBM)
1958	Machining centre (Kearney & Trecker)
1959	First industrial robot (Unimation)
1960	Adaptive controlled milling machine (Bendix)
1963	Computer-aided design (Sutherland)
1968	DNC (direct numerical control) (Molins, etc.)
1970	First robot welding line (GM)

Robotic Industry Environment

During the 1970s, synergistic developments in microprocessors, controls, software, sensors, and computers combined to produce the intelligent adaptive robots that are now available. Today industrial robots are being used throughout the United States, Western Europe, and Japan. Approximate distribution by location (1983) is as follows:

Country	Robots
Japan	14,000
USA	4,000
Germany	700
Great Britain	700
Sweden	500
Italy	300
Other	300

Initial applications of robots were primarily in the automotive and electronic industries. The current trend is toward wider deployment, as indicated by the following data from the Japanese Industrial Robot Association (JIRA).

Industry	1976	1977	1978	1979	1980
Automotive	30	34	39	38	29
Electronic, electrical	21	23	24	18	36
Precision machinery	1	1	1	3	2
Construction machinery	1	1	1	1	1
Metal-working machinery	5	6	4	3	2
Other machinery	2	2	—	1	1
Bicycles, industrial vehicles	2	2	1	1	—
Plastics, molded products	13	10	10	11	10
Chemicals, petroleum, coal products	2	1	1	—	1
Ceramics	1	—	—	—	1
Iron, steel	6	6	3	4	1
Nonferrous metals	2	2	2	2	3
Metal-working, boilers, engines, turbines	7	3	8	9	5
Textiles	1	1	—	3	1
Others	6	8	3	6	5

36.2 CLASSIFICATION

Robots can be classified by three major components, namely, type of manipulator, controller, and prime mover. These three basic elements or components combine to define the capabilities of a robot. In this section, the basic types of components are defined and related to capability.

Types of Manipulators

Manipulators can be classified according to their mechanical configuration. Six common arrangements of robot manipulators are shown in Fig. 36.1. The first is a jointed arm that can rotate about a vertical axis and also about three horizontal axes. The second is a cartesian or x-y-z arm that moves

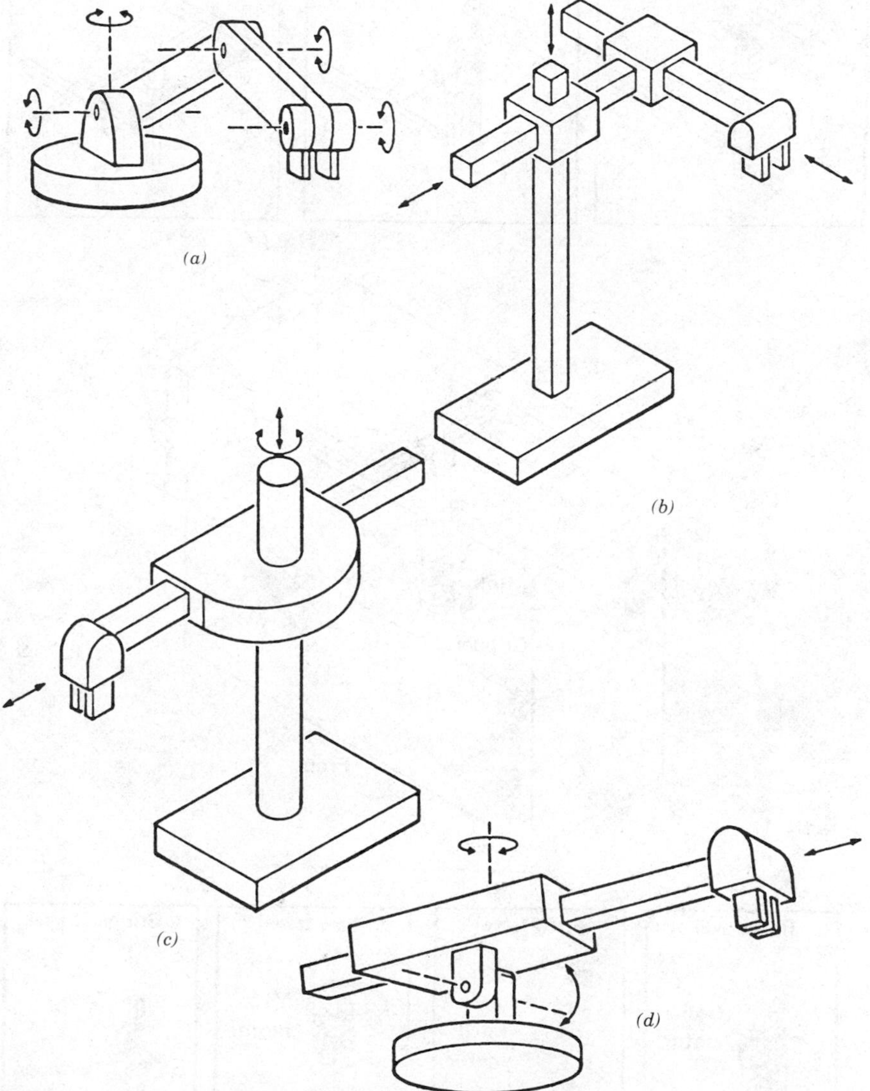

Fig. 36.1 Six common arrangements of robot manipulators: (*a*) Jointed arm, (*b*) Cartesian or *x*-*y*-*z* arm, (*c*) cylindrical arm, (*d*) spherical, (*e*) gantry, (*f*) selective compliance assembly robot arm (SCARA).

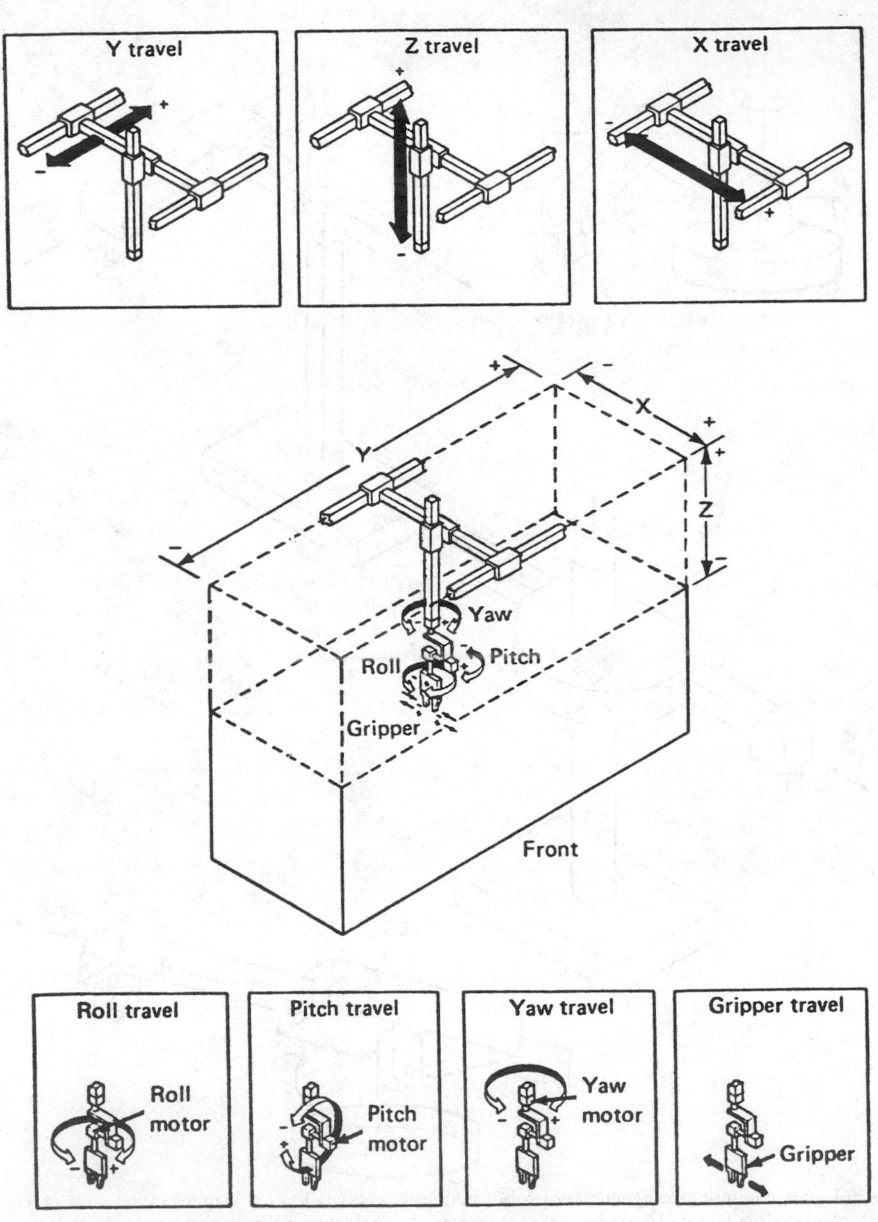

(e)

Fig. 36.1 (*Continued*)

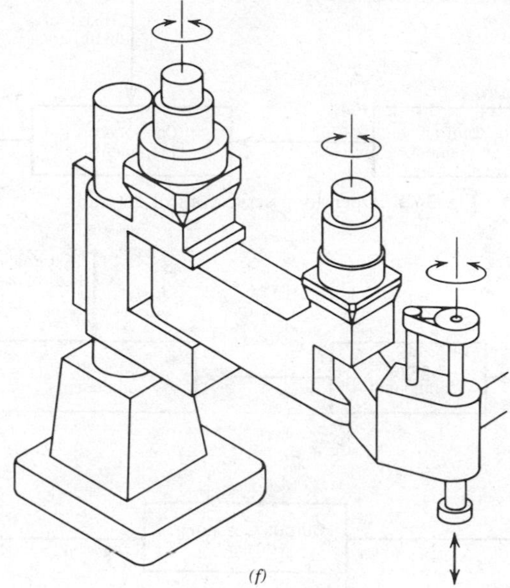

(f)

Fig. 36.1 (*Continued*)

horizontally with cantilevered joints. The third is a cylindrical arm that can rotate about a vertical axis and also move radially. The fourth is a spherical arm that rotates about the vertical and horizontal axes and also moves radially. The fifth is a box-frame or gantry-type robot that moves horizontally in x, y, and z with additional articulation provided at the end of the suspended arm for rotation of roll, pitch, and yaw. The sixth is a jointed arm with two joints that rotate about vertical axes, a third joint that moves up and down, and the ability to spin the gripper about a vertical axis. This configuration is commonly referred to as a selective compliance assembly robot arm (SCARA) because of the inherent capability of the arm to move slightly in the horizontal plane to accommodate misalignment of parts without allowing the parts to rotate, which could cause wedging.

Types of Controllers

Controllers for industrial robots include the following, arranged in ascending order of capability:

1. Manual control with push buttons or other.
2. Relay or pneumatic logic with fixed sequence.
3. Programmable controller.
4. Microprocessor with servo control capability.
5. Computer control of motion with logic capability.
6. Computer control with motion, logic, data processing, sensing, and communications capability.

The first three are usually programmed with relatively simple circuit and logic diagrams ("ladder diagrams"). Many of these systems would operate with an open-loop control, as sketched in Fig. 36.2.

At the fourth level a closed-loop control system, as shown in Fig. 36.3, is more likely. Programming may be at the assembler level. Above the fourth level a robotic programming language is usually required.

The three most common types of servo control are position, velocity, and force (pressure). Typical schematic and/or control-loop diagrams for these types of servo systems are shown in Figs. 36.4, 36.5, and 36.6, respectively. For a discussion of automatic control theory as used in the analysis and design of these systems, see Chapters 33, 34, and 35.

Important performance parameters usually include static accuracy, repeatability, speed, and stability. Factors affecting performance include hysteresis, dynamic response, mass, structural stiffness, friction, lost motion or backlash in the drive system, transducer repeatability, resolution, linearity, and dynamic response.

A summary of performance characteristics is given in Table 36.1.

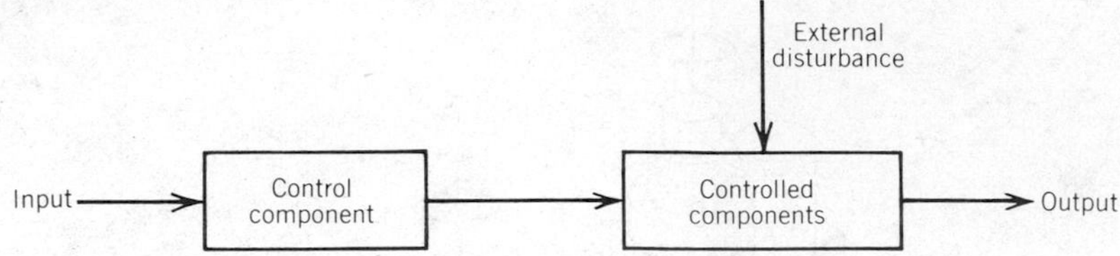

Fig. 36.2 Open-loop servo control diagram.

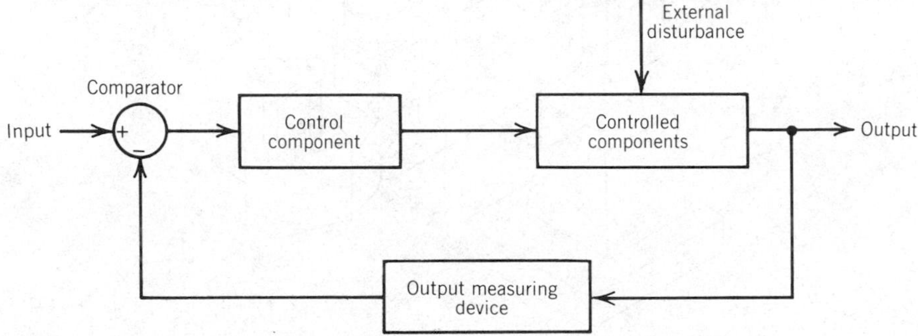

Fig. 36.3 Closed-loop servo control diagram.

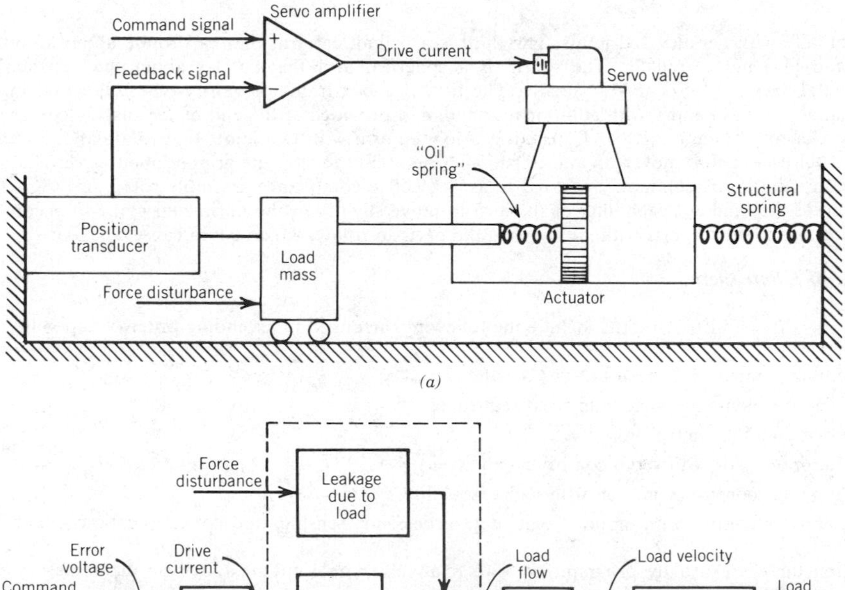

(*a*)

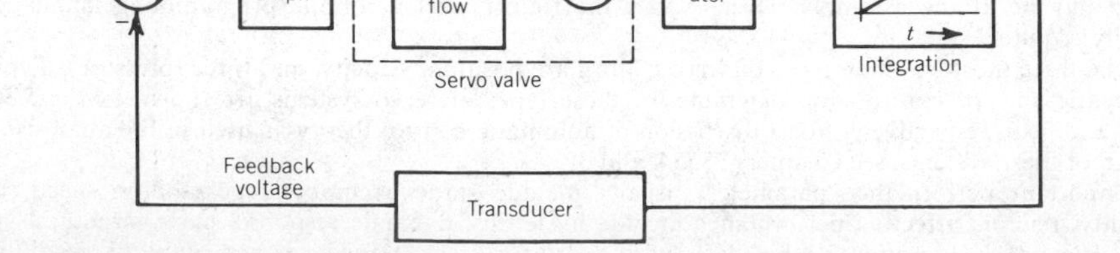

(*b*)

Fig. 36.4 Servo schematic (*a*) and control diagram (*b*) for position control.

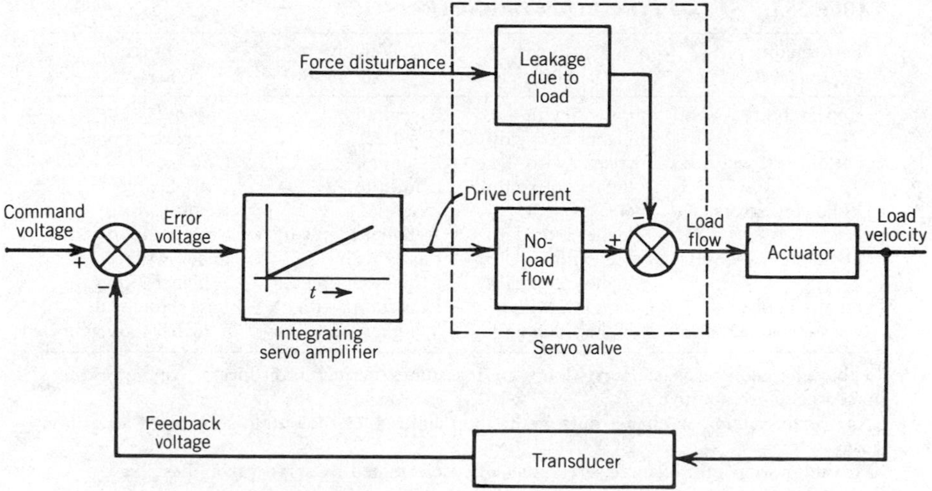

Fig. 36.5 Servo control diagram for velocity control.

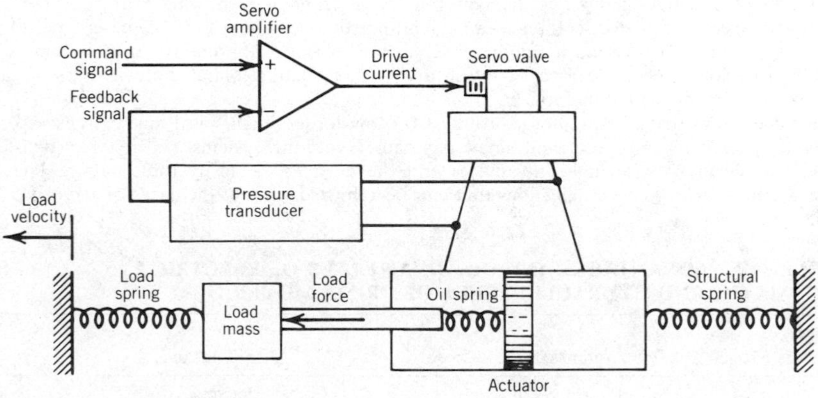

Fig. 36.6 Servo schematic (*a*) and control diagram (*b*) for force control.

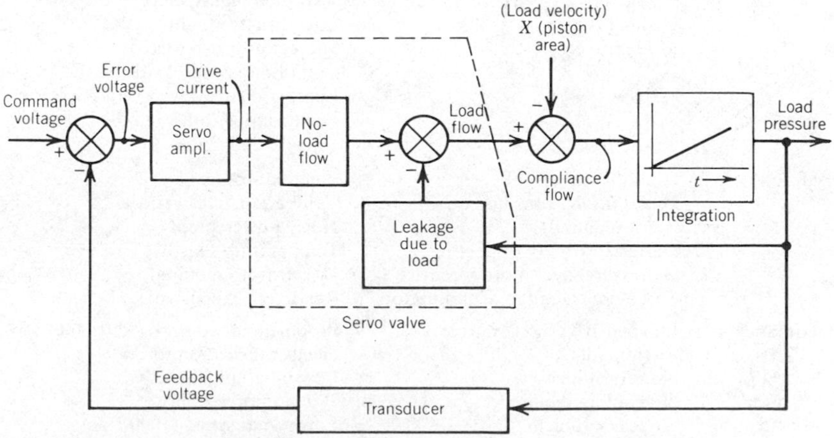

TABLE 36.1. SERVO CONTROL CHARACTERISTICS[a]

Performance Parameter	Position Servo[b]	Velocity Servo	Force Servo[c]
Repeatability	Improves with higher loop gain	Essentially perfect	Improves with higher loop gain
Output stiffness	Increases with higher loop gain	Essentially infinite	Not applicable
Following errors	Increases with higher loop gain	Essentially higher loop gain	Reduced with higher loop gain
Dynamic response	Faster with higher loop gain	Faster with higher loop gain	Faster with higher loop gain
Damping of load resonance	Less stable with higher loop gain	Less stable with higher loop gain	Less stable with higher loop gain

[a] Table characterizes basic servo loops and assumes perfect transducers. Any transducer inaccuracies are additive.

[b] Any drive system compliance outside the loop will reduce the output stiffness of a position servo.

[c] Maximum loop gain of force servo is usually determined by servo valve phase lag.

Types of Prime Movers

The motion of a robot manipulator is powered by a prime mover, which provides the energy necessary for acceleration, deceleration, and payload positioning. There are three basic types of prime movers: electric, pneumatic, and hydraulic. Selection of the most appropriate drive system depends on required payload, speed, accuracy, control, and cost. Advantages and disadvantages of each of these systems are summarized in Table 36.2.

The table shows that electric motors are best for low-torque, high-speed applications with a low duty cycle, near constant speed, and small payload. Hydraulic systems are best for high-torque, low-speed applications with high duty cycles and the need for a steady force output. Pneumatic systems are best for low-cost applications that can be achieved without the need for servoed position control.

TABLE 36.2. ADVANTAGES AND DISADVANTAGES OF ELECTRIC, PNEUMATIC, AND HYDRAULIC TYPES OF PRIME MOVERS

Type of Prime Mover	Advantages	Disadvantages
Electric	Suitable for clean room Convenient technology High resonance (easy to control) Cabling is easy Moderate cost	Electric hazard Can't use in explosive environment Needs transmission (low torque, high speed) Low thrust/weight Subject to brush wear Can't hold steady load Limited duty cycle (motor cooling) Power supply must handle peak load
Hydraulic	High thrust/weight Direct drive (transmission not required) Design flexibility (higher pressure gives more force) Can store energy in accumulator	Potential for leaks Can't go in "clean rooms" Noisy power supply Temperature sensitive (warm-up required) Relatively expensive
Pneumatic	Inexpensive power source (shop air) No return lines No cooling required No leak problem Can store or buffer energy Low cost	Proportional control is difficult Needs lubrication for air Low inherent stiffness due to compressibility Low signal speed (1000 ft/s) Low force levels (120 psi vs. 3000 psi) Little programming flexibility

Levels of Capabilities

A convenient way to classify robots in terms of performance capability is as follows:

Classification	Typical Application
Pick and place	Simple parts transfer
Point-to-point	Machine load and unload
Continuous path	Process (spraying, etc.)
Intelligent	Assembly, test, inspect

36.3 SOFTWARE

Requirements

Robots can be programmed at the machine level, the assembler level, or a higher level. The most sophisticated languages resemble a natural language with commands such as MOVE, IF, THEN, ELSE, WRITE (to disk), READ (from disk), PRINT, and SENSE. A full-function language will include motion control, logic, sensing, and some data-processing capability.

Motion control is the primary requirement of a robot-programming language. Each joint of a robot experiences acceleration, motion at constant velocity, and deceleration with each move command. These can be achieved by having the joints move in series (one joint moves, then the second moves, etc.), by allowing each joint to move independently (may be unacceptable since the path of such a move may be unpredictable), or by coordinating the motion of all the joints, that is, starting all joints at the same time and having each joint reach its goal at the same time. The last is the most sophisticated approach and is obviously not a trivial task. It normally requires control of the trajectory in addition to definition of the starting point and the goal for each joint.

Motion can be programmed in joint, world, or robot coordinate systems. Speed scaling is a desirable characteristic of motion control.

Decision making or logic is a second major requirement of a robot-programming language. This normally is achieved through instructions for looping, branching, position indexing, and so on. Operation in a real-time mode may be required to achieve this capability.

Interfacing is also an important requirement. Typically a robot control language should be able to control digital outputs and read digital inputs. This allows the robot to control the process around it and to respond to changes in its environment.

Data-processing capability is required to control complicated robot applications. Such capability would include the normal functions of reading data from storage, saving programs by writing to storage, displaying messages for operators, and others.

If the robot is to be used in an application where adaptability is required, the control language must include integrated sensing. This allows the robotic system to recover from errors due to missing or out-of-tolerance parts, empty feeders, positional drift, operator errors, and so on. Optical sensing, vision, and tactile sensing can be used to meet this requirement.

Built-in diagnostic capability is a desirable feature of robot control languages. Safety features can also be built into the language. For example, tolerance limits can be established in software for the position of each joint during a move. If any joint is sensed to be out of the tolerance limit, the robot can be shut down automatically.

Communication with other robots and/or a host computer may be required. The architecture of the language can allow this communication to take place while the robot is operating or can require batch transmission of data files with the robot in a nonoperating mode. Concurrent robot operation and data communication would require multitasking capability in the control processor. Communication with the operator can be achieved through a terminal, an operator panel, or a "teach" pendant.

Languages

A number of commercial robot languages have been introduced in the past decade. These include the following:

Language	Source
AL	Stanford University
AML	IBM Corporation
HELP	General Electric
MCL	McDonnell Douglas
RAIL	Automatix
VAL	Unimation

Details of these robot languages can be found in the *Allegro Operator's Manual*, the *AML Reference Manual*, Mujtaba and Goldman, the *RAIL Software Manual*, and the *User's Guide to VAL*. In addition to the explicit robot control languages listed, nonrobot languages such as BASIC, PASCAL, and APT can be used to program robots.

Sample Program

The following example is representative of the type of programming that is available. This program, written in AML, will execute the simple task of moving to a feeder, approaching a part, grasping the part, withdrawing, moving to a fixture, approaching the assembly point, releasing the part, withdrawing, and returning to the feeder. Several of the basic commands of AML are used in this program to generate the various subroutines. To expand this sample program to an actual application, additional subroutines would be needed for calibration, monitoring of sensors, and error recovery.

36.4 SPECIFICATIONS

Approximately 230 companies manufacture and market robots worldwide. In the United States, Cincinnati Milicron, Westinghouse/Unimation, GMF, Prab, General Electric, IBM, US Robots,

```
ASSEMBLY: SUBR;
-- ***************************************
-- ***** BASE POINTS                 *****
-- ***************************************
HOME_:       NEW
<4.96093,-23.5437,2.22167,80.6694,-8.81979,5.84205,.02095>;
-- ***************************************
-- ***** POINTS RELATIVE TO HOME_    *****
-- ***************************************
FEEDER_:    NEW
<-4.81933,1.87207,1.78459,0.,-31.627,0.,0.>
       + HOME_ ;
FIXTURE_:  NEW
<-8.30561,1.87207,1.78459,0.,6.37118,0.,0.>
       + HOME_ ;
-- ***************************************
-- ***** MAIN LINE PROGRAM           *****
-- ***************************************
          DISPLAY(EOP,'USE THE PENDANT TO POSITION THE',
                    ' ARM TO A SAFE POSITION.',EOL,
                   'HIT THE END-BUTTON TO START ',
                   'PROGRAM: ASSEMBLY EXECUTION.',EOL);
          PRINT(2,<'ENABLED',EOL>); -- MESSAGE TO PENDANT
          GUIDE(ARM);
          PRINT(2,<' ',EOL>);
ST1 :   REACH(FEEDER_);
ST2 :   APPROACH(1.00000);
ST3 :   GRASP(500.000);
ST4 :   WITHDRAW(1.00000);
ST5 :   POSITION(FIXTURE_);
ST6 :   APPROACH(1.00000);
ST7 :   SETGRIPPER(2.00000);
ST8 :   WITHDRAW(1.00000);
ST9 :   REACH(FEEDER_);
--   ********************************************************
--   *                     VERB SUBROUTINES                 *
--   *                                                      *
--   ********************************************************
   GRASP: SUBR(HOWHARD) ;
                  MID: NEW <INT,INT> ;
            IF ?JG EQ 0 THEN RETURN;       -- NO-OP IF NO GRIPPER
            MID = MONITOR(<SLP,SRP>,3,-50,HOWHARD) ;
            CLEANUP($UND);        -- INSURE ENVIRONMENT RESTORED.
                   UND: SUBR;
                        ENDMONITOR(MID);
                        END;
            MOVE(JG,0,MID,<.08>); -- MAKE MOTION SLOW & MONITOR.
            END;
-----------------------------------------------------------
   SETGRIPPER:  SUBR(INCHESOPEN);  -- 0.0 TO 3.25
                IF ?JG EQ 0 THEN RETURN;  -- NO-OP IF NO GRIPPER
                TRMOVE(JG,INCHESOPEN);
                END;
-----------------------------------------------------------
   WITHDRAW: SUBR(DISTANCE);
           WAITMOVE;
           TRMOVE(<JX,JY,JZ>,QGOAL(<JX,JY,JZ>)+DISTANCE*BOXTRANS()(2,3));
           END;
-----------------------------------------------------------
```

```
    APPROACH: SUBR(DISTANCE);
              WITHDRAW(-DISTANCE);
              END;
--------------------------------------------------------------------------------
    POSITION:  SUBR(POINTVALUE);
       J: NEW SELECT(<1,1,1,?JR,?JP,?JW>,<1,2,3,4,5,6>);
            -- J IS ALL JOINTS BUT GRIPPER AVAILABLE ON SYSTEM
               TRMOVE(J,POINTVALUE(J)) ;
               END;
--------------------------------------------------------------------------------
    REACH:     SUBR(POINTVALUE);
               TRMOVE(ARM,POINTVALUE(ARM));
               END;
--------------------------------------------------------------------------------
    TRMOVE:    SUBR(JS,GS);
               AMOVE(JS,GS);
               END;
--------------------------------------------------------------------------------
---------- 4/13/82 ----- TROPS1 -----------------------------------
END; --ASSEMBLY SUBR;
```

Intelledex, and numerous other companies provide a variety of robotic systems. Asea, Electrolux, Olivetti, and Renault, Trallfa, Binks, and Volkswagen are leading suppliers in Europe. Dainichi Kiko, Fujitsu Fanuc, Hitachi, Kawasaki, Mitsubishi, Sankyo Seiki, and Yasakawa are among the leaders in Japan.

Performance specifications include number of axes, coordinate system, drive system, payload capacity, motion type, programming method, number of digital inputs and outputs, repeatability, and price. Representative specifications are given in Figs. 36.7 through 36.11.

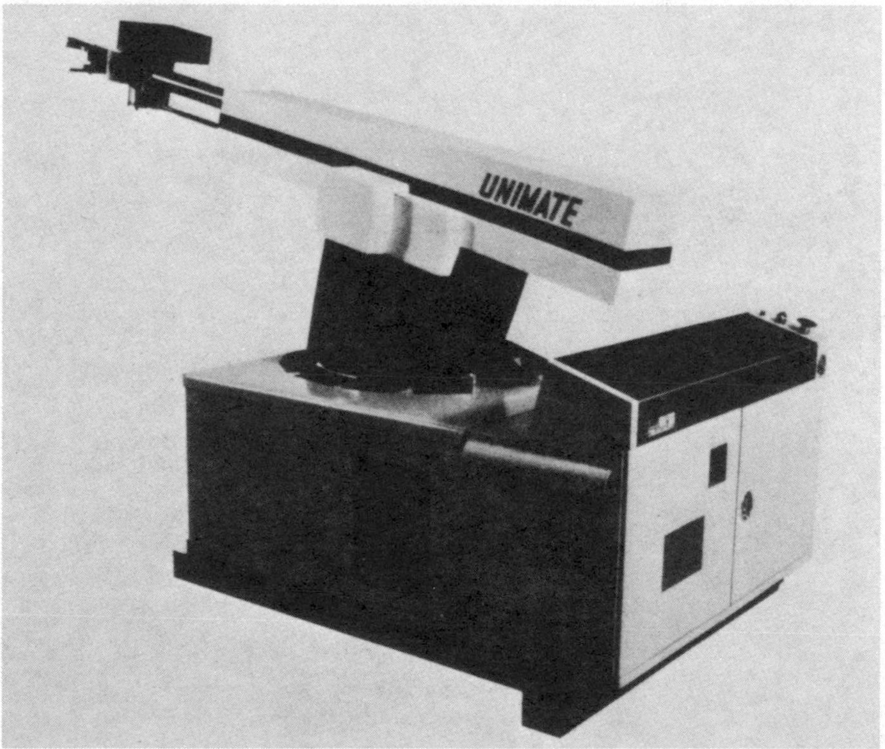

Fig. 36.7 Westinghouse/Unimation robot. Number of axes: up to six. Coordinate system: polar. Drive system: hydraulic. Payload: 5, 16, 100, 450 lb. Control: closed-loop servo. Motion type: point-to-point. Programming method: teach box or VAL language. Repeatability: ±0.050 to 0.080 in. by model.

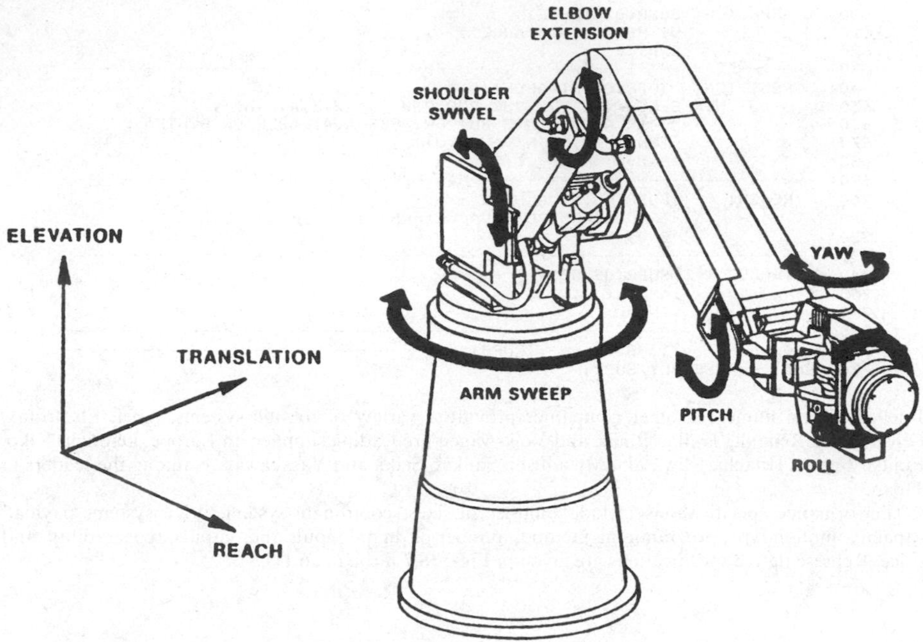

Fig. 36.8 Cincinnati Milicron robot.

Fig. 36.9 Prab robot. Number of axes: five. Coordinate system: polar. Drive system: hydraulic. Payload: 50, 75, 100, 125 lb. Control: nonservo. Motion type: point-to-point. Programming method: limit switches and adjustable stops. Repeatability: ± 0.016 to 0.060 in. (axis dependent).

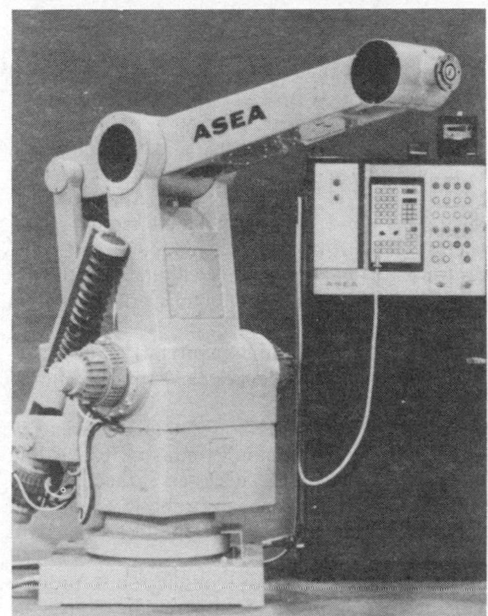

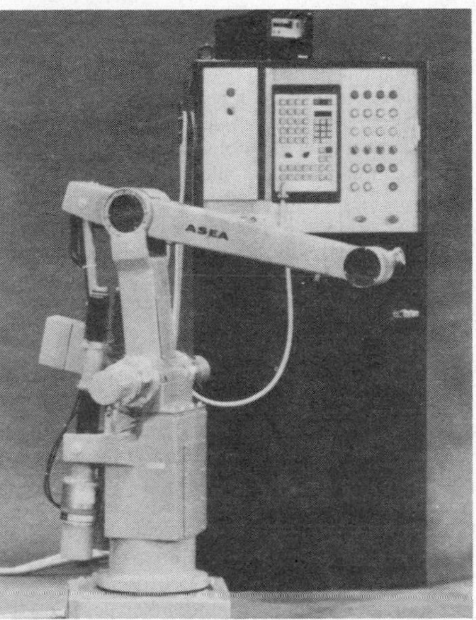

Fig. 36.10 Asea robot. Number of axes: six. Coordinate system: jointed arm. Drive system: DC servo. Payload: 13, 16 lb. Control: closed-loop servo. Motion type: point-to-point. Programming method: teach box. Repeatability: ±0.008 to 0.016 in. by model.

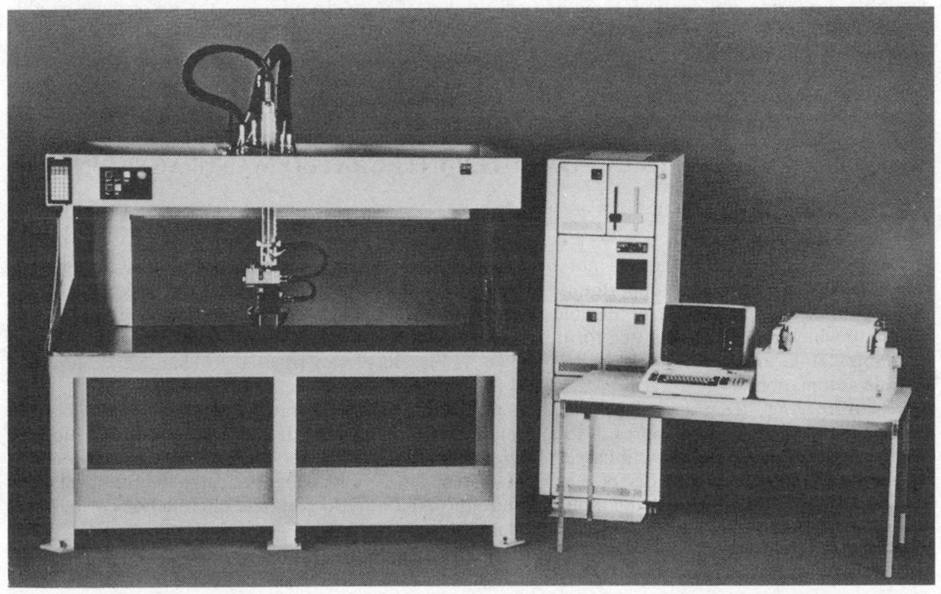

Fig. 36.11 IBM robot. Number of axes: six, plus gripper. Coordinate system: Cartesian. Drive system: hydraulic. Payload: 5, 13 lbs. Control: closed-loop servo. Motion type: point-to-point. Programming method: teach pendent and "AML" language. Repeatability: ±.005 in.

36.5 APPLICATIONS

Robot applications can be found throughout the factory. They range from stand-alone stations to completely integrated lines. A typical application would be machine load/unload. The handling of die castings is an example of an ideal application for a spherical robotic configuration. Loading/unloading a machine tool such as a lathe or milling machine is an application that can be done with a jointed arm type manipulator. Many machine load/unload robots can be justified on the basis of improved efficiency of the machine tool.

Material handling is a second generic type of robot application. Examples include interfacing with conveyors and automatic storage and retrieval systems, palletizing, kitting, and packaging. The introduction of group technology and other techniques for classification of parts will facilitate the use of robots for material handling. Economic justification can be based on reduction of work in progress and improved inventory control.

Fabrication represents another general application area. Pilot hole drilling in aircraft manufacturing, grinding, investment casting, deburring, and routing are typical applications. Improvements in quality often result from use of robots for fabrication.

Approximately one half of the initial applications for robots were in spotwelding. A variety of configurations are necessary for this application, particularly in the automotive industry for welding car bodies. Spherical, cylindrical, cartesian, and jointed arms provide the necessary articulation to reach the required locations. Arc welding has the potential to be an important robotic application area. Technical problems are centered on the requirement for following less than perfect seams and compensating for variations in parts.

Spray painting by means of robots requires smooth motion control. Most painting robots provide a significant savings in amount of paint used by reducing overspraying. Their use can reduce the need for expensive environmental controls for safety and health.

Inspection and testing can also be performed by robots. Trend analysis can be particularly valuable. In addition, certain data-driven tests such as inspection for shorts and/or continuity in electronic assembly are ideally suited to robot use.

Assembly requires speed and accuracy to a greater degree than do other applications. In addition, assembly may require a higher level of communications in that cells, lines, and integrated manufacturing are usually required to assemble a product. Gantry-type robots provide uniform accuracy throughout their work envelope. Selective compliance arms are also well suited to assembly, since they naturally accommodate uncertainty. The keys to successful use of robots in assembly are a product that has a minimum number of parts, assured access by the robot, a minimum number of assembly directions, automatic feeding and orientation of parts. In addition, the robot must be able to accommodate the uncertainty associated with part geometry and placement as well as the inherent inaccuracy of its own motion and that of any ancillary tooling. Sensing and logic requirements tend to be more demanding for assembly applications, and either the tooling must be extremely accurate or the robot must have enough intelligence to recover from errors.

36.6 ROLE OF ROBOTS IN COMPUTER-INTEGRATED MANUFACTURING

Flexible Automation

Heretofore much of industry has not been suitable for fixed automation. For this reason it is important to appreciate the opportunity presented by the advent of intelligent robots, which now make possible the utilization of flexible automation as an alternative to manual operations. The degree to which robots can be used for flexible automation can be defined in terms of a combination of manipulation and control. The matrix shown in Fig. 36.12 can be used to characterize levels of flexible automation.

"Leapfrogging" from a totally manual operation to a fully integrated operation is probably not practical, and a reasonable approach to automation is to define the desired levels of automation for each area and then to develop a phased plan for moving toward those levels.

A sketch of the architecture for a control system for a fully integrated manufacturing plant is shown in Fig. 36.13. Obviously this would require the capability to combine the functions of business planning, product design, operations, and manufacturing.

Application Selection and Development

The process of finding and developing suitable robotic applications in industry requires a systematic approach if maximum benefits are to be obtained. The flow chart shown in Fig. 36.14 is suggested as a guideline for this process.

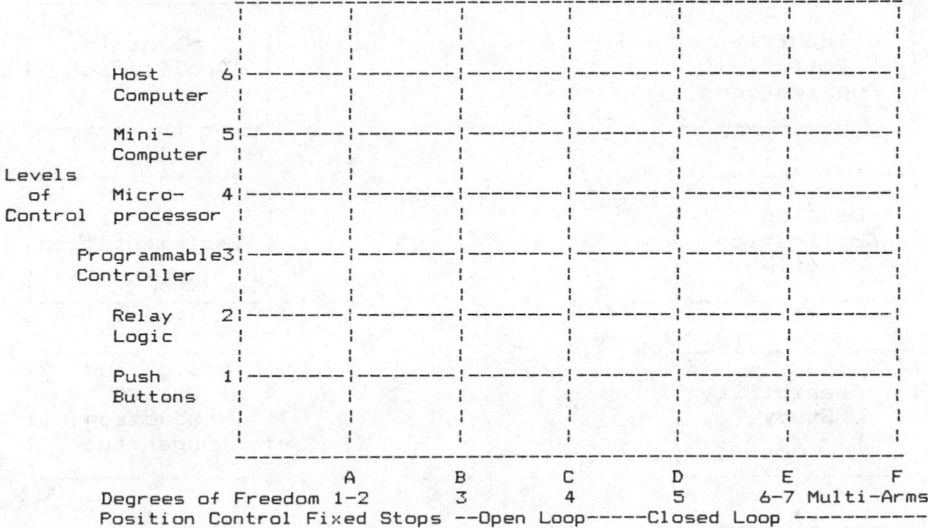

Fig. 36.12 Flexible automation.

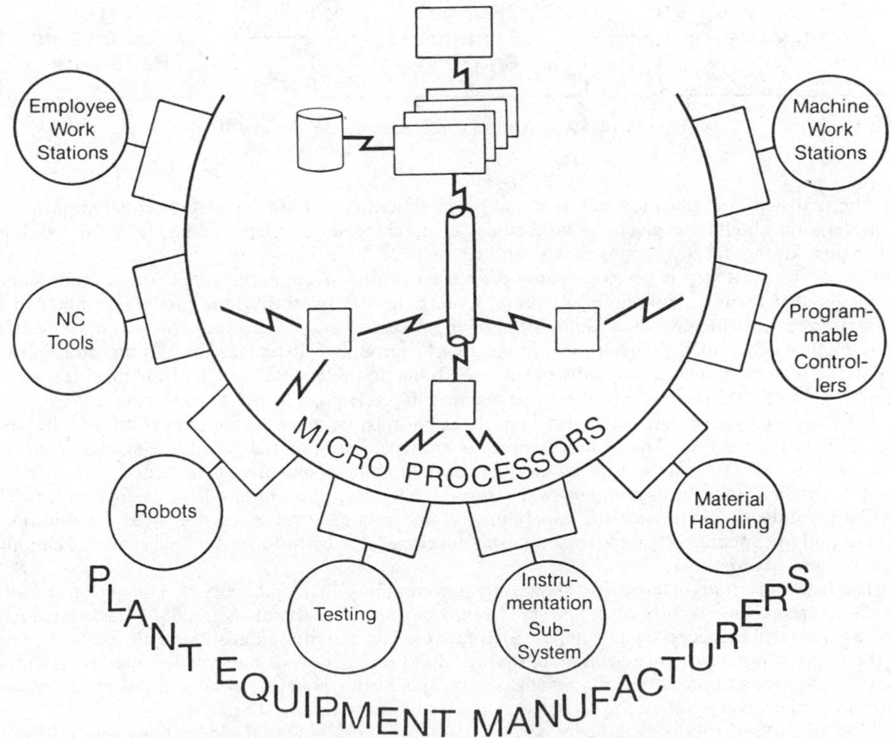

Fig. 36.13 Computer-integrated manufacturing.

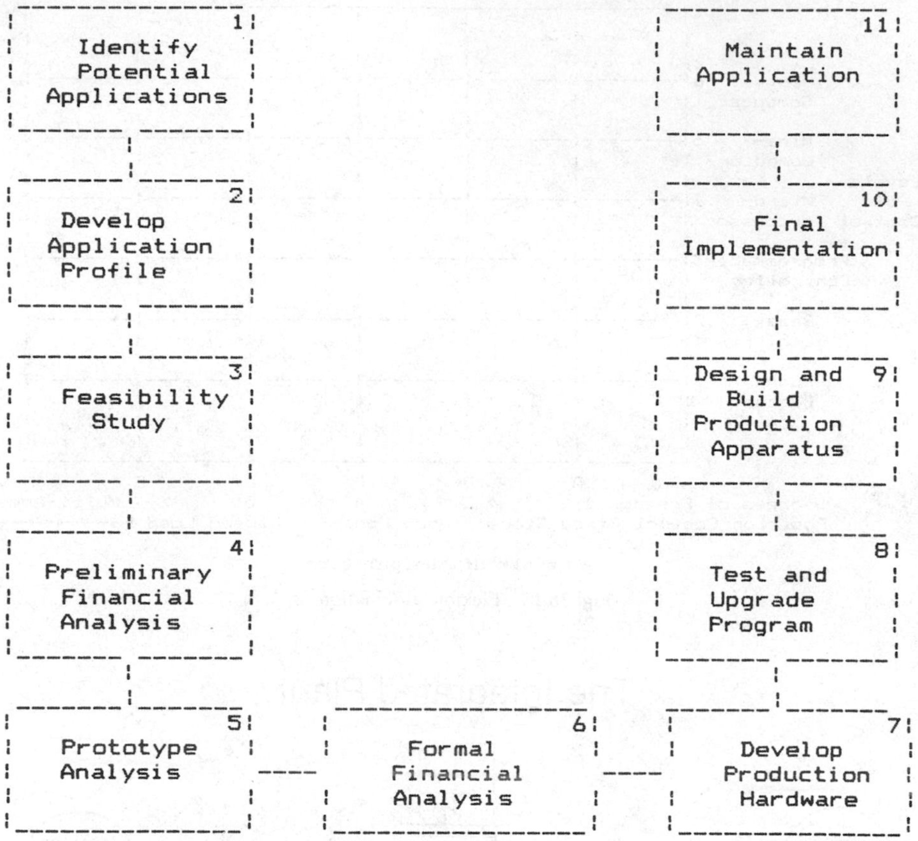

Fig. 36.14 Application selection and development.

The first step in application selection and development is identification of potential applications. Consideration should be given to tasks such as machine loading/unloading, material handling, palletizing, kitting, fabrication, assembly, and testing.

The next logical step is development of a detailed profile of the application. Applications should be sought that require a minimum of special tooling, have a relatively long production life, can be characterized as belonging to a family of similar processes, and have production volumes ranging from small batches to high quantities. If the robotic process will replace an existing process, it is important that the previous and subsequent operations be understood and that the need for special handling due to the environment be documented both from a process viewpoint and from an operator exposure viewpoint. Obviously consideration must be given to the impact of the robot as a source of contamination. This would require an understanding of the possible emissions from the robot such as oil mist from lubricated mechanical, hydraulic, or pneumatic elements.

A feasibility study is appropriate when several potential applications have been profiled. The feasibility study should review the capabilities of the robot(s) and assess the need for additional tooling and/or apparatus. Concepts should be developed for end-of-arm tooling, material handling systems, and fixtures.

The feasibility study is logically followed by a preliminary financial analysis. The objective of this step is to ensure the viability of the project from a business standpoint. Normally a discounted cash flow analysis will be necessary to compare alternatives. This requires an accurate estimate of the costs of the entire robotic system. In addition, savings due to reduction in direct labor, quality assurance benefits, documentation cost containment, safety and health benefits from reduced exposure, and other potential savings should be included in the preliminary financial analysis.

Since the use of robots in industry is relatively new, many potential applications will need to be analyzed at a prototype level before further steps are taken. A layout of the work area and a list of the process sequence will be required. Then prototype end-of-arm tooling and fixtures can be used with the robot to simulate the final process. The process can be optimized, timed, and videotaped.

A formal financial analysis can now be performed. In addition to the costs of the robot, consideration must be given to the costs of tooling, additional hardware or apparatus, special maintenance and test equipment, installation and site preparation, personnel retraining, engineering, and development. The formal analysis will incorporate unit costs and production schedules as well as tax credits and depreciation. The usual result is an estimate of return on investment and payback period. A formal proposal is often required to obtain approval for the capital expenditures necessary to proceed. Most proposals contain a statement of the problem, a method of solution, the rationale, a list of available and needed resources, personnel, schedule, and budget.

After the project is funded, development of the production hardware and software will require definition of major interfaces from both functional and process standpoints. Sensors, diagnostic routines, error recovery, documentation, and safety and health issues must be resolved at this time. In addition, consideration should be given to operator training and physical planning.

Designing and building the production tooling, fixtures, feeders, and material handling systems require special skills. Attention to detail can mean the difference between success and failure.

Final implementation involves floor space allocation and relocations, provision of service drops for air and/or electricity, and integration of all tooling, feeders, material handling systems, and so on.

Maintenance of the application requires trained maintenance personnel, spare parts, and adequate buffers.

It can be seen that application selection and development is a combination of technical and nontechnical considerations. Technical factors often can be expressed in numbers, for example, robot capabilities, cycle times, return of investment, tooling design, and material handling. Nontechnical issues related to the process involve concerns for the intangible impact of the robot. Furthermore, it can be seen that implementation of a robotic manufacturing system will involve participation from numerous individuals and organizations. For this reason, the trend has been for most robotic installations to be provided by systems integrators who can select the required robots and design the tooling necessary to deliver a turn-key system.

Economic Justification of Robots

Traditionally, economic justification of capital equipment has been based on a discounted cash flow method of analysis, using equations that reflect the time value of money. For example, the present worth of series of uniform future savings can be calculated as follows:

$$P = \frac{R[(1+i)^n - 1]}{(1+i)^n i} \tag{36.1}$$

where P = present worth
i = rate of return
n = number of compounding periods
R = amount of each uniform future saving

The corresponding cash flow diagram for Eq. (36.1) is shown in Fig. 36.15. Note that it is assumed that the uniform savings R are made at the end of each compounding period.

Since the first cost or present worth of the capital equipment is generally known, and the resulting uniform future savings R can be estimated for the life of the project in years n, the only unknown in Eq. (36.1) is the rate of return i.

A second factor used to compare alternatives is the payback period, which is defined as the time required for net accumulated savings to equal accumulated expenditures.

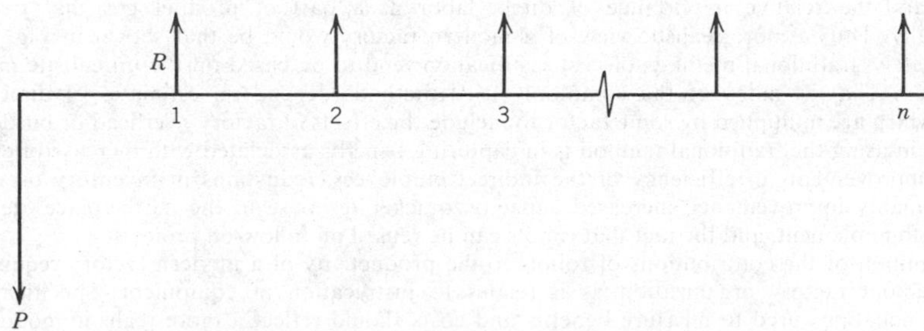

Fig. 36.15 Cash flow diagram for economic analysis.

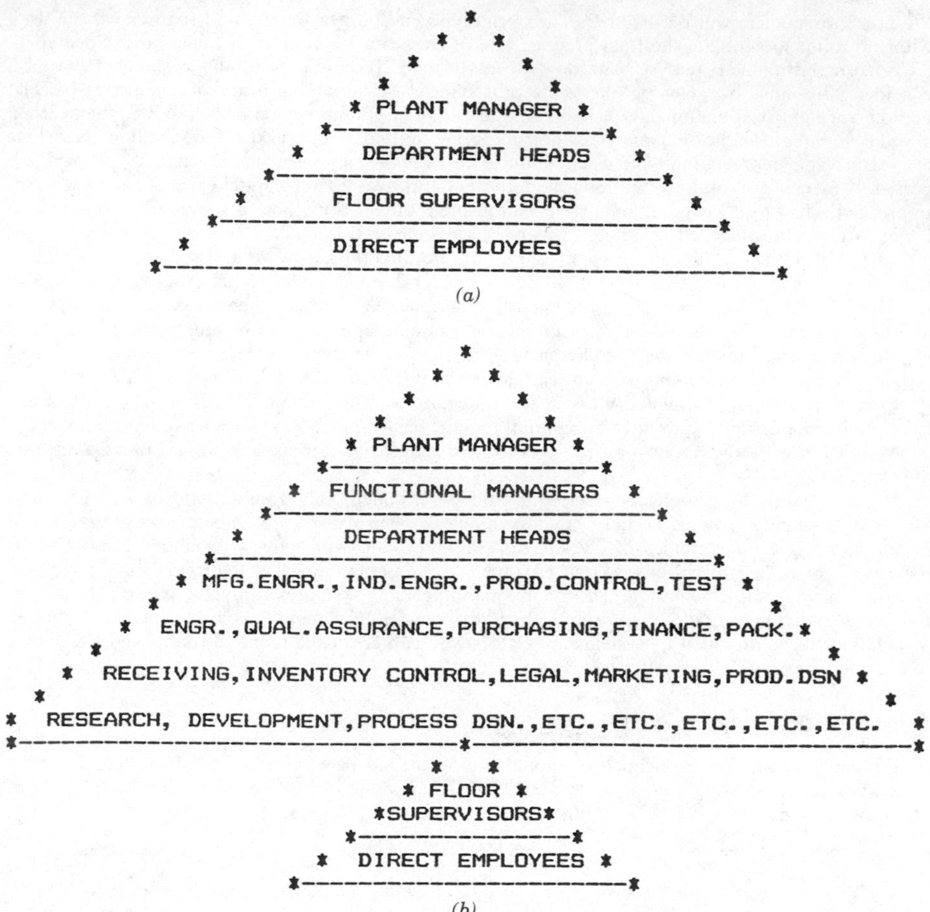

Fig. 36.16 Factory organization: (*a*) Traditional view, (*b*) more realistic view.

A combination of these two factors has traditionally been sufficient to prioritize projects that are competing for funding. In other words, the project having the highest projected rate of return and the shortest payback period has the lowest risk, and hence is the "best" from a business standpoint.

The difficulty with using traditional methods to compare alternatives to robotics is that the benefits associated with enhanced flexibility are seldom easy to quantify, particularly if existing methods rely on measurements of direct labor savings only.

In the past, product cost was dominated by direct labor, and factory organization had a basically pyramid structure, as shown in Fig. 36.16*a*. However, as a result of directed engineering efforts over many years, the relative importance of direct labor as a part of product cost has decreased dramatically. Thus a more realistic view of a modern factory would be that shown in Fig. 36.16*b*. Unfortunately, traditional methods of cost justification tend to be based on the unrealistic model of Fig. 36.16*a*. That is, many of the commonly used methods depend on estimates of direct labor savings, which are multiplied by some factor to include the effects of factory overhead or burden. The challenge in using the traditional method is in capturing benefits associated with increased flexibility, such as improvements in efficiency of the indirect employees, reductions in inventory or work in process, quality improvements, increased capacity, quicker response to the marketplace, decreased lead time to implement, and the fact that robots can be reused on follow-on projects.

Recognition of the contributions of robots to the productivity of a modern factory requires new thinking about factory organization as it relates to justification of equipment. Specifically, the economic equations used to measure benefits and costs should reflect a more realistic model of the factory, as shown in Fig. 36.16*b*. When robots are deployed in sufficient numbers, it must be

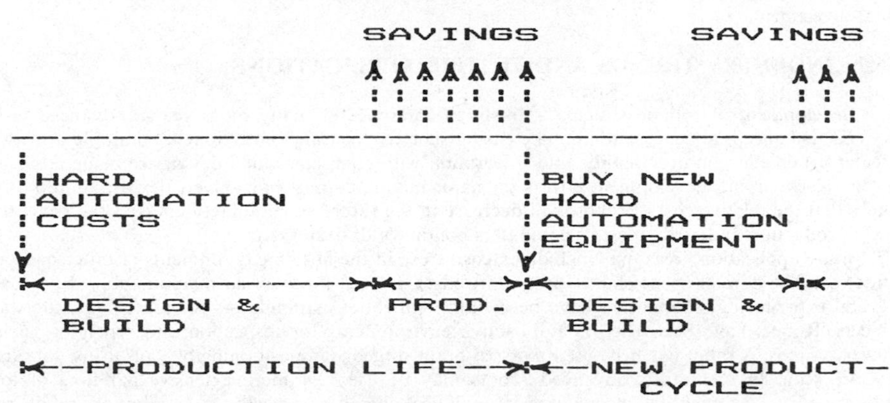

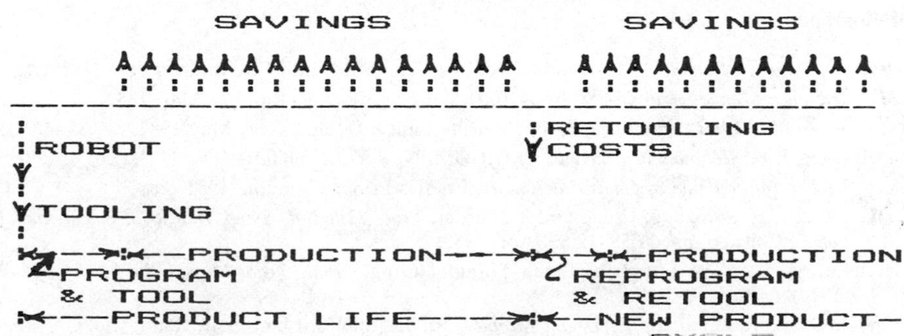

Fig. 36.17 Comparison of economic alternatives.

recognized that they will change the way many "information" or indirect employees work in manufacturing, in much the same way that the invention of the transistor and the microprocessor changed the way electrical and electronic engineers and computer programmers worked. Methods of equipment justification must accurately reflect the improved efficiency of these indirect employees.

In addition to the changing nature of the factory organization, an important characteristic of modern manufacturing is the compression of production cycles. International competition coupled with advances in technology have reduced the time allowed for product development as well as manufacturing process development. Furthermore, the actual production life of the product may be relatively short. In addition, the time between introduction of a new product and the need for high-volume production has become a more critical factor in business management. All of these factors inhibit to some degree the opportunity to utilize traditional fixed automation. However, these very factors enhance the opportunity for use of robots and other flexible methods.

This is illustrated in the cash flow diagrams shown in Fig. 36.17. The upper diagram is representative of the use of traditional fixed automation. The process begins with the decision to build the required equipment. After a relatively long period, the equipment is built and put into production. In order for the savings to provide a satisfactory return on investment, the equipment may have to be utilized for a relatively long period. However, if the market for the product does not materialize or competitive products are introduced rapidly, the equipment can become obsolete prematurely.

Now consider the lower half of Fig. 36.17, which represents the cash flow diagram for a robotic manufacturing system. Major advantages include a shorter lead time with corresponding higher proportion of production being automated. In addition the robots can be retooled and reprogrammed at the end of the product's life, thereby making possible continued utilization of a major part of the total investment.

36.7 INDUSTRY TRENDS AND FUTURE APPLICATIONS

The development of robotic systems is likely to continue for many more years. Advanced vision systems, enhanced languages, additional sensing such as "hearing" and "touch," multiple arms with greater articulation, more mobility, and integration with computer-aided design are being addressed by the research and development efforts of major manufacturers and universities. In addition, it is likely that the relative cost of robots will decrease in the future as manufacturers gain the benefits of mass production and as the cost of computers continues to decrease.

Future application areas may include extensive use in the hostile environments of outerspace and undersea. The current space shuttle program utilizes a robot for external manipulation of payloads. Several experimental submarines have been built with robot manipulators for retrieving underwater objects. Robots have been used in radioactive environments for inspection and repair of nuclear power stations. A robot has been used to assist brain surgeons in positioning of tools using data from a CAT scan. In the future, advanced robots may be used for more extensive handling of toxic substances such as might be encountered in analytical chemistry procedures. In addition, robots may be used in "clean room" applications where the presence of people tends to contaminate the process. Perhaps in the future robots will even be able to perform some routine household tasks, such as cleaning and cooking. However, it is unlikely that this will occur for many years, since it depends on significant technological development with corresponding cost reductions.

Bibliography

Allegro Operator's Manual (A12 Assembly Robot), General Electric Co., Bridgeport, CN, 1982.

AML Reference Manual, 2nd ed., SC34-0410, IBM Corp., Boca Raton, FL, Mar. 1982.

Ayres, R. and S. Miller, "Industrial Robots on the Line," *Technol. Rev.*, May 1982, pp. 35–46.

Engelberger, J. F., *Robotics in Practice*, AMACOM, New York, 1980.

Fleck, J., *The Introduction of Industrial Robots*, Francis Printer, London, 1983.

Gruver, W. A., B. Soroka, J. Craig, and T. Turner, *Proc. 13th Intl. Symp. Industrial Robots and Robots 7*, Chicago, pp. 12.58–12.83, Apr. 1983.

Merchant, M. E., "The Future of Batch Manufacturing," *Phil. Trans. Royal Soc.* **275A**:356–372 (1976).

Mujtaba, S. and R. Goldman, *AL User's Manual*, 3rd ed., Stanford University, Palo Alto, CA, Report No. STAN-CS-81-889, Dec. 1981.

Nevins, J. L. and D. E. Whitney, "Computer Controlled Assembly," *Sci. Am.* **238**(2):62–74 (Feb. 1978).

Program Robotics by Example User's Guide, #08009, IBM Corp., Boca Raton, FL, Dec. 1982.

RAIL Software Manual, Rev. 3.0, MN-RB-07, Automatix Inc., Burlington, MA, Jan. 1982.

Summers, P., R. Taylor, and J. Meyer, "AML: A Programming Language for Automation," *COMPSAC 81*, Chicago, IL, Nov. 18, 1981.

Tanner, W., *Industrial Robots*, Society of Manufacturing Engineers, Dearborn, MI, 1978.

Tanner, W. R., "A Users' Guide to Robot Applications," SME Technical Paper MR76-601, 1976.

User's Guide to VAL, Version 12, #398-H2A, Unimation Inc., Danbury, CN, Jun. 1980.

Will, P. M., and D. D. Grossman, "An Experimental System for Computer-controlled Assembly," *IEEE Trans. Comput.* **C-24**:879–888 (Sept. 1975).

Wood, B. O. and M. Fugelso, *Proc. 13th Intl. Symp. Industrial Robots and Robots 7*, Chicago, pp. 12.84–12.96, Apr. 1983.

CHAPTER 37
SYSTEMS MEASUREMENTS

RICHARD G. COSTELLO

The Cooper Union for the Advancement of Science and Art
New York, New York

37.1 INTRODUCTION TO THE MEASUREMENT OF SYSTEM STATES

The fundamental concept of control systems involves a comparison between the controlled variables, which usually include the system output, and the desired reference values for these variables. To make this comparison, the controlled variables must be measured in units corresponding to those used by the system controller. For example, in an electronic feedback amplifier where both input and output are voltages, an actual measurement device is not required. Only a straightforward wired connection that serves to tap off a measurement of the system output is used. This is the usual case in an all-electronic system.

In systems that have a mechanical (or hydraulic or pneumatic) output with an electrical controller, the mechanical (or hydraulic or pneumatic) output must be converted to an electrical signal. For example, in mechanical systems various electromechanical devices are used, such as potentiometers, tachometers, gyroscopes, accelerometers, and specialized transducers. These devices are discussed in Sections 37.6 through 37.12.

In some self-contained specialized control systems, electrical signals and components are absent and the system output is measured hydraulically, pneumatically, or mechanically. A common example of this is the force-balance pressure regulator. It drops and regulates the pressure output from a high-pressure input source, which is generally a gas or liquid. Such regulators are hidden under the streets of major cities to regulate gas and water distribution line pressures and are seen on oxygen tanks and other high-pressure gas tanks used in hospitals or in welding shops.

Float regulators, which keep the level of a fluid constant, are present in almost every automobile carburetor. This purely hydromechanical control system performs its measurement mechanically as a consequence of a moving float. Such systems, which do not use electrical signals at all, will not be discussed specifically in this chapter. Most of the discussion centers on methods of measurement that have a relationship to electrical or electronic control. Section 37.2, which deals with the requirements a measurement must meet, and Section 37.3, which deals with the scaling of a measurement, are general and apply to nonelectrical as well as electrical measurements.

37.2 MEASUREMENT REQUIREMENTS

To be useful, a measurement must be accurate (precise), be repeatable, possess adequate resolution, and have an adequate dynamic range and response time. The advent of the digital display, the digital computer, and assorted digital instruments, transducers, and signal conditioners has made it imperative to distinguish between digits displayed and "garbage." It is not uncommon to see a computer printout of 10 or more digits, when only the first three or four digits have any significance. The remaining six or seven digits are quite literally "garbage" and should be thrown out, if it is known that the original data source was only accurate to one tenth of 1%, or to 1 part in a thousand. This is typically the case for most control system components. Only high-cost high-precision measurement devices have an inherent accuracy exceeding one tenth of 1%.

Before we proceed further, several measurement or metrological terms will be defined and discussed fairly completely from an engineering point of view.

1. Accuracy. The accuracy of an engineering measurement is the validity or quantitative correctness of the measurement, usually expressed in terms of the largest possible error in the measurement. The largest possible error is often called the worst-case error. There are at least four common formats for an accuracy specification:

(a) Reading $\pm N\%$ of reading, where $N =$ percent accuracy, say, 1%. Worst-case error for (a) $= \pm N\%$ of reading.

(b) Reading $\pm M\%$ of full-scale reading, where $M =$ percent of full-scale accuracy, not equal to N. One percent N is superior to $1\%M$, at low-scale readings. The two are identical at full scale. Worst-case error for (b) $= \pm M\%$ of full-scale reading.

(c) Reading $\pm K$, where $K =$ constant, often the quantizing error in an analog-to-digital conversion. For an N-bit converter, the quantizing error K is given by the full-scale reading divided by 2^N for an unsigned conversion, or divided by 2^{N-1} for a signed conversion. Worst-case error for (c) $= \pm K$.

(d) Reading $\pm f(\Upsilon)$, where $f(\Upsilon)$ is some function of a variable Υ, which is usually temperature but can be other variables such as pressure or time. $f(\Upsilon)$ may be an absolute function, such as ± 2.5 μV/°C times Υ°C, or $\pm 2.5\Upsilon\mu$V. $f(\Upsilon)$ may also be a percentage function, such as $\pm 0.0001\Upsilon\%$ times the reading, or the full-scale reading. Worst case error for (d) $= \pm f(\Upsilon)$.

For example, a typical specification for a digital transducer, such as an angle-measuring potentiometer, readout by a digital voltmeter, might be

$$\text{Error} = \pm 0.01\% \text{ of full-scale reading} \pm 1 \text{ digit} \pm 0.01 \times \Delta T$$
$$\text{(temperature in degrees Centigrade) digits}$$

If for the sake of illustration the full-scale reading was 360 units of angular degrees with a display resolution of 1 degree per least significant digit and the temperature change was 30°C, the maximum (or worst-case) error in any angular measurement would be

Max. error $= \pm 0.01\%(360) \pm 1 \pm 0.3$
Max. error $= \pm(3.6 + 1 + 0.3)$
Max. error $= \pm 4.9$ digits $= \pm 4.9$ angular degrees (which would round to ± 5 displayed digits)

Thus a device nominally accurate to one tenth of 1% of full scale could produce very large percentage errors at small readings. In the example just given, a reading of 10°(angular) would correspond to a measurement of $10 \pm 4.9°$, or a possible error of 49%. This is not an unusual situation. Fortunately, feedback control systems do not require extremely accurate absolute measurements if the system error can be directly measured from the input and output, which is sometimes the case. This is true because, at least for closed-loop systems, the error is driven to zero. Therefore all that is needed is a measurement transducer that is linear and has a nearly zero output for a zero input. Any nonzero output, called a DC offset or zero drift, will produce an equivalent output error.

However, if the system error cannot be directly measured but instead must be obtained by measuring the input and output and then subtracting the second from the first, an accurate measurement system is required. Any error in the measurement will be reflected by an error in the system output.

2. Resolution. The resolution of a measuring device refers to the smallest quantity that the device can identify, resolve, or distinguish. Thus for a meter stick divided into millimeters, the resolution is 1 mm, since this is the smallest measurement that can be made (using the meter stick's divisions). For a yardstick divided into sixteenths of an inch, the resolution would be one sixteenth of an inch. For an 8-bit digital transducer, which divides its full-scale measurement into $2^8 - 1 = 255$ parts, the resolution is 1 part in 255, or roughly 0.4% of full scale. For the human eye, the resolving power or resolution is approximately 1 to 1.5 arc-min. For a typical inertial navigation system's angular resolver, the resolution is approximately 1 arc-s. For a typical camera lens, the resolving power ranges from approximately 100 to 400 lines per millimeter. For a commercial US television set, the resolving power is approximately 300 lines per picture height, whereas some video display monitors have a resolution of 2000 lines. The resolution of a 3-digit digital voltmeter is 1 part in 999, while the resolution of a 3.5-digit digital voltmeter is 1 part in 1999. On the lowest or most sensitive scale, the resolution of the same digital voltmeter might be 1 mV (0 to 1999 mV), or even 1 μV (0 to 1999 μV), depending on the meter's construction and ranges. Resolution may be relative, for example, 1 part in N parts, or it may be absolute, such as 1 mV or 1 mm. Thus there are many different types of resolution specifications. The resolution of a measuring device is the smallest change it can detect or display. That change may not be accurate; for most digital devices, the accuracy is almost always significantly less than the resolution. For example, a digital angle-measuring device might have a resolution of 1° and a worst-case accuracy of ±4.9° (as discussed previously in the definition of accuracy). Thus a 30° angle might be measured as 34°, and a 31° angle measured as 35°. This measuring device resolves a 1° input change from 30° to 31°, but in this example it has an absolute error of +4°. Therefore it produces a 1° change in output measurement, from 34° to 35°.

3. **Precision.** Precision is usually a synonym for accuracy. A precision of four decimal digits, or simply four-digit precision, means an accuracy of at least 1 part in 9999. It is sometimes taken to mean that each of the four digits is precise or accurate, and that the maximum error is $\pm 1/2$ the least significant digit.

4. **Repeatability.** It is very desirable that a measurement system be consistent and give repeatable results, even though the results are somewhat inaccurate. For example, let us assume that a 360° full-scale angle-measuring device has a maximum worst-case error of 4.9°, and it produces a measurement of 34° when the actual angle is 30°. Then if the angle is varied up to 270° and back to 30°, a repeatable device would give the same measurement as before: 34°. Note that the accuracy specification would be met by a measurement of 26°, and certainly by an exact measurement of 30°. However, neither of these would satisfy the requirement of repeatability.

5. **Dynamic range.** Dynamic range is the range or ratio of the smallest resolvable measurement to the largest possible measurement. For electronic transducers, the smallest resolvable signal or measurement is determined by the total magnitude of the various electronic and thermal noises present in the transducer and its associated electronics. For real-time measuring instruments that do not employ convolution or elaborate signal-processing techniques, the smallest resolvable signal or measurement must exceed the noise level. Any smaller signal will be "hidden" in the noise. At the other extreme, the largest possible measurement is limited by either hard constraints (i.e., the physical limits of undamaged transducer operation) or by the onset of unacceptable nonlinearities. The term "dynamic range" is often applied to individual components of the measurement system, such as the transducer element, or an associated electronic instrumentation amplifier. The dynamic range of a particular instrumentation amplifier might be 1 to 1000 mV, referred to the input terminals. This range is commonly expressed in decibels (dB) by employing the relation

$$\text{Dynamic range} = 20 \log\left(\frac{\max}{\min}\right) \tag{37.1}$$

For this illustration

$$\text{Dynamic range} = 20 \log\left(\frac{1000 \text{ mV}}{1 \text{ mV}}\right)$$

or

$$\text{Dynamic range} = 20 \log(1000) = 20(3) = 60 \text{ dB}$$

Dynamic range is related to resolution by Eq. (37.1). A 60-dB dynamic range corresponds to a resolution of 1 part in 1000, and an 80-dB dynamic range corresponds to a resolution of 1 part in 10,000 (e.g., $10^{80/20} = 10^4 = 10{,}000$).

If the noise level of the measurement system should increase with age, the size of the smallest resolvable signal would increase and both resolution and dynamic range would decrease markedly. For example, let us assume that the instrumentation amplifier, which had a dynamic range of 1 to 1000 mV, should become noisy, so that the minimum clearly discernible signal is 2 mV. Then the resolution would become 2 parts in 1000 (which is equivalent to 1 part in 500), and the dynamic range would be lowered from 60 dB to $20 \log(1000/2) = 20 \log(500) = 20(2.7) = 54$ dB. Note that what seems like a small increase in the noise level, in this case 1 mV, produces a much larger change in resolution and dynamic range.

For a digital measurement system there is a relation between the number N of digital bits used and the dynamic range. The resolution of a positive unsigned measurement is 1 part in $2^N - 1$. Applying Eq. (37.1) yields

$$\text{Dynamic range} = 20 \log \frac{2^N - 1}{1}$$

$$\text{Dynamic range} \cong 20N \log 2, \qquad \text{if } 2^N \gg 1 \tag{37.2}$$

$$\text{Dynamic range} \cong 20N(0.30) = 6N \text{ dB}$$

where N = digital word length, in bits.

Thus if a dynamic range of 60 dB (or 1 part in 1000) is required, the digital word length must be $6N = 60$, or $N = 10$ bits. If a dynamic range of 80 dB is required, the digital word length must be $6N = 80$, or $N = 13.3$, which must be rounded up to the nearest integer, or $N = 14$ bits. Since the physical world is analog, with continuously varying quantities, any digital measuring device must contain an analog-to-digital converter.

In the early 1980s, the commonly available off-the-shelf analog-to-digital converters have word lengths of 8, 10, and 12 bits which correspond to dynamic ranges of 48, 60, and 72 dB, respectively.

If signed measurements are required, so that half the measurement range is positive and half the range is negative, the positive range is halved or decreased by a power of two. This is equivalent to removal of 1 digital bit from the word length. This removed bit becomes the sign bit of the measurement, and Eq. (37.2) becomes

$$\text{Signed dynamic range} = 20 \log 2^{N-1} = 20(N-1)\log 2 \tag{37.3}$$

$$\text{Signed dynamic range} = 20(N-1)(0.30) = 6(N-1) \quad \text{dB}$$

where N = digital word length. Thus if a positive and negative measurement is required with a resolution of 1 part in 1000, or 60 dB for each range, the required digital word length N is given by

$$6(N-1) = 60, \quad \text{or} \quad N - 1 = 10, \quad \text{or} \quad N = 11 \text{ bits}$$

Recall that for a single positive range with a resolution of 1 part in 1000 or 60 dB, only a 10-bit word length was required.

6. Response time. The response time of a measuring device is the time that elapses between the change in the physical input parameter and the appearance of the final output measurement signal or display. One convenient way to measure this time delay is to apply a step change, whose rise time is negligible compared with the response time, as a test input to the measurement device. When the output measurement signal reaches 90% of its final value, that time is noted. The response time is then defined as the time difference between the application of the step input and the time to reach the 90% response output point. If the measurement device contains an analog-to-digital converter, a sample-and-hold unit, or a digital-to-analog converter, each of these devices will add significantly to the response time.

Response time is particularly important in the measurement of control system variables, because any delay in obtaining such measurements decreases the control system's relative stability. In most instances the response time of electronic measurement devices is measured in microseconds, which is usually much shorter than the delay time that would lead to system instability. As long as the measurement response time is two or more orders of magnitude less than the control system's time constants or oscillatory periods, little adverse effect will be produced on the control systems by the response-time delay.

37.3 SCALING OF MEASUREMENTS

In practice, most measurements require a change of scale before they can be used, owing to several factors. Four of these factors are that the range or variation of the physical input being measured is too large or too small or the output signal of the measurement device is too large or too small for the purpose intended.

Scaling of Rotational Quantities

Scaling of rotational quantities is very important from a practical viewpoint. Rotational physical inputs (typically obtained from a rotating shaft) are often scaled using a set of gears. When a "big gear" is meshed with a "small gear," the big gear turns slower, turns through a smaller angle, experiences a larger torque, and experiences a smaller acceleration than the meshing smaller gear.

The scale factor N for these four effects is given by

$$N = \frac{\text{teeth on large gear}}{\text{teeth on small gear}} > 1 \tag{37.4}$$

Referring to Fig. 37.1:

	Large Gear_1	Small Gear_2
Angle	$\begin{cases} \theta_1 \\ \theta_2/N = \theta_1 \end{cases}$	$\begin{matrix} N\theta_1 = \theta_2 \\ \theta_2 \end{matrix}$
Angular velocity	$\begin{cases} \omega_1 \\ \omega_2/N = \omega_1 \end{cases}$	$\begin{matrix} N\omega_1 = \omega_2 \\ \omega_2 \end{matrix}$
Angular acceleration	$\begin{cases} \alpha_1 \\ \alpha_2/N = \alpha_1 \end{cases}$	$\begin{matrix} N\alpha_1 = \alpha_2 \\ \alpha_2 \end{matrix}$
Torque	$\begin{cases} T_1 \\ NT_2 = T_1 \end{cases}$	$\begin{matrix} T_1/N = T_2 \\ T_2 \end{matrix}$

When more than two gears are used, there is a scale factor N for each pair of gear meshes. As long as the effect of each gear pair is similar, either a speed up or a slow down, the overall gear ratio or scale factor is given by the product of the individual gear ratios. Note that torque is scaled in the opposite sense to the angular quantities. This is a consequence of the law of the conservation of energy. For frictionless gears, power in, P_1 = power out, P_2. For a rotating shaft, power is given by the product of torque T and angular velocity ω, or $P = T\omega$. Thus equating P_1 to P_2:

$$P_1 = P_2 \quad \text{or} \quad T_1\omega_1 = T_2\omega_2 \tag{37.5}$$

$T_1/T_2 = \omega_2/\omega_1 = N$ = gear ratio of Eq. (37.4). Equation (37.5) shows the inverse relation between the scaling of torque and the scaling of angular velocity (or angle) for a pair of gears with the gear ratio N.

A set of gears may also be employed to scale or change the apparent magnitude of a rotational movement of inertia J, a rotational viscous damper B, and/or a rotational spring K. In each of these three cases the scale factor is N^2, with the bigger gear getting the effective bigger inertia, damper, or spring. The bigger gear experiences a bigger torque, N times that of the small gear, and the bigger gear turns through a smaller angle, $1/N$ times that of the smaller gear. The rotational spring constant K is given by

$$K = \frac{T}{\theta}$$

For the bigger gear:

$$K_1 = \frac{T_1}{\theta_1}$$

while for the small gear:

$$K_2 = \frac{T_2}{\theta_2}$$

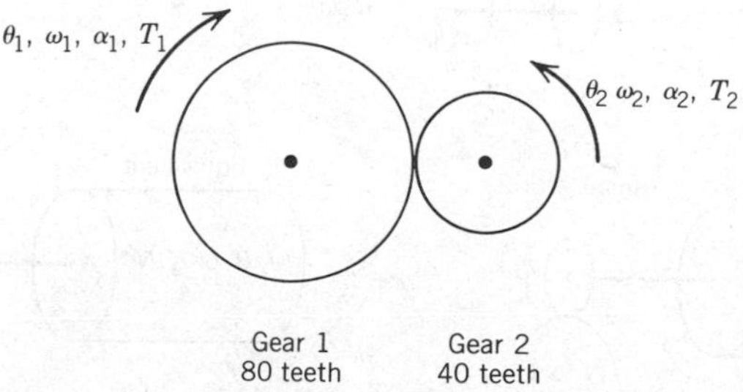

Fig. 37.1 Scaling of rotational systems using gearing. Gear ratio N = teeth on large gear/teeth on small gear = 80/40 = 2.0 where θ = gear or shaft angle, rad; $\omega = \dot{\theta} = d\theta/dt$ = angular velocity, rad/s; $\alpha = \ddot{\theta} = d^2\theta/dt^2$ = angular acceleration, rad/s^2; T = torque, N-m or lb-ft (previously ft-lb). [1 N = 0.224809 lb-force (1 kg · 1 m/s^2), 1 m = 3.28084 ft, 1 N-m = 0.224809 lb · 3.28084 ft = 0.737562 lb-ft. Pound-feet (or pounds-foot) is used to differentiate the torque unit from the energy unit of foot-pounds.]

TABLE 37.1. SCALING OF ROTATIONAL SPRING, DAMPER, AND MOMENT OF INERTIA BY GEAR MECHANISM, WITH GEAR RATIO $N > 1$

	Bigger Gear_1	Smaller Gear_2	Torque Equation, Either Gear
Rotational spring K	K_1	$K_1/N^2 = K_2$	$T = K\theta$
	$N^2K_2 = K_1$	K_2	
Rotational damper B	B_1	$B_1/N^2 = B_2$	$T = B\dot{\theta} = B\omega$
	$N^2B_2 = B_1$	B_2	
Rotational moment of inertia J	J_1	$J_1/N^2 = J_2$	$T = J\ddot{\theta} = J\alpha$
	$N^2J_2 = J_1$	J_2	

However, $T_1 = NT_2$ and $\theta_1 = \theta_2/N$. Inserting these relations into the expression for K_1:

$$K_1 = \frac{T_1}{\theta_1} = \frac{NT_2}{\theta_2/N} = N^2\frac{T_2}{\theta_2} = N^2K_2 \tag{37.6}$$

Thus the spring constant experienced by the bigger gear K_1 is N^2 times the spring constant experienced by the smaller gear. Since angle, angular velocity, and angular acceleration are scaled identically, the expressions for scaling a rotational damper $B = T/\dot{\theta} = T/\omega$ and a rotational moment of inertia $J = T/\ddot{\theta} = T/\alpha$ are all the same, namely the scale factor is N^2. This scaling is summarized in Table 37.1.

The scaling of K, B, and J by a factor of N^2, using a pair of gears with a tooth ratio of N, is often called reflecting K, B, or J to the other side of the gear set. This is summarized in Fig. 37.2.

Scaling of Linear Quantities

The scaling of linear quantities also is very important in practice. Linear positional inputs are often scaled with levers or linkages. Unlike gear systems, lever systems can operate only over a limited range and are often limited to an even smaller range if the lever system scale factor N_{lev} is to be

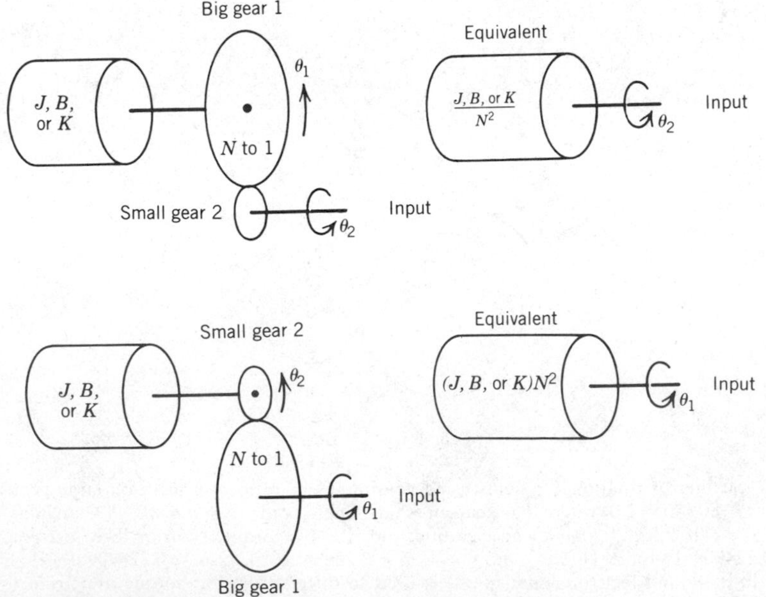

Fig. 37.2 Reflecting or scaling of an inertia J, a rotational viscous damper B, and/or a rotational spring K by a gear set of gear ratio N, where N is always greater than 1.

constant. The nonlinearity in a lever system, if it exists, is usually due to a straight-line motion actuating a pivoting lever which moves in an arc. The arc and the straight-line motion are nearly identical for small angles of arc, growing divergent as the arc angle increases. The scale factor N_{lev} is illustrated in Fig. 37.3, for a simple lever.

Conversion of Linear Motion to Angular Motion

Linear position, velocity, and acceleration are often converted in practice into equivalent angular units by use of a flexible string, wire, belt, tape, or band which is wrapped about a pulley or drum. One revolution of a drum of radius R corresponds to a linear motion of a circumferential string, wire, band, or belt equal to $2\pi(R + r)$ units, where r is the radius of the string or wire, or one half the thickness of a band, belt, or tape. This is illustrated in Fig. 37.4. Usually the radius of the drum R is much greater than the radius of the string or wire r, so that $2\pi(R + r)$ is approximately equal to $2\pi R$. Such a combination of scaling (by $2\pi R$ for one revolution, or by R for 1 rad) and the resultant conversion of linear motion to angular motion is used in many common devices such as the servo feedback mechanisms of x-y plotters, digital drum plotters, and magnetic recording-head positioning systems, and electronically controlled positioning of the printing mechanisms for low-speed line printers. The basic drum mechanism is shown in Fig. 37.4, along with the conversion factors for a drum of radius R:

Linear Quantity	Angular Quantity
Length L, m	$\theta = L/R$, rad
Velocity $V = \dot{L}$, m/s	$\omega = \dot{\theta} = \dot{L}/R = V/R$, rad/s
Acceleration $A = \ddot{L}$, m/s^2	$\alpha = \ddot{\theta} = \ddot{L}/R = A/R$, rad/s^2

Scaling of Electrical Quantities

Electrical scaling of measurements is the preferred choice whenever scales (or scale factors) must be changed repeatedly or over a large range, or whenever the transducer of choice provides an electrical output. The simplest and most commonly used scaling mechanism is the resistive divider network, which consists of two series resistors R_1 and R_2, as shown in Fig. 37.5. In many cases R_1 and R_2 are

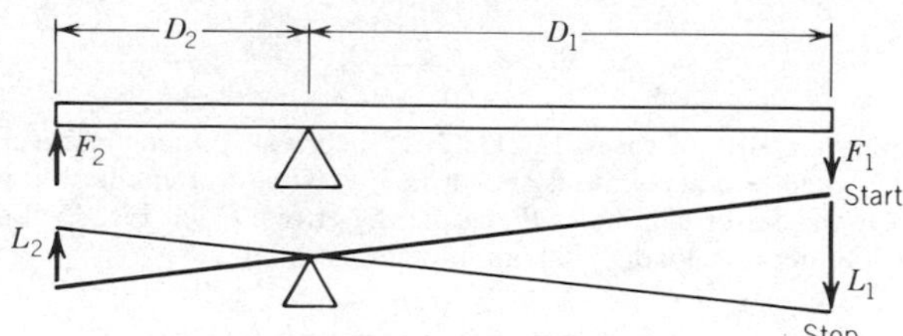

Fig. 37.3 Scaling of linear quantities using levers.

The scale factor N_{lev} for a simple lever is

$$N_{lev} = \frac{\text{longer lever arm distance } D_2}{\text{shorter lever arm distance } D_1} > 1$$

For small motions, where $L_2 \ll D_2$, $L_1 \ll D_1$, and $N = N_{lev}$

	Long Arm$_1$	Short Arm$_2$
Position	L_1	$L_1/N = L_2$
L, m	$NL_2 = L_1$	L_2
Velocity	V_1	$V_1/N = V_2$
V, m/s	$NV_2 = V_1$	V_2
Acceleration	A_1	$A_1/N = A_2$
A, m/s^2	$NV_2 = A_1$	A_2
Force	F_1	$NF_1 = F_2$
F, N	$F_2/N = F_1$	F_2

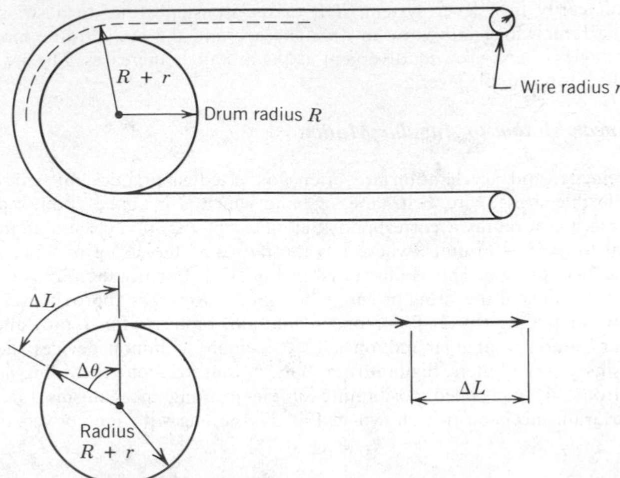

Fig. 37.4 Conversion of linear motion to angular motion. Effective radius of drum and wire = $R + r$. L = length or position, m. ΔL = change in L. In radians, $\Delta L = (R + r)\Delta\theta$, or $\Delta\theta = \Delta L/(R + r)$. If $R \gg r$, then $\Delta\theta = \Delta L/R$, or $\theta = L/R$, and $\omega = \dot{\theta} = \dot{L}/R, \omega = V/R$ and $\alpha = \ddot{\theta} = \ddot{L}/R, \alpha = A/R$ where θ = angle, rad; $\dot{\theta} = \omega$ = angular velocity, rad/s; $\ddot{\theta} = \alpha$ = angular acceleration, rad/s^2; $V = \dot{L}$ = linear velocity, m/s; $A = \ddot{L}$ = linear acceleration, m/s^2.

the two portions of a single divided (or tapped) resistor. If the tap-off point is movable in response to a physical rotary or linear adjustment, the resistive divider is often called a variable resistor, a variable attenuator, a potentiometer, or a pot. An ideal potentiometer is used in a bridge circuit and has identical current flow through its two resistors, with no current flow out of the tap connection (theoretically). Almost every variable scaling device called a potentiometer (or a pot) is not a true pot, because the currrent flow through R_1 is not exactly equal to the current flow through R_2; this produces a scaling error. As shown in Fig. 37.5, the ideal potentiometer scale factor N_{pot} is

$$N_{\text{pot}} = \frac{R_2}{R_1 + R_2} \tag{37.7}$$

In the overwhelming majority of cases, Eq. (37.7) is incorrect. Either the effect of the finite load resistance R_L must be taken into account, as shown in Fig. 37.6, or a calibration procedure must be employed that compensates for or eliminates these loading errors. From Fig. 37.6, the scale factor for an actual resistive divider network loaded with an impedance R_L is

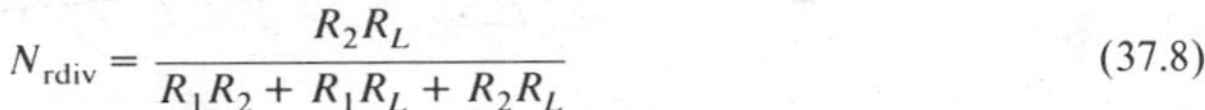

$$N_{\text{rdiv}} = \frac{R_2 R_L}{R_1 R_2 + R_1 R_L + R_2 R_L} \tag{37.8}$$

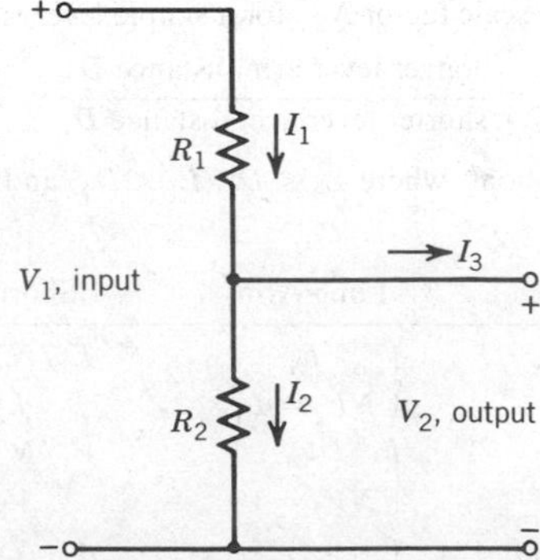

Fig. 37.5 Ideal resistive divider network. True potentiometer, $I_1 = I_2$, $I_3 = 0$. Scale factor $N_{\text{pot}} = R_2/(R_1 + R_2)$. I_3 will equal zero if a voltage of $V_1[R_2/(R_1 + R_2)]$V is connected to the V_2 terminals, or if an infinite impedance is connected to the V_2 terminals.

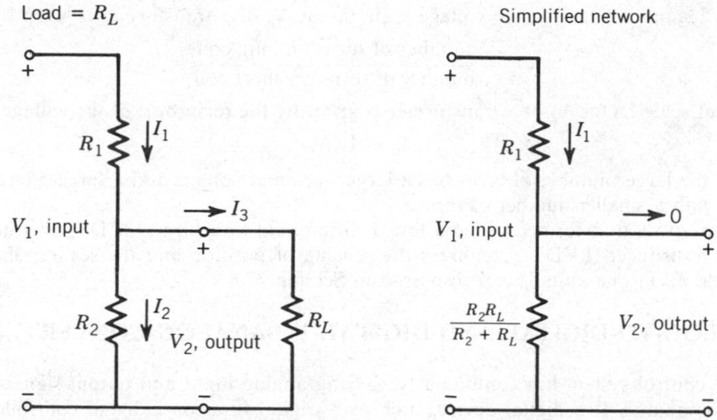

Fig. 37.6 Actual resistive divider Network with a load R_L. Loaded resistive divider scale factor, N_{rdiv}, is

$$N_{\text{rdiv}} = \frac{R_2 R_L/(R_2 + R_L)}{R_1 + [R_2 R_L/(R_2 + R_L)]} = \frac{R_2 R_L}{R_1 R_2 + R_1 R_L + R_2 R_L}$$

Observe that as R_L approaches infinity, the $R_1 R_2$ term becomes negligible, and

$$N_{\text{rdiv}} \cong \frac{R_2 R_L}{R_1 R_L + R_2 R_L} = \frac{R_2}{R_1 + R_2} = N_{\text{pot}} \quad \text{if} \quad R_L \to \infty$$

If the load is a complex impedance Z_L rather than a resistive load R_L, Eq. (37.8) is easily altered by replacing R_L with Z_L.

In many scaling applications, the potentiometer is set by applying a known calibration voltage V_1 and observing the loaded output voltage V_2 with a device that has an input impedance much greater than Z_L (or R_L). Hence the device that measures V_2 produces a negligible change in the desired scale factor. That is, all the error in the simple scaling equation $N_{\text{pot}} = R_2/(R_1 + R_2)$ is produced by R_L (or Z_L), and none (or very little) is produced by the input impedance of the device that measures V_2. This procedure is employed to set the coefficient pots in analog computers, to calibrate electronic transducers and measuring instruments, and to calibrate electronic consumer products.

Scaling a measurement by the application of a simple resistive divider network is the procedure most commonly employed for signals ranging in frequency from 0 Hz (DC) to several megahertz. This is the frequency range for almost all control systems. At higher frequencies, shunt capacitance becomes important, and the resistors R_1, R_2, and R_L of the scaling Eq. (37.8) must be replaced by Z_1, Z_2, and Z_L, respectively, where each of these contains a capacitive component (in the usual case). The net result is that a continuously variable control becomes replaced by a multiposition switch that selects various Z_1 and Z_2 elements. Each Z_1 or Z_2 usually consists of a resistor in parallel with a capacitor. Therefore the real and imaginary parts of the scaling Eq. (37.8) can produce a frequency-independent real constant. In effect, capacitors are added in parallel with R_1 and R_2 to compensate for the shunt capacity of Z_L. For example, these capacitors are used on the vertical scale factor control of the electronic oscilloscope, which is common to almost every electronics laboratory.

Another common method of scaling electrical measurements involves the use of transformers. Since transformers do not operate at DC or zero frequency, the electrical signal being scaled must not contain any DC or zero-frequency component (since the DC component will not be transmitted by the transformer). Transformers that contain magnetic cores have a frequency response that can range from a few hertz to a few megahertz (with a typical frequency range much less than this). Transformers that have no core or have an air core typically operate from 1 MHz up to the gigahertz region (which is generally well above the frequency band of most control systems). There is a class of control system that uses a modulated alternating current (AC) carrier to power a two-phase servo motor. The modulation can be provided by a phase-shifting device or by a control transformer that looks like a small motor and contains a rotating winding. The angular position of the movable winding determines the magnitude and phase of the modulated AC carrier signal. The typical operating frequency of the carrier signal is 400 Hz. Transformers are widely used to scale control signals in such systems.

Within its region of linearity, the voltage scale factor N_v of a transformer is given by

$$N_v = \frac{\text{number of turns on long coil}}{\text{number of turns on short coil}} \tag{37.9}$$

The current scale factor N_I of a transformer is given by the reciprocal of the voltage scale factor:

$$N_I = 1/N_v \tag{37.10}$$

The coil with the larger number of turns has a larger terminal voltage and a smaller terminal current than the coil with a smaller number of turns.

A special purpose transformer, called a linear differential transformer (LDT) or a linear variable displacement transducer (LVDT), combines the sensing of position and the scaling of the resultant electrical signal all in one unit. This is discussed in Section 37.6.

37.4 ANALOG-TO-DIGITAL AND DIGITAL-TO-ANALOG CONVERTERS

Almost every control system has continuously varying analog input and output signals because the "real world" is analog. If a digital device, such as a digital filter or a digital controller, is to be a component of a control system with analog inputs and outputs, an analog-to-digital (A/D) converter must precede the digital device, and a digital-to-analog (D/A) converter must follow the digital device. An A/D converter has one analog input line and N digital output lines. Typical values of N are 8, 10, 12, and 14. Values of N smaller than 8 and larger than 14 are occasionally used, with $N = 18$ being the 1981 upper limit for an A/D converter in the commercial market. Because many microcomputers have an 8-bit word length, the 8-bit A/D converter is very commonly used. As discussed in Section 37.1, the resolution of an N-bit A/D converter is one part in 2^N. For an 8-bit converter, the resolution is then one part in $2^8 = 256$. If the range of the analog input signal is from 0 to some maximum value, the increment corresponding to each digital step is:

$$\text{Analog increment} = \frac{\text{maximum value}}{2^N} \tag{37.11}$$

$$\text{Maximum number of increments} = 2^N - 1 \tag{37.12}$$

The -1 term in Eq. (37.12) accounts for the fact that the first digital count corresponds to a zero analog value, so that one digital count is effectively lost, and is used simply to begin the range. This is exactly analogous to measuring with a ruler. A 10-cm ruler has 11 marks on it, at 0, 1, 2, 3, 4, 5, 6, 7, 8, 9, and 10. Thus 8 digital bits can produce $2^8 = 256$ marks on an imaginary digital ruler, and these marks can measure 255 units. Thus by convention, if an 8-bit A/D converter is used to measure a 10-V range, the 10-V range will be divided into 256 steps of 39.1 mV ($10/256 \simeq 0.0391$) and the digital output will range from 0 to 255 of these steps. The last step corresponding to exactly 10.0 V does not exist. The converter stops at one step ($2^N - 1 = 255$) below the maximum value, or 9.9609 V.

It is possible to define the analog increment to take into account the missing step, so that the A/D converter will measure the full maximum range. In this case the analog increment would be defined as

$$\text{Analog increment} = \frac{\text{maximum value}}{2^N - 1}$$

$$\text{Maximum number of increments} = 2^N - 1$$

This procedure is not commonly used. When the number of digital bits N is reasonably large, say 10 or 12 bits, there is very little difference between 2^N and $2^N - 1$, and the effect of the missing last analog step is not significant.

To emphasize this effect, and to illustrate the quantizing uncertainty error of plus or minus one half of the analog increment, a 2-bit 10-V A/D converter is described by Fig. 37.7. This figure relates the input analog voltage to the digital code output. The analog increment is $10/2^N = 10/4 = 2.5$ V for each of the digital states 00, 01, 10, 11. This counting procedure is called straight binary code. The nominal analog voltage corresponding to each of these four digital states is 0, 2.5, 5.0, and 7.5 V. The quantizing uncertainty (error) is $\pm 1/2(2.5) = \pm 1.25$ V. Therefore the four digital states actually correspond to the voltages 0 ± 1.25, 2.5 ± 1.25, 5.0 ± 1.25, and 7.5 ± 1.25 V. These voltage ranges for each digital code are shown in Fig. 37.7 as rectangles, containing the digital code for each voltage range. The analog increment corresponds to the nominal voltage assigned to the least significant digital bit, often referred to as simply the least significant bit (LSB).

A similar illustration is shown in Fig. 37.8 for a 3-bit ± 5 V bipolar A/D converter.

Thus the quantizing error is often given as $\pm 1/2$ LSB. What is actually meant is $\pm 1/2$ of the nominal analog voltage corresponding to the least significant bit. Table 37.2 is a glossary of commonly used terms for A/D (and D/A) converters. Table 37.3 summarizes the resolution and theoretical maximum accuracy of A/D (and D/A) converters.

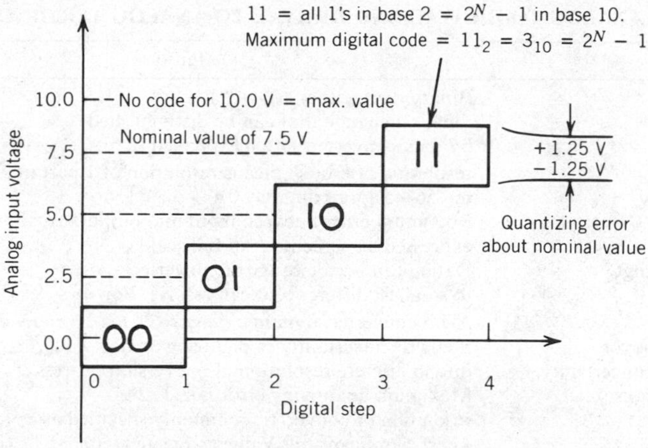

Fig. 37.7 Input-output response of a 2-bit ($N = 2$) 10-V analog-to-digital (A/D) converter, using straight binary coding. Analog increment = maximum value/2^N = $10/2^2$ = 10/4 = 2.5 V = least significant bit = (LSB). Quantizing uncertainty (error) = analog increment/2 = ±2.5/2 = ±1.25 V. Maximum nominal voltage coded = $(2^N - 1)$ analog increment = $(2^N - 1)$ maximum value/2^N = $(2^N - 1)/2^N$ maximum value = (3/4)10 = 7.5 V.

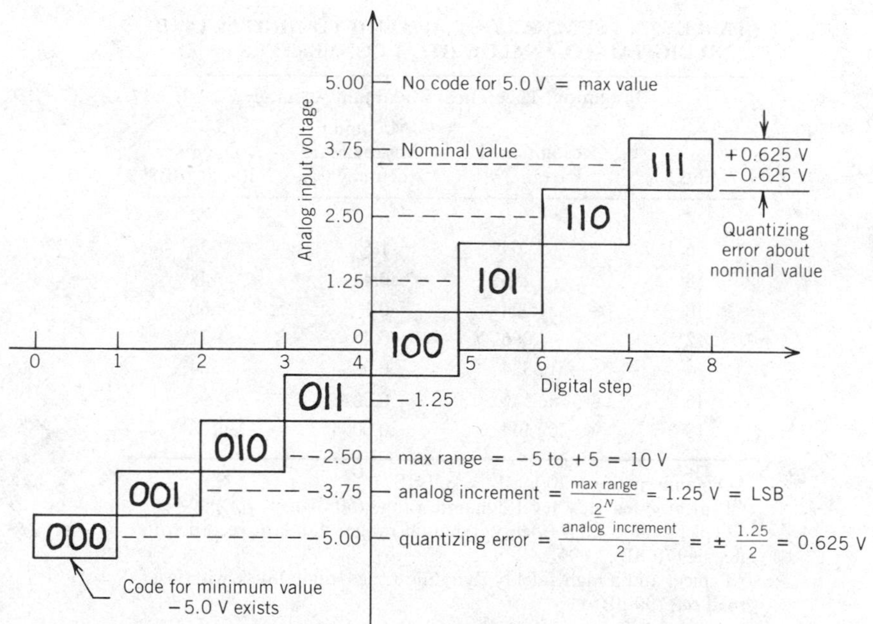

Fig. 37.8 Input-output response of a 3-bit ($N = 3$) ± 5-V Bipolar analog-to-digital (A/D) converter, using offset binary coding.

Code	Voltage
000	− 5.00
001	− 3.75
010	− 2.50
011	− 1.25
100	0.0
101	1.25
110	2.50
111	3.75

TABLE 37.2. ANALOG-TO-DIGITAL AND DIGITAL-TO-ANALOG TERMINOLOGY

Term	Definition
Bit	Binary digit, either "0" or "1"
Resolution	Smallest change that can be distinguished by data converter. Often expressed in bits. "A resolution of 8 bits" means resolution of 1 part in $2^8 = 1$ part in 256, or approximately 0.4%
Accuracy	Maximum error between input and output, usually expressed as percentage of full scale
Dynamic range	Ratio of full-scale range of converter to smallest difference it can resolve. For N-bit converter, dynamic range $= 20 \log 2^N \cong 6N$ dB
Quantizing error (quantizing uncertainty, quantizing noise)	Stepwise uncertainty in digitizing analog value, due to discrete resolution of conversion process. Maximum quantizing error is $\pm 1/2$ of resolution of converter, commonly specified as $\pm 1/2$ LSB
LSB	Least significant bit. Value associated with change of 1 bit. Generally identical to resolution of data converter. For 10-bit converter with 5-V range, 1 LSB $= 5/2^{10} = 5/1024 = 0.0048828$ V

TABLE 37.3. SUMMARY OF ANALOG-TO-DIGITAL (A/D) AND DIGITAL-TO-ANALOG (D/A) CONVERTERS

Resolution, Theoretical Maximum Accuracy			
Bits	Resolution, 1 Part in 2^N	Maximum Approximate Accuracy (%)	Dynamic Range (dB)[a]
4	16	6	24
6	64	1.5	36
8	256	0.4	48[b]
10	1,024	0.1	60
12	4,096	0.25	72[c]
14	16,384	0.006	84
16	66,536	0.0015	96[d]
18	262,144	0.0004	108

[a] Dynamic range $= 20 \log 2^N \cong 6N$ dB.
[b] Typical video gray level dynamic range (50 dB).
[c] Typical audio high fidelity dynamic range, standard record with needle (70 dB).
[d] Typical audio high fidelity dynamic range, digital disk with laser read-out (90 dB).

Various codes are used by A/D and D/A converters.[1] Some of the common codes are summarized in Table 37.4.

D/A Converters

A D/A converter has N digital input lines and one analog output line. The resolution and coding schemes of A/D and D/A converters are comparable. In general, D/A converters are simpler, more accurate, and less costly than A/D converters. Because of this, a D/A converter is often used to construct an A/D converter. This can be achieved in the following manner.

TABLE 37.4. CODES USED BY ANALOG-TO-DIGITAL (A/D) AND DIGITAL-TO-ANALOG (D/A) CONVERTERS

Code	Comments
Straight binary	For unipolar converters, as used in Fig. 37.7
Complementary binary	Complement of straight binary. All zeros changed to ones and all ones to zeros
Offset binary	For bipolar converters. 0 count = -analog full-scale voltages; midcount = 0 analog voltage; $2^N - 1$ or max count = maximum analog voltage − 1 LSB. 3-bit offset binary A/D converter is illustrated in Fig. 37.8
Complementary offset binary	Complement of offset binary
2's Complement	Positive and negative voltages represented by 2's complement binary code. Positive voltages have a most significant bit of 0, and negative voltages have a most significant bit of 1 and are in 2's complement form
Sign magnitude binary	Positive and negative voltage magnitudes represented by straight binary. Added most significant sign bit (0 or 1) differentiates positive and negative voltages. Sign magnitude binary has two distinct representations for 0 V: plus zero and minus zero
Binary coded decimal	BCD code, used for 4-, 8-, 12-, or 16-bit converters. Each group of 4 bits codes one decimal digit of analog input voltage, ranging from $0_{10} = 0000_2$ to $9_{10} = 1001_2$. Six binary states from 1010 to 1111, which correspond to digits greater than 9, are not used. Hence BCD is inefficient code, using only 10 out of possible 16 states for each 4 binary bits. BCD code is commonly used for digital panel meters and other decimal display applications, where human factor is very important and code inefficiency is tolerable
Complementary BCD	Complement of binary coded decimal

TABLE 37.5. TYPICAL CONVERSION (OR DELAY) TIMES FOR ANALOG-TO-DIGITAL (A/D) AND DIGITAL-TO-ANALOG (D/A) CONVERTERS

Type of Converter	Relative Speed	Conversion or Delay Time 8 Bit	10 Bit	12 Bit	16 Bit
Dual slope integrating A/D	Slow	20 ms	30 ms	40 ms	20 ms
	Medium	1 ms	5 ms	20 ms	—
	Fast	0.5ms	1 ms	4 ms	—
Successive approximation A/D	Slow	30 μs	40 μs	50 μs	400 μs
	Medium	10 μs	15 μs	20 μs	100 μs
	Fast	0.5 μs	1 μs	2 μs	35 μs
	Video/flash[a]	0.02 μs	—	—	—
D/A	Slow	3 μs	3 μs	3 μs	20 μs
	Medium	1 μs	1 μs	1 μs	—
	Fast	0.2 μs	0.2 μs	0.2 μs	—
	Video/flash	0.02 μs	—	—	—

[a] A "flash" A/D converter checks every possible digital output code simultaneously, by using 2^N (or $2^N - 1$) electronic comparator circuits. An 8-bit flash A/D converter thus contains $2^8 = 256$ (or 255) electronic comparators, each with a different analog trigger level, and each with a different 8-bit digital output.

A binary counter drives the D/A converter, and the resultant analog output of the D/A converter is compared with the analog input signal. When the two analog signals match, the binary counter contains the desired digital code (which is the output of the A/D converter). Such an A/D converter is relatively slow but also relatively inexpensive. At the present time (1982), the National Bureau of Standards (NBS) uses a 20-bit D/A converter to calibrate high-resolution (18-bit) A/D converters. A 20-bit converter has a resolution of 1 part in $2^{20} = 1,048,576$. As with A/D converters, the most common commercially available D/A converters are 8, 10, and 12 bits wide, although 14-, 16-, and 18-bit D/A converters are available. In most control applications, the delays of the A/D and D/A processes are negligible compared with the time constants of the control system plant. Typical conversion times are given in Table 37.5. A summary of the resolution and theoretical maximum accuracy of D/A (and A/D) converters is given in Table 37.3. These tables have been compiled from various manufacturers' specification sheets and catalogs as well as from manufacturers' handbooks. One of the best of these is put out by Datel/Intersil.[2]

37.5 CONVERSION BETWEEN LINEAR AND ANGULAR MEASUREMENTS

Mechanical Methods

Systems using a rotational prime mover, such as an electric motor, to produce a linear motion are quite common (e.g., raising an elevator). In such a system the need often arises to convert a linear motion into an angular measurement, or vice versa. In a practical elevator control system, one common approach is to have a slender measurement cable attached to the elevator. The measurement cable is then wrapped around a drum, located in the elevator control room. As the elevator rises, the drum turns and produces an angular measurement related to the elevator's linear position. The rotation of the control drum then can be utilized to measure the linear position, velocity, and acceleration of the elevator, in angular units, suitable for motor control.

The motor shaft (or hoisting drum shaft) is not the best choice for this measurement task, because the variable weight of the elevator, its load, and the weight of the elevator hoisting cables causes the hoisting cables to stretch. Therefore a unique relationship between linear position and motor shaft angle does not exist. Any stretch in the loaded elevator cable is compensated for in the measurement system. This illustrates a common practice in control system design, namely, the separation of the measurement components from the control components. Therefore, variations in the controlled components are automatically compensated for by the measurement loop.

The drum and cable method of converting angular and linear coordinates appears in many variations, both in measurement and in output motion applications. The basic process was discussed in Section 37.3 and illustrated in Fig. 37.4. The cable may be replaced by a flat belt or by a band made of steel, fabric, or composite material. Devices that utilize such a linear-to-angular or angular-to-linear conversion process include

1. Elevators.
2. Robots.
3. Remote manipulators.
4. Computer disk drive head mechanisms, particularly floppy disks.
5. Digital plotters, drafting machines, X-Y plotters.

When the magnitude of the linear or angular motion is relatively small, a crank and arm or linkage may be employed to convert between angular and linear units. In almost all such situations, the absolute maximum angular limit is $\pm 90°$ or $\pm \pi/2$ rad, or one half of a revolution of the rotating shaft. Some practical limits may make this smaller. Typical applications include sensing of the angular position of the various control surfaces of aircraft, ships, and submarines or of any other control device that moves or rotates a limited amount.

Linkages are generally attached to the control surfaces and then to a crank attached to the angular transducer's shaft. The transducer's shaft could theoretically be simply connected directly to the shaft of the control surface. In practice this may be mechanically difficult owing to the presence of gears, support bearings, drive mechanisms, motors, or in some cases, a stationary shaft about which the control surface rotates. Figure 37.9 illustrates a crank having a swing radius R connected by means of an arm of length A to a slider constrained to move in a radial direction perpendicular to the axis of the crank. The angle θ describes the angular position of the crank as it rotates about its shaft axis. When θ is zero, the slider is furthest from the crank's shaft; when θ is π radians, the slider is closest to the crank's shaft. This distance is denoted by L. As θ increases, L decreases, as illustrated

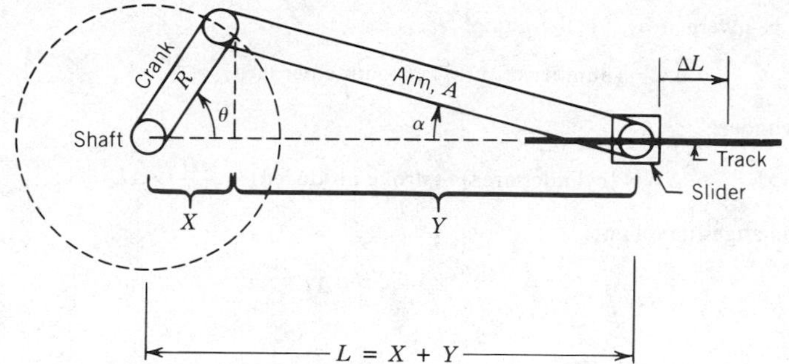

Fig. 37.9 Conversion of linear to angular measurements, and vice versa, with a crank shaft. $X = R\cos\theta$; $Y = A\cos\alpha$; $L = X + Y = R\cos\theta + A\cos\alpha$. If $A \gg R$, then $\alpha \to 0$ and $\cos\alpha = 1$. $L \cong R\cos\theta + A$; $dL/d\theta \cong -R\sin\theta$; $\Delta L \cong -R\sin\theta \cdot \Delta\theta (A \gg R)$.

in Fig. 37.9. The conversion between linear and angular measurements, for the geometry shown, is given by

$$\Delta L \cong -R\sin\theta \cdot \Delta\theta \qquad (A \gg R) \tag{37.13}$$

This relationship holds only for the simplified case where the connecting arm length A is much longer than the crank's swing radius R. Observe that Eq. (37.13) is nonlinear owing to the nonlinear function $\sin\theta$.

Another common mechanical conversion method employs a rack and pinion gear. A rack is a long straight bar into which gear teeth have been cut or otherwise formed. The pinion gear is a standard disk-shaped round gear that meshes with and rolls along the rack. Quite often the axis of the gear simply rotates as the rack translates back and forth. This conversion process is similar to that of a drum and belt. Angular units (θ) are converted to linear units (L), and vice versa, in accordance with the relation

$$\Delta L = R\Delta\theta \tag{37.14}$$

where θ is in radians and R and L are identical linear units (m, mm, cm, in.). R represents the effective radius or pitch radius of the drum or the pinion gear. A rack and pinion gear cannot slip and can be manufactured to a very high accuracy. A numerically controlled lathe, milling machine, or jig borer may employ a rack and pinion gear with a linear tolerance of a few ten thousandths of an inch or a few hundredths of a millimeter.

A related mechanical conversion scheme, used in numerically controlled machines, robots, and positioning systems, is the lead screw and precision nut. This is simply an oversized nut and machine screw. As the screw is turned, the nut, which does not turn, moves in a linear manner along the axis of the screw. In some precision applications, the screw has semicircular grooves rather than a standard V groove, and the mating nut becomes a recirculating ball nut. The nut contains a helix of ball bearings, which causes it to roll up on the rotating screw. If the lead screw has N threads per inch (or threads per millimeter or threads per centimeter), the conversion between angular units of θ (in radians) and linear units of L (measured in inches or millimeters or centimeters) is governed by

$$\Delta L = \frac{\Delta\theta}{2\pi} \cdot \frac{1}{N} = (\text{number of turns}) \cdot (\text{length moved per turn}) \tag{37.15}$$

where ΔL = linear motion, in., mm, or cm
$\Delta\theta$ = angular motion, rad
N = screw pitch, threads/in., /mm, or /cm

A recently designed microprocessor-controlled x-y plotter uses a stepper motor driving a lead screw to position the pen, in accordance with Eq. (37.15).

Hydraulic Methods

A positive displacement rotary hydraulic pump/motor, in conjunction with a hydraulic cylinder, converts angular units θ at the pump to linear units L at the cylinder. If the pump/motor displaces a volume V_T per revolution and the hydraulic cylinder has a diameter D (where V and D are measured in the same units and θ is measured in radians), then the conversion between angular units θ and linear units L is governed by, for the pump:

$$\Delta V = (\text{number of turns}) \cdot (\text{volume per turn}) = \frac{\Delta\theta}{2\pi} \cdot V_T$$

and for the cylinder:

$$\Delta V = (\text{cylinder area}) \cdot (\text{stroke produced}) = \frac{\pi D^2}{4} \cdot \Delta L$$

Equating the changes in volume:

$$\frac{\Delta\theta}{2\pi} V_T = \frac{\pi D^2}{4} \Delta L$$

Solving for ΔL:

$$\Delta L = \frac{2V_T}{\pi^2 D^2} \Delta\theta \tag{37.16}$$

where V_T = pump displacement/turn
D = cylinder diameter
$\Delta\theta$ = turns of pump, rad
ΔL = piston linear movement
V, D, L = measured in identical units (mm, in., cm, etc.)

Errors will be introduced into Eq. (37.16) if the pump has any leakage, the connecting hoses expand slightly in volume, or the total volume of fluid changes for any reason (e.g., temperature variations). Unlike hydraulically powered systems that contain fluid reservoirs or accumulators, this measurement conversion system can tolerate no leaks (since any leak directly produces an error). A practical system employing this procedure would require periodic calibration to compensate for the inevitable leaks or fluid volume changes.

Electrical Methods

Electrical methods of converting linear and angular units generally involve scaling the voltage or current that represents the measurement in accordance with the scale factors of the linear and angular measurements themselves. Thus a slidewire sensing a linear motion L might have a scale factor of K_{lin} volts per meter. A motor, driving a ball screw to produce the linear motion, might have a mechanical scale factor of K_m motor turns per meter of ball screw extension. A 100-turn potentiometer and gearing might have an electrical scale factor of K_{ang} volts per motor shaft revolution. The angular position of the motor shaft θ is thus related to the linear motion of the ball screw L. Therefore the electrical measurement of angle E_A may be converted to the electrical measurement of linear position E_L in terms of the given constants. Thus

$$\Delta E_L = K_{\text{lin}} \times \Delta L = \frac{\text{volts}}{\text{meter}} \times \text{meters (for linear slide wire)}$$

$$\Delta E_A = K_{\text{ang}} \times \frac{\Delta\theta}{2\pi} = \frac{\text{volts}}{\text{turn}} \times \text{turns (for angular pot)}$$

$$\Delta L = \frac{\Delta\theta/2\pi}{K_m} = \frac{\text{turns of motor}}{\text{turns/meter of screw}} \text{ (screw extension)}$$

Inserting ΔL into ΔE_L:

$$\Delta E_L = K_{\text{lin}} \times \Delta L = K_{\text{lin}} \frac{\Delta\theta/2\pi}{K_m}$$

For the ratio $\Delta E_L/\Delta E_A$:

$$\frac{\Delta E_L}{\Delta E_A} = \frac{K_{\text{lin}}}{K_{\text{ang}} K_m} = K_{\text{scale}} \tag{37.17}$$

Thus in this simple hypothesized analog scaling system, the output of the angular measurement transducer E_A is related to the output of the linear measurement transducer E_L by a constant scale factor, K_{scale}, consisting of an appropriate function of the various system scale factors [given by Eq. (37.17)]. Only one transducer is actually necessary, since the linear transducer output can be obtained from the angular transducer, and vice versa, by using Eq. (37.17).

The important conclusion to be drawn is that if angular and linear measurements are linearly related to voltages (or currents) in a linear control system, then angular and linear unit conversions may be produced electrically by a single constant multiplication. If the constant is less than unity, an analog conversion may be produced by a simple attenuator, voltage divider, or potentiometer. If the constant is greater than unity, an amplifier (generally an operational amplifier, op amp) will be needed to produce the constant multiplication factor. This is shown in Fig. 37.10.

Very often current signals are employed, and the output of a particular transducer is a current that is proportional to the physical quantity being measured. If a voltage signal is desired, the current may be "summed" or converted into a voltage by use of a differential operational amplifier containing a feedback resistor R_f connected from the amplifier output to the negative input terminal. The current input signal is also connected directly to the negative input terminal. The positive terminal is grounded, either directly or by a compensation network. A current-to-voltage scaling network is shown in Fig. 37.11*a*. Certain transducers may require negligible loading, in which case a high-input impedance buffer amplifier is required. Inverting and noninverting buffer amplifiers are also shown in Fig. 37.11*b*, *c*. Some specialized transducers may have a charge output (Q, coulombs, out per unit physical input), in which case a charge-to-voltage scaling is required. The necessary op amp circuit is shown in Fig. 37.11*d*.

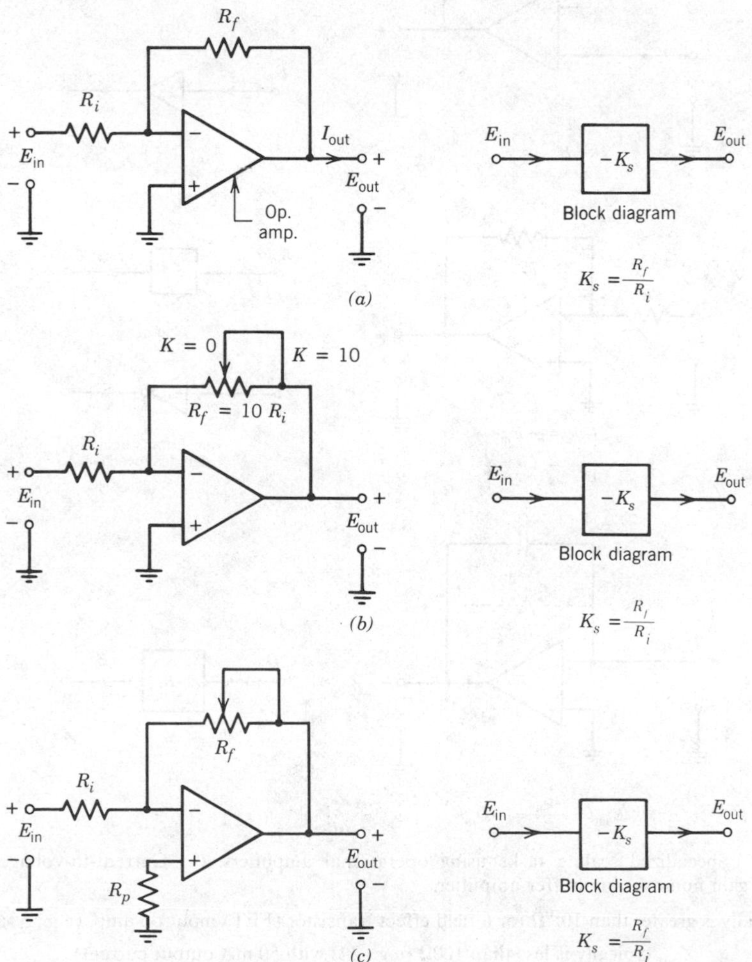

Fig. 37.10 Scaling voltages using operational amplifiers. In all cases $E_{out} = E_{in}(R_f/R_i) = -E_{in}K_s$, $K_s = R_f/R_i$. (*a*) Scaling a voltage measurement with an operational amplifier by the factor $K_{scale} = K_s$. (*b*) To vary K_{scale} from 0 to 10, use a variable resistor for R_f, where R_f, max $= 10R_i$. (*c*) To compensate for input bias currents, add a resistor of value R_p in series with the grounded positive input. The typical value of R_p chosen is $R_p = R_iR_f/(R_i + R_f)$. If R_f varies widely, a common alternate choice of R_p is $R_p = R_i$.

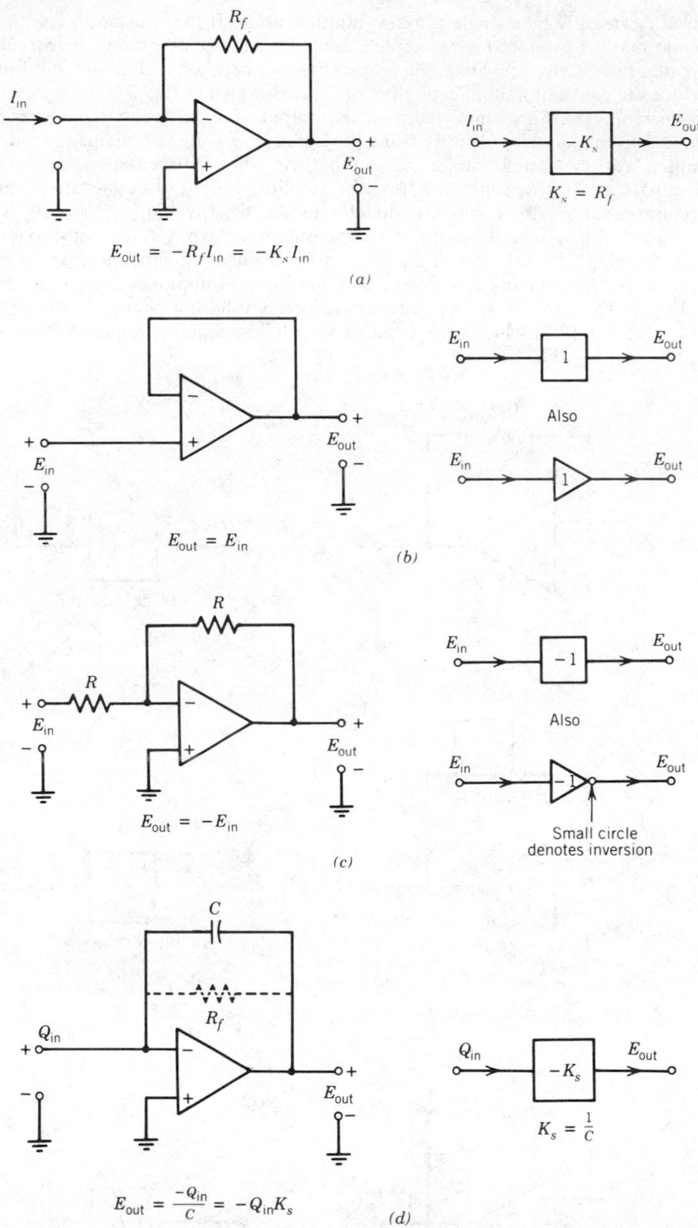

Fig. 37.11 Specialized scaling tasks using operational amplifiers. (*a*) Current-to-voltage scaling. (*b*) Unity gain noninverting buffer amplifier.

Z_{in} typically is greater than $10^{10}\Omega$ for a field effect transistor (FET) input op amp, (e.g., $5 \times 10^{11}\Omega$).

Z_{out} typically is less than 100Ω (e.g., $2\ \Omega$ with 50 mA output current).

(*c*) Unity gain inverting buffer amplifier. $Z_{in} \cong R$, typically 10^4 to $10^6\ \Omega$. $Z_{out} \cong$ output impedance of op amp, typically $< 100\Omega$. Note that for an FET input stage op amp, Z_{in} for the inverting buffer amplifier is many orders of magnitude smaller than Z_{in} for the noninverting buffer amplifier. The FET may be a metal oxide semiconductor (MOS) or complementary metal oxide semiconductor (CMOS). (*d*) Charge-to-voltage scaling. To discharge the amplifier in (*c*), a practical circuit will add a resistor R_f in parallel with it. This produces a -3-dB lower cutoff frequency of $1/2\pi R_f C$.

TABLE 37.6. SUMMARY OF OPERATIONAL AMPLIFIER PERFORMANCE AND LIMITATIONS

Factor	Performance Limitation
E_{out}[a]	E_{out} is limited by amplifier and power supplies. Typical ranges are ± 5 to ± 15 V
I_{out}	I_{out} is limited by amplifier and power supplies. Typical ranges are ± 10 to ± 100 mA
E_{in}	Magnitude of E_{in} is constrained by highest frequency component of E_{in} and frequency response of amplifier. For 10-V op amp with slew rate of SR (V/s) and gain band width product of GB (Hz), maximum input voltage $E_{in,max}$ at full-power frequency is given by $E_{in,max} = E_{out,max} \dfrac{SR}{20\pi GB}$ *High speed op amp example*: $E_{out,max} = \pm 10$ V, SR = 1000 V/μs = 10^9 V/s, GB = 100 MHz = 10^8 Hz $E_{in,max} = \pm 10 \dfrac{10^9}{20\pi \times 10^8} = \pm \dfrac{10}{2\pi} = \pm 1.6$ V peak to peak at full-power frequency If this level is exceeded, amplifier goes into slew rate limit, at full-power frequency (FPF), of $FPF = \dfrac{SR}{2A\pi} = \dfrac{SR}{20\pi} (A = \pm 10 \text{ V})$ FPF = maximum frequency for $\pm A$ or ± 10 V output $FPF = \dfrac{10^9}{20\pi} \cong 16$ MHz Since most control systems operate at frequencies well below 16 MHz, they do not require high-speed op amps and there is no slew rate limit constraining $E_{in,max}$ for typical control system applications. Under such situations, E_{in} is limited to a value that causes E_{out} to approach $E_{out,max}$, namely $E_{in,max} = \dfrac{E_{out,max}}{R_f/R_i} = \dfrac{E_{out,max}}{\lvert gain \rvert}$
Input bias currents and offset voltages	Op amps also suffer from small, but not necessarily negligible, input bias currents and input offset voltages. Compensation networks are often used to null out these errors which produce finite amplifier output voltage when input voltage is zero. Quite often identical resistance is inserted in series with both positive and negative input op amp terminals, to compensate for input bias currents. Typical input bias current for field effect transistor input-stage high-speed op amp is 100 to 1000 pA (10^{-12} A), while typical input offset voltage ranges from 1 to 10 mV (10^{-3} V). Input impedance typically ranges from 1×10^{10} to 100×10^{10} Ω, and output impedance typically ranges from 1 to 100 Ω

[a] Refer to Fig. 37.9 for E_{in}, E_{out}, R_f, and R_i symbol definitions.

Operational amplifiers suffer from various limitations, which can affect the performance of a control system. In general, the amplifier's frequency response limits are not critical, since most control systems act as low-pass filters (because of their mechanical component inertias). Therefore the system's physically controlled plant limits frequency response to a value far below the frequency limits of the op amps. On the other hand, DC drift, input-offset voltage, and input-bias currents of an op amp can produce a finite signal output when the op amp's input voltage is zero. This can result in a shift of the control system's output and introduce errors in its calibration. In critical cases, chopper-stabilized op amps must be used. A chopper-stabilized op amp periodically shuts off its DC amplifier, grounds the input, and automatically adjusts a bias or compensation network to produce an isolated internal zero output. The properly zeroed DC amplifier is then switched back into the external circuit.

In the early 1950s, vacuum tube operational amplifiers were used where the switching was provided by an electromechanical oscillating switch, similar in concept to an electric door buzzer. The oscillating or vibrating switch was called a chopper, hence the name "chopper stabilized operational amplifier." These days, solid state integrated circuit switches are incorporated within the amplifier chip.

A summary of op amp performance and limitations is contained in Table 37.6, which is based on manufacturers' literature.[2]

37.6 DEVICES FOR ANALOG MEASUREMENT OF SYSTEM ANGULAR OR LINEAR POSITION

Any device whose electrical characteristics vary in response to mechanical motion may be directly converted into an angular or linear position transducer. Depending on the application, the transducer either may be used directly or may require signal processing or signal conditioning.

Angular Transducers: Potentiometers

A rotary potentiometer or variable resistor has a resistance that varies with the angular position of the potentiometer's control shaft. The resistance may vary linearly or nonlinearly with angle, depending on the intended application of the potentiometer. Such a potentiometer, with no alterations, can be used as an analog angular transducer over the range of angle through which the potentiometer's input shaft can turn. This range varies from less than 1 turn, perhaps 300° for the common potentiometer, to various integral numbers of turns. The 10-turn potentiometer is often used. Potentiometers of more than 10 turns, such as 20 turns, are available but are far less common.

In its most common application the potentiometer produces a variable voltage output, as a function of mechanical shaft angle input. The varying resistance is used to convert a constant voltage into a variable voltage, $V(\theta)$, which is a function of shaft angle θ, as shown in Fig. 37.12.

An angular transducer constructed in this manner suffers from the following limitations:

1. The maximum angle is limited to typically 1 to 10 turns (0 to 360° or 0 to 20π rad). Gearing extends this range by perhaps a factor of 100, but the resolution and accuracy decrease by the same factor [or more, if the gears exhibit backlash or "play" (dead zones)].

2. The moving wiper arm of the potentiometer produces frictional wear on the resistance material. This wear is accelerated at high rotational speeds, due to heating effects. Thus a potentiometer might be limited to 1 million turns, at a rate not to exceed 100 r/min. Such a device could not be used to determine the shaft angle of a turbine turning at 100,000 r/min.

3. Precision wire-wound potentiometers produce tiny discrete resistance changes as the wiper arm moves over each wire. These discrete changes produce a small electrical noise, which tends to increase as the potentiometer wears or becomes dirty. The maximum resolution of a wire-wound potentiometer is given by the total angle through which the potentiometer shaft turns, divided by the number of turns of resistance wire wrapped within this angle. If a simple potentiometer has a total shaft angle excursion of 300°, and 1000 turns of resistance wire are wrapped around a form within this 300°, the resolution of the potentiometer is

$$\text{Resolution} = \frac{\text{total shaft angle range}}{\text{turns of resistance wire}} = \frac{300°}{1000 \text{ turns}}$$

$$\text{Resolution} = 0.3\,°$$

A potentiometer utilizing a composition-based resistance element has a theoretical resolution of infinity.

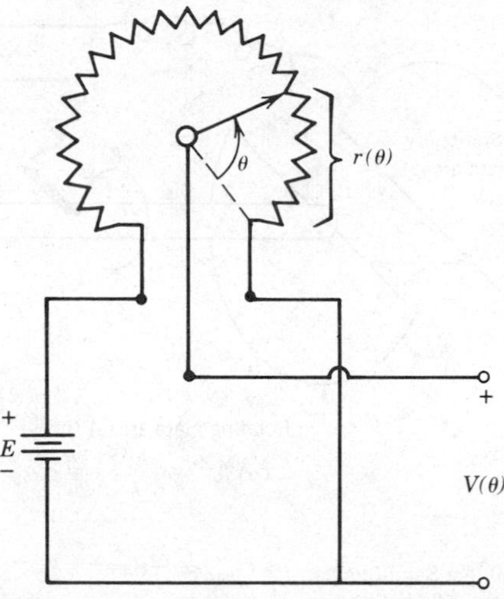

Fig. 37.12 Potentiometric angular transducer. $V(\theta) = (r(\theta)/R)E$, where R = total resistance of potentiometer. If $r(\theta)$ is a linear function of angle, then $r(\theta) = R(\theta/\theta \text{ max})$ and $V(\theta) = (\theta/\theta_{max})E = (E/\theta_{max})\theta = K_1\theta$, where K_1 = transducer constant. $K_1 = E/\theta_{max}$ = volts/degree or volts/radian.

Linear Transducers: Slide Wires

Conceptually, the linear counterpart of the potentiometer is produced by "straightening" out the potentiometer's circular resistance track. The central shaft is removed and replaced by a sliding contact that moves a finite linear amount bounded by the length of the resistance element. Slide wires using this type of construction are used in analog x-y plotters and in control mechanisms that have limited movement. Linear slide-wire transducers have all the mechanical drawbacks and limitations that angular potentiometers have. Potentiometers (angular) and linear slide wires are conceptually easy to understand. The conversion of a resistance variation to a voltage or current variation is also conceptually direct and involves no additional circuitry. Some other transducers that require additional circuitry and signal processing will be briefly described.

Rotary Variable Capacitance Transducer

A variable capacitor can be constructed to sense either angular or linear movement. The semicircular plate capacitor, shown in Fig. 37.13*a*, produces a capacitance variation from maximum to minimum for one half of a rotation of the movable semicircular plate. The change in capacitance must be converted by additional circuitry into a usable electrical signal. Possible conversion schemes include charge balance methods and frequency-to-voltage conversion. In the latter technique, the variable capacitor is used to vary the frequency of an oscillator, which serves as the input to a frequency-to-voltage converter. Frequency-to-voltage converters often utilize phase-locked loops to produce the desired output voltage, whose magnitude varies with the applied input frequency.

The rotary variable capacitance transducer enjoys the following advantages:

1. It is a noncontact device and can operate at very high rotational rates (limited only by dynamic balance problems and bearing problems).

2. Unlike variable resistance transducers, it is essentially a nonwearing device. The mechanical support or bearings are the life-limiting elements.

3. It can tolerate very high G forces, because of its simple mechanical design.

The disadvantages of the rotary variable capacitance transducer include the following:

1. Relatively elaborate electronics are required to convert the capacitance variation into a voltage or current signal.

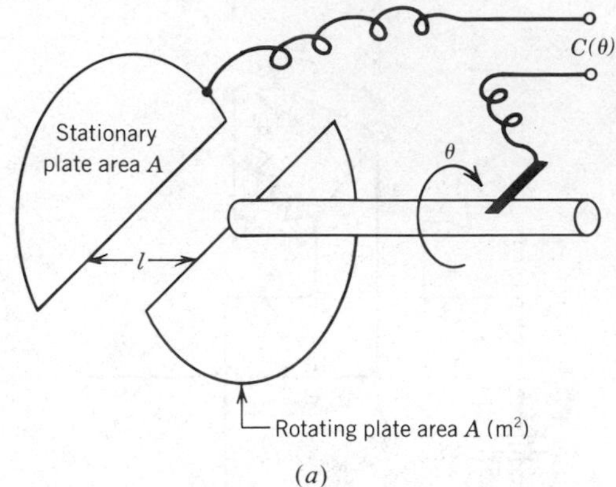

(a)

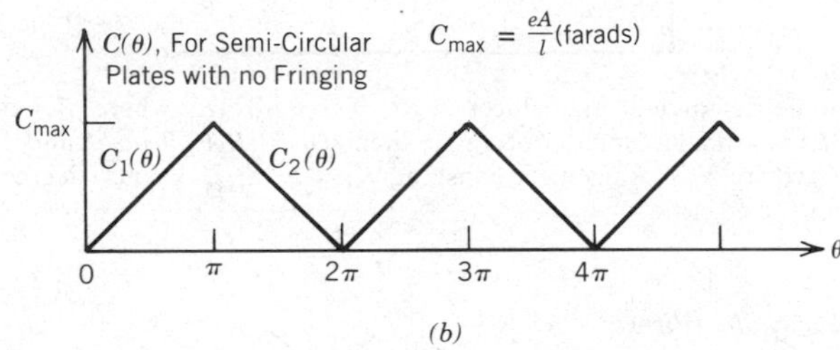

(b)

Fig. 37.13 Angular transducer using a rotating plate capacitor: (*a*) Diagram. A = plate area, m^2; l = plate separation, m; e = permittivity, F/m; $e = K_D e_0$, where K_D = dielectric constant and e_0 = permittivity of vacuum $= 8.85 \times 10^{-12}$ F/m. (*b*) Capacitor variation as a function of angle θ for semicircular plates with no fringing. $C_1(\theta) = (C_{max}/\pi)\theta = K_2\theta, 0 \leqslant \theta \leqslant \pi$, and $C_2(\theta) = C_{max}/\pi(2\pi - \theta) = K_2(2\pi - \theta), \pi \leqslant \theta \leqslant 2\pi$.

2. It has a limited operating range, which may be increased by levers, linkages, or gears. A rotary variable capacitance transducer is periodic and produces a unique capacitance value only within a one half revolution range.

Linear Variable Capacitance Transducer

The linear variable capacitance transducer enjoys advantages (2) and (3) of the rotary variable capacitance transducer. It is illustrated in Fig. 37.14. The linear variable capacitance transducer, as shown in this figure, can be configured in two ways. If the plates always present a constant area to one another and the distance between the plates is varied in accordance with the position to be sensed, a nonlinear capacitance variation results.

Referring to Fig. 37.14*a*:

$$C(l) = K_3 \frac{1}{l}$$

where K_3 = scale constant, F/m
$K_3 = eA = K_d e_0 A$
l = plate separation, m
$C(l)$ = variable capacitance, F
K_d = dielectric constant (1 for vacuum; 1.0006 for air; 2 to 10 for oil, glass, mica, most plastics)
e_0 = permittivity of free space, 8.85×10^{-12} F/m
A = plate area, m^2
e = permittivity of dielectric $= K_d e_0$, F/m

If instead the plate separation is kept fixed while the plates slide over one another (so that the area common to both plates varies in accordance with the position to be sensed), then a linear capacitance variation results. Referring to Fig. 37.14*b*:

$$C(l) = K_4 l$$

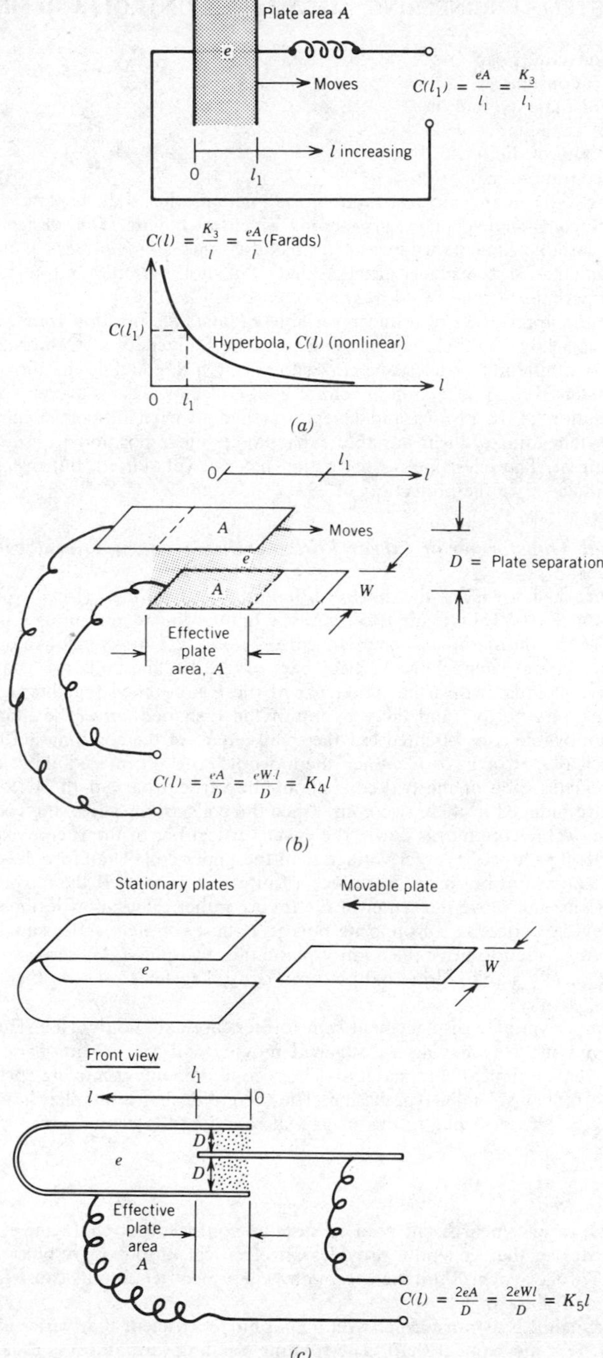

Fig. 37.14 Position transducer using a parallel plate capacitor: (*a*) Variable plate separation, constant area. A = plate area, m^2; l = plate separation, m (a variable); e = permittivity, F/m; $e = K_D e_0$, where K_D = dielectric constant (1 for air or vacuum) and e_0 = permittivity of vacuum $= 8.85 \times 10^{-12}$ F/m. (*b*) Constant plate separation, variable area. l = effective plate length, m (a variable) and W = plate width, m. (*c*) Constant plate separation, variable area, shielded U plate. W = plate width, m; D = plate separation, m; l = length of plate overlap, m; A = effective area $= W \times$ l, m^2; e = permittivity, F/m.

where K_4 = scale constant, F/m
$K_4 = ew/D$ = constant
l = length of plate overlap, m
$C(l)$ = variable capacitance, F
e = permittivity of dielectric
D = plate separation, m

To reduce fringing effects and decrease the noise pickup, the sliding plate capacitor can be configured as a folded U-shaped plate surrounding a central I plate. This design is shown in Fig. 37.14*c*. The linear variable capacitance transducer has the same disadvantages as those listed for the rotary variable capacitance transducer, except that it is not periodic but instead is limited in operating range to perhaps 5 mm or 1/4 in., and often much less.

An example of the application of a linear variable capacitance position transducer concerns the adaptive focusing of a large multiple-mirror telescope. The University of California has planned to construct a 400-in. multiple-mirror telescope consisting of 36 hexagonal 80-in. mirrors, each of which can be separately focused onto the common central light collector. This is accomplished by means of the coordinated motion of 18 pistons and levers attached to each of the 36 mirrors. The mirror position control system utilizes linear variable capacitance linear position transducers to sense the position of each mirror. The telescope's construction has not yet started, but over 1 million dollars has already been allocated to the project (as of 1981).

Linear Differential Transformer or Linear Variable Displacement Transducer

A special transformer, known as the linear differential transformer (LDT), or linear variable displacement transducer (LVDT), is used to sense the limited linear motion of a movable coil. It is illustrated in Fig. 37.15. The stationary magnetic core of an LVDT looks just like the capital letter E, with a coil wrapped about each of the vertical bars of the E. The coils are wrapped in opposite directions, and the coil ends nearest the center bar of the E are joined together, leaving one end of each coil free at the very bottom and the very top of the E-shaped core. The short movable coil is wrapped around a movable core, oriented like the small letter i in the combination Ei. The movable i core can move vertically, E_i or Ei or E^i. When the movable core is centered about the central bar of the E, equal flux excites each of the two coils about the vertical bars of the E coil, and equal but opposite voltages are induced in these two coils. Since the coils are in series, the voltages add, with a zero result. If the movable core moves down, the lower vertical bar of the E core experiences a larger flux and the lower coil produces a larger voltage than the upper coil. Therefore the opposing induced voltages no longer cancel out but instead produce a finite sum voltage. If the movable coil should be displaced an equal amount above the center of the E core rather than below it, the same sum voltage is produced. However, it contains an opposite polarity, phase, or sign. After suitable amplification, this sum voltage can be used to drive the control phase of a two-phase AC servomotor. Such systems are used in gyroscopically stabilized inertial platform control systems and in other types of positioning control systems.

Two types of linear variable displacement transformers are commonly used. The first operates as discussed, with a moving core having an attached moving coil and a stationary E core with two opposed coils. Ferrite is a typical core material. The second has only a moving core; all the coils are stationary and of a hollow-solenoid-type design. The second design is simpler, but the reluctance of the magnetic circuit is greater owing to the air gap surrounding the central core.

Synchros

There are many types of synchros, all used in systems controlling a shaft angle. A synchro is an electromechanical device that generally provides an electrical output in response to a mechanical shaft angle input. Torque synchros are the exception, where electrical inputs cause the output shaft to assume a commanded shaft angle.

Electrically, a synchro is a transformer with a coupling coefficient that varies as one transformer winding is rotated by a mechanical shaft. The rotating winding (armature) is usually a single-phase distributed coil, and the outer surrounding stator windings are generally of three space-phase coil construction, with a physical angle of 120° between the electrical center of each coil. However, all three windings are in phase electrically. The three-space-phase stator coils are wye-connected internally, with only the three leads from each of the ends of one branch of the wye brought out. The two leads from the single-phase rotor are brought out by means of slip rings.

The exception to the single-phase rotor occurs for the case of the control differential synchro. This synchro has a three-phase rotor and a three-phase stator. It is used to offset, adjust to zero, add, or subtract an angle electrically from the three-phase synchro representation of a mechanical shaft

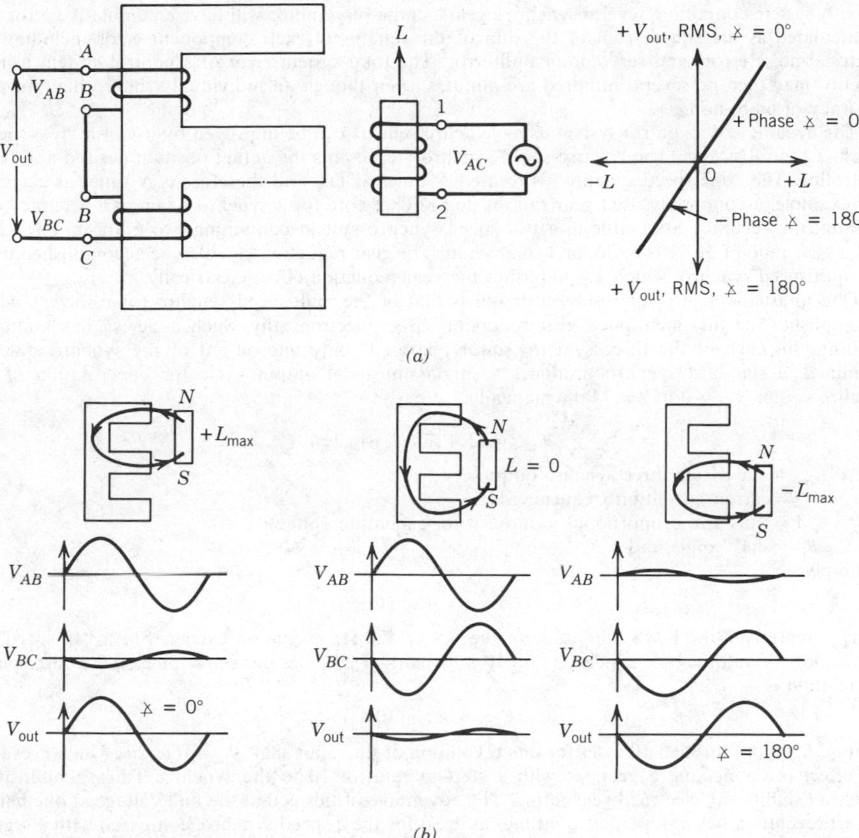

Fig. 37.15 Linear differential transformer (LDT) or a linear variable displacement transducer (LVDT). (*a*) L = linear motion of coil 1, 2 in meters. Coil AB is wound in the opposite sense of coil BC. V_{AC} = reference input voltage, ∡0°. (*b*) Flux paths at the instant when V_{AC} produces a maximum flux.

angle. The three-space-phase windings have electrical signals that are in time phase. No three-phase AC is used in a synchro.

Synchros are generally excited by either 60 or 400 Hz AC, and their mechanical shaft rotation rates should be kept very small compared with these rates. This ensures that the modulation frequency produced by the rotating synchro armature is much less than the carrier or excitation frequency. In addition, since the synchro contains coils rotating in a magnetic field, considerable voltages can be added to the normal transformer voltages of the synchro, owing to the generator effect of a rapidly rotating synchro armature. However, the synchro is supposed to act as a variable transformer, not as a generator. It is important to recognize that any significant speed-generated voltages can lead to stability problems. For these reasons it is common practice to limit the synchro's maximum speed to approximately one hundreth of the excitation frequency. For a synchro excited at a frequency F containing P poles, the maximum shaft speed is then

$$\left(\frac{F}{100}\ \text{cycle/s} \times 60\ \text{s/min} \times 2\ \text{poles/pair}\right)/P \cong \text{rpm max}$$

$$1.2F/P \cong \text{rpm max}$$

For a two-pole 60-Hz synchro, the maximum revolution per minute is, therefore, $1.2(60)/2 = 36$ rpm. For a two-pole 400-Hz synchro, the corresponding maximum is $1.2(400)/2 = 240$ rpm. In the great majority of applications, synchros turn at much slower rates, and the very slow speed (or nearly static case) is quite common.

At slow speeds the accuracy of a typical synchro may be 0.1° or 6 arc-min. A compensated precision synchro may have an accuracy 10 times better than this, or 0.6 arc-min, which equals 36 arc-s. A system containing several synchros, gears, and a servomotor will have a cumulative error that is calculated as the square root of the sum of the squares of each component error including the electrical noise errors and servomotor null error. The total system error of a control system using a synchro may then be several hundred arc-minutes, even though an individual synchro may have an accuracy of 6 arc-min.

The accuracy of a control system using synchro sensors can be improved by use of a "two-speed" synchro control system. The "coarse-speed synchro" represents the actual position desired and being controlled. The "fine-speed synchro" is rotated N times as fast and therefore is N times as accurate. For example, a commonly used gear ratio is 36 : 1. Therefore for a synchro having an accuracy of 6 arc-min, the accuracy attainable in a two-speed synchro system containing two 6 arc-min synchros and a gear ratio of 36 : 1 is 6/36, or 1/6 arc-min. The gear reduction can also be accomplished using the multispeed synchro, which accomplishes the gear reduction effect electrically.

One unusual synchro (or resolver) design is that of the multispeed synchro (or resolver), which accomplishes the just mentioned gear reduction effect electronically. Such a device has multipole windings for each of the three synchro stator phases. If only one output of the synchro stator is examined, a standard synchro produces a single sinusoidal output cycle for one rotation of the synchro's rotor, $\theta_{in} = 0$ to 2π. Mathematically

$$E_{out} = \sqrt{2}\, A \sin(\theta_{in}) \sin 2\pi f t$$

where E_{out} = one of the three synchro outputs, V
f = carrier excitation frequency, Hz
A = constant, proportional to input rotor excitation voltage
θ_{in} = shaft angle, rad

or simply

$$E_{out} = A \sin(\theta_{in})$$

E_{out} represents the RMS amplitude of the 60- or 400-Hz excitation carrier. For a "10-speed" or "tenth-order" multiple pole synchro (with 10 pole pairs or 20 poles per stator phase), the corresponding relation is

$$E_{out} = A \sin(10\theta_{in})$$

That is, E_{out} goes through 10 cyles for one revolution of the input shaft $\theta_{in} = 0$ to 2π, which is exactly the effect of connecting a gear set with a step-up ratio of 10 to the synchro. Thus, a multispeed synchro exhibits an "electrical gear ratio." The advantage of this is that the null voltage at one half of a shaft revolution has a slope that is n times as great for the n speed synchro, compared with a regular synchro. This results in a decrease in error in the single-speed synchro (say 10 min of angle) by a factor of n for the n-speed synchro (to 10 min$/n = 1$ min of angle for $n = 10$).

Tables 37.7 and 37.8 list the common types of synchros and briefly describe them. Figure 37.16 illustrates the internal windings of each device in a schematic manner. Figure 37.16*e* illustrates the connections for a torque-transmitting synchro pair, which acts as an electrically connected flexible shaft (unfortunately with a shaft angle error that increases with increasing transmitted torque).

Synchros were used extensively during the years 1940 to 1970, primarily in military, aerospace, and radar applications. Synchros are still used in many systems today. However, the advent of the microprocessor and the digital shaft angle encoder has seriously eroded the synchro applications base. Prior to development of the microprocessor, synchros were used, in conjunction with resolvers and tachometers, to solve elaborate differential or trigonometric equations, such as fire control problems, or the transformation of three-dimensional coordinates.

Resolvers

A resolver looks like a synchro. Its stator, however, has two windings at right angles to each other and has four leads brought out. Ninety degrees between the coil axes produce the 90° shift between a sine- and a cosine-related voltage output. A schematic representation of a resolver is shown in Fig. 37.17. A resolver resolves a mechanical shaft angle input into its sine and cosine components, which are represented electrically. This is accomplished by rotating a single-phase armature, excited with a reference voltage $E_R\sqrt{2}\sin(2\pi f t)$, denoted hereafter simply as E_R, within a stator which is so designed that the coupling coefficient between rotor and stator coils varies as the sine (or cosine) of the rotor position. If the shaft angle input is denoted by θ_{in}, then the output of each resolver stator coil is

$$E_{out_1} = KE_r \cos(\theta_{in})$$

$$E_{out_2} = KE_r \sin(\theta_{in})$$

TABLE 37.7. COMMON SYNCHRO TYPES

Name	Use	Input	Output
Synchro control transmitter (abbreviated CX). Excitation of 60 or 400 Hz AC applied to rotor, E_R, V. n-transformation ratio = maximum stator rms output voltage divided by E_R	Translates physical shaft angle into three stator synchro voltages whose magnitude varies with shaft angle. Used as command or input device, to apply reference input signal	Rotor shaft angle θ (single-phase rotor is excited by E_R, but E_R is constant, not variable)	Three stator voltages in time-phase, measured between stator terminals S_1, S_2, and S_3. Three stator voltages are $E_{S13} = nE_R\sin(\theta + 0°)$, $E_{S32} = nE_R\sin(\theta + 120°)$, $E_{S21} = nE_R\sin(\theta + 240°)$
Synchro control transformer (abbreviated CT). A CX and a CT are used in pairs to develop an error voltage related to the difference between the two shaft angles. A CX and a CT could be physically identical devices, although a CX often has a lower impedance than a CT, since one CX may drive several CTs in parallel	Translates applied three stator synchro voltages into error voltage output taken from rotor. Error output varies as sine of difference between shaft angles of CX and CT	Three stator voltages from a CX (electrical input); rotor shaft angle ϕ, driven by controlled system output; (mechanical input); both are variable	Single-phase error voltage, developed by rotor coil, $E_{R_{out}}$. $E_{R_{out}} = K\sin(\theta - \phi)$. $E_{R_{out}}$ is rms magnitude of AC voltage of 60 or 400 Hz. It is used to drive control winding of two-phase servo motor, after suitable amplification
Synchro control differential transmitter (abbreviated CDX). CDX is inserted between three wires joining CX to CT. CDX has three space-phase stator input and three space-phase rotor output	Adds or subtracts angle electrically from three-phase synchro voltage representation of shaft angle. Error voltage from CX-CT pair is $E_{R_{out}} = K\sin(\theta - \phi)$; for CX-CDX-CT set	Mechanical shaft angle D of CDX; three synchro stator voltages from CX, applied to three-phase stator of CDX	Three CDX rotor voltages, from three-phase rotor of CDX. Output of CDX is connected to three stator coils of CT. CT error output voltage then depends on sine of difference of input CX

TABLE 37.7. (*Continued*)

Name	Use	Input	Output
	error voltage is $K_{R_{out}} = K \sin(\theta - \phi + D)$, where D = differential angle		and output CT shaft angles, offset by shaft angle of CDX
Synchro torque transmitter (abbreviated TX). Basically identical to CX, except used without amplifcation and without control system or servo motor. Usually of lower impedance than CX. Excitation of 60 or 400 Hz applied to rotor	Used to drive remote dials and pointers connected to synchro torque receiver TR, also called synchro repeater. TX is always used with TR. Together they act as electrical equivalent of flexible shaft	Mechanical rotor shaft angle θ. (Rotor of TX is excited by reference voltage E_R, but this is constant and not an input variable)	Three stator voltages in time-phase, occurring between stator terminals S_1, S_2, and S_3. These voltages are not measured or amplified but are directly applied to stator of synchro torque receiver
Synchro torque receiver (abbreviated TR). Basically identical to CT. TR and TX could be physically identical devices, although TX often has lower impedance, since one TX may drive several TRs. Torque synchros are usually much less accurate than control synchros. Larger the delivered torque, larger the shaft angle error. Unlike CT, TR has excitation applied to its single phase rotor	Used to indicate position of shaft of TX and to drive pointers or dials. When shaft of TX is turned, shaft of paired TR turns equal amount. Rotor windings of TX and TR are excited, in parallel, by same reference source, typically 60 or 400 Hz, at 26 or 115 V AC	Three stator voltages from TX. Stator of TX is connected directly to stator of TR, S_1 to S_1, S_2 to S_2, and S_3 to S_3. (Rotor of TR is excited by reference voltage E_R, but this is constant and not input variable)	Rotor shaft angle ϕ of synchro torque receiver TR. Ideally, ϕ of TR = θ of TX. Actually, $\phi = \theta - f(t)$ where $f(t)$ is function of torque developed by TR. If developed torque is small, as is case in driving pointer where only bearing torques occur, then $\phi \cong \theta$

Synchro torque differential transmitter (abbreviated TDX). Basically identical to CDX. TDX has three-phase stator input and three-phase rotor output	Adds or subtracts shaft angle or offset from three-phase synchro voltage representation of shaft angle	Mechanical shaft angle of TDX; three synchro stator voltages from TX, applied to three-phase stator of TDX	Three CDX rotor voltages from three-phase CDX rotor. Three CDX rotor output voltages are applied as inputs to three-space-phase stator of TR
Multispeed control transmitter, multispeed control transformer. These devices have N pole pairs per phase, rather than standard one pole pair per phase. One mechanical shaft rotation produces N electrical output cycles, rather than one cycle produced by standard synchro	Used to improve accuracy of standard single-speed CX or CT by factor of N, where N is "speed," "electrical gear ratio," or "order" of multispeed synchro	Identical to that for CT or CX	Identical to that of CT or CX equipped with step-up gear ratio of N to 1

TABLE 37.8. SELECTED SYNCHRO AND RESOLVER SPECIFICATIONS

Variable	Specification				
Size	Size of synchro or resolver is specified by integer 10 times nominal maximum outside diameter of device measured in inches. Identical size code is also employed for two-phase servo motors, torquers, and stepper motors, although actual exact measurements may differ. For example size-25 servo may measure 2.478 in. outside diameter, while size-25 stepper motor may measure 2.500 in. Length of servo or resolver is independent of size specification. For example, size-25 resolver may be 3.47 in. long, while size-25 "pancake" resolver may be 1.01 in. long. Any external cooling fins are not counted in size specification. For example, size 20-stepper motor may have cylindrical body measuring 2.000 in. for mounting purposes, covered with fins of diameter 2.8 in.				
Accuracy per Size	Size	Maximum Housing Diameter (in.)	Typical Synchro Accuracy (arc-min)	Resolver Accuracy[a]	Stepper Motor Holding Torque (in.-oz)
	5	0.500	± 10	± 10	—
	8	0.750	± 5, ± 7, ± 10	± 3, ± 5, ± 7	0.5
	11	1.062	± 5, ± 7, ± 10	± 3, ± 5, ± 7	3
	12	1.188	± 5, ± 7, ± 10	—	—
	15	1.437	—	± 40	8
	18	1.750	—	—	15
	25	2.478	± 2.5	± 20, ± 30	—
	28	2.734	—	± 20, ± 30	—
	35	3.500	—	—	90
Voltage (line-to-line, volts rms)	Common input excitation voltages: 11.8, 26, 115 Other input excitation voltages: 28.7, 40, 57.5 Common output voltages: 11.8, 22.5, 90 Other output voltages: 2, 2.5, 4, 12.6, 18, 18.2, 22, 23.5, 57.3				
Synchro resolver frequencies (hertz)	Common operating frequencies: 60, 400 Other operating frequencies: 800, 1600, 10,000				

[a] All in arc-minutes except sizes 15, 25, and 28, which are in arc-seconds.

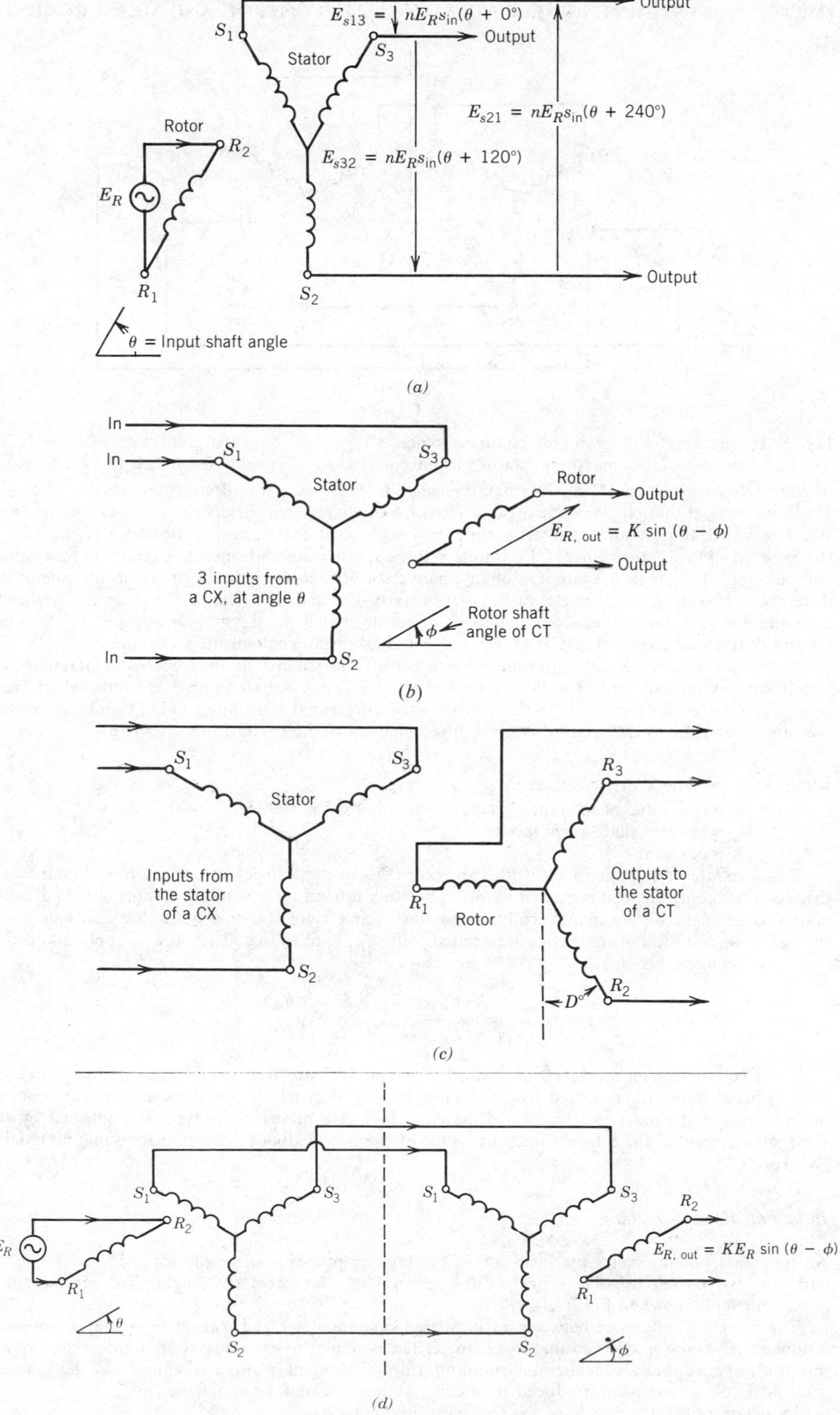

Fig. 37.16 (*See next page for legend.*)

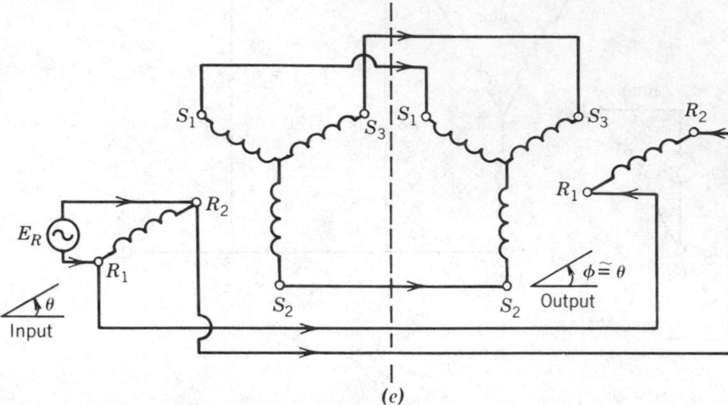

Fig. 37.16 Schematic diagrams of common synchro types. (*a*) Synchro control transmitter (CX). n = transformation ratio, maximum stator output voltage/E_R. Typically 0.2 to 2.0; E_R = rms value of rotor reference voltage, $\sqrt{2}\, E_R \sin(2\pi f t)$, typically 26 V; f = excitation frequency, usually 60 or 400 Hz; θ = electrical shaft angle, angle of rotor shaft for single pole pair device; E_{S13} = voltage rise from stator terminal 1 to stator terminal 3. Input = θ = physical shaft angle. Output = $E_{S13}, E_{S32}, E_{S21}$. (*b*) Synchro control transformer (CT). Inputs = three stator voltages from a CX (control) transmitter with an input shaft angle θ. Output = single-phase error voltage $E_{R,\text{out}}$ proportional to the sine of the difference between the shaft angles of the two synchros CX and CT. At null, the error $E_{R,\text{out}}$ goes to zero and $\theta = \phi$, or the output shaft angle ϕ is driven to match the input shaft angle, θ. (*c*) Synchro control differential transmitter (CDX). $D°$ = differential angle, angle of the CDX shaft. (*d*) CX–CT error detector pair. (*e*) Synchro torque transmitter (TX)–synchro torque receiver (TR) pair. E_R = reference excitation, applied to both rotors in parallel. Input = shaft angle θ. Output = shaft angle ϕ, and $\phi = \theta$ + error terms $\cong \theta$. The synchro torque differential transmitter (TDX) (not shown) has the same schematic as the synchro control differential transmitter (CDX) shown in (*c*).

where $E_{\text{out}_{1,2}}$ = stator output voltages, V_{rms}
E_r = rms value of reference excitation at either 60 or 400 Hz
θ_{in} = resolver shaft angle input
K = constant

Resolvers are often constructed with two rotor coils at right angles to one another. For the case just described, only one coil is excited by E_R. The other unused rotor winding is short circuited. Since coils at right angles do not magnetically couple, the excited rotor coil does not induce currents in the shorted rotor coil. If both rotor coils are excited, with E_{R1} applied to coil 1 and E_{R2} applied to coil 2, the output voltages become

$$E_{\text{out}_1} = K(E_{R1}\cos\theta_{\text{in}} + E_{R2}\sin\theta_{\text{in}})$$
$$E_{\text{out}_2} = K(E_{R1}\sin\theta_{\text{in}} - E_{R2}\cos\theta_{\text{in}})$$

These equations, and most data in Sections 37.6 and 37.7, are abstracted from specification sheets and technical publications issued by the manufacturers of synchros, resolvers, and related components. Some of the most comprehensive material, in fact a model of its type, is published by the Keargott Division of the Singer Company.[3] One of the best textbooks covering the same material is Del Toro's.[4]

Induction Potentiometer

An induction potentiometer produces an AC voltage output with an amplitude that varies linearly with angle, over an angular range of less than 180°. A schematic diagram of an induction potentiometer is shown in Fig. 37.18.

The principal difference between an induction potentiometer and a resolver is that a resolver's output varies as the sine (or cosine) of an angle. The windings of an induction potentiometer are not uniformly spaced, but are deliberately nonuniformly distributed in order to cancel out the inherent sinusoidal voltage variation produced by a coil rotating in a uniform magnetic flux.

Induction potentiometers have the following advantages:

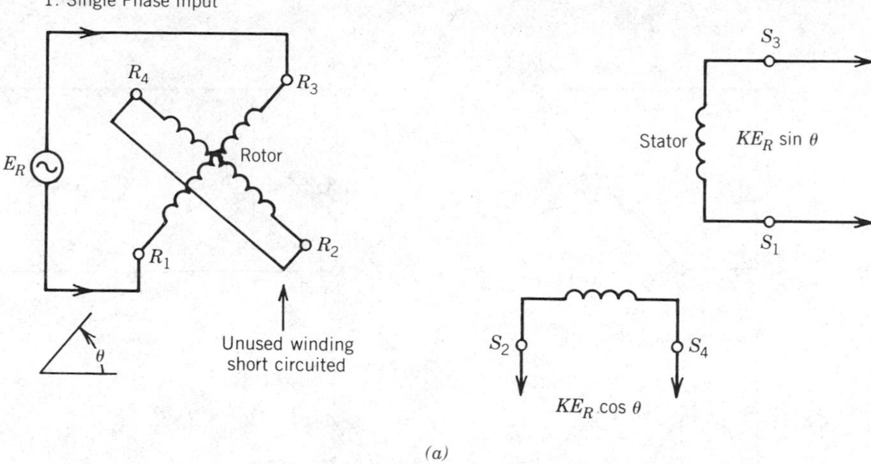

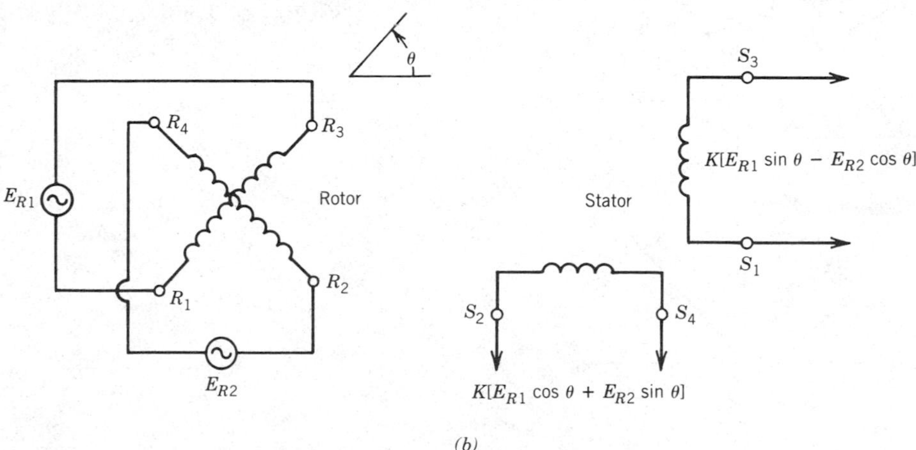

Fig. 37.17 Schematic diagrams of a resolver. (*a*) Single-phase input. θ = shaft angle; K = constant, typically 0.2 to 2.0 V/degree; E_R = reference voltage, typically 26 V at 400 Hz. E_R ranges from 11.8 to 115 V_{AC}. (*b*) Two-phase input.

1. They have no sliding or wiping contacts and hence experience very little wear.

2. Having no wiping contacts, they produce very little frictional torque, and hence can be used in applications where minimal loading torques can be tolerated (e.g., gyroscope angular pickoffs).

3. Their angular resolution is theoretically infinite.

4. Having no wiping contacts, they are immune to contact-generated noise, which can plague resistance potentiometers.

Induction potentiometers have the following disadvantages:

1. They are limited to a range of at most 180° or ±90° and may be linear over a significantly smaller range.

2. They produce a varying amplitude of AC voltage which is ideal for AC carrier-type control systems but which introduces complications into the typical noncarrier control system. The complications include providing the AC carrier, then rectifying, and filtering the resultant output.

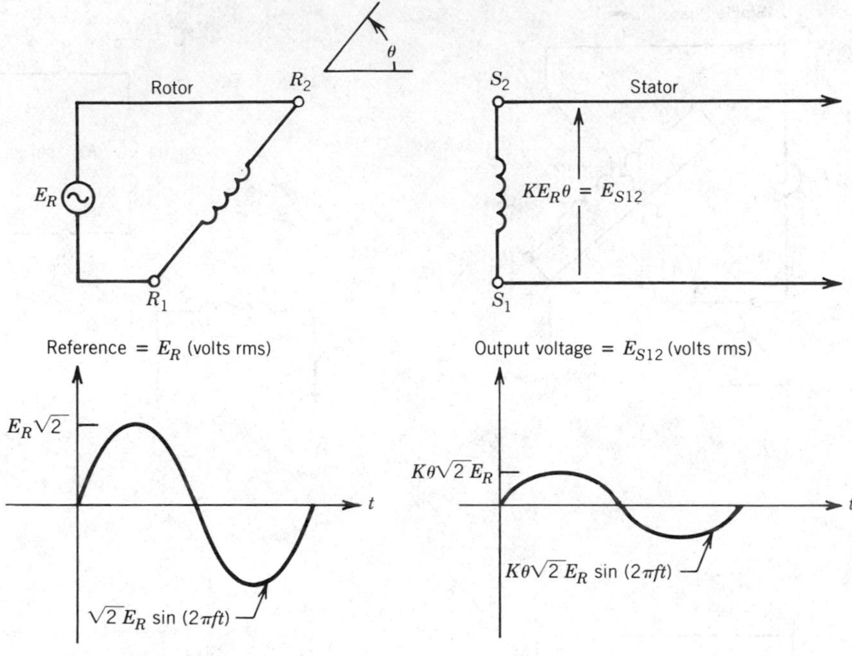

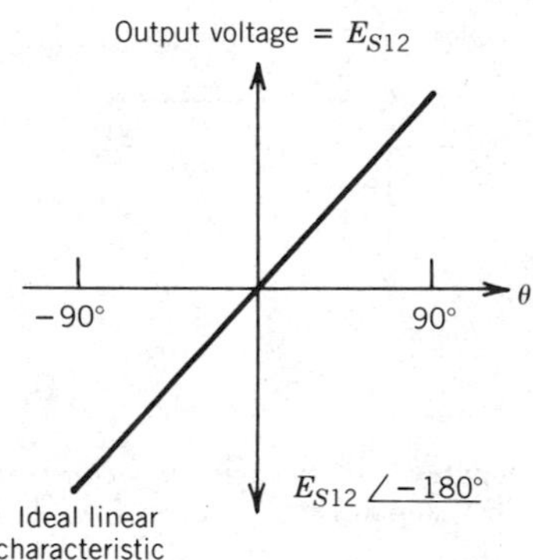

Fig. 37.18 Schematic diagram of an induction potentiometer. K = constant = maximum stator rms output voltage$/E_R\theta_{max}$; f = Excitation frequency, typically 400 Hz.

37.7 SUMMARY OF ANALOG MEASUREMENTS OF SYSTEM ANGULAR OR LINEAR POSITION

Although digital control systems and microprocessors have made large inroads into angular control systems, the angles themselves must still be measured. Synchros, resolvers, and induction potentiometers still represent some of the most accurate angular transducers, and are used in the most sophisticated control systems currently being designed (e.g., NASA's space shuttle robotlike loading arm).

37.8 DIGITAL CONTROL SYSTEM TRANSDUCERS

The applicability and usefulness of synchros have been extended into the digital control era by various synchro-to-digital converters, processors, angle displays, and electronic modules which

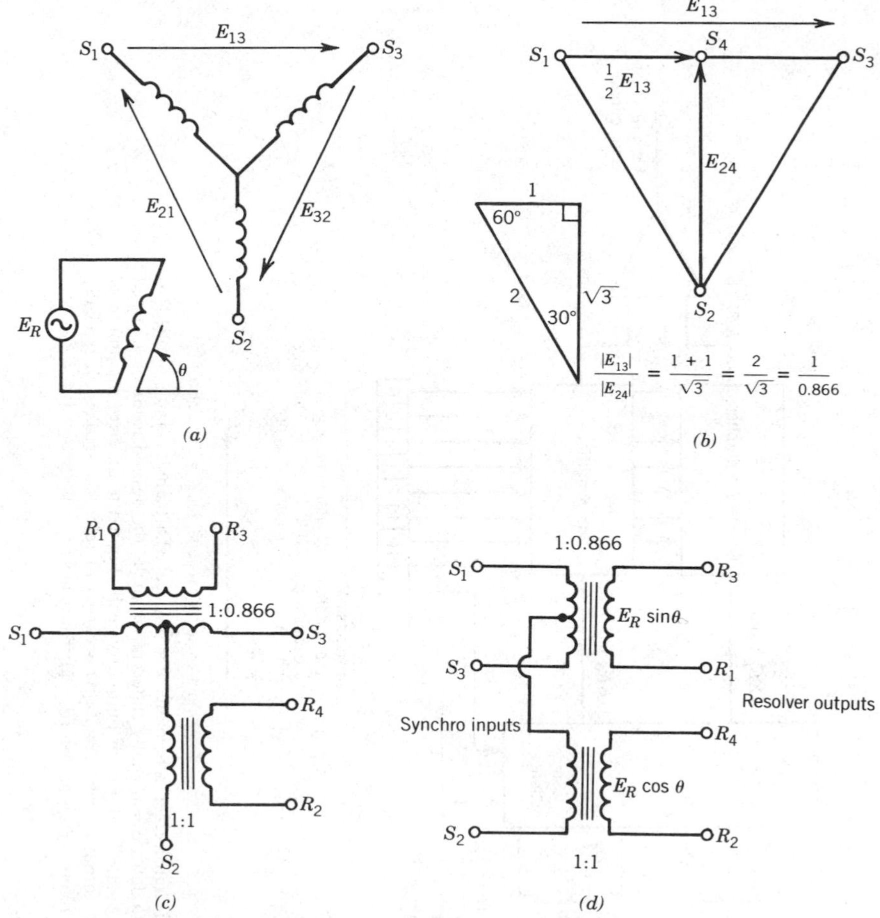

Fig. 37.19 Scott "T" transformer. Conversion of three-wire three-space-phase synchro outputs $E_R\sin(\theta + O°)$, $E_R\sin(\theta + 120°)$, $E_R\sin(\theta + 240°)$, into four-wire two-space-phase resolver information. θ = shaft angle of synchro; E_R = synchro reference voltage; S_1, S_2, S_3 = stator terminals of synchro; $R_1 + R_3$ = resolver phase 1; $R_2 + R_4$ = resolver phase 2. The voltage E_{13} is perpendicular or 90° out of space phase with E_{24}. Note that the center tap S_4 produces $1/2\ E_{13}$. (*a*) Synchro outputs. (*b*) Scott T geometry. (*c*) Scott T transformer connections. (*d*) Redrawn connections.

perform the functions of various synchro components. These include:

1. **Solid state control transformers.** The shaft angle input to the electromechanical synchro is replaced by a 10- to 14-bit digital shaft angle code. The analog signals from an electromechanical synchro control transmitter are applied, just as usual, or these three-space-phase angle signals may be converted to two-space-phase signals via a Scott "T" transformer (Fig. 37.19). The output is an amplitude-modulated error signal given by $A \sin 2\pi ft(\sin\theta - \phi)$, where θ represents the three-wire synchro input angle (or four-wire two-phase equivalent resolver angle), ϕ represents the 10- to 14-bit parallel binary input angle, and $A \sin 2\pi ft$ is the excitation signal of amplitude A and frequency f. Typically the maximum error $(\theta - \phi)$ is limited to much less than 90°, say 10 to 15°. A schematic of a solid state control transformer is given in Fig. 37.20.
2. **Synchro-to-digital converters.** The analog inputs are either three-wire three-phase synchro or four-wire two-phase resolver at the standard synchro or resolver levels of 115, 90, 26, or 11.8 V rms line to line at 400 or 60 Hz. The single-phase reference signal of 26 or 115 V rms at 60 or 400 Hz must also be applied. Miniature transformers are sometimes built right into the hybrid

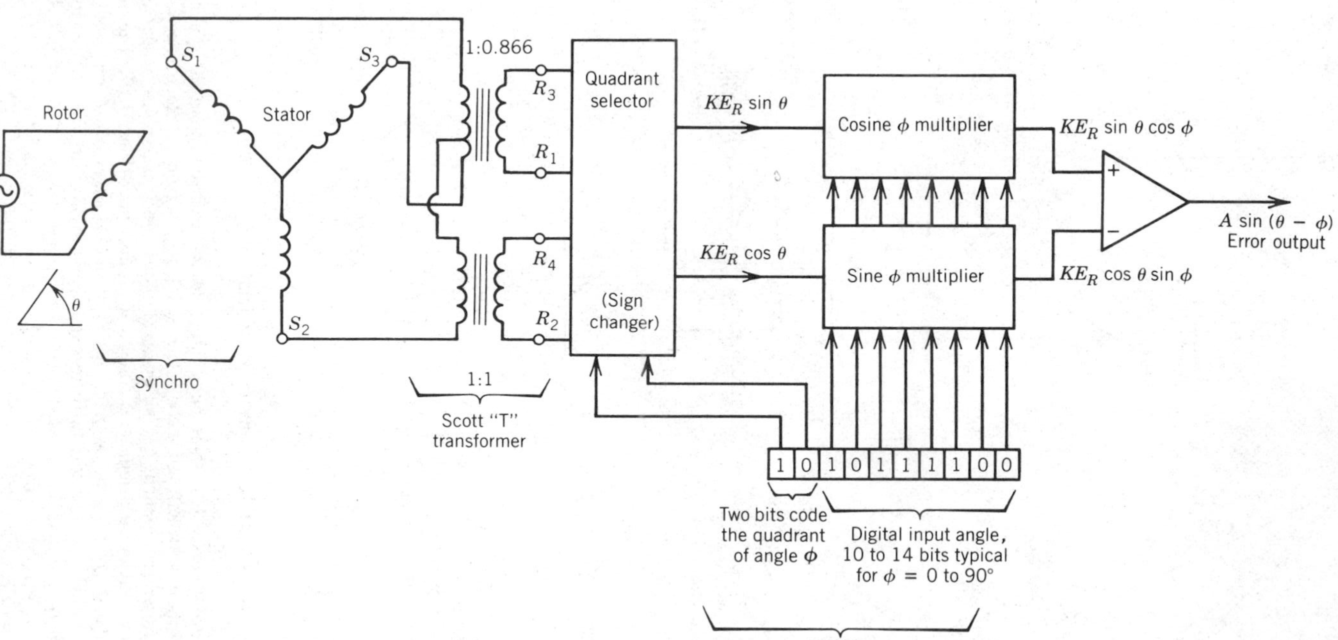

Fig. 37.20 Solid state control transformer schematic. θ = physical shaft angle input, applied to synchro control transmitter, 0 to 360°. ϕ = digitally coded input angle, the desired value of θ, coded as four quadrants of 0 to 90°. $A \sin(\theta - \phi)$ = error output, typically limited to an error maximum of 15° or so. $A = A_0 \sin \omega t$ = error output carrier; $\omega = 2\pi f$; f = carrier of reference frequency, typically 60 or 400 Hz. $E_R = E_0 \sin \omega t$ = reference input. $\sin(\theta - \phi) = \sin\theta\cos\phi - \cos\theta\sin\phi$. $A = KE_R$.

solid-state converter modules to convert voltages; sometimes, at higher voltage or lower frequency, external transformers must be used. Scott T transformers (Fig. 37.19) are used to change three-phase signals to two-phase signals. Finally, the resulting output is typically 10- to 14-bit parallel digital straight binary code, representing the syncho shaft angle from 0 to 360°. This represents a shaft angle resolution that ranges from

Bits	Resolution
10	$\frac{360°}{2^{10}} = \frac{360°}{1024} = 0.3516° = 21$ arc-min
12	$\frac{360°}{2^{12}} = \frac{360°}{4096} = 0.0879° = 5.3$ arc-min
14	$\frac{360°}{2^{14}} = \frac{360°}{16{,}384} = 0.0220° = 1.3$ arc-min

Typically, an additional DC voltage output is provided, which is linearly proportional to $d\theta/dt$ or $-d\theta/dt$.

3. **Synchro-to-linear DC converter.** The inputs are either three-wire synchro or four-wire resolver. The reference excitation signal is that specified for the synchro-to-digital converter at the standard synchro levels also specified. The output is a DC level that is linearly proportional to the input angle. A typical output voltage range is ± 10 V DC for an input angle range of $\pm 180°$. Typically, an additional DC voltage output is provided, which is proportional to $d\theta/dt$ or $-d\theta/dt$ and can be directly employed for derivative feedback system compensation (tachometer feedback).
4. **Digital-to-synchro and digital-to-resolver converters.** The input is a 10- to 16-bit parallel straight binary angle code at standard TTL† levels, and the outputs are amplitude-modulated AC in either three-wire three-phase synchro format or four-wire two-phase resolver format. The AC reference signal determines the maximum amplitude and the frequency of the AC outputs, which can be any of the standard synchro levels specified in paragraph 2. However, since solid-state devices usually dissipate less power than mechanical devices, the typical digital-to-synchro converter provides a relatively small volt-ampere output, generally on the order of 1 to 2 VA.
5. **Two-speed synchro digital converter, processor, and angle display modules.** The inputs are the AC reference signal used by both the coarse and the fine synchro control transmitter and the three stator output signals generated by both the coarse and the fine synchro. A typical module is designed to handle the common speed or gear ratios of 36 : 1, 18 : 1, and 9 : 1 used by older two-speed synchro systems. Newer two-speed synchros use binary gear ratios of 64 : 1, 32 : 1, 16 : 1, and 8 : 1, which are handled by slightly modified and simpler modules. The output comprises a combination of parallel bits derived from both the coarse and the fine synchros. Up to 19 bits of parallel binary angular output code can presently be produced in this fashion. The coarse synchro, which directly measures the desired angle, must be capable of determining that angle to better than 90° on the fine shaft. The fine synchro, which measures the desired angle multiplied by the step-up gear ratio, must be capable of resolving an angle smaller than the backlash angle of the gears on the fine shaft. The ultimate limit of accuracy becomes the "play" or backlash of the two-speed synchro gearing.

Table 37.8 presents a summary of selected synchro and resolver specifications, abstracted from Singer–Kearfott data sheets, with effective dates ranging from 1966 to 1981.‡

37.9 DIGITAL MEASUREMENT OF SYSTEM POSITION OR ANGLE

Two basic approaches are used to obtain digital measurements of system position or angle. The first approach involves conversion of a basic analog transducer into a digital measuring system, by use of an analog-to-digital converter and the associated necessary control logic added to the basic analog transducer. The second approach is use of a digital transducer that provides digital outputs directly.

If a single-signal analog output measurement is linear, a straightforward analog-to-digital converter may be employed to produce the desired digital output. The digital output consists of N digital output lines resulting from the single analog input line. Analog-to-digital converters are discussed in Section 37.4.

†TTL = transistor-transistor logic, nominally 0 to 5 V. Logic 0 is taken as 0 to 0.4 V, and logic 1 as greater than 2.4 V.

‡Singer Company, Kearfott Division, Little Falls, NJ.

If the analog measurement is not linear, or if several signals code the analog output (e.g., a three-wire synchro or four-wire resolver), a specialized digital converter must be employed. A common example is the synchro-to-digital converter discussed in Section 37.6. The three-amplitude modulated AC synchro outputs are electronically processed by the synchro-to-digital converter to produce an N-bit digital output. With TTL (transistor-transistor logic), typical voltage levels are 0 (= 0 to 0.4 V) and 1 (= 2.4 to 5.5 V). In this case the inputs are three AC signals and a reference AC voltage, and the outputs are N digital lines. The overall accuracy of such a system must be less than the basic accuracy of the analog transducer, which determines the required number of bits, N, of the digital output. Providing more bits will provide greater resolution (and will result in greater complexity and cost) but will not result in greater accuracy.

For a synchro accurate to 6 arc-min, the maximum resolution needed is

$$\text{Resolution} = \frac{\text{smallest accurate measurement}}{\text{maximum range}} = \frac{6 \text{ min}}{360° \times 60 \text{ min}/°} = \frac{1}{3600}$$

The corresponding number of bits of resolution N needed is

$$\text{Resolution} = \frac{1}{2^N}, \text{ or } 2^N = 3600, \text{ or } N = \frac{\log 3600}{\log 2} = 11.8 \cong 12$$

Thus 12 bits is the largest digital resolution required for the given analog accuracy. However, the inevitable conversion errors will lower the final accuracy. Ten bits might be a readily obtainable resolution in this case. Commonly available synchro-to-digital converters have resolution of 10, 12, 14, and 16 bits for 360° input.

Any of the analog methods discussed in Section 37.7 for measurement of system position or angle can be adapted to produce a digital result by using the procedures just discussed. Several examples are shown in Fig. 37.21.

Direct Digital Measurements

Position- and angle-measuring transducers, which produce N digital outputs directly, are often called encoders. The most common is the digital shaft-angle encoder. A schematic of a straight 3-bit binary shaft-angle encoder is shown in Fig. 37.22*a*. The particular device illustrated uses photoelectric sensing of the code disk. Other sensing methods are often used such as magnetic sensing, brush-type wiping electrical contacts, or any other switching scheme amenable to the physical requirements. Photoelectric shaft-angle encoders are common, because the code disks can be produced rapidly, accurately, and relatively inexpensively by photographic methods.

Referring to Fig. 37.22, the photoelectric shaft-angle encoder operates as follows. The rotating disk contains N concentric tracks. Each track is viewed on one side by a photosensitive device, which is usually a photodiode or a phototransistor. On the other side of the disk a single light source illuminates all N tracks with a narrow, collimated beam of light. Whenever a given track is clear, the corresponding photodiode or phototransistor is illuminated, and whenever the track is opaque, there is no illumination. This masking of light provides the digital switching action. Care must be taken to ensure that a particular photoelement receives light from one track and one track only. The opacity patterns of each track follow the binary truth table patterns for the particular code to be encoded. In gray code, only 1 bit at a time changes between successive states, whereas in straight binary code, all the bits change as the count goes from maximum to zero, and various combinations of bits change at intermediate counts. Therefore the gray code device is subject to less error. Figure 37.22*a* and *b* show the truth table patterns for 3-bit straight binary and 3-bit gray code, respectively.

The advantages of a digital shaft-angle encoder are as follows:

1. Absolute code outputs, directly produced, with no calibration procedure. This is a marked advantage over incremental pulse-counting shaft-angle indicators, which lose calibration if the electronics package loses a pulse. The absolute indication is limited to one revolution of the encoder. For larger angles, electronics must keep track of total number of revolutions.

2. High-speed long-life operation (thousands of revolutions per minute and hundreds of millions of revolutions) because of the noncontact optical switching method employed. Brush-type sliding-contact encoders experience contact wear, and consequently their maximum rotational rate is usually limited to 100 to 200 rpm and a few million revolutions.

3. Comparatively simple interfacing requirements, such as a single pull-up resistor to each digital output line from the shaft-angle encoder.

4. Relatively high angular resolution. A 12-bit digital shaft-angle encoder is an off-the-shelf item, and a three-stage cascaded V-code encoder can give 20 or more bits of resolution. The disk pattern on a V-code encoder looks like the letter V, and the first stage enables or switches the second stage, which enables or switches the third stage.[5]

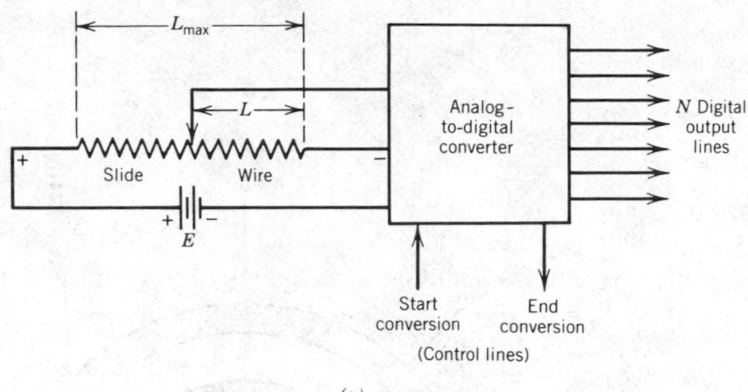

(a)

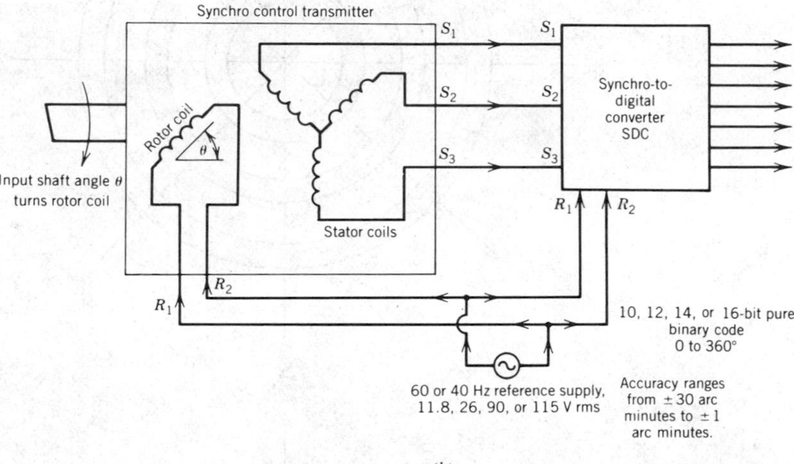

(b)

Fig. 37.21 Digital measurement systems using analog transducers. (*a*) Measurement of position. L_{max} = maximum length of slidewire, L = length to be measured, $E_{in} = (L/L_{max})E$. Typical values of $E = 5$ to 10 V; typical values of $N = 8$, 10, 12, and 16. The 3-bit unipolar straight binary code output for a 10-V digital-to-analog converter with E = 10 V, is

Position (L_{max})	E_{in} (V)	Digital Code	Step
$0 \pm 1/16$	0 ± 0.625	0 0 0	0
$1/8 \pm 1/16$	1.25 ± 0.625	0 0 1	1
$1/4 \pm 1/16$	2.50 ± 0.625	0 1 0	2
$3/8 \pm 1/16$	3.75 ± 0.625	0 1 1	3
$1/2 \pm 1/16$	5.00 ± 0.625	1 0 0	4
$5/8 \pm 1/16$	6.25 ± 0.625	1 0 1	5
$3/4 \pm 1/16$	7.50 ± 0.625	1 1 0	6
$7/8 \pm 1/16$	8.75 ± 0.625	1 1 1	7

(*b*) Measurement of angle. Accuracy ranges from ±30 to ±1 arc-min. The first few outputs for this 10-bit synchro-to-digital (SDC) converter are

Step	Nominal Angle (°)	Digital Code		
0	0.00000	00	0000	0000
1	0.351563	00	0000	0001
2	0.703125	00	0000	0010
3	1.054688	00	0000	0011
⋮	⋮		⋮	

Least significant bit (LSB) = $360°/2^{10} = 360°/1024 = 0.351563°$.

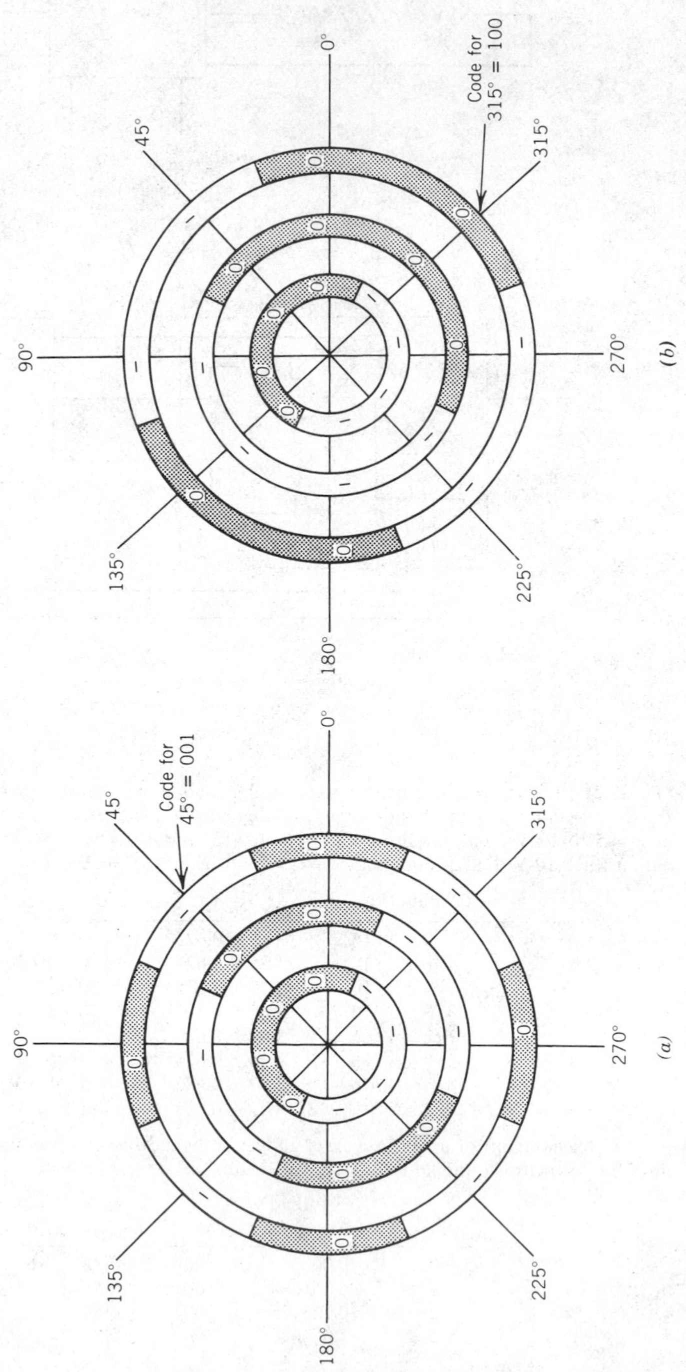

0°
45°
90°
135°
180°
225°
270°
315°
Code for
45° = 001
(a)
Code for
315° = 100
(b)

Fig. 37.22 Three-bit binary codes for shaft angle encoders.

(*a*) Example of a 3-bit digital shaft-angle encoder disk, using straight binary code:

Angle (°)	Step	Code		
0 ± 22.5	0	0	0	0
45 ± 22.5	1	0	0	1
90 ± 22.5	2	0	1	0
135 ± 22.5	3	0	1	1
180 ± 22.5	4	1	0	0
225 ± 22.5	5	1	0	1
270 ± 22.5	6	1	1	0
315 ± 22.5	7	1	1	1
		Inner track		Outer track

1 = clear, 0 = opaque.

(*b*) Example of a 3-bit digital shaft-angle encoder disk, using gray code. Only one bit changes between successive codes.

Angle (°)	Step	Code		
0 ± 22.5	0	0	0	0
45 ± 22.5	1	0	0	1
90 ± 22.5	2	0	1	1
135 ± 22.5	3	0	1	0
180 ± 22.5	4	1	1	0
225 ± 22.5	5	1	1	1
270 ± 22.5	6	1	0	1
315 ± 22.5	7	1	0	0
		Inner track		Outer track

For optical encoder, 1 = clear = light; 0 = opaque = no light. *For brush-type encoder*, 1 = metal track = unshaded; 0 = insulator = shaded track.

The disadvantages of a digital shaft-angle encoder are as follows:

1. Relatively high cost, compared with incremental pulse-counting shaft-angle indicators.
2. N output lines required, plus power supply, ground, and shields (if used). An incremental, pulse-counting shaft-angle indicator has only one or two output lines. This results in a relatively small cable for the incremental device, and a relatively large cable for the digital shaft-angle encoder. This is a major problem in aircraft and aerospace vehicles where cabling represents significant weight, volume, and cost of the total system design.

The same basic concept employed in a digital shaft-angle encoder may be adapted to linear position measurements by use of a set of N parallel-coded switching or optical tracks, rather than N circular tracks. This is illustrated in Fig. 37.23. The N parallel optical tracks would be photographically printed on a transparent digital "ruler" attached to the element whose position is to be determined. Mechanical sliding-brush switching tracks consist of a metal layer laid down on an insulating track. Photographic resists may be employed to produce the desired metallic pattern by an etching procedure.

This type of N-bit linear position encoder is infrequently seen, compared with the N-bit digital shaft-angle encoder, which is comparatively common. A linear pulse-counting relative-position indicator is seen much more frequently. Pulse-counting position and angle indicators are discussed next.

Incremental Pulse-Counting Measurement Systems

In a pulse-counting measurement system, a single-track pattern capable of producing a switching action is attached to either a disk or a linear rulerlike element. This corresponds to the most rapidly alternating track shown in Fig. 37.22 and Fig. 37.23. The other tracks are not used. As the switching

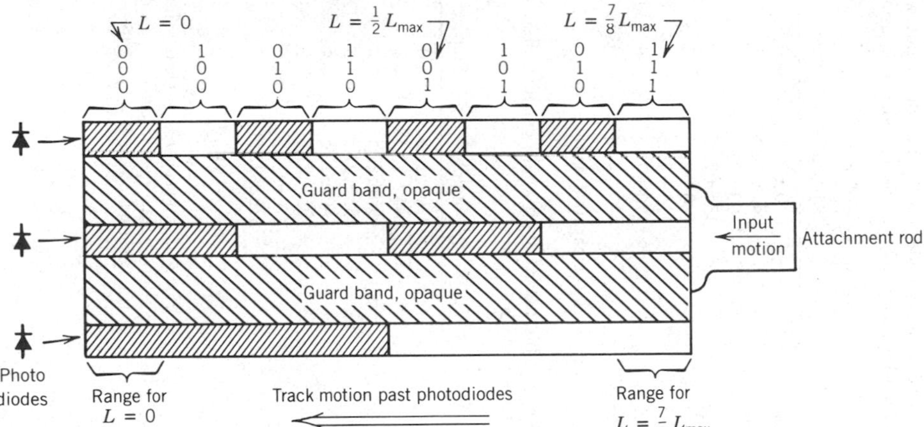

Fig. 37.23 Direct digital encoding of position. Three-bit straight binary code example. This optical encoder track layout results in the following straight binary code:

Position	Step	Code
$0 \pm 1/16$	0	0 0 0
$1/8 \pm 1/16$	1	0 0 1
$1/4 \pm 1/16$	2	0 1 0
$3/8 \pm 1/16$	3	0 1 1
$1/2 \pm 1/16$	4	1 0 0
$5/8 \pm 1/16$	5	1 0 1
$3/4 \pm 1/16$	6	1 1 0
$7/8 \pm 1/16$	7	1 1 1

Note that the length of the track for a 3-bit encoder is L_{max}, but that the measurement range is 0 to $[(2^N - 1)/2^N]L_{max} = \frac{7}{8} L_{max}$. The missing increment of $1/2^N = \frac{1}{8} L_{max}$ appears as a tolerance of $\pm \frac{1}{16} L_{max}$ on either side of $L = 0$ and $L = \frac{7}{8} L_{max}$.

track moves relative to the switching element, the switch element produces a digital 101010 . . . pattern. Each transition corresponds to one half period of the switching track pattern. If the pattern is an optical one, with alternatively opaque and clear bands spaced 1 mm apart, a pulse transition (1 to 0 or 0 to 1) occurs for every millimeter of optical track movement. By counting the pulses and knowing the direction of motion a priori, it is possible to determine the total motion of the moving element. Note that this pulse-counting scheme, by itself, cannot determine the direction of motion. If two tracks are used, with the code 00, 01, 11, 10, 00, 01, 11, 10, 00, . . . , position and direction can both be determined, since the bit-switching sequence is different for forward and reverse rotation. For example, consider the code 11. If the next code is 10, a forward rotation has occurred, while if the next code is 01, a reverse rotation has occurred. This is a 2-bit gray code.

Magnetic pulse tracks, or variable reluctance tracks, are commonly used. The magnetic tracks can be produced and read out using magnetic recording principles. The angular position of a magnetic disk in a computer disk memory is often determined in this fashion, with use of a magnetically recorded pulse train called the clock track. On the other hand, floppy magnetic memory disks are often punched with a sequence of equally spaced holes, with one pair of double holes. When read out by a photoelectronic sensing system, these holes determine the position of the rotating disk in terms of so-called disk sectors. The pair of double holes provides a point of initialization. Such a disk is called an optically hard-sectored disk, each sector being indicated by a hole. A soft-sectored floppy disk has only one hole, to indicate the point of initialization. The remaining sectors are determined by a magnetic track.

Variable reluctance pulse-counting schemes are extremely rugged. In certain applications an ordinary gear, constructed of high-permeability material such as iron or steel, is used as the variable reluctance element. Gears of plastic, brass, or other nonmagnetic material cannot be used. The gear is often part of the existing system. The variable reluctance sensor is simply a coil of wire wrapped around a high-permeability core, such as ferrite, iron, or steel. A magnetic flux is produced within this core, either by a permanent magnet or by a DC-powered electromagnet. The sensor's core must be smaller in diameter than the gear teeth spacing, at the end nearest the gear teeth. The core can be pencil shaped, with the pointed end closest to the gear teeth. The magnetic flux passes from the sensor core across a small air gap and into the ferrous gear. When a gear tooth is directly opposite the sensor, the air gap and the reluctance of the magnetic circuit are minimum and the magnetic flux is a maximum in accordance with

$$\phi = K \frac{\text{MMF}}{R}$$

where ϕ = magnetic flux, Wb
MMF = magnetomotive force produced by the permanent magnet, or a DC current in an electromagnet
R = reluctance of magnetic circuit
K = constant of proportionality

When the gear turns by one half the tooth spacing, the sensor moves from inline with a tooth to midway between the gear teeth, and the air gap between the sensor core and the gear teeth becomes a maximum. This increases the reluctance of the magnetic circuit and decreases the flux ϕ. The flux ϕ thus cycles from a maximum to a minimum and back to a maximum as each gear tooth passes directly under the sensor. The coil of wire wrapped around the sensor core has a voltage induced in it by this changing flux:

$$V = n \frac{d\phi}{dt}$$

where ϕ = flux through coil, Wb
n = number of turns in coil
V = voltage, V

Thus the variable reluctance sensor produces a voltage pulse every time a gear tooth passes the sensor. Electronic counting of these pulses allows the angular position of the shaft to be determined (assuming that no pulses are lost and that the shaft does not reverse its direction of rotation). Identical pulses are produced for either direction of rotation if the gear teeth are symmetrical, which is almost always the case. Hence the direction of rotation cannot be determined from the electrical pulses produced by a variable reluctance sensor. If two sensors are used and one sensor is shifted by one quarter of the tooth spacing, the relative phase of the resulting two pulse trains can be used to determine the direction of the rotation.

Gear teeth are ideal for sensing the position of relatively slow-moving shafts, since the magnitude of the sensor pulse depends on the angular velocity ω of the shaft multiplied by the number of gear teeth. At relatively high speeds (e.g., 10,000 rpm and up), the rate of change of flux will be adequately large for a single-lobed cam or metallic pin attached to the shaft. This is akin to a gear with one tooth. Gas turbines turn at high speeds (up to 100,000 rpm and more), and their angular position can be determined by a variable reluctance sensor producing only one pulse per revolution. Since the speed of the turbine is effectively constant over the time period of a single revolution, the interval between two sensor pulses can be electronically divided into, for example, 360 parts, and each interval will correspond very closely to 1° of turbine revolution (measured from the absolute reference point that generates the original sensor pulse). A similar result can be produced by attaching a permanent magnet to the rotating shaft, so that the magnet passes by a stationary magnetic field sensor as the shaft rotates. The magnetic field sensor can be the same coil of wire on a ferrous core as that used for the variable reluctance sensor. The magnetic field sensor can also be a solid-state Hall-effect device, which produces a variable voltage output in response to a changing magnetic field input. The permanent magnet could be a gearlike disk, with each tooth an alternate north or south pole. No actual teeth or protrusions are necessary, simply an alternating magnetic field pattern around the periphery of the rotating disk. This disk could be perfectly smooth aluminum, for example, with one or more permanent magnets embedded in its periphery (along the outer circumference).

Such pulse-counting schemes are widely used to determine position or angle, in conjunction with digital processing circuitry and a digital controller. Numerically controlled machines employ almost all of the methods discussed here. There are more exotic control systems which employ digital pulse-counting position transducers. Probably the most accurate digital position-measurement device utilizes a photooptical sensor to count the passage of light and dark laser interferometric fringes. A mirror, which is one arm of the interferometer, is attached to the object whose position is to be measured. The reflected light interferes with a reference light beam and a bright constructive interference band is produced when the two light beams are in phase. When the two light beams are 180° out of phase, a dark destructive interference band is produced. Since the light reflected from the mirror travels out and then back, motion of the mirror produces a double change in the light-path length. Thus a mirror movement corresponding to one half a wavelength of the laser light will produce a full wavelength shift of 360° and a complete cycle of the interference band pattern: from light to dark and back to light. If this pattern is projected onto a slit which is viewed by a phototransistor or photodiode, an electrical pulse will be produced for every half-wavelength motion of the mirror, since the mirror motion will cause a cycling in the light intensity falling on the slit—as a consequence of the changing interference pattern. Thus each electrical pulse will correspond to a motion of one half wavelength of light. Suitable counting circuits can keep track of the total motion, as long as the motion is slow enough to produce a pulse repetition frequency within the operating bandwidth of the electronic circuits.

Assume the laser is a helium-neon device that emits red light with a wavelength of 630 nm ($1 \text{ nm} = 1 \text{ m}\mu = 10^{-9} \text{ m} = 10 \text{ Å} = 10 \times 10^{-10} \text{ m}$). An electrical pulse will be produced by the interferometer for a one half wavelength motion of $1/2 \times 630 \text{ nm} = 315 \text{ nm} = 315 \times 10^{-9} \text{ m}$. Assuming that the electronics can count up to 10^7 pulses per second, then the maximum allowable mirror velocity is

$$\text{Maximum velocity} = \text{maximum pulse counting rate} \times \frac{\text{wavelength}}{2} = 10^7 \frac{\text{pulses}}{\text{s}} \times \frac{630 \times 10^{-9}}{2} \frac{\text{meters}}{\text{pulse}}$$

$$= 315 \times 10^{-2} \cong 3 \text{ m/s}$$

Thus for a 10-MHz (10^7) counter and a red light laser, the maximum allowable velocity is 3 m/s for interferometric pulse-position sensing. Three meters per second is about 9.8 ft/s, or 6.7 miles/hr, which is about double walking velocity. This relatively small maximum velocity and the relatively complex interferometer limit the application of this position-measuring technique to relatively rare systems requiring the highest possible accuracy (e.g., a diffraction grating ruling engine or optical measuring systems).

37.10 ANALOG MEASUREMENT OF SYSTEM VELOCITY

The most common device used to measure analog velocity is the tachometer or tachometer generator, often called a tach (pronounced tack). The tachometer converts angular velocity to a voltage whose magnitude and sign (or phase) varies almost linearly with the magnitude and direction of the applied angular velocity. The tachometer may be of either DC or AC construction and may have a permanent magnet field or a separately excited field requiring power input. Figure 37.24 illustrates the characteristics of AC and DC tachometers.

A tachometer is basically a voltage generator, especially designed to produce a voltage output that varies linearly with the applied shaft speed. A DC tachometer produces a DC output voltage with a superimposed ripple frequency, which increases with the applied angular velocity. The ripple is caused by the switching effect of the commutators used in DC generators. This effect is minimized by maximizing the number of commutator bars, and the ripple voltage can be readily held down to a few percent of the DC output voltage. A typical DC tachometer, utilizing a permanent magnet field, has the following characteristics: measurement of 1/2 to 2 in. in diameter; 1 to 3 in. long; 8 to 18 commutator bars; linear within 5 to 1/3% over a speed range of 600 to 4000 rpm; weight of 1/4 to 1/2 lb; and produces 6.5 V DC for each 1000 rpm of shaft input angular velocity. If the direction of rotation is reversed, the polarity of the DC voltage reverses. Although the minimum speed specified is 600 rpm, the DC tachometer operates at speeds below 600 rpm, but its output departs increasingly from the linear relation given by Eq. (37.18), both at low speeds and at very high speeds:

$$V_{out} = K_1 \cdot \omega \text{ (DC tach)} \qquad \text{V DC} \tag{37.18}$$

$$V_{out} = K_2 \cdot V_{in} \cdot \omega = K_2 \cdot 115\sqrt{2}\, \sin 2\pi ft \cdot \omega = K_3 \cdot \omega \text{ (AC tach)} \qquad \text{V AC rms} \tag{37.19}$$

where K_1, K_2, K_3 = constants
ω = shaft angular velocity, rad/s
V_{in} = AC tach reference voltage (e.g., $115\sqrt{2}\sin 2\pi ft$ for $V_{in} = 115$ V)
f = frequency of reference voltage, Hz

A similar departure from nonlinearity occurs for the AC tachometer at low speeds and at very high speeds. Thus DC tachometers are specified to be linear over a limited rpm range, although they will work at all speeds from zero up to some relatively high speed limited by early bearing failure, insulation failure, or mechanical destruction due to rotational forces. AC two-phase induction tachometers have a performance limited by upper speed limits depending on the frequency of the applied excitation voltage. The output of an AC two-phase introduction tachometer is a sinusoidal voltage in phase with the reference voltage for a positive angular velocity, and 180° out of phase for a negative angular velocity. A positive angular velocity is usually a counterclockwise rotation, when viewed from the shaft end, although the reverse definition (clockwise = positive) is also used. The magnitude of this AC voltage output varies with the magnitude of the angular velocity input but, unlike that of ordinary AC generators, the output frequency of the two-phase AC induction tachometer is constant. To minimize the phase shift between the reference voltage and the output voltage, the maximum AC tachometer rpm must be much less than the synchronous speed corresponding to the excitation frequency, in order to preserve the desired phase relationship of 0 or 180°. For a two-pole two-phase AC induction tachometer, the synchronous speed is $60 \cdot f$ rpm, where f is the excitation frequency in hertz. For $f = 60$ Hz, the synchronous speed is 3600 rpm, and the typical AC two-phase tach is limited to at most one half this rate, or 1800 rpm. Even so, the desired phase relationships will vary by several degrees, perhaps tens of degrees, over the specified speed range (which might be 500 to 1750 rpm), as shown in Fig. 37.24*h*. A two-pole 400-Hz tachometer has a synchronous speed of $60 \times 400 = 24{,}000$ rpm, which allows production of phase-error limit of a few degrees through the limiting of shaft speed to one quarter of synchronous speed, or 6000 rpm. This undesirable phase shift is caused by the varying frequency f_r of the induced rotor currents which act upon the rotor impedance $Z_R = R_r + j2\pi f_r L_r$, where R_r is the effective rotor resistance and L_r is the effective rotor inductance.[4] R_r is made large compared with L_r to minimize the effect of changing frequency f_r.

At low speeds, the induced rotor frequency f_r is large and approaches the excitation frequency as speed approaches zero. Hence the phase-shift problem is large at low speeds.

At high-shaft speed, f_r is small and approaches zero at synchronous speed, as given by Eq. (37.21). Note the unusual behavior of the frequency of the induced-rotor currents. At low shaft speed the frequency is high, and at high shaft speed the frequency is low.

The rotor frequency f_r produces a rotating magnetic field of speed $60 \cdot f_r$ rpm. The sum of this speed plus the shaft speed ω is constant and equal to the synchronous speed $60 \cdot f$ produced by the excitation voltage of frequency f. This constant-speed rotating field induces constant-frequency voltages in the output winding of the AC tachometer, independent of shaft speed ω.

Unit Conversions

$$\omega = \frac{\text{rad}}{\text{s}} \times \frac{60}{1}\frac{\text{s}}{\text{min}} \times \frac{1\text{ r}}{2\pi\text{ rad}} = \omega\frac{60}{2\pi}\frac{\text{r}}{\text{min}} = \text{rpm}$$

$$\text{or} \quad \omega\frac{\text{rad}}{\text{s}} \times \frac{60}{2\pi} = \text{rpm}, \quad \text{or} \quad \omega = \frac{\text{rad}}{\text{s}} = \text{rpm}\frac{2\pi}{60} \tag{37.20}$$

Induced rotor frequency

Rotor field rotation + rotor rotation = synchronous speed

$$60 \cdot f_r \text{ rpm} + \omega\frac{60}{2\pi}\text{ rpm} = 60 \cdot f \text{ rpm}$$

$$f_r + \frac{\omega}{2\pi} = f$$

$$f_r = f - \frac{\omega}{2\pi} \tag{37.21}$$

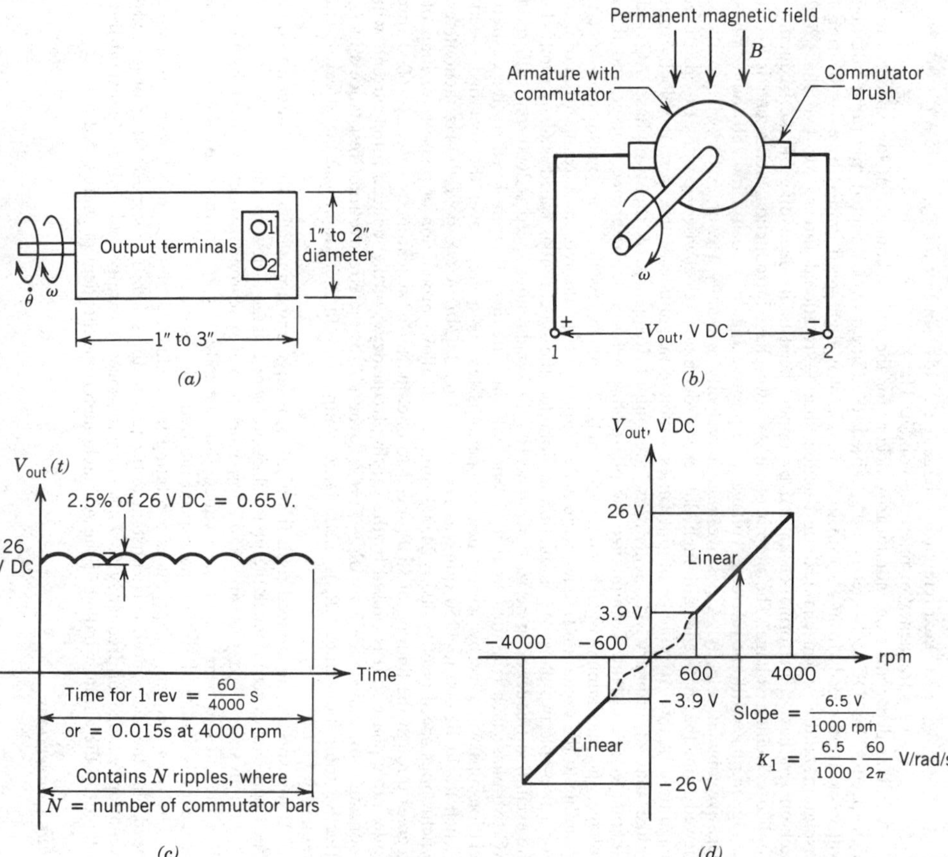

Output terminals
1
2
1" to 2" diameter
1" to 3"
$\dot{\theta}$ ω
(a)
Permanent magnetic field
B
Armature with commutator
Commutator brush
ω
+
−
1
2
V_{out}, V DC
(b)
$V_{out}(t)$
2.5% of 26 V DC = 0.65 V.
26 V DC
Time
Time for 1 rev = $\frac{60}{4000}$ s
or = 0.015s at 4000 rpm
Contains N ripples, where
N = number of commutator bars
(c)
V_{out}, V DC
26 V
Linear
3.9 V
−4000
−600
600
4000
rpm
−3.9 V
Slope = $\frac{6.5 \text{ V}}{1000 \text{ rpm}}$
$K_1 = \frac{6.5}{1000} \frac{60}{2\pi}$ V/rad/s
Linear
−26 V
(d)

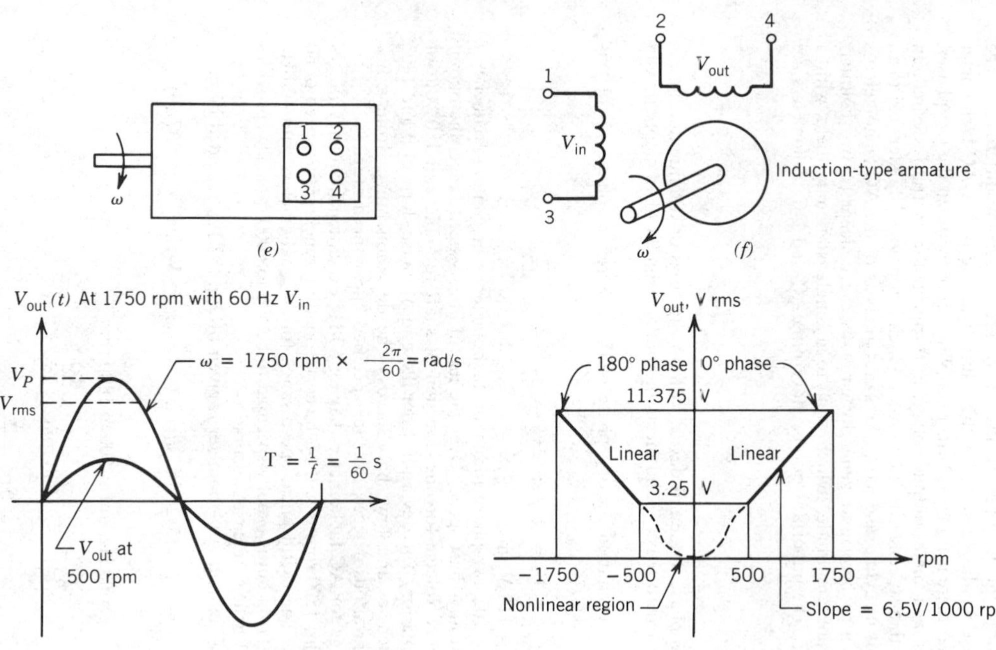

Fig. 37.24 DC and AC tachometers. (*a*)–(*d*) *DC tachometer with permanent magnetic field.* $\omega = d\theta/dt$ = angular velocity, rad/s. If ω reverses direction, V_{out} reverses polarity. $V_{out} = K_1\omega$ in the linear region, V DC. K_1 = tachometer constant = V/rad/s; typical value = 0.06 V/rad/s. (*a*) Generalized outline. (*b*) Electrical schematic. (*c*) Ripple voltage, maximum at 4000 rpm or $\omega = 4000 \times 2\pi/60$ rad/s. (*d*) V_{out} V/s rpm. (*e*)–(*h*) *AC tachometer*, which requires an AC reference, V_{in}, typically 115 V rms at 60 or 400 Hz. Thus $V_{in} = 115\sqrt{2}\sin(2\pi 60t)$ or $115\sqrt{2}\sin(2\pi 400t)$. The output voltage, V_{out}, is given by $V_{out} = K_2 V_{in}\omega = K_3\omega$ in the linear region. Note that ω, the angular velocity, modulates the amplitude of V_{in} but does not change the phase or frequency. When ω reverses, V_{out} undergoes a 180° phase shift. A typical value of K_3 is K_3 = 6.5 V/1000 rpm × 60/2π = V/r/s. The phase angle of V_{out} is measured with respect to V_{in}, the reference. (*e*) *Generalized outline drawing.* (*f*) *Electrical schematic.* (*g*) $V_{out}(t)$, maximum and minimum. V_{rms} = 6.5 V/1000 rpm × 1750 rpm = 11.375 V. $V_p = \sqrt{2}\, V_{rms}$ = 16.08 V. (*h*) V_{out} v.s. rpm.

where ω = shaft speed, rad/s
rpm = shaft speed, r/min
f_r = rotor frequency, Hz
f = excitation frequency, Hz (60 or 400 typical)

At synchronous speed:

$$\omega = 2\pi \cdot f \text{ rad/s}$$
$$\text{rpm} = 60 \cdot f \text{ r/min}$$

For AC tachometers, the region of linear operation with small phase shift resides at speeds well above zero and well below synchronous speed. Conceptually, there is no difference between an AC two-phase induction tachometer and an AC two-phase servomotor, except that the load turns the tachometer, whereas the motor turns the load. In reality, there are significant differences. The effective axis of the excitation-phase winding and the output-phase winding of the AC tachometer must be at exactly 90°, to ensure that there is no transformer coupling action between the two phases, which would result in an erroneous tachometer output voltage at 0 rpm. If low inertia is desired, an AC tachometer may be designed with a so-called drag cup rotor which is topologically identical to an open-top empty tin can, spun about its cylindrical axis.[6] An exact analysis of the AC tachometer is far more involved than that presented here, and generally employs the concept of two counter-rotating magnetic fields of one half amplitude, which sum to a single stationary field with a sinusoidally varying unit amplitude. The important points concerning AC and DC tachometers are summarized as follows.

AC Tachometer Characteristics

1. The AC tachometer is a two-phase induction device with no slip rings or commutator, and is spark free. The rotating "drag cup" armature is of solid metal and hollow (similar to an empty tin can or cup).

2. The AC tachometer requires excitation on one phase, with a reference voltage V_R of rms amplitude A (volts) and frequency f (hertz)

$$V_R = A\sqrt{2}\sin 2\pi ft$$

Typically, $f = 60$ or 400 Hz and $A = 115$ V rms.

3. The output voltage from an AC two-phase induction tachometer is of contant frequency f equal to the excitation frequency at any shaft speed. Over the useful linear speed range, the output voltage is in phase with the reference voltage for one direction of shaft rotation, and 180° out of phase for the reverse direction of rotation. The deviation from these ideal phase shifts of 0 and 180° is greater for low excitation frequencies and low shaft speeds. The deviation ranges from a few degrees to a few tens of degrees for various devices, over the specified linear-speed range.

4. The useful linear-speed range for AC tachometers having 60-Hz excitation frequency ranges from approximately 500 to 1800 rpm; 1800 rpm is half of the corresponding synchronous speed of 60 cycles/s × 60 s/min = 3600 r/min. The useful linear-speed range for AC tachometers with 400-Hz excitation ranges from 500 to 12,000 rpm but is usually specified to be 500 to 6000 rpm. This provides a smaller phase-shift deviation, and hence increases accuracy.

5. The output voltage of an AC induction tachometer is given by Eq. (37.19), repeated here:

$$V_{\text{out}} = K_2 \cdot V_{\text{in}} \cdot \omega = K_2 \cdot 115\sqrt{2}\sin 2\pi ft \cdot \omega = K_3 \cdot \omega \qquad \text{V rms} \tag{37.19}$$

Some typical values are:

a) $f = 60$ Hz, linear 500 to 1800 rpm

$$K_3 = 6.5 \frac{\text{V}}{1000 \text{ rpm}} \times \frac{60}{2\pi}$$

$$K_3 \cong 0.06 \frac{\text{V}}{\text{rad/s}}$$

b) $f = 400$ Hz, linear 500 to 6000 rpm

$$K_3 = \frac{2.0 \text{ V}}{1000 \text{ rpm}} \times \frac{60}{2\pi}$$

$$K_3 \cong 0.02 \frac{\text{V}}{\text{rad/s}}$$

where ω = shaft velocity, rad/s = rpm × $2\pi/60$
V_{in} = tachometer reference voltage, V rms ($V_{\text{in}} = 115\sqrt{2}\sin 2\pi ft$ is typical)

f = frequency of reference voltage ($f = 60$ or 400 Hz is typical)
K_2 = constant, 1/rad/s
K_3 = constant, V rms/rad/s $\left(K_3 = \frac{V}{\text{rpm}} \times \frac{60}{2\pi} = \frac{V}{\text{rad/s}} \right)$

Thus at 1800 rpm (or approximately 180 rad/s), the voltage output for a typical 60-Hz AC tachometer might be approximately 12 V rms. This is the useful maximum voltage output, which is approximately one tenth the typical excitation voltage input of 115 V rms. A similar voltage is produced by a 400-Hz tachometer at a higher speed.

6. Some typical AC tachometer specifications are:[7]

Variable	Specification
Size	2″ diameter × 2-$\frac{1}{2}$″ long
Excitation power input	6 W
Excitation current	60 mA
Excitation voltage	115 V rms at 60 Hz
Inertia	0.01 in. oz^2
Weight	3/4 lb
Operating temperature	− 55 to + 55°C
Linearity	± 0.5% (500 to 1750 rpm for 60 Hz)
Output impedance	5000 Ω
Output resistance	3000 Ω
Phase shift	1 to 10°
Residual voltage at 0 rpm	0.05 V

7. The hollow metal "drag-cup" armature has thin cuplike walls and contains no windings or squirrel-cage bars, as do standard induction or DC machines. Hence the inertia of the drag-cup armature is low compared with that of the solid iron armature (with windings) of a comparable DC tachometer. Recent advances in DC machines have produced hollow cup-shaped or flat disk-shaped DC windings without rotating iron cores, which enable a DC machine to be constructed with a low-inertia armature. Such construction is often used in DC motors, but not at present in tachometers.

DC Tachometer Characteristics

1. The DC tachometer is a miniature DC generator, usually containing a permanent-magnet field. The rotating armature consists of an iron core wound with many windings terminated on a rotating cylindrical, multisegmented commutator. Each winding is terminated by two commutator bars, usually made of copper. Typically the windings are interconnected in a closed-loop pattern, called a lap winding or a wave winding. The stationary rubbing connections to the rotating copper-segmented commutator are usually constructed of compressed carbon and are called carbon brushes. These carbon brushes can produce a small carbon arc if excessive current flows (generally not the case for a DC tachometer). However, the possibility of a spark is always present, and DC tachometers should not be used in explosive or combustible atmospheres. Brushless DC motors have been developed, which utilize solid-state switching elements to replace the commutator switch and hence cut down on sparking and commutator/brush wear. Such a procedure has recently been applied to DC tachometers.

2. Unlike an AC tachometer, a DC tachometer requires no excitation voltage or power input when constructed with a permanent-magnet field (the typical case). The permanent magnet is usually alnico (aluminum, nickel, cobalt, iron alloy), which can produce a magnetic flux density of up to 1 T = 1 Wb/m^2 = 10,000 gauss. As a point of comparison, the earth's magnetic field is approximately 0.7 gauss. If miniaturization is a necessity, high-strength samarium cobalt plus iron (rare earth magnets) can be used. Samarium cobalt magnets provide flux densities slightly less than those of alnico but can withstand four to ten times greater demagnetizing current, compared with alnico magnets. Thus the actual energy product of $\text{samarium}_2\text{cobalt}_{17}$(240 kJ/m^3) greatly exceeds that for alnico_5(30 kJ/m^3).

3. The output voltage of a DC tachometer is positive for one direction of shaft rotation and negative for the reverse direction of rotation. The DC voltage is subject to a small ripple that is at most a few percent of the DC output, for a typical tachometer. The frequency of the ripple voltage component increases linearly with shaft speed. The DC voltage output is given by Eq. (37.18) (repeated here):

$$V_{out} = K_1\omega \tag{37.18}$$

where ω = shaft speed, rad/s
K_1 = constant, V/rad/s (typical value: $K_1 = 0.06$ V/rad/s)

4. The speed range for linear operation (±3% average quality, ±1/3% good quality) typically varies from several hundred to several thousand revolutions per minute (300 to 3000 or 600 to 4000 rpm; or up to 15,000 rpm and beyond with decreased linearity). There is no fundamental upper limit to the speed range of a DC tachometer, as there is for an AC tachometer. Material problems limit the speed of DC tachometers.

5. Some typical DC tachometer specifications are:[7]

Variable	Specification
Size	2″ diameter × 2-$\frac{1}{2}$″ long
Inertia	0.08 in.-oz^2
Weight	1/2 lb.
Operating temperature	− 55 to +55°C
Linearity	± 0.33% (600–4000 rpm)
Output resistance	100 Ω
Commutator bars	18
Ripple	2.5%
Friction torque	0.5 in. oz
Field type	Alnico permanent magnet

Other Velocity Measurements Using Tachometers

When the velocity to be measured is not a shaft rotational velocity, tachometers are still often used. The velocity to be measured is then converted mechanically to a shaft velocity, which the tachometer measures. Thus linear motion is often converted to a shaft rotation using a rack and pinion gear, a drum and cable, or any of the methods given in Section 37.5, where the conversion of linear to angular measurements was discussed.

Another conversion process involves attaching a turbine, rotor, or propeller to the tachometer shaft. This enables the tachometer to measure fluid velocity or stream velocity. The fluid may be a gas, a liquid, or a particulate mixture. An anemometer (air velocity), a ship's log (water velocity), and a turbine-flow rate meter all operate on this principle.

Nontachometric Velocity Measurements

Doppler Frequency Shift.[8] When an emitted traveling wave is reflected from a moving object, the frequency of the reflected wave is shifted up if the object is moving toward the measurement system and down if the object is moving away from the system. This is known as the Doppler frequency shift. When a measurement system is moving toward a fixed source of traveling waves, the perceived frequency f is shifted up from the source frequency f_0; when the motion is away from the source, the frequency f is shifted down from the source frequency f_0. These relations are expressed by Eq. (37.22a, b) for the Doppler shift for sound waves, radio waves, microwaves, light waves, and radar, with a fixed source and with an observer moving with a velocity V m/s:

Moving toward source

$$f = f_0 \frac{C + V}{C} \tag{37.22a}$$

Moving away from source

$$f = f_0 \frac{C - V}{C} \tag{37.22b}$$

where C = velocity of wave, m/s
$C = 335$ m/s (sound)
$C = 3 \times 10^8$ m/s (EM wave)
f = observed frequency, Hz
f_0 = source frequency, Hz
where V is much less than C.

The Doppler shift for sound waves, radio waves, microwaves, light waves, and radar, with a fixed source and observer, colinear, with reflector moving at a velocity V m/s, is (the reflected beam travels the path twice over, which doubles the shift):

Reflector moving toward source

$$f = f_0 \frac{C + 2V}{C} \tag{37.23a}$$

Reflector moving away from source

$$f = f_0 \frac{C - 2V}{C} \tag{37.23b}$$

where V is much less than C.

A special case applies to sound waves and not to light waves, since the speed of light or other electromagnetic wave (e.g., radar or microwaves) is believed to be independent of the motion of the source or the medium.

The Doppler shift, for sound only with a fixed observer, and a moving sound source with velocity V m/s:

Source moving toward observer

$$f = f_0 \frac{C}{C - V} \tag{37.24a}$$

Source moving away from observer

$$f = f_0 \frac{C}{C + V} \tag{37.24b}$$

where V is much less than C.

Many different velocity-measuring systems utilize the Doppler-shift principle, which is analog in nature. The electronics that produce a velocity signal from the varying frequency shift may be analog or digital in nature. A few Doppler shift velocity-measuring devices are enumerated in the following. Since all of these systems employ velocities much less than the speed of light, the relativistic Doppler equation (the "red shift" equation of astronomy)

$$f = f_0 \frac{1 - V/C}{\sqrt{1 - (V/C)^2}} \tag{37.24c}$$

is never needed.

Note that Eq. (37.24c) reduces to Eq. (37.22b) if $C \gg V$, or equivalently $V/C \to 0$; e.g.:

$$f = f_0 \frac{1 - V/C}{\sqrt{1 - 0}} = f_0 \frac{1 - V/C}{1} = f_0 \frac{C - V}{C}$$

1. **Doppler shift velocity measurement systems**
   (a) Aircraft true speed indicator (radar frequency).
   (b) Police car speed radar (radar frequency).
   (c) Doppler fluid flow (sonic frequency).
2. **Differential pressure velocity measurement.** Fluid flow rates are often determined by measuring the pressure drop across an orifice or Venturi tube placed in the pipe, or by measuring the static and dynamic pressure of the flowing fluid via a Pitot tube. These pressures are measured as "heads," which represent the linear displacement of a differential manometer connected to the

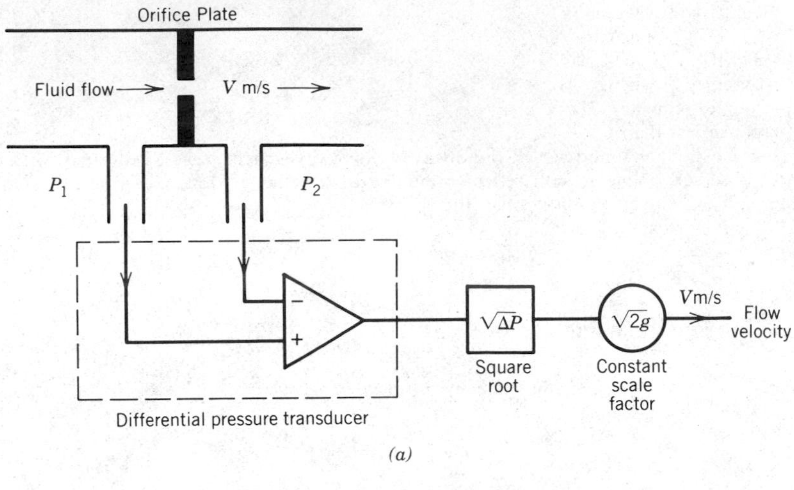

(a)

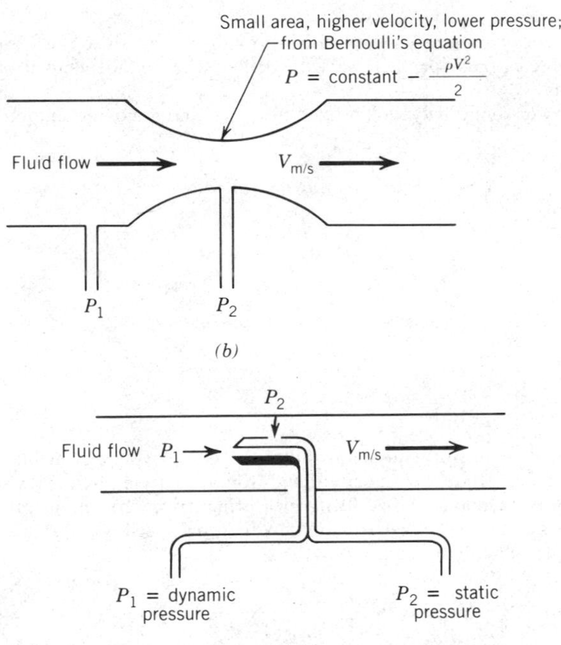

(c)

Fig. 37.25 Three differential pressure velocity measurement approaches. (*a*) Orifice Plate differential pressure sensor for fluid flow velocity measurements. P = pressure head, m; V = flow velocity, m/s; g = acceleration of gravity. (*b*) Venturi tube. ρ = density of fluid, P = static pressure, $V = f(P_1 - P_2)$, f = nonlinear function. (*c*) Pitot tube.[12] For air, $V = V_A$ = ft/min, $V_A = 1096.7\sqrt{h/d}$ ft/min or $V_A = 4004.4\sqrt{h}$, for $d = 0.075$ lb/ft^3 dry air at 70°F, 29.92 in. mercury (Hg). For other temperatures and barometric pressures, $d = 1.325\ P_b/T$ lb/ft^3. P_b = barometric pressure, in. Hg, T = absolute temperature, °R = 459° + temperature °F. $h = P_2 - P_1$ = velocity head or pressure, in. water, and d = density of fluid (air).

two pressure sources. A sketch of these three differential pressure velocity measurement approaches is given in Fig. 37.25. The velocity of the flow is proportional to the square root of the differential pressure ΔP, multiplied by the constant factor of the square root of $2g$, where g is the acceleration of gravity:[10]

$$V_F = \sqrt{2g} \cdot \sqrt{\Delta P} \tag{37.25}$$

where V_F = velocity of flow, m/s
g = acceleration of gravity, 9.9 m/s^2
ΔP = pressure difference head, m

For real fluids with various Reynolds (Re) numbers and densities or viscosities which vary with temperature, pressure, or process, the square root relation of Eq. (37.25) may fail to hold exactly. Calibration and correction are required. Viscosity independence can be improved by introducing a disturbance, such as a short-radius bend, elbow, or tee, which develops turbulent flow. Such correction is required for a sharp-edged orifice plate at low Reynolds numbers ($3000 < \text{Re} < 5000$). Certain flow-restricting orifice designs can be used for Reynolds numbers as low as 300. The Reynolds number is a dimensionless quantity, given by $\text{Re} = \rho Vl/\mu$, where ρ is the fluid density, V is the fluid velocity, l is the length of the flow-disturbing body, and μ is the viscosity of the fluid. Low Reynolds numbers correspond to viscous flow; high values to inviscid flow. For Reynolds numbers below 2300 (approximately), liquid flow is laminar; above 2300, flow is turbulent. For air flow, Reynolds numbers can exceed 1 million.[11]

3. **Sonic transit time velocity measurement.** Two sound beams are transmitted to two receivers, one upstream and the other downstream. The transmitted sound waves are carried along by the fluid flow, resulting in an earlier time of arrival for the downstream receiver. The transit time difference is proportional to the flow velocity.
4. **Hydrodynamic oscillators.** An impediment in a fluid stream, such as fixed blades or a blunt body, can produce vortices that are shed periodically from the impediment. The vortex-shedding rate is related to the fluid flow rate. The discrete vortex swirls are detected by an electronic transducer, which produces pulses proportional to the fluid flow rate.
5. **Magnetic flow rate measurement.** When the flowing fluid is conductive, its velocity can be measured using Faraday's law for induced voltages: voltage = BLV, where B is a uniform magnetic flux density in teslas perpendicular to both L and V, L is the length in meters of the moving conductor (which in this case will be the pipe diameter for pickoffs placed diametrically opposite), and V is the fluid flow velocity in meters per second.[12] For a fixed configuration, B and L are constant and the generated voltage is proportional to the flow velocity V and is independent of viscosity, pressure, temperature, Reynolds number, mach number, or specific gravity. In a real system, B and V will not be uniform across the pipe. The linear relationship between voltage and velocity will still hold, but an additional scale factor will be required. If the velocity profile varies or if turbulence occurs, the linear relation will not hold.
6. **Hot-wire anemometer or hot-thermo sensor velocity transducer.** An element whose electrical properties are a function of temperature is heated by the application of a suitable voltage and placed in a fluid stream moving with velocity V. If V is zero, the temperature is a maximum. As V increases, the temperature drops. The changing temperature causes a change in electrical properties of the heated element, which can be correlated to fluid flow rate. The most common variable electrical property is resistance, which can have a positive temperature coefficient. Therefore resistance increases with temperature, as it does for a hot wire. The resistance can also have a negative temperature coefficient, so that resistance decreases with temperature (as it does for most thermistors).

Most of these velocity-measuring techniques are employed in process control, either to monitor the process or to form part of a closed-loop control system. Volume flow rate, mass flow rate, and total fluid flow can be automatically calculated from the fluid-flow rate with a knowledge of system parameters and dimensions. In the case of total fluid flow, an integration is required.

37.11 ANALOG MEASUREMENT OF SYSTEM ACCELERATION

Acceleration is measured by devices called accelerometers. All common accelerometers are based on an application of Newton's Second Law of Motion:

$$\text{Acceleration} = \frac{\text{force}}{\text{mass}} \quad \text{or} \quad a = \frac{F}{M} \tag{37.26}$$

$$\text{Weight} = W = M \cdot g \tag{37.27}$$

Figure 37.26 illustrates gravity acting on a mass from the viewpoints of different systems of units.

The fundamental design of an accelerometer is based on the use of a mass m which, when accelerated, produces a reaction force $F = Ma$ in accordance with Newton's Second Law. This force is converted to an output measurement of acceleration a. A conceptually simple accelerometer simply restrains the mass with a set of springs, with spring constant K, as shown in Fig. 37.27. The spring-restraining force F is given by $F = K\Delta X$, whereas the equal (but opposite) force accelerating the mass is given by $F = Ma$. Equating these forces yields $Ma = K\Delta X$ or $a = (K/M)\Delta X$. The acceleration input a is thus converted to a linear displacement or movement ΔX which is proportional to a. A simple potentiometer converts this displacement ΔX into a voltage output. The bridge-type potentiometer circuit, shown in Fig. 37.28, produces a zero voltage output for zero acceleration, and a positive (or negative) voltage output for a positive (or negative) voltage output for a positive (or negative) acceleration. The final accelerometer output equation, derived in Fig. 37.27, is[13]

$$V_{\text{out}} = aK_2 \tag{37.28}$$

where V_{out} = voltage or output, V
a = acceleration
K_2 = acceleration constant, V/m/s^2

Practical accelerometers are not built in the fashion illustrated by Fig. 37.27 for the following reasons:

1. The mass is sensitive to accelerations applied in any direction, but the output equation yields correct results only for accelerations applied colinearly with the springs, as illustrated. In particular, the acceleration of gravity would tend to cause the mass to "fall down" into the potentiometer, as shown in Fig. 37.27. To prevent this, the mass would have to be somehow restrained by, perhaps, placing it inside a horizontal tube or floating it in a dense fluid.
2. If a large range of acceleration is to be sensed with good resolution at low accelerations, a large excursion of the mass is required. Real springs are not perfectly linear for large excursions, and they will introduce nonlinearities into the accelerometer output.
3. The mass physically moves the sliding arm of a potentiometer, and hence the mass must first overcome the frictional forces of the potentiometer before any voltage change can take place (or any acceleration can be sensed). This sets a lower limit on the acceleration that can be sensed. In addition, a sliding contact potentiometer eventually wears out.
4. The dynamics of the mass and spring configuration shown in Fig. 37.27 represent an undamped or slightly damped second-order system, which will oscillate at a frequency of $\omega = \sqrt{K/M}$ rad/s whenever a sudden change in acceleration occurs.

To overcome these limitations, the practical accelerometer is usually of the force-balance pendulum type, with a noncontact position transducer used to sense the motion of the pendulous mass. A sketch of a force-balance accelerometer is given in Fig. 37.28. The pendulum may be vertical or horizontal—the concept remains the same. The acceleration input causes the pendulous mass to move relative to its housing. This motion is sensed and converted to an electrical signal, which is then amplified. The amplified signal provides an output current that flows through a coil located in a magnetic field. A magnetic reaction force is generated, which produces a torque that drives the pendulum back to its null position.

The input acceleration determines the amount of force-balance torque required, which is, in turn, proportional to the torquer current. Hence the DC torquer current is proportional to the input acceleration.

The value of the current can be sensed by a small resistance placed in the torquer current paths.

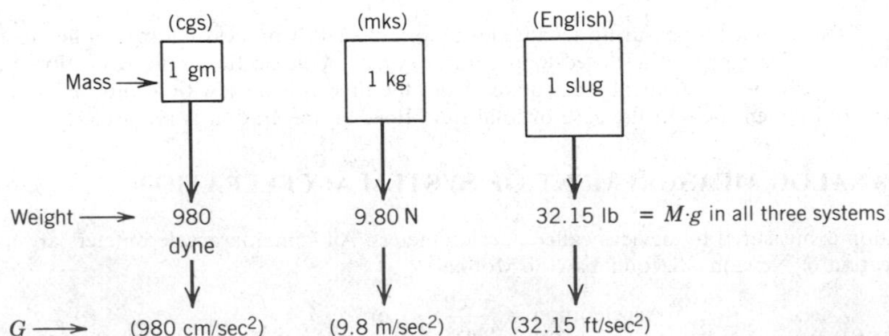

Fig. 37.26 *Gravitational force acting on a mass.* Weight = force exerted on the mass by the acceleration of gravity. a = acceleration, m/s/s; F = force, N (1 N = 0.224 lb); M = mass, k (1 kg = 2.205 lb); G = acceleration of gravity = 9.8 m/s^2 = 980 cm/s^2 = 32.15 ft/s^2; Slug = 32.15 lb = 14.6 kg; Dyne = 2.24×10^{-6} lb = g-cm/s^2; Newton = 0.224809 lb (force) = Kg-m/sec^2; cgs = centimeter-gram-second system; mks = meter-kilogram-second system; English = foot /slug/second system; W = weight, N = mass × Acceleration of gravity; GAL = acceleration unit = 0.01 m/s^2 (980 GAL = 1 g). To easily remember the conversion factor between newtons and pounds, simply remember the story concerning a falling apple prompting Isaac Newton to investigate gravity. One large apple weighs a newton, or about 1/4 to 1/5 pound.

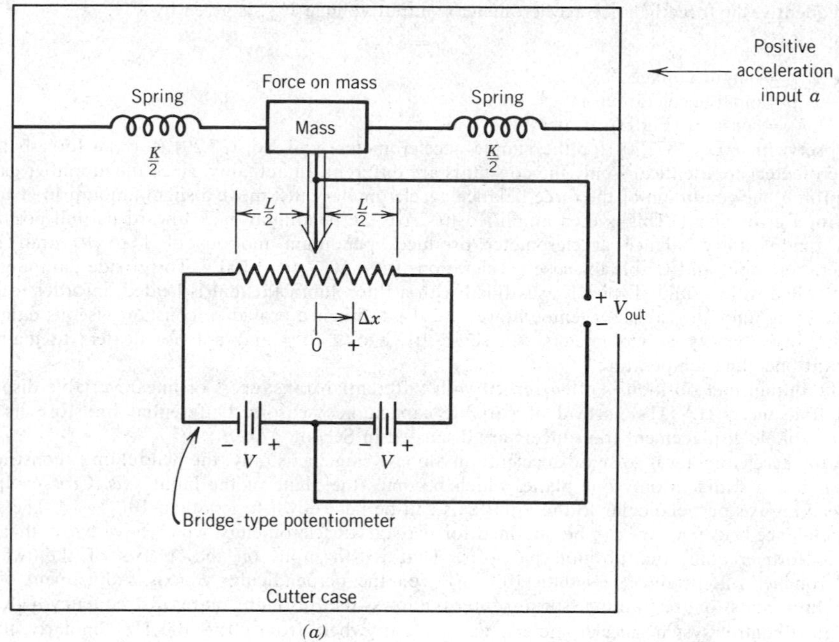

(a)

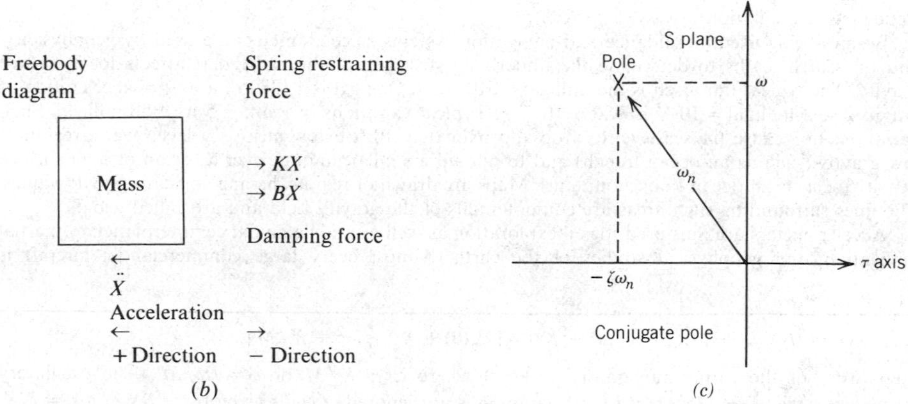

(b) (c)

Fig. 37.27 (*a*) Simple accelerometer. 1. The equivalent spring constant is $(K/2) + (K/2) = K$ and the force exerted on the mass by the springs is $F = K\Delta X$, where ΔX is the change in the spring length, measured from the equilibrium position. $F = Ma$ (force in springs, exerted by mass). $F = K\Delta X$ (force on mass, exerted by springs). Then $Ma = K\Delta X$ and $a = (K/M)\Delta X$. For the bridge-type potentiometric circuit, $V_{out} = 0$ for $\Delta X = 0$ and $V_{out} = \Delta X/(L/2)V$, where L is the total length of the resistance element, V is the battery voltage, and $L/2$ is the length of the positive (or negative) voltage segment of the resistance element. Thus $V_{out} = \Delta X\,[V/(L/2)]$, $\Delta X = a(M/K)$, and $V_{out} = a(M/K)\,[V/(L/2)] = aK_2$, where $K_2 = MV/KL/2$. V_{out} = output voltage, V; a = input acceleration, m/s^2; K_2 = accelerometer constant, V/m/s^2; m = mass, kg; V = battery voltage, V; K = equivalent spring constant, N/m; $L/2$ = half length of resistance element. (*b*) Freebody diagram. If the damping produced by the potentiometer sliding contact is ignored, the oscillatory frequency of the accelerometer in response to an initial displacement $X(0)$ or an initial velocity input $X(0)$ is really evaluated. Steps in solving:

1. Ignore damping. $B = 0$, sum of the forces acting on M.
2. $F = Ma$. $F = -KX$, since F acts in the negative $\ddot{X} = a$ direction.
3. $-KX = M\ddot{X}$, Laplace transform.
4. $-KX(s) = M[s^2X(s) - sX(0) - X(0)]$. Simplify.
5. $MsX(0) + M\dot{X}(0) = KX(s) + Ms^2X(s)$.

Consequently, the force balance accelerometer's output voltage V_{out} is given by

$$V_{out} = aK_3 \tag{37.29}$$

where V_{out} = output voltage, V
a = input acceleration, m/s^2
K_3 = constant (Fig. 37.27, item 4), $V/m/s^2$

Observe that Eq. (37.28) (for the simple accelerometer) and Eq. (37.29) (for the force-balance accelerometer) are identical—only the constants are different. In actuality, since the amplifier gain is not infinite, the pendulum of the force-balance accelerometer must move a slight amount to generate a position error signal. This is then amplified to drive the pendulum back toward its null position.

A quality force-balance accelerometer produces pendulum motions of 1 to 10 μrad/g of acceleration input and is able to sense accelerations from 10^{-7} to ± 100 g. To provide damping, the case is filled with a fluid. Typically it is filled with a fluorolubricant that is heated, in order to keep the damping fluid at a constant temperature (± 1°F) and hence provide a constant viscous damping effect. Lower quality accelerometers will sense 10^{-5} to ± 20 g and not use heaters to maintain constant operating temperature.[13]

The output null position is often sensed with differential transducers or linear variable displacement transducers (LVDTs) instead of variable capacitance sensors. Differential transformers and linear variable displacement transducers are discussed in Section 37.6.

If the accelerometer is to sense acceleration along a single axis only, the pendulum is constructed so that it can swing in only one plane, which becomes the plane of the input axis. Cross-coupling between an axis perpendicular to the input axis can be held down to less than 10^{-5} g/g. The same independence between axes can be obtained for two-axis accelerometers, which have two orthogonal sets of torquer coils and position pick-offs. That is, an input on the A axis of the two-axis accelerometer will produce less than 10^{-5} g/g on the perpendicular B axis. Adjustment of the pendulum mass or restraining spring forces allows the resonant natural frequency ($\sqrt{K/M}$ of typical control system accelerometers) to be set anywhere from 60 to 100 Hz. Similarly, use of various fluorolubes allows the damping constant ξ to be adjusted from 0.7 to 200 times critical damping (e.g., $\xi = 1$ for critical damping; 200ξ is overdamped with real roots and no resonant frequency or oscillation).

Besides being used in guidance and navigation systems, accelerometers are used by geophysicists and oil search teams to determine the underlying structure of the earth that affects local surface gravity. The typical unit used is the milligal = 10^{-3} gal. One gal is 1/980 of a g (g = acceleration of gravity), so 1 milligal = 10^{-3} $g/980 \cong 10^{-6}$ g. Typical variations in g are ± 5 to ± 50 milligal. Such variations in g cause the sea level to drop approximately 10 ft below mean sea level over a region of low gravity (such as an ocean trench) and to pile up a similar amount over a region of high surface density (such as an ocean floor mountain). Maps are drawn of regions having constant gravity values. The lines surrounding such areas are equipotentials of the gravity field and are called geoids.

Accelerometers are thus used for oil exploration as well as missile guidance, interplanetary inertial navigation, and geophysical studies of the earth. Almost every large commercial jet aircraft is

---

6. $X(s) = [MsX(0) + M\dot{X}(0)/(Ms^2 + K)] = [sX(0) + \dot{X}(0)/(s^2 + K/M]$

Step (6) is of the form: numerator$/(s^2 + \omega^2)$ where $\omega^2 = K/M$ or $\omega = \sqrt{K/M}$. The oscillatory frequency is then $\omega = \sqrt{K/M}$ rad/s. If damping is not ignored, $X(s)$ is given by $-BX - KX = M\ddot{X}$, or $X(s) = [(S + B/M)X(0) + \dot{X}(0)/(S^2 + BS + K/M]$ (assuming that damping is proportional to velocity, $\dot{X}$).

The damped oscillatory frequency ω is then given by the imaginary part of the roots of $S^2 + BS + K/M = 0$. If this term is written in standard form:

$$S^2 + BS + K/M = S^2 + 2\zeta\omega_n S + \omega_n^2$$

where ω_n = undamped natural frequency $= \sqrt{K/M}$, rad/s; ζ = damping constant $B/2\sqrt{M/K}$, ω = observed frequency $= \omega_n\sqrt{1 - \zeta^2} = \sqrt{K/M}\,\sqrt{1 - \zeta^2}$, rad/s. Then ω_n, ζ, and ω are related by the Pythagorean theorem as follows:

$$\omega_n^2 = \omega^2 + (\zeta\omega_n^2)$$
$$\omega^2 = \omega_n^2 - (\zeta\omega_n^2) = \omega_n^2(1 - \zeta^2)$$

or

$$\omega = \omega_n\sqrt{1 - \zeta^2} = \text{observed damped frequency of oscillation of the accelerometer}$$

equipped with an automatic pilot that contains accelerometers. All accelerometers are basically a "box" containing a mass m, which is restrained by a force F, which is in turn related to the input acceleration a by Newton's Second Law, $F = Ma$. The accelerometer's electronics produces an output voltage proportional to this force F, so that the output voltage is linearly related to Ma; $V_{out} = K \cdot F = K \cdot Ma$. Since M is constant, the output voltage is linearly related to the acceleration a, or $V_{out} = K_2 \cdot a$ [as originally given by Eq. (37.28)].

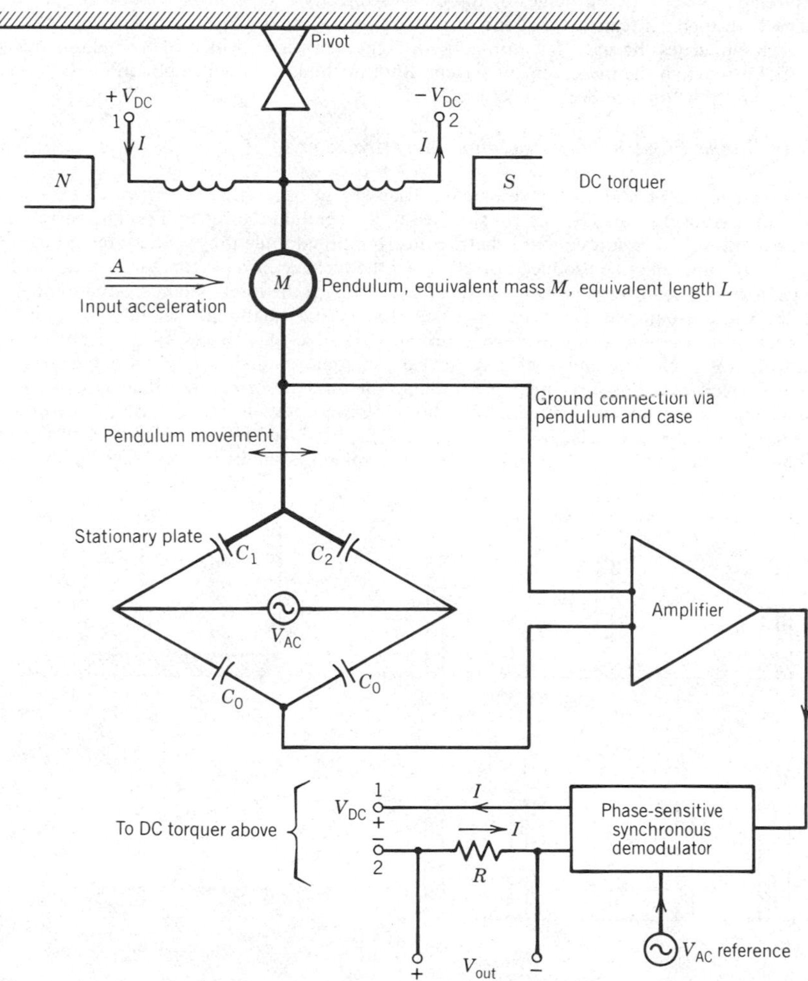

Fig. 37.28 Force-balance pendulous accelerometer with variable capacitance position transducer. V_{AC} = excitation reference voltage, V_{AC}; V_{DC} = DC torquer voltage, produced by phase-sensitive synchronous demodulator; I = torquer current, A_{DC}; $V_{out} = A(K_3)$ = DC output voltage, proportional to acceleration A. C_0 is a fixed capacitance. C_1 and C_2 comprise a noncontact capacitance transducer. If C_1 increases, C_2 decreases, and vice versa. M is the equivalent pendulous mass, including the two torquer coils and the variable capacitance plates. N and S indicate the poles of the torquer's permanent magnet structure.

Operation: Input acceleration A produces a force F on pendulous mass M, $F = MA$, which in turn produces a torque $T_A = LMA$. L is the length from the pivot point to the center of the pendulous mass M. The pendulum moves in response to T_A, which unbalances the capacitance bridge. The unbalanced bridge produces an AC signal that is converted to a DC voltage V_{DC} by the demodulator. V_{DC} is applied to the torquer coils, resulting in a current I and a reaction torque $T = KI$. At null, these two torques are equal and opposite, $|T| = |T_A|$ or $KI = LMA$ or $I = A(LM/K)$. The output voltage V_{out} is developed across resistor R, or $V_{out} = RI = A(RLM/K)$ or simply $V_{out} = A(K_3)$. Thus the output voltage V_{out} is proportional to the input acceleration A.

37.12 DIGITAL MEASUREMENT OF SYSTEM ACCELERATION

Many control systems (e.g., spacecraft and guided missiles) rely on digital control because of the added accuracy possible with digital controllers.[14] As discussed in Section 37.9, there are two general approaches to digital measurements. The first approach is to make an analog measurement and then employ an analog-to-digital converter to obtain a digital output. The second approach is to employ a measurement method that provides digital outputs directly.

The first approach can use existing, highly developed analog measurement devices. The second approach eliminates the need for an analog-to-digital converter, which in turn eliminates a possible source of error from the measurement system. Both methods are used to obtain digital measurements of angle and position (see Section 37.9).

Digital Output From Analog Pendulous Accelerometer

The force-balance pendulous accelerometer, discussed in Section 37.11, provides an analog voltage output proportional to the DC current supplied to the pendulum torquer. This DC current I produces a force that results in a torque that exactly cancels out the torque produced by the input acceleration a. A digital output may be produced by changing the torquer current from a constant direct current I to a pulse train $I(t)$ that has an average value equal to I. If the force-balance accelerometer, shown in Fig. 37.28, is overdamped ($\xi > 1$, or more than 100% critical damping, see item 9 of Fig. 37.27), the pendulum will respond to the average value of a set of torque pulses and will not jump or move erratically for each tiny pulse of torque. For smooth gradual pendulum movement, the pulse repetition frequency should be much greater than the resonant or natural frequency of the pendulum. Under these circumstances, the pendulum cannot respond to each torque pulse. Instead it responds to the average value of the pulse train of torque pulses, produced by the pulse train of current $I(t)$. A sketch of several pulse trains $I(t)$ and their average value I is shown in Fig. 37.29.

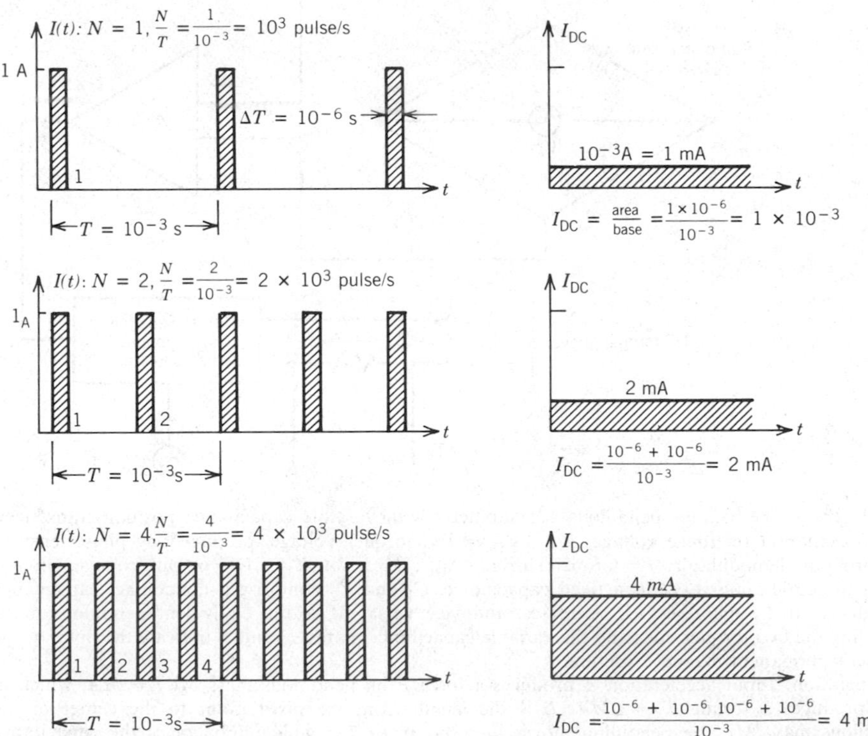

Fig. 37.29 Average value of a constant current pulse train $I(t)$. $I(t)$ = the current pulse train of constant current pulses, of magnitude 1A and pulse width 10^{-6} s. I_{DC} = DC or average value of $I(t)$, or $I_{DC} = 1/T\int_0^T I(t)dt$ = area of pulses divided by averaging period T. For this example $T = 10^{-3}$ s and is constant. ΔT = pulse width = 10^{-6} s = constant. The area of one pulse is 1 A × 10^{-6} s = 1 × 10^{-6} A · s = constant. N = number of pulses per averaging period, N/T = pulses per second = pulse rate.

The expression relating acceleration a to average current I, developed in Fig. 37.28, remains unchanged, except that I is replaced by the average value of $I(t)$. From Fig. 37.28:

$$I = a\frac{LM}{K} \tag{37.30}$$

where L = distance from pivot to pendulum center of mass M
M = mass of pendulum, kg
K = torquer constant, N · m/A
a = input acceleration, m/s^2
I = DC torquer current, A

When I is replaced by a pulse train $I(t)$, the average value of $I(t)$ is equated to the DC current I. Thus

$$I = \text{average } I(t) = \frac{1}{T}\int_0^T I(t)\,dt = \frac{\text{area of pulses}}{\text{base period}} \tag{37.31}$$

where T = 1 or more periods of $I(t)$, s
area = total area under $I(t)$ curve, from $t = 0$ to $t = T$, A · s
$I(t)$ = pulse train, A
I = average value of I, A

From Fig. 37.29 it can be seen that by choosing T to be much longer than the period of $I(t)$, T may be held constant even though the period varies. Whether T is 10 or 100 or 37 periods of $I(t)$, it has no effect on the evaluation of the average value, since the multiplicative factor (10 or 100 or 37) appears in both the numerator and denominator of Eq. (37.31) and cancels out. A possible source of error occurs if T is a fractional number of periods, such as 37.1 or 100.3 periods. Choosing T to be large minimizes this fractional error. That is, 100.3 periods differs from the correct 100 periods by 0.3%.

The area, or numerator, term of Eq. (37.31) is given by

$$\text{Area} = \int_0^T I(t)\,dt = N\Delta T \tag{37.32}$$

where N = number of pulses in averaging period T
T = averaging period $\gg \Delta T$, s
ΔT = constant pulse width, s
$\dfrac{\Delta T}{T}$ = fixed constant $\ll 1$

See Fig. 37.29.

Substituting Eq. (37.32) into Eq. (37.31):

$$I = \frac{\text{area}}{T} = \frac{1}{T}\int_0^T I(t)\,dt = \frac{N\Delta T}{T} = N\frac{\Delta T}{T} \tag{37.33}$$

Thus the average value I of the pulse train $I(t)$ depends on the number of pulses N within the constant averaging period T. As N (the number of pulses per time interval) increases, the average current I increases in direct proportion.

Inserting Eq. (37.33) into Eq. (37.30):

$$N\frac{\Delta T}{T} = I = a\frac{LM}{K}$$

or

$$\frac{N}{T} = a\frac{LM}{K}\frac{1}{\Delta T} \tag{37.34a}$$

where M = mass of pendulum, kg
L = distance from pivot to center of gravity of pendulum, m
K = torquer constant, N · m/A
T = averaging period of pulse train, s
a = input acceleration, m/s^2
N = output pulse count, N pulses per averaging period T

Writing Eq. (37.34a) in a more compact form:

$$\frac{N}{T} = aK_4 \tag{37.34b}$$

where K_4 = accelerometer constant
$K_4 = (LM/K)(1/\Delta T)$
N/T = number of pulses/s = pulse rate

Thus the number of pulses per second is proportional to acceleration, and the sum of the individual pulses is proportional to velocity, as shown in the following:

$$\text{Velocity} = \int_0^{T_f} a\,dt = \int_0^{T_f} \frac{dv}{dt}\,dt = \int_0^{T_f} dv = V \tag{37.35}$$

$$\int_0^{T_f} \frac{N}{T}\,dt = \sum_0^{T_f} (\text{total pulse output}) = N_{\text{total}} \tag{37.36}$$

where T_f = final time, s
N/T = pulse rate, pulses/s
N_{total} = total pulse output, summed up
V = velocity, m/s
a = acceleration, m/s^2
K_4 = accelerometer constant

From Eq. (37.34), integrating

$$\frac{N}{T} = aK_4$$

$$\int_0^{T_f} \frac{N}{T}\,dt = \int_0^{T_f} aK_4\,dt$$

Substituting Eqs. (37.35) and (37.36):

$$N_{\text{total}} = VK_4 \tag{37.37}$$

Thus a digital system can determine velocity by simply summing up the individual pulse outputs of the digitally pulsed torque force-balance accelerometer. Sign must be taken into account, and a negative acceleration must produce a negative N count.

Notice that no integration of acceleration takes place in hardware to determine velocity from acceleration. The effect of integration is produced by summing up pulse counts. Each pulse corresponds to an increment in velocity, with the size of the increment depending on the accelerometer constant K_4.

Direct Digital Accelerometers

The pitch of an electric guitar "string" (it is actually a steel wire) is a function of the tension in the wire. As the tension is increased, the pitch increases, in accordance with[13]

$$f^2 = \frac{T}{4WL^2} \tag{37.38}$$

where T = wire tension, N
W = wire mass per unit length, kg/m
L = wire length, m

The pitch is sensed by a permanent magnet and coil-type variable reluctance pickup, which produces an electrical voltage output of frequency equal to the vibrating frequency of the guitar's steel string. If another coil of wire were placed near the steel string, a pulse of current sent through this coil would exert magnetic force on the steel string and produce the effect of plucking the string electronically. It is thus possible to electrically sense the frequency of vibrations of the steel string. The vibrating steel string accelerometer is constructed on these principles. A mass is suspended midway between two steel wires in tension as shown in Fig. 37.30. At rest the tensions in both wires are equal, $T_2 = T_1$. Under acceleration in the direction shown in Fig. 37.30, T_2 increases and T_1 decreases, producing a change in the vibrational frequency of the two wires f_2 and f_1. The larger the acceleration, the larger the tension change and the frequency change. Analytically, from Fig. 37.30, the input acceleration a acting on the mass M produces a tension change $T_2 - T_1$ given by

$$F = Ma = T_2 - T_1 \tag{37.39}$$

T_2 and T_1 may be related to the frequencies f_2 and f_1, by applying Eq. (37.38) twice, yielding

$$Ma = T_2 - T_1 = 4WL^2f_2^2 - 4WL^2f_1^2 \tag{37.40}$$

Factoring:

$$\frac{Ma}{4WL^2} = f_2^2 - f_1^2 = [(f_2 - f_1)(f_2 + f_1)]$$

Simplifying:

$$\frac{Ma}{4WL^2(f_2 + f_1)} = (f_2 - f_1)$$

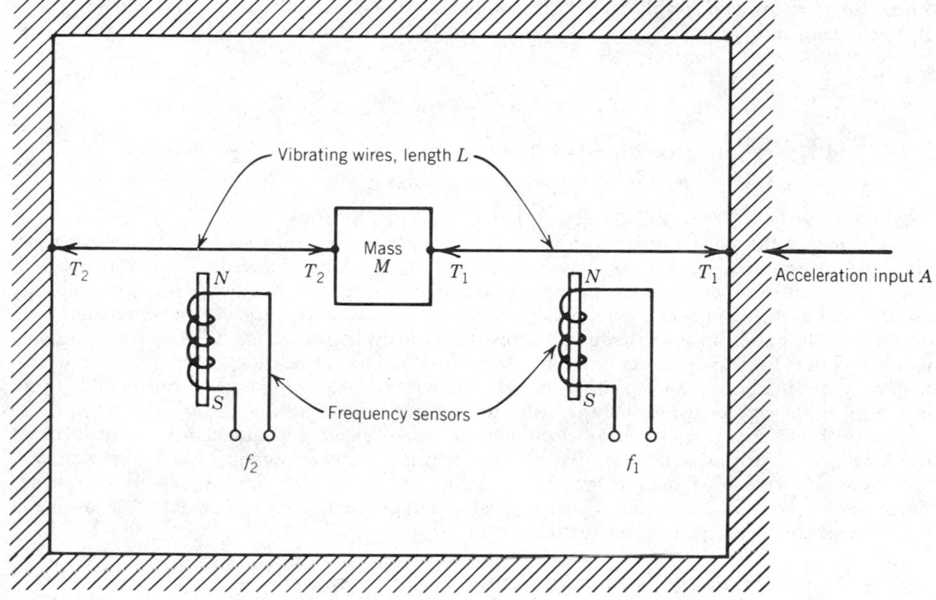

Fig. 37.30 Vibrating steel string accelerometer. Mass M is supported by two vibrating steel wires of length L, with frequencies f_1 and f_2 and wire tensions T_1 and T_2, respectively. The frequencies f_1 and f_2 are measured by variable reluctance frequency sensors, analogous to electric guitar pickups. For increasing acceleration A in the direction shown, T_2 and hence f_2 increase, while T_1 and hence f_1 decrease.

Forcing $f_2 + f_1$ to be a constant:

$$aK_5 = f_2 - f_1 \tag{37.41}$$

where $K_5 = M/[4WL^2(f_2 + f_1)]$ = accelerometer constant
W = wire mass per unit length, kg/m
M = accelerometer test mass, kg
L = wire length, m
f_1, f_2 = transverse vibrational frequency of wires, Hz

The difference between the output frequencies, $f_2 - f_1$, is thus proportional to the input acceleration a, as long as the sum frequency $f_1 + f_2$ is held constant. Normally $f_1^2 + f_2^2$ is constant, as may be seen by considering Eq. (37.38), repeated here:

$$f^2 = \frac{T}{4WL^2} = KT \tag{37.38}$$

where $K = 1/4WL^2$
W = wire mass per unit length, kg/m
T = wire tension, N
L = wire length, m

The frequency f_2 is produced when T is increased by ΔT, and f_1 is produced when T is lowered by ΔT (where ΔT is the change in wire tension produced by the acceleration of the test mass, and T is the constant base tension in both wires).

$$\begin{aligned} f_2^2 &= K(T + \Delta T) \\ f_1^2 &= K(T - \Delta T) \end{aligned} \tag{37.42}$$

Adding:

$$f_2^2 + f_1^2 = 2KT = \text{constant} \tag{37.43}$$

Since $f_2^2 + f_1^2$ is constant for all changes in tension ΔT, $f_2 + f_1$ is not exactly constant. For a 25% change in tension, $f_2 + f_1$ changes by 0.8%. For example, if $T = 4$ and $\Delta T = 1$, then $f_2 = \sqrt{4+1} = \sqrt{5}$ and $f_1 = \sqrt{4-1} = \sqrt{3}$:

$$f_2 + f_1 = \sqrt{5} + \sqrt{3} = 3.968$$

For $\Delta T = 0$, $f_2 = \sqrt{4+0} = 2$ and $f_1 = \sqrt{4-0} = 2$:

$$f_2 + f_1 = 2 + 2 = 4.000$$

and $(4.000 - 3.968)/4.000 = 0.8\%$ change in $f_2 + f_1$ for $\Delta T/T = 1/4 = 25\%$.

To eliminate the small variations in the sum frequency $f_2 + f_1$, such as the 0.8% deviation just discussed, the sum frequency $f_2 + f_1$ may be held constant by means of a feedback control loop with a reference oscillator input. It is therefore possible to obtain a frequency output $f_2 - f_1$ that is directly proportional to the input acceleration a as given by Eq. (37.41). Each cycle can be represented by a digital count or pulse. Since acceleration is proportional to cycles per second, velocity is proportional to the integral of cycles per second or to the sum of the individual cycle count. Each cycle of different frequency corresponds to an increment in velocity, just as each current pulse corresponds to an increment in velocity for the digitally pulsed torqued force-balance accelerometer.

Although the driving and sensing electronics are complicated, and although it is possible that metal fatigue or stretch could shift the natural frequency of the vibrating wires, vibrating wire accelerometers have been used in guidance systems. However, they are not off-the-shelf items. Force-balance accelerometers with digital or analog output are available as off-the-shelf items and are the most common type of accelerometer.

Digital Readout Pendulous Integrating Gyroscopic Accelerometer

A rather complicated accelerometer can be constructed from a gyroscope with an added mass.[13] The mass is added to the gyroscope support cradle, in line with the spin axis, to unbalance the gyro support. The gyro wheel itself remains balanced about its spin axis. A sketch of the gyroscopic pendulous mass accelerometer is given in Fig. 37.31, along with other sketches describing the gyroscope torque equation in terms of the precession rate vector ω and the angular momentum vector $\mathbf{H}$ of the gyro wheel. From Fig. 37.31, an acceleration a at the pendulous mass M results in a force $F = Ma$, which acts on a moment arm of length R to produce a torque $T_A = MaR$. An angular transducer senses any rotation about the torque axis and produces a signal that, when amplified, drives a rotary torquer to produce a precession rate ω. The precession rate ω produces a gyroscopic reaction torque $\mathbf{T}_r = \omega \times \mathbf{H}$ which opposes the original acceleration torque. This closed-loop feedback system reaches a null when the two torques are equal but opposite. Mathematically

$$\mathbf{T}_R = \mathbf{T}_A$$

Using magnitudes only

$$\omega H = MaR$$

or

$$\omega = \frac{MR}{H} a \tag{37.44}$$

where $T_R = \omega \times H$ = gyro reaction torque, N · m
T_A = pendulous mass acceleration torque, N · m
H = gyro angular momentum, $kg \cdot m^2/s$
a = input acceleration, m/s^2
R = pendulous mass moment arm, m
ω = angular velocity about the precession axis, rad/s
V = input linear velocity, m/s

Integrating both sides of Eq. (37.44), noting $\omega = d\theta/dt$, $a = dv/dt$

$$\int \omega\, dt = \frac{MR}{H} \int a\, dt$$

$$\theta = \frac{MR}{H} V \tag{37.45}$$

Thus the angle θ through which the gyroscope precesses is a direct measure of the integral of the input acceleration a. θ is read out by a digital angular transducer and is directly proportioned to the input velocity V. An actual integrator is not required to integrate acceleration to obtain velocity information. This simplicity is obtained at the expense of the complication of a gyroscope.

An excellent background in gyro dynamics is given by Cochin[15] and Pitman.[16]

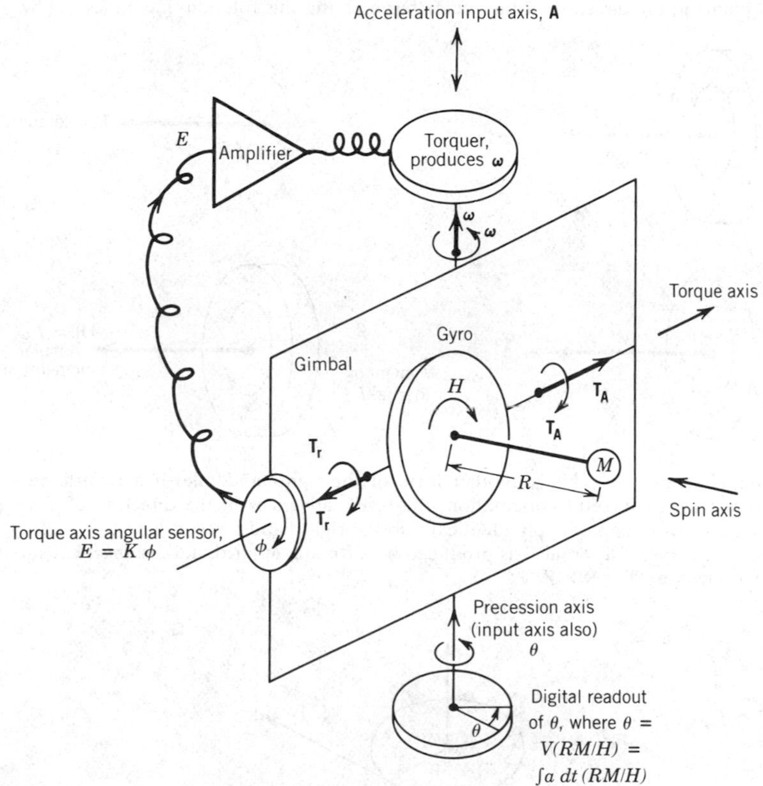

Fig. 37.31 Digital readout pendulous integrating gyroscopic accelerometer. Gyro-restoring torque $= \mathbf{T_r} = \boldsymbol{\omega} \times \mathbf{H}$. Torque due to acceleration input $= \mathbf{T_a} = \mathbf{R} \times \mathbf{F} = RMa$ in magnitude, using $\mathbf{F} = M\mathbf{a}$. At null, $|\mathbf{T_r}| = |\mathbf{T_a}|$ or $|\omega \mathbf{H}| = RMa$, from which $|\omega| = a(RM/H)$. Integrating yields $|\theta| = V(RM/H)$. $|\theta|$ is thus linearly related to velocity input V, or the integral of the acceleration input **a**, as long as R, M, and **H** are constant. R is the radius arm length to M, M is the pendulous mass, and **H** is the angular momentum of the gyro wheel, or $\mathbf{H} = I\omega_{\text{spin}}$, where I is the moment of inertia of the gyro wheel and $\boldsymbol{\omega}_{\text{spin}}$ is the spin rate of the gyro wheel, typically greater than 100,000 rad/s. ω is the precession rate of the gyro, typically zero or a few tenths of a radian per second. $|\theta| = \int \omega dt$ is the angle of rotation about the precession axis. $\mathbf{V} = \int \mathbf{a} dt$ is the velocity component along the input or precession axis and is equal to the integral of the acceleration input **a** along the input axis. Units: $T = \text{N/m}$, $\omega = \text{rad/s}$, $H = \text{kgm}^2/\text{s}$, $R = \text{m}$, $F = \text{N}$, $a = \text{m/s}^2$, $V = \text{m/s}$, $\theta = \text{rad}$, $m = \text{kg}$. Operation: Any rotation ϕ about the torque axis, due to the acceleration of the pendulous mass M, is converted to a voltage $E = K\phi$ by the torque axis angular sensor. This voltage E is amplified and fed to the torquer, which produces the precession rate ω. The precession rate ω produces a gyro reaction torque $\mathbf{T_r}$, which opposes the input torque $\mathbf{T_a}$, and nulls ϕ. Thus the mass M does not fall—instead the whole gyro gimbal frame rotates about the precession axis.

Notes on torque, angular momentum and cross-product vectors.

1. Right-hand rule convention. If the fingers of the right hand curl in the direction of rotation, then the thumb points in the direction of the vector representing this rotation. Examples follow.

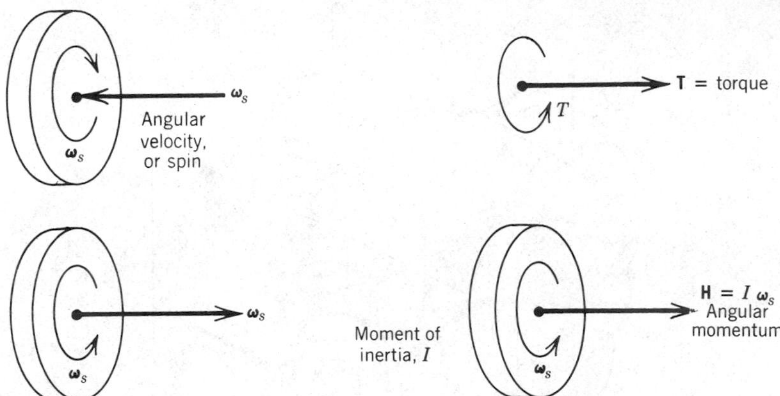

2. Right-hand screw rule. This is another form of the right-hand rule. If a right-hand (ordinary) screw is turned in the direction of rotation, the screw advances in the direction of the equivalent vector. The previous examples apply identically to the right-hand screw rule.

3. Torque expression. If a torque **T** is produced by a force **F** applied at a distance **R** from an axis of rotation, **T** is given by $\mathbf{T} = \mathbf{R} \times \mathbf{F}$

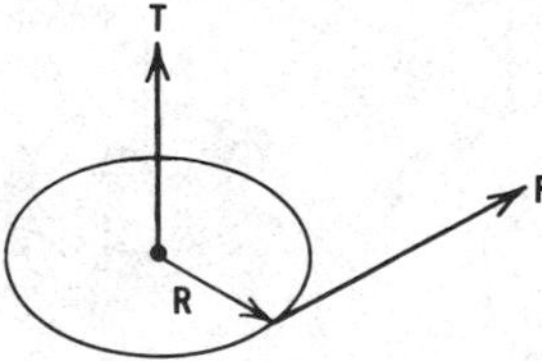

To find the direction of **T**, rotate **R** into **F** and apply the right-hand rule or the right-hand screw rule Also:

$$|\mathbf{T}| = |\mathbf{R} \times \mathbf{F}| = |R| \cdot |F| \cdot \sin \theta_{RF}$$

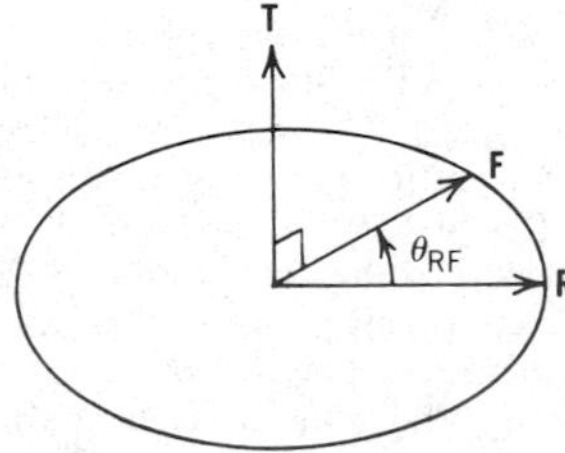

where θ_{RF} is the angle between R and F.

Fig. 37.31 (*Continued*)

4. Gyroscopic reaction torque

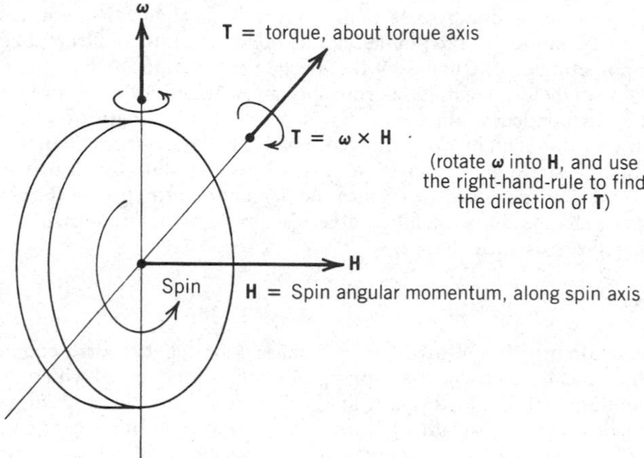

Precession axis, Rotation rate = ω rad/s (all three axes are mutually perpendicular)

If a spinning gyro wheel has a constant angular momentum magnitude $|\mathbf{H}|$, and the wheel is rotated at a rate ω about a perpendicular axis called the precession axis, the torque $\mathbf{T}$ necessary to produce ω is given by

$$\mathbf{T} = \omega \times \mathbf{H}$$

Similarly if a torque $\mathbf{T}$ is applied, the resulting precession rate ω is given by

$$\omega = \frac{\mathbf{H} \times \mathbf{T}}{|H|^2}$$

The previous expression (rather than $\mathbf{T}/\mathbf{H}$) is needed to preserve the vector directions. It is obtained as follows:

$$\mathbf{T} = \omega \times \mathbf{H}$$
$$\mathbf{H} \times \mathbf{T} = \mathbf{H} \times \omega \times \mathbf{H} = \mathbf{W}|\mathbf{H}|^2$$
$$\omega = \frac{\mathbf{H} \times \mathbf{T}}{|\mathbf{H}|^2}$$

(Using the right-hand rule, it can be shown that if ω and $\mathbf{H}$ are perpendicular, then the double cross product with $\mathbf{H}$ reduces to an ordinary multiplication by $|H|^2$.)

If only the magnitude of ω is needed:

$$|\omega| = |\omega| = \frac{|\mathbf{H}|\,|\mathbf{T}|\sin 90°}{|H|^2} = \left|\frac{T}{H}\right|$$

It is a property of gyroscopes that they "fall" or precess at right angles to the applied torque. The precession turns the angular momentum vector toward the torque vector, in accordance with $\omega = (\mathbf{H} \times \mathbf{T})/(|\mathbf{H}|^2)$, or $|\omega| = |\mathbf{T}/\mathbf{H}|$.
As long as only magnitudes are considered:

$$T = \omega H = \dot{\theta} H \quad \text{and} \quad \omega = \dot{\theta} = \frac{T}{H} \qquad (T, H, \omega \text{ mutually perpendicular})$$

Thus a gyroscope converts a torque T into an angular rate $\dot{\theta}$, or it converts an angular rate $\dot{\theta}$ into a torque T. The gyro equations are reversible, and all the variables are at right angles to each other.

Fig. 37.31 (*Continued*)

37.13 MEASUREMENT OF WEIGHT, FORCE, TORQUE, PRESSURE, AND TEMPERATURE

One common method employed to measure weight, force, torque, pressure, or temperature involves conversion of the input variable into a mechanical motion. The mechanical motion then indicates the desired measurement directly or it can be converted into an electrical signal by means of a variable resistance, capacitance, inductance, reluctance, or transformer action. For example, force or weight can stretch a spring inside a spring scale and produce a mechanical measurement by moving a pointer. The mechanical motion of the pointer can be converted into an electrical signal by means of a linear or rotary potentiometer actuated by the mechanical motion. If the spring is made very stiff and constructed of a solid steel bar that deforms only a few thousandths of an inch, the deformation can be measured by a bonded strain gauge,[17] which is either a length of variable-resistance wire thinner than a hair or a thin metallic foil arranged in various flat geometric patterns on a thin flexible planar substrate. Such a strain gauge looks like a small postage stamp with two wires attached. It is glued or bonded to the deforming member. Since the percent deformation of the strain gauge is small, the percent resistance change is also small, and very sensitive high-gain measurement techniques are required to measure the resistance change.

Weight and Force

The strain gauge is almost always used in an electrical bridge network, coupled to a high-gain amplifier. Quite often a calibrated steel bar, spring, or cantilever equipped with a bonded strain gauge is packaged as a unit and called a load cell. A load cell, with the matching strain-gauge amplifier, can be used to electrically measure weight of force. For example, railroad cars can be weighed[18] by placing a length of rail on load cell supports and then simply rolling the railroad cars onto the instrumental rail. A scale to directly indicate to the pilot of an airplane the takeoff weight of the aircraft can be constructed by incorporating a load cell into the landing gear supports of the airplane. Automatic mixing equipment can use load-cell/strain-gauge instrumentation of a mixing hopper to provide closed-loop valve control of various raw material streams. This concept is readily applied to batch processes, such as mixing of cement, mixing of bread dough, or brewing of beer.

One possible mixing process operates as follows: A control sequencer specifies that W_y pounds of material y be placed in the hopper. An electronic or microprocessor controller reads the present weight of the hopper and its contents and adds W_y to this result. The valve controlling the flow of material y is opened by the controller, which continually checks the hopper weight indicated by the strain gauge. When the commanded weight is reached, the controller closes the valve controlling the flow of material y. A block diagram of this type of weight-measuring mixture control system is shown in Fig. 37.32.

Pressure and Torque

Since pressure is force per unit area and torque is force times moment arm length, fundamental techniques for measuring force are equally applicable to the measurement of pressure or torque. A typical pressure-measuring device consists of a diaphragm that is deformed by the applied pressure. The diaphragm may be equipped with a strain gauge whose resistance change is measured, amplified, and then processed to produce an electrical signal proportional to pressure. Specialized integrated circuits have been constructed in one or two hybrid units which contain the diaphragm, the strain gauge, and the strain-gauge amplifier.† For all strain-gauge applications, whether the measured quantity is force, torque, or pressure, the following two assumptions are made:

1. Strain-gauge resistance is a single-valved function of strain. For a limited range of strain, this variation is linear. For larger strain variations, nonlinear corrections are required. Strain is the relative change in length $\Delta L/L$, where L is the length at zero stress and ΔL is the change in length due to the applied stress, which is colinear with L. Strain is dimensionless.

2. Stress, which is force per unit area, is assumed to be linearly related to strain. This assumption is termed Hooke's law, after Robert Hooke who in 1660 discovered by experiment that it is true for real materials over a limited range. The constant of proportionality is termed the modulus of

† National Semiconductor, LX1602G; sensor die on one chip, signal conditioner on a second chip. 15-V supply; 12.5 V = full scale, 2.5 V = 0 psi. This popular device has a full scale of 15 psi. Some are used to measure blood pressure. The average systolic blood pressure ranges between 100 and 150 mm Hg, which is equivalent to 2 to 3 psi. Other devices, LX1600, LX1601, LX1603, LX1470A, etc., cover a wide range of pressures, up to 5000 psi.[19]

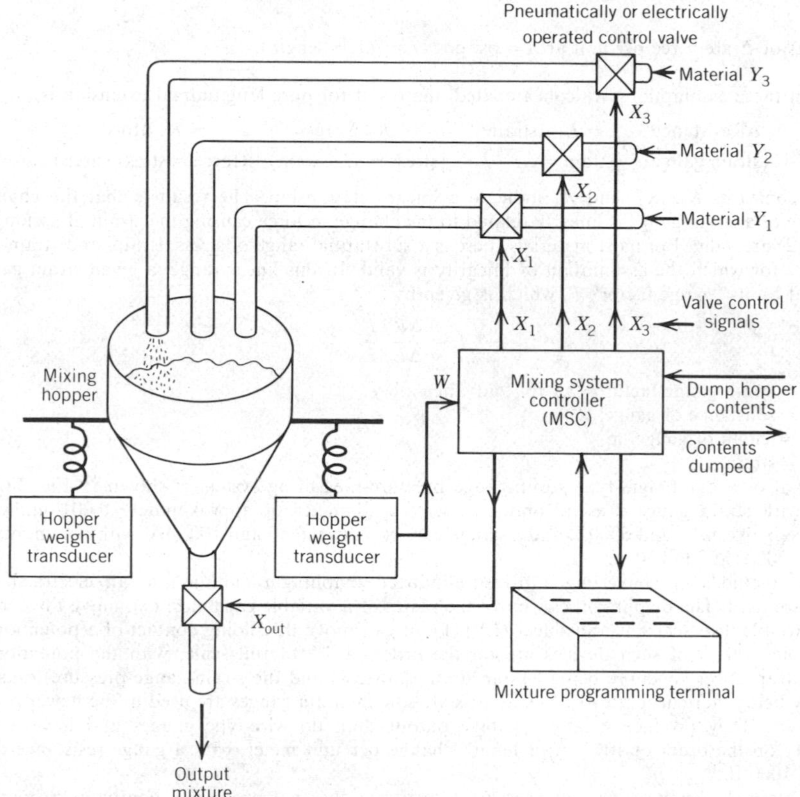

Fig. 37.32 Weight-measuring mixture control system. Y_1, Y_2, Y_3 = raw materials to be batch mixed; X_1, X_2, X_3 = on/off control signals for valves controlling the raw materials; W = signal proportional to weight of hopper plus contents; X_{out} = on/off control signal for hopper output; tare weight = weight of hopper before any of material Y_n is added. W-tare = 0 before Y_n is added. $n = 1, 2, 3$. *Operational Steps*:

1. The desired weights of materials Y_1, Y_2, Y_3 are programmed.
2. The Mixing System Controller (MSC) checks for an empty hopper, sets the tare weight, opens valve Y_1 by enabling X_1, and monitors weight W.
3. When weight W reaches tare plus weight Y_1, the MSC closes valve X_1, resets the tare weight, opens X_2, and monitors W.
4. When W increases by the desired weight of Y_2, the MSC closes X_2, resets the tare weight, opens X_3, and monitors weight W.
5. When weight W increases by the desired weight of Y_3, the MSC closes X_3 and enters a standby mode.
6. The external central controller commands X_{out} to open and dump the hopper contents when the process system is ready for a new batch of input materials. When the hopper weight decreases to its minimum valve (minimum tare weight), the MSC sends a "contents dumped" reply signal back to the central system controller. If the mixing process is not yet completed, the MSC holds the X_{out} command until mixing is completed and then opens the X_{out} valve.

elasticity, or Young's modulus E, which is given by

$$E = \frac{\text{longitudinal stress}}{\text{resultant longitudinal extension}} = \frac{\text{force/area}}{\Delta L/L} = \frac{\text{stress}}{\text{strain}}$$

The units of E are force per unit area = psi or N/m^2 (L = length).

When these assumptions are concatenated, the result for pure longitudinal extension is

$$\underset{\text{(strain gauge)}}{\Delta\text{Resistance}} = \underset{(\text{strain} = \Delta L/L)}{K_1\,\Delta\text{strain}} = \underset{(\text{stress} = E \cdot \text{strain})}{K_2\,\Delta\text{stress}} = \underset{(\text{force} = \text{stress} \cdot \text{area})}{K_3\,\Delta\text{force}}$$

The constants K_1, K_2, and K_3 indicate a linear relationship. The result is that the change in resistance of a strain gauge is linearly related to the change in force causing the strain, if assumptions (1) and (2) are valid. For most materials there is a substantial range of stress, strain, and strain-gauge resistance for which the assumption of linearity is valid. In this linear range, a given strain gauge is described by its "gauge factor" F, which is given by

$$F = \frac{\Delta R/R}{\Delta L/L}$$

where F = strain gauge factor = 2.1 for "advance alloy"
R = resistance of gauge, Ω
L = length of gauge, m
$\Delta L/L$ = strain

A diagram of a diaphragm-type strain-gauge pressure-measuring system is shown in Fig. 37.33. A typical wire strain gauge uses a conductor with a diameter of approximately 0.001 in., with a resistance between 60 and 6000 Ω and a gauge current between 10 and 100 mA, with common values of 100 to 200 Ω at 20 to 30 mA.

Other methods of converting a pressure-induced diaphragm motion into an electrical signal motion are used. The diaphragm can move the plates of a variable capacitor, can move the core of a linear variable displacement transducer (LVDT), or can move the sliding contact of a potentiometer. Overall, accuracies of such devices are on the order of 1% of full scale, with the potentiometric pressure transducer generally being less accurate (2 or 3%) and the strain-gauge pressure transducer generally being more accurate (1/2%). Solid-state silicon strain gauges are used in the newer pressure transducers. They produce a larger voltage output than do wire-type gauges and have a strain sensitivity on the order of 10^{-6} m of length change per unit meter, with a gauge resistance on the order of 1000 Ω.[20]

An extremely common type of pressure transducer does not produce a continous proportional electrical output but rather operates a switch when a preset pressure is reached. Such pressure-actuated switches generally consist of a pressure-sensing diaphragm whose motion is opposed by a spring. The spring force is usually adjustable, which adjusts the switching pressure. When the applied pressure generates a sufficient diaphragm force to overcome the spring force, the diaphragm moves and trips the electrical switch.

Almost every electrically powered air compressor is equipped with such a "pressure-regulating switch" that comprises the entire pressure-control system. The natural mechanical hysteresis of such a switch provides the differential pressure between the higher pressure that switches the compressor off and the lower pressure that switches it on. Other pressure switches, such as those used on domestic steam boilers, have an adjustable differential pressure or hysteresis, provided by another spring and an added lever (or two).

Pressure sensors of this nature which produce on-off control may appear simple, yet an exact analysis of a system utilizing them can be complex. Under certain assumptions, such as limited power availability and a desire for a minimum system-response time, on-off controllers are the optimum choice. Although the growth of microprocessors and electronic sensors has been explosive during the 1980s, the predicted demise of mechanical and pneumatic sensors has not occurred. The principal reasons given for the continued use of mechanical and pneumatic sensors are the following: relatively low cost; relatively high noise immunity; good vibration, shock, and overload immunity; relative simplicity; and proven reliability.

Temperature

Temperature can be measured mechanically, by means of the differential expansion of liquids, solids, or gases. The mechanical motion can move a pointer, trip a switch, or operate an electrical transducer. Temperature can also be measured electrically, by means of an electrical device that is

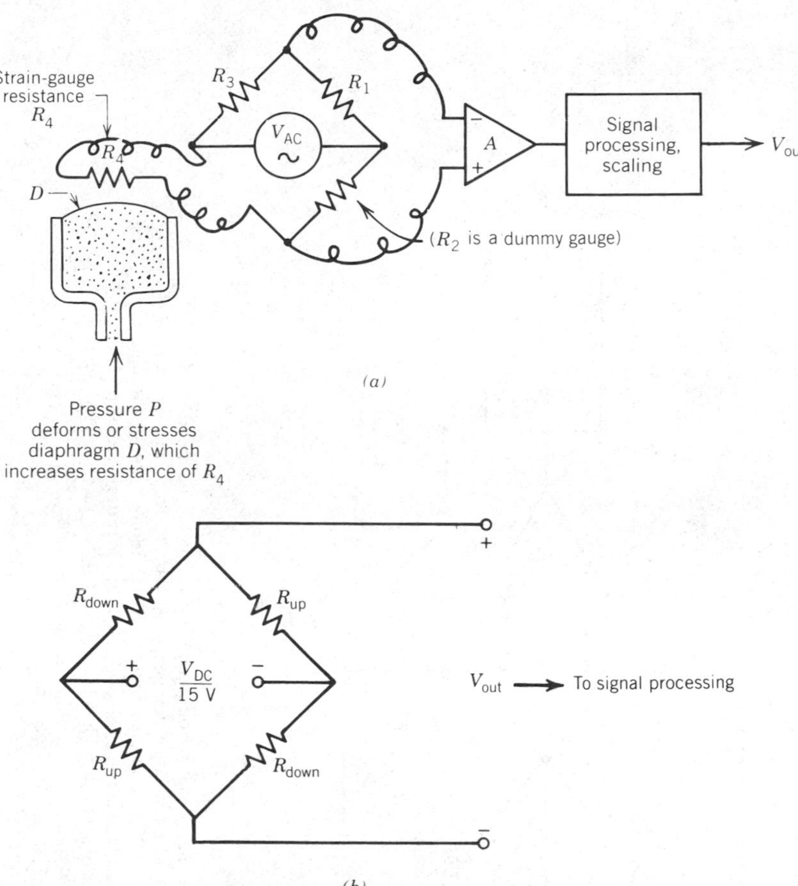

Fig. 37.33 (*a*) Diaphragm-type strain-gauge pressure transducer. Typical accuracy is 1/2% of full scale, with temperature compensation. The strain gauge may be directly bonded to the diaphragm or may be unbonded and connected to the diaphragm via levers. In solid-state units, the diaphragm may be crystalline, with a diffused piezoresistor and with all signal-processing circuitry contained on a single chip. To compensate for temperature-induced changes in the strain gauge R_4, R_2 may be an identical dummy gauge.
(*b*) A single chip solid-state four-strain-gauge bridge. R_4 is replaced by a single chip with four strain-sensitive resistors. R_{up} and R_{down} vary oppositely with increasing strain. This effect is produced by changing element orientation on the chip.

sensitive to temperature changes. Examples of these are thermocouples that generate a voltage; nonmetallic solid semiconductor resistors called thermistors whose resistance decreases nonlinearly with increasing temperature; and wire-wound resistors, typically a small coil of fine platinum wire whose resistance increases with increasing temperature. These wire-wound devices are called RTDs (resistance temperature dependence) or resistance thermometers. The platinum resistance wire element is usually constructed with three leads, two of which are attached to one end of the coil and the third attached to the other end of the coil. A thermistor, which may be the size of a pea, is provided with two leads. A thermocouple, which is the junction of two dissimilar metals, is also provided with two leads. As mentioned, an RTD is usually provided with three leads, although sometimes two or four leads are used. When more than two leads are provided for an RTD, the extra leads are used to compensate for any lead resistance changes, so that only the resistance change of the RTD's temperature-sensing coil affects the temperature determination. A diagram of a three-lead RTD and its associated bridge circuitry is shown in Fig. 37.34. The temperature is often measured by sensing of a change in gas pressure or volume. Many thermostats used in refrigeration control or process control operate on this principle. The gas, or liquid gas mixture, is contained within a sealed reservoir or

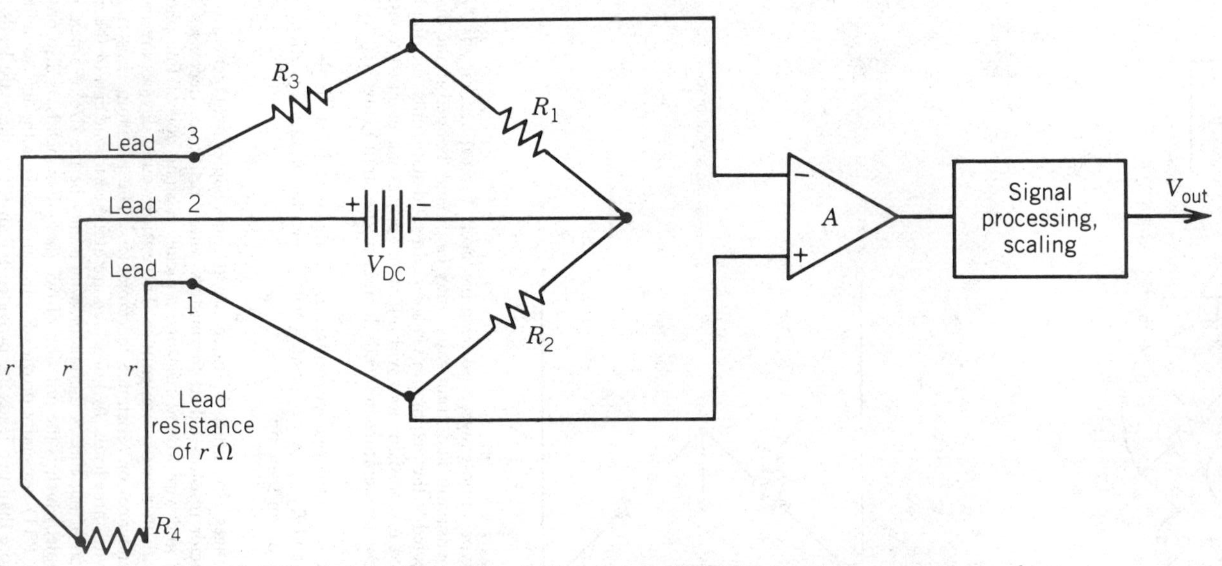

Fig. 37.34 RTD three-lead resistance thermometer bridge (RTD = resistance temperature dependence). $V_{out} = K \cdot$ temperature of R_4. Lead resistance compensation current from the source V_{DC} flows into node 2, through one lead to the RTD, and then either to R_3 via another lead or through R_4 and another lead to point 1. In either case, the current passes through two identical leads, one pair of leads in series with R_3 and one pair in series with R_4. Thus both R_3 and R_4 are increased by identical resistances of value $2r$.

TABLE 37.9. SUMMARY OF TEMPERATURE-MEASURING TRANSDUCERS[20]

Device	Typical Readout Accuracy	Typical Minimum Temperature	Typical Maximum Temperature
Thermistor[a] (solid non-metallic resistor)	High, ±0.25°C over a limited range	−100°C	400°C
RTD[b] (resistance temperature dependence) (coil of wire)	High, often used as a standard; ±0.1°C over a limited range	−264°C	600°C (platinum) (900°C at reduced accuracy) 400°C (type T, Copper-Constantan. Color Code: Blue plus, Red Minus)
Thermocouple[c] (two dissimilar wires)	High, ±0.3°C over a limited range; ±1°C over an extended range	−212°C	1000°C (type K, Chromel-Alumel. Color Code: White plus, Red Minus) 1600°C (type R, platinum, platinum + 13% rhodium Color Code: Black plus, Red Minus)
Vapor pressure (sensing bulb plus mechanical bellows liquid + vapor)	Medium, ±1°C	−30°C	+120°C (sulfur dioxide)
Gas pressure (no liquid, linear scale)	High, used as a calibration standard	−90°C	+450°C (nitrogen)
Bimetallic	Low, ±1% full scale	−40°C	+430°C[20]

[a] A thermistor has a negative temperature coefficient of resistance and displays a nonlinear resistance with a nonconstant coefficient of exponential resistance variation. Over a limited range: $R \cong R_0 e^{b/T}$, where R = resistance, Ω at absolute temperature T °Kelvin; b = constant, but only over a limited temperature range; and R_0 = resistance at reference temperature, Ω.

[b] An RTD has a positive temperature coefficient of resistance (as do all "good conductors"). The resistance increases with temperature in accordance with $R(t) = R_0(1 + at + bt^2 + ct^3 + \cdots)$ where R_0 = calibrated resistance at $t = t_0$, typically 0°C and t = temperature, °C. In many cases only the first coefficient a is used. The coefficient a is referred to as the temperature-resistance coefficient and is 0.00392 for platinum, the most common RTD element. (European RTD calibration curves for platinum use 0.00385 for a.)

[c] A thermocouple generates a small potential by the Seebeck effect (Thomas Seebeck, German scientist, ca. 1821). This potential ranges from about 1 mV at room temperature to 50 mV at 1500°F (816°C) for an iron-constantan-type thermocouple.

bulb, attached by means of a thin tube or capillary to an expansion chamber. The expansion chamber may be a movable diaphragm or a metallic bellows. When the temperature of the bulb increases, the temperature of the gas inside increases and the gas expands, causing the diaphragm or bellows to move. This mechanical motion is utilized to operate an electrical switch, a continuously variable electrical transducer, or a mechanical device such as a fluid control valve. The temperature calibration can be adjusted by variation of a restraining spring force applied to the bellows or diaphragm. A complete temperature control or temperature regulator system can be constructed in this manner, requiring no source of power or other control elements. A common example is the automobile thermostat, used to control the operating temperature of a water-cooled internal combustion engine.

The automobile thermostat consists of a temperature-operated mechanical water valve of fixed temperature range. The valve is directly operated by a sealed expandable bulb, usually containing a fluid suspension of a metallic dust or powder. Older thermostats employ bimetallic springs, which bend or expand with increasing temperature and thus open the water valve. When the water valve is opened, engine cooling water is allowed to circulate through a cooling device or heat exchanger. (This is incorrectly called a radiator. Automotive "radiators" cool primarily by conduction and convection; the actual radiation effect is secondary.)

A summary of temperature-measuring devices is presented in Table 37.9.

37.14 ERROR DETECTION

The error in a feedback control system (e.g., servomechanism, process controller) is defined as the difference between the desired output signal or state and the actual output signal or state. For a single-input, single-output control system with an input reference signal $r(t)$ and an output controlled variable $c(t)$, the error $e(t)$ is given by

$$e(t) = r(t) - c(t) \tag{37.46}$$

In simple cases both $r(t)$ and $c(t)$ correspond to the same type of physical signal (e.g., they are both pneumatic, hydraulic, mechanical, or electrical). For these cases it is then possible to subtract $c(t)$ from $r(t)$ by a suitable pneumatic, hydraulic, mechanical, or electronic device.

Actual physical devices that produce error signals typically consist of:

Input Signal Type	Basis of Operation	Output Signal Type
Pneumatic	Differential diaphragm or bellows	Mechanical
Pneumatic	Above, plus flapper valve	Pneumatic
Hydraulic	Differential cylinders	Mechanical
Hydraulic	Above, plus spool valve	Hydraulic
Mechanical position	Levers	Mechanical position
Mechanical angle	Differential gears	Mechanical angle
Any of the above	Above, plus electrical transducer	Electrical
Electrical	Differential amplifier	Electrical
Thermal	Thermal expansion	Mechanical position or angle
Thermal	Above, plus valve	Pneumatic or hydraulic
Thermal	Above, plus electrical transducer	Electrical
Force	Force balance, plus valve	Pneumatic or hydraulic
Force	Above, plus electrical transducer	Electrical

The actual construction of these listed error-sensing devices is often ingenious, compact, and sometimes complicated by the incorporation of compensating or limiting devices within the error-sensing unit. It is not unusual for the error-sensing unit to be inseparable from the control unit. This occurs, especially, in relatively simple regulators of pressure, temperature, position, or angle. Examples of such devices are thermostats, float valves, and gas-pressure regulators, of either single-stage or dual-stage design.

The vast majority of complex control systems require a conversion of signal type in order to form an error signal. For example, a velocity control system may have an angular output rate delivered by a rotating motor shaft. The shaft angular velocity ω, measured in radians per second, may have to be converted to an electrical signal measured in volts; that is, related to ω by a scale factor K having the units of volts per radian per second. A tachometer is commonly used for such purposes. The error signal is then produced by applying an equivalent reference command signal $r(t)$ in electrical voltage units and subtracting the shaft angular output velocity ω, also expressed in electrical units. The error signal that results is expressed in volts, whereas the physical output is a mechanical shaft's angular velocity.

Many such conversions may exist in a complex control system. Pneumatic, hydraulic, thermal, and mechanical outputs may have to be converted to each other, to electrical signals, and vice versa.

Whenever an error is calculated, however, the two signals being subtracted must be of the same physical type. The resulting error signal may then be converted to a different type for purposes of system control. Whenever complex control strategies are employed, electrical errors are the principal choice because of the relative simplicity of electronic computations compared with pneumatic, hydraulic, or mechanical computations or signal processing.

When a system has multiple inputs and multiple outputs, a state variable approach is often used, where the signals $\mathbf{r}(t)$ and $\mathbf{c}(t)$ become vectors of n and m components, respectively. Their transforms, $\mathbf{R}(s)$ and $\mathbf{C}(s)$, are

$$\mathbf{R}(s) = \begin{bmatrix} R_1(s) \\ R_2(s) \\ \vdots \\ R_n(s) \end{bmatrix} \qquad \mathbf{C}(s) = \begin{bmatrix} C_1(s) \\ C_2(s) \\ \vdots \\ C_m(s) \end{bmatrix} \tag{37.47}$$

A block diagram for such a multidimensional system is shown in Fig. 37.35.

In the case where not all the outputs may be involved in determining the system error ($n \neq m$), the transfer feedback matrix $\mathbf{H}(s)$ operates on the output state vector $\mathbf{C}(s)$, and the error definition is changed to reflect this fact:

$$\mathbf{E}'(s) = \mathbf{R}(s) - \mathbf{H}(s)\mathbf{C}(s) \tag{37.48}$$

where $\mathbf{E}'(s)$ = modified error vector. In the case of an equal number of inputs and outputs ($n = m$), with unity feedback $H(s) = I(s)$ (the identity matrix), Eq. (37.48) reduces to

$$\mathbf{E}(s) = \mathbf{R}(s) - \mathbf{C}(s)$$

where $\mathbf{E}(s)$ = error vector

$$\begin{bmatrix} E_1(s) \\ E_2(s) \\ \vdots \\ E_n(s) \end{bmatrix} = \begin{bmatrix} R_1(s) \\ R_2(s) \\ \vdots \\ R_n(s) \end{bmatrix} - \begin{bmatrix} C_1(s) \\ C_2(s) \\ \vdots \\ C_m(s) \end{bmatrix}, n = m \tag{37.49}$$

In this situation the error is a state vector, with n components.

For a satellite attitude control system, the components of the error vector might be pitch, roll, and yaw angle along with pitch, roll, and yaw rate. Some selected error detection systems are shown in Fig. 37.36. Both Murphy[22] and Cochin[23] contain many examples and drawings of error-sensing devices.

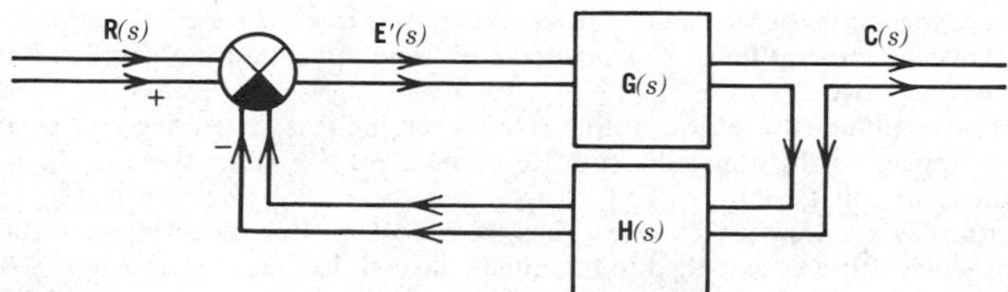

Fig. 37.35 Block diagram of a multiple input, multiple output control system. $\mathbf{G}(s)$ = controlled plant transfer matrix, $\mathbf{H}(s)$ = feedback transfer matrix, $\mathbf{C}(s)$ = output vector, $\mathbf{R}(s)$ = input vector, $\mathbf{E}'(s)$ = modified error vector.

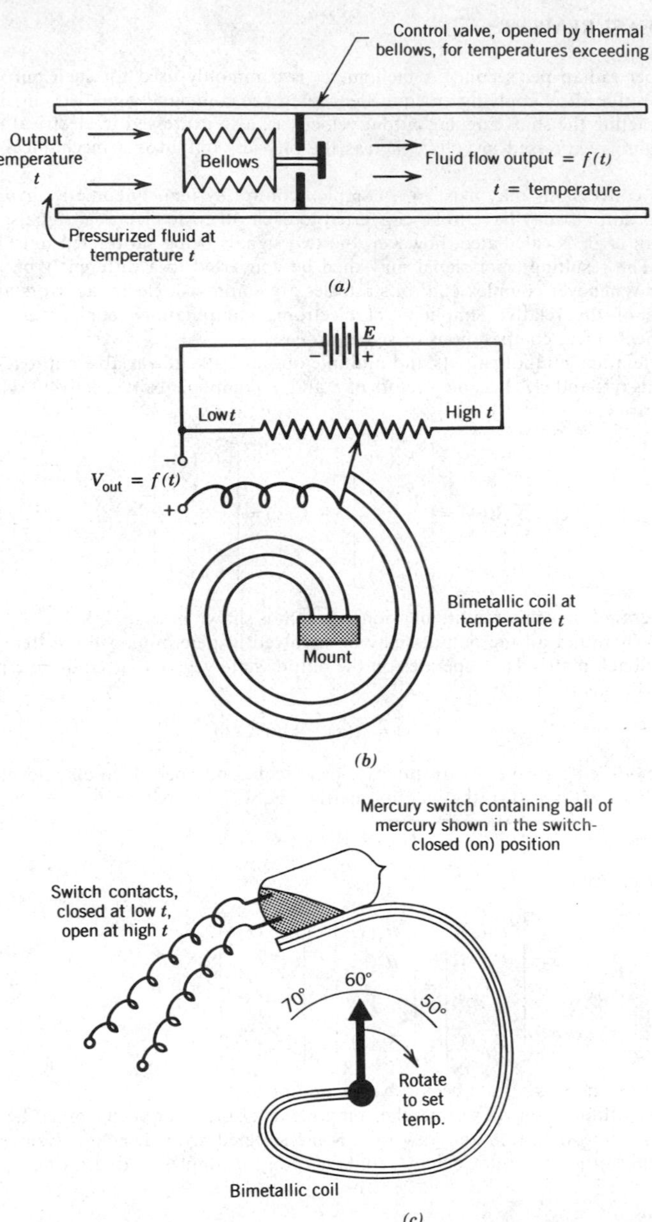

Fig. 37.36 Error detection systems.[20]
(a) Thermal error, pneumatic or hydraulic output (thermostat in car). Reference input r is set by the bellows design. Output temperature C is associated with fluid temperature, t degrees. Resultant error $e = r - c$ operates the valve.
(b) Thermal error, continuous electrical output. Reference input is set by the coil position. Output temperature C is applied to the bimetallic coil. Resultant error $e = r - c$ operates the potentiometer and produces the error voltage V_{out}.
(c) Thermal error, electrical switch closure (house thermostat). Reference input is set by the coil position. Output temperature C is applied to the bimetallic coil. Resultant error $e = r - c$ expands the coil and tips the mercury switch
(d) Electrical error (electrical output and inputs). A = high-gain differential input operational amplification. $V_{out} = R_2/R_1[V_2(1 + R_1/R_2)R_4/(R_3 + R_4) - V_1]$
If $R_1 = R_2$ and $R_3 = R_4$, then $V_{out} = V_2 - V_1$. Reference input r is set equal to V_2. Controlled output C is set equal to V_1. Error e is set equal to V_{out}, resulting in $e = r - c$.

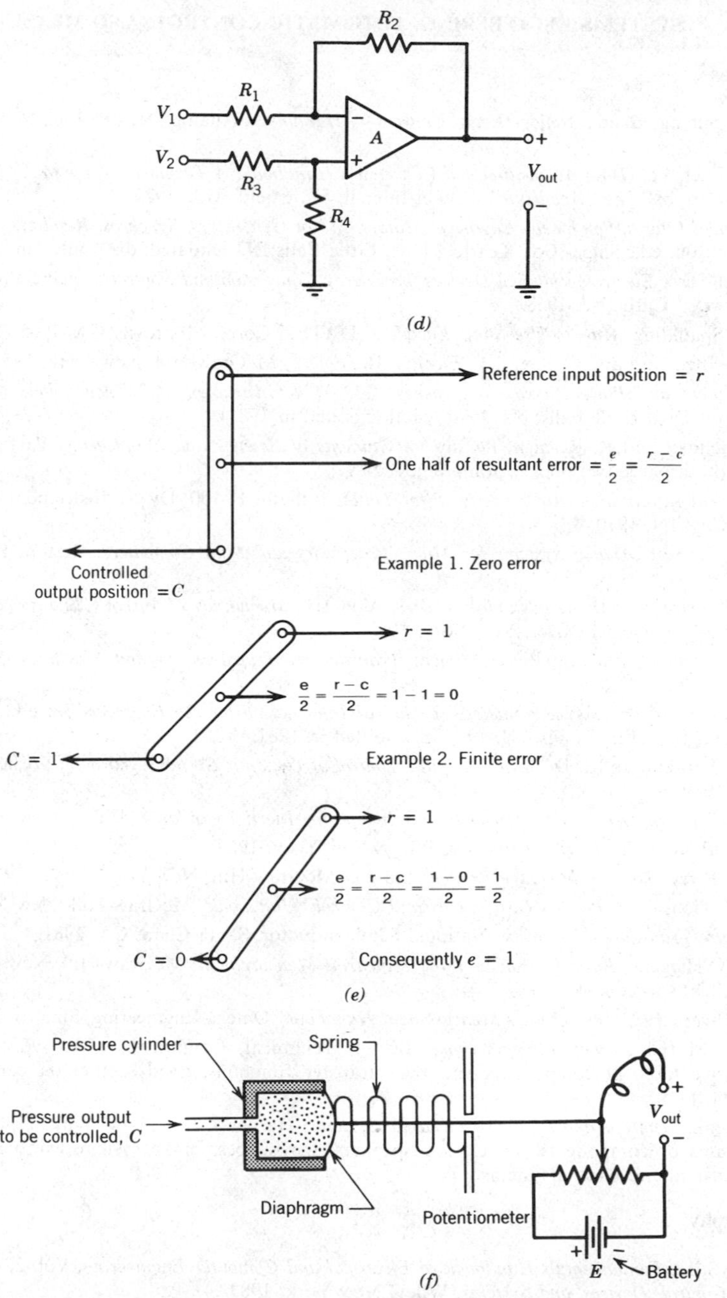

(*e*) Mechanical error/mechanical output.

(*f*) Pneumatic Error/Electrical Output. The reference pressure r is set by the spring force, divided by the area of the diaphragm, pressure = force/area. The output pressure C is applied to the pressure cylinder. The resultant error $e = r - c$ expands (or contracts) the spring-restrained diaphragm, which adjusts a potentiometer producing a voltage output V_{out} proportional to the pressure error e.

Pneumatic output may be produced by replacing the potentiometer with a pneumatic valve. Electrical switched output may be produced by replacing the potentiometer with an electrical switch. Hydraulic devices operate on the same principle. The oil pressure "idiot light" in an automobile is operated by a pressure error sensor of the design pictured (with the potentiometer replaced by a single-contact normally closed switch).

References

1 D. H. Sheingold, ed., *Analog-Digital Conversion Handbook*, Analog Devices, Inc., Norwood, MA, 1972.

2 E. L. Zuch, ed., *Data Acquisition and Conversion Handbook, A Technical Guide to A/D and D/A Converters and Their Applications*, Datel-Intersil, Mansfield, MA, 1979.

3 *Technical Information for the Engineer: Motors, Motor Generators, Synchros, Resolvers, Electronics, Servos*, 10th ed., Singer Co., Kearfott Div., Little Falls, NJ, undated, distributed in 1981.

4 V. Del Toro, *Electromechanical Devices for Energy Conversion and Control Systems*, Prentice-Hall, Englewood Cliffs, NJ, 1968.

5 C. P. Spaulding, *How to Use Shaft Encoders*, DATEX Corp., Monrovia, CA, 1965.

6 A. E. Fitzgerald and C. Kingsley, *Electric Machinery*, McGraw-Hill, New York, 1952.

7 *Instrument and Power Servo Components, G72-0474, Instrument AC Tachometers*, Singer Co., Kearfott Div., Little Falls, NJ, 1974 (catalog issued in 1981).

8 D. Halliday and R. Resnick, *Physics for Students of Science and Engineering*, Part II, 2nd ed., 1962 (both parts, two books). John Wiley, N.Y.

9 *Technical Information, Air Velocity, Pitot Tubes*, Bulletin H-100, Dwyer Instrument Co., Michigan City, IN, 1980.

10 P. H. Garrett, *Analog Systems for Micro Processors and Mini Computers*, Reston, Reston, VA, 1978.

11 W. F. Durand, ed., *Aerodynamic Theory*, Vol. III, *Mechanics of Viscous Fluids, Mechanics of Compressible Fluids*, Dover, New York, 1963.

12 G. R. Slemon, *Magnetoelectric Devices; Transducers, Transformers, and Machines*, Wiley, New York, 1966.

13 *GYROS, Platforms, Accelerometers, Technical Information for the Engineer*, 7th ed., Singer Co., Kearfott Div., Little Falls, NJ, 1967, distributed in 1981.

14 G. F. Franklin and J. D. Powell, *Digital Control of Dynamic Systems*, Addison Wesley, Reading, MA, 1980.

15 I. Cochin, *Analysis and Design of the Gyroscope for Inertial Guidance*, Wiley, New York, 1963.

16 G. R. Pitman, Jr., *Inertial Guidance*, Wiley, New York, 1962.

17 C. C. Perry, *The Strain Gauge Primer*, 2nd ed., McGraw-Hill, New York, 1962.

18 D. M. Considine, *Process Instruments and Controls Handbook*, McGraw-Hill, New York, 1957.

19 *Pressure Transducer Databook*, National Semiconductor, Santa Clara, CA, 1981.

20 P. J. O'Higgens, *Basic Instrumentation, Industrial Measurement*, McGraw-Hill, New York, 1966. (Excellent sourcebook.)

21 *The Omega 1982 Temperature Measurement Handbook*, Omega Engineering, Stamford, CN, 1982.

22 G. J. Murphy, *Control Engineering*, Boston Technical, Cambridge, MA, 1965. (Relatively extensive table of components and their transfer functions, good section on carrier control systems.)

23 I. Cochin, *Analysis and Design of Dynamic Systems*, Harper & Row, New York, 1980. (Excellent diagrams of components, valves, loudspeakers, gyroscopes, brake systems, auto and aircraft suspension, and control systems.)

Bibliography

Chang, S. S. L., *Fundamentals Handbook of Electrical and Computer Engineering*, Vol. 2, *Communication, Control, Devices, and Systems*, Wiley, New York, 1983.

Doebelin, E. O., *Measurement Systems Application and Design*, 3rd ed., McGraw-Hill, New York, 1983. (Excellent sourcebook.)

Grabbe, E. M., S. Ramo, and D. E. Wooldridge, *Handbook of Automation, Computation, and Control*, Vol. 3, *Systems and Components*, Wiley, New York, 1961.

Kuo, B. C., *Automatic Control Systems*, 4th ed., Prentice-Hall, Englewood Cliffs, NJ, 1982. (Good section on sensors and encoders, Section 4.5.)

Ogata, K., *Modern Control Engineering*, Prentice-Hall, Englewood Cliffs, NJ, 1970. (Has excellent treatment of pneumatic controllers, good diagrams; see Sections 5.2, 5.3.)

Shinners, S. M., *Modern Control System Theory & Application*, Addison-Wesley, Reading, MA, 1972. (Has good discussion of hydraulic devices, see Section 3.5. Also in second edition of 1978.)

CHAPTER 38
MODELING AND SIMULATION

RICHARD G. COSTELLO

Cooper Union for Advancement of Science and Art, New York

38.1 ANALOG COMPUTER SIMULATION

The large general-purpose analog computer has almost faded from the simulation scene, having been replaced by the digital computer (beginning in the late 1960s). However, the concepts and techniques of analog simulation are still widely used in many electronic and control system devices.

Basic analog simulation consists of solving a high-order differential equation by the repeated integration of an analog variable that is equated with or related to the actual problem variable. In the electronic analog computer, the analog variable is usually the volt. For example, a problem in heat flow would typically relate British thermal units (Btu's) to volts and Btu's per second to volts per second by use of a scaling constant chosen to match the magnitude of the problem variables to the available voltage of the analog computer. This is termed magnitude scaling.

Scaling

If the simulated problem proceeds too rapidly or too slowly, time scaling of the variables becomes necessary. The necessity for time scaling is indicated by an unscaled differential equation whose coefficients increase or decrease monotonically by an order of magnitude or more with increasing or decreasing order of differentiation. Examples of magnitude scaling and time scaling are illustrated as follows:

Magnitude Scaling Alone (Mechanical Simulation).

Unscaled Differential Equation

$$2000\ddot{X} + 4000\dot{X} + 1000X = 2\sin(3t)$$

where X = position, m
$\dot{X}$ = velocity, m/s
$\ddot{X}$ = acceleration, m/s^2

Use the scale factors:

$$\frac{X_s}{X} = \frac{1\text{ V}}{1000\text{ m}}, \qquad \frac{\dot{X}_s}{\dot{X}} = \frac{1\text{ V}}{1000\text{ m/s}}, \qquad \frac{\ddot{X}_s}{\ddot{X}} = \frac{1\text{ V}}{1000\text{ m/s}^2}$$

that is, the magnitude scale factor = 1/1000; $X_s = X/1000$; $X = 1000X_s$.

Scaled Differential Equation

$$2\ddot{X}_s + 4\dot{X}_s + X_s = 2\sin(3t)$$

where X_s = scaled position, V
$\dot{X}_s$ = scaled velocity, V
$\ddot{X}_s$ = scaled acceleration, V

Time Scaling (with Later Magnitude Scaling).

Unscaled Differential Equation

$$20\ddot{X} + 400\dot{X} + 1000X = 2\sin(3t)$$

where t = time, s
t_s = scaled time, s
n = time scale factor
$\dot{X} = dX/dt$
$\ddot{X} = d^2X/dt^2$

Change from dot notation to derivative notation:

$$20\frac{d}{dt}\frac{dX}{dt} + 400\frac{dX}{dt} + 1000X = 2\sin(3t)$$

Let $t = nt_s$. Then $dt = d(nt_s) = ndt_s$:

$$20\frac{d}{ndt_s}\frac{dX}{ndt_s} + 400\frac{dX}{ndt_s} + 1000X = 2\sin(3nt_s)$$

[The computer operates with scaled time t_s. If n is greater than one, the solution is speeded up

$$20(1/n^2)(d^2X/dt_s^2) + 400(1/n)(dX/dt_s) + 1000X = 2\sin(3nt_s)]$$

Multiply by n^2 and change back to dot notation:

$$20\ddot{X} + 400n\dot{X} + 1000n^2X = n^2 2\sin(3nt_s)$$

Choose n to minimize coefficient variations; let $n = 1/10$.

$$20\ddot{X} + 40\dot{X} + 10X = 0.02\sin(0.3t_s)$$

Divide by 10.

Time-Scaled Differential Equation

$$2\ddot{X} + 4\dot{X} + X = 0.002\sin(0.3t_s)$$

Increasing the input driving function $0.002\sin(0.3t_s)$ by a factor of 1000 is equivalent to magnitude scaling $\ddot{X}$, $\dot{X}$, and X by a factor of 1000; the observed results will be 1000 times too large and hence each result must be scaled down by a factor of 1000 in magnitude, as shown in the following:

Magnitude and Time-Scaled Differential Equation

$$2\ddot{X}_s + 4\dot{X}_s + X_s = 2\sin(0.3t_s)$$

$$\frac{\ddot{X}_s}{\ddot{X}} = \frac{1\text{ V}}{1000\text{ m/s}^2}, \qquad \frac{\dot{X}_s}{\dot{X}} = \frac{1\text{ V}}{1000\text{ m/s}}, \qquad \frac{X_s}{X} = \frac{1\text{ V}}{1000\text{ m}}$$

where desired solution time $t = nt_s = (1/10)t_s$, where t_s is the scaled time frame in which the computer actually operates; that is, $2\sin(0.3t_s)$ can be supplied by a signal generator of frequency $2\pi f = 0.3$ or $f = 0.3/2\pi$ Hz.

To summarize, magnitude scaling requires each coefficient of the homogeneous differential equation to be divided (multiplied) by a scale factor n, whereas time scaling requires the coefficient of the sequential derivative terms to be divided (multiplied) by $1, n, n^2, \ldots, n^K$. The desired goal is a scaled differential equation that has all its coefficients bounded by the range 0.1 to 10.0. The practical limits for a typical analog simulation with 1% accuracy are coefficients bounded by the range 0.01 to 100.0. Although gains or coefficients of 1000 or 10,000 are readily produced by cascading gains of 10, the percent error of the resultant simulation will generally exceed 1% by an unacceptable amount.

Analog Computer Components

The principal analog computer components are:

1. **Inverting integrator/summers.** These have typical fixed gains of 1, 2, 5, and 10 and provision for application of an initial condition (which is often inverted). One to five inputs are typically provided.
2. **Inverting amplifier/summers.** These also have typical fixed gains of 1, 2, 5, and 10. One to five inputs are usually provided.
3. **Noninverting attenuators (potentiometers).** These can provide gains or coefficients of 1.00 to 0. Usually gains of less than 0.1 are not used because of resolution and noise problems. A pot is used to multiply a function of time by a constant between zero and one.
4. **Electronic multipliers.** These allow multiplication of two functions of time, such as a sine wave times a cosine wave. If both functions can change sign, a four-quadrant multiplier is needed. If only one function changes sign, a two-quadrant multiplier is needed. If neither function changes sign, a single-quadrant multiplier is needed. Essentially, two single-quadrant multipliers can be combined into one two-quadrant multiplier, and two two-quadrant multipliers can be combined into one four-quadrant multiplier.

The operational equations describing the four principal analog computer components are given in Table 38.1.

Analog Integrator/Summer. An analog integrator is constructed from a high-gain operational amplifier by the addition of capacitive feedback. If long-term DC stability is required (zero output for zero input), a chopper-stabilized operational amplifier (op amp) is used. A chopper-stabilized op amp consists of separate parallel AC and DC amplifiers. The AC amplifier operates continuously, but the DC amplifier's input is periodically grounded by a switching circuit (chopper) and its output is automatically zeroed by suitable offset-bias compensating circuitry. This occurs hundreds or thousands of times each second.

The basic performance of an electronic analog integrator and electronic analog summer are developed in Figs. 38.1 and 38.2, respectively, on the basis of the following two assumptions concerning the operations of an op amp:

1. The differential input voltage to an op amp is zero. (If the maximum DC output is 10 V and the open-circuit gain of the amplifier is 10^6, a reasonable value, the maximum input voltage is $10/10^6 = 10^{-5}$ V, which is effectively zero.)

2. The input impedance of the op amp is infinite. (A field effect transistor may have an input impedance greater than 10^{10} to 10^{14} Ω.)

Quarter Square Multiplier. The single-quadrant quarter square multiplier, or analog multiplier, is based on the following mathematical development:

$$(X + Y)^2 = X^2 + 2XY + Y^2$$

$$-(X - Y)^2 = -X^2 + 2XY - Y^2$$

$$(X + Y)^2 - (X - Y)^2 = 0 + 4XY + 0$$

or

$$XY = \tfrac{1}{4}[(X + Y)^2 - (X - Y)^2]$$

This form is advantageous, because the operation of squaring can be fairly well approximated by the voltage-current characteristic of a forward-biased diode, as shown in Fig. 38.3.

The operation of multiplying a signal X by a signal Y is thus performed by adding X to Y, squaring the resultant sum, subtracting Y from X, and squaring the resultant difference. These two intermediate results are then subtracted and the final result is attenuated by one quarter. All of these operations are readily carried out by operational amplifiers, and a complete quarter square multiplier is available as a single integrated-circuit package.

A greatly simplified form of a squaring circuit is shown in Fig. 38.4, with bias circuits and attenuators omitted.

TABLE 38.1. ANALOG COMPUTER COMPONENTS

Component	General Device Symbol	Output Relation
Integrator	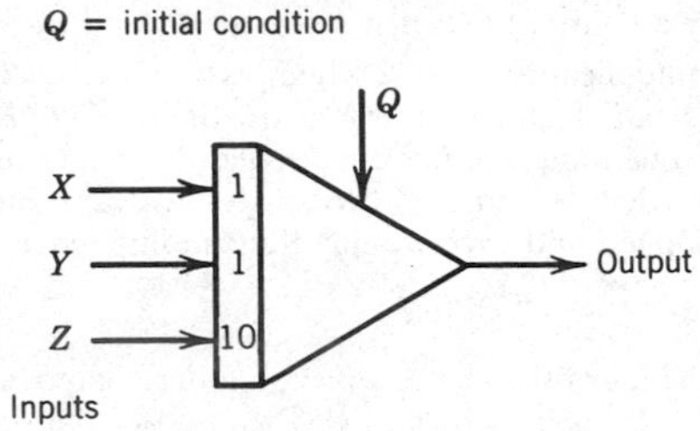 Q = initial condition; inputs X (1), Y (1), Z (10); Inputs; Output	X, Y, Z = time functions, inputs Q = constant $-\int_0^t (X + Y + 10Z)\,dt - Q$
Summer	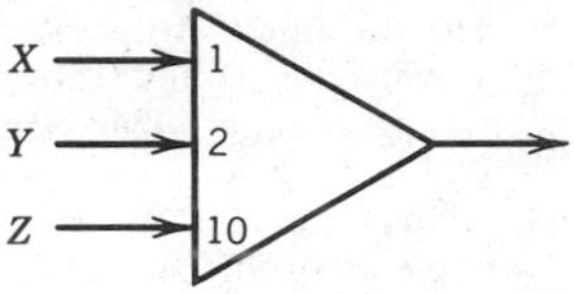 inputs X (1), Y (2), Z (10)	$-(X + 2Y + 10Z)$
Multiplier	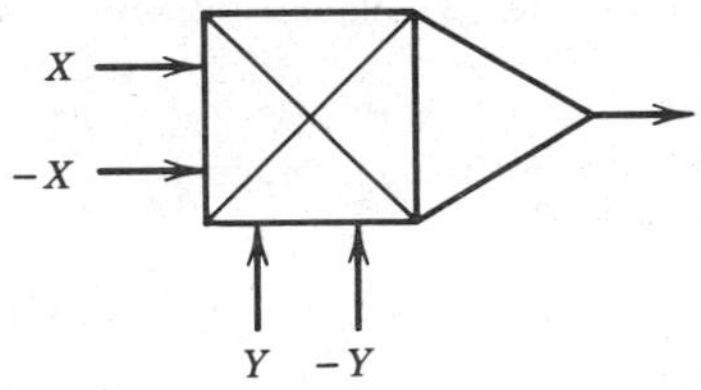 inputs X, $-X$, Y, $-Y$	$+\dfrac{(X)(Y)}{V}$ $V = 10$ for a ten volt multiplier
Coefficient pot	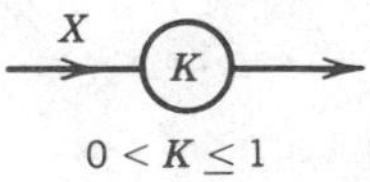 X → K → $0 < K \leq 1$	$+(KX)$ $0 < K \leqslant 1$

$F(s)/f(t)$ Notation	Dot/Derivative Notation
$F(s)$, $f(t)$, 1, $\frac{-1}{s}F(s)$, $-\int_0^t f(t)dt$	$\dot{X}$, $\frac{dx}{dt}$, 1, $-X$, $-X$
$F(s)$, $f(t)$, 1, $-F(s)$, $-f(t)$	$\dot{X}$, $\frac{dx}{dt}$, 10, $-10\dot{X}$, $-10\frac{dx}{dt}$
$F(s)$, $f(t)$, $G(s)$, $g(t)$, $[F(s)G(s)]/V$, $[f(t)g(t)]/V$	$\dot{X}$, $\ddot{Y}$, $(\dot{X} \cdot \ddot{Y})/V$, V = scale constant
$F(s)$, $f(t)$, K, $KF(s)$, $kf(t)$, $0 < K \leq 1$	$\dot{X}$, $\frac{dx}{dt}$, K, $K\dot{X}$, $K\frac{dx}{dt}$, $0 < K \leq 1$

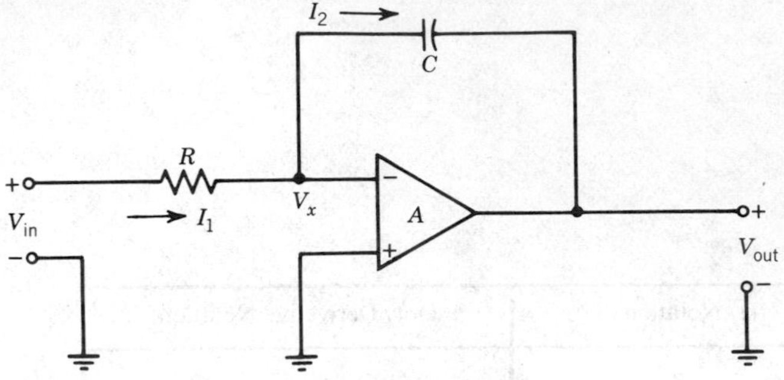

Fig. 38.1 Electronic analog integrator circuit. From assumption 1 (see text), $V_x = 0$. From assumption 2, $I_1 = I_2$. Solving for I_1 and I_2 in the Laplace domain:

$$I_1 = \frac{V_{\text{in}} - V_x}{R} = \frac{V_{\text{in}}}{R}$$

$$I_2 = \frac{V_x - V_{\text{out}}}{1/SC} = -V_{\text{out}}SC$$

$$I_1 = I_2, \frac{V_{\text{in}}}{R} = -V_{\text{out}}SC$$

$$V_{\text{out}} = \frac{-1}{RC}\left(\frac{V_{\text{in}}}{S}\right)$$

Transforming from the Laplace domain to the time domain, using the Laplace transform pair:

$$X(s) = \frac{Y(s)}{s}, \qquad X(t) = \int_0^t Y(t)\,dt \text{ (with zero initial conditions)}$$

and identifying $X(s)$ with V_{out} and $Y(s)$ with V_{in}, therefore $V_{\text{out}}(t) = (-1/RC)\int_0^t V_{\text{in}}(t)\,dt$.

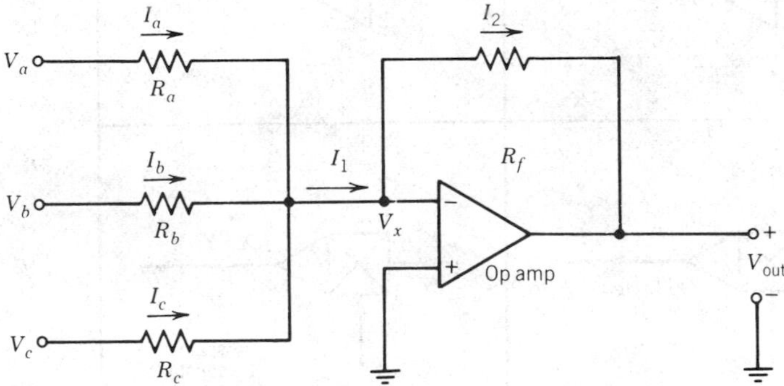

Fig. 38.2 Electronic analog summation. From assumption 1 (see text), $V_x = 0$. From assumption 2, $I_1 = I_2$.

$$I_1 = I_a + I_b + I_c = \frac{V_a - V_x}{R_a} + \frac{V_b - V_x}{R_b} + \frac{V_c - V_x}{R_c}$$

$$I_1 = \frac{V_a}{R_a} + \frac{V_b}{R_b} + \frac{V_c}{R_c}$$

$$I_2 = \frac{V_x - V_{\text{out}}}{R_f} = \frac{-V_{\text{out}}}{R_f}$$

$$I_1 = I_2, \qquad \frac{V_a}{R_a} + \frac{V_b}{R_b} + \frac{V_c}{R_c} = \frac{-V_{\text{out}}}{R_f}$$

$$V_{\text{out}} = -\left(V_a\frac{R_f}{R_a} + V_b\frac{R_b}{R_b} + V_c\frac{R_f}{R_c}\right)$$

If $R_a = R_b = R_c = R_f$, then $V_{\text{out}} = \sum V_{\text{in}} = -(V_a + V_b + V_c)$.

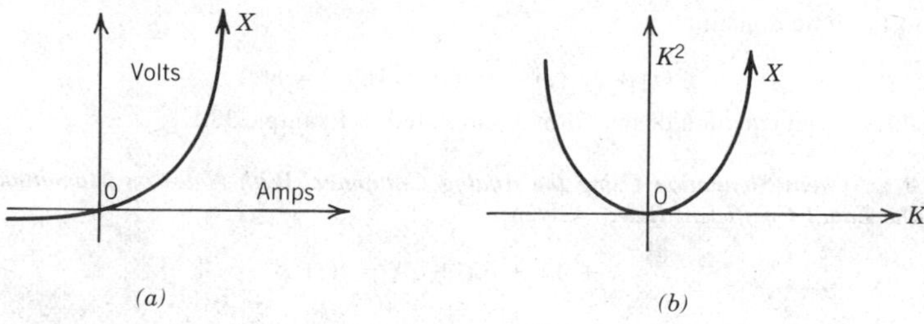

Fig. 38.3 Diode voltage-current characteristic (*a*) and plot of K^2 v. K (*b*). Note that the curves are almost identical from 0 to X.

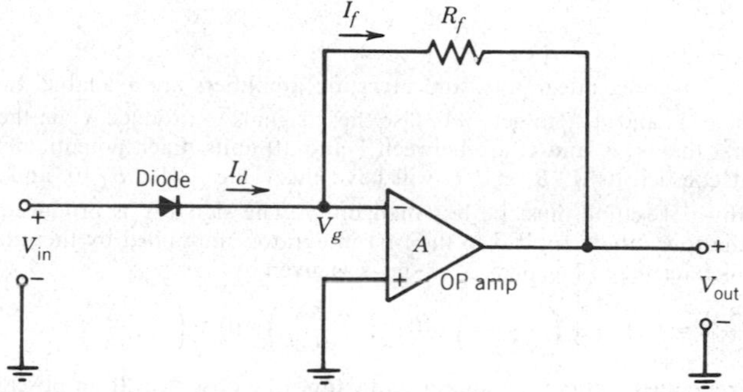

Fig. 38.4. Diode operational amplifier squaring circuit. V_g = virtual ground $\cong 0$ V, since $V_g = |V_{out}|/A = 10_{max}/10^5 = 10^{-4}$, for typical op amps. $I_d \cong I_f$ because of the very high input impedance of the op amp (10^{10}–10^{14} Ω for field effect transistors. Over a certain range, say 0.1 to 1.0 V, $I_d \cong V_{in}^2$ = diode characteristic. $V_0 = -I_f R_f = -I_d R_f = -V_{in}^2 R_f$. $V_0 \cong -V_{in}^2 \cdot R_f$ $(0.1 \leqslant V_{in} \leqslant 1.0$ V). In practice, many diodes comprising a diode function generator are used to obtain an accurate square law response.

Other Analog Computer Elements. A large variety of specialized linear and nonlinear analog computer elements exist. These have been adapted to other uses, which include:

1. **Adjustable limiters or clamping circuits.** These nonlinear devices exhibit a constant gain until the output reaches the clamping level. At this point, the output is held constant for all larger input signals. Basically a limiter consists of a resistor, a diode, and a voltage source for the limiting level.
2. **Diode function generators.** These devices use bias circuits, adjustable resistors, and diodes to produce a multisegment straight-line approximation to a nonlinear function [e.g., sin(X) or log(X)]. In other contexts they are called shaping circuits and can, for example, change a triangular waveform into a sinusoid. This approach is used in many contemporary "function generators," which basically are square wave generators followed by an analog integrator which produces triangular waves followed by a diode function generator which rounds off the top of the triangles into sinusoids.
3. **Comparators and electronic switches.** A comparator compares two input signals X and Y. It produces one output for $X > Y$ and a different output for $X < Y$. The different comparator outputs may be used as digital control signals or they may (internally or externally) control an analog electronic switch. For example, the electronic switch may connect its output to either one of two inputs A and B applied to the analog switch. Either A or B will be selected depending on whether X is greater than or less than Y. Thus four analog signals are inputs: X and Y, which are compared, and A and B, which are switched. The single analog output signal is switched from A to B as the inputs X and Y vary.

For illustrative purposes, consider the transfer function

$$\frac{C(s)}{R(s)} = \frac{1}{s^3 + 2s^2 + 3s + 4}$$

Cross multiplying:

$$C(s)(s^3 + 2s^2 + 3s + 4) = R(s)$$

Inverting to the time domain:

$$\dddot{C}(t) + 2\ddot{C}(t) + 3\dot{C}(t) + 4C(t) = r(t)$$

A general differential equation of this form is simulated in Example 38.1.

Example 38.1. System Simulation Using the Analog Computer, With Notes on Magnitude and Time Scaling and the Equal Coefficient Rule. Given:

$$\dddot{X} + A\ddot{X} + B\dot{X} + CX = f(t)$$

where

$$\dddot{X} = \frac{d^3X(t)}{dt^3}, \qquad \ddot{X} = \frac{d^2X(t)}{dt^2}, \qquad \dot{X} = \frac{dX(t)}{dt}$$

1. Solve for $\dddot{X}$: $\dddot{X} = f(t) - A\ddot{X} - B\dot{X} - CX$.
2. Assume that $\dddot{X}$, inverting integrators, and inverting amplifiers are available. Integrate $\dddot{X}$ to get $-\ddot{X}$, $-\ddot{X}$ to get $+\dot{X}$, and $+\dot{X}$ to get $-X$. Use these signals to produce $\dddot{X}$ via the relation of step 1. Assume at first that A, B, and C are between 1 and 10 units. Each potentiometer (pot) used to set the constant coefficients A, B, and C will have the value $A/10$, $B/10$, and $C/10$, since the magnitude of any pot setting must be less than unity. The signal $\dddot{X}$ is produced by the internal summation of the four inputs applied to the first integrator, multiplied by the integrator gains for each signal. Thus referring to Fig. 38.5, $\dddot{X}$ of step 1 is given by

$$\dddot{X} = f(t)\cdot 1 + \left(\frac{-CX}{10}\right)\cdot 10 + \left(\frac{-B\dot{X}}{10}\right)\cdot 10 + \left(\frac{-A\ddot{X}}{10}\right)\cdot 10$$

Note that $\dddot{X}$ is generated within integrator 1 and cannot be observed. If an observation of $\dddot{X}$ were desired, an additional summation amplifier would have to be added (preceding integrator 1), and an additional inverter would be needed to counteract the sign change introduced by the added summation amplifier.

Note 1. Every loop has an odd number of sign inversions (integrators + summers + amplifiers) for a bounded response. There are $1, 3, 5, \ldots, 2n-1$ active elements in any loop.

Note 2. If the input $f(t)$ is doubled (with zero initial conditions), the output $X(t)$ is doubled. Thus if the sign of $f(t)$ is changed to $-f(t)$, the need for the last inverter in Fig. 38.5 can be eliminated, since for linear systems an input of $K\cdot f(t)$ produces an output of $K\cdot X(t)$.

Note 3. For the determination of system response, only the loop gains are important. The individual gains making up the loop gains can be traded off at will, as long as the total loop gain is kept constant. Changing the forward path gains does not affect the shape or frequency response, it merely changes the magnitude scale—that is, instead of calculating $X(t)$ you may calculate $KX(t)$; K = extra added forward-path gain (with a compensating $1/K$ gain added elsewhere in the loop).

Note 4. The procedure of Note 3 is the basis of magnitude scaling. All the coefficients of an equation are divided by a single constant factor, to achieve a scaled equation with coefficients ranging from 0.1 to 10.0, if possible.

Note 5. The "equal coefficient rule" allows the maximum value of X, $\dot{X}$, $\ddot{X}$, and so on to be estimated, usually within a factor of two.

Example: If $1\dddot{X} + 2\ddot{X} + 4\dot{X} + 10X = 25u(t)$ where $u(t)$ = unit step, then

$$1\dddot{X}_{max} = 25; \qquad \dddot{X}_{max} \cong 25$$

$$2\ddot{X}_{max} = 25; \qquad \ddot{X}_{max} \cong 25/2 = 12.5$$

$$4\dot{X}_{max} = 25; \qquad \dot{X}_{max} \cong 25/4 = 6.25$$

(*Special Case, no derivative*)

$$10X_{max} = \text{twice } 25 = 50; \; X_{max} \cong 50/10 = 5$$

An easy way to remember the special case ($KX_{max} = 2\cdot$ input magnitude) is to recall that a step response can exhibit overshoot, up to twice as high as the input step (100% overshoot $= 2\cdot$ input magnitude).

Note 6. If the coefficients of an equation increase or decrease markedly (by a factor of 10 or 100 or 1000) for each derivative of X, time scaling will be required. The scaled time t_s will be real time for the analog computer. To time scale, the gain of *every* integrator is changed by a

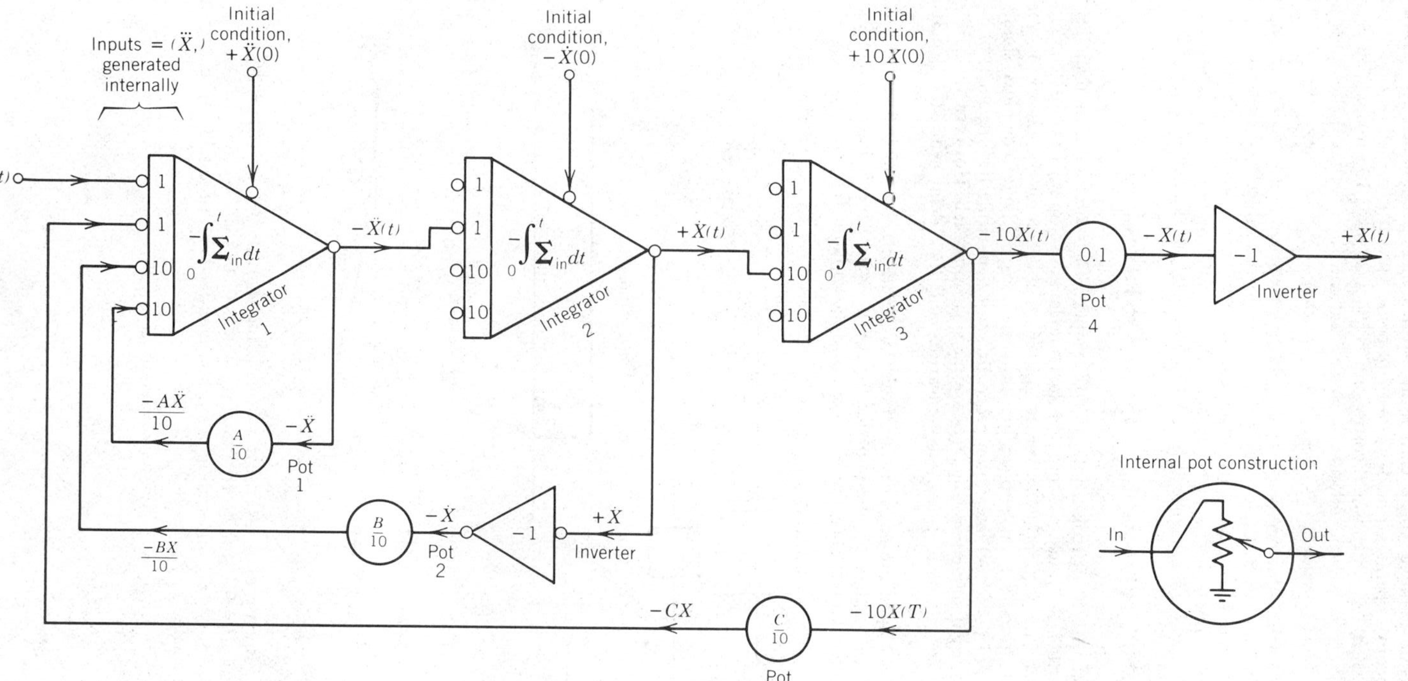

Fig. 38.5. Circuit for Example 38.1. Scale factors: 1 V = 1 unit; X scaled/X given = 1. Real time; t scaled/t given = 1. It is poor procedure to generate $-10X$ and then attenuate it to $-CX$. The larger-than-necessary signal may cause a needless overload of the integrator. However, if amplifier overload is not a problem, this procedure saves the two amplifiers needed to generate a separate gain of +10.

factor n. The original problem time t becomes n times real time t_s. If $n > 1$, the solution is speeded up. For example, for a certain time-scaled analog computer, $n = 500$ in the "rep-op" (repetitive operation) mode. Thus a response viewed in real time on an oscilloscope might last 20 ms, which would correspond to 500×20 ms $= 10$ s in unscaled problem time. Similarly, if $n < 1$, the solution is slowed but the same equation holds, problem time $= n$ times real time, or $t = nt_s$.

Many of the standard introductory linear control theory texts contain a section dealing with analog computer system simulation. Ogata[1] devotes 12 pages to the subject, D'Azzo and Houpis[2] devote four or five pages, and Kuo[3] devotes five or six pages. Texts such as Korn and Korn's[4] dealing exclusively with analog computers tend to be decades old. Shannon's[5] book dealing with various system simulation approaches is more current, as is Kleijnen's,[6] Hausner's,[7] James's,[8] and others.[9–14]

38.2 DIGITAL COMPUTER SIMULATION OF SYSTEMS

The digital computer has effectively replaced the analog computer for system simulation studies. This change began in 1955 when Selfridge, working with machine language on an IBM 701 computer at the US Naval Ordnance Test Station in Inyokern, California, published an account of his work on system simulation.[15] He proposed and applied the basic methods of digital computer simulation, namely, the application of numerical integration procedures to evaluate the time response of a given differential equation. Current simulation languages use the same approach. However, they are easier to apply and can handle variations, such as nonlinearities and time delay with standard built-in functions.[16]

The basic concept of digital simulation is illustrated by simulating the simple first-order system shown in Fig. 38.6, which is subjected to a unit step input. The system block diagram, along with a plot of the input and exact analytic output response, are also shown in the figure.

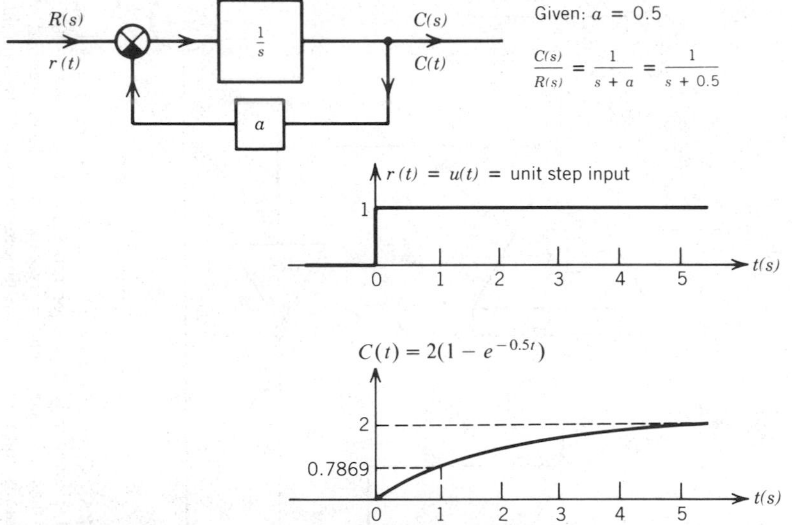

Fig. 38.6. Unit step response of a first-order system. Note that at $t = 1$ s, $C(t) = 0.7869$. When system is simulated with $\Delta t = 0.1$ s, the corresponding value is 0.8025, a 2% error; with $\Delta t = 1$ s, the value is 1.000, a 27% error. Euler's integration algorithm, described in Section 38.3, is used. The 2% and 27% errors are calculated in Tables 38.2 and 38.3, respectively.

Analytic input:

$$R(s) = \frac{1}{s}$$

Analytic output:

$$C(s) = \frac{1}{s}\,\frac{1}{s + a}$$

$$C(s) = \frac{1}{s}\,\frac{1}{s + 0.5}$$

$$C(t) = 2(1 - e^{-.5t})$$

From Fig. 38.6:

$$\frac{C(s)}{R(s)} = \frac{G(s)}{1 + G(s)H(s)} = \frac{1/s}{1 + (1/s)a} = \frac{1}{s + a}$$

where $a = \frac{1}{2}$, $R(s) = 1/s$.

$$C(s) = R(s)\frac{1}{s+a} = \frac{1}{s}\,\frac{1}{s+a} = \frac{1/a}{s} + \frac{-1/a}{s+a}$$

$$c(t) = \frac{1}{a} - \frac{1}{a}e^{-at} = \frac{1}{a}(1 - e^{-at}) = 2(1 - e^{-0.5t})$$

The results of the digital simulation will be compared with the exact analytic value of $C(t)$ obtained from this equation.

A step-by-step procedure for simulating the system $C(s)/R(s) = 1/(s + a)$ by means of digital techniques follows.

1. Find the system differential equation relating output to input.

$$C(s)(s + a) = R(s)$$

$$\frac{dc(t)}{dt} + ac(t) = r(t)$$

2. Solve for the highest order derivative of the system output $c(t)$.

$$\frac{dc(t)}{dt} = r(t) - ac(t)$$

3. Substitute in a discrete step or sampled time value $t = n\Delta\text{T}$, where ΔT is the step size or increment in time and n is the step number.

$$\frac{d}{dt}c(n\Delta\text{T}) = r(n\Delta\text{T}) - ac(n\Delta\text{T})$$

4. The solution starts at $t = 0$, or $n = 0$.

$$\frac{d}{dt}c(0) = r(0) - ac(0)$$

The input function $r(0), r(\Delta\text{T}), r(2\Delta\text{T}), \ldots,$ is given, as is $c(0)$, the initial condition of the output.

5. Solve for the next value of $c(t) = c(\Delta\text{T})$ by integrating the previous derivative of

$$c(t) = (d/dt)c(0).$$

$$\int_0^{\Delta\text{T}} \frac{d}{dt}c(0)\,dt = \int_0^{\Delta\text{T}} dc(0) = c(\Delta\text{T})$$

This is not done analytically but rather numerically, using one of the many possible numerical integration algorithms. This step is the essence of digital system simulation. For mth-order systems, the mth derivative is integrated to yield the $(m - 1)$th derivative, which is then integrated to yield the $(m - 2)$nd derivative; and the process is continued until the first derivative is integrated to yield the desired system output, $c(n\Delta\text{T})$. For this example the Euler integration algorithm is used.

6. Euler's algorithm[17] is

$$c[(n + 1)\Delta\text{T}] = c(n\Delta\text{T}) + \Delta\text{T}\frac{d}{dt}c(n\Delta\text{T})$$

Usually the ΔT in the argument is dropped, so that $c(2)$ is understood to be $c(2\Delta\text{T})$. This results in

$$c(n + 1) = c(n) + \Delta\text{T}\frac{d}{dt}c(n)$$

It is also common to employ state variables and dot notation for derivatives. In this case there is only one state variable and one derivative, c dot$(n) = \dot{c}(n) = (d/dt)c(n)$.

$$c(n + 1) = c(n) + \Delta\text{T}\dot{c}(n)$$

7. To illustrate the process for the given example, choose ΔT to be much smaller than the system time constant of $1/a = 1/0.5 = 2$ s (e.g., let $\Delta\text{T} = 0.1$ s). For the given simulation, $c(0) = 0$ and $r(0) = r(1) = r(2) = \cdots = 1$ = unit step. Then from step 4:

$$n = 0, \quad \text{and} \quad \frac{d}{dt}c(0) = \dot{c}(0) = r(0) - ac(0)$$

$$\dot{c}(0) = 1 - 0.5(0) = 1$$

This derivative is then integrated by substituting $\dot{C}(0) = 1$ into the integrating algorithm, for $n = 0$:

$$n = 0, \qquad \text{and} \quad c(n+1) = c(n) + \Delta T\dot{c}(n)$$
$$c(1) = 0 + 0.1(1) = 0.1$$

This new value of $c(t)$ is then used to calculate a new value of the derivative $\dot{c}(t)$, given in step 3 and repeated here. The process then repeats in a circular, iterative, or closed-loop fashion.

Derivative

$$\dot{c}(n) = r(n) - ac(n) = 1 - 0.5c(n)$$

Next c

$$c(n+1) = c(n) + \Delta T\dot{c}(n) = c(n) + 0.1\dot{c}(n)$$

A table of calculated values (Table 38.2) is shown for $\Delta T = 0.1$ s. $c(n)$ is the digitally simulated result.

To illustrate the importance of step size, the digital simulation is repeated with $\Delta T = 1$ s. Again $c(n)$ is the digitally simulated step response output sequence. For this tabular calculation:

Derivative

$$\dot{c}(n) = r(n) - ac(n) = 1 - 0.5c(n)$$

Next c

$$c(n+1) = c(n) + \Delta T\dot{c}(n) = c(n) + 1 \cdot \dot{c}(n)$$

The results are shown in Table 38.3. Observe that for $\Delta T = 1$ s, the error at step 1 is $(1.0000 - 0.7869)/0.7869 = 0.27 = 27\%$, as shown in line two of Table 38.3, whereas for $\Delta T = 0.1$ s, the error at step 1 is $(0.1000 - 0.0975)/0.0975 = 0.0256 = 2.6\%$, as shown in line two of Table 38.2. In addition, at $t =$

TABLE 38.2. DIGITALLY SIMULATED STEP RESPONSE OF $C(s) = R(s)/(s + a)$ USING EULER'S INTEGRATION ALGORITHM[a]

n = Step	Exact Value $2(1 - e^{-0.5t})$	$t = n\Delta T$ (s)	$r(n)$	$C(n)$	$\dot{C}(n)$	Error (%)	$C(n+1)$
0	0.0000	0	1	0.0000	1.0000	0	0.1000
1	0.0975	0.1	1	0.1000	0.9500	2.6	0.1950
2	0.1903	0.2	1	0.1950	0.9025	—	0.2853
3	0.2785	0.3	1	0.2850	0.8574	—	0.3710
4	0.3625	0.4	1	0.3710	0.8145	—	0.4524
5	0.4424	0.5	1	0.4524	0.7738	—	0.5298
6	0.5184	0.6	1	0.5298	0.7351	—	0.6033
7	0.5906	0.7	1	0.6033	0.6983	—	0.6732
8	0.6594	0.8	1	0.6732	0.6634	—	0.7395
9	0.7247	0.9	1	0.7395	0.6302	—	0.8025
10	0.7869	1.0	1	0.8025	0.5987	2.0	0.8624
11	0.8461	1.1	1	0.8624	0.5987	—	0.9193
12	0.9024	1.2	1	0.9193	0.5688	—	0.9733
13	0.9559	1.3	1	0.3733	0.5404	—	1.0247
14	1.0068	1.4	1	1.0734	0.5133	—	1.0734
15	1.0553	1.5	1	1.0734	0.4877	—	1.1197

[a] For $a = 0.5$ and where $\Delta T = 0.1$ s = step size; $r(n)$ = system input = unit step function; $c(n)$ = digitally simulated system output response; error = [(exact value − simulated value)/exact value] × 100%; $\dot{c}(n)$ = approximation to $(d/dt)c(n)$; $\dot{c}(n) = 1 - 0.5c(n)$; $c(n+1)$ = next $c = c(n) + 0.1\dot{c}(n)$.

TABLE 38.3. DIGITALLY SIMULATED STEP RESPONSE OF $C(s) = R(s)/(s + a)$ USING EULER'S INTEGRATION ALGORITHM[a]

n = Step	Exact Value $2(1 - e^{-0.5t})$	$t = n\Delta T$ (s)	$r(n)$	$C(n)$	$\dot{C}(n)$	Error (%)	$C(n+1)$
0	0.0000	0	1	0.0000	1.0000	0	1.0000
1	0.7869	1	1	1.0000	0.5000	27	1.5000
2	1.2642	2	1	1.5000	0.2500	—	1.7500
3	1.5537	3	1	1.7500	0.1250	—	1.8750
4	1.7293	4	1	1.8750	0.0625	—	1.9375
5	1.8358	5	1	1.9375	0.0313	—	1.9688
6	1.9004	6	1	1.9688	0.0156	—	1.9844
7	1.9396	7	1	1.9922	0.0078	—	1.9922

[a] For $a = 0.5$ and where $\Delta T = 1.0$ s = step size; $r(n)$ = system input = unit step function; $c(n)$ = digitally simulated system output response; error = [(exact value − simulated value)/exact value] × 100%; $\dot{c}(n)$ = approximation to $(d/dt)c(n)$; $\dot{c}(n) = 1 - 0.5c(n)$; $c(n+1)$ = next $c = c(n) + 1.0\dot{c}(n)$.

1 s for $\Delta T = 0.1$ and $n = 10$, the error is $(0.8025 - 0.7869)/0.7869 = 0.0198 = 2.0\%$. A 2% error may be acceptable, but a 27% error is usually intolerable.

This illustrates the principal problem with digital simulation. For reasonable accuracies, small step sizes must be used. Other integration algorithms produce much smaller errors at the expense of more complicated calculations. The gain in accuracy more than offsets the complications, however, allowing a larger step size to be used, which decreases the total number of individual arithmetic operations for a given simulation accuracy. Several common integration algorithms are listed in Section 38.3. For example, the fourth-order Runge–Kutta algorithm typically can use a 10-times larger step size than the Euler algorithm for a given accuracy. For this reason, the Runge–Kutta algorithm is greatly preferred to the relatively simple Euler algorithm.

38.3 INTEGRATION ALGORITHMS

The heart of any continuous system simulation process utilizing a digital computer is the numerical integration algorithm. As discussed in Section 38.2, the numerical integration algorithm operates on a function specified by a set of sampled numerical values to produce an approximation of the integral of that function. The following algorithms are analyzed in this section:

Euler's integration algorithm
Modified Euler's integration algorithm
Adam's four-point predictor-corrector algorithm
Milne's second-difference predictor-corrector algorithm
Milne's fourth-difference predictor-corrector algorithm
Runge–Kutta fourth-order integration algorithm

Euler's Integration Algorithm[†]

Euler's algorithm approximates the slope of the function to be integrated, and then multiplies this slope $f(t_n, Y_n)$ by the step size T to obtain one increment of the integral.

From Fig. 38.7:

$$Y_1 = Y_0 + \int_{t_0}^{t_1} \dot{Y}(t)\,dt \tag{38.1}$$

Approximating the integral by

$$(t_1 - t_0)\dot{Y}(t_0) = \mathrm{T}f(t_0, Y_0) = T \cdot \text{slope}$$

$$Y_1 \cong Y_0 + \mathrm{T}f(t_0, Y_0) = Y_0 + \mathrm{T}\dot{Y}(t_0) = Y_0 + \mathrm{T}\dot{Y}_0 \tag{38.2}$$

[†] R. W. Southworth and S. L. Deleeuw, *Digital Computation and Numerical Methods*, McGraw-Hill, New York, 1965. All the algorithms given are adapted from this text, with changes in notation. System simulation variables are generally a function of time, or $Y = f(t)$. Southworth and Deleeuw consider the problem $Y = f(x)$. Thus the variable x, its integral, and its derivatives, were changed to t.

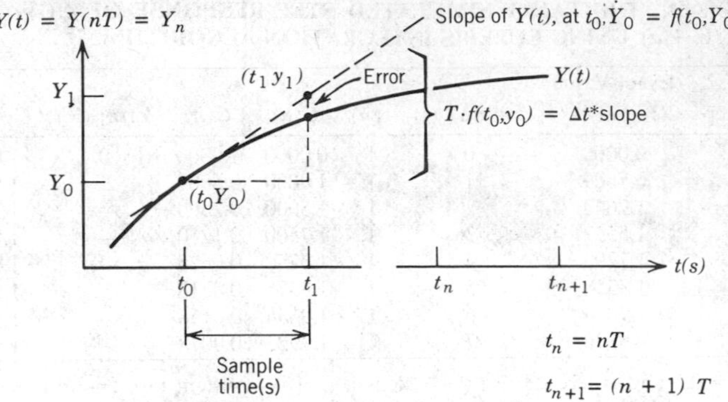

Fig. 38.7. Euler's integration algorithm.

In general:

$$Y_{n+1} = Y_n + \mathrm{T}f(t_n, Y_n) = Y_n + \mathrm{T}\dot{Y}(t_n) = Y_n + \mathrm{T}\dot{Y}_n \tag{38.3}$$

where

$$\dot{Y}(t_n) = \frac{dY(t_n)}{dt} = f(t_n, Y_n) = \text{slope} = \dot{Y}_n$$

$$\mathrm{T} = t_1 - t_0 = \text{step size, } \Delta t, \text{ s}$$

$$n = \text{step or time index}$$

Modified Euler's Integration Algorithm

The error produced by the Euler algorithm may be decreased by evaluating the slope of $Y(t)$ at two points and taking an average of the two slopes. Denoting the slope dY_0/dt by $f(t_0, Y_0)$, and replacing $f(t_0, Y_0)$ by

$$\frac{dY_0}{dt} = \frac{f(t_0, Y_0) + f(t_1, Y_1)}{2} \tag{38.4}$$

the resulting algorithm is

$$Y_{n+1}^{(r+1)} = Y_n + \frac{\mathrm{T}}{2}\left[f(t_n, Y_n) + f\left(t_{n+1}, Y_{n+1}^{(r)}\right)\right] \tag{38.5}$$

where $f(t_n, Y_n) = dY(t)/dt$ at $t = t_n$ = slope of $Y(t)$ and r = the number of iterations of Eq. (38.5) for a single increment of n; the iteration index.

Equivalently:

$$Y_{n+1}^{(r+1)} = Y_n + \frac{\mathrm{T}}{2}\left[\dot{Y}_n + \dot{Y}_{n+1}^{(r)}\right] \tag{38.6}$$

Although the modified Euler algorithm slope of Eq. (38.4) appears simple and direct, this is not the case. The reason is that the value of Y_1 is needed to calculate Y_1, a seemingly impossible requirement. The procedure is to either use Y_0 for Y_1 or use the unmodified Euler algorithm to calculate a first approximation for Y_1, called $Y_1^{(1)}$. With this first approximation a slope can be calculated at $Y_1^{(1)}$, which is then averaged with the slope at Y_0 to obtain a new more accurate slope via Eq. (38.4). This averaged slope is then used to calculate an improved second approximation for Y_1, called $Y_1^{(2)}$, using Eq. (38.5). The process is repeated r times, to obtain $Y_1^{(r+1)}$. The process stops when no further changes (to the desired degree of accuracy) occur in $Y_1^{(r+1)}$. Thus $(r + 1)$ iterations are employed to calculate one point, Y_1. Y_2 is then calculated by the same iterative process, continuing on to Y_n and Y_{n+1}.

Adam's Four-Point Predictor-Corrector Algorithm

The Adam's four-point predictor-corrector algorithm is similar to the basic Euler algorithm; it extrapolates forward-using slopes calculated at previously evaluated points. It is used once for each new point.

Predictor (used once)

$$Y_{n+1}^{(0)} = Y_n + \frac{T}{24}(55\dot{Y}_n - 59\dot{Y}_{n-1} + 37\dot{Y}_{n-2} - 9\dot{Y}_{n-3}) \tag{38.7}$$

The simple predicted value for Y_{n+1}, called $Y_{n+1}^{(0)}$, is then iteratively corrected $r+1$ times, in a fashion analogous to that discussed for the modified Euler algorithm.

Corrector (used r + 1 times)

$$Y_{n+1}^{(r+1)} = Y_n + \frac{T}{24}\left[9\dot{Y}_{n+1}^{(r)} + 19\dot{Y}_n - 5\dot{Y}_{n-1} + \dot{Y}_{n-2}\right] \tag{38.8}$$

where T = step size = $t_{n+1} - t_n$
$\dot{Y}_n = dY(t_n)/dt = f(t_n, Y_n)$ = slope
r = number of iterations of Eq. (38.8) for a single point Y_{n+1}. Iterations start with $Y_{n+1}^{(0)}$ obtained from the predictor equation and end with $Y_{n+1}^{(r+1)}$. If $r = 0$, only one correction is made and Eq. (38.8) is used only once.

Historical Note. According to Scarborough,[18] "The method of replacing the derivative of a function by a polynomial and integrating that polynomial over an interval was used by J. C. Adams as early as 1855." In fact, many of the most popular numerical integration methods used for computer simulations date back to the 1800's. The Runge–Kutta method, one of the most popular integration algorithms used today, was developed by Runge in the years 1894–95[19] and was extended by Kutta a few years later.[20]

Milne's Second-Difference Predictor-Corrector Algorithm

Milne's method, devised in 1926[21,22] integrates ahead by extrapolating an average slope from the last three values of slope, $\frac{1}{3}(2\dot{Y}_n - \dot{Y}_{n-1} + 2\dot{Y}_{n-2})$. Y_{n+1} is then found from Y_{n-3} by adding on a four-step-size extrapolation of 4T times the average slope.

Predictor (used once)

$$Y_{n+1}^{(0)} = Y_{n-3} + \tfrac{4}{3}T(2\dot{Y}_n - \dot{Y}_{n-1} + 2\dot{Y}_{n-2}) \tag{38.9}$$

Corrector (used r + 1 times)

$$Y_{n+1}^{(r+1)} = Y_{n-1} + \frac{T}{3}(\dot{Y}_{n+1}^{(r)} + 4\dot{Y}_n + \dot{Y}_{n-1}) \tag{38.10}$$

If the corrector equation is not repeatedly iterated but is instead used only once, then $r = 0$. Refer to the discussion concerning the Adam's predictor-corrector algorithm for the definition of terms.

Milne's Fourth-Difference Predictor-Corrector Algorithm

Milne's fourth-difference method is essentially similar to the second-difference method except that the slope is extrapolated from the last five values to yield the average slope $\frac{1}{20}(11\dot{Y}_n - 14\dot{Y}_{n-1} + 26\dot{Y}_{n-2} - 14\dot{Y}_{n-3} + 11\dot{Y}_{n-4})$. Y_{n+1} is then found from Y_{n-5} by adding on a six-step-size extrapolation of 6T times the average slope. The fraction 6T/20 is reduced to 3T/10 in the formula.

Predictor (used once)

$$Y_{n+1}^{(0)} = Y_{n-5} + \frac{3T}{10}(11\dot{Y}_n - 14\dot{Y}_{n-1} + 26\dot{Y}_{n-2} - 14\dot{Y}_{n-3} + \dot{Y}_{n-4}) \tag{38.11}$$

Corrector (used r + 1 times)

$$Y_{n+1}^{(r+1)} = Y_{n-3} + \frac{2T}{45}\left[7\dot{Y}_{n+1}^{(r)} + 32\dot{Y}_n + 12\dot{Y}_{n-1} + 32\dot{Y}_{n-2} + 7\dot{Y}_{n-3}\right] \tag{38.12}$$

If the corrector equation is not iterated but is used only once, then $r = 0$ [as previously observed for Eq. (38.10)]. Similarly, refer to the discussion concerning the Adam's predictor-corrector algorithm for the definition of terms. The fraction 2T/45 in Eq. (38.12) is the reduced value of 4T/90, where 90 is the sum of all the slope coefficients in the corrector equation. Y_{n+1} is then found from Y_{n-3} by adding on a four-step-size extrapolation of 4T times the average slope.

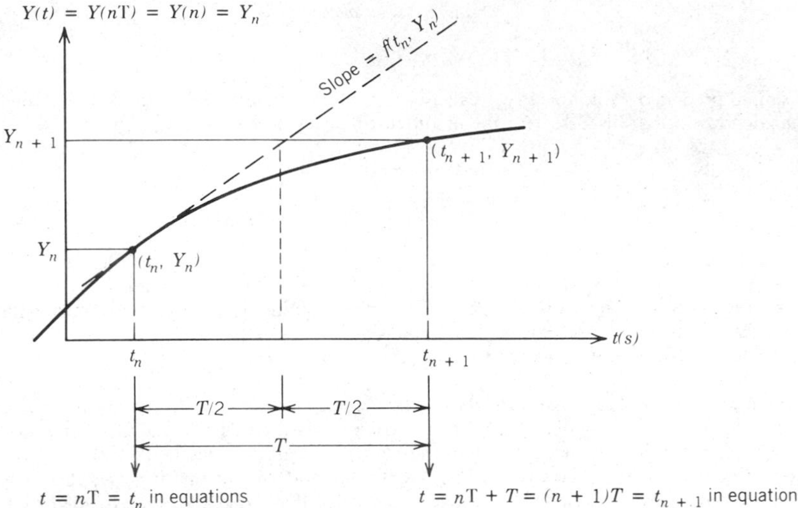

Fig. 38.8. Runge–Kutta fourth-order integration algorithm. $(d/dt)Y(t) = Y'(t) = \dot{Y}(t) = f(t, Y)$ = slope of the curve approximating $Y(t)$ at the point (t, Y).

Runge–Kutta Fourth-Order Integration Algorithm

The Runge–Kutta method is unlike any of the predictor-corrector methods or most other numerical integration methods, in that calculations for the first step or increment are exactly the same as those for any other increment. The Runge–Kutta method is often used as a starter for predictor-corrector methods. It calculates four slopes at points determined by the previous slope calculation, so that each calculation tends to correct the previous one. The final increment from $Y(n)$ to $Y(n+1)$ is given by a weighted average of these four slopes times the step size T.

The Runge–Kutta algorithm is

$$Y(n+1) = Y(n) + \tfrac{1}{6}(K_0 + 2K_1 + 2K_2 + K_3) \tag{38.13}$$

where $K_0 = \mathrm{T}f(t_n, Y_n)$ = slope at starting point, times T

$$K_1 = \mathrm{T}f\left(t_n + \frac{\mathrm{T}}{2}, Y_n + \frac{K_0}{2}\right) = \text{slope at extrapolated center point, times T}$$

$$K_2 = \mathrm{T}f\left(t_n + \frac{\mathrm{T}}{2}, Y_n + \frac{K_1}{2}\right) = \text{slope at reextrapolated center point, times T}$$

$$K_3 = \mathrm{T}f(t_n + \mathrm{T}, Y_n + K_2) = \text{slope at extrapolated end point, times T}$$

A diagram of $Y(t)$ vs. t is given in Fig. 38.8.

38.4 SPECIALIZED SIMULATION LANGUAGES

Specialized simulation languages were developed so that an individual interested in simulating a large complex system would not be required to spend months writing and debugging a unique simulation program. Instead, the simulation language developers spent months and years developing their generalized simulation programs, which allow a user to simulate a large system in hours or days.

The principal ingredient in a simulation language is an integration algorithm, which can be applied sequentially, just as an analog computer integrator is applied, to solve a differential equation. Simulation languages usually have simplified input and output statements, free of the format requirements of the underlying procedural language, which is often FORTRAN. Many of the common simulation languages developed as a compiler or preprocessor that would produce a FORTRAN program as their output. The user would specify a block diagram in terms of interconnections and transfer functions. Alternatively, the user would specify the differential equations describing the system. These specifications served as the input to the simulation language, which would then produce a FORTRAN program that would be run in the usual manner. There were, of course, a great many other little details that the user had to specify, such as the time increment for the

TABLE 38.4. CHRONOLOGY OF SIGNIFICANT SIMULATION LANGUAGES

Acronym or Name	Decoding of Acronym or Name	Approximate Date	Author(s) and Affiliations	Comments
No name	No name	1955	R. G. Selfridge, US Naval Ordnance Test Station, Inyokern, CA	Used machine language on IBM 701; FORTRAN was not developed until 1957
DEPI to DEPI-4	Differential equations pseudocode interpreter	1958	H. F. Lesh, Jet Propulsion Laboratory, Pasadena, CA	Used machine language and the fourth-order Runge–Kutta integrating algorithm on Datatron 204 decimal fixed-point machine
ASTRAL	Analog schematic translator to algebraic language	1958	M. L. Stein, J. Rose, and J. B. Parker, Convair Astronautics, CA	Compiled input statements that modeled analog computer into FORTRAN program. FORTRAN program was then run as usual. ASTRAL also sorted operations so that variables were calculated in order of need
DAS	Digital analog simulator	1963	R. A. Gaskil, J. W. Harris, and A. L. McKnight, Martin Co., Orlando, FL	Used rectangular integration algorithm on IBM 7090, which was modified (hence MIDAS) to fifth-order predictor-corrector algorithm for MIDAS
MIDAS	Modified DAS	1963	R. T. Harnett, F. J. Sanson, and H. E. Petersen, Wright-Patterson AFB	Written for IBM 7090/7094, it used sorting scheme similar to that of ASTRAL and improved integration algorithm (see DAS). Program was distributed at no charge and became very popular in its day
PACTOLUS	King Midas bathed in River Pactolus, to wash away golden touch	1964	R. D. Brennan, IBM, San Jose, CA	Written for IBM 1620, small computer that users could individually operate, like analog computer being replaced. Big 7090/7094 computers were not operated by ordinary users. PACTOLUS was forerunner of DSL-90 and CSMP.

TABLE 38.4. (*Continued*)

Acronym or Name	Decoding of Acronym or Name	Approximate Date	Author(s) and Affiliations	Comments
MIMIC	Related to or MIMIC's MIDAS	1965	H. E. Peterson, F. J. Sanson, and L. M. Warshawsky, Wright-Patterson AFB	Improvement on MIDAS, changing fifth-order predictor-corrector integration algorithm to variable-step-size fourth-order Runge–Kutta algorithm
DSL-90	Digital simulation language for IBM 7090/7094	1965	W. M. Lyn and D. G. Wyman, IBM, San Jose, CA	DSL-90 translates block diagrams of analog computer simulations or differential equations into FORTRAN IV subroutine which is compiled and run with any one of four integration algorithms. Format-free input and output was provided
CSMP	Continuous system modeling program	1967	R. D. Brennan, IBM Application Program Number 360A-CX-16X	Developed for IBM 1130 and IBM 360 series, CSMP incorporates features of DSL-90 and PACTOLUS and adds option of console interaction. CSMP is still being used in early 1980s on machines capable of running FORTRAN IV. It provides eight integrating algorithms
CSSL	Continuous system simulation language	1967	Simulation Software Committee of Simulation Councils, Inc.	CSSL is preprocessor that converts differential equation description of system into procedural language such as ALGOL, PL/1, or FORTRAN IV. It is extension of MIMIC, which was based on MIDAS, which was improvement of ASTRAL. CSSL, usually FORTRAN IV version, is available today for most major computers
GPSS	General purpose simulation systems	1967	IBM Application Program Numbers 360A-CS-17X(OS) and 360A-CS-19X(DOS)	GPSS is intended to model discrete transactions that move through system, such as parts moving along assembly line or people queueing for service at bank. It does not solve differential equations as all previous models do. It solves queueing problems, using storage, delay, switches, Boolean algebra, and regular algebra

TABLE 38.5. OTHER SIMULATION LANGUAGES

Acronym or Name	Decoding of Acronym or Name	Applications and Comments
ECAP	Electrical circuit analysis program	Available for most major computers. Used to design and analyze electronic circuits. Steady-state and transient response available, along with nonlinear and element models
STRESS	Structural engineering system solver	Available for most major computers. Used to model elastic statically loaded framed structures. Developed at MIT
STRUDL I & II	Structural design language	Advancement of STRESS, STRUDL II handles dynamic loads for discrete or continuous systems. STRUDL was developed by Robert D. Logcher at MIT Civil Engineering Systems Labs in 1968
COGO	Coordinate geometry	Available for most major computers. Used by surveyors to produce maps and evaluate property areas
SAAM-27	Simulation, analysis, and modeling	Available for most major computers. Developed by National Institutes of Health and used by it for biomedical simulation, SAAM is being promoted for all simulation applications
DYNAMO	Dynamic Models	Developed by Jay Forrester and Alexander Pugh at MIT to model economic production and world dynamics. Unique set of symbols is used but they basically describe coupling coefficients and rates that are integrated
SIMSCRIPT II.5		Extension of GPSS. Handles queueing and scheduling simulations. Available for most major computers. In SIMSCRIPT, system is described by "entities" that have "attributes" that interact with activities described by "event routines"

integration process, which integration algorithm to use if a choice existed, which variables to print out, and whether to sort the sequence of calculations.

Table 38.4 contains a chronologic sampling of significant simulation languages. It is interesting to note that each language built upon and evolved from an earlier one. Table 38.5 lists several specialized simulation languages used in electronics, civil engineering, economics, and queuing theory.

38.5 GENERAL COMMENTS ON THE REAL-TIME PROBLEM

Many simulation problems can be carried out on the computer, with the results being a list of numeric values that can be examined or plotted. In such a situation it would not matter if 10 minutes were needed to calculate 10 seconds of simulated response data. On the other hand, consider the case of a pilot-training simulator. In this situation the simulator solves the differential equations in real time, describing the behavior of a jet aircraft in response to the pilot's various control inputs. The outputs of the simulator are displayed on gauges, readout panels, and a simulated viewport or window. In this case the simulation must respond in real time so that the pilot can see the results of his control actions now, not 10 minutes later. The digital computer must be able to calculate 10 seconds of simulated response data in 10 seconds or less. If the sample time is 1/100 of a second and if 1000 lines of code requiring an average of 10 operations must be executed to calculate each point, then 1 million operations are needed for 1 second of data. However, all the available time cannot be allocated to computing. The display process, with its digital-to-analog conversion, requires significant time, as does the sampling of the pilot's inputs. The net result is that the 1 million computer operations must be carried out in a fraction of a second, if one second of data is to be displayed in one second of real time. This requires a fast computer with fast input-output devices, equipped with a so-called real-time clock. The real-time clock synchronizes the computer's output with real time at every one hundredth of a second for the example discussed. In general, a real-time simulation is far more difficult than an ordinary non-real-time simulation. For the case discussed, the real-time simulation requires specialized input-output devices and a computer capable of executing several million floating-point operations per second. Stephenson[15] and Kochenburger[23] touch on this problem.

References

1 K. Ogata, *Modern Control Engineering*, Prentice-Hall, Englewood Cliffs, NJ, 1970.

2 J. D'Azzo and C. Houpis, *Linear Control System Analysis and Design, Conventional and Modern*, 2nd ed., McGraw-Hill, New York, 1981.

3 B. Kuo, *Automatic Control Systems*, 4th ed., Prentice-Hall, Englewood Cliffs, NJ, 1982.

4 G. Korn and T. Korn, *Electronic Analog Computers*, McGraw-Hill, New York, 1956.

5 R. Shannon, *Systems Simulation: The Art and Science*, Prentice-Hall, Englewood Cliffs, NJ, 1975.

6 J. Kleijnen, *Statistical Techniques in Simulation*, Marcel Dekker, New York, Part I, 1974, Part II, 1975.

7 A. Hausner, *Analog and Analog/Hybrid Computer Programming*, Prentice-Hall, Englewood Cliffs, NJ, 1971.

8 M. L. James, G. M. Smith, and J. C. Wolford, *Analog Computer Simulation of Engineering Systems*, Intext, Scranton, NJ, 1971.

9 J. E. Stice and B. S. Swanson, *Electronic Analog Computer Primer*, Blaisdell, Waltham, MA, 1965.

10 G. A. Bekey and W. J. Karplus, *Hybrid Computation*, Wiley, New York, 1968.

11 R. C. Weyrick, *Fundamentals of Analog Computers*, Prentice-Hall, Englewood Cliffs, NJ, 1969.

12 A. S. Jackson, *Analog Computation*, McGraw-Hill, New York, 1960.

13 A. Durling, *Computational Techniques: Analog, Digital and Hybrid Systems*, Intext, New York, 1974.

14 W. J. Karplus, *Analog Simulation: Solution of Field Problems*, McGraw-Hill, New York, 1968.

15 R. E. Stephenson, *Computer Simulation for Engineers*, Harcourt Brace Jovanovich, New York, 1971.

16 A. M. Law and W. D. Kelton, *Simulation Modeling and Analysis*, McGraw-Hill, New York, 1981.

17 W. E. Grove, *Brief Numerical Methods*, Prentice-Hall, Englewood Cliffs, NJ, 1966.

18 J. B. Scarborough, *Numerical Mathematical Analysis*, 2nd ed., Johns Hopkins, Baltimore, MD, 1950.

19 C. Runge, *Mathematische Annalen* **46**, 1895, cited by Scarborough.[18]

20 W. Kutta, *Zeitschrift Math. Phys.* **46**, 1901, cited by Scarborough.[18]

21 W. E. Milne, "Numerical Integration of Ordinary Differential Equations," *Am. Math. Mon.* **33**:455–460 (1926).

22 A. A. Bennett, W. E. Milne, and H. Bateman, *Numerical Integration Differential Equations*, Dover, New York, 1956.

23 R. J. Kochenburger, *Computer Simulation of Dynamic Systems*, Prentice-Hall, Englewood Cliffs, NJ, 1972.

Bibliography

Dodes, I. A., *Numerical Analysis for Computer Science*, Elsevier North-Holland, New York, 1978.

Glorioso, R. M. and F. C. C. Osorio, *Engineering Intelligent Systems*, Digital, Bedford, MA, 1980. (Wide-spectrum text, primarily collection of models and modeling methods suitable for computer simulation or solution. Covers adaptive control systems, robotics, reliability, stochastic automata, and pattern recognition.)

Southworth, R. W. and S. L. Deleeuw, *Digital Computation and Numerical Methods*, McGraw-Hill, New York, 1965.

CHAPTER 39
RELIABILITY

AMRIT L. GOEL

Syracuse University
Syracuse, New York

JOSEPH J. NARESKY (DECEASED)

Formerly Chief, Reliability and Compatibility Division
Rome Air Development Center
Griffiths Air Force Base
Rome, New York

39.1 INTRODUCTION

An important attribute of a unit or system is the degree to which it can be relied on to perform its intended function. A commonly used quantifiable measure of this attribute is called reliability. The great technological advances of recent decades have led to the development and deployment of more and more complex equipment and devices. This, in turn, has been responsible for an increased concern with the measurement, prediction, and improvement of equipment reliability. The military services, because they have the most complex systems and the most acute problems, have provided the impetus to development of the discipline of reliability engineering. They were instrumental in developing mathematical models for reliability and design techniques to permit quantitative specification, prediction, and measurement of reliability. Research was significantly accelerated by World War II and the Korean War. The growth of this field continues owing to continued emphasis on military preparedness.

This chapter describes basic concepts and relevant details associated with the term "reliability." Its intent is to provide theoretical and practical information sufficient for most readers. References are provided for those who want more details or whose problems are complex.

References 1 through 4 give details about basic concepts in reliability. Probability distributions in reliability are well covered in Ref. 5 and their applications are illustrated in Ref. 6. Details about the models and computations of nonrepairable and repairable systems are given in Refs. 1, 2, 7, and 8. Reference 4 thoroughly discusses reliability estimation, while Refs. 6, 9, and 10 give examples of reliability data analyses. The relatively new subject of software reliability and hardware-software reliability is covered in Refs. 11 through 13. Reliability engineering is thoroughly covered in Ref. 6;

some relevant military documents are Refs. 14 through 16. For a discussion of recent developments see Refs. 17 through 19. The details of Bayesian aspects in reliability assessment, not covered in this chapter, can be found in Refs. 19 and 20.

39.2 BASIC CONCEPTS AND DEFINITIONS

To perform its intended function, a device must possess certain properties. A deviation from such properties is termed a *fault*; the manifestation of a fault is called a *failure*. Not all faults have the same significance, but for our purposes we consider all failures of approximately the same importance. Since failure can be caused by a variety of faults, the time to failure is a random variable T. Associated with T are the probability density function (pdf) $f(t)$ and the cumulative distribution function (cdf) $F(t)$, defined as follows:

$$F(t) = \text{probability of a failure occurring by } t,\ P(T \leqslant t)$$

and

$$F(t) = \int_0^t f(x)\,dx$$

where x is a dummy variable. It is easy to see that

$$f(t) = \frac{dF(t)}{dt}$$

Reliability. Reliability at time t, $R(t)$, is defined as the probability of failure-free operation during a specified time, also called mission time, t. Thus

$$R(t) = P(X > t) = 1 - F(t)$$

or

$$R(t) = \int_t^\infty f(x)\,dx$$

Mean Time to Failure (MTTF). This concept applies to nonrepairable systems. It is defined as the expected value of time to failure and is given by

$$\text{MTTF} = \int_0^\infty t f(t)\,dt$$

It can be shown that MTTF is related to $R(t)$ as follows:

$$\text{MTTF} = \int_0^\infty R(t)\,dt$$

Mean Time Between Failures (MTBF). This concept applies to repairable systems in which failed elements are replaced on failure. It is a commonly used measure of performance and represents the mean of all times over an infinite length of time during which the unit or system is operational prior to entering a failed state. It is given by the ratio of the total operational time to the number of failures, that is,

$$\text{MTBF} = \frac{\text{total operating time}}{\text{number of failures}}$$

Hazard Rate. This is a very useful term in reliability engineering which makes it possible to distinguish between different failure models on the basis of physical considerations. The hazard rate is defined as

$$h(t) = \frac{f(t)}{1 - F(t)}$$

The hazard rate also has a probabilistic interpretation, namely, $h(t)\Delta t$ represents the probability that a device of age t will fail in the interval $(t, t + \Delta t)$, or

$$h(t) = \lim_{\Delta t \to 0} \left\{ \frac{\text{P}_\text{r}\,[\text{device will fail in } (t, t + \Delta t) \mid \text{device survived up to } t]}{\Delta t} \right\}$$

or

$$h(t) = \lim_{\Delta t \to 0} \left\{ \frac{F(t + \Delta t) - F(t)}{\Delta t[1 - F(t)]} \right\} = \frac{f(t)}{1 - F(t)} = \frac{f(t)}{R(t)}$$

On the basis of physical considerations, one can choose an appropriate functional form for $h(t)$. Once this is done, both $f(t)$ and $R(t)$ can be obtained from $h(t)$ using the following formulas:

$$f(t) = h(t)\exp\left[-\int_0^t h(x)\,dx\right]$$

$$R(t) \equiv 1 - F(t) = \exp\left[-\int_0^t h(x)\,dx\right]$$

Hazard rates for commonly used distributions are shown in Table 39.1.

Bathtub Curve

To assist in the choice of $h(t)$, three types of failures are generally recognized as having a time-dependent characteristic. These constitute the so-called bathtub curve, as shown in Fig. 39.1.

1. Initial failures. These appear shortly after $t = 0$ and gradually decrease in frequency during the initial period of operation.

2. Chance failures. These occur during the period in which a device exhibits a constant failure rate, generally at a lower rate than that during the initial period. Such failures are usually caused by severe and unpredictable environmental conditions.

TABLE 39.1. HAZARD RATES FOR COMMONLY USED DISTRIBUTIONS

Distribution	Hazard Rate	Nature of Hazard Rate
Binomial; Poisson	$\dfrac{p(x)}{\sum_{j=x}^{\infty} p(j)}$	IFR IFR
Normal	$\dfrac{f(t)}{1 - F(t)}$	IFR
Lognormal	$\dfrac{f(t)}{1 - F(t)}$	DFR for long life
Exponential	λ	Constant
Gamma	$\dfrac{t^{k-1}\lambda^k}{(\Gamma k)\sum_{j=0}^{k-1}\left\{(\lambda t)^j \dfrac{1}{\Gamma(j+1)}\right\}}$	DFR for $k < 1$ Constant for $k = 1$ IFR for $k > 1$
Weibull	$\dfrac{\beta}{\eta}\left(\dfrac{t}{\eta}\right)^{\beta-1}$	DFR for $\beta < 1$ Constant for $\beta = 1$ IFR for $\beta > 1$

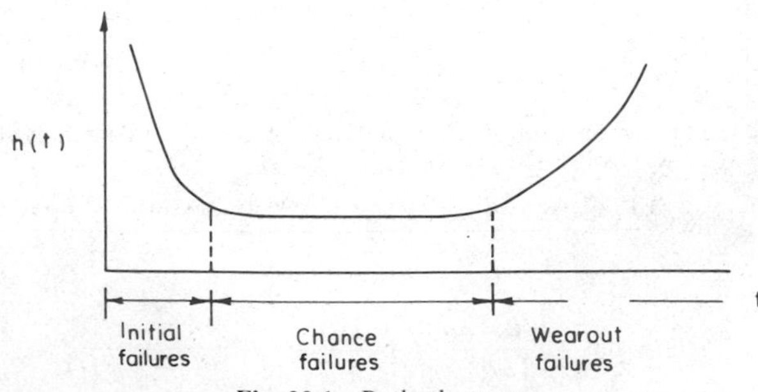

Fig. 39.1 Bathtub curve.

3. Wearout failures. These occur due to gradual depletion of a material or due to an accumulation of shocks, fatigue, and so on. The rate of their occurrence increases rapidly with time.

Important Terms and Concepts

System Effectiveness. The probability that the system can successfully meet an operational demand within a given time when operated under specified conditions. For a one-shot device such as a missile, it is the probability that the system (missile) will operate successfully (kill the target) when called on to do so under specified conditions.

Mission Reliability. The ability of an item to perform its required functions for the duration of a specified mission profile.

Operational Readiness. The ability of an item (military unit) to respond to its operational plan(s) on receipt of an operating order (total calendar time is the basis for computation of operational readiness).

Availability. A measure of the degree to which an item is in an operable and committable state at the start of a mission when the mission is called for at an unknown (random) time (includes operating time, active repair time, administrative time, and logistic time, but excludes mission time).

Intrinsic Availability. The probability that the system is operating satisfactorily at any time when used under stated conditions, where the time considered is operating time and active repair time.

Maintainability. The measure of the ability of an item to be retained in, or restored to, specified condition when maintenance is performed by personnel having specified skill levels and using prescribed procedures and resources, at each prescribed level of maintenance and repair.

Repairability. The probability that a failed system will be restored to operable condition within a specified active repair time.

Serviceability. The degree of ease or difficulty with which a system can be repaired.

39.3 PROBABILITY DISTRIBUTIONS USED IN RELIABILITY MODELS

A number of probability distributions have been found to be good descriptors of failures in components; some are described here. For reliability applications, an important characteristic of a distribution is its hazard or failure rate, which can be a decreasing (DFR) constant or an increasing (IFR) function of time depending on the value of the parameter (see Table 39.1).

Binomial Distribution

This distribution arises in reliability applications when we are interested in the probabilities of the number of failures in Bernoulli trials. Bernoulli trials are repeated independent trials of an experiment with two possible outcomes at each trial, success or failure, with a constant probability of each outcome.

Let X = number of failures in n trials and p = probability of success at each trial. Then

$$p(x) = P[X = x] = \binom{n}{x} p^x (1-p)^{n-x}, \qquad x = 0, 1, \ldots, n$$

where

$$\binom{n}{x} = \frac{n!}{x!(n-x)!}$$

It is a two-parameter (n, p) distribution with mean $E(X) = np$ and variance $V(X) = np(1-p)$.

Poisson Distribution

This distribution is used quite frequently in reliability analysis. It can be obtained as a limiting form of the binomial distribution with $n \to \infty$ and np = constant. If failures are Poisson distributed, they occur at a constant rate of λ per unit time and the number of failures x in a given time t has a probability

$$P(X = x) = \frac{e^{-\lambda t}(\lambda t)^x}{x!}, \qquad x = 0, 1, \ldots, n$$

This is a one-parameter (λ) distribution with $E(X) = \lambda t$ and $V(X) = \lambda t$.

As an example of the use of this distribution, consider a minuteman launch console that has an average lamp failure rate (λ) of 0.001/hr. Then the reliability for a 500-hour mission, if the allowable number of lamp failures cannot exceed 2, is obtained as follows:

$$\lambda = 0.001 \qquad t = 500, \qquad x \leqslant 2 \qquad \lambda t = 0.5$$

$$R(500) = P(X \leqslant 2) = \sum_{x=0}^{2} \frac{(0.5)x_e^{-0.5}}{x!}$$

or

$$R(500) = e^{-0.5} + 0.5e^{-0.5} + \frac{(0.5)^2 e^{-0.5}}{2}$$

or

$$R(500) = 0.986$$

Normal (or Gaussian) Distribution

There are two principal applications of the normal distribution in reliability. One application deals with the analysis of items that exhibit failure due to wear, such as mechanical devices. The wearout failure distribution is frequently sufficiently close to normal to make the use of this distribution for predicting or assessing reliability valid.

The other application deals with the analysis of the ability of manufactured items to meet specifications. No two parts made to the same specifications are exactly alike. The variability of parts leads to a variability in systems composed of these parts. The design must take this part variability into account, or the system may not meet specifications owing to the combined effects of part variability. Yet another aspect of this application is in quality control procedures.

The basis for the use of normal distribution in this application is the central limit theorem, which states that the sum of a large number of identically distributed random variables, each with finite mean and variance, is normally distributed. Thus the variations in values of electronic component parts due to manufacturing, for example, are considered normally distributed.

The failure density function for the normal distribution is

$$f(t) = \frac{1}{\sqrt{(2\pi)}\,\sigma} \exp\left[-\left(\frac{t-\mu}{\sqrt{2}\,\sigma}\right)^2\right]$$

where μ = the population mean and σ = the population standard deviation, which is the square root of the variance.

For most practical applications, probability tables for the standard normal distribution are used (see Naresky[6]). The standard normal distribution density function with $z = (t-\mu)/\sigma$ is given by

$$f(z) = \frac{1}{\sqrt{2\pi}\,\sigma} \exp\left(\frac{-z^2}{2}\right)$$

where $\mu = 0$, $\sigma^2 = 1$. Normal is a two-parameter (μ, σ^2) distribution with $E(X) = \mu$ and $v(X) = \sigma^2$. Plots of the shape and other characteristics of the normal distribution are shown in Fig. 39.2.

Expressions for hazard function and reliability are best shown via an example. Suppose the failures of a microwave-transmitting tube have been observed to follow a normal distribution with $\mu = 5000$ hours and $\sigma = 1500$ hours. Then the reliability of such a tube for a mission time of 4100 hours and its hazard rate at age 4400 hours are obtained as follows:

$$R(4100) = P\left(Z > \frac{4100 - 5000}{1500}\right) = 0.73$$

and

$$h(4400) = \frac{f(4400)}{R(4400)}$$

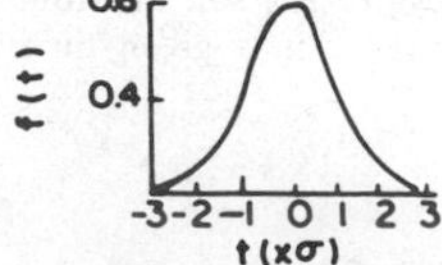

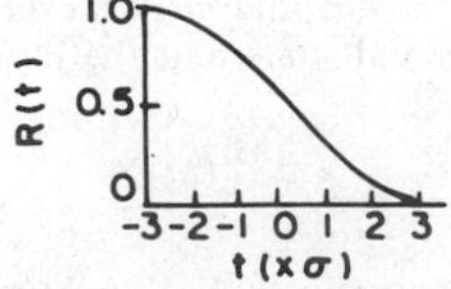

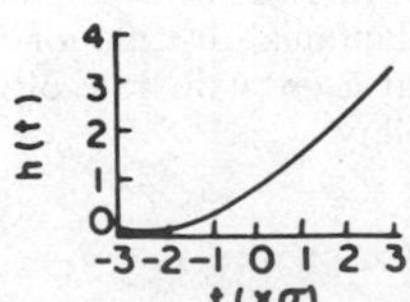

Fig. 39.2 Failure density, reliability, and hazard function for the normal distribution.

Now

$$f(4400) = \frac{1}{\sqrt{2\pi}\,(1500)} \exp\left[-\frac{1}{2(1500)^2}(4400 - 5000)^2 \right]$$

or

$$f(4400) = 0.00025$$

and

$$R(4400) = 0.65$$

so that

$$h(4400) = \frac{0.00025}{0.65} = 0.00038 \text{ failures/hr}$$

Lognormal Distribution

The lognormal distribution is used in reliability analyses of semiconductors and fatigue life of certain types of mechanical components. Its most frequent application, however, is in maintainability analysis of time-to-repair data. It is the distribution of a random variable whose natural logarithm is distributed normally. In other words, it is the normal distribution with ln t as the variate. The density function is

$$f(t) = \frac{1}{\sigma t (2\pi)^{\frac{1}{2}}} \exp\left[-\left(\frac{\ln t - \mu}{\sigma} \right)^2 \right], \qquad t \geqslant 0$$

where μ and σ are the mean and standard deviation of ln t. For this distribution, the mean = $\exp[\mu + (\sigma^2/2)]$ and the standard deviation = $[\exp(2\mu + 2\sigma^2) - \exp(2\mu + \sigma^2)]^{\frac{1}{2}}$.

Exponential Distribution

This is probably the most important distribution in reliability work and is used almost exclusively for reliability prediction of electronic equipment. It describes the situation wherein the hazard rate is constant, which can be shown to be generated by a Poisson process. Some of the advantages of this distribution are that it has only one parameter, which can be easily estimated, and that its underlying mathematics is very tractable. Some particular applications of this model include items whose failure rate does not change significantly with age, complex and repairable equipment without excessive amounts of redundancy, and equipment for which the early failures or "infant mortalities" have been eliminated by "burning in" the equipment for some reasonable time period.

The pdf, cdf, and reliability for $t > 0$ are, respectively,

$$f(t) = \lambda \exp(-\lambda t)$$

$$F(t) = 1 - \exp(-\lambda t)$$

and

$$R(t) = \exp(-\lambda t)$$

It is easy to see that the hazard rate is a constant equal to λ:

$$h(t) = \frac{f(t)}{R(t)} = \frac{\lambda \exp(-\lambda t)}{\exp(-\lambda t)} = \lambda$$

The mean of the exponential distribution is $1/\lambda$ and the variance equals $1/\lambda^2$. Plots of the important characteristics of this distribution are shown in Fig. 39.3.

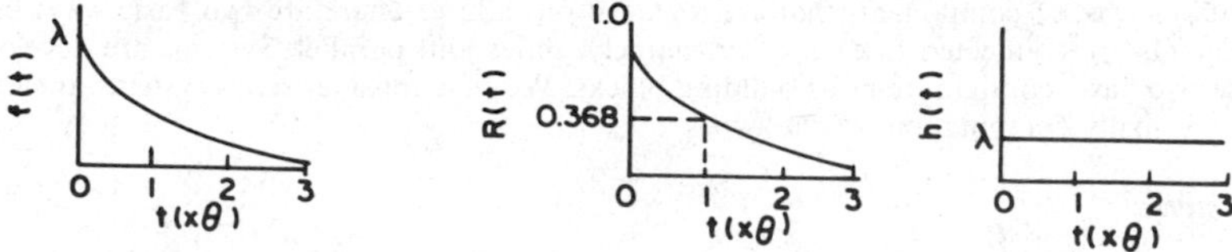

Fig. 39.3 Plots of failure density, reliability, and hazard rate for the exponential distribution.

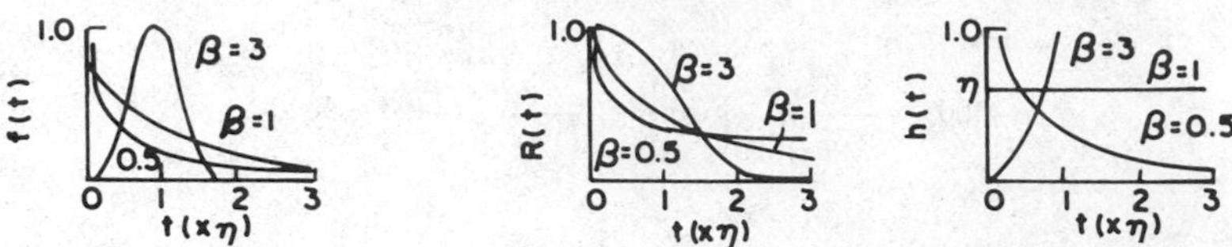

Fig. 39.4 Plots of density, reliability, and hazard function of a Weibull distribution.

Gamma Distribution

The gamma distribution is used in reliability analysis for cases where partial failures can exist, that is, when a given number of partial failures must occur before an item fails (e.g., redundant systems). It is a flexible distribution and can be used to model various kinds of observed failures. The gamma pdf is

$$f(t) = \frac{\lambda^k}{\Gamma k} t^{k-1} \exp(-\lambda t), \qquad k, t \geqslant 0, \qquad \lambda > 0$$

where Γk $(= \int_0^\infty e^{-y} y^{k-1}\, dy)$ is the gamma function, $1/\lambda$ is the scale parameter, and k is the shape parameter.

Note that when k is an integer, gamma distribution is also called an Erlang distribution; when $k = 1$, it reduces to an exponential distribution. Also its mean is k/λ and its variance is k/λ^2.

Weibull Distribution

The Weibull distribution is particularly useful in reliability work since it is very flexible and, with adjustment of its parameters, can be used to model a wide range of life distribution characteristics of different classes of engineering items. Specifically, it can be used to model decreasing, constant, or increasing failure rates. The Weibull failure density function is

$$f(t) = \frac{\beta}{\eta}\left(\frac{t-\gamma}{\eta}\right)^{\beta-1} \exp\left[-\left(\frac{t-\gamma}{\eta}\right)^{\beta}\right], \qquad t \geqslant 0$$

where β = shape parameter
η = scale parameter or characteristic life (life at which 63.2% of the population will have failed)
γ = minimum location parameter or minimum life

In most practical reliability situations, γ is zero (failure assumed to start at $t = 0$) and the failure density function becomes

$$f(t) = \frac{\beta}{\eta}\left(\frac{t}{\eta}\right)^{\beta-1} \exp\left[-\left(\frac{t}{n}\right)^{\beta}\right]$$

The corresponding reliability and hazard functions are

$$R(t) = e^{-(t/\eta)^{\beta}}$$

$$h(t) = \left(\frac{\beta}{\eta}\right)\left(\frac{t}{\eta}\right)^{\beta-1}$$

Note that for the Weibull distribution, $h(t)$ is a decreasing function of t for $\beta < 1$, is constant for $\beta = 1$, and is increasing for $\beta > 1$. Also, for $\beta = 1$ the Weibull distribution reduces to an exponential distribution. The mean and variance of this distribution are $\eta\Gamma[(\beta+1)/\beta] + \gamma$ and $\eta^2\{\Gamma[(\beta+2)/\beta] - \Gamma^2[(\beta+1)/\beta]\}$, respectively.

Plots of the Weibull distribution for several parametric values are shown in Fig. 39.4.

39.4 NONREPAIRABLE SYSTEMS

We now consider the reliability of systems composed of various interconnected nonrepairable components, that is, of components that are replaced on failure. There are two basic ways in which the components are connected to each other, namely, series and parallel. Systems are developed by using these two basic configurations as building blocks. We first consider series systems, then parallel systems, and finally combinations of the two.

Series Systems

Consider a system composed of n independent components or subsystems, as shown in Fig. 39.5. This is a series system and has the property that it operates only when all its subsystems operate. This is

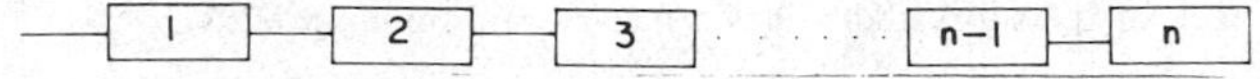

Fig. 39.5 Series system.

the simplest and perhaps the most commonly occurring configuration in reliability modeling. In this case, since the components are independent of each other, the reliability of the system R_S is equal to the product of the reliabilities of each of them, that is,

$$R_S = R_1 \cdot R_2 \cdot R_3 \cdot \ \ldots \ \cdot R_{n-1} \cdot R_n$$

or

$$R_S = \prod_{i=1}^{n} R_i$$

If the reliabilities are time dependent, then

$$R_S(t) = \prod_{i=1}^{n} R_i(t)$$

It can be shown that R_S is less than or equal to the minimum of reliabilities of its subsystems. Series systems are therefore sometimes called the weakest-link systems.

In the time-dependent case, if each subsystem i has a hazard rate of $h_i(t)$,

$$R_S(t) = \prod_{i=1}^{n} \exp\left[- \int_0^t h_i(x)\,dx \right]$$

If each component i exhibits a constant hazard λ_i:

$$R_S(t) = \prod_{i=1}^{n} \exp\left[- \int_0^t h_i(x)\,dx \right] = \exp\left(- \sum_{i=1}^{n} \lambda_i t \right)$$

If each component exhibits a linear hazard rate, $h_i(t) = K_i t$,

$$R_S(t) = \exp\left(- \sum_{i=1}^{n} K_i \frac{t_i^2}{2} \right)$$

Parallel Systems

The next most commonly occurring configuration encountered in reliability mathematical modeling is the parallel configuration, as shown in the reliability block diagram of Fig. 39.6. In this case, for the system to fail all of the components would have to fail. The purpose of configuring components this way is to achieve a reliability greater than that of the individual components. Such arrangements are also called redundant configurations.

Letting $Q_i = 1 - R_i = 1 - e^{\lambda_i t}$ be the probability of failure (or unreliability) of each component, the unreliability of the system is given by

$$Q_S = Q_1 \cdot Q_2 \cdots Q_n = \prod_{i=1}^{n} Q_i$$

and the reliability of the system is

$$R_S = 1 - Q_S$$

since $R_S + Q_S = 1$. If reliabilities are time dependent, then

$$R_S(t) = 1 - \prod_{i=1}^{n} [1 - R_i(t)]$$

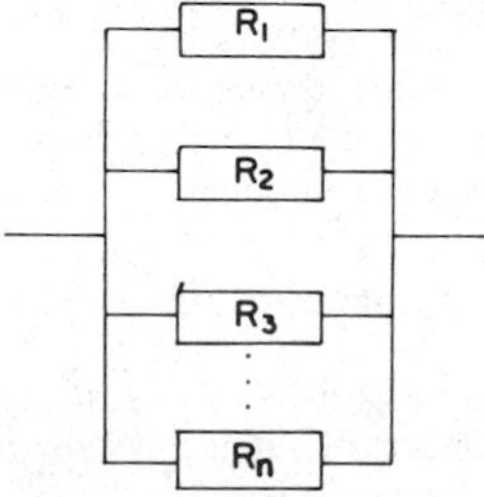

Fig. 39.6 Parallel system.

As before, if each of the subsystems i has a hazard rate function $h_i(t)$,

$$R_S(t) = 1 - \prod_{i=1}^{n} \left\{ 1 - \exp\left[-\int_0^t h_i(x)\,dx \right] \right\}$$

Consider a parallel system composed of five parallel components, each with a reliability of 0.99. Then

$$Q_i = 1 - R_i = 1 - 0.99 = 0.01$$

$$Q_S = (0.01)^5 = 10^{-10} = 0.0000000001$$

$$R_S = 1 - Q_S = 0.9999999999$$

Thus we see that it is possible to obtain a very high system reliability by using a parallel configuration. Of course this is a very simple concept; practical implementations are much more complicated.

MTBF. Consider two components connected in parallel, with hazard rates λ_1 and λ_2. Then

$$R_S(t) = 1 - (1 - e^{-\lambda_1 t})(1 - e^{-\lambda_2 t})$$

and

$$\text{MTBF} = \int_0^\infty R_S(t)\,dt = \frac{1}{\lambda_1} + \frac{1}{\lambda_2} - \frac{1}{\lambda_1 + \lambda_2}$$

Similarly, for n components in parallel with failure rates $\lambda_1, \lambda_2, \ldots, \lambda_n$,

$$\text{MTBF} = \sum_{i=1} \frac{1}{\lambda_i} - \sum_{i \neq j} \frac{1}{\lambda_i + \lambda_j} + \sum_{1 \neq j \neq k} \frac{1}{\lambda_i + \lambda_j + \lambda_k} - \ldots + (-1)^{n-1} \frac{1}{\sum_{i=1}^{n} \lambda_i}$$

If the n components in parallel are identical, that is, if $\lambda_1 = \lambda_2 = \ldots = \lambda_n = \lambda$, then the system MTBF is

$$\text{MTBF} = \sum_{i=1}^{n} \frac{1}{i\lambda}$$

Comparison of Series and Parallel Systems

Several comparisons can be made between the basic series and parallel systems. Two such comparisons follow.

Reliability. The reliability of a parallel system composed of n independent components is greater than that of a series system composed of the same n components. If $0 < R_i(t) < 1, i = 1, \ldots, n$ represent the component reliabilities, then

$$\underbrace{1 - \prod_{i=1}^{n} [1 - R_i(t)]}_{\text{Reliability of parallel system}} > \underbrace{R_j(t)}_{\text{Individual component reliability}} > \underbrace{\prod_{i=1}^{n} R_i(t)}_{\text{Reliability of series system}}$$

Thus we can improve system reliability by adding components in parallel. As stated earlier, this technique is commonly called redundancy.

MTBF. Let, for $i = 1, 2, \ldots, n$, t_i = failure time for ith component, t_{ss} = failure time for series system composed of n independent components with failure times t_i, and t_{ps} = failure time for parallel system composed of n independent components with failure times t_i. Then it can easily be seen that

$$t_{ss} = \min_{1 < i < n} t_i \leqslant \max_{1 \leqslant i \leqslant n} t_i = t_{ps}$$

because a series system fails as soon as one of its components fails and a parallel system fails only on failure of all of its components. Therefore the series system never has a larger MTBF than the corresponding parallel system.

To obtain the MTBF of a series or a parallel system, recall that

$$\text{MTBF} = \int_0^{\infty} R(t)\, dt$$

so that

$$\text{System MTBF} = \int_0^{\infty} R_S(t)\, dt$$

Calculation of this quantity is quite straightforward for series systems but gets messy for parallel systems, as seen earlier.

k-Out-of-n Systems

Another class of systems that are commonly used are the so-called k-out-of-n systems, sometimes denoted as (k, n) systems. In this configuration there are n subsystems in parallel, out of which at least $k(< n)$ must be good for the system to operate.

Let $R(t)$ be the reliability of each subsystem and $R_S(t)$ be the system reliability. The probability of x subsystems being good at some time t is given by the binomial distribution and $R_S(t)$ is given as

$$R_S(t) = \sum_{x=k}^{n} \binom{n}{k} \cdot [R(t)]^x [1 - R(t)]^{n-x}$$

If each subsystem has a constant failure rate λ, then

$$R_S(t) = \sum_{x=k}^{n} \binom{n}{x} (e^{-\lambda t})^n \cdot (1 - e^{-\lambda t})^{n-x}$$

Also the MTBF of the (k, n) system is

$$\text{MTBF}(k, n) = \frac{1}{\lambda} \sum_{j=k}^{n} \left(\frac{1}{j}\right)$$

General expressions for the reliability and MTBF of (k, n) systems when the components do not have identical reliability expressions are very cumbersome. Explicit expressions for given structures can be obtained, however.[7]

Series-Parallel Systems

A series-parallel system of order (m, n) is a system consisting of m identical parallel systems with n components each, arranged in series, as shown in Fig. 39.7. If the components are independent and $R_i(t), i = 1, 2, \ldots, n$ is the reliability of the ith component, then the reliability of each parallel subsystem at time t is $\{1 - \prod_{i=1}^{n} [1 - R_i(t)]\}$. Because the system consists of m such subsystems in series, the system reliability is

$$R_S = \left\{1 - \prod_{i=1}^{n} [1 - R_i(t)]\right\}^m$$

In case each component has a constant failure rate

$$R_S = \left[1 - \prod_{i=1}^{n} (1 - e^{-\lambda t})\right]^m$$

and the system MTBF is

$$\text{MTBF} = \int_0^{\infty} R_S(t)\, dt = \frac{1}{\lambda} \int_0^1 \frac{(1 - x^n)^m}{(1 - x)}\, dx$$

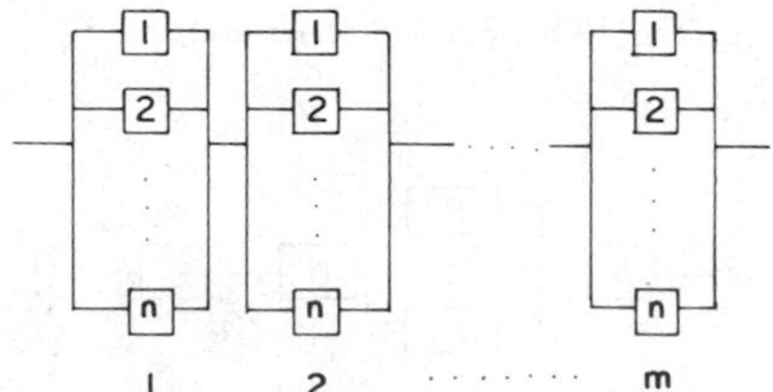

Fig. 39.7 Series-parallel system.

Parallel-Series Systems

A parallel-series system of order (m, n) consists of m identical series systems with n components each, arranged in parallel as shown in Fig. 39.8. If the components are independent, the reliability of each series subsystem at time t is $[R_i(t)]^n$. Because there are m such subsystems in parallel, the system reliability is

$$R_S(t) = 1 - Q_S(t) = 1 - [1 - R_i^n(t)]^m$$

For the special case when each component has a constant failure rate λ:

$$R_S(t) = 1 - (1 - e^{-n\lambda t})^m$$

and the system MTBF is

$$\text{MTBF} = \frac{1}{n} \sum_{j=1}^{m} \left(\frac{1}{j} \right)$$

In general, systems are a mixture of the series-parallel and parallel-series cases, and the reliability has to be calculated by first principles.

Consider a system composed of five subsystems connected as shown in Fig. 39.9. Each fails exponentially and the MTBFs of A, B, C, D, and E are 20, 10, 15, 25, and 30 hours, respectively. The reliability and the MTBF of the system for a period of two hours are obtained as follows.

$$R_S(2) = R_{\text{ADE}}(2) \cdot R_{\text{BC}}(2) = 0.752$$

where

$$R_{\text{ADE}}(2) = \exp\left[-\left(\frac{1}{20} + \frac{1}{25} + \frac{1}{30} \right) 2 \right] = 0.781$$

$$R_{\text{BC}}(2) = \exp\left(-\frac{2}{10} \right) + \exp\left(-\frac{2}{15} \right) - \exp\left(-\frac{2}{10} \right) \exp\left(-\frac{2}{15} \right) = 0.978$$

and

$$\text{MTBF} = \int_0^2 0.752\, dt = 1.504 \text{ hr}$$

Standby Redundancy with Perfect Switching

In the simplest case there is one primary component or subsystem and $(n - 1), n \geqslant 2$, cold (unpowered) components. The primary component operates until it fails, at which time one standby component is switched on. It is assumed that the switching is perfect and that the standby systems have zero dormancy failure rates. The system is considered failed when all n components have failed.

Consider a system with one primary and one standby component. Let their times to failure, pdf's, and cdf's be T_i, $f_i(t)$, and $F_i(t), i = 1, 2$, respectively. Then the probability of system failure by time t

PARALLEL SERIES SYSTEMS

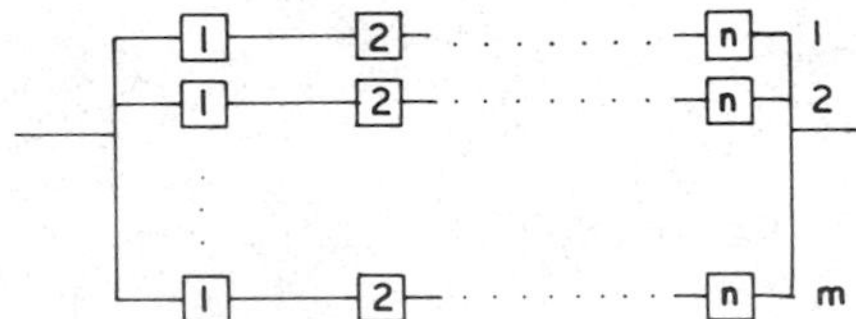

Fig. 39.8 Parallel-series system.

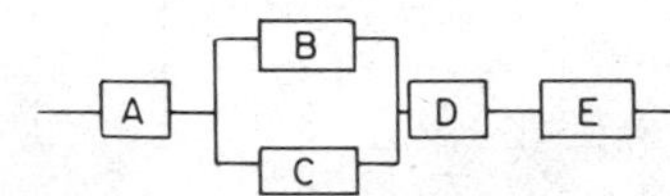

Fig. 39.9 Mixture of series and parallel systems.

is

$$Q_S(t) = P[\text{standby fails in } (t-u)/\text{primary fails in } (u, u+du)]$$

or

$$Q_S(t) = \int_0^t F_2(t-u) \cdot f_1(u)\, du$$

Note that we are really asking a question about the total life of two independent random variables T_1 and T_2, that is,

$$Q_S(t) = P(T_1 + T_2 \leqslant t)$$

or

$$Q_S(t) = \int_0^t F_2(t-u)\, dF_1(u) = F_1 * F_2$$

$F_1 * F_2$ is called the convolution of F_1 and F_2.

For the case where there are $(n-1)$ subsystems in standby and the cdf's of the n subsystems are $F_j(t)$, $j = 1, 2, \ldots, n$, we have

$$Q_S(t) = P(T_1 + T_2 + \cdots + T_n \leqslant t) = F_1 * F_2 * \ldots * F_n$$

and the system reliability is

$$R_S(t) = 1 - Q_S(t)$$

where $T_1, T_2, \ldots, T_n$ represent the failure times of the n subsystems.

If the T_i's are exponentially distributed with failure rates λ_i's, $i = 1, 2, \ldots, n$,

$$R_S(t) = \frac{\lambda_2\lambda_3\lambda_4 \ldots \lambda_n e^{-\lambda_1 t}}{(\lambda_2-\lambda_1)(\lambda_3-\lambda_1)\ldots(\lambda_n-\lambda_1)} + \frac{\lambda_1\lambda_3\lambda_4 \ldots \lambda_n e^{-\lambda_2 t}}{(\lambda_1-\lambda_2)(\lambda_3-\lambda_2)\ldots(\lambda_n-\lambda_2)}$$
$$+ \cdots + \frac{\lambda_1\lambda_2\lambda_3 \ldots \lambda_{n-1} e^{-\lambda_n t}}{(\lambda_1-\lambda_n)(\lambda_2-\lambda_n)\ldots(\lambda_{n-1}-\lambda_n)}$$

and

$$\text{MTBF} = \int_0^\infty R_S(t) = \sum_{i=1}^{n} \left(\frac{1}{\lambda_i}\right)$$

If each subsystem is identical with a constant failure rate λ,

$$R_S(t) = e^{-\lambda t} \sum_{i=0}^{n-1} \frac{(\lambda t)^i}{i!}$$

and

$$\text{MTBF} = \frac{n}{\lambda}$$

Reliability Improvement

As seen, the reliability of a system can be improved by adding either components in parallel or components in standby redundancy. Some points relevant to improving reliability this way are summarized here.

1. Suppose p is the reliability of a single component and we add identical components in parallel, one at a time. Then it can be shown that the greatest incremental improvement in reliability is obtained when $p = 1/(1+i)$, $i = 1, 2, \ldots, N-1$.

2. Given $(N-1)$ redundant components, the overall improvement in reliability is the greatest if $p = 1 - (1/N)^{1/(N-1)}$.

3. A system designer may have the option of adding parallel elements or using improved elements in a nonparallel configuration to improve reliability. We compare these two choices, assuming that effectiveness, cost, weight, maintenance, and other related limiting factors are equivalent. Consider a system with two components each with failure rate λ. Then for time t the reliability of a parallel system is

$$R_S = 2e^{-\lambda t} - e^{-2\lambda t}$$

The failure rate of the improved element will have to be $(2/3)\lambda$ to get the same reliability.

39.5 REPAIRABLE SYSTEMS

In this section we discuss the performance measures of systems that consist of units that can be repaired on failure. Such systems operate for a time in what is called an up state. If a unit fails, it undergoes repair and the system continues to function in an up state if a redundant unit can be switched on. Otherwise the system fails and is said to go into a down state. The failure and repair times are random variables and hence the system behavior is also a random variable and is generally described by a stochastic process. System performance measures are obtained by analyzing the appropriate stochastic processes.

We first present some relevant terms and concepts. Next we obtain the performance measures of selected system configurations. Analysis of complicated configurations gets very involved and general expressions are not easily obtainable.

Stochastic Processes

A stochastic process $[X(t); t \in T]$ is a family of random variables $X(t)$. The variable t is often interpreted as time, and hence $X(t)$ is used to represent the state of the process at time t.

A Markov process is a stochastic process with the property that given the value of $X(t)$, the probability of $X(s + t)$ where $s > 0$ is independent of the previous values of $X(t)$. More formally, the process $[X(t), t \in T]$ is said to be a Markov process if

$$P[X(t) \leqslant x | X(t_1) = x_1, X(t_2) = x_2, \ldots, X(t_n) = x_n] = P[X(t) \leqslant x | X(t_n) = x_n],$$
$$\text{where} \quad t_1 < t_2 < \cdots < t_n < t$$

If T is a countable set, then the process is said to be a discrete time process. If T is an open or closed interval on the real line, the process is said to be a continuous time process.

A discrete-time Markov process with a finite or countable state space is said to be a Markov chain. A continuous time process with a finite or countable state space is said to be a continuous-parameter Markov chain.

System Maintainability

System maintainability is a characteristic of design and installation which is expressed as the probability that an item will be restored to specified conditions within a given time when a maintenance action is performed in accordance with prescribed procedures and resources. The total downtime is a random variable and is referred to as the repair time. If the total downtime T has probability density function $f(t)$, the probability of a failed system returning to service by some time t is given by

$$P[T \leqslant t] = \int_0^t f(x)\, dx$$

Most often the downtime distribution is described by the exponential or lognormal distribution. The mean downtime or mean time to repair (MTTR) is given by

$$\text{MTTR} = \int_0^\infty t f(t)\, dt$$

If $f(t)$ is an exponential distribution with parameter μ, that is, $f(t) = \mu e^{-t\mu}$, then $\text{MTTR} = 1/\mu$.

System Availability

System availability is a measure of the "availability" of the system and can be expressed as point availability, interval availability, and inherent availability. Each of these is described below.

Point Availability. Point availability, $A(t)$, is the probability that the system will be able to operate within the tolerance at a given instant of time.

$$\text{Let } Z(t) = \begin{cases} 1 & \text{if the system is up at time } t \\ 0 & \text{if the system is down at time } t \end{cases}$$

Then

$$A(t) = P[Z(t) = 1] = 0 \times P[Z(t) = 0] + 1 \times P[Z(t) = 1]$$

or

$$A(t) = E[Z(t)]$$

Interval Availability. Interval availability, $\overline{AV}(a,b)$, is the expected fraction of a given interval of time that the system will be able to operate within the tolerance.

$$\overline{AV}(a,b) = \frac{1}{b-a}\int_a^b A(t)\,dt, \qquad \text{for time interval } (a,b)$$

Inherent Availability. Inherent availability (limiting interval availability) is defined as the expected fractional amount of time in a continuum of operating time that the system is in an up state, that is, the expected fraction of time in the long run that the system operates satisfactorily. It is more commonly denoted by UTR (uptime ratio) and can be written as

$$\text{UTR} = \overline{AV}(+\infty) = \lim_{t\to\infty} \overline{AV}(t)$$

It can be shown that

$$\text{UTR} = \frac{\text{MTBF}}{\text{MTBF} + \text{MTTR}}$$

An additional term used occasionally is the downtime ratio, denoted by DTR, which is defined as

$$\text{DTR} = 1 - \text{UTR}$$

or

$$\text{DTR} = \frac{\text{MTTR}}{\text{MTBF} + \text{MTTR}}$$

The downtime ratio thus represents the probability of the system being down at any time over a sufficiently large time period.

One-Unit System

Suppose we are given a system with constant failure rate λ and constant repair rate μ. At any instant the system must be in either one of the following mutually exclusive states:

State 0. The system is operating.

State 1. The system has failed and repairs have begun.

Let $P_i(t)$, $i = 0, 1$, denote the probability that at time t the system is in state i. The performance measures of interest are the probabilities $P_0(t)$, $P_1(t)$ and availabilities $\overline{AV}(t)$ and $\overline{AV}(\infty)$. To obtain expressions for these quantities we consider the transition diagram of the system, which gives the probabilities of going from state 0 to state 1 and back to state 0. It also gives the probabilities of the system staying in state 0 or state 1. The required expressions are then obtained by solving the defining equations under the specified boundary conditions. For details of such derivations, see Refs. 1 and 12.

By following the previous procedure one can obtain the performance measures for a one-unit repairable system:

$$P_0(t) = \frac{1}{\lambda+\mu}\{\mu - [\mu - P(\mu+\lambda)]e^{-(\mu+\lambda)t}\}$$

and

$$P_1(t) = \frac{1}{\lambda+\mu}\{\lambda + [\mu - P(\mu+\lambda)]e^{-(\mu+\lambda)t}\}$$

Note that $P_0(t) + P_1(t) = 1$, that is, the system has to be either operating or in repair so that in this case

$$A(t) = P_0(t)$$

Also

$$\overline{AV}(t) = \frac{\mu}{\lambda+\mu} + \frac{\mu - P(\mu+\lambda)}{t(\lambda+\mu)^2}(1 - e^{-(\mu+\lambda)t})$$

and as $t \to \infty$,

$$\text{UTR} = \overline{AV}(\infty) = \frac{\mu}{\lambda+\mu}$$

This can also be obtained directly from the definition of UTR as

$$\text{UTR} = \frac{\text{MTBF}}{\text{MTBF} + \text{MTTR}} = \frac{1/\lambda}{1/\lambda + 1/\mu} = \frac{\mu}{\lambda+\mu}$$

In most practical situations the MTBF is much smaller than the MTTR, and we get

$$\overline{AV}(\infty) \approx 1 - \frac{\lambda}{\mu}$$

Two-Unit Standby System

Consider a system consisting of two units each with constant failure rate λ and constant repair rate μ. Both units are originally operable; one component is actually in operation and the other component is in inactive standby. We assume that switching and sensing are perfect and that no warm-up is required.

At any time t the system must be in one of the following three states with probability, say, $P_i(t)$, $i = 0, 1, 2$.

State 0. Both components are operable but only one is operating.

State 1. One component has failed and the other is operating.

State 2. Both components have failed.

By using the theory of stochastic processes, the probabilities $P_i(t)$, the reliability $R_S(t)$, and the system MTBF can be obtained in terms of λ and μ and are as follows:

$$P_0(t) = \frac{\lambda + \mu + s_1}{s_1 + s_2} e^{s_1 t} - \frac{\lambda + \mu + s_2}{s_1 - s_2} e^{s_2 t}$$

$$P_1(t) = \frac{\lambda}{s_1 - s_2}(e^{s_1 t} - e^{s_2 t})$$

$$R(t) = P_0(t) + P_1(t) = \frac{s_1 e^{s_2 t} - s_2 e^{s_1 t}}{s_1 - s_2}$$

and

$$\text{MTBF} = \frac{2\lambda + \mu}{\lambda^2}$$

where

$$s_1 = \frac{-(2\lambda + \mu) + (4\lambda\mu + \mu^2)^{1/2}}{2}$$

and

$$s_2 = \frac{-(2\lambda + \mu) - (4\lambda\mu + \mu^2)^{1/2}}{2}$$

Recall that the MTBF without repair (i.e., $\mu = 0$) is given by MTBF $= 2/\lambda$. The ratio of these two MTBF values yields $1 + \mu/2\lambda$, which denotes the measure of the improvement in system MTBF due to repair.

Two-Unit Parallel System

In this case our basic assumptions are the same as in the standby system except that we assume that both units are operable and we redefine state 0 to be that state in which both components are operating. With use of the theory of stochastic processes, the system performance measures are obtained to be

$$P_0(t) = \frac{\lambda + \mu + s_1}{s_1 - s_2} e^{s_1 t} - \frac{\lambda + \mu + s_2}{s_1 - s_2} e^{s_2 t}$$

$$P_1(t) = \frac{2\lambda}{s_1 - s_2}(e^{s_1 t} - e^{s_2 t})$$

$$P_2(t) = 1 + \frac{s_2 e^{s_1 t} - s_1 e^{s_2 t}}{s_1 - s_2}$$

$$R(t) = \frac{s_1 e^{s_2 t} - s_2 e^{s_1 t}}{s_1 - s_2}$$

and

$$\text{MTBF} = \frac{3\lambda + \mu}{2\lambda^2}$$

where

$$s_1 = \frac{-(3\lambda + \mu) + (\lambda^2 + 6\lambda\mu + \mu^2)^{1/2}}{2}$$

and

$$s_2 = \frac{-(3\lambda + \mu) + (\lambda^2 + 6\lambda\mu + \mu^2)^{1/2}}{2}$$

The system MTBF without repair (i.e., $\mu = 0$) is given by $3/2\lambda$, and so the measure of improvement provided by allowing repair is $1 + \mu/3\lambda$.

Standby System with Parallel Repair

Let us consider a standby system consisting of n units each with the same failure rate λ and repair rate μ. Suppose $r \leqslant n - 1$ repair facilities are provided and that we initially have one unit in operation and the other units as standby. We assume that a unit is not subject to failure while waiting in standby, and that the time to switch from a failed unit to a standby unit is negligible. When a unit is repaired, it is returned to a standby condition. Thus a unit can be in one of three possible states: operating in the system, waiting in standby, or waiting for or receiving repair service. Let i denote the system state in which exactly i ($i = 0, 1, \ldots, n$) units have failed and are either waiting for or receiving repair. In the steady state, that is, as $t \to \infty$, we have

$$P_0 = \left[1 + \sum_{i=1}^{r} \frac{1}{i!}\left(\frac{\lambda}{\mu}\right)^i + \sum_{i=r+1}^{n} \frac{1}{r!r^{i-r}}\left(\frac{\lambda}{\mu}\right)^i\right]^{-1}$$

$$P_i = \begin{cases} \dfrac{1}{i!}\left(\dfrac{\lambda}{\mu}\right)^i P_0, & \text{if } 1 \leqslant i \leqslant r \\[2ex] \dfrac{1}{r!r^{i-r}}\left(\dfrac{\lambda}{\mu}\right)^i P_0, & \text{if } r + 1 \leqslant i \leqslant n \end{cases}$$

and

$$\text{UTR} = 1 - P_n$$

In particular, when $n = 2$ and $r = 1$,

$$\text{UTR} = \frac{\mu(\mu + r)}{\mu^2 + \lambda\mu + \lambda^2}$$

and when $n = 2$ and $r = 2$

$$\text{UTR} = \frac{2\mu(\lambda + \mu)}{\lambda^2 + 2\mu\lambda + 2\mu^2}$$

The measure of improvement in uptime ratio due to having two repair facilities instead of one is

$$\frac{2\mu^2 + 2\lambda\mu + 2\lambda^2}{\lambda^2 + 2\lambda\mu + 2\mu^2} = 1 + \frac{\lambda^2}{(\lambda + \mu)^2 + \mu^2}$$

Two Components in Series and One Repair Facility

Consider the simple system where two equipments are connected in series such that if either fails the system fails. We assume that each equipment fails at the same rate λ and that each can be repaired at the same rate μ.

At any time t the system must be in one of the three following states with probabilities, say, $P_i(t)$, $i = 0, 1, 2$.

State 0. Both equipments are operating.

State 1. One equipment is operating and the second is under repair.

State 2. Both equipments are under repair.

In the steady state the above probabilities are

$$P_2 = \frac{2\lambda^2}{\mu^2 + 2\lambda\mu + 2\lambda^2}$$

$$P_1 = (\mu/\lambda)P_2$$

$$P_0 = 1 - P_1 - P_2$$

and

$$\overline{AV}(\infty) = \frac{\mu^2}{\mu^2 + 2\lambda\mu + 2\lambda^2}$$

n Components in Series and One Repair Facility

In general it can be shown that if $X = \mu/\lambda$, the following recurrence relation holds:

$$P_n = X^0 P_n$$

$$P_{n-1} = X^1 P_n$$

$$P_{n-2} = X^2 P_n$$

$$\dots$$

$$P_0 = (X^n/n!)P_n$$

Since $P_0 + P_1 + P_2 + \dots + P_n = 1$, we have

$$P_n = 1/\sum_{j=0}^{m} (x^j/j!)$$

and

$$\overline{AV}(\infty) = A(\infty) = P_0 = \frac{X^n}{n! \sum_{j=0}^{n} (X^j/j!)}$$

The expected number of equipments that will be down at any time can be found from

$$E(n) = \sum_{j=0}^{n} n_j P_j = \frac{\sum_{j=0}^{n} [jX^{n-j}/(n-j)!]}{\sum_{j=0}^{n} (X^j/j!)}$$

where $X \leqslant 1$.
The variance of the distribution of the number of equipments that will be down at any time is

$$\sigma_2(n) = \frac{\sum_{j=0}^{n} [j^2 X^{n-j}/(n-j)!]}{\sum_{j=0}^{n} (X^j/j!)} - \left[\frac{\sum_{j=0}^{n} jX^{n-j}/(n-j)!}{\sum_{j=0}^{n} X^j/j!} \right]^2$$

Two Components in Series and Two Repair Facilities

Proceeding as we did just previously we have

$$P_2 = \frac{\lambda}{2\mu} P_1$$

$$P_1 = \frac{2\lambda}{\mu} P_0$$

$$\overline{AV}(\infty) = A(\infty) = P_0 = \mu^2/(\lambda + \mu)^2$$

n Components in Series and n Repair Facilities

If as before we let $X = \mu/\lambda$, we see that the following recurrence relations hold:

$$P_n = \binom{n}{n} X^0 P_n$$

$$P_{n-1} = \binom{n}{n-1} X^1 P_n$$

$$P_{n-2} = \binom{n}{n-2} X^2 P_n$$

$$\cdots$$

$$P_0 = \binom{n}{n-n} X^n P_n$$

$$\overline{AV}(\infty) = A(\infty) = P_0 = \frac{X_n}{\sum_{j=0}^{n} \binom{n}{n-j} X^j} = \frac{\mu^n}{(\lambda + \mu)^n}$$

This result was to be expected, since for each equipment the existence of a repairman, who works independently of the rest of the repairmen, defines a set where each equipment's availability is independent of the availability of others. Thus if the availability of any one equipment is the ratio $\mu/(\lambda + \mu)$, the probability that all n are available is simply the compound event $[\mu/(\lambda + \mu)]^n$.

The expected value and the variance of the number of equipments down at any time are

$$E(n) = n\left(\frac{1}{1+X}\right)$$

and

$$\sigma^2(n) = n\left(\frac{1}{1+X}\right)\left(\frac{X}{1+X}\right)$$

Thus if there are four equipments and $\lambda = \mu = 1$, we would expect two equipments to be down at any time and the square root of the variance would be one equipment.

39.6 RELIABILITY ESTIMATION

In most practical cases the parameters of the failure model or even the model itself are not known. These parameters have to be estimated from the available data from life tests. Various methods are available for obtaining estimates of the parameters of the failure model, for fitting a specified failure model to a given set of data, and for fitting a model to an observed hazard function. The estimated parameters are then used for reliability assessment. The commonly used methods for the stated purpose are:

Method of maximum likelihood.
Method of matching moments.
Method of least squares.
Probability paper plotting.

Methods of estimating the parameters of the failure model can be for point estimation and for internal estimation. Also we use the appropriate goodness of fit tests for judging the adequacy of the fitted failure model as a descriptor of the observed failure data.

In this section we describe the listed methods for estimating the model parameters and two methods for goodness of fit tests. For a detailed treatment of this subject, see Mann and colleagues[4] and Hahn and Shapiro.[9]

Relevant Terms

Point Estimation. The choice of a single number, called a statistic, for which there is some expectation or assurance that it is "reasonably close" to the parameter it is supposed to estimate.

Unbiased Estimator. A statistic $\hat{\lambda}$ is said to be an unbiased estimator of the parameter λ if

$$E(\hat{\lambda}) = \lambda$$

Interval Estimation. The process of choosing two statistics $\hat{\lambda}_1, \hat{\lambda}_2$ such that

$$P(\hat{\lambda}_1 < \lambda < \hat{\lambda}_2) = 1 - \alpha$$

After the values are chosen, $(\hat{\lambda}_1, \hat{\lambda}_2)$ is called a 100 $(1 - \alpha)\%$ confidence interval for λ.

Method of Maximum Likelihood

Consider $t_1, t_2, \ldots, t_n$ as a random sample of lifetimes with a pdf $f(t)$. Let λ be the parameter of interest of this distribution; to reflect this, $f(t)$ can be written $f(t, \lambda)$. Then the joint probability of the occurrence of $t_1, t_2, \ldots, t_n$ can be written as

$$f(t_1, t_2, \ldots, t_n; \lambda) = f(t_1, \lambda), f(t_2; \lambda) \ldots f(t_n; \lambda)$$

since $t_1, t_2, \ldots, t_n$ are independent observations. Now if the observations have been made, we might be interested in obtaining an estimate of λ. Given $t_1, t_2, \ldots, t_n$ as observed lifetimes, we can write $l(\lambda | t_1, t_2, \ldots, t_n)$ as the likelihood function for λ.

The method of maximum likelihood consists of obtaining the value $\hat{\lambda}$ of λ such that the described likelihood function is maximized.

To see how a maximum likelihood estimate (mle) is obtained, we consider the gamma distribution discussed in Section 39.3. The parameters of the distribution are k and λ, so that for observed values of failure times $t_1, t_2, \ldots, t_n$

$$l(\lambda, k | t_1, t_2, \ldots, t_n) = \left(\frac{\lambda^k}{\Gamma k}\right)^n \left(\prod_{i=1}^{n} t_i\right)^{k-1} \exp\left(-\lambda \prod_{i=1}^{n} t_i\right)$$

The mle's $\hat{\lambda}$ and $\hat{k}$ of λ and k, respectively, are obtained as solutions of

$$\frac{\partial \ln l}{\partial \lambda} = 0 \quad \text{and} \quad \frac{\partial \ln l}{\partial k} = 0$$

where l represents the likelihood function. The expressions for $\hat{\lambda}$ and $\hat{k}$ are

$$\hat{\lambda} = \frac{\hat{k} - \frac{1}{2} + \frac{1}{24}(\hat{k} - \frac{1}{2})}{\left(\prod_{i=1}^{n} t_i\right)^{1/n}}$$

$$\hat{k} = \hat{\lambda} \cdot \sum_{i=1}^{n} t_i / n$$

Method of Matching Moments

This is an easy-to-use method for estimating the parameters of the failure model from observed lifetimes. However, it is generally less efficient than, say, the method of maximum likelihood. In this method the sample mean and the sample variance are equated to the theoretical mean and theoretical variance of the failure model. These equations are then solved to obtain the estimates of the parameters.

For example, suppose that the time to failure of an electrical generator is represented by a gamma distribution. Seven similar generators were observed to fail at 100, 110, 150, 175, 185, 200, and 220 hours. For these data the sample mean = 162.9 and the sample variance = 2032. The theoretical mean of a gamma distribution is k/λ and the theoretical variance is k/λ^2, so that $k/\lambda = 162.9$ and $k/\lambda^2 = 2032$. This gives $\hat{k} = 13.06$ and $\hat{\lambda} = 12.47$.

Method of Least Squares

To understand the general method, let us suppose that we have n paired outcomes of an experiment $x_1, x_2, \ldots, x_n$ and $y_1, y_2, \ldots, y_n$, where x's are the independent and y's the dependent variables. We are interested in determining a relationship between x's and y's for prediction purposes.

Let us postulate the model

$$y = ax + b + \epsilon$$

where a and b are the constants and ϵ is the random error in prediction. The least squares method consists of obtaining those estimates of a and b that minimize the error sum of squares $\sum_{i=1}^{n} \epsilon_i^2$. On the basis of this criterion, it can be shown that the least squares estimates of a and b are

$$\tilde{a} = \frac{\sum_{i=1}^{n} y_i \left(x_i - \sum_{i=1}^{n} x_i/n \right)}{\sum_{i=1}^{n} \left(x_i - \sum_{i=1}^{n} x_i/n \right)^2}$$

and

$$\tilde{b} = \sum_{i=1}^{n} y_i/(n - \tilde{a}) \sum_{i=1}^{n} x_i/n$$

Probability Paper Plotting

This is a graphical method of estimating parameters and of subjectively checking how well a distribution fits the given failure data. The procedure consists of plotting the ungrouped or grouped data on the probability paper for the chosen distribution. If the plotted points fall on a straight line, the fit is assumed to be good. For details of the procedure see Hahn and Shapiro.[9]

Goodness of Fit Tests

We describe two commonly used goodness of fit tests for checking the adequacy of fitted models.

Kolmogorov-Smirnov Test. This test is used when relatively small amounts of data are available and it is desired to determine if the fitted distribution is adequate. The procedure is illustrated here by an example concerning time to failure data.

1. Twenty observed times to failure, $t_1, t_2, \ldots, t_{20}$, are given in Table 39.2. The failure times are ordered according to their numerical value.
2. On the basis of historical information, engineering judgment, and so on, is taken to be a Weibull distribution.

TABLE 39.2. TIME TO FAILURE DATA AND KOLMOGOROV-SMIRNOV TEST

t	$\hat{F}(t)$	$F(i)$	$\lvert\hat{F}(t) - F(i)\rvert$
92	0.03	0.05	0.02
130	0.05	0.10	0.05
233	0.12	0.14	0.02
260	0.14	0.19	0.05
320	0.18	0.24	0.06
325	0.19	0.29	0.10
420	0.26	0.33	0.07
430	0.27	0.38	0.11
465	0.30	0.43	0.13
518	0.34	0.48	0.14
640	0.43	0.52	0.09
700	0.48	0.57	0.09
710	0.49	0.62	0.13
770	0.53	0.67	0.14
830	0.57	0.71	0.14
1010	0.68	0.76	0.08
1020	0.68	0.81	0.13
1280	0.80	0.86	0.06
1330	0.82	0.90	0.08
1690	0.91	0.95	0.04

3. The shape and scale parameters are estimated and the estimates are $\hat{\beta} = 1.50$ and $\hat{\eta} = 930.84$.
4. The computed values of the estimated cumulative distribution function $\hat{f}(t)$ are given in Table 39.2. The values are computed at each failure time.
5. The percentile for each of the (i) failure times is computed from $F(i) = i/(n+1)$. Table 39.2 gives the values of $F(i)$ and $|\hat{F}(t) - F(i)|$.
6. $\max|\hat{F}(t) - F(i)|$ is compared with the critical value of the Kolmogorov-Smirnov statistic for the given n and the specified significance level.

For the example, $n = 20$ and the significance level is, say, 0.05. From the tables (see Bury[5]), the critical value is 0.29; from Table 39.2 the maximum value of $|\hat{F}(t) - F(i)|$ is 0.14. Therefore in this case the hypothesis that the failure data came from a Weibull distribution cannot be rejected at the 5% level of significance.

χ^2 Goodness of Fit Test. This test is generally used to check the adequacy of the fitted distribution when relatively large samples are available.

1. The number of cycles to failure for a group of 50 relays on a life test is given in Table 39.3.
2. The failure distribution is assumed to be two-parameter Weibull and the estimated parameters are $\hat{\beta} = 1.2$ and $\hat{\eta} = 2539.55$.
3. The cycles to failure are divided into intervals; and the number of failures in each interval, denoted by f_i, is shown in Table 39.4.
4. The expected number of failures F_i in each interval is obtained from

$$P(\text{failure in an interval}) = F(x_n) - F(x_{n-1})$$

where $F(x) = 1 - \exp(-x/\eta)^{\beta}$.

5. The χ^2 statistic is computed from

$$\chi^2 = \sum_{i=1}^{k} \frac{(f_i - F_i)^2}{F_i}$$

TABLE 39.3. NUMBER OF CYCLES TO FAILURE FOR GROUP OF 50 RELAYS

1283	4865	8185	13210	28946
1887	5147	8559	14833	29254
1888	5350	8843	14840	30822
2357	5353	9305	14988	38319
3137	5410	9460	16306	41554
3606	5536	9595	17621	42870
3752	6499	10247	17807	62690
3914	6820	11492	20747	63910
4394	7733	12913	21990	68888
4398	8025	12937	23449	73473

TABLE 39.4 TABLE FOR χ^2 GOODNESS-OF-FIT TEST BASED ON DATA IN TABLE 39.3

Interval	f_i	$F(x)$	$[F(x_n) - F(x_{n-1})]$	F_i	$(f_i - F_i)^2/F_i$
0–4,000	8	0.16	0.16	8	0
4,000–7,200	10	0.30	0.14	7	1.29
7,200–13,000	12	0.52	0.22	11	0.11
13,000–18,000	7	0.66	0.14	7	0.0
18,000–25,000	3	0.80	0.14	7	2.29
25,000–∞	10	1.00	0.20	10	0
Total	50			50	$\chi^2 = 2.69$

where k = number of intervals. This χ^2 statistic is compared with the tabulated value to determine the adequacy of fit.

The degrees of freedom for this case are $(k-1)-v=(6-1)-2=3$. The value from the χ^2 table for 3 degrees of freedom and the 0.05 level of significance is 7.815. Since 3.69 does not exceed the tabled value, the hypothesis that these data came from a Weibull distribution cannot be rejected at the 5% level of significance.

39.7 SOFTWARE RELIABILITY

As mentioned earlier, an important quality attribute of any computer system is the degree to which it can be relied on to perform its intended function.† Measurement, prediction, and improvement of this attribute have concerned computer designers and users from the early days of computer evolution. Until the late 1960s, attention was almost solely on the hardware-related performance of the system. In the early 1970s, software also became a matter of concern, primarily due to a continuing increase in the ratio of software-to-hardware costs, in both the development and operational phases of the system. Software, also called a program, is essentially an instrument for transforming a discrete set of inputs into a discrete set of outputs. Since to a large extent software is produced by humans, the finished product is often imperfect. It is imperfect in the sense that a discrepancy exists between what the program can do and what the user of the computing environment wants it to do. These discrepancies are called software errors.

Even if we know that software contains errors, we generally do not know their exact identity. Currently there are two approaches for exposing software errors: program proving and program testing. Program proving, though formal and mathematical, is still an imperfect tool for verifying program correctness. Program testing is more practical and somewhat heuristic. It consists of a symbolic or physical execution of a set of test cases with the intent of exposure of any errors embedded in the program. However, it too is an imperfect approach for guaranteeing program correctness. Owing to the imperfectness of these approaches in guaranteeing a correct program, a quantifiable metric is needed which reflects the degree of program incorrectness. Such a measure is also useful in planning and controlling additional resources needed for enhancing software quality. One such measure that has been in use for the past decade is software reliability.

There are a number of conflicting views as to what software reliability is and how it should be quantified. For example, some people believe that the measure should be binary in nature, so that an incorrect program would have zero reliability while a perfect program would have a reliability of one. Others, however, feel that the measure should be a percentage of the time that the program works as intended by the user. Following the latter approach we define software reliability as follows: Let E be a class of errors, defined arbitrarily, and let T be a measure of relevant time, the units of which are dictated by the application at hand. Then the reliability of the program with respect to the class of errors E and with respect to metric T is the probability that no error of the class occurs during the execution of the program for a prespecified period of relevant time.

A number of models have been proposed during the last 10 years for estimating software reliability. Most of them are based on the error history of software. They can be classified as follows according to the nature of the failure process studied.

1. **Time Between Failures Models.** The random variable X_i, time between the $(i-1)$st and the ith failures, is assumed to follow a distribution whose parameter(s) generally depends on the number of remaining errors during this interval. Estimates of software reliability and other performance measures use these models based on the times between the first k failures, $x_1, x_2, \ldots, x_k$.
2. **Failure Counting Models.** The random variable $[N(t); t \geqslant 0]$, the number of failures to time t_i, is assumed to follow a stochastic process generally based on a time-dependent (discrete or continuous) failure rate. Appropriate performance measures are then estimated from the observed values of $N(t_1), N(t_3), \ldots, N(t_i)$ or of $x_1, x_2, \ldots, x_k$.
3. **Error Seeding Models.** A known number of errors are seeded in a program that is assumed to have an unknown number of indigenous errors. After the program is run, the exposed seeded and indigenous errors are counted and the resulting numbers are used to obtain an estimate of the error content of the program. Estimates of software reliability are then obtained from this value.
4. **Input Domain-Based Models.** The basic approach taken here is to generate a set of test cases from an input distribution that is assumed to be representative of the operational usage of the program. Because of the difficulty in obtaining this distribution, the input domain is partitioned into

† This is especially true for a computer system that in many cases is responsible for life-critical functions.

a set of equivalence classes, each of which is usually associated with a program path. An estimate of program reliability is obtained from the failures observed during physical or symbolic execution of the test cases sampled from the input domain.

For a detailed discussion of these models, their limitations, and their applicability, see Goel.[11] For models dealing with both the hardware and software failures, see Goel and Soenjoto.[12]

39.8 RELIABILITY ENGINEERING

Reliability engineering is the technical discipline of estimating, controlling, and managing the probability of failure in devices, equipment, and systems. In a sense it is engineering in its most practical form, since it consists of two fundamental aspects: paying attention to detail and handling uncertainties. This section describes a "bag of tools" that the design engineer needs to design a system to meet a specified reliability requirement. The earlier sections laid the theoretical mathematical foundation for the reliability engineering discipline; this section emphasizes practical approaches to the solution of reliability engineering problems.

Reliability Specification

The first step in the reliability engineering process is specification of the reliability that the equipment/system must be designed to achieve. The essential elements of a reliability specification are:

A quantitative statement of the reliability requirement.

A full description of the environment in which the equipment/system will be stored, transported, operated, and maintained.

The time measure or mission profile.

A clear definition of what constitutes failure.

A description of the test procedure with accept/reject criteria that will be used to demonstrate the specified reliability.

Methods of Specifying the Reliability Requirement. To be meaningful, a reliability requirement must be specified quantitatively. There are four basic ways in which a reliability requirement may be defined:

1. **"Mean life" or mean time between failures, MTBF.** This definition is useful for long-life systems in which the form of the reliability distribution is not too critical or where the planned mission lengths are always short relative to the specified mean life. Although this definition is adequate for specifying life, it gives no positive assurance of a specified level of reliability in early life, except as the assumption of an exponential distribution can be proven to be valid.
2. **Probability of survival for specified time *t*.** This defintion is useful for defining reliability when high reliability is required during the mission period but mean time to failure beyond the mission period is of little tactical consequence except as it influences availability.
3. **Probability of success, independent of time.** This defintion is useful for specifying the reliability of one-shot devices. It is also used for items that are cyclic, such as the flight reliability of missiles, the launch reliability of launchers, and the detonation reliability of warheads.
4. **"Failure rate" over specified period of time.** This definition is useful for specifying the reliability of parts, units, and assemblies whose mean life is too long to be meaningful or whose reliability for the time period of interest approaches unity.

The reliability requirement may be specified in either of two ways: as a nominal value with which the customer would be statisfied, on the average, or as a minimum value below which the customer would find the system totally unacceptable and not to be tolerated in the operational environment—a value based on operational requirements. Of the two methods, the first is by far the best, since it automatically establishes the design goal at or above a known minimum.

As an example, consider a complex radar which has both search and track functions. The search function can be operated in both a low and a high power mode. The reliability requirement for this system could be expressed as follows:

The reliability of the system shall be at least:

case I—High power search—28 hours MTBF;

case II—Low power search—40 hours MTBF;

case III—Track—0.98 probability of satisfactory performance for 1/2 hour.

Description of Environment and/or Use Conditions. The reliability specification must cover all aspects of the use environment to which the item will be exposed and that can influence the probability of failure. The specification should establish in standard terminology the "use" conditions under which the item must provide the required performance. These conditions refer to all known use conditions under which the specified reliability is to be obtained, including the following:

Temperature	Penetration/abrasion
Humidity	Ambient light
Shock	Mounting position
Vibration	Weather (wind, rain, snow)
Pressure	Operator skills

Time Measure or Mission Profile. Time is vital to the quantitative description of reliability. It is the independent variable in the reliability function. The system usage from a time standpoint in large measure determines the form of the reliability expression, of which time is an integral part. For those cases where a system is not designed for continuous operation, total anticipated time profile or time sequences of operation should be defined either in terms of duty cycles or profile charts.

Clear Definition of Failure. A clear unequivocal definition of "failure" must be established for the equipment or system in relation to its important performance parameters. Succcssful system (or equipment) performance must be defined. It must also be expressed in terms that will be measurable during the demonstration test.

Parameter measurement will usually include both go/no-go performance attributes and variable performance characteristics. Failure of go/no-go performance attributes such as channel switching, target acquisition, motor ignition, and warhead detonation are relatively easy to define and measure to provide a yes/no decision boundary. Failure of a variable performance characteristic, on the other hand, is more difficult to define in relation to the specific limits outside of which system performance is considered unsatisfactory. The limits of acceptable performance are those beyond which a mission may be degraded to an unacceptable level.

Description of Method(s) for Reliability Demonstration. It is not enough merely to specify the reliability requirement. One must also delineate the test(s) that will be performed to verify whether the specified requirement has been met. In essence, the element of reliability specification should answer the followng questions:

How the equipment/system will be tested: the specified test conditions, for example, environmental conditions, test measures, length of test, equipment operating conditions, accept/reject criteria, test reporting requirements, and so on.

Who will perform the tests: contractor, government, independent organization.

When the tests will be performed: development, production, field operation.

Where the tests will be performed: contractor's plant, government organization.

Reliability Apportionment/Allocation

The first step in the design process is to translate the overall system reliability requirement into reliability requirements for each of the subsystems. This process is known as reliability apportionment (or allocation).

The allocation of system reliability involves solving the basic inequality:

$$f(\hat{R}_1, \hat{R}_2, \ldots, \hat{R}_n) \geqslant R_S$$

where $\hat{R}_i$ = allocation reliability parameter for ith subsystem
R_S = system reliability requirement parameter
f = functional relationship between subsystem and system reliability

As an example, for a simple series system in which the R's represent probability of survival for t hours, the equation becomes

$$\hat{R}_1(t) \cdot \hat{R}_2(t) \cdot \ldots \cdot \hat{R}_n(t) \geqslant R_S(t)$$

Theoretically, this equation has an infinite number of solutions, assuming no restrictions on the allocation. The problem is to establish a procedure that yields a unique or limited number of solutions by which consistent and reasonable reliabilities may be allocated. For example, the allocated reliability for a simple subsystem of demonstrated high reliability should be greater than for a complex subsystem whose observed reliability has always been low.

The allocation process is approximate. The reliability parameters apportioned to the subsystems are used as guidelines to determine design feasibility. If the allocated reliability for a specific subsystem cannot be achieved at the current state of technology, the system design must be modified and the allocations reassigned. This procedure is repeated until an allocation is achieved that satisfies the system level requirement and all constraints and results in subsystems that can be designed within the state of the art.

In the event that it is found that even with reallocation some of the individual subsystem requirements cannot be met within the current state of the art, the designer must use one or any number of the following approaches (assuming that they are not mutually exclusive) to achieve the desired reliability:

Find more reliable component parts to use.

Simplify the design by using fewer component parts, if this is possible without degrading performance.

Apply component derating techniques to reduce the failure rates below the averages.

Use redundancy for those cases where the first three approaches do not apply.

It should be noted that the allocation process can, in turn, be performed at each of the lower levels of the system hierarchy, for example, equipment, module, component.

Next we describe six different approaches to reliability allocation. These approaches differ in complexity, depending on the amount of subsystem definition available and the degree of rigor to be employed.

Equal Apportionment Technique. In the absence of definitive information on the system other than the fact that n subsystems are to be used in series, equal apportionment to each subsystem would seem reasonable. In this case the nth root of the system reliability requirement would be apportioned to each of the n subsystems. The equal apportionment technique assumes a series of n subsystems, each of which is to be assigned the same reliability goal. A prime weakness of the method is that the subsystem goals are not assigned in accordance with the degree of difficulty associated with achievement of the goals. For this technique the model is

$$R_i^* = (R_S)^{1/n} \qquad \text{for } i = 1, 2, \ldots, n$$

where R_S is the required system reliability and R_i^* is the reliability requirement apportioned to subsystem i.

As an example, consider a proposed communication system that consists of three subsystems (transmitter, receiver, and coder), each of which must function if the system is to function. Each of these subsystems is to be developed independently. Assuming each to be equally expensive to develop, the apportioned subsystem requirements are found as:

$$R^*T = R^*R = R^*C = (R_S)^{1/n} = (0.729)^{1/3} = 0.90$$

Then a reliability requirement of 0.90 should be assigned to each subsystem.

Agree Apportionment Technique. A method of apportionment for electronics systems is outlined in the AGREE report.[14] This technique takes into consideration both the complexity and the importance of each subsystem. It assumes a series of k subsystems, each with exponential failure distributions. The apportioned reliability goal is expressed in terms of MTBF.

A concept of module is used in this technique for three purposes: (1) so that the relative complexity inherently required can be taken into account, (2) so that the minimum acceptable reliability figures will not be grossly inconsistent, and (3) so that the reliability requirements will be dynamic and state-of-the-art changes can be incorporated as they occur. A module is designated as the basic electronic building block and is considered a group of electronic parts.

ARINC Apportionment Technique. This method assumes series subsystems with constant failure rates, such that any subsystem failure causes system failure and that subsystem mission time is equal to system mission time. This apportionment technique requires expression of reliability requirements in terms of failure rate.

Feasibility-of-Objectives Technique. This technique was developed primarily as a method of allocating reliability without repair for mechanical-electrical systems. In this method, subsystem allocation factors are computed as a function of numerical ratings of system intricacy, state of the art, performance time, and environmental conditions. These ratings are estimated by the engineer on the basis of experience.

Minimization-of-Effort Algorithm. This algorithm considers minimization of total effort expended to meet system reliability requirements. It assumes a system comprising n subsystems in series. Certain assumptions are made concerning the effort function. It assumes that the reliability of each subsystem is measured at the present stage of development or is estimated and it apportions reliability such that greater reliability improvement is demanded of the lower reliability subsystems.

Let $R_1, R_2, \ldots, R_n$ denote subsystem reliabilities. The system reliability R_S would be given by

$$R_S = \prod_{i=1}^{n} R_i$$

Let R_S^* be the required reliability of the system, where $R_S^* > R_S$. It is then required to increase at least one of the values of R_i to the point that the required reliability R_S^* will be met. To accomplish such an increase takes a certain effort, which is to be allotted in some way among the subsystems. The amount of effort would be some function of number of tests, amount of engineering manpower applied to the task, and other factors. The algorithm assumes that each subsystem has associated with it the same effort function $G(R_i, R_j^*)$, which measures the amount of effort needed to increase the reliability of the ith subsystem from R_i to R_i^*.

Dynamic Programming Approach. If all subsystems are not equally difficult to develop, dynamic programming provides an approach to reliability apportionment with minimum effort expenditure when the subsystems are subject to different but identifiable effort functions. The preceding minimization-of-effort algorithm requires that all subsystems be subject to the same effort function.

Additional Comments on Dynamic Programming Allocation Method. The dynamic programming approach can be most useful because it can be implemented with a simple algorithm that consists of only arithmetic operations. Some advantages of the dynamic programming approach are:

1. Large problems can be solved with a minimum number of calculations (this "minimum" may be very large for a complex system).

2. There is always a finite number of steps required in computing an optimum solution.

3. There are no restrictions of any kind on the form of the functional expression for computing reliability or the form of the cost-estimating equations. Nonlinear functions can be used if required.

The dynamic programming algorithms provide a guide through the maze of possible alternate calculations that may arise when big systems are being analyzed. The dynamic programming approach also can be applied to the problem of reliability optimization of redundant systems with repair. The use of the dynamic programming algorithm does not in any way remove the requirement for computing the reliability and cost for each system configuration. However, it minimizes the total number of calculations by rejecting those configurations that would result in decreasing reliability or in costs exceeding the cost constraints, and so on.

The dynamic programming optimization technique has application potential in other areas of reliability analysis. For example, useful models have been developed for determining an optimal number of redundant units (subsystems) subject to restraints such as weight, cost, volume, and opposing failure modes. Also a dynamic programming model has been developed for providing a systems approach to test planning, that is, planning for an optimal number of tests.

The important point to remember is that the dynamic programming approach can be readily computerized, and a number of computer models are available.

Reliability Prediction

Reliability prediction is the process of quantitatively assessing whether a proposed or actual equipment/shelter design will meet a specified reliability requirement. The real value of the quantitative expression lies in the information conveyed with this value and in the use which is made of the information. Predictions do not, themselves, contribute significantly to system reliability. They do, however, constitute decision criteria for selecting courses of action that affect reliability.

The primary objective of reliability prediction is the provision of guidance relative to the expected inherent reliability of a given design. Reliability predictions are most useful and economical during the early phase of system design and acquisition, before hardware is constructed and tested.

During design and development, predictions serve as quantitative guides by which design alternatives can be judged for reliability. Basically, the purpose of reliability prediction includes feasibility evaluation; comparison of alternative configurations; identification of potential problems during design review, logistics support planning, and cost studies; and determination of data deficiencies, tradeoff decisions, and allocation of requirements. It also provides criteria for reliability growth and demonstration testing.

Some important uses of reliability prediction include:

1. Establishment of firm reliability requirements in planning documents, preliminary design specifications, and requests for proposals, as well as determination of the feasibility of a proposed reliability requirement.

2. Comparison of the established reliability requirement with state-of-the-art feasibility for guidance in budget and schedule decisions.

3. Provision of a basis for uniform proposal preparation and evaluation and ultimate contractor selection.

4. Evaluation of reliability predictions submitted in technical proposals and reports in precontract transactions.

5. Identification and ranking of potential problem areas and the suggestion of possible solutions.

6. Allocation of reliability requirements among subsystems and lower-level items.

7. Evaluation of the choice of proposed parts, materials, units, and processes.

8. Conditional evaluation of the design for prototype fabrication during the development phase.

9. Provision of a basis for tradeoff analysis.

Thus reliability prediction is a key to system development and allows reliability to become an integral part of the design process. To be effective, the prediction technique must relate engineering variables to reliability variables.

In general there is a hierarchy of reliability prediction techniques available to the designer, depending on the depth of knowledge of the design and the availability of historical data on equipment and component part reliabilities. As the system design proceeds from the conceptual stage through full-scale development to the production phase, data describing the system design evolves from a qualitative description of systems functions to detailed specifications and drawings suitable for hardware production. Therefore a hierarchy of reliability prediction techniques has been developed to accommodate the different reliability study and analysis requirements and the availability of detailed data as the system design progresses. These techniques can be roughly classified in five categories, depending on the type of data or information availability for the analysis:

Similar Equipment Techniques. The equipment under consideration is compared with similar equipments of known reliability in estimating the probable level of achievable reliability.

Similar Complexity Techniques. The reliability of a new design is estimated as a function of the relative complexity of the subject item with respect to a "typical" item of similar type.

Prediction by Function Techniques. Previously demonstrated correlations between operational function and reliability are considered in obtaining reliability predictions for a new design.

Part Count Techniques. Equipment reliability is estimated as a function of the number of parts, in each of several part classes, to be included in the equipment.

Stress Analysis Techniques. The equipment failure rate is determined as an additional function of all individual part failure rates, with consideration of part type, operational stress level, and derating characteristics of each part.

Failure Mode and Effects Analysis

Failure mode and effects analysis (FMEA) is a reliability procedure that documents all possible failures in a system design within specified ground rules. It determines, by failure-mode analysis, the effect of each failure on system operation and identifies single failure points, that is, those failures critical to mission success or crew safety. It may also rank each failure according to the criticality category of failure effect and probability occurrence. This procedure is the result of two steps: the failure mode and effects analysis (FMEA) and the criticality analysis (CA). In the analysis, each failure studied is considered the only failure in the system (single failure analysis). The FMEA can be accomplished without a CA, but a CA requires that the FMEA has previously identified critical failure modes for items in the system design. When both steps are done, the total process is called a failure mode, effects, and criticality analysis (FMECA).

The principles of FMEA are straightforward and easy to grasp. The practice of FMEA is tedious and very time consuming. It is best done in conjunction with cause-consequence and fault tree analysis. The bookkeeping aspects, namely, the keeping track of each item and its place in the hierarchy, are very important, because mistakes are easy to make.

References

1 R. E. Barlow and F. Proschan, *Mathematical Theory of Reliability*, Wiley, New York, 1965.

2 M. L. Shooma, *Probabilistic Reliability: An Engineering Approach*, McGraw-Hill, New York, 1968.

3 D. K. Lloyd and M. Lipow, *Reliability: Management, Methods and Mathematics*, Prentice-Hall, Englewood Cliffs, NJ, 1962.

4 N. R. Mann, R. E. Schafer, and N. O. Singpurwalla, *Methods for Statistical Analysis of Reliability and Life Data*, Wiley, New York, 1974.

5 K. V. Bury, *Statistical Models in Applied Science*, Wiley, New York, 1975.

6 J. J. Naresky, *Electronic Reliability Design Handbook*, Rome Air Development Center, Rome, NY, 1982.

7 J. G. Rau, *Optimization and Probability in Systems Engineering*, Van Nostrand Reinhold, New York, 1970.

8 J. E. Arsenault and J. A. Roberts, eds., *Reliability and Maintainability of Electronics Systems*, Computer Science Press, 1980.

9 G. Hahn and S. Shapiro, *Statistical Methods in Engineering*, Wiley, New York, 1967.

10 ARINC Research Corporation, *Reliability Engineering*, Prentice-Hall, Englewood Cliffs, NJ, 1969.

11 A. L. Goel, *A Guidebook for Software Reliability Assessment*, Rome Air Development Center, Rome, NY, 1983.

12 A. L. Goel and J. Soenjoto, "Models for Hardware–Software System Operational Performance Evaluation," *IEEE Trans. Reliability* **R-30**(3):232–238 (Aug. 1981).

13 P. D. T. O'Conner, *Practical Reliability Engineering*, Heyden, Philadelphia, 1981.

14 Reliability of Military Electronic Equipment, Advisory Group on Reliability of Electronic Equipment (AGREE), Office of the Assistant Secretary of Defense (Research and Engineering), Washington, DC, Jun. 1957.

15 *Maintainability Verification/Demonstration, Evaluation*, MIL-STD-471, Department of Defense, Washington, DC, 1966.

16 *Reliability Design Qualification and Production Acceptance Tests: Exponential Distribution*, MIL-STD-781C, Department of Defense, Washington, DC, 1977.

17 R. S. Barlow and F. Proschan, *Statistical Theory of Reliability and Life Testing*, Holt, Rinehart & Winston, New York, 1975.

18 J. D. Kalbflusch and R. L. Prentice, *The Statistical Analysis of Failure Time Data*, Wiley, New York, 1980.

19 H. F. Marts and R. A. Waller, *Bayesian Reliability Analysis*, Wiley, New York, 1982.

20 A. L. Goel and A. M. Joglekar, *Reliability Acceptance Sampling Plans Based Upon Prior Distribution*, RADC-TR-76-294(A033576), 1976.

PART 8

MEDICAL APPLICATIONS OF ELECTRONICS

CHAPTER 40

INTRODUCTION TO HUMAN PHYSIOLOGY

CLIFFORD BOGEN

New York Institute of Technology, New York

40.1 INTRODUCTION

All life systems, from apparently primitive single cells to the most complex multicellular organisms including man, are characterized by a unique highly organized state of specific biomolecules (proteins, lipids, nucleic acids, and carbohydrates) suspended in a water-based system of electrolytes (Na^+, K^+, Cl^-) and small soluble molecules (carbon dioxide, oxygen, amino acids, glucose, fatty acids and their derivatives). This system (protoplasm) is essentially the same in all cells but is marked by subtle differences, mostly in the proteins and nucleic acids. These differences lead to the variations in populations of living things that cumulatively are expressed in speciation and evolution.

All life systems originated from and are physically and chemically part of the energy fields and matter that make up the universe. All are subject to laws governing the nature of the universe, including the second law of thermodynamics that suggests that the universe in general is characterized by an increasing loss of organization and in fact seems to be moving toward total disarray (high entropy).

The unique highly organized states of living systems represent islands of low entropy that are sustained and maintained by the controlled transduction of environmental energy. The controlling substances are catalytic proteins (enzymes) and the processes involved are collectively the metabolism of the organism.

The high degree of order and organization that characterizes life infers the maintenance of a steady state (dynamic equilibrium) in the face of the high entropy that prevails in the rest of the universe. An interacting series of physical, chemical, and biochemical signals are processed by each life system (biofeedback) to maintain this steady state (homeostasis).

In single cells, metabolism is controlled by enzymes. In multicellular organisms, metabolism is coordinated by biochemical messengers (hormones) and electrochemical charges (nerve impulses, muscle contractions).

Figure 40.1 shows the transduction of energy from the high entropy of the environment to the low entropy of homeostasis. It is based on the most common ecosystem (interaction between life and its environment) functional today. Track A shows radiant light energy trapped by chlorophyll and converted to the chemical bond energy of the biomolecule C–C, C–H. The trapped energy splits the H_2O, producing molecular O_2 and making free H_2 (photolysis), which will be used to fix (reduce the carbon of) CO_2 and produce CH_2O (carbohydrate). Track B shows how enzymatically controlled oxidations with or without O_2 convert the carbohydrate back into CO_2 and H_2O and convert its bond

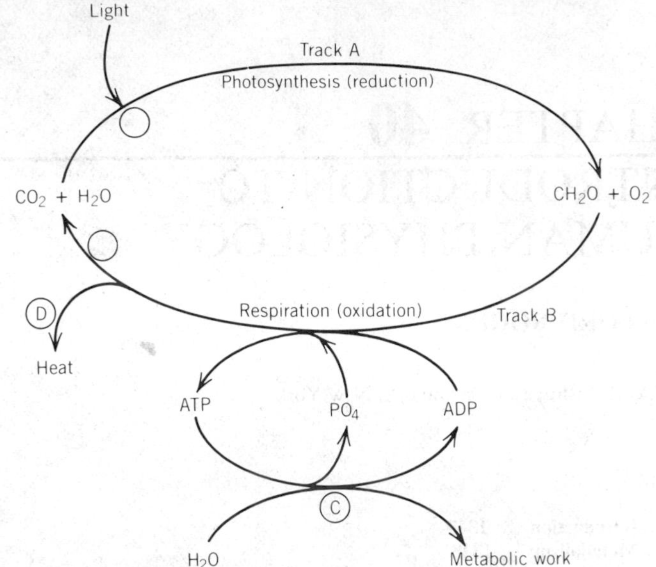

Fig. 40.1 Transconduction of energy from high entropy to homeostasis (low entropy). In track A, radiant light energy is trapped by chlorophyll and is converted to chemical bond energy. The trapped energy splits the H_2O, producing molecular O_2 and free H_2. The H_2 is used to produce CH_2O. In track B, enzymatically controlled oxidations convert CH_2O back into CO_2 and H_2O and convert its bond energy into a high-energy bond that links two phosphate radicals in the form of adenosine triphosphate (ATP). At C, hydrolysis of ATP releases energy available to do metabolic work. At D, the heat lost during the transductions is released. ADP, adenosine diphosphate.

energy into the high-energy bond (pyro bond) linking two phosphate radicals in the form of adenosine triphosphate (ATP: A–P~P~P). This is ultimately the only form of metabolic energy in life systems. At C, hydrolysis of the P~P bond releases metabolic energy that will be transduced to perform metabolic work, including nerve impulses, muscle contraction, biochemical synthesis, and active transport associated with absorption and excretion. Point D emphasizes the fact that all metabolic transductions are inefficient and not capable of being expressed as useful metabolic work. The difference between potential bond energy and work done is released in all life systems in the form of heat (high entropy).

40.2 METABOLISM

Metabolism is the enzymatically controlled intake, transduction, and utilization of environmental energy to maintain organization (homeostasis) of a life system. In this context, metabolism and nutrition are essentially synonymous.

Single Cell Systems

Single cells interact with their environment (H_2O, electrolytes, soluble molecules) to carry out certain functions, the sum of which is the metabolism. These functions are:

1. **Ingestion.** Essentially the external transport of nutritional substances to within the biochemical control of the cell.
2. **Digestion.** Extracellular biochemical conversion of nutritional substances into soluble absorbable form.
3. **Absorption.** Transport of nutritional substances from environmental fluids into cellular cytoplasm across a selective living membrane.
4. **Circulation.** Transport of substances (metabolites) within the cell to specific regions (organelles) for processing.
5. **Respiration.** Catabolic processing; oxidation of some metabolites to produce potential metabolic energy (ATP).

6. Assimilation. Anabolic processing; reduction of metabolites leading to synthesis of proteins (protoplasm). Most of the newly formed proteins will be used metabolically for cellular maintenance and repair. When conditions lead to the accumulation of additional protoplasm (increased cellular volume), assimilation equals growth. When cellular surfaces (membranes) reach critical maximum areas, new surfaces must be developed, leading to an increase in the number of cells (Fig. 40.2). At this level assimilation equals reproduction.

7. Excretion. Transport of excess metabolites and metabolic wastes from protoplasm back to environmental fluids across cell membranes.

8. Elimination. Unprocessed ingested substances returned to the environment beyond cellular biochemical control.

9. Irritability. Sensing and processing of environmental signals (stimuli) to locate nutritional substances and/or potential danger.

10. Motion. Responses to environmental stimuli toward nutritional substances and away from potential danger.

Multicellular Systems

In multicellular organisms, including man, most cells are highly specialized (only a portion of the genetic potential is expressed). This allows for more efficient functioning in the carrying out of certain metabolic activities but causes loss of capability in independent functioning. Therefore these specialized cells act as part of aggregates (tissues, organs, systems) that collectively manage the metabolism of the organism.

Metabolic activity of multicellular forms results in homeostasis of the extracellular fluid (ECF) that makes up the internal environment of the organism. The ECF therefore is more or less homogeneous even though it is compartmentalized (blood, plasma, lymph, interstitial fluids, cerebrospinal fluids). This fluid ultimately bathes all cells and supplies them with their metabolic needs as well as accepts their metabolic byproducts (wastes, excess metabolites, hormones).

The parts of the organ system of multicellular organisms and their contribution to metabolism are as follows. (It should be understood that while each tissue, organ, and organ system is highly specialized to perform a specific metabolic rôle, its activity also contributes directly or indirectly to the biofeedback that coordinates the homeostasis of the whole organism).

1. Digestive system. The gastrointestinal tract including liver, pancreas, and salivary glands are involved in the ingestion, digestion, and absorption of food (biomolecules, water, electrolytes) and the elimination of nondigested ingested substances (defecation).

2. Respiratory system. The air passageways and the lungs are involved in gaseous interchange, that is, the absorption of O_2 into the blood, and the excretion of CO_2 out of the blood.

3. Circulatory system. The heart, blood vessels, and lymphatic system manage the ECF, distribute nutritional substances to cells, and remove metabolic excesses and waste from cells. They also are involved in immunity (the prevention of infection or invasion of the ECF by pathogens (disease-causing microparasites) and by toxic foreign substances).

4. Excretory system. The kidneys and urinary tract as well as the sweat glands maintain homeostasis of the ECF and are ultimately responsible for removing metabolic waste from, and regulating the volume of, the ECF.

5. Nervous system. The central nervous system, autonomic nerves, peripheral nerves, and sense organs are a complex system of highly specialized cells (neurons) that sense, transmit, and process stimuli both from the external and the internal (ECF) environment. In coordination with various hormones produced by the endocrine glands they control and regulate the metabolism of the organism.

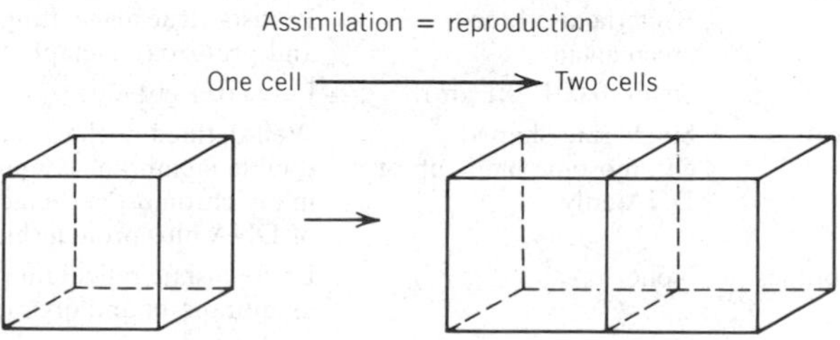

Fig. 40.2 Formation of a new cell.

6. Skeletomuscular system. This system allows movement in response to regulatory signals. Movement may involve locomotion (movement from place to place), distribution of the ECF, or other kinds of displacement.

7. Integuments. The skin and subcutaneous tissues conserve the ECF, prevent loss of fluids, protect against invasion by pathogens, maintain the temperature of the body, and carry out other similar functions.

40.3 CELLULAR STRUCTURE AND FUNCTION

Before we examine the nature of human structure and function, it would be useful to review the nature of cellular structure and function. There is no question that understanding of the biochemical and physiological events of the individual cells directly leads to a full understanding of human physiology.

It is possible, although perhaps somewhat simplistic, to develop a hierarchy of energy states from the most random to the most highly organized (Table 40.1). These states encompass most of the natural universe that could conceivably develop into a functional life system.

Collectively all cells have similar, if not identical, biochemical and physiological characteristics. However, it is quite apparent that on dimensional, morphological, and anatomical consideration, all cells are of one of two basic types: procaryotic or eucaryotic. Table 40.2 compares the two forms. Extant species can be arranged into taxonomic categories on the basis of function and anatomy. This arrangement (Table 40.3) gives insight into relationships and evolution.

Eucaryotic Cell Structure and Function

A number of investigators into the origin of the eucaryotic cell are supporting a polyphyletic origin based on three invasions by viruses and procaryotic cells into a primitive eucaryotic one over a long period of time. This theory is supported by biochemical, metabolic, and dimensional factors. It

TABLE 40.1. HIERARCHY OF ENERGY STATES

Entropic Level	Energy State	Examples
Highest[a]	Radiant energy	Electromagnetic spectrum, heat
↓	Particulate energy	Electron, proton, neutron
	Atoms	H, O, C, N
	Ions and simple molecules	H_2, O_2, CO_2, CH_4, NH_3, Na^+, Cl^-
	Simple biomolecules	Amino acids, fatty acids ↓ ↓
	Complex biomolecules	Proteins, phospholipids ↓ ↓
	Coacervates	Cytoplasm, membranes
Lowest	Cells	

[a] As entropy decreases, stability increases.

TABLE 40.2. COMPARISON OF PROCARYOTIC AND EUCARYOTIC CELL TYPES

Characteristic	Procaryotic	Eucaryotic
Distribution	Bacteria and blue-green algae	Protista (true algae, fungi, and protozoa), metaphyta, metazoa
Size	Small (0.001–0.1 μm)	Large (0.1–10 μm)
Genome (genes)	Single ring-shaped chromosome made up of DNA only	Well-defined nucleus enclosed by double membrane containing two to many chromosomes made up of DNA plus protein (histone)
Internal membranes	None	Endoplasmic reticulum plus membrane-bound organelles
Ribosomes	60 S	80 S

TABLE 40.3. TAXONOMIC CATEGORIES OF SPECIES

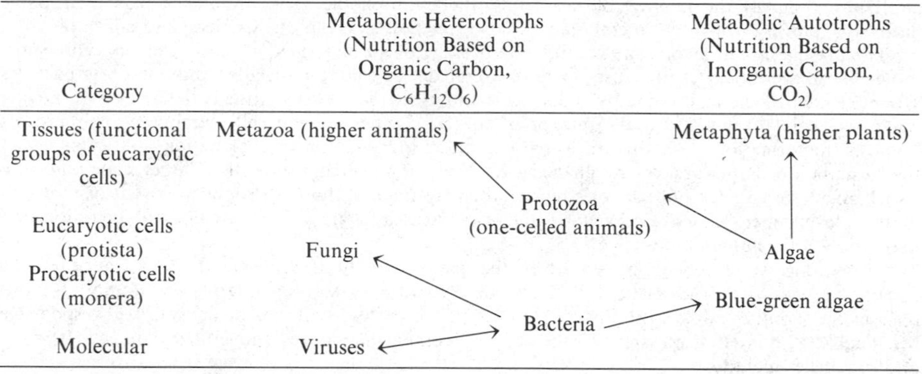

Category	Metabolic Heterotrophs (Nutrition Based on Organic Carbon, $C_6H_{12}O_6$)	Metabolic Autotrophs (Nutrition Based on Inorganic Carbon, CO_2)
Tissues (functional groups of eucaryotic cells)	Metazoa (higher animals)	Metaphyta (higher plants)
Eucaryotic cells (protista)	Fungi; Protozoa (one-celled animals)	Algae
Procaryotic cells (monera)	Bacteria	Blue-green algae
Molecular	Viruses	

considers the following facts:

1. Ribosomes are virus particles that have completely integrated into the metabolism of the cell and carried out protein synthesis.
2. Mitochondria are aerobic bacteria that initially were parasitic but have become responsible for aerobic respiration and have retained their ability to reproduce independently of the host.
3. Chloroplasts are symbiotic blue-green algae that, like mitochondria, retain their ability to reproduce but have taken over the cell's ability to carry out photosynthesis.

To best understand the nature of the eucaryotic cell, it must be realized that it is as complex structurally and functionally as any multicellular form including the human organism. Therefore it should be studied systematically.

The cell is made up of four separate but highly integrated systems: the membrane system, the cytoplasm, the nucleoplasm, and the semiautonomous organelles.

Membrane System. Eucaryotic cells have two systems of membranes, which act to enclose and partition the various parts of the cell but at the same time allow for communication and integration between parts. All cell membranes have the same structural and biochemical makeup. They are all essentially phospholipid bilayers with neutral lipid interiors (Fig. 40.3) and are strongly complexed with a varying number of proteins that function in different ways. The proteins act to make the membrane selective or semipermeable.

Figure 40.3 displays a molecular diagram of the cell membrane based functionally on the currently favored crystal model. This model suggests that the lipid portion of the membrane is more or less fixed in position but is fluid enough to maintain the integrity of the membrane by adjusting to the varying conformational changes of the membrane proteins. The phospholipid bilayer allows the

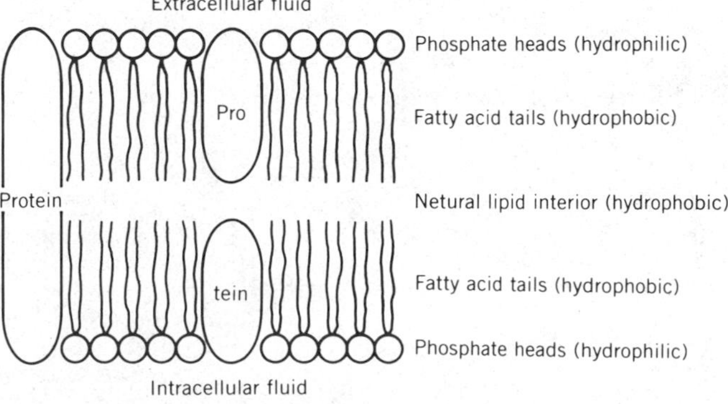

Fig. 40.3 Molecular diagram of the cell membrane.

surfaces of the membrane to be hydrophilic and, in combination with neutral lipids (triglycerides and cholesterol), makes the interior of the membrane hydrophobic. The proteins act as hydrophilic pathways (pores) through the membrane but also function as receptors, carriers, and enzymes.

The plasma or cell membrane controls the passage of all materials in and out of the cells. Small molecules and electrolytes are absorbed and excreted while large molecules and aggregate particles are moved across the membrane by endocytosis and exocytosis. Endocytosis is the process of moving macromolecular particles into the interior of the cell by-passing the cell membrane. This process involves the formation of small vesicles by membrane invagination or evagination (pinocytosis) and the formation of larger vesicles by phagocytosis (Fig. 40.4). (Phagocytosis is a larger scale pinocytosis.) Pinocytosic vesicles and vacuoles represent environmental fluids surrounded by cut off portions of the plasma membranes. Endocytosis and exocytosis and are cellular analogs of ingestion and defecation in the human body.

The endoplasmic reticulum represents the major structural and metabolic advance of the eucaryotic over the procaryotic cell. It is an elaborate system of internal membranes that is responsible for all the lipid synthesis and almost all the protein synthesis of the cell. It is responsible for much of the internal transport of the cell and compartmentalizes and unitizes the cell. Its parts and organelles include:

Smooth endoplasmic reticulum. Region of lipid synthesis.

Rough endoplasmic reticulum. Surface studded with heavy concentration of ribosomes; responsible for protein synthesis.

Peroxisomes. Small vesicles budded off from reticulum that contain enzymes that manage peroxides. Peroxides are a byproduct of aerobic respiration. They are highly toxic to a number of metabolic activities in the cell and must be eliminated.

Cisterna. Large membrane-bound vesicles formed by the reticulum. They may be used for storage or act as reservoirs for certain electrolytes.

Nuclear membrane. A double membrane that separates the nucleoplasm from the cytoplasm. It represents a number of cisternae that collect around the daughter nuclei after nuclear reproduction. These flatten and then fuse almost completely, leaving a number of nuclear pores that allow for communication between the nucleoplasm and the cytoplasm. Figure 40.5 illustrates this process.

Golgi apparatus. A highly elaborated part of the reticulum that varies in appearance from cell to cell. It is most evident in cells that are actively secreting various products. It contains, in various patterns, all of the parts of the reticulum (both rough and smooth reticulum): vesicles, cisternae, and membranes in series.

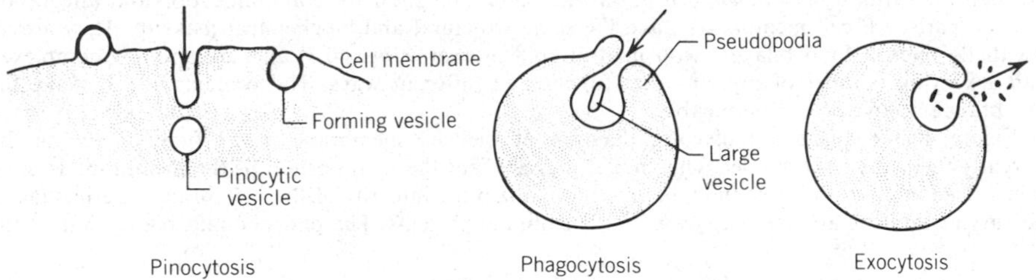

Endocytosis

Fig. 40.4 Endocytosis and exocytosis in the cell.

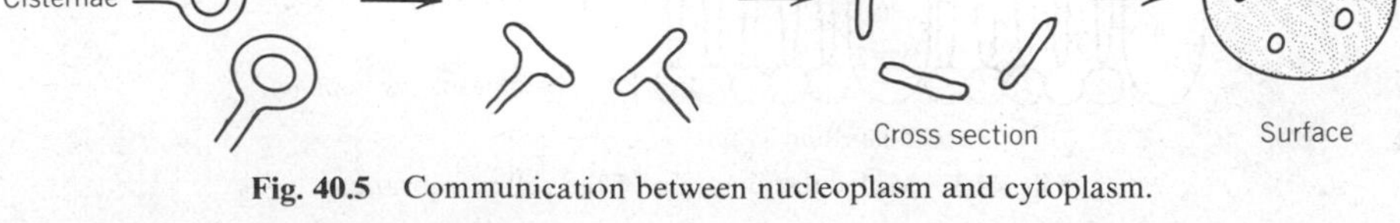

Cisternae

Nuclear pores

Cross section

Surface

Fig. 40.5 Communication between nucleoplasm and cytoplasm.

Lysosomes. Small membraned-bound packages of proteolytic enzymes formed by Golgi apparatus. Since the food of one cell is the product of or part of another cell, there really is no difference between food and cytoplasm; certainly enzymes of digestion cannot make such a distinction. Therefore all digestion must take place in the ECF either outside the cell or in food vacuoles inside the cell. The lysosomes fuse either with the cell membrane or the vacuolar membrane and release their enzymes into the ECF of the vacuole or outside the cell. Either directly or indirectly as a result of exocytosis of the vacuole or of the lysosome, the fusion of the lysosome membrane with the plasma membrane increases the dimensions of the plasma membrane. Therefore the growth and repair of the plasma membrane is a function of the reticulum via the golgi apparatus and the lysosome.

Cytoplasm. The bulk of the cell is made up of cytoplasm, which is a two-part system comprising a basic fluid ground material and a protein matrix suspended in it. The ground material is a water solution of electrolytes (ions) and metabolites (small soluble molecules). The cytoplasmic matrix is a colloidal suspension of proteins that acts as an endoskeleton as well as being responsible for a series of cellular events dealing with contractility and motility. The structure of the cytoplasm varies between a sol, that is, a continuous fluid with dispersed protein, and a gel, which is composed of a continuous protein with dispersed fluid.

The cytoplasm also contains biomolecular inclusions that act as storage depots. These molecules are either precipitates or are hydrophobic. Glycogen is a precipitate utilized for short-term storage. Triglycerides are hydrophobics used for long-term storage.

The proteins in the cytoplasmic matrix are of essentially two kinds: They are globular, with unique surfaces that allow them to interact with specific substrates such as enzymes and antibodies, or they are filamentous, which generally are structural proteins that give physical organization to the cell.

The basic structural protein unit is the microfilament. Microfilaments are frequently found to exist in highly organized arrays that are responsible for a number of cellular events (especially contractility). Thick (myosin) filaments are interspersed with thin (actin) filaments. When biochemically activated, the thin filaments slide over the thick filaments, shortening (contracting) the array of filaments. Occasionally the array may be microscopically visible, as in striated skeletal and cardiac muscle.

Thirteen microfilaments in the form of an elongated hollow cylinder form a microtubule. Microtubules occur singly or in specifically organized groups. They are the basic structural units of a number of cytoplasmic organelles, all of which are responsible for various cellular activities relating to motility. Single microtubules form spindle fibers. These fibers collect in arrays to form spindles. Spindles are tapered groups of fibers that develop during nuclear reproduction and are responsible for the separation and distribution of daughter chromosomes prior to the formation of daughter nuclei.

Patterned groups of microtubules are bundles of microtubules, essentially cylindrical in appearance, that in cross section show a 9 + 0 or 9 + 2 pattern (Fig. 40.6). The 9 + 0 organelles include basal bodies and centrioles. Basal bodies are shortened cylinders that are continuous with the base of the flagella and cilia. Centrioles are typical of animal and fungal cells and organize the spindle in these cells. The 9 + 2 organelles include flagella and cilia. Flagella are long whiplike extensions of the cytoplasm, while cilia are short hairlike extensions. Flagella and cilia protrude but do not break through the cell membrane. They are therefore covered by the plasma membrane. Basal bodies and centrioles are within the normal boundary of the plasma and are not covered with the cell membrane.

Nucleoplasm. Nucleoplasm is the content of the nucleus. The fluid part of the nucleoplasm is really an extension of the ground material of the cytoplasm by way of the nuclear pores. The major difference between the cytoplasm and the nucleoplasm is the replacement of the protein matrix of the cytoplasm by highly structured combinations of deoxyribonucleic acid (DNA) and special proteins called histones in the nucleoplasm. This complex of nucleoproteins makes up the nuclear organelles, the chromosomes. Chromosomes, except in special cells, are only visible during nuclear reproduction

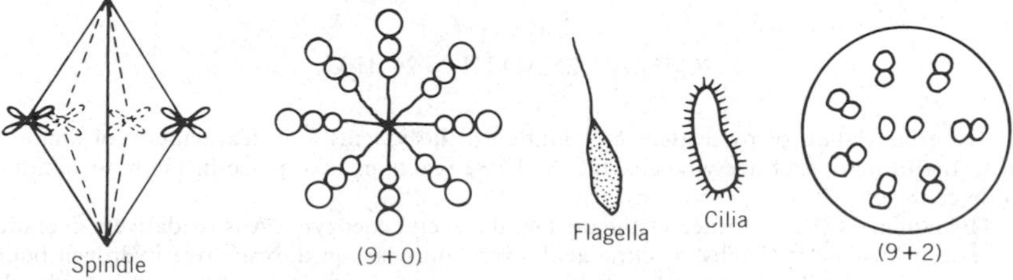

Fig. 40.6 Patterns of microtubules.

when they are highly condensed. For the most part they exist in a diffuse state. In the preparation of cells for microscopic viewing, various nuclear stains are used. Condensed chromosomes and/or their condensed portions stain darkly (heterochromatin); diffuse chromosomes stain lightly (euchromatin). The chromosomes are really linear arrays of genes (units of heredity). Biochemically a gene is made up of a molecule of DNA that has two unique biological functions: replication and transcription. Replication is the ability to duplicate identically (clone). This is the basic reproductive function in all living systems. In fact, all of the different types of reproduction are sequelae of replication where the duplicated genes are distributed and recombined.

Transcription is the synthesis of ribonucleic acid (RNA). There are three kinds of RNA, all of which are involved in protein synthesis. [The synthesis of protein represents the physical expression of a gene (phenotype). The stored information of genes is indicated by the genotype.]

Ribosomal RNA (rRNA) in combination with cytoplasmic proteins makes up ribosomes. Ribosomes are the organelles of the cytoplasm where protein synthesis occurs. Messenger RNA (mRNA) transmits information for a given protein from genes in the nucleus to ribosomes in the cytoplasm. Transfer RNA (tRNA) is amino-acid specific and can translate the information of mRNA into a sequence of amino acids that can make up the fundamental structure of a protein.

Replication and transcription take place only in euchromatin. There are four kinds of genes in eucaryotic cells; the most common can replicate but do not ever transcribe. Therefore they have no known function. The other three can all replicate, and each can specifically transcribe one of the three RNAs. There are a number of genes for transcribing rRNA. They are located in clusters on one or more specific chromosomes. The region on a given chromosome where these genes are operative is called the nucleolar organizing site. When the cell is active metabolically, a nucleolus forms in the nucleoplasm abutting the nucleolar organizing site. Initial assembly of ribosomes take place in the nucleoli. This assembly involves the migration of specific proteins from the cytoplasm in conjunction with the transcription of rRNA. The ribosome produced therefore is a molecular aggregate of nucleic acid and protein. From a biochemical and dimensional viewpoint, the ribosome falls within the range of virus particles. The assembled ribosome migrates from the nucleoplasm back to the cytoplasm but is modified and broken up into two parts. When protein synthesis is initiated, the two subunits reassemble in conjunction with the rRNA if the Mg^{2+} concentration is high enough.

Semiautonomous organelles. These are relatively large double-membraned bodies with single ring-shaped chromosomes without protein. Dimensionally they fall within the range of procaryotic cells. There are two types:

1. Mitochondria. These are generally oval shaped with a binding outer membrane and a highly folded inner membrane. Each fold is called a crista. Cristae have arrays of oxidative enzymes on their inner surface. Mitochondria are responsible for terminal oxidation of hydrogen in aerobic respiration. All aerobic cells degrade either glucose or acetyl coenzyme A into CO_2 and H_2O along metabolic pathways that lead to the synthesis of ATP. These pathways are:

(a) Glycolysis. With glucose as a substrate, partial oxidation produces 2 M of pyruvate, 2 M of H_2 (bound to carrier molecule NAD), and 2 M of ATP. (This reaction does not require free oxygen and is anaerobic.) Subsequently the pyruvate is reduced by the hydrogen released in its formation into ethyl alcohol, in some plant cells, or into lactic acid, in most animal cells. Glycolysis is a cytoplasmic function.

Glycolysis

$$C_6H_{12}O_6 \rightarrow 2C_3H_4O_3 + 2NADH_2 + 2ATP$$

Fermentation

$$2C_3H_4O_3 + 2NAD \rightarrow 2C_2H_5OH + 2CO_2$$

Lactic acid

$$2C_3H_4O_3 + 2NAD \cdot H_2 \rightarrow 2C_3H_6O_3$$

(b) Decarboxylation of pyruvate. *b*-Oxidation of fatty acids and deamination of amino acids all lead to the formation of acetyl coenzyme A. These reactions take place in the mitochondrion in eucaryotic cells.

(c) Oxidation. On the surface of the cristae, the acetyl coenzyme A is oxidatively degraded by a cyclic series of reactions (Krebs, or citric acid, cycle) into carbon dioxide, free hydrogen bound to carriers, and ATP. Finally, the bound hydrogen from any of the reactions discussed is oxidized by a series of redox steps to produce large amounts of ATP. The low-energy hydrogen resulting from redox is terminally oxidized by oxygen, forming water. This terminal oxidation of hydrogen is carried out by a number of enzymes and coenzymes collectively called the cytochrome sink.

The steps listed in (a), (b), and (c) can be summarized as follows:

Glycolysis

$$C_6H_{12}O_2 \rightarrow 2C_3H_4O_3 + 2NAD \cdot H_2 + 2ATP$$

Synthesis of acetyl coenzyme A

$$2C_3H_4O_3 + 2H_2O \rightarrow 2CO_2 + 2NAD \cdot H_2 + 2CH_3COOH$$

Citric acid cycle

$$2CH_3COOH + 4H_2O \rightarrow 4CO_2 + 6NAD \cdot H_2 + 2FAD \cdot H_2 + 2ATP$$

A summary of steps 1, 2, and 3 is:

$$C_6H_{12}O_6 + 6H_2 + 10NAD \cdot H_2 + 2FAD \cdot H_2 + 4ATP$$

(Every mole of $NAD \cdot H_2$ passed through the cytochrome sink and terminally oxidized yields 3 mol of ATP and 1 mol of H_2O. Every mole of $FAD \cdot H_2$ yields 2 mol of ATP and 1 mol of H_2O.) The overall reaction for aerobic respiration is given by

$$C_6H_{12}O_6 + 6H_2O + 6O_2 \rightarrow 6CO_2 + 12H_2O + 38ATP$$

2. Chloroplasts. These are found in eucaryotic plant cells only. Like the mitochondrion, they have an outer binding membrane and an inner membrane elaborated into layers that anastomose with one another. The series of connected layers enclose the stroma. Embedded in the stroma are stacks of flattened disks (thylakoid). The stacks of thylakoids make up the granum, which is the actual structure of the chloroplast responsible for photosynthesis. The photosynthetic equations add up both qualitatively and quantitatively to the reverse of aerobic respiration:

$$6CO_2 + 12H_2O \xrightarrow[\text{chlorophyll}]{\text{light}} C_6H_{12}O + 6H_2 + 6O_2$$

40.4 CELLULAR BIOCHEMISTRY

At this point a brief review of cellular biochemistry and biofeedback would be useful because, as stated earlier, human physiology is simply the sum of the metabolism of the 70 trillion cells that make up the human organism.

The biochemistry of the cell takes place in a water-based solution of electrolytes (basically Na^{2+}, Cl^-, and K^+). This water solution is the major ingredient of the ground material of the cell and is the foundation of the intracellular fluid (ICF). The ICF contains many other solute particles, both ionic and molecular.

The ionic particles are trace elements (mostly plus-two metallic ions) including Ca^{2+}, Mg^{2+}, Fe^{2+}, Zn^{2+}, Co^{2+}, Mn^{2+}, and Cu^{2+}, all of which act as cofactors, that is, inorganic chemicals that activate or enhance enzyme function; dissociation products of water and carbon dioxide (H^+, or OH^-, HCO_3^-), all of which interact to set up an optimum pH level (acid-base) for enzymatic activity; and various radical groups (PO_4^{3-}, SO_4^{2-}, NO_3^{1-}, among others). These particles play various roles in the cell relating to water balance and biosynthesis.

The molecular particles are metabolites and metabolic wastes. Metabolites are the basic molecular units in biochemistry. With the exception of O_2 and CO_2, the metabolites are all organic and essentially water soluble. They consist of:

1. Metabolic gases. Oxygen and carbon dioxide. In heterotrophic cells, oxygen is the metabolite and carbon dioxide is the waste. In autotrophic cells, carbon dioxide is the metabolite and oxygen is the waste.

2. Glucose. Glucose ($C_6H_{12}O_6$) is the basic hexose. In human cells all nutritional carbohydrates are ultimately processed into glucose and then reprocessed into derivative biochemicals or used as the basic source of metabolic energy by respiratory pathways. The derivative molecules include some coenzymes (vitamin C). Polysaccharides are polymers of glucose. They are used for short-term energy storage (glycogen) or are incorporated with other metabolites used in mechanical tissues as filling (collagen). In polymeric form they are essentially removed from the ongoing metabolism of the cells by enzymatic reduction of their solubility (polymerization). In polymeric form they may either be suspended (colloidal) or precipitated in the body fluids. The glucose may be recovered or recycled for metabolism by digestion. The chemistry is

$$nC_6H_{12}O_6 \underset{\text{digestion}}{\overset{\text{polymerization}}{\rightleftharpoons}} (C_6H_{10}O_5) \cdot H_2O + (n-1) \cdot H_2O$$

3. Fatty or organic acids (RCOOH). In the fatty acids, R represents a hydrocarbon chain. In the simplest fatty acid, acetic acid, the R is the methyl group (CH_3) and the formula is CH_3COOH. Acetic acid is too reactive to exist metabolically within the cell and is usually found in the acetyl group that is combined with a coenzyme as acetyl coenzyme A. The whole class of nonpolar biochemicals called lipids are really derivatives of acetyl coenzyme A.

Triglycerides are esters of fatty acids and glycerine. Fats are animal triglycerides characterized by saturated hydrocarbon chains in fatty acid residues. Oils are plant triglycerides with unsaturated hydrocarbon chains. The triglycerides are hydrophobic molecules that act as long-term energy storage depots in cells. In the long term, all organic molecules that are in excess within the cell are converted into fats and oils. The chemistry is

$$3RCOOH + C_3H_5(OH)_3 \underset{\text{digestion}}{\overset{\text{synthesis}}{\rightleftharpoons}} C_3H_5(OOCR)_3 + 3H_2O$$

Isoprene derivatives from unsaturated fatty acids include a number of coenzymes including vitamin A, vitamin K, α-tocopherol (vitamin E in rats).

Steroid derivatives of saturated fatty acids include cholesterol. These derivatives, along with phospholipids, are important in membrane structure. Vitamin D is a steroid, as are the hormones produced by the gonads (ovaries and testes) and the adrenal cortex. These hormones include the sex hormones (estrogen and progesterone in the female; testosterone in the male). Aldosterone and cortisol are produced by the adrenal cortex.

4. Amino acids ($NH_2RCHOOH$). These are the monomers of the fundamental structure of proteins (polypeptide chain). Proteins are by far the most complex molecules that we know. They serve every metabolic role. Essentially a living system is a minimum number of proteins that supply a structural framework for the functioning of metabolic proteins (enzymes). Most proteins are made up of two or more polypeptide chains complexed with other organic molecules or plus-two cations and include:

(a) Metalloproteins—usually enzymatic.
(b) Lipoproteins—usually in membranes or as lipid carriers.
(c) Glucoproteins—mostly structural.
(d) Mucoproteins—coenzymatic or enzymatic.
(e) Nucleoproteins—structural.
(f) Polypeptide chains—enzymatic or hormonal.
(g) Antibody—immune processes.
(h) Interferon—antiviral.

Amino acids are derivatives of acetic acid. There are 20 amino acids, differing from one another only in the exact nature of the hydrocarbon chain (R). The simplest amino acid is glycine, which is derived from acetic acid by amination:

$$NH_3 + H-\underset{H}{\overset{H}{C}}-COOH \underset{\text{deamination}}{\overset{\text{amination}}{\rightleftharpoons}} NH_2-\underset{H}{\overset{H}{C}}-COOH + H_2$$

Polymerization of amino acids biochemically is identical to that discussed earlier for polysaccharides. That is,

$$nNH_2RCHCOOH \underset{\text{digestion}}{\overset{\text{synthesis}}{\rightleftharpoons}} (NHRCHCO)_n \cdot H_2O + (n-1) \cdot H_2O$$

Derivatives of amino acids serve a number of functions. Some are hormones (thyroxine), others are coenzymatic (vitamin B complex). The nitrogen bases (purines and pyrimidines), important elements in the chemistry of the nucleotides, are derived from amino acid.

40.5 HOMEOSTASIS, BIOFEEDBACK, AND DYNAMIC EQUILIBRIUM

The cell's essential activity is to establish a biochemical dynamic equilibrium to maintain the concentration of metabolite. Failure of cells to maintain the metabolite pool is life threatening.

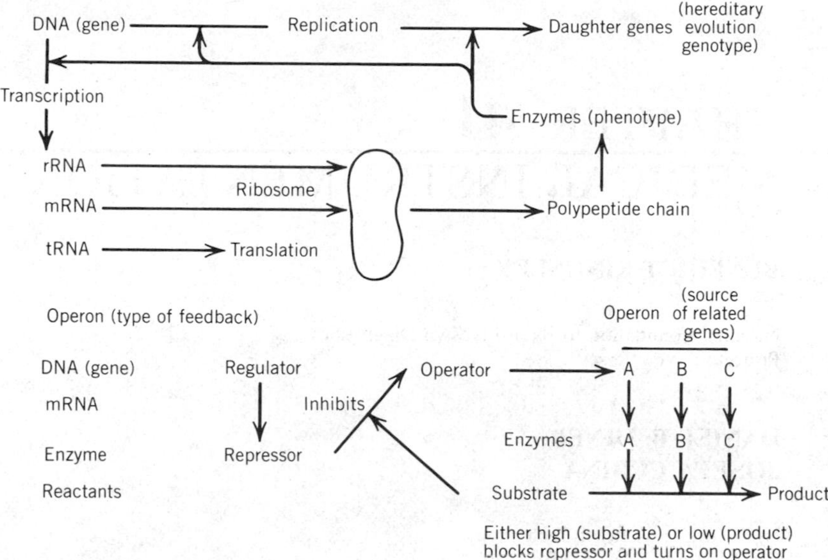

Fig. 40.7 Flow sheet of informational molecules. rRNA, ribosomal RNA; mRNA, messenger RNA; tRNA, transfer RNA.

If a given metabolite concentration is not maintained nutritionally, enzymatic conversion of one metabolite into the one in short supply may occur. The depot molecules will be digested, first the polysaccharides and then the triglycerides. When all the depot molecules are exhausted, the cell as a final response will convert its own proteins into metabolites. When the protein level falls below the minimal necessary amount, the metabolism fails and the cell loses its functional organization and dies.

The homeostatic mechanisms are cellular regulation and gene control. For maintenance of the metabolite balance necessary for cellular function, there must be a series of built-in feedback mechanisms. Ultimately, feedback must lead to maintaining the right amount of the right enzyme in the right place at the right time. As the flow sheet in Fig. 40.7 suggests, genes are informational molecules that represent a blueprint for specific proteins. A number of feedback mechanisms turn on or off a specific gene, as need dictates.

Bibliography

Ayers, C. J., *Basic Cell Biology*, 2nd ed., Prindle, Weber & Schmid D., Boston, 1982.

DeRoberts, E. D. P., and G. M. F. DeRoberts, *Essentials of Cell and Molecular Biology*, Saunders, Philadelphia, 1981.

Dyson, R. D., *Cell Biology, A Molecular Approach*, 2nd ed., Allyn & Bacon, Boston, 1978.

Guyton, A. C., *Textbook of Medical Physiology*, 6th ed., Saunders, Philadelphia, 1981.

Holzman, *Cells and Organelles*, 3rd ed., Saunders, Philadelphia, 1984.

Molecules to Living Cells, readings from *Scientific American*, Freeman, San Francisco, 1980.

CHAPTER 41
MEDICAL INSTRUMENTATION

BENEDICT KINGSLEY

National Foundation for Non-Invasive Diagnostics
Princeton, New Jersey

DANIEL B. DINER
JOSEFA CUBINA

New York Institute of Technology
Old Westbury, New York

JEFF TOSK
JULIUS SIMON

Laboratory for Chromatography
Flushing, New York

41.1 ULTRASONOGRAPHY

Benedict Kingsley

41.1-1 Introduction

Echo sounding has been used to measure distance by animals such as bats, dolphins, and certain birds long before man began to use it shortly after World War I. During 1920 to 1945 it was used for depth sounding and the location of submarines and schools of fish or seaweed.

After 1945, echo sounding found extensive use and applications in nondestructive testing of materials and in medical diagnosis. Actually, the plotting of ultrasonic echoes has been found useful in a number of fields including the testing of homogeneous metals and the charting of ocean beds. The technique is now being applied to specific aspects of medical diagnosis.

The principle involved is relatively simple. When an electric current is introduced into a piezoelectric crystal, the crystal vibrates and produces sound waves. If a very rapid, fluctuating current is employed, the sound produced is well beyond the range of human hearing in the realm of ultrasound. Each time a beam of this energy crosses the boundary or interface between two structures or tissues of different density, some energy is reflected off the interface and the echoes are picked up by the same crystal and amplified as an electrical signal. The depth or position of a large number of reflecting structures can be plotted, as is done in charting the bed of the ocean. These echoes are immediately visible for interpretation on an oscilloscope screen and may provide information for medical diagnosis.

41.1-2 Historical Development

In 1880 the Curie brothers detected electrical potentials on the surface of a specially cut piece of quartz when it was subjected to mechanical stress, and the name piezoelectric was applied.

In the following year the same workers were able to demonstrate vibrations in a piezoelectric crystal when it was subjected to an oscillating electrical potential, the reverse piezoelectric effect. These mechanical vibrations are transmitted to the surrounding medium in the form of sound waves. If the sound waves have a frequency greater than approximately 20,000 Hz, they lie above the audible range and are termed ultrasonic. Thus a single piezoelectric crystal can act both as a generator and as a receiver for ultrasound waves. The Curies' discovery of the double property of piezoelectric crystals such as quartz, Rochelle salts, or barium titanate is the basis for high-frequency diagnostic ultrasound. However, in all probability the first recorded description of ultrasound was by Gordon in 1883 when he described a device that served as the forerunner for the jet generators used in industry.

The first practical application of acoustic vibrations of high frequencies was developed by Langevin in France in 1916 when he invented a method of underwater communication. As a matter of fact, Langevin was commissioned by the French government during World War I to find a means of locating enemy submarines that were attacking French vessels. His patent proposed an assembly of mosaic quartz crystals cemented between steel plates to be used for generation and reception of ultrasonic waves. Although this device was not employed to any extent, it led to the development of sonar during World War II.

It is of historical interest that Gordon's description of ultrasound preceded the discovery of x-rays by 12 years. X-rays found comparatively immediate application, but ultrasound did not come into widespread use until half a century later. It is usually to Sokolov of Leningrad that credit is given for the development of ultrasound. In 1937 he worked intensively in this field and studied the propagation of ultrasonic waves in both solids and liquids and developed the technique for the detection of internal flaws in metals which was the first ultrasonic method to visualize internal structures.

Credit for the first medical application of ultrasound is given to an Austrian by the name of Dussik, who in 1937 used the Sokolov transmission method in attempting to demonstrate intracranial processes. This method, which he called hyperphonography, assumed that a different absorption of ultrasound in brain matter and cerebral spinal fluid would get a picture of the form and position of the various structures within the brain and thus enable tumors to be detected by inference. The transmission method, in fact, proved of little clinical value for the detection of intracranial space-occupying lesions, because a skull filled with water gave pictures very similar to those reported to represent various pathologic lesions within the brain. During the Second World War, Firestone developed a more accurate method of detecting internal flaws in solid structures by studying the reflections of ultrasonic energy from the flaws. His instrument, called a reflectoscope, utilized an echo method in which the emitter crystal, which was piezoelectric quartz, and the receiver crystal, piezoelectric Rochelle salts, were both on the same side of the object under examination. The high-frequency sound waves were emitted in pulses of short duration and relatively long intervals. The reflected sound waves from interfaces lying in the path of the beam were picked up by the receiver crystal and displayed quantitatively on an oscilloscope screen. This pulse-echo method ultimately led to the first successful attempt to apply ultrasound for medical purposes.

Until 1952 the use of ultrasound in medical diagnosis had been confined to unidimensional echography, that is, transmission and pulse-echo methods, which can be compared to needle biopsy. By utilizing a 15-MHz pivotal crystal mounted in a water chamber closed by a rubber membrane, Wild and Reid employed a simple linear movement of the probe at right angles to the emersion beam to produce the first two-dimensional echogram. Using this two-dimensional method, they examined many palpable tumors of the breast and were able to diagnose preoperatively 26 of 27 malignant tumors and 43 of 50 benign tumors confirmed by microscopic diagnosis.

This two-dimensional technique, though geometrically different, had a common limiting feature, namely, that a beam reached any one point in a tissue from a single probe or transducer position. This defect was alleviated by Howry and co-workers when they developed the first practical instrument, called a tomograph, which employed a compound scan. In 1952, Howry and associates, using their original ultrasonic instrument (a "sonoscope") with a scanning method of lower frequency and lower power and with lens-focused ultrasonic pulses, produced echograms of high technical quality. The area to be examined was immersed in water to facilitate the propagation of sound; scanning a complete 360° produced a picture that presented a cross section of excellent definition. Their ultrasound pictures of neoplastic breast masses and cross sections through the forearm of one of the investigators were the first produced by pulse-echo method that showed the interior construction of solid human tissue.

Another ultrasonic examining technique developed within the last decade is based on the Doppler principle. By transmitting and receiving ultrasound, the instrument utilizing Doppler techniques detects the motion of organs and blood within the body. The transducer containing both a transmitting and a receiving crystal is placed against the patient's chest or abdomen. A beam of low-intensity ultrasound is transmitted into the body as a continuous beam, and part of that sound is reflected back from the internal structures. Ultrasound received from motionless structures has precisely the same frequency as the transmitted sound and is not heard. Ultrasound received from moving organs or flowing blood is shifted slightly in frequency from the transmitted sound and this difference in frequency is converted to an audible signal. For example, several distinctive Doppler sounds of clinical importance can be identified from the pregnant uterus. Most easily heard is that of the fetal pulse. It is produced by the blood flow through the fetal artery. Depending on the structure through which the ultrasonic beam is directed, a distinctive placental sound is also easily identified and is used to localize the placenta.

On the basis of the five major stepping stones in the historical development of diagnostic ultrasound, namely, piezoelectric effect, transmission method, pulse-echo method, scanning techniques, and Doppler techniques, many medical applications have been developed. Diagnostic ultrasound is a safe, painless, harmless, and risk-free technique.

The average power levels employed in this type of test are very low, 0.01 to 0.04 W/cm^2. Ultrasonic examinations are totally free of discomfort, completely external, repeatable as often as indicated, and of course do not expose the human body to ionizing radiation and, therefore, can be used in medical situations where x-ray examination cannot be applied.

41.1-3 Physical Principles

Sound is sometimes defined as an audible sensation, but from the physicist's viewpoint, sound consists of mechanical vibration at any frequency, audible or otherwise. In short, ultrasound does not differ from audible or subaudible sound in any way other than frequency.

To define the important concepts of frequency, a sine wave should be visualized. The frequency of the sine is called F. The period of the sine wave is defined as the smallest interval of time for which the wave form repeats itself. The period T in our example is equal to 0.5 s. The frequency of this wave is defined as a reciprocal of the period, or $1/T$. For our example, the frequency is 1 divided by 0.5 s or 2 Hz.

The length from any point on the sine curve to the corresponding point on the next cycle is defined as wavelength and is usually symbolized by the Greek letter lambda (λ). For sake of simplicity, the corresponding points may be taken either as crest to crest or trough to trough.

Sound waves may be categorized in terms of frequency as follows:

Subsonic or infrasonic waves fall into a category from 0 to 20 Hz.

Audible waves are categorized in a frequency region from 20 to 20,000 Hz.

Ultrasonic waves are defined as waves that exist at a frequency higher than 20,000 Hz.

Thus a sound wave does not have to be audible, just as light waves are not necessarily visible if they fall into the infrared or ultraviolet part of the spectrum.

Fluids, solids, or gases can move in various fashions. They can undergo translation, such as blood flowing in an artery or a mass of air being moved by blowing wind. They can also undergo rotational motion, as in a whirlpools of blood at an arterial obstruction or in a tornado. However, the type of motion that is of greatest concern in biological applications, vibratory motion, can produce shape distortions that travel outward. Such a moving shape distortion is called a wave. For production of a sound wave, an initial disturbance in an elastic medium is required. If the initial disturbance in correcting itself disturbs a neighboring region in a fashion similar to the original disturbance, the wave is propagated.

If the disturbed particles in a medium move back and forth with respect to an equilibrium position in the direction parallel to the wave motion, resulting in a series of alternate compressions and rarefications in the propagating medium, the wave is called longitudinal. Each particle vibrates back and forth about an equilibrium position and transmits its energy to an adjoining particle, causing it to execute a motion similar to that of the exciting particle. The particle itself does not travel far in the direction of wave motion but only vibrates and then returns to its original position.

Longitudinal waves can exist in fluids, solids, or gases. Since the propagation of longitudinal waves depends on alternate compressions and rarefications of the medium, these waves will only travel in an elastic medium. Longitudinal waves are sometimes also referred to as compressional or elastic waves. In short, any sound wave, whether audible or ultrasonic, is produced by actual mechanical motion of the medium through which it passes.

The velocity of the sound wave V_w is an entirely different parameter from the particle velocity V_p and depends on the density and elasticity of the medium in which it is traveling as well as on the type of wave, for example, longitudinal or transverse. Since mammalian tissue can be considered primarily a fluid, a transverse component of any sound wave propagated in tissue can be considered negligible as compared with a longitudinal component.

From a practical viewpoint, therefore, only longitudinal waves are considered in terms of human tissues. The wave velocity V_w of a longitudinal wave depends on the density and elastic contents of the fluid in which it is traveling and is given by $V_w = 1/\rho B$, where ρ is equal to the density of fluid and B is equal to bulk compressivity of fluid.

In the literature, the following values for the velocity of sound in human tissues are given: refrigerated muscle, 1568 m/s; refrigerated liver tissue, 1570 m/s; fatty tissue, 1476 m/s. Therefore the average velocity for longitudinal sound waves in human tissues at body temperature is usually taken as 1540 m/s. The velocity of sound in the human skull bone at body temperature is 3360 m/s.

Wave velocity is independent of the frequency of the sound wave. However, the frequency F, wave velocity V_w, and wavelength λ of a sound wave are interrelated as follows:

$$F = V_w/\lambda$$

This means that a wave velocity of sound in human tissue at body temperature remains constant at 1540 m/s no matter whether the sound frequency is 200 or 2×10^6 Hz. However, the wavelength λ will differ considerably; 7.7 m λ for 200 Hz and 0.77 mm λ for 2 MHz. In short, the higher the frequency of a sound wave the smaller the wavelength, and vice versa.

The characteristic impedance R is equal to the product of density ρ times the wave velocity V_w.

The acoustic resistance is the parameter that is associated with the dissipation of acoustic energy. This parameter is very useful in quantifying various physical characteristics of a sound wave. The intensity of a sound wave (I), which is analogous to loudness, is measured in terms of the quantity of energy passing each second through 1 cm^2 of area, the area being perpendicular to the direction of the propagation of the sound. Acoustic intensity I is expressed in equation form as follows:

$$I = \frac{\rho V_w V_p^2}{2}$$

where ρV_w = acoustic impedance
V_p = particle velocity
V_w = wave velocity
ρ = density of medium

The average acoustic intensity level used for diagnostic ultrasound procedures is 0.04 W/cm^2. Intensities from 1 to 3 W/cm^2 are used for physiotherapy using therapeutic ultrasound. Surgical applications entailing the destruction of tissues or welding of a detached retina require intensities ranging as high as 1500 W/cm^2.

When a sound wave strikes an interface between two media of different characteristic impedance or acoustic resistance, the sound wave is partly reflected and partly transmitted through the second medium. If the sound wave strikes the interface at right angles, the reflection coefficient K_r, which represents a proportionality constant relating the reflected incident sound intensities, is given by

$$K_r = \frac{\rho_1 V_{w1} - \rho_2 V_{w2}}{\rho_1 V_{w1} + \rho_2 V_{w2}}$$

where ρ_1 = density of medium 1
ρ_2 = density of medium 2
V_{w1} = wave velocity of sound in medium 1
V_{w2} = wave velocity of sound in medium 2

The reflected part or portion of the sound wave is called an echo. The greater the difference in acoustical resistance between the two media, the greater becomes the reflection coefficient, resulting in a greater size or amplitude of the echo. At normal incidence, the greatest amount of sound energy or the largest echo is reflected. When a sound wave strikes an interface between two media obliquely, the reflection and refraction characteristics of the sound wave are governed by Snell's law. As the sound wave traverses an interface between the two media of different acoustic impedance, reflection and refraction occur.

A sound wave meeting an interface in a medium may not be reflected at all. The French physicist Fresnel was able to show in 1818 that whether a wave was reflected or not depended on the comparative size of the wavelength λ and the obstacle. When an obstacle is the size of the wavelength or less, the wave is not reflected but instead is bent around the obstacle. This physical phenomenon is called diffraction. If the obstacle is considerably larger than the wavelength, the wave is reflected. This means that for a wavelength of approximately 0.5 mm of the incident ultrasonic beam, any obstacle in the tissue larger than about 2 mm should give rise to detectable reflections or echoes.

As a sound wave passes through a homogeneous medium, its intensity becomes progressively diminished as the result of absorption. Absorption is usually caused by viscosity or inner friction of the medium as well as by heat conduction. The absorption coefficient is proportional to the square of the frequency. However, it has been shown experimentally that for soft tissue the absorption coefficient is directly proportional to frequency, which probably is due to the complex nature of tissue protein. It has been shown that 80% of the absorption by tissue is caused by tissue protein.

41.1-4 Diagnostic Ultrasound Equipment

A diagnostic ultrasonic system has the capability to measure the distance or range from a transducer to the biological structure of interest. The basic diagnostic ultrasound system consists of a transducer, a transmitter, a receiver, and a display unit. This system operates by transmitting a particular type of waveform, ultrasonic impulses, and detecting the nature of the echo signals.

The transmitter generates a short electrical pulse which is converted by the transducer into a 1-μs burst of ultrasonic energy having a frequency usually of 2 to 7.5 MHz. Echoes of varying amplitude and range are received whenever the beam encounters an interface of different acoustic impedance. These echoes are amplified in the receiver for display.

The optimum ultrasonic frequency represents a compromise between several factors and usually depends on the specific application. For example, in cardiology work the usual frequency ranges from 2 to 2.5 MHz. A frequency of 15 MHz would be desirable because it would allow the use of a narrow ultrasonic beam, resulting in good resolution of structural detail. However, this feature would be obtained at the expense of decreased penetration because of increased absorption at higher frequencies, particularly in bone. Frequencies below 2 MHz would lack beam sharpness and resolution.

The transducer is usually positioned directly on the patient's chest or abdomen or wherever the best approach to the organ of interest occurs. A liquid or cream coupling fluid must be used between the transducer probe and the skin of the subject; an air film between the transducer and the skin is nearly a perfect reflector of sound waves and would interfere with transmission into the tissue.

Whenever the transmitted ultrasonic pulses encounter an interface composed of biological structures of different acoustic impedance, a portion of the transmitted energy contained in each pulse is reflected. The acoustic impedance, as has been previously discussed, depends on both the density of structures ρ and the wave velocity of sound V_w in the particular biological structure. Immediately after emission of each pulse the transducer becomes inactive and mechanically excited by the returning echoes. These echoes are converted into voltage pulses and are then detected and amplified by the receiver for display.

The purpose of display is the presentation of information contained in the echo signals in a form suitable for human interpretation. The most common form of echo display is visual and usually is accomplished via a cathode ray tube. However, if information is obtained at a rate greater than the human equipment operator can assimilate, for example, at a rate of 1000 pulses per second, it may become necessary to display only the condensed information and to utilize automatic data processors, such as digital computers, to digest and interpret the echo data.

The most basic display on the oscilloscope is the A presentation, which consists of deflections of an electron beam on the screen caused by the presence of a target. The deflections are commonly referred to as blips. This display can easily be photographed by an instant picture camera, showing the range of the target from the transducer and the amplitude of the echoes.

The B presentation is electronically generated from the A mode; it displays echoes as dots along the horizontal axis, the relative echo amplitudes being indicated by brightness variations of the dots. The M-mode presentation is obtained by connecting a slow sweep unit to the display oscilloscope. This unit moves the trace across the oscilloscope screen at a given rate with markers recorded every 0.5 s in time and 1 cm in range. This presentation also provides intensity modulation for the echo amplitudes. In short, the M-mode presentation recorded by instant photography or strip chart displays time and range against a background of amplitude measured by the degree of brightness.

The distance traveled by the echo signals, or the range, may be read directly from a calibrated scale on the face of the oscilloscope. This is accomplished by superimposing calibration markers on the cathode ray tube sweep. These markers appear as regularly spaced vertical deflections. The distance or range R for the target is determined by measuring the time taken by the pulse to travel to the target and return. This relationship is expressed quantitatively by

$$R = V_w \times \frac{T}{2}$$

where V_w is the velocity of sound waves in tissue and T is the time duration for sound waves to travel out to the target and back.

The wave velocity of sound in tissue at body temperature is usually taken as 1540 m/s. Once the pulse is emitted by the transducer, a sufficient amount of time must elapse to allow any echo signals to return and be detected before the next pulse may be transmitted. Therefore the rate at which the pulses may be transmitted is determined by the longest range at which targets are expected. If the pulse repetition frequency is too high, echo signals from some targets might arrive after the transmission of the next pulse and ambiguity in measuring range might result. For example, assuming that the longest possible range for any target of interest in echocardiography is 20 cm, the highest permissible pulse rate is 154,000 cm/s divided by 40 cm, which is equal to 3850 pulses per second.

Just as with frequency, the pulse rate selected for diagnostic ultrasound represents a compromise. The higher the pulse rate, the greater the resolution and measurement accuracy, especially for fast-moving targets. However, a higher pulse rate increases the average amount of acoustic power of the ultrasonic beam entering the body. The pulse rates utilized in diagnostic ultrasound systems currently available vary from 200 pulses per second to 2000 pulses per second. Most diagnostic ultrasound units currently available operate at average acoustic power levels of 0.01 to 0.04 W/cm^2. These levels have been shown to be completely safe.

In learning to operate the equipment, the engineer/physician should become familiar with the proper adjustment of the available controls. All commercial systems have gain controls which electronically increase the echo amplitude. Some units have time control gain adjustments which electronically control the gain as a function of range. Thus the echoes from deep-seated structures receive more gain than do structures closer to the transducers to compensate for larger attenuation and absorption losses as the ultrasonic pulses travel through greater distances in the tissue.

When the ultrasonic beam intercepts interfaces of small differences in acoustic resistance, many low amplitude noise echoes will be generated. However, increasing the gain will increase not only the amplitude of the echo signal but also the amplitude of the noise echoes. For this reason, most commercial units contain a reject control which raises the baseline of the A display to eliminate the extraneous noise echoes.

The disadvantages of a reject control are that it is of no value if the echo signal is buried in noise and that it results in a measurement error of range due to sloping of the anterior edge of the echo. In cases when echoes close to the transducer must be measured, a different type of noise problem exists. Since the transducers are used for transmitting and receiving, a portion of the energy of each transmitted pulse leaks over into the receiver causing a cluster of echoes following the transmitted pulse called the main bang. Any echo within 2 cm of the transducer is usually affected by the main bang. A damping control provides electronic filtering in the receiver to reduce the noise in the main-bang region. The damping and near gain controls are frequently very useful in isolating the echo from, for example, the anterior wall of the right ventricle in the heart. However, it must be noted that use of these controls will reduce the sensitivity of the receiver so that weak echoes may not be recorded at all.

The most important component of a diagnostic ultrasound system is probably the transducer. By definition, the transducer is a device that converts one form of energy into another form. In the case of diagnostic ultrasound, the transducer converts electrical to sound energy and vice versa. The transducer usually consists of a piezoelectric crystal, ceramic, barium titanate, lithium sulfate, or lead zirconate or titanate. Crystal size varies because there is a predictable relationship between the dispersion of the sound beam and the width in wavelengths of the crystal face. Crystals from about 4 to 25 mm in diameter have been used, the most common diameter of commercial diagnostic crystals being about 13 mm. A standard diagnostic ultrasound transducer is usually a barium titanate crystal.

All crystal transducers operate on the basic principle of the piezoelectric effect. When an electric current is applied to certain natural crystals such as quartz, a mechanical strain is induced. If the frequency of the alternating electric current is equal to the resonant frequency of the crystal, the crystal will mechanically vibrate at this frequency, for example, 2.25 MHz. The crystal is mechanically designed to resonate at the desired frequency. When the exciting current is removed, the mechanical vibrations are damped out very quickly. Conversely, when a mechanical stress is applied in a specified direction through the crystal, an electrical field is created. Thus a piezoelectric crystal is able to act as both a generator of ultrasonic vibrations and as a converter of such vibrations into electrical current.

The piezoelectric crystal of the ultrasonic transducer must have electrodes in contact with its front and back surfaces. A layer of plastic is placed on the front of the crystal for protective purposes. In the immediate vicinity of a disk-shaped transducer, the ultrasonic vibrations are transmitted as a beam of a diameter similar to the diameter of the disk. Within the region close to the transducer, the near zone, the sound intensity distribution is not homogeneous because of interference. The length of the near zone L is given by

$$L = R^2/\lambda$$

where R is the radius of the transducer and λ is the wavelength. The angle of divergence α for the beam is derived from the formula

$$\alpha = \frac{0.61 \times \lambda}{R}$$

Thus it can be seen that the higher the frequency or the smaller the wavelength of the beam in relation to the radius of the transducer surface, the smaller will be the spread of the sound, resulting in better resolution of the targets crossing the beam.

The main limitation of a stationary ultrasonic transducer is a loss of visualization of a significant number of interfaces in a plane not parallel to the surface of the body. To alleviate this limitation as well as others connected with visualization of geometrically complex structures, use has been made of the so-called scanning diagnostic ultrasound system. In sector scanning, the transducer is rocked back and forth or through a multiplicity of angles. Inasmuch as for each angle of inclination of the transducer to the surface of the body there will be a corresponding change in angle of beam incidence in the interfaces, the chance of striking the interface at an angle suitable for visual detection is increased. In addition the transducer may be made to move around in a circle, so that the ultrasonic beam is always directed to the center of the circle. This technique is known as circular scanning. A combination of sector and circular scanning represents compound scanning. Compound scanning makes it possible to follow the natural contours of body parts like extremities or the head.

Although compound scanning has some advantages, it also requires much more complex equipment than does simple scanning. It is especially useful in obtaining ultrasonic pictures of the kidney, liver, spleen, and uterus.

The earliest two-dimensional real-time cross-sectional scanning system was a linear-array scanner. This system utilizes a linear multielement transducer from which the sound beam is transmitted and received by the elements along the array and the entire array is scanned at a high rate. The multielement array transducer is usually 6 to 8 cm in length and contains 20 to 60 individual elements. The ultrasonic beam is generated by triggering sequentially the individual elements in the array in rapid succession. Thus an ultrasonic beam is formed that is electronically "swept" along the length of an array and in a direction perpendicular to the line of crystals. This linear multielement array generates large rectangular images.

Phased-array electronic sector scanners also utilize a linear multielement transducer in which the ultrasound beam is electronically "swept" by varying the delay of the transmitted and received pulse. The resultant image is a two-dimensional cross-sectional sector scan of cardiac structures within the beam. The phased-array transducer is about the same size as a conventional transducer but generally consists of 30 individual elements spaced over a total distance of approximately 15 mm. Each element in the array is triggered sequentially, with a small delay line interposed in the excitation path of each crystal. Variations in the degree of delay result in phase shifts which electronically steer or swing the beam through a sector.

The number of lines in the image is determined by the pulse repetition frequency and the number of frames used in the study. Since the maximum pulse repetition rate without interference of transmitted and received pulses is about 4000/s, each frame contains approximately 133 lines when a frame rate of 30/s is used (Taylor, 1978). The real-time image displays a high density of information when the sector angle is relatively small (e.g., 30°) and the individual lines are closely spaced. It is important to note that at wide sector angles (e.g., 80°) the spaces between the data lines are wide, since a fixed number of available pulses must be divided more often and spread over a wider area. The cost of a phased-array system is generally high (approximately two to six times higher than a linear-array system), mainly because the hardware for producing delays is very expensive and the beam-steering technique can only be accomplished by computers.

Mechanical sector scanning, as the name implies, is based on the concept of a transducer being driven through a selected arc from a fixed spot on the patient's chest. In this system the transducer is rapidly moved through a small arc. Each angular excursion of the crystal produces a frame with every beam sweep so that a full cycle of angulation produces two frames as the crystal is angled and then returned to the starting position. The same relationships exist between sector angle, frame rate, and line density as have been described previously for phased arrays.

41.1-5 Clinical Applications

For many years fairly simple pulse-echo instruments have been used for determining midline shift of the brain following injury or disease. The name given to the procedure is echoencephalography. This technique is particularly valuable in accident cases in which there is neither time nor facilities for angiographic studies and a decision has to be made about emergency surgical intervention. Reportedly, echoencephalographic equipment now finds a place in the specially equipped ambulances used for motor accident work in Japan, where emergency treatment, including surgery, is undertaken at the site of the accident because of the difficulty of getting patients to the hospital through the intensely heavy traffic.

We can understand the principle of the echoencephalography procedure if we imagine a needle passing from one side of the skull through to the other side. The resistance felt when the needle penetrates the tissue surface is indicated by a mark on the needle corresponding to the amount of resistance. The needle thus gives a one-dimensional report on the state of the path the needle has followed through the head. This principle is realized when a sound pulse, instead of a needle, is sent into the head as a plane-wave packet. Interfaces offer resistance to the sound pulse, which is partly reflected. The returning ultrasonic echoes are conducted to a display unit which converts the time differences between transmitter pulse and echo received and displays them on a latitudinal axis, thereby creating a one-dimensional display that we call an A scan.

The best frequency for routine echoencephalography is 2 MHz. In the brain, the wavelength of 0.75 mm is short enough to provide adequately accurate data about the site of the various interfaces at which the echoes originate. Furthermore, the area of amplification is large enough to pick up all reflections coming from the inside of the adult skull. A transducer at the frequency of 2 MHz and a diameter of 15 mm may be regarded as the optimum transducer because it is able to solve nearly every echoencephalographic problem. This probe produces a rectified ultrasonic beam of 80 mm in the direction of propagation so that a structure in the center of the brain can be easily localized.

Furthermore, the manipulation of the light transducer is very simple. A suitable application point in the temporal area where the surface of the 15-mm probe can make complete contact may be found in nearly every patient, while the 24-mm probe presents coupling problems.

For special problems of investigation, for example, the demonstration of subdural hematomas on the side of the hemorrhage of infantile hydrocephalus, a transducer with the frequency of 4 MHz and a diameter of 10 mm has proved to be very useful, because the depth of ultrasound penetration required in such cases is only slight.

Since ultrasound of a frequency of 4 MHz is attenuated much more in bone than in soft tissue, this condition is fulfilled. The inside of the skull as far as the origin of the midline echo can be evaluated by relating the midline echo to the anatomy of the brain, because the initial echo is much shorter here than with the 2-MHz transducer and a distortion free production of the echoes on the oscilloscope screen is obtained from the outset. To achieve this with low-frequency transducers, special measures must be employed for the amplifier gain, time voltage controls, and reject controls.

Sometimes obtaining a clear midline echo in older patients is impossible with use of a normal transducer. The difficulty can often be avoided by using a 1-MHz transducer with a diameter of 10 mm, which penetrates bone much better. For perfect reception of ultrasonic echoes, there must be good contact between the transducer and the skull. A coupling medium such as water, paraffin, or some special ultrasonic jelly is best for this. Air bubbles between the probe and the skull must be avoided under all circumstances.

The first step in the echoencephalography procedure is the positioning of the patient. Patients who are able to walk are best seated in a chair with a headrest facing the physician. After the patient is positioned correctly, the theoretical midline of the patient's head must be determined. This can be done by relating the midline echo to the anatomy of the brain. Two transducers are used. Each one is applied to the unshaved skull slightly above each ear, facing each other through the skull. At the point corresponding to half the diameter of the head, a single reflection appears because the ultrasonic beam covers the distance from one transducer to the other only once. With the oscilloscope A presentation, a high reflection may be seen at the left edge of the screen before the transducer is applied to the surface. This is the previously mentioned transmitter pulse or main bang. The variations in electrical voltage which stimulate the crystal are responsible for this reflection. It is also called the crystal artifact. The transmitter pulse or main bang is followed by a smooth baseline. Reflected quantities of ultrasonic energies or echoes are then received and displayed as positive blips on this baseline.

Echo sounding is first carried out from the left temple, then from the right. Once a comprehensive, satisfactory A trace has been obtained by moving the transducer and adjusting and regulating the cathode ray picture, the trace can be photographed with an instant camera. It has proved useful to project all traces, including that from the theoretical midline obtained by means of the through transmission approach and those from the left and from the right, on a single triple-exposure photograph. This can be done either by elevating the cathode ray tube exposure or by projecting the trace in the center of the cathode ray tube for the first picture and then turning the tube upside down for the second picture. The latter method has the advantage that the displacements of the midline echo can be recognized more easily. On the other hand, it has the disadvantage that the amplitude of reflection, which is often important in the interpretation of the picture, is poorly recorded because only half of the screen is available for its projection.

The amount of time required to determine the midline echo in a healthy subject is only one to two minutes. If there are pathological findings present, on the other hand, the investigation takes four to five minutes, including the taking of the picture. This time expenditure is certainly justified even in acute cases, since the findings obtained with echoencephalography not infrequently indicate immediate surgery.

Lack of alignment of the left and right echo traces or midline echoes signifies a shift of midline structures and thus the presence of an expanding process and in some cases a pulling of midline structures toward a shrunken hemisphere. The degree of shift from the midline equals half the distance between the left and right echoes and is expressed in millimeters. According to most workers, a shift of 3 mm either to the right or to the left is considered the upper limit of normal variation. Any shift beyond that, for example, a shift of 5 mm to the right, signifies some type of space-occupying lesion in the left hemisphere of the brain. A shift of, say, 7 mm to the left probably signifies a space-occupying lesion in the right hemisphere of the brain.

Echoencephalography complements the electroencephalography procedure in the most advantageous way because, once a focal lesion is detected by the electrical method, the ultrasonic technique adds, in many cases, the necessary differentiation between an expanding and a stationary process and very frequently can produce a precise diagnosis. Echoencephalography augments the diagnostic information without hazard or inconvenience to the patient.

The human eye is an ideal organ for ultrasonic examination. It consists of a cornea, an anterior chamber, a lens, a vitreous space filled with homogeneous fluid, a retina, and a posterior wall. The ultrasonic transducer usually operates at a frequency of 5 to 15 MHz. Of course, the higher frequency is more desirable because it is associated with better resolution. However, since it is associated with decreased penetration, the normal sensor used in eye work is usually a transducer at a frequency of about 7.5 MHz.

The transducer is placed either directly on the cornea after the cornea has been anesthetized with a few drops of anesthetic or it is placed directly on the lid of the closed eye with some ultrasonic jelly underneath it to prevent air from generating noise in the ultrasonic beam. The ultrasonic pulses emanate from the transducer face and pass through the eye. Whenever a pulse meets an interface of different acoustic impedance, a portion of the energy is reflected as an echo. Residual energy continues to the next interface, which in turn reflects an echo, and these echoes establish a pattern. The anterior chamber, the lens, and the vitreous cavity are normally acoustically homogeneous and produce no echoes.

The echogram of the normal eye usually shows an average diameter from the most anterior to the most posterior echo of 22 to 26 mm. The distance representing the anterior chamber is usually approximately 2 mm and the echo representing the lens normally measures 4 mm. The vitreous cavity has a diameter of about 18 mm.

One of the most significant areas for the A-mode application in eye work is the measurement of the dimensions of the globe. Changes associated with certain diseases like glaucoma or myopia can be demonstrated. A particularly interesting application of diagnostic ultrasound is the analysis of the vitreous cavity. Pathology in the vitreous cavity can present a variable acoustic response. Both the nature of the target lesion and the state of the surrounding vitreous humor are important in interpreting the echo configurations that are obtained. The greater the difference in their impedances the more pronounced will be the echo response. Fresh hemorrhages in the vitreous cavity differ from those of several days standing in that they usually reflect clear sharp echoes. The latter frequently become quite acoustically homogeneous with the degenerative vitreous reaction and may not reflect any patterns at all.

If the hemorrhage is well separated from the surrounding vitreous humor, the echoes reflected may be of such caliber as to mimic a foreign body, and localization becomes quite difficult. However, frequently there is destruction of the normal vitreous architecture, which leads to eventual dispersion of the blood and prevents the formation of sharp interfaces with confusing patterns.

One of the most important applications of diagnostic ultrasound in ophthalmology is the detection of retinal detachment. The detachment may be serous or fluid, where the subretinal fluid is acoustically homogeneous and reflects no echoes, or may be solid, where echoes are reflected in terms of multiple blips disrupting the baseline and extending posteriorly to also involve the wall of the globe. When a single high-amplitude spike is obtained, the sonar beam is probably striking the retina perpendicular to its surface. This is most easily demonstrated in an area of flat detachment. The evaluation of solid detachments may be carried out by ultrasonic techniques even in the absence of a fundus view. The A mode is capable of identifying the presence of a solid detachment. It may also identify tumors attached to the retina or located behind the retina in the choroid region. Ultrasound is also useful in foreign body localization within the eye as well as in extraction of this body by a surgical technique.

Whenever it is necessary, desirable, or convenient to terminate pregnancy prior to the onset of labor, there is need for accurate estimation of the gestational age. Avoidance of excessive prematurity is of prime importance. A significant error in the estimate of fetal weight or expected date of delivery imposes an additional hazard on the fetus and increases the likelihood of prenatal death. Unfortunately, no clinical criteria sufficiently predict fetal maturity.

Additional evidence to support the clinical estimation of the gestational age can be obtained by diagnostic ultrasound. The measurement of the biparietal diameter of the fetal head within the uterus is a practical procedure. From this measurement one can make the usual prediction about birth date and to a certain extent about gestational age. More important, the biparietal diameter indicates minimum birth weight. Armed with this information the physician can with greater certainty avoid the problems associated with premature delivery.

Ultrasonic fetal cephalometry is a painless, safe, and accurate method of measuring the biparietal diameter of the fetal head. Measurements correlate significantly with other measurements made after birth. There is also a statistically significant correlation between ultrasonic measurements and birth weight. These measurements have proved valuable to clinical obstetricians for a wide range of medical and obstetric complications that call for early termination of pregnancy. They are more valuable as indicators of the minimum birth weight that might be expected than of the exact weight of an infant. Their proper use is to supplement rather than to supplant the patient's history and clinical data. When so employed, they are useful in clarifying the question of fetal maturity.

The procedure is simple, safe, and reliably accurate. After the fetal head is located by palpation, the transducer of the diagnostic ultrasonic equipment is coupled to the abdominal wall either with water or lubricating jelly and is then moved over the surface of the abdomen until a satisfactory echo pattern appears on the oscilloscope screen. If the transducer is perpendicular to the fetal skull at the biparietal diameter, three echoes appear. Echo one is returned from the abdominal wall. Echoes two and three are returned from the near and far skull walls and are approximately the same height but are placed higher than the first echo. The biparietal diameter measurement is obtained by counting the markers on a superimposed centimeter scale; posterior and anterior diameters can produce a similar pattern. These can be recognized because they are considerably larger than the biparietal diameter. Occasionally it is possible to obtain an echo from the third ventricle, which is an additional confirmation that the biparietal diameter has been located.

Once the fetal head is engaged in pelvic delivery it is usually not possible to obtain a satisfactory measurement because of interfering echoes from the mother's pelvic bones. This is no disadvantage because the biparietal diameter is seldom needed following engagement. The pregnant abdomen has proved particularly ideal for pulse-echo ultrasonic techniques both because of its anatomical configuration and the fact that the uterus is filled with fluid, which makes good sonic contrast for depicting intrauterine structures. Pregnancy represents a condition for which a new diagnostic technique would be particularly valuable, since radiation hazards limit the use of x-ray and isotope techniques for study of various abnormalities of pregnancy and for following fetal development over a long period of time.

Ultrasonic scanning, providing a B-scan presentation, can be achieved utilizing a compound contact scanner. In this method the transducer, which contains a lead zirconate crystal, moves mechanically 30° to each side of the perpendicular while at the same time it is moved by the operator across the pregnant abdomen. A storage oscilloscope tube stores on the oscilloscope screen the echo information reflected from the tissue interfaces. The skin surface is coated liberally with mineral oil to provide sonic contact between the tissues and the transducer. Each cross-sectional scan requires approximately one minute. The transducer carriage is then moved up and down the uterus in both horizontal and sagital planes to provide cross-sectional presentations at approximately 2-cm intervals. The entire examination requires approximately 30 minutes.

By reviewing the pictures taken at 2-cm intervals both horizontally and sagitally in a composite fashion, the viewer can plot out graphically the gross location and boundaries of the placenta. The placenta pattern should be seen in three horizontal and one to three sagital films. Thus ultrasound can provide specific localization in any part of the uterus, which makes it particularly useful prior to amniocentesis or intrauterine transfusion. In contrast to other diagnostic techniques, ultrasonic placentography visualizes the placenta and its relationship to the uterine wall, the fetus, and the maternal abdomen. Echoes from the placenta form an ultrasonic pattern peculiar to that tissue and thus can outline precise boundaries for the placental margin.

The examining technique causes the patient no particular discomfort and thus can be used to follow the patient serially during pregnancy. Up to the present no untoward reactions have been reported secondary to use of ultrasound for diagnostic examination in either mother or infant.

Since the advent of cardiac surgery, the need for accuracy of the cardiovascular diagnosis preceding the operation has increased tremendously. As a result, several new diagnostic techniques have been developed, such as measurement of the pressure in the heart chamber by heart catheterization and various methods of angiocardiography. Nearly all of the methods are complicated, require a relatively large medical staff, and involve some risk to the patient. They are also uncomfortable for the patient and therefore cannot be repeated frequently for control purposes.

In 1954 Edler and Hertz in Sweden showed that heart structures reflect ultrasound and that echoes could be obtained from the heart by placing a 2.5-MHz transducer from commercial diagnostic ultrasound equipment designed for material testing externally on the chest of the patient. The pattern of movement of echo signals varies according to the location of the crystal on the chest and the direction of the beam. If the heart is enlarged, pulsating echo signals can be obtained over a significantly larger area than in normal cases. By applying the 2.5-MHz ultrasonic transducer to the third or fourth left interspace 1 to 4 cm from the left sternal border, echo signals are obtained which pulsate in position and size synchronously with the heartbeat of the patient. Since the lung tissue surrounding the heart has a very high absorption coefficient for ultrasound, with a frequency of about 2 MHz, no echoes from the body structures are obtained. On the other hand, the heart can be located by this method only from a relatively small area under the thorax where no lung tissue lies between the heart and the chest wall.

Since the movement of echoes represents the movement of heart structures, it was clear in the beginning that echocardiography could eventually be used for diagnostic purposes. In the diseased heart we would expect that these typical heart movements would be greatly altered.

Before the various echoes from the heart structures recorded on an echocardiogram can be identified and analyzed, the basic function of the human heart must be understood. The contracting ventricles of the human heart constitute a unique type of pump. The source of energy is a repetitive contraction initiated through electrochemical reactions. As a pump, the human heart has two inlet valves, the mitral and tricuspid, and two outlet valves, the aortic and pulmonary. During a cardiac cycle, two heart sounds are generated that consist mainly of two separate energy bursts of vibrations occurring after closure of the two sets of valves. Partial obstruction of any valve or blood leakage due to imperfect closure results in turbulence of blood flow and concurrent audible sounds called murmurs. In heart disease diagnosis, such murmurs are often symptomatic of one or more malfunctioning valves.

Echocardiograms as measured by the ultrasonic echoing technique contain characteristic patterns for the valve motion and structural detail of normal and diseased valves. The ultimate objective of analysis is to establish operational signatures of normal and diseased natural valves, and for normal and diseased structures in the heart, such as walls or implanted prosthetic valves. This includes the detection of particular valve faults either by interpretation of the echocardiogram itself or by analysis of the correlations between the ultrasonic echoes and heart sound recordings.

Obstruction in any heart valve is called stenosis, while a leak in the heart valve is referred to in medical nomenclature as insufficiency or regurgitation. The echoes on the echocardiogram may be depicted in the A presentation (A mode), which simply shows the depth of penetration and the amplitude of the reflected ultrasound energy. The depth-of-penetration scale is calibrated for an average speed of sound and tissue at body temperature of 1540 m/s. On this scale, precise measurements of dimensions from the chest wall to the posterior wall of the left ventricle can be made and of left and right ventricular sizes or the thickness of the myocardium or heart muscle itself.

The M-mode presentation permits the study of valvular or myocardial motion by depicting not only the depth of the structure but also the changing position of the echoes as though they were viewed from above. This presentation provides intensity modulation for the large number of echo amplitudes and may be recorded photographically on instant film or on a strip recorder.

The clinical application of ultrasound in cardiovascular disease is well established and has gained widespread acceptance for the diagnosis of valvular disease. The waveform is specific for the scarred, calcified, and immobile mitral valve leaflet. The range of motion or amplitude of the recording by the M-mode presentation is characteristically reduced. Echocardiography serves as an important guide in defining the precise structural abnormality of the mitral valve. Severe calcification or fibrosis is shown by increased brightness of the echoes on the B presentation. Marked calcification causing a relatively immobile leaflet is also demonstrated by reduced amplitude of motion of the echo.

Echocardiography has also been very valuable in detecting blood clots or tumors in the heart. Additionally, it has been helpful in detecting pericardial effusion or fluid accumulating in the pericardial sack, the sack that holds the entire heart. Combined M-mode and real-time two-dimensional echocardiography is the noninvasive technique of choice for detecting segmental or global disease of the heart muscle. Abnormalities in cardiac wall motion as imaged by ultrasound provide reliable indications of coronary artery or intrinsic cardiac muscle disease. The echocardiogram also yields anatomic and hemodynamic information so that most indices for left ventricular function, including cardiac output and ejection fraction, can be derived.

Bibliography

Feigenbaum, H., *Echocardiography*, 3rd ed., Lea & Febiger, Philadelphia, 1981.

Hobbins, J. C., *Diagnostic Ultrasound in Obstetrics*, Churchill Livingstone, New York, 1979.

Kingsley, B., J. W. Linhart, and P. Kantrowitz, *Advances in Non-Invasive Diagnostic Cardiology*, Slack, Thorofare, NJ, 1976.

Kotler, M. N. and B. L. Segal, *Clinical Echocardiography*, Davis, Philadelphia, 1978.

McDicken, W. N., *Diagnostic Ultrasonics; Principles and Uses of Instruments*, 2nd ed., Wiley, New York, 1981.

Nanda, N. C. and R. Gramiak, *Clinical Echocardiography*, Mosby, St. Louis, MO, 1978.

Purnell, E. W., *Ultrasound in Ophthalmological Diagnosis*, Plenum, New York, 1966.

Schiefer, W., E. Kazner, and St. Kouze, *Clinical Echoencephalography*, Springer Verlag, New York, 1968.

Taylor, K. J. W., *Atlas of Gray Scale Ultrasonography*, Churchill Livingstone, New York, 1978.

Wells, P. N. T., *Biomedical Ultrasonics*, Academic, New York, 1977.

41.2 COMPUTERS IN MEDICINE: A LOOK AT COMPUTED TOMOGRAPHY AND NUCLEAR MAGNETIC RESONANCE IMAGING

Daniel B. Diner

41.2-1 Introduction

Medicine, like many other aspects of life, is undergoing profound changes as the age of computers dawns. These changes include both the elegant handling of the massive medical data collections and the development of new medical equipment and techniques.

We would like to focus attention on two of these techniques: computed tomography and nuclear magnetic resonance imaging.

Computed tomography (also called computerized axial tomography, CAT scanning, EMI scanning, x-ray tomography, or simply CT) was invented in 1971 by Dr. Godfrey N. Hounsfield of EMI Ltd. Working in cooperation with Dr. James Ambrose at Atkinson Morley's Hospital, Wimbledon, England, he successfully scanned a human brain with surprising clarity and detail. This opened the door for further development of the CT system, and led to Dr. Hounsfield's sharing of the 1979 Nobel Prize in Physiology and Medicine.

Nuclear magnetic resonance imaging (also called NMR imaging or NMR zeugmatography) was first performed by Dr. Paul Lauterbur of the State University of New York in 1973. However, not until 1978 was what is believed to be the first NMR head scan presented. It was made by Dr. Hugh Clow and Dr. Ian Young of EMI Central Research Laboratories, London. (NMR spectroscopy, incidentally, was first described in 1946 by Dr. Felix Block of Stanford University, Palo Alto, California, and by Dr. Edward Purcell of Harvard University, Boston. For this they won the 1952 Nobel Prize in Physics.)

Thus although NMR spectroscopy and x-ray photography existed for quite some time, it took the power of modern-day computers to make possible the medical imaging techniques that we have today.

41.2-2 Computed Tomography

Method

Anyone who looks through an anatomy or a pathology textbook will be aware of the value of a clear cross-sectional image (a slice) of anatomical tissue. Such a slice at the proper level of, say, the brain, could reveal the presence of a tumor, the blockage or rupture of a blood vessel, or any number of structural or disease-related problems.

Thus the problem faced by the inventors of CT was to take the information available in the two-dimensional x-ray shadow of, say, the brain, and from this develop cross-sectional images of any desired "thin slice" through the brain. Suppose, for example, one wanted to image the details of a 1-cm thick horizontal slice of the brain located just above the eyes. One would first pass a narrow beam of x-rays (say, 1-cm thick) through the horizontal slice of interest and then measure the transmitted beam by a scintillation detector on the other side. Let this beam start at the left front of the brain, for example, pass perpendicular to the midline, and exit at the right front of the brain. The beam and the scintillation detector are then moved 1 cm back and a second measurement is made exactly parallel to the first. This is repeated until the entire slice of interest has been scanned.

One would then rotate the head by 1°, and take a second set of parallel measurements across the entire slice of interest. This is repeated until the head has been rotated 180° and one has measured every possible line through the head at each of the 180° orientations. (Actually, the head is not rotated. The x-ray tube and the scintillation detector are rotated instead. Not only does this ensure greater patient comfort, but it improves the accuracy of the system.)

Consider the slice of interest as a two-dimensional grid, with an independent x-ray absorption coefficient for each grid element. Imagine each grid element to be 1 cm^3 in size.

Provided that one has been able to make sensitive enough measurements, one is now able to determine mathematically, within some level of accuracy, the x-ray absorption coefficient of each grid element in the slice of interest. This is true because the absorption coefficient is defined as

$$\text{Absorption} = \log\left(\frac{\text{intensity of x-rays}}{\text{detector reading}}\right)$$

Thus the absorption for each measurement across the slice of interest is actually the sum of the x-ray absorption coefficients of the grid elements that the beam has passed through. Because the brain is much less than 180 cm in diameter and we have taken 180 sets of measurements, all at different angles, it is easy to see that we have more independent equations than variables. Because only n independent equations are needed to solve for n variables, the problem is solved. The mathematical method most widely used is called the filtered back-projection algorithm.

Once the x-ray absorption coefficient of each grid element has been computed, a two-dimensional cross-sectional image of the slice of interest is generated and displayed on a computer screen. Each grid element is shaded in proportion to the value of its x-ray coefficient. Normally several cross-sectional images are made. For example, several slices of tissue, one above the other, can be scanned to get a three-dimensional analysis of that region of tissue. Another possible configuration would be three scans of slices that are mutually perpendicular to each other and that pass through a point within a region of interest. These three images would be combined to approximate a three-dimensional analysis of that region.

The detail and clarity of the original CT images were immediately recognized to be of extreme medical value. In actuality, things are a bit more complex. The complexity arrives from several factors, including human tolerance of x-rays, possible patient motion during the scanning time, and the statistical nature of x-ray detection. Each of these will be discussed later. Suffice it to say that the technique works and, for the first time, remarkable internal images of living human brains and organs have been made without surgical intervention.

Many improvements on the original system followed rapidly. The translate-rotate system (just described) was superseded by the rotate system and the stationary-circular-detector-array system. In the rotate system, the set of parallel measurements taken at each angle was replaced by a fan of measurements (from the x-ray tube up to 300 detectors) at each angle. The x-ray tube and the 300 detectors rotate together through 180°. In the stationary-circular-detector-array system, 700 to 1000 fixed detectors surround the patient and only the x-ray tube rotates. All three methods take approximately the same pattern of readings.

Resolution

The grid element size (the resolution) of commercial CT systems has been reduced to as low as 0.5 mm. Slice thickness has been reduced to as low as 2 mm. The 180° are still measured but in 0.33° steps, and about 1.5 million readings can be taken in as few as three seconds to give a 320×320 image grid.

The fast scans (three seconds) minimize errors due to patient motion. The narrower slice thickness cuts down on the radiation exposure for the patient. Radiation exposure is also minimized by use of the weakest x-ray beams that still give a clear picture.

This leads us to the statistical nature of the x-ray detection. Because only a limited number of photons arrive at the detectors, there is a statistical spread between readings. The standard deviation, according to Hounsfield, is about 0.5% on tissue (on a 320×320 matrix). This yields a picture noise (or grain) that is inherent in the system. Again, because radiation exposure must be limited, there is a limit on the improvement of grain reduction one can expect from CT.[1]

In industrial applications, however, there is no radiation exposure problem, and thus the accuracy of the system can be increased.[1]

Medical Impact

Many malignant tumors can be seen clearly on CT[1] scans. Particularly important is the location of brain tumors. The resolution of CT enables us for the first time to guide and align radiotherapy beams while irradiating brain tumors. CT also enables us to determine the optimal strength of the radiotherapy beams and even to measure the effect of radiotherapy on the tumor.

CT has greatly reduced the need for exploratory surgery and tracer-dye injections, some of which are life threatening.

The ability to very accurately measure the x-ray absorption coefficients of tissues in vivo promises great advances in the study of biological tissues.

For the first time, soft tissues, such as the liver and kidneys, can be distinguished by x-rays.

The greatest impact of CT on medicine, however, will come from the ability to see inside the living body without surgery. This particular aspect of CT is still being developed.

One system of particular interest is the dynamic spatial reconstructor at the Mayo Clinic, Rochester, Minnesota.[2] This apparatus gives a three-dimensional moving image of an internal organ or of any part of the internal organ from any view desired. The image can be life size.[2]

By using (eventually) 28 x-ray tubes that rotate around the patient every four seconds on a circular gantry 4.5 m in diameter, the dynamic spatial reconstructor sends data for 15,000 cross sections per second to the computer. Each cross section is 0.9 mm thick.[2]

The computer can display the images as moving three-dimensional reconstructions on a television screen. For example, a three-dimensional image of a heart can be called up, rotated, sliced through any plane desired, and inspected for the internal structures. If one desires to see the arteries but muscle is obscuring the view, the muscle can be "dissolved" electronically.[2]

Eventually, all this will be done almost instantaneously while the patient is in the apparatus. The limiting factor now is the processing speed of the computers.[2]

An additional goal is development of the display device, a problem already being addressed. The display device will be circular with a "drumheadlike flexible mirror" which vibrates at high speed. The mirror will reflect the images from the television screen. The reflected images will be life size and the picture will resemble a hologram. Besides being able to change the position of the image on the screen, one will be able to walk around the image and inspect it from any angle. One will also be able to watch the organ move, to speed up or slow down the image motion, or to call for an instant replay.[2]

If the system is successful, it will be like a "magic window" through which one can watch the human body's internal functions.[2]

Problems

The main problem with CT is the x-ray exposure. X-rays can cause changes in cellular DNA which might lead to cancer.

A second problem is that of body motion. For brain scans this is not a serious problem, because the head can be stabilized. However, for study of the lungs or the heart, this problem can be serious enough to cause tumors to be overlooked.

A third problem is that of artifacts. Owing to the high difference in x-ray attenuation coefficients of tissues (bone vs. lung, for example), linear artifacts are quite common. Thus a CT scan may indicate the presence of something when nothing is there.

Another problem is that bone is opaque to x-rays. Thus tissue near bone or surrounded by bone cannot be seen clearly. This is particularly important in scans of the posteria fossa of the brain or if one wishes to study bone marrow.

Despite these problems, however, CT stands as a remarkable advance in medical science.

41.2-3 Nuclear Magnetic Resonance Imaging

Method

The invention of CT was rapidly followed by the development of nuclear magnetic resonance imaging. NMR imaging was first performed by Dr. Paul Lauterbur of the State University of New York in 1973.

To understand the more elaborate uses of NMR, one must understand quantum physics. However, the classical physics approach is sufficient to understand clinical NMR imaging. We shall deal only with classical physics here.

Certain nuclei, when placed in a magnetic field, can be caused to resonate by the application of a particular radio frequency (RF). The nuclei absorb energy from the RF. If the RF is then turned off, the nuclei will continue to resonate in the magnetic field at that same frequency, thus emitting signals and giving up the absorbed energy. The emitted signals can be observed by a receiver coil.[3] The emission demonstrates an approximately exponential time decay.

Let us step back a moment and visualize what occurs.

The nuclei that can be so excited are those that have an odd number of protons, an odd number of neutrons, or both.[4] The net rotation or spin of their charge distribution generates a magnetic field. Thus each such nucleus acts like a magnetic dipole.[4]

Before the magnetic field is applied, the magnetic dipoles of the nuclei are randomly distributed. The application of a uniform static magnetic field causes the dipoles to align such that their net magnetization vector aligns with the lines of induction of the magnetic field. Let us call this direction the z axis.[4]

For a given magnetic field strength, the nuclei of an NMR-sensitive element can absorb energy only in response to an RF pulse of a particular frequency. This frequency is called the Larmor frequency, for that nucleus and that magnetic field strength.[4]

The RF pulse acts like a smaller magnetic field that rotates about the net magnetization vector in the x, y plane. When the Larmor frequency is applied, the nuclei experience a torque, and this changes the direction of the net magnetization vector away from the direction of the static magnetic field (the z axis).[4]

If the RF pulse is left on just long enough for the net magnetization vector to be entirely in the x, y plane and is then turned off, the net magnetization vector will continue to rotate in the x, y plane, thus generating a signal. This signal is the maximum signal the net magnetization vector can generate, because only its x, y component can generate a signal. Clearly if the vector were pointing in the z or $-z$ direction, no signal would be generated, as no x, y component would be present.[4]

When the RF pulse is turned off, signal emission begins. This implies a loss of energy from the resonating nuclei, which causes the precessional motion of the net magnetization vector to decay. Eventually the net magnetization vector will point, once again, in the direction of the z axis. The emitted signals diminish accordingly.

The emission amplitude is a function of the density of the NMR-sensitive resonating nuclei present. The emission relaxation (decay) time has two components: one longitudinal to the magnetization field, which is called $T1$, and the other transversal to the magnetic field, which is called $T2$. $T1$ is related to the nuclear environment of the resonating nuclei and is thus called the spin-lattice relaxation time. It measures the rate of return of the net magnetization vector to the z axis. $T2$ is related to the interactions of the resonating nuclei and is therefore called the spin-spin relaxation time. It measures the dephasing of the resonating nuclei (in the x, y plane) before they lose their absorbed energy. $T2$ is clearly never longer than $T1$, and is often much shorter.

$T1$ and $T2$ are actually defined as the time constants of their respective (approximately) exponential decays. Thus after three $T1$'s, 95% of the longitudinal decay is complete.

The three parameters (nuclei density, $T1$, and $T2$) are sufficient to distinguish between many human tissues. Precisely how each is measured, and what each distinguishes, will be discussed later.

Now let us discuss localization. Everything so far described takes place within a volume—the volume of the static magnetic field. We wish to image cross-sectional slices of this volume. This can be done in the following ways.

The volume can be reduced to a plane by tailoring the RF bandwidth to one specific frequency. This is because, for only one strength of the magnetic field (found in only one plane perpendicular to the direction of the magnetic field), will the RF be the Larmor frequency of the nuclei of interest. In biological tissue, the nuclei of interest are usually free hydrogen ions; that is, protons. Other excitable nuclei exist, but their Larmor frequencies are such that we can tailor the magnetic field to prevent their excitation. This is called the method of selective irradiation.

A second way to isolate one plane is by oscillating the magnetic field over time such that only one null plane does not vary, and averaging all measurements over time. By this technique we limit our measurements to one plane. This is called the method of the time-dependent gradient. In either case, the one plane corresponds to the slice of interest in CT scanning.

If we can reduce the data collection to lines, we can use the filtered back-projection algorithm to get the cross-sectional images that we desire. To reduce the collection to lines, the following technique is used.

After the RF pulse is turned off and the nuclei in the plane are rotating at their Larmor frequency, a small transverse magnetic gradient is introduced across the plane. This causes the Larmor frequency to vary across the plane, thus causing the resonating nuclei to adjust their frequency of resonance to match the new Larmor frequency at each location. The signal frequency emitted by each resonating nucleus will, of course, be its new Larmor frequency. Thus each line of equal magnetic field strength will emit only one frequency. The amplitude of this frequency will tell us how many resonating nuclei are present.

The receiving coil measures all the emitted frequencies mixed together. However, by use of Fourier analysis, the amplitudes of each frequency can be computed. Thus we can measure the nuclear magnetic resonance for a set of parallel lines across a plane of interest.

By rotating the transverse magnetic gradient 1° at a time and repeating the measurements, we can obtain 180 sets of parallel lines of information. Now we can use the filtered back-projection algorithm to get the NMR image.

Let us go back now to the three parameters of NMR imaging: nuclear density, $T1$, and $T2$.

There are many NMR scanning techniques. However, three in particular stand out in the literature: the saturation-recovery (SR) technique, inversion-recovery (IR) technique, and the spin-echo (SE) technique.[5]

In the SR technique, a 90°-RF pulse is applied (one that rotates the net magnetization vector

90°), the transverse magnetic gradient (the "read" gradient) is then applied, and data are collected. If a time much longer than $T1$ is waited before the next 90°-RF pulse is applied, full recovery is possible, and the second signal read will be comparable to the first, that is, strong. Thus tissue with nuclei with a short $T1$ will give strong signals and appear light. However, if a time comparable to $T1$ is waited before the next RF pulse is applied, only partial recovery of the net magnetization vector will have occurred and the RF pulse will shift a reduced magnetization vector 90°, thus resulting in a smaller measured signal. Thus tissue with nuclei with a long $T1$ will give a weak signal and appear dark. Blood has a long $T1$, as does cerebrospinal fluid. Flowing blood, however, appears light. This is because the new blood will not have been magnetized and thus will appear fully recovered, that is, as if with a short $T1$. For this reason the SR technique shows flowing blood and blood vessels very clearly.

Bone, incidentally, appears dark, but that is due to its low concentration of free hydrogen nuclei.

In the IR technique, a 180°-RF pulse is applied, a period of time is waited, and then a 90°-RF pulse is applied. After the 90°-RF pulse is shut off, the "read" gradient is applied and data are taken. If $T1$ is short with respect to the waiting time, full recovery will have taken place and the 90° pulse will leave the net magnetization vector in the x, y plane. Thus nuclei with a short $T1$ will give strong signals and appear light on the scan. However, if $T1$ is comparable to the wait time, a reduced signal will be measured after the 90°-RF pulse. Thus nuclei with a long $T1$ will give weaker signals and appear dark on the scan. This technique yields a marked differentiation between gray and white matter, because in gray matter the protons are mostly in water, that is, they are free, but in white matter they are usually found in lipids.

In the SE technique, a 90°-RF pulse is applied, a short period of time is waited, a 180° pulse is applied, an equally short period of time is waited, and then the "read" gradient is applied and data are taken. This is repeated for various short waiting periods. The data signals typically decay very rapidly but the peaks are plotted against the waiting periods to determine $T2$. Although this sounds confusing, it can be easily understood.

Short waiting periods are necessary because $T2$ is much shorter than $T1$ for many interesting solid tissues. The 180° pulse and the repeated short waiting period are necessary because the transversal decay time one measures otherwise is a combination of spin-spin dephasing and dephasing caused by inhomogeneities in the magnetic field. By shifting all the components of the net magnetization vector 180°, and allowing them to continue to spin an equal amount of time, the inhomogeneities which earlier caused dephasing now cause rephasing. An "echo" of the signal appears precisely when the dephasing due to the inhomogeneities has been canceled out by the rephasing. At this instant, the attenuation of the echo signal is essentially due to spin-spin interactions and thus reflects $T2$ decay. After this instant, the rephasing due to the inhomogeneities has passed the point of recovery and becomes dephasing. Thus only the peak of the measured signal, that is, the full echo, is used to determine $T2$.

The measurement is repeated for various waiting times simply to better define the $T2$ exponential decay.

SE scans show differences in soft-tissue detail.

In general, for water and simple fluids, $T1$ and $T2$ are about equal and range from tenths of seconds to several seconds. This time is temperature dependent. For solids, however, $T1$ can be many seconds and $T2$ can be milliseconds or shorter.[4,5]

Resolution

NMR is a noise-limited technique. The noise is the thermal noise of the body. Resolution (grid element size) can be on the order of 2 to 3 mm and can be improved by longer acquisition times. However, acquisition time is on the order of one minute, and thus body motion can become a problem. The long acquisition times come from the necessary waiting for measuring $T1$. The inducting RF pulses are usually quite short.

The accuracy depends on amount of subject movement, nonlinearity of the read gradient, phase errors, and nonuniformity of the RF and the static magnetic field.

Some fast-scan methods already exist, and the advances in this field are so rapid that one hesitates to say anything definitive about future NMR resolution or accuracy. However, one can expect rapid progress.

Medical Impact

NMR imaging may well revolutionize medicine. One must remember that NMR is essentially a chemical and physiological imaging technique and not a structural technique, such as radiography, CT, or sonography. Chemical and physiological changes often precede changes in histology and anatomy. Thus our basic approach to medicine may change.

A variety of other nuclei can be traced with NMR imaging. One very interesting one is phosphorus-31, which can reveal much about metabolism and metabolic impairment. For example, myocardial and cerebrovascular ischemia and infarction may be easily detected and better understood by phosphorus NMR imaging. Fluorine-19, sodium-23, carbon-13, nitrogen-23, and potassium-39 are also possible candidates. There is also the possibility of infusion of the patient with paramagnetic agents.[6]

Anatomic advances can be expected in the study of blood flow, diffusion rates, and soft-tissue structure and even in gene identification. NMR also presents the possibility of *in vivo* analysis of various body functions. For example, bone is not opaque to NMR, and thus the possibility exists to visualize bone marrow to study hematology.

Diagnostic possibilities include disease states in which the water content of tissue is changed, cancer, and even demyelination diseases. The lesions of multiple sclerosis (MS) show up as distinctly white in NMR scans, probably owing to their greater water content. In any case, for the first time one can measure the progress of MS independent of symptoms.

And finally, NMR presents treatment possibilities. Perhaps, for example, one could find a way to excite cancer cells until they self-destruct.

Considering that NMR is essentially noninvasive and that, unlike with CT, limits of human tolerance do not directly hamper its widespread and frequent use, all of the above discussed possibilities and many more are within reason.[7]

Problems

The main problem with NMR lies in the magnets. They are expensive to buy and to maintain. Many advances have been made in this direction. Suffice it to say that the uniformity of the magnetic field, safe from outside disturbances, and the limits of tolerance of the human body are all manageable problems.[6,8]

Artifacts also present problems, but these also are not very large. Small central artifacts are often seen. Large linear artifacts, however, are not usually seen except at skin surfaces.

There is a mathematical problem in that the IR, SR, and SE images do not truly represent the respective variables of nuclei density, $T1$, and $T2$, but mixtures of each. However, combinations of these images allow each of these variables to be isolated, and pure nuclear density, $T1$, and $T2$ images can be formed. These combined images are noisier but can be made when needed.

Medical problems with NMR are patient motion and the tradeoff between resolution and acquisition times. Both of these problems are being worked on, and we must patiently wait to see what progress will be made.

41.2-4 CT vs. NMR Imaging

CT measures one major variable, density, and one minor variable, atomic number. (Atomic number can be measured using the photoelectric effect. For example, iodine can be discriminated by subtracting two CT pictures taken at different x-ray levels.) NMR imaging measures three variables: proton density, spin-lattice decay time, and spin-spin decay time. Thus the two must be understood as complementary techniques, each capable of different measurements. CT measures form, and NMR imaging measures chemical composition.

At present, CT has a greater picture resolution and a shorter scan time than NMR imaging. These are extremely important medical considerations.

NMR, however, is noninvasive and can measure gray and white matter and flowing blood, very clearly. NMR also has the distinct advantage that it can see through bone.

Thus in the near future, we can expect the two systems to complement each other. However, it is reasonable to expect that NMR will eventually replace most CT applications.

References

1 G. N. Hounsfield, "Computed Medical Imaging," *Science* **210**:22–28 (1980).

2 P. Gunby, "At Mayo Clinic, The Latest Word in Computed Tomographic Systems," *JAMA* **244**:2393–2395 (1980).

3 D. Halliday and R. Resnick, *Physics*, Wiley, New York, 1966.

4 I. L. Pykett, J. H. Newhouse, F. S. Buonanno, et al., "Principles of Nuclear Magnetic Resonance Imaging," *Radiology* **143**:157–168 (1982).

5 I. R. Young, M. Burl, G. J. Clarke, et al., "Imaging the Posterior Fossa," *Am. J. Roentgenology* **137**:895–901 (1981).

6 A. E. James Jr., C. L. Partain, G. N. Holland, et al., "Nuclear Magnetic Resonance Imaging: The Current State," *Am. J. Roentgenology* **138**:201–210 (1981).

7 W. H. Oldendorf, "NMR Imaging: Its Potential Clinical Impact," *Hospital Practice* **17**(9):114–128 (1982).

8 A. E. James Jr., C. L. Partain, F. D. Rollo, et al., "Nuclear Magnetic Resonance (NMR) Imaging: The Potentials and the Technic," *S. Med. J.* **74**(12):1514–1519 (1981).

41.3 CARDIOPULMONARY MEASUREMENTS

Josefa Cubina

41.3-1 Introduction

The human body is a very delicate machine composed of a variety of organs arranged in systems. Each system performs a specific function essential for the survival of the body. The sum total of the functions of all the systems represents the function of the human body. It is, perhaps, the greatest wonder of life that so many parts could be integrated as intimately as in the human body.

As with every system on Earth, function in the body represents work, and work in turn is translated into energy expenditure. Every cell in the body is an energy-generating device; the energy is supplied in food in the form of chemical bond energy. Each cell has the ability to break down chemical bonds. This process is facilitated by the great number of enzymes located in cells. These biological catalysts are characterized by an astounding specificity, so that chemical reactions within the cell take place not only at a surprisingly rapid rate but with great efficiency. The energy thus produced is utilized to drive the various cell functions that make up total body function.

In all energy conversion processes a certain amount of energy is lost. Cells are no exception to this rule. However, even this inevitable energy drain is put to good use by the cell. The ultimate and least useful form of energy is heat. The heat produced in cell metabolism is utilized to maintain a core temperature optimal for enzyme action.

All energy conversion systems require a primary energy source to ignite the system and keep it going. In cells this "fuel" is oxygen. On this planet, oxygen is produced by green plants. It is thus plentiful in the atmosphere, but must be brought inside the body to reach the cells. The system that brings oxygen into the body is the respiratory system. Distribution of the oxygen to every corner of the body is accomplished by the cardiovascular system, with the help of the most important fluid in the body: the blood.

41.3-2 The Cardiovascular System

The cardiovascular system is composed of the heart, which acts as a pump, and the blood vessels. These vessels are a system of elastic conduits of varying caliber and length, distributed profusely throughout the body so that every area is assured a supply of blood. The density of the vasculature in any area is directly proportional to the metabolic activity of that area.

The principle of blood flow is based on the ability of the heart not only to develop sufficient energy to propel the blood through its velocity of ejection but to do this several thousand times in a day at a constant rate.

There are two basic requirements of the cardiovascular system: it must maintain a constant blood flow and it must adjust the magnitude of flow to the needs of the body. Since the heart plays a pivotal role in this function, nature has endowed it with the mechanism necessary for doing so. The heart is a mass of specialized muscle, termed cardiac muscle. It is striated and involuntary muscle and, like all muscle, capable of contracting. The heart is a hollow structure and is filled with blood. Most of the time, on contraction, the heart muscle exerts pressure on the blood within. When this pressure reaches a certain limit, movement of the blood takes place.

All cell functions, of which contraction is one, represent a response of the cell to some type of stimuli. The stimulus quite frequently is in the form of an external electrical disturbance—the nerve impulse. The heart is unique in that it need not wait for an electrical stimulus, since it is capable of initiating its own impulses. Discrete areas within the heart contain cardiac muscle cells of extreme sensitivity, capable of self-excitation. This excitation results in a change in the polarity of the cell membrane, a phenomenon called depolarization (Fig. 41.1). This term suggests a loss in the polarity of the membrane, which is misleading, since the polarity only changes from some value at the resting level to a different value in the excited state. A detailed discussion on cell membrane polarity is not possible here, but the reader is urged to seek further information on this subject, which is of paramount importance in the understanding of cell function.

The areas within the heart where the electrical impulse is initiated are situated at some distance

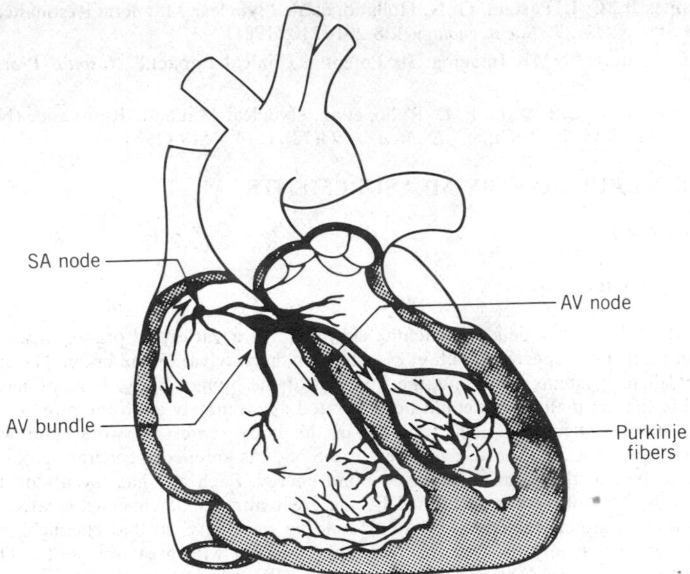

Fig. 41.1 Depolarization. The wave of depolarization is initiated in the SA node and transmitted in concentric waves through the atrial mass. When it reaches the AV node in the anatomical area between the atria and the ventricles, it continues into the ventricles using a set pathway: the AV bundle and the Purkinje system.

from each other, separated by cells which, while not normally involved in self-depolarization, are capable of transmitting the wave of depolarization.

In the resting state, heart cells can be considered negatively charged in relation to their immediate environment, owing to their ionic composition. As depolarization occurs, a sudden influx of positive ions causes the cells to become positive in relation to the environment. This state cannot be maintained for long periods of time, and once the stimulus ceases the cells return to their resting state: negative in relation to the environment. This return is called repolarization. This electrical activity can be recorded as deflections in relation to a baseline on a recording apparatus, and is traditionally referred to as the electrocardiogram (ECG). It can be measured directly by applying electrodes to the heart, a method that is rarely used for obvious reasons. Indirectly the ECG can be measured by positioning the electrodes on appropriate areas on the surface of the body. Since all body tissues and fluids represent electrical conducting systems, any electrical disturbance will be transmitted throughout the body in the form of concentric waves; these waves may be picked up by external sensors placed on the skin and recorded by appropriate monitoring equipment.

Understanding of electrocardiography therefore requires understanding of the distribution of electrical currents through a three-dimensional conductor, represented by the various tissues located between the source of the potential, in this case the heart, and the point of measurement, the skin.

The passage of the electrical current from the heart to the skin is a complex phenomenon, a result of vectorial addition of an infinite number of dipoles of different strength and orientation. It is only logical that measurement of these vectors be made at different angles, as represented by a variety of skin positions, or lead positions.

Traditionally, positioning of the leads is determined by the Einthoven triangle (Fig. 41.2). In this lead system, the vector sum of all electrical activity in the heart at any moment is assumed to lie in the center of an equilateral triangle surrounding the heart and formed by the right and left shoulder and the pubic region.

Einthoven's law (Fig. 41.3) states that if the electrical potential of two of the three leads is known, the third one can be calculated by simply summing the first two. For convenience, the electrodes are positioned on the right and left arms and the left leg, considering these areas as extensions of the basic triangle. Additional intersecting lines may be obtained by placing additional electrodes (Fig. 41.4).

The normal ECG obtained in this manner is shown in Fig. 41.5. As the positive wave of depolarization moves toward a positive skin electrode, a positive deflection is recorded. Where the *P* wave represents atrial depolarization, the QRS complex represents ventricular depolarization, and the

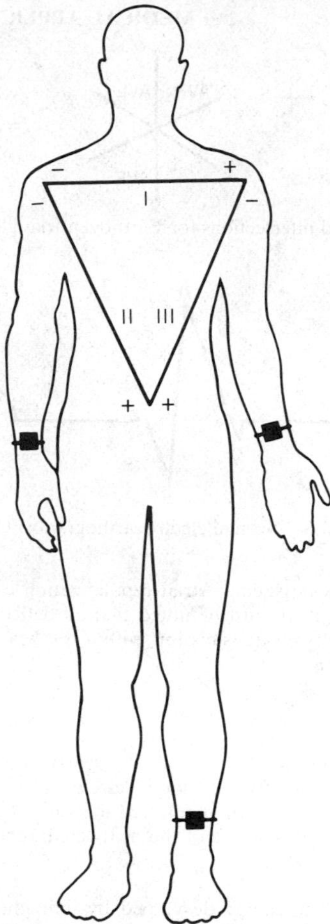

Fig. 41.2 Positioning of test leads per the Einthoven triangle.

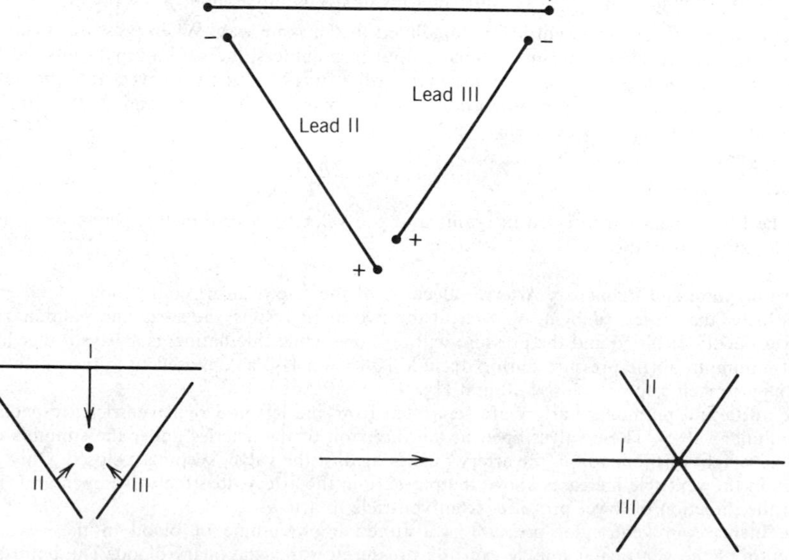

Fig. 41.3 Einthoven's law.

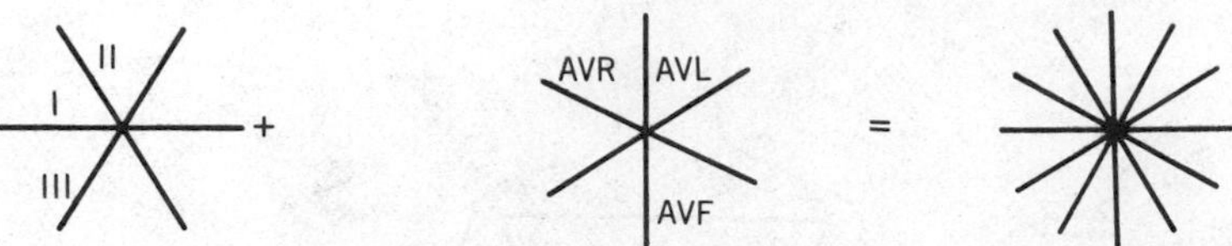

Fig. 41.4 Obtaining additional intersections for Einthoven triangle; six intersecting limb leads.

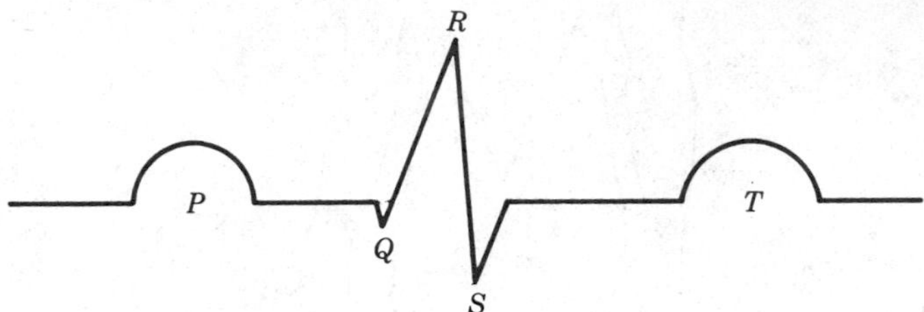

Fig. 41.5 Normal electrocardiogram (ECG).

T wave represents ventricular repolarization. Atrial repolarization cannot be detected, since it falls under the QRS and is masked by it. It must be noted that all deflections are positive under normal circumstances, because the pathway of transmission within the heart for depolarization is different from the pathway for repolarization.

Stroke Volume

As stated previously, the heart contracts as a result of depolarization. This contraction propels the blood from the heart cavity into the efferent blood vessels, called arteries. The amount of blood pumped by the heart during each contraction is referred to as the stroke volume and is a function of a number of factors, among them the strength of the heart contraction and the pressure in the aorta and pulmonary artery.

Strength of Heart Contraction. The energy developed by individual muscle fibers is transformed into work and utilized for two tasks. One of them is the raising of the blood pressure from the low level in the veins to the high level in the arteries. This is called potential energy and can be calculated as

$$\text{stroke volume output} \times (\text{left ventricular mean ejection pressure} - \text{left atrial pressure})$$

The work output of the right ventricle is calculated in the same way. When pressure is expressed in dynes per square centimeter and the volume output in milliliters, the work output units are ergs.

The rest of the energy developed by the myocardial fibers is utilized to accelerate the blood. The work output of each ventricle is proportional to the mass of the blood and to the square of its velocity, and represents the kinetic energy:

$$\text{kinetic energy} = \frac{mv^2}{2}$$

When the blood mass is expressed in grams and the velocity in centimeters per seconds, the work output is expressed in ergs.

Pressure in Aorta and Pulmonary Artery. Because of the rate of heart contraction, blood enters the arteries faster than it leaves them. As a result the two main vessels, the aorta and pulmonary artery, are never devoid of blood and the pressure within them, while fluctuating, is always rather high. The normal minimum aortic pressure during diastole is 80 mm Hg, as opposed to intracardiac pressure, which might reach a low of close to 0 mm Hg.

The aorta and pulmonary artery are separated from the left and right ventricle, respectively, by the semilunar valves. These valves open in the direction of the arteries under the stimulus of blood flow. The weight of the blood in the artery pushes against the valve, keeping it closed. Only after the pressure in the ventricle increases above the pressure in the artery does the valve open and the blood flow in the direction of lower pressure: from ventricle to artery.

The increase in ventricular pressure is a function of volume of blood in the ventricle and contraction of the ventricular muscle exerting pressure on this stationary blood. The point at which the valve opens is a function of the magnitude of pressure in the aorta and the ability of the ventricle to contract against that pressure. As the pressure increases, as in untreated hypertension, it becomes increasingly difficult for the ventricle to develop sufficient energy to open the valve, and stroke volume decreases. If pressure continues to increase, the ventricle might reach the point where it can no longer maintain an adequate stroke volume, the condition known as cardiac failure.

Intracardiac Pressure: Cardiac Catheterization and Angiocardiography

While it is somewhat easy to conceptualize the existence of pressure within the heart chambers as well as its fluctuation as a function of myocardial contractility, the actual measurement of these parameters is not easy. Two basic techniques exist for measurement which can be classified as invasive and noninvasive.

Invasive techniques involve the entering of the area of the heart to be studied by a catheter and the connection of the catheter to a pressure transducer that records pressure changes. Noninvasive techniques involve the detection of sounds within the heart and vessels (sonocardiography) or the measurement of reflected waves from tissues that have been exposed to ultrasonic treatment (diagnostic ultrasound). Cardiac catheterization, while an invasive technique that carries considerable risk, is still the most accurate method for diagnosis of cardiac distress, and is used prior to any heart surgery.

Since the cardiovascular system is basically a closed system, theoretically it is possible to enter it at any point and follow its course completely. It is impossible to manufacture a catheter small enough to fit through all the divisions of the cardiovascular system (the diameter of some capillaries being 3–5 μm), but it is quite easy to enter one of the larger vessels and follow its course to the heart. If a vein is entered, the right side of the heart will be the ultimate site; if an artery, the left side will be reached. The progress of the catheter is followed with the help of a fluoroscope. A trained physician would have no difficulty in positioning the tip of the catheter, which is opened to face the blood, in the desired area. If the other end of the catheter is connected through a pressure transducer to an oscilloscope or any other pressure-recording apparatus, measurement of pressure fluctuations can be made with great accuracy.

Figure 41.6 shows the relationship between the pressure tracings obtained from the aorta, the left ventricle, and the left atrium. Their relationship to the electrocardiogram can also be seen. Notice that the onset of left ventricular pressure increase coincides with the R segment of the ECG; this also represents the point of maximal ventricular blood volume. Notice also that the ventricular pressure peaks following the S segment and begins to fall during ventricular repolarization (T segment).

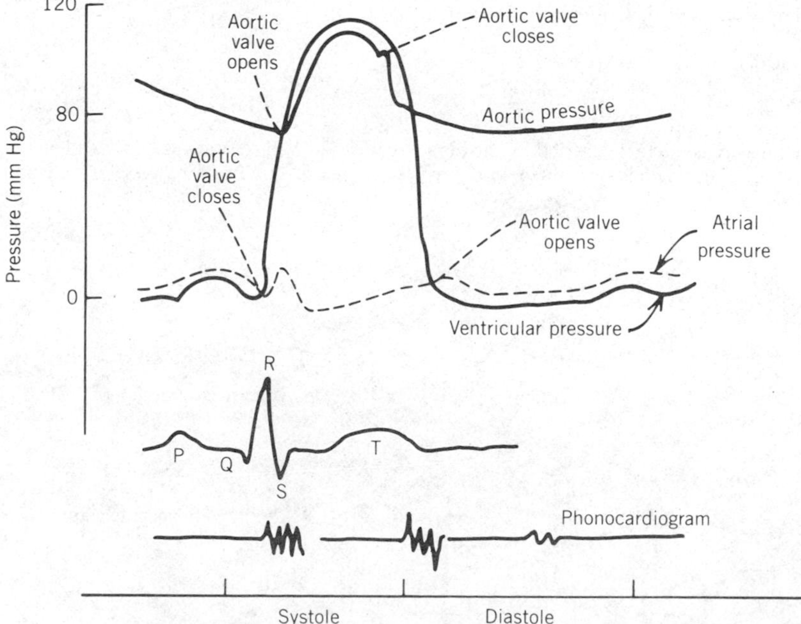

Fig. 41.6 Pressure tracings from aorta, left ventricle, and left atrium.

If at the time that the catheter is positioned in the area to be studied a radioopaque dye is injected, the flow of the dye can be followed and delineation of the vessel can be observed with the aid of a fluoroscope. This method, referred to as angiocardiography, can be applied to the study of any vessel in the body.

Intracardiac Pressure: Sonocardiography and Echocardiography

The series of events that precede and follow heart contraction are referred to as the cardiac cycle. The cycle extends from the end of one heart contraction to the end of the next contraction. During this time the valves within the heart open and close, following pressure changes and serving to control blood flow. This valve activity produces vibrations in the audible range that can be detected with the help of a stethoscope and can be recorded with the use of electronic amplification, in the form of a phonocardiogram. The ECG and the phonocardiogram are compared in Fig. 41.7.

In recent years the use of ultrasound as a diagnostic tool has gained popularity. In addition to its being a noninvasive and hence relatively safe procedure, it yields a volume of information considerably greater than other conventional methods.

Ultrasonic waves can be produced inside any material by placing a piezoelectric element in close proximity to the surface of the material and pulsating the element with appropriate voltage. The piezoelectric element converts electric energy to mechanical energy. A couplant material situated between the piezoelectric element and the material to be studied allows propagation of the ultrasonic waves into the material. The ultrasonic energy requires a given period of time to be propagated from one point to another within the material being tested; this time is a function of the nature of the material and the distance covered. If the wave speed for a given material is known, the arrival time of the ultrasonic wave can be used to measure the distance between two points.

The type of equipment selected will depend on the needs of the experimenter. Equipment to fit all degrees of sophistication is available. The basic features of any ultrasonic arrangement are a transducer, an intermediate signal-processing stage, and a display mode. The amplitude-time signal can be displayed with the aid of a cathode ray tube (CRT), which is a basic component of all systems. The basic oscilloscope is described in many textbooks.

Blood Pressure Measurement

Blood flow through a vessel is determined by the interplay of two factors:

1. The pressure gradient between the two ends of the vessel.
2. The impediment to flow offered by the vessel.

Mathematically this can be expressed as

$$Q = \frac{\Delta P}{R}$$

Blood pressure is the force exerted by the blood against the walls of the vessel. It can be measured directly by inserting a canula or catheter into any vessel; the pressure from the vessel is transmitted to

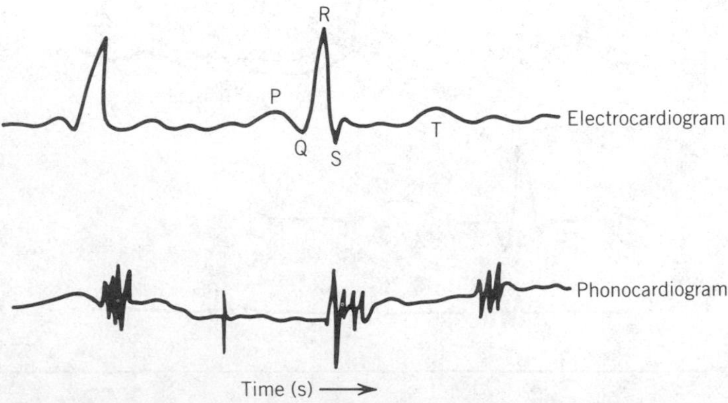

Fig. 41.7 Comparison of electrocardiogram (ECG) and phonocardiogram.

a mercury manometer where displacement of the mercury is measured. Unfortunately the inertia of the mercury is so great that the mercury cannot rise and fall rapidly, and pressure changes cannot be measured accurately by this method. When rapid changes in pressure are to be measured, a pressure transducer is indicated. This transducer converts pressure changes into electrical signals, which are then recorded on a high-speed electrical recorder.

A variety of pressure transducers are available commercially, but they all use a very thin and highly stretched metal membrane that forms a wall of a fluid chamber connected to a catheter inserted into the vessel. Changes in pressure within the vessel are transmitted to the fluid in the chamber and hence to the metal membrane. These changes can be measured by the appropriate electronic equipment.

Pressure transducers make possible the direct measurement of pressure in specific areas of the cardiovascular system; they are thus valuable and even indispensable tools in diagnostic studies. This type of measurement, however, though highly accurate, carries the danger of all invasive techniques and is never to be used unless the need for accuracy compensates for the risk. With pressure transducer measurements, graphs depicting pressure changes as a function of time can be developed for any area of the cardiovascular system (Fig. 41.8).

One of the most useful of all such graphs is the aortic pressure curve (Fig. 41.9). We must bear in mind that the aorta, being directly attached to the left ventricle, receives the full impact of the stroke volume in each cardiac cycle and is consequently subjected to maximal abuse. It also plays a pivotal role in the maintenance of blood flow.

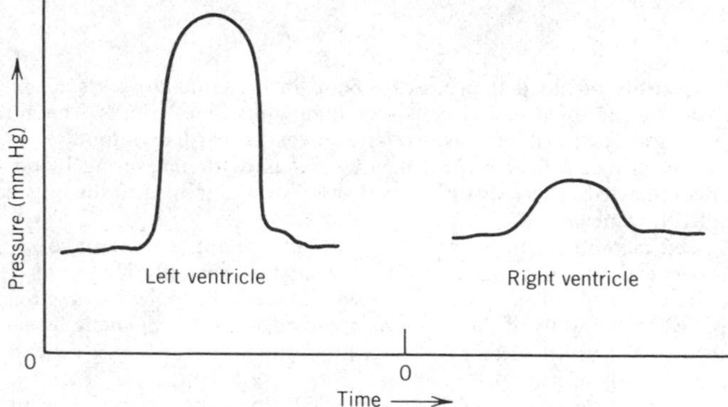

Fig. 41.8 Pressure changes vs. time for two areas of the heart. The magnitude of the pressure change varies with each area of the heart; it is a function of the strength of the muscle in that area. The rate of change, however, is the same throughout the heart; it is a function of the rate of depolarization.

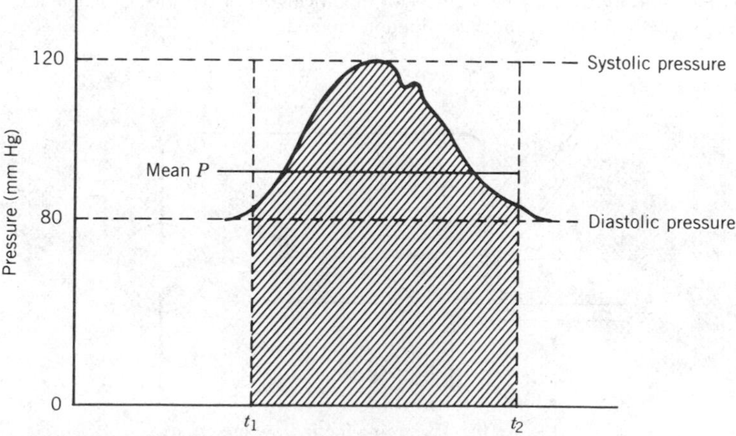

Fig. 41.9 Aortic pressure curve. $P_a = (\int_{t_1}^{t_2} P_a\,dt)/(t_2 - t_1)$.

Lying between the heart and the lesser vessels, the aorta acts as a hydraulic filter, and as such maintains the peripheral blood flow during diastole and minimizes the work load of the heart. It is in the aorta where the greatest magnitude of blood pressure is developed; this pressure, the direct result of left ventricular contraction, is the driving force for systemic blood flow. It is difficult to measure the pressure in any portion of the aorta at any given time, but such a measurement is unimportant, since the parameter that has physiological significance is the mean arterial pressure, which is defined as the average pressure in the aorta and all its branches. This value can be calculated with great accuracy from the aortic pressure tracing by integrating the area under the curve (Fig. 41.9).

Under most circumstances the blood pressure is measured indirectly by means of a sphygmomanometer. This instrument consists of an inflatable cuff containing a bag. The cuff is wrapped around an extremity directly above the artery to be compressed. Inflation of the bag to a pressure higher than the systolic pressure causes occlusion of blood flow in the artery. No pulse can be perceived at this point by palpation or auscultation. The pressure in the bag is released slowly; when it reaches the systolic pressure, blood begins to flow through the artery and a slight pulse is felt. As the pressure in the bag continues to drop, the pulse is heard loudly; the sounds are called Korotkoff sounds and represent vibrations caused by the blood from the occluded artery as it hits a static column of blood immediately underneath. When the pressure in the cuff reaches that of the diastolic pressure, the sounds cease, since at that point the flow is continuous. The mean arterial pressure can be approximated satisfactorily according to the formula

$$\overline{P}_a = P_d + \tfrac{1}{3}(P_s - P_d)$$

Blood Flow

Blood flow is the quantity of blood that passes a point in the circulatory system in any measured length of time. It can be expressed in any units of volume units time^{-1} units. The normal value in a resting adult human is about 5 l/min. This is referred to as the cardiac output.

Many mechanical or mechanicoelectrical devices are used for measuring blood flow. In some instances these are connected in series with a blood vessel or are applied to the outside of the vessel. All are simply called flowmeters.

A commonly used method for measuring flow without opening a vessel involves the use of the electromagnetic flowmeter, which works on a principle similar to the production of electricity by the electric generator (Fig. 41.10). The vessel is placed between the poles of a strong magnet and electrodes are applied to the walls of the vessel, perpendicular to the magnetic lines of force. When the blood flows through the vessel, electrical voltage proportional to the flow is generated between the electrodes, and can be easily measured.

Another type of flowmeter often used is the so-called ultrasonic Doppler flowmeter. It involves the use of a minute piezoelectric crystal. This crystal, when energized, transmits sound downstream along the flowing blood. The sound waves are reflected back by the red cells at a lower frequency (the Doppler effect), and measurement of these sound waves is possible.

The methods previously described provide highly accurate measurements but unfortunately carry the dangers of all invasive techniques and are rarely used in humans. In human beings, flow is measured indirectly using the Fick principle. Two modifications of the Fick principle are commonly used: the oxygen method and the indicator dilution method.

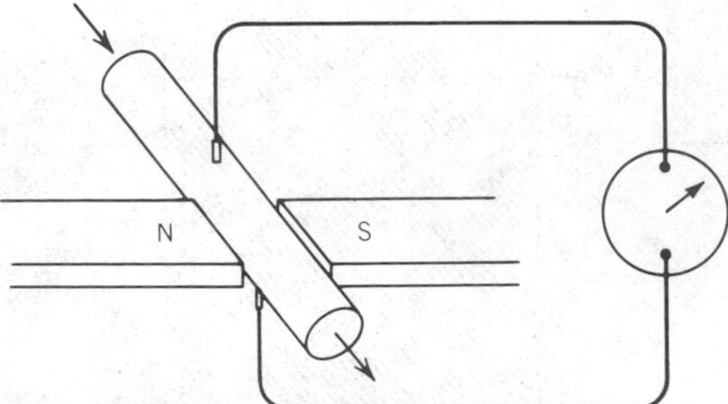

Fig. 41.10 Electromagnetic flow meter.

Oxygen Method (Fig. 41.11). The blood leaving the right ventricle has a low O_2 content, while the blood entering the left atrium has a much higher content, a result of having been exposed to the O_2 in the lungs.

Three highly accurate measurements are required for determining cardiac output by this method:

1. The rate of oxygen absorption by the lungs, measured by a respirometer.
2. Venous oxygen concentration, measured in the right ventricle.
3. Arterial oxygen saturation, measured in any large artery or in the left atrium.

$$\text{Cardiac output (l/min)} = \frac{O_2 \text{ absorbed by the lungs (ml/min)}}{\text{arteriovenous } O_2 \text{ difference (ml/l blood)}}$$

Indicator Dilution Method (Fig. 41.12). A small amount of indicator is injected into a large vein and its appearance and concentration change are measured in a peripheral artery. A curve showing concentration of dye in arterial blood vs. time is developed.

$$\text{Cardiac output} = \frac{\text{amount of dye injected (mg)}}{\begin{array}{l}\text{average concentration of dye} \\ \text{in each ml blood for duration of} \\ \text{curve}\end{array} \times \begin{array}{l}\text{duration of} \\ \text{curve (s)}\end{array}}$$

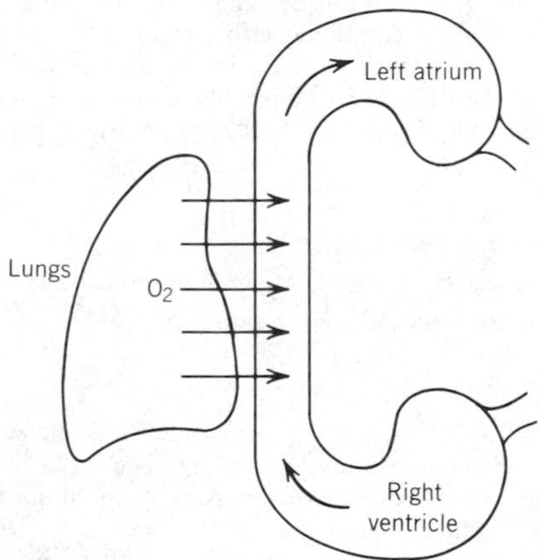

Fig. 41.11 Measuring flow by the "oxygen" method. The rate of absorption of oxygen by the lungs is a function of alveolar ventilation. The amount of this oxygen that will be accepted by the pulmonary blood is a measure of the magnitude of the pulmonary blood flow and the hemoglobin content of the blood.

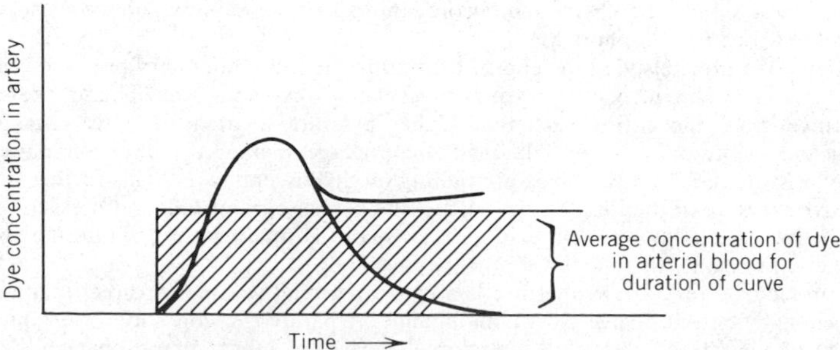

Fig. 41.12 Measuring flow by the "indicator dilution" method.

41.3-3 Pulmonary System

In the previous section we discussed the absolute need of the tissues for oxygen and the role that the circulatory system plays in making that oxygen available to each cell in the body. Oxygen, although plentiful in the atmosphere, must enter the body and come into contact with the blood, and it must do so on a constant basis and in amounts proportional to the needs of the body.

The system in the human body responsible for bringing the atmospheric oxygen into contact with the blood is the pulmonary system. Since this system works very closely with the cardiovascular system, the two are often grouped together into what physiologists call the cardiopulmonary system. The respiratory system can be loosely divided into two parts, the airway passages and the lungs, even though the lungs themselves consist of millions of extremely small passages profusely invested with capillaries.

The airway passages are open to the atmosphere at one end via the nose and mouth. When the mouth is open, the entire pulmonary tree is in equilibrium with the atmosphere and the pressure within the system is equal to atmospheric pressure. The architecture of the respiratory system is such that the passages increase in number and decrease in diameter as they progress toward the lungs, so that the system starts with one single tube of considerable diameter, the nasal passage, and ends in hundreds of thousands of very fine tubes in the alveoli.

Ventilation is the name given to the cyclic phenomenon by which fresh air from the environment enters the tracheobronchial tree and moves into the alveoli (inspiration). This is followed by the exit of an equal amount of air from the alveoli to the environment (expiration). In this process the composition of the air changes. The air leaving the lungs has a higher concentration of CO_2 and a lower concentration of O_2 than the air entering it. The increased CO_2 is a result of cell metabolism; the decreased O_2 is a measure of O_2 absorption by the pulmonary blood.

Inspiration is brought about by the combined effort of the inspiratory muscles. Contraction of these muscles leads to a three-dimensional increase in the size of the thorax and a proportional increase in the lung volume. Inspiration is an active phenomenon requiring energy expenditure to overcome the resistances of the structures in the chest cage and the lungs. These resistances are very real, and include:

1. Elastic resistance of the chest cage and lungs.
2. Nonelastic frictional resistance of the tissues of the lungs and thorax.
3. Nonelastic frictional resistance opposed by the lining cells of the air passages and the humidity within the passages.

As with all elastic structures, the energy utilized to overcome the elastic resistance is stored within the elastic structures in the form of potential energy. This energy is utilized for elastic recoil to the resting expiratory level. Unless some pathologic condition causes an abnormal increase in nonelastic resistance, expiration is a passive phenomenon.

Anatomically the lungs lie very close to the chest cage, and both structures are endowed with elastic tissue. Under normal circumstances each of these two structures is affected by the pull of the other. The two opposing forces come into play—the elastic forces of the chest cage, which tend to expand it, and the elastic forces of the lungs, which tend to collapse them. As a result, a negative pressure develops in the interphase between the thorax and lungs. The interphase consists of the pleura, one layer of which faces the lungs (visceral) and the other layer faces the thorax (parietal). The space between the two layers is called the potential intrapleural space and exists at a negative pressure at all times. Should the chest cage become punctured, air rapidly enters this space, creating the condition known as pneumothorax.

As was mentioned previously, at the end of expiration the intrapulmonary pressure is equal to the atmospheric pressure. The alveolar pressure must be lower than atmospheric pressure during inspiration, since gases move from areas of higher pressure to areas of lower pressure. Active contraction of the inspiratory muscles enlarges the chest cage, making it pull on the parietal layer of the pleura at the same time that the lungs are pulling on the visceral layer. This further decreases the already negative pressure in the intrapleural space. The decrease in intrapleural pressure enlarges the alveoli, and the increase in volume results in a decrease of the total pressure within the alveoli to less than atmospheric pressure. Air now flows into the alveoli.

The most immediate result of ventilation is a change in lung volume. A direct relationship exists between muscle activity and degree of volume change. A pathologic condition often quickly affects the magnitude of volume change. It is therefore logical that lung volume changes be used as a criterion of pulmonary function.

Traditionally, four volumes and four capacities are analyzed. The volumes are:

1. **Tidal volume (TV).** Volume of gas inhaled or exhaled during a normal breathing cycle.
2. **Inspiratory reserve volume (IRV).** Maximal volume of gas that can be inspired from the end-inspiratory position.
3. **Expiratory reserve volume (ERV).** Maximal volume of gas that can be expired from the end-expiratory position.
4. **Residual volume (RV).** Volume of air remaining in the lungs after a forced expiration.

The capacities are:

1. **Total lung capacity (TLC).** Amount of gas in the lungs after a maximal inspiration.
2. **Vital capacity (VC).** Maximal volume of gas that can be expelled from the lungs by forceful expiration following a maximal inspiration.
3. **Inspiratory capacity (IC).** Maximal volume of gas that can be inspired from the end-expiratory position.
4. **Functional residual capacity (FRC).** Volume of gas remaining in the lungs from the resting level.

Some of these volumes and capacities can be measured with the use of a gas volume recorder called a spirometer. Most spirometers have a pen that writes on a rotating drum. As the subject breathes into the drum, the drum moves against the pen plotting gas volume vs. time. Spirometers also measure the capacity for dynamic changes of lung volume (Fig. 41.13). FRC, RV, and TLC cannot be measured with a spirometer, since these are volumes that cannot be expired. Three indirect methods can be used to calculate them.

1. Closed-circuit method. This method involves dilution of an inert, insoluble, foreign gas (usually He) as it mixes with the gas in the lungs. The subject breathes into a sealed spirometer of known volume containing a known concentration of He. As the gas in the spirometer mixes with the gas in the lungs, the He concentration decreases. The volume of gas in the lungs can be calculated:

$$V_1 = \left(V_{sp} \times \frac{\mathrm{He}_{initial}}{\mathrm{He}_{final}} \right) - V_{sp} + C$$

where V_1 = initial lung volume
V_{sp} = spirometer volume
$\mathrm{He}_{initial}$ = initial concentration of He in spirometer
He_{final} = final concentration of He in both spirometer and lungs
C = correction factor that cancels volume change

When this method is used, the initial lung volume is FRC.

$$\mathrm{FRC} - \mathrm{ERV} = \mathrm{RV}$$
$$\mathrm{FRC} + \mathrm{IC} = \mathrm{TLC}$$

2. Open-circuit N_2 washout method. This method involves washing out the lung N_2 with O_2 and measuring the quantity of N_2 that was contained in the lung volume (FRC). Since the air has about 80% N_2, if we measure the amount of N_2 in the lungs we can calculate the total lung volume.

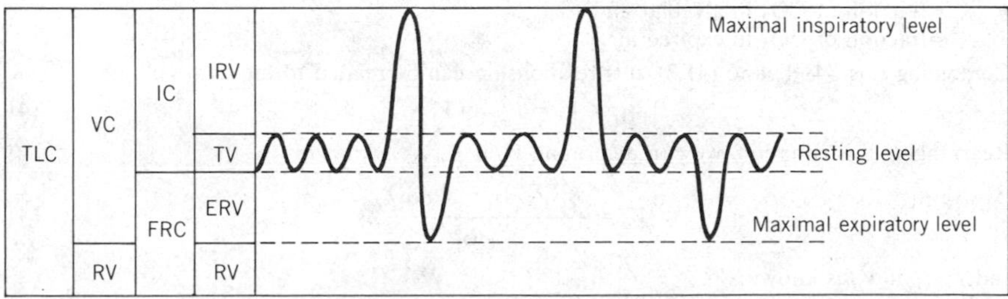

Fig. 41.13 Typical spirometer record. TLC, total lung capacity; VC, vital capacity; RV, residual volume; IC, inspiratory capacity; FRC, functional residual capacity; IRV, inspiratory reserve volume; TV, tidal volume; ERV, expiratory reserve volume.

3. Body plethysmography. This method is based on Boyle's law, $P_1V_1 = P_2V_2$ if temperature is maintained constant. The subject is placed within a gas-tight chamber and is allowed to breathe fresh air through a mouthpiece. After humidity and temperature within the chamber have stabilized, the breathing line is blocked by a pressure transducer, and the subject breathes against it. The volume changes within the box result in corresponding P changes that can be measured by a second transducer and recorded. Enlargement of the thorax as the subject breathes against the obstruction compresses the air around the patient, increasing the plethysmographic pressure.

The measured P' and V' changes in the lungs, together with the known initial ambient pressure, permit calculation of the initial FRC, since this value would be the same or slightly larger than the thoracic volume (V) measured by the body plethysmograph.

$$PV = P'V' = P'(V + \Delta V); \qquad V = \frac{P'\Delta V}{P - P'}$$

Alveolar Ventilation

While ventilation refers to the breathing in of fresh air and the breathing out of stale air, it is a term that does not convey the full idea of the efficiency of the respiratory system. We must bear in mind that the ultimate goal of the ventilatory process is to bring O_2 into the alveoli and to allow it to come into contact with the pulmonary blood. This fact gives rise to the concept of alveolar ventilation, also termed effective ventilation, and defined as the volume of fresh air that reaches the gas exchange regions of the lungs per unit of time.

The airway passages are always filled with gas. A portion of this gas has already been involved in gas exchange with the blood, and another portion is fresh air. This last portion of gas is the only one that will be available to ventilate the alveoli. The amount of air that fills the airway passages is referred to as dead space. Therefore

$$V_a = (V_t - V_d)f \tag{41.1}$$

where V_a = alveolar air volume
V_t = tidal volume
V_d = dead space volume
f = frequency

The relationship between frequency and dead space ventilation is almost linear. This guarantees an adjustment in alveolar ventilation in response to changes in metabolism by an increase only in breathing rate and not in tidal volume.

While the function of the respiratory system is always thought of as one of O_2 acquisition, a second function is of equal importance, namely, removal of CO_2 produced by the tissues. This function is so important that the level of CO_2 in the blood serves as the main stimulus for control of ventilatory rate. An increased CO_2 level stimulates the respiratory center in the medullary portion of the central nervous system, leading to hyperventilation. The converse is true when the CO_2 level drops.

Under normal circumstances the rate of CO_2 removal can be calculated from either alveolar or expired gas analysis:

$$V_{CO_2} = F_{eCO_2} \times V_e = F_{aCO_2} \times V_a \tag{41.2}$$

where V_e = expired air volume
V_a = alveolar air volume
V_d = dead space volume
F_{aCO_2} = fraction of CO_2 in alveolar air
F_{eCO_2} = fraction of CO_2 in expired air

Combining Eqs. (41.1) and (41.2), this relationship can be related to the tidal volume:

$$F_{eCO_2} \times V_t = F_{aCO_2}(V_t - V_d) \tag{41.3}$$

Rearranging this equation, we can determine V_d if F_{aCO_2} is known:

$$V_d = \frac{V_t(F_{aCO_2} - F_{eCO_2})}{F_{aCO_2}} \tag{41.4}$$

or find F_{aCO_2} if V_d is known:

$$F_{aCO_2} = \frac{F_{eCO_2} \times V_t}{V_t - V_d} \tag{41.5}$$

Equation (41.5) was developed by Bohr to calculate the fraction of alveolar gas that is CO_2, on the basis of an assumed value for dead space. Haldani later modified the equation to calculate dead space on the basis of measured expired CO_2.

To convert the fraction of alveolar gas to the partial pressure of the same gas:

$$_{p}CO_2 = F_{aCO_2}(P_B - 47)$$
$$_{p}O_2 = F_{aCO_2}(P_B - 47)$$

where P_B is the atmospheric pressure and 47 is the water vapor pressure in the lungs.

Gas Concentration in Lungs: Gas Analyzer

On an average day, at sea level, the composition of atmospheric air is as follows:

Component	Partial Pressure (mm Hg)	Percent in Air
N_2	597.0	78.62
O_2	159.0	20.84
CO_2	0.3	0.04
H_2O	3.7	0.50

This is the composition of air as it enters the airway passages. Once in the passages, the air becomes humidified by the H_2O present in the secretions of the cells lining the passages. Since the total pressure of all the gases has to add up to 760 mm Hg (atm), the partial pressure of the gases will proportionately decrease to accommodate the H_2O.

As the air moves toward the alveoli, it mixes with the stale air in the dead space. Once it is in the alveoli, O_2 diffuses from it across and into the blood and CO_2 diffuses from the blood into the alveolar air, causing pCO_2 to rise and pO_2 to drop. Since no concentration gradient exists for N_2, the partial pressure of this gas changes little. The pN_2 remains unchanged.

Once the gas exchange is completed and the air begins to exit from the alveoli, it once more mixes with the gas in the dead space and its composition changes again: pCO_2 decreases, and pO_2 increases. A comparison of the air composition in the different compartments is as follows:

Component	Humidified Air		Alveolar Air		Expired Air	
N_2	563.4 mm Hg	74.09%	56.9 mm Hg	74.9%	566.0 mm Hg	74.5%
O_2	149.5 mm Hg	19.67%	104.0 mm Hg	13.6%	120.0 mm Hg	15.7%
CO_2	0.3 mm Hg	0.04%	40.0 mm Hg	5.3%	27.0 mm Hg	3.6%
H_2O	47.0 mm Hg	6.20%	47.0 mm Hg	6.20%	47.0 mm Hg	6.2%

The O_2 in the alveoli reflects a balance between the rate of O_2 absorption by the blood and alveolar ventilation (Fig. 41.14). As the O_2 absorption increases, the ventilatory process must proportionately increase to maintain the same alveolar pO_2.

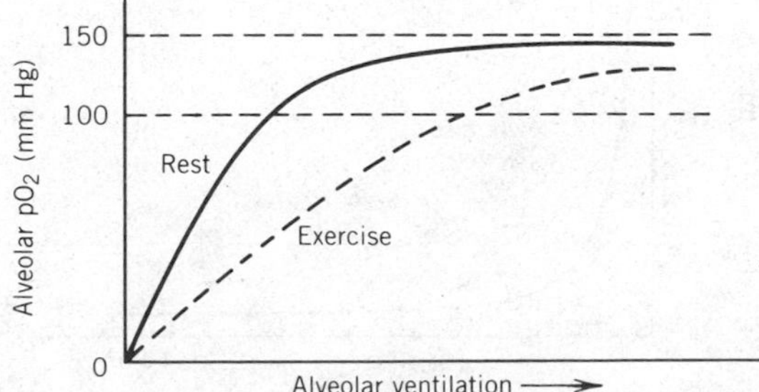

Fig. 41.14 Blood pO_2 level vs. alveolar ventilation. The maximal pO_2 that the blood can achieve is that of the humidified air (150 mm Hg). The average determined pO_2 is 104 mm Hg. To maintain this level, the ventilation rate must be increased in times of stress, as in exercise.

The pCO_2 in the alveoli is a function of the rate of diffusion of CO_2 from the blood into the alveoli and the rate of removal of CO_2 from the alveoli by the ventilatory process (Fig. 41.15). The composition of the expired air is in between that of the alveoli and that of the atmospheric air. The air at the beginning of expiration is quite similar in composition to humidified air. As more air exits, its composition resembles more and more that of alveolar air.

One method for analyzing alveolar air is to collect the last portion of expired air and to place it through a gas analyzer. CO and CO_2 may be analyzed by infrared absorption, N_2 by emission spectroscopy, and a great variety of gases by mass spectrometry.

Ventilation-Perfusion Ratio

Gas exchange in the lungs is contingent on interaction between the gas in the alveoli and the blood perfusing it. Perfusion describes the amount of blood flowing through the pulmonary capillary bed per unit of time. Ventilation has already been defined elsewhere in this chapter. The relationship between ventilation and perfusion has great physiological significance and is often expressed as $\dot{V}_a/\dot{Q}_c$.

Blood flow in the pulmonary capillaries is subjected to the same rules that govern flow through any area in the body. As has been explained in previous sections, ventilation of the alveoli is affected by the nature of the lungs and the general behavior of gases. Under the best of circumstances, distribution of air throughout the lungs is uneven, owing to the position of the lungs in the body and the density of the gases. The degree of unevenness varies from moment to moment as the individual shifts body position. Nature has given us a mechanism that compensates for this change in ventilation, which consists of the capability of blood flow to rapidly shift from poorly ventilated areas to well ventilated areas. What controls this blood flow shift is not clear, but the shift helps in the maintenance of optimum $\dot{V}_a/\dot{Q}_c$ at all times.

Measurement of the magnitude of flow for calculation of the ventilation-perfusion ratio can be made in the same manner as described in the section on systemic blood flow. Both the direct Fick method and the indicator dilution method are frequently employed.

Blood Gases

The ability to transport gases is often referred to as the respiratory function of the blood.

O_2 Transport. Oxygen is transported in two ways:

1. Dissolved as O_2 in plasma. About 1% of the total O_2 is transported in this manner.
2. Combined with hemoglobin in the form of oxyhemoglobin. The bulk of the O_2 is transported in this manner.

Hemoglobin is the main component of the red cell, and each molecule contains four binding sites for molecular O_2. Therefore each gram of hemoglobin can combine chemically with a maximum of about 1.36 ml [STPD (standard temperature and pressure-dry)] of O_2. One hundred milliliters of blood contain about 15 g of hemoglobin and can combine with $15 \times 1.36 = 20.4$ ml (STPD) of O_2.

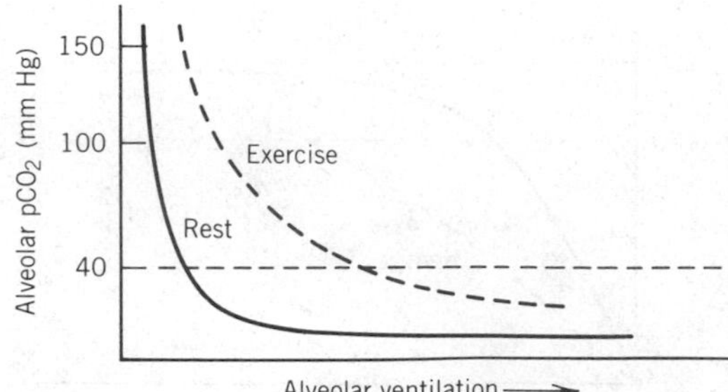

Fig. 41.15 Relation of alveolar pCO_2 to alveolar ventilation. When the metabolic rate increases, as in exercise, the ventilatory rate must be increased to maintain the normal alveolar pCO_2 of 40 mm Hg. This will increase the O_2 consumption of the respiratory muscles, which limits the capabilities of these muscles.

The amount of O_2 that combines with hemoglobin is not linearly related to pO_2, but it is described by an S-shaped curve, Fig. 41.16. The graph shows a nonlinear relationship between arterial oxygen saturation and pO_2. As can be seen, the hemoglobin will hold tighter to the oxygen at both extremes of the curve. Under normal circumstances the conditions that prevail will be those at the middle of the curve. Oxyhemoglobin dissociation is then controlled by fluctuating values of blood pCO_2 and pH.

The amount of gas, either O_2 or CO_2, present in a blood sample is measured in a number of ways; the manometric method of Van Slyke and Neil gives accurate, though somewhat slow, results. For rapid determination, as during cardiac catheterization, oximeters, instruments that measure O_2 tension of both the spectrophotometric and the reflectance type, are often used even though the results obtained are not as accurate as those obtained by manometry. In any case, great care must be given to preserve the samples under strict anaerobic conditions. The Lukas method, combining Van Slyke gas extraction techniques with the gas chromatographic method of analysis, provides a rapid and highly accurate determination of blood gases. Careful selection of the chromatograph adsorption column permits simultaneous analysis of several gases in one blood sample. pO_2 can also be measured directly with the help of an O_2 electrode. Dissolved O_2 can then be calculated for each patient.

CO_2 Transport. CO_2 is transported in the blood in three different ways:

1. Dissolved in plasma. The amount of CO_2 transported in this manner is minimal and is proportional to the pCO_2 to which it is exposed.
2. Combined with the amino group in the amino acid of plasma protein to form carbamino acid compounds.

$$R - NH_2 + CO_2 \longleftrightarrow R - NHCOO^- + H^+$$

   CO_2 reacts rapidly to form these compounds.
3. Transported in the form of bicarbonate ion. The bulk of CO_2 in the blood is transported in this way. In the erythrocyte, in the presence of the enzyme carbonic anhydrase, CO_2 reacts rapidly with H_2O to give carbonic acid:

$$CO_2 + H_2O \longleftrightarrow H_2CO_3 \longleftrightarrow H^+ + HCO_3^- \longleftrightarrow H^+ + CO_3^{2-}$$

This reaction is fully reversible; obviously the equilibrium will shift depending on the concentration of CO_2. Under normal circumstances formation of HCO_3^- is favored, and with its formation of H^+, whose concentration will determine the pH of the blood.

According to the Henderson–Hansselbach equation:

$$pH = pk + \log \frac{[HCO_3^-]}{[CO_2]} \tag{41.6}$$

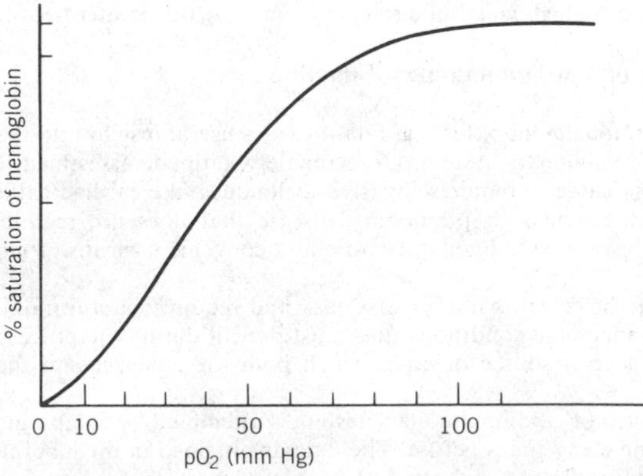

Fig. 41.16 Relation between arterial oxygen saturation and pO_2.

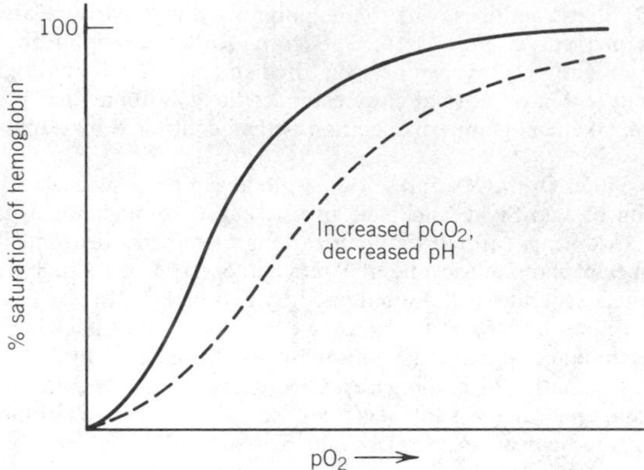

Fig. 41.17 Effect of increased pCO_2 and decreased pH on oxyhemoglobin. The normal oxyhemoglobin dissociation curve is affected by varying concentrations of CO_2 and concurrent fluctuations in pH. When the pCO_2 increases and the pH decreases, the curve shifts to the right and more O_2 is made available to the cells.

The pk for this system is 6.1. Substituting in Eq. (41.6) the pK and the measured blood pH of 7.40:

$$7.40 = 6.1 + \log \frac{[HCO_3^-]}{[CO_2]}$$

$$\frac{[HCO_3^-]}{[CO_2]} = \frac{20}{1}$$

Since the blood pH must be maintained close to 7.40 for proper body function, this ratio of HCO_3^- to CO_2 must be strictly kept. It is normally maintained by pulmonary ventilation. Whenever arterial pCO_2 tends to increase, the medullary respiratory center is stimulated and hyperventilation results.

This explanation applies only to those hydrogen ions that are of respiratory origin. Under pathological conditions when the H^+ might originate from other sources, for example, from metabolism, this picture is upset and the ventilatory system is incapable of coping with the increase in $[H^+]$. In metabolic acidosis, which often appears with diseases like diabetes mellitus, the patient adapts an abnormal breathing pattern in a futile attempt to hyperventilate and remove the excess H^+. This pattern often helps in the early diagnosis of metabolic distress.

The pH of the blood can be easily measured by the use of a pH meter. This measurement, however, must be accompanied by simultaneous measurements of total blood pCO_2.

Blood pH and pCO_2 greatly affect the rate of dissociation of oxyhemoglobin and hence the availability of O_2 to the tissues (Fig. 41.17). At conditions of high pCO_2, low pH, and high temperature, oxyhemoglobin will dissociate and free the O_2. It is remarkable that these are the conditions that characterize high metabolic rate and consequently greater need for O_2.

41.3-4 Assessment of Cardiopulmonary Function

Careful analysis of cardiopulmonary testing data provides useful insight into the function of both the cardiovascular and the pulmonary system. Unfortunately, ultimate assessment in critical times, as in preparation for cardiac surgery, requires invasive techniques like cardiac catheterization. However, for patients with cardiovascular or pulmonary disease that does not require surgery, pulmonary testing and blood analysis provide highly accurate and convenient means of monitoring the progress of the disease.

Data interpretation, however, is not an easy task and requires careful training on the part of the interpreter. The strict anaerobic conditions that must prevail during the process of sample collection and analysis provide a great source of error which both the analyzer and the interpreter must be quick in recognizing.

Diagnosis on the basis of cardiopulmonary testing is attempted by comparing results with those of correlation studies done many times before. The data as obtained in the laboratory does not indicate the presence or absence of any disease but rather suggest the most likely clinical syndrome compatible with the observed pattern. We must keep in mind that every human being is a unique

TABLE 41.1. TESTS USEFUL IN ASSESSING CARDIOPULMONARY DYSFUNCTION

Measurement	Test
Lung volume	Spirometry, body plethysmography
Pulmonary ventilation	
Volume	Measurement of lung volume, alveolar or arterial pCO_2; calculation of alveolar ventilation from measured anatomic dead space
Distribution	Single-breath N_2 testing, volume of trapped air (body plethysmography)
Pulmonary circulation	
Pressure, flow, resistance, volume	ECG; cardiac catheterization; pulmonary capillary blood volume and flow
Distribution	Pulmonary angiography; arterial O_2 saturation; alveolar pCO_2
Alveolar capillary diffusion	Diffusing capacity; arterial O_2 saturation at rest and during exercise
Arterial O_2, CO_2, and pH	Arterial O_2 content, capacity, and saturation; arterial pCO_2; hemoglobin content
Mechanical factors of breathing	Spirometry, pulmonary compliance, airway resistance

machine, similar but not identical to other human beings. Deviation from reported norms does not clearly or necessarily indicate pathology.

Table 41.1 displays tests that can be used to assess cardiopulmonary dysfunction. Which test to order for a given patient is a decision to be made by the physician. In any case, the clinical history and radiological and physical findings must precede any further testing.

Bibliography

Cardiovascular System

Aranjo, J. and D. S. Lukas, "Interrelationships among Pulmonary Capillary Pressure, Blood Flow and Value Size in Mitral Stenosis. The Limited Regulatory Effects of the Pulmonary Vascular Resistance," *J. Clin. Invest.* **31**:1082 (1952).

Armour, J. A. and W. C. Randal, "Structural Basis for Cardiac Function," *Am. J. Physiol.* **218**:1517 (1970).

Attinger, E. O., *Pulsatile Blood Flow*, McGraw-Hill, New York, 1964.

Attinger, E. O., H. Sugawara, A. Navarro, et al., "Pressure-Flow Relations in Dog Arteries," *Circ. Res.* **19**:230 (1966).

Bader, H., The Anatomy and Physiology of the Vascular Wall, in W. E. Hamilton and P. Dow, eds., *Handbook of Physiology*, sec. 2, *Circulation*, vol. 2, American Physiological Society, Washington, DC, pp. 865–889, 1963.

Baum, G., ed., *Fundamentals of Medical Ultrasonography*, Putnam, New York, 1975.

Beeler, G. W. Jr. and H. Reuter, "The Relation between Membrane Potential, Membrane Currents and Activation of Contraction in Ventricular Myocardial Fibers," *J. Physiol.* (*London*) **207**:211 (1970).

Bergel, D. H., ed., *Cardiovascular Fluid Dynamics*, Academic, New York, 1972.

Bergel, D. H. and D. L. Schultz, "Arterial Elasticity and Fluid Dynamics," *Prog. Biophys. Mol. Biol.* **22**:1 (1971).

Berne, R. M. and M. N. Levy, *Cardiovascular Physiology*, Mosby, St. Louis, 1972.

Boom, H. B. R., *Elasticity of the Heart: Instantaneous Pressure-Volume Relationship of the Left Ventricle Throughout the Cardiac Cycle*, Bronderoffset, Rotterdam, 1971.

Brecher, G. A., *Venous Return*, Grune & Stratton, New York, 1956.

Bruce, T. A. and J. E. Douglas, Dynamic Cardiac Performance, in E. D. Frohlich, ed., *Pathophysiology*, Lippincott, Philadelphia, 1976.

Carlin, B., *Ultrasonics*, 2nd ed., McGraw-Hill, New York, 1960.

Cheng, E. K., ed., *Artificial Cardiac Pacing: Practical Approach*, Williams & Wilkins, Baltimore, 1978.

Chengcharoen, D., "Genesis of Korotkoff Sounds," *Am. J. Physiol.* **207**:190 (1964).

Dobrin, P. B., "Mechanical Properties of Arteries," *Physiol. Rev.* **58**:397 (1978).

Dow, P., "Estimations of Cardiac Output and Central Blood Volume by Dye Dilution," *Physiol. Rev.* **36**:77 (1956).

Dubin, D., *Rapid Interpretations of EKGs*, Cover, Florida, 1977.

Freeman, I., *Sound and Ultrasonics*, Random House, New York, 1973.

Ghista, D. N., et al., eds., *Theoretical Foundations of Cardiovascular Processes*, Karger, New York, 1979.

Goodman, A. H., et al., "A Television Method for Measuring Capillary Red Cell Velocities," *J. Appl. Physiol.* **37**:126 (1974).

Green, H. D., Circulation: Physical Principles, in O. Glasser, ed., *Medical Physics*, Yearbook, New York, 1944.

Grodins, F. S., "Integrative Cardiovascular Physiology: A Mathematical Synthesis of Cardiac and Blood Vessel Hemodynamics," *Q. Rev. Biol.* **34**:93 (1959).

Gross, J. F. and A. Popel, eds., *Mathematics of Microcirculation Phenomena*, Raven, New York, 1980.

Grossman, W., ed., *Cardiac Catheterization and Angiography*, Lea & Febiger, Philadelphia, 1980.

Guyton, A. C., "Determination of Cardiac Output by Equating Venous Return Curves with Cardiac Response Curves," *Physiol. Rev.* **35**:123 (1955).

Guyton, A. C., *Circulatory Physiology: Cardiac Output and Its Regulation*, Saunders, Philadelphia, 1963.

Guyton, A. C., *Arterial Pressure and Hypertension*, Saunders, Philadelphia, 1980.

Guyton, A. C., *Textbook of Medical Physiology*, 6th ed., Saunders, Philadelphia, 1981.

Guyton, A. C., et al., "Pressure-Volume Curves of the Entire Arterial and Venous Systems in the Living Animal," *Am. J. Physiol.* **184**:253 (1956).

Haft, J. I. and M. S. Horowitz, *Clinical Echocardiography*, Futura, Mt. Kisco, NY, 1978.

Hallock, P. and I. C. Benson, "Studies on the Elastic Properties of Isolated Human Aorta," *J. Clin. Invest.* **16**:595 (1937).

Hamilton, W. E., Measurement of the Cardiac Output, in W. E. Hamilton and P. Dow, eds., *Handbook of Physiology*, sec. 2, *Circulation*, vol. 3, American Physiological Society, Washington, DC, pp. 1875–1886, 1965.

Harlan, J. C., et al., "Pressure-Volume Curves of Systemic and Pulmonary Circuit," *Am. J. Physiol.* **213**:1499 (1967).

Hech, H. H., "Some Observations and Theories Concerning the Electrical Behavior of Heart Muscle," *Am. J. Med.* **30**:720 (1961).

Hoffman, B. F. and P. F. Cranefield, *Electrophysiology of the Heart*, McGraw-Hill, New York, 1960.

Hoffman, B. F. and P. F. Cranefield, "The Physiological Basis of Cardiac Arrhythmias," *Am. J. Med.* **37**:670 (1964).

Huang, C. and A. L. Copley, eds., *Biorrheology*, American Institute of Chemical Engineers, New York, 1978.

James, D. G., ed., *Circulation of the Blood*, University Park Press, Baltimore, 1978.

Josephson, M. E. and S. F. Seides, *Clinical Cardiac Electrophysiology Techniques and Interpretations*, Lea & Febiger, Philadelphia, 1979.

King, D. L., *Diagnostic Ultrasound*, Mosby, St. Louis, 1974.

Kotler, M. N. and B. L. Segal, eds., *Clinical Echocardiography*, Davis, Philadelphia, 1978.

Kremkau, F. W., *Diagnostic Ultrasound; Physical Principles and Exercises*, Grune & Stratton, New York, 1980.

Langer, G. A., "Ion Fluxes in Cardiac Excitation and Contraction and Their Relation to Myocardial Contractility," *Physiol. Rev.* **48**:708 (1968).

Lear, M. W., *Heartsounds*, Simon & Schuster, New York, 1979.

Lighthill, M. J., "Physiological Fluid Dynamics: A Survey," *J. Fluid Mech.* **52**:475 (1972).

Lukas, D. S., *Cyanosis in Signs and Symptoms*, Lippincott, Philadelphia, 1970.

Lukas, D. S., J. Aranjo, and I. Steinberg, "The Syndrome of Patient Ductus Arteriosus with Reversal of Flow," *Am. J. Med.* **17**:298 (1954).

Massie, E. and T. J. Walsh, *Clinical Vectorcardiography and Electrocardiography*, Year Book, Chicago, 1960.

McDonald, D. A., *Blood Flow in Arteries*, Edward Arnold, London, 1960.

McKusick, V. A., *Cardiovascular Sound in Health and Disease*, Williams & Wilkins, Baltimore, 1958.

Milnor, W. R., D. H. Bergel, and J. D. Bargainer, "Hydraulic Power Associated with Pulmonary Blood Flow and Its Relation to Heart Rate," *Circ. Res.* **19**:467 (1966).

Mitchell, J. H., D. N. Gupta, and R. M. Payne, "Influence of Atrial Systole on Effective Ventricular Stroke Volume," *Circ. Res.* **17**:11 (1965).

Moss, A. J., "Prediction and Prevention of Sudden Cardiac Death," *Annu. Rev. Med.* **31**:1 (1980).

Noble, M. I. M., "The Contribution of Blood Momentum to Left Ventricular Ejection in the Dog," *Circ. Res.* **23**:663 (1968).

O'Rourke, M. F. and M. G. Taylor, "Input Impedance of the Systemic Circulation," *Circ. Res.* **20**:365 (1967).

Page, E., "The Electrical Potential Difference across the Cell Membrane of Heart Muscle," *Circulation* **26**:582 (1962).

Patel, D. J., et al., "Hemodynamics," *Annu. Rev. Physiol.* **36**:125 (1974).

Pedley, T. J., *The Fluid Mechanics of Large Blood Vessels*, Cambridge University Press, New York, 1979.

Rappaport, E., ed., *Current Controversies in Cardiovascular Disease*, Saunders, Philadelphia, 1980.

Remington, J. W. and L. J. L. O'Brien, "Construction of Aortic Flow-Pulse from Pressure Pulse," *Am. J. Physiol.* **218**:437 (1970).

Resnekov, L., "Hemodynamic Effects of Acute Myocardial Infarction," *Med. Clin. North Am.* **57**:243 (1973).

Roach, M. R., "Biophysical Analysis of Blood Vessel Walls and Blood Flow," *Annu. Rev. Physiol.* **39**:51 (1977).

Samet, P. and N. El-Sherif, eds., *Cardiac Pacing*, 2nd ed., Grune & Stratton, New York, 1979.

Scher, A. M., Excitation of the Heart, in W. E. Hamilton and P. Dow, eds., *Handbook of Physiology*, sec. 2, *Circulation*, vol. 1, American Physiological Society, Washington, DC, pp. 287–322, 1962.

Schneck, D. J. and D. L. Vawter, eds., *Biofluid Mechanics*, Plenum, New York, 1980.

Smith, J. J. and J. P. Kampine, *Circulatory Physiology: The Essentials*, Williams & Wilkins, Baltimore, 1979.

Starling, E. H., *The Linacre Lecture on the Law of the Heart*, Longmans, Green, London, 1918.

Stehbens, W. E., ed., *Hemodynamics and the Blood Vessel Wall*, Thomas, Springfield, IL, 1978.

Sugimoto, T., et al., "Effect of Maximal Workload on Cardiac Function," *Jap. Heart J.* **14**:146 (1973).

Taylor, M. G., "Hemodynamics," *Annu. Rev. Physiol.* **35**:87 (1973).

Wells, P. N. T., *Biomedical Ultrasonics*, Harcourt Brace Jovanovich, New York, 1977.

Wells, P. N. T., *Ultrasonics in Clinical Diagnosis*, 2nd ed., Longman, New York, 1977.

Wells, P. N. T., *Physical Principles of Ultrasonic Diagnosis*, Academic, New York, 1969.

Wenger, N. K., ed., *Exercise and the Heart*, Davis, Philadelphia, 1978.

Willis, J., ed., *The Heart: Update*, McGraw-Hill, New York, 1979.

Wolf, S., et al., eds., *Structure and Function of the Circulation*, Plenum, New York, 1979.

Respiratory System

Agostoni, E., Thickness and Pressure of the Pleural Liquid, in A. P. Fishman and H. H. Hecht, eds., *The Pulmonary Circulation and Interstitial Space*, University of Chicago Press, Chicago, 1969.

Altose, M. D., *The Physiological Basis of Pulmonary Function Testing*, vol. 31, no. 2, *Clinical Symposia*, Ciba Pharmaceutical, Summit, NJ, 1979.

Bartels, H. and R. Baumann, "Respiratory Function of Hemoglobin," *Int. Rev. Physiol.* **14**:107 (1977).

Bauer, C., et al., eds., *Biophysics and Physiology of Carbon Dioxide*, Springer-Verlag, New York, 1980.

Bradley, G. W., "Control of the Breathing Pattern," *Int. Rev. Physiol.* **14**:185 (1977).

Briscoe, W. A., R. E. Forster, and J. H. Comroe Jr., "Alveolar Ventilation at Very Low Tidal Volumes," *J. Appl. Physiol.* **7**:27–30 (1954).

Bruley, D. F., "Mathematical Considerations for Oxygen Transport Tissue," *Adv. Exp. Med. Biol.* **37**:149 (1973).

Cherniack, N. S. and A. P. Fishman, "Abnormal Breathing Patterns," *D. M.: Disease-a-Month* (Jul. 1975).

Cherniack, R. M., Ventilation, Perfusion, and Gas Exchange, in E. D. Frohlich, ed., *Pathophysiology*, 2nd ed., Lippincott, Philadelphia, 1976.

Cohen, M. I. and J. L. Feldman, "Models of Respiratory Phase-Switching," *Fed. Proc.* **36**:2367–2374 (1977).

Comroe, J. H., Jr., et al., *The Lung: Clinical Physiology and Pulmonary Function Tests*, 2nd ed., Year Book, Chicago, 1963.

Cumming, G., Alveolar Ventilation: Recent Model Analysis, in *MTP International Review of Science: Physiology*, vol. 2, University Park Press, Baltimore, p. 139, 1974.

Domm, B. M. and C. L. Vassallo, "Pulmonary Function Testing," *Clin. Anesth.* **9**:191 (1973).

DuBois, A. B., et al., "A Rapid Pethysmographic Method for Measuring Functional Residual Capacity in Normal Subjects," *J. Clin. Invest.* **35**:322–326 (1956).

Ebert, R. V., "Small Airways of the Lung," *Ann. Intern. Med.* **88**:98–103 (1978).

Engel, L. A. and P. T. Macklem, "Gas Mixing and Distribution in the Lung," *Int. Rev. Physiol.* **14**:37 (1977).

Fenn, W. O. and H. Rahn, eds., *Handbook of Physiology*, sec. 3, *Respiration*, vols. 1 and 2, American Physiological Society, Bethesda, MD, 1964–1965.

Fishman, A. P., *Assessment of Pulmonary Function*, McGraw-Hill, New York, 1980.

Forster, R. E., Pulmonary Ventilation and Blood Gas Exchange, in W. A. Sodeman, Jr. and W. A. Sodeman, eds., *Pathologic Physiology: Mechanisms of Disease*, 5th ed., Saunders, Philadelphia, 1974.

Forster, R. E. and E. D. Crandall, "Pulmonary Gas Exchange," *Annu. Rev. Physiol.* **38**:69 (1976).

Guyton, A. C., "Measurement of the Respiratory Volumes of Laboratory Animals," *Am. J. Physiol.* **150**:70 (1947).

Guyton, A. C., et al., "An Arteriovenous Oxygen Difference Recorder," *J. Appl. Physiol.* **10**:158 (1957).

Hart, M. C., M. M. Orzalesi, and D. D. Cook, "Relation Between Anatomic Respiratory Dead Space and Body Size and Lung Volume," *J. Appl. Physiol.* **18**:519–522 (1963).

Hawker, R. W., *Notebook of Medical Physiology: Cardiopulmonary, with Aspects of Clinical Measurement and Monitoring*, Longman, New York, 1979.

Horsfield, K., "The Regulation Between Structure and Function in the Airways of the Lung," *Br. J. Dis. Chest* **68**:145 (1974).

Hyatt, R. E. and L. F. Black, "The Flow-Volume Curve. A Current Perspective," *Am. Rev. Resp. Dis.* **107**:191 (1973).

Jones, N. L., *Blood Gases and Acid-Base Physiology*, Decker, New York, 1980.

Kao, F. F., *An Introduction to Respiratory Physiology*, American Elsevier, New York, 1972.

Macklem, P. T., "Relationship Between Lung Mechanics and Ventilation Distribution," *Physiologist* **16**:580–588 (1973).

Macklem, P. T., "Respiratory Mechanics," *Annu. Rev. Physiol.* **40**:157 (1978).

Mead, J., "Respiration: Pulmonary Mechanics," *Annu. Rev. Physiol.* **35**:169 (1973).

Meyer, B. J., A. Meyer, and A. C. Guyton, "Interstitial Fluid Pressure in the Lungs," *Circ. Res.* **22**:263–271 (1968).

Morris, J. F., A. Koski, and L. C. Johnson, "Spirometric Standards for Healthy Nonsmoking Adults," *Am. Rev. Respir. Dis.* **108**:57–67 (1971).

Murray, J. F., *The Normal Lung*, Saunders, Philadelphia, 1976.

Nadel, J. A., et al., "Control of Mucus Secretion and Ion Transport in Airways," *Annu. Rev. Physiol.* **41**:369 (1979).

Nagaishi, C., *Functional Anatomy and Histology of the Lung*, University Park Press, Baltimore, 1973.

Nunn, J. F., *Applied Respiratory Physiology*, 2nd ed., Butterworth, London, 1977.

Otis, A. B., Quantitative Relationships in Steady-State Gas Exchange, in W. O. Fenn and H. Rahn, eds., *Handbook of Physiology*, sec. 3, *Respiration*, vol. 1, Williams & Wilkins, Baltimore, 1964.

Piiper, J. and P. Scheid, "Respiration: Alveolar Gas Exchange," *Annu. Rev. Physiol.* **33**:131 (1971).

Rahn, H., et al., "The Pressure-Volume Diagram of the Thorax and Lung," *Am. J. Physiol.* **146**:161 (1946).

Rahn, H. and L. E. Farhi, Ventilation, Perfusion, and Gas Exchange—The V_a/Q Concept, in W. O. Fenn and H. Rahn, eds., *Handbook of Physiology*, sec. 3, *Respiration*, vol. 1, Williams & Wilkins, Baltimore, 1964.

Rahn, H. and W. O. Fenn, *A Graphical Analysis of Respiratory Gas Exchange*, American Physiological Society, Washington, DC, 1955.

Reichel, G. and M. S. Islam, "Measurement of Static Lung and Thorax Compliance in Health and Pulmonary Diseases," *Respiration* **29**:507–515 (1972).

Riley, R. L. and S. Permutt, The Four-Quadrant Diagram for Analyzing the Distribution of Gas and Blood in the Lung, in W. O. Fenn and H. Rahn, *Handbook of Physiology*, sec. 3, *Respiration*, vol. 2, Williams & Wilkins, Baltimore, 1965.

Secker-Walker, R. H. and R. G. Evens, "The Clinical Application of Computers in Ventilation-Perfusion Studies," *Prog. Nucl. Med.* **3**:166 (1973).

Siggard-Anderson, O., *The Acid-Base Status of the Blood*, 4th ed., Williams & Wilkins, Baltimore, 1974.

Sorensen, S. C., "The Chemical Control of Ventilation," *Acta. Physiol. Scand. (suppl.)* **361**:1–72 (1971).

Stonin, N. B. and L. H. Hamilton, *Respiratory Physiology*, 4th ed., Mosby, London, 1981.

Thurlbeck, W. M., "Structure of the Lungs," *Int. Rev. Physiol.* **14**:1 (1977).

Wagner, P. D., "The Oxyhemoglobin Dissociation Curve and Pulmonary Gas Exchange," *Sem. Hematol.* **11**:405 (1974).

Wagner, P. D., "Diffusion and Chemical Reaction in Pulmonary Gas Exchange," *Physiol. Rev.* **57**:257 (1977).

Wagner, P. D., "Ventilation-Perfusion Relationships," *Annu. Rev. Physiol.* **42**:235–247 (1980).

Weibel, E. R., "Morphological Basis of Alveolar Capillary Gas Exchange," *Physiol. Rev.* **53**:419 (1973).

Weibel, E. R. and H. Backofen, Structural Design of the Alveolar Septum and Fluid Exchange, in A. P. Fishman and E. M. Renkin, eds., *Pulmonary Edema*, Waverly, Baltimore, 1979.

West, J. B., *Ventilation/Blood Flow and Gas Exchange*, 2nd ed., Lippincott, Philadelphia, 1970.

West, J. B., *Respiratory Physiology—The Essentials*, Williams & Wilkins, Baltimore, 1974.

West, J. B., "Pulmonary Gas Exchange," *Int. Rev. Physiol.* **14**:83 (1977).

41.4 MEDICAL LABORATORY AUTOMATION

Jeff Tosk and Julius Simon

41.4-1 Introduction

Laboratory automation was introduced and developed in response to an ever-increasing volume and diversity of laboratory services requested and offered. This is best exemplified by the virtually explosive growth in the demand for clinical chemical measurements, which prompted an extensive research and development effort. For the design engineer faced with the task of automating a laboratory procedure, an examination of the methodology and design philosophies developed for automating clinical instrumentation is useful, if not unavoidable. Indeed the state of the art in laboratory automation has been largely defined by developments in clinical chemical analysis.

Automation of a given laboratory procedure, in whole or in part, is considered when the sample throughput demand approaches the throughput capability of the available instrumentation. Laboratory automation draws such diverse disciplines as electronics, robotics, computer science, and chemistry, to name a few.

Laboratory automation is a complex and often confusing topic, made so by the extent to which it is interdisciplinary. For example, in laboratory automation economics, psychology, and business management interact extensively with more obvious technically based fields. Considering also the rapid state of flux to which many of these areas are subject, one begins to realize that laboratory automation is difficult to even define. The definition of "automate" suggested by the International Union of Pure and Applied Chemistry (IUPAC) excludes most laboratory systems marketed as automatic in that IUPAC states "automate is to replace human manipulative effort and facilities in the performance of a given process by mechanical and instrumental devices which are regulated by a feedback of information so that the apparatus is self-monitoring or self-adjusting." The requisite feedback system is often not encountered in commercially available "automatic" laboratory equipment. Stockwell used a simpler definition of automation. It is "the use of any facility, either electronic or mechanical, which eliminates some aspect of manual interaction and improves the efficiency of the analytical process."

With Stockwell's simplified definition in mind, let us now proceed to an overview of just how laboratory automation is currently implemented. It will be seen that in general, the rudiments of practical automation are quite similar to those first developed. While such aspects of instrumentation as detector technology; electronics; and data acquisition, reduction, and reporting have advanced with time, the fundamental modes of automation are apparently here to stay for the foreseeable future. This may be due in part to the inherent efficiency of the three major methods of automating a laboratory procedure. These methods are continuous-flow, centrifugal, and discrete analysis, into which most automated instrumentation may be classified. A brief description of each with examples is presented.

41.4-2 Continuous-Flow Analysis

Continuous-flow analysis (CFA) was introduced by Skeggs.[1,2] A typical CFA system comprises a sampler, a multichannel peristaltic pump, one or more mixing/reaction delay coils, and a detector, usually of the spectrophotometric variety. The samples are introduced into the system by the sampler functional block. The samples are separated by an air bubble. Once inside the system, samples, added reagents, and solvents flow through tubing until the effluent exits the instrument. The sample is mixed with appropriate reagents which are admitted by the peristaltic pump, usually employing one pump channel per reagent. The sample reagent mixture traverses a given length of tubing wherein mixing and reaction occur. Reaction time is established by the length of the tubing, which is coiled to conserve space while providing a sufficient time delay before additional reagents are added or the tubing enters the detector block.

41.4-3 Centrifugal Analysis

The centrifugal analyzer was developed at the Oak Ridge National Laboratory by Anderson. The sample and reagents are added near the center of a centrifuge rotor and are rapidly mixed by centrifugal action. The reactions occur in discrete compartments in the form of cuvettes which are photometrically monitored as each one passes through the detector once during each revolution. This is particularly useful for kinetic analysis of an enzyme where absorbance changes as a function of time. Since the detector is virtually multiplexed by rapid rotation of the cuvettes, which permit each to traverse the optical detector once in each revolution, in a short time numerous readings per sample are acquired. The data may then be presented as the change of absorbance vs. time. A wash cycle may be included when the analysis is completed.

41.4-4 Discrete Analysis

Discrete analyzers are designed to simulate the function or procedure as carried out by the technician performing the manual version of an analysis. They offer the potential to automate an existing manual method. In general, sample and reagents are metered by motor- or pneumatic-driven syringes. The sample and reagents are added to individual reaction vessels, usually in the form of a cuvette or test tube. Mixing is achieved via mechanical agitation, while temperature control is implemented with a water bath or an incubator. After a suitable reaction time, the reaction vessel is mechanically introduced into the detector for quantitation. Additional reagents may be added during the analysis, and many discrete analyzers are equipped with a wash cycle to remove used reagent-analyte mixtures and prepare the reaction vessel for the next sample. While it readily lends itself to adaptation of existing techniques, the discrete analyzer is often mechanically complex and subject to failure modes associated with mechanical systems where operating characteristics vary with time due to friction, stress, fatigue, and the associated need for adjustment and parts replacement. The DuPont ACA system is offered as a currently available instrument employing discrete analysis.

The Technicon autoanalyzer CFA system was among the first commercially available instrumentation for automated clinical chemistry. Although modular in design such that the system could be switched from one assay to another by changing the module between the pump and the detector, the autoanalyzer was usually dedicated to the batch analysis of one or two serum components. The Technicon SMA 12/50 system incorporated a 12-channel assay manifold within one CFA system. This made the CFA assay of up to 12 serum parameters simultaneously available. The Technicon

SMAC biochemical analyzer was yet another improvement employing CFA, permitting 20 simultaneous analyses at a rate of 150 samples per hour.

The air bubble serves to contain and confine the sample and reagents used in each analysis. This permits the chemistry performed in each analysis to be viewed as though the segments were discrete reaction vessels. That is, once the segmented reagent-sample bolus has been formed, no segment can react with any other. An unsegmented version of the basic technique is known as flow-injection analysis (FIA).[3] Detailed differences between CFA and FIA are discussed in the literature, to which the reader is referred. The salient feature that distinguishes the two techniques is generally considered to be the reaction time. FIA systems have reaction times in seconds, while CFA systems may have reaction times in minutes. It is thought that FIA may become the choice for many automated and semiautomated assays.

41.4-5 Laboratory Computer and Automation

Laboratory automation increases sample throughput, thereby necessitating the acquisition, storage, processing, and reporting of proportionally more data. The laboratory computer is therefore becoming increasingly visible. It can assist with each stage of a laboratory procedure. Samples can now be identified on receipt with a computer-generated/computer-readable bar code accession number. Once assigned, the bar-encoded sample number may be referred to at any stage of sample processing. In this way samples may even be submitted to analysis out of conventional numerical order on a routine basis. This is helpful when some samples are given priority. Manual handling of this situation is tedious and error prone. Computer-assisted sample management is a logical, if not essential, enhancement of laboratory automation. In addition to bar-encoded sample recognition, computerized voice recognition may be employed to identify, describe, or otherwise characterize samples. Hardware and software for this is available in readily useable form.[4] Results that must be recognized by eye may be dictated via microphone for acquisition by a laboratory computer. An example is the reading of thin-layer chromatography plates. The technologist determining the result can verbally identify the sample via alphanumerics, and report the results as well. Sufficiently large computer-recognizable vocabularies have been developed to the extent that in the analytical setting, for example, results may be requested by voiced commands.

41.4-6 Electronics

With the advent of the integrated circuit, especially large-scale integration (LSI) and very-large-scale integration (VLSI), increasingly complex electronic functions may be implemented with fewer parts and heretofore unavailable functions may be obtained. State-of-the-art electronics as applied to laboratory automation is a subject best discussed at length elsewhere. However, some recent developments in electronics are appropriately mentioned here. Two developments worthy of brief consideration as they may be applied to laboratory automation are costorn and semicostorn logic arrays and digital signal-processing (DSP) integrated circuits.

The design engineer may now routinely and cost effectively design his or her own "logic array," which is then reduced to monolithic form at a suitably equipped facility employing computer-assisted design/computer-assisted manufacture (CAD/CAM) technology. Such services are offered by several semiconductor manufacturers. Few if any design engineers thought that such a facility would become available. Now, in addition to the availability of an ever-increasing variety of off-the-shelf logic, it is possible to specify unique and complex logic functions and with a rapid turn-around time receive ready-to-install integrated circuits so specified. The primary advantage to this capability is that the instrumentation manufacturer can keep proprietary essential segments of circuitry and thus have a significant advantage in the marketplace.

Digital signal processing (DSP) is now available on single VLSI chips. DSP technology affords the capability of "cleaning up" a noisy signal from an instrument operated near the limits of detection. A variety of techniques have been available for such purposes, but now such methods as the Fourier transform, convolution integral, digital correlation, and digital error correction are available on a single chip. An entire minicomputer with an array processor was needed in some cases to achieve what is now available at a fraction of the cost and software overhead. Most notable of these DSP chips is the Texas Instruments 320 DSP integrated circuit. At an initial unit cost of about $300, this device can in many cases replace 10 to 100 times its cost in hardware alone. Such related devices assist in implementing error correction and enhancing signal-to-noise ratios. To the end user this translates into vastly improved data integrity, quality control, and assurance which are otherwise difficult to obtain when processing automated laboratory procedures at the upper limit of an instrument's capability.

41.4-7 Image Processing

Thousands of petri plates, often with hundreds to thousands of bacterial colonies, must be examined and the colonies counted. Such a task has fortunately yielded to automation. The Fisher automated count-all counter is a commercially available instrument for counting not only bacterial colonies but grain in microscopic autoradiographs as well. The circuitry is proprietary but the available techniques for digital image processing are well developed and a typical counter of the aforementioned description may be implemented in several ways. Some of these ways are hardware intensive, such as the Fisher counter, and some are software intensive, such as a computerized colony counter developed by one of the authors of this chapter. They share in common a vidicon tube with raster scouring such as is used in television cameras. The composite video signal is digitized via an analog-to-digital (A/D) converter and stored in digital form as a memory map in random access memory (RAM). The Fisher counter uses a hardware-intensive approach wherein the digitized, memory-mapped image is divided into individual picture elements (pixels) and the light pixels are distinguished from the dark, that is, colonies are distinguished from background agar, with subsequent conversion of pixels into the form of a histogram. Once the colony size parameters are set via calibration, the histogram may be used to perform a calculation of the number of colonies. The Fisher counter is fast and accurate. However, it costs nearly $10,000. The device is dedicated with respect to its counting function, the said function being easily implemented with a microcomputer-based system comprising a video A/D board and a television camera in addition to a 5-100-8080 based microcomputer, which may be used as a stand-alone computer. The cost of this microcomputer-based system is $7,000; it can function as a laboratory word processor in addition to being an image-processing system capable of many functions besides colony counting. The algorithm employed[4] stores in the memory map of the pixels and searches for adjacent pixels until a dark (background) pixel is encountered. This process is repeated line by line (as derived from the digitized raster scour image of the plate being counted). Light areas surrounded by dark are counted, yielding the number of colonies. The point to be made is that in certain cases a dedicated instrument is simply not cost effective, however satisfactory the results obtained with it.

Another important application of image processing in laboratory automation is the recently developed scouring photodiode array spectrophotometer offered as a liquid chromatography detector. The image to be processed in this case is the response of the individual photodiodes to wavelengths of light, which transilluminate the sample cell containing the chromatographic effluent. The rate of the scan and subsequent processing is sufficiently fast to enable work at several wavelengths in real time and in some cases to scan a complete spectrum in real time during the chromatographic separation.[5] Some procedures employed in discrete analyzers lend themselves to video photography with subsequent manipulation of the sample on the basis of data reduced from the processed video image. This is an example of robotics at work in the automated laboratory. In such cases a charge-coupled device (CCD) often replaces the vidicon tube. The CCD is more easily adapted to digital image-processing techniques than is the vidicon tube. Image processing will play an increasingly important role in laboratory automation, virtually replacing the human eye in many instances.

41.4-8 Automated Sample Preparation

One of the most time-consuming and critical tasks in the laboratory is sample preparation prior to analysis. Many samples, particularly of biological fluids such as blood, serum, and urine, contain a variety of substances that can easily foul or even destroy delicate instrumentation. In other settings the samples are crude, unmeasured, and contaminated by interfering substances known and unknown. Currently available automated sample preparation (ASP) functional blocks are still rather unsophisticated, offering routinely accurate measurement by volume (autopipette and syringe) or by weight.[6] Chemical modification of samples has been automated to the extent that a sample may be derivatized, pH adjusted, or buffered. Such functions as grinding, mixing, homogenizing, dissolution, dialysis, filtering, and evaporation are also available. (In the case of pharmaceutical quality control analysis, we find automated crushing of tablets and their dissolution prior to chemical analysis.) The fact is that not all sample preparations can or should be automated. It is clear that extensive automated sample preparation is as yet unavailable for a wide variety of needs.

References

1 L. T. Skeggs, "New Dimensions in Medical Diagnosis," *Analyt. Chem.* **38**(6):31A–44A (May 1966).

2 L. T. Skeggs, *Am. J. Clin. Pathol.* **28**:311–322 (1957).

3 K. K. Stewart, "Flow Injection Analysis," *Analyt. Chem.* **55**(9):931A–940A (Aug. 1983).

4 S. B. Tilden and N. B. Denton, "Advanced Software Concepts for Employing Micro-Computers in the Laboratory," *J. Autom. Chem.* **1**(5):128–134 (Oct. 1979).

5 A. D. Mills, I. Mackensie, and R. J. Dolphin, "The Use of the Micro-Computer for Flexible Automation of a Liquid Chromatograph," *J. Autom. Chem.* **1**(5):134–140 (Oct. 1979).

6 D. A. Burns, "Automated Sample Preparation," *Analyt. Chem.* **53**(12):1403–1418A (Oct. 1981).

Bibliography

Dessy, R. E., "Local Area Networks: Part 1," *Analyt. Chem.* **54**(11):1167A–1184A (Sep. 1982).

Haeckel, R. and O. Sonntag, "An Evaluation of the Kodak Ektachen System for the Determination of Glucose and Urea," *J. Autom. Chem.* **1**(5):273–281 (Oct. 1979).

CHAPTER 42
PATIENT MONITORING

STEVEN G. EPSTEIN

JOHN DAWSON

LOUIS R. M. DEL GUERCIO

New York Medical College, Valhalla, New York

42.1 INTRODUCTION

Patient monitoring has in recent years progressed from infancy to adolescence as a result of technologic advances and a better medical understanding of the critically ill patient. The acute-care setting requires continuous monitoring of selected physiologic parameters, in contrast to the hospital ward or the physician's office, where routine measurements are obtained. The acute setting might be the operating room (OR), intensive care unit (ICU), or neonate intensive care unit (NICU). The rate or interval at which measurements are made depends both on the parameter being monitored and on the setting in which it is being monitored. In the OR, minute-by-minute knowledge of a patient's temperature is necessary to detect a condition known as malignant hyperthermia, a rapid and often fatal rise in temperature due to the effects of anesthesia. In the ICU, temperature measurements are required but less frequently to detect the onset of fever due to the passage of bacteria into the blood. Obviously, a sophisticated method of temperature measurement is required in the OR, while in the ICU a simple mercury thermometer suffices. It is also evident that the sampling frequency or interval is a critical component of patient monitoring, and that the interval time (t) must be short enough to detect any perturbation in the physiologic parameter being monitored.

Another example of degrees of monitoring is the use of the electrocardiogram (ECG). In adults an ECG might be obtained every year in the physician's office for detection of minor changes in heart rhythm, while in the ICU the ECG must be continuous for detection of life-threatening disturbances in rhythm secondary to myocardial infarction. Thus $t = 1$ yr in the first case and 10 ms in the second.

Although a physician is always present in the OR, this is not always the situation in the ICU, where a critical care nurse is responsible for the immediate care of the patient. It is not always easy for a nurse to differentiate between machine malfunction and patient abnormality. Monitoring devices must remain simple in both access (transducer input) and retrieval of information (output), and should contain alarms that indicate when measured parameters are out of the acceptable range. Already in use are microprocessors, which are better able to evaluate both input and output of data and which increase the reliability of the instrumentation.

Which physiologic parameters require monitoring in the acute setting is often debated. The decision is often based on financial considerations as well as technologic limitations. The technologic limitations exist both in transduction of a physiologic event to an electromechanical event and in processing of the information. The current trend in transduction is toward noninvasive methods, that is, methods that do not violate the integrity of the body. No technology yet exists for continuous monitoring of the absolute blood pressure noninvasively. It remains necessary to place an indwelling catheter in an artery for this measurement. Invasive techniques create a portal of entry to the body from the nonsterile environment, and thus pose the risk for infection. Catheters act as a foreign body and cause inflammation at their point of entry into the body. They may also cause thrombosis and subsequent emboli when placed in situ.

The instrumentation discussed in this chapter is that which the authors consider important in the critical care unit or operating room; it is by no means exhaustive. Devices may differ according to the condition of the patient in the ICU. All patients entering the ICU have their temperature, blood pressure, heart action, and respiratory rate monitored routinely. Blood pressure may be monitored by invasive methods or by auscultation, that is, continuously or intermittently. The heart rate is monitored by observing the ECG and manually counting the pulse rate. The flow-directed balloon catheter (discussed later) is employed when fluid balance or cardiac insufficiency is of concern. Transcutaneous oxygen monitoring is necessary in the neonate because of its delicate respiratory status. The infrared carbon dioxide detector is important in the pulmonary ICU to detect impending respiratory failure or to aid in the adjustment of automatic ventilators. Intracranial pressure monitors find their use in the trauma unit following head trauma, where intracranial pressure may rise due to cerebral edema or bleeding.

These are just a few of the multitude of devices in use and being developed. Automated therapeutic response to the output of these devices is becoming reality. The ultimate decisions will always be in the hands of professional health care personnel and the engineers who create the devices.

42.2 CARDIOVASCULAR MONITORING

Pressure systems provide the bulk of the clinical information extracted in the critical care unit. The cardiovascular pressures of interest may be divided into the low-pressure venous system (-5 to 60 mm Hg) and the high-pressure arterial system (0 to 300 mm Hg). Motive force is provided by a four-chambered lift pump—the heart. Ancillary force is lent by the negative intrathoracic pressure during inspiration and by limb muscle contraction compressing the peripheral vasculature.

Arterial blood pressure may be obtained in the conventional manner by auscultation. Another method is placement of an indwelling catheter into an artery. This method employs a catheter, connecting tubing, a transducer, an amplifier and a filter, and a display system. The same method can be used to measure venous pressures via insertion of the catheter into the central venous system. The equipment should be capable of measuring changes from -5 to 300 mm Hg with a frequency of 0 to 5 Hz and harmonics to 100 Hz. The transducer may be of the piezoceramic, strain gauge, or resistance type. A balance vernier to provide null level at atmospheric pressure with the transducer vent open at the patient's midchest is available to the user. Additional gain, range, filter, and alarm limit settings may also be present. Data output is by cathode ray tube (CRT) display of the pressure waveform and digital or analog readout. A nonpermeable diaphragm and fluid interface physically decouples the patient's bloodstream and transducer (Fig. 42.1). Stopcocks and constant infusion devices are often interposed along the fluid path. These provide useful therapeutic ingress or assure line patency, but also throttle direct fluid column transmittance and harbor entrained or outgassed bubbles and thus markedly affect fidelity and precision.

To reduce resonance effects within the fluid system, the resonant frequency must be well above expected physiologic frequencies. For practical purposes, a 2-Hz mean with tenth harmonic reproduction is adequate. The resonant frequency of the fluid-transmitting system is given by $f_0 = (D/4)(1/L) \times (P/V)$, where D = internal diameter, L = column length, P = pressure, and V = system volume. Short thick-walled small internal diameter (< 1.5 mm) catheters provide the best combination of low volumetric compliance and resonance-to-damping ratio. Maximum fidelity and reproduction may be obtained with microtransducer-tipped needles introduced directly into the bloodstream. However, owing to limitations imposed by sterility, cost, and difficulty in zeroing and calibration, these are not practical for clinical use.

The system just described is used to measure arterial pressures. By changing the type of catheter and its anatomic position from the arterial system to the pulmonary vascular system, central pulmonary artery pressures may be measured. These pressures are essential in calculating the amount of fluids a patient requires. The catheter employed is the balloon-tipped flow-directed catheter developed by Swan and Ganz. This catheter is also capable of measuring cardiac output by the thermodilution method and a microcomputer. The device is a multilumened catheter that provides a direct path to either chamber of the right heart and to the pulmonary artery. One lumen of the catheter terminates 2 cm proximal to the catheter tip and is enclosed by a latex balloon. A second

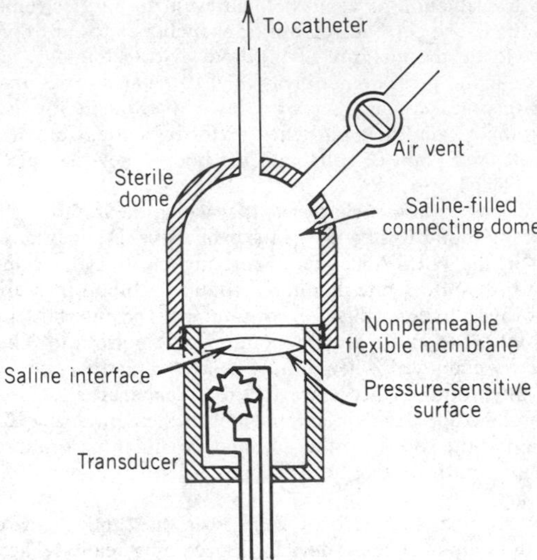

Fig. 42.1 Pressure transducer.

lumen terminates at the tip and provides pressure measurements via a fluid column. A third lumen terminates 30 cm inside the tip and is also fluid filled. The fourth lumen encapsulates two flexible leads to a thermistor bead implanted on the outer surface of the catheter just proximal to the balloon (Fig. 42.2).

After a sufficiently large peripheral or central vein has been entered, the balloon is inflated with 1.25 cc of air or CO_2 and the device is advanced through the vasculature toward the right side of the heart. Venous flow carries the catheter through the right heart to the pulmonary artery. Monitoring of the distal tip waveform allows the physician to determine the catheter's position, each section of the venous system, right heart, and pulmonary artery having a distinctive waveform and typical pressures. The final position of the catheter is in the pulmonary artery such that when the balloon is inflated it obstructs all fluid flow upstream of the balloon. The definitive position of the catheter is determined by either fluoroscopy or radiography.

Owing to the small diameter of the catheter (2 mm) and its flexibility, valvular incompetence is negligible. The position of the distal tip allows sampling of the mixed gas partial pressures in the pulmonary artery, which give information about cardiac output and heart shunts. When the balloon is inflated, the pressure measured is that reflected by the left atrium, also representative of the body's fluid status.

One of the most important hemodynamic variables amenable to monitoring is the cardiac output. The flow-directed catheter is perfectly adapted for this measurement. The Stewart-Hamilton dilution method is the technique utilized; in this instance it is the dilution of a cool liquid in the body's warmer blood. An aliquot of fluid of known temperature is injected into the right atrium (30-cm lumen). It mixes with the blood and flows past the distal thermistor. The rate of change of temperature is measured at the thermistor and a characteristic curve is obtained (Fig. 42.3). The

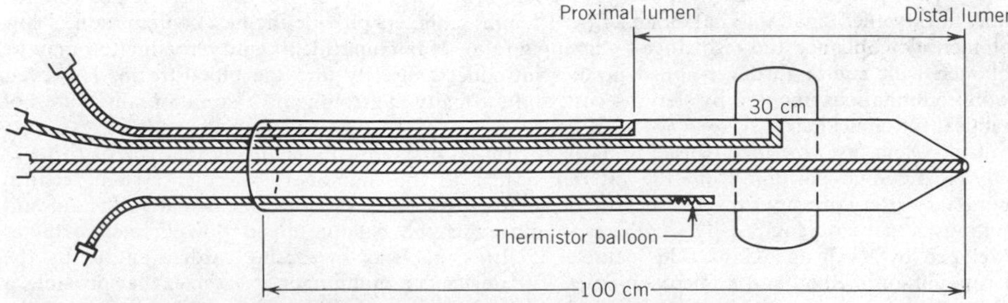

Fig. 42.2 Four-lumen flow-directed balloon catheter.

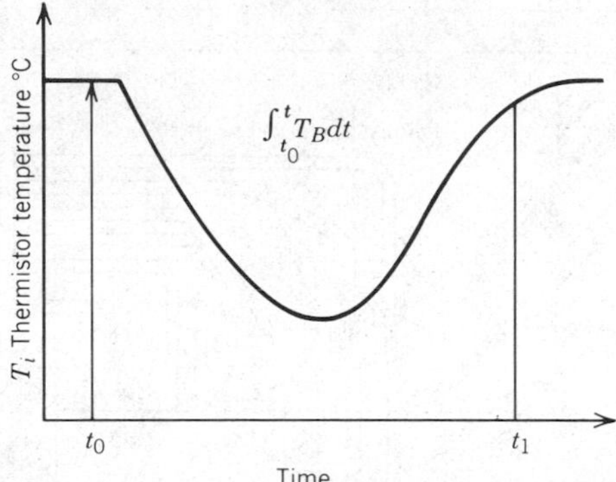

Fig. 42.3 Cardiac output thermodilution curve. t_0 = injection of coded fluid, t_1 = completion of dilution, T_i = initial blood temperature before injection.

adapted Stewart-Hamilton equation is given by

$$\text{CO} = \frac{k(T_B - T_I)}{k_1} \int_0^t T_B(t)\,dt$$

where CO = cardiac output
k = manufacturer's constant
T_B = blood temperature
T_I = injectate temperature
k_1 = manufacturer's constant
$\int_0^t T_B(t)\,dt$ = area under the time-temperature dilution curve

Each manufacturer has established its own criteria for determining when dilution is complete, marking the end of the curve: k and k_1 are related to these criteria.

42.3 RESPIRATORY MONITORING

Monitoring of respiratory parameters in the intensive care setting and in the operating room has always been important but until recent years has been limited to simple mechanical or electrical monitoring of respiratory rate, apnea spells, or airway pressures. Cumbersome equipment was needed to measure gas concentrations.

In recent years many attempts have been made to monitor arterial oxygen and carbon dioxide concentrations (gas partial pressures) with noninvasive methods. Current invasive methods either require use of an indwelling catheter or serial arterial puncture and subsequent in vitro analysis of the sample obtained for pH, partial arterial oxygen pressure (P_aO_2), and partial arterial carbon dioxide pressure (P_aCO_2). Although direct sampling is the most precise and reliable method, it creates technical problems and risks, such as placement of the arterial catheter, sepsis (infection), embolism, and thrombosis of the blood vessel. Recently, engineered devices such as the transcutaneous oxygen monitor ($P_{tc}O_2$) and exhaled gas monitors of the infrared and spectrometer type have alleviated many of the problems of invasive sampling or monitoring. It is of note that the results obtained by these monitors do not replace those provided by in vitro analysis with use of refined electrochemical techniques of an actual blood sample. Monitors may supply actual values but at times may only indicate a changing trend rather than the absolute blood value. Unlike the EKG or blood pressure monitor, where malfunction is easy to detect and then verify by direct measurement (i.e., a finger on the pulse), these instruments may yield an erroneous value that has lost its correlation with the actual in vivo value. A high index of suspicion and awareness on the part of the staff are essential and a prerequisite to the use of gas monitors.

Because they are noninvasive instruments, these devices are electrically safe; there is no direct connection to the patient except by air columns.

Transcutaneous oxygen monitoring was introduced into clinical use in the early 1970s by two independent groups—Huch, Huch, Menzer and Lubbers, and Eberhard and colleagues. The present popularity of such monitoring is centered in the neonate intensive care unit. In this setting the correlation between $P_{tc}O_2$ and P_aO_2 has been demonstrated to be 0.95 in most instances.

The basic system consists of a Clark electrode mounted within a housing containing an electrolyte solution and separated from the patient's skin by a thin membrane. Within this housing a heating element and thermistor are provided to increase the temperature of the skin (Fig. 42.4).

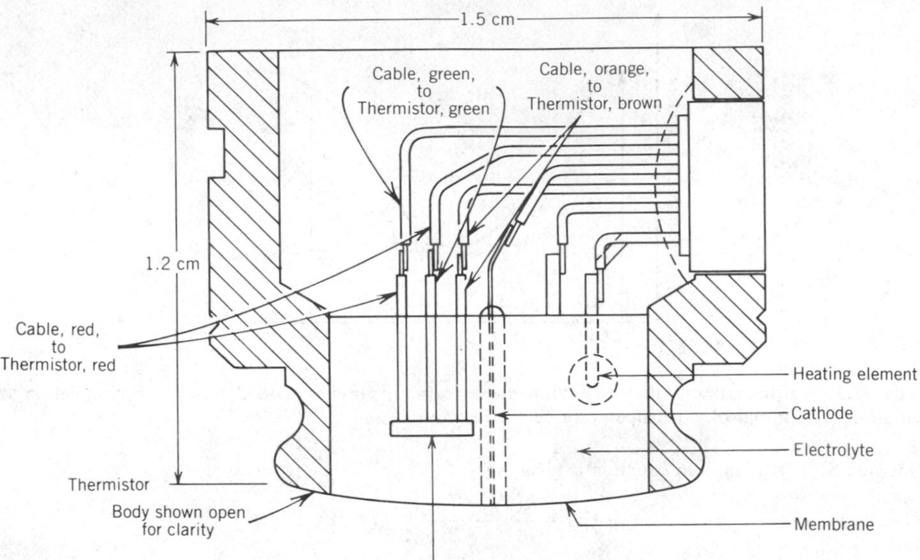

Fig. 42.4 Transcutaneous oxygen monitor. Used with permission of Biochem International Inc.

Application of the $P_{tc}O_2$ monitor is based on the diffusion of oxygen from the capillaries below the epidermis to the electrode. Oxygen is supplied to the epidermis via a complex system of capillaries that underlie the epidermal layer of the skin and lie within the dermis. The thickness of the epidermal layer and the capillary density vary with their anatomic location. The epidermis is thickest on the back, palms, and soles of the feet. It ranges in thickness from 30 to 90 μm in the adult, and is considerably thinner in the neonate. Capillary density is maximal on the face. Blood flow through a given capillary bed is a function of cardiac output and the intrinsic resistance to flow within the capillary bed, which is governed by the arterioles at the entrance to the bed. The diameter of these arterioles is regulated by both nervous influx and circulating humoral factors. The oxygen concentration in the blood is a function of the oxygenating ability of the lungs. The amount of oxygen delivered to the tissue is the product of the flow to that region multiplied by the difference between the arterial inflow oxygen concentration and the venous outflow concentration.

In an attempt to reduce the number of parameters controlling oxygen delivery to the skin underlying the transducer, a heating element is placed within the electrode housing. Increasing the temperature of the skin causes local vasodilatation, which aids in maintaining a constant flow through the capillaries. The temperature of the skin is raised to about 43°C and is maintained constant by use of a thermistor in a feedback loop with the heating element. The amount of current or energy required to maintain this temperature is displayed on the control panel of the instrument. This variable is physiologically important because an increase in current may reflect a decrease in ambient or body temperature. All neonates are kept in a temperature-controlled environment. The high correlation of the $P_{tc}O_2$ and P_aO_2 in the neonate is a result of the thin epidermis and the controlled environment. If the cardiac output of the infant should fall, this correlation decreases.

The instrument should contain high- and low-current temperature alarms and should display both the temperature on the surface of the skin and the amount of power required to maintain that temperature. Because of the heat applied to the surface of the skin, the transducer position must be changed at regular intervals or a second-degree burn may result, especially in neonates. The reliability of the heating circuitry is also critical for the same reason.

Stability, calibration, and response time of the transducer are the main problems encountered both in the design of the transcutaneous oxygen monitor and in its clinical use. Dehydration of the electrolyte solution causes its permeability to O_2 to change, resulting in drift and necessitating recalibration. This problem has been resolved in part by the addition of ethylene glycol to the electrolyte solution to slow dehydration. Unfortunately, the ethylene glycol reduces the sensitivity of the electrode to oxygen. Oxygen diffusion from the skin to the electrode is also affected by the material and thickness of the membrane interposed between the skin and the electrode. Among the synthetic materials used are Teflon, Mylar, and polypropylene, each with its own permeability characteristic and response time. These membranes vary from 0.5 to 1.0 thousandth of an inch in thickness. Although in vitro testing has demonstrated quantitative differences between these materials, in vivo evaluation has not borne out these differences. It has been shown that when anesthetic gases are circulating in the bloodstream, there is interference between the anesthetic agent and oxygen, and that this interaction may be reduced by using a membrane of low oxygen permeability.

Recently introduced into clinical application is infrared detection and analysis of carbon dioxide concentration in expired respiratory gases. The instruments are based on the property of carbon dioxide to absorb infrared radiation. Carbon dioxide has absorption spectra at 15, 4.26, 2.77, and 2.69 μm. It is the peak at 4.26 μm that is used in commercially available analyzers. The expired CO_2 is passed through a cuvette between an infrared source and a filter/detector. The filter is an interference filter at 4.2 μm and the detector is a germanium-arsenide photoelectric sensor. Electronic stabilization is achieved by mechanically chopping the infrared source and demodulating the signal after its detection. The CO_2 concentration is inversely proportional to the amount of radiation received by the sensor while the gas is passing through the cuvette, increasing concentration causing increased absorption of the infrared source.

The concentration of CO_2 at the termination of the expiratory phase of the respiratory cycle is normally equal to the alveolar concentration of CO_2. The end expiratory CO_2 (and alveolar CO_2) is a reflection of the adequacy of ventilation (rate and volume) to clear the blood of carbon dioxide. When rate and volume are inadequate, the end expiratory CO_2 rises (as well as the P_aCO_2). The rise may be detected by sampling the expiratory gases from the oral pharynx or ventilator tubing and analyzing the CO_2 concentration with the infrared analyzer. This reduces the need for frequent arterial blood gas samples when ventilatory adequacy is in question.

These instruments are not, however, free of operational or design difficulties. During anesthesia, other gases may interfere with the absorption spectra of CO_2. Nitrous oxide (N_2O), an anesthetic gas, has an absorption peak at 3.9 μm and interferes with the CO_2 absorption peak. To avoid this problem, the interference filter should have a narrow bandwidth.

Water condensation within the sensor head results in a change in sensitivity. This problem is resolved by heating the sensor or backflushing the sampling tubing with warm dry air. Nevertheless, excess accumulation of detritus such as blood or sputum may occur, resulting in inaccurate concentration readings.

Mass spectrometry is also being used to analyze the concentration of inspired and expired respiratory gases. This is done by aspirating gases from the oral pharynx or ventilator tubing, as with the infrared detector, and then continuously leaking these gases from the low-pressure aspiration pump into the high vacuum chamber of the spectrometer. Each gas has its own particular mass spectrum and is analyzed accordingly. This system may be used to monitor more than one patient at a time by multiplexing the sampling tubing from each patient to the spectrometer. The tubing must be of the same length from each bed to the spectrometer so that sampling aspiration times are identical. Each patient may be monitored continuously or at intervals determined by the number of beds being monitored. Flow rates within the sampling tubing are of the order of 100 ml/min. Each patient should be sampled for approximately 15 to 30 seconds to obtain an adequate respiratory profile.

As with the infrared analyzer, the sampling tubing must be kept free of any foreign material that might obstruct the flow of gas. This tubing may be as long as 30 m from patient to spectrometer and may be contained within wall conduit, which makes it extremely difficult to clean should it become obstructed.

42.4 INTRACRANIAL PRESSURE MONITORING

The brain is contained within the skull, which in the adult is an enclosed and rigid compartment. Certain pathologic processes, such as bleeding from trauma or a cerebrovascular accident or cerebral edema, cause the pressure within this compartment to increase. The increase in pressure, if not treated, ultimately results in death of the nerve cells. Monitoring of intracranial pressure (ICP) has been employed in neurosurgical intensive care units since 1960, when Lundberg continuously monitored ICP by means of an indwelling cerebral ventricular catheter connected to an external strain gauge transducer. This technique had its limitations, such as infection and difficult insertion of the catheter into the ventricles which were compressed owing to the increased pressure.

The normal ICP ranges from 0 to 11 mm Hg (or 0–150 cm H_2O), varies slightly during sleep, and is dependent on the position of the head relative to the body (sitting or prone). Episodic waves are imposed on normal static pressures as a result of respiration and heartbeat. Beyond these normal variations, rises in pressure are ominous findings; when clinical evidence of raised intracranial pressure is present, rapid medical or surgical intervention is mandatory.

The cellular material of the brain is enclosed within the bony skull and a tissue known as the dura mater, which lies just beneath the bone. Drilling of a hole in the skull and incising of the dura permit the underlying brain tissue to come into contact with the exterior. The pressure within the cranium may be measured by placing a hollow screw in the hole and connecting it to a strain gauge transducer via fluid-filled tubing, which transmits the hydrostatic pressure within the skull to the transducer.

This device may be inserted in the intensive care unit; its use is less invasive than cannulation of the ventricles. It cannot be used in an infant because the skull is too thin to hold the bolt. Small air leaks occur in all transducer systems and require addition of fluid to the tubing. This must be done with great care to prevent increase of the ICP. Tissue or debris may become entrapped at the mouth of the screw, damping changes in the transmitted pressure and necessitating flushing the system. Less than 1 cc may artificially raise the ICP. One of the disadvantages of this system is the risk of meningitis because of the open dura. Absolute zeroing of the transducer is possible with this system.

The fiberoptic-pneumatic device introduced in the mid 1970s alleviated many of the problems with open intracranial monitoring. This device consists of a mirror mounted on a membrane within a closed chamber which is implanted through a hole in the skull but above an intact dura. An afferent fiberoptic light source transmits light to the mirror, which is then reflected onto efferent fiberoptic threads. Changes in ICP cause the membrane to displace the mirror, which causes light to be reflected back via the efferent threads; the difference of intensity between the afferent light and the efferent light is measured by a differential amplifier. The output of the differential amplifier is fed to a bellows, which increases the pressure inside the sensor, counteracting the increase in ICP and returning the mirror to a neutral position (Fig. 42.5). The sensor is zeroed when it is free from all contact. Calibration is electronic.

This device has been used both intracranially and extracranially. In the neonate, the anterior fontanel is soft and bulges with increasing ICP. The fiberoptic pressure transducer may be applied to the fontanel on the scalp; the pressure measured is not absolute and is influenced by the method of fixation of the sensor to the scalp. The initial pressure measured is the sum of two unknown pressures, fixation and ICP. The device then measures changes in the ICP. The situation is similar when the sensor is placed intracranially between the dura and skull. The device is implanted as gently as possible so that the pressure of fixation is minimal. Because the dura is intact, the risk of meningitis is reduced (Fig. 42.6).

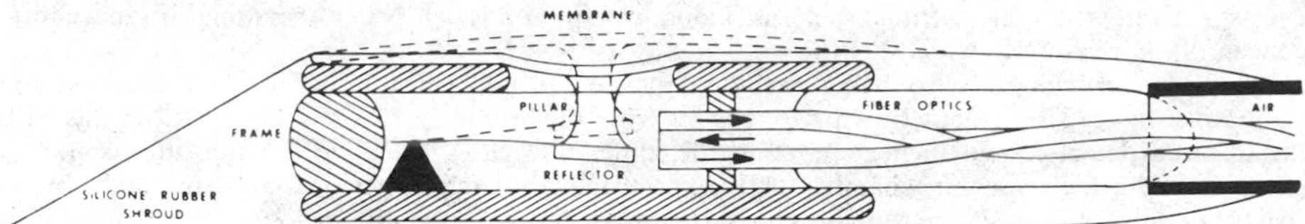

Fig. 42.5 Cross section of intracranial portion of pressure sensor. Reproduced from A. B. Levin, *Neurosurgery* **1(3)**:267 (1977), by permission of Ladd Research Industries, Inc.

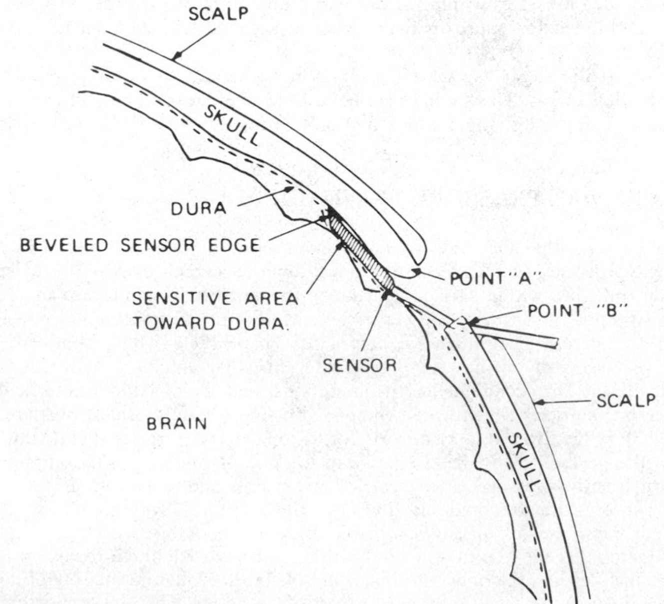

Fig. 42.6 Placement of intracranial pressure sensor. Reproduced from A. B. Levin, *Neurosurgery* **1(3)**:267 (1977), by permission of Ladd Research Industries, Inc.

42.5 TEMPERATURE MONITORING

Temperature is the simplest and most time-honored parameter monitored. Interpretation of trended data reveals the onset of sepsis, loss of humoral or metabolic control, specification of infection type and organ system, and efficacy of treatment.

The transducers used include mercury-in-glass, thermistor, and reverse-biased diode. Infrequently, thermocouple pairs may be employed. Monitoring of the anesthetized or hypothermic patient is usually accomplished with a rectal or esophageal probe coupled with a nontrending digital or analog meter. Nonacute patients are easily managed with under-the-tongue or rectal mercury thermometers. Many institutes have converted to the more rapid and less fragile hand-held digital probes. For accurate assessment of the critically ill, core temperature measurement via the thermistor on the distal end of the balloon-directed catheter (discussed previously) is indispensable. Owing to great disparity between skin and core temperature due to constriction or dilatation of peripheral vessels, the self-adhering surface contact probes are of limited use in the critical care setting but have found use in vascular research. Despite wide availability of well-constructed and accurate automated equipment, the measurement, recording, and hand trending of temperature data remain a prerogative of the nursing staff.

42.6 COMPUTERS

Computer utilization in the intensive care setting initially involved machine analysis of EKG patterns and patient information retrieval. Gradually, trend analysis of several physiologic parameters was incorporated and investigational disease diagnostic programs were developed. Most computerized units of this nature are based on a central time-sharing mainframe or a dedicated micro. Consequently the expense of purchasing, maintaining, and operating such a system limits use to well-endowed major medical centers. With the advent of the microcomputer, automation of even the smallest ICU is now possible. The relative ease of operation of the micro, the versatility of peripheral devices, and the amenability of the micro to custom programming make patient-computer interfacing attractive. A multitude of uses have been found for microcomputers, ranging from the setting up of pharmacopeia data bases to instructional programming for employees. Clinically, many dedicated units are utilized for data reduction, analysis, storage, and retrieval in conjunction with every form of monitoring device. Fully programmable devices allow customization of clinical programs on the basis of local need and have eliminated many previously time-consuming chores. An example of monitor-computer interfacing is the Automated Physiologic Profile, a widely employed clinical monitoring scheme. Data obtained from pressure monitoring, blood gas analysis, right heart catheterization, and hematologic indices is collated and entered at the bedside. Derived hemodynamic and pulmonary function indices are calculated and a bar-graph printout against a background of normal values is obtained. Comparison of serial printouts provides visualization of clinical trends previously unavailable (Fig. 42.7).

42.7 SAFETY

Owing to the adverse electrical and physical environment of the typical hospital, design safety for the protection of operator and patient is imperative. Within the physical plant, unequal ground resistances and ground loops exist on power-line runs. Frame grounding of beds and equipment is dubious, and spillage of conductive fluids is commonplace.

Microshock is the major hazard encountered, with the potentially fatal complication of ventricular fibrillation the result. A current of 10 to 20 μA at 50 to 60 Hz is sufficient to induce ventricular fibrillation if applied directly to the myocardium. Patients undergoing open-heart surgery leave the operation with pacemaker wires attached to the epicardium and passing through the skin of the thorax with uninsulated terminations. The lethal potential of these wires is evident.

Patients with indwelling catheters for pressure measurement often have salt-containing solutions filling the tubing connecting the catheters to the transducers. Improper grounding with leakage current along the tubing may result in current injection to the bloodstream. Front-end isolation may be achieved with optical coupling, inductive isolators, or resistor-diode bridges. A similar situation occurs with externally applied macroshock, the result of ground loops or induced currents. Here, however, the threshold levels for ventricular fibrillation are increased, to approximately 10 mA. Burns may occur at levels of only 100 mA. There are few survivors when the delivered current exceeds 5 A. As a patient's skin resistance may vary from several ohms to greater than 10 kΩ, depending on the current path and humidity on the skin, the voltage needed to produce these currents may be from several volts to several thousand volts.

Unavoidable wiring deficiencies and poor shielding are present in most multistoried hospital environs. In addition to inherent construction and design problems, abuse, poor maintenance, and

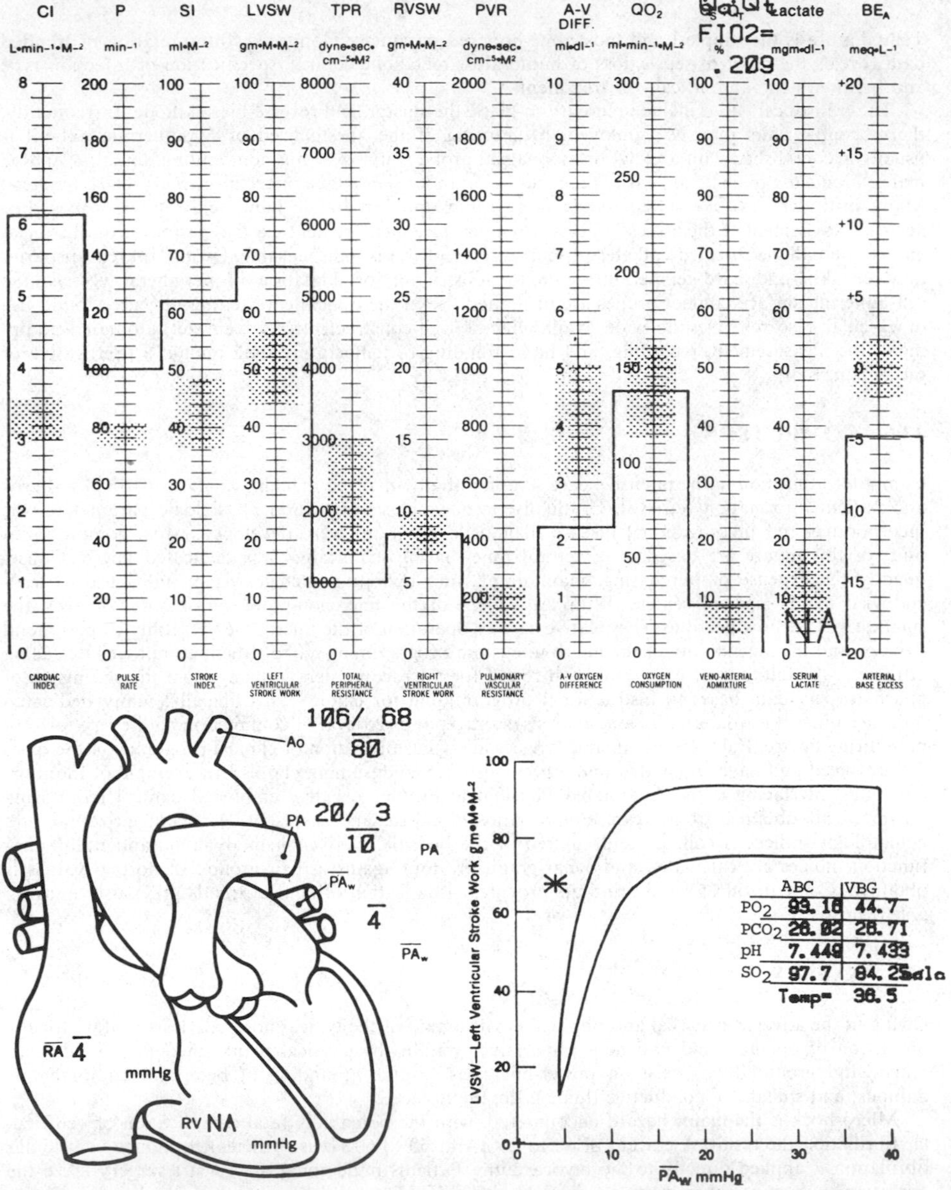

Fig. 42.7 Automated physiologic profile.

ignorance of equipment faults may lead to disaster. Rigid adherence to common isolation and grounding practice is required. Ground-fault interrupters of the inductive variety should be employed. Regular inspection schedules and education of personnel in the proper use of equipment and grounding faults are a necessity.

Apart from safety considerations, poor grounding and/or shielding may result in decreased signal-to-noise ratios and readout interference. A variety of portable equipment (x-ray, ultrasound) can engender large current surge-voltage spike transients. Survival of poorly engineered equipment in this setting is low. Interruption of microprocessor function is frequent. Line input filters and constant voltage transformers combined with surgistor insertion across the line will help alleviate the problem.

Bibliography

Automated Physiological Profile-User's Manual, Life Sciences Inc., 1981.

Baker, L. E., "A Rapidly Responding Narrow-Band Infrared Gaseous CO_2 Analyzer for Physiological Studies," *IRE Trans. Biomed. Electr.* **24**:16–24 (1961).

Blackburn, J. P. and T. R. Williams, "Evaluation of the Datex CD-101 and Codart Capnograph Mark II Infra-Red Carbon Dioxide Analysers," *Br. J. Anaesth.* **52**:551–555 (1980).

Blumenfeld, W., et al., "On-Line Respiratory Gas Monitoring," *Computers Biomed. Res.* **6**:139–149 (1973).

Bruner, J. M. R., L. J. Krenis, J. M. Kunsman, and A. P. Sherman, "Comparison of Direct and Indirect Methods of Measuring Arterial Blood Pressure," *Med. Instrumen.* **15**:11–21 (Jan. /Feb. 1981).

Bruner, J. M. R., L. J. Krenis, J. M. Kunsman, and A. P. Sherman, "Comparison of Direct and Indirect Methods of Measuring Arterial Blood Pressure, Part II," *Med. Instrumen.* **15**:97–101 (Mar./Apr. 1981).

Bruner, J. M. R., L. J. Krenis, J. M. Kunsman, and A. P. Sherman, "Comparison of Direct and Indirect Methods of Measuring Arterial Blood Pressure, Part III," *Med. Instrumen.* **15**:182–188 (May/Jun. 1981).

Collier, C. R., et al., "Continuous Rapid Infrared CO_2 Analysis," *J. Lab. Clin. Med.* **45**:526–539 (1955).

Eberhard, P. and W. Mindt, "Reliability of Cutaneous Oxygen Measurement by Skin Sensors with Large Size Cathodes," *Acta Anaesth. Scand.* (*Suppl.*) **68**:20–27 (1978).

Eberhard, P. and W. Mindt, "Interference of Anaesthetic Gases at Skin Surface Sensors for Oxygen and Carbon Dioxide," *Crit. Care Med.* **9**:717–720 (1981).

Ferris, C. D., "Gaseous Oxygen Monitor and Instrument for Evaluating Polarographic O_2 Electrodes," *ISA Biomed. Sci. Instrum.* **18**:99–102 (1982).

Ferris, C. D., *Introduction to Bioinstrumentation*, Humana, Clifton, NJ, 1978.

Ferris, C. D. and D. N. Kunz, "Design Considerations for Oxygen-Sensing Electrodes," *ISA Biomed. Sci. Instrum.* **17**:103–108 (1981).

Finer, N. N., "Newer Trends in Continuous Monitoring of Critically Ill Infants and Children," *Ped. Clin. N. America* **27**:553–566 (1980).

Finial, B. R., "Indirect Monitoring of Blood Pressure," *Anesth. Analges.* **49**:204–210 (1970).

Flitter, M. A., Techniques of Intracranial Pressure Monitoring, in M. Weiss, ed., *Clinical Neurosurgery*, Williams & Wilkins, Baltimore, 1981.

Fowler, K. T., "The Respiratory Mass Spectrometer," *Phys. Med. Biol.* **14**:185–199 (1969).

Graham, G. and M. A. Kenny, "Performance of a Radiometer Transcutaneous Oxygen Monitor in a Neo Natal Intensive Care Unit," *Clin. Chem.* **26**:629–633 (1980).

Guide to Physiological Pressure Monitoring, Hewlett-Packard Co., 1977.

Hebrank, D. R., "Noninvasive Transcutaneous Oxygen Monitoring," *J. Clin. Eng.* **6**:41–47 (1981).

Hill, A. and J. J. Volpe, Measurement of Intracranial Pressure Using the Ladd Intracranial Pressure Monitor," *J. Pediatr.* **98**:974–976 (1981).

Kerr, D. R. and I. V. Malhotra, "Electrical Design and Safety in the Operating Room and Intensive Care Unit in Anesthetic Considerations in Setting Up a New Medical Facility," *Internat. Anesth. Clin.* **19**:27–47 (1981).

Levin, A. B., "The Use of a Fiberoptic Intracranial Pressure Transducer in the Treatment of Head Injuries," *J. Trauma* **17**:767–773 (1977).

Lubbers, D. W., "Theoretical Basis of the Transcutaneous Blood Gas Measurements," *Crit. Care Med.* **9**:721–733 (1981).

Murray, I. P., "Complications of Invasive Monitoring," *Med. Instrum.* **15**:85–89 (1981).

Olsson, S. G., et al., "Clinical Studies of Gas Exchange during Ventilatory Support—A Method Using the Siemens-Elema CO_2 Analyzer," *Br. J. Anaesth.* **52**:491–499 (1980).

Osborn, J. J., "Cardiopulmonary Monitoring in the Respiratory Intensive Care Unit," *Med. Instrum.* **11**:278–282 (1977).

Peabody, J. L., et al., "Clinical Limitations and Advantages of Transcutaneous Oxygen Electrodes," *Acta Anesth. Scand.* (*Suppl.*) **68**:76–82 (1978).

Powner, D. J. and J. V. Snyder, "In Vitro Comparison of Six Commercially Available Thermodilution Cardiac Output Systems," *Med. Instrum.* **12**:122–127 (Mar./Apr. 1978).

Rafferty, T. D., et al., "In Vitro Evaluation of a Transcutaneous CO_2 and O_2 Monitor: The Effects of Nitrous Oxide Enflurane and Halothane," *Med. Instrum.* **15**:316–318 (1981).

Russel, R. O. Jr. and C. E. Macklet, *Hemodynamic Monitoring in a Coronary Intensive Care Unit*, Futura, Mount Kisco, NY, 1974.

Swan, H. J., W. Ganz, et al., "Catheterization of the Heart in Man with the Use of a Flow-Directed Balloon Tipped Catheter," *N. Engl. J. Med.* **283**:447–451 (1970).

Tremper, K. K. and W. C. Schowmaker, "Transcutaneous Oxygen Monitoring of Critically Ill Adults, with and without Low Flow Shock," *Crit. Care Med.* **9**:706–709 (1981).

Turney, S. Z., et al., "Automatic Respiratory Gas Monitoring," *Ann. Thorac. Surg.* **14**:159–172 (1972).

Vale, R. J., Monitoring of Temperature During Anesthesia in Anesthetic Considerations in Setting Up a New Medical Facility, *Internat. Anesth. Clin.* **19**:61–83 (1981).

Wilson, F., ed., *Critical Care Manual*, Upjohn, 1977.

CHAPTER 43
PROSTHETICS

HAROLD Z. HAUT

Medical Systems Corporation, Greenvale, New York

HARRY LEVITT

State University of New York, New York

WILLIAM LEMBECK

New York University Post Graduate Medical School, New York

43.1 IMPLANTS

Harold Z. Haut

43.1-1 Introduction

A wide variety of medical prostheses are being implanted in hundreds of thousands of people worldwide each year. These devices perform many roles, from relieving pain, restoring functions lost due to injury or disease, to ultimately extending life.

Nearly half a million people are alive today because their deficient heart is aided by a heartpacer. Close to 300,000 people who would be crippled and in constant pain from arthritis can walk with ease owing to their artificial hip joints. Patients with cataracts can see because of implanted intraocular lenses, and victims of arteriosclerosis have their circulation improved by vascular grafts.

Other implants are being developed to prevent fatal arrhythmias, to replace the entire human heart, to help some deaf persons to hear, to help the paralyzed move muscles, to pump drugs where they are most needed, to stimulate the growth of bone, to help heal difficult fractures.

Most of these implants were practically unknown before 1950. With the exception of dental implants, few were used before 1960. New developments in materials science, along with the electronics revolution of the last few decades, have made these devices possible.

This chapter discusses some electronic implants that have seen widespread human use. Table 43.1 lists several of the more successful implants and their applications.

TABLE 43.1. SOME ELECTRONIC IMPLANTS AND THEIR USE

Device Name	Application
Automatic implantable defibrillator (AID)	Defibrillate arrhythmic heart
Bladder stimulator	Restore lost bladder control
Bone growth stimulator	Heal difficult fractures
Cardiac pacer	Correct arrhythmia in heart
Cerebellar stimulator	Control spasticity in cerebral palsy; control seizures in epilepsy; control severe psychosis
Deep brain stimulator	Control pain
Diaphragm pacer	Restore lost breathing control
Peripheral nerve stimulator	Control pain
Peroneal nerve stimulator	Improve gait in hemiplegics
Scoliosis stimulator	Improve scoliotic curve
Spinal cord stimulator	Control pain Control spasticity Control systems of multiple sclerosis

43.1-2 Basic Principles

System Organization

The basic rationale for the implantation of an electronic device is the performance of some function within the body without the need for bringing wires through the skin. All implant systems with few exceptions have an external means of control. Command information and in some cases operating power are transmitted across the intact skin to an implanted electronic package. Radio-frequency energy, magnetic fields, and ultrasonic acoustic energy are used as carriers. Modulation schemes include analog pulse-amplitude, pulse-width, and a variety of digital pulse-code schemes. In some sophisticated implants, telemetry of implant system status or of physiologic parameters is provided from the implant to the outside world. Figure 43.1 depicts a block diagram of a typical implant system.

The majority of electronic implants work by delivering an electrical stimulus to excitable tissue such as nerve or muscle. The stimulus with few exceptions is a rectangular pulse of current that is capacitively coupled from a pulse generator to tissue-interface electrodes. Although there is some question about the energy efficiency of a rectangular pulse as opposed to, for example, an exponentially increasing waveform, the ease with which rectangular pulses can be electronically generated has made them almost standard.

Electrophysiology of Stimulation

The mechanism by which an electrical stimulus affects muscle or nerve tissue is depolarization of the target tissue cell membrane. Muscle or nerve cells normally maintain a resting potential difference across their cell membrane of approximately -70 to -90 mV (intracellular to extracellular). When the muscle cell contracts or the nerve cell fires, this potential difference is momentarily brought toward zero, concomitantly with the firing. These cells can be artificially induced to fire by changing the potential difference across the membrane to a threshold level, usually a value closer to 0 than the negative resting potential. This is normally accomplished with an extracellular electrode which, on receiving a cathodic pulse, floods its vicinity, the target tissue, with negative ions. The presence of extracellular negative ions counterbalances the normal intracellular negative potential to reduce the transmembrane potential to the threshold, firing the cell. Which cells in a nerve or muscle are fired depends on their proximity to the electrode and their position in the electric field created by the electrodes. For a detailed discussion of this subject see Ranck.[1]

Stimulus pulse parameters, such as pulse duration, pulse amplitude, and pulse timing, vary depending on application. The pulse width can vary from 50 μs to several ms, the amplitude from less than 1 mA to nearly 30 mA, and the repetition rate from 1 Hz (cardiac pacing) to 1500 Hz (spinal cord stimulation for spasticity). Some applications also require that the stimulus be turned on and off for equal or unequal periods lasting seconds, minutes, or hours. A typical capacitively coupled waveform is depicted in Fig. 43.2.

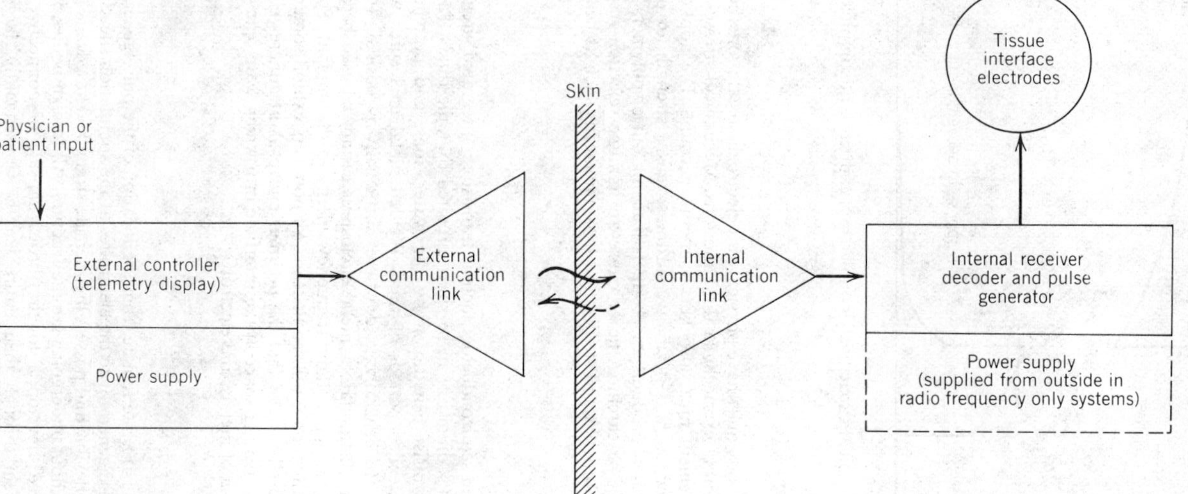

Fig. 43.1 Block diagram of implant system with external control and telemetry.

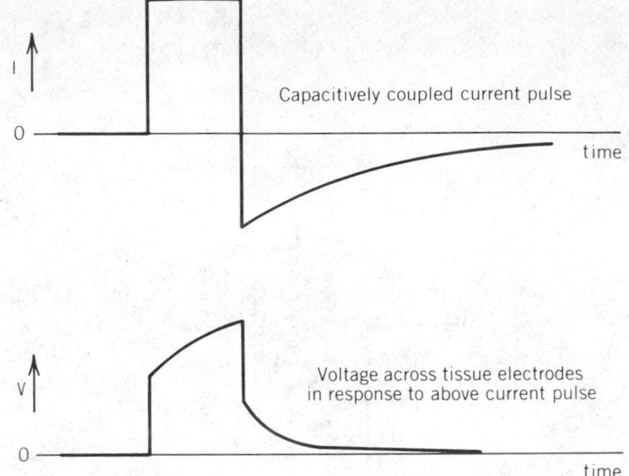

Fig. 43.2 Typical capacitively coupled constant current stimulus pulse.

Materials Considerations

Materials used within the body must have physical and electrochemical properties able to withstand a harsh corrosive environment of extracellular fluid and ions. Mechanical stresses associated with body movement or organ movement must also be taken into account. The application will determine physical properties required, such as flexibility, hardness, tensile strength, or permeability to water. Metal corrosion or dissolution products, or products leaching out of polymers, must not be toxic to tissue. They must also not be carcinogenic or cause an immune rejection response. All implanted components must be able to withstand sterilization by some means such as heat, ethylene oxide gas, or irradiation.

Packaging

Electronic circuits based on semiconductors such as transistors, integrated circuits, or hybrids can be destroyed by ions such as sodium or can be short circuited by ionic fluids. Packaging materials for electronic implants must resist penetration by the water, ions, and proteins that make up the extracellular fluid, for extended periods of time. Welded metal enclosures with glass-metal feed-throughs for conductors are generally used where hermeticity is desired. In externally powered implants, such as some neurostimulators, metal enclosures cannot be used because the metal would excessively attenuate the transfer of radio-frequency energy to the implant. Polymeric encapsulants such as epoxy or silicone rubber are utilized instead. All polymers exhibit some permeability to water (see Table 43.2). Care must be taken that the polymer encapsulant adheres closely to the encapsulated electronics, to avoid creating space at the interface into which water can pool and cause shunt conduction paths, corrosion of metals, and electrolysis.

Leads and Electrodes

Stimulation is usually done at a site remote from the location of the pulse generator. Lead wires must be used to connect the pulse generator to the electrodes at the stimulation site. Subcutaneously placed lead wires must be highly flexible and possess high tensile strength. These leads must move freely with repetitive body motion without undergoing fatigue fracture. All foreign bodies, including leads placed in the subcutaneous space, develop a tough fibrous capsule of connective tissue which tightly envelops and anchors them through their entire length. Consequently, large tensile stresses can develop in leads with extreme bending or stretching of the body.

Endocardial cardiac pacing leads, on the other hand, must have a minimum stiffness factor, which will allow them to be advanced into the heart through a venous route. They must have a conductor and insulation that can withstand some 100,000 flexions a day, or one for each heartbeat.

Lead wire conductors must exhibit good biocompatibility in the event that the insulation that normally isolates them from extracellular fluid is broken. Passive corrosion and the tissue toxicity of the corrosion products cause the exclusion of conductors such as silver and copper from chronic use

TABLE 43.2. MOISTURE VAPOR TRANSMISSION OF SEVERAL POLYMERS

Polymer (2-mil Film)	Moisture Vapor Transmission (Measured at 30.8°C and 90–98% RH) (g-mil/100 in.^{2}-24 hr)
Polymonochlorotrifluoroethylene	0.02
Ethylene/vinyl acetate	0.20
Polyvinylidene chloride	1
Paraxylylene	1
Epoxy	2
Polypropylene	0.7–3.0
Polytetrafluoroethylene	3
Polyurethanes[a]	2–9
Polycarbonate	10
Polyethylene	21
Polymethyl methacrylate	35
Polyethylene terephthalate	48
Polystyrene	120
Cellophane	134
Silicone rubber	170

Source. From M. Szycher and W. J. Robinson, *Synthetic Biomedical Polymers*, Technomic, Lancaster, PA, 1981, p. 193, with permission.
[a] Depends on type.

in implanted leads. Presumably lead, which is a component of solder, would also have toxic corrosion products. Gold, platinum, iridium, tantalum, stainless steel, and platinum-iridium alloys are all considered nontoxic as passive implants. Under pulsed biphasic (capacitively coupled) stimulation conditions, platinum, iridium, and rhodium were found to be more corrosion resistant than gold, which was more resistant than stainless steel. Platinum and various stainless-steel alloys are widely used as conductors. Pure platinum has excellent biocompatibility but is a soft, malleable metal. It is usually alloyed with 10 to 20% iridium to improve mechanical properties when used in leads or electrodes. Conductors are constructed as large bundles of fine stainless-steel fibers, braids, or spring coils wound around a central polymeric tensile element. Silicone rubber, polyurethane polyester, and Teflon have been used as insulation in flexible leads.

Tissue electrodes form the interface that conducts current from metal conductors to extracellular fluid, which bathes all target tissue. Current flow within metal conductors is by the movement of electrons, in tissue by the movement of ions. The transfer of charge between a metallic conductor and an ionic conductor can occur by a capacitive mechanism of charging, a so-called Helmholtz double layer, or by oxidation reduction reactions, which may or may not be reversible. Reversible reactions are those in which the reaction products are immobilized at the electrode surface. Irreversible reactions generate new, potentially toxic chemical species that leave the electrode. To prevent tissue damage due to electrochemical byproducts, irreversible reactions must be avoided.

The impedance of the electrode in tissue can be modeled by an *RC* network, as shown in Fig. 43.3. The value of the lumped parameter of each component of the model depends on the electrode material, the electrode surface area, the spacing of the electrodes, and the character of the tissue near the electrodes.

Only balanced biphasic waveforms, which have no net direct current component, can achieve current flow utilizing primarily reversible reactions. Lilly[2] and Brummer and Turner[3] studied the electrochemical limitations of the most common electrode material, platinum. They reported the gassing limit for cathodically pulsed platinum electrodes as being a current density of approximately 400 μC per real square centimeter per phase. This is the current density above which electrolysis and gas evolution occur. Real surface area of an electrode is the geometric surface area multiplied by a factor that takes into account the surface roughness. For moderately polished platinum electrodes, real surface area is approximately 1.4 times geometric surface area. The limit for the production of oxidation products of Cl^- ions is even higher than the gassing limits. Most stimulation is done at charge densities well below the stated gassing limit. Some dissolution of platinum metal into platinum

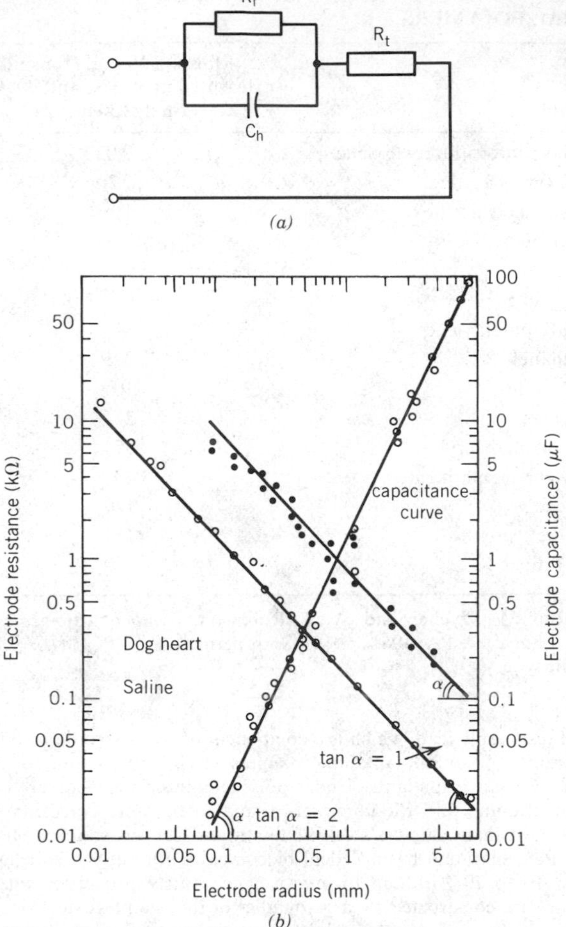

Fig. 43.3 (*a*) Simplified lumped parameter model of a tissue electrode interface equivalent circuit. R_t is the tissue resistance, C_h the Helmholtz capacitance, and R_f the Faraday resistance. R_f is usually large and can be ignored. C_h and R_t are related to the surface area of the electrode. (*b*) Plot of the C_h and R_t values as a function of the radius of a hemispherical cardiac pacing electrode in vivo and in vitro. Courtesy of Pacesymp, Montreal, Quebec, Canada.

ions occurs at charge densities as low as 10 μC per real square centimeter per phase. It is not clear whether this has any long-term harmful effects.

43.1-3 Cardiac Pacing

The cardiac pacer is the most successful and best known of the implantable electronic devices. Hundreds of thousands of pacers, manufactured by over two dozen companies, are implanted annually. In 1980 the worldwide pacer market was $600 million; it is expected to grow to $757 million by 1985. The modern era of implantable cardiac pacers started in 1958 when the first nickel cadmium battery-powered pacer was implanted in Sweden by Elmquist and Senning.[4] Schechter[5] has reviewed the history of electrical stimulation of the heart leading to the development of cardiac pacing.

A normal heart-pumping cycle starts with the filling of the right and left atria with blood from the vena cava and pulmonary vein, respectively. The atria then contract in response to a stimulus originating from the sinoatrial (SA) node, a natural pacemaker located at the top of the right atrium. Blood is ejected from the atria into the right and left ventricles through one-way valves. It is the contraction of the thicker-walled more muscular ventricles that provides most of the blood pressure required for circulation.

The timing of ventricular contraction is controlled by an intracardiac neural conduction system. During atrial contraction, the SA node timing stimulus is carried to another neural node on the opposite side of the atrium called the atrioventricular (AV) node. The AV node gives rise to a bundle of nerve tissue called the bundle of Hiss, which travels to the ventricles and subdivides into bundle branches to innervate both ventricles. It is the AV node, bundle of Hiss, and bundle-branch system that synchronize the ventricular contraction to occur at the correct time with respect to the atria, so that pumping efficiency is maximized. This cycle is repeated for each heartbeat. Figure 43.4 illustrates the natural intracardiac pacer and conduction system.

The generation of the natural pacing stimulus, the conduction of that stimulus, and the contraction of cardiac muscle in response to the stimulus are all events that are marked by an electrical depolarization of muscle or nerve cell membranes. Cardiac muscle and nerve are composed of thousands of individual cells, many of which depolarize at a given time in the cardiac cycle. This mass multicell depolarization results in an electrical event that, when carried to the skin surface through the volume conductor of ionic tissue fluid, results in a recordable surface electrocardiogram (ECG). The ECG is composed of components that are related to the various cardiac timing events. It is routinely used for diagnosis of heart disease. Figure 43.5 illustrates an ECG and its relation to cardiac cycle events.

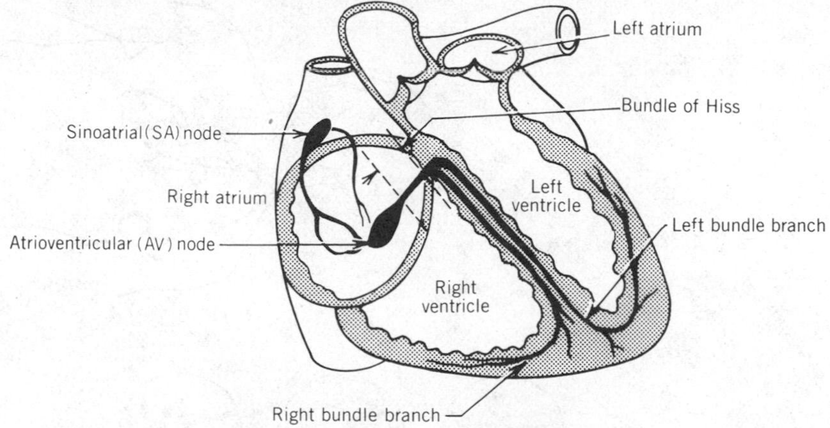

Fig. 43.4 Normal cardiac conduction system.

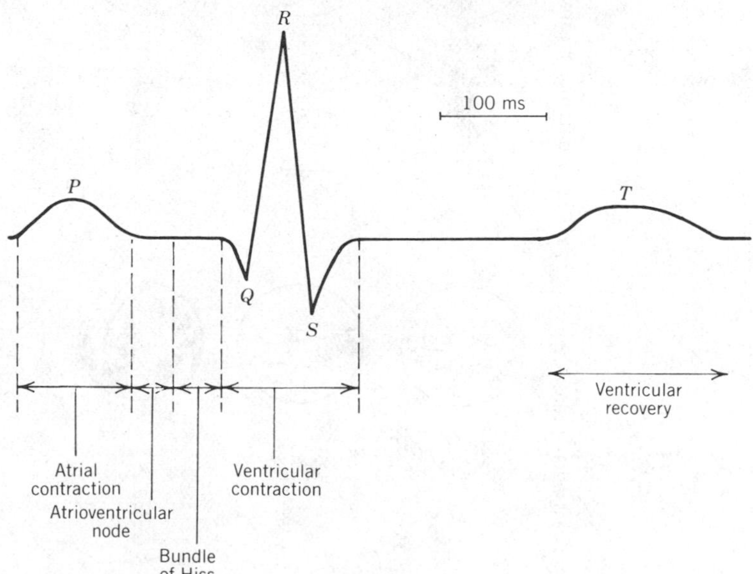

Fig. 43.5 Surface electrocardiogram and its relationship to the timing of cardiac events.

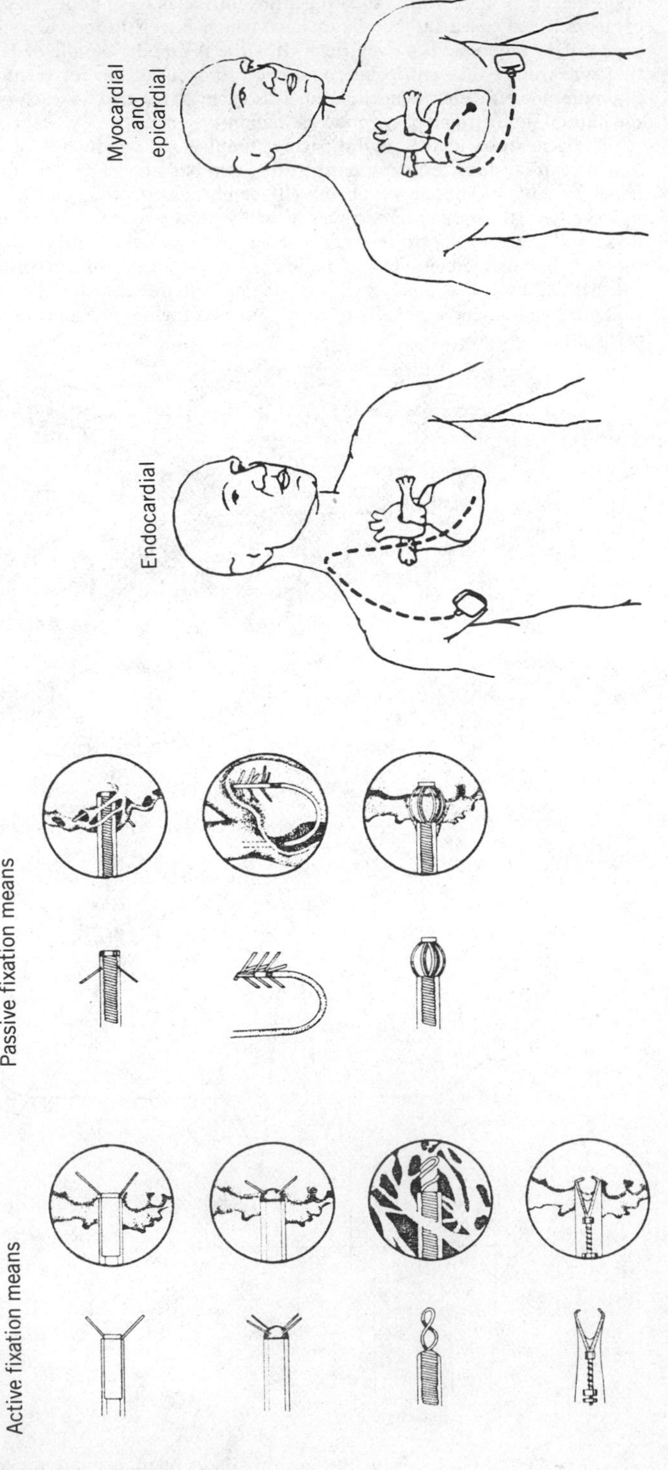
Active fixation means
Passive fixation means
Endocardial
Myocardial
and
epicardial

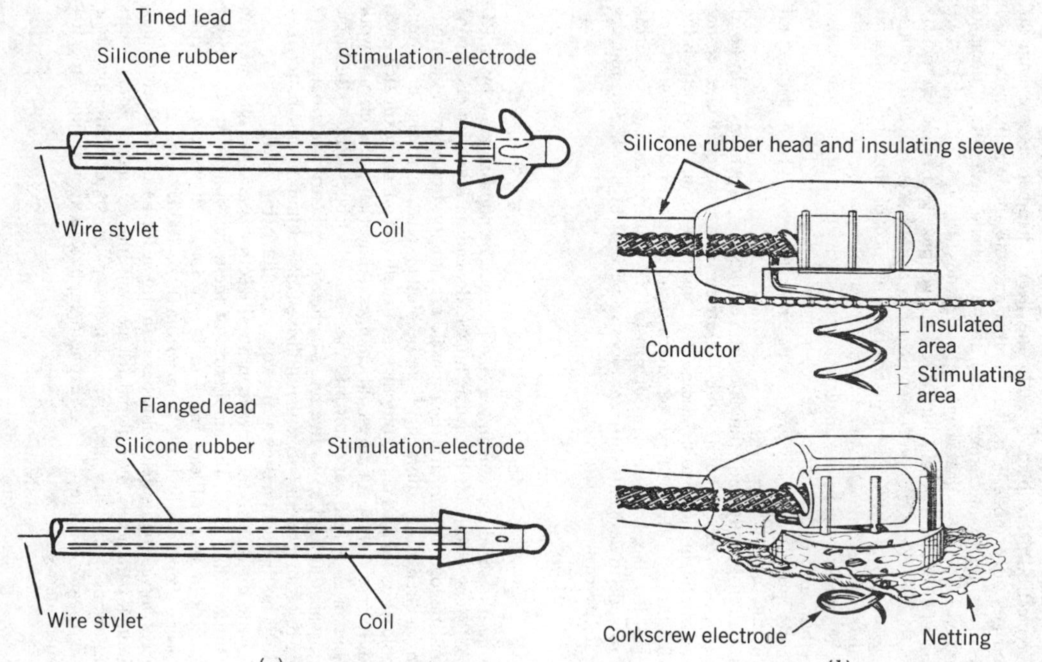

Fig. 43.6 Various styles of cardiac pacing electrodes and their position during use. Courtesy of Pacesymp, Montreal, Quebec, Canada.

Electronic cardiac pacers are primarily used to treat defects in the natural pacer or its conduction system. Electronic stimuli are substituted for irregular or absent natural stimuli. One such defect, for which pacers saw some of their earliest use, is atrioventricular block, or Stokes-Adams syndrome. This condition is characterized by normal SA node operation and concomitant atrial contraction but irregular conduction of the stimulus to the ventricles. In this case the electronic pacer delivers a substitute stimulus to an electrode located at the ventricle. Other heart rhythm defects have also been treated with heart pacing, but a discussion of these is beyond the scope of this chapter.

Pacing electrodes or leads can be surgically attached to the surface of the heart or more conveniently can be placed through a large vein into the heart chamber that is to be paced. Today the majority of cases are done using the latter, or endocardial, technique. Bipolar configurations are used, where both output or sensing electrodes are placed within the heart chamber, as are unipolar configurations, where one electrode, the cathode, is in the chamber and the anode is the pacer case itself. Bipolar electrodes are less sensitive to interference when used as sensing electrodes in demand cardiac pacer applications. Figure 43.6 depicts some pacing electrode leads.

Early cardiac pacers were simple fixed-rate oscillators that were coupled to a single ventricular electrode. In the later pacers a form of feedback control is utilized to adapt pacer function to more subtle rhythm defects. Sensing amplifiers are used to detect the presence or absence of natural cardiac events by measuring the ECG correlate of these events from intracardiac electrodes. In a simple demand cardiac pacer, the pacer would not stimulate unless it sensed an absent ventricular signal (QRS in Fig. 43.5), a certain time interval after the occurrence of the last QRS. Therefore, if natural rhythm is slower than the preset pacer rhythm, the pacer rhythm prevails; if the natural rhythm is faster, the natural rhythm prevails.

Other complex pacers sense atrial activity and stimulate the ventricles in synchrony with atrial contraction. Still other "physiologic" pacers stimulate both the atrium and ventricle while sensing activity in these two chambers and utilizing logic to determine the appropriate stimulus based on the sensed information. The proliferation of different types of pacers from different manufacturers and different pacing modalities, has led the Intersociety Commission for Heart Disease Resources (ICHD) to formulate a standard code to designate different types of pacers.[6] Figure 43.7 reproduces the recommended ICHD code for pacer nomenclature.

Circuits

Pacer circuits must deliver a suprathreshold stimulus into a load that can be approximated by an RC network, as shown in Fig. 43.3. Pacing thresholds in cardiac muscle depend on many factors, among them electrode surface area and biologic factors affecting muscle excitability. For a given electrode size and physiologic conditions, a strength-duration curve can be plotted, relating voltage or current and pulse duration to threshold (Fig. 43.8). Pacing outputs are usually set at approximately 50% above the measured threshold, to safely allow for threshold changes. Almost all pacer circuits work by the switching and discharging of a charged capacitor into the tissue electrodes (Fig. 43.9). In the case of early fixed-rate pacers, which utilized up to five series-connected mercury oxide batteries (7 V), a single capacitor was utilized. More modern pacers that utilize a single lithium-iodide battery (2.8 V) use output circuits with voltage multiplication such as the voltage doubler (Fig. 43.10).

Sensing circuits must be able to distinguish the P wave in atrial electrodes and the QRS wave in ventricular electrodes. In the intracardiac ECG, the amplitudes of these component waveforms are typically 2 to 6 mV for the P wave in an atrial lead and 7 to 14 mV for the R wave in ventricular electrode. The passband of sensing amplifiers typically ranges from 20 to 100 Hz.

In complex automated pacers, logic circuits determine pacer firing on the basis of sensed signals. Programmability of many stimulus parameters by external telemetry is also available. Some pacers utilizing complex microprocessor-based circuitry also can telemeter pacer status and physiologic parameters to an external interrogator. Figure 43.11 is a photograph of a typical multiprogrammable pacer with its programmer interrogator.

Power Sources

The first implanted pacer was a simple transistor blocking oscillator powered by a nickel-cadmium rechargeable battery. The battery was charged monthly with an external coil placed over the pacer implantation site. Aside from the need to recharge, these batteries were plagued by reliability problems. In the early 1960s, Greatbach and Chardack[7] reported the development of a pacer powered by mercury oxide cells that could last from two to four years. This development overshadowed the use of nickel-cadmium rechargeable batteries and other power sources, so that from 1960 to 1973 more than 95% of pulse generators used the mercury oxide batteries.

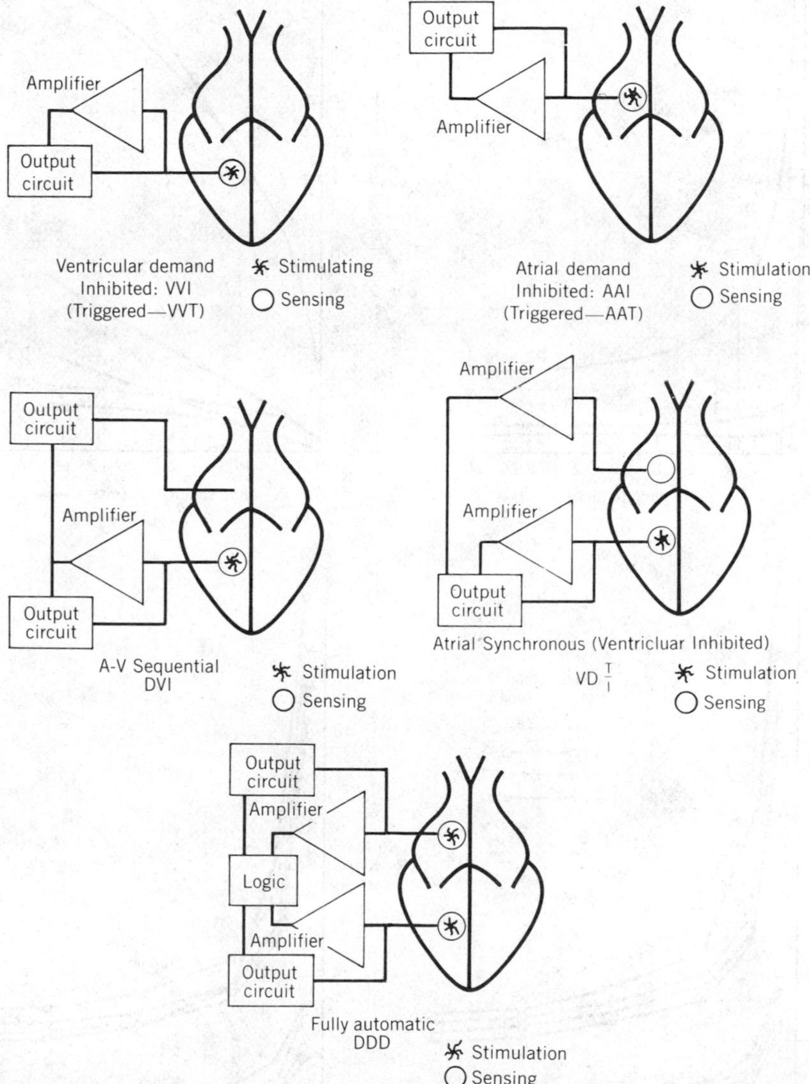

Fig. 43.7 Intersociety Commission for Heart Disease Resources (ICHD) code for standard pacemaker terminology to describe different pacer functional types. Reproduced, with permission, from the draft AAMI Pacemaker Standard, developed by the Association for the Advancement of Medical Instrumentation under contract with the US Food and Drug Administration (Contract no. 223-74-083).

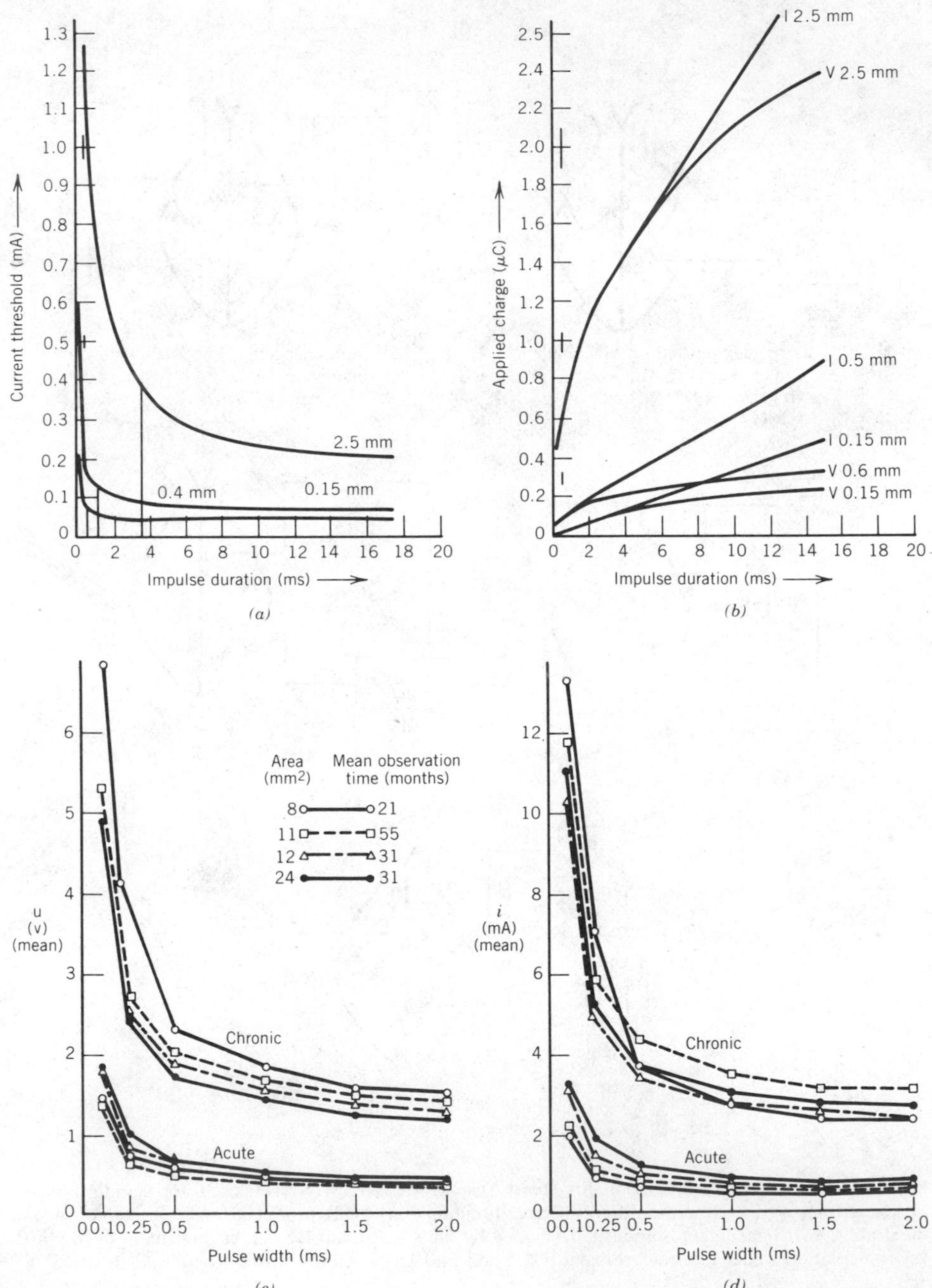

Fig. 43.8 Typical strength-duration curves for applied current (*a*) and voltage and charge (*b*), as a function of the surface area of cardiac-pacing electrodes. The change (increase) in threshold as a function of duration of implantation is shown in (*c*) and (*d*). Courtesy of Pacesymp, Montreal, Quebec, Canada.

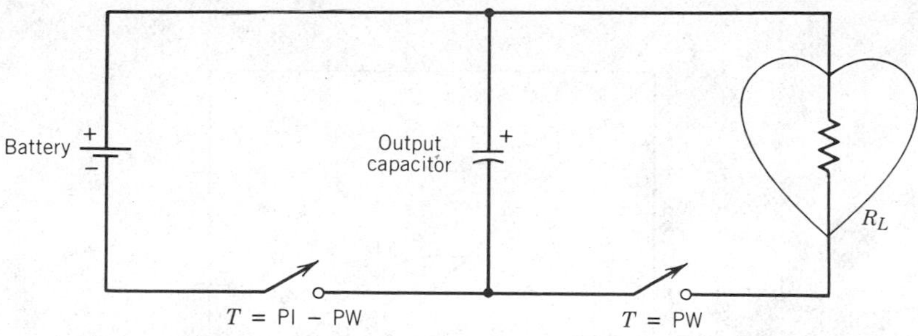

Fig. 43.9 Simple cardiac pacer output circuit based on the changing of a capacitor and its discharge across the tissue electrodes.

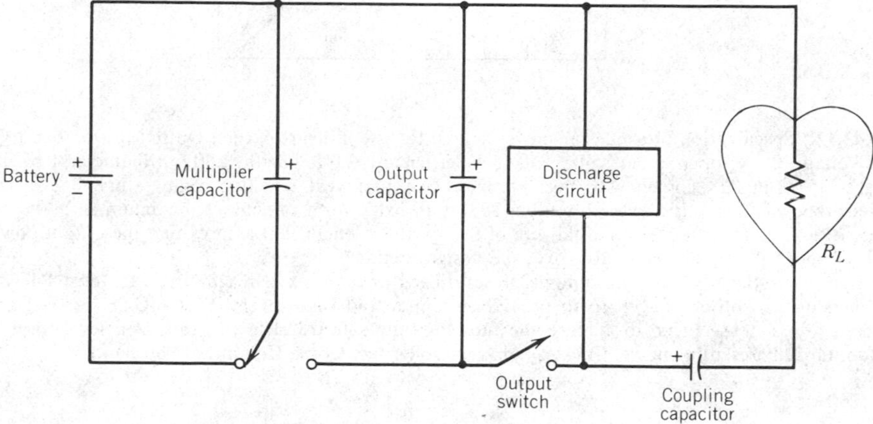

Fig. 43.10 Typical capacitive voltage-doubler cardiac pacer output circuit, used to boost output voltage in cardiac pacers utilizing a single battery cell such as a lithium-iodide cell. (R_L represents load.)

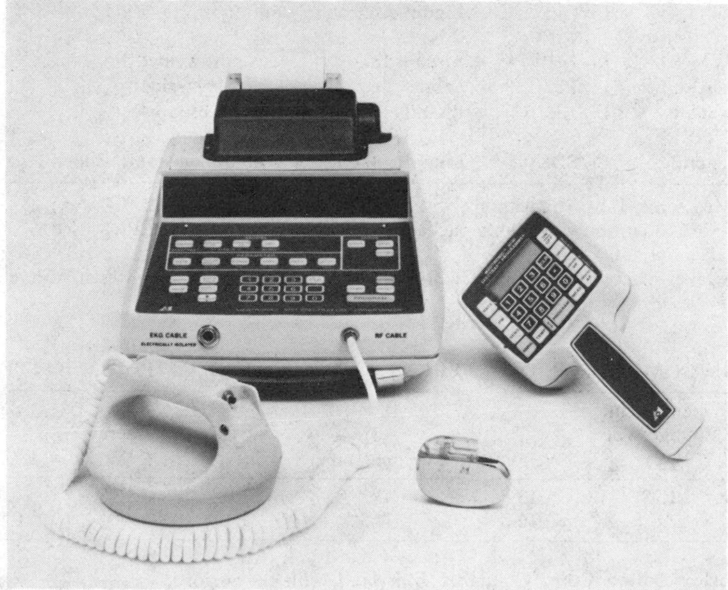

Fig. 43.11 Typical modern multiprogrammable heart pacer and its programmers. Courtesy of Medtronic, Inc., Minneapolis.

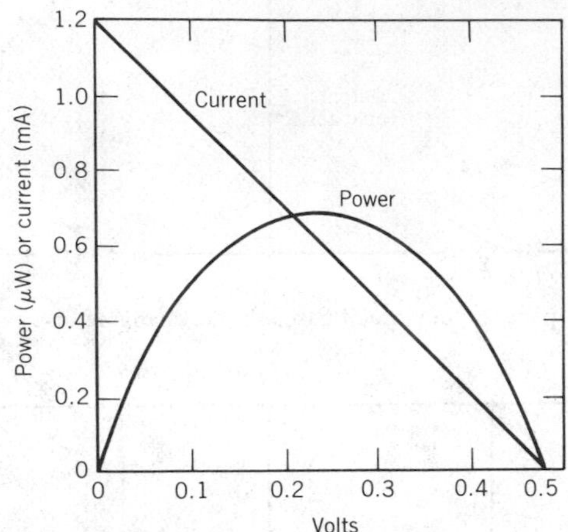

Fig. 43.12 Specification of a nuclear power source for use in heart pacers. Output power, 600 μW; load voltage, 0.3 V; open circuit voltage, 0.48 V; efficiency, 1.0%; length, 1.80 in.; diameter, 0.67 in.; weight, 35 g; Pu-238 content 140 mg; lifetime, 10 years; heat rejection temperature, 37°C. The radioisotope employed as fuel is plutonium-238 in an oxide form to achieve maximum stability and safety of use. Plutonium-238 has a half-life of 89.7 years, a length that assures that the output power of the device will remain very stable over the design lifetime.

High-purity (i.e., $> 90\%$) plutonium-238 is utilized and the oxygen employed to form PuO_2 is enriched in the isotope oxygen-16 to provide minimum radiation levels. The PuO_2 is pressed and sintered into a dense pellet to achieve the radiologically safest and lowest radiation level form of plutonium-238 presently known. Courtesy of Nuclear Battery Corp., Columbia, Maryland.

TABLE 43.3. SEVERAL LITHIUM ANODE BATTERY SYSTEMS USED IN PACEMAKERS

Cathode	Symbol	Cathode Type	Package Type[a]	Voltage[b] (V)
Bromine	$Li/Br_2(PVP)$	Viscous fluid	Not available	3.5
Copper sulfide	Li/CuS	Liquid electrolyte	Button; crimp seal	2.1[c]
				1.8
Iodine (PVP)	$Li/I_2(PVP)$	Viscous fluid	Custom; welded	2.8
Lead iodide, sulfide	Li/PbI_2, PbS	Solid state	Cylinder; welded	1.9
Silver chromate	$Li/AgCrO_4$	Liquid electrolyte	Button; crimp seal	3.2[c]
				2.4
Thionyl chloride	$Li/SOCl_2$	Liquid cathode	Cylinder, custom; welded	3.6

Source. Medtronic, Inc., Minneapolis, with permission.
[a] The shape of the currently available package, followed by the type of sealing used.
[b] Nominal open-circuit voltage.
[c] Discharge takes place in two sequential stages, each characterized by a different voltage.

TABLE 43.4. COMPARISON OF A TYPICAL MERCURY AND A LITHIUM BATTERY

Battery[a]	Open Circuit Voltage, Voc (V)	Weight (g)	Volume (cm^3)	Capacity, C (AH)	Energy, E (WH)	E/cm^3	E/kg
Hg	1.38	13.6	3.2	1.0	1.38	0.43	101
LI	2.8	16.0	4.5	1.5	4.0	0.89	250

Source. Nuclear Battery Corp., Columbia, Maryland, with permission.
[a] Hg, mercuric oxide silver Mallory certified cell 317827; LI, lithium-iodide catalyst research 903.

Many power source types have been considered in the attempt to extend pacer lifetime. One type utilized continuous radio-frequency inductive coupling through the intact skin to supply energy to the pacer. Nickel-cadmium rechargeable systems with improved reliability and service life were built by Fischell.[8] Systems utilizing a biogalvanic process or the products of metabolism in a fuel cell were built with the hope of producing electric energy for the life of the patient. All of these systems proved to be either limited in output, unreliable, or nonbiocompatible.

A pacer powered by a radioisotope-based battery was first implanted in 1970. Nuclear power sources for pacers have been built utilizing two different technologies. One type utilizes plutonium 238 as the fuel and a series-connected array of thermocouples to produce electric power by the thermoelectric effect. The other utilizes promethium 147 to bombard a silicon *pn* photojunction with electrons to produce electricity by the beta-voltaic principle. Figure 43.12 describes the performance and specifications of a nuclear power source based on plutonium-238.[9] Although these nuclear power sources can produce electric power using a simple reliable process for a lifetime (the half life of plutonium 238 is 86 years), their utilization has been limited by cost and stringent governmental regulations regarding the dispensing, use, and disposal of nuclear materials.

Probably the most widely used energy sources today are primary batteries based on solid lithium anodes. These batteries, based on cells developed in 1968, have proven to be highly reliable high energy density power sources. Several types of cathode materials have been utilized, and new chemistries are being investigated in search of improved performance. Table 43.3 lists some lithium battery cathode materials that have been utilized in pacing batteries and their nominal open-circuit voltages. Table 43.4 compares the specifications of a widely used lithium-iodide battery and a mercuric oxide-silver battery.

Pacer lifetimes of ten years or more can be expected, not only because lithium batteries have larger energy densities than do mercury cells (see Table 43.4) but also because their chemistries allow them to be hermetically sealed. On the other hand, mercury oxide batteries evolve hydrogen gas, and consequently pacers utilizing them have to be encapsulated in a gas-permeable container such as epoxy. Modern pacers utilize low-power-consumption high-impedance circuitry such as the complementary metal oxide semiconductor (CMOS). Hermetic encapsulation is essential for reliable long-term operation.

43.1-4 Automatic Implantable Defibrillator (AID)

Ventricular fibrillation of the heart is an arrhythmic condition characterized by rapid, shallow, ineffective contractions of the ventricles. If unchecked, it is followed by a decline in blood pressure leading to death. Defibrillation can be achieved by application of a large discharge current across the heart tissue. This is often done with an external defibrillator, which can deliver as much as 400 J to paddle electrodes held across the chest.

The automatic implantable defibrillator (AID) is an electronic device programmed to monitor the heart continuously, recognize ventricular fibrillation, and when necessary, deliver corrective defibrillatory discharges to implanted cardiac electrodes. Its purpose is to protect selected patients from sudden death whenever and wherever they are stricken by lethal arrhythmia. The AID has the unique advantage of being permanently available to the patient at risk with no need for specialized personnel or facilities.

The first clinical version of the AID was encased in a titanium hermetic package weighing 250 g and having a volume of 145 ml. The electrodes were made from silicone rubber and titanium. One of them is located on a catheter pervenously introduced into the superior vena cava. The second electrode, in the form of a cup, is placed extrapericardially over the apex of the heart. The outside surface of the apical electrode is insulated to achieve optimum current distribution.

The AID delivers truncated exponential pulses of 25 J about 15 seconds after the onset of the arrhythmia, and recycles as many as three times if the first shock is ineffective; the energy of the third and fourth shocks can be increased to 35 J. The delivered energy is independent of the differing heart-electrode resistances. After the fourth shock, about 35 seconds of normal rhythm is required to reset a counter and to allow a full series of pulses to be delivered again. The device is powered by lithium batteries, which have a projected monitoring life of three years or a discharge capability of approximately 100 shocks.

The sensing system monitors the probability density function of the slope of ventricular electrical activity, determining the fraction of time spent by the differentiated input electrogram between two amplitude limits located close to the isoelectric line. The sensor recognizes ventricular fibrillation by the striking absence of isoelectric potential segments. This approach has a "passive mode of failure." A sensing malfunction would result in passive behavior of the device rather than in an unwanted shock resulting from a false-positive diagnosis of ventricular filbrillation. Figure 43.13 depicts the AID in position.

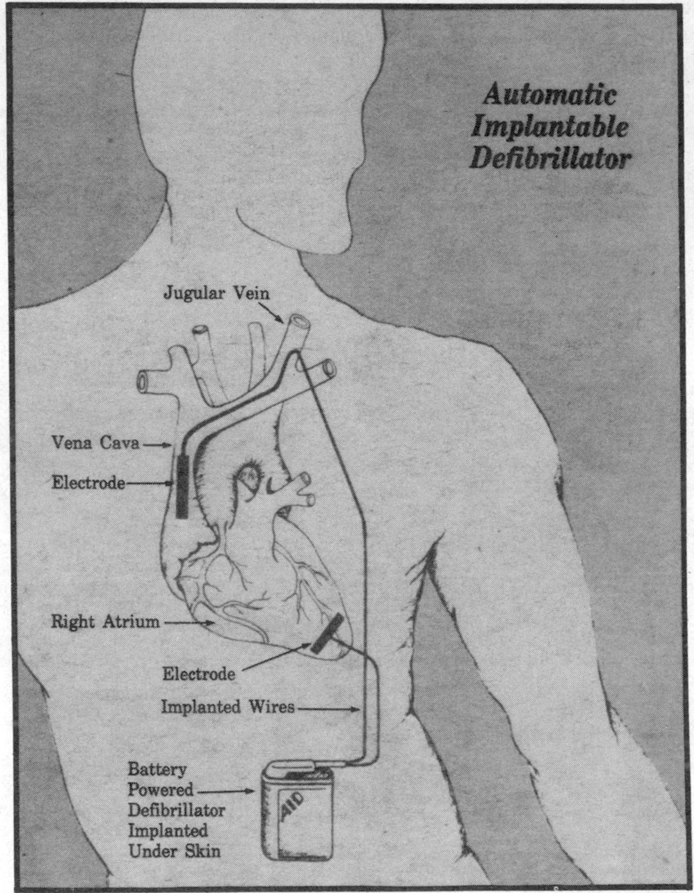

Fig. 43.13 Automatic implantable defibrillator. Courtesy of Newsday Magazine.

43.1-5 Control of Respiration by Electrical Stimulation-Diaphragm Pacer

The application of externally supplied electrical stimulation to the phrenic nerve as a means of restoring ventilation is not a new idea. As early as 1756 Shechter[10] caused diaphragmatic contractions in a dog whose phrenic nerve was excited with an electrostatic induction machine. In 1818 Ure[11] did the first recorded human demonstration by stimulating with a voltaic battery the exposed phrenic nerve of a "criminal who had hung on the gallows for one hour," and demonstrated diaphragmatic movement. The modern era of diaphragm pacing was inaugurated in 1948, when Sarnoff and associates[12,13] stimulated the phrenic nerve for prolonged periods with electrode wires inserted through the skin. Bringing wires through the skin for prolonged periods of time is impractical owing to the problem of infection. The first practical long-term diaphragm pacer was built in the early 1960s by Glenn[13] by adopting some of the radio-frequency-coupled technology previously developed for cardiac pacing. The diaphragm pacer (DP) in use today has evolved from the Glenn technique. It has been applied to chronically support ventilation in patients who lose neural control over the muscles of breathing. This includes patients with congenital defects or who have suffered trauma, stroke, or tumor in the brain stem respiratory control center. One such group of pacing candidates are those with Ondine's curse, or sleep apnea, a condition in which breathing is inadequate only during sleep. Another group that benefits from DPs are those with a spinal cord injury that disrupts the neural connection between the brain respiratory control center and the phrenic nerve that innervates the diaphragm muscle. In all cases, both the phrenic nerve and the diaphragm muscle must be intact for pacing to be usable.

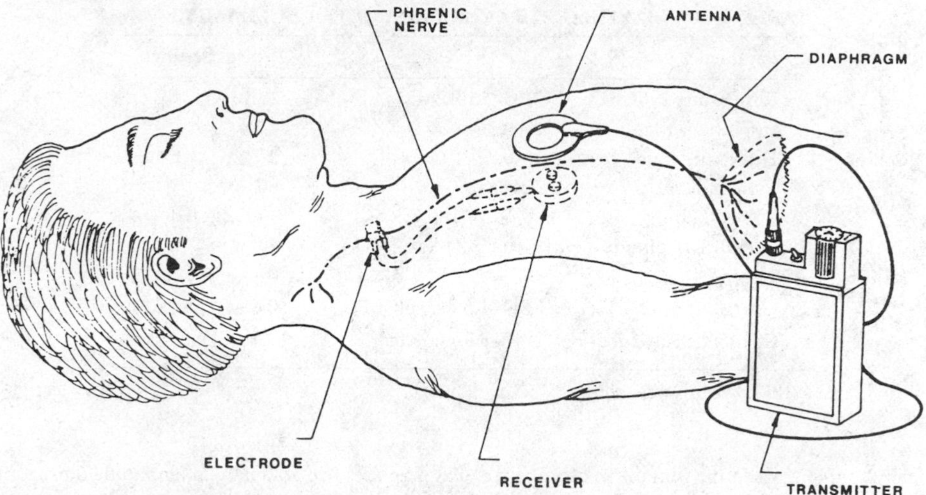

Fig. 43.14 Unilateral diaphragm pacer. Courtesy of Avery Laboratories, Inc., Farmingdale, New York.

The DP utilizes partially implanted and partially external radio-frequency-coupled technology. Figure 43.14 depicts a unilateral DP schematically. The external transmitter contains all the controls to adjust the stimulus delivered to the phrenic nerve, thereby allowing regulation of tidal volume, respiratory rate, and other parameters after implantation is completed.

Rhythmic contractions of the diaphragm are caused by delivery of timed trains of electrical-stimulating pulses to electrodes placed on the phrenic nerves. When stimulating pulses are delivered to the phrenic nerve, inhalation results. Expiration is passive and occurs due to elastic recoil of the

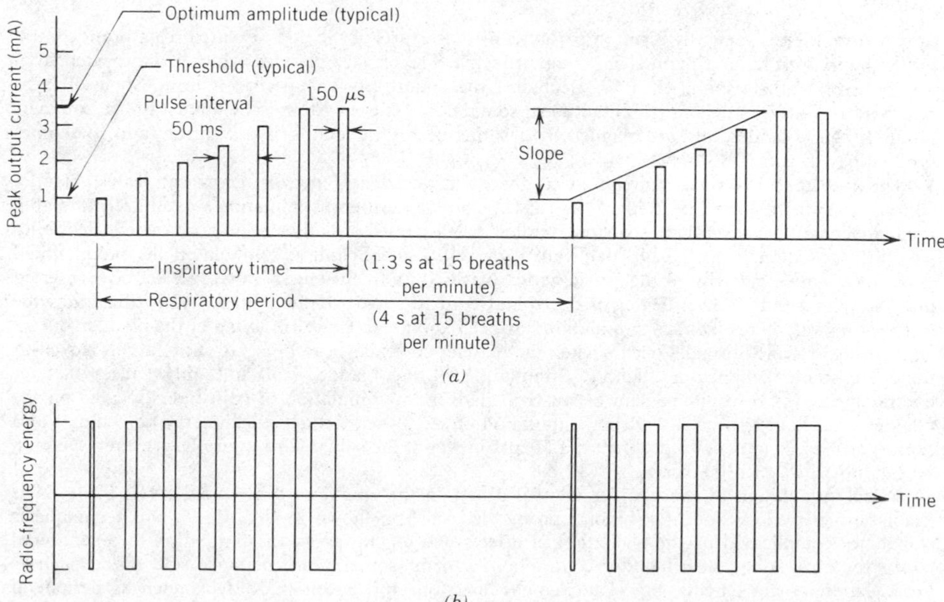

Fig. 43.15 Output waveforms of the diaphragm pacer receiver (*a*) and transmitter (*b*).

TABLE 43.5. DIAPHRAGM PACER PARAMETER SUMMARY

Parameter	Range	Typical Setting
Respiratory rate	5–55 BPM[a]	12 BPM
Inspiratory time	0.36–1.35 s	1.3 s
Expiratory time	0.72–10.65 s	2.7 s
Pulse interval	40–180 ms	60 ms
Pulse width	—	150 μs fixed
Stimulus amplitude	0–10 mA (0–peak)	4.0 mA
Electrode surface area	—	0.17 cm^2
Charge transfer	0–1.5 μC/phase	0.6 μC/phase
Charge density/phase	0–8.8 μC/cm^2	3.5 μC/cm^2

[a] BPM, breaths per minute.

chest and lung when the stimulus is off. The stimulating pulses rise gradually in amplitude during the inspiratory phase so that the diaphragm contracts in a graded quasiphysiologic manner.

A platinum electrode molded in a silicone rubber substrate is surgically implanted on one or both phrenic nerves. Each electrode is connected by silicone rubber-coated stainless-steel lead wires and connectors to a radio-frequency receiver, which is implanted in a subcutaneous thoracic pocket. The implanted receiver is a passive electronic package containing no batteries.

An external transmitter supplies both electrical power and stimulus information to the implanted receiver by radio-frequency electromagnetic coupling across the intact skin. A transmitting loop antenna is maintained over the receiver site for this purpose. The transmitter delivers a pulse-width-modulated 2.05 MHz carrier into the antenna. A tuned receiving coil in the implant links the transmitted energy to a hybrid circuit, which decodes the transmittted pulse width to a stimulus amplitude. Implant power supply energy is derived from the same transmitted pulse. The time occurrence of the stimulus is determined by the time occurrence of the transmitted signal. Figure 43.15 summarizes both the transmitted and the decoded waveforms. Table 43.5 summarizes stimulation parameters.

43.1-6 Control of Bladder Function

Spinal cord injuries generally result in paralysis of those parts of the body isolated from brain control by the injury, resulting in paraplegia or quadriplegia. The urinary bladder, which is innervated from the lower part of the spinal cord, is affected in the majority of spinal cord injuries. Some of these patients can void by utilizing still-intact local spinal cord reflexes. Many others must utilize catheters to empty their bladder, and are plagued by chronic urinary infections, hypertension, and sometimes fatal kidney disease.

The fact that electrical stimulation to the spinal cord and pelvic nerves can cause bladder contractions was reported by Budge[14] in 1854. In the modern era, stimulation was utilized to empty the bladder through several approaches. Bradley[15,16] described a bladder stimulator, where electrodes were attached directly to the bladder wall. Brindly[17,18] and Schmidt[19] stimulated the sacral spinal nerve roots which give rise to the pelvic nerve. Nashold and colleagues[20–23] used electrodes inserted into the micturition control center of the spinal cord to evoke voiding. Figure 43.16 illustrates the different points where electrical stimulation has been applied to cause emptying of the bladder.

Voiding of urine normally occurs when the bladder contracts in response to stimulation. However, there are some complicating factors. Stimulation of the bladder wall not only causes bladder contraction but also results in pain because of inadvertent stimulation of pain fibers located on the bladder wall. Pain also occurs during stimulation of sacral nerve roots, which carry both motor and sensory nerves. Pain is not a problem in patients in whom the spinal cord injury has severed the pain sensation pathways to the brain.

Direct stimulation of the bladder control center in the spinal cord at spinal levels S2 to S4 is accomplished by insertion of a bipolar prong electrode, as shown in Fig. 43.17. This technique is sometimes complicated by the elicitation of effects due to the spread of stimulation to other spinal cord structures. Along with bladder contraction, patients sometimes experience side effects such as lower extremity movements, bowel movement, erection, and autonomic effects such as peripheral vasodilation.

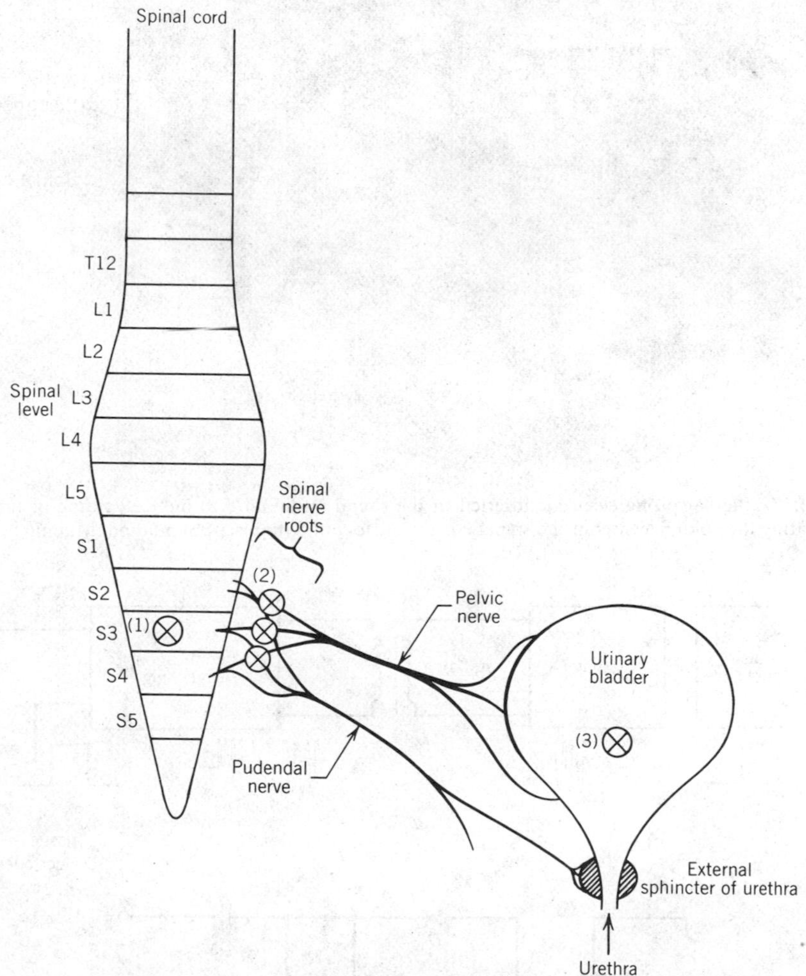

Fig. 43.16 Innervation of the urinary bladder and the location of stimulating electrodes (⊗) for spinal cord (1), spinal nerve root (2), and direct bladder (3).

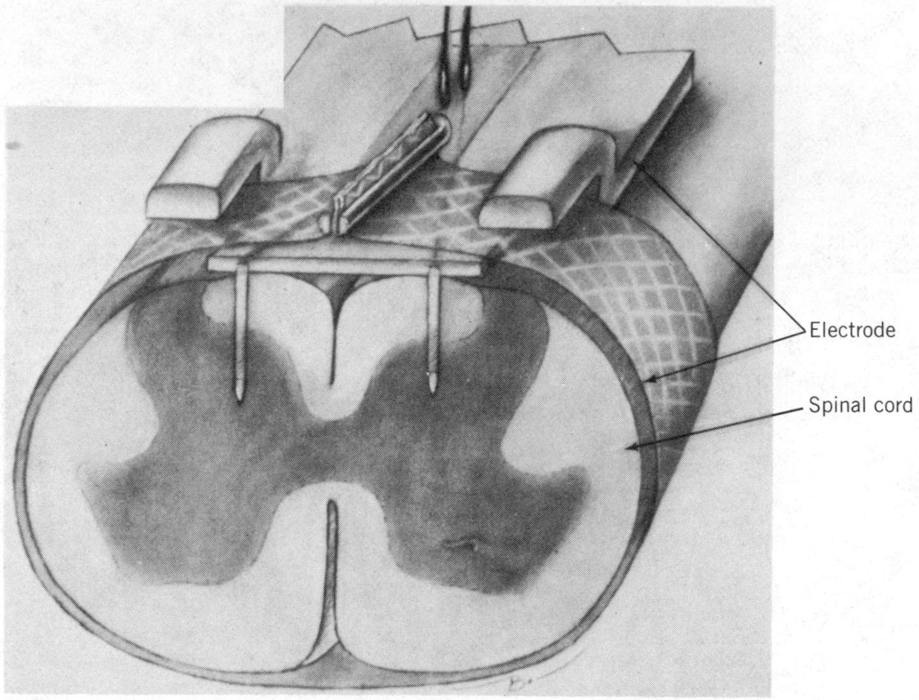

Fig. 43.17 Bipolar prong electrode inserted in the sacral spinal cord to induce voiding of urine, by stimulating the voiding center of the spinal cord. Courtesy of Symposia Foundation, Miami.

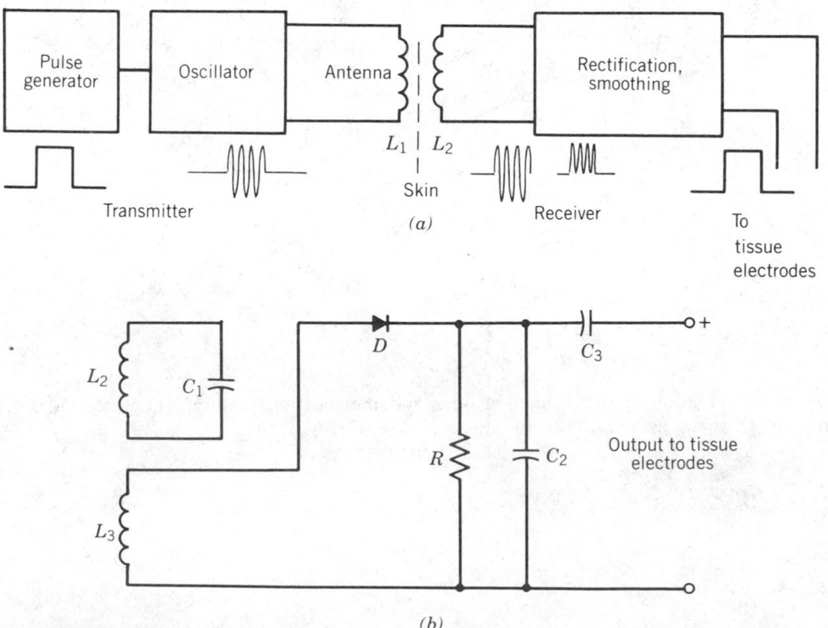

Fig. 43.18 Block diagram (*a*) and schematic (*b*) of a radio-frequency-coupled and a radio-frequency-energized implant.

TABLE 43.6. SPECIFICATIONS OF DIFFERENT BLADDER STIMULATION SYSTEMS

Parameter	Direct Bladder	Sacral Nerve Roots	Spinal Cord
Frequency	20 Hz	15–30 Hz	10–50 Hz
Pulse width	1–5 ms	200 μs	200 μs
Pulse amplitude	5–30 V	1–14 V	5–14 V

Two types of voiding and stimulation patterns have emerged in the spinal cord approach. In the first, patients use stimulation for 60 to 90 seconds and begin to void to completion 15 to 30 seconds after the start of the stimulation. In the second pattern, the patient voids at the end of a stimulation period, with urine continuing to flow after the current is turned off. The second pattern is seen primarily in males and results from simultaneous stimulation of urethral sphincters. When stimulation begins, both the sphincter and bladder muscles contract. Even though bladder pressure rises, urine does not flow because the urethra is kept closed by the sphincter. The bladder muscle is a visceral smooth-type muscle and therefore has a slower response time than the skeletal-type sphincter muscle. When stimulation stops, both the sphincter and the bladder muscle begin to relax. The sphincter relaxes faster, and therefore the urethra is opened before bladder pressure is fully dissipated. Urine flow thus occurs.

The stimulation equipment used for bladder control has been based on transcutaneous radio-frequency coupling. An external transmitter energizes a loop antenna with pulsed radio-frequency energy. This antenna is placed on the skin surface over an implanted receiver. The receiver is a simple tuned circuit, followed by an AM detector. The detected rectangular pulse is then capacitively coupled to the appropriate electrode. Figure 43.18 illustrates a typical receiver schematic. Table 43.6 lists stimulation parameters used in the different types of bladder stimulators.

43.1-7 Electrical Stimulation to Correct Scoliotic Curvature

Scoliosis, or lateral curvature of the spine, is a deforming disease that starts in childhood, proceeds most rapidly during the pubertal growth spurt, and then slows drastically at skeletal maturity. A curve of less than 45° measured at maturity usually progresses very slowly during later life, requiring no further treatment. Conversely, a curve greater than 45° progresses relentlessly during later life, leading to cardiopulmonary complications and increased morbidity and mortality rates.

Treatment of scoliosis is designed to prevent moderate curves from progressing to become a serious medical problem for the patient. This means that curves more than 45° must be straightened

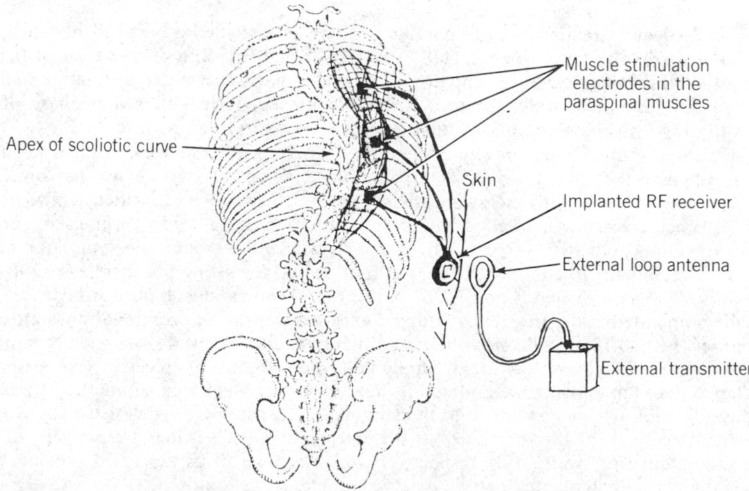

Fig. 43.19 Partially implanted stimulator for the correction of scoliotic curvature.

TABLE 43.7. STIMULATION PARAMETERS IN SCOLIOSIS CORRECTION

Parameter	Value
Amplitude	0–15 V
Pulse duration	220 μs
Frequency	30 Hz
On timing	1–5 s
Off timing	5–25 s
Use	Nighttime only

mechanically and the spine fused rigid, and that curves less than 45° must be braced aggressively to prevent their progression to the stage where they require stabilization by fusion.

Treatment by fusion or bracing has significant associated problems. Correction and stabilization of the curved spine by fusion usually require the use of internal metal rods or cables attached to the spine to correct the curve (at least partially) and hold it rigidly until the bony fusion grows strong. At this point the curve will no longer progress, but in return the patient is left with a rigid spine. All the forces associated with the motion of the vertebral levels are now transmitted through the few unfused levels that may remain above or below the fused region. Premature invertebral disk degeneration is one of several problems associated with these abnormally high stresses generated on these disks.

Bracing (usually using the Milwaukee brace) has only limited (70%) success, is costly, and in 19% of patients leads to psychologic problems and brace rejection.

Bobechko[24] first described a method to correct scoliotic curves by electrically stimulating the paraspinal muscles on the convex side of the scoliotic curve with an implanted stimulator. His technique was later adopted for human use.[25] Three platinum stimulating electrodes are surgically placed in the paraspinal muscles around the convex side of the curve. The electrodes are connected to an implanted batteryless radio-frequency receiver similar to that shown in Fig. 43.18. Power and stimulus programming is transmitted through the skin from an antenna taped on the skin over the implant site. The antenna is plugged into a battery-operated transmitter that has been adjusted to provide the desired cycling time and amplitude. The patient uses the equipment at night only, leaving the stimulator on at bedside while he or she sleeps. In the morning it is shut off, the antenna is removed, and the patient has unrestricted activity all day. Figure 43.19 illustrates the scoliosis correction system in place, and stimulation parameters are given in Table 43.7.

43.1-8 Peroneal Nerve Stimulation for Footdrop

Stroke patients frequently retain a partial paralysis of one side of the body called hemiplegia. Often the stroke victim finds that he or she can stand on the affected foot and can move it sufficiently for walking, but that the gait is seriously impaired because of paralysis of the anterior tibialis muscle group. During the attempt to walk, the foot is not dorsiflexed during the swing phase of gait but rather tends to drag, causing a condition called equinovarous dropfoot.

The tibialis anterior muscle group is innervated by the peroneal nerve, which runs near the surface of the skin just behind the knee. Liberson[26] and Vodovnik[27,28] described a device that stimulated the peroneal nerve in synchrony with the swing phase of gait. A switch was located in the heel of the patient's shoe. When the heel was lifted, a stimulator connected to electrodes on the skin surface over the peroneal nerve was activated. When the heel was placed on the ground the stimulator was turned off. This caused the tibialis to contract, lifting the foot during the swing phase, releasing it after heel strike, and improving gait. Waters[29] and Medtronics, Inc., Minneapolis, built a similar device based on a partially implanted radio-frequency-coupled receiver-transmitter combination similar to that described earlier (Fig. 43.18). A silicone rubber and platinum cuff electrode is surgically implanted on the peroneal nerve and connected to an implanted radio-frequency receiver. An external loop antenna is taped over the implant site and connected to a battery-operated stimulating transmitter. A heel switch is also radio-frequency linked to the stimulating transmitter, which is usually worn on the patient's belt. When a heel liftoff or heel strike occurs, the heel switch transmitter activates or deactivates the stimulating transmitter, respectively, to improve gait as described previously. Figure 43.20 depicts the partially implanted peroneal nerve stimulator for footdrop.

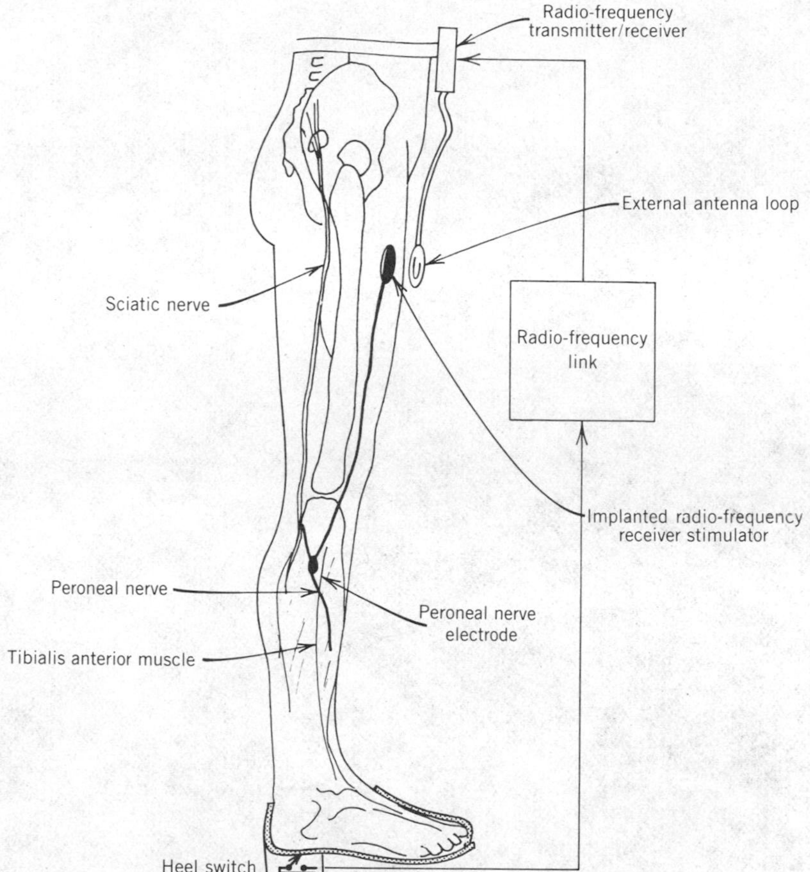

Fig. 43.20 Peroneal nerve stimulator for footdrop (partially implanted).

43.1-9 Healing of Bone Fractures by Electrical Stimulation

Interest in the effects of electric current on bone was stimulated when Yasuda[30] demonstrated new bone formation in the vicinity of a cathode when microamp DC current was continuously applied to a rabbit femur for three weeks. He also described a piezoelectric effect in bone, where the electric potentials were generated in bone in response to mechanical stress. When bending stress was applied to bone, the side under compression became electronegative, and the side under tension became electropositive.

Similar findings were later independently reported. Friedenberg and Brighton[31] reported that in unstressed bone, a steady-state negative potential was generated in the areas undergoing active growth and repair, compared with less active areas. These findings stirred interest in the possible use of electric current to aid bone healing. Friedenberg and colleagues[32] reported the healing of bone nonunions, or fractures that do not heal normally, by use of direct-current stimulation. Basset and co-workers[33,34] healed nonunions by inducing electric currents in bone with a strong alternating electromagnetic field. Many forms of electricity have been applied to bone growth and repair in both animals and humans. Electric stimulation has proven to be very effective in healing difficult nonunions but has not accelerated the healing of normal fractures.

Two competitive techniques have evolved for clinical use, noninvasive electromagnetic induction[35] and direct current applied by electrodes inserted in the fracture site. Although the noninvasive technique does not require surgery, the equipment is bulky and not portable due to the large currents required in the excitation coil. The direct-current technique has been implemented in a totally

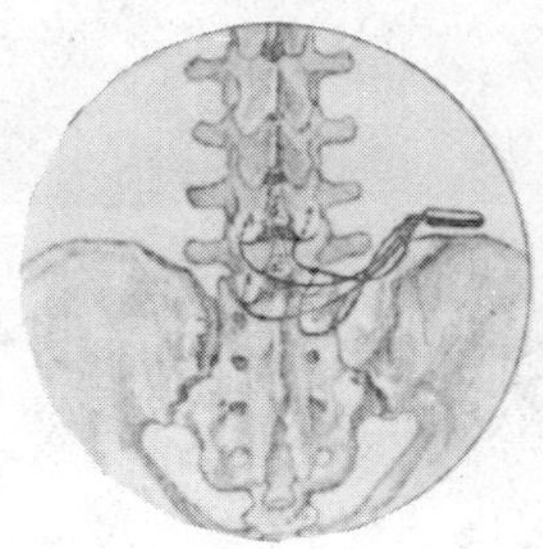

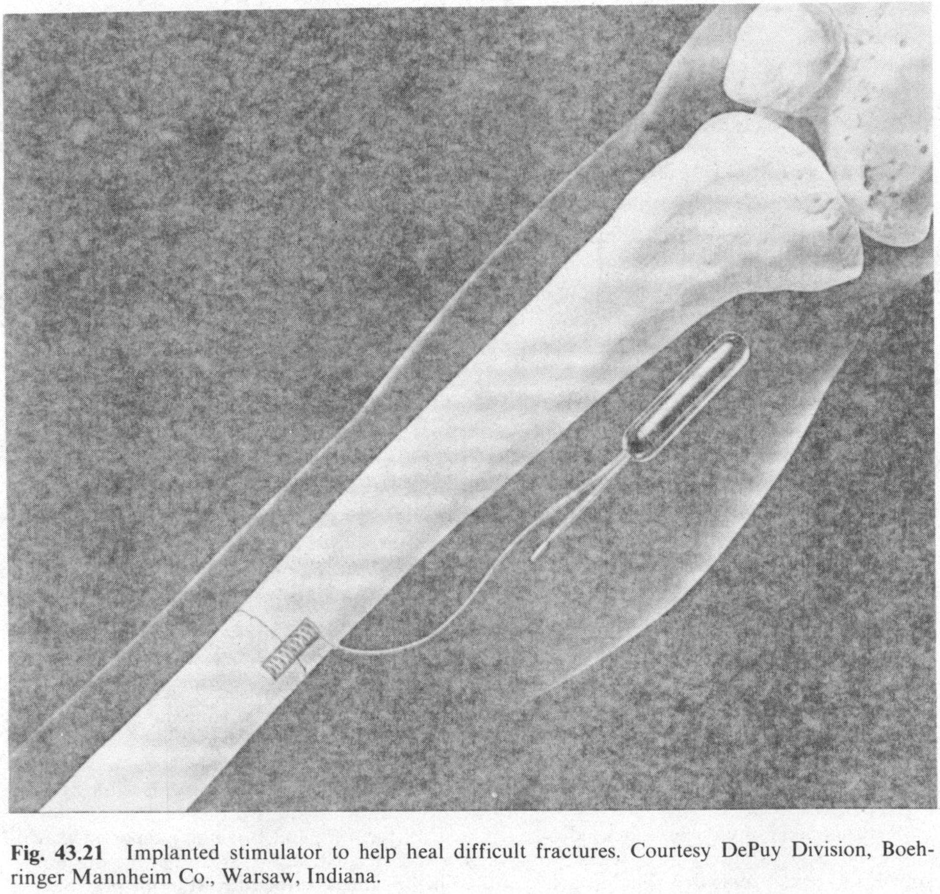

Fig. 43.21 Implanted stimulator to help heal difficult fractures. Courtesy DePuy Division, Boehringer Mannheim Co., Warsaw, Indiana.

implanted device. One type of implant consists of a silver-oxide alkaline zinc battery and a current regulator set to deliver 20 μA to a load ranging from 0 to 100 kΩ. The device is encapsulated in a bullet-shaped epoxy and titanium package and has an active life of 16 to 24 weeks. The cathode consists of a triple-stranded titanium wire, polyethylene insulated for the first 15 cm and bare for another 25 cm. The uninsulated part of the cathode is coiled and surgically inserted at the fracture site. The current source and an 80-mm^2 platinum anode are implanted in soft tissue near the fracture site. Healing usually takes three to four months, after which the implant is removed. Figure 43.21 depicts a bone stimulator in situ.

43.1-10 Control of Pain by Electrical Stimulation

As early as 47 AD Scribonius Largus described the case of a patient who was cured of the pain of gout (arthritis) when he accidentally came in contact with a live electric ray, or torpedo fish, while strolling on the beach. Scribonius recommended torpedo-fish therapy for headache as well as gout and other assorted afflictions. Interest in electrotherapy grew in the seventeenth century with the advent of controlled production of electricity and has continued until today. The modern era of electrotherapy for pain was ushered in with the publication by Melzack and Wall of the "gate control theory" to explain the effect.[36] The theory stipulates that electrical pain relief begins when large A-delta-type sensory nerve fibers are electrically stimulated. These nerve fibers, which normally carry light touch-type sensation, cause a "chemical gate" to be closed, preventing pain sensation normally carried by smaller C-type nerve fibers from reaching conscious levels of the brain. Melzack and Wall did not indicate the exact location of this "gate" in the nervous system, but stipulated that it may exist in the spinal cord or at various levels of the brain that integrate sensory stimuli. Their theory served to unify not only pain relief due to electrical stimulation but also acupuncture, another ancient pain therapy.

Implantation of a stimulator for pain relief was first performed by Shealy.[37] Motivated by the gate theory, he surgically placed a set of platinum electrodes on the dorsal surface of the spinal cord. The dorsal portion of the spinal cord carries primarily large sensory A-delta-type nerve fibers. The technique proved successful and was soon followed by the implantation of stimulating electrodes on large peripheral nerves (see Fig. 43.22) and into pain-related structures within the brain (see Fig. 43.23).

The technique utilized by Shealy consisted of placement of a silicone rubber pad embedded with a pair of platinum electrodes on the surface of the spinal cord through an extensive surgical procedure called a laminectomy. Cook[38,39] modified the technique by utilizing a thin catheterlike electrode with an uninsulated tip, which was placed on the dorsal epidural surface of the spinal cord through a percutaneous needle inserted between the vertebra. In both cases a subcutaneous radio-frequency receiver-external transmitter combination is used to stimulate the electrodes (Fig. 43.24).

Stimulation of deep brain structures is accomplished by placing a wirelike electrode through a skull burr hole, using a target-positioning device called a stereotaxic apparatus. Once the electrode is in position, a lead from it is tunneled under the scalp and through a subcutaneous neck tunnel, to a radio-frequency receiver usually located in a subclavicular skin pocket (Fig. 43.23).

Only patients with severe intractable pain of organic, not psychic, origin are implanted with electrical stimulators. The patient usually stimulates using an external transmitter and a loop antenna taped over the receiver implant site. Stimulation is done on an as-needed basis. Some patients need to stimulate continuously; others can obtain pain relief for several hours after a few minutes of stimulation. Stimulation parameters used for pain control are summarized in Table 43.8.

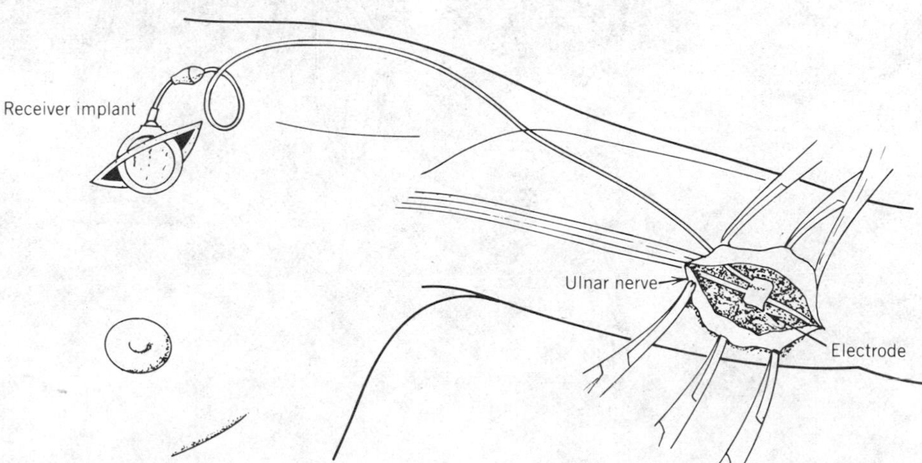

Fig. 43.22 Peripheral nerve stimulator for pain control implanted on the ulnar nerve. Courtesy of Trent Wells, Inc., South Gate, California.

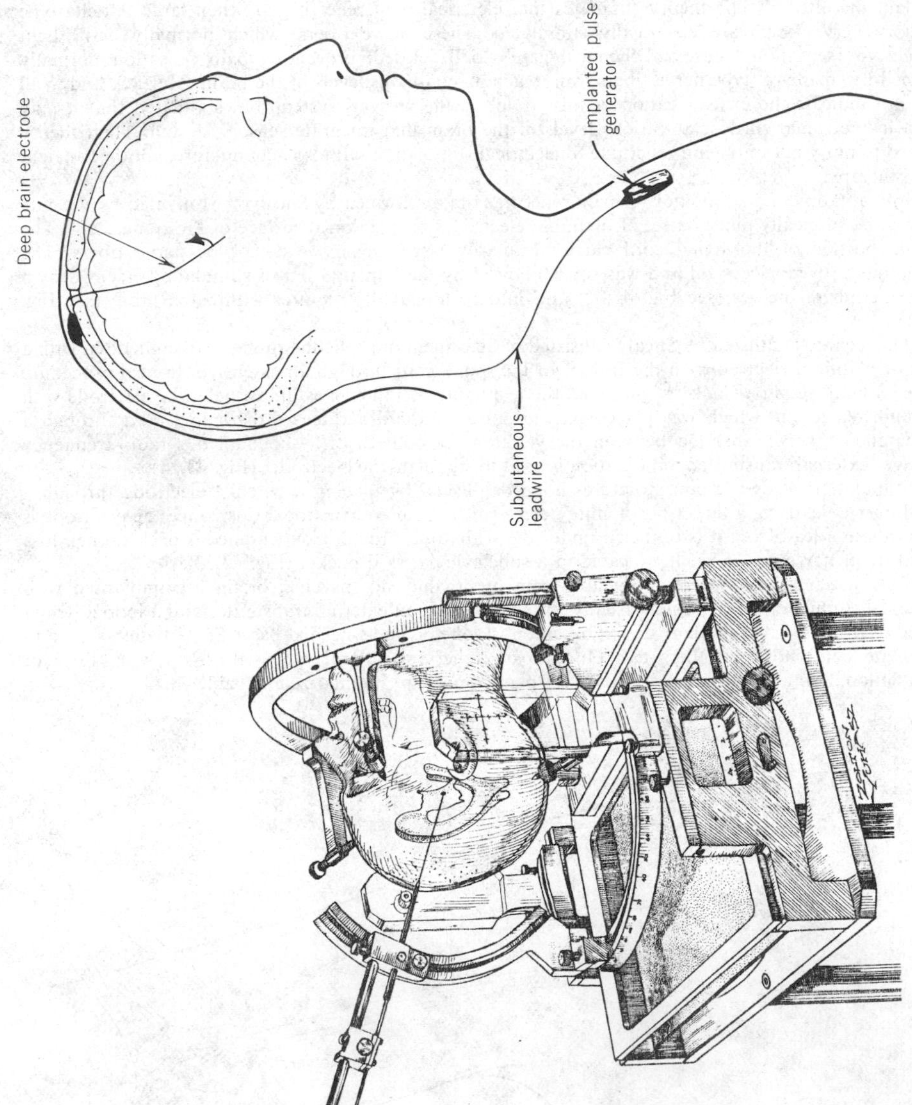

Fig. 43.23 Deep brain stimulation system and procedure.

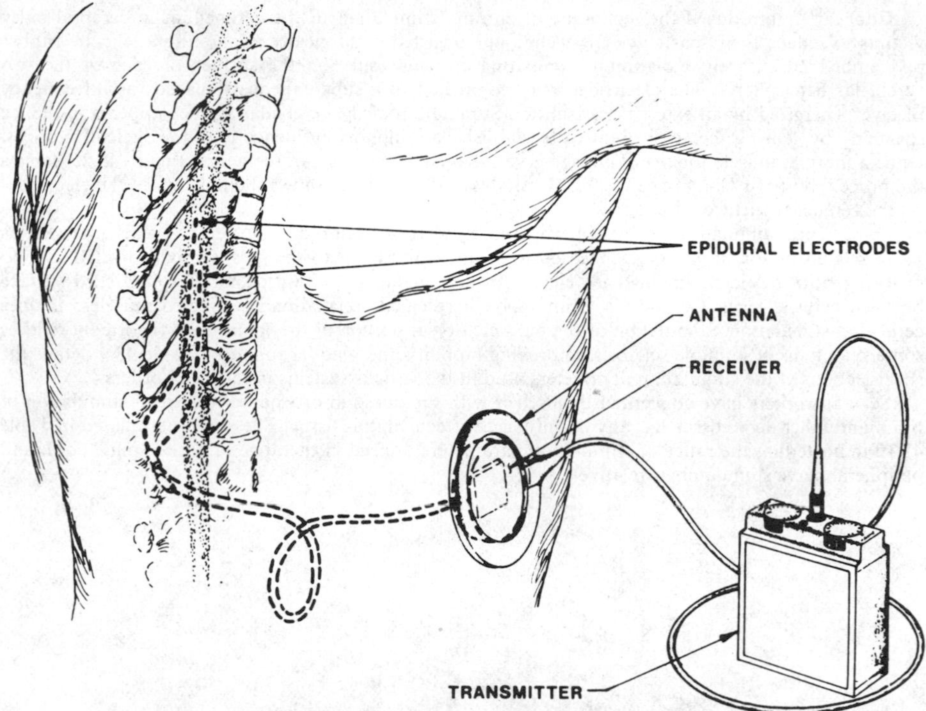

Fig. 43.24 Spinal cord stimulator for pain control in position. Courtesy of Avery Laboratories, Inc., Farmingdale, New York.

TABLE 43.8. STIMULATION PARAMETERS USED IN PAIN RELIEF

Parameter	Range
Amplitude	0–14 mA
Pulse width	50–400 μs
Pulse rate	7–200 Hz
Timing	As needed
Load (resistive component)	500–1500 Ω
Load (capacitive component)	Approx. 1.0 μF

43.1-11 Control of Spasticity and Movement Disorders

The cerebellum is the portion of the brain that coordinates voluntary muscle activity. It serves to organize the contractions and relaxations of the many muscles necessary to perform even simple normal movement. The way the cerebellum performs this function is by selectively inhibiting and facilitating the contractions of the different muscles. The overall effect of the cerebellum is inhibitory. It has been shown that activation of the cerebellum by electrical stimulation reduces decerebrate rigidity of muscles. When the cerebellum is damaged, by anoxia or other means, its influence is reduced and muscle hypertonia results.

Cerebral palsy is a disease that often occurs due to anoxic injury to the fetal brain. It is characterized by intractable muscle hypertonia, or spasticity; exaggerated reflexes; and occasionally involuntary movements called athetosis. Affected patients have difficulty ambulating or performing any voluntary movement because of the overriding involuntary muscle tone or movements.

Cooper[40,41] introduced the technique of chronic stimulation of the cerebellum in cerebral palsy victims to reduce their spasticity. His technique consisted of surgically implanting a silicone rubber pad embedded with eight platinum stimulating electrode sites on the surface of each of the two cerebellar hemispheres. The electrodes were connected to a subcutaneously placed radio-frequency receiver, energized by an external transmitter. Stimulation of the cerebellar surface appears to reduce spasticity by causing the cerebellum to reestablish its inhibitory influence over the skeletal muscles. Once abnormal muscle tone is diminished, voluntary movements can reemerge. Figure 43.25 depicts the procedure for implanting cerebellar electrodes, and Fig. 43.26 shows the position of the electrodes on the cerebellar surface.

Spinal cord stimulation has also been used to treat the symptoms of cerebral palsy, as well as some related motor disorders such as dystonia musculorum deformans, spasmodic torticollis, and posttraumatic spasticity. In this case electrodes are implanted on the posterior spinal cord surface between cervical levels C1 and C4. Stimulation here causes relaxation of affected muscles, as does cerebellar stimulation. Stimulation of the thoracic dorsal portion of the spinal cord reportedly relieves some symptoms of multiple sclerosis, improving mobility and bladder function. Table 43.9 details the specifications of the stimulation parameters used in the various systems in motor disorders.

Several workers have concerned themselves with the question of whether electrical stimulation of brain and other nerve tissue has any harmful side effects on this tissue. The studies tabulated in Table 43.10 indicate that the range of stimulation parameters utilized in therapeutic brain, spinal cord, and peripheral nerve stimulators is relatively safe.

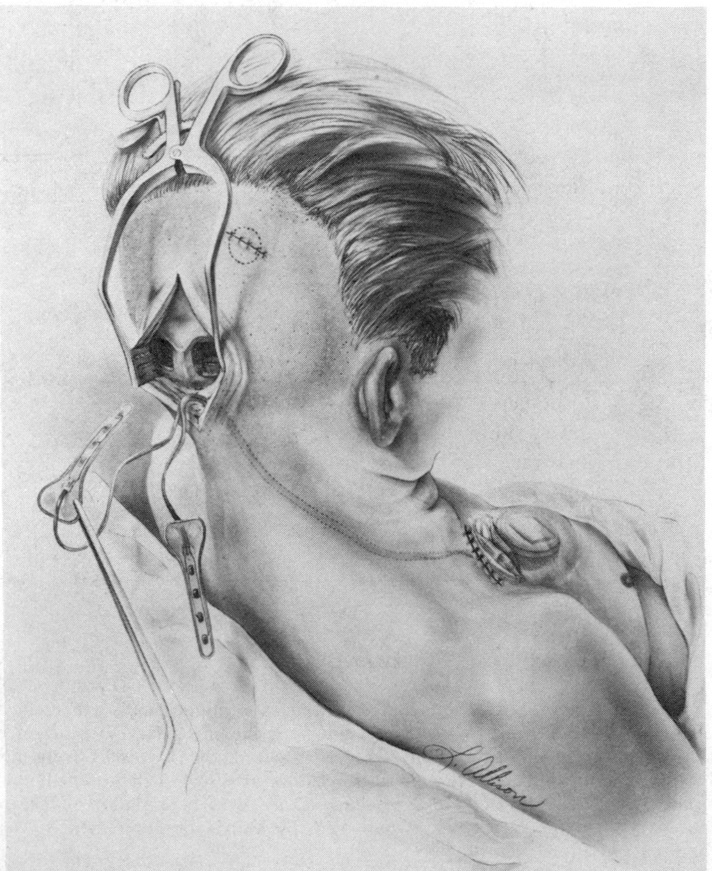

Fig. 43.25 Surgical procedure used to implant cerebellar-stimulating electrodes for the treatment of movement disorders. Courtesy of Avery Laboratories, Inc., Farmingdale, New York.

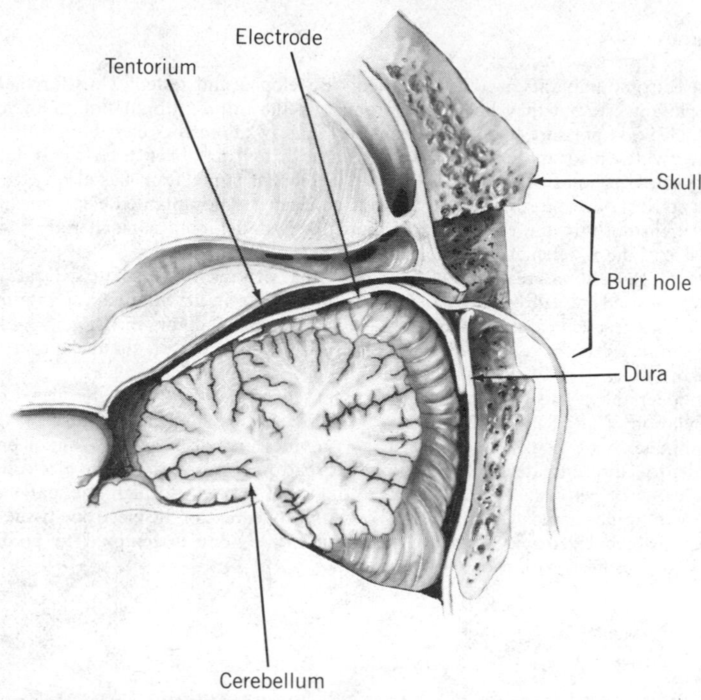

Fig. 43.26 Stimulating electrodes located on the surface of the cerebellum for spasticity reduction. Courtesy of Avery Laboratories, Inc., Farmingdale, New York.

TABLE 43.9. STIMULATION PARAMETERS USED FOR MOTOR DISORDERS

Author	Condition	Amplitude	Rate (Hz)	Pulse Width	Timing (On/Off) (min)
Cooper[41]	Cerebral palsy	0–14 V	200	0.5–1.0 ms	1–8
Cooper[42]	Epilepsy	0–14 V	10	0.5–1.0 ms	1–8
Davis[43]	Cerebral palsy	0.3–1.0 mA	200	0.5 ms	4–8
Gildenberg[44]	Torticolis	0–14 V	50–1500	100 μs	
Waltz[45]	Spasticity	0–14 V	100–1400	100 μs	

TABLE 43.10. STUDIES OF TISSUE DAMAGE FROM STIMULATION WITH PLATINUM ELECTRODES

Author	Tissue	Time and Stimulus	Result
Kim[46]	Human, phrenic nerve	2 yr, 4.3 μC/cm^2/pH	No damage
Gilman[47]	Macaques, cerebellum	205 hr, 30 μC/cm^2, 2.4 μC/pH	
		2.4 mA, 1 ms, 10 Hz	
		Continuous, 7–10 hr/day	Damage
		intermittent, 8 min on/off	Less damage
Brown[48]	Monkey, cerebellum	205 hr, 10 Hz, 1 ms, 0.5 mA	
		7.4 μC/cm^2	No damage
		35 μC/cm^2	Some damage
		70 μC/cm^2	Definite damage
Agnew[49]	Cat, cerebral cortex	36 hr, 4 mA, 0.25 ms, 50 Hz	
		10 μC/cm^2	No damage
		32 μC/cm^2	Some damage
		50 μC/cm^2	Definite damage

43.1-12 Future

Many other electronic implants are currently being developed and tested. Thus far these have seen only limited patient use, but they hold great promise in the future. Stimulation of the carotid sinus nerve to control blood pressure has been attempted[50] but awaits further elucidation of the barostatic mechanism for widespread application. Cerebellar stimulation has been used to modulate emotion and to reduce the symptoms of severe psychosis.[51] Cerebellar stimulation has also been used for the abolition and control of seizures in certain types of epilepsy.[42] An implanted electronically controlled pump to meter insulin into the circulation of diabetics is undergoing clinical trials.[52] A number of devices have been built to telemeter physiologic parameters outside the body.[53]

One of the more exciting areas of development is that of sensory prosthetics. Deaf patients have had electrodes inserted into the cochlea or the auditory nerve itself. Stimulation with an implanted pulse generator has resulted in an encouraging degree of auditory comprehension.[54] Additional work is required to improve the signal-processing algorithm, the number of channels, and the coupling electrodes before more useful hearing is achieved.

Blind volunteers have had electrode arrays implanted on the surface of the visual cortex of the brain.[55] Stimulation of an element of this array is sensed by the subject as a flash of light in space, called a phosphene. A 64-element electrode array provides enough spatially distinct phosphenes to allow the patient to discriminate the braille alphabet read to him or her via the electrode phosphene map. This work was performed with an electrode cable passed through the patient's skin and connected to a computer-controlled stimulator. With improvement in the electrode tissue face and the current rate at which electronic elements are miniaturized, one is tempted to predict a totally implanted visual prosthesis within one or two decades.

References

1 J. B. Ranck, Jr., "Which Elements Are Excited in Electrical Stimulation of Mammalian Central Nervous System: A Review," *Brain Res.* **98**:417–440 (1975).

2 J. C. Lilly, in D. E. Sheer, ed., *Electrical Stimulation of the Brain*, Texas University Press, Dallas, pp. 60–64, 1961.

3 S. B. Brummer and M. J. Turner, "Electrochemical Considerations for Safe Electrical Stimulation of the Nervous System with Platinum Electrodes," *IEEE-BME Trans.* 59–63 (1977).

4 R. Elmquist and A. Senning, in C. N. Smyth, ed., *Medical Electronics*, Proc. 2nd Int. Conf. Medical Electronics, Hiffe, London, 1960.

5 D. C. Schechter, "Background of Clinical Cardiac Electrostimulation," *N. Y. State J. Med.* **71**, **72** (1971–1972).

6 V. Parsonnet, S. Furman, and N. P. Smyth, "Implantable Cardiac Pacemakers Status Report and Resource Guideline. Pacemaker Study Group," *Circulation* **50**:A21–A35 (Oct. 1974).

7 W. Greatbach and W. M. Chardack, "A Transistorized Implantable Pacemaker for the Long-Term Correction of Complete Atrioventricular Block," *Med. Electron. NEREM* **8**:73 (1959).

8 R. E. Fischell, K. B. Lewis, J. H. Schulmann, and J. W. Love, in W. Schaldach and S. Furman, eds., *Advances in Pacemaker Technology*, Springer-Verlag, New York, p. 357, 1975.

9 W. E. Matheson, in M. Schaldach and S. Furman, eds., *Advances in Pacemaker Technology*, Springer-Verlag, New York, p. 401, 1975.

10 D. C. Schechter, "Background of Clinical Cardiac Electrostimulation," *N.Y. State J. Med.* **71**, **72** (1971, 1972).

11 A. Ure, "Experiments Made on the Body of Criminal Immediately after Execution, with Physiological and Philosphical Observations," *J. Sci. Arts* **12**:1 (1818); *Ann. Chim. Phys.* **14**:350 (1828).

12 S. J. Sarnoff, E. Hardenberg, and J. L. Whittenberger, "Electrophonic Respiration," *Am. J. Physiol.* **155**:1 (1948).

13 W. W. L. Glenn, W. G. Holcomb, J. F. Hogan, et al., "Diaphragm Pacing by Radio-frequency Transmission in the Treatment of Chronic Ventilatory Insufficiency," *J. Thor. Cardiovasc. Surg.*, **66**:505 (1973).

14 J. Budge, "Ueber den Einfluss des Nervensystems auf die Bewegung der Blase," *Z. Ration. Med.* **21**:1 (1864).

15 W. E. Bradley, L. E. Wittmers, S. N. Chou, and L. A. French, "Use of a Radio Transmitter Receiver Unit for the Treaτment of Neurogenic Bladder: A Preliminary Report," *J. Neurosurg.* **19**:782–786 (1962).

16 W. E. Bradley, L. E. Wittmers, and S. N. Chou, "An Experimental Study of the Treatment of the Neurogenic Bladder," *J. Urol.* **5**:575–582 (1963).

17 G. S. Brindley, "An Implant to Empty the Bladder or Close the Urethra," *J. Neurol. Neurosurg. Psychiat.* **40**:358–369 (1977).

18 G. S. Brindley, "Emptying the Bladder by Stimulating Sacral Ventral Roots," *Proc. Physiol. Soc.* **9**:15P, 16P (Nov. 1973).

19 R. A. Schmidt, H. Bruschini, and E. A. Tanagho, "Feasibility of Inducing Micturition Through Chronic Stimulation of Sacral Roots," *J. Urol.* **12**(4):471–477 (Oct. 1978).

20 B. S. Nashold, Jr., H. Friedman, and S. Boyarsky, "Electrical Activation of Micturition by Spinal Cord Stimulation," *J. Surg. Res.* **11**:144–147 (1971).

21 B. S. Nashold, Jr., H. Friedman, J. P. Glenn, et al., "Electromicturition in Paraplegia," *Arch. Surg.* **104**:195–202 (1972).

22 B. S. Nashold, Jr., H. Friedman, and J. Grimes, "Electrical Stimulation of the Conus Medullaris to Control the Bladder in the Paraplegic Patient. A 10-Year Review," *Appl. Neurophysiol.* **44**:225–232 (1981).

23 B. S. Nashold, Jr., H. Friedman, J. Grimes, and R. Avery, in *Neural Organization and Its Relevance to Prosthetics*, Symposia Specialists, Miami, pp. 349–367, 1973.

24 W. P. Bobechko, Electrical Stimulation in Scoliosis, in T. E. Hambrechet, ed., *Functional Neuromuscular Stimulation*, National Academy of Sciences, Washington, DC, p. 115, 1972.

25 W. P. Bobechko, M. A. Herbert, and H. C. Friedman, "Electrospinal Instrumentation for Scoliosis: Current Status," *Orth. Clin. No. Am.* **10**(4) (1979).

26 W. T. Liberson, H. J. Holmquest, D. Scot, and M. Dow, "Functional Electrotherapy: Stimulation of the Peroneal Nerve Synchronized with the Swing Phase of the Gait of Hemiplegic Patients," *Arch. Phys. Med.* **42**:101–105 (1961).

27 L. Vodovnik, M. R. Dimitrijevic, T. Prevec, and M. Logar, "Electronic Walking Aids for Patients with Peroneal Palsy," *World Med. Electron.* **4**:58–61 (1966).

28 L. Vodovnik, U. Stanic, A. Kral, R. Acimovic, F. Gracanin, S. Grobelnik, P. Suhel, C. Godec, and S. Plevnick, in F. T. Hambrecht and J. B. Peswick, eds., *Functional Electrical Stimulation: Applications in Neural Prostheses*, Decker, New York, pp. 465–477 (1977).

29 R. Waters and D. McNeal, Task No. 1.76—Effectiveness of Surface Stimulation. Annual Reports of Progress, Rancho Los Amigos Hospital, Downey, CA, 1976–1978.

30 I. Yasuda, "Fundamental Aspects of Fracture Treatment," *J. Kyoto Med. Soc.* **4**:395–406 (1953).

31 Z. B. Friedenberg, and C. T. Brighton, "Bioelectric Potentials in Bone," *J. Bone Joint Surg.* **48-A**:915–923 (Jul. 1966).

32 Z. B. Friedenberg, M. C. Harlow, and C. T. Brighton, "Healing of Nonunion of the Medial Mallcolus by Means of Direct Current. A Case Report," *J. Trauma* **11**:883–885 (1971).

33 C. A. L. Bassett and R. J. Pawluk, "Noninvasive Methods for Stimulating Osteogenesis," *J. Biomed. Mater. Res.* **9**:371–374 (1975).

34 C. A. L. Bassett, R. J. Pawluk, and A. A. Pilla, "Augmentation of Bone Repair by Inductively Coupled Electromagnetic Fields," *Science* **184**:575–577 (1974).

35 C. A. L. Bassett, S. N. Mitchell, L. Norton, et al., Electromagnetic Repairs of Nonunions, in C. T. Brighton, J. Black, and S. R. Pollack, eds., *Electrical Properties of Bone and Cartilage: Experimental Effects and Clinical Applications*, Grune and Stratton, New York, pp. 605–630, 1979.

36 R. Melzack and P. D. Wall, "Pain Mechanisms: A New Theory," *Science* **150**(3699) (Nov. 19, 1965).

37 C. N. Shealy, J. T. Mortimer, and J. R. Reswick, "Electrical Inhibition of Pain by Stimulation of the Dorsal Columns. Preliminary Clinical Report," 1967.

38 A. W. Cook, "Percutaneous Trial for Implantable Stimulating Devices," *J. Neurosurg.* **44**:65–651 (1976).

39 A. W. Cook, A. Tahmouresis, A. Oygar, et al., "Epidural Electrical Stimulation of Spinal Cord for Intractable Pain and Other Abnormal Conditions," *Acupunc. Electrotherapeut. Res. Int. J.* **2**:259–270 (1977).

40 I. S. Cooper, "Chronic Stimulation of Palaeocerebellar Cortex in Man," *Lancet* **I**:206 (1973).

41 I. S. Cooper, M. Riklan, I. Amin, et al., "Chronic Cerebellar Stimulation in Cerebral Palsy," *Neurology* **26**:744–753 (1976).

42 I. S. Cooper, I. Amin, M. Riklan, et al., "Chronic Cerebellar Stimulation in Epilepsy," *Arch. Neurol.* **33**:559–570 (1976).

43 R. Davis, M. Barolat-Romana, and H. B. Engle, "Spasticity; Chronic Cerebellar Stimulation for Cerebellar Palsy, Five-Year Study," *Acta. Neurochirurgica*, Suppl. 30, 317–332, 1980.

44 P. L. Gildenberg, "Treatment of Spasmodic Torticollis with Dorsal Column Stimulation," *Appl. Neurophysiol.* **41**:113–121 (1978).

45 J. M. Waltz, L. O. Reynolds, and M. Riklan, "Multi-Lead Spinal Cord Stimulation for Control of Motor Disorders," *Appl. Neurophysiol.* **44**:244–257 (1981).

46 J. H. Kim, E. E. Manuelidis, W. W. L. Glenn, and T. Kaneyuki, "Diaphragm Pacing: Histopathological Changes in the Phrenic Nerve Following Long-Term Electrical Stimulation," *J. Thorac. Cardiovasc. Surg.* **4**:602–608 (1976).

47 S. Gilman, G. W. Dauth, V. M. Tennyson, and L. T. Kremzner, "Chronic Cerebellar Stimulation in the Monkey," *Archs. Neurol.* **32**:474–477 (1975).

48 W. J. Brown, T. L. Babb, H. V. Soper, J. P. Lieb, C. A. Ottino, and P. H. Crandall, "Tissue Reactions to Long-Term Electrical Stimulation of the Cerebellum in Monkeys," *J. Neurosurg.* **47**:366–379 (1977).

49 W. F. Agnew, R. H. Pudenz, L. A. Bullara, T. G. H. Yuen, and D. B. Jacques, Progress Rep. Contract No. N01-NS-0-2275, pp. 1–24, Bethesda, MD, NINCDS, 1978.

50 N. R. Hagfors and S. I. Schwarts, "Implantable Electronic Carotid Sinus Nerve Stimulators for Reducing Hypertension," *Proceed. 19th Annual Conference on Engineering in Medicine and Biology* **8**:36 (Nov. 14, 1966).

51 D. C. Schechter, "Background of Clinical Cardiac Electrostimulation," *N.Y. State J. Med.* **71**, **72** (1971–1972).

52 P. J. Blackshear, et al., "Control of Blood Glucose in Experimental Diabetes by Means of a Totally Implantable Insulin Infusion Device," *Diabetes* **28**:634–639 (1979).

53 T. B. Fryer, H. A. Miller, and H. Sandler, eds., *Biotelemetry III*, 1976.

54 R. L. White, S. A. Shamma, and N. E. Cotter, "Development of Multi-Channel Electrodes for an Auditory Prosthesis," Stanford Electronics Laboratory, NIH Contract No. N01-NS-7-2366, 1977.

55 G. S. Brindley and W. S. Lewin, "The Sensations Produced by Electrical Stimulation of the Visual Cortex," *J. Physiol.* (*Lond.*) **196**:479–493 (1968).

43.2 SIGNAL PROCESSING FOR COMMUNICATIVELY HANDICAPPED

Harry Levitt

Prosthetic aids that incorporate modern signal-processing techniques can vastly improve communication for hearing-impaired, visually impaired, and speech-impaired individuals. Mass production of highly sophisticated but inexpensive signal-processing systems is opening new possibilities for aids for the handicapped and shifting the emphasis in research and development work. The challenge no longer is one of engineering development; rather it is to apply existing sophisticated technology to the solution of fundamental problems.

Sophisticated technologies, which potentially can provide breakthroughs for handicapped people, also add to the burdens and challenges they face. Telecommunications technology has been revolutionized during the last 20 years, for example, but even today deaf telephone users must rely on cumbersome and slow telecommunications devices and they are still largely isolated from hearing users of the same system.

43.2-1 Extent of Communicative Handicaps

The communicatively handicapped include the hearing impaired, visually impaired, and speech impaired. The three groups are not mutually exclusive: A hearing-impaired person can also be visually impaired and, quite commonly, a person with a severe hearing impairment also has speech problems, although the reverse is not usually true. Signal-processing prosthetic aids have been used most extensively by the hearing impaired, the largest severely impaired group, and it is for this group that these aids have the greatest potential.

It is estimated that in the United States alone more than 15 million people—approximately 8% of the adult population and 1.2% of the population under 18 years of age—have a hearing impairment of some kind. More than 3 million are moderately impaired and have difficulty understanding speech without the help of a hearing aid. Severely hearing-impaired people have problems understanding speech even with the help of a hearing aid. Profoundly hearing-impaired people cannot understand even amplified speech. There are more than 1 million severely or profoundly hearing-impaired people in the United States. Although people with this degree of hearing impairment are often called deaf, only a small proportion of severely hearing-impaired people have no hearing at all.

The impact of a hearing impairment is very damaging, the most serious problems occurring in young children who are either hearing impaired at birth or who become hearing impaired before their language has developed to a significant degree. The prelingually deafened child—the prevalence of this condition is about 0.1%—has considerable difficulty in acquiring speech and language and usually lags well behind normal-hearing children in educational development. The vast majority of children at schools for the deaf are prelingually hearing impaired.

While about 50% of the population have minor visual impairments (those that can be compensated for adequately with eyeglasses or contact lenses), there are far fewer people with severe or profound visual impairments than with severe or profound hearing impairments. Roughly 1.25 million people in the United States suffer from severe visual impairment; about 300,000 are classified as legally blind.

The incidence of speech impairments is about 1% for the population as a whole, but it is much higher for children. Surveys show that teachers recommend speech therapy for approximately 9% of noninstitutionalized 6-year-old children. This percentage decreases with age, falling to about 1% for 17-year-olds. Institutionalized children have a high incidence of speech disorders.

Serious speech disorders often accompany neuromotor diseases, the most common of which are cerebral palsy (0.5% in children), stroke (1.3% in adults), and Parkinson's disease (0.2% in adults). Serious speech disorders also accompany deafness and mental retardation, which together afflict 1 to 3% of the population. There are also about 1.5 million stutterers in the United States and over 200,000 people with serious orofacial anomalies, such as cleft palates. The incidence of laryngeal carcinoma that results in surgical removal of the larynx is roughly 9000 cases per year. Language impairment, which is closely related to developmental speech problems, is present in 2 to 3% of 3-year-old children and in 1% or less of children entering school.

Most speech disorders do not seriously impede communication, but a small proportion of severe disorders are extremely damaging to communication ability. Many of these serious disorders either result from or are accompanied by other major diseases or impairments.

43.2-2 Acoustic Amplification for Hearing Impaired

The ordinary hearing aid, the most commonly prescribed prosthetic aid that uses electronic signal processing, is basically a conventional amplifier whose output is limited so as not to overload the auditory system.

Several factors make the design and prescription of hearing aids difficult:

1. The demographics of the hearing-impaired population have changed substantially since the 1940s. Then the majority of hearing-aid users suffered from some form of middle-ear impairment. With the growth of preventive medicine and the development of medical and surgical procedures to ameliorate this condition, people with middle-ear impairments are no longer the primary hearing-aid users. But the results of classic early studies, which primarily involved subjects with middle-ear problems, still have a profound influence on the design and prescription of modern hearing aids.

2. Cosmetic factors play a very large role in the design of new hearing aids. Many modern technologic advances have been directed toward making hearing aids smaller and less noticeable rather than toward improving their overall effectiveness.

3. Hearing loss is not simply a loss in auditory sensitivity, but rather is a loss in the overall ability to process speech sounds. Even if sounds are amplified, hearing-impaired individuals are unable to detect changes in the sounds' characteristics, and they therefore cannot distinguish one word from another.

4. The transmission path from sound source to eardrum is extremely complicated; head and body baffle effects, reflections off the pinna, and room reverberation must all be taken into account. The transmission path also changes continually as a result of head movements and changes in the location of the sound source.

Although the most basic measure of hearing impairment, the audiogram, shows loss in hearing sensitivity as a function of frequency, this is not the primary difficulty in hearing loss. A loss in hearing sensitivity per se is easily corrected by means of frequency-selective amplification in which the hearing loss at each frequency is restored by a matching acoustic amplification. The enormous amount of gain needed at the high frequencies to make sounds audible also makes the more intense sounds uncomfortably loud and potentially hazardous to the patient's remaining hearing. Some method of reducing the intensity range of the speech signals is thus necessary. A fundamental problem is how to accomplish this without reducing intelligibility.

With purely conductive impairment such as otosclerosis, in which a bony growth develops, the sound transmitted through the middle ear is simply attenuated. Most impairments of this type can be treated medically or surgically, thus obviating the need for a hearing aid altogether, but the vast majority of candidates for acoustic amplification have some sensory and/or neural impairment that cannot be cured medically.

A characteristic of sensorineural hearing impairment is that the dynamic range—the range from threshold of audibility to discomfort level—is reduced. Although the auditory threshold may be raised substantially, there is no corresponding increase in the level at which sound becomes uncomfortably loud. In fact, the discomfort level for a sensorineurally impaired person is quite often lower than for someone with normal hearing. Within this limited dynamic range, the frequency resolution of the impaired ear is reduced. In particular, a hearing-impaired person usually has much more difficulty discriminating speech in a noisy background than does a person whose hearing is normal.

Signal-processing techniques can be useful in automatically adjusting the frequency-gain characteristic of hearing aid amplifiers so as to place as much as possible of the information-bearing content of the speech signal within the available region of residual hearing. Researchers at the Central Institute for the Deaf in St. Louis, MIT, and City University of New York[1,2,3] have independently identified similar sets of electroacoustic characteristics that need to be taken into account to do this. The characteristics are determined individually for each hearing-impaired person. The implementation of such an individualized amplification system in a practical instrument, which could be worn either behind the ear or in the ear canal, is a challenging problem that would benefit from the intelligent application of modern signal-processing techniques.

Still being explored is the extent to which the electroacoustic characteristics of the hearing aid should be adjusted automatically to best match the time-varying characteristics of speech. There is little doubt that automatic gain control is helpful, but the most effective form of signal-dependent adjustment of a hearing aid's characteristics is not known. Edgar Villchur of the Foundation for Hearing-Aid Research, Woodstock, New York, believes that a multichannel system with independent automatic gain control in each channel is a promising approach.[4] Data from preliminary experimental evaluations of this approach have not shown significant advantages, but there are strong theoretical reasons to believe that the system will work well with certain classes of hearing impairments, such as those exhibiting a very narrow dynamic range.

Modern signal-processing techniques can also improve hearing aids by enhancing signals in the presence of noise. Ample experimental evidence shows that the deleterious effects of noise on speech intelligibility are much worse for the hearing impaired than for those with normal hearing. In everyday face-to-face communication, we tend to raise our voices in the presence of noise, thereby increasing the signal-to-noise ratio by an amount that is usually sufficient for a normal-hearing listener. For hearing-impaired listeners, however, these everyday signal-to-noise ratios are very difficult to handle. Using signal-processing techniques to improve the effective signal-to-noise ratio would be of immense value, but this improvement must not be achieved by causing distortion of the speech signal and reduced intelligibility.

While improving signal-to-noise ratio is difficult, certain factors contribute to a positive solution. In face-to-face communication it is quite common for the speech and noise sources to occupy spatially separate locations. As a result, there are important interaural differences between the speech and noise that the normal binaural auditory system uses to good effect. Hearing-impaired listeners, however, do not appear to use these interaural differences efficiently; several studies show little or no binaural advantage for the hearing impaired.[5] For these listeners, some degree of interaural processing before the signals are delivered can effectively increase the signal-to-noise ratio. A two-channel or multichannel preprocessor, unlike present hearing aids, need not be restricted to microphone inputs at the two ears (an array of microphones could be used, for example). Such systems have the potential to improve speech intelligibility to the point where it is actually better than that for normal-hearing listeners with their binaural advantage.

Excessive room reverberation also reduces speech intelligibility, and the reduction is far greater for hearing-impaired than for normal-hearing persons. Although signal-processing techniques for reducing the effects of room reverberation have been developed, practical implementation of these techniques in a wearable hearing aid has not been achieved.[6]

Relatively simple solutions to the dual problem of increasing signal-to-noise ratio and reducing reverberation have been obtained under limited conditions. In one approach, which has been tried in theaters, churches, and other large gathering places, the signal from the speaker is transmitted throughout the room by means of frequency-modulated (FM) infrared light. The hearing-impaired members of the audience wear lightweight FM receivers, with the outputs coupled acoustically to the two ears through ear plugs. Extraneous noises generated within the room are eliminated, and the signals generated on the stage or pulpit are delivered directly to the ears with little or no reverberation.

These FM systems have been received enthusiastically by many hearing-impaired persons. Cosmetic factors do not appear to be of great consequence in this case, possibly because the devices, although larger than conventional hearing aids, are worn in comfortable anonymity in a darkened theater or auditorium, with someone else as the center of attention.

Acoustic amplification works extremely well with the mildly hearing impaired, moderately well with the moderately hearing impaired, and less well with the severely or profoundly hearing impaired. For this last group, the hearing aid provides auditory cues to supplement lipreading as well as cues for monitoring of speech production. It is also possible to recode the speech signal for the profoundly hearing impaired, bringing important speech cues within their limited range of residual hearing.

43.2-3 Speech-Analyzing Aids for Hearing Impaired

Speech-analyzing aids are devices that recode the speech signal to convey speech information in an alternative form. In the case of severe or profound sensorineural hearing impairment, it is quite common to find some hearing in the low frequencies (up to about 1000 or 2000 Hz) and very little, if any, hearing above 2000 Hz. A form of acoustic recoding known as frequency transposition shifts the inaudible speech energy in the high frequencies into the low-frequency region, where it produces distinctly audible cues.[7]

Transposition works well with voiceless fricative sounds. Since these sounds have very little low-frequency energy, the superposition of coded signals representing the high-frequency structure of the sound destroys very little low-frequency information. The success of transposition in recoding sounds that contain information in both the low and high frequencies has not been proved.

Experimental evaluations of several different transposition systems show improvements in speech and sound reception under restricted conditions, such as in single-word speech identification tasks or in the use of speech-training aids designed to improve production of the fricative sounds.[8] Evaluations under more general conditions have yielded conflicting results.[9,10] Some studies show small improvements in comparison with conventional acoustic amplification.[11] A major difficulty in using and evaluating frequency transposers that radically alter the speech signal is that the users must learn a new code. It could be argued that the negative experimental results are due primarily to the subjects' not having yet learned the transposed speech code.

It has also been argued that acquisition of a radically different speech code is best done during childhood, when children are beginning to learn the sounds of speech.[12] This hypothesis is very difficult to test, and there are ethical questions involved in doing so. Conceivably, a hearing-impaired child could acquire speech and language much more readily if all the important speech cues were transposed to his or her region of residual hearing. If such an experiment failed, however, the child's acquisition of speech and language could be retarded further than it would have been otherwise. The difficulty of reliably measuring the amount of residual hearing in a hearing-impaired infant makes this kind of experiment even riskier. Children with a fair degree of residual hearing can acquire reasonably good speech and language skills with conventional acoustic amplification. Radical distortions of the speech signal could cause irreparable damage to the normal processes of development in such children.

Many researchers argue that the auditory system has inherent speech-feature detectors that severely restrict the ability to learn any radically different speech codes.[12] They argue further that the information capacity of a severely impaired auditory system is extremely limited and that to be intelligible, the information content of recoded speech signals necessarily exceeds this capacity.

In an important experiment, currently under way at MIT, Louis Braida and his colleagues are attempting to determine whether an intelligent normal-hearing person can learn a new speech code in which all of the acoustic cues lie in the low-frequency region.[13] The new code is designed to maximize the perceptual distance between the different sounds of speech. The experiment should demonstrate whether it is feasible to learn a radically different low-frequency speech code under ideal conditions. If results are positive, there remains the major problem of developing a practical speech-processing device that will translate speech into such a code.

43.2-4 Direct Electrical Stimulation

Direct electrical stimulation of the auditory system is the newest approach to exploiting residual hearing. This surgical technique bypasses the nonfunctioning sensory cells and provides a useful input to the auditory system at the neural level. It benefits the profoundly sensorially impaired, but not the neurally impaired. It is not clear how many people qualify for direct electrical stimulation, but they represent only a small proportion of the total number of the profoundly hearing impaired.

Extracochlear electrical stimulation, in which a point peripheral to the cochlea is stimulated, is by far the safest direct stimulation approach.[14] The operation is reversible, as the stimulator can be removed without damaging the auditory system. The major limitation of extracochlear stimulation is that only one channel of limited bandwidth is available. It does provide the psychologic benefit of allowing people to hear again, albeit very little, and it has proved valuable in supplying voice-pitch information to profoundly deaf individuals. The cues about pitch and intonation are a useful supplement to lipreading and provide important feedback for monitoring one's own voice. Methods of processing speech signals so as to make most effective use of extracochlear stimulation are currently being investigated.

Inserting electrodes into the cochlea is another method of direct electrical stimulation.[15] Each region of the cochlea is sensitive to different frequency components in the speech signal; for transmission of intelligible speech, several frequency regions must be stimulated independently. Independent electrodes can be used in a cochlear implant, thus providing the possibility of multichannel stimulation. Since the electrical fields generated by intracochlear stimulation are fairly diffuse, a sophisticated combination of skilled surgery and precise engineering is required to achieve a true multichannel cochlear implant.

Several prominent research groups both in the United States and abroad are working on this problem. Graeme Clark of the University of Melbourne, Australia, was among the first to insert a multichannel electrode array in the cochlea and obtain improvements in speech reception.[16] The improvements were impressive, but they fell far short of the goal of making speech fully intelligible. Several other research groups have recently achieved impressive success with multiple-channel implants.[15] At least one major US corporation, the 3M Company, has invested in the development of cochlear implants.

Research currently is focused on how to process the speech signal for multichannel implants so as to maximize intelligibility. Despite the much heralded publicity about cochlear implants, researchers are still asking the same fundamental question: Can the speech signal be recoded so that it is intelligible to an impaired auditory system of limited channel capacity? This question is not very different from that posed with respect to the low-frequency recoding of speech. (It is also essentially the same question being addressed, but in a different format, in current research on visual and tactile displays of speech.)

Researchers have also experimented with the signal-channel cochlear implant, but this has proved to be among the least promising implants largely because it combines all the dangers of cochlear implantation with all the disadvantages of single-channel stimulation. There is no evidence, for example, that single-channel cochlear implants provide benefits superior to those obtained with the much safer extracochlear stimulation. In addition, any cochlear implant does irreversible damage to the very delicate structures within the cochlea, including the destruction of any remaining intact sensory cells. In some cases, evidence shows that a cochlear implant may interfere with the sense of balance as a result of the close proximity of the semicircular canals to the cochlea. Despite its obvious disadvantages and the availability of a much safer alternative, the single-channel cochlear implant procedure is being performed at an increasing rate.

A third form of electrical stimulation involves implanting electrodes in the auditory nerve. Blair Simmons of Stanford University used multiple electrodes in his pioneering work with this approach.[17] Results similar to those obtained with other cochlear implants have been reported. A disadvantage of the nerve implant is that the electrode array cannot be located precisely enough to allow consistent tonotopic stimulation (that is, each electrode stimulating a specified frequency region). Because of the well-organized layout of auditory nerve fibers in different frequency regions of the cochlea, however, consistent tonotopic stimulation can be obtained with a multielectrode cochlear implant.

Stimulating the auditory cortex itself is a fourth approach. A group of researchers under the direction of William Dobelle, then at the University of Utah, placed electrodes on the surface of the auditory area of the brain to study direct electrical stimulation of hearing.[18] This approach involves extremely complex problems of electrode placement and signal processing and is too difficult and dangerous to warrant consideration as an imminent aid to communication.

43.2-5 Nonauditory Aids for Hearing Impaired

Supplementing or bypassing the impaired sensory system by means of another modality is a very attractive option in the development of sensory aids for the communicatively handicapped. For the hearing impaired, the use of visual and/or tactile cues to facilitate communication offers many benefits.

Initial attempts at such an approach simply displayed the speech waveform tactilely or visually. A person could separate weak from strong sounds and determine whether speech was present, but not much else. More sophisticated methods, which provide either spectrum or speech-feature displays of the speech signal, have since been developed. Although some very useful results have been obtained with these displays, the goal of effective speech communication through another modality remains elusive.

During the course of these nonauditory developments, several distinct schools of thought have emerged. One group argues that speech is a special code and the auditory system a unique decoder.[19] It is unrealistic, according to this view, to expect either the visual or tactile system to substitute effectively for an impaired auditory system. This argument is used to account for the relatively modest gains obtained with nonauditory techniques. Proponents of this view feel that visual or tactile displays have served as useful supplements to lipreading or as speech-training aids, but not as substitutes for the auditory system.

A counterargument to this view is advanced by Robert Houde of the Center for Communications Research, Rochester, New York.[20] He contends that the gains realized in communicating speech through alternative modalities have been modest because of such factors as the poor resolution of early experimental devices, the use of inappropriate training strategies, and exposure to displays for limited amounts of time. In his view, a person can understand speech through a display if information is properly presented and the person is thoroughly trained. Victor Zue of MIT has demonstrated that a person can read and understand a spectrographic display of speech.[21] Though the process is very time-consuming and impractical as a means of communication for the deaf, Zue's methods could prove extremely useful in designing more effective displays.

There may be critical periods in a child's development when the ability to learn speech is at a peak. If effective displays were provided to a child during these early stages of speech and language development, such displays would be far more effective as communication aids later.

A third school of thought believes that regardless of whether speech is special, it is possible to communicate effectively using visual or tactile symbols that reflect the features of speech in a well-organized way.[22] One can understand speech, for example, through the printed word or through lipreading, even though there is little or no auditory input. In the somewhat cumbersome Tadoma method of communication, deaf-blind people pick up speech cues by touching speakers' faces and feeling their articulatory movements.

The success of these methods raises questions about whether more effective displays could be designed for transmitting only the important features of speech. Several experimental aids have been developed that provide supplemental cues to reduce the ambiguity that occurs in lipreading.[23–27] Positive results have been obtained with many of these devices, usually in the form of increased speech reception skills, but, as in the case of implants, the size of the improvements are relatively small compared with the overall goal of making speech fully intelligible.

Automatically extracting important features of speech is a major technical difficulty in creating effective displays. This problem is essentially the same as that underlying automatic speech recognition. If a reliable automatic speech-recognition device operating on unconstrained continuous speech could be developed, it would then be a simple matter to convert such a device to a reliable speech-feature indicator, or vice versa.

The Sensory Aids Foundation, Palo Alto, California, has initiated research that will apply the technology of automatic speech recognition to the problem of developing a practical visual speech recognition system for the deaf.[28] An ongoing research project at the City University of New York is attempting to reach a middle ground between visual speech displays that are easy to generate automatically but are very difficult to read by humans (the spectrogram, for example) and displays that are easy to read but are very difficult to generate automatically (such as the printed output of an automatic speech recognition device).[29]

Meanwhile, the hearing impaired are increasingly using aids that take advantage of the printed word, such as captioned television or teletypewriters. Such methods can be extremely effective ways to communicate, but they are not without serious shortcomings. An automatic speech recognition device would go a long way toward solving one major problem with captioned television—the time and cost involved in producing the captions.

While reliable automatic speech recognition of unconstrained material is unlikely to be achieved in the near future, semiautomatic systems already exist. One system, TOMCAT, is currently being evaluated as a teaching aid for deaf students at the National Technical Institute for the Deaf, Rochester, New York.[30] A stenographer types out a verbatim transcription of a lecture, as it occurs, on a shorthand typewriter whose keys are wired to a computer. The computer translates the incoming signals into English text, which is displayed within seconds on a television screen. A deaf lawyer arguing a landmark case before the Supreme Court has used the TOMCAT system. Other possible applications include more efficient preparation of television captions or facilitation of communication between deaf and hearing telephone users.

Telephone aids for the hearing impaired could benefit enormously from the insightful application of modern signal-processing technology. In many cases, inexpensive, minor modifications to existing computer systems would suffice. Many inexpensive pocket computers have an audio output used for recording programs and data on audio casettes. With a slight change in frequency, these audio signals could be used by the hearing impaired to communicate over the telephone. A device of this type was recently awarded first prize in the Johns Hopkins First National Search in Personal Computing to Aid the Handicapped.[31]

Inexpensive personal computers have several advantages as communication devices: They cost much less than conventional teletypewriters and they have both memory and logic. Word processors designed specifically for the hearing impaired, for example, could store frequently used words and phrases in an easily accessible format and then print them out at the touch of a button, thus reducing considerably the time spent typing out messages in current telecommunications devices.

Because hearing-impaired people might perceive computers as difficult to use, it is important that computers programmed to operate as telecommunications devices be at least as easy to use as conventional teletypewriters.

Most telecommunications devices for the hearing impaired use the relatively slow Baudot code, while most personal computers use the ASCII code for telecommunications. This incompatibility can be solved by providing deaf users with devices that can communicate in either code or with telephone access to a centrally located code converter. A common code must be agreed on or, alternatively, inexpensive code converters must be developed if the hearing impaired are to benefit fully from the use of personal computers as communication devices.

43.2-6 Speech-Training Aids

Most hearing-impaired children have a doubly severe handicap: They have difficulty understanding speech, and they also have difficulty producing intelligible speech. This is because they lacked auditory input during the critical early years when speech is acquired and because they are unable to monitor their own speech production effectively.

Speech-training aids have been relatively successful in improving the speech of the hearing impaired. These devices provide visual and/or tactile displays of the speech features that supplement auditory cues available through acoustic amplification. Researchers are currently attempting to determine which speech features should be displayed and how they should be displayed. The displays that have been most successful show the fundamental frequency contour, the degree of nasalization, and the spectral characteristics of fricative sounds. An early computer-based speech-training system embodying many of these features was developed by Raymond Nickerson and Kenneth Stevens at Bolt, Beranek, and Newman, Cambridge, Massachusetts.[32] Positive results obtained with this system have led to the development of a relatively inexpensive speech-training aids using personal computers.

More effective speech-training aids await the development of reliable procedures for automatic extraction of speech features. Because only one speech feature at a time need be extracted and displayed, the task is not as difficult as that involved in developing automatic speech-recognition devices. In addition, for speech training, miniature sensors can be placed on the speaker to obtain articulatory information that would be extremely difficult to extract automatically from the acoustic speech signal.

The size of speech-training aids also limits their effectiveness. Most of the devices are desk-mounted, requiring the child to come to the device, which severely limits access to the aid. Wearable devices, which would provide continuous feedback on specific features from the child's own speech production (as well as good examples from the teacher), would do much to facilitate the speech-learning process.

The pedagogical problem of integrating speech-analyzing aids into the speech-training curriculum is being evaluated at the Lexington School for the Deaf, Queens, New York, where a relatively inexpensive minicomputer speech-training system is part of the general speech-training curriculum.[33] The computer serves as sensory aid, diagnostic tool, and record keeper for the teacher.

43.2-7 Controversies

The oralism vs. manualism controversy extends well beyond the issues surrounding prosthetic aids per se. Complex issues are involved, and there are significant shades of opinion within each camp.

The oralists believe that maximum use should be made of residual hearing and that through acoustic amplification and effective speech and auditory training, including lipreading, hearing-impaired individuals can acquire the skills they need to function effectively in a hearing world. They emphasize speech and auditory training during a child's early years, because the hearing-impaired child will be at a substantially greater disadvantage in acquiring oral language later. They point to the fact that the communication skills of children who become hearing impaired after they have acquired speech and language are generally superior to those of the prelingually impaired.

The manualists believe that sign language is the natural language of the deaf and spoken language is of little importance. Consequently, they feel that attempting to teach speech and auditory communication skills to a profoundly hearing-impaired child is bound to end in failure, which in turn will lower the child's self-esteem and could result in psychological damage. While some profoundly hearing-impaired children will never acquire a modicum of proficiency in spoken language, the oralists counter, some can and have acquired oral communication skills and are able to function effectively in a hearing world.

At bottom, the philosophical battle lines between the two groups are drawn over the issue of integrating the hearing impaired into the larger, hearing world. Most oralists believe that integration is both necessary and desirable; many manualists feel that it is neither.

The educational philosophy of total communication adopts the compromise position that profoundly hearing-impaired children should be taught both to speak and to use sign language. Though most of the schools for the deaf in the United States have ostensibly embraced this point of view, there are vast differences among them in the importance they assign to speech training. Despite their educational label, many of the total-communication schools follow the manualist philosophy de facto. Another compromise position is cued speech, which advocates that the speaker use hand gestures to reduce ambiguity in sounds that are visually identical when lip-read.

Each group advocates the development of prosthetic aids that will harmonize with its own philosophy. The manualists, a small but very influential segment of the hearing-impaired population, consist mostly of profoundly hearing-impaired individuals, many of whom grew up with sign language as their first language. They are primarily interested in sensory aids that facilitate communication by means of the printed word and aids that facilitate the learning of sign language. Interestingly, they have not yet shown great interest in devices that would permit communication by sign language over the telephone using low-bandwidth video channels.

The oralist philosophy is embraced by most of the moderately hearing impaired and severely hearing impaired and a small proportion of the profoundly hearing impaired. The oralists are interested in all types of prosthetic aids except those involving sign language, but they are particularly concerned with auditory aids, such as conventional hearing aids, aids that supplement lipreading, and aids for speech training.

The total-communication group consists primarily of severely or profoundly hearing-impaired people. Since many American schools for the deaf have recently switched to the total-communication philosophy, the proportion of hearing-impaired adults who subscribe to it is steadily growing. Their interest in sensory aids is eclectic; any device that will help them communicate more effectively, either orally or manually, is well received.

The cued-speech group is very small, because both speaker and listener need to learn special hand cues. If a practical, automatic cueing device is developed, the size and importance of this group is likely to increase dramatically.

43.2-8 Aids for Visually Impaired

The philosophical disputes that have impeded the development of aids for the hearing impaired are not as intense or as numerous in the visually impaired community. Because blindness does not preclude the acquisition of language skills, its effects are less damaging to a child's development than severe or profound deafness. Educational issues do not loom as large and battle lines are drawn much more pragmatically.

Corrective lenses, such as eyeglasses or contact lenses, are by far the most common visual aid. While they do not in themselves involve electronic signal processing, their prescription and preparation could benefit from the application of signal-processing techniques. The manual methods still used to prescribe eyeglasses and lenses are quite effective, but they are unnecessarily time-consuming and not wholly accurate; computer-aided techniques would do much to improve both efficiency and precision. To implement these techniques, however, would require overcoming the inertia engendered by years of tradition.

Aids for the severely visually impaired or blind fall into two broad groups: reading aids and mobility aids.

The most common reading aids are low-vision aids, designed to improve a poorly functioning visual system, and sensory-substitution aids in which another sensory modality—touch or hearing—is used in place of the impaired visual system. Low-vision aids typically are simple devices that provide optical magnification. Well-designed optical fiber systems have the potential to inexpensively provide a flat-field image without the spherical aberrations typical of the more common magnifying glasses.

The growing use of video terminals in the workplace and at home opens up new possibilities for low-vision aids. Closed-circuit television systems are often used to aid the partially sighted by projecting magnified text and other material on a television screen. With the growing use of computers there is also the need to improve display formats on computer terminals for the partially sighted.

Reading aids that depend on other sensory modalities range from the technologically simple to the highly sophisticated. Braille is perhaps the most well known of the sensory-substitution aids, although it is not as widely used as is commonly thought. The conventional Braille alphabet is difficult to learn and the preparation of materials is fairly expensive. Several simplified alphabets have been developed, but these are of limited scope. Current developments in computer-assisted typesetting, however, make it feasible to prepare Braille materials more economically. Connecting relatively inexpensive word processors to Braille embossers is an exciting possibility for low-cost preparation of these materials.

An audio recording of printed material is another simple sensory-substitution aid, although preparation of such talking books is expensive. Costs could be reduced if the preparation were automated using computer-generated speech. Practical text-to-speech converters have been developed, and applications of that technique to aids for the visually impaired should be pursued.

With any sensory-substitution system, the rate of information transfer is usually far slower in the substituted modality than in an unimpaired system. This fundamental problem is not as severe in the case of talking books, which increase the playback speed of the speech, thereby increasing the rate at which the spoken message is received. Since intelligibility breaks down very rapidly as playback speed is increased, methods of processing speeded speech to increase intelligibility have been developed. Lowering the frequency components of the speeded speech signal is the most common method and can be done quite easily using digital techniques.

Photoelectric scanning devices that, like Braille, provide a tactile representation of the printed word are more sophisticated. The Optacon, developed by James Bliss and John Linville at Stanford University, is the most successful of these devices.[34] A miniature camera scans the printed page, and the image is electronically transferred to a matrix of tiny vibrators, which stimulate the fingertips of the user. With training it is possible for users to tactilely recognize facsimiles of letters and other printed symbols. A serious impediment to the effective use of tactile photoelectric scanning devices is that the rate of information transfer is relatively slow. One of the reasons for the success of the Optacon is that reading rates as high as 50 words per minute can be obtained with intensive training. These rates make the device practical for highly skilled readers, but they are slower than normal rates of speech communication (100 to 200 words per minute) and reading rates of sighted people (in excess of 400 words per minute).

The development of a practical text-to-speech converter has been the most significant advance in sensory-substitution aids in the last decade. Creating a practical device meant solving a two-part problem: converting graphemes to phonemes and phonemes to spoken output. The pioneering work at the Haskins Laboratories of New Haven on text-to-speech conversion led to the development of practical procedures for automatic speech synthesis, and research conducted at Bell Laboratories and elsewhere demonstrated that intelligible speech could be generated by computer. Raymond Kurzweil exploited the constraints imposed by the use of limited, well-defined fonts of printed text with the statistical constraints of English to derive a practical recognition strategy. Combining automatic speech synthesis with his method for optical character recognition, he developed the Kurzweil reading machine, which is now used in many libraries and institutions for the blind. Though they are an important breakthrough, the machines are far from perfect. Current research is directed toward reducing errors in character recognition and in improving the unnatural, machinelike quality of the synthesized speech.

Attempts to develop practical electronic mobility aids have involved either ultrasonic probes or optical probes, which convey information by auditory or tactile displays of objects in the environment. Two significant developments are the laser cane and the binaural sonar sensor. The cane uses optical signals to convey information similar to that provided by a conventional cane, but it also uses tactile and auditory signals to give an early warning of nearby objects or discontinuities in the terrain. The sonar sensor, developed by Leslie Kay of Christchurch, New Zealand, provides a binaural sound image of objects in the immediate vicinity. The device is remarkably sensitive and can tell the user not only the spatial location of large objects, but also their texture. A blind person could distinguish between a hard, smooth surface like a wall and a nonsmooth surface like a curtain, for example.

Devices designed to replace the conventional cane and guide dog may not offer significant new advantages. Using modern signal processing techniques to provide information that is not easily obtained by conventional means—for example, using radar to distinguish between moving and stationary objects or to identify potentially dangerous objects that are moving toward the user—might prove a fruitful direction for research.

43.2-9 Aids for Speech Impaired

The design of prosthetic aids for the speech impaired is critically dependent not only on how the disorder manifests itself, but also on its underlying causes and on the other sensory, motor, or mental abilities that have been impaired.

The most common speech impairments are relatively minor problems of deviant articulation or improper phonation, occurring mainly in children. Prosthetic aids are seldom used for these problems, although simple speech-training devices similar to those used to teach speech to the profoundly hearing impaired are frequently employed. In these aids, signal-processing techniques are used to extract important speech features (voicing, nasality, friction) and display them visually or tactually.

In cases where automatic extraction of the speech features is extremely difficult to obtain from the acoustic speech signal, special sensors are used to monitor the relevant articulators. The palatograph, for instance, is a very fine plastic sheet that fits on the palate and contains tiny sensors that monitor tongue placement. Another very useful instrument, the laryngograph, very precisely measures the opening and closing of the vocal cords by monitoring changes in electrical capacitance across the throat. In addition to their value in diagnosing speech problems, these and similar instruments are also useful for obtaining basic information on the speech production process.

The use of speech-analyzing instruments as both speech-training aids and diagnostic tools is growing steadily. Although a trained human ear is a remarkably sensitive instrument for detecting and identifying speech problems, it is far from infallible. Those aspects of impaired speech that the human ear is relatively poor at diagnosing happen to be well suited to instrumental evaluation (for example, improper control of nasality, shape of the fundamental frequency contour), and many aspects of impaired speech that are very difficult for a machine to identify (for example, phoneme substitutions) can be identified relatively easily by ear. There is a need for researchers in speech pathology to explore the merits of both these modalities and to develop diagnostic procedures that most efficiently combine the two.

Signal processing aids are frequently used in the case of severe speech impairment. These aids either augment an impaired system or substitute for it. One of the earliest electronic aids designed specifically for the severely speech impaired is the artificial larynx, which provides pulsive acoustic stimulation to replace the stimulation that had been provided by vibration of the vocal cords. Other augmentative aids amplify a weak voice or provide timing cues to a person with poor control of speech rhythm.

Most substitution aids permit the generation of messages by motor activities other than speaking. People who cannot speak as a result of paralysis, nervous disorder, or reduced mental functioning but are still capable of a limited number of actions commonly use a device in which they can communicate a short message by means of typewriter keys, buttons labeled with special symbols or diagrams, or a limited set of body movements. In more recent versions, a speech synthesizer producing the desired message may replace or augment printed output. Although very slow, these devices are invaluable in allowing speech-impaired people to convey emergency messages or other important information. The special symbol sets used with these devices have also opened up a new world for people who cannot communicate through words but who can convey concepts. A major constraint of these devices is the limited number of diagrams they provide, but computer technology could be used to expand the range and speed of diagram generation.

For people with good manual dexterity but no useful speech output, an electronic speech synthesizer, such as the widely used Votrax, can be extremely useful, although keying in phoneme sequences by hand can be time-consuming. Communication can be speeded up by using abbreviations for common words and phrases or a shorthand typewriter that feeds the message into a speech synthesizer. A convenient computer-based retrieval system could enhance the speed of input and output. Extensive training may be necessary, however, for a person to reach rates of communication comparable to those of normal speech.

Ingenuity is required to match the speech synthesis device to the capabilities and needs of the user, and improved methods of coding the input to the devices is needed. The coding methods must reflect the inherent structure of speech in a concise, learnable format.

References

1 M. W. Pascoe, "An Approach to Hearing Aid Selection," *Hearing Instruments* **29**:12–16, 36–37 (1978).

2 L. D. Braida, N. I. Durlach, R. P. Lippman, B. L. Hicks, W. M. Rabinowitz, and C. M. Reed, *Hearing Aids—A Review of Past Research of Linear Amplification, Amplitude Compression and Frequency Lowering*, ASHA Monograph No. 19, Rockville, MD; American Speech-Language-Hearing Association, 1979.

3 H. Levitt, A. Neuman, R. Mills, and T. Schwander, "A Digital Master Hearing Aid," *Journal of Rehabilitation Research and Development*, to appear, 1985.

4 E. Villchur, "Signal Processing to Improve Speech Intelligibility in Perceptive Deafness," *Journal of the Acoustical Society of America* **53**:1646–1657 (1973).

5 H. Levitt, J. M. Pickett, and R. A. Houde, eds., *Sensory Aids for the Hearing Impaired*. See introduction to Part III, IEEE Press, New York, 1980.

6 J. B. Allen, D. A. Berkley, and J. Blauert, "Multimicrophone Signal-Processing Technique to Remove Room Reverberation from Speech Signals," *Journal of the Acoustical Society of America* **62**:912–915 (1977).

7 B. Johannson, "The Use of the Transposer for the Management of the Deaf Child," *International Audiology* **5**:362–373 (1966).

8 N. Guttman and J. R. Nelson, "An Instrument that Creates Some Artificial Speech Spectra for the Severely Hard of Hearing," *American Annals of the Deaf*, **113**:295–302 (1968).

9 D. Ling, "Speech Discrimination by Profoundly Deaf Children Using Linear and Coding Amplifiers," *IEEE Transactions Audio and Electroacoustics* **AU-17**:298–303 (1969).

10 M. Mazor, H. Simon, J. Scheinberg, and H. Levitt, "Moderate Frequency Compression for the Moderately Hearing Impaired," *Journal of the Acoustical Society of America* **62**:1273–1278 (1977).

11 K. O. Foust and R. W. Gengel, "Speech Discrimination by Sensorineural Hearing Impaired Persons Using a Transposer Hearing Aid," *Scandinavian Audiology* **Z**:161–170 (1973).

12 H. Levitt, J. M. Pickett, and R. A. Houde, eds., *Sensory Aids for the Hearing Impaired*. See introduction to Part IV, IEEE Press, New York, 1980.

13 K. K. Foss, *Identification Experimentation on Low Frequency Artificial Codes as Representation of Speech*, B.S. dissertation, Massachusetts Institute of Technology, 1983.

14 E. Douek, A. J. Fourcin, B. C. J. Moore, and G. P. Clark, "A New Approach to the Cochlear Implant," *Proceedings of the Royal Society of Medicine* **70**:379–383 (1977).

15 C. W. Parkins and S. W. Anderson, eds., *Cochlear Prostheses: An International Symposium, Annals of the New York Academy of Sciences* **405**, New York, 1983. A description of the different approaches used by various research groups throughout the world.

16 G. M. Clark, Y. C. Tong, R. Black, I. C. Forster, J. F. Patrick, and D. J. Dewhurst, "A Multiple-Electrode Cochlear Implant," *Journal of Laryngology and Otology*, 1979.

17 F. B. Simmons, "Electrical Stimulation of the Auditory Nerve in Man," *Archives Otolaryngology* **84**:2–54 (1966).

18 W. H. Dobelle, S. S. Stensaas, M. G. Mladejovsky, and J. B. Smith, "A Prosthesis for the Deaf Based on Cortical Stimulation," *Annals of Otology, Rhinology, and Laryngology* **82**:445–462 (1973), with discussion by H. Davis, *et al.*, 462–463.

19 A. M. Liberman, F. S. Cooper, D. P. Shankweiler, and M. Studdert-Kennedy, "Why Are Spectrograms Hard to Read?" *American Annals of the Deaf* **113**:127–133 (1968).

20 R. A. Houde, "Visual and Tactile Aids for Speech Reception: Spectrum and Temporal Displays," in H. Levitt, J. M. Pickett, and R. A. Houde, eds., *Sensory Aids for the Deaf*, IEEE Press, New York, pp. 243–246, 1980.

21 V. W. Zue and R. A. Coles, *Proceedings of ICASSP—1979*, IEEE Press, New York, 1979, pp. 116–119.

22 H. Levitt, "Speech Processing Aids for the Deaf: An Overview," *IEEE Transactions in Audio and Electroacoustics* **AU-21**:269–273 (1973).

23 H. Upton, "Wearable Eyeglass Speech Reading Aid," *American Annals of the Deaf* **113**:222–229 (1968).

24 W. O. Beadles and B. Wilson, "Research on the Autocuer," *Gallaudet Research Conference on Speech Processing Aids for the Deaf*, J. M. Pickett, ed., Gallaudet College, Washington D.C., 1977.

25 A. J. Goldberg, "A Visible Feature Indicator for the Severely Hard of Hearing," *IEEE Transactions Audio and Electroacoustics* **AU-20**:16–23 (1972).

26 M. Rosenstein, "A Vowel Meter and Its Application to Constrained Speech," *IEEE Transactions Audio and Electroacoustics* **AU-13**:135–141 (1965).

27 M. Rosenstein, "Feasibility of Supplying High-Frequency Speech Information via Visual Channels," *Journal of the Acoustical Society of America* **46**:82(A) (1969).

28 W. R. Huggins and R. A. Houde, personal communication, 1984.

29 M. Weiss and P. Chien, personal communication, 1984.

30 R. Stuckless, "Real-Time Graphic Display and Language Development for the Hearing Impaired," *Volta Review* **83**:291–300 (1981).

31 H. Levitt, "Use of a Pocket Computer as a Communication Aid for the Deaf," *American Annals of the Deaf* **128** (1983).

32 R. S. Nickerson and K. N. Stevens, "Teaching Speech to the Deaf: Can a Computer Help?" *IEEE Transactions Audio and Electroacoustics* **AU-21**:445–455 (1973).

33 N. McGarr, J. Head, M. Friedman, A. M. Behrman, and K. Youdelman, "Phonatory Problems in Hearing-Impaired Children: Systematic Speech Training and Sensory Aids," *Journal of Rehabilitation Research and Development*, to appear (1985).

34 J. C. Bliss and J. G. Linvill, "A Direct Translation Reading Machine," in R. Dulton, ed., *Proceedings of the International Conference on Sensory Devices for the Blind*, St. Dunstans, London, 1967.

43.3 ELECTRICAL PROSTHESES (UPPER LIMB)

William Lembeck

43.3-1 History

While there are sketches and notes describing powered assistance for amputees that go back to the nineteenth century, workable schemes in electrically powered limbs did not appear until the 1920s.

The first electronically controlled device was demonstrated in Germany about 1944; it used vacuum tubes and a huge power supply. A flurry of activity occurred after the Russian demonstration of a myoelectrically controlled (muscle-generated signals) hand that used germanium transistors, at the Belgium World's Fair in 1958. This was the beginning of myoelectric control (what the layman would call brain-activated) of prosthetic articulations performed by battery-powered DC motor-driven mechanisms.

In essence, the 1950s and 1960s were a time of experiment during which the experimenter had little pressure to produce efficient and reliable devices that would upgrade the level of the amputee's function. The pressure increased radically in the late 1960s, when many Thalidomide-deformed babies where born with serious upper-limb deficiencies. The governments of Germany, Sweden, England, and Canada, who had allowed the use of this drug, now had a responsibility to provide funds and technologic help.

At first, pressurized CO_2 cylinders driving various fluid actuators were used, but by the early 1970s batteries powering tiny but powerful DC motors emerged as the leading source of external power and activation.

Despite the state of advanced technology in which we live, it would be misleading to assume that the use of externally powered prostheses is common and widespread. Presently, the prosthetic profession still relies heavily on conventional, body-powered devices.

43.3-2 Body-Powered Devices

Prior to development of electric devices, body power had been the sole powering and controlling mechanism not only for hands or hooks but for all functional upper-limb prostheses.

The typical body-powered system uses a flexible cable attached at one end to a harness on the shoulder girdle and at the other end to the moving member of the prehension device. The device is operated by a pulling movement. Relaxation of the body returns the device to its normal state. Speed, force, and position of the prehension "fingers" is directly related to the motion of the body musculature and skeletal structure around the shoulder girdle.

As such, body-powered devices can be considered proportionally controlled with some degree of proprioceptive feedback. These positive attributes, coupled with the mechanical simplicity and low cost of body-powered devices, make them reliable mechanisms that will continue in the prosthetic marketplace for the forseeable future.

It is the disadvantages of the body-powered devices, however, that are responsible for the development of externally powered prostheses. The large demands on the body's energy, particularly with congenital amputees who may have additional deformities; the inadequate excursion and force that the high-level amputee experiences; and/or the painful pressures on the scarred and sensitive trunk caused by the harness strapping on an acquired amputee are the major impediments to body-powered prosthesis wear.

43.3-3 Why Electric Energy?

The energy source for an externally powered prosthesis must provide reliable and sufficient energy for a patient's daily functions. The source must be very portable and preferably rechargeable at the patient's home during the hours of sleep.

TABLE 43.11. MAJOR JOINTS OF UPPER LIMB, RANGE OF MOTION, AND TYPES OF PROSTHESES AVAILABLE

Bones Involved in Joint Movement	Normal Anatomical Range of Motion	Common Nomenclature For Various Amputation Levels[a]	Available Prosthetic Joints	Types of Prostheses Used for Various Subcatagories of Amputations[b]: Long Stump	Short Stump	Very Short Stump
Metacarpophalangeal	Flexion: to 90° Active extension: to 40° Passive extension: to 90°	Partial hand	Partial hand prosthesis	P-√√; B-√; E-×		
Wrist (carpals pivot from radius)	Extension: to 85°	Wrist disarticulation	Prehensile hook	P-×; B-√√; E-√		
	Flexion: to 85°		Prehensile hand	P-√; B-√√; E-√		
			Wrist flexor	P-√; B-×; E-×		
Forearm (radius rolls over ulna)	Pronation: 85°	Below elbow	Prehensile hook or hand	P-×; B-√√; E-√	P-×; B-√√; E-√√	P-×; B-√√; E-√
	Rotation (palm down)		Wrist rotator	P-√√; B-×; E-×	P-√√; B-√; E-√	P-√√; B-×; E-×
	Supination: 90° Rotation (palm up)		Wrist flexor	P-√; B-×; E-×	P-√; B-×; E-×	P-×; B-×; E-×
Elbow (ulna pivots from humerus)	Extension: 0°	Above elbow	Prehensile hook	P-×; B-√√; E-√	P-×; B-√√; E-√	P-×; B-√√; E-√
	(women & children can		Prehensile hand	P-√; B-√√; E-√	P-√: B-√√; E-√	P-√; B-√; E-√
	hyperextend to 10°)		Wrist rotator	P-√√; B-×; E-√	P-√√; B-×; E-×	P-√√; B-×; E-×
	Active flexion: to 145°		Wrist flexor	P-√; B-×; E-×	P-√; B-×; E-×	P-×; B-×; E-×
	Passive flexion: to 160°		Elbow (includes passive humeral rotation)	P-×; B-√√; E-√	P-×; B-√√; E-√√	P-√; B-√; E-√√

Shoulder (spherical surface of humeral head moves three axes within glenoid cavity. Additional movements occur at acromioclavicular, scapulothoracic, and sternoclavicular joints)	Humeral rotation (along longitudinal axis of the humerus): Lateral rotation: 80° Medial rotation: 90° Humeral abduction (upper limb moves away from body): 180° (vertically upward) Humeral adduction (upper limb moves toward the body): 30° if arm moves in front of the body Humeral flexion: 180° (arm vertically upward) Humeral extension: 50° (arm to rear of body) Scapula adduction and abduction provide about 15 cm of medial to lateral displacement	Shoulder disarticulation (forequarter amputations use the same prosthetic joints with slight variations in controlling activation of these joints)	Prehensile hook or hand	P-√; B-√√; E-√√
			Wrist rotator	P-√√; B-×; E-√
			Wrist flexor	P-√; B-×; E-×
			Elbow	P-√; B-√√; E-√√
			Shoulder flexor	P-√√; B-×; E-×
			Shoulder abductor	P-√; B-×; E-×
			Shoulder abductor and flexor	P-√√; B-×; E-×

[a] Levels arranged from least loss to greatest loss; each succeeding level requires prosthetic joint for loss of anatomical joint(s) in same and previous rows.
[b] P = passive and/or friction joint; B = body-powered joint; cable actuated; E = electrically powered joint; either switch or myoelectric control.
Note. √√ = most often used; √ = sometimes used; × = used sparingly and, in some categories, never.

The nickel-cadmium battery answers all of these requisites, and does so for close to 1000 rechargeable cycles. An efficient adult hand, for example, uses between 3 and 6 W-hr of energy for a wearing period of 12 hours. Five AA cells, costing $1.50 each, can power this hand for three years, with only a $1.00 additional cost for the entire three years of nightly recharging energy. High energy-to-weight and energy-to-volume ratios are still the key factors that could use greater improvement. If the lithium cell were rechargeable it could quadruple these energy ratios and would be the ideal prosthesis energy source. Since it is not rechargeable, however, use of nonrechargeable lithium cells would cost $4 to $8 per day.

43.3-4 Why Only Upper-Limb Prostheses?

If the nickel-cadmium battery can adequately supply the energy needs for upper-limb prostheses, one may wonder why there is not more research and development in lower-limb prostheses, since by far the majority of patients are lower-limb amputees.

Aside from the much greater energy demand for proper prosthetic function for lower limbs, many upper-limb amputees would have no function at all without external power, while present lower-limb prostheses or wheelchairs provide sufficient function without the use of electrical power. Thus the needs of the severely handicapped upper-limb amputee are the ones that are best served by electrical power and control. As energy sources become more efficient and compact, it is likely that commercial applications of external power will be applied to lower-limb amputees as well.

43.3-5 Which Anatomical Joint?

Table 43.11 describes the major joints of the upper limb and their average range of motion. Obviously, to get food to the mouth or bring an object close to the eyes, a grasping (prehension) device (traditionally called a terminal device whether it be a hand or a hook) and an arm-flexing device (anatomically and prosthetically called an elbow) are essential. Less essential to function is wrist rotation (called pronation and supination), which interestingly does not occur at the wrist, as is commonly believed, but over the entire length of the forearm. Although the shoulder, a triple-axis joint, is essential for spatial placement of the hand, it does not have to be supplied with power to perform the rudimentary functions a high-level amputee needs for everyday living. Bending the wrist is the least essential amputee function. Table 43.11 also lists the relative importance of powering the joints of the upper limb. For example, an amputee with a short above-elbow stump would most probably be fitted with a body-powered hand or hook, a passive friction wrist rotator, and either a body-powered or electrically powered elbow.

Ideally, if sufficient electrical energy were available, all of the upper-limb joints could be powered. In reality, powering a joint requires suitable control of that joint. This is presently the limiting factor in the selection of a powered joint. Insufficient control sites coupled with the brain's inadequate ability to learn to control the devices smoothly and synchronously have, in most cases, limited the number of electrically powered joints to two on each side of the amputee, with three powered joints being the maximum practical number at present.

Another factor that determines the choice of a powered joint is the age of the amputee. Young children do not require the manipulative skills of their elders, and passive friction joints or spring assists are often adequate. Powered elbows and hooks, in combination with conventional body-powered prosthetic components (called hybrid systems), are more suitable for amputees from ages 5 to 12.

Teenagers, especially, feel the need for cosmesis, and use of powered hands is becoming more prevalent in this age group.

Adults usually make the choice of hands or hooks, whether powered bodily or electrically, depending on their desire for either cosmesis or function. Generally the hook is considered more functional than the cosmetic hand, since the hand's many fingers obstruct visibility, and the cosmetic hand is awkward to use.

Other considerations in the adult include whether the impairment is unilateral or bilateral and such other factors as cost, reliability, ease of fitting, comfort, and weight. Most often a hybrid system that uses both body and electrical power is used.

43.3-6 Prehension Devices (Hands and Hooks)

The human hand, far more complex than most machines, is the most difficult to synthesize in a prosthetic limb. Prehension, whether it be a fine or crude grasp, is the basic function of the hand. In the past, the hand amputee would either choose between finely controlled prehension, using a split-finger hook, or grossly controlled grasp (although greatly improved cosmesis) using an artificial

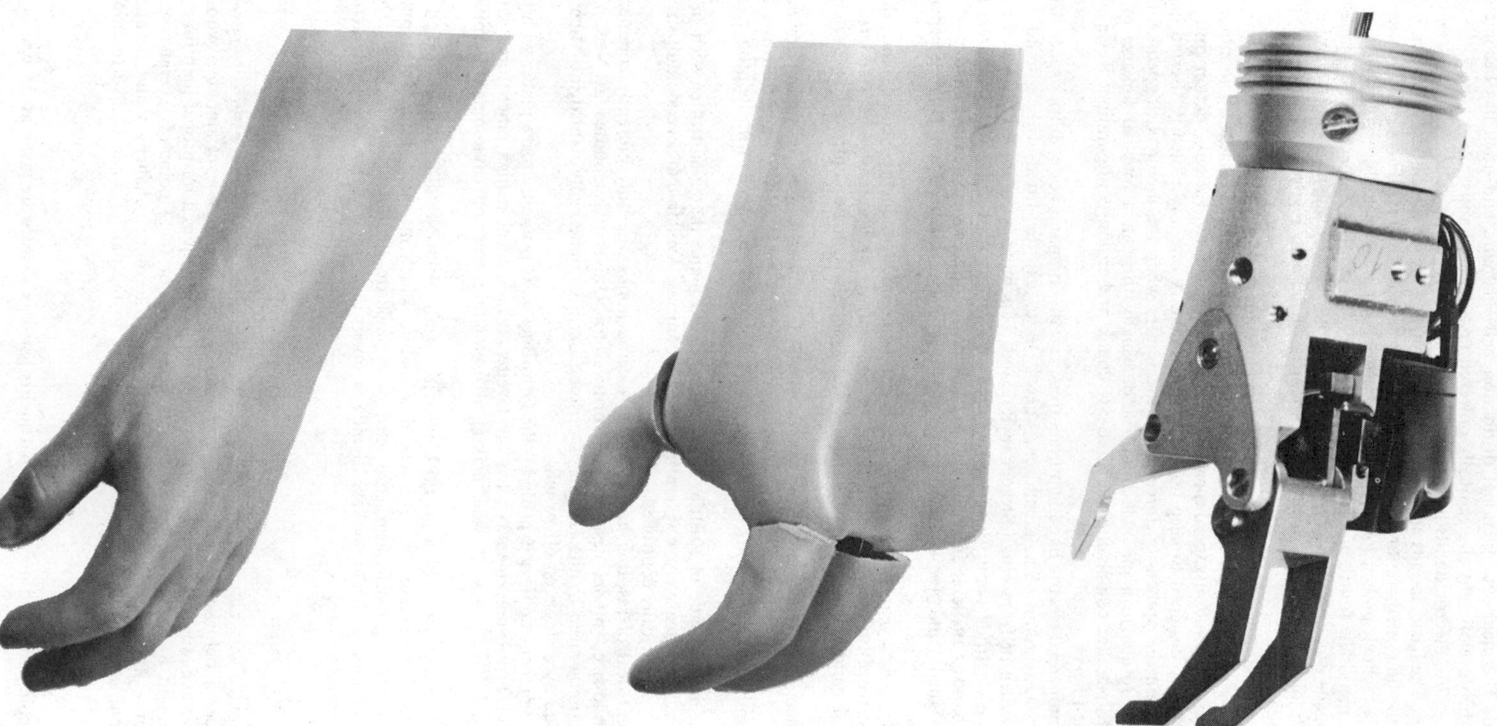

Fig. 43.27 Views of the Swedish myoelectrically controlled child-size hand showing three stages of enclosure: the outer cosmetic glove, the inner cosmetic coverings, and the internal mechanism.

glove-covered hand. The clear-cut functional distinction between hand and hook has become less polarized through the development of electrically powered myoelectric hands.

Since the early 1970s, electric hands have become commercially available in many sizes and for amputees of all ages. The basic designs and operating mechanisms are similar but they differ in motor selection, speed-reduction method, or kinematic coupling to the fingers. Their grasp pattern is classified as three-jaw chuck prehension, with the thumb opposing and usually nesting between the second and third fingers. The fourth and fifth fingers are passive flexible members usually made of a rubberlike filling inside a cosmetic outer glove. The thumb and its opposing two-finger member are rigid structures with joints that merely pivot around the metacarpal axis. Figure 43.27 shows a typical design of these hands.

The powering mechanism is a DC permanent-magnet motor, usually made by Micro Mo Electronics,† a company that supplies most of the motors used in electric prostheses. The motor has both its torque increased and its speed proportionately decreased by various friction and gear mechanisms, ingeniously designed to keep noise to a minimum. A one-way clutch is inserted in the mechanism to maintain the prehension force on all the fingers when the motor is stopped. A final output gear supplies torque to a movable thumb and, usually through a four-bar linkage, to the opposing two-finger member completing the three-jaw chuck prehension configuration. Reversal of the motor opens the fingers.

Of the many commercial electric hand manufacturers, certain companies or countries specialize in particular sizes or age groups. Small children, ages 2 to 6, can be fitted with a hand from Systemteknik in Sweden or from Hugh Steeper in England. This hand is controlled myoelectrically, using on-off surface electrode control. Simple modifications can convert this hand to microswitch control. A similar hand is available from the Ontario Crippled Children's Center in Canada.

A similarly designed hand, for children ages 5 to 9, is also available from Hugh Steeper in England.

For the amputee over 9 years of age, there are several companies producing electric hands. Four sizes are made by the Otto Bock Company, whose components are presently the standards for electrically powered hand prostheses. The hands of this company can be controlled by switch or myoelectrically.

Another myoelectrically controlled hand, made by Fidelity Electronics, is an outgrowth of combined research with the Veterans Administration Prosthetic Center (VAPC) and Northwestern University. The hand is for adults only and is very responsive to input signals with its proportional control.

The Viennatone hand, made in Austria, is an older reliable hand that is not often seen in the United States but is used extensively in several European countries. Available in several adult sizes, it can be switch or myoelectrically controlled.

Although a great deal of research has gone into electric hooks, the only presently commercial realization is a child-sized model designed at Northwestern University, modified at New York University, and manufactured by the Hosmer/Dorrance Corp. Commercially named the Michigan hook, it is controlled by a single on-off switch.

Another electric prehension device, made by the Otto Bock Co., is an adult-size pincer with jaws that remain parallel during movement. Called the Greifer, it can develop high pinch force with a manual override and can open very wide, making it useful in a working environment despite its uncosmetic appearance.

A recent development from New York University is an electric pulling device, called the prehension actuator, that operates like a powerful miniature winch mounted inside an amputee's hollow forearm. This pulling mechanism, controlled by a single on-off switch or myoelectrode, can actuate any of the multitude of body-powered hands or hooks that are available. It is manufactured by the Hosmer/Dorrance Corp.

43.3-7 Elbows

The high-level amputee with elbow or shoulder joint loss has more need for an externally powered prosthesis than does a below-elbow amputee. Unfortunately for the high-level amputee, who has fewer muscular sites available for control, more powered joints are required for proper limb function. Thus the improvement in the prosthetic elbow has been a major focal point of several research centers, both because it presents to the creative designer extraordinary challenges and because it offers great personal satisfaction to the designer as well as to the patient when lifelike function is achieved.

† Addresses of all companies and university departments listed as sources are given at the end of this section.

At first glance, a prosthetic researcher sees the elbow as a simple hinged joint with a range of 140°. A look at the power requirements, however, reveals that to flex the elbow fully in 1 second to lift a nominal 4-lb load held in the hand at a distance of 12 in. from the elbow flexing axis, requires 9 W. Considering the practical weight and size restrictions for an elbow prosthesis, and given the most powerful DC motor within these restrictions, the efficiency and torque requirements of the speed reduction mechanism approach unattainable values. Thus unorthodox approaches and imaginative solutions have been found to approximate anthropomorphic movement. The following descriptions of five elbow designs that have attained commercial status are cursory examinations of the unique features of these devices.

The Ontario Crippled Children's Center (OCCC) elbow, manufactured by Variety Village Electro-Limb Production Centre, is the most conventional approach taken to provide elbow function. The drive mechanism is housed in a round-bottomed cylindrical module, mounted at the lower end of the humeral section, with the shaft supplying the output torque aligned along the natural elbow-flexing axis. A pancake-shaped DC motor, located at the upper end of this module, outputs to a self-locking worm drive and then, through a series of spur gears, transmits the required speed and torque to the drive shaft.

Designed for children, the overall live-lift specifications are slightly lower than for adult elbows (see Table 43.12 for all the specifications of the elbows described). What the design lacks in efficiency and quietness it makes up in its small size, simplicity, comparatively low cost, and a few other nonessential but nonetheless very much appreciated features. These include free swing of the forearm in the extended position, an overload clutch in case of an accidental fall on the arm, dimensional compatibility with standard upper-limb prosthetic components, and an optional integral control switch activated by a cable-pulling technique familiar to above-elbow amputees wearing a conventional body-powered prosthesis. Controlled mostly by microswitches, this elbow is popular throughout North America.

The New York University-Hosmer/Dorrance Corp. elbow is used by amputees from age 7 onward. Such use is accomplished by the provision of two sizes of elbow housings that span all age groups, while utilizing the same internal mechanism. Basically a 5-V system, the substitution of a 6.3-V interchangeable battery pack increases the live-lift capabilities to adult proportions.

The drive mechanism, encased in a humeral section module, is dimensionally compatible with conventional components. A MicroMo 2233 DC motor, mounted along the elbow flexing axis, supplies the driving torque through a 650 : 1 speed reduction mechanism. Almost noiseless operation is accomplished using a special O-ring pulley drive coupled to three pairs of precision spur gears, after passing through an overload clutch. Positional self-locking is assured by a fail-safe mechanical lock driven by a separate motor-gearbox cam mechanism.

Using a highly efficient, noiseless drive mechanism, this elbow provides a compact, lightweight, and reliable device at the low-cost end of the prosthetic elbow spectrum.

The Fidelity elbow, designed by the Veterans Administration Prosthetic Center, is another example of a complete elbow mechanism encased in a housing fitted to the end of the humeral section. This unit fits adult patients only.

Fidelity's 8.5-V switch-controlled elbow uses a modified 2233 MicroMo motor that drives a harmonic speed reducer directly. This attempt at state-of-the-art mechanism design has unfortunately fallen short of its anticipated goal because of the excessive noise associated with one-step harmonic gear reduction. Although commercially produced, it is rarely used because of this noise problem.

The first successful attempt at myoelectric control of an elbow system has been the Boston arm joint venture between the Liberty Mutual Insurance Co. and the Massachusetts Institute of Technology. Since then the major redesign of this elbow has occurred at Liberty Mutual's research center and the system is now known as the Liberty Mutual Boston elbow.

Designed for adults, this heavy-duty unit has its drive mechanism mounted in a humeral module, while the 12-V battery and electronics are housed in the empty space in the amputee's forearm. The drive incorporates one-way clutching, planetary gearing, and a final harmonic-gear reduction to supply high-output torque in a compact space. While the system includes most of the desired features for an elbow, especially a proportional, myoelectric control, the cost of the system prevents its widespread use in the prosthetic field.

The most advanced elbow system is the Utah artificial arm, a product of the University of Utah and Motion Control, Inc. Its structural, mechanical, and electronic systems utilize advanced state-of-the-art design.

Structurally, injection-molded carbon and glass-fiber reinforced nylon are used for the exoskeletal construction. Mechanically, a low-noise timing belt, combined with two subsequent high-efficiency Evoloid helical gearing stages, provides relatively low 323 : 1 speed reduction, permitting the forearm to free swing through the transmission. A torque strain-gauge load-cell, a velocity-feedback tachometer, an implanted thermistor in the drive motor, and a separate motorized lock mechanism all provide feedback to the sophisticated electronic control system.

TABLE 43.12. PERFORMANCE SPECIFICATIONS FOR FIVE ELECTRIC ELBOWS

Feature	Variety Village	NYU-Hosmer	Fidelity	Liberty-Boston	University of Utah
Battery system	12 V, nickel-cadmium, 10 cells @ 500 mAh	5 V, nickel-cadmium, 4 AA cells @ 500 mAh	8.4 V, nickel-cadmium, 7 cells @ 450 mAh, 1 hr charge rate with thermal sensor	12 V, nickel-cadmium, 10 cells @ 450 mAh	12 V, nickel-cadmium, 10 cells @ 450 mAh
Weight					
Elbow alone	370 gm	420 gm	383 gm	1,130 gm	950 gm
Battery pack	312 gm	140 gm	175 gm	(complete system)	(complete system)
Range of travel	0 to 135°	5 to 135°	5 to 140°	0 to 135°	5 to 140°
Live lift					
Maximum torque to stall	5.4 N-m	5.7 N-m	6.1 N-m	6.1 N-m	5 N-m
Time for full flexion of unloaded arm	2.5 s	1.5 s	Approx. 1 s	Approx. 1 s	0.9 s
Maximum torque load before clutch slips	16 to 23 N-m	20 to 24 N-m	68 N-m maximum (no clutch)	70 N-m	95 N-m maximum (no clutch)
Dimensions of elbow	5.6 cm diameter, 7.0 cm elbow axis to end of socket	Medium: 6.0 cm diameter; large: 7.1 cm diameter; both: 6.4 cm elbow axis to end of socket	7.5 cm diameter, 5.7 cm elbow axis to end of socket	7.5 cm diameter, 5 cm elbow axis to end of socket, 21 cm minimum forearm length	7.0 cm diameter, 4.8 cm elbow axis to end of socket, 17 cm minimum forearm length
Free swing capability	At full extension	at full extension	External mechanism at full extension	Limited to 30°	Through transmission; adjustable damping
Noise level (subjective)	Always audible, increasing in intensity at greater loads	Barely audible at all speeds and loads	Very loud, often annoying	Audible, but not objectionable	Very low
Control	Options: internal double function pull switch; external pull or push switch; three-state myoelectric, on-off control	Double-function pull or push switch; double function myoelectric on-off control	Double-function pull switch	Proportional myoelectric or on-off microswitch	Proportional myoelectric

Electronically, a 12-V battery (which, since this is the heaviest component, is located in the lower humeral section) powers a proportional control utilizing myoelectric signals. Hybrid circuitry is used for minimum weight, volume, and power consumption. The circuitry can be adapted to control other arm functions as well as those of the elbow. It is the overall dynamic response of this system that sets it apart from all others and gives it the appearance of a real moving elbow.

Space does not permit a complete discussion of the multiple design achievements of this device. Along with its advanced performance is a cost factor an order of magnitude higher than that of the first three elbow systems.

43.3-8 Wrists and Other Powered Devices

As previously explained, elbow motion and prehensile grasp are the primary articulations for minimal upper-limb function. Lack of proper control sites generally limits the number of devices that can be activated to these two categories. Therefore the commercial development of other electric devices has been limited, and further work has only proceeded in wrist rotators.

Otto Bock and Viennatone have led the field, while others are developing their own wrist rotators. The usual procedure is to drive a worm-gear output by a subminiature Micro Mo motor-gearbox combination, since high torque and speed are not essential. What is essential is the positioning of the prehension device. Thus wrist rotator drives can be compact, lightweight, and low powered. Either switch or myoelectric controls provide activation for these devices.

Many other powered devices are either on the drawing board or will soon be forthcoming. These include devices for wrist flexion, humeral rotation, and kinematically coupled elbow and shoulder movement and, of course, a whole new generation of devices for prehension.

43.3-9 Control Systems

Although intimately associated with and, in many cases, specifically designed for the devices providing motion, prosthetic controls can be classified as two types: on-off and proportional control. Each type can be further subdivided into mechanical switch and myoelectric activation. In practice, however, switches are only used for on-off control, while myoelectrics are equally divided between on-off and proportional control.

The simplest type of control is the mechanical or myoelectric switch providing a single on-off (SPST) function for a unidirectional motor. As previously described, the "Michigan hook" and the "prehension actuator" are activated by such a single-function control.

The next level of on-off activation is the double-function control, which can be two SPDT switches either mounted independently or mounted in a single push- or pull-actuated enclosure, for example, the Bock, NYU-Hosmer, or OCCC switch. Activation usually appears in the sequence off—1 on—off—2 on. The same double-function action can be provided by two separately located myoelectrodes, for example, the Bock myoelectrodes, or by a three-level myoelectric sensor, for example, Bock and University of New Brunswick electrodes. The double-function control is used to activate the OCCC, NYU-Hosmer, and Fidelity elbows, as well as the Systemteknik, Viennatone, and Bock prehension devices.

A third level of on-off activation is triple-function control, which has an operational form off—1 on—off—2 on—off—3 on. NYU-Hosmer provides this triple function in a single pull-activated enclosure to control its combined elbow and prehension devices. Figure 43.28 shows an electrical schematic of switch-controlled prosthetic devices often used for above-elbow amputees.

Endless combinations of single- and double-function on-off controls can activate any combination of electric devices. Despite the many worldwide producers of electric prosthetic devices, it is encouraging to see how interchangeable the control systems have become.

Proportionally controlled prostheses are a closer step toward natural arm movements. Myoelectric control provides the best means of attaining variable motion, although a few linear transducers are available to produce positional response.

Basically, cutaneous myoelectric signals, usually preamplified at the source, are processed at a control stage to produce a command signal for a pulse-width-modulated power amplifier. This is the general means of providing variable speed to a DC drive motor. Figure 43.29 shows the overall electronic schematic for the Liberty Mutual Boston elbow system. Aside from more natural motion, proportional control provides high speed for rapid positioning and slow speed for precise location.

At the present, four proportionally controlled systems are available for prostheses. They are the Liberty Mutual Boston system for its elbow, the Fidelity Electronics system for its hand, the University of Utah system for its elbow and for the Bock hand, and the Northwestern University system, which is adaptable to several prosthetic devices.

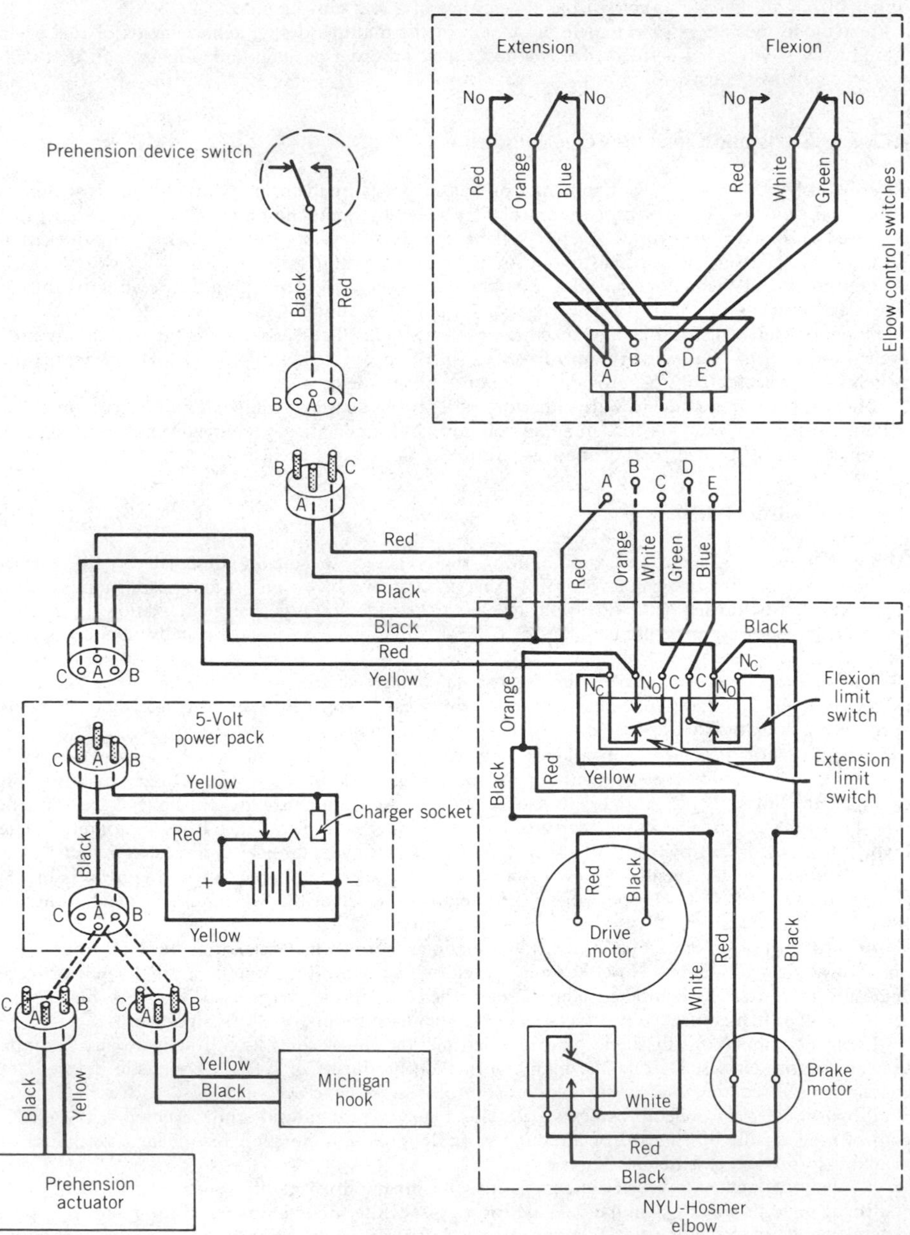

Fig. 43.28 Electrical schematic of the New York University-Hosmer/Dorrance Corp. on-off switch-controlled prosthetic arm components.

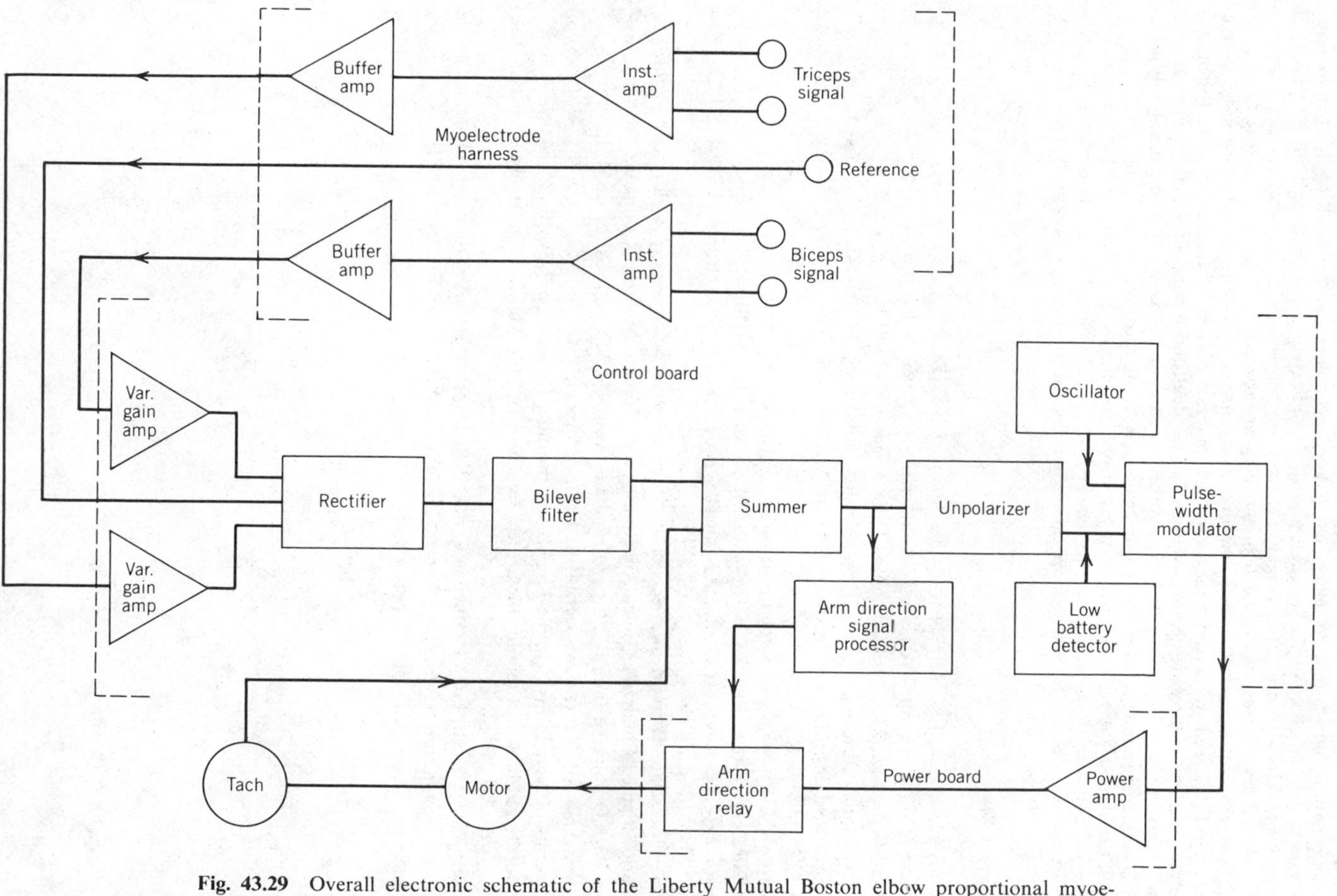

Fig. 43.29 Overall electronic schematic of the Liberty Mutual Boston elbow proportional myoelectric system.

43.3-10 Future Goals

The problem with producing prosthetic upper limbs with performance that compares favorably with that of a natural limb is that, at the present state of technology, we must still apply compromising solutions which fall short of the long-range goal of duplicating the amputated limb. The kinds of compromises, however, are ever changing as technology progresses. The space and electronic eras have allowed developments unthinkable a mere 20 years ago.

The mechanical drive mechanisms, improving constantly, are awaiting a breakthrough in interactive control and feedback systems that will allow multiple degrees of freedom beyond the present limitation to two or three degrees of freedom. Independent and natural five-finger motion, proportional wrist flexion and rotation, powered triple-axis shoulder movements, and possibly all movements occurring simultaneously can be implemented with an improved control procedure.

In the distant future, most probably the prosthetic arm will become an electromechanical-biological arm with direct attachment to the skeletonervous system. On the basis of the directions genetic and microbiological sciences are taking, the prosthetic arm may eventually evolve into the natural, biological arm.

Source List

Fidelity Electronics, Ltd., 8800 N.W. 36 St., Miami, FL 33178.

Hosmer/Dorrance Corp., P.O. Box 37, Campbell, CA 95008.

Hugh Steeper Ltd., 237-239 Roehampton Lane, London SW15 4LB, England.

Liberty Mutual Research Center, 71 Franklin Rd., Hopkinton, MA 01748.

Massachusetts Institute of Technology, Dept. of Mechanical Engineering, 77 Mass Ave., Cambridge, MA 02139.

Micro Mo Electronics, 742 Second Ave. So., St. Petersburg, FL 33701.

Motion Control, Inc., 1005 So. 300 West, Salt Lake City, UT 84101.

New York University Post Graduate Medical School, Prosthetics & Orthotics, 317 E. 34. St., New York, NY 10016.

Northwestern University Medical School, Prosthetics Research Laboratory, Chicago, IL 60611.

Ontario Crippled Children's Center, 350 Rumsey Rd., Toronto M4G 1R8, Canada.

Otto Bock Orthopedic Industry, Inc., 610 Indiana Ave. No., Minneapolis, MN 55422.

Systemteknik AB, Vasavagen 76, S-181-41 Lidingo, Sweden.

University of New Brunswick, Bio-Engineering Institute, Fredericton, New Brunswick, Canada.

University of Utah, Institute for Biomedical Engineering, 3168 Merrill Engineering Bldg., Salt Lake City, UT 84112.

Variety Village ElectroLimb Production Centre, 3701 Danforth Ave., Scarborough, Ontario, Canada.

Veterans Administration Prosthetic Center, 252 Seventh Ave., New York, NY 10001.

Viennatone, Vienna, Austria; Dist. by Hosmer/Dorrance Corp.

PART 9

SOUND AND VIDEO RECORDING AND REPRODUCTION

CHAPTER 44
CHARACTERISTICS OF SOUND

ROBERT B. NEWMAN
AND STAFF

Bolt Beranek and Newman, Inc.
Cambridge, Massachusetts

STANLEY A. GELFAND

Veterans Administration Medical Center
East Orange, New Jersey

44.1 ACOUSTICS

Robert B. Newman and Staff of Bolt Beranek and Newman, Inc.

44.1-1 Introduction

This section should help the building designer to understand the basic principles of architectural acoustics and to design buildings in which later "correction" will be unnecessary. The proper acoustic environment for any kind of activity in a building can be determined in advance, and the necessary provisions can be made during the design.

Often acoustics problems are not recognized explicitly by the designer or owner of a building. Everyone knows, for example, that special attention to acoustics is required for an auditorium or a school of music, but too few people realize that every motel and apartment house, every office building and hospital have important acoustical problems. Many of these problems can be handled with little added expense for acoustics itself. Each element of the design and construction of a building has some influence on its acoustical characteristics, and unless all of the factors involved are clearly understood and properly incorporated during the design of the building, satisfactory results will seldom be achieved. The important thing is to understand how much and what kind of influence these various elements have.

As will be shown in this section, acoustics should influence not only the choice of finish materials in rooms but also the basic disposition of the elements of the building—locating noisy rooms far away from quiet rooms, for example. Much expense for special noise-isolating construction can be avoided simply by use of good common sense in design.

Reprinted in part by permission from J. H. Callender, ed., *Time-Saver Standards for Architectural Design Data*, McGraw-Hill, New York, 1974.

Source, Path, Receiver

Almost every acoustical situation can be described in terms of a source of sound, a path for transmission of sound, and a receiver of the sound. Sometimes the source strength can be increased or reduced, the path can be made less or more effective, and the receiver can be made more attentive, either by removal of distraction or by being made more tolerant to disturbance.

If, for example, a noisy air-conditioning unit (source) bothers the occupant of an office (receiver), the problem must be analyzed in terms of what can be done about reducing the noise at the source (selection of the quietest available equipment, proper mountings, etc.), what can be done about reducing the transmission (path) by way of structure and ducts (resilient separation, absorbent lining, etc.), and what can be done to get the receiver to tolerate a bit of noise. Attack of only a single aspect of the problem may result in overdesign or an unsatisfactory solution.

Basic Problems and Criteria

Acoustics is one of the many aspects of the environment in which we live. Sounds can distract us; they can make us happy or sad. The quantity of sound we hear and its context determine the overall effect. The loud noise of an airplane flying overhead may interfere with a telephone conversation or may bring on a feeling of fear. Laughter in an adjoining classroom may prove quite distracting to students attending a lecture: The knowledge that something funny is going on in the next room will distract the students, although the level of sound transmitted will not interfere with the audibility of speech.

The basic purpose of architectural acoustics is to provide a satisfactory acoustic environment for the use for which the space is intended. In the office building the designer may wish to provide freedom from distraction or privacy for conversation. In the concert hall he or she may wish to provide maximum communication between the performers and the listeners, allowing the room itself to enhance the quality of the musical sounds. In almost any situation the designer can determine just what the environmental requirements are and then proceed to design the building to satisfy them.

Factors Influencing Acoustic Environment

Qualities that characterize the desired acoustic environment vary widely depending on how the space is to be used, how fussy the users may be, and how the space relates to other parts of the building. A library reading room, for example, should certainly be free from distraction. This freedom can be achieved either by having the reading room quiet (forbidding all sorts of disturbing sound) or by allowing the room to have a moderate, continuous background sound level of an unobtrusive, unrecognizable character that hides or masks the many minor intrusions that inevitably come along (people entering and leaving, books being delivered from the stacks, typewriters in operation, etc.). The latter is usually the more realistic approach. In a large business office one might be able to accept even more noise, but here again there are limits beyond which workers would find it difficult to perform their tasks. People usually tolerate a noise conveying no information better than they do one that tells them something about activities in an adjoining space. An expected noise is often more tolerable than an unexpected one of the same magnitude.

In addition to describing the magnitude and dynamic characteristics of the background sound, we should also describe the character of the occupied space. If a room is finished in materials that are highly sound reflective, sounds will persist for a long time and will seem to come from all directions; the space will probably be less pleasant than one that has a moderate amount of sound-absorptive finish. Everyone knows the experience of going into an empty house before the furniture has been put in place—how much more pleasant it is after rugs, curtains, and upholstered chairs have been moved in. A room can be too "dead," however, and therefore quite oppressive. There are optimum ranges for reverberation time in occupied spaces. All these matters must be thought through carefully when planning a building.

An important aspect of the acoustic environment, often overlooked, is the opportunity to introduce a sequence of sound qualities as one goes from space to space in a building. A uniform acoustic environment throughout a building can be just as monotonous as a uniform lighting environment. It is pleasant to go from a reverberant space where this quality adds a sense of monumentality to a "dead" space where, perhaps, communication is important, or where one may merely wish to sit down and read or experience a feeling of enclosure and quiet. Both kinds of space gain by contrast. It is certainly to be hoped that the old specification of "acoustical tile on all ceilings" is now out-of-date!

Factors Influencing Hearing Conditions

If the environment is to be favorable to good hearing conditions:

1. It must be completely quiet.
2. The desired sounds must be sufficiently loud.
3. The sounds must be well distributed through the room to give a desirable degree of acoustic uniformity and to avoid disturbing echoes, focusing, or "islands" of low intensity.
4. The reverberation time must be long enough to give proper blending of sounds and yet be short enough so that there is no excessive overlapping and confusion.

These simple criteria, if satisfied, will result in good hearing conditions in any space. Sometimes we can use the natural sounds in the space and, by proper design of the enclosing surfaces, achieve all of the requirements for even the most weak-voiced speakers. In large or noisy spaces it may be necessary to use a carefully designed electronic sound reinforcement system. But whatever the requirements or the space, good hearing conditions can be achieved for any type of use. The important thing, as in all aspects of acoustics, is to recognize the problems in advance and solve them in the design stage of the project, not after it is finished.

44.1-2 BASIC TERMINOLOGY AND DEFINITIONS

To deal effectively with acoustics problems in building design, the architect must be familiar with some of the basic acoustical concepts and terminology. The intelligent evaluation of a product often hinges on simple matters of acoustical terminology. Obviously it is not possible to include here all of the terms and concepts that will be encountered, but it is hoped that those outlined will cover many practical situations. (For more detailed information, refer to bibliography.)

Terms Dealing With Character of Sound

Frequency (f). Frequency is the rate of repetition of a periodic phenomenon. Sound waves are basically periodic phenomena (for example, in air they consist of a series of compressions and rarefactions of air particles moving outward from some vibrating source, and this determines the "pitch" of a sound). Frequency is basic to the description of sounds and materials to control sound. The frequency is the reciprocal of the time period, or the time necessary for the phenomenon to repeat. The unit is the cycle per second (cps) or the hertz (Hz).

The frequency range for the human ear extends from about 20 to about 20,000 Hz for young persons with acute hearing. Some musical instruments encompass almost this entire range, notably the pipe organ. The range of human speech that is most important for understanding extends from about 600 to about 4000 Hz. On the other hand, if we are concerned with the annoyance of speech sounds, as we might be in an office privacy situation, the lower frequency of the speech range may also be important, and this may extend down to about 200 Hz.

Pure Tone. A pure tone is the simplest kind of sound because it is composed entirely of sound waves of a single frequency. A pure tone can be generated by striking a tuning fork, but very few of the sounds around us are this "pure."

Musical Tone. A musical tone is actually a combination of many pure tones. For example, the striking of middle C on the piano (256 Hz) gives rise to a tone composed of this fundamental frequency plus integral multiples of this frequency, called harmonics. These harmonics are what determine the quality or "timbre" of the musical tone.

Common Sounds (Speech, Music, Noise). In real life the sounds that surround us are much more complex than the simple pure tone or musical tone just discussed. These more complex sounds include speech, music, and a much wider range of sounds that we call noise if they are sounds we do not want to hear. Figure 44.1 shows graphically these various kinds of sounds as well as a pure tone and a musical tone.

Frequency Band. For the measurement and specification of matters pertaining to sound, it is often convenient to divide the audible frequency range into sections. One common division is into octave bands, which divide the frequency range into sections centered at the following frequencies (Hz): 31.5, 63, 125, 250, 500, 1000, 2000, 4000, 8000. Further breakdowns of the frequency range are used for more detailed analyses of sound problems. These include one half octave bands, one third octave bands, and even smaller divisions of the frequency range.

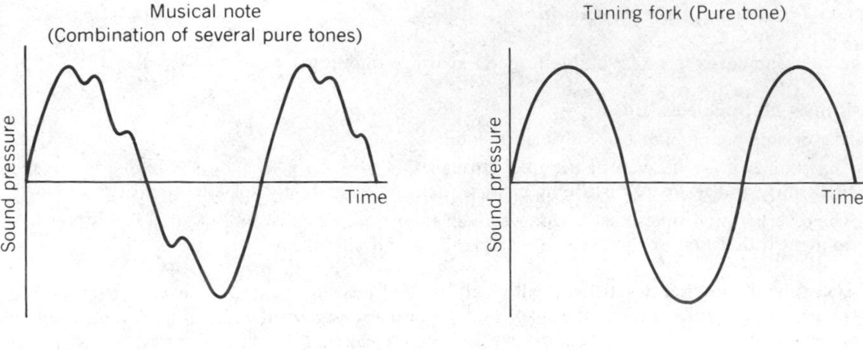

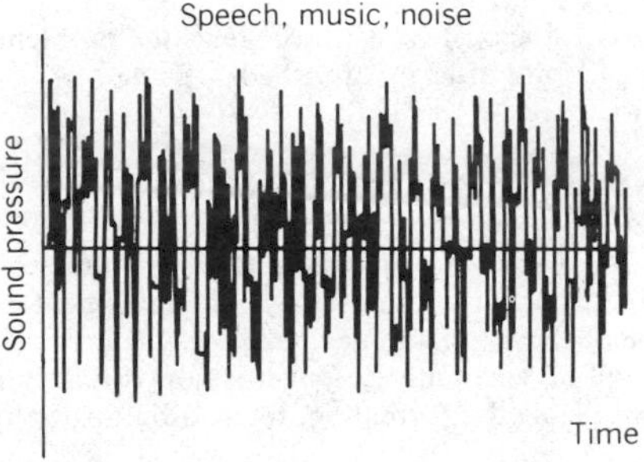

Fig. 44.1 Schematic representations of a pure tone (tuning fork) (*a*), a musical note (combination of several pure tones) (*b*), and more complex sounds (speech, music and noise) (*c*), showing the variation of sound pressure with time.

Velocity of Sound (*c*). A sound wave travels at a velocity that depends primarily on the elasticity and density of the medium. In air at normal temperature and pressure the sound velocity is approximately 1100 ft/s. This is extremely slow when compared with the velocity of light, which is 186,000 mi/s.

Wavelength (λ). Knowing the velocity of sound one can calculate at any frequency the wavelength of the sound (i.e., the distance that the sound wave travels in one cycle) by the following expression:

$$\lambda = \frac{c}{f}$$

where λ = wavelength, ft
c = velocity of sound, fps
f = frequency of sound, Hz

A few simple calculations will reveal that high-frequency sounds are characterized by short wavelengths and low-frequency sounds by long wavelengths. For example, at 100 Hz the wavelength of sound in air is about 11 ft, while at 1000 Hz the wavelength is only 1 ft.

Terms Dealing With Magnitude of Sound

Sound Power (*W*). Sound power in watts describes the energy of the sound source. This power may be (1) the total power radiated by the source over its entire frequency range, (2) the power radiated in a limited frequency range, or (3) the power radiated in each of a series of frequency bands. Obviously the frequency range of sound power (or any of the other quantities dealing with the magnitude of sound discussed hereafter) should be clearly specified.

Sound Intensity (*I*). The sound intensity is the power radiated in a specified direction through unit area normal to this direction; for example, watts per square foot or watts per square centimeter. This term is analogous to light intensity.

Sound Pressure (*p*). Under certain circumstances the sound pressure of a sound wave is equivalent to the sound intensity. The unit is the microbar (1 dyne/cm^2). Most equipment for measuring sound is pressure sensitive, and it is usually easier to measure pressure fluctuations than intensities.

Decibels (dB). The decibel is a dimensionless unit for expressing the ratio of two numerical values on a logarithmic scale. It is convenient to use decibels in dealing with sound power, sound intensity, or sound pressure because of the tremendous range of values of these quantities that can be perceived by the ear. For example, the range of sound intensities that can be perceived by the normal ear extends all the way from the faint rustle of leaves up to the roar of a jet engine, which encompasses a ratio of sound intensities from 1 million-million to 1. The number of decibels is ten times the logarithm to the base 10 of the numerical ratio of the two quantities. For example, let W_1 and W_2 designate two powers, or I_1 and I_2 designate two sound intensities, and p_1 and p_2 designate two pressures. Then the corresponding number of decibels (M in each case) is

$$M(\text{sound power}) = 10 \log \frac{W_1}{W_2} \quad \text{dB}$$

$$M(\text{sound intensity}) = 10 \log \frac{I_1}{I_2} \quad \text{dB}$$

$$M(\text{sound pressure}) = 10 \log\left(\frac{p_1}{p_2}\right)^2 \quad \text{dB}$$

Addition of Decibels. Decibels, since they are logarithmic units and not like ordinary units such as feet and pounds cannot be added directly. One must convert them back to power, intensity, or pressure; add these quantities; and finally convert the total back to decibels. In other words, 50 dB plus 50 dB is not 100 dB but 53 dB. A simplified chart for adding quantities in decibels is given in Fig. 44.2.

Sound-Power Level (PWL). Sound-power level is the designation in decibels of the ratio of two sound powers. The reference power is usually taken to be 10^{-12} W. Therefore

$$\text{PWL} = 10 \log\left(\frac{W}{10^{-12}}\right) \quad \text{dB}$$

Sound-Intensity Level (IL). Sound-intensity level is the designation in decibels of the ratio of two intensities. The reference intensity is usually taken to be 10^{-16} W/cm^2. Therefore

$$\text{IL} = 10 \log\left(\frac{I}{10^{-16}}\right) \quad \text{dB}$$

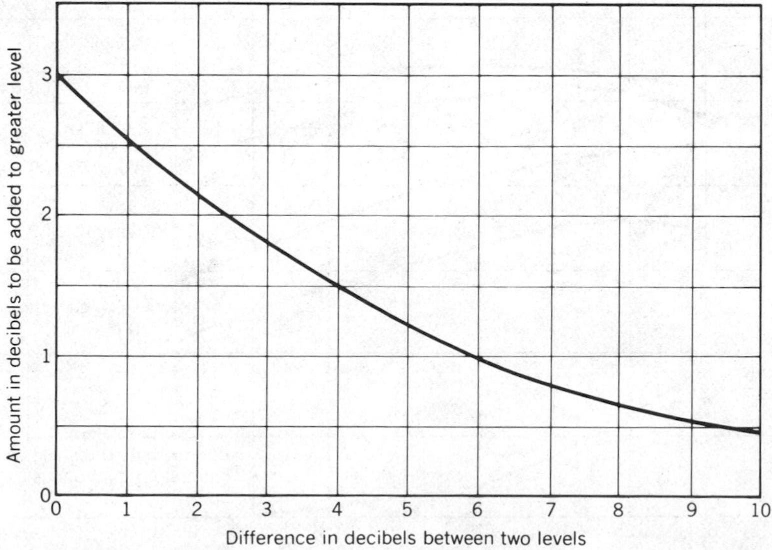

Fig. 44.2 Chart for adding two sound-pressure levels (SPL) or intensity levels (IL).

Sound-Pressure Level (SPL). Sound-pressure level is the designation in decibels of the ratio of two pressures squared. The reference value is always taken to be 0.0002 dyne/cm^2. Therefore

$$\text{SPL} = 10\log\left(\frac{p}{0.0002}\right)^2 \qquad \text{dB}$$

As already stated, under some circumstances sound pressure can be taken to be equivalent to sound intensity. For most architectural acoustics problems, then, the sound-pressure level can be considered equivalent to the sound-intensity level. Some typical measured sound-pressure levels are shown in Figs. 44.3 and 44.4.

Sound Level. Simple sound-measuring devices are available to record the physical magnitude of sound in terms of single numbers. Single-number sound levels are defined as the quantity read on a standard sound-level meter with an appropriate frequency-weighting network. A commonly used system of single numbers are A-scale readings which, for relatively low overall sound levels, correspond to the way our ears respond to the sound (i.e., they weight or "ignore" the low-frequency end of the sound spectrum). The frequency-weighting network must always be known in order to evaluate single-number readings. Figure 44.5 shows the range of some common sounds measured in terms of average levels obtained from sound-level meter readings, using standard A-, B-, and C-scale frequency-weighting networks.

Noise Reduction (NR). Noise reduction is the difference in decibels of the sound-pressure levels or the sound-intensity levels at two points along a sound path. Alternatively, it is the difference in decibels of the sound-pressure levels or sound-intensity levels existing at a single point before and after a change of acoustical treatment to a space. Therefore the following expressions are often used:

$$\text{NR} = \text{IL}_1 - \text{IL}_2 \qquad \text{dB}$$

$$\text{NR} = \text{SPL}_1 - \text{SPL}_2 \qquad \text{dB}$$

Attenuation. Attenuation is often used in the same sense as noise reduction, as already described.

Terms Dealing With Sound Under Free-Field Conditions

Inverse Square Law. Under free-field conditions of sound radiation (i.e., no reflecting surfaces around the sound source), the sound intensity is reduced by one fourth each time the distance

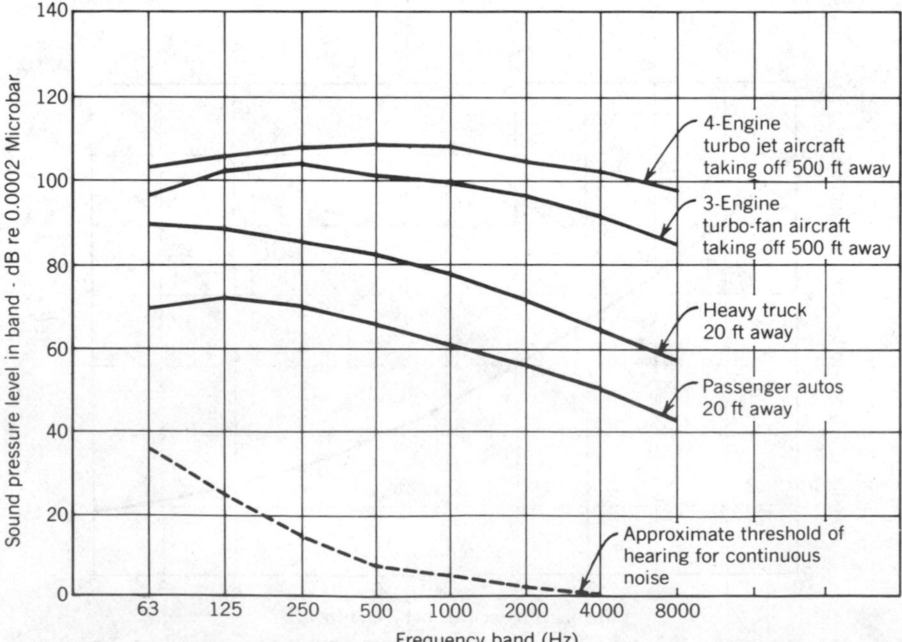

Fig. 44.3 Sound-pressure levels for some noise sources, measured outdoors.

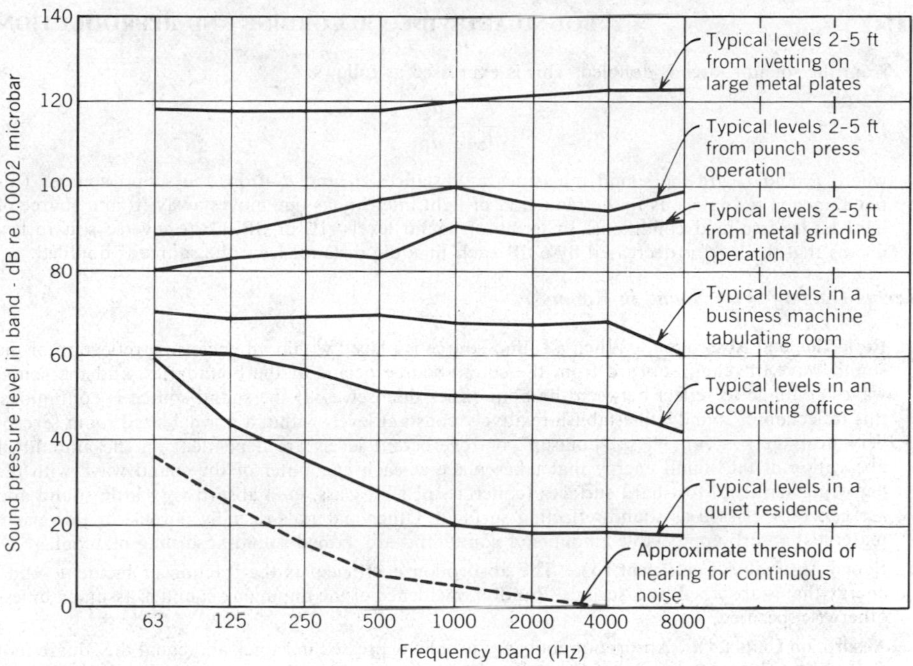

Fig. 44.4 Sound-pressure levels for some noise sources, measured indoors.

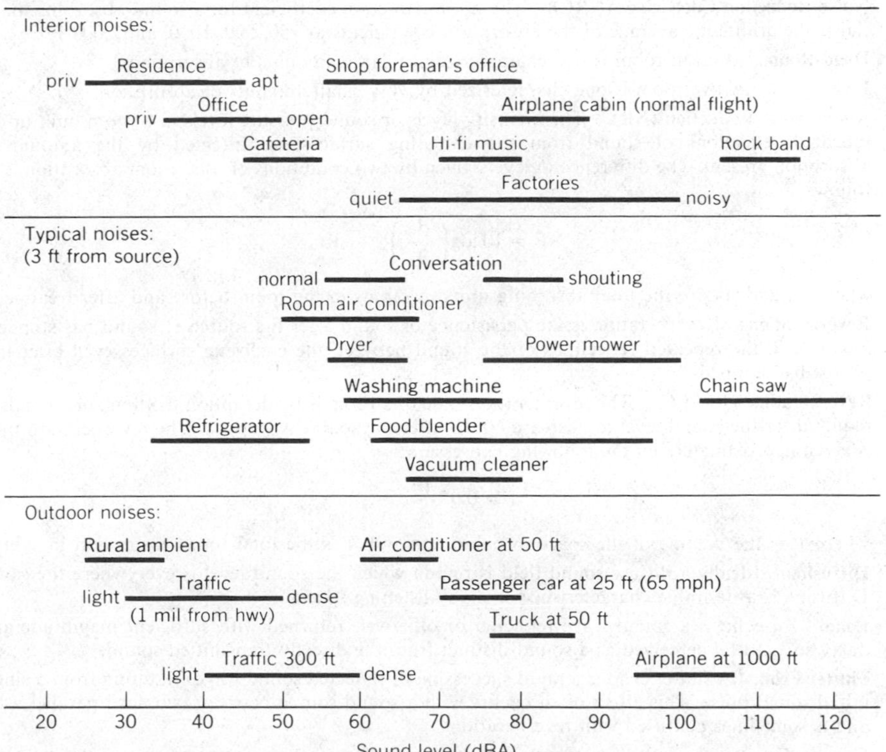

Fig. 44.5 Range of sound levels in decibel absorption units found in typical environments from a variety of sources. Single-number sound-level recordings must be used with caution when analyzing noise problems.

from the sound source is doubled. This is expressed as follows:

$$\frac{I_1}{I_2} = \frac{d_2^2}{d_1^2}$$

where I_1 and I_2 are the sound intensities at distances d_1 and d_2 from the sound source. This phenomenon is analogous to the reduction of light intensity as one moves away from a source of light under free-field conditions. In terms of sound levels (IL or SPL), the inverse square law means that the level is decreased by 6 dB each time the distance from the source is doubled.

Terms Dealing With Sound in Rooms

Reflection and Absorption. When a sound source is placed within an enclosure, reflection of the sound wave traveling outward from the sound source occurs at the boundaries, and the sound waves continue to reflect between the boundaries themselves. If the sound source is continuous, this reflection of sound will establish relatively constant levels within a normal-sized room (except very near the source). These "built-up" or reverberant levels are dependent on the amount of absorption of the sound energy that takes place at each encounter of the sound wave with the enclosing surface. Most hard surfaces (concrete, plaster, glass, etc.) absorb very little sound and are generally classed as sound-reflecting surfaces. Other materials (usually porous, or thin panel materials) absorb appreciable amounts of sound and are termed sound-absorbing material.

Sound-Absorption Coefficient (α). The absorption coefficient is the fraction of incident sound energy that is absorbed by a surface. Random incidence of the impinging sound is assumed unless otherwise specified.

Absorption Units (A). Absorption units are usually expressed in sabins and equal the square foot area of a surface S times its absorption coefficient α. Usually several kinds of surfaces or materials are included in a room, and the total absorption is the sum of the areas times their absorption coefficients.

Noise-Reduction Coefficient (NRC). The noise-reduction coefficient for a sound-absorbing material is the arithmetic average of the absorption coefficients at 250, 500, 1000, and 2000 Hz.

Dead Room. A dead room is one characterized by large amounts of absorption.

Live Room. A live room is one characterized by very small amounts of absorption.

Room Noise Reduction (NR). The intensity levels or sound-pressure levels in a room built up by repeated reflections of sound from the enclosing surfaces are affected by the amount of absorption present. The difference in levels given by two conditions of total room absorption is as follows:

$$\text{NR} = 10 \log\left(\frac{A_2}{A_1}\right) \quad \text{dB}$$

where A_1 and A_2 are the total absorbing units in sabins in the room before and after treatment.

Reverberation. Reverberation is the persistence of sound after the source of sound has stopped. It is due to the repeated reflections of the sound between the enclosing surfaces even after this source has stopped.

Reverberation Time (T). The reverberation time of a room is by definition the time, in seconds, s required for the sound level to decrease 60 dB after the source is stopped. The reverberation time is given approximately by the following expression:

$$T = 0.05 \frac{V}{A} \quad \text{s}$$

where V is the volume of the space in cubic feet and A is the total room absorption in sabins.

Diffusion. Ideally a diffuse sound field is one in which the sound level is everywhere the same. Diffusion is a desirable characteristic for many listening spaces.

Echo. An echo is a sound wave reflected or otherwise returned with sufficient magnitude and delay so as to be perceived as a sound distinct from the directly transmitted sound.

Flutter Echo. A flutter echo is a rapid succession of reflected sound waves resulting from a single initial sound pulse. This effect often occurs with a sound source between two hard parallel walls and is sometimes confused with reverberation.

Creep. Creep is the reflection of sound along a curved surface. It occurs when a sound source is located close to surfaces such as domes, vaults, and so on so that the reflected sound energy is conserved and can be heard distinctly at some point further along the surface; for example, a "whispering gallery."

Focusing. Focusing occurs when sound waves are reflected from concave surfaces and build up the reflected sound levels at some point or area away from the reflecting surface.

Standing Waves. In small rooms, in particular, sound waves can build up nodes and antinodes that are characterized by regions of maximum sound-pressure level and minimum sound-pressure level. Standing-wave effects are usually restricted to the low-frequency range.

Terms Dealing With Sound Transmission Between Rooms

Transmission Coefficient (τ). The transmission coefficient is the fraction of incident energy that is transmitted through a barrier. Thus

$$\tau = \frac{W_2}{W_1}$$

where W_1 is the sound energy in watts incident on the barrier and W_2 is the sound energy in watts transmitted.

Transmission Loss (TL). The transmission loss of a barrier is given by the following expression:

$$\text{TL} = 10 \log \frac{1}{\tau} \quad \text{dB}$$

It is a basic property of a barrier and varies with the frequency of the impinging sound, the surface weight, stiffness, and edge-mounting condition of the barrier.

Effective Transmission Loss (Eff TL). The effective transmission loss of a barrier consisting of two or more different materials is given by the following expression:

$$\text{Eff TL} = 10 \log \frac{\Sigma S}{\Sigma \tau S} \quad \text{dB}$$

where ΣS is the sum of the areas of the parts of the barrier in square feet and $\Sigma \tau S$ is the areas of materials times their respective transmission coefficients.

Sound Transmission Class (STC). This is a single number rating assigned to a measured transmission-loss curve obtained by comparison with a standard curve in accordance with recommended rules given in ASTM E90 (see also discussion in section on sound isolation).

Room-to-Room Noise Reduction (NR). The difference in sound-intensity levels (IL) or sound-pressure levels (SPL) between a "source" room and a "receiving" room is given approximately by the following expression:

$$\text{NR} = \text{TL} - 10 \log \frac{S}{A_2} \quad \text{dB}$$

where TL is the effective transmission loss of the barrier in decibels, S is the total area of the barrier in square feet, and A_2 is the total absorption in the receiving room in sabins.

Attenuation Factor. Attenuation factor is sometimes used to describe the room-to-room noise reduction of a particular construction. For example, the noise reduction values through suspended ceiling configurations over two adjacent test rooms are reported in terms of attenuations in decibels (see section on sound isolation).

Impact Transmission. Impact transmission usually occurs when an impulsive sound source acts directly on the structure and causes radiation of airborne sound on the other side of the structure in question.

Structure-Borne Transmission. Structure-borne transmission refers to sound waves transmitted within a structure. These structure-borne waves may be induced by airborne sound impinging on a barrier or by direct-impact sound sources. If the structure-borne waves are of sufficient magnitude, they may be reradiated into a space as airborne sound.

Background Noise. Background noise refers to the ambient or all-encompassing noise associated with a given environment and is usually due to a composite of sounds from many sources near and far.

Masking Noise. Masking noise usually refers to the ability of background noise to cover up some other specific intruding sound. This is also referred to as acoustical perfume.

Speech Privacy. Speech privacy is a condition of sound isolation of a room in which the occupant feels he has sufficient freedom from intruding speech sounds so that he can conduct his work in an undisturbed manner. This is usually achieved when the speech levels transmitted to the space are unintelligible, or very nearly so.

44.1-3 Criteria for Acoustic Environment

Before a designer can begin actual engineering work on a building, he or she must establish the criteria. Basically, a satisfactory acoustic environment is one in which the character and magnitude of all sounds are compatible with the satisfactory use of the space for its intended purpose. While this is a reasonable objective, it is not always easy to express it in quantitative terms. In talking about the thermal environment, for example, one cannot simply say that 70°F is comfortable but must also talk about humidity, air movement, and so on. In lighting, one cannot simply say that 100 fc is an ideal light intensity but must also talk about other factors in the luminous environment, such as specular glare, color, and continuity. Similarly, in acoustics we cannot just say how much noise we want but rather we must specify what kind of noise; what pitch; whether it is continuous, expected, or contains information; and so on.

Human beings are highly adaptable to the various physical phenomena of heat, light, and sound, and their sensitivity varies widely. The human ear can detect the sound-intensity levels of less than 10 dB (the gentle rustle of leaves) and yet can survive without permanent hearing damage the powerful roar of a jet engine—as much as 120 dB, a million-million times the intensity of the leaf rustle sound.

Psychologists and research workers in acoustics have laid a good deal of the groundwork that enables us to understand how much and what kind of noise will affect speech communication, annoyance, and fatigue. Work has also been done on the problem of hearing damage due to high-intensity noise levels, but this is seldom an important problem in ordinary buildings. Although the research results are far from complete, they do lead us to certain generally accepted specifications on the noise environment in many of the kinds of spaces we design.

Buildings are often built on sites near airports or other loud noise sources. Recent work on acoustical criteria has included consideration of allowable levels of intruding sound from these sources into occupied spaces and also methods of predicting "acoustic suitability" of sites. In critical spaces, infrequent intrusion of noise can be permitted to be as much as 5 dB above the normal background without degradation of the environment. If the space is not one that requires a very low background, occasional intrusions of up to 10 dB above the normal background may be acceptable. More than 10 dB of intrusion is normally unacceptable and causes major complaints or even legal action.

Some recommended noise criteria for various types of occupancy are indicated in Table 44.1.

As one might expect, spaces where listening is important require low background levels, and business offices and factories where speech communication is restricted to short distances can have higher background levels.

Background Noise Criteria

Although single-number sound level readings (dBA) give us some idea of how much noise we have, a much better method of specifying continuous background noise level is the use of noise criteria (PNC) curves (see Fig. 44.6). The PNC curves are a further refinement of the formerly used NC curves, taking more nearly into account the noise spectrums actually observed in a great many

TABLE 44.1. RECOMMENDED NOISE CRITERIA FOR VARIOUS USES

Type of Space	Recommended Noise Criteria (Range or Maximum)	Approximate Sound Level (dB)
Broadcast and recording studios, concert halls	PNC[a] 10–20	20–30
Legitimate theaters (no amplification), churches	PNC 20	30
Large conference rooms (for 50 or so), small auditoriums, music rehearsal rooms, motion picture theaters	PNC 30	38
Classrooms, conference rooms (for 20 or so)	PNC 35	42
Bedrooms (hotels, apartment houses, hospitals, residences)	PNC 25–40	34–47
Private or semiprivate offices, living rooms, libraries	PNC 30–40	38–47
Sports coliseums (amplification)	PNC 35–40	42–47
Restaurants, stores	PNC 35–45	42–52
General offices (typing, etc.)	PNC 40–50	47–56
Factories	PNC 50–75	56–80

[a] PNC, preferred noise criteria.

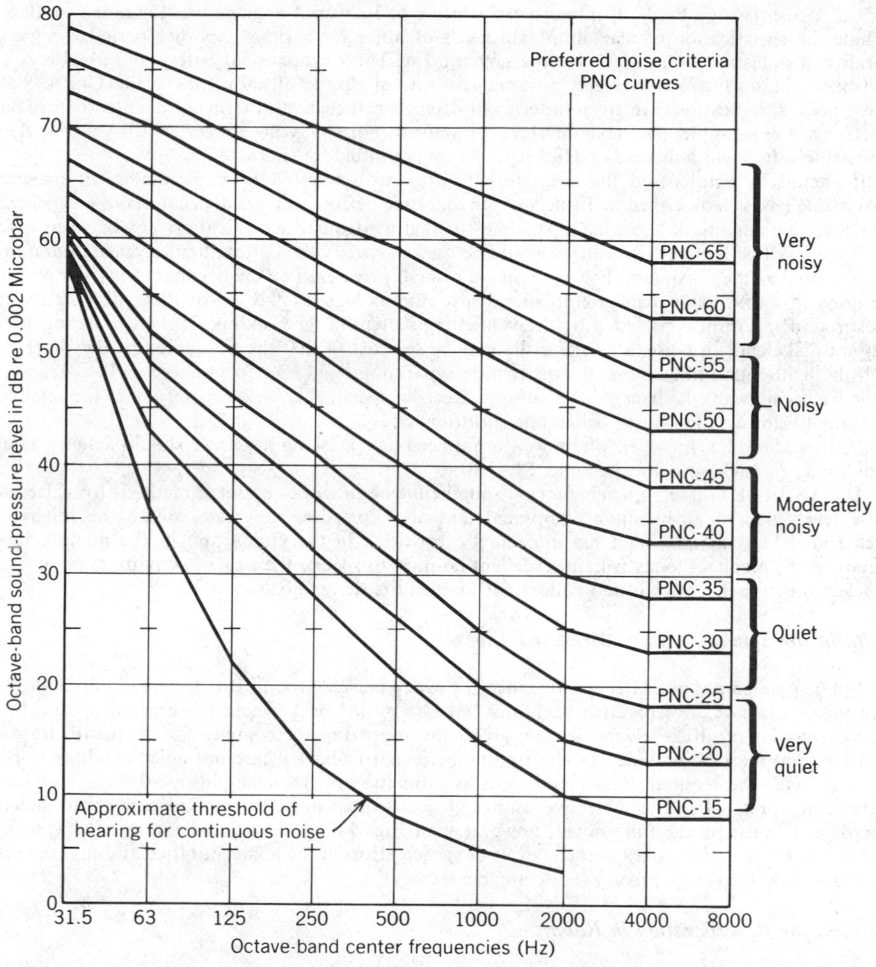

Fig. 44.6 Preferred noise criteria (PNC) curves.

building situations. The PNC curves provide a system of rank ordering various noise spectrums in terms of specified sound pressure levels in each of nine octave frequency bands. Octave bands are convenient for dividing the audible frequency range into segments for purposes of measurement. The PNC numbers are the arithmetic average of the sound pressure levels in the 500-, 1000-, and 2000-Hz bands. These frequency bands are very closely related to the important frequencies for speech intelligibility. Thus the presence of noise in these bands can interfere with speech, and most people rate the noisiness of the environment with speech interference.

Also shown in Fig. 44.6 is the subjective evaluation that a listener might give to a particular acoustic environment that has octave band sound levels approximating these curves. Below PNC-25, for example, most people would judge a space "very quiet"; above PNC-55, "very noisy"; and between these extremes, "quiet," "moderately noisy," and "noisy." In terms of speech communication, a background sound spectrum of PNC-30 would permit understanding of speech at normal voice levels at distances up to about 20 ft. A PNC-40 spectrum would permit a raised voice to be understood at 20 ft but would only permit normal voice communication at distances up to about 6 ft. With a background spectrum of PNC-50, a raised voice would be required to be heard clearly at more than 3 ft between speaker and listener. Even higher levels than PNC-50 would be permissible in a factory, where speech communication and annoyance are not too important. However, if the continuous background noise levels exceed PNC-70, it is impossible to use the telephone; and with spectrum levels as high as PNC-80, there may be a possibility of permanent hearing damage after long exposure.

The Walsh-Healey Public Contracts Act relating to industrial hygiene was amended in 1969 to include the specification of maximum safe levels of noise for various exposure periods for people working in industries doing work for the government. The Occupational Safety and Health Act of 1970 incorporates the Walsh-Healey requirements and applies to all industries in the United States. These noise specifications are given in terms of decibels rather than in terms of detailed octave band spectrums. For example, the maximum level of ambient noise may not exceed 90 dBA if a worker is exposed to it for eight hours a day. Higher levels are permitted for shorter exposures.

It should be emphasized that we are often just as interested in minimum as in maximum permissible levels as described in Table 44.1. In an office or even in a residence it may be desirable to have a certain amount of "acoustical perfume" to assure adequate acoustical privacy between spaces, ant it may well be that our PNC curve would be used to specify the bottom limit of background noise levels as well as the maximum. For example, in a small private office the occupant may not object to the noise of a continuous background sound spectrum as high as PNC-35 or PNC-40. Often such a background spectrum is provided by the ventilating system in the building. If the ventilating system will not make enough noise, a noisier grille may be selected so that the adequate privacy is achieved with the lightweight, movable wall construction separating one office from another. In other words, while we do not want the background noise to exceed a specified criterion spectrum, at the same time we do not want it to fall much below this spectrum.

In an auditorium, however, where we do not need any masking noise, we should achieve as near inaudibility of background as possible, PNC-10 to -20.

The ASHRAE Guide, in its chapter on sound control, describes in detail methods for calculating noise levels due to air-handling equipment and also considers in greater detail the criteria for background level design. With the information provided in the Guide and in the manufacturers' literature, the architect today is in an excellent position to specify the acoustic environment and then to select the proper materials and equipment to meet the design goals.

Criteria for Speech Privacy Between Offices

As will be pointed out in a later section, speech privacy between rooms is determined not only by the transmission loss of the separating partitions but also by the background noise in the spaces. In the special situation of office privacy, we can give some approximate recommendations for the transmission loss that must be provided by the partitions for particular background noise conditions. These criteria, giving the required transmission loss as a function of frequency in one-third octave bands, are shown in Fig. 44.7. The subjective ratings of background noise correspond with those shown in Fig. 44.6. It must be carefully noted, however, that Fig. 44.7 gives approximate criteria values for transmission loss and assumes average levels of speech effort in the adjacent office and that the office occupant expects average privacy (not complete secrecy).

Criteria for Reverberation in Rooms

Every occupied space has some sort of reverberation characteristic. It may be live or it may be dead, but there is always a certain amount of return of energy reflected from distant surfaces to the listener that gives him or her a "feel" of the space. There are no criteria for exactly what reverberation time one should have in a house or in an office, but experience shows that unless such occupied spaces are furnished or finished with reasonable amounts of sound-absorbing material, they will not be comfortable. The next section discusses the procedure for calculating reverberation time in a space as well as the effect of the amounts of sound-absorbing treatment in a room on the general sound level in the room. However, the quality of the sound in the space is usually a more important factor than the absolute level, at least within the ordinary range of interest. Sound-absorbing treatment in a space reduces the spreading of sound and localizes the direction of sources: sounds do not seem to come from everywhere, and the annoyance level from all sorts of noise sources is less. In a restaurant, sound-absorbing materials to reduce reverberation time are very important in making an acceptable acoustic environment. With hard surfaces in such spaces, the din often becomes unbearable, and corrective measures must be taken later. In the house with an open plan and sparse furnishings, some sound-absorbing material on floor or ceiling surfaces can do a great deal to improve the acoustic environment.

In larger rooms where hearing is important, we can put definite numbers on the range of reverberation times that seem to most listeners to be satisfactory. A small conference room or lecture room used primarily for speech needs a relatively low reverberation time to achieve high articulation and separation of successive sounds for maximum audibility. At the other end of the scale is the large cathedral church, where liturgical music is of major importance and where the audibility of a sermon can be handled with a carefully designed sound reinforcement system and where maximum blending of the musical sounds must be the criterion for design. Between these extremes lies the whole range of

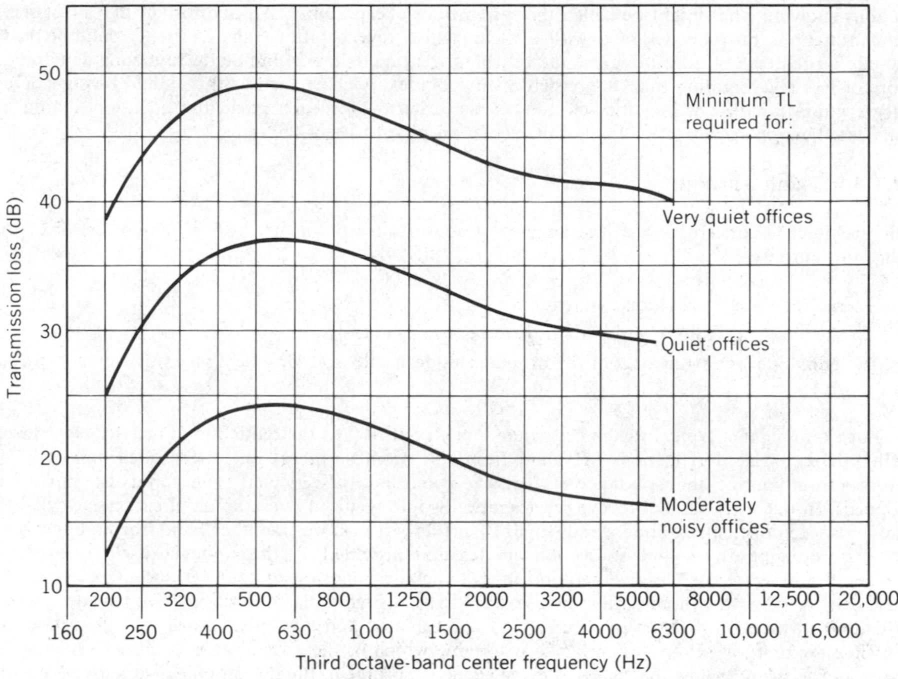

Fig. 44.7 Approximate transmission-loss criteria for speech privacy between offices.

types of performance. In general, the reverberation time of an auditorium of any size should lie somewhere between 1 and 2 s for best results. In general, the larger the hall, the longer the reverberation time must be for satisfactory listening conditions. But for every situation there is considerable latitude in the choice of a design reverberation time that will give satisfaction. In Fig. 44.8 we show the range of reverberation times generally considered acceptable for various types of use. The preferred range for most uses is shown as a black bar for the given function, the dotted

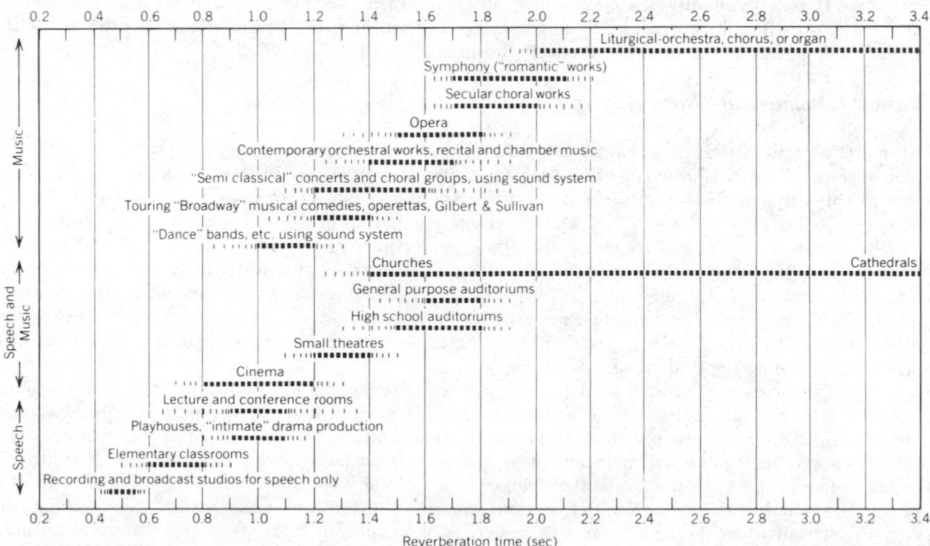

Fig. 44.8 Optimum reverberation (500–1000 Hz) for auditoriums and similar facilities.

sections showing what might be called the extremes of acceptability. An auditorium that must serve many functions may be designed with a compromise reverberation time. A better solution is to provide large areas of adjustable sound-absorbing treatment to accommodate functions demanding more or less reverberation than the basic design provides. A chart such as this should be considered only a guide for design, and the selection of the criterion for each particular situation should be considered very carefully in the light of the actual proposed uses of the space.

44.1-4 Sound Absorption

The architect is concerned with the amount of sound absorption in a space for any or all of the following purposes:

1. To reduce noise levels (noise control).
2. To shorten or prolong reverberation (reverberation control).
3. To eliminate echo (echo control) or other undesirable sound reflections (focusing or flutter control).

For example, in a typical school classroom, noise control (i.e., the reduction of activity noise levels built up by repeated reflections of sound from the room surfaces) may be just as important as reverberation control (the avoidance of excessive persistence of reflected sound after the source has stopped). In the latter case, excessive reverberation would result in overlapping of successive syllables making speech difficult to understand. In a concert-hall design, on the other hand, the architect may avoid introducing any sound absorption besides that provided by the audience itself, in order to achieve the longest possible reverberation time. Even the inclusion of sound-absorbing material for echo control may prove undesirable in a concert hall. A preferable solution may be to redesign any offending surfaces so that the reflected sound is redirected and no longer heard as a discrete echo.

Thus the first job of the designer is to determine which of these requirements must be met in the space and to what extent; that is, as in every acoustics problem, the design objective must be clearly defined. Success in meeting these objectives depends largely on the designer's knowledge and skill in the selection and use of materials.

It should be emphasized that the principal uses of sound-absorbing materials are for the control of sound *within* a space and not for the control of sound transmission between spaces. There is much confusion on this point and much disappointment has resulted from the misapplication of sound-insulating materials to room surfaces or in the stud space of some partition constructions. Such use of sound-absorbing material often makes no significant difference in the sound transmission between spaces. For the most part, sound-absorbing materials, especially porous lightweight ones, offer little resistance to sound transmission. There are, however, some new specially designed materials (suspended acoustical tile ceilings in particular) that act as both sound absorbers and sound-reducing barriers. (These combination materials will be discussed in the section on sound isolation.) However, in general, one should view cautiously any cure-alls that promise to solve all acoustics problems with a single homogeneous material of practically no thickness.

Basic Mechanism of Sound Absorption

All materials and objects in a space where a sound field exists absorb some of the sound incident on their surfaces. Porous, fibrous materials such as carpets, draperies, upholstered furniture, and clothing and specially designed sound-absorbing materials are capable of appreciable sound absorption (that is, they do not reflect very much of the sound energy that strikes them). Impervious, thin, flexible panels (plywood, etc.) absorb sound also, but their effectiveness is usually limited to the low-frequency range of the audible spectrum. By contrast with these absorbers, most common building materials (brick, concrete, glass, plaster, etc.) are very poor sound absorbers and most often absorb less than 5% of the incident sound energy in the frequency range of interest. In fact, these latter materials may be classified as sound-reflecting materials and can be used effectively in auditorium design to distribute the desired sounds properly.

Sound absorption results when the impinging sound energy is converted to heat energy in the body of the absorber (although the amount of heat is quite small). In the porous type of sound absorber, this occurs as the pressure of the air increases and decreases with the arrival of successive sound waves causing the air molecules near the porous surface to migrate into the labyrinth of capillary-like tunnels in rapid to-and-fro motion. Part of the acoustical energy is thus converted to heat by frictional drag. The amount of friction, and hence absorption, provided by the material is, of course, determined by the actual physical properties of the porous layer: thickness, density, porosity, orientation of the fibers or passageways, and, of prime importance, resistance the material offers to the passage of air. The careful design and control of these parameters is the domain of the acoustical researcher and manufacturer. However, it is important for the architect to have some basic understanding of what is involved in absorption by the porous materials that are most commonly used. Only then can he or she understand why a very thin layer of a so-called acoustical paint cannot possibly absorb sound efficiently where, on the other hand, a carefully designed perforated thin-sheet material such as that used for sound-absorbing luminous ceiling applications can be an effective sound absorber. But even with these latter materials the absorption depends on the material in combination with the enclosed air volume behind, and in this context they may actually be thought of as thick materials.

With thin, impervious, flexible panels, absorption results when the surface is set in to-and-fro flexural motion by the alternating pressure of the impinging sound wave, and part of the sound energy is converted to heat through internal viscous damping. Few panel absorbers are manufactured and marketed as such, and the architect must rely on the information published in acoustical texts for the effect such materials may have on the room design. Broadly speaking, however, the absorption of thin panels is confined primarily to the low-frequency range.

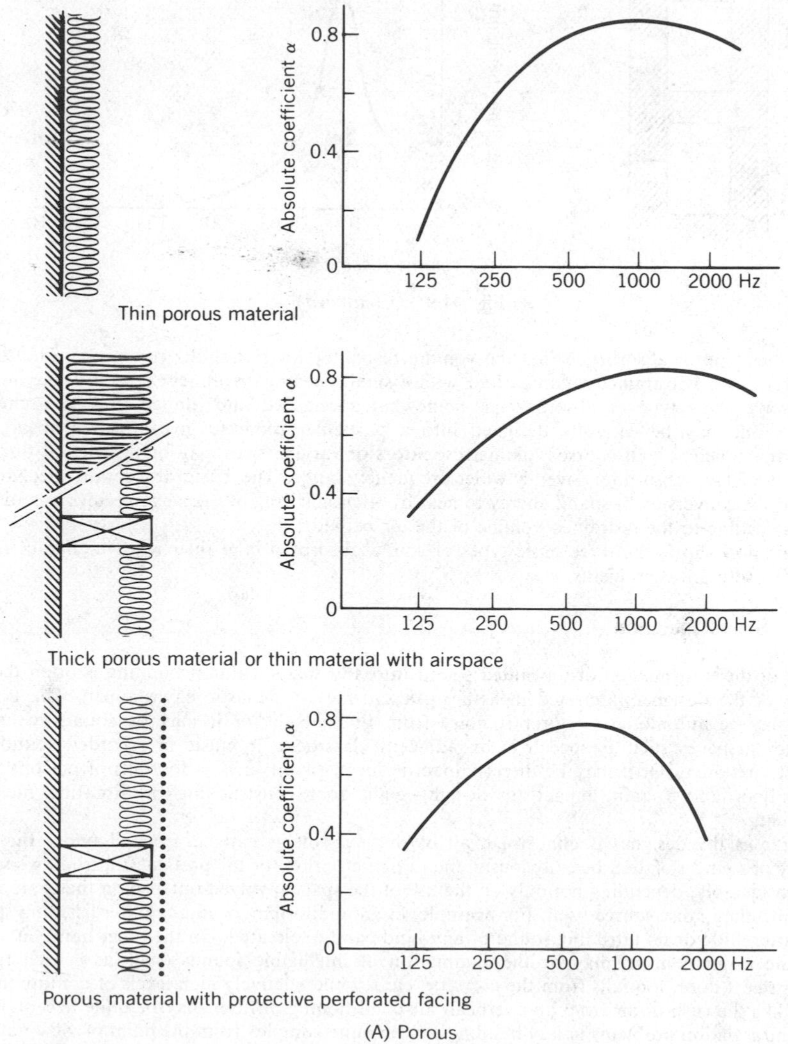

Fig. 44.9 Basic types and relative efficiencies of sound-absorbing materials: (*a*) porous, (*b*) vibrating panel, (*c*) volume resonator.

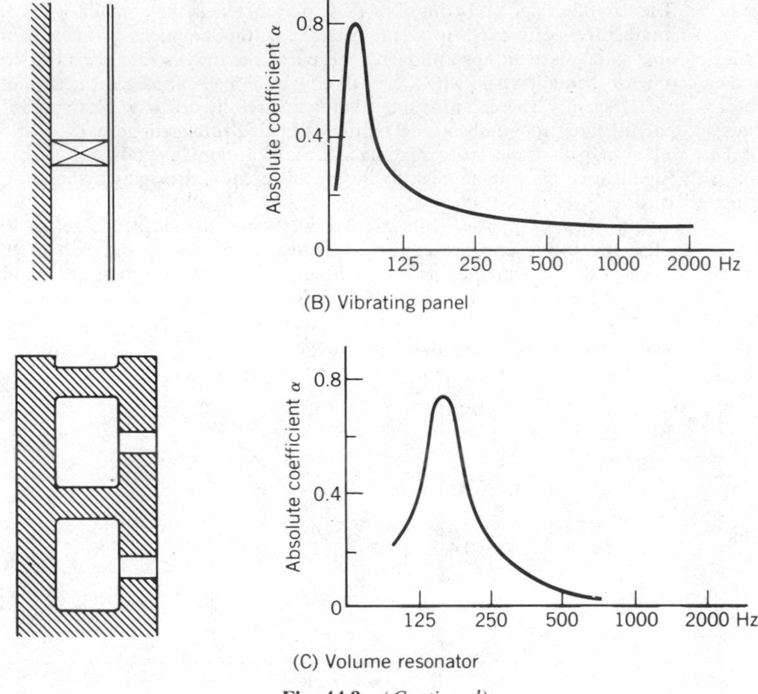

Fig. 44.9 (*Continued*).

Another type of absorber, called the volume resonator (or Helmholtz resonator in honor of its discoverer), uses a restrained volume of air with a small opening or tunnel exposed to the impinging sound wave. This type of absorber has somewhat specialized and limited uses in architectural acoustics but may be carefully designed into a particular room to give effective absorption at particular frequencies. Of course, volume resonators of varying sizes may be distributed throughout the room to give absorption over a wider frequency range. The basic absorption mechanism is, however, the conversion of sound energy to heat by frictional drag of the air molecules in and around the neck leading to the restrained volume of the air behind.

Figure 44.9 shows the three basic types of sound absorption and their relative characteristics in terms of absorption coefficients.

44.1-5 Sound Isolation

Control of the transmission of unwanted sound into any space within a building is often the major concern of the designer interested in assuring a satisfactory acoustic environment. The undesired sound may be automotive or aircraft noise from the outside, or it may be sound generated in surrounding spaces such as speech in an adjacent classroom or music or recorded sounds in an adjacent apartment. Or it may be direct impact-induced sound such as footfalls of persons walking on the floor above, rain impact on a lightweight roof construction, or vibrating mechanical equipment.

As far as the designer is concerned, all of these problems can be grouped under the general category of sound isolation but, obviously, the design criterion for the particular intruding sound will vary considerably, depending not only on the use of the space involved but also on the characteristics of the intruding noise source itself. For example, in an auditorium, or in any other listening space for that matter, little or no intruding sound of any kind can be tolerated. On the other hand, in a private office the major concern may be the elimination of intelligible sounds (such as speech from the occupant next door, footfalls from the corridor, etc.), while relatively high levels of continuous bland sounds like the rush of air from an overhead air-conditioning diffuser may be quite acceptable.

Sound-isolation problems in any building can be quite complex from the point of view not only of analyzing the many potential sources of intruding sound but also of evaluating what levels of intruding sound can be tolerated by the occupants of the space. The designer must not only have some fundamental knowledge of the general aspects of the analysis of a sound-isolation problem but also some understanding of the important physical characteristics of barriers and how these can best be used to isolate a given space from both airborne and structure-borne sounds.

The problems of sound isolation are usually considerably more complicated than problems of sound absorption and involve reductions of sound level that are of greater orders of magnitude than can be achieved by either absorption or separation of the noise sources from the listener. These large reductions of sound level from one space to another can be achieved only by continuous and massive impervious barriers and, if the problem involves structure-borne sound as well, it may be necessary to introduce discontinuities or resilient layers into the barrier also.

The significant point is that sound-absorbing materials and sound-isolating materials are used for entirely different purposes. Just as one does not expect much sound absorption from an 8-in. concrete wall, there is not much reason to expect high sound isolation from a porous, lightweight material that may be applied to the surfaces of a room. As was mentioned previously, some materials have been developed to fill a demand for materials that will perform both tasks simultaneously. However, the basic mechanisms of sound absorption and sound isolation are quite different. This point cannot be emphasized too strongly, since much confusion still exists among architects and building designers.

Simple Case of Room-to-Room Sound Transmission

To illustrate a number of the important variables in any sound-isolation problem, we will try to visualize a simple case of airborne sound transmission between two rooms separated by a common barrier (Fig. 44.10). One of the rooms (called the source room) contains a continuously operating noise source, and the other (called the receiving room) has a listener. To keep the situation simple, we

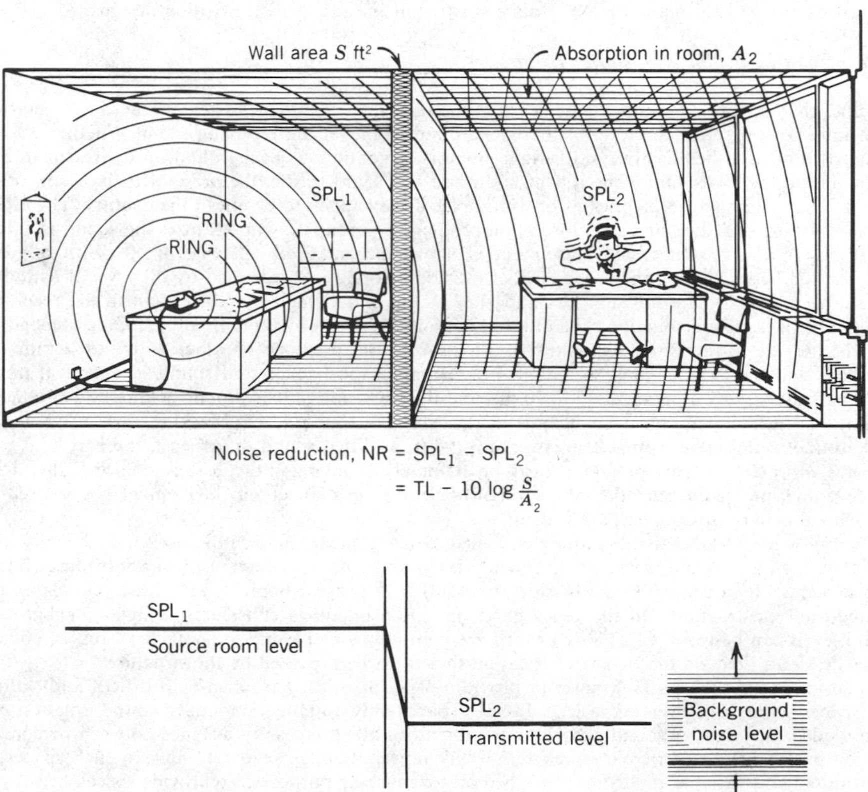

Fig. 44.10 Illustration of the simple case of airborne sound transmission between adjacent rooms through a common barrier. With a sound source in one room, the transmitted sound level is dependent not only on the transmission loss of the barrier but also on the area of the barrier and the receiving room absorption. The actual background "masking" noise levels determine whether the transmitted sound will be heard.

will assume that the only way for sound to get to the receiving room is through the common wall, which is completely airtight so that all the sound has to go through the material itself. The source room sound, which is relatively uniform throughout the room except very near the source, impinges on the barrier at many angles of incidence. In essence, when the sound impinges, it tries to move the barrier and to the extent that it does move it, sound is reradiated by the barrier to the receiving room. The level of transmitted sound in the receiving room (not very near the wall) is dependent primarily on three factors: (1) the transmission loss (TL) of the wall, (2) the area of the wall, and (3) the amount of absorption in the receiving space. This can be expressed approximately by the following equation:

$$\mathrm{NR} = \mathrm{TL} - 10 \log \frac{S}{A_2} \quad \mathrm{dB}$$

where NR = noise reduction, dB (difference in reverberant sound levels between two spaces in question, in this case $\mathrm{SPL}_1 - \mathrm{SPL}_2$)

TL = transmission loss of wall, dB

S = area of wall, ft^2

A_2 = total sound absorption of receiving room, sabin (sum of areas of various materials in room, ft^2 times their respective sound-absorption coefficients)

The transmission loss accounts for the largest part of the room-to-room noise reduction but, as can be seen, the area of the wall and the amount of sound-absorbing material in the receiving space also have some effect. The larger the wall area is, the more sound energy will be transmitted. The more sound-absorbing material there is in the receiving space, the lower the reverberant sound level will be. In most practical situations, the correction term that accounts for the area of the wall and the receiving room absorption usually affects the room-to-room noise reduction by not more than about ± 5 dB. However, this amount may be quite significant in many sound-isolation designs.

Role of Masking Noise in Sound Isolation

Whether the transmitted sound will be heard in the receiving room depends on another factor that we have thus far neglected: the level of the background sound in the receiving room. Whether we are aware of it or not, there is always a certain amount of continuous background noise present in any space due to the air-conditioning system, the noise of distant traffic, the noise of activities in other parts of the building, or even crickets or wind noise if we happen to be out in the country. The effect of this masking sound on any sound-isolation problem is perhaps as important as the sound-isolating properties of the barrier itself. This effect is shown schematically in Fig. 44.10. With a given construction between two spaces, the level of intruding sound is determined by the level of sound in the source room, the transmission loss and area of the barrier, and the absorption in the receiving space. The background sound, on the other hand, may vary considerably in any building, depending on whether the air-conditioning system is on or off; the presence of other activities within the building; and the exterior noise situation. For example, the background sound in a typical office building may vary by as much as 15 to 20 dB, and this variation in background sound level can mean that it will fall either above or below the transmitted sound level. Masking occurs when the background sound either completely covers up the transmitted sound or, at least, the part of it that conveys information. A dripping faucet can be extremely annoying in the deathly silence of the night. However, during the daytime, the same level of noise from this faucet may be completely obscured by the general activity sounds that are present.

In any space whose activities require extreme quiet, the background sound itself must be very low, and thus the barriers are called on to provide large amounts of reduction of any intruding sound. This explains why a concert hall, a broadcast studio, or a special laboratory may need very elaborate, double-wall construction. On the other hand, in office buildings considerably higher background sound levels can be tolerated as long as they are continuous and bland in character. This, of course, places less of a demand on the sound isolation that must be provided by the structure.

The importance of masking noise in providing speech privacy in open-plan offices and "office landscapes" cannot be overemphasized. In such spaces only a limited amount of sound isolation can be provided by extensive sound-absorbing treatment on all surfaces, by distance between personnel, and by partial height barriers or screens. Activity noise, although relatively high in such spaces, is intermittent and variable and cannot be relied on for masking purposes. To provide speech privacy in open-plan offices, a carefully designed electronic background noise system is almost always required. Such a system normally consists of a solid-state noise generator and amplifier plus filters and controls for adjusting the character and level of the noise. Loudspeakers are distributed in or above the ceiling and provide a uniform, bland, innocuous sound throughout the space, ensuring maximum speech masking and minimum annoyance.

Much research is under way to try to arrive at a better understanding of the masking effects of background sound on various kinds of intruding noise. For speech-isolation problems, these effects are well understood, and it is possible to achieve a good balance between partition selection and the background noise to solve speech privacy problems.

The precise effect of masking on other intermittent, intruding sounds, such as mechanical equipment noise, music, and so on, are less well understood and, for the time being, the designer must provide some safety factor in the sound-isolation design so that the transmitted sound will be reduced somewhat below the lowest background noise levels that will actually exist in the space.

44.1-6 Sound Reinforcement Systems

In many situations, to obtain adequate loudness and good distribution of sound it is necessary to augment the natural transmission of sound from source to listener by means of a sound system. In large sports arenas, in airport terminal buildings, and in other noisy locations it is almost always necessary to provide sound reinforcement. Even in rooms where most strong-voiced speakers can be heard clearly, the weaker voices must be amplified, and there is often the need to reproduce recorded material or movie sound. In all cases, however, the design of the sound reinforcement system must be carefully integrated with the design of the room and with its acoustical characteristics.

There are two principal types of sound reinforcement systems: central and distributed. The preferred type in most situations is the central system in which a loudspeaker (or cluster of loudspeakers) is located directly above the actual source of sound. Only one loudspeaker position is used in a system of this sort, and it is capable of giving maximum realism. Listeners with their two ears are readily able to localize the direction of the source of sound, and if the amplified signal comes from the same direction as the original sound, they get an impression merely of increased loudness or clarity but not of artificial "amplified" sound (Fig. 44.11).

The other principal type of sound reinforcement system is the distributed type. In this system, one uses a large number of loudspeakers distributed uniformly over the audience areas. With its loudspeakers located overhead, this type of system operates much like downlighting. We cover the room with small "pools" of sound, each listener receiving sound from only the closest loudspeakers. This type of system is used in any situation where the ceiling height is inadequate to use a central system or where all listeners cannot have "line-of-sight" on a central loudspeaker. It is also used in such spaces as large convention rooms, where there must be a very flexible arrangement of the space for amplifying sources of sound in any position in the hall. It is the logical system for most airport terminal buildings, where the amplified signal usually must be somewhat higher in level to override the high background noise levels due to aircraft operations. The distributed system is a flexible system, and while it does not give maximum realism in reinforcing live activities, it can be made to provide high intelligibility in many difficult situations (Fig. 44.12).

In spaces with very high ceilings or when other considerations will not permit mounting the loudspeakers in the ceiling, loudspeakers in a distributed system can be installed within the audience areas. They can be attached to the backs of the seats in conference rooms, mounted in the desks in assembly rooms or legislative chambers, or installed in the backs of church pews.

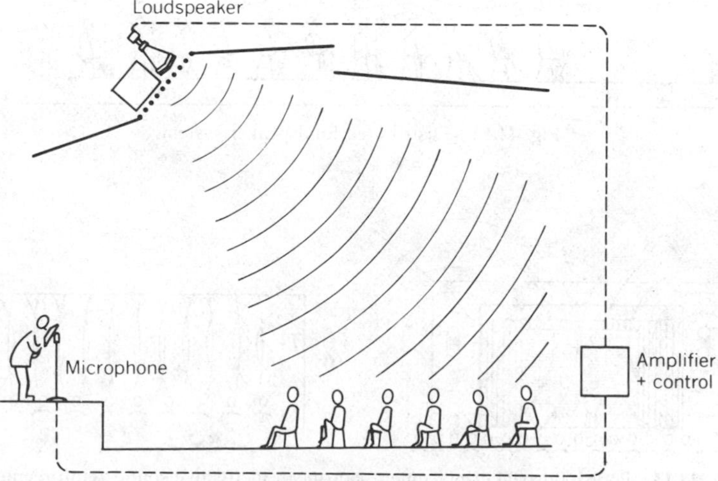

Fig. 44.11 Central loudspeaker system.

Loudspeakers should never be located at the two sides of the proscenium opening, nor should they be distributed along the two sides of the room or in the four corners of a large reverberant space. This never works well, and the hearing conditions in a space can usually be improved by shutting the loudspeakers off (see Fig. 44.13).

Central Systems

The loudspeaker for a central system usually consists of a cluster of directional horns, some of which handle the high-frequency end of the audible spectrum, and larger loudspeakers that handle the low-frequency end of the spectrum. The high-frequency horns are usually exponential, multicell, or radial horns and are arranged in clusters to give coverage of specific areas of the seating. It is important that the horn arrays be designed to have proper directional characteristics and that the level of sound from the several units be individually adjustable. One cannot achieve high-quality sound amplification without loudspeakers with carefully controlled directional characteristics. If a loudspeaker system is to be used only for speech purposes, the system need not have any low-frequency loudspeakers and can be housed in a smaller space than a full-frequency-range system (used for music). Usually, a speech system is cut off at approximately 300 Hz (i.e., these loudspeakers do not amplify sounds below that frequency). This results in no loss in realism and actually improves intelligibility in rooms with "boomy" characteristics.

The designer of an auditorium incorporating a loudspeaker system must realize that the system will take a great deal of space and that it cannot be tucked conveniently into a 1-ft slot. The grille in front of the loudspeaker must be completely transparent to sound and must contain no large-scale elements (see Fig. 44.14). Every listener in the room must have line-of-sight to the loudspeaker; one cannot count on reflection of sound from room surfaces to fill in any areas not covered by direct line-of-sight.

The operator of the sound system should be located toward the rear of the seating area where he or she can hear the system as it is heard by the audience. The operator should not be behind a glass window in a booth receiving sound only from a monitor loudspeaker. The power amplifiers can be in any convenient location, but the actual control must be "in the room."

Microphones must be placed near the sources of sound. If there are to be many sources, as in a play, there must be a sufficient number of microphones provided within the acting area, concealed in the scenery, so that the actors are always relatively close to these pickup devices.

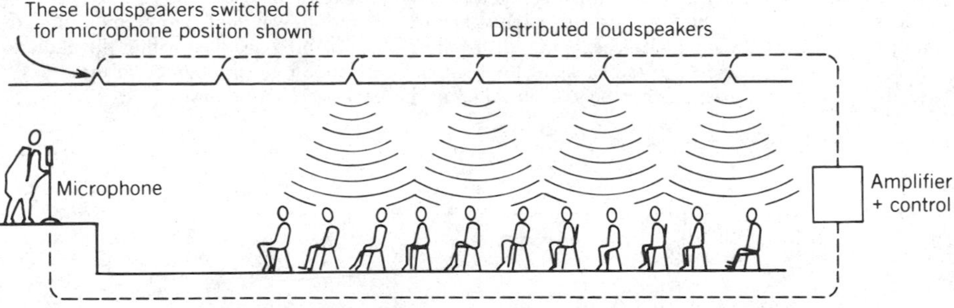

Fig. 44.12 Distributed loudspeaker system.

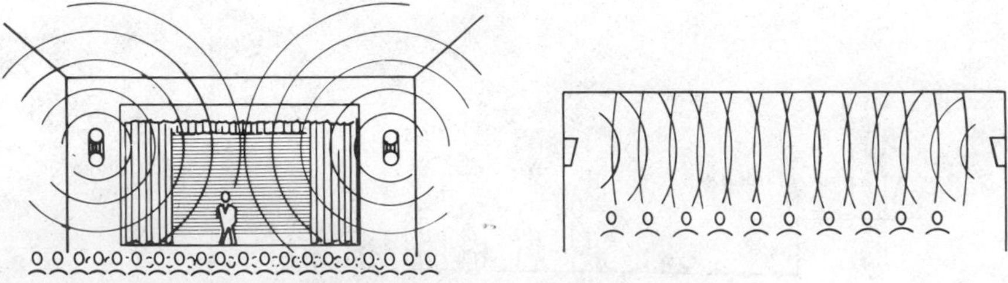

Fig. 44.13 Poor loudspeaker placement can mean ineffective sound reinforcement.

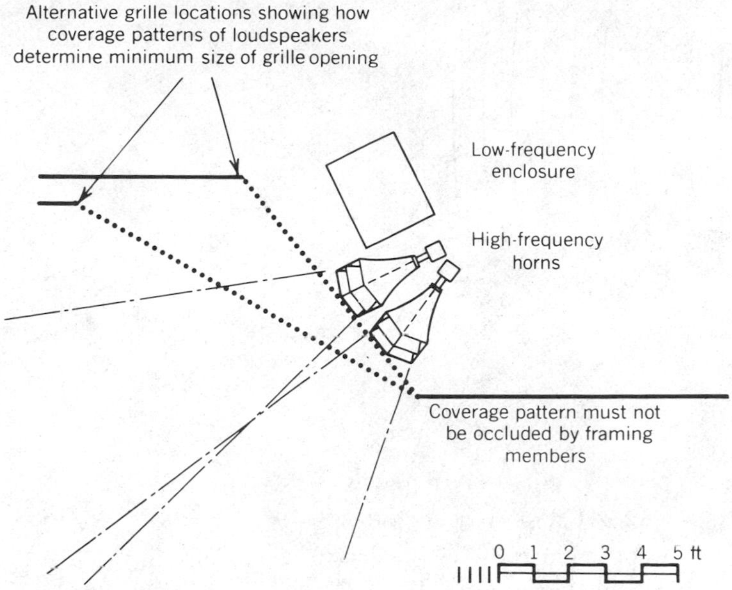

Fig. 44.14 Loudspeaker grille sizes are determined by the coverage patterns of the loudspeakers behind them.

There is also the important problem of feedback of sound energy from loudspeaker to microphone, and the relative locations of microphones to the loudspeakers must be carefully considered to avoid the familiar squealing or howling of a poorly designed and operated system. This is a matter for detailed consideration by the designer of the system and is not primarily an architectural question except insofar as relative location of loudspeaker to microphone is concerned. In the large arena or sports building, the central cluster of loudspeakers is usually suspended without any attempt at concealment, and some central position can usually be found giving everyone clear line-of-sight on the unit.

In some situations, line-source loudspeakers are preferable to radial or multicellular horns (see Fig. 44.15*a*). These units, made up of a series of cone-type loudspeakers, also take space, and this must be carefully considered in the design of the building.

Distributed Systems

In this type of sound-amplifying system it is extremely important to have an adequate number of loudspeaker units. They are generally placed in the ceiling, facing down and sounding through appropriate grillage. Each loudspeaker unit is considered to cover between 60 and 90°, depending on the type selected. Even the highest quality units with the most suitable grilles do not cover more than 90° adequately; when concealed behind fake diffusers, they are far less satisfactory (Fig. 44.15*b*). Unfortunately, many loudspeakers employed in such systems beam high-frequency energy rather sharply. Uniform, high speech intelligibility is not possible with such systems; also they must be operated at an uncomfortably high level to permit listeners between loudspeakers to hear properly.

To prevent feedback, a flexible system is usually provided, with switching arrangements so that certain loudspeakers can be shut off when a source of sound is to be placed immediately under one of the units in a space for flexible use.

As mentioned earlier, loudspeakers should never be placed along the side walls of a room, producing "cross firing." This cross firing always causes the listener to hear from many loudspeakers at the same time, with multiple time delays reducing speech intelligibility. In a church, the loudspeakers might be located in the bottoms of chandeliers (over the heads of the worshippers) or in the backs of the pews; there are many ways in which loudspeakers can be located properly for assistance. Sometimes when a central loudspeaker system is used for an auditorium where people are seated on the platform with the speaker, participating perhaps in a panel or merely serving as background, a few loudspeakers operated at low level can be placed for the convenience of these listeners so that they hear more than merely the reverberant sound from the hall itself.

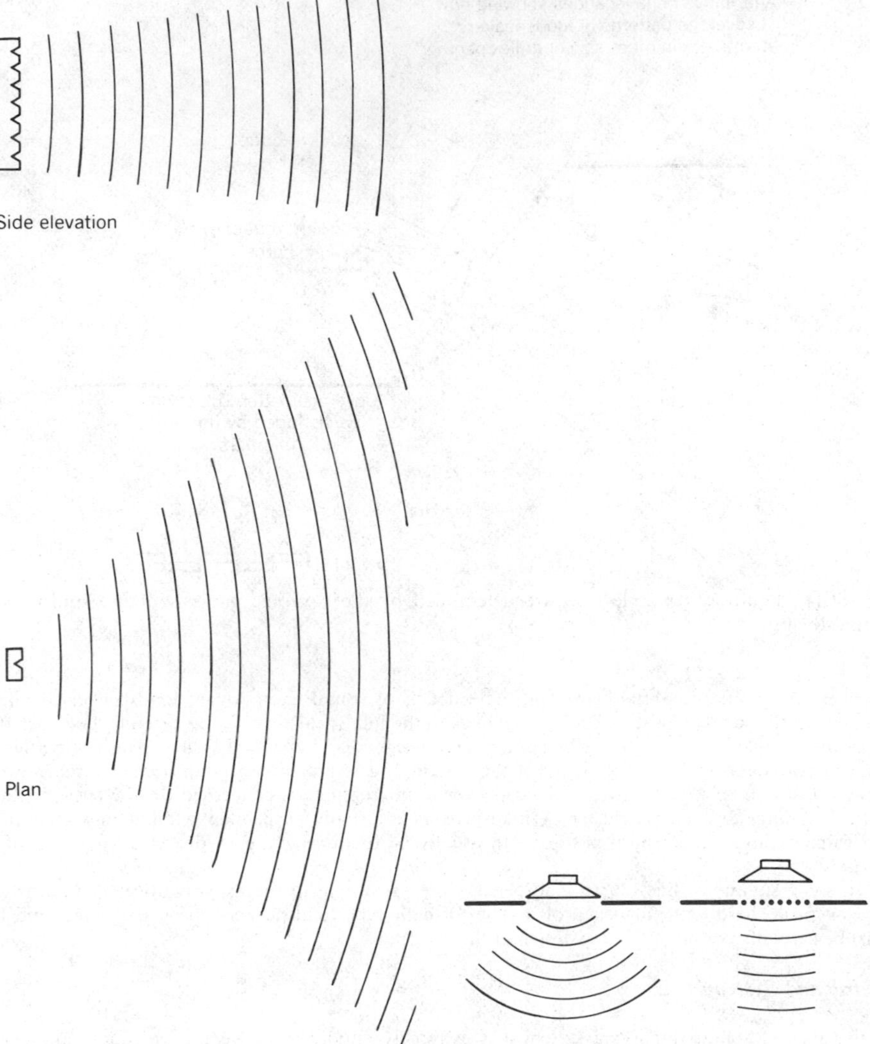

Fig. 44.15 (*a*) Line-source or column loudspeaker and its coverage characteristics. (*b*) Poorly chosen loudspeaker grille materials can markedly affect the sound distribution from any loudspeaker.

Time Delay

Artificial echo is often a problem, particularly in distributed systems. Consider a listener at the rear of a long auditorium that uses a distributed loudspeaker system for any of the reasons discussed. This listener will hear amplified sound almost instantaneously from the nearest loudspeaker, while the natural sound will arrive at some later time, depending on the distance to the platform. This delay in the arrival of the natural sound (due to the fact that sound in air travels at the rate of about 1120 ft/s compared with the very much greater speed of the electrical signal between the microphone and the loudspeaker) may be sufficient to cause a discrete echo if the delay is on the order of 65 ms or more; if the delay is somewhat less than this, the echo can appear as simply a muddying effect on the sound heard by the listener. To resolve this problem, it is necessary to introduce a delay device in the electrical circuit that, in effect, delays the loudspeaker sound so that it arrives at approximately the same time as the natural sound. For a very long room, two or more delay circuits may be required, serving several zones along the length of the room. These devices have been used most successfully where distributed loudspeaker systems supplement a central system (e.g., under a deep balcony overhang), but, of course, they mean additional initial expenditure and maintenance costs. Time delay is usually a last resort and may not ever be required if the sound system design is considered in the early stages of space design.

Bibliography

Acoustical and Insulating Materials Association, *Sound Absorption Coefficients of Architectural Acoustical Materials*, Bulletin XXX (1970).

Beranek, L. L., *Noise Reduction*, McGraw-Hill, New York, 1960.

Beranek, L. L., *Music Acoustics and Architecture*, Wiley, New York, 1963.

"Better Architecture for the Performing Arts," *Architectural Record* (Dec. 1964).

Harris, C. M., *Handbook of Noise Control*, McGraw-Hill, New York, 1957.

Knudsen, V. O. and C. M. Harris, *Acoustical Designing in Architecture*, Wiley, New York, 1950.

Newman, R. B., ed., "Design for Hearing," *Progressive Architecture* (May 1959).

Parkin, P. H. and H. R. Humphreys, *Acoustics, Noise and Buildings*, Faber and Faber, London, 1958.

Schultz, T. J., "Acoustics of the Concert Hall," *IEEE Spectrum* (Jun. 1965).

44.2 PSYCHOACOUSTICS

Stanley A. Gelfand

44.2-1 Introduction

The perception of sound becomes of interest to the engineer whenever an application involves a signal that must be heard. Typical applications include communication systems and channels, electroacoustic transducer design, commercial and entertainment recording systems, alarm systems, and computer-generated acoustic signals, to name but a few general areas of concern. The engineer must also have a fundamental knowledge of how sound is perceived in order to address engineering problems in the areas of reverberation and room acoustics, noise measurement and control, and similar applications. Further, the design of hearing aids and other prosthetic aids for the hearing impaired (particularly future designs that might employ signal-processing approaches) necessitate an understanding of hearing on the part of the engineer.

In this section we will examine a number of fundamental concepts of sound perception, or psychoacoustics. As the term suggests, "psychoacoustics" is concerned with how sound is perceived. It represents a category of the more inclusive science of psychophysics, which deals with the representation, perception, and so on of the physical world by the various sensory and perceptual modalities. We will discuss sound perception in terms of simple acoustic signals (tones, noise) rather than from the standpoint of speech, which is an extremely complex signal that varies continuously over time.

44.2-2 Scales of Measurement

The psychoacoustician attempts to quantify the manner in which sound is perceived by applying appropriate methods and scales of measurement. Four such scales may be considered classical (nominal, ordinal, interval, and ratio scales). Other hybrid scales are beyond our scope here. They have been described in detail by Stevens (1975).

The lowest level of scaling involves simply categorizing observations into categories on the basis of some attribute. However, no assumptions are made about the relationships between categories except that the categories are different. Thus we could categorize subjects according to sex or eye color. There is no inherent ordering or hierarchy of any kind between the sex or eye-color groups other than the fact that they happen to be different. These are *nominal* scales. On the other hand, if one could establish an ordering among the categories (such as relative prices of automobiles), the groups would constitute an *ordinal* scale. In this case there is some orderly progression from group to group, but the spacing between groups is different.

We may now appreciate the impact of scaling differences. With nominal scales (where $A \neq B \neq C \ldots$) we can do little more than count the number of observations in each category and establish modes. Ordinally scaled material (where $A > B > C \ldots$) may be described in terms of medians, percentiles, and so on so that more information is available to describe the data.

If the distances between categories or data points are the same (e.g., hours on a clock), an *interval* scale is said to exist. Here most mathematical operations may be performed on the data, so that averages (means) may be calculated. However, ratios among the data may not be expressed because a true zero is not assumed. *Ratio* scales have all the characteristics of interval scales plus a true origin. Hence ratios among such data are possible as well as the use of logarithms and decibels. In the physical realm, temperature in Kelvin represents a ratio scale.

44.2-3 Methods of Measurement

To study some attribute of interest we must be able to make appropriate (valid) and repeatable (reliable) measurements. Physical measurements are fundamental to the engineer and need not be discussed in this context other than to draw an analogy to the measurement of psychoacoustic parameters. However, in psychoacoustics we must consider more than an appropriate metric measurement (e.g., weight in grams or length in millimeters) and allowances for physical tolerances: To measure the parameters of what an individual hears (in response to a known physical stimulus) we must, in fact, ask him or her what he or she has heard as opposed to making a direct measurement of the sensation itself. This indirect aspect of psychoacoustic measurements adds a dimension to what the subject hears ("sensory capability"), which is how he responds ("response proclivity"). The latter represents a myriad of biases that must be excluded from the results of a psychoacoustic measurement.

Three general methods of measurement have been employed in classical psychophysics. The first approach is the method of limits. Here the experimenter controls the level (or some other attribute) of the stimulus and varies it in a fixed direction until the subject detects a change. Table 44.2 shows two runs of a threshold experiment. The stimulus may be increased from below the expected threshold until the subject reports that it is now audible (ascending run) or from above the expected threshold until the subject reports no longer hearing the stimulus (descending run). Each run ends when there is a change (from − to + or from + to −), and the threshold is taken as the average of the cross-over points.

In the method of adjustment, the subject is asked to control the stimulus attribute. In the threshold example the subject increases (decreases) a level control until the stimulus becomes audible (inaudible). Biases in responding are avoided to some extent by use of a dial that turns continuously without markings, and often by insertion of a second attenuator under the experimenter's control. "Bracketing," or increasing and decreasing level alternately until a change point is reached, represents a commonly employed modification of the methods of adjustment and of limits.

The method of constants is the third classical measurement method. It differs from the other two approaches in that an equal number of stimuli are randomly presented at each of a predetermined number of stimulus levels. Here the method of constants is not a sequential method because one presentation does not depend on the response to the previous one. Since an equal number of stimuli are presented at each level, the experimenter may examine the responses to determine the level at which there is a 50% probability (or any other reasonable proportion) of a response, which may then be taken as the threshold value.

Same/different, louder/softer, and other judgments may also be made using these methods, as in differential sensitivity experiments (discussed later).

Space does not permit in-depth coverage of all types of measurement methods; however, at least a brief discussion of adaptive procedures is indicated. These are sequential methods much like the method of limits but are considered a separate category because the strategy by which stimuli are presented depends on the previous response in such a way that observations tend to concentrate in the vicinity of, and converge on, the target value (Levitt, 1971; Taylor and Creelman, 1967).

44.2-4 Auditory Domain

We may conceive of the auditory domain as the range of sounds that are audible and tolerable. In particular, we are interested in the range of frequencies that can be heard, the softest audible sounds,

TABLE 44.2. EXAMPLE OF THE METHOD OF LIMITS

Stimulus Level in dB:	40	38	36	34	32	30	28	26	24	22	20	18	16	14
Ascending run					+	−	−	−	−	−	−	−	−	−
Descending run	+	+	+	+	+	+	−							

and the most intense sounds that are tolerable. We would also like to know how brief a sound can be and still be audible.

Figure 44.16 shows the sound pressure levels of the softest audible sounds as a function of frequency, obtained in several studies. These threshold data are shown in two forms. Results obtained with subjects listening through earphones, minimum audible pressures (MAP), are about 6 to 10 dB less sensitive than data obtained listening to a speaker placed in a sound field. The latter are minimum audible fields (MAF). This difference between MAP and MAF (the "missing 6 dB") has been the subject of considerable controversy since it was reviewed in detail by Sivian and White (1933). However, it now appears that the difference is more apparent than real, being accounted for by such factors as physiologic noise masking stimuli presented under phones, differences between real ear and electroacoustic coupler-refined measures, and other technical factors (Killion, 1978; Rudmose, 1982).

Figure 44.17 reveals that threshold sensitivity is far from uniform across the frequency range. These curves show that human hearing is quite sensitive between approximately 100 and 10,000 Hz, being most acute in the 2000- to 5000-Hz range. Hearing sensitivity becomes poorer as frequency increases and decreases above and below this range. (The frequency scale between 10,000 and 20,000 Hz is expanded to show this phenomenon more clearly.)

A particularly important application of MAP data is in the specification of normal hearing levels for audiometers used in clinical hearing evaluations. Here the number of dB SPL (sound pressure level) required to reach the threshold of audibility of normal hearing persons is expressed as 0 dB hearing level (HL, ANSI-1969). For example, the reference levels for 0 dB HL would be 7 dB SPL at 1000 Hz and 9 dB SPL at 2000 Hz, using TDH-39 receivers. The relationship between SPL and HL is shown graphically in Fig. 44.17, where the curved normal threshold function (circles) for SPL in (*a*) is seen as a straight line in decibels HL (*b*). The triangles in the two figures are for the same noise-induced hearing loss, expressed in decibels SPL and decibels HL.

Figure 44.16 shows the upper limits of usable hearing, which are expressed in the form of thresholds for unpleasant sensations, generally of a tactile rather than an auditory nature. These levels fall between 120 and 140 dB SPL, depending on the nature of the sensation. If uncomfortable loudness is of concern, the upper levels would be approximately 100 dB SPL (Hood and Poole, 1966, 1970). Regardless of the sensation (auditory or tactile), the upper limits curves are essentially flat across frequency, in contrast to the threshold sensitivity curves.

If the duration of a sound is decreased well below 1 s, it becomes necessary to increase the stimuli level to overcome the effect of reduced on-time. Sounds longer than roughly 200 to 300 ms require the same intensity to be heard. However, shorter durations must be overcome by an increase in signal level such that a decade duration change is offset by a 10-dB level adjustment (Zwislocki, 1960), a

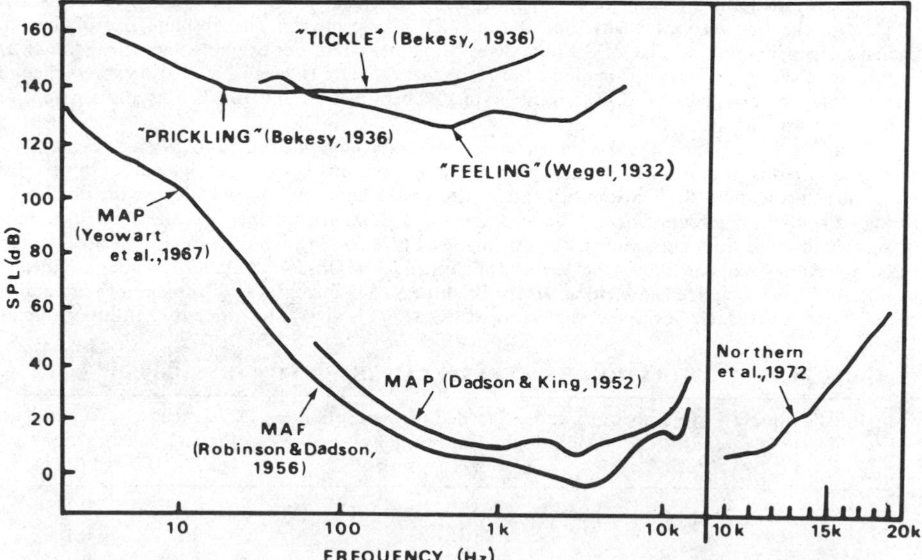

Fig. 44.16 Minimum audible pressures (MAP) and fields (MAF) and highest tolerable levels from various sources. From Gelfand, 1981, with permission.

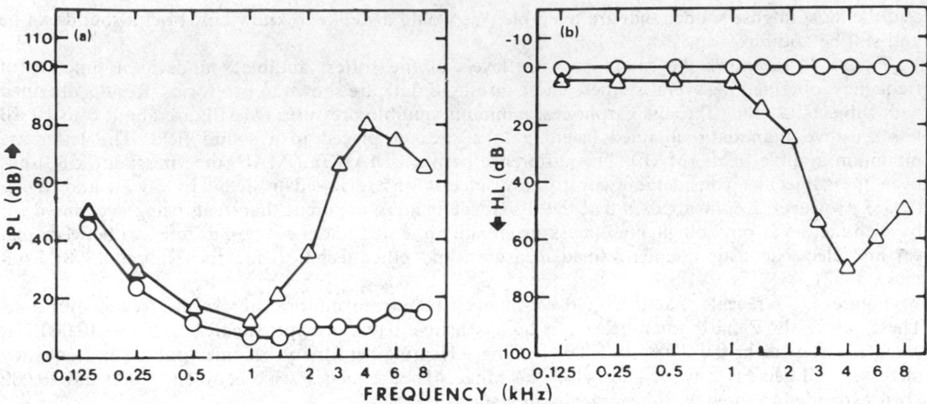

Fig. 44.17 Audiograms expressed in decibels for sensitivity to pressure (SPL) (*a*) and hearing level (HL) (*b*) for normal individuals (circles) and for a patient with hearing loss due to noise exposure (triangles). Increasing level is read downward on the clinical audiogram. From Gelfand, 1981, with permission.

phenomenon called temporal summation (integration). Hence a decrease in duration of a signal from 500 to 50 ms must be counteracted by an increase in its level by 10 dB to maintain its audibility (or loudness, for that matter), whereas a decrease in duration of a signal from 5 s (5000 ms) to 500 ms does not affect its threshold.

Another interesting effect of duration relates to the sensation of tonality (Doughty and Garner, 1947). For frequencies above about 1000 Hz, a signal must be on for about 10 ms for a sensation of tonality to be obtained. Below 1000 Hz the tonality threshold increases with decreasing frequency (e.g., 15 ms at 500 Hz and 60 ms at 50 Hz).

44.2-5 Differential Sensitivity

We may now direct our attention to how sounds are perceived within the audible range. The logical first step is to ask what is the smallest perceptible difference between two sounds that can be heard, again in terms of frequency, intensity, and duration. In absolute terms, such a minimally perceptible difference is called a just noticeable difference (jnd) or difference limen (DL), also written as Δf, ΔI, and Δt for the frequency, intensity, and duration DLs, respectively. The absolute size of the DL changes depending on the value of the parameter being measured. For example, the actual size of Δf is different if the measurement is made at 100 Hz vs. at 3000 Hz. Hence we may also express the jnd relatively as, for example, $\Delta I/I$. This ratio is the Weber fraction and is taken as the measure of difference sensitivity.

Let us refer to Hirsh's (1952) classic illustration in which the DL and Weber fraction are expressed in terms of how many candles must be added to a number of existing candles to perceive the smallest perceptible difference in brightness. The first column in Table 44.3 shows the number of candles already lit in a room (I), and the second column shows the number needed if there is to be a noticeable increase in brightness over the starting level ($I + \Delta I$). The third column indicates the DL (ΔI) and the last column shows the Weber fraction ($\Delta I/I$). Observe that the absolute number of candles that must be added to yield a DL for brightness (ΔI) increases with the number of candles originally present (I). However, the relative increase ($\Delta I/I$) is always the same (0.1 in this example).

TABLE 44.3. ILLUSTRATION OF WEBER'S LAW (BASED ON HIRSH, 1952)

Initial Number of Candles (I)	Number of Candles Needed for Perception of More Light Than Initially ($I + \Delta I$)	Difference Limen (ΔI)	Weber Fraction ($\Delta I/I$)
10	11	1	0.1
100	110	10	0.1
1,000	1,100	100	0.1
10,000	11,000	1,000	0.1
100,000	110,000	10,000	0.1

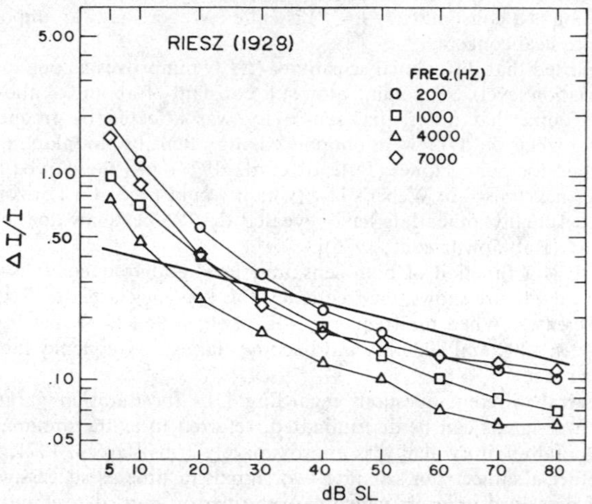

Fig. 44.18 Weber fraction for intensity ($\Delta/I/I$) as a function of sensation level (dB SL). The straight line shows the data of Jesteadt et al. (1977); the symbols show Reisz's (1928) results. From Jesteadt et al., 1977, with permission.

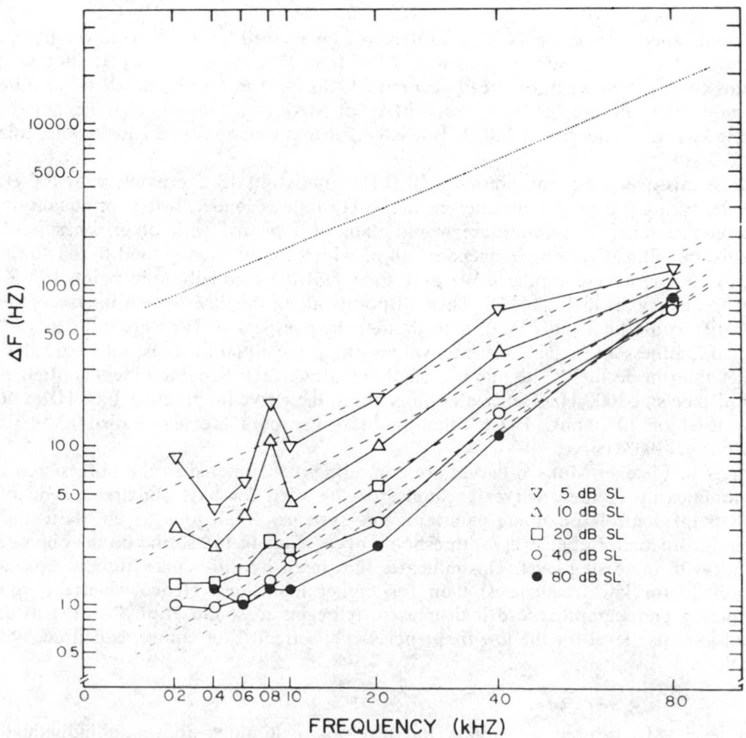

Fig. 44.19 Frequency difference linear (DL) (ΔF) as a function of frequency, with sensation level (SL) as the parameter. From Wier et al., 1977, with permission.

That the Weber fraction is a constant ($\Delta I/I = k$) is called Weber's law, an important and frequently encountered psychophysical concept.

In 1928 Riesz reported that differential sensitivity ($\Delta I/I$) improves as one increases the test level above threshold (sensation level),† becoming almost a constant of about 0.3 above about 40 dB (Fig. 44.18). While Riesz found that differential sensitivity was affected by frequency (shown by the symbols in the figure) when the DLs were obtained using intensity modulation, this does not occur when DLs are obtained for pulsed tones (Jesteadt et al., 1977). [As the curved line clearly shows in Fig. 44.20, there is a "near miss" to Weber's law (which would predict a horizontal line, since $\Delta I/I$ would be a constant). Actually, other data have revealed that Weber's law does hold, at least between about 10 and 40 dB SL (Rabinowitz et al., 1976)].

The frequency DL is a function of both sensation level and frequency, as shown in Fig. 44.19 (Wier et al., 1977). As the figure shows, the frequency DL becomes larger as frequency increases and as sensation level decreases. When the frequency DL is converted to Weber fractions, we find that $\Delta f/f$ is smallest between 1000 and 2000 Hz and becomes larger as frequency increases and decreases above and below this range.

One may ask several different questions regarding DLs for duration parameters. The shortest interval over which two signals can be discriminated, referred to as the minimal integration time or the threshold of temporal auditory acuity, is approximately 2 ms (Green, 1971, 1973). We may view this measure of whether a subject can separate two signals in time as successiveness, as opposed to being an estimate of perceived order, in which case the person must determine which of two signals (e.g., "low" and "high" frequency tones) came first (Hirsh, 1959). In the latter case, a separation of about 20 ms is needed. Finally, we may ask what the DL is for signal duration, which is of course the analogous parameter to the frequency and intensity DLs just covered. The duration DL (Δt) increases as overall duration (t) increases, from approximately 0.5 ms at 50 ms to roughly 50 ms for a duration of about 1 s.

44.2-6 Loudness and Pitch

Loudness and pitch refer to the perceptual correlates generally associated with the physical parameters of intensity and frequency, respectively. While loudness does tend to increase with intensity and pitch similarly rises with frequency, the manner in which the psychologic and physical correlates relate is far from linear.

A commonly encountered aspect of how intensity is perceived has to do with whether a particular signal level at one frequency sounds louder or softer than that same intensity at another frequency. Recall in this context that we have already examined this issue at threshold, where we observed that differing amounts of SPL are needed to reach MAP or MAF as a function of frequency (Fig. 44.16). We may now ask, what intensity at 100 Hz is needed for that tone to sound equally loud to a 1000-Hz tone of 40 dB SPL?

To address this query, we may present a 1000-Hz tone at 40 dB alternately with 100-Hz tones of various levels, asking the subject whether each 100-Hz tone is louder, softer, or equally loud to the 1000-Hz reference tone. This procedure would then be repeated with other tones until we have obtained a list of SPLs at various frequencies, all of which sound equally loud to the 40-dB 1000-Hz tone (and, of course, to one another). We may then plot these equally loud points in dB SPL as a function of frequency, as in Fig. 44.20. Thus all points along this line (which intersects the 1000-Hz point at 40 dB) would be equally loud, even though their respective SPLs are different. So as not to confuse equal loudness and equal intensity, values along the equal-loudness curve are designated in phons rather than in decibels. The number of phons along each equal-loudness contour is equal to the level in decibels at 1000 Hz. Hence all points along the curve intersecting 1000 Hz at 40 dB have a loudness level of 40 phons. These equal-loudness contours are also called phon curves and Fletcher-Munson (1933) curves.

References to Fletcher-Munson curves are commonly encountered in the hi-fi/stereo and literature; the nonlinearity of these curves accounts for the need for bass and treble (and often other response-shaping) controls on home entertainment systems. Analogous to the flattening we have already seen for uncomfortable level vs. threshold curves (Fig. 44.16), so too do the phon curves tend to flatten out with increasing level. This indicates that intensity differences are greater among lower levels (especially for low frequencies) than for higher intensities. Hence when the volume of a natural-sounding phonograph record is decreased, it begins to sound "tinny" owing to the relative reduction of loudness level for the low frequencies. This effect is, of course, equalized by use of the bass control.

†Sensation level (SL) is used to express the level of a stimulus above an individual's hearing threshold. For example, if a person's threshold for some signal is 32 dB SPL and a measurement is made at 50 dB SPL, the sensation level would be 50 − 32 = 18 dB SL.

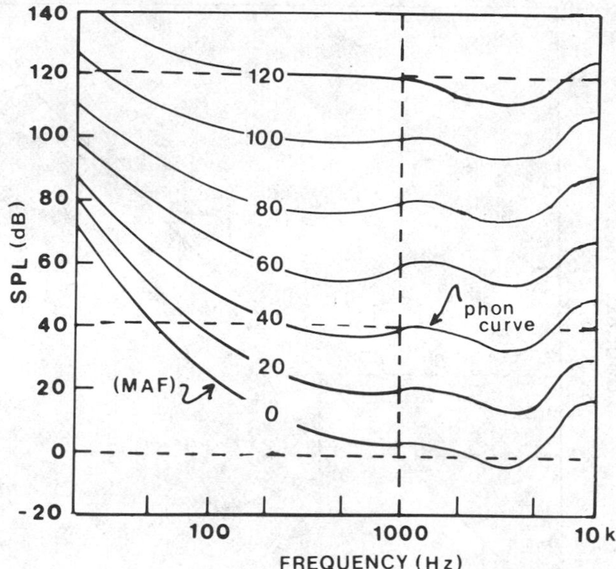

Fig. 44.20 Idealized equal loudness (or phon or Fletcher–Munson) curves, adapted from various sources. Dotted lines show equal sensitivity to pressure levels (SPLs) (0, 40, and 120 dB) for comparison. Adapted from Gelfand, 1981.

Other direct engineering applications of equal-loudness contours are encountered in frequency-response-weighting networks of sound-level meters. The A- and B-weighting responses are rough equivalents of the 40- and 70-phon curves, respectively, while the C-weighting response attempts to approximate equal loudness at high intensities. (The use of A-weighting networks in noise measurements at high levels is really inconsistent with the actual equal-loudness contours at these levels, grossly underestimating low-frequency effects). Weighting formulas for the noisiness of wide-band signals are beyond the scope here but may be found in work by Kryter (1970) and Stevens (1972).

Phon curves reveal which sounds are as loud as others but not how loudness is related to intensity. To generate a scale of loudness as a function of intensity, we might ask a subject to listen to a reference tone (called a modulus) and then to adjust another sound to be twice as loud as the modulus, and so on. Techniques such as these, pioneered by Stevens (1959), are direct-scaling methods. If a 1000-Hz tone at 40 dB is the reference intensity, and we call its loudness 1 sone (Fig. 44.21), we may then generate a scale so that the loudness of a tone sounding twice as loud as 40 dB would be called 2 sones, half as loud 0.5 sone, and so on. Figure 44.21 shows that an increase of 9 to 10 dB results in a doubling of loudness (e.g., 1 to 2 sones, 2 to 4 sones, etc.). We also observe that the sone scale is a straight line on the log-log coordinates, indicating that loudness (L) is a power function of stimulus level (I); or $L = kI^e$, where k is a constant and e is an exponent equal to 0.67. This relationship is Stevens' power law (1957). Like loudness, other percepts such as that of brightness, increase with stimulus intensity with exponents less than 1.0. On the other hand, exponents greater than 1.0 are encountered for percepts that increase faster than the physical stimulus level (e.g., electric shock).

Two interesting facets of loudness are its relationships to stimulus duration and to bandwidth. As for threshold, stimulus level and duration interact for signals shorter than about 200 to 300 ms, so that a decade-time/10 dB time-intensity trade, as previously defined, also exists for loudness (Small et al., 1962).

Increasing the bandwidth of a signal (or measuring the separation between two tones) does not affect the loudness of a noise (or the tonal complex) until the bandwidth (separation) exceeds a certain critical bandwidth. However, once this critical bandwidth is exceeded, loudness increases with increasing bandwidth even though the overall signal level is the same (Scharf, 1970).

Using scaling techniques like those just described for the sone scale, we may develop a pitch scale as a function of frequency such that the pitch of a 1000-Hz tone (at 40 phons) is called 1000 mels. The tone sounding twice as high would be 2000 mels; half as high, 500 mels; and so on (e.g., Stevens and Volkman, 1940; Stevens et al., 1937). Such a scale is represented in Fig. 44.22. Note that the entire audible range of roughly 20 kHz is "compressed into a pitch" range of only about 3500 mels, and that the relation of pitch to frequency follows a sigmoid function. Hence a pitch doubling from

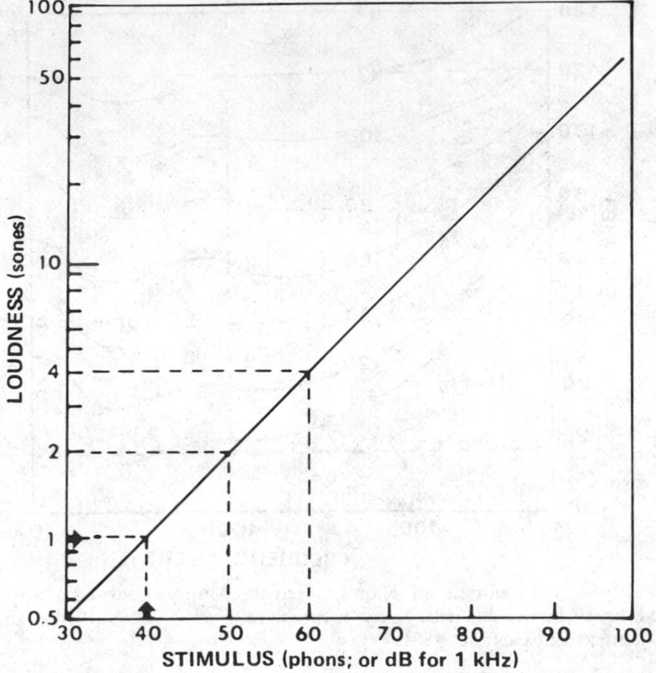

Fig. 44.21 Idealized sone (loudness) scale. From Gelfand, 1981, with permission.

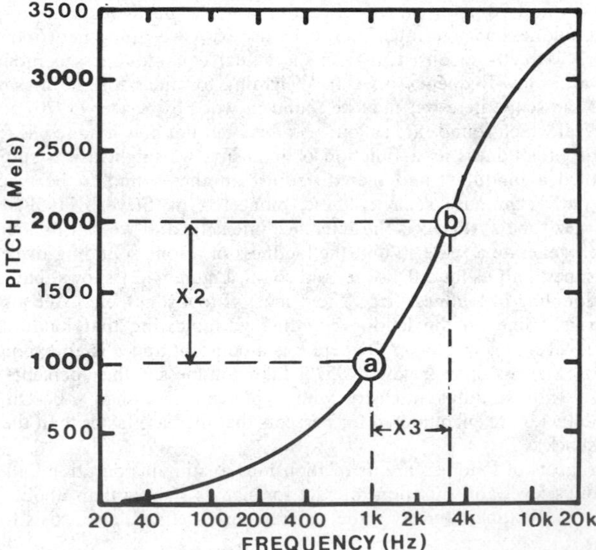

Fig. 44.22 Idealized mel (pitch) scale modified after work of Stevens and colleagues.

1000 to 2000 mels corresponds to tripling frequency from 1000 to 3000 Hz. The critical bandwidth corresponds to approximately 150 mels (Scharf, 1970).

The psychoacoustic pitch scale does not quite agree with that of musical pitch, in that, for example, the two ranges 100 to 200 Hz and 1000 to 2000 Hz are both musical octaves, but not so in mels (see Fig. 44.22). This discrepancy is beyond the current scope; however, the interested reader might consult Terhardt (1974) for an insightful approach to this apparent paradox. It appears that pitch is not substantially affected by intensity (Cohen, 1961).

44.2-7 Masking and Temporary Threshold Shift (TTS)

Masking is interference with the perception of one sound (signal) because of the presence of a second sound (masker), generally in the sense that the signal is rendered inaudible by the presentation of the masker. In the typical masking experiment one simply obtains the threshold of a signal (usually a tone), adds a masker (usually a noise), and then measures the threshold of the tone, this time in the presence of the noise. For example, if the threshold of a signal is 50 dB when it is presented alone and the level must be raised to 65 dB when a second sound (masker) is added, we may say that:

1. Fifty decibels is the unmasked threshold of the signal.
2. The masker caused the signal to be masked at 50 dB.
3. The masked threshold of the signal is 65 dB.
4. Fifteen decibels of masking has been produced by the masker.

Hence we may describe the event of masking in a variety of ways, so that the manner of presenting masking data should be consistent with the intended application.

The masking patterns in Fig. 44.23 show the effects of masker intensity and spectrum on the audibility of tones (Wegel and Lane, 1924; Ehmer, 1959; Small, 1959). In each case the abscissa shows the frequency of the signal (tone) and the ordinate shows the amount of masking compared with the unmasked thresholds. The frequencies indicated in each frame are the center frequencies of narrow-band noise maskers, and the small numbers indicate the levels (in decibels) at which the maskers were presented. Thus each curve shows the amount of masking produced by a particular level of a narrow-band noise as a function of frequency.

Several points are illustrated by these masking audiograms. First, the amount of masking

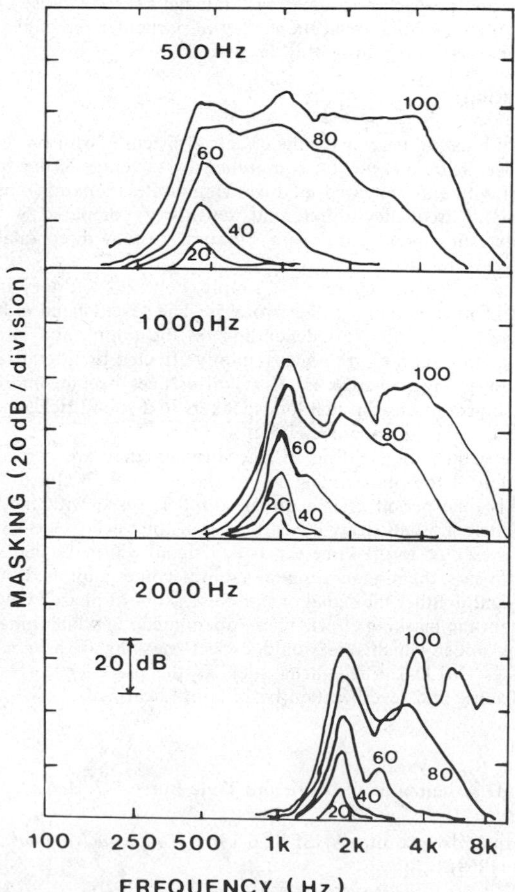

Fig. 44.23 Typical masking patterns. Adapted from Ehmer, 1959.

increases with masker level. In fact, the amount of masking has been shown to be a linear function of masker level (Hawkins and Stevens, 1950). Second, more masking is produced at frequencies close to that of the masker. Third, masking patterns tend to be narrow and symmetrically distributed around the masker frequency at lower masker levels but become wider with increasing masker intensity, extending asymmetrically upward in frequency. Fourth, masking patterns tend to be restricted to narrower (high) frequency regions for high-frequency maskers, but are much wider for low-frequency maskers. Finally, peaks in the masking patterns tend to occur roughly at multiples of the masker frequency, though not precisely so.

These masking patterns constitute an auditory representation of the traveling wave activity in the cochlea, in which the displacement pattern meanders upward from base (higher frequencies) to apex (lower frequencies), with a sharp decline of activity on the apical side of the wave peaks.

An interesting point has to do with the amount of masking and the bandwidth of the masker. For a constant-spectrum level, widening the bandwidth of a noise will increase the amount of masking it will produce (for a tone at its center frequency). However, widening this band beyond a certain bandwidth results in no further masking of that signal (Hawkins and Stevens, 1950). The masked threshold occurs when noise (N_0) and signal (S) power are equal in the critical band (C), or $S = C(N_0)$. Hence this critical band for masking is actually a critical ratio: $C = S/N_0$. Multiplying masking critical ratios by 2.5 yields critical bandwidths obtained from loudness and other data (Scharf, 1970).

If one measures a subject's threshold to a tone, then presents an intense noise for some time, and subsequently again measures the threshold, a shift from the original sensitivity will be noted. This "after the fact" effect is called temporary threshold shift (TTS) or poststimulatory fatigue and is related to the level of the fatiguing stimulus. The maximal effect is generally in the 4000- to 6000-Hz region, and the TTS tends to decrease with the logarithm of time after exposure to the stimulus. The extent to which the amount of TTS (generally measured two minutes post exposure) relates to permanent hearing loss, an approach to estimating "damage risk" for noise exposure, is commonly encountered in acoustical engineering applications. See in particular the paper by Kryter (1973) and the commentary that follows it for an insightful discussion.

44.2-8 Binaural Hearing

At least some mention of binaural hearing (or the effects of listening with two ears as opposed to one) should be mentioned. [See Gelfand (1981) for comprehensive coverage of this topic.] Besides resulting in better threshold sensitivity and increased loudness (binaural summation), binaural hearing results in improved intensity and frequency differential sensitivity compared with monaural hearing. However, the most interesting points deal with binaural fusion, directional hearing, precedence effects, and masking level differences.

Binaural fusion refers to the perception of a simple, coherent auditory image when earphones present similar but not identical stimuli to the two ears. The fused signal is heard within the head, lateralized along a plane between the ears depending on the timing and intensities of the signals. Below about 1500 Hz, lateralization right or left is mainly affected by interaural time differences (the sooner-arriving signal governing), whereas intensity differences become more important at higher frequencies. When the subject listens through loudspeakers in a sound field, the fused image is heard extracranially, which causes the stereophonic effect.

If two signals arrive within about 40 ms of one another, they are perceived as a single image coming from the direction of the first-arriving sound (Haas, 1949; Wallach et al., 1949). Hence an echo arriving within this time period after a direct sound is fused with the direct sound, but one arriving after approximately a 50-ms delay is heard as a distinct echo.

Recall that a noise presented to the same ear as is a signal will mask that signal. However, if the noise is presented to both ears, the monaural signal again becomes audible. If a signal and a noise are presented to both ears, and if either the signal or the noise is out of phase between the two ears, there will also be a release from the masking effect. These phenomena in which binaural hearing permits a release from conditions under which one would expect masking of a signal to occur are called masking level differences (MLDs). Phenomena such as the precedence (Haas) effect and MLDs underlie the squelch of noise and reverberation by binaural hearing.

Bibliography

Abel, S. M., "Duration Discrimination of Noise and Tone Bursts," *J. Acoust. Soc. Am.* **51**:1219–1223 (1974).

American National Standards Institute, ANSI S3.6-1969, *American National Standard Specifications for Audiometers* (1969).

Bekesy, G., *Experiments in Hearing*, McGraw-Hill, New York, 1960.

Cohen, A., "Further Investigation of the Effects of Intensity upon the Pitch of Pure Tones," *J. Acoust. Soc. Am.* **33**:1363–1376 (1961).

Dadson, R. S. and J. H. King, "A Determination of the Normal Threshold of Hearing and Its Relation to the Standardization of Audiometers," *J. Laryngol. Otol.* **46**:366–378 (1952).

Doughty, J. M. and W. R. Garner, "Pitch Characteristics of Short Tones: I. Two Kinds of Pitch Threshold," *J. Exp. Psychol.* **37**:351–365 (1947).

Ehmer, R. H., "Masking Patterns of Tone," *J. Acoust. Soc. Am.* **31**:1115–1120 (1959).

Fletcher, H. and W. A. Munson, "Loudness, Its Definition, Measurement and Calculation," *J. Acoust. Soc. Am.* **5**:82–105 (1933).

Gelfand, S. A., *Hearing: An Introduction to Psychological and Physiological Acoustics,* Marcel Dekker, New York, 1981.

Green, D. M., "Temporal Auditory Acuity," *Psych. Rev.* **78**:540–551 (1971).

Green, D. M., "Temporal Auditory Acuity as a Function of Frequency," *J. Acoust. Soc. Am.* **54**:373–379 (1973).

Haas, H., "The Influence of Single Echoes on the Audibility of Speech," *Library Communication 363*, Dept. Sci. Indust. Rest., Garston, Watford, England (1949).

Hawkins, J. E. and S. S. Stevens, "The Masking of Pure Tones and of Speech by White Noise," *J. Acoust. Soc. Am.* **22**:6–13 (1950).

Hirsh, I. J., *The Measurement of Hearing*, McGraw-Hill, New York, 1952.

Hirsh, I. J., "Auditory Perception of Temporal Order," *J. Acoust. Soc. Am.* **31**:759–767 (1959).

Hood, J. D. and J. P. Poole, "Tolerable Limits of Loudness: Its Clinical and Physiological Significance," *J. Acoust. Soc. Am.* **40**:47–53 (1966).

Hood, J. D. and J. P. Poole, "Investigations in Hearing upon the Upper Physiological Limit of Normal Hearing," *Int. Audiol.* **9**:250–255 (1970).

Jesteadt, W., C. C. Wier, and D. M. Green, "Intensity Discrimination as a Function of Frequency and Sensation Level," *J. Acoust. Soc. Am.* **61**:169–177 (1977).

Killion, M. C., "Revised Estimate of Minimum Audible Pressure: Where is the 'Missing 6 dB'?" *J. Acoust. Soc. Am.* **63**:1501–1508 (1978).

Kryter, K. D., *The Effects of Noise on Man*, Academic, New York, 1970.

Kryter, K. D., "Impairment to Hearing from Exposure to Noise," *J. Acoust. Soc. Am.* **53**:1211–1234 (1973). (Also comments by A. Cohen, H. Davis, B. Lempert, W. D. Ward, and reply by Kryter, *J. Acoust. Soc. Am.* **53**:1235–1252, 1973.)

Levitt, H., "Transformed Up-Down Methods in Psychoacoustics," *J. Acoust. Soc. Am.* **49**:467–477 (1971).

Northern, J. L., M. A. Downs, W. Rudmose, et al., "Recommended High-Frequency Audiometric Threshold Levels (8000-18,000 Hz)," *J. Acoust. Soc. Am.* **52**:585–595 (1972).

Rabinowitz, W. M., J. S. Lim, L. D. Braida, and N. I. Durlach, "Intensity Perception: VI. Summary of Recent Data on Deviations from Weber's Law for 1000-Hz Tone Pulses," *J. Acoust. Soc. Am.* **59**:1505–1509 (1976).

Riesz, R. R., "Differential Intensity Sensitivity of the Ear for Pure Tones," *Physiol. Rev.* **31**:867–875 (1928).

Robinson, D. W. and R. S. Dadson, "A Re-determination of the Equal Loudness Relations for Pure Tones," *Brit. J. Appl. Phys.* **7**:166–181 (1956).

Rudmose, W., "The Case of the Missing 6 dB," *J. Acoust. Soc. Am.* **71**:650–659 (1982).

Scharf, B., Critical Bands, in J. V. Tobias, ed., *Foundations of Modern Auditory Theory*, vol. 1, Academic, New York, 1970.

Sivian, L. J. and S. D. White, "On Minimal Audible Sound Fields," *J. Acoust. Soc. Am.* **4**:288–321 (1933).

Small, A. M., "Pure Tone Masking," *J. Acoust. Soc. Am.* **31**:1619–1625 (1959).

Small, A. M., J. F. Brandt, and P. G. Cox, "Loudness as a Function of Signal Duration," *J. Acoust. Soc. Am.* **34**:513–514 (1962).

Stevens, S. S., "On the Psychophysical Law," *Psych. Rev.* **54**:153–181 (1957).

Stevens, S. S., "On the Validity of the Loudness Scale," *J. Acoust. Soc. Am.* **31**:995–1003 (1959).

Stevens, S. S., "Perceived Level of Noise by Mark VII and Decibels (E)," *J. Acoust. Soc. Am.* **51**:575–601 (1972).

Stevens, S. S., *Psychophysics*, Wiley, New York, 1975.

Stevens, S. S. and J. Volkman, "The Relation of Pitch to Frequency: A Revised Scale," *Amer. J. Psychol.* **53**:329–353 (1940).

Stevens, S. S., J. Volkman, and E. B. Newman, "A Scale for the Measurement of the Psychological Magnitude Pitch," *J. Acoust. Soc. Am.* **8**:185–190 (1937).

Taylor, M. M. and C. D. Creelman, "PEST: Efficient Estimates of Probability Functions," *J. Acoust. Soc. Am.* **41**:782–787 (1967).

Terhardt, E., "Pitch Consonance and Harmony," *J. Acoust. Soc. Am.* **55**:1061–1069 (1974).

Wallach, H., E. B. Newman, and M. R. Rosensweig, "The Precedence Effect in Sound Localization," *Am. J. Psych.* **62**:315–336 (1949).

Wegel, R. L., "Physical Data and Physiology of Excitation of the Auditory Nerve," *Ann. Otol.* **41**:740–779 (1932).

Wegel, R. L. and C. E. Lane, "The Auditory Masking of One Pure Tone by Another and Its Probable Relation to the Dynamics of the Inner Ear," *Physiol. Rev.* **23**:266–285 (1924).

Wier, C. C., W. Jesteadt, and D. M. Green, "Frequency Discrimination as a Function of Frequency and Sensation Level," *J. Acoust. Soc. Am.* **61**:178–184 (1977).

Yeowart, N. S., M. Bryan, and W. Tempest, "The Monaural MAP Threshold of Hearing at Frequencies from 1.5 to 100 c/s," *J. Sound. Vibr.* **6**:335–342 (1967).

Zwislocki, J., "Theory of Auditory Summation," *J. Acoust. Soc. Am.* **32**:1046–1060 (1960).

CHAPTER 45
AUDIO RECORDING AND PLAYBACK

KATSUAKI TSURUSHIMA

Sony Corporation of America, Park Ridge, New Jersey

ELECTRONIC DEFENSE LABORATORIES

Sylvania Electric Products, Inc., Mountain View, California

ALBERT B. GRUNDY

Audio Research Institute, New York, New York

45.1 SONY COMPACT DISC SYSTEM

Katsuaki Tsurushima

45.1-1 Introduction

CD is the abbreviation for the Compact Disc Digital Audio System, announced in June 1980, as the result of a cooperative venture between Sony and Philips. In this joint research program, Philips investigated the basic operating principles and designed the hardware. Sony's contribution centered mainly on the development of software, including the signal-processing method.

Interchangeability with video disk systems was not included in the design goals of the CD system. Instead the system offers a number of advantages that far outweigh this video interchangeability. Noteworthy among them are the small disk diameter of 12 cm, long playing time of more than 60 minutes on one side, easy handling, lower production costs, and the possibility of making the player system also very compact. These merits have been achieved by adopting Pulse-Code-Modulation (PCM) direct digital recording as well as entirely new modulation and error-correction systems.

45.1-2 Operating Principles

The CD system employs a noncontact signal-readout system using a semiconductor laser. The noncontact design offers a definite advantage over contact systems, such as mechanical and variable

Reprinted by permission of Sony Corporation of America.

capacitance methods. That is, neither the stylus nor the disk ever wear out, no matter how many times the disk is played, assuring long-term high quality and reliability.

It was not at all easy, however, to incorporate the laser pickup in a compact unit. Various technical difficulties were also encountered in designing rotational CLV (constant linear velocity), focus, and tracking servo systems, since these systems involve complex optical devices within their servo loops. To overcome these problems and make the CD system a true practical reality, innovative technology and creative engineering were required.

Disk Structure and Specifications

The main specifications and structure of the disk are shown in Table 45.1 and Fig. 45.1, respectively. As seen in Fig. 45.1*b*, the laser beam is applied from below, passes through a 1.2-mm-thick transparent layer, and focuses on the signal surface. When viewed from the incident laser beam, the signal surface is close to the upper side of the disk, which is coated with an extremely thin

TABLE 45.1. SPECIFICATIONS OF THE SONY COMPACT DISC PLAYER

Variable	Specification
Disk	
Playing time	Approx. 60 min on one side
Rotation	Counterclockwise when viewed from readout surface
Rotational speed	1.2–1.4 m/s
Track pitch	1.6 μm
Diameter	120 mm
Thickness	1.2 mm
Center hole diameter	15 mm
Recording area	46–117 mm
Signal area	50–116 mm
Material	Any transparent material with 1.5 refractive index, such as acryl-plastic
Minimum pit length	0.833–0.972 μm (1.2–1.4 m/s)
Maximum pit length	3.05–3.56 μm (1.2–1.4 m/s)
Pit depth	Approx. 0.11 μm
Pit width	Approx. 0.5 μm
Optical System	
Standard wavelength	$\lambda = 780$ nm (7800 Å)
Focal depth (λ/NA $\leqslant$ 1.75 μm; NA, numerical aperture)	$\pm$ 2 μm
Signal Format	
Number of channels	Two (four-channel recording will also be possible at twice the present rotational speed)
Quantization	16-bit linear quantization
Quantizing timing	Concurrent for all channels
Sampling frequency	44.1 kHz
Channel bit rate	4.3218 Mb/s
Data bit rate	2.0338 Mb/s
Data-to-channel bit ratio	8:17
Error correction code	CIRC (with 25% redundancy)
Modulation system	EFM

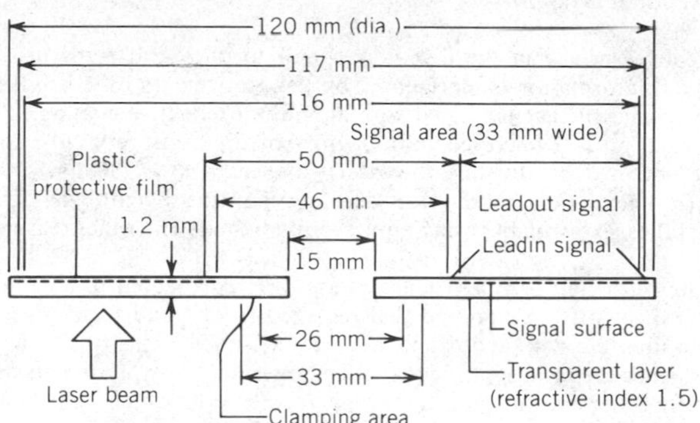

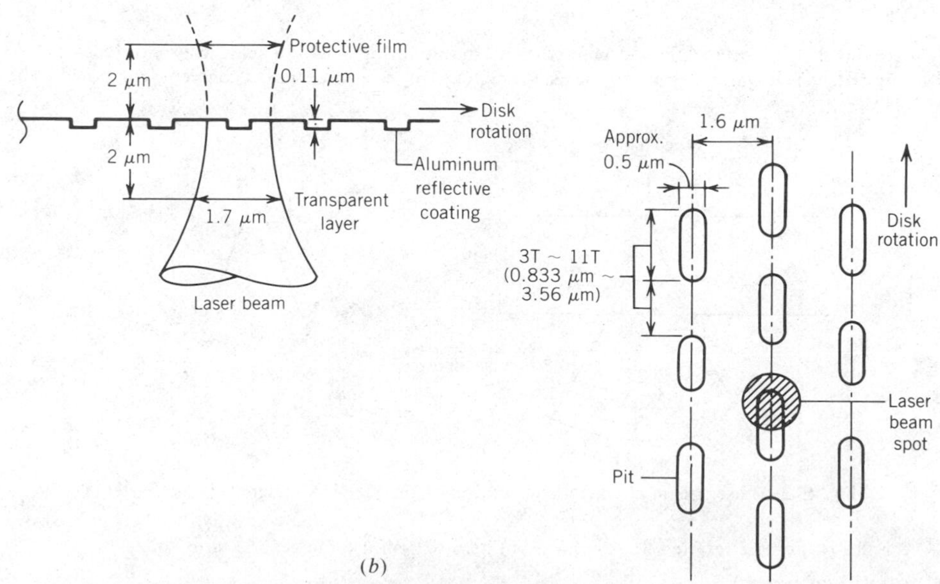

Fig. 45.1 Sony compact disk player: (*a*) unit, (*b*) diagram of disk structure.

(10–30-μm) plastic protection film. Consequently the possibility of the signal surface being damaged physically or chemically is very slight.

The 1.2-mm transparent layer has a very important function, as shown in Fig. 45.2. Its refractive index (n) is 1.5. The spot size of the laser beam on the disk surface is approximately 0.8 mm in diameter, but refracted by the transparent layer becomes as small as 1.7 μm at the signal surface. This means that a bit of dust or a scratch on the disk surface is actually only a millionth in size on the signal surface. In fact, any dust or scratch less than 0.5 mm becomes insignificant and causes no error in signal readout. Accordingly the disk does not need as much delicate handling as the conventional analog disk.

On the signal surface beneath the transparent layer, the disk contains a series of tiny pits† impressed outward from the inner circumference at a pitch of 1.6 μm. The 1.6-μm pitch means that there are over 20,000 tracks composed of such pits in a signal area 33-mm wide. To be more concrete, it is possible to pack 60 tracks in a conventional analog record's track pitch of 100 μm, and approximately 30 tracks in a piece of thin hair. The precision of the disk's track pitch actually equals that of large-scale-integration pattern drawing.

The unusually narrow track pitch of the compact disk naturally requires extremely critical center-hole precision and track eccentricity. For this reason, highly accurate cutting and stamping techniques had to be developed. In addition, an advanced tracking servo system had to be devised to assure precise tracking by compensating for the slight deviation (within 0.1 mm) in the center hole size.

Optical Signal Readout System

The pits and reflective surface in the disk correspond to ones and zeros in digital data. In other words, various digital information is represented by the presence and the length of pits on the signal surface. The whole signal surface is treated with aluminum reflective coating.

The height of each pit is a microscopic 0.11 μm. This figure is very close to but slightly smaller than the quotient obtained by dividing the laser wavelength ($\lambda = 780$ nm) by 4 and then by the refractive index ($n = 1.5$). One reason for deciding on this height was to make the tracking error-detection signal easily available using a push-pull method. The other reason can be explained as follows.

Assume that the pit height is $\lambda/4$. Then a phase difference equal to $\lambda/2$ or 180° is engendered between the beam reflected from a pit and that reflected from an adjacent reflective surface, causing these two beams to interfere with and cancel each other.

In other words, since the laser spot is equal to or larger than any pit dimension, the reflected beam is modulated in proportion to the ratio of the pit in the spot, which can then be detected by the photodiode and converted into an electrical signal (Fig. 45.3).

Optical Pickup

One example of optical pickup is structured as illustrated in Fig. 45.4. To achieve a sharp focus of 1.7 μm in diameter, it was necessary to use as a luminous source a laser that is coherent, composed of a

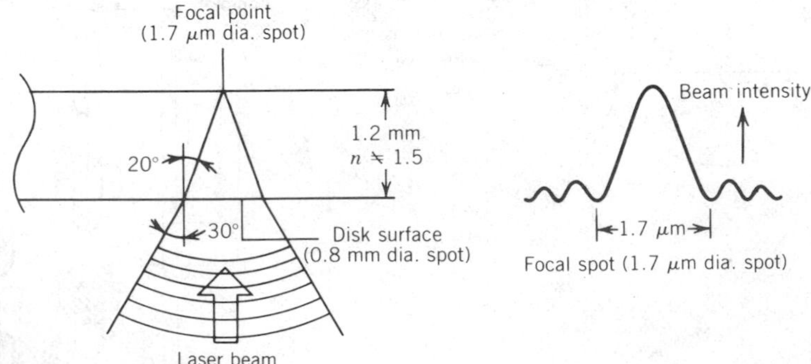

Fig. 45.2 Laser beam characteristics of Sony Compact Disc Digital Audio System.

†Technically they are referred to as bumps, but here we will use the general term "pits."

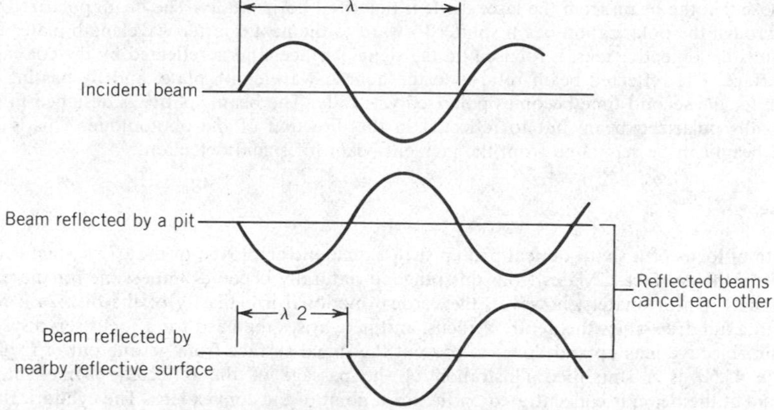

Fig. 45.3 Beam cancellation.

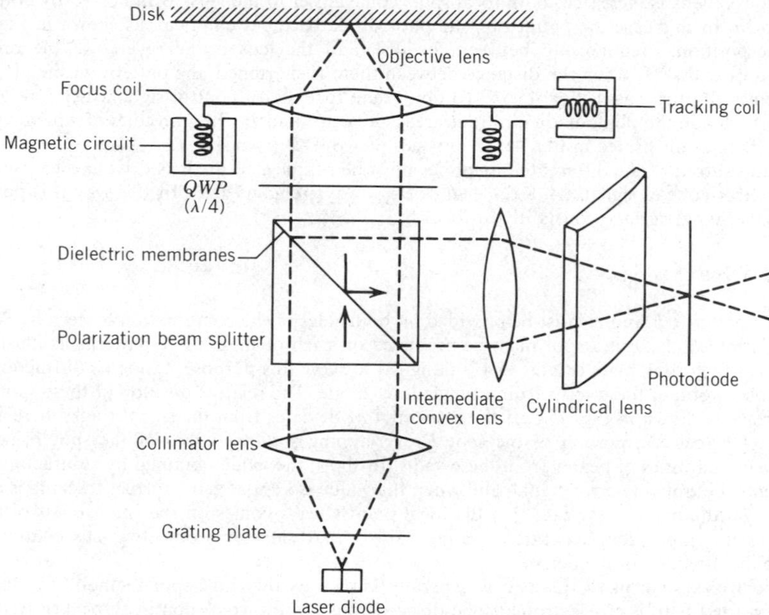

Fig. 45.4 Optical pickup of Sony Compact Disc Digital Audio System.

single frequency, and of high intensity. The laser diode that emits the laser beam is placed at the focal point of a collimator lens with a relatively long focal distance.

The laser beam coming through the collimator lens becomes a parallel beam, which is then brought into focus on the signal surface by passing through the subsequent objective lens. The objective lens used here is a convex type and is characterized by approximately the same diameter and focal distance of 4 mm. Despite its small size, however, the lens is as precise as those used in the most advanced microscopes. Extra advantages include outstanding brightness. Aberration is also nonexistent.

Between the collimator lens and the objective lens is the polarization beam splitter. This is a kind of prism but incorporates dielectric membranes. The polarization beam splitter serves to direct the laser beam emitted from the laser diode to the signal surface and the reflected beam to the photodiode.

Suppose that the beam from the laser diode is polarized horizontally. The beam polarized this way can go through the polarization beam splitter forward to the next quarter-wavelength plate. Then the beam enters the objective lens, is focused on the signal surface, and is reflected by the coating on the signal surface. The reflected beam returns to the quarter-wavelength plate, and by passing through this plate for the second time becomes polarized vertically. The beam splitter is designed to not pass the vertically polarized beam but to reflect it in the direction of the photodiode. This is how the reflected beam can be separated from the incident beam for signal detection.

Focus Servo

The depth of focus of a sharp optical pickup such as the one employed in the CD system is only ± 2 μm. If the signal surface deviates from this range, it naturally becomes impossible for the pickup to detect signals. Unfortunately, however, the vertical motional irregularity of the disk as it rotates is more than a hundred times the depth of focus, and here arises the need for a focus servo system that moves the objective lens up and down to prevent the signal surface from getting out of focus.

Figure 45.5*a* is a simplified illustration of the passage of the reflected beam. The vertical component of the beam is concentrated on the focal point of the convex lens. The cylindrical lens has no effect on this beam. The horizontal component of the beam, on the other hand, is refracted as it passes through the cylindrical lens and makes a focal spot at a shorter distance. If the photodiode is placed at the point where the vertical component of the beam intersects with the diffused horizontal component of the beam, a circular pattern can be obtained on it. When the distance between the disk and the convex lens is increased, both focal spots come closer to the lens, as indicated by dotted lines in Fig. 45.5*a*. In this case the pattern on the photodiode becomes elliptical, as shown in Fig. 45.5*b*. When the positional relationship between the disk and the lens is the reverse of the previously mentioned case, that is, when the distance between them is shortened, the pattern on the photodiode also becomes elliptical, as in Fig. 45.5*b*. To obtain the focus error-control signal, therefore, it is first necessary to divide the photodiode into four areas (A, B, C, and D, the sum total of which represents the RF signal), as illustrated in Fig. 45.5*c*, and subtract $B + D$ from $A + C$.

Next the output of the differential amplifier must be supplied to the lens drive mechanism so that the focus error-control signal, $(A + C) - (B + D)$, always remains zero. In this way it is possible to maintain the signal surface constantly in focus.

Tracking Servo

First, extra patterns E and F must be provided on both sides of the combination pattern A, B, C, and D on the photodiode, and a spot must be produced on each of these three patterns as illustrated in Fig. 45.5*b*. The grating plate in Fig. 45.4 is designed to serve this purpose. Using its diffraction effect, it is possible to obtain three spots from a single laser diode. The relative position of these spots on the signal surface is shown in Fig. 45.6. If the main spot M deviates from the signal track, there arises an imbalance between the portion of the spot E overlapping a pit and that of the spot F, causing a difference in quantities of beams from these spots. In short, the value obtained by subtracting F from E represents the tracking-error signal, and when this value is kept at zero, correct tracking is ensured. Therefore, on the basis of the fact that the focal point always comes on the central axis of the lens, the CD system employs an accurate drive mechanism that moves the objective lens bilaterally with respect to the direction of the beam.

The control system just explained is generally known as the three-spot method. The lens drive mechanism used here is an electromagnetic device similar to the voice coil in a speaker system. The drive mechanism as a whole is also called a two-axis device, since it is capable of adjusting the lens position in two directions, up-down and right-left.

CLV Servo

CLV stands for constant linear velocity and signifies a state in which a uniform relative velocity is maintained between the disk and the optical pickup.

Conventional LP records rotate at a constant angular velocity of 33 1/3 rpm. Compared with that of the LP, the rotation of the compact disk varies depending on the position of the pickup for the purpose of assuring a constant linear velocity of 1.25 m/s. The turntable platter of the CD system rotates at a speed of 500 rpm when the pickup is tracing the track close to the inner circumference. As the pickup moves outward, the rotational speed gradually decreases to a minimum of 200 rpm. The CLV servo is designed to control the speed by synchronizing the frame sync impressed in the disk and the frequency of the quartz crystal oscillator built in the player.

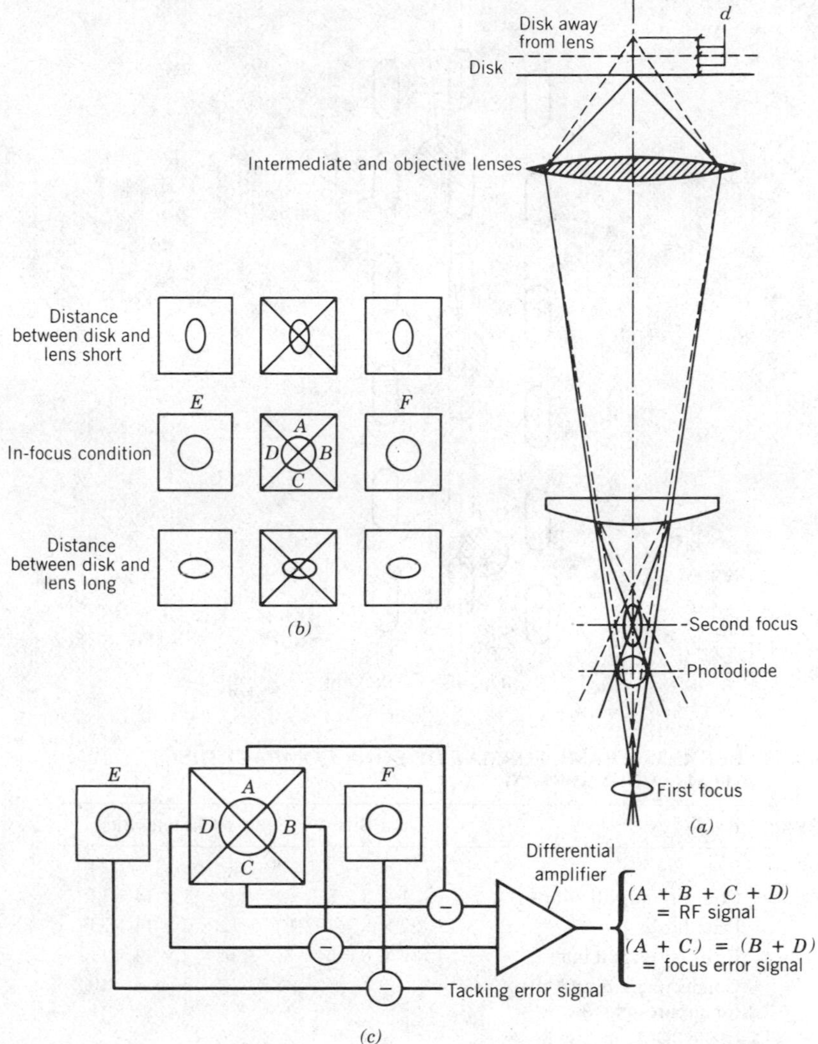

Fig. 45.5 Focusing action of Sony Compact Disc Digital Audio System: (*a*) passage of reflected beam, (*b*) pattern on photodiode, (*c*) computation of signals.

The three servo systems so far explained are very important for the CD system. Only when they operate properly does it become possible to detect digital signals correctly.

EFM (Eight to Fourteen Modulation) System

Listed in Table 45.2 is the frame format. The L and R channel audio signals are first sampled at a frequency of 44.1 kHz and then converted into binary representations by undergoing a 16-bit linear quantization process. A frame is composed of six such sampled signals. Accordingly, each frame cycle is 7.35 kHz, or 136 μs. With the EFM system, the sampled signal is first divided into two symbols, one consisting of the lower 8 bits and the other the higher 8 bits, and modulation thereafter is carried out on the basis of each symbol. EFM literally implies that each symbol of 8 bits is converted into 14 channel bits.

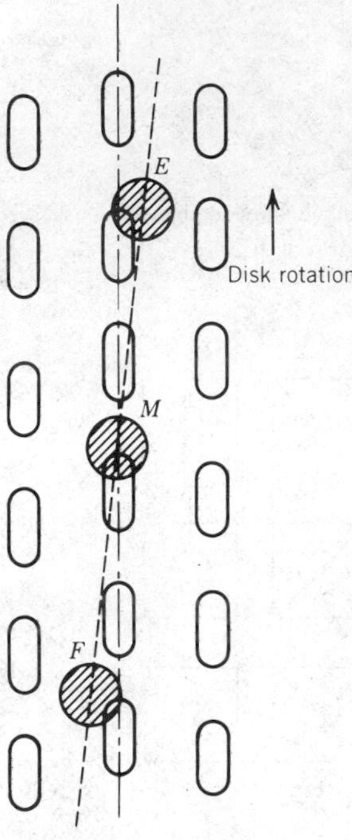

Fig. 45.6 Tracking.

TABLE 45.2. FRAME FORMAT OF SONY COMPACT DISC DIGITAL AUDIO SYSTEM

Bits	Data Bits	Channel Bits
Sync bits		24
Control/indication bits	$1 \times 8 = 8$	$1 \times 14 = 14$
Data bits	$12 \times 2 \times 8 = 192$	$12 \times 2 \times 14 = 336$
Error correction bits	$4 \times 2 \times 8 = 64$	$4 \times 2 \times 14 = 112$
Connecting bits and bits for suppressing low frequencies		$34 \times 3 = 102$
Total	261	588

As a result, the frame configuration becomes as illustrated in Fig. 45.7. This is the best compromise between the highest possible recording density and the easiest possible clock bit extraction.

In binary notation, 8 bits can offer 256 possible code combinations. Similarly, 14 bits are capable of making 2^{14}, that is, 16,384, different code patterns. Assume that we select from these patterns only those in which more than 2 but less than 10 zeros appear continuously. Then only 267 patterns satisfy this condition. With these 267 patterns, in other words, the minimum inversion width of a NRZ signal that inverts at 1 is limited by the recording density, and the maximum inversion width is limited in terms of clock bit extraction. Figure 45.8 indicates an example of an 8-bit code converted into a code of 14 channel bits fulfilling this requirement. Notice that the original time interval T is extended to 1.5 T in the converted code. The two channel bits marked X in the diagram are connecting bits inserted between code patterns. These bits in this case correspond to zeros.

In actual application, one more bit is added to each code pattern. The binary number, either 1 or 0, to be assigned to this extra bit depends on the code contents of the preceding and succeeding

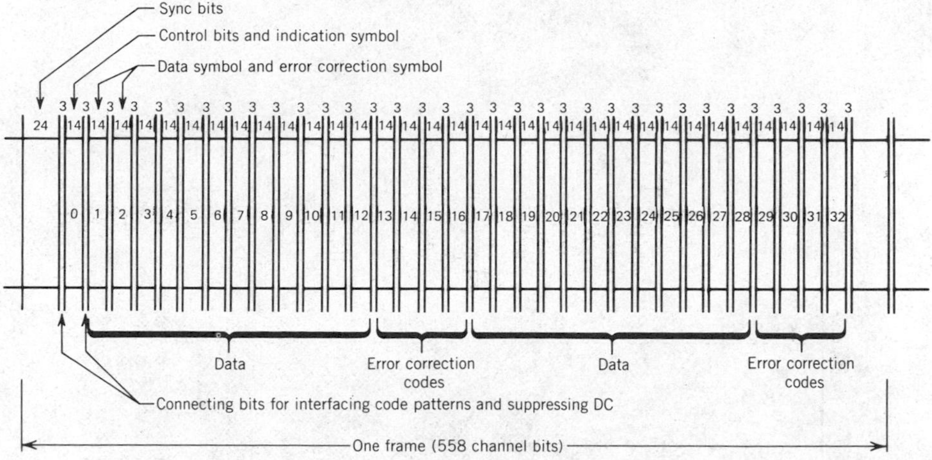

Fig. 45.7 Frame configuration of Sony Compact Disc Digital Audio System.

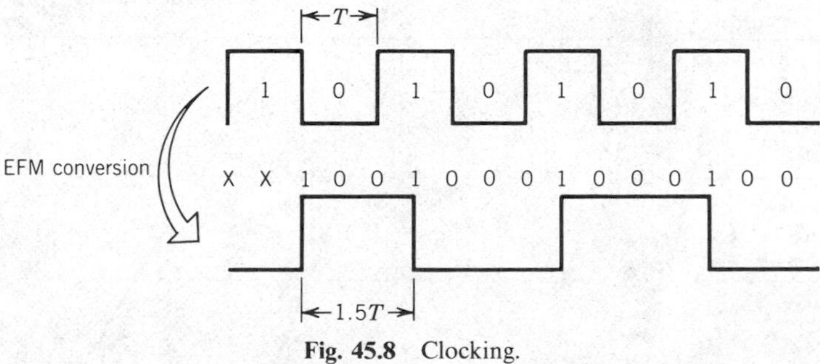

Fig. 45.8 Clocking.

patterns. That is to say, it is determined so that it can suppress the DC component that affects the signal-to-noise ratio of the servo error signal.

As you have seen, a symbol of 8 bits is transformed into 17 bits in the end. In demodulation, however, the additional 3 bits are ignored and the remaining 14 data bits are subjected to demodulation.

These are the basic operating principles of EFM. In practice, 11 code patterns of 267 are further eliminated, and thus the code patterns memorized in the ROM (read-only memory) for use in conversion and reconversion total 256.

CIRC Error Correction

An effective error correction scheme is an essential part of the design of a digital disk system. Roughly categorized, there are two types of code errors. The first type is random errors, caused in disk production by inaccurate photoresist coating, cutting, and so on. The second type is burst errors, resulting from scratches and fingerprints on the disk surface.

The CIRC (cross interleave Reed Solomon code), originally developed by Sony, can cope perfectly with both these errors. It can even correct an extensive error of over 4000 bits. Moreover, the CD system employs a strong code restoration method called super strategy. Its correction capability is graphed in Fig. 45.9.

In the CD system, three LSIs comprise the EFM demodulator and CIRC circuits, and these, together with a 16 K random access memory (RAM), comprise a black box. Figure 45.10 is the block diagram of the CD system.

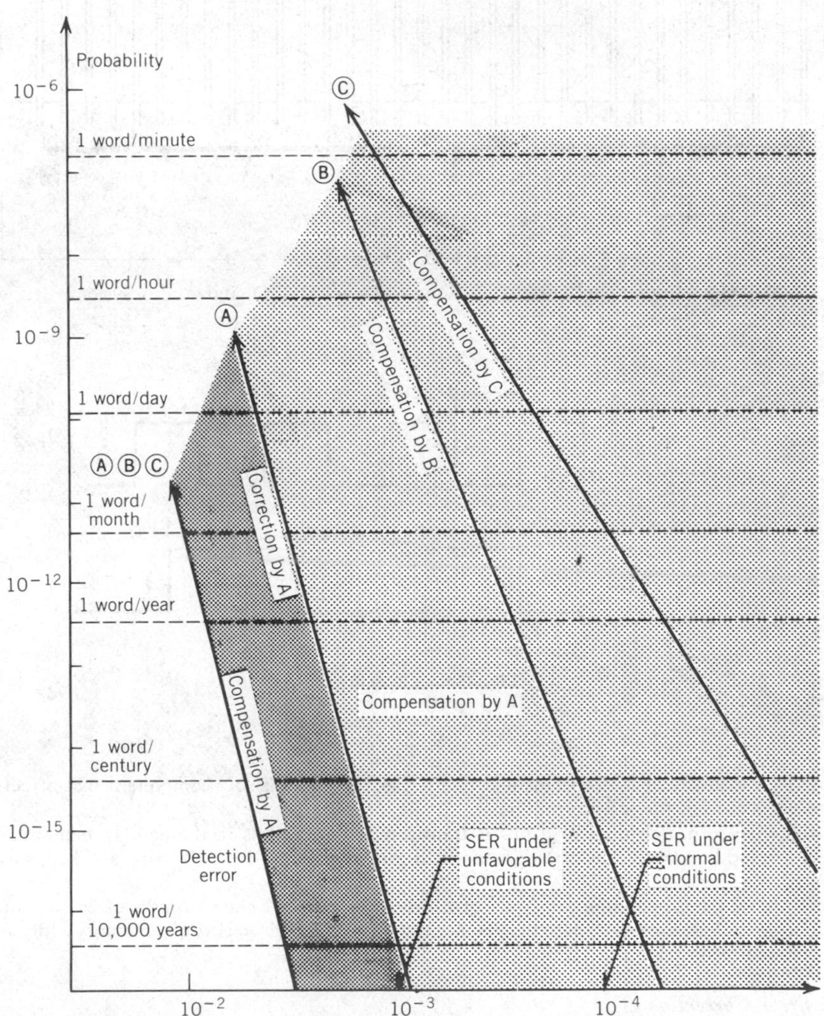

Fig. 45.9 Correction capability of "super strategy," code restoration method developed by Sony and used in the Sony Compact Disc Digital Audio System. A, superstrategy; B, standard error-correction system; C, simplified error-correction system. SER, error probability per symbol (8 bits) before being corrected.

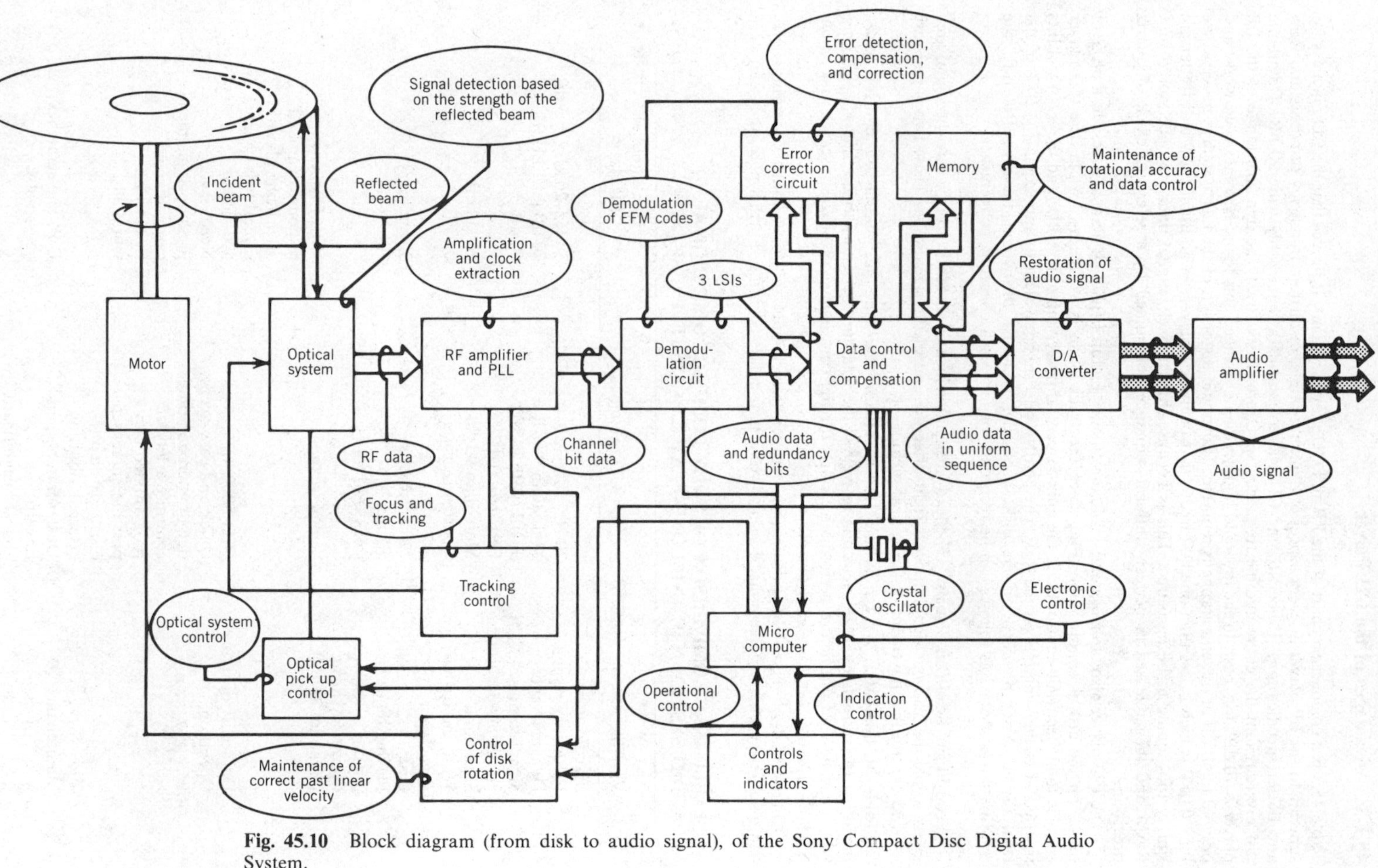

Fig. 45.10 Block diagram (from disk to audio signal), of the Sony Compact Disc Digital Audio System.

45.1-3 Advantages of the CD Player

Table 45.3 is a comparison chart of the LP and CD player systems. It proves that the CD player is far superior to the LP player in every aspect of dynamic range, distortion, frequency response, and wow and flutter specifications. Especially, the CD player exhibits a remarkably wide 90 dB of dynamic range throughout the entire audible frequency spectrum. In contrast, the dynamic range of the LP player is 70 dB at its best. In the low- and high-frequency ranges, this figure becomes even worse and only 40 to 50 dB of dynamic range is reported. Harmonic distortion of the CD system is never worse than 0.01%, which is less than one hundredth that of the LP player. In addition, wow and flutter are simply too minute to be measured. This is because, in playback, digital data are once stored in a RAM and then released in perfect, uniform sequence determined by a reference clock of quartz precision.

The CD player also offers a number of convenience features. Though manual access to a piece of music on the disk is simply impossible, since the track pitch is as fine as 1.6 μm, the CD player enables total electronic control using the built-in central processing unit (CPU) and the 8 control bits inserted after the sync bits on the disk. The control bits are equivalent to approximately 2.7% of the total digital codes recorded on the disk. Presently, with use of two of these control bits, the information described as follows can be recorded for increased operating convenience.

Lead-In and Lead-Out Signals. The lead-in signal is located just in front of where the signal area starts, while the lead-out signal appears right after the end of the signal area. These two signals are used to control the movement of the optical pickup.

Table of Contents. Written in the lead-in area is the time information on control codes, including the start time of each selection as well as the total number and playing time of selections, which can be read out before performance starts for the convenience of program search and so on. Such information can also be displayed.

Control Codes. These codes distinguish between two-channel and four-channel recording. They can also detect whether preemphasis is given to a particular recording. Furthermore, they can even

TABLE 45.3. COMPARISON OF SONY COMPACT DISC DIGITAL AUDIO SYSTEM (CD) AND LONG-PLAY (LP) SYSTEM

Variable	CD System	Conventional LP Player
Specifications		
Frequency response	20 Hz–20 kHz ± 0.5 dB	30 Hz–20 kHz ± 3 dB
Dynamic range	More than 90 dB	70 dB (at 1 kHz)
Signal-to-noise	90 dB (with MSB)	60 dB
Harmonic distortion	Less than 0.01%	1–2%
Separation	More than 90 dB	25–30 dB
Wow and flutter	Quartz precision	0.03%
Dimensions		
Disk	12 cm (dia.)	30 cm (dia.)
Playing time (on one side)	60 min (max. 74 min)	20–25 min
Operation/Reliability		
Durability of disk	Semipermanent	High-frequency response degraded after being played several tens of times
Durability of stylus	Over 5000 hr	500–600 hr
Operation	Quick and easy access due to microcomputer control, variety of programmed play possible, increased resistivity to external vibration	Needs stylus pressure adjustment, easily affected by external vibration
Maintenance	Dust, scratches, and fingerprints almost insignificant	Dust and scratches cause noise

change the circuit connections of the player automatically according to the detected conditions of the disk.

Music Start Flag. The music start flag is provided in the blank space between selections. By counting this flag, therefore, it is possible to locate any desired selection on the disk. Even when the end of a selection fades into the beginning of the next, the flag can be inserted for a minimum of two seconds in front of the second selection so that such selections can be recognized not as a single piece of music but as two separate selections.

Track Number and Index. Selections recorded on the disk can be numbered from 1 through 99. Each bar in a piece of music can also be addressed up to 99.

Time Code. With the time code function, you can tell the time lapse from the beginning of each selection in minutes, seconds, and 1/75 of a second. During the blank space between selections, the time is counted down.

As indicated previously, the control bits can be used in a variety of ways. They not only let you locate any portion of music exactly and quickly but also allow random playback in any order you set. And despite the long playing time of an hour on one side, access to the desired point can be achieved almost instantly. This is one of the greatest advantages of the CD system that cannot be found in any other system.

45.1-4 Future of the CD System

The compact disk and the player were marketed simultaneously at the end of 1982. For the time being, the player will be quite expensive. Before long, however, as the optical parts industry matures, it will surely come down into a price range everyone can afford. As to the disk, it is expected that the cost per playing time will be equal to that of the LP.

You may remember that with the present CD system, only one fourth of the control bits are utilized and 6 bits still remain unused. This suggests that redundancy bits can be used to record, for example, the title and text of the music for the extra enjoyment of singing to the accompaniment of CD music.

Besides use as a home audio component, the CD system is suitable for portable and in-car applications, since it incorporates precise servo systems and thus is highly resistant to vibration. In fact, with just a little more sophistication in its optical design, commercialization of a CD Walkman is possible. Unlike conventional record players, moreover, the CD system never suffers from howling even when it is placed on a speaker box.

45.2 MAGNETIC TAPE RECORDING

Staff of Electronic Defense Laboratories of Sylvania Electric Products, Inc.

45.2-1 Magnetic Recording

Three basic elements are required to make a magnetic recording and later to reproduce it: (1) a device that can respond to an electrical signal and create a magnetic pattern in a magnetizable medium, (2) a magnetizable medium that will conform to and retain the magnetic pattern, and (3) a device that can detect such a magnetic pattern and convert it once again to the original electrical signal. These three elements take the physical form of the record head, the magnetic tape, and the reproduce head. With the addition of some electronic amplification and a mechanical tape handler, a basic magnetic recorder results.

Direct Record / Reproduce

A record head is similar to a transformer with a single winding. Signal current flows in the winding, producing a magnetic flux in the core material. So that it performs as a record head, the core is made in the form of a closed ring, but unlike a transformer core, the ring has a short nonmagnetic gap in it. When the nonmagnetic gap is bridged by magnetic tape, the flux detours around the gap through the tape completing the magnetic path through the core material. Magnetic tape is simply a ribbon of plastic on which tiny particles of magnetic material have been uniformly deposited. When the tape is moved across the record head gap, the magnetic material, or oxide, is subjected to a flux pattern that

---

Reprinted by permission of General Telephone and Electronic Corporation, Mountain View, California.

is proportional to the signal current in the head winding. As it leaves the head gap, each tiny particle retains the state of magnetization that was last imposed on it by the protruding flux. Thus the actual recording takes place at the trailing edge of the record head gap. A simplified diagram of the recording process is shown in Fig. 45.11.

To reproduce the signal, the magnetic pattern on the tape is moved across a reproduce head. Again a small nonmagnetic gap in the head core is bridged by the magnetic oxide of the tape. Magnetic lines of flux are shunted through the core, and are proportional to the magnetic gradient of the pattern on the tape that is spanned by the gap. At this point, analysis of the reproduce function is divided into two possible alternatives, that is, the use of current or voltage amplifiers in the reproduce electronics. Each method is treated separately here, but for both the same example is used to allow for easier comparison.

Suppose the signal to be recorded on the tape is a sine wave voltage described by $A \sin(\omega t)$. Both the current in the record head winding and the flux ϕ through the record head core will be proportional to this voltage.

Voltage Amplifier. The induced voltage in the head winding follows the law of electromagnetic induction: $e_s = Nd\phi/dt$. It is important to note that the reproduced voltage is not proportional to the magnitude of the flux, but to its rate of change (see Fig. 45.12). If the tape retains this flux pattern and regenerates it in the reproduce head core, the voltage in the reproduce head winding will be

$$e_{\text{repro}} \propto \frac{d\phi}{dt}$$

where

$$\begin{aligned} \frac{d\phi}{dt} &= \frac{d}{dt} A \sin(\omega t) \\ &= \omega A \cos(\omega t) \end{aligned}$$

Thus the reproduce head acts as a differentiator and the reproduced signal is actually the derivative of the recorded signal and not the signal itself. This fact imposes two well-known limitations on the direct-reproduce process. The output of the reproduce head is proportional to the signal frequency, and for maintenance of amplitude fidelity, a 6-dB-per-octave rise in the head output must be compensated for in the reproduce amplifier by a process known as equalization.

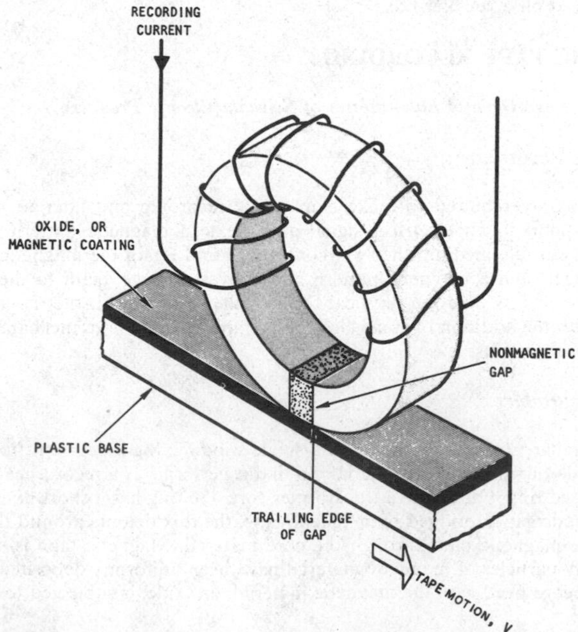

Fig. 45.11 Simplified diagram of the magnetic recording process.

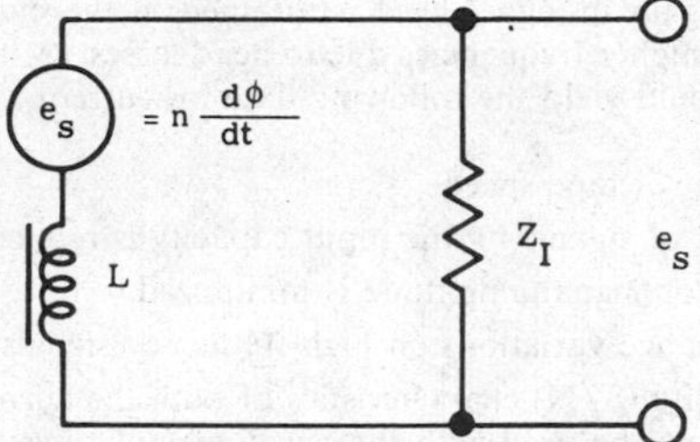

Fig. 45.12 Voltage equivalent head model.

Current Amplifier. The induced current in the head winding follows the equation $i_s = n\phi/L$ (see Fig. 45.13). The corresponding current in the reproduce head winding is

$$i_{\text{repro}} \propto \frac{\phi}{L}$$

where

$$\phi = A \sin(\omega t)$$

Since this eliminates the 6-dB-per-octave rise in the head output (see Fig. 45.14), the required equalization circuitry need only compensate for the normal high frequency losses encountered in tape recording. The resultant signal attenuation is much greater for the short wavelengths (high frequencies) but is flat for the lower frequencies. An L-C boost circuit after the preamplifier can be employed in the higher frequency regions to, in effect, duplicate the voltage head circuit and thus make the two

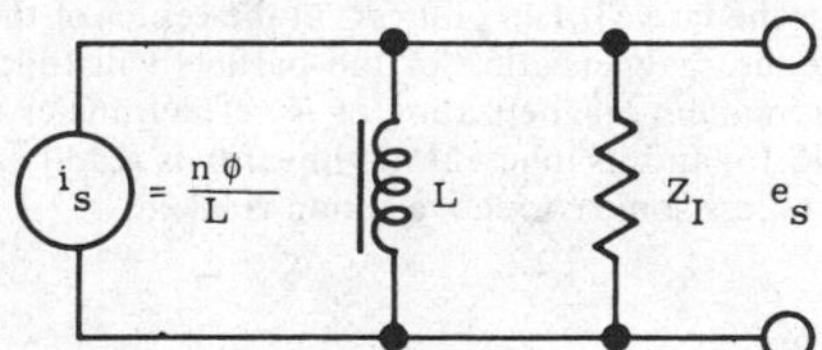

Fig. 45.13 Current equivalent head model.

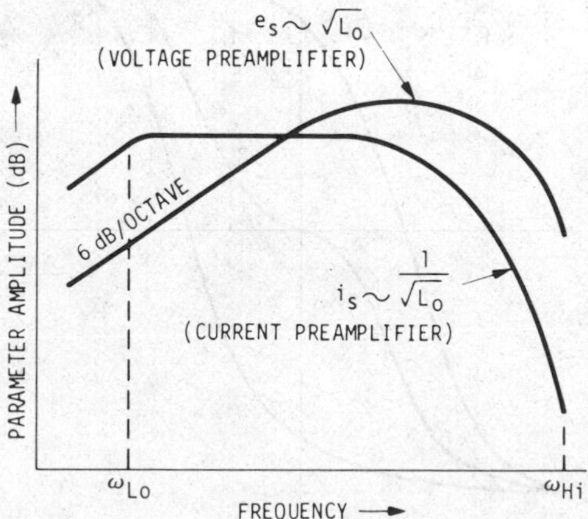

Fig. 45.14 Intensity vs. voltage curves.

systems equivalent. A slight gain in signal level is attained in the current model because we have eliminated attenuation of the higher frequencies due to head losses.

An overall comparison would yield the following data on current amplifiers:

1. The output is independent of tape speed.
2. The bandwidth limitation of signals by the input capacity is reduced.
3. The effect of head losses on signal amplitude is minimized.
4. The effect of head inductance variations on high-frequency signals is removed.
5. Although the signal-to-noise (S/N) characteristics of both the current and voltage amplifiers are the same, high Q leads may be used with the current amplifiers to achieve higher S/N ratios.

For both types of amplification, another limitation occurs as the recorded frequency approaches zero. At some point the output voltage from the reproduce head falls below the inherent noise level of the overall recording system. So there is a low-frequency limit in the direct record process, below which reproduction cannot be made.

Bias

Up to now our discussion has assumed that the magnetizable medium responds linearly to the magnetizing force of the record head. As might be expected, the perversity of nature asserts itself and the assumption is in error. Like other magnetic materials, the particles on the tape exhibit a very nonlinear characteristic when exposed to a magnetizing force. A typical magnetization curve, or hysteresis loop, is shown in Fig. 45.15.

H is the magnetizing force and is determined by the number of turns and the current in the record head winding. B is the resultant induced magnetization on the tape.

As a demagnetized particle on the tape approaches the record head gap, it carries no residual magnetism. (Point 0, at the origin, Fig. 45.15). Assuming that a cycle of the recorded signal along the tape is very long compared with the gap length, the particle will pass through an essentially constant magnetizing force created by the recording current. Referring to the curve of Fig. 45.15, such a force (H_R) will carry the particle up the curve $0A$ to point R, at the center of the gap. As the particle leaves the gap, H falls to zero, but the magnetization of the particle will follow a minor hysteresis loop, RB_R, retaining a residual or remanent magnetization of B_R. The transfer characteristic of this process is shown graphically in Fig. 45.16, and its inherent nonlinearity is readily apparent. High distortion in the reproduced signal results unless some corrective action is taken.

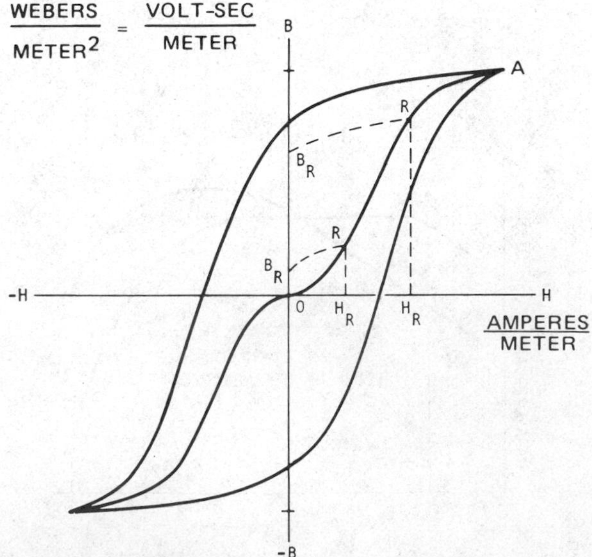

Fig. 45.15 Typical magnetization curve, or hysteresis loop.

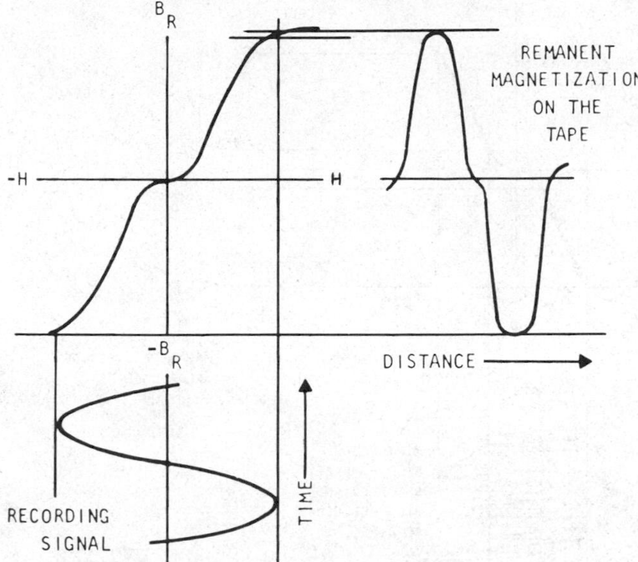

Fig. 45.16 Head-to-tape transfer characteristic with no bias.

Fortunately there are two fairly linear segments in the transfer characteristic curve, one on each side of the origin with its center about half way to saturation. (See Fig. 45.16.) If the recording can be confined to one (or both) of these straight sections, low distortion can be realized. Some method of "biasing" the recording function into the linear transfer region must be used. Early recorder designers went naturally to a DC bias produced simply by adding a constant DC current to the signal, and obtained operation in one or the other of the two linear ranges. With the relatively limited range thus available, DC-biased recorders give a very restricted dynamic range, but they are quite improved over zero-bias recording. If both linear sections of the curve are to be used, some means of rapidly switching from one to the other must be devised. This is exactly what a high-frequency AC bias does. There are several theories about how an AC bias performs this function, and no one of them really accounts for all aspects. One of the older and still widely accepted theories, however, is that since the bias itself is not seen at the output of the reproduce head, because of its high frequency, its switching function is not detectable and the gap between the two linear sections disappears.

Figure 45.17 shows graphically how a low-distortion magnetic signal is thus recorded. Some further reduction in nonlinearity is also obtained in the same way as in push-pull amplifier operation, since nonlinearities are symmetrically disposed around the origin.

Several features of AC bias operation are worth noting:

1. The bias and the signal are linearly mixed (or added) together. It is not a modulation process.

2. The proper amplitude for the bias is dependent on the exact transfer characteristic of the tape and should be adjusted to reach from center to center of the linear regions. Too much bias will greatly reduce the high-frequency response, while inadequate bias will cause increased distortion of the lower frequencies.

3. Bias frequencies are usually not critical, but should be at least 3.5 times the highest frequency to be recorded to minimize interaction with harmonics of the signal.

4. Care must also be taken to provide a harmonically pure sine wave bias current to minimize distortion products.

In practice, bias currents from 1.0 to 20.0 mA are common, and they may be from 5 to 30 times the signal current, depending on the tape and head characteristics.

High-Frequency Response

Several factors combine to limit the high-frequency response of tape recorders, but before these are discussed we should understand what is meant by recorded wavelength, resolution, and packing density.

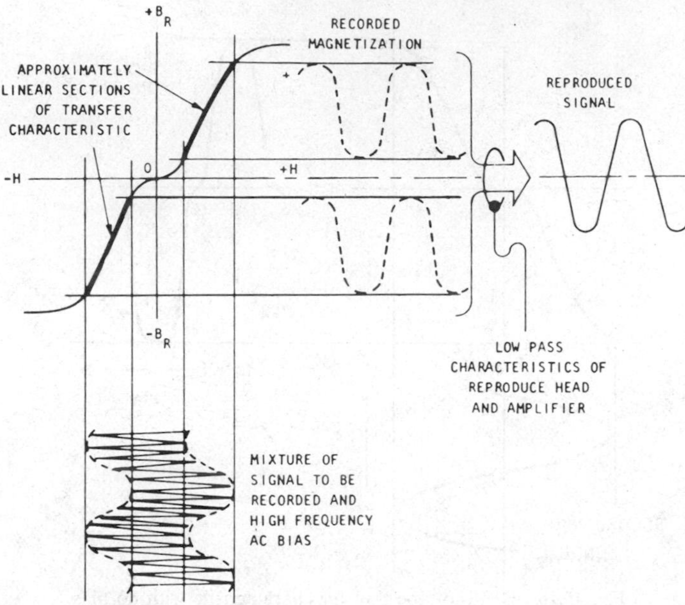

Fig. 45.17 Graphic representation showing how AC bias alternately transfers the signal from one linear section of the curve to the other.

If a sine wave signal is recorded, the magnetic intensity of the recorded track will vary sinusoidally. The distance along the tape required to record a complete cycle is called the recorded wavelength, or λ, and is directly proportional to tape speed and inversely proportional to signal frequency. For example, a particular recorder quotes 60 kHz response at 15 ips. There are several other ways to describe this response. Dividing 60 kHz by 15 in. shows the machine is capable of a *packing* density of 4000 cycle/in. Such a signal has a wavelength of 0.00025 in., which is the limit of the machine's *resolution*. Both packing density and resolution can be used to describe a recorder's response independent of tape speed, and thus are more definitive of a recorder's capability than is just a frequency specification at a given speed.

Seven factors contribute to the high frequency limitation of tape recorders: (1) gap effect, (2) recording demagnetization, (3) self-demagnetization, (4) penetration losses, (5) head losses, (6) separation loss, and (7) azimuth misalignment.

1. **Gap Effect.** As shown in Fig. 45.18, the reproduce head output increases with frequency up to a point and then decreases rapidly to zero. The decrease is primarily the result of gap effect and occurs as the recorded wavelength (λ) becomes shorter and shorter until it eventually equals the reproduce gap dimension itself. At this point there is no magnetic gradient spanned by the gap and thus no output voltage. This is the most serious, single restriction on a tape recorder's high-frequency response. Figure 45.18 shows a graphical representation of the gap effect.
2. **Recording Demagnetization.** This causes a decrease in the reproduced signal level at the shorter wavelengths and, as the name implies, it occurs in the recording process. Normally, with the longer wavelengths the particles on the tape are driven through large symmetrical hysteresis loops by the AC bias while in the influence of the recording field. These loops are offset by the much smaller recording signal; as the particle leaves the field the loops collapse, leaving the particle magnetized in proportion to the signal. For shorter wavelengths the recording signal may vary considerably as the particle is leaving the field, and a corresponding reduction in the remanent magnetization will result.
3. **Self-Demagnetization.** This occurs in the magnetic medium itself when the external magnetizing force is removed; it is most pronounced when magnetic poles are crowded closer and closer (shorter wavelengths). Actually, self-demagnetization is probably the least important of the high-frequency limitations.
4. **Penetration Losses.** These losses are wavelength dependent and cause another reduction in the reproduced signal level. The full depth (or thickness) of the magnetic coating on the tape

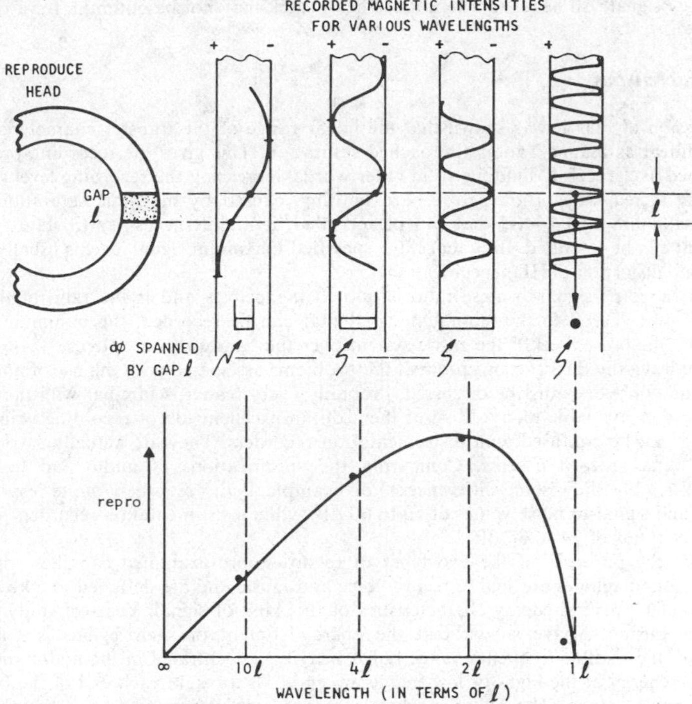

Fig. 45.18 Graphic representation of gap effect.

becomes magnetized at long wavelengths, but as the wavelength decreases the depth of magnetization is reduced and, at very short wavelengths, only the surface particles are effectively magnetized. Thus the shorter wavelengths influence fewer particles, there is less intensity in the recorded magnetic pattern, and the reproduced output falls off.

5. **Head Losses.** Unlike the limitations already discussed, head losses are not wavelength dependent, but like any AC-driven ferromagnetic material are related strictly to frequency. Both core and winding losses act to reduce the effective recording current at high signal frequencies. Hysteresis and eddy-current losses in the core material and the distributed capacity of the windings are the major contributors and, of course, they increase with frequency.
6. **Separation Loss.** This effect is seen in the reproduced signal as very short and random reductions (or dropouts) in signal level. Actually, dropouts are not caused by the recorder but result from imperfect tape-to-head contact. Modern tape technology has reduced dropout problems by several orders of magnitude but has not eliminated them. In audio recording, dropouts cause very little problem because the human ear tends to integrate short amplitude variations, and with this smoothing process they are not discernible.
7. **Azimuth Misalignment.** Short wavelength losses occur when the reproduce head gap is not precisely parallel to the record head gap. These losses are defined by the equation

$$\text{Loss (dB)} = 20\log\frac{\sin[\pi W(\tan A)/\lambda]}{\pi W \tan A/\lambda}$$

where W = track width
A = angle of misalignment, rad
λ = wavelength

For example, with a track width of 37 mils and a wavelength of 0.25 mil, a 3-dB loss occurs for a 1/6 degree (0.003 rad) misalignment. Obviously proper head alignment is important in reducing azimuth losses; however, there can be other contributors such as oversized or worn tape guides, subnormal tape widths, or improper capstan pressure roller alignment. On extremely wide band recorders, azimuth adjustment screws are built into the reproduce head stack mounting base. Optimum reproduce head azimuth can be obtained by "tweeking" this adjustment to maximize the reproduced high-frequency signal. All heads in a reproduce head stack may not be optimum for a given azimuth setting.

Other Characteristics

In the discussion of bias it was shown that the linear range of the transfer characteristic gradually became nonlinear as magnetization approached saturation. This gives the recording process what is aptly described as "graceful" limiting, or in other words, increasing the recording level above normal will gradually increase distortion before hard limiting, caused by magnetic saturation, occurs. To define the maximum signal level that can be recorded, it is thus necessary to state the maximum distortion that can be tolerated. In practice, the specified maximum signal level is usually tied to a 1% total harmonic distortion (THD) specification.

Dynamic range or signal-to-noise ratio is quoted in decibels and is the ratio of the maximum signal (for a given THD) to the minimum signal that can be recorded, the minimum signal being determined by the noise level of the entire system over the bandwidth of interest.

Before we leave the direct record process, the problems associated with the use of audio recorders for instrumentation work must be discussed. Probably every reader is familiar with the large variety of audio recorders available today. Most of them do an excellent job of recording voice and music, but they must not be confused with instrumentation recorders. They are actually a very specialized case of the direct record process. Comparing the specifications of audio and instrumentation recorders will readily show some differences. For example, audio recorders quote response down to 40 or 50 Hz and signal-to-noise values of 50 to 60 dB, while instrumentation recorders stop at 50 Hz and have S/N ratios of 24 to 38 dB.

Only when the purposes of the two types of recorders are examined can these differences be reconciled. Audio machines are sold to record voice and music and are designed to take advantage of the rather peculiar spectral-energy characteristics of this kind of signal. Years of study, primarily in the telephone business, have shown that the energy content of such signals is not distributed uniformly over the audible frequency range but is heavily concentrated in the mid-audio band, with relatively little energy at the high- or low-frequency ends. As a result, each end of the band requires very little dynamic range. The audio machine's record amplifier boosts, or preemphasizes, these band-edge signals as they are put on the tape. When the tape is played back, the audio reproduce amplifier operates with reduced gain at the band edges to equalize the signal and restore fidelity. This factor also improves the overall noise level of the audio recorder because it is at the band edges that the system noise level is highest. With reduced gain in these regions, the cumulative noise within the passband is greatly reduced. This fancy footwork has gained the audio recorder a lower frequency response and a lower noise level in the reproduced signal, at the cost of less dynamic range (or earlier saturation) at each end of the audible spectrum, that is, at 50 Hz or 14,000 Hz an audio recorder may saturate with a 12- to 15-dB signal, yet at 400 Hz where S/N is measured, the machine may handle a 55-dB signal. This type of preemphasis and postequalization has been widely accepted in audio recorders, both for home and professional use, and standard audio equalization curves have been adopted by the National Association of Broadcasters (NAB).† They are similar to the well-known RIAA equalization used in the record-making industry.

Instrumentation recorders cannot favor any portion of their recorded bandwidth because of the nature of the signals they must accommodate. Their low-frequency response is quoted at 50 cps so that the added noise contribution below that point can be eliminated. Of course the added noise from the higher frequency response does add to the cumulative noise and reduces the S/N figure. Also, in striving for the higher frequency response much narrower gaps are used in the reproduce heads, with a corresponding reduction in signal power. The definition of maximum record level also works against the instrumentation machine, that is, that level that produces 1% THD, while for audio recorders it is that level that produces 3% THD.

As you can see, there are many differences between audio recorders and instrumentation recorders. Some are obvious, some are very subtle. There are very few occasions when an audio machine will record instrumentation data without compromise. Be careful if you select one for your recording job.

Response Standards for the Direct Record Process

As a result of the need to ensure compatibility when recorded data were to be exchanged between machines at different locations or of different origin, some form of standardization was required.

†The NAB is composed of members involved in and concerned with all phases of commercial radio broadcasting: AM, FM, and TV. NARTB is the older term for the same organization—it stands for the National Association of Radio and Television Broadcasters.

Standards have been established for four groups of direct record parameters. The IRIG Standards, first established in the early 1950s, are applicable only to 1/2 and 1-in. tape systems, while technologic advances have enabled manufacturers to offer 4-, 7-, and 8-channel instrumentation recording capability on 1/4-in. tape. The lack of IRIG standards for 1/4-in. tape has not prevented the incorporation of the applicable high-performance specifications to the newer recorders. There are many industrial, medical, government, and other users who do not require the interchange standards of IRIG, and who should be aware of the price-performance package these 1/4-in. recorders can provide.

FM Record/Reproduce

Following World War II, the limitations of the direct record process severely restricted the use of tape recorders for general instrumentation work. By 1950 the growing aeromissile business and the several military test ranges had to record an increasing amount of DC and very-low-frequency test data. To serve in these areas, magnetic recording had to somehow provide DC response, good DC linearity, and better signal-to-noise ratios.

The recording industry filled this need with the development of FM recording. The technique was widely accepted and the majority of recorders in use in the 1950s were equipped for FM operation. Still in wide use today, the first FM scheme used center frequencies and deviations adapted to 80- to 100-kHz recorders and put a single carrier on each track, recorded to saturation without bias.

Data recording using a frequency-modulated carrier is accomplished by deviating the carrier frequency in response to the amplitude of a data signal, and recording it. A DC signal of one polarity increases the carrier frequency, and the opposite polarity decreases it. An AC signal alternately increases and decreases the carrier above and below its center frequency at a rate equal to the data-signal frequency. In the reproduce process, the carrier's amplitude instability is essentially wiped out by limiting and the data signal is reconstructed by detecting zero crossings. Residual carrier signal and out-of-band noise are removed by a low pass filter. But FM recording has a problem of its own. It is extremely sensitive to tape speed fluctuations (flutter), since in either the record or reproduce mode, tape speed variations produce unwanted modulation of the carrier (or noise).

The increased noise level in the reproduced signal and corresponding reduction in dynamic range is the first-order effect of flutter. A second-order effect is the actual time-base variation of the data signal, much the same as the direct recording process.

A detailed technical description of FM theory can be found in textbooks, but a brief discussion of a few of the more pertinent factors is included here to assist the reader in a cursory understanding of the FM recording process. First a few definitions:

f_c	carrier center frequency
Δf	maximum carrier deviation from f_c
f'	data signal frequency
f' max	highest data signal frequency
$\dfrac{\Delta f}{f' \text{ max}}$	deviation ratio, or modulation index
$100 \dfrac{\Delta f}{f_c}$	percentage deviation

Deviation ratio is one of the most important factors in any FM process. Basically, the higher the deviation ratio, the more immunity the system has to noise. In FM recording, however, there are practical limits to deviation ratio, since Δf is restricted by recorder bandwidth limitations and f' max must be kept high to accommodate the data signals. Common deviation ratios in use today range from 5 in the telemetry FM subcarriers to 0.675 in wideband FM recording. FM broadcasting, which enjoys an excellent noise immunity, uses a deviation ratio of 5 ($\Delta f = 75$ kHz, f' max = 15 kHz).

The percentage deviation, 100 $\Delta f/f_c$, is another factor in FM recording. When recording low-percentage deviation systems, such as the FM telemetry subcarriers (7.5 and 15%), the effect of flutter is essentially multiplied with a corresponding increase in noise. For instance, if a 7.5% frequency deviation corresponds to a 100% input signal, a 1% deviation caused by flutter will appear as 100/7.5 = 13.3% noise signal. The same flutter imposed on a 40% deviation system will cause only 100/40 = 2.5% noise signal. The higher percentage deviation systems are thus less influenced by tape speed variations, but circuit design limitations make 70 to 75% a practical limit.

A well-designed FM carrier recording system will give reasonably good amplitude accuracy, DC response, good DC linearity, and low distortion. The price paid for this improvement over direct record performance is greatly reduced frequency response for a given tape speed, added complexity and cost in the record/reproduce electronics, and a much greater need for constant tape speed (low flutter).

There are many uses for FM recording. The obvious advantages, of course, are the DC response and stability. It also is a very handy tool for time expansion and compression techniques. As an example, 5-kHz data recorded at 15 ips on a 27-kHz carrier can be reproduced at 15/32 ips with the data frequency spectrum reduced by a factor of 32. Combinations of changing tape speeds and rerecording can provide time base change and frequency shift factors to well over 200. Data reduction, spectrum analysis, hard copy output, and other laboratory chores can be simplified with such expansion and compression techniques.

FM Recording "Standards"

As long as recorder response was limited to about 100 kHz at 60 ips, there were fairly well-defined standards for FM recording. Tape speeds 1 and 7/8 to 60 ips were assigned FM carrier frequencies from 1688 Hz to 54 kHz, all harmonically related, that is, double the tape speed and double the FM carrier frequency. Percentage deviation was established as ±40% and a deviation ratio of approximately 2 was selected. The major recorder manufacturers adhered to the standards, and FM recording proceeded in an orderly fashion.

From 1958 to 1965, however, tape recorder response moved up by an order of magnitude, and many extensions and modifications to the old standards came into being. By 1965 there was considerable confusion in the industry and among users regarding FM recording. Such terms as wideband FM, single-carrier FM, standard-response FM, extended-response FM, and even double-extended-response FM were used to describe the many new kinds of FM recording, but truly authoritative definitions were lacking. Finally, in 1965, an updated version of the IRIG document, "Telemetry Standards" (IRIG document no. 106-65) was published and has since been revised several times, the latest being 106-73.

FM Recording Systems

"Single carrier FM recording" is the name now used to describe the original FM technique that used saturation recording. First called wideband FM to distinguish it from telemetry subcarrier recording, it has relinquished this name to the new schemes that use fractional deviation ratios and are truly wideband.

The majority of the single-carrier FM recording systems in use today still record the carrier to saturation on the tape without bias, using FM electronics provided by the recorder manufacturer. In this manner the maximum possible voltage is obtained from the reproduce head. Direct recording (with AC bias) may also be used with voltage-controlled oscillators and discriminators external to the recorder.

Carrier frequencies are proportional to tape speed and have been selected near the middle of the recorder's response to keep distortion products above the passband of the head/amplifier combination. FM record amplifiers can usually be "tuned" to any of the standard carrier frequencies by selection of the proper plug-in or switchable tuning unit. Similarly the FM reproduce amplifier is set for a specific carrier by selection of the proper frequency-determining element and low-pass filter.

Single-carrier FM recording uses ±40% deviation with a deviation ratio of approximately two. Performance will usually equal or exceed the following:

Characteristic	Performance
Frequency response	± 0.5 to 1.0 dB over specified band
RMS signal-to-noise ratio	40 to 54 dB if full deviation is used and flutter specification is 0.25 to 0.7% peak-to-peak
DC drift	Less than 1% of full deviation over 24-hr period
DC linearity	Within ± 1.0% of zero-based straight line

If electronic flutter compensation is used, the first-order effects of flutter or flutter-induced noise can be reduced to a point where other noise contributors set the noise level. At best, a signal-to-noise ratio of 50 to 56 dB may be obtained at tape speeds of 60 or 120 ips.

Rotary head video recorders, which appeared about 1958, also use a type of FM recording. Primarily designed for TV broadcast use, instrumentation versions of these machines can now provide outstanding performance. The cost of such a machine certainly restricts its usage, but the fact that such performance can be obtained in a production machine is an outstanding tribute to its developers.

There are two other techniques of FM recording that put multiple subcarriers on a single track. The direct recording process is used to minimize distortion and avoid cross talk between carriers. Basically these schemes provide a means of frequency-division multiplexing of many low-frequency data signals on a single-tape track.

The first and oldest scheme uses the FM/FM telemetry subcarrier oscillators. Their availability and standardization long ago made their use a natural for FM subcarrier recording. Based on a constant percentage deviation (7.5 or 15%), these units provide increased bandwidth with each higher center frequency. As discussed earlier, their low percentage deviation makes them highly susceptible to flutter-induced noise. Electronic flutter compensation must be used if data amplitude accuracy is important. Unlike with the single-carrier FM system, however, flutter compensation for multiple subcarrier recording is complicated by the characteristics of the bandpass filters, which are used to separate the subcarriers in the reproduce process. Space has been allowed in the IRIG subcarrier frequency assignments for a 17-kHz reference tone to be recorded between channels 13 and 14. Usually this reference frequency is used for tape speed compensation of the tape drive mechanism. A 100- or 200-kHz reference tone is also often recorded for use by the electronic flutter-compensation equipment.

The constant percentage deviation, or proportional bandwidth system, fell short in supplying the need for a large number of channels having relatively high frequency response. To fill this requirement, several telemetry ground equipment manufacturers provided "constant bandwidth" subcarrier systems. Deviation ratios of from 1 to 5 can be used; however, 2 is most common. Percentage deviation varies with the subcarrier frequency. A typical system might consist of a baseband of subcarriers, perhaps five in number. Their center frequencies and deviations might be 24 ± 2 kHz, 32 ± 2 kHz, 40 ± 2 kHz, 48 ± 2 kHz, and 56 ± 2 kHz. If additional channels are required, another baseband of five subcarriers is generated and translated in frequency to be recorded at 64 kHz $\pm$ 2 kHz, 72 kHz $\pm$ 2 kHz, and so on. Several basebands of subcarriers can be recorded in this manner if appropriate translation frequencies are used for each.

45.2-2 Tape Recorder Heads

The key to the success of any recorder's performance are its heads. There are record heads, reproduce heads, and erase heads. In rare cases, such as in some FM carrier recorders, a single head is used for both the record and reproduce function, but the great majority of instrumentation recorders use separate record and reproduce heads.

Tape recorder heads are electromagnetic transducers. A record head converts an electrical signal to a magnetic signal suitable to impose on the tape, and the reproduce head converts the magnetic signal of the tape into an electrical signal. Erase heads are a sort of specialized record head designed to saturate the tape at a specific frequency; however, they are seldom used on instrumentation recorders. When tapes must be erased, bulk erasure is most commonly used because it provides a more thorough erasing job and is less likely to be done accidentally.

The most carefully guarded proprietary secrets in the recording industry are those concerned with head manufacture and assembly. Many aspects of the assembly of high-resolution record and reproduce heads are closer to being an art than a science. Successful head production is more often related to the experience and know-how of key individuals, rather than to established production techniques.

The basic construction of both record and reproduce heads is similar, consisting of a magnetic core on which is wound a number of turns of wire (see Fig. 45.19). The core consists of two "C"-shaped half sections that are made from a number of bonded laminations of thin high-permeability ferromagnetic material. The surfaces of the half sections, which interface with each other, are lapped and polished very carefully and the gap material is deposited on one. The two cores and their windings are then joined to form a head.

One head is used for each track of the tape. In multitrack operation, several heads must be assembled together with intertrack shields to form a headstack. Extreme precision is necessary in aligning the heads in a headstack, since the gaps of all heads must fall within a 100 μin. band. After alignment, the heads are potted. The headstack is finished by contouring and lapping for optimum head-to-tape contact and then it is mounted on a precision-machined mounting base by which it is secured to the tape transport. The complete head assembly thus includes a base with a mounting surface whose plane is perpendicular to the gap line at the contact surface of the headstack. It is because of the precision mounting base that headstacks may be removed for cleaning and new

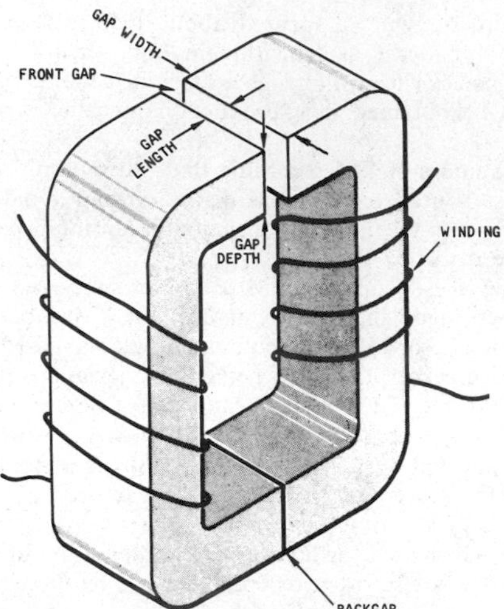

Fig. 45.19 Head core showing dimension definitions.

headstacks installed without factory adjustment while maintaining minimum losses due to azimuth misalignment (see previous discussion).

Record Heads

It was explained earlier that the record process takes place at the trailing edge of the record head gap. Because of this, the record head gap length has little effect on a machine's frequency response. Most machines use a record gap length of 0.2 to 0.5 mil. The accuracy of the gap edge, however, is very important to the recorder's high-frequency capability, and extreme care is taken to get a sharp well-defined gap edge.

The ideal record head/record amplifier combination will place an equal-amplitude flux pattern on the tape for a given signal level throughout the recorder's frequency range. This would truly be a "constant flux" recording. Actually, performance closely approximating this ideal condition is obtained in modern recorders, through the use of recording preemphasis. Both head losses and to some extent the recording demagnetization losses are made up for by the use of a high-frequency "boost" characteristic in the record amplifier. Additional preemphasis is not used, however, since the constant flux condition is optimum for best dynamic range and minimum distortion products.

Reproduce Heads

The high-frequency response of a tape recorder is determined by the gap length and the quality of its reproduce heads. For a machine to reproduce a 2-MHz signal at 60 ips, it must "see" recorded wavelengths on the tape as short as 30 μin., and this requires a reproduce head gap length of approximately 10 μin. In addition, the gap must be sharply defined and have adequate depth to withstand the abrasive effect of the tape. The gap material and the core material must have similar wear characteristics for the gap to maintain its definition. To produce several such heads in a single headstack in production quantities is one of the greatest industrial achievements in the United States today. The price of such a high-resolution headstack may run several thousand dollars.

The push for shorter and shorter gap lengths has one disadvantage. Unfortunately, as the gap length decreases, the output voltage from the head winding also decreases, thus lowering the signal-to-noise ratio of the recorder. One of the first symptoms of excessive head wear is increased head output at lower signal frequencies. This results from the lengthening of the gap. Even with such adverse conditions, however, present-day recorders obtain 38 dB or better signal-to-noise ratios while realizing packing densities of 30,000 to 33,000 cycles per inch.

Head Care and Head Life

The care and the life of tape recorder heads are completely interdependent. Factors that can affect the life of the head during normal use include the following:

1. Cleanliness of the tape, the transport, and the environment in which the equipment is operated.
2. Maintenance procedures that involve the checking of tape tension, tracking, and so on.
3. The abrasiveness of the tape being used.
4. Solvents used for cleaning the heads.

Cleanliness in and around the head area is of utmost importance in all instrumentation machines. Not only can the dirt particles become a serious threat to the data "take" in terms of spacing loss but they can also become minute scrapers, gougers, and cutters to the head and tape surfaces when dragged between them. High tape-to-head pressures are necessary to keep the spacing loss low, and such pressures not only increase the abrasive effects of the tape, but any piece of dirt is crushed that much harder into the head or tape material.

Maintenance procedures involving nearly all aspects of the mechanical part of the recorder will affect the head area in one way or another. Tape tensions must be kept as specified to ensure the optimum compromise between wear and performance. Care must be taken not to touch the heads with any metallic or hard object for fear of scratching, gouging, or magnetizing the heads.

The recording/reproducing of shorter and shorter wavelengths demands intimate contact between the oxide particles and the head gap to eliminate spacing losses. Higher tape tensions and more abrasive tape surfaces have resulted, and both tend to increase head wear. Gamma ferric oxide particles are very sharp and hard and resemble extremely fine sandpaper particles. The tape binders used on wideband tapes are also smoother and harder and form a firm base for each scratching particle. If high packing densities are not necessary, do not use premium tapes.

Clean heads are necessary if good recording is to be done. However, it is not safe to use just any solvent that appears to dissolve the residue left by the tape. The material used to hold the head cores in the head assembly is in some cases softened by such solvents as toluene, methyl ethyl ketone, or xylene. If one is in doubt about the head construction, use only alcohol, naptha, freon TF, gasoline, or even jet fuel! Freon TF is probably the best all-around cleaner and is now available in an aerosol can for convenient storage and use. Most head cleaners will also dissolve lubricating greases and tape binders and should be used carefully, especially around bearings and the tape. A cotton-tipped applicator makes a good disposable cleaning tool for the majority of cleaning requirements.

Clean all transport parts that come in contact with the tape. Oxide buildup can cause degradation of high-frequency response, distorted recording and playback, and tape and head damage. Clean tape heads with cotton-tipped applicators dampened in head cleaner by rotating the applicator against the head surface and noting the discoloration of cotton. Use as many applicators as are required to avoid wiping head surfaces with dirty applicators. When the applicator comes away clean, the head surface is clean. Also use applicators to clean tape guides, rollers, capstan, and pinch roller. Take particular care to clean the undercut edges of the guides and the roller grooves. Accumulation of wear products in corners of guides or roller grooves will damage the tape.

Magnetized heads produce unpredictable results in the reproduced data. *Under no circumstance should continuity of a head be checked with an ohmmeter*. High second-harmonic distortion results from a magnetized record head; the demagnetizing effects of the AC bias used in the direct record process, however, reduce the chance that record heads will be magnetized unless they are subjected to large unidirectional current surges (such as a continuity check) or close proximity to high-intensity magnetic fields. It is unwise, therefore, to remove record electronics cards from the recorder while it is in the record mode. The last current surge as one removes an amplifier may serve to magnetize the record head.

Magnetization of the reproduce head is more prevalent and affects its performance in an unpredictable fashion, though generally increased noise levels result. Demagnetization of recorder heads is accomplished by using any of the commercially available head degaussers. The recorder should always be turned off when degaussing heads, and on some machines removal of the head assembly simplifies the job and ensures more complete degaussing. Proper procedures can usually be found in the recorder's instruction manual. Be sure to understand the complexity of head removal before removing the head assembly, though, as head assembly removal and replacement always requires extreme care and attention to detail.

45.2-3 Tape Transport Mechanisms

Development

The sole purpose of a tape transport is to move the tape by the heads at a constant speed and to provide the various winding modes of operation required for tape handling, without straining, distorting, or wearing of the tape. To accomplish this, a transport must guide the tape past the heads with extreme precision and maintain the proper tension within the head area to obtain adequate tape-to-head contact. Spooling or reeling of the tape must be done smoothly so that a minimum of perturbations are reflected into the head area. Takeup torque must be controlled so that a good tape pack results on the takeup reel. It is also the job of the transport to move the tape from one reel to the other quickly in the fast-forward or rewind mode. Even with fast speeds, the tape must be handled gently and accurately so that a good tape pack is maintained on each reel. In going from a fast mode to a stop (or vice versa), precise control of the tape must be maintained so that undue slack or stress is not incurred by the tape. Most of these functions are provided in modern instrumentation tape recorders, but improvements in the area of uniform tape motion are constantly being sought.

The earliest instrumentation recorders were an outgrowth of the best audio recorders of the time. They used the open-loop transport design (see Fig. 45.20) where the tape in the head area has only one end controlled by the capstan, the other end being tensioned directly by a reeling function. As instrumentation demands increased, audio machines were refined and specialized electronics were built with the open-loop transport predominating until the mid 1950s. It was basically a sound design, both simple and reliable; however, its flutter performance left something to be desired. There were two inherent sources of unwanted tape speed variations in the head area. One came from the perturbations caused by the supply reel, its drive motor or braking mechanism, and the other resulted from erratic vibrations in the tape because of the long unsupported tape length in the head area. Vibrations were also induced by friction between the tape and the heads or fixed guides.

The closed-loop transport design shown in Fig. 45.21 would seem to solve both these problems. The unsupported tape length is halved and, if both ends of the tape in the head area are under positive control of the capstan, the reeling function perturbations would be eliminated. Actually these characteristics are only partly achieved.

If there were no slippage or creepage between the tape and the capstan at either point of contact, any initial tension in the loop would be maintained. But the only elements that can apply tension to the tape are outside the loop, that is, the reeling functions. Thus there has to be some creep between the tape and the capstan for the reeling functions to maintain tension within the loop. So even in the closed-loop transport design, motional perturbations in the reeling functions still cause uneven tension variations in the head area and corresponding tape speed variations, but the effects are greatly reduced when compared with those of the old open-loop designs.

By 1958 nearly all manufacturers were using closed-loop transports, but still striving for improved performance. About this time one manufacturer developed a new capstan drive design that provides a differential action. It is basically a capstan with two discrete diameters (only a few thousandths of an inch difference). As shown in Fig. 45.22, special contoured pinch rollers are used and tape entering the head area is forced against the smaller capstan diameter while tape leaving the head area is forced against the larger capstan diameter. Thus tension is generated within the loop because of the speed differential between the tape entering and the tape leaving the loop. In this design, less tape tensioning is necessary from the reeling functions and more isolation is obtained in the head area from reeling perturbations. However, the drive is not suitable for bidirectional operation.

Yet another type of transport configuration is now widely used. This is the two-capstan transport shown in Fig. 45.23 and called dual capstan or differential capstan. Some designs turn both capstans at the same peripheral speed and rely on the reeling functions to establish the tape tension in the head area, much the same as the conventional closed-loop transport. However, most two-capstan machines turn the capstans at slightly different peripheral speeds and establish the tape tension in the head area with this differential action. Less tension external to the head area is required and considerable isolation from reeling function perturbations is thus obtained.

In the mid 1960s the open-loop transport reappeared in a refined form that competes quite well with the more sophisticated closed-loop designs. A major effort in one manufacturer's design, shown in Fig. 45.24, was devoted to the use of swinging damper arms with a high-inertia high-torque capstan. The significant reduction in the length of unsupported tape allowed for a substantial improvement in high-frequency flutter.

All in all, the new open-loop design retains all the simplicity of the old, yet provides greatly improved performance at considerably less cost than most closed-loop machines.

Another feature that appeared in transport designs is the elimination of the pinch rollers. These drive systems use a capstan with high surface friction and large tape-wrap angles. When they are used with the dual (differential) capstan drive, reasonable isolation is obtained from reeling perturbations. One design, however, is similar to the closed-loop transport, with the high surface friction capstan replacing the turn-around idler. In this machine, tape tension in the head area and isolation from reeling perturbations is provided by vacuum-controlled tape chutes before and after the record and reproduce heads. Flutter and skew performance of this machine are said to be excellent.

With the elimination of the pinch roller, one more source of tape speed irregularity and tape guiding is removed. Pinch rollers with surface deformations, slick or sticky spots, or noisy bearings can be large contributors to a machine's flutter and dynamic skew.

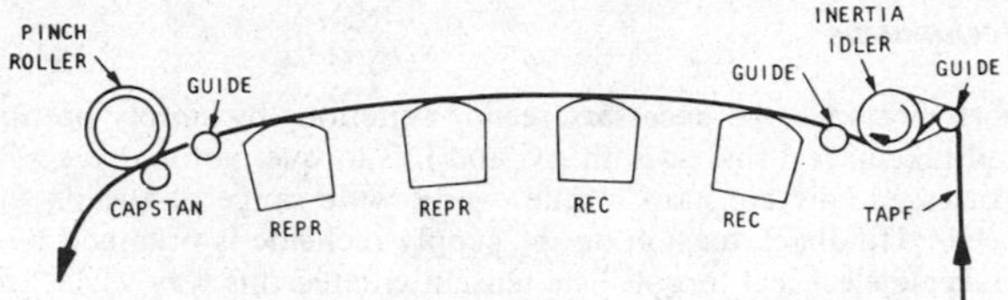

Fig. 45.20 Open-loop drive.

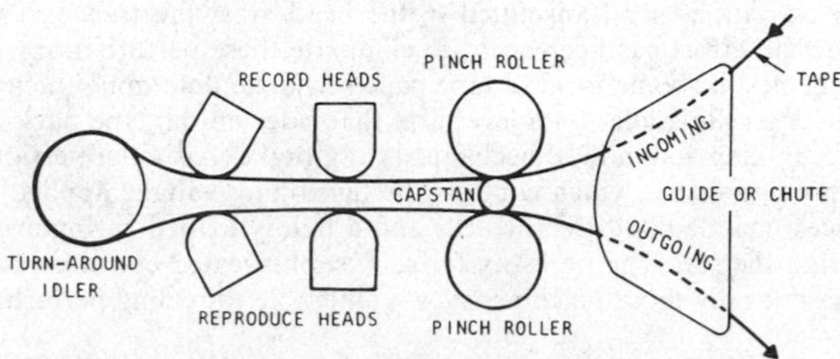

Fig. 45.21 Closed-loop drive.

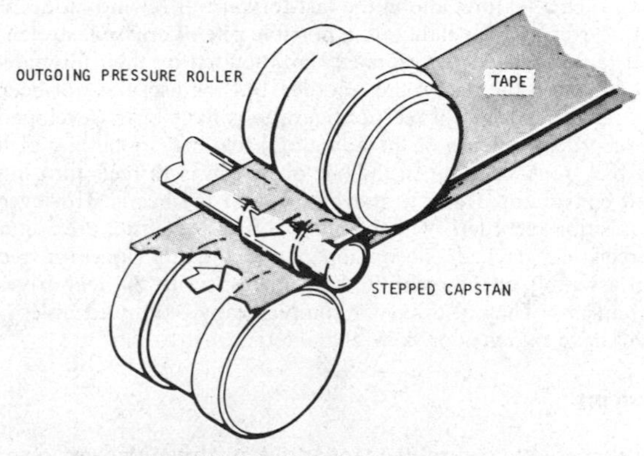

Fig. 45.22 Two-diameter capstan.

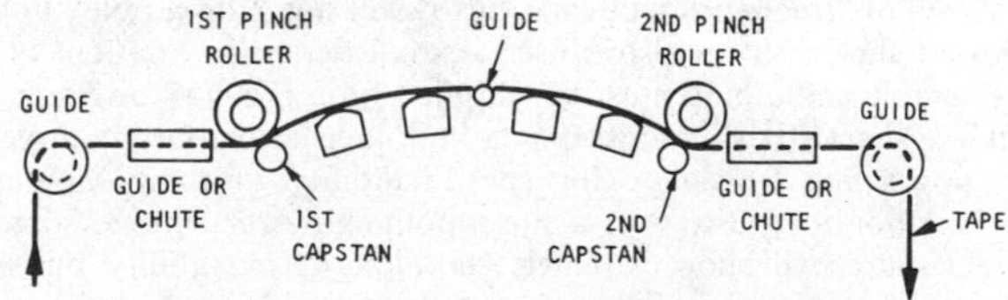

Fig. 45.23 Dual-capstan closed-loop drive.

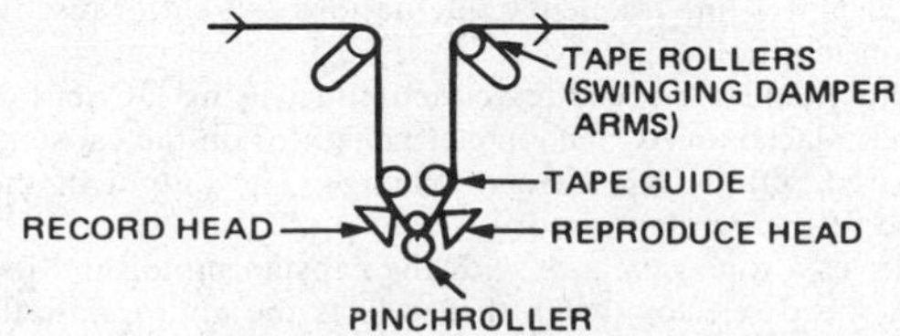

Fig. 45.24 Refined open-loop drive.

Tape-Reeling Mechanisms

Early tape transports provided the necessary reeling functions by simply putting a torque motor on both the supply and takeup reel shafts. Both AC and DC torque motors have been used, and they are designed to give an essentially constant torque over a wide range of speeds without overheating at extremely slow speeds. Holdback tension on the supply reel side is obtained by a simple mechanical drag brake or by reverse electrical torque. The tension created this way varies from full reel to empty reel, however, because of the changing radius of the tape pack. Minimum tension is provided with a full reel and maximum tension with one almost empty. Both schemes also contribute short-term tension variations because of irregular braking action or cogging pulsations in the torque motor. When such tension variations are transmitted to the head area, the transport's flutter performance will suffer. Much design effort has been spent to eliminate these perturbations, with varying degrees of success. Most schemes use some form of tape pack sensor to determine the amount of tape on the reel. One method uses a rolling idler on a lever arm that rides on the tape pack. As the arm moves, a linkage controls the braking action of a mechanical drag brake. A similar version uses the same lever arm to position a potentiometer, which controls the tensioning voltage applied to the torque motor. Other sensing devices include photoelectric cells and a lightly loaded spring arm that tracks the exit point of the tape from the reel. The necessity for such sophisticated controls is, of course, dependent on the transport design; only those machines most vulnerable to reeling perturbations must use these techniques.

There is one highly desirable feature that probably should be included in every tape recorder reeling mechanism. This is a fail-safe function in the form of a band or disk brake mounted on the reel shafts to bring the tape to a safe controlled stop in the event of a power failure. Some older machines do not have such a feature, and in the fast-forward or rewind mode, a power failure can put several hundred feet of irreplaceable data into a horrible pile of creased, stretched, and mangled tape.

The packaging of tape recorders is to some extent dictated by their ultimate use. The applications have imposed severe limits on the size of the recorder, but the user has not been willing to reduce the recording time. As a result, a variety of reel configurations have been developed, often at the expense of simplicity of the recorder and ease of threading and loading/unloading of the tape.

The side-by-side reel transport, that is, the one on which both reels turn in the same plane, is the most convenient and conventional way to move tape past the heads. However, space recorders and some other data acquisition recorders, where volume is at a premium, use a concentric reel transport, one on which the reels are stacked one on top of the other. Concentric reel machines require a complex mechanical assembly at the reel hub, both in respect to the reel drive and tape-reel loading and unloading mechanisms. They also have a complex tape path, and unless great care is taken in their design, they will have rather poor skew characteristics.

Speed-Control Systems

Since the capstan is supposed to control the tape speed, its driving power in early tape recorders was chosen for constant speed. The most popular choice was the hysteresis-synchronous motor. Such a motor runs phase locked to the AC power frequency, and its long-term speed stability is thus as good as the power line frequency stability if it is operated with a constant load. With a varying load its phase relation with its supply frequency will vary, but it will not "slip a pole" unless it is overloaded. The pole construction of these motors also causes a small periodic variation, or "cogging," in their torque output. Considerable smoothing must be supplied by a flywheel and sometimes the elasticity of a drive belt. Such a system is best described as a high-inertia low-torque drive.

Dependency on power-line frequency for speed stability in many applications is a strong disadvantage. Even in laboratory use, with a metropolitan power source, some instability can be expected. Such a power source will show extremely good long-term stability, but short-term variations as high as $\pm 0.25\%$ are not uncommon. These factors long ago led to the development of precision 60-Hz power supplies consisting of very stable 60-Hz frequency sources (tuning forks or counted-down crystal oscillators) and power amplifiers capable of 50 to 150 W. Capstans powered by this means are independent of power-line frequency fluctuations but suffer disadvantages in size, weight, and amount of power consumption.

Several quite satisfactory constant speed drive techniques using DC motors have been developed. One of the earlier methods placed an AC tachometer generator on the capstan motor shaft. Its output was rectified and compared with a DC reference voltage. The output of the comparator was then used to control the speed of the capstan motor.

Another technique drives a tone generator with the capstan motor and its output goes to a series of frequency dividers. A speed selector switch then selects the appropriate divider output for phase comparison with a highly stable reference frequency. The output of the phase comparator is then used to control the field current of the shunt-wound DC capstan motor. When both signals are locked in phase, the speed of the motor is constant. This control method, in addition to being independent of line frequency, has several advantages over the AC drive. One relates to the ease of changing tape speeds. Simply switching one more divide-by-two circuit into the divider circuit will double the motor speed. Another advantage is the reduced size, weight, and power consumption, since a high-power, low-efficiency amplifier is not required.

Early DC-powered capstans were also relatively high inertia systems but with somewhat more torque than the AC drives. There is increasing usage of DC permanent magnet motors, which use a printed circuit rotor with many poles. This device has made possible a low-inertia high-torque drive system, which several manufacturers have incorporated in their recorders.

Standard tape speeds of 120, 60, 30, 15, 7 1/2, 3 3/4, 1 7/8, 15/16, and 15/32 in./s are available, and most instrumentation recorders provide easy selection of three to six of them. Some of the earlier recorders required drive belt changes if a speed change greater than 2 : 1 was required.

Any of these methods of capstan drive control gives reasonably accurate tape speed in the record operation. If precise reproduction of recorded data is required, however, servo speed control is necessary in the reproduce mode. Several techniques are used for this; they all operate from a reference signal recorded on the tape with the data signals.

If the recording machine is driven by a hysteresis-synchronous capstan motor, its precision 60-Hz supply is used as the reference signal and is recorded by modulating a 17-kHz carrier. If a DC capstan motor (other than the new low-inertia versions) powers the recording machine, the 60-Hz reference is generated from the speed-determining reference oscillator and is recorded in the same manner. Thus either system can provide a reference signal that can be used by either system. Servo control of a hysteresis-synchronous reproduce machine is accomplished by recovering the reproduced 60-Hz reference signal and phase comparing it with the local precision 60-Hz reference voltage. The output of the comparator is a DC or very-low-frequency AC signal, which controls the output of a 60-Hz voltage-controlled oscillator. This output is then amplified and used to drive the capstan motor. The servo action effectively locks the recorded reference signal to the local precision reference signal, and a length of tape equivalent to 1 cycle of the recorded reference is passed through the machine for each cycle of the local reference. Servo control of a reproduce machine with a DC capstan drive is usually accomplished by replacing the local tone generator signal with the reproduced reference signal from the tape.

Both methods are phase comparison systems and thus are basically positional servos. Pull-in range is limited by the high mass of the capstan drives, but the higher torque capability of the DC motor gives that system a small edge. Pull-in ranges from 1/2 to 3 or 4 cycles are probably the limit of their response.

The appearance of the low-inertia high-torque DC motors for use in capstan drives brought a totally different reproduce speed control into existence. The "snappy" response of these devices made it possible to eliminate practically all long- and short-term speed variations up to 100 Hz or more. The majority of these systems use a dual servo control system embodying a rate control function and a positional control function. For recording, there is an AC tachometer generator attached to the capstan shaft that produces several hundred cycles for each capstan rotation. Frequency variations in this tachometer signal are sensed by a discriminator and become a DC control voltage for the capstan motor. Basically a rate servo, this control is used to bring the capstan to the proper speed. The tachometer signal is also phase compared with a local reference oscillator after the proper speed is reached and phase-locked operation is provided. During recording, the signal from the reference oscillator is also recorded on the tape. For servo control of the tape speed in the reproduce operation, the recorded reference is used to replace the tachometer generator and essentially the same operation just described takes place.

Flutter specifications are very similar for both the high-inertia and the low-inertia capstan drive systems, as far as percentage flutter is concerned. There is, however, a marked difference in the spectral components of the flutter in the two systems. The high-inertia system exhibits greater amplitudes of low-flutter frequencies, and only the usual higher frequency flutter caused by tape scrape and vibration. The low-inertia systems, when servo controlled from the tape, practically eliminate the low-frequency flutter, but in being so responsive they actually add to the normal high-frequency flutter by a form of spectrum spreading.

Even though both systems seem to show approximately the same total flutter percentage, there is a strong point favoring the low-inertia system: the dramatic improvement in time base error. The specified TBE on some machines is $\pm 0.5\ \mu s$ absolute. This means that between two points anywhere on the tape, the timing error will be no greater than $\pm 0.5\ \mu s\ \pm$ the crystal reference tolerance.

Tape Motion Irregularities

Ideal tape motion may be simply defined. The tape must move across the heads with an absolutely uniform, precisely known velocity. Actually, no tape transport will ever attain this ideal motion, though improvements are continuously being made. Medium or long-term deviations from the desired average speed can be corrected by servo means, as previously discussed. In fact, the low-inertia high-torque capstans are snappy enough to correct some of the short-term variations, but in general, all short-term variations cannot be eliminated in this manner. As a result, they must be considered among the basic characteristics of a recorder.

Flutter. Short-term speed variations that are uniform across the tape can be caused in many ways in a tape transport mechanism. Some of these are pulsations of the torque motors, reel eccentricities, irregularities in the tape pack or tape physical characteristics, vibrations in the tape caused by friction as it passes over fixed guides or heads, mechanical runout of rotating parts, slight cogging of the capstan drive motor, power-line voltage transients that may affect the motors, pinch rollers with surface deformations, and sticky bearings. The problem is further compounded by reels and reel drive assemblies that have varying velocities and a mass that is constantly changing.

Velocity variations that are uniform across the tape have been variously described as flutter and drift. Flutter denotes variations in speed that occur at frequencies above 0.10 Hz, and drift (or tape speed accuracy) is used for those frequencies below 0.1 Hz. As applied to instrumentation recorders, common usage has broadened the definition of the term "flutter" to include all variations 0.1 Hz to 10 kHz. "Time displacement error" (TDE), "time base error" (TBE), and "jitter" are terms used to describe the same tape-speed variations from a different point of view. These terms are sometimes improperly used, and confusion can result in trying to describe the ability of a recorder to reproduce a signal with its original time relationships. It should be remembered that TDE, TBE, and jitter figures must state the time over which they were measured, and flutter must be quoted in either rms or peak-to-peak values over a specific band of frequencies.

The flutter spectrum of a well-designed machine is made up of a combination of small, discrete sinusoidal components and a more or less uniformly distributed noise signal. Because of the noise involved, a clear well-defined measurement of flutter is difficult. The flutter signal itself can be obtained by recording an extremely stable reference sine wave and passing the reproduced signal through an FM discriminator. The signal thus generated would be zero if the tape speed were exactly the same for reproduction as for recording. Any variations between the two speeds deviates the frequency of the reference tone and produces an output from the discriminator. It is customary to measure flutter components to at least 10 kHz in instrumentation recorders. The flutter meter is the most common device used to make the peak-to-peak flutter measurements when verifying flutter specifications.

One common form of flutter specification is "cumulative flutter" measurement. This can be made by passing the flutter signal through a variable cutoff low-pass filter and measuring the filter's output for increasing values of cutoff frequency.

Since the noise contribution in the flutter signal is essentially uniform, the shape of the cumulative flutter curve rises with frequency. At each point where some rotating component produces a discrete sinusoidal contribution, there is a small step function in the cumulative curve. Many manufacturers publish curves of this type, but they are usually the average results of testing many transports and do not show the extremes that may be found in individual machines.

An rms measurement for flutter has long been used for audio machines, but the peak-to-peak measurement is more useful for instrumentation machines. Actually a true rms value of flutter is almost impossible to attain, since the flutter signal contains a DC component, a noise component, and many sine wave components. As a very rough approximation, the rms value can be assumed to be 1/6 to 1/4 of the peak-to-peak value.

Many techniques are used for measuring time base perturbations caused by flutter. Basically they all involve comparison between a precise electronic time delay and the time base represented by some length of tape. A reproduced pulse from the tape is used to initiate the electronic delay, and some period of time later a second pulse from the tape is compared with the electronically delayed pulse for time coincidence. Time mismatch between the two pulses than represents a time base error attributable to tape speed variations. Equipment to make such measurements accurately is costly and not readily available.

Skew. The term "skew" is used to describe the fixed and variable time differences between the several tracks of a single headstack. It implies that the tape is moving in some manner other than longitudinal as it passes the heads, that is, that it is skewing or yawing. Fixed, or static, skew contributes a constant relative timing difference; dynamic skew produces a variable timing difference between tracks.

Fixed skew is usually caused by misalignment of head to tape, misalignment of individual heads in a headstack, and misguiding, which produce fixed differences in tension distribution as the tape crosses the heads. Dynamic skew is produced when there is uneven tension distribution across the tape. Such flutter producers as tape scrape and vibration will initiate dynamic skew.

Fixed skew and some forms of dynamic skew produce relative timing errors between the tracks that are proportional to the track spacing. Some dynamic skew, however, is caused by tape dimension irregularities and random flutter components, and the timing errors so produced are not correlated.

As with flutter, skew errors are imposed by both the recording operation and the reproducing operation. Tapes reproduced on the same machine they were recorded on will have less skew error than those reproduced on different machines.

Tape transport specifications usually show skew, dynamic skew, or total interchannel time base error (static and dynamic). Figures are given for different tape speeds, and some of the modern machines can provide ± 0.5 μs or better at 120 ips between adjacent tracks on the same headstack.

Stretching. Another source of timing error occurs between odd- and even-numbered tracks owing to the spacing between the two headstacks. If the physical dimensions of the tape vary due to environment or stretching after the recording is made, a corresponding timing error will occur when it is reproduced. As an exaggerated example, suppose there is to be a 1% change in part of a tape's length. This represents 15 mil for a distance of 1.5 in. and might represent many wavelengths of recorded signal. The less the spacing distance between the headstacks, the less loss of signal will be observed for the same change in tape length. Differences in tape tension between the record and reproduce machines can also contribute to this type of timing error.

Note. Gross timing errors will result if data are recorded on one machine and reproduced on a machine with a different track format.

Effects of Flutter. Flutter has several harmful effects on recorded data. Perhaps best known is the noise it produces in FM carrier recordings. A constant-frequency carrier will be frequency modulated by tape speed variations; when the carrier is reproduced and discriminated, the expected DC output will also include the noise of the unwanted modulation. When the carrier is modulated with a data signal, this noise is added with the data in the demodulation process. Thus flutter's first-order effect is to increase the noise level of the reproduced signal with a corresponding reduction in dynamic range. For a given percentage flutter, the ability of the flutter noise to interfere with the data signal is dependent on the deviation ratio of the FM carrier system being recorded.

In the direct-record mode, flutter primarily perturbs the time base of the reproduced signal. Akin to this but somewhat more subtle, the reproduced waveform will show broadening of its spectral components in a spectral analysis.

45.2-4 Electronics

The electronic circuitry of an instrumentation magnetic recorder performs two distinctly different tasks: tape speed control and data handling or signal conditioning. Each performs rather independently of the other, with an interdependency being realized in the quality of the reproduced data. Tape speed control systems were discussed in the previous section and the remaining electronics will be covered here. Each type of record/reproduce operation requires its own rather specialized electronics, but only the direct and single carrier FM modes are described. Flutter compensation circuitry will be briefly introduced with the FM electronics.

Direct Record/Reproduce

The direct record/reproduce operation requires an amplifier head driver, a bias oscillator, a reproduce head preamplifier, and a reproduce equalizer-amplifier. The amplifier head driver is more commonly called a record amplifier and it provides several functions. It should present a nominally high impedance to the data signal to minimize loading of the signal source. A first approximation of its gain/frequency characteristics would suggest a constant current output for all frequencies. Unfortunately, the impedance of the record head will change quite drastically over the frequency range covered by modern wideband machines. Also, as mentioned previously, a constant flux recording (equal magnetic intensity on the tape for all frequencies) is ideal, yet head losses and recording demagnetization are a function of frequency. For these reasons it is customary for the record amplifier to provide increased output, or preemphasis, at the higher frequencies. This should not be confused with the exaggerated preemphasis used in audio machines: Only the high-frequency boost required to approximate a constant flux recording is used. As an example, one wideband machine provides 8 to 10 dB of preemphasis at 1.5 MHz. Another function of the record amplifier is to add the bias signal to the data signal while providing a buffer for the bias oscillator.

Somewhere in the recording circuitry a monitor point is often provided to allow observation of the record head current waveform. Usually the monitor level is about 10 Ω above ground and presents a very low-level signal. Considerable gain is required in the oscilloscope for proper viewing. Waveforms at this point, especially on the wideband machines, may appear quite distorted, and the machine's instruction book should be consulted for details on their use. Recording levels should be set to obtain an optimum matching between the dynamic ranges of the data signal and the recorder. A well-designed record amplifier will not saturate before the tape does, and it should be remembered that saturation of the tape is a somewhat gradual process with very quick recovery. The distortion that results when the tape is overrecorded can sometimes be tolerated in exchange for the slightly increased dynamic range.

A record level meter is often provided on each record amplifier but is of little use if the data signal is not sinusoidal. Low duty-cycle pulse trains can be drastically saturating the tape with little or no indication on the conventional rms meter. Peak reading meters have been offered by some manufacturers as an accessory but they are, unfortunately, not in wide use.

The bias oscillator must operate at several times (3.5 to 5.0 or more) the highest frequency the recorder can handle. A pure sinusoidal waveform is required, since any distortion will be reflected by the recorded signal. A single-bias oscillator is used to eliminate sync or "beating" problems.

A buffer amplifier, usually located in the record amplifier, provides the necessary bias drive for each track and at the same time eliminates the possibility of cross talk between channels from occurring through the bias distribution wiring. Integral with this buffer amplifier, there is sometimes provided a convenient means for monitoring bias current on the record level meter.

The output of a reproduce head can be a very small signal, that is, a very few microvolts in a wideband machine. At these levels the signal is subject to all kinds of noise pickup and ground loop problems. The output of the reproduce head is immediately applied to a preamplifier, through a carefully shielded cable of minimum length. To a large extent this preamplifier sets the system noise level, and great care must be taken in its design to provide low noise operation and optimum impedance matching with the head. The ideal design will probably result with integrated circuit preamplifiers embedded in the reproduce headstack.

The next operation on the reproduced signal is accomplished by a reproduce amplifier. This is the unit that equalizes the reproduce head's output/frequency characteristic. When a voltage amplifier is used, a 6-dB-per-octave rolloff must be inserted from the lowest frequency to the mid-frequency peak of the head's output curve. At this point, regardless of amplifier type, an increasing amount of gain (up to 12 or 18-dB per octave) must be added to compensate for the drooping head output characteristic and, to some extent, other high-frequency losses. A different equalizer network is required for each tape speed to properly match the different head output curves, and each network has two to four adjustments associated with it. Each adjustment affects a specific portion of the response curve and when properly set can provide the machine's specified response. Particularly in the wideband machines, one of the adjustments is usually associated with phase response. It is used to provide phase equalization, or in other words to ensure that the various spectral components arrive at the output with the proper time relation. Phase equalization is especially desirable when reproducing pulse-type data signals to minimize overshoot in the reproduced waveform.

Selection of the proper equalizer network for the tape speed in use varies from one type of machine to another. In some, selection is automatic with speed selection; in others, switching with a front panel control is required; others require replacement of plug-in units when speeds are changed.

The last function of the reproduce amplifier is to bring the signal to a standard output voltage and impedance level and perhaps to provide for switching to a meter for monitoring.

FM Record / Reproduce

Single-carrier FM recording electronics are supplied by the recorder manufacturer as plug-in units that can be interchanged with the direct record/reproduce units. An FM record amplifier contains an amplifier, an oscillator whose frequency varies with the data signal, and an output amplifier used primarily as an impedance match for the record head. As indicated earlier, the signal is recorded to saturation on the tape at a constant amplitude. Each tape speed requires a different center frequency; in the record amplifier this is provided for with plug-in or electrically switchable frequency-determining units. Typical oscillator circuits are multivibrators or a modified phantastron design.

An FM reproduce amplifier contains an amplifier to raise the level of the head output, a limiter, an FM demodulator, a low-pass filter, and an output amplifier to bring the signal to a standard output voltage and impedance level. The design of the first amplifier is not critical, since the saturated signal already enjoys an excellent signal-to-noise ratio on the tape, and the center frequency is located to minimize head output variations over the frequencies used. A hard limiting operation is used to eliminate amplitude variations and provide a standardized signal to the demodulator. Several types of demodulators are used. The simplest method is nothing more than a zero-crossing detector, which triggers a constant energy pulse for each crossing. A one-shot multivibrator may be used, with the real challenge lying in developing the constant energy pulse. Both pulse amplitude and duration must be precisely constant, since these are the basis for the accuracy of the reproduce unit. The resulting pulse train is then passed through a low-pass filter to remove noise and residual carrier components, and is amplified to provide a standardized output. Another demodulation technique sometimes used consists of a phase-locked oscillator whose frequency is phase compared with the reproduced signal. Differences in the two frequencies produce an output used to drive the local oscillator to obtain a null. The driving voltage thus is a duplicate of the data signal. Again a low-pass filter is used in the output. As with the FM record amplifier, plug-in or electrically switchable elements are used to set center frequencies and filter characteristics.

The harmful effects of flutter cause serious deterioration in an FM recorded signal. If the utmost resolution is to be obtained, a method of flutter compensation must be used. One of the simplest forms of compensation is provided by demodulating a reference signal that was recorded with the data (on a separate track) and thus generating the flutter signal itself. This signal is then simply subtracted from the many data channels as they are demodulated. Great care must be taken to match the phase and gain characteristics of both the reference and data demodulators. In a similar system, the flutter signal is similarly recovered from the recorded reference but is used to control the duration of the constant energy pulses being generated in the data demodulator.

Other FM recording techniques include the wideband FM, constant bandwidth FM subcarrier, and IRIG proportional bandwidth subcarriers. The equipment for using these techniques is almost totally supplied by the telemetry ground equipment manufacturers, and their signals are all recorded by the direct record mode. They are not peculiar to tape recorders and will not be discussed here.

Other Features

Remote Control. Nearly all operating modes of magnetic recorders are controlled by relays or electronic switching. When using remote control the operating mode is controlled by momentary contact switches, either by closing, or in the case of the stop circuit by opening, appropriate circuits. Remote controls permit the activation of all transport modes from another position, with the exception of turning power on and off. Other indications of the recorder's mode of operation often presented at the remote control include a tape-break indicator, a tape-remaining meter, an indication of tape speed, and a capstan servo sync indicator. With today's emphasis on interface standards among electronic measuring devices, the ability to interface the remote capabilities of instrumentation recorders further increases the machine's flexibility.

Bidirectional Recording. This capability has been made possible by the dual capstan transports and improved reel servo systems. At least one manufacturer has used this technique to greatly extend the recording time of a few data channels. If two, four, or seven data signals are all that are to be recorded, a special sequencing circuit may be used in the following manner. The recorder is put into operation with the desired number of tracks being recorded. At the end of each pass, the recorder senses the end of usable tape, reverses the tape's direction, and shifts the data signals to the next set of tracks. This operation continues until all tracks are used, at which time the recorder turns itself off.

Overlap. Overlap or recorder-to-recorder sequencing allows two recorders to be operated as one, continuously with no loss of data. As the running recorder reaches the "end-of-usable" tape, the waiting one starts and is up to speed as the running one shuts down. Tape is rethreaded on the now-waiting machine, making it ready for the sequence signal from the running machine as it reaches the "end-of-usable" tape. Data signals are usually connected in parallel to both machines.

Tape Loop Adapter. Tape loops are used for special instrumentation purposes. Loops can be associated with any type of capstan drive and can be a tensioned reversing-idler loop, a random bin loop, or a lubricated-tape circular loop. Loop adapters for conventional machines are available from some manufacturers. Storage from 5 to 150 ft is commonly available in most loop configurations. Tape loop adapters are ideal where data analysis problems require continual replay of data into the analysis equipment or for any application that requires recording and continuously reproducing data without the necessity of stopping and rewinding tape.

Signal Actuation. As the name implies, a signal starts the recording operation by exceeding some preset threshold level. In the event the recorder cannot get up to speed in time to catch the signal, a tape loop machine may be used also. With the loop operating constantly in the record, reproduce, erase sequence, and the loop delay between the record head and the reproduce head, the sudden appearance of a signal allows the loop-delay time for getting the data recorder up to speed. The signal is then dubbed to the data recorder from the reproduce head of the loop machine. Following the reproduce head on the loop machine, the tape is erased and the operation continues.

45.2-5 Magnetic Tape

About 30 years ago recording tape was a paper ribbon coated with a crude red oxide material similar to red barn paint. Today's tape is a thin ribbon of plastic, usually polyester (Mylar), on which an emulsion of highly refined magnetic oxides is placed. The binder, necessary to retain the oxides, also contains a variety of chemicals to reduce friction, improve the wetting characteristic of the oxide particles, reduce static charge, lubricate the tape, and so on. The tape backing or web is manufactured in wide (usually 24–26 in.) continuous rolls, and after the binder (containing the oxide) is coated on the web and cured, is slit into the desired tape widths by precision slitting machines.

An instrumentation tape must have a very uniform overall thickness and a very close tolerance on width. Normally the width tolerance (also known as the slitting tolerance) is +0.000 and −0.004 in. The coating should not rub off onto the heads or stick to other portions of the tape transport mechanism, even at high tape speeds. Oxide particle distribution must be constant throughout the thickness of the oxide layer as well as the length of the tape. Oxide particle size and magnetic characteristics must be extremely well controlled for tapes that are to be used in very short wavelength reproduction.

The mechanical characteristics of a tape are governed by the base material. A tape base of polyester will exhibit the following properties:

1. Its moisture absorption is the lowest of all base materials and, therefore, it will not exhibit layer-to-layer adhesion in moist climates.
2. Its fungus resistance is excellent, while that of a cellulose compound is exceedingly poor.
3. It is usable over an extremely wide temperature range, from −60° to 150°C.
4. Its dimensional stability makes it far superior to any of the other base materials.

The major disadvantage of polyester is that it will stretch rather than break when stressed beyond its yield strength. If this should occur, the entire portion of tape and the data it includes must be discarded. Present-day tape transports handle tapes very gently, however, and stretching is quite uncommon unless the transport is out of adjustment. As a result, polyester bases are almost universally used on instrumentation machines because the cost differential is quite small in relation to the value of the data to be recorded.

Types of Tape

The question of which tape to use for any specific application has consistently posed a problem for the users of instrumentation tape recorders. Optimum performance for any machine is usually specified for a particular type of tape, and the use of other tape can seriously affect the machine's ability to operate at the manufacturer's specifications. IRIG Document 106-73 covers the standard for 1/2- and 1-in. magnetic tape.

The problems of tape selection have been compounded with the advent of 1/4-in. instrumentation tape recorders. Unlike the 1/2- and 1-in. tapes, which are made mostly for instrumentation applications, the majority of 1/4-in. tapes are made for use in audio equipment. Given the list of all available 1/4-in. tapes, where a high-quality instrumentation tape can cost ten times that of a lower quality audio tape, the selection of the proper tape for a particular application can become an economic as well as a performance decision. The variables in the tape itself and how they can affect performance are combined in the following guidelines for ease of reference.

1. **Oxide coating.** Differences in oxide characteristics and coating thickness can cause up to an 8-dB peak or dip in the frequency response of a recorder. The use of those types of tape for which the recorder was designed is the best method to minimize the problem.

2. **Frequency response.** For medium bandwidth recorders (up to 0.25 mil wavelength), quality audio tapes can be used with satisfactory results. However, variation in frequency response characteristics and output levels will cause performance of different types of tape to vary greatly. The audio tapes that do perform consistently well compared with instrumentation tapes are as expensive as the instrumentation tapes; they offer the advantage of being readily available at local retail stores. A significant disadvantage of audio tapes is that the important tape parameters are not specified or guaranteed by the manufacturer. The continued existence of a particular type of audio tape is a doubtful assumption in such a dynamic industry.

3. **Effect on head wear.** Some of the cheaper audio tapes do not use an adequate method to bind the oxide coating to the tape base material. The oxide that flakes off as a result causes excessive wear to both tape heads and guides. In light of the cost of replacing recorder heads, the expense of higher quality tape may well reduce the lifetime cost of heavily used machinery.

The majority of tapes use a ferrous oxide particle, while a harder more permeable chromium dioxide is used where higher frequency response is the main objective. The use of this chromium dioxide tape on machines having mu-metal heads rather than ferrite heads will cause a rapid increase in the rate of head wear.

Table 45.4 lists some tape types, their characteristics, and how they perform in comparison with each other. Although the higher quality tapes command a premium price, they are not the total answer to every measurement problem. Instrumentation recorders that specify audio tape to meet published specifications are commercially available and should be considered in any analysis that is undertaken.

Signal Dropouts

A universal problem in direct record instrumentation recording is signal dropouts. A dropout is defined as a 50% (or greater) amplitude reduction in the reproduced data. While generally caused by poor head-to-tape contact during the record or reproduce process, they may also be caused by poor oxide particle distribution within the binder. The effects of poor head-to-tape contact may be reduced by cleaning the heads, guides, and parts of the recorder that the tape contacts, but there is nothing the user can do about a tape that has poor oxide distribution or a rough surface.

The signal attenuation resulting from poor contact (or a rough tape surface) between tape and the head is most pronounced in the shorter wavelengths, while absence of magnetic material in the binder affects all frequencies. The effect of poor head-to-tape contact is illustrated by the following formula and is termed spacing loss.

$$\text{Drop in playback level (dB)} = 54\frac{d}{\lambda}$$

where d is separation of the tape from the head in inches and λ is the recorded wavelength in inches.

Figure 45.25 shows attenuation as a function of the separation-to-wavelength ratio. This is a

TABLE 45.4. TAPE CHARACTERISTICS

Type	Audio		Instrumentation		
Parameters	Low Cost	High Cost	Medium Band	Wide Band	Chromium Dioxide
Head wear characteristics	Highly abrasive	Low	Low	Very low	High
Minimum recorded wavelength (λ)[a]	> 0.25 mil only	> 0.25 mil only	> 0.25 mil only	> 0.0625 mil	> 0.0625 mil
Signal to noise @ $\lambda > 0.25$ mil	Poor	Good	Good	Poor to good	Excellent
Signal to noise @ $\lambda < 0.25$ mil	Poor	Poor	Poor	Good	Good
Reliability of tape performance	Poor	Poor	Good	Good	Good

[a] Frequency = tape speed/λ.

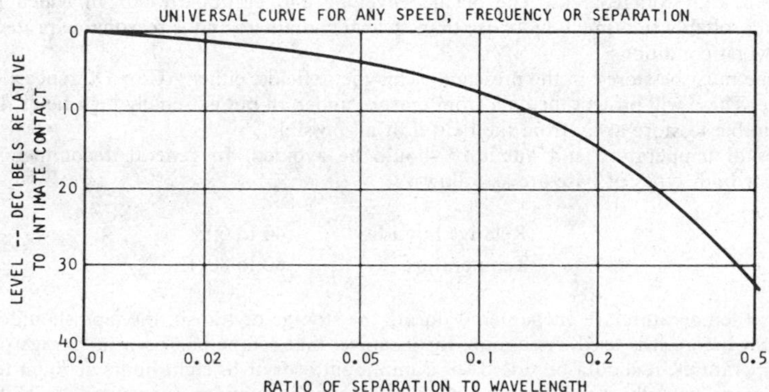

Fig. 45.25 Signal attenuation caused by poor tape-to-head contact in playback.

universal curve, applicable to any speed, frequency, and separation. It should be noted that this figure illustrates the loss in signal level in playback only, assuming that the head and tape make adequate contact during recording. Obviously a similar signal loss will occur if there is tape/head separation at the record head during the record process. In addition to such frequency-dependent losses, very high-bias frequencies are perturbed by spacing loss to the extent that all frequency components of a data signal will suffer attenuation or distortion.

In the playback of a low-frequency signal of 15 mil wavelength, a separation of 1 or 2 mils affects the level only slightly, but at a high frequency of 1-mil wavelength, even a half mil spacing results in a drop of more than 20 dB. (A 1-mil wavelength results when one records 15,000 Hz at 15 ips, 60,000 Hz at 60 ips, etc.) In instrumentation recording, a 6-dB or 50% drop in signal is considered critical. With a 1-mil wavelength, this takes place with 111 μin. (0.111 mil head-to-tape separation). Since a dust particle might easily approach this size, the possibility of spacing loss and the importance of keeping the recorder extremely clean are obvious. When the heads are not cleaned for extended periods of time, complete loss of high frequencies can easily occur, and every other part of the recorder may be blamed before the heads are finally cleaned.

Handling of Magnetic Tape

When tape is handled, as during splicing, the operator's hands should be thoroughly clean to prevent contamination of the tape by body oils and salts, which will cause the tape to pick up foreign particles. The use of sticky masking tape or cellulose ("Scotch cellophane") tape as splicing or tail-end holddown is strongly not recommended. Small deposits of the adhesive will stick to the tape.

Heads and guides should be cleaned to remove accumulations of foreign matter each time a tape is placed on a recorder. The machine manufacturer's recommended cleaning procedure should be followed. If extreme importance is placed on the data to be recorded, it should be remembered that a tape that has been run through the transport several times will have fewer dropouts than a new one, owing to the resulting polishing or smoothing action.

Cleaning

If there are signal dropouts arising from contamination by dust, carefully wipe the surface and backing of the tape with a lintfree cloth, such as a very soft chamois. To get rid of contamination that does not brush off easily, use a cloth slightly moistened with Freon TF. Aliphatic hydrocarbon-type solvents (heptane, gasoline, naptha) can also be used, but care should be exercised because they are flammable. Freon TF is nontoxic and nonflammable. Do not use carbon tetrachloride, ethyl alcohol, trichloroethylene, or other unknown cleaning agents because they may soften the oxide, deform the backing, or both.

Storage

When not in use, tapes should be placed on a precision reel, uniformly wound at a moderate tension, and then given protected storage. Recommended takeup tension for most instrumentation recorders using 1-mil tape is 4 to 5 oz for each 1/4 in. of tape width. The best method of storage is to place the reel of tape in a self-sealing plastic case and to store it on edge in a storage bin equipped with partitions between each reel. The plastic case protects from dust and sudden humidity and temperature changes. It also guards both tape and reel from damage in handling when the tape is transported between work and storage areas. The plastic envelope and cardboard box in which the tape is supplied will probably find much more use than special containers and is reasonable protection under all but the worst conditions.

If the tape must be stored in the presence of magnetic fields, either AC or DC, special containers are available. These will protect the data from erasure under all but extremely high fields. However, it is more desirable to store away from the field if at all possible.

Extremes of temperature and humidity should be avoided. In general, recommended storage conditions for both types of base are as follows:

Relative humidity	40 to 60%
Temperature	60 to 80°F

If extremes in temperature are encountered during the storage or transit, the tape should be brought to equilibrium before it is used. Assuming, for instance, that a tape has been in storage or transit at subzero temperatures, it should be stored for a minimum of four to eight hours at room temperature before it is used. Actually it will not regain complete equilibrium for approximately 16 hours. This time can be shortened by raising the ambient temperature, but the temperature should not exceed 100°F; otherwise condensation will form on the tape, which may prove to be a problem. Avoid using direct heat, such as from lamps or other spot heaters, to warm up a tape.

Temperature and humidity conditions during shipment are unpredictable, especially when shipment is made by air. Temperature and humidity vary so rapidly on takeoff that the stresses across (or through) the tape pack do not reach equilibrium before these conditions reverse during landing. Uneven stretching across the width of the tape results in edge ruffles when the tape is reproduced. Thermally insulated plastic containers, taped at the edges, reduce the effect of temperature and humidity shock and, depending on the time of the flight, the tape should be allowed to reach temperature equilibrium as described in the previous paragraph. Procurement of hermetically sealed tape shipping cartons may be necessary if the ultimate in tape safety is required.

If the tape has been subjected to combinations of temperature and humidity that cause between-layer sticking, it may be beyond saving. Do not place the recorder in fast forward when unrolling a reel suspected of abnormal storage or transit conditions; instead use a low-speed drive mode. All tapes that have been stored for long periods of time should be unrolled and rewound to ensure normal tape pack on the supply reel.

Physical Distortion

While most signal dropouts in instrumentation recordings are caused by specks of dust and other contaminants, which lift the tape away from the head, two other significant causes are dents and creases in the base material. Dents can be caused either by foreign particles becoming wound up tightly in the roll or by roughness in the surface of the hub on which the tape is wound. These may cause a permanent dent or crease in many layers of the tape, which cannot be stretched out flat as the tape passes over the head. Stresses in the roll that are sufficient to stretch the backing 5% will generally leave a permanent impression. Stresses below the 5% point are not normally permanent. Creases usually are caused by tape handling (i.e., threading, splicing, removing the tape from the guides, etc.) or by damage to the edges of the tape because of uneven winding.

Most causes of distortion of the base material can be eliminated by the use of a precision reel. A typical precision reel has straight or tapered flanges that are accurately machined and spaced to minimize the scattering of turns during winding. The flange design also affords greatly increased protection against dust and crushing of the tape edges. The hub has no threading slots that cause distortion of the inner turns. Instead it is often covered by a neoprene friction band to aid in threading. This ring acts as a cushion for the innermost tape layers and tends to minimize distortion from winding pressure and expansion-contraction stresses.

Erasure or Saturation

Magnetic properties of instrumentation tapes are stable indefinitely. Magnetic retentivity is permanent unless altered by magnetic means. It may be altered, for example, by magnetic fields from permanent magnets or electromagnets. These, very likely, will cause partial erasure if placed within a few inches of the tape.

This principle is used in the bulk-erasing process in which a whole reel of tape is demagnetized without unwinding. The fields necessary to produce complete erasure, however, are so intense that it is not likely that stray magnetic fields will cause trouble of this kind. Complete erasure (considered for purposes of this discussion to be reduction of signal to a point below the noise level of the system) does not usually take place unless the field is strong enough to exert a noticeable attraction for the tape. Slight erasure can occur, however, without any noticeable attraction or vibration.

Figure 45.26 illustrates the relation between field intensity and erasure as shown in experiments conducted with a typical AC bulk eraser. Some erasure is noticeable at a field intensity of only 100 Oersted. A 6-dB loss is generally considered critical because it represents a 50% reduction in signal strength. In some applications a loss of 1 dB might be critical.

Both unrecorded and recorded tapes should be kept away from electromagnetic bulk erasers and storage cabinets with magnetic latches. Unrecorded tapes should not be placed near DC magnetic fields, such as traveling-wave tubes or magnetron magnets, because they may become heavily biased or even create gross distortion in the record process (i.e., resultant signal-to-noise ratio will be reduced).

If parts of the recorder become magnetized, they can cause tape erasure, possible tape saturation, and signal degradation. As a preventive measure, periodic demagnetization of critical parts, particularly heads, is recommended.

To guard against accidental erasure of recorded tape during shipment, tape can be packed with bulk spacing (such as wood) between the tape and its shipping carton. Bulk spacing is effective in reducing the possibility of accidental erasure by fields encountered during transit because field

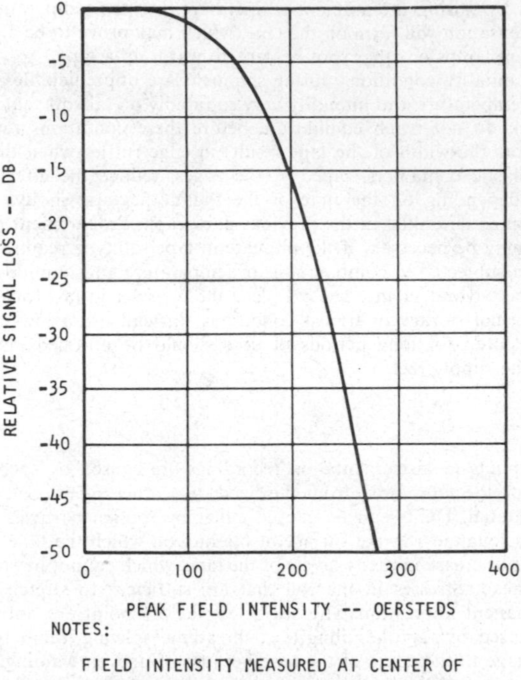

Fig. 45.26 Erasure as a function of field intensity.

strength varies inversely with the square of the distance. Assuming that no field strength greater than 1000 Oersted would be encountered during shipment (this is unverifiable but a reasonable assumption), 3 in. of bulk spacing would give adequate protection.

The special shielded container described in the paragraph on storage may be used in shipment if large stray magnetic fields are expected during shipment. Experience, however, indicates that standard shipping cartons are usually satisfactory.

Bibliography

Bannerjee, S., *Audio Cassettes: The User Medium*, UNESCO, 1977.

Bauer, B. B., F. A. Comerci, E. J. Foster, and A. J. Rosenbeck, "Audibility of Tape Dropouts," *JAES* **15**:147–151 (Apr. 1967).

Dolby, R. M., *Audio Noise Reduction System*, Dolby Labs, London, April 25, 1967.

Everest, A., *Handbook of Multichannel Recording*, Tab Books, Blue Ridge Summit, PA, 1975.

Jorgenson, F., *The Complete Handbook of Magnetic Recording*, Tab Books, Blue Ridge Summit, PA, 1980.

Levine, I. and D. E. Daniel, "Determination of the Recording Performance of a Tape from Its Magnetic Properties," *JAES* **7**:181–188 (Oct. 1959).

McKnight, J. G., "Erasure of Magentic Tape," *JAES* **11**:223–233 (Jul. 1963).

McKnight, J. G., "Mechanical Damping in Tape Transports," *JAES* **12**:140–146 (Apr. 1964).

McKnight, J. G., "Tape Reproducer Response Measurements with a Reproducer Test Tape," *JAES* **15**:152–156 (Apr. 1967).

Morris, A. H., "A New Magnetic Tape with Greater Dynamic Range," *JAES* **12**:36–39 (Jan. 1964).

Runstein, R. E., *Modern Recording Techniques*, Howard W. Sams, Indianapolis, IN, 1974.

Salm, W. G., *Cassette Tape Recorders, How They Work*, Tab Books, Blue Ridge Summit, PA, 1973.

Sarati, A. A., "A High Resolution Stereo Magnetic Head for Four Track Application," *JAES* **8**:243–245 (Oct. 1960).

45.3 LOUDSPEAKERS

Albert B. Grundy

45.3-1 Introduction

Understanding loudspeakers involves at least three separate fields of knowledge: (1) transducers, that is, the principles behind the elements that generate the sound, (2) enclosures, that is, how enclosures and baffles influence the sound of the system, and (3) room acoustics, that is, the influence the room or hall has on the response of the loudspeaker. We must know how to measure the response of a loudspeaker both anechoically and in a real room, how to interpret the measurements, and how to correct properly for deficiencies of response.

45.3-2 Transducer Types

Dynamic Loudspeaker

Cone Loudspeaker. Most transducers are electrodynamic motors. The most common of these is often called the PM speaker, which has become almost a generic name. We prefer the name "cone loudspeaker," to distinguish it from its nearest relative, the "compression driver."

The cone loudspeaker is constructed almost exactly like a moving coil microphone, in that they both depend on coils moving within a magnetic field. Referring to Fig. 45.27, a cone loudspeaker may be anywhere from about 1 in. in diameter, to as much as 30 in. The cone (or diaphragm) is bound to a soft outer material, called the surround, which flexes but also helps to act as a centering spring. At the center, the diaphragm is attached to the "voice coil." The "spider" keeps the voice coil centered in the magnetic gap and prevents it from scraping the sides. It also provides the elastic restoring force to return the cone to rest position.

Although the voice coil is free to travel forward and backward, its excursion should stay within the flux lines of the magnet or nonlinear distortion may result. One approach to guarantee linearity is to make the voice coil longer than the gap, so that the same amount of copper is always immersed in the field within a given range of excursion. Efficiency is sacrificed with this approach.

Some cone loudspeakers (called coaxial) have more than one diaphragm about a single axis. Although the circular cone is most common, rectangular and irregular shapes have also been used in high-quality cone loudspeakers. Cones were traditionally made of paper or pulp material, but increasing numbers are now formed of synthetics such as polypropylene or bextrene, or of light metals, usually aluminum. Two advantages of alternative materials over paper include lower susceptibility to humidity (the hygroscopic paper cone is more sluggish in humid environments) and increased stiffness without high weight.

In consumer applications, cone loudspeakers are used for both low-frequency and high-frequency reproduction, but usually they have not proved rugged or efficient enough for professional high-frequency applications. Under demanding conditions (e.g., high sound pressure levels) a compression

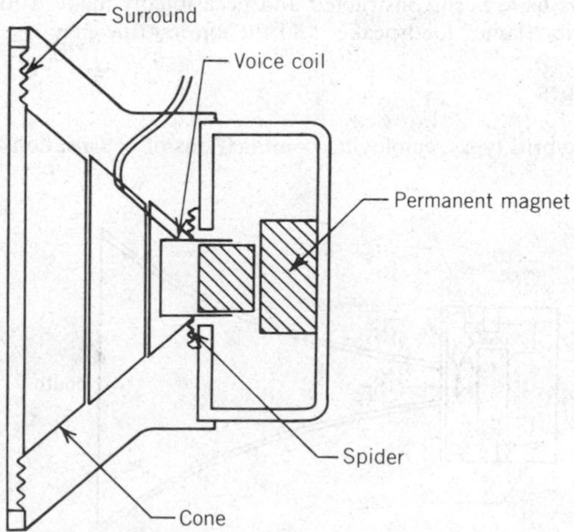

Fig. 45.27 Cone loudspeaker.

driver/horn combination is usually used instead of a direct radiator cone loudspeaker. However, with the increasing demand for high fidelity and lower distortion in the control room, we may find more rugged cone speakers replacing the traditional compression driver. A cone loudspeaker can also be mated to a horn to produce greater output.

Compression Driver. As a transducer, the compression driver is a cousin of the cone type. The primary difference is that the cone is replaced by a smaller diaphragm, usually made of aluminum. A compression driver is always mated to the throat of a specially matched horn, which couples the driver's acoustical impedance to the air (Fig. 45.28).

Magneplanar and Ribbon. Used exclusively in the home, the Magneplanar loudspeaker is a unique design employing a low-mass polyester diaphragm like the electrostatic loudspeaker, except that a static magnetic field is produced by many tiny bar magnets. Wires that act as a "voice coil" run through the thin polyester panel. The total thickness of this bipolar radiator is 2 in.

The ribbon loudspeaker is the analog of the ribbon microphone. It produces excellent high-frequency response, but at limited sound pressure levels. Although present-day technology permits ribbon tweeters and upper mid-range speakers, no one has produced a practical ribbon capable of reproducing low frequencies.

Electrostatic Loudspeaker

The electrostatic loudspeaker is the analog of the condenser or capacitor microphone. Two stationary plates operate on a push-pull basis to minimize second harmonic distortion. The two electrodes are composed of fine mesh to let sound pass (Fig. 45.29). Since each electrode is oppositely polarized, the inert thin diaphragm (often composed of clear polyester) is attracted alternately to either the front or back electrostatic pole. To prevent the diaphragm from breaking through the dielectric or air gap and "sticking" to either pole, a series of tiny spacers keep the diaphragm centered between the electrodes.

The dynamic range of the electrostatic loudspeaker is determined by the polarization voltage, which must be limited to about 1000 to 2000 V to prevent arcing. A relatively low dynamic range and a bipolar radiation pattern usually restrict the electrostatic loudspeaker to home use.

Piezoelectric Transducer

Analogous to the ceramic microphone, the piezoelectric speaker uses a transducer element usually consisting of polycrystalline ceramic materials. The diaphragm of a piezoelectric speaker is connected by a mechanical linkage to the transducer (Fig. 45.30). A horn then usually couples to the air. The minimal excursion capabilities of piezoelectric transducers have generally limited their use to high-frequency low-level applications. However, many sound reinforcement firms have produced practical sound columns made up of multiple piezo transducers. An economical price and a high internal impedance (enabling parallel connections) have made this professional application viable.

Other Types

Experimental transducers have been constructed and occasionally make it to the consumer market. These include the esoteric "flame" loudspeaker and the ion loudspeaker.

45.3-3 Enclosure Types

Most loudspeakers are hybrid types, employing combinations of or variations on the enclosure types described as follows.

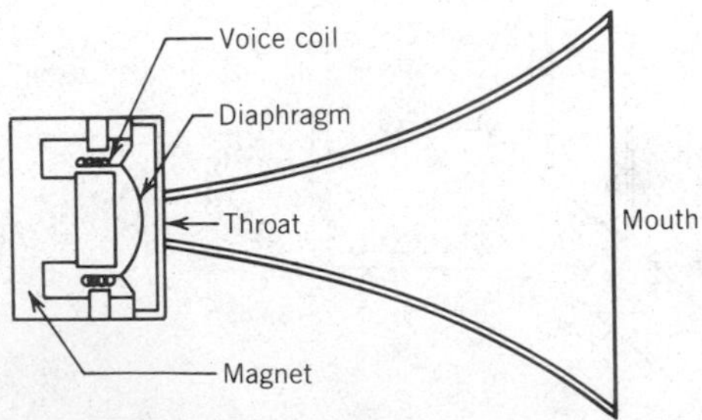

Fig. 45.28 Compression driver and horn.

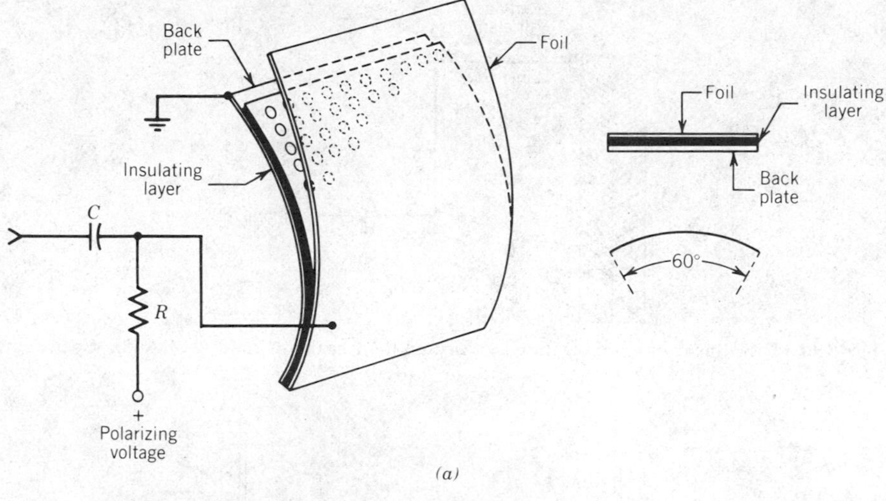

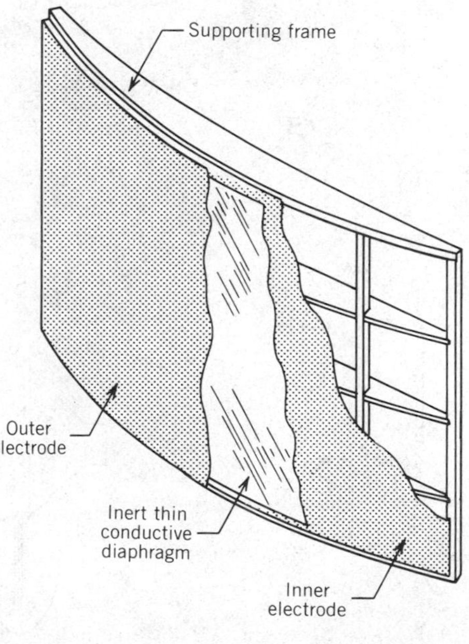

Fig. 45.29 (*a*) Electrostatic loudspeaker. (*b*) Cutaway of electrostatic loudspeaker showing internal construction. Courtesy of Pickering and Co. Inc.

Baffle

Referring to Fig. 45.31, we see that a loudspeaker in a free air environment must by necessity have a low-frequency cutoff, determined by signal wavelengths and the path length difference between the back and front of the driver. The addition of the baffle lengthens the path difference between front and back and lowers the frequency of cancellation. The ultimate baffle is, of course, the infinite baffle, such as a speaker mounted in a wall between two large rooms.

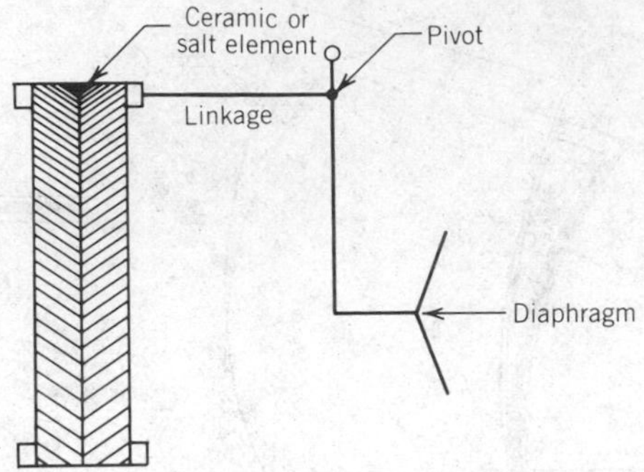

Fig. 45.30 Drive mechanism of a ceramic or crystal loudspeaker. Courtesy of Audio Cyclopedia.

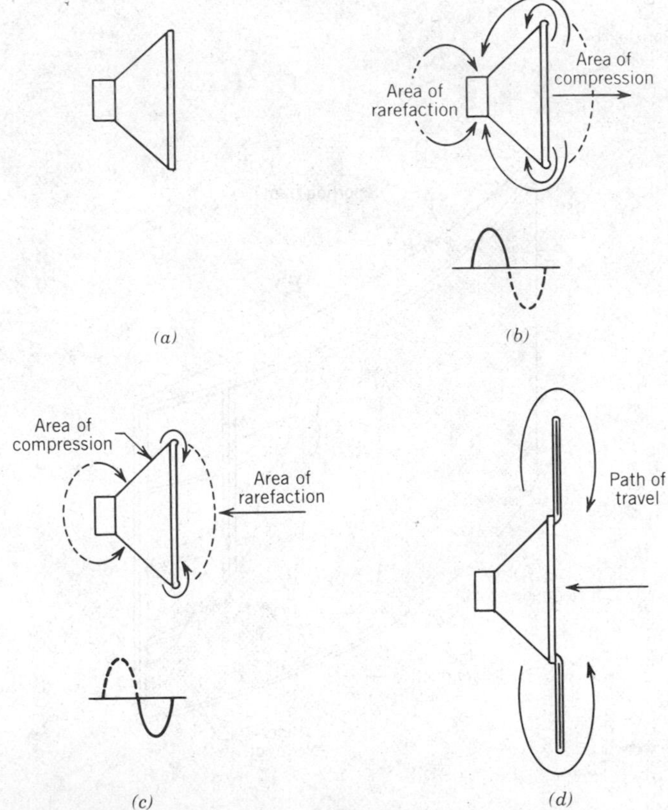

Fig. 45.31 Loudspeaker operated without baffle, showing how the front and rear sound waves cancel at the lower frequencies. (*a*) Diaphragm at rest. (*b*) Diaphragm moves forward. (*c*) Diaphragm moves backward. (*d*) Speaker mounted in baffle, showing the increased length of travel caused by the baffle. Courtesy of Audio Cyclopedia.

Horn Loading

Even before Rice and Kellogg invented the dynamic loudspeaker in 1925, various types of horns provided acoustical augmentation of what were essentially telephone receivers. Modern variants on the horn loudspeaker can be found in all professional systems designed to produce high sound-pressure levels with high efficiency. With increased knowledge of the physical principles involved, the overall distortion and resonant quality of horn loudspeakers have been considerably reduced. In fact, engineers are noting that the sound qualities of the large professional-type horn-loaded systems and of the smaller home-type direct radiator loudspeakers are actually converging. For example, while the efficiency and power handling of the home units are increasing, the frequency response and distortion of the professional units have also been improving.

Figure 45.28 shows an exponential horn. An exponential flare rate will provide a sharp rise in throat resistance at the lower frequencies for a given horn length. Other types of flare rates include parabolic, conical, and hypex. The factors that determine the length of an exponential horn are the flare rate, the intended low frequency of cutoff, the velocity of sound in air, and the throat area of the horn where the driver meets the horn.

Horn resonance, which causes frequency response aberrations and distortion, is one of the prime considerations of the engineer. If the bell of the horn is made large with respect to the lowest frequency to be reproduced, both resonance and distortion are reduced. It is also the responsibility of the designer to place an electrical crossover before the unit to limit the frequencies of input to within the design limits of the acoustic horn and driver.

Reentrant, or folded, horns follow much the same principle as many brass musical instruments, that is, they have a larger total area and a lower cutoff frequency but reduced physical length. (See Fig. 45.32.)

Vented Box

Other names for the vented box are bass reflex and ported enclosure. First introduced around 1937, the vented box enclosure was designed to increase low-frequency response without sacrificing efficiency, according to the principles of the Helmholtz resonator. Before about 1971 the design of these systems was actually more of a "black art" than a science. However, in that year, A. N. Thiele published a series of milestone papers on vented box loudspeaker design.[1,2] His work was further refined in 1972 and 1973 by R. H. Small.[3,4] Before Thiele and Small, the sealed box was thought to be inherently more linear than the vented box, although it is now known that both systems can be made with linear frequency response. Therefore, present-day speaker designers make the choice of design principles on the basis of empirical factors, including efficiency, port size, cabinet volume, driver resonance, low frequency cutoff, and harmonic distortion. Most of these factors can now be modeled on a computer long before the wood for the box is cut. Thanks to Thiele and Small, a properly designed vented box system no longer merits the nickname "boom box."

The principle of the port is to act as a resonator, extending the low-frequency response of the loudspeaker below the cone resonance frequency (Fig. 45.33). Usually the port is a simple hole of calculated dimensions cut into the front or side of the cabinet. Interestingly, the ported speaker can even be considered a subcategory of the horn-loaded speaker, for if we start with a ported design and construct an acoustical labyrinth within the enclosure whose cross section increases gradually and terminates at the "mouth" of the port, we have created a low-frequency folded horn!

Sealed Box

Said to be developed in the mid 1950s by Edgar Vilchur, the acoustic suspension enclosure is known for a deep well-damped low-bass response down to a limit defined by the interior volume of the

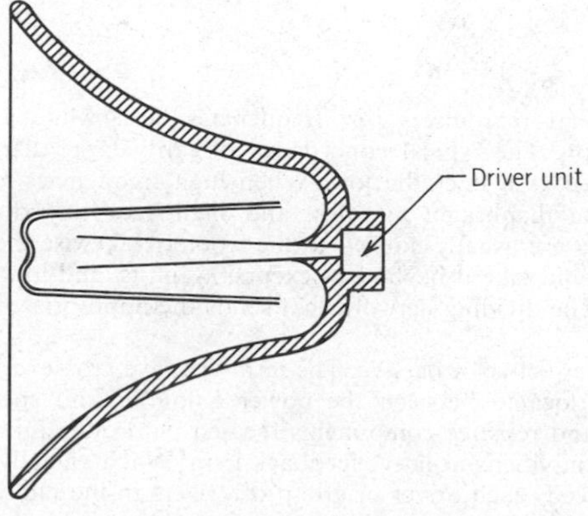

Fig. 45.32 Folded horn.

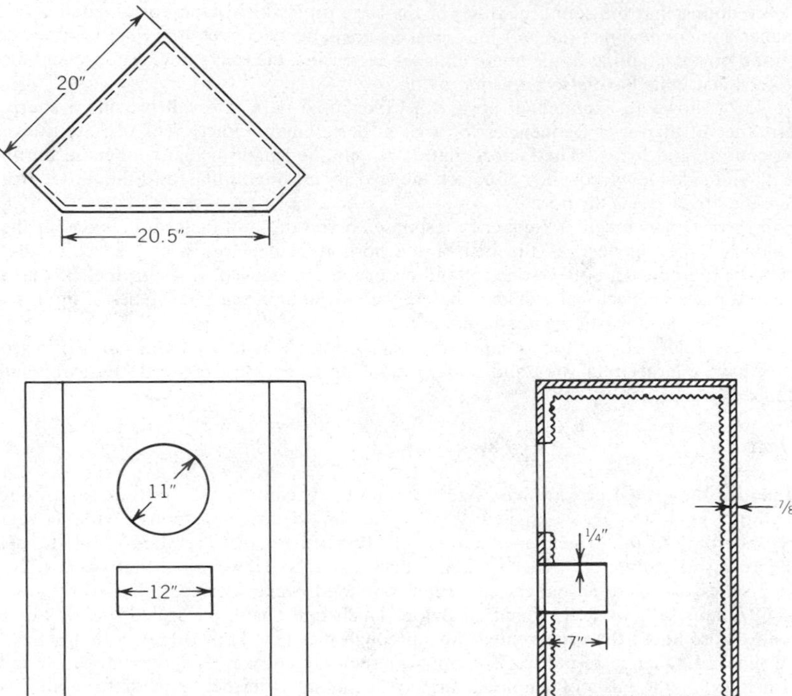

Fig. 45.33 Corner-vented baffle.

enclosure. It is a relatively low-efficiency device that only became popular when high-powered amplifiers became available. An acoustic suspension speaker uses the combination of a damped, sealed cabinet of a given volume with a speaker that has a very compliant surround.

This more compliant speaker is designed for almost a subsonic free-air resonant frequency, therefore it is almost completely undamped. Even so, the addition of a sealed, damped cabinet (with wool or fiberglass) acts as a spring, in fact a more linear spring than would be provided by an unaided low-compliance speaker. It has been claimed that the linear spring provided by the air within the sealed cabinet produces lower bass harmonic distortion than does the vented box system, but recent advances now make the two systems entirely competitive. The acoustic suspension speaker has little or no impedance or frequency response rise and a rapid falloff below the low-frequency cutoff. Efficiency is measured around 0.5 to 2%.

45.3-4 Crossovers

A woofer is a speaker that reproduces low frequencies best, while a tweeter is designed for high-frequency reproduction. The typical cone driver of fairly large diameter can reproduce high frequencies, but not as well as it does the low. When high frequencies are reproduced by a large driver, only a part of the diaphragm vibrates, and high-frequency dispersion is compromised. Therefore, high frequencies are usually not fed to the woofer. Likewise the tweeter can be damaged by low frequencies that would take it beyond its excursion limits, and low frequencies are prevented from driving the tweeter. The dividing network that sends the appropriate frequencies to each driver is called a *crossover*.

Crossovers may be either active or passive. The term "passive crossover" usually refers to reactive and resistive components located between the power amplifier and speakers. "Active crossover" usually refers to reactive and resistive components located before the input of a power amplifier or (rarely) within the power amplifier's negative feedback loop. When an active crossover is employed, a separate power amplifier feeds each driver or group of drivers in the multiple speaker system.

The slope rates and frequency responses of crossovers must be carefully calculated to produce the best overall response from the total system and guarantee that the speakers will not be overdriven. Designers have used slopes from 6 dB per octave to 18 dB per octave and occasionally beyond.

45.3-5 Polar Patterns

Figure 45.34 shows polar patterns of typical speaker drivers. When multiple drivers are used, these polar patterns can be rather complex. Designers must take steps to obtain a uniform frequency response within a prescribed horizontal and vertical listening area.

45.3-6 Measurement Techniques

Early designers and some of today's engineers often measure the frequency response of a loudspeaker outdoors, in order to remove the negative influence of reflections within a listening room. Many even bury the speaker cabinet until its face is flush with the earth, thus eliminating most audible and measurable diffraction effects. The anechoic chamber is designed for the same purpose, but most chambers of affordable or practical size are not linear at low frequencies owing to standing waves.

The response curves taken within the anechoic chamber rarely seem to correlate with what the ear hears, even within an "excellent" listening room. Many speakers that exhibit linear high-frequency response within the anechoic chamber sound very bright or shrill when played in a normal room. At least one reason for this is that the ear/brain combination integrates to some extent the direct sound from a loudspeaker with the sounds reflected off nearby walls and floor. If, for example, the off-axis high-frequency response of the loudspeaker does not parallel its anechoic response, rolling off much faster, the speaker will probably sound bright to the listener.

This illustrates only one of many problems associated with the measurement of loudspeaker responses and correlation of these with what the ear interprets as "correct." Also in controversy is the type of measurement microphone that should be used, for no omnidirectional microphone can have both perfectly flat on-axis and off-axis responses. Each speaker designer has necessarily arrived at his or her own set of standard correction curves that seem to correlate with the dispersion of the loudspeakers, the type of microphone used, and the reflection characteristics of the listening room. Despite recent advances in measurement techniques, including time delay spectrometry and fast Fourier analysis, an unequivocal scientific determination of "correct" frequency response lies somewhere in the future.

In fact, computerized measurement techniques now permit us to learn much more about loudspeakers than we can fully interpret. Some measurements that are more easily available to us include analysis of the vibrational nodes of drivers, energy-time curve (ETC) of a system, and phase response. Although even "good sounding" loudspeakers measure imperfectly on various empirical scales, we have not yet defined audibility acceptance windows for such parameters as ETC, phase error, and IM distortion.

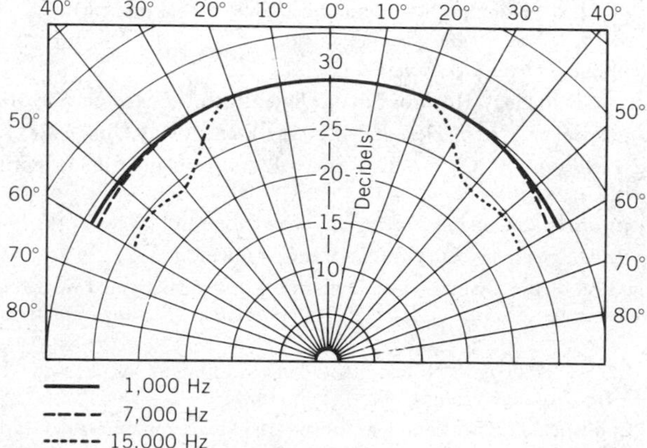

Fig. 45.34 Polar curves of a typical high-quality wide-range loudspeaker in an enclosure. Courtesy of Audio Cyclopedia.

References

1 A. N. Thiele, "Loudspeakers in Vented Boxes, Part I," *JAES* (May 1971).

2 A. N. Thiele, "Loudspeakers in Vented Boxes, Part II," *JAES* (Jun. 1971).

3 R. H. Small, "Closed-Box Loudspeaker Systems, Part I," *JAES* (Jun. 1972).

4 R. H. Small, "Closed-Box Loudspeaker Systems, Part II," *JAES* (Dec. 1972).

Bibliography

Belkin, B. G., "The Need for Transient Distortion Measurements in Loudspeakers," *Sov. Phys.-Acoust.* **18**(2) (Oct.–Dec. 1972).

Beranek, L. L., *Acoustics*, McGraw-Hill, New York, 1954.

Brittain, F. H., "Metal Cone Loudspeaker," *Wireless World* (Nov. 1952).

Broadhurst, A. D., "Loudspeaker Enclosure to Simulate an Infinite Baffle," *Acustica* **39** (1978).

Capetanopoulos, C. D., "Measurement of the Directivity Characteristics of Loudspeakers and Microphones in a Reverberant Enclosure," *Trans. Audio Electroacoust.* **AU-20**(2) (Jun. 1972).

Collems, M., *High Performance Loudspeakers*, Wiley, New York, 1978.

de Boer, E., "Theory of Motional Feedback," *IRE Trans. Audio* **AU-9**(1) (Jan.–Feb. 1961).

Ewaskio, C. A. and O. K. Mawardi, "Electroacoustic Phase Shift in Loudspeakers," *J. Acoust. Soc. Am.* **22**(4) (Jul. 1950).

Frankort, F. J. M., "Vibration Patterns and Radiation Behaviour of Loudspeaker Cones," *Philips Tech. Rev.* **36**(1) (1976).

Gilford, C. L., "The Acoustic Design of Talk Studios and Listening Rooms," *Proc. IEE* **106**, Part B(27) (May 1959).

Hanna, C. R., and J. Slepian, "The Function and Design of Horns for Loudspeakers—Reprint," *J. Audio Eng. Soc.* **25**(9) (Sep. 1977). "Discussion: The Function and Design of Horns for Loudspeakers—Reprinted from *Trans. AIEE*, Feb. 1924," *J. Audio Eng. Soc.* **26**(3) (Mar. 1978).

Harwood, H. D., "New BBC Monitoring Loudspeaker: Design of the Low-Frequency Unit," *Wireless World* (Mar. 1968).

Harwood, H. D., "Loudspeaker Developments," *Br. Acoust. Soc. Electroacoust. Air Water* (Jan. 1970).

Harwood, H. D., "Testing High-Quality Loudspeakers, Part II," *Audio* (Sep. 1971).

Harwood, H. D., "Some Factors in Loudspeaker Quality," *Wireless World* (May 1976).

Heyser, R. C., "Acoustical Measurements by Time Delay Spectrometry," *J. Audio Eng. Soc.* **15**(4) (Oct. 1967).

Holdaway, H. W., "Design of Velocity-Feedback Transducer Systems for Stable Low-Frequency Behaviour," *IEEE Trans. Audio* **AU-11**(5) (Sep.–Oct. 1963).

Hunt, F. V., *Electroacoustics*, Wiley, New York, 1954.

Klaassen, J. A. and S. H. de Koning, "Motional Feedback with Loudspeakers," *Philips Tech. Rev.* **29**(5) (1968).

Klein, S., "The Ionophone," *Onde Electrique* **32** (1952).

Klipsch, P. W., "A Low-Frequency Horn of Small Dimensions," *J. Acoust. Soc. Am.* **13** (Oct. 1941).

Linkwitz, S. H., "Loudspeaker System Design," *Wireless World* (May, Jun. 1978).

McLachlan, N. W., *Loudspeakers*, Oxford University Press, England, 1934, reprinted Dover, New York, 1960.

Olson, H. F., *Elements of Acoustical Engineering*, D. Van Nostrand, New York, 1940.

Olson, H. F., *Acoustical Engineering*, D. Van Nostrand, Princeton, 1957.

Rajkai, G., "Investigation of the Distortion of Dynamic Loudspeakers at Low Frequencies," *Seventh International Congress on Acoustics, Acoustical Commission of Hungarian Academy of Sciences, Budapest*, 1971.

Rice, C. W. and E. W. Kellogg, "Notes on the Development of a New Type of Hornless Loudspeaker," *J. Am. Inst. Elec. Engrs.* **44**(9) (Sep. 1925).

Shorter, D. E. L., "Loudspeaker Transient Response—Its Measurement and Graphical Representation," *BBC Q.* **1** (1946).

Shorter, D. E. L., "A Survey of Performance Criteria and Design Considerations for High-Quality Monitoring Loudspeakers," *IEE* Paper No. 2604 (Apr. 1958).

Small, R. H., "Efficiency of Direct-Radiator Loudspeaker Systems," *J. Audio Eng. Soc.* **19**(10) (Nov. 1971).

Small, R. H., "Performance Limitations and Synthesis of Direct-Radiator Loudspeaker Systems," *Proc. IREE* **34**(8) (Aug. 1973).

Stroh, W. R., "Phase Shift in Electroacoustic Transducers," *Acoust. Res. Lab., Harvard Univ. Tech. Mem.* **42** (18 Mar. 1958).

Thiele, A. N., "Loudspeakers, Enclosures and Equalisers," *Proc. IREE* **34**(11) (Nov. 1973).

Walker, P. J., "Wide Range Electrostatic Loudspeakers," *Wireless World* (May, Jun., Aug. 1955).

Werner, R. E., "Loudspeakers and Negative Impedances," *IRE Trans. Audio* **AU-6**(4) (Jul.–Aug. 1958).

CHAPTER 46
VIDEO RECORDING AND PLAYBACK

RCA CORPORATION

Consumer Electronics Technical Training, Indianapolis, Indiana

46.1 FUNDAMENTALS OF VIDEO TAPE RECORDING

46.1-1 Introduction

Most technologists are somewhat familiar with the way in which audio information is recorded on magnetic tape. The objective in recording video is essentially the same—to convert the picture and synchronization information into electrical energy and store it on a tape in the form of varying magnetic fields impressed in a metallic coating on the tape. It would appear to be possible to record video with the same technique as used in recording audio, until we consider the different characteristics of the two signals and the recording mechanism.

The objective with video is to record and play back video frequencies. Also certain other criteria must be considered during the design of a system. These basically include:

1. Good video resolution.
2. Acceptable record/play time (minimum tape usage).
3. Simplified operation (assuming use as a home entertainment product).
4. Reliable, exacting mechanisms.
5. Capability for playback through a television receiver.

In audio, we are dealing with signals in a frequency range from about 20 to 20,000 Hz. In video, the frequency range is from 30 Hz to 4 MHz (Fig. 46.1). This difference presents several problems when we consider the characteristics of a recording head and the associated system. We will find the head gap imposes the greatest limitation (influence) on a system for recording the video spectrum.

Figure 46.2 shows that the relation between the width of the gap in the record head and the length of the wave on the tape is reflected in the voltage output, and indicates that output is maximum when the gap is one half the wavelength, and that there is a point beyond which the system is ineffective. This becomes more meaningful when we apply the formula

$$\lambda = \frac{V}{F}$$

where λ = recording wavelength, in./cycle
V = tape speed, in./s
F = frequency of signal, cycle/s

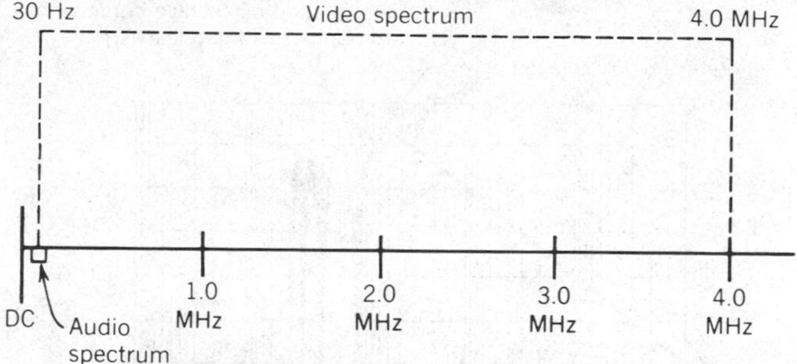

Fig. 46.1 Video frequency spectrum.

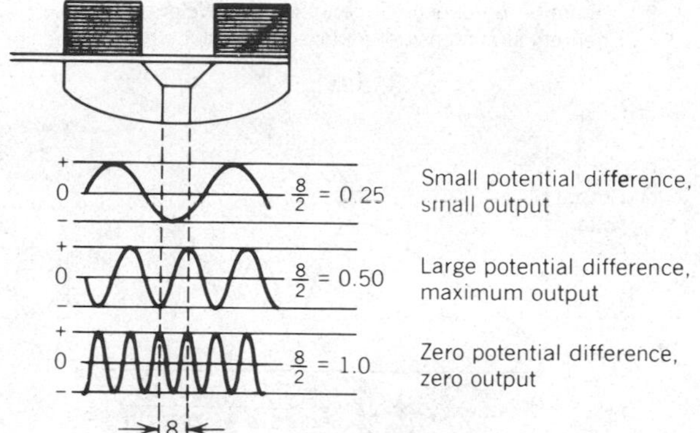

Fig. 46.2 Relationship of gap to wavelength.

If we consider a hypothetical audio situation using a tape speed of 15 in./s and a desired frequency response to 20,000 cycle/s, we find $\lambda = 15/20{,}000 = 0.00075$, or $\lambda = 0.75$ mil. In this situation, a system using a head gap of 0.75 mil would develop no output at 20,000 cycles/s, while a system using a head gap one half that size (0.375 mil) would develop maximum output at that frequency. A gap width of 0.3 mil is common in audio tape recorders.

If we apply this same principle to a video signal of a conservative 3.0 MHz, we find that we need a record head with a gap width of 0.0025 mil at 15 in./s for maximum output at 3.0 MHz.

However, if we use a 0.3-mil head and increase the tape speed sufficiently to record 3 MHz (which is possible), we would use 900 in. of tape per second, and tape supply becomes a serious problem.

A better way to increase the head-to-tape speed (and solve part of the problem) is to move both the head and the tape, rather than just moving the tape past a stationary head. A rotary head system is employed to increase the relative speed between head and tape. In conjunction with a selected head gap (narrowed), it is possible to record and play back the high video frequencies.

This technique can provide the necessary head-to-tape speed and still provide an acceptable length of recording time without consuming an impractical amount of tape.

Just recording higher frequencies is fine, but recall that we need to record a wide frequency range. This problem becomes evident when we examine the theoretical 6-dB per octave curve and the output characteristics, as represented in Fig. 46.3. Since we experience a 6-dB rise in output for every octave and the video range covers about 18 octaves, we find a 100-dB difference in output between the low end and the high end of the video range. This is too wide a range to accommodate, and simply increasing the tape speed would not accomplish our objective. It becomes evident that some other method is needed.

One way to solve the output problem is to move the necessary 4-MHz bandwidth up the frequency spectrum, selecting a range for acceptable output over our needed bandwidth. For example, 20 Hz to 4 MHz represents 18 octaves, but 4 to 8 MHz is still a 4-MHz spread and only represents one octave and a 6-dB output differential. (An octave is considered a relationship of one given frequency to another in a 2:1 ratio. Here if we choose 20 to 40 Hz as octave one, 40 to 80 Hz is considered the next octave, 80 to 160 Hz octave three, and the 2:1 relationship (octave) extends up the frequency spectrum each time the frequency is doubled.)

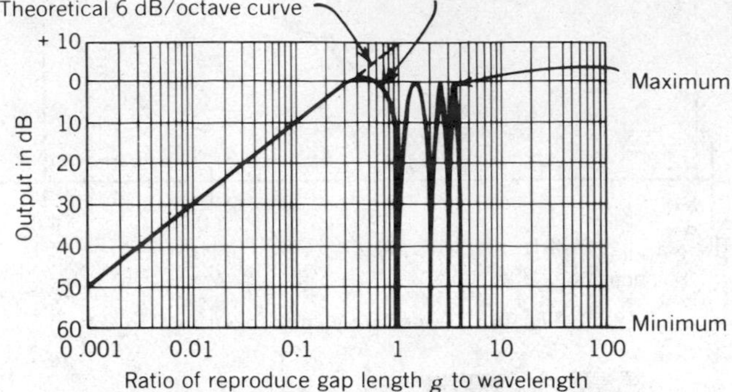

(*a*)

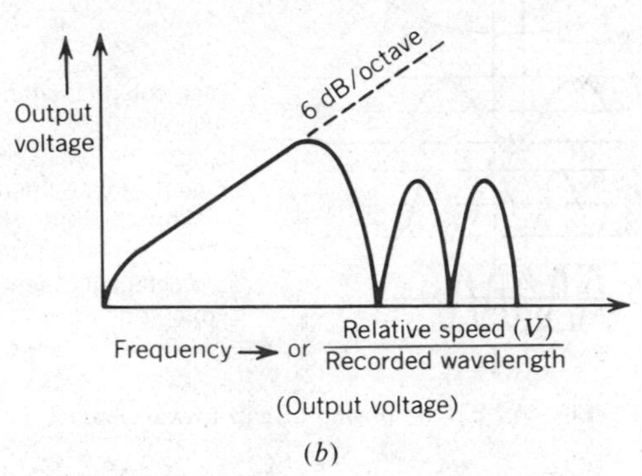

(*b*)

Fig. 46.3 (*a*) Theoretical head response. (*b*) Simplified output of a 6-dB per octave slope.

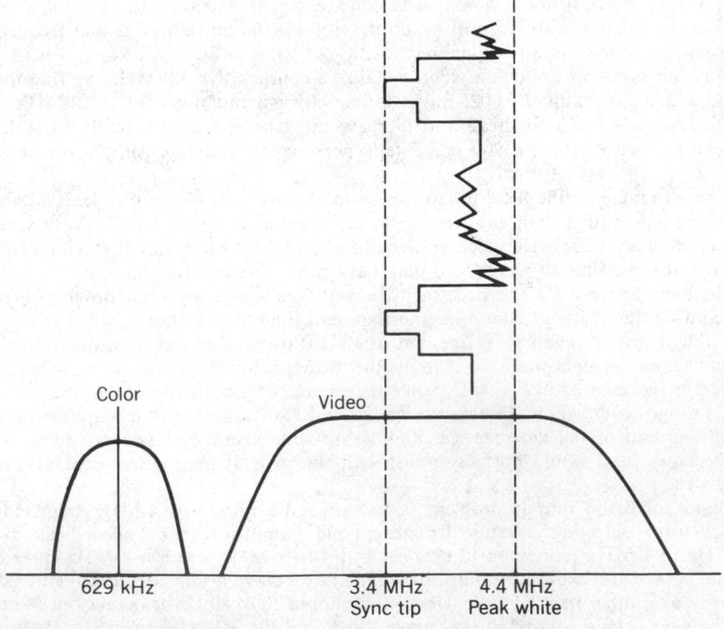

Fig. 46.4 Typical video spectrum recorded.

The method most often used (presently) is to impress a carrier on the tape and modulate the frequency of the carrier with the video signal (see Fig. 46.4). Carriers are generally in the 2-MHz range. The one shown in Fig. 46.4 uses a carrier frequency of 3.4 MHz, with sync tips appearing at 3.4 MHz and peak whites at 4.4 MHz. Figure 46.4 also shows a fairly common color record system. Chroma information is down converted to a band in the 629-kHz area of the spectrum before recording.

The basic system just discussed solved the problem of output differential for frequency response in the 4-MHz video range.

46.1-2 Mechanical System

The first practical video tape recorder was developed for the broadcast industry. It was designed to very tight specifications, and after a few "growing-pains" developed into a very satisfactory system which is still in use today. Much of the programming we see on television is coming from this type of machine. The system uses four record heads mounted on a wheel that rotates so that the heads scan tracks across the tape (similar to simplified paths shown in Fig. 46.5). It uses a 2-in.-wide tape, traveling at a speed of 15 in./s. While this system does an excellent job, it is much too complex and expensive for home use.

The quad-head machine records about 16 lines of video on each track, and 16 tracks are required for one complete field of a TV picture (262 plus lines). For home use, it would be much more desirable to have a method that would record one complete field, on one track. Slowing the headwheel down would be unsatisfactory, however, for the head-to-tape speed would be too slow to provide adequate frequency response.

Thus the helical scan system was developed for institutional and home use. Instead of having four heads traveling across the tape at almost right angles, the tape is wrapped around a rotating drum containing one or two heads and traveling at an angle to the direction of tape travel (see Fig. 46.6). By selecting a cylinder head diameter, degree of tape wrap, tape speed, and cylinder head speed, it is possible to record a long-enough track (across the tape at an angle) to hold a complete TV field and still maintain adequate head-to-tape speed. One helical scanning concept used a "full" tape wrap around the drum (see Fig. 46.7), and used one record head; most subsequent machines use a half-wrap with two heads (Fig. 46.8). Several different methods are used to form the head-to-drum tape path and/or tracking. Supply and take-up reels can be placed at different heights, with the drum level; the cylinder head drum itself can be slanted (angle mounted), with the tape reels parallel. This slant-drum arrangement is probably the most common, offering parallel "loading" and several other advantages.

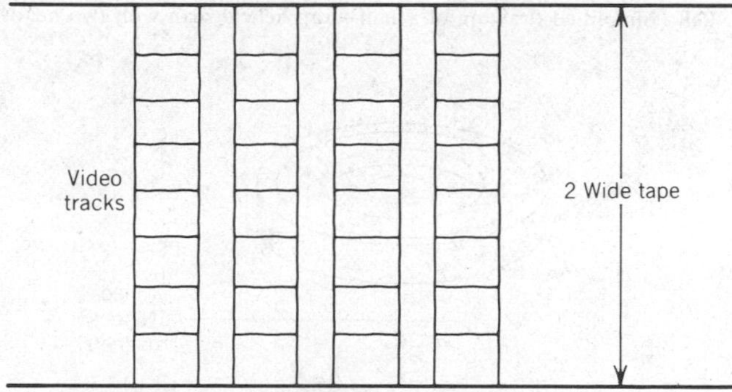

Fig. 46.5 Track path of a quad-head recorder.

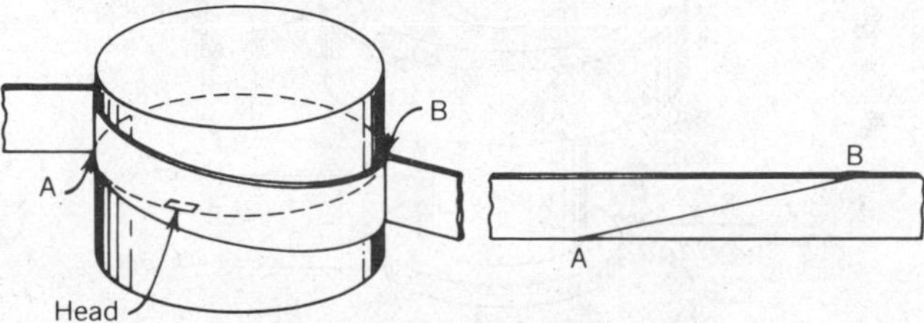

Fig. 46.6 Helical scan recording track.

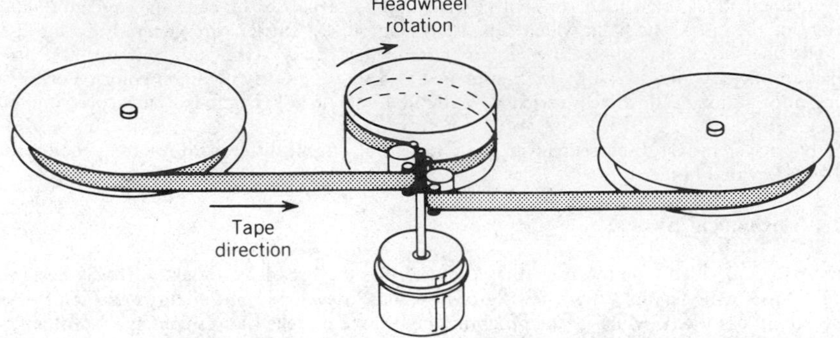

Fig. 46.7 Simplified drawing illustrating full-wrap helical scanning concept.

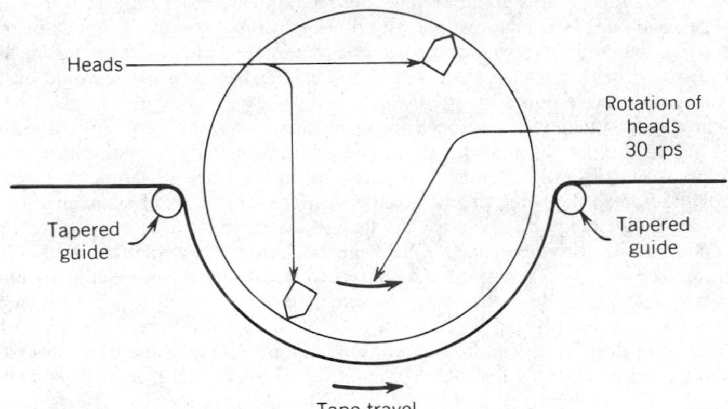

Fig. 46.8 Simplified drawing of a half-wrap helical scan with two heads.

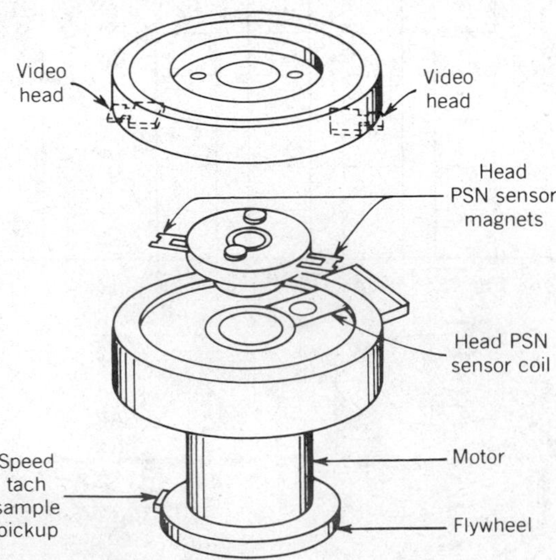

Fig. 46.9 Partially exploded view of a typical head drum assembly.

A typical cylinder head drum arrangement is shown in Fig. 46.9. The record heads are positioned 180° apart and are made to protrude slightly beyond the surface of the drum. The energy is transmitted to (or from) the heads through the use of a rotary transformer as an integral part of the drum assembly. The wheel rotates at 30 rps (Hz) so that each head records a complete field, and one complete rotation of the headwheel records a full frame (2 fields) of TV information (525 lines).

The resultant recording is positioned on the tape as shown in Fig. 46.10*a*. Each track encompasses one TV field, which consists of $262\frac{1}{2}$ horizontal lines including all associated information such as luminance signals, chroma signals, burst pulses, sync pulses, and any other material that may have been transmitted on the signal except the audio. The sound carrier is demodulated in the system and is recorded on the tape as audio, as shown in Fig. 46.10*b*. The control track (shown in Fig. 46.10*b*) consists of pulses generated and recorded by the VCR system and used to keep the video heads properly aligned with the recording on the tape during playback. The various heads needed to accomplish all of this are positioned as shown in Fig. 46.11. As in audio tape recording, the tape passes over a full erase head to remove any residual magnetism which may be on the tape. The tape next passes the rotating video heads where the picture information is recorded and then moves on to the audio head assembly (two) where audio and control track information is recorded. In the playback mode, the same video, audio, and control heads are used to recover the signals from the tape.

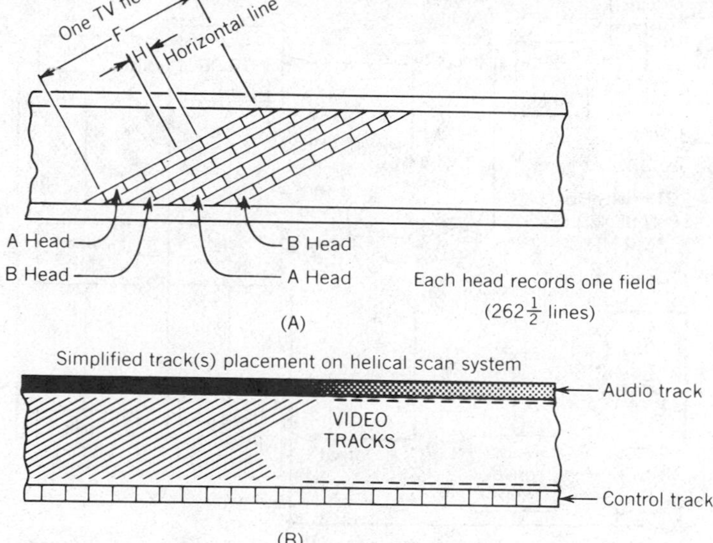

Fig. 46.10 Three tracks are generally recorded: audio, video, and control.

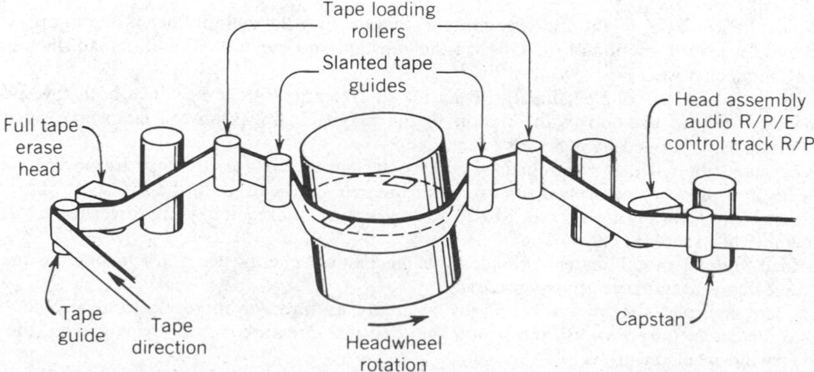

Fig. 46.11 Typical head arrangement in a helical scan VCR.

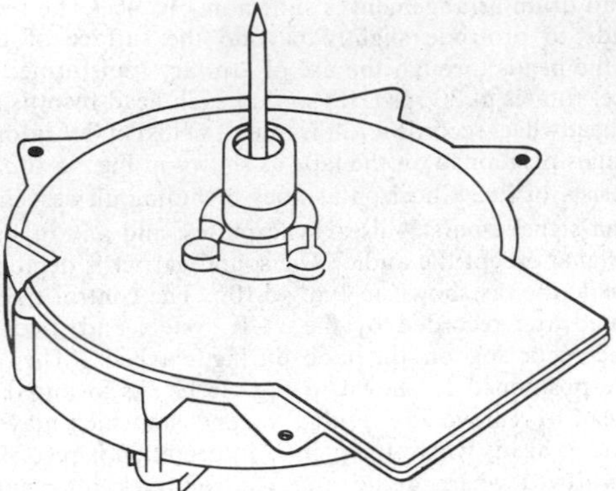

Fig. 46.12 Direct drive (DD) capstan motor.

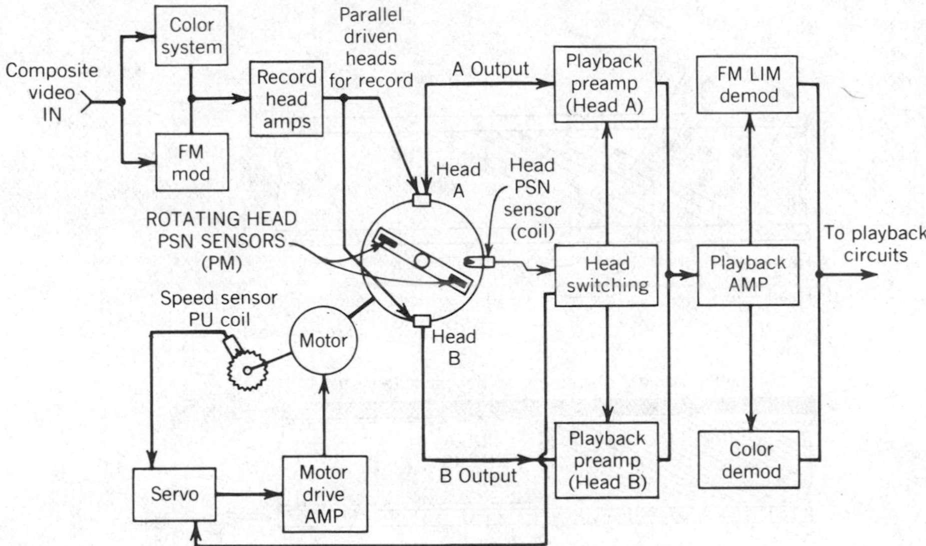

Fig. 46.13 Diagram illustrating typical interface of play/record electronics.

A typical helical VCR might use three motors: one to drive the cylinder head, one to provide the forward and/or reverse motion of the tape via the capstan, and one to load and unload the tape into or out of the mechanism.

The cylinder head is coupled directly to its motor (refer to Fig. 46.9). Since both the speed of rotation and the head position (with relation to the playback tracks on the tape) are critical, the cylinder motor is controlled by a servo system.

The capstan motor is also controlled by a servo system that causes the tape to move across the video heads at a precisely controlled rate. As with the cylinder head, both the speed of rotation and the position of the capstan are critical. Most capstan motors utilized today are direct-drive (DD), as in the cylinder head motor (Fig. 46.12).

The load/unload motor merely provides the mechanical energy necessary to operate the load /unload mechanism and is relatively uncritical.

A modern helical VCR may also employ as many as three or more electronically controlled solenoids for such purposes as activating the capstan pressure roller or engaging the fast-forward/rewind mechanisms.

A simplified block diagram of the electronics system in a two-head VTR (Fig. 46.13) indicates some of the complexities involved in accomplishing this objective. As previously mentioned, the heads serve a dual purpose, both record and playback.

46.1-3 Recording System

The heart of the VCR is the video record/playback heads. As mentioned before, the head must be capable of recording and reproducing signals in the 4-MHz region. In spite of the increased head-to-tape speed that the helical scan system provides, the ability to reproduce those frequencies requires a rather small head gap.

Modern video heads are made of ferrite material formed something like the one shown in Fig. 46.14. Typical heads are made of material about 10 mils thick and have a gap of 40 to 50 μm.

The video information is impressed on the tape in the form of a frequency modulated squarewave carrier. Figure 46.15 shows how the magnetic impression would look when the carrier is modulated by a horizontal sync pulse. The frequency at which the pulses occur is a function of the intensity of

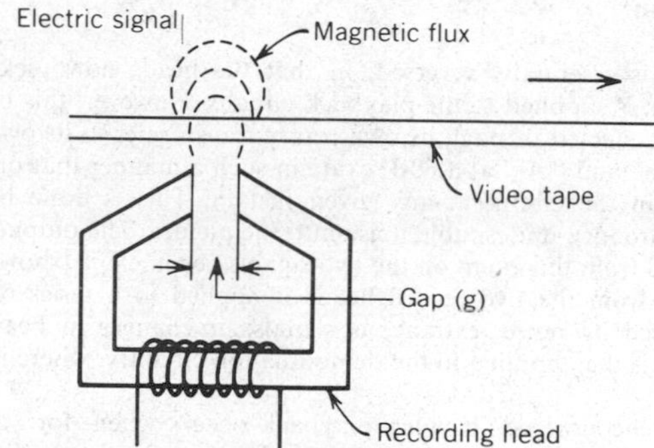

Fig. 46.14 Sample of video head form.

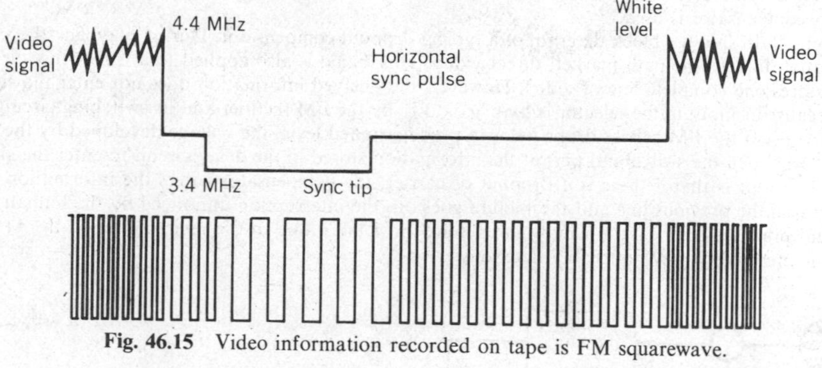

Fig. 46.15 Video information recorded on tape is FM squarewave.

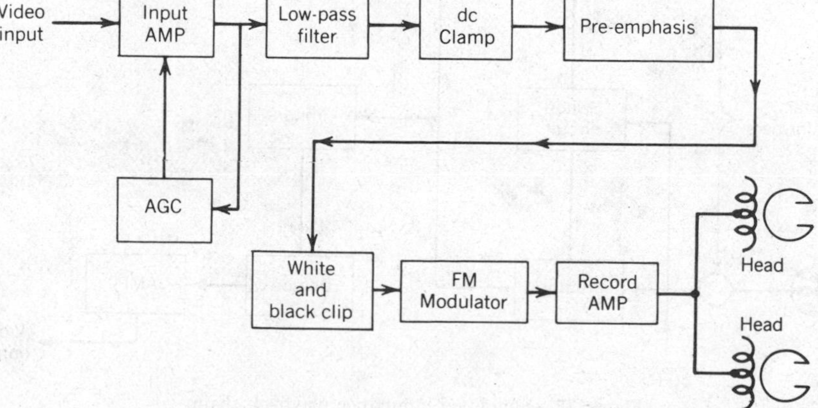

Fig. 46.16 Simplified block diagram of a luminance recording chain.

the video information. Whites are at the highest frequency (4.4 MHz) and sync tips (blacks) are at the lowest frequency (3.4 MHz), as shown in Fig. 46.4. The period of time a given repetitive rate lasts is a function of the horizontal width (or frequency) of the video information.

Figure 46.16 is a simplified block diagram of the processing to which the video information is subjected in preparation for recording. A standard video signal is applied to the input amplifier. The gain of this stage is controlled to assure a stable output. The signal then passes through a lowpass filter to remove any information above 3.8 MHz, and the clamp circuit restores the DC level of the signal, referencing it to sync tips. A preemphasis circuit then emphasizes the high-frequency response and a clipping circuit limits the amount of overshoot developed at both extremes of the spectrum. The signal is applied to an FM modulator and then to a record amplifier that drives (in parallel) the record heads through rotary transformers.

46.1-4 Playback System

In playback, the process is essentially reversed, in that the heads now pick up the information recorded on the tape and it is supplied to the playback circuits. However, this time the heads are not in parallel. Each head is connected through its own rotary transformer to its own preamplifier. These amplifiers are switched "on" and "off" at a 60-Hz rate in such a manner that only the head that is in contact with the tape is in the circuit at any given instant. This is done because the head not contacting the tape will introduce undesirable noise into the picture. The outputs of the preamplifiers are connected together and from this point on the two signals become one (shown in Fig. 46.17).

The combined output from the two record heads is applied to a series of limiters, where any amplitude variations caused by noise, extraneous signals, or changes in head-to-tape contact are removed. The clean signal is then applied to the demodulation circuitry, where it is converted back to video.

One of the problems encountered in video playback occurs when, for some reason, the head breaks contact with the metal-oxide coating on the tape. This may be due to dirt on the tape or from minute defects in the coating. The resultant phenomenon is called a dropout, and if ignored can create unacceptable flashing in the picture. To compensate for this situation, a circuit called a dropout compensator is used.

Figure 46.18 shows a block diagram of a typical dropout compensator. During playback the video information follows the path marked direct video. This signal is also applied to a time delay circuit, which stores one complete line of video. However, the delayed information does not enter the signal path because its input to the selector is switched "off" by the FM rectifier and its switching circuit.

In the event the FM carrier drops below a predetermined level, the voltage developed by the FM rectifier will open the switch and permit the information stored in the delay circuit to enter the signal path. The result is that if there is a dropout of carrier, the compensator enters the information that appeared on the previous line and the picture goes on, the interruption unnoticed by the human eye. In actual practice, compensation may go on for 3 or 4 lines, due to the recirculation path. At that point picture quality may noticeably deteriorate.

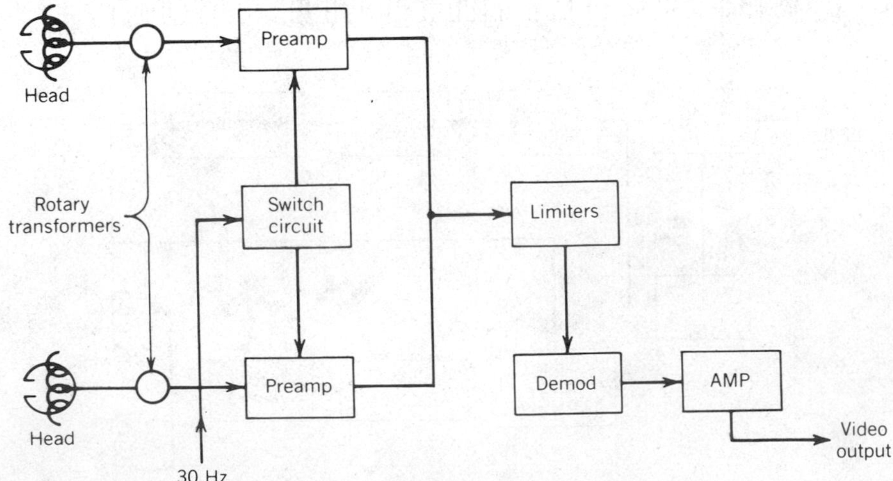

Fig. 46.17 Simplified luminance playback chain.

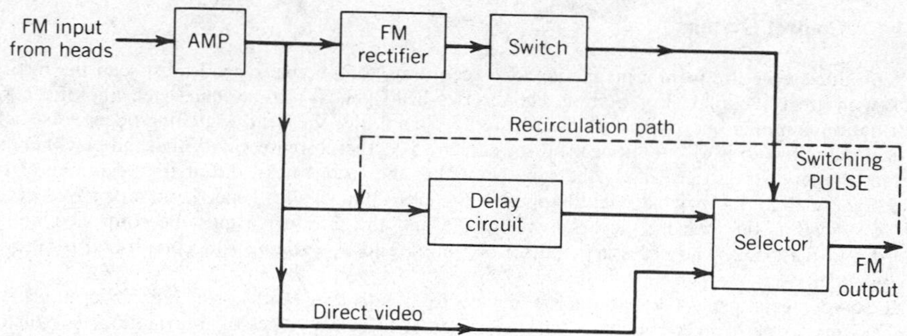

Fig. 46.18 Block diagram of a dropout compensator.

46.1-5 Color System

Most helical-scan video recording systems use some method of separating color information from luminance information before the signal is recorded on the tape. A common method is to convert the chroma to some other frequency. The system depicted in Fig. 46.19 converts the 3.58-MHz subcarrier to 629 kHz, with the color sidebands appearing around that point. In this system steps must be taken to limit the low-frequency excursion of the luminance FM signal to prevent interference patterns in the color portions of the picture.

The procedure for converting the 3.58 MHz signal to 629 kHz is shown in the block diagram in Fig. 46.20. The chroma is separated from the composite video signal before it enters the luminance recording chain. The 3.58-MHz signal is then beaten against a 4.20-MHz oscillator, and the resultant 629-kHz frequency, containing all the chroma information, is recombined with the luminance FM signal and recorded on the tape.

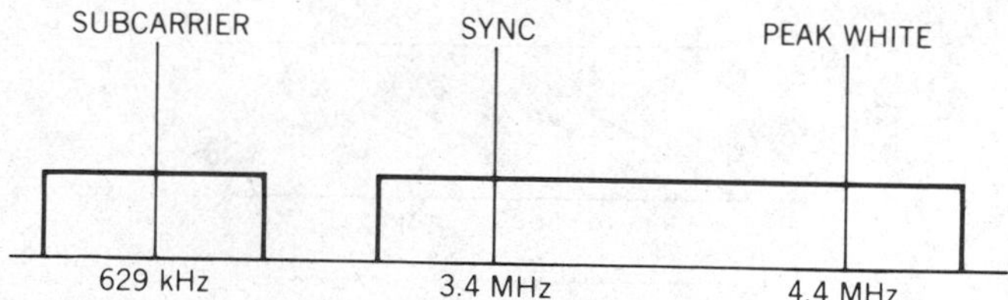

Fig. 46.19 Color carrier frequencies on tape.

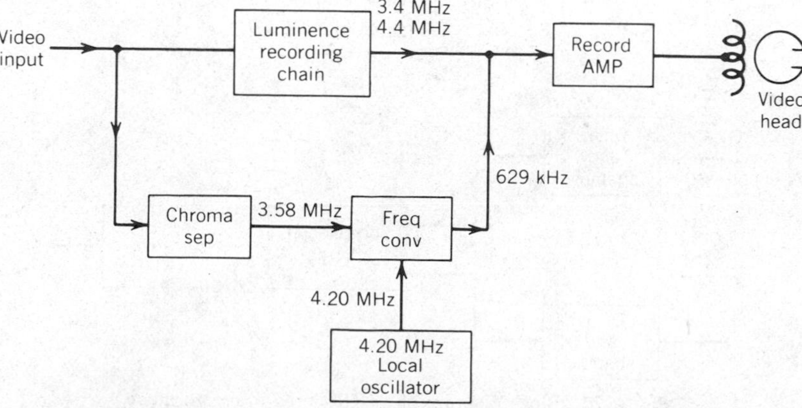

Fig. 46.20 Principle of recording with the converted subcarrier method.

46.1-6 Control System

To reproduce a picture from a magnetic video recording, the playback head must scan the recorded track with great accuracy. For example, in the two-head helical scan system, each line of recorded information contains one complete field of the TV signal and the head switching occurs just before vertical blanking (so it will not appear on the screen). It is therefore necessary that the playback head contact the tape (track) in exactly the same spot that the record head did at the beginning of each track. Figure 46.21 shows what might happen in the event that the head speed and tape speed become out of sync. For this reason, the headwheel drive and the tape drive must be controlled both on record and playback. The necessary control of timing and positioning is accomplished by use of a servo system.

A servo mechanism is an automatic device for controlling and correcting performance of a mechanism. To perform its function, a servo needs at least two types of information: (1) an input from the machine to determine how it is performing (feedback) and (2) a constant (reference) source to determine the degree to which the machine deviates from the desired performance.

In record mode, the control system starts with the vertical sync pulse of the incoming video signal (see Fig. 46.22); the pulse is processed and becomes the reference input of a comparison circuit. The other input of the comparison circuit receives a feedback signal developed from the drive motor. The comparison circuit compares the two inputs and develops an output that is representative of the difference between the two. This output is used to control the drive motor, and therefore any change in the relation will cause a corrective change in drive. This arrangement could be used to control either the headwheel or the tape drive (capstan). In either case, the other drive would be at a predetermined constant speed.

At the same time the signal is being recorded and the vertical sync pulse is controlling one drive motor, that same sync pulse is applied to a frequency divider that produces 30-Hz pulses. These are used to trigger a circuit that records 30-Hz pulses on the control track of the tape (Fig. 46.23).

In playback mode, the control system is essentially the same except that the reference input is now the 30-Hz pulses taken from the control track. This time the processing circuit provides for some external control (variable tracking) of the timing of the pulses before they are applied to the comparison circuit (see Fig. 46.24). Theoretically, a tracking control should not be necessary, but in

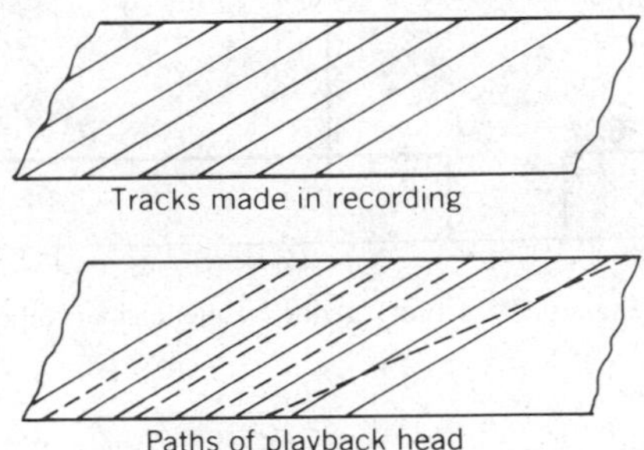

Fig. 46.21 Tracking in a VCR.

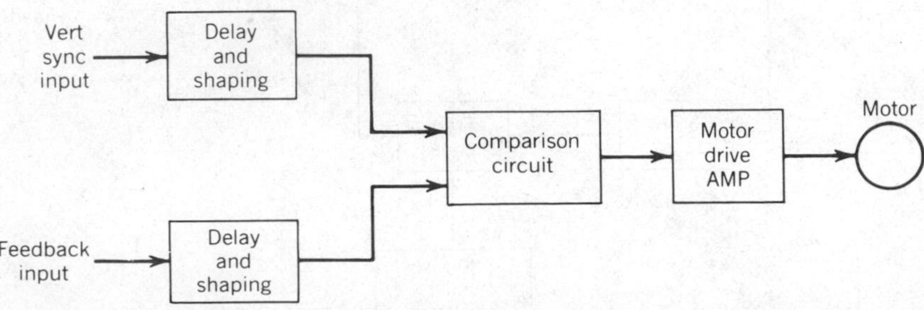

Fig. 46.22 Block diagram of a VCR servo in record mode.

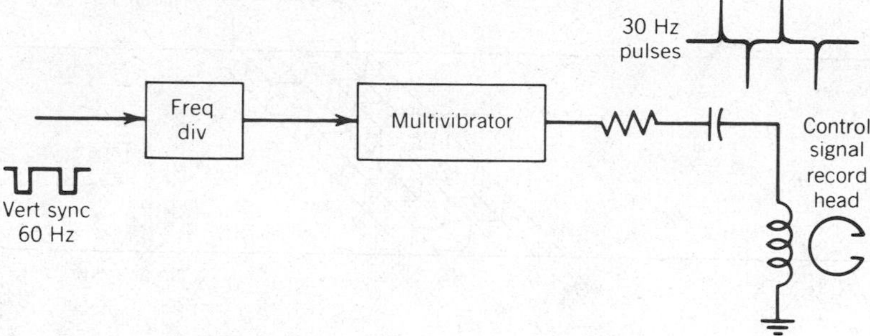

Fig. 46.23 Block diagram of a control pulse recording.

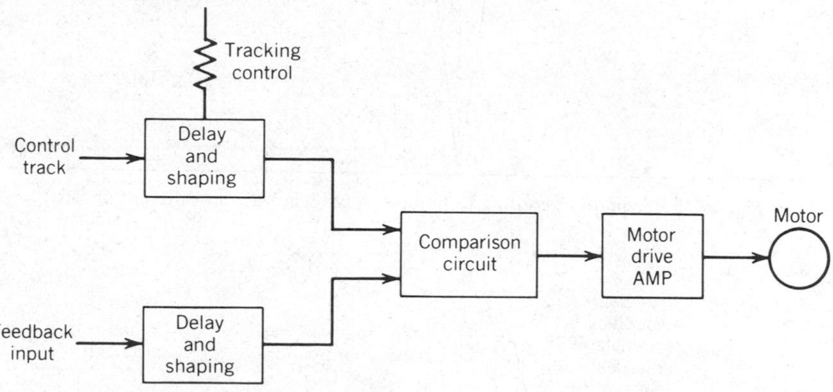

Fig. 46.24 Block diagram of a VCR servo in play mode.

practice it is sometimes desirable owing to tape stretching or when reproducing a recording made on another machine (tracking is generally adjustable to the user via a front-panel control).

A problem develops on playback when a head does not scan the recorded track perfectly. If the head deviates from the track, it will deliver reduced output, but can also pick up signals from an adjacent track causing cross talk. One of the ways of minimizing cross talk is to put a guardband between the tracks. This is very effective, but requires more space on the tape and therefore reduces the recording time capability. Another recording method that eliminates the guardband has been devised. It is called azimuth recording.

46.1-6 Azimuth Recording

In most magnetic recordings it has been customary to position the gap of the record head at right angles to the direction of scan. In playback, it is important that the pickup head gap be at that same precise angle, or reduced frequency response and reduced output will result.

In azimuth recording, the two heads of a two-head system are deliberately placed at different angles with relation to the direction of scan. The system depicted in Fig. 46.25 places the head gaps at 84° and 96°, respectively. This makes it possible to record with no guardband between tracks because, if a head deviated over the wrong track, the pickup head would be at a 12° angle from the recorded signal and cross talk would be minimal. This does complicate the servo system, however, because not only must a head scan a track accurately, it must scan the correct track accurately. The Video Home System (VHS) uses the azimuth recording technique and this makes it possible to record four hours of program material in a relatively small tape cassette.

The way in which the tracks are recorded on the tape is shown in Fig. 46.26. You will notice that there is a guardband when recording in the two-hour or standard speed mode. However, when the machine is operated in the LP (four-hour) mode, the tracks actually overlap with no undesirable effect.

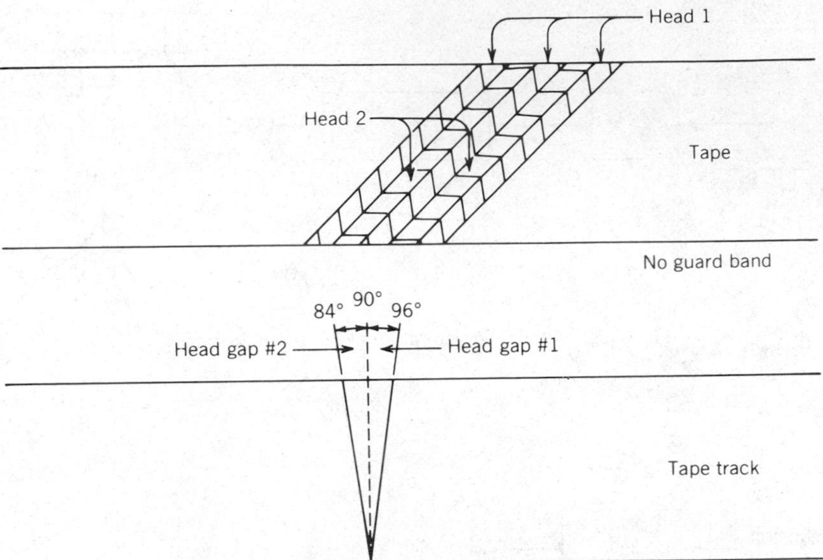

Fig. 46.25 Azimuth recording technique places head gap at different angles.

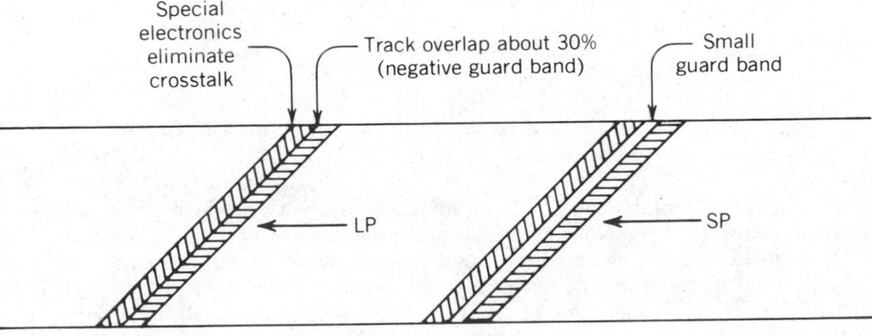

Fig. 46.26 Track recording in RCA VCR instrument (Video Home System).

46.1-7 Summary

The information just discussed was intended to present only a brief overview of video tape recorders.

Video recording techniques and systems have been improved at a rapid pace. Helical scanning, azimuth recording (head gaps at different angles), FM modulation of luminance information, and direct recording of down-converted color information (usually rotary phase) are common factors.

Various input and output jacks serve for external interface of other instruments, accessories, signals or controls; cameras are among the accessories normally considered optional.

Bibliography

Bensinger, C., *The Home Video Handbook*, 3rd ed., Howard Sams, Indianapolis, 1982.

Chesire, D., *The Video Manual*, Van Nostrand, New York, 1982.

Clason, W., *Dictionary of Television and Video Recording*, Elsevier, New York, 1975.

Effrein, J., *Video Tape Production and Communication Techniques*, Tab Books, Blue Ridge Summit, PA, 1970.

Effrein, J., *Home Video Yearbook*, Knowledge Industry, While Plains, NY, 1981.

Kortman, P., *Handbook of Video Cassette Recording*, Van Nostrand, New York, 1978.

Kybett, H., *Video Tape Recorders*, 2nd ed., Howard Sams, Indianapolis, 1978.

Lanzendorf, P., *The Video Taping Handbook*, Harmony, Prospect, KY, 1983.

Lenk, J. D., *Complete Guide to Video Recorder Operation and Service*, Prentice Hall, Englewood Cliffs, NJ, 1983.

McGinty, G. P., *Video Cassette Recording, Theory and Servicing*, McGraw-Hill, New York, 1979.

Robinson, R., *The Video Primer: Equipment Production and Concepts*, Putnam, New York, 1974.

46.2 VIDEODISC PLAYER

46.2-1 Introduction

The RCA CED (capacitance electronic disk) VideoDisc player represents a major new home entertainment product that provides the consumer with an economical means of viewing prerecorded program material. Capacitance pickup technology, which has been researched and developed for many years by RCA, is utilized in the VideoDisc player. The capacitance system was selected for development after careful evaluation of the advantages and disadvantages of alternate approaches. The capacitance pickup technique was found to provide the best combination of manufacturing simplicity, features, and performance. The RCA VideoDisc player provides such features as visual search (allows the VideoDisc program to be scanned at 16 times normal speed) and rapid access (making it possible to quickly and easily access a particular program segment anywhere on the video disk). Also a full "stereo" feature is provided in selected models to allow the user to play back stereo or bilingual encoded disks. Five-function infrared remote control has also been incorporated into some models, allowing the user to remotely access the player functions.

Playing time was also a major factor in the selection of the capacitance system. The RCA CED system provides two hours of program material per video disk, or one hour of program material per side.

The VideoDisc player itself is a very compact instrument measuring $5\frac{3}{4}$ in. high, 17 in. wide, and $15\frac{1}{2}$ in. deep. The player is also lightweight, weighing in at under 20 lb. Power consumption of the VideoDisc player is only 35 W.

The player is also very simple to operate. The video disk is stored in a protective plastic sleeve or "caddy" which is inserted into the player when the machine is in the load position. The caddy ensures that the disk is always correctly loaded as well as protected when not in the player. Once a disk is loaded, the machine is placed in the play mode.

At that time, a system control microcomputer assumes control of the player. At the end of the program, the user again places the machine lever in the load position and reinserts the caddy into the player to remove the video disk. The caddy now serves as a safe, convenient storage container for the video disk.

46.2-2 Customer Controls

Customer controls on the RCA VideoDisc player include the function lever (found on the SFT075/100 and SGT100/200), which controls the off, play, and load modes of operation. The newer SGT250 has a load/unload and a power on/off button, which eliminates the need for the mechanical function lever. The rapid access forward and reverse buttons, visual search forward and reverse buttons, and a pause button are also located on the front of the player.

The SGT250 SelectaVision VideoDisc player represents another major step in VideoDisc player technology in that the SGT250 not only offers the true "stereo" capability that was introduced in the SGT200 but also the added convenience of full-function IR (infrared) remote control along with electronic "soft-touch" operation.

Except for the addition of some new switches, the "basic" mechanical configuration and its operation remain essentially unchanged. These new switches and their respective functions are covered in detail later.

The basic signal-processing circuitry of the SGT250 is very similar to that found in the SFT100 and SGT200. The primary areas that are new and different consist of the "soft-touch" mechanism control and the remote control circuitry.

Customer operation remains very simple and straightforward, and is very similar to that of the earlier models. In many ways the SGT250 is easier to operate than previous players. For example, all the operator must do to place the machine in the play mode is to press the power on button, insert the disk into the machine, then withdraw the empty caddy. Once the empty caddy is withdrawn, the machine automatically places itself in the play mode.

As for remote operation, the customer may operate the unit in rapid access forward and reverse, visual search forward and reverse, and also cause the player to go into the pause mode by simply pressing the appropriate button on the IR remote hand unit.

Located on the back of the VideoDisc player is the channel selector switch for the RF output signal. The RF output of the VideoDisc player can be switched to channel three or four with this switch. Also, on the "stereo" models the audio selector switch, stereo output jacks, and video output jack (1-V p-p output) are located on the rear of the instrument.

The audio selector switch should be placed in the "normal" position except for playing disks with bilingual sound tracks. When using the stereo output jacks, use the 5-ft audio cables to connect from the player output jacks to a stereo amplifier. (Do not connect stereo output jacks directly to a set of speakers.) The stereo output connections to a stereo amplifier should represent an input impedance of 10 kΩ or greater.

The manual function lever (SFT100, SGT075/100/200) controls the power to the player as well as the loading and unloading operation for the caddy and disk. The function lever also connects the external antenna input directly to the RF output connector when the player is "off." On the newer models having the automatic load (soft-touch) feature, the function drive motor replaces the manual lever. This motor performs essentially the same mechanical functions as the manual lever.

To operate the manual VideoDisc player, the function lever is first placed in the load position. This applies B + to the electronic circuits in the player and also opens the caddy entry door at the front of the VideoDisc player, allowing the caddy to be inserted into the player. After the caddy has been removed from the player, leaving the disk and spine inside, the function lever is placed in the play position. This causes the stylus to be lowered onto the video disk, allowing the player to begin detecting the signals on the disk and generating a display on the receiver screen. Approximately a six-second delay occurs between placement of the function lever in the play position and the development of video and audio signals.

Three indicator devices (four on the "stereo" models—SGT200/250) are located on the front panel of the VideoDisc player. One is a two-digit LED display that indicates the elapsed time of the VideoDisc program in minutes. The other two indicators are the side one and side two indicator LEDs. When side one of the video disk is being played, the side one LED illuminates, and likewise with side two. When the caddy is to be inserted, the LED display flashes the letter "L," indicating that the player is now in the load mode. When the function lever is placed in the play position to begin playing a video disk, the LED display shows two dashes (--) until a video signal is recovered from the disk, at which time the LED display shows the elapsed time of the program in minutes.

The fourth and newest indicator is the "stereo" LED, which is illuminated when a stereo video disk is played and the audio selector switch is in the "normal" position.

The rapid access forward and reverse buttons allow the user to rapidly move in both forward and reverse directions across the disk. This is to enable quick location of a particular segment of the video disk. Using the rapid access buttons, the entire side of the disk can be scanned in less than 30 seconds. Pressing either button causes the stylus to be lifted from the disk and the signal circuits to be "squelched." This blanks the video display, preventing noise from being developed as the servo moves the pickup assembly rapidly across the disk.

While the rapid access feature is operating, the LED display provides an approximate (within one minute) indication of the elapsed program time. In the rapid access mode, the stylus is lifted; therefore the exact time information that is encoded on the video disk is not available. The approximate time is developed by an optical switch connected to the pickup arm drive gear.

The visual search forward and reverse buttons allow the user to scan the VideoDisc program at approximately 16 times normal speed. During visual search, the video display is present on the screen, allowing program material to be viewed in "fast" motion. The audio portion of the signal, however, is muted during the visual search feature. Audio signals processed at 16 times normal rate would be garbled and unintelligible. Like the rapid access feature, the visual search feature is operational in both the forward and reverse directions; thus the user can scan forward and reverse to any portion of the VideoDisc program while observing the video signal.

Stylus kicker electronics, which make the visual search features possible, are also responsible for preventing the stylus from becoming "locked" into a groove. A video disk-locked groove produces an effect similar to a locked groove in an audio system, in that the video and audio signals in that groove are repeated over and over. Since eight vertical fields are contained in a revolution of the disk groove, those eight video fields and audio signals are repeated. The system control electronics, by checking a special code on the video disk, causes the kicker circuits to move the stylus forward when a locked groove is detected.

The RCA VideoDisc player has a pause feature. When the pause button is pressed, the stylus is raised from the video disk. At the same time the electronic circuits in the VideoDisc player are placed in the squelch mode, preventing noise from occurring on the video display. When the pause button is pressed, the LED display shows the letter "P," which flashes at a one-second rate. Once placed in the pause mode, the player remains in this mode until the pause button is again pressed or until the rapid access or visual search buttons are pressed. When the pause button is pressed to exit the pause mode, the stylus is lowered on the video disk and video signal is again generated at approximately the same point in the program as when the pause button was initially pressed. When either the rapid access or visual search buttons are pressed to exit the pause mode, the player immediately enters either rapid access or lowers the stylus and enters visual search, respectively.

At the end of the program on each side of the disk, a special code is recorded that causes the player to go into the end mode. When this code is detected, the stylus is lifted and the LED display blinks "E" at a one-second rate. The player remains in the end mode until the disk is removed or until the rapid access reverse or reverse visual search buttons are pressed. The forward buttons are deactivated in the end mode. The blinking LED display in the load, pause, and end modes indicates that the player remains in these modes until a customer control is actuated.

46.2-3 Capacitance Pickup Theory

The RCA VideoDisc player utilizes capacitance pickup technology to detect the video and audio signals placed in the grooves on a capacitance electronic disk (CED). The CED is somewhat like a typical audio record in that the signal information is placed in grooves. There are some striking differences, however, in the video disk as compared with an audio disk. One such difference is the density of the grooves, as illustrated in Fig. 46.27. In an audio disk, the grooves are spaced approximately 0.004 in. apart. In the video disk, however, the grooves are much closer. As many as 40 grooves can be placed in the space between the grooves of an audio disk. This means that approximately 10,000 grooves are placed in a 1-in. radius of the video disk.

The FM-modulated video and audio carriers are placed on the disk by varying the depth of the groove with the carrier signals, as shown in Fig. 46.28. The groove is very small, only 2.5 μm wide. Therefore the modulation of the groove depth is necessarily small, approximately 850 Å peak to peak maximum for video.

In a typical audio disk system, the signal is detected by the movement of a magnet in a small coil that is modulated by a stylus riding in the grooves of the record. This generates a signal that varies in both frequency and amplitude, thus recovering the original audio signal. Because of the small size of both the video disk grooves and the depth modulation, detection of both amplitude and frequency

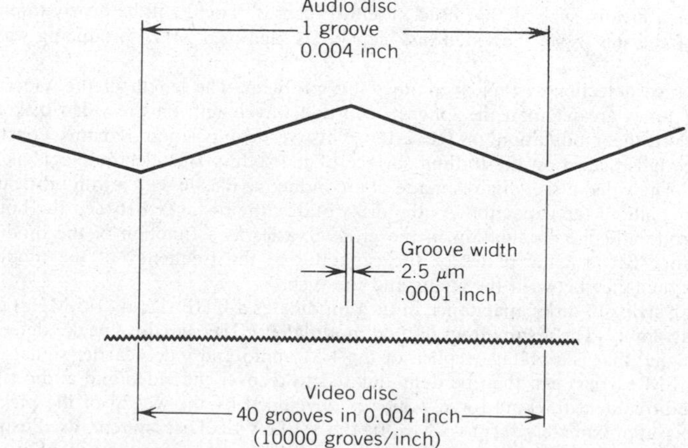

Fig. 46.27 Video/audio disk groove density comparison.

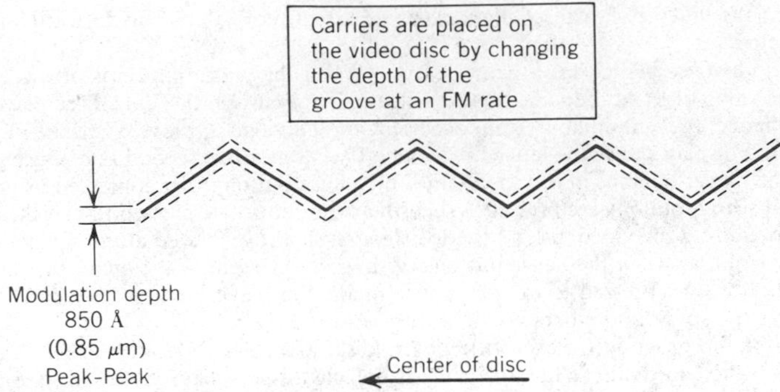

Fig. 46.28 Video disk modulation.

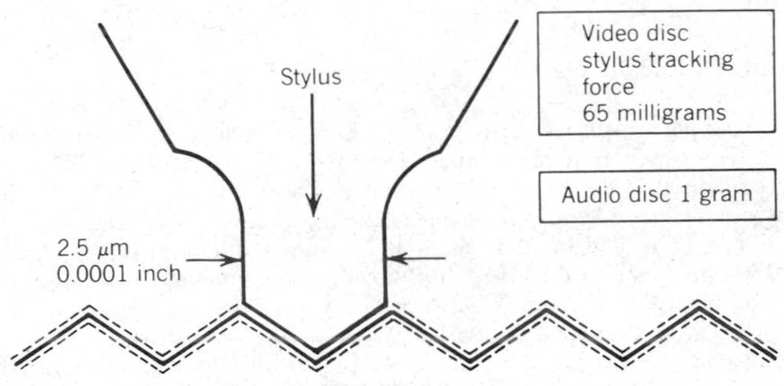

Fig. 46.29 Video disk stylus tracking force.

changes by this method would be virtually impossible. Therefore the capacitance pickup system was developed.

In the CED (capacitance electronic disk) system, the audio and video signals are placed on the disk via FM carriers. This eliminates the need for precise detection of amplitude variations, since the video signal amplitude is now represented by the frequency deviation of the FM carrier signal. Therefore it becomes necessary only to detect the frequency of the signal on the video disk. Doing this by mechanical means, such as that used in audio records, would still be nearly impossible, since the frequency of the signals on the video disk can be as high as 7 MHz, producing wavelengths as short as 0.5 μm.

The capacitance detection system overcomes this problem. The length of the VideoDisc pickup stylus is several times greater than the longest recorded wavelength on the video disk. Thus as the stylus travels over the modulation in the groove, its vertical position remains constant. A thin metalized electrode is placed on the trailing surface of the stylus. This electrode acts as one plate of the "capacitor." The video disk, which is made of a conductive plastic with a thin lubricating coating, acts as the other plate of the capacitor. As the disk rotates, the distance between the bottom edge of the stylus electrode and the modulation in the groove varies as a function of the modulation. This varies the distance between the plates of the capacitor at the frequency of the modulation, thus changing the capacitance between the stylus and the disk.

The changing stylus-to-disk capacitance in turn modulates a UHF signal (915 MHz) in the pickup arm resonator assembly. The resultant amplitude-modulated UHF signal is "peak" detected, generating an output signal that is a voltage replica of the FM audio and video carrier signals recorded on the disk. These FM carriers can then be demodulated to recover the video and audio signals.

A typical audio system tracking force, which is developed by the weight of the pickup cartridge and tone arm, is approximately 0.5 to 2.5 g. In the CED VideoDisc system, as illustrated in Fig. 46.29, the tracking force is only 65 mg (0.065 g). Stylus tracking force for the VideoDisc is developed by the very fine stylus flylead and the light weight of the pickup stylus. This flylead also connects the stylus electrode to the resonator electronics in the pickup arm assembly.

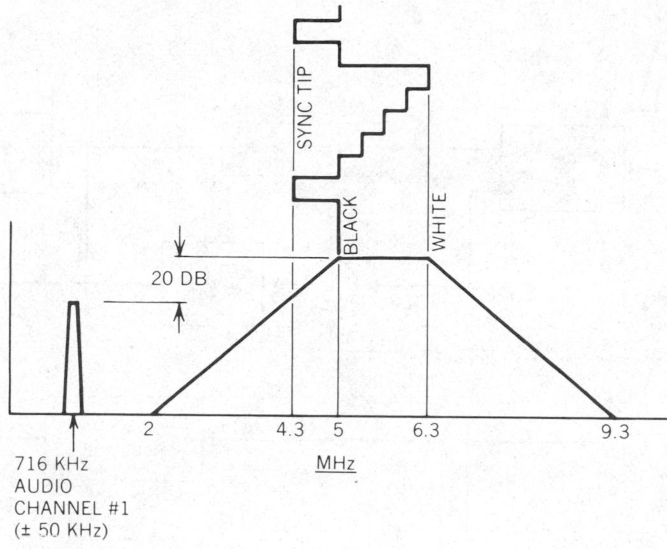

Fig. 46.30 Video disk frequency spectrum (signal recorded on disk).

46.2-4 Video Disk Signals

Unlike audio disks, in which the audio signals are recorded directly on the disk, the video and audio signals for the video disk are used to frequency modulate two carrier signals that are, in turn, then recorded on the disk. This reduces the dynamic range of the signals that are to be recorded and also allows the placement of more than one signal on the video disk by utilization of two or more different carrier frequencies.

The video signal recorded on the disk is a frequency-modulated 5-MHz video carrier. Black level of the video signal causes zero deviation of the carrier, or a frequency of 5 MHz. Sync tips cause the video carrier frequency to deviate to 4.3 MHz. Peak white in the video signal causes the video carrier signal to deviate to 6.3 MHz. The sidebands generated from the FM modulation of the 5-MHz carrier extend from 2 to 9.3 MHz.

Audio information must also be placed on the video disk at the same time as the video information. The audio signals are placed at a different carrier frequency than the video signals. A channel of audio information is placed on an FM carrier frequency of 716 kHz. The audio signal generates a frequency deviation of ±50 kHz. The video disk spectrum thus contains an audio carrier at 716 kHz and a video FM carrier from 4.3 to 6.3 MHz with sidebands from 2 to 9.3 MHz (Fig. 46.30).

Prior to modulation of the 5-MHz video carrier, the 3.58-MHz chrominance subcarrier and resultant sidebands are down converted to 1.535625 MHz (1.53 MHz). This produces several positive effects. First, down converting the chroma information allows the use of a relatively low video carrier frequency (5 MHz). Normal sidebands of a 3.58-MHz subcarrier would exceed 4 MHz, thus increasing the inherent system noise level. Second, the down-converted chrominance allows the shortest wavelength recorded on the video disk to be comparatively long. This facilitates the recording of one hour of program material on each side of the disk.

Down-converted chrominance is developed by heterodyning the 3.58 MHz chrominance with a 5.115170-MHz oscillator signal. The resultant 1.535625-MHz (1.53 MHz) chrominance subcarrier is then sideband limited to ±500 kHz. Luminance information is then added to the down-converted chrominance to generate a composite video signal with the chrominance signals completely within the luminance bandwidth. Down conversion of the chroma information to the middle of the luminance spectrum is sometimes referred to as a buried subcarrier or buried chroma system.

The resultant composite video signal modulates the 5-MHz video carrier generating the 4.3- to 6.3-MHz video FM carrier signal. Video carrier sidebands extend from 2 to 9.3 MHz. The frequency-modulated audio carrier of 716 ±50 kHz is then added to the video FM carrier and applied to the cutter head of the video disk mastering machine. The video disk master is made much the same as an audio record master, using a mechanical assembly to "cut" the modulation in the disk grooves. An overall block diagram of the video disk recording process is shown in Fig. 46.31.

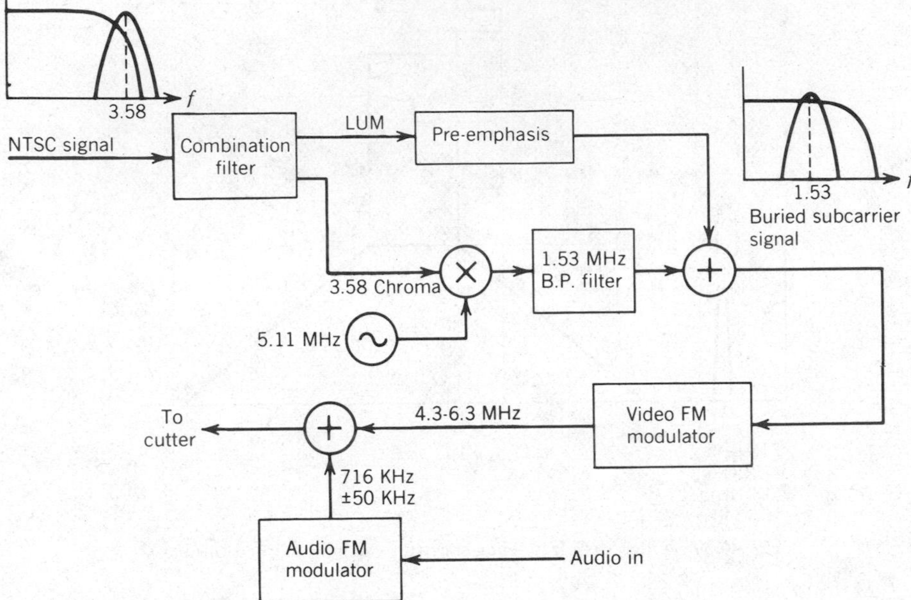

Fig. 46.31 Simplified video disk recording block diagram.

46.2-5 Caddy Operation

Owing to the extremely small size of the video disk grooves, accumulations of dust, grit, and other debris on the disk surface could cause undesirable noise in the detected video and audio signals. To protect the disk surface from collecting such impurities, the disk is contained in a plastic housing known as a caddy. When the disk is outside the player, it resides in the plastic caddy. When the video disk is to be played, the caddy is inserted in the player, which removes the disk and spine assembly from inside the caddy. The empty caddy is then removed from the player to allow the disk to be played. After the disk has been played, the empty caddy is reinserted into the player to recover the disk and spine assembly. The caddy with disk and spine is then removed, thus protecting the disk from direct exposure to the outside atmosphere.

The caddy also prevents the user from touching the disk, as oils and acids from the fingertips could cause degradation of the disk lubricating surface, thus possibly causing loss of signal. In addition to protecting the video disk from contamination and scratches, the caddy serves as a handy storage case for the disk. A soft cloth seal in the caddy opening forms a tight, dust-resistant seal, preventing contamination from entering the caddy.

46.2-6 General System Overview

The majority of the electronic circuits in the RCA VideoDisc player can be separated into two basic categories—system control and signal-processing circuits. A microcomputer is the heart of the system-control electronics. The computer receives inputs from the customer function switches and, in turn, controls the operation of the player. The system-control microcomputer is also responsible for decoding the digital auxiliary information (DAXI) code supplied from the video disk to develop the time indication. The majority of the system-control electronics are located on the PW 500 board, which is mounted above the video disk turntable.

The signal-processing circuits, which are primarily contained on the PW 3000 board with some located in the arm assembly, are responsible for detecting the video information placed on the video disk, demodulating this information, processing it through a comb-filter circuit, and modulating it onto channel 3 or channel 4. The modulated television RF signal may then be applied to a standard NTSC television receiver. The complete signal-processing system is made up of six integrated circuits, which provide most of the signal-processing functions. A "monaural" VideoDisc system operational block diagram is illustrated in Fig. 46.32.

All of the electronics necessary for the processing of stereo on applicable models are located on the PW 4000 board. The PW 5900 board contains all the electronics necessary to accommodate the remote control and "soft-touch" features found in certain models.

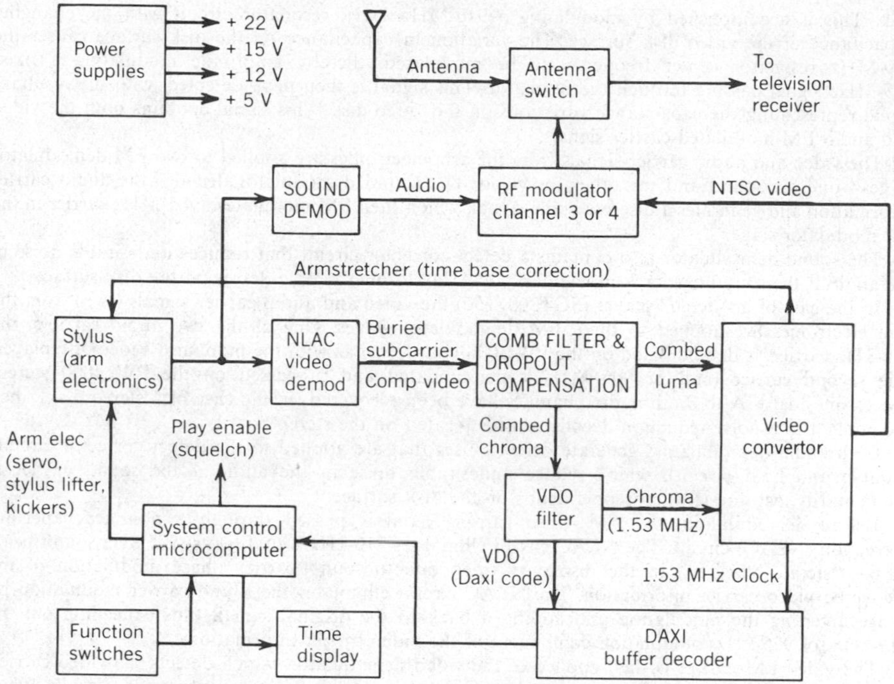

Fig. 46.32 RCA VideoDisc player block diagram.

Functional Operation

The operation of the VideoDisc player is controlled totally by the system-control circuitry. Customer functions of play, rapid access forward and reverse, visual search forward and reverse, pause, and load are input to the microcomputer through the customer-function switches. The microcomputer decodes these input commands and, in turn, controls the electronics to carry out those functions. The state of all the signal-processing circuits is controlled by the "not squelch" output of the microcomputer. When the not squelch line goes to a logic "lo" state, all the electronic circuits are disabled.

The system-control microcomputer also has direct control over the pickup arm assembly. This control involves the servo operation, which moves the arm forward during normal play; the stylus lifter operation, which raises and lowers the stylus as the various functions are initiated; and the kicker circuits, which enable the system to provide the visual search features in forward and reverse. The microcomputer also controls the direction of the servo system. In the rapid access reverse and visual search reverse modes of operation, the microcomputer instructs the servo system to operate in the reverse mode rather than in the normal forward mode of operation.

The system-control microcomputer is also responsible for the time display. The time display information is developed from a DAXI signal that is recorded on the video disk during manufacture. The DAXI code, which appears on line 17 of each vertical field, contains a field identification number that is decoded by the system-control microcomputer to display the elapsed time of the program in minutes.

In the rapid access forward and reverse modes of operation, the DAXI code is not available owing to the stylus being lifted from the disk. In this mode of operation, time display must be maintained so the approximate elapsed time of the program can be tracked while the arm is moved in either direction across the disk. This is accomplished by a "photointerrupter" circuit, which provides approximate elapsed time by tracking the relative position of the arm with respect to the disk.

The electronics located in the pickup arm assembly are responsible for detecting the video information on the disk. The arm also contains electronics responsible for providing the features of visual search forward and reverse as well as locked groove protection. These features are carried out by activation of the "stylus kicker" coils, which when activated cause the stylus to skip two grooves of the video disk. Also located in the pickup arm assembly is the "armstretcher" transducer, which corrects for timebase variations in the recovered chrominance and luminance signals due to disk warpage, eccentricity, and/or changes in the rotational speed of the turntable.

The primary function of the pickup electronics is to detect the video signals placed on the video disk. This is accomplished by modulating a 910-MHz UHF resonator circuit with the changing capacitance of the video disk surface. The variation in capacitance on the disk surface causes the 910-MHz resonator center frequency to be modulated, thereby amplitude modulating a fixed 915-MHz signal passing through the resonator. This signal is then peak detected, with the resultant signal representing the capacitance variations on the video disk. This signal contains both the video and audio FM-modulated carrier signals.

The video and audio carrier signals from the arm electronics are applied to two FM demodulator stages—one for sound and the other for video. The sound demodulator decodes the audio carrier information and generates a discrete audio signal, which then FM modulates a 4.5-MHz carrier in the RF modulator stage.

The sound demodulator also contains a defect corrector circuit that reduces undesirable noise in the audio if the sound carrier is momentarily lost due to microscopic debris on the disk surface.

In the case of a "stereo" player (SGT200/250) the video and audio carrier signals (two) from the arm electronics are applied to three FM demodulator stages. One of the two audio carriers, the 716-kHz carrier, is demodulated on the PW 3000 board, just as with the monaural VideoDisc player. The second carrier (and/or 905 kHz) is demodulated and processed on the PW 4000 stereo-processing board. After both audio channels have been recovered (in the case of a stereo disk), they are applied to a noise-reduction decoder that is located on the stereo board.

Both audio demodulators generate defect pulses that are applied to the defect corrector circuits (sample and hold circuit), which reduce undesirable noise in the audio if the sound carrier is momentarily lost due to microscopic debris on the disk surface.

Before demodulation, the FM video carrier signal is passed through a nonlinear aperture correction (NLAC) circuit. The NLAC circuit eliminates 716-kHz sound beats (905-kHz sound beats in the "stereo" versions) in the displayed video due to sound carrier phase modulation of the recovered video carrier information. The NLAC circuit eliminates the sound carrier modulation by phase inverting the modulation and adding it back to the original signal, thus cancelling out the 716-kHz (or 905-kHz) modulation component of the video carrier information.

The video FM carrier is then applied to the video demodulator, which detects the video carrier. The video demodulator also contains a defect correction or dropout compensation circuit, which allows the previous horizontal line to be inserted when a defect caused by loss of carrier occurs.

The output of the video demodulator, which is composite video with buried chroma subcarrier, is applied to a comb-filter circuit. The comb filter dynamically separates chrominance and luminance information from the composite video signal. Efficient chroma/luminance filtering is necessary, as the chroma subcarrier information is "buried" within the luminance frequency bandwidth.

The output of the comb filter is "combed" chrominance and "combed" luminance. The combed chrominance output signal also contains low-frequency luminance information as well as the DAXI signal, which is transmitted with each vertical field. Bandpassing recovers the 1- to 2-MHz chroma signal; the two remaining signals are separated by lowpass filters. The low-frequency luminance information is recombined with the combed luminance information to provide the luminance output. The DAXI signal is coupled through a DAXI buffer/decoder to the system-control microcomputer.

After processing by the comb-filter circuitry, the luminance and chrominance information is coupled to the video converter circuits. The video converter up converts the 1.53-MHz chrominance information back to 3.58-MHz. The 3.58-MHz chroma and the luminance information are then combined. The composite video signal is applied to the RF modulator, where the audio FM carrier is added and an RF signal on channel 3 or channel 4 is developed for output to a standard NTSC television receiver.

Also developed in the video converter stage is the drive signal for the armstretcher time base corrector circuitry. The correction signal is developed by comparing the 3.58-MHz chroma information developed in the video converter with a 3.58-MHz reference. Any phase or frequency difference between the two signals develops an error signal, which is applied to the armstretcher circuitry. This circuitry operates a solenoid that moves the stylus to maintain a constant disk-to-stylus velocity. The armstretcher circuitry output is also coupled to the converter oscillator to maintain phase lock between the up-converted 3.58-MHz color signal and the 3.58-MHz reference oscillator.

Another board found in the newer VideoDisc players is the pulse interference corrector (PIC) board. The purpose of the pulse interference corrector circuit is to prevent radar and other strong RF pulses in the 900-MHz range from interfering with the operation of the VideoDisc player. The PIC circuit detects the presence of such pulses and instructs the defect corrector in IC 3301 to substitute the previous line of video information.

AC and DC Power Supplies

All the electronic circuits in the RCA VideoDisc player are isolated from the power line, that is, they are cold grounded. Referring to Fig. 46.33, AC input is applied to the PW ACIN board on which

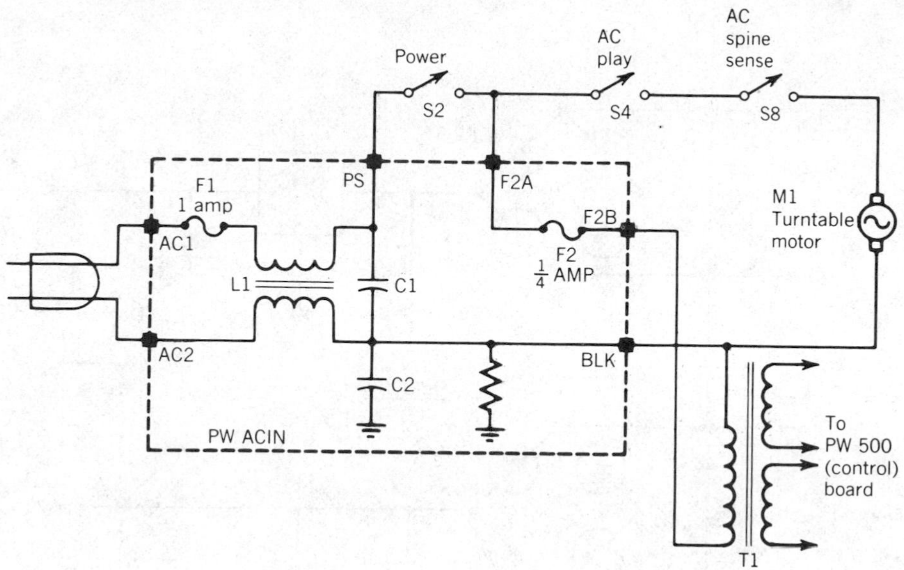

Fig. 46.33 AC input and turntable power.

initial protection is provided by a 1-A fuse. The AC power switch, S2, is controlled by the function lever. AC power is applied to power transformer T1 when the function switch is in the load or play position. Power transformer T1 is also protected by F2, a $\frac{1}{4}$-A fuse.

The secondary of T1 contains two windings, one developing 9 V rms, used to generate a 5-V supply for the microcomputer control system; the other developing 18 V rms, which provides a 22-V DC supply.

AC power from S2 is also applied to the AC play switch, S4. The AC play switch is controlled by the function lever and is closed only in the play position. The AC play switch is open in the load position.

AC play switch S4 connects to AC spine sense switch S8, which is in series with the turntable motor. The AC spine sense switch is closed only when a caddy has been inserted into the VideoDisc player and then removed, leaving the spine and disk in the player. Closing of the power switch, the AC play switch, and the AC spine sense switch then applies power to the turntable motor.

The turntable motor is a two-pole shaded-pole AC motor that drives the turntable at 450 rpm. Synchronization between the turntable and the power line is maintained by two magnetic poles attached to the motor, which drive a 16-pole magnetic ring located inside the turntable base. The magnetic poles on the motor driving the magnetic ring around the turntable base ensures that the turntable maintains a constant phase relationship with the 60-Hz AC input. This results in a very stable 60-Hz video field rate.

As illustrated in Fig. 46.34, 9 V rms developed between terminals 5 and 6 of T1 is applied to bridge rectifier diodes CR 2, CR 3, CR 4, and CR 5 located on the PW 500 board. The bridge rectifier provides a DC voltage that is filtered by capacitor C9, a 2200-μF electrolytic capacitor. The filtered DC is then applied to a three-terminal 5-V regulator (U4) that develops an output voltage of 5 V, which is then filtered by C6. The 5-V supply is used to power the system-control microcomputer, LED display, and other system-control circuitry.

Eighteen volts AC from terminals 3 and 4 of transformer T1 is applied to bridge rectifier diodes CR 6, CR 7, CR 8, and CR 9 on the PW 500 board. The output of that bridge rectifier is applied to capacitor C10 and C11—two 1500-μF electrolytics in parallel. The result is a well-filtered 22-V B + supply, which is utilized by the servo and stylus kicker driver circuits.

The 22-V DC supply developed on the PW 500 board is also supplied to the PW 3000 board (terminal N), where it is applied to the input terminal of U801, a 15-V three-terminal regulator (Fig. 46.35). Regulated 15-V output from U801 can be measured at terminal K. A 12-V regulated supply is developed from the 15-V supply by U401 and Q801. A divider network consisting of R801 and R802 provides approximately 12 V at the positive input of an operational amplifier. The output of the operational amplifier drives the base of 12-V regulator transistor Q801. The regulated 12-V output is then coupled back to the negative input of the operational amplifier, maintaining a constant output

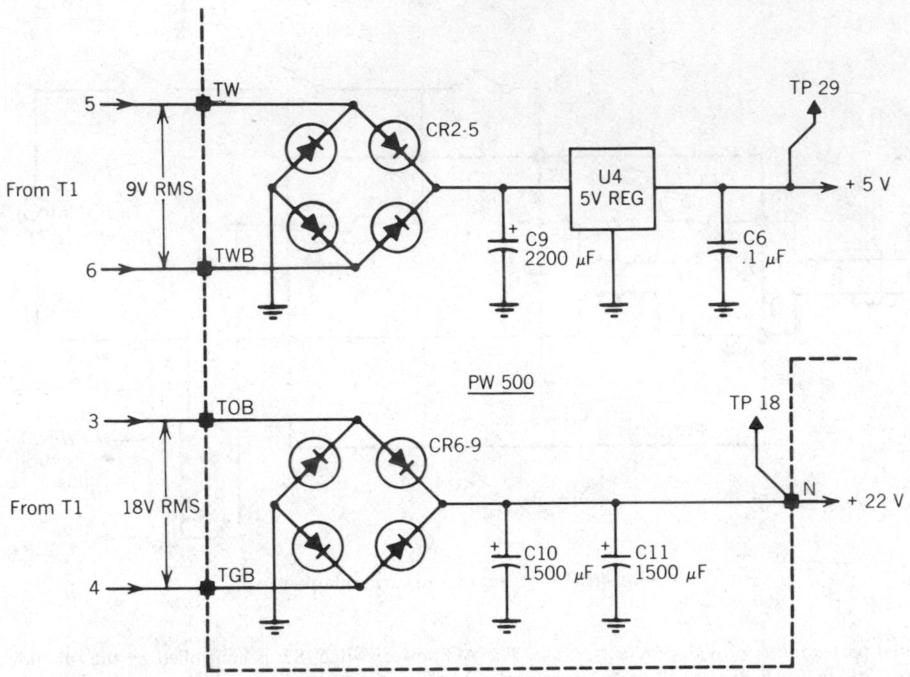

Fig. 46.34 Development of a +5- and +22-V B+ supplies.

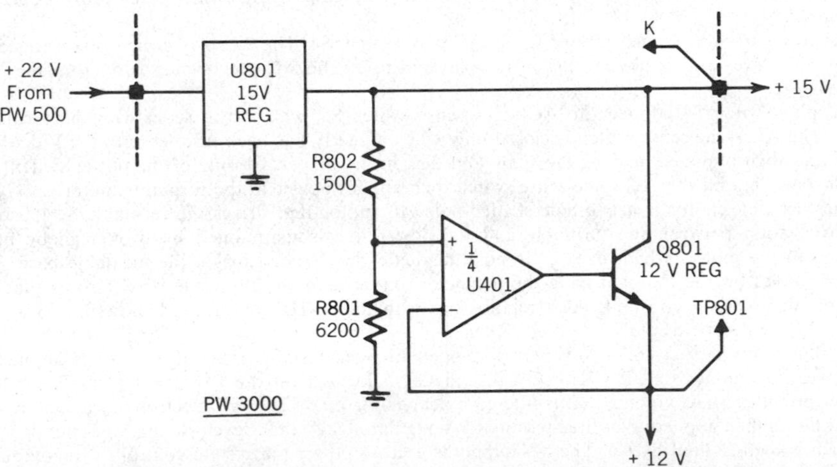

Fig. 46.35 Development of a +12- and +15-V power supplies.

of 12 V at TP 801. The operational amplifier is part of U401, which is a quad-operational amplifier package.

Most of the electrical components in the VideoDisc player are located on one of several large subassemblies. The PW 3000 board contains all the signal-processing circuitry as well as the RF modulator stages. This board is attached to the bottom of the VideoDisc player and can easily be placed in the service position by removing the mounting screws and folding the board out to the side of the player. At this point, all assemblies are accessible from the top of the VideoDisc player.

The PW 500 board, which contains all of the system-control components, is mounted to the top of the player with the components facing down toward the turntable. The PW 500 board can be removed and placed at the side of the player or inverted and inserted into a slot at the top of the player. Both fuses, as well as the AC input circuits, are located on the PW ACIN board, which is located directly under the right side of the turntable.

The pickup arm assembly contains the 915-MHz resonator circuit, preamplifier, and AFT circuits, as well as the armstretcher, stylus kicker, and servo sensor circuits. The replaceable stylus cartridge is also contained in the pickup arm assembly. The stylus cartridge is easily replaceable through a door above the arm assembly.

The servo motor and reduction gear assembly is located toward the rear of the player. A gear rack (part of the pickup arm assembly) is driven by the servo motor, which moves the arm across the video disk. The servo reduction gear assembly incorporates a clutch that is operated by the function lever. The clutch disconnects the servo drive gears to allow the arm to be moved easily during disk removal.

Bibliography

Bensinger, C., *The Home Video Handbook*, 3rd ed., Howard Sams, Indianapolis, 1982.

Chesire, D., *The Video Manual*, Van Nostrand, New York, 1982.

Clason, W., *Dictionary of Television and Video Recording*, Elsevier, New York, 1975.

Clason, W., *Home Video Yearbook*, Knowledge Industry, White Plains, NY, 1981.

Ennes, D., *Television Broadcasting; Tape and Disc Recording Systems*, Howard Sams, Indianapolis, 1982.

Robinson, R., *The Video Primer: Equipment Production and Concepts*, Putnam, New York, 1974.

Sigal, E., et al., *Video Discs: The Technology, the Applications and the Future*, Van Nostrand, New York, 1980.

PART 10
COMMUNICATIONS

CHAPTER 47
COMMUNICATIONS CONCEPTS

YENG S. KUO
KWEI TU

Lockheed Engineering and Management Services Company, Inc., Houston, Texas

RODGER E. ZIEMER

University of Colorado–Colorado Springs, Colorado Springs, Colorado

WILLIAM H. TRANTER

University of Missouri–Rolla, Rolla, Missouri

47.1 CONCEPTS OF MODERN COMMUNICATIONS

Yeng S. Kuo

47.1-1 Introduction

The purpose of communication is to convey information from one point to another. The era of electrical communication was born when Morse transmitted the first telegraph message over a 16-km wire. Ten years after Hertz verified Maxwell's theory in 1877, Marconi successfully demonstrated and patented a complete wireless telegraph system. During World War II, radar and microwave systems were developed, along with improved electronics and statistical methods for analyzing signal extraction problems. The foundations for modern communications were solidly laid during this period.

An inherent characteristic of electrical communication is the presence of uncertainty, which is due in part to the inevitable presence of undesired signal perturbations, commonly referred to as noise. Noise has been an ever-present problem since the early days of electrical communication. It was long thought that noise placed an inescapable restriction on communication accuracy. Due to the stochastic nature of noise, probabilistic analysis procedures were used by Rice to analyze communications operating in the presence of noise. Weiner and Kolmogoroff applied statistical methods to signal-detection problems.

In 1948 Shannon published his famous paper "A Mathematical Theory of Communications," in which he recognized the stochastic nature of information. Shannon posed this problem: Given a set of possible messages a source can produce in a random fashion, how shall the messages be represented or coded so as best to convey the information over a given system? Shannon proved that for a given information source transmitting over a given communication channel, there exists a coding technique such that the error can be made arbitrarily small despite the presence of noise as long as the information transmission rate is less than the channel capacity. His work was the beginning of modern communications. Although Shannon has proved that error-free transmission over a noisy channel is possible with proper coding, his theory did not reveal what kind of coding technique should be employed. Numerous coding theories were developed during the 1950s, during which period emphasis was on the code construction, block decoding, weight structure, and bounds on distance. Convolutional coding with sequential decoding by Fano and maximum likelihood decoding by Viterbi were developed theoretically in the 1960s. Also, analyses on more efficient signal design and on modulation and demodulation techniques were treated during this period. With the rapid development of advanced solid-state electronics technologies, such as the ultra-high-speed very-large-scale integrated circuit (VLSI) chip and the powerful microprocessor in the 1970s, many complex coding/decoding and modulation/demodulation techniques were realized. Electrical communication has since entered a new era. Its potential applications are bounded only by our needs, aspirations, and imagination.

47.1-2 Typical Communication System Model

A typical model for a communication system is shown in Fig. 47.1. As illustrated, we define a communication system as the process of transferring source information from one point to another. The functional elements are information source and destination, transmitter and receiver, and the channel.

Information Source

The information source produces a message or a sequence of messages to be conveyed to the receiving terminal or destination. There are many kinds of information sources, and thus the messages appear in a variety of forms. Generally, we may roughly classify the messages into three different types:

1. Analog (continuous waveform) signal, which may be modeled as functions of continuous-time variables. Voice, music, television, and temperature measurements are good examples of analog signals.
2. Digital signal, which consists of discrete symbols such as the output of a digital computer, digitized voice, and digitized television.
3. Pulse signal, which consists of a sequence of narrow pulses such as those used in radar or ranging applications.

The message (signal) produced by a source is not necessarily in electrical form and a transducer may be required to convert the message into a form more suitable for transmission. For the purposes of this section it is assumed that messages are in electrical form.

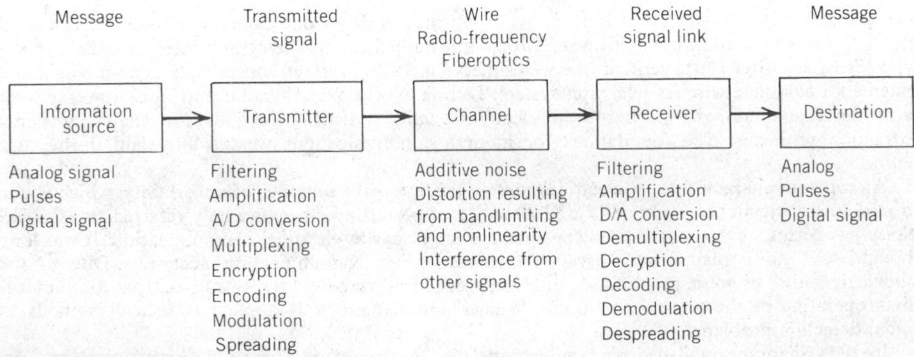

Fig. 47.1 Generic communication system model.

Transmitter

The purpose of the transmitter is to convert the message into a form suitable for transmission over the channel. This process includes one or more of the operations defined as follows.

Filtering. To limit the bandwidth of the information source or to shape the waveform.

Amplification. To increase the amplitude of the signal to a level appropriate for processing and transmission.

Modulation. To produce a waveform suitable for the channel.

In addition to the three basic operations, many transmitters are also able to perform special operations, defined as follows.

Analog-To-Digital Conversion. To convert a continous waveform (analog signal) into digital form.

Multiplexing. To combine two or more separate information signals into a single transmitted signal.

Encryption. To protect data from tampering or eavesdropping.

Encoding. To encode digital data for error detection and correction.

Spectrum Spreading. To expand the radio frequency spectrum of the transmitted signal for purposes of controlling access and reducing spectral density. (See Section 48.5.)

Channel

The channel is the transmission medium for the communication system. It may consist of a free-space link (with antennas), a pair of wires, a cable, or an optical fiber. A communication channel has the following characteristics.

Attenuation. A function of distance traveled by the signal (see Section 48.2).

Distortion. Caused by fading (see Section 48.7), interference, nonlinearity, or bandlimiting.

Noise. Generally modeled as a channel characteristic, but does not necessarily originate in the channel. (See Section 47.2.)

All of these channel characteristics tend to degrade the quality of the signal arriving at the receiver and introduce the possibility that the signal that was transmitted has not been received with complete fidelity.

Receiver

The receiver recovers the signal from the channel, along with the attending noise, distortion, and interference, and reconstructs as accurately as possible the original information signal.

47.1-3 Noise and Distortion

Noise and distortion are the two principal causes of performance degradation in communication systems. Noise is a general name given to a variety of phenomena that are discussed in detail in Section 47.2. The presence of noise in communication systems is the motivation for most research into the theory of communications, for without noise it would be possible to build a perfect (error-free) communication system. Information theory, coding theory, and statistical estimation and detection theory all deal with the problem of extracting a desired signal from a noise background.

Distortion is the alteration of the signal due to imperfect response of the communication system. The most common types of distortion in a communication system are the following.

Amplitude distortion. Caused by nonuniform frequency response in the band of interest.

Phase distortion. Caused by nonuniform phase response in the band of interest.

Harmonic distortion. Caused by nonlinear response.

Intermodulation distortion. Caused by a nonlinear combination of multiple frequency components of the signal.

Yet another type of distortion is that caused by interference from other communication systems, whether inadvertent or intentional (jamming). In modern communication systems, techniques have been developed for reducing the effects of this type of distortion, as discussed in Section 48.5.

Most techniques for counteracting the effects of distortion and noise are based on the principle of increasing the bandwidth of the transmitted signal and the complexity of the signal processing in the receiver. With reference to Fig. 47.1, this means that while it is desirable to keep the bandwidth of the message as small as possible to reduce the ultimate effects of noise, it is possible and even desirable to occupy a much larger bandwidth in the communications channel to combat the effects of noise, distortion, and interference.

47.1-4 Measures of Quality

An ideal communication system is free of distortion and provides error-free transmission. In other words, an identical replica of the transmitted signal can be recovered at the receiver, with possible scaling and time delay. Unfortunately, noise in the system is unavoidable, and distortion usually occurs in implementation. Thus we must examine the effects of noise and distortion on the performance of the system and employ some quantitative measures to describe the quality of the recovered signal.

Analog Signals

A simplified block diagram of a communication system is shown in Fig. 47.2. In the case of an analog system, the input message $x(t)$ is a continuous (analog) waveform; $y(t)$ is the sum of the received message, attenuated by passing through the channel; and the noise added in the channel, $n(t)$, is the channel noise and $x'(t)$ is the recovered message. It is assumed that $x(t)$ is bandlimited and has a maximum upper frequency W, that the noise is gaussian (see Section 47.2), and that the channel is distortion free.

The signal-to-noise ratio at the input of the receiver can be represented as

$$(S/N)_i = \frac{\overline{y^2(t)}}{\overline{n^2(t)}} = \frac{S_r}{N_0 B}$$

where S_r = received signal power
N_0 = noise power spectral density
B = receiver bandwidth

and the overbar indicates an average value.

The signal-to-noise ratio of the recovered message is then

$$(S/N)_0 = \frac{\overline{x'^2(t)}}{N_0 W} = \frac{S_0}{N_0 W}$$

where S_0 is the signal power at the output of the receiver.

The $(S/N)_0$ is the usually accepted measure of quality of the received signal and can be expressed as a function of the input signal-to-noise ratio, $(S/N)_i$. In a linear modulation system, $(S/N)_0 \leqslant (S/N)_i$, while in a nonlinear modulation system, such as FM, performance can be improved by increasing the transmission bandwidth (see Section 48.1). Figure 47.3 shows a comparison of $(S/N)_0$ vs. $(S/N)_i$ for AM and FM with two values of modulation index, β.

Linear distortion in a communication system can sometimes be reduced by the use of equalization devices. These devices have a transfer function that, when multiplied by the system transfer function, produces a composite response without distortion. Such techniques require a knowledge of the actual system transfer function or some adaptive techniques to determine the transfer function. Nonlinear distortion, on the other hand, is curable only in specific cases, such as when the nonlinearity is intentionally introduced. Such a case is the use of a compander (for compressor/expander) to improve the dynamic range of a system. The conventional measurements of nonlinear distortion are

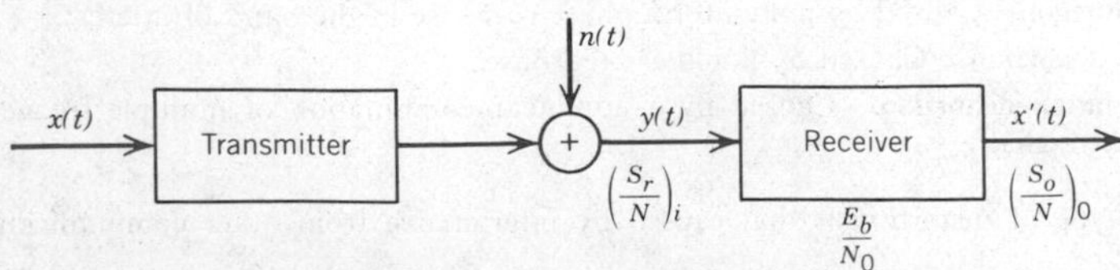

Fig. 47.2 Analytical model of a communication system.

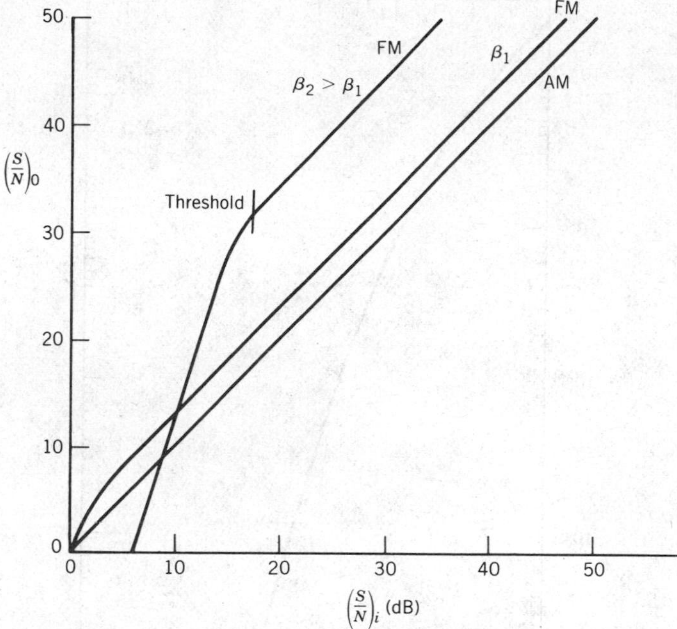

Fig. 47.3 Output signal-to-noise ratio for FM and AM systems.

the following:

$$\text{Total harmonic distortion} = \left(\frac{\text{total harmonic power}}{\text{total fundamental power}}\right)^{1/2} \times 100\%$$

$$\text{Total intermodulation distortion} = \left(\frac{\text{total intermodulation power}}{\text{total fundamental power}}\right)^{1/2} \times 100\%$$

Digital Signals

The objective of a digital communication system is to transmit a discrete message in a prescribed amount of time with a minimum number (or minimum probability) of error. The accepted measure of quality in a digital communication system is the probability of error, either for a message, a symbol, or a single binary digit (bit). In the analytical model of Fig. 47.2, the input message $x(t)$ is a sequence of binary messages, and the input signal-to-noise ratio is represented by the ratio of symbol (or bit) energy to noise spectral density, E_b/N_0. Figure 47.4 shows the average probability of error in detecting a single bit in the presence of additive white gaussian noise (explained in Section 47.2). The curve shown is for an optimum binary transmission system. Performance can be improved (lower probability of error for a given E_b/N_0) by using nonbinary symbols and by advanced techniques, discussed in Section 48.5.

Bibliography

Rice, S. O., Mathematical Analysis of Random Noise, in N. Wax, ed., *Selected Papers on Noise and Stochastic Processes*, Dover, New York, 1954.

Schwartz, M., *Information Transmission, Modulation, and Noise*, McGraw-Hill, New York, 1959.

Schwartz, M., W. R. Bennett, and S. Stein, *Communication Systems and Techniques*, McGraw-Hill, New York, 1966.

Taub, H. and D. L. Schilling, *Principles of Communication Systems*, McGraw-Hill, New York, 1971.

VanTrees, H. L., *Detection, Estimation, and Modulation Theory, Part 1*, Wiley, New York, 1968.

Viterbi, A., *Principles of Coherent Communications*, McGraw-Hill, New York, 1966.

Wozencraft, J. and I. Jacobs, *Principles of Communication Engineering*, Wiley, New York, 1965.

Ziemer, R. E. and W. H. Tranter, *Principles of Communications*, Houghton Mifflin, Boston, 1976.

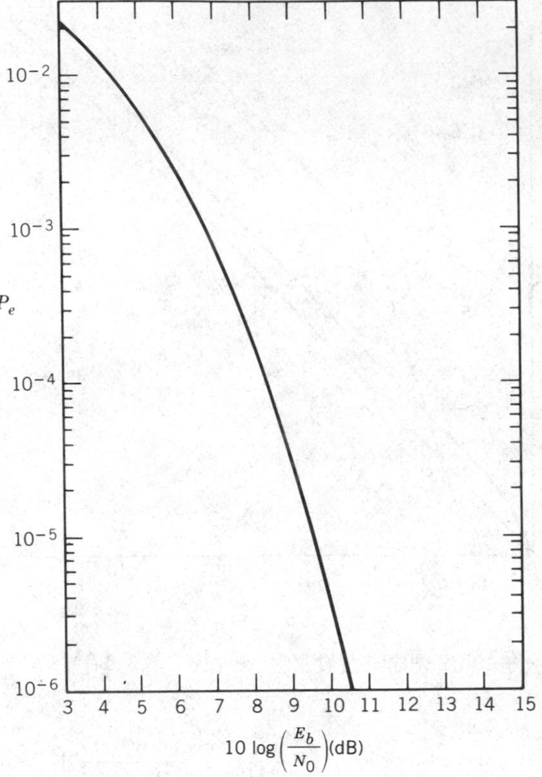

Fig. 47.4 Probability of error for an optimum binary system.

47.2 NOISE THEORY

Kwei Tu

47.2-1 Sources of Noise

Noise and interference from other communications systems are two factors limiting the performance of all communications systems. Noise occurs from a number of different sources. The most important are (1) noise from external sources including atmospheric noise, galactic noise, and man-made noise, (2) thermal noise generated from dissipative losses in the receiving transmission line system, and (3) noise from sources internal to the receiving system. The level of noise is usually expressed as temperature in degrees Kelvin (K) or as decibels (dB) relative to a standard noise temperature of 290 K.

External Sources

Atmospheric Noise. This noise is produced mostly by lighting discharges in thunderstorms. The noise level thus depends on frequency, time of day, weather, season of the year, and geographic location. Subject to variations due to local stormy areas, this noise generally decreases with increasing latitude on the surface of the earth. Noise is particularly severe during the rainy seasons in areas such as the Caribbean, East Indies, equatorial Africa, and northern India. However, note from Fig. 47.5 that this atmospheric noise is not a significant contribution to the system noise for ultra-high frequency (UHF) and higher frequencies. Noise due to atmospheric absorption can be of importance above 1 GHz when low-noise amplifiers are employed at the receiver. The sky temperature for an infinitely sharp beam is given in Fig. 47.6. For an elevation angle of 0.1 rad, the typical effective sky temperature is 13 K for 1 GHz and 25 K for 10 GHz, increasing to 80 K at 20 GHz.

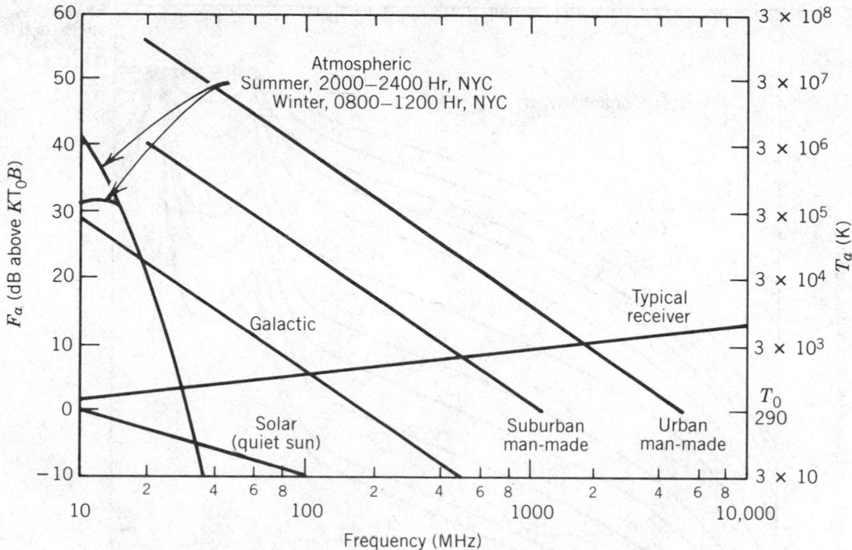

Fig. 47.5 Median values of average noise power expected from various sources. (Source: CCIR 10th Plenary Assembly, "World Distribution and Characteristics of Atmospheric Radio Waves," Report No. 322, Geneva, 1963.)

Galactic Noise. Galactic noise may be defined as the noise at radio frequency caused by disturbances that originate outside the earth or its atmosphere. The chief causes of such noise are the sun and a large number of discrete radio sources distributed chiefly along the galactic plane. Galactic noise reaching the surface of the earth ranges from about 15 MHz to 100 GHz, being limited by ionospheric absorption at the low end of the spectrum and by atmospheric absorption at the high end. In practice, the importance of galactic noise is restricted to frequencies no lower than about 18 MHz by atmospheric noise and to frequencies not higher than 500 MHz by receiver noise temperature and antenna gain, as shown in Fig. 47.5. However, with a high-gain receiving antenna directed at the sun, the antenna noise temperature may exceed 290 K at frequencies as high as 10 GHz. Figure 47.7 shows the level of galactic noise in decibels relative to a noise temperature of 290 K when the antenna is a half-wave dipole. The noise levels shown in this figure assume no atmospheric absorption, and refer to the following sources of galactic noise.

Galactic plane. Galactic noise from the galactic plane in the direction of the center of the galaxy. The noise levels from other parts of the galactic plane can be as much as 20 dB below the levels given in Fig. 47.7.

Quiet sun. Noise from the "quiet" sun; that is, solar noise at times when there is little or no sunspot activity.

Disturbed sun. Noise from the disturbed sun. The term "disturbed" refers to times of sunspot and solar-flare activity.

Cassiopeia. Noise from a high-intensity discrete source of cosmic noise known as Cassiopeia. This is one of more than a hundred known discrete sources, each of which subtends an angle at the earth's surface of less than half a degree.

The levels of cosmic noise received by an antenna directed at a noise source may be estimated by correcting the relative noise levels with a half-wave dipole (from Fig. 47.7) for the receiving antenna gain realized on the noise source. Since the galactic plane is an extended nonuniform noise source, free-space antenna gains cannot be realized, and 10 to 15 dB is approximately the maximum antenna gain that can be realized here. However, on the sun and other discrete sources of galactic noise, antenna gains of 50 dB or more can be obtained.

Man-Made Noise. The amplitude of man-made noise decreases with increasing frequency and varies considerably with location. It is chiefly due to electric motors, neon signs, power lines, and ignition systems located within a few hundred yards of the receiving antenna; certain high-frequency medical appliances and high-voltage transmission lines may, however, cause interference at much

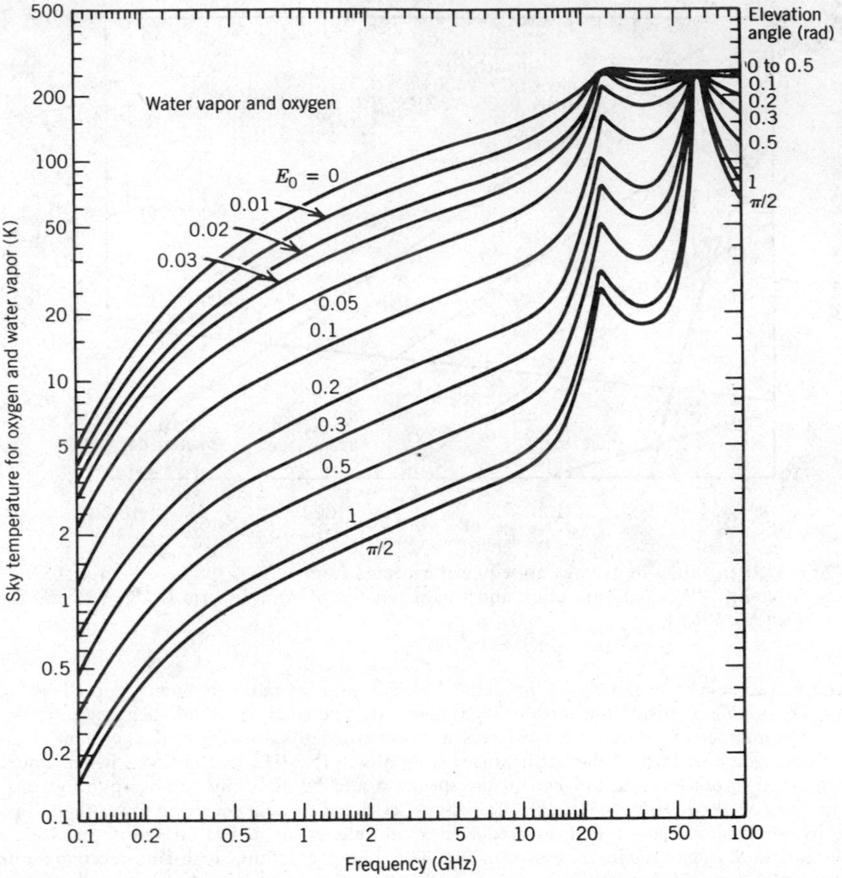

Fig. 47.6 Sky noise temperature due to reradiation by oxygen and water vapor. (Source: CCIR 10th Plenary Assembly, "World Distribution and Characteristics of Atmospheric Radio Waves," Report No. 322, Geneva, 1963.)

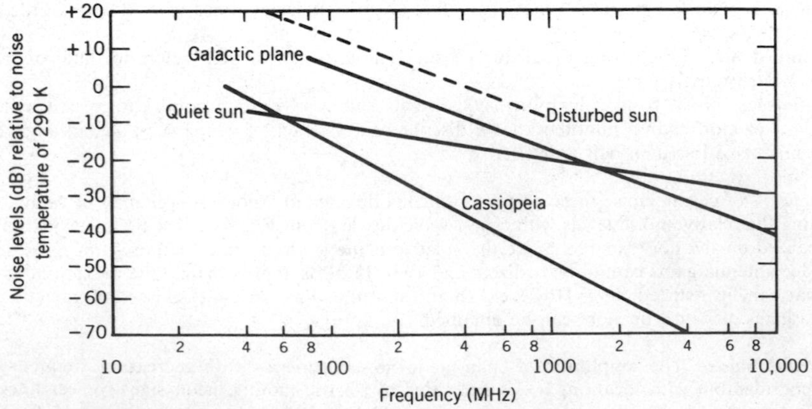

Fig. 47.7 Galactic noise levels for a half-wave-dipole receiving antenna.

greater distances. The average level of man-made noise power can be 16 dB or more higher in urban than in suburban areas in the United States; in remote rural locations the level may be 15 dB below that experienced in a typical suburban site. In quiet remote locations, the noise level from man-made sources will usually be below galactic noise in the frequency range above 10 MHz. Propagation of man-made noise is chiefly by transmission over power lines and by groundwave; however, it may also be by ionospheric reflection at frequencies below about 20 MHz. Measurements indicate that the peak level of man-made noise is not always proportional to bandwidth for bandwidth greater than about 10 kHz. According to the best available information, the peak field strengths of man-made noise (except diathermy and other narrow-band noise) increase as the receiver bandwidth is increased, substantially for bandwidths greater than 10 kHz, as shown in Fig. 47.8.

Internal Sources

Thermal Noise. Thermal noise is caused by the random motion of free electrons in a conductor excited by thermal agitation. All passive components such as waveguides, coaxial cables, couplers, orthomode transducers, and so on of the receiving system generate thermal noise due to a dissipative loss (ohmic loss). Let R = resistive component in ohms of an impedance Z. The mean square value of thermal-noise voltage is given by

$$E^2 = 4RkTB_n$$

where k = Boltzmann's constant (1.38×10^{-23} J/K)
T = absolute temperature, K
B_n = bandwidth, Hz
E = root-mean-square noise voltage

This equation assumes that thermal noise has a uniform distribution of power through the bandwidth B_n. In case two impedances Z_1 and Z_2 with resistive components R_1 and R_2 are in series at the same temperature, the square of the resulting root-mean-square voltage is the sum of the squares of the root-mean-square noise voltages generated in Z_1 and Z_2:

$$E^2 = E_1^2 + E_2^2 = 4(R_1 + R_2)kTB_n$$

In case the same impedances are in parallel at the same temperature, the resulting impedance Z is calculated as is usually done for alternating current circuits, and the resistive component R of Z is then determined. The root-mean-square noise voltage is the same as it would be for a pure resistance R. It is customary in temperate climates to assign to T a value such that $1.38T = 400$, corresponding to about 17°C or 63°F. Then $E^2 = 1.6 \times 10^{-20} RB_n$.

Shot Noise. Shot noise is caused by the random and discrete arrival of electrons in an amplifier. This noise can also come from the current flow in a circuit that includes a thermionic diode or a semiconductor junction. In such a circuit each charge carrier contributes to current flow during the course of the time it is crossing from cathode to anode, or crossing the junction. The average junction current determines the average interval that elapses between the times when two successive carriers enter the junction. However, the exact interval that elapses is subject to random statistical fluctuations. This randomness gives rise to a type of noise called shot noise. This noise consists of pulses of identical shape but of random amplitudes and random times of occurrences. The mean squared value

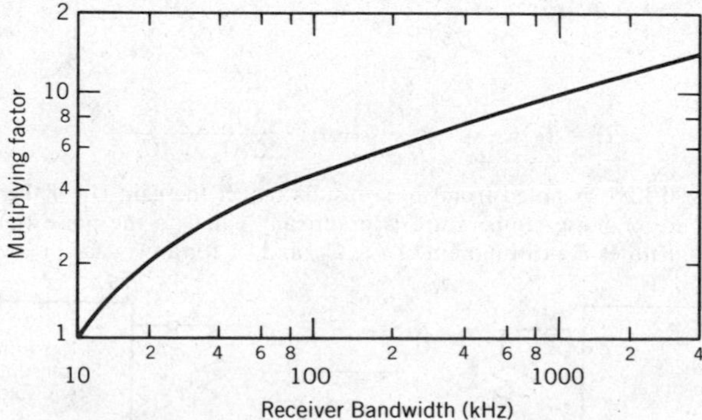

Fig. 47.8 Bandwidth factor for man-made noise.

of shot noise current is given by

$$I^2 = 2eI_{DC}B_n$$

where e is the charge on an electron ($= 1.6008 \times 10^{-19}$ C) and I_{DC} is the average (direct) current.

47.2-2 Parameters and Characteristics

Noise Figure (Noise Factor)

Noise figure is usually used to describe the quality of a receiver as far as noise is concerned. In a linear two-port transducer, the noise figure F is defined by

$$F = \frac{S_i/N_i}{S_O/N_O} = \frac{N_O}{kTB_nG}$$

where S_i = available input signal power
N_i = available input noise power = kTB_n
S_O = available output signal power
N_O = available output noise power
B_n = equivalent noise bandwidth of receiver
G = average available power gain

By definition, the input noise power is equivalent to the thermal noise power provided by a resistor matched to the input terminals of the transducer at a temperature of $T_0 = 290$ K, and hence is equal to kT_0B_n. This quantity is frequently referred to as the theoretical ambient noise power. Noise figure is commonly expressed in decibels where $F_{dB} = 10 \log F$.

Noise Temperature

For noisy receivers ($F > 1$) the concept of noise figure is adequate to describe performance. However, with the advent of low-noise receivers, such as masers and parametric amplifiers, a more useful measure of receiver noise is the effective input noise temperature of the receiver. This is defined (for a two-port transducer) as that temperature T_r of a fictitious passive resistor at the input of an ideal noise-free receiver that would generate the same output noise power as that of the actual transducer connected to a noise-free termination, as illustrated in Fig. 47.9, where $\Delta N = kT_rB_nG$ and $T_r = (F-1)T_0 = (F-1)290$ K. Figure 47.10 plots the relationship between noise temperature and noise figure.

Noise in Cascaded Networks

For simplicity, consider two networks in cascade, each with the same noise bandwidth B_n but with different noise figures (or noise temperatures) and available gains. The overall noise figure F_{12} of the two networks in cascade is $F_{12} = F_1 + (F_2 - 1)/G_1$ and the equivalent input noise temperature of the combination is $T_e = T_1 + T_2/G_1$, where F_1, G_1, and T_1 are the noise figure, available gain, and noise temperature, respectively, of the first network and F_2, G_2, and T_2 are the same values for the second network, as shown in Fig. 47.11. Similarly, the noise figure and equivalent input noise temperature of n networks in cascade are given by

$$F = F_1 + \frac{F_2 - 1}{G_1} + \frac{F_3 - 1}{G_1G_2} + \cdots + \frac{F_n - 1}{G_1G_2 \ldots G_{n-1}}$$

and

$$T_e = T_1 + \frac{T_2}{G_1} + \frac{T_3}{G_1G_2} + \cdots \frac{T_n}{G_1G_2 \ldots G_{n-1}}$$

where $T_i = (F_i - 1)290$ K. It is noted from these results that if the gain G_1 of the first stage is high, the overall noise figure or noise temperature is practically equal to the noise figure F_1 of the first stage. Figure 47.12 illustrates F as a function of F_1, F_2, and G_1 for a two-stage network.

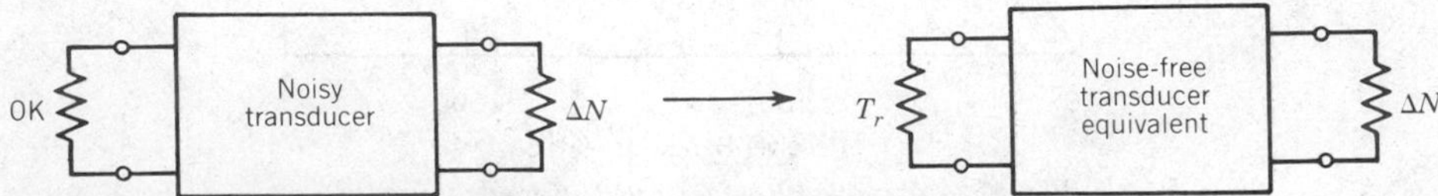

Fig. 47.9 Equivalent noisy transducer.

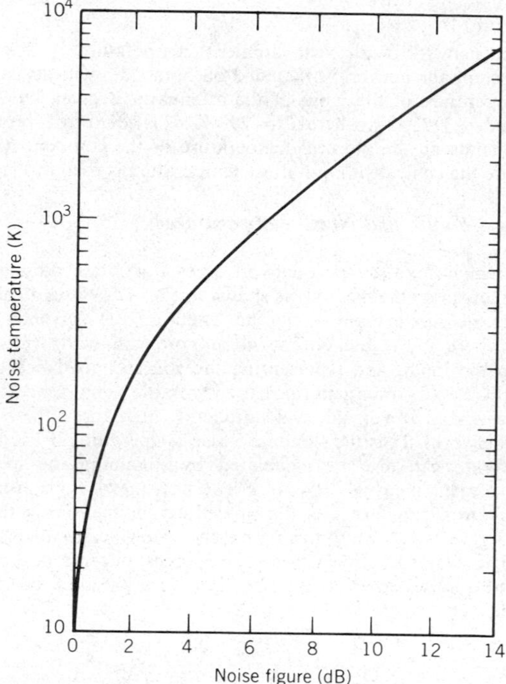

Fig. 47.10 Relationship between noise temperature and noise figure.

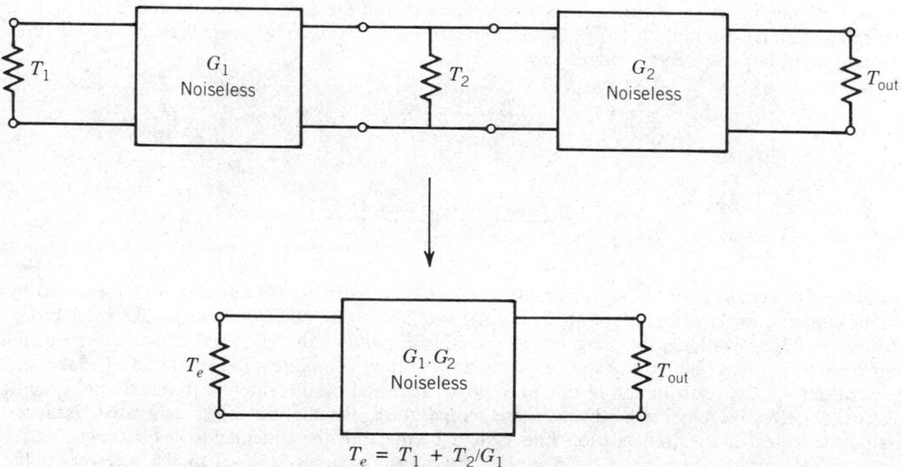

Fig. 47.11 Two amplifiers in cascade (equivalent noise temperature).

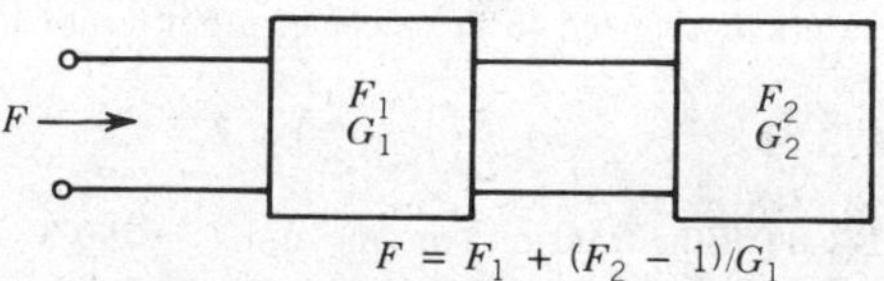

Fig. 47.12 Two amplifiers in cascade.

Noise Due to Lossy Networks

Consider an attenuating network of physical (ambient) temperature T_p and loss factor $L(>1)$, as shown in Fig. 47.13. Assume the network is matched on both sides with its characteristic impedance. The effective noise temperature of the input of the attenuator is given by $T_e = (L-1)T_p$ and the noise figure is $F = 1 + (L-1)T_p/T_0$, where $T_0 = 290$ K. F is equal to L provided $T_p = T_0$. Since F for a lossy network depends on the physical temperature of the component, it is not an especially useful concept. Therefore the equivalent input noise temperature is used instead.

Measurements of Noise Figure and Noise Temperature

The noise figure measurement for a two-port network is usually obtained by measuring the total noise power delivered by the network. One method, as shown in Fig. 47.14 uses a calibrated standard signal generator at the input to measure experimentally the function $G(f)$. F is then calculated according to $F = N_0/kT_0\int G(f)\,df$, where N_0 is the output noise power in watts (measured with the signal generator connected to the input, and representing the source impedance but not generating any signal) and $G(f)$ is the available power gain (measured with the signal generator and the power meter at all essential frequencies, and at a signal level sufficiently high above the noise level but below the saturation level of the network). T_0 is the standard noise temperature 290 K and k is the Boltzmann constant. The noise figure can also be calculated by obtaining the measured effective noise temperature T_E. The measurement of T_E for a two-port network is usually by means of the "Y-factor" method. Figure 47.15 illustrates the procedure for measuring the effective input noise temperature and noise figure. As shown in this figure, two known calibrated ("hot" and "cold") noise sources are required. By switching between these two sources, one can determine a Y-factor using a precision attenuator and a noise power meter. The Y-factor is the ratio of noise power delivered by the hot source to that delivered by the cold source, that is,

$$Y = \frac{T_{\text{hot}} + T_E}{T_{\text{cold}} + T_E}$$

where T_{hot} and T_{cold} are effective noise temperatures of hot and cold sources, respectively, and T_E is the effective noise temperature of the network under test. The measurement is as follows: (1) Connect the hot noise generator to the network under test and adjust gain for convenient meter reference, (2) switch the cold noise generator; adjust the attenuator for the same meter reference, (3) note the change in attenuator setting, ΔA (dB), (4) calculate $Y = 10^{-\Delta A/10}$, and (5) calculate the noise figure or effective input noise temperature given by

$$T_E = \frac{T_{\text{hot}} - YT_{\text{cold}}}{Y-1}$$

and

$$F = \frac{T_{\text{hot}}/T_0 - YT_{\text{cold}}/T_0}{Y-1} + 1$$

System Noise Temperature. The total noise power in the receiving system may be represented by a system temperature T_s of such level that the total available noise power referred to the receiver input terminal is kT_sB_n, with B_n denoting the receiver noise bandwidth. The total effective system noise temperature T_s may be considered to consist of three components: (1) effective space noise temperature T_a, the contribution of the noise power received by the antenna from external radiating sources, (2) effective passive hardware noise temperature, the thermal noise generated because of dissipative losses in the transmission line system connecting the antenna to the receiver, and (3) effective RF amplification noise temperature, noise from sources internal to the receiver itself. It should be noted here that this division of the receiving system into three components is rather arbitrary, since very often the antenna system contains the transmission line, thus eliminating the second component. The effective noise temperature of each of the three noise sources shown schematically in Fig. 47.16 are represented by T_a, T_t, and T_r, respectively.

The system noise temperature T_s referred to the receiver input terminal is given by

$$T_s = \frac{T_a}{L} + T_t\left(1 - \frac{1}{L}\right) + T_r$$

where L is the loss factor ($L > 1$) of the transmission line and $T_t = 290$ K, the reference temperature, for the receiver noise figure measurement. Note that the system noise temperature referred to the

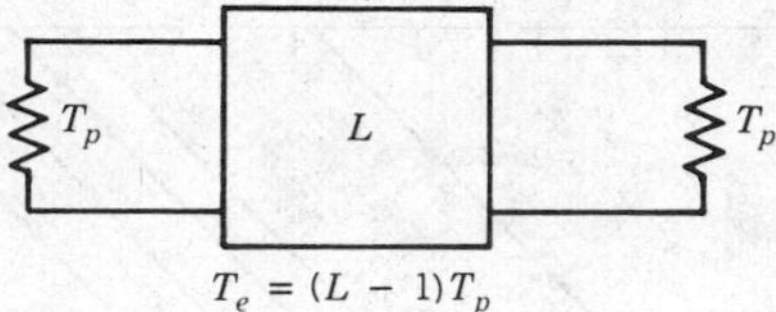

Fig. 47.13 Noise generated in a lossy network.

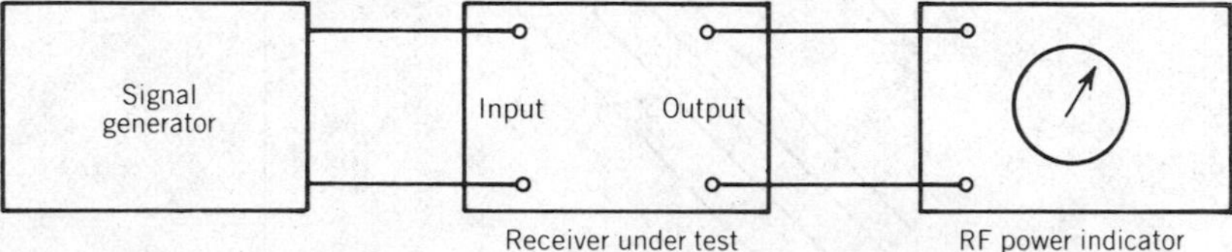

Fig. 47.14 Measurement of noise figure.

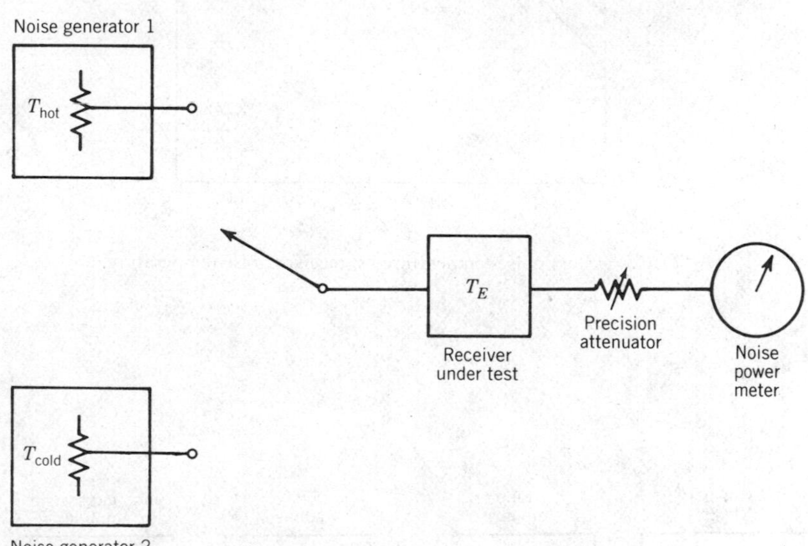

Fig. 47.15 Effective noise temperature measurement of true receiver.

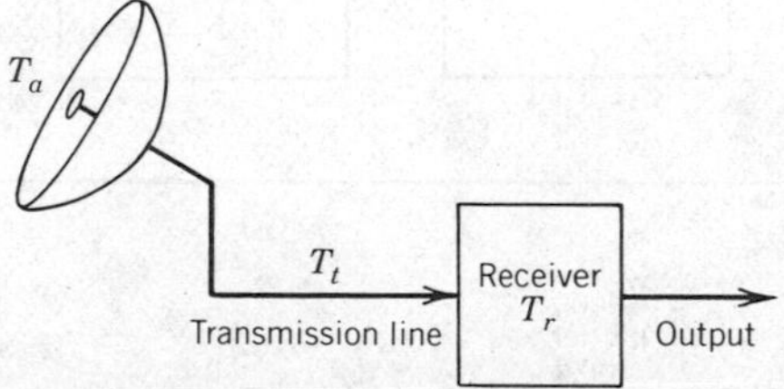

Fig. 47.16 Typical receiving system.

input of the transmission line is given by

$$T_s' = T_a + (L - 1)T_t + LT_r$$

The importance of minimizing the loss between the antenna and the receiver input is illustrated in Fig. 47.17, which contains a plot of the system noise temperature T_s vs. receiver noise temperature T_r for a given antenna temperature of $T_a = 100$ K and for various values of L (in decibels). The system noise temperature for a commonly used typical receiving system is given in Fig. 47.18.

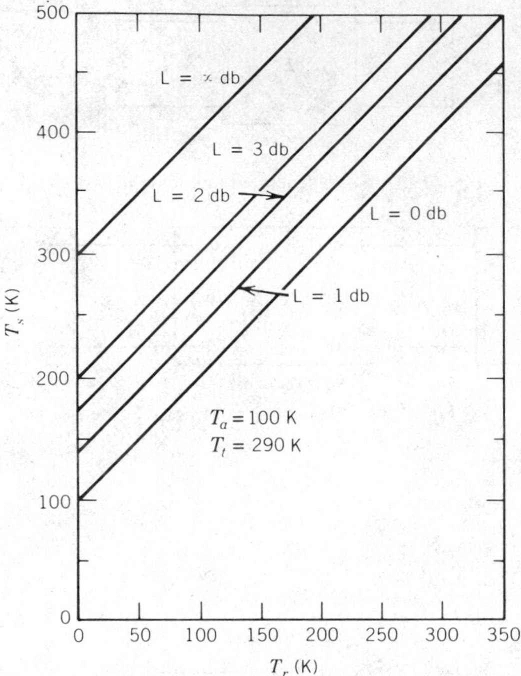

Fig. 47.17 System noise temperature vs. receiver noise temperature.

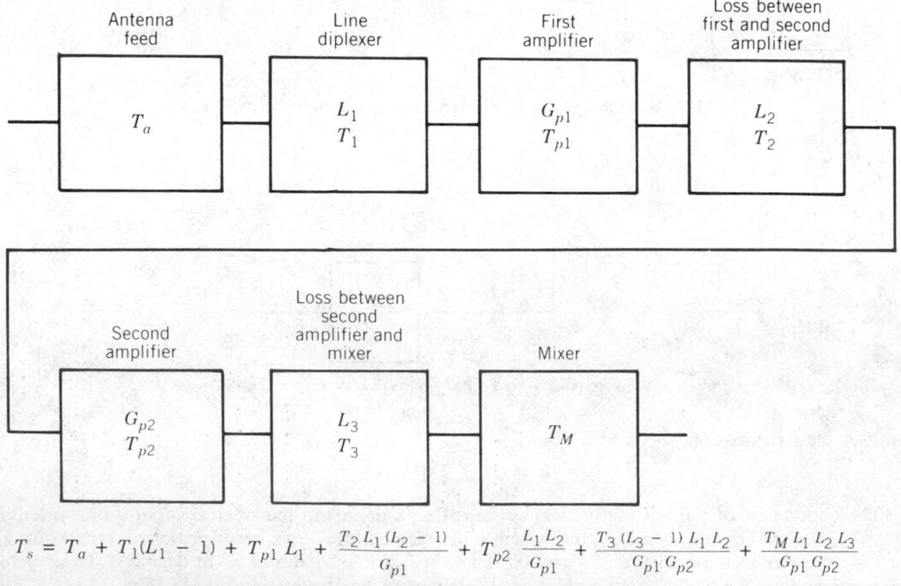

$$T_s = T_a + T_1(L_1 - 1) + T_{p1} L_1 + \frac{T_2 L_1 (L_2 - 1)}{G_{p1}} + T_{p2} \frac{L_1 L_2}{G_{p1}} + \frac{T_3 (L_3 - 1) L_1 L_2}{G_{p1} G_{p2}} + \frac{T_M L_1 L_2 L_3}{G_{p1} G_{p2}}$$

Fig. 47.18 System noise temperature for a typical receiving system.

Antenna Noise Temperature. The antenna receives noise that may originate from celestial or atmospheric sources, from the ground or nearby objects, and from man-made sources such as ignition interference. The latter may have peculiar characteristics and is not always present, but the other sources are in general present, as shown in Fig. 47.19. The noise power received by an antenna from all natural external sources is customarily represented by an effective noise temperature T_a defined by

$$T_a = \frac{P_a}{kB_n}$$

where P_a is the noise power available from the antenna over the frequency range B_n. Antenna noise temperature is the temperature at which the equivalent antenna radiation impedance must be maintained to generate the amount of thermal noise power received by the antenna from its external surroundings. Antenna temperature then gives a measure of the amount of noise power received with the signal, and as such becomes a fundamental sensitivity limitation on any communication system. It should be emphasized here that the antenna noise temperature is an equivalent or fictitious noise temperature equal to that of the antenna. Thus the antenna noise temperature T_a is the temperature that the antenna "sees" rather than the physical temperature that the antenna "feels." As an example, the antenna structure may be in a cold environment of 100 K, and yet T_a may be at several thousand K. On the other hand, the antenna structure may measure 300 K, while T_a may be only 30 K. The antenna noise temperature T_a depends both on the direction in which the antenna is pointed and on the pattern characteristics of the antenna itself. The effective antenna noise temperature T_a is the sum of noise contributions from various sources: the transmission line that feeds the antenna; resistive losses in the antenna; sky noise due to the galaxy, sun, and moon; absorption by atmospheric gases and precipitation; radiation from earth into backlobes of the antenna; and interference from man-made radio sources. The suppression of antenna sidelobe levels in low-noise systems is of great importance because, while the main beam may be looking at a cold region of the sky, one of the sidelobes may be looking at a hot region such as the sun to produce a large value of T_a. For the total antenna noise, the equivalent diagram of the complete path is illustrated in Fig. 47.20. With respect to the reference port, all noise sources are combined into one and the corresponding noise temperature

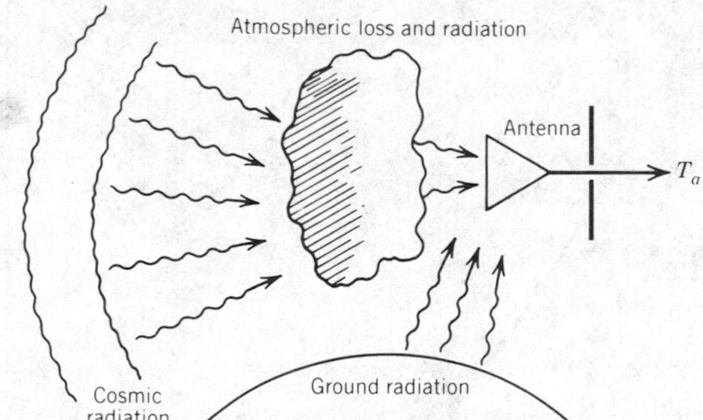

Fig. 47.19 External elements contributing to antenna noise temperature.

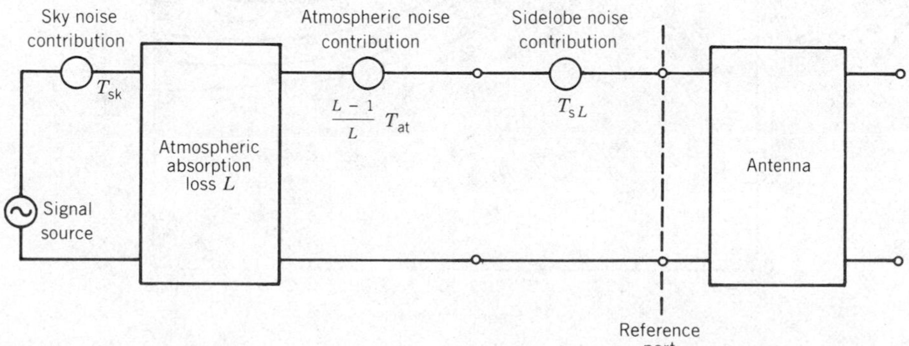

Fig. 47.20 Equivalent circuit for calculation of antenna noise temperature.

is calculated:

$$T_a = \frac{T_{sk}}{L} + \frac{(L-1)T_{at}}{L} + T_{sL}$$

where T_{sk} = background sky temperature
L = loss factor of medium ($= e^{\alpha l}$), α the absorption coefficient for the atmosphere
T_{at} = atmosphere temperature
T_{sL} = antenna noise contributed by all sidelobes

It is of practical interest in many instances, while making system calculations, to take antenna noise into account in a general way without the necessity of a detailed computation. Typical antenna noise curves derived by Blake are reproduced in Fig. 47.21. For many applications the accuracy provided by these curves will be adequate. The curves assume a surface-based antenna and are applicable neither to airborne radar antenna nor to ground-based antennas at extremely high altitudes. The result of this assumption is to include the effect of the entire earth's atmosphere between the antenna and extraterrestrial noise sources. The dashed curves represent maximum and minimum cosmic and atmospheric noise.

Measurement of Antenna Noise

The effective antenna noise temperature T_a can also be measured by means of the Y-factor method using a known calibrated noise source, as shown in Fig. 47.22.

$$Y = \frac{T_p + T_E}{T_a + T_E}$$

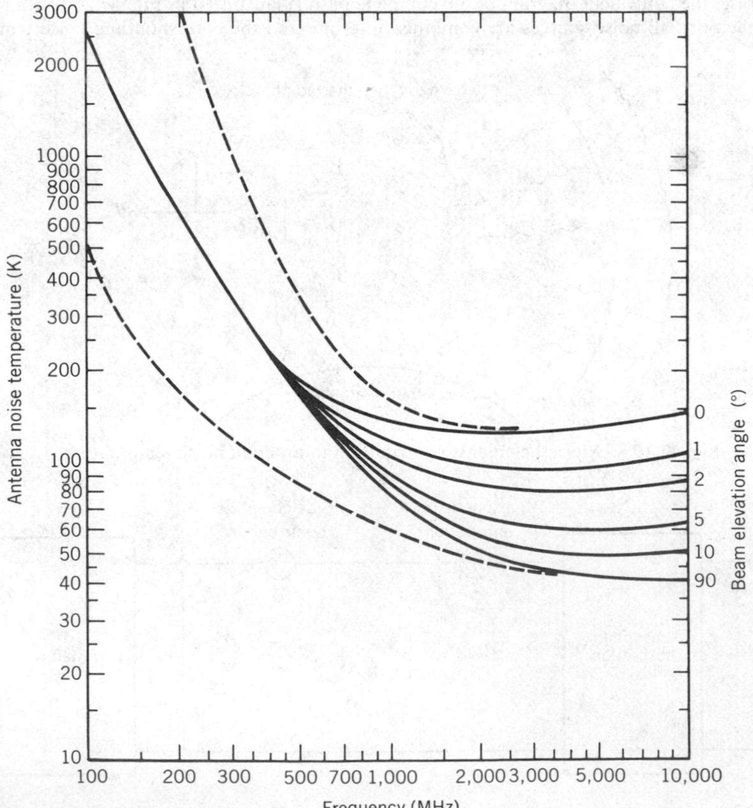

Fig. 47.21 Noise temperature of a typical directive antenna for representative environmental conditions.

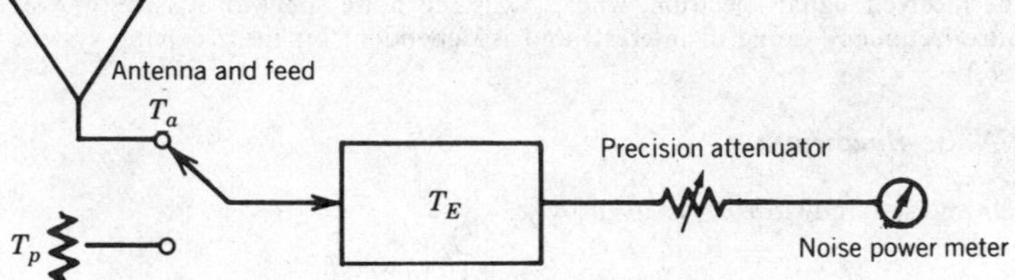

Fig. 47.22 Effective noise temperature measurement of the source.

and

$$T_a = \frac{T_p + T_E}{Y - T_E}$$

where T_p is the effective noise temperature of the calibrated noise surface and T_E is the effective noise temperature of the receiver measured using the Y-factor method.

Additive White Gaussian Noise (AWGN)

Thermal noise has a power spectral density that is quite uniform up to frequencies of the order 10^{13} Hz. Shot noise has a power spectral density that is reasonably constant up to frequencies of the order of the reciprocal of the transit time of charge carriers across the junction. Other noise sources like atmospheric and galactic noise sources similarly have very wide spectral ranges. These kinds of noise are usually referred to as additive white gaussian noise. White noise is noise whose power spectral density is uniform over the entire frequency range of interest. The term "white" is used in analogy with white light, which is a superposition of all visible spectral components. "Additive" refers to noise that is added to the signal to be received by the communication system. "Gaussian" refers to noise whose probability density function is gaussian. In very many communication systems and in a wide variety of circumstances, the assumption of a gaussian density function is justifiable. On the other hand, such an assumption is hardly universally valid. For example, if gaussian noise is applied to the input of a rectifier circuit, the output is not gaussian. Similarly, it may well be that the noise encountered on a telephone line or on other channels consists of short, pulse-type disturbances whose amplitude distribution is decidedly not gaussian. Figure 47.23 shows a typical AWGN model, where $s(t)$ is the signal, $n(t)$ is the channel noise, and $y(t)$ is the received signal plus noise. Figure 47.24

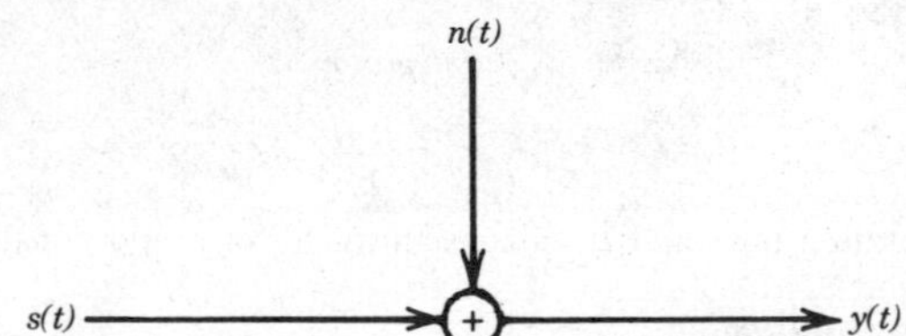

Fig. 47.23 Additive White Gaussian Noise (AWGN) model.

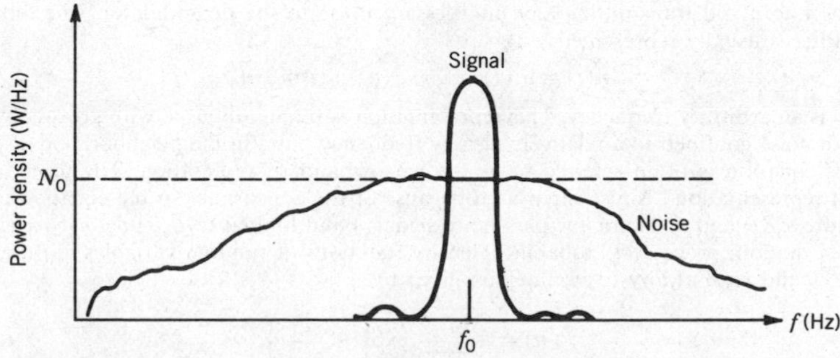

Fig. 47.24 Received signal spectrum.

illustrates the received signal spectrum, where N_0 is the noise spectral density (one-sided, constant over the entire frequency range of interest) and is dependent on the receiving system temperature (i.e., $N_0 = kT_s$).

Equivalent Noise Bandwidth

The equivalent noise bandwidth B_n is given by

$$B_n = (1/2\pi)\int_{-\infty}^{\infty} \frac{|H(jw)|^2\, dw}{|H(jw_0)|^2}$$

where $H(jw)$ is the transfer function of the network and w_0 denotes the angular frequency of maximum response. B_n is therefore equal to the bandwidth of an equivalent rectangular filter of height $H(jw_0)$, whose noise power output is equal to the total output noise from the network.

Effect of Filtering

Rectangular (ideal) low-pass filter

$$P_n = N_0 B_n$$

where P_n is the output noise power.

RC low-pass filter

$$P_n = \frac{\pi N_0 f_c}{2}$$

where f_c is 3-dB frequency of the low-pass filter.

Rectangular bandpass filter

$$P_n = N_0(f_2 - f_1)$$

where $(f_2 - f_1)$ is the filter bandwidth.

Differentiating filter (followed by rectangular low-pass filter)

$$P_n = \frac{4\pi^2 N_0 \tau^2 B_n^3}{3}$$

where τ is a constant factor of proportionality of the differentiating filter.

Integrator

$$P_n = \frac{N_0 T}{2\tau^2}$$

where T is the integration time, and τ is a constant factor of proportionality.

Bandpass Noise Representation

To minimize the noise power that is presented to the demodulator of a receiving system, a filter before the demodulator is usually employed. The bandwidth B_n of the filter is made as narrow as possible so as to avoid transmitting any unnecessary noise to the demodulator. The output of this bandpass filter is usually represented by:

$$n(t) = n_c(t)\cos 2\pi f_0 t + n_s(t)\sin 2\pi f_0 t$$

in which f_0 is an arbitrary frequency. This representation is frequently used with great convenience in dealing with noise confined to a relatively narrow frequency band in the neighborhood of f_0. For this reason, this equation is often referred to as the narrowband representation. The term "quadrature component representation" is also often used because of the appearance in the equation of sinusoids in quadrature. $n_c(t)$ and $n_s(t)$ are low-pass noise signals band limited to B_n. Like $n(t)$, $n_c(t)$ and $n_s(t)$ are gaussian random processes. Probability density functions of random variables n, n_c, and n_s [that is, $n(t)$, $n_c(t)$, and $n_s(t)$ at any fixed time] are given by

$$P(n) = \frac{1}{\sqrt{2\pi\sigma_n^2}} \exp\left(-\frac{n^2}{2\sigma_n^2}\right)$$

and

$$P(n_c) = \frac{1}{\sqrt{2\pi\sigma_{n_c}^2}} \exp\left(-\frac{n_c^2}{2\sigma_{n_c}^2}\right)$$

$$P(n_s) = \frac{1}{\sqrt{2\pi\sigma_{n_s}^2}} \exp\left(-\frac{n_s^2}{2\sigma_{n_s}^2}\right)$$

where $\sigma_n^2 = \sigma_{n_c}^2 = \sigma_{n_s}^2 = N_0 B_n$ = total noise power = mean square values of $n(t)$, $n_c(t)$, and $n_s(t)$, respectively. $n(t)$ can also be represented in the polar form

$$n(t) = R(t)\cos[2\pi f_0 t + \theta(t)]$$

where

$$R(t) = \sqrt{n_c^2(t) + n_s^2(t)}$$

and

$$\theta(t) = \tan^{-1}\left[\frac{n_s(t)}{n_c(t)}\right]$$

Since both $n_c(t)$ and $n_s(t)$ are very slowly varying signals, it follows that $R(t)$ and $\theta(t)$ are also slowly varying signals. The bandpass noise signal with a narrowband spectrum has the appearance of a sinusoidal signal with amplitude and phase varying slowly. The envelope of this signal is given by $R(t)$ and the phase is $\theta(t)$.

Bibliography

Benoit, A., "Signal Attenuation Due to Neutral Oxygen and Water Vapour, Rain and Clouds," *Microwave J.*, pp. 73–80 (Nov. 1968).

Blake, L. V., "Antenna and Receiving-System Noise-Temperature Calculation," *NRL Rep. 5668* (Sep. 19, 1961).

CCIR 10th Plenary Assembly, "World Distribution and Characteristics of Atmospheric Radio Waves," Report No. 322, Geneva, 1963.

CCIR 11th Plenary Assembly, vol. 2, Report No. 234, Oslo, 1966.

Crane, R. K., "Propagation Phenomena Affecting Satellite Communication Systems," *Proc. IEEE*, p. 178 (Feb. 1971).

Crichlow, W. Q., et al., "Special Report on Characteristics of Terrestrial Radio Noise," International Scientific Radio Union (URSI), Commission IV, August 1960.

Dijk, J., M. Jeuken, and F. J. Maanders, "Antenna Noise Temperature," *Proc.* IEE, **115**(10):1403–1410 (Oct. 1968).

Fredrick Research Corp., *Handbook of Radio-Frequency Interference*, vols. 1 to 4, Wheaton, MA, 1962.

Grimm, H. H., "Noise Temperature in Passive Circuits," *Microwave J.*, pp. 52–54 (Feb. 1960).

Hogg, D. C., "Millimeter-Wave Communication Through the Atmosphere," *Science* **159**(3810): 39–46 (5 Jan. 1968).

Johnson, J. B., "Thermal Agitation of Electricity in Conductors," *Phys. Rev.* **32**:97–109 (Jul. 1928).

MacDonald, D. K. C., *Noise and Fluctuations: An Introduction*, Wiley, New York, 1962.

Mumford, W. W., and E. H. Scheibe, *Noise Performance Factors in Communication Systems*, Horizon House-Microwave, Delham, MA, 1968.

Powsey, J. L. and R. N. Bracewell, *Radio Astronomy*, Clarendon Press, Oxford, England, 1955.

Rusch, W. V. T. and P. D. Potter., *Analysis of Reflector Antennas*, Academic, New York, 1970.

Staras, H., "The Propagation of Wideband Signals Through the Ionosphere," *Proc. IRE* **49**:1211 (Jul. 1961).

Steinberg, J. L. and J. Lequeux, *Radio Astronomy*, McGraw-Hill, New York, 1963.

Stephenson, R. D., External Noise, in A. V. Balakrishnan, ed., *Space Communications*, Ch. 6, McGraw-Hill, New York, 1963.

Taub, H. and D. L. Schilling, *Principles of Communication Systems*, Chs. 7, 8, 9, McGraw-Hill, New York, 1971.

Watt, A. D., R. M. Coon, E. L. Maxwell, and R. W. Plush, "Performance of Some Radio Systems in the Presence of Thermal and Atmospheric Noise," *Proc. IRE* **46**:1914–1923 (Dec. 1958).

47.3 INFORMATION THEORY

Rodger E. Ziemer and William H. Tranter

Information theory presents us with the performance characteristics of an *ideal*, or optimum, communication system. The performance of an ideal system provides a meaningful basis for comparing the performance of the realizable systems. It illustrates the gain in performance that can be obtained by implementing more complicated transmission and detection schemes.

Motivation for the study of information theory is provided by Shannon's coding theorem, sometimes referred to as Shannon's second theorem, which can be stated as follows: If a source has an information rate less than the channel capacity, there exists an encoding procedure such that the source output can be transmitted over the channel with an arbitrarily small probability of error.

This is a truly surprising statement. We are being led to believe that transmission and reception can be accomplished with *negligible* error, even in the presence of noise. An understanding of this process called coding, and an understanding of its impact on the design and performance of communication systems, require an understanding of several basic concepts of information theory.

47.3-1 Basic Concepts

Consider a hypothetical classroom situation occurring early in a course at the end of a class period. The professor makes one of the following statements to the class:

I shall see you next period. (Statement A)

My colleague will lecture next period. (Statement B)

Everyone gets a grade of A in the course, and there will be no more class meetings. (Statement C)

What is the relative information conveyed to the student by each of these statements, assuming that there had been no previous discussion on the subject? Obviously there is little information conveyed by statement (A), since the class would normally assume that their regular professor would lecture, that is, the probability of the regular professor lecturing, $P(A)$, is nearly unity. Intuitively, we know that statement (B) contains more information, and the probability of a colleague lecturing, $P(B)$, is relatively low. The third statement, (C), contains a vast amount of information for the entire class, and most would agree that such a statement has a very low probability of occurrence in a typical classroom situation. It appears that the lower the probability of a statement, the greater is the information conveyed by that statement. Stated another way, the students' surprise on hearing a statement seems to be a good measure of the information contained in that statement.

We shall now define information mathematically in such a way that the definition is consistent with the preceding intuitive example.

Information

Let x_j be an event that occurs with probability $p(x_j)$. If we are told that event x_j has occurred, we say that we have received

$$I(x_j) = \log_a \frac{1}{p(x_j)} = -\log_a p(x_j) \tag{47.1}$$

units of information. This definition of information is consistent with the previous example, since $I(x_j)$ increases as $p(x_j)$ decreases.

The base of the logarithm in Eq. (47.1) is quite arbitrary and determines the units by which we measure information. Hartley, who first suggested the logarithmic measure of information in 1928, used logarithms to the base 10, and the measure of information was the hartley. Today it is standard to use logarithms to the base 2, and the unit of information is the binary unit or bit. Logarithms to the base e are sometimes utilized, and the corresponding unit is the nat.

There are several reasons for us to be consistent in using the base 2 logarithm to measure information. The simplest random experiment that one can imagine is one with two equally likely

---

Reproduced from R. E. Ziemer and W. H. Tranter, *Principles of Communications*, Houghton Mifflin, Boston, 1976, with permission.

outcomes, such as the flipping of an unbiased coin. Knowledge of each outcome has associated with it one bit of information. Also, since the digital computer is a binary machine, each logical 0 and each logical 1 has associated with it 1 bit of information, assuming that each of these logical states are equally likely.

Example 47.1. Consider a random experiment with 16 equally likely outcomes. The information associated with each outcome is

$$I(x_j) = -\log_2 \tfrac{1}{16} = \log_2 16 = 4 \text{ bits}$$

where j ranges from 1 to 16. The information is greater than 1 bit, since the probability of each outcome is much less than one half.

Entropy

In general the *average information* associated with the outcome of an experiment is of interest rather than the information associated with each particular event. The average information associated with a discrete random variable X is defined as the entropy $H(X)$. Thus

$$H(X) = E\{I(x_j)\} = -\sum_{j=1}^{n} p(x_j)\log_2 p(x_j) \tag{47.2}$$

where n is the total number of possible outcomes. Entropy can be regarded as average uncertainty, and therefore should be maximum when each outcome is equally likely.

Example 47.2. For a binary source $p(1) = \alpha$ and $p(0) = 1 - \alpha = \beta$, derive the entropy of the source as a function of α and sketch $H(\alpha)$ as α varies from zero to one.

From Eq. (47.2)

$$H(\alpha) = -\alpha \log_2 \alpha - (1 - \alpha)\log_2(1 - \alpha) \tag{47.3}$$

This is sketched in Fig. 47.25. We should note the maximum. If $\alpha = \frac{1}{2}$, each symbol is equally likely and our uncertainty is a maximum. If $\alpha \neq \frac{1}{2}$, one symbol is more likely to occur than the other, and we are less uncertain as to which symbol appears on the source output. If α or β is equal to zero, our uncertainty is zero, since we know exactly which symbol will occur.

From the preceding example we concluded, at least intuitively, that the entropy function has a maximum, and the maximum occurs when all probabilities are equal. This fact is of sufficient importance to warrant a more complete derivation.

Assume that a random process has n possible outcomes and that p_n is a dependent variable depending on the other probabilities. Thus

$$p_n = 1 - (p_1 + p_2 + \cdots + p_k + \cdots + p_{n-1}) \tag{47.4}$$

where p_j is concise notation for $p(x_j)$. The entropy associated with the process is

$$H = -\sum_{i=1}^{n} p_i \log_2 p_i \tag{47.5}$$

To find the maximum value of entropy, the entropy is differentiated with respect to p_k, holding all probabilities constant except p_k and p_n. This gives a relationship between p_k and p_n that yields the

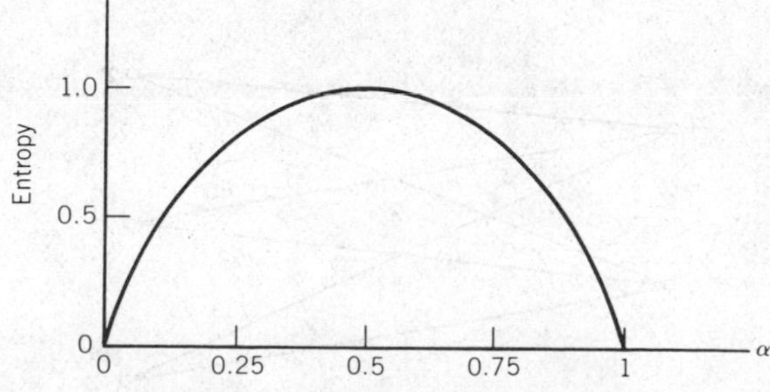

Fig. 47.25 Entropy of a binary source.

maximum value of H. Since all derivatives are zero except the ones involving p_k and p_n

$$\frac{dH}{dp_k} = \frac{d}{dp_k}(-p_k \log_2 p_k - p_n \log_2 p_n) \tag{47.6}$$

By using Eq. (47.4) and

$$\frac{d}{dx}\log_a u = \frac{1}{u}\log_a e \frac{du}{dx}$$

we obtain

$$\frac{dH}{dp_k} = -p_k \frac{1}{p_k}\log_2 e - \log_2 p_k + p_n \frac{1}{p_n}\log_2 e + \log_2 p_n$$

or

$$\frac{dH}{dp_k} = \log_2 \frac{p_n}{p_k}$$

which is zero if $p_k = p_n$. Since p_k is arbitrary

$$p_1 = p_2 = \cdots = p_n = \frac{1}{n} \tag{47.7}$$

To show that the above condition yields a maximum and not a minimum, note that when $p_1 = 1$ and all other probabilities are zero, entropy is zero. From Eq. (47.5), the case where all probabilities are equal yields $H = \log_2 n$.

Channel Representations

Throughout most of this chapter the communication channel will be assumed to be memoryless. For such channels, the channel output at a given time is a function of the channel input *at that time* and is not a function of previous channel inputs. Memoryless discrete channels are completely specified by the set of conditional probabilities that relate the probability of each output state to the input probabilities. An example illustrates the technique. A diagram of a channel with two inputs and three outputs is illustrated in Fig. 47.26. Each possible input-to-output path is indicated along with a conditional probability p_{ij}, which is concise notation for $p(y_j \mid x_i)$. Thus p_{ij} is the conditional probability of obtaining output y_j given that the input is x_i, and is called a channel transition probability.

We can see from Fig. 47.26 that the channel is completely specified by the complete set of transition probabilities. Accordingly, the channel is often specified by the matrix of transition probabilities $[P(Y \mid X)]$ where, for the channel of Fig. 47.26,

$$[P(Y \mid X)] = \begin{bmatrix} p(y_1 \mid x_1) & p(y_2 \mid x_1) & p(y_3 \mid x_1) \\ p(y_1 \mid x_2) & p(y_2 \mid x_2) & p(y_3 \mid x_2) \end{bmatrix} \tag{47.8}$$

Since each input to the channel results in some output, each row of the channel matrix must sum to unity.

The channel matrix is useful in deriving the output probabilities given the input probabilities. For example, if the input probabilities $P(X)$ are represented by the row matrix

$$[P(X)] = \begin{bmatrix} p(x_1) & p(x_2) \end{bmatrix} \tag{47.9}$$

then

$$[P(Y)] = \begin{bmatrix} p(y_1) & p(y_2) & p(y_3) \end{bmatrix} \tag{47.10}$$

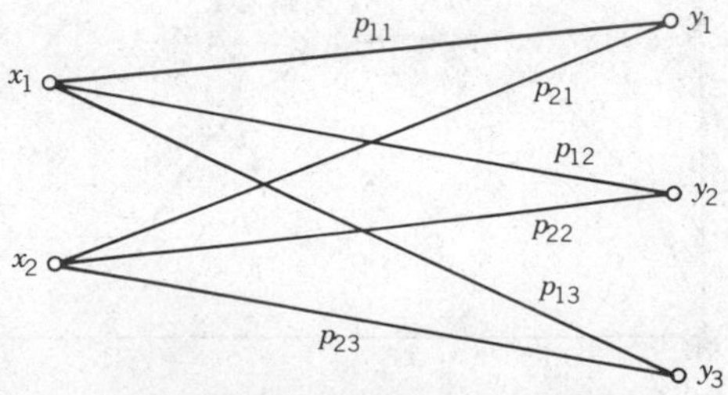

Fig. 47.26 Channel diagram.

which is computed by

$$[P(Y)] = [P(X)][P(Y \mid X)] \tag{47.11}$$

If $[P(X)]$ is written as a diagonal matrix, Eq. (47.11) yields a matrix $[P(X, Y)]$. Each element in the matrix has the form $p(x_i)p(y_j \mid x_i)$, or $p(x_i, y_j)$. This matrix is known as the joint probability matrix, and the term $p(x_i, y_j)$ is the joint probability of transmitting x_i *and* receiving y_j.

Example 47.3. Consider the binary input-output channel shown in Fig. 47.27. The matrix of transition probabilities is

$$P[Y \mid X] = \begin{bmatrix} 0.7 & 0.3 \\ 0.4 & 0.6 \end{bmatrix}$$

If the input probabilities are $P(x_1) = 0.5$ and $P(x_2) = 0.5$, the output probabilities are

$$P(Y) = [0.5 \quad 0.5]\begin{bmatrix} 0.7 & 0.3 \\ 0.4 & 0.6 \end{bmatrix} = [0.55 \quad 0.45]$$

and the joint probability matrix is

$$P[X, Y] = \begin{bmatrix} 0.5 & 0 \\ 0 & 0.5 \end{bmatrix}\begin{bmatrix} 0.7 & 0.3 \\ 0.4 & 0.6 \end{bmatrix} = \begin{bmatrix} 0.35 & 0.15 \\ 0.2 & 0.3 \end{bmatrix}$$

Joint and Conditional Entropy

If we use the input probabilities $p(x_i)$, the output probabilities $p(y_j)$, the transition probabilities $p(y_j \mid x_i)$, and the joint probabilities $p(x_i, y_j)$, we can define several different entropy functions for a channel with n inputs and m outputs. These are

$$H(X) = -\sum_{i=1}^{n} p(x_i)\log_2 p(x_i) \tag{47.12}$$

$$H(Y) = -\sum_{j=1}^{m} p(y_j)\log_2 p(y_j) \tag{47.13}$$

$$H(Y \mid X) = -\sum_{i=1}^{n}\sum_{j=1}^{m} p(x_i, y_j)\log_2 p(y_j \mid x_i) \tag{47.14}$$

and

$$H(X, Y) = -\sum_{i=1}^{n}\sum_{j=1}^{m} p(x_i, y_j)\log_2 p(x_i, y_j) \tag{47.15}$$

Another useful entropy, $H(X \mid Y)$, which is sometimes called equivocation, is defined as

$$H(X \mid Y) = -\sum_{i=1}^{n}\sum_{j=1}^{m} p(x_i, y_j)\log_2 p(x_i \mid y_j) \tag{47.16}$$

These entropies are easily interpreted. $H(X)$ is the average uncertainty of the source, while $H(Y)$ is the average uncertainty of the received symbol. Similarly, $H(X \mid Y)$ is a measure of our average uncertainty of the transmitted symbol after we have received a symbol. The function $H(Y \mid X)$ is the average uncertainty of the received symbol given that X was transmitted. The joint entropy $H(X, Y)$ is the average uncertainty of the communication system as a whole.

Two important and useful relationships, which can be obtained directly from the definitions of the various entropies, are

$$H(X, Y) = H(X \mid Y) + H(Y) \tag{47.17}$$

and

$$H(X, Y) = H(Y \mid X) + H(X) \tag{47.18}$$

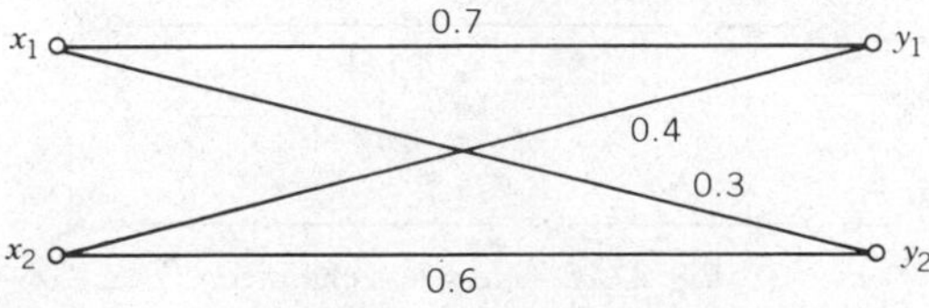

Fig. 47.27 Binary channel.

Channel Capacity

Consider for a moment an observer at the channel output. The observer's average uncertainty concerning the channel input will have some value before the reception of an output and his or her average uncertainty of the input will usually decrease when the output is received. In other words, $H(X \mid Y) \leqslant H(X)$. The decrease in the observer's average uncertainty of the transmitted signal when the output is received is a measure of the average transmitted information. This is defined as transinformation, or mutual information $I(X; Y)$. Thus

$$I(X; Y) = H(X) - H(X \mid Y) \tag{47.19}$$

It follows from Eqs. (47.17) and (47.18) that Eq. (48.19) can also be written

$$I(X; Y) = H(Y) - H(Y \mid X) \tag{47.20}$$

It should be observed that transinformation is a function of the source probabilities as well as the channel transition probabilities.

The channel capacity C is defined as the maximum value of transinformation, which is the maximum average information *per symbol* that can be transmitted through the channel. Thus

$$C = \max[I(X; Y)] \tag{47.21}$$

The maximization is with respect to the source probabilities, since the transition probabilities are fixed by the channel. However, the channel capacity is a function of only the channel transition probabilities, since the maximization process eliminates dependence on the source probabilities. Several examples will illustrate the method.

Example 47.4. Find the channel capacity of the noiseless discrete channel illustrated in Fig. 47.28. We start with

$$I(X; Y) = H(X) - H(X \mid Y)$$

and write

$$H(X \mid Y) = -\sum_{i=1}^{n} \sum_{j=1}^{n} p(x_i, y_j) \log_2 p(x_i \mid y_j)$$

For the noiseless channel, all $p(x_i, y_j)$ and $p(x_i \mid y_j)$ are zero unless $i = j$. For $i = j$, $p(x_i \mid y_j)$ is unity. Thus $H(X \mid Y)$ is zero for the noiseless channel, and

$$I(X; Y) = H(X)$$

We have seen that the entropy of a source is maximum if all source symbols are equally likely. Thus

$$C = \sum_{i=1}^{n} \frac{1}{n} \log_2 n = \log_2 n \tag{47.22}$$

Example 47.5. Find the channel capacity of the binary symmetric channel illustrated in Fig. 47.29.

This problem has considerable practical importance in the area of binary digital communications. We will determine the capacity by maximizing

$$I(X; Y) = H(Y) - H(Y \mid X)$$

where

$$H(Y \mid X) = -\sum_{i=1}^{2} \sum_{j=1}^{2} p(x_i, y_j) \log_2 p(y_j \mid x_i)$$

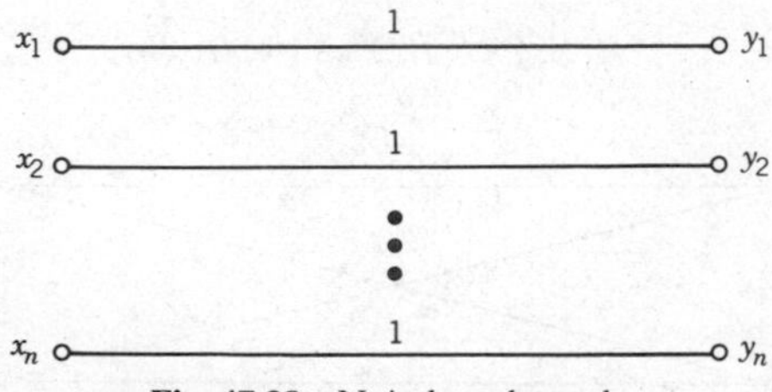

Fig. 47.28 Noiseless channel.

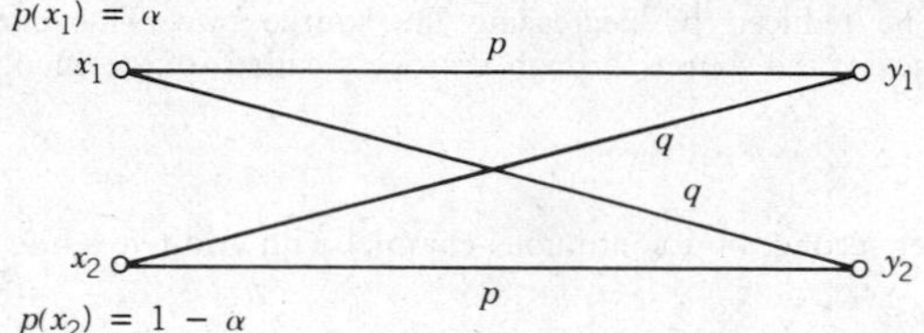

Fig. 47.29 Binary symmetric channel.

Using the probabilities defined in Fig. 47.29, we obtain

$$H(Y \mid X) = -\alpha p \log_2 p - (1-\alpha)\, p \log_2 p - \alpha q \log_2 q - (1-\alpha) q \log_2 q$$

or

$$H(Y \mid X) = -p \log_2 p - q \log_2 q$$

Thus

$$I(X; Y) = H(Y) + p \log_2 p + q \log_2 q$$

which is maximum when $H(Y)$ is maximum. Since the system output is binary, $H(Y)$ is maximum when each output has a probability of $\frac{1}{2}$ and is achieved for *equally likely inputs*. For this case $H(Y)$ is unity, and the channel capacity is

$$C = 1 + p \log_2 p + q \log_2 q = 1 - H(p) \tag{47.23}$$

where $H(p)$ is defined in Eq. (47.3).

The channel capacity for a binary symmetric channel is sketched in Fig. 47.30. As expected, if $p = 0$ or 1, the channel output is completely determined by the channel input, and the capacity is 1 bit per symbol. If p is equal to $\frac{1}{2}$, an input symbol yields either output with equal probability, and the capacity is zero.

The error probability P_E of a binary symmetric channel is easily computed. From

$$P_E = \sum_{i=1}^{2} p(e \mid x_i) p(x_i) \tag{47.24}$$

where $p(e \mid x_i)$ is the error probability given input x_i, we have

$$P_E = qp(x_1) + qp(x_2)$$

Thus

$$P_E = q \tag{47.24a}$$

which states that the unconditional error probability P_E is equal to the conditional error probability $p(y_j \mid x_i)$, $i \neq j$.

P_E is a decreasing function of the energy of the received symbols. Since the symbol energy is the received power multiplied by the symbol period, it follows that, if the transmitter power is fixed, the

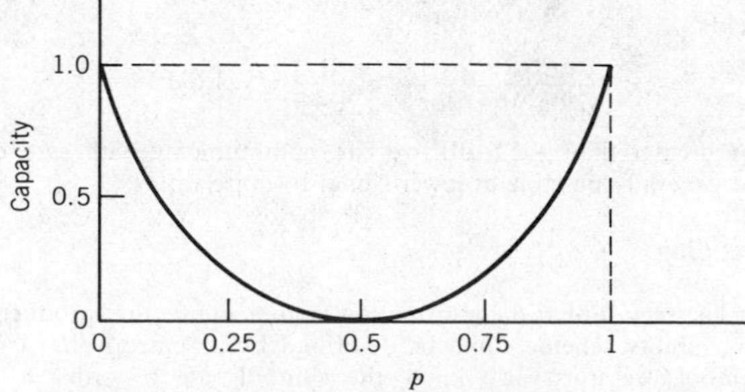

Fig. 47.30 Capacity of a binary symmetric channel.

error probability can be reduced by decreasing the source rate. This can be accomplished by removing the redundancy at the source through a process called source encoding.

Continuous Channel

The capacity, in bits per second, of a continuous channel with additive white gaussian noise is given by

$$C_c = B \log_2\left(1 + \frac{S}{N}\right) \tag{47.25}$$

where B is the bandwidth in hertz and S/N is the signal-to-noise ratio. This particular formulation of channel capacity is referred to as the Shannon–Hartley law. The subscript is used to distinguish Eq. (47.25), which has units of bits per second, from Eq. (47.21), which has units of bits per symbol.

The tradeoff between bandwidth and signal-to-noise ratio can be determined from the Shannon–Hartley law. For the noiseless case, infinite signal-to-noise ratio, the capacity is infinite for any nonzero bandwidth. However, as we shall show, the capacity cannot be made arbitrarily large by increasing bandwidth if noise is present.

For a noisy channel, it is interesting to compute the signal-to-noise ratio that still permits transmission at a rate equal to the channel capacity. First Eq. (47.25) is written as

$$C_c = B \log_2\left(1 + \frac{S}{N_0 B}\right) \tag{47.26}$$

where S is the signal energy per time unit, that is, power, and N_0 is the noise power spectral density. A useful bound can be determined by taking the limit as B approaches infinity. Before taking the limit, it is convenient to write Eq. (47.26)

$$C_c = \frac{S}{N_0} \log_2\left[\left(1 + \frac{S}{N_0 B}\right)^{N_0 B/S}\right] \tag{47.27}$$

We can then use

$$\lim_{x\to 0} (1 + x)^{1/x} = e$$

to write

$$\lim_{B\to\infty} C_c = \frac{S}{N_0} \log_2 e \tag{47.28}$$

The signal power ($S = E/T$) for M-ary signaling can be written in terms of the rate by recognizing that

$$R = \frac{\log_2 M}{T} \tag{47.29}$$

if all M signals are equiprobable. Thus the signal power is

$$S = \frac{ER}{\log_2 M} \tag{47.30}$$

and Eq. (47.28) becomes

$$\lim_{B\to\infty} C_c = \frac{ER}{N_0 \log_2 M} \log_2 e \tag{47.31}$$

which, for $R = C$, yields

$$\frac{E}{N_0 \log_2 M} = \frac{1}{\log_2 e} \cong \frac{1}{1.44} \cong -1.6 \text{ dB} \tag{47.32}$$

Thus for $E/N_0 \log_2 M$ greater than -1.6 dB, we can communicate with zero error, while reliable communication is not generally possible at lower signal-to-noise ratios.

47.3-2 Source Encoding

We determined in the last section that the information from a source that produced different symbols according to some probability scheme could be described by its entropy, $H(X)$. Since entropy has units of bits per symbol, we must also know the symbol rate in order to specify the source information rate in bits per second. In other words, the source information rate R_s is given by

$$R_s = rH(X) \qquad \text{bits per second} \tag{47.33}$$

where $H(X)$ is the source entropy in bits per symbol and r is the symbol rate in symbols per second.

Let us assume that this source is the input to a channel with capacity C bits per symbol or SC bits per second, where S is the available symbol rate for the channel. An important theorem of information theory, the noiseless coding theorem, sometimes referred to as Shannon's first theorem, is stated: Given a channel and a source that generates information at a rate less than the channel capacity, it is possible to encode the source output in such a manner that it can be transmitted through the channel. A proof of this theorem is beyond the scope of this introductory treatment of information theory and can be found in any of the standard information theory textbooks. We shall, however, demonstrate the theorem by a simple example.

An Example of Source Encoding

Consider a discrete binary source that has two possible output symbols, A and B (Fig. 47.31). Assume that the source symbol rate r is 2.66 symbols per second. The output of the source is assumed to be the input to a channel that can transmit a binary 0 or a binary 1 at a rate S of 1.5 symbols per second with negligible error. Thus from Example 47.5 with $p = 1$, the channel capacity is 1 bit per symbol or

$$SC = 1.5 \text{ bits per second}$$

Clearly the source *symbol* rate exceeds the channel *symbol* rate so that the source symbols cannot be directly input to the channel.

Shannon's first theorem, however, requires that the source *information* rate be compared to the channel capacity to determine whether transmission through the channel is possible. To carry out this calculation, two sets of probabilities are assumed. These are illustrated in Table 47.1. For the first set $P(A) = 0.9$ and $P(B) = 0.1$ and for the second set $P(A) = 0.6$ and $P(B) = 0.4$. A code is defined by a mapping of the source symbols (A, B) onto a sequence of code symbols (0, 1). In order to determine whether transmission is possible, the source entropy must first be computed. For the first set of probabilities the entropy is

$$\begin{aligned} H_1(X) &= -0.9 \log_2 0.9 - 0.1 \log_2 0.1 \\ &= 0.4690 \text{ bits per symbol} \end{aligned}$$

For a source symbol rate of 2.66 symbols per second, this yields an information rate of

$$rH_1(X) = 1.248 \text{ bits per second}$$

This is less than the channel capacity of 1.5 bits per symbol so that transmission is possible.

For the second set of probabilities the source entropy is

$$\begin{aligned} H_2(X) &= -0.6 \log_2 0.6 - 0.4 \log_2 0.4 \\ &= 0.9710 \text{ bits per symbol} \end{aligned}$$

This yields an information rate of

$$rH_2(X) = 2.583 \text{ bits per second}$$

which exceeds the channel capacity. Thus transmission is not possible.

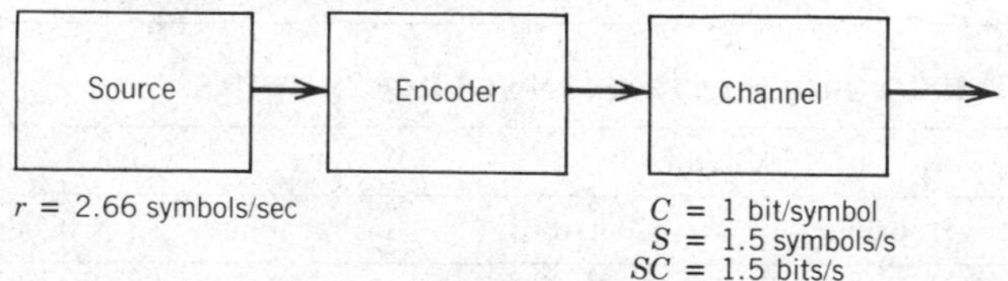

Fig. 47.31 Transmission scheme.

TABLE 47.1. SOURCE OUTPUTS

Source Symbol	Code 1		Code 2	
	Probability	Codeword	Probability	Codeword
A	0.9	0	0.6	0
B	0.1	1	0.4	1
		$\bar{L} = 1$		$\bar{L} = 1$

With the knowledge that the source symbols can be transmitted through the channel with the first set of probabilities, we search for a suitable coding scheme. One scheme is to group the source symbols by pairs and assign the shortest codeword to the most likely pair of source symbols and the longest codeword to the least likely pair of source symbols. This is shown in Table 47.2. The symbol pairs (AA, AB, BA, and BB) are known as the outputs of the second-order source extension. The average word length $\overline{L}$ is given by

$$\overline{L} = \sum_{i=0}^{2^n} p_i l_i$$

where p_i is the probability of the ith output of the extended source and l_i is the length of the ith codeword. The parameter n denotes the order of the source extension, which in this case is 2. The symbol rate at the encoder output is $r\overline{L}/n$. Thus for Code 1, the symbol rate at the encoder output is 1.716, which still exceeds the allowable symbol rate for the channel. The symbol rate has been reduced and we therefore try again.

The third-order source extension is illustrated in Table 47.3. With this scheme $\overline{L}$ is 1.598 and the symbol rate at the encoder output is $r\overline{L}/3 = 1.418$ symbols per second, which the channel can accept.

The behavior of these schemes is illustrated in Fig. 47.32. It can be seen that the symbol rate approaches the information rate $rH(X)$ as n grows large. This is a fundamental result of information theory. In Fig. 47.32, the heavy dots represent source symbol rates in symbols per second.

In this example the codewords were assumed. We now turn our attention to the problem of determining codewords.

Several Definitions

Before we discuss in detail the method of deriving codewords, we should pause to make a few definitions that will clarify our work.

Each codeword is constructed from an *alphabet*, which is a collection of symbols used for communication through a channel. For example, a binary codeword is constructed from a two-

TABLE 47.2. SECOND ORDER EXTENSION SOURCE OUTPUTS

Source Symbol	Code 1 Probability	Code 1 Codeword	Code 2 Probability	Code 2 Codeword
AA	0.81	0	0.36	00
AB	0.09	10	0.24	01
BA	0.09	110	0.24	10
BB	0.01	111	0.16	11
		$\overline{L} = 1.290$		$\overline{L} = 2$
		$\frac{1}{2}\overline{L} = 0.645$		$\frac{1}{2}\overline{L} = 1$

TABLE 47.3. THIRD ORDER EXTENSION SOURCE OUTPUTS

Source Symbol	Code 1 Probability	Code 1 Codeword	Code 2 Probability	Code 2 Codeword
AAA	0.729	0	0.216	00
AAB	0.081	100	0.144	010
ABA	0.081	101	0.144	011
BAA	0.081	110	0.144	100
ABB	0.009	11100	0.096	101
BAB	0.009	11101	0.096	110
BBA	0.009	11110	0.096	1110
BBB	0.001	11111	0.064	1111
		$\overline{L} = 1.598$		$\overline{L} = 2.944$
		$\frac{1}{3}\overline{L} = 0.533$		$\frac{1}{3}\overline{L} = 0.981$

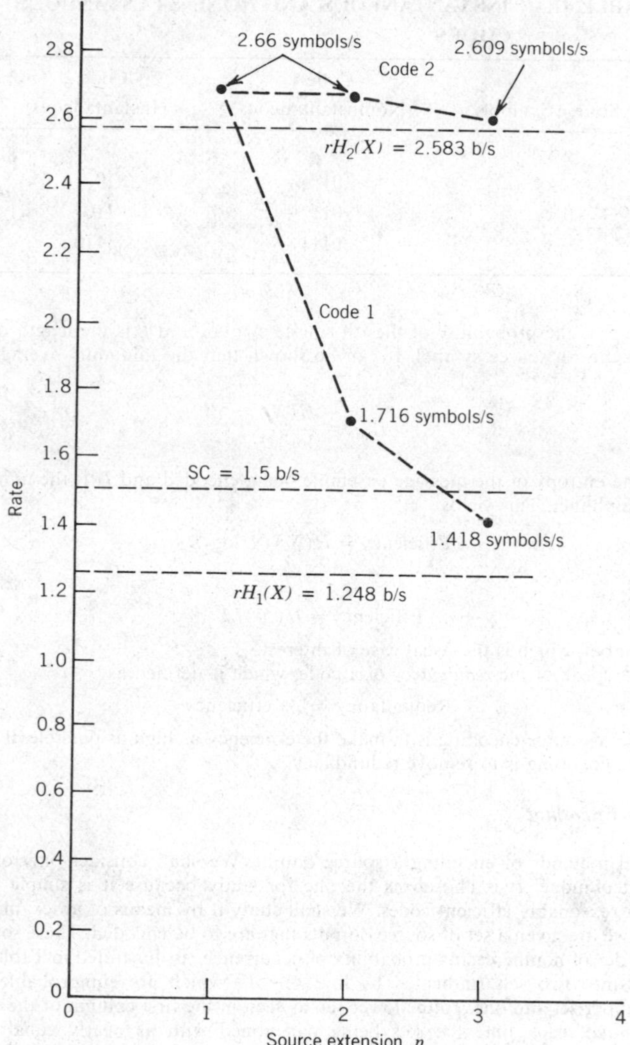

Fig. 47.32 Information rates as a function of source extension.

symbol alphabet, wherein the two symbols are usually taken as the zero and the one. The *wordlength* of a codeword is the number of symbols in the codeword.

There are several major subdivisions of codes. For example, a code can be either *block* or *nonblock*. A block code is one in which each block of source symbols is encoded into a fixed-length sequence of code symbols. A uniquely decipherable code is a block code in which the codewords may be deciphered without using spaces. These codes can be further classified as instantaneous or noninstantaneous according to whether or not it is possible to decode each word in sequence without reference to succeeding code symbols. Alternatively, noninstantaneous codes require reference to succeeding code symbols, as illustrated in Table 47.4. It should always be remembered that a noninstantaneous code can be uniquely decipherable.

A useful measure of goodness of a source code is the *efficiency*, which is defined as the ratio of the minimum average wordlength of the codewords, $\bar{L}_{\min}$, to the average wordlength of the codeword, $\bar{L}$. Thus

$$\text{Efficiency} = \frac{\bar{L}_{\min}}{\bar{L}} = \frac{\bar{L}_{\min}}{\sum_{i=1}^{n} p_i l_i} \tag{47.34}$$

TABLE 47.4. INSTANTANEOUS AND NONINSTANTANEOUS CODES

Source Symbols	Code 1 (Noninstantaneous)	Code 2 (Instantaneous)
x_1	0	0
x_2	01	10
x_3	011	110
x_4	0111	1110

where, as before, p_i is the probability of the ith source symbol and l_i is the length of the codeword corresponding to the ith source symbol. It can be shown that the minimum average wordlength is given by

$$\bar{L}_{\text{min}} = \frac{H(X)}{\log_2 D} \tag{47.35}$$

where $H(X)$ is the entropy of the message ensemble being encoded and D is the number of symbols in the encoding alphabet. This yields

$$\text{Efficiency} = H(X)/\bar{L}\log_2 D \tag{47.36}$$

or

$$\text{Efficiency} = H(X)/\bar{L} \tag{47.37}$$

for a binary alphabet, which is the usual case of interest.

Sometimes we speak of the *redundancy* of a code, which is defined as

$$\text{Redundancy} = 1 - \text{efficiency} \tag{47.38}$$

Since the purpose of source encoding is to make the efficiency as high as possible, it follows that the purpose of source encoding is to remove redundancy.

Shannon–Fano Encoding

There are several methods of encoding a source output. We shall consider only one of these, the Shannon–Fano technique. It is chosen as the one for study because it is simple to perform and usually results in reasonably efficient codes. We shall study it by means of an example.

Assume that we are given a set of source outputs that are to be encoded. These source outputs are first ranked in order of nonincreasing probability of occurrence, as illustrated in Table 47.5. The set is then partitioned into two sets (indicated by line $A - A'$) which are equiprobable, and zeros are assigned to the upper set and ones to the lower set, as seen in the first column of the codewords. This process is continued, each time the sets being partitioned with as nearly equal probabilities as possible, until further partitioning is not possible. This scheme will give a 100% eff[illegible] if the

TABLE 47.5. SHANNON–FANO ENCODING

Source Words	Probability	Codeword	(Length) (Probability)
x_1	0.2500	00	2 (0.25) = 0.50
x_2	0.2500	01	2 (0.25) = 0.50
		A----A'	
x_3	0.1250	100	3 (0.125) = 0.375
x_4	0.1250	101	3 (0.125) = 0.375
x_5	0.0625	1100	4 (0.0625) = 0.25
x_6	0.0625	1101	4 (0.0625) = 0.25
x_7	0.0625	1110	4 (0.0625) = 0.25
x_8	0.0625	1111	4 (0.0625) = 0.25
			Average wordlength = 2.75

partitioning always results in equiprobable sets; otherwise the code will be less efficient. For this particular example,

$$\text{Efficiency} = \frac{H(X)}{\bar{L}} = \frac{2.75}{2.75} = 1$$

since equiprobable partitioning is possible.

A procedure suggested by Huffman gives a code with the shortest average wordlength. Such a code is termed an optimum code. His procedure is slightly more difficult to apply than the Shannon–Fano code, so it will not be explained here.

47.3-3 Reliable Communications in the Presence of Noise

We shall now turn our attention to methods for achieving reliable communication in the presence of noise by combating the effects of that noise. We undertake our study with a promise of considerable success from Claude Shannon. Shannon's theorem, sometimes referred to as the fundamental theorem of information theory, is stated: Given a discrete memoryless channel (each symbol is perturbed by noise independently of all other symbols) with capacity C and a source with positive rate R, where $R < C$, there exists a code such that the output of the source can be transmitted over the channel with an arbitrarily small probability of error.

Thus Shannon's theorem predicts essentially error-free transmission in the presence of noise. Unfortunately, the theorem tells us only of the existence of codes and tells nothing of how to construct these codes.†

†Editor's comment: Shannon's theorem proves the existence of coding techniques for reliable communication. The development of practical coding techniques has resulted in significant advances in the reliability of communications. The most popular of these techniques are discussed in Section 48.5.

Bibliography

Abramson, N., *Information Theory and Coding*, McGraw-Hill, New York, 1963.

Carlson, A. B., *Communication Systems*, McGraw-Hill, New York, 1968.

Gallager, R. G., *Information Theory and Reliable Communication*, Wiley, New York, 1968.

Lathi, B. P., *Communication Systems*, Wiley, New York, 1968.

Lin, S., *An Introduction to Error-Correcting Codes*, Prentice-Hall, Englewood Cliffs, NJ, 1970.

Peterson, W. W., *Error-Correcting Codes*, MIT, Cambridge, MA, 1961.

Reza, F. M., *An Introduction to Information Theory*, McGraw-Hill, New York, 1961.

Sakrison, D., *Communication Theory: Transmission of Waveforms and Digital Information*, Wiley, New York, 1968.

Schwartz, M., *Information Transmission, Modulation, and Noise*, 2nd ed., McGraw-Hill, New York, 1970.

Shannon, C. E., "A Mathematical Theory of Communications," *Bell Sys. Tech. J.* **27**:379–423, 623–656 (Jul. 1948).

Shannon, C. E. and W. Weaver, *The Mathematical Theory of Communication*, University of Illinois, Urbana, IL, 1963.

Taub, H. and D. L. Schilling, *Principles of Communication Systems*, McGraw-Hill, New York, 1971.

Wiener, N., *Extrapolation, Interpolation, and Smoothing of Stationary Times Series with Engineering Applications*, MIT, Cambridge, MA 1949.

Woodward, P. M., *Probability and Information Theory with Applications to Radar*, Pergamon, New York, 1953.

CHAPTER 48
COMMUNICATIONS TECHNIQUES

JACK W. SEYL

NASA Johnson Space Center, Houston, Texas

GEORGE W. RAFFOUL

Lockheed Engineering and Management Services Company, Inc., Houston, Texas

MATTHEW J. QUINN, JR.

College of Technology, University of Houston, Houston, Texas

K. K. CHOW

Lockheed Missiles and Space Company, Palo Alto Research Laboratory,
Palo Alto, California

BERNARD SKLAR

Aerospace Corporation, Los Angeles, California

PETER MONSEN

P. M. Associates, Stow, Massachusetts

E. T. DICKERSON

University of Houston—Clear Lake, Houston, Texas

48.1 MODULATION TECHNIQUES

Jack W. Seyl

Modulation can be defined as the alteration or changing of some characteristic of a known signal or waveform, usually called a carrier, as a function of some unknown signal or waveform that conveys information. In radio-frequency communication systems, the carrier is almost universally a sinusoid, and there are several methods of altering or modulating the carrier. These include linear modulation, angle (or exponential) modulation, and various types of pulse modulation. Each will be treated in more detail in the following paragraphs, but general characteristics are described first.

A sinusoidal carrier is described by the following function:

$$e_c(t) = E_c \sin(\omega_c t + \phi_0) \tag{48.1}$$

where E_c = amplitude, V
ω_c = frequency, rad/s
ϕ_0 = initial phase, rad

The argument of the sine function in Eq. (48.1) is called the phase angle, or instantaneous phase angle, of the carrier. The time domain and frequency domain characteristics of the sinusoidal carrier are described in Fig. 48.1.

As seen in the sinusoidal time expression of Eq. (48.1), there are two parameters or characteristics of the carrier that can be varied in accordance with an information signal $g(t)$. These parameters are the amplitude and the phase angle. In linear modulation, the amplitude of the carrier signal is varied

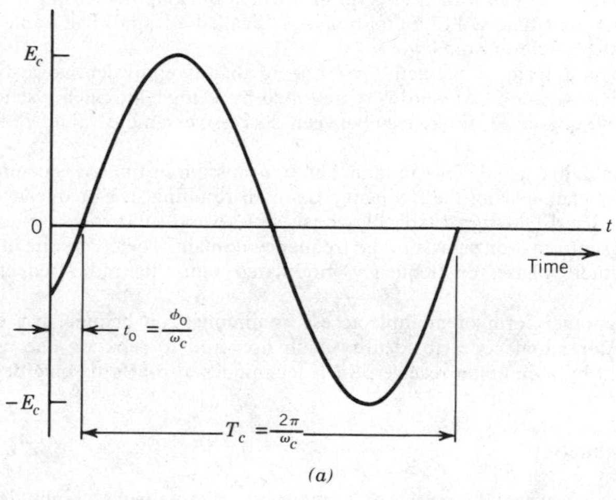

(a)

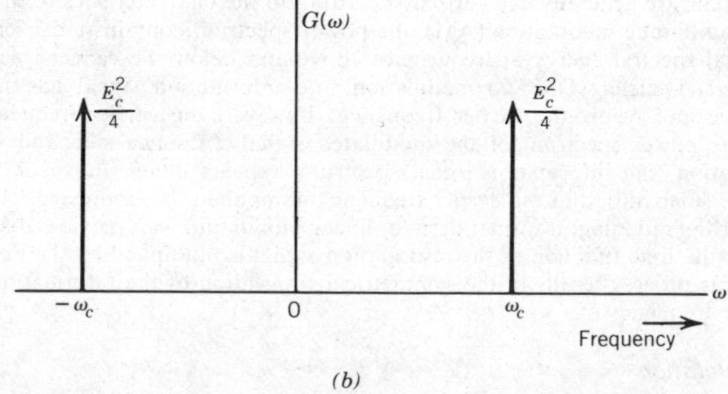

(b)

Fig. 48.1 Characteristics of sinusoidal carrier signal. (*a*) Time function. (*b*) Two-sided power spectral density.

in proportion to the information signal. Three forms of linear modulation are commonly encountered. They are amplitude modulation (AM), double sideband suppressed-carrier (DSBSC) modulation, and single sideband suppressed-carrier (SSBSC) modulation. In each of these forms the modulating process is linear and equivalent to a multiplication of the information signal and the carrier. The resulting modulated carrier effectively translates the energy of the information signal to a band of frequencies (sidebands) immediately adjacent to the sinusoidal carrier frequency, allowing radio-frequency transmission of the information signal.

In angle modulation, also called exponential modulation, the argument of the sinusoidal carrier is varied by modulating either the phase θ or the frequency ω. When the phase is varied in direct proportion to the information signal $g(t)$, then $\theta(t) = [\omega_c t + \phi(t) + \phi_0]$, which represents phase modulation (PM). Alternately, when the frequency ω of the carrier is varied in direct proportion to the information signal, then $\theta(t) = [\omega(t)t + \phi_0]$, which represents frequency modulation (FM).

In addition to the modulation of continuous waves, a technique of sampling information signals in time can be utilized to generate a pulse-modulated signal. Although the pulsed signal can be used as the information carrier directly, it is commonly used to modulate a continuous-wave carrier for transmission. Time sampling can also be used to combine or multiplex several information signals together for transmission on a single communication carrier or channel. This process of sampling and combining in time is referred to as time division multiplexing (TDM). Pulse parameters that can be modulated or varied by an information signal include amplitude, duration, and position, resulting in pulse amplitude modulation (PAM), pulse duration modulation (PDM), and pulse position modulation (PPM), respectively.

A more complex form of pulse modulation that is widely used for high-capacity information channels is pulse code modulation, or PCM. In PCM systems, the individual information signals are sampled in time and quantized in amplitude. The sampled amplitude is then transmitted as a coded binary word representing the quantization interval into which the amplitude samples fall. Each word is transmitted in a particular time slot of a group of words called a frame. The frame starting time is identified with a unique synchronization word.

Delta modulation is a form of predictive quantizing that is equivalent to a 1-bit pulse code modulation system, that is, each PCM word is represented by a single bit. Such systems are based on the transmission of the quantized differences between successive sample values rather than of the samples themselves.

In addition to combining signals for simultaneous transmission in time, it is common to separate information signals by channelizing the frequency domain, resulting in a frequency-division multiplexed (FDM) signal. The FDM signal typically consists of several subcarriers (sinusoids) that serve to separate the various information signals in the frequency domain. The composite of the subcarriers is then either amplitude, phase, or frequency modulated onto the radio-frequency carrier for transmission.

In recent years, another form of multiple-access communication channel has emerged. These channels utilize pseudorandom codes to identify each user and to separate one user signal from another when received by a common receiver. Such techniques also afford some degree of security and multipath protection.

48.1-1 Linear Modulation

Linear modulation of a sinusoidal carrier results when its instantaneous amplitude is varied as a linear function of the information signal. Linear-modulated signals can be transmitted in several formats. The signals are generally categorized according to the characteristics of their power spectra. For the case of amplitude modulation (AM), the power spectrum contains a carrier component with information signal spectral energy at frequencies above and below the carrier frequency. In double sideband suppressed-carrier (DSBSC) modulation, the information signal has spectral energy at frequencies above and below the carrier frequency; however, the carrier frequency component is suppressed in the power spectrum of the modulated signal. In single sideband suppressed-carrier (SSBSC) modulation, the information signal spectrum appears either above or below the carrier frequency (single sideband), and the carrier frequency component is suppressed.

Regardless of the modulation format desired, linear modulation can be viewed as a multiplication process in which the time function of the information signal is multiplied by the time function of the carrier signal. This process results in the symmetrical translation of the information signal spectrum about the carrier frequency.

Amplitude Modulation

In amplitude modulation, the instantaneous amplitude of the carrier (sinusoid) is varied in proportion to the amplitude of the modulating signal. In the general case the time domain description of an AM signal is

$$e_{AM}(t) = [1 + mg(t)]E_c\sin(\omega_c t + \phi_0) \tag{48.2}$$

where $g(t)$ = information signal
E_c = unmodulated carrier peak amplitude, V
ω_c = carrier frequency, rad/s
ϕ_0 = initial phase angle of carrier, rad
m = modulation index

Figure 48.2 depicts the instantaneous time function for the special case when $g(t)$ is a sinusoidal information signal with amplitude normalized to unity.

$$g(t) = \sin(\omega_m t + \phi_m)$$

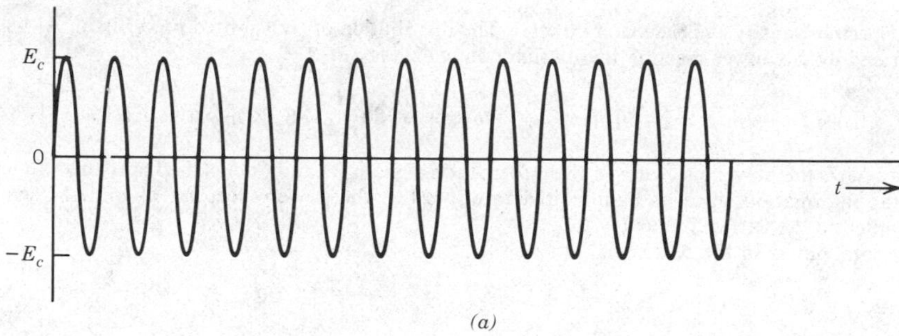

(a)

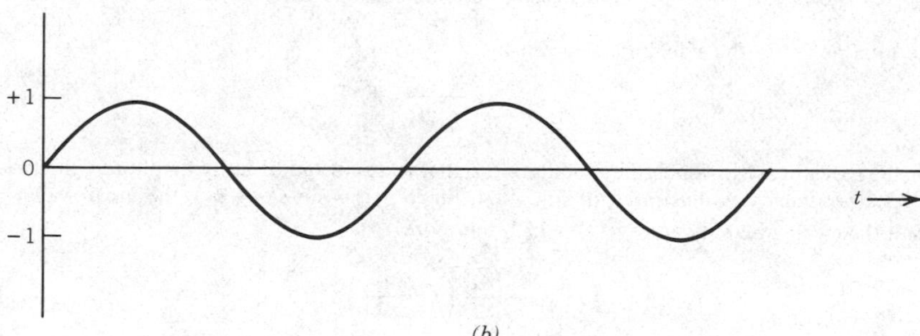

(b)

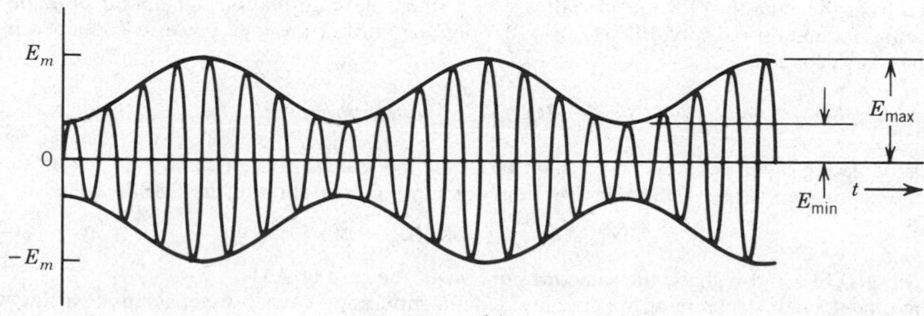

(c)

Fig. 48.2 Amplitude modulation time functions. (*a*) Carrier signal, $e_c(t) = E_c \sin \omega_c t$ (assuming that $\phi_0 = 0$). (*b*) Information signal, $g(t) = \sin \omega_m t$ (assuming that $\phi_m = 0$). (*c*) Amplitude-modulated signal, $e_{AM}(t) = E_c[1 + m \sin \omega_m t]\sin \omega_c t$.

where ω_m is the information signal frequency in radians per second and ϕ_m is the information signal phase in radians.

The modulation index of the AM signal determines the amount of energy contained in the signal sidebands. In the AM signal $e_{AM}(t)$ of Fig. 48.2, the modulation index m is defined as

$$m \quad \% = \left(\frac{E_{max} - E_{min}}{E_{max} + E_{min}} \right) \times 100 \tag{48.3}$$

where E_{max} is the maximum (peak) of the modulated signal amplitude and E_{min} is the minimum (valley) of the modulated signal amplitude.

In most practical AM systems, the modulation index is constrained to be no more than 100% ($m = 1.0$) to avoid nonlinear distortion of the information signal and generation of harmonics of the transmitted carrier.

Power Spectral Density and Sideband Power. The distribution of transmitted power in an AM signal is described by the power spectral density function, which is

$$G_{AM}(\omega) = [1 + 2m\overline{g(t)}]\frac{E_c^2}{4}[\delta(\omega - \omega_c) + \delta(\omega + \omega_c)] + \frac{E_c^2}{4}[G_g(\omega - \omega_c) + G_g(\omega + \omega_c)] \tag{48.4}$$

where $G_g(\omega)$ is the power spectrum of the information signal $g(t)$ and $\delta(\omega)$ is the Dirac delta function.

If the information signal is bandlimited to ω_m rad/s, the power spectrum of the AM signal is bandlimited to the carrier frequency $\pm\omega_m$.

The total power in the AM signal is

$$P_{AM} = \left[1 + 2\,m\,\overline{g(t)} + m^2\,\overline{g^2(t)}\,\right] E_c^2/2 \tag{48.5}$$

where m = modulation index

$\overline{g(t)}$ = mean value of $g(t)$

$\overline{g^2(t)}$ = mean squared value of $g(t)$

Figure 48.3 shows the power spectral density functions of the information signal and of the AM signal.

The total power in the sidebands is given by

$$P_{SB} = \left[m^2\,\overline{g^2(t)}\,\right] \frac{E_c^2}{2} \tag{48.6}$$

For the special case in which the modulation signal is a sinusoid of unity amplitude, the two-sided power spectral density is illustrated in Fig. 48.4. Since $g(t) = \sin(\omega_c t + \phi_0)$, the mean value $g(t)$ is zero, and the mean squared value $\overline{g^2(t)}$ is 1/2, therefore

$$P_{AM} = \frac{E_c^2}{2}\left(1 + \frac{m^2}{2}\right) \tag{48.7}$$

and

$$P_{SB} = m^2\frac{E_c^2}{4} \tag{48.8}$$

As seen from the equations for the special case of a sinusoidal information signal, and assuming the maximum modulation index of 100% ($m = 1.0$), only one third of the total power is contained in the information sidebands.

Double Sideband Suppressed-Carrier (DSBSC) Modulation

If the modulating signal contains no constant term, the modulation format changes to that of double sideband suppressed carrier (DSBSC). The time domain representation of the DSBSC signal is

$$e_{DSB}(t) = mg(t)\sin(\omega_c t + \phi_0) \tag{48.9}$$

where m, $g(t)$, ω_c, and ϕ_0 have the same meaning as in the case of AM.

The time-domain structure of the double sideband suppressed-carrier wave, assuming a sine wave modulating signal $g(t)$, is shown in Fig. 48.5. Each time the information signal passes through zero value, the phase of the carrier signal is shifted 180° (i.e., a phase reversal occurs). This feature of the DSBSC wave is important in digital transmission systems, since it allows generation of a special digital phase-modulation technique referred to as phase shift keying (PSK). The characteristics of PSK or digital DSBSC modulation are described in detail in the section on digital phase modulation.

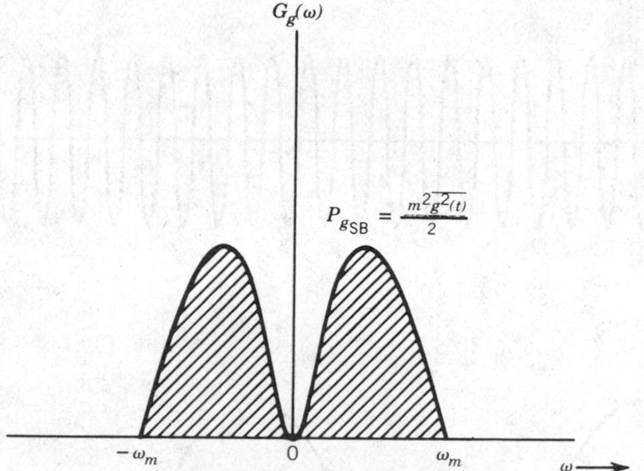

(a) Two-sided Power Spectral Density of Bandlimited Information Signal

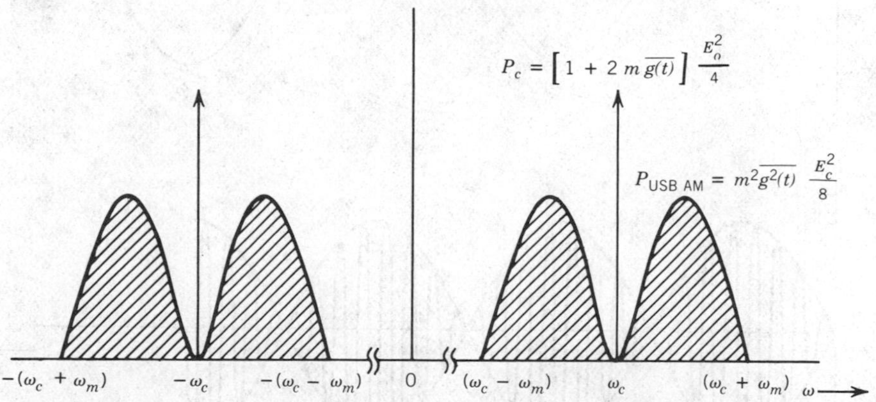

(b) Two-sided Power Spectral Density of Amplitude Modulated Signal

Fig. 48.3 Power spectral density, AM modulation with bandlimited information signal. (*a*) Two-sided power spectral density of bandlimited information signal. (*b*) Two-sided power spectral density of amplitude-modulated signal.

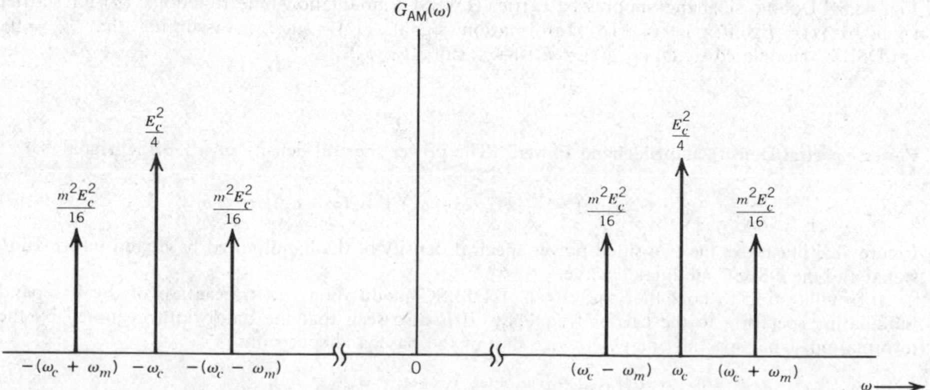

Fig. 48.4 Power spectrum of an amplitude-modulated signal with sinewave (tone)-modulating signal $[e_m(t) = \sin \omega_m t]$.

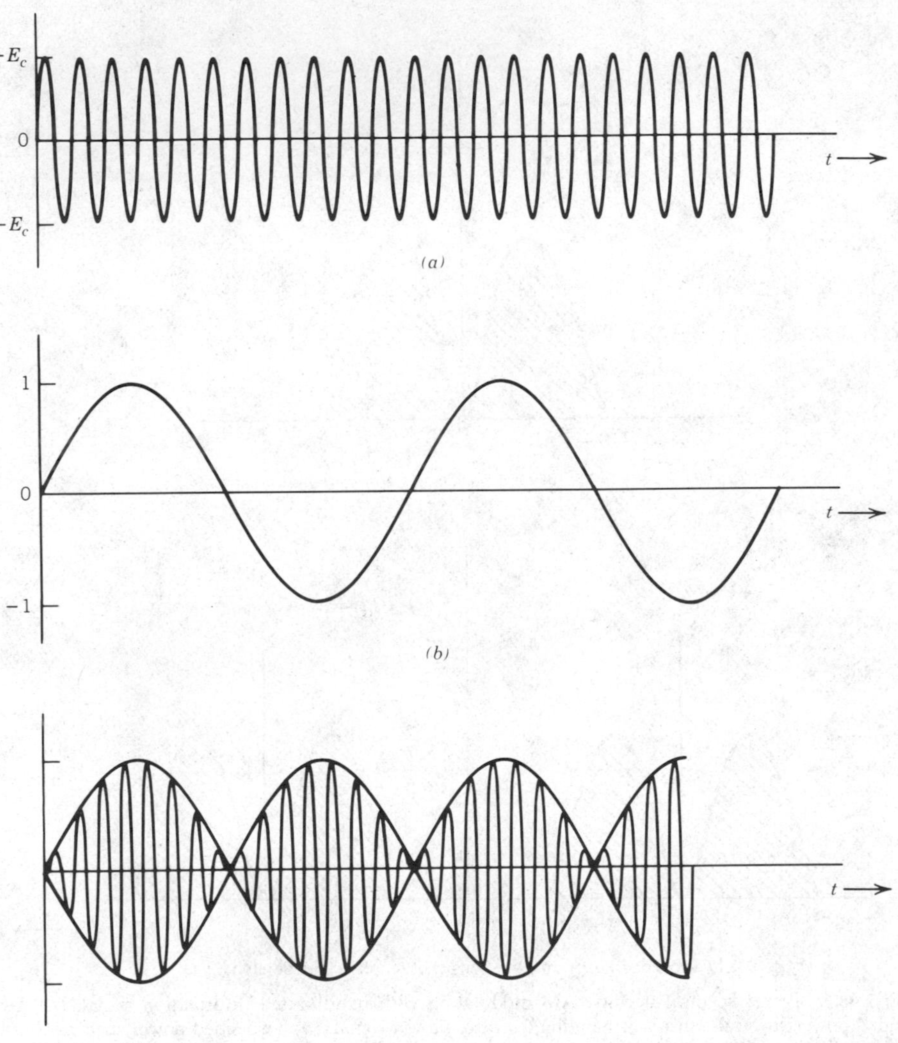

Fig. 48.5 Double sideband suppressed-carrier (DSBSC) modulation time functions. (a) RF carrier signal, $e_c(t) = E_c \sin(\omega_c t) + \phi_0$. ($b$) Information signal, $g(t) = \sin \omega_m t$ (assuming that $\phi_m = 0$). (c) DSBSC-modulated signal $e_{\text{DSB}}(t) = E_c \sin \omega_m t \sin(\omega_c t + \phi_0)$.

Power Spectral Density and Sideband Power. The power spectral density of a DSBSC signal is

$$G_{\text{DSB}}(\omega) = \frac{E_c^2}{4} [G_g(\omega - \omega_c) + G_g(\omega + \omega_c)] \tag{48.10}$$

Figure 48.6 illustrates the two-sided power spectral density of the bandlimited baseband information signal and the DSBSC modulated wave.

It is interesting to note that the effect of DSBSC modulation is a translation of the baseband modulating spectrum to the carrier frequency. It is also seen that the bandwidth occupied by the information signal spectral energy is twice that of the baseband spectrum:

$$BW_{\text{DSB}} = [(\omega_c + \omega_m) - (\omega_c - \omega_m)] = 2\omega_m \tag{48.11}$$

This effective doubling of the baseband bandwidth was also true for the AM signal spectrum. The

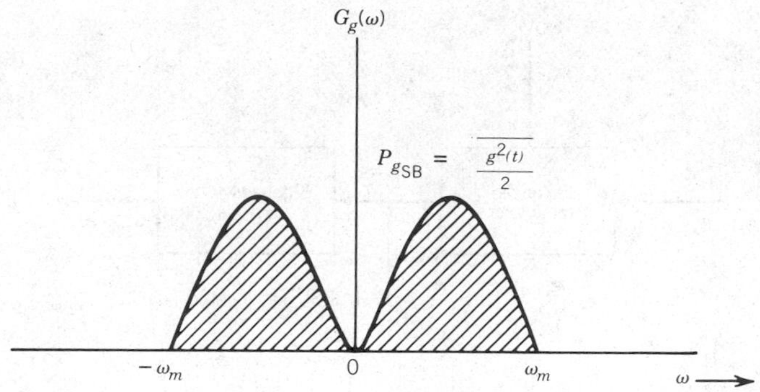

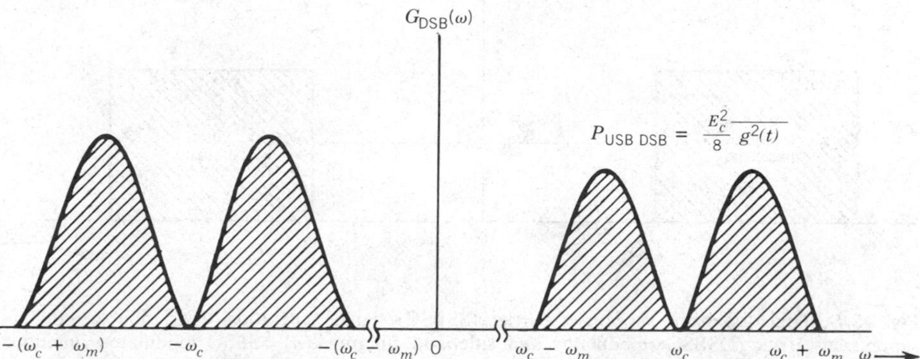

Fig. 48.6 Power spectral density spectra: (*a*) of bandlimited information signal, (*b*) of double sideband suppressed-carrier signal.

total power of the DSBSC signal is

$$P_{DSB} = \frac{E_c^2}{2} \overline{g^2(t)} \tag{48.12}$$

where $\overline{g^2(t)}$ is the mean squared value or total power of the information signal.

The total sideband power of the DSBSC is

$$P_{SB.(DSB)} = \frac{E_c^2}{2} \overline{g^2(t)} \tag{48.13}$$

Thus all of the power of a DSBSC wave is contained in the information signal spectrum, resulting in a more efficient transmission process than AM. However, the demodulation process for suppressed-carrier modulated signals is considerably more complex than for the case of AM, and thus there is an economic consideration of receiver complexity and cost vs. transmitter efficiency.

Single Sideband Suppressed-Carrier (SSBSC) Modulation

A single sideband suppressed-carrier signal can be generated by filtering (passing only one of the modulation sidebands) or by phase-shifting techniques during the modulation process. If a double sideband suppressed-carrier wave is passed through a filter that completely eliminates all signals below the carrier, as shown in Fig. 48.7, a single sideband suppressed-carrier wave is obtained.

A similar result can be achieved by use of the so-called phase discrimination technique, which requires the addition of a quadrature version of the double sideband suppressed-carrier modulated signal to the original DSBSC wave (Panter, 1965). Figure 48.8 illustrates this technique of SSBSC modulation.

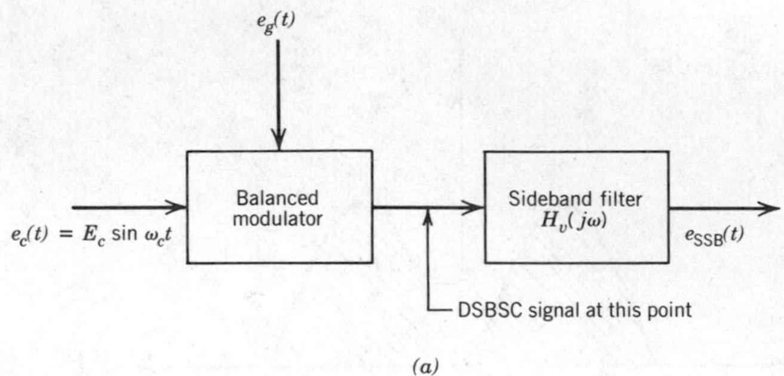

(a)

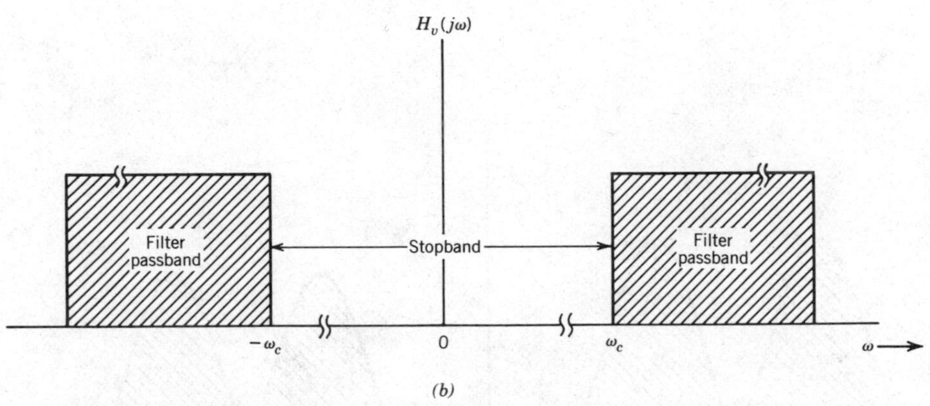

(b)

Fig. 48.7 Single sideband suppressed-carrier (SSBSC) signal generation using double sideband suppressed-carrier (DSBSC) modulator and sideband filtering. (*a*) SSBSC modulator functional diagram. (*b*) Sideband filter characteristics (shown for upper sideband case).

The resulting time function representation for SSBSC modulation can be most compactly expressed in the carrier quadrature form

$$e_{SSB}(t) = g(t) \cos \omega_c t - U(t) \sin \omega_c t$$

where $U(t)$ is a quadrature version of $g(t)$ obtained by introducing a constant 90° phase shift at all frequencies of $g(t)$.

Power Spectral Density and Sideband Power. The power spectrum for the SSBSC wave is essentially that of the DSBSC signal with one of the information sidebands eliminated. Figure 48.9 illustrates the SSBSC spectrum when the upper sideband has been selected for transmission.

The power density spectrum is given by

$$G_{SSB}(\omega) = \begin{cases} \dfrac{E_c^2}{4}\left[G_g(\omega - \omega_c) + G_g(\omega + \omega_c)\right] & \text{for} \quad |\omega| > \omega_c \\ 0 & \text{for} \quad |\omega| < \omega_c \end{cases} \tag{48.14}$$

The total power of the SSBSC wave is

$$P_{SSB} = \frac{E_c^2}{4}\,\overline{g^2(t)} \tag{48.15}$$

and the sideband power is

$$P_{SB} = \frac{E_c^2}{4}\,\overline{g^2(t)} \tag{48.16}$$

As was the case for the DSBSC signal, the total power and the information signal sideband power are equal for the SSBSC signal. However, the total available power is one half that for the case of the DSBSC modulated signal, if the carrier amplitude, E_c, is the same.

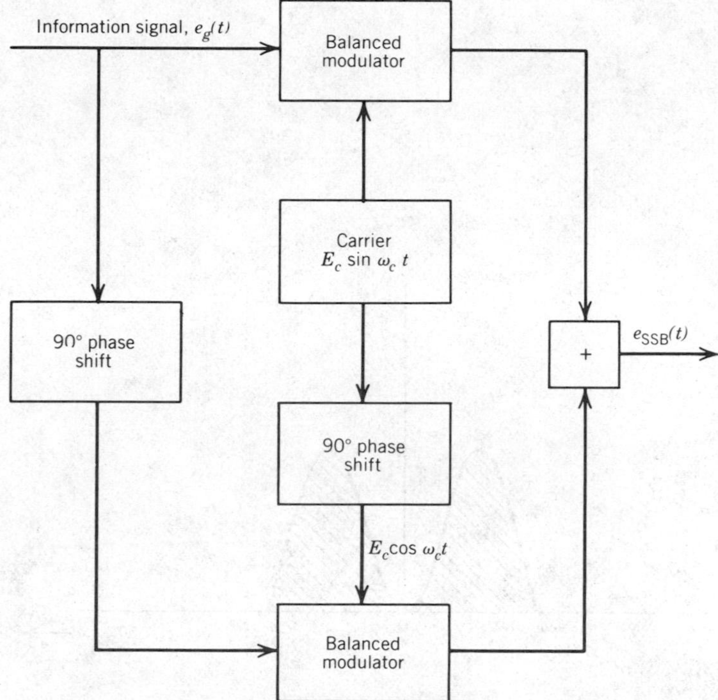

Fig. 48.8 Phase shift technique for single sideband suppressed-carrier (SSBSC) modulation generation.

Detection of Linear Modulation

The transmitted spectra of the different forms of linear-modulated signals vary in both magnitude and frequency distribution. Recovery of the information from the received carrier signal requires a frequency conversion and detection process.

In double sideband and single sideband suppressed-carrier detection, the carrier must be restored at the receiver before demodulation can occur. This form of detection is generally referred to as coherent or synchronous detection, since a carrier signal must be obtained at the receiver, a carrier that is phase coherent or synchronous with the transmitted carrier signal. For conventional AM signals, synchronous detection is unnecessary, since the information signal is completely contained in the envelope of the carrier and not in the phase. This results in a decided advantage in complexity and cost for conventional AM receivers over DSBSC and SSBSC receivers, since the detection system can be substantially simpler if carrier reconstruction is not required.

A comparison of the performance of linear modulation and demodulation techniques can be made on the assumption that the available carrier power prior to modulation, $P_c = E_c^2/2$, is constant. The results are summarized in Table 48.1. In the table the sideband power is shown in terms of the available carrier power. In the case of SSBSC, the sideband power is half that for AM and DSBSC, as was previously discussed.

Using the two-sided frequency domain model, the required receiver bandwidth prior to detection is 4 ω_m for AM and DSBSC but only 2 ω_m for SSBSC. If the receiver is in practice restrained to this bandwidth, the predetection signal-to-noise ratio, $(S/N)_i$, is exactly the same for SSBSC as for DSBSC and AM. Consequently, the overall performance as measured by the output signal-to-noise ratio, $(S/N)_o$, is equal for all three cases.

In practical SSBSC systems, it is common to increase the peak power output so that the average output power (sideband power) is the same as would be the case in DSBSC. The output signal-to-noise ratio for SSBSC is then twice that of AM or DSBSC.

Table 48.2 shows a similar comparison, but with the output signal-to-noise ratio normalized to the total received signal power, P_T. Since AM requires the transmission of a carrier, only one third of the total signal power is in the sidebands, and therefore the overall performance for a fixed transmission

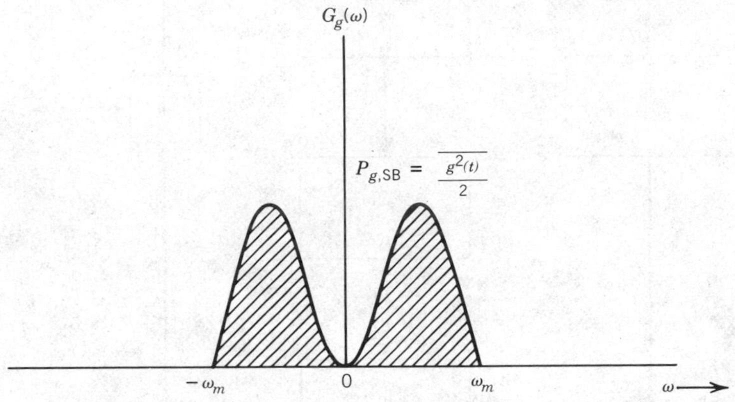

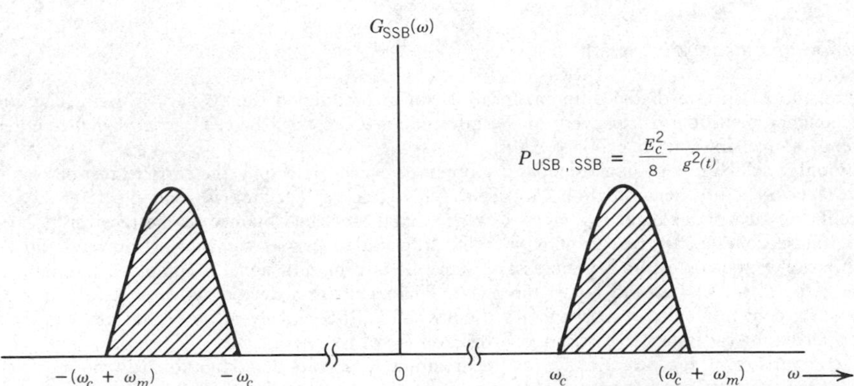

Fig. 48.9 Power spectral density for (*a*) two-sided information signal baseband spectrum, assuming bandlimiting at $\pm\omega_m$, and (*b*) two-sided single sideband suppressed-carrier (SSBSC) modulated-wave spectrum.

TABLE 48.1. COMPARISON OF LINEAR MODULATION/DEMODULATION TECHNIQUES (SIDEBAND POWER CONSTRAINED)

	Modulation Type		
	Amplitude Modulation (AM)[a]	Double Sideband Suppressed-Carrier (DSBSC)	Single Sideband Suppressed-Carrier (SSBSC)
Sideband power (W)	$\overline{g^2(t)}\, P_c$ [b]	$\overline{g^2(t)}\, P_c$	$\overline{g^2(t)}\, P_c/2$
Two-sided predetection bandwidth (Hz)	$4\omega_m$	$4\omega_m$	$2\omega_m$
Noise power spectral density (W/rad/s)	$N_0/2$	$N_0/2$	$N_0/2$
Predetection signal-to-noise $(S/N)_i$	$\dfrac{\overline{g^2}(t) P_c}{2N_0\omega_m}$	$\dfrac{\overline{g^2}(t) P_c}{2N_0\omega_m}$	$\dfrac{\overline{g^2}(t) P_c}{2N_0\omega_m}$
Output signal-to-noise $(S/N)_o$ [c]	$\dfrac{\overline{g^2}(t) P_c}{P_n}$	$\dfrac{\overline{g^2}(t) P_c}{P_n}$	$\dfrac{\overline{g^2}(t) P_c}{P_n}$

[a] Modulation index = 1.0.
[b] $P_c = \alpha E_c^2/2$ = available carrier power at detector. α is an attenuation constant.
[c] $P_n = 2N_0\omega_m$.

TABLE 48.2. COMPARISON OF LINEAR MODULATION/DEMODULATION TECHNIQUES (NORMALIZED TO TOTAL SIGNAL POWER)

	Modulation Type[a]		
	Amplitude Modulation (AM)[b]	Double Sideband Suppressed-Carrier (DSBSC)	Single Sideband Suppressed-Carrier (SSBSC)
Total signal Power P_T, W	$3P_c/2$	$P_c/2$	$P_c/4$
Sideband power[c] P_{SB}	$P_c/2 = P_T/3$	$P_c/2 = P_T$	$P_c/4 = P_T$
Predetection signal-to-noise $(S/N)_i$	$P_T/6N_0\omega_m$	$P_T/2N_0\omega_m$	$P_T/N_0\omega_m$
Output signal-to-noise $(S/N)_o$[d]	$P_T/3P_n$	P_T/P_n	$2P_T/P_n$

[a] Sine wave modulation.
[b] $m = 1.0$.
[c] $P_c = \alpha E_c^2/2$.
[d] $P_n = 2N_0\omega_n$.

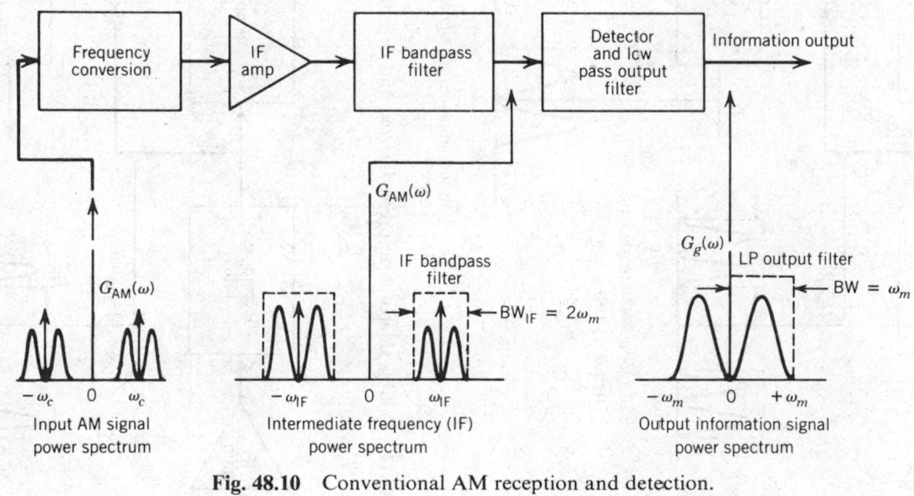

Fig. 48.10 Conventional AM reception and detection.

power is poorer than for DSBSC or SSBSC. In Table 48.2, SSBSC appears to have a 2 : 1 (3-dB) advantage over DSBSC, because SSBSC requires only half the predetection bandwidth. This advantage is realized if the *total received power* is equal in both cases. If the *total sideband powers* are equal, then the results of Table 48.1 hold.

Figures 48.10, 48.11, and 48.12 illustrate the reception/detection process for AM, DSBSC, and SSBSC, respectively.

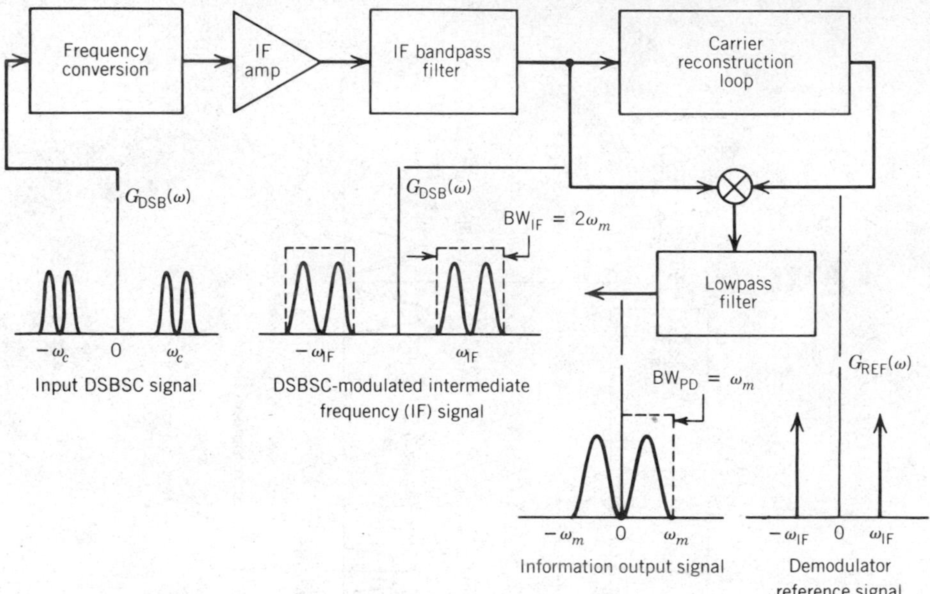

Fig. 48.11 Double sideband suppressed-carrier and detection.

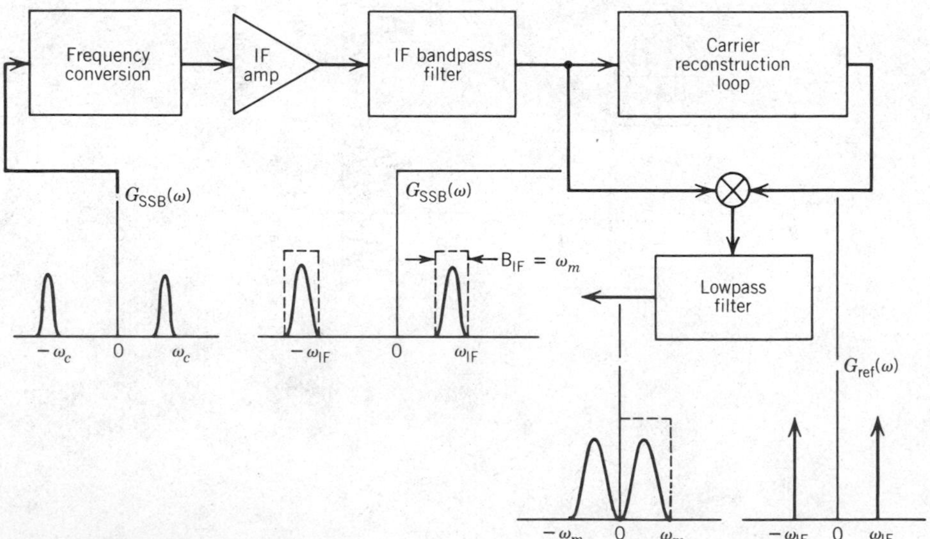

Fig. 48.12 Single sideband suppressed-carrier (SSBSC) reception and detection.

48.1-2 Angle Modulation/Demodulation

As noted earlier, one parameter of the carrier that can be modulated or varied to convey information is the phase angle.

If a term is included in the argument of the sinusoidal function that varies in proportion to the modulating or information signal, the result is phase modulation (PM). Thus for PM the time domain equation is

$$e_{\mathrm{PM}}(t) = E_c \sin[\omega_c t + \theta(t) + \phi_0] \tag{48.17}$$

where $\theta(t) = \Delta\theta g(t)$
$\Delta\theta$ = phase modulation index, rad
$g(t)$ = information signal time function
ω_c = carrier frequency, rad/s
ϕ_0 = arbitrary initial phase angle, rad

If the argument is such that the difference in the instantaneous frequency $\omega_i(t)$ and the carrier frequency is proportional to the modulating or information signal, the signal is frequency modulated (FM). Thus for FM, the frequency varies in proportion to the information signal $g(t)$.

$$\omega_i(t) - \omega_c = \Delta\omega g(t)$$

Since $\omega_i(t) = d\theta(t)/dt$, the FM time domain waveform equation is

$$e_{\mathrm{FM}}(t) = E_c \sin[\omega_c t + \theta(t) + \phi_0] \tag{48.18}$$

where $\theta(t) = \Delta\omega \int_{-\infty}^{t} g(t)\,dt$.

$\Delta\omega$ is the peak frequency deviation in radians per second. Thus in general the angle-modulated sinusoid can be expressed in complex form as the phasor

$$\overrightarrow{E_\theta(t)} = E_c \exp[j(\omega_c t + \phi_0)] \exp[j\theta(t)] \tag{48.19}$$

where $j = \sqrt{-1}$. The phasor expression reveals that for the case of angle modulation the carrier $E_c \exp[j(\omega_c t + \phi_0)]$ has been modified by an exponential function of the message $\theta(t)$. Owing to this characteristic, the terms "exponential modulation" and "angle modulation" have become synonymous. As seen from the time structure of the angle-modulated carrier, the signal amplitude remains constant when modulation is applied. Figure 48.13 illustrates an angle-modulated carrier when the information signal is a single sine wave $g(t) = \Delta\theta \sin \omega_m t$. As seen in the illustration, the resulting PM and FM signals differ only in that the phase variation for the PM case is a sine function and for the FM case a cosine function.

To demodulate or recover the information signal, the angle-modulated carrier must be processed by the receiving system to obtain an output signal proportional to its instantaneous phase or frequency. For the case of PM, the detection can be accomplished using product demodulations where the multiplying reference signal is in phase quadrature with the transmitted carrier. FM detection requires the use of a frequency discriminator, which in its simplest form is a network that first converts the frequency modulation to amplitude modulation and then performs envelope detection of the AM signal.

Phase Modulation

Because of the nonlinear characteristics of the PM process, the power spectrum of the resulting modulated carrier is much more complex than in the case of linear modulation. Thus for PM only, a few selected types of modulating signal formats will be considered in discussing the power spectrum of the PM wave.

For the special case when $g(t) = \sin \omega_m t$, the PM-modulated-wave time domain function is

$$e_{\mathrm{PM}}(t) = E_c \sin[\omega_c t + \Delta\theta \sin \omega_m (t)] \tag{48.20}$$

which can be expanded (Giacoletto, 1947) to

$$e_{\mathrm{PM}}(t) = E_c \Bigg(J_0(\Delta\theta) \sin \omega_c t + \sum_{n=1}^{\infty} J_{2n}(\Delta\theta)[\sin(\omega_c + 2n\omega_m)t + \sin(\omega_c - 2n\omega_m)t] + \sum_{n=1}^{\infty} J_{2n-1}(\Delta\phi)\{\sin[\omega_c + (2n-1)\omega_m]t - \sin[\omega_c - (2n-1)\omega_m]t\} \Bigg) \tag{48.21}$$

where $J_x(\Delta\theta)$ is the Bessel function of the first kind and order x, with argument equal to the PM modulation index $\Delta\theta$.

Thus the power spectrum for a sinusoidally modulated PM wave is a series of impulse functions

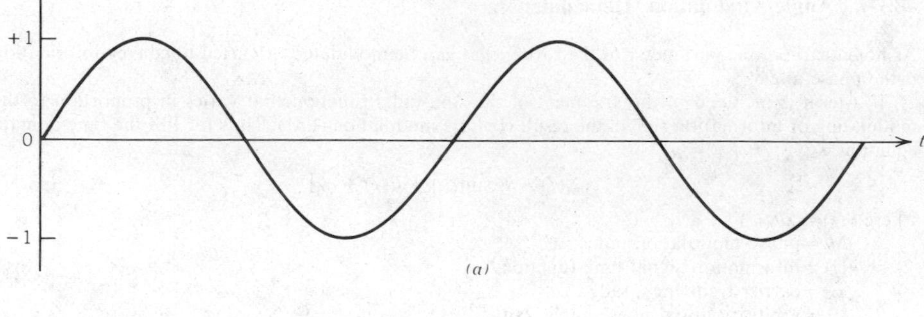

(a)

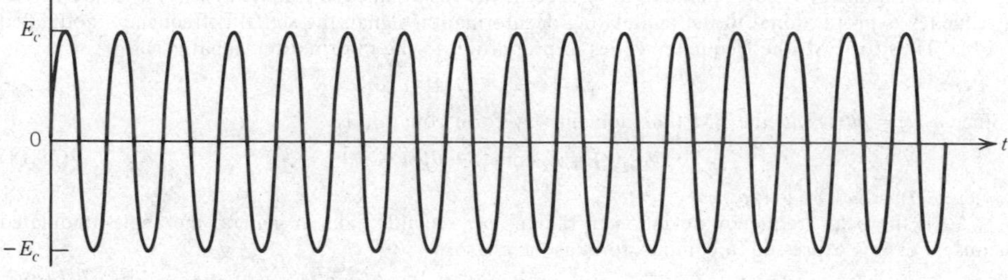

(b)

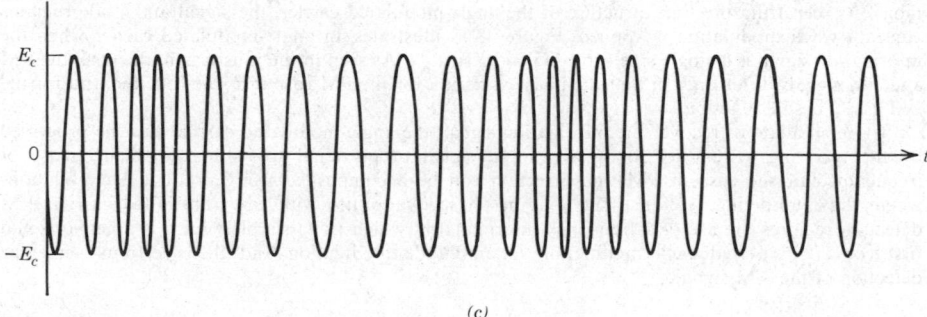

(c)

Fig. 48.13 Time domain illustration of angle modulation by a single sinewave information signal. (*a*) Information signal, $g(t) = \sin \omega_m t$. (*b*) RF carrier signal, $e_c(t) = E_c \sin \omega_c t$. (*c*) Angle-modulated RF carrier signal, $e(t)_{\mathrm{PM}} = E_c \sin[\omega_c t + \Delta\phi \sin \omega_m t]$; $e(t)_{\mathrm{FM}} = E_c \sin\{\omega_c t + [(\Delta\omega/\omega_m)\cos \omega_m t]\}$.

spaced symmetrically about the carrier frequency ω_c at multiples of the modulation frequency ω_m. The weight of each impulse is determined by the Bessel function of appropriate order for that particular sideband.

Theoretically, an infinite number of sidebands appear in the PM spectrum. However, the Bessel function characteristics are such that the sideband magnitudes decrease rapidly as the order increases. Figure 48.14 illustrates the Bessel coefficients for the first nine-order sidebands of a PM or FM wave vs. its argument $\Delta\theta$ or β.

Phase modulation removes power from the carrier signal and deposits it in the sidebands. For single-tone modulation, the power spectrum is illustrated in Fig. 48.15 assuming a constant phase modulation index $\Delta\theta$ and an increasing modulating signal frequency ω_m.

The total power in the sidebands and the carrier components of the PM signal can be shown to be (Painter and Hondros, 1965)

$$P_{\mathrm{PM}} = \frac{E_c^2}{2}\left[\sum_{-\infty}^{\infty} J_n^2(\Delta\theta)\right] \tag{48.22}$$

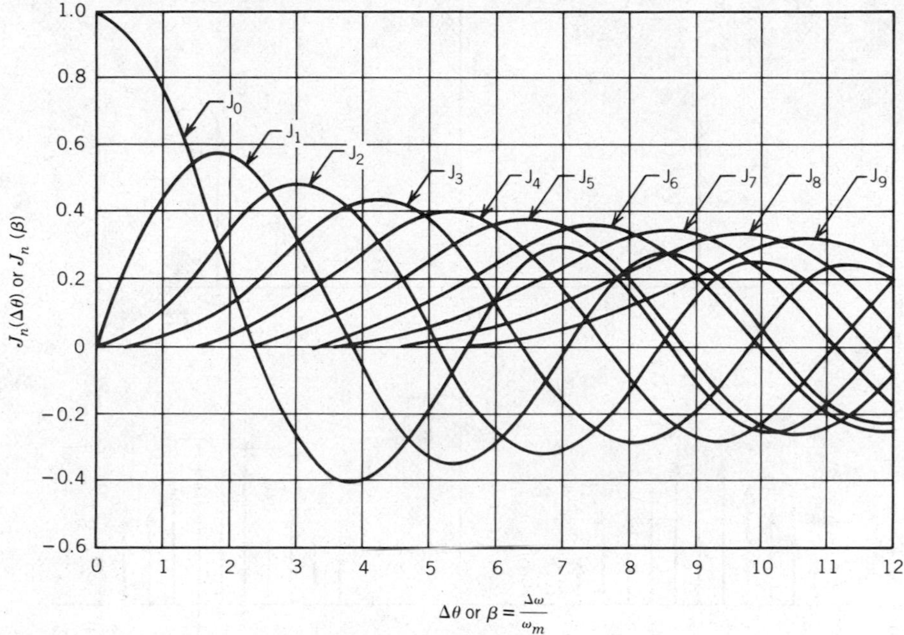

Fig. 48.14 Bessel coefficients for phase and frequency modulation (PM and FM) spectral components.

Since the quantity $\sum_{-\infty}^{\infty} J_n^2(\Delta\theta) = 1$, the total power of the modulated wave is equivalent to that of the unmodulated carrier $E_c^2/2$. If more than one sine wave signal is used to phase modulate the same carrier, the resulting spectrum will be a complex interrelation of the two signals' frequencies and modulation indices. In fact, if the two modulation signals are not coherently or harmonically related, the power spectrum will consist of impulses at frequencies around the carrier equivalent to the sum and differences of each of the modulating signals and their harmonic frequencies. In the general case of PM modulation by K sine waves, the time domain expression for the PM wave becomes (Giacoletto, 1947)

$$e_{\mathrm{PM}}(t) = E_c \sum_{n_i=-\infty}^{\infty} \cdots \sum_{n_k=-\infty}^{\infty} \prod_{i=1}^{K} [J_{ni}(\Delta\theta_i)] \left\{ \begin{matrix} \sin \\ \cos \end{matrix} \right\} \left[\omega_c t + \sum_{i=1}^{K} n_i(\omega_i t + \phi_i) \right] \tag{48.23}$$

where $\left\{ \begin{smallmatrix} \sin \\ \cos \end{smallmatrix} \right\}$ denotes either sine or cosine functions and $\prod_{i=1}^{K}$ denotes the product of terms $i = 1$ to K.

The addition of more modulating tones to a phase-modulated carrier takes power from the existing modulating tones as well as from the carrier. Therefore, before use of frequency division multiplexing techniques (subcarriers) to transfer information on a PM channel, the effects of power distribution among the various subcarriers when modulation indices and the number of subcarriers are selected must be considered. An illustration of the power distribution of the carrier term and the first sideband (J_1) terms of the power spectral density for the case of two PM subcarriers is given in Fig. 48.16.

PM Demodulation

Demodulation of PM signals generally employs a carrier recovery and product detection process similar to that depicted in Fig. 48.17 (Painter and Hondros, 1965).

Since the process of multiplication is employed in PM detection, the results can be viewed simply as a translation of the PM-modulated spectrum to baseband, as illustrated in Fig. 48.18. Then the ω_m (J_1 term) sideband can be filtered with a lowpass filter as illustrated in the figure and the subcarrier signal recovered. The power in the recovered signal is equal to the total PM signal power multiplied by the square of the first-order Bessel term if there is no phase error between the quadrature

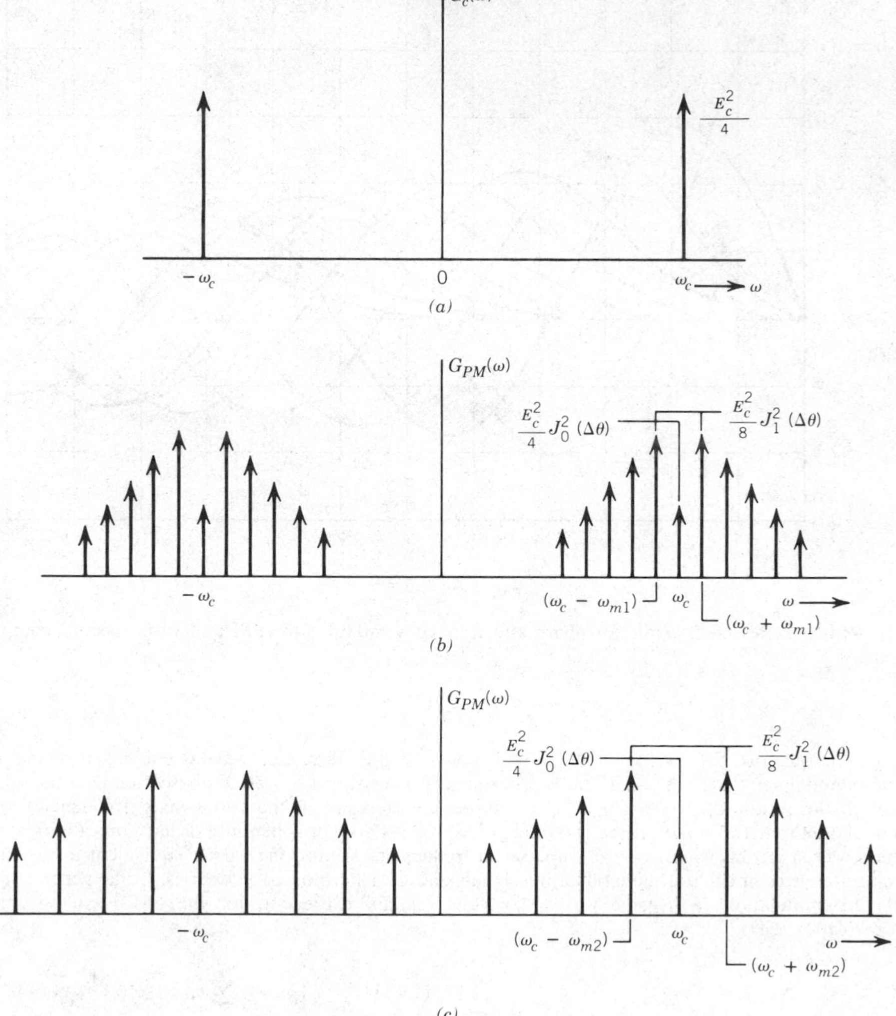

Fig. 48.15 Power spectra for single-tone phase-modulated wave ($\Delta\theta$ constant): (*a*) Two-sided, unmodulated carrier, (*b*) two-sided, PM-modulated signal ($\omega_{m_1} = 10$ kHz; $\Delta\theta = 1$ rad), (*c*) two-sided, PM-modulated signal ($\omega_{m_2} = 20$ kHz; $\Delta\theta = 1$ rad).

reference carrier and the received carrier, that is (Painter and Hondros, 1965),

$$e_{o,\mathrm{PM}}(t) = KE_c J_1(\Delta\theta)\sin\omega_m t$$

where $K \doteq \cos\Delta\theta_c$. Then

$$P_{\mathrm{SC}} = \frac{E_c^2}{2} J_1^2(\Delta\theta)$$

which is the power contained in the first sideband pair of the PM-modulated wave.

The ratio of the power in the first sideband pair to the total power of a PM-modulated wave for a sine wave subcarrier signal is referred to as the modulation loss

$$\mathrm{ML} = \frac{2J_1^2(\Delta\theta)(E_c^2/2)}{E_c^2/2} = 2J_1^2(\Delta\theta)$$

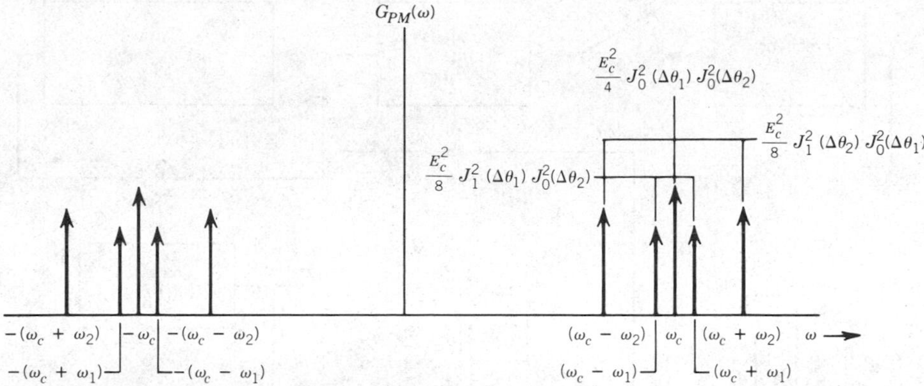

Fig. 48.16 Partial power spectrum resulting from two subcarriers at frequencies ω_1 and ω_2, with modulation indices $\Delta\theta_1$ and $\Delta\theta_2$, respectively.

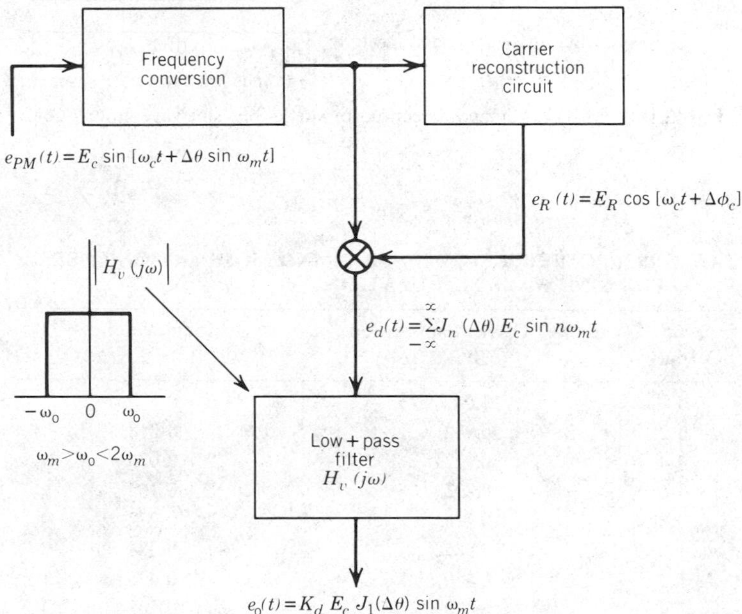

Fig. 48.17 Detection process for a phase-modulated carrier modulated by a single sine wave of frequency ω_m.

The performance of the carrier recovery (carrier-tracking) loop is affected by the number of modulating tones and their respective modulation indices. This factor is referred to as the carrier suppression of the PM wave, which for single-tone modulation is given by

$$\text{CS} = J_0^2(\Delta\theta)$$

Table 48.3 lists the carrier suppression $J_0^2(\Delta\theta)$ and the modulation loss $2J_1^2(\Delta\theta)$ for a single sine wave subcarrier channel as a function of the modulation index $\Delta\theta$.

Since it is only necessary to recover the first sidebands, that is, $J_1(\Delta\theta)\sin \omega_m t$ terms, of the modulated signal for single-tone subcarrier modulation, the receiver IF bandwidth requirement is equivalent to that of a single-tone amplitude-modulated wave.

The receiver input noise is assumed to be white and Gaussian with two-sided noise spectral density $N_0/2$. The relation between output signal-to-noise ratio and input (IF) signal-to-noise ratio

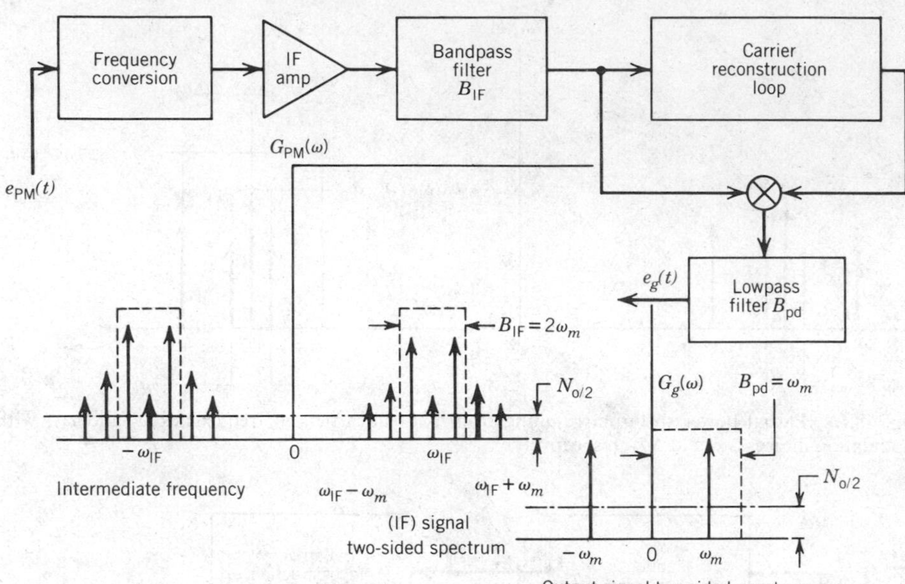

Fig. 48.18 PM reception and detection for single-tone sinewave modulation.

TABLE 48.3. CARRIER SUPPRESSION AND MODULATION LOSS VS. $\Delta\theta$ FOR SINE WAVE MODULATION

$\Delta\theta$ (rad)	Carrier Suppression (CS) = $J_0^2(\Delta\theta)$	Modulation Loss (ML) = $2J_1^2(\Delta\theta)$
0	1.0	0
0.2	0.980	0.018
0.4	0.922	0.076
0.6	0.832	0.162
0.8	0.716	0.272
1.0	0.585	0.388
1.2	0.450	0.496
1.4	0.320	0.588
1.6	0.207	0.650
1.8	0.115	0.676
2.0	0.050	0.666
2.2	0.017	0.618
2.4	6.3×10^{-6}	0.542
2.6	0.0094	0.442
2.8	0.034	0.336
3.0	0.068	0.230
3.2	0.102	0.136
3.4	0.133	0.064
3.6	0.153	0.018
3.8	0.162	0.0003
4.0	0.158	0.009

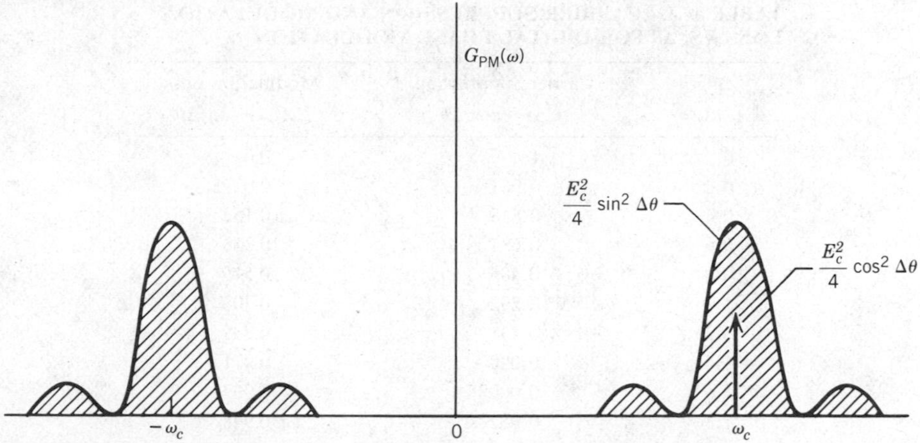

Fig. 48.19 Power spectral density of digitally phase-modulated wave [nonreturn-to-zero (NRZ) binary information signal].

(considering only the fundamental J_1 terms as input signal power) is

$$(S/N)_{o,\text{PM}} = \frac{E_c^2 J_1^2(\Delta\theta)}{2N_0\omega_m} \tag{48.24}$$

which is equivalent to the 100% AM, DSBSC, and SSBSC performance based on constant sideband power. However, because of the coherent detection process inherent in PM demodulation, performance under poor input signal-to-noise ratios, that is, $(S/N)_{\text{in}} \ll 10$ dB, is significantly improved over that for conventional noncoherent AM detection.

Digital Phase Modulation

A carrier signal phase modulated by a digital (binary) signal can be expressed in the time domain as

$$e_{\text{PM}}(t) = E_c \sin[\omega_c t + \Delta\theta\, s(t)] \tag{48.25}$$

where $s(t)$ is defined as a digital switching function of time with values ± 1.

If the transition from $+1$ to -1 and vice versa is instantaneous (i.e., infinite bandwidth is assumed), the expression can be expanded to give (Panneton, 1968)

$$e_{\text{PM}}(t) = E_c[\cos\Delta\theta \sin\omega_c t + s(t)\sin\Delta\theta \cos\omega_c t] \tag{48.26}$$

The power spectrum of a digitally phase-modulated signal is illustrated in Fig. 48.19. The amount of power distributed in the sidebands representing the information signal is

$$P_{\text{SB}} = \frac{E_c^2}{2}\sin^2\Delta\theta \tag{48.27}$$

Since the total power of the PM wave is $E_c^2/2$, the modulation loss can again be defined for a digital modulating signal as

$$\text{ML} = \sin^2\Delta\theta \tag{48.28}$$

where $\Delta\theta$ is the PM modulation index as defined previously.

The amount of carrier power remaining in the modulated wave is

$$P_c = \frac{E_c^2}{2}\cos^2\Delta\theta \tag{48.29}$$

Then the carrier suppression can also be defined as

$$\text{CS} = \cos^2\Delta\theta \tag{48.30}$$

Table 48.4 lists the values of ML and CS as a function of $\Delta\theta$. When $\Delta\theta = \pi/2$ radians, the carrier is totally suppressed

$$\text{CS} = \cos^2\frac{\pi}{2} = 0$$

TABLE 48.4. CARRIER SUPPRESSION AND MODULATION LOSS VS. $\Delta\theta$ FOR DIGITAL PHASE MODULATION

$\Delta\theta$ (rad)	Carrier Suppression (CS) = $\cos^2\Delta\theta$	Modulation Loss (ML) = $\sin^2\Delta\theta$
0	1	0.0
0.2	0.961	0.0395
0.4	0.848	0.152
0.6	0.681	0.318
0.8	0.485	0.515
1.0	0.292	0.708
1.2	0.131	0.869
1.4	0.029	0.971
1.6	0.00085	0.999
1.8	0.052	0.948
2.0	0.173	0.827
2.2	0.346	0.654
2.4	0.544	0.456
2.6	0.734	0.266
2.8	0.888	0.112
3.0	0.980	0.019

and no modulation loss occurs.

$$\text{ML} = \sin^2\frac{\pi}{2} = 1$$

Hence for $\pi/2$ radians modulation index, assuming a digital modulating function $s(t)$, the carrier signal is totally suppressed and all the power appears in the information sidebands. The resulting time function then becomes

$$e_{\text{PSK}}(t) = E_c s(t)\cos(\omega_c t + \phi_0) \tag{48.31}$$

This special case of digital PM (i.e., where $\Delta\theta = \pm\pi/2$ radians) is referred to as phase shift keying (PSK) modulation. Since $e_{\text{PSK}}(t)$ and $e_{\text{DSBSC}}(t)$ time functions are of the same form, a PSK signal can be generated with a linear product modulator configured for DSBSC modulation. The resulting modulated-signal spectrum is simply a translation in frequency of the digital information spectrum to the RF carrier frequency band. Figure 48.20 illustrates a PSK signal spectrum when the digital data rate is $\omega_R = 2\pi/T$.

Digital Phase Demodulation

Since the digital PM signal is equivalent to DSBSC modulation, detection can be accomplished as illustrated in Fig. 48.21, utilizing a quadrature (90° phase-shifted) reconstructed carrier signal as the demodulator reference (Lindsey, 1972). As noted previously, the output of the demodulator can thus be viewed simply as the IF signal spectrum translated to baseband. The signal power contained in the information signal sidebands, assuming no phase error in the detection process or power loss due to IF bandwidth limitations, is $E_c^2\sin^2\Delta\theta/2$, where $\Delta\theta$ is the PM modulation index.

For $\Delta\theta = \pm\pi/2$ radians (PSK modulation), the information signal power is equal to the total received power. Assuming receiver noise as being white and Gaussian with spectral density equal to $N_0/2$, the output signal-to-noise ratio in the data rate bandwidth ω_R is

$$(S/N)_{o,\text{PSK}} = \frac{E_c^2\sin^2\Delta\theta}{2N_0\omega_R} \tag{48.32}$$

For a practical PSK system, the PM channel receiver IF bandwidth required to pass 90% or more of the signal power must range from 1.5 to 6 times the data rate ω_R, depending on the digital code format. Figures 48.22 and 48.23 (Batson, 1968) illustrate the percentage of signal power passed by an IF filter of bandwidth B_{IF} assuming a PSK-modulated carrier signal. Data are given for both nonreturn-to-zero (NRZ) and biphase (bi-ϕ) digital data formats.

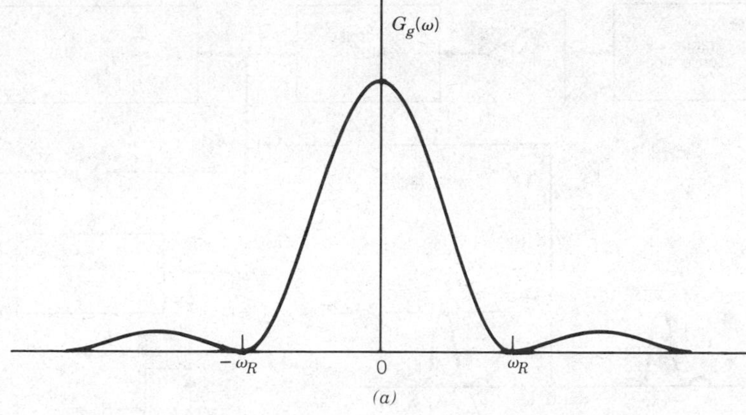

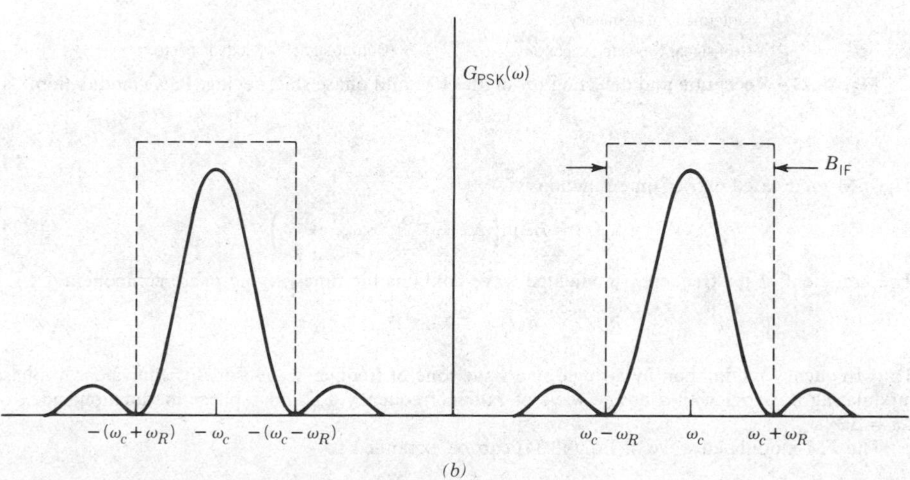

Fig. 48.20 Digital phase shift keying (PSK) modulation spectra assuming nonreturn-to-zero (NRZ) information signal: (*a*) Two-sided power spectrum of baseband NRZ digital information signal, (*b*) two-sided power spectrum of PSK-modulated wave.

Frequency Modulation

For the case of continuous-wave frequency modulation, the time domain expression is

$$e_{FM}(t) = E_c \sin\left[\omega_c t + \Delta\omega \int_{-\infty}^{t} g(t)\,dt + \phi_0\right] \tag{48.33}$$

As for the case of PM, the FM modulation process is nonlinear and results in a spectrum of infinite bandwidth. Owing to the complexity of establishing the general FM spectrum characteristics and transmission bandwidth requirements, only a few selected types of modulating signal functions are discussed.

Sine Wave Frequency Modulation. For the special case when the modulating signal is single sine wave

$$g(t) = \Delta\omega \sin \omega_m t$$

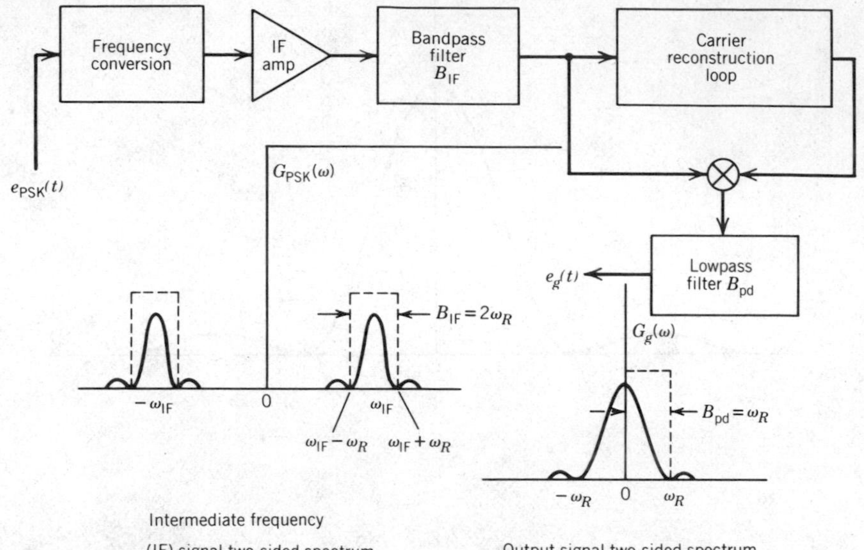

Fig. 48.21 Reception and detection for digital PM and phase shift keying (PSK) modulation.

The FM modulated carrier time function is

$$e_{FM}(t) = E_c \sin\left(\omega_c t + \frac{\Delta\omega}{\omega_m} \cos \omega_m t + \phi_0\right) \tag{48.34}$$

The argument of the frequency-modulated wave contains the time-varying phase component

$$\theta(t) = \frac{\Delta\omega}{\omega_m} \cos \omega_m t \tag{48.35}$$

Thus frequency modulation by a single sine wave tone of frequency ωm signal is equivalent to phase modulating a carrier with a cosine wave of radian frequency ω_m and a phase modulation index of $\Delta\theta = \Delta\omega/\omega_m$.

The FM-modulated wave of Eq. (48.34) can be expanded to

$$\begin{aligned} e_{FM}(t) = E_c\bigg[& J_0\left(\frac{\Delta\omega}{\omega_m}\right) \sin \omega_c t \\ & + \sum_{n=1}^{\infty} (-1)^{n+1} J_{2n-1}\left(\frac{\Delta\omega}{\omega_m}\right) \{\cos[\omega_c + (2n-1)\omega_m]t + \cos[\omega_c - (2n-1)\omega_m]t\} \\ & + \sum_{n=1}^{\infty} (-1)^n J_{2n}\left(\frac{\Delta\omega}{\omega_m}\right) \{\sin[\omega_c + 2n\omega_m]t + \sin[\omega_c - 2n\omega_m]t\}\bigg] \end{aligned} \tag{48.36}$$

The power spectrum of the FM wave with sinusoidal modulation is an infinite series of impulses spaced symmetrically about the carrier with spacing equal to the modulating signal frequency ω_m and its harmonics. The weight of each impulse is dependent on the appropriate-order Bessel function for that particular sideband term and the frequency modulation index $\beta = \Delta\omega/\omega_m$. Since the practical case does not allow for infinite transmission bandwidth, some filtering occurs at the transmitter and/or receiver. The transmission bandwidth required for the sinusoidal modulated FM carrier is dependent on the modulation index and the frequency of the modulating sine wave. It has been shown (Painter and Hondros, 1965) that the bandwidth encompassing all the sidebands of significant amplitude for this special case is equivalent to

$$BW_{FM} = 2(\Delta\omega + \omega_m) \tag{48.37}$$

which is often referred to as Carson's rule. However, application of Carson's rule for frequency modulation with complex information signals will not necessarily result in optimum system performance.

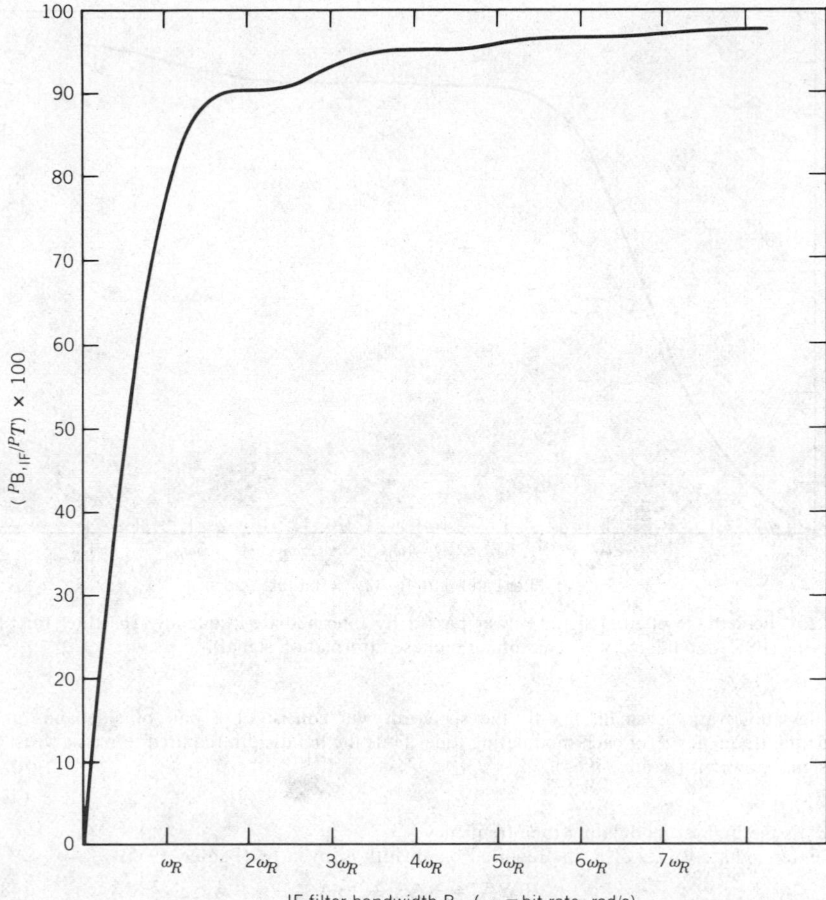

Fig. 48.22 Percentage of total signal power passed by intermediate-frequency (IF) filter for phase shift keying (PSK)-modulated signal [assuming nonreturn-to-zero (NRZ) information signal].

Figure 48.24 illustrates an FM wave modulated by a single sine wave, as the modulating frequency ω_m is varied and the frequency deviation $\Delta\omega$ is held constant.

For the case of multiple sine wave modulation (not harmonically related), the spectrum is of infinite bandwidth, as it is for multitone PM. Spectral components occur at frequencies corresponding to the sum and difference of each modulating tone and its harmonics with all other modulating tones and their harmonics. The multitone modulated-wave time domain expression is (Giacoletto, 1947)

$$e_{FM}(t) = E_c \sum_{n_i=-\infty}^{\infty} \cdots \sum_{n_k=-\infty}^{\infty} \prod_{i=1}^{K} \left[J_{n_i}(\beta_i) \left\{ \begin{matrix} \sin \\ \cos \end{matrix} \right\} \right] \left[\omega_c t + \sum_{i=1}^{K} n_i(\omega_i t + \phi_i) \right] \qquad (48.38)$$

The bandwidth requirements for noncoherent multitone FM can be obtained by evaluation of the Bessel functions. Since the power of an FM wave is constant and equal to the carrier power, it is obvious that the power in each sideband component decreases as the number of modulating tones increases. Beyond a certain frequency range from the carrier, the sideband power becomes negligible. If β_i is large for each tone, the bandwidth required for passing 90% or more of the sideband power at the receiver IF is approximately (Panter, 1965)

$$\mathrm{BW}_{FM} = 2 \sum_{i=1}^{K} \Delta\omega_i \qquad (48.39)$$

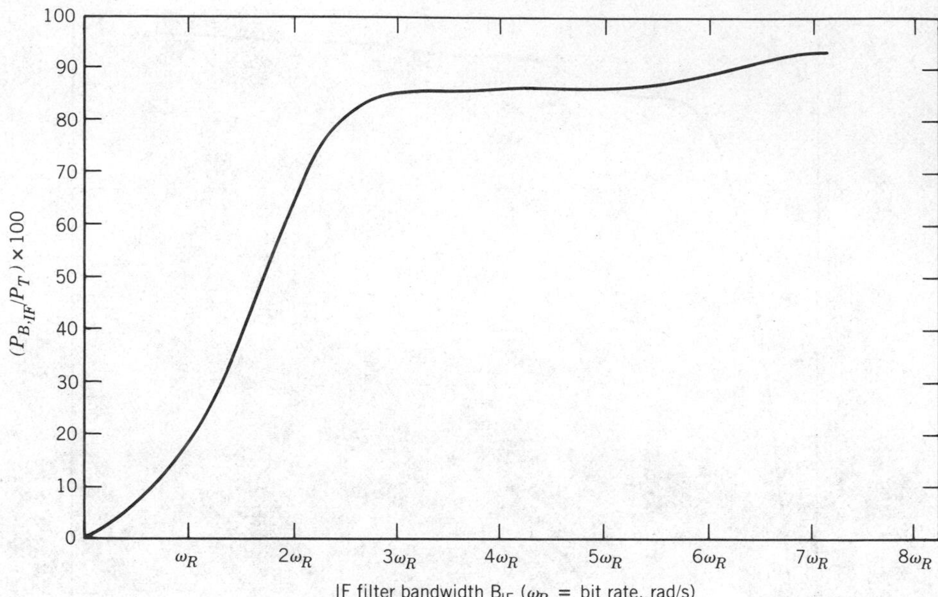

Fig. 48.23 Percentage of total signal power passed by intermediate-frequency (IF) filter for phase shift keying (PSK)-modulated wave (assuming biphase information signal).

If the deviation ratio is small, $\beta_i \ll 1$, the spectrum will consist of a pair of sidebands at the fundamental frequency ω_i of each modulating tone. Thus the bandwidth required to pass at least 90% of the signal power is (Panter, 1965)

$$\text{BW}_{\text{FM}} = 2\omega_J \tag{48.40}$$

where ω_J is the highest modulating tone frequency.

For intermediate values of β, the required bandwidth is given by (Panter, 1965)

$$\text{BW}_{\text{FM}} = 2(\Delta\omega_p + 2\omega_J) \tag{48.41}$$

where $\Delta\omega_p$ is the composite peak frequency deviation of the system and ω_J is the highest modulating tone frequency.

Digital Frequency Modulation. Two cases of digital FM of interest are continuous-phase FM, which is analogous to the FM modulation process discussed thus far, and frequency shift keying (FSK) modulation, which is equivalent to switching between oscillators of frequencies ω_1 and ω_2 depending on the digital information signal to be transmitted.

Continuous-phase modulation with a square wave or binary digital process can be expressed as

$$e_{\text{FM}}(t) = E_c \sin[\omega_c t + \Delta\omega s(t)] \tag{48.42}$$

where $s(t)$ represents the function that takes on only values $+1$ or -1, depending on the information symbol being transmitted.

The determination of the spectrum requires knowledge of the coefficients in the Fourier series expansion of $e_{\text{FM}}(t)$. If $s(t)$ is a square wave or repetitive sequence of binary 1's and 0's, then the FM waveform can be expressed as (Panter, 1965)

$$e_{\text{FM}}(t) = \sum_{-\infty}^{\infty} |\bar{F}_n| \cos(\omega_c t + n\omega_0 t) \tag{48.43}$$

where ω_0 is the fundamental frequency of the modulating square wave and

$$\bar{F}_n = \frac{1}{2}\left\{ \frac{\sin[(\beta - n)\pi/2]}{(\beta - n)\pi/2} + (-1)^n \frac{\sin[(\beta + n)\pi/2]}{(\beta + n)\pi/2} \right\} \tag{48.44}$$

where $\beta = \Delta\omega/\omega_0$.

The modulated spectrum is illustrated in Fig. 48.25 as the deviation ratio is increased by maintaining a constant $\Delta\omega$ and reducing ω_0. As seen, the spectral energy tends to concentrate at the

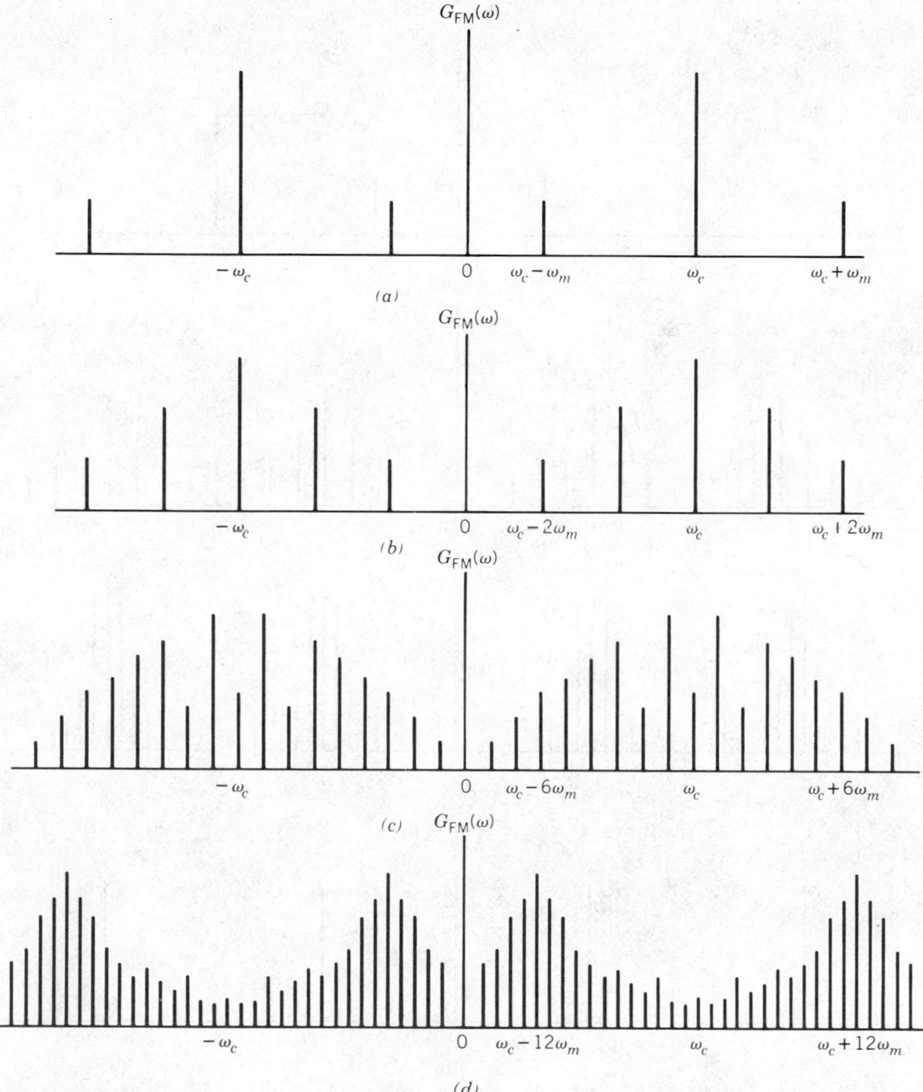

Fig. 48.24 FM modulation power spectra for single-tone modulating signal ($\Delta\omega$ constant): (*a*) $\Delta f = 24$ kHz, $f_m = 48$ kHz, $\beta = 0.5$; (*b*) $\Delta f = 24$ kHz, $f_m = 24$ kHz, $\beta = 1.0$; (*c*) $\Delta f = 24$ kHz, $f_m = 4$ kHz, $\beta = 6.0$; (*d*) $\Delta f = 24$ kHz, $f_m = 2$ kHz, $\beta = 12$.

frequency equivalent to the frequency deviation $\Delta\omega$. The FM spectral distribution tends to follow the modulating signal time distribution. The longer the signal remains at a particular carrier frequency location, the more spectral energy appears in that band of the modulated signal spectrum. As β approaches ∞ with a fixed $\Delta\omega$, the information rate approaches zero in the limit. Thus when $\beta = \infty$, the modulating signal is at $+\Delta\omega$ and $-\Delta\omega$ for an infinitely long period of time, resulting in half the power located at $\omega_c + \Delta\omega$ and half at $\omega_c - \Delta\omega$.

FSK Signaling. In the special case of frequency shift keying, the modulated signal is formed by switching between two oscillators at frequencies ω_1 and ω_2. Since the oscillators are not coherent, the phase of the resulting modulated wave is not continuous, that is, a discontinuity occurs at the switching time. Figure 48.26 illustrates the technique of FSK modulation. The resulting signal spectrum consists of two impulse functions at the symbol frequencies modulated by the random binary information process illustrated in Fig. 48.27.

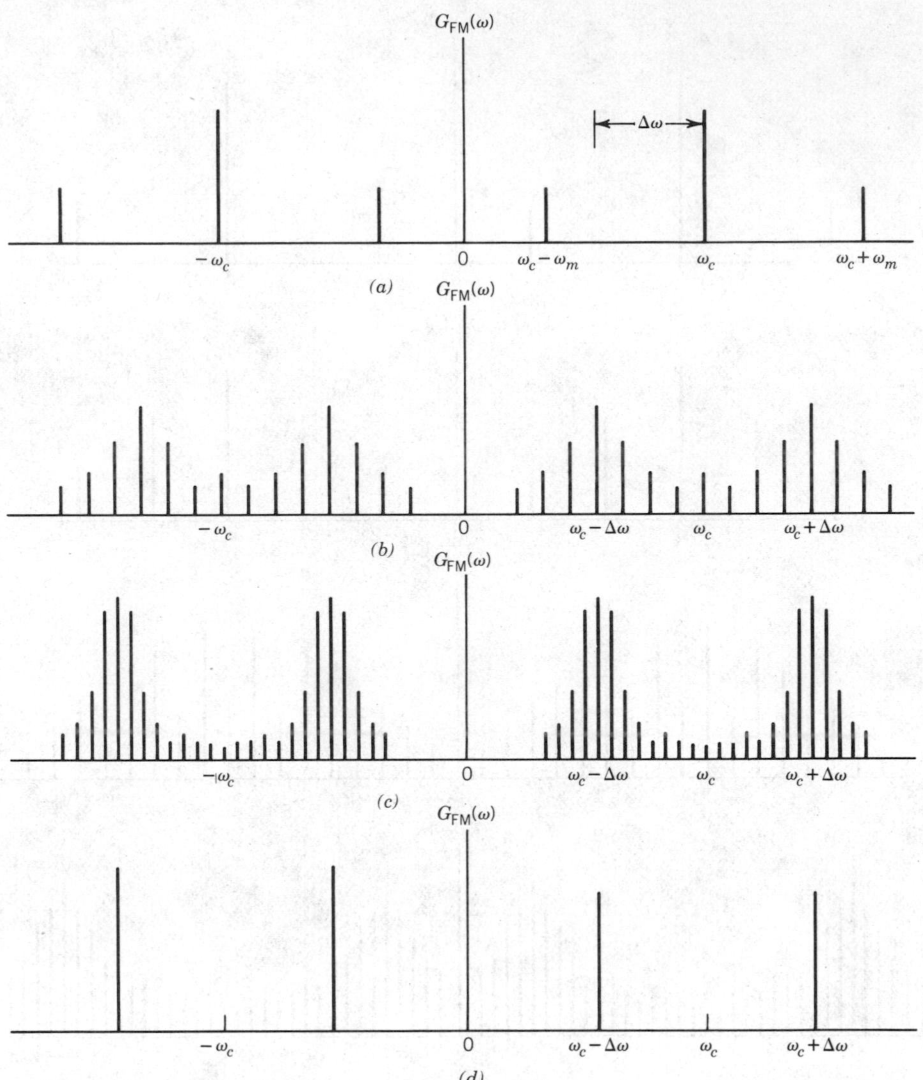

Fig. 48.25 FM modulation power spectra for square-wave modulation signals ($\Delta\omega$ constant): (*a*) $\beta < 1$, (*b*) $5 > \beta < 10$, (*c*) $10 > \beta < 25$, (*d*) $\beta = \infty$.

Since the FSK signal is not continuous in phase, it can be recovered by individually filtering each of the symbol frequencies at the receiver. Hence the actual bandwidth required for reception of FSK is less than $\omega_1 + \omega_2$, as would be inferred from Fig. 48.27.

FM Demodulation. Figure 48.28 illustrates FM reception and detection. If the IF noise at the output of the bandpass filter B_{IF} is white and Gaussian with two-sided noise spectral density $N_0/2$, and the signal (carrier)-to-noise ratio is large (i.e., $C/N > 10$ dB), the output noise power spectrum (Fig. 48.29) is given by

$$G_{n0}(\omega) = \begin{cases} \dfrac{K_d^2\omega^2}{4\pi^2 E_c^2} N_0 & |\omega| < 2\pi\left[\dfrac{B_{IF}}{2}\right] \\ 0 & |\omega| > 2\pi\left[\dfrac{B_{IF}}{2}\right] \end{cases} \tag{48.45}$$

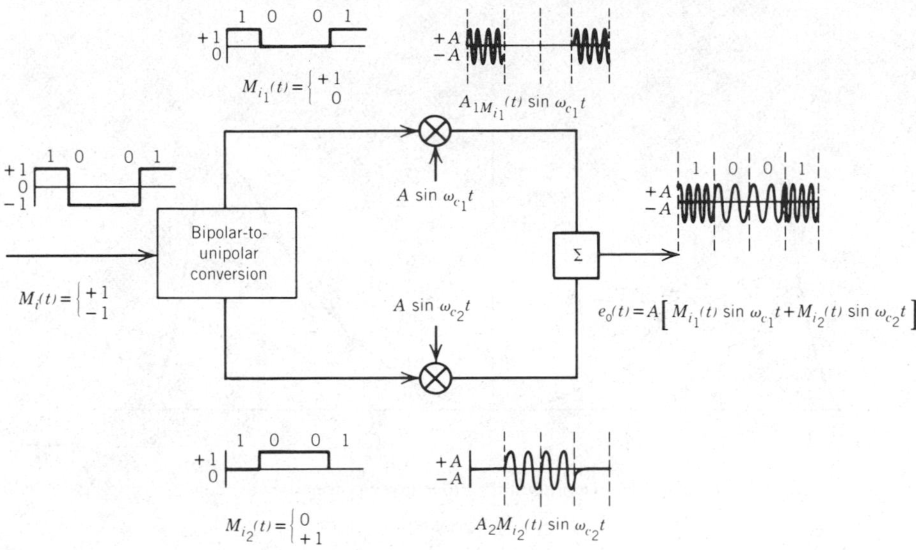

Fig. 48.26 Functional representation of orthogonal (coherent) frequency shift keying (FSK) modulation [assuming $M_i(t)$ is a nonreturn-to-zero (NRZ) digital signal].

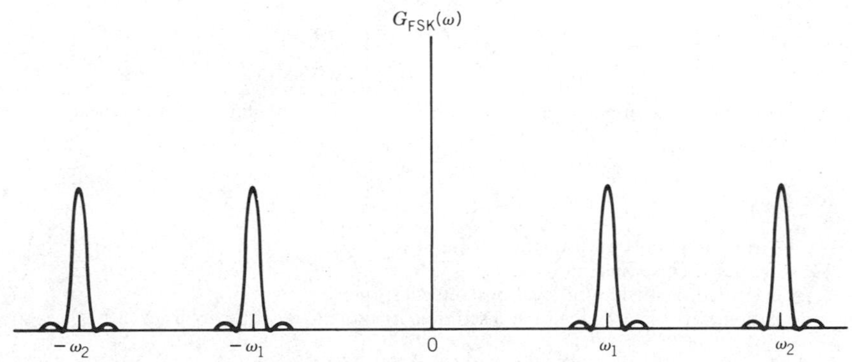

Fig. 48.27 Power spectral density, frequency shift keying (FSK) signal.

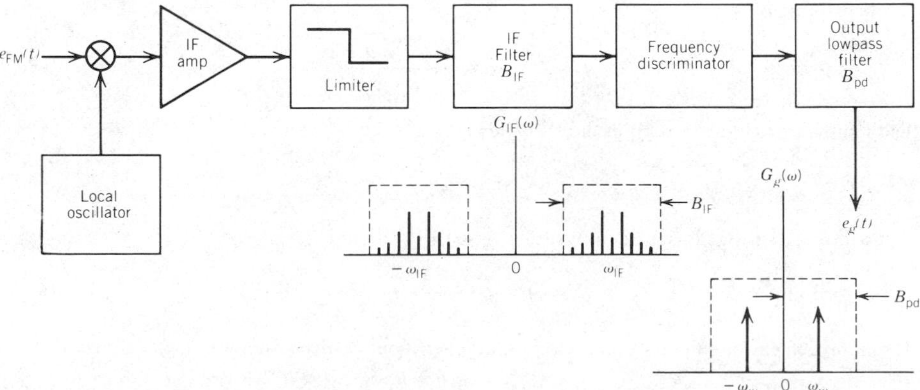

Fig. 48.28 Typical FM reception and detection system (single-tone modulating signal).

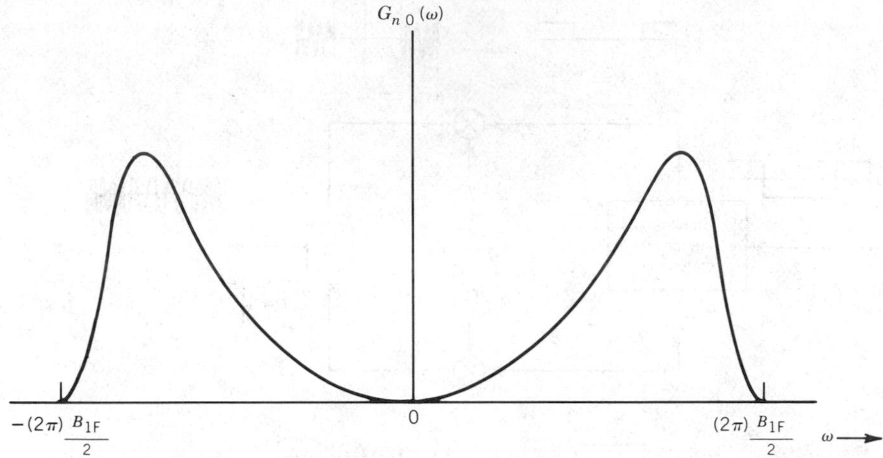

Fig. 48.29 FM discriminator output noise power spectral density.

Note that the output noise spectrum is dependent on ω^2, resulting in a parabolic form. The FM baseband noise output is then obtained by integrating over the transfer function of the postdetection lowpass filter. Assuming the discriminator sensitivity is given by K_d and the output filter is an ideal lowpass filter with response

$$|H(j\omega)|^2 = \begin{cases} 1 & |f| < B_{\text{pd}} \\ 0 & |f| > B_{\text{pd}} \end{cases} \tag{48.46}$$

the output noise power, assuming that the input carrier is not modulated by an information signal, that is, $g(t) = 0$, is

$$P_n = \left(\frac{K_d}{E_c}\right)^2 \frac{2N_0 B_{\text{pd}}^3}{3} \tag{48.47}$$

where E_c = carrier amplitude at discriminator input
$N_0/2$ = two-sided noise spectral density
B_{pd} = one-sided bandwidth of ideal postdetection filter

If the information signal is $g(t)$, then the discriminator output signal power is given by

$$P_s = \left(\frac{K_d}{2\pi}\right)^2 \Delta\omega^2\left[\overline{g^2(t)}\right] \tag{48.48}$$

and the signal-to-noise ratio out of the discriminator for IF carrier-to-noise ratios greater than 10 dB can be expressed as

$$(S/N)_{o,\text{FM}} = \frac{3E_c^2 \Delta f^2 \overline{g^2(t)}}{2N_0(B_{\text{pd}})^3} \tag{48.49}$$

The input carrier-to-noise ratio in the IF bandwidth is given by

$$\text{CNR}_i = \frac{E_c^2}{2N_0 B_{\text{IF}}} \tag{48.50}$$

If the information signal is a sine wave, then $\overline{g^2(t)} = 1/2$ and

$$(S/N)_{o,\text{FM}} = \frac{3}{2}\frac{\Delta f^2 B_{\text{IF}}}{B_{\text{pd}}^3}\text{CNR}_i \tag{48.51}$$

The improvement in output signal-to-noise ratio over input carrier-to-noise ratio is defined as

$$I_{\text{FM}} = \frac{3}{2}\Delta f^2 \frac{B_{\text{IF}}}{B_{\text{pd}}^3} \tag{48.52}$$

If the output lowpass filter is ideal and the bandwidth is equivalent to the modulating signal frequency, ω_m, and if the IF bandwidth is made equal to twice the modulating signal frequency, 2 ω_m, then a direct comparison with the AM signal-to-noise performance is possible. For these conditions, and assuming operation is above discriminator threshold, the improvement factor is

$$I_{\mathrm{FM}} = 3\left(\frac{\Delta f}{f_m}\right)^2 \tag{48.53}$$

and since $\Delta f/f_m$ was previously defined as the modulation index β, the FM improvement over AM is often defined as $3\beta^2$. This is only true for the very special conditions of postdetection and IF filter bandwidths equivalent to the AM channel requirements.

When the modulation signal is composed of one or more sinusoidal subcarriers, the postdetection filtering employed is generally that of bandpass selection of the individual subcarrier signals, as illustrated in Fig. 48.30. In this case the signal-to-noise improvement factor, when operating above discriminator threshold, is

$$I_{\mathrm{FM,BPF}} = \frac{3}{2}\,\frac{\Delta f^2 B_{\mathrm{IF}}}{(f_U^3 - f_L^3)} \tag{48.54}$$

where f_U is the upper half power cutoff frequency of the postdetection bandpass filter (BPF) and f_L is the lower half power cutoff frequency of the postdetection BPF.

The FM improvement relationships defined thus far have been arrived at using approximations only valid under high IF carrier-to-noise ratio conditions ($\mathrm{CNR}_i > 10$ dB). Below a certain critical value of CNR_i, referred to as the threshold level, the FM system performance deteriorates rapidly. Analysis of below-threshold performance is somewhat complicated. The resulting output signal-to-noise expression is (Shilling and Taub, 1971)

$$(S/N)_{o,\mathrm{FM}} = \frac{3\rho^2\,\Delta f^2\left(B_{\mathrm{IF}}/B_{\mathrm{pd}}^3\right)\mathrm{CNR}_i}{\sqrt{3}\,(B_{\mathrm{IF}}/B_{\mathrm{pd}})^2\mathrm{CNR}_i\left(1 - \operatorname{erf}\sqrt{\mathrm{CNR}_i}\right) + 1} \tag{48.55}$$

In this expression, ρ is the peak-to-rms ratio of the modulating signal and $\operatorname{erf}\sqrt{\mathrm{CNR}_i}$ is the Gaussian error function. Note that for a sine wave modulating signal $\rho = 1/\sqrt{2}$ and for a strong CNR_i, the error function approaches unity. Then the equation becomes

$$(S/N)_{o,\mathrm{FM}} = \frac{3}{2}\Delta f^2\,\frac{B_{\mathrm{IF}}}{B_{\mathrm{pd}}^3}\,\mathrm{CNR}_i \tag{48.56}$$

as discussed earlier. The baseband signal-to-noise ratio transfer function characteristics of a typical FM discriminator showing the threshold effect is given in Fig. 48.31. The threshold level of CNR_i is defined as the CNR_i where the output signal-to-noise ratio $(S/N)_{o,\mathrm{FM}}$ has diverged 1 dB from the

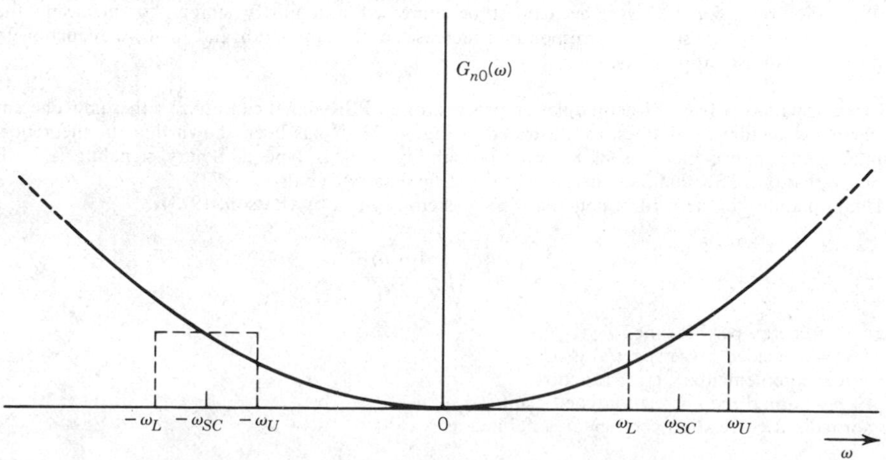

Fig. 48.30 Bandpass postdetection filtering of FM discriminator.

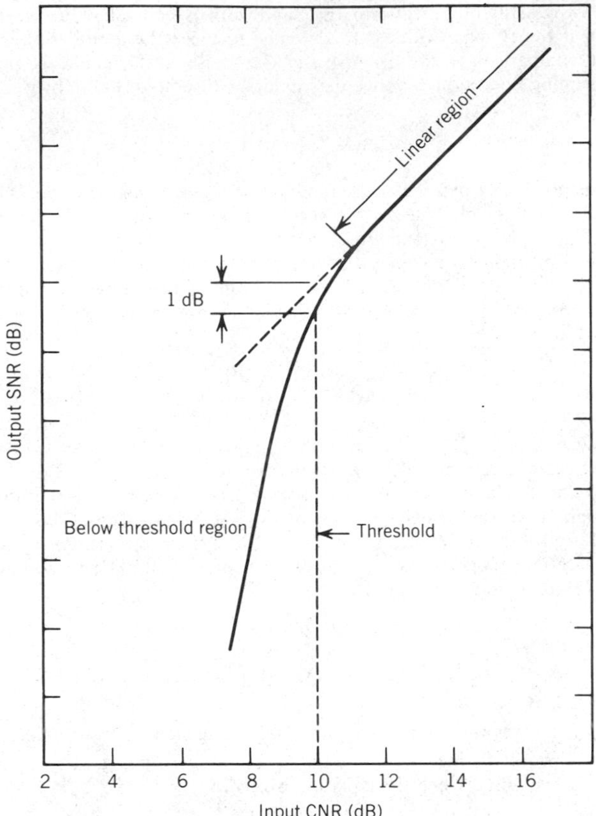

Fig. 48.31 Typical output signal-to-noise ratio (SNR) vs. input carrier-to-noise ratio (CNR) for FM discriminator, illustrating threshold phenomenon.

straight-line projection of the strong-signal FM improvement curve. The baseband output noise becomes impulsive as the threshold condition occurs, producing spikes in the output noise signal. These spikes or impulses are often referred to as clicks and account for the rapid degradation in output signal-to-noise ratio that occurs below threshold. This threshold effect is why the output signal-to-noise ratio for FM systems cannot be increased indefinitely simply by increasing the frequency deviation Δf, since a corresponding increase in IF bandwidth and hence a reduction in input carrier-to-noise ratio also occurs.

FSK Demodulation. The FM demodulation process for an FSK signal can utilize either noncoherent or coherent detection techniques, as illustrated in Fig. 48.32. It has been shown that the theoretical optimum performance in terms of bit-error probability for this type of binary signaling is 3 dB inferior to that of a PSK channel operating at the same data rate (Batson, 1973).

The probability of error for a coherent FSK system is given by (Batson, 1973)

$$P_e = \frac{1}{2} \operatorname{erfc} \sqrt{\frac{(1-\rho)E_b}{2N_0}} \tag{48.57}$$

where E_b = energy per bit time
N_0 = one-sided noise spectral density
erfc = complementary error function
ρ = normalized correlation coefficient

The normalized correlation coefficient is defined as

$$\rho = \int_0^T S_1(t) S_2(t)\, dt \tag{48.58}$$

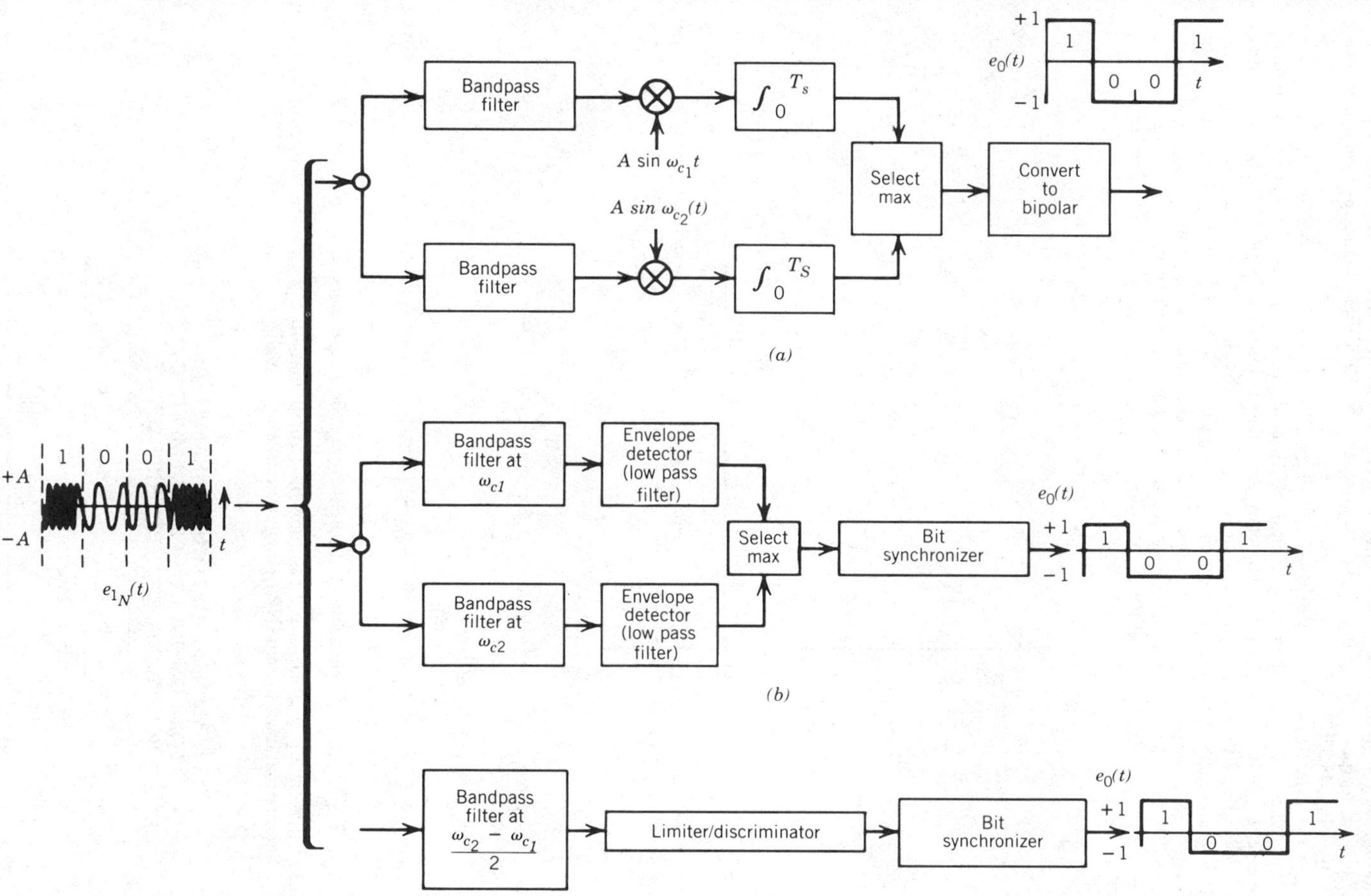

Fig. 48.32 Functional illustration of coherent and noncoherent demodulation processes: (*a*) coherent detection, (*b*) envelope (noncoherent) detection, (*c*) discriminator (noncoherent) detection.

where $S_1(t)$ and $S_2(t)$ are the transmitted symbols of unit energy. It can be shown that for the case of continuous-phase FM transmission, the optimum correlation coefficient is not zero (Seyl et al., 1976). The correlation coefficient assuming a nonreturn-to-zero (NRZ) digital modulating signal is shown graphically in Fig. 48.33.

As seen from the figure, the peak frequency deviation for orthogonal performance (i.e., $\rho = 0$) is $\Delta f = 0.25R$, where R is the data rate of the NRZ modulation. However, for best (optimum) performance, $\rho = -0.22$ and the peak frequency deviation required is $\Delta f = 0.35R$. Thus optimum performance only 2.2 dB worse than optimum PSK is achievable.

Another parameter that affects optimum performance is the receiver IF filter bandwidth. Simulations and numerical results have been used to establish the optimum ratio of IF filter bandwidth $\mathrm{BW_{IF}}$ to data rate for continuous-phase FM digital signaling when the optimum Δf is employed. The optimum $\Delta f/R$ and $\mathrm{BW_{IF}}/R$ ratios for NRZ and biphase data formats are given in Table 48.5.

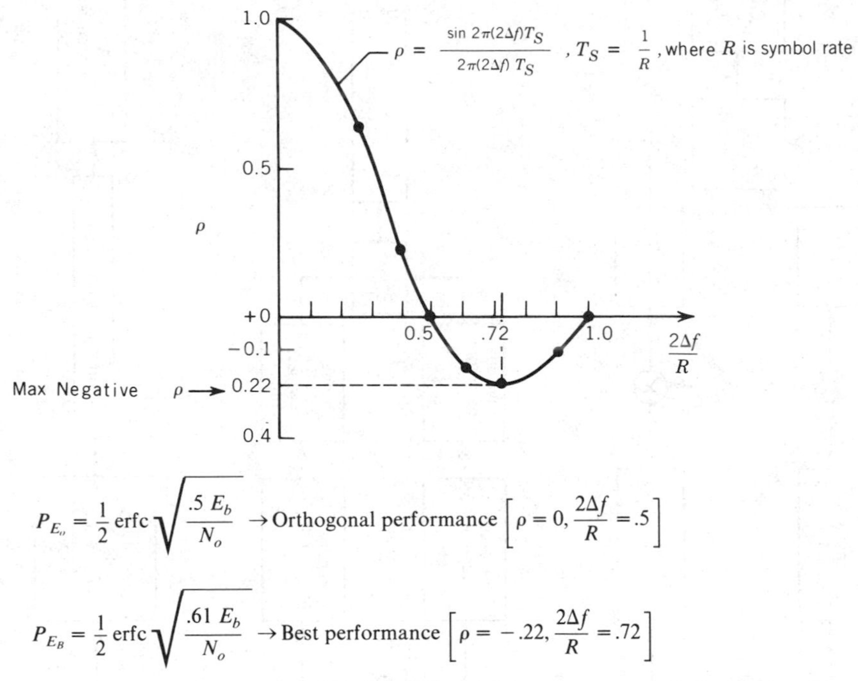

$$P_{E_o} = \frac{1}{2}\,\mathrm{erfc}\sqrt{\frac{.5\,E_b}{N_o}} \rightarrow \text{Orthogonal performance}\left[\rho = 0, \frac{2\Delta f}{R} = .5\right]$$

$$P_{E_B} = \frac{1}{2}\,\mathrm{erfc}\sqrt{\frac{.61\,E_b}{N_o}} \rightarrow \text{Best performance}\left[\rho = -.22, \frac{2\Delta f}{R} = .72\right]$$

$P_{E_B} \rightarrow$ Represents performance bound on coherent FSK 2.2 dB worse than coherent PSK

Fig. 48.33 Correlation coefficient and resulting bit-error probabilities for continuous-phase digital FM modulation [nonreturn-to-zero (NRZ) digital modulating signal].

TABLE 48.5. OPTIMUM PARAMETERS FOR CONTINUOUS-PHASE FM WITH DIGITAL INFORMATION SIGNALS

Data Format	$\Delta f/R^a$	B^b/R
Nonreturn-to-zero (NRZ)	0.36	1.4
Biphase	0.62	2.7

Source. From Seyl, Batson, and Smith, 1976, with permission.
[a] R, information signal data rate.
[b] B, receiver intermediate-frequency (IF) bandwidth.

48.1-3 Pulse Modulation

In pulse modulation systems, data are represented by a series of periodically recurring pulses. The information signal $g(t)$ is conveyed by varying some parameter associated with the pulse train. Since several parameters are available for variation, a large number of modulation methods are feasible. The simplest method of pulse modulation is pulse amplitude modulation (PAM), where samples of the information signal are used to modulate the amplitudes of the successive carrier pulses. Instead of being used to modulate the pulse amplitude, the samples could be used to modulate the pulse width or duration and thus to obtain pulse width modulation (PWM) or pulse duration modulation (PDM). In this case the information is conveyed in the leading- and trailing-edge zero crossings of the pulses. A further utilization would be pulse position modulation (PPM), in which the pulse time is varied with respect to the regularly spaced pulse repetition (reference) time. Finally, the frequency of the pulse train can be varied to obtain PFM, pulse frequency modulation.

A more complex approach to pulse modulation incorporates a technique of coding individual samples of the information signal amplitude and transmitting the coded information to represent each sample size. The process of sampling and coding is referred to as pulse code modulation (PCM). PCM information signal samples must be converted to preassigned discrete number values; hence each sample must be stored in the code bin that most closely represents its actual value. The process of digitizing an information signal is referred to as quantization, and results in some additional noise (quantization noise) in the recovered data. Another form of pulse modulation is delta modulation, in which discrete symbols relative to previous sample approximations are transmitted according to a particular algorithm. A similar algorithm is applied at the receiver to recover the original information signal. Delta modulation is often used for voice digitizing where data transmission rates are limited. Since in all pulse modulation schemes the information signal must be sampled at regular time intervals, the concepts of signal sampling and the required sampling rate are important.

Sampling Theorem

The sampling theorem specifies the number of time samples of the information time function $g(t)$ required for complete reproduction of $g(t)$ without ambiguities. If $g(t)$ is an aperiodic function with a Fourier transform $G_g(\omega)$ that is bandlimited, as shown in Fig. 48.34, then

$$g(t) = \int_{-\infty}^{\infty} G_g(\omega)e^{j\omega t}\, d\omega = \int_{-W}^{W} G_g(\omega)e^{j\omega t}\, d\omega \tag{48.59}$$

and $G_g(\omega)$ is expanded as a Fourier series that converges for $|\omega| < W$

$$G_g(\omega) = \sum_{n=-\infty}^{\infty} C_n e^{j2\pi n(f/2W)} \tag{48.60}$$

where

$$C_n = \frac{1}{2W}\int_{-W}^{W} G_g(\omega)e^{-j\omega(n/W)}\, d\omega \tag{48.61}$$

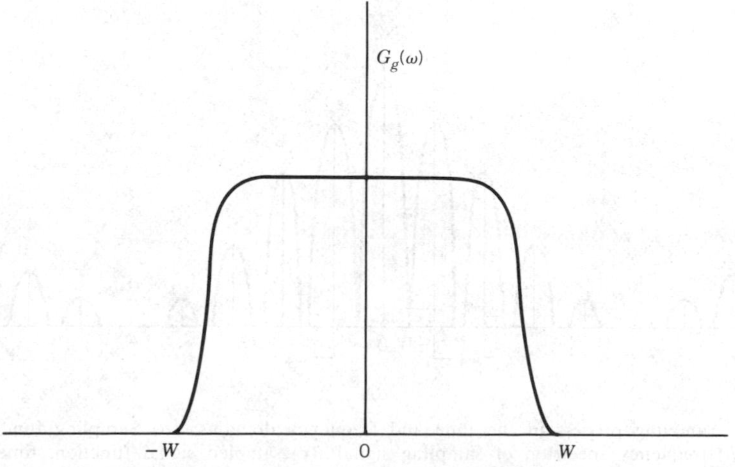

Fig. 48.34 Ideally bandlimited spectrum of $g(t)$.

Thus if $g(t)$ is sampled at regularly spaced time intervals of $t = -(n/2W)$, each sample will determine a Fourier coefficient, and if t goes to ∞ then the Fourier transform $G_g(\omega)$ is defined. This in turn defines the time waveform, which is merely the inverse Fourier transform of $G_g(\omega)$. Therefore, sampling $g(t)$ at time intervals $1/2W$ is sufficient to determine the function $g(t)$ for all time. Figure 48.35 illustrates the sampling process in both the time and the frequency domain. $S(t)$, the sampling function, is usually a train of periodic pulses of width d and constant amplitude A. The sampled signal $m(t)$ is formed by the product of $S(t)$ and $g(t)$, resulting in a frequency spectrum $G_m(\omega)$ that is

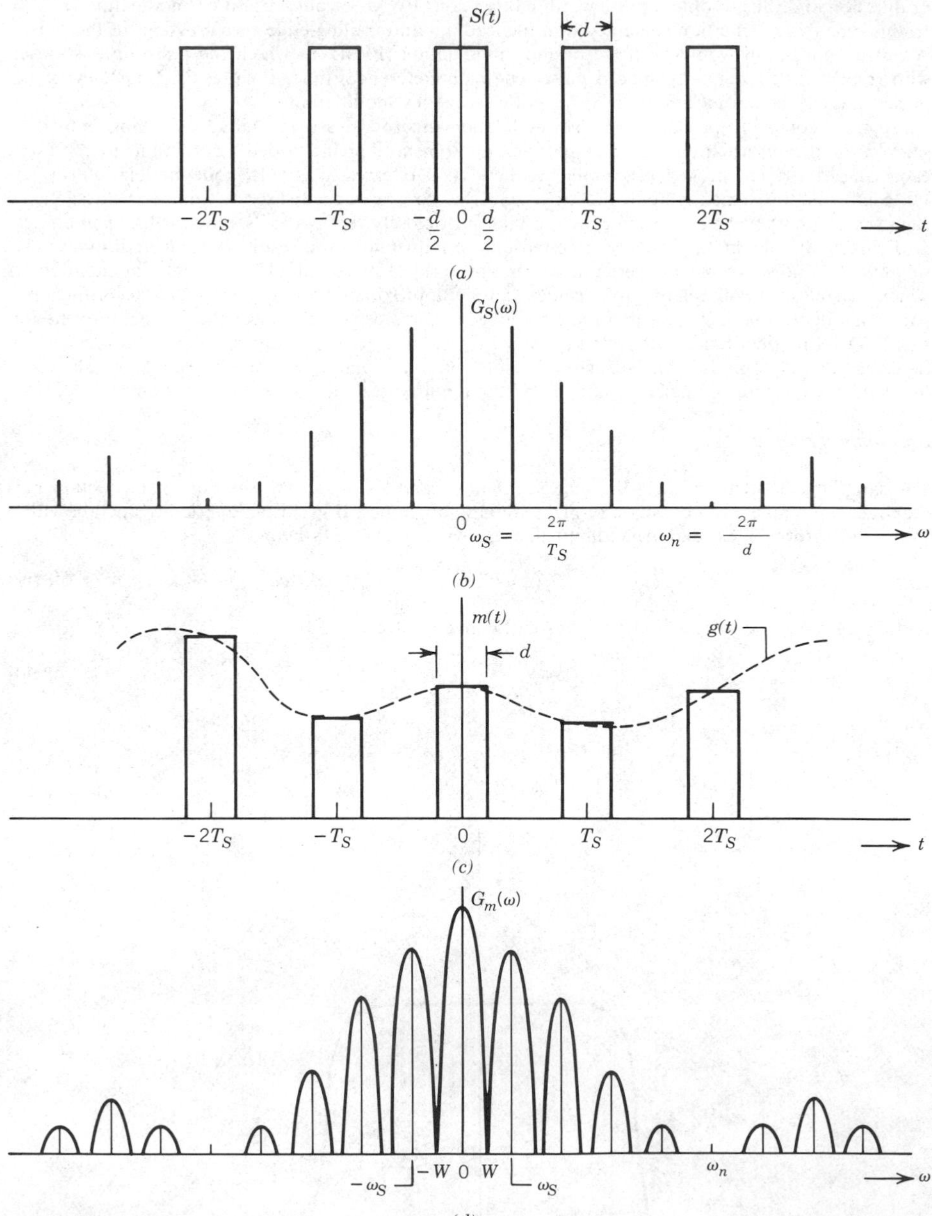

Fig. 48.35 Sampling process in the time and frequency domains: (*a*) Sampling function, time domain; (*b*) frequency spectrum of sampling signal; (*c*) sampled signal function, time domain; (*d*) frequency spectrum of sampled signal.

the convolution of $G_g(\omega)$ and $G_s(\omega)$. For the ideal bandlimited information signal spectrum of Fig. 48.34, the sampled signal spectrum of $G_g(\omega)$ will not overlap if

$$\frac{1}{T_S} > 2W \tag{48.62}$$

If the sampling rate is twice the highest frequency component of the information signal, there will be no overlapping of the information signal in the sampled signal. This sampling rate ($2W$) is referred to as the Nyquist rate and is theoretically sufficient to recover the orginal information signal $g(t)$ undistorted (Panter, 1965).

To recover $g(t)$, the sampled signal can be passed through an "ideal" lowpass filter whose transfer function is defined by

$$|H(j\omega)|^2 = \begin{cases} 1 & |\omega| < W \\ 0 & |\omega| > W \end{cases} \tag{48.63}$$

The frequency spectrums at the filter input and output are illustrated in Fig. 48.36.

Thus the frequency spectrum of the output of the ideal lowpass filter is that of the information signal, and hence its time function is $g(t)$. If the sampling rate is less than the Nyquist rate (i.e., $\omega_S < 2W$), then some sampled spectrum overlapping occurs that introduces an aliasing error in the recovered signal, as seen in Fig. 48.37.

Since an ideally bandlimited spectrum exists only in theory, there will always be some aliasing and resulting distortion. However, the distortion can be reduced to acceptable levels by increasing ω_S, the sampling rate.

Pulse Amplitude Modulation (PAM)

If the sampled signal $m(t)$ defined in Fig. 48.35 is transmitted directly, the resulting pulse train is that of a pulse amplitude-modulated (PAM) signal. If the average power of the unmodulated pulse train is

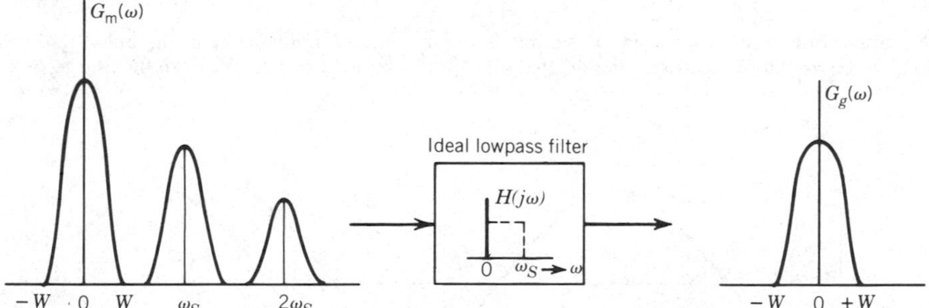

Fig. 48.36 Frequency spectra of sampled signal and recovered signal after lowpass filtering.

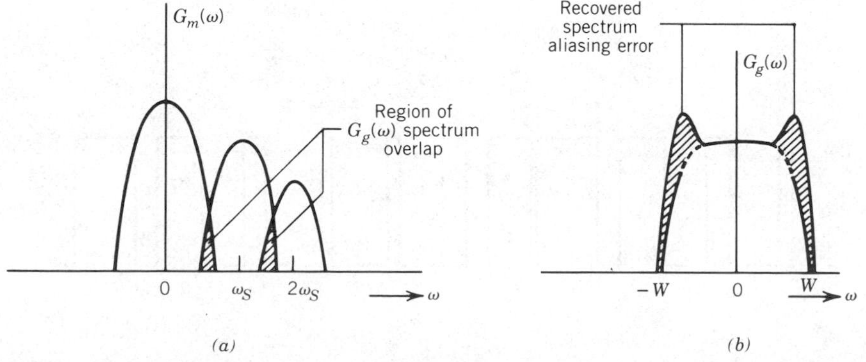

Fig. 48.37 Aliasing due to insufficient sampling rate ($\omega_S < 2W$): (*a*) Sampled signal spectrum, (*b*) recovered signal spectrum.

denoted as P_p, then when modulation is introduced, as shown in Fig. 48.35, the signal power is $m_p^2\,\overline{g^2(t)}\,P_p$ where m_p is the modulation index or modulation factor of the pulse train. The signal-to-noise output, assuming an ideal lowpass filter, can then be shown to be (Panter, 1965)

$$(S/N)_{o,\mathrm{PAM}} = \frac{m_p^2\,\overline{g^2(t)}\,(\tau/T_S)P_p}{N_0 W} \tag{48.64}$$

where W = highest frequency of the bandlimited information signal
$N_0/2$ = two-sided noise spectral density
τ = pulse width of sampling signal
T_S = pulse rate of sampling signal

Pulse Duration Modulation

Another pulse modulation technique is to vary the time of occurrence of the leading edge or the trailing edge of the pulse, which results in varying pulse width or duration. Figure 48.38 illustrates a pulse duration modulated signal and the information or modulating time function.

If the longest and shortest pulses are denoted as T_L and T_X, respectively, and the message information is regarded as residing directly in the pulse duration, then the mean pulse duration $\overline{T}$ will represent a zero-information signal amplitude sample and is

$$\overline{T} = \tfrac{1}{2}(T_L + T_X) \tag{48.65}$$

The positive and negative full-scale excursion of the information signal is then represented by the time duration

$$(\overline{T} - T_X) = (T_L - \overline{T}) = \tfrac{1}{2}(T_L - T_X) \tag{48.66}$$

Thus for direct-time measurement demodulation of PDM signals, the output signal power, assuming a continuous information signal $g(t)$ of unity peak amplitude, is

$$S_{o,\mathrm{PDM}} = \tfrac{1}{4}(T_L - T_X)^2\,\overline{g^2(t)} \tag{48.67}$$

The output noise in a PDM system is manifested as jitter in the trailing edge of the pulse. Since the leading edges of the pulses are not modulated and occur at a regular rate, they can be used to derive

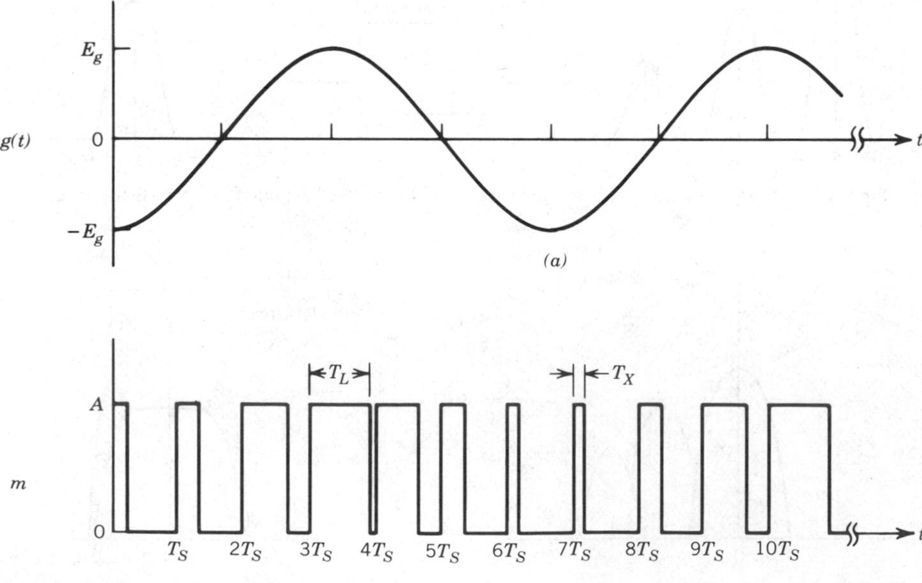

Fig. 48.38 Pulse-duration modulation signal time function. (*a*) Information signal, $g(t) = E_g \sin \omega_g t$. (*b*) Pulse duration modulated signal, $m_{\mathrm{PDM}}(t)$.

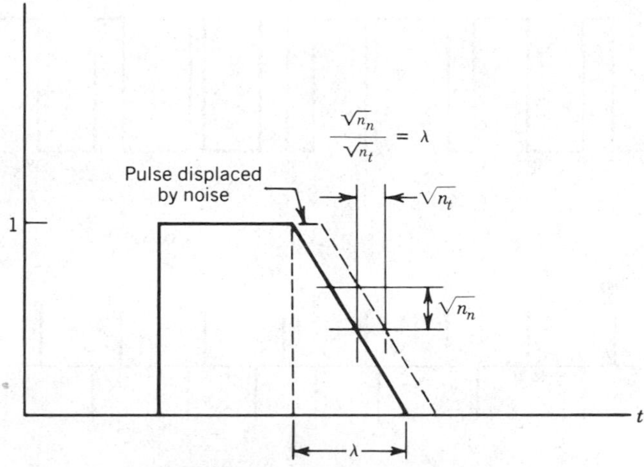

Fig. 48.39 Variation in pulse width due to system broadband noise with standard deviation $\sqrt{n_n}$.

a noiseless timing reference. Figure 48.39 illustrates the noise characterization of the PDM system. The noise power output of the time demodulation process is

$$n_t = \frac{n_n}{\lambda^2} \tag{48.68}$$

where $\sqrt{n_n}$ is the standard deviation of the baseband noise (rms noise voltage), and λ is the slope of the pulse trailing edge.

The output signal-to-noise can be expressed as (Downing, 1964)

$$(S/N)_{\text{out,PDM}} = \frac{\lambda^2(T_L - T_X)^2\,\overline{g^2(t)}}{4n_n} \tag{48.69}$$

Note that the output signal-to-noise is dependent on the ratio λ^2/n_n and hence a tradeoff between these quantities with respect to baseband noise filtering must be made. Since the maximum slope of the step response for a lowpass filter of a given type is directly proportional to its bandwidth, the presence of λ in the numerator argues for the widest possible baseband filter bandwidth; however, as the baseband filter is increased, the noise power increases monotonically, and this quantity in the denominator weighs in favor of the narrowest possible baseband filter. In AM systems, where the noise power spectral density is uniform, the baseband noise power increases as the first power of the baseband filter bandwidth. Thus the best system performance would be achieved if no baseband filtering was employed.

Pulse Position Modulation

If the pulse duration modulated signal is differentiated and another pulse train of constant amplitude and constant duration pulses is generated coincident with the negative spikes of the differentiated wave, then a pulse position-modulated (PPM) signal is obtained. Figure 48.40 illustrates the process of PPM by differentiation of a PDM signal.

A PPM system can be considered analogous to an angle-modulated system, since the information is contained in the position of the pulse leading edges relative to the mean or average position of the edges. Figure 48.41 illustrates the effects of system noise on the recovered PPM signal.

If T_o is the maximum offset from the mean pulse occurrence time in both positive and negative directions, and $\sqrt{nt}$ is the time error or noise jitter due to the system noise with standard deviation $\sqrt{n_n}$, then the output signal-to-noise ratio is a function of the leading edge slope λ of the pulse train where

$$\lambda = \frac{\sqrt{nt}}{\sqrt{n_n}} \tag{48.70}$$

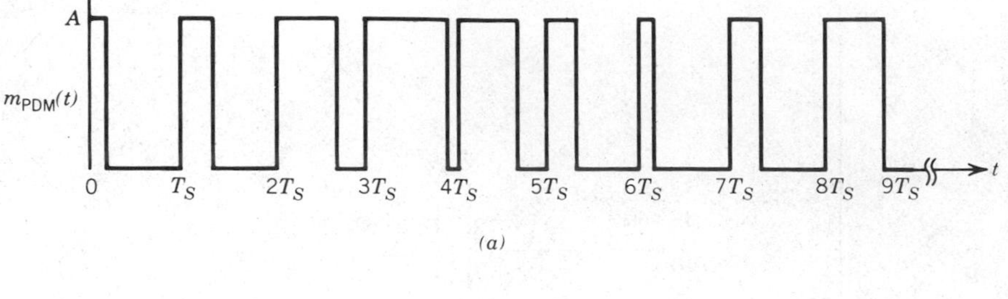

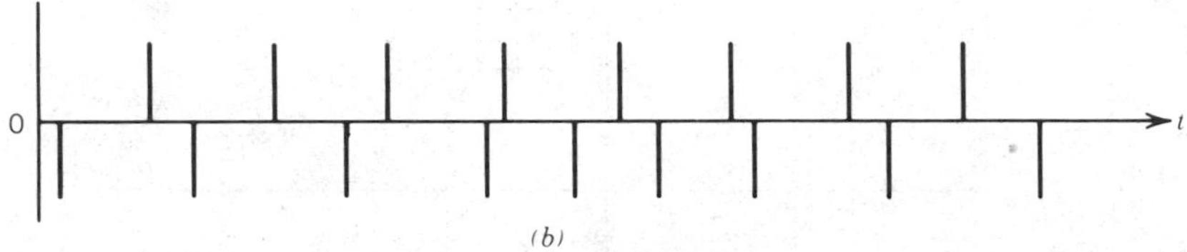

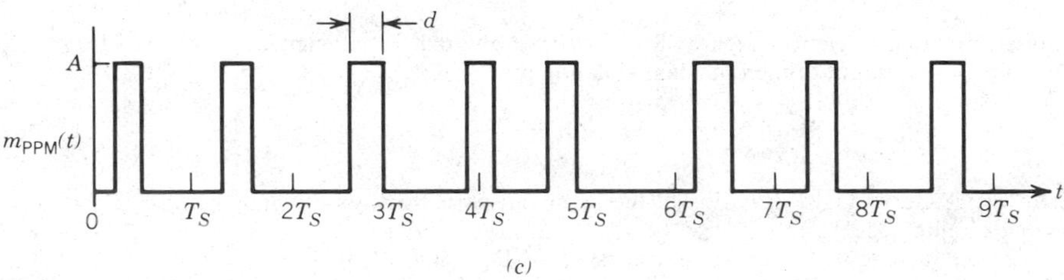

Fig. 48.40 Pulse position modulation (PPM). (*a*) Pulse duration modulated (PDM) signal, time function. (*b*) Differentiated PDM signal, time function. (*c*) PPM signal, time function.

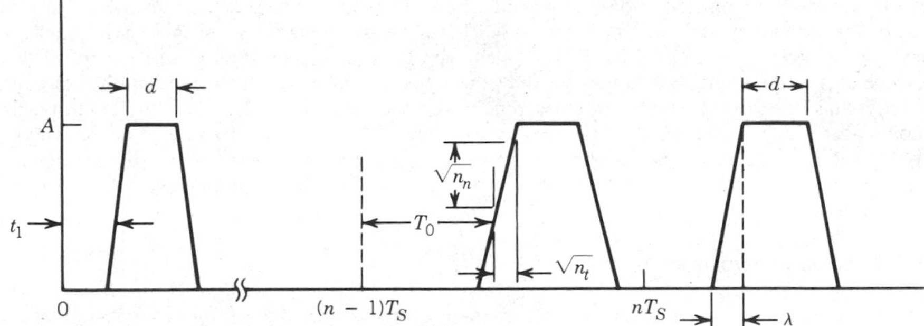

Fig. 48.41 Variation in pulse position due to broadband system noise with standard deviation $\sqrt{n_n}$.

or

$$nt = \frac{n_n}{\lambda^2} \tag{48.71}$$

The output signal power, assuming direct-time demodulation is employed, is

$$S_{P,\text{PPM}} = kT_o^2\,\overline{g^2(t)} \tag{48.72}$$

The output signal-to-noise ratio is then (Panter, 1965)

$$(S/N)_{\text{out,PPM}} = \frac{T_o^2\,\overline{g^2(t)}\,\lambda^2}{n_n} \tag{48.73}$$

Again it is obvious that increasing the slope λ of the pulse leading edges will improve the output signal-to-noise ratio providing that the output noise power spectrum is flat. Thus a wide bandwidth criterion similar to that for PDM modulation systems would produce the optimum SNR performance for the recovered information signal.

Pulse Code Modulation

The previously considered methods of pulse modulation represented signals by sampling analog functions. It is also common practice to represent continuous-information signals in a digital coded form. To accomplish this digitizing process, the information signal $g(t)$ must be sampled in time and each sample quantized in amplitude. Figure 48.42 illustrates a sampled information signal quantized in amplitude.

The quantization interval Q is dependent on the resolution required and the level of quantization noise that can be tolerated by the information channels. In Fig. 48.42, at time t_1 the system would transmit $\gamma_q = 3$; at time t_2, $\gamma_q = -1$; and at time t_3, $\gamma_q = -2$; and so on. The information thus actually transmitted is the symbol representing the individual quantization number rather than the exact information signal's sampled amplitude. The number of quantization intervals (QIs) is related to the resolution desired and the peak-to-peak amplitude of the information signal.

If the peak-to-peak range of $g(t)$ is E_g and the desired resolution is R, then the quantization interval size is $Q = RE_g$. The number of quantization intervals required would be

$$N_q = \frac{E_g}{Q} = \frac{1}{R} \tag{48.74}$$

Quantization Noise. Figure 48.43 illustrates noise effects resulting from the quantization process. At time t_1 the transmitter will transmit the symbol representing $\gamma_q = 3$ and at the receiver the recovered sample of $g(t_1)$ will be assigned the value corresponding to the center of the quantization number $\gamma_q = 3$. Thus an error (q) in the amplitude of the recovered information signal exists. q is a random variable representing the difference between $g(t)$ and the quantization number mean value. If the random variable of q is assumed to be uniformly distributed over the quantization interval Q, then the quantization error probability distribution is uniform, as shown in Fig. 48.44.

The noise power resulting from this random error process is (Downing, 1964)

$$P_Q = \frac{Q^2}{12} = \frac{RE_g^2}{12} = \frac{E_g^2}{12\gamma_q^2} \tag{48.75}$$

If the information signal power is given by $\overline{g^2(t)}$, then the signal-to-quantization noise becomes

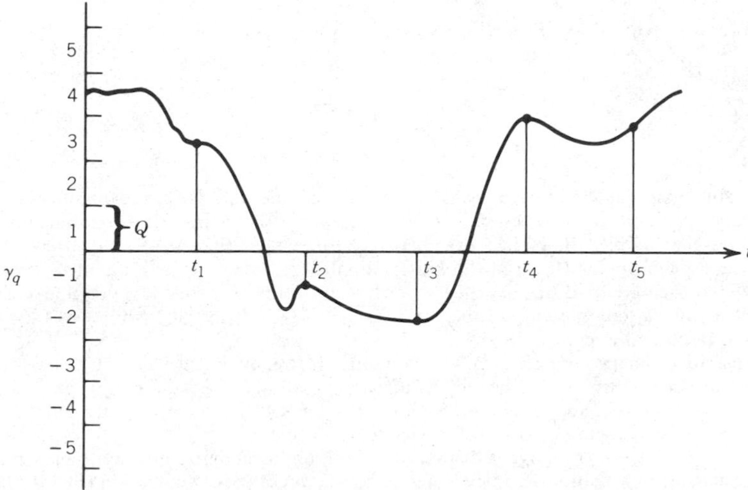

Fig. 48.42 Quantization of a continuous waveform.

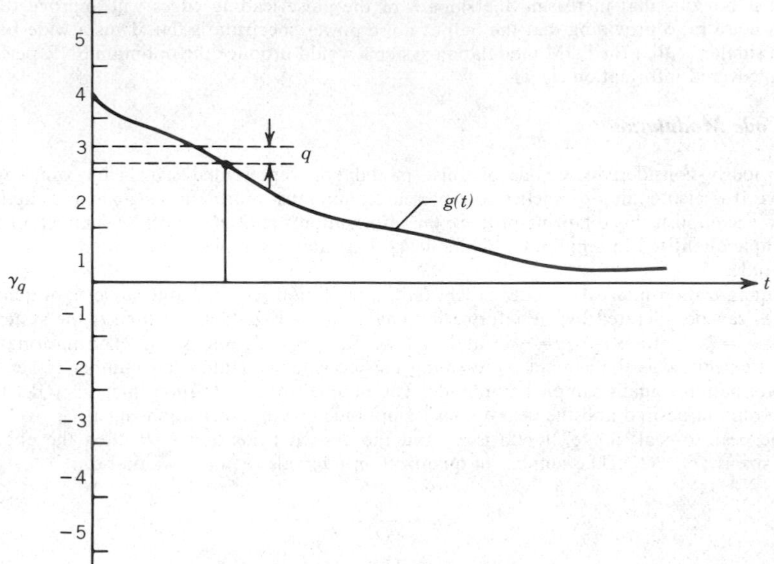

Fig. 48.43 Quantization noise.

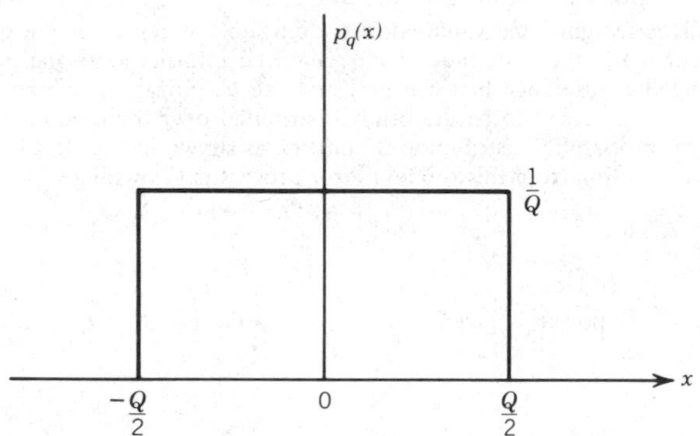

Fig. 48.44 Probability distribution of quantization error (q).

(Downing, 1964)

$$(S/N)_q = \frac{12\, \overline{g^2(t)}\, \gamma_q^2}{E_g^2} \tag{48.76}$$

Digitizing. Once the signal amplitudes have been sampled and quantized, the quantization intervals can be coded digitally and the digital code representing the appropriate quantization interval transmitted at the sampled time. For example, if $N_q = 8$, then Fig. 48.45*a* illustrates this binary coding process. Each one and zero in the binary number representing each quantization interval is called a bit, and each group of bits is called a word. With words of N bits, it is possible to distinguish between 2^N quantization intervals. Table 48.6 lists word sizes in bits, number of quantization intervals, and resolution in percent.

The transmitted binary symbols form a continuous-bit stream, and the rate of this continuous-bit stream must be sufficiently high to accommodate the initial digitizing (sampling) rate multiplied by the binary word lengths (number of bits per word). Figure 48.45*b* illustrates the final transmitted binary time waveform, assuming an NRZ binary data format.

The system in which the message is quantized in both time and amplitude and binary symbols are used to transmit the information sample is referred to as a pulse code modulation (PCM) system. The minimum PCM data rate required is equal to the word size divided by the sampling rate. Thus if the amplitude samples are taken at regularly spaced intervals of 1 ms and the PCM word size is 8 bits, a minimum PCM data rate of 8 kb/s (kilobits per second) would be required.

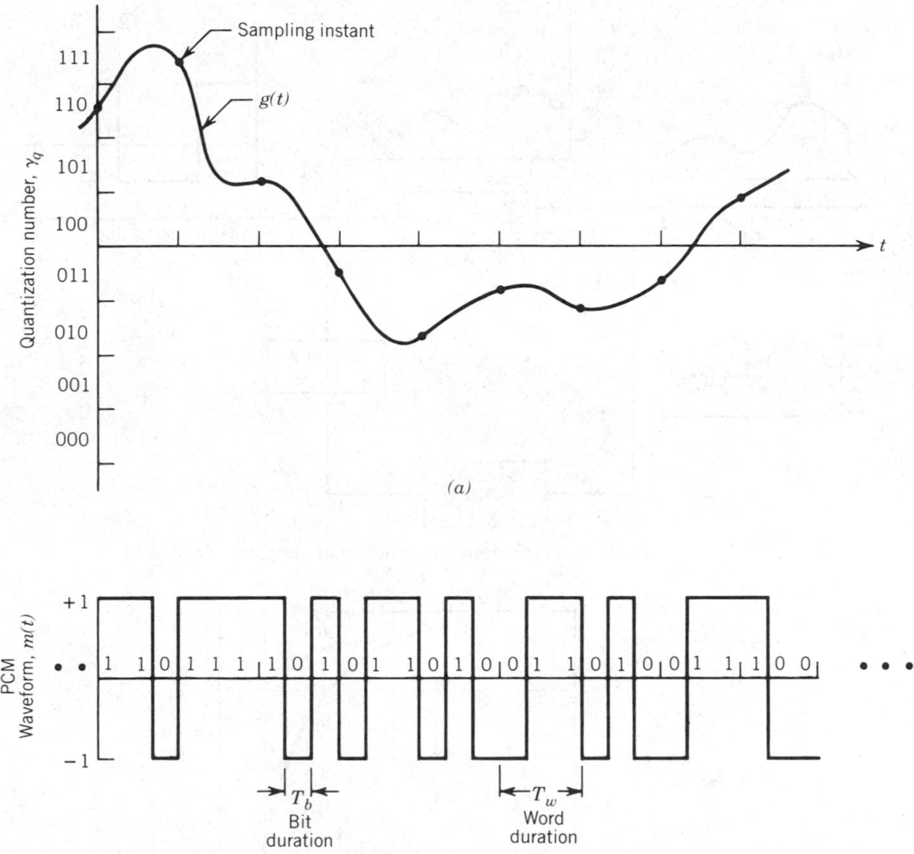

Fig. 48.45 Quantization example [pulse code modulation (PCM)]. (*a*) Binary coding process. (*b*) Transmitted binary time waveform, assuming nonreturn-to-zero (NRZ) binary data format.

TABLE 48.6. PULSE-CODE MODULATION WORD SIZE VS. RESOLUTION REQUIRED AND NUMBER OF QUANTIZATION INTERVALS

Word Size N (No. of Bits)	Number of Quantization Intervals $\gamma_q = 2^N$	Resolution, $R = 1/\gamma_q$ (%)
1	2	50
2	4	25
3	8	12.5
4	16	6.25
5	32	3.125
6	64	1.563
7	128	0.78125
8	256	0.391
9	512	0.195
10	1024	0.098

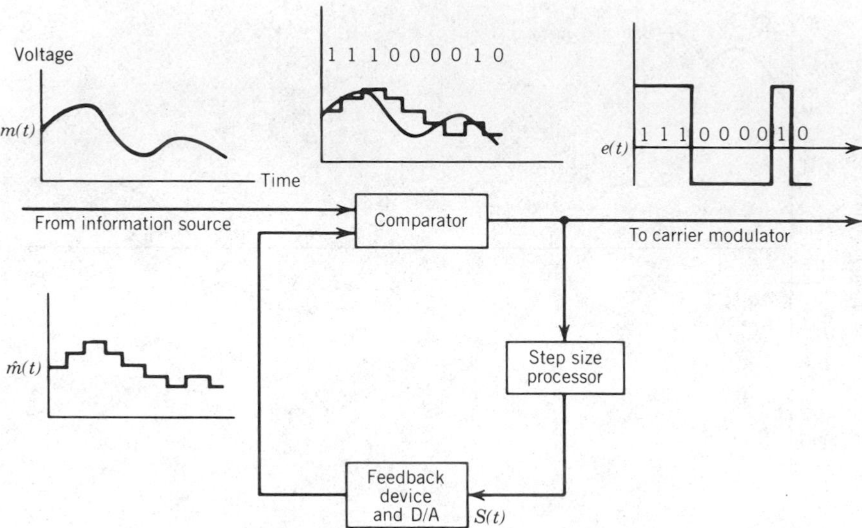

Fig. 48.46 Delta modulation, functional diagram.

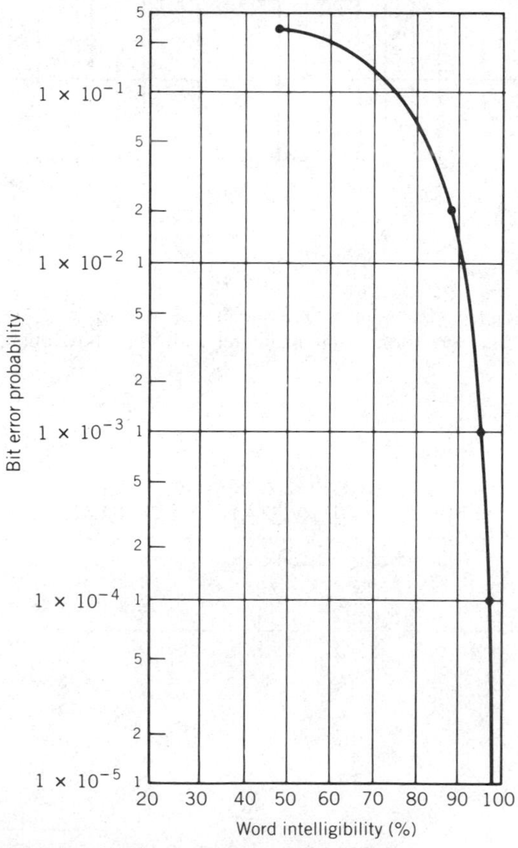

Fig. 48.47 Word intelligibility vs. bit error probability for shuttle delta modulation system. From Seyl, J. W. et al, June 1983 with permission.

Delta Modulation

Delta modulation is a type of predictive quantizing system equivalent to a one-digit differential pulse code modulation system. Such systems are based on transmission of the quantized difference between successive sample values rather than of the samples themselves. Both the modulators and demodulators make an estimate or prediction of the input signal value on the basis of the previously transmitted sequence.

There are two types of delta modulation, linear and adaptive. In linear delta modulation, the value of the input signal at each sample time is predicted to be a particular linear function of the past values of the signal. The effect is to make the difference between any two consecutive predictor values equal to the quantizing level. In adaptive delta modulation, the value of the input signal at each sample time is predicted to be a nonlinear function of the past values of the signal. Introducing nonlinear prediction into delta modulation provides a useful means of extending the range over which the system yields its optimum performance.

A simplified block diagram of a delta modulator is shown in Fig. 48.46. The input is an analog information signal whose instantaneous value is digitized by delta modulation techniques. In the operation of the modulator, the feedback-device analog output is compared with the input analog signal. The result of this comparison, which is a digital signal, causes the feedback device to either increase or decrease its analog output magnitude to approach the magnitude of the input signal. The rate at which the modulator changes its magnitude is controlled by a clock signal. The digital-output bit stream runs at the clock rate.

The operation of the demodulator is identical, in that demodulation occurs at the output of the feedback device. The demodulator output is then an analog reproduction of the input signal predicted by the modulator. If bit errors occur in the transmission process, the demodulator bit stream is not identical to that of the modulator output, and hence the demodulator analog output will not precisely reproduce the modulator input signal. In general, however, delta modulation has a high tolerance to bit errors and hence a graceful degradation in the presence of noise can be expected. For

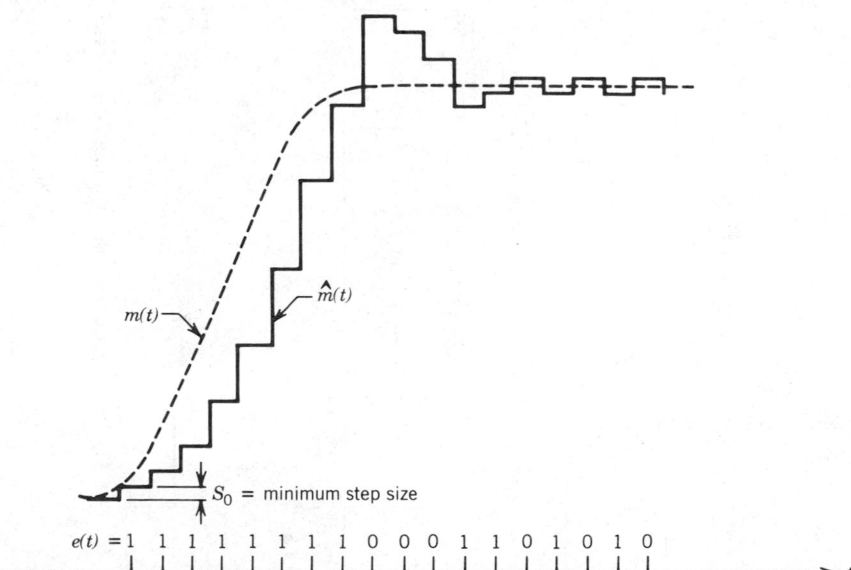

Fig. 48.48 Adaptive delta modulation algorithm used for space shuttle digital voice links.

$$S(t+1) = \begin{cases} [|S(t)| + S_0]v(t) & \text{when } e(t) = e(t-1) \text{ and } |S(t)| < 8S_0 \\ |S(t)|v(t) & \text{when } e(t) = e(t-1) \text{ and } |S(t)| = 8S_0 \\ |S_0|v(t) & \text{when } e(t) \neq e(t-1) \end{cases}$$

$$v(t) = \begin{cases} +1, & e(t) = 1 \\ -1, & e(t) = 0 \end{cases}$$

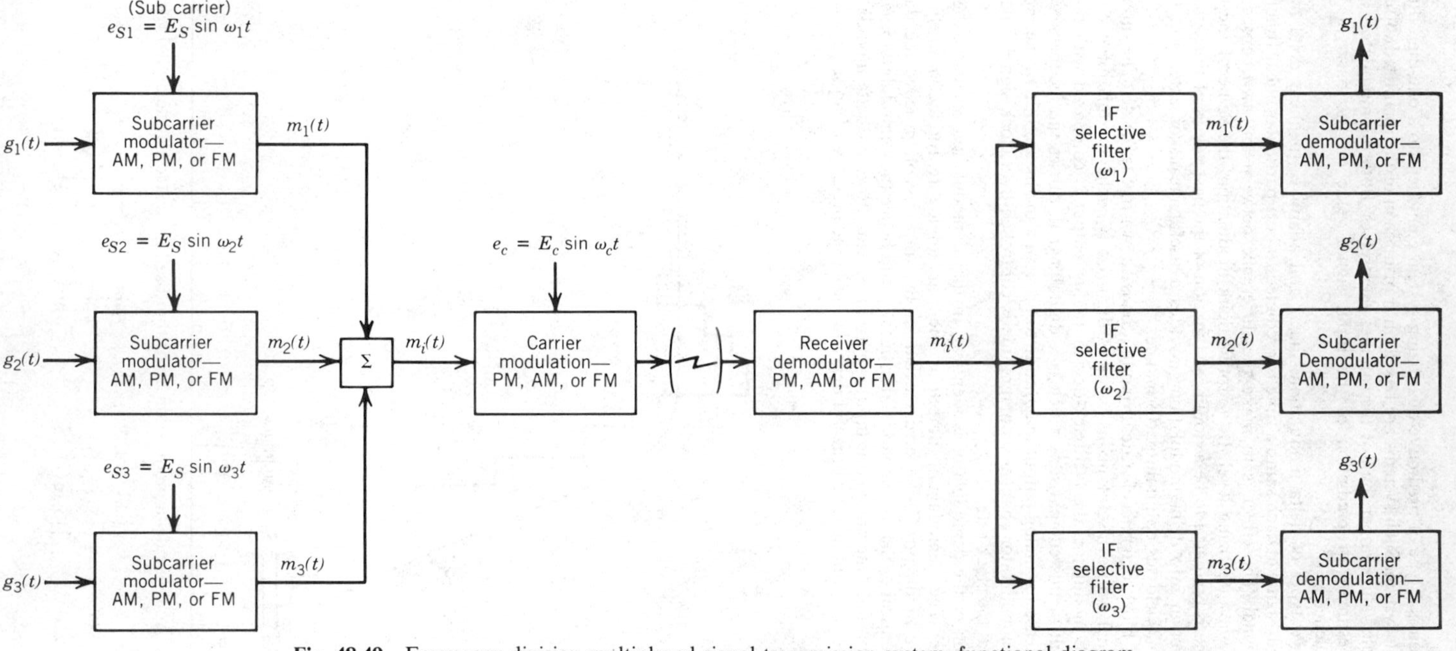

Fig. 48.49 Frequency-division multiplexed signal transmission system, functional diagram.

the case when the information signal is voice, the intelligibility factor of the voice channel does not significantly degrade until the error rate is on the order of one error in 10 bits, or 1×10^{-1}. This is far below the usable range for most digital PCM-type systems. Figure 48.47 shows a graph of word intelligibility (based on phonetically balanced words) vs. bit error rate for the particular adaptive delta modulation algorithm (Fig. 48.48) used on the space shuttle audio channels. These channels operate at a delta modulation clock rate of 32 kb/s (Seyl et al., 1983).

Multiplexing and Demultiplexing

To transmit more than one information signal simultaneously over the same communication channel, some method of combining or multiplexing the information signals must be devised. Two domains exist in any communication channel where multiplexing can be accomplished: One is the time domain, resulting in time division multiplexing (TDM), and the second is the frequency domain, resulting in frequency division multiplexing (FDM). In addition, special forms of multiplexing have been devised, such as code division multiplexing, where each information signal is identified by a unique pseudorandom code, and quadrature carrier multiplexing, which utilizes a carrier signal with quadrature components. Each of these forms of multiplexing will be discussed in the following section.

Frequency Division Multiplexing (FDM). As the term implies, the FDM system combines multiple information channels by combining subcarrier signals of different frequencies at the transmitter and filtering the recovered signals to select the individual information channel at the receiver. Figure 48.49 illustrates an FDM system.

If the information signal spectra are ideally bandlimited and double sideband suppressed-carrier modulation is employed for the subcarrier modulation method, then the subcarrier frequencies can be selected to prevent information signal spectrum overlap, as seen in Fig. 48.50.

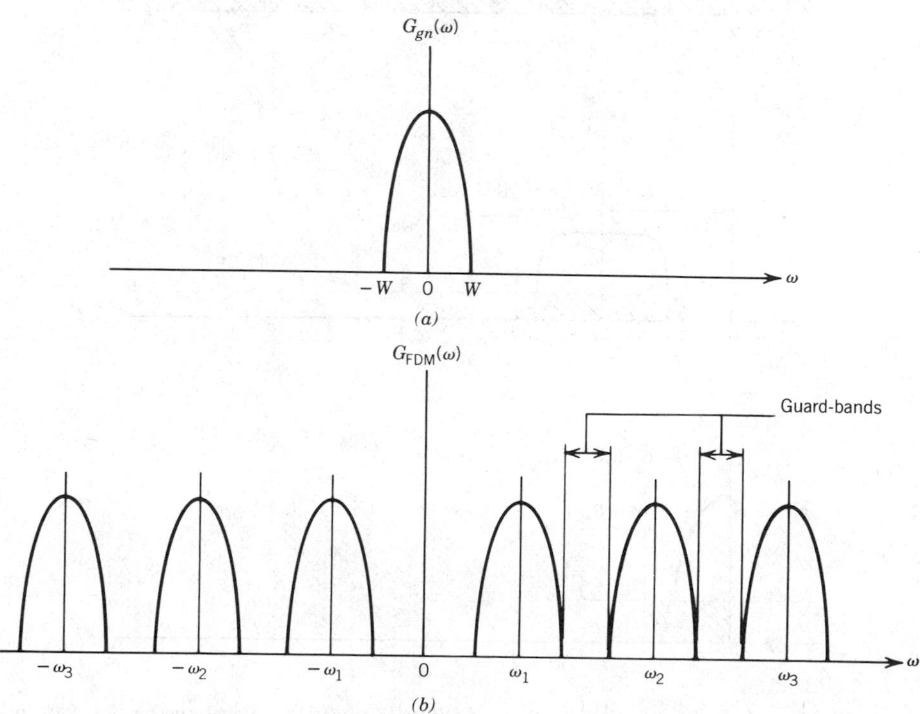

Fig. 48.50 Baseband frequency spectra for information signal and frequency division multiplexed signal: (*a*) Information signal spectrum (bandlimited to $\pm W$ for all channels), (*b*) frequency division multiplexed signal spectrum, assuming double sideband suppressed-carrier (DSBSC) modulation for each subcarrier.

Assuming a DSBSC subcarrier modulation format, the baseband signal spectrum thus consists of the individual subcarrier frequencies with the appropriate information signal spectrum translated to each subcarrier frequency. The spacing between the upper sideband spectral components of channel one and the lower sideband spectral components of channel two is normally referred to as the guardband. Thus for no overlap of the information spectrum, the subcarrier frequency spacing required is $2W$ plus the guardband. The combined FDM signal spectrum in turn modulates the carrier signal. The carrier modulation can be either AM, PM, or FM. For AM and PM carrier modulation, the baseband-detected signal spectrum would be as seen in Fig. 48.51.

Since the noise spectrum at the receiver output is flat for all frequencies, use of equal amplitude or phase modulation factors for each subcarrier channel would result in the same signal-to-noise performance for each channel, assuming equal information bandwidths. However, for FM carrier modulation this is not the case, owing to the parabolic amplitude nature of the FM receiver output noise. Figure 48.52 illustrates an FDM signal output of an FM channel.

The composite modulating signal is $m(t)$, where

$$m(t) = \sum_{i=1}^{V} K_i m_i(t) \tag{48.77}$$

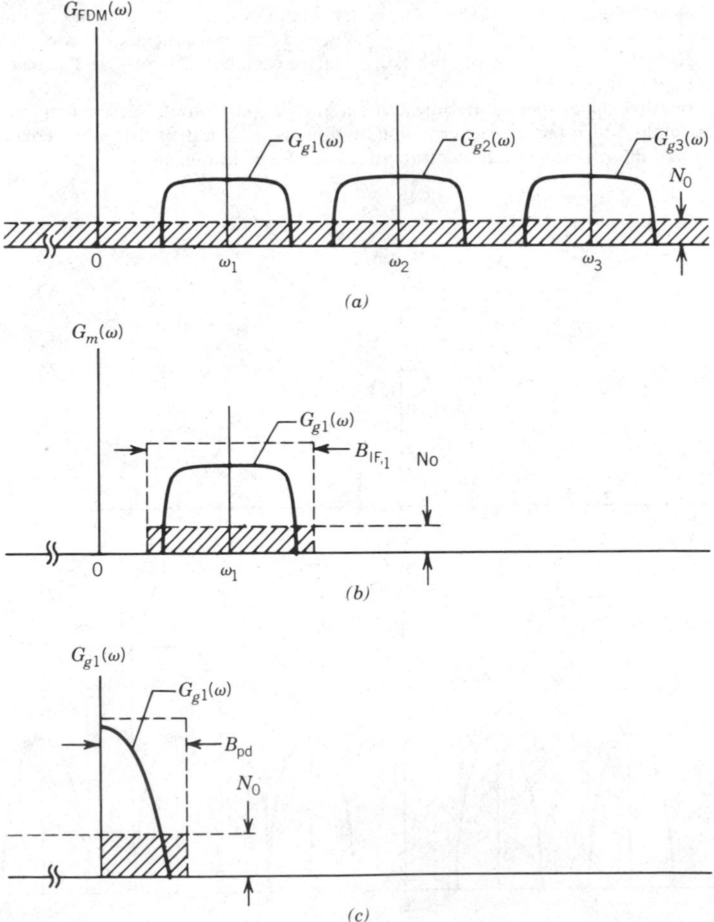

Fig. 48.51 Typical detected frequency division multiplexing (FDM) signal spectra for PM- or AM-modulated channel: (*a*) Detected baseband spectrum plus system noise, (*b*) selective bandpass filter (ω_1) output signal spectrum plus system noise, (*c*) post-detection signal spectrum for channel 1 plus noise.

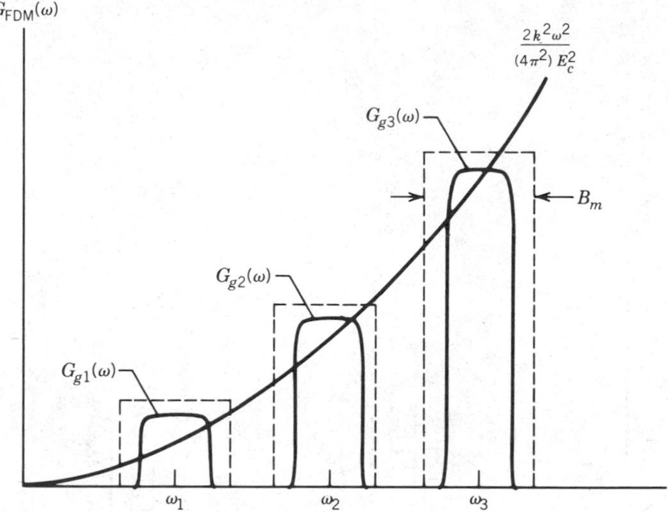

Fig. 48.52 Frequency division multiplexed (FDM) output signal for frequency-modulated communication channel.

K_i represents the individual subcarrier channel weighting factor that determines its modulation index in the composite signal. To ensure that the carrier frequency modulator is not overmodulated, it is necessary to select the peak frequency deviation on the basis of these weighting factors and the probability of the composite signal exceeding a specified value. If many subcarriers are present, and they are not coherently related in frequency or phase, then a Gaussian probability distribution is a good assumption, and Δf can be established using the 3σ value as the peak deviation of the composite FDM signal (Downing, 1964).

$$\Delta f^2 = 9 \sum_{i=1}^{V} K_i^2 \overline{m_i^2(t)} \tag{48.78}$$

If $\overline{m_i^2(t)}$ is normalized to unity, then

$$\Delta f^2 = 9 \sum_{i=1}^{V} K_i^2$$

with probability 99.7%. Assuming an output baseband noise power spectral density, as shown in Fig. 48.52, the output SNR for the ith channel would be

$$(S/N)_{o,\mathrm{FDM}}^{(i)} = \frac{E_c^2 K_i^2}{2k^2 f_i^2 B_m} \tag{48.79}$$

Thus for equal information channel SNR peformance, the weighting factor K_i must be proportioned to the channel subcarrier frequency f_i. The proportionality depends on the number of channels V, the information bandwidth B_m, and the carrier peak modulation constraint.

Qualitatively then, equal-channel SNR performance requires that the higher frequency channels deviate the carrier more than the lower frequency channels. Thus some form of preemphasis is usually employed for setting the weighting factors in FDM-FM multiplexing systems.

Time Division Multiplexing. As the term implies, time division multiplexing accomplishes information signal combining in the time domain. As discussed in the section of pulse modulation, information signals can be sampled in time and reconstructed at the receiver from the transmitted time samples. If several information signals are sampled at different time intervals and the resulting sampled signals are combined in a synchronous method, the time waveform is referred to as a time division multiplexed (TDM) signal. Figure 48.53 illustrates TDM signaling using PAM, PDM, and PCM techniques.

Figure 48.54 illustrates sampling using a rotary switch as a commutator and decommutator. The

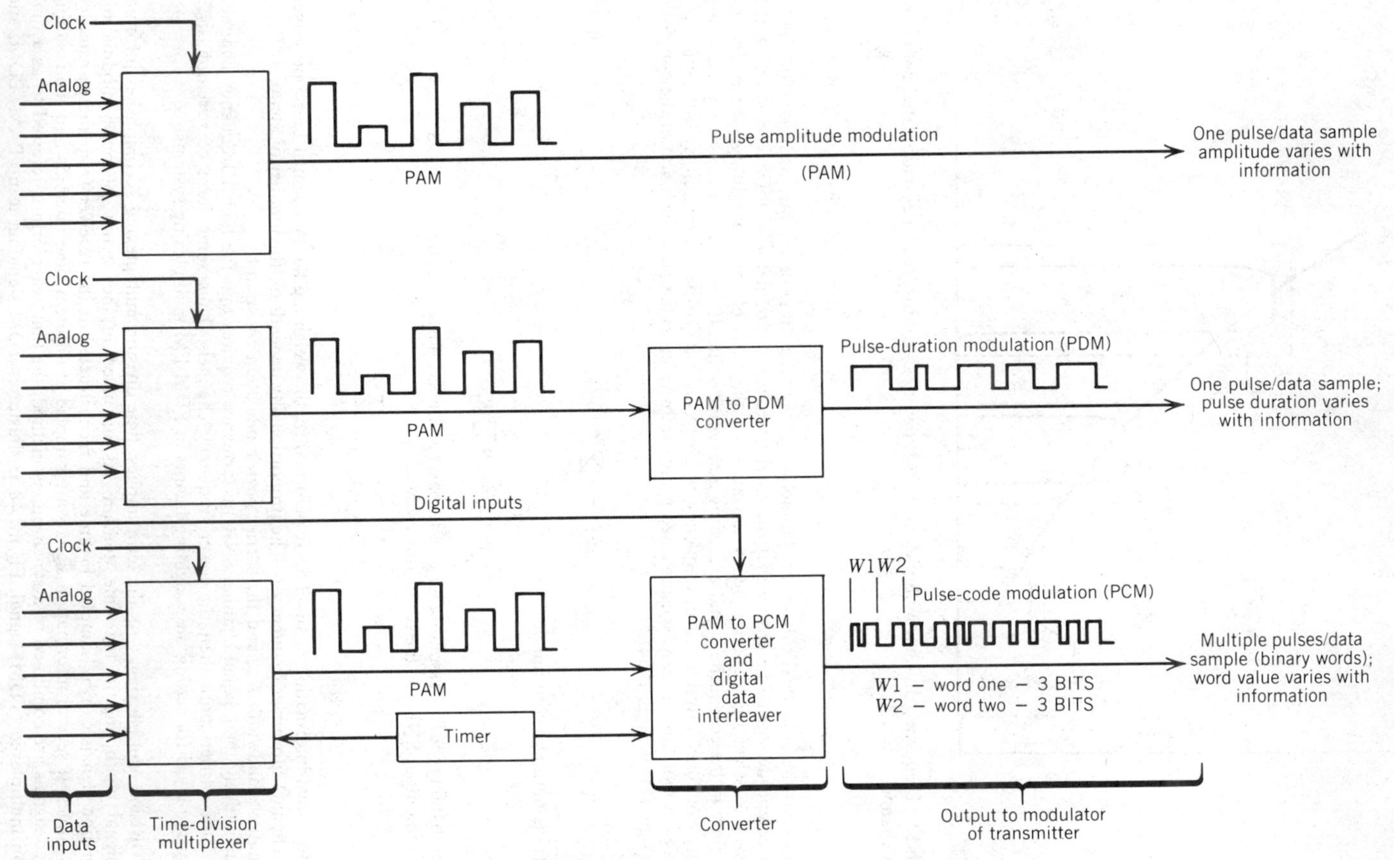

Fig. 48.53 Time division multiplexing (TDM), functional diagram.

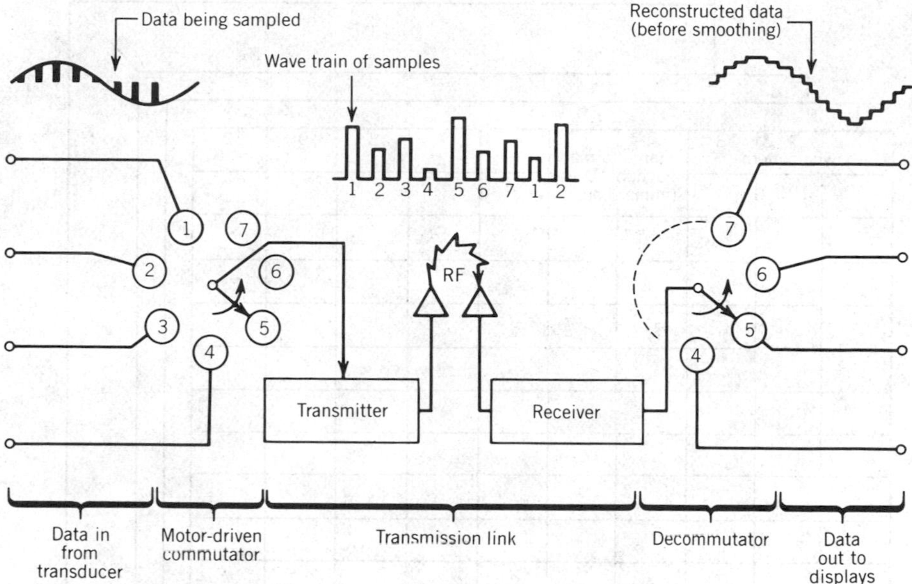

Fig. 48.54 Time division multiplexing using a rotary switch as a commutator.

process requires that the commutator and decommutator rotate synchronously in time to ensure that the proper information signal sample appears at the desired output terminal. One method of synchronization commonly used is the reconstruction of a clock signal by the receiving system, which can be used to ensure that the decommutator rotates at the same rate as the commutator. However, in addition to the basic timing established by the clock, some form of channel identification must be provided to establish a reference channel location. This can be accomplished by devoting one channel in the system to a definitive signal or marker pulse. When pulse code modulation of TDM formats is employed, the commutation and decommutation process is accomplished through the formation of data frames of information. Each frame contains data words that are located in time slots relative to a synchronizing word that identifies the beginning of the data frame. In addition, subcommutated frames of data can be included by identifying each frame with a binary counter (word). Figure 48.55 illustrates a typical telemetry (TDM) data frame format. The sample rate and word length of the PCM system determine the TDM data rate that must be transmitted over the communications channel.

Quadrature Carrier Multiplexing. Although not as commonly used as FDM and TDM systems, a quadrature carrier signal can be used to transmit two channels of information simultaneously. Quadrature signaling has been widely discussed in terms of so-called quadrature phase modulation systems in which the information channel is a single digital data stream that is reduced in rate by a factor of two, with one half of the information being modulated on the in phase and one half on the quadrature phase of the carrier. The demodulated data from each carrier component are recombined to obtain the original data stream. The obvious advantage in this case is the reduced transmission data rate, allowing for higher rate data to be transferred through a bandwidth-limited channel. It is not necessary, however, that both data inputs be synthesized from one another in any sense. In this case, the quadrature and in-phase signals would serve as separate carriers of the same frequency for simultaneous transmission of two information channels. Figure 48.56 illustrates a conceptual mechanization of quadrature carrier multiplexing and Fig. 48.57 depicts a typical modulation power spectrum. The spectra $G_I(\omega)$ and $G_q(\omega)$ are the resulting data modulation spectra for the in-phase and quadrature signals, respectively (Lindsey, 1976).

A more sophisticated form of multiplexing that has recently evolved is referred to as code division multiplexing. In this method of multiplexing, information signals are normally transmitted simultaneously from separate transmitting sources and received by a single receiving element. The system is illustrated in Fig. 48.58. Each information signal transmitter is modulated with a pseudorandom (PRN) code sequence prior to transmission. At the receiver, the combined coded signals are

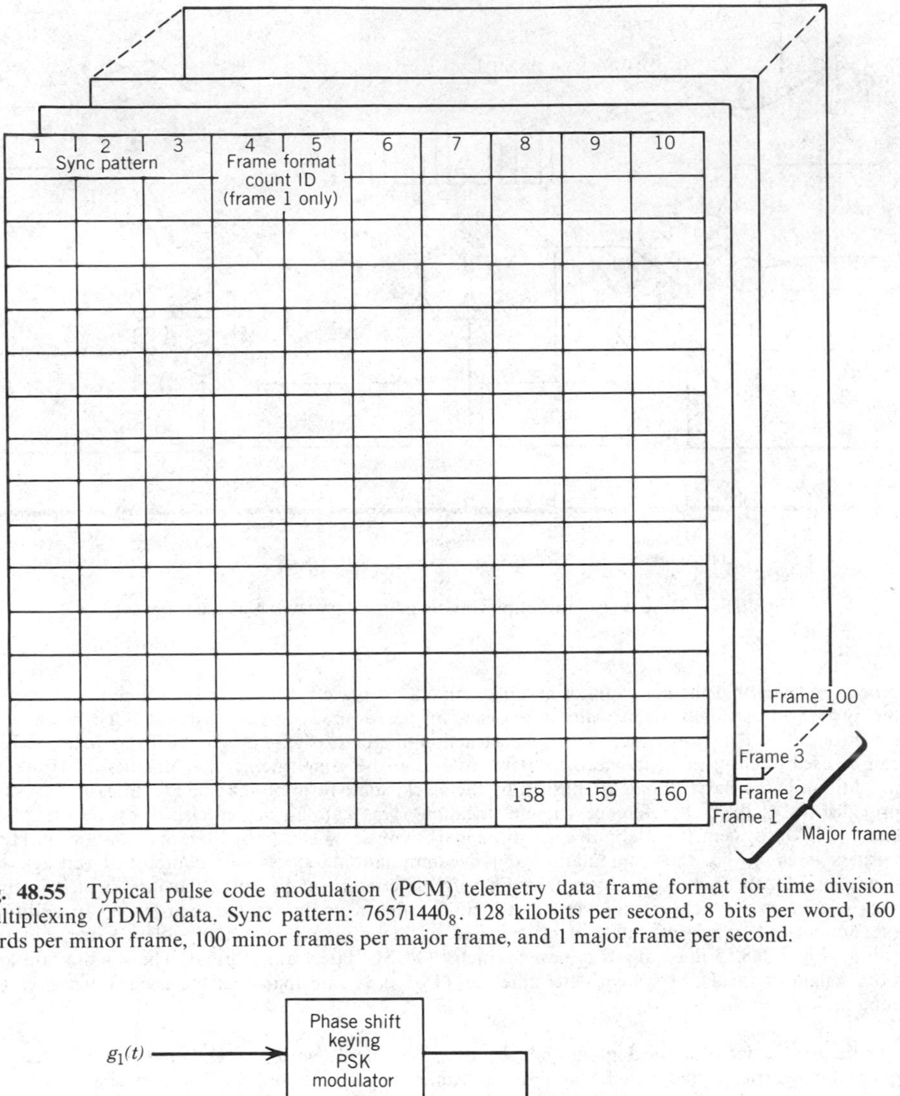

1 2 3 4 5 6 7 8 9 10

Sync pattern

Frame format count ID (frame 1 only)

158 159 160

Frame 100

Frame 3

Frame 2

Frame 1

Major frame

Fig. 48.55 Typical pulse code modulation (PCM) telemetry data frame format for time division multiplexing (TDM) data. Sync pattern: 76571440_8. 128 kilobits per second, 8 bits per word, 160 words per minor frame, 100 minor frames per major frame, and 1 major frame per second.

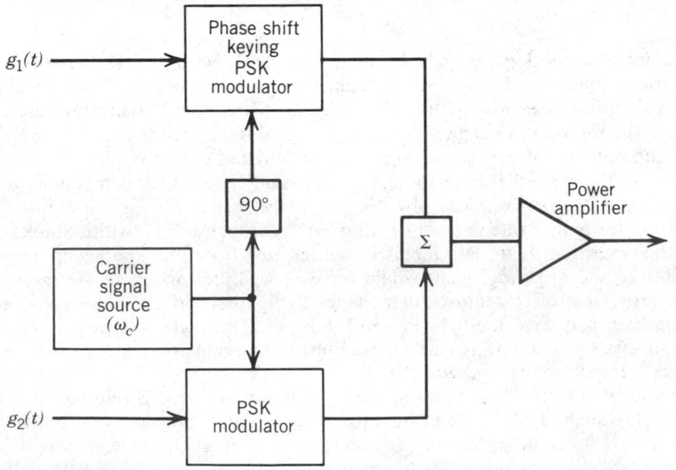

Fig. 48.56 Quadrature carrier multiplexing, functional illustration.

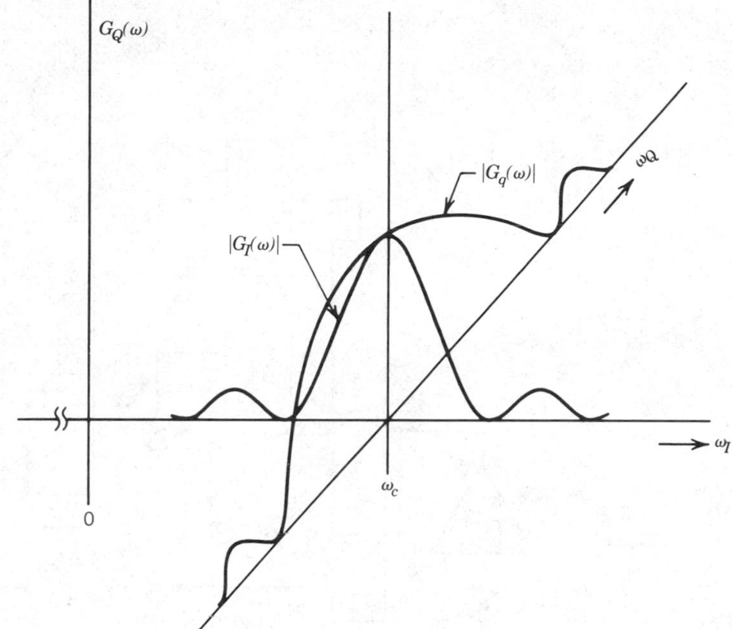

Fig. 48.57 Quadrature carrier multiplexed frequency spectra.

converted to an IF signal with bandwidth sufficiently wide to pass the coded spectra. Each of the codes is then detected and synchronized with the receiver replica of the transmitted PRN sequence. The code replica is then used to demodulate the IF signal that produces the received version of that code's information signal at IF. The other coded signals remain as additional background noise or interference, and hence an upper limit on the total number of coded signals that can be received simultaneously is established. The coded channel is then demodulated to obtain the information signal $g(t)$.

Code-division multiplexing is an example of spread spectrum techniques. These are described in more detail in Section 48.5, Advanced Techniques for Digital Communications.

Bibliography

Batson, B. H., *An Analysis of the Relative Merits of Various PCM Code Formats*, JSC Internal Note MSC-EB-R-68-5, Nov. 1968, pp. 3-1–4-11.

Batson, B. H., "Performance of Binary FSK Data Transmission Systems," Johnson Space Center Report, JSC 08097, Jul. 1973.

Downing, J. J., *Modulation System and Noise*, Prentice-Hall, Englewood Cliffs, NJ, 1964, pp. 113–177.

Giacoletto, L. J., "Generalized Theory of Multitone Amplitude and Frequency Modulation," *Proc. IRE*, pp. 680–693 (Jul. 1947).

Hopkins, P. M. and M. H. Kapell, "A Correction Factor for Filter Degradation in Frequency Modulation System Performance Models," Johnson Space Center Report, JSC-09059, Jun. 1974.

Lindsey, W. C., *Synchronization Systems in Communication and Control*, Prentice-Hall, Englewood Cliffs, NJ, 1972.

Lindsey, W. C., "Power Division Analysis of a Three Channel Unbalanced QPSK Signal Out of a Band Pass Filter," LinCom Corp. TR No. 04-7604-8, Apr. 1976.

Painter, J. H. and G. Hondros, "Unified S-Band Telecommunication Techniques for Apollo," Vol. II, *Mathematical Models and Analysis*, NASA, #1184, Sep. 1965.

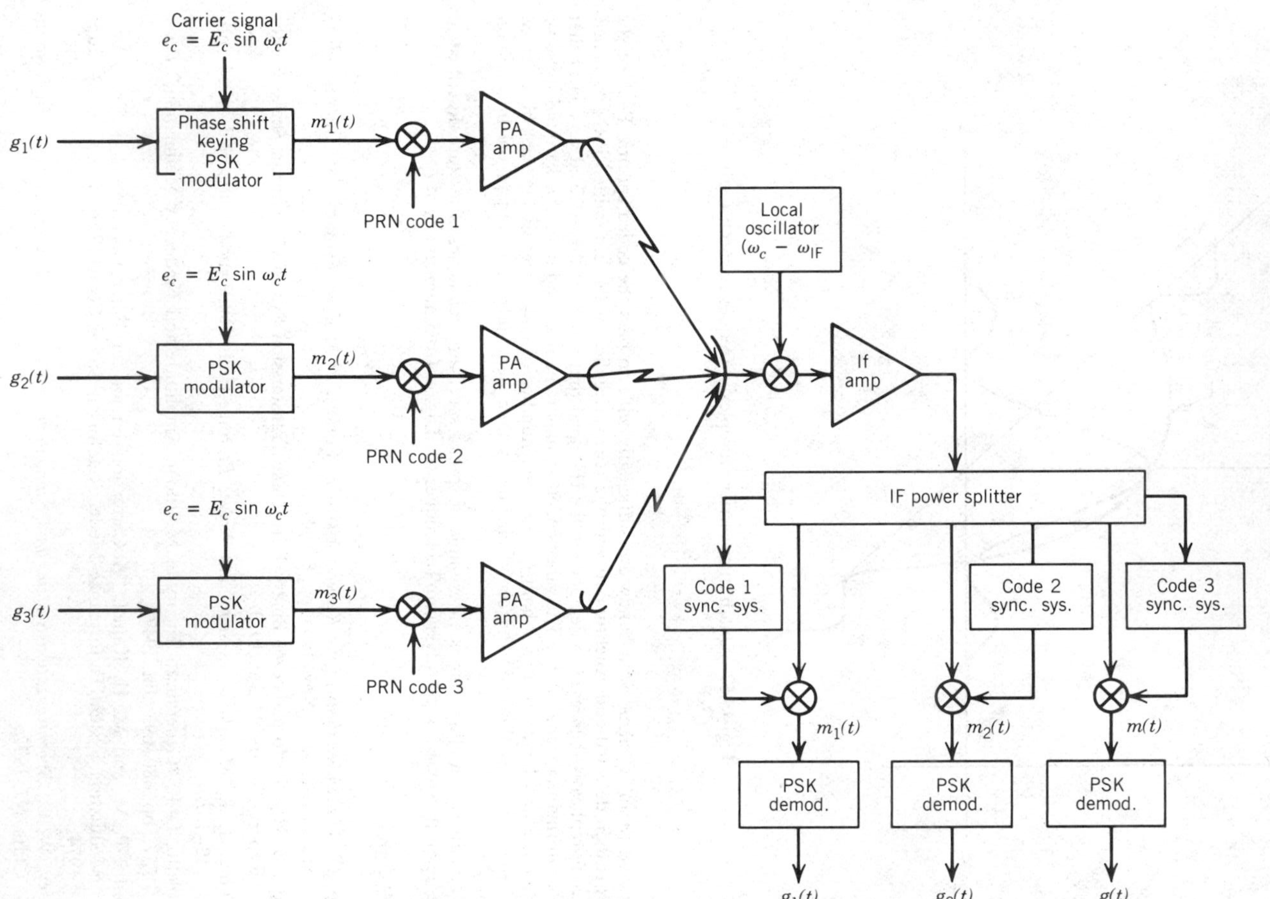

Fig. 48.58 Code division multiplexing system, functional diagram.

Panneton, R. J., *Analysis of Reduction in the 1.024 MHz Telemetry Subcarrier Modulation Index*, TRW, Project Technical Report Task E-94, Aug. 1968.

Panter, P. F., *Modulation, Noise, and Spectral Analysis*, McGraw-Hill, New York, 1965.

Platkin, S. C., "FM Bandwidth as a Function of Distribution and Modulation Index," *IEEE Communications Technology* issue, Jun. 1967.

Schwartz, M., *Information Transmission, Modulation, and Noise*, McGraw-Hill, New York, 1959.

Seyl, J. W., B. H. Batson, and B. G. Smith, "Experimental Result for FSK Data Transmission Systems Using Discriminator Detection," Proceedings, National Telecommunication Conference, Dec. 1976.

Seyl, J. W., et al., "Shuttle S-Band Communications Technical Concepts," Proceedings, Space Shuttle Technical Conference, Johnson Space Center, Jun. 1983.

Shilling, D. and M. Taub, *Principles of Communications Systems*, McGraw-Hill, New York, 1971.

Titsworth, R. C. and L. R. Welch, "Power Spectra of Signals Modulated by Random and Pseudo Random Sequence," Technical Report 32-140, Jet Propulsion Laboratory, Pasadena, CA, Oct. 1961.

Viterbi, A. J., *Principles of Coherent Communications*, McGraw-Hill, New York, 1966, pp. 185–215.

48.2 RADIO COMMUNICATIONS

George W. Raffoul

48.2-1 Introduction

Radio communication is a process by which intelligence or information at one location can be relayed to another distant location without the burden of connecting cables between source and destination. The transmission of intelligence, such as audio, video, or telemetry signals, is accomplished by modulating either the amplitude, frequency, or phase (or a combination thereof) of a higher frequency carrier. The modulated carrier is transformed into a propagating electromagnetic wave that travels through a propagating medium (normally the atmosphere) to reach the intelligence destination. At the destination the wave is intercepted and transformed back into a modulated carrier signal, is amplified, and is then demodulated to recover the original intelligence.

Thus the radio communication process involves a series of sequential steps in transporting the intelligence from source to destination. Step one consists of the modulation of the carrier at the source location by the intelligence signal. This function is performed by the transmitter. Step two involves the translation of the modulated carrier into a propagating electromagnetic wave by the transmitting antenna. Step three consists of the propagation of the electromagnetic wave through the atmosphere until it reaches the intended destination. Step four involves the conversion of the received electromagnetic wave at the destination into a modulated carrier signal by the receiving antenna.

Finally step five consists of the demodulation or extraction of the intelligence from the modulated carrier after its having been sufficiently amplified. Amplification is performed by the receiver, whose ability to recover the original intelligence exactly is hampered by the addition of noise in the channel. The overall performance of the radio communication link is a function not only of the noise content but also of the type of modulation and coding techniques used. In an analog communication system, the performance is expressed in terms of the postdetection signal-to-noise ratio (SNR) at the receiver. However, in a digital communication system, where the object is to decide whether a 1 or a 0 has been transmitted, the performance is usually described in terms of the probability of bit error, P_B, at the output of the detector.

In this section an overview of each of the four components of a radio communication link is given along with a brief link performance analysis. On the topic of radio wave propagation, special emphasis is placed on atmospheric effects. The current trend toward millimeter wave frequencies will make the impact of absorption by water vapor and oxygen significant enough to be included in the calculation of the link margin.

In the antenna domain, current development efforts are directed toward electronically scanned phased arrays, which may include advanced features such as adaptive control and conformance to curved surfaces. Also considerable development effort is directed toward satellite multibeam antennas, low-profile microstrip antennas, and millimeter wave antennas.

On the subject of transmitters and receivers, the discussion in this section is limited mostly to classical analog systems like AM (amplitude modulation), FM (frequency modulation), and SSB (single sideband). Digital communication systems are currently receiving the most attention in the research arena and in the marketplace. More detailed treatment of digital communication techniques is found in Section 48.5.

The final topic covered in this section deals with radio communication link analysis. A formula for the link margin is given and its application in link budget calculation is illustrated by an example.

48.2-2 Radio Wave Propagation

Propagation in Free Space

Free space is defined ideally as a homogeneous medium with no conductive currents or charges present, and with no objects that absorb or reflect radio energy. The concept of free space is used because it simplifies the approach to wave propagation and because propagating conditions sometimes do approximate those of free space, particularly at frequencies in the upper ultrahigh-frequency region. The extent of radio-wave propagation in free space is limited only by signal attenuation as the wave moves from the source of radiation. The transmission or path loss is given by the inverse square law of optics applied to radio transmission. The path attenuation between two isotropic antennas is given by

$$L_p = \left(\frac{4\pi d}{\lambda}\right)^2 \tag{48.80}$$

where L_p = path loss, or numeric ratio of transmitted power to received power (P_t/P_r)
λ = wavelength
d = path length
and where physical quantities are in the same units. A more convenient form of the path loss expression is

$$L_p = 32.5 + 20\log f + 20\log d \tag{48.81}$$

where L_p = path loss, dB
f = frequency, MHz
d = path length, km

Electromagnetic Spectrum

The speed of electromagnetic wave propagation in free space is of fundamental importance. This value is equal to the speed of light in free space, designated by the symbol c. The value of c is 186,283 statute miles per second, or 299,793 km/s, rounded off for most purposes to 186,000 mi/s or 3×10^8 m/s. In other than free space, the velocity of the electromagnetic wave is generally less, depending on the characteristics of the propagation medium involved. Another parameter of the electromagnetic wave is the wavelength, which is equal to the distance that the wave travels during one cycle period. It is denoted by λ_0 in free space and is related to the velocity by the relation

$$\lambda_0 = \frac{c}{f} \tag{48.82}$$

where the wavelength λ_0 is in meters, the velocity c is in meters per second, and the frequency f is in cycles per second, or hertz. Since the various phenomena of wave propagation are largely dependent on the frequency of the wave, the complete electromagnetic spectrum is shown in Table 48.7. Note the subdivision of the radio spectrum into several bands in accordance with accepted practice. The frequencies and wavelengths are shown in customary radio units.

Atmospheric Effects

The concept of free space transmission assumes that the atmosphere is perfectly uniform and nonabsorbing, and that the earth is either infinitely far away or that its reflection coefficient is negligible. In practice, during propagation near the earth, waves are reflected by the ground, mountains, and buildings; they are refracted as they pass through layers of the atmosphere that have different densities or different degrees of ionization. Also electromagnetic waves may be diffracted around large intervening obstacles and may even interfere with each other, as when two waves from the same source meet after having traveled by different paths. Waves may also be absorbed by various atoms and molecules found in the atmosphere. Some of these effects are desirable, to a certain extent, in the case of very low frequency (VLF), low frequency (LF), medium frequency (MF), high frequency (HF), and troposcatter links, because these aid in the establishment of a communication path between transmitter and receiver. However, in a line-of-sight communication link, the effects represent hindrance to the communication function and result in temporal fading or simply attenuation of the carrier level. We will now examine briefly each of these atmospheric effects.

TABLE 48.7. ELECTROMAGNETIC SPECTRUM

Band Designation	Frequency Range	Free-Space Wavelength Range
Extremely low frequency (ELF)	< 3 kHz	> 100 km
Very low frequency (VLF)	3–30 kHz	10–100 km
Low frequency (LF)	30–300 kHz	1–10 km
Medium frequency (MF)	300 kHz–3 MHz	100 m–1 km
High frequency (HF)	3–30 MHz	10–100 m
Very high frequency (VHF)	30–300 MHz	1–10 m
Ultrahigh frequency (UHF)	300 MHz–3 GHz	10 cm–1 m
Superhigh frequency (SHF)	3–30 GHz	1–10 cm
Extremely high frequency (EHF)	30–300 GHz	1–10 mm
Infrared	8×10^{11}–4×10^{14} Hz	80–400 μm
Visible light	4×10^{14}–7.5×10^{14} Hz	40–80 μm
Ultraviolet light	7.5×10^{14}–10^{16} Hz	1.2–40 μm
X-rays, gamma rays	10^{16}–10^{20} Hz	0.6 mμ–1.2 μm
Cosmic rays	$> 10^{20}$ Hz	< 0.6 mμ

Atmospheric Absorption. Of the main gases in the atmosphere, water vapor and oxygen absorb energy from electromagnetic waves, the former because of its electric dipole moment and the latter because of its magnetic dipole moment. Fortunately, the atmospheric absorption at frequencies below 10 GHz is quite insignificant. As shown in Fig. 48.59, absorption by both the oxygen and the water vapor content of the atmosphere becomes significant at that frequency and then rises gradually. Because of various molecular resonances, however, certain peaks and valleys of attenuation exist. For example, frequencies such as 60 and 120 GHz are not recommended for long-distance communication in the atmosphere. Also it is best not to use 23 or 180 GHz except in dry-air environments. On the other hand, Fig. 48.59 shows windows or frequency bands where absorption is greatly reduced, as in the case of frequencies around 33 and 110 GHz. Note also that the absorption level due to the water vapor content is based on a medium or standard humidity level. If the latter is increased or if there is fog, rain, or snow, this form of absorption is increased tremendously, and scattering due to rainwater drops may also develop.

Atmospheric Refraction. Atmospheric refraction phenomena observed in radio propagation are due to variations of the refractive index of air as height increases. Under normal conditions, the refractive index of the atmosphere decreases slightly but linearly with height. As a result of the slight refraction that takes place, waves bend down somewhat instead of traveling strictly in straight lines, and the radio horizon is thereby increased. Basically, what happens is that the bottom of the wavefront travels in denser atmosphere than the top of the wavefront and therefore travels slower, so that the wavefront is bent downward.

Uniform bending may be represented by straight-line propagation, but with the radius of the earth modified so that the relative curvature between the radio rays and the earth remains unchanged. The new radius of the earth is called the effective earth radius, and the ratio of effective to actual earth radii is designated by K. The average value of K is about $4/3$. The distance to the radio horizon over smooth earth, when the height h is very small compared with the earth radius, is given by

$$d_h = \left(\frac{3Kh}{2} \right)^{\frac{1}{2}} \tag{48.83}$$

where d_h is the distance to the radio horizon in miles and h is the altitude above the earth's surface in feet.[2]

Diffraction by Obstacles. Signals propagated on a line-of-sight basis may be received behind tall buildings, mountains, and other similar obstacles as a result of diffraction. The diffraction phenomenon is best explained by application of the notions introduced by Fresnel in optics. Let a broad ridge be situated at distances d_1 and d_2 from the transmitter T and the receiver R, respectively, at a height

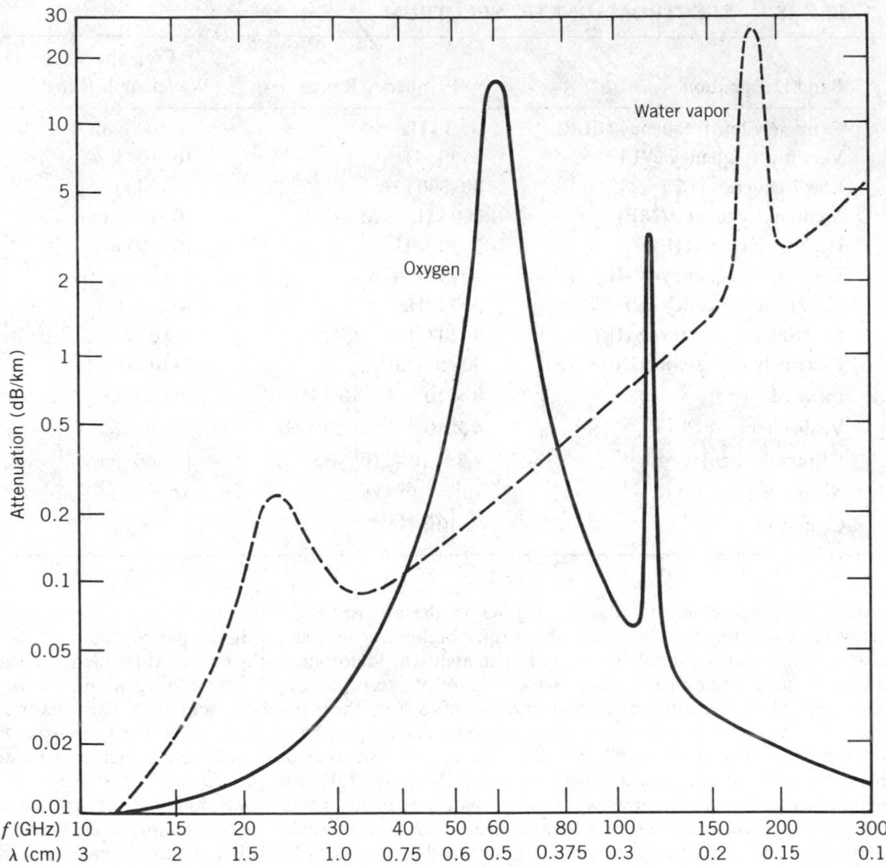

Fig. 48.59 Atmospheric absorption of electromagnetic waves. From Kennedy,[1] with permission.

below or above the optical line of sight, as shown in Fig. 48.60. Regions of space corresponding to increased path lengths of multiples of $\lambda/2$ with respect to $TR = d_1 + d_2 = d$, are ellipsoids with foci at T and R. The nth Fresnel surface is that for which the sum of distances between the transmitter and receiver and a point on the surface of the ellipsoid of revolution (dotted line in Fig. 48.60) exceeds by $n(\lambda/2)$ the distance between the transmitter and the receiver:

$$S = (r_1 + r_2) - (d_1 + d_2) = \frac{n\lambda}{2} \tag{48.84}$$

When h is small compared with d_1 and d_2, the distance S may be written as

$$S = \frac{h^2}{2}\left(\frac{1}{d_1} + \frac{1}{d_2}\right) = \frac{n\lambda}{2} \tag{48.85}$$

Corresponding to each integer value n there exists a value h_n known as the radius of the nth Fresnel zone and given by

$$h_n = \left(\frac{n\lambda}{1/d_1 + 1/d_2}\right)^{\frac{1}{2}} \tag{48.86}$$

A criterion to determine whether the ridge is sufficiently removed from the radio line of sight between transmitter and receiver to allow mean free-space propagation conditions to apply is to have the first Fresnel zone clear all obstacles in the path of the rays. This first Fresnel zone clearance is given by

$$h_1 = \left(\frac{\lambda}{1/d_1 + 1/d_2}\right)^{\frac{1}{2}} \tag{48.87}$$

Figure 48.61 shows the effect of path clearance on radio transmission.

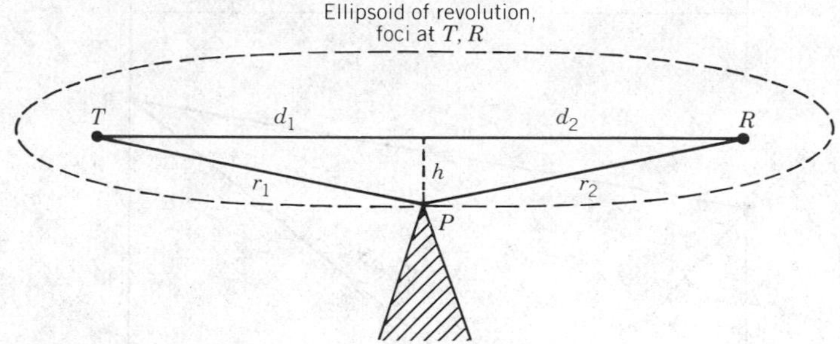

Fig. 48.60 Diffraction effect of a wide ridge. T, transmitter; R, receiver.

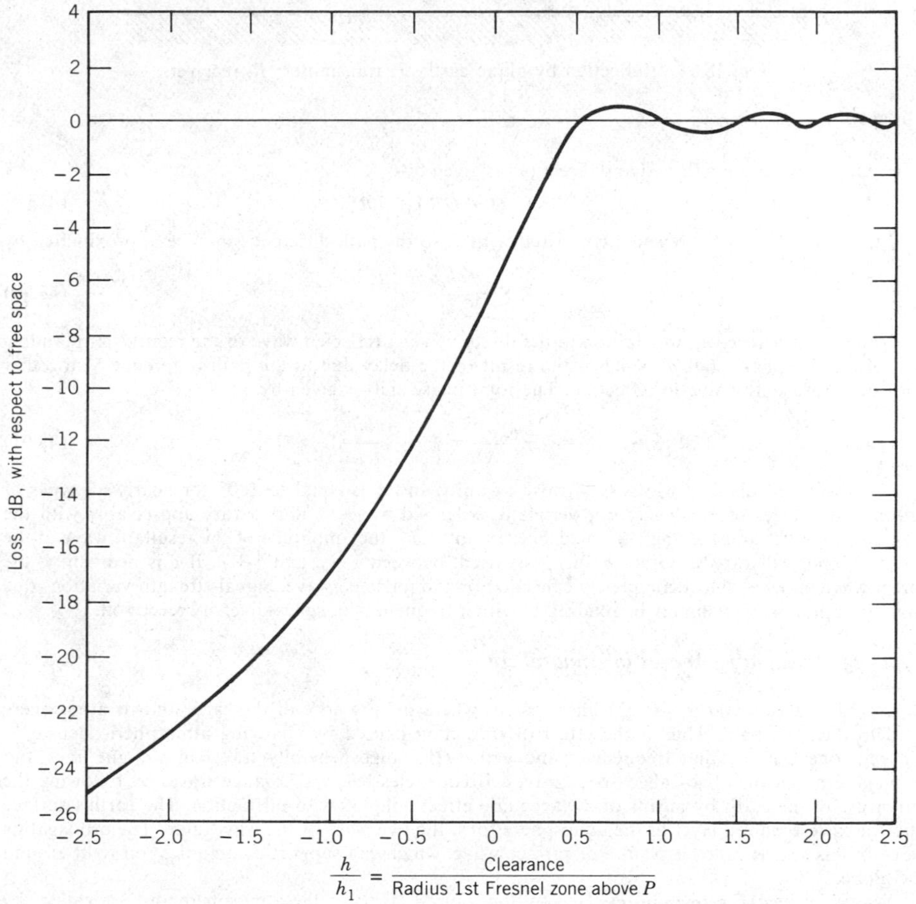

Fig. 48.61 Effect of path clearance on radio transmission, assuming a knife-edge diffraction. From Panter,[3] with permission.

Ground Reflections. Reflections from the earth can seriously affect the level of the received signal. The simplest case to consider is the reflection from a plane earth, as shown in Fig. 48.62. The resultant wave at the receiver R consists of the direct ray TR and the reflected ray OR. Reflection causes attenuation and phase shift along the indirect ray, which may be accounted for by the reflection coefficient expressing the ratio of the reflected field strength to the incident field strength

$$\Gamma = \rho e^{j\psi} \tag{48.88}$$

where ρ and ψ are the magnitude and phase of the reflection coefficient, respectively. The path

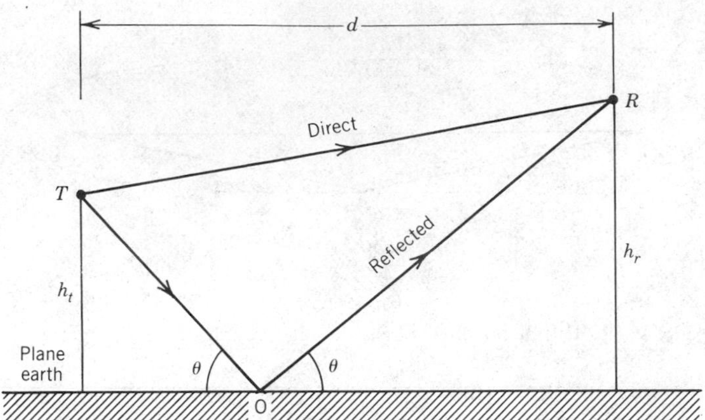

Fig. 48.62 Reflection by plane earth. T, transmitter; R, receiver.

difference between the reflected and direct ray is given by

$$S = (TO + OR) - TR \tag{48.89}$$

In practice, h_t and h_r are small compared with d, so the path difference may be approximated by

$$S \simeq \frac{2h_t h_r}{d} \tag{48.90}$$

Thus the receiver receives, in addition to the direct wave, a reflected wave of the relative amplitude ρ and of relative phase shift α, which is the result of the delay due to the path difference S, together with the phase shift ψ due to reflection. The total phase shift is given by

$$\alpha = \psi - \frac{2\pi S}{\lambda} = \psi - \frac{4\pi h_t h_r}{\lambda d} \tag{48.91}$$

For horizontally polarized waves, ρ is close to unity and ψ is equal to 180° for nearly all types of ground and angles of incidence. For vertically polarized waves, ρ and ψ vary appreciably with the ground properties and the angle of incidence. In any case, the amplitude of the resultant wave at the receiver input will vary, as either h_t or h_r is varied, between $1 - \rho$ and $1 + \rho$. If ρ is near unity, the direct wave may be almost completely cancelled by the reflected wave. Signal-strength variations due to multipath transmission may be reduced by either frequency- or space-diversity reception.

Types of Propagating Waves in Atmosphere

In general, waves travel in straight lines, except where the presence of the earth and its atmosphere tend to alter the path. Thus if the effect of refraction caused by changing atmospheric density is ignored, one can say that frequencies above the HF range generally travel in straight lines and propagate by means of so-called space waves. Frequencies below this range travel by following the curvature of the earth by means of a waveguide effect coupled with diffraction. The earth's surface and the lowest ionized layer of the atmosphere form the two walls of the waveguide. The propagating wave in this case is called a ground or surface wave, which can support communication right around the globe.

Waves in the HF range are reflected by the ionized layers of the atmosphere and are called sky waves. Signals in the HF band are transmitted into the sky, where they get reflected by the ionospheric layers and return to the ground well beyond the horizon. A long-distance communication, popularly called shortwave, is possible when these waves are alternately reflected again and again by the ground and the ionosphere to reach receivers on the opposite side of the earth.

Two other methods of beyond-the-horizon propagation are tropospheric scatter and geosynchronous satellite communication. The special character of HF propagation deserves special treatment and thus will be described in some detail next.

HF Propagation. The ionosphere is a sparse region that extends from around 50 to around 400 km in altitude. During daylight hours it is bombarded by solar radiation, so that the rarefied air becomes ionized by the ultraviolet and x-ray emissions. Different parts of the solar spectrum are absorbed at different altitudes, so that several ionized layers are formed. The upper layers reflect radio waves, while the lower layers attenuate the waves passing through them, as illustrated in Fig. 48.63.

The radio refractive index μ of an ionized layer is given approximately by

$$\mu = \left(1 - \frac{f_p^2}{f^2}\right)^{1/2} \tag{48.92}$$

where f_p is the plasma frequency and f is the frequency of the wave. For any given density of ionization N (electrons per cubic meter), the refractive index decreases as the frequency f decreases. At $f = f_p = 81N$, the refractive index goes to zero. A radio wave at a frequency below f_p and normally incident on an ionized layer penetrates only as far as the point where μ falls to zero and is then totally reflected. The plasma frequency at the very top of the layer is called the critical frequency f_c. Normally incident waves at frequencies above f_c are transmitted right through the layer.

At frequencies above the critical frequency, reflection may still take place if the wave is obliquely incident. The highest frequency that can be totally reflected at a given angle of incidence θ is called the maximum usable frequency, or MUF, and given by

$$\text{MUF} = f_c \sec\theta \tag{48.93}$$

The MUF may be several times greater than the critical frequency, and its expression is only approximate because it is based on a flat earth and a flat reflecting layer.

The attenuation loss depends mainly on the product of the ionization density and the frequency of collisions between the free electrons and the heavier particles. When the electrons are set in motion by a radio wave, they acquire oscillatory energy from the wave that is converted to heat on collision. Since the energy is extracted from the wave, the wave is attenuated. The rate of attenuation A is given approximately by

$$A = \frac{1.16 \times 10^{-6} N\nu}{f^2} \quad \text{dB/m} \tag{48.94}$$

where ν is the electron collision frequency. Thus the attenuation level determines the lowest usable frequency, while the MUF sets the upper limit.

Since the ionization density of the upper atmosphere can be estimated from the time, season, latitude, and sunspot number, it is possible to forecast critical frequencies and attenuations for most practical purposes except during the peak of the 11-year sun cycle. At this critical period, solar activity is very intense and a complete loss of HF communication over the whole band may result. Figure 48.64 gives the altitudes of the various ionospheric layers with their regular variations. The D layer is strictly an attenuating layer. E, F_1, and F_2 are all reflecting layers during the day. At night, the D and E layers disappear and the F_1 and F_2 layers combine into the F layer located at the altitude of the F_1 layer.

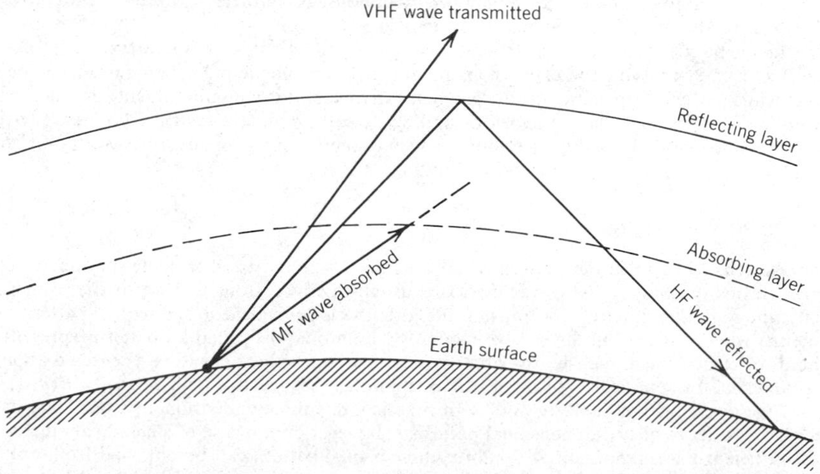

Fig. 48.63 Reflection of high-frequency waves by the ionosphere.

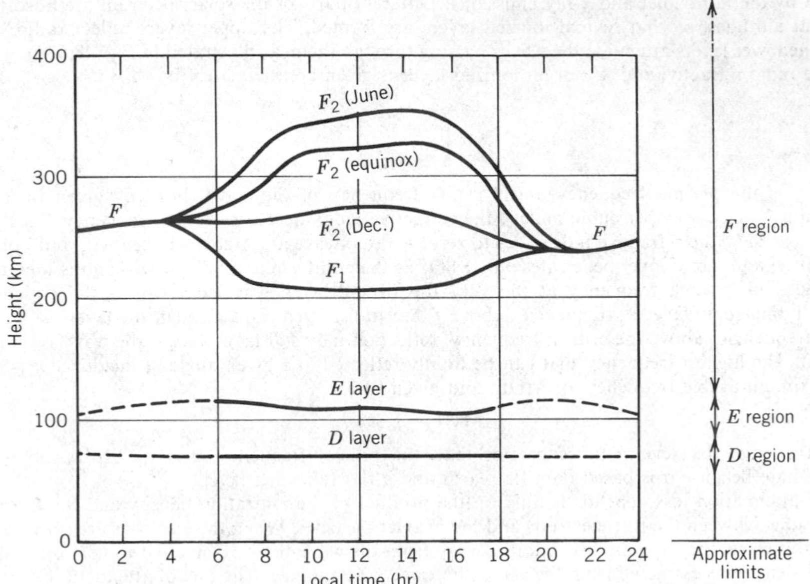

Fig. 48.64 Ionospheric layers and their regular variations.

48.2-3 Antenna Fundamentals

Antenna Types

Antennas can be very broadly classified either by the frequency spectrum in which they are commonly applied or by their basic mode of radiation. In the frequency spectrum classification, antenna types follow the usual band designations and are described as VLF, LF, MF, HF, VHF, UHF, and microwave antennas according to their frequency of operation. In the radiation classification, antennas can be divided into four groups: elemental current antennas, traveling-wave antennas, array antennas, and aperture antennas. The four groups can be distinguished by the size of the antenna measured in wavelengths, which in turn can be related to the region of the spectrum in which the antennas are commonly applied, as shown in Fig. 48.65. Table 48.8 lists examples of antennas that can be allocated to each of the four groups. The classification of antenna types into these four groups is only an approximation, with numerous exceptions. Nevertheless, it provides a convenient form of organizing the subject of antenna fundamentals.

In the mathematical analysis of antennas, the correct choice of coordinate system is often an important factor in simplifying the expressions for the electromagnetic fields and currents associated with the antenna system. Depending on the geometry involved, it is common practice to make use of the conventional cartesian, polar, cylindrical, and spherical coordinate systems. Figure 48.66 illustrates the components of the radiated electric field, at some point in space, due to a point source located at the origin in terms of spherical coordinates.

Antenna Parameters

Radiation Pattern. The radiation pattern of an antenna is the most fundamental parameter of the antenna, since many of the other parameters are usually derived from it. Due to the principle of reciprocity, the radiation pattern of a transmitting antenna is equivalent to the receive pattern of the same antenna when it is used in the receive mode. By definition, the radiation pattern represents the variation of the electric field intensity over the surface of a large sphere of radius r centered about the radiating antenna. In spherical coordinates, it is a plot of the electric field intensity $E(\theta, \phi)$ as a function of the directional variables θ and ϕ. In practice, this three-dimensional pattern is measured and recorded in a series of two-dimensional patterns. However, in the case of single-beam directional antennas, sufficient information about the three-dimensional pattern can be obtained from only two two-dimensional plane patterns that include the major beam maximum direction. These plane patterns are called the principal-plane patterns of the antenna, such as the xy plane ($\theta = 90°$) and the

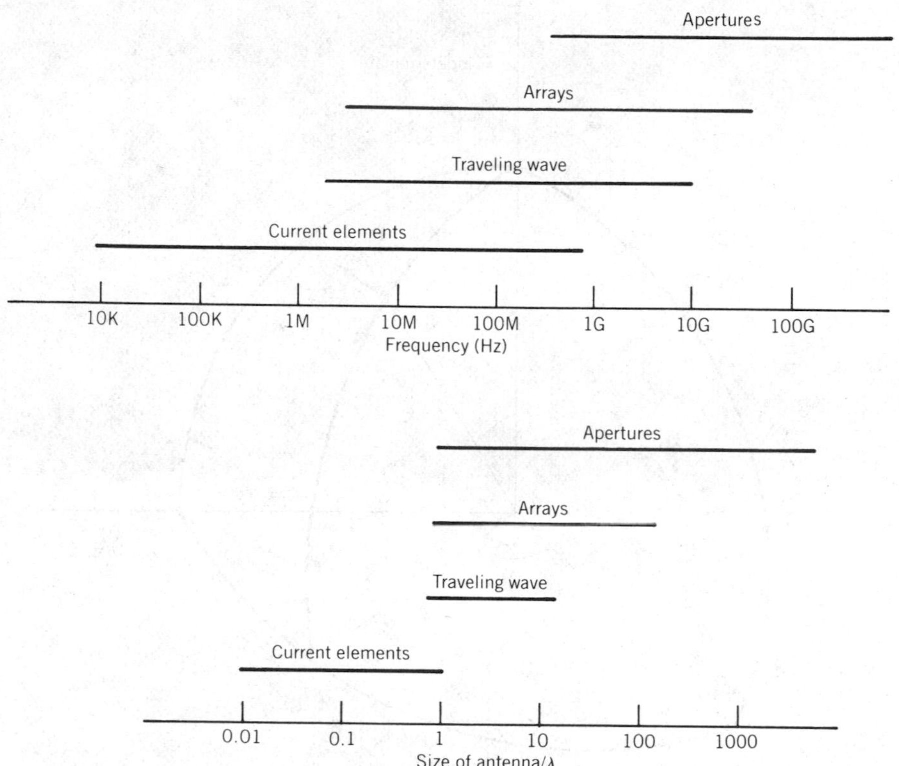

Fig. 48.65 Classification of antennas on the basis of mode of radiation. From Rudge et al.,[4] with permission.

TABLE 48.8. TYPES OF ANTENNAS

Current Element	Traveling Wave	Array	Aperture
Monopole	Line source	Broadside	Reflector
Dipole	Long wire	Endfire	Horn
Loop	Rhombic	Linear	Lens
Slot	Slotted waveguide	Planar	Backfire
Biconical	Spiral	Circular	Short dielectric rod
Notch	Helix	Conformal	Parabolic horn
Spheroidal	Log periodic	Log periodic	
Disk	Slow wave	Signal processing	
Microstrip	Fast wave		
	Leaky wave		
	Surface wave		
	Long dielectric rod		

xz plane ($\phi = 0°$) shown in Fig. 48.66. For a linearly polarized antenna, the principal-plane patterns may also be called the E-plane and H-plane patterns, provided one plane contains the E-field vector and the other contains the H-field vector.

The radiation pattern may be drawn graphically in a variety of formats. The most common are two-dimensional polar or cartesian coordinate plots. The relative amplitude of the radiated energy may be recorded as a relative power pattern (P/P_{max}), a relative field pattern (E/E_{max}), a

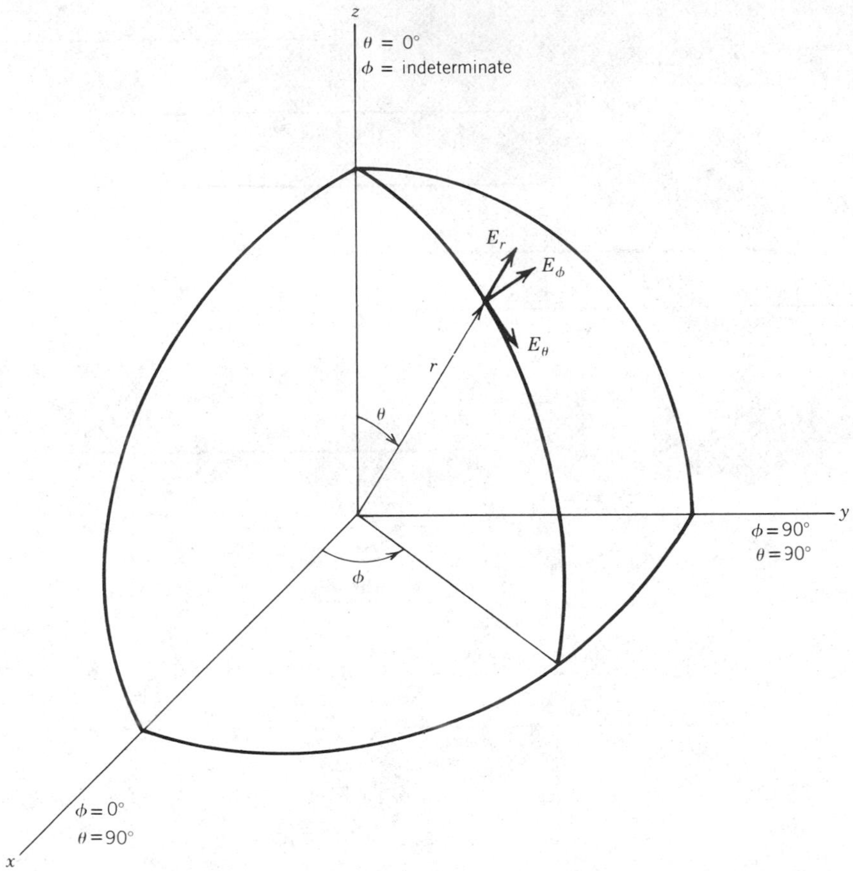

Fig. 48.66 Spherical coordinate system used in antenna analysis.

logarithmic power pattern $[10\log(P/P_{max})]$, or a logarithmic field pattern $[20\log(E/E_{max})]$. Figure 48.67 illustrates a typical relative field pattern plotted in both polar and rectangular form.

The radiation pattern can be used to obtain the beamwidth of the main beam and the sidelobe level. The beamwidth is specified as the angular difference between the two points on the radiation pattern where the power has fallen to one half of the peak value, or -3 dB on the decibel scale. The sidelobe level represents the level of the largest minor lobe as a fraction of the main beam level and is usually specified in decibels (e.g., -20 dB).

The space surrounding an antenna is usually subdivided into three regions or zones: (1) reactive near-field, (2) radiating near-field (Fresnel), and (3) far-field (Fraunhofer) regions. The reactive near-field region exists very close to the antenna, where the reactive components of the electromagnetic fields are very large with respect to the radiating fields. The radiating near-field region is the region located between the reactive near-field region and the far-field region where radiation fields predominate and where the angular field distribution is dependent on the distance from the antenna. The far-field region is defined as the region where the angular field distribution is essentially independent of the distance from the antenna. Its inner boundary is taken to be the radial distance $R = 2D^2/\lambda$, where D is the antenna's largest dimension and λ is the wavelength, while the outer boundary is located at infinity. It is in this region that the radiation pattern of the antenna is measured and/or calculated.

Directivity, Gain, and Efficiency. The directivity of an antenna is a measure of its directional properties or its ability to concentrate the radiated power in various directions. Usually the directivity is specified with reference to an isotropic radiator. The latter is a hypothetical antenna that radiates

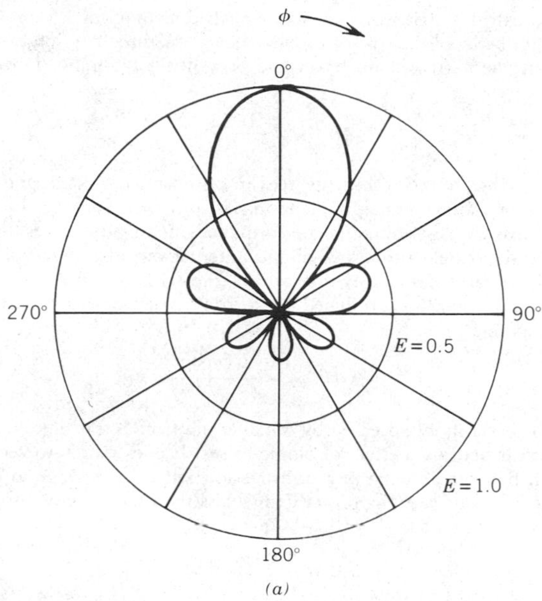

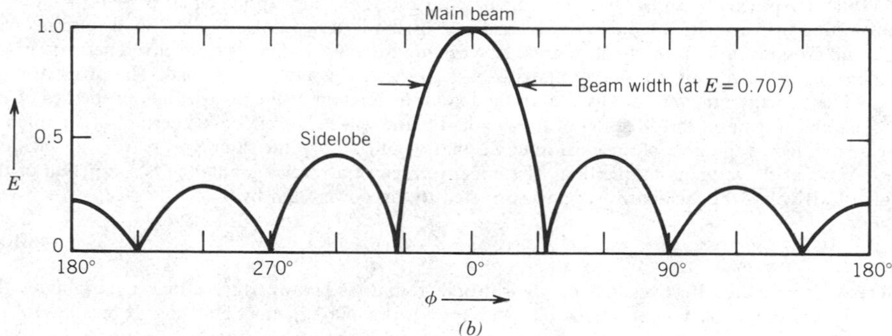

Fig. 48.67 Comparison of plane relative E-field pattern plotted in polar and rectangular form: (a) Polar plot, (b) rectangular plot.

uniformly in all directions. Then the directivity $D(\theta, \phi)$ in a given direction (θ, ϕ) is given by the ratio of the radiation intensity of the antenna in the direction (θ, ϕ) to the radiation intensity produced by an isotropic radiator.

$$D(\theta, \phi) = \frac{U(\theta, \phi)}{U_0} = \frac{U(\theta, \phi)}{(P_{\text{rad}}/4\pi)} \tag{48.95}$$

where $U(\theta, \phi)$ = radiation intensity of antenna in direction (θ, ϕ), W/unit solid angle
U_0 = radiation intensity of isotropic source, W/unit solid angle
P_{rad} = total power radiated by antenna, W

In terms of the far-field electric field intensity $E(\theta, \phi)$, the directivity can be expressed as

$$D(\theta, \phi) = \frac{|E(\theta, \phi)|^2}{1/4\pi \int_0^{\pi} \int_0^{2\pi} |E(\theta, \phi)|^2 \sin\theta \, d\theta \, d\phi} \tag{48.96}$$

Although directivity can be specified in any direction, it is usual to refer to the peak value that is associated with the direction of the main beam radiated by the antenna. Hence in any reference to the directivity of an antenna, the peak figure D_0 is normally implied. For example, a short linear current element (also called hertzian dipole), has a peak directivity of 1.5, or 1.76 dB, relative to an isotropic source, while the larger half-wave dipole has a peak directivity of 1.64, or 2.14 dB.

For antennas with one narrow main beam and very negligible minor lobes, the peak directivity can be approximated by[5]

$$D_0 \simeq \frac{41{,}253}{\theta_{1d}\theta_{2d}} \tag{48.97}$$

where θ_{1d} is the half-power beamwidth in degrees in one principal plane and θ_{2d} is the half-power beamwidth in degrees in another principal plane that is at a right angle to the first.

Another useful measure describing the performance of an antenna is the gain. It takes into account not only the directional properties of the antenna but also its efficiency. The power gain $G(\theta, \phi)$ in a given direction is defined as the ratio of the radiation intensity of the antenna in that direction to the radiation intensity produced by a lossless isotropic source having the same total power input as the antenna.

$$G(\theta, \phi) = \frac{U(\theta, \phi)}{(P_{in}/4\pi)} \tag{48.98}$$

where P_{in} is the total power input accepted by the antenna from a transmitter. Unlike directivity, the power gain expression includes the effect of ohmic losses. It does not, however, include losses due to impedance mismatch between antenna and transmission line or losses due to polarization mismatch at a receiving antenna. In any case, the power gain is less than the directivity by a factor equal to the radiation efficiency η of the antenna.

$$\eta = \frac{G(\theta, \phi)}{D(\theta, \phi)} \tag{48.99}$$

Both gain and directivity may be referenced to any standard antenna, such as half-wave dipole or horn, instead of the isotropic radiator used here.

While reciprocity ensures that the calculated values of gain apply equally well to either a transmitting or receiving antenna, the performance of the latter can also be described in terms of a receiving cross section or an effective area. A receiving antenna will collect effective energy from an incident plane wave and, if properly matched, will transfer this power to a load. The proportion of the incident energy that will find its way to the load is a function of the polarization properties of the antenna and its gain in the direction of the incident plane wave. The effective aperture of an antenna can be defined as the area of an ideal antenna that would absorb the same power from an incident plane wave as the antenna in question. The effective area of a receiving antenna is a function of the angle of arrival of the incident wave and is related to the power gain by

$$A_{eff}(\theta, \phi) = \frac{\lambda^2}{4\pi} G(\theta, \phi) \tag{48.100}$$

where $\lambda^2/4\pi$ is the effective area of an isotropic radiator. The aperture efficiency evaluates the effective aperture as a fraction of the physical aperture of the antenna. This idea is meaningful for antennas that have a well-defined collecting aperture, as in the case of aperture antennas (see Table 48.8 for examples).

$$\eta_{aperture} = \frac{A_{eff}}{A_{phy}} \tag{48.101}$$

For example, the aperture efficiency of reflector antennas falls in the range of 50 to 70%.

Bandwidth. The term "bandwidth" is used to describe the frequency range over which an antenna will operate satisfactorily. There is no unique definition for satisfactory performance, because such performance is dependent on the antenna application. Usually one can distinguish between a bandwidth dictated by pattern considerations and a bandwidth dictated by impedance considerations. Associated with pattern bandwidth are gain, sidelobe level, beamwidth, polarization, and beam direction, while input impedance and radiation efficiency are related to impedance bandwidth.

In practice usually one or more of the antenna parameters is more sensitive to frequency change than the others and thus may become the factor-limiting bandwidth. This is so only if the variation of such parameters is constrained by performance limits imposed by the application.

For broadband antennas the bandwidth is normally expressed as the ratio of the upper-to-lower frequencies in the acceptable band (e.g., 10 : 1). For narrowband antennas, the bandwidth is expressed as a percent of the band center frequency (e.g., 5%). The physical design factors limiting the bandwidth vary from antenna to antenna. In monopoles, dipoles, slots, and microstrip elements the structures are resonant at particular frequencies and the bandwidth is determined by the impedance characteristics at the input terminals. On the other hand, horn radiators are bandlimited by the modal nature of the wave propagation in the waveguiding structure.

Polarization. The polarization of an electromagnetic wave at a single frequency describes the shape of the locus of the extremity of the instantaneous electric field vector as a function of time at a fixed location in space, and the sense in which the locus is traced as observed along the direction of propagation. A single current element oriented along the x-axis will radiate a linearly polarized wave with an electric field vector oriented along the x direction. A more complicated antenna may radiate a wave where the electric field vector has both x and y components. If the two components E_x and E_y differ in phase by 0° or 180°, the wave will still be linearly polarized. If the two components have equal magnitude and $\pm 90°$ phase difference, the resultant electric field vector at a given point in space will rotate at an angular rate ω in such a way that its extremity will trace out a circle. The wave in this case is said to be circularly polarized. In general, if the two components have arbitrary amplitudes and phase difference, the instantaneous electric field will trace out an ellipse and the wave is called elliptically polarized. Moreover, when the rotation around the ellipse or circle is clockwise, the polarization is termed right-hand; conversely, when the rotation is counterclockwise, the polarization is termed left-hand.

If the polarization of the incident wave does not match the polarization of the receiving antenna, a polarization loss occurs owing to the mismatch. This polarization loss must always be taken into account in the link calculations design of a communication system, especially in power-limited applications.

Input Impedance. An antenna must be coupled to a transmitter by means of a transmission line or waveguide in order to be excited and produce radiation. The antenna input impedance presented to the feed line constitutes an important parameter whose value is required in the design of efficient coupling networks that will assure maximum power transfer. The antenna input impedance will generally have both a resistive and a reactive component:

$$Z_A = R_A + jX_A \tag{48.102}$$

The reactive component arises from the near-zone induction fields, because these fields give rise to a reactive energy storage in the region surrounding the antenna. The resistive component of the input impedance has contributions from all the various elements that lead to a loss of energy from the antenna. In the case of an antenna in free space, where mutual coupling from other sources is nonexistent, the antenna resistance can be thought of as the sum of the radiation resistance R_r and the ohmic resistance R_L:

$$R_A = R_r + R_L \tag{48.103}$$

The radiation resistance is defined as the equivalent resistance that would dissipate a power equal to the radiated power when the current through the resistance is equal to the current at the antenna input terminals. The ohmic resistance R_L accounts for the losses due to a finite conductivity on the antenna structure. For an efficient antenna, the radiation resistance must be much larger than the ohmic resistance. For example, a practical, thin half-wave dipole has a radiation resistance of 73 Ω and an ohmic resistance of about 2 Ω.

Measurement of input impedance at high frequencies is usually done by measuring the reflection coefficient and the voltage standing wave ratio (VSWR). The latter is related to the magnitude of the reflection coefficient Γ by the equation

$$\text{VSWR} = \frac{1 + |\Gamma|}{1 - |\Gamma|} \tag{48.104}$$

The antenna impedance Z_A is given by

$$Z_A = Z_0\left(\frac{1 + \Gamma}{1 - \Gamma}\right) \tag{48.105}$$

where Z_0 is the characteristic impedance of the transmission line.

Current Element Antennas

Current element antennas are the most fundamental radiators in use today. They may be of the electric current variety, such as dipole and loop, or of the magnetic current variety, such as a slot. We will limit our discussion here to the short- and half-wavelength dipoles.

Consider a short, infinitely thin dipole positioned along the z-axis, as shown in Fig. 48.68. In the far-field region, the radiation consists of a transverse electromagnetic wave propagating away from

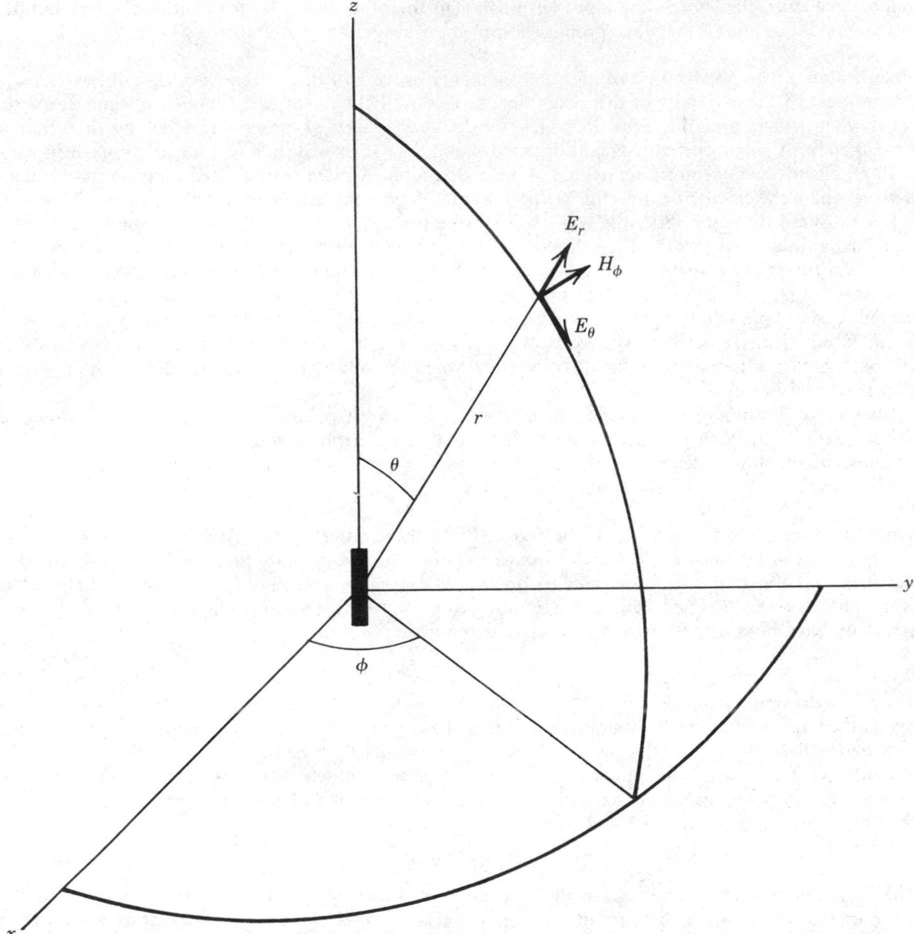

Fig. 48.68 Calculation of the radiated fields of a short dipole.

the dipole and given by

$$E_\theta = j60\pi \frac{I\Delta z}{r\lambda} e^{-jKr} \sin\theta \tag{48.106}$$

$$H_\phi = \frac{E_\theta}{\eta_0}$$

where I = constant current on dipole, A
Δz = dipole length, m
λ = wavelength, m
$K = 2\pi/\lambda$ = propagation constant, rad/m
$j = \sqrt{-1}$
η_0 = 377 ohms = intrinsic impedance of free space

If the length of the dipole is increased to a half of a wavelength ($\lambda/2$), the current distribution is no longer uniform but equal to $I(Z) = I_0 \cos KZ$ to a first approximation. The far-field is then given by

$$E_\theta = j60I_0 \frac{e^{-jKr}}{r} \frac{\cos[(\pi/2)\cos\theta]}{\sin\theta} \tag{48.107}$$

The elevation plane patterns for both the short and the $\lambda/2$ dipoles are compared in Fig. 48.69. In the

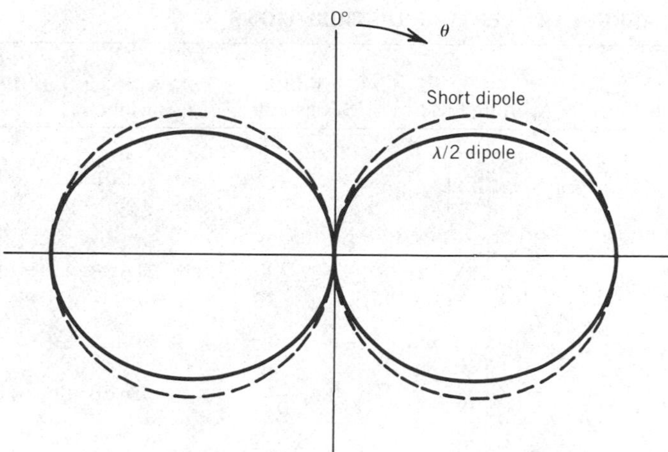

Fig. 48.69 Elevation plane patterns of a vertical dipole.

azimuth plane (xy plane), the radiation of either antenna is omnidirectional and the pattern is a circle.

Aperture Antennas

There are a large number of antenna types for which the radiated electromagnetic field can be considered to emanate from a physical opening or aperture. Antennas that fall into this category include several types of reflectors, lenses, and horns, to name a few.[6] In general, the aperture consists of a finite opening in an infinite plane. The electric and/or magnetic fields in the aperture region are first determined by approximate methods; then the radiated fields are calculated, using as sources the fields in the aperture surface alone.

Table 48.9 lists the pattern characteristics of a circular aperture of diameter D under various tapering distributions. For each aperture distribution, the 3-dB beamwidth (in radians), the sidelobe level, and the position of the first pattern zero are given. The uniform distribution has a $\sin x/x$ far-field pattern and a -13-dB sidelobe level. The tapered distributions reduce the sidelobe level at the expense of some beam broadening and reduction in gain. The Taylor distribution is an optimized distribution in which the sidelobe level is low and the main beam is still reasonably narrow.

Array Antennas

Array antennas are made up of a group of radiating elements arranged in some geometrical and electrical fashion. The total pattern for the array is obtained by multiplying the pattern of an individual element (element factor) and the pattern due to an array of isotropic point sources (array factor). The array factor generally depends on the geometrical orientation of the array, the spacing between elements, and the excitation of the elements. Usually the individual radiating elements have low directivities and so the array pattern is largely determined by the array factor, although there may be exceptions to this rule. Typical array elements include dipoles, monopoles, waveguide slots, open-ended waveguides, and microstrip radiators. The choice of a given element for a particular application depends on various factors such as frequency, power handling capability, polarization, feeding arrangement, and the like. The array is called linear when the elements are arranged in a straight line, planar when the elements are arranged in a plane, and conformal when the elements adhere to some geometrical shape such as a circle or a sphere.

One of the attractive features of an array is the ability to scan the radiated beam electronically by controlling the excitation phase of the individual elements. When used in this fashion, the array is termed a phased array. Advanced arrays of considerable current interest are the signal processing and adaptive arrays. In the latter, the array radiation pattern adapts itself under computer control to a particular situation. In particular, the null steering array steers nulls in the radiation pattern to coincide with the direction of an unwanted or interfering signal.

TABLE 48.9. CIRCULAR APERTURE DISTRIBUTIONS

Distribution	Aperture Field	3-dB Beamwidth	Level of 1st Sidelobe	Angular Position of 1st Zero
$0 \leqslant r \leqslant 1$ Uniform	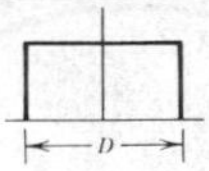	$1.02\,\frac{\lambda}{D}$	− 17.6 dB	$1.22\,\frac{\lambda}{D}$
Tapered to zero at edge $(1 - r^2)$	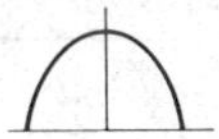	$1.27\,\frac{\lambda}{D}$	− 24.6 dB	$1.63\,\frac{\lambda}{D}$
Tapered to zero at edge $(1 - r^2)^2$	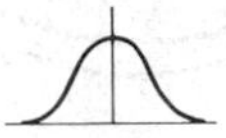	$1.47\,\frac{\lambda}{D}$	− 30.6 dB	$2.03\,\frac{\lambda}{D}$
Tapered to 0.5 at edge $[0.5 + (1 - r^2)^2]$	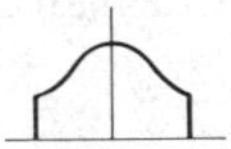	$1.16\,\frac{\lambda}{D}$	− 26.5 dB	$1.51\,\frac{\lambda}{D}$
Taylor distribution	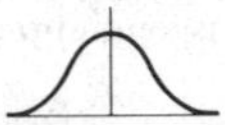	$1.31\,\frac{\lambda}{D}$	− 40.0 dB	

Source. From Rudge et al.,[4] with permission.

48.2-4 Transmitter Fundamentals

General

Transmitters can readily be classified in terms of the modulation technique on which they are based. These techniques include amplitude modulation (AM), frequency modulation (FM), phase modulation (PM), and single sideband (SSB). The choice of a given transmitter type normally depends on the intended application, the frequency of operation, and the host of rules and regulations governing radio transmission as promulgated by the FCC, the CCITT, the ITU, and other international regulatory bodies.

In this subsection are described, at the block diagram level, the most basic transmitter types. For an explanation of the modulation theory involved in each transmitter type, the reader is referred to Section 48.1.

AM Transmitter

The usual structure of an AM transmitter is a master oscillator followed by power amplifiers and/or frequency multipliers in numbers sufficient to obtain the desired frequency and power output. The output from this final stage is passed through a bandpass filter to attenuate harmonics and is then fed to the antenna. Modulation may be applied to any of the power amplifiers. Depending on the stage of application, AM transmitters are classified as employing high-level or low-level modulation. Figure 48.70 is a block diagram of a typical high-level AM transmitter, where the modulation is applied to the final power amplifier. The master oscillator is used to generate the carrier frequency. However, if the operating frequency is high, better stability is obtained by designing the oscillator for some lower frequency and then using frequency multipliers to obtain the desired frequency value. Where ability to change the operating frequency is desired, crystal oscillator and frequency synthesizer outputs can be used as inputs to a mixer whose carrier output becomes a function of the digital control input to the synthesizer.

The advantage of high-level modulation lies in the fact that all the intermediate power amplifiers can be operated class C for maximum efficiency. This is possible since the signal is unmodulated. The number of stages used depends on the power required to drive the final power amplifier. One disadvantage of high-level modulation is the large modulating signal power required to produce modulation.

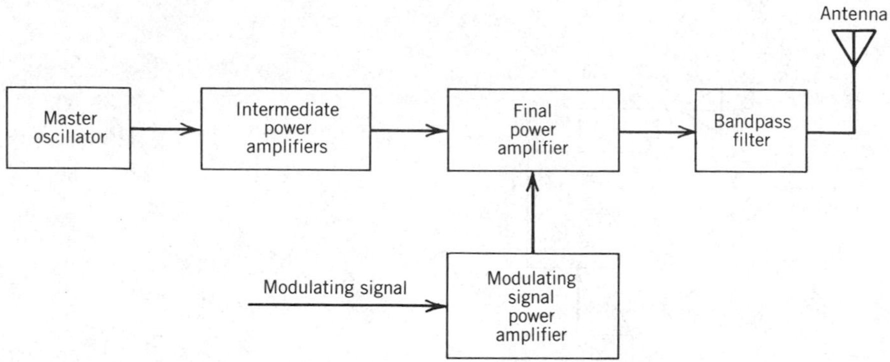

Fig. 48.70 Block diagram of an AM transmitter.

In a low-level modulation transmitter, the modulating signal is applied to one of the intermediate power amplifiers. The earlier the stage, the lower the carrier power level, and the modulating signal power requirements decrease. However, since the RF signal is now modulated, all subsequent stages must handle sideband power as well as carrier power and must have sufficient bandwidth for the sideband frequencies. Furthermore, all these stages must be operated in a linear class, A or B, with an attendant loss in efficiency.[7]

FM Transmitter

Transmitters for frequency modulation service fall into two broad categories, depending on the technique used to obtain the frequency deviation. In the "direct-modulation" technique, the frequency of the oscillator is made to vary in accordance with the modulating signal. In the "indirect-modulation" technique, frequency modulation is obtained indirectly after phase modulating the carrier.

In each of these categories, FM can further be classified in terms of peak frequency deviation as "wideband" and "narrowband." Wideband FM, with peak deviation of ± 75 kHz, is used when high-fidelity signals are to be transmitted, such as in FM broadcasting and television sound. Narrowband FM, with peak deviations of ± 15 kHz, is employed by the so-called FM mobile communication services, such as police, ambulances, taxicabs, and so on.

A typical FM transmitter based on the direct-modulation concept is shown in Fig. 48.71. The preemphasis filter, preceding the peak limiter, forces the higher audio frequencies to be limited first, thus giving the same peak deviation for all audio frequencies. The voltage-controlled oscillator (VCO), which is phase-locked to the automatic frequency control (AFC) crystal oscillator, generates the FM-modulated carrier. The bandpass filter, located between the power amplifier and the antenna, serves to attenuate any spurious signals that may result from coupling of signals from other nearby transmitters.

The configuration of a phase modulation (PM) transmitter is similar to that of the FM transmitter, with two specific differences. The preemphasis filter is not used in the PM transmitter, and the filtered modulation signal is applied to the VCO circuit in such a way that the phase, rather than the frequency of the oscillator is varied. The FM transmitter of Fig. 48.71 could be made into a PM transmitter (conceptually) by replacing the preemphasis filter with a differentiator.

SSB Transmitter

There are basically three practical methods for generating single sidebands: the filter method, the phase-shift method, and the Weaver method. All three use the balanced modulator to suppress the carrier, but each uses a different technique for removing the unwanted sideband.

The filter method is the simplest of the three, in which the unwanted sideband is removed by a bandpass filter located at the output of the balanced modulator. The filter may be LC, crystal, or mechanical, with the mechanical type possibly the most popular. A block diagram of a filter-type SSB transmitter is shown in Fig. 48.72. The balanced mixer in the figure serves a twofold purpose. On the one hand it raises the frequency to a value desired for transmission, and on the other hand it makes it possible to recover the sum frequency (of sideband filter output and oscillator ♯2) with a tuned filter.

The phase-shift method makes use of two balanced modulators and two phase-shifting networks, as shown in Fig. 48.73. One modulator receives the oscillator voltage (shifted by 90°) and the

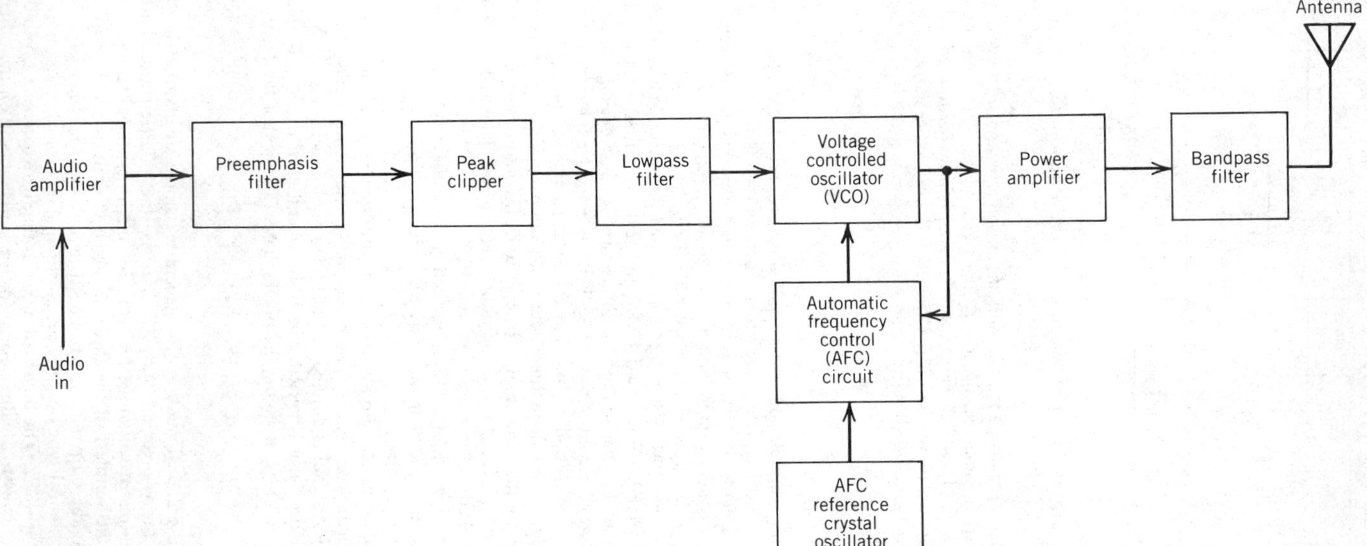

Fig. 48.71 Block diagram of a typical FM transmitter using the direct-modulation concept.

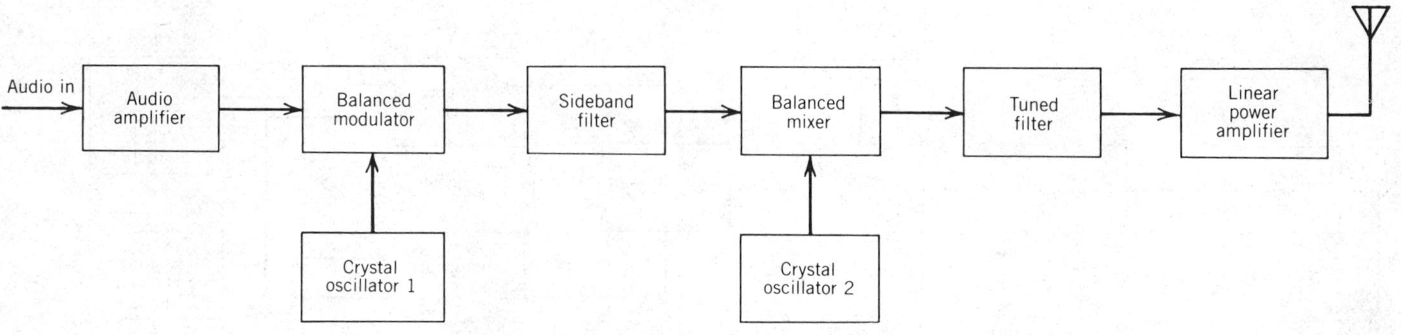

Fig. 48.72 Block diagram of a filter-type single-sideband (SSB) transmitter

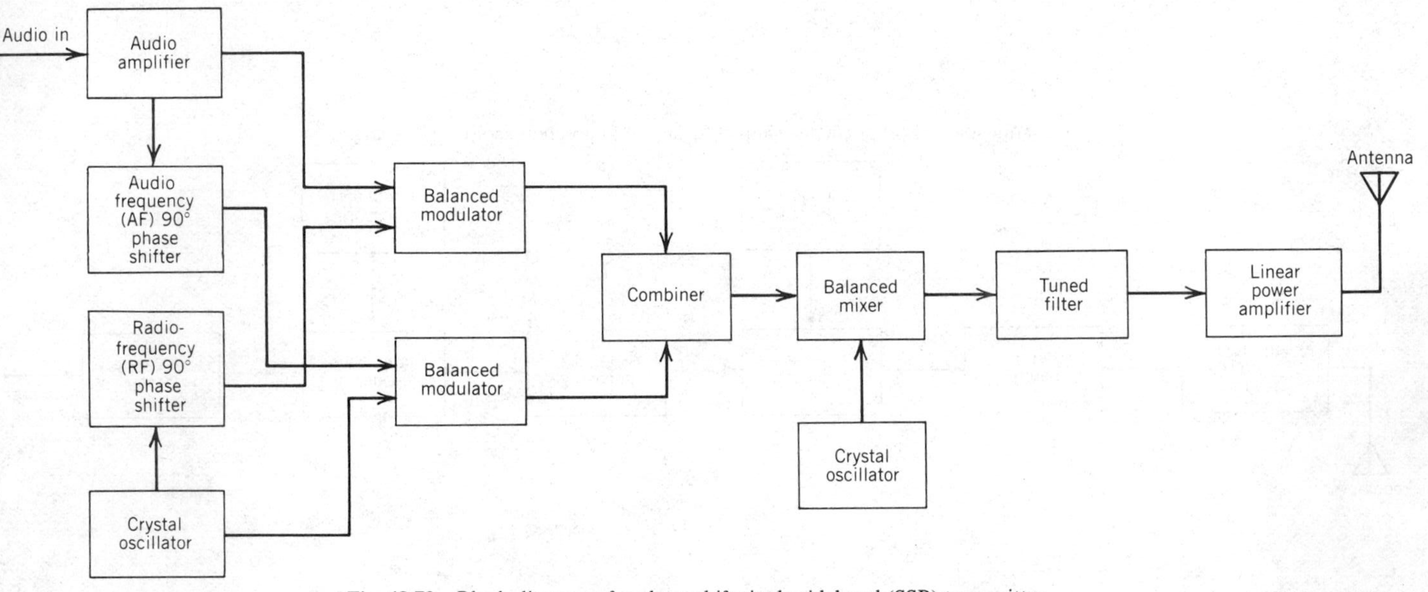

Fig. 48.73 Block diagram of a phase-shift single-sideband (SSB) transmitter.

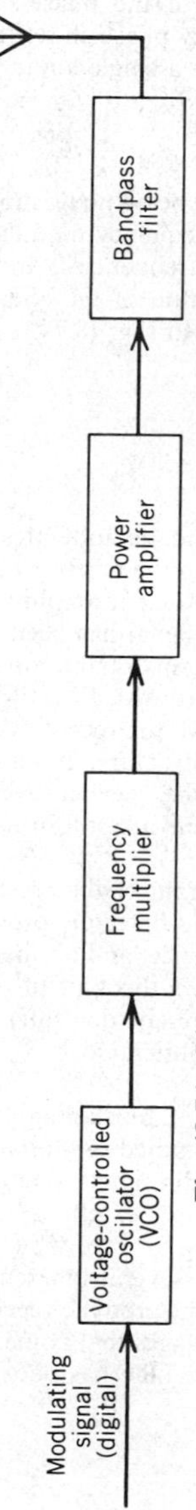

Fig. 48.74 Block diagram of a frequency shift keying (FSK) transmitter.

modulating voltage, while the other modulator receives the modulating voltage (shifted by 90°) and the oscillator voltage. Both modulators produce an output consisting only of sidebands. However, referenced to the oscillator voltage, both upper sidebands lead the reference by 90° while only one of the lower sidebands does this, the other lagging it by 90°. The two lower sidebands are thus 180° out of phase, and when added in the combiner they cancel each other out. The upper sidebands are in phase and therefore will add at the combiner, resulting in a SSB signal where the lower sideband has been canceled. If generation of the lower sideband is desired, one modulator must have both inputs shifted by 90° while the other modulator must have only unshifted inputs.

The Weaver method is an extension of the phase-shift method. It removes the cumbersome requirement of a wideband audio-frequency phase shifter and replaces it by two additional balanced modulators, one audio-tone generator, and a single audio-frequency phase shifter.[1]

FSK Transmitter

Frequency shift keying (FSK) is one of the best-known modulating schemes currently in use for radio telegraphy. FSK may be thought of as a frequency-modulation system in which the carrier frequency is midway between the mark and space frequencies, and in which the modulating waveform is a digital signal. As a result, the block diagram of an FSK transmitter is a shortened version of the normal FM transmitter diagram, as shown in Fig. 48.74.

48.2-5 Receiver Fundamentals

Classification of Receivers

A receiver is any device that accepts and demodulates an RF signal to extract the transmitted information. Since the signal at the input to the receiver ordinarily has extremely low level, it must be amplified by the receiver before it is of sufficient amplitude for practical use. The amplification may be accomplished before and/or after the signal has been demodulated.

Receivers are normally classified in terms of the modulation in the signal they are designed to receive and demodulate. Therefore there are AM, FM, PM, and various other receivers. Furthermore, within one modulation scheme such as AM, the receivers may be further subdivided according to the type of equipment used in the receiver structure. In this sense one can make distinction between crystal detector, tuned RF, superheterodyne, regenerative, and superregenerative receivers. Of these, only the tuned RF and the superheterodyne are still in use today.

Crystal Detector Receiver. A block diagram of the crystal detector receiver is shown in Fig. 48.75. The bandpass filter, located ahead of the detector, provides some receiver selectivity. The crystal detector is a small-signal or square-law device, and all the amplification in the receiver is provided by the audio amplifier. The single advantage in this type of receiver concerns its hardware simplicity (no local oscillator, no tuned amplifiers, square-law detector). Among its disadvantages are poor sensitivity and the requirement of high audio amplification.

Tuned Radio-Frequency (TRF) Receiver. A block diagram of the tuned RF receiver is shown in Fig. 48.76. Several cascaded RF amplifiers, all tuned to the same frequency, amplify the received signal to a level suitable for detection. Then the audio signal is detected, amplified, and fed to the loudspeaker. Without loss in generality, the detected signal can be a video or telemetry signal that is fed to a video monitor or some recording device.

The TRF receiver is characterized by several inherent advantages and disadvantages, compared with the nowadays more customary superheterodyne receiver. Simplicity of design and the absence of image frequency problems are two advantages that come readily to mind. On the other hand, the list of disadvantages is much larger. First, the TRF has poor selectivity at high frequencies, owing to the

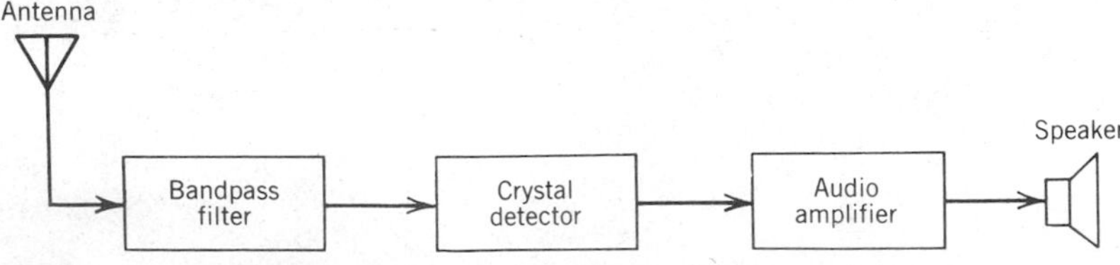

Fig. 48.75 Block diagram of a crystal detector receiver.

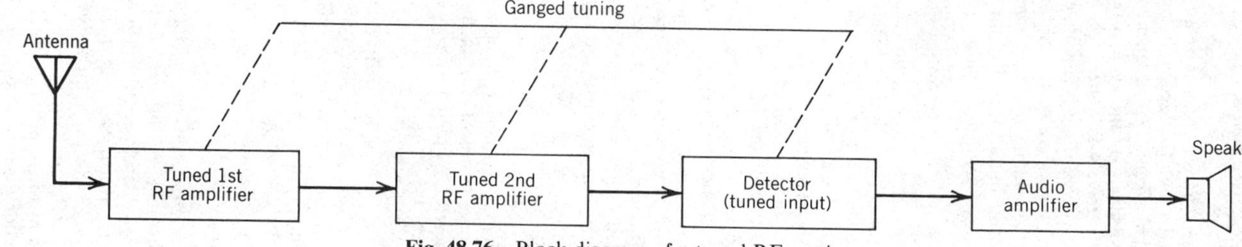

Fig. 48.76 Block diagram of a tuned RF receiver.

enforced use of single-tuned circuits. Poor selectivity in turn results in adjacent channel interference. Second, for optimum gain all the tank circuits in the various RF stages must be tuned to the same frequency throughout the tuning range. Such exactness or perfect "tracking" is not possible in practice. Therefore a variation in gain or sensitivity as a function of tuning frequency results. Third, the risk of instability is quite real at high frequency whenever high gain is being achieved by a multistage amplifier. In this case positive feedback due to stray capacitance or stray paths can put the amplifier into oscillation and render the receiver useless.

Superregenerative Receiver. A superregenerative receiver is an RF amplifier or detector having sufficient positive feedback to cause oscillation. The receiver is made to go in and out of oscillation by a control signal known as the "quench" signal. Typical quench-signal frequencies are between 10 kHz and 1 MHz. Very high gains are possible with this type of receiver.

The radiation of superregenerative receivers is quite high, so problems may arise if several receivers must operate in the vicinity of one another. This type of receiver is not well suited for normal reception of several channels, but may serve quite well for small beacons, transponders, and remote-control applications. During a quenching cycle, a sample of the signal is detected. A modulated carrier will result in a constant output signal if the quench frequency is equal to the modulation frequency. Otherwise the output frequency will equal the modulation frequency minus the quench frequency.[8]

Superheterodyne Receiver. Practically every form of radio receiver in use today is based on the superheterodyne concept. It can be found in such diverse applications as communication, television, and radar receivers, with only slight modifications in principle. Although the superheterodyne receiver is more complex than other types, it generally enjoys greater sensitivity and selectivity.

A block diagram of a superheterodyne receiver is shown in Fig. 48.77. The basic principle is the conversion of the RF signal to an intermediate frequency by heterodyning the RF signal with a local oscillator whose frequency differs from that of the RF signal by an amount equal to the desired intermediate frequency. Tuning the receiver consists of simultaneously tuning the local oscillator, mixer, and RF amplifier so that the intermediate frequency remains constant.

The IF amplifier generally uses a large number of double-tuned circuits operating at a constant frequency to provide most of the gain and bandwidth requirements of the receiver. As a result of the lower frequency and constant tuning of the IF amplifier, the sensitivity and selectivity are essentially constant throughout the tuning range of the receiver. The RF section is used mainly to trap the desired frequency, reject "image frequency" interference, and reduce receiver noise figure.

Receiver Performance

Irrespective of the particular application of a receiver, four main factors govern its performance: sensitivity, selectivity, fidelity, and noise figure. These factors depend on the combination of amplification, phase and amplitude linearity, frequency response, stability, and noise in the various stages of which the receiver is composed.

Sensitivity. The sensitivity of a radio receiver is its ability to amplify weak signals. Sensitivity is defined in terms of the voltage that must be applied to the receiver input terminals to give a standard output power, measured at the output terminals. A sensitivity of about 50 μV is typical of many AM broadcast-band receivers. Quality communication receivers may have sensitivities below 1 μV in the HF band. The sensitivity level must be chosen larger than the noise level generated within the receiver. Otherwise the receiver output will contain more noise than intelligence, and thus part of the amplification is wasted.

Selectivity. The selectivity of a receiver is its ability to reject all unwanted signals. It is a function of the frequency response of the tuned circuits ahead of the detector. Graphically it can be represented by a curve that shows the increase in signal level needed to maintain the standard output for various input frequencies on either side of the receiver resonant frequency. Such a selectivity curve is shown in Fig. 48.78. Caution must be exercised to not make the selectivity too sharp and thereby threaten the integrity of the sideband energy of the desired signal.

Selectivity, in general, is determined by the response of the IF section, with the RF section playing a small but important role. The RF section determines the front-end selectivity of the receiver, which is necessary for rejection of the "image frequency." The latter frequency is given by

$$f_{s1} = f_s + 2f_i \tag{48.108}$$

where f_s is the desired signal frequency and f_i is the IF frequency. An interference signal of frequency

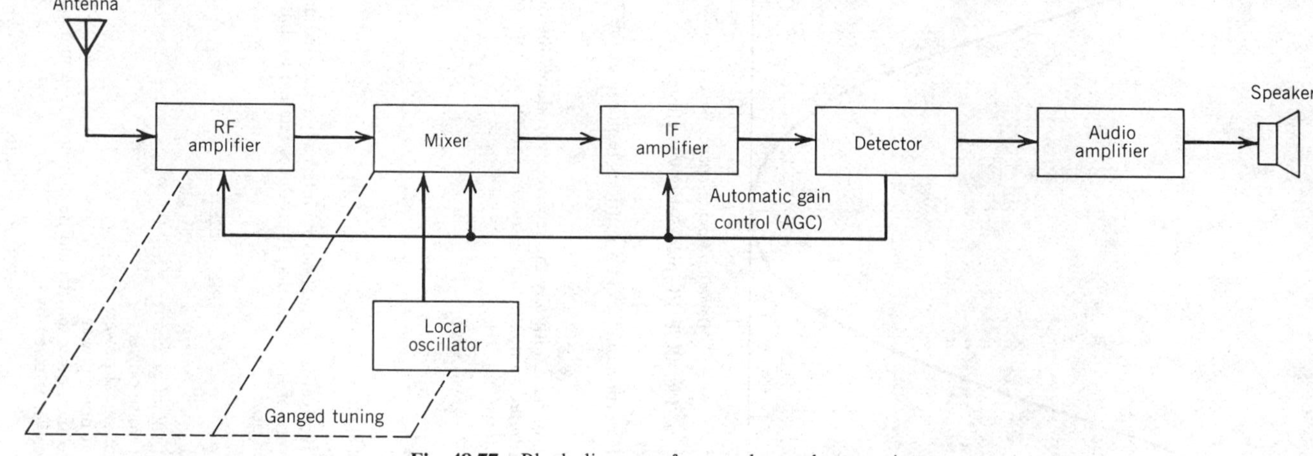

Fig. 48.77 Block diagram of a superheterodyne receiver.

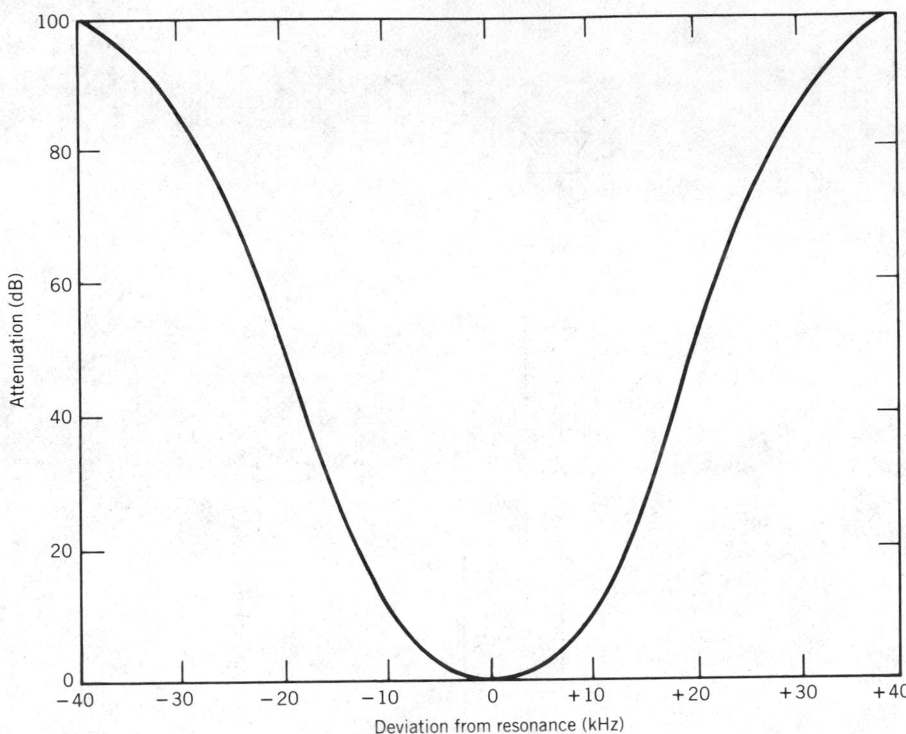

Fig. 48.78 Typical selectivity curve.

f_{s1}, if unchecked by the RF portion, will yield a spurious response at the IF stage and compromise receiver operation.

Fidelity. This is a measure of how faithfully the receiver reproduces the original intelligence signal. Loss of fidelity can be attributed to nonlinearities in the various sections of the receiver, which can cause undesired effects, such as

1. Undesired modulation.
2. Modulation distortion.
3. Cross modulation.
4. Generation of spurious signals.

Noise Figure. An "ideal" receiver is one that has no internally generated noise. The noise figure of a practical receiver is a measure of the noise that it produces relative to that of an ideal receiver. The noise figure may be defined as

$$F = \frac{S_{in}/N_{in}}{S_{out}/N_{out}} \tag{48.109}$$

where S_{in} = available input signal power
N_{in} = available input noise power
S_{out} = available output signal power
N_{out} = available output noise power

"Available power" refers to the power that would be delivered to a matched load. The noise power, N_{in}, at the input to the receiver is given by

$$N_{in} = KT_0B \quad \text{watts} \tag{48.110}$$

where $K = 1.38 \times 10^{-23}$ J/K° = Boltzmann's constant
$T_0 = 290$ K = ambient temperature
B = noise bandwidth

Using the receiver gain definition $G = S_{out}/S_{in}$ and substituting the noise power input N_{in} from Eq. (48.110) into Eq. (48.109) yields

$$F = \frac{N_{out}}{KT_0BG} \tag{48.111}$$

From Eq. (48.111), the noise figure may be interpreted as the ratio of the actual available output noise power to the noise power that would be available if the receiver had no internally generated noise. The noise figure may also be written as

$$F = 1 + \frac{\Delta N}{KT_0BG} \tag{48.112}$$

where ΔN is the noise power introduced by the receiver itself. The effective noise temperature of the receiver is defined as that temperature T_e at the input to the receiver that would account for the added noise ΔN at the output. Therefore

$$\Delta N = KT_eBG$$

and

$$F = 1 + \frac{T_e}{T_0} \tag{48.113}$$

or

$$T_e = (F - 1)T_0 \tag{48.114}$$

For an ideal receiver, $F = 1$ (0 dB); $T_e = 0$ K. For $F = 2$ (3 dB); $T_e = 290$ K.

AM Receivers

AM receivers, in use today, have basically the same structure as the superheterodyne receiver already described. Certain aspects of the block diagram in Fig. 48.77 will now be treated in the AM modulation context.

In standard-broadcast AM receivers, the carrier tuning range extends from 540 to 1650 kHz with a constant intermediate frequency of 455 kHz. For the usual case of local oscillator frequency above carrier frequency, the corresponding frequency range of the local oscillator is between 995 and 2105 kHz, giving a ratio of maximum-to-minimum frequencies of 2.2 : 1, which can be achieved by variable-capacitor tuning.

To recover the intelligence signal, an AM receiver uses an envelope detector. In its simplest form, the envelope detector has the circuit diagram shown in Fig. 48.79. The half-wave rectified signal at the output of the diode is loaded by the parallel *RC* circuit. At each positive peak of the RF cycle, *C* charges up to the peak signal voltage E_s. In the short time between RF peaks, the capacitor *C* discharges slightly into resistor *R* only to be recharged on the next positive peak. The result is the voltage E_0, which reproduces the modulating voltage accurately except for the small amount of RF ripple. Note that the *RC* time constant must be long to keep the RF ripple small and short enough to follow rapid variations in the modulating signal. Envelope detectors, used in practice, incorporate features to eliminate the DC component and the small RF ripple from the detected waveform.

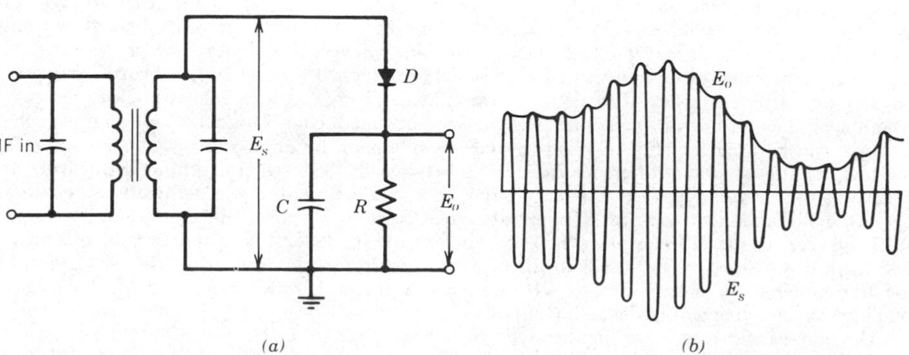

Fig. 48.79 Simple envelope detector: (*a*) Circuit diagram, (*b*) input and output voltages.

Another aspect of receiver design that must be mentioned at this point is the concept of automatic gain control. AGC, as it is called, is a technique by which the overall gain of a radio receiver is varied automatically with the changing strength of the received signal, to keep the output substantially constant. This is accomplished by means of a DC bias voltage, derived by the detector, and applied to a selected number of RF, IF, and mixer stages. AGC allows the tuning of stations having great disparity in carrier signal strengths without appreciable change in the level of the output signal. Also it helps to smooth out the rapid fading that may occur with long-distance short-wave reception and prevents the overloading of the last IF amplifier stage in the case of excessively strong received signals.

FM Receivers

FM communication channels are found throughout the frequency spectrum. Regardless of frequency, an FM receiver has a superheterodyne structure similar to its AM counterpart. A number of differences with the AM receiver exist, as can be seen from the block diagram in Fig. 48.80. Unique features of the FM receiver include amplitude limiting, deemphasis, a different demodulation technique, and a different method to generate AGC. Generally the IF frequency chosen is a function of the carrier frequency band and the class of communication service. For example, standard-broadcast receivers operating in the frequency range of 88 to 108 MHz use an IF frequency of 10.7 MHz.

All FM demodulators are preceded by an amplitude limiter that eliminates amplitude variations in the IF signal. These variations, if not removed, can lead to distortion in the demodulator output.

The basic function of an FM demodulator is as a frequency-to-amplitude converter. It translates the frequency deviation of the incoming IF signal into an amplitude variation that is a reproduction of the intelligence signal. The balanced slope detector, the Foster-Seeley discriminator, and the ratio detector are three of the best known FM demodulators.

The balanced slope detector splits the FM signal into two 180° out-of-phase components and feeds each one to a separate tuned circuit. The two tuned circuits have resonances above and below the center frequency, respectively. Their outputs are passed through separate diode detectors and then summed together to yield the final demodulator output.

The Foster-Seeley discriminator employs two tuned circuits that are both resonant at the center frequency. The frequency deviation in this case is translated into a phase difference between the outputs of the two tuned circuits. Because of its dependence on phase relations, this demodulator is also known as a phase discriminator. It is much easier to align and provides better linearity than the balanced slope detector.

The ratio detector operates on the same principle as the Foster-Seeley discriminator but has the advantage of possessing an inherent voltage-limiting capability, which renders prior limiting unnecessary.

SSB Receivers

SSB receivers are often required to operate in crowded bands where the attenuation of adjacent channel interference is of paramount importance. For this reason they are usually of the double-conversion variety, incorporating two mixers and generating two intermediate frequencies. The high first IF pushes the image frequency further away from the signal frequency, and therefore permits much better attenuation of it. The low second IF, on the other hand, has all the advantages of a low fixed operating frequency, particularly sharp selectivity and hence good adjacent-channel rejection.

To demodulate the SSB signal at the output of the second IF amplifier, SSB receivers generally use one of two methods: the product detector and the balanced modulator. The product detector resembles an ordinary mixer in which the SSB signal is multiplied with the nominal carrier frequency or the pilot-carrier frequency, if the latter is transmitted. Then the low-frequency audio is separated from the product terms by lowpass filtering. The balanced modulator is normally used in transceivers where it can be used for both modulation and demodulation by appropriate switching. As in the product detector, sum and difference frequencies between the SSB signal and a nominal carrier are generated and then only the low-frequency terms, or audio, are retained and delivered to the output.

A block diagram of the SSB suppressed-carrier receiver is shown in Fig. 48.81. The upper-sideband receiver uses a frequency synthesizer to generate stable local oscillators for the two mixers and the product detector. The oscillator input to the first mixer is in steps of 1 kHz, which requires the transmitter to be channeled at 1 kHz frequencies for proper reception. The AGC voltage is obtained by rectifying part of the audio output.

The pilot-carrier receiver is another type of SSB receiver in which the recovered pilot carrier serves two purposes. First, it is used as a reference for an automatic frequency control circuit to ensure good overall frequency stability. Second, since the pilot carrier amplitude varies with the strength of the input signal, a suitable AGC signal can be obtained from it by simple rectification.

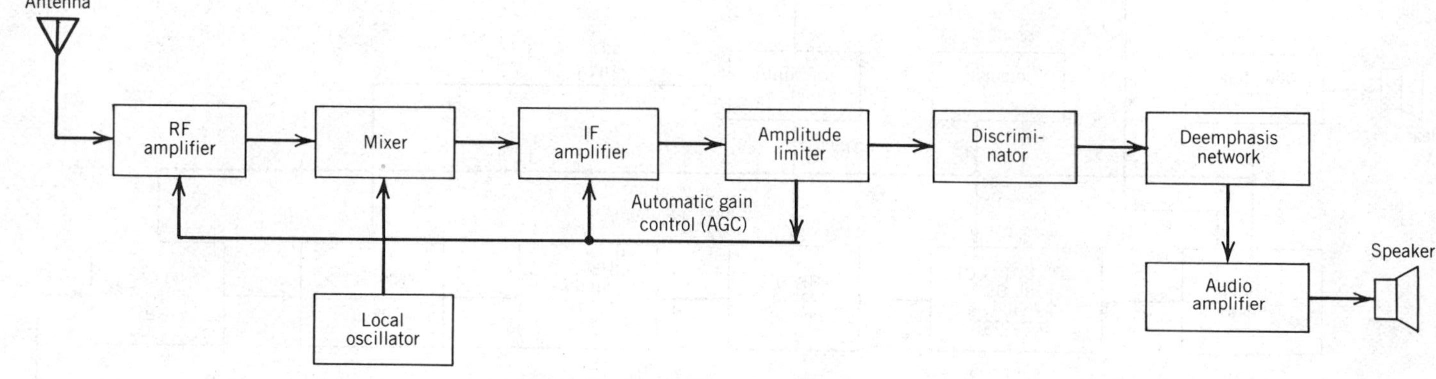

Fig. 48.80 Block diagram of an FM receiver.

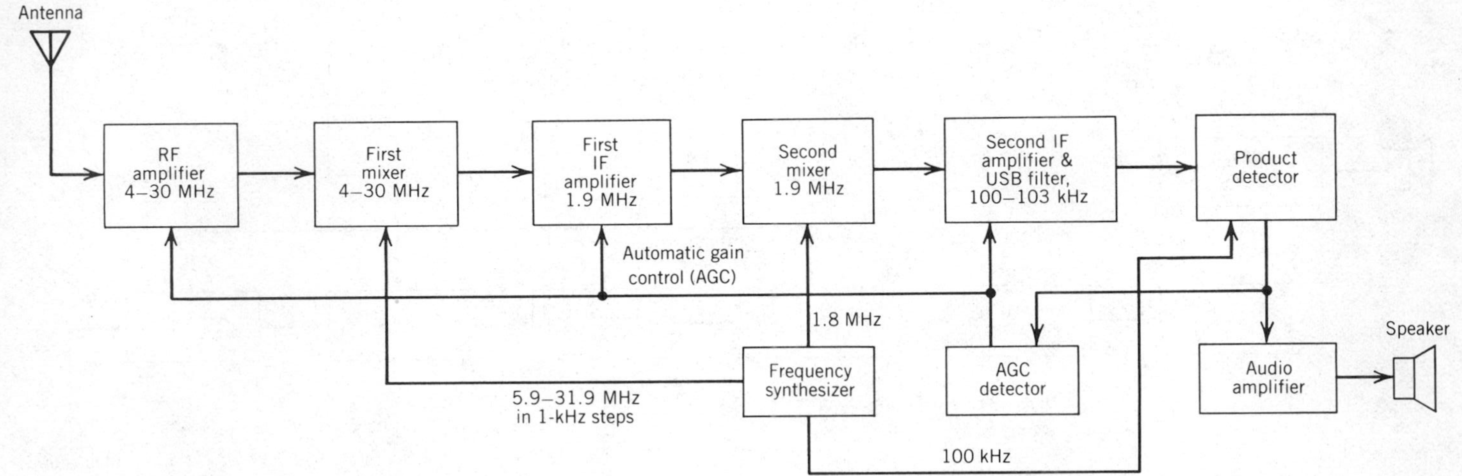

Fig. 48.81 Block diagram of a single sideband suppressed-carrier (SSBSC) receiver.

48.2-6 Total Radio Communication Link

Communication Link Analysis

The end-to-end performance of any communication link can be expressed in terms of the signal-to-noise ratio (SNR) evaluated in an appropriate bandwidth at some point in the receiving system. In an analog system in which noise bandwidth is generally greater than signal bandwidth, we often speak of the average carrier power-to-noise ratio or (P_r/N) as the SNR of particular interest.

$$\frac{P_r}{N} = \frac{\text{EIRP}(G_r/T_e)}{L_p L_{oe} L_{oi} KB} \tag{48.115}$$

The various parameters in Eq. (48.115) can be defined with the help of the link model given in Fig. 48.82,

where P_r = received signal power at detector input, W
$N = KT_eB$ = thermal noise power at detector input, W
K = Boltzmann's constant, 1.38×10^{-23} J/K
T_e = equivalent system noise temperature, K
B = detector input bandwidth, Hz
EIRP = P_tG_t = equivalent isotropically radiated power, W
P_t = transmitted power, W
G_t = transmitter antenna gain, dimensionless
G_r = receiver antenna gain, dimensionless
G_r/T_e = figure of merit gain-to-equivalent system noise temperature ratio, K^{-1}
$L_p = (4\pi d/\lambda)^2$ = free space loss as defined earlier
L_{oe} = other external losses = atmospheric loss + antenna polarization loss + antenna pointing loss
L_{oi} = other internal losses = transmit circuit loss + receive circuit loss + intermodulation noise loss

In a digital system in which the signal bandwidth is taken to be equal to the noise bandwidth, link performance is expressed in terms of the received signal power-to-noise spectral density ratio (P_r/N_0).

$$\frac{P_r}{N_0} = \frac{\text{EIRP}(G_r/T_e)}{L_p L_{oe} L_{oi} K} \tag{48.116}$$

where $N_0 = N/B$ = noise spectral density in watts per hertz.

If we assume that all the received power stems from the modulating signal (suppressed carrier), then we can evaluate the link performance in terms of the bit energy-to-noise spectral density ratio. We can write

$$\frac{P_r}{N_0} = \left(\frac{E_b}{N_0}\right)R \tag{48.117}$$

Then

$$\frac{E_b}{N_0} = \frac{\text{EIRP}(G_r/T_e)}{L_p L_{oe} L_{oi} KR} \tag{48.118}$$

where R is the information data rate in bits per second. If the carrier power is not negligible we can still use Eq. (48.118) provided we reflect the carrier power as a loss within the parameter L_{oi}. In decibels, Eq. (48.118) can be expressed as follows:

$$\frac{E_b}{N_0}(\text{dB}) = \text{EIRP(dBW)} + \frac{G_r}{T_e}(\text{dB/K}) - L_p(\text{dB}) - L_{oe}(\text{dB}) - L_{oi}(\text{dB}) - K(\text{dBW/K} - \text{Hz}) - R(\text{dB Hz}) \tag{48.119}$$

The E_b/N_0, as defined in Eq. (48.118) refers to the required E_b/N_0 necessary for the achievement of a certain bit error probability, P_b, at the detector output. In practice, the actual E_b/N_0 is related to the required E_b/N_0 by a safety factor M, which is commonly known as the link margin.

$$\left(\frac{E_b}{N_0}\right)_{\text{actual}} = M\left(\frac{E_b}{N_0}\right)_{\text{req}} \tag{48.120}$$

In decibels, the link margin is simply the difference between the actual and required values of E_b/N_0.

$$M(\text{dB}) = \left.\frac{E_b}{N_0}\right|_{\text{actual}}(\text{dB}) - \left.\frac{E_b}{N_0}\right|_{\text{req}}(\text{dB}) \tag{48.121}$$

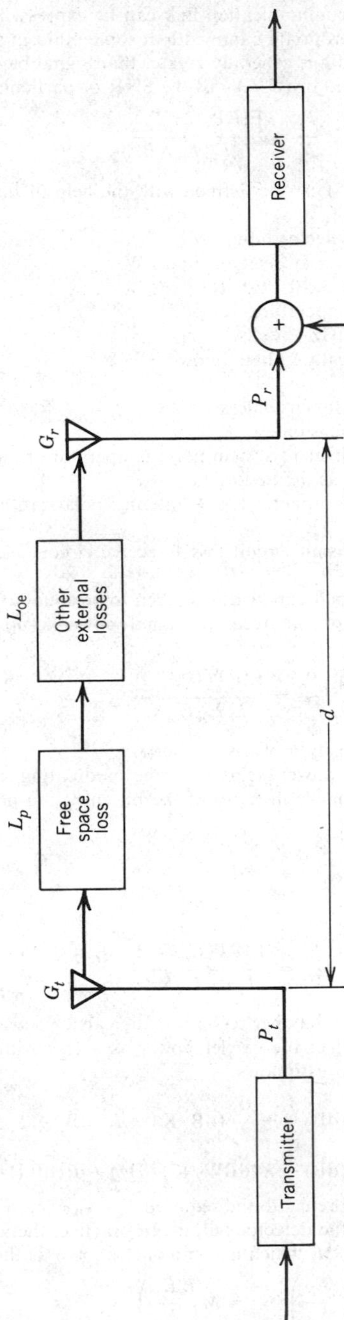

Fig. 48.82 Simplified communication link model.

For a given bit error probability, the required E_b/N_0 is a function of the modulation and coding techniques used on the link. The amount of link margin to use depends to a large extent on the statistical nature of the link and on our ability to predict all sources of gain, loss, and noise phenomena. Link margin values ranging from 0 to 6 dB have been used in practice. This reflects the wide variation in the level of certainty with which the various communication links can be modeled.

Example of Link Budget Calculation

As an illustration of a link budget calculation, consider the candidate space station (SS) to tracking and data relay satellite (TDRS) communication system shown pictorially in Fig. 48.83. Initial concepts call for the system to support three separate communication links at three frequency bands (S, Ku, and W). We shall restrict our discussion here to the Ku-band communication links, as an example.

The space station is planned to be put into service in a low earth orbit about 270 nautical miles above the surface of the earth. It will communicate, among other things, with the TDRS, which is situated in a geosynchronous orbit at about 22,000 nautical miles above the earth's surface. The TDRS will act as a relay station between the space station and satellite earth stations on the ground. Figure 48.84 illustrates this relay function and separates the Ku-band system into forward (ground to space station) and return (space station to ground) links.

Forward Link. The link budget for the forward link is given in Table 48.10, using the parameter definitions and link equations presented previously. The calculation serves as an illustrative example of link budget calculations. There are three partial sums included (shown underlined in the table). The strategy is to determine the first partial sum (item 10), then the second partial sum (item 12), and

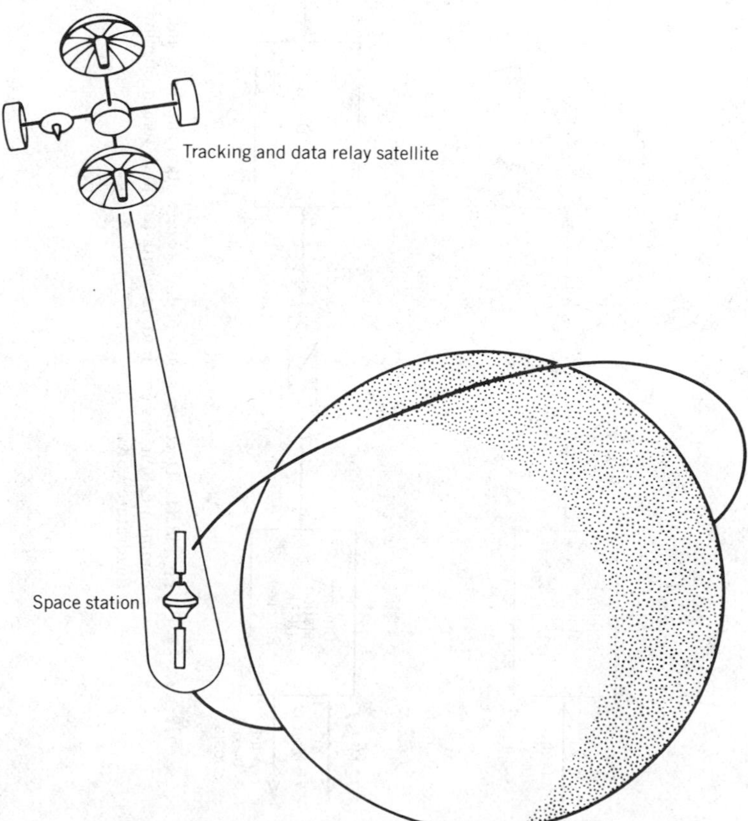

Fig. 48.83 Proposed space station to tracking and data relay satellite communication system.

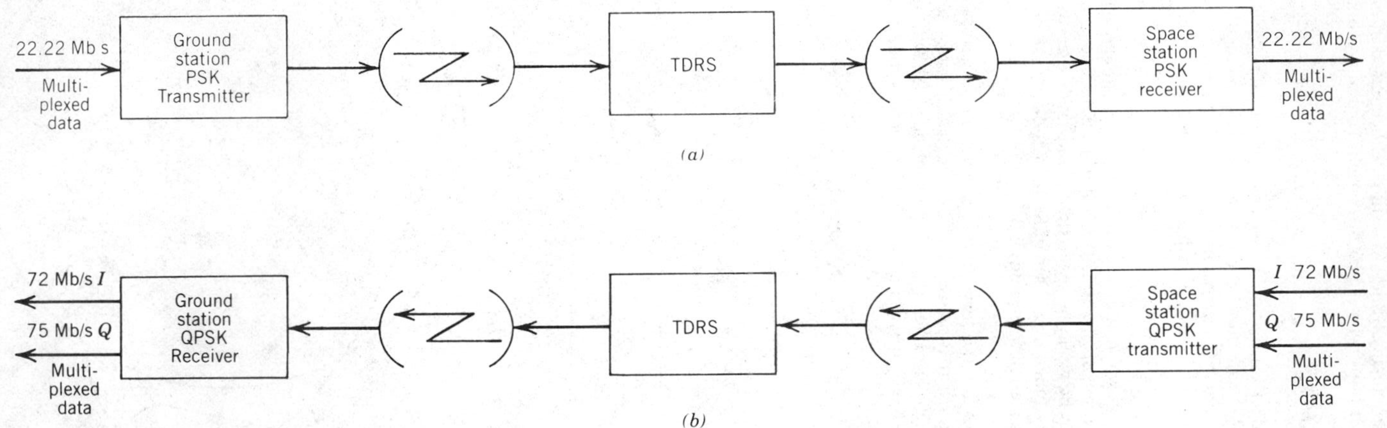

Fig. 48.84 Decomposition of the space station/tracking and data relay satellite (TDRS) Ku-band communication system into forward (*a*) and return (*b*) links. PSK, phase shift keying; QPSK, quadrature PSK.

TABLE 48.10. FORWARD KU-BAND LINK BUDGET OF THE CANDIDATE SPACE STATION (SS) TO TRACKING AND DATA RELAY SATELLITE (TDRS) COMMUNICATION SYSTEM

Item	Variable	Sum
1	TDRS EIRP	46.5 dBW
2	Free space loss L_p ($f = 13.775$ GHz, $d = 38,000$ km)	206.8 dB
3	Antenna pointing loss	0.5 dB
4	Antenna polarization loss	0.2 dB
5	SS receive antenna gain G	G dB
6	SS receive system noise temperature T_e (440 K)	26.4 dBK
7	SS G/T_e	$G - 26.4$ dBK
8	Boltzmann's constant K	-228.6 dBW/K-Hz
9	SS receive circuit loss	2.5 dB
10	SS P_r/N_0	$38.7 + G$ dBHz
11	Bit rate R (22.22 Mbs)	73.5 dBHz
12	Actual E_b/N_0	$-34.8 + G$ dB
13	Required E_b/N_0 (for 10^{-5} BER[a])	12.3 dB
14	Link margin M[b]	$-47.1 + G$ dB

[a] BER, bit error rate.

[b] For $M = 3$ db, $G = 50.1$ dB.

then the final sum (item 14) in this order. But before a partial sum can be computed, its various elements must be computed (if need be) and/or converted to decibel units. For the first partial sum (item 10), the first element to be calculated is "free space loss" (item 2). We have from Eq. (48.81):

$$L_p(\text{dB}) = 32.5 + 20 \log f + 20 \log d = 206.8 \text{ dB}$$

The second item to be calculated is the "SS receive system noise temperature" (item 6), which is specified as 440 K corresponding to a receiver with 4-dB noise figure.

$$T_e(\text{dB/K}) = 10 \log(440) = 26.4 \text{ dB/K}$$

The next element is the "SS G/T_e" (item 7), which when converted to decibels gives

$$10 \log(G/T_e) = 10 \log G - 10 \log T_e$$

$$G/T_e(\text{dB/K}) = G_{\text{dB}} - 26.4 \text{ dB/K}$$

Next we have to evaluate the Boltzmann's constant (item 8) in decibels. We have

$$K = 1.38 \times 10^{-23} \text{J/K} = -228.6 \text{ dBW/K/Hz}$$

Now we are ready to calculate the first partial sum, "SS P_r/N_0" (item 10). We have:

$$\begin{aligned} \text{SS } P_r/N_0 \text{ (dBHz)} &= \text{TDRS EIRP (dBW)} - \text{free space loss (dB)} \\ &\quad - \text{antenna pointing loss (dB)} - \text{antenna polarization loss (dB)} \\ &\quad + \text{SS } G/T_e \text{ (dB/K)} - \text{Boltzmann's constant (dBW/K/Hz)} \\ &\quad - \text{SS receive circuit loss (dB)} \qquad (48.122) \\ &= 46.5 - 206.8 - 0.5 - 0.2 + (G - 26.4) - (-228.6) - 2.5 = 38.7 + G \end{aligned}$$

The second partial sum requires the evaluation of the bit rate in decibels before it can be calculated. The bit rate R is given as 22.22 Mb/s on the forward link, or

$$R(\text{dBHz}) = 10 \log(22.22 \times 10^6) = 73.5 \text{ dBHz}$$

The second partial sum (difference in reality), "actual E_b/N_0" (item 12) is then given by

$$\begin{aligned} \text{Actual } E_b/N_0 \text{ (dB)} &= \text{SS } P_r/N_0 \text{ (dBHz)} - R \quad \text{dBHz} \qquad (48.123) \\ &= 38.7 + G - 73.5 = -34.8 + G \text{ (dB)} \end{aligned}$$

The final sum (actually difference) gives the link margin M (item 14).

$$M(\text{dB}) = \text{actual } E_b/N_0 \text{ (dB)} - \text{required } E_b/N_0 \text{ (dB)} \qquad (48.124)$$

$$M(\text{dB}) = -34.8 + G - 12.3 = -47.1 + G \text{ (dB)}$$

TABLE 48.11. RETURN KU-BAND LINK BUDGET OF THE CANDIDATE SPACE STATION (SS) TO TRACKING AND DATA RELAY SATELLITE (TDRS) COMMUNICATION SYSTEM

Item	Variable	Sum
1	SS transmit power P_t	P_t dBW
2	SS transmit circuit loss	2.5 dB
3	SS transmit antenna gain G	50.1 dB
4	SS EIRP	$47.6 + P_t$ dBW
5	Free space loss L_p ($f = 15.0034$ GHz, $d = 38,000$ km)	207.6 dB
6	Antenna pointing loss	0.5 dB
7	Antenna polarization loss	0.2 dB
8	TDRS G/T_e	23.1 dB/K
9	Boltzmann's constant K	− 228.6 dBW/K-Hz
10	TDRS intermodulation and other internal losses	7.0 dB
11	TDRS P_r/N_0	$84.0 + P_t$ dBHz
12	Bit rate R (75 Mb/s)	78.8 dBHz
13	Actual E_b/N_0	$5.2 + P_t$ dB
14	Required E_b/N_0 (for 10^{-5} BER[a])	13.5 dB
15	Link margin M[b]	$-8.3 + P_t$ dB

[a] BER, bit error rate.

[b] For $M = 3$ dB, $P_t = 11.3$ dBW (13.5 W).

If a link margin of 3 dB is specified, the unknown "SS receive antenna gain," G, can be determined.

$$M = 3 \text{ dB} = -47.1 + G$$

and

$$G = 3 + 47.1 = 50.1 \text{ dB}$$

Return link. The link budget for the return link is given in Table 48.11. Basically four partial sums are to be evaluated, as delineated by underlining in the table. The first partial sum is the "SS EIRP" (item 4) which is given by

$$\begin{aligned} \text{SS EIRP (dBW)} &= \text{SS transmit power } P_t \text{ (dBW)} - \text{SS transmit circuit loss (dB)} \\ &\quad + \text{SS transmit antenna gain (dB)} \\ &= P_t - 2.5 + 50.1 = 47.6 + P_t \quad \text{dBW} \end{aligned} \tag{48.125}$$

where the assumption is made that the SS transmit and receive antenna gains are equal to 50.1 dB, as was determined in the forward link budget.

The remainder of the return link budget calculation follows the same steps outlined for the forward case. In the return calculation, however, the link margin is calculated in terms of the unknown "SS transmit power," P_t. By selecting a link margin of 3 dB, the value of P_t can be determined:

$$M = 3 \text{ dB} = -8.3 + P_t$$

$$P_t = 3 + 8.3 = 11.3 \text{ dBW (13.5 W)}$$

References

1 G. Kennedy, *Electronic Communication Systems*, McGraw-Hill, New York, 1970.

2 *Reference Data for Radio Engineers*, 5th ed., Howard W. Sams, Indianapolis, 1968.

3 P. F. Panter, *Communication Systems Design*: *Line-of-Sight and Tropo-Scatter Systems*, McGraw-Hill, New York, 1972.

4 A. W. Rudge, K. Milne, A. D. Olver, and P. Knight, eds., *The Handbook of Antenna Design*, vol. 1, Peter Peregrinus, London, 1982.

5 C. A. Balanis, *Antenna Theory Analysis and Design*, Harper & Row, New York, 1982.

6 L. V. Blake, *Antennas*, Wiley, New York, 1966.

7 J. J. DeFrance, *Communications Electronics Circuits*, Holt, Rinehart and Winston, New York, 1966.

8 L. J. Giacoletto, *Electronics Designers Handbook*, McGraw-Hill, New York, 1977.

48.3 WIRE COMMUNICATIONS

Matthew J. Quinn, Jr.

Modern data communications involves the communication of data between computers and between elements of data networks. (See Section 49.3, Network Communications.) Data communications are essential to the operation of distributed processing networks, which have become more and more common with the introduction of smaller and more economical computers. Since the nationwide telephone system interconnects approximately 125 million telephones in the United States, use of this system to interconnect computers and data networks is logical. The techniques and devices for using the telephone system as a data communications network are the subject of this section. Since the telephone system was once interconnected totally by wires, these techniques are commonly referred to as wire communications, even though microwave links and satellites are now common to the system.

48.3-1 Definitions

A modem is a device used to connect a data source or receiver to a telephone line. The name "modem" is derived from the names of the two functions the device performs, modulation and demodulation. (A general treatment of modulation and demodulation techniques is found in Section 48.1.) Modems are discussed in more detail later in this section.

Data communications links may be classified as simplex, full-duplex, or half-duplex mode. In the simplex mode, there is a single one-way link between transmitter and receiver, that is, communication is in one direction only. In the full-duplex mode, simultaneous communication is possible between the two ends of the link. Half-duplex operation allows communications in both directions but not simultaneously. Figure 48.85 illustrates the three modes of operation. Since the telephone network is normally a full-duplex network, most data communications links are full duplex.

Links may be further classified as synchronous or asynchronous, according to how data are transmitted and synchronized on the lines. In the case of synchronous transmission, the transmitter and receiver are synchronized so that blocks of data of arbitrary length may be sent without interruption. Synchronous transmission is typically used on high-speed data links, such as computer-to-computer links. In these cases the transmitting modem transmits the data at constant speed and the receiving modem synchronizes to the incoming data stream to generate a local data clock signal. Once synchronized, the data transfer may continue for an extended time. Such links can be operated very efficiently (as defined in a later paragraph). Often an error-detecting code, such as a cyclic redundancy code (CRC), is added to check that the data have been transmitted without errors.

Asynchronous transmission is typical of the link between a simple terminal, for example, teletype, without buffering, that sends characters one at a time with an arbitrary amount of time between characters. Since the characters do not constitute a continuous data stream or even a long data block, the receiving modem must achieve synchronization on each character individually. Figure 48.86 shows a typical asynchronous data format. Each character begins with a start bit (zero) and continues with the character code of six, seven, or eight characters, depending on the particular code used. The transmission terminates with a stop pattern of 1, $1\frac{1}{2}$, or 2 bits that are usually ones.

One advantage of asynchronous transmission is that the transmitting terminal can be a simple design. Only a single character buffer is required in most cases. In particular, low-cost terminal devices are usually built for asynchronous transmission.

Encoding

Both synchronous and asynchronous links use ASCII (American Standard Code for Information Interchange) codes. The 7-bit code yields 128 possible combinations, as shown in Tables 48.12 and 48.13. There are 96 printable characters and 32 control characters. IBM equipment normally uses EBCDIC (extended binary coded decimal interchange code), an 8-bit code yielding 256 possible

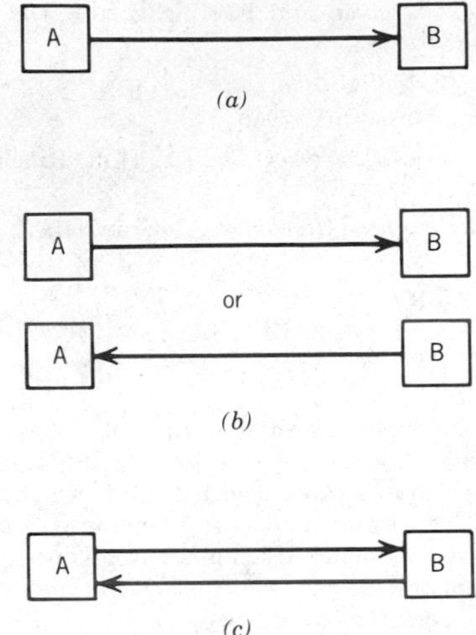

Fig. 48.85 Simplex and duplex operation of data communications links: (*a*) Simplex (one-way), (*b*) half-duplex (two-way, not simultaneous), (*c*) full duplex (two-way, simultaneous).

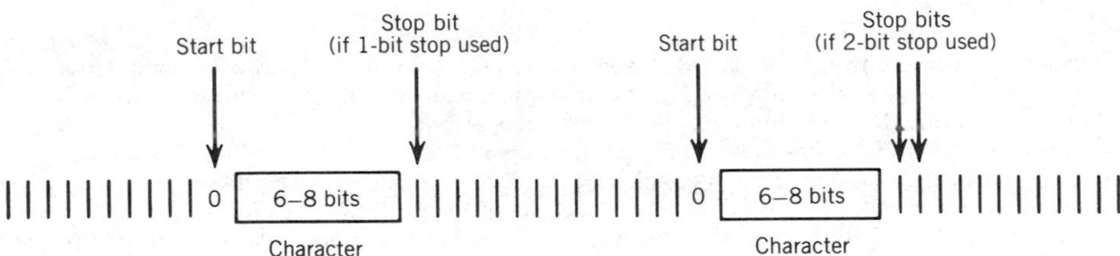

Fig. 48.86 Asynchronous data format.

combinations, as shown in Table 48.14. Baudot code is an early teletype coding scheme using 5 bits that yields only 32 possible combinations. One character is used as a shift, so that 64 possible combinations can be obtained to print the 26 alphabet and 10 number characters. A Baudot code is shown in Table 48.15.

Efficiency

The efficiency of a data transmission scheme can be defined as the fraction (or percentage) of each transmission sequence or block that is actually data. For example, a 6-bit asynchronous character code, with 1 start bit and 1 stop bit, has maximum efficiency of

$$\eta = (6 \text{ data bits})/(6 \text{ data bits} + 2 \text{ sync bits}) = 0.75$$

This maximum efficiency occurs when the characters are transmitted in a continuous sequence. If 2 stop bits are used instead of 1, the efficiency is reduced to $6/9 = 0.67$.

Synchronous transmission can be made more efficient than asynchronous transmission because the data can be sent in long blocks of characters. Block lengths of 1000 bits are common, with synchronization sequences of as few as 24 bits. In this case, the efficiency is

$$\eta = 1000/1024 = 0.976$$

Therefore transmission efficiencies of greater than 90% can be achieved with synchronous transmission. If error control techniques, such as those described in Section 48.5, are used, the apparent efficiency will be smaller, but in that case, efficiency as defined in this section is not the primary concern.

TABLE 48.12. 7-BIT ASCII (AMERICAN STANDARD CODE FOR INFORMATION INTERCHANGE) CODE SET

Bits				b_7 →	0	0	0	0	1	1	1	1
				b_6 →	0	0	1	1	0	0	1	1
				b_5 →	0	1	0	1	0	1	0	1
b_4 ↓	b_3 ↓	b_2 ↓	b_1 ↓	Column Row ↓ →	0	1	2	3	4	5	6	7
0	0	0	0	0	NUL[a]	DLE	SP	0	@	P	`	p
0	0	0	1	1	SOH	DC1	!	1	A	Q	a	q
0	0	1	0	2	STX	DC2	"	2	B	R	b	r
0	0	1	1	3	ETX	DC3	#	3	C	S	c	s
0	1	0	0	4	EOT	DC4	$	4	D	T	d	t
0	1	0	1	5	ENQ	NAK	%	5	E	U	e	u
0	1	1	0	6	ACK	SYN	&	6	F	V	f	v
0	1	1	1	7	BEL	ETB	'	7	G	W	g	w
1	0	0	0	8	BS	CAN	(	8	H	X	h	x
1	0	0	1	9	HT	EM	)	9	I	Y	i	y
1	0	1	0	10	LF	SUB	*	:	J	Z	j	z
1	0	1	1	11	VT	ESC	+	;	K	[	k	{
1	1	0	0	12	FF	FS	,	<	L	\	l	|
1	1	0	1	13	CR	GS	—	=	M	]	m	}
1	1	1	0	14	SO	RS	.	>	N	^	n	~
1	1	1	1	15	SI	US	/	?	O	___	o	DEL

[a]NUL = all zeros.
SOH = start of heading.
STX = start of text.
ETX = end of text.
EOT = end of transmission.
ENQ = enquiry.
ACK = acknowledgement.
BEL = bell or attention signal.
BS = backspace.
HT = horizontal tabulation.
LF = line feed.
VT = vertical tabulation.
FF = form feed.
CR = carriage return.
SO = shift out.
SI = shift in.
DLE = data link escape.
DC1 = device control 1.
DC2 = device control 2.
DC3 = device control 3.
DC4 = device control 4.
NAK = negative acknowledgement.
SYN = synchronous/idle.
ETB = end of transmitted block.
CAN = cancel (error in data).
EM = end of medium.
SUB = start of special sequence.
ESC = escape.
FS = information file separator.
GS = information group separator.
RS = information record separator.
US = information unit separator.
SP = space.
DEL = delete.

TABLE 48.13. ASCII (AMERICAN STANDARD CODE FOR INFORMATION INTERCHANGE) CODE FORMATS

	8-Bit			7-Bit			Even Parity			Odd Parity	
	Binary	Hex		Binary	Hex		Binary	Hex		Binary	Hex
:	10 111 010	BA	:	0 111 010	3A	:	00 111 010	3A	:	10 111 010	BA
;	10 111 011	BB	;	0 111 011	3B	;	10 111 011	BB	:	00 111 011	3B
<	10 111 100	BC	<	0 111 100	3C	<	00 111 100	3C	<	10 111 100	BC
=	10 111 101	BD	=	0 111 101	3D	=	10 111 101	BD	=	00 111 101	3D
>	10 111 110	BE	>	0 111 110	3E	>	10 111 110	BE	>	00 111 110	3E
?	10 111 111	BF	?	0 111 111	3F	?	00 111 111	3F	?	10 111 111	BF
@	11 000 000	CO	@	1 000 000	40	@	11 000 000	CO	@	01 000 000	40
[	11 011 011	DB	[	1 011 011	5B	[	11 011 011	DB	[	01 011 011	5B
\	11 011 100	DC	\	1 011 100	BC	\	01 011 100	BC	\	11 011 100	DC
]	11 011 101	DD	]	1 011 101	ED	]	11 011 101	DD	]	01 011 101	ED
ˆ	11 011 110	DE	ˆ	1 011 110	EE	ˆ	11 011 110	DE	ˆ	01 011 110	EE
—	11 011 111	DF	—	1 011 111	SF	—	01 011 111	SF	—	11 011 111	DF
\			\	1 100 000	60	\	01 100 000	60	\	11 100 000	70
{	11 111 011	FB	{	1 111 011	7B	{	01 111 011	7B	{	11 111 011	FB
:	11 111 100	FC	:	1 111 100	7C	:	11 111 100	FC	:	01 111 100	7C
}	11 111 101	FD	}	1 111 101	7D	}	01 111 101	7D	}	11 111 101	FD
~	11 111 110	FE	~	1 111 110	7E	~	01 111 101	7E	~	11 111 110	FE
ACK	10 000 110	B6	ACK	0 000 110	O6	ACK	00 000 110	O6	ACK	10 000 110	86
BEL	10 000 111	B7	BEL	0 000 111	O7	BEL	10 000 111	87	BEL	00 000 111	O7
BS	10 001 000	B8	BS	0 001 000	O8	BS	10 001 000	88	BS	00 001 000	O8
CAN	10 011 000	98	CAN	0 011 000	18	CAN	00 011 000	18	CAN	10 011 000	98
CR	10 001 101	8D	CR	0 001 101	0D	CR	10 001 101	8D	CR	00 001 101	0D
DC1	10 010 001	91	DC1	0 010 001	11	DC1	00 010 001	11	DC1	10 001 001	91
DC2	10 010 010	92	DC2	0 010 010	12	DC2	00 010 010	12	DC2	10 010 010	92
DC3	10 010 011	93	DC3	0 010 011	13	DC3	10 010 011	93	DC3	00 010 011	13
DC4	10 010 100	94	DC4	0 010 100	14	DC4	00 010 100	14	DC4	10 010 100	94
DEL	11 111 111	FF	DEL	1 111 111	7F	DEL	11 111 111	FF	DEL	01 111 111	7F
DLE	10 010 000	90	DLE	0 010 000	10	DLE	10 010 000	90	DLE	00 010 000	10
EM	10 011 001	99	EM	0 011 001	19	EM	10 011 001	99	EM	00 011 001	19
ENQ	10 000 101	85	ENQ	0 000 101	05	ENQ	00 000 101	05	ENQ	10 000 101	85
EOT	10 000 100	84	EOT	0 000 100	04	EOT	10 000 100	84	EOT	00 000 100	04
ESC	10 011 011	9B	ESC	0 011 011	1B	ESC	00 011 011	1B	ESC	10 011 011	9B
ETB	10 010 111	97	ETB	0 010 111	17	ETB	00 010 111	17	ETB	10 010 111	97
ETX	10 000 011	83	ETX	0 000 011	03	ETX	00 000 011	03	ETX	10 000 011	83
FF	10 001 100	8C	FF	0 001 100	0C	FF	00 001 100	0C	FF	10 001 100	8C
FS	10 011 100	9C	FS	0 011 100	1C	FS	10 011 100	9C	FS	00 011 100	1C
GS	10 011 101	9D	GS	0 011 101	1D	GS	00 011 101	1D	GS	10 011 001	99
HT	10 001 001	89	HT	0 001 001	09	HT	00 001 001	09	HT	10 001 001	89
LF	10 001 010	8A	LF	0 001 010	0A	LF	00 001 010	0A	LF	10 001 010	8A
NAK	10 010 101	95	NAK	0 010 101	15	NAK	10 010 101	95	NAK	00 010 101	15
NUL	10 000 000	80	NUL	0 000 000	00	NUL	00 000 000	00	NUL	10 000 000	80

TABLE 48.13. (*Continued*)

	8-Bit			7-Bit			Even Parity			Odd Parity	
	Binary	Hex		Binary	Hex		Binary	Hex		Binary	Hex
RS	10 011 110	9E	RS	0 011 110	1E	RS	00 011 110	1E	RS	10 011 110	9E
SI	10 001 111	8F	SI	0 001 111	0F	SI	00 001 111	0F	SI	10 001 111	8F
SO	10 001 110	8E	SO	0 001 110	0E	SO	10 001 110	8E	SO	00 001 110	0E
SOH	10 000 001	81	SOH	0 000 001	01	SOH	10 000 001	81	SOH	00 000 001	01
STX	10 000 010	82	STX	0 000 010	02	STX	10 000 010	82	STX	00 000 010	02
SUB	10 011 010	9A	SUB	0 011 010	1A	SUB	10 011 010	9A	SUB	00 011 010	1A
SYN	10 010 110	96	SYN	0 010 110	16	SYN	10 010 110	96	SYN	00 010 110	16
US	10 011 111	9F	US	0 011 111	1F	US	10 011 111	9F	US	00 111 111	1F
VT	10 001 011	8B	VT	0 001 011	0B	VT	10 001 011	8B	TV	00 001 011	0B

	8-Bit			7-Bit			Even Parity			Odd Parity	
	Binary	Hex		Binary	Hex		Binary	Hex		Binary	Hex
A	11 000 001	C1	A	1 000 001	41	A	01 000 001	41	A	11 000 001	C1
B	11 000 010	C2	B	1 000 010	42	B	01 000 010	42	B	11 000 010	C2
C	11 000 011	C3	C	1 000 011	43	C	11 000 011	C3	C	01 000 011	43
D	11 000 100	C4	D	1 000 100	44	D	01 000 100	44	D	11 000 100	C4
E	11 000 101	C5	E	1 000 101	45	E	11 000 101	C5	E	01 000 101	45
F	11 000 110	C6	F	1 000 110	46	F	11 000 110	C6	F	01 000 110	46
G	11 000 111	C7	G	1 000 111	47	G	01 000 111	47	G	11 000 111	C7
H	11 001 000	C8	H	1 001 000	48	H	01 001 000	48	H	11 001 000	C8
I	11 001 001	C9	I	1 001 001	49	I	11 001 001	C9	I	01 001 001	49
J	11 001 010	CA	J	1 001 010	4A	J	11 001 010	CA	J	01 001 010	4A
K	11 001 011	CB	K	1 001 011	4B	K	01 001 011	4B	K	11 001 011	CB
L	11 001 100	CC	L	1 001 100	4C	L	11 001 100	CC	L	01 001 100	4C
M	11 001 101	CD	M	1 001 101	4D	M	01 001 101	4D	M	11 001 101	CD
N	11 001 110	CE	N	1 001 110	4E	N	01 001 110	4E	N	11 001 110	CE
O	11 001 111	CF	O	1 001 111	4F	O	11 001 111	CF	O	01 001 111	4F
P	11 010 000	D0	P	1 010 000	50	P	01 010 000	50	P	11 010 000	D0
Q	11 010 001	D1	Q	1 010 001	51	Q	11 010 001	D1	Q	01 010 001	51
R	11 010 010	D2	R	1 010 010	52	R	11 010 010	D2	R	01 010 010	52
S	11 010 011	D3	S	1 010 011	53	S	01 010 011	53	S	11 010 011	D3
T	11 010 100	D4	T	1 010 100	54	T	11 010 100	D4	T	01 010 100	54
U	11 010 101	D5	U	1 010 101	55	U	01 010 101	55	U	11 010 101	D5
V	11 010 110	D6	V	1 010 110	56	V	01 010 110	56	V	11 010 110	D6
W	11 010 111	D7	W	1 010 111	57	W	11 010 111	D7	W	01 010 111	57
X	11 011 000	D8	X	1 011 000	58	X	11 011 000	D8	X	11 011 000	58
Y	11 011 001	D9	Y	1 011 001	59	Y	01 011 001	59	Y	11 011 001	D9
Z	11 011 010	DA	Z	1 011 010	5A	Z	01 011 010	5A	Z	11 011 010	DA
a	11 100 001	E1	a	1 100 001	61	a	11 100 001	E1	a	01 100 001	61
b	11 100 010	E2	b	1 100 010	62	b	11 100 010	E2	b	01 100 010	62
c	11 100 011	E3	c	1 100 011	63	c	01 100 011	E3	c	11 100 011	E3
d	11 100 100	E4	d	1 100 100	64	d	11 100 100	E4	d	01 100 100	64
e	11 100 101	E5	e	1 100 101	65	e	01 100 101	65	e	11 100 101	E5
f	11 100 110	E6	f	1 100 110	66	f	01 100 110	66	f	11 100 110	E6
g	11 100 111	E7	g	1 100 111	67	g	11 100 111	E7	g	01 100 111	67
h	11 101 000	E8	h	1 101 000	68	h	11 101 000	E8	h	01 101 000	68
i	11 101 001	E9	i	1 101 001	69	i	01 101 001	69	i	11 101 001	E9
j	11 101 010	EA	j	1 101 010	6A	j	01 101 010	6A	j	11 101 010	EA
k	11 101 011	EB	k	1 101 011	6B	k	11 101 011	EB	k	01 101 011	6B
l	11 101 100	EC	l	1 101 100	6C	l	01 101 100	6C	l	11 101 100	EC

TABLE 48.13. (*Continued*)

	8-Bit			7-Bit			Even Parity			Odd Parity	
	Binary	Hex		Binary	Hex		Binary	Hex		Binary	Hex
m	11 101 101	ED	m	1 101 101	6D	m	11 101 101	ED	m	01 101 101	6D
n	11 101 110	EE	n	1 101 110	6E	n	11 101 110	EE	n	01 101 110	6E
o	11 101 111	EF	o	1 101 111	6F	o	01 101 111	6F	o	11 101 111	EF
p	11 110 000	F0	p	1 110 000	70	p	11 110 000	F0	p	01 110 000	70
q	11 110 001	F1	q	1 110 001	71	q	01 110 001	71	q	11 110 001	F1
r	11 110 010	F2	r	1 110 010	72	r	01 110 010	72	r	11 110 010	F2
s	11 110 011	F3	s	1 110 011	73	s	11 110 011	F3	s	01 110 011	73
t	11 110 100	F4	t	1 110 100	74	t	01 110 100	74	t	11 110 100	F4
u	11 110 101	F5	u	1 110 101	75	u	11 110 101	F5	u	01 110 101	75
v	11 110 110	F6	v	1 110 110	76	v	11 110 110	F6	v	01 110 110	76
w	11 110 111	F7	w	1 110 111	77	w	01 110 111	77	w	11 110 111	F7
x	11 111 000	F8	x	1 111 000	78	x	01 111 000	78	x	11 111 000	F8
y	11 111 001	F9	y	1 111 001	79	y	11 111 001	F9	y	01 111 001	79
z	11 111 010	FA	z	1 111 010	7A	z	11 111 010	FA	z	01 111 010	7A
0	10 110 000	B0	0	0 110 000	30	0	00 110 000	30	0	10 110 000	B0
1	10 110 001	B1	1	0 110 001	31	1	10 110 001	B1	1	00 110 001	31
2	10 110 010	B2	2	0 110 010	32	2	10 110 010	B2	2	00 110 010	32
3	10 110 011	B3	3	0 110 011	33	3	00 110 011	33	3	10 110 011	B3
4	10 110 100	B4	4	0 110 100	34	4	10 110 100	B4	4	00 110 100	34
5	10 110 101	B5	5	0 110 101	35	5	00 110 101	35	5	10 110 101	B5
6	10 110 110	B6	6	0 110 110	36	6	00 110 110	36	6	10 110 110	B6
7	10 110 111	B7	7	0 110 111	37	7	10 110 111	B7	7	00 110 111	37
8	10 111 000	B8	8	0 111 000	38	8	10 111 000	B8	8	00 111 000	38
9	10 111 001	B9	9	0 111 001	39	9	00 111 001	39	9	10 111 001	B9
SP	10 100 000	A0	SP	0 100 000	20	SP	10 100 000	A0	SP	00 100 000	20
!	10 100 001	A1	!	0 100 001	21	!	00 100 001	21	!	10 100 001	A1
˜	10 100 010	A2	˜	0 100 010	22	˜	00 100 010	22	˜	10 100 010	A2
#	10 100 011	A3	#	0 100 011	23	#	10 100 011	A3	#	00 100 011	23
$	10 100 100	A4	$	0 100 100	24	$	00 100 100	24	$	10 100 100	A4
%	10 100 101	A5	%	0 100 101	25	%	10 100 101	A5	%	00 100 101	25
&	10 100 110	A6	&	0 100 110	26	&	10 100 110	A6	&	00 100 110	26
′	10 100 111	A7	′	0 100 111	27	′	00 100 111	27	′	10 100 111	A7
(	10 101 000	A8	(	0 101 000	28	(	00 101 000	28	(	10 101 000	A8
)	10 101 001	A9	)	0 101 001	29	)	10 101 001	A9	)	00 101 001	29
*	10 101 010	AA	*	0 101 010	2A	*	10 101 010	AA	*	00 101 010	2A
+	10 101 011	AB	+	0 101 011	2B	+	00 101 011	2B	+	10 101 011	AB
,	10 101 100	AC	,	0 101 100	2C	,	10 101 100	AC	,	00 101 100	2C
-	10 101 101	AD	-	0 101 101	2D	-	00 101 101	2D	-	10 101 101	AD
.	10 101 110	AE	.	0 101 110	2E	.	00 101 110	2E	.	10 101 110	AE
/	10 101 111	AF	/	0 101 111	2F	/	10 101 111	AF	/	00 101 111	2F

TABLE 48.14. HEXADECIMAL REPRESENTATION OF CHARACTER CODES [IBM BINARY SYNCHRONOUS COMMUNICATIONS (BSC)]

ASCII Character Assignments				EBCDIC[a] Character Assignments (as Defined for IBM 3270[b])				6-Bit Transcode Character Assignments	
Character	Hex	Character	Hex	Character	Hex	Character	Hex	Character	Hex
A	41	"	22	A	C1	*	5C	A	01
B	42	#	23	B	C2	%	6C	B	02
C	43	$	24	C	C3	@	7C	C	03
D	44	%	25	D	C4	(	4D	D	04
E	45	&	26	E	C5	)	5D	E	05
F	46	'	27	F	C6	—	6D	F	06
G	47	(	28	G	C7	.	7D	G	07
H	48	)	29	H	C8	+	4E	H	08
I	49	*	2A	I	C9	;	5E	I	09
J	4A	+	2B	J	D1	>	6E	J	11
K	4B	,	2C	K	D2	=	7E	K	12
L	4C	-	2D	L	D3	|	4F	L	13
M	4D	.	2E	M	D4	¬	5F	M	14
N	4E	/	2F	N	D5	?	6F	N	15
O	4F	:	3A	O	D6	"	7F	O	16
P	50	;	3B	P	D7	SBA	11	P	17
Q	51	<	3C	Q	D8	EUA	12	Q	18
R	52	=	3D	R	D9	IC	13	R	19
S	53	>	3E	S	E2	RA	3C	S	22
T	54	?	3F	T	E3	DLE	10	T	23
U	55	@	40	U	E4	EM	19	U	24
V	56	[	5B	V	E5	ENQ	2D	V	25
W	57	\	5C	W	E6	ETB	26	W	26
X	58	]	5D	X	E7	EOT	37	X	27
Y	59	∧	5E	Y	EB	ESC	27	Y	28
Z	5A	¬	5F	Z	E9	ETX	03	Z	29
a	61	\	60	a	81	FF	OC	0	30
b	62	{	7B	b	82	PT	05	1	31
c	63	|	7C	c	83	DUP	1C	2	32
d	64	}	7D	d	84	SF	1D	3	33
e	65	~	7E	e	85	FM	1E	4	34
f	66	BEL	07	f	86	ITB	1F	5	35
g	67	BS	08	g	87	NAK	3D	6	36
h	68	CAN	18	h	88	NL	15	7	37
i	69	CR	0D	i	89	NUL	00	8	38
j	6A	DC1	11	j	91	ESC	27	9	39
k	6B	DC2	12	k	92	SOH	01	Space	1A
l	6C	DC3	13	l	93	Space	40		0B
m	6D	DC4	14	m	94	STX	02	.	2B
n	6E	DEL	7F	n	95	SUB	3F	$	1B
o	6F	DLE	10	o	96	SYN	32	#	3B
p	70	EM	19	p	97			,	0C
q	71	ENQ	05	q	98			+	1C
r	72	EOT	04	r	99			/	21
s	73	ESC	1B	s	A2			%	2C
t	74	ETB	17	t	A3			@	3C
u	75	ETX	03	u	A4			-	20

TABLE 48.14. (*Continued*)

ASCII Character Assignments				EBCDIC[a] Character Assignments (as Defined for IBM 3270[b])				6-Bit Transcode Character Assignments	
Character	Hex	Character	Hex	Character	Hex	Character	Hex	Character	Hex
v	76	FF	0C	v	A5			&	10
w	77	FS	1C	w	A6			BEL	0D
x	78	GS	1D	x	A7			DEL	3F
y	79	HT	09	y	A8			DLE	1F
z	7A	LF	0A	z	A9			EM	3E
0	30	NAK	15	0	F0			ENQ	2D
1	31	NUL	00	1	F1			EOT	1E
2	32	RS	1E	2	F2			ESC	2A
3	33	SI	0F	3	F3			ETB	0F
4	34	SO	0E	4	F4			ETX	2E
5	35	SOH	01	5	F5			HT	2F
6	36	STX	02	6	F6			NAK	3D
7	37	SUB	1A	7	F7			SOH	00
8	38	SYN	16	8	F8			STX	0A
9	39	US	1F	9	F9			SUB	OE
Space	20	VT	0B	&	50			SYN	3A
!				/	61			US	1D
				$	5B				
				¢	4A				
				!	5A				
					6A				
				:	7A				
				#	7B				
				,	6B				
				.	4B				
				<	4C				

[a] EBCDIC, extended binary coded decimal interchange code.
[b] Other terminals will have functions defined for them that do not appear here, e.g., hex 04 for IBM 2780 is PF. Similarly, other terminals may have different names for functions defined here, ie., hex 05 for 2780 is HT.

TABLE 48.15. BAUDOT CODE FORMAT

Character		Mark Position				
Lower Case	Upper Case					
A	—	|	|			
B	?	|			|	|
C	:		|	|	|	
D	$	|			|	
E	3	|				
F	!	|		|	|	
G	&		|		|	|
H	#			|		|
I	8		|	|		
J	′	|	|		|	
K	(	|	|	|	|	
L	)		|			|
M	.			|	|	|
N	,			|	|	
O	9				|	|
P	0		|	|		|
Q	1	|	|	|		|
R	4		|		|	
S	Bell	|		|		
T	5					|
U	7	|	|	|		
V	;		|	|	|	|
W	2	|	|			|
X	/	|		|	|	|
Y	6	|		|		|
Z	″	|				|
Letters (shift to lower case)		|	|	|	|	|
Figures (shift to upper case)		|	|		|	|
Space				|		
Carriage return					|	
Line feed			|			
Blank						

48.3-2 Data Modems

A data modem is a device that accepts digital data from a computer or terminal and transforms that digital signal into a more appropriate form for transmission over the telephone network. On the receive end, another modem restores the data to its original form. Modems are often called data sets and dataphones by the telephone companies.

Prior to January 1, 1969, only telephone company modems or dataphones could be used on the public telephone network. At that time the Carterfone decision took effect, and foreign attachments were allowed on the telephone network. Nonetheless, a Bell-supplied DAA (data access arrangement) had to be connected to protect the telephone system. Since then the FCC has certified many other modems, which allows them to be connected to the telephone system without the DAA.

48.3-3 Modulation Techniques for Modems

There are three ways to modulate a carrier signal—by varying its amplitude, by varying its phase, or by varying its frequency. The various ways refer to AM, PM, and FM. For binary data these types of modulation are referred to as ASK, PSK, and FSK (amplitude shift keying, phase shift keying, and frequency shift keying).

ASK. Amplitude shift key is normally only used in fiber optics links.

FSK. Frequency shift keying is normally used on asynchronous links because it can be demodulated asynchronously. It is the simplest system to build, and the cheapest. Therefore it makes the ideal system for low-speed operation. It is a very robust system, which is important when it is used in operational computer networks. The disadvantage of FSK is the relatively large amount of bandwidth used.

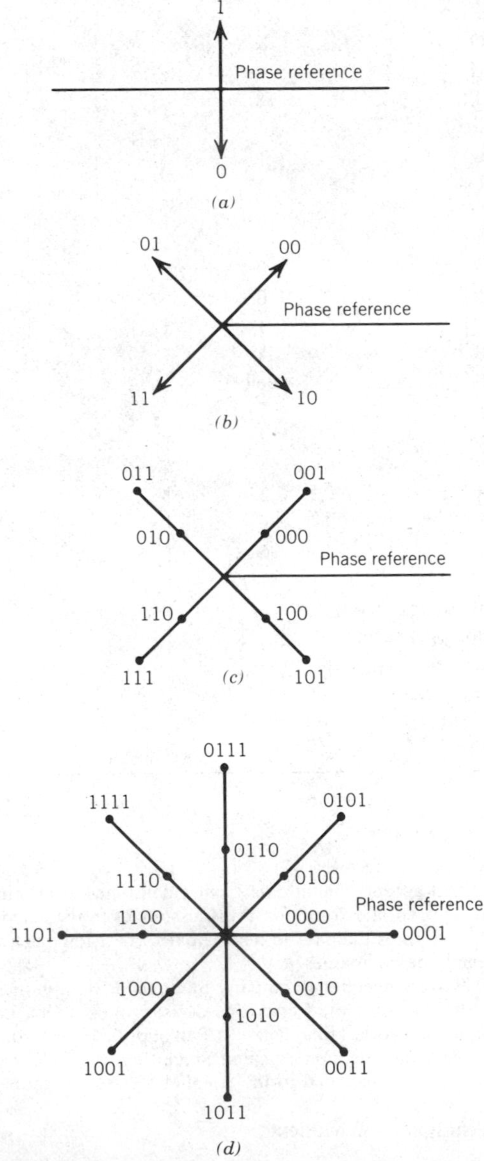

Fig. 48.87 Phase modulation schemes for data modems: (*a*) Binary phase shift keying (BPSK), (*b*) quadrature PSK (QPSK), (*c*) four-phase quadrature amplitude modulation (QAM), (*d*) eight-phase QAM.

PSK. Phase shift keying is the most efficient method of transmitting binary data in the presence of noise. The major disadvantage of PSK is the complexity of the transmitter and receiver, as compared with those for FSK. The PSK signal must be synchronously demodulated, and therefore, PSK is an ideal candidate for the synchronous modem. The bandwidth requirements are excellent if QPSK (quadrature PSK) is used. Sometimes DPSK (differential PSK) is used to simplify the receiver.

QAM. Quadrature amplitude modulation is a very efficient method of getting the maximum efficiency out of a limited bandwidth. It is a combination of phase and amplitude shift keying.

The PSK transmitter uses two phases to send a one and zero, as shown in Fig. 48.87*a*. To gain more efficiency, PSK transmitters can send four phases to send the four numbers 00, 01, 10, 11, as shown in Fig. 48.87*b*. The signals are encoded using Gray code.

Further efficiencies are gained by transmitting eight phases to send the eight numbers 000, 001, 010, 011, 100, 101, 110, 111. QAM is commonly used for such transmissions, as diagrammed in Fig. 48.87*c*. In Fig. 48.87*d*, eight phases and two amplitudes characterize eight-phase QAM, which has 16 signal assignments, each corresponding to a 4-bit number. The assignment of signal phase and amplitude according to a Gray code ensures that there is only a 1-bit difference between adjacent phases. Therefore if a phase detection error is made, the likelihood is that only 1 bit will be in error.

48.3-4 Data Transmission Rates

The transmission rates of modems begin as low as 50 b/s (bits per second) and go up to the megabits per second (Mb/s) range. This section will discuss only the common voiceband, or voice grade, modems capable of being connected to the telephone network. Normally these modems are limited to 9600 b/s and use about 3 kHz of bandwidth. With a signal-to-noise ratio (SNR) of about 30 dB and a bandwidth of about 3 kHz, data transmission rates in the neighborhood of 30 kb/s (kilobits per second) are theoretically possible (see Section 47.3, Information Theory). However, the telephone channel is not an example of the familiar AWGN (additive white Gaussian noise) channel. The telephone channel is characterized as a burst error channel, because the network is affected by switching transients on the voice channel, is subject to a great deal of cross talk, and is vulnerable to weather disturbances because of the microwave links in the telephone system. As a result, the highest data rate in common use for modems is 9600 b/s.

After nearly 30 years of demand for modems and 12 years of competition in the voiceband modem field, a family of compatible modems has survived. The rates give the user a reasonable selection of types and costs, to allow cost effectiveness to dictate which modem to use in a particular application. Modems are available with transmission rates of 300, 1200, 1800, 2400, 4800, and 9600 b/s.

In the case of low-rate data modems, the data transmission rate (b/s) is usually the same as the keying rate [symbols per second (s/s)]. This ratio of 1 bit per symbol is usable up to 2400 b/s, although the basic telephone voice channel bandwidth of 2700 Hz is narrower than the transmitted power spectrum at 2400 b/s. The amount of distortion on the received data symbols at the receiver has been found to be acceptable at 2400 b/s (Davey, 1972). For low-rate data communications, the typical full-duplex modem uses FSK, with transmission and reception frequencies arranged as shown in Fig. 48.88.

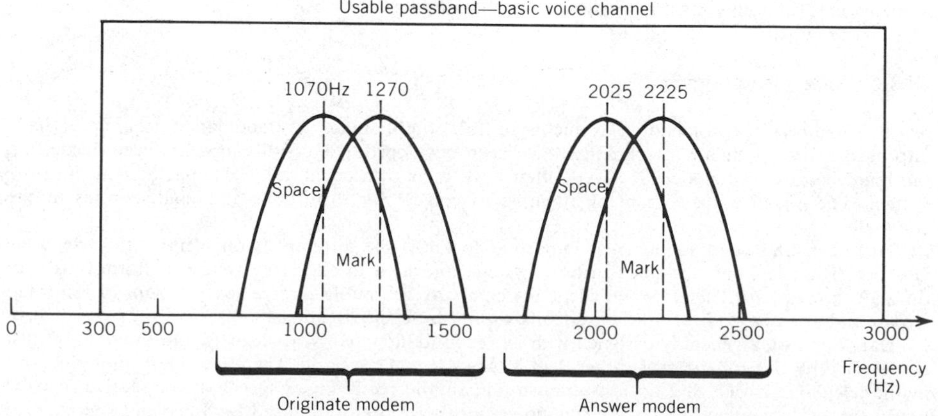

Fig. 48.88 Frequency spectrum for full-duplex modem.

TABLE 48.16. TYPICAL DATA TRANSMISSION CHARACTERISTICS

Type of Modulation	Bits/Symbol	Bits/Second in Voice Channel	Mode	Line Requirement
Binary FSK[a]	1	1200	ASYNC	DDD[b]
Binary PSK	1	2400	SYNC	DDD/PL
Four-phase PSK	2	4800	SYNC	DDD/PL
Four-phase QAM	3	7200	SYNC	PL
Eight-phase QAM	4	9600	SYNC	PL

[a] FSK, frequency shift keying; PSK, phase shift keying; QAM, quadrature amplitude modulation.

[b] DDD, direct distance dialing; PL, private line.

For data transmission rates higher than 2400 b/s, more complex modulation techniques, such as those shown in Fig. 48.87, are required. These schemes normally use coherent phase modulation techniques. If four-phase PSK is employed, there are 2 bits per symbol, and a data rate of 4800 b/s is possible. Similarly, if eight-phase QAM is used (see Fig. 48.87*d*), there are 16 signal states, or 4 bits per symbol. With this modulation format, 9600 b/s can be transmitted through a telephone voice channel. Table 48.16 summarizes the achievable data rates for the various modulation formats. For the higher data rates, nonswitched private lines are required.

48.3-5 Types of Telephone Service

Several types of telephone service are available for data transmission. One is the private line. The interface for this connection is very simple, because it is a full-period connection. There is no need to perform many of the housekeeping functions of a normal dial-up call. There is no need to access the line, hook onto the line, and then go off the hook when the operation is completed. All that is required are three signals: clear to send, request to send, and carrier detect.

A private line is the simplest type of circuit, but the full-period connection can be expensive. If the circuit is used all the time, the expense is worth the luxury of having a permanent connection. However, if the circuit is used only a few hours a day, a dial-up, or DDD (direct distance dialing), modem is more economical. No specific amount of usage is necessary to justify the full-period vs. the dial-up decision. Each time the tariffs change, the user must reassess the toll charges vs. the cost of the full-period connection.

With a DDD modem, the user must be concerned with automatic answering, automatic disconnect, and the problem with originate/answer designations. In certain full-duplex modems, the originate modem transmits a mark (one) at 1270 Hz and a space (zero) at 1070 Hz and receives a mark at 2225 Hz and a space at 2025 Hz. The answer modem must obviously handle the opposite frequencies. At times a modem will be used interchangeably as the originate modem and the answer modem. This capability must be designed into the modem.

48.3-6 Line Conditioning

Another important consideration is the need for conditioning. As modulation techniques have improved and demodulation techniques have been developed, less conditioning has been needed on the lines, because the modem receivers often have their own equalization equipment built into the system. The equalization equipment attempts to correct for distortions and nonlinearities in the channel.

To understand conditioning, it is important to know the different parameters of the telephone network. The telephone network is a huge system, and as with any large system, uniformity is very difficult to maintain. Therefore the telephone company has published the specifications of a normal circuit and attempts to keep all their circuits within these specifications.

Data signals are generally designed with less redundancy than voice signals, and the signals must be detected by electronic means rather than by the ear and the brain. Therefore simple imperfections in the telephone circuit that do not seriously impair the reception of speech at the receiver can be detrimental to reception of data. Telephone circuits have been optimized for voice and speech, and many of the filters that have very sharp cutoffs to ensure a minimum of cross talk between adjacent channels adversely affect the transmission of data because their phase characteristics are very poor. Before the large demand for data circuits, the telephone company did not worry about the poor phase characteristics, because the human ear is relatively insensitive to phase distortion. Therefore, in providing a flat amplitude response over the frequency band, phase response was not carefully controlled.

During transmission of a pulse through a communications circuit, the effect of phase distortion (delay distortion) is that the many frequency components of the pulse experience different amounts of delay through the circuit. At the receiver the components do not reassemble to form a well-defined pulse. The result is a distorted pulse, particularly in the leading and trailing edges of the pulse. The pulse, which occupied a distinct time interval when transmitted, becomes smeared out in time. Because of the time smearing, the pulses tend to interfere with each other in the receiver. This phenomenon is called intersymbol interference and can be the limiting performance effect in a seriously bandlimited system.

Line conditioning, through the use of compensating devices called equalizers, is employed to reduce the effects of amplitude and phase distortion on telephone lines used for data transmission. A standard selection of conditioned circuits is summarized in Table 48.17. In the table, the basic voice grade private line (type 3002) is shown for comparison.

TABLE 48.17. TELEPHONE LINE CONDITIONING FOR DATA TRANSMISSION

	Attenuation Distortion (Frequency Response) Relative to 1004 Hz		Envelope Delay Distortion	
Channel Conditioning[a]	Frequency Range (Hz)[b]	Variation (dB)[d]	Frequency Range (Hz)	Variation (μs)
Basic	500–2500	−2 to +8	800–2600	1750
	300–3000	−3 to +12		
C1	1000–2400[c]	−1 to +3	1000–2400[c]	1000
	300–2700[c]	−2 to +6	800–2600	1750
	300–3000	−3 to +12		
C2	500–2800[c]	−1 to +3	1000–2600[c]	500
	300–3000[c]	−2 to +6	600–2600[c]	1500
			500–2800[c]	3000
C3 (access line)	500–2800[c]	−0.5 to +1.5	1000–2600[c]	110
	300–3000	−0.8 to +3	600–2600[c]	300
			500–2800[c]	650
C3 (trunk)	500–2800[c]	−0.5 to +1	1000–2600[c]	80
	300–3000[c]	−0.8 to +2	600–2600[c]	260
			500–2800[c]	500
C4	500–3000[c]	−2 to +3	1000–2600[c]	300
	300–3200[c]	−2 to +6	800–2800[c]	500
			600–3000[c]	1500
			500–3000[c]	3000
C5	500–2800[c]	−0.5 to +1.5	1000–2600[c]	100
	300–3000[c]	−1 to +3	600–2600[c]	300
			500–2800[c]	600

[a] C conditioning applies only to attenuation and envelope delay characteristics.
[b] Measurement frequencies will be 4 Hz above those shown. For example, the basic channel will have −2 to +8 dB loss between 504 and 2504 Hz.
[c] Tariffed items.
[d] (+) means loss and (−) means gain with respect to 1004 Hz.

48.3-7 Conclusions

The systems implemented using modems should have certain performance objectives. These objectives are summarized in Table 48.18. First, the data transmission speed is the most important choice. We are interested in minimizing the delay in most cases. The faster transmission rates offer shorter delays, and normally the requirement is expressed in some average response time. In most real time systems, reasonable response time is about five seconds to keep a dialogue with a person. Also the throughput is important and the typical measurement is some peak traffic level.

Second, the availability of service is equally important. The availability is normally specified in requirements. Availability is defined as

$$\text{Availability} = \frac{\text{MTBF}}{\text{MTBF} + \text{MTTR}}$$

where MTBF is the mean time between failures and MTTR is the mean time to restore. For example, a system that has a 1000-hour MTBF and a 10-hour MTTR has an availability of

$$\text{Availability} = \frac{1000}{1000 + 10} = 99\%$$

On the other hand, a system that has a 1000-hour MTBF and a 100-hour MTTR has an availability of only

$$\text{Availability} = \frac{1000}{1000 + 100} = 91\%$$

Data integrity is an important measure; it is rated as BER (bit error rate). A normal modem channel should have a BER of no more than 10^{-6}. This means that one of every million bits is in error. With an average BER of 10^{-6}, message loss rate will be low, depending on the size of the message and the bursty nature of the bit errors. Since burst errors may bunch up in one message, the message loss rate may be less on telephone channels than on the ideal additive white Gaussian noise channels with randomly occurring errors.

Finally, data security is an important measure to ensure that messages will not be delivered to the wrong destination. This is particularly important in network communications.

A significant amount of research and development has been devoted to the design, development, and perfection of modems and related devices to enable digital data to be communicated over an essentially analog system. Concomitantly, equally significant amounts of time have been devoted to the development of techniques, devices, and systems to replace large segments of the analog telephone system with digital systems. Figure 48.89 shows a simplified diagram of multiplexed digital communications by way of a T1 carrier system. It is now common for a user's digital data to be converted to a digitally modulated analog carrier, which is then converted to digital form for transmission. A digital receiver converts the signal back to a modulated carrier, which is then demodulated by the receiving modem to restore the original data. While this may seem to be a circuitous route to data communications, it does allow the simultaneous use of the vast resources of the nationwide telephone system by data users and voice channel users.

In summary, the use of the telephone network for data communications has proven to be remarkably effective. Computer networks, remote telemetering systems, and subscriber data bases for computer users are just three examples of current practices that would be extravagantly expensive, if not impossible, without the widespread use of "wire" communications.

TABLE 48.18. PERFORMANCE OBJECTIVES OF WIRE COMMUNICATION SYSTEMS USING MODEMS

Component	Parameter	Typical Measure
Speed	Delay	Average response time
	Throughput	Peak traffic level
Service	Availability	% time down
	Data integrity	Bit error rate
	Message integrity	Message loss rate
	Security	Message misdelivery rate

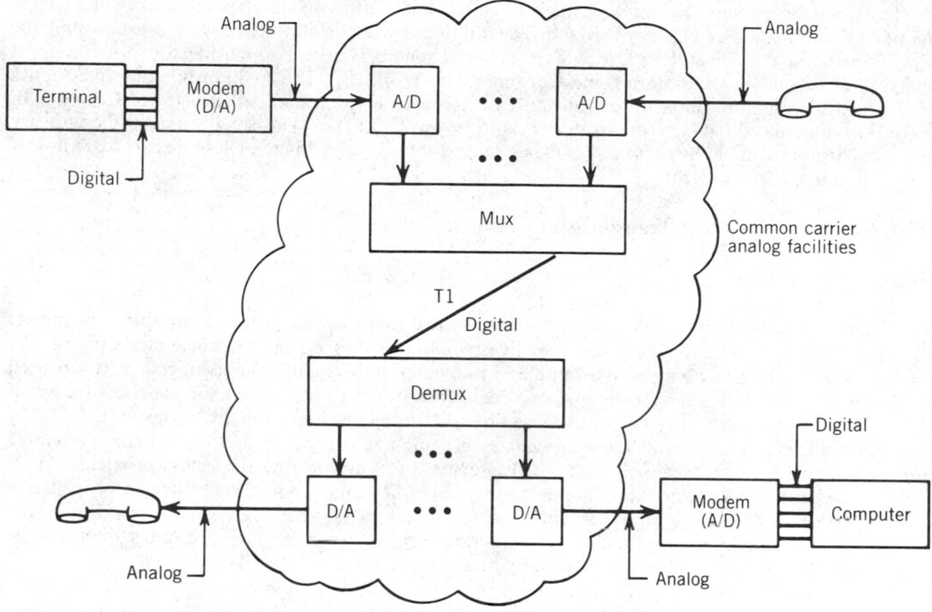

Fig. 48.89 Digital transmission within the telephone network.

Bibliography

Bell Telephone Labs, *Engineering and Operations in the Bell System*, Indianapolis, 1977.

Davenport, W. P., *Modern Data Communication*, Hayden, Rochelle Park, NJ, 1971.

Davey, J. R., "MODEMS," *Proc. IEEE* **60**(11):1284–1292 (Nov. 1972).

Lenkhurt Electric Co., *The Lenkhurt Demodulator*, San Carlos, CA, 1966.

Martin, J., *Telecommunications and the Computer*, Prentice-Hall, Englewood Cliffs, NJ, 1976.

Martin, J., *Communication Satellite Systems*, Prentice-Hall, Englewood Cliffs, NJ, 1978.

Schwartz, M., *Computer-Communication Network Design and Analysis*, Prentice-Hall, Englewood Cliffs, NJ, 1977.

Sinnema, W., *Digital, Analog and Data Communication*, Reston, Reston, VA, 1982.

48.4 FIBER OPTIC COMMUNICATION SYSTEMS

K. K. Chow

48.4-1 Early Development of Fiber Optics

The trapped propagation of light inside a transparent medium has been known for well over a century now. Tyndall, for instance, demonstrated light propagation in a water jet before the Royal Society in 1870.[1] However, serious work on the practical utilization of guided light waves did not begin until the early 1950s. The centers of activity then appeared to be Holland (van Heel) and England (Hopkins, Kapany, and Hirschowitz), where the propagation of light on uncoated, plastic-coated, and glass-coated fibers was studied for image transmission.[2] As a matter of fact, it was Kapany et al.[3] who coined the term "fiber optics" in 1956.

Since then, fiber optics has developed into a separate branch of communications. The potential of using guided optical waves was obvious to those working in telecommunications. This, coupled with the rapid advent of heterojunction light-emitting devices in the mid 1960s (for a review of early development, see Panish[4]), encouraged laboratory research for telephony applications. However, with a fiber loss typically in the 1000 dB/km range, practical application of fiber optics for long-distance communications was rather dubious. During the period 1966 to 1969, Kao and co-workers[5–8] at the Standard Telephone Laboratories pointed out that if purer materials were used, optical losses could be substantially reduced. In addition, they devoted much effort to developing measurement techniques for ultra low-loss glasses. This spurred additional work in different laboratories around the world, including the British Post Office, Telefunken, Siemens Halske, Nippon Electric, Nippon Sheet Glass, Bell Telephone Laboratories, and Corning Glass Works.[9–12] The credit of bringing fiber optics into practical communication utilization must be given to Kapron and colleagues[12] of Corning: In 1970 they announced having reduced fiber loss to below 20 dB/km. This so encouraged the industry that much more competitive effort was expended; within 3 years fibers of less than 6 dB/km loss could be purchased regularly.

48.4-2 Fiber Types and Transmission Characteristics

Fiber Types

From the historic and propagation point of view, optical fibers can be grouped into three categories: (1) fiber bundles, (2) multimode (step-index or graded-index) fibers, and (3) single-mode fibers. The fiber bundle, as the name implies, is a bundle of individual fibers lightly bonded together. It was used in practically all of the early experiments. Each constituent fiber, however, comprises a core of relatively high refractive index (n_1) surrounded by a cladding of lower refractive index (n_2). As better optoelectronic components and coupling devices became available, a single fiber became preferable for telecommunications, because it eliminated interstitial loss and relative dispersion problems. These single fibers are also known as multimode fibers, since they all support a multitude of propagating optical modes. A multimode fiber can either be step index or graded index, as shown in Fig. 48.90. The step-index fiber has a constant refractive index core[8]; the core of a graded-index fiber, on the other hand, varies approximately parabolically across its diameter.[10,11]

According to Snell's law, light rays striking the core/cladding boundary at an angle below the critical angle θ will be totally internally reflected, as depicted in Fig. 48.90. Thus there is an acceptance cone angle 2θ for any given fiber. This cone angle is also related to the numerical aperture (discussed next). As depicted in Fig. 48.90, under the most perfect conditions in the graded-index fiber the accepted light rays will be refocused in a periodic fashion to preserve the relative phase relationship among rays, while in the step-index fiber the rays tend to go out of step after a distance. The graded-index fiber therefore has less modal dispersion and is preferred for telecommunications. Modal dispersion is typically rated in either bandwidth distance products (MHz-km) or time per unit distance (ns/km); typical values are given in Fig. 48.90. A fiber that has now come into prominence is the single-mode fiber (Fig. 48.90*c*), in which the core is so restricted (3–5 μm diameter) that essentially only the axial rays can propagate. This fiber, having the least modal dispersion, is most preferred for long-distance communications.[13] Unfortunately, because of their small diameter, alignment of single-mode fibers is difficult, so that low-loss demountable connectors and fiber splices are not as readily obtainable as those for a multimode fiber.

Numerical Aperture (NA)

An important parameter characterizing fibers is the numerical aperture (NA), which gives a measure of light acceptance into the fiber. For step-index and single-mode (which is basically step-index) fibers, the NA is defined as[14]

$$\text{NA(step index)} = n_1 \sin\theta = \sqrt{n_1^2 - n_2^2} \tag{48.126}$$

where θ = internal half cone angle of acceptance
n_1 = index of core
n_2 = index of cladding

For a graded-index fiber, the NA is a function of radius r at which the light is injected into the fiber:[15]

$$\text{NA}(r) = n(r)\sin\theta = \left[n^2(r) - n_0^2\right]^{1/2} = \left\{2n_1^2[1 - (r/a)^g]\Delta\right\}^{1/2} \tag{48.127}$$

where $n(r)$ is the refractive index profile as a function of r, given by $n_1[1 - \Delta(r/a)^g]$, and a is the radius of the graded-index core.

For axial injection, that is, at $r = 0$, NA(0) of a graded-index fiber reduces to NA (step index).

Fiber Loss and Material Dispersion

The loss of a fiber is wavelength dependent, the limiting value being the Raleigh scattering loss, which has a λ^{-4} dependency. However, the impurities contained in a fiber (notably OH^- radicals and various metal ions) contribute significantly to resonant peaks in the loss-vs.-wavelength curve, an

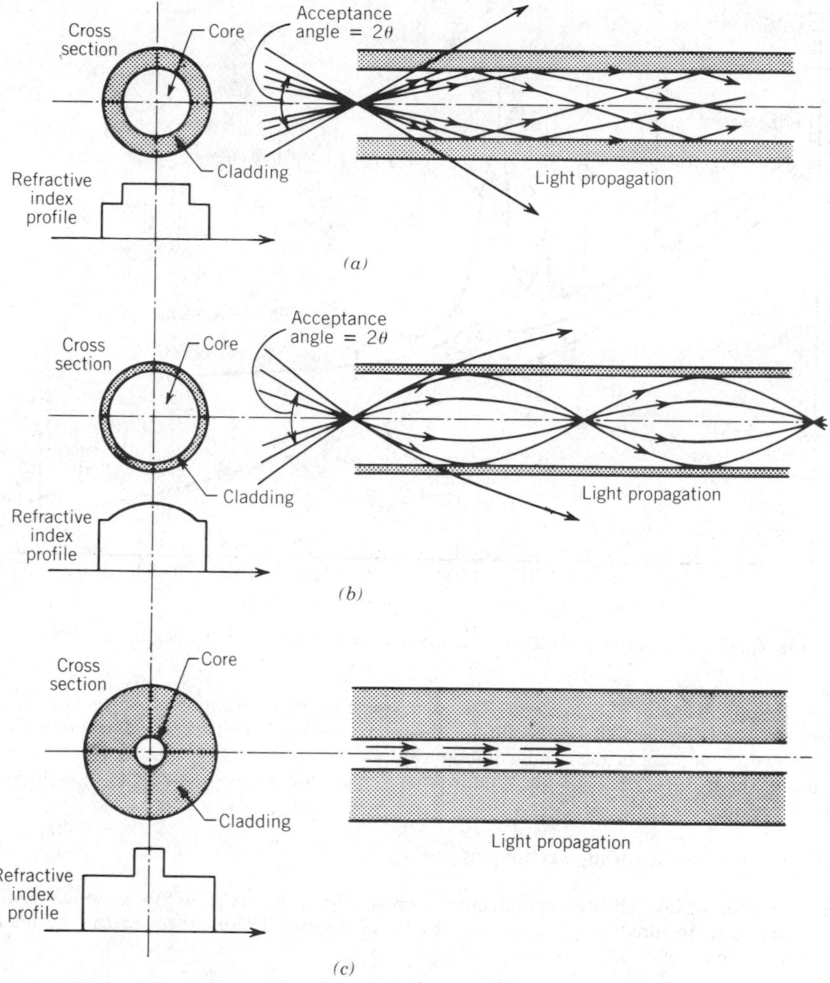

Fig. 48.90 Diagrammatic view of commonly used fibers: (*a*) step-index fiber, 20 to 100 MHz-km; (*b*) graded-index fiber, 400 to 1000 MHz-km; (*c*) single-mode fiber, up to 40 GHz-km.

example of which is shown in Fig. 48.91. The curves in the figure are composite curves showing the state of the art of fiber attenuation from about 1977. Initially the major transmission window was in the 800- to 900-nm range. Later, however, through continuing work on process control and other aspects, "double-window" fibers having low attenuation in the 800- to 900-nm as well as in the 1.1- to 1.6-μm ranges became commercially available. In the laboratory, fibers having attenuations of 0.2 dB/km at 1.55 μm have been reported.[16] It has been generally held that fibers of this low loss must be made from ultrapure (99.999%) starting materials; however, later experiments indicate that less pure starting materials (99.17%) would give attenuation of 0.6 dB/km at 1.55 μm. This could greatly reduce the manufacturing cost of low-loss fibers in the future.

In addition to the modal dispersion discussed earlier, fibers also exhibit material dispersion properties that are wavelength dependent. For multimode fibers, material dispersion is generally small compared with modal dispersion. For single-mode fibers, on the other hand, material dispersion is the dominant factor.[17] Since material dispersion and modal dispersion go in opposite directions over a wide range of wavelength, it is possible to select material, core/cladding ratio, and so on, for a single fiber so that the two dispersions essentially cancel each other over a spectral range of 1.35 to 1.67 μm.[18,19] For such fibers, the product of bandwidth times distance is well above 200 GHz-km for a source spectral width of less than 5 nm.[20]

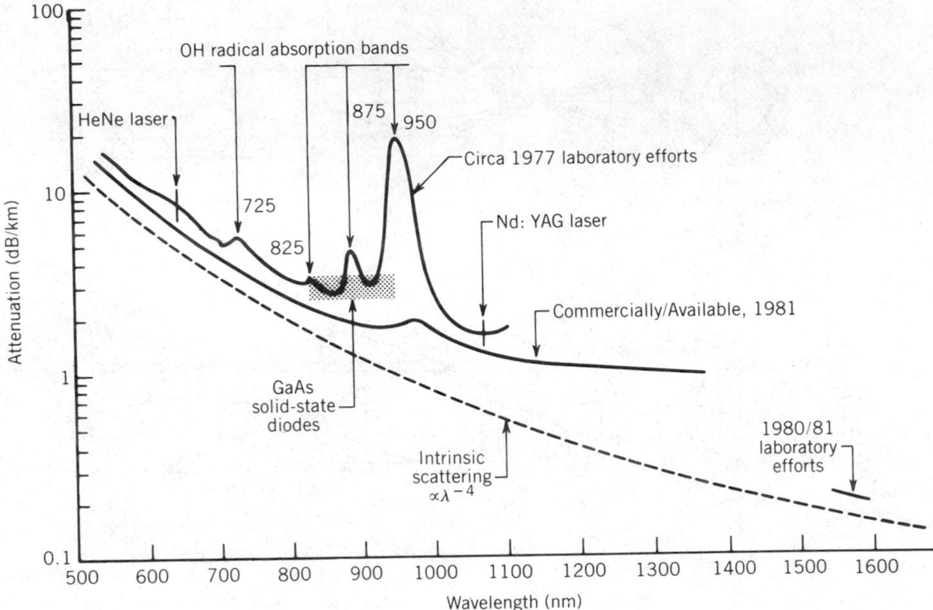

Fig. 48.91 Composite spectral attenuation curves of low-loss high-silica optical fibers.

From this discussion of fiber loss and dispersion, it is clear that utilization of single-mode fibers in the 1.3- to 1.6-μm region will be the future direction for long-haul telecommunications. For the present, however, because of the relative immaturity of sources, detectors, connectors, and so on for single mode fibers in the 1.3-to-1.6 μm window, many operational systems still tend to be in the 0.8- to 0.9-μm region.

48.4-3 Fiber Manufacturing Techniques

For minimization of loss, all fiber manufacturing must start with very pure raw materials. Different raw materials are required for different processes, of course. These processes can generally be grouped into five categories:

1. Double-crucible method.
2. Outside vapor-phase oxidation.
3. Modified chemical vapor deposition.
4. Plasma chemical vapor deposition.
5. Vapor-phase axial deposition.

The Double-Crucible Method

The starting materials for this method are the highly purified glasses, with total transition metal impurities on the order of 10 parts per billion. The core and cladding glasses are melted in concentric crucibles; the melts are drawn from the nozzles situated at the bottom of the crucibles so as to form a cladded fiber. This method is shown schematically in Fig. 48.92*a*. During the melting and drawing processes, care must be taken to prevent contamination either from the environment or from the crucibles. For graded-index fibers, the index profile is obtained by allowing the diffusion of mobile ions across the core/cladding interface. In general, this method is suitable for the rapid drawing of high NA fibers.[21]

Outside Vapor-Phase Oxidation (OVPO)

In the OVPO method, high-purity vapor reactants (e.g., $SiCl_4$, $GeCl_4$, BCl_3O_2, etc.) are burnt to produce small particles of glass of proper composition. These particles, known as soot, are deposited on a rotating mandrel that also travels axially back and forth. The soot is therefore deposited on the mandrel, layer on layer, eventually to form a porous cylindrical preform that is then slipped off the mandrel. Depending on the starting material and control, preforms for either step-index or graded-index fibers can be made. The preforms are first sintered to form solid glass blanks, following which they are treated with chlorine to remove the hydroxyl (OH^-) ions, which give high spectral absorption peaks in fibers. The glass blank is then heated in a furnace to a proper viscosity and a continuous fiber is drawn. During the process of drawing, the central hole of the blank, left from the withdrawing of the mandrel, is collapsed to render a solid fiber. This method is shown diagrammatically in Fig. 48.92*b*.

This process commercially produces highly concentric continuous fibers having attenuations of less than 3 dB/km and bandwidth times length products of 1 GHz-km. Up to 10 km of fiber of 125-μm diameter can readily be drawn from one of the blanks.[22]

Modified Chemical Vapor Deposition (MCVD)

MCVD is almost the inverse of OVPO in the sense that the starting vapors are caused to react inside a high-purity quartz tube. The soot so obtained is deposited inside the quartz tube, which is rotated to

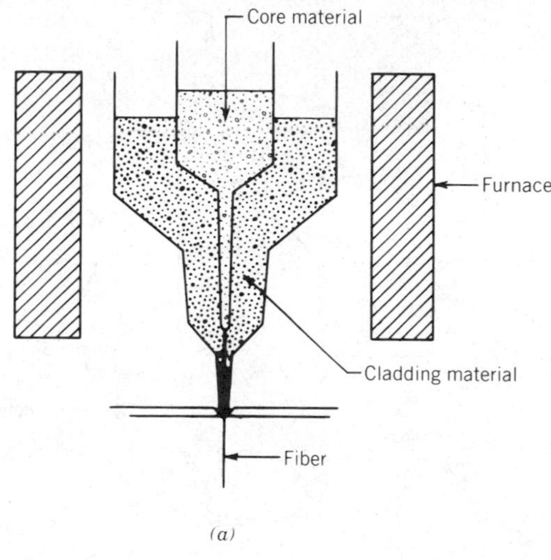

(*a*)

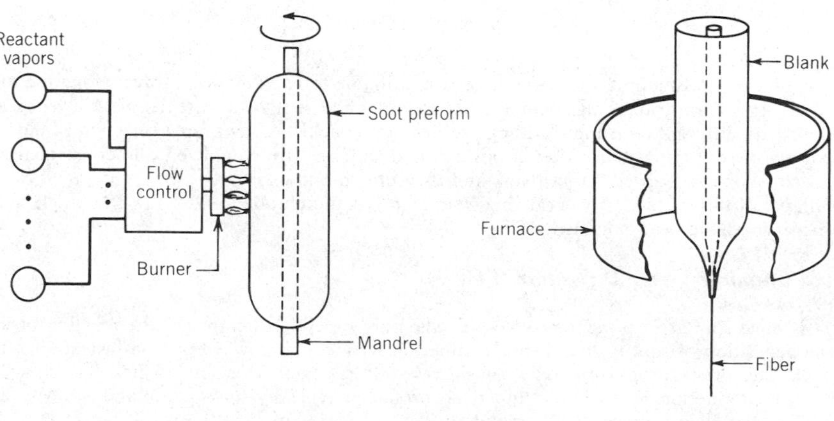

(*b*)

Fig. 48.92 Schematic view of different fiber-drawing processes: (*a*) double-crucible process; (After N. S. Kapany. Courtesy Academic Press) (*b*) outside vapor-phase oxidation process: preform making (left) and fiber drawing (right); (After P. C. Schultz, © 1980 IEEE) (*c*) modified chemical vapor deposition method; (After J. B. MacChesney, © 1980 IEEE) (*d*) vapor-phase axial deposition. (After T. Izawa et al, © 1980 IEEE).

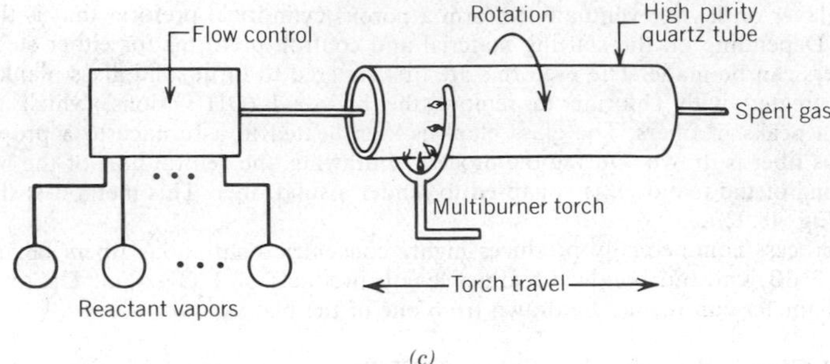

(c)

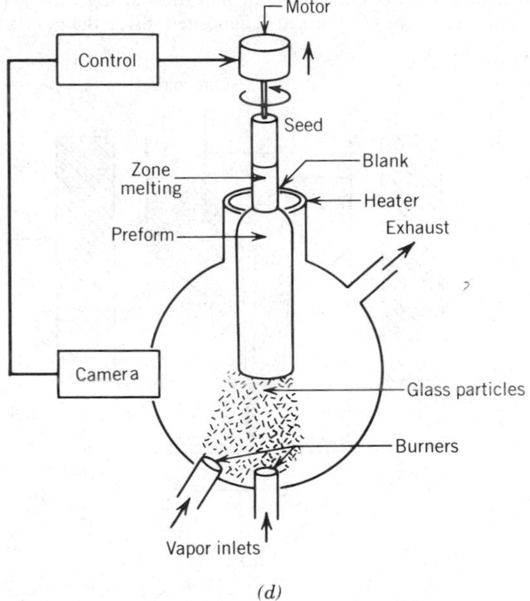

(d)

Fig. 48.92 (*Continued*)

ensure uniform deposition. At the same time, a multiflame torch is made to travel along the tube so as to heat it uniformly to a high temperature to sinter the soot to a vitreous glass layer (see Fig. 48.92*c*). Again different core and cladding profiles can be achieved by proper selection of the vapors and control of soot deposition. After deposition and sintering are completed, the entire quartz tube with its contents is collapsed into a blank and then drawn into a fiber.

With use of this method, fibers ranging from high NA multimode to ultra low-loss single mode to zero dispersion have been obtained.[23]

Plasma Chemical Vapor Deposition (PCVD)

PCVD is quite similar in principle to MCVD, the only exception being that the reaction of vapors and the deposition of soot is done through either an RF or a microwave plasma instead of a flame. Although this process was originally considered "slow," later work at Philips has shown that commercial production is feasible.[24] Fibers so produced typically have attenuation of less than 4 dB/km and bandwidth times length product of 0.5 to 1 GHz-km.

Vapor-Phase Axial Deposition (VAD)

The VAD process of growing preforms is rather similar to that of crystal growing: A starting silica rod is used as the seed on which glass soot is deposited to form the porous preform. The soot is generated by burning the raw vapors in an enclosed vessel while the silica rod is rotated and pulled upward. Zone heating near the top of the reaction chamber changes the porous preform to a glassy blank (see Fig. 48.92*d*). To maintain cylindrical geometry and constant diameter, the preform/vapor interface must be fixed in position and maintained in accuracy. This is generally done by a viewing TV camera and elaborate electronic control circuitry.[25] This process avoids the central hole required in the OVPO, MCVD, and PCVD processes and has produced both the high NA and the ultra low-loss single-mode fibers.

48.4-4 Fiber Optic Transmission Systems and Components

System Considerations

Optical fibers, by virtue of their low-loss, low-dispersion, and low-pickup properties, are ideal for long-distance wide-band transmissions. In addition, since low-loss and low-dispersion propagation is possible over a wide wavelength range (in fact, several wavelength ranges for the new fibers), and since emission spectra of light sources can be made very narrow (e.g., laser diodes having linewidths of the order of 2 nm), it is possible to multiplex a number of wavelengths onto a single fiber so that each wavelength can carry its own full communication channel capacity. Various wavelength division multiplexing (WDM) techniques are receiving increasing attention in the laboratory as well as being used in prototype systems. Therefore the most general fiber optic system is as shown in Fig. 48.93, in which a number of low bandwidth signals (e.g., telephone channels or video data links) are electronically multiplexed together to form a single constituent optical communication channel, a number of which are then optically multiplexed to form the final multiwavelength optical channel transmitted over a single fiber. At the receiver end, the inverse occurs and each low bandwidth signal is finally retrieved. These individual signals can be either digital or analog and can have bandwidths ranging from a few kilohertz to tens of megahertz.

From Fig. 48.93 it is clear that to design an efficient fiber optic transmission system one must first group the individual signals intelligently so as to minimize problems in electronic multiplexing. The complex electronic signal so obtained is then used to modulate the light source (e.g., wavelength λ_1) at the transmitter. Although electronic multiplexing, modulation, demodulation, and so on are

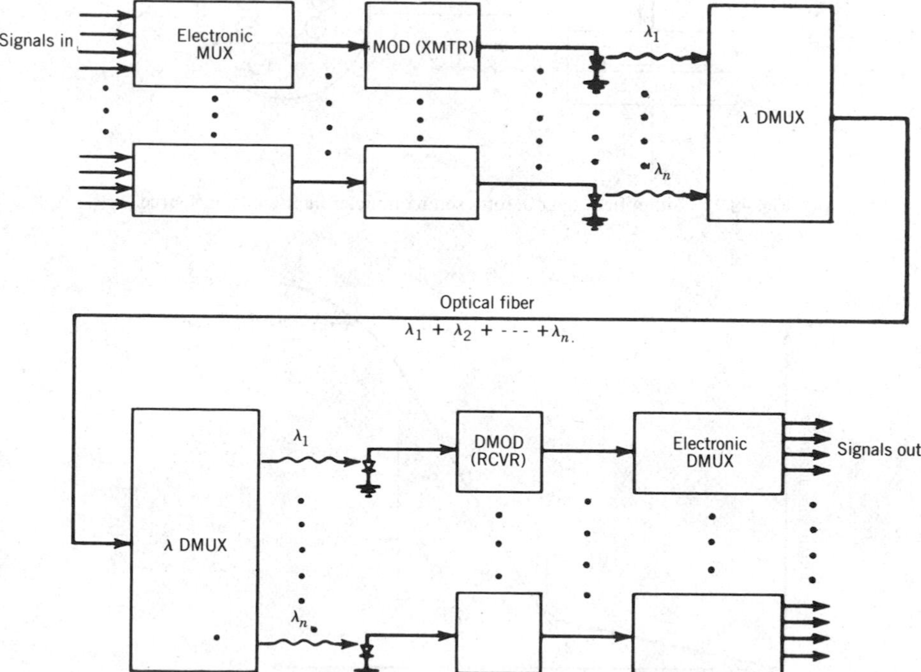

Fig. 48.93 General communication system configuration. MUX, multiplexer; MOD (XMTR), electronic modulator-transmitter; λMUX, wavelength multiplexers; λDMUX, wavelength demultiplexers; DMOD (RCVR), electronic demodulator-receiver; DMUX, demultiplexer.

integral parts of a fiber optic system, they are electronic in nature and are discussed in Section 48.1. In this section we are concerned only with optoelectronic and optical components.

Light Sources: Light-Emitting Diodes and Lasers

Although in theory many different types of light sources can be used in fiber optic communication systems, only semiconductor devices radiating in the 0.8- to 0.9 and 1.06- to 1.55-μm regions are of practical importance, because of their compactness and simplicity, as well as ease of modulation and match to the fiber transmission windows. Therefore this section presents these devices in some detail.

Basically, all the emitters considered here are constructed from materials doped differentially during growth to form a junction between p-type and n-type materials (pn junction). Typically, the junction is of the order of 1 μm in thickness and 10 to 15 μm in width. In operation, an electrical current is passed through the device so that electrons and holes are injected into the junction. These carriers recombine radiatively in the junction to emit photons. This process is shown diagrammatically in Fig. 48.94. When the photons are emitted incoherently, the device operates as a light-emitting diode (LED). When the front and the back faces of the device are well cleaved so that the device forms its own Fabry-Perot cavity, laser action occurs when the drive current is sufficiently high. With use of different host materials, dopants, and dopant concentrations, the emitter can be tailored precisely as to its radiation wavelength.

Figure 48.94 clearly suggests that varying the magnitude of the drive current I changes the optical output power accordingly. This is in fact true, as shown in Fig. 48.95 for a laser diode. At low drive currents, there is not enough positive feedback to cause laser action, and the device behaves as an LED. As the current is increased past the threshold, lasing occurs and the optical power output vs. drive current curve is essentially linear. At a higher drive current, saturation occurs. For an LED, a similar curve is obtained except there is not a pronounced "knee" corresponding to laser action.

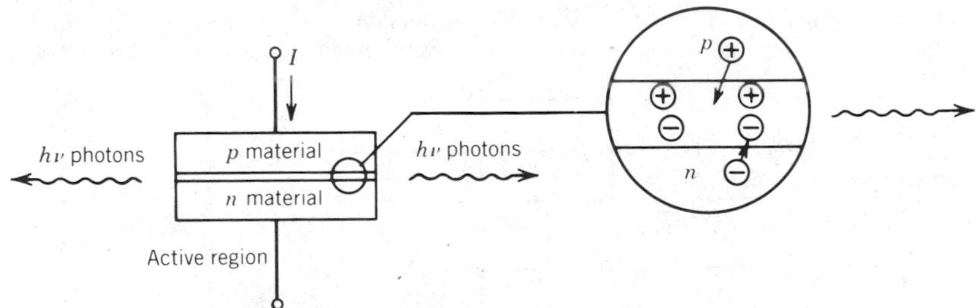

Fig. 48.94 Simplified diagram of a semiconducter light-emitting source.

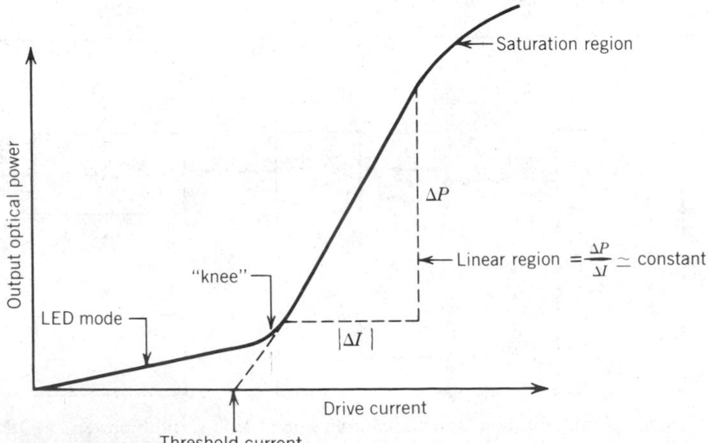

Fig. 48.95 Typical power vs. drive current curve for a semiconductor laser.

Obviously, for analog system applications, the emitter must be biased at the middle of the linear region and the modulating signal swing must be under the linear span. For digital applications, biasing can be made at cutoff, and higher drive currents are permitted.

LEDs, being noncoherent devices, have high output spectral widths (e.g., 50–60 nm at 820 nm), wide emission angles, and lower modulation bandwidths (~150 MHz). Thus they are typically used in low-data-rate short-haul systems employing large NA fibers. Lasers, being coherent, have low spectral bandwidths (e.g., 2–4 nm at 820 nm), narrow emission angles, and high modulation bandwidths (several GHz) and are more suitable for high-data-rate long-haul systems using small core fibers.

Since inception of the semiconductor emitters in the early 1960s, much work has been done by a number of authors, such as Burrus, Kressel, Lee, Paoli, Ripper, to name only a few. Among the material systems studied, $Ga_xAl_{1-x}As$ on a GaAs substrate is probably the most well understood. By tailoring the dopants, this system emits anywhere in the 750- to 920-nm region.[26] For the 1.06- to 1.55-μm region, lower bandgap energy systems are needed; presently the double-mixed crystal system $In_xGa_{1-x}As_yP_{1-y}$ is under intensive study.[27]

The most significant development in the entire semiconductor light emitter area is perhaps the invention of the double heterostructure (DH) device,[28–30] typified by the gallium-aluminum arsenide ($Al_xGa_{1-x}As$-GaAs-$Al_xGa_{1-x}As$) system shown schematically in Fig. 48.96. The double heterostructure derives its name from the fact that there are two distinct doping profiles (*p-p* and *n-n*) on either side of the light-emitting region. Proper arrangement of the layer thicknesses and dopant levels allows bandgap energies and refractive indices to be obtained to result in effective carrier and optical confinement in the *z*-direction (in the GaAs active region). In addition, if the contact to the emitter is made in a stripe geometry so that the injected current is further restricted in the *y*-dimension, very

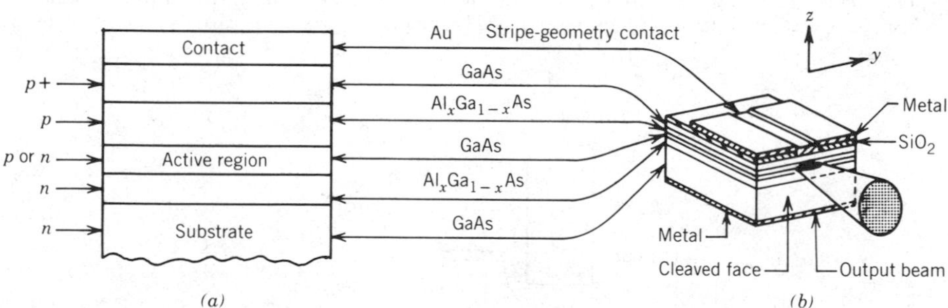

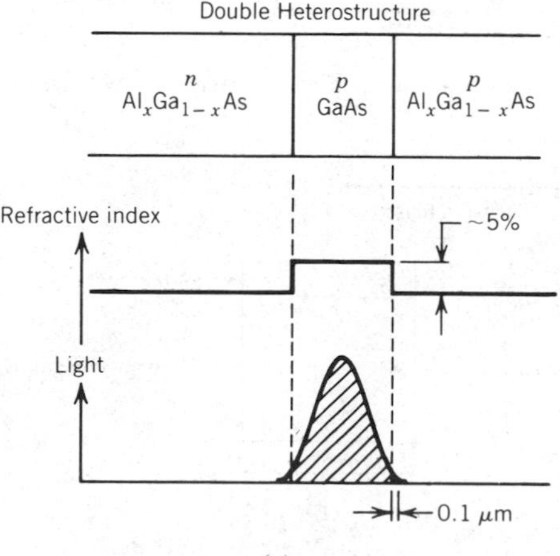

Fig. 48.96 Schematic view of a stripe geometry double-heterostructure (DH) device: (*a*) Material composition, (Courtesy M. B. Panish) (*b*) device configuration, (After T. P. Lee et al, © 1973 IEEE) (*c*) optical confinement. (Courtesy M. B. Panish)

low threshold current for lasing results. Invention and improvement of such stripe-geometry DH devices made possible CW lasing action at room temperature and firmly established the use of semiconductor devices as light sources for fiber optic communication systems. To date, the stripe-geometry DH configuration remains the basis of commercial lasers.

Modern LEDs are constructed along a similar line, using DH structures. However, since no Fabry-Perot resonator is required for operation, the emitting face can be shaped at will for best optical coupling. To this end, the Burrus diode,[31] shown in Fig. 48.97, remains the most popular basis for commercial LEDs. In such a diode, a hemisphere is etched well into the substrate to reduce absorption; a single fiber is brought close to the emitting area and is permanently cemented with an index-matching epoxy. The DH structure and the stripe geometry are also clearly shown in the figure. Many other variations have been proposed and tested, among them planar geometry and domed geometry.[32–34]

Photodetectors

As in the case of emitters, the most appropriate photodetectors for fiber optic communication systems are semiconductor devices such as PIN (P-Intrinsic-N) photodiodes and avalanche photodiodes (APD). For operation down to the 1.06-μm region, silicon devices constitute the best choice; for operation in the 1- to 1.6-μm region, other materials such as germanium and indium-gallium arsenide-phosphide are commonly used. Typically, since the output from a fiber is multimode (or at least tends to change modes), heterodyne detection is impractical. Thus all detectors are employed in the direct detection (photon-counting) mode in practical schemes.[26,35,36]

The operation of a PIN diode is schematically shown in Fig. 48.98, in which an absorbed photon is shown to generate an electron-hole pair, which is then separated in the depleted intrinsic region. The large reverse bias typically employed (e.g., 10–90 V) creates a high electric field in the drift region, thereby reducing the drift time and junction capacitance, both of which are essential for high-speed operation. External current i is induced in the load R_L while the carriers are drifting in the

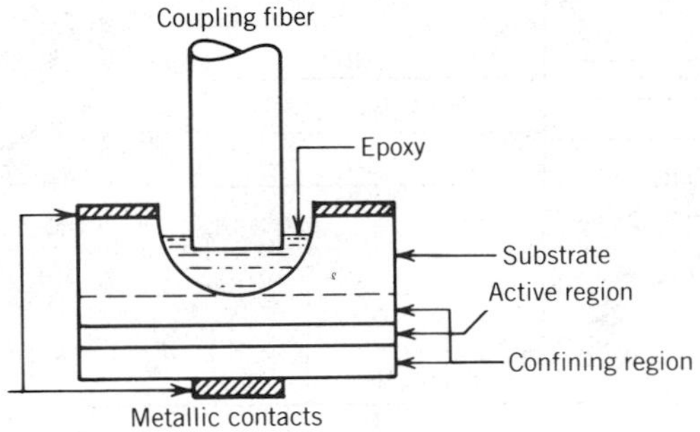

Fig. 48.97 Schematic view of a Burrus diode. (After S. E. Miller et al. © 1973 IEEE)

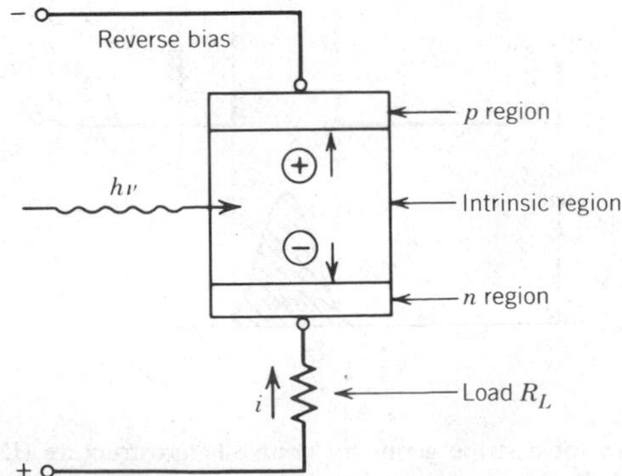

Fig. 48.98 Principle of operation of a PIN diode. (After R. G. Smith, © 1980 IEEE)

depletion region. Thus i is linearly proportional to the incident intensity until saturation is reached. For a small-area PIN diode under high bias, operation over 1 GHz is easily obtained. The principal noise contribution in a silicon PIN diode is the shot noise arising from dark current and surface leakage current. However, in a normal situation, thermal noise due to R_L and the first amplifier stage dominates.

In an avalanche photodiode (APD), the pn^+ junction is highly reverse biased, as shown in Fig. 48.99, so that the photon-induced carriers create new electron-hole pairs through impact ionization. The secondary carriers continue the ionization process, resulting in avalanche multiplication. Therefore, the APD effectively combines photon counting with internal gain and is useful in low received-energy situations. However, since the avalanche multiplication is a statistical process, there is a basic fluctuation in the current gain, so that the APD is fundamentally more noisy than is a PIN diode. This added noise is typically specified by an excess noise factor $F(M)$ defined by

$$\frac{d}{df}\langle i^2\rangle = 2qI\langle M\rangle^2 F(M) \tag{48.128}$$

where $d/df\langle i^2\rangle$ = noise spectral density of multiplied photocurrent
q = electronic charge
I = total circuit current
$\langle M\rangle^2$ = mean square avalanche gain

$F(M)$ can also be defined in terms of $\langle M\rangle$ by the following equation:

$$F(M) = \langle M\rangle^x \tag{48.129}$$

For silicon (Si), $x \approx 0.3$, while for germanium (Ge), $x \simeq 1$.

In earlier APDs, uniformity of the avalanche region was difficult to obtain so that premature breakdowns, known as microplasmas, frequently occurred. This would decrease the gain and increase the noise. By better processing, use of guard rings, field shaping, and other techniques, uniform avalanche gain has been obtained. Other considerations that must be taken into account in using the APD are higher bias voltage (~150–200 V) and sensitivity to temperature and voltage variations. Many commercial APDs now provide means of temperature/voltage feedback control so that these effects can be minimized.

Optical Switches

Most optical switches described in the recent literature are integrated optic devices in which the two optical waveguides are deposited as thin layers and coupled through an electrooptic effect under an applied voltage.[37–41] Soref and colleagues,[42] however, did describe a three-port device that was made from an $LiNbO_3$ crystal, shown in Fig. 48.100. Switching is effected by inducing channel waveguides A and B (each 220 μm wide) in the $LiNbO_3$ crystal via the electrooptic effect. In one switch position, a voltage V_a is applied to surface electrode A, increasing the refractive index under it and creating channel waveguide A. To change to the other switch position, the voltage V_b is applied, creating channel B and removing light from channel A by a directional coupler type action. Total inherent losses are about 4 dB and the switch length is about 1.7 cm. To use this switch, polarizers and lenses (not shown) are required. To date, these switches remain as laboratory devices and are not commercially available.

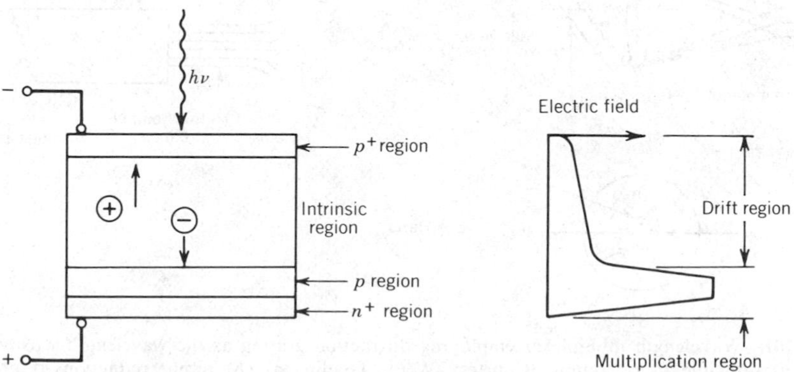

Fig. 48.99 Principle of operation of an avalanche photodiode (APD). (After S. E. Miller et al. © 1973 IEEE)

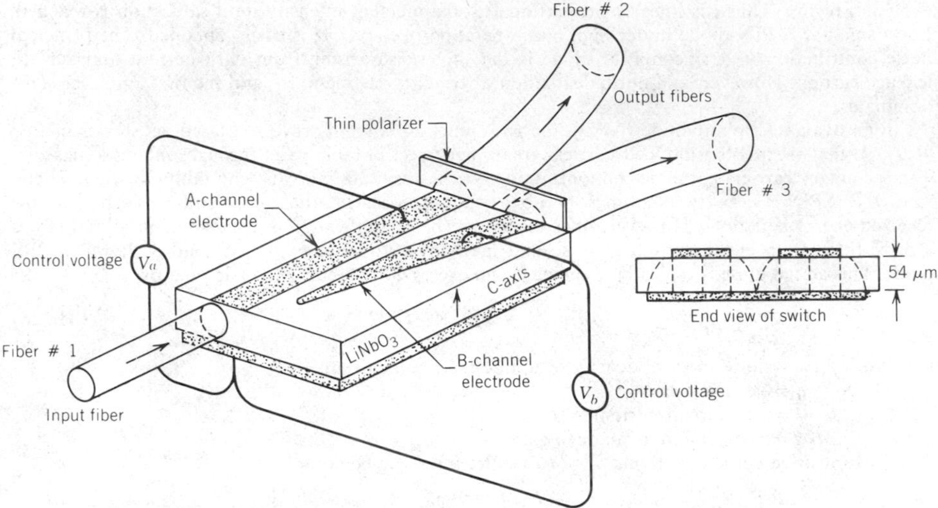

Fig. 48.100 Three-port multimode directional coupler switch. (Courtesy R. A. Soref)

The most readily available commercial optical switches are of the acoustooptic type using the Bragg diffraction principle. A traveling acoustic wave, excited by an RF transducer, is launched into a crystal to create a periodic variation of its refractive index. This effectively creates a diffraction grating in the crystal. When a laser light is incident on this virtual grating at the proper angle, deflection occurs. Optical switching of an input beam between two output locations is typically accomplished by electronically switching between two crystal oscillators in the RF driver. Insertion loss is typically about 2 dB. Although this type of switch tends to be bulky, it is the most mature in terms of technology and is thus widely used. In actual configuration, no polarizer is required and the use of lens is optional, depending on the system requirements.

Wavelength Multiplexers and Demultiplexers

Wavelength multiplexers are typically of two types: gratings and dielectric filters. Several grating multiplexers have been reported in the literature;[43–46] they generally utilize a lens-grating combination, as shown in Fig. 48.101*a*. Light beams from the input fibers (λ_1, λ_2, and λ_3) are collimated by

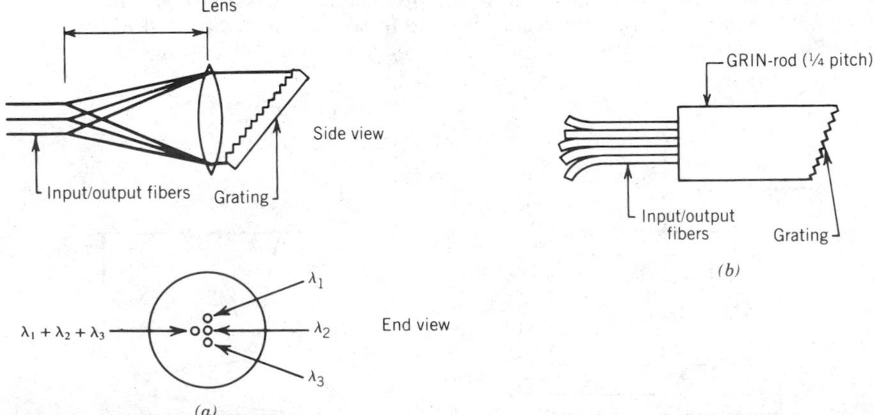

Fig. 48.101 Wavelength multiplexer employing diffraction grating as the wavelength-sensitive element: (*a*) principle of operation, (Courtesy W. J. Tomlinson) (*b*) recent reduction to practice. (Courtesy A. H. Fitch et al.)

the lens onto the brazed grating and are diffracted according to their wavelengths. They are refocused on the output fiber as the multiplexed output. In practice, a graded refractive index (GRIN)-rod lens is frequently used, because it offers advantages of planar input and output faces, coaxial orientation of fibers, easy matching of NA, minimum aberration, convenience of mounting, and others. A recent configuration using GRIN-rod/grating combination is shown in Fig. 48.101*b*. Typical insertion loss is about 2.5 dB, and isolation between channels 20 nm apart is about 24 dB.

The use of multilayer dielectric interference filters as multiplexers and demultiplexers has also been investigated extensively.[47–49] Since dielectric filters generally exhibit high polarization sensitivity, they are usually used in a near-normal incidence configuration in conjunction with GRIN rods, as shown in Fig. 48.102. Although the general feeling has been that the interference devices give low transmission and poor isolation, a properly designed device such as that shown in Fig. 48.102*a* offers an average insertion loss of about 2 dB and isolation between channels (20 nm apart) of about 34 dB.[45]

Since these devices are reciprocal, a multiplexer can also be used as a demultiplexer. However, greater optical output can be achieved in a demultiplexer if the output fibers have cores larger than the core of the input fiber.

Fiber Optic Couplers: Directional, Tee, and Star

In a multiterminal system (e.g., a data bus), various optical couplers are required. Three of the most basic ones are discussed here.

Directional Couplers. A directional coupler is a three-port device, shown schematically in Fig. 48.103*a*. Ideally, power from Port A goes only to Port C, that from Port C goes only to Port B, and there is infinite isolation between A and B. Thus such a coupler is used in a full-duplex system. In practice, one successful construction is shown in Fig. 48.103*b*: Two identical fibers are twisted together, heated, and drawn to form a "biconical taper."[50] Commercial devices so constructed typically have isolation between A and B of greater than 50 dB and an insertion between the throughput ports of about 4 dB.

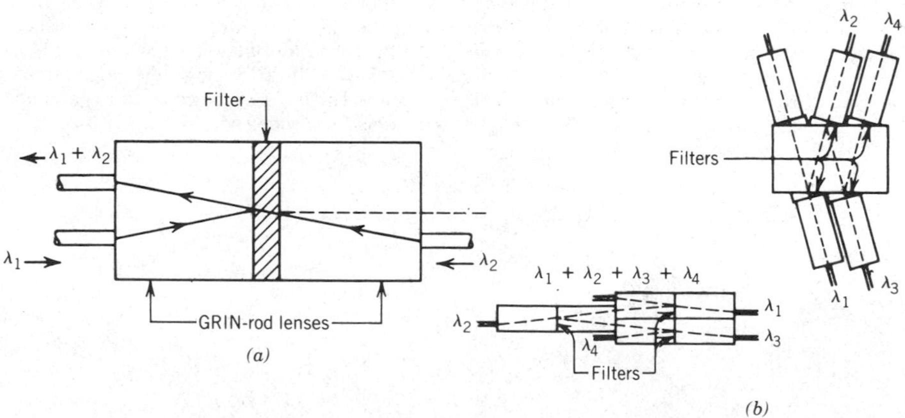

Fig. 48.102 Wavelength multiplexer employing multilayer dielectric interference filters as the wavelength-sensitive element: (*a*) single stage, (Courtesy A. H. Fitch et al.) (*c*) cascade configuration. (Courtesy W. J. Tomlinson)

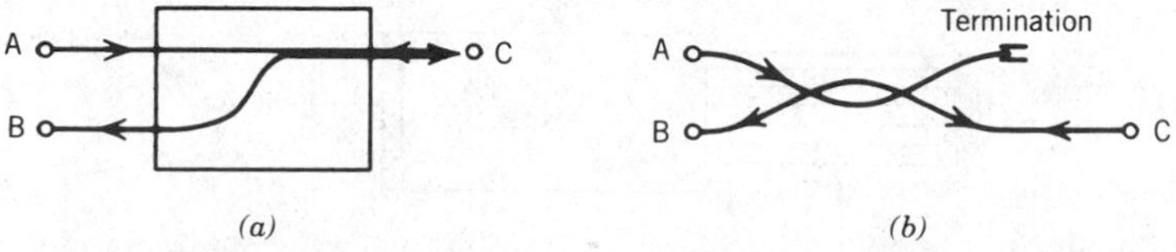

Fig. 48.103 Directional coupler: (*a*) principle of operation, (*b*) reduction to practice.

Tee Couplers. The tee coupler (also known as a tap) is also a three-port device, shown in Fig. 48.104. The main power flow is from A to C, with a prescribed portion (the tap ratio) of the power coming out of Port B. Although a simple construction can be obtained by using a fiber-filter combination much in the same way as in the multiplexer (see Fig. 48.102*a*), recent commercial devices again employ the biconical taper construction. Excess loss $[(P_A - P_B - P_C)/P_A]$ of such a device is typically about 1 dB.

Star Couplers. A star coupler is an *n*-port device in which a signal from one fiber is distributed evenly among the other fibers, as shown in Fig. 48.105. That is, the input power P_{in} from any one fiber is allowed to fill out the entire scrambler rod, which is typically a large step-index fiber, and then reflected back to uniformly illuminate the fibers. Each fiber thus ideally receives P_{in}/N of the signal. Practical devices usually have uniformity among ports of the order 1 to 2 dB, and an excess loss $(P_{\text{in}} - \sum P_{\text{out}})$ of the order of 4 dB.

Connectors and Splices

When two optical fibers are connected together, they can be joined either in a demountable fashion by the use of connectors or permanently in a splice. Since fibers typically have very small diameters (angular and linear), alignment is a serious consideration. Figure 48.106 shows the calculated losses for step-index fibers under various misalignment and end-separation conditions, and shows that substantial losses can be obtained if care is not exercised in making the connections.

Manufacturers today employ two basic approaches in their designs: (1) use of mechanical devices (grooves, alignment sleeves, watch-jewel bearings, etc.) to align the outer diameter of the fibers and (2) use of optical elements (lenses, Selfoc rods, virtual lenses, etc.) to obtain a larger diameter, collimated output beam to be received by the matching input optical system. In the first approach the dominant loss factors are axial displacement and separation, while in the second, one basically trades the axial displacement problem for angular misalignment. Since angular alignment is easier to obtain if one can maintain mechanical rigidity over a long distance, and since a large-diameter beam is less sensitive to dust inclusion, scratches, and so on, the optical approach generally gives superior results, especially after multiple mating and demating.

Figure 48.107 shows the principles of two designs currently in use. Figure 48.107*a* shows a multiple groove approach used by Bell Telephone Laboratories. Twelve grooves are precision etched in a silica wafer, between two wafers 12 fibers are embedded in epoxy, and a total of 12 layers are stacked together (144 fibers). The block so formed is then polished for butt joining to a mating block. Insertion loss of 0.2 to 0.3 dB has been reported.[15,51] Figure 48.107*b* shows the principle of a lens-type of commercially available demountable connector. The heart of the connector is a molded plastic element into whose conical chambers the bare fibers protrude; optical fluid is used in the

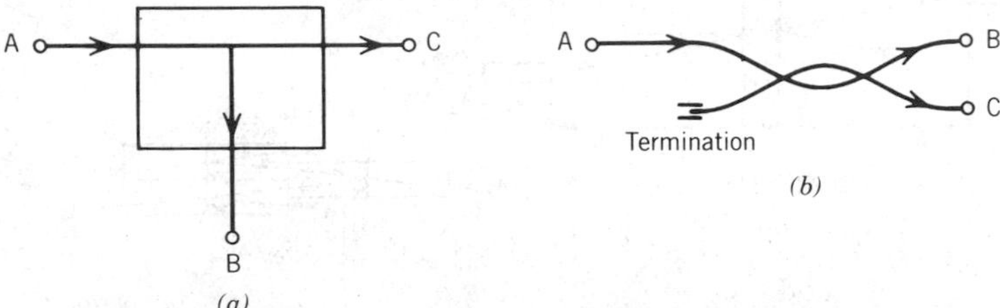

Fig. 48.104 Tee coupler: (*a*) principle of operation, (*b*) reduction to practice.

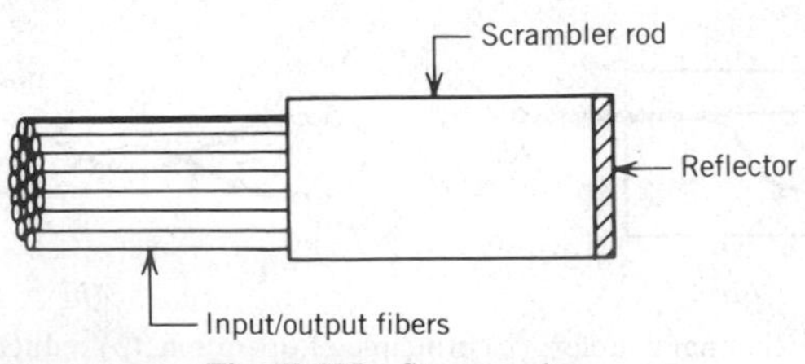

Fig. 48.105 Star coupler.

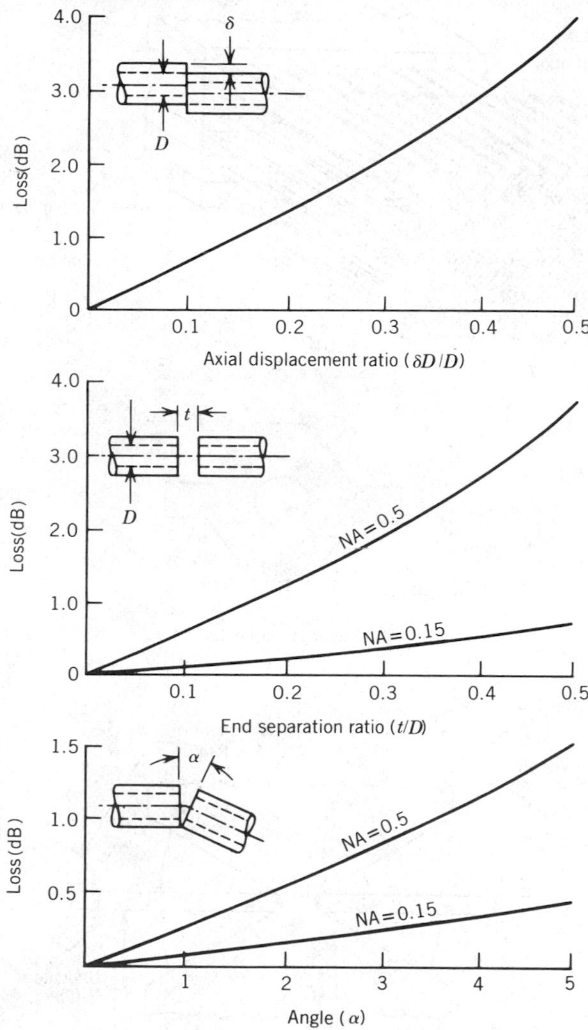

Fig. 48.106 Calculated losses of a multimode step-index fiber splice due to axial misalignment, end separation, and angular offset. (Courtesy F. L. Thiel et al.)

conical chambers for matching and lensing. To minimize insertion loss, the lens parameters are selected to optimize coupling of specific fibers and the biconical section is precision molded for transverse alignment. Insertion loss of about 1 dB can generally be maintained even after many matings.

Another technique, known as fusion splice, is now in common use. The fibers to be spliced together are stripped, the ends are cleaved and butted, and the fibers are aligned under a microscope by means of micromanipulators. An electric arc is then struck to fuse the ends together to form a permanent bond. With practice so that square, well-cleaved, and clean ends are obtained, fusion splices generally have loss of 0.1 to 0.2 dB and are mechanically stable and strong. An external support structure is required after splicing to maintain mechanical integrity.

48.4-5 Nuclear Radiation Effects

Optical fibers, being made from silica and silicates with various dopants, are not intrinsically hard in the face of ionizing radiations. That is, radiation will cause damaged or broken bonds between atoms

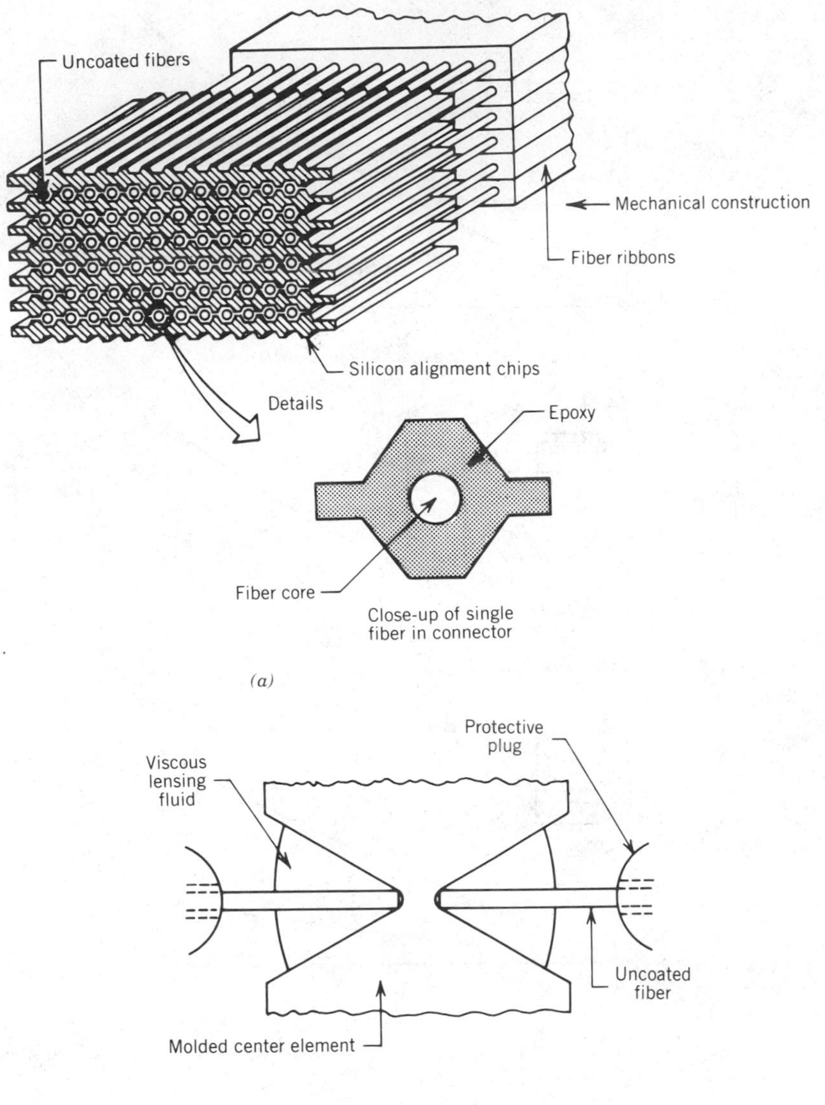

Fig. 48.107 Two types of fiber optic connectors: (*a*) array connection, (After J. F. Dalgleish, © 1980 IEEE) (*b*) single-fiber connector. (Courtesy The Deutsch Co.)

and thus set up charge trapping sites in the glass matrix. These sites create color centers in the fiber, so that attenuation is greatly increased. The total amount of damage increases with total dose received in a highly nonlinear fashion. Such damage, fortunately, is not permanent and usually anneals out with time, particularly at elevated temperatures. The annealing processes are complex and depend not only on the fiber host composition and dopant concentration, among other factors, but also on the fiber-drawing process and previous history of irradiation. Although such annealing is continuous with time, the recovery rate is not spectrally uniform because absorption in different spectral regions may be caused by different defects so that recovery mechanisms could be quite different. An optically induced annealing effect has been reported and this could improve fiber hardening, especially when temperature annealing is not available.

From this brief discussion it is quite clear that fiber response to a pulse of radiation is quite different from that to steady-state radiation. In the pulse (transient) case, the peak transient loss depends on the damage generation rate (dose rate) and the continuous recovery rate, while in the steady-state case, total damage is the net result of two steadily ongoing processes and thus depends only on the total dose. In addition, under transient irradiation, pulse emission of light usually occurs with a pulse width generally equal to that of the ionizing radiation. Thus transient response of a fiber is typically as shown in Fig. 48.108. The origin of this light emission is not clearly understood: One school of thought attributes it to spontaneous emission, while the other attributes it to Cerenkov radiation. In any event, the light emitted tends to be in the short wavelength region and can be efficiently filtered out if the system transmits in the near infrared region (e.g., at 820 nm).

Because of the nonlinear response mechanisms and the strong dependence of recovery on environmental parameters, comparison of test data is difficult. Strictly speaking, comparison can only be made if all the parameters are identical: This includes not only the test and measurement conditions but also preirradiation and postirradiation storage and handling. However, with the various testing data now available it is possible to indicate the following trends for the three most popular classes of fibers (i.e., doped silica, polymer clad silica (PCS), and doped silicate glass) in the 0.8- to 0.9-μm spectral range.

Steady-State Radiation. This is typically done using a cobalt (Co)-60 source. At low doses (under 1000 rad si), doped silica fibers tend to exhibit low loss. At high doses (1000 to 10,000 rad si), PCS fibers usually show lower attenuation. Silicate glass fibers, unfortunately, usually show high attenuation at all levels.

Transient Radiation Response. Normally such irradiation tests are done in a pulsed x-ray machine or electron source. Again, PCS fibers tend to give low peak transient loss as well as fast recovery at room temperature, with doped silica fibers generally trailing behind.

At the longer wavelengths (e.g., 1.3–1.5-μm range), measurements to date indicate that radiation hardening is highly enhanced for doped silica and more so for PCS. Thus with proper choice of wavelength, fiber material, dopants and dopant concentration, controlled fiber processing, operation temperature, and handling and storage considerations, optical fibers can be used for various transient and steady-state environments, as is evident from the number of prototype systems for nuclear power stations and military installations.

The discussion so far has concentrated on the response of optical fibers, as they are the "new" or "unknown" elements in a fiber optic transmission system. Nuclear hardened electronics design handbooks and parts lists are available from the Department of Defense and Department of Energy. As a matter of fact, a large body of literature exists. Basically the electronics are shielded wherever possible and are designed to give minimum system response to a nuclear radiation-induced stimulus. Therefore one usually sacrifices sensitivity for nuclear hardness. In the case of a high transient radiation after which the peak transient loss of the fiber may be considerable, one may have to specify an acceptable "outage" time for the system to recover from this high attenuation state. During the outage time, data transmitted may be lost entirely or may be stored in a hardened memory to await recovery.

48.4-6 Simple Design Examples

In designing a fiber optic system, the first consideration is the system bandwidth, because it determines the basic choice of the optical source (LED vs. laser). The next consideration is the link

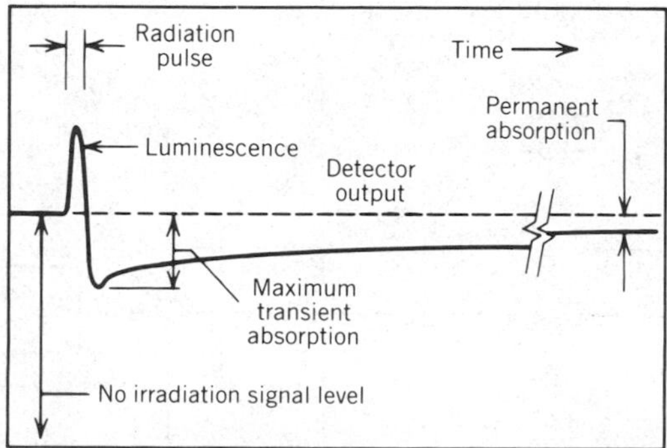

Fig. 48.108 Typical response of an optical fiber under irradiation by a short ionizing pulse. (Courtesy: B. E. Kincaid et al.)

length, because it determines what type of fiber should be used and what overall fiber attenuation is expected. The latter, together with various coupling, connection, splice, and tap-off losses, determines whether a laser diode should be used regardless of the bandwidth, what type of optical detectors should be used, and whether repeaters are needed. If a nuclear radiation environment is to be encountered, the choice of fiber will be primarily determined by the radiation hardening requirements and allowable outage time. Receiver sensitivity, in this case, will also be reduced. In the following, we shall illustrate these points with two simple examples.

Wide Bandwidth, Long Distance, Point-to-Point Link

Consider the case of a 200-MHz 3.2-km analog link, with source at point A and switchable sink either at B or C, as shown in Fig. 48.109. The system bandwidth requires the use of a good laser diode; the length and bandwidth require, as a minimum, a good graded-index fiber having a bandwidth times length product greater than 600 MHz-km. Since commercial fibers typically come in 1-km lengths, three splices are indicated. For a good commercial laser diode with a pigtail, a modulated linear output of 1 mW (0 dBm) from the pigtail may be expected. Typically, the diode will also be biased at 1 mW DC output. Thus we can compute the following link budget:

Source power		0 dBm
Link losses		
3 connectors at 1 dB ea	3.0 dB	
3 splices at 0.2 dB ea	0.6 dB	
3.2 km of fiber at 4 dB/km	12.8 dB	
1 optical switch at 2 dB	2.0 dB	
Coupling loss at photodiode	1.0 dB	
Aging, misalignment	3.0 dB	
Total link loss	22.4 dB	
∴ Power incident on photodiode		− 22.4 dBm

Using a PIN diode of responsivity 0.5 mA/mW, the photocurrent i from the diode is 2.9 μA, which means the PIN diode will be shot-noise limited. Since the shot-noise power is given by $2qi\Delta f$, while the signal power is given by i^2, the signal-to-noise ratio at the photodiode $(S/N)_i$ is $i/2q\Delta f$. Now $i = 2.9 \times 10^{-6}$ A, $q = 1.6 \times 10^{-19}$ C (the charge of one electron), and $\Delta f = 200$ MHz $= 2 \times 10^8$ Hz, giving

$$(S/N)_i = \frac{2.9 \times 10^{-6}}{2 \times 1.6 \times 10^{-19} \times 2 \times 10^8} = \frac{2.9 \times 10^{-6}}{6.4 \times 10^{-11}} = 4.5 \times 10^4$$

or

$$(S/N)_i = 46.5 \text{ dB}$$

This, of course, is not the system signal-to-noise ratio, since the first stage (postdetector) amplifier noise will contribute to additional noise. The exact contribution of noise from the first amplifier depends on the detection scheme, circuit design, and other factors and cannot be arbitrarily assigned. However, assuming a conservative 5-dB noise figure for a 23-dB amplifier, the output signal-to-noise ratio $(S/N)_o$ is reduced to 41.5 dB and this should essentially be the system signal-to-noise ratio.

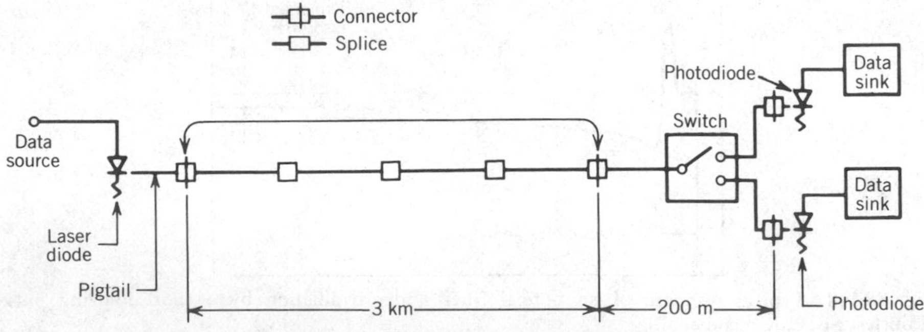

Fig. 48.109 Design example: analog data link.

If the required signal-to-noise performance is 35 dB, then the present link has a margin of 6.5 dB, which is adequate for most applications but not necessarily a comfortable one. For more margin or higher signal-to-noise ratio requirement, an APD should be used which would typically give another 13 dB of signal.

Medium Bandwidth Multiterminal Link

Consider a simple N-terminal data bus situation in which a master station broadcasts to the rest $(N-1)$ slave stations. Furthermore, for simplicity, a star coupler is used so that each station is equidistant from the star. This situation is shown in Fig. 48.110.

Typically such a data bus will transmit at about 10 Mb/s, biphase encoded, at a station-to-star separation of 100 m or less, and will have about 32 terminals in the system. These parameters indicate that an LED and a large NA step-index fiber may be used, giving a coupled input into the fiber of about 100 μW (-10 dBm) average. Again an optical power budget may be computed:

Source power		-10 dBm
Link losses between master and any one slave station		
4 connectors at 1 dB ea	4.0 dB	
200 m of fiber at 4 dB/km	0.8 dB	
Division loss $= 1/N = 1/32$	15.0 dB	
Excess loss of star	2.0 dB	
Coupling loss at photodiode	0.5 dB	
Aging, misalignment	3.0 dB	
Total station-to-station loss	25.3 dB	
∴ Power incident on photodiode		-35.3 dBm

Again using the same PIN diode of responsivity of 0.5 mA/mW, we find the photocurrent is about 0.15 μA, so that the detector is still shot-noise limited. The bandwidth of a biphase-encoded 10 Mb/s signal is conservatively estimated at 40 MHz, that is, $\Delta f = 4 \times 10^7$ Hz.

$$\therefore (S/N)_i = \frac{i}{2q\Delta f} = \frac{1.5 \times 10^{-7}}{2 \times 1.6 \times 10^{-19} \times 4 \times 10^7} = \frac{1.5 \times 10^{-7}}{1.28 \times 10^{-11}} = 1.17 \times 10^4$$

or

$$(S/N)_i = 40.7 \text{ dB}$$

Again the system signal-to-noise ratio will be reduced by the noise of the postdetector amplifier so that an overall system $(S/N)_o$ of about 37 dB may be expected. The exact noise contribution and the final system performance again depend on the modulation/detection scheme and cannot be assigned here. As a rule of thumb, for a digital signal a signal-to-noise ratio of about 20 dB will typically give a bit error rate of less than 10^{-9}, which is a generally accepted performance goal. Thus this link should have a margin of about 17 dB.

References

1 N. S. Kapany, *Fiber Optics, Principles and Applications*, Academic Press, New York, 1967.

2 N. S. Kapany, "Fiber Optics. V. Light Leakage Due to Frustrated Total Reflection," *J. Opt. Soc.*

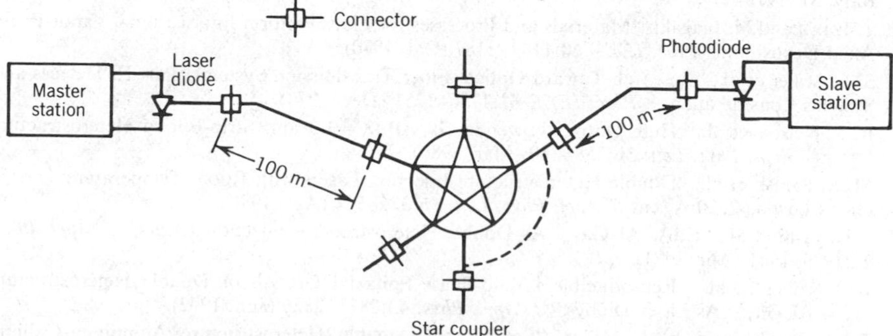

Fig. 48.110 Design example: data bus employing a star coupler.

Am. **49**:770–778; "Fiber Optics. VI. Image Quality and Optical Insulation," 779–787 (Aug. 1959).

3 N. S. Kapany, "An Introduction to Fiber Optics"; N. S. Kapany et al., "Fiber Optics—Image Transfer on Static and Dynamic Scanning with Fiber Bundles," *J. Opt. Soc. Am.* **47**:117 (Jan. 1957).

4 M. B. Panish, "Heterostructure Injection Lasers," *Proc. IEEE* **64**:1512–1540 (Oct. 1976).

5 K. C. Kao and G. A. Hockham, "Dielectric-Fibre Surface Waveguides for Optical Frequencies," *Proc. IEE* **113**:1151–1158 (Jul. 1966).

6 K. C. Kao and T. W. Davies, "Spectrophotometric Studies of Ultra Low Loss Optical Glasses. I: Single Beam Method," *J. Sci. Instr.* **1** (Series 2):1063–1068 (1968).

7 M. W. Jones and K. C. Kao, "Spectrophotometric Studies of Ultra Low Loss Optical Glasses. II: Double Beam Method," *J. Sci. Instr.* **2** (Series 2):331–335 (1969).

8 R. D. Maurer, "Glass Fibers for Optical Communications," *Proc. IEEE* **61**:452–462 (Apr. 1973).

9 S. E. Miller and L. C. Tillotson, "Optical Transmission Fiber Guide," *Appl. Opt.* **5**:1538–1549 (Oct. 1966).

10 T. Uchida et al., "A Light Focusing Fiber Guide," *IEEE J. Quant. Elec.* **QE-5**:331 (Jun. 1969).

11 S. Kawakami and J. Nishizawa, "An Optical Waveguide with the Optimum Distribution of Refractive Index with Reference to Waveform Distortion," *IEEE Trans. MTT* **MTT-16**:814–818 (Oct. 1968).

12 F. P. Kapron et al., "Radiation Losses in Glass Optical Waveguides," *Appl. Phys. Lett.* **17**:423–425 (Nov. 1970).

13 S. Shimada, "Systems Engineering for Long Haul Optical-Fiber Transmission," *Proc. IEEE* **68**:1304–1309 (Oct. 1980).

14 S. E. Miller et al., "Research Toward Optical Fiber Transmission Systems," *Proc. IEEE* **61**:1703–1751 (Dec. 1973).

15 S. E. Miller and A. G. Chynoweth, eds., *Optical Fiber Telecommunications*, Academic Press, New York, 1979.

16 T. Miya et al., "Ultimate Low-Loss Single-Mode Fibre at 1.55 μm," *Elec. Lett.* **16**:106–108 (Feb. 1979).

17 D. Marcuse, "Pulse Distortion in Single Mode Fibers," *Appl. Opt.* **19**:1653–1660 (May 1980).

18 H. Tsuchiya and N. Imoto, "Dispersion-Free Single-Mode Fibre in 1.5 μm Wavelength Region," *Elec. Lett.* **15**:476–478 (Jul. 1979).

19 K. Okamoto et al., "Dispersion Minimization in Single Mode Fibres over a Wide Spectral Range," *Elec. Lett.* **15**:729–731 (Oct. 1979).

20 T. Li, "Structures, Parameters and Transmission Properties of Optical Fibers," *Proc. IEEE* **68**:1175–1179 (Oct. 1980).

21 K. J. Beales et al., "Multicomponent Glass Fibers for Optical Communications," *Proc. IEEE* **68**:1191–1194 (Oct. 1980).

22 P. C. Schultz, "Fabrication of Optical Waveguides by the Outside Vapor Deposition Process," *Proc. IEEE* **68**:1187–1190 (Oct. 1980).

23 J. B. MacChesney, "Materials and Processes for Preform Fabrication—Modified Chemical Vapor Deposition and Plasma Chemical Vapor Deposition," *Proc. IEEE* **68**:1181–1184 (Oct. 1980).

24 J. G. J. Peelen, *Technical Digest of 4th European Conf. Optical Fiber Communication*, Genoa, Italy, **37** (1978).

25 T. Izawa and N. Inagaki, "Materials and Processes for Fiber Preform Fabrication—Vapor-Phase Axial Deposition," *Proc. IEEE* **68**:1184–1187 (Oct. 1980).

26 S. E. Miller et al., "Research Toward Optical Fiber Transmission Systems: Part II, Devices and Systems Considerations," *Proc. IEEE* **61**:1726–1751 (Dec. 1973).

27 R. J. Nelson et al., "High Output Power in GaAsP ($\lambda = 1.3$ μm) Strip-Buried Heterostructure Lasers," *Appl. Phys. Lett.* **36**:358–360 (Mar. 1980).

28 M. B. Panish et al., "Double Heterostructure Injection Lasers with Room Temperature Threshold as Low as 2300 A/cm^2," *Appl. Phys. Lett.* **16**:326–327 (Apr. 1970).

29 I. Hayashi et al., "GaAs-Al_xGa_{1-x}As Double Heterostructure Injection Lasers," *J. Appl. Phys.* **42**:1929–1941 (Apr. 1971).

30 B. I. Miller et al., "Reproducible Liquid-Phase-Epitaxial Growth of Double Heterostructure GaAs-Al_xGa_{1-x}As Laser Diodes," *J. Appl. Phys.* **43**:2817–2826 (Jun. 1972).

31 C. A. Burrus and B. I. Miller, "Small Area, Double Heterostructure Aluminum-Gallium

Arsenide Electroluminescent Diode Sources for Optical Fiber Transmission Lines," *Opt. Comm.* **4**:307–309 (Dec. 1971).

32 T. Yamaoka et al., "GaAlAs LEDs for Fiber-Optical Communication Systems," *Fujitsu Sci. Tech. J.* **14**:133–146 (Mar. 1978).

33 T. P. Lee et al., "A Stripe Geometry Double-Heterostructure Amplified Spontaneous-Emission (Superluminescent) Diode," *IEEE J.Q.E.* **QE-9**:820–828 (Aug. 1973).

34 Zh. I. Alferor et al., "High Efficiency Heterojunction LED with Spherical Radiating Surface," *Sov. Tech. Phys. Lett.* **3**:293–294 (Aug. 1977).

35 R. G. Smith, "Photodetectors for Fiber Transmission Systems," *Proc. IEEE* **68**:1247–1253 (Oct. 1980).

36 H. W. Ruegg, "An Optimized Avalanche Photodiode," *IEEE Trans. Electron. Dev.* **ED-14**:239–251 (May 1967).

37 H. Kogelnik, "An Introduction to Integrated Optics," *IEEE Trans. Microwave Theory Tech.* **MTT-23**:2–16 (1975).

38 R. V. Schmidt and H. Kogelnik, "Electro-Optically Switched Coupler with Stepped $\Delta\beta$ Reversal Using Ti Diffused $LiNbO_3$ Waveguides," *Appl. Phys. Lett.* **28**:503–506 (May 1976).

39 R. V. Schmidt and L. L. Buhl, "Experimental 4×4 Optical Switching Network," *Elec. Lett.* **12**:575–577 (Oct. 1976).

40 M. Papuchon et al., "Electrically Switched Optical Directional Coupler: Cobra," *Appl. Phys. Lett.* **27**:289–291 (Sep. 1975).

41 R. A. Steinberg et al., "Polarization-Insensitive Integrated Optical Switches: A New Electrode Design," *Appl. Opt.* **16**:2166–2170 (Aug. 1977).

42 R. A. Soref et al., "Multimode Achromatic Electro-Optic Waveguide Switch for Fiber-Optic Communications," *Appl. Phys. Lett.* **28**:716–718 (Jun. 1976).

43 W. J. Tomlinson and G. D. Aumiller, "Optical Multiplexer for Multimode Fiber Transmission Systems," *Appl. Phys. Lett.* **31**:169–171 (Aug. 1977).

44 W. J. Tomlinson, "Wavelength Multiplexing in Multimode Optical Fibers," *Appl. Opt.* **16**:2180–2194 (Aug. 1977).

45 R. Watanabe et al., "Optical Demultiplexer Using Concave Grating in 0.7–0.9 μm Wavelength Region," *Elec. Lett.* **16**:106–108 (Jan. 1980).

46 K. Kobayashi and M. Seki, "Micro-Optic Grating Multiplexers and Optical Isolators for Fiber-Optic Communications," *IEEE J. Quantum Electron.* **QE-16**:11–22 (Jan. 1980).

47 W. J. Tomlinson, "Applications of GRIN-Rod Lenses in Optical Fiber Communication Systems," *Appl. Opt.* **19**:1127–1138 (Apr. 1980).

48 "OD-8601-8604 Optical Directional Coupler," Nippon Electric Co. Ltd., Data Sheet DPR-047E, 1978.

49 A. H. Fitch et al., "Four Channel Fiber Optic Link Employing Wavelength Division Multiplexing and Full Duplex Transmission," Topical Meeting on Optical Fiber Communications, Phoenix, AZ, Apr. 1982.

50 B. S. Kawasaki and K. O. Hill, "Low Loss Access Coupler for Multimode Optical Fiber Distribution Networks," *Appl. Opt.* **16**:1794–1795 (Jul. 1977).

51 J. F. Dalgleish, "Splices, Connectors and Power Couplers for Field and Office Use," *Proc. IEEE* **68**:1226–1232 (1980).

48.5 ADVANCED TECHNIQUES FOR DIGITAL COMMUNICATIONS

Bernard Sklar

48.5-1 Introduction

An impressive assortment of communications signal-processing techniques has arisen during the past two decades. This section presents an overview of some of these techniques, particularly as

---

A version of Section 48.5 has appeared as a two-part series in the IEEE Communications Magazine (August and October 1983) under the title "A Structured Overview of Digital Communications—A Tutorial Review." That paper was selected for the 1984 Communications Society Prize Paper Award for the Communications Magazine.

they relate to digital satellite communications. The material is developed in the context of a structure used to trace the processing steps from the information source to the information sink. Transformations are organized according to functional classes: formatting and source coding, modulation, channel coding, multiplexing and multiple access, frequency spreading, encryption, and synchronization.

Satellite communication has two unique characteristics: the ability to cover the globe with a flexibility that cannot be duplicated with terrestrial links, and the availability of bandwidth exceeding anything previously available for intercontinental communications.[1] Most satellite communication systems to date are analog in nature. However, digital communication is becoming increasingly attractive because of the ever-growing demand for data communication and because digital transmission offers data-processing options and flexibilities not available with analog transmission.[2]

This section presents an overview of digital communications in general; for the most part, however, the treatment is in the context of a satellite communications link. The key feature of a digital communication system (DCS) is that it sends only a finite set of messages, in contrast to an analog communication system, which can send an infinite set of messages. In a DCS, the objective at the receiver is not to reproduce a waveform with precision; it is instead to determine from a noise-perturbed signal which of the finite set of waveforms has been set by the transmitter. An important measure of system performance is the average number of erroneous decisions made, or the probability of error (P_E).

Figure 48.111 illustrates a typical DCS. Let there be M symbols, or messages $m_1, m_2, \ldots, m_M$, to be transmitted. Let each symbol be represented by a corresponding waveform $s_1(t), s_2(t), \ldots, s_M(t)$. The symbol (or message) m_i is sent by transmitting the digital waveform $s_i(t)$ for T seconds, the symbol period. The next symbol is sent over the next period. Since the M symbols can be represented by $k = \log_2 M$ binary digits (bits), the data rate can be expressed as

$$R = (1/T)\log_2 M = k/T \qquad \text{bits per second (b/s)}$$

Data rate is usually expressed in bps whether or not binary digits are actually involved. A binary symbol is the special case characterized by $M = 2$ and $k = 1$. A digital waveform is taken to mean a voltage or current waveform representing a digital symbol. The waveform is endowed with specially chosen amplitude, frequency, or phase characteristics that allow the selection of a distinct waveform for each symbol from a finite set of symbols. At various points along the signal route, noise corrupts the waveform $s(t)$ so that its reception must be termed an estimate $\hat{s}(t)$. Such noise and its deleterious effect on system performance are treated in Section 48.5-3.

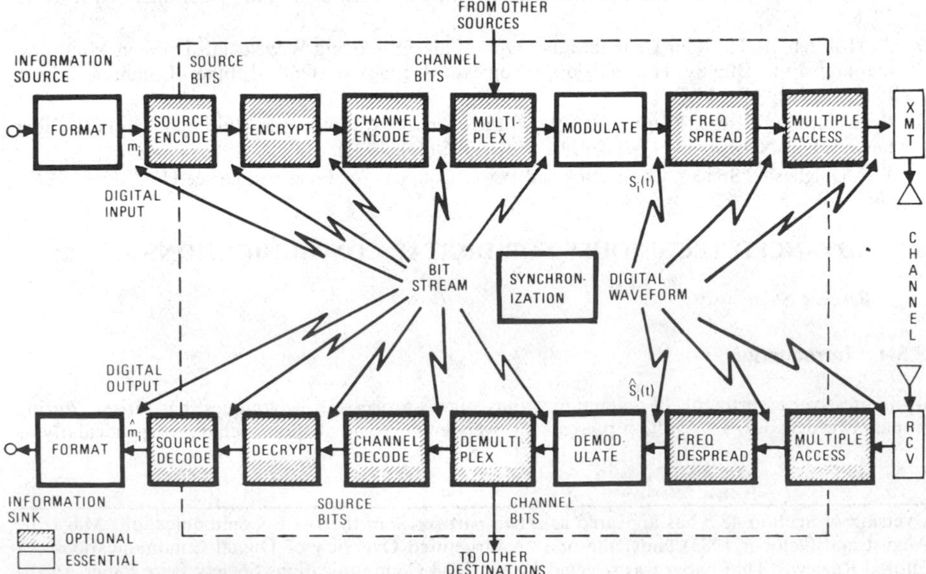

Fig. 48.111 Block diagram of a typical digital communication system.

48.5-2 Signal-Processing Steps

The functional block diagram shown in Fig. 48.111 illustrates the data flow through the DCS. The upper blocks, which are labeled format, source encode, encrypt, channel encode, multiplex, modulate, frequency spread, and multiple access, dictate the signal transformations from the source to the transmitter. The lower blocks dictate the signal transformations from the receiver back to the source; the lower blocks essentially reverse the signal-processing steps performed by the upper blocks. The blocks within the dashed lines initially consisted only of the modulator and demodulator functions, hence the name "modem." During the past two decades, other signal-processing functions were frequently incorporated within the same assembly as the modulator and demodulator. Consequently, the term "modem" often encompasses the processing steps shown within the dashed lines of Fig. 48.111. When this is the case, the modem can be thought of as the "brains" of the system and the transmitter and receiver as the "muscles." While the transmitter consists of a frequency up-conversion stage, a high-power amplifier, and an antenna, the receiver portion is occupied by an antenna, a low-noise front-end amplifier, and a down-converter stage, typically to an intermediate frequency (IF).

Of all the signal-processing steps, only formatting, modulation, and demodulation are essential for all DCSs; the other processing steps within the modem are considered design options for various system needs. Source encoding, as defined here, removes information redundancy and performs analog-to-digital (A/D) conversion. Encryption prevents unauthorized users from understanding messages and from injecting false messages into the system. Channel coding can, for a given data rate, improve the P_E performance at the expense of power or bandwidth, reduce the system bandwidth requirement at the expense of power or P_E performance, or reduce the power requirement at the expense of bandwidth or P_E performance. Frequency spreading renders the signal less vulnerable to interference (both natural and intentional) and can be used to afford privacy to the communicators. Multiplexing and multiple access combine signals that might have different characteristics or originate from different sources.

The flow of the signal-processing steps shown in Fig. 48.111 represents a typical arrangement; however, the blocks are sometimes implemented in a different order. For example, multiplexing can take place prior to channel encoding, prior to modulation, or—with a two-step modulation process (subcarrier and carrier)—it can be performed between the two steps. Similarly, spreading can take place anywhere along the transmission chain; its precise location depends on the particular technique used. Figure 48.111 illustrates the reciprocal aspect of the procedure; any signal-processing steps that take place in the transmitting chain must be reversed in the receiving chain. The figure also indicates that, from the source to the modulator, a message takes the form of a bit stream, also called a baseband signal. After modulation, the message takes the form of a digitally encoded sinusoid (digital waveform). Similarly, in the reverse direction, a received message appears as a digital waveform until it is demodulated. Thereafter it takes the form of a bit stream for all further signal-processing steps.

Figure 48.112 shows the basic signal-processing functions, which may be viewed as transformations from one signal space to another. The transformations are classified into seven basic groups: formatting and source coding, modulation, channel coding, multiplexing and multiple access, spreading, encryption, and synchronization. The organization has some inherent overlap, but nevertheless provides a useful structure for this overview. The text by Lindsey and Simon[3] is an excellent reference for the modulation, coding, and synchronization transformations treated here. The comprehensive books by Spilker[4] and Bhargava et al.[5] specifically address digital communications by satellite. The seven basic transformations will now be treated individually, in the general order of their importance rather than in the order of the blocks shown in Fig. 48.111.

Formatting and Source Coding

The first essential processing step, formatting, renders the communicated data compatible for digital processing. Formatting is defined as any operation that transforms data into digital symbols. Source coding means data compression in addition to formatting. Some authors consider formatting a special case of source coding (for which the data compression amounts to zero), instead of making a distinction between the two. The source of most communicated data (except for computer-to-computer transformations already in digital form) is either textual or analog in nature. If the data consist of alphanumeric text, it is character encoded with one of several standard formats, such as American Standard Code for Information Interchange (ASCII), Extended Binary Coded Decimal Interchange Code (EBCDIC), or Baudot, and is thereby rendered into digital form. If the data are analog, the (bandlimited) waveform must first be sampled at a rate of at least $2f_m$ Hz (the Nyquist frequency), where f_m is the highest frequency contained in the waveform. Such sampling ensures perfect reconstruction of the analog signal; undersampling results in a phenomenon called aliasing, which introduces errors. However, the minimum sampling rate can be less than $2f_m$ if the lowest

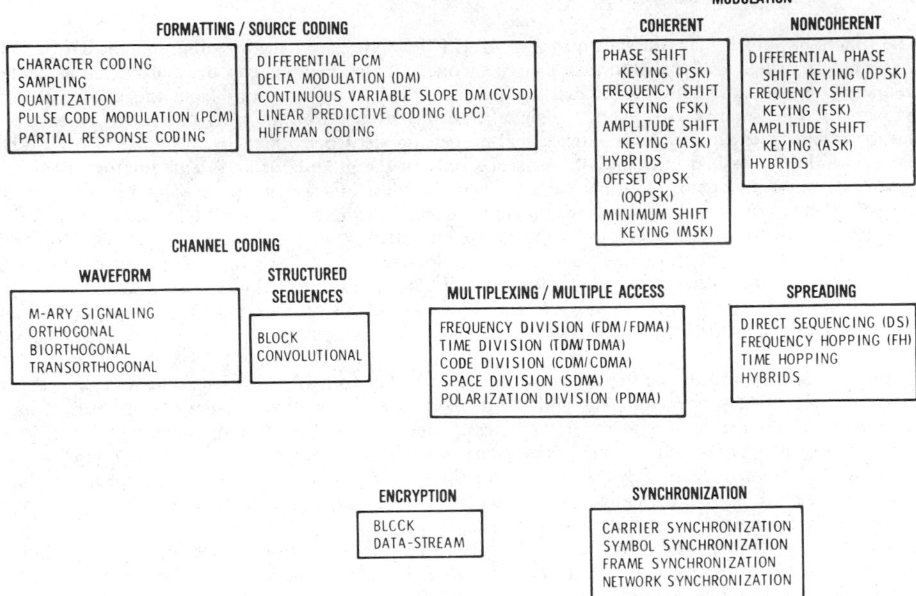

Fig. 48.112 Basic digital communication transformations.

signal frequency contained in the waveform is nonzero.[6] Quantization of the time samples allows each sample to be expressed as a level from a finite number of predetermined levels; each such level can be represented by a digital symbol. After quantization, the analog waveform can still be recovered, but not precisely; improved reconstruction fidelity of the analog waveform can be achieved by increasing the number of quantization levels (requiring increased transmission bandwidth).

Pulse code modulation (PCM), the classical and most widely used digital format, converts the quantized samples into code groups of two-level pulses using fixed amplitudes. Each pulse group represents a quantized amplitude value expressed in binary notation. There are several PCM subformats (such as nonreturn to zero, Manchester, and Miller), each providing some special feature, such as self-clocking or a compact spectral signature.[3] Duobinary, or partial response coding (also called correlative coding), is a formatting technique that improves bandwidth efficiency by introducing controlled interference between symbols. The technique also provides error-detecting capabilities without introducing redundancy into the data stream.[7–9]

Both source encoding and formatting mean encoding the source data with a digital format (A/D conversion); in this sense alone, the two are identical. However, the term "source encoding" has taken on additional meaning in DCS usage. Besides digital formatting, "source encoding" has also come to denote data compression (or data rate reduction). With standard A/D conversion using PCM, data compression can only be achieved by lowering the sampling rate or by reducing the number of quantization levels per sample, each of which increases the mean squared error of the reconstructed signal. Source-encoding techniques accomplish rate reduction by removing the redundancy that is indigenous to most message transmissions, without sacrificing reconstruction fidelity. A digital data source is said to possess redundancy if the symbols are not equally likely or if they are not statistically independent. Source encoding can reduce the data rate if either of these conditions exists. A few descriptions of common source coding techniques follow.

Differential PCM (DPCM) utilizes the differences between samples rather than their actual amplitude. For most data, the average amplitude variation from sample to sample is much less than the total amplitude variation; therefore, fewer bits are needed to describe the difference. DPCM systems actually encode the difference between a current amplitude sample and a predicted amplitude value estimated from past samples. The decoder utilizes a similar algorithm for decoding. Delta modulation (DM) is the name given to the special case of DPCM where the quantization level of the output is taken to be 1 bit. Although DM can be easily implemented, it suffers from "slope overload," a condition in which the incoming signal slope exceeds the system's capability to follow the analog source closely at the given sampling rate. To improve performance, whenever slope overload is detected, the gain of the system can be varied according to a predetermined algorithm known to the receiver. If the system is designed to adaptively vary the gain over a continuous range, the modulation is termed continuous variable slope delta (CVSD) modulation, or adaptive delta modulation (ADM). Speech coding of good quality has been demonstrated with a CVSD modulation at bit rates less than 25 kb/s, a notable data rate reduction when compared with the 56-kb/s PCM used with commercial telephone systems.[10]

Another example of source coding is linear predictive coding (LPC). This technique is useful when the waveform results from a process that can be modeled as a linear system. Rather than samples of the waveform being encoded, significant features of the process are encoded. For speech these include gain, pitch, and voiced or unvoiced information. Whereas in PCM each sample is processed independently, in a predictive system such as DPCM a weighted sum of the n-past samples is used to predict each present sample; the "error" signal is then transmitted. The weights are calculated to minimize the average energy in the error signal that represents the difference between the predicted and actual amplitude. For speech, the weights are calculated over short waveform segments of 10 to 30 ms and thus change as the speech statistics vary. The LPC technique has been used to produce acceptable speech quality at a data rate of 2.4 kb/s and high quality at 7.2 kb/s.[11–13] For current perspectives in digital formatting of speech, see Crochiere and Flanagan.[14]

Some source-coding techniques employ code sequences of unequal length so as to minimize the average number of bits required per data sample. A useful coding procedure, called Huffman coding,[15,16] can be used for effecting data compression on any symbol set, provided the a priori probability of symbol occurrence is known and not equally likely. Huffman coding generates a binary sequence for each symbol so as to achieve the smallest average number of bits per sample, for the given a priori probabilities. The technique involves assigning shorter code sequences to the symbols of higher probability and longer code sequences to those of lower probability. The price paid for achieving data rate reduction in this way is a commensurate increase in decoder complexity. In addition, there is a tendency for symbol errors, once made, to propagate for several symbol periods.

Digital Modulation Techniques

Digital Modulation Formats. Modulation, in general, is the process by which some characteristic of a waveform is varied in accordance with another waveform. A sinusoid has just three features that can be used to distinguish it from other sinusoids—phase, frequency, and amplitude. For the purpose of radio transmission, modulation is defined as the process whereby the phase, frequency, or amplitude of a radio-frequency (RF) carrier wave is varied in accordance with the information to be transmitted. Figure 48.113 illustrates examples of digital modulation formats: phase shift keying (PSK),

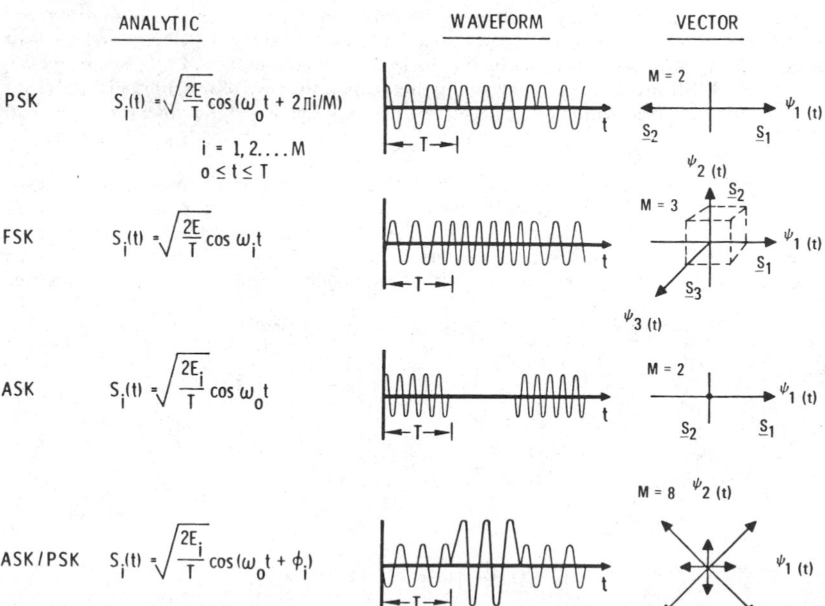

Fig. 48.113 Digital modulation formats. Phase shift keying (PSK); frequency shift keying (FSK), amplitude shift keying (ASK); hybrid combination of ASK/PSK.

frequency shift keying (FSK), amplitude shift keying (ASK), and a hybrid combination of ASK and PSK, sometimes called quadrature amplitude modulation (QAM). The first column lists the analytic expression, the second is a pictorial of the waveform, and the third is a vectorial picture. In the general M-ary signaling case, the processor accepts k source bits at a time and instructs the modulator to produce one of an available set of $M = 2^k$ waveform types. Binary modulation, where $k = 1$, is just a special case of M-ary modulation. For the binary PSK (BPSK) example in Fig. 48.113, M equals two waveform types (2-ary). For the FSK example, M equals three waveform types (3-ary); note that this $M = 3$ choice for FSK has been chosen to emphasize the mutually perpendicular axes. In practice, M is usually a nonzero power of two (2, 4, 8, 16, . . .). For the ASK example, M equals two waveform types; for the ASK/PSK example, M equals eight waveform types (8-ary). The vectorial picture for each modulation type (except FSK) is characterized on a plane whose polar coordinates represent signal amplitude and phase. Signal sets that can be depicted with opposing vectors (phase difference equals 180°) on such a plane, for example BPSK, are called antipodal signals. In the case of FSK modulation, the vectorial picture is characterized by cartesian coordinates, such that each of the mutually perpendicular axes represents a different transmission frequency. Signal sets that can be characterized with such orthogonal axes are called orthogonal signals. Modulation techniques are discussed in greater detail in Section 48.1.

Modulation was defined as that process wherein a carrier or subcarrier is varied by a baseband signal; the hierarchy for digital modulation is shown in Fig. 48.112. When the receiver exploits knowledge of the carrier wave's phase reference to detect the signals, the process is called coherent detection; when it does not have phase reference information, the process is called noncoherent. In ideal coherent detection, prototypes of the possible arriving signals are available at the receiver. These prototype waveforms exactly replicate the signal set in every respect, even RF phase. The receiver is then said to be phase locked to the transmitter. During detection the receiver multiplies and integrates (correlates) the incoming signal with each of its prototype replicas. Under the heading of coherent modulation (see Fig. 48.112) PSK, FSK, and ASK are listed as well as hybrid combinations.

Noncoherent modulation refers to systems designed to operate with no knowledge of phase; phase-estimation processing is not required. Reduced complexity is the advantage over coherent systems, and increased P_E is the tradeoff. Figure 48.112 shows that the modulation types listed in the noncoherent column almost identically replicate those in the coherent column. The only difference is that there cannot be "noncoherent PSK" because noncoherent means without using phase information. However, there is a "pseudo PSK" technique termed differential PSK (DPSK) that utilizes RF phase information of the prior symbol as a phase reference for detecting the current symbol (described under demodulation later in this section).

Two digital modulation schemes, of special interest for use on nonlinear bandlimited channels, are called staggered (or offset) quadraphase PSK (SQPSK or OQPSK) and minimum shift keying (MSK). Both techniques retain low spectral sidelobe levels while allowing efficient detection performance. The generation of both can be represented as two orthogonal, antipodal binary systems with the symbol timing in the two channels offset by one half of a symbol duration. OQPSK uses rectangular pulse shapes, and MSK uses half-cycle sinusoid pulse shapes. Because of the sinusoidal pulse shaping in MSK, it can be viewed as continuous-phase FSK with a frequency deviation equal to one half the bit rate.[17,18]

Demodulation. The analysis of all coherent demodulation or detection schemes involves the concept of distance between an unknown received waveform and a set of known waveforms. Euclideanlike distance measurements are easily formulated in a signal space described by mutually perpendicular axes. It can be shown[19] that any arbitrary finite set of waveforms $s_i(t)$, where $s_i(t)$ is physically realizable and of duration T, can be expressed as a linear combination of N orthonormal waveforms $\phi_1(t), \phi_2(t), \ldots, \phi_N(t)$, such that

$$s_i(t) = \sum_{j=1}^{N} a_{ij}\phi_j(t) \tag{48.130}$$

where

$$a_{ij} = \int_0^T s_i(t)\phi_j(t)\,dt \qquad \begin{array}{l} i = 1, 2, \ldots, M \\ j = 1, 2, \ldots, N \\ N \leq M \end{array} \qquad 0 \leq t \leq T \tag{48.131}$$

and

$$\begin{aligned} \int_0^T \phi_i(t)\phi_j(t)\,dt &= 1 \qquad \text{for} \quad i = j \\ &= 0 \qquad \text{otherwise} \end{aligned} \tag{48.132}$$

Additive white Gaussian noise (AWGN) can similarly be expressed as a linear combination of orthonormal waveforms

$$n(t) = \sum_{j=1}^{N} n_j \phi_j(t) + \tilde{n}(t) \tag{48.133}$$

where

$$n_j = \int_0^T n(t)\phi_j(t)\,dt$$

For the signal-detection problem, the noise can be partitioned into two components

$$n(t) = \hat{n}(t) + \tilde{n}(t) \tag{48.134}$$

where

$$\hat{n}(t) = \sum_{j=1}^{N} n_j \phi_j(t)$$

is taken to be the noise within the signal space, or the projection of the noise components on the signal axes $\phi_1(t), \phi_2(t), \ldots, \phi_N(t)$, and

$$\tilde{n}(t) = n(t) - \hat{n}(t)$$

is defined as the noise outside the signal space. In other words, $\tilde{n}(t)$ may be thought of as the noise that is effectively tuned out by the detector. The symbol $\hat{n}(t)$ represents the noise that will interfere with the detection process; it will henceforth be referred to simply as $n(t)$. Once a convenient set of N orthonormal functions has been adopted [note that $\phi(t)$ is not constrained to any specific form], each of the transmitted signal waveforms $s_i(t)$ is completely determined by the vector of its coefficients

$$\mathbf{s}_i = (a_{i1}, a_{i2}, \ldots, a_{iN}) \qquad i = 1, 2, \ldots, M$$

Similarly, the noise $n(t)$ can be expressed by the vector of its coefficients

$$\mathbf{n} = (n_1, n_2, \ldots, n_N)$$

where $\mathbf{n}$ is a random vector with zero mean and Gaussian distribution.

Since *any* arbitrary waveform set, as well as noise, can be represented as a linear combination of orthonormal waveforms we are justified in using Euclideanlike distance in such an orthonormal space as a decision criterion for the detection of *any* signal set in the presence of AWGN.

Detection in Presence of AWGN. Figure 48.114 illustrates a two-dimensional signal space, the locus of two noise-perturbed prototype binary signals ($\mathbf{s}_1 + \mathbf{n}$) and ($\mathbf{s}_2 + \mathbf{n}$), and a received signal $\mathbf{r}$. The

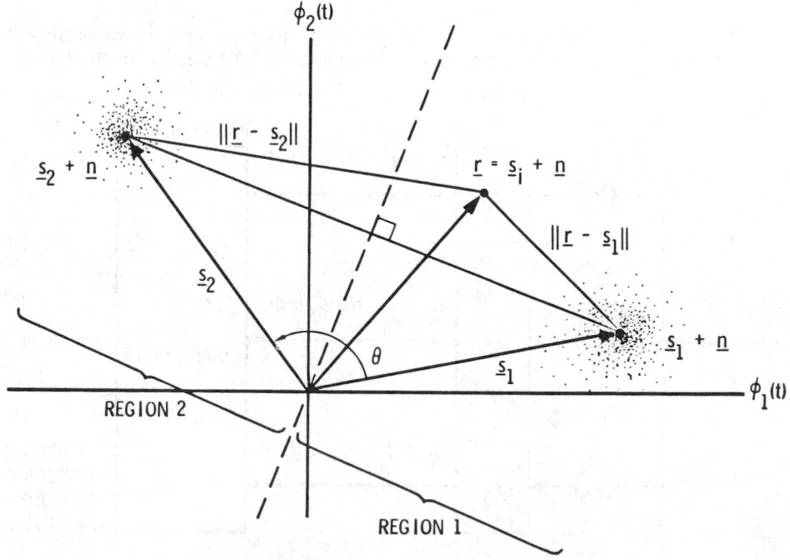

Fig. 48.114 Two-dimensional signal space.

received signal, in vector notation, is $\mathbf{r} = \mathbf{s}_i + \mathbf{n}$, where $i = 1$ or 2. This geometric or vector view of signals and noise facilitates the discussion of digital signal detection. The vectors $\mathbf{s}_1$ and $\mathbf{s}_2$ are fixed, since the waveforms $s_1(t)$ and $s_2(t)$ are nonrandom. The vector or point $\mathbf{n}$ is a random vector; hence $\mathbf{r}$ is also a random vector.

The detector's task after receiving $\mathbf{r}$ is to decide whether signal $\mathbf{s}_1$ or $\mathbf{s}_2$ was actually transmitted. The method is usually to decide on the signal classification that yields the minimum P_E, although other strategies are possible.[20] For the case where M equals two signal classes, with $\mathbf{s}_1$ and $\mathbf{s}_2$ being equally likely and with the noise being AWGN, the minimum-error decision rule turns out to be: Whenever the received signal $\mathbf{r}$ lands in region 1, choose signal $\mathbf{s}_1$; when it lands in region 2, choose signal $\mathbf{s}_2$ (see Fig. 48.114). An equivalent statement is: Choose the signal class such that the distance $d(\mathbf{r}, \mathbf{s}) = \|\mathbf{r} - \mathbf{s}_i\|$ is minimal, where $\|\mathbf{x}\|$ is called the "norm" of vector $\mathbf{x}$ and generalizes the concept of length.

Detection of Coherent PSK. The receiver structure implied by the previous rule is illustrated in Fig. 48.115. There is one product integrator (correlator) for each prototype waveform (M in all); the correlators are followed by a decision stage. The received signal is correlated with each prototype waveform known a priori to the receiver. The decision stage chooses the signal belonging to the correlator with the largest output (largest z_i). For example, let

$$s_1(t) = \sin \omega t$$

$$s_2(t) = -\sin \omega t$$

$$n(t) = \text{a random process with zero mean and Gaussian distribution}$$

Assume $s_1(t)$ was transmitted, so that

$$r(t) = s_1(t) + n(t) \qquad \text{and} \qquad z_i = \int_0^T r(t)s_i(t)\,dt \qquad i = 1, 2$$

The expected values of the product integrators, as illustrated in Fig. 48.115, are found as follows:

$$E[z_1(t = T)] = E\left[\int_0^T \sin^2\omega t + n(t)\sin \omega t\,dt\right] = T/2$$

$$E[z_2(t = T)] = E\left[\int_0^T - \sin^2\omega t + n(t)\sin \omega t\,dt\right] = -T/2$$

where E is the statistical average.

The decision stage must decide which signal was transmitted by measuring its location within the signal space. The decision rule is to choose the signal with the largest value of z_i. Unless the noise is large and of a nature liable to cause an error, the received signal is judged to be $s_1(t)$. In the presence of noise this process is statistical; the optimal detector is one that makes the fewest errors on the average. The only strategy that the detector can employ is to "guess," using some optimized decision rule.

Figure 48.116 shows the detection process with the signal space in mind. It represents a coherent four-level (4-ary) PSK coherent or quadraphase shift keying (QPSK) system. In the terms we used

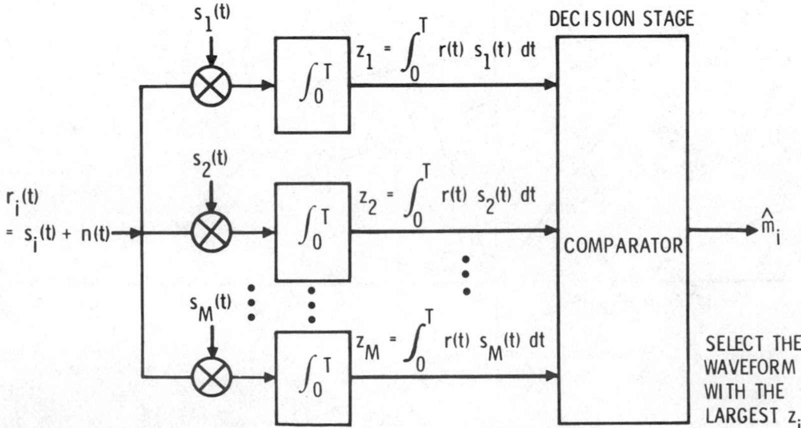

Fig. 48.115 Product integrators (correlators) and decision stage.

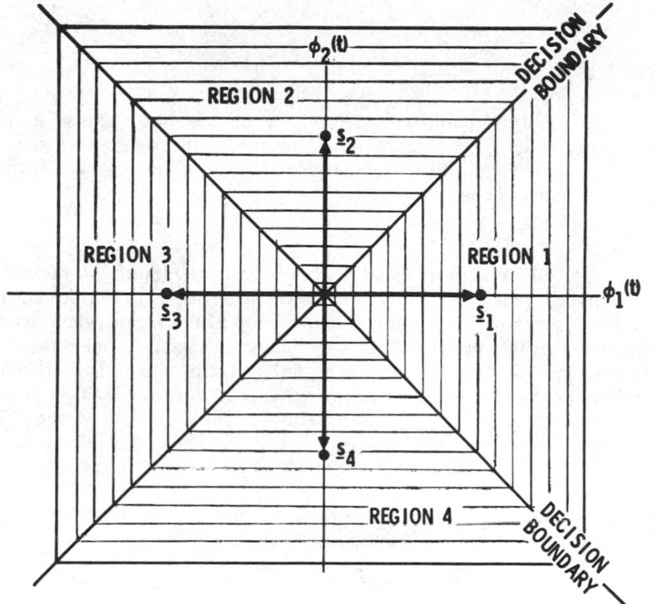

Fig. 48.116 Signal space and decision regions for quadraphase shift keying (QPSK) detection.

earlier for M-ary signaling, $k = 2$ and $M = 2^2 = 4$. Binary source digits are collected two at a time, and for each symbol interval the two sequential digits instruct the modulator as to which of the four waveforms to produce. In general, for coherent M-ary PSK (MPSK) systems, $s_i(t)$ can be expressed as

$$s_i(t) = \sqrt{E/T}\cos(\omega_0 t - 2\pi i/M) \qquad \text{for} \quad 0 \leqslant t \leqslant T \quad \text{where} \quad i = 1, 2, \ldots, M$$

Here E is the energy content of $s_i(t)$, and ω_0 is an integral multiple of $2\pi/T$. We can choose a convenient set of orthogonal axes scaled to fulfill Eq. (48.132) as follows

$$\begin{aligned} \phi_1(t) &= \sqrt{2/T}\cos\omega_0 t \\ \phi_2(t) &= \sqrt{2/T}\sin\omega_0 t \end{aligned} \tag{48.135}$$

Now $s_i(t)$ can be written in terms of these orthogonal coordinates, giving

$$s_i(t) = \sqrt{E}\cos(2\pi i/M)\phi_1(t) + \sqrt{E}\sin(2\pi i/M)\phi_2(t) \tag{48.136}$$

The decision rule for the detector (see Fig. 48.116) is to decide that $s_1(t)$ was transmitted if the received signal point falls in region 1, that $s_2(t)$ was transmitted if the received signal point falls in region 2, and so forth. In other words, the decision rule is to choose the ith waveform with the largest value of correlator output z_i (see Fig. 48.115).

Detection of Coherent FSK. FSK modulation is characterized by the information being contained in the frequency of the carrier wave. A typical set of signal waveforms is described by

$$\begin{aligned} s_i(t) &= \sqrt{2E/T}\cos\omega_i t \qquad \text{for} \quad 0 \leqslant t \leqslant T \qquad i = 1, 2, \ldots, M \\ &= 0 \qquad \text{otherwise} \end{aligned}$$

where E is the energy content of $s_i(t)$ and $(\omega_{i+1} - \omega_i)$ is an integral multiple of $2\pi/T$. The most useful form for the orthonormal coordinates $\phi_1(t), \phi_2(t), \ldots, \phi_N(t)$ is

$$\phi_j(t) = \sqrt{2/T}\cos\omega_j t \qquad j = 1, 2, \ldots, N$$

and, from Eq. (48.131)

$$a_{ij} = \int_0^T \sqrt{2E/T}\cos\omega_i t\sqrt{2/T}\cos\omega_j t\, dt$$

Therefore

$$a_{ij} = \sqrt{E} \qquad \text{for } i = j$$
$$= 0 \qquad \text{otherwise}$$

In other words, the ith signal point is located on the ith coordinate axis at a displacement $\sqrt{E}$ from the origin of the signal space. Figure 48.117 illustrates the signal vectors (points) and the decision regions for a 3-ary coherent FSK modulation ($M = 3$). In this scheme, the distance between any two signal points $\mathbf{s}_i$ and $\mathbf{s}_j$ is constant

$$d(\mathbf{s}_i, \mathbf{s}_j) = \|\mathbf{s}_i - \mathbf{s}_j\| = \sqrt{2E} \qquad \text{for} \quad i \neq j$$

As in the coherent PSK case, the signal space is partitioned into M distinct regions, each containing one prototype signal point. The optimum decision rule is to decide that the transmitted signal belongs to the class whose index number is the same as the region where the received signal was found. In Fig. 48.117, a received signal point $\mathbf{r}$ is shown in region 2. Using the decision rule, the detector classifies it as signal $\mathbf{s}_2$. Since the noise is a random vector, there is a probability greater than zero that the location of $\mathbf{r}$ is due to some signal other than $\mathbf{s}_2$. For example, if the transmitter sent $\mathbf{s}_2$, then $\mathbf{r}$ is the sum of $\mathbf{s}_2 + \mathbf{n}_a$, then the decision to choose $\mathbf{s}_2$ is correct; however, if the transmitter actually sent $\mathbf{s}_3$, then $\mathbf{r}$ must be the sum of $\mathbf{s}_3 + \mathbf{n}_b$ (see Fig. 48.117), and the decision to select $\mathbf{s}_2$ is an error.

Detection of DPSK. With noncoherent systems, no provision is made to phase synchronize the receiver with the transmitter. Therefore, if the transmitted waveform is

$$s_i(t) = \sqrt{2E/T}\cos(\omega_0 t + \phi_i) \qquad i = 1, 2, \ldots, M$$

the received signal can be characterized by

$$r(t) = \sqrt{2E/T}\cos(\omega_0 t + \phi_i + \alpha) + n(t)$$

where α is unknown and is assumed to be a uniformly distributed random variable between zero and 2π.

For coherent detection, product integrators (or their equivalents) are used; for noncoherent detection, such use is generally inadequate because the output of a product integrator is a function of

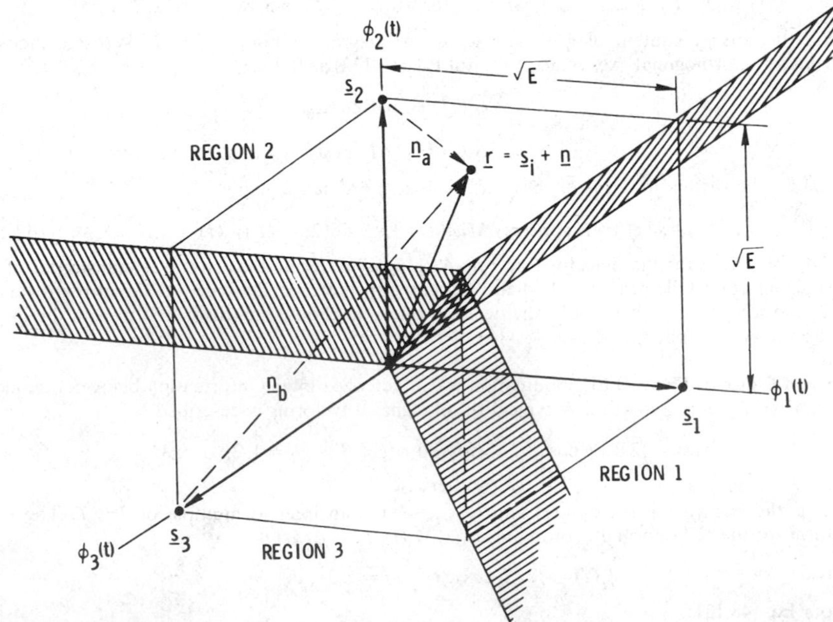

Fig. 48.117 Signal space and decision regions for 3-ary coherent frequency shift keying (FSK) detection.

the unknown angle α. However, if we assume that α varies slowly enough to be considered constant over two period times $(2T)$, the relative phase difference between two successive waveforms is independent of α, that is,

$$(\phi_1 + \alpha) - (\phi_2 + \alpha) = \phi_1 - \phi_2$$

This is the basis for DPSK modulation. The carrier phase of the previous signaling interval is used as a phase reference for demodulation. Its use requires differential encoding of the message sequence at the transmitter since the information is carried by the difference in phase between two successive waveforms. To send the ith message $(i = 1, 2, \ldots, M)$, the current signal waveform must have its phase advanced by $2\pi i/M$ radians over the previous waveform. The detector can then calculate the coordinates of the incoming signal by product integrating it with the locally generated waveforms $\sqrt{2/T}\cos\omega_0 t$ and $\sqrt{2/T}\sin\omega_0 t$. In this way it measures the angle between the current and the previously received signal points (see Fig. 48.118).[19]

One way of viewing the difference between coherent PSK and DPSK is that in the former, the received signal is compared with a clean reference; in the latter, however, two noisy signals are compared with each other. Thus we might say there is twice as much noise in DPSK as in PSK. Consequently, DPSK manifests a degradation of approximately 3 dB when compared with PSK; this number decreases rapidly with increasing signal-to-noise ratio. In general, the errors tend to propagate (to adjacent period times) due to the correlation between signaling waveforms. The tradeoff for this performance loss is reduced system complexity.

Detection of Noncoherent FSK. A noncoherent FSK detector can be implemented with correlators, such as those shown in Fig. 48.115. However, the hardware must be configured as an energy detector, without exploiting phase measurements. For this reason it is implemented with twice as many channel branches as the coherent detector. Figure 48.119 illustrates the in-phase (I) channels and quadrature (Q) channels used to detect the signal set noncoherently. Another possible implementation uses filters followed by envelope detectors; the detectors are matched to the signal envelopes and not to the signals themselves. The phase of the carrier is of no importance in defining the envelope; hence no phase information is used. In the case of binary FSK, the decision whether a 1 or a 0 was transmitted is made on the basis of which of the two envelope detectors has the largest amplitude at the moment of measurement. Similarly, for a multifrequency shift keying (M-ary FSK, or MFSK) system, the decision as to which of the M signals was transmitted is made on the basis of which of the M envelope detectors has maximum output.

Probability of Error. The calculations for probability of error (P_E), which can be viewed geometrically (see Fig. 48.114), involve finding the probability that given a particular signal, say $\mathbf{s}_1$, the noise vector $\mathbf{n}$ will give rise to a received signal falling outside region 1; all P_E calculations have this goal. For the general M-ary signaling case, the probability of making an incorrect decision is termed the probability of symbol error, or simply P_E. It is often convenient to specify system performance by the probability of bit error (P_B), even when decisions are made on the basis of symbols for which $k > 1$. P_E and P_B are related as follows, for orthogonal signals,[21]

$$P_B/P_E = (2^{k-1})/(2^k - 1)$$

For nonorthogonal schemes, such as MPSK signaling, one often uses a binary-to-M-ary code such

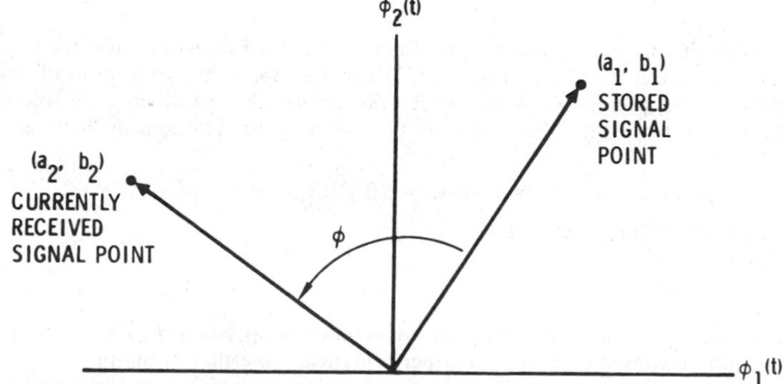

Fig. 48.118 Signal space for differential phase shift keying (DPSK) detection.

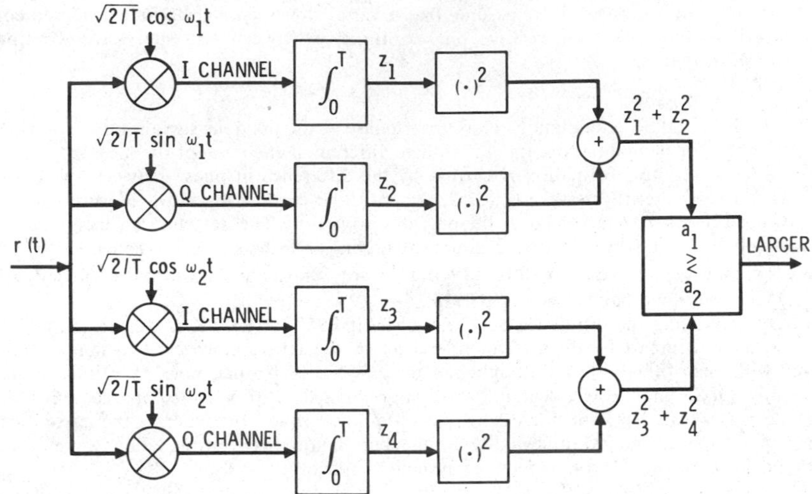

Fig. 48.119 Quadrature receiver for noncoherent FSK detection.

that binary sequences corresponding to adjacent symbols (phase shifts) differ in only one bit position; one such code is the Gray code. When an M-ary symbol error occurs, it is more likely that only one of the k input bits will be in error. For such signals[3]

$$P_B \cong P_E/\log_2 M = P_E/k \qquad \text{for} \quad P_E \ll 1$$

For convenience, the discussion in this section is restricted to BPSK ($k = 1$, $M = 2$) modulation. For the binary case, the symbol error probability equals the bit error probability. Assume that signal $s_1(t)$ has been transmitted and that $r(t) = s_1(t) + n(t)$. Assuming equally likely signals, and recalling that the decision of region 1 vs. region 2 depends on the product integrators and the decision stage (see Fig. 48.115), we can write

$$P_E]_{\text{binary}} = P_B = \Pr\left[\int_0^T r(t)s_2(t)\,dt > \int_0^T r(t)s_1(t)\,dt \mid r(t) = s_1(t) + n(t)\right]$$

for $0 \leqslant t \leqslant T$. The solution for the P_B expression is, for the case of additive white Gaussian noise (see Section 47.1)

$$P_B = 1/\sqrt{2\pi}\int_{\sqrt{E_b/N_0(1-\cos\theta)}}^{\infty}\exp(-u^2/2)\,du$$

where E_b is the signal energy per bit in joules, N_0 is the noise density at the receiver in watts per hertz, and θ is the angle between $\mathbf{s}_1$ and $\mathbf{s}_2$ (see Fig. 48.114). When $\theta = \pi$, the signals are termed antipodal, and the P_B becomes

$$P_B = 1/\sqrt{2\pi}\int_{\sqrt{2E_b/N_0}}^{\infty}\exp(-u^2/2)\,du \equiv Q\left(\sqrt{2E_b/N_0}\right) \tag{48.137}$$

The same kind of analysis is pursued in finding the P_B expressions for the other types of modulation. The parameter E_b/N_0 in Eq. (48.137) can be expressed as the ratio of average signal power to average noise power, S/N (or SNR). By arbitrarily introducing the baseband signal bandwidth W, we can write the following identities, showing the relationship between E_b/N_0 and SNR:

$$E_b/N_0 = ST/N_0 = S/RN_0 = SW/RN_0W = (S/N)(W/R) \tag{48.138}$$

where S = average modulating signal power
T = bit time duration
$R = 1/T$ = bit rate
$N = N_0W$

The dimensionless ratio E_b/N_0 (required to achieve a specified P_B) is uniformly used for characterizing digital communications system performance. Note that optimum digital signal detection implies a correlator (or matched filter) implementation, in which case the signal bandwidth is equal to the noise bandwidth. Often we are faced with a system model for which this is not the case (less than optimum); in practice we just reflect a factor into the required E_b/N_0 parameter that accounts for the suboptimal detection performance. Therefore, required E_b/N_0 can be considered a metric that characterizes the performance of one system vs. another; the smaller the required E_b/N_0, the more efficient the system modulation and detection process.

The P_B expressions for the binary modulation schemes discussed are listed in Table 48.19 and are graphically compared in Fig. 48.120. At large SNRs, it can be seen that there is approximately a 4-dB difference between the best (coherent PSK) and the worst (noncoherent FSK). In some cases, 4 dB is a small price to pay for the implementation simplicity gained in going from a coherent PSK to a noncoherent FSK; however, for some applications even a 1-dB saving is worthwhile. There are other

TABLE 48.19. PROBABILITY OF BIT ERROR (P_B) FOR SELECTED BINARY MODUALTION SCHEMES

Modulation	P_B[a]
Coherent PSK	$Q\left(\sqrt{\frac{2E_b}{N_0}}\right)$
Noncoherent DPSK	$1/2\exp(-E_b/N_0)$
Coherent FSK	$Q\left(\sqrt{\frac{E_b}{N_0}}\right)$
Noncoherent FSK	$1/2\exp[-1/2(E_b/N_0)]$

[a] Here $E_b/N_0 =$ (energy/bit)/noise density $= S/N_0R$ = signal power received/(noise density × bit rate) and $Q(X) = (1/\sqrt{2\pi})\int_x^\infty \exp(-u^2/2)\,du$; $Q(X) \cong (1/x\sqrt{2\pi})\exp(-x^2/2)$, for $x > 3$.

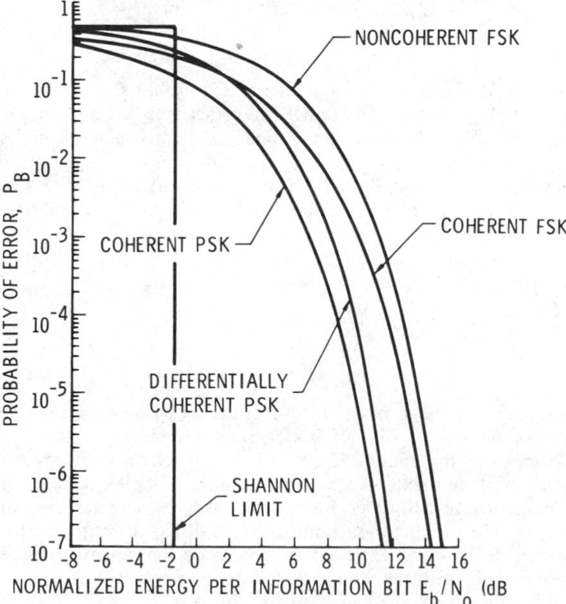

Fig. 48.120 Probability of bit error for selected binary modulation schemes. Frequency shift keying (FSK); phase shift keying (PSK).

considerations besides P_B and system complexity; for example, in some cases (such as randomly fading propagation conditions), a noncoherent system is more robust and desirable because there may be difficulty in establishing a coherent reference.

An exception to Table 48.19 and Fig. 48.120 is worth mentioning, in light of today's bandwidth efficient modulation schemes. MSK modulation, which can be regarded as coherent FSK, manifests error-rate performance equal to BPSK when detected with the appropriate receiver.[18]

Channel Coding

Channel encoding (see Figs. 48.111 and 48.112) refers to the data transformation, performed after source encoding but prior to modulation, that transforms source bits into channel bits. Channel coding is partitioned into two groups, waveform coding and structured sequences (see Fig. 48.112). Waveform (or signal design) coding is herein defined to mean any source data transformation that renders the detection process less subject to errors and thereby improves transmission performance. It can best be viewed as a transformation that demonstrates improved performance in an overall or "gestalt" sense, because the encoding produces a set of signals with better distance properties than those of the original signal set. The structured sequences category, by comparison, improves performance by embedding the data with structured redundancy, which may then be used to detect and correct transmission errors.

Waveform Coding. M-ary signaling was described as a modulation or coding scheme that processes k bits at a time. The system directs the modulator to choose one of its $M = 2^k$ waveforms for each k-bit sequence, where M is the symbol-set size and k is the number of binary digits that each symbol represents. M-ary signaling alone, for the case in which $k > 1$, can be regarded as a waveform coding procedure that affects system performance. Orthogonal signals manifest improved P_B at the expense of bandwidth as k increases; nonorthogonal signals manifest improved bandwidth efficiency (R/W) at the expense of power or P_B performance. R/W is measured in bits per second or per hertz (b/s/Hz); for binary signaling, the typical value of bandwidth efficiency is approximately 1 b/s/Hz. However, present-day multiphase shift keying (MPSK) and quadrature-amplitude modulation (QAM) systems frequently have bandwidth efficiencies of 3 b/s/Hz and higher.[22,23]

Another example of waveform coding is the use of an improved signal set as replacement for the original data symbols. The most popular of these codes are referred to as orthogonal and biorthogonal signal sets. The orthogonal signal set $s_i(t)$, where $i = 1, 2, \ldots, M$, is said to be orthogonal if and only if

$$
\begin{aligned}
z_{ij} = 1/E\int_0^T s_i(t)s_j(t)\,dt &= 1 \qquad \text{for} \quad i = j \\
&= 0 \qquad \text{otherwise}
\end{aligned}
\tag{48.139}
$$

where it is assumed that all M signals have equal energy E and that T is the symbol duration. Just as M-ary signaling with an orthogonal modulation format (such as MFSK) improves the P_B performance with increasing k, so too coding with an orthogonally constructed signal set, prior to MPSK modulation, produces the same improvement.

A biorthogonal signal set can be obtained from an orthogonal set of $M/2$ signals by augmenting it with the negative of each signal. The biorthogonal set consists of a combination of orthogonal and antipodal signals. Since antipodal signal vectors have better distance properties than orthogonal ones (see Fig. 48.113), biorthogonal codes perform slightly better than orthogonal ones. With respect to z_{ij} of Eq. (48.139), biorthogonal codes can be characterized as follows:

$$
\begin{aligned}
z_{ij} &= 1 \qquad \text{for} \quad i = j \\
&= -1 \qquad \text{for} \quad i \neq j, \quad |i - j| = M/2 \\
&= 0 \qquad \text{for} \quad i \neq j, \quad |i - j| \neq M/2
\end{aligned}
$$

For completeness, a code generated from an orthogonal set by deleting the first digit of each code word is called a transorthogonal or simplex code. Such a code represents the minimum energy equivalent (in the P_B sense) of the equally likely orthogonal set. In comparing the performance of orthogonal, biorthogonal, and simplex codes, we can state that simplex coding requires the minimum SNR for a specified bit-error rate. However, for large k all three schemes are essentially identical in performance as they approach the Shannon limit of -1.6 dB.[5,24] Biorthogonal coding requires half the bandwidth of the others. However, for each of these codes bandwidth requirements (and system complexity) grow exponentially with the value of k.

Structured Sequences (Linear Block Codes). Channel coding with structured sequences represents a method of inserting structured redundancy into the source data so that transmission errors can be identified. Structured sequences are partitioned into two important subcategories: block coding and convolutional coding (see Fig. 48.112). With block coding, the source data is first segmented into blocks of k data bits each; each block can represent any one of $M = 2^k$ distinct messages. The encoder transforms each message block into a larger block of n digits. This set of 2^k coded messages is called a code block. The $(n - k)$ digits, which the encoder adds to each message block, are called redundant digits; they carry no new information. The ratio of data bits to total bits within a block, k/n, is called the code rate. The code itself is referred to as an (n, k) code.

Suppose that $\mathbf{c} = (c_1, c_2, \ldots, c_n)$, where $c_j = 1$ or 0, is the transmitted code word, from a set of $i = 1, \ldots, M$ code words, and that $\mathbf{r} = (r_1, r_2, \ldots, r_n)$ is the sequence received at the decoder input. If all code words have equal likelihood of being transmitted, the optimum decoding scheme to use is called maximum likelihood decoding; it is similar to the optimum demodulation scheme under similar a priori assumptions. The decoder computes the conditional probability $P(\mathbf{r} \mid \mathbf{c}_i)$ for all 2^k code words. The code word $\mathbf{c}_t$ is identified as the transmitted word if $P(\mathbf{r} \mid \mathbf{c}_t)$ is the maximum of the computed probabilities.

The performance improvement possible with channel coding can be illustrated with the following example of a (15, 11) single error-correcting code. The notation (15, 11) means that each block of 15 bits comprises 11 data bits and 4 redundant bits. Consider the following uncoded transmission. Assume BPSK modulation: $S/N_0 = 43{,}776$; data rate $R = 4800$ b/s. P_B^U and P_M^U represent the uncoded probabilities of bit error and message error, respectively. P_B^C and P_M^C represent the coded probabilities of bit error and message error, respectively.

Without Coding. From Eq. (48.138)

$$E_b/N_0 = S/RN_0 = 9.12 \ (= 9.6 \text{ dB})$$

$$P_B^U = Q\left(\sqrt{2E_b/N_0}\right) = Q\left(\sqrt{18.24}\right) = 1.02 \times 10^{-5} \tag{48.140}$$

where

$$Q(x) \cong \left(1/x\sqrt{2\pi}\right)\exp(-x^2/2) \qquad \text{for} \quad x > 3$$

$$P_M^U = 1 - (1 - P_B^U)^{11} = 1.12 \times 10^{-4} \tag{48.141}$$

With Coding. The coded bit rate R_C is 15/11 times the data bit rate.

$$R_C = 4800 \times 15/11 \cong 6545 \text{ b/s}$$

$$E_b/N_0 = S/R_C N_0 = 6.688 \ (= 8.25 \text{ dB})$$

The E_b/N_0 for the coded bit is a little less than for the uncoded bit because the bit rate has increased but the transmitter power is assumed to be fixed.

$$P_B^C = Q\left(\sqrt{2E_b/N_0}\right) = Q\left(\sqrt{13.38}\right) = 1.36 \times 10^{-4} \tag{48.142}$$

It can be seen by comparing Eq. (48.140) and Eq. (48.142) that the bit error rate has degraded; more bits must be detected during the same time interval and with the same available power. The performance improvement due to the coding is not yet apparent. We now compute the coded message error-rate P_M^C, as follows

$$P_M^C = \sum_{k=2}^{15} \binom{15}{k} (P_B^C)^k (1 - P_B^C)^{15-k}$$

The summation is started with $k = 2$, since the code corrects all single errors within a block of $n = 15$ bits. A good approximation is obtained by using only the first term of the summation. For P_B^C we use the value computed in Eq. (48.142)

$$P_M^C \cong \binom{15}{2} (P_B^C)^2 (1 - P_B^C)^{13} = 1.94 \times 10^{-6} \tag{48.143}$$

Equation (48.143) yields the message error-rate for the block of 15 coded bits. In this typical example, it can be seen by comparing Eq. (48.141) with Eq. (48.143) that the probability of message error has improved by a factor of 58 through the use of a block code.

Most research on block codes has been concentrated on a subclass of linear codes known as cyclic codes. A cyclic code word, after any number of end-around cyclic shifts, has the property of remaining a valid code word from the original set of code words. Cyclic codes are attractive because they can be easily implemented with feedback shift registers. The decoding methods are simple and efficient and generally use the same feedback shift registers as are employed for encoding.[25,26]

Structured Sequences (Convolutional Coding). A convolutional encoder convolves an input data sequence of bits with an encoding function, as shown in Fig. 48.121. The encoder is mechanized with a K stage shift register and n modulo-2 summers, where K is called the constraint length. The source is represented by the data bit sequence $U = (u_1, u_2, \ldots, u_i, \ldots)$. At the ith unit of time, data bit u_i is shifted into the first stage; all bits in the register are shifted one stage to the right, and the output of the n summers is sequentially sampled and transmitted. Since there are n code bits per data bit, the code rate is $1/n$ in this case. For the general case, k bits at a time are shifted into the register, and the code rate is k/n. The n code bits occurring at time i constitute the ith branch of the code word $Y_i = (y_{1i}, y_{2i}, \ldots, y_{ni})$. The code word Y consists of the sequence of branches $Y = (Y_1, Y_2, \ldots)$.

Let the state of the encoder at time i be defined as $X_i = (u_{i-1}, u_{i-2}, \ldots, u_{i-K+1})$, which is the contents of the right-most $K-1$ stages of the shift register. The ith code word branch Y_i is completely determined by state X_i and the present input bit u_i; thus the state X_i represents the history of the encoder in determining the encoder output. The encoder state is said to be Markov, in the sense that the probability $P(X_{i+1} \mid X_i, X_{i-1}, \ldots, X_0)$ of being in state X_{i+1}, given all previous states, depends only on state X_i, that is, the probability is equal to $P(X_{i+1} \mid X_i)$. One simple way to represent an encoder is by a state diagram, as shown in Fig. 48.122 for the encoder with $K = 3$. The states of the diagram represent the possible contents of the right-most $K-1$ stages of the encoder shift

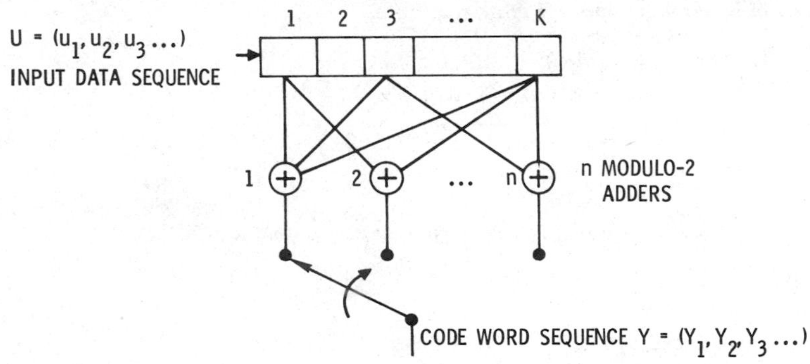

Fig. 48.121 Convolutional encoder with constraint length K and rate $1/n$.

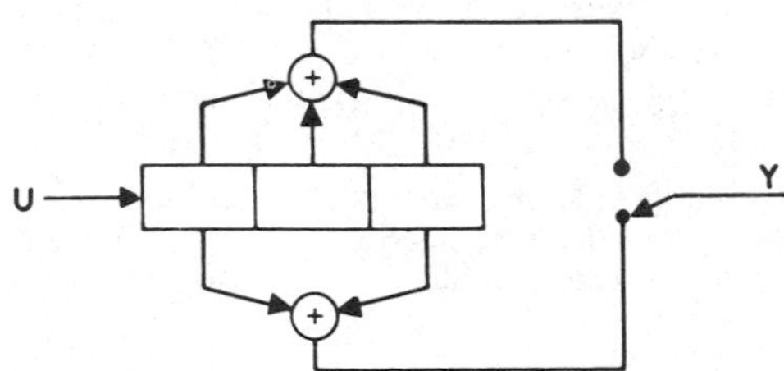

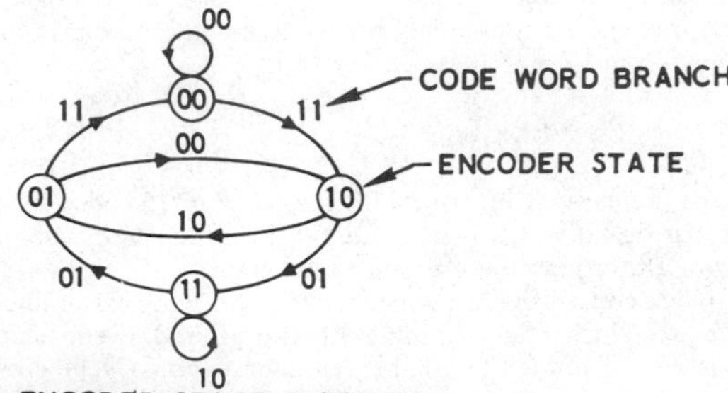

Fig. 48.122 Characterization of encoder by state diagram.

register. There are two transitions from each state, corresponding to the two possible input bits, and there is a code word branch associated with each transition. The state diagram can be used to obtain a transfer function, which in turn can be used to derive error probability bounds.[27]

A convenient way of incorporating encoder time history into the state diagram is through the trellis diagram shown in Fig. 48.123. At each time unit the trellis shows all possible transitions between states. There are two possible paths leaving each state, corresponding to the two possible values of the input data bit. By convention, a dashed line in the trellis corresponds to input data 1 and a solid line to input data 0. The output code-word branches corresponding to the transitions are also shown for the encoder of Fig. 48.122; they appear as labels on the trellis branches.

An optimal decoder makes the maximum likelihood estimate of the transmitted code word Y, given the observation Z. The decoder chooses code word $\hat{Y}$ if $P(Z \mid \hat{Y}) = \max_Y P(Z \mid Y)$. Since the noise is assumed to be independent

$$P(Z \mid Y) = \prod_{i=1}^{\infty} P(Z_i \mid Y_i) = \prod_{i=1}^{\infty} \prod_{j=1}^{n} P(z_{ji} \mid y_{ji})$$

where Y_i is the ith branch of code word Y, Z_i is the ith branch of the received sequence Z, z_{ji} is the jth code bit of Z_i, and y_{ji} is the jth code bit of Y_i, each branch comprising n coded bits. The decoder problem consists of choosing a path through the trellis of Fig. 48.123 (each possible path defines a code word) such that $\prod_{i=1}^{\infty}\prod_{j=1}^{n} P(z_{ji} \mid y_{ji})$ is maximized.

Binary channels are characterized by the need to make hard decisions (two-level decisions) on the received code bits. Continuous channels are characterized by the ability to make soft decisions (multilevel decisions). A multilevel decision can be thought of as a decision with a confidence factor attached. The ability to carry soft decisions along during the decoding process results in better decoding performance (approximately 2 dB) than if hard decisions are made. Ultimately, for a digital system all decisions must be converted to hard decisions, that is, 1 or 0.

A brute-force maximum-likelihood decoder calculates the likelihood of the received data on all the paths through the trellis. The number of paths for an L-bit information sequence is 2^L; the brute-force method becomes impractical as L increases. The Viterbi algorithm essentially performs maximum likelihood decoding; however, it reduces the computational load by taking advantage of the special structure in the code trellis. The advantage of Viterbi decoding (compared with brute-force) is that the decoder complexity is a linear rather than an exponential function of L.[28] The algorithm involves calculating a metric (measure of similarity) between the received signal (at time t_i) and all the trellis paths entering each state (at time t_i), where $i = 1, 2, \ldots$. In the event that two paths terminating on a given state are redundant, the one having the largest metric is stored (the surviving path). This selection of survivor is performed for all paths entering each of the other states. The decoder continues in this way to advance deeper into the trellis, making decisions by eliminating the least likely paths. Surviving paths need to be stored over an interval of about five constraint lengths to allow for coding delay. Storage requirements grow exponentially with constraint length; the present state of the art limits Viterbi decoders to about $K = 10$. Viterbi decoders are very cost effective for moderate error rates but cannot achieve very low error rates effectively. On the other hand, they are capable of very high speeds, where sequential decoders become uneconomical.[29]

A sequential decoder works by generating hypotheses about the transmitted data sequence; it

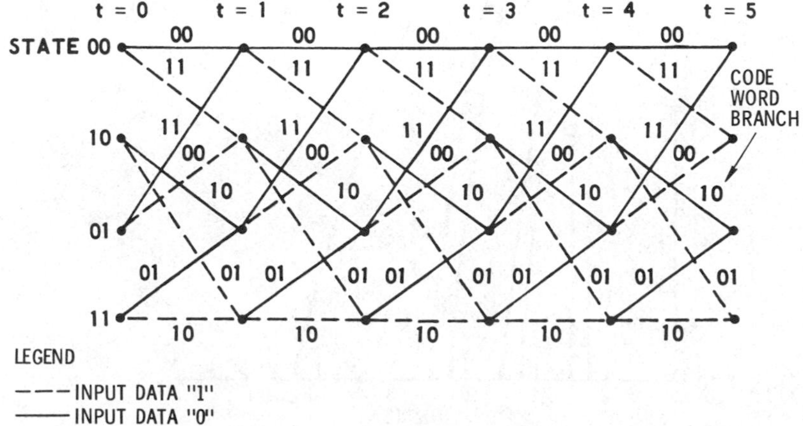

Fig. 48.123 Encoder trellis diagram.

generates a metric between its hypotheses and the received signal. It goes forward as long as the metric remains within tolerance; otherwise it goes backward, changing hypotheses until it finds an improved metric through a trial-and-error search. It can be implemented to work with hard or soft decisions, but soft decisions are usually avoided because they greatly increase the required storage and computations. Decoder complexity is relatively insensitive to the code constraint length; hence constraint lengths are generally made very large ($K = 40$), which is an important factor in providing such low P_B performance. The number of poor hypotheses and backward searches are a function of the SNR; with greater noise, more hypotheses must be generated. Because of this variability in computational load, buffer storage must be provided. Occasionally these buffers will overflow, leaving a section of data in its encoded form. Therefore an important part of a sequential decoder specification is the probability of buffer overflow.

The performance of these two popular solutions to the decoding problem, Viterbi decoding and sequential decoding, is illustrated in Fig. 48.124. The curves compare Viterbi decoding (rates 1/2 and 1/3 hard decision) with Viterbi decoding (rates 1/2 and 1/3 soft decision) and with sequential decoding (rates 1/2 and 1/3). Figure 48.124 illustrates that for $P_B = 10^{-5}$, coding gains of approximately 7 dB can be achieved with sequential decoders. Since Shannon's work foretold the potential of 11.2 dB of coding gain,[5] it appears that the major portion of what is theoretically possible has already been accomplished.

Interleaving. Codes used for satellite channels are designed to combat independent errors; they are called random-error-correcting codes. There are also channels (e.g., telephone lines, magnetic tape storage, troposcatter links, and sometimes satellite channels) on which the disturbances introduce errors that are clustered together in bursts. Use of an interleaver is a way of enhancing the random-error-correcting capabilities of a code so that it is also useful in a burst-noise environment. The interleaver shuffles the encoded bits over a span of several block lengths (for block codes), and several constraint lengths (for convolutional codes). The span length required is determined from the need for error protection over some specified burst duration. The details of the bit redistribution pattern must be known to the receiver as well as to the transmitter in order for the bit stream to be

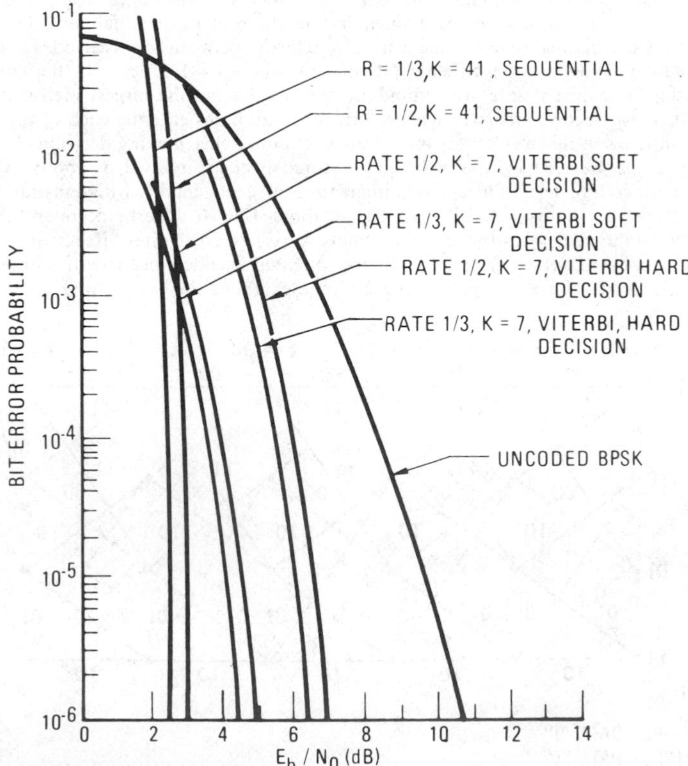

Fig. 48.124 Performance of Viterbi decoders and sequential decoders. Binary phase shift keying (BPSK).

deinterleaved before being decoded. The overall result is to "spread out" the effect of burst noise so that induced errors appear to be independent (thereby matching the code's error-correcting capabilities).

Ramsey[30] discusses interleaver configurations that reorder a sequence of symbols so that no contiguous sequence of n_2 symbols in the reordered sequence contains any pair of symbols that were separated by fewer than n_1 symbols in the original ordering. He also shows that one such configuration is optimum in the sense of minimum possible coding delay and storage capacity.

Multiplexing and Multiple Access

The terms "multiplexing" and "multiple access" both refer to the sharing of a communication resource (CR) (see Figs. 48.111 and 48.112). There is a subtle difference between them. With multiplexing, the system controller (which may be a human, an algorithm, or even a wired logic board, either centralized or distributed) has instantaneous knowledge of all users' requirements or plans for CR sharing. No overhead is needed to organize the resource allocation, and the process is usually considered to take place within the confines of a local site (e.g., an assembly or a circuit board). Multiple access usually involves the remote accessing of a resource; additionally there may be a nonzero amount of time required for the controller to become aware of each user's CR needs. Such time constitutes an overhead impact to the system utilization.

There are fundamentally two approaches to improving the ability of a satellite to support communications traffic. One way is to seek technological improvements toward increasing EIRP (effective radiated power referenced to isotropic), or to provide more bandwidth (there is great interest in developing the 30/20-GHz band for satellite communications). The second approach is to make the allocation of the CR more efficient. This second approach is the domain of communications multiple access.

The problem is to efficiently allocate portions of the satellite's fixed CR to a large number of users who seek to communicate digital information to each other at a variety of bit and message rates, and with various traffic requirements. A mechanism must be employed whereby the multiple signals can access the CR without creating interference to each other in the detection process. The avoidance of such interference requires that signals on one CR channel do not increase the probability of error in another channel. It should be obvious that orthogonality of the signals on separate channels suffices to avoid interference between users. Two signals are orthogonal if they can be described in the time domain by Eq. (48.139). Similarly, they are orthogonal if they can be described in the frequency domain by

$$\int_{-\infty}^{\infty} S_i(f) S_j(f)\, df = \begin{cases} 1 & \text{for } i = j \\ 0 & \text{otherwise} \end{cases} \tag{48.144}$$

where the functions $S_i(f)$ are the Fourier transforms of some signal waveforms $s_i(t)$. Channelization characterized by Eq. (48.139) is called time division multiple access (TDMA), and that characterized by Eq. (48.144) is called frequency division multiple access (FDMA). In general, orthogonality can be achieved using code division multiple access (CDMA), a method involving both the time and frequency domains. In practice, CDMA offers user access flexibility but requires more complicated signal processing than either TDMA or FDMA.[31]

A diagram of the time-frequency resource is shown in Fig. 48.125. We assume there are M users

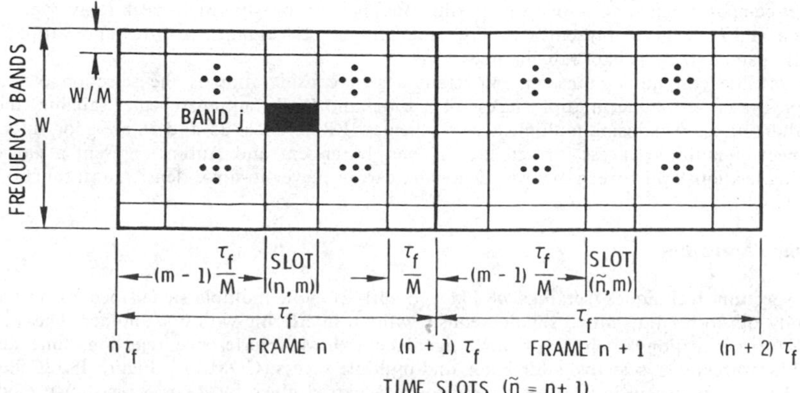

Fig. 48.125 Communication resource: time-frequency channelization.

and that the total frequency bandwidth available is W Hz. M frequency bands of width W/M Hz are available and intervals of time duration τ_f/M seconds can be envisioned. We assume that the channel is time synchronized so that periodic time intervals, called slots, are available. The slots are defined by

$$\text{Slot}\,(n,m) = [t : n\tau_f + (m-1)\tau_f/M \leqslant t \leqslant n\tau_f + m\tau_f/M]$$

where n and m are integers. The time interval $[n\tau_f, (n+1)\tau_f]$ is called a frame and has duration τ_f seconds. The domain of the unit signal is the intersection of the slot (n,m) and "band j" in Fig. 48.125. For any channelization of the CR, we assume that a modulation-coding system is chosen so that the full bandwidth W of the CR can support R bits per second as the total available CR bit rate. In any subchannel having bandwidth W/M Hz, the associated bit rate will be R/M bits per second.

Two additional access schemes useful for satellite communications are space division multiple access (SDMA) and polarization division multiple access (PDMA). For production of SDMA, the signals in different channels (allowed to occupy the same frequency band) are transmitted by use of spot beam antennas. The spot beams produce orthogonality by physically separating the signals so they can be collected with physically separated receivers. For production of PDMA, the antennas are orthogonally polarized to separate the electromagnetic fields. A flexible implementation of SDMA, called satellite-switched TDMA (SS/TDMA), uses a microwave switch matrix in the satellite. The switching sequence of the matrix is controlled according to a programmable memory; the TDMA signals are cyclically interconnected among different antenna spot beams in rapid sequence. An earth station in the network communicates with those in other beams by transmitting TDMA bursts in proper timing to the sequence.[32]

The multiple-access schemes discussed thus far would be termed fixed assignment for the case in which a user has access to the channel independent of the actual message traffic. By comparison, dynamic assignment schemes, sometimes called demand assignment multiple access (DAMA), give the user access to the channel only when he or she has a message to send. If the traffic from users tends to be burstlike or intermittent, great efficiencies can be gained by use of DAMA procedures to access the CR. The Intelsat IV satellite implemented a DAMA scheme called single channel per carrier access-on-demand equipment (SPADE) in the early 1970s. At each terminal the SPADE subsystem responds to service requests by allocating an unused carrier frequency to the user; it then notifies the other terminals of the frequency use through a common signaling channel. The initiating terminal requests a frequency pair at random; such random selection makes it unlikely that two terminals will simultaneously request the same channel unless very few channels remain. SPADE utilizes QPSK modulation at 32 ks/s (kilosymbols per second), using a signal bandwidth of 38 kHz for each 64-kb/s digitized voice channel ($R/W = 1.68$ b/s/Hz). This first important commercial satellite use of a DAMA scheme has resulted in more efficient use of power and bandwidth per channel than any of the fixed multiple-access schemes used earlier.[33]

The development of packet-switching techniques represents an important breakthrough in communications resource sharing. In circuit-switched networks such as the telephone network, calls and message routing are set up prior to the commencement of message transmission. Once the route has been established, the message is transmitted on the dedicated circuit; after completion of the call, the circuit is disconnected. In packet communications, messages are packetized (partitioned into modular groups, each containing an address header). Each packet may be regarded as moving autonomously through the network, queuing at specific nodal points together with packets from other traffic. The key feature of a packet-switching system is the potential for very efficient utilization of a communications or computer network, especially in the presence of bursty (high peak-to-average) traffic. Bhargava and colleagues[5] present a concise summary of performance features and various access methods characterizing packet satellite networks.

For satellite communications, an important design consideration is the selection of signaling techniques suitable for the multiple access of a wideband hard-limiting repeater satellite; there are many alternatives in choosing multiple access schemes.[34] References 35 and 36 are good tutorials on the subject of multiple access for satellite systems. Nirenberg and Rubin[35] present a particularly interesting relationship between message delay and carrier power-to-noise density, as a function of bit error rate.

Frequency Spreading

Spread spectrum techniques (see Figs. 48.111 and 48.112) allow multiple signals occupying the same RF bandwidth to be transmitted simultaneously without interfering with one another. The technique is used for applications such as privacy, signal covertness, interference rejection, time delay or ranging measurements, selective addressing, and multiple access (CDMA).[37] Figure 48.126 illustrates a spread spectrum system in its most general form. A carrier given by $A \cos \omega_0 t$ is shown modulated

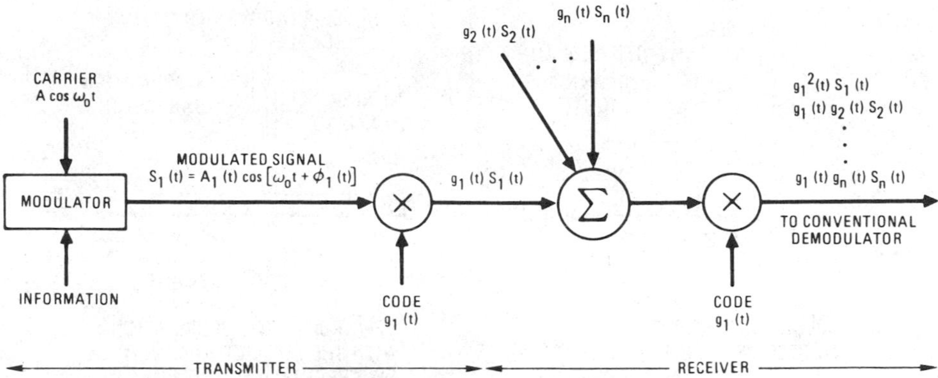

Fig. 48.126 Spread spectrum system.

with information to produce a signal $s_1(t)$, where

$$s_1(t) = A_1(t)\cos[\omega_0 t + \phi_1(t)]$$

No restriction has been placed on the type of modulation that can be used. $s_1(t)$ is now multiplied by some code function $g_1(t)$. (Frequently each code function is kept secret, and its use is restricted to a community of authorized users.) The resulting signal $g_1(t)s_1(t)$ is transmitted over the channel. At the same time, other users have multiplied their signals by other code functions. The signal present at the receiver is the linear combination of the emanations from each user

$$g_1(t)s_1(t) + g_2(t)s_2(t) + \cdots + g_n(t)s_n(t) \tag{48.145}$$

Multiplication of $s_1(t)$ by $g_1(t)$ produces a signal whose spectrum is the convolution of the spectra of the two component signals. Thus if the signal $s_1(t)$ is relatively narrowband compared with the code or spreading signal $g_1(t)$, the product will have nearly the bandwidth of $g_1(t)$. Assume that the receiver is configured to receive messages from user number 1. The first stage of the receiver multiplies the incoming signal of Eq. (48.145) by $g_1(t)$. The output of the multiplier will yield the following terms

Wanted signal

$$g_1^2(t)s_1(t)$$

Unwanted signals

$$g_1(t)\,g_2(t)s_2(t) + g_1(t)\,g_3(t)s_3(t) + \cdots + g_1(t)\,g_n(t)s_n(t) \tag{48.146}$$

If the code functions $g_i(t)$, where $i = 1, 2, \ldots, n$, are chosen with orthogonal properties similar to Eq. (48.139), then the desired signal can be extracted perfectly, and the unwanted signals yielding zero terms are easily rejected.

Figure 48.127*a* illustrates the wideband input to the receiver; it consists of wanted and unwanted signals, each spread by its own code, with code rate R_p, and each having a spectrum of the form $(\sin^2 x)/x^2$. Figure 48.127*b* illustrates the spectrum after correlation with the code $g_1(t)$ (despreading). The unwanted signals of Eq. (48.146) remain effectively spread by $g_1(t)g_i(t)$, where $i = 2, 3, \ldots, n$. Only that portion of the spectrum of the unwanted signals falling in the information bandwidth of the receiver will cause interference to the wanted signal.

If there is any jamming signal at the receiver (intentional or otherwise), the spreading signal will affect it just as it did the original signal at the transmitter. Thus even a narrowband jamming signal in the middle of the information band will be spread to the bandwidth of the spreading signal; call it W. If the power of the jamming signal is J watts, its average density can be treated as wideband noise $J_0 = J/W$ watts per hertz. If $s_1(t)$ has power S watts and if the data rate is R bits per second, the received energy per bit is $E_b = S/R$ watt-seconds and, similar to Eq. (48.138), the parameter E_b/J_0 that dictates the bit error rate performance in the presence of wideband noise jamming can be written

$$E_b/J_0 = (S/J)(W/R)$$

Thermal noise is present also, but we will assume that the jamming noise is so much greater than the

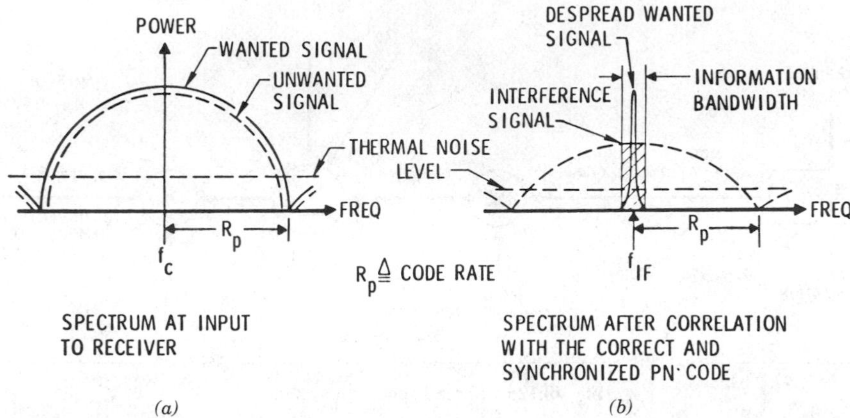

Fig. 48.127 Spread spectrum detection: (*a*) Wideband input to the receiver, with wanted and unwanted signals; (*b*) spectrum after despreading.

thermal noise that we can neglect the latter. Hence the ratio of jamming power to signal power is

$$J/S = (W/R)/(E_b/J_0) \tag{48.147}$$

This illustrates that if E_b/J_0 is the minimum ratio of bit energy to jamming noise density needed to support a given bit error rate, and if W/R is the ratio of spread bandwidth to the original data rate, also called the processing gain, then J/S is the maximum tolerable ratio of jamming power to signal power. It is commonly used as a figure of merit to describe a system's vulnerability to jamming; the larger the J/S, the greater the resistance against jamming. Another way of describing the relationship in Eq. (48.147) is as follows: An adversary would like to employ a jamming strategy so that the effective E_b/J_0 is as large as possible. He may try to employ pulse, tone, or partial band jamming rather than wideband noise jamming. A large E_b/J_0 would cause a small J/S for a fixed processing gain, or it would force the communicator to employ a larger processing gain for some desired J/S. The system designer strives to choose the waveform so that the jammer can gain no special advantage with a jamming strategy other than wideband noise.

There are two popular techniques for spectrum spreading (see Fig. 48.112). The first is called direct sequencing or pseudonoise spread spectrum. Spreading is achieved through multiplication of the data by a binary pseudorandom sequence (discussed next) whose symbol rate is many times the data rate. The second technique uses a frequency-hopping carrier. The carrier remains at a given frequency for a duration and then hops to a new frequency somewhere in the spreading bandwidth W. Frequency hopping is generally classified as slow or fast hopping. In the case of slow hopping, there are typically several symbols per hop, and the bandwidth of the transmitted signal is equal to that of the data signal. In the case of fast hopping, there are typically several hops per symbol, and the bandwidth of the transmitted signal is equal to the reciprocal of the hopping duration. Figure 48.127, the spectral illustration of the spreading-despreading phenomenon, is an accurate rendition for each of the spreading techniques described. One important difference between direct sequencing and frequency hopping signals is that the former can be coherently demodulated. However, with frequency hopping, phase coherence is difficult to maintain; hence it is usually demodulated noncoherently. The performance of spread spectrum systems is treated in detail in Refs. 38 through 41.

Encryption

Two reasons for using cryptosystems in communications (see Figs. 48.111 and 48.112) are as follows: (1) privacy, to prevent unauthorized persons from extracting information from the channel, and (2) authentication, to prevent unauthorized persons from injecting information into the channel. The message, or plaintext P, is encrypted with an invertible transformation E_k that produces the ciphertext $C = E_k(P)$. The ciphertext is transmitted over an insecure or public channel. When an authorized receiver obtains C, he decrypts it with the inverse transformation $D_k = E_k^{-1}$ to obtain

$$D_k(C) = E_k^{-1}[E_k(P)] = P$$

the original plaintext message.

E_k is chosen from a family of cryptographic transformations, frequently regarded as public information. The parameter k, or the key that selects the individual transformation within the family, is safeguarded; in typical cryptosystems, anyone with access to the key can both encrypt and decrypt messages. The key is transmitted to the community of authorized users over a secure channel (in some cases, a courier), and generally remains unchanged for a considerable number of transmissions. The goal of an eavesdropper or adversary (cryptanalyst) is to produce an estimate of the plaintext $\hat{P}$ by analyzing the ciphertext obtained from the public channel, without benefit of the key.

Encryption schemes fall into two generic categories: block encryption and data-stream (or simply stream) encryption. With block encryption, the plaintext is segmented into blocks of fixed size; each block is encrypted independently from the others. A particular plaintext block will therefore be encrypted into the same ciphertext block each time it appears (as with block encoding). In general, however, the properties desired in a block cipher are quite different from those desired in an error-correcting code. For example, with encryption, plaintext data should never appear directly in the ciphertext; also, changing even a single bit of either the plaintext or the key should cause approximately 50% of the ciphertext bits to change.

Cryptosystems have their roots in Shannon's work[42] connecting cryptography with information theory. Shannon introduced the terms "confusion" (substitution) and "diffusion" (permutation). He suggested a method of using both of these in concert (a product cipher) to build a stronger encryption system than either method alone could produce. Figure 48.128 shows an example of a nonlinear substitution transformation. In general, n input bits are first represented as one of 2^n different characters (binary to octal translation in this example). A substitution is then made to one of the other characters from the set of 2^n characters. The character is then converted back to an n-bit output. It is easily shown that there are $(2^n)!$ different substitution or connection patterns possible. The cryptanalyst's task becomes computationally unfeasible as n gets large; say $n = 128$, then $2^n = 10^{38}$ and $(2^n)! =$ an astronomical number. We recognize that for $n = 128$, this substitution box (S box) represents the ideal encryption device. However, although we can identify the S box with $n = 128$ as ideal, its implementation is not feasible because it requires a unit with $2^n = 10^{38}$ connections.

Figure 48.129 represents an example of data permutation (a linear operation). Here the input data are simply rearranged or permuted (P box). The technique has one major disadvantage when used alone; it is vulnerable to trick messages. Such a message is illustrated in Fig. 48.129. A single 1 at the input and all the rest 0's quickly reveal one of the internal connections; similar messages can be used to reveal each of the remaining connections. The product cipher of Fig. 48.130 illustrates the combination of substitution and permutation transformations originally suggested by Shannon. It represents the compromise solution to the difficulties with the S box or P box alone; the combination of S and P boxes yields a more powerful system than would either one alone. The basic product cipher concept was used by IBM in developing its Lucifer system. In 1977, the National Bureau of Standards adopted a modified Lucifer system as the national data encryption standard.[43]

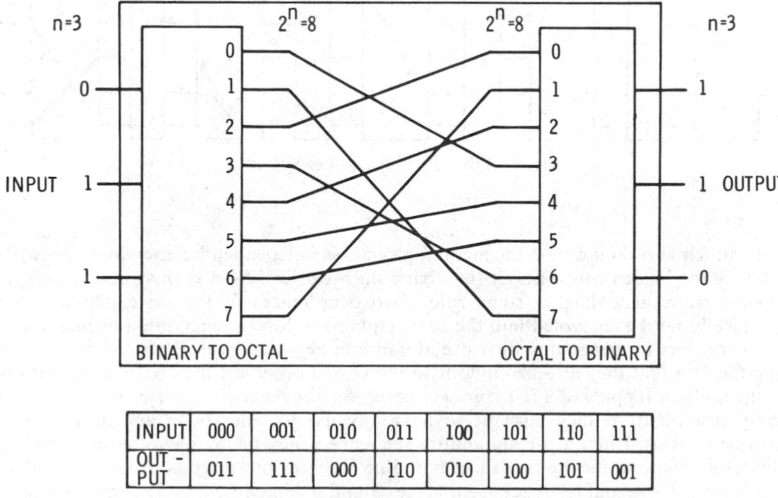

INPUT	000	001	010	011	100	101	110	111
OUT-PUT	011	111	000	110	010	100	101	001

Fig. 48.128 Substitution box.

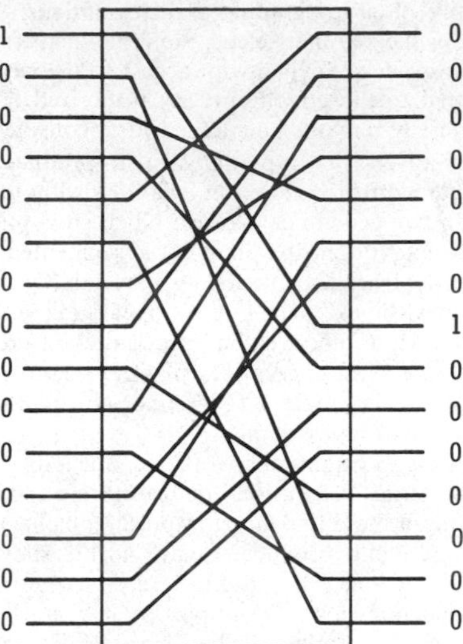

Fig. 48.129 Permutation box.

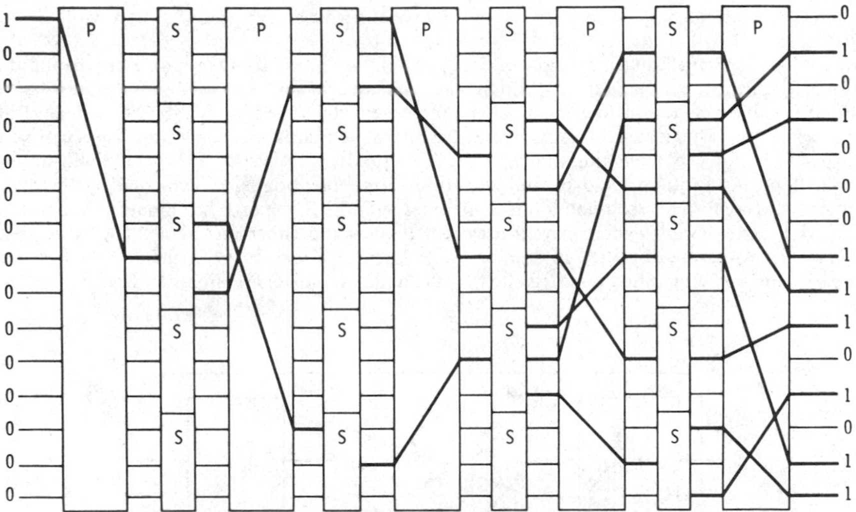

Fig. 48.130 Product cipher system.

Data-stream ciphers do not treat the incoming symbols independently; encryption depends on the internal state of the implementing device (feedback shift register). After each symbol is encrypted, the device changes state according to some rule. Two occurrences of the same plaintext input will therefore typically not be encrypted into the same ciphertext. Stream encryption techniques generally employ shift registers for generating their pseudorandom key sequences. Such sequences derive their name from the fact that they appear random to the casual observer; they have statistical properties similar to the random flipping of a fair coin. However, the sequences, of course, are not random; they are precisely structured, as they must be to have any use for encryption and decryption. A shift register can be converted into a pseudorandom sequence generator by inclusion of a feedback loop that computes a new term for the first stage based on the previous n terms. An example is shown in Fig. 48.131, where $n = 4$, and feedback from stages 3 and 4 is modulo-2 added and returned to stage 1. If the initial stage of the register is 1 0 0 0, then the succession of states triggered by clock pulses

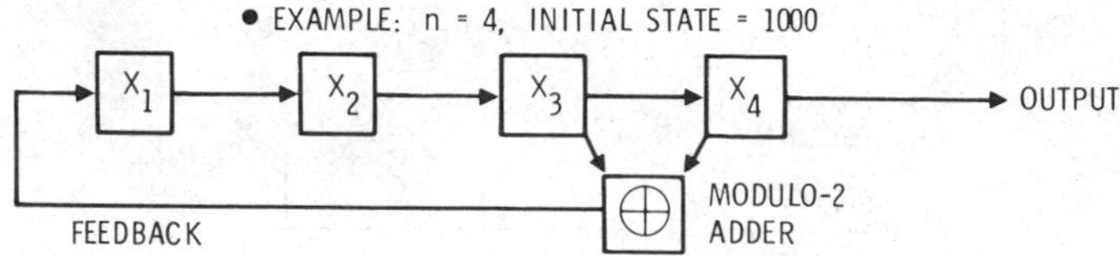

Fig. 48.131 Shift register sequence generator.

would be 1 0 0 0, 0 1 0 0, 0 0 1 0, 1 0 0 1, 1 1 0 0, ... The output sequence is made up of the bits shifted out from the fourth register, that is, 0 0 0 1 0 0 1 1 0 1 0 1 1 1 1. Given any linear feedback shift register of degree n, the output sequence is ultimately periodic. Any output sequence achieving the maximum possible period, $p = 2^n - 1$, is called a maximum-length shift register sequence.[25] These sequences have the following randomness properties:

1. **Balance property.** In each period of the sequence, the number of ones differ from the number of zeros by at most 1.
2. **Run property.** Among the runs of ones and zeros in each period, one half the runs of each kind are of length 1, one fourth are of length 2, one eighth are of length 3, and so on.
3. **Correlation property.** If a period of the sequence is compared term by term with any cyclic shift of itself, the number of agreements differs from the number of disagreements by at most 1.

Maximum-length linear codes are used not only for compact encryption-decryption keys but also for many similar purposes, such as code sequences for CDMA schemes and for other frequency-spreading techniques. A linear shift register is very vulnerable to attack by a cryptanalyst. Even if the feedback taps are not known, the analyst needs only $2n$ bits of plaintext and ciphertext to learn the feedback taps, the initial state of the register, and the entire sequence of the code.[44] The use of nonlinear feedback in the shift register makes the cryptanalyst's task much more difficult, if not computationally impracticable.

Diffie and Hellman[44] provided a thorough treatment of encryption and decryption fundamentals in their 1979 paper; their bibliography and references constitute an excellent resource for further reading in the field. Also Kahn,[45] in his 1967 book, presents a comprehensive and fascinating history of cryptography from ancient to modern times. The *IEEE Communications Magazine* special issue on communications privacy is a further invaluable primer.[46]

Synchronization

Synchronization can be defined as the alignment of time scales of spatially separated periodic processes. In the context of digital communications it involves the estimation of both time and frequency. Here we have space only to list and generally describe the synchronization requirements for digital systems. Throughout the signal-processing discussions, we have dealt with the operation on a digital symbol m_i or a digital waveform $s_i(t)$ during the time interval $[nT, (n+1)T]$, where $n = 0, 1, 2, \ldots$ is an integer indexing each symbol time duration T, and $i = 1, \ldots, M$, is an integer indexing individual symbols or waveforms from a finite set. An implicit assumption for each one of the processing steps (see Fig. 48.111) has been that the system is synchronized with respect to time and frequency; that the demodulator "knows" when to start accumulating energy for the decision-making process; and that it "knows" when to stop accumulating, when to make its decision, and when to repeat the operation. Any error in timing or frequency will result in lost energy, which effectively reduces the received E_b/N_0 and therefore degrades P_B performance.

The hierarchy of system synchronization levels (see Fig. 48.112) is as follows: Carrier synchronization refers to the restoration of the carrier (accurate with respect to frequency and phase) from a carrier-suppressed waveform. It is needed only for the demodulation of phase-coherent systems. Symbol (or bit) synchronization is needed for determining when the modulation may be changing state. Word, frame, or packet synchronization is needed for the proper reconstruction of the data. Network synchronization is needed for synchronizing channel access time among several users sharing the CR.

Efficient signal design dictates that any discrete component of carrier or clock signal be suppressed; transmitted power is devoted exclusively to data. In this case the system must recover the carrier and clock from a signal that contains neither one in explicit form. A survey by Franks[47] reviews some popular sychronizers and analysis techniques for implementing carrier and symbol synchronization. The survey demonstrates the application of maximum-likelihood estimation to

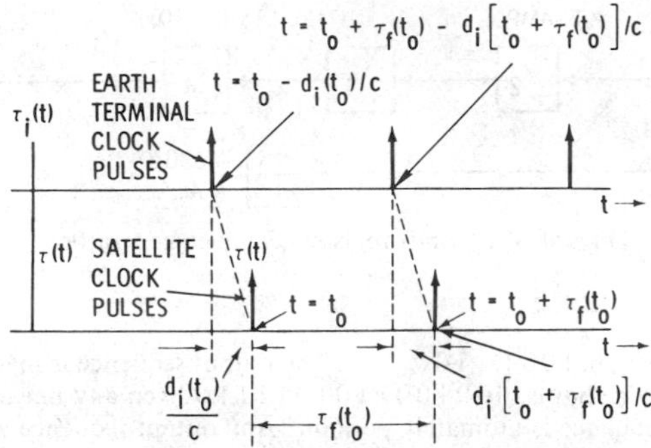

Fig. 48.132 Timing at satellite and at slaved terminal.

practical signals; it also discusses commonly used circuits for approximating maximum-likelihood estimators. Other detailed treatments for implementing carrier and symbol synchronization can be found in Refs. 3 and 48.

Figure 48.132 illustrates the general satellite and earth terminal considerations for maintaining a terminal clock slaved to the satellite clock (network synchronization). The ith terminal must adjust its clock pulses and transmissions so that they arrive in synchronism with the satellite clock pulses. Therefore any transmission from the ith terminal must be initiated early (with reference to the satellite clock) by a factor of $d_i(t)/c$, where $d_i(t)$ is the propagation distance from the satellite to the ith terminal and c is the speed of light.

Since range generally varies with time (either the satellite or the terminal may be moving with respect to one another), $\tau_i(t)$, the clock pulse sequence at the ith terminal, cannot be maintained at a fixed pulse rate; therefore the frame duration $\tau_f(t)$ varies with time. For many cases, the terminal clock needs to be slaved continually to ensure that it is in synchronism with the satellite clock. A good reference for synchronizing time between a satellite and an earth station is Ref. 49.

48.5-3 Communications Link Analysis

The communications system link budget is a balance sheet of gains and losses. It is composed of the detailed apportionment of transmission and reception resources, noise sources, and signal attenuators measured from the modulator and transmitter, through the channel, up to and including the receiver and demodulator (see Fig. 48.111). The budget is mainly derived from the calculation of received useful power. Some of the budget parameters are statistical, for example, RF propagation fades due to meteorological events.[50,51] Link budget analysis is therefore an estimation technique for evaluating communications system performance.

The propagating medium or electromagnetic path connecting the transmitter and receiver is called the channel. The concept of free space assumes a channel region free of all objects that might affect RF propagation by absorption, reflection, or refraction. It further assumes that the atmosphere in the channel is perfectly uniform and nonabsorbing, and that the earth is infinitely far away or its reflection coefficient is negligible. The RF energy arriving at the receiver is assumed to be a function of distance from the transmitter (simply following the inverse square law of optics). In practice, of course, propagation in the atmosphere and near the ground results in refraction, reflection, and absorption that modify the free space transmission.[52]

SNR is a convenient measure of performance at various points in the link. The definition is

$$\text{SNR} = \frac{\text{power in desired waveform}}{\text{power in interfering waveform}}$$

The desired waveform can be an information signal, a baseband waveform, or a modulated carrier. The SNR can degrade in one of two ways: (1) through the attenuation of desired waveform power relative to interfering waveform power or (2) through the increase of interfering waveform power relative to desired waveform power. These degradations are termed "loss" and "noise," respectively. Losses occur when, by some mechanism, a portion of the signal is diverted, scattered, or reflected from its intended route. Noise occurs when unwanted signal energy is injected into the link, or thermal noise is generated within the link. There are two main types of noise, thermal noise and intermodulation noise. Thermal noise is radiated into the antenna by oxygen and water vapor molecules in the atmosphere; it is also introduced by the first stages of the receiver. Intermodulation noise is caused by nonlinearities in the system; its deleterious effects are generally grouped quantitatively, with other system losses, under the heading of a loss parameter. We will restrict our discussion of noise to thermal noise, in which case the power spectral density is assumed to be flat up through the gigahertz range; the thermal noise process in communication receivers is generally accepted to be an AWGN process.[53] A well-known physical model[54] for thermal or Johnson noise generated in dissipative components consists of a noise generator with open-circuit mean squared voltage equal to $4\kappa T^\circ W\mathscr{R}$

where κ = Boltzmann's constant, 1.38×10^{-23} J/K
T° = temperature, K
W = bandwidth, Hz
$\mathscr{R}$ = resistance, Ω

It can be shown that the maximum available thermal noise power N, coupled from the noise generator into the front end of an amplifier, is[53]

$$N = \kappa T^\circ W \qquad \text{W}$$

and the noise density N_0 is simply

$$N_0 = N/W = \kappa T^\circ \qquad \text{W/Hz} \tag{48.148}$$

Development of the fundamental link power relationships assumes an omidirectional RF source transmitting uniformly over 4π steradians (isotropic radiator). The power density on a hypothetical sphere at a distance d from the source is related to the transmitter power P_t by

$$p(d) = P_t/4\pi d^2$$

The power extracted with the receiving antenna can be written

$$P_r = p(d)A_{er} = P_t A_{er}/4\pi d^2 \tag{48.149}$$

where the parameter A_{er} is the absorption cross section (effective area) of the antenna defined by

$$A_{er} = \frac{\text{total power absorbed}}{\text{incident power flux density}}$$

The receiving antenna's effective area and physical area are related by the efficiency parameter η, as follows

$$A_{er} = \eta A_{pr}$$

which accounts for the fact that the total power is not absorbed; some of it is lost through reradiation, scattering, or spillover. Typical values for η are 0.55 for a dish and 0.75 for a horn.

A common antenna parameter that relates the power output (or input) to that of an isotropic radiator is the antenna gain G, where

$$G = \frac{\text{maximum power intensity in some fixed direction}}{\text{average power intensity over } 4\pi \text{ steradians}}$$

Antenna gain, unlike that of an electronic amplifier, is the result of concentrating the RF flux in some restricted region less than 4π steradians; thus the effective power radiated with respect to an isotropic source (EIRP) is defined as

$$\text{EIRP} = P_t G_t$$

where G_t is the transmitter antenna gain. To find received power for the general case in which the transmitting source manifests antenna gain over isotropic, we replace P_t with EIRP in Eq. (48.149)

$$P_r = \text{EIRP}\, A_{er}/4\pi d^2 \tag{48.150}$$

The relationship between antenna gain and antenna effective area is[55]

$$G = 4\pi A_e/\lambda^2 \qquad \text{for} \quad A_e \gg \lambda^2 \tag{48.151}$$

where λ is the wavelength of the radiation. Similar expressions apply at the transmitter and receiver antennas by the reciprocity theorem.[55] Since the effect of an antenna can be expressed as a gain or area, we can replace A_{er} in Eq. (48.150) with $G_r\lambda^2/4\pi$ from Eq. (48.151), as follows

$$P_r = \text{EIRP}\, G_r\lambda^2/(4\pi d)^2 = \text{EIRP}\, G_r/L_s \tag{48.152}$$

In Eq. (48.152) the parameters $(4\pi d/\lambda)^2$ have been replaced by the term L_s, the space loss or path loss. Path loss characterizes the decrease in received power as a function of distance and frequency; it is a *definition* predicated on the use of an isotropic receiving antenna ($G_r = 1$). Hence path loss is an abstraction that cannot be measured; it represents a hypothetical received-power loss that *would occur if the receiving antenna were isotropic*. In a radio communications system, path loss accounts for the largest loss in signal power. In satellite systems, the path loss for a C-band (6-GHz) link to a synchronous satellite is typically 200 dB.

In evaluating system performance, the quantity of greatest interest is not the received power P_r but the SNR. This is because the basic system constraint is our ability to detect the signal, with an acceptable P_B, in the presence of noise. Since the desired signal here is a modulated carrier waveform, we often speak of the average carrier power-to-noise ratio (C/N, or CNR) or (P_r/N) as the SNR of particular interest. Into Eq. (48.152) we introduce P_r/N,

$$\frac{P_r}{N} = \frac{\text{EIRP}\, G_r/N}{L_s} \tag{48.153}$$

Equation (48.153) applies to any one-way satellite RF link. In analog systems, noise bandwidth is generally greater than signal bandwidth, and P_r/N is the main parameter for measuring signal detectability and performance quality. In digital receivers, however, correlators or matched filters, where signal bandwidth is taken to be equal to noise bandwidth, are usually used. Rather than consider input noise power, a common formulation for digital links is to replace noise power with noise density. We can use Eq. (48.148) for rewriting Eq. (48.153) as follows

$$\frac{P_r}{N_0} = \frac{\text{EIRP}\, G_r/T^\circ}{\kappa L_s L_o} \tag{48.154}$$

where the system effective temperature T° is a function of the thermal noise radiated into the antenna and the thermal noise generated within the first stages of the receiver.[56–58] We have introduced a term L_o in Eq. (48.154) to represent all degradation factors due to various losses and noise sources; this term L_o represents "other losses" not specifically addressed by the other terms of Eq. (48.153). It allows for a large assortment of different losses and noise sources (e.g., intermodulation noise), which have been catalogued in detail.[59] Equation (48.154) summarizes the key parameters of any link analysis, which are the received signal power-to-noise density (P_r/N_0), the magnitude of transmitted power (EIRP), the sensitivity of the receiver (G_r/T°), and the losses ($L_s L_o$).

If we assume that all the received power stems from the modulating signal (suppressed carrier), then we can write, from Eq. (48.138)

$$P_r/N_0 = S/N_0 = (E_b/N_0)R \tag{48.155}$$

If some of the received power is lodged in the carrier (a signal power loss), we can still employ Eq. (48.155), but we additionally represent the carrier power as a loss (within the parameter L_o of Eq. (48.154).

Until now we have referred only to one kind of E_b/N_0, that value of bit energy-to-noise density *required* to yield a specified P_B. Now, to facilitate calculating a margin of safety factor M, we need to differentiate between the required E_b/N_0 and the actual or *received* E_b/N_0. From this point on we will refer to the former as $(E_b/N_0)_{\text{reqd}}$ and to the latter as $(E_b/N_0)_r$. We can rewrite Eq. (48.155), introducing the link margin parameter M, as follows

$$P_r/N_0 = (E_b/N_0)_r R = M(E_b/N_0)_{\text{reqd}} R \tag{48.156}$$

The difference in decibels between $(E_b/N_0)_r$ and $(E_b/N_0)_{\text{reqd}}$ yields the link margin. Consider a system specified to operate at an $(E_b/N_0)_{\text{reqd}} = 10$ dB, with $P_B = 10^{-4}$. Suppose we require a link margin of 4 dB (let us assume that the commensurate P_B for an E_b/N_0 of 14 dB is 10^{-6}). We can look on this margin in one of two ways:

1. We can state that we have 4 dB more E_b/N_0 than we actually need to meet our required P_B of 10^{-4}.

2. We can state that we are operating at an E_b/N_0 of 14 dB and therefore that the actual operating P_B of the system is 10^{-6}, a margin of 100 times better error probability performance than we require.

The parameter $(E_b/N_0)_{\text{reqd}}$ reflects the differences from one system to another; these might be due to differences in modulation or coding schemes. A larger than expected $(E_b/N_0)_{\text{reqd}}$ may be due to a suboptimal RF system, which manifests larger timing errors or allows more noise into the detection process than does an ideal matched filter.

Combining Eqs. (48.154) and (48.156) and solving for the link margin M yields

$$M = \frac{\text{EIRP}\, G_r/T^\circ}{(E_b/N_0)_{\text{reqd}} R \kappa L_s L_o} \tag{48.157}$$

Since link budget analysis is typically calculated in decibels, we can express Eq. (48.157) as follows:

$$\begin{aligned} M(\text{dB}) = {} & \text{EIRP(dBW)} + G_r(\text{dBI}) - (E_b/N_0)_{\text{reqd}}(\text{dB}) \\ & - R(\text{dB} - \text{b/s}) - \kappa T^\circ(\text{dBW/Hz}) - L_s(\text{dB}) - L_o(\text{dB}) \end{aligned} \tag{48.158}$$

Transmitted signal power is expressed in decibel-watts (dBW), noise density is in decibel-watts per hertz (dBW/Hz), antenna gain is in decibels referenced to isotropic gain (dBI), data rate is in decibels referenced to b/s (dB-b/s), and all other terms are in decibels (dB). The values of the parameters in Eq. (48.158) constitute the link budget, a useful tool for allocating communications resources. In an effort to maintain a positive margin, we might trade off any parameter with any other parameter; we might choose to reduce transmitter power by giving up excess margin, or we might elect to increase the data rate by reducing the required E_b/N_0 (through the selection of improved modulation and coding). Any one of the decibels in Eq. (48.158), regardless from which parameter it stems, is just as good as any other decibel. It should be noted, however, that as requirements become more constrained, it may not be possible to trade or yield on some items. For example, even though binary PSK modulation outperforms binary FSK (in the P_B sense), requirements to operate in a scintillating environment would dictate the avoidance of PSK and the choice of the more robust FSK. Also certain coverage requirements may constrain antenna dimensions, so that one might *not* have the freedom of trading off or selecting any antenna gain one desires.

How Much Link Margin is Required

The question of how much link margin should be designed into the system is asked frequently. The answer is that if all sources of gain, loss, and noise have been rigorously detailed, and if the link parameters with large variances (e.g., fades due to weather) match the statistical requirements for link availability, very little margin is needed. For satellite communications at C-band, where the parameters are well known and fairly well behaved, it should be possible to design a system with only 1 dB of link margin. Receive-only television stations operating with 16-ft-diameter dishes at C-band are frequently designed with only a fractional decibel of margin. However, telephone communications via satellite using standards of 99.9% availability require considerably more margin; some of the Intelsat systems have 4 to 5 dB of margin. Designs using higher frequency (e.g., 14/12 GHz) generally call for larger margins because atmospheric losses increase with frequency and are highly variable. It should be noted that a byproduct of the attenuation due to atmospheric loss is greater antenna noise. When extra margin is allowed for weather loss, additional margin should simultaneously be added to compensate for the increase in antenna temperature (a function of thermal noise radiated into the antenna). With low-noise amplifiers, small weather changes can result in increases of 40 to 50 K in antenna temperature.

Satellite Repeaters

Satellite repeaters retransmit the messages they receive (with a translation in carrier frequency). A regenerative (digital) repeater regenerates, that is, demodulates and reconstitutes the digital information embedded in the received waveforms; however, a nonregenerative repeater only amplifies but does not transform the signal to its baseband format. A nonregenerative repeater, therefore, can be used with many different modulation formats (simultaneously), but a regenerative repeater is usually designed to operate with only one or a very few modulation formats. Link analysis for a regenerative satellite repeater treats the uplink and downlink as two separate point-to-point analyses. To estimate the performance of a regenerative repeater link, it is necessary to determine separately the bit error probability on the uplink and downlink. The overall error rate is obtained by simply summing the individual rates.[60,61]

By comparison, link analysis for a nonregenerative repeater generally treats the entire "round trip" (uplink transmission to the satellite and downlink retransmission to an earth terminal) as a single analysis. To estimate performance of a nonregenerative repeater link, the uplink and downlink values of E_b/N_0 (or P_r/N_0) are combined as follows, in the absence of intermodulation noise[60]

$$(E_b/N_0)_U^{-1} + (E_b/N_0)_D^{-1} = (E_b/N_0)_R^{-1}$$

where the subscripts U, D, and R indicate uplink, downlink, and resultant values of E_b/N_0, respectively.

Most conventional commercial satellites in use today are the simple nonregenerative kind. However, it seems clear that digital satellite systems of the future, which will require on-board processing, switching, or selective message addressing, will start with a regenerative repeater to transform the received waveforms to message bits. Besides the potential for sophisticated data processing, one of the principal advantages of regenerative compared with nonregenerative repeaters is the decoupling of the uplink and downlink so that the uplink noise power is not retransmitted on the downlink. There are significant performance improvements in terms of reducing the E_b/N_0 values needed on the uplinks and downlinks relative to the values needed for the conventional transponder designs in use today. Improvements as much as 5 dB on the uplink and 6.8 dB on the downlink (using coherent QPSK modulation, with $P_B = 10^{-4}$) have been demonstrated.[60]

Power is severely limited in most satellite communications systems, and the inefficiencies associated with linear power amplification stages are intolerable. For this reason many satellite repeaters employ highly nonlinear power amplifiers; the main feature here is that efficient power amplification is obtained at the cost of nonlinear distortions. The major undesirable effects of the repeater nonlinearities are

1. Intermodulation (IM) noise owing to the multiplicative action of different carriers. The harm caused is twofold: Useful power can be lost from the channel as IM energy and spurious IM products can be introduced into the channel as interference.[62,63]
2. AM-to-PM conversion is a phase-noise phenomenon occurring in nonlinear devices such as traveling-wave tubes (TWT). Fluctuation in operating level (amplitude modulation) produces phase variations that impact the P_B performance for systems using an MPSK modulation format.[64,65]
3. Signal suppression of weak signals by stronger signals,[62] by as much as 6 dB.

Conventional nonregenerative repeaters are generally operated "backed off" from their highly nonlinear saturated region, to avoid appreciable IM noise and to thus allow efficient utilization of the system's entire bandwidth. However, backing off to the linear region is a compromise; some level of IM noise must be accepted to achieve a useful level of output power.

One set of features, unique to nonregenerative repeaters, is worth describing here—it is the dependence of the downlink SNR on the uplink SNR and the sharing of the repeater downlink power in proportion to the uplink power from each of the various uplink signals and noise. Henceforth, reference to a repeater or transponder will mean a nonregenerative repeater, and for simplicity we will assume the transponder is operating in its linear range.

A satellite transponder is limited in transmission capability by its downlink power, the earth terminal's uplink power, satellite and earth terminal noise, and channel bandwidth. One of these usually represents the dominant performance constraint; most often the downlink power or the channel bandwidth proves to be the major system limitation. Figure 48.133 illustrates the important link parameters of a linear satellite repeater channel. The repeater transmits all arriving uplink messages (or noise, in the absence of messages) without any processing beyond frequency translation. Let us assume that the multiple uplinks within the receiver's bandwidth W are separated from one another through the use of a multiple-access scheme, such as FDMA or CDMA. The satellite downlink power P_{sat} is constant, and since we are assuming a linear transponder, P_{sat} is shared among the multiple uplink signals (and noise) in proportion to their respective power levels.

The transmission starts from a ground station (bandwidth $< W$), for instance terminal number 1, with an $\text{EIRP}_{t1} = P_{t1}G_{t1}$. Simultaneously, other signals are being transmitted to the satellite (from

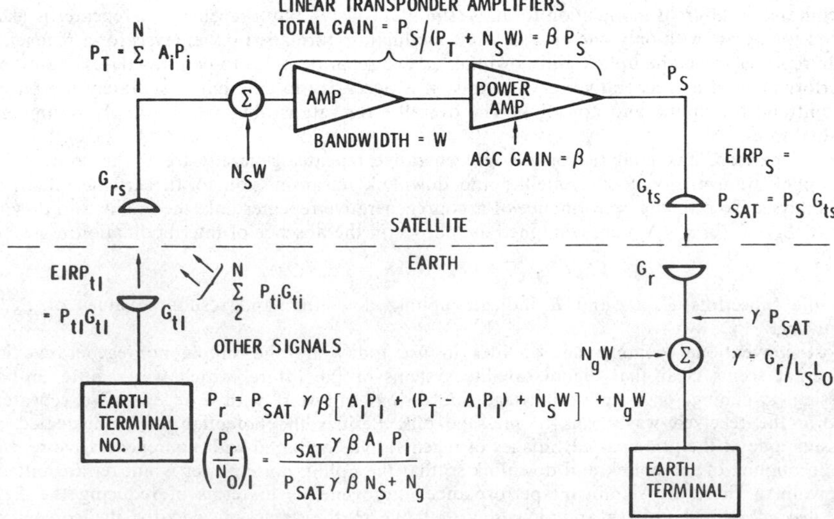

Fig. 48.133 Nonregenerative satellite repeater.

other terminals). At the satellite, a total signal power $P_T = \sum A_i P_i$ is received, where the A_i reflect the various propagation losses the different signals experience on arrival at the satellite. Noise power $N_s W$ is also received at the satellite, where N_s is the noise density due to thermal noise radiated into the satellite antenna and thermal noise generated in the satellite receiver. The total satellite downlink $\text{EIRP}_s = P_{\text{sat}}$ can be expressed with the following identity:[4]

$$P_{\text{sat}} = P_{\text{sat}} \beta [A_1 P_1 + (P_T - A_1 P_1) + N_s W]$$

where $\beta = 1/(P_T + N_s W)$ is the AGC gain and P_T has purposely been written as $A_1 P_1 + (P_T - A_1 P_1)$ to separate signal number 1 power from the remainder of simultaneous signals in the transponder. Using Eq. (48.154) we can write the $(P_r/N_0)_{1i}$ for signal number 1 arriving at the ith terminal, as follows:[4]

$$\left(\frac{P_r}{N_0}\right)_{1i} = \frac{P_{\text{sat}} \gamma_i \beta A_1 P_1}{P_{\text{sat}} \gamma_i \beta N_s + N_g} \tag{48.159}$$

where $\gamma_i = G_r / L_s L_o$ for the ith terminal and N_g is the receiver noise density for the ith terminal.

When the satellite receiver noise dominates, that is, when $P_T \ll N_s W$, the link is said to be uplink limited, and most of the downlink P_{sat} is wastefully allocated to noise power. When this is the case, and when $\gamma_i P_{\text{sat}} \gg N_g W$, we can rewrite Eq. (48.159) as

$$\left(\frac{P_r}{N_0}\right)_{1i} \cong \frac{\gamma_i P_{\text{sat}} A_1 P_1 / N_s W}{(\gamma_i P_{\text{sat}} / W) + N_g} \cong \frac{A_1 P_1}{N_s} \tag{48.160}$$

Equation (48.160) illustrates that, in the case of an uplink limited channel, the resultant P_r/N_0 ratio essentially follows the uplink SNR. The more common situation is the downlink-limited channel, in which case $P_T \gg N_s W$, and the satellite EIRP is limited. In this case, Eq. (48.159) can be rewritten as[4]

$$\left(\frac{P_r}{N_0}\right)_{1i} \cong \frac{\gamma_i P_{\text{sat}} A_1 P_1 / P_T}{N_g}$$

The power of the transponder is then shared primarily among the various uplink transmitted signals; very little uplink noise is transmitted on the downlink. In this downlink-limited case, the resultant P_r/N_0 of the repeater is constrained only by the downlink parameters.

References

1 W. L. Pritchard, "Satellite Communication—An Overview of the Problems and Programs," *Proc. IEEE* **65**:294–307 (Mar. 1977).

2 M. P. Ristenbatt, "Alternatives in Digital Communications," *Proc. IEEE* **61**:703–721 (Jun. 1973).

3 W. C. Lindsey and M. K. Simon, *Telecommunication Systems Engineering*, Prentice-Hall, Englewood Cliffs, NJ, 1973.

4 J. J. Spilker, Jr., *Digital Communications by Satellite*, Prentice-Hall, Englewood Cliffs, NJ, 1977.

5 V. K. Bhargava, D. Haccoun, R. Matyas, and P. P. Nuspl, *Digital Communications by Satellite*, Wiley, New York, 1981.

6 C. B. Feldman and W. R. Bennett, "Band Width and Transmission Performance," *BSTJ* **28**:594–595 (1949).

7 A. Lender, "The Duobinary Technique for High Speed Data Transmission," *IEEE Trans. Commun. Electron.* **82**:214–218 (May 1963).

8 E. R. Kretzmer, "Generalization of a Technique for Binary Data Communication," *IEEE Trans. Commun. Tech.*, pp. 67–68 (Feb. 1966).

9 S. Pasupathy, "Correlative Coding: A Bandwidth Efficient Signaling Scheme," *IEEE Commun. Mag.*, pp. 4–11 (Jul. 1977).

10 N. S. Jayant, "Digital Coding of Speech Waveforms: PCM, DPCM, and DM Quantizers," *Proc. IEEE* **62**(5):611–632 (May 1974).

11 B. S. Atal and S. L. Hanover, "Speech Analysis and Synthesis by Linear Prediction of the Speech Wave," *J. Acoust. Soc. Am.*, **50**:637–655 (1971).

12 J. W. Bayless, S. J. Campanella, and A. J. Goldberg, "Voice Signals: Bit by Bit," *IEEE Spectr.* **10**:28–39 (Oct. 1973).

13 J. L. Flanagan, *Speech Analysis, Synthesis, and Perception*, 2nd ed., Springer-Verlag, New York, 1972.

14 R. E. Crochiere and J. L. Flanagan, "Current Perspectives in Digital Speech," *IEEE Commun. Mag.* **21**(1):32–40 (Jan. 1983).

15 D. Huffman, "A Method for Constructing Minimum Redundancy Codes," *Proc. IRE* **40**:1098–1101 (May 1952).

16 R. G. Gallager, *Information Theory and Reliable Communication*, Wiley, New York, 1968.

17 S. A. Gronemeyer and A. L. McBride, "MSK and Offset QPSK Modulation," *IEEE Trans. Commun.* **COM-24**(8):809–820 (Aug. 1976).

18 S. Pasupathy, "Minimum Shift Keying: A Spectrally Efficient Modulation," *IEEE Commun. Mag.* pp. 14–22 (Jul. 1979).

19 E. Arthurs and H. Dym, "On the Optimum Detection of Digital Signals in the Presence of White Gaussian Noise—A Geometric Interpretation of Three Basic Data Transmission Systems," *IRE Trans. Commun. Sys.* **CS-10**:336–372 (Dec. 1962).

20 H. L. Van Trees, *Detection, Estimation, and Modulation Theory, Part I*, Wiley, New York, 1968.

21 A. J. Viterbi, *Principles of Coherent Communication*, McGraw-Hill, New York, 1966.

22 K. Feher, *Digital Communications: Microwave Applications*, Prentice-Hall, Englewood Cliffs, NJ, 1981.

23 P. R. Hartman, "Digital Radio Technology: Present and Future," *IEEE Commun. Mag.* **19**(4):10–14 (Jul. 1981).

24 J. M. Wozencraft and I. M. Jacobs, *Principles of Communication Engineering*, Wiley, New York, 1965.

25 S. Golomb, ed., *Digital Communications with Space Applications*, Prentice-Hall, Englewood Cliffs, NJ, 1964.

26 V. K. Bhargava, "Forward Error Correction Schemes for Digital Communications," *IEEE Commun. Mag.* **21**(1):11–19 (Jan. 1983).

27 A. J. Viterbi, "Convolutional Codes and Their Performance in Communication Systems," *IEEE Trans. Commun. Tech.* **COM-19**:751–772 (Oct. 1971).

28 J. A. Heller and I. M. Jacobs, "Viterbi Decoding for Satellite and Space Communication," *IEEE Trans. Commun. Tech.* **COM-19**:835–848 (Oct. 1971).

29 G. Forney, Jr., "Coding and Its Application in Space Communications," *IEEE Spectr.*, pp. 47–58 (Jun. 1970).

30 J. L. Ramsey, "Realization of Optimum Interleavers," *IEEE Trans. Infor. Theory* **IT-16**(3):338–345 (May 1970).

31 I. L. Lebow, K. L. Jordan, Jr., and P R. Drouilhet, Jr., "Satellite Communications to Mobile Platforms," *Proc. IEEE* **59**(2):139–159 (Feb. 1971).

32 T. Muratani, "Satellite-Switched Time-Domain Multiple Access," *Record IEEE Electron. Aerosp. Syst. Conv. (EASCON)*, October 7–9, 1974, pp. 189–196.

33 J. G. Puente and A. M. Werth, "Demand-Assigned Service for the Intelsat Global Network," *IEEE Spectr.*, pp. 59–69 (Jan. 1971).

34 J. W. Schwartz, J. M. Aein, and J. Kaiser, "Modulation Techniques for Multiple Access to a Hard-Limiting Satellite Repeater," *Proc. IEEE* **54**(5):763–777 (May 1966).

35 L. M. Nirenberg and I. Rubin, "Multiple Access System Engineering—A Tutorial," *IEEE WESCON/78 Professional Program*, Modern Communication Techniques and Applications, Session 21, Los Angeles, September 13, 1978.

36 G. D. Dill, "TDMA, The State-of-the-Art," *Record IEEE Electron. Aerosp. Syst. Conv. (EASCON)*, Sept. 26–28, 1977, pp. 31-5A to 31-5I.

37 R. C. Dixon, *Spread Spectrum Analysis*, Wiley, New York, 1976.

38 M. P. Ristenbatt and J. L. Daws, Jr., "Performance Criteria for Spread Spectrum Communications," *IEEE Trans. Commun.* **COM-25**:756–761 (Aug. 1977).

39 S. W. Houston, "Modulation Techniques for Communication, Part 1: Tone and Noise Jamming Performance of Spread Spectrum M-ary FSK and 2, 4-ary DPSK Waveforms," *IEEE 1975 Nat. Aerosp. Electron. Conf.*, June 10–12, 1975, pp. 51–58.

40 G. K. Huth, "Spread Spectrum Techniques," *IEEE WESCON/78 Professional Program*, Modern Communication Techniques and Applications, Session 15, Los Angeles, September 13, 1978.

41 J. K. Holmes, *Coherent Spread Spectrum Systems*, Wiley, New York, 1982.

42 C. E. Shannon, "Communication Theory of Secrecy Systems," *BSTJ* **28**:656–715 (Oct. 1949).

43 National Bureau of Standards, "Data Encryption Standard," *Federal Information Processing Standard (FIPS)*, Publication No. 46, January 1977.

44 W. Diffie and M. E. Hellman, "Privacy and Authentication: An Introduction to Cryptography," *Proc. IEEE* **67**(3):397–427 (Mar. 1979).

45 D. Kahn, *The Codebreakers*, Macmillan, New York, 1967.

46 Special Issue on Communications Privacy, *IEEE Commun. Mag.* **16**(6) (Nov. 1978).

47 L. E. Franks, "Carrier and Bit Synchronization in Data Communication—A Tutorial Review," *IEEE Trans. Commun.* **COM-28**(8):1107–1121 (Aug. 1980).

48 W. Lindsey, *Synchronization Systems in Communications and Control*, Prentice-Hall, Englewood Cliffs, NJ, 1972.

49 P. P. Nuspl, K. E. Brown, W. Seenaart, and B. Ghicopoulos, "Synchronization Methods for TDMA," *Proc. IEEE* **65**:434–444 (1977).

50 R. K. Crane, "Prediction of Attenuation by Rain," *IEEE Trans. Commun.* **COM-28**(9):1717–1733 (Sep. 1980).

51 L. M. Schwab, "World-Wide Link Availability for Stationary and Critically Inclined Orbits Including Rain Attenuation Effects," *Lincoln Laboratory, Project Report DCA-9*, January 27, 1981.

52 P. L. Bargellini, "Principles and Evolution of Satellite Communications," *Acta Astron.* **5**:135–149 (1978).

53 R. Gagliardi, *Introduction to Communication Engineering*, Wiley, New York, 1978.

54 H. Nyquist, "Thermal Agitation of Electric Charge in Conductors," *Phys. Rev.* **32**:110–113 (Jul. 1928).

55 R. E. Collin and F. J. Zucker, *Antenna Theory, Part I*, McGraw-Hill, New York, 1969.

56 P. F. Panter, *Communications Systems Design: Line-of-Sight and Tropo Scatter Systems*, McGraw-Hill, New York, 1972.

57 D. C. Hogg and T-S. Chu, "The Role of Rain in Satellite Communications," *Proc. IEEE* **63**(9):1308–1331 (Sep. 1975).

58 D. C. Hogg and W. W. Mumford, "The Effective Noise Temperature of the Sky," *Microw. J.*, pp. 80–84 (Mar. 1960).

59 B. Sklar, "What the System Link Budget Tells the System Engineer," *Proc. Int. Telemetering Conf.* **15** (1979).

60 S. J. Campanella, F. Assal, and A. Berman, "Onboard Regenerative Repeaters," *Int'l. Conf. Communications*, Chicago, 1977, vol. 1, pp. 6.2-121 to 6.2-125.

61 K. Koga, T. Muratani, and A. Ogawa, "On-Board Regenerative Repeaters Applied to Digital Satellite Communications," *Proc. IEEE* **65**(3):401–410 (Mar. 1977).

62 J. J. Jones, "Hard Limiting of Two Signals in Random Noise," *IEEE Trans. Infor. Theory*, pp. 34–42 (Jan. 1963).

63 F. E. Bond and H. F. Meyer, "Intermodulation Effects in Limiter Amplifier Repeaters," *IEEE Trans. Commun. Tech.* **COM-18**(2):127–135 (Apr. 1970).

64 O. Shimbo, "Effects of Intermodulation, AM-PM Conversion, and Additive Noise in Multicarrier TWT Systems," *Proc IEEE* **59**:230–238 (Feb. 1971).

65 P. Jain, T. C. Huang, K. T. Woo, et al., "Detection of MPSK Signals Transmitted Through a Nonlinear Satellite Repeater," *NTC '77 Conference Record*, December 5–7, 1977.

48.6 FADING CHANNEL COMMUNICATIONS

Peter Monsen

48.6-1 Introduction

Two radio propagation channels for beyond-the-horizon communications, troposcatter, and HF are currently being reexamined. In the past, transmission over these radio channels had been considered unreliable due to fading effects. Recently, conversion from analog to digital transmission and the use of new adaptive signal-processing techniques have offered promise of acceptable network communication quality. In addition, over-the-horizon radio provides economic and/or security advantages relative to satellite, cable, or line-of-sight terrestrial microwave links. There is renewed interest in the use of high-frequency (HF) and troposcatter communications for networks carrying digital traffic.

HF radio uses frequencies in the range of 2 to 30 MHz. At these frequencies, communications beyond line-of-sight is achieved through refractive bending of the radio wave in the ionosphere from ionized layers at different elevations. In most cases more than one ionospheric "layer" causes the return of a refracted radio wave to the receiving antenna. The impulse response of such a channel exhibits a discrete multipath structure. The time between the first arrival and the last arrival is the multipath delay spread. Changes in ion density in individual layers due to solar heating cause fluctuations in each multipath return. This time-varying multipath characteristic produces alternately destructive and constructive interference. The resulting fading can produce complete loss of signal, as those who have heard replays of Winston Churchill's radio talks during World War II can attest.

Troposcatter radio transmission was discovered only after World War II when it was noted that microwave signals from beyond-the-horizon radars were much stronger than predicted from diffraction calculations over the earth's surface. A commonly held theory of this phenomenon, developed by Tatarski,[1] holds that random fluctuations in the dielectric constant in the troposphere divert some small fraction of the impinging energy back to the receiver. The name "tropospheric scatter," or "troposcatter," derives from this concept of random redirection of the incident wave by the troposphere. Significant scatter returns occur from a "common volume," defined by the receiving and transmitting antenna beam patterns. Scatter returns from different points within the common volume have different path delays. Signals scattered from points separated by more than the decorrelation distance of the fluctuations in the dielectric constant are not correlated. Thus as in the HF example, the impulse response characterizing the channel has a time-varying multipath structure with delay spread but without the discrete layers. Troposcatter systems have been widely used in military applications for beyond line-of-sight communications up to about 600 miles. The frequency range for this application extends from about 400 to 5000 MHz. Reliable communications require redundant transmission paths provided through the use of multiple frequencies, antennas separated in space, or scatter at two different beam angles. The several redundant paths are referred to as diversity paths, and the number of paths is termed the order of diversity.

The multipath delay spread limits the channel capacity that can be achieved in present analog systems. Only transmission bandwidths less than the reciprocal of this multipath delay can be achieved. Signals of larger bandwidths become distorted owing to the multipath dispersion. In FM systems this dispersion causes intermodulation noise after detection.

With the introduction of satellite communication systems, which do not suffer from extensive multipath fading, the future of HF and troposcatter systems appeared to be limited. Economic and security factors have altered this assessment, particularly for digital transmission. With digital signal formats, adaptive methods can be devised to measure the multipath structure and exploit it as an extra form of diversity to improve performance. Unlike the capacity of analog systems, the capacity of digital systems is not restricted by the multipath delay spread. From a network viewpoint, fades in tandem digital links do not have a cumulative effect because the signal can be regenerated at each node.

Adaptive troposcatter systems have been demonstrated that are efficiently able to detect digital signals perturbed by a fading channel medium while tracking the fading variations. If receiver adaptation requires significantly less signal-to-noise ratio per bit than receiver detection of the digital signal, the receiver decisions can be effectively used as the estimate of the transmitted signal to achieve what is referred to as *decision-directed* adaptation. Such systems can be operated without transmission of special pilot or reference signals for channel tracking.

In troposcatter communication, adaptive techniques have increased the digital rate capability by at least an order of magnitude. An adaptive equalizer modem[2,3] developed for military troposcatter links has been successfully field tested at digital rates up to 12.6 Mb/s in a 15 MHz channel allocation. In HF systems, adaptive signal-processing techniques are now being considered with goals of digital rates on the order of 5 kb/s in a 3-kHz channel. In both these examples, the digital symbol period is of the same order as the channel multipath delay spread.

Applicability of adaptive signal-processing techniques is critically dependent on whether the rate of fading is slower than the rate of signaling. As discussed later in this section, both HF and troposcatter radio links can be considered slow-fading multipath channels.

48.6-2 Slow-Fading Multipath Channels

For digital communication over beyond-the-horizon radio links, an attempt is made to maintain transmission linearity, that is, the receiver output should be a linear superposition of the transmitter input plus channel noise. This is accomplished by operation of the power amplifier in a linear region, or with saturating power amplifiers, by use of constant-envelope modulation techniques. For linear systems, multipath fading can be characterized by a transfer function of the channel $H(f; t)$. This function is the two-dimensional random process in frequency f and time t that is observed as carrier modulation at the output of the channel when sine wave excitation at the carrier frequency is applied to the channel input. For any continuous random process, we can determine the minimum separation required to guarantee decorrelation with respect to each argument.

For the time-varying transfer function $H(f; t)$, let t_d and f_d be the decorrelation separations in the time and frequency variables, respectively. If t_d is a measure of the time decorrelation in seconds, then

$$\sigma_t = \frac{1}{2\pi t_d} \quad \text{Hz}$$

is a measure of the fading rate or bandwidth of the random channel. The quantity σ_t is often referred to as the Doppler spread, because it is a measure of the width of the received spectrum when a single sine wave is transmitted through the channel. The dual relationship for the frequency decorrelation f_d in Hertz suggests that a delay variable

$$\sigma_f = \frac{1}{2\pi f_d} \quad \text{s}$$

defines the extent of the multipath delay. The quantity σ_f is often referred to as the multipath delay spread, because it is a measure of the width of the received process in the time domain when a single impulse function is transmitted through the channel.

Typical values of these spread factors for HF and troposcatter communication are

HF	Troposcatter
$\sigma_t \sim 0.1$ Hz	$\sigma_t \sim 1$ Hz
$\sigma_f \sim 10^{-3}$ s	$\sigma_f \sim 10^{-7}$ s

where the symbol $\sim$ denotes "on the order of."

The spreads can be defined precisely as moments of spectra in a channel model,[4] which assumes wide-sense stationarity (WSS) in the time variable and uncorrelated scattering (US) in a multipath delay variable. This WSSUS model and the assumption of Gaussian statistics for $H(f; t)$ provide a statistical description in terms of a single two-dimensional correlation function of the random process $H(f; t)$.

This characterization has been quite useful and accurate for a variety of radio link applications. However, the stationarity and Gaussian assumptions are not necessary for the utilization of adaptive signal-processing techniques on these channels. What is necessary is first that sufficient time exists to "learn" the channel characteristics before they change, and second that decorrelated portions of the frequency band be excited such that a diversity effect can be realized. These conditions are reflected in the following two relationships in terms of the previously defined channel factors, the data rate R, and the bandwidth B:

Learning constraint

$$R(\text{b/s}) \gg \sigma_t \quad \text{Hz}$$

Diversity constraint

$$B(\text{Hz}) \gtrsim f_d \quad \text{Hz}$$

where the symbol $\gtrsim$ denotes "on the order of or greater than."

The learning constraint ensures that sufficient signal-to-noise ratio (SNR) exists for reliable communication at rate R over the channel. Clearly if $R \sim \sigma_t$, the channel would change before significant energy for measurement purposes could be collected. When $R \gg \sigma_t$, the received data symbols can be viewed as the result of a channel-sounding signal and appropriate processing can generate estimates of the channel character during that particular stationary epoch. The signal-processing techniques in an adaptive receiver do not necessarily need to measure the channel directly in the optimization of the receiver, but the requirements on learning are approximately the same. If only information symbols are used in the sounding signal, the learning mode is referred to as decision directed. When digital symbols known to both the transmitter and receiver are employed, the learning mode is called reference directed. An important advantage of digital systems is that in many adaptive communications applications, adaptation of the receiver with no wasted power for sounding signals can be accomplished using the decision-directed mode. This is possible in digital systems because of the finite number of parameters or levels in the transmitted source symbols and the high likelihood that receiver decisions are correct.

Diversity in fading applications is used to provide redundant communications channels so that when some of the channels fade, communication will still be possible over the others that are not in a fade. Some of the forms of diversity employed are space with use of multiple antennas and angle of arrival with use of multiple feedhorns, polarization, frequency, and time. These diversity techniques are sometimes called explicit diversity because of their externally visible nature. An alternate form of diversity is termed implicit diversity because the channel itself provides redundancy. To capitalize on this implicit diversity for added protection, receiver techniques have to be employed to correctly assess and combine the redundant information. The potential for implicit frequency diversity arises because different parts of the frequency band fade independently. Thus while one section of the band may be in a deep fade, the remainder can be used for reliable communications. However, if the transmitted bandwidth B is small compared with the frequency decorrelation interval f_d, the entire band will fade and no implicit diversity can result. Thus the second constraint $B \gtrsim f_d$ must be met if an implicit diversity gain is to be realized. In diversity systems a little decorrelation between alternate signal paths can provide significant diversity gain. Thus it is not necessary for $B \gg f_d$ to realize implicit frequency diversity gain, although the implicit diversity gain clearly increases with the ratio $B : f_d$. Note that the condition $R \ll B \gtrsim f_d$ does not preclude the use of implicit diversity because a bandwidth expansion technique can be used in the modulation process to spread the transmitted information over the available bandwidth B. We shall distinguish between these low-data rate and high-data rate conditions, because the appropriate receiver structures take on somewhat different forms.

The implicit diversity effect described here results from decorrelation in the frequency domain in a slow-fading ($R \gg \sigma_t$) application. This implicit frequency diversity can in some circumstances be supplemented by an implicit *time* diversity effect, which results from decorrelation in the time domain. In fast-fading applications ($R \gtrsim \sigma_t$), redundant symbols in a coding scheme can be used to provide time diversity provided the code word spans more than one fade epoch. In our slow-fading application this condition of spanning the fade epoch can be realized by interleaving the code words to provide large time gaps between successive symbols in a particular code word. The interleaving process requires the introduction of signal delay longer than the time decorrelation separation t_d. In many practical applications that require transmission of digitized speech, the required time delay is unsatisfactorily long for two-way speech communication. For these reasons there is more emphasis on implicit frequency diversity techniques in practical systems. The receiver structures to be discussed next are applicable to situations where the implicit frequency diversity applies.

48.6-3 Adaptive Receiver Structures

We consider a pulse amplitude modulation system wherein the sample set $\{a_k\}$ is to be communicated over the channel using a modulation technique that forms a one-to-one correspondence between the sample a_k and the amplitude of a transmitted pulse. Independent modulation of quadrature carrier signals (i.e., $\sin 2\pi f_o t$ and $\cos 2\pi f_o t$, f_o = carrier frequency) is included in this class. An important example with optimum detection properties is quadrature phase shift keying (QPSK), which transmits the sample set $\{a_k = +1 + j\}$ by changing the sign of quadrature carrier pulses in accordance with the sign of the real and imaginary parts of the source sequence $\{a_k\}$.

Receivers for Channels with Negligible Intersymbol Interference

If each sample a_k can be one of M possible amplitudes ($M = 4$ for QPSK), the transmitted data rate is

$$R = [\log_2(M)]/T$$

where $1/T$ is the transmitted symbol rate.

Most terrestrial over-the-horizon channel applications utilize media that are signal-to-noise-ratio limited rather than bandwidth limited. To maximize signal detectability, only a few amplitudes are usually employed in these applications. When the symbol period T is much greater than the total width of the multipath dispersion of the channel, only a small portion of adjacent symbols interfere with the detection of a particular symbol. For the slow-fading application, the diversity constraint requires that the signal bandwidth B be on the order of or larger than the frequency decorrelation interval f_d. Conditions for negligible intersymbol interference (ISI) and adaptive processing to obtain implicit diversity are then

$$T \gg 2\pi\sigma_F = \frac{1}{f_d}$$

$$B \gtrsim f_d$$

When the number of amplitudes M is small, these conditions imply a low-data-rate system relative to the available bandwidth, that is, a bandwidth expansion system. The low-data-rate condition for implicit frequency diversity can be expressed in terms of the data rate and bandwidth as

$$R \ll B \log_2(M)$$

In the absence of intersymbol interference, it is well known[5] that the optimum detection scheme contains a noise filter and a filter matched to the received pulse shape. The optimum noise filter has a transfer function equal to the reciprocal of the noise power spectrum $K(f)$. When the additive noise at the receiver input is white, that is, its spectrum is flat over the frequency band of interest, the noise filter can be omitted in the optimum receiver. The optimum receiver for a fixed channel transfer function $H(f)$ then contains a cascade filter with component transfer functions

$$R(f) = \frac{1}{K(f)} H^*(f)$$

where the * denotes complex conjugation. In practical applications, signal delay must be introduced to make these filters realizable.

In general, K and H change with time and the adaptive receiver must track these variations. The tapped-delay-line (TDL) filter is an important filter structure for such channel-tracking applications. The TDL filter shown in Fig. 48.134 consists of a tapped delay line with signal multiplications by the tap weight w_i for each tap. For a bandpass system of bandwidth B, the sampling theorem states that any linear filter can be represented by parallel TDL filters operating on each quadrature carrier component with a tap spacing of $1/B$ or less. The optimum receiver can then be realized by a cascade of two such parallel TDL quadrature filters: one with tap weights adjusted to form the noise filter, the second with tap weights adjusted to form the matched filter. Since the cascade of two bandlimited linear filters is another bandlimited filter, in some applications it is more convenient to employ one TDL to realize $R(f)$ directly. In practice, signals cannot be both time and frequency limited, so that these TDL filters can only approximate the ideal solution. One advantage of the TDL filter is the convenience in adjusting the tap weight control voltage as a means of tracking the channel and noise spectrum variations.

The optimum receiver requires knowledge of the noise power spectrum $K(f)$ and the channel transfer function $H(f)$. When $K(f)$ is not flat over the band of interest, the input noise process contains correlation that is to be removed by the noise filter. Techniques to reduce *correlated noise effects* include: (1) prediction of future noise values and cancellation of the correlated component, (2) mean square error filtering techniques using an appropriate error criterion, and (3) noise excision techniques where a fast Fourier transform (FFT) is used to identify and excise noise peaks in the frequency domain. The problem of noise filtering is usually important in bandwidth expansion systems because of interference from other users as well as hostile jamming threats. For realization of the matched filter, a RAKE TDL filter using the concepts developed by Price and Green[6] can be used to adaptively derive an approximation to $H^*(f)$. A RAKE filter is so named because it acts to "rake" all the multipath contributions together. This can be accomplished using the TDL filter shown in Fig. 48.135, where the TDL weights are derived from a correlation of the tap voltages with a common test sequence, that is, $S(t)$. This correlation results in estimates of the equivalent TDL channel tap values. By proper time alignment of the test sequence, the RAKE filter weights become estimates of the channel tap values but in inverse time order, as required in a matched-filter design. For adaptation of the RAKE filter, the test sequence may be either a known sequence multiplexed with the modulated information or receiver decisions used in a decision-directed adaptation.

An alternate structure for realizing the matched filter is a recirculating delay line that forms an average of the received pulses. This structure was proposed as a means of reducing complexity in a RAKE filter design[7] for a frequency shift keying (FSK) system. For a pulse amplitude modulation (PAM) system, the structure would take the form shown in Fig. 48.136. An inverse modulation operation between the input signal and local replica of the signal modulation is used to strip the signal modulation from the arriving signal. The recirculating delay line can then form an average of

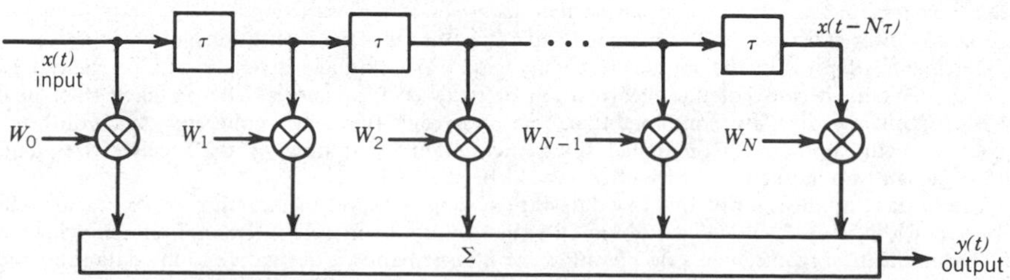

Fig. 48.134 Tapped-delay-line (TDL) filter.

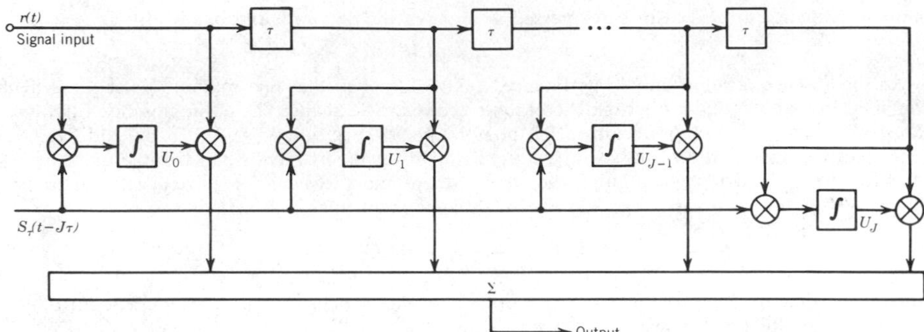

Fig. 48.135 RAKE filter.

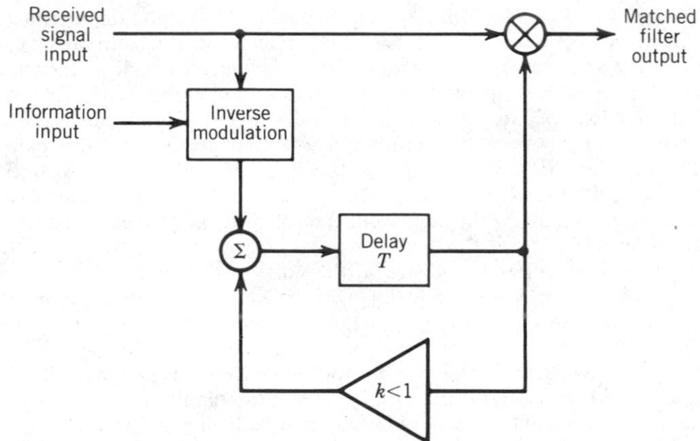

Fig. 48.136 Correlation filter for pulse amplitude modulation (PAM) system.

the received pulse, which is used in a correlator to produce the matched-filter output. This correlation filter is considerably simpler than the RAKE TDL filter shown in Fig. 48.135.

In both the RAKE TDL and the correlation filter, an averaging process is used to generate estimates of the received signal pulse. Because this signal pulse is imbedded in receiver noise, the measurement process must realize sufficient signal-to-noise ratio. This fundamental requirement is the basis for the learning constraint

$$R\ (\text{b/s}) \quad \gg \sigma_t \qquad \text{Hz}$$

introduced earlier. If the signal rate R from which adaptation is being accomplished is not much greater than the channel rate of change σ_t, the channel will change before the averaging process can build up sufficient signal-to-noise ratio for an accurate measurement. This requirement limits the application of adaptive receiver techniques with implicit frequency diversity gain to slow-fading applications relative to the data rate. Fortunately many channels have fade rates on the order of a few hertz and data requirements thousands of times larger.

The receiver for this small-ISI example has, in general, a noise filter to accentuate frequencies where noise power is weakest and a matched-filter structure that coherently recombines the received-signal elements to provide the implicit diversity gain. The implicit diversity can be viewed as a frequency diversity because of the decorrelation of received frequencies. The matched filter in this view is a frequency diversity combiner that combines each frequency coherently according to its received strength. Without the matched filter, incoherent combining of the received frequencies would occur and no implicit diversity effect would be realized.

An important application of this low-data-rate system is found in jamming environments where excess bandwidth is used to decrease jamming vulnerability. In more benign environments, however, most communication requirements do not allow for a large bandwidth relative to the data rate, and if implicit diversity is to be realized in these applications, the effect of intersymbol interference must be considered.

High-Data-Rate Receivers

When the transmitted symbol rate is on the order of the frequency decorrelation interval of the channel, the frequencies in the transmitted pulse will undergo different gain and phase variations resulting in reception of a distorted pulse.

Although there may have been no intersymbol interference (ISI) at the transmitter, the pulse distortion from the channel medium will cause interference between adjacent samples of the received signal. In the time domain, ISI can be viewed as a smearing of the transmitted pulse by the multipath, thus causing overlap between successive pulses. The condition for ISI can be expressed in the frequency domain as

$$T^{-1} \quad \gtrsim f_d \quad \text{Hz}$$

or in terms of the multipath spread

$$T \quad \lesssim 2\pi\sigma_f \quad \text{s}$$

Since the bandwidth of a PAM signal is at least on the order of the symbol rate T^{-1} hertz, there is no need for bandwidth expansion under ISI conditions in order to provide signal occupancy of decorrelated portions of the frequency band for implicit diversity. However, it is not obvious whether the presence of the intersymbol interference can wipe out the available implicit diversity gain. Within the last decade it has been established that adaptive receivers can be used that cope with the intersymbol interference and in most practical cases wind up with a net implicit diversity gain. These receiver structures fall into three general classes: correlation filters with time gating, equalizers, and maximum likelihood detectors.

Correlation Filters. These filters approximate the matched-filter portion of the optimum no-ISI receiver. The correlation filter shown in Fig. 48.136 would fail to operate correctly when there is intersymbol interference between received pulses because the averaging process would add overlapped pulses incoherently. When the multipath spread is less than the symbol interval, this condition can be alleviated by transmitting a time-gated pulse whose "off" time is approximately equal to the width of the channel multipath. The multipath causes the gated transmitted pulse to be smeared out over the entire symbol duration but with little or no intersymbol interference. The correlation filter can then be used to match the received pulse and provide implicit diversity.[8] In a configuration with both explicit and implicit diversity, moderate intersymbol interference can be tolerated because the diversity combining adds signal components coherently and ISI components incoherently. Because the off time of the pulse can not exceed 100%, this approach is clearly data-rate limited for fixed multipath conditions. In addition, the time gating at the transmitter results in an increased bandwidth, which may be undesirable in a bandwidth-limited application. The power loss in peak power-limited transmitters due to time gating can be partially offset by using two carrier frequencies with independent data modulation.[9]

Adaptive Equalizers. Adaptive equalizers are linear filter systems with electronically adjustable parameters that are controlled in an attempt to compensate for intersymbol interference. Tapped-delay-line filters are a common choice for the equalizer structure, because the tap weights provide a convenient adjustable parameter set. Adaptive equalizers have been widely employed in telephone channel applications[10] to reduce ISI effects due to channel filtering. In a fading multipath channel application, the equalizer can provide three functions simultaneously: noise filtering, matched filtering for explicit and implicit diversity, and removal of ISI. These functions are accomplished by adapting a tapped-delay-line equalizer (TDLE) to force error measure to a minimum. By designing the error measure to include the degradation due to correlated noise, ISI, filtering, and improper diversion combining, the TDLE will minimize their combined effects.

A linear equalizer (LE) is defined as an equalizer that linearly filters each of the N explicit diversity inputs. An improvement to the LE is realized when additional filtering is performed on the detected data decisions. Because it uses decisions in a feedback scheme, this equalizer is known as a decision-feedback equalizer (DFE).

The operation of a matched-filter receiver, an LE, and a DFE can be compared from examination of the received pulse train example of Fig. 48.137. The binary modulated pulses have been smeared by the channel medium, producing pulse distortion and interference from adjacent pulses. Conventional detection without multipath protection would integrate the process over a symbol period and decide a $+1$ was transmitted if the integrated voltage is positive and a -1 if the voltage is negative. The pulse distortion reduces the margin again in that integration process. A matched filter correlates the received waveform with the received pulse replica, thus increasing the noise margin. The intersymbol interference arises from both future and past pulses in these radio systems, since the multipath contributors near the mean path delay normally have the greatest strength. This ISI can be

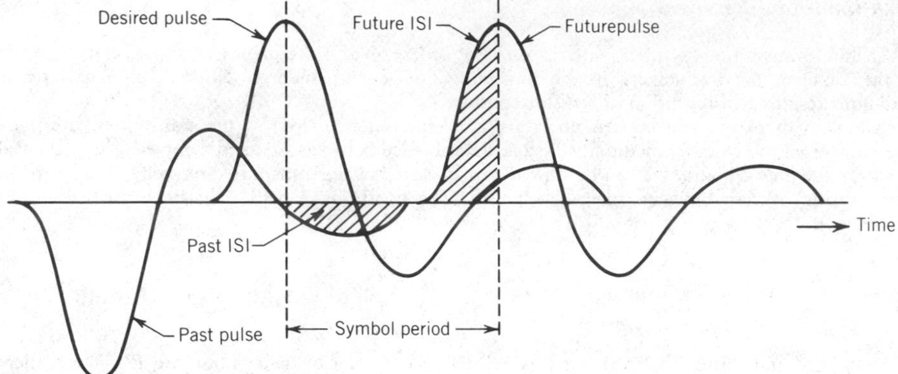

Fig. 48.137 Received pulse sequence after channel filtering. ISI, Intersymbol interference.

compensated for in a linear equalizer by using properly weighted time-shifted versions of the received signal to cancel future and past interferers. The DFE uses time-shifted versions of the received signal only to reduce the future ISI. The past ISI is canceled by filtering past-detected symbols to produce the correct ISI voltage from these interferers. The matched-filtering property in both the LE and DFE is realized by spacing the taps on the TDLE at intervals smaller than the symbol period.

The DFE is shown in Fig. 48.138 for an Nth order explicit diversity system. A forward filter (FF) TDLE is used for each diversity branch to reduce correlated noise effects, provide matched filtering and proper weighting for explicit diversity combining, and reduce ISI effects. After diversity combining, demodulation, and detection, the data decisions are filtered by a backward filter TDLE to eliminate intersymbol interference from previous pulses. Because the backward filter compensates for this "past" ISI, the forward filter need only compensate for "future" ISI.

An automatic gain control (AGC) amplifier is shown for each diversity branch to bring the fading signal into the dynamic range of the TDLE. A decision-directed error signal for adaptation of the DFE is shown as the difference between the detector input and output. Qualitatively one can see that if the DFE is well adapted, this error signal should be small. Reference-directed adaptation can be accomplished by multiplexing a known bit pattern into the message stream for periodic adaptation.

When error propagation due to detector errors is ignored, the DFE has the same or smaller mean square error than the LE for all channels.[11] The error propagation mechanism has been examined by a Markov chain analysis[12] and shown to be negligible in practical fading channel applications. Also in an Nth-order diversity application, the total number of TDLE taps is generally less for the DFE

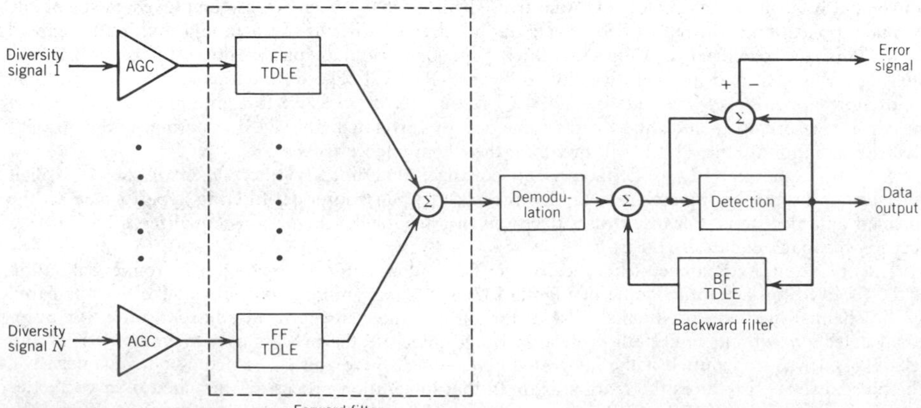

Fig. 48.138 Decision-feedback equalizer, Nth-order diversity. AGC, automatic gain control; FF TDLE, forward filter tapped-delay-line equalizer; BF TDLE, backward filter TDLE.

than for the LE. This follows because the former uses only one backward filter after combining of the diversity channels in the forward filter.

The performance of a DFE on a fading channel can be predicted[13-15] using a transformation technique that converts implicit diversity into explicit diversity and that treats the ISI effects as a Gaussian interferer. As an example, the average probability of error vs. the total received bit energy (E_b) relative to the noise spectral density (N_0) is shown in Fig. 48.139 for a quadruple diversity system. The dashed line represents the zero multipath spread ($\sigma_f = 0$) performance and the solid lines show performance for different DFE configurations (N = number of forward filter taps and Δ = normalized tap spacing) and ISI conditions when the ratio of multipath spread to symbol period T is 0.25. The no-ISI conditions are performance bounds determined by setting the ISI components to zero. When $\sigma/T = 0.25$, performance would be to the right of the dashed line if adaptive signal processing were not employed. The equalizer is seen to remove this degradation and also provide an implicit diversity gain, which is measured by the difference between the solid line $N = 3$, $\Delta = 0.5$ curve and the dashed line. The difference between the $N = 3$, $\Delta = 0.5$ curve and the next curve labeled No ISI is the intersymbol interference penalty. With the filter parameters $N = 3$, $\Delta = 0.5$, no technique for removing ISI can do better than this curve. The small ISI penalty in this typical example is a strong argument for the use of the DFE vs. more powerful ISI techniques. Finally the leftmost solid line approximates the very best that can be done, as results show negligible improvement as the number of taps is increased further. The small difference exhibited shows that a DFE with only a modest number of forward filter taps performs almost as well as an ideal DFE with an infinite number of taps.

A DFE modem has been developed[3] with data rates up to 12.5 Mb/s for application on troposcatter channels with up to four orders of diversity. This DFE modem uses only a three-tap forward filter TDLE and a three-tap backward filter TDLE. Extensive simulator and field tests[3,13] have shown that implicit diversity gain is realized over a wide range of actual conditions, while ISI effects are mostly eliminated. Thus operation at data rates near the frequency decorrelation distance is possible with no large intersymbol interference penalty. Measured results agree well with the predicted performance for which Fig. 48.139 is a typical example.

Maximum Likelihood Detectors. Since the DFE minimizes an analog detector voltage, it is unlikely that it is optimum for all channels with respect to bit error probability. By considering intersymbol interference as a conventional code defined on the real line (or complex line for bandpass channels),

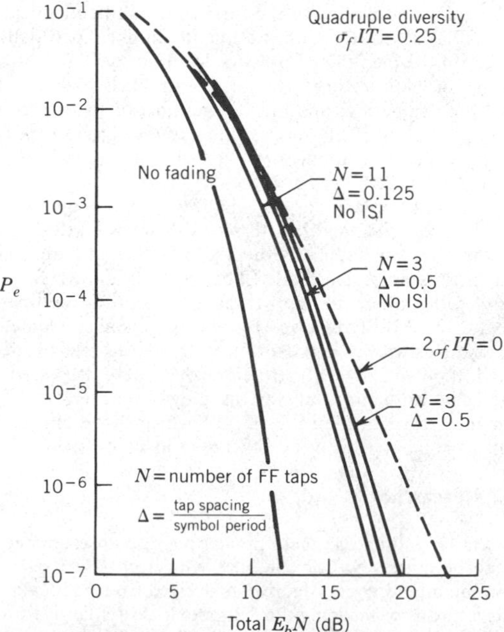

Fig. 48.139 Decision-feedback equalizer (DFE) performance, quad diversity. ISI, intersymbol interference; FF, fast forward. Pe, Probability of Bit Error.

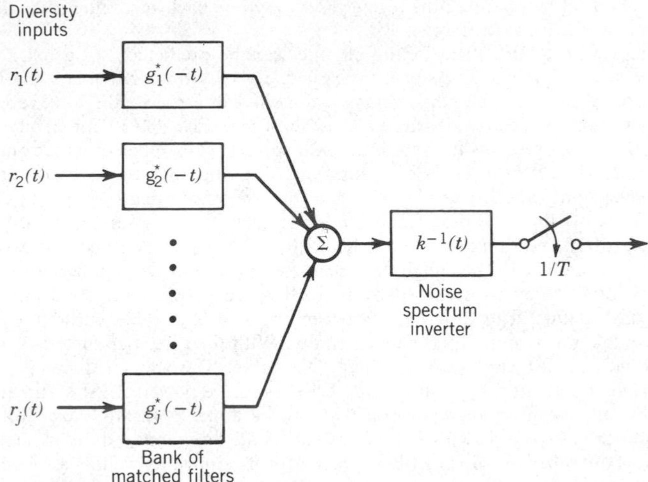

Fig. 48.140 Diversity combiner for maximum-likelihood sequence estimator (MLSE) receiver.

maximum likelihood sequence estimation algorithms have been derived[16,17] for the PAM channel. These algorithms provide a decoding procedure for receiver decisions that minimize the probability of sequence error. A maximum-likelihood sequence estimator (MLSE) receiver still requires a noise filter and matched filters for each diversity channel. After these filtering and combining operations, a trellis decoding technique is used to find the most likely transmitted sequence. Figure 48.140 illustrates the filtering, combining, and sampling functions which precede the MLSE.

The MLSE algorithm works by assigning a state for each intersymbol interference combination. Because of the one-to-one correspondence between the states and the ISI, the maximum-likelihood source sequence can be found by determining the trajectory of states.

If some intermediate state is known to be on the optimum path, then the maximum-likelihood path originating from that state and ending in the final state will be identical to the optimal path. If at time n each of the states has associated with it a maximum-likelihood path ending in that state, it follows that sufficiently far in the past the path history will not depend on the specific final state to which it belongs. The common path history is the maximum-likelihood state trajectory.[16]

Since the number of ISI combinations and thus the number of states is an exponential function of the multipath spread, the MLSE algorithm has complexity that grows exponentially with multipath spread. The equalizer structure exhibits a linear growth with multipath spread. Also the requirement for diversity combining and matched filtering in the MLSE receiver requires about the same circuit and adaptation implementation complexity as an equalizer for this requirement alone. A comparison of Figs. 48.138 and 48.140 for the DFE and MLSE receiver shows that the systems are similar except for the replacement of the backward filter in the DFE by the decoding algorithm in the MLSE receiver. The backward filter is an L tap TDL filter, whereas the MLSE decoding algorithm has computational complexity with exponential growth as a function of multipath spread. In return for this additional complexity, the MLSE receiver results in a smaller (sometimes zero) intersymbol interference penalty for channels with isolated and deep frequency-selective fades. However, in many applications in which high orders of diversity are employed, these deep selective-frequency fades do not occur frequently enough to significantly affect the average error probability. This result is illustrated in the performance curve given in Fig. 48.139, which shows only a small ISI penalty for the DFE with just three taps as compared with the DFE with as many taps when there is no ISI.

48.6-4 New Areas of Research

Present adaptive equalizers for fading channel applications use an estimated gradient algorithm for channel tracking that can be quite slow for channels with deep selective-frequency fades or in the presence of hostile electronic interference. Algorithms derived from the Kalman estimation equations have been suggested[18] as a means of realizing the full potential of adaptive tracking capability. Faster tracking would provide an impetus for HF equalization where digital data rates are not always many orders of magnitude greater than the channel rate of change. Adaptive receivers using multiple antennas in a fading multipath channel environment can provide antijamming protection from antenna nulling in addition to implicit diversity gain. Faster tracking algorithms would increase system flexibility to a wider range of jamming threats.

References

1 V. I. Tatarski, "The Effects of the Turbulent Atmosphere on Wave Propagation," Israel Program for Scientific Translation, Jerusalem, Israel, 1971.

2 P. Monsen, "High Speed Digital Communication Receiver," U.S. Patent 3 879 664, Apr. 22, 1975.

3 D. R. Kern and P. Monsen, "Megabit Digital Troposcatter Subsystem (MDTS)," GTE Sylvania, Needham, MA, and SIGNATRON, Lexington, MA, Final Rep. ECOM-74-0040-F.

4 P. A. Bello, "Characterization of Randomly Time-Variant Linear Channels," *IEEE Trans. Commun. Syst.* **CS-11**:360–393 (Dec. 1963).

5 L. A. Wainstein and V. D. Zubakov, *Extraction of Signals from Noise*, Prentice-Hall, Englewood Cliffs, NJ, 1962, ch. 3.

6 R. Price and P. E. Green, Jr., "A Communication Technique for Multipath Channels," *Proc. IRE* **46**:555–569 (Mar. 1958).

7 S. M. Sussman, "A Matched Filter Communication System for Multipath Channels," *IRE Trans. Inform. Theory* **IT-6**:367–372 (Jun. 1960).

8 M. Unkauf and O. A. Tagliaferri, "An Adaptive Matched Filter Modem for Digital Troposcatter," *Conf. Rec., Int. Conf. Commun.* (June 1975).

9 M. Unkauf and O. A. Tagliaferri, "Tactical Digital Troposcatter Systems," in *Conf. Rec., Nat. Telecommun. Conf.*, vol. 2, pp. 17.4.1–17.4.5 (Dec. 1978).

10 R. W. Lucky, J. Salz, and E. J. Weldon, Jr., *Principles of Data Communication*, McGraw-Hill, New York, 1968, ch. 6.

11 P. Monsen, "Feedback Equalization for Fading Dispersive Channels," *IEEE Trans. Inform. Theory* **IT-17**:56–64 (Jan. 1971).

12 P. Monsen, "Adaptive Equalization of the Slow Fading Channel," *IEEE Trans. Commun.* **COM-22** (Aug. 1974).

13 P. Monsen, "Theoretical and Measured Performance of a DFE Modem on a Fading Multipath Channel," *IEEE Trans. Commun.* **COM-25**:1144–1153 (Oct. 1977).

14 M. Schwartz, W. R. Bennett, and S. Stein, *Communications Systems and Techniques*, McGraw-Hill, New York, 1966, ch. 10.

15 P. Monsen, "Digital Transmission Performance on Fading Dispersive Diversity Channels," *IEEE Trans. Commun.* **COM-21**: 33–39 (Jan. 1973).

16 G. D. Forney, Jr., "Maximum-Likelihood Sequence Estimation of Digital Sequences in the Presence of Intersymbol Interference," *IEEE Trans. Inform. Theory* **IT-18**:363–377 (May 1972).

17 G. Ungerboeck, "Adaptive Maximum-Likelihood Receiver for Carrier-Modulated Data Transmission Systems," *IEEE Trans. Commun.* **COM-22**:624–636 (May 1974).

18 D. Godard, "Channel Equalization Using a Kalman Filter for Fast Data Transmission," *IBM J. Res. Develop.*, pp. 267–273 (May 1974).

48.7 COMPARATIVE ANALYSIS OF COMMUNICATION TECHNIQUES

E. T. Dickerson

A number of communication techniques have been introduced in Sections 47.2, 47.3, and 48.1. Two classes of communication systems were discussed: analog techniques, in which the modulating processes are continuous functions of time, and digital techniques, in which the modulating processes are in the form of a sequence of discrete data symbols.

In this section, various modulation techniques that belong to these two broad classes will be compared. In the course of this comparative analysis we will see that different criteria of comparison are required for analog and digital modulation techniques.

48.7-1 Criteria for Comparison

Selecting the "best" communication system design for a particular application requires a definition of "best" for that application. There are several widely accepted criteria for evaluating quantitatively the performance of a communication system. These include (1) the bandwidth required, (2) the probability of error (in a digital system), and (3) the received signal-to-noise ratio (in an analog system).

Other quality measures may be applicable to a particular case *as defined by the user*. That is, what is best in one case may not be best in another. Evaluation of performance is subject to constraints that relate to the *suitability* of a candidate system for a particular application. These constraints may include one or more of the following:

1. Complexity.
2. Cost.
3. Available power.
4. Skill of the operator.
5. State of the art.

The list is, of course, not exhaustive but is illustrative of the practical constraints that must be considered. Clearly there is no single best communication system, or otherwise there would not be such a wide variety to choose from.

Analog vs. Digital

An important separation of communication systems, and one that will be used in this section, is between continuous waveform modulation (analog) systems and discrete waveform modulation (digital) systems. There is no simple yet general choice between analog and digital implementation. For some applications, such as the communication of computer data already stored in digital form, a digital system seems to be indicated clearly. On the other hand, broadcast-quality television demands an analog system, by most reasonable criteria. Therefore the evaluation approach taken in the following discussion is to consider analog and digital communication systems separately. To avoid lengthy discussions in the comparisons, tabular and graphic techniques will be employed.

48.7-2 Comparison of Continuous Waveform (Analog) Techniques

Two categories of analog modulation techniques are considered here, namely, linear modulation and angle (or exponential) modulation (nonlinear modulation). To compare a number of linear and nonlinear modulation schemes it is necessary to define the criterion from which the systems are to be judged. The relation between the predetection signal-to-noise ratio, SNR_i, and the postdetection signal-to noise ratio, SNR_o, is used here as a measure of "goodness" in comparing the various analog modulation techniques. To establish this relation mathematically requires a brief and condensed discussion of information theory.

Information theory, of course, applies to both digital and analog communication systems. However, the application to analog systems is not as apparent as it is to digital systems (see Section 47.3, Information Theory). This is because the principal results of information theory are expressed in terms of discrete measures (bits and b/s). These principles and measures can be applied to analog communication systems by considering the equivalent discrete representation of the analog information.

The sampling theorem of Nyquist guarantees exact reproduction of a sampled bandlimited analog process if samples are taken at a rate equal to twice the highest frequency contained in the signal. If the number of quantization levels per sample is chosen to represent the signal with sufficient accuracy, then the product of samples/second and bits/sample represents to a good approximation the information rate in bits/second. For example, if a voice signal is bandlimited to 4 kHz, sampled at 8 ks/s, and quantized at 64 levels (6 b/s), then the approximate information rate is 48 kb/s. From a strict application of information theory, the actual information rate may be less than the binary symbol rate, because the measure of information depends on the probability distribution of the signal. From the designer's point of view, however, the binary symbol rate is a significant parameter which is practically equivalent to the information rate.

The signal-to-noise ratio in the predetection bandwidth, SNR_i according to the Shannon-Hartley law, determines the maximum rate at which information may arrive at the detector. Mathematically this channel capacity C_i is given as

$$C_i = B_i \cdot \log_2[1 + (\mathrm{SNR})_i]$$

where C_i = maximum information rate at system input, b/s
B_i = predetection bandwidth, Hz
$(\mathrm{SNR})_i$ = ratio of signal power to noise power in bandwidth B_i

In an ideal communication system, the information rate is constant throughout the system. In particular, the information rate at the receiver input and output are equal, which implies negligible loss of information due to errors, which in turn implies $SNR_i \gg 1$. The definition of "ideal" may be further extended to stipulate that the information rate is equal to the channel capacity defined above. Under these conditions, it has been shown (Panter, 1965, p. 605) that the input and output signal-to-noise ratios are exponentially related. That is,

$$SNR_o \cong (SNR_i)^{\alpha}$$

or

$$SNR_o \cong (P/\alpha N_0 W)^{\alpha}$$

where P = signal power at receiver input, W
N_0 = one-sided noise power spectral density, W/Hz
C_0 = maximum information rate at output
W = signal bandwidth, Hz
α = bandwidth expansion factor, $\alpha W = B_i$

Figure 48.141 gives plots of postdetection SNR_o as a function of P/N_0W for various values of the bandwidth expansion factor. For values of α greater than 22, little improvement is realizable by continuing to expand the transmission bandwidth B_i.

The performance of several practical analog modulation-demodulation systems operating in the presence of additive Gaussian white noise will now be discussed. Figure 48.142 is also a plot of SNR_o vs. P/N_oW. Included in this figure are detection gain curves for several analog modulation schemes. Ideal detection of amplitude modulation (AM), double sideband (DSB), and quadrature DSB modulation would result in a detection gain curve that corresponds to a bandwidth expansion factor of two. However, as shown, the coherent detection of these waveforms falls short of the ideal case. The detection gain curve for FM, with a bandwidth expansion factor of 22, is also shown. Although

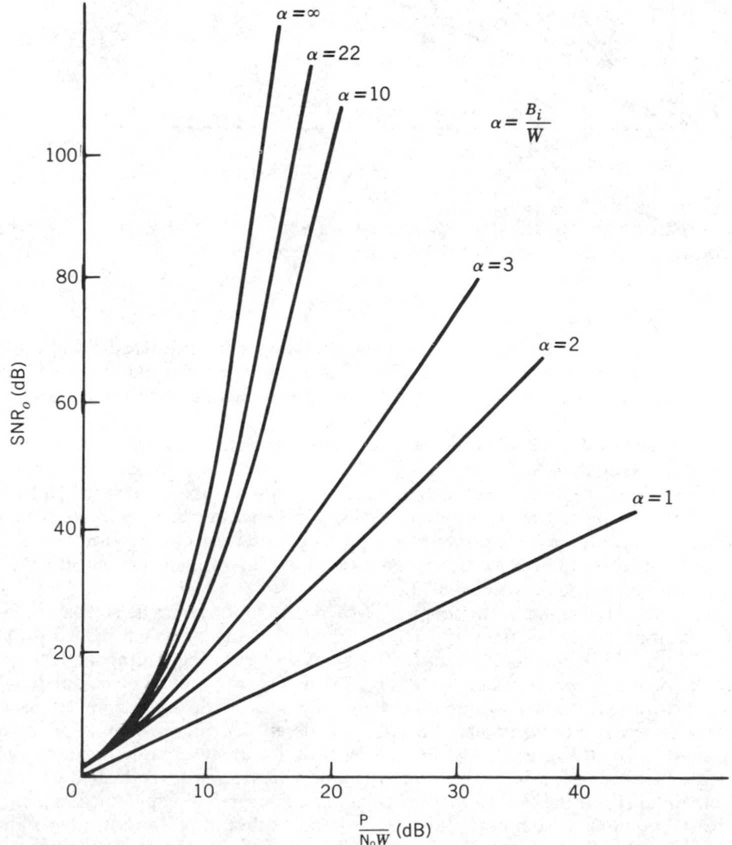

Fig. 48.141 Performance curves for "ideal" modulators and detectors as a function of the bandwidth expansion factor.

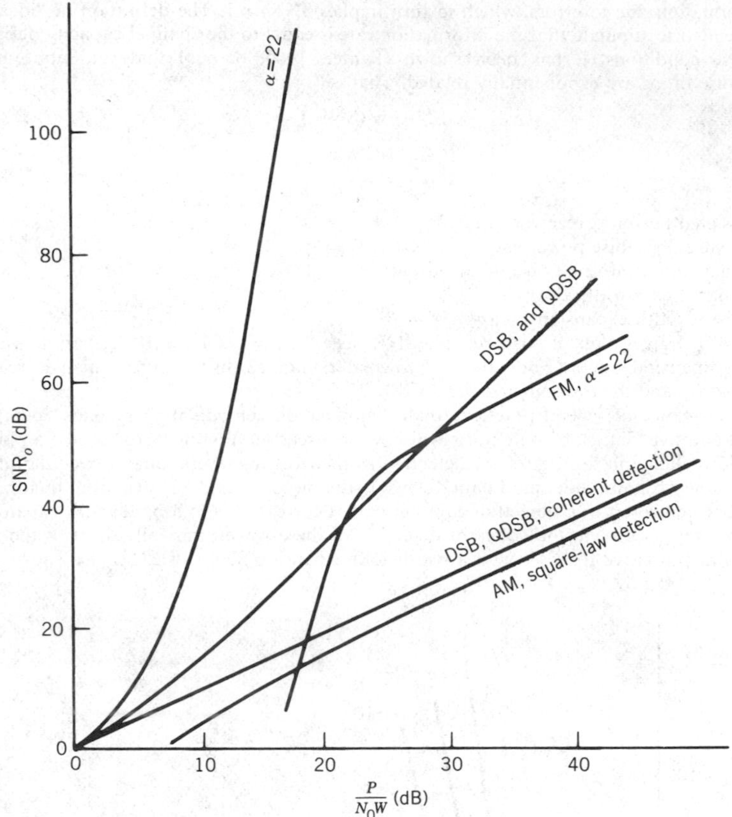

Fig. 48.142 Comparison of realizable analog schemes with the "ideal" modulation scheme. DSB, double sideband; QDSB, quadrature DSB; AM, amplitude modulation.

the FM system is not ideal, it is a widely used and versatile system. The threshold effect for the FM system is well defined. The curves also indicate that FM operation for signal-to-noise ratios less than 10 dB is marginal owing to the threshold effect. A general discussion of threshold effect is included later.

For completeness, Table 48.20 gives a relative comparison of a number of linear and nonlinear analog modulation techniques. In the analysis of modulation techniques presented here, it is assumed that the noise component of the received signal is additive white Gaussian noise with a power spectral density of N_0 watts per hertz. The expression for SNR_i can be written as the ratio of the signal power P at the receiver input and the noise power $N = N_0B_i$ in the predetection bandwidth B_i. The ratio of the input signal-to-noise ratio SNR_i and the signal-to-noise ratio after signal detection SNR_o is used as one criterion for comparison in Table 48.20.

As indicated in the table, the linear suppressed-carrier techniques, single sideband (SSB), double sideband (DSB), and vestigial sideband (VSB), are "better" than AM when judged on the basis of detection gain ($\mathrm{SNR}_o/\mathrm{SNR}_i$). Also SSB and DSB are not subject to the threshold effect that degrades the performance of AM when envelope detection is used. The wide range of modulation techniques discussed here permit user groups with diverse requirements to obtain a communications system that is best for their needs. For example, the FCC (Federal Communications Commission) limits transmitter power to 1000 W average and to 2000 W peak for amateur radio operators. The spectral occupancy is also greatly restricted for the amateur radio operator, to several narrow bands sprinkled throughout the radio spectrum. For all these reasons the amateur radio operator is willing to pay the price of increased equipment complexity to achieve bandwidth and power advantages characteristic of the SSB system. Another advantage of SSB is that the carrier heterodynes (tone interference) inherent in the AM system are eliminated.

TABLE 48.20. COMPARISON OF ANALOG MODULATION SCHEMES

Modulation Type	Transmission Bandwidth[a]	Detection Gain SNR_o/SNR_i	Bandwidth Expansion Factor, B_i/W[b]	DC Response	Equipment Complexity	Popular Applications
Double sideband (DSB)	W	1	2	Yes	Moderate: Coherent demodulation required; receiver can be simplified by transmission of pilot tone	Analog data
Amplitude (AM)	$2W$	1/3	2	No	Minor: Simple envelope detection used; modulator is simple; DC isolation required at receiver	Broadcast radio
Single sideband (SSB)	$2W$	1	1	No	Major: Coherent demodulation required; modulator requires complex sideband filtering	Point-to-point, voice communications, amateur radio
Vestigial sideband (VSB)	$2W$	1	2	Yes	Major: Coherent demodulation required; complex symmetric filters required	Digital and analog, wideband transmission
VSB plus carrier (VSB + C)	$2W$	1/3	2	No	Moderate: Envelope detector used; symmetric filtering required	Wideband signals, video for broadcast TV
Frequency (FM)	$2(D+1)W$	$3/2D^2$	$2(D+1)$	Yes	Moderate: Transmitter complex receiver utilizes simple frequency discriminator	Broadcast radio, point-to-point, microwave relay, broadcast TV audio
Phase (PM)	$2(D+1)W$	K_p^2	$2(D+1)$	Yes	Moderate: Simple modification of FM receiver required	Data transmission voice
Baseband	W	1	1	Yes	Minor: No modulation or frequency translation required	Intercomputer communications, Short point-to-point communications

[a] W, baseband signal bandwidth; D, deviation constant.
[b] B_i, transmission bandwidth.

The AM system is attractive to a very different user group. This group requires transmission to a large number of users for a minimum equipment cost. The AM system using an envelope detector meets these requirements.

The linear analog techniques discussed thus far require linear amplification after the modulation process has taken place. It is often desirable to utilize efficient techniques for power amplification. For example, a large number of spaceborne applications utilize a traveling-wave tube (TWT) power amplifier that must operate in the saturated (nonlinear) mode for efficient operation. This application requires that the modulated signal have a constant envelope and that the information be contained in the phase angle. Of all the techniques mentioned thus far, FM and PM meet this requirement. Compared with other analog modulation techniques, FM provides the best detection gain. This improved performance is accomplished at the expense of bandwidth.

As indicated in Table 48.20, the FM system offers a considerable savings in power for equivalent performance, measured in terms of detection gain, over the other analog schemes discussed thus far. The VSB and SSB schemes are the most complex of the schemes, while the AM system is the simplest to realize in terms of hardware. This analysis of analog modulation schemes has been brief and has only included the basic techniques. A number of communication systems, such as color TV, require a combination of these basic techniques. This complex signal utilizes vestigial sideband plus carrier (VSB + C), frequency (FM), and quadrature double sideband (QDSB) modulators and frequency division (FDM) and time division (TDM) multiplexers, respectively.

48.7-3 Comparison of Discrete Waveform (Digital) Techniques

Probability of Error

As indicated in Section 48.7.1, the principal criterion for comparison of digital modulation techniques is the probability that the receiver will make an error when detecting the received signal in the presence of additive white Gaussian noise. In general, the probability of an incorrect decision is viewed as the probability of a symbol error. For the binary system, one symbol usually contains 1 bit of information.

For ease of comparison, only binary modulation techniques will be considered initially, that is, the cases where the symbol error probabilities and bit error probabilities are equal. However, the relation between binary system error rates and M-ary system error rates is given here for completeness. For the orthogonal signaling case,

$$P_E(\text{binary}) = P_E(M\text{-ary}) \cdot [2^{k-1}/(2^k - 1)]$$

where P_E(binary) is the probability of error for a binary signaling scheme and $P_E(M$-ary) is the probability of error for M-ary signaling schemes.

For nonorthogonal signals, assuming only 1 bit per symbol, the error probabilities are related as follows:

$$P_E(M\text{-ary}) = P_E(\text{binary}) \log_2 M = P_E(\text{binary}) \cdot k$$

where M is equal to 2^k bits per symbol.

Since the signals considered here are deterministic and the additive noise is a Gaussian distributed random process, the signal plus the noise will exhibit a Rician distributed-probability density function. The Rician density function is approximated by the Gaussian density function for large signal-to-noise ratios.

Assuming that the received signal is an equiprobable independent sequence of bits and that a correlation receiver is employed, the probability of error P_E for a binary phase-shift-keyed (PSK) signal can be expressed as

$$P_E = \frac{1}{\sqrt{2\pi}} \int_{\sqrt{(E_b/N_0)(1-\cos\theta)}}^{\infty} \exp(-u^2/2)\, du$$

where E_b = signal energy per bit, J
N_0 = one-sided noise power spectral density, W/Hz
θ = angle between signaling waveforms in signal space

The ratio E_b/N_0 is used to describe system performance for various digital modulation techniques. This quantity is easily related to the signal-to-noise ratio in the bit rate bandwidth at the receiver, providing the signaling rate R and the bit rate bandwidth W are known. The following equation describes this relation.

$$P/N_0 W = (E_b/N_0)(R/W)$$

Figure 48.143 compares a number of the modulation schemes on the basis of probability of a bit error. These plots of probability of error vs. E_b/N_0 indicate that coherent schemes give generally

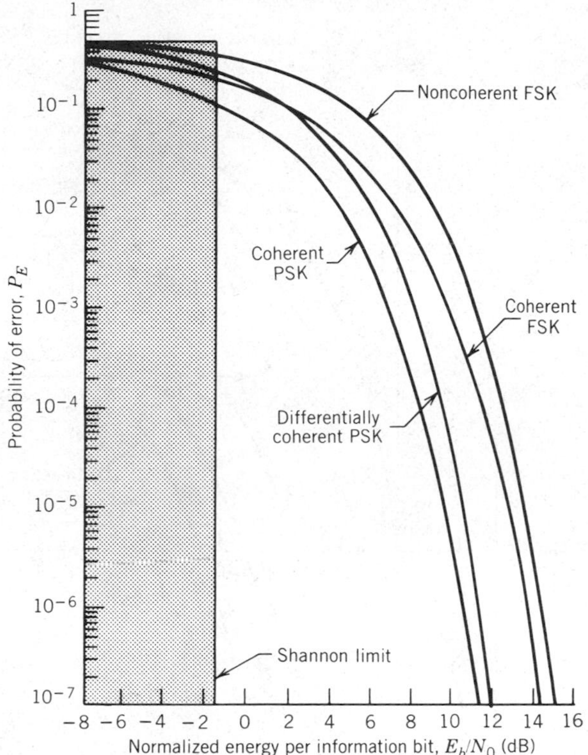

Fig. 48.143 Probability of bit error for binary modulation techniques. FSK, frequency shift keying; PSK, phase shift keying.

better results than noncoherent schemes and that PSK generally performs better than FSK. For example, for a bit error probability of 10^{-4}, noncoherent FSK requires that E_b/N_0 be approximately 4 dB greater than that required for coherent PSK. Therefore the tradeoff between performance and complexity is clear. In a power-limited situation, as encountered in spaceborne systems, the justification for the increased cost due to increased system complexity is evident. As indicated by Sklar (1983), there are other considerations besides P_E and system complexity when choosing a communication scheme. For example, in some cases (such as randomly fading propagation conditions), a noncoherent system is more robust and desirable because there may be difficulty in establishing a carrier reference for the coherent schemes.

Bandwidth-Efficiency Plane

The bandwidth-efficiency plane is an effective pictorial description for comparing a number of digital signaling techniques. Figure 48.144 and parameter discussion have been adapted from Sklar (1983). The Shannon-Hartley theorem can be written mathematically in terms of E_b/N_0 and W/R, assuming that the information rate input to the receiver and the information rate out of the receiver are equal. The following equation is derived using this law.

$$E_b/N_0 = W/R(2^{R/W} - 1)$$

This equation has been plotted on the R/W vs. E_b/N_0 plane in Fig. 48.144. The ordinate R/W is a measure of how much data can be transmitted in a specified bandwidth within a given time; it therefore reflects how efficiently the bandwidth resource is utilized. The abscissa is the metric E_b/N_0 in decibels. For the case when the channel capacity C and the signaling rate R are equal in the equation just given, the hashed curve represents a boundary that separates potential error-free communication from regions where such communication is not possible. On the bandwidth-efficiency plane of Fig. 48.144 are plotted the operating points for minimum PSK (MPSK) and minimum FSK (MFSK) modulation, with $P_E = 10^{-5}$. For MPSK modulation, R/W increases with increasing M;

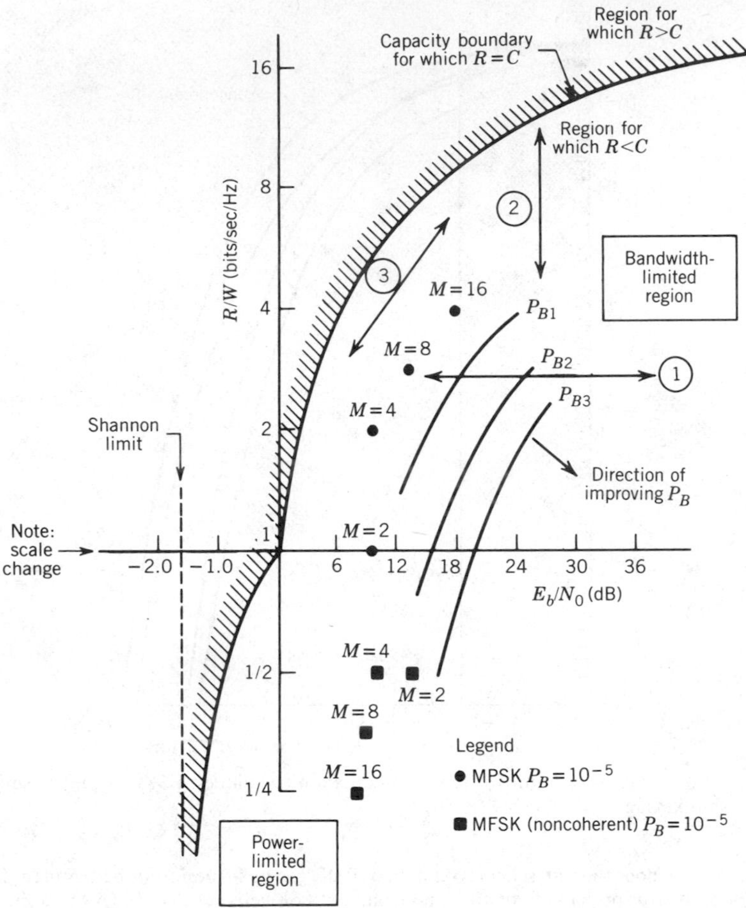

Fig. 48.144 Bandwidth-efficiency plane.

$$R/W \leqslant C/W = \log_2(1 + S/N) = \log_2[1 + E_b/N_0(C/W)].$$

Any digital signaling scheme that transmits $\log_2 M$ bits in T seconds, using bandwidth W (Hz), always operates at an efficiency of $R/W = \log_2 M/WT$ (b/s/Hz). Capacity boundary: locus of minimum E_b/N_0 vs. R/W for reliable transmission. Shannon limit: limit as $C/W \to 0$ of $E_b/N_0 = -1.6$ dB. MPSK: multiphase shift keying, MFSK: multiple frequency shift keying. From Sklar (1983), reproduced with permission.

however, for MFSK modulation, R/W decreases with increasing M. The location of the MPSK points indicate that binary PSK (BPSK) ($M = 2$) and quadrature PSK (QPSK) ($M = 4$) require the same E_b/N_0. That is, for the same value of E_b/N_0, QPSK has a bandwidth efficiency of 2b/s/Hz, compared with 1 b/s/Hz for BPSK. This unique feature stems from the fact that QPSK is effectively a composite of two BPSK signals, transmitted on waveforms orthogonal to one another and having the same spectral occupancy. Also plotted on the bandwidth-efficiency plane are the operating points for noncoherent MFSK modulation $P_E = 10^{-5}$. The position of the MFSK points indicates that binary FSK, (BFSK) ($M = 2$) and quadrature FSK (QFSK) ($M = 4$) have the same bandwidth efficiency, even though the former requires greater E_b/N_0 for the same error rate. Bandwidth efficiency varies with the modulation index.

The theoretical information capacity boundary is shown on the bandwidth-efficiency plane. This boundary is the theoretical limit for the rate at which information can be transmitted error free through a channel. This rate provides a goal for the designer when specifying the desired performance for a particular communications system application. Curves of equal error probability for various modulation and coding schemes are also shown. The curves labeled P_{B1}, P_{B2}, and P_{B3} are hypothetical constructions for some arbitrary modulation and coding scheme; the P_{B1} curve

TABLE 48.21. COMPARISON OF DIGITAL MODULATION TECHNIQUES

Scheme	Symbols, $S_1(t), S_2(t)$	Bit Rate Bandwidth, W (Hz)	Probability of Error, P_E	S/N for $P_E = 10^{-4}$ (dB)	Equipment Complexity	Popular Usage
Amplitude shift keying (ASK), coherent	$S_1(t) = A\cos\omega_c t$,[a] $S_2(t) = \theta$	$2W$	$Q(\sqrt{E_b/2N_0})$[b]	11.4	Moderate	Rarely used
ASK, noncoherent	Same as coherent ASK	$2W$	$1/2\exp(-E_b/2N_0)$	12.3	Minor	Requires high peak power and is sensitive to channel variations
Frequency shift keying (FSK), coherent	$S_1(t) = A\cos(\omega_c - \omega_d)t$, $S_2(t) = -A\cos(\omega_c + \omega_d)t$	$> 2W$	$Q(\sqrt{E_b/2N_0})$	10.6	MAJOR	Seldom used
FSK, noncoherent	Same as coherent ASK	$> 2W$	$1/2\exp(-E_b/2N_0)$	12.3	MINOR	Used for slow speed data transmission
Phase shift keying (PSK), coherent	$S_1(t) = A\cos\omega_c t$, $S_2(t) = -A\cos\omega_c t$	$2W$	$Q(\sqrt{2E_b/N_0})$	8.4	Major	Transmission best overall performance used for high-speed data
Differential PSK (DPSK)	Same as coherent PSK with differential coding	$2W$	$1/2\exp(-E_b/N_0)$	9.3	Moderate	Medium-speed data transmission

[a] A, carrier amplitude; ω_c, carrier frequency, rad/s.
[b] E_b, energy per bit, J; N_0, noise power spectral density, W/Hz; $Q(x)$, complementary error function with argument x.

represents the largest error probability of the three curves and the P_{B3} curve represents the smallest. The general direction in which the curves move for improved P_B is indicated in the figure.

Tradeoffs between P_E and E_b/N_0 can be viewed using the bandwidth-efficiency plane. Such potential tradeoffs are changes in operating point in the direction shown by the arrows. Movement of the operating point along line 1 can be viewed as trading P_E vs. E_b/N_0 performance, with R/W fixed. Similarly, movement along line 2 is seen as trading P_E vs. W (or R/W) performance, with E_b/N_0 fixed. Finally, movement along line 3 illustrates trading W (or R/W) vs. E_b/N_0 performance, with P_E fixed. Movement along line 1 is effected simply by increasing or decreasing the available E_b/N_0. Movement along line 2 or line 3 is effected through appropriate changes to the system by modification of the modulation or coding scheme.

System performance has been described for a number of digital signaling schemes. This performance has been assessed by the probability of error as a function of such parameters as noise power spectral density, signaling rate, and signal power. The principal metric used was E_B/N_0, which is related to the three system parameters just mentioned. Table 48.21, as adapted from Shanmugam (1979), summarizes the digital modulation schemes described in this section. The parameters used for the comparative analysis in this table will now be discussed.

Bandwidth. Bandwidth requirements for a particular digital communication system are of paramount importance. Channel bandwidth is usually limited for communication channels by the FCC or other governmental agencies. Therefore signaling schemes, such as FSK, that utilize bandwidth inefficiently are generally not considered for high-data-rate applications. Bandwidth also determines the noise power level in the receiver predetection filter, B_i.

Probability of error. The designer must also decide what error rate the communication requirement can tolerate. The equations listed in Tables 48.20 and 48.21 can be used to calculate the probability of a bit error for the signaling schemes listed. These calculations require knowledge of the complementary error function $Q(x)$ and the metric E_b/N_0. $Q(x)$ can be calculated using the following integral expression:

$$Q(x) = \frac{1}{\sqrt{2\pi}} \int_x^{\infty} \exp(-u^2/2)\, du$$

Numerous tables are available that list $Q(x)$ for a wide range of the argument x. When using these tables to obtain values for the complementary error function $Q(x)$, the designer may have to compensate for the various definitions used.

Signal-to-noise ratio. As mentioned on a number of occasions throughout this section, P_E, bit error probability, is the most widely used metric when comparing different digital communication schemes. The metric E_b/N_0 was the independent variable in the equations for P_E. The predetection signal-to-noise ratio SNR_i is directly related to this quantity. SNR_i is an excellent basic function to evaluate system performance. This quantity is listed in the table for the various signaling schemes at a fixed $P_E = 10^{-4}$. Using SNR_i as a metric ($P_E = 10^{-4}$), it is observed from the values in the table and from Fig. 48.143 that coherent PSK requires the least amount of power of the modulation schemes listed. The others in ascending order of power required are DPSK, coherent FSK, noncoherent ASK, coherent ASK, and noncoherent FSK. The most widely used of these schemes are PSK, DPSK, and noncoherent FSK. From the table it is observed that DPSK requires approximately 1 dB more power than PSK and that noncoherent FSK requires between 6 and 7 dB more power than PSK.

Equipment complexity. The major differences with respect to equipment complexity for the digital signaling schemes under discussion are in the demodulation process. Among the most frequently used modulation techniques, the coherent PSK receiver is the most complex. DPSK is the next in descending degree of complexity, and noncoherent FSK is the least complex of the popular signaling schemes.

Application tradeoffs. It is clear that none of the digital signaling schemes mentioned is the best for all applications. In satellite communications, power is usually at a premium. Coherent PSK is the optimum choice for the power-limited system, despite the major equipment complexity. On the other hand, when power and data rates are not the principal factors, such as in a multiuser environment, noncoherent FSK is frequently a suitable choice.

48.7-4 Digital vs. Analog Modulation Techniques

An answer to the question of whether to implement a communications system employing analog or digital techniques is applications specific. In the previous sections, a comparison of analog and digital modulation techniques was outlined. Here the question of digital vs. analog techniques is analyzed and discussed.

It was shown that the output of the information source will usually determine the type of modulation that is most easily implemented. However, in some cases it is desirable to transmit analog information over a digital channel. In this case the analog information must be accurately converted to digital information. Nyquist's sampling theorem specifies the number of samples required for exact reconstruction of the analog signal as being two samples per cycle at the highest frequency component contained in the bandlimited analog process. These sample values are then coded into some specific digital format (pulse code modulation, for example). The designer may now take advantage of all the signal-processing techniques available for digital communications systems. This digital transmission of analog information increases the complexity of the equipment as well as increases the required transmission bandwidth.

One such application, where the technique might warrant the cost in complexity and bandwidth, would be in a spaceborne link in which repeaters or transponders are employed. In this application the information would be demodulated, error corrected, and then remodulated, thereby eliminating the repeater input noise. TV is an example of an analog signal that is often impractical to transmit by digital means, because of the required bandwidth and the clock rates.

48.7-5 Practical Considerations

Synchronization

Synchronization is an important concept in both analog and digital systems. System complexity often will determine the degree of synchronization required.

Both analog and digital coherent demodulation schemes require at least one level of synchronization. Carrier synchronization for both digital and analog communications systems requires an accurate knowledge of the carrier phase reference. This carrier reference can be obtained in a number of ways. If the modulation signal spectrum contains a carrier component, carrier coherence is obtained using a simple phaselock loop. Also a pilot tone coherently related to the carrier can be transmitted with the information and recovered for use as a carrier reference at the receiver. If no carrier component is contained in the modulation signal spectrum, alternate techniques such as a squaring loop or a Costas loop can be used to recover the carrier for coherent demodulation. Other levels of synchronization are required for more complex schemes. Some of these are (1) bit synchronization, (2) word synchronization, and (3) code synchronization.

Threshold

An important characteristic of some communication systems is threshold effect. Noncoherent FM is a good example of a practical system that has a very definite threshold below which performance is degraded rapidly. (See Fig. 48.142.) The FM system is also an excellent example of a practical system that can be used for a wide range of communication requirements so long as there is sufficient power to maintain operation above threshold. If sufficient power is not available, some of the more complex and costly coherent techniques must be employed or the FM system complexity must be increased for threshold extension.

48.7-6 Summary

This section has provided a brief review of the principles of comparative analysis of communication techniques. These analytical techniques are useful in the conceptual design and planning stages of the development of a communication system. More detailed discussions can be found in the bibliography.

Bibliography

Bennett, W. R., "Methods of Solving Noise Problems," *IRE* **44**:609–638 (1956).

Bennett, W. R., "Envelope Detection of a Unit-Index Amplitude Modulated Carrier Accompanied by Noise," *IEEE Trans. Inform. Theory* **IT-20**:723–728 (Nov. 1974).

Berlekamp, E. R., *Algebraic Coding Theory*, McGraw-Hill, New York, 1968.

Black, H. S., *Modulation Theory*, Van Nostrand, New York, 1953.

Carlson, A. B., *Communication Systems*, McGraw-Hill, New York, 1968.

Feher, K., *Digital Communications, Microwave Applications*, Prentice-Hall, Englewood Cliffs, NJ, 1981.

Feller, W., *An Introduction to Probability Theory and Its Applications*, vol. 1, Wiley, New York, 1957.

Gagliardi, R. M., *Introduction to Communication Engineering*, Wiley, New York, 1978.

Gardner, F. M., *Phaselock Techniques*, Wiley, New York, 1966.

Gregg, W. D., *Analog and Digital Communications*, Wiley, New York, 1977.

Lee, Y. W., *Statistical Theory of Communications*, Wiley, New York, 1964.

Lindsey, W. C. and M. K. Simon, *Telecommunication Systems Engineering*, Prentice-Hall, Englewood Cliffs, NJ, 1973.

Lucky, R. W., "A Survey of the Communication Theory Literature: 1968–1973," *IEE Trans. Inform. Theory* **IT-19**:725–739 (Nov. 1973).

Middleton, D., *An Introduction to Statistical Communication Theory*, McGraw-Hill, New York, 1960.

Panter, P. F., *Modulation, Noise, and Spectral Analysis*, McGraw-Hill, New York, 1965.

Panter, P. F., *Communication Systems Design: Line-of-Sight and Troposcatter Systems*, McGraw-Hill, New York, 1972.

Papoulis, A., *The Fourier Integral and Its Application*, McGraw-Hill, New York, 1962.

Papoulis, A., *Probability, Random Variables, and Stochastic Processes*, McGraw-Hill, New York, 1965.

Rice, S. O., "Mathematical Analysis of Random Noise," *Bell Sys. Tech. J.* **24**:46–156 (Jan. 1945).

Sakrison, D. J., *Communication Theory: Transmission of Waveform and Digital Information*, Wiley, New York, 1968.

Schwartz, M., *Information Transmission, Modulation, and Noise*, 2nd ed., McGraw-Hill, New York, 1980.

Schwartz, M., W. R. Bennett, and S. Stein, *Communication Systems and Techniques*, McGraw-Hill, New York, 1966.

Schwartz, R. J. and B. Friedland, *Linear Systems*, McGraw-Hill, New York, 1965.

Shanmugam, K. S., *Digital and Analog Communication Systems*, Wiley, New York, 1979.

Shannon, C. E., "A Mathematical Theory of Communications," *Bell Sys. Tech. J.* **27**:379–423, 623–656 (Jul. 1948).

Sklar, B., "A Structured Overview of Digital Communications," *IEEE Commun. Mag.* (Aug. 1983).

Skolnik, M. I., ed., *Radar Handbook*, McGraw-Hill, New York, 1970.

Spilker, J. J., *Digital Communications by Satellite*, Prentice-Hall, Englewood Cliffs, NJ, 1977.

Stark, H. and F. B. Tuteur, *Modern Electrical Communications Theory and Systems*, Prentice-Hall, Englewood Cliffs, NJ, 1979.

Taub, H. and D. L. Schilling, *Principles of Communication Systems*, McGraw-Hill, New York, 1971.

Weiner, N., *The Extrapolation, Interpolation, and Smoothing of Stationary Time Series with Engineering Applications*, Wiley, New York, 1949.

Wozencraft and Jacobs, *Principles of Communication Engineering*, Wiley, New York, 1965.

Ziemer, R. and W. H. Tranter, *Principles of Communications, Systems, Modulations, and Noise*, Houghton Mifflin, Boston, 1976.

CHAPTER 49
COMMUNICATION SYSTEMS

RAY W. NETTLETON

AMECOM Division of Litton Industries, Inc., College Park, Maryland

JEFFERSON F. LINDSEY III

Southern Illinois University at Carbondale, Carbondale, Illinois

SURYA V. VARANASI

Lockheed Engineering and Management Services Company, Inc., Houston, Texas

DANIEL F. DiFONZO

COMSAT Laboratories, Clarksburg, Maryland

49.1 TERRESTRIAL SYSTEMS

Ray W. Nettleton

49.1-1 Introduction

This section is concerned with the transmission of information from one point to another, where both points are earthbound and no conductive path exists between the two. The use of extraterrestrial mechanisms such as satellites is not covered here; see Section 49.4 on satellite communications.

49.1-2 Signal Propagation

Propagation of a radio signal between two earthbound points relies on the interaction of the radio wave with the earth's atmosphere, the earth's surface, and any obstructions and inhomogeneities that may be present. It is possible to identify five principal mechanisms: (1) line-of-sight propagation, (2) groundwave propagation, (3) ionospheric propagation, (4) tropospheric scattering, and (5) local scattering[1,2] (see Fig. 49.1). Each of these mechanisms creates a channel with its own characteristics that vary with frequency, link distance, time of day and year, and solar and atmospheric conditions. Often several mechanisms are present together, but usually the signal strength due to one of them

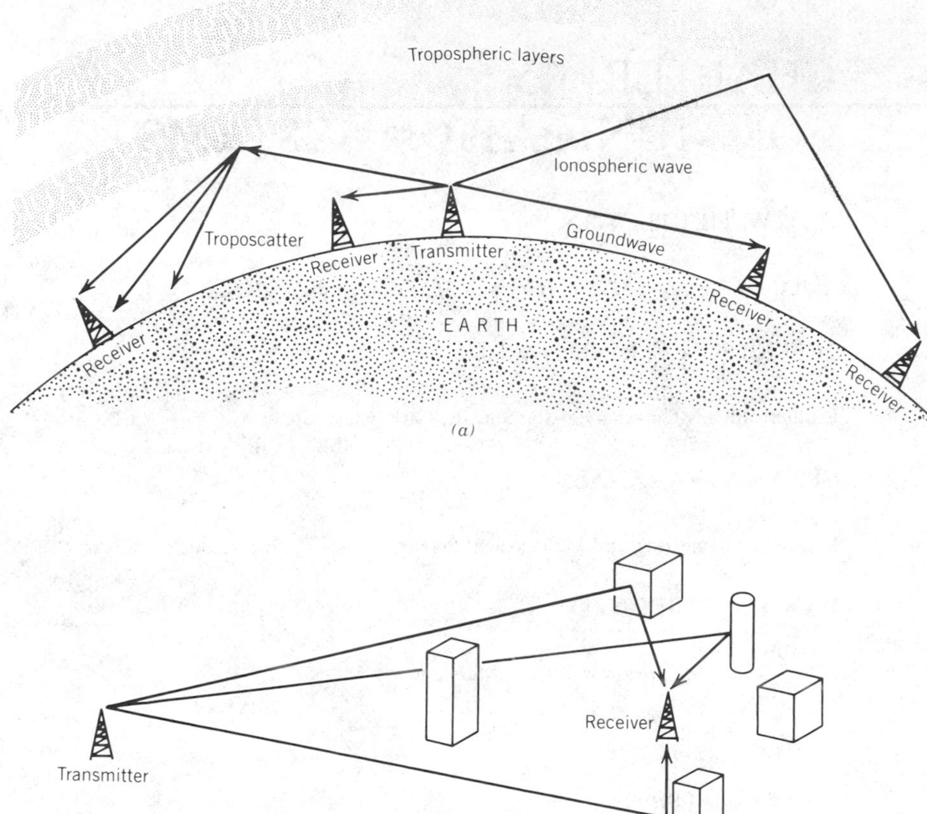

Fig. 49.1 Principal propagation mechanisms: (*a*) Atmospheric and terrestrial propagation, (*b*) local scattering propagation.

dominates the others and the channel can be modeled as if only one mechanism were operating. The dominant transmission mechanism may change from location to location (particularly if the path length is significantly changed), and sometimes from time to time in the same location.

The design of a communication link depends on many factors, some of which are discretionary to the designer (such as the cost/performance tradeoff and the modulation method) and some of which are outside the designer's control. In the case of unguided terrestrial links, the most significant uncontrolled design factor is the nature of the channel through which transmission must take place. Thus in this section the nature of each channel is considered, and then some typical applications are described for fixed point-to-point links and for mobile radio links.

Line-of-Sight Propagation

A line-of-sight (LOS) channel can be established whenever there is a straight-line path between transmitter and receiver that is free of significant obstructions. The term "significant" here means any obstruction that is large in dimension compared with a wavelength. For radio waves this usually means that small but visible obstructions (such as the leaves of a tree) may be ignored, so that a LOS path might exist even though an observer located at the transmitter cannot actually see the receiver. In addition, the presence of significant obstructions near the transmission path can give rise to multipath distortion of the signal, and so most LOS links use narrow antenna patterns focused on each other to prevent excessive multipath problems. This also permits the multiple use of the same frequency band in independent but nearby systems. To provide narrow-beam antennas with reasonable physical size, small wavelengths are necessary. Thus most LOS links use microwave, millimeter, and optical frequencies, even though such links can theoretically be established at any frequency.

In spite of careful design and location of LOS link equipment, however, the earth and the atmosphere frequently conspire to provide multipath reception. Figure 49.2 shows two of the possibilities. The first is present under otherwise perfect conditions; a part of the transmitted wavefront is reflected from the surface of the earth (and/or other fixed obstructions) and impinges on the receiver antenna. Since the direct and reflected waves have traveled different distances, they may not be in phase; indeed if they are of the same magnitude, they may even cancel to give a zero net received field. Thus there is an optimum receiver antenna height for any link, corresponding to the first Fresnel zone clearing from the transmitter and depending on the transmitter-earth-receiver geometry and the phase shift at the reflection (i.e., higher is not necessarily better; notice how microwave antennas are often mounted only part way up a tower or tall building rather than at the top).

The first multipath mechanism is relatively easy to account for, since it is not time varying. The second is more difficult: refraction of part of the transmitted wave from regions of the atmosphere that have different refractive indices. This occurs due to thermal layering and to turbulence, and so the phenomenon changes with time of day and with the weather. At one time the received signals may be in phase and at another they may partially or completely cancel. The time frame of the phenomenon varies widely; in microwave links it typically takes minutes or hours to cycle, and the effect is a "fade." In optical links it may happen many times per second and the result is called scintillation.

The most common way of combating these fades is to provide space diversity, that is, multiple receiving antennas located some wavelengths apart so that the fades encountered by each tend to be statistically uncorrelated. In the microwave case this means two antennas mounted at different heights up the tower. For optical frequencies, simply increasing the physical size of the photodetector surface usually suffices, so that the detector output corresponds to the summation of several received optical modes.

A second major problem for the LOS link designer is the presence of precipitation in the propagation path. Rain, snow, and fog all cause attenuation of radio waves. The effect is greater for denser precipitation, for larger precipitation particles (e.g., snow is worse than fog, and hail is often the worst of all), and at higher frequencies. At optical frequencies even dust and other atmospheric particles (aerosols) are problematic. Specific loss curves are given in the sections on microwave and optical links.

Groundwave Propagation

Since the earth is an imperfect conductor, radio waves penetrate some distance into its surface, diminishing rapidly in field strength with depth, and propagating more slowly than in air. This gives rise to a "dragging" or surface-wave effect and causes the wave just above the surface to follow the curvature of the earth. The process is inherently lossy and so is useful only for relatively low frequencies and for relatively short-distance communication beyond the horizon [order of 200 km for medium frequency (MF); order of 2000 km for low frequency (LF) and very low frequency (VLF)].

Groundwave propagation is often present with other forms of propagation and causes interference patterns that cause fading phenomena. It is useful during the daytime when the strongly absorbing D layer of the ionosphere makes reflection from the upper layers very weak at low and medium frequencies. It is also used for marine and submarine communications at LF, VLF, and extremely low frequency (ELF) (< 300 kHz).[3]

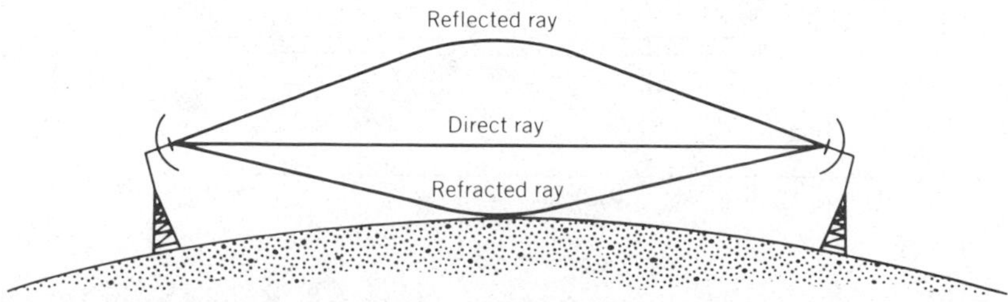

Fig. 49.2 Multipath propagation in line-of-sight links.

Ionospheric Propagation

The ionosphere is a region in the upper atmosphere that contains layers of electrically charged atmospheric gases. The ionization of the gases is due to the action of solar radiation (both electromagnetic radiation and the so-called solar wind, a stream of ionized particles ejected at high velocity from the sun). Thus the composition of the ionosphere depends strongly on solar activity, particularly sunspots, and on the time of day (diurnal variation), season of the year (seasonal variation), and geographic latitude.

Figure 49.3 shows the diurnal variation in the location of the principal ionospheric layers. During the day there are four distinct layers: D, E, F_1, and F_2. During the night the D layer disappears and the F layers merge. The density of the E layer is reduced, permitting greater penetration into the F layer. In addition to the more predictable layers, small cloudlike layers of increased ion density roam around the E layer, permitting brief enhanced communication, usually in the upper HF band. The clouds are referred to as sporadic E.

The ionosphere is also affected by the earth's own magnetic field. The presence of this field makes the ionosphere birefringent, breaking up an incident radio wave into two components, one polarized along and one normal to the magnetic lines of force. The resultant reflected waves are called the ordinary and extraordinary waves, and they may have quite different characteristics. Solar corona activity also causes magnetic storms in the ionosphere, particularly at the poles, which often manifest in the celebrated and spectacular aurora borealis (northern lights). Such storms wreak havoc in ionospheric radio channels, often causing complete blackouts of communication.

Ionized particles vibrate physically in response to the presence of a radio-frequency field, and the energy absorption resulting from ion collisions causes attenuation of the radio wave. Absorption reaches a broad maximum at a particular resonant frequency peculiar to each molecular species (the gyromagnetic frequency),[4] generally occurring at frequencies below 1 kHz for atmospheric gases (ELF band). However, the most important ion for HF propagation is the free electron, which has a gyromagnetic resonance at approximately 1.4 MHz. Thus absorption is a maximum about in the center of the MF band (300 kHz–3 MHz). Absorption is also dependent on ion density and is a maximum in the dense D layer. When the D layer disappears at night, some propagation does occur in the MF band. However, attenuation is gradually less at frequencies above and below the resonant region, providing more useful reflection effects at LF and VLF frequencies (< 300 kHz) and in the high frequency (HF) range (3–30 MHz and sometimes beyond) below the current maximum usable frequency (MUF) for the layer being used.

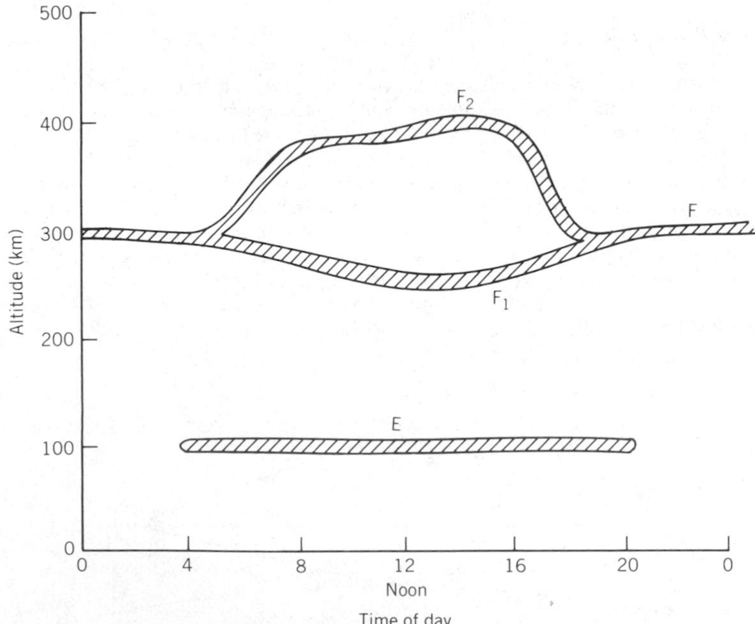

Fig. 49.3 Diurnal variations in the ionospheric layers. Adapted from National Bureau of Standards Circular 462, "Ionospheric Radio Propagation."

The "critical frequency" of an ionospheric layer is the highest frequency that will be reflected from that layer at a normal (vertical) angle of incidence. Critical frequency depends strongly on diurnal, seasonal, and sunspot cycle variations. Figure 49.4 shows the diurnal variations in and the effect of the sunspot cycle on critical frequency.

Propagation by reflection occurs above the critical frequency at oblique angles of incidence. The maximum usable frequency (MUF) at an incident angle ϕ is given by

$$\text{MUF} = f_c \sec \phi$$

where f_c is the critical frequency.

Since absorption begins to increase as frequency is reduced below the MUF, the operating frequency is commonly chosen just below the MUF, leaving a suitable margin for errors due to changing conditions and inaccurate ionospheric forecasting. Tables of data for optimum operating frequency as a function of geographic location, time of day, and date are issued by the National Bureau of Standards of the U.S. Department of Commerce[5] and by other agencies in the United States and elsewhere. More immediate reports of current propagation conditions are issued, for example, every hour at 14 minutes past the hour (Greenwich mean time) by radio station WWV, Fort Collins, Colorado, at 2.5, 5, 10, 15, 20, and 25 MHz. The same information can be obtained from WWV by telephone at (303) 499-7111 (USA).

By these arguments it also follows that communication between earth and space (including satellite earth-to-earth relays) must use frequencies *above* the MUF at the relevant slant angle, since such signals must pass through all the layers of the ionosphere.

A direct consequence of the common practice of choosing an operating frequency between the critical frequency and the MUF is that the transmitter is surrounded by an approximately annular "skip zone" in which reception is difficult or impossible. Moving outward from the transmitter, the zone begins where groundwave field strength has fallen below the reception threshold and ends where the first ionospheric wave has returned to the earth. Figure 49.5 illustrates the notion of "skip distance," the width of the zone measured radially.

A secondary consequence of the effect of radio waves on the ionosphere is that a sufficiently intense radio field can cause measurable perturbation of the ionosphere itself. The most celebrated phenomenon in this regard is the so-called Luxemburg effect, in which the signal from a distant station is observed to have been modulated by the message from a nearby, powerful AM radio station.

Fading is a particularly troublesome effect in many instances of ionospheric propagation. It is produced by several mechanisms. The first and most common problem is the nonstationarity of the ionospheric layers themselves. Physical movement, both horizontally and vertically, and changes in density occur constantly as a result of celestial and meteorologic phenomena. Thus long-term

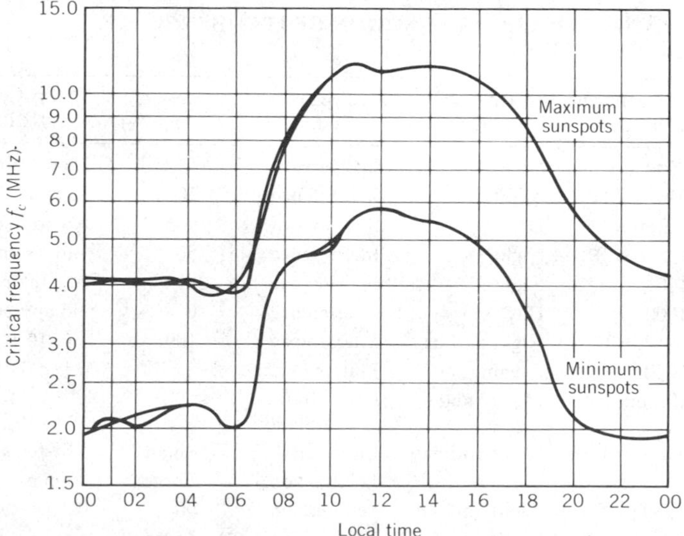

Fig. 49.4 Diurnal variation of critical frequency. Adapted from National Bureau of Standards Circular 462, "Ionospheric Radio Propagation."

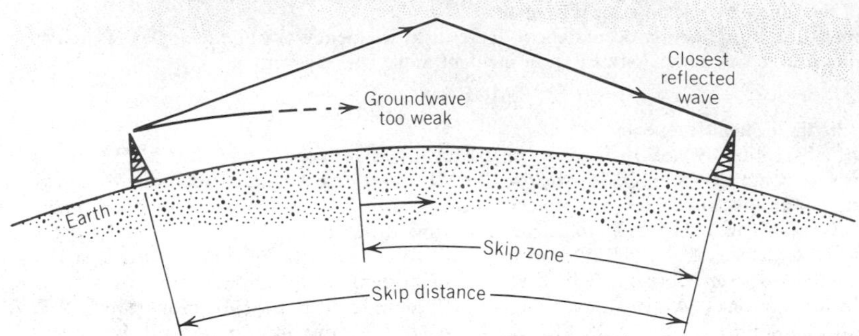

Fig. 49.5 The "skip" phenomenon in ionospheric propagation.

changes in reflection characteristics are the rule rather than the exception. This type of fading is particularly noticeable at dawn and dusk when the ionospheric structure is changing radically and the boundary of a skip zone may oscillate from one side of a receiver to another. Such fading is referred to as MUF fading.

Short-term changes are more noticeable, however, and occur due to multipath reception. The received wave may be the sum of several incident waves, arriving from slightly different angles and via several different path lengths. The result is interference-type fading with Rayleigh or Rician statistics and a field strength that varies over relatively short periods of time. Table 49.1 shows the fading characteristics at several frequencies and wavelengths.

Tropospheric Scatter

The region of the atmosphere in which we live is called the troposphere. It is characterized by a temperature that generally decreases with altitude, and by alternating layers of laminar and turbulent air flow at significantly different temperatures. These layers have sufficiently different indices of refraction that some weak scattering back to the earth is experienced in the very-high-frequency (VHF) and ultrahigh-frequency (UHF) bands, up to the frequencies where atmospheric gas and vapor absorption begins to cause severe attenuation (molecular dipole resonance). This phenomenon

TABLE 49.1. FADING SPEEDS AS FUNCTION OF FREQUENCY (ORDER OF MAGNITUDE)

f	Transmission Distance (km)	Time of Day	Type of Fading	Depth of Fading (%)	Period
20 kHz	7000	Night	Interference	5–25	30–60 min
20 kHz	0	Night	Interference	50	10 min
20 kHz	Variable	Day	Interference	0	No appreciable fading
100 kHz	0	Night	Interference	50	1 to 5 min
1MHz	1000	Night	Interference	85	0.1–1.0 min
1MHz	1000	Day	Interference	0	No appreciable fading
10MHz	0–10,000	Night and day	Interference	100	1 to 10 s
10MHz	3000	Evening	Flutter	?	0.1 to 0.01 s
100MHz	Oblique	Mostly night	Radio star scintillation		10 to 100 s
100MHz	Moon echoes	Night and day	Libration	Small	1 to 100 s
			Polarization	Large	30 min
20 MHz	Satellite	Night and day	Scintillation	90	0.2 s
20 MHz	Satellite	Night and day	Polarization	Large	1 s

Source. K. Davies.[4]

(troposcatter) has a usable range of about 30 MHz to 10 GHz, fortuitously taking over about where ionospheric propagation leaves off.

Compared with ionospheric propagation, troposcatter is far less prone to changes in solar or seasonal conditions, though it varies quite strongly with geographical location, being more useful in the temperate zones than in the tropics. Since the scattering occurs at lower altitudes, the range is more limited (order of several hundred kilometers). Troposcatter also suffers from multipath and fading phenomena. The number of discernible paths for the received wave is usually very large, resulting in the well-known Rayleigh probability density function for the received field strength:

$$p(E) = (E/P)\exp(-E^2/2P)$$

where E is the magnitude of the received electric field and P is the short-term mean received power. In turn, the short-term mean received power varies over longer periods of time and is log normally distributed:

$$p(P_d) = (2\pi S)^{-1/2}\exp\left[-(P_d - Q)^2/2S\right]$$

where P_d = short-term mean, dB
Q = global mean, dB
S = square of standard deviation of distribution, typically about 2 to 5 dB.

The short-term fading due to multipath and turbulence is called fast fading, and the fading of the short-term mean (due to diurnal and meteorologic variations in the layer structure of the troposphere) is called slow fading.

To reduce the probability that the received signal will fall below the reception threshold during a deep fade, a "fading margin" is included in the computation of the required transmitted power. If satisfactory performance *without* slow fading can be achieved with a minimum transmitted power of T dBW, then the probability of unsatisfactory performance *with* slow fading when the transmitted power is $T + mS$ dBW is $\frac{1}{2}\text{erfc}(m/\sqrt{2})$, where $\text{erfc}(\cdot)$ is the complementary error function. For example, for a standard deviation of 3 dB, a probability of satisfactory performance of 99.7% will require $m = 3$ or a fading margin of 9 dB.

Fast fading can also be combated by use of diversity. Space diversity with antenna spacing on the order of 100 wavelengths is usually sufficient. Because of the multipath nature of the received signal, the full gain of a large antenna is unlikely to be realized, since different portions of the received wavefront will be out of phase over a large receiving area (incoherent wavefront). Thus with appropriate electronics, two or more small antennas are likely to do at least as well as, and at lower cost than, a single large antenna in troposcatter receivers.

Local Scattering Propagation

This propagation mechanism does not belong to the group of "classical" channels already described, and it will not be mentioned in most other handbooks of this type. Its characteristics have been fully investigated only recently, and yet it has existed as long as radio communication. It is typified by the urban mobile radio channel,[2,6] but other examples such as the submarine acoustic channel[7] have very similar properties.

When the link path is relatively short, it need not rely on high-altitude atmospheric effects or on groundwave propagation, even when there is no line-of-sight path between transmitter and receiver. When the receiver is surrounded by obstructions, it is the obstructions themselves—buildings, lampposts, trees, hills—that provide a multitude of paths for the radio signal.

The received signal from this channel arrives from many different directions and via several mechanisms. In the case of large obstructions (large compared with a wavelength), waves may be reflected from surfaces and/or diffracted around edges. In the case of small obstructions, scattering will occur, each particle acting as a weak omnidirectional radiator. Fortunately, it is usually unnecessary to consider the specific mechanisms present in a particular channel of this type. The net effect is as if the receiving antenna were surrounded by a large number of randomly disposed scatterers, each acting as a source of the same signal but with random phase, delay, and amplitude. The resulting received field has a Rayleigh-distributed amplitude and a uniformly distributed phase. Its statistics are essentially independent for samples taken several wavelengths apart in space (at or beyond the lateral "coherence distance"), for samples taken at two sufficiently spaced frequencies (separated by at least the channel "coherence bandwidth"), and, whenever movement occurs, either in the transmitter, receiver, or scatterers, for samples taken sufficiently far apart in time (by at least the channel "coherence time").

Occasionally a single one of the paths will produce a field strength that dominates the others in magnitude. This may occur at a relatively "open" location such as near a park or lake, where a LOS path may be accompanied by many scattered paths; or when a large surface causes a particularly strong reflection. Such dominant paths are called specular paths and are similar to the well-known radar phenomenon called glint. In this case the resultant field is the phasor sum of the specular and the scattered field and has a Rician-distributed probability density function

$$p(E) = (E/S)\exp[-(E^2 + P^2)/2S]I_0(EP/S)$$

where $I_0(\cdot)$ = Bessel function of first kind and of order zero
E = magnitude of received electric field
P = short-term mean of specular received amplitude
S = short-term mean of total received power

How short a time the mean is averaged over depends on how long the channel continues to exhibit quasistationary statistics. In the case of a fast-moving car in a mobile radio channel, this may be only a few seconds. Such a channel is said to experience Rician fading, which, as one would expect, is less severe than Rayleigh fading.

Other Propagation Phenomena

Occasionally propagation phenomena other than the five described may be responsible for reception of signals. Some of these are as follows.

Meteor Burst Propagation. This occurs due to the bombardment of the atmosphere by meteors, which causes brief but dense columns of ionized particles in the ionosphere, particularly the E layer. VHF and UHF signals are reflected from such columns and may be received as far away as 1000 km. Such channels, though short lived, may be utilized for practical communication of so-called bursty or low-duty-cycle messages, particularly during periods of meteor storms in the spring and fall.

Ducted Propagation. This occurs when atmospheric layers are so disposed as to produce an effective "waveguide" in the space between them. Ducts are believed to form between layers of the ionosphere, giving enhanced long-distance HF propagation, within the troposphere, and between the earth itself and the bottom of the D layer (in the daytime) or E layer (at night) to provide near-global coverage of VLF and ELF waves.

Propagation by Diffraction. This occurs whenever a wave passes over the edge of an obstruction. The classical theory of diffraction is predicated on the "infinite sharpness" of the edge, and some correction is needed if the edge is not sufficiently sharp. Attempts were made early in the development of radio communication to explain propagation beyond the horizon by means of the diffraction theory, and lengthy discussions of it still appear in many handbooks. The fact is that although diffraction does occur in most instances of propagation, only in exceptional cases does it account predominantly for the field strength present at a receiver. Probably the most common occurrence is in diffraction over the top ridge of a mountain to a "shadow area" behind. Paradoxically, diffraction is important in LOS propagation where the ray path is close to the earth (see the discussion of microwave LOS links).

Orbital Dipole Belts. These are probably more of historical interest than anything else. In 1963 the West Ford Experiment launched a belt of small dipoles some 1.8-cm long into orbit about 3700 km above the earth in an attempt to provide a microwave reflecting medium for long-distance communication. Transmission experiments demonstrated the idea to be marginally feasible, but a furor in scientific circles, particularly (and understandably) among radio astronomers, as well as the success of active communication satellites put the experiment to rest. It is doubtful that we shall see a repetition of any such experiment.

49.1-3 Point-to-Point Radio Links

The following sections describe the specific problems, design considerations, and practices of radio links in which the transmitting and receiving terminals are located at fixed points on the earth. If the sections have anything at all in common, it is that ordinarily no significant Doppler shift occurs in the link; but nevertheless, most such links still have characteristics that are time variant. The discussion is organized by ascending order of frequency. Since the frequency bands at MF and below are used for broadcasting, mobile, navigational, and frequency standard applications, rather than for communications, we begin at the HF band.

HF Radio Links

The HF band (3–30 HMz) is used for medium- and long-range links where satellite or land line coverage is unavailable or uneconomical. Typical applications are telegraphy, telephony, and facsimile and data transmission. The band is also used for broadcasting, mobile communications, and standard frequency transmissions. There are several amateur allocations in the band, and the region 21.85 to 21.87 MHz is reserved by international treaty for radio astronomy.[1]

Propagation in the HF band is primarily by groundwave for short distances beyond the horizon (farther during the day and at lower frequencies) and is ionospheric for long-range links, particularly at night. Because of diurnal and other variations, more than one frequency allocation is often made for a single link so that communication can be maintained as conditions vary. Typically two frequencies are used on a given day, one during the day and one at night, but the frequency pair might also change from season to season and year to year according to the current ionospheric forecasts.

HF links usually suffer from severe noise conditions, primarily from the atmosphere. Electrical storm activity occurs constantly in the earth's atmosphere, and the impulsive noise generated by thunderstorms is propagated to the vicinity of the link by the ionosphere. Atmospheric noise has a power spectral density that is inversely proportional to frequency over almost the entire range of radio frequencies. In the HF band its level is substantially higher than that of galactic and other noise sources, with the possible exception of severe man-made noise in urban areas.[8] It is more prevalent in tropical regions and less so at the poles, and varies seasonally, in both cases following the incidence of electrical storms.[4] Thus calculation of required transmitted power must take these factors into account to provide sufficient signal-to-noise ratio at the receiver. (See Chapter 47.)

The most severe constraint on HF link design, however, is the degree of multipath distortion that frequently occurs. The presence of several ionospheric layers at different altitudes makes possible the reception of several reflected signals at significantly different delay times. In addition, each reflection from a layer may be subject to a smaller range of delays because of inhomogeneities in the layer. The overall delay spread may reach as high as 8 ms and is not infrequently on the order of 1 ms. Unless elaborate adaptive channel-compensation systems are used, this will severely constrain the range of frequencies (bandwidth) that can be transmitted without distortion in an analog modulation scheme, and equivalently the rate at which symbols can be transmitted in a digital scheme.

The coherence bandwidth of a channel is the smallest frequency separation between two signals that are statistically uncorrelated. In the case of a multipath channel, coherence bandwidth is inversely proportional to the time spread of the channel impulse response. For a 1-ms time spread it would be on the order of 1 kHz, the exact value depending on the shape of the channel impulse response. An uncompensated HF link using analog modulation would experience severe distortion if its bandwidth greatly exceeded the channel coherence bandwidth, and so the narrowest band modulation scheme available is usually chosen. This helps to explain, for example, the popularity of single sideband (SSB) AM for HF voice links; intelligible speech can be transmitted with a bandwidth of around 3 kHz.

In a digital transmission, the delay spread itself is a more useful measure of channel characteristics. The spread of signal arrival times determines the extent to which a given transmitted symbol overlaps its neighbors on either side (intersymbol interference). Thus the symbol rate (number of symbols per second) of an uncompensated HF digital link cannot exceed some fraction of the reciprocal of the delay spread, which restricts most HF links to less than 1000 symbols per second. For applications that require data rates higher than 1 kb/s (kilobit per second), therefore, source encoding is necessary so that each transmitted symbol represents more than 1 bit of information (see Sections 47.3 and 48.5). Extension of bit rate by source encoding is a somewhat limited technique, however, since the resultant bandwidth increases exponentially as the number of bits per symbol increases linearly.[9] One of the most popular techniques is QPSK (quadraphase shift keying), since it doubles the bit rate without increasing bandwidth (see Sections 48.1, 48.5, and 48.7).

A more successful approach to high-data-rate HF communication lies in the use of channel compensation techniques in conjunction with channel sounding.[10] If the delay sequence of the channel is known, a filter mechanism can be introduced into the receiver to compensate for the multipath distortion.[11] Since the channel characteristics change over time, a mechanism must also be included to "sound" the channel from time to time and adjust the compensation filter accordingly.

Ionospheric sounding may be performed as an integral part of the signaling technique or as an entirely separate operation. Separate sounding techniques involve transmitting a specially designed signal (pulse, chirp, pseudorandom sequence) into the ionosphere and observing the response. Vertical sounding (ionosonde) is a technique that is performed regularly as a part of the ionospheric forecasting procedure. Sounding at oblique angles is possible if there is a suitably placed receiver and a synchronization network (or if atomic clocks are used at transmitter and receiver); or if there is enough back-scattered energy to be received at the transmitter site.

Integral channel sounding can be performed by the periodic interspersion of a special sounding signal within the message signal. Such signals are often called learning sequences and may also be used for synchronization. Another technique is to multiply each bit of the message by one period of a pseudorandom bit sequence (see Section 48.5), so that the time resolution of the signal is increased to where each path delay can be individually resolved (i.e., the autocorrelation function of the pseudorandom sequence has a central peak that is narrower than a typical delay interval). This technique is referred to as RAKE.[11]

The use of adaptive receivers and channel-sounding techniques results in an increase of signaling rate in the HF band of at least an order of magnitude, with an accompanying improvement in error performance.

Performance improvements are well known to be available also through the use of diversity, but the ionospheric channel does not lend itself well to diversity techniques. Space diversity is difficult to use because the required spacing—several hundred wavelengths—translates to rather more real estate than is available to the average receiver. Frequency diversity *is* feasible from a technical point of view—the required frequency spacing is often only a few kilohertz—but the HF band is very crowded and regulatory agencies are usually loath to allocate multiple channels for simultaneous use in one link. Finally, time diversity is usually not feasible because the coherence time is too long (several seconds even with fast fading).

Troposcatter Links

Troposcatter radio links operate in the VHF to UHF bands (30 MHz–3 GHz) and beyond, with link distances up to 1000 km. Because the scattering region is more compact than the ionosphere, multipath delay spreads are significantly smaller, giving coherence bandwidths on the order of 500 kHz that decrease with increasing distance. Thus wider transmitted bandwidths and faster data rates are possible than in the HF band.

The VHF and UHF bands are used for long-distance telephony and telegraphy, particularly in multiplexed form; for FM and television broadcasting; for radiolocation and instrument landing systems; and for satellite communication systems. In addition there are allocations for radio astronomy, space research, and amateur radio. Atmospheric and man-made radio noise are significantly less than at lower frequencies, declining inversely with frequency and often falling below the galactic background in the region of 1 GHz and above.[12] Attenuation due to precipitation is negligible except in the most extreme storms at the higher frequencies (see Fig. 49.6).

A further benefit of the higher frequencies is the ability to use more directional antennas of reasonable size, resulting in better received signal-to-noise ratios and the possibility of reusing the same frequency for different services in a given geographic area. In addition, there is more spectral space available: 2.07 GHz of bandwidth compared with 2.7 MHz in the HF band. These factors contribute to the growing obsolescence of the HF band as currently utilized.

Whether the use of a troposcatter link is cost effective compared with a LOS link depends on the communication traffic to be carried and the geographic topology of the link path. For a given path length, a LOS link requires at least one repeater for each radio-horizon distance; the troposcatter link would certainly require fewer and probably none. Thus a long-distance LOS link would invariably cost more. But LOS links usually have very wide coherence bandwidths and so can handle much more traffic. If there is sufficient demand, the revenues generated by the LOS link will usually overcome the original capital cost disadvantage. The troposcatter link also requires substantial amounts of transmitted power (order of 1 kW), so that frequency utilization is poorer than with LOS links because of interference constraints.

Troposcatter is attractive in areas where repeater site development would be unreasonably expensive, such as in jungle or mountainous regions. Thus troposcatter links are found commonly in the less developed countries of the world. (Note that links from island to island beyond the horizon *must* use troposcatter in this band, since the use of repeaters is impractical. The only alternatives are undersea cable or satellite link.)

Many of the transmission-rate extension techniques developed for HF links have been adapted for troposcatter links. Thus spread-spectrum transmission and antimultipath techniques are often used together to provide links at megabit-per-second rates. Diversity schemes are also common; the short wavelengths permit space diversity to be used with reasonable antenna spacing (100 wavelengths translates to 30 m at 1 GHz, for example). Space diversity has the advantage that no extra transmitted power or bandwidth is required, and the design of transmitting equipment is unaffected by its use. Frequency diversity is also feasible if the two frequency allocations needed, separated by several times the coherence bandwidth, can be obtained. Time diversity is unlikely to be useful in most applications because the coherence time of the channel, on the order of minutes, is too long.

Microwave LOS Links

At microwave frequencies (>1 GHz) radio waves begin to exhibit somewhat more of the properties of optical radiation than at the lower radio frequencies. The term "line of sight" is thus more often associated with these frequencies than elsewhere. Microwave LOS links require that neither the earth nor any other significant obstruction block the path between transmitter and receiver, but it would be

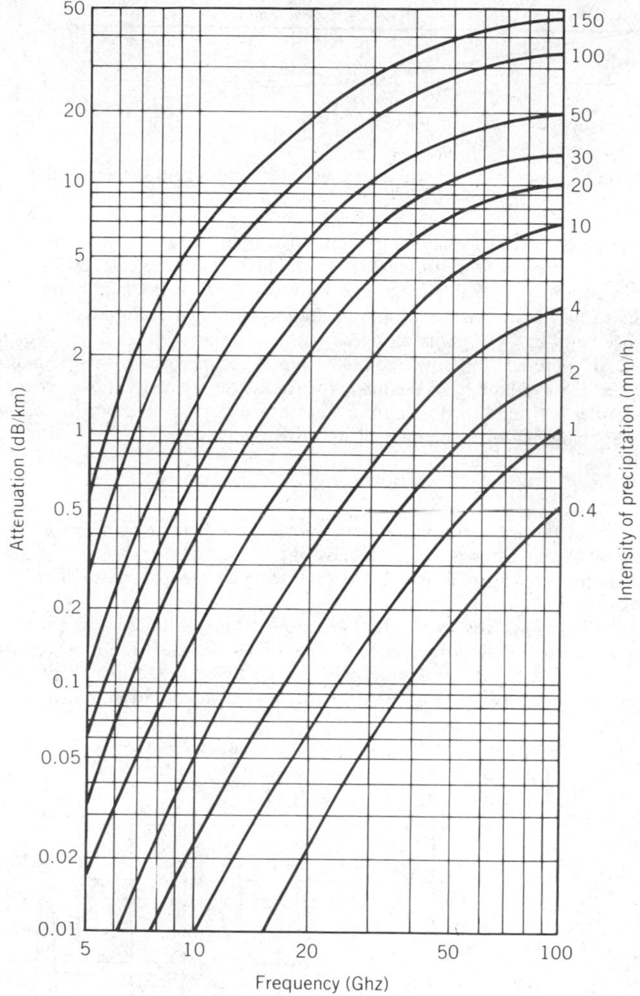

Fig. 49.6 Attenuation of microwave signals due to precipitation. Adapted from CCIR XIIIth Plenary Assembly, Vol. V, Report 233-3, Geneva, 1974.

improper to assume that all that is required is to establish an unobstructed straight-line path between them.

Because the density of the atmosphere tends to vary with altitude, refraction occurs and the "direct" wave is bent. The effect is the same as if no refraction occurs but the radius of the earth is different from its true value. In typical conditions the resultant "effective earth radius" is about four thirds the true mean value of 6378 km. However, in various atmospheric conditions the ratio of effective-to-true radius may vary from 0.6 to 5.0. Thus to allow for the worst condition, the clearance between the earth and the closest straight-line ray path must be increased from that arrived at using a simple geometric model.

Figure 49.7 shows the geometry of a LOS link. To determine if a point P near the ray path has sufficient clearance from the ray, we denote b as the distance from transmitter to P, c as the distance from P to receiver, and d as the distance from transmitter to receiver. Then the radius R of the first Fresnel zone from P, measured in a plane perpendicular to the ray, is given by

$$R^2 = \frac{\lambda bc}{d}$$

where λ is wavelength and all measurements are in the same units.

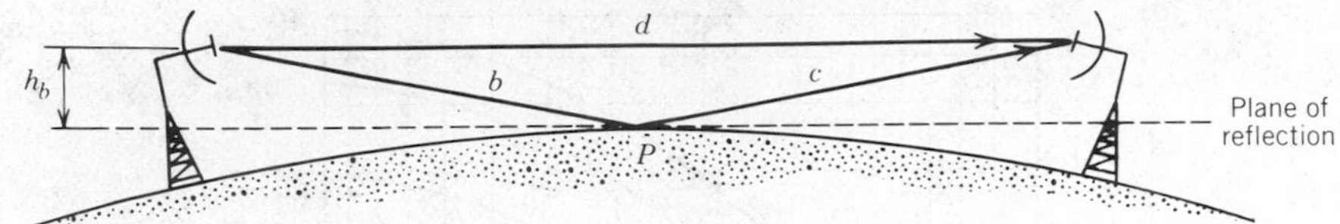

Fig. 49.7 Interference between direct wave and wave reflected from the earth in a microwave link.

An obstruction is considered to have sufficient clearance from the ray if it is outside the first Fresnel zone measured from the ray's nearest pass. The resultant calculations can be used to determine the required antenna heights for a given link distance, or the maximum link distance for a given pair of antenna heights, when the profile of the intervening terrain is known.[1]

Multipath also occurs in LOS paths as described in Section 49.1-2. Both frequency and space diversity are used in practical microwave LOS links,[12] sometimes in combination. Regulatory agencies have limited the availability of channels for frequency diversity, however, so it is likely that space diversity schemes will tend to dominate in the future. Space diversity to combat the problem requires a vertical spacing between antennas of approximately

$$d_s = \frac{\lambda d}{4h_t}$$

where h_t is the height of the transmitting antenna above a plane tangent to the earth at the point of reflection (see Fig. 49.7) and all measurements have the same units.

Time diversity is not used in microwave LOS links, since the coherence time is at least on the order of hours under normal conditions.

Microwave LOS links also suffer from attenuation due to precipitation and from absorption in the oxygen and water molecules of the atmosphere. Water molecule absorption has a peak at just over 2 GHz and oxygen at 60 GHz, so these frequencies are usually avoided in link designs. Precipitation attenuation curves are given as a function of frequency, with precipitation density as a parameter, in Fig. 49.8.

These links have the largest coherence bandwidth of all (order of 10 to 100 MHz) with the exception of optical links, permitting data transmission at rates of up to 300 Mb/s. Direct

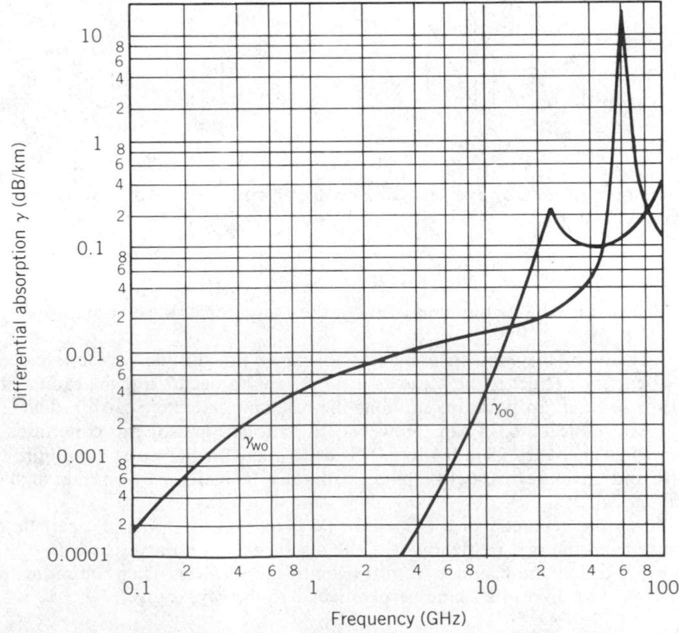

Fig. 49.8 Atmospheric absorption vs. frequency. Adapted from CCIR XIIIth Plenary Assembly, Vol. V, Report 233-3, Geneva, 1974.

transmission of television signals and very large groups of telephone links in multiplexed form is possible without equalization. Thus the revenue-generating capability of the link compares favorably with its large capital cost in most cases, even when several repeater stations are needed to complete a long-haul link.

Multiplexed transmission methods presently in use are mostly analog, with FM-FM and SSB-FM being the most popular frequency division multiplexing methods. (See Section 48.1.) Recently most new systems being constructed use digital methods, with both frequency division and time division multiplex systems being developed. The latter are more popular where saturating repeaters are used, and its seems likely that time division multiple access (TDMA) methods will dominate in the future.

Optical LOS Links

Optical LOS links operate at frequencies in excess of 30 THz (3×10^{13} Hz) and offer the widest coherence bandwidth of any frequency band (order of 1–10 GHz). Thus the potential exists to transmit large multiplexed groups of television signals and huge groups of telephone and data links on a single channel. (See also Section 48.4.) Unfortunately, optical atmospheric links are restricted to line of sight in the most literal sense, are subject to severe attenuation in the presence of precipitation and atmospheric aerosols, suffer severe absorption from atmospheric gas molecules, and are very susceptible to refraction due to thermal and turbulent atmospheric effects. These factors combine with the practical difficulties of mechanical alignment of "antennas" with very narrow beams, and the technical success and low cost of optical fiber links, to render the LOS optical link a most unattractive alternative for any but the shortest links. Thus we may expect to see such links used only for short-haul (a few hundred meters) high-density traffic in urban areas and in areas where fiber-link installation is costly or impossible, such as to nearby islands or to satellites. (Probably the most promising application for LOS optical links is in deep space, where none of these problems pertains.)

Molecular absorption occurs at fixed points in the spectrum owing to the presence of atmospheric gases and vapors. The absorption structure is too detailed to show in a graph that covers the entire infrared-to-ultraviolet spectrum; for example, Table 49.2 shows some important infrared laser sources that are strongly absorbed by atmospheric water.[13] Similar tables can be compiled for many other atmospheric molecules, including some strong absorbers that occur only in trace amounts. Thus the atmosphere is characterized by many "windows" in the spectrum, often very narrow, alternating with almost opaque regions in which propagation is almost impossible. The selection of an operating frequency, and hence a source laser, is of critical importance.

Molecular scattering also occurs, depending on the gas density in the propagation path and hence on altitude. This phenomenon is known as Rayleigh scattering (it is the phenomenon that gives the sky its blue color), and the resulting attenuation is proportional to the fourth power of frequency.

TABLE 49.2. LASER LINES IN INFRARED REGION STRONGLY ABSORBED BY WATER VAPOR MOLECULES

Laser Source	Wavelength (μm)
Atomic krypton	1.7843
Atomic krypton	1.9211
Tm^{+3}-$CaWo_4$	1.911
Tm^{+3}-$CaWo_4$	1.916
U^{+3}-SrF_2	2.472
U^{+3}-CaF_2	2.511
Atomic krypton	2.5234
U^{+3}-BaF_2	2.556
U^{+3}-CaF_2	2.613
CO	5.2–7
Cesium	7.1821
Atomic neon	18.3040
Atomic neon	20.351

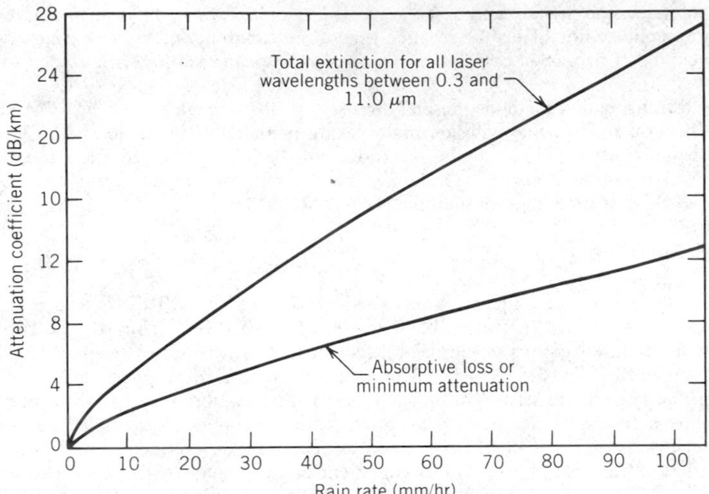

Fig. 49.9 Optical attenuation for precipitation vs. rain rate.

The effect of aerosols is the most variable of the attenuation mechanisms, depending on the type and density of aerosols in the propagation and thus varying with location, altitude, weather patterns, industrial and volcanic activity, and other factors. For example, Fig. 49.9 shows a graph of attenuation rate vs. precipitation intensity.

In addition to absorption effects, turbulence in the atmosphere breaks up the plane wavefront of the optical beam and causes losses due to scattering and lack of coherence. The coherence distance of the received wave is frequently less than the diameter of the receiver aperture, resulting in an "effective aperture" that may be significantly smaller than the physical size.[14] Thus the full theoretical power gain of the receiver aperture cannot be obtained if the aperture is large compared with the channel coherence distance (a factor that also limits the useful size of optical astronomical telescopes for use on the earth). If the receiver aperture contains several zones that have statistically independent signal strengths (optical modes), the resultant powers add, rather than the resultant amplitudes. This is the equivalent of space diversity in radio systems, and helps to eliminate "scintillation" due to turbulence.[15]

For modulation methods, see the section on optical fiber communications systems (48.4).

49.1-4 Mobile Radio Links

There are frequency allocations for mobile radio in almost every band of the regulated electromagnetic spectrum. The band chosen depends on the application; marine and submarine long-distance systems use the HF bands and lower; the maritime services use VHF for short-haul service; aeronautical services use the HF band and higher for communication and the MF and LF bands for navigation; and land-mobile systems use the VHF and UHF bands in most cases. Table 49.3 shows the primary mobile radio allocations in the United States. Mobile communications are probably the most challenging of all applications, mainly because the channel is time varying, owing to vehicle motion, and its nature is uncontrollable, owing to the randomness of the vehicle's location.

In most cases of interest, the mobile radio channel behaves as the local scattering channel, described in the section on propagation. In many senses this channel is a "worst case," so that a system design that performs satisfactorily in such a channel is likely to perform well in almost any other situation.

Most mobile services currently deployed in civilian applications use analog modulation techniques, most often narrowband frequency modulation. AM-double sideband (DSB) is used in the so-called citizen's band in the 27-MHz region, and AM-SSB is being deployed in some congested bands in an effort to improve the availability of the system. Digital techniques are beginning to find their way into use, but regulatory and technologic constraints have often restricted these to baseband tone-keying signals that are simply analog modulated onto a carrier. This severely restricts the

TABLE 49.3. SOME IMPORTANT PUBLIC MOBILE RADIO ALLOCATIONS IN UNITED STATES[a]

35.19–35.69 MHz
43.19–43.69 MHz
152.0–152.255 MHz
152.495–152.855 MHz
157.755–158.115 MHz
158.475–158.715 MHz
454.0–455.0 MHz
459.0–460.0 MHz
470.0–512.0 MHz
806.0–821.0 MHz
825.0–845.0 MHz[b]
851.0–866.0 MHz
870.0–890.0 MHz[b]

[a] For a full list see Ref. 1.
[b] Cellular allocations.

maximum bit rate to a few kilobits per second, but is still useful in such applications as remote keyboard terminals for data base access. Common and widespread use of medium- or high-rate digital mobile radio still appears to be some way off. Since its technical feasibility has been demonstrated and its cost is not high, the principal obstacle appears to be the regulatory process at the present time.

Dedicated Land Mobile Services

Dedicated land mobile radio (LMR) services include all services that use fixed allocations for a single purpose and a single user or group of users. These services include business, transportation, industrial, public safety, and public uses where a given user needs to have the exclusive use of a channel or set of channels in a local sense.

Modulation methods for such applications are almost exclusively FM at the present time, with baseband tone modulation for digital links. In almost all cases the transmitted bandwidth is smaller than the channel coherence bandwidth, so that these transmissions are considered to suffer "flat fading," that is, non-frequency-selective fading, with Rayleigh field strength statistics. The effect of fading is not usually accounted for in receiver design except for the occasional use of space diversity. Fading margins on the order of 10 dB are included in the transmitter power rating in an effort to reduce the probability of outage as the receiver moves about the service area. If an outage does occur there is a high probability that satisfactory performance can be had by moving the receiver antenna just a few wavelengths away from the "dead spot," so that the use of sophisticated techniques is rarely necessary.

Most LMR links of this type are directly between a mobile unit and some centrally located base station or between two mobile units. Since the service area is often very large (such as an entire city), base stations are often located at as high an elevation as possible in a carefully selected central location, and transmitter powers are large enough to provide coverage out to the fringes of the service area. If mobile transmitter power is inadequate for mobile-to-mobile communication directly, the base station may act as a relay either by "patching" the two links or by verbally relaying the message.

Where extended coverage is required, especially in hilly terrain, the technique known as simulcasting is employed. This technique involves transmitting the same carrier and modulation from two or more separately located base stations and results in a received signal that closely resembles a severe multipath signal (see Fig. 49.10). Care must be taken to synchronize carriers and to compensate for propagation delays in the multiple transmission, or the result will be a severely distorted received signal.

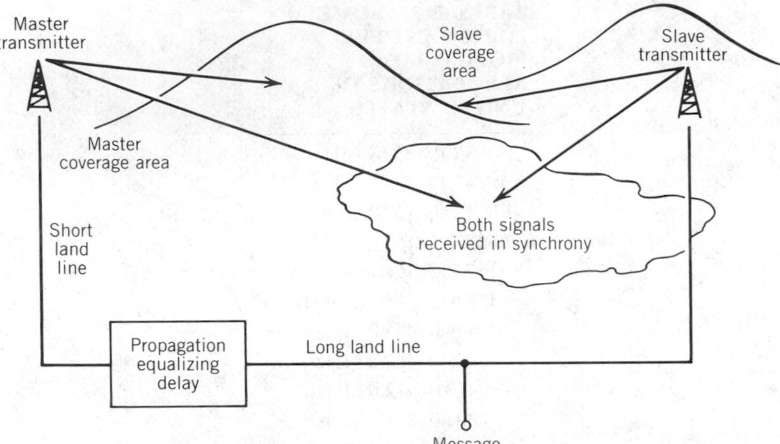

Fig. 49.10 Principle of simulcasting.

Common-Carrier LMR Systems

Mobile public radiotelephone, paging, and some private communication systems are provided by companies that have channel allocations sufficient to provide accessible service to a large population of subscribers, each with a small average call time per day (say 5% or less). These companies are called common carriers and are often identified with local wired-service telephone companies (e.g., the Bell system). Utilization of each channel is determined along similar principles to a line or "trunk" of a long-distance telephone network. If a channel is available, the call may be completed; if none is available, "blocking" occurs and a busy signal is heard.

Most common-carrier services currently offered use a single base station per service area, so that each channel may be used only once at any given time in the service area (and for some distance beyond the service area wherever large transmitted powers are required). Thus the number of simultaneous calls per service area is limited to one half the available number of channels, since two channels are required for one full-duplex telephone conversation. In addition, not all common-carrier channels may be allocated to a single service area, because neighboring cities requiring service may be too close to permit the same channels to be reused. The result has been severe congestion in many metropolitan areas and a large backlog of potential users of the system waiting for an operating license. The use of cellular formats to improve the congestion situation is now being put into use.

Narrowband frequency modulation, with channel spacings in the 15- to 30-kHz region, is the standard for voice transmission on almost all common-carrier LMR systems. Space diversity is occasionally offered as an option on receiving equipment to reduce the effects of fading.[2] Data transmission, which includes the control channels used to establish calls in a radiotelephony system, use baseband tone-keying modulation. Time diversity is commonly used on data channels, since the data rate is not high and the coherence time for mobile radio channels is typically 0.1 s for mobile units at a standstill and significantly less when the unit is moving.

Recently the use of AM-SSB, possibly including analog companding techniques, has been proposed as an interim solution to the congestion problem.[17] It is proposed to use 5-kHz spacing of channels to yield an immediate increase in the number of channels available. But it has yet to be shown if performance in the presence of fading, co-channel, and adjacent-channel interference will be satisfactory; whether the signal distortion caused by amplitude and frequency companding will be acceptable; and whether equipment with the necessary close frequency tolerances can be produced cost effectively.[18] If these problems can be addressed satisfactorily, the SSB idea may have a short-term future. However, the problems of digital transmission on such links, their compatibility with an increasingly digital wired-telephone service, and their suitability for use in cellular schemes are unlikely to be solvable.

Cellular LMR Systems

A cellular LMR system is a scheme in which the service area is divided into smaller "miniservice areas" called cells, each of which is provided with its own base station. In the literature the cells are often represented as squares or rectangles, but in a real service area the boundaries of the cells will be located where signal strength from adjacent base stations is about equal. Communication between a mobile and a base station is set up by means of a short radio link between the mobile and its nearest base station, or at least the base station with the best signal-to-noise ratio measured at the mobile, which will *most probably* be the nearest one. Communication between one mobile and another is set up by means of two such radio links, one in the cell of each mobile, coupled with a land line or other nonradiative link between the two base stations; see Fig. 49.11. Thus the required transmitted power is only that needed to reach across the largest link distance, which does not exceed the "radius" of each cell. When a mobile unit moves from one cell to another, both links of the message are rerouted via the base station of the mobile's new cell. (This implies that the central controller must keep track of the location of all mobile units at all times.[20]) This feature provides a simple means to limit and control the interference between users who occupy the same portion of the spectrum simultaneously, and thus opens the way to more efficient use of the spectrum.

Two distinct means of achieving the efficiency goal in a cellular LMR system have been proposed and are described as follows.

Narrowband Frequency-Reuse Schemes. These schemes use narrowband FM or some other type of modulation in which the transmitted bandwidth does not greatly exceed the message bandwidth. Each channel pair is assigned to more than one cell in the service area, but cells with the same assignments are separated by "rings" of cells that have different assignments; see Fig. 49.12. Mobile units must use the channels assigned to them (via a dedicated control channel) by the central control system. When a mobile moves from one cell to another, the central controller issues a new channel allocation so that the new cell's allocations are not violated. At the same time the signals for the call are rerouted, as described previously. Thus a complete call might use several base stations and channels at different times as the mobile moves around the service area. The smallest group of cells that together uses the entire set of available channels is called a "cluster"; in Fig. 49.12 each cluster contains seven cells.

The number of cells in a cluster is determined by the reuse distance of the system, which is the minimum permissible distance between centers of cells with the same channel assignment. A system composed of clusters of N hexagonal cells has a reuse distance of $R\sqrt{3N}$, where R is the cell "radius," that is, the distance from center to any corner. Thus the reuse distance in Fig. 49.12 is $\sqrt{21} = 4.58$ cell radii.

Field strength in a city tends to fall off more rapidly than the inverse-square law for free-space propagation would predict; the field strength equation is more commonly of the form

$$P_r = KP_t d^{-\alpha}$$

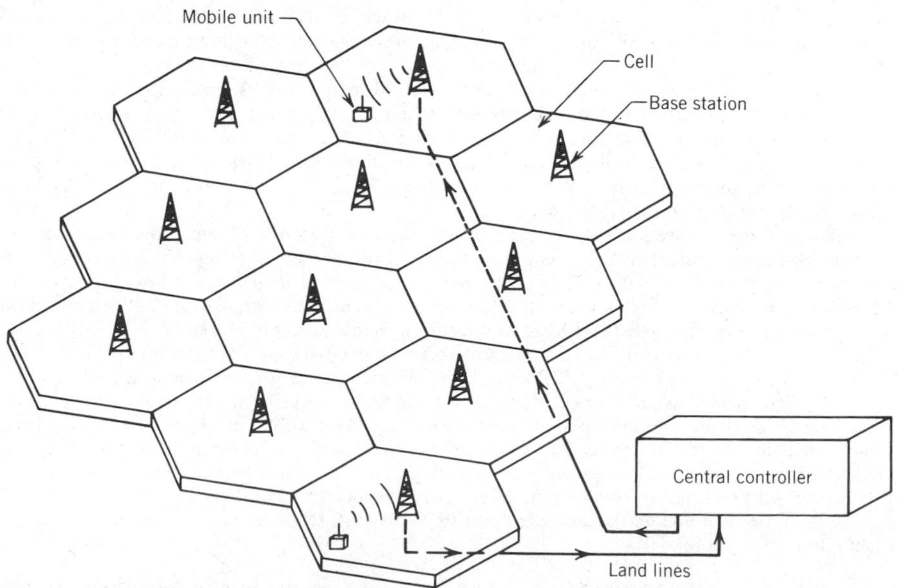

Fig. 49.11 Cellular land mobile radio (LMR) system concept.

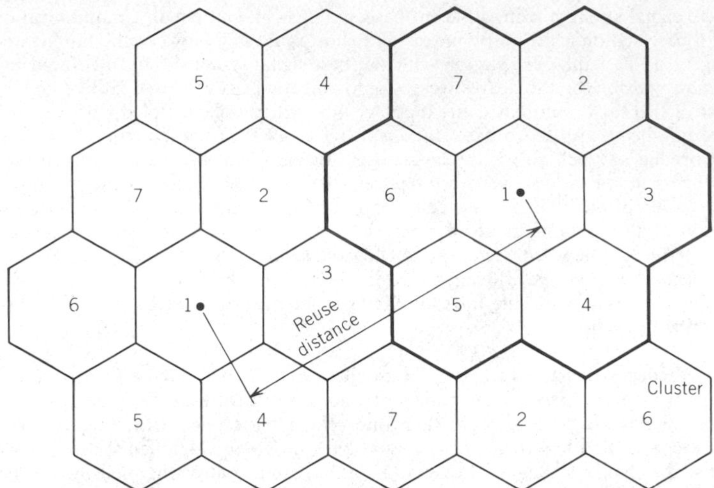

Fig. 49.12 Narrowband frequency-reuse system with seven cells per cluster. Numbers refer to the subset of channels allocated to the cell.

where P_t = transmitted power
P_r = *average* received power (with effects of fading removed)
d = distance from base station
K = a constant that depends on antenna design, operating frequency, and environmental factors

The exponent α is in the region 3.0 to 4.0, with the lower value pertaining to shorter distances (i.e., to smaller cells). A typical value is around 3.5.

Because of this propagation characteristic, the average signal-to-interference ratio for a base station receiving signals from a mobile in its own cell at distance D, and co-channel interference from a mobile in a nearby cell at distance d, is simply given in decibels as $\text{SIR} = 10\alpha \log(d/D)$ dB.

The important point to note about this equation is that because the distance ratio is independent of cell size, only the cell geometry determines the interference characteristics. If α remains fixed, the amount of traffic handled by a given cell is independent of its size. Thus an increase in traffic demand can be met by reducing the size of the cell rather than by allocating more channels to the system.

The essential design tradeoff of such a system is between spectral efficiency and communication reliability. To get an efficient system, the smallest cluster size N is required so that each cell has the maximum possible number of channels. But signal-to-interference ratio is poorer for small N because of the smaller reuse distance. Thus the cluster size is a critical design parameter. So too is the choice of modulation methods; a system with some processing gain (ratio of output to input SNR) will enable smaller reuse distances to be used.

To illustrate the connection between cluster size N, reuse distance T, and propagation factor α, consider the following simplified example. A seven-cell-per-cluster system, with only two clusters, operates in a city with $\alpha = 4.0$. A mobile situated at the edge of its cell, on a line joining its base station with the other base station using the same channels, receives signals from both base stations; see Fig. 49.13. The average ratio of desired to interfering signal power is $10\alpha \log(T-1) = 40 \log(3.58) = 22.2$ dB approximately. If a similar system were to be deployed in a city with $\alpha = 3.0$, the same mobile would experience an average SIR of 16.6 dB. To restore the performance to the same level as that of the first system would require a cluster size of about 14 cells, which in our example is the entire system! Thus there would be no channel reuse, and the traffic capacity of the second system would be half that of the first. This illustrates the sensitivity of the narrowband channel-reuse design to the parameter α, which unfortunately is environmental and not amenable to engineering control.

A demonstration system of the type described has been established by the Bell system in Chicago.[19] The system uses the parameters given in Table 49.4. Other similar systems are planned for Washington, DC, and for Tokyo.

Spread-Spectrum Cellular LMR. This is a scheme in which the available spectrum is divided into only two very wide "channels," one for base-station-to-mobile transmission and the other for

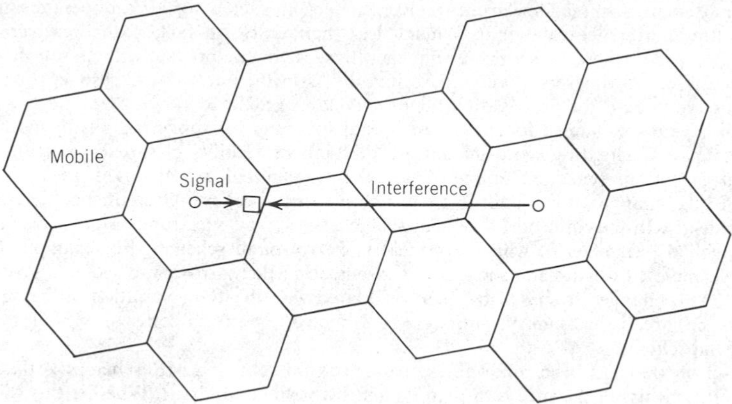

Fig. 49.13 Worst-case interference geometry for a cellular land mobile radio (LMR) system.

mobile-to-base-station transmission. All system users occupy the appropriate part of the spectrum in its entirety for the duration of a call. The different signals are distinguished by assigning a different signature sequence or "code" to each of the users in the system (such systems are referred to as code division multiple access, or CDMA).

The signature sequence is used as the spreading function for the spread-spectrum transmissions; see Section 48.5 on spread spectrum. Any spreading function that is an appropriate choice for use in CDMA applications would be suitable for the scheme, but frequency-hopping functions have been mentioned as suitable candidates in two separate proposed schemes in the literature.[21,22] The high dimensionality of frequency-hopped signals makes possible the assignment of a unique signature sequence to every user on the system, thus permitting private communication and simplifying the addressing procedure. This feature is not an inseparable part of the scheme, however, and a smaller signal set such that there is only one sequence for all *simultaneous* users (a much smaller number than all users) would be satisfactory if each receiver were capable of receiving instructions to use any sequence in the set.

Instead of assigning only part of the available band to each cell, the spread-spectrum system uses all of the band in every cell. The inevitable interference that occurs between users is controlled in two ways. First, the transmitted power from each mobile is controlled dynamically so that the received power at any base station is the same for every mobile unit in that station's cell. Second, the design of the set of signature sequences is such that the cross correlation between every pair of sequences in the

TABLE 49.4. OPERATING PARAMETERS FOR CHICAGO DEMONSTRATION OF CELLULAR MOBILE RADIO SYSTEM

Cluster size, N	12
Operating frequencies	825–890 MHz
Voice modulation method	FM
Channel spacing	30 kHz
Number of channels	800 duplex
Probability of blocking	0.02
Data modulation method	Frequency shift keying with 8-kHz spacing
Data rate	10 kb/s
Data coding	Manchester
Error correction	(40, 28) BCH code
Data time diversity	5 repetitions
Diversity wait time	10 ms
Space diversity	Optional

Source. Special issue of *Bell System Technical Journal*,[19] with permission.

set is uniformly small. A spread-spectrum receiver "tuned" to such a signal can operate satisfactorily with a signal-to-interference ratio that is much less than unity; in fact, if the receiver operating threshold SIR is $1/M$, where M is a positive integer, the system can operate with an interference level that is the equivalent of M users at every base station. As in the narrowband system, this number is independent of cell size if the propagation factor α remains fixed.

The main advantage claimed for the use of spread spectrum in mobile radio is its insensitivity to environmental parameters. In general this includes the inherent ability of spread-spectrum signals to operate well in the presence of fading. Since the transmitted signal spans several coherence bandwidths of the channel, it has built-in frequency diversity because different spectral components fade independently. In the context of a cellular scheme, the spread-spectrum system is also insensitive to the propagation parameter α, which is critical in narrowband schemes. For example, the traffic capacity of a simple 14-cell system, such as was discussed in the narrowband section, would vary by less than 30% as α changed from 4.0 to 3.0, as compared with the 100% variation of the narrowband system. Thus the spread-spectrum system might be expected to perform better in difficult terrain than the narrowband scheme.

Another claim that has been made for spread-spectrum cellular LMR schemes is that they will make more efficient use of the spectrum than the narrowband counterpart. Whether this will prove to be the case is difficult to predict; the advantage may be only small with large-cell schemes in which α is fairly controllable and quite large. It appears, however, that when small-cell (high-density) schemes become necessary, spread spectrum will prove to be superior in this respect.

References

1 International Telephone and Telegraph Corporation, *Reference Data for Radio Engineers*, 6th ed., Howard W. Sams, Indianapolis, 1975.

2 W. C. Jakes, Jr., ed., *Microwave Mobile Communications*, Wiley, New York, 1974.

3 A. D. Watt, *VLF Radio Engineering*, Pergamon, Elmsford, NY, 1967.

4 K. Davies, *Ionospheric Radio Waves*, U.S. Department of Commerce, National Bureau of Standards, Boulder, CO, 1965.

5 M. Leftin, ed., *Ionospheric Predictions*, U.S. Department of Commerce, Institute for Telecommunications Sciences, Boulder, CO, 1980.

6 William C. Y. Lee, *Mobile Communications Engineering*, McGraw-Hill, New York, 1982.

7 A. V. Venetsanopoulos, Modelling of the Sea Surface Scattering Channel and Undersea Communications, in J. Skwirtzynski, ed., *Communication Systems and Random Process Theory*, Sijthoff & Noordhoff, Amsterdam, 1978.

8 E. N. Skomal, *Man-Made Radio Noise*, Van Nostrand Reinhold, New York, 1978.

9 J. M. Wozencraft and I. M. Jacobs, *Principles of Communication Engineering*, Wiley, New York, 1965.

10 M. Darnell, Channel Evaluation Techniques for Dispersive Communications Paths, in J. Skwirtzynski, ed., *Communication Systems and Random Process Theory*, Sijthoff & Noordhoff, Amsterdam, 1978.

11 R. Price and P. E. Green, Jr., "A Communication Technique for Multipath Channels," *Proc. IRE* **46**:555–570 (Mar. 1958). Reprinted in B. Goldberg, ed., *Communication Channels: Characterization and Behavior*, IEEE Press, New York, 1976.

12 K. Feher, *Digital Communications: Microwave Applications*, Prentice-Hall, Englewood Cliffs, NJ, 1981.

13 CRC Handbook of Lasers with Selected Data on Optical Technology, Chemical Rubber Company, 1971.

14 Advisory Group for Aerospace Research and Development, *Optical Propagation in the Atmosphere*, AGARD Conference Proceedings No. 183, 1976. Distributed in United States by National Technical Information Service, Springfield, VA.

15 R. M. Gagliardi and S. Karp, *Optical Communications*, Wiley, New York, 1976.

16 V. E. Benes, *Mathematical Theory of Connecting Networks and Telephone Traffic*, Academic, New York, 1965.

17 B. Lusignan, "Single-Sideband Transmission for Land Mobile Radio," *IEEE Spectr.*, pp. 33–37 (Jul. 1978). Condensed from report, "Spectrum Efficient Technology for Voice Communications," submitted to UHF Task Force of Office of Plans and Policy, Federal Communications Commission, Feb. 1978. Includes demonstration tape. Available from National Technical Information Service, Springfield, VA.

18 Report of Electronics Industries Association TR-8 Ad Hoc Committee, Communications Division, for Spectrum-Efficient Technology. On file at Land Mobile Communications Council, Washington, DC. Summary can be found in *IEEE Commun. Mag.* **17**(2):25–28 (Mar. 1979). Report concluded that recommendations in Ref. 17 were "unsound."

19 "Advanced Mobile Phone Service," Special Issue of *Bell Sys. Tech. J.* **58**(1) (Jan. 1979).

20 E. N. Skomal, *Automatic Vehicle Location Systems*, Van Nostrand Reinhold, New York, 1981.

21 G. R. Cooper and R. W. Nettleton, "A Spread-Spectrum Technique for High-Capacity Mobile Communications," *IEEE Transactions on Vehicular Technology*, VT-27(4), November 1978.

22 D. J. Goodman, P. S. Henry, V. K. Prabhu, "Frequency—Hopped Multilevel FSK for Mobile Radio," *Bell Syst. Tech. J.* **59**(9) (Sept. 1980).

49.2 BROADCAST SYSTEMS

Jefferson F. Lindsey III

Standard broadcasting refers to the transmission of voice and music received by the general public in the 535- to 1605-kHz frequency band. Amplitude modulation is used to provide service ranging from that needed for small communities to higher power broadcast stations needed for larger regional areas. The primary service area is defined as the area in which the groundwave is not subject to objectionable interference or objectional fading. The secondary service area refers to an area served by skywaves and not subject to objectional interference. Intermittent service refers to an area receiving service from a groundwave but beyond the primary service area and subject to some interference and fading.

49.2-1 Standard Broadcasting (Amplitude Modulation)

Frequency Allocations

The carrier frequencies for standard broadcasting in the United States are designated in the Federal Communications Commission (FCC) *Rules and Regulations*, Volume 3, Part 73. A total of 107 carrier frequencies are allocated from 540 to 1600 kHz in 10-kHz intervals. Each carrier frequency is required by FCC rules to deviate no more than ±20 Hz from the allocated frequency, to minimize heterodyning from two or more interfering stations. Double sideband full carrier modulation, commonly called amplitude modulation (AM), is used in standard broadcasting for sound transmissions. Typical modulation frequencies for voice and music range from 50 Hz to 16 kHz. Each channel is generally thought of as 10 kHz in width, and thus the frequency band is designated from 535 to 1605 kHz; however, when the modulation frequency exceeds 5 kHz, the radio frequency bandwidth of the channel exceeds 10 kHz and adjacent channel interference may occur.

Channel and Station Classifications

In standard broadcast (AM), channels are classified according to the range of coverage. Clear channel stations are used to render service over a wide area and are cleared of objectionable interference by license controls and international agreements. The maximum power output of a clear channel station within the United States is limited to 50 kW. Regional channels use a maximum power output of 5 kW and are limited by the field intensity that is produced from other regional channel stations. The third type of channel is the local channel, which uses a maximum power output of 1 kW in the daytime and 250 W at night, except for stations in the southern part of Florida where the power is limited to 250 W both day and night. Maximum output powers outside the United States range as high as 2 MW.

Of the 107 standard broadcast channels, 62 are designated clear channels, and stations on these frequencies may be classified as I-A or I-B. For class I-A stations, the output power is 50 kW and the primary service area is free from interference on the same and adjacent channels. Also the secondary service area is free from interference on the same channel but not from that on adjacent channels. The primary service area is defined by the level of the electric field intensity of the groundwave. This level varies from 0.1 mV/m in rural areas to 50 mV/m in a city. Secondary service is considered to be delivered in areas where the skywave for 50% or more of the time exceeds 0.5 mV/m. A clear channel station classified as I-B has a transmitter output power ranging from 10 to 50 kW, and more than one station within the United States may operate on the same frequency subject to co-channel and adjacent channel limitations. For the class I-A stations, only one station within a 750-mile radius is permitted to operate at night; thus co-channel interference is avoided.

Clear channel stations may also be classified as class II, in which case the output power may be as low as 250 W depending on the particular subclassification (i.e., class II-A, II-B, or II-C). The class II stations have a maximum output power of 50 kW. Class II stations are considered secondary to class I stations and are designed to render service to a primary service area subject to and limited by interference from class I stations. Whenever necessary, a directional antenna is used to reduce interference. Also the nighttime power may be reduced to avoid interference.

Of the 107 standard broadcast channels, 41 are designated for use as regional channels. Several stations may operate with transmitter output power of 5 kW or less on the same frequency subject to the limitation imposed by the field intensity contour and the consequence of interference. A class III station operates to render service to a population center and the rural area contiguous thereto. With class III there are two subclassifications—class III-A, in which output power ranges from 1 to 5 kW inclusive, and class III-B, in which nighttime power ranges from 0.5 to 1 kW inclusive and maximum daytime power is 5 kW.

The third channel classification, local, is for class IV stations, which are designed to render service to a city or town and the suburban and rural areas contiguous thereto. Such stations operate with a maximum power of 1 kW in the daytime and 250 W at night except for stations in Florida south of the 28° north latitude and between meridians 80° and 82° west longitude, in which case the power is limited to 250 W both day and night. Some stations in this class have been previously licensed to operate with an output power of 100 W.

Field Strength

The field strength produced by a standard broadcast station is a key factor in determining primary and secondary service areas and interference limitations of possible future radio stations. The field strength limitations are specified as field intensities by the FCC, with the units volts per meter; however, measuring devices may read volts or decibels referenced to 1 mw (dBm), and a conversion may be needed to obtain the field intensity. The power received may be measured in dBm and converted to watts. Voltage readings may be converted to watts by squaring the root mean square (rms) voltage and dividing by the field strength meter input resistance, which is typically on the order of 50 or 75 Ω. Additional factors needed to determine electric field intensity are the power gain and losses of the field strength receiving antenna system. Once the power gain and losses are known, the effective area with loss compensation of the field strength receiver antenna may be obtained as

$$A_{\text{eff}} = G\frac{\lambda^2}{4\pi}L$$

where A_{eff} = effective area including loss compensation, m^2
G = power gain of field strength antenna, W/W
λ = wavelength, m
L = mismatch loss and cable loss factor, W/W

From this calculation, the power density in watts per square meter may be obtained by dividing the received power by the effective area, and the electric field intensity may be calculated as

$$E = \sqrt{\mathcal{P} \times Z_{\text{fs}}}$$

where E = electric field intensity, V/m
$\mathcal{P}$ = power density, W/m^2
$Z_{\text{fs}} = 120\pi$ Ω, impedance of free space

The field intensities necessary to render primary service for business and factory areas in a city range from 10 to 50 mV/m; for residential areas in a city intensities range from 0.1 to 0.5 mV/m except in southern areas in the summer, when the range is from 0.25 to 1 mV/m.

The protected service contours and permissible interference contours for standard broadcast stations are given in Table 49.5. The field intensities given in this table, along with a knowledge of existing broadcast stations, may be used in determining the potential for establishing new standard broadcast stations.

Propagation Characteristics

One of the major factors in the determination of field strength is the propagation characteristic, which is described by the change in electric field intensity with an increase in distance from the broadcast station antenna. This variation depends on a number of factors including frequency, distance, surface dielectric constant, surface loss tangent, polarization, local topography, and time of day. Generally speaking, groundwave propagation occurs at shorter ranges both during day and night periods. Skywave propagation permits longer ranges and occurs during night periods, and thus some

TABLE 49.5. PROTECTED SERVICE SIGNAL INTENSITIES AND PERMISSIBLE INTERFERING INTENSITIES FOR STANDARD BROADCASTING (AM)

Class of Station	Class of Channel Used	Permissible Power	Signal Intensity Contour of Area Protected From Objectionable Interference[a]		Permissible Interfering Signal on Same Channel	
			Day[b]	Night	Day[b]	Night[c]
I-A	Clear	50 kW	SC 100 uV/m AC 500 uV/m	SC 500 uV/m (50% skywave)[d] AC 500 uV/m[b]	5 uV/m	25 uV/m[d]
I-B	Clear	10 to 50 kW	SC 100 uV/m AC 500 uV/m	SC 500 uV/m 50% skywave AC 500 uV/m[b]	5 uV/m	25 uV/m
II-A	Clear	0.25 to 50 kW (daytime) 10 to 50 kW (nighttime)	500 uV/m	500 uV/m[b]	25 uV/m	25 uV/m
II-B and II-D	Clear	0.25 to 50 kW	500 uV/m	2500 uV/m[b,e]	25 uV/m	125 uV/m
II-B and II-D[f]	Clear	0.25 to 1 kW[f]	500 uV/m	10,000 uV/m[f]	25 uV/m	500 uV/m[f]
III-A	Regional	1 to 5 kW	500 uV/m	2500 uV/m[b]	25 uV/m	25 uV/m
III-B	Regional	0.5 to 1 kW (night), and 5 kW (day)	500 uV/m	4000 uV/m[b]	25 uV/m	200 uV/m
IV	Local	0.25 kW (night), and 0.25 to 1 kW (day)	500 uV/m	Not prescribed	25 uV/m	Not prescribed

Source. From Federal Communications Commission, *Rules and Regulations*, Part 73.183.
Note. SC, same channel; AC, adjacent channel.
[a]When a station is already limited by interference from other stations to a contour of higher value than that normally protected for its class, this contour shall be the established standard for such station with respect to interference from all other stations.
[b]Groundwave.
[c]Skywave field strength for 10% or more of the time.
[d]Class I-A stations on channels reserved for the exclusive use of one station during nighttime hours are protected from co-channel interference on that basis.
[e]Values are with respect to interference from all stations except class I-B, which stations may cause interference to a field strength contour of higher value. However, it is recommended that class II stations be so located that the interference received from class I-B stations will not exceed these values. If the class II stations are limited by class I-B stations to higher values, then such values shall be the established standard with respect to protection from all other stations.
[f]Applies only to nighttime operations of class II-B stations coming within §73.21(a)(2)(ii)(C), and to the operation of limited-time class II-D stations during nighttime hours other than those during which they were authorized to operate as of June 1, 1980.

stations must reduce power or not operate at night to avoid interference. Propagation curves in the broadcast field are frequently referred to a reference level of 100 mV/m at 1 mile; however, a more general expression of groundwave propagation may be obtained by using the Bremmer series (Bremmer, 1949). Typical groundwave propagation curves for the standard broadcast band are shown in Fig. 49.14.

The effective radiated power (ERP) refers to the effective power output from the antenna in a

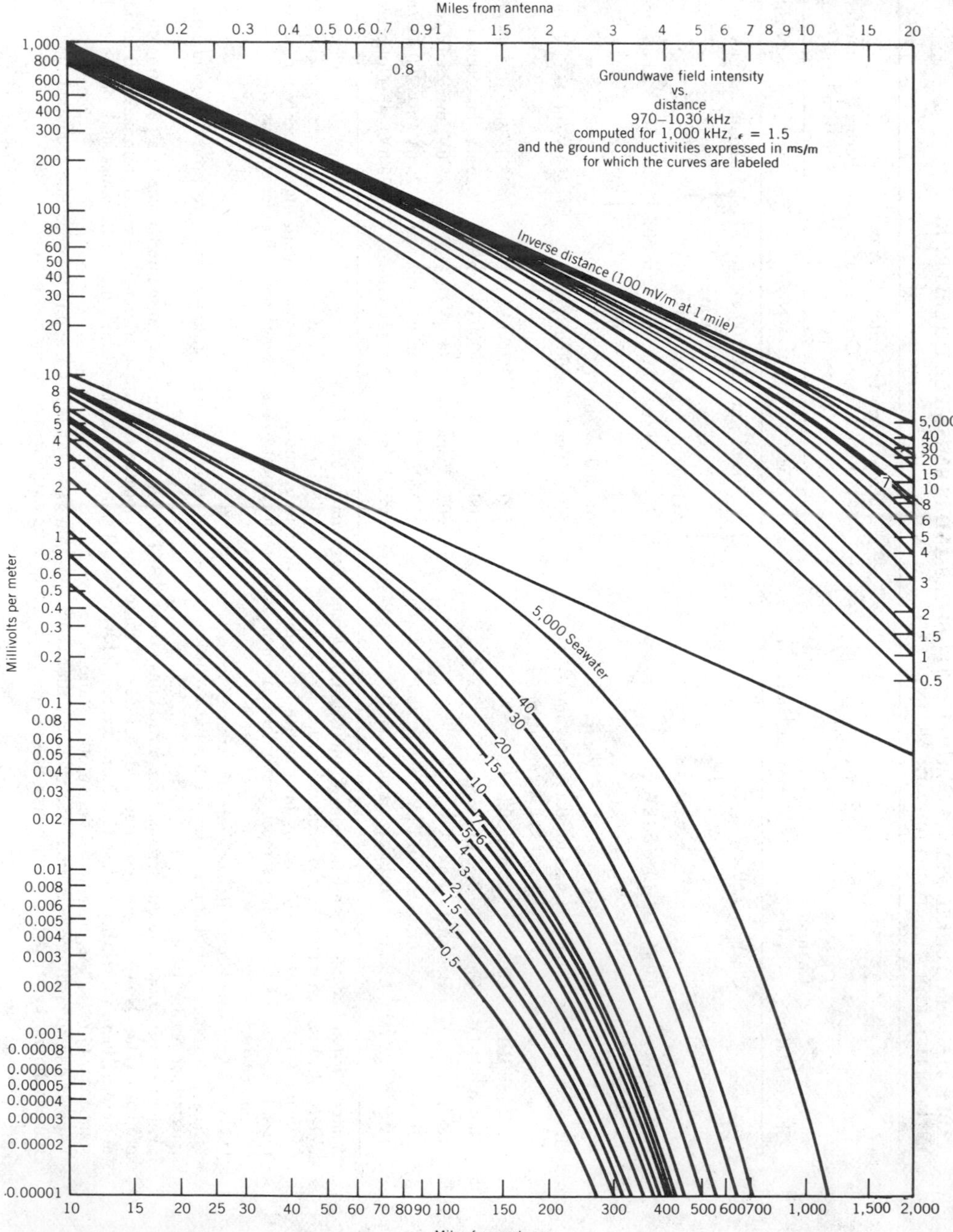

Fig. 49.14 Groundwave propagation curves. From FCC, *Rules and Regulations*, Vol. III, part 73, Oct. 1982, p. 74.

specified direction and includes the transmitter power output, transmission line losses, and antenna power gain. The ERP in most cases exceeds the transmitter output power. For a hypothetical perfect isotropic radiator, the ERP is found to be

$$\mathrm{ERP} = \frac{E^2 r^2}{30}$$

where E is the electric field intensity in volts per meter and r is the distance in meters.

For a distance of 1 mile (1609.34 m), the ERP required to produce a field intensity of 100 mV/m is found to be 863.3 W. Since the field intensity is proportional to the square root of the power, field intensities may be determined at other powers.

Skywave propagation necessarily involves some fading and less predictable field intensities and is most appropriately described in terms of statistics or the percentage of time a particular field strength level is found. Figure 49.15 shows skywave propagation for 1 MHz for latitudes from 36° to 50°.

Transmitters

Standards that cover AM broadcast transmitters are given in the Electronic Industry Association (EIA) Standard TR-101A, *Electrical Performance Standard for Standard Broadcast Transmitters.* Parameters and methods for measurement include the following: carrier output rating, carrier power output capability, carrier frequency range, carrier frequency stability, carrier shift, carrier noise level, magnitude of radio-frequency harmonics, normal load, transmitter-output-circuit adjustment facilities, RF and audio interface definitions, modulation capability, audio input level for 100% modulation, audio frequency response, audio frequency harmonic distortion, rated power supply, power supply variation, and power input.

Standard AM broadcast transmitters range in power output from solid-state 250-W type to vacuum-tube 50-kW units. A block diagram of a typical 50-kW broadcast transmitter is shown in Fig. 49.16.

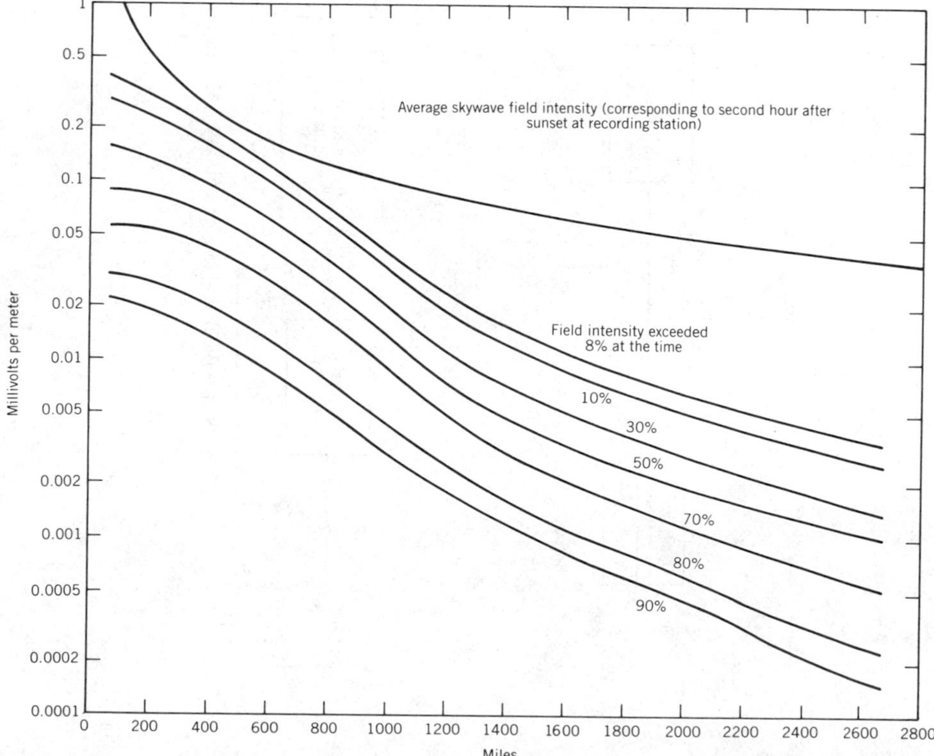

Fig. 49.15 Skywave propagation curves. The curves are not considered sufficiently accurate for practical use for distances less than about 250 miles. From FCC, *Rules and Regulations*, Vol. III, part 73, Oct. 1982, p. 90.

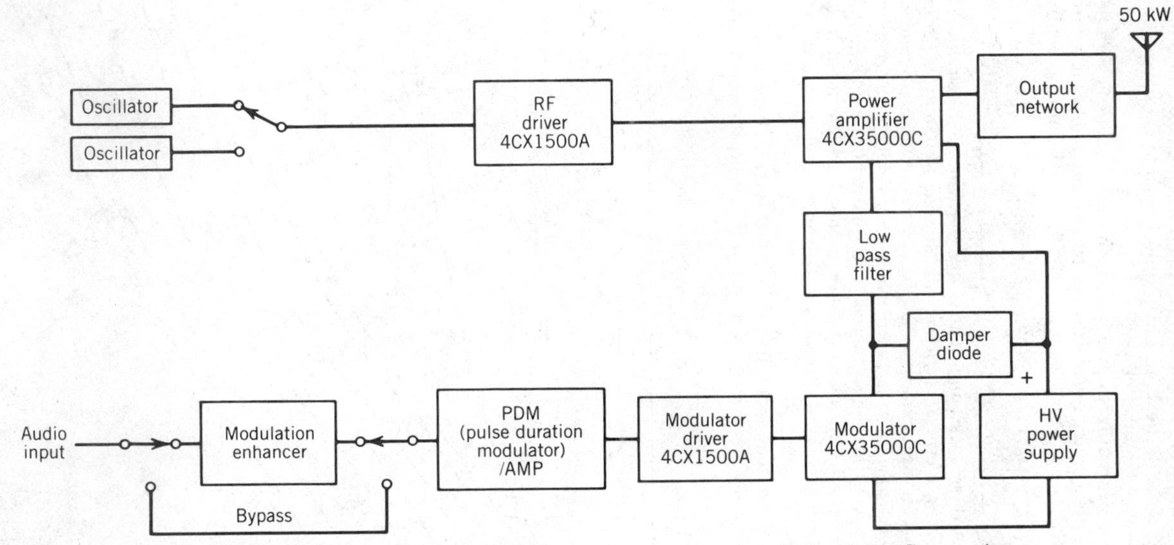

Fig. 49.16 Block diagram of a 50-kW AM transmitter. Courtesy of the Harris Corporation.

Antenna Systems

The antenna systems for standard AM broadcast typically consist of a group of quarter-wave vertical elements that are combined in a phased array to produce desired electric field intensity contours. For example, if a transmitter and an antenna system are located north of the population center they serve, the radiated energy may be redistributed to provide better coverage to the south of the transmitter. The antenna system must be well matched to the transmitter with a voltage standing wave of less than 1.2 : 1, to avoid excessive power being reflected back to the transmitter. Typical gains for broadcast antenna systems range from 3 to 8 dB over a perfect linear isotropic. Typical heights of each vertical range from 200 to 500 ft.

49.2-2 Frequency Modulation

Frequency modulation (FM) broadcasting refers to the transmission of voice and music received by the general public in the 88- to 108-MHz frequency band. Frequency modulation is used to provide higher fidelity reception than is available with standard broadcast AM. In 1961 stereophonic broadcast was introduced with the addition of a double sideband suppressed carrier for transmission of a left-minus-right difference signal. The left-plus-right sum channel is sent with use of normal frequency modulation. Some FM broadcast systems also include a subsidiary communication authorization (SCA) subcarrier for private commercial uses. Frequency modulation broadcast is typically limited to line-of-sight ranges. As a result, FM coverage is localized to a range of approximately 75 miles depending on the antenna height and effective radiated power.

Frequency Allocations

The 100 carrier frequencies for FM broadcast range from 88.1 to 107.9 MHz and are equally spaced every 200 kHz. The channels from 88.1 to 91.9 MHz are used for noncommercial broadcasting and those from 92.1 to 107.9 MHz for commercial broadcasting. Each channel has a 200-kHz bandwidth, which includes a 25-kHz guardband that allows a maximum frequency swing of ± 75 kHz. The carrier frequency is required to be maintained with ± 2000 Hz.

Station Classifications

In FM broadcast, stations are classified as class A, class B, or class C. Class A stations serve a relatively small city or town and the surrounding area and are limited to a maximum effective radiated power of 3 kW. Also the class A stations are limited to 20 frequencies between 92.1 and 107.1 MHz inclusive. Classes B and C stations render service to sizeable cities or towns and the surrounding area. Class B stations are permitted to operate to a maximum effective radiated power of 50 kW with an effective antenna height of 500 ft. For heights over 500 ft above existing terrain, the power must be reduced. Class C stations operate to 100 kW effective radiated power and heights to 2000 ft without power reductions.

Field Strength and Propagation

The field strength produced by an FM broadcast station depends on the effective radiated power, antenna heights, local terrain, tropospheric scattering conditions, and other factors. From a statistical point of view, however, an estimate of the field intensity may be obtained from Fig. 49.17. A factor in the determination of new licenses for FM broadcast is the separation in miles from existing co-channel and adjacent-channel stations, the class of the station, and the antenna heights. Typical guidelines are given in Table 49.6. In addition, field strength contours are estimated for 3.16-mV/m (70-dBμ) and for 1-mV/m (60-dBμ) levels. These may be determined from graphs similar to that of Fig. 49.17. Propagation at FM frequencies (88–108 MHz) is generally thought of as line of sight; however, larger effective radiated powers along with the effects of diffraction, refraction, and tropospheric scatter allow coverage slightly beyond the line of sight.

Transmitters

FM broadcast transmitters typically range in power output from solid-state 300 W to tube-type 100 kW. A block diagram of a dual FM transmitter is shown in Fig. 49.18. This system consists of two 25-kW transmitters that are operated in parallel and that provide increased reliability in the event of a failure in either the exciter or transmitter amplifier.

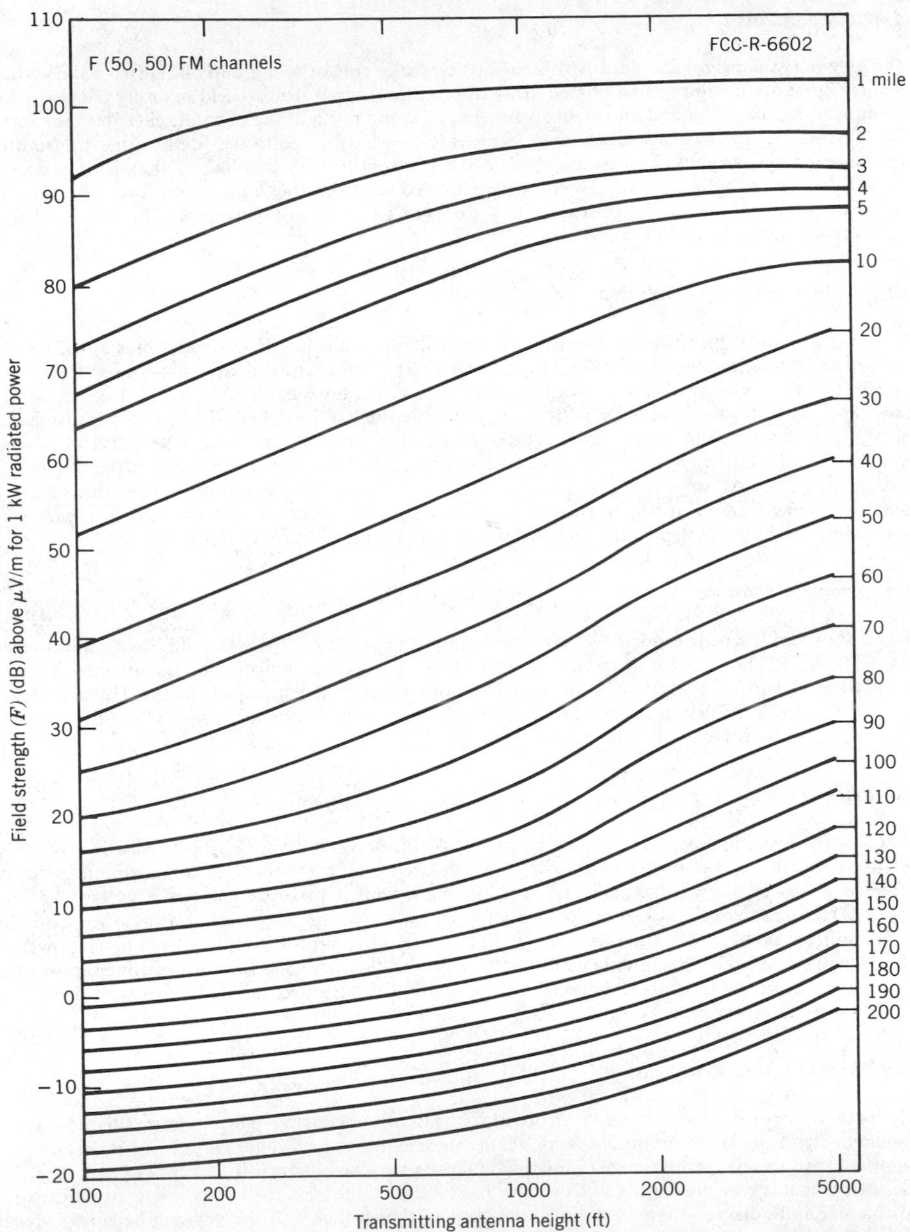

Fig. 49.17 Estimated FM field strength for a receiving antenna height of 30 feet. $F(50, 50)$, field strength for 50% of time at 50% of locations. From FCC, *Rules and Regulations*, Vol. III, part 73, Oct. 1982, p. 150.

Antenna Systems

FM broadcast antenna systems are required to have a horizontal polarization component. Most antenna systems, however, are circularly polarized, having both horizontal and vertical components. Typically the antenna system concentrates the energy in the horizontal plane, which provides an increased effective radiated power resulting from power gains up to approximately 10 dB. This means

TABLE 49.6. GUIDELINES FOR FM STATION MINIMUM SPACINGS

	Separation in Miles		Facilities Authorized	
Class of Station	Co-channel	First Adjacent	Power (kW)	Antenna Height (ft)
A to A	45 to 65		3	300
A to A	40 to 44		2	300
A to A	Less than 40		1	300
A to B		50 to 65	3	300 class A
			50	500 class B
A to B		40 to 49	3	300 class A
			20	500 class B
A to B		Less than 40	3	300 class A
			10	500 class B
A to C		80 to 105	3	300 class A
			100	2000 class C
A to C		60 to 79	3	300 class A
			50	2000 class C
A to C		Less than 60	3	300 class A
			20	2000 class C
B to B	125 to 150	80 to 105	50	500
B to B	100 to 124	65 to 79	20	500
B to B	75 to 99	50 to 64	10	500
B to B	Less than 75	Less than 50	5	500
B to C	140 to 170	110 to 135	50	500 class B
			100	2000 class C
B to C	110 to 139	85 to 109	20	500 class B
			50	2000 class C
B to C	90 to 100	60 to 84	10	500 class B
			20	2000 class C
B to C	Less than 90	Less than 60	5	500 class B
			10	2000 class C
C to C	150 to 180	125 to 150	100	2000
C to C	120 to 149	95 to 124	50	2000
C to C	100 to 119	75 to 94	20	2000
C to C	Less than 100	Less than 75	10	2000

Source. Federal Communications Commission, *Rules and Regulations*, Part 73, pp. 128–129 (Oct. 1982).

that a 5-kW transmitter could have an effective radiated power of 50 kW. In addition, a directional pattern in the horizontal plane may be permitted with up to 15 dB variation to emphasize coverage in predetermined areas.

Preemphasis

Preemphasis is employed in an FM broadcast transmitter to improve the received signal-to-noise ratio. The preemphasis upper limit shown in Fig. 49.19 is based on a time constant of 75 μs as required by the FCC for FM broadcast transmitters. Audio frequencies from 50 to 2120 Hz are essentially transmitted with normal FM, whereas audio frequencies from 2120 Hz to 15 kHz are emphasized with a larger modulation index. There is significant signal-to-noise improvement when the receiver is equipped with a matching deemphasis circuit. The lower limit on preemphasis is also given in Fig. 49.19.

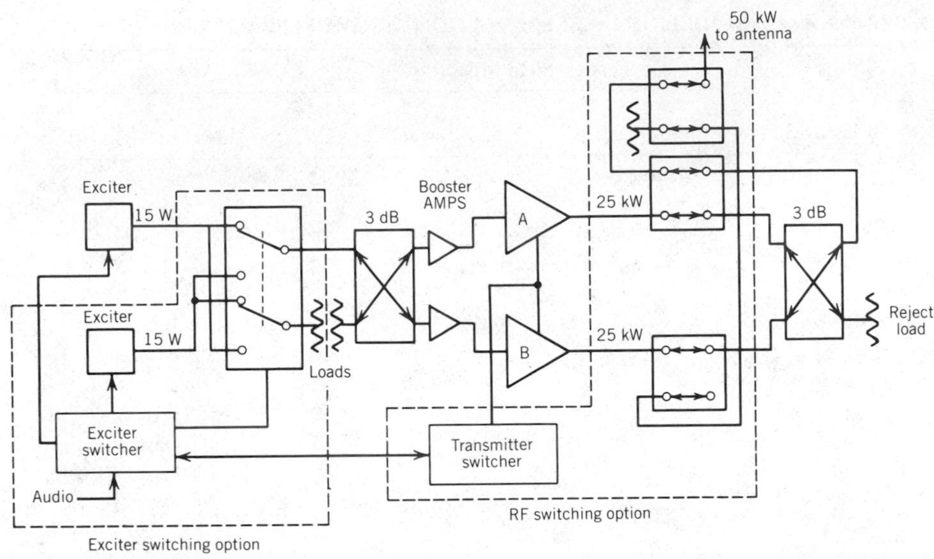

Fig. 49.18 Block diagram of dual FM transmitter. Courtesy of the Harris Corporation.

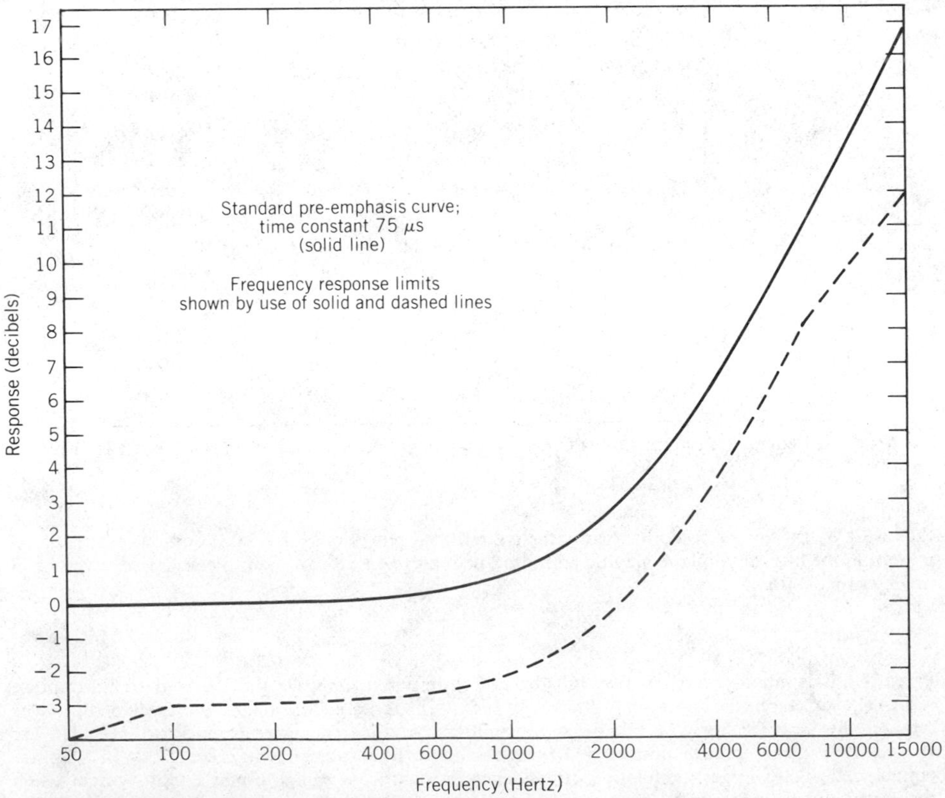

Fig. 49.19 Preemphasis limits based on a 75-μs time constant. From FCC, *Rules and Regulations*, Vol. III, part 73, Oct. 1982, p. 153.

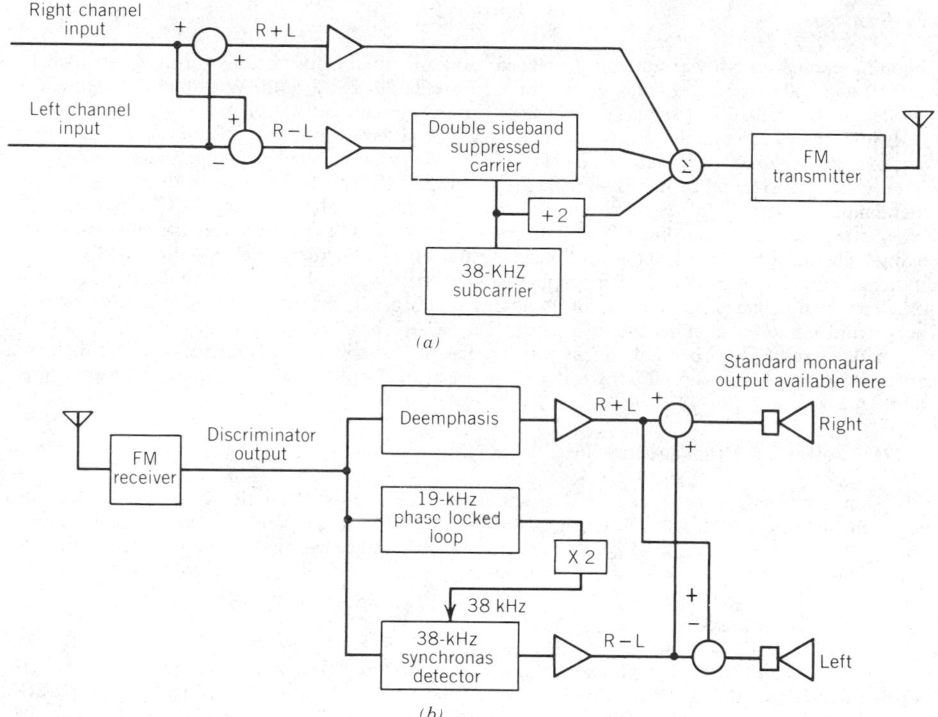

Fig. 49.20 Block diagram of an FM stereophonic system: (*a*) Transmitter, (*b*) receiver. R, right; L, left.

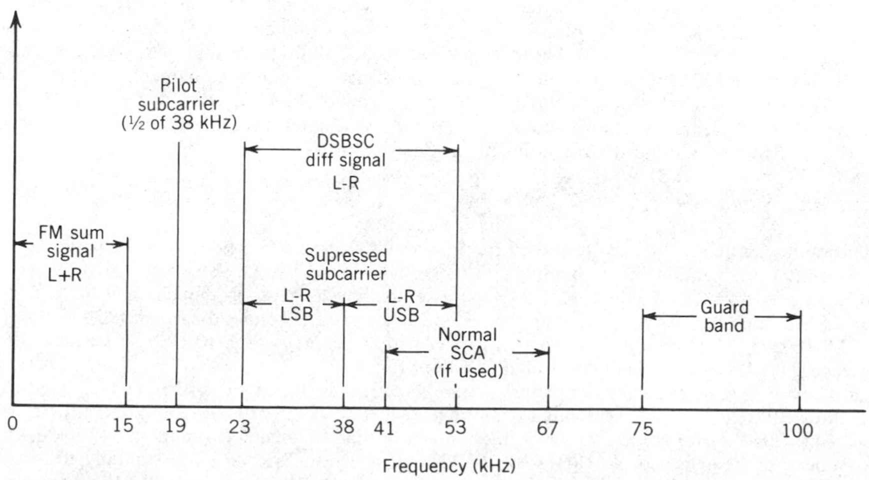

Fig. 49.21 Spectrum locations for FM broadcast using stereo and subsidiary communication multiplexed subcarriers (SCA). DSBSC, double sideband suppressed carrier; L, left; R, right; LSB, lower sideband; USB, upper sideband.

FM Spectrum

The monophonic system was initially developed to allow sound transmissions for audio frequencies from 50 to 15,000 Hz to be contained within a ±75-kHz RF bandwidth. With the development of FM stereo, the original FM signal (consisting of a right-plus-left channel) is transmitted in a smaller bandwidth to be compatible with a monophonic FM receiver, and a right-minus-left channel is frequency multiplexed on a subcarrier of 38 kHz using double sideband suppressed carrier. An unmodulated 19-kHz subcarrier is derived from the 38-kHz subcarrier to provide a synchronous demodulation reference for the stereophonic receiver. A functional block diagram of a stereophonic FM system is shown in Fig. 49.20. The synchronous detector at 38 kHz recovers the left-minus-right channel information, which is then combined with the left-plus-right channel information in sum and difference combiners to produce the original left channel and right channel signals. In addition, a subsidiary communication authorization multiplexed subcarrier, designated SCA, is used to provide background music for commercial subscribers. The SCA subcarrier limits range between 20 and 75 kHz, with 41 and 67 kHz being the most popular. The SCA modulation is limited to 30% of the main carrier for monophonic and 10% for stereo. The spectrum for a typical FM broadcast using stereo and SCA is shown in Fig. 49.21.

49.2-3 Television Broadcasting

Television broadcasting consists of video and sound transmissions to the general public in the very-high-frequency (VHF) and ultrahigh-frequency (UHF) bands. The composite system utilizes separate carriers for video, sound, and color information, designed so that compatible reception may be obtained on a black and white or color receiver.

Frequency Allocations

In the United States, the frequency allocations are divided into 68 channels in the VHF and UHF bands. For VHF, channels 2 through 13 include 54 to 72, 76 to 88, and 174 to 216 MHz. For UHF, channels 14 through 69 include 470 to 806 MHz and channels 70 through 83 include 806 to 890 MHz. Each channel consists of a 6-MHz frequency band that contains video, audio, color, and possibly a special modulation used by hearing-impaired viewers. The channel and frequency band designations are given in Table 49.7. Some of the channels are shared with other services, including biomedical telemetry, mobile radio, and radio astronomy.

Field Strength and Propagation Characteristics

The field strength of a television broadcast transmitter depends on distance, antenna heights, local terrain, frequency, and effective radiated power. A plot of field strength for channels 2 through 6 is given in Fig. 49.22. Propagation characteristics of other channels are given in the FCC *Rules and Regulations*, Part 73.699. Licensing of new stations depends on minimum separation of co-channel assignments and ranges from 170 to 220 miles for channels 2 through 13 and 155 to 205 miles for channels 14 through 83, depending on the location. For adjacent channel assignments, the minimum distance is 60 miles for channels 2 through 13 and 55 miles for channels 14 through 88. This minimum separation requirement does not always provide protection from interference caused by the grant of a new station or authority to modify an existing station.

Transmission Standards

A television transmitter is required to have a minimum of 100 W visual power in the horizontal plane with horizontal polarization; the maximum power is determined by the values in Table 49.8. The visual carrier is normally located 1.25 MHz above the lower limit of the channel. The sound subcarrier is located 4.5 MHz higher than the visual carrier and the color subcarrier is 63/88 times 5 MHz (3.57954545 . . . MHz) above the visual carrier with a tolerance of ±10 Hz and a rate of drift not to exceed 0.1 Hz/s. A typical spectrum is shown in Fig. 49.23.

The number of scanning lines per picture (frame) is 525, interlaced two-to-one in successive fields (262.5 lines per field) with a horizontal scanning frequency of 15,750 Hz for monochrome and 15,734.264 ± 0.044 Hz for color. The color horizontal scanning frequency is obtained by dividing the sound subcarrier frequency of 4.5 MHz by 286. The aspect ratio is 4 units horizontally to 3 units vertically. The synchronizing waveforms for color transmissions are given in Fig. 49.24 and for black and white in Fig. 49.25.

The vertical scanning frequency is 60 Hz for monochrome and 59.94 Hz for color. The vertical scanning frequency for color may be obtained by dividing the horizontal scanning frequency of

TABLE 49.7. TELEVISION CHANNEL AND FREQUENCY BANDS

Channel No.	Frequency Band (MHz)	Channel No.	Frequency Band (MHz)
2	54–60	43	644–650
3	60–66	44	650–656
4	66–72	45	656–662
5	76–82	46	662–668
6	82–88	47	668–674
7	174–180	48	674–680
8	180–186	49	680–686
9	186–192	50	686–692
10	192–198	51	692–698
11	198–204	52	698–704
12	204–210	53	704–710
13	210–216	54	710–716
14	470–476	55	716–722
15	476–482	56	722–728
16	482–488	57	728–734
17	488–494	58	734–740
18	494–500	59	740–746
19	500–506	60	746–752
20	506–512	61	752–758
21	512–518	62	758–764
22	518–524	63	764–770
23	524–530	64	770–776
24	530–536	65	776–782
25	536–542	66	782–788
26	542–548	67	788–794
27	548–554	68	794–800
28	554–560	69	800–806
29	560–566	70	806–812
30	566–572	71	812–818
31	572–578	72	818–824
32	578–584	73	824–830
33	584–590	74	830–836
34	590–596	75	836–842
35	596–602	76	842–848
36	602–608	77	848–854
37	608–614	78	854–860
38	614–620	79	860–866
39	620–626	80	866–872
40	626–632	81	872–878
41	632–638	82	878–884
42	638–644	83	884–890

Source. Federal Communications Commission, *Rules and Regulations*, Part 73, pp. 171–172 (Oct. 1982).

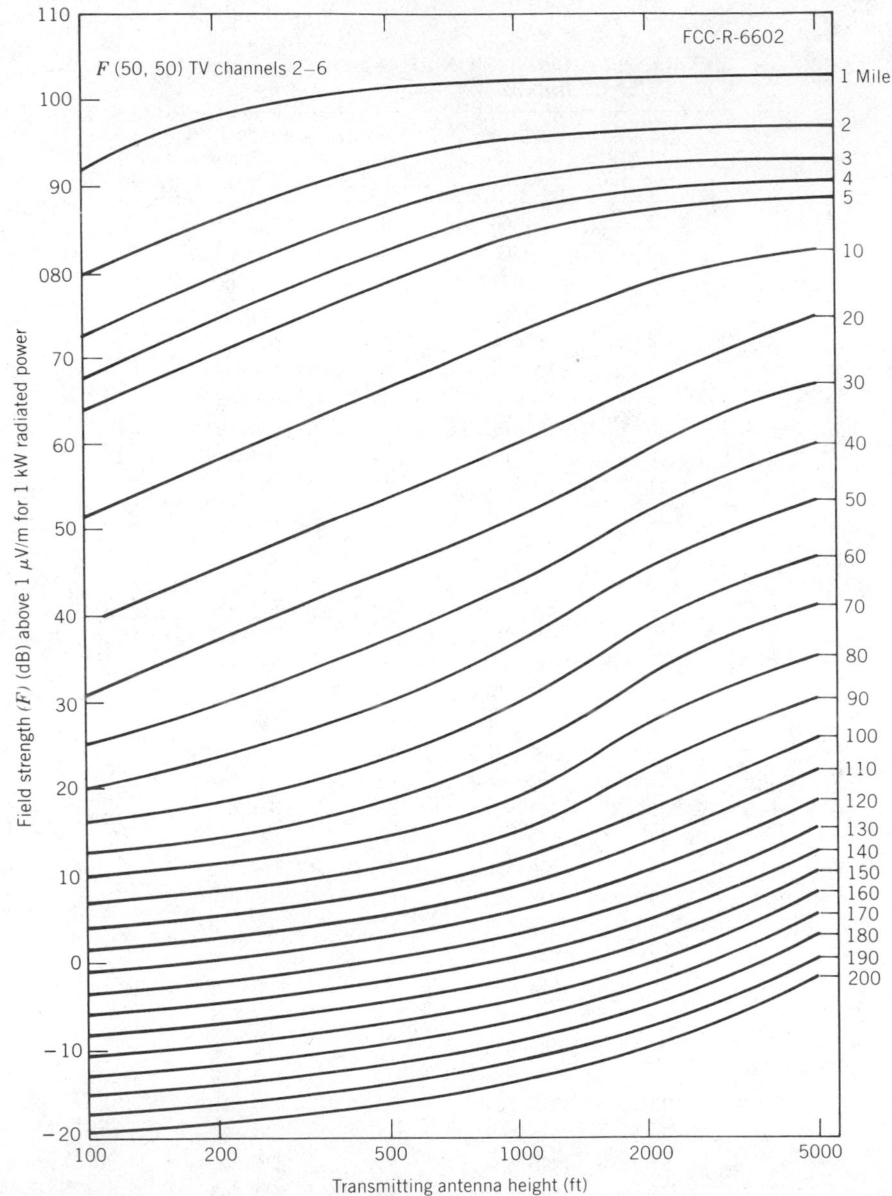

Fig. 49.22 Field strength for channels 2 through 6 for a receiving antenna height of 30 ft. $F(50, 50)$, field strength for 50% of time at 50% of locations. From FCC, *Rules and Regulations*, Vol. III, part 73, Oct. 1982, p. 226.

15,734.264 by number of lines per field, 262.5. The blanking level is transmitted at 75 ± 2.5% of the peak carrier level and the reference black level is separated from the blanking level by 7.5 ± 2.5% of the video range from the blanking level to the reference white level. A decrease in initial light intensity causes an increase in radiated power. The modulation of the video is a form of amplitude modulation (AM) in which the lower sideband is significantly attenuated and is referred to as vestigial modulation. This permits transmission of a video signal within approximately one half the RF bandwidth required with conventional AM. Attenuation of the lower sideband begins 0.75 MHz below the picture carrier; therefore, the lower sideband is not completely attenuated.

TABLE 49.8. TELEVISION MAXIMUM POWERS

Channel Nos.	Maximum Visual Effective Radiated Power (dB) Above One kW (dBk)
2 to 6	20 dBk (100 kW)
7 to 13	25 dBk (316 kW)
14 to 83	37 dBk (5000 kW)[a]

Source. Federal Communications Commission, *Rules and Regulations*, Part 73, p. 183 (Oct. 1982).

[a] Maximum visual effective radiated power of television broadcast stations operating on channels 14–83 within 250 miles of Canadian–United States border may not be in excess of 30 dBk (1000 kW).

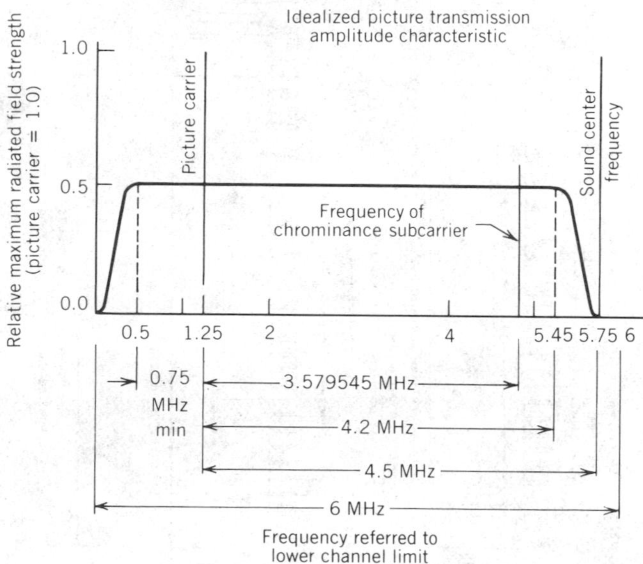

Fig. 49.23 Television spectrum (not drawn to scale).

The sound is transmitted on a subcarrier that is 4.5 MHz above the video carrier. The modulation used is FM with preemphasis to improve the received signal-to-noise ratio.

For color transmission, the picture information is processed to provide separate signals for picture content without color (luminance) and for the color content (chrominance). A functional diagram of the standard premodulation processing of color television signals is shown in Fig. 49.26.

A color television camera produces three simultaneous signals, one each in the primary colors red, blue, and green. These three signals are combined in accordance with industry standards ("NTSC Signal Specifications," 1954) to produce a luminance signal (designated the Y signal) and two chrominance signals (I and Q). The I and Q signals modulate two components of the 3.58-MHz chrominance subcarrier that are in phase quadrature. Double sideband suppressed-carrier (DSBSC) modulation is used (see Section 49.1), and the outputs of the two modulators are added together and to the luminance signal.

In the composite video signal, which modulates the picture carrier, the luminance signal is applied directly to the modulator. This component of the signal is compatible with black and white television receivers. The instantaneous phase of the chrominance subcarrier signal is a function of the color of

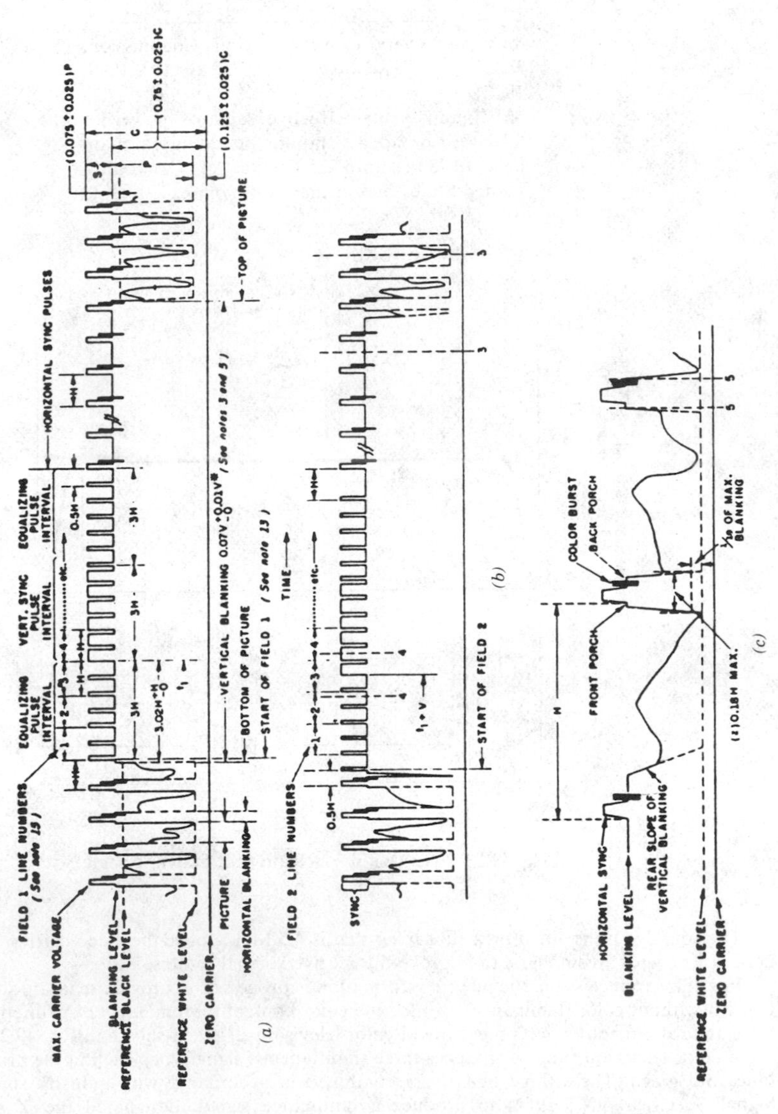

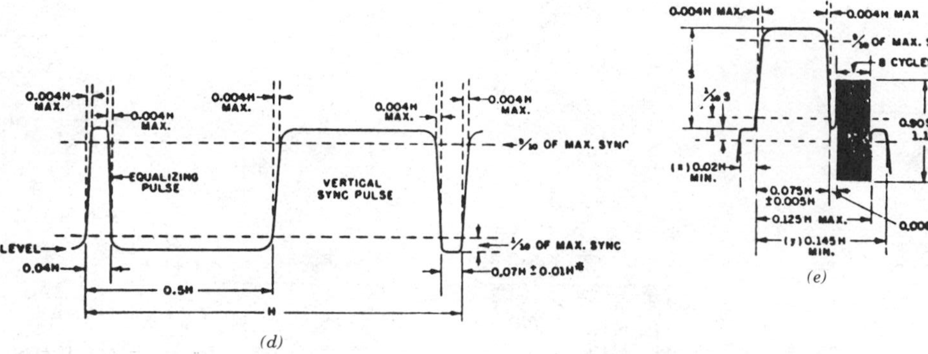

Fig. 49.24 Color synchronizing waveforms: (*a*) Field 1, (*b*) field 2, (*c*) detail between 3—3 in (*b*), (*d*) detail between 4—4 in (*b*), (*d*) detail between 5—5 in (*c*). Horizontal dimensions not to scale in (*a*), (*b*), (*c*). H, time from start of one line to start of next; V, time from start of one field to start of next; P, peak excursion of luminance signal from blanking level (does not include chrominance signal); S, sync amplitude above blanking level; C, peak carrier amplitude. Asterisk indicates that tolerances given are permitted only for long time variations and not for successive cycles.

The leading and trailing edges of vertical blanking should be complete in less than $0.1H$. The leading and trailing slopes of horizontal blanking must be steep enough to preserve minimum and maximum values of $(x + y)$ and (z) under all conditions of picture content.

The equalizing pulse area is between 0.45 and 0.5 of the area of a horizontal sync pulse. A color burst follows each horizontal pulse but is omitted following the equalizing pulses and during the broad vertical pulses. Color bursts are also omitted during monochrome transmission. The burst frequency is 3.579545 MHz, with a tolerance of ± 10 cycles, the maximum rate of change of frequency not to exceed $1/10$ Hz per second. The horizontal scanning frequency is $2/455$ times the burst frequency. The dimensions specified for the burst determine the times of starting and stopping the burst, but not its phase. The color burst consists of amplitude modulation of a continuous sine wave.

The start of field 1 is defined by a whole line between the first equalizing pulse and the preceding H sync pulses. The start of field 2 is defined by a half line between the first equalizing pulse and the preceding H sync pulses. Field 1 line numbers start with the first equalizing pulse in field 1; field 2 line numbers start with the second equalizing pulse in field 2.

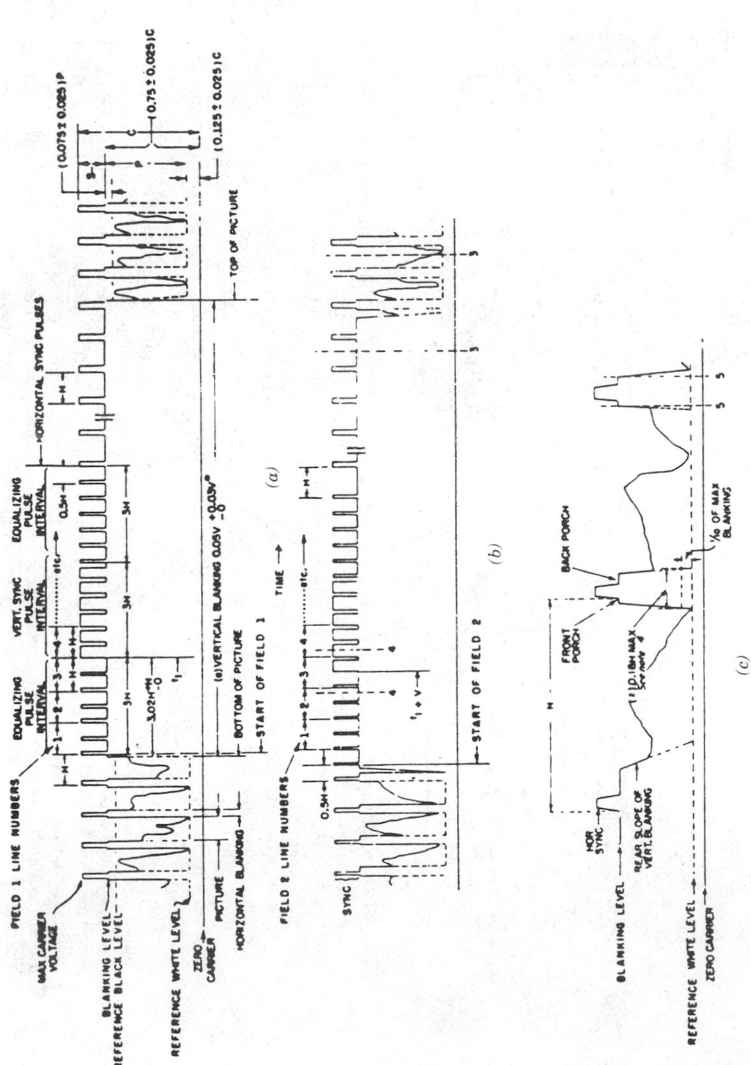

FIELD 1 LINE NUMBERS
MAX CARRIER VOLTAGE
EQUALIZING PULSE INTERVAL
VERT. SYNC PULSE INTERVAL
EQUALIZING PULSE INTERVAL
HORIZONTAL SYNC PULSES
BLANKING LEVEL
REFERENCE BLACK LEVEL
REFERENCE WHITE LEVEL
ZERO CARRIER
PICTURE
HORIZONTAL BLANKING
3.02H +H -0
(e) VERTICAL BLANKING 0.05V +0.03V -0
BOTTOM OF PICTURE
START OF FIELD 1
TOP OF PICTURE
(0.075 ± 0.025) P
(0.75 ± 0.025) C
(0.125 ± 0.025) C
TIME →
FIELD 2 LINE NUMBERS
SYNC
START OF FIELD 2
HOR SYNC
BLANKING LEVEL
REAR SLOPE OF VERT. BLANKING
FRONT PORCH
BACK PORCH
(f) 0.18H MAX See note 4
REFERENCE WHITE LEVEL
ZERO CARRIER
1/10 OF MAX BLANKING

(a)

(b)

(c)

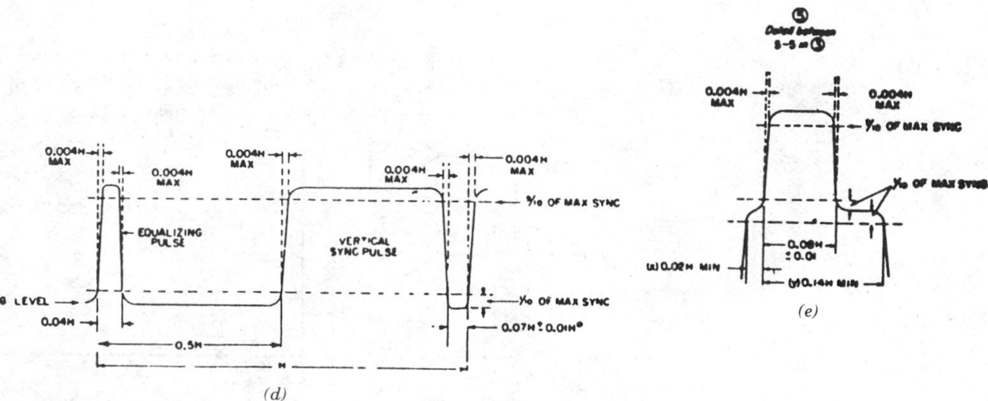

Fig. 49.25 Black and white synchronizing waveforms: (*a*) Field 1, (*b*) field 2, (*c*) detail between 3—3 in (*b*), (*d*) detail between 4—4 in (*b*), (*e*) detail between 5—5 in (*c*). Horizontal dimensions not to scale in (*a*), (*b*), (*c*). *H*, time from start of one line to start of next; *V*, time from start of one field to start of next; *P*, peak excursion of luminance signal from blanking level; *S*, sync amplitude above blanking level; *C*, peak carrier amplitude. Asterisk indicates that tolerances given are permitted only for long time variations and not for successive cycles.

The leading and trailing edges of vertical blanking should be complete in less than $0.1H$. The leading and trailing slopes of horizontal blanking must be steep enough to preserve minimum and maximum values of $(x + y)$ and (z) under all conditions of picture content. The equalizing pulse area is between 0.45 and 0.5 of the area of horizontal sync pulse.

The start of field 1 is defined by a whole line between the first equalizing pulse and the preceding H sync pulses. The start of field 2 is defined by a half line between the first equalizing pulse and the preceding H sync pulses. Field 1 line numbers start with the first equalizing pulse in field 1; field 2 line numbers start with the second equalizing pulse in field 2.

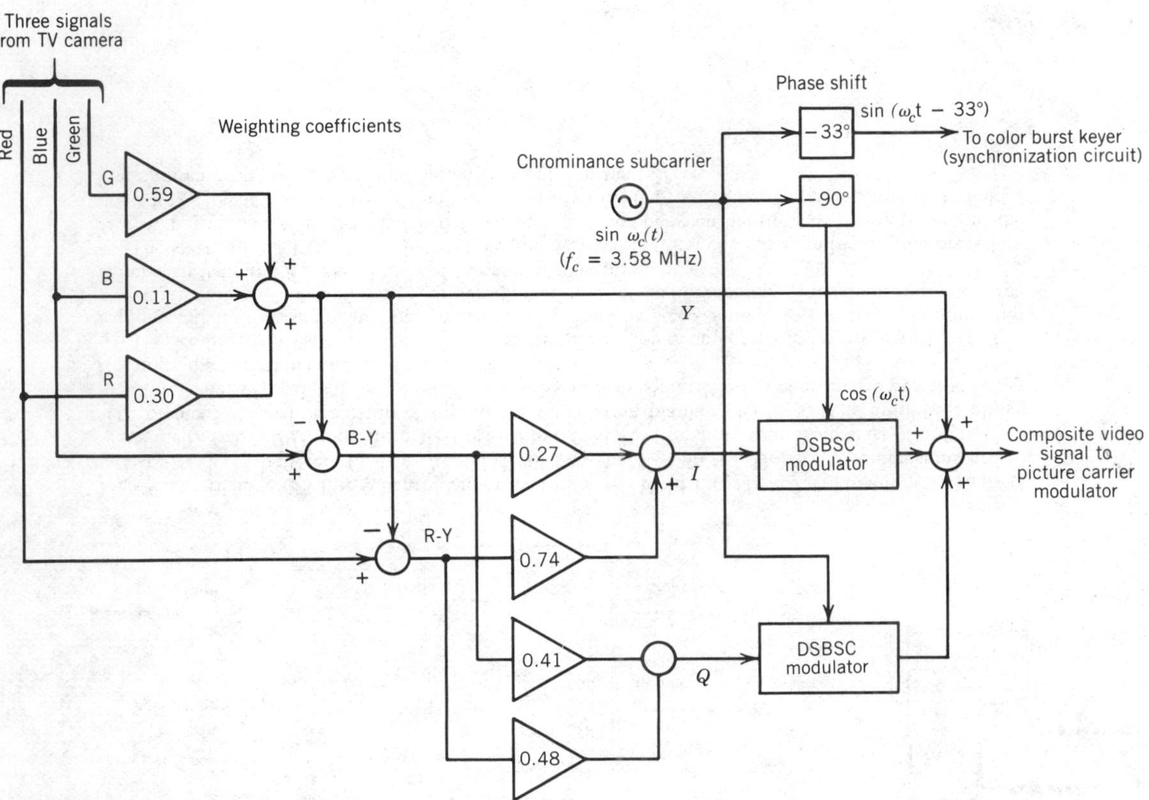

Fig. 49.26 Standard premodulation processing for color television signals. DSBSC, double sideband suppressed carrier; *Y*, luminance signal; *I* and *Q*, chrominance signals.

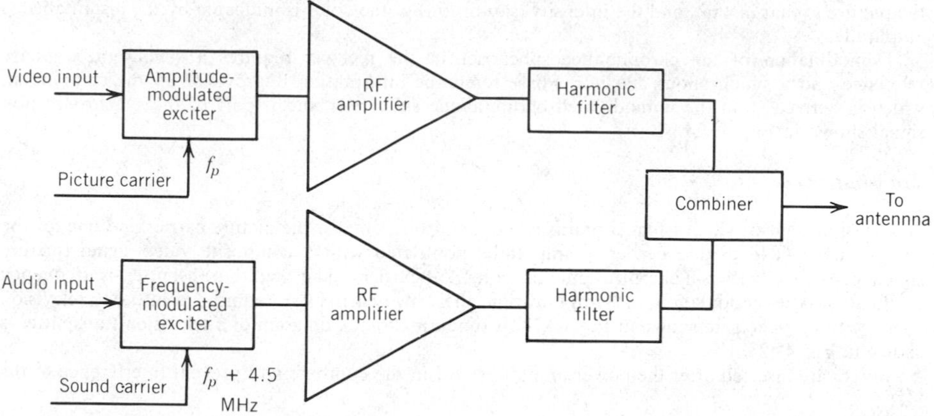

Fig. 49.27 Functional block diagram of a television transmitter. Courtesy of the Harris Corporation.

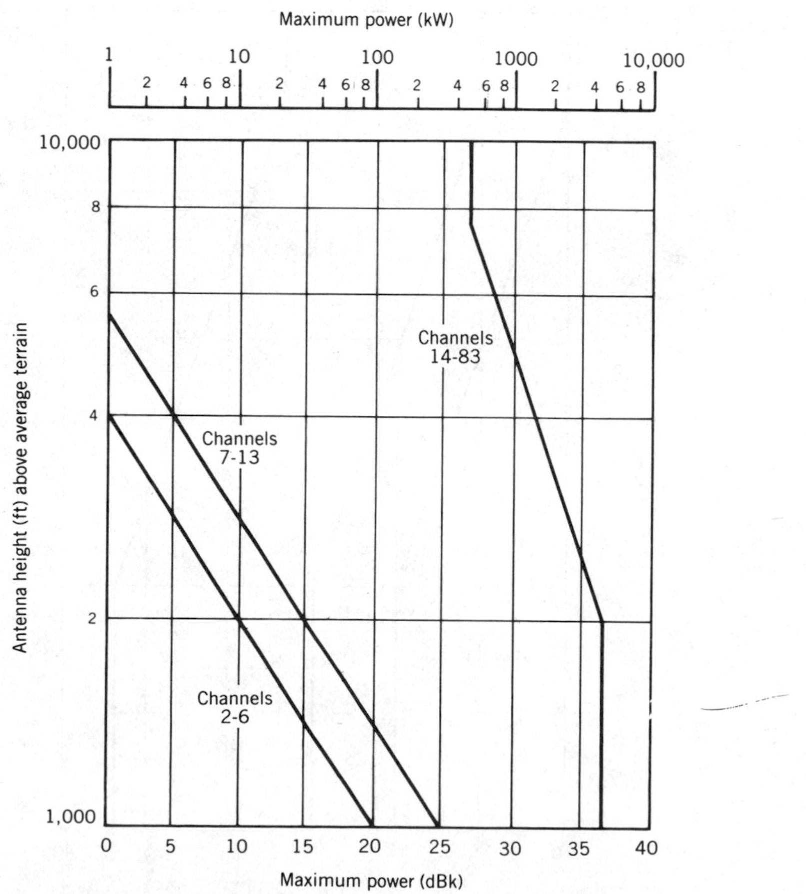

Fig. 49.28 Antenna height and power limitations for zone I.

the picture at that instant, and the intensity (saturation) of the color is indicated by the instantaneous amplitude.

Demodulation of the chrominance subcarrier in the receiver requires a synchronized phase reference and a synchronous detector. Phase reference information is provided by the color burst, which is derived from the unmodulated chrominance subcarrier and is part of the synchronization signal shown in Fig. 49.24.

Transmitters

A television transmission system contains two transmitters, one for the picture carrier and one for the sound carrier. The picture carrier is amplitude modulated with a composite video signal (picture signal combined with synchronization and color subcarrier). The sound transmitter is frequency modulated. After modulation and amplification, these two carriers are combined into one television signal with the spectrum shown in Fig. 49.23. A functional block diagram of a television transmitter is shown in Fig. 49.27.

Filters are inserted after the power amplifiers and in the combiners to prevent interference to the

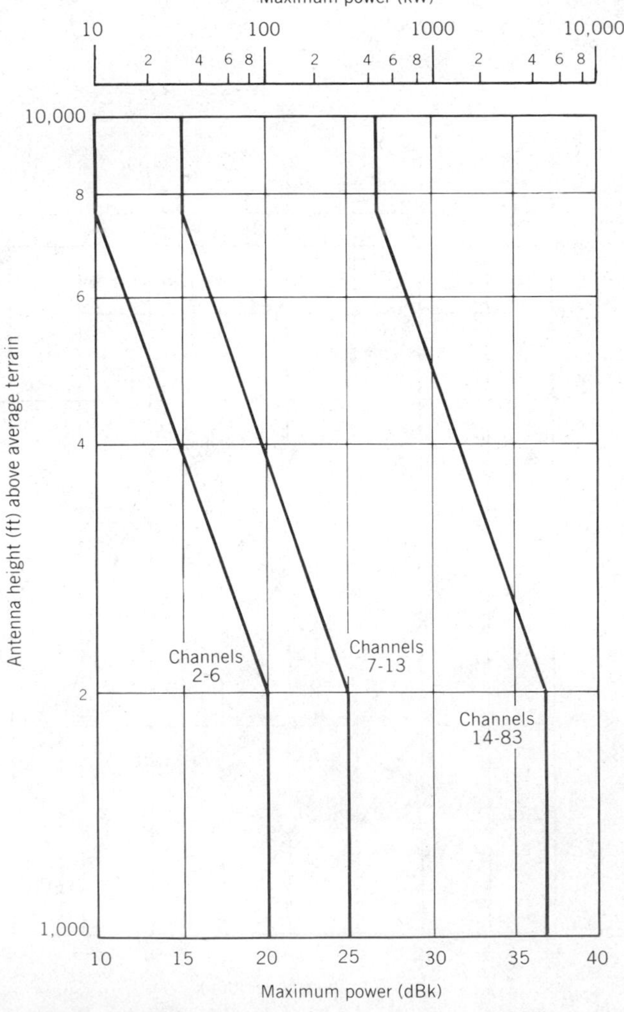

Fig. 49.29 Antenna height and power limitations for zones II and III.

sound signal by the picture signal and vice versa. In addition, the filters suppress harmonics of the radio-frequency carriers.

Vestigial sideband filtering can be accomplished at low power levels if a linear power amplifier is used for the picture carrier. If a class C RF amplifier is used for high efficiency in the final power amplifier, the vestigial sideband filter must be placed after the final power amplifier. The latter scheme provides the highest power efficiency in the final amplifier, but requires that the vestigial sideband filter dissipate large amounts of power.

Antenna Systems

The antenna systems for television are generally designed to provide omnidirectional coverage in the horizontal plane. The polarization is required to have a horizontal component; however, the use of circular polarization is becoming more popular because of the increased coverage that may be obtained particularly from the reduction of polarization fading and increased effective radiated power. The power gains vary from 6 to 20 dB, depending on the frequency and the number of bags. Elements or bags may be composed of dipoles, slots, and helices as well as of other less well-known types. Tower height and wind loading are major factors in the design. The FCC regulates the effective radiated power for antenna heights above 1000 ft for zone I VHF stations and 2000 ft for zone I UHF stations. These limitations are shown in Figs. 49.28 and 49.29. Detailed descriptions of the zones are given in the FCC *Rules and Regulations*, Part 73.609.

Bibliography

Barghausen, A. F., "Medium Frequency Sky Wave Propagation in Middle and Low Latitudes," *IEEE Trans. Broad.* **12**:1–14 (Jun. 1966).

Bartlett, G. W., ed., *National Association of Broadcasters Engineering Handbook*, 6th ed., The National Association of Broadcasters, Washington, DC, 1975.

Bremmer, H., *Terrestrial Radio Waves: Theory of Propagation*, Elsevier, Amsterdam, 1949.

Electronic Industries Association, Standard TR-101A, *Electrical Performance Standards for AM Broadcast Transmitters*, 1948.

Federal Communications Commission, *Rules and Regulations*, Vol. III, Parts 73 and 74 (Oct. 1982).

"NTSC Signal Specifications" and "NTSC Color Standards," *Proc. IRE* (Jan. 1954).

49.3 NETWORK COMMUNICATION SYSTEMS

Surya V. Varanasi

49.3-1 Introduction

Data communications is the fastest growing segment of telecommunications. A major reason for the rapid advancement of data communications is the increasing use of electronic data processing and computers. Data networks consist of either a network of computers or a set of terminals connected to one or more computers. Networks are capable of handling information in many forms, including written words, symbols, graphics, or just bit sequences. Airline and train reservations, banking transactions, air traffic control, and remote text composition for newspapers and magazines are only a few examples of the applications and use of data communication networks in everyday life today.

49.3-2 Network Design Considerations

Most communication networks consist of nodes, hosts, terminals, and transmission links. A node refers to a computer whose primary function is to switch data. Hosts are those computers used primarily for functions other than switching data. Terminals are user-network interfacing devices, and transmission links join this collection of subnet elements to form a communication network. A data network consists of transmission links, nodes, and the essential control software.

There are two general ways to organize a data network: centralized and distributed. A basic two-terminal data communication network is shown in Fig. 49.30. Figures 49.31 and 49.32 demonstrate a centralized and a distributed data network, respectively. In a centralized data network with

---

I wish to thank Dr. Roger Freeman and Dr. Udo Pooch for permitting me to use materials from their textbooks.

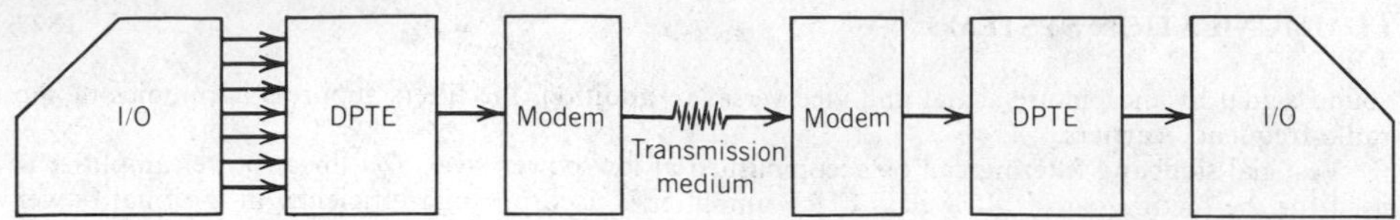

Fig. 49.30 Basic two-terminal data communications system, I/O, input-output device, DPTE, data processing terminal equipment.

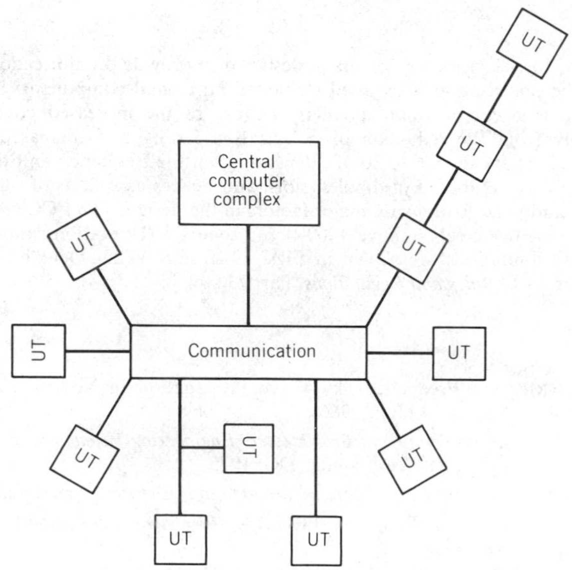

Fig. 49.31 Centralized data network. UT, user terminal.

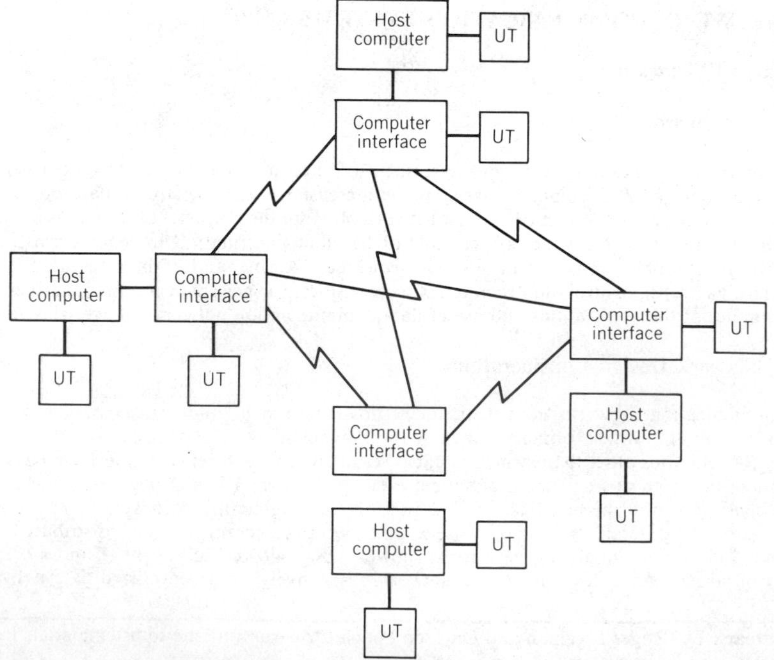

Fig. 49.32 Distributed data network. UT, user terminal.

TABLE 49.9. COST TO TRANSMIT 1 MB OF GRAPHIC INFORMATION 1400 MILES

Medium[a]	Relative Cost to Index	Comments[b]
Telegram	1650	Daytime rate 30 bits/word, 100-word messages
Night letter	280	Overnight delivery, 30 bits/word
Telex	100	50 b/s
DDD (103A)	11	Direct-distance dial, 300 b/s, daytime (dial-up)
Autodin	4	Full use during working hours (US military network)
DDD (202)	1.75	Data sets (dial-up), 2000 b/s
Letter (airmail)	1.15	30 bits/word, 250 words/page
Western Union broadband	1.00	Full duplex, 2400 b/s
WATS	0.75	Special ATT telephone service, 200 b/s per working day
Leased line (201)	0.28	Full duplex, 2000 b/s, commercial
Leased line (303)	0.11	Full duplex, 50 kb/s, commercial
Mail DEC tape	0.10	Airmail, 2.5-Mb tape
Mail IBM tape	0.017	Airmail, 100-Mb tape

[a] Numbers in parentheses identify Bell System modems used in the indicated service.
[b] Mb, megabits.

only one main processing location, all traffic occurs between the remote terminals and the single central processing unit (CPU). In contrast, in a distributed data network the major data-processing capabilities are located in more than one central location.

In the design of a data network, a delicate tradeoff between capabilities and cost is made to satisfy network demand. A properly designed network must provide reliable, error-free communication within a reasonable time. Data network design considerations include (1) the network organization, (2) tariffs and tariff structures, (3) reliability, (4) type of communication services, that is, switched, leased, or private lines or a combination thereof, (5) line routings, (6) types of terminal equipment used at remote sites, (7) protocols, that is, location and types of communication-control procedures, (8) error control procedures, (9) economy, (10) simple user access, and (11) security as required by the user. The design must also take into account the user's business application requirements in addition to technical specifications such as (1) number and locations of processing sites, (2) number and locations of remote terminals, (3) types of transactions to be processed, (4) traffic intensities for each type of transaction by the type of terminal, (5) urgency of information to be transmitted (timeliness), (6) patterns of traffic flow, (7) acceptable error rates, and (8) required availability of the system.

It is not always necessary for a cost-efficient data network design to incorporate all the technical specifications just listed. Instead it may have only those essential for a specific application. Table 49.9 demonstrates the relative cost to transmit a megabit of graphic information in the United States, assuming leased equipment and use of the service for eight hours per working day. The distance factor is taken to be 1400 statute miles (2200 km) and a cost index of 100 is set for standard 50-baud telex service.

49.3-3 Data Terminals

The seven types of terminals in use are: (1) teleprinter, (2) alphanumeric cathode ray tube (CRT), (3) graphic CRT, (4) remote batch, (5) data preparation, (6) point of sale (POS), and (7) industrial data collection.

A teleprinter can serve for input, output, or both. Teleprinters may be made up of the following configurations: (1) receiving only (RO), providing printed copy, (2) keyboard send-receive (KSR), and (3) automatic send-receive (ASR), with keyboard, paper tape reader, paper tape perforator, and a hard-copy printer. Teleprinters are generally asynchronous (start-stop), operating at a rate of 100 words per minute. Alphanumeric CRTs are functionally analogous to teleprinters except the hard-

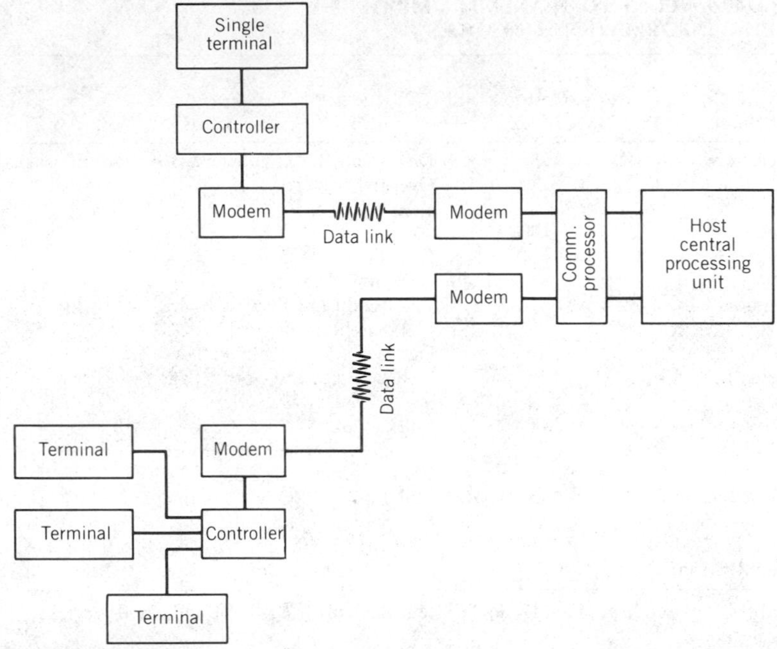

Fig. 49.33 Stand-alone or single terminal and terminal cluster accessing a host computer.

copy printer is replaced by a CRT with the same capabilities. The graphic CRT provides direct display and input of pictorial data representation. Remote data batch terminals provide such input-output (I/O) facilities as card reader (and card punch), perforated tape reader, and magnetic tape units to provide batch computer processing input from a remote terminal. Line operation is a serial bit stream at rates of 1200 to 9600 b/s. The common term for this type of terminal is "remote job entry" (RJE).

Data preparation usually implies traditional key punching, that is, IBM card preparation facilities, but can also extend from keyboard to tape or disk systems where the operator works from source documents.

A point of sale (POS) system is a terminal system designed for retail store application. Each transaction is either directly entered into a computer or placed in off-line storage for later input. A POS terminal resembles an enhanced cash register. Other inputs include badge readers, credit cards, and encoded merchandise tags. Many POS terminals have internal processing to calculate taxes, perform arithmetic operations, verify check digits, and so on, and these functions can be performed either on line or off line.

Industrial data collection terminals are primarily found in such industrial areas as storerooms, receiving departments, tool stations, assembly lines, and quality control (QC) areas. Data input is obtained either optically, magnetically, or by card reader/keyboard entry.

Programmable data terminals perform certain communication interface functions and are increasingly alleviating the CPU of certain preprocessing functions. The programming capability is provided by a built-in microprocessor, which can help the user acquire, edit, sort, update, file, calculate, and manipulate source data off line. Typical memory capacity of a programmable terminal is 2 to 8 kilobytes.

Cluster terminals are now in common use, where a cluster accesses the communication link via a common data link controller. The cluster concept is illustrated in Fig. 49.33.

49.3-4 Network Configurations and Data Switching

This section briefly discusses centralized and distributed data networks from the point of view of network data processing and examines the configurations of various data networks.

Network Configurations

A centralized data network has the structure of a "star network," with one central computer and the following characteristics:

1. Its computing and switching facilities are centrally located at one site.
2. It has a treelike appearance, although in some cases it can be a ring or a loop.
3. There is only one unique communication path between the terminal and the CPU.
4. It is a terminal-oriented system.
5. The traffic flow is between the terminals and the CPU.

A centralized data network configuration is shown in Fig. 49.31. In the ring network configuration shown in Fig. 49.34, the data input-output points form a ring structure. In ring network, the traffic can be unidirectional or bidirectional, in which case the reliability of the network is considerably increased. The processors can be simple or complex. In the simple configuration, only one pair of stations can communicate at a time, whereas in complex configurations, using time division multiplexing (TDM) bit streams in both directions, all stations can intercommunicate quasisimultaneously.

A multipoint system is one in which two or more terminals share a dedicated or leased line. A point-to-point connection is any connection between a source-sink pair. Figures 49.35 and 49.36 illustrate these configurations, respectively.

Data Switching

The two approaches to data switching are (1) message switching, also known as store-and-forward message switching, and (2) packet switching.

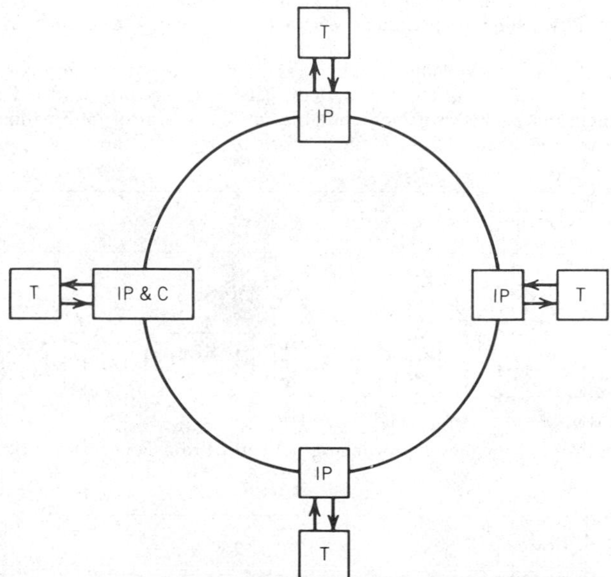

Fig. 49.34 Ring network configuration. IP, interface processor; T, data terminal or computer; C, ring controller.

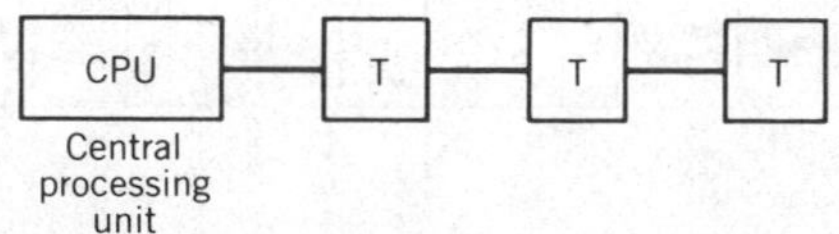

Fig. 49.35 Multipoint network system. T, terminal.

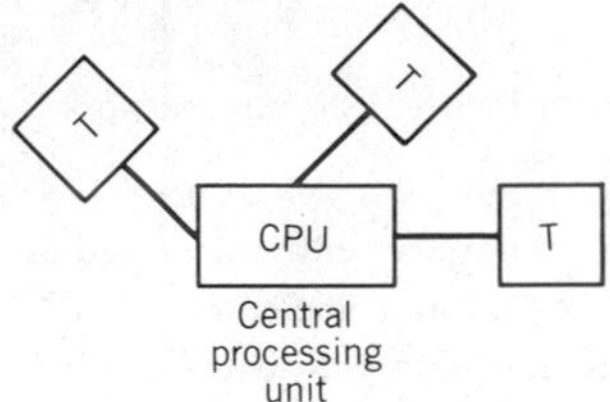

Fig. 49.36 Single-point network system. T, terminal.

Message Switching. A message can be defined as a logical unit of information for purposes of communication. A message is composed of (1) a "headline" (header), which contains information suitable for network control operations, (2) the "body," or text, which contains the information to be transferred, and (3) a "trailer," which contains fields that signify the end of message. Telegrams, programs, and data files are examples of messages. Basically, store-and-forward message switching stores data messages at switching nodes and forwards that traffic to the next node or addressee(s) when a circuit becomes available. Messages within a message-switching network are transferred between switches on a message-by-message basis. This means that a message, sometimes broken into blocks of data, must be either transmitted and received in its entirety or canceled across a link before the next message can be transmitted. Each block of a message must be transmitted in its proper sequence so the receiving switch can rebuild the message and verify to the sending switch that it has received it. Figure 49.37 shows a typical format for a message-switching network. Note that in the figure only the initial block has sufficient control information for further routing. "Cybernet," designed and implemented by Control Data Corporation (CDC), is an example of a message-switching network. Message switching has the following disadvantages: (1) expensive switch costs, (2) long message delays, (3) less efficiency in the utilization of network resources, and (4) less flexibility in adjusting to traffic conditions. For a regular user, however, message switching can be less expensive than circuit switching, since circuit costs are divided among the users sharing the system.

Packet Switching. In packet switching, a data message is broken down into parts called packets. These packets could be called short little messages, each with its own header. Unlike in message switching, each packet in packet switching contains sufficient control information to transmit the packet across a network independent of all other packets belonging to the same message. Figure 49.38

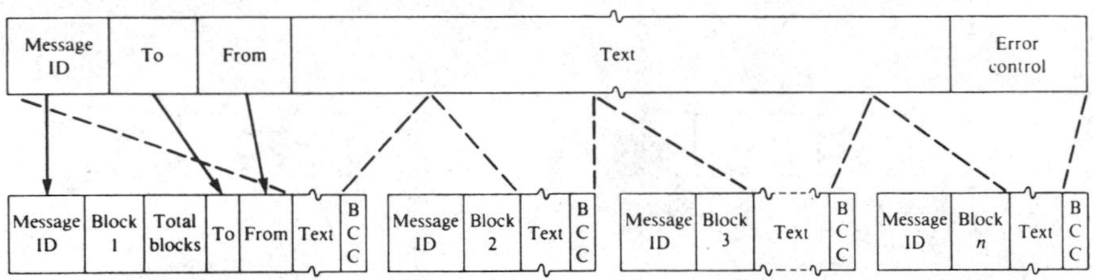

Fig. 49.37 Typical format for message-switching network. From Pooch et al., 1983, reprinted with permission.

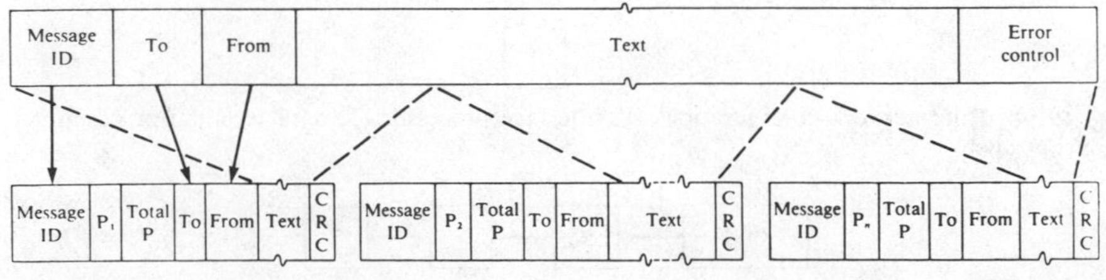

Fig. 49.38 Typical format for packet switching network. From Pooch et al., 1983, reprinted with permission.

shows the format for packet switching. For a more detailed description of a packet-switching network, see Kleinrock (1970). A popular and well-known packet-switching network is the advanced research projects agency network (ARPANET).

49.3-5 Circuit Optimization and Effective Data Transfer

Circuit Optimization

Data networks, whether packet switched or message switched, have source sinks with varying requirements regarding the "quantity" of data and its urgency. It would be uneconomical to underutilize expensive lines that can support 2400, 4800, or even 9600 b/s if urgency were not a consideration in a particular application. Hence some method of optimizing data communication links is required. Two basic techniques of data transmission are used when terminals are connected to a computer system: multiplexing and concentration.

The multiplexing process combines at one end of the data communication channel lower-speed subchannels (i.e., low-speed data streams) and demultiplexes higher-speed data into the original lower-speed subchannels. Two multiplexing techniques are frequency division (FDM) and time division (TDM). FDM subdivides the communication line's bandwidth into narrower individual channels guarding against mutual interface or cross talk. TDM uses a time-synchronization signal to assign a specific time slot to each channel. The rate at which data can enter or leave terminals, when multiplexed, cannot exceed the data rate of the communication channel.

Data concentration, although similar to multiplexing, assembles low-speed data into characters and blocks with appropriate code conversion, error checks, and compression. To be more exact, data are combined into a complex composite signal, which contains more information per unit time. On the other end of the high-speed communication line, the data are deconcentrated in a front-end processor similar to the demultiplexing process. Because of data communication (or compression), more economical use of high-speed communication lines is made.

Effective Data Transfer

In a large data network, suppose a link were established from source to sink operating at 2400 b/s. How efficient is the circuit? At this rate, in one hour 8.64 Mb (2400 × 60 × 60 bits) can be delivered. But the question is, how many bits in a time period are really useful to the CPU at the sink? The American National Standards Institute recommends the use of the term "transfer rate of information bits" (TRIB) to qualify the net data transfer rate. TRIB is defined as the ratio of number of information bits accepted by the sink to total time required to get those bits accepted. The formula used by Doll (1978), assuming block transmission, is

$$\text{TRIB} = \frac{K_1(M - C)}{N_t(M/R) + \Delta T}$$

where K_1 = information bits per character
M = message block length, characters
R = line transmission rate, characters/s
C = average number of noninformation characters per block
N_t = average number of transmissions required to get block accepted at sink
ΔT = time between blocks, seconds

If P is the probability of having to retransmit a block, then N_t can be expressed as

$$N_t = \frac{1}{1 - P}$$

Then

$$\text{TRIB} = \frac{K_1(M - C)(1 - P)}{(M/R) + \Delta T}$$

In this formula there is no direct reference to error rate; it is implied in the term $(1 - P)$. It follows that there is an optimum block length (given channel error performance) that will optimize data transfer, at least as far as M is concerned, all other terms remaining constant. Also it should be noted that the term C can reduce the TRIB considerably. Other delays may be (1) dial-up time, (2) satellite channel propagation delay, (3) modem synchronization delay, and (4) type of automatic repeat request (ARQ), whether stop and wait or a continuous running on full duplex. A data link can be so overburdened with inefficiencies that, for example, a 2400 b/s data link may only afford 100 b/s of TRIB source to sink.

49.3-6 Operational Considerations

A multipoint structure is one of the earliest techniques established to decrease the cost of transmitting data to remote terminals. Instead of attaching a single terminal to each remote line, several terminals are attached, each with one or more addresses unique to that terminal, as shown in Fig. 49.39. Thus traffic received at a terminal, but not addressed to that terminal, is ignored. Multipoint configurations, furthermore, frequently have the capability of "group or broadcast addressing," in which more than one terminal on a line can receive the same message without the sender needing to transmit the message more than once. A form of multipoint, frequently used for short-distance loops in which modems are not required, involves the use of a synchronous transmission data stream. Each message is preceded by the address of the destination terminal; all other terminals ignore this transmission. Synchronization pulses are transmitted to keep all terminals synchronized. Synchronous multipoint networks are very popular where detection of failed elements in the network is essential to its continued and efficient operation. Other procedures, namely, contention and polling, are used if more than one source of data is connected to the line.

Contention

Contention access protocol procedures, popular because of their simplicity, will function adequately only if the user has sufficient understanding of the demands made of the network. If there were only two or three terminals, the operation could simply be on a first-come, first-served basis. But more sophisticated schemes require that transmission begin only on predetermined instants, thereby lowering the likelihood of interference.

The contention-access procedure requires a mechanism to acknowledge the receipt of a signal. If this acknowledgment does not occur, the originator assumes that the transmitted data were interfered with and retransmits until such acknowledgment does occur. Contention access is applicable to both land-line and radio media, but is more useful in radio communications where frequency allocation constraints require users to share frequencies.

An example of a network well known for its use of the contention-access procedure is the "Aloha" network that connects terminals in Hawaii with various locations in the mainland. The Aloha packet radio system uses a contention-access scheme that allows users to transmit information randomly on two 100-kHz channels assigned for use by the net. A maximum channel utilization of 18.4% can be achieved using Aloha contention-access procedure (Abramson, 1970), which represents a poor utilization of a very valuable resource, the RF spectrum. The slotted Aloha contention-access protocol provides an improvement in efficiency over the simple Aloha procedure. In the slotted Aloha contention scheme, packets can be transmitted only at the start of a predefined clock interval (Kleinrock and Lam, 1973). Whenever two packets try to occupy the channel at the same time, there is a collision. Transmission of packets only at the start of a predefined clock interval limits the time in

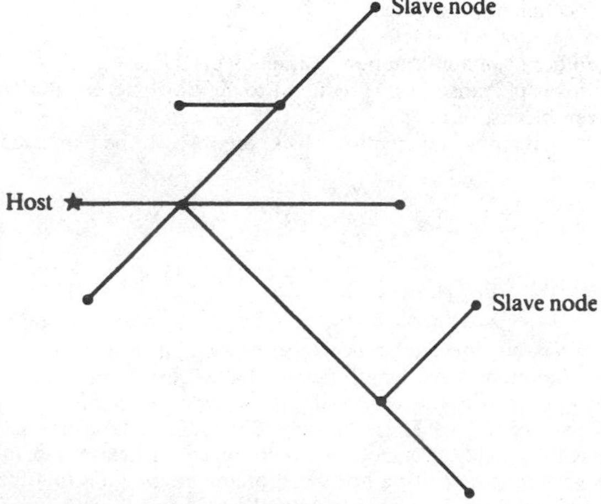

Fig. 49.39 Multipoint network. From Pooch et al., 1983, reprinted with permission.

which collisions can occur. If a collision is going to occur, it will always occur at the beginning of a clock interval. Thus random collisions do not occur in a transmission once it has started. The slotted Aloha contention-access scheme functions properly only if all of the terminals in the network are synchronized. This requires a very accurate clocking mechanism and a procedure for implementing the synchronization. The maximum utilization rate, using load-factor analysis, for a slotted Aloha channel is 36.8%.

Polling

Polling involves the central control of all nodes in the network. The ring or loop network may also use polling. It is basically a master-slave(s) operation in which the master queries each slave to determine if it has anything to say. If the answer is affirmative, the slave is either given permission to transmit or scheduled to transmit at a later time. Two types of polling are now in use, hub polling and roll-call polling.

With hub polling, a network of terminals can operate only if the network is properly configured. Each slave node is serially connected to another slave node until a path is completed back to the master node, thus forming a "hub" configuration as in Fig. 49.40. The master node initiates a polling request to the first slave node on the hub, which passes the request to the next slave node if it has nothing else to communicate. This process continues until the request has completed the hub cycle. The cycle is broken and must be restarted anytime a slave node needs to communicate with the master node. Whenever a slave node on the hub fails, the hub is broken and must be reconfigured to continue operation. The additional complexity for recovering from such a failure must be shared by each slave node on the hub, and thus increases expense. These disadvantages make the hub-polling procedure less popular than other schemes, even though a cost savings in communications circuitry may be realized.

The concept of "roll-call" polling, while simple, adds a dimension of flexibility for meeting changing demands more typical of most data-processing environments. The terminal node need only be cognizant that it is a slave node of a network and be able to appropriately respond when requested by the master node. The roll-call terminal is considerably less complex than the hub terminal, which must also be able to reconfigure itself in the event that an adjacent slave node fails. However, the required complexity of the master can be greatly increased depending on the particular roll-call scheme used. A simple roll-call network configuration is shown in Fig. 49.41.

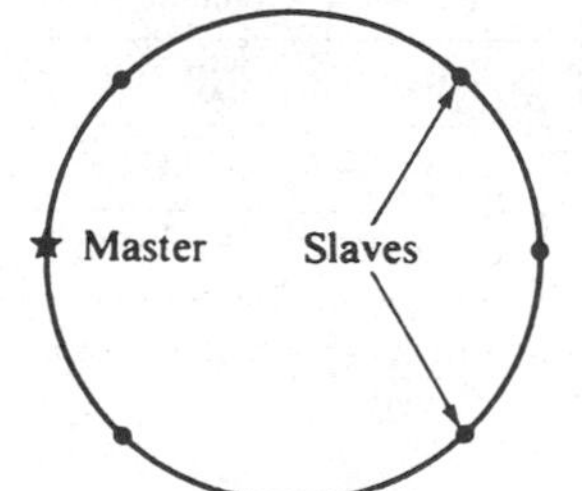

Fig. 49.40 Hub polling network. From Pooch et al., 1983, reprinted with permission.

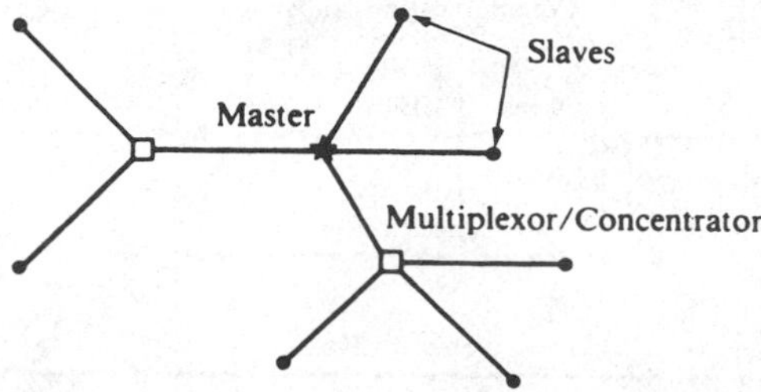

Fig. 49.41 Roll-call polling network. From Pooch et al., 1983, reprinted with permission.

The simplest roll-call procedure is to establish a predefined sequence of nodes that will ensure that each slave node has an opportunity to communicate with the master node, and only then to proceed to another cycle. The interval of delay between sequential transmissions is constant. Suppose, however, that some slave nodes require communications more frequently than other nodes, or perhaps less frequently but with a higher priority. Both of these conditions warrant special consideration by allocating more opportunities for communications than would otherwise be typical. This, of course, decreases the opportunities that the remaining terminals have to communicate, and so a careful balance between the needs of all the users and the available resources must be made. This additional alternative allows a slave node, for example, node A in Fig. 49.42, to receive a roll call more frequently than the other slave nodes. Changes to the roll-call algorithm occur only in the master node.

49.3-7 Protocols

Protocols are common tools for controlling information transfer between computer systems. In a distributed network involving three or more computers, large or small, communicating to transfer a message accurately between two or more hosts, not only must the integrity of the host information be retained, but messages must be properly formatted for transmission within the network and then efficiently transferred across each link along the selected network path. Each of these processes requires a different procedure and in most cases is under control of different elements within the network. The logical control of the allocation of resources in a complex network is achieved by the use of host-to-host, end-to-end, and link-to-link protocols, as shown diagrammatically in Fig. 49.43.

Host-to-host protocols are also referred to as transport protocols because of the basic function they perform. For example, consider a data base containing elements of information that must be transmitted to a remotely located data base. The originating host knows the form in which the destination must receive the data and configures the individual data elements accordingly. The servicing communications subsystem node would, however, become confused if given the data in this basic form. The originating host must collect the data, format them into a message, and attach the appropriate header and trailer information for intelligent interpretation by the servicing node.

The additional information required for the header and trailer varies from network to network. The header consists essentially of those fields that will allow the communications subsystem to deliver the message in an accurate and timely manner. As a minimum, the host attaches data fields such as

	Roll-call sequence
1	ABCDE ABCDE ABC . . .
2	ABCADEABCADEA . . .
3	ABCADE . . . BCDBEA . . .

Fig. 49.42. An alternate roll-call algorithm. From Pooch et al., 1983, reprinted with permission.

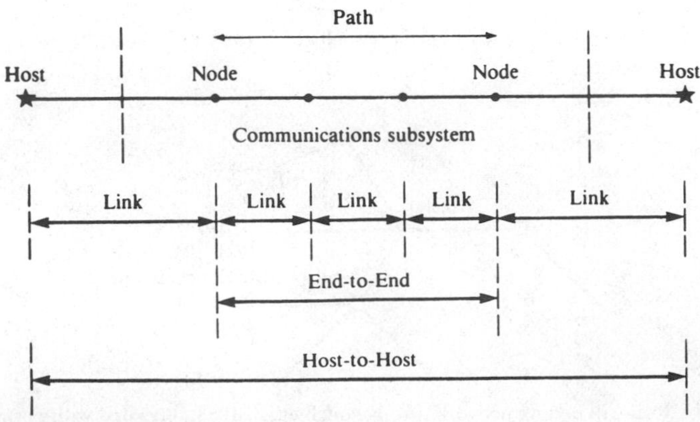

Fig. 49.43 Hierarchical protocol structure. From Pooch et al., 1983, reprinted with permission.

those illustrated in Fig. 49.44. The need for the destination field is obvious, but the requirement for priority and classification fields will vary depending on the characteristics of the data transmitted. Precedence allows a user to identify the urgency for delivering a message, perhaps at the cost of delaying other traffic in the network. The user may desire that delivery be inhibited if the receiving host is unknowingly not authorized to receive proprietary information. The originator is customarily notified in some manner. The message number, while sufficient for some interval of time (usually 24 hours), is limited by the size of the field. However, if the message is associated with the date and time of origination (DTG) and the originating host, it is (forever) uniquely identified. The origination field is used by the destination, so that receipt of the data can be acknowledged and the integrity of the input can be tested.

Information contained in the trailer is usually basic, consisting of some error control mechanism and an indication that the text has ended. Additional fields may or may not be included, depending on the particular needs of the environment.

End-to-end protocols are those protocols dedicated to transmitting a message from one end of the communications subsystem to the other, that is, between each of the servicing nodes. The message generated by the host is usually in a format that requires the least effort on the part of the host. This message must then be formatted for network transmission by the node that serves the originating host.

Messages received from a host are usually too long to allow efficient error-free transmission across a network. At a packet-switched node, for example, the message is subdivided so that portions of the message received in error can be retransmitted without the entire message having to be retransmitted. The servicing nodes, using established length, subdivide the message into blocks of data called packets. Each packet of a message traverses the network independently of other member packets of the same message. Packets are sequentially numbered as data are stripped from an originating message. This process allows the destination node to reorder packets into a message even when the packets are received out of sequence. The arrival sequence is very much dependent on any delay experienced by each packet as it travels its own path across the network. This implies that each packet must contain essentially the same control information so that it can be treated as a separate entity. The originating node must, therefore, attach to each packet header information that will assure proper delivery.

Packet headers do little more than duplicate information contained in the message header, provided by the host, which has now become the text of one or more packets in the packet stream. The significant differences are: (1) Data elements are reformatted to equivalent bit-oriented versions of the character data typically contained in a message header, (2) each packet contains an ID that uniquely identifies it from other packets in the network, and (3) some method for tracing the route taken by a packet is placed in the header so that looping can be detected and/or prevented. Trailer information for a packet is similar to the information contained in the message header. Packet-switching networks usually employ a very complex and powerful error-control mechanism to guarantee a high probability of error detection. Figure 49.45 shows a simple sequence of events that might occur in the control of packets by the end-to-end protocol hierarchy. Release of packets to the destination node does not, nor is it intended to, indicate the particular route taken by each packet. Such release implies, however, that certain protocol communications occur between the hosts and their servicing nodes.

Link protocols allow adjacent nodes to carry on organized communication under normal circumstances. Link protocols, at the lowest level in the protocol hierarchy, are far more complex than those at other levels. Link-protocol procedures can be divided into four subprocedures: one for normal communications, another for "no-notice" link-failure conditions, a third for prenotice link failure, and a fourth to establish the link. Each of the procedures is further subdivided as shown in Fig. 49.46.

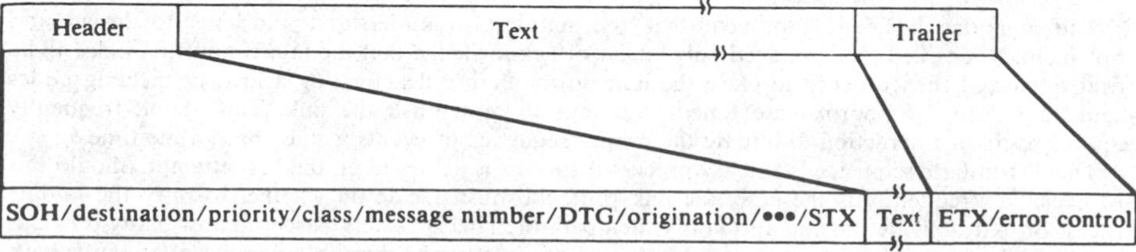

Fig. 49.44 Data field provided by host. SOH, start of heading; DTG, date and time of origination; STX, start of text; ETX, end of text. From Pooch et al., 1983, reprinted with permission.

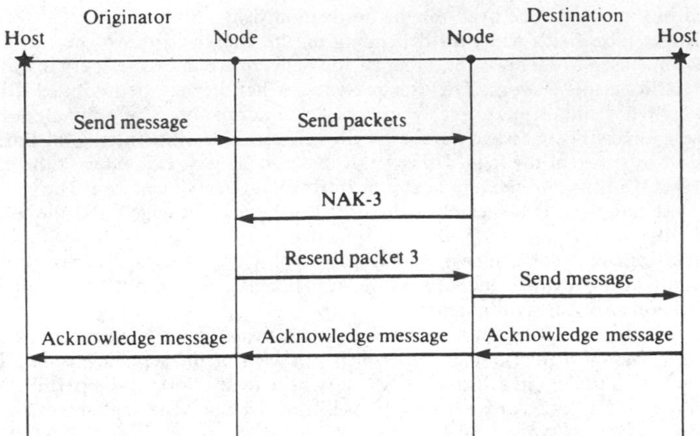

Fig. 49.45 End-to-end packet-switching protocol sequence. From Pooch et al., 1983, reprinted with permission.

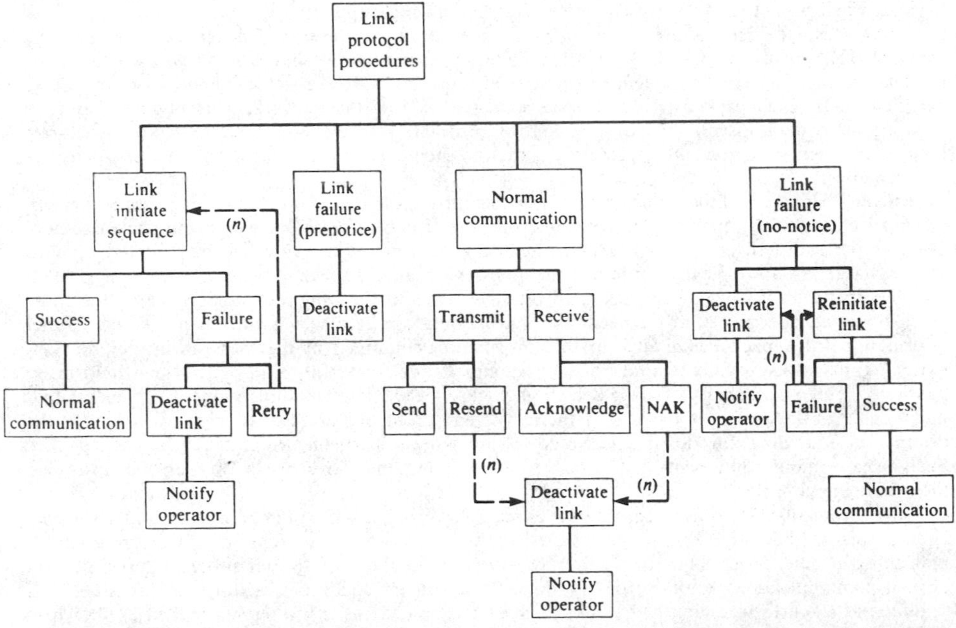

Fig. 49.46 Categories of link-protocol procedures. From Pooch et al., 1983, reprinted with permission.

The prenotice link-failure procedure is the simplest to understand and thus to design and implement. Nodes tied to a link need only advise or acknowledge that the interconnection link will be deactivated and then proceed to close the link down. Before deactivating a link, connecting nodes should agree on the approximate time to attempt to reestablish the link. This action frequently requires operator interaction to initiate the proper sequence of events at the appropriate time.

The link-initiate sequence is an easy process if success is achieved on the first attempt. Should this not occur, as frequently is the case, the link protocol must decide on whether to retry the initiate process or to deactivate the link and notify the operator. This is usually based on how many previous unsuccessful attempts have been made. After some predetermined threshold number of attempts have failed, little else can be done automatically and the human operator should be brought into action. If success has been achieved, the link goes into the normal communications protocol procedure.

Another method of classifying different types of protocol is by the message-framing techniques used. There are character-oriented, byte-oriented, and bit-oriented types of protocol. A typical character-oriented protocol is IBM's BISYNC or binary synchronous protocol. Character-oriented protocol uses special characters to indicate such events as "start of heading" (SOH), "start of text" (STX), and "end of text" (EOT). Byte-oriented protocol also uses character-sequence delimiters in the header similar to character-oriented protocol and includes a "count" that indicates the number of data characters in the message. For a more detailed treatment of procedures and protocols, the reader is directed to Pouzin and Zimmermann (1978).

Bibliography

Abramson, N., "The Aloha System—Another Alternative for Computer Communications," *Proc. FJCC*, pp. 281–285 (1970).

Doll, D. R., *Data Communication Facilities, Networks and Systems Design*, Wiley, New York, 1978.

Freeman, R. L., *Telecommunications Systems Engineering, Analog and Digital Network Design*, Wiley, New York, 1980.

Kleinrock, L., "Analytical and Simulation Methods in Computer Network Design," *Proc. AFIPS JJCC*, pp. 569–679 (1970).

Kleinrock, L. and S. Lam, "Packet Switching in a Slotted Satellite Channel," NCC, *AFIPS Conf. Proceed.* **42** (1973).

Pooch, U. W., W. H. Green, and G. G. Moss, *Telecommunications and Networking*, Little, Brown, Boston, 1983.

Pouzin, L. and H. Zimmermann, "A Tutorial on Protocols," *Proc. IEEE* **66**(11):1346–1370 (Nov. 1978).

49.4 SATELLITE COMMUNICATIONS SYSTEMS

Daniel F. DiFonzo

49.4-1 Introduction and Overview

Communications satellites, acting as radio-frequency (RF) repeaters in orbit, have revolutionized the telecommunications industry since their inception in the 1960s. Most communications satellites are in geostationary orbit, a circular orbit that lies in the equatorial plane at an altitude of 36,000 km. The orbit period is one sidereal day, so that the satellite appears stationary to an observer on the earth. From this altitude more than one third of the earth's surface is visible. Figure 49.47 illustrates the use of a satellite to relay signals among widely separated earth locations A and B by means of direct line-of-sight microwave links between each station and the satellite. A system of three geostationary satellites spaced approximately 120° apart in longitude could provide nearly full global coverage and interconnectivity. Communications satellite features include wide bandwidth for high channel capacity for voice communications as well as to provide digital data and TV; multiple access so that a large number of users can communicate via the same satellite; multiple destination and point-to-point traffic; direct broadcasting to large geographic regions; communication with mobile users such as ships and aircraft; links between satellites in different orbits for data relay; and intersatellite links to improve system interconnectivity.

A satellite communications system consists of a space segment and an earth segment. The space segment includes the active and spare satellites. The earth segment includes the earth stations that constitute communications nodes for the system, as well as special earth stations for telemetry, tracking, command, and monitoring (TTC & M). These stations are used to control the satellite orbital position, monitor its health, and reconfigure its circuits if necessary. In some systems the earth segment can also include terrestrial distribution networks and the interfaces to other communications systems. In the INTELSAT global system the TTC & M facilities are considered part of the space segment for cost allocation purposes.

This section describes the evolution and basic features of communications satellite systems, as well as orbit geometry and launch vehicles, communications link parameters, modulation and access, satellite and earth terminal subsystems, representative satellite systems, coordination issues, and future trends.

History

In 1945 Arthur C. Clarke first proposed geostationary satellites for worldwide communications.[1] He described "active" repeaters or transponders that would receive signals from the earth and retransmit

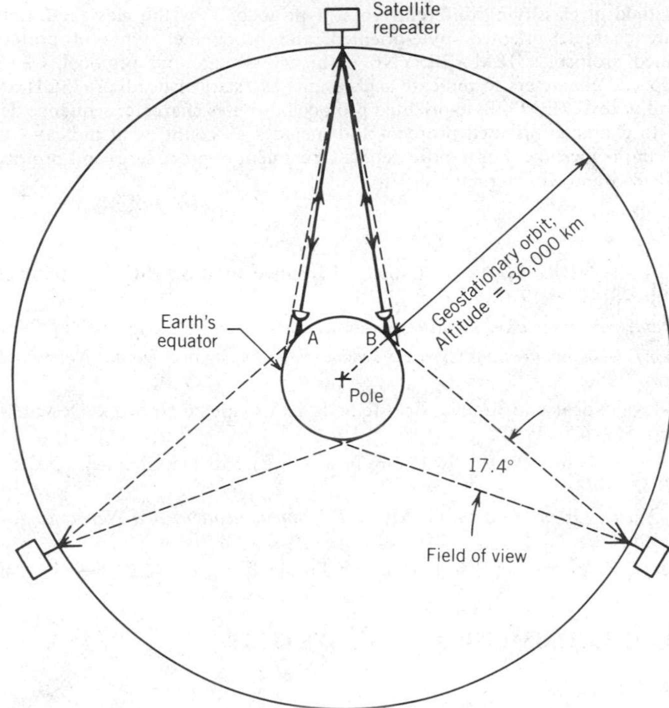

Fig. 49.47 Communications satellite is a radio-frequency repeater in orbit. Three geostationary satellites can cover most of the earth's surface.

them. J. R. Pierce later described passive and active satellite concepts for signal relay.[2,3] The launch of SPUTNIK by the Soviet Union on October 4, 1957, and of Explorer I by the United States on January 1, 1958, stimulated space activity. On August 12, 1960, Project ECHO resulted in the launch of a 30-m diameter metallized balloon into an earth orbit with an apogee (highest altitude) of 1688 km and an orbital period of approximately two hours. This passive reflector could relay signals among widely separated earth antennas as long as it was in their common field of view. On July 10, 1962, TELSTAR, an active repeater satellite built by AT&T's Bell Telephone Laboratories, was launched by NASA's Thor Delta vehicle into an orbit with an apogee of 5640 km, a perigee (lowest altitude) of 949 km, a period of 158 minutes, and an inclination of 44.8° relative to the equatorial plane. It weighed 80 kg, was less than 1 m in diameter, and could receive at 6390 MHz and retransmit at 4170 MHz signals having a bandwidth of 50 MHz at a transmitter power level of 2 W. It carried the first live international TV transmission.[3]

The period of a geosynchronous (or synchronous) orbit is a multiple (usually equal to 1) of the earth's rotation period. If the orbit plane is inclined relative to the equatorial plane, the satellite's earth track, that is, the subsatellite point, traces a figure-eight pattern whose maximum north-south latitude excursion is equal to the orbit inclination in degrees. The ideal geostationary orbit is a special case of the geosynchronous orbit with an inclination of zero degrees. The subsatellite point is then stationary on the equator. SYNCOM I, built by Hughes Aircraft Company for NASA, was launched February 14, 1963, but because of a rocket failure did not achieve its intended orbit. SYNCOM II, launched on July 26, 1963, achieved a nearly perfect synchronous orbit inclined at 33° with apogee and perigee of 35,887 km (22,300 miles) and period of 23 hours, 55.9 minutes. It received up-link signals from the earth at 7360 MHz and transmitted to the ground at 1815 MHz (down link). SYNCOM III, launched on August 19, 1964, into a geostationary orbit, successfully relayed the eighteenth Olympiad opening ceremonies from Tokyo to California.[4,5]

The Communications Satellite Corporation (COMSAT) was incorporated on February 1, 1963, to establish satellite communications on a commercial basis. COMSAT was the U.S. signatory and the first manager of the International Telecommunications Satellite Organization (INTELSAT), formed on August 20, 1964. It originally represented 11 countries that would share the ownership of the space segment, consisting of the satellites and TTC&M facilities, and would share responsibility for launch arrangements for a worldwide system. Individual members of INTELSAT own their earth stations and collect revenues in their countries.

The Early Bird satellite (later named INTELSAT I), launched on June 28, 1965, demonstrated the commercial viability of a geostationary satellite system. Prior to that time, a system of medium-altitude (and therefore not stationary) satellites had been envisioned. These would require earth station antennas that could track the satellites from horizon to horizon and "hand over" from one satellite to another. However, because the geostationary system proved to be much more cost effective, most commercial systems now use geostationary satellites. The INTELSAT system has grown to the point where it carries a significant amount of international telecommunications traffic. As of November, 1984, INTELSAT had 109 member countries. Its space segment consisted of 13 satellites in geostationary orbit over the Atlantic, Indian, and Pacific Oceans. The earth segment included 834 operational antennas at 662 sites representing 172 users. The system provided 1318 communications pathways among earth stations carrying more than 35,000 voice and data circuits as well as TV traffic.

INTELSAT is not the only international satellite communications entity. In 1971, INTERSPUTNIK was formed to provide communications among a number of Eastern Bloc countries. As of 1982 there were 14 signatories to INTERSPUTNIK. Space segment capacity is leased from the U.S.S.R.'s GORIZONT geostationary satellite.[6]

The International Maritime Satellite Organization (INMARSAT) was formed in 1979 for international mobile communications (e.g., ships at sea).[7] Many regional and domestic satellite systems have also come into being with the result that, as of April 1983, there were more than 350 active, spare, or planned satellites for the geostationary orbit,[7] and the cumulative market for communications satellites as of the end of 1983 was more than $3000 million.[8]

System Overview

Figure 49.48 illustrates the essential features of a microwave repeater satellite system. An end user signal, such as a telephone call, enters an earth station via terrestrial facilities. There it may be combined with other signals, and these signals modulate an up-link microwave radio-frequency carrier, for example, at 6 GHz. Table 49.10 depicts the prevalent frequency bands allocated for satellite communications.[9] The factors affecting the choice of radio frequency will be discussed later. For the "C-band" satellite illustrated in Fig. 49.48, the up-link signal in the frequency band 5.925 to 6.425 GHz is received, amplified, and down converted (linear translation) to the down-link band of 3.7 to 4.2 GHz. In the frequency division multiple access (FDMA) satellite configuration depicted

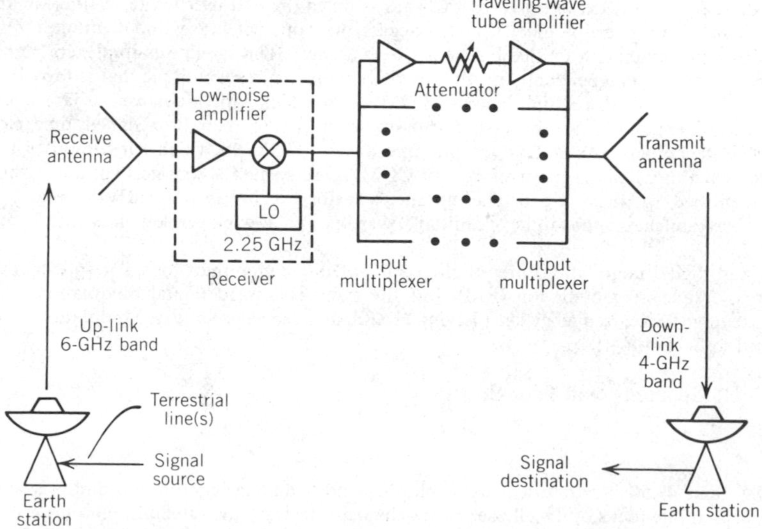

Fig. 49.48 "C-Band" frequency division multiple access (FDMA) satellite receives 6-GHz up-link signals and transmits 4-GHz down-link signals.

TABLE 49.10. PARTIAL LIST OF SATELLITE COMMUNICATIONS FREQUENCY BANDS

Down-Link Frequency (GHz)	Up-Link Frequency (GHz)	Band	Typical Uses and Systems[a]
1.535–1.5425	1.635–1.645	L	Maritime mobile, INMARSAT, MARISAT
2.5–2.655	2.655–2.69	S	FSS, mobile INSAT, ARABSAT (broadcast-community reception)
3.4–4.8	5.85–7.075		FSS, INTELSAT, most domestic satellites
3.7–4.2	5.925–6.425	C	(pre-WARC 1979)
4.5–4.8			
2.25–7.75	7.9–8.4	X	Mobile, military
10.7–12.7	12.75–13.25		FSS, INTELSAT, SBS
10.95–11.2		K_u	INTELSAT (pre-WARC 1979), SBS, ANIK
11.45–11.7			
11.7–12.2	14.0–14.5		
12.2–12.7	17.3–17.7		Broadcast
17.7–21.2	27.0–31.0	K_a	FSS, ACTS, INTELSAT, JAPAN experimental
40.0–41.0	50.0–51.0	V	FSS
41.0–43.0		Q	Broadcast
54.25–58.2			Intersatellite link
59.0–64.0			

[a]INMARSAT, International Maritime Satellite Organization; MARISAT, Maritime Satellite; FSS, Fixed Satellite Service; WARC, World Administrative Radio Conference; INTELSAT, International Telecommunications Satellite Organization; SBS, Satellite Business Systems; ACTS, Advanced Communications Technology Satellite.

here, the input multiplexer separates the signals by frequency into individual transponder channels where they are amplified to a level appropriate for down-link transmission. The output multiplexer combines the individual channels onto a common output waveguide for transmission by the down-link antenna.

At the receiving earth station, the signal is amplified by a low-noise receiver, down-converted to an intermediate frequency, demultiplexed, and passed on to the end user via terrestrial facilities.

The FDMA system permits many users to access the same satellite without undue interference with each other by assigning a unique frequency to each user. The satellite multiplexers permit each transponder to perform independently of the others. Another access technique that is rapidly gaining prominence is time-division multiple access (TDMA), in which individual users share a common frequency assignment but transmit in short bursts according to their preassigned time slots. Yet another access technique is code-division multiple access (CDMA), which superimposes a unique waveform on each user's message waveform. In CDMA, users share a common band and individual users are identified by decoding of the unique waveform. This is a form of spread-spectrum communications and has application to military systems and low-power low-data-rate commercial systems.

Analog and digital modulations are employed in satellite communications. The prevalent analog modulation is frequency modulation (FM), but the trend is toward digital modulations including phase shift keying (PSK), frequency shift keying (FSK), and their derivatives. Modulation techniques are discussed in Section 48.1.

49.4-2 Orbit Geometry and Launch

Orbits

Figures 49.49 and 49.50 illustrate a general elliptical orbit and geocentric coordinate system. The earth's center is at one focus of the ellipse, F. Six elements describe the satellite's motion: eccentricity, e; length of semimajor axis, a; inclination, i; right ascension of the ascending node, Ω; angle (in orbit plane) of perigee axis, ω; and eccentric anomaly (i.e., angle), E. The true anomaly, v, is the actual angle from perigee axis to the satellite. The mean anomaly, $M = E - e \sin E$, is the angle that would

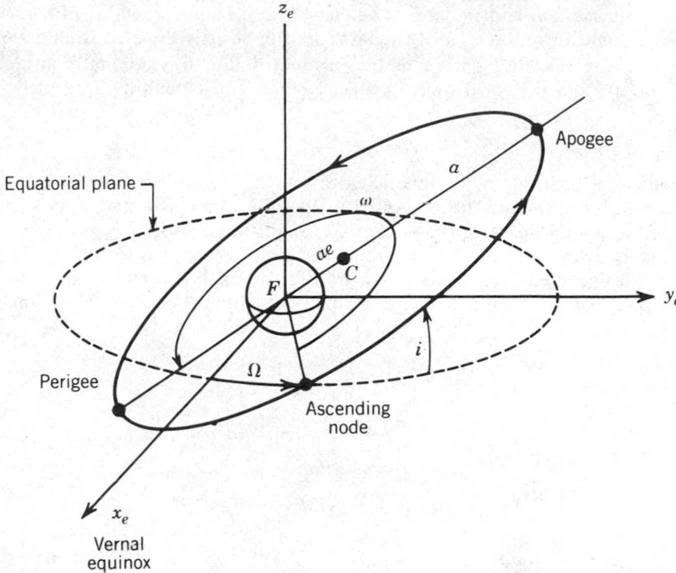

Fig. 49.49 Elliptical orbit in geocentric coordinate system. The center of the earth is at one focus of the ellipse, *F*. The center of the ellipse is at *C*.

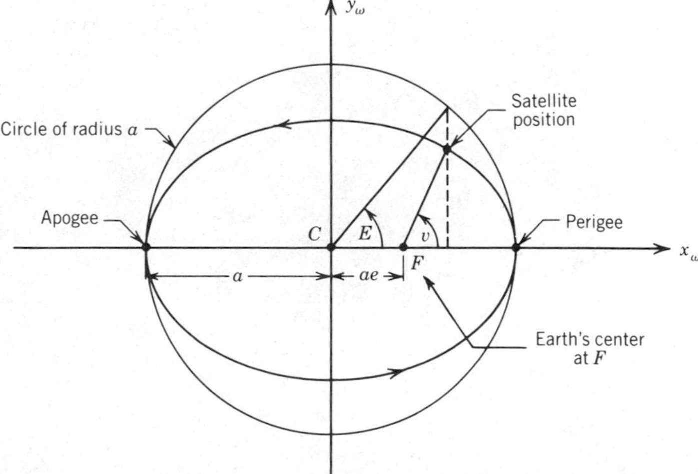

Fig. 49.50 Orbit plane (x_ω, y_ω) depicting the true anomaly v and the eccentric anomaly E.

be traversed if the satellite were moving at its mean angular velocity. Figure 49.50 illustrates the true anomaly v and the mean anomaly E. Miya,[10] Baker and Makenson,[11] and Herrick[12] give formulas for calculating satellite position from the orbital elements. The right ascension and declination (angle north from equatorial plane) of the satellite can be determined from the elements, and then related through standard trigonometric formulas to the range distance and azimuth and elevation angles from specific earth locations.[13,14] However, because of the nonideal nature of orbits, the practical problem of pointing an earth station to the satellite is more complicated and may require semiempirical formulas.[15]

Most quantities of interest for satellite communications can be obtained by considering the special case of circular orbits with zero inclination. Then $e = 0$, $i = 0$, and $M = E = v$.

A satellite having mass m and moving at velocity v in a circular path at radius r (from earth's center) experiences centrifugal force mv^2/r equal to and oppositely directed from gravitational force GMm/r^2, where $GM = 398{,}601.2\ \text{km}^3/\text{s}^2$ is the product of the gravitational constant, G, and the mass of the earth, M, and is known more accurately than either quantity separately. Equating the forces yields

$$v = (GM/r)^{1/2} = r(2\pi/T)$$

where the period, T, of the orbit is the time taken to traverse 2π radians. For an orbit period of one sidereal day, $T = 86{,}164.091$ s and the geosynchronous orbit radius is found to be 42,164.2 km. The radius normalized to that of the earth ($r_e = 6378.153$ km at the equator) is $r_0 = r/r_e = 6.61072$. Table 49.11 summarizes the relevant quantities for circular orbits.

For some orbit calculations, for example, for transfer orbits, it is useful to know the velocity and period of an elliptical orbit. For a satellite at radius r, in elliptical orbit with semimajor axis a, the velocity is[16]

$$v = \left[2GM\left(\frac{1}{r} - \frac{1}{2a}\right)\right]^{\frac{1}{2}} \tag{49.1}$$

and the period is

$$T = \frac{2\pi a^{\frac{3}{2}}}{\sqrt{GM}} \tag{49.2}$$

Geometric Relationships and Pointing Angles

Figure 49.51 depicts the satellite geometry relative to an earth station at point p having latitude ψ and east longitude l_e. For $i = 0$, the orbit normal is the $+z_e$ direction. The satellite is on the x_e axis at normalized distance $r_0 = 6.6107$. The quantity λ is equal to $l_e - l_s$, where l_s is the east longitude of the subsatellite point (directly beneath the satellite). The arc γ is the spherical angle between the subsatellite point and p. The vector relations between the earth-centered coordinates (x_e, y_e, z_e) and spacecraft-centered coordinates (x_s, y_s, z_s) yield the normalized distance S, and pointing angles *from* the satellite axis *to* an earth station:

$$S = |\mathbf{r}_s| = \sqrt{(1 + r_0^2 - 2r_0\cos\psi\cos\lambda)} \tag{49.3}$$

$$\cos\theta_s = \frac{r_0 - \cos\psi\cos\lambda}{S} \tag{49.4}$$

$$\theta_s = \tan^{-1}\frac{\sqrt{(1 - \cos^2\psi\cos^2\lambda)}}{(r_0 - \cos\psi\cos\lambda)} \tag{49.5}$$

$$\phi_s = \tan^{-1}\frac{\sin\lambda}{\tan\psi} \tag{49.6}$$

$$\alpha_s = \tan^{-1}\frac{\cos\psi\sin\lambda}{(r_0 - \cos\psi\cos\lambda)} \tag{49.7a}$$

$$\alpha_s \approx \theta_s\sin\phi_s \tag{49.7b}$$

$$\beta_s = \sin^{-1}\frac{\sin\psi}{S} \tag{49.8a}$$

$$\beta_s \approx \theta_s\cos\phi_s \tag{49.8b}$$

where (θ_s, ϕ_s) are the satellite coordinate polar angles and (α_s, β_s) are "azimuth" (east-west) and "elevation" (north-south) pointing angles from the satellite to the earth location.

Figure 49.52 illustrates the earth station antenna's azimuth and elevation-pointing angles to the satellite (α_e, β_e). The azimuth angle, α_e, is the angle clockwise from north that the great circle arc, γ, makes with the local meridian. To determine α_e, first compute the auxiliary quantity, A:

$$A = \tan^{-1}\frac{\tan\lambda}{\sin\psi}$$

The azimuth angle is then found from Table 49.12. The elevation angle, β_e, measured from the local horizon, is $\beta_e = 90° - \gamma - \theta_s$, where $\cos\gamma = \cos\psi\cos\lambda$. This can be shown to be

$$\beta_e = \cos^{-1}(r_0\sin\theta_s) \tag{49.9}$$

TABLE 49.11. QUANTITIES FOR CIRCULAR GEOSTATIONARY ORBITS

Parameter	Measurement
Earth's radius at equator, r_e	$r_e = 6378.153$ km
Mass of earth, M	$M = 5.9734 \times 10^{24}$ kg
Gravitational constant, G	$G = 6.673 \times 10^{-20}$ km^3/kg · s^2
$G \times$ mass of earth, GM	$GM = 3.986012 \times 10^{14}$ m^3/s^2
Orbit radius (equatorial plane), r	$r = 42164.2$ km
Orbit radius/earth radius, r/r_e	$r/r_e = r_0 = 6.61072$
Height of orbit from equator, h	$h = 35786.04$ km $= 5.61072$ earth radii
Velocity, $v = \sqrt{GM/r}$	$v = 3.074662$ km/s
Period, $T = 2\pi r^{3/2}/(GM)$	$T = 86164.091$ s
Acceleration, GM/r^2	$GM/r^2 = 0.2242079 \times 10^{-3}$ km/s^2
Angular velocity, v/r	$v/r = 72.92115 \times 10^{-6}$ rad/s

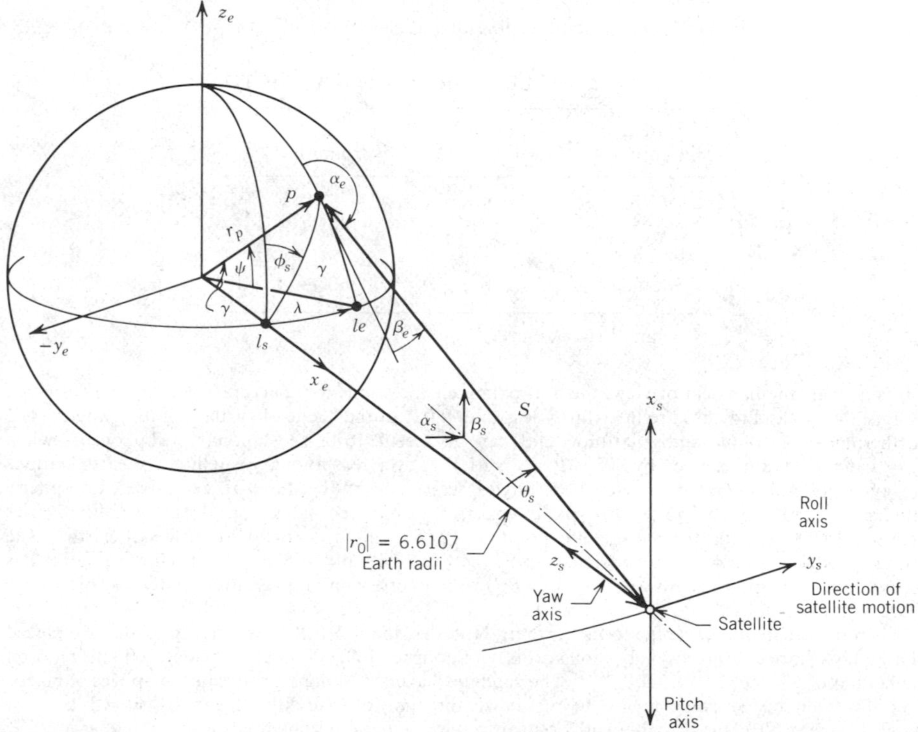

Fig. 49.51 Geostationary orbit geometry. $\hat{r}_p = \cos\varphi \cos\lambda \hat{x}_e + \cos\varphi \sin\lambda \hat{y}_e + \sin\varphi \hat{z}_e$. $r_0 = 6.6107\hat{x}_e$. $\mathbf{r_s} = \mathbf{r_p} - \mathbf{r_0}$.

Figure 49.53 provides a graphic means to determine earth station azimuth and elevation.

If the earth station must be pointed along the horizon to access the satellite, its elevation angle, β_e, is zero. For this case a ray from the satellite is tangent to the earth's surface and the polar angle, θ_s, from the satellite axis is

$$\theta_s = \theta_{s,\max} = \sin^{-1}(1/r_0) = 8.7°$$

This limits the view of the earth from geostationary orbit to a full cone angle of 17.4°. Because

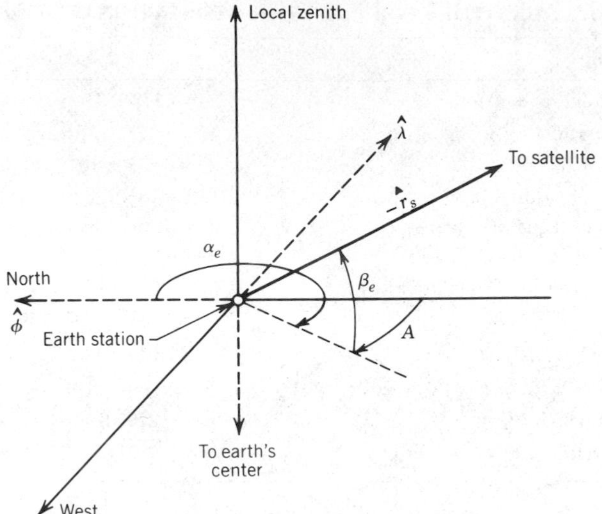

Fig. 49.52 Earth station azimuth and elevation angles (α_e, β_e).

TABLE 49.12. RULES FOR FINDING AZIMUTH

Station Latitude	Sign of λ ($\lambda = l_e - l_s$)	Azimuth (°)
North	+	$180° + A^a$
North	−	$180° + A$
South	−	A
South	+	$360° + A$

$^a A = \tan^{-1} (\tan \lambda / \sin \psi)$.

this is a small angular field of view, the approximate expressions for spacecraft pointing angles given by Eqs. (49.7b) and (49.8b) are in error by less than 0.02° at the "edge of earth" pointing angles. The northernmost or southernmost latitude that can be "seen" from geostationary orbit occurs when $\beta_e = 0$ and $\lambda = 0$ and is given by $90° - \theta_{s,\max} = 81.3°$. For a geostationary satellite, the earth stations are typically not operated at elevation angles less than 5° because of excessive atmospheric attenuation and thermal noise. The path length, Sr_e, between satellite and earth station varies between 36,000 km for the subsatellite point to 41,680 km for a point at "edge of earth." The round-trip delay time for a radio signal varies between 240 and 278 ms. If an elliptical orbit has eccentricity, e, the path length will vary $\pm e(Sr_e)$ with a corresponding variation in delay. This affects TDMA transmissions.

To provide satellite coverage to its northern latitudes, the U.S.S.R.'s Molniya satellites are placed in a highly elliptical nongeosynchronous orbit with apogee of 40,000 km, perigee of 500 km, an orbit inclination of 63 to 65°.[17,18(Report 207-5)] The satellites have a 12-hour period and the apogee is placed over the intended coverage region to maximize the time of visibility. It can be shown that for $i = 63.4°$, the orbit plane is stable with respect to the gravitational potential of the oblate earth.[10]

Orbit Variations

Orbit variations are caused by initial orbital errors, the oblateness of the earth, the gravitational pull of the sun and moon, and perturbations caused by correction maneuvers. An orbit plane inclination of $i°$ causes the subsatellite point to trace a figure eight every 24 hours with maximum declination (measured from earth's center) equal to the inclination and the angular "width" of the figure eight (i.e., the right ascension variation) of approximately $i^2/229°$.[13,14] An earth station antenna on the earth's surface would see similar variations, but the exact angles would be slightly different from the right ascension and declination because of parallax relative to earth's center. It could be deduced from Eq. (49.2) that a reduction of 0.78 km in the radius of the geostationary orbit from ideal would

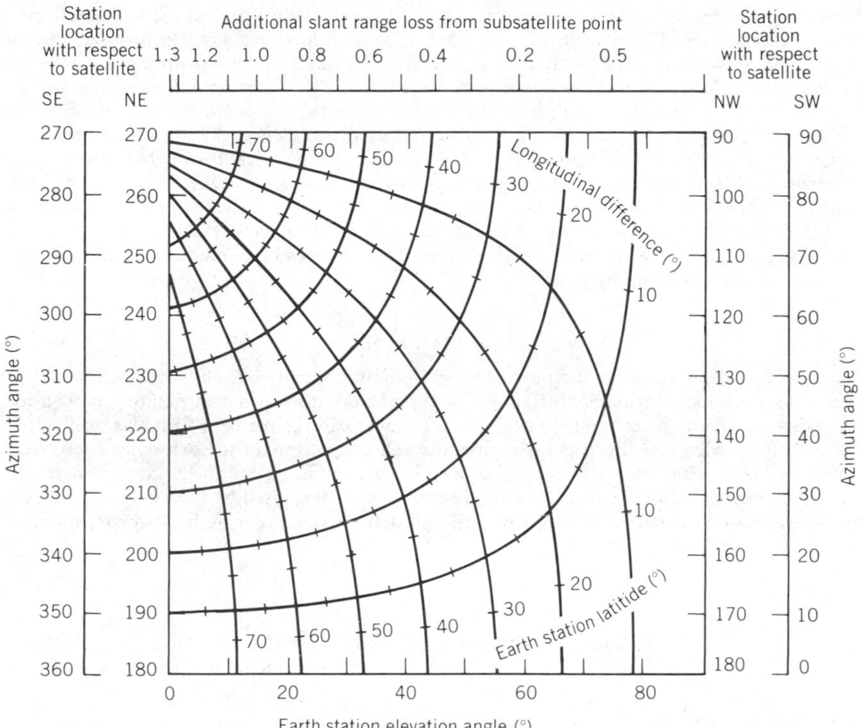

Fig. 49.53 Azimuth and elevation angles from an earth station to a geostationary satellite. Courtesy of Communications Satellite Corp.

result in an eastward drift of 0.01° per day in longitude. The gravitational pulls of the sun and moon affect the radius but primarily disturb the orbit plane inclination. These effects occur over an 18-year period owing to lunar precession, in addition to a 6-month seasonal variation and a 14-day period because the moon's orbit is not in the equatorial plane. In the 1970 to 1988 time frame, the orbit inclination change varies between 0.73° and 0.93°/year with mean of approximately 0.85°/year.[19]

Variations in longitude and inclination owing to all causes require that small thrusters on board the satellite be periodically fired to maintain the longitude and orientation of the orbit plane. Correction of longitude is referred to as east-west stationkeeping, and inclination correction is called north-south stationkeeping. The thruster firings to control inclination consume the most fuel on board the satellite. Orbit corrections require changes to the velocity vector by the *vector* addition of a velocity increment, ΔV ("delta-vee"). For satellite mass m, velocity increment dv, and thruster fuel relative exhaust velocity v_e directed opposite to the satellite velocity vector, conservation of momentum requires $mdv = -v_e dm$. For constant v_e, the relative mass of fuel used to effect a velocity increment ΔV m/s is

$$\frac{\Delta m}{m} = 1 - e^{(\Delta V/gI_{sp})} \tag{49.10}$$

where $g = 9.8$ m/s^2 is the sea level gravitational acceleration and $I_{sp} = v_e/g$ is the specific impulse of the thruster, having units of seconds. I_{sp} depends on the chemicals used for the fuel and has typical values of 220 to 300 s for hydrazine and 290 to 450 s for liquid bipropellants. It can take on very large values (e.g., 2500 s, although at low thrust) for ion propulsion.[20] Typical values of ΔV are: $\Delta V \approx 54$ m/s per degree of change in inclination, and $\Delta V \approx 2.8$ m/s per change of 1°/day in longitude drift rate. Typical ΔV requirements for east-west stationkeeping are 3 to 6 m/s/yr, and for north-south 50 m/s/yr.[21] Typically, east-west and north-south stationkeeping for satellites is maintained to ±0.1°.

Attitude

Figure 49.51 depicts the satellite roll, R, pitch, P, and yaw, Y, axes. The roll axis points "east" in the direction of motion, the yaw axis points "down" toward the earth's center, and the pitch axis is directed "south" parallel to the orbit normal. These axes are rigidly fixed to the spacecraft body. In the ideal case they are aligned with y_s, $-x_s$, and z_s axes, respectively, but in practice they depart from these axes by small amounts. If the satellite antenna beam directions are fixed relative to the body, the antenna beam-pointing direction will vary. Figure 49.54 depicts one quadrant of the earth in units of satellite-pointing angle linearized by using the small angle approximations, Eqs. (49.7b) and (49.8b), which give satellite-pointing angles azimuth α_s and elevation β_s to a point p, which will appear to move to point p_3 owing to R, P, and Y errors. A roll error causes the beam to move in the north-south direction; a pitch error causes an east-west movement. A yaw error produces a rotation around the subsatellite points, for example, the electric field of a linearly polarized wave directed toward the subsatellite point would be rotated by the amount of the yaw error.

For small errors in R, P, Y, expressed in degrees, the changes in azimuth and elevation-pointing angles may be approximated by

$$|\Delta\alpha| \approx P + (Y/57.3)\,\beta_s \tag{49.11}$$

$$|\Delta\beta| \approx R + (Y/57.3)\alpha_s \tag{49.12}$$

The location of p_3 must be computed for all permutations of $\pm R$, $\pm P$, $\pm Y$, and the locus of points defines the apparent motion of p. Where extreme accuracy in beam pointing is required, the exact excursions should be computed using standard spherical trigonometry formulas. In practice, the errors are not independent. For example, yaw and roll errors convert to each other every six hours (one fourth of the orbital period). The errors have a deterministic part and a random part, and the exact statistics depend on the particular attitude-control system utilized by the satellite. These attitude errors necessitate that antenna coverages be appropriately enlarged to include all apparent motions of the earth coverage locations.

Attitude Control

The two predominant stabilization methods are spin stabilization and three-axis stabilization (Fig. 49.55). A spin-stabilized spacecraft uses a spinning drum in which the spin and pitch axes are

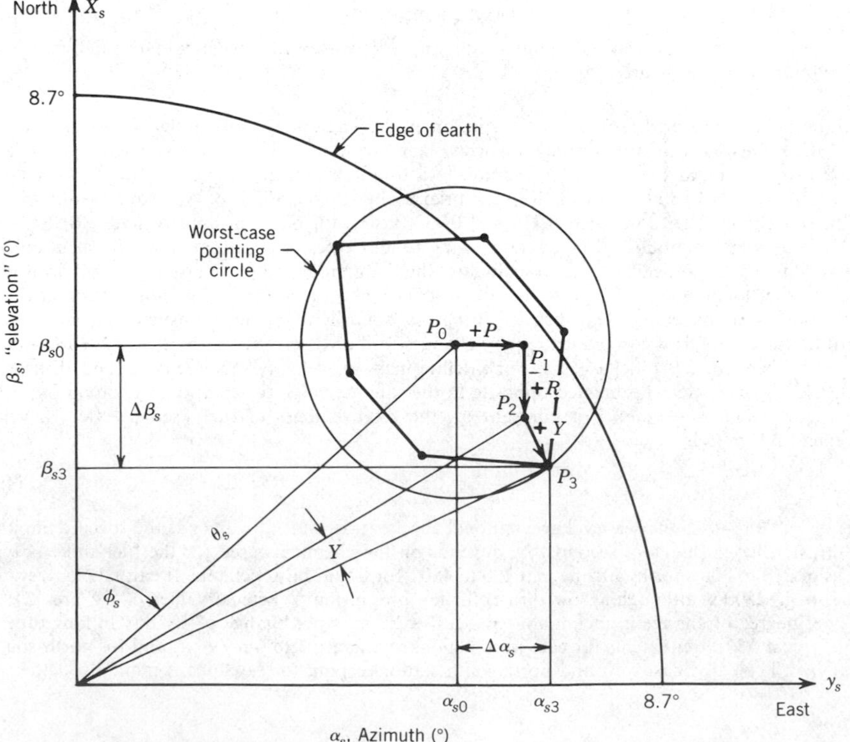

Fig. 49.54 Angular map of earth seen from satellite. Attitude errors are exaggerated to illustrate apparent motion of an earth point p: From p_0 to p_3 given (+) pitch followed by (+) roll followed by (+) yaw. The polygon vertices correspond to other permutations.

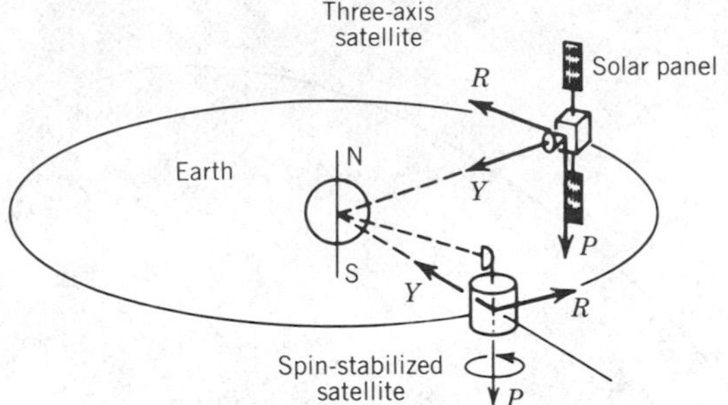

Fig. 49.55 Spin-stabilized and three-axis stabilized satellites. For the "spinner," the communications antennas must be despun to face the earth.

aligned; the ratio of the moment of inertia about this axis to that about either of the other axes is typically between 1.1 and 1.3, and the spin rate is approximately 100 r/min.[16] The "spinner" has solar cells around the surface of the drum and the antennas must be despun to face the earth. In the Hughes "despun platform" class of spinners, such as the INTELSAT VI satellite, the entire RF package is despun relative to the drum that contains the other satellite subsystems.

Three-axis satellites maintain attitude by exchanging torques among a number of spinning momentum wheels inside the satellite body. The solar cells are usually in the form of panels extending out the north and south faces. These panels must rotate to face the sun as the satellite proceeds in its orbit. The satellite must have a programmed pitch rotation to maintain the antennas pointed to the earth as the satellite moves around the earth.

For both types of satellites, infrared sensors are used to detect the limb of the earth against the cold sky. In addition, sun sensors and rate-integrating gyroscopes are used to refine the attitude determination. The state of the art is continually improving with respect to attitude control, and there is an ongoing debate as to the relative merits of spin vs. three-axis stabilization.[22] State-of-the-art attitude control systems permit roll and pitch to be maintained at $\pm 0.1°$ and yaw to be maintained at less than 0.25° for both kinds of systems. In a spinner, pitch errors can be corrected by momentum exchange without fuel being consumed, but roll and yaw corrections require that thrusters be fired. For a three-axis satellite, the number and orientation of momentum wheels determine whether thrusters must be fired.

Launch Vehicles and Launch Sequence

Table 49.13 summarizes some of the capabilities of several launch vehicles. The most prominent of the expendable vehicles for the 1980s are the Delta[23] and Ariane[24] rockets. Each is available with a

TABLE 49.13. LAUNCHERS

			Payload Bay Envelope (mm)	
Launcher[a]	Weight to Transfer Orbit (kg)	Weight to Geostationary Orbit (kg)	Diameter	Height
Ariane 4 (ESA)	1852–4135	1080–2410	3650	8600
			3650	9600
Atlas-Centaur	2358	1093	2820	8540
Delta 3910/PAM	1111	600	2184	4368
Delta 3920/PAM	1247	680	2184	4368
N-II (Japan)		350	2200	
STS/PAM-D	1247	665	2184	2543
			2921	2543
STS/PAM-DII		983	2184	2543
			2921	2543

[a] ESA, European Space Agency; PAM, Payload Assist Module; STS, space transportation system.

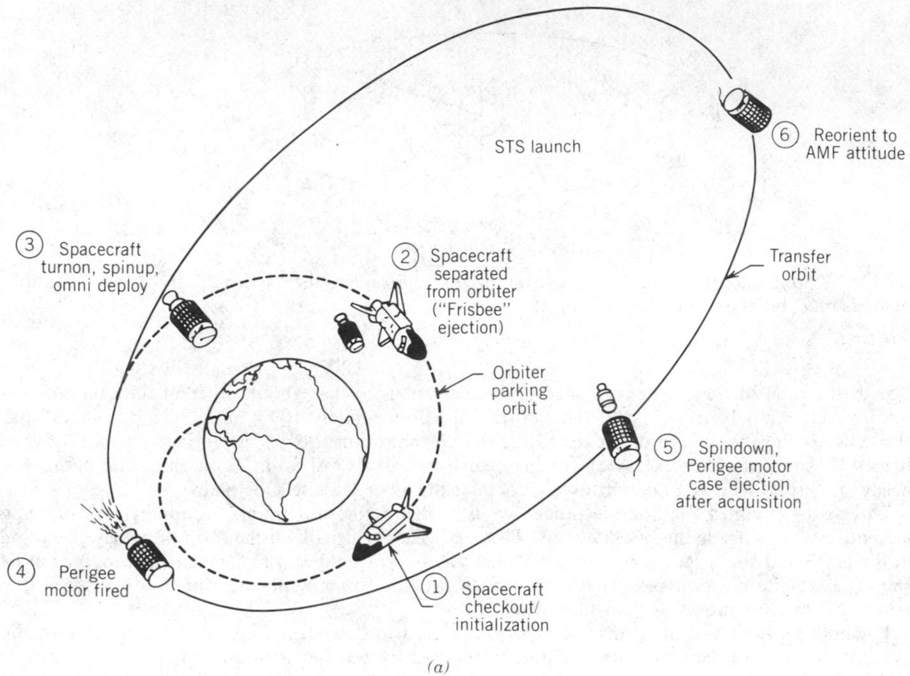

(a)

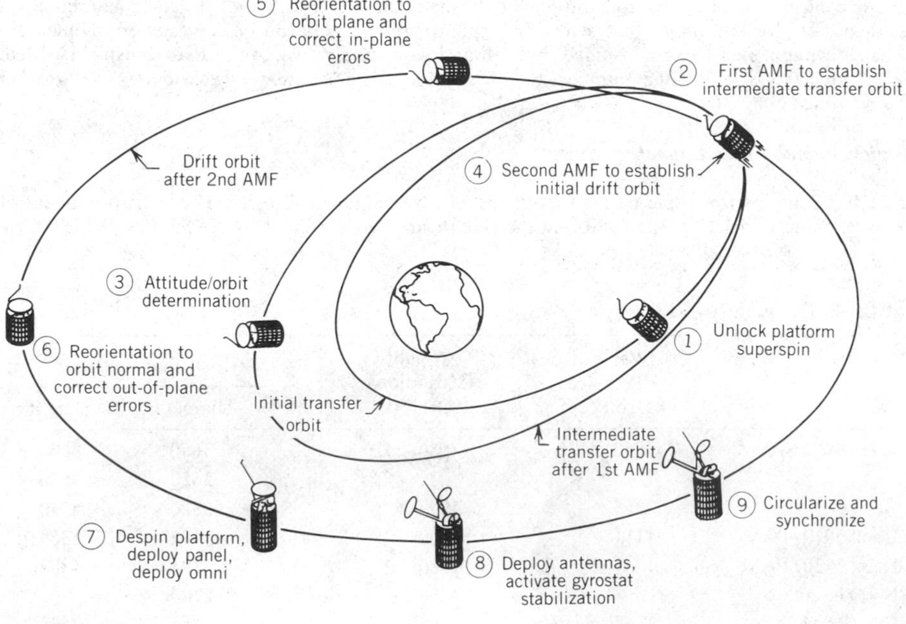

(b)

Fig. 49.56 INTELSAT VI STS (space transportation system) launch sequence: (*a*) Launch and transfer orbit, (*b*) transfer orbit to geosynchronous orbit. Courtesy of Hughes Aircraft Co.

variety of options. The NASA Space Transportation System (STS) is a reusable vehicle that places a spacecraft into a circular orbit of altitude 296 km and inclination 28.5°.[25] Figure 49.56 illustrates the sequence for the INTELSAT VI satellite. The STS is launched into low earth orbit. The spacecraft is ejected from the shuttle after having been mechanically spun. After the spacecraft drifts away from the STS, it is oriented and the perigee motor is fired to put the spacecraft into the transfer orbit, which is an elliptical orbit of altitude 296 × 35,786 km and inclination 27°. At apogee, the apogee kick motor (AKM) is fired to circularize the orbit and to apply sufficient ΔV to bring inclination to 0°. From Eq. (49.1) and Table 49.2, with $r = r_e + 35{,}786 = 42{,}164$ km and $a = (42{,}164 + r_e + 296)/2 = 24{,}419$ km, it is easy to show that the velocity at apogee is 1608 m/s. Since geostationary velocity is 3074 m/s (Table 49.11), the required $\Delta V = 1796$ m/s, obtained by using the cosine law to account for the vector addition of velocities and inclination change of 27° to achieve geostationary orbit. Generally the net satellite payload weight deliverable to geostationary orbit is about 58% of the weight delivered to transfer orbit. Costs for launches vary with the launch system and depend on the year of launch. The STS can place up to 29,484 kg into its 296-km parking orbit; its payload bay is 4.6 m diameter × 18.3 m long. Costs for shuttle launch are a function of the fraction of weight and bay length occupied.[25] Typically several users share the bay.

49.4-3 Communications Link Parameters

Link Equations

The starting point for satellite link calculations is the power transfer relationship between two antennas separated by distance R (Fig. 49.57). The following relations will be developed for a general link. They can then be specialized to the up-link and down-link cases. For a transmitting antenna having gain g_t (numerical ratio) and radiating p_t watts, the power flux density at distance R is

$$\psi = \frac{p_t g_t}{4\pi R^2} \cdot \frac{1}{l_i} \qquad \text{W/m}^2 \tag{49.13}$$

where l_i represents atmospheric path attenuation due to dissipative loss or scattering. If a receiving antenna having effective area A_r (not necessarily equal to its physical area) intercepts this flux, the received power is

$$p_r = \psi A_r \gamma \qquad \text{W}$$

where γ is the dimensionless polarization coupling factor ($\gamma \leqslant 1$). The received power is not explicitly dependent on frequency. That apparent dependence arises from the ratio of an antenna's effective area to its gain, which is a universal constant for all types of antennas and is given by

$$\frac{A}{g} = \frac{\lambda^2}{4\pi} \tag{49.14}$$

where λ is the wavelength. For $g = 1$, this ratio is the effective area of an isotropic antenna, A_1, that decreases as the square of frequency, since $\lambda = c/f$, where $c = 2.998 \times 10^8$ m/s is the velocity of light and f is the frequency in hertz. Its reciprocal, $4\pi/\lambda^2$, is the gain of an antenna with $A = 1$ m^2. For a passive reciprocal antenna, the gain, area, and radiation pattern do not depend on whether the antenna is transmitting or receiving.

The quality of a radio link depends on the ratio of received signal power, p_r, to the system noise power

$$n = kTb_{\text{RF}} \qquad \text{W}$$

where k = Boltzmann's constant = 1.38×10^{-23} J/K
b_{RF} = RF bandwidth, Hz
T = system noise temperature, K

Then

$$\frac{p_r}{n} = p_t g_t \cdot \frac{1}{4\pi R^2} \cdot \frac{\lambda^2}{4\pi} \cdot \frac{g_r}{T} \cdot \frac{1}{k} \cdot \frac{1}{b_{\text{RF}}} \cdot \gamma \cdot \frac{1}{l_i} \tag{49.15}$$

It is common to express all quantities in decibels. The carrier-to-noise ratio (C/N) is then

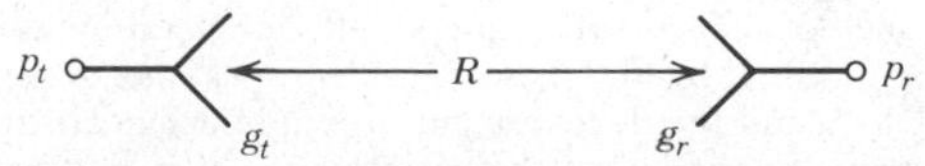

p_t — g_t ← R → g_r — p_r

Fig. 49.57 RF link between two antennas.

$C/N = 10\log(p_r/n)$:

$$C/N = \text{EIRP} - L_s + A_1 + G_r/T + 228.6 - B + \Gamma - L_i \quad \text{dB} \tag{49.16}$$

where each term in Eq. (49.16) is the expression in decibels, that is, 10 log( ), of the corresponding parenthesized term. The term EIRP (dBW) is called the equivalent isotropically radiated power. It is the decibel sum of the antenna gain, $10\log(g_t)$ dBi (decibels relative to isotropic) and the transmit power $10\log(p_t)$ dBW. L_s (dB · m^2) is the free space spreading factor; A_1 (dB · m^2) is the area of an isotropic antenna; G_r/T (dB/K) is a figure of merit for the receive system; $-10\log$ (k) = 228.6 dBW/K · Hz; B (dBHz) is the RF bandwidth; Γ (dB) is the polarization factor; L_i (dB) is the path attenuation.

For a geostationary orbit, the range distance, $R = Sr_e$, is between 36,000 and 41,680 km, and the corresponding spreading factor, L_s, ranges between 162.1 and 163.4 dB · m^2, with 163 dB · m^2 usually taken as typical for link calculations. Sometimes L_s and A_1 are combined and called the free space loss, a term that depends on frequency. At 4 GHz, for example, $\lambda = 0.075$ m so that $A_1 = -33.5$ dB · m^2 and the free space loss is $L_s - A_1 = 196.5$ dB. The RF bandwidth B depends on the modulation parameters. To permit link evaluation independent of this parameter, it is common to normalize Eq. (49.16) to unit bandwidth. The result is the carrier-to-noise-density ratio,

$$C/N_0 = (C/N) + 10\log B \quad \text{dBHz}$$

For digital modulation, the quantity of interest is the dimensionless ratio of energy per bit to the noise power density:

$$E_b/N_0 = (C/N_0) + 10\log(t_b) \quad \text{dB} \tag{49.17}$$

where t_b is the time duration of 1 bit in seconds and $1/t_b$ is the transmission bit rate in bits per second.

Since most parameters of a link such as frequency and power level are fixed by system constraints, the ratio (G/T), actually calculated as $10\log(g_r) - 10\log(T)$, is a figure of merit determining the link quality. For an antenna connected to a receiver, the system noise temperature, T, is

$$T = T_a + (l_g - 1)T_g + l_g T_r \quad \text{K} \tag{49.18}$$

where T_a is the antenna noise temperature, $l_g (\geqslant 1)$ and T_g are the loss factor and physical temperature, respectively, of the nonradiating parts of the feed system and connecting waveguide, and T_r is the equivalent noise temperature of the receive system:

$$T_r = T_0 \,(\text{nf} - 1) \tag{49.19}$$

Here $T_0 = 290$ K and nf is the receiver noise figure expressed as a power ratio. The antenna temperature results from noise power received from such sources as the earth, the sky, the sun, and the galaxy.

The foregoing applies to any link. For complete evaluation of an earth-station-to-earth-station path via satellite, the preceding equations must be applied separately to the up-link and the down link and then the results must be properly combined.

The satellite transponder gain is a fixed value designed to drive the output amplifier to a specified saturation output power for a specified value of up-link "saturation flux density." Some satellites have commandable attenuators in the signal path, as shown in Fig. 49.48, to permit several possible values of saturation flux density to be chosen to suppress up-link noise. The output amplifiers, whether they are traveling-wave tubes (TWTs) or solid-state power amplifiers (SSPAs), do not have a linear input-output transfer characteristic near saturation. Figure 49.58 illustrates a typical transfer curve applicable to many TWTs regardless of the absolute output power. A reduction in input drive level, called the input backoff, BO_i, results in a generally different value of reduction in output power, called the output backoff, BO_o. In the figure, a 19-dB input backoff yields a 12-dB output backoff. Nonlinearities produce intermodulation products when two or more carriers are present, with the third-order intermodulation product predominant. This increases the noise level and degrades overall C/N_0 for the link. The definition of "linear" operation is somewhat arbitrary but is usually specified in terms of a given ratio of carrier-to-intermodulation noise power, $(C/N)_{\text{im}}$. For a given bandwidth, a typical range of acceptable values for $(C/N)_{\text{im}}$ is 20 to 30 dB.

The drive level strongly affects the efficiency of transponder operation. Figure 49.58 implies that to achieve $(C/N)_{\text{im}} > 20$ dB for two equal carriers, $\text{BO}_i \approx 13$ dB and $\text{BO}_o \approx 7.5$ dB. A 20-W TWT, for example, would then be supplying only 3.5 W total (7.5 dB below 20 W), or 1.75 W per carrier. A tube with DC-to-RF efficiency of 35% at saturation would be operating at an efficiency of only 6%, since the DC power consumption is fixed. These considerations encourage the use of single-carrier access schemes such as TDMA and the development of amplifier linearization techniques. It should be noted that backoff is generally applied to the earth station transmitters also. If the station must

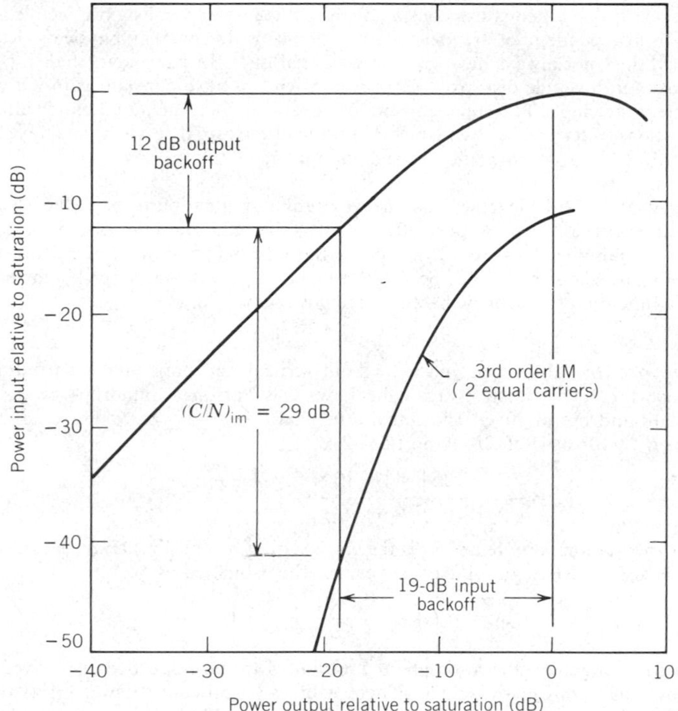

Fig. 49.58 Typical traveling-wave tube (TWT) power transfer response is not linear near saturation. IM, intermodulation.

transmit many carriers over the entire satellite bandwidth, the backoff can be more substantial than for the satellite.

The up-link flux density, the transponder transfer gain, the output amplifier saturated output power, the output losses, and the satellite down-link antenna gain determine the down-link EIRP. The down-link range distance, path attenuation, and earth station figure of merit (G_{re}/T_{re}) are then incorporated into Eq. (49.16) to calculate $(C/N_0)_d$, the down-link carrier-to-noise density. The overall link $(C/N_0)_t$ is related to the individual contributions by

$$\left(\frac{C}{N_0}\right)_t = \left[\left(\frac{C}{N_0}\right)_u^{-1} + \left(\frac{C}{N_0}\right)_{\text{im}}^{-1} + \left(\frac{C}{N_0}\right)_d^{-1}\right]^{-1} \tag{49.20}$$

It might appear that the overall C/N_0 can be increased simply by maximizing the earth station EIRP; however, Eq. (49.20) indicates that, for multiple carriers, there is an "optimum" value for the satellite TWT backoff to maximize $(C/N_0)_t$. This limits the maximum attainable $(C/N_0)_t$ and hence the communications capacity of the transponder.[26] The capacity also depends on the modulation and access method. Examples of link calculations may be found in Sections 48.2 and 48.5 and in Refs. 27 and 28.

Link Performance Factors

Frequency. For a free space link, the total path loss is, from Eq. (49.15):

$$\frac{p_r}{p_t} = k_1 \frac{g_r g_t}{f^2}$$

where k_1 includes all other parameters. If both antennas have fixed gain vs. frequency, the loss increases as f^2, suggesting the use of the lowest practical frequency. If both have fixed aperture area, the numerator is proportional to f^4 and the loss decreases as f^2, suggesting the use of the highest practical frequency. If one antenna has constant gain and the other a constant effective area, the free space link loss is independent of frequency. An antenna with constant physical area, such as a reflector or array, does not always have gain proportional to f^2. In particular, shaped-beam satellite reflector antennas for coverage of a fixed earth region tend to have constant gain. Of course many other factors affect the choice of frequency band for a satellite link, including bandwidth, worldwide frequency allocations, interference, frequency sharing with terrestrial services, available power, state of the art in hardware components, and propagation factors.

Propagation. Figure 49.59 depicts the total one-way zenith attenuation through the atmosphere as a function of frequency (Ref. 18, Report 205-4). For other elevation angles, $5° \leqslant \beta_e \leqslant 90°$, the ordinate should be multiplied by cosec (β_e), which is the approximate ratio of the path length at β_e to that at zenith. Figure 49.60 shows the specific attenuation γ_R in decibels per kilometer for various rain rates R as a function of frequency, based on the power-law relation

$$\gamma_R = kR^{\alpha} \qquad \text{dB/km}$$

where the curves are for values of k and α for both vertical and horizontal polarizations obtained from CCIR Report 721-1,[28] which assumes the Laws and Parsons[29] raindrop-size distribution for oblate spheroidal raindrops at 20°C. The actual attenuation is $A = \gamma_R L_s$, where L_s is the slant path length given for $\beta_e \geqslant 10°$ as (Ref. 28, Report 564-2):

$$L_s = \frac{5.1 - 2.15 \log[1 + 10^{(\psi - 27)/5}] - h_0}{\sin \beta_e} \qquad \text{km} \tag{49.21}$$

where ψ is the earth station latitude and h_0 is the earth station height (km) above mean sea level. The attenuation exceeded for 0.01% of an average year can be estimated as

$$A_{0.01} = \gamma_R L_s \left[\frac{90}{90 + 4L_s \cos(\beta_e)} \right] \qquad \text{dB} \tag{49.22}$$

In addition to attenuation, the atmosphere contributes to the noise power received at an earth station. Contributions to this noise include galactic sources (significant below 1 GHz) and the signal absorption of water vapor, oxygen, and hydrometers. Figure 49.61 depicts the atmospheric noise

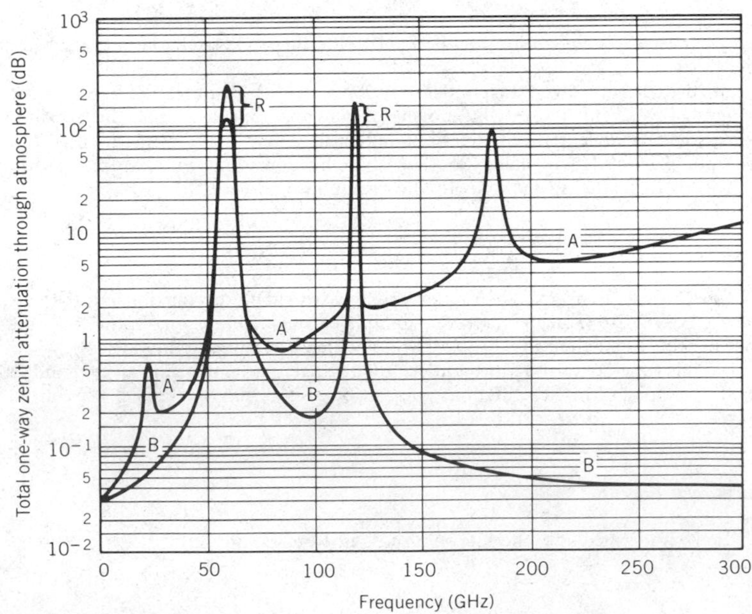

Fig. 49.59 Total one-way zenith attenuation through the atmosphere. Curve A, moderate humidity (moisture 7.5 g/m^3 at ground). Curve B, dry atmosphere (moisture 0 g/m^3). R, range of values due to fine structure. From Recommendations and Reports of the CCIR,[18] Report 208-4, with permission.

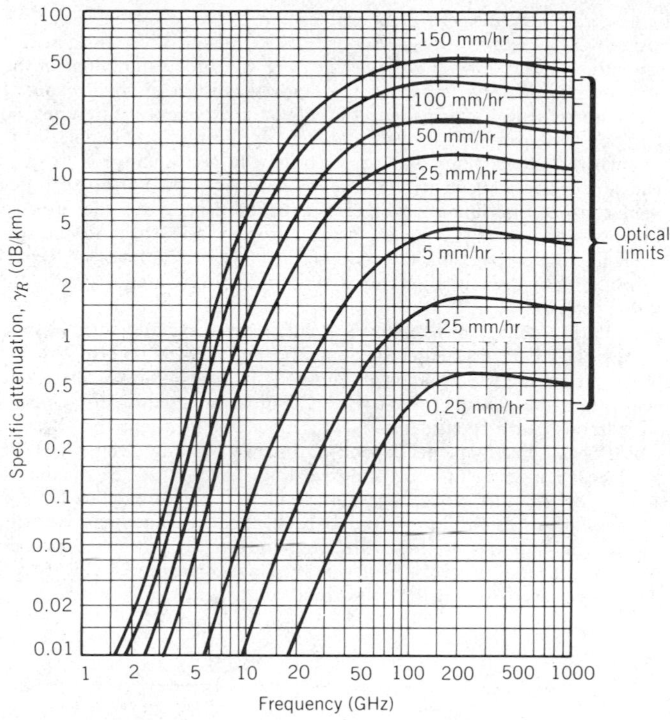

Fig. 49.60 Specific attenuation γ_R due to rainfall. Total attenuation $A = \gamma L_s$, where L_s = slant path length. From Recommendations and Reports of the CCIR,[28] Report 721-1, with permission.

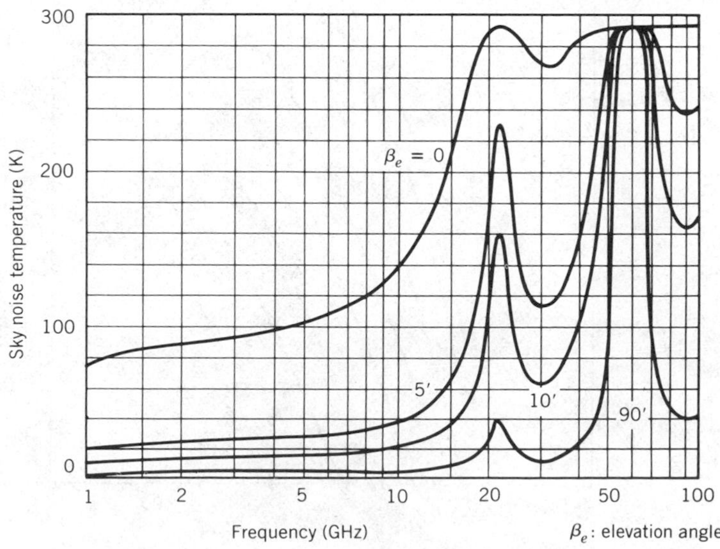

Fig. 49.61 Clear sky noise temperature (K). From Miya,[10] with permission.

temperature for moderate humidity (10 g/m^3 water vapor), 1 atmosphere of surface pressure, and surface temperature = 20°C (Refs. 10; 28, Report 720-1). Below 1 GHz, galactic noise increases rapidly and dominates the atmospheric noise. Thus there is a broad minimum in the range 1 to 10 GHz between the cosmic noise and the absorption/radiative effect of water vapor. The orthogonal components of electromagnetic waves passing through rain experience differential attenuation and differential phase shift, as shown in Figs. 49.62 and 49.63.[30] The resultant depolarization of the waves introduces cross-polarized components that can degrade the performance of frequency reuse links that attempt to achieve increased capacity via simultaneous use of dual orthogonal polarizations. The resultant cross-polarization discrimination (XPD) statistics depend on the effective path length, dropsize distributions, and canting angle of the raindrops with respect to the incident polarizations.[31,32] Other significant propagation effects described in greater detail in Refs. 10, 31, and 32 include atmospheric refraction, scintillation (rapid fading), Faraday rotation, and noise due to rain.

Frequency Reuse. Early satellites had relatively small solar arrays and their channel capacity was limited by the available RF power. Later satellites, beginning with INTELSAT IV[33] in the early 1970s, had sufficient power and were "bandwidth limited" in their ability to carry increasing numbers of voice channels within the allocated 500-MHz "C-band" bandwidth. To achieve greater system capacity, several methods can be considered. New frequency allocations may be sought, but this is not always practical because assignments depend on international agreement and are changed only infrequently. Also, heavy investment in existing facilities encourages the maximum utilization of existing frequency bands. Another method is to utilize more than one satellite for a given system, as shown in Fig. 49.64. Earth station sites A and B each have two antennas operating in the same frequency bands and the pathways are independent as long as the power radiated by antenna A in the direction of satellite B and vice versa is acceptably low (typically 28–30 dB below the desired signal). The interference levels depend on the orbital arc separation, λ, between the satellites, the

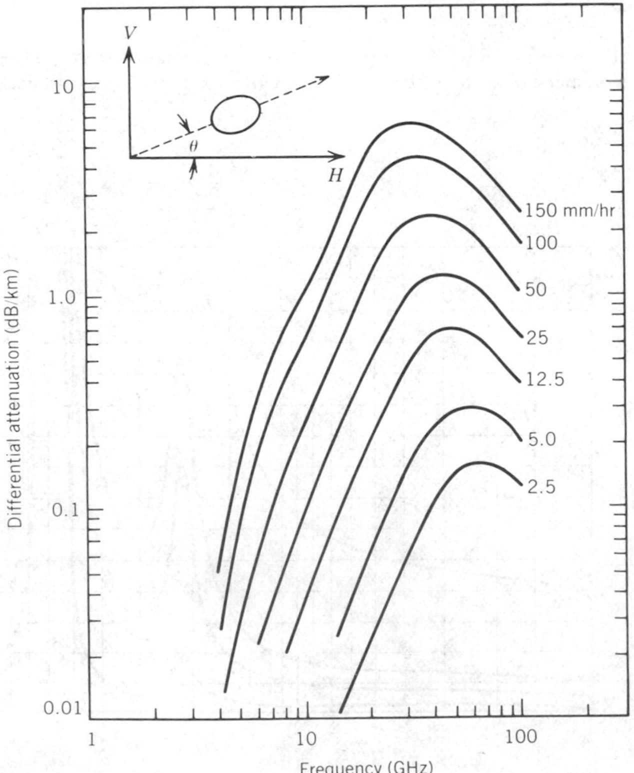

Fig. 49.62 Rain-induced differential attenuation between polarizations I and II for various rain rates. From Hogg and Chu,[30] with permission.

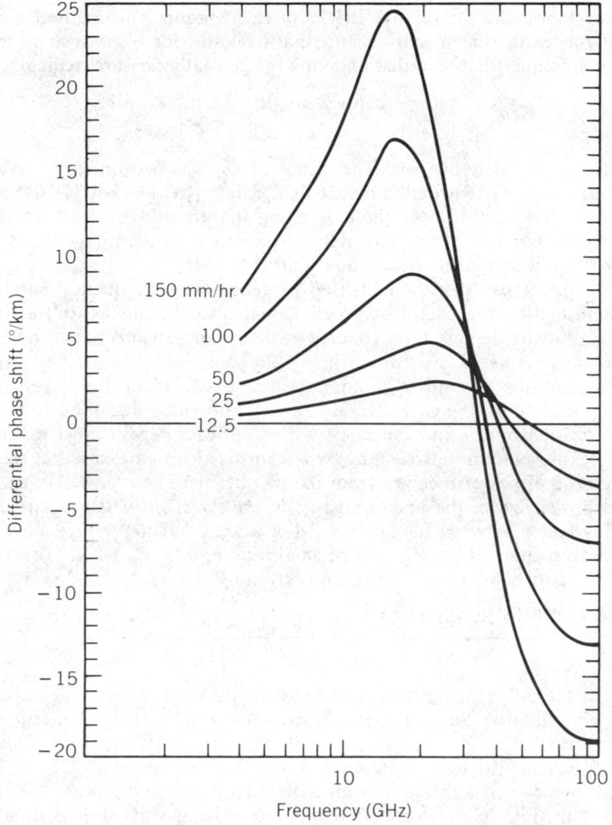

Fig. 49.63 Rain-induced differential phase shift between polarizations I and II for various rain rates. From Hogg and Chu,[30] with permission.

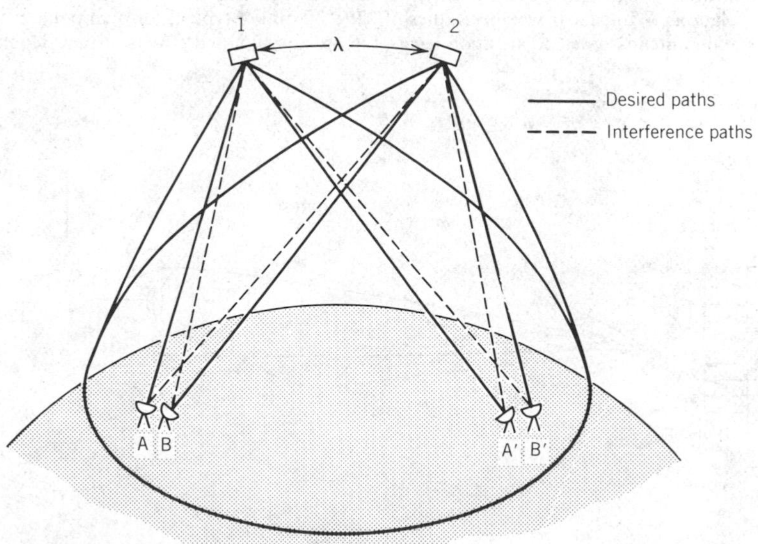

Fig. 49.64 Interfering path geometry with two satellites. Both satellites cover all stations.

antenna sidelobe envelopes, and the relative EIRPs of each station. This method is sometimes called frequency reuse at the earth. Earth station antenna sidelobes for most systems must maintain a prescribed level of maximum sidelobe radiation along the geostationary arc, typically defined as

$$G(\theta) \leqslant \begin{cases} G_0 - 25 \log \theta & \text{dBi} \quad 1 \leqslant \theta \leqslant 48° \\ -10 & \text{dBi} \quad \theta \geqslant 48° \end{cases}$$

where θ is the off-axis angle in degrees. The value of G_0 was 32 for early systems, but as the geostationary arc became more crowded, a recent FCC ruling (CC Docket 81-704) specified $G_0 = 29$ at 6 GHz for all sidelobes within 9° of the axis along the arc and $G_0 = 19$ for cross-polarization response at 6 GHz. The purpose of these tighter restrictions is to permit satellites eventually to be spaced as close as 2° apart along the geostationary arc.

Satellite system capacity can also be increased by reuse of the frequency bands at the satellite. This can be accomplished by the use of dual orthogonal polarizations as well as multiple spatially isolated beams. Figure 49.65 depicts both concepts where four satellite beams are actually present. The "west" region is illuminated by two orthogonally polarized beams A_V and A_H. Similarly, orthogonally polarized beams B_V and B_H illuminate the "east" region. Both west beams must have low sidelobes everywhere within the east coverage region, and vice versa. Also both must maintain a high degree of polarization purity and the earth stations in each region must maintain polarization orthogonality. The beam isolation between any two beams is always measured at the receive location and is given by the ratio of power received from the two beams. These power levels, in turn, depend on the transmitted power levels, the antenna gains *in the direction of the receive antenna*, and the polarization coupling factor between the antennas. For a wave having voltage axial ratio r_W and tilt angle τ_W and a receiving antenna having voltage axial ratio r_A $(0 \leqslant r_A, r_W \leqslant 1)$ and tilt angle τ_A, the polarization coupling factor γ noted in Eq. (49.15) is given by

$$\gamma = \frac{(1 + r_W^2)(1 + r_A^2) \pm 4 r_W r_A + (1 - r_W^2)(1 - r_A^2)\cos 2\delta}{2(1 + r_W^2)(1 + r_A^2)} \tag{49.23}$$

where $\delta = \tau_W - \tau_A$ is the difference in tilt angles and the $(-)$ sign is used for opposite polarization sense. INTELSAT typically uses dual circular polarizations ($r \approx 1$); many domestic systems use dual linear polarizations ($r \approx 0$).

Typical values for beam isolation between any two beams are in the 27- to 33-dB range. Of course, total interference power is the sum of all individual interfering powers.[34] For the example of Fig. 49.65 there are three interference entries. The interference appears similar to noise and must be added on a power basis to the thermal and intermodulation noise levels, further reducing the total carrier-to-noise density ratio of Eq. (49.20). Note that since isolation is measured at the receive location, the up-link beam isolation is measured at the satellite.

The implications regarding satellite antenna patterns are slightly different for up-link and down-link isolations. While clear weather values of 27 to 30 dB are typical for dual-polarized satellite and earth station antenna cross-polarization levels, the previously noted effects of rain depolarization

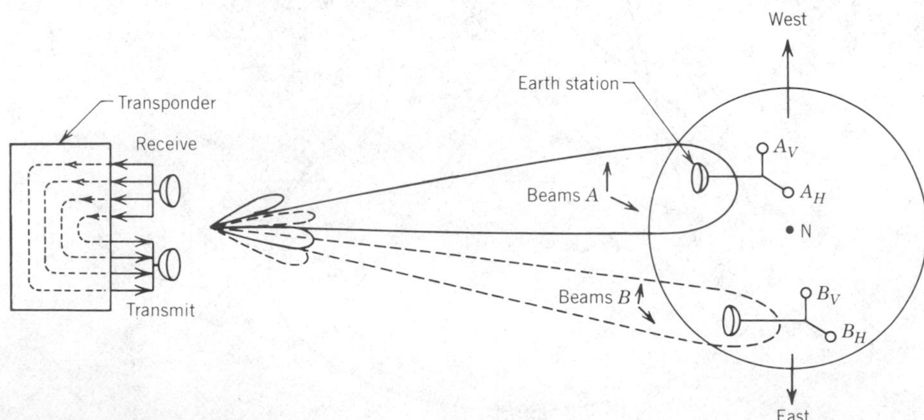

Fig. 49.65 Frequency reuse by dual polarization and spatial beam separation for four satellite down-link and four up-link beams.

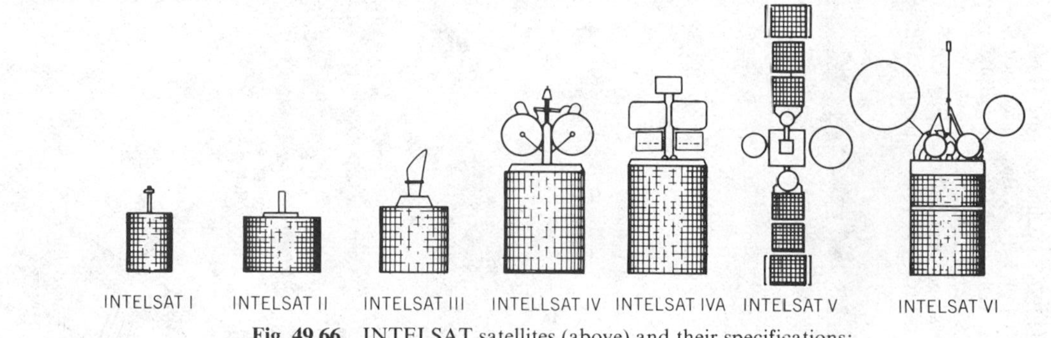

Fig. 49.66 INTELSAT satellites (above) and their specifications:

	INTELSAT I	INTELSAT II	INTELSAT III	INTELSAT IV	INTELSAT IVA	INTELSAT V	INTELSAT VI
Year of first launch	1965	1967	1968	1971	1975	1980	1986
Dimensions (cm)							
Diameter	72.1	142	142	238	238	1560 (wingspan)	360 (drum)
Height	59.6	67.3	104	282 (drum) 528 (overall)	282 (drum) 590 (overall)	640	1163 (overall)
Mass (kg)							
At launch	68	162	293	1385	1469	1928	3740
In orbit	38	86	152	700	790	1037	2225
Launch vehicle	Thor-Delta	Improved Thor-Delta	Long-tank Thor-Delta	Atlas Centaur	Atlas Centaur	Atlas Centaur	Ariane-STS
Primary power (W)	40	75	120	400	500	1200	2100
Transponders	2	1	2	12	20	27	48
Bandwidth per transponder (MHz)	25	130	225	36	32, 36	40, 80, 240	40, 80, 160
Coverage	Northern hemisphere	Global	Global	Global, spot beams	Global, spot beams	Global, regional, spot beams	Global, regional, spot beams
EIRP (dBW)	11.5	15.5	23	22.5 (global) 33.7 (spot beam)	22 (global) 29 (spot beam)	22.29 (4 GHz) 44 (11 GHz)	23.31 (4 GHz) 41.44.4 (11 GHz)
Number of telephone circuits	240 (no multiple access)	240	1200	4000 (avg.)	6000 (avg.)	12,000	30,000
Lifetime (yr)	1.5	3	5	7	7	7 +	10
Satellite cost/circuit year ($K)	30	10	2	1	1	0.9	0.5

can seriously degrade isolation. This has motivated the study of adaptive polarization compensation networks.[35] The depolarization due to rain is worst for circular polarization and for linear polarization with raindrop canting angle of 45° relative to the wave. Degradation is much less severe for linear polarization if the wave components are oriented parallel and perpendicular to local vertical and horizontal at the earth.[30] Seen from the satellite, lines of "local earth vertical" are radial from the subsatellite point and loci of constant local horizontal are concentric circles. The angle ϕ_s in Figs. 49.51 and 49.54 defines a satellite polarization tilt angle relative to x_s that will be aligned with the local gravity vector at earth location p.

49.4-4 Representative Satellite Systems

The steady and rapid growth of satellite systems is typified by the INTELSAT satellites. Figure 49.66 depicts the major features of the INTELSAT satellites.[18,33,36,37] INTELSAT I (Early Bird), launched in 1965, had two 25-MHz transponders at 6/4 GHz and the solar cells on the spinning drum produced only 40 W prime power. Each succeeding generation provided a substantial growth in power and channel capacity. INTELSAT IV[33] had twelve 36-MHz transponders, fully utilizing the available 500-MHz C Band. The INTELSAT IV-A satellite,[36–38] was the first to employ frequency

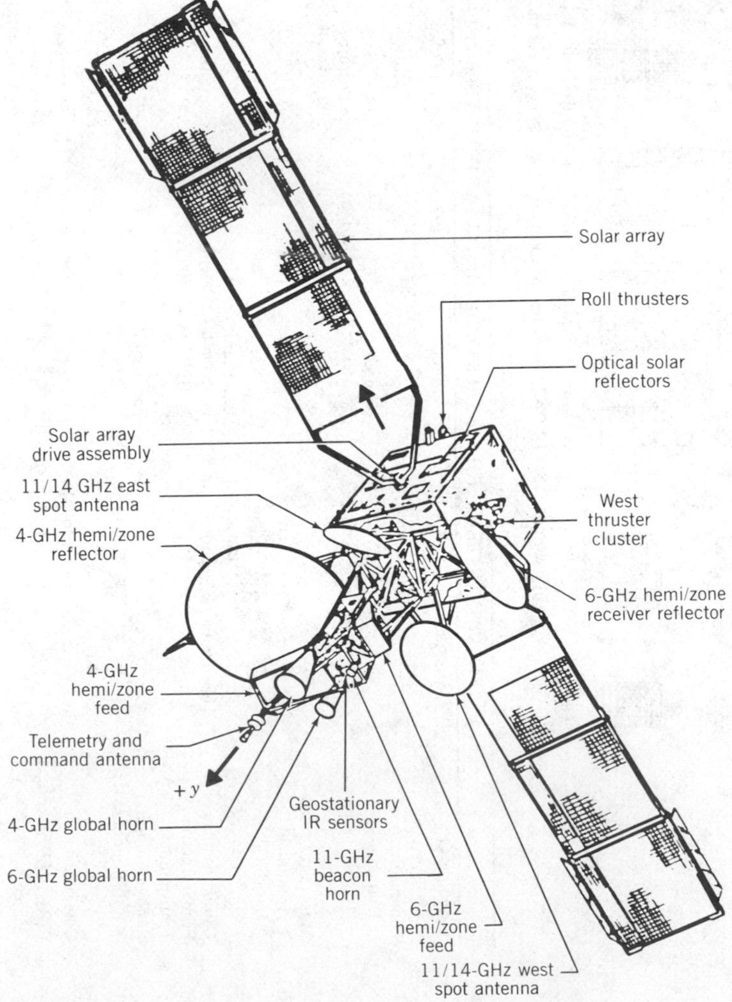

Fig. 49.67 INTELSAT V satellite has a total length of more than 15 m. Courtesy of Ford Aerospace and Communications Corp.

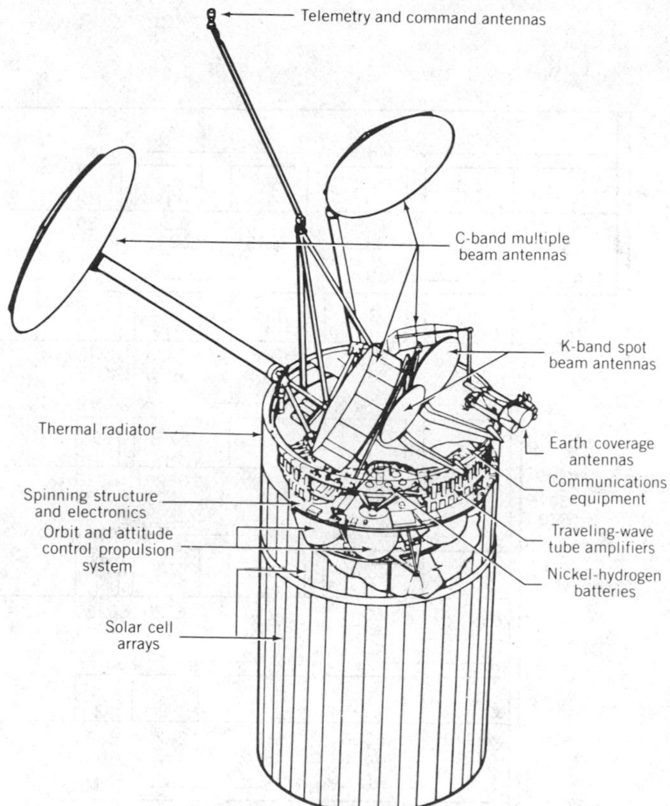

Fig. 49.68 INTELSAT VI satellite. Courtesy of Hughes Aircraft Co.

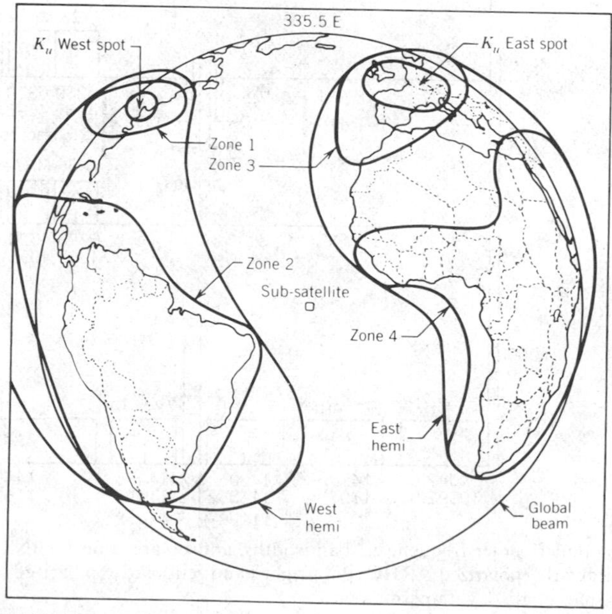

Fig. 49.69 INTELSAT VI antenna coverage contours for the Atlantic Ocean region (AOR) from primary satellite location. Global, hemi, zone beams are at 6/4 GHz and east and west spot beams are at 14/11 GHz.

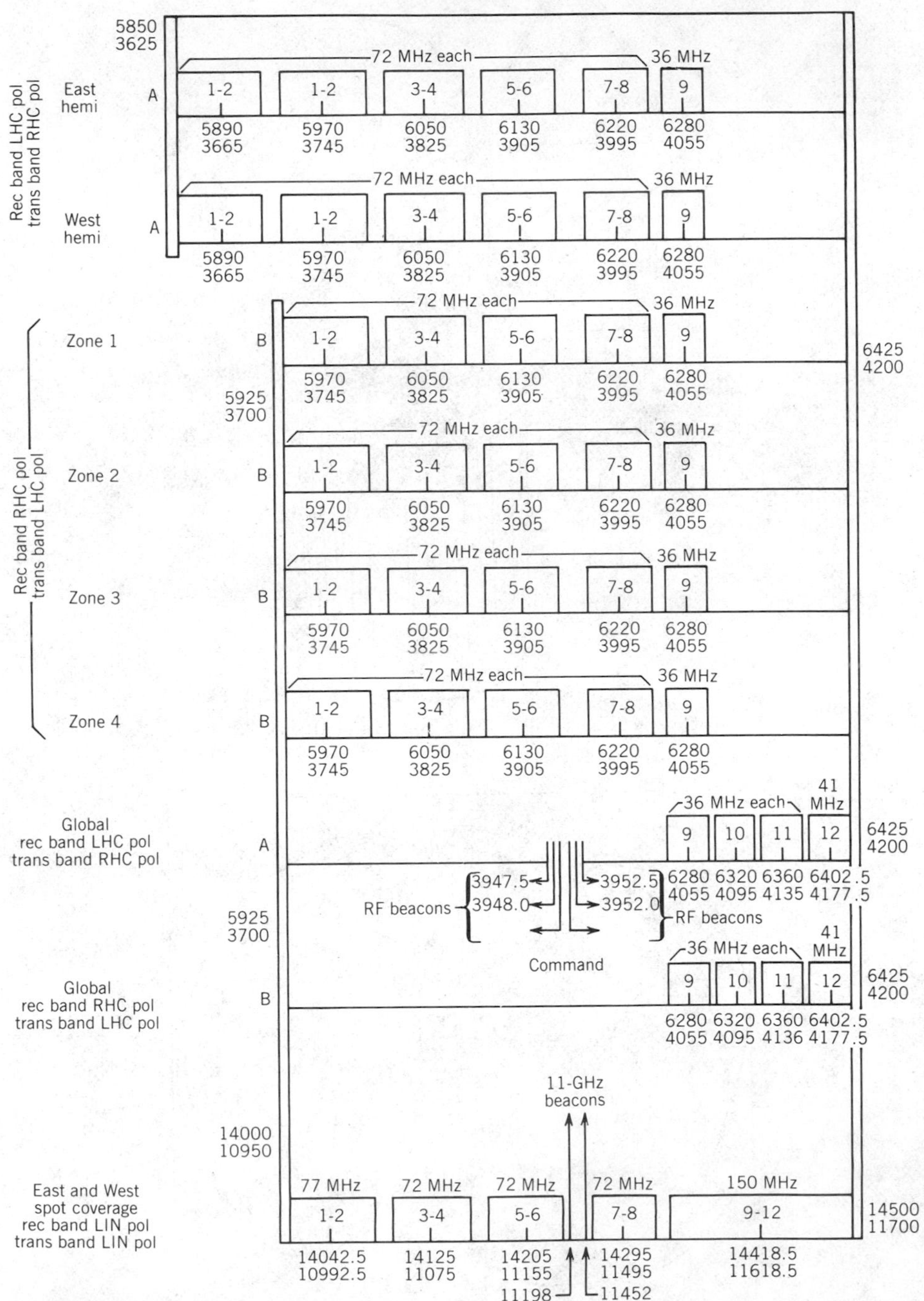

Fig. 49.70 Transponder center frequencies, bandwidths, and polarization for INTELSAT VI. LHC pol, left-hand circularly polarized; RHC pol, right-hand circularly polarized; lin pol, linearly polarized. From Ghais et al.,[40] with permission.

TABLE 49.14. PROPERTIES OF SOME SATELLITE SYSTEMS

	GSTAR[42]	ANIK-C[43]	ARABSAT[44]
Launch date	1984	1982, 1984	1984
Launch vehicle	STS,[a] Ariane 3, or Delta 3920	STS or Delta 3910	Ariane, STS
Owner	GTE Corp., USA	TELESAT, Canada	ARABSAT
Applications	Digital 60–90 Mb/s, voice, video	Message, TV, digital 91 Mb/s	Communications, community TV at 2.5 GHz
Frequency bands, up/down (GHz)	14–14.5/11.7–12.2	14–14.5/11.7–12.2	5.925–6.425/3.7–4.2; 2.5
Coverage	Contiguous United States, Alaska, Hawaii	Southern Canada	Northern Africa and Middle East
Number of transponders	16 (8 on each polarization)	16	25 at 6/4, 1 at 2.5 GHz
Transponder bandwidth (MHz)	54	54	33
Polarizations	Dual linear, isolation = 33 dB	Dual linear	Dual circular polarization
EIRP (dBW)	42 (contiguous United States)	47	31 at 4 GHz; 39 at 2.5 GHz
Output amplifier power	5 at 20 W, 2 at 30 W	15	8.5 at 4 GHz; 50 W at 2.5 GHz
Manufacturer	RCA	Hughes Aircraft	Aerospatiale/Ford
Stabilization	Body	Spin	Body
Mass in orbit (kg)	1250	567	1195
Solar array power (W)	1900	900	1285

[a] STS, space transportation system.

reuse via spatially isolated beams shaped to cover the western and eastern hemispheres as seen from orbit. It permitted a 50% increase in the channel capacity. INTELSAT V,[39] the first INTELSAT body-stabilized satellite, further reused the 6/4-GHz bands by overlaying an orthogonally polarized "zone" beam on the northern part of each hemisphere beam. Figure 49.67 illustrates the major components of INTELSAT V. INTELSAT VI,[40] scheduled for 1986 launch and shown in Fig. 49.68, contains 48 transponders operating at C and K_u bands and can utilize satellite-switched TDMA (SS-TDMA)[27,33] to yield a capacity of more than 30,000 voice circuits.

As mentioned previously, many systems use more than one satellite in a particular region. The INTELSAT system uses several satellites in the Atlantic region spaced over the orbital arc from approximately 325° to 340° east longitude. The satellite that interconnects all users is called the primary. A second satellite, interconnecting those heavy traffic sites that must install a second earth station antenna, is called the major path satellite. The primary and major-path satellite systems are independent. There may be situations that call for the systems to be interconnected. This is now done at the earth stations, but intersatellite links[18,41] are being considered to improve sensitivity. Each additional major-path satellite requires a set of earth station antennas.

The INTELSAT VI Atlantic Ocean region coverage shown in Fig. 49.69 provides a sixfold reuse of the 6- and 4-GHz bands at the satellite as well as spot beams at the K_u band. There are four co-polarized spatially isolated zone-coverage shaped beams (beam shaping is described in Section 49.4-5) overlayed on two hemispheric ("hemi") beams that are co-polarized and spatially isolated from each other but polarized orthogonal to the zone beams. Circular polarizations are used with the down-link zones left-hand circularly polarized (LHCP) and the down-link hemi beams right-hand circularly polarized (RHCP). Up-link beams have senses opposite to their down-link counterparts. The K_u-band spot beams are linearly polarized. Interbeam isolations are specified to be 27 to 30 dB. Transponder characteristics are functions of each specific beam and include (G/T) of -2 to -14 dB/K, saturation flux densities from -67 to -87 dBW/m^2, depending on the settings of gain step attenuators (as in Fig. 49.47). Down-link EIRPs range from 23.5 to 44.4 dBW. Figure 49.70 depicts the transponder frequency plan. Design life for the satellite is seven years, stationkeeping is intended to be $\pm 0.1°$ for longitude and latitude, and the attitude control subsystem is expected to maintain pointing at $\pm 0.10°$ in roll and pitch and to 0.11° in yaw. The 2-kW solar array is backed up with nickel-hydrogen batteries having more than 2.2 kWhr capacity.

Table 49.14 lists parameters for a few[42–44] of the many systems in use around the world. The GSTAR[42] shaped-coverage EIRP contours for the contiguous United States (CONUS) and Hawaii are shown in Fig. 49.71. This satellite is typical of the trend toward the use of K_u band, shaped beams, and frequency reuse via dual linear polarizations. The pattern shapes are achieved with multiple feed horns illuminating offset paraboloids. The transponder block diagram for GSTAR, Fig. 49.72, illustrates the channelization, receiver redundancy, and antenna beam-forming networks. Variable power dividers are used to enable several discrete coverage shapes to be obtained for a number of channels, for example, for coverage of only one half of CONUS. The "dual mode networks," which will be discussed later, permit contiguous transponder channels to be combined in one antenna, while the multiplexers need only accommodate alternate channels (e.g., odd and even channel numbers), thereby simplifying the multiplexer design.

Just as systems moved toward the K_u band as the C band became congested, future trends will be toward even higher frequencies, for example, the 30/20-GHz bands, and the use of spot beams for greater capacity will continue to develop.[45] Military satellite systems use multiple beams for protection against jamming[46] and there is also a trend toward higher frequencies.[47] Other satellite systems, such as INMARSAT[48,49] serve special needs such as providing L-band ship-to-shore communications to mobile users. The tracking and data relay satellite (TDRSS)[50] provides data relay between low-orbit satellites and the ground via geostationary orbit. Direct broadcast satellites could provide transmission of TV, data, and other services directly to the home.[51]

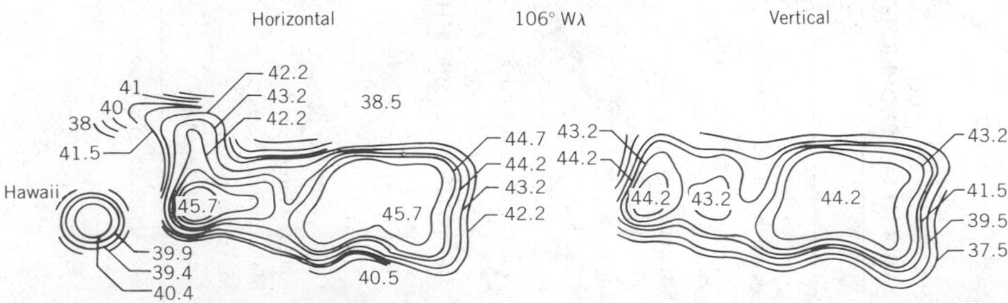

Fig. 49.71 GSTAR EIRP contours for contiguous United States (CONUS) and Hawaii. From Lovell and Cuccia,[45] with permission.

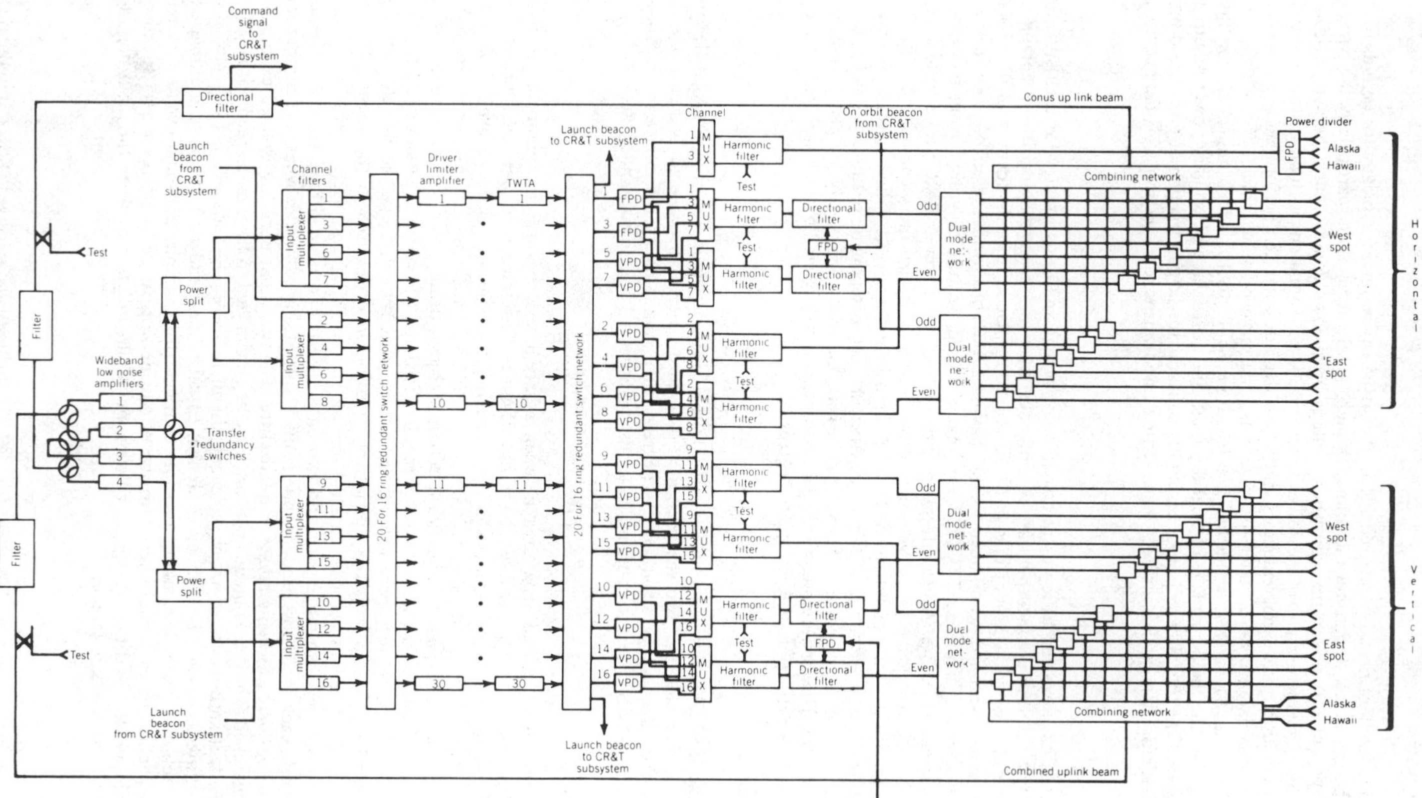

Fig. 49.72 GSTAR transponder block diagram. D, diplexer; FPD, fixed power divider; VPD, variable power divider; MUX, multiplexer; TWTA, traveling wave tube amplifier. From Napoli,[42] with permission.

49.4-5 Subsystems

Satellite Subsystems

Communications satellites must operate reliably in the space environment, typically for periods of 7 to 10 years. The spacecraft design must account for such factors as vacuum [particularly the partial vacuum ($\approx 10^{-6}$ mm Hg) at transfer orbit where voltage breakdown could occur]; thermal environment, which causes temperature extremes of as much as 200°C; radiation, including electrons and protons, which could damage semiconductors and which degrade solar cell performance; ultraviolet radiation, which can damage some materials; meteroids; and the geomagnetic field.[10,52]

The principal subsystems for a satellite are listed in Table 49.15, which indicates, for the INTELSAT VI satellite, the percentage of total dry mass and percentage of primary power required by each subsystem. The transponder output amplifiers consume the largest fraction of power, and the communications subsystem, which includes the antennas and transponder electronics, is the heaviest.

Antennas

Satellite antennas have evolved to represent a significant weight and design factor for each system. This is primarily a result of the incorporation of frequency reuse and beam shaping. Beam shaping is illustrated in the simplified diagram of Fig. 49.73. Each feed displaced from the focus, f, of the offset paraboloid produces a scanned "component beam" having a 3-dB beamwidth, $\theta_c \approx 65\lambda/D$, where D is the projected aperture diameter. Offset-fed antennas are used because they are free of blockage, which would degrade sidelobes. A beam-forming network such as a power divider coherently sums the beams to form a composite beam, shown in cross section and also on a contour basis in the figure. With many feeds it is possible to build up desired beam shapes to follow desired coverage contours, such as a country boundary. For larger aperture, the component beams are smaller, and more are required to fill in the coverage, thereby increasing hardware complexity but also permitting better pattern control. The minimum coverage gain is approximately related to the beam contour area around the -4- to -5-dB contour, Ω in square degrees, as

$$g = \frac{k}{\Omega}$$

where k is typically 13,000 but can vary from 10,000 to 15,000, depending on the particular design and the reflector diameter. For two shaped beams that must be spatially isolated, a useful rule of thumb is that the edge-to-edge separation, σ degrees, that maintains 27 dB interbeam isolation between adjacent -4-dB coverage contours that must be at least 1.5 θ_c. This implies that

$$\sigma \approx \frac{100}{(D/\lambda)} \quad \text{degrees} \tag{49.24}$$

For a system with specified separation between adjacent spot beams, the minimum antenna diameter is determined by Eq. (49.24). This result assumes that the beams are not substantially degraded by off-axis scan aberrations, which in turn implies reflectors or lenses with $f/D \geqslant 1$. Finding the optimum feed excitations in amplitude and phase to simultaneously produce high-coverage gain and low sidelobes is a significant design problem. Beam coverages are optimized with computer-aided techniques. Polarization control requires feeds that have good pattern symmetry for circular polarization. To achieve pure linear polarizations in offset reflectors, the reflector surface may be composed of parallel grids or, alternatively, gratings may be used in the aperture plane. Antenna designs for satellites received much attention in the 1970s and early 1980s.[34,53] If the network in Fig. 49.73 has two input ports and two output ports and is an orthogonal (i.e., lossless) multiport network, each input port produces nominally identical shaped beams, where the component beams may be considered as being added in quadrature. The two input ports are isolated as long as the feeds are matched. This "dual mode" network may be generalized to have many feed ports and is useful in satellite transponders in which adjacent channels are closely spaced in frequency and contiguous multiplexers would be difficult to design. In such transponders, two sets of multiplexers can be used, each multiplexer containing alternate channels, for example, odd- and even-numbered channels. Connection of each multiplexer to one of the network ports produces nearly identical beams for each "mode," and the orthogonal network maintains isolation between the input ports to prevent adjacent channel interference.

Transponder

The transponder subsystem, as shown in Fig. 49.72, includes the receiver, filters, redundancy switches, input multiplexer, amplifiers and attenuators, output amplifier, output multiplexer, and all switches for appropriate interconnections of up-link and down-link beams. Key transponder compo-

TABLE 49.15. INTELSAT VI MASS AND POWER SUBSYSTEM PERCENTAGES OF TOTAL

Subsystem	Mass (% of 1570 kg)	Power (% of 2000 W)
Antennas	18	0
Repeaters	19	74
Tracking, command, and ranging	3	3
Attitude	5	1
Propulsion	8	1
Electric power	18	
Power electronics		1
Battery charge		13
Losses		3
Load uncertainty		2
Thermal control	4	2
Structure	17	0
Integration	8	0

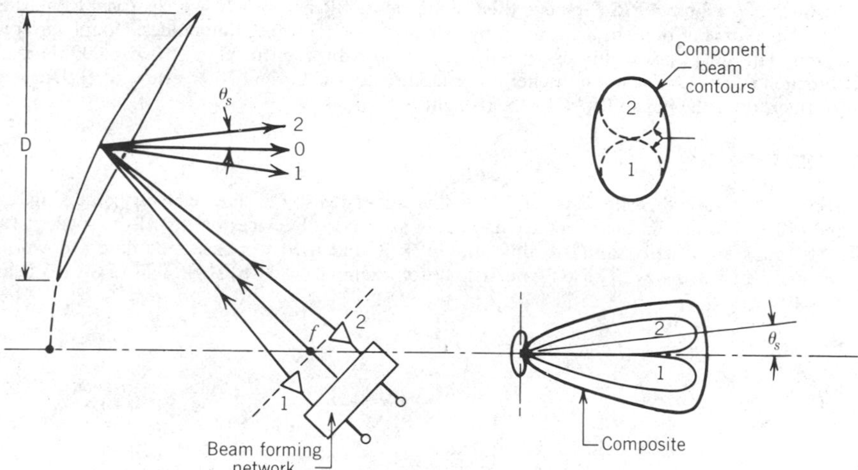

Fig. 49.73 Beam shaping with two feeds and offset paraboloid.

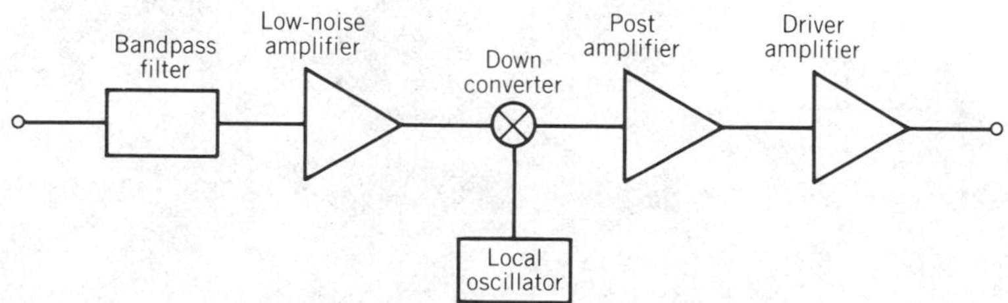

Fig. 49.74 Single-conversion receiver.

nents are described in detail in Ref. 36. A typical single-conversion receiver diagram, shown in Fig. 49.74,[41] could use a tunnel diode amplifier for the C-band low-noise amplifier (LNA), but GaAs field effect transistor (FET) amplifiers are gaining prominence, especially at higher frequencies.[36] The gain of the LNA must be sufficiently high, typically at least 10 to 15 dB, to establish the system noise temperature. A typical overall receiver gain is about 70 dB, and the typical overall gain of a satellite transponder, including output amplifier, is in the 100- to 130-dB range. Since the satellite antenna is typically about 300 K because it views the earth, there is a limit on G/T improvement obtainable by reducing satellite LNA noise figure. An earth station, in contrast, "sees" mostly cold sky, so its antenna temperature is lower and its LNA should have a correspondingly lower noise temperature.

The RF filters must limit the noise bandwidth of the system while not distorting the carrier signal spectrum due to amplitude nonlinearity response and group delay. Group-delay distortion can sometimes be compensated by including an equalizer in the signal path. Filter degradation for digital carriers can result in intersymbol interference.[41] Other parameters that must be controlled include spurious outputs from the local oscillator (LO) frequency stability of the LO, and spurious outputs, adjacent channel multipath in multiplexers, intermodulation products in the output amplifiers, and for digital carriers, pulsed transient responses.

Typically TWTs are used for the output amplifier, but solid-state amplifiers are also being used at C band. TWTs are available for many of the frequency bands of interest for satellite communications and can provide a wide range of output power levels. Typical C-band TWT performance includes[54] efficiencies (DC to RF) of 35 to 45%, output power of 1 to 15 W, bandwidths of 1 to 700 MHz, and gains of 30 to 55 dB. Higher gains and TWTs with higher output power at higher frequencies are available. Typical K_u-band satellite TWTs operate at 20 to 50 W and a direct broadcast satellite[50] uses a 200-W TWT at 12.25 to 12.75 GHz. System performance degradations from TWTs stem primarily from their nonlinear amplitude response and their phase shift near saturation. Impairments include AM-to-PM conversion, indicated by the tube's AM/PM conversion coefficient, which is typically 2 to 7°/dB, and tube phase shift, which can range from 15 to 50° at saturation.[16]

The input and output multiplexers constitute a significant fraction of the volume and weight of the transponders. Early designs were based on Tchebycheff responses, but significant breakthroughs[55] in the synthesis of dual-mode elliptical function filters for satellites promise improved selectivity and weight reduction.[56] Figure 49.75 depicts a dual-mode cavity filter in which two electrical cavities are realized in the volume of one physical cavity by using orthogonal waveguide modes. Coupling screws provide important couplings within each cavity to permit realization of high Q ($>$ 10,000), and the elliptic function response. The use of higher order modes can yield even higher values of Q. Dielectric resonator filters offer the possibility of further weight reductions.

Other Subsystems

Van Trees[36(p. 451)] describes the remaining satellite subsystems. For the telemetry, tracking, and command (TT&C) function, the telemetry data from sensors and devices on board the spacecraft to monitor its status are usually sampled, time multiplexed, and used to phase modulate a down-link carrier at assigned frequency. The bit rate is low, for example, 4800 b/s for INTELSAT VI, and

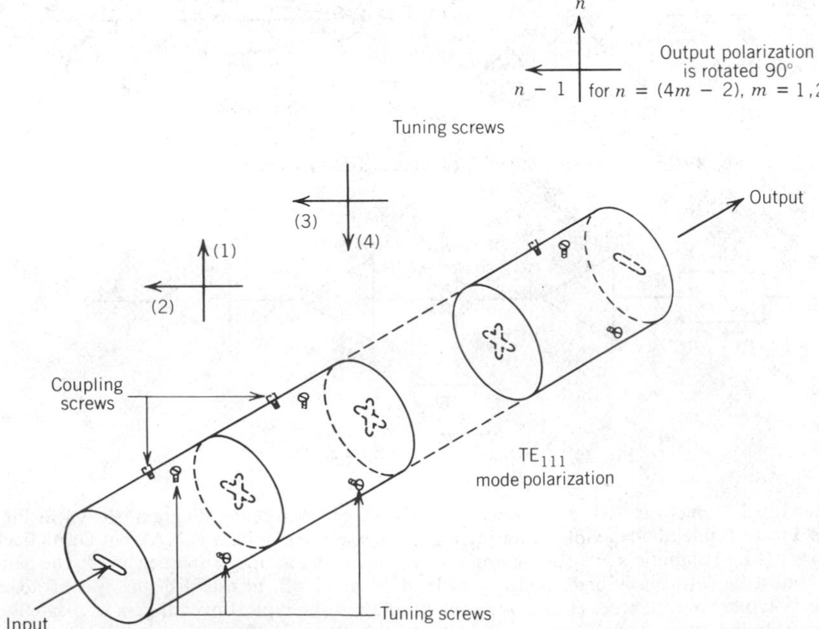

Fig. 49.75 Longitudinal dual-mode cavity filter. From Atia and Williams,[55] with permission.

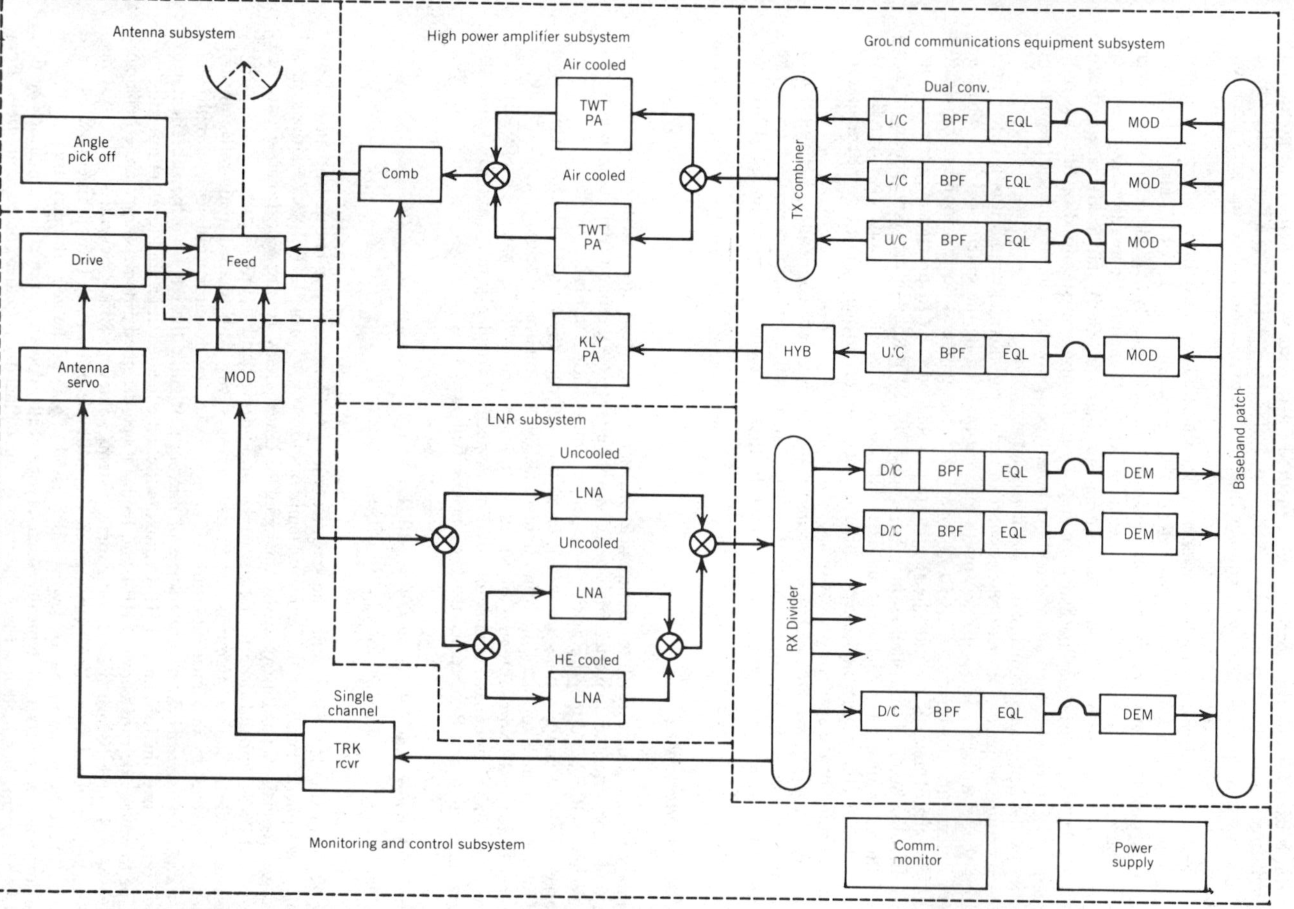

Fig. 49.76 Typical ground station block diagram. HPA, high power amplifier; GCE, grand communications equipment; TWT, traveling wave tube; PA, power amplifier; KLY PA, Klystron power amplifier; LNA, low noise amplifier; LNR, low noise receiver; U/C, up converter; BPF, bandpass filter; EQL, equalizer; HYB, hybrid; MOD, modulator; DEM, demodulator; D/C, down converter. From E. R. Walthall, "Earth Station Technology," in Van Trees,[36] pp. 580–587, with permission.

lower for less complicated spacecraft. Low-bit-rate FM command signals are received in the up-link band and must be reliably decoded. The command link should have high noise immunity to prevent errors and the possibility of false commands. Complicated satellites must distinguish several hundred commands. Several omnidirectional antennas, usually bicones, are used for transfer orbit, and usually a small horn is used for on-orbit operation. One method for ranging is to modulate the command carrier with three related tones (e.g., 283.4, 3968, and 27,777 Hz for INTELSAT VI), allowing a unique range distance to be determined. Another ranging technique uses pulses and the measured time delay yields the range.

The electric power subsystem supplies power from the solar cells and also provides regulation and conditioning of bus voltages. The solar power available is approximately 1.35 kW/m^2, and typical cell efficiencies are 13%. The space environment electron flux will typically degrade cell output over a seven-year lifetime by about 30 to 40%.[16] Batteries must be able to maintain operation during eclipse, which can occur for as much as 72 min/day during the equinoxes. Body-stabilized spacecraft can have the full solar panel surface follow the sun's daily motion, but a spin-stabilized satellite has only about one third of its cells exposed at any time. For both types, between equinox and solstice the sun angle varies by the 23.5° ecliptic inclination, further reducing power.

The temperature in space of a small black sphere of radius r in thermal equilibrium is determined by equating the solar flux of 1.35×10^3 W/m^2 to the heat radiated by the sphere: $4\pi r^2 \sigma T^4$ where σ, the Stefan-Boltzmann constant, is 5.67×10^{-8} $W/m^2/K^4$. The temperature T is found to be ≈ 280 K. In the absence of any solar input, the temperature would be 4 K. In practice, the temperature difference between sunlight and shadow may be around 200 K. Radiation and conduction are the only mechanisms available for heat removal. Thermal control subsystems include passive reflective materials to shield components and may include thermal heaters to maintain narrow temperature excursions for some of the spacecraft electronics.

Reliability considerations dictate choices for materials, designs, and redundancy. Reliability of a component or system is the probability that it will be functional after some prescribed time. The INTELSAT VI is designed for reliability of 0.996 for the first two months of operation and 0.781 over its seven-year lifetime.

Earth Station Subsystems[10,17,57]

The primary elements of a satellite communications earth station are antenna, low-noise receiver (LNR)/down converter, up converter, high-power amplifier(s) (HPA), ground communication terrestrial equipment (GCE), and monitor and control equipment, as shown in Fig. 49.76.

A large variety of earth stations are in existence, ranging from the large INTELSAT stations to small shipboard terminals for maritime communications. Direct broadcast satellite systems will utilize home receiver terminals with antennas of less than 1 m operating at K_u band. The INTELSAT standard A, B, and C antennas are characterized by a specified frequency range and G/T. For the 6/4-GHz standard A and B antennas, the G/T is specified as

$$\text{Standard A: } G/T \geqslant 40.7 + 20 \log f/4 \quad \text{dB/K}$$
$$\text{Standard B: } G/T \geqslant 31.7 + 20 \log f/4 \quad \text{dB/K}$$

For the K_u-band standard C antenna, the specification of G/T is more complicated because of the effect of atmospheric noise on the antenna temperature.[10(p. 341)] The INTELSAT stations must equalize group-delay distortion caused by the satellite. Linear delay equalization ranges from ± 10 ns/MHz for 1.25-MHz bandwidth to ± 1 ns/MHz for 36-MHz bandwidth.

Most large earth station antennas are Cassegrain systems with shaped main-reflector and subreflector surfaces to maximize aperture efficiency. Also, many large stations are built with "beam-waveguide" systems. These are a series of mirrors that guide the RF from the subreflector to the feed horn. Hence, the feed and all HPA equipment can be located at ground level, minimizing the losses due to waveguide runs between HPA and feed. In well-designed antennas, aperture efficiencies in excess of 70% have been achieved, with some systems achieving 80%.[53] Several satellite systems must access more than one satellite from a given site. Multiple-beam earth station antennas, such as the Torus,[57] have been developed to view as much as 50 to 60% of the geostationary arc.

The low-noise receivers had typically been cryogenically cooled parametric amplifiers in the INTELSAT system, but in many cases they are being replaced by Peltier-cooled amplifiers. For smaller systems, for example, for TV distribution, GaAs FET amplifiers are used, having noise temperatures of 70 to 100 K.

The HPA, either a TWT or a Klystron with power up to 10 kW, must be substantially backed off from saturation for multicarrier operation. One approach to HPA implementation is to utilize a number of smaller amplifiers and an output multiplexer.

49.4-6 Modulation and Access

The techniques of analog and digital modulations and access techniques have been described in Sections 48.1 and 48.5. Only the highlights of those aspects related to satellite communications will be described here; further details may be found in the extensive literature on these subjects.[10,16,17,19,26,27,36,41]

Analog modulation, particularly FM, has been prevalent in satellite systems. However, the trend is clearly toward digital modulation. Also, frequency division multiple access (FDMA) is by far the most predominant, but time division multiple access (TDMA) systems are coming into use.

For FDMA, users are assigned a unique carrier frequency. "Users" are earth stations that may frequency division multiplex (FDM) many individual voice channels onto a given carrier, for example, for telephone traffic multiplexed according to group and supergroup hierarchies.[58] These signals can frequency modulate a carrier, and each carrier defines an access. With FM, this scheme is referred to as FDM-FM-FDMA. The transponder utilization efficiency, voice channels per transponder, decreases as the number of accesses is increased. For a simple carrier, the satellite TWT may be operated near saturation, maximizing the power for each voice channel. As the number of carriers in the transponder increases, each carrier contains fewer voice channels. Also the spectrum of each carrier is approximately Gaussian and there must be some guardband between carriers. Furthermore, the satellite amplifier must be backed off to avoid intermodulation products among the carriers. A typical effect is to decrease the number of voice channels per 36-MHz transponder, from 1000 for a

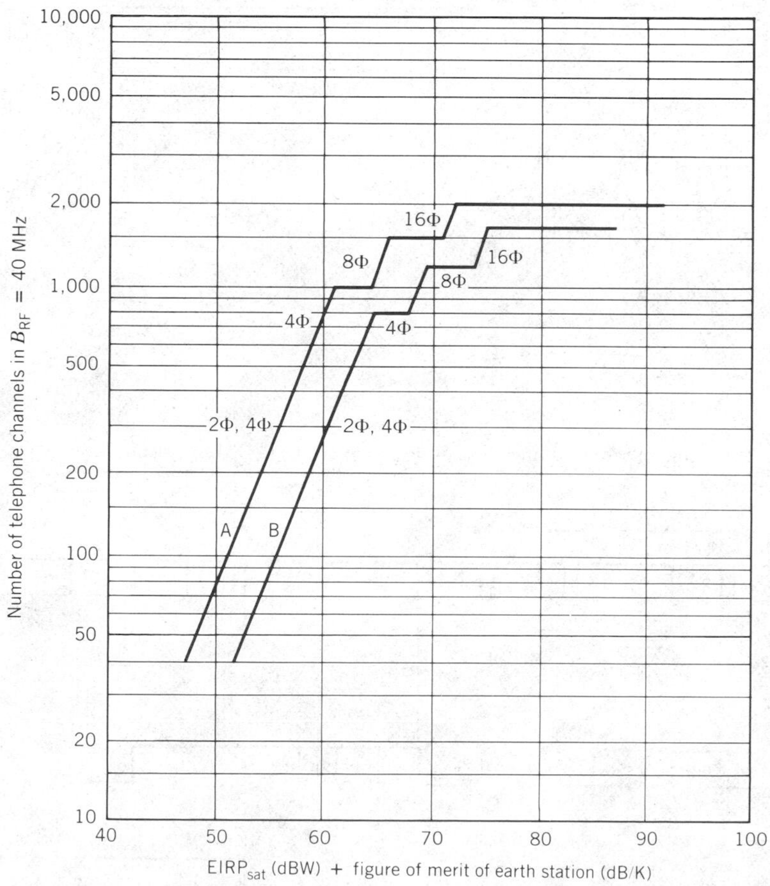

Fig. 49.77 Transponder utilization for multiple access, phase shift keying. Curves A: TDM-PCM-CPSK-TDMA. Curves B: TDM-PCM-CPSK-FDMA. TDM, time division multiplex; PCM, pulse code modulation; CPSK, coherent phase shift keying; TDMA, time division multiple access; FDMA, frequency division multiple access.

single broadband carrier to ≈400 for 15 carriers. A 36-MHz transponder normally would carry one TV channel, but some systems carry two channels per transponder.

Digital modulation offers the possibility of improved utilization. Various combinations of multiplexing, modulation, and access can be envisioned. For example, a baseband signal may be pulse code modulated (PCM) and time division multiplexed (TDM) with other signals, and these may phase shift key (PSK) modulate an RF carrier in a TDMA system. The overall scheme would be referred to as PCM-TDM-PSK-TDMA. This particular scheme offers high utilization that does not degrade significantly as the number of accesses increases. The single-channel-per-carrier (SCPC) approach assigns one voice channel per carrier. This is typically PCM encoded, for example, 8 bits[6] at a rate of 8 kb/s for a 64 kb/s rate per channel. SCPC permits a constant utilization efficiency. For analog systems, signal fidelity is the measure of performance and is related to demodulated signal-to-noise ratio. For FM, the RF carrier-to-total-noise ratio must be typically greater than 13 dB for good signal quality. The criterion for digital signals is the probability of error in the demodulation of a bit. For each type of digital modulation it is related to the ratio E_b/N_0 (Sections 49.4-3 and 48.7). For a given acceptable probability of error, the throughput, or utilization efficiency, can be obtained for various digital modulations. Figure 49.77 illustrates the utilization of a 40-MHz transponder for multiple access with n-phase PSK signals where ϕ refers to the number of phases.[18(Report 708)] The abscissa is the decibel sum of satellite EIRP and earth station G/T, which is directly proportional to E_b/N_0. Reference 18, Report 708, describes the assumptions for these curves, which indicate a slight advantage for TDMA compared with FDMA.

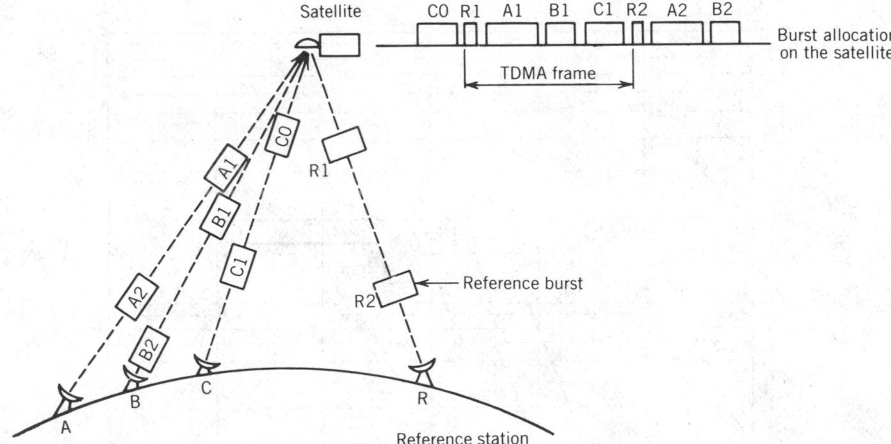

Fig. 49.78 Basic configuration of time division multiple access (TDMA) system. From Miya,[10] with permission.

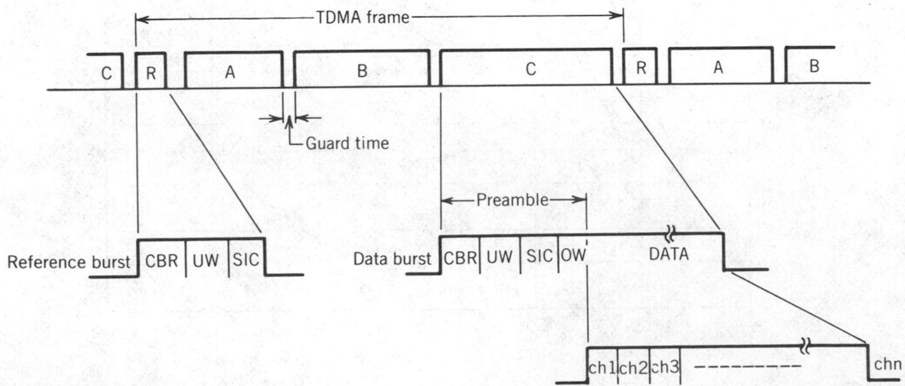

Fig. 49.79 Example of time division multiple access (TDMA) frame and burst formats. CBR, carrier and clock recovery; UW, unique word; SIC, station identification code; OW, order wire; DATA, communication channel. From Miya,[10] with permission.

TABLE 49.16. U.S. DOMSAT COSTS (MILLIONS OF DOLLARS) TRANSPONDER[7]

Procurement Year	Satellite	Spacecraft +	Launch =	Orbiting Spacecraft	No. of XPRDRS[a] Per spacecraft	Design Life	Cost Per XPRDR-yr	Cost Per XPRDR-yr (1982$)
1972	WESTAR 1, 2, 3	11	11	22	12	9	0.20	0.50
1973	COMSTAR 1, 2, 3, 4	16	21	37	24	7	0.22	0.52
1974	SATCOM 1, 2, 3	23	15	38	24	8	0.23	0.50
1978	SBS 1, 2, 3, 4	23	24	47	11	7	0.61	0.93
1980	TELSTAR 1, 2, 3	44	30	74	24	10	0.31	0.39
1980	WESTAR 3, 4	25	30	55	24	9	0.25	0.31
1981	GSTAR 1, 2, 3	44	30	74	24	10	0.31	0.34
1981	SATCOM 3R, 4	24	30	54	24	10	0.23	0.26
1981	Galaxy 1, 2, 3	37	30	67	24	8.5	0.33	0.37
1982	Spacenet 1, 2	40	30	70	36	7	0.28	0.28
	Average			54	23	9	0.30	0.44

Source. Lovell and Fordyce,[8] reprinted with permission.

[a]XPRDR, transponder.

TDMA using satellites is conceptually illustrated in Fig. 49.78. Traffic bursts in preassigned time slots are sent from several stations. The bursts are assigned a specified time slot and the satellite retransmits them to the ground in their prescribed order. Figure 49.79 shows a typical TDMA frame and burst format. Critical factors include carrier recovery, bit timing clock recovery, unique word detection, synchronization of bursts, and acquisition. This imposes precise ranging requirements for the system. For multiple satellite spot beams, a switch matrix on board the satellite can be used to interconnect the up-link and down-link beams on a frame-by-frame basis. This is done simulta-

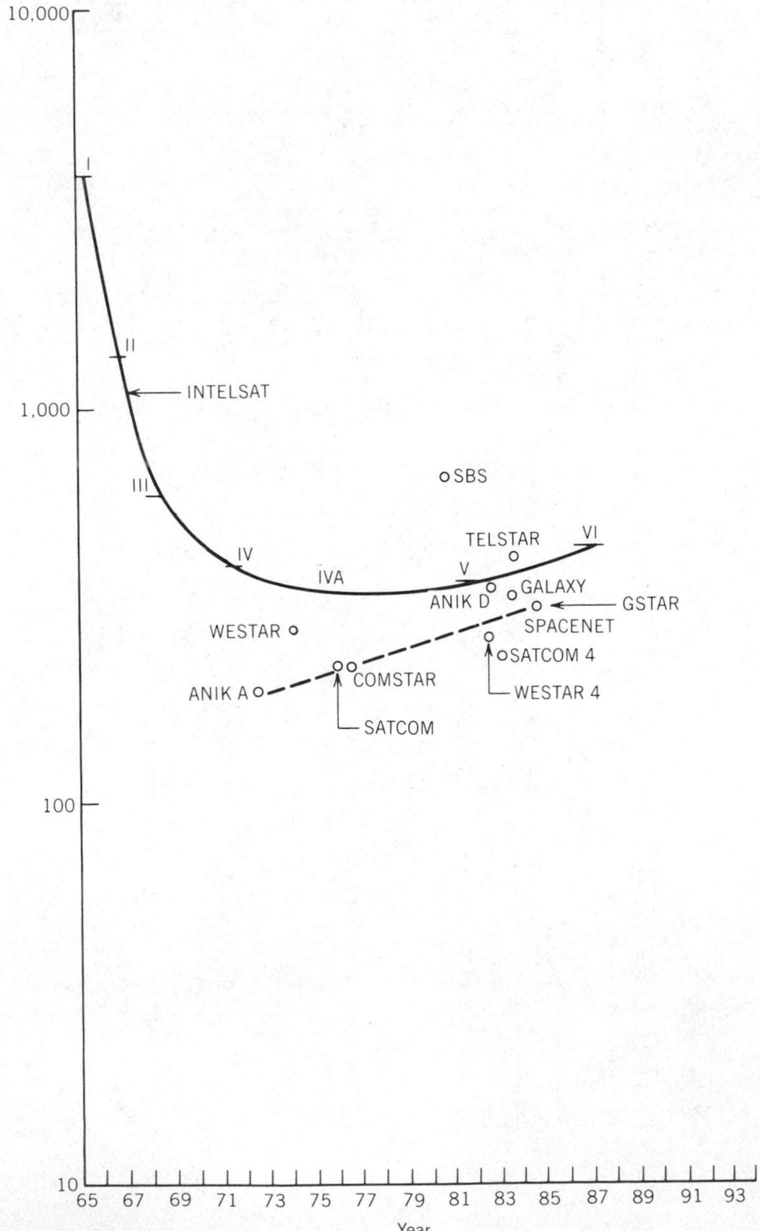

Fig. 49.80 Space segment investment year cost per equivalent transponder (36-MHz bandwidth) of INTELSATS and North American DOMSATS.

neously for all beams, so for N up-link and down-link beams an $N \times N$ switch matrix is used. The switch must be able to operate at the TDMA frame rates. Reference 59 contains an excellent description of satellite TDMA systems. Reference 60 also describes a variety of access methods not mentioned here.

49.4-7 Miscellaneous Topics

Satellite System Costs

As noted for the INTELSAT satellites, improvements in technology have allowed the costs of satellite services to steadily decrease from the earliest system to the present generation. Lovell and Fordyce[8] have conducted a survey of satellite systems, including launch costs, and have defined a cost figure of merit that is the on-orbit investment cost per equivalent 36-MHz transponder per year. Table 49.16[8] displays spacecraft and launch costs for several U.S. domestic satellites and, for the given spacecraft lifetimes, shows that the average cost per transponder per year (1982 dollars) is \$0.44M. Figure 49.80[8] displays the space segment investment cost per transponder for the DOMSATs and the INTELSAT satellites in real-year dollars. The figure of merit for direct broadcast satellites (DBS) is expected to be much higher, perhaps up to \$2M per transponder per year.

Orbit Utilization and Policy

The geostationary orbit and the frequency spectrum are finite natural resources that cannot be increased. The extensive worldwide growth of satellite systems has placed increasing emphasis on technical, policy, and regulatory matters relating to efficient utilization of the orbit. Reference 61 contains a thorough discussion of these issues. The International Telecommunications Union (ITU) with advice from the CCIR regulates orbit and spectrum resources.

While Fig. 49.64 shows two satellites in orbit, it may be generalized to include many adjacent satellites and other earth station locations. The lower limit on orbital spacing in degrees of longitude is limited by interference between systems which, in turn, relates to power levels, modulation, and antenna polarization and sidelobe levels. Moreover, an additional interference source may be the sharing of frequency bands with terrestrial networks, as in the case of the 6/4-GHz bands. CCIR Report 455-2 treats sharing constraints and the International Radio Regulations establishes requirements for administrations to coordinate with each other with respect to new services to avoid mutual interference.

CCIR Recommendation 466 describes the "increase in equivalent noise temperature" (ΔT) method for estimating interference effects. For the case in which potentially interfering systems share the same frequency band, this method involves the following steps. First, it is necessary to compute the flux density, I_u (W/Hz) of interference at the satellite receiver due to an interfering earth station. The gain of the earth station, its spectral power density (W/Hz), and the satellite antenna gain response must be incorporated with the link equations. Then the flux density is converted to an increase in satellite noise temperature, T_s, according to $10\log(k\Delta T_s) = I_u$. For the down link, similar reasoning is used to compute the ΔT_e at the earth station due to an interfering satellite using similar reasoning. The change in noise temperature for the link is $\Delta T/T = \gamma\Delta T_s + \Delta T_e$, where γ is the transmission gain of the satellite link, evaluated from the wanted satellite receive antenna to the wanted earth receive antenna. If $\Delta T/T$ is less than 4%, interference is considered acceptably low. If it is greater, coordination must be effected.

References

1 A. C. Clarke, "Extra-Terrestrial Relays," *Wireless World* **51**(10):303–308 (Oct. 1945).

2 J. R. Pierce, "Orbital Radio Relays," *Jet Propulsion* **25**:153–157 (Apr. 1955).

3 J. R. Pierce and R. Kompfner, "Transoceanic Communication by Means of Satellite," *Proc. IRE* **47**:372 (Mar. 1959).

4 L. Jaffe, *Communications in Space*, Holt, Rhinehart and Winston, New York, 1966.

5 J. R. Pierce, *The Beginnings of Satellite Communications*, San Francisco Press, San Francisco, 1968.

6 T. Pirard, "INTERSPUTNIK: The Eastern Brother of INTELSAT," *Satellite Commun.* **6**(8):38–40, 42, 44 (Aug. 1982).

7 W. Morgan, "Satellite Notebook #31," *Satellite Commun.* **7**(13):28–31 (Dec. 1983).

8 R. R. Lovell and S. W. Fordyce, "A Figure of Merit for Competing Communications Satellite Designs," *Space Commun. Broadcast.* **1**(1):53–63 (Apr. 1983).

9 Final Acts of the World Administrative Radio Conference, Geneva, 1979.

10 K. Miya, ed., *Satellite Communications Technology*, KDD Engineering and Consulting, Tokyo, 1981.

11 R. M. L. Baker, Jr. and M. W. Makenson, *An Introduction to Astrodynamics*, Academic Press, New York, 1960.

12 S. Herrick, *Astrodynamics*, Vols. 1 and 2, Van Nostrand Reinhold, New York, 1971.

13 H. E. Rowe and A. A. Penzias, "Efficient Spacing of Synchronous Communication Satellites," *Bell Syst. Tech. J.*, pp. 2379–2433 (Dec. 1968).

14 D. V. Z. Wadsworth, "Longitude Reuse Plan Doubles Communication Satellite Capacity of Geostationary Arc," AIAA 80-0507, pp. 198–204, 1980.

15 V. J. Slabinski, "Expressions to the Time-Varying Topocentric Direction of a Geostationary Satellite," *COMSAT Tech. Rev.* **5**(1):1–14 (1975).

16 E. Fthenakis, *Manual of Satellite Communications*, McGraw-Hill, New York, 1984.

17 R. G. Gould and Y. F. Lum, *Communications Satellite Systems: An Overview of the Technology*, IEEE Press, New York, 1976.

18 *Recommendations and Reports of the CCIR*, Vol. IV, Part 1, Geneva, 1982.

19 J. J. Spilker, *Digital Communications by Satellite*, Prentice-Hall, Englewood Cliffs, NJ, 1977.

20 R. G. Jahn, *Physics of Electric Propulsion*, McGraw-Hill, New York, 1968.

21 D. E. Fritz, R. L. Sackheim, and H. Macklis, "Trends in Propulsion Systems for Communications Satellites," *Space Commun. Broadcast.* **1**(2):173–188 (Jul. 1983).

22 G. W. Durling, "High Power and Pointing Accuracy From Body-Spun Spacecraft," *Space Commun. Broadcast.* **1**(1):65–71 (Apr. 1983).

23 *PAM-D/PAM-DII User's Manual*, McDonnell Douglas Astronautics, Huntington Beach, CA, Feb. 1983.

24 *Ariane 4 User's Manual*, Arianespace, Evry, France, Apr. 1983.

25 *Space Transportation System User's Handbook*, NASA, Washington, DC.

26 P. S. Bargellini and S. J. Campanella, "Overview of Satellite Transmission Techniques," in J. T. Manassah, ed., *Innovations in Telecommunications*, Academic, New York, 1982.

27 K. Feher, *Digital Communications-Satellite/Earth Station Engineering*, Prentice-Hall, Englewood Cliffs, NJ, 1983.

28 CCIR, *Recommendations and Reports of the CCIR*, Vol. V, Geneva, 1982.

29 J. O. Laws and D. A. Parsons, "The Relation of Raindrop Size to Intensity," *Trans. Am. Geophys. Un.* **24**:452–460 (1943).

30 D. C. Hogg and T. S. Chu, "The Role of Rain in Satellite Communications," *Proc. IEEE* **63**:1308–1331 (Sep. 1975).

31 W. L. Flock, "Propagation Effects on Satellite Systems at Frequencies Below 10 GHz," NASA Reference Publication 1108, 1983.

32 L. J. Ippolito, R. D. Kaul, and R. G. Wallace, "Propagation Effects Handbook for Satellite Systems Design—A Summary of Propagation Impairments on 10- to 100-GHz Satellite Links with Techniques for Systems Design," NASA Reference Publication 1082(03), 1983.

33 P. Bargellini, ed., "The INTELSAT IV Communications System," *COMSAT Tech. Rev.* **2**(2): 437–572 (Fall 1972).

34 R. W. Kreutel et al., "Antenna Technology for Frequency Reuse Satellite Communications," *Proc. IEEE* **65**(3):370–378 (Mar. 1977).

35 D. DiFonzo, A. E. Williams, and W. S. Trachtman, "Adaptive Polarization Control for Satellite Frequency Reuse Systems," *COMSAT Tech. Rev.* **6**(2) (Fall 1976).

36 H. L. Van Trees, ed., *Satellite Communications*, IEEE Press, New York, 1979.

37 E. I. Podraczky and J. L. Dicks, "International System Development," reprinted in J. T. Manassah, ed., *Innovations in Telecommunications*, Part B, Academic Press, New York, 1982, pp. 603–662.

38 J. L. Dicks and M. P. Brown, Jr., "INTELSAT IV—A Satellite Transmission Design," *COMSAT Tech. Rev.* **5**(1):73–104 (Spring 1975).

39 R. J. Rusch, J. T. Johnson, and W. Baer, "INTELSAT V Design Summary," AIAA Paper 78-528, AIAA 7th Communications Satellite Systems Conference, San Diego, CA, Apr. 24–27, 1978, pp. 8–20.

40 A. Ghais et al., "Summary of INTELSAT VI Communications Performance Specifications," *COMSAT Tech. Rev.* **12**(2):413–429 (Fall 1982).

41 R. M. Gagliardi, *Satellite Communications*, Lifetime Learning Publications, Belmont, CA, 1984.

42 J. Napoli, "GSTAR—A High Performance Ku-Band Satellite for the 80's," AIAA 9th Communications Satellite Systems Conference, May 1982, pp. 436–447.

43 W. Morgan, "Satellite Notebook #16," *Satellite Communications* **6**(7):62 (Jul. 1982).

44 A. Al-Mashat, "The Arab Satellite Communications System," AIAA 9th Communications System Conference, March 1982, pp. 187–191.

45 R. R. Lovell and C. L. Cuccia, "Type-C Communication Satellites Being Developed for the Future," *Microwave Sys. News*, pp. 58–81 (Mar. 1984).

46 I. S. Haas and A. T. Finney, "The DSCS III Satellite: A Defense Communication System for the 80's," AIAA 7th Communications Satellite Systems Conference, April 24–28, 1978.

47 J. B. Schultz, "Milstar to Close Dangerous C^3I Gap," *Defense Electr.* **15**(3):46–59 (Mar. 1983).

48 T. Takahashi, "The INMARSAT System and Its Development," AIAA 9th Communications Satellite Systems Conference, March 1982, pp. 202–209.

49 D. W. Lipke et al., "MARISAT—A Maritime Satellite Communications System," *COMSAT Tech. Rev.* **7**(2):351–391 (Fall 1977).

50 D. W. Harris and W. O. Macoughtry, "Tracking and Data Relay Satellite System (TDRSS): A Worldwide View From Space," AIAA Paper No. 1AF-83-78, October 1983.

51 L. M. Keane, ed., "A Direct Broadcast Satellite System for the United States," *COMSAT Tech. Rev.* **11**(2):195–265 (Fall 1981).

52 G. E. Mueller and E. R. Spangler, *Communication Satellites*, Wiley, New York, 1964.

53 A. W. Rudge et al., eds., *The Handbook of Antenna Design*, P. Peregrinus, New York, 1982–83.

54 R. Strauss, J. Bretting, and R. Metirier, "Traveling Wave Tubes for Communications Satellites," *Proc. IEEE* **65**:387–400 (Mar. 1977).

55 A. E. Atia and A. E. Williams, "New Types of Waveguide Bandpass Filters for Satellite Transponders," *COMSAT Tech. Rev.* **1**(7):21–43 (Fall 1971).

56 A. E. Atia and A. E. Williams, "Waveguide Filters for Space—Smaller, Lighter—and Fewer Cavities," *Microwave Sys. News* **6**(4):75–77 (Aug./Sep. 1976).

57 G. Hyde, R. W. Kreutel, and L. V. Smith, "The Unattended Earth Terminal Multiple Beam Torus Antenna," *COMSAT Tech. Rev.* **4**(2):231–262 (Fall 1974).

58 *Transmission Systems for Communications*, Bell Telephone Laboratories, Inc., 1982.

59 S. J. Campanella and D. Shaefer, Time Division Multiple Access System (TDMA), in K. Feber, ed., *Digital Communications Satellite Earth Station Engineering*, Prentice-Hall, Englewood Cliffs, NJ, 1981.

60 J. Martin, *Communications Satellite Systems*, Prentice-Hall, Englewood Cliffs, NJ 1978.

61 D. M. Jansky and M. C. Jeruchim, *Communication Satellites in the Geostationary Orbit*, Artech House, Dedham, MA, 1983.

PART 11

RANGING, NAVIGATION, AND LANDING SYSTEMS

CHAPTER 50
PRINCIPLES OF NAVIGATION

EMIL R. SCHIESSER

NASA Johnson Space Center, Houston, Texas

Navigation may be defined as the process of directing the movement of a vehicle from one position to another. Modern navigation systems use a variety of electronic devices and systems to perform the required tasks. These systems vary greatly in complexity and expense, but the basic principles and functions are common.

50.1 FUNCTION DEFINITIONS

In-travel functions performed in support of the navigation process include state estimation, guidance, control, prediction, and targeting.

State Estimation. State estimation is the maintenance of a knowledge of vehicle position and of other parameters that adequately describe the situation or state for navigation purposes. The navigation state includes essential vehicle translational, rotational, and environmental elements and vehicle characteristics such as position, velocity, acceleration, orientation, atmospheric drag, and navigation sensor measurement biases. State estimation includes the determination of the state through the use of observations from navigation aids and the propagation or extrapolation in time of this information in the absence of such observations. This function is sometimes called "navigation" in a narrow sense.

Guidance. Guidance is a related function by which a determination is made of the action required to direct the vehicle along the desired trajectory. Guidance commands are based on a comparison of projected and desired target conditions.

Control. Control is the determination and execution of detailed command instructions to forcing devices on the basis of the results of the guidance function and the status of the forcing devices. These commands alter forces acting on the vehicle through changes in orientation of control surfaces, or the firing of rotational control jets and translational rocket motors.

State Prediction. This is the prediction of future state on the basis of current state estimates and the expected effect of forces acting on the vehicle. State prediction is often called trajectory prediction when the primary parameters are position and velocity.

Targeting. This is the definition of the travel objective—for example, the destination position—in terms useful for vehicle guidance.

50.2 STATE PROPAGATION

State propagation is the backbone of state estimation and navigation. It was implemented as "dead reckoning" for early ship navigation, and it takes the form of "inertial navigation" in modern navigation systems such as those for airplanes and missiles.

Dead Reckoning. In dead reckoning navigation, the position of the vehicle (a ship) is determined by using the last known position, course heading, speed, and elapsed time since the last position (Bowditch, 1977).

Inertial Navigation. Inertial navigation is associated with the use of an inertial measurement unit or device to aid state propagation (Pitman, 1962). Such a device operates on the principle that if a mass is experiencing no force, it will remain stationary or will continue to move at a constant speed and direction until force is applied.

Inertial measurement units (IMUs) contain accelerometers, which behave as pendula and indicate displacement from a rest position when disturbed. An accelerometer is required for each direction of possible motion. IMUs also provide measurements related to orientation. This is accomplished with gyroscopes, of either the traditional electromechanical type or the newer laser type. Additional information on IMUs may be found in Pitman (1962) and also in Section 52.3 on ship's inertial navigation systems (SINS).

Accelerometers do not measure gravitational forces; therefore, state propagation with use of IMUs requires modeling of the gravitational forces for some applications, such as missile flights. For example, a rocket hovering just above the surface of the earth experiences a rocket force roughly equal to but opposite in direction to the gravitational force. An accelerometer properly aligned along the vertical would measure a velocity change of about 32 $\mathrm{ft/s^2}$ caused by the rocket contact force, even though no such velocity change is occurring.

Propagation of position and velocity are performed in on-board computers by numerical integration of measurements of velocity changes obtained from IMUs and computed accelerations from the model of gravitational forces.

State estimation and navigation are based on state propagation alone when observation measurements to external features are not available. Inertial systems and state propagation in general accumulate errors in time, since the initial state, IMU measurements, and force models are imperfect. To compensate for these natural error tendencies, the state is periodically adjusted using data from other sources, such as the electronic navigation aids that are discussed in following chapters.

50.3 STATE DETERMINATION

One or more parameters that change with vehicle position changes are measured to determine position and velocity. The ideal would be to measure directly, cheaply, accurately, continuously, and reliably the desired state elements. Most systems actually measure parameters different from but related to the state elements of interest, and these parameters relate only to a portion of the state at any given time. Consequently, a time history of measurements is required to determine position and velocity.

The state correction function is to adjust position, velocity, and other state element estimates so that a resulting curve of the state as a function of time is consistent with a set of observations. The curve is defined through the state propagation process. State propagation and correction can be viewed as the extraction of a signal from a noisy background by use of a filter.

The curve-fitting process can be performed over a group of observations, as with a least-squares fit of a fixed data set. Sequential filtering techniques with one observation at a time or with sequential groups of observations are also used. There are two basic sequential techniques, one that involves an inversion of a matrix of the same dimensions as the state, and another, the Kalman filter, that operates without explicit matrix inversion if measurements taken at the same time are uncorrelated. Otherwise a matrix of the size of the locally correlated observations must be inverted. Both techniques are concerned with determination of the portion of the state observed and the amount to correct that portion on the basis of present uncertainty of the state and quality of the observation. Determination of the entire state usually requires the use of a series of observations over some time interval so that a sufficient amount of physical geometry and data are available to determine a relatively smooth curve through the generally noisy measurements. Modified Kalman filtering techniques are widely used for the computation of state correction in on-board computers.

50.4 OBSERVATION AND MEASUREMENT

Observation geometry, type, and availability along with measurement quality, time-tag correlation, and reliability are critical to state determination. Observations may be classified according to their relationships to the vehicle state elements, their dependence on external devices and features, whether transmission of energy is required, and whether the measurement is obtained on board or external to the vehicle, for example, on the ground. External features include natural and artificial ones and the measurement devices can be passive, supportive, or cooperative. Typical observations and measurements are summarized in Table 50.1 along with relationships to the state elements.

Measurement Types

"Range" is the length of a path between two points, in this case between the vehicle and an external measurement point. The straight-line distance is called a slant range, the measurement of which is influenced by refraction of radio waves in the medium of propagation. These effects are discussed in Section 48.2.

The "range rate" is the instantaneous rate of change of range. The rate is approximated by measuring the change in range over small intervals of time through the use of Doppler shift of the radio frequency of the ranging signal. The range rate so derived is accurate enough for most navigation applications, but the measurement is viewed and used as range change for difficult navigation tasks. A number of systems produce an accumulated Doppler frequency count, so that the user can select the time interval for processing the range change measurement.

"Angle" measurements include the angle of arrival of a received signal and the angle between two line-of-sight directions, for example, one to a star and one to the nearest point on the horizon.

Timing Considerations

Navigation observations are valid only for the instant at which they are measured, and usually must be appropriately time tagged to be useful. That is, a value for the time at which the measurement was made is determined and that time is associated with the measurement. The navigation process depends on the ability to maintain an adequate timekeeping and measurement data synchronization.

50.5 NAVIGATION SYSTEMS

In general, a system capable of supporting the navigation process includes vehicle motion sensors, vehicle control equipment, information input and output equipment, processing and interfacing equipment, and manual controls. Figure 50.1 is a functional block diagram of a typical navigation

TABLE 50.1. OBSERVATION AND STATE ELEMENT RELATIONSHIPS

	State Elements				
	Translational			Rotational	
Observation Type	Position	Velocity	Acceleration	Orientation	Rate
Range	x				
Range change	x	x			
Velocity change		x	x		
Angle					
Near object	x				
Far object				x	
Angle change					
Near object	x	x			
Far object					x
Static pressure	x				
Dynamic pressure	x	x			
Gravity gradient	x				
Occultation	x	x			

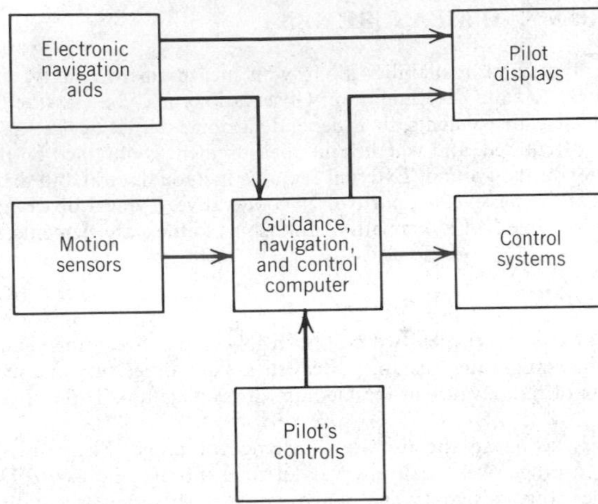

Fig. 50.1 Functional block diagram of a modern navigation system.

system, showing only major groups of functions. Electronic navigation aids, which are treated in detail in the next four chapters, along with the on-board motion sensors and inputs from the pilot (if the vehicle is manned), provide data for a processing subsystem often called the guidance, navigation, and control (GN & C) computer. In a simple system, the GN & C computer may be a hand-held calculator or a special-purpose slide rule. However implemented, the purpose of the computer is to provide information to the pilot, in the form of displays, charts, indicators, and so on, and to provide commands to the control systems.

The composition of a specific navigation system depends on the application. A high-speed airplane, for example, places different demands on a navigation system than does a slow-moving ship navigating in coastal waters. There are many ways of classifying navigation systems; some of these are shown in Table 50.2. These classifications are not mutually exclusive, since a particular system

TABLE 50.2. CLASSIFICATION OF NAVIGATION SYSTEMS

Classification Parameter	Example	Special Features and Requirements
Vehicle type	Airplane	High speed, long range
	Surface ship	Low speed, long duration
	Submarine	Low speed, long duration, difficult communications
	Spacecraft	Extreme speed and range
Area of travel	Air traffic lanes	Extensive support from ground
	Remote land areas	Limited access to ground-based systems
	Open sea	No landmarks, less accuracy required
	Deep space	Long-term accuracy required, difficult geometry
	Coastal zone	Accuracy required, many aids accessible
Navigation base	On board	State estimation performed on board
	Ground based	Requires communications link
	Mixed or shared	Updates transmitted from ground
Phase of travel (aircraft example)	Takeoff/ascent	Rapid change of state
	Cruise	Small changes from planned trajectory
	Landing	Precision required, rapid change of state

may be applicable to several of the example classifications. Moreover, a particular system may be required to operate in several different situations. For example, an aircraft navigation system is required to operate in takeoff, cruise, and landing phases.

50.6 SUMMARY

Modern navigation is a process that includes a variety of measurements, computations, and activities, specialized to meet the particular needs of the application. Of particular interest in this handbook are the electronic navigation aids that provide the parameter measurements and many of the display functions that are an integral part of the navigation process. In the next four chapters, many of those systems and devices will be described in detail.

Bibliography

Bowditch, N., *American Practical Navigator*, Defence Mapping Agency Hydrographic Center, Publication No. 9, Vol. 1, 1977.

Pitman, G., *Inertial Guidance*, Wiley, New York, 1962.

Proceedings of the IEEE, special issue on global navigation systems, IEEE Press, New York, Oct. 1983.

CHAPTER 51
RADAR

FRANK R. CASTELLA

Johns Hopkins University, Laurel, Maryland

51.1 THE NAVIGATION PROBLEM

Surface navigation is the process of directing the movement of a ship from one point to another on the earth's surface. Piloting is navigation in restricted waters and involves frequent determination of position relative to geographic points with a high degree of accuracy. In navigation, radar can serve not only for position finding but as an anticollision aid. It is a very important aid to navigation because it provides the same information during the night and in fog as in more favorable conditions. A great advantage of radar is that it does not require the cooperation of other stations, unlike other navigation aids.

In the open ocean the primary navigation systems are LORAN, OMEGA, and the navigation satellite system. In LORAN or OMEGA, the time or phase differences at a navigator's receiver are measured on transmissions from geographically distributed stations. From this information the navigator's position can be determined. In the navigation satellite system, position is determined from the Doppler frequency history of a received signal from a satellite. The role of the radar on a ship in the open ocean (i.e., a long distance from land) is primarily that of collision avoidance with other ships, icebergs, islands, and so on. In proximity to land the radar can function as a precision sensor for location accuracy in addition to its normal collision-avoidance role. With maps available, the image on the radar display device (the plan position indicator, or PPI) can be compared with the map to determine ship location. However, the radar image in many cases will not have map quality owing to factors such as sensor resolution, terrain shielding, and scattering properties of targets, so human judgment is generally necessary to identify coastlines, mountain peaks, buoys, and other landmarks. Conceivably in future equipments a map-matching procedure can be performed whereby internally stored maps are automatically correlated with the radar PPI information to determine ship location. In such a system an operator may only need to indicate his or her approximate location and then the successive search and map correlation function can be performed by a computer. The final output would be the ship coordinates (i.e., latitude and longitude) and its location on the map coordinate grid.

Collision avoidance is an area in which radar is mandatory. In fog-bound environments it is the only successful method for such avoidance. The operator at present determines the presence of other ships or obstacles and from the time history of the PPI determines if they represent a collision threat.

If a collision is predicted for some future time, a maneuver is effected that puts the ship on a noncollision course. The situation is continuously monitored until physical separation and relative velocities indicate no further danger. The process of automatic target tracking and the performance of collision avoidance calculations are presently being implemented in devices called collision avoidance aids. They will provide information to the ship captain or navigator, and the human alone will be able to use this information as desired.

51.2 RADAR PRINCIPLES

A radar operates as follows: A transmitter generates a sequence of very short sinusoidal pulses that are radiated through an antenna with a narrow beamwidth. These pulses propagate through space with approximately the velocity of light and interact with objects (targets) such as buoys, ships, and aircraft. A portion of the energy incident on these objects is backscattered in the direction of the radar, detected by the radar receiver, and displayed on a cathode ray tube (CRT). Typical parameters of a radar system employed for navigation purposes and illustrated in Fig. 51.1 are:

Pulse duration	0.1×10^{-6} s or 0.1 μs
Interval between pulses	10^{-3} s or 1.0 ms
Radio frequency of radiated pulses	10^{10} Hz or 10 GHz
Antenna mechanical rotation rate	15 rpm

The time separation between transmission of a pulse and reception of the backscattered or echo energy from the object indicates the range or distance between the antenna and the object. If the object distance is R, the time separation T is given by the formula

$$T = \frac{2R}{C}$$

where C is the velocity of light. The maximum range for unambiguous detection of an object (i.e., the echo energy is returned prior to transmission of the next pulse) is

$$R_{\max} = \frac{C}{2} T_{\max}$$

Since $T_{\max}$ corresponds to the interval between pulses, then

$$R_{\max} = \frac{3 \times 10^8}{2} \text{ m/s} \times 10^{-3} \text{ s} = \frac{3}{2} \times 10^5 \text{ m} = 150 \text{ km}$$

An object at 160 km that gives an echo energy large enough to be detected in the radar receiver will be displayed at a range of 10 km on the CRT. This is called a range ambiguous detection.

The CRT display, also called a PPI, operates such that the object echo is displayed along a radius vector from the center of the tube. For an object at 75 km, with a full-scale deflection corresponding

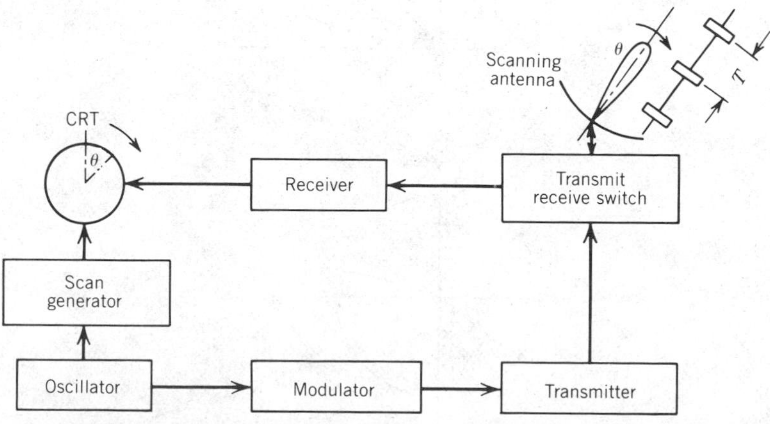

Fig. 51.1 Block diagram of a radar system. CRT, cathode ray tube.

to 150 km, the object echo will appear halfway along the radius vector. Thus if the tube diameter is 35 cm, the echo will appear at 8.75 cm from the center of the tube.

The azimuth angle of the radius vector is electromechanically linked to the pointing direction of the antenna. If the PPI presentation is such that the north direction corresponds to the top of the tube, then an object detected at 45° east of north is displayed at angle $\theta = 45°$. The radius vector rotates about the center of the tube at the antenna mechanical rotation rate.

Since the radar transmits high-power pulses for short periods of time (i.e., low duty cycle), the average transmitted power can be quite low. With the previously stated parameters and a peak pulse power of 10^6 W (1 MW), the average transmitted power is $10^6 \times 10^{-7}/10^{-3} = 100$ W. Echoes that are visible on the screen may be radar returns from desired and undesired objects. It is the function of the human observer studying the CRT to sort the desired returns from the undesired ones. This detection process can be automated if pointlike targets only, such as ships, airplanes, and buoys, are of interest. If mapping of harbor entrances and land clutter is desired, the human operator is the most sophisticated detector.

51.3 RADAR TYPES

Radars can be classified by their coverage characteristics and the type of waveform they transmit. They can also be classified as monostatic or bistatic, the former referring to transmitting and receiving antennae at the same geographic site (and usually the same antenna), the latter referring to transmitting and receiving antennae at separated sites. Radars can be of the three-dimensional (3D) or two-dimensional (2D) surveillance type, where 360° is covered in azimuth on a periodic basis. Three-dimensional radars have narrow beamwidth in the azimuth and elevation coordinates, so that with the range coordinate all three coordinates of a target can be determined. Elevation coverage is typically about 45° (Fig. 51.2). Two-dimensional radars have a narrow fanbeam in azimuth and thus cannot determine target elevation angle (Fig. 51.3). A typical time period to cover 360° in azimuth, 45° in elevation, and ranges out to 200 nm is 4 s. The coverage in azimuth can be provided more cheaply with a mechanically scanning antenna, while the more expensive but more flexible phased array with multiple faces can do the scanning electronically. A three-faced electronically steered phased array can cover 360° in azimuth if each face is called on to steer its beam ±60° in azimuth. The coverage in elevation by a 3D surveillance radar is usually accomplished by scanning of a narrow elevation beam as the antenna rotates or is steered in azimuth. This narrow elevation beam can be scanned by applying a phase gradient across the antenna aperture in the vertical direction. The magnitude of the gradient determines the elevation angle to which the beam is directed. The function of a 2D or 3D surveillance radar is to detect targets over a large field of view. Tracking of the detected targets can be accomplished by an automatic track-while-scan system, whereby the detections of the surveillance radar serve as inputs to a computer and targets are tracked via some mathematical algorithm.

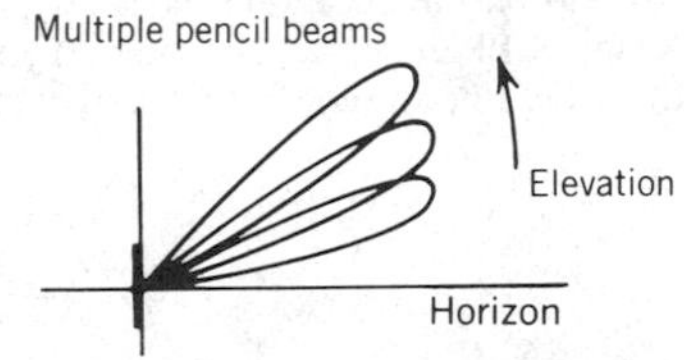

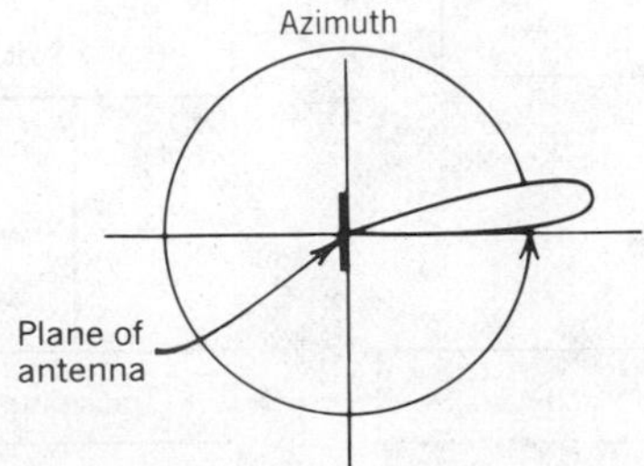

Fig. 51.2 Three-dimensional radar coverage.

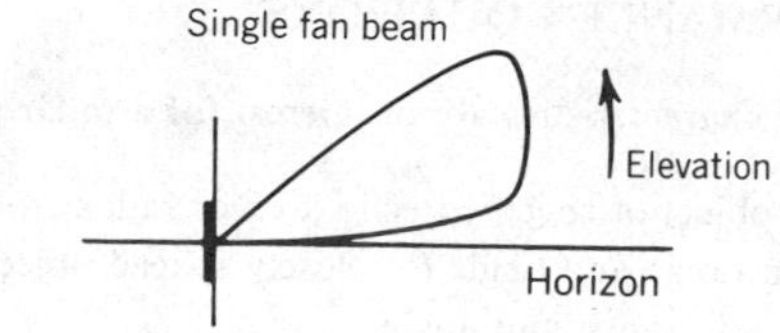

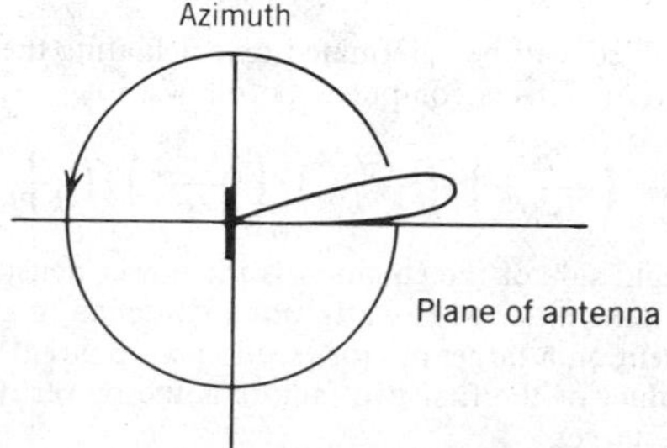

Fig. 51.3 Two-dimensional radar coverage.

A tracking radar is a single-target-dedicated 3D radar with narrow beamwidths in azimuth and elevation. While a surveillance radar may only revisit a target every few seconds, a track radar is continuously locked on to a desired target and thus can more accurately determine target coordinates. Typical track radar beamwidths are 1° in azimuth by 1° in elevation (Fig. 51.4). To acquire the desired target with a narrow beam such as this requires a designation from a surveillance radar or other source and a search of a specified volume in space about that designation point. Tracking the desired target in angle is generally accomplished by angle-tracking techniques referred to as conical scan or monopulse.

Radars may also be classified by the type waveform they transmit. A continuous-wave (CW) or frequency-modulated (FM) radar transmits 100% of the time and generally needs separated transmit and receive antennae for maximum detection sensitivity. These radars work on the Doppler effect (see Section 51.8), which is obtained with moving targets. With CW transmission target range cannot be determined, while with FM transmission target range can be determined. These radars are capable of detecting targets in very heavy clutter. Moving target indicator (MTI) and pulse Doppler radars can also detect moving targets in clutter. Generally these types of radar employ pulsed transmissions, reject clutter somewhat less effectively than CW or FM radars, but determine the range more accurately than these radars.

The most conventional radar type employs pulsed transmissions and does not exploit the Doppler effect for moving targets. These radars are referred to as noncoherent radars and utilize only the amplitude characteristics of the target echo. Radars exploiting the Doppler effect utilize both the phase and amplitude characteristics of the target echo.

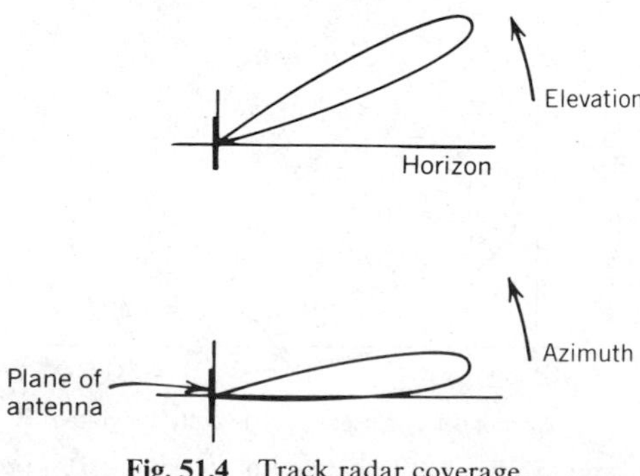

Fig. 51.4 Track radar coverage.

51.4 RADAR PERFORMANCE EQUATIONS

The following performance characteristics are of interest for a radar system.

1. Detection range on an object of specified echo area or radar cross section.
2. Resolution capability in range and angle for closely spaced objects.
3. Object location accuracy in range and angle.
4. Ability to detect moving targets in sea and land clutter of specified density.
5. Radar coverage envelope (i.e., range azimuth, elevation limits).

The detection range of an object can be determined by calculating the signal-to-noise ratio (SNR) in the receiver of the target return. This is computed as follows:

$$\text{SNR} = \left(\frac{P_t G_t}{4\pi R^2}\right)\cdot\left(\frac{\sigma}{4\pi R^2}\right)\cdot\left(\frac{G_R \lambda^2}{4\pi}\right)\cdot\left(\frac{1}{kTB}\right)$$

The first quantity on the right side of the equation is the power density in watts per square meter at a range of R meters from a transmitter of P_t watts and an antenna of gain G_t. The second factor is the fraction of the power incident on a target of cross section σ (apparent area), which is reradiated at a range of R meters. The product of the first two factors is the power density reflected back to the point of transmission from the target.

The third quantity on the right side of the equation is the effective area of the receiving antenna, expressed in terms of the gain, G_R, and the wavelength, λ, in meters. Antenna area is expressed in square meters. The product of the first three terms is the total signal power received back at the receiver, given that P_t watts were transmitted.

This radar cross section is a function of the material composition, shape, and size of the object and also of the wavelength λ and polarization of the radar transmissions. These transmissions may be vertically, horizontally, or circularly polarized. A reference value for radar cross section is $\sigma_0 = 1\ \text{m}^2$.

The last quantity in the denominator is the receiver noise, which competes with the signal in the detection process. k is Boltzmann's constant, equal to 1.37×10^{-23} J/K, and T is the noise temperature in Kelvin. B is the receiver bandwidth in hertz and is approximately equal to the reciprocal of the transmitted pulse duration. F is the receiver noise figure, a factor greater than one, which gives a measure of the noise inherently introduced by the amplifiers and mixers in the receiver.

The signal-to-noise ratio must be large enough to reliably detect an object. This required value is a function of the following quantities: (1) number of pulses N returned from a target as the antenna sweeps by the target, (2) desired probability of detection, and (3) desired probability of false alarm. Item 1 influences the detection process, since multiple pulses integrate to larger intensity values on the CRT owing to the integrating capability of the phosphor that coats the inner surface of the CRT. For a 1° azimuth beamwidth, 90°/s antenna rotation rate, and 1000 pps pulse repetition frequency, the number of pulses returned by a point object during its illumination time is $N = 1000 \text{ pps} \times 1°/90° \text{ s} \approx 11$ pulses. Since the interfering noise has random properties, the process of target detection is a statistical event. Figure 51.5 is a plot of signal-to-noise ratio in decibels (i.e., 10 log SNR) required for detection vs. number of pulses N. The probability of detection, P_d, is equal to 50% and the probabilities of false alarm, P_{fa}, are 10^{-6}, 10^{-4}, and 10^{-2}. A value of $P_{\text{fa}} = 10^{-4}$ indicates

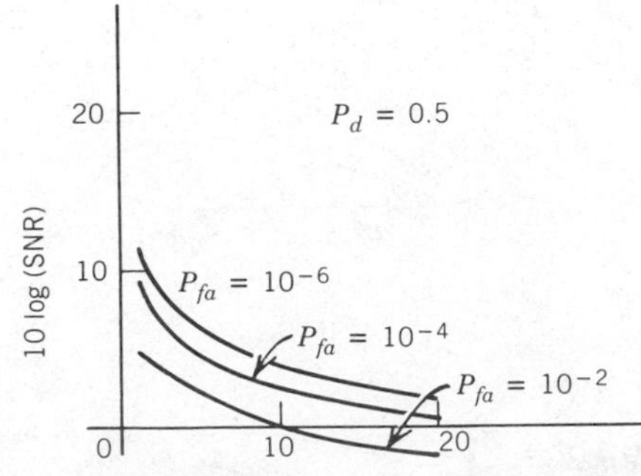

Fig. 51.5 Signal-to-noise ratio (SNR). P_d, probability of detection; P_{fa}, probability of false alarm.

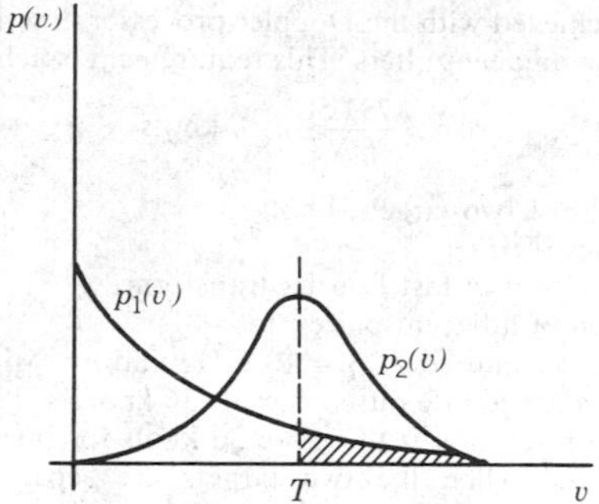

Fig. 51.6 Probability density of detector voltage. $p_1(v)$, no target; $p_2(v)$, target present.

that noise will be falsely declared as a target once in 10,000 opportunities. It can be seen that the required signal-to-noise ratio increases for decreasing P_{fa} and decreases as more pulses are integrated.

An illustration of the statistical nature of the detection decision is shown in Fig. 51.6. Curves 1 and 2 represent the probability density functions of the receiver voltage (v) after the detector. The effect of integration on the CRT phosphor has been modeled as the summation of the amplitudes of a succession of N pulses, as would be done by an automatic detector. The function $p_1(v)$ is the probability density of v when noise alone is present at the input. $p_2(v)$ is probability density of v when signal plus noise is present at the input. It can be seen that $p_2(v)$ has a larger mean value than $p_1(v)$. If a threshold value of T volts is established as the detection threshold, the P_d and P_{fa} values are given by

$$P_d = \int_T^\infty p_2(v)\,dv$$

$$P_{fa} = \int_T^\infty p_1(v)\,dv$$

In this manner results as in Fig. 51.5 can be determined.

51.5 IDEAL RADAR RESOLUTION

Resolution is the ability of the radar to distinguish two closely spaced targets. Targets can be resolved in any two of the following four coordinates: slant range, azimuth angle, elevation angle, and range rate. The first three of these radar coordinates are illustrated in Fig. 51.7, in which the radar is located at the origin of the x, y, z coordinate system. Range rate resolution is available only to radars employing Doppler processing, while elevation angle resolution is available only to 3D radars. In a conventional noncoherent processor with a fanbeam antenna (2D), resolution of closely spaced targets can be achieved only with the range and azimuth angle (i.e., bearing angle) coordinates.

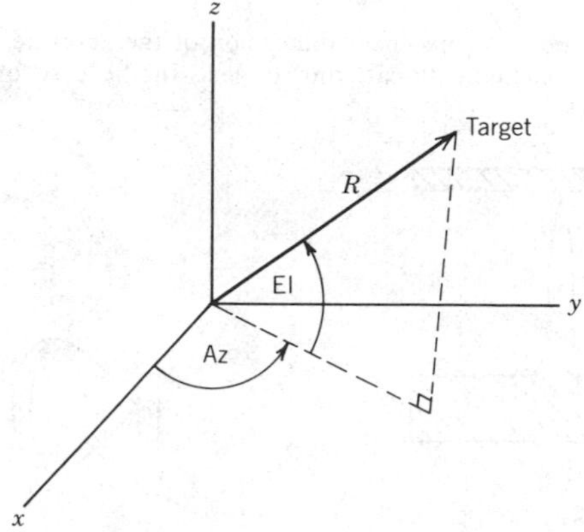

Fig. 51.7 Radar coordinates. Az, azimuth angle; El, elevation angle, R, range.

Range rate resolution can be achieved with the Doppler processor described in Section 51.8 if the two targets give responses that are in adjacent filters. This requirement is achieved when

$$\Delta \dot{r} > \frac{291.58}{f_0 NT} \quad \text{knots}$$

where $\Delta \dot{r}$ = range rate difference of two targets, knots
f_0 = transmitter frequency, GHz
N = number of pulses utilized in fast Fourier transform
T = interpulse separation of adjacent pulses, ms

Thus if a 1-kHz pulse train is transmitted at $f_0 = 10$ GHz and an 8-point fast Fourier transform is employed, then two targets with range rate differences of 36 knots will give filter responses separated by one filter width. Range rate resolution is therefore 36 knots for this case.

Range resolution is achieved when the two targets are separated in range by about two pulsewidths. Since a pulsewidth τ gives a response over a range interval Δr for a point target where

$$\Delta r = \frac{C}{2} \tau$$

the range resolution is given by

$$\text{Range resolution} \approx 2\Delta r = C\tau$$

With a transmitted pulsewidth of 0.1 μs, the range resolution is 3×10^8 m/s $\times 0.1 \times 10^{-6}$ s = 30 m. Thus targets separated by 30 m in range can be resolved with 0.1-μs transmissions. Narrow effective pulsewidths can also be achieved by encoding (modulating) longer duration pulses. For instance, if pulses of duration τ are transmitted and the modulated bandwidth of each pulse is Δf, then the effective pulse duration after decoding in the receiver is

$$\tau_{\text{eff}} = \frac{\tau}{\tau \Delta f} = \frac{1}{\Delta f}$$

In this manner, if each pulse has a modulation bandwidth equal to 10 MHz, then the effective pulsewidth after decoding in the receiver is 0.1 μs. This encoding can be achieved with linearly frequency-swept pulses (i.e., chirp) and phase- or frequency-coded modulations. With peak power limitations of the transmitter, higher average power can be achieved in this manner. The disadvantage of encoding the individual pulses is that they have to be decoded in the receiver, which makes the receiver more complicated. Also the longer duration transmissions increase the minimum range at which targets can be detected, since the receiver is turned off when the transmitter is on.

Angular resolution of a radar system is limited by the beamwidth of the antenna pattern. This beamwidth is related to the physical size in wavelengths of the antenna aperture (i.e., radiating area), as illustrated in Fig. 51.8. If the aperture dimensions are L_2 meters in width by L_1 meters in height, then the azimuth 3-dB beamwidth (i.e., between half-power points of the transmit pattern) is approximately λ/L_2 while the 3-dB elevation beamwidth is approximately λ/L_1 radians. Thus a 10λ aperture gives a beamwidth of 1/10 radian or 5.6°. A separation of two beamwidths is generally assumed as the angle resolution limit.

Theoretical resolution limits are somewhat smaller than the values given here, which are "engineering value" limits. For instance, the Rayleigh limit for angular resolution is

$$\Delta\theta = 1.22 \frac{\lambda}{D}$$

where Δ is the wavelength and D is the linear dimension of the aperture. This, however, applies to equal-strength targets and noncoherent scattering targets. In the case of nonequal scatterers, the

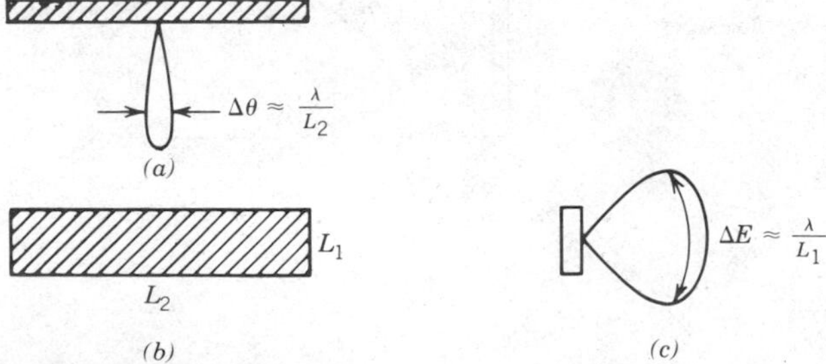

Fig. 51.8 Antenna beamwidths for rectangular aperture: (*a*) Top view, (*b*) front view, (*c*) side view.

sidelobe response of one scatterer may hide the mainlobe response of the weaker target. A similar situation exists in range resolution, especially when coded pulses are used. On decoding of these pulses, time sidelobes result, which interfere with the resolution process. Range rate resolution limits smaller than the value given previously can be achieved if one exploits the knowledge of the filter transfer functions. Amplitude information from responses in adjacent filters can be utilized to determine where the target is in the Doppler interval. As we try to refine these resolution limits the presence of noise creates a lower, impenetrable bound.

51.6 SYSTEM LIMITATIONS ON MEASUREMENT ACCURACY

In Section 51.5, resolution was defined as the ability to distinguish two closely spaced targets. This section deals with the accuracies achievable by a radar system when measuring the coordinates of a single target: slant range, azimuth angle, elevation angle, and range rate. The accuracies are generally a fraction of the resolution values and are a function of the measurement method, the transmitted waveform, and the amount of noise present. We will also discuss in this section the accuracies achievable when tracking is performed. Tracking refers to the combination of successive measurements from a target in order to more accurately estimate the target trajectory. Improved performance can be obtained in this manner, since the effects of noise is reduced when multiple measurements are combined.

Measurement accuracy for a single measurement is inversely proportional to the square root of the signal-to-noise power ratio. Examples of this will be given for both Doppler frequency measurement accuracy and angular position measurement accuracy. The in-phase and quadrature (I and Q) components of a signal combined with noise are

$$I = a \cos\theta + \eta_I$$

$$Q = a \sin\theta + \eta_Q$$

where a = signal amplitude
θ = signal phase
η_I, η_Q = two components of noise

When the target has a Doppler frequency f_d, the signal phase is given by

$$\theta = 2\Pi f_d t + \theta_0$$

where θ_0 is an arbitrary, initial phase. A discriminator or frequency measuring circuit estimates the Doppler frequency $\hat{f}_d$ as

$$\hat{f}_d = \frac{1}{2\Pi T}\left[\tan^{-1}\left(\frac{Q_2}{I_2}\right) - \tan^{-1}\left(\frac{Q_1}{I_1}\right)\right]$$

where T = signal duration, during which Doppler frequency is constant
I_1, Q_1 = quadrature components of input at time t_1
I_2, Q_2 = quadrature components at time $t_2 = t_1 + T$

When the standard deviation of this expression is determined to see how accurate the estimate is, we obtain

$$\sigma_{\hat{f}_d} = \frac{1}{2\Pi T\sqrt{a^2/2\sigma_n^2}} = \frac{1}{2\Pi T\sqrt{\text{SNR}}} = \frac{K}{T\sqrt{\text{SNR}}} \quad \text{Hz}$$

where $a^2/2\sigma_n^2$ is the input signal-to-noise ratio, SNR. Thus the Doppler frequency measurement accuracy is inversely proportional to the signal duration T and inversely proportional to the square root of the input signal-to-noise ratio. All frequency measurement techniques have the same functional relationship to SNR, the only difference being the value of K, which is $1/2\pi$ for this example.

A similar result can be derived for a phase comparison monopulse method for measuring direction of arrival of a signal wavefront. Monopulse means that this measurement can be made with a single pulse. In this method, shown in Fig. 51.9, two antennae separated by a distance D and with the same angular response pattern $f(\theta)$ receive the following signals:

$$S_1 = f(\theta)\cos(\omega_0 t + \psi) + n_1$$

$$S_2 = f(\theta)\cos(\omega_0 t - \psi) + n_2$$

where $\psi = (2\pi/\lambda)D\sin\theta$
n_1, n_2 = noise components superimposed on each signal
λ = wavelength of radiation
θ = direction of signal off boresight

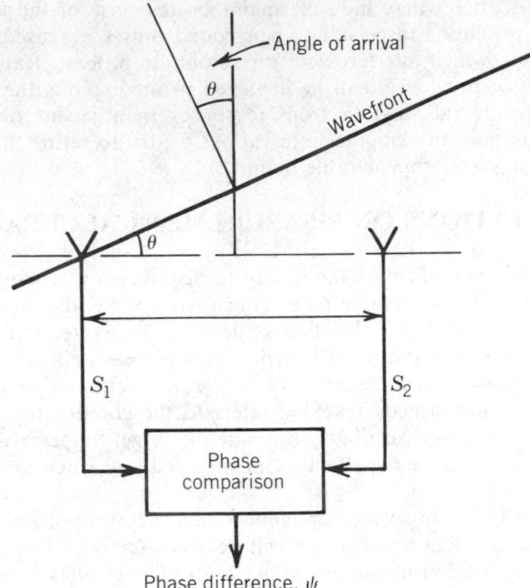

Fig. 51.9 Monopulse phase measurement.

The boresight direction is perpendicular to the line connecting the two antennae. From the measurement of phase angle, ψ, the angle of arrival of the signal can be estimated.

$$\hat{\theta} = \sin^{-1} \frac{\lambda\psi}{2\pi D}$$

The mean error in this measurement is zero, and the standard deviation is

$$\sigma_\theta = \frac{\lambda}{2\pi D\sqrt{2 \cdot \text{SNR}}}$$

Thus the accuracy of angular determination is inversely proportional to the square root of the input signal-to-noise ratio. Thus two dissimilar measurements, such as Doppler frequency and angle of arrival, lead to the same dependence on input signal-to-noise ratio. Similar dependence can also be derived when estimating the target range.

The previous result for angle of arrival accuracy assumes an interferometer-type measurement. In a scan radar system, in which the antenna pattern sweeps by the target and the angular position of the target is estimated by the centroid of the individual pulse responses, results are not as good. Angle estimation by this technique is considered highly accurate when σ_θ is 10% of λ/D. The same can be said for range measurements. When the range measurement $\hat{r}$ has a standard deviation equal to 10% of the pulse duration [i.e., $\sigma_{\hat{r}} = 0.1\ (C\tau/2)$], the measurement is considered highly accurate.

When successive measurements over a time interval are combined to estimate more accurately the target coordinates, target motion during the time interval must be considered. If target motion is negligible during the time in which N measurements are gathered, the resultant accuracy σ_N is given by

$$\sigma_N = \frac{\sigma_1}{\sqrt{N}}$$

where σ_1 is the standard deviation for a single measurement.

When target velocity is significant during the measurement period and it is necessary to estimate this velocity as well as the range, the standard deviation of the range error is

$$\sigma_N = \frac{2}{\sqrt{N}} \sigma_1$$

The estimate $\hat{R}$ in this case is

$$\hat{R} = \Sigma \alpha_i [R_0 + V_0(t_i - t_0) + n_i]$$

where α_i are selected to minimize the error in a least-squares sense. Extrapolating these results to the case where p parameters of motion have to be estimated (i.e., $p = 1$ for position, $p = 2$ for position and velocity, $p = 3$ for position, velocity, and acceleration), it can be derived that

$$\sigma_N = \frac{p}{\sqrt{N}} \sigma_1$$

Thus the accuracy continuously degrades as more parameters are estimated. These are the accuracies achievable by a pth-order tracker utilizing N measurements.

51.7 PROPAGATION PATH EFFECTS

The primary factors influencing the propagation of radar waves are multipath, refraction or bending, and attenuation and scattering by atmospheric gases.

Multipath

Multipath results when a target is illuminated by both a direct ray and a surface-reflected ray, as shown in Fig. 51.10. The resultant electric field strength at point T is a function of the path length difference between the direct and indirect paths, the reflection coefficient of the surface, and the antenna directional pattern and pointing angle. For a broad-beam antenna pattern in elevation angle, as for a two-dimensional radar measuring range and azimuth only, the antenna pattern effects are small. If there is a phase shift Δ at the surface and a path length difference equal to ϵ and if the magnitude of the reflection coefficient is ρ, then the resultant field strength at T is proportional to

$$\gamma = \sqrt{1 + \rho^2 + 2\rho \cos[\Delta + (2\pi/\lambda)\epsilon]}$$

The round-trip power back at the receiver is proportional to γ^4, so that the received power for a target at range R and radar cross section σ is

$$P_{\text{REC}} = \frac{P_t G^2 \lambda^2 \sigma}{(4\pi)^3 R^4} \left[1 + \rho^2 + 2\rho \cos\left(\Delta + \frac{2\pi}{\lambda} \epsilon \right) \right]^2$$

As shown in Fig. 51.10, if h_R and h_T are the heights of the radar and target above the reflecting surface, and d_0 is their separation along the horizontal plane, then the path length difference can be approximated as

$$\epsilon = 2 \frac{h_T h_R}{d_0}$$

If the magnitude of the reflection coefficient is 1 and its phase Δ is equal to 180°, which is the normal situation over seawater for horizontal polarization, it can be shown that

$$\gamma^4 \approx 16 \sin^4\left(\frac{2\pi}{\lambda} \cdot \frac{h_T h_R}{d_0} \right)$$

This gives a doubling of the target detection range, as compared with that obtained with the radar in free space, for targets where the argument of the $\sin^4$ function is as follows

$$\frac{2\pi}{\lambda} \cdot \frac{h_T h_R}{d_0} = \left(N + \frac{1}{2} \right) \pi$$

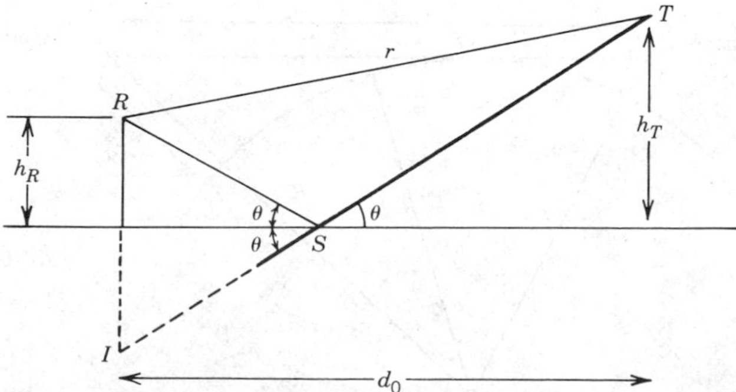

Fig. 51.10 Multipath geometry. R, radar location; T, target location; I, radar image location; S, point of reflection; h_R, height of radar antenna; h_T, height of target; r, slant range.

The first maximum occurs at $N = 0$ and successive detection maxima at $N = 1, 2, 3 \ldots$. A null, or zero response, is obtained in the multipath fade zone where the argument of the $\sin^4$ function is a multiple of π so that

$$\frac{2\pi}{\lambda} \cdot \frac{h_T h_R}{d_0} = N\pi$$

Thus as an airborne target flies on an incoming radial trajectory of a specified altitude toward the radar, it will alternately traverse fade zones and enhancement zones as a function of time.

The actual multipath effect will be less severe than calculated here because of the following factors:

1. Spherical scattering surface, owing to the spherical earth.
2. Roughness of the scattering surface, owing to waves of a specified rms wave height.
3. Directivity of the radar antenna pattern, which illuminates the sea surface with less intensity for the multipath ray as compared with the direct ray.

Refraction

The radiation from radar transmitters undergoes refraction or bending as these waves propagate through the atmosphere. This is due to the gradual change in refractive index of the atmosphere with increasing height, due in turn to changes in humidity, pressure, and temperature in the atmosphere. As a result, radar waves propagate about 15% beyond the geometric horizon. The calculation of these curved paths for the radar waves is quite complex and thus it is approximated by assuming a four-thirds earth radius, as shown in Fig. 51.11. Thus for a radar antenna mounted at a height h_R and a target at height h_T, the radar horizon range is given by

$$R_H = \sqrt{2R_e h_R} + \sqrt{2R_e h_T}$$

where R_e is the equivalent earth radius given by 4/3 times its actual value R_0. The 4/3 factor assumes a specific temperature and pressure at sea level and a certain decrease of these quantities with height. This is referred to as a standard atmosphere. A constant relative humidity is also assumed. When variable atmospheric factors over the globe are considered, R_e can be as large as $2R_0$, which leads to an even greater horizon range. The formula for R_H does not include the multipath effects discussed previously. Superrefraction refers to the case in which the rays are bent downward more than in a standard atmosphere. This condition generally occurs when a warm air mass, after passing over a warm land mass, flows over a relatively cold sea. In the Mediterranean Sea, which is almost completely surrounded by land, superrefraction is observed 90% of the time in spring and summer.

If the radar rays are bent down still more, they may be repetitively reflected and curved down again, as shown in Fig. 51.12. This phenomena is referred to as ducting, the rays appearing to be confined to a duct much as in a waveguide. With this phenomena, a radar transmission and its echo can be propagated very long distances (i.e., many times the radar ambiguous-range interval) before they escape from the duct. A radar PPI display can be quite confusing under these circumstances.

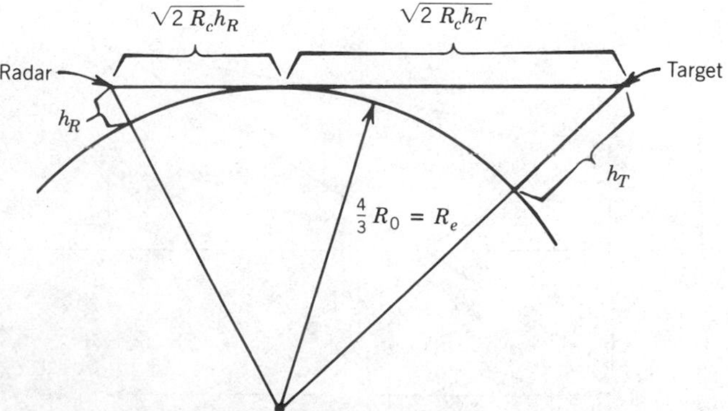

Fig. 51.11 Radar horizon range geometry.

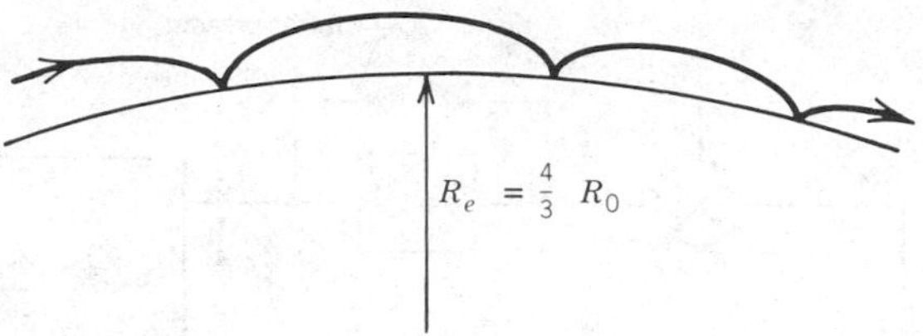

Fig. 51.12 Ducting geometry.

Attenuation and Scattering

In a clear atmosphere, radar energy is diminished by absorption by oxygen and water vapor. A part of the incident energy on the molecules is absorbed and lost as heat. This is primarily a quantum mechanical effect in which the molecular state goes from one energy level to a higher energy level. Fog, clouds, snow, hail, or rain absorb and scatter incident radar energy also. The amount of scattered radiation depends on the dielectric properties of the scatterers, the polarization of the incident radiation, and the relative size of the scatterer to the wavelength of the radiation. When the particles are much smaller than the wavelength, the scattered energy is near isotropic (i.e., uniformly scattered in angle). This is the usual case for radar waves interacting with atmospheric components. Clouds and precipitation can give echoes at the receiver owing to this scattering, and these are visible on the screen. If the echoes are intense enough they will interfere with normal target detection. If the target is moving with sufficient radial velocity with respect to this weather clutter, its presence can be determined by Doppler processing or MTI in the receiver.

51.8 DOPPLER RADAR AND VELOCITY MEASUREMENTS

Doppler radars have the ability to detect moving targets in the presence of heavy clutter, such as that returned from sea, land, and weather formations. The motion of the target relative to the radar transmitter and receiver produces a frequency shift in the received signal that is not present on the return signal from the clutter. Thus the Doppler radar can discriminate between signals from moving targets and signals from stationary objects.

If a radar target has a radial velocity $\dot{r}$ with respect to the receiver and transmitter (here assumed to be at the same location), then the frequency of the received signal will be

$$f_R = f_0(1 - 2\dot{r}/C) = f_0 - 2\dot{r}/\lambda$$

where f_0 = transmitter frequency
C = speed of light
λ = wavelength = $f_0 C$

The quantity added to the transmitted frequency f_0 is referred to as the Doppler frequency f_d. Thus

$$f_d = -2\dot{r}/\lambda$$

If the target has a component of motion radial outbound (inbound), then the Doppler frequency is negative (positive). For each knot of radial target motion at $f_0 = 10$ GHz, the received signal frequency changes by 34.29 Hz.

The target radial velocity can be measured nearly instantaneously with a surveillance radar if a signal processor, as shown in Fig. 51.13, is utilized. In this processor, called a pulse Doppler processor, the signal spectrum is analyzed. A phase-coherent pulse train of N pulses is transmitted with a constant interpulse period T between pulses (i.e., constant PRF). The received signal, which has induced phase modulation $\phi(t)$ owing to target motion, is combined with interfering noise plus clutter, indicated by $n(t) + C(t)$. The input signal is mixed with two reference signals in phase quadrature (i.e., the two reference signals are 90 electrical degrees out of phase with each other) in separate channels. The low-frequency components of this mixing operation are referred to as I and Q, the in-phase and quadrature components of the signal. An MTI operation is then performed on I and Q to reduce the effects of low-velocity clutter. The simplest MTI operation of this type is obtained by subtracting two successive samples of I and Q at the same range so that the moving target indicator (MTI) outputs are

$$I'(t) = I(t) - I(t - T)$$
$$Q'(t) = Q(t) - Q(t - T)$$

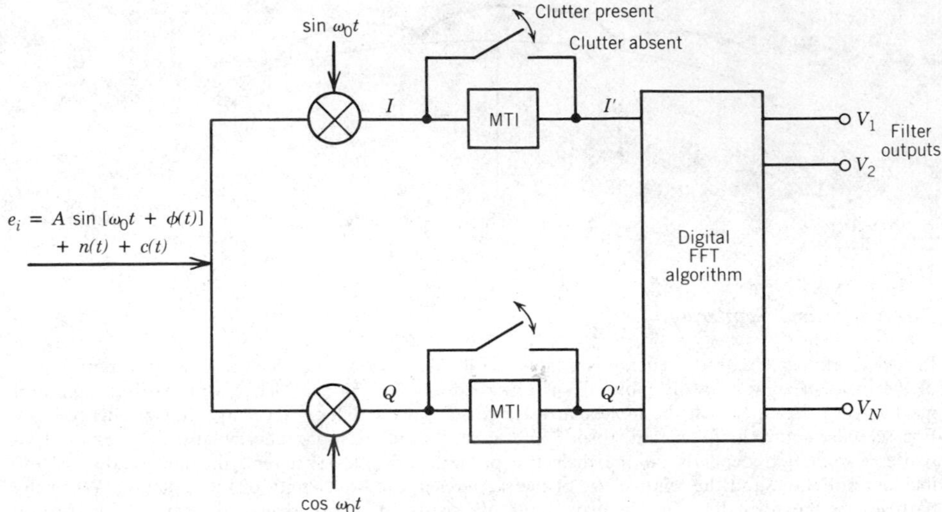

Fig. 51.13 Doppler filter processing. FFT, fast Fourier transform; MTI, moving target indicator.

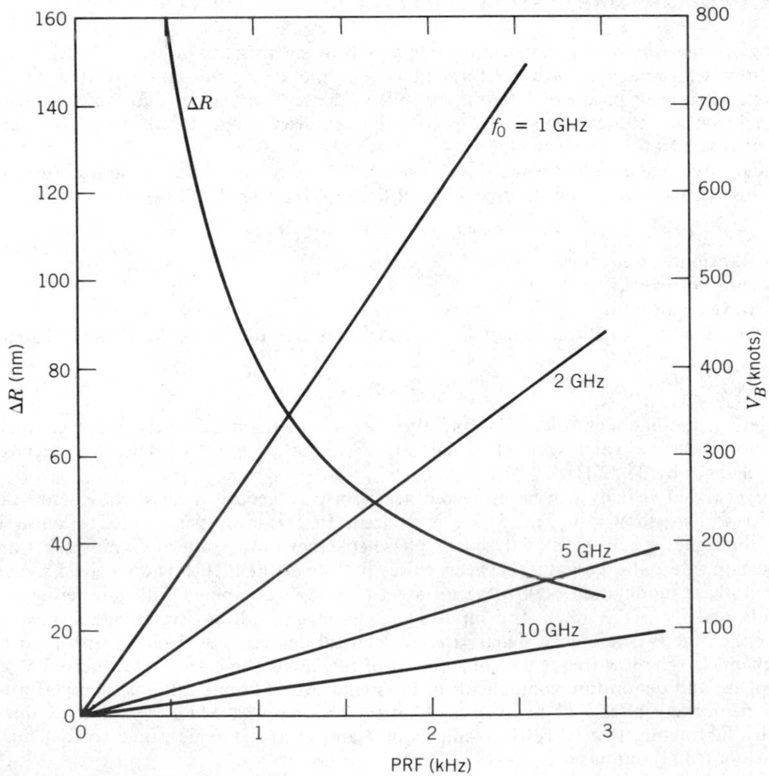

Fig. 51.14 First blind speed for MTI or pulse Doppler radar. Also shown is maximum unambiguous range, ΔR in nm. f_0, transmitter frequency.

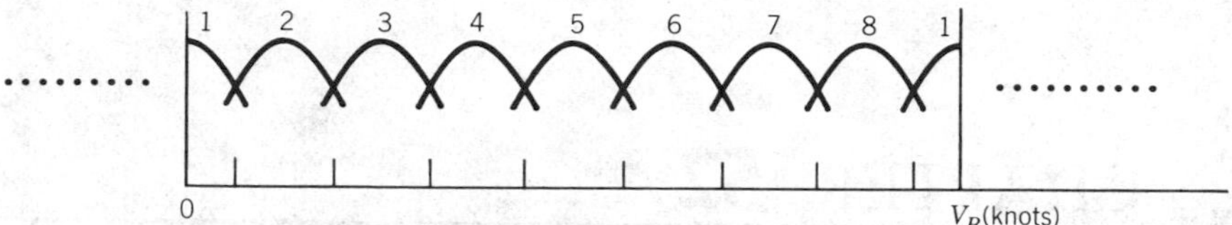

Fig. 51.15 Doppler filter bank. V_B, first blind speed.

where T is the interpulse period. MTI operations of this kind are sensitive to the target radial velocity. In fact, target responses are canceled if $f_d T = M$, where M is an integer. When $M = 1$, the value of $\dot{r}$ corresponding to $1/T$ is referred to as the first blind speed. This value of $\dot{r}$, defined as V_B in knots, is plotted in Fig. 51.14 as a function of PRF $= 1/T$ and transmitted frequency f_0. The relation is given as

$$V_B = 291.58\left[\frac{\text{PRF (kHz)}}{f_0\,\text{(GHz)}}\right] \quad \text{knots}$$

Thus for PRF = 1 kHz and $f_0 = 1$ GHz (i.e., 10^9 Hz), the value of V_B is 291.58 knots. Thus targets with range rates equal to 291.58 knots will have zero response out of the MTI. Also shown in this figure is the radar unambiguous range ΔR for a fixed PRF waveform. This relationship is given by $\Delta R = 81/\text{PRF (kHz)}$ nm.

Finally, the I' and Q' components enter a digital FFT processor which analyzes the signal spectrum in a Doppler filter bank with coverage as shown in Fig. 51.15. This filter bank results when an 8-point FFT is employed. The filter responses as shown are periodic with a period equal to V_B, the first blind speed. Thus if $V_B = 400$ knots, each filter handles a Doppler interval of $400/8 = 50$ knots. A 200-knot target will excite the central filter. Targets with range rates of 600, 1000, or 1400 knots will also excite this central filter. This Doppler ambiguity can be resolved by tracking the target range with time or transmitting other pulse batches at a different PRF. (This technique also eliminates possible range ambiguities.) For this latter case a different filter will be excited. With proper radar parameters the target range rate can be determined instantaneously to an accuracy better than 5 knots. With separated sites employing this method, the velocity vector of the target can be determined. If targets are detected in regions where clutter is absent, the MTI operation can be bypassed by closing the switches in Fig. 51.13. Targets at multiples of the blind speed can now be detected.

The pulse Doppler processor can be used with a scanning surveillance radar, if there is sufficient illumination time on the target, or with a dedicated tracking radar. Other modulations such as continuous wave (CW) or interrupted continuous wave (ICW, on-off CW with a high repetition rate and duty cycle) modulation can also be used. The returned signal can be spectrum analyzed in a digital or analog Doppler filter bank. Swept-frequency modulation is required to determine target range for CW or ICW transmissions. This can be accomplished by linearly or sinusoidally modulating the transmitter frequency. Range determination in this manner is generally more coarse than that achievable with a pulse Doppler processor.

51.9 ADVANCED TECHNIQUES—MILLIMETER WAVES

The development of millimeter wave radar systems will enhance the resolution capability of navigation systems in the future. With the utilization of frequencies exceeding 30 GHz, angular resolution will improve at least threefold for similar sized antenna apertures. Narrower pulsewidths will improve the resolution in the range dimension also. These systems will have limited range capabilities owing to atmospheric propagation characteristics at these frequencies, so they will be useful for close-in applications only. A dual-frequency system will be required (keeping the present X-band) if long-range coverage is required.

Systems are coming into production in the near future that will adequately serve as collision avoidance aids.[5] Measurement of the signature of signal returns will allow targets to be identified and automatically tracked with the rejection of clutter and unwanted echoes. With a millimeter wave system it is easy to envision the utilization of map-matching techniques within a computer to compare a radar-generated image with stored maps in order to plot a position in the map coordinate system.

Laser ranging systems can also be envisioned as an aid in performing the docking procedure. Measured relative position and rates can be utilized to automatically position the transmitting vehicle for this critical close-in procedure.

CHAPTER 52
NAVIGATIONAL AIDS

FREDERICK B. POGUST

Eaton Corporation, Farmingdale, New York

STEPHEN C. MARTIN

Lockheed Missiles and Space Company, Palo Alto, California

THOMAS J. CUTLER

United States Naval Academy, Annapolis, Maryland

52.1 INTERNATIONAL STANDARDS AND CONVENTIONS IN AIR NAVIGATION

Frederick B. Pogust

52.1-1 Early History

In the early days of aviation, visual contact with the ground was a necessity, and navigation was accomplished by techniques more akin to those associated with the auto than with the ship. Government involvement and regulation were equally primitive, and of course international coordination was unnecessary.

The evolution to the complex system of local, national, and international rules, standards, conventions, and procedures that govern air navigation today began with the need to fly under conditions of visual obscurity. Towns, farm buildings, roads, and rail lines (the iron compass) were no longer suitable guides to the aviator. During the 1920s, when aviation became a business, segments of both the industry and the government recognized the urgency of providing order to what was fast becoming a chaotic and dangerous situation in the air. In 1926 the Congress passed the first Air Commerce Act, and the federal government formally entered into activities that, insofar as navigation is concerned, constitute a partnership between those who fly and those charged with the responsibility to encourage flight in a safe and efficient manner.

The navigation systems that are currently employed in air commerce are survivors of hundreds of concepts that have been proposed, developed, tested, and used to achieve the dual objectives of guided flight: the pilot's need for information to assure that he or she is on the desired track between points A and B and the air traffic control authority's need for information to assure that plane Y is sufficiently distant from plane Z to avoid a collision. It is these two related but separate objectives that have led to the organization of most commercial flight into a system of designated and demarked airways.

The first airways consisted of ground markings along the route. Eventually rotating light beacons that flashed a beam along the airway were added. By 1928 the first radio range stations generating four aural courses were being operated by the Department of Commerce on the New York-Cleveland airway. To utilize this new form of radio navigation, the aircraft had to be equipped with proper reception equipment. The government supplied a standardized signal from the ground and the user needed equipment that could interpret that signal. The requirement for standards and conventions was born.

The institutionalization of the systems employed in air navigation proceeded at a fast pace throughout the 1930s. In 1935 the industry formed a coordinating group called the Radio Technical Committee for Aeronautics (RTCA) that would play a leading role in defining the systems that would be standardized. In 1938 the Congress passed a new Civil Aeronautics Act creating the Civil Aeronautics Authority (CAA) in the Department of Commerce. The CAA led the post-war evolution into our present system. World War II had the contradictory effects of halting any changes to the in-place civilian systems and of greatly accelerating the technology that would eventually replace them.

By the beginning of the 1940s, air commerce as carried on by the commercial airlines was almost totally performed by the DC-3. The United States and other developed parts of the world were laced with airways laid out along the courses of radio ranges. Aircraft of this era were equipped with voice radio for air-to-ground communications and with radio direction finders that would automatically orient their antennas toward a tuned station, providing a radio compass reading. These medium- and high-frequency receivers could also tune in a radio range. Flying an airway was accomplished by steering the aircraft to achieve a steady tone in the earphones. This indicated alignment with one of the four radio range legs. Straying off-airway would result in a Morse-coded "A" or "N" in the pilot's ears.

Overseas or international air commerce was confined to a handful of flying boats. They would find their way with the help of professional navigators, who took celestial fixes with hand-held sextants as had their mariner brothers for hundreds of years.

The CAA provided the system of radio ranges, nondirectional beacons, and fan markers that were standard to that era. It also had new systems under development: (1) the visual omnirange (VOR), which would provide any selected course to or from the range station, (2) distance measuring equipment (DME), which would provide slant range distance from a ground station, and (3) the instrument landing system (ILS), consisting of a localizer that would provide precise alignment with a runway, a glide path that would direct the correct descent rate, and marker beacons to give an indication of distance to the runway. These systems are discussed in detail in Section 52.2.

52.1-2 International Standardization

The end of the war set the stage for the transformation of the entire commercial air system. Aircraft were now capable of easily flying over the ocean and from one country to another. The technology needed to guide them had been perfected, but many of the systems that resulted from this technology were not compatible with each other. The few technologically advanced nations differed on the techniques that best suited the new era. In 1947 the first meeting of an international agency to deal with these matters was convened, and soon the International Civil Aviation Organization (ICAO), a specialized agency of the United Nations, became a permanent institution of air commerce with a staff and headquarters in Montreal, Quebec. Today ICAO has among its many responsibilities the ratification of all international aviation standards and conventions.

ICAO's role is to assure that nations participating in international air commerce provide standard services that can be utilized by suitably equipped aircraft. This means that aircraft can fly from country to country with a complement of equipments to perform the required procedures. ICAO has no official role in domestic flight, though pressure to make domestic systems conform to ICAO standards is obvious.

An interesting exception to this norm is the Soviet Union, which equips Aeroflot aircraft in international commerce with ICAO standard instruments and has installed ICAO navigation aids to serve the handful of its international gateway airports. All internal systems, however, were developed and deployed before the Soviet Union joined ICAO and these are totally nonconforming. Efforts are now under way to eventually convert the internal systems to ICAO standards.

The ICAO functional documentation for achieving its objective of standardization is the *International Standards and Recommended Practices (SARPS), Aeronautical Telecommunication*, Annex 10 to the Convention on International Civil Aviation. Volume 1 of Annex 10 includes Part 1, *Equipment and Systems*, and Part 2, *Radio Frequencies*. Volume 3 covers communications procedure. These volumes have been periodically updated since initially becoming applicable in 1950. Currently volume 1 is in its third edition, dated July 1972. At least 63 amendments have become applicable since its original publication. Specifications for the following radio navigation aids are included:

1. Instrument landing system (ILS)
2. Precision approach radar (PAR) systems
3. VHF omnidirectional radio range (VOR)
4. Nondirectional radio beacon (NDB)
5. UHF distance measuring equipment (DME)
6. Enroute VHF marker beacons (75 MHz)
7. Console system characteristics
8. Secondary surveillance radar (SSR)
9. Airborne ADF (automatic direction finder) receiver systems
10. LORAN-A system
11. Microwave landing system (angle guidance) (MLS)

These are the current international standardized systems. (The precision DME that will be used with the MLS is in process of being standardized.) There are other non-ICAO standard systems in use for air navigation throughout the world, and even in the United States. There is an entirely separate structure of standardization promoted by the Federal Aviation Administration (FAA), and a system of binding Federal Air Regulations (FARs) that also define the usable air navigation systems. In general, where ICAO and FAA standards overlap, they are in complete agreement.

52.1-3 Standardization in the United States

Within the United States the FAA is charged with the development of air navigation and air traffic control systems. The military operates its own components of these systems in conformance with FAA standards. When the CAA was the aviation authority, it did not have jurisdiction over military systems and there was divergence. A dispute between civilian and military agencies over the form of the short-range navigation system for domestic use was an element in the demise of the CAA and the formation of the FAA in the Federal Aviation Act of 1958.

Short-Range Navigation

The expressed unwillingness of Congress to fund separate military and civilian systems, as well as the possible hazards in such an uncoordinated policy, led to a study by the RTCA Special Committee 31 that recommended a common military–civilian system of air traffic control and navigation. The military proposed that its TACAN (tactical air navigation) system be adopted as this common navigation system, while the CAA had been proceeding with the implementation of the VOR and its own version of a DME. The civilian DME and the military TACAN, which includes a VOR-type signal and a DME, occupied the same band in the radio spectrum. The resolution of this conflict resulted in the VORTAC system. A VORTAC station, which is the standard short-range navigation aid operated by the FAA, includes a VHF ICAO standard VOR azimuth system (mostly used by civilian aircraft), an L-band (UHF) non-ICAO standard azimuth system (mostly used by military aircraft), and an L-band ICAO standard DME (used by both). While in practice few, if any, civilian aircraft are equipped for flying the L-band azimuth system, nothing prevents them from so doing in U.S. airspace. (Cost is the inhibitor.)

Landing Systems

The FAA also maintains a system of approach and landing aids called the instrument landing system (ILS) at over 1000 runways at qualifying civilian public-use airports. This system, while a guaranteed ICAO standard through at least 1995, is scheduled to be replaced by a new ICAO standard system called the microwave landing system (MLS) in future years. Many U.S. and foreign military air fields do not have ILS, but instead use a form of "talk-down" system employing an ICAO standard precision approach radar (PAR), an element of a ground controlled approach (GCA) system. The FAA at one time operated some PARs but no longer does. However, PARs are found at some foreign airports. Aircraft landing systems are discussed in Chapter 53.

Long-Range Navigation Systems

In the field of long-range navigation, the situation is even more complex and confused. Long-range navigation systems are generally used for over-water flying. This is true because long-range systems utilize frequencies in the lower regions of the radio spectrum that propagate further (but not as well over land as over water) than the higher, line-of-sight frequencies used by VORTAC. Because the continental United States is covered by over 800 VORTACS, the more complex long-range systems are seldom required. However, helicopter operators who often fly too low to receive continuous VORTAC coverage have been interested in having the FAA standardize a low-frequency system for their use.

The most common long-range navigation system that was developed during the war was LORAN-A, which is an ICAO standard. CONSUL is another that was favored by the British. These systems did not provide worldwide coverage, nor were they as accurate as is sometimes desired.

The U.S. Navy and Coast Guard had need of a system that was global in coverage. They developed and have installed the OMEGA system. This system, though wanting in accuracy for many air navigation functions (long-range low-frequency navigation systems are also used by ships, as discussed in Section 52.3), does provide worldwide coverage from only eight stations.

A more accurate version of LORAN, at a lower frequency, called LORAN-C, has had extensive deployment for coverage over the more heavily traveled over-ocean routes. LORAN-A stations have mostly been decommissioned.

Neither OMEGA nor LORAN-C have been standardized by ICAO. Instead the accuracy and reliability of so-called self-contained navigation systems (Doppler radar and inertial navigation) have predominated in recent over-ocean airline usage. This has resulted in the retirement of most of the human navigators. Self-contained systems require no ground stations and result in sufficient accuracy to feed ocean-crossing airplanes directly into the short-range navigation systems when they are within line of sight of land.

The result of this profusion of systems and techniques is that ICAO's early standardization of long-range systems has not been updated. The need has not arisen. From the ICAO point of view, self-contained systems need not conform to rigid compatibility specifications (there is nothing to be compatible with) so long as they are reliable enough to assure that aircraft in international flight can report their position with the required accuracy.

FAA's Role in Policy

The FAA's role in the specification and regulation of navigation systems differs from ICAO's in that (1) the FAA provides a direct service at federal expense and (2) the FAA is charged with assuring that aircraft are operated safely within the U.S. territory. The FAA therefore must assure itself that, insofar as it supplies the service, it limits its expense to only those systems it perceives as required to support the aviation community. Any navigation system used by any aircraft also must not, in any way, present a hazard. The FAA currently operates only the VORTAC and the ILS systems at civilian airports. The long-range navigation systems, that is, OMEGA and LORAN-C, are operated by the Navy and the Coast Guard. Equipment to use these latter systems is commercially available to aircraft owners, and indeed is quite common among the business class of aircraft. The FAA has the responsibility to regulate the quality of these products to ensure that they are airworthy and to certify that these systems are adequate to support the air traffic control system.

Plans for the Future

The Department of Transportation, of which the FAA and the Coast Guard are a part, has been charged by Congress to be the coordinator of a national navigation policy. Air navigation, because it is the most demanding, tends to dominate this policy. Current studies for this policy are centered on the eventual effect of the emergence of the satellite navigation system called global positioning system (GPS)/NAVSTAR. (See Chapter 54.)

GPS/NAVSTAR is a Department of Defense development that promises highly accurate, worldwide coverage to any vehicle suitably equipped to receive the signals from at least four satellites of an 18- or 24-satellite galaxy. The implication of such a system is that it could replace OMEGA, LORAN-C, and VORTAC, which are all currently maintained by the U.S. government, as well as all other short- and long-range navigation systems. To do so, however, it would also have to become an ICAO standard. The process by which a dedicated U.S. military system could become a civilian international standard is not obvious. The internationalization of a satellite navigation system in the future has many attractive technical features, even if the political and economic thorns may keep it at arm's length from implementation.

52.1-4 The Process of Standardization

ICAO Process

The process by which a system becomes an international standard has not been standardized. In 1950 when Annex 10 was initiated, the United States proposed the standardization of the systems that were being implemented at the time in this country. There was some controversy, mainly because the British proposed that their DECCA navigation system be the short-range system. The basic U.S. systems were eventually adopted.

Most of the amendments through the years have been the result of experience within the United States, where air traffic volume is the highest and where congestion puts the greatest demands on the air navigation systems. Therefore the SARPS, when written and ratified, usually reflect the results of extended real-life experience.

The adoption of a new system, or the dropping of an old one, is a laborious and prolonged process. Over 130 nations can participate, and any change can have grave financial consequences to large nations as well as small undeveloped ones. Special technical panels are formed by ICAO to study all the implications of a change, and any lingering doubts tend to prolong the process. ICAO, when setting a standard, guarantees that it will remain a standard for an extended period of time to assure the nations that equipment they buy will not be rendered quickly obsolete, and to assure aircraft owners that they can safely invest in airborne equipment.

The cost and difficulty in adding to the list of systems is illustrated by the process by which the MLS has recently been made an ICAO standard. The United States can no longer be sure that a vote in the ICAO assembly will support its position. On the other hand, even the United States cannot afford to develop a new system, install it domestically, amass the volume of data that would prove its suitability, and then risk possible rejection by ICAO. Therefore the development of the MLS was undertaken as an international effort. The vehicle the FAA chose to spearhead the development of a new landing system was the RTCA.

The RTCA is a joint government–industry coordinating group that welcomes international participation. In 1963 a similar organization was founded in Europe called the European Organization for Civil Aviation Electronics (EUROCAE). The two organizations do parallel work and effectively coordinate their efforts. Their role is to provide technical consultation to the governments that must make the implementing decisions. The procedure calls for the identification of a technical or operational problem and then the formation of a committee to explore the question. Its report is basically a recommendation which, as such, has no legal force.

The RTCA's role in the development of a new landing system was strictly advisory. After three years of deliberation, a large special committee recommended that one of two types of system be further developed. It reached this conclusion after study of the operational requirements as well as the technical alternatives.

ICAO also recognized the interest in replacing the ILS and instituted its own studies in an All-Weather Operations Panel (AWOP) concerned with documenting operational requirements. ICAO membership is limited to governments. Its panels are made up of individuals who have been designated by their governments to participate. Many of these same individuals participate in RTCA and EUROCAE activities, since those bodies welcome both government and industry membership.

The actual proposal to amend the SARPS can only be made by an ICAO member government. Five nations in all proposed specific landing system designs to the ICAO Air Navigation Panel as replacements for the ILS. A 12-member council was designated to review these systems and to recommend one to the ICAO Assembly. This body has representation from all member states.

France, Great Britain, Germany, Australia, and the United States made proposals for the MLS. The last two countries proposed the same system. The decision-making process was both technical and political. The Soviet Union endorsed the U.S.–Australia proposal, and after some accommodation for their concepts, Germany joined this group and the time reference scanning beam (TRSB) concept was selected.

The process of formal ratification and drafting of the exact language of the SARPS took about three years. The MLS has been incorporated by amendment to Annex 10 of the Convention. Implementation of the system will take many more years. Users are assured that the ILS will remain standard at international airports until at least 1995, and neither availability of MLS signals nor MLS equipment in aircraft will be required for many more years.

Process in the United States

While the RTCA's role when a major issue such as the MLS is debated receives some public notice, the real work of the organization goes on in its special committees. At any one time, as many as 15 such committees are active. Since 1935 there have been over 150 special committees. The work they do that bears most directly on the standards that apply to navigation is the issuing of minimum operational performance standards (MOPS) for various systems and equipments. These documents are issued as guidelines to builders of aviation equipments. They are often incorporated totally or by reference into FAA regulations. RTCA official documents, issued with distinctive green covers, represent the best possible consensus of technical requirements that can be worked out between government, equipment manufacturers, and equipment users, all of whom participate in the special committees.

The FAA implements and enforces navigation standards on domestic U.S. aviation by several processes. First, it is customary for the ground portions of a navigation system to be owned by the federal government. While the commissioning of a nonfederal aid is possible, such an aid must conform to FAA standards if it is to be open to public use. Several states or localities have, from time to time, bought and installed their own navigational aids, but always under FAA standards. It is also possible to operate a private navigation aid in the United States provided it is barred to the public and provided that the FAA is convinced that it represents no hazard to the user or the public. The FAA issues a special technical certificate for such private use.

A further control over the use of navigational facilities is provided by the FAA's requirements for an aircraft to receive air traffic control services. Under some weather conditions, all aircraft are barred from operating without air traffic control clearance, and some classes of air-carrier aircraft must always operate with air traffic control clearance. The FAA can and does require that such aircraft have adequate and approved navigation facilities.

VOR, DME, and ILS are the navigation systems that most easily meet this requirement (communications equipment and air traffic control radar beacon equipment are also usually required). However, other systems such as LORAN-C or OMEGA have been proposed as possible qualifiers. The FAA studies such proposals on an individual basis. Use of these nonstandard systems within domestic air space is rare.

Finally, the FAA in its certification authority does authorize the use of each specific item of equipment for aircraft. The design and manufacture of airborne units must be accomplished in accordance with FAA standards before the units are distributed to the aircraft-owning public. The FAA issues a technical standard order (TSO) for each item it approves. It is forbidden by federal air regulation to sell or use a navigation device for air traffic compliance that has not been issued a TSO. As issuance of a TSO is contingent on extensive tests, the TSO represents an enforced standardization. The equipment must perform its intended purpose.

The evolution of air navigation has been slow and steady since the earliest days of flight. There is no reason to believe that this trend will not continue. So far, the four-course radio range that pre-dated ICAO standardization is the only system that has totally disappeared. It is likely to be decades before any other system follows it into oblivion. On the other hand, future trends seem to point to self-contained navigation systems or earth satellite-based systems. These will no doubt bring forth new and imaginative methods to assure proper standards within workable conventions.

Bibliography

Federal Aviation Regulations, Part 37, "Technical Standard Order Authorizations," Department of Transportation, Federal Aviation Administration, Washington, DC.

Federal Radio Navigation Plan, Department of Transportation, Federal Aviation Administration, Washington, DC, 1980.

International Standards and Recommended Practices, Aeronautical Telecommunications, Annex 10 to the Convention on International Civil Aviation," volume 1, 3rd ed., July 1972 (amended).

Kent, R. J., Jr., "Safe, Separated and Soaring; A History of Federal Civil Aviation Policy, 1961–1972," US Department of Transportation, Federal Aviation Administration, Washington, DC, 1980.

Komins, N. A., "Bonfires to Beacons, Federal Civil Aviation Policy Under the Air Commerce Act, 1926–1938," US Department of Transportation, Federal Aviation Administration, Washington, DC, 1978.

Rochester, S. I., "Takeoff at Mid-Century; Federal Civil Aviation Policy in the Eisenhower Years, 1953–1961," US Department of Transportation, Federal Aviation Administration, Washington, DC, 1976.

Wilson, J. R. M., "Turbulence Aloft, the Civil Aeronautics Administration Amid Wars and Rumors of Wars, 1938–1953," US Department of Transportation, Federal Aviation Administration, Washington, DC, 1979.

52.2 NAVIGATION AIDS FOR AVIATION

Stephen C. Martin

52.2-1 Radio Aids to Aircraft Navigation

The ability of commercial, military, and private aircraft to navigate and land through almost all weather and visibility conditions is now taken for granted. This performance is achieved through the use of many different radio aids, most of which are the culmination of over 40 years of development. As an example, Fig. 52.1 shows all the antenna locations on a modern airliner required for the varied

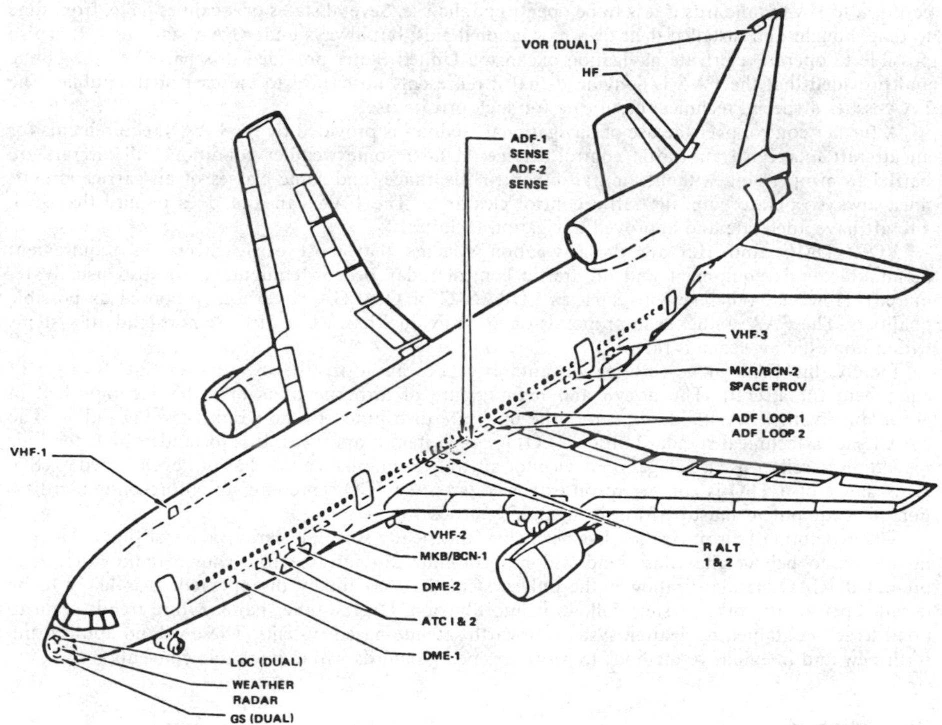

Fig. 52.1 Antenna locations on aircraft. From Lockheed California Company, with permission.

Antenna	Use
VHF (very high frequency) HF (high frequency)	Communications
VOR (VHF Omnidirectional radio range) ADF, loop and sense (automatic direction finder) DME (distance measuring equipment)	Short- and medium-range navigation
ATCRBS (air traffic control reporting beacon system)	Ground air traffic control surveillance
LOC (localizer) ADF (compass locator) GS (glide scope) MKR/BCN (75-MHz marker beacon) R ALT (radar altimeter)	Instrument landing

communication, navigation, surveillance, and instrument landing functions. This is the minimum complement required for continental navigation. For flight in oceanic and remote regions, additional long-range radio navaids or inertial navigation are used.

In modern large transport aircraft, management of these different navigational aids is assisted by flight management computers. These are able to select and tune the different systems automatically and to generate a single navigational display based on combinations of inputs. The latest generation of aircraft present these data to the crew using multifunction color cathode ray tube displays, integrated with other aircraft data such as heading, airspeed, and roll and pitch angles plus data from onboard weather detection equipment.

Discussed here will be radio aids for enroute navigation (short and long range) and for weather detection. Radio aids for instrument landing are discussed in Chapter 53.

52.2-2 Radio Aids for Short- to Medium-Range Navigation

Major systems providing service up to 200 nmi (nautical miles) from the ground station (depending on aircraft altitude) are summarized in Table 52.1. Of these systems, all except DECCA (a British system) and the azimuth measurement portion of TACAN have their parameters standardized by ICAO (International Civil Aviation Organization) agreement.

The low-frequency four-course radio range (Ref. 2, Ch. 2), which was the first navaid to be deployed extensively starting in the 1920s, is now obsolete. The VHF marker beacon, originally developed for use with this system, has survived for enroute use but is now mostly used as part of the instrument landing system (ILS).

Secondary surveillance radar (SSR), known in the United States as the air traffic control reporting beacon system (ATCRBS), is not an aircraft navigation system per se but rather an aid to ground surveillance of air traffic. As such it is a vital part of the air traffic system and is undergoing extensive further development as a data link for the transfer of standardized ATC messages and as the basis for an automated collision avoidance scheme.

Very-High-Frequency Omnirange (VOR) System

The very-high-frequency omnirange (VOR) system is the primary short- and medium-range radio navigation aid used by civil aircraft. It was originally adopted as an international standard by ICAO in 1949. VOR ground stations are to be found all over the world, although few countries except the United States have yet achieved almost 100% VOR coverage of domestic airspace.

VOR is a point source navaid, which gives the user a measurement of azimuthal bearing from the station referenced to local magnetic north. Unlike airborne direction finding or "homing" systems, VOR does not give a reading as a function of the heading of the aircraft. The ground station radiates

TABLE 52.1. MAJOR SHORT- TO MEDIUM-RANGE NAVAIDS

System	Frequency Range	Usable Coverage	Projected Number of Users in 1985		
			US Civil	US Military	World Total
VOR (VHF omni-directional range)	108–117.95 MHz	Line of sight	196,000	12,000	270,000
DME (distance measuring equipment	962–1213 MHz	Line of sight	70,000	13,000	130,000
TACAN (tactical air navigation)	962–1213 MHz	Line of sight	—	13,000	16,000
NDB/ADF (nondirectional beacon/automatic direction finding)	200–1650 kHz	50–200 nmi	100,000	24,000	170,000
DECCA	70–130 kHz	$\simeq$210 nmi	—	—	1,000
Marker beacon	75 MHz	$<$ 6 nmi	77,000	12,000	150,000
SSR (secondary surveillance radar)	1030 and 1090 MHz	Line of sight	180,000	24,000	250,000

Source. Federal Radio Navigation Plan.[1]

a VHF carrier simultaneously modulated by two 30-Hz signals. One 30-Hz component is of fixed phase and is called the reference signal. The phase of the second, or variable, signal varies as a function of azimuth from the station. Phase difference between the two signals then gives the observer a measure of azimuth from the station. Since the phasing of the signals is adjusted such that zero phase difference is observed at a location due magnetic north of the station, the actual phase difference (0–360°) measured by an aircraft is equal to its azimuth angle from the station.

In a conventional VOR the variable phase signal is generated by application of an unmodulated carrier to an antenna system with a cardioid azimuth pattern rotating at 30 r/s. From this component the user observes double sideband (DSB) amplitude modulation of the carrier at 30 Hz. The reference phase signal is transmitted from an omnidirectional antenna. The carrier is first DSB amplitude modulated with a 9960-Hz subcarrier. This subcarrier is sinusoidally frequency modulated at a 30-Hz rate with ±480-Hz peak deviation. The detected phase of this 30-Hz modulation provides the fixed reference.

Superimposed on the carrier of the reference signal are amplitude-modulated Morse-coded identifying signals using a 1020-Hz tone. In addition, at some locations voice signals for identification, weather broadcast, or communications may be added by amplitude modulation. The resultant composite spectrum is shown in Fig. 52.2.

VOR operating frequencies are in the band from 108 to 118 MHz. Channels have been assigned for 50-kHz spacing. At present only channels with frequencies in 100-kHz increments are actually used (e.g., 112.10 MHz, 114.30 MHz). Reception distance is essentially line of sight with little or no contamination by skywave propagation. Horizontal polarization is used. In initial tests of prototype systems[3] this orientation was believed to give smaller multipath propagation errors than did vertical polarization, although this conclusion was not universally accepted.

The modulation specifications are controlled by ICAO[4] and U.S. standards.

Station Design. An important element in radiating the horizontally polarized VOR signals is the Alford horizontal loop antenna.[6] This has a radiation pattern similar to a vertical dipole, but with horizontal polarization. It is a horizontal square loop with sides approximately 0.25 λ in length. For the VOR transmitter, four such elements are arranged in a closely packed square. When two diagonally opposite loops are fed equally with the signal, their field pattern is a horizontal figure eight. The field pattern of the other loop pair is a figure eight oriented at 90° to the first. Feeding each pair of loops with two signals that are amplitude modulated at a 30-Hz rate but are 90° out of phase results in a composite field pattern that is a horizontal figure eight rotating at 30 r/s. When this pattern is combined at the receiver with the carrier component of the reference signal that is being radiated from an omnidirectional antenna, the resultant pattern becomes a rotating cardioid.

In most VOR stations existing today, the two 30-Hz phase-shifted signals are generated by feeding carrier energy into a capacitative goniometer, a mechanical phase shifter, which is rotated at an accurate 1800 rpm by a synchronous motor.[7] A block diagram of the conventional VOR station employing mechanical modulation is shown in Fig. 52.3. The 9960 ± 480-Hz signal is derived from a tone wheel driven off the same shaft as the goniometer. This wheel has 332 teeth with cyclically staggered spacing. The signal is inductively picked off from this wheel, summed with the voice and identification signals, and the composite used to amplitude modulate the transmitter. The goniometer is fed by a portion of this signal that has had its modulation stripped off by a limiter. This arrangement preserves the phase relationship between the two signals. Combination of the reference- and variable-phase components occurs in two RF bridges.

Receiver Design. Figure 52.4 is the overall block diagram of a VOR receiver. The composite signal is recovered by the AM detector. Separate filters extract the voice, the 30-Hz variable-phase signals,

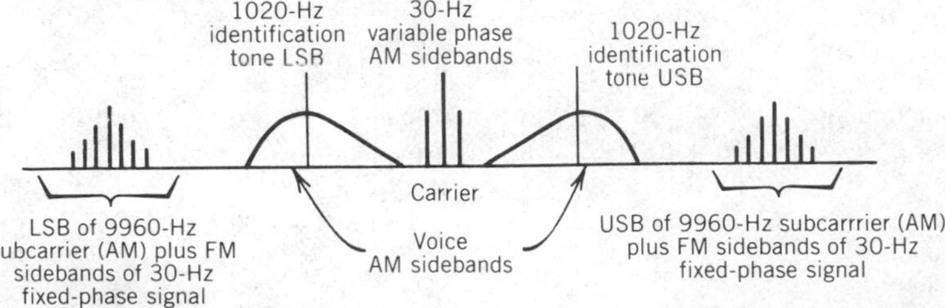

Fig. 52.2 Spectrum of a conventional VOR (VHF omnidirectional radio range) signal. LSB, lower sideband; USB, upper sideband.

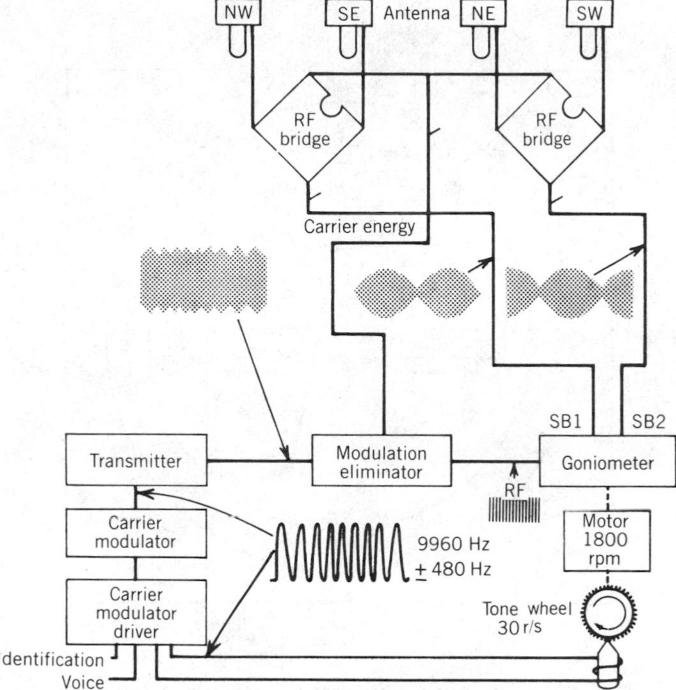

Fig. 52.3 Conventional VOR (VHF omnidirectional radio range) with four-loop antenna. From R. B. Flint, "VOR Past, Present, Future," presented at Institute of Navigation National Aerospace Symposium, Atlantic City, NJ, April 1978, used with permission.

and the frequency-modulated 9960-Hz subcarrier. The 30-Hz reference-phase signal is recovered from this subcarrier by a limiter–discriminator. To maintain system accuracy, close phase tracking of the two 30-Hz bandpass filters is essential.

Aircraft antennas range from simple horizontal V-dipoles to variations on the Alford loop. A mounting high on the aircraft vertical stabilizer is preferred to achieve omnidirectional coverage.

For the pilot's display, although phase reading indicators are available that give a direct readout

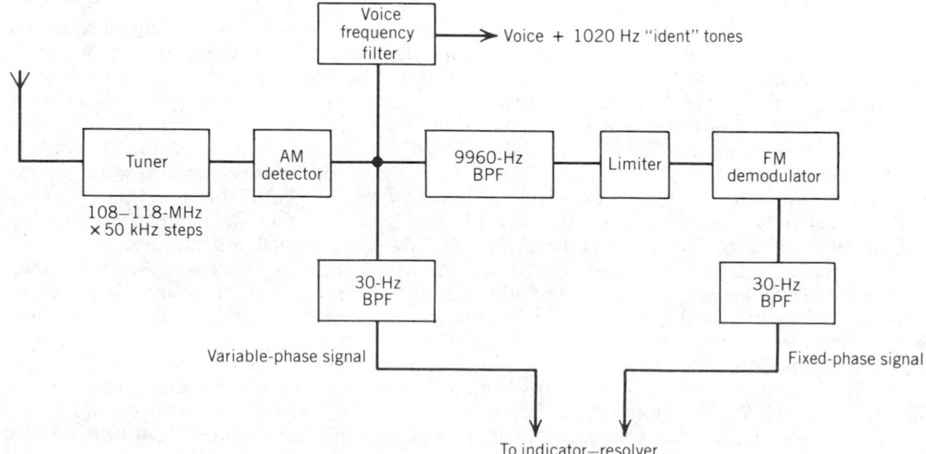

Fig. 52.4 VOR (VHF omnidirectional radio range) receiver. BPF, bandpass filter.

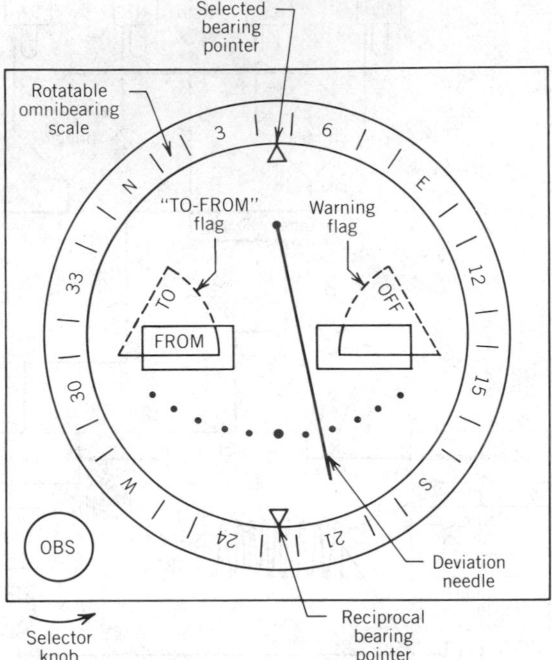

Fig. 52.5 Course deviation indicator (CDI). OBS, omni bearing selector.

(analog or digital) of the actual azimuth or "radial" from the selected station, a more common display is the course deviation indicator (CDI), sketched in Fig. 52.5. The rotatable omnibearing scale, called the omnibearing selector (OBS), is connected to a 30-Hz phase shifter. This shifts the reference phase signal by the number of degrees shown on the "selected bearing" pointer away from 0° (north). The deviation needle is driven by a resolver that compares the phase of this shifted reference with the variable-phase 30-Hz signal. Usually full-scale deviation indicates 10° bearing error. The needle centers when the aircraft is on the selected bearing from the station, or on its reciprocal. To resolve this ambiguity, one of the 30-Hz signals is shifted 90° and compared with the other in a second high-gain resolver, which drives a "TO–FROM" ambiguity resolution flag. When the 30-Hz signal levels become too weak for reliable use, a warning flag appears in a second window in the instrument.

By virtue of its being installed in about 90% of U.S. single and light twin-engined aircraft, the simple CDI is the most common VOR indicator; however, on large aircraft an indicator similar to the horizontal situation indicator (HSI), shown in Fig. 52.6, will be found. Indications are superimposed on a rotating compass card slaved to the aircraft's master compass system. The course arrow points to the desired VOR "radial" selected by the course set knob (290° in the illustration). The course deviation bar moves to show the position of the selected azimuth radial with respect to the symbolic aircraft fixed in the center of the display. When the bar is aligned with the course arrow, the aircraft is on the selected azimuth. A pair of direction arrows show whether the VOR station is in front of or behind the aircraft, and correspond to the "TO FROM" flags of the CDI. In addition, warning flags will appear prominently on the face of the display when the signal strength is inadequate.

Many variations of the basic HSI exist. Data from other navigation sources are usually included, for example, from the automatic direction finding (ADF) system (described later) and the instrument landing system (ILS) (see Chapter 53). For the next generation of transport aircraft, the symbology will be computer generated and displayed on a cathode ray tube.

Minimum performance characteristics for VOR receivers have been set by the Radio Technical Committee for Aeronautics (RTCA)[8] and standards for airline equipment by Aeronautical Radio, Inc. (ARINC). A VOR receiver with indicator for light aircraft would weigh about 2 kg, have a volume of 1750 cm^3, and consume 20 W, while for an airline unit these figures could be 6 kg, 5300 cm^3, and 70 W.

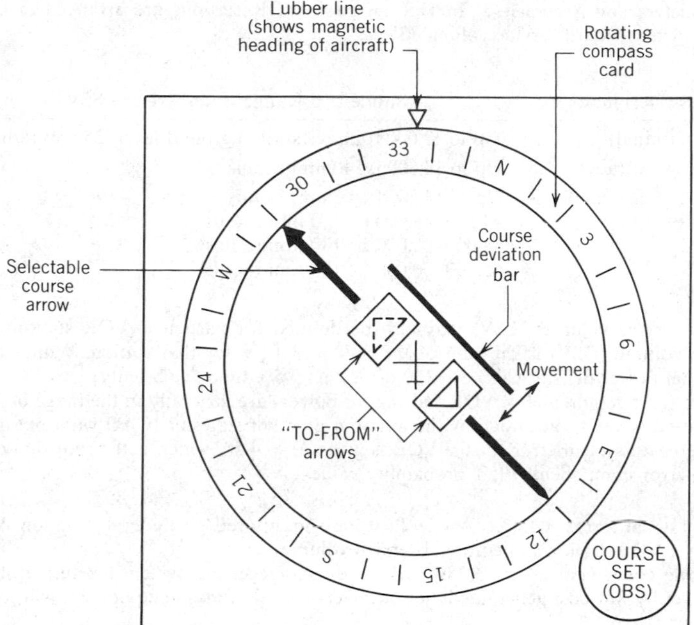

Fig. 52.6 Horizontal situation indicator (HSI). OBS, omni bearing selector.

Site Errors. The conventional VOR station requires careful siting for avoidance of angular azimuth errors caused by multipath reflections from nearby structures and terrain. Typically the antennas are placed about one-half wavelength above a circular metallic counterpoise some 7 m in diameter, which is itself elevated about 5 m above the ground. Much larger and higher counterpoises are sometimes necessary. For more difficult sites, the Doppler VOR was developed.

Doppler VOR. The Doppler VOR[9] is able to reduce multipath effects partly by employing a much larger antenna aperture than the conventional station. A circular array of 50 Alford loop elements on a diameter of 13.4 m is placed above a counterpoise of nominally 45.7 m diameter. With f_0 the nominal station frequency, a carrier at $f_0 + 9960$ Hz is commutated sequentially to each antenna element at a rotation rate of 30 r/s. An observer at a distance will see an apparent sinusoidal 30-Hz frequency modulation owing to the Doppler effect. The spacing and diameter of the antenna ring are such that the peak frequency deviation is ± 480 Hz, corresponding to that of the conventional VOR. The phase of this 30-Hz modulation is a function of the observer's azimuth from the station. A reference phase is generated by DSB amplitude modulating with a fixed 30-Hz tone an omnidirectionally radiated carrier of frequency f_0. Thus for the Doppler VOR, the sources of the "variable" and "reference" 30-Hz signal modulations are reversed with respect to the conventional VOR. However, since the aircraft receiver measures only their phase difference, its performance is the same as would be obtained with a conventional transmitting station.

Compared with the spectrum of the conventional VOR signal, the Doppler VOR generates only the upper sideband components of the frequency-modulated 9960-Hz subcarrier. Some VOR receivers (mainly the smaller, less expensive units) have shown sensitivity to this effect, producing larger than allowable phase errors. To compensate for this effect, double sideband Doppler VOR has been tested. Here a carrier at $f_0 - 9960$ Hz is simultaneously commutated with that at $f_0 + 9960$ Hz, but with connections physically 180° away around the ring of antennas, which inserts the lower sideband. VORs of this type have only recently begun to be operationally installed.

In Europe a similar double-sideband technique known as alternating sideband Doppler VOR has been implemented.[10] This uses 39 loops in the circle. Instead of exciting diametrically opposed loops simultaneously with the upper and lower sideband signals, it excites them alternately. Power shaping is used so that as the signal power is increasing at one antenna element, it is decreasing at the appropriate loop on the opposite side of the circle.

System Parameters and Accuracies. In U.S. airspace, VOR stations are arranged in three classes, each with its own standard service volume (SSV).[5]

Class Designator	Altitude and Range Boundaries of SSV
T (terminal)	Up to 12,000 ft above station ground level; 25 nmi radius
L (low altitude)	Up to 18,000 ft; 40 nmi radius
H (high altitude)	Up to 14,500 ft; 40 nmi radius 14,500–18,000 ft; 100 nmi radius 18,000–45,000 ft; 130 nmi radius 45,000–60,000 ft; 100 nmi radius

Expanded service volumes (ESV) may also be defined for particular VOR stations. The operational service volume (OSV) is the sum of the SSV and ESV for that station. Within the OSV the minimum signal power density will be $-120\ \mathrm{dBW/m^2}$ (95% time availability).

To achieve these requirements, VOR transmitter powers are generally in the range of 10 to 200 W.

Methods used to define accuracy by the different member states of ICAO vary, but in the United States the aggregate system error for the VOR is given as $\pm 4.5°$, which is the root sum of squares of the following error components (95% probability values):

1. **Radial signal error: $\pm$1.4°.** A tolerance limit monitored by Federal Aviation Agency flight checks within a VOR station's operational service volume.
2. **Airborne component error: $\pm$3.0°.** Error due to receiving system (exclusive of indicators). This value is set by low-cost general aviation receivers. Modern airline equipment will reduce this by a factor of four.
3. **Instrumentation error: $\pm$2.0°.** Errors due to indicator (including setting error) and auto-pilot coupling, when used. It is significantly lower for digital displays.
4. **Flight technical errors: $\pm$2.3°.** Error due to piloting inaccuracy. It is significantly reduced by auto-pilot coupling.

Distance Measuring Equipment (DME)

The FAA and ICAO standard DME is a pulse-ranging system. Each aircraft interrogates a ground transponder and measures the round-trip delay time between the interrogation and the returned reply pulses. It was originally developed as the ranging element of the TACAN (tactical air navigation) system; however, the DME functions are capable of totally independent implementation.[11]

Aircraft interrogations are made on one of 126 frequencies spaced 1-MHz apart from 1025 to 1150 MHz. The ground beacon reply is on a frequency 63 MHz offset from the interrogation. When the reply is 63 MHz lower, the arrangement is called an X channel, and when higher, a Y channel. There are thus 252 reply frequencies ranging from 962 MHz to 1213 MHz. Both air and ground signals are radiated with vertical polarization, and the aircraft and ground antenna patterns are made as near omnidirectional as possible. Each interrogation and response consists of a pair of pulses. The pulse spacings are as follows.

Channel	Airborne Interrogator Pulse Spacing	Ground Reply Spacing
X	12 μs	12 μs
Y	36 μs	30 μs

Tolerance on spacing is 0.25 μs. Each pulse is Gaussian shaped in order to minimize spectral occupancy, with a half-amplitude width of $3.5 \pm 0.5\ \mu$s (Ref. 4, Vol. 1, Sec. 3.5). The ground transponder is designed to introduce a fixed delay of 50 μs between correct decoding of a pulse pair and retransmission of a pulse pair. This fixed delay is subtracted by the airborne equipment in its estimation of round-trip time.

At present only the X channels are actually used by civil aircraft, although use of Y channels has been authorized by the US Department of Transportation and by ICAO, and most civil aircraft DME installations already have this capability.

Acquisition and Tracking. Each ground DME transponder beacon will, of course, be operating simultaneously with many aircraft within its area of coverage. Since each aircraft interrogating that beacon is using the same frequency and signal format, and the aircraft interrogations are not in any way synchronized, each aircraft has the problem of sorting the replies to its own interrogations from those to the interrogations of other aircraft. For this purpose, the interrogation rate of each aircraft is deliberately jittered in a random manner. The DME receiver then searches for a pattern of returned pulse pairs occurring at a fixed delay from the aircraft interrogation time. When the aircraft's search gate is correctly positioned relative to the time of interrogation, a reply will be detected in response to most of the "own aircraft" interrogations. Replies from other aircraft interrogations will vary randomly with respect to this gate time.

For initial search, the aircraft interrogates at a mean rate of about 150 interrogations per second. The range gate is moved outward from zero range delay until a high proportion of reply pairs is detected. This stops the search and initiates the track mode. In track, the mean interrogation rate is reduced to a lower value of around 20 interrogations per second. Now the gate is continually repositioned, with use of a predictive tracking loop, to maintain the returned pulses in the center of the gate as the aircraft's range changes. The prediction loops allow track to be continued even when a high proportion of the reply pulses are missing. Interrogation rates as low as three per second are practicable with modern airborne equipment.

DME measures slant range between the aircraft and the beacon, rather than ground range. The practical effect of this difference is negligible except when the aircraft is within a few nautical miles of the ground station.

Ground Beacon Implementation. The ground transponding beacon receives and transmits on a pair of fixed frequencies separated by 63 MHz. The beacon antenna generally consists of vertically stacked dipoles, producing an omnidirectional pattern in the horizontal plane. Gain is at least 4 dB over a single dipole; values up to 9 dB sometimes are used. Following each successful interrogation, the ground receiver is inhibited for a delay of about 60 μs. This is intended to prevent triggering by delayed multipath echoes of the direct interrogation. In severe, mountainous terrain, this 60-μs delay may have to be increased to as much as 150 μs.

Beacons are designated to operate at a reply rate of 2700 pulse pairs per second. This corresponds to the handling of 100 aircraft, 5% of which are in the search mode (150 interrogations per second) and 95% in the track mode (average of 20 interrogations per second). This pulse-pair rate is maintained even when fewer or no interrogations are present, by automatically increasing the ground receiver gain until the beacon triggers on its own front-end noise. Gain is dynamically adjusted so that the total number of transmitted pulses due to interrogations and random noise averages 2700 pulse pairs per second. This arrangement allows (1) the transmitter to operate at a constant duty cycle, (2) the aircraft receiver automatic gain control to have a steady number of pulses to integrate over, even before commencing search interrogations, and (3) graceful degradation in the event of saturation by too many aircraft: interrogations from the most distant aircraft are the ones that are disregarded first.

The range of beacon gain variation is 50 dB or greater.

Identification. DME beacons are required by FAA and ICAO specification to multiplex audible identification codes along with the navigation signals. At intervals of about 30 s transmission of reply and random pulse pairs is suppressed, while two or three Morse-coded identification characters are transmitted. Symbols are 0.125 s long for dot and 0.375 s for dash and are sent as pulse pairs at a steady rate of 1350 pulse pairs per second. In the aircraft this signal is detected by a tuned circuit and fed to the audio distribution system as a 1350-Hz Morse-coded tone. In the spaces between the dots and dashes, regular replies and random pulses are interspersed. Transmission of each identification code group lasts less than 5 s.

Airborne Implementation. Figure 52.7 is a block diagram of typical airborne DME. Transmit and receive frequencies are always 63 MHz apart, so use of a 63-MHz intermediate frequency allows a single frequency source to act as transmitter excitation and receiver local oscillator. Aircraft antennas are vertical quarter wave stubs, which at DME frequencies are about 7 cm in length. Aircraft transmitter peak pulse powers range from 100 W for the simplest units to 1 kW or more for airline equipment.

In all modern equipment, round-trip delay is measured by digitally counting down a precise clock signal. After subtraction of the fixed 50-μs beacon delay time, each nautical mile corresponds approximately to 12 μs round-trip delay. Although maximum usable range is set by line-of-sight limitations, the maximum measurement range of the equipment is generally 200 nmi, although some units designed for high altitude aircraft can measure out to 400 nmi.

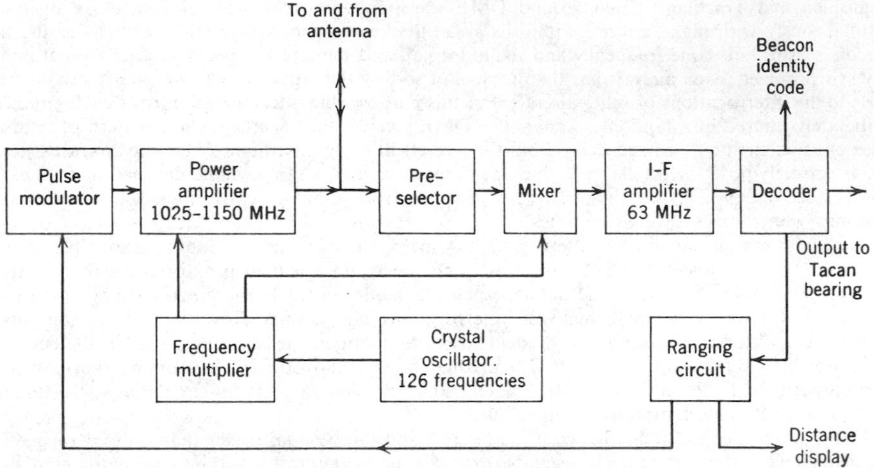

Fig. 52.7 Block diagram of airborne distance measuring (DME). From Kayton and Fried, eds.,[12] with permission.

Displays present a digital readout of slant range to the selected beacon in nautical miles and tenths. A second digital display is derived from the velocity tracking loops and gives range rate in knots. When the aircraft is tracking directly to or from the beacon (the usual case for civil aircraft), this is a close approximation to ground speed. A third digital display is usually provided that gives "time to go to station" directly in minutes, whenever range rate is negative (i.e., during flight toward the beacon).

Minimum performance standards for airborne DME have been set by RTCA.[13] A basic DME for light aircraft would weigh 1.5 kg, occupy 1650 cm^3, and consume about 30 W. For airline units these figures would be closer to 7.5 kg, 11,000 cm^3, and 55 W.

System Parameters and Accuracies. Operational service volumes (standard plus extended) are defined for DME beacons (Ref. 5, Ch. 4) in exactly the same way as for VOR stations. Usable volumes are defined as being those where the power density at the aircraft antenna is as follows:

Above 18,000 ft above ground level	-91.5 dBW/m^2
Below 18,000 ft above ground level	-86.0 dBW/m^2

These are averages of the peaks of the pulses, and are 95% time availability numbers. To achieve these values, DME ground beacon peak powers range from 10 kW for long-range enroute service to as low as 1 kW for beacons designed for terminal-only service (30 nmi radius or less) or for beacons installed in association with an instrument landing system (ILS) (see Chapter 53).

At maximum gain, the beacon receiver sensitivity specification requires 70% reply efficiency when the incident peak power density from an aircraft interrogation is -101.5 dBW/m^2.

Required DME accuracies are given in ICAO and FAA specifications. The error allocation due to all functions of the ground beacon is a maximum of 0.1 nmi (about 1 μs delay error). The combination (root sum squared) of this error with that attributable to the airborne equipment is the total error in distance, which as displayed to the pilot must not exceed 0.5 nmi or 3% of the total distance, whichever is greater. In practice the actual error is, to 95% confidence, ± 0.2 nmi or 0.25% of slant range, whichever is greater. A compatible DME, but one that has significantly higher accuracy for short ranges, is being developed as part of the microwave landing system (MLS) (see Chapter 53).

Tactical Air Navigation (TACAN)

Tactical air navigation (TACAN) is a point source (rho-theta) navigation system used by the US military, NATO, and other forces of the Western world. Its development dates from 1948 and its production from 1953.[14]

A TACAN ground (or shipborne) beacon consists of a DME beacon, as described previously, with amplitude modulation added to its reply and randomly transmitted pulses in order to provide azimuthal information. The formats and frequencies of the distance-measuring pulses are the same as defined by the FAA and ICAO specifications for DME (i.e., 126 X channels and 126 Y channels). Specifications of the azimuth-measuring functions are given in the US TACAN military standard document.[15]

Azimuth Signal Format. The transmitted pulses are amplitude modulated by a composite of two synchronized waveforms generated by an antenna system rotating at 15 r/s. The first component is sinusoidal amplitude modulation at 15 Hz (once per revolution) and the second is sinusoidal AM at 135 Hz (nine times per revolution). The rotation rate is clockwise, viewed from above the beacon. When the beacon is at a magnetic azimuth angle γ from an observer, the envelope of detected pulses seen by that observer is given by

$$y(t) = 1 + A\sin(2\pi ft - \gamma) + B\sin(18\pi ft - 9\gamma)$$

where $y(t)$ = normalized composite detected envelope
A = modulation of 15-Hz component = 0.21 ± 0.09
B = modulation of 135-Hz component = 0.21 ± 0.09 (sum of A and B not to exceed 0.55)
f = pattern rotation frequency = 15 Hz $\pm$ 0.23%
t = elapsed time from some defined "reference time"

Azimuth position is computed by measuring, in $y(t)$, the phase of both the 15-Hz and 135-Hz signals. Unlike VOR, in which 1° of electrical phase corresponds to 1° of azimuth, for TACAN 9° of electrical phase of the 135-Hz signal corresponds to 1° of azimuth. Since the 135-Hz signal has nine ambiguities per revolution (at 40° intervals), the phase of the 15-Hz component is needed to resolve these.

Two sets of identifying signals define the "reference time" from which t is measured in the equation given. The first set is called the main reference group and is transmitted as the peak of the 15-Hz modulation is pointing due east of the station (this is also known as the "north" reference signal). The second signal set is transmitted each time the peak of the 135-Hz modulation points east and is called the auxiliary reference group. There are thus nine auxiliary groups to each main group, except that, since the peaks of the two modulations are synchronized when the 15- and 135-Hz modulation peaks are both pointing east, the auxiliary reference signals are then suppressed in favor of the main reference signal. Figure 52.8 shows the pulse envelope waveform that would be seen by a user magnetically due east of the TACAN station.

Main and auxiliary reference groups are generated by groups of pulses with specific pulse spacings, as defined in Ref. 15. During transmission of main or auxiliary reference group pulses, the random, identity, and distance reply pulses are suppressed.

Beacon Configuration. A TACAN ground beacon is a DME beacon plus a modulating antenna system and circuits to generate the reference pulse groups. Figure 52.9 shows the configuration. The majority of TACAN beacons in use have mechanically rotating antenna systems. Two sets of

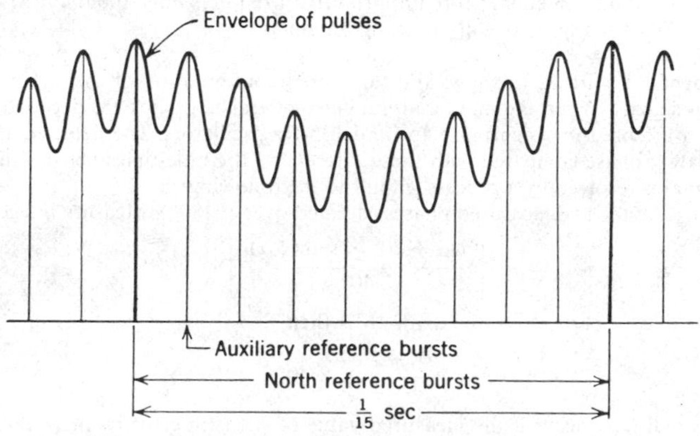

Fig. 52.8 Tactical air navigation (TACAN) signal (east of station).

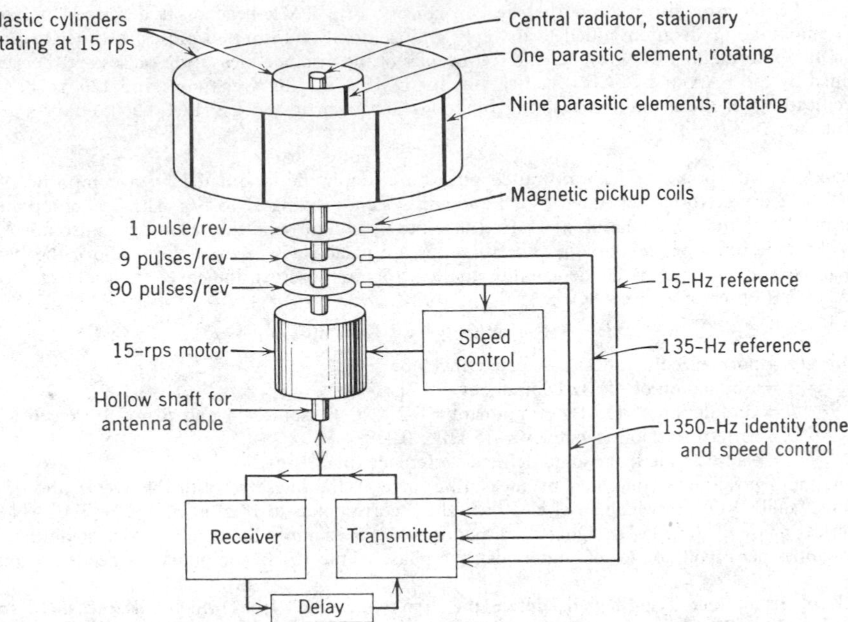

Fig. 52.9 Tactical air navigation (TACAN) ground beacon. From Kayton and Fried, eds.,[12] with permission.

parasitic elements are rotated about the central radiator at 15 r/s to produce the composite amplitude modulation. The diameters of the inner and outer plastic cylinders carrying these elements are approximately 15.25 and 91 cm, respectively. Trigger pulses for the main reference, auxiliary, and 1350-Hz identity pulse groups are derived from three disks rotating on the same shaft. For shipboard use the TACAN antenna is stabilized mechanically or electronically to compensate for ships' heading and roll, so that the signals in space accurately define compass bearings referenced to the ship.

Solid-State Antennas. Electronically modulated solid-state antennas have been developed and are in service. One version (Model AT-100, Rantec, Division of Emerson Electric) uses 36 vertical arrays arranged in a circle of 1.5 m diameter. Each array is 3.15 m high and consists of 12 cavity radiators. A modulation generator and azimuth feed network produce correctly phased drive currents to generate the rotating nine-lobe pattern. This assembly weighs 350 kg, as opposed to the over 1000 kg of the conventional rotating antenna. More importantly, it requires only about 100 W drive power as opposed to the 5 kW of the mechanically rotating antennas.

Airborne Equipment. Figure 52.10 shows the functional configuration of the equipment to decode TACAN azimuth signals. When the main and auxiliary reference groups are decoded, sample pulses are generated that are used to synchronize 15- and 135-Hz oscillators. The detected 15- and 135-Hz signals are separately phase compared with these references. the nine-times-per-revolution ambiguity in the 135-Hz signal is resolved by a process similar to the following:

Let ϕ_{15} and ϕ_{135} be the two measured phases (defined over $\pm 180°$). Measure integer K as

$$\left\lfloor \frac{\phi_{15} + 20° \times \text{sign}(\phi_{15})}{40°} \right\rfloor$$

where $\lfloor\ \rfloor$ denotes integer part. Measured azimuth is then

$$\frac{\phi_{135}}{9} + K \times 40°$$

Ambiguity errors will only occur if the measured value of ϕ_{15} is in error by more than $\pm 20°$.

Airborne Displays. The basic azimuth display for TACAN is a rotating pointer driven from the selected TACAN station's signal over $0 \rightarrow 360°$. When the indicator is slaved to the aircraft's gyrocompass, the card and pointer rotate together and show (1) aircraft heading and (2) TACAN

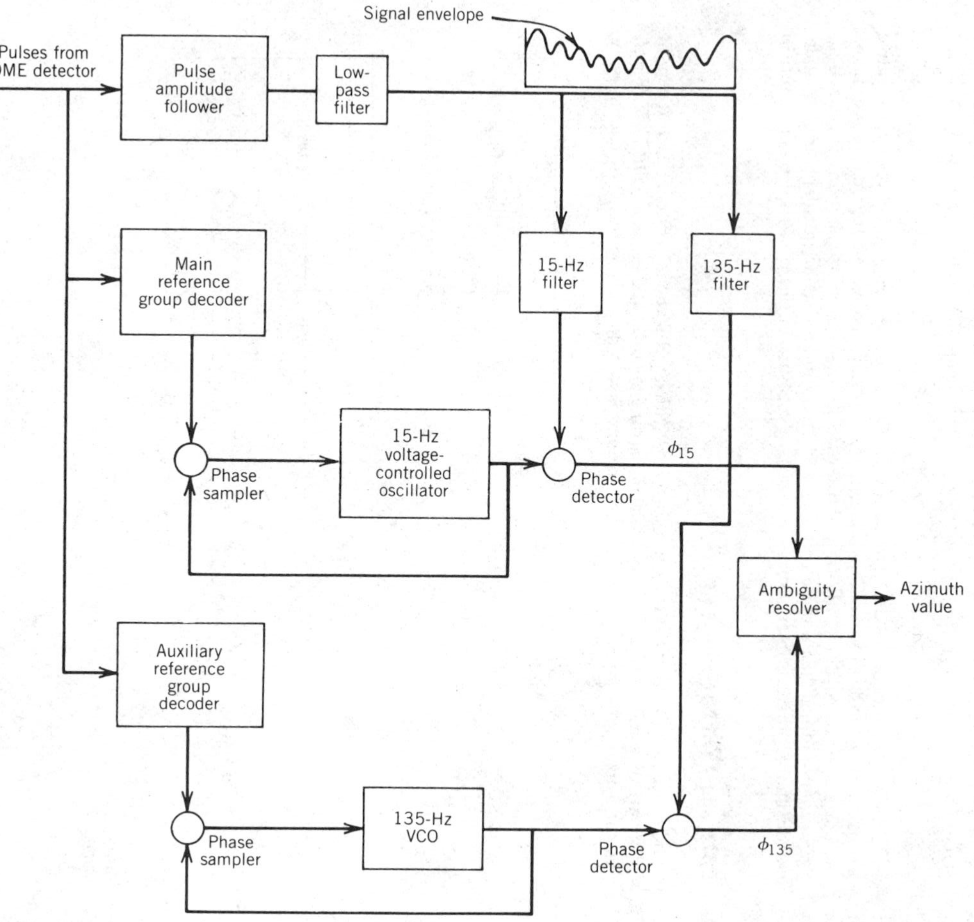

Fig. 52.10 Airborne tactical air navigation (TACAN) azimuth measurement. DME, distance measuring equipment; VCO, voltage-controlled oscillator.

azimuth. This instrument is known as a radio magnetic indicator (RMI). Modern military aircraft, using navigation flight computers, use data acquired simultaneously from several TACAN stations (within range), combine this with other sources (e.g., inertial), and produce a geographical display. This could be latitude and longitude coordinates, bearing and distance to any selected point, or a pictorial moving map.

Signal Parameters and Accuracies. Because of the commonality of TACAN azimuth signals with DME signals, service volumes and required power densities are as specified for DME-only service. The U.S. National Standard covering TACAN specifies $\pm 3°$ (95% probability) as the allowable error, including site, decoding, and display errors.[5] In practice, azimuth accuracy is significantly better. Total system errors of less than $\pm 0.5°$ have been demonstrated using well-sited ground stations. The ground station bearing tolerance is $\pm 1°$ for the 135-Hz component and $\pm 4.5°$ for the 15-Hz component.

Air-to-Air TACAN. A special TACAN mode exists for air-to-air ranging between cooperative military aircraft. Characteristics can be found in Ref. 15. Transmit and receive frequencies are still 63 MHz apart but are arranged differently from those used in air-to-ground operation. The air-to-air reply is a single pulse rather than a pulse pair. No intrinsic azimuth data are available in this mode, except that directional aircraft antennas can be used to measure angles of arrival of the returned pulses.

VOR/DME and VORTAC

DME ground beacons for civil enroute use are colocated with VOR stations (VOR/DME). When the DME is also part of a military TACAN beacon, the arrangement is known as a VORTAC. Figure 52.11 shows the usual configuration. The TACAN antenna is mounted above the VOR counterpoise,

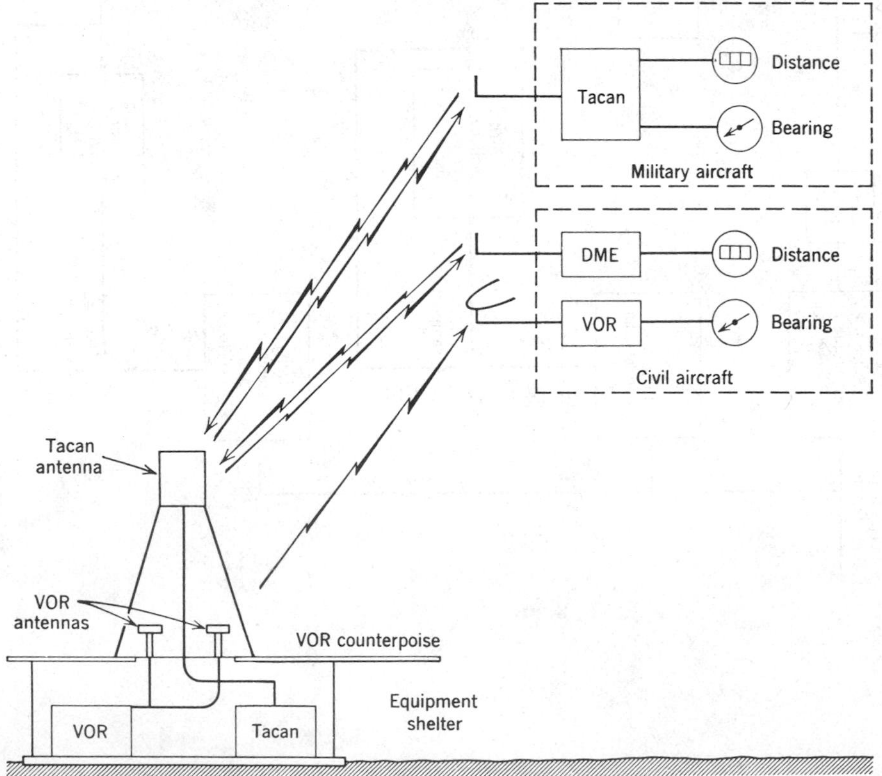

Fig. 52.11 VORTAC configuration. TACAN, tactical air navigation; VOR, VHF omnidirectional radio range; DME, distance measuring equipment. From Kayton and Fried, eds.,[12] with permission.

centrally located with respect to the four Alford loop VOR antennas. A cone-shaped radome then covers the assembly. This arrangement minimizes the distortion of the VOR radiation patterns by the TACAN antenna. If the VOR is of the Doppler type, the coaxial arrangement gives excessive VOR signal distortion. In this case the VOR and TACAN antennas are separated by up to 80 m. This offset between the VOR and DME origins is negligible in practice.

The utilization of a VORTAC station by civil and military aircraft to obtain range and bearing is shown in Fig. 52.11. The TACAN amplitude modulation of the DME pulses is ignored by civil aircraft DME receivers. Colocation allows the two sets of users, with different navigational systems, to fit into a common air traffic control route structure.

Channeling. Every VOR frequency is "paired" with a DME or TACAN channel number. When the pilot selects a particular VOR frequency, the DME will usually be automatically tuned to the appropriate frequency associated with that VOR. These frequency pairings have been set by international agreement and are promulgated in the appropriate U.S. standard.[5] There are more TACAN/DME channels than assigned VOR frequencies. Some TACAN/DME channels are paired with VHF channels used by the instrument landing system (ILS) (see Chapter 53).

The use of two different navigational aids for bearing measurement, with similar coverage and accuracy, might seem an unnecessary duplication. The shorter wavelength of TACAN allows the ground beacon to be much smaller, and more easily relocatable, than does the VOR beacon with its requirement for extensive site preparation and leveling. VOR beacons are also not suitable for ship use. Thus TACAN fulfills the special requirements of the military forces. The aircraft VOR receiver, on the other hand, can be a much simpler and cheaper instrument than the most basic TACAN receiver (particularly at the time of initial deployment of the two systems in the early 1950s). As well as providing accurate navigational data to air transport aircraft, the VOR/DME system allows large numbers of smaller, general aviation aircraft to use the airways and air traffic control systems at affordable cost.

Nondirectional Beacons (NDBs): Automatic Direction Finders (ADFs)

A nondirectional beacon (NDB) is a ground transmitter radiating an essentially unmodulated signal using an omnidirectional antenna. The aircraft uses a directional antenna system to track this signal and to produce an indication of the bearing of the NDB relative to the nose of the aircraft. The modern version of this airborne equipment is known as an automatic direction finder (ADF). Unlike VOR or TACAN, the NDB/ADF system does not give a direct indication of the aircraft's azimuth relative to the beacon; this is obtained by combining the ADF reading with that of the aircraft's magnetic compass. Also no range information is available.

In the United States, NDBs are seldom used as the primary aid for enroute navigation, except in Alaska. In some parts of the world, they may be used in combination with VOR. In developing countries, NDBs are likely to be the only ground aids to aircraft navigation. NDBs are also used as an adjunct to the instrument landing system (ILS). When used in this way they are known as compass locator beacons.

In the United States, NDBs are allocated frequencies in the 200- to 415-kHz band. The ICAO standard allows frequencies in the range of 200 to 1750 kHz.[4] Most airborne ADF equipment will cover this range. This allows also the use for navigation of marine radio beacons (285–325-kHz band) as well as commercial broadcast stations in the standard AM band. Each NDB is required to transmit an identifying two or three letter Morse-coded signal, repeated at least every 30 seconds. This is usually achieved by amplitude modulating the carrier with a keyed audio tone of 1020 Hz (occasionally 400 Hz). Outside the United States, some low-power NDBs transmit identifying codes by on-off keying the unmodulated carrier. Many NDB transmitters are also used for simultaneous voice transmission of aviation weather information.

Ground Equipment. The ground transmitter radiates the signal using a vertically polarized, omnidirectional antenna. Frequency stability is specified by ICAO as 0.01% (0.005% for some high-powered stations). This value is easily exceeded with modern equipment. Transmitter power varies from 20 W to 5 kW, depending on the grade of service.

Airborne Equipment. The operation of the direction-finding equipment relies on two phase-matched antennas, the "loop" antenna and the "sense" antenna. The loop antenna was originally just that, a loop of several turns of wire that could be rotated about a vertical axis and oriented for a "null" in the selected signal. A pointer, slaved to the rotation of the loop, indicated the signal direction. Because of the figure-eight horizontal pattern of the loop antenna, the sense antenna, which has an omnidirectional pattern, was introduced to resolve the 180° null ambiguities. The resultant pattern of properly phased combined loop and sense antennas is a cardioid, which has only a single null.

In modern equipment, the loop antenna consists of a pair of fixed wound ferrite slab antennas mounted along the aircraft's transverse and longitudinal axes. These are called the sine loop and cosine loop antennas, because the voltages induced in them are proportional to the sine and cosine of the angle of arrival of the signal. Again an arbitrary phase angle must be resolved using a sense antenna. This is either a long wire from the fuselage to the top of the fin or, on high-speed aircraft, an E-field plate antenna mounted usually on the bottom of the fuselage in a flush or almost flush configuration. Of the various techniques for processing the signals, one method is shown in Fig. 52.12. The outputs from the two loop antennas are modulated by separate components of a 32-Hz square wave, 90° apart in phase. (It is +90° or −90° depending on whether the antenna is mounted above or below the fuselage.) These two signals are then summed with a scaled and phase-compensated signal from the sense antenna. The resultant is a signal phase modulated by a 32-Hz square wave, the phase of which modulation, relative to the original 32-Hz switching signal, is proportional to the angle of arrival of the incoming signal. After phase detection, filtering, and limiting, this 32-Hz signal is phase compared with the two original 32-Hz components. This produces two DC voltages, which after passing through a nonlinear shaping network are proportional to sine and cosine of the angle of arrival. These signals are fed to a DC resolver, and the signal direction is shown by a pointer attached to the resolver shaft.

Figure 52.13 shows a simple version of an ADF indicator of the radio magnetic indicator (RMI) type. The angle of the needle shows the station bearing relative to the nose of the aircraft. The actual bearing from magnetic north of the NDB from the aircraft's position is given by the value on the compass card pointed at by the needle. In all but light aircraft, the ADF indication is usually combined with that of other navaids such as VOR or TACAN bearing and presented on the horizontal situation indicator (HSI) display (Fig. 52.6).

Inevitably the aircraft structure distorts the patterns of the loop and sense antennas. This results in significant errors of up to 20° at certain angles. These "quadrantal errors" are compensated for each specific aircraft type by the voltage-shaping circuits in the ADF receiver.

Minimum performance standards for airborne ADF receivers are set by RTCA.[16] An ADF receiver for light aircraft weighs about 3 kg, has a volume of 2500 cm^3, and consumes less than 10 W. For an airline unit, equivalent numbers would be 5 kg, 3800 cm^3, and 25 W.

System Parameters and Accuracy. By its ability to operate in the LF and MF bands, the NDB/ADF system does not have line-of-sight limitations. However, for aeronautical use the usable coverage is considered to be (among other criteria) wherever the ratio of vertically polarized ground wave to total skywave power is 10 dB or greater. ICAO recommends that within the rated coverage

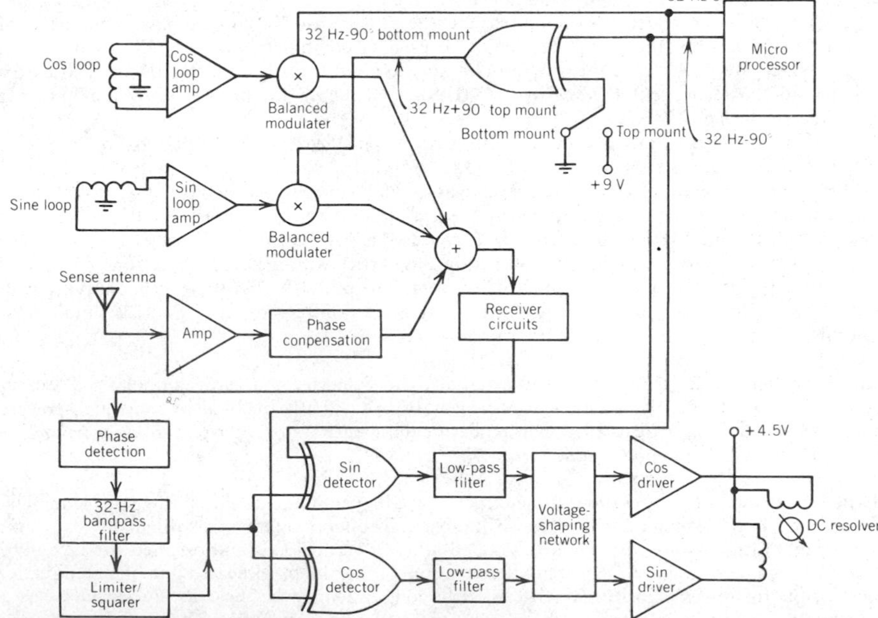

Fig. 52.12 Pointing system KR87 automatic direction finder (ADF). From King Radio, Olathe, Kansas, with permission.

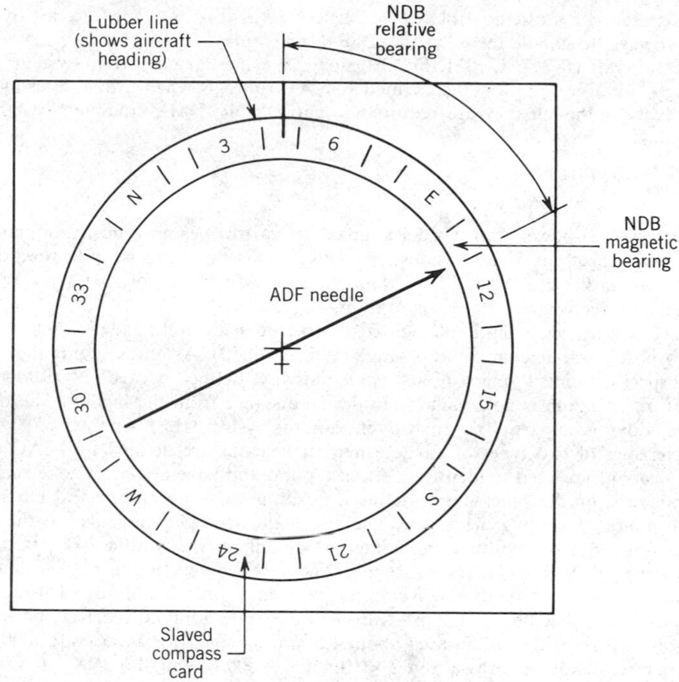

Fig. 52.13 ADF (automatic direction finder) indicator. NDB, nondirectional beacon.

area of the NDB, the signal strength be no less than 70 μV/m, except in tropical areas, where 120 μV/m is preferred. Specific requirements depend on such factors as surface conductivity and background noise from urban areas. A detailed discussion of these factors may be found in attachment C to Ref. 4.

In the United States, three classes of NDBs are defined for enroute service:

Class Code	Power (W)	Usable Radius (nmi)
MH	Under 50	25
H	50–1999	50
HH	2000 or more	75

Bearing accuracy of properly calibrated airborne equipment is of the order of $\pm 3°$. Within an NDB's usable radius, the short-term errors due to terrain and propagation effects are required by the FAA to not exceed $\pm 10°$. At night, during periods of anomalous propagation and particularly in the vicinity of magnetic storms, the accuracy of the NDB/ADF system is likely to be significantly degraded.

Area Navigation

The present system of overland airways in the United States and elsewhere is defined by straight lines joining point source navigational aids of the rho-theta type (VOR/DME, NDB). In many cases this rigid structure introduces unnecessary deviations into the route between two points. In recent years a series of airborne computing aids have become available, known as area navigation (R-NAV) systems. These are now available even for light single-engine aircraft. Using these, a pilot, when within coverage of two or more VOR/DME stations, is able to define "phantom" VOR/DME sites, or "waypoints" along the desired route of flight. The pilot then flies directly between these waypoints using standard flight instrumentation. In the United States, a number of special area coverage (R-NAV) routes have been defined and charted for use by aircraft in instrument flight conditions. However, they are not available in many areas of high traffic density. This is mainly due to the inability of the present air traffic control system, structured as it is to handle aircraft flying on fixed point-to-point airways, to handle these "random" R-NAV routes.

Many larger aircraft (B-747, L-1011, DC-10) now have inertial navigation systems as standard equipment. These are able to follow the defined R-NAV routings when available using the inertial system for guidance, although they are required to carry VOR/DME equipment as their primary navigation system.

Marker Beacons

Marker beacons are low-power transmitters coupled to narrow-beam vertically oriented antennas. They are provided at locations where it is desired to give an aircraft a precise indication of its passage over that point. All marker beacons operate, worldwide, on a frequency of 75.0 MHz, relying on low power and antenna directivity for signal separation.

Marker beacons were originally implemented to provide a method of identifying the center and legs of the now obsolete four course radio range (Ref. 2, Ch. 2). As enroute aids they may still be found defining specific points ("intersections") along airways defined between NDB facilities, where, of course, DME information is not available to define distance from the facility. The major use for marker beacons today is as part of the instrument landing system (ILS) (Chapter 53).

The characteristics of two types of enroute marker beacons are defined by ICAO (Ref. 4, Sec. 3.6). Z-marker beacons are used to identify a specific point and have a vertical cone-shaped antenna pattern. Fan markers, on the other hand, define a position along an airway and have an elliptical beam, with the minor axis along the airway and the major axis perpendicular to the airway. The 75-MHz carrier frequency is amplitude modulated at a nominal 95% with a 3000-Hz tone, which is on-off keyed to provide Morse identifying signals. The area defined by the marker is a function of transmitted radiated power, aircraft receiver sensitivity, and aircraft altitude. The nominal signal strength for detection is specified at 1.5 mV/m. With correctly adjusted receiver sensitivities, the Z marker generates this intensity over an area about 1.5 nmi diameter for an aircraft 3000 ft above the station. A fan marker defines an area about 2×10 nmi at 3000 ft, and about 18×6 nmi at 13,000 ft above the station.

Ground Equipment. Z-marker transmitters have a power of around 4 W. The antenna is a "turnstile" of four horizontal half-wave elements placed one-quarter wavelength above a counterpoise. Correct phasing of the elements generates the required vertical conical beam, with horizontal, circular polarization. Fan-marker transmitter power is typically 100 W. One type of antenna array consists of four colinear half-wave elements, centerfed, with matching stubs between elements. The whole is placed one-quarter wavelength above the counterpoise. This produces a vertical elliptical beam, with the narrow dimension perpendicular to the line of antenna elements.

Airborne Equipment. Light aircraft use a simple half-wave wire antenna mounted on spacers about 8 cm below the fuselage, parallel to the centerline. Larger and high-speed aircraft use resonant cavity antennas mounted flush or almost flush with the bottom of the fuselage. The aircraft receiver is a fixed 75-MHz frequency crystal controlled unit. Sensitivity beyond the 1.5 mV/m figure is not desirable. When the signal exceeds the threshold, the detected 3000-Hz identification signals are switched into the aircraft's audio system. The detected voltage across a 3000-Hz resonant circuit is used to activate a white signal light on the cockpit display, which flashes as the 3000-Hz modulation is keyed on and off.

DECCA

The DECCA system, now primarily a marine navigational aid (described in Section 52.3) is also used by fixed and rotary-wing aircraft. In the past the system was proposed to ICAO for adoption as the primary worldwide short- to medium-range aircraft navigational aid. In 1959, however, the VOR/DME system was chosen by ICAO for a variety of political and technical factors, one being the difficulty at that time of providing a small, inexpensive DECCA receiver and display suitable for light aircraft.

As an airborne navigational aid, the DECCA system is now primarily used by helicopters. The area coverage of DECCA is particularly suitable for the random routes flown by such aircraft. In addition, since DECCA is not limited by line-of-sight effects and has accuracies in the tens of meters in most usable coverage areas, the system is of great value in assisting helicopters to make descents into unprepared landing sites not served by other navigational aids. In the North Sea of Europe, DECCA is extensively used by helicopters serving the many oil drilling platforms in that area.

In Europe, where DECCA coverage is almost complete, the receiving system is installed on aircraft of British Airways for domestic and European service, and is used on many corporate fixed-wing aircraft. Helicopters and some fixed-wing aircraft of the British Army and Royal Air Force are equipped with DECCA receivers.

In airborne DECCA receivers[17] there is no multiplication of the received component frequencies, but rather all components are divided down (by 5, 6, 8, or 9) to the common frequency, f, for the DECCA chain in use. Position is measured as fractions (typically 1/1024) of "zones," rather than "lanes" as for marine receivers. Ambiguities within the zone caused by the division process are resolved by multipulse lane identification signals. Combination of these allows direct phase comparisons to be made at f and also $0.2f$. Ambiguities can then be automatically resolved within a zone and within groups of five zones.

The most significant element of DECCA for aircraft navigation was the development of the moving map pictorial navigation display, or flight log. In its basic form this uses the zone counts and fractions for the two "colors" in use as the rectangular $X - Y$ coordinates to a roller-map display. Since the original coordinates are hyperbolic, this gives sometimes a very distorted display using specially prepared charts. However, in practice this does not significantly affect the ability of the pilot to fly along any precharted track or to identify specific checkpoints. A second form of DECCA flight log uses a digital computer to perform a hyperbolic-to-rectangular coordinate conversion, which allows largely undistorted charts to be used.

Many combinations of equipment are possible, but a modern airborne DECCA installation (Mk 19) might consist of a receiver and antenna amplifier, the digital computer, the flight-log display, a zone identification meter (for identifying gross errors), and two control units. This system weighs approximately 19 kg and consumes about 240 W. Chart information is coded into a 9-bit binary optical track on the side of the plastic chart roll. The output from this is fed to the associated computer to provide closed-loop control of chart position. A similar track governs the left-right position of a cursor.

Secondary Surveillance Radar / ATCRBS

Primary radar suffers from several problems when used as a surveillance tool by ground air traffic control (ATC). The signals are attenuated by precipitation, the radar cross section of light aircraft is too small to generate consistent radar returns, and most importantly, primary radar is not capable of directly identifying a return as being from a specific aircraft. Secondary surveillance radar (SSR) eliminates these problems almost completely. In the United States, SSR is known as the air traffic control radar beacon system (ATCRBS).

Developed from the World War II Mark X IFF (identification friend or foe) system, SSR relies on a transponder in each aircraft that replies when interrogated by coded signals from the ground station. Using a directional antenna and measuring round-trip delay, the ground terminal measures bearing and range to each target. At the present state of development, each equipped aircraft is able to send back a unique identifying code plus readout of its barometric altitude. Extensions to this system are now being developed in the United States: the discrete address beacon system (DABS) and the beacon-based collision avoidance system (B-CAS). DABS allows exchange of limited air traffic control messages between the ground and specific aircraft; B-CAS will provide a hierarchy of aircraft detection capabilities for collision avoidance acting both autonomously and under ground supervision.

Basic System Description. System parameters are defined by U.S.[18] and ICAO (Ref. 4, Sec. 3.8) specifications. All ground-to-air transmissions are made on a frequency of 1030 MHz. The reply frequency in all cases is 1090 MHz. Interrogation and reply signals consist of pulse groups coded by pulse spacing. Each interrogate pulse is nominally 0.8 ± 0.1 μs wide and each of the reply pulses is 0.45 ± 0.1 μs wide. Controlled rise and fall times are used to minimize spectral occupancy.

The six modes of military and civil interrogation are shown in Fig. 52.14. Civil aircraft transponders are designed generally to recognize only modes 3/A and C. To mode 3/A interrogations the aircraft replies with its uniquely assigned code; to mode C it replies with a code indicating barometric altitude. The ground antenna used to radiate P_1 and P_3 in Fig. 52.14 has a narrow azimuth beam (typically 3°). Usually this antenna is mounted on, and rotates with, the antenna of the ATC primary radar with which the facility is associated. This allows correlation between primary and secondary radar returns. The interrogation rate is a function of the antenna rotation rate; generally it is set so as to obtain 4 to 5 sequential replies from each aircraft in the 3° beamwidth, up to a maximum of 450 per second. A typical civil interrogation code sequence is two mode 3/A followed by one mode C.

In civil modes, another pulse, P_2, is transmitted in order to suppress aircraft replies to sidelobe interrogations, or multipath delayed interrogations. P_2 follows P_1 by 2 μs and is radiated using an omnidirectional antenna such that its effective radiated power (ERP) is less than that of P_1 and P_3 in the main beam of their antenna but higher than in the sidelobes and back lobes. The airborne

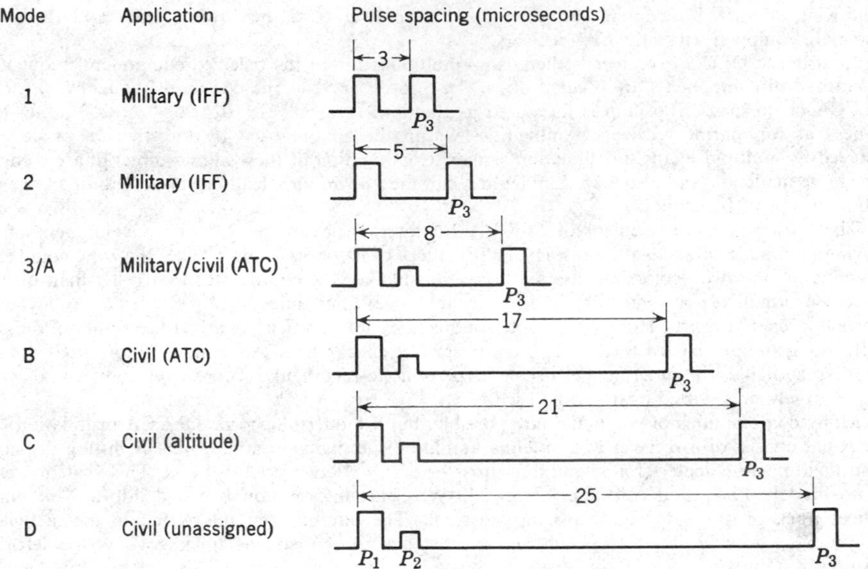

Fig. 52.14 Interrogation pulse modes used in secondary surveillance radar. IFF, identification—friend or foe; ATC, air traffic control. From Kayton and Fried, eds.,[12] with permission.

transponder will reply only if the following conditions are met:

1. The spacing of P_1 to P_3 is correct within $\pm 1\ \mu s$ for the selected mode.
2. The amplitude of P_3 is equal to the amplitude of P_1 (+3 dB, −1 dB).
3. The amplitude of P_2 is more than 9 dB below the amplitude of P_1.

The transponder reply consists of up to 15 pulses, as shown in Fig. 52.15. Every reply contains the two framing pulses 20.3 μs apart. To mode 3/A interrogations the identification reply contains one of 4096 discrete codes (0000 through 7777 octal) selected manually by the pilot in response to ATC instruction (usually set only once, before takeoff). The value of the code is sent by presence or absence of the A, B, C, D pulses in accordance with their arithmetical values, as indicated in Fig. 52.15, for example, code $4034 = C_1 + C_2 + A_4 + D_4$. Certain fixed codes are assigned for special circumstances, for example, 7700 = emergency, 7600 = communications failure, and 7500 = hijack in progress! To transmit altitude data, the aircraft requires an encoding altimeter generating binary data in a special 11-bit code. The 11 bits are switched to the A, B, C, D pulse encoders of the transponder (D_1 is not used) when replying to mode C interrogations. The pressure reference for the encoding altimeter is always fixed at 29.92 in. Hg (1013.2 millibars). Corrections for the difference between this and the actual local reference pressure are made in the ground equipment after the reply is decoded.

To positively identify an aircraft and to guard against blunders in setting in the identification code, the special position identification pulse of Fig. 52.15 is used. This is transmitted with all replies for about 30 seconds following the pilot's pressing an "ident" button when requested to do so by air traffic control. This pulse generates a special mark on the controller's display associated with that aircraft.

Ground Processing and Display. Because of the use of a common frequency pair (1030 and 1090 MHz), an aircraft will reply to all interrogations within its line of sight. The resulting interference on the display at one ground station caused by replies by an aircraft in its area of coverage to interrogations from a different ground station is known as "fruit." Since this fruit is not synchronized with the primary ground station interrogations, it is removed by accepting for display only those replies that correlate in range over two or more successive interrogations. "Garble" is interference caused by two aircraft with similar slant ranges and azimuths being triggered by the same interrogation and producing overlapping replies. Although difficult to remove, it is relatively easy to detect.

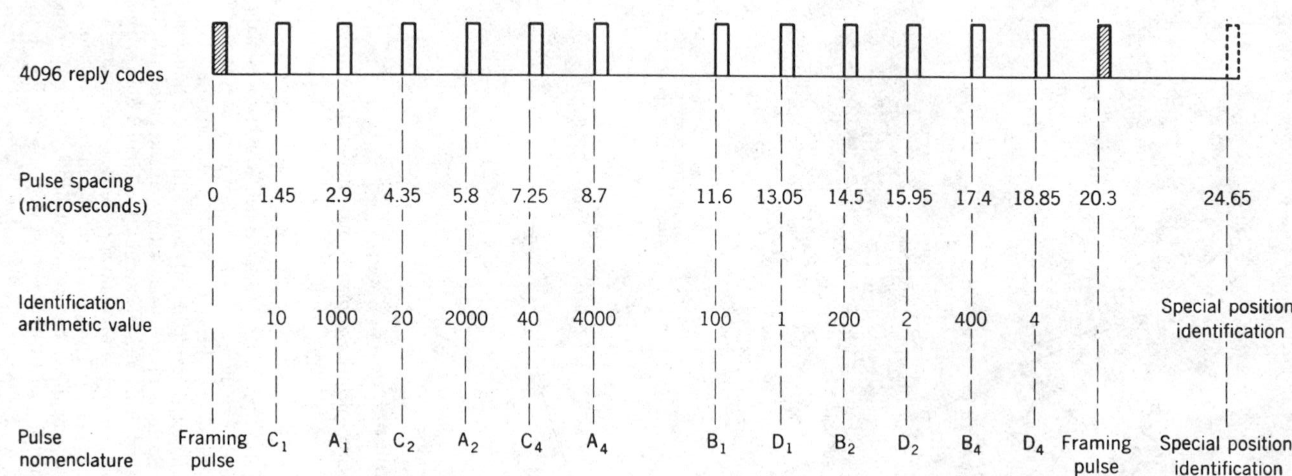

Fig. 52.15 Transponder reply codes used in secondary surveillance radar. From Kayton and Fried, eds.,[12] with permission.

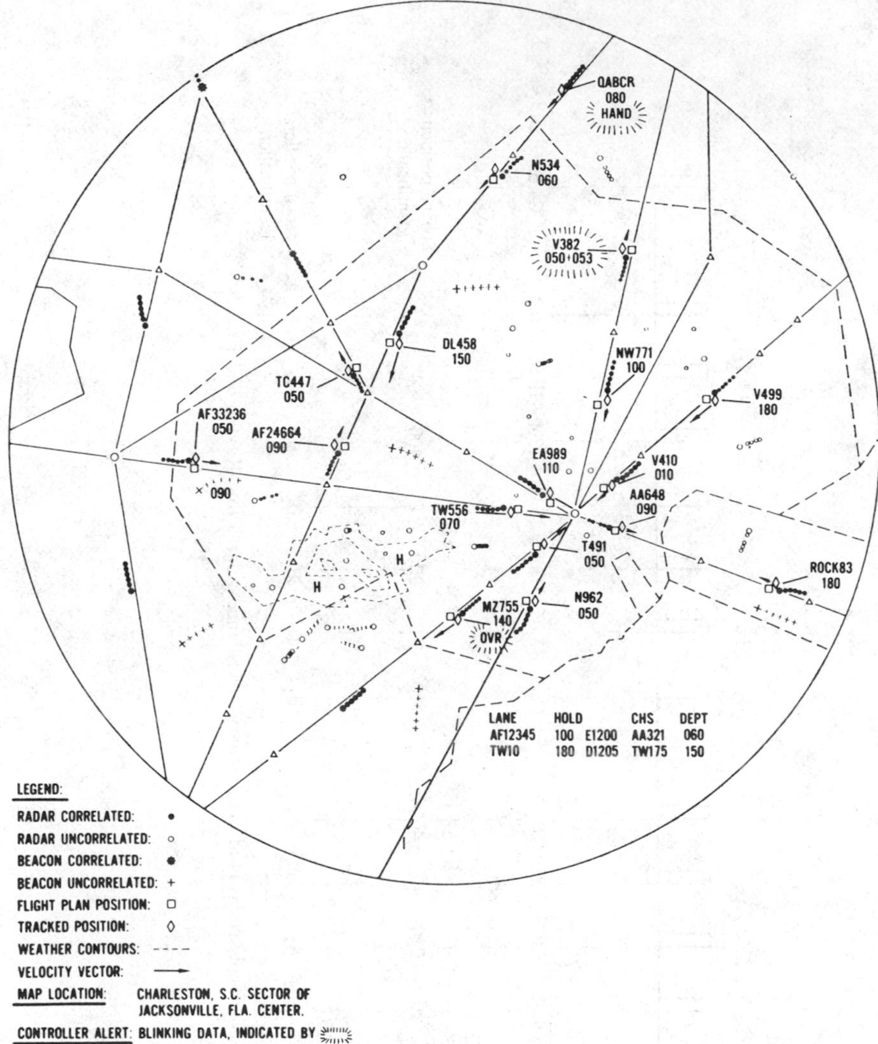

Fig. 52.16 Example of automatic pictorial display of national airspace system. From Kayton and Fried, eds.,[12] with permission.

Figure 52.16 shows a typical controller display associated with the U.S. national airspace system (NAS) stage A,[19] now implemented at all air route traffic control centers (ARTCC) handling enroute traffic. Tracking of "radar" (primary), "beacon" (SSR), and correlated targets is possible, with alphanumeric tagging of each target aircraft's identity and altitude (in hundreds of feet) for the SSR returns. For terminal areas, a Sperry-Univac system known as the automated radar terminal system III (ARTS-III) has been installed at 64 airports around the United States. This system has similar display capability, plus an ability to perform ground speed measurement, low-altitude warning, track extrapolation during loss of signal or garbling, and limited conflict prediction. Similar enroute and terminal processing and display capability has been installed by many national airspace agencies throughout the world, including the Soviet Union.

Airborne Equipment. The airborne transponder is a small and, in these days of integrated microcircuits, relatively simple unit. The fixed transmit and receive frequencies simplify the RF design. Vertical quarter-wave stub antennas are used, with an effectively horizontal azimuth pattern. In the cockpit the pilot interface consists of four thumbwheel switches for selecting the assigned code plus the "ident" button for transmission of the special position identification pulse. In many installations a "reply" light on the panel flashes in response to successful interrogations by the ground. Minimum performance standards for airborne SSR transponders are set by RTCA.[20]

The parameters of the SSR system are as follows:

Operational range (from interrogator)	200 nmi
Operational altitude	100,000 ft
Maximum interrogator ERP (P_1 and P_3)	52.5 dBW (500 W–1.5 kW actual transmitter power)
Aircraft receiver triggering level (minimum for 90% reply probability)	− 101 dBW
Aircraft transponder peak pulse power	21–27 dBW (125–500 W)
Ground receiver sensitivity	− 115 dBW

Discrete Address Beacon System (DABS)

The DABS system[21,22] is being developed for the FAA by the Lincoln Laboratories of the Massachusetts Institute of Technology. It is an evolutionary development from ATCRBS and incorporates a two-way data link capacity for ground-air-ground messages. In ATCRBS, interrogation is spatial; all aircraft in the ground antenna beam reply. DABS assigns each aircraft a unique code, and has the capability to elicit replies only from selected aircraft. It therefore also has the ability to schedule interrogations on the basis of predicted aircraft range, thus reducing interference caused by overlapping replies (garble) and considerably increasing the capacity of the system over ATCRBS. The discrete addressing capability also eliminates the problem of fruit caused by unwanted replies to distant interrogations. As well as improved surveillance capability, the data link function of DABS will provide the communication capability for the projected automatic traffic advisory and resolution service (ATARS) of the air traffic control system. It is also intended to serve as the prime element of the proposed beacon-based collision avoidance system (B-CAS).

DABS uses the same frequencies and largely the same signal formats as ATCRBS. DABS- and ATCRBS-equipped aircraft will operate within a common structure until all ATCRBS aircraft upgrade to DABS equipment. The specifications for the next generation of airline SSR transponder equipment require ATCRBS and DABS capability.[23]

Five signal types are used in the DABS system.

1. **ATCRBS/DABS all-call interrogation.** All ATCRBS and DABS transponders in the interrogator beam reply with surveillance data (mode A or mode C, as requested).
2. **ATCRBS only all-call interrogation.** Same as (1) but ignored by DABS aircraft.
3. **ATCRBS reply.** In response to (1) for ATCRBS and DABS and to (2) for ATCRBS.
4. **DABS only interrogation.** Used to solicit surveillance data from a specific or all DABS aircraft in the antenna beam, or to send or solicit communications messages.
5. **DABS reply.**

Several variations on these exist. For example, specific DABS aircraft may be instructed to ignore "all-call" interrogations (lockout), and for communications three different types of messages have been so far defined.

The "all-call" interrogation is the same as an ATCRBS interrogation (mode A or mode C), except that an additional pulse, P_4, of length 1.6 or 0.8 μs follows 2 μs after the P_3 pulse. ATCRBS transponders reply normally and ignore P_4. If P_4 is 1.6 μs long, DABS transponders also reply with an ATCRBS-type response. When P_4 is only 0.8 μs long, however, all DABS transponders are designed to ignore the interrogation.

A DABS-only interrogation has the form shown in Fig. 52.17. It consists of a two-pulse preamble (P_1, P_2) followed by a data block (P_6) that contains a synchronizing signal. Within the data block, information is transmitted at a 4-megabit rate using binary differential phase shift keying (DPSK). The preamble P_1 and P_2, transmitted in the main interrogator beam, serves to suppress ATCRBS transponders, since they will interpret P_2's being received at the same level as P_1 as indicating a sidelobe interrogation. At 1.25 μs after the start of the DABS data block, a 180° phase reversal always occurs. This synchronization signal must be recognized by the DABS transponder in order to start processing subsequent data. The sidelobe suppression (SLS) pulse, P_5, is transmitted by the DABS interrogator, using an omnidirectional antenna. If the transponder is not in the main lobe of the narrow beam interrogator, the higher received level of P_5 will mask the synchronizing phase reversal and will cause the reply to be suppressed.

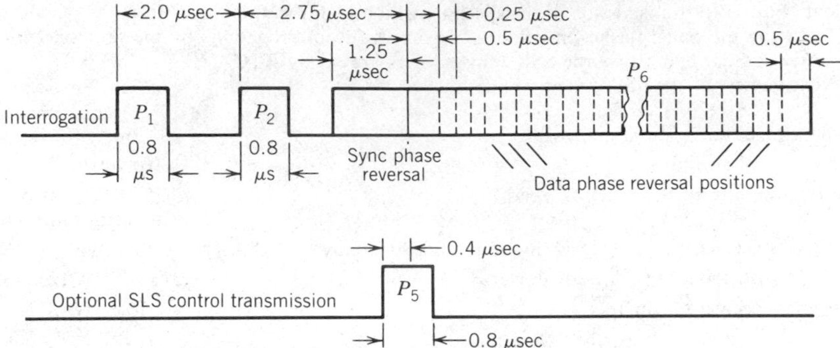

Fig. 52.17 DABS (discrete address interrogation system) interrogation. SLS, sidelobe supression. From Orlando and Drouilhet,[22] with permission.

Each data block is either 56 or 112 bits long. Of these, 24 bits define the aircraft's unique address. This would be assigned by permanent tail number or, for airline use, by company and flight number. A DABS aircraft only replies when it recognizes its own address in the interrogation, except that an "all-ones" code is used for the "all-call" DABS address. The format of the messages are defined in Ref. 21. Provision is made for transmission of extended length messages (ELM) in which multiple blocks are sent to the same aircraft before a reply is triggered. The DABS signal format has been adopted by ICAO, and is referred to as SSR mode S.

The format of the DABS reply is shown in Fig. 52.18. It again consists of a preamble plus a 56- or 112-bit data block. Now, however, pulse position modulation is used for data at a 1 Mb/s rate using 0.5-μs pulses. The four-pulse preamble allows the DABS reply to be easily distinguished from any overlapping ATCRBS responses. A 24-bit parity check code is included in the DABS uplink and downlink messages for error detection and correction.

Performance characteristics of DABS are summarized in Table 52.2. The σ-azimuth figure in it is achieved by use of monopulse antenna processing of the replies at the ground interrogator. A number of ATCRBS interrogators now operational are provided with antennas suitable for monopulse processing, but the technique is not yet being used. Experimental airborne DABS units have been in use since early 1980, being tested both with experimental DABS interrogators and in a standard ATCRBS environment. Operational DABS installations will use dual aircraft antennas (above and below the fuselage) with diversity combining to improve message transfer reliability.

Automatic Air Traffic Advisory and Resolution Service (ATARS). As part of the future DABS system, ATARS is intended to provide a pilot-oriented ground-based collision avoidance aid. At each DABS interrogator site, the ATARS generates track history files for all aircraft within coverage—DABS equipped, ATCRBS equipped, and primary radar (nontransponder-equipped) targets. For DABS aircraft and those aircraft with ATCRBS altitude-encoding capability (mode C), this is a three-dimensional track. Using the ground-air data link capability, ATARS transmits to DABS-equipped aircraft advisories concerning the presence of other tracked aircraft in its vicinity (for the non-mode C and primary radar-only traffic, this can only be done for the horizontal plane). When the ATARS algorithms detect a potential conflict, the data link sends maneuvering advisories (turn left, turn right, climb, descend) to the appropriate DABS-equipped aircraft. The ATARS also sends alert messages to the responsible air traffic control facility whenever a conflict-resolution advisory is sent to a controlled aircraft.

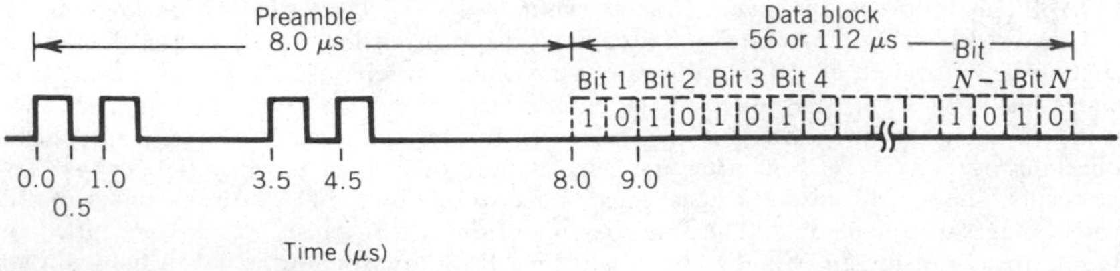

Fig. 52.18 DABS (discrete address interrogation system) reply. From Orlando and Drouilhet,[22] with permission.

TABLE 52.2. DISCRETE ADDRESS BEACON SYSTEM (DABS) PERFORMANCE SUMMARY

Surveillance	
Capacity	Up to 700 aircraft per interrogator
σ-Azimuth	0.06° ± 0.033° bias
σ-Range	50 ft ± 150 ft bias
Update interval	4 s terminal coverage; 5–10 s enroute coverage
Data Link	
Capacity	$\simeq$ 3% duty cycle for total message requirements identified to date
Reliability	$>$ 0.99 in 4 s for short messages
Undetected error rate	$< 10^{-7}$

Beacon-Based Collision Avoidance System (B-CAS). B-CAS allows equipped aircraft an autonomous capability to detect and to maneuver to avoid other DABS and mode C ATCRBS-equipped aircraft when outside ATARS coverage.[24] It is a further evolutionary development to DABS, for which experimental equipment is already being tested in high-density traffic areas.

A DABS aircraft with B-CAS capability periodically interrogates all DABS and ATCRBS mode C aircraft in its vicinity, and measures air-to-air range and altitude differences from their replies. On-board track files are maintained for each aircraft considered to represent a potential conflict. To assist in acquisition, all DABS aircraft periodically spontaneously transmit (squitter) an identifying transmission. Between DABS-equipped aircraft, the air-to-air communications data link allows exchange of aircraft speed and altitude rate capabilities to avoid continual reinterrogation of distant aircraft that do not pose a threat. There is no angle-of-arrival measurement, so discrimination must be performed on the basis of range and altitude rates only. Similarly, B-CAS evasive maneuvers are performed only in the vertical plane. When a B-CAS aircraft encounters a basic DABS or ATCRBS threat, it is the B-CAS aircraft that performs the evasive maneuver if such is needed. Between two B-CAS aircraft, the air-to-air data link is used automatically to coordinate the maneuvering of one or both aircraft.

Plans for the eventual implementation of ATARS and B-CAS are still being developed. In another recent concept,[25] the use of ground-based facilities is deemphasized and airborne capability is enhanced. In this threat alert and collision avoidance system (T-CAS), fully equipped aircraft will have directional interrogation and response antennas and will be able to compute and perform evasive maneuvers in the horizontal as well as the vertical plane.

Joint Tactical Information Distribution System (JTIDS)

The joint tactical information distribution system (JTIDS) is a full-scale tactical control and command system under joint development by the U.S. Armed Forces. As well as having a communication function, it has an important navigational capability. JTIDS operates in the same band used for TACAN (960 to 1215 MHz) using pulse techniques that allow simultaneous JTIDS and TACAN operations without significant degradation due to mutual interference. Although performance details are largely classified, an overview of the system and its signal structure may be found in Refs. 26 and 27.

Each user (which may be a ship or ground station as well as an aircraft) interrogates other users within line of sight and determines range by the two-way round-trip delay. Each user, when requested, sends an estimate of its position and of position error, which is based on all the navaids that user may be employing. The original requestor is thus able to make an "optimal" estimate of its own position using measured ranges to other users and weighted estimates of their positions. For example, if one source is a ground unit navigating by global positioning system (GPS), this input would be weighted heavily relative to another source, such as an aircraft, navigating autonomously by Doppler radar.

Antijam resistance and message security are provided by a combination of frequency hopping and direct sequence (pseudonoise) spectral spreading. Within each pulse, information is transmitted by continuous phase shift keying (CPSK). Both code shift keying and Reed–Solomon block coding, based on a 32-ary alphabet, are used to improve message error rate. As part of its signal structure, JTIDS will incorporate functions simulating TACAN, eliminating the future need for JTIDS-equipped aircraft to carry TACAN equipment as well.

52.2-3 Long-Range Aircraft Radio Navigation Systems

The earliest radio aids used by aircraft for long distance and oceanic navigation were of the low- to medium-frequency type in the 200- to 400-kHz band. Prominent among these was the four-course radio range,[2] now obsolete, which was successfully used in pioneering flights between the U.S. west coast and Hawaii in 1927. This system also guided aircraft across the North Atlantic during World War II. A similar radio aid in the 300-kHz band was CONSOL, developed in World War II by Germany as SONNE, and continued in operation, primarily for marine use. These radio aids were employed intermittently to give cross-check "fixes" on position during navigation primarily by dead reckoning. Radio aids in these frequency bands suffered from atmospheric noise due to thunderstorm activity, large course errors at night, or when the signals crossed a land–water boundary, and from cochannel interference from distant stations, especially at night.

On-board Doppler radar (Section 51.8), operating in the 10- to 20-GHz frequency band, initially provided air transport aircraft with an electronic aid to dead reckoning. By virtue of requiring no ground stations, this system had the capability of providing data anywhere in the world; however, it was still necessary to provide a means for taking occasional independent fixes. Star sightings provided this capability in many cases. Doppler radar has now been almost entirely supplanted by inertial navigation for oceanic airline service. However, it is still used by many long-range corporate aircraft and extensively by the military.

LORAN-C (described in Section 52.3) is used extensively by U.S. and NATO military forces for enroute navigation. Civilian users are mainly helicopters operating offshore, particularly in the North Sea and U.S. Gulf Coast regions for oil rig service. It is used by some commercial fixed-wing aircraft for North Atlantic navigation where coverage is continuous, and also for flights between the U.S. east coast and points in the Caribbean.

The OMEGA navigation aid (described in Section 52.3) is being increasingly used for oceanic navigation by aircraft of both the airline and corporate classes.[28] Aircraft OMEGA receivers are also being designed to use certain very stable VLF communications signals that are broadcast from various points throughout the world.

Rather than specify a particular system for oceanic navigation, ICAO has recommended minimum navigation performance specifications (MNPS). These MNPS requirements are expressed in terms of the lateral track error as follows:

Standard deviation of error (1σ): 6.3 nm (11.7 km)

Error exceeds 30 nm (55.6 km) for less than 1 hour per 1900 flight hours.

Error is between 50 and 70 nm (92.6–129.6 km) less than 1 hour per 8000 flight hours.

Operation with equipment meeting the MNPS is now mandatory in certain high-density traffic regions, such as the "preferred" routes across the North Atlantic. The estimated number of users of these long-range radio navaids is given in Table 52.3.

Equipment Descriptions

OMEGA/VLF. The principles of OMEGA operation are described in Section 52.3. As well as the standard phase difference (hyperbolic) mode, aircraft OMEGA receivers are usually arranged to work in a direct-ranging mode, using a stable on-board frequency reference. The phase change of each received station relative to a known starting position represents change of range to that station. Navigation is possible in this relative mode using only two received OMEGA stations (rho-rho mode), but accuracy progressively deteriorates with time owing to reference oscillator drift. With three stations (rho-rho-rho mode), limited correction for this reference clock drift can be made. In addition to using the eight projected OMEGA transmitters, many airborne receivers are designed to operate with stable VLF transmitters in various parts of the world, such as those maintained by the

TABLE 52.3. PROJECTED 1985 LONG-RANGE RADIO NAVAID USERS

System	Projected Number of Airborne Users		
	U.S. Civil	U.S. Military	World Total
OMEGA/VLF	800[a]	1700[a]	4000
LORAN C/D	1000[a]	1100[a]	4000
Doppler radar	1000	1500	5000

[a] From Ref. 1.

U.S. Navy for worldwide data transmission. The seven stations most commonly utilized are:

Station	Location	Frequency (kHz)
NWC	Australia	22.3
NDT	Japan	17.4
GBR	England	16.0
NAA	Maine, USA	17.8
NPM	Hawaii	23.4
NSS	Maryland, USA	21.4
NLK	Washington State, USA	18.6

Transmission power of these stations is of the order of 1 MW. In one implementation, the airborne equipment deactivates reception of the 11.33-kHz OMEGA channels and superimposes the seven VLF signals in a fixed sequence in the time slots ordinarily occupied by the 11.33-kHz signals. Thus each transmission is sampled once per 10 seconds. The position fix is a composite of OMEGA and VLF signals. Since the VLF transmitters have a single frequency with arbitrary phase, they cannot be used for unambiguous position determination but only in a relative navigation mode. This relative navigation mode, however, starting from a known position, actually gives better short-term navigation errors than does pure OMEGA (hyperbolic navigation). Between occasional VOR/DME updates a relative OMEGA/VLF system operating over land can perform as an area navigational aid with position errors of the order of 1.5 nmi. On long overwater flights, where updates are not available, the relative mode error increases without limit. Operation in the pure OMEGA mode typically will bound the errors to around 4 nmi. To assist in tracking and to allow extrapolation through periods of signal dropout, airborne OMEGA/VLF equipment usually is designed to accept airspeed and heading data from the aircraft's air data system.

Minimum performance specifications for airborne OMEGA systems are given in Ref. 29. Some significant parameters are as follows:

Minimum operational field strength	10 μV/m
Minimum signal-to-noise ratio (SNR) for acquisition	0 dB in 100 Hz BW
Minimum SNR for tracking	− 12 dB in 100 Hz BW
Static accuracy (95% probability)	$\pm$5 centilanes (CEL) (approx. $\pm 5 \times 0.08$ nmi at 10.2 kHz on baseline)
Automatic propagation correction error (95% confidence)	$<$ 30 centicycles (CEC) (approx. 30×0.16 nmi on baseline)
Dynamic track error (95% probability) due to 180° turn at 3°/s or to step velocity change of 200 knots:	$<$ 10 nmi
Initial synchronization time (at min. SNR)	$<$ 6 min

A typical airborne system consists of a receiver–processor unit, a control–display unit, and an antenna assembly. Figure 52.19 shows the panel and functions of one commercial control–display unit. As well as output of latitude-longitude coordinates, the unit can give distance to go and cross-track error on a course between two selected waypoints. Left-right steering signals are provided to drive a meter of the course deviation indicator (CDI)-type or an autopilot.

H-field loop antennas are preferred for OMEGA reception. A typical antenna housing has the dimensions 16.5×35.5 cm and protrudes 4.75 cm from the mounting surface. However, loop antennas are very susceptible to 400-Hz interference from aircraft AC power systems, so that careful skin mapping of each aircraft type is required to find the quietest location. E-field antennas of the capacitive, cavity type are also used. In this case care must be taken to minimize the effects of precipitation static. For some aircraft installations, use of the sense antenna of the automatic direction finding (ADF) system proves satisfactory for OMEGA reception.

LORAN-C. Characteristics of LORAN-C are given in Section 52.3. Aircraft LORAN-C receivers generally require a minimum of operator control, following chain selection. Acquisition and lock to the third cycle of the pulse of the groundwave signal is achieved automatically. At long ranges,

CONTROL DISPLAY UNIT

The CDU provides comprehensive displays of navigation and steering data, while remaining simple to operate. One variation of the CMA-771 CDU includes a remote data loading capability. In the Test position of the display selector switch, various self test functions can be monitored as well as signal and station conditions, external system inputs, position error estimate, and maintenance checks while the system continues to navigate.

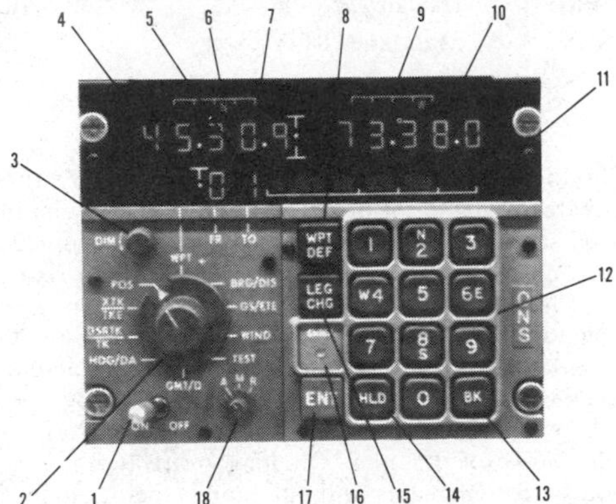

1 ON/OFF switch;
2 DISPLAY MODE SELECTOR SWITCH — Greenwich Meantime/Date (GMT/D), Heading/Drift Angle (HDG/DA), Desired Track/Actual Track (DSRTK/TK), Cross Track Distance/Track Angle Error (XTK/TKE), Position Lat. Long. (POS), Waypoint (WPT), Bearing/Distance (BRG/DIS), Groundspeed/Estimated Time to Waypoint (GS/ETE), Wind Direction and Speed (WIND) Test.
3 DIMMER CONTROL (DIM)
4 WAYPOINT NUMBER DISPLAY (WPT)
5 FROM/TO WAYPOINT DISPLAY (FR TO)
6 ANNUNCIATORS — (Left or Right, North or South).
7 LEFT DISPLAY
8 WAYPOINT DEFINE KEY (WPT DEF)
9 ANNUNCIATORS — (Left or Right, East or West).
10 RIGHT DISPLAY
11 ANNUNCIATORS — System failure (SYS), Dead Reckoning Mode of Operation (DR), Position Ambiguity (AMB), Omega Synchronization status (SYN), System in VLF mode (VLF).
12 DATA KEYBOARD.
13 BACK KEY (BK)
14 HOLD POSITION KEY (HLD).
15 LEG CHANGE KEY (LEG CHG).
16 ENTER INDICATOR.
17 ENTER KEY (ENT)
18 AUTOMATIC/MANUAL LEG CHANGE SWITCH (AUTO/MAN) — A variation of the CDU incorporates an Automatic/Manual/Remote Switch.

Fig. 52.19 Control–display unit CMA-771 OMEGA. Courtesy of Canadian Marconi Co. Avionics Division, Montreal, Quebec.

operation on skywave, with degraded accuracy, can be selected manually. Recommended performance standards for airborne LORAN receiving equipment are defined by RTCA in Ref. 30. Some of these are tabulated here.

Operating signal range	30 dB above 1 μV/m to 120 dB above 1 μV/m
Differential dynamic range	80 dB (between any two usable stations of a chain)
Maximum auto acquisition time	450 s (90% probability)
Time difference error standard deviation	1 μs

In addition, equipment must be capable of tracking time difference rates appropriate for the type of aircraft to be used (up to 8 μs/s change at 2500 knots). Typically, 100 ns static and 300 ns dynamic time difference accuracies can be achieved.

A LORAN airborne receiving system weighs on the order of 8.5 kg and consumes 50 W of power. This is for a relatively simple unit displaying to the pilot sets of time differences in microseconds. The primary output, however, is a serial data stream feeding these data to an on-board navigation computer, for mix with other on-board sensor outputs. The aircraft automatic direction finding (ADF) sense antenna is commonly used for receiving LORAN signals.

LORAN-D. LORAN-D is a tactical version of LORAN-C operated by the U.S. Air Force. The ground stations are air transportable, allowing relocation of a chain in less than 48 hours. LORAN-D pulses are sent in groups of 16 pulses, 500 μs apart. The phasing within each pulse can be coded by cryptographic devices to prevent unauthorized use of the signals. LORAN-D receivers often have the capability to handle both C and D types of signals. During processing of LORAN-D signals, correlation processing over many pulses is sometimes used to reduce substantially the effective processing bandwidth.

Doppler Navigation. Principles of Doppler radar for navigation are given in Section 51.8. Minimum performance standards have been set by RTCA.[31] Early systems used one- or two-axis stabilized antennas to compensate for aircraft pitch and roll, but fixed antenna systems with compensation in the signal processing are now more common. Doppler radar generates signals proportional to along-track and across-track rates, and also vertical velocity in some cases. Derivation of position by integration of these rates uses an external accurate heading source. Typical velocity error is of the order of 0.3% (2 sigma). Typical along-track error is 0.5% of distance flown (95% probability) and drift angle measurement error 0.75°, excluding heading source error contribution.

The AN/APN 200-V unit is approximately 41 × 41 × 13 cm and weighs 16.5 kg. Each separate transmit and receive antenna is a planar array of slotted waveguides, producing three beams with widths of 5° in the depression plane and 11° in the broadside plane, with a depression angle of 67°. The radiated signal is continuous wave at a frequency of around 13.3 GHz with a power of 100 mW. The Doppler frequency outputs from the three receiving channels are audio frequencies that are processed to provide signals proportional to along-track, cross-track, and vertical velocities. Total power consumption is about 220 W for this unit.

In the simplest configuration, the Doppler data are combined with aircraft heading inputs to give a display of integrated distance and bearing from a known starting point. For airline and military systems, the Doppler signals are fed to a navigational computer and combined with heading, altitude, and other air data. The display is then in the form of left-right steering commands and distance-to-go to a selected waypoint, or geographic coordinates to a "moving map" type of display. Doppler navigation is extensively used on military and civil helicopters. With systems optimized for these vehicles' speed ranges, the Doppler information can be used for hover control.

52.2-4 Weather Detection Equipment

Airborne Radar

Many military aircraft are equipped with ground-mapping radar for use as a navigational aid. However, airborne radar is not generally used for this purpose by civil aircraft. Rather, airborne radar is used as a weather-avoidance or penetration aid. Installation of some form of weather avoidance aid is required by the FAA for all airline and most commercial U.S. aircraft.

Airborne weather radars detect returns from raindrops and hail. In thunderstorm and severe turbulence regions, the drop size and intensity are much larger than in fog or stable stratiform clouds. Since the radar return, based on Rayleigh scattering, varies as the sixth power of drop diameter, the appearance of strong returns correlates well with regions of turbulence. Many radars operate in the "contour" mode, in which the display is presented in terms of the gradient of the return intensity with distance. A steep gradient is a good indicator of turbulence.

Performance standards for airborne weather radars have been set by RTCA.[32] Equipment operates at around 5 GHz, or more commonly now at around 9.35 GHz. Pulse widths range from 2 to 5 μs. Peak powers range from 3 to 5 kW for equipment designed for light twin-engine aircraft, up to 50 kW or more for airline units. Airline equipment now under development uses much lower peak powers, but with longer pulse lengths to maintain sensitivity and with correlation processing to preserve range resolution. Antennas vary from 30 cm in diameter to around 80 cm, depending on the aircraft size. For larger installations, the antenna is stabilized against aircraft pitch and roll. Scan width is typically ±60° horizontally about the aircraft nose at a rate of around 12 to 20 scans per minute. Usually the angle of the pencil beam antenna may be tilted above or below the horizon up to ±15° under manual control of the pilot to investigate the vertical extent of the precipitation returns. At low depression angles, the radar can also be used for ground mapping as an aid to navigation. Typical airline equipment weighs 40 kg and consumes 250 W; a system for small aircraft could weigh less than 10 kg and consume less than 100 W. Maximum usable ranges vary from 120 up to 300 nmi for airline equipment.

Although many weather radar systems still in use provide a rho-theta sector scan CRT display, modern weather radars perform storage and scan conversion to give a steady X-Y type raster display. Color is now being used to depict different return intensities. The trend now is also to use the weather radar CRT to combine weather with other navigation data in one display, and also to use it for the presentation of alphanumerics such as checklists and emergency messages.

Lightning Detection and Ranging

Studies by the U.S. Air Force and NASA have shown significant correlation between turbulent conditions associated with thunderstorms and the position and intensity of electrical discharges. As well as visible lightning, these discharges include many nonvisible phenomena that can be detected using low-frequency radio receivers. A commercial device designed to detect turbulence by measuring these effects is the "Stormscope" (3M-Ryan Corp., Columbus, Ohio). It is a receive-only system with direction of arrival of discharge energy being measured using an antenna similar to that for the ADF. Selected frequencies in a band around 50 kHz are monitored. Measurement of range is performed by a proprietary, patented technique. Signal intensity, pulse rise and fall times, and differences between the E and H field components are measured and compared with sets of prestored "fingerprint" models representing various range conditions. The rho-theta coordinates of each detected discharge are presented as a dot on a storage CRT, giving a 360° display in azimuth about the aircraft. Close groupings of dots indicate severe activity and hence probable turbulence. The range accuracy is approximately 10%, with maximum ranges out to 250 nmi. Although originally developed as a weather aid for those light aircraft for which weather radar installation was not practicable, the Stormscope is gaining increased usage in larger aircraft and has been certified by the FAA as an approved weather avoidance aid for certain U.S. commercial operations.

References

1 *Federal Radio Navigation Plan*, US Department of Defense and US Department of Transportation, DOT-TSC-RSPA-80-16, Vols. 1–4, Jul. 1980.

2 P. C. Sandretto, *Electronic Avigation Engineering*, International Telephone and Telegraph Corp., 1958.

3 H. C. Hurley, S. R. Anderson, and H. F. Keary, "The CAA VHF Omnirange," *Proc. IRE* **39**:1506–1520 (Dec. 1951).

4 *Aeronautical Telecommunications*, Annex 10 to the Convention on International Civil Aviation, ICAO, Montreal, Canada, 3rd ed., Jul. 1972.

5 "United States National Aviation Standard for the Very High Frequency Omnidirectional Radio Range (VOR)/Distance Measuring Equipment (DME)/Tactical Air Navigation (TACAN) Systems," FAA Advisory Circular, AC-00-31, 10 Jun. 1970. Proposed revised version released 10 March 1981.

6 A. Alford, A. G. Kandoian, "Ultra High Frequency Loop Antenna," *Electr. Commun.* **18**(5) (Apr. 1940).

7 S. R. Anderson, "VHF Omnirange Accuracy Improvements," *IRE Trans. Aero. Nav. Electr.* **ANE-12**(1):26–35 (Mar. 1965).

8 *Minimum Performance Standards—Airborne VOR Equipment*, DO-153A, RTCA, Washington, DC, 2 Nov. 1978.

9 S. R. Anderson and R. B. Flint, "The CAA Doppler Omnirange," *Proc. IRE* **47**(5, Pt. 1):808–821 (May 1959).

10 W. J. Crone and E. L. Kramar, "Development of the Doppler VOR in Europe," *IEEE Trans. Aero. Nav. Electr.* **ANE-12**(1):36–40 (Mar. 1965).

11 S. H. Dodington, "Development of 1000 MHz Distance Measuring Equipment," *IEEE Trans. Aero. Electr. Syst.* **AES-16**(4) (Jul. 1980).

12 M. Kayton and W. R. Fried, eds., *Avionics Navigation Systems*, Wiley, New York, 1969.

13 *Minimum Performance Standards—Airborne DME*, DO-151A, RTCA, Washington, DC, 2 Nov. 1978.

14 P. C. Sandretto, "Development of TACAN at Federal Telecommunication Laboratories," *Electrical Communication* **33**(1):4–10 (Mar. 1956).

15 "Standard Tactical Air Navigation (TACAN) Signal," US Military Standard MIL-STD-291B, Dec. 1967.

16 *Minimum Performance Standards—Airborne ADF Equipment*, DO-142, RTCA, Washington, DC, 8 Jan. 1970.

17 "The Decca Navigator System and Its Uses," Pub. K20/3, Decca Navigator Co., Ltd., London, 1980.

18 "Selection Order: U.S. National Standard for the IFF Mark X (SIF) Air Traffic Control Radar Beacon System (ATCRBS) Characteristics," DOT/FAA Order 1010.51, 10 Oct. 1968.

19 "Design for the National Airspace Utilization System," FAA, Washington, DC, Jun. 1962.

20 *Minimum Performance Standards—Airborne ATC Transponder Equipment*, D0-150, RTCA, Washington, DC, 17 Mar. 1972.

21 "Discrete Address Beacon System (DABS) National Standard," FAA Order #6365.1, 9 Dec. 1980.

22 V. A. Orlando and P. R. Drouilhet, "Discrete Address Beacon System (DABS) Functional Description," MIT Lincoln Labs, FAA-RD-80-41, Apr. 1980, NTIS AD-A085169.

23 Characteristic 718-2, "Mark 3 Air Traffic Control Transponder (ATCRBS/DABS)," Aeronautical Radio Inc. (ARINC), Annapolis, MD, 8 Aug. 1980.

24 J. D. Welch and V. A. Orlando, "Active Beacon Collision Avoidance System—Functional Overview," MIT Lincoln Labs, FAA-RD-80-127, Dec. 1980, NTIS AD-A094177.

25 "Briefing Spurs Carrier Issue of Beacon," *Aviat. Week Space Technol.*, 3 Aug. 1981.

26 R. L. Eisenberg, "JTIDS System Overview," in C. T. Leondes, ed., *Principles of Precision Position Determination Systems*, AGARDograph No. 245, Jul. 1979.

27 G. I. Palattucci, "JTIDS Signal Structure," in C. T. Leondes, ed., *Principles of Precision Position Determination Systems*, AGARDograph No. 245, Jul. 1979.

28 "Long Range Navigation Market Thriving," *Aviat. Week Space Technol.*, 5 Jun. 1978.

29 *Minimum Performance Standards—Airborne Omega Receiving Equipment*, DO-164A, RTCA, Washington, DC, 21 Sep. 1979.

30 *Minimum Performance Standards—Airborne LORAN Equipment*, DO-159, RTCA, Washington, DC, 17 Oct. 1975.

31 *Minimum Performance Standards—Airborne Doppler Radar Navigation Equipment*, DO-158, RTCA, Washington, DC, 17 Oct. 1975.

32 *Minimum Performance Standards—Airborne Weather and Ground Mapping Pulsed Radars*, DO-134, RTCA, Washington, DC, 16 Feb. 1967.

52.3 NAVIGATION AIDS FOR MARITIME USE

Thomas J. Cutler

52.3-1 Introduction

The science of finding one's location at sea (navigation) is many faceted and ranges from the very simple to the extremely complex. The one common denominator to nearly every system of navigation, however, is the "line of position" (LOP). Simple piloting while within sight of land is accomplished by observing a known charted object in conjunction with a compass to determine a straight-line visual bearing. That bearing, when transferred to a chart, is an LOP. The rather complex science of celestial navigation utilizes the angle of altitude of astronomical bodies (obtained by the sextant) to determine a "circle of equal altitude," which is simply another form of LOP.

The LOP by itself does not reveal position to the navigator but instead provides a *clue* as to location. The navigator knows that he or she is somewhere on that LOP but cannot determine exactly where with any degree of precision without more information. But if the navigator is able to obtain two or more LOPs, he knows that where they intersect is his location, which he calls a fix.

Electronic navigation is no different. Despite the degrees of complexity and sophistication, nearly all systems have only one basic purpose: to provide the navigator with LOPs. How this is accomplished varies considerably over a wide spectrum of systems and equipment designs. Some of the electronic systems used by mariners to aid in navigation, specifically radar, sonar, and satellite systems, are addressed in other sections of this handbook because their applications range far beyond the science of navigation. There are, however, a significant number of systems that exist solely for the purpose of navigation and these will be addressed in this section.

It should be emphasized that no true navigator relies on only one source of information for fixes. All methods of obtaining LOPs have limitations. Electronic systems are subject to power failures or fluctuations, atmospheric conditions, and operator error. Celestial navigation can be constrained by clouds that mask the heavenly bodies from view, and coastal piloting is worthless in fog. Further, the utter importance of what he is doing and the consequences of failure dictate to the navigator that he must never be satisfied until he has tapped all available sources of information to satisfy himself that his vessel is not standing into danger.

Electronic Systems

The electronic systems unique to maritime navigation are radio direction finding (RDF), long-range navigation (LORAN), OMEGA, DECCA, and CONSOL. Table 52.4 provides a comparison of these systems. The ranges and accuracies portrayed are theoretical ideals and would not necessarily be actually encountered except under the best circumstances; they are provided to facilitate comparison of the systems. "Hyperbolic" and "radial" are explained in a later subsection.

Another navigational system, ships inertial navigation system (SINS), also addressed in a later subsection, is not included in Table 52.4 because its uniqueness does not permit a meaningful comparison with the other systems.

Frequencies. All of the systems to be discussed (except SINS) rely on the transmission of radio signals through the earth's atmosphere and therefore occupy portions of the radio frequency spectrum. Table 52.5 illustrates the relationship of marine navigation usage of the spectrum to the usage by other electronic systems.

It can be seen that electronic navigation is confined to the lower portion of the RF spectrum (i.e., MF and below) except for satellite navigation.

Very high frequencies (VHF) and above are limited in range to line of sight. Medium and high frequencies (MF and HF) are vulnerable to high propagation losses when traveling over land, and propagation predictability can be erratic if man-made conductive structures whose size is a significant fraction of the wavelength being used are in the path of transmission. Low frequency (LF) transmissions are much more stable and therefore predictable and because of their large wavelength are suitable for accurate time and phase comparisons. At longer ranges, however, skywaves begin to return to the earth's surface and cause interference with the groundwave, necessitating some means of discerning between the two (Fig 52.20). Very low frequencies (VLF) travel for tremendously long distances in the natural waveguide created between the earth's surface and the ionospheric layers of the atmosphere. The range is of great advantage, but the necessity for real-time knowledge of ionospheric conditions complicates the usage of these frequencies.

It can be seen from Table 52.5 that the LF portion of the radio-frequency spectrum has been the most popular for navigation systems. LORAN-A started in the MF region, but its improved successor LORAN-C opted for the more stable LF portion. Only OMEGA has appeared in the VLF region; the desire to create a worldwide navigation system with a minimum number of required transmitting sites was sufficient to overcome the drawbacks of cost magnification and design complexity.

Radial Systems. Early attempts at incorporating electronics into the realm of navigation employed shore-based direction-finding antennae that could take directional bearings from ships transmitting their radios at sea. The shore station could then send this information to the ship, thus providing the vessel with an LOP. Two or more of these shore stations could provide a fix.

Later development reversed the roles by placing the direction-finding antenna aboard the vessel and the shore stations became the transmitting sites. The result was the same but permitted the vessel to operate in a receive-only, or passive, mode and did not require communications to be established. This is radio direction finding (RDF) and is explained in more detail in a later subsection.

The principle of obtaining a straight-line bearing from a single object makes this a radial system, since the bearing is obtained from a transmitter radiating signals in all directions.

Hyperbolic Systems. Hyperbolic systems are created by placing two transmitting stations at a considerable distance apart and having them transmit a recognizable signal. A theoretical line between the two stations is called the baseline and the line continued beyond the stations is the baseline extension (Fig. 52.21*a*). If a vessel is equipped to receive the signals and to discern any difference in the time that each signal is received, an indication of the vessel's position can be obtained (i.e., an LOP). For example, if the vessel receives both transmitted signals simultaneously (i.e., no time difference), then the vessel must be somewhere along the perpendicular bisector of the baseline (indicated by "Zero" in Fig. 52.21*a*). The lack of a time difference indicates to the vessel's navigator that he is located at some point equidistant from the two transmitting stations. As soon as the vessel is no longer exactly the same distance from each of the stations, one station's signal will arrive before the other (there will be a discernible time difference in the receipt of the signals). Once again this indicates to the navigator that he is somewhere along a definable line. But because we are dealing with differences in the mathematical sense, the resulting line is not straight but hyperbolic in

TABLE 52.4. COMPARISON OF MARITIME NAVIGATION SYSTEMS

	RDF	LORAN-A	LORAN-C	OMEGA	DECCA	CONSOL
Frequency (kHZ)	285–325	1850–1950	90–110	10.2–13.6	70–130	190–370
Area Covered	Most coastal areas out to 175 nm	Japan	Most coastal areas out to 2000 nm	Virtually worldwide	Eastern Canada; Western Europe; Japan; Persian Gulf; portions of Africa, India, Australia	USSR; Europe (Norway to Spain); United States (Nantucket, San Francisco, Miami)
Type	Radial bearing measurement	Hyperbolic time difference	Hyperbolic time difference	Hyperbolic phase	Hyperbolic phase	Hyperbolic/radial
Maximum Range (nmi)	175	800	1500	Virtually unlimited	240	1400 (min 25–30)
Best Accuracy (nmi)	Variable	0.5	0.25	1	0.1	1
Comments	Least expensive	Phasing out; only Japan phaseout undetermined	Recognized as primary U.S. electronic navigation system	Most expensive; newest	Privately owned	"CONSOL" in Europe; "CONSOLAN" in U.S.; USSR has comparable but different system

[a] RDF, radio direction finding; LORAN, long-range navigation; nmi, nautical miles (1 nmi = 2000 yd = 1828.8 m).

TABLE 52.5. FREQUENCY SPECTRUM USAGE FOR MARITIME NAVIGATION

Band	Frequency	Marine Navigation Use[a]	Other Uses
EHF	30–300 GHz		Microwave communication
SHF	3–30 GHz		Microwave communication; Radar
UHF	300–3000 MHz		Radar; NAVSTAR; TV; Satellite navigation
VHF	30–300 MHz		Radar; TV; Satellite navigation
HF	3–30 MHz		
MF	300–3000 kHz	LORAN-A; RDF	AM radio
LF	30–300 kHz	RDF; CONSOL; LORAN-C; DECCA	
VLF	3–30 kHz	OMEGA	

[a] LORAN, long-range navigation; RDF, radio direction finding.

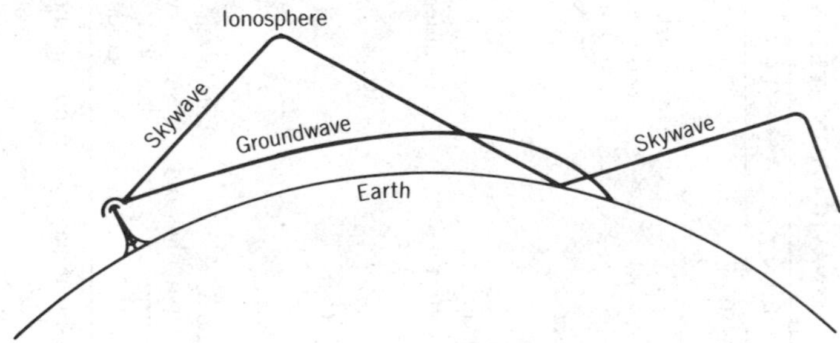

Fig. 52.20 Groundwave and skywave propagation.

form. Each combination of time differences (measured in microseconds) results in a unique hyperbola (Fig. 52.21*b*). Using a nautical chart (map) with the appropriate hyperbolae superimposed, the navigator can measure the time difference between receipt of each station's signals to identify the appropriate hyperbola. This process yields a single LOP. To obtain a fix, another set of stations with their own unique hyperbolae must be used (Fig 52.21*c*). Where the hyperbolae intersect is the fix.

Since synchronization between the two stations is so critical, original designs used a "master" and "slave" concept, the master station not only transmitting its own signal but providing the triggering to the slave to ensure simultaneity; hence the "M" and "S" designations for the stations. Slave has evolved into "secondary" with the advent of modern, extremely accurate clocks that obviate the need for dependent triggering.

The advantages of the hyperbolic system are significant. A directional antenna is not required; a simple wire antenna is all that is needed. Although highly sophisticated, the equipment required on board is relatively small and requires no transmitting capability as does a radar set. The range and accuracy of a hyperbolic system is far superior to any simple azimuthal system.

Although not readily evident to the human eye, the lines appearing on the nautical chart are not true hyperbolae because of distortions caused by the spheroidal shape of the earth.

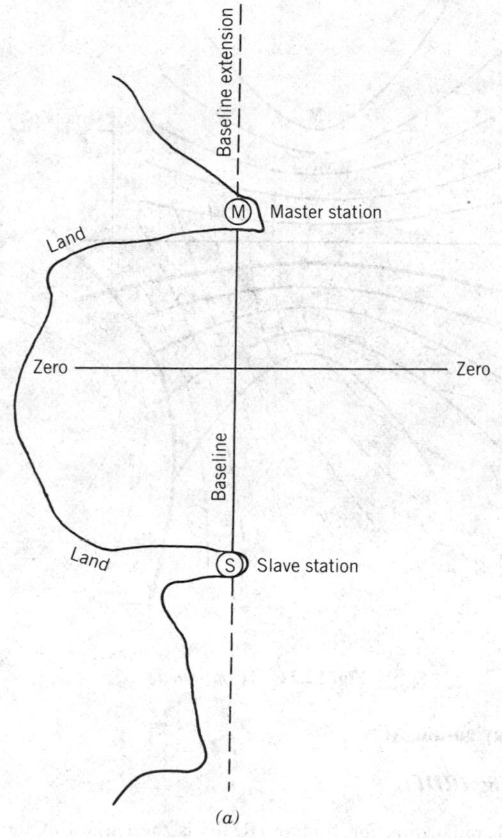

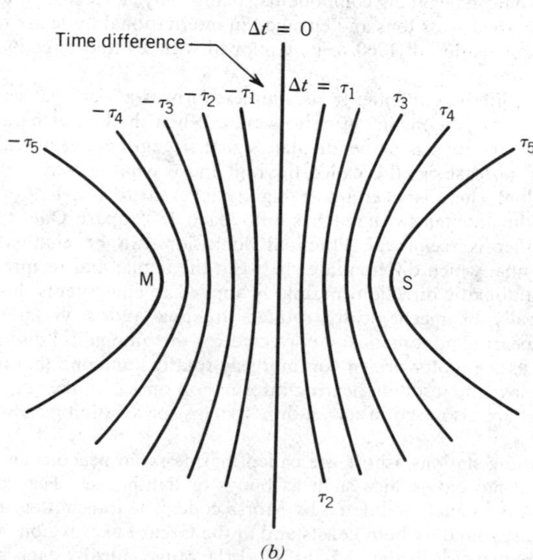

Fig. 52.21 Hyperbolic system of navigation: (*a*) basic geometry, (*b*) hyperbolic pattern, (*c*) intersecting hyperbolic systems.

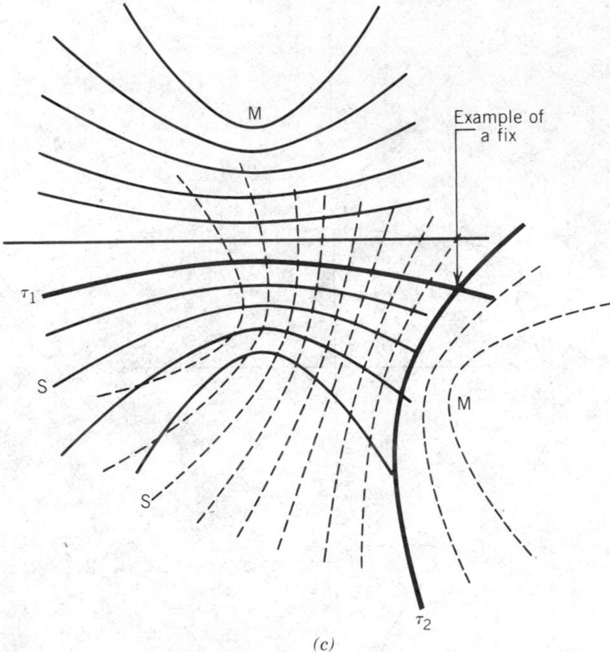

(c)

Fig. 52.21 (*Continued*)

52.3-2 Electronic Navigation Aids

Radio Direction Finding (RDF)

As previously discussed, radio direction finding (RDF) is the earliest of the methods of electronic navigation and works on the azimuthal, or radial, principle. The LOPs obtained are simple straight-line bearings easily plotted on almost any nautical chart. RDF equipment is the least expensive of any electronic navigation components, being only a receiver and a direction-finding antenna. All vessels over 1600 gross tons and engaged in international trade are required by the safety of life at sea (SOLAS) convention of 1960 to be equipped with a radio direction finder.

Antennae. The direction-finding antenna in its simplest form is a loop that varies the strength of a received signal by its angle to the incoming radio waves. When the loop is parallel to the waves, the signal is strongest. When the loop is perpendicular, signal strength is at its weakest or is completely indiscernible. This point of least signal is called the null and is what is used to determine the bearing to the transmitting station. There is a certain ambiguity inherent to a system of this type because in rotating through 360°, the antenna twice reaches nulls, each 180° apart. One is the actual bearing to the station and the other is reciprocal. Potential confusion can be eliminated by attaching an additional "sense" antenna, which differentiates between the actual and reciprocal bearings.

The term "ADF" (automatic direction finding) is applied to equipments that have the additional features of an electrically or mechanically rotated antenna and a visual display that gives a continuous readout of bearing (automatically corrected for the ambiguity) discussed before. Modern naval RDF equipment uses a motor-driven continuously rotating antenna that monitors comparative signal strength and displays the resulting bearing information on a cathode ray tube (CRT) indicator. RDF and ADF systems are also used in navigation systems for aviation (see Section 52.2).

Stations. The transmitting stations ashore are called marine radio beacons and are often combined with conventional visual navigation aids such as buoys or lighthouses. They normally broadcast a low-power continuous-wave signal modulated by Morse code. The transmitters are maintained by the U.S. Coast Guard and are found on both coasts and in the Great Lakes region. Marine radio beacons are assigned specific frequencies in the 285- to 325-kHz range. Ideally each station operates on a discrete frequency, but where this is not possible the stations are placed with enough geographic separation to preclude interference. In some cases such separation also is not possible, and a time-sharing system is utilized instead.

Besides the radio beacon stations created specifically for marine navigational use, there are other sources of RDF information that the mariner can use. Commercial radio stations can be utilized provided that the location of the transmitting antenna is shown on the nautical chart being used. Aircraft beacons can also provide reliable bearing information; although these beacons are gradually changing to higher frequencies which are beyond the range of most marine RDF units, a significant number are still receivable.

In some countries stations still exist that employ the old method of RDF, where the ship does the transmitting and the shore station interprets the direction and sends it to the vessel. These stations are called radio compass stations.

Publications. Information as to location, frequency, range, and identifying characteristic (i.e., Morse code identifier of marine radio beacons located in the United States) can be found in the U.S. Coast Guard publication *Light List* which, despite its rather restrictive title, contains information on radio navigational aids, fog signals, buoys, lightships, and daybeacons as well as lights. It is divided into five separate volumes according to region and is continuously updated through the "Notice to Mariners" system.

RDF information for foreign waters is contained in Defense Mapping Agency Publication 117, *Radio Navigational Aids* (formerly HO117 when the U.S. Navy Hydrographic Office had publication responsibility). There are two separate volumes, 117A and 117B, which apply to the Atlantic-Mediterranean and Pacific-Indian Ocean areas, respectively.

Accuracy. A radio signal traveling over land is subject to some degree of bearing distortion. Therefore the farther inland a station is located, the greater is the loss in accuracy. The mariner must be aware of this anomaly when selecting stations because many of the commercial broadcast and aircraft beacon stations are located well inland, which means there is potential for significant error.

Structural components of the vessel itself can create interference that will detrimentally affect accuracy. Incoming signals can be received and reradiated by metallic objects such as the vessel's standing rigging (wire mast supports, etc.). The existence of such spurious signals in close proximity to the antenna adversely affects the receiver's ability to resolve direction.

A polarization effect is evident during night hours, particularly near the times of sunrise and sunset, which creates some inaccuracy not encountered in daylight.

Other factors that can affect the accuracy of RDF bearings are signal strength (the weaker the signal, the less accurate the bearing) and the ability of the mariner to use the equipment (practice and experience greatly improve results).

Omni. Designed as an aircraft navigation system, Omni (short for omnidirectional) is used by some mariners who have the financial capability to own the considerably more expensive receiving equipment required. The system operates in the VHF (108–118 MHz) range and is small, lightweight, and simple to use. There are a significant number of stations located close enough to the coastline to enable vessels to use them, but coverage is by no means continuous for U.S. waterways.

Omni can be used to obtain bearings (hence LOPs) but is most effective when used as a "homing" system, that is, the mariner picks a station located at or near his destination and then uses the Omni receiver to keep him headed toward the transmitter. Because of the near line-of-sight characteristic of VHF signals, range is rather limited for surface vessels. Section 52.2 contains a more detailed discussion of Omni.

LORAN. The word "LORAN" is an acronym for the term "long range navigation." It is one of the hyperbolic navigation systems and has both maritime and aeronautical applications as well as some current research into land use (possible utilization for transit scheduling, emergency vehicle monitoring, train locations, etc.).

Development. LORAN was developed by the Massachusetts Institute of Technology (MIT) during World War II. The original version was called LORAN-A and continues in use to a limited degree even today. A subsequent version, LORAN-B, was a pulse and cycle type but never succeeded because of unsatisfactory propagation tolerances. LORAN-C, a design improvement of the original LORAN-A, is the primary LORAN system in use today and has been designated by the U.S. government as the principal civil aid to navigation in the "coastal confluence zone" [CCZ, defined as 50 miles (92.6 km) offshore or out to the edge of the continental shelf, whichever is greater] until the year 2000.

A system of transportable stations developed by the U.S. Air Force, characterized by relatively low-power and resistance to electronic jamming, were designated LORAN-D.

The need to extend the range of LORAN-A stimulated research that eventually led to the development of LORAN-C. Lowering the frequency to increase the range and stabilize the propagation characteristics required higher power output at the transmitter sites, and the resulting extended ranges created new problems in distinguishing between groundwaves and skywaves. LORAN-C emerged employing methods of phase comparison as well as time-difference measurement.

The first LORAN-C stations began operating in 1957, and LORAN-A began phasing out in 1977. LORAN-A is scheduled for complete phaseout during the early 1980s; the one exception appears to be Japan where scheduled termination is listed as "undetermined" in Defense Mapping Agency Publication 117B (*Radio Navigation Aids*).

Referring to Table 52.4, the improvements in range and accuracy are evident. The ranges listed in the table can be more meaningfully analyzed as follows (1 nmi = 2000 yd = 1828.8 m):

	Groundwave Range (nmi)	One-Hop Skywave Range (nmi)	Two-Hop Skywave Range (nmi)
LORAN-A	450–800	1400	—
LORAN-C	1200	2300	4000

Being a hyperbolic system, LORAN requires a two-station arrangement for each LOP to be generated. The possible distance between stations increased from a maximum of 500 nmi for LORAN-A to 1500 nmi with LORAN-C. This increase in the baseline distance gives greater flexibility in station site selection as well as provides a much broader area of coverage by the resulting hyperbolae.

Stations, Chains. Although a minimum of two stations are required, the normal arrangement is a chain comprising a master and two, three, or four secondary stations (master designated M and secondaries designated W, X, Y, Z). Configurations are determined by the coverage desired and the availability of station sites, but adherence to the basic patterns illustrated in Fig. 52.22 is attempted wherever possible to optimize efficiency.

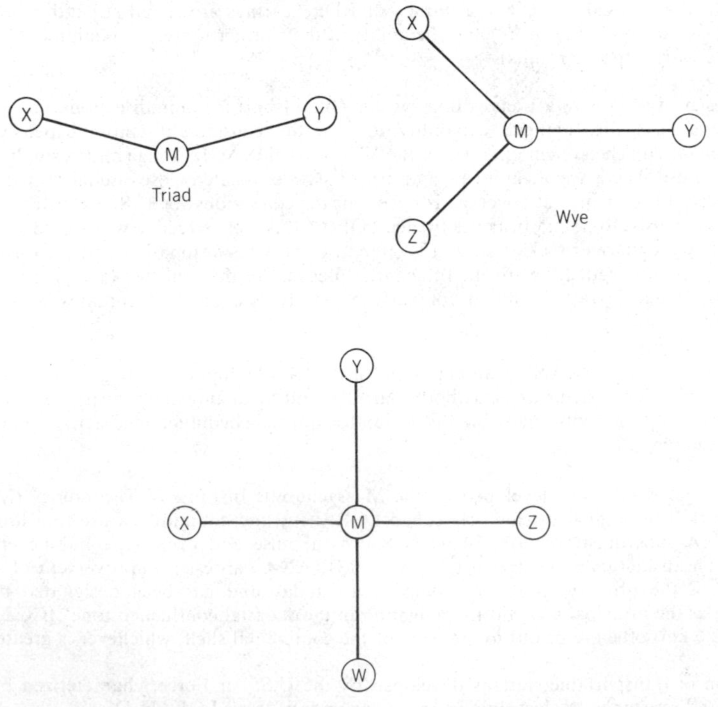

Fig. 52.22 LORAN station arrangement patterns. M, master station; X, Y, Z, secondary stations.

There are currently thirteen LORAN-C chains worldwide consisting of 41 stations. A similar system is owned by the Soviet Union and consists of eight stations.

Within each chain there are one or two monitoring stations to monitor and control the operation of the chain, ensuring proper format and continuity of signals.

The sequence of transmission starts with a pulse from the master station followed by a pulse from each of the secondaries sequentially and ends with the start of another master pulse. The span of time from master pulse to master pulse is termed a group repetition interval (GRI) and is unique for each chain. A chain is identified by its GRI in microseconds less the final zero; for example, a chain having a GRI of 49,900 μs would have an identifier of 4990 and the hyperbolic lines on the nautical chart (map) would appear with this identifier as well as a station designator (W, X, Y, or Z) and a time difference indicator (also in microseconds). With the information gained from the receiver and the corresponding line information on the chart, the mariner is able to select the appropriate line (LOP) from those represented, or can interpolate as necessary to refine his position.

The radiated power of the various stations ranges from 165 to 1800 kW. It is interesting to note that the combination of high power and low frequency permits groundwave signals to actually penetrate the sea to some extent, enabling submarines to receive them while submerged.

Signals. LORAN signals are pulse modulated, transmitted in multipulse groups of comparatively low-power pulses. The use of pulsed emissions of this type enhances the receiver's ability to discriminate between groundwaves and skywaves and also provides more signal energy for the receiver to process without having to substantially increase the power output at the transmitting sites.

The incoming signals are compared for time difference by the receiver as the primary means for obtaining LOP information, but an additional feature of LORAN-C includes a phase comparison as a further refinement of accuracy.

Potential Errors. The receiving antenna aboard the vessel must be properly grounded and care must be exercised in selecting a location that is optimally high in the vessel and away from metal objects insofar as is feasible.

Interference may be generated by engines or auxiliary equipments, and television sets installed too close to a LORAN receiver can produce detrimental signals that can reduce accuracy substantially or even render the equipment useless.

To ensure minimal distortion to incoming signal shape, LORAN-C receivers have a relatively wide bandwidth (20–30 kHz normally). This characteristic, however, makes the equipment vulnerable to extraneous noise. To counter this adversity, manufacturers often include notch filters in their equipments. These filters can be *internal* (set by the manufacturer) or can be *tunable* (set by the operator).

Despite their precision, timing intervals can be off between master and slave. This "synchronization error" is normally kept within ± 0.2 μs (corresponding to ± 60 m error in distance) and is considered tolerable. However, should the error increase to greater than 0.2 μs, an automatic warning system goes off within the transmitter and alters the outgoing signal sufficiently to warn users and to alert monitoring stations.

Equipment Variations. Receiver types vary in sophistication from a manual matching of signals on a cathode ray tube to a computerized conversion of time differences to a readout of latitude and longitude. Remote readouts are available for those who desire access to the navigational information in more than one location aboard their vessel, and "track-plotters" have been developed that record the vessel's track directly onto a chart.

OMEGA. OMEGA is the fairly recent culmination of research into usage of the very-low-frequency (VLF) range to create an extremely long-range system that virtually provides worldwide coverage. The formation of a natural waveguide between the earth's surface and the ionosphere at the chosen OMEGA frequency of 10 to 14 kHz permits propagation of the signal over extremely long distances. These very low frequencies are also quite stable, and phase shifts are relatively predictable. This latter characteristic is particularly important because OMEGA does not depend on time-difference measurement, as does LORAN, but is a phase-comparison system. The receiver measures the difference in phase between signals transmitted continuously and simultaneously by two or more stations to obtain position information.

Stations. Unlike with LORAN, there is no master-slave (or secondary) relationship. Any two stations can be combined to obtain an LOP.

The OMEGA design calls for eight permanent stations located in North Dakota, Hawaii, Liberia, Norway, Japan, Argentina, Madagascar, and Australia. A temporary station was located in Trinidad while the Australian station was being completed.

Typical baseline length (distance between stations) is 5000 to 6000 miles. This extremely long baseline results in a hyperbolic pattern with far less divergence problem as distance increases from the baseline.

Normal radiated power is 10,000 W (10 kW) at each station. Once again, the high-power low-frequency combination permits signal penetration below the ocean's surface, so that OMEGA is available to submerged submarines as well as surface vessels.

Because of the extremely long wavelength of the VLF signal, the antenna spans at these stations are approximately 3500 m tower to tower and weigh about 17 tons. In fact, the antenna in Norway spans a fjord!

Frequencies. OMEGA stations transmit on 10.2, 11.05, 11.22, and 13.6 kHz. Each of the eight stations transmits on each of the four frequencies in such a manner as to have only four stations transmitting at any given instant and each on a separate frequency.

Receiver. The OMEGA receiver must be capable of discerning the OMEGA signal in the midst of normal atmospheric noise and of accurately determining the phase of the received signal.

Shipboard receiving antennae are normally tunable, vertical-whip type, approximately 3 m in height. Special antenna configurations are available for aircraft use.

Lanes. Since the transmissions travel in repetitive sine waves, the phase comparison yields position information that is good only within one-half wavelength (half since the receiver establishes phase position relative to the nearest zero contour, which occurs twice each cycle). These half-wavelength corridors are depicted on nautical charts (maps) and identified as lanes. It is therefore imperative that the mariner be aware of which lane he is in at any given time. OMEGA equipments normally include a lane-counting function that registers a change every time one lane is departed from and a second lane is entered. The prudent mariner of course would also keep track manually by plotting the expected track on the basis of the courses and speeds the vessel has been using during the voyage. A technique known as broad lane locating, which basically creates an artificial larger lane for comparing with a normal lane, permits the resolution of ambiguity in most cases and allows the mariner to determine which lane the vessel is currently occupying.

The average receiver sensitivity permits accuracy to approximately 1/100th of a lane, called a centilane. This of course yields 100 hyperbolic LOPs per lane. The appropriate half wavelength for the OMEGA frequency yields an average lane width of 8 nmi (approximately 15 km).

Differential OMEGA. A one-mile accuracy in the vast reaches of the open sea is acceptable to the mariner, but on approaching coastal areas a greater accuracy is required. To accomplish this a system known as differential OMEGA was devised.

Since radio propagation conditions are the major source of OMEGA error, and because these propagation conditions are relatively constant in a given local area, a monitoring station, capable of measuring errors and determining compensatory corrections, broadcasts these corrections to users in the area. The resultant accuracy is within 0.25 nmi—a decided improvement!

DECCA. The DECCA Navigator System (usually simply called DECCA) is unique in that it is privately owned. The Racal-Decca Company finances the operation of the system through the sale and leasing of the receiving equipment. DECCA is used by aircraft as well as surface vessels at sea.

Although invented in 1937, DECCA first received notable recognition in 1944 when it was used to guide minesweepers and landing craft onto the Normandy beaches during the D-Day invasion. The first commercial stations were set up in southeastern England in 1946.

Characteristics. DECCA uses an unmodulated continuous wave (CW) signal in the 70- to 130-kHz frequency range. Unlike LORAN and OMEGA, which are considered open ocean systems, DECCA is essentially a coastal aid to navigation with a nominal range of 240 nmi (444 km). The limited range capability is somewhat compensated for by the high degree of accuracy. DECCA is similar to OMEGA in that it uses phase comparison rather than time difference to establish position.

Another major advantage is DECCA's ease of operation. No particular skill is required to read off the information by the dials (called decometers) on the front of the receiver, and DECCA charts have lattice grids that are color coded to match the color of the decometer being used.

Stations. A "chain" of stations normally consists of the master station surrounded by three slaves arranged in a pattern similar to the LORAN "wye" pattern (see Fig. 52.22) with approximately 120° spacing. The baseline length is generally between 60 and 120 nmi (111–222 km). Neighboring chains often share common slave sites.

The slaves are color coded to match the colors provided on the receiver's decometers (phase meters). The colors used—red, green, and purple—are standard throughout the system. The resulting hyperbolae from the master and the red slave are printed *in red* on the DECCA nautical chart (map). The others are correspondingly colored as well. Each chain operates on a unique fundamental frequency between 14 and 14.33 kHz. Exactly which frequency is what determines or identifies the particular chain.

Receiver. The DECCA receiver is essentially four separate receivers housed within one unit. All four are set to the appropriate frequency simply by selecting the "fundamental frequency" for the particular chain. The fundamental frequency (f) is not actually radiated. Instead a series of harmonics is used, as depicted in Table 52.6. Within the receiver the incoming signals are frequency multiplied to arrive at a simple comparison frequency that provides compatibility in phase so that proper comparison can take place.

Lanes. Because of the considerable difference in frequency, a single lane (distance corresponding to one-half wavelength) of DECCA is small: only 357 m (compared with an OMEGA lane of 14,816 m).

There is an additional feature of DECCA that permits lane identification by the transmission of a separate lane identifier signal by each master and slave once each 20 seconds. This identifier is only 0.6 second in duration but is essential to the proper usage of the system, since lane identification would be virtually impossible without it.

CONSOL. Originally named "SONNE," CONSOL was developed by the Germans during World War II and later refined as well as renamed by the British. CONSOL is an improved version of RDF that is hyperbolic in that it uses multiple stations to generate hyperbolic lines of position, but is more closely associated with radial systems since the extremely short baseline (distance between stations) creates a system of "collapsed hyperbolae" which appear as *radial* lines when any significant distance from the stations. The principle here is that when a hyperbolic line is extended beyond a distance of approximately 12 times the length of the baseline, the line becomes a straight one (called an asymptote in mathematical terms). Since the baseline for CONSOL is normally 4 to 5 km, the lines become "radial" at a relatively short distance from the stations.

Used by both aircraft and surface vessels, CONSOL has the unique advantage of not requiring any specialized equipment. CONSOL LOPs can be obtained using an RDF or standard communications receiver.

Stations. CONSOL uses equidistant stations placed in a line, the separation being approximately three times the wavelength of the transmitted frequency. Station complexes are located in European waters from Norway to Spain.

Receiver. Any receiver that is 250 to 370 kHz capable can be used with CONSOL. If a communications receiver is employed, automatic gain control (AGC) should be off and a beat-frequency oscillator (BFO) is desirable. The system provides an audible series of dots and dashes that reveal bearing (and therefore LOP) information.

Range. Because of the "collapsed hyperbolic" nature of the system, CONSOL has a *minimum* effective range of 25 to 50 nmi (46–93 km). The maximum range is anywhere from 500 to 1400 nmi (926–2593 km), depending on equipment capability and atmospheric conditions.

Accuracy. Because of pattern distortion, there is an unusable arc of approximately 30° on either side of the baseline extension, centered on the nearest station (see Fig. 52.21*a*). This results in a usable sector of approximately 240° centered about the bisector of the baseline. Accuracy is greatest at the

TABLE 52.6. DECCA FREQUENCY USAGE

Signals To Be Compared	Harmonic	Multiplier	Resulting Comparison Frequency
Master	$6f^a$	4	$24f$
Red slave	$8f$	3	
Master	$6f$	3	$18f$
Green slave	$9f$	2	
Master	$6f$	5	$30f$
Purple slave	$5f$	6	

[a] f, fundamental frequency (always between 14.00 and 14.33 kHz).

perpendicular bisector of the baseline and diminishes with proximity to the baseline extension. With a *best* accuracy of 1.0 nmi, CONSOL is classed as an open-ocean system rather than a coastal one.

Versions. In the United States a similar system is in use in three locations: Nantucket Island, San Francisco, and Miami. This American version of CONSOL is called CONSOLAN, uses only two antennae instead of three, and functions at a higher power output level, in the 190- to 194-kHz range.

The Soviet Union also has a version of CONSOL called BPM5, which uses a chain of five stations in a cross pattern. This modification provides narrower dot and dash sectors and therefore refines the accuracy somewhat.

Ship's Inertial Navigation System. The ship's inertial navigation system (SINS) is a unique electronic navigation system in a number of significant ways:

1. It requires no transmitting stations and is therefore neither hyperbolic nor radial.
2. It provides a fix instead of a series of LOPs.
3. It requires specially qualified personnel to operate and maintain and is extremely expensive.
4. It is self-contained within the vessel and requires no outside electronic source to function (periodic updates for correction, however, are necessary and must be obtained from some external source such as navigational satellites).

SINS was originally developed for use in aircraft and spacecraft and later found application in the Polaris submarine program. It has subsequently been extended to use in surface vessels of the U.S. Navy.

Principles of Operation. Two devices that sense accelerations in relation to the surface of the earth (termed accelerometers) are stabilized by a system of gyroscopic components to ensure a continuous north-south and east-west orientation. Movements by the vessel, and therefore by the stabilized accelerometers, are sensed and, by application of Newton's second law of motion, are integrated to determine velocity and distance with respect to time. A computer component of the system converts the resulting rectangular coordinates to spherical equivalents to compensate for the curvature of the earth's surface. The result is a continuous readout of latitude and longitude as well as ship's heading (direction) and speed.

Errors. The daily rotation of the earth, minimal but nonetheless extant; friction; and gravitational misinterpretation all contribute very small but cumulative errors to the SINS system, which must be periodically purged. This is accomplished by updating the actual position of the vessel through some other source of reliable information.

Bibliography

Admiralty Manual of Navigation, Vol. 1, Her Majesty's Stationery Office, London, 1964.

Appleyard, S. F., *Marine Electronic Navigation*, Routledge & Kegan Paul, London, 1980.

Beck, G. E., *Navigation Systems: A Survey of Modern Electronic Aids*, Van Nostrand Reinhold, London, 1971.

Bowditch, N., *American Practical Navigator*, Defense Mapping Agency Hydrography Center, 1977.

Cotter, C. H., *The Elements of Navigation and Nautical Astronomy*, Brown, Son & Ferguson, Glasgow, 1977.

Davies, H. N., *The Navigators' Guide to Hyperbolic Navigation*, Brown, Son & Ferguson, Glasgow, 1949.

Defense Mapping Agency, *Radio Navigational Aids*, Publication 117A & B, Defense Mapping Agency Office of Distribution Services, 1981.

Hobbs, R. R., *Marine Navigation 2: Celestial and Electronic*, U.S. Naval Institute, Annapolis, MD, 1981.

Institute of Navigation, *Navigation: Journal of the Institute of Navigation*, Washington, DC (quarterly).

Lenk, J. D., *Electronic Navigation Made Easy*, John F. Rider, New York, 1964.

Litton Systems Inc., *An Introduction to OMEGA*, Litton Aero Products, Canoga Park, CA, 1977.

Maloney, E. S., *Dutton's Navigation & Piloting*, U.S. Naval Institute, Annapolis, MD, 1978.

Moody, A. B., *Navigation Afloat: A Manual for the Seaman*, Van Nostrand Reinhold, New York, 1980.

Paushin, D. A., *Termination of LORAN-A*, Oregon State University, Corvallis, OR, 1977.

Time-Life Library of Boating, *Navigation*, Time-Life Books, Alexandria, VA, 1981.

U.S. Coast Guard, *Light List CG 158*, Department of Transportation, Washington, DC, 1980.

U.S. Coast Guard, *LORAN-C User Handbook*, U.S. Government Printing Office, Washington, DC, 1969.

United States Naval Institute, *Proceedings*, Annapolis, MD (monthly).

U.S. Naval Oceanographic Office, *Navigation Dictionary*, U.S. Government Printing Office, Washington, DC, 1969.

CHAPTER 53
AIRCRAFT LANDING SYSTEMS

STEPHEN C. MARTIN

Lockheed Missiles and Space Company, Palo Alto, California

53.1 RADIO AIDS TO AIRCRAFT LANDING

From the earliest days of aviation it was quickly recognized that the most critical phase of flight was the final approach and landing. As soon as enroute navigation in clouds had been facilitated by gyroscopic flight instruments and the early radio ranges, considerable effort was devoted to providing aids for landing guidance. An account of early U.S. government work in this area can be found in Ref. 1. In 1929 James H. Doolittle (later General) performed a fully blind landing using the low-frequency radio range for lateral guidance and a sensitive altimeter for vertical reference. These experiments showed that for reliable operation close to the ground, a precise vertical guidance signal would be necessary. Further work in the United States and Europe led to the development of the instrument landing system (ILS).

Radio landing aids are not normally used to guide the aircraft to touchdown but rather to provide guidance on the approach until visual contact is made with the runway, from which point a normal landing can be completed. The capability for reliable fully automatic landing by a commercial carrier is comparatively recent; the first automatic landing by a commercial carrier with fare-paying passengers was made by British Airways in 1965.

Precision and Nonprecision Instrument Approaches

In a nonprecision instrument approach, lateral guidance is provided by one or more of the short-range radio navaids described in Section 52.2 [e.g., very-high-frequency (VHF) omnidirectional radio range (VOR), distance measuring equipment (DME), tactical air navigation (TACAN), nondirectional beacons (NDB)]. The pilot makes timed descents to specific altitudes after passing a sequence of one or more radio fix positions. Each approved approach procedure is peculiar to a specific airport or runway and is individually charted. Legal minimum descent altitudes before achievement of visual contact for landing vary considerably with terrain conditions but are seldom lower than 500 ft above the ground. Nonprecision instrument approaches are flown routinely by private, military, corporate, and smaller commercial aircraft at thousands of airports throughout the world.

A precision instrument approach is one in which the pilot receives continuous vertical guidance by electronic means. For commercial aircraft there are presently three such precision approach systems: conventional instrument landing system (ILS); microwave landing system (MLS), now in advanced development; and precision approach radar (PAR), a "talkdown" system that, when used by the military, is an element of a ground-controlled approach (GCA). Characteristics of these systems are defined by International Civil Aviation Organization (ICAO) agreements. There are also a considerable number of special landing systems that have been developed for military requirements, such as aircraft carrier landing and tactical operation of helicopters and vertical takeoff and landing (VTOL) aircraft.

For the conventional ILS, different categories of service have been defined, as shown in Table 53.1. These are now often applied to the performance of precision landing systems in general. Operation in category III conditions clearly requires some form of automatic landing capability.

53.2 INSTRUMENT LANDING SYSTEM (ILS)

The conventional ILS consists of three elements: the localizer, which gives lateral guidance; the glide slope, which gives vertical guidance; and one or more marker beacons that define specific points along the course. The whole defines a straight course, about 10 nautical miles (nmi) long, descending at a fixed angle, usually 3°, to the near end of the runway. The localizer operates on one of 40 VHF channels between 108.1 and 111.95 MHz [interspersed with VHF omnirange (VOR) channels]. Minimum channel spacing is 50 kHz. The glide slope occupies one of 40 ultrahigh-frequency (UHF) channels between 329.15 and 334.7 MHz (minimum spacing 150 kHz), each of which is "paired" with a localizer channel (Ref. 2, Sec. 3). The marker beacons operate at 75 MHz, the same as the enroute beacons described in Section 52.2.

Demonstrated first in 1939, the present configuration of the ILS dates from 1942 and was used extensively in World War II. By 1985 there are expected to be over 1200 ground installations worldwide, with over 150,000 equipped aircraft.

Localizer

The localizer signal is amplitude modulated by two signals at 90 and 150 Hz and radiates continuously with a directive, multielement, phased antenna system using horizontal polarization. The aircraft uses an omnidirectional receiving antenna. When "on course," equal levels of the demodulated 90- and 150-Hz signals are received. When the aircraft is left of course, the 90-Hz signal is stronger, and the 150-Hz signal predominates when it is right of course. Figure 53.1 shows the typical radiation patterns for the two sets of localizer signals, the "carrier" and "sideband" signals. The carrier signal is actually double sideband amplitude modulated with equal levels (normally 20% modulation) of the 90- and 150-Hz signals. The sideband signal contains DSB suppressed-carrier modulation at 90 and 150 Hz. On the left side of course (lower part of Fig. 53.1) the 90-Hz modulation is in phase with that of the carrier signal and the 150-Hz is out of phase. Right of course these transmitted phases are reversed. Thus in the aircraft receiver, the composite received waveform favors 90 or 150 Hz when left or right of course, respectively.

The transmitted modulation and patterns are adjusted such that the received difference in depth of modulation (DDM) varies linearly with angle off course over a range of approximately $\pm 2.5°$ (depending on installation). At the extremes of coverage, DDM is ± 0.175. Aircraft equipment gives "full scale" left or right signals at this DDM. Recommended localizer coverage volumes are defined

TABLE 53.1. ICAO[a] DEFINITIONS OF LANDING VISIBILITY CATEGORIES

Category	Runway Visual Range (RVR)	Decision Height (Continue Visually or Go Around)
I	800 m	60 m (200 ft)
II	400 m	30 m (100 ft)
III(a)	200 m	0
III(b)	50 m	0
III(c)	0	0

Source. International Standards and Practices.[2]
[a]ICAO, International Civil Aviation Organization.

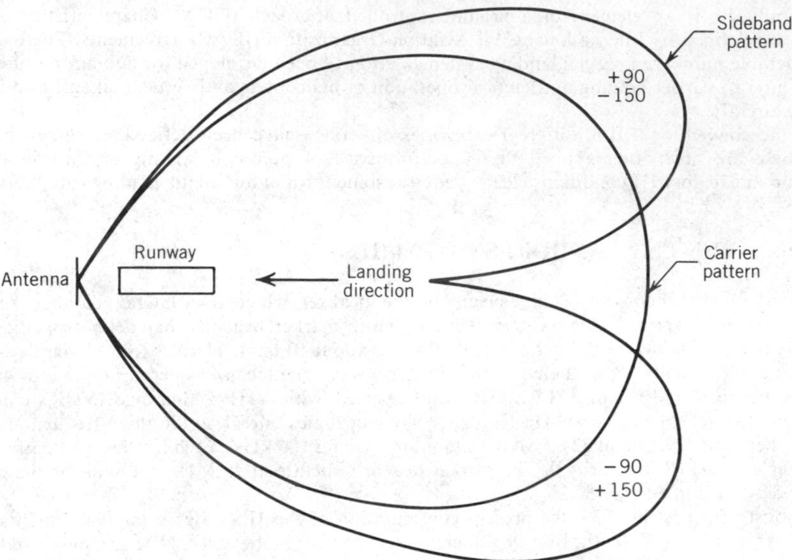

Fig. 53.1 Localizer radiation patterns.

by ICAO as follows:

±10° from course line	25 nmi distance
10°–35° from course line	17 nmi distance
Greater than 35° from course line	10 nmi distance
Vertical coverage	0° to 7° above horizon

The minimum field strength within this volume is to be 100 μV/m. Outside the linear course region (±2.5°) but within the coverage region, the aircraft must receive essentially full-scale indication with correct sensing, known as a "clearance" signal. Localizer radiated powers are in the 10- to 15-W range. For identification, Morse-coded 1020-Hz tones are simultaneously transmitted with a modulation depth of 10%. At some sites, voice transmissions from the control tower can also be superimposed.

Many different types of localizer antenna array and array elements are in use. The standard U.S. localizer uses eight elements in a linear array spanning 14 m. Details of spacing and signal phasing can be found in Ref. 3. The array is mounted about 300 m beyond the stop end of the runway being served, broadside to the extended centerline. The antenna elements may be omnidirectional Alford loops (see Ref. 6, Section 52.2) or directive types when sidelobe radiation is to be minimized. Among these latter are the loop plus folded dipole (V-Ring) element[5] and the traveling-wave antenna.[6] The folded log-periodic dipole element[4] is another approach with advantages of reduced height (2.1 m) and resistance to the effects of ice coverage.

Use of elements that individually have an omnidirectional pattern provides the advantage that a usable back course may exist, reciprocal to the primary approach direction. This can provide overflight guidance and a guidance signal for opposite-direction approaches where, however, the sense of the left-right information is reversed. On the other hand, this wide-angle radiation can lead to severe course bends or "scalloping" caused by multipath reflections from buildings or higher terrain parallel to the approach path.

The ICAO requirements[2] are that the 2-sigma excursion from the mean localizer course shall not exceed 0.36° beyond 4 nmi, decreasing to 0.06°, 2-sigma, when 1050 m from runway threshold. At sites subject to multipath, much wider aperture localizer antennas are used to give a narrower radiation pattern. A typical array uses 14 elements over a width of 26 m.[5] Other types of wide-aperture localizer include parabolic reflecting arrays and slotted waveguides of up to 60 m in length.[3] The narrowness of the radiation pattern often means that adequate clearance signals are not provided at wide angles. For this purpose, a separate narrow aperture array is often provided that radiates guidance signals on a frequency about 8 kHz shifted from the main signals. When the aircraft is close to the on-course path, the high-accuracy signals have higher power and capture the aircraft receiver. At wide angles it is the clearance signal that predominates and that gives correct sensing information.

Glide Slope

The UHF glide slope uses amplitude modulations of 90 and 150 Hz, as does the localizer. The antenna system has vertical directivity such that 90-Hz modulation is stronger when "too high" and 150-Hz stronger when "too low." Horizontal polarization is used. The linear region extends 0.7° above and below the nominal glide path, which is usually set at 3° above horizontal. At the extremes of the linear region, the difference in depth of modulation (DDM) is ± 0.175. Horizontal coverage is a minimum of $\pm 8°$, about the centerline, for a distance of 10 nmi. Within this coverage area the minimum field strength is to be 400 μV/m. Glide slope transmitter power is around 4 W.

The great majority of glide slope antenna systems are of the image type. Figure 53.2 illustrates the pattern formation for the most common "null-reference" system. Two horizontal dipole antennas are mounted on a vertical mast. The lower transmits the carrier signal and the upper, which is twice as high, transmits the sideband signals only. Each antenna, together with its image in the ground plane formed by the earth, forms a vertical lobe structure, as shown on the right of Fig. 53.2. The lobe structure for the sideband antenna is twice as fine as for the carrier antenna. For the 330-MHz glide slope frequency, antenna heights of about 4.25 and 8.5 m produce a peak in the carrier pattern and a null in the sideband pattern at the desired deviation angle of 3°. The ground reflection causes the phasing of the sideband signals in the second lobe to be inverted relative to those in the first lobe, giving correct sensing about the on-course null point. Since the lobe structure continues to higher elevation angles, another stable false glide slope appears at 15°. This is avoided, in practice, by intercepting the correct glide path in level flight from below. The antenna mast is placed about 120 m to the side of the runway and set back about 300 m from the approach end so that the glide path center has an elevation of 15 ± 3 m crossing the runway end.

ICAO specifications require that course bends not exceed 0.15° (2σ) at 4.5 nmi decreasing to 0.1° (2σ) inside 1050 m from touchdown. The imaging array requires that all the terrain in the first Fresnel zone (see Section 48.2) be suitably flat and uncluttered, to achieve correct patterns. At smaller airfields and in hilly terrain this is very difficult to achieve. The sideband reference system[7] is a different image-type array that is less sensitive to terrain effects. At sites with an up slope in front of the glide slope antenna, the "capture-effect" array can give improved results.[5] In this configuration, the radiation patterns of the separate antennas are shaped, with use of reflecting screens, so as to tilt the resulting patterns above the horizon. With use of a third antenna element, a low-power clearance signal consisting only of "fly-up" information is transmitted with a broad pattern on a carrier frequency about 8 kHz shifted from the primary radiation. Close to "on course" the main signal is stronger, but below the linear course region the clearance signal is stronger and captures the aircraft receiver, giving a strong fly-up command.

At especially difficult sites, nonimaging glide slope antennas have been used. Vertical, slotted 330-MHz waveguides up to 35 m long have been tested,[5] tilted at 3° from vertical to generate the

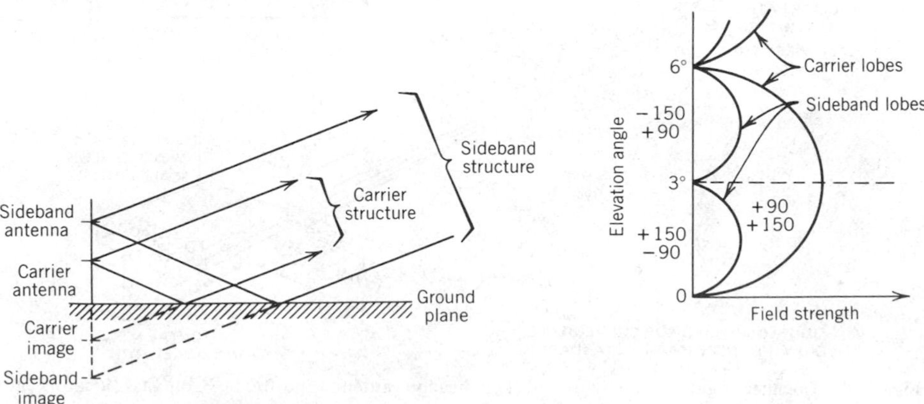

Fig. 53.2 Glide path formation for "null-reference" system.

required beam slope. A 20-m antenna of this type is installed at LaGuardia, New York, where the ground plane is actually the sea surface and cannot be used to reflect an image because of 3-m tidal variations.

Considerable research has been conducted on nonimaging end-fire arrays.[6,8] One configuration in use has two radiating sources about 200 m apart, each offset 60 m from the runway centerline and about 1.25 m above ground. Each is fed with carrier and both sidebands in phase quadrature. In spite of these different glide slope systems, it is still true that for many small airports and for those in mountainous regions, provision of an acceptable ILS glide slope is not feasible.

Marker Beacons

Low-power 75-MHz marker beacons (see Section 52.2) are used with the ILS to define specific points along the course. As the aircraft passes over the vertically radiating beacon, the pilot receives a few seconds of aural and visual indication. An outer marker beacon is installed at a distance of 3.5 to 6 nmi from the end of the runway, usually at the point where the level-flight portion of the designated approach procedure intercepts the glide slope. The outer marker transmitter is 95% amplitude modulated with a sequence of 400-Hz tone dashes at two dashes per second. A middle marker beacon is placed about 1050 m from the end of the runway, at the point where the glide slope is 60 m above ground (decision height for category I approaches). Its transmitter is modulated with alternate dots and dashes using a 1300-Hz tone. At runways certified for category II and III landings, an inner marker is also provided. This is between 75 and 450 m from the end of the runway, at the point where the glide slope height is 30 m (category II decision height). It is modulated with dots at six per second using a 3000-Hz tone.

ILS marker beacon transmitters use vertical fan beam antennas with horizontal polarization, and radiate less than 2.5 W. Figure 53.3 shows the relative coverage of the localizer, glide slope, and two usual marker beacon signal patterns.

ILS-DME

At locations where installation of marker beacons is not practical (such as approaches over water), distance measuring equipment (DME) is usually provided (see Section 52.2). The DME transponder is colocated with the localizer antenna and operates on a channel that is "paired" with that of the VHF frequency of the localizer.[2] Some airborne DME equipments are designed to recognize this as

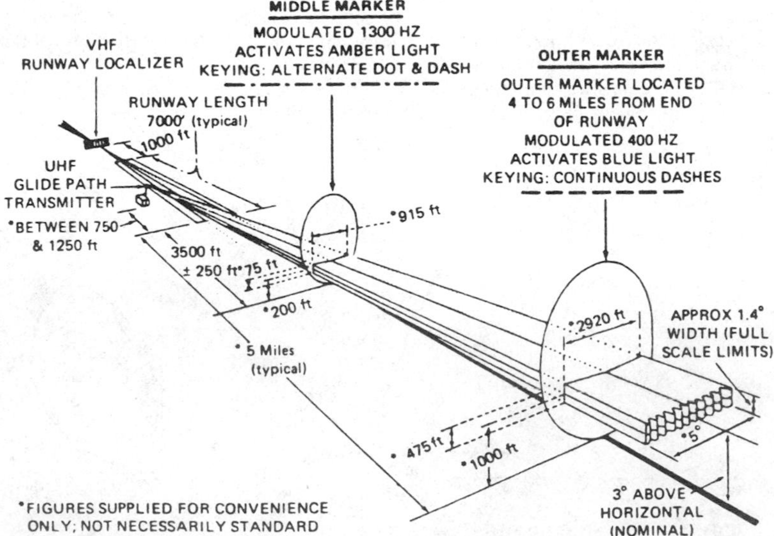

Fig. 53.3 Localizer, glide slope, and marker beacon antenna positions. Compass locators are sometimes colocated with outer and middle markers.

being an ILS-assigned channel, and will then maintain a relatively high interrogation rate ($\simeq$ 50–100/s) in the tracking mode. This provides smoother data to the autopilot.

Categories II and III ILS

The vast majority of installed ILS facilities provide service to category I standard (60-m decision height). At most major airports in the United States and Europe, a category II ILS (30-m decision height) serves at least one runway. Category III service (0 decision height) is still relatively rare. In the United States there are little more than half a dozen category III systems presently commissioned. In Europe progress is rather more advanced, and considerable experience has been obtained with fully automatic category III operations.[9] The latest generation of air transport aircraft is being produced with the special flight control systems needed for category III operation incorporated as standard equipment (e.g., L-1011, Concorde, A-300, Boeing 757).

In many cases the difference between a ground ILS facility being certificated for category II or category III operation lies not in the inherent quality of guidance but rather in the provision and degree of monitors and of duplicated transmitting equipment. These are needed to ensure that an almost instantaneous switchover can be made in the event of an out-of-tolerance guidance signal being detected.

Airborne ILS Equipment

The aircraft localizer and glide slope receivers each tune the selected channel pair, AM demodulate the signals, and separately compare the 90- and 150-Hz modulation amplitudes. Although the system is conceptually simple, care is needed in each receiver to maintain gain balance between the two tone channels to ensure that zero difference in modulation depth accurately gives an "on-course" signal.

On smaller aircraft and in earlier types of airline equipment, the localizer functions are incorporated within the VOR receiver (see Section 52.2). The VOR antenna similarly serves to receive the localizer signals. A separate glide slope receiver is used in this case, with tuning slaved to that of the VOR receiver. The glide slope antenna is a U-shaped UHF folded dipole mounted on the aircraft nose, or a flush cavity antenna. Modern airline equipment uses a localizer receiver separate from the VOR, combined with the glide slope in a common "ILS receiver" unit. Also a separate localizer antenna, not shared with VOR, is provided.

ICAO standards (Ref. 2, Part 1, Attachment C) require selectivities of 60 dB at ± 50 kHz for localizer receivers and 60 dB at ± 300 kHz for glide slope receivers. Other minimum performance standards have been set by the Radio Technical Committee for Aeronautics (RTCA).[10] Typical sensitivities are 3 μV for the localizer receiver and 5 μV for the glide slope receiver. A complete airline ILS receiver typically occupies 5500 cm^3, weighs 5.5 kg, and consumes 26 W. The separate glide slope receiver used with the existing VOR/LOC (localizer) receiver on a smaller aircraft would occupy about 1400 cm^3, weigh 1 kg, and consume 7 W.

The simplest ILS cockpit display is the two-needle crosspointer, as sketched in Fig. 53.4. This type of display is usually combined with the course deviation indicator (CDI) (see Section 52.2), but this basic cross-pointer form is usually found only on smaller civil aircraft and in some military installations. On commercial and airline aircraft the localizer signal is switched to the course deviation bar of the horizontal situation indicator (HSI), as shown in Section 52.2. An adjacent vertical pointer scale then shows glide slope deviation. Separate lateral and vertical error pointers are often also incorporated into the pilot's attitude indicator display (artificial horizon). For commercial and airline operations, nearly all ILS approaches are flown down to decision height in the autopilot coupled mode, with the deviation indicators used only for pilot monitoring purposes.

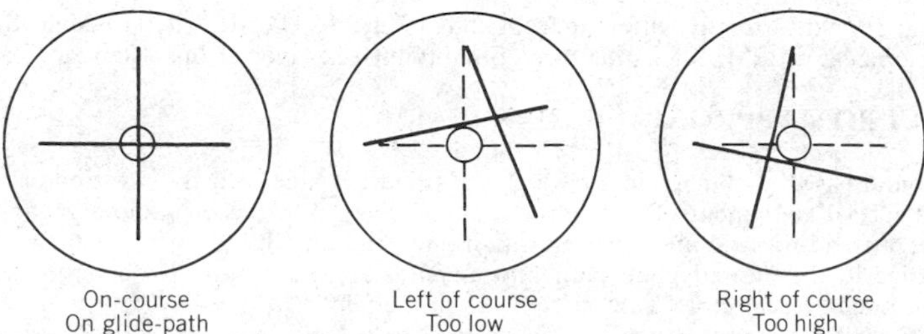

Fig. 53.4 Instrument landing system (ILS) cross-pointer indications.

Future Plans

The conventional ILS system, by international agreement, will be supported as a standard system until at least 1995. It will be replaced eventually by the scanning-beam microwave landing system (MLS), which is described later. The U.S. military forces plan to start the use of MLS in 1985 for land and naval use, with a gradual phaseout of ILS by 1995.[11]

Radio Altimeter

The radio altimeter is a low-power downward-looking radar used by civil aircraft to determine height above ground in the final stages of the landing approach. It is required equipment for aircraft certified to make approaches in category II or category III conditions. Civil radio altimeters operate in the 4.2- to 4.4-GHz band. Minimum performance standards are set by RTCA.[12] Both pulse and modulated continuous-wave (CW) systems are in use. The modulated CW types use triangular or sinusoidal modulation, measuring height by means of the difference frequency between transmitted and received signals (Ref. 13, Sec. 3.3). A typical modulation deviation is 100 MHz peak to peak at a 100-Hz rate.

Civil radar altimeters operate to a maximum height of around 750 m (2500 ft), but the information below 150 m (500 ft) is of major interest. The pilot's radio altitude display has an expanded scale in this region showing height to a resolution of around 3 m (10 ft). As the aircraft descends through the decision height for the category of approach being flown, the radar altimeter can be preset to trigger a visual and audible warning signal. For automatic category III landings, the vertical guidance of the aircraft is transferred progressively from the ILS glide slope to the radio altimeter signal as the aircraft descends to within 30 m (100 ft) of the ground.

Pulse-type altimeters mostly use a single antenna for transmit and receive, transmit 5 to 10 W peak pulse power, and have a minimum range (set by pulsewidth considerations) of around 12 m. FM-CW altimeters use two separate antennas, radiate about 400 mW, and are usable down to zero height. Typical accuracy is ± 0.5 m or 2% of height when below 150 m. A pulse system for a small aircraft would typically occupy 1400 cm^3, weigh 1 kg, and consume 6 W. For an airline unit of the FM-CW type, these figures would be 9000 cm^3, 5.5 kg, and 50 W.

Military radar altimeters may operate on different frequencies from the civil 4300-MHz band. As well as for landing, they are used for enroute measurement of height above terrain, to altitudes of 6000 m or more. For an aircraft equipped with a Doppler navigation system (see Section 52.2), radio altitude can often be computed directly from the data used for navigation.

Interim Standard Microwave Landing System (ISMLS)

This is a microwave (5-GHz) system, using step scanned beams for azimuth and vertical guidance, which has been installed at about 12 sites in the United States starting in 1975. It was developed to provide ILS-type guidance at sites where installation of conventional ILS was too difficult. Azimuth of $\pm 40°$ is covered by eight switched beams, and elevation of 0 to 6° with four beams. The cycle rate for each set is about 600 Hz. Each beam is pulsewidth modulated with sample values of a composite 90/150-Hz waveform, with relative amplitude ratios appropriate to the beam distance away from centerline or 3° glide path. The aircraft receiver amplitude detects the appropriate azimuth and vertical signal frequencies and obtains two sets of proportional 90/150-Hz signals which are a function of error angle, just as for ILS. The aircraft receiver in fact uses conventional VHF-UHF localizer-glide slope receivers [with minor automatic gain control (AGC) modifications] preceded by a single down-conversion stage from 5 GHz. The two 5-GHZ signals are transmitted with the appropriate offset ($\simeq 200$ MHz) so that the "localizer" and "glide slope" down-converted signals have one of the standard ILS frequency pairings.

This signal format is totally different from that of the ICAO MLS described in Section 53.4. Further development of ISMLS has now been discontinued in favor of this "standard" MLS.

53.3 PRECISION APPROACH RADAR (PAR)

PAR is a ground-based guidance aid by which an aircraft is able to make low approaches without carrying any special equipment other than a radio receiver. A precision ground radar tracks each incoming aircraft and uses simultaneous elevation and azimuth CRT displays to show the aircraft's position relative to the desired glide path and azimuth angles. A trained operator, viewing these displays, then radios verbal correction instructions to the pilot.

ICAO has set PAR performance specifications for civil aviation use (Ref. 2, Sec. 3.2). These require that the radar be capable of tracking a 15-m^2 target (equivalent to a small twin-engined aircraft) at a range of 9 nmi. Position measurement accuracy is to be 9 m laterally, 6 m vertically, and

30 m in range. PAR can provide guidance down to ILS category I minimums (60-m decision height). It is, however, now hardly used at all by civil aircraft. PAR was once provided as a backup at many large airports to monitor aircraft making ILS approaches, but most such installations have now been removed.

The military has continued to use PAR since its development in World War II.[14] A military installation consisting of a surveillance radar for acquisition and the PAR for final guidance is known as a ground-controlled approach (GCA) system. The system is much more suited to forward bases than is ILS, which has its requirements for extensive site preparation and leveling. PAR under many circumstances is used to guide military aircraft down to ILS category II equivalent minimums (30-m decision height).

PAR typically operates at 9 to 10 GHz. Many systems in use employ two mechanically scanned orthogonal fan beam antennas to make separate measurements in azimuth and elevation. An example of a modern system is the AN/TPN-19 precision radar. This uses a single 824-element phased-array reflector antenna 2.8 m wide $\times$ 3.55 m high to generate a spot beam 1.4° in azimuth $\times$ 0.75° in elevation. This can be raster scanned for search or operated as a monopulse tracker over a $\pm 10°$ azimuth $\times$ 15° elevation sector. Data rates are 22 scans per second in search and 2 scans per second in track. With peak power of 320 kW, the operating range in 50 mm/hr rainfall is 15 nmi. Angular tracking accuracy is 0.143° azimuth and 0.072° elevation. Using a 1-μs pulse with 120 MHz of chirp frequency compression, a ranging resolution of around 1.5 m is claimed.

A PAR system in which the operator's voice commands are replaced by a surface-to-air data link, the AN/SPN-42, is used by the U.S. Navy for aircraft carrier landings. This is discussed later.

53.4 MICROWAVE LANDING SYSTEM (MLS)

The MLS has been adopted by ICAO as the system that will eventually supplant the conventional ILS sometime after 1995. Coexistence of both systems is envisaged for some considerable time. The MLS will provide many operational benefits over conventional ILS. Among these are:

1. Guidance for precision instrument approaches at sites where the terrain is too difficult for ILS installation.
2. Guidance for curved-approach profiles to reduce airspace usage and to avoid noise-sensitive areas on the ground.
3. Variable glide-path angles to suit each aircraft's performance.
4. Increased number of operating channels.
5. Reduced ground installation and maintenance costs.

Prototype MLS ground and airborne equipment is being tested by the United States, England, France, Japan, Australia, and the Soviet Union. The United States is presently conducting a service test and evaluation program (STEP) using ground stations at Atlantic City, Philadelphia, and Washington, DC.[15]

The U.S. military services are supporting development of a compatible joint tactical microwave landing system (JTMLS). This program will provide a manpack transportable ground station and suitable avionics for testing of joint Army, Navy, Marine, and Air Force procedures using both fixed-wing aircraft and helicopters.

The MLS angle guidance scheme is known as time-referenced scanning beam (TRSB). This technique was selected by ICAO in 1978 after study and evaluation of several competitive guidance systems. These were commutated antenna (Doppler) scanning beam, multilateration using precision distance measuring equipment (DME), and frequency-coded scanning beam methods.

With the TRSB concept, angular data are provided by orthogonally scanned fan beams. The narrow width of each beam in its critical dimension allows the airborne processor to perform significant discrimination of correct signals from those caused by multipath reflections from terrain or ground structures.[16] It is because of this ability that the MLS will be usable at landing sites for which conventional ILS is not feasible.

Angle guidance and data transmission functions at a given site are time multiplexed onto a single frequency channel at C band. There are 200 assigned channels at 300-kHz spacings from 5031.0 to 5090.7 MHz. When distance measurement functions are provided at an MLS site, these use the conventional L-band DME frequencies and formats except that pulse spacing and processing are modified at close range for increased precision. Provisional specifications for the angle guidance and data functions have been published by ICAO (Ref. 2, Part 1, Attachment I). Similar specifications for the precision DME were adopted in 1984.

Conventional ILS generates a single straight guidance path, fixed in azimuth and elevation. MLS, on the other hand, transmits "signals in space" from which the approaching aircraft is able to measure its azimuth and elevation from the runway centerline over a volume that (typically) will extend ±40° in azimuth, 0 to 10° in elevation, and 20 nmi from touchdown. Within this volume the airborne equipment can compute its deviations from any preselected approach path (which could be curved) and preselected descent angle. Data transmitted by the MLS ground station provide (among other things) warning of obstructions in azimuth and elevation.

Signal Format

The azimuth and elevation fan beams are scanned electronically at precisely controlled rates. Figure 53.5 illustrates the principle for azimuth measurement. The vertically oriented narrow fan beam makes a pair of lateral scans across the coverage area called the "TO" and "FRO" scans. The aircraft's azimuth, θ, is found from the time difference between detection of these two beam passages:

$$\theta = \frac{T_0 - t}{K}$$

where $\theta°$ = required angle
t = measured interval, μs
T_0 = known constant corresponding to delay when $\theta = 0$, μs
K = known scaling constant, μs/°

Elevation angle is similarly defined using a lateral fan beam, scanned vertically to and fro. For azimuth, T_0 is 6800.0 μs, and for elevation, 3466.7 μs. K in both cases is 100 μs/°.

The azimuth and elevation functions form the basic MLS format. Further capabilities can be added on a modular basis. These include a separate "missed approach" or back-azimuth lateral guidance beam and a separate high-precision elevation guidance beam covering the region where the aircraft begins its "flare," or round out just before landing. Each function is transmitted sequentially on the assigned frequency channel, as illustrated in Fig. 53.6. Each separate function consists of a preamble followed by a single TO-FRO scan pair. The preamble is formed by 400 μs of carrier for acquisition, followed by a 5-bit sync word, 5 bits identifying the function, and 2 parity bits plus, for azimuth, a further bit to key on and off in the aircraft an audible Morse-code identity signal. These data are transmitted using differential phase shift keying at 15 kb/s. Each function is processed independently by the aircraft receiver. Transmission of each guidance/data function may therefore be interleaved with other functions in any sequence. The minimum repetition rates for each of the

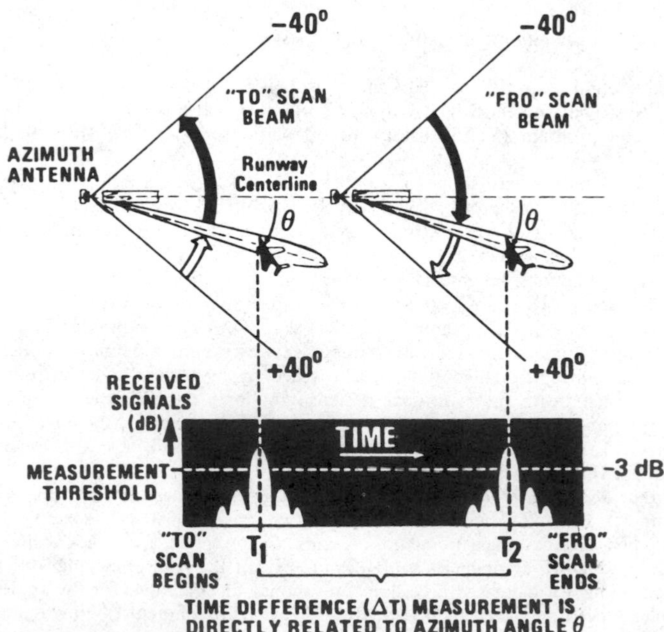

Fig. 53.5 Microwave landing system (MLS) angle measurement. From Cox and Shirey,[15] with permission. © 1980 IEEE.

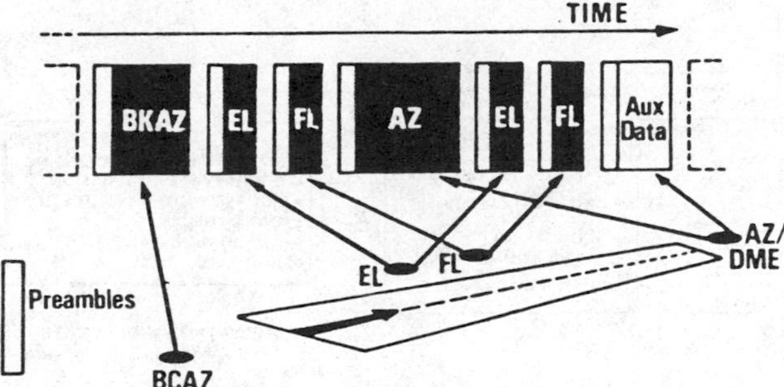

Fig. 53.6 Microwave landing system (MLS) signal format. BKAZ (BCAZ), back azimuth; EL, elevation; FL, flare; AZ, azimuth; DME, distance measuring equipment. From Cox and Shirey,[15] with permission. © 1980 IEEE.

functions, as specified by ICAO,[2] are

Function	Repetition Rate (TO/FRO Scans per Second)
Azimuth	13.5
Elevation	40.5
Back azimuth	6.75
Flare elevation	40.5

Provision is made in the format for future addition of a separate 360° azimuth function. The scan formats include test pulses plus, for the azimuth scan, out-of-coverage indication (OCI) pulses. These are equivalent to the clearance patterns of the conventional ILS, and their relative received amplitudes are used to give a warning to an aircraft when it is outside the region of proportional angle guidance.

Ground-to-air data transmission is an inherent part of the MLS signal. These data are transmitted in blocks in the time-shared format using differential phase shift keying (DPSK), and are radiated from the basic azimuth site using an antenna pattern that fills the guidance coverage volume. Formats for these data are not yet completely defined. Basic data will be used by all aircraft. These include

Performance category
Equipment status
Antenna beamwidths
Azimuth coverage limits
Minimum selectable glide path angle
Associated DME channel number (if used)

Following transmission of basic data, auxiliary data will be transmitted for use by suitably equipped aircraft, such as for the sending of messages in alphanumeric characters.

Airborne Processing

Figure 53.7 shows a typical airborne architecture[17] for MLS angle and data processing. It is implemented using two microprocessors, one to perform the angle measurement and the other to use this information plus site data to generate position and steering outputs. Analog outputs are provided for cross-pointer-type displays, and digital outputs to interface with an autopilot. The I/O processor also receives inputs from distance measuring equipment (DME) and from the control unit on which the pilot selects the required flight track and descent angle.

Significant multipath rejection is possible in the angle measurement processing. The split-gate signal processor (SGSP)[17] determines the time of arrival of the peaks of the TO and FRO beams by essentially differentiating the video response and identifying the zero crossing of this derivative. The

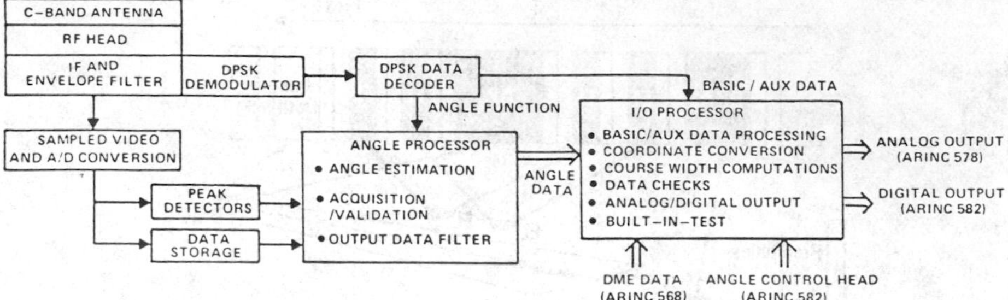

Fig. 53.7 Angle receiver processing architecture. DPSK, differential phase shift keying; DME, distance measuring equipment. From Kelly and Skudrna,[17] with permission. © 1981 IEEE.

position of this zero crossing is relatively insensitive to multipath distortion of the overall pulse shape. The SGSP performs successive interpolations over sample sequences of each pulse response to determine the zero crossing in a manner that is less noise sensitive than conventional differentiation.

When a separate flare elevation guidance signal is provided for automatic landing, a technique known as single-edge processing (SEP) is being used to obtain the extreme accuracy needed for measuring these signals.[17] In this, the time between TO and FRO passage is measured between points on the leading edge of each detected pulse at which the slope reaches a selected value. Multipath, being a delayed signal, will not significantly distort the first part of the detected pulse shape.

Guidance Accuracy

For automatic landing, ICAO requirements for MLS require ±6 m (20 ft) lateral and ±0.6 m (2 ft) vertical guidance accuracies. These are 2-sigma numbers and are defined for the touchdown zone of the runway. For long runways of 3000 to 3500 m, MLS has demonstrated these accuracies using azimuth and elevation antenna beamwidths of 1°, with glide slope elevations, in testing, as low as 1°.[16] For shorter runways and where full automatic landing capability is not required, the antenna beamwidths can be relaxed. For example, an MLS ground system developed for short-runway small community airports uses 3° azimuth and 2° elevation beamwidths, but has still demonstrated ICAO guidance accuracies down a standard 3° glide path angle.

Precision Distance Measuring Equipment (PDME)

PDME is used to define distance to touchdown for MLS approaches. Although not part of the most basic (Az-El) MLS format, it is expected to be installed at the great majority of sites. Originally conceived as being a C-band system, like MLS angular guidance, it was later shown that PDME requirements could be satisfied using an L-band system that was compatible to a large extent with the enroute DME, described in Section 52.2. Current PDME range performance requirements are

Fixed-wing aircraft	± 30 m (100 ft), 2σ
Helicopters	± 12 m (40 ft), 2σ

These specifications must be met in the presence of strong multipath reflections both in the vertical plane, caused by ground reflections, and laterally, such as those from airport hangars. Reference 18 discusses a number of the techniques available for the achievement of PDME performance. The proposed interrogation rates are high (40 per second) compared with the aircraft dynamics. This allows considerable data filtering to be performed in the aircraft to reduce the perturbations caused by long-period multipath.

For multipath delays that are short compared with the DME pulse lengths, the effect is to distort the leading edges of the pulses and to cause large pulse-to-pulse variations in received amplitudes. A technique used to normalize arrival time measurements under these conditions is the delay-and-compare process illustrated in Fig. 53.8. The pulse detector does not use a fixed threshold. Each pulse is delayed an amount τ and is then compared with a threshold formed by an attenuated (but undelayed) version of the input. As shown in Ref. 18, the threshold crossing time t_0 is, to the first order, independent of input signal amplitude and rise time. The appropriate delay τ and attenuation A are functions of pulse length and signal-to-noise ratio.

The PDME scheme proposed to ICAO for incorporation into the MLS specification is a two-pulse two-mode technique.[19] When the aircraft is more than about 7 nmi from the landing site, it

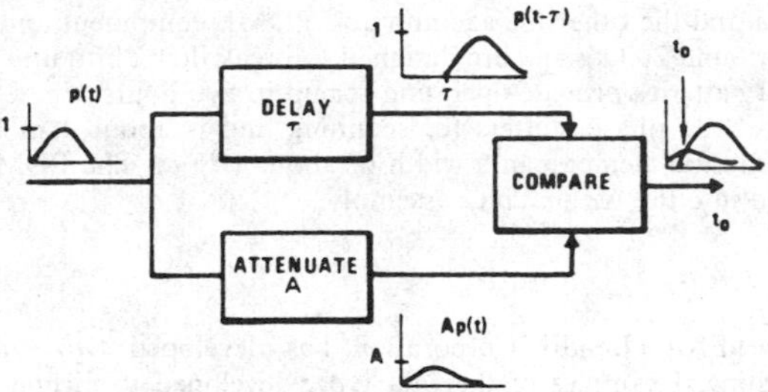

Fig. 53.8 Delay and compare signal processing. From Kelly and LaBerge,[18] with permission.

interrogates the ground PDME transponder with pairs of 12-μs-spaced pulses, and the ground beacon replies with pairs of 12-μs-spaced pulses in the ICAO standard DME X-channel mode. Inside 7 nmi, the aircraft switches to high-rate interrogations ($\simeq 40$ per second) using 18-μs-spaced pulse pairs. The beacon replies also using 18-μs-spaced pulses. Replies to near and distant aircraft are interleaved, using the appropriate pulse spacings. In the close-in mode, the air and ground receivers use delay-and-compare processing with an attenuation (A in Fig. 53.8) of 17 dB. Also a pulse approximating a cosine-cosine^2 shape is used, which has an 800-ns rise time. Static testing of prototype MLS PDME has demonstrated range errors of around 5.5 m (18 ft) rms.[18]

MLS Ground Equipment

The basic MLS azimuth and elevation antennas will be sited similarly to the localizer and glide slope antennas of the conventional ILS. The added antenna for precise flare elevation data will be about 100 m to the side of the runway and about 900 m from the threshold. The back-azimuth guidance antenna will be 100 to 200 m in front of the runway threshold, on the centerline.

Specific design of ground equipment to meet MLS performance requirements varies throughout the various countries active in the program. The United States is developing two classes of basic azimuth-elevation (Az-El) systems. The equipment for long runways at major airports uses 1° beamwidth antennas, each a 116-element phased array. The guidance and control electronics are housed in a separate shelter with fully redundant equipment. Short runways are to be served by smaller 2° elevation and 3° azimuth beamwidth antennas. An integrated packaging concept for these is used; each antenna and its relevant electronics are combined in a single unit. Further, the precision DME antenna and transponder electronics are integrated with the azimuth guidance assembly. For both versions of the equipment, 20-W transmitter power is presently used. Tests show that for operational use, 10 W should be adequate.

Figure 53.9 shows the two ground elements developed for the U.S. military forces under the joint tactical microwave landing system program (JTMLS). One unit houses the elevation scanning

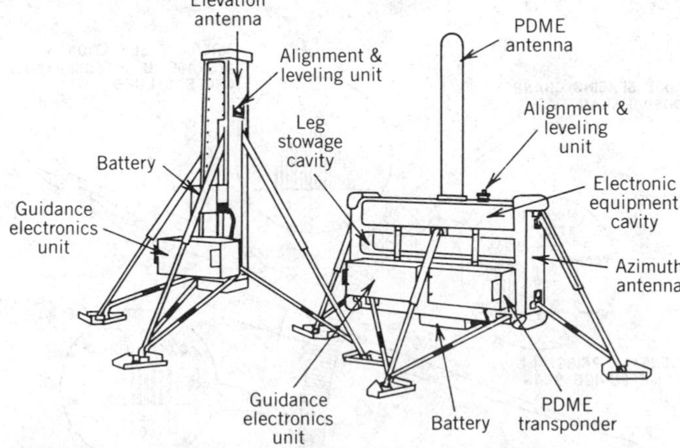

Fig. 53.9 Joint tactical microwave landing (JTMLS) ground equipment. PDME, precision distance measuring equipment. From Cox and Shirey,[15] with permission. © 1980 IEEE.

electronics and antenna and the other the azimuth and PDME equipment and antennas. These have a combined weight of around 200 kg and are designed to break down into units suitable for manpack transportation. Internal batteries provide operation for up to two hours. The 2° El guidance antenna uses 40 elements with diode phase shifters for scanning and is about 2 m long, while the 3° Az guidance phased array has 36 elements in a width of about 1.25 m. The PDME, with its 1.3-m-high antenna, is incorporated into the Az guidance assembly.

Airborne Equipment

Under contract to the FAA, Bendix Corporation has developed two types of MLS airborne equipment for U.S. testing. Examples of the first type, developed to airline (ARINC) format and specification, have been in use in various forms since 1977. The latest version (the phase III receiver) weighs around 6 kg and is designed to operate in conjuction with a standard ARINC configuration DME airborne unit. A second type of airborne unit has been developed for eventual use in small aircraft. This operates using angle data only and has a weight of around 3.2 kg. Minimum operating performance standards for airborne MLS equipment have been set by the RTCA.[20]

53.5 SPECIAL-PURPOSE MILITARY LANDING SYSTEMS

The armed services have developed many different systems to solve their particular problems in landing helicopters and fixed-wing aircraft. For tactical use, a landing system is expected to provide accurate guidance in all weathers into a relatively small and unimproved landing field. For naval operations, such a "field" may be the deck of an aircraft carrier or a ship with a helicopter landing pad. The ground components of a tactical landing system need to be small, easily transported, and capable of being set up and operating within hours or even minutes of arrival at a new site. The various systems developed, therefore, all operate at microwave frequencies, so that the narrow search and scanning beams required for accurate guidance can still be achieved with small antennas.

It is the policy of the U.S. Armed Services that the ICAO microwave landing system (MLS) signal format will eventually be used to satisfy all tactical landing requirements.[11] Common airborne equipment will then serve for landings at tactical fields as well as at major military and civilian airports. To this end, prototypes of ground equipment for the joint tactical microwave landing system (JTMLS) have been developed. However, the various special-purpose landing systems now in existence will continue in operation for some time to come.

AIL Landing Systems

The AIL Division of Eaton Corporation in Farmingdale, New York, has developed a family of landing systems using a common signal format primarily for military customers, but also with some limited commercial uses. These systems all operate in the 15.4- to 15.7-GHz range. Guidance is provided by pairs of pulse-coded narrow fan beams, which are orthogonally swept to cover the region of interest using mechanically scanned antennas.

Figure 53.10 illustrates the principle for elevation guidance. The fan beam, narrow in elevation,

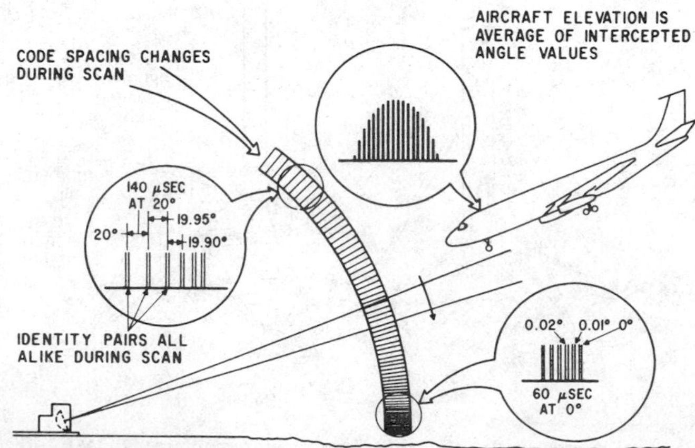

Fig. 53.10 AIL angle measurement. Courtesy of AIL, Division of Eaton Corporation, Farmingdale, NY.

transmits identity pulse pairs, with a spacing between pairs that varies linearly with beam elevation, from 140 μs at 20° to 60 μs at 0° for this example. The aircraft finds its elevation angle by measuring the average pulse-pair spacing in its observed sequence of pulses. A beam shape-fitting procedure can be used to obtain increased measurement accuracy at low elevation angles.[21] For azimuth guidance using the horizontally swept beam, the principle is the same except that a different fixed spacing is used for the "identity pairs." Elevation and azimuth guidance signals are time multiplexed on a common frequency channel, with an update rate of 5 scans per second. With some configurations the aircraft can interrogate the ground station on the same frequency, within this time-multiplexed format, to obtain distance information from the round-trip delay of the reply.

These scanning beam methods all descend from the FLARESCAN system, developed by AIL for the U.S. FAA in 1962. This was the earliest practical demonstration of the microwave scanning beam concept for low angle elevation guidance.

C-Scan. Supplied to the U.S. Navy starting in 1966, this version of the AIL landing system provides guidance to aircraft making carrier landings. The shipboard guidance system (designated AN/SPN-41) uses a 2° azimuth beam and a 1.3° elevation beam to scan a volume ±20° laterally and 0 to 20° vertically in the approach direction. The elevation antenna is mechanically stabilized against ship's roll and pitch. With 2 kW of peak power, the usable range is about 20 nmi. The airborne equipment presents a cross-pointer display to the pilot similar to that of conventional ILS equipment. The correct glide path angle is preset into each aircraft's elevation guidance equipment at a value appropriate to that type of aircraft. Accuracy is ±0.1° elevation and ±0.2° azimuth. C-scan is either used alone, in which case the pilot transitions to visual guidance for the last 200 ft of descent, or in conjunction with the AN/SPN-42 system (described later) for a fully automatic landing capability. The AN/TRN-28 is a truck-mounted version of C-scan, used by the U.S. Navy and Marine Corps for aircraft guidance at some shore bases.

Space Shuttle Landing System (MSBLS). The microwave scanning beam landing system (MSBLS) version of the AIL system is being used by NASA to provide fully automatic approach and landing guidance for the Space Shuttle Orbiter vehicle.[22] Following its initial intercept of MSBLS signals at around 14,000 ft altitude, the Shuttle receives guidance throughout the remaining 90 seconds of flight. For MSBLS, the azimuth scan coverage is ±20° and elevation coverage is increased to 30° to accommodate the 21° to 24° glide slope angle of the Shuttle. Azimuth and elevation accuracies are 0.05° and 0.03° (1σ), respectively, with integral ranging (DME) accuracy of better than 30 m. To enhance system reliability, dual hardware capability is provided, together with continuous field monitoring of the transmitted signals. Five MSBLS ground stations have so far been procured to serve the planned civil and military landing sites of the Space Shuttle.

TILS (Tactical Instrument Landing System). This is a readily transportable version of the AIL system which is being deployed by the Swedish and Finnish Air Forces to provide landing guidance to their SAAB Viggen and Draken aircraft at dispersed, remote landing sites. Over 50 ground stations and 200 sets of airborne equipment have been supplied.

A-Scan (AN/TRQ-33). This is a version of the basic system developed for the U.S. Army primarily for providing steep-angle wide-azimuth coverage for helicopter guidance. It provides multiple landing paths, ground obstacle warning signals, and precision DME.

Co-Scan. This is a commercial version of the AIL system intended for guidance of short- and vertical-takeoff and landing (STOL, VTOL) aircraft. It has been extensively tested in Canada by a commuter airline providing service to city centers using small restricted-operation downtown airfields, at which it provided guidance accuracies equivalent to ILS category I.

AN/SPN-42

As part of the U.S. Navy's all-weather carrier landing system (AWCLS), the AN/SPN-42 landing control system has been developed by Bell Aerospace Company to provide the capability for fully automatic landings on aircraft carriers. The SPN-42 is essentially an automated shipborne precision approach radar. The incoming aircraft is tracked by a Ka-band (32-GHz) radar mounted on the ship's "island" superstructure. The position and rate of the aircraft are compared with the desired track and correction steering signals derived in a shipborne computer. The AN/SPN-42 uses a digital computer. In its predecessor system, the AN/SPN-10, and its land-based equivalent, the AN/GSN-5,[23] analog computation was performed. The computation requires knowledge of the dynamic coefficients of the aircraft. The correct set for each particular aircraft must therefore be selected before the approach. The steering commands are sent to the aircraft using a standard UHF Navy data link and are displayed to the pilot using the same cross-pointer instrument as used for

AN/SPN-41 guidance. On suitably equipped aircraft, coupling to the autopilot allows achievement of a fully automatic landing capability.

Transfer of guidance from SPN-41 to SPN-42 takes place at about 4 nmi range. Following switchover, SPN-41 angle measurements are continued in order to monitor the performance of the SPN-42. The transmitted guidance signals are computer compensated for ship's pitch roll and heave. In the last few seconds of guidance this requires algorithms to predict the position and rate of the deck. The aircraft's path is corrected to minimize the relative vertical velocities at touchdown.

Microwave Aircraft Digital Guidance Equipment (MADGE)

Several versions of this system have been developed in England by the MEL Division of Philips Electronic and Associated Industries, Ltd. It was originally developed to satisfy a military NATO requirement for a tactical portable approach aid, and was adopted as a NATO standard in 1971. Since that time, MADGE has also been certificated for civilian use by operators of helicopters serving North Sea oil rigs.[24]

MADGE operates on several frequency channels in the 5- to 5.25-GHz band. The ground or shipborne equipment does not radiate until interrogated by a user aircraft, a feature that has obvious security advantages in a tactical situation. On receiving an interrogation, the ground station measures the angular position of the aircraft, using two orthogonal (Az-El) fixed interferometer antenna arrays. Two modes of operation are then possible. In the "air-derived" mode, these angle data are retransmitted to the aircraft. From these and an estimate of range obtained by measuring the round-trip transponding delay, the airborne equipment is able to calculate steering corrections needed to bring the aircraft on to the pilot-selected desired azimuth and glide slope.

The "ground-controlled" mode is used when fixed-path approach procedures are being used, particularly for approaches to landings at offshore oil platforms. Deviations are computed in the ground station and transmitted to the aircraft in the form of off-course deviation signals. This mode is particularly appropriate for offset approach courses that do not intersect the "origin" represented by the physical location of the guidance antennas.

The air-to-ground interrogation in the normal air-derived mode consists of a 3-μs marker pulse followed by 25 bits of data including aircraft and ground station address codes. These are transmitted at a 1-MHz rate using pulse amplitude modulation with about 250 W of radiated power. This is repeated at a 50-Hz rate. The ground station reply consists of a marker pulse plus 60 bits of data in the same format. For the ground control mode, the air and ground data messages are longer, and a 100-Hz interrogation rate is used. The digital aircraft and ground codes allow up to 25 ground systems to operate in close proximity on the same frequency, although 12 different frequency groups are available.

The ground azimuth measurement antenna is a horizontal interferometer array of seven horn elements with unequal spacing over a width of about 1.5 m. The elevation antenna uses coarse and fine interferometers of four and eight elements respectively over a height of about 2 m. A third ground unit is the reply transponder, which has a 1-m-high antenna, radiating pulses at 150-W peak power. The total ground station weight in a tactical configuration is around 100 kg.

Coverage dimensions are typically $\pm 45°$ horizontally and 25° vertically from the ground station, with an operating range of 15 nmi. Guidance accuracy is given as 0.1° rms in angle and ± 30 m in range. MADGE has been certificated for landing approaches down to 30 m (100 ft) minimum descent altitude and 400 m visual range, equivalent to ICAO ILS category II standards. It is now in production to equip aircraft carriers and VTOL Harrier aircraft of the British Royal Navy.

Marine Remote Area Approach and Landing System (MRAALS)

This system has been developed by the Singer–Kearfott Company of New Jersey under contract to the U.S. Marine Corps. MRAALS is designed to provide helicopters of the Marine Corps with an all-weather landing capability at remote sites in a tactical environment. It is a microwave (15-GHz) scanning beam system, with signal format identical to the AIL C-scan (AN/SPN-41) system described earlier. Using mechanically scanned antennas, the single-unit ground station (designated AN/TPN-30) provides guidance signals over a volume $\pm$ 20° in azimuth, 0° to 20° in elevation, and 10 nmi in range. Accuracy is $\pm 0.1°$ azimuth, $\pm 0.05°$ elevation, and ± 30 m in range using an associated TACAN-compatible distance measuring system. The total ground station weight is around 50 kg.

References

1 F. G. Kear, "Instrument Landing at the National Bureau of Standards," *IRE Trans. Aeronaut. Nav. Electr.* (special issue on instrument landing systems) (Jun. 1959).

2 *International Standards and Practices, Aeronautical Telecommunications*, Annex 10 to the Convention on International Civil Aviation, ICAO, Montreal, Canada, 3rd ed., Jul. 1972.

3 W. E. Jackson, "Improvements on the Instrument Landing System," *IRE Trans. Aeronaut. Nav. Electr.* (Jun. 1959).

4 "Localizer-Mark 1D System," Northrop-Wilcox Corp., Kansas City, MO.

5 H. H. Butts and R. H. McFarland, "New Developments in Instrument Landing System," *IEEE Trans. Aero. Electr. Syst.* **AES-2**(6) (Supplement) (Nov. 1966).

6 J. J. Battistelli, "The Conventional ILS—So What's New?," *Navigation*, Journal of The Institute of Navigation, **21**(2), (Summer 1974).

7 F. W. Iden, "Glide Slope Antenna Arrays for Use under Adverse Siting Conditions," *IRE Trans. Aeronaut. Nav. Electr.* (Jun. 1959).

8 R. H. McFarland, "Application of End-Fire Arrays at Contemporary Glide-Slope Problem Sites," *IEEE Trans. Aero. Electr. Syst.* **AES-17**(2) (Mar. 1981).

9 "Automatic Landing—Now a Matter of Course," *Interavia* (Jul. 1980).

10 *Minimum Performance Standards—Airborne ILS Receiving Equipment*, DO-131A (localizer), DO-132A (glide slope), RTCA, Washington, DC, 2 Nov. 1978.

11 *Federal Radio Navigation Plan*, DOT-TSC-RSPA-80-16, Department of Defense/Department of Transportation, Jul. 1980 (4 vols.).

12 *Minimum Performance Standards—Airborne Low-Range Radar Altimeters*, DO-155, RTCA, Washington, DC, 11 Jan. 1974.

13 M. I. Skolnik, *Introduction to Radar Systems*, McGraw-Hill, New York, 1962.

14 H. R. Ward et al., "GCA Radars: Their History and State of Development," *IEEE Proc.* **62**(6) (Jun. 1974).

15 R. M. Cox and J. M. Shirey, "MLS—A New Generation Landing System is Here," presented at IEEE Position Location and Navigation Symposium (PLANS 80), Atlantic City, NJ, Dec. 1980.

16 R. J. Kelly, H. W. Redlien, and J. L. Shagena, "Landing Aircraft under Poor Conditions," *IEEE Spectr.* (Sep. 1978).

17 R. J. Kelly and J. T. Skudrna, "Joint Tactical Microwave Landing System (JTMLS) Airborne Signal Processing," presented at 4th AIAA/IEEE Digital Avionics Systems Conference, St. Louis, MO, Nov. 1981.

18 R. J. Kelly and E. F. LaBerge, "Guidance Accuracy Considerations for the Microwave Landing System Precision DME," *Navigation*, Journal of The Institute of Navigation, **27**(1) (Spring 1980).

19 Final Report, ICAO All-Weather Operations Panel, Working Group M, ICAO, Paris, Sep. 1981.

20 *Minimum Operational Performance Standards—MLS Airborne Receiving Equipment*, DO-177, RTCA, Washington, DC, Jul. 1981.

21 G. Blazek and A. Charych, "Beam-Fit Decoding Technique for Increasing Elevation Measurement Accuracy from Scanning Beam Systems at Low Elevation Angles," AIAA Guidance and Controls Conference, Colorado Springs, CO, Aug. 1979, AIAA 79-1710.

22 G. Blazek and L. M. Carrier, "Microwave Landing System Selected for Space Shuttle Program," *ICAO Bull.* (Aug. 1974).

23 F. D. Powell, "An Automatic Landing System," *IRE Trans. Aeronaut. Nav. Electr.* (Jan. 1959).

24 D. A. Brown, "MADGE Certification Expected in March," *Aviat. Week Space Technol.* (Feb. 25 1980).

CHAPTER 54
SATELLITE NAVIGATION

JOHN H. PAINTER

Texas A & M University
College Station, Texas

54.1 PRINCIPLES OF SATELLITE RADIO NAVIGATION

The basic mathematical ideas underlying satellite radio navigation are no different from those underlying any other kind of navigation employing independent measurements to determine position. First a coordinate system is chosen by means of which the navigator's position is described. For the present purpose, a right-hand Cartesian system using components x, y, and z may be taken with origin the center of the earth. Such a system is called earth-centered. It is usual practice to take the z axis through the North Pole with a positive sense. Then the x and y axes lie in the plane of the equator. Such a coordinate frame is called equatorial, as opposed to the ecliptic coordinate frame used in astronomy and celestial navigation.

If the x axis is projected from the center of the earth toward a fixed point in space at an infinite distance, such as the vernal equinox on the celestial sphere, the coordinate frame does not rotate with the earth. If the x axis is projected through a point on the earth's surface, such as the zero-latitude zero-longitude point, the frame is called earth-centered earth fixed (ECEF). We shall deal in this chapter with either type of earth-centered Cartesian coordinate frame, as the need arises.

Let us now define the navigator's position in the x-y-z frame as a three-vector, or 3×1 column matrix, as

$$\mathbf{P} = \begin{bmatrix} x \\ y \\ z \end{bmatrix} \tag{54.1}$$

If the coordinate frame for **P** is ECEF, there exist well-known transformations for expressing **P** in latitude, longitude, and altitude, for instance, on a perfect spherical earth. Corrections for the true nonspherical earth are more difficult and are discussed in Section 54.3.

Navigation in three space is accomplished by obtaining three "independent" measurements of quantities known to be functionally related to the navigator's position. Given a knowledge of the functional mathematical relationships connecting the position components and the measurements, the position is then "solved" for, from the measurements. This solution exists providing the measurements are suitable to the navigation problem. For instance, a navigator's distance measurements to three known points yield a finite number of solutions[1] provided the three known points are not colinear. The ambiguous solution is rejected by the navigator's a priori knowledge of his or her position.

Let the three measurements be denoted as m_1, m_2, and m_3. In a well-posed navigation problem, there exist three scalar functions, $f_1(\ ,\ ,\ )$, $f_2(\ ,\ ,\ )$, and $f_3(\ ,\ ,\ )$ such that

$$\begin{aligned} m_1 &= f_1(x, y, z) \\ m_2 &= f_2(x, y, z) \\ m_3 &= f_3(x, y, z) \end{aligned} \tag{54.2}$$

Now denote the measurements and functions by three vectors, as

$$\mathbf{m} = \begin{bmatrix} m_1 \\ m_2 \\ m_3 \end{bmatrix} \qquad \mathbf{f} = \begin{bmatrix} f_1(\ ) \\ f_2(\ ) \\ f_3(\ ) \end{bmatrix} \tag{54.3}$$

Then we have a compact notation for the functional relationship as

$$\mathbf{m} = \mathbf{f}(\mathbf{P}) \tag{54.4}$$

where $\mathbf{f}(\ )$ is a vector-valued function of vector argument, $\mathbf{P}$.

The navigation problem is, "Given the structure of the function $\mathbf{f}(\ )$, and the measurements $\mathbf{m}$, solve for $\mathbf{P}$." If, for instance, the relationship between $\mathbf{m}$ and $\mathbf{P}$ were linear and invertible, the problem would be simple. Unfortunately, most navigation problems have no inverse function for $\mathbf{f}$ that can be applied straightaway to solve for $\mathbf{P}$.

As an example, in the problem of measuring distances from the navigator's position to three known points described by the vectors $\mathbf{P}_1, \mathbf{P}_2, \mathbf{P}_3$, the components of the function $\mathbf{f}(\ )$ are given by

$$\mathbf{f}_i(\mathbf{P}) = \|\mathbf{P}_i - \mathbf{P}\| \qquad i = 1, 2, 3 \tag{54.5}$$

That is, the function is the square root of the dot product, or the Euclidian norm. This function is nonlinear and has no inverse.

In cases where the function $\mathbf{f}(\ )$ has no inverse, a so-called perturbation solution is generally used, based on "linearizing" the navigation problem. This is done as follows.

Assume there exists a reasonably precise estimate of the navigator's position, $\mathbf{P}_0$. (This might be a dead-reckoning position, for instance.) From this assumed position $\mathbf{P}_0$, calculate the corresponding "predicted" measurement $\mathbf{M}_0$ using the known function $\mathbf{f}(\ )$. That is,

$$\mathbf{M}_0 = \mathbf{f}(\mathbf{P}_0) \tag{54.6}$$

Now $\mathbf{M}_0$ is the first term of a vector Taylor series for $\mathbf{M}$, expanded about the "point" $\mathbf{P}_0$. Calculate the coefficient of the linear second term of the series, which is the Jacobian matrix H, given by

$$H = \left.\frac{\partial \mathbf{f}(\ )}{\partial \mathbf{P}}\right|_{\mathbf{P}=\mathbf{P}_0} \qquad \text{a } 3 \times 3 \text{ matrix} \tag{54.7}$$

Then write

$$\mathbf{M} = \mathbf{M}_0 + H(\mathbf{P} - \mathbf{P}_0) \tag{54.8}$$

Equation (54.8) is a two-term (linear) approximation to $\mathbf{M}$. If a navigation solution exists, H will be invertible with inverse H^{-1}, and we may solve for $\mathbf{P}$ as

$$\mathbf{P} = \mathbf{P}_0 + H^{-1}(\mathbf{M} - \mathbf{M}_0) \tag{54.9}$$

Such a linearized perturbation solution for the navigator's position is common practice today in computer-based navigation systems.

In a satellite-based radio navigation system, it is the function of the navigator's radio receiver to provide the measurements m_1, m_2, and m_3, comprising $\mathbf{M}$. In the Transit system, for instance, these measurements are of the Doppler offsets of the satellite's transmitted radio frequency, due to satellite velocity. In the Global Positioning System (GPS), the measurements are of distances, or ranges, from the satellites to the navigator's receiver. In both the GPS and the Transit system, the navigation computer requires knowledge of the satellite position at the time of the measurement. Thus both systems also incorporate data messages from satellite to navigator in the radio transmission. These data messages contain the Keplerian orbital elements from which the satellite orbit can be computed in the navigation processor.

To be highly accurate, satellite radio navigation systems must contain means for correcting errors induced by passage of the radio signals through the ionosphere. The orbital altitudes for both the Transit system and the GPS are sufficiently great that the radio signal paths to a navigator near the earth's surface traverse a significant portion of the ionosphere. The effects of ionospheric passage on the radio signal are more pronounced the lower the signal frequency and the lower the elevation angle of the satellite as viewed from the navigator's position. Both the Transit system and the GPS employ techniques to correct ionospherically induced positioning errors.

Radio navigation systems such as Transit and GPS, which employ satellite radio transmissions, are termed passive in that the navigation system is a receive-only system. No transmissions are required on the part of the navigator, as in tactical air navigation (TACAN)-distance measuring equipment (DME). These space-born communication channels operate in a line-of-sight mode only and employ relatively low-power transmissions from the satellites. Transmission power of 20 W or less are the rule. Thus the signal power levels available to the receiver are quite small, being of the order of -160 dBW for Transit and GPS. With receiver noise figures on the order of 3 dB, ratios of available signal power to noise power spectral density on the order of 40 to 50 dBHz imply the need for efficient signal processing.

The following sections treat in more detail some of the elements of satellite radio navigation that are common to all systems. Examples are satellite orbital characterization and ionospheric effects. Following those expositions, both the Transit system and the GPS are described in some depth.

54.2 SATELLITE ORBIT CHARACTERIZATION—KEPLER'S EQUATION

It is not intended to give a complete treatise on orbital mechanics here. For that the reader is referred to one of the standard texts, such as Battin.[2] However, some familiarization is required, since contemporary satellite radio navigation receiver-processors, such as for the Transit system and the GPS, perform satellite orbit calculations internally. Furthermore, the satellite data messages contain the so-called orbital elements for the satellite being received. Thus a knowledge of these elements and their use is necessary for understanding of the system itself.

The form of a stable satellite orbit is an ellipse when only the satellite and earth are considered. That is, considering the satellite and the earth to form a "two-body problem" in celestial mechanics, the solution for the satellite trajectory is an ellipse with the earth (center of mass) at one focus. The elliptic orbit is fundamentally described by its semilatus rectum, eccentricity, and time of nearest approach to earth. These three quantities are called orbital elements. To relate this elliptic orbit to the coordinate system in use requires three more elements, classically taken as the Euler angles (defined later).

Figure 54.1 shows several of the orbital elements used to describe the orbits of contemporary radio navigation satellites. The following quantities are identified with respect to Fig. 54.1:

$\angle AFP = V$: true anomaly
$\angle ACQ = E$: eccentric anomaly
$\overline{AC} = a$: semimajor axis
$\overline{FP} = r$: orbit radial distance
$0 \leqslant e \leqslant 1$: orbit eccentricity
P : satellite position on orbit
A : perigee point
F : ellipse focal point (position of earth center of mass)
C : center of circumscribing circle
μ : earth universal gravitational parameter

$n_0 = \dfrac{2\pi}{T} = \sqrt{\dfrac{\mu}{a^3}}$: mean angular motion

T : orbital period
t_{oe} : time of ephemeris
M : mean anomaly

The orbit radial distance, $r = \overline{FP}$, is given by

$$r = a(1 - e\cos E) \tag{54.10}$$

Let x'' and y'' denote a local Cartesian coordinate system in the plane of the ellipse with the x'' axis passing through the focus and perigee point. The y'' axis is orthogonal to x'' and passes through the focus. The coordinates of the point P on the elliptic orbit are then given by

$$\begin{aligned} x'' &= a(\cos E - e) = r\cos V \\ y'' &= a(1 - e^2)^{1/2}\sin E = r\sin V \end{aligned} \tag{54.11}$$

The eccentric anomaly E is determined by solving Kepler's equation, given as

$$n(t - t_{oe}) = M = E - e\sin E \tag{54.12}$$

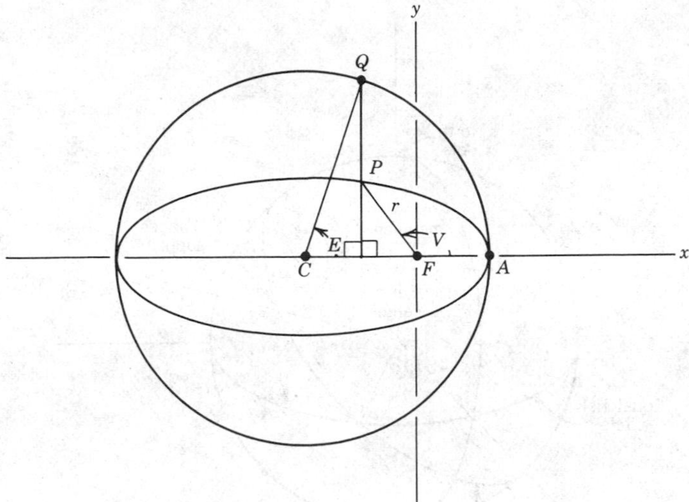

Fig. 54.1 Elliptic orbit parameters.

In Eq. (54.12), the quantity t is the time at which the value of E (and x'' and y'') is to be determined. The values of t and t_{oe} must be measured with respect to the same clock.

Equations (54.11) and (54.12) define the oribit position as a function of time, referenced to the orbital plane, major axis, and perigee time. These equations define a right-hand, earth-centered Cartesian coordinate system x''-y''-z'', for which z'' is identically zero.

Now there are two other right-hand, earth-centered, Cartesian systems that are of interest and in which the oribit must be defined. Both of the systems are known as equatorial, since in both the x and y axes lie in the plane of the equator. The z axis in both passes through the true North Pole. The first coordinate system is called inertial (relative only to the earth), since the x axis is directed at a fixed point in the firmament. That point is the first point of Aries, or vernal equinox. It is the point at which the sun rises above the plane of the equator. The second coordinate system is called earth fixed, since the x axis passes through the Greenwich meridian of longitude (0°).

In a two-body problem with an earth of spherically symmetric mass density and no air friction or powered flight, the elliptic orbit is fixed relative to the inertial coordinate system. Thus if we denote the inertial system as x'-y'-z', these new coordinates may be obtained from the x''-y''-z'' by three successive simple (one-axis) rotations about the origin (the center of the earth).

To aid in visualization of the required rotations, Fig. 54.2 is given. In the figure the orbital plane is shown inclined to the equatorial plane by an inclination angle i. Also the orbital major axis, which passes through the perigee point, is displaced in the orbital plane from the intersection point with the equatorial plane by an angle, ω, called the argument of perigee. The intersection point between the two planes, with the satellite z coordinate increasing, is called the ascending node crossing. The intersection of the two planes, called the line of nodes, is angularly displaced eastward of the x' axis in the equatorial plane by an amount Ω. This latter angle is called the right ascension of the ascending node. These three angles are the Euler angles.

From this view of the angular relationships between the elliptic orbit in x'', y'', z'' and the earth-centered inertial frame in x', y', z', it is clear what rotations are needed to transform from the former to the latter. The rotation order is ω, i, and Ω. The compound transformation is given in vector-matrix form as

$$\begin{bmatrix} x' \\ y' \\ z' \end{bmatrix} = \Lambda \cdot \begin{bmatrix} x'' \\ y'' \\ z'' \end{bmatrix}$$

$$= \begin{bmatrix} (\cos\omega\cos\Omega) - \sin\omega\cos i\sin\Omega & -\ (\sin\omega\cos\Omega) + \cos\omega\cos i\sin\Omega & -\ (\sin i\sin\Omega) \\ (\cos\omega\sin\Omega) + \sin\omega\cos i\cos\Omega & -\ (\sin\omega\sin\Omega) - \cos\omega\cos i\cos\Omega & -\ (\sin i\cos\Omega) \\ (\sin\omega\sin i) & (\cos\omega\sin i) & (\cos i) \end{bmatrix} \tag{54.13}$$

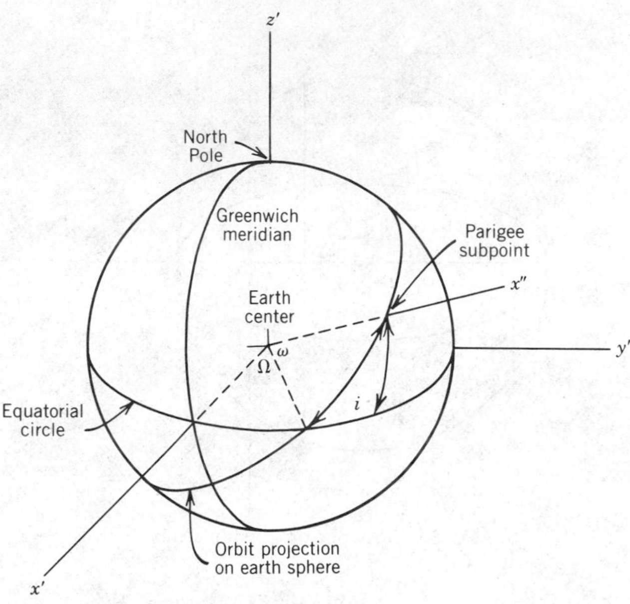

Fig. 54.2 Orbit coordinate conversion. i, inclination angle; ω, argument of perigee; Ω, right ascension of ascending node. i, ω, and Ω = Euler angles.

TABLE 54.1. SATELLITE ORBIT COMPUTATIONS

Computation	Explanation
$\mu = 3.986008 \times 10^{14}$	WGS-72[a] gravitational parameter, m^3/s^2
$\dot{\Omega}_e = 7.292115147 \times 10^{-5}$	WGS-72 earth rotation rate, rad/s
$C_{rc}, C_{rs}, C_{uc}, C_{us}, C_{ic}, C_{is}$	Zonal harmonic coefficients, given
t_{oe}	Time of ephemeris, given
e	Eccentricity, given
a	Semimajor axis, given
$n_0 = (\mu/A^3)^{1/2}$	
$n = n_0 + \Delta n$	Δn given
$M = M_0 + n(t - t_{oe})$	M_0 given
$M = E - e \sin E$	Solved implicity for E
$\cos v = (\cos E - e)/(1 - e\cos E)$	
$\sin v = (1 - e^2)^{1/2} \sin E/(1 - e\cos E)$	
$v = \arctan(\sin v/\cos v)$	
$\phi = v + \omega$	ω given
$u = \phi + \delta u$	$\delta u = C_{uc}\cos 2\phi + C_{us}\sin 2\phi$
$r = a(1 - e\cos E) + \delta_r$	$\delta_r = C_{rc}\cos 2\phi + C_{rs}\sin 2\phi$
$x'' = r\cos u$	
$y'' = r\sin u$	
$i = i_0 + \delta_i$	$\delta_i = C_{ic}\cos 2\phi + C_{is}\sin 2\phi$, i_0 given
$\Omega = (\Omega_0 - \dot{\Omega}_e \cdot t_{oe}) + (\dot{\Omega} - \dot{\Omega}_e)\cdot(t - t_{oe})$	Ω_0, $\dot{\Omega}$ given
$\begin{bmatrix} x \\ y \\ z \end{bmatrix} = \Lambda(\omega, \Omega, i) \cdot \begin{bmatrix} x'' \\ y'' \\ 0 \end{bmatrix}$	$\Lambda(\cdot)$ given by Eq. (54.13)

[a] WGS-72, world geodetic system of 1972.

The transformation Λ brings the satellite into the earth-centered inertial coordinate system x', y', z' by a rotation about the earth's rotational axis by an angle Ω, the right ascension of the ascending node. To bring the position into the ECEF system, x, y, z where x projects through the Greenwich meridian at the equator, can be done by redefining Ω and using the transformation Λ.

Let Ω now be redefined to be the longitude of the ascending node, measured positively westward from the Greenwich meridian. The angle Ω is now time varying, owing to the rotation of the earth and the precession of the orbital plane in inertial space. Thus define

$$\Omega = \left(\Omega_0 - \dot{\Omega}_e \cdot t_{oe}\right) + \left(\dot{\Omega} - \dot{\Omega}_e\right)(t - t_{oe}) \tag{54.14}$$

In Eq. (54.14), the first parenthetical expression gives Ω_0, in longitude of the ascending node at $t = 0$, corrected by the easterly rotation rate of the earth $\dot{\Omega}_e$, acting up to $t = t_{oe}$. The second parenthetical expression then corrects Ω to the time t for earth rotation and for precession of the orbital plane at the rate $\dot{\Omega}$, positive westerly. The parameter t_{oe} is called epoch time and is aligned with time of perigee passage.

Because a two-body problem is not sufficiently accurate, the earth's gravitational field is not spherically symmetric, and other reasons, the parameters of the orbit are not stable. Thus a host of real-time corrections is needed for precise orbit calculations. In particular, the mean motion n, inclination i, right ascension of the ascending node Ω, and mean anomaly M are all specified at t_{oe} and corrected at subsequent times using rate factors. Also the "argument of latitude," $\phi = v + \omega$, as well as satellite orbit radius r and inclination i, are all corrected for the second zonal harmonics of the gravitational field.

The order of computations, with corrections, is given in Table 54.1.

54.3 USER COORDINATES AND GEODESY

Although satellite navigation systems such as Transit and GPS operate in a right-hand orthogonal Cartesian (ECEF) coordinate system, the user usually does not. The most common user coordinate system is latitude, longitude, and altitude (LLA), a system that is approximately differentially orthogonal within a local neighborhood of the user. The chief uses of the LLA system are in locating the user on a navigation chart or with respect to a known point. Maritime users generally do not worry about altitude. Aeronautical users are generally concerned with ranges and bearings from their position to a known point.

There are two complicating factors in satellite navigation with respect to locating the user position in the LLA system on a navigation chart. The first problem is that the earth is not a sphere. The model that is universally accepted for the earth is that of an ellipsoid of revolution, symmetric about the axis of rotation of the earth. This ellipsoid is a "flattened" sphere, with the flattening taking place in the polar regions. A flattening coefficient f is defined that is related to the eccentricity e of the generating ellipse by

$$e^2 = 2f - f^2 \tag{54.15}$$

The current universal earth model is the world geodetic system of 1972 (WGS-72). This is an ellipsoid that provides a "best fit" to the actual earth geoid. The WGS-72 flattening coefficient is

$$f = 1/298.26 \tag{54.16}$$

A consequence of the flattening of the earth spheroid model is that altitude and latitude must now be measured with respect to the WGS-72 ellipsoid rather than to a spheroid. Longitude remains the same, theoretically. Latitude and altitude, however, are now computed as "geodetic" quantities.

In the case of a perfect spheroid with a spherically symmetric mass distribution, the radius vector to earth center, the surface normal vector, and the gravitational vertical all are in the same direction, independent of position on the surface. For an ellipsoid of revolution, these three directions do not necessarily coincide, except at the poles and on the equator. Thus latitude at a given position is measured between the line intersection of the equatorial and meridianal planes and a position vector normal to the ellipsoid surface having a length that is the radius of curvature at the surface. The angle so subtended is called geodetic latitude λ. The surface-normal position vector does not have origin at the ellipsoid center. This geodetic latitude is not the same as latitude on a spheroid nor is it equal to astronomical latitude, which is based on the local gravitational vertical.

A closed-form relationship exists between the ECEF position vector $[x, y, z]^T$, and the user geodetic LLA vector $[\lambda, \phi, h]^T$. These relations are set down in Table 54.2.

Longitude λ, positive westerly, can be solved directly from the model as

$$\lambda = \arctan\left(\frac{-y}{x}\right) \tag{54.17}$$

Latitude ϕ and altitude h, however, require implicit solution, analogous to that for solving Kepler's equation [Eq. (54.12)]. One particular method is given in Ref. 1.

TABLE 54.2. RELATIONS BETWEEN EARTH-CENTERED EARTH-FIXED (ECEF) SYSTEM AND LATITUDE, LONGITUDE, AND ALTITUDE (LLA) SYSTEM

Computation	Comments
$a = 6378135$ m	WGS-72[a] semimajor axis
$e^2 = 2f - f^2$	$f = 1/298.26$
$r_c = a/(1 - e^2\sin^2\phi)^{1/2}$	Radius of curvature
$x = (r_c + h)\cos\phi\cos\lambda$	
$y = (r_c + h)\cos\phi\sin\lambda$	
$z = [r_c(1 - e^2) + h]\sin\phi$	

[a]WGS-72, world geodetic system of 1972.

The second problem in establishing user position on a navigational chart is that of the reference datum. For many years various regions of the world have used particular ellipsoids fitted to the local region. Thus there are many localized reference datum systems used for mapping purposes. In the United States, the North American datum is used. In Europe, the European datum is used. In Japan, the Tokyo datum is used, and so forth. For a partial listing see Ref. 3 (p. 76).

There exists a set of standard equations for translating ECEF positions from some other datum into the WGS-72 datum. These are the so-called Molodensky formulas.[4] These formulas may be used either for defining WGS-72 positions from local survey (maps) or locating WGS-72 positions on local maps.

It would seem that with the advent of satellite navigation, all world geodesy might soon gravitate to the WGS datum, whether in its 1972 form or future forms, such as that of 1982. As a matter of fact, with satellite positioning accuracies approaching the submeter range, the most accurate world geodesy will soon be satellite derived.

54.4 NAVIGATION DATA ACQUISITION

It has been seen in the previous sections that a wealth of subsidiary data is required in a satellite navigation receiver. Most of these data concern the orbits for the various satellites and the calculations thereof. Much data consist of parameters whose values are continuously changing. Because of the accuracy required in determining satellite positions, it is necessary to know the orbital parameters quite accurately each time a user position is calculated.

The simplest way for a user to obtain this time-variable data is from the satellite itself. As the satellites are tracked by their ground control stations, the orbits are continuously calculated to great accuracy. Tables of Keplerian orbital parameters and corrections are then predicted ahead in time and transmitted to the satellites for storage. At a particular time, a satellite transmits in a broadcast mode the data table pertaining to that particular satellite at that particular time. Any user receiving signals from a particular satellite also receives that satellite's current orbital elements as part of a "data-link" message. The data-link signal is usually incorporated into the signal used to make the navigation measurement itself. The exact mechanizations of data-link and measurement signals for the Transit system and the GPS are detailed in a subsequent section.

It is important to realize that the satellite navigation techniques detailed here and embodied in the Transit system and the GPS are entirely passive on the part of the user. That is, the user does not transmit to the satellite, as in a radar-transponder navigation system. The user need only receive satellite signals in order to accumulate sufficient measurements and auxiliary data for solution of the user position. This passive receive-only mode of radio navigation has great benefits in terms of user equipment size, cost, and reliability. The tradeoff against active transmit-receive methods is in the amount of computation required in the user receiver. The Transit system and the GPS are essentially user–computer-based systems.

54.5 IONOSPHERIC EFFECTS

The ionosphere is the interface between our atmosphere and outer space. It is a region of ionized atmospheric gases and free electrons that have not recombined with the atmospheric ions from whence they came. Historically the ionosphere was described as being composed of "layers" that were thought to give rise to "reflection" of radio waves from the earth. Now it is known that the ionosphere is described by a density of electrons per cubic centimeter that has a rapid onset in the neighborhood of 50 km altitude, has a maximum density in the neighborhood of 250 to 450 km, and decreases more gradually to zero in the neighborhood of 800 to 3000 km.

A radio wave passing upward into the ionosphere is subject to deleterious effects whose magnitude is inversely proportional to the square of the transmission frequency. For suitably low frequencies, such as in the range 15 to 1500 kHz, during the daytime when the ionic density is greatest, signals are subject to absorption in the lowest ionospheric region, the D region. Higher frequency signals that pass through the D region are subject to smooth refraction, or gradual bending of their ray paths, by the higher regions, called E, F_1, and F_2, in order of increasing altitude. If the electron density is sufficiently great with respect to signal frequency, the wave may be sufficiently refracted to return to earth. Such behavior is called skip.

The presence of the ionosphere and its density are due to solar bombardment and radiation of the atmospheric molecules. Thus the ionosphere is most dense around local noon, in the daytime, at latitudes nearest the sun, and at the peak of the 11-year sunspot cycle. At night the density decays to just a single F layer with peak electron density less than the daytime maximum by 1 to 1.5 orders of magnitude.

Satellite radio navigation frequencies are sufficiently high so that there is no marked refraction of the ray paths or absorption. However, because of the accuracies desired, the small refractive effects are significant and must be accounted for in precision navigation.

The significant effects of ionospheric passage on the radio wave are two. First, the wave propagates at an infinitesimally smaller velocity in the ionosphere than in vacuum. Second, as the wave propagates through a region of changing density the wave is refracted smoothly, according to the gradient of the density. This means that from a point above the ionosphere to a point below the ionosphere, the wave travels via a curved path that is longer than the straight path connecting the two points. This causes different effects on a Doppler system, such as the Transit system, than on a ranging system, such as the GPS.

Because the ray path from satellite to user is not straight, the range measured by a ranging system is always a little long. Since the curved ray path always lies above the straight-line path and therefore sees a greater projection component of satellite velocity, the Doppler frequency measured by a Doppler system is always a little greater in magnitude. However, this is a very small effect compared with the effect on a Doppler system of just the excess propagation delay of the curved path. The effects on the Transit system and GPS are explained as follows.

Given that the ionosphere creates an excess delay over that created by a straight-line path, the question is how to calculate the excess delay.

The bending of the ray path is a function of refractive index μ or rather of its gradient along the path. μ is a function of frequency F (in hertz) and electron density per cubic centimeter N, given as

$$\mu(N) = \left(1 - \frac{80.6N}{F^2}\right)^{1/2} \tag{54.18}$$

When μ becomes imaginary, the gradient becomes infinite and a radio wave cannot penetrate the ionosphere further. However, for satellite navigation frequencies and relatively large peak electron densities, μ is very near 1 and small bending of the ray path occurs. For values of frequency of 400 and 1575 MHz, respectively, the μ's are of the order of

$$\begin{aligned} \mu_{400} &= 1 - 8\pi \times 10^{-11} \\ \mu_{1575} &= 1 - \frac{\pi}{2} \times 10^{-11} \end{aligned} \tag{54.19}$$

Likewise the angular separation of the straight and curved paths is very small. For a satellite-to-user range of 2000 km (Transit system) and an excess delay of 10 m, the path separation angle at the satellite is of the order of $\theta = 4 \times 10^{-3}$ rad, or 14 min of arc.

A first-order model of excess range is

$$\Delta R = \frac{K}{F^2}(\text{TEC}) \qquad \text{m} \tag{54.20}$$

where $K = \dfrac{e^2}{4\pi^2 m\epsilon_0} = 80.6 \dfrac{\text{m}^3}{\text{electron} \cdot \text{s}^2}$

e = electron charge = 1.6×10^{-19}C/electron

m = electron mass = 9.1×10^{-31}kg/electron

ϵ_0 = permittivity = $10^{-9}/36\pi \dfrac{\text{C}^2 \cdot \text{s}^2}{\text{m}^3 \cdot \text{kg}}$

F = frequency, cycles/s or Hz

TEC = total integrated electron content in column of 1 m^2 cross section lying along path from user to satellite, electron/m

The value for TEC is obtained by integrating the electron density per cubic meter along the ray path. Because of the slight difference between the curved and straight paths, it is sufficient to integrate along the straight path. Typical values of TEC range from 10^{16} to 10^{19}.

The electron density model as a function of r, the geocentric radius to the satellite, is taken from Yip and Von Roos[5] as

$$N(h) = N_{max}\exp\left(\tfrac{1}{2}\right)\left\{1 - z + \int_{-\infty}^{z} \frac{\exp(-x)\,dx}{\left[1 - (y_0^2 + z_0^2)/(Hx + R_e - h_{max})^2\right]^{1/2}}\right\} \tag{54.21}$$

where N_{max} = peak electron density $(\mathrm{m}^3)^{-1}$, at height h_{max}

$z = \dfrac{r - R_e - h_{max}}{H}$ = normalized height

r = geocentric radius

R_e = earth radius

H = scale height of ionosphere varies from about 8 at 100 km altitude to 110 at 700 km altitude relative linearly

x = solar zenith angle

y_0 = user coordinates in x_0, y_0, z_0 system where $z_0 = [X_0, 0, 0]$ is solar subpoint

Applying a correcting factor to Eq. (54.21) for satellite elevation angle (γ) and combining with Eq. (54.20), we obtain the integral for excess path delay as

$$\Delta R = \left(\frac{e^2}{4\pi^2 m\epsilon_0 F^2}\right)\int_s N_{max}\cdot\exp\left\{\frac{1 - z + \int_{-\infty}^{z} \dfrac{\exp(-x)\,dx}{\left[1 - (y_0^2 + z_0^2)/(Hx + R_e + h_{max})^2\right]^{1/2}}}{2\left[1 - (R_e^2\cos^2\gamma)/r^2\right]^{1/2}}\right\}dr \tag{54.22}$$

This integral requires numerical evaluation by computer using some such technique as Simpson's rule, for instance.

Graphs of results are easily plotted with satellite elevation angle γ as abscissa and excess range ΔR as ordinate. Plots may be made for fixed solar zenith angle χ, or local hour angle ϕ, with the other serving as parameter for a family of plots. Figure 54.3 shows an example for two values of N_{max} for a local hour angle of +45°, which is roughly 15:00 hours, local time. A family of curves for solar zenith angle χ from 0° to 90° is shown.[6]

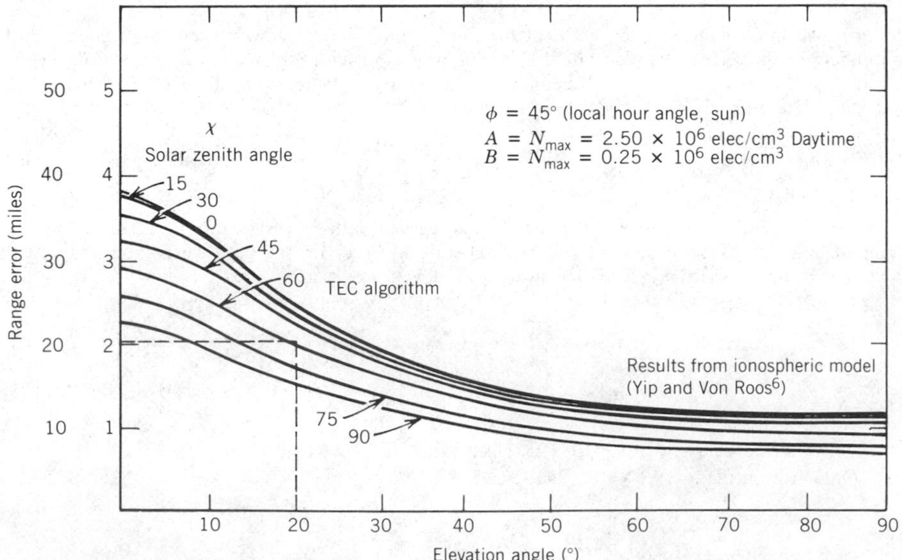

Fig. 54.3 Ionospheric range error. TEC, columnar electron density.

An interesting feature of the example given is that the excess delay does not increase markedly for lower elevation angles, as has been held previously. The increase from 90° to 0° elevation angle is seen to be only by a factor of 2 to 3. The absolute value of ΔR is seen to be linearly dependent on $N_{\max}$.

Given the existence of excess propagation delay, two questions are (1) what is the effect on the navigation solution, and (2) what can be done about it? The effect will be addressed first.

In both the Transit system and the GPS, the effect of ΔR is to bias the measured position away from the satellite. If three satellites of approximately the same elevation angle but separated in azimuth by 120° can be quickly received, then the ΔR biases will tend to cancel in the latitude-longitude coordinates. The effect on altitude will remain, however, with the tendency being to measure lower than the true altitude. For surface navigation, this is not a problem.

With respect to correcting the effect, there are several possibilities. The first is to use two separate transmission frequencies to correct the excess delay effects. This technique takes advantage of the fact that excess delay is inversely proportional to the square of frequency.

Suppose that two frequencies f_1 and f_2 are used for transmission, with $f_1 < f_2$, and the frequencies are phase-coherently generated from the same oscillator source so that

$$\frac{f_2}{f_1} = (K)^{\frac{1}{2}} \tag{54.23}$$

where K is known and stable. Now denote the ranges measured at frequencies f_1, f_2, as, respectively,

$$\begin{aligned} R_1 &= R + \Delta R_1 \\ R_2 &= R + \Delta R_2 \end{aligned} \tag{54.24}$$

where R is true range and $\Delta R_1, \Delta R_2$ are the excess ranges at frequenices f_1 and f_2. There is a constant C_1 such that

$$\Delta R_1 = \frac{C_1}{f_1^2}, \qquad \Delta R_2 = \frac{C_1}{f_2^2} \tag{54.25}$$

Thus

$$\Delta R_1 = \left(\frac{f_2}{f_1}\right)^2 \cdot \Delta R_2 \tag{54.26}$$

and

$$R_1 - R_2 = \Delta R_1 - \Delta R_2 \tag{54.27}$$

From these equations it follows that

$$R = \frac{R_2(f_2/f_1)^2 - R_1}{(f_2/f_1)^2 - 1} = \frac{KR_2 - R_1}{K - 1} \tag{54.28}$$

where K is given by Eq. (54.23).

Thus from Eq. (54.28) the true range R may be determined by measuring apparent ranges R_1 and R_2 at two known frequencies, f_1 and f_2, whose ratio is a stable constant, $\sqrt{K}$.

In the case of Doppler, let the Doppler of frequency f be denoted by D, where

$$D = \frac{\nu_R}{C} \cdot f \tag{54.29}$$

and ν_R is radial velocity component (signed) with C the speed of light. Taking time derivatives of Eq. (54.28), we obtain

$$\frac{\dot{R}}{C} f_1 = \frac{D_2(f_2/f_1) - D_1}{(f_2/f_1)^2 - 1} = \frac{\sqrt{K}\, D_2 - D_1}{K - 1} \tag{54.30}$$

The quantity $(\dot{R}/C)f_1$ is the true Doppler at frequency f_1. D_1 and D_2 are the apparent Dopplers measured at frequencies f_2 and f_1, respectively.

Equations (54.28) and (54.30) show the two-frequency method for correcting ranging (GPS) or Doppler (Transit system) measurements for the effects of ionospherically induced excess range.

There is another possibility for correcting for ΔR, not using multiple frequencies. As is shown later with respect to Transit and GPS, there must be as many independent measurements of satellite data as there are navigation variables to be solved. Thus to compute user position x, y, z, requires three independent measurements. In the case of the user's clock being biased from satellite time, a fourth measurement is required to solve for clock bias. Given estimates of user position, satellite position, and sun's position (all calculable), the only unknown in the solution for ΔR in Eq. (54.22) is N_{max}, peak electron density. If five independent satellite measurements were taken with ray paths all passing through an ionospheric region of common density (the same N_{max}) then theoretically N_{max} could also be solved for. This solution would be computation intensive but not impossible. This method has possible future potential.

54.6 TRANSIT, A LOW-DYNAMICS DOPPLER SYSTEM

The Transit satellite system is a system employing multiple satellites in circular polar orbits of 107-minute period. Orbital altitude is a nominal 1075 km. The system was originally implemented by the U.S. Navy for use by the Department of Defense only. The system became operational in 1964 and was released for civil use in 1967. The satellites on orbit are backed up by spare spacecraft of the original type which are in storage. Also a Transit improvement program has resulted in an upgraded Transit-compatible satellite called NOVA which is now available. The new satellites employ orbit control to maintain precession of the orbital plane (right ascension of the ascending node) at negligible values. Orbital precession of the original five Transit satellites was a problem that gave unacceptably long waiting times between fixes in some cases (Ref. 3, p. 48).

Although originally designed for four satellites with planes separated by 45° of longitude (ascending nodes separated by 90°), uncontrolled orbital precession soon made a fifth satellite highly desirable. Recently the same effect has caused planning for a sixth satellite also. The inability to control orbital precession has caused waiting times, even for six satellites, to exceed four hours, with 10% probability. For six satellites the mean time between fixes is 81 minutes at worst. Both worst cases occur at the equator.

A Transit fix is available every time a single satellite is visible, which is somewhere between 80 and 100 minutes average for the worst case of low latitudes, depending on whether five or six satellites are active. Thus Transit is a position-fixing system and not a continuous-navigation system. For maritime users with good dead-reckoning capability and low maneuvering dynamics, Transit is a very useful system.

Each fix is obtained from one satellite as it passes the user location. A useful satellite pass will be visible between 10 and 18 minutes. During this period, the user receiver must take Doppler measurements from the satellite several times. Each period of Doppler measurement is essentially one of the several independent measurements that must be made to resolve the various navigational uncertainties. Since each satellite Doppler measurement can correct the user's estimated position only in the line-of-sight direction of the satellite, it can be seen that several measurements are necessary as the satellite changes position in azimuth with respect to the user. Since user altitude is not important in the maritime world, two good measurements at orthogonal azimuths are sufficient to resolve longitude and latitude. A third measurement is also required to resolve uncertainty in the user's Doppler reference oscillator frequency, which later enters into processing of the measurements. In practice, as many independent measurements as possible in the available satellite times are taken to give smoothing of the calculated position. Modern equipment takes from 20 to 40 sets of measurements.

Because the various Doppler measurements are taken at different times, the user's motion must be predicted to have an estimated position at the time of the Doppler measurement, which can be corrected by the Doppler measurement. Thus dead reckoning is required during the measurement interval. Needless to say, this requirement constrains the user in dynamic maneuvering during this period.

Transit Signal and System Structure

The Transit satellites transmit two signals, one at 399.968 MHz and one at 149.988 MHz, each with a long-term stability of 1 part in 10^{11} during a satellite pass time. Each frequency is offset by 80 parts per million below 400.00 MHz and below 150.00 MHz, respectively. This ensures that the greatest positive Doppler will not result in received frequencies greater than 400.00 and 150.00 MHz, respectively, which are the user–receiver Doppler reference measuring frequencies, nominally.

The purpose of two transmitted frequencies is for ionospheric correction, as detailed in Section 54.5. A precision Transit receiver employs dual channels for reception of these two frequencies. A standard single-channel receiver uses only the 400-MHz signal.

The transmitted signal powers are of the order of a watt, leaving significant margin for most channel conditions, using state-of-the-art low-noise receivers. The carrier signals are phase shift keyed in quadraphase. A three-level waveform is placed in the sine phase to convey the binary data of the satellite message. This waveform, shown in Fig. 54.4, represents a binary "one" as shown. For a binary "zero," the waveform is inverted. The phase deviation of the data waveform is $\pm 60°$. A two-level clock waveform is placed in the cosine phase of the signal. The clock runs at twice the bit

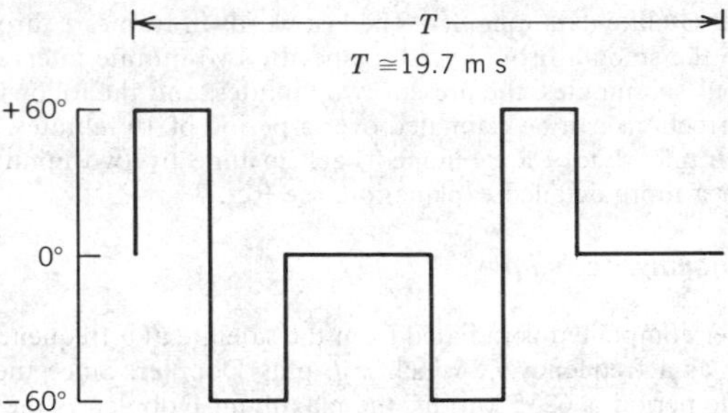

Fig. 54.4 Transit signal waveform.

rate and is used for synchronizing the receiver bit detector. The phase deviation of the carrier is such that a residual unmodulated carrier component remains, which contains 56.25% of the total signal power. This enables the receiver to employ a standard phase-locked loop for tracking the unmodulated carrier component. The loop oscillator then provides demodulation reference sinusoids for recovering the sine phase and cosine phase data bits and clock, respectively, as well as the Doppler signal.

The powers allotted to data and clock are 37.5% and 6.25%, respectively (Ref. 3, p. 48). Figure 54.5 shows a representative receiver implementation.

The Transit navigation cycle is organized around a basic two-minute interval, which is the duration of the transmitted navigation data messages. The message consists of 6103 binary digits of data organized into 26 frames, followed by a final 19 bits. Each frame consists of six 39-bit words. The final 25 bits of each two-minute message block are a synchronization word denoting the time mark between two-minute messages. Each message begins and ends precisely at the start of an even minute in Transit system time (satellite time).

The orbital parameters are located in the sixth word of frames 1 through 22 inclusive. That is, the orbit elements occur in the sixth word in each frame from 1 through 22. The last words in frames 9 through 22 are the elements defining a smooth elliptical orbit over the period of roughly 12 hours

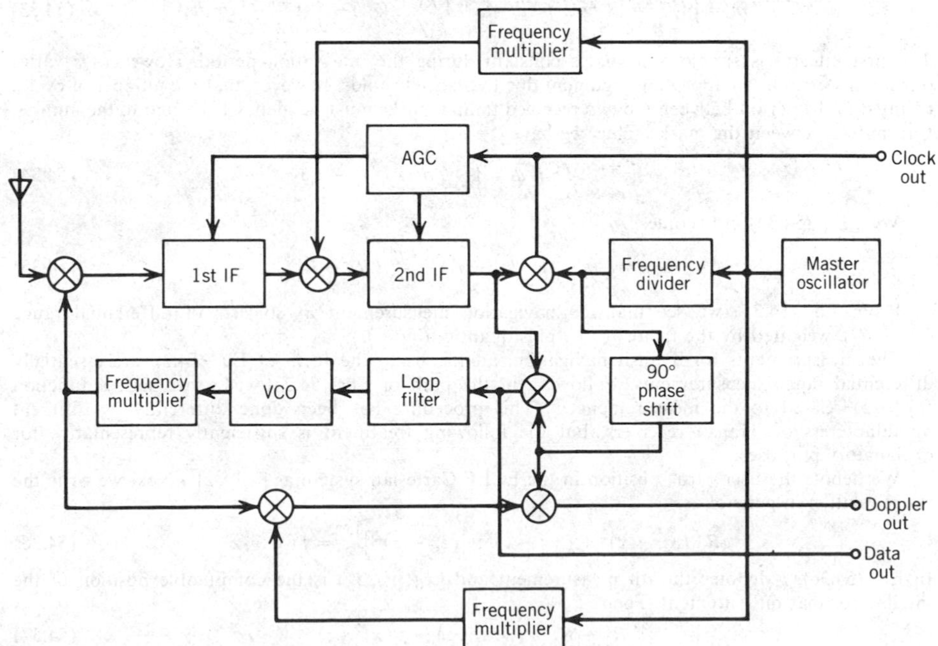

Fig. 54.5 Transit receiver implementation. AGC, automatic gain control; VCO, voltage-controlled oscillator.

between up-loads of the satellite data memory. The last words in frames 1 through 8 inclusive are the real-time corrections to the smooth orbit, keyed to specific two-minute intervals. That is, corrections are given for the previous six minutes, the present two minutes, and the following eight minutes. Thus by interpolation the corrections can be estimated over a period of 16 minutes. Every two minutes the entries in word six of frames 1 to 9 are bumped back in time by two minutes and a new word is entered in frame 9. For a more detailed explanation, see Ref. 3.

Transit Doppler Positioning Technique

The unmodulated carrier component is radiated from the satellite at a frequency $f_t = 399.968$ MHz. It is received by the user as a frequency f_r, which is f_t plus Doppler. Since the orbital velocity for a satellite with 107-minute period is 6235 km/hr, the maximum Doppler as the satellite crosses the 0° horizon is ± 7057 Hz on a 400-MHz carrier.

Rather than measuring the Doppler frequency directly, the Doppler zero crossings are counted for 23 seconds, under the timing control of the timing marks inherent in the telemetry data. Because of the frequency offset of $8 \times 10^{-6} f_t$ in the satellite ($\sim$32 kHz), the frequency f_r that is counted always reflects a positive Doppler bias, even when the satellite is going away from the user. This offset plus any long-term inaccuracy in the user's oscillator frequency must be solved for, along with the user's position coordinates.

The user-receiver oscillator forms a reference frequency $fg = 400.00$ MHz. The Doppler frequency output of the receiver is then $fd = fg - fr$. This Doppler frequency is then counted from time t_1 to time t_2, a period of exactly two minutes at the satellite. The resulting count m is given as

$$m = \int_{t'_1}^{t'_2} fd\,dt = \int_{t'_1}^{t'_2} (fg - fr)\,dt \tag{54.31}$$

Now the times t'_1 and t'_2 are signal event times measured by the user's clock. They may be related to satellite signal event times t as

$$t' = t + \frac{R_{t,t'}}{C} \tag{54.32}$$

That is, a satellite signal event seen at time t (such as a data-timing mark) is seen at the receiver at time t^1. $R_{t_1t'}$ is the range that the signal traveled between t and t^1.

Thus we may write the Doppler count in the form

$$m = \int_{t_1 + R_1/C}^{t_2 + R_2/C} (fg - fr)\,dt = \int_{t_1 + R_1/C}^{t_2 + R_2/C} fg\,dt - \int_{t_1 + R_1/C}^{t_1 + R_2/C} fr\,dt \tag{54.33}$$

The first integral is simple, since fg is constant during the integration period. However, fr varies during this period. An appealing argument due to Stansell[3] holds, however, that the number of cycles counted in the period between the two received timing marks must be identically equal to the number transmitted between the marks. Thus we have

$$\int_{t_1 + R_1/C}^{t_2 + R_2/C} fr\,dt = \int_{t_1}^{t_2} ft\,dt = f_t(t_2 - t_1) \tag{54.34}$$

With Eq. (54.34) m becomes

$$m = (fg - ft)\cdot(t_2 - t_1) + \frac{fg}{C}(R_2 - R_1) \tag{54.35}$$

From Eq. (54.35) we see that the navigation measurement m consists of differential range $(R_2 - R_1)$, weighted by the factor fg/C, plus quantity $(fg - ft)\cdot(t_2 - t_1)$.

The measurements for Transit navigation, analogous to the form of Eq. (54.2), are essentially differential range measurements. Following in the path of Eqs. 54.2 to 54.9, we seek a function $f(x, y, z)$ related to the measurement m. This procedure has been done differently by different manufacturers of Transit receivers, but the following treatment is sufficiently representative for explanatory purposes.

We denote the user's true position in the ECEF Cartesian system as $[x, y, z]^T$. Next we write the range relation in general as

$$R_i\left[(x_{si} - x)^2 + (y_{si} - y)^2 + (z_{si} - z)^2\right]^{1/2} = f_i(x, y, z) \tag{54.36}$$

In Eq. (54.36), i denotes the ith measurement and (x_{si}, y_{si}, z_{si}) is the computable position of the satellite for that measurement. Then

$$R_2 - R_1 = f_2(x, y, z) - f_1(x, y, z) \tag{54.37}$$

We then have

$$m = (fg - ft)(t_2 - t_1) + \frac{fg}{C}[f_2(x, y, z) - f_1(x, y, z)] \tag{54.38}$$

Viewing the quantity $(fg - ft)(t_2 - t_1)$ as an unknown differential range bias b, we may also choose to solve for it, in addition to the ECEF coordinates x, y, z. Note that fg, being derived from the user's basic frequency reference, may be assumed known. But $(fg - ft)$ is not a priori known, since ft is not available for measurement with respect to fg. Thus we may incorporate b into the navigation state and rewrite Eq. (54.38) as

$$m = f(x, y, z, b) \tag{54.39}$$

Now the usual next step is the linearization of the problem by assuming a position (x_0, y_0, z_0, b_0), perhaps by dead reckoning. This position is taken in the midpoint of the user course presumably traversed during the full satellite measurement period.[7] The linearization is written as

$$\begin{aligned} e &= m - f(x_0, y_0, z_0, b_0) = m - m_0 \\ &= \left.\frac{\partial f}{\partial x}\right]_{x_0} \cdot (x - x_0) + \left.\frac{\partial f}{\partial y}\right]_{y_0} \cdot (y - y_0) + \left.\frac{\partial f}{\partial z}\right]_{z_0} \cdot (z - z_0) + \left.\frac{\partial f}{\partial b}\right]_{b_0} \cdot (b - b_0) \end{aligned} \tag{54.40}$$

Four such measurements may be taken, resulting in

$$\begin{bmatrix} e_1 \\ e_2 \\ e_3 \\ e_4 \end{bmatrix} = \begin{bmatrix} \left.\frac{\partial f_1}{\partial x}\right]_{x_0} & \left.\frac{\partial f_1}{\partial y}\right]_{y_0} & \left.\frac{\partial f_1}{\partial z}\right]_{z_0} & \left.\frac{\partial f_1}{\partial b}\right]_{b_0} \\ \left.\frac{\partial f_2}{\partial x}\right]_{x_0} & \cdot & \cdot & \cdot \\ \left.\frac{\partial f_3}{\partial x}\right]_{x_0} & \cdot & \cdot & \cdot \\ \left.\frac{\partial f_4}{\partial x}\right]_{x_0} & \cdot & \cdot & \left.\frac{\partial f_4}{\partial b}\right]_{b_0} \end{bmatrix} \cdot \begin{bmatrix} x - x_0 \\ y - y_0 \\ z - z_0 \\ b - b_0 \end{bmatrix} \tag{54.41}$$

In Eq. (54.41), the f_i are the versions of Eq. (54.38) computed for the satellite position applicable in the ith measurement time. Equation (54.41) may be written compactly as

$$(\mathbf{m} - \mathbf{m}_0) = H(\mathbf{x} - \mathbf{x}_0) \tag{54.42}$$

where $\mathbf{m}$ is the four vector of measurements, $\mathbf{x}$ is the four vector of coordinates, and H is a 4×4 matrix of partial derivatives of the transformation, known generally as the Jacobian matrix.

For a Transit satellite pass, the matrix H is invertible with inverse H^{-1}. Thus Eq. (54.42) may be solved as

$$\mathbf{x} - \mathbf{x}_0 = H^{-1}(\mathbf{m} - \mathbf{m}_0) \tag{54.43}$$

Now each solution requires only four 23-second Doppler-count measurements. It is possible to make from 20 to 40 measurements during a nominal pass. Thus the capability exists to smooth, or filter, the positions. If the user is at rest, this can be done by simple averaging of the coordinates. If the user is in motion, a sequential filter of the Wiener or Kalman type may be used. It is even possible to operate a sequential filter on individual measurements, since each residual, e_i, may be "projected" onto all four coordinates through the appropriate elements of the H^{-1} matrix.

This theoretical explanation has been for the purposes of illumination only. The computations carried out in a manufacturer's receiver may vary in detail within the scope of the basic differential ranging technique. Also some manufacturers compute latitude, longitude, and bias directly from the residuals, transforming to x, y, z only for the purpose of computing the estimated differential ranges. Various refinements can be made to increase the accuracy of a moving user, but these are beyond the scope of the present discussion.

Transit Performance and Error Characteristics

The accuracy of a typical Transit navigational fix depends on the type of receiver used, the motion of the user, and the geometry of the given satellite pass. Stansell[3] quotes typical accuracies for a stationary user as 27 to 37 m rms for a dual-channel (ionospherically corrected) set and 80 to 100 m rms for a single-channel set. The rule of thumb quoted for users in motion is 370 m of position error for each knot of unknown velocity of the user.

For a dual-channel stationary user, most of the error is systematic rather than instrumentation induced. Stansell[3] quotes a 1973 error budget attributed to the Applied Physics Laboratory, which assigned 15 to 30 m rms to uncertainties in satellite position computation and only 3 to 6 m rms to measurement noise. For a single-channel daylight user, ionospherically induced errors at 400 MHz sometimes were as great as 200 to 500 m, mostly in longitude. Errors due to tropospheric refraction at elevation angles greater than about 20° typically were quoted at 4 to 8 m. For sets that do not calculate altitude as part of the navigation solution, unknown altitude errors propagate into the navigation solution principally as longitude error. For such a set, altitude must be known with respect to the WGS-72 geoid, rather than sea level. It is known that the physical ocean surface exhibits "hills" and "valleys" of ± 80 to 100 m!!

Because of its sensitivity to unknown user motions and its mean time between fixes of 81 minutes, the Transit system has been characterized as a position-fixing system for low-dynamics users. This is not a derogatory categorization, since Transit advanced the science of navigation and geodesy by an order of magnitude after it became operational in 1964. The state of the art in miniaturized computing hardware has advanced sufficiently since then, however, to make possible another order of magnitude advance, NAVSTAR.

54.7 NAVSTAR GPS, A HIGH-DYNAMICS RANGING SYSTEM

The NAVSTAR global positioning system, or GPS, is a system employing ultimately 18 satellites in 12-hour orbits of 55° inclination. The system is being implemented by the Department of Defense for military use. However, it has a "clear access" (C/A) channel that is available for general civil use. The GPS development program grew from two 1960s programs, the Air Force's 621-B program and the Navy's Timation program, which were merged in 1973. GPS is a second-generation satellite navigation system that applies the pseudonoise (PN) ranging technology developed by NASA[8] in the 1960s to the satellite navigation technology embodied in the Transit system.

The GPS employs satellites that are precisely controlled in their orbital positions. Indeed, knowledge of a set of orbital elements, or ephemerides, over one year old is sufficient to predict satellite visibility times within five minutes from a known earth position. Like the Transit system, the GPS is a passive navigation system on the part of the user, in that only reception of satellite-transmitted signals is used by the navigator to compute position. Unlike Transit, GPS uses simultaneous or near-simultaneous reception of signals from four satellites to compute three coordinates of position and one of time difference, due to error between satellite and user clocks. Thus given visibility of four satellites, GPS provides the capability for continuous navigation processing rather than isolated position fixes interconnected with dead reckoning. As will be described, measurement integration times are much less with GPS than with Transit. Thus positions may be computed on the order of every second. Hence GPS provides the capability to "track" vehicles characterized by high-dynamic maneuvering.

The positions of the GPS satellites are uniformly distributed about the earth, three per orbit, in six orbits.[†] With the full orbital constellation of 18 satellites, more than four are thus normally visible at any given time anywhere on earth. This leads to a problem, discussed later, of selecting the four satellites having the "strongest" navigation geometry. Also if a user vehicle should lose line of sight with one GPS satellite, because of, say, blockage of signal by vehicle super structure, the possibility exists of immediately "acquiring" another satellite to replace the one just lost. This procedure is the same as if one of four satellites has set below the usable horizon. The latter occurrence can be predicted, of course, and acquisition of a replacement satellite planned ahead.

The GSP navigation-processor operates from measurements of range between satellite and user. Actually the measurement is one of elapsed time between the time of satellite transmission of a known signal reference and the time of user reception of that same reference. Given the speed of light, the elapsed time measurement equates to a measured range. Now the satellite transmission time is measured according to its clock, which is precisely set to GPS time, with an error of the order of 3 ns. However, the user reception time is measured with respect to the user clock, based on a user reference frequency oscillator. The user oscillator is generally of much lower quality than the satellite frequency standards by many orders of magnitude.

The GPS demonstration satellites, the so-called NDS series, have carried both the cesium and rubidium frequency standards with basic long-term frequency stabilities of the order of 10^{-12} to $10^{-13}(\Delta f/f_0)$. This is truly amazing stability when one realizes that the frequency offset due only to relativity effects is of the order of 4 parts in 10^{-10}.[9] Thus the satellites are able to maintain a system reference time to within several nanoseconds over a 12-hour period, which equates to about 1 m of range. The user frequency reference oscillators (for a supposed reasonable cost civil user) have long-term frequency stabilities of the order of 10^{-7} to 10^{-9}. This stability yields clock drifts of 1 to 100 ns/s, equivalent to 0.3 to 30 m of range per second. Thus it is necessary for the inclusion of a fourth coordinate in the user's position, this coordinate being user clock bias (in meters).

Because the measured range is always in error by the clock bias, the measured quantity is called pseudorange. Pseudoranges without correction are processed directly, as detailed later, to solve for the four user coordinates.

[†] According to plans revealed by the U.S. Air Force Space Division as of the time of this writing.

GPS Signal and System Structure

GPS operates with two available satellite signal frequencies, as does the Transit system, for the purpose of removing the unknown ionospheric additional delay time. These frequencies are 1575.420 MHz, called L1, and 1227.600 MHz, called L2. These two carrier-wave frequencies are phase-coherently generated by frequency multiplication of the same basic standard frequency of 10.23 MHz. The multiplication factors are 154 for L1 and 120 for L2. Actually, the satellite basic reference is lowered from 10.23 MHz by $4.45 \times 10^{-10} \times 10.23$ MHz to equalize the relativity effect, but this is not significant to our consideration of signal structure.

The L1 transmitted carrier frequency carries two signals in phase quadrature. One phase carries the clear access (C/A) pseudonoise (PN) ranging signal and a 50-baud data link. The orthogonal phase carries a precision PN ranging signal (the P code) and the same 50-baud data link. The exact phase relationship is

$$\text{C/A carrier phase} = \text{P carrier phase} + 90° \tag{54.44}$$

The L2 transmitted carrier may carry either the C/A or the P code.

The ranging signals and data-link signal are digital, of varying baud rates. The P-code rate is 10.23 megabaud. The C/A code rate is 1.023 megabaud. The data-link rate is 50 baud. The data-link bits are combined with the code bits using modulo-two addition (exclusive OR). The composite code-data bit stream is then modulated onto the proper carrier phase using full binary phase shift keying ($\pm 90°$), which leaves no unmodulated carrier residual in the transmitted spectrum. That is, no carrier phase reference is transmitted for the various digital signals. Thus it is to be expected that some of the chief problems in receiving these signals have to do with achieving signal synchronization.

The PN ranging codes are special examples of maximal-length linear-shift register sequences,[10] called Gold codes after their inventor, Robert Gold. The structures of the particular GPS Gold codes are explained in detail in Ref. 11. The unambiguous length, or repetition period, of the C/A codes is 1023 chips, or 1.0 ms, exactly. This basic time period is also related to the data-bit clock period, since there are exactly 20 C/A code epochs per data bit. Since the C/A code is only 1.0 ms long, it can resolve ranges unambiguously only within multiples of about 300 km. The resolution of the ambiguity is performed during the initial satellite acquisition process as a part of obtaining the first fix.

There is really only one P code, which is the product of two PN codes whose lengths in chips are relatively prime numbers (that is, with no common divisors). These two codes are of lengths 15,345,000 and 15,345,037 chips, respectively. Since a chip period is 100 ns, the length of the product code is more than 38 weeks. P codes for individual satellites are taken as nonoverlapping one-week segments of the 38-week-long code. The long code is restarted at midnight Saturday-Sunday (Greenwich time), every week.

The digital data transmitted by each satellite contain all the Keplerian orbital parameters for the satellite position computations shown in Table 54.1. Additionally, reduced accuracy orbital elements are contained in one satellite's data message for rough computation of the other satellites' positions. The precise orbital information is called ephemeris, while the less precise data are called almanac.

The satellite data transmission uses 30-bit words, with 10 words per each six-second subframe. Five subframes complete one satellite data message. With 20 ms/b, each word is 0.6 second in duration, each subframe takes 6 seconds, and a complete message requires 30 seconds. Every 30-second data message contains almanac for one of the 18 possible satellites. Thus to acquire almanac for all satellites requires 9 minutes.

Each 6-second subframe begins with a 30 bit telemetry word (TLM). The first 8 bits of the TLM word is the hexadecimal "8B." Owing to the way the data are detected in a receiver, the bits may be complemented. The final 2 bits of the TLM word should be zeros. If they are ones, the data are inverted. Thus a search for "8B" or its complement establishes subframe synchronization.

The second 30-bit word in each subframe is a hand-over word (HOW). This word is a number (z count), which when multiplied by 4 indicates the number of 1.5-second epochs (X1 epochs) that will have occurred since the beginning of the GPS week, at the beginning of the *next* subframe (TLM word). The purpose of this word is to enable rough synchronization of the P code (for the first time) at the beginning of the next subframe. The HOW word is an acquisition aid for P code. The structure of the 30-bit data message is indicated in Table 54.3.

Subframe 1 contains data block I, which contains four parameters for making a quadratic correction to the indicated satellite clock time. Also in data block I are eight parameters for making a rough correction for ionospheric delay for those users not equipped with dual-frequency (L1-L2) receivers. Data block II occupies both subframes 2 and 3 and contains the complete, accurate ephemeris for the satellite being received. Subframe 4 makes provision for special broadcast messages. Subframe 5 contains the rotating almanacs for all satellites.

Of interest besides the structure of the transmitted signals is the signal-to-noise ratio environment in which the signals will be received. In designing or analyzing the performance of satellite-to-earth

TABLE 54.3. STRUCTURE OF 30-BIT DATA MESSAGE

Subframe	Type of Word		Structure
1	TLM[a]	HOW[b]	Data block I—clock correction
2	TLM	HOW	Data block II—"ephemeris"
3	TLM	HOW	Data block II, cont'd—"ephemeris"
4	TLM	HOW	Message block
5	TLM	HOW	Data block III—"almanac"

[a] TLM, telemetry word.
[b] HOW, hand-over word.

links, it is usual to formulate the ratio of received signal power divided by the value of the white noise power spectral density effective in the receiver. This is the so-called C/N_0 ratio.

The received signal power C is equal to that transmitted, multiplied by various gain and loss factors that affect the link. Chief among these is the "space loss," which is just the attenuation of power in the transmitted electromagnetic wave due to spherical spreading of the wavefront with distance from the source. This loss factor, L_s, is given by

$$L_s = (V_c/4\pi FR)^2 \tag{54.45}$$

where V_c = velocity of light, m/s
R = range, m
F = frequency, Hz
The units for V_c and R must be compatible.

It is seen that the space loss varies inversely as the square of frequency and of the range. Thus $0 < L_s < 1$, and L_s will always diminish the received power over that transmitted. Since L_s varies with range, there is a maximum and minimum value for L_s, depending on the elevation angle of the satellite as seen from the user's position, assumed near the surface of the earth. The maximum range (at 0° elevation angle) and the minimum range, respectively, are

$$\begin{aligned} R_{\max} &= 25{,}231 \text{ km} \\ R_{\min} &= 19{,}652 \text{ km} \end{aligned} \tag{54.46}$$

assuming a spherical earth of mean radius 6371 km. Thus we have maximum and minimum values for L_s at frequencies L1 and L2, as given in Table 54.4. In this table both the scalar value and the decibel value of the losses are given, where X dB = 10 log(X).

The amount of noise effective in the receiver is, for GPS, essentially the noise generated in the radio frequency preamplifier, which is of special low-noise design. This assumes that the low-noise preamplifier has sufficient gain (20–30 dB) to override the noise generated in the first stages of the receiver itself. In this case the white noise spectral density, N_0, is given by

$$N_0 = KT_s \tag{54.47}$$

where T_s is the "system noise temperature" in Kelvin. T_s is the noise temperature of the preamplifier itself, plus any contributions from cable and connector losses between antenna and preamplifier. Great pains are usually taken to mount the preamplifier as near the antenna as possible. The constant K is Boltzmann's constant given by

$$K = 1.38 \times 10^{-23} \text{ W/Hz/K} \tag{54.48}$$

The system noise temperature T_s may be specified according to the standard "noise figure," F, for the

TABLE 54.4. SPACE LOSS FOR L1 AND L2 CARRIERS

	L1	L2
$R_{\min}$	5.9460×10^{-19} (−182.3 dB)	9.7926×10^{-19} (−180.1 dB)
$R_{\max}$	3.607×10^{-19} (−184.4 dB)	5.9408×10^{-19} (−182.3 dB)

system by

$$T_s = 290°(F - 1) \tag{54.49}$$

A handy device for determining the values of available C/N_0 for the various cases is a design control table. Such a table enters the various parameters of the link, such as transmitted powers, gains, losses, noise spectral density, and so on, to arrive at a value for available C/N_0. This value is then used to calculate performance levels of various parts of the receiver.

An example table for the C/A channel at L1 is given as Table 54.5. The following comments are made with respect to this table.

Modulation loss accounts for the fact that the total satellite transmitter power is apportioned between C/A and P signals at L1 and the signal at L2. The power proportions are

$$\text{C/A (L1)} : \text{P (L1)} : \text{L2} = 4 : 2 : 1 \tag{54.50}$$

Tolerances account for uncertainties in specifications, changes with age or temperature, or variations in the geometric relation between user and satellite. For a user with an antenna that is nominally omnidirectional over the upper hemisphere, there is a large change in gain between the zenith direction and the horizon. The tolerance reflects this. Also an omnidirectional antenna that is circularly polarized at the zenith becomes elliptical at lower elevation angles. Hence the tolerance for polarization loss.

The nominal value for C/N_0 in the table is 48.4 dB, and the tolerance is −12.9 dB. Thus the value available is +48.4 dB in the best case and perhaps as small as +35.5 dB in the worst case. Experience has shown that the adverse tolerances hardly ever add linearly, except for those that reflect correlated effects. Three such correlated tolerances are those for space loss, antenna polarization loss, and antenna gain for receiving. The sum of these three correlated tolerances is −8.5 dB, which can be expected to occur at the low elevation angles below, say, 10°.

The availability of 48.4 dB for C/N_0 in the best case sets the upper limit for the receiver-processing possibilities in the example shown. Using a rule of thumb that 10 dB or greater signal-to-noise ratios are required for good processing indicates that the widest final processing bandwidth can be no greater than about 7 kHz. Being more conservative, in view of the adverse tolerances, might bring this final processing bandwidth down to, say, 3 kHz. The actual bandwidths used for final processing will depend, of course, on more detailed considerations. But at least the value of available C/N_0 sets the scale for further thought.

GPS Pseudonoise (PN) Ranging Technique

The GPS PN ranging technique is a direct lineal descendent of the original scheme developed by NASA's Jet Propulsion Laboratory during the 1960s.[12] Although PN ranging was considered during

TABLE 54.5. DESIGN CONTROL TABLE, CLEAR ACCESS CHANNEL, L1 FREQUENCY

Parameter	Nominal Value (dB)	Tolerance (dB)
1. Transmitter power	13.0 dBW	0
2. Modulation loss, power split	−2.4	0
3. Circuit loss, transmitter	−0.3	−0.1
4. Antenna gain, transmit	+13.7	−2.2
5. Antenna pointing loss, transmit	0	0
6. Space loss; $F = 1575.4$ MHz, $R = 19{,}652$ km	−182.3	−2.1
7. Antenna polarization loss	0	−1.4
8. Antenna gain, receive (omni)	+3.0	−5.0
9. Antenna pointing loss, receiver	0	0
10. Circuit loss, receiver	−0.3	−0.2
11. Net link loss (add 2 through 10)	−168.6	−11.0
12. Total received power, C (1 + 11)	−155.6 dBW	−11.0
13. System noise spectral density, N_0		
Preamp noise figure 3.0 dB $\left\{\begin{array}{l}+1.0\\-0.0\end{array}\right.$		
Loss temperature +20 K, nominal	−204.0 dBW/Hz	+1.9
14. Received C/N_0 (12 − 13)	+48,4	−12.9

the 1950s,[13] it was perfected by JPL and applied in the "heyday" of spaceflight to such projects as Pioneer, Mariner, Surveyor, Viking, Apollo, and Voyager. It continues today in Shuttle and others.

The basic idea behind PN ranging is the following. Two identical PN codes are involved. One code propagates between satellite and user. The other code is maintained in the user equipment. Since a PN code is completely deterministic, the sequence of ones and zeros in the code is completely known. Both codes are started up in synchronism at a particular known time, t_0. The code that propagates between satellite and user is received by the user, delayed by the propagation time. The user then shifts his or her own local version of the code, keeping track of the amount of time delay injected into the "local code," until its sequence exactly matches the sequence being received from the satellite. When the user observes that the two versions of the same PN code were "matched" or synchronized at the observation time, t_r, he notes that the range-delay time, ΔT, at time t_r was just that amount of time delay that the user injected into the local code to cause it to match the received code.

The matching of the local code with the received code is essentially matching the zero-one sequences and then "exactly" matching the leading edges and trailing edges of each bit (or chip as they are called in ranging) within the sequences. Any error in matching the codes in time delay translates into an error in determining the range separating the satellite and user. The conversion between time delay error ΔT and range error ΔR, using a velocity of light, $V_c = 2.99739 \times 10$ m/s, is

$$\frac{\Delta R}{\Delta T} = 0.983 \text{ ft/ns} \tag{54.51}$$

To make the code-matching problem yield very precise time measurements, the time duration of each code chip is made very small. For the C/A code, the chip duration is 0.9775 μs. For P code, the duration is 97.75 ns. Because of the method used for matching code chip to code chip, the resolution in the delay time is even smaller than the chip durations themselves.

The use of submicrosecond code chip durations means that the modulated bandwidth of the ranging signal is of the order of megahertz. For the case of $\pm 90°$ phase shift keying, which is used in GPS to place the code modulation on the radio frequency carrier, the modulated signal may be written in the form

$$s(t) = AC(t)\cos(2\pi f_0 t) \tag{54.52}$$

where A = signal amplitude
f_0 = carrier frequency, Hz
$C(t)$ = analog code waveform

The analog code waveform is either +1 or −1, that is, a digital 0 chip gives an analog ±1 and a digital 1 chip gives an analog −1. Since $C(t)$ characteristically is a rectangular waveform, the frequency power spectrum has the characteristic form

$$S(f) \sim \frac{\sin^2\left[\frac{\pi}{R}(f-f_0)\right]}{\left[\frac{\pi}{R}(f-f_0)\right]} \tag{54.53}$$

where f = frequency, Hz
f_0 = carrier frequency, Hz
R = code chip rate, chips/s

The graph of Eq. (54.53) is given as Fig. 54.6.

It is known that 92% of the power in the PN signal spectrum resides between the first nulls on either side of the carrier frequency. Therefore most receivers employ bandwidths of $2R$ or greater to pass the modulated signal. Now with an available C/N_0 of 48.4 dB/Hz, the signal-to-noise ratio for the C/A signal in a bandwidth of $2R = 2.046$ MHz is −14.7 dB. Thus the signal chips are not observable in such a bandwidth, owing to the overriding effect of the accompanying noise.

Because of the impossibility of synchronizing the received and local codes on a bit-by-bit basis, a more indirect method is used, which can operate in a smaller bandwidth. This method uses the "correlation" properties of PN codes.

Suppose we ignore for the time being the fact that the received PN code exists as modulation on a sinusoidal carrier waveform. Let us consider the simpler problem of synchronizing two identical PN code waveforms that exist in the ±1 analog format. Let us assume that the directly received code is input to one port of an analog muliplier as $C(t)$. Let the local code be input to the multiplier's second port as $C(t+\tau)$, denoting a slight desynchronization of amount τ seconds. Let the output of the analog multiplier be processed by a time averager, such as a lowpass filter. This operation is shown in Fig. 54.7.

The device depicted in Fig. 54.7 is called a time-average cross correlator. Let us now examine its operation on two relatively delayed versions of the same PN code. For this, let us also view Fig. 54.8. In this figure are shown several representative chips of the direct PN code and local PN code when

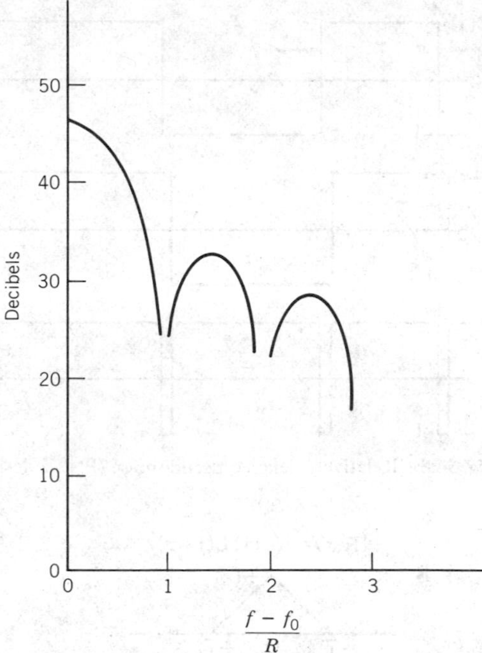

Fig. 54.6 Pseudonoise (PN) signal spectrum.

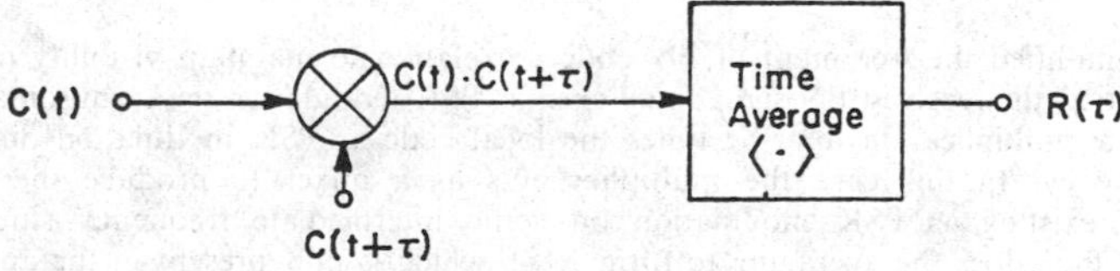

Fig. 54.7 Cross correlator.

they are near synchronism. The upper two graphs are the codes themselves, while the lower drawing is of the code product, $C(t + \tau) \cdot C(t)$.

It is clear by inspection of Fig. 54.8 that the product, $C(t + \tau) \cdot C(t)$, is $+1$ most of the time, with quick excursions to the -1 state during those short periods of duration, τ, when $C(t)$ and $C(t + \tau)$ are of opposite sign. It is also clear that the time average of the product is positive and nearly $+1$. When $\tau = 0$, or the codes are exactly synchronized, the average is exactly $+1$. It can be shown that the output $R(\tau)$, of the average varies linearly with the offset, or relative delay, τ.[12] Also for a very long PN code, the average is essentially zero when $C(t)$ and $C(t + \tau)$ are desynchronized by more than one chip period, T. This function, $R(\tau)$, which is the autocorrelation function of the basic PN code, is shown in Fig. 54.9.

The correlation properties of PN codes allow the synchronization of the received and local PN codes to be accomplished using a smaller bandwidth than that of the code itself. The lesser bandwidth is that of the time-averaging lowpass filter, employed in the cross correlator. For example, if we wished to observe $R(\tau)$ with only $\pm 5\%$ noise with an available C/N_0 of 48.4 dB, a signal-to-noise ratio out of the averager of 26 dB would be required. This would, in turn, require the bandwidth of the averaging lowpass filter to be 174 Hz.

Going on with the example, one next asks the interesting question: If $R(\tau) = 1$ signifies code synchronization and it is known that the output of the correlator has $\pm 5\%$ noise, what is the possible ranging error incurred in accepting any single measurement of $R(\tau)$ within a $\pm 5\%$ neighborhood of $+1$? From Fig. 54.9, with $T = 293$ m equivalent, it can be easily determined that the ranging strategy might incur an error of ± 29.3 m. Many other questions might be asked at this point, but they will be deferred to following sections.

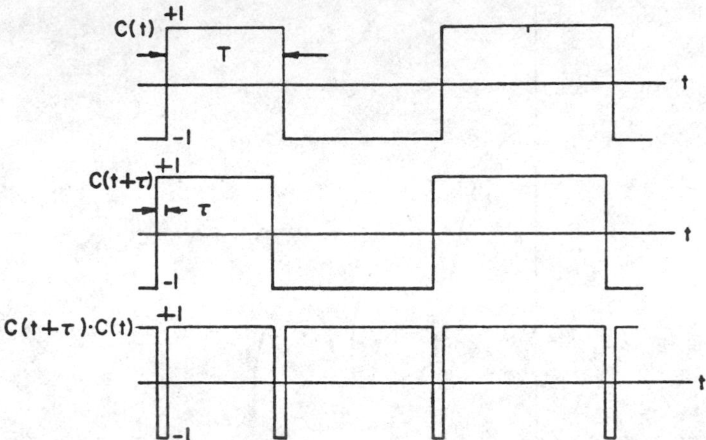

Fig. 54.8 Relatively delayed pseudonoise (PN) codes.

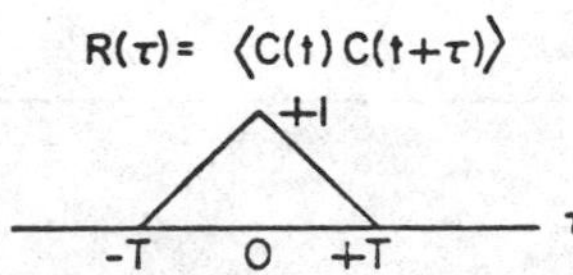

Fig. 54.9 Pseudonoise (PN) autocorrelation function. Upper two graphs show direct and local PN codes, lower graph shows code product.

We have simplified the treatment of PN code correlation to maintain visibility of the essential results. In practice, the received PN signal may exist as PSK modulation on a sinusoidal carrier, as in Eq. (54.52). The multiplication may be times the local code as PSK modulation on a sinusoid of different frequency. In this case the multiplier acts as a mixer to produce the code product $C(t) \cdot C(t + \tau)$ existing as PSK modulation on some intermediate frequency sinusoid. The IF bandpass filter then does the averaging to form $R(\tau)$, which is now present in the amplitude of the output IF sinusoid. There are many such ways to perform the correlation.

The essential parts of the PN ranging operation are the forming of $R(\tau)$ and the adjusting of the delay of the local code to achieve an "estimate" that $R(\tau) = 1$. The error in achieving this estimate translates into the rms error in the range measurement. The speed with which a range measurement can be made is directly proportional to the time required to search out the $R(\tau) = 1$ synchronization point. Speed can be traded for accuracy of the range estimate, and this tradeoff is considered in particular user receiver designs. The question of search strategies for "correlation" is treated in the next section.

Phase-Coherent and Noncoherent Receiver Techniques

In forming $R(\tau)$ and searching out the local code delay corresponding to the center of the triangle of Fig. 54.9, there are several possibilities. We may seek to implement a feedback servo device in the receiver that will automatically adjust the local code delay so as to "track" the $R(\tau) = 1$ point. Otherwise we may choose to sweep the local code past the synchronization point with the received code, observe $R(\tau)$ during the sweep, and determine after the sweep the local code delay that yielded $R(\tau) = 1$.

The choice between the two ranging methods just described depends on several other factors. One of these is whether the receiver is "multichannel." This means whether a separate analog channel in the receiver can be devoted to each of the four required satellites. Several of the early military GPS receivers were designed in this manner. In such a case, four separate feedback systems can be employed to drive four separate local code generators to simultaneously track the four received satellite range codes.

The alternative to a multichannel receiver simultaneously tracking four satellite range codes is a "sequential" receiver that alternately makes range measurements between the four satellites. In this case a single code generator and correlator is sequentially switched to synchronize with the incoming satellite range codes, one at a time. An early military demonstration unit for such a concept yielded a minimum switching time between satellites of 1.2 seconds. A sequential receiver can be made to perform like a multichannel receiver, however, if the switching can be made sufficiently rapid. Such a receiver employs a high level of digital processing and is called a multiplex receiver.[14]

Let us consider now the received signal as it might exist at the output of a hypothetical code correlator that is implemented by a mixer in the formation of the last intermediate frequency in the receiver. Figure 54.10 shows the circuitry. In this figure, the input signal to the mixer correlator has frequency f_1. A digitally controlled oscillator produces a cosine of frequency f_2, which is phase shift keyed by the local code, $C(t + \tau)$. The mixer forms the code product and the bandpass filter averages it while also selecting the IF frequency $f_i = f_1 - f_2$. The bandpass filter bandwidth, B_f, is small enough to average the code product but large enough to pass the data modulation, $d(t)$.

The output of the correlator may now be processed in one of several ways. If recovery of the data is desired, the IF signal may be applied to the input of a Costas loop[15] (Fig. 54.11). The Costas loop regenerates a reference sinusoid which is phase coherent with the suppressed IF frequency f_i. The reference is applied to a synchronous amplitude detector (product detector) which is also driven by the received signal. A 90° phase-shifted version is applied to a similar detector. The two detectors demodulate the IF signal and produce two signal components called the I component (in phase) and Q component (quadrature phase). These components are given as

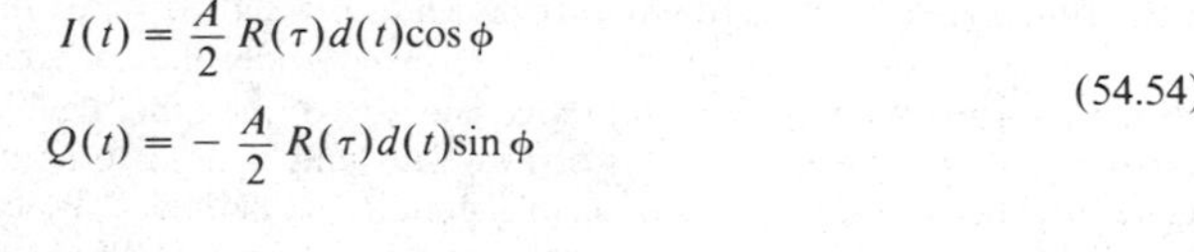

$$I(t) = \frac{A}{2} R(\tau)d(t)\cos\phi$$
$$Q(t) = -\frac{A}{2} R(\tau)d(t)\sin\phi \tag{54.54}$$

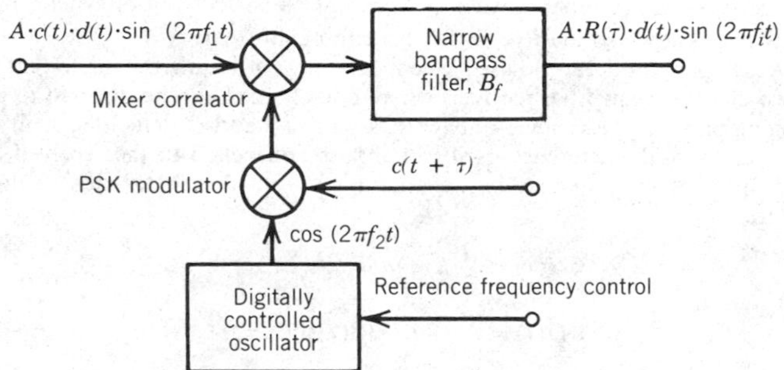

Fig. 54.10 Intermediate frequency (IF) correlator. PSK, phase shift keying.

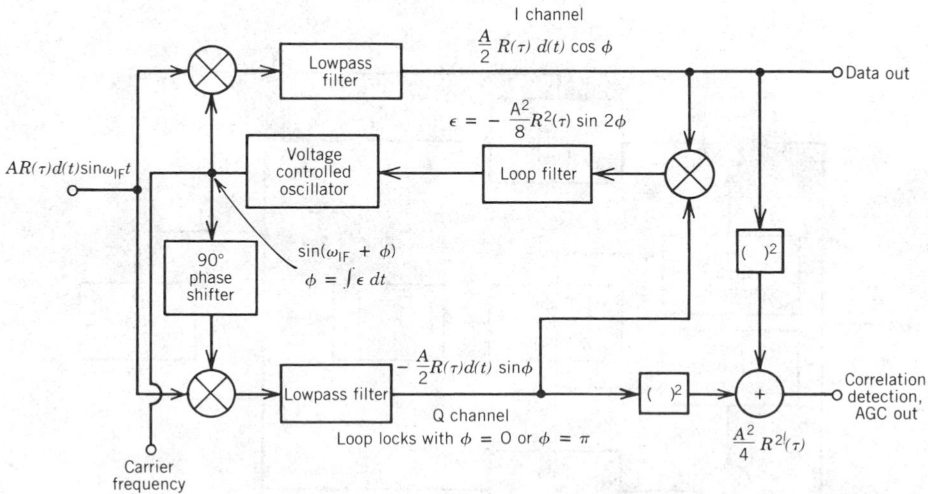

Fig. 54.11 Costas loop.

In Eq. (54.54), ϕ is the error (tracking) in phase between the regenerated IF sinusoid and the received IF sinusoid. For proper operation of the loop $\phi = 0$ or π and the I channel contains the data waveform, weighted in amplitude by $\pm R(\tau)$. Note that if $\phi = \pi$, then the data in the I channel are inverted. Taking the absolute value of the I channel waveform just gives $(A/2)|R(\tau)| = (A/2)R(\tau)$, since $d(t) = \pm 1$. Thus the I channel may be used to derive both data and $R(\tau)$.

A Costas loop synchronizes and tracks the phase (and frequency) of the IF signal sinusoid and produces an output proportional to $R(\tau)$. Two Costas loops may be employed in a feedback loop configuration to synchronize and track the clock frequency of the received code. Such a configuration is called an early-late clock loop. This scheme employs two correlation mixers and two local code generators, as shown in Fig. 54.12.

The early-late loop works in the following way. The two local code generators are driven by the same code clock generator (oscillator). However, one code generator is advanced slightly in delay with respect to the other. Thus the code from one generator is slightly "early" with respect to the code from the other generator (say, 1 bit early, for example). As both code generators are brought near the synchronization point, the "early code" passes the synchronization point and the "early Costas loop" locks, producing an "early $R(\tau)$," called $R_E(\tau)$. When the early code is $\frac{1}{2}$ bit past the synchronization point, the "late code" is $\frac{1}{2}$ bit before the synchronization point, the "late Costas loop" comes into lock, and produces a "late $R(\tau)$," called $R_L(\tau)$.

Referring again to Fig. 54.12, we see that $R_L(\tau)$ is subtracted from $R_E(\tau)$, forming a "tracking error function" of τ. This function is shown in Fig. 54.13. When the early and late codes are exactly $\frac{1}{2}$ bit early and late with respect to the received code, then the tracking error signal is exactly zero. When the received code tends to move one way or the other from the bracketed position, the tracking error signal increases or decreases from zero and is used to adjust the frequency of the code clock generator in the proper direction, to maintain "code lock." When the loop is stably locked, neither the early or late code is exactly synchronized, or "prompt." Each are out of synchronization by $\frac{1}{2}$ bit, but the error is exactly known so the $R(\tau) = 1$ delay can be inferred.

Another processing method can be used when data are not required. This might be the case, for instance, in a two-channel sequential receiver, where one channel can be devoted to getting data and the other to ranging only. In the ranging-only case, a pair of I and Q detectors, as in the Costas loop, can be used with a sinusoidal reference oscillator that is not locked to the received carrier phase. If this oscillator is different in frequency by an amount Δ from the IF frequency, then the I and Q channel signals are

$$\begin{aligned} I(t) &= \frac{A}{2} R(\tau) d(t) \cos(2\pi\Delta t) \\ Q(t) &= -\frac{A}{2} R(\tau) d(t) \sin(2\pi\Delta t) \end{aligned} \tag{54.55}$$

The two signals in this equation may be squared and summed to produce a signal

$$I^2(t) + Q^2(t) = \frac{A^2}{4} R^2(\tau) \tag{54.56}$$

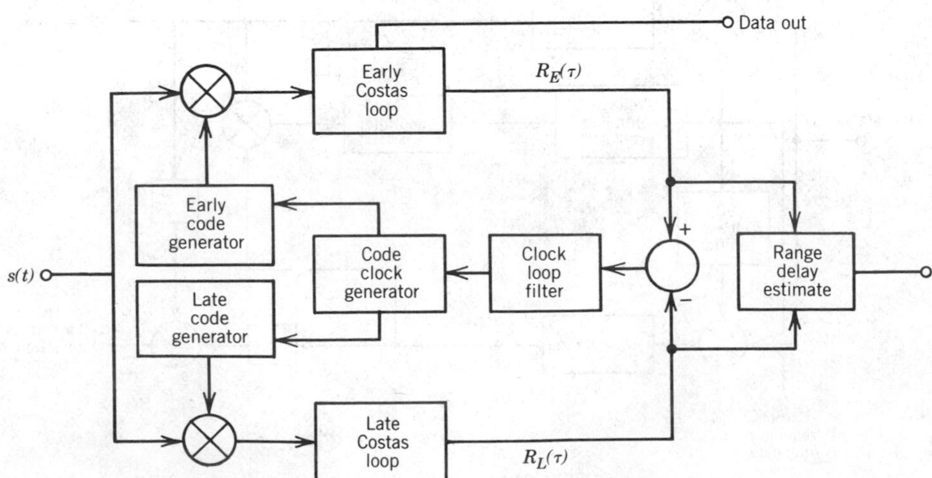

Fig. 54.12 Early-late code synchronizer.

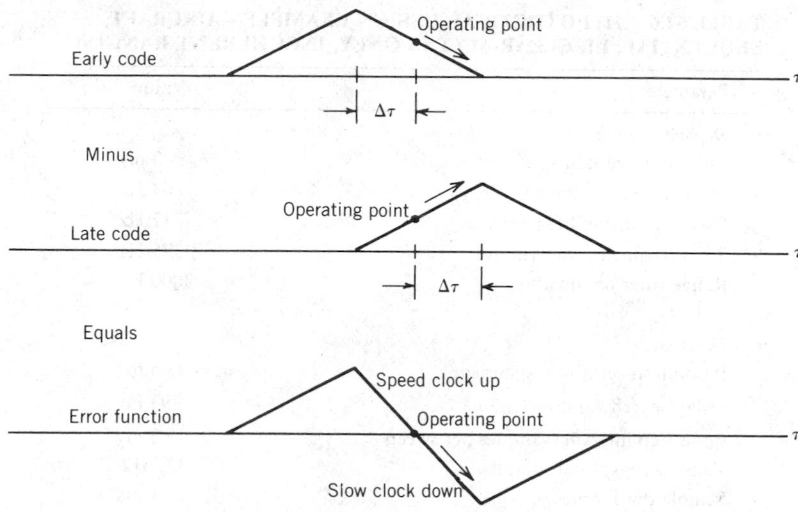

Fig. 54.13 Tracking error signal.

As the single local code generator is swept past the synchronization point, the signal $R^2(\tau)$ may be used to estimate the $R(\tau) = 1$ delay, since $0 \leqslant R(\tau)$ and a one-to-one relationship exists between $R(\tau)$ and $R^2(\tau)$. Since this latter processing method does not depend on having a phase or frequency lock on the IF sinusoid, the method is called noncoherent. The Costas loop method is called coherent.

Because of Doppler effects on the received signal, both the coherent and noncoherent ranging methods require a search, not just in delay τ, but also in frequency. Referring back to Fig. 54.10, the digitally controlled oscillator (DCO) must be adjusted to offset the Doppler effects. It is necessary that the IF signal be accurately centered in the narrowest processing bandwidth encountered in the remainder of the receiver. In acquiring the first satellite signal, there may be a considerable frequency uncertainty range to be searched. During normal navigation the uncertainty should be minimal.

For either coherent or noncoherent range processing, the DCO is preset and then the delay uncertainty region is searched. If no evidence of correlation is observed, the DCO is reset at a different frequency and delay searched again. In this manner the two-dimensional frequency-delay uncertainty region is searched.

A hypothetical example of a system design is given in Table 54.6 to illustrate some of these points. This design is for an aircraft user, with access only to the C/A code, on frequency L1, employing an incoherent (phase)-ranging technique. First some assumptions are made about the accuracies of velocity and position provided by the navigation filter. The standard deviations of velocity error and position error are taken as 3 m/s σ_v and 48.6 m σ_p, respectively. From σ_v is calculated the resulting uncertainty in received frequency due to user velocity error. It is assumed that navigation is nominal and that satellite velocity computations contribute no uncertainty. The search range due to Doppler uncertainty is taken, conservatively, as $\pm 2\sigma_v$, or 63 Hz. Next the frequency uncertainty at the IF due to user's oscillator instability is computed as 312 Hz. The total frequency uncertainty is thus 375 Hz. Theoretically, the bandwidth for estimating the $R(\tau) = 1$ delay, using the incoherent technique, need only be 375 Hz. However, since a sampled-data digital processor is used, the bandwidth is taken as the much greater value of 3 kHz, to allow fast sampling. Three times the reciprocal of this bandwidth is just the period of one epoch of the C/A code.

Next the delay search window is taken as 293 m, or one chip width, which is $\pm 3\sigma_p$ for the position uncertainty. The $R(\tau)$ signal is repeatedly swept 32 times, collecting 16 samples per sweep, for a total of 512 samples of $R(\tau)$. The delay resolution is 19.5 m, but the large number of samples taken reduces the quantization error to negligible proportions. With a dwell time of 1 ms per sample, the time to compute one satellite range measurement is 0.512 second. One complete sequence of four satellites takes 2.048 seconds.

Next the performance of the ranging estimator is determined. The best available C/N_0 is taken from the previous example as 48.4 dB/Hz. In a 3-kHz bandwidth, the resulting signal-to-noise ratio is 13.6 dB. The resulting estimation accuracy is determined by Monte Carlo simulation of the range-processing algorithms to be $\sigma_r = 10.8$ m.

Finally an error budget is developed to compute the position uncertainty, σ_p, previously assumed. The performance of the navigation algorithms adds 5.8 m to the range estimator error of 10.8 m. Daytime ionospheric delay is taken at an additional 20-m error in final position. Tropospheric error is

TABLE 54.6. HYPOTHETICAL DESIGN EXAMPLE—AIRCRAFT, SEQUENTIAL, L1, CLEAR-ACCESS ONLY, INCOHERENT RANGING

Parameter	Value
Doppler Search	
Velocity uncertainty	$\sigma_v = 3$ m/s
Doppler search range ($\pm 2\sigma_v$)	63 Hz
Oscillator instability ($\pm \sigma_{IF}$)	312 Hz
Total frequency uncertainty	375 Hz
Range filter bandwidth	3000 Hz
Delay Search	
Position uncertainty, spherical	$\sigma_p = 48.6$ m
Delay search window ($\pm 3\sigma_p$)	293 m
Code step interval/samples per sweep	19.5 m/16
Total sweeps/total samples	32/512
Sample dwell time	1 ms
Satellite ranging time	0.512 s
Fix time (four satellites)	2.048 s
Range Estimator Signal-to-Noise Ratio (SNR)	
Available C/N_0 (best case)	48.4 dB/Hz
Required bandwidth	34.8 dB Hz
Available ranging SNR	13.6 dB
Ranging accuracy	$\sigma_r = 10.8$ m
Error Budget	
Range estimator	10.8 m
GDOP-NAV filter contribution (GDOP[a] = 2.0)	5.8 m
Ionosphere contribution	20.0 m
Satellite ephemeris contribution	11.0 m
Troposphere contribution	1.0 m
Total position uncertainty	$\sigma_p = 48.6$ m

[a] GDOP, geometric dilution of precision.

taken as 1.0 m. Worst case satellite ephemeris error of 11.0 m is taken. The sum total position uncertainty is then 48.6 m.

It is seen from this hypothetical example that a design case is an iterative process. Since some of the design performance can be determined only by computer simulation, the design must be "fine tuned" until all entries are consistent. The design is obviously optimized for one particular value of C/N_0.

Navigation Processing: Linearization and Geometric Dilution of Precision (GDOP)

The navigation processor in a GPS user-receiver is a major component. In this and the following sections, we attempt to show some of the computations that must be made. Naturally the processor embodies one or more digital computers of the microprocessor variety. Studies[16] have shown that a digital word length of 40 bits (32-bit precision) is a necessity to maintain computational roundoff at a low value.

The purpose of the navigation processor is to accept each pseudorange measurement as it is obtained and to apply it to the solution for the user's coordinate vector, consisting of x, y, z, and b (clock bias) in the ECEF system. In so doing, the processor must solve the Keplerian equations to determine the position vector for each satellite being received. The processor also determines which satellites are visible and, of those, which four to choose for best navigation geometry. The processor usually incorporates a filter that minimizes the statistical average of the sum of the squares of errors due to system noise and user dynamic maneuvers. Moreover, this filter automatically adjusts its parameters to accommodate sensed changes in the signal-to-noise environment and navigation geometry.

Referring back to Eqs. (54.2) through (54.6), the measurements in GPS are pseudoranges, R_i, for $i = 1, 2, 3, 4$. We have

$$R_i = f_i(x, y, z, b) = \sqrt{(x_i - x)^2 + (y_i - y)^2 + (z_i - z)^2} + b \tag{54.57}$$

where the ith satellite coordinates are x_i, y_i, z_i, the user coordinates are x, y, z, and user clock bias in meters is b.

As in Eq. (54.7), the problem is linearized by assuming a user position and clock bias $\hat{x}, \hat{y}, \hat{z}, \hat{b}$ approximately equal to the true quantities, and the corresponding $\hat{R}_i$, as calculated from Eq. (54.57). Then we make the first-order linear expansion as

$$R_i \cong = \hat{R}_i + \frac{\partial f_i(\ )}{\partial x}\bigg]_{\hat{x}} \cdot (x - \hat{x}) + \frac{\partial f_i(\ )}{\partial y}\bigg]_{\hat{y}} \cdot (y - \hat{y}) + \frac{\partial f_i(\ )}{\partial z}\bigg]_{\hat{z}} \cdot (z - \hat{z}) + \frac{\partial f_i(\ )}{\partial b}\bigg]_{\hat{b}} \cdot (b - \hat{b}) \tag{54.58}$$

We may write the equations for four satellites in vector matrix form as

$$\begin{bmatrix} R_1 - \hat{R}_1 \\ R_2 - \hat{R}_2 \\ R_3 - \hat{R}_3 \\ R_4 - \hat{R}_4 \end{bmatrix} = H \cdot \begin{bmatrix} x - \hat{x} \\ y - \hat{y} \\ z - \hat{z} \\ b - \hat{b} \end{bmatrix} \tag{54.59}$$

where H is the square Jacobian matrix of ordered partial derivatives, as given in Eq. (54.58). Now, for example, we have

$$\frac{\partial f_i(\ )}{\partial x} = \frac{(x - x_i)}{\sqrt{(x_i - x)^2 + (y_i - y)^2 + (z_i - z)^2}} \qquad i = 1, 2, 3, 4 \tag{54.60}$$

$$\frac{\partial f_i(\ )}{\partial b} = 1 \qquad i = 1, 2, 3, 4$$

and so the elements of the matrix H are direction cosines for the directed distance from user position to satellite position (see Fig. 54.14). Let α, β, γ be the angles of $\bar{R}_i$, the directed distance, with respect

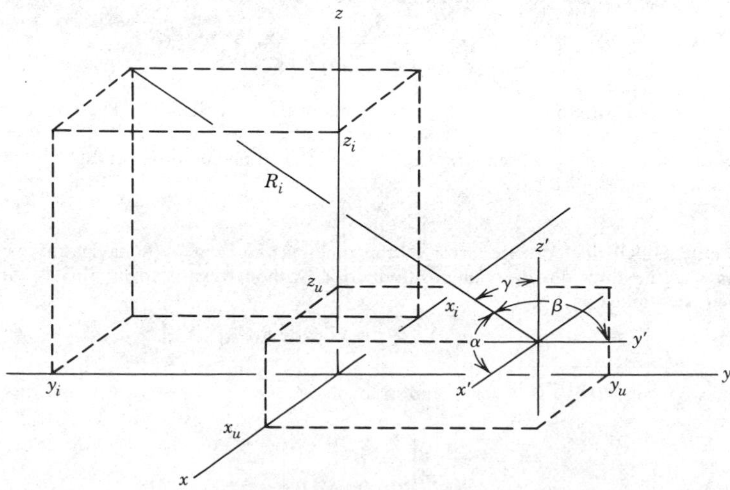

Fig. 54.14 Relative geometry.

to the x, y, z axes, respectively. Then

$$H = \begin{bmatrix} \cos\alpha_1 & \cos\beta_1 & \cos\gamma_1 & 1 \\ \cos\alpha_2 & \cos\beta_2 & \cos\gamma_2 & 1 \\ \cos\alpha_3 & \cos\beta_3 & \cos\gamma_3 & 1 \\ \cos\alpha_4 & \cos\beta_4 & \cos\gamma_4 & 1 \end{bmatrix} \tag{54.61}$$

Observing the structure of the H matrix we can see two cases in which H would not be invertible. One would be if the user was colinear with two satellites. The other is if the four satellites were all on the surface of a cone.

For the nominal case, when H is invertible, we may solve for the user's coordinate "vector" as

$$\begin{bmatrix} x \\ y \\ z \\ b \end{bmatrix} = \begin{bmatrix} \hat{x} \\ \hat{y} \\ \hat{z} \\ \hat{b} \end{bmatrix} + H^{-1} \cdot \left\{ \begin{bmatrix} R_1 \\ R_2 \\ R_3 \\ R_4 \end{bmatrix} - \begin{bmatrix} \hat{R}_1 \\ \hat{R}_2 \\ \hat{R}_3 \\ \hat{R}_4 \end{bmatrix} \right\} \tag{54.62}$$

We may define position and range vectors and their differential forms formally as

$$\mathbf{X} = \begin{vmatrix} x \\ y \\ z \\ b \end{vmatrix}, \qquad \mathbf{R} = \begin{vmatrix} R_1 \\ R_2 \\ R_3 \\ R_4 \end{vmatrix}; \quad \begin{matrix} \boldsymbol{\delta}\mathbf{x} = \mathbf{x} - \hat{\mathbf{x}} \\ \boldsymbol{\delta}\mathbf{R} = \mathbf{R} - \hat{\mathbf{R}} \end{matrix} \tag{54.63}$$

Then from Eq. (54.62) we have

$$\boldsymbol{\delta}\mathbf{x} = H^{-1}\boldsymbol{\delta}\mathbf{R} \tag{54.64}$$

which is the differential form of the linearized solution in vector-matrix notation.

Suppose now that we assume that the differential range vector, $\boldsymbol{\delta}\mathbf{R}$, is subject to some random statistical error (noise). Let us assume that each element of $\boldsymbol{\delta}\mathbf{R}$ is subject to a random noise of standard deviation $\sigma_{\delta R}$ and that the four noises are statistically independent. The validity of this assumption depends on the mechanisms for generating the pseudoranges, R_i, and their linearizing estimates, $\hat{R}_i$.

Under these assumptions we may form the covariance matrix for the differential solution, $\boldsymbol{\delta}\mathbf{x}$, as

$$\operatorname{var}\{\boldsymbol{\delta}\mathbf{x}\} = \operatorname{var}\{H^{-1}\boldsymbol{\delta}\mathbf{R}\} = \sigma_{\delta R}^2 \cdot [H^T H]^{-1} \tag{54.65}$$

Now let us define the standard deviation of the differential solution as

$$\sigma_{\delta P} = \sqrt{\sigma_{\delta x}^2 + \sigma_{\delta y}^2 + \sigma_{\delta z}^2 + \sigma_{\delta b}^2} \tag{54.66}$$

The elements $\sigma_{\delta x}^2$, $\sigma_{\delta y}^2$, $\sigma_{\delta z}^2$, and $\sigma_{\delta b}^2$ are the main diagonal elements of the covariance matrix. Thus we have

$$\sigma_{\delta p} = \sigma_{\delta R}\sqrt{\operatorname{trace}\left[(H^T H)^{-1}\right]} \tag{54.67}$$

where trace $(\cdot)$ is the matrix operator that sums the main diagonal elements. The quantity under the radical has an important name in GPS. It is called GDOP, for geometric dilution of precision. It is the quantity that relates error in pseudorange to error in position solution, through the goemetry as expressed in the H matrix. We have

$$\sigma_{\delta p} = \text{GDOP} \cdot \sigma_{\delta R} \tag{54.68}$$

The quantity GDOP includes the error in the time-bias element b. A navigator may be more interested in just the three-dimensional position error, without regard to b. This is the so-called PDOP, which is obtained as

$$\sigma(\delta_x, \delta_y, \delta_z) = \text{PDOP} \cdot \sigma_{\delta R} \tag{54.69}$$

where $\text{PDOP} = \sqrt{\operatorname{trace}\{D[H^T H]^{-1} D\}}$ and

$$D = \begin{bmatrix} 1 & 0 & 0 & 0 \\ 0 & 1 & 0 & 0 \\ 0 & 0 & 1 & 0 \\ 0 & 0 & 0 & 0 \end{bmatrix}$$

The GDOP factor is used in selecting the four best satellites for the navigation problem from those in view. The optimum navigation performance is obtained for the minimum value of GDOP, which is always positive. The theoretical minimum GDOP occurs for four satellites, each separated from all others by an angle of 120°. This value is 1.62. Unfortunately, for an earth-surface navigator, three of the optimum satellites would be below the horizon from the navigator's point of view.

Surprisingly enough, adding the constraint of satellite minimum visibility does not increase GDOP much. For the case of three satellites with 0° elevation angles, separated in azimuth by 120°, and a fourth overhead satellite, GDOP = 1.74.

It is obviously necessary that satellites be selected to minimize GDOP, and this is one task of the navigation processor. The question arises as to how best to do the selection with minimum computational load on the processor. The early "brute-force" method was to compute the number of visible satellites, n, and then compute GDOP for all sets of four satellites out of n visible. The number of such computations is

$$N \text{ comp} = \binom{n}{4} = \frac{n!}{(n-4)!\,4!} \tag{54.70}$$

For example, if $n = 7$, then N comp = 35.

It turns out that there are more practical selection algorithms which still yield good GDOP.[16] For instance, suppose the navigator can choose satellites on track, cross track, and overhead (three-orthogonal) and take the fourth above the horizon in the direction of the vector resultant of the first three. For this combination, GDOP = 1.67. This algorithm requires $4(n-1)$ computations, which is 24 for seven satellites, and the required computations are not as complex as in the "optimal" method. In computer simulation for a practical case, the average GDOP over six hours was 2.64 for the optimum selection algorithm and 2.92 for the latter algorithm.[17]

Kalman Filtering: Standard and Adaptive

Given four simultaneous pseudorange measurements, four computed satellite positions at the same instant, and the required estimated coordinate vector and pseudorange vector, the navigator's position may be determined from Eq. (54.62) on a point-by-point basis. There are practical difficulties, of course, in having all of these necessary pieces of data at the same instant every time a position solution is desired. But these computational practicalities are not dealt with here, in order to not obscure the basic principles involved.

The main question, at this point, is whether Eq. (54.62) yields sufficiently accurate results or whether some filtering or smoothing of the solutions is necessary. The answer depends entirely on the accuracy desired. In the C/A design example of Table 54.6, using an unfiltered solution would increase the position uncertainty only an additional 10%, for a GDOP of 2.0. However, the example was for the very best signal-to-noise ratio case and better than average GDOP. Applying the worst case C/N_0 tolerance of -12.9 dB from Table 54.5 would increase the C/A-only position uncertainty to the neighborhood of 650 m without filtering and for the average GDOP = 3.0. Such performance would probably be unacceptable to most navigational users.

A navigational filter is very useful for a vehicle that spends much of its time in uniform motion (unaccelerated). In this case a filter may be devised that reduces the effects of noise on the pseudorange measurement almost without limit. However, when the vehicle is subject to acceleration and higher-order motion, as in a turn, the situation is not as good. The filter output will not necessarily track a nonuniform motion input without error. An "optimum" filter is designed to "split the difference" between errors due to measurement noise and errors due to filter mistracking of nonuniform vehicle motion. That is, an optimum filter design attempts to equalize some measure of the errors due to the two different sources. The measure usually is the statistical average of the squared total error.

If a vehicle were subject only to motions whose graphs were describable by finite-order polynomials, theoretically a filter could be devised to track such motion without error. Unfortunately, motion as simple as a circular turn is described by sinusoids, whose higher derivatives never terminate, rather than finite polynomials. Thus, errorless filters are not possible.

A popular approach to the filtering problem is to describe the vehicle motion as a sample function from some family of random functions. The ranging noise is similarly described. In the case where both vehicle motion and noise are describable as Gaussian random functions, there exists a linear filter that minimizes, over all such functions, the statistical average of the square of the instantaneous total tracking error. This filter is variously called a Kalman filter or a Wiener filter. The Kalman filter is optimum from the instant it is activated. The Wiener filter, which is historically older, is optimum only in the steady state, after the "turn-on transient" has died away.

Both the polynomial filters and the Kalman filters require that the class of inputs for which the filter is expected to be optimum shall be mathematically modeled. The filter design is then based on that specific input model. In the polynomial case, if the input is describable as a fifth-order polynomial, then the filter order (number of poles) is made commensurate to ensure optimum dynamic response to the function being tracked. In the Kalman case, a generating filter is actually specified for generating the class of inputs to be tracked. Elements of the generating filter end up being incorporated into the Kalman filter structure.

The starting place for modeling the class of inputs is to define the "state vector" to be tracked. Heretofore we have defined the "position" vector of the user as being composed of the elements x, y, z, and b. But this is not sufficient for modeling purposes. To generate a higher order input model that will result in a higher order filter, we will also define time derivatives of the position components. In GPS receivers it is common practice to model position, velocity, and acceleration for x, y, and z and to model b and $\dot{b}$ for clock bias. The reason for limiting the model for b to just its first derivative is that b results from a clock oscillator whose frequency is inaccurate. There exists little acceleration in b. The same is not true for x, y, and z.

The order in which the various elements are inserted into the 11-state vector is not arbitrary and has computational consequences. However, one particular order lends itself more easily to explanation. The two popular definitions of the state vector are

$$\mathbf{x}^T = [x, y, z, \dot{x}, \dot{y}, \dot{z}, \ddot{x}, \ddot{y}, \ddot{z}, b, \dot{b}] \tag{54.71a}$$

$$\mathbf{x}^T = [x, \dot{x}, \ddot{x}, y, \dot{y}, \ddot{y}, z, \dot{z}, \ddot{z}, b, \dot{b}] \tag{54.71b}$$

The vector-matrix generating model for x, y, z, b is, for either state-vector definition, given canonically as

$$\begin{aligned} \dot{\mathbf{x}}(t) &= A\mathbf{x}(t) + B\mathbf{w}(t) \\ [x, y, z, b]^T &= \Gamma\mathbf{x}(t) \end{aligned} \tag{54.72}$$

For the state-vector definition of Eq. (54.71a), the coupling matrix A will be upper triangular, and this will result in some computational advantages in terms of 11×11 matrices. For the state-vector definition of Eq. (54.71b), the generally 11×11 vector-matrix equation of Eq. (54.71a) breaks apart into four independent equations, three of which are 3×3 and the one involving b is 2×2. That this is true is shown as follows.

The driving function, $\mathbf{w}(t)$, in Eq. (54.72) is a vector

$$\mathbf{w}^T(t) = [w_x(t), w_y(t), w_z(t), w_b(t)] \tag{54.73}$$

where the four components are independent of each other and in time, are of zero mean and known variance, and are Gaussian. The structure of the input matrix B is

$$B = \begin{bmatrix} 0 & & & \\ 0 & & 0 & \\ b_x & & & \\ & 0 & & \\ & 0 & & \\ & b_y & & \\ & & 0 & \\ & & 0 & \\ 0 & & b_z & \\ & & & 0 \\ & & & b_b \end{bmatrix} \tag{54.74}$$

The structure of the coupling matrix A is

$$A = \begin{bmatrix} 0 & 1 & 0 & & & & & & & & \\ 0 & 0 & 1 & & & & & & & & \\ 0 & 0 & 0 & & & & & & & & \\ & & & 0 & 1 & 0 & & & & & \\ & & & 0 & 0 & 1 & & & & & \\ & & & 0 & 0 & 0 & & & & & \\ & & & & & & 0 & 1 & 0 & & \\ & & & & & & 0 & 0 & 1 & & \\ & & & & & & 0 & 0 & 0 & & \\ & & & & & & & & & 0 & 1 \\ & & & & & & & & & 0 & 0 \end{bmatrix} \tag{54.75}$$

From Eqs. (54.74) and (54.75), it can be seen that Eq. (54.72) reduces to four equations

$$\begin{aligned}\dot{\mathbf{X}}_x(t) &= A_x\mathbf{X}_x(t) + \mathbf{b}_x W_x(t)\\ \dot{\mathbf{X}}_y(t) &= A_y\mathbf{X}_y(t) + \mathbf{b}_y W_y(t)\\ \dot{\mathbf{X}}_z(t) &= A_z\mathbf{X}_z(t) + \mathbf{b}_z W_z(t)\\ \dot{\mathbf{X}}_b(t) &= A_b\mathbf{X}_b(t) + \mathbf{b}_b W_b(t)\end{aligned} \tag{54.76}$$

where each of the A matrices and $\mathbf{b}$ vectors are of the canonical form

$$A = \begin{bmatrix} 0 & 1 & 0\\ 0 & 0 & 1\\ 0 & 0 & 0\end{bmatrix}, \qquad \mathbf{b} = \begin{bmatrix} 0\\ 0\\ b\end{bmatrix} \tag{54.77}$$

For x, y, z, A is 3×3 and $\mathbf{b}$ is 3×1. For the coordinate b, A is 2×2 and $\mathbf{b}$ is 2×1. The coefficient b in $\mathbf{b}$ should not be confused with the coordinate b. Shortly we will do away with this possibility for confusion.

What makes the decomposition of Eq. (54.76) possible is that the coupling matrix A in Eq. (54.75) shows no coupling between the states of the coordinates x, y, z, and b. That is, each of those coordinates is generated independently of the other. Now this is an assumption. We know that in reality it is generally violated. For example, in a turn in the x-y plane, $x(t)$ is proportional to a sine function and $y(t)$ is proportional to a cosine. In this case the x and y coordinates are certainly not independent. The problem is that in accepting the Kalman assumption that $x(t)$ and $y(t)$ are random, we have no a priori knowledge of their joint correlation, or coupling, unless we have separate aiding information, such as from an inertial measurement unit. Thus we simply assume zeros for the cross-coupling coefficients in the A matrix of Eq. (54.75) and the decomposition of Eq. (54.76) occurs.

Now we may restrict our attention to a reduced order 3×3 system for any one of the coordinates, say x. We have

$$\begin{bmatrix}\dot{x}(t)\\ \dot{v}(t)\\ \dot{a}(t)\end{bmatrix} = \begin{bmatrix} 0 & 1 & 0\\ 0 & 0 & 1\\ 0 & 0 & 0\end{bmatrix}\begin{bmatrix} x(t)\\ v(t)\\ a(t)\end{bmatrix} + \begin{bmatrix} 0\\ 0\\ 1\end{bmatrix} j(t) \tag{54.78}$$

$$\dot{\mathbf{x}} = A \cdot \mathbf{x} + \mathbf{b} \cdot j$$

where now x denotes position, v denotes velocity, a denotes acceleration, and j denotes jerk, the first derivative of acceleration.

The Kalman filter is implemented in sampled-data form. Therefore the vector-matrix differential equation [Eq. (54.78)] must be converted to a discrete-time difference equation. This is done by assuming that the input $j(t)$ is sampled every t seconds and directed to the input of the generating filter through a zero-order hold circuit. The output of the filter is then observed only at sampling times. With this transformation the discrete-time generating filter equations become

$$\begin{bmatrix} x(k+1)\\ v(k+1)\\ a(k+1)\end{bmatrix} = \mathbf{\Phi} \cdot \begin{bmatrix} x(k)\\ v(k)\\ a(k)\end{bmatrix} + \gamma j(k) \tag{54.79}$$

where now k is sample number and $\mathbf{\Phi}$ and γ are given by

$$\begin{aligned}\mathbf{\Phi} &= \mathbf{exp}(\mathbf{A}T)\\ \gamma &= \int_0^T \mathbf{exp}(\mathbf{A}q)\, dq \cdot \mathbf{b}\end{aligned} \tag{54.80}$$

In Eq. (54.80), the quantity $\exp(Aq)$ is the exponent function of a 3×3 matrix and is itself a 3×3 matrix. Since the matrix A is a Jordan-block matrix, the exponent form is

$$\mathbf{exp}(\mathbf{A}T) = \begin{vmatrix} e^{\lambda T} & Te^{\lambda T} & \frac{T^2}{2}e^{\lambda T}\\ 0 & e^{\lambda} & Te^{\lambda T}\\ 0 & 0 & e^{\lambda T}\end{vmatrix} \tag{54.81}$$

where λ is the single repeated eigenvalue of the matrix A. Since $\lambda = 0$ by inspection, we have

$$\begin{aligned}\mathbf{\Phi} = \exp(AT) &= \begin{vmatrix} 1 & T & T^2/2\\ 0 & 1 & T\\ 0 & 0 & 1\end{vmatrix}\\ \gamma &= \begin{vmatrix} T^3/6\\ T^2/2\\ T\end{vmatrix}\end{aligned} \tag{54.82}$$

We may now write Eq. (54.79) in coupled scalar-equation form as

$$\begin{aligned} x(k+1) &= x(k) + T \cdot v(k) + T^2/2 \cdot a(k) + T^3/6 \cdot j(k) \\ v(k+1) &= v(k) + T \cdot a(k) + T^2/2 \cdot j(k) \\ a(a+1) &= a(k) + T \cdot j(k) \end{aligned} \tag{54.83}$$

Under the standard Kalman assumptions, the data input to the filter, $z(k)$, consists of coordinate to be tracked $x(k)$ plus noise $n(k)$. It is assumed that the noise $n(k)$ is independent of $j(k)$ and is zero mean, white, Gaussian, and of known variance σ_n^2. Then the data form is

$$z(k) = x(k) + n(k) \tag{54.84}$$

The Kalman filter equations are given here together with the generator equations and data form, for completeness.

Generator

$$\mathbf{x}(k+1) = \mathbf{\Phi x}(k) + \gamma j(k)$$
$$x(k) = \boldsymbol{\lambda}^T \mathbf{x}(k);\ \boldsymbol{\lambda}^T = [1, 0, 0]$$

Data

$$z(k) = x(k) + n(k) \tag{54.85}$$

Filter

$$\hat{x}(k) = \boldsymbol{\lambda}^T \hat{\mathbf{x}}(k)$$
$$\hat{\mathbf{x}}(k) = \mathbf{g}(k)[z(k) - \boldsymbol{\lambda}^T \mathbf{\Phi} \hat{\mathbf{x}}(k-1)] + \mathbf{\Phi} \hat{\mathbf{x}}(k-1)$$

In Eq. (54.85), the quantity $\mathbf{g}(k)$ is a gain vector, defined by

$$\mathbf{g}^T(k) = [g_x(k), g_v(k), g_a(k)] \tag{54.86}$$

$\mathbf{g}(k)$ can be time varying. Its method of calculation is detailed as follows.

To obtain understanding of the operation of the filter, the vector-matrix equations of Eq. (54.85) are written in coupled scalar form as

$$\begin{aligned} \hat{x}(k) &= \hat{x}(k-1) + T \cdot \hat{v}(k-1) + T^2/2 \cdot \hat{a}(k-1) + g_x(k) \cdot \epsilon(k) \\ \hat{v}(k) &= \hat{v}(k-1) + T \cdot \hat{a}(k-1) + g_v(k) \cdot \epsilon(k) \\ \hat{a}(k) &= \hat{a}(k-1) + g_a(k) \cdot \epsilon(k) \\ \epsilon(k) &= z(k) - [\hat{x}(k-1) + T\hat{v}(k-1) + T^2/2\hat{a}(k-1)] \end{aligned} \tag{54.87}$$

Figure 54.15 shows the filter structure.

It is seen from Eq. (54.87) and the figure that each new estimate consists of a prediction, based on the old estimates, plus a correction, based on the new data. The correction is a weighted-error term where the error is data minus predicted position. The weighting factors are the "Kalman gains." The error term, $\epsilon(k)$, is called filter residual or innovation.

In the standard Kalman filter, the gain vector, $\mathbf{g}(k)$, is computed from a set of coupled vector-matrix equations given as follows. These equations are based on the covariance matrix of the filtering error, where

Filtering error

$$\tilde{\mathbf{x}}(k) = \mathbf{x}(k) - \hat{\mathbf{x}}(k)$$

Covariance matrix

$$\boldsymbol{v}(k) = E[\tilde{\mathbf{x}}(k)\tilde{\mathbf{x}}^T(k)] \tag{54.88}$$

where $E[\cdot]$ is the statistical average and under the assumption that $\mathbf{x}(k)$ and $\hat{\mathbf{x}}(k)$ both have zero mean values. The gain equations are

$$\begin{aligned} \boldsymbol{v}(k|k-1) &= \Phi \boldsymbol{v}(k-1)\Phi^T + \sigma_j^2 \gamma\gamma^T \\ \mathbf{g}(k) &= \boldsymbol{v}(k|k-1)\boldsymbol{\lambda}[\sigma_n^2 + \boldsymbol{\lambda}^T \boldsymbol{v}(k|k-1)\boldsymbol{\lambda}]^{-1} \\ \boldsymbol{v}(k) &= [I - \mathbf{g}(k)\boldsymbol{\lambda}^T]\boldsymbol{v}(k|k-1) \end{aligned} \tag{54.89}$$

In Eq. (54.89), $\boldsymbol{v}(k)$ is covariance matrix computed using the $\hat{\mathbf{x}}(k|k-1)$ one-step predicted estimate.

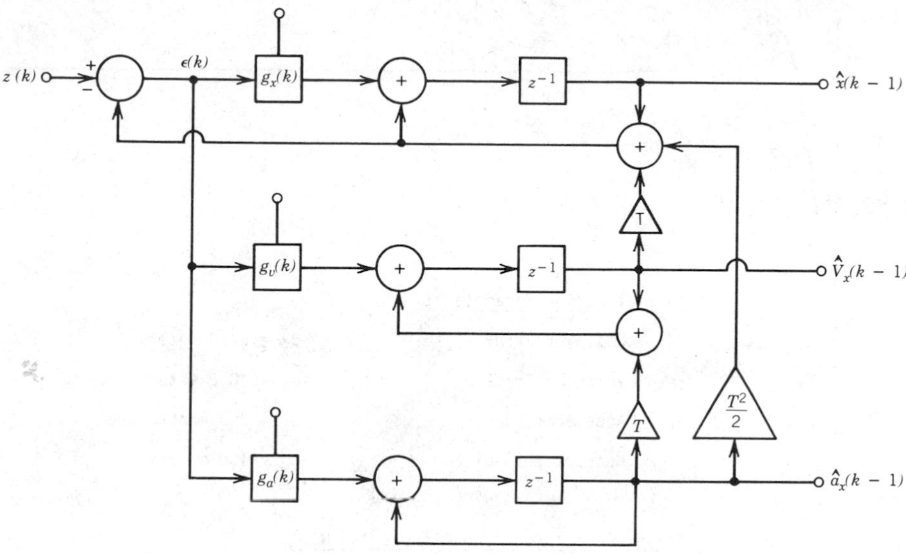

Fig. 54.15 Kalman filter.

σ_j^2 and σ_n^2 are the known variances of the modeled jerk and additive white noise processes, respectively.

The $\mathbf{g}(k)$ computation is recursive and is initialized with a value $\mathbf{v}(k=0)$. Usually this is the steady-state variance matrix for the process, $\mathbf{x}(k)$. However, in the present navigation case, with the given structure of the $\mathbf{\Phi}$ matrix, the variance of $\mathbf{x}(k)$ grows without bound. Thus some other ad hoc initialization should be used for $\mathbf{v}(k)$.

All the gain elements of $\mathbf{g}(k)$ rapidly approach steady-state values. In the steady state, the Kalman filter is just the optimum Wiener filter. The steady-state gains are a function of T, the sampling period, and the ratio $\sigma_j^2/\sigma_n^2 = \text{SNR}$, which is a signal-to-noise ratio. The greater the ratio, the greater the gains, and vice versa. The greater the gains, the more responsive the filter is to abrupt changes (maneuvering) in the input, but the more white measurement noise is accepted by the filter. The lower the gains, the more sluggish the filter becomes, but the less noisy becomes the output. Figures 54.16 and 54.17 show examples of the evolution of the gains and of the steady-state dependence on signal-to-noise ratio.

A problem with using a Kalman filter in the navigation problem is that the SNR ratio changes drastically, dependent on vehicle maneuvering. During unaccelerated motion, SNR$\rightarrow 0$ and the optimum steady-state gains are very small, resulting in little noise on the filter output. During accelerated motion, SNR increases and the optimum steady-state gains are larger, resulting in more white noise acceptance but lower dynamic filter error.

For example, a 300-knot aircraft subject to ± 3 m normal control excursions with a 30-second period and 10.8-m ranging noise results in SNR $= 6.50 \times 10^{-6}$. During a standard rate two-minute turn, SNR $= 1.54 \times 10^{-3}$, a +30-dB change. The standard Kalman gain computation has no knowledge of these radically different environments for filter operation. Thus to use the filter effectively, an "adaptive" gain algorithm must be employed.

There exists no well-developed body of theory for adaptive-gain filters comparable to Kalman-Wiener theory. Thus developments in this area are all ad hoc and for special cases. The only basic requirement is that the filter residual, $\epsilon(k)$, be used to provide information for adjusting the filter gains. The gains may not be adjusted independently. Rather they must be adjusted consistent with results that are produced by solution of the gain equations. Otherwise filter instability may result.

Extended Kalman Filtering for GPS

There is one final problem in applying the Kalman filter to the GPS navigation problem: the noisy navigation measurements, $R_i(k)$, do not constitute the correct data form for the Kalman filter. The filter requires noisy measurements of the position coordinates. The pseudorange measurements, however, are nonlinear transformations of the position coordinates. An approach to this problem is to

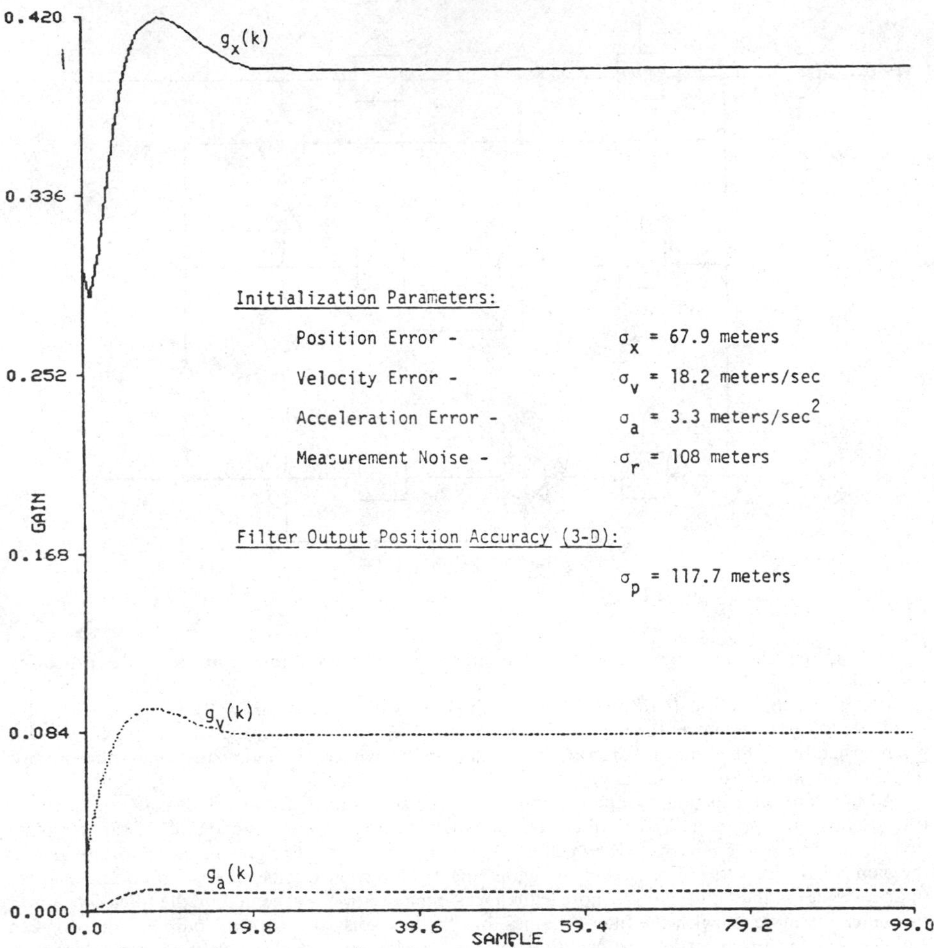

Fig. 54.16 Evolution of gains. From "The NAVSTAR Global Positioning System—A User's Approach to Understanding," a short course by Altair Corporation, College Station, Texas, with permission.

use the linearization developed previously in this chapter and embodied in Eq. (54.64). That relation is expanded here as

$$\mathbf{x}(k) - \hat{\mathbf{x}}(k) = H^{-1}[\mathbf{R}(k) - \hat{\mathbf{R}}(k)] \tag{54.90}$$

where $\mathbf{x}(k)$ = true position vector

$\mathbf{x}^T(k) = [x, y, z, b]$

$\hat{\mathbf{x}}(k)$ = position assumed for purposes of computing H^{-1}

$\mathbf{R}(k)$ = vector of four measured pseudoranges

$\hat{\mathbf{R}}(k)$ = prediction of $\mathbf{R}(k)$ computed using $\hat{\mathbf{x}}(k)$ and nonlinear function of Eq. (54.57)

Because $\mathbf{R}(k)$ has additive white noise, Eq. (54.90) will add white noise to $\mathbf{x}(k)$ through the linear transformation $\overline{H}$. However, Eq. (54.90) gives a differential transformation from $\mathbf{R}$ to $\mathbf{x}$, rather than a global transformation. So the question remains, how to use Eq. (54.90) to provide the input data for the Kalman filters.

Suppose that in Eq. (54.90) at sample number k, the assumed position vector $\hat{\mathbf{x}}(k)$ is made up of the Kalman filter predicted position components, which are all of the same form as that for the x component, given as

$$\hat{\mathbf{x}}(k|k-1) = \boldsymbol{\lambda}^T \Phi \hat{\mathbf{x}}(k-1) \tag{54.91}$$

Now we may rewrite Eq. (54.90), showing explicitly the additive noise contribution due to $H^{-1}\mathbf{R}(k)$,

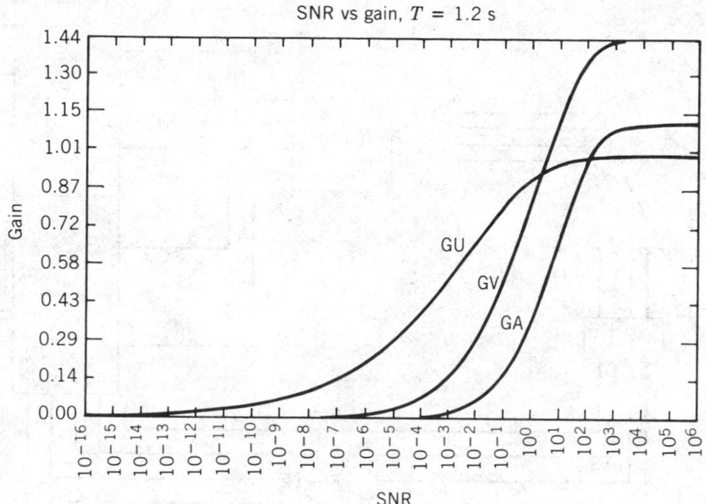

Fig. 54.17 Signal-to-noise ratio (SNR) dependence of gains. From "The NAVSTAR Global Positioning System—A User's Approach to Understanding," a short course by Altair Corporation, College Station, Texas, with permission.

as

$$\begin{aligned}\mathbf{x}(k) + \mathbf{n}(k) - \hat{\mathbf{x}}(k|k-1) &= H^{-1}[\mathbf{R}(k) - \hat{\mathbf{R}}(k|k-1)] \\ &= z(k) - \hat{\mathbf{x}}(k|k-1) = H^{-1}[\mathbf{R}(k) - \hat{\mathbf{R}}(k|k-1)] \\ &= \boldsymbol{\epsilon}(k) = H^{-1}[\mathbf{R}(k) - \hat{\mathbf{R}}(k|k-1)]\end{aligned} \tag{54.92}$$

That is, we may identify the quantity $H^{-1}[\mathbf{R}(k) - \hat{\mathbf{R}}(k|k-1)]$ as the four-vector error term in an 11-state Kalman filter.

Because of our choice of the coupling matrix, Φ, in the generating model, that model is reduced from an 11-state model to four independent models having 3, 3, 3, and 2 states, respectively, based on the state vector description of Eq. (54.71b). The Kalman filter may be similarly composed of four independent sections, each driven by one of the four elements of the error vector, $\boldsymbol{\epsilon}(k)$. The extended Kalman filter, so realized, is diagrammed in Fig. 54.18.

Although the structure of the Kalman filter is successfully decoupled so far as computation is concerned, the position estimates are not independent, since errors in x, y, z, and b are cross coupled by $\mathbf{f}(\ )$ and H^{-1}. Another effect is that the white noise variances in the inputs to the x, y, z, and b filters are not necessarily equal or fixed, since they are dependent on H^{-1}, which is variable with the geometry.

$$H^{-1} = \begin{bmatrix} lx_1 & lx_2 & lx_3 & lx_4 \\ ly_1 & ly_2 & ly_3 & ly_4 \\ lz_1 & lz_2 & lz_3 & lz_4 \\ lb_1 & lb_2 & lb_3 & lb_4 \end{bmatrix} \tag{54.93}$$

Let the additive noise in $\mathbf{R}(k)$ be explicitly indicated by

$$\mathbf{R}(k) = \begin{bmatrix} R_1(k) \\ R_2(k) \\ R_3(k) \\ R_4(k) \end{bmatrix} + \begin{bmatrix} n_1(k) \\ n_2(k) \\ n_3(k) \\ n_4(k) \end{bmatrix} \tag{54.94}$$

Then the noises effective in x, y, z, and b are given by

$$\begin{aligned} n_x(k) &= \sum_{i=1}^{4} l_{xi} n_i(k) \\ &\vdots \\ n_b(k) &= \sum_{i=1}^{4} l_{bi} n_i(k) \end{aligned} \tag{54.95}$$

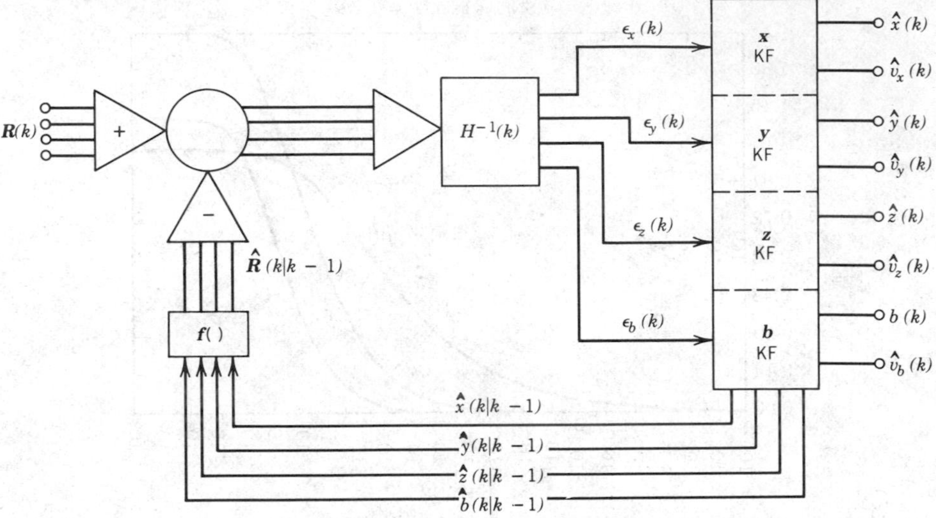

Fig. 54.18 Extended Kalman filter (KF). From "The NAVSTAR Global Positioning System—A User's Approach to Understanding," a short course by Altair Corporation, College Station, Texas, with permission.

It is assumed that the range measurement noises, $n_i(k)$ are zero mean, independent, and of equal variance σ_r^2. Then the variances of $n_x(k)$, $n_y(k)$, $n_z(k)$, and $n_b(k)$ are

$$\sigma_{nx}^2 = \sigma_r^2 \cdot \sum_{i=1}^{4} l_{xi}^2$$
$$\vdots \tag{54.96}$$
$$\sigma_{nb}^2 = \sigma_r^2 \cdot \sum_{i=1}^{4} l_{bi}^2$$

This development of the extended Kalman filter has assumed that all four pseudorange measurements are available at the same instant and that the total filter is cycled with four available pseudoranges. This is, of course, not the case with a sequential receiver. Let us examine Eq. (54.91) in more detail.

$$\begin{bmatrix} x-\hat{x} \\ y-\hat{y} \\ z-\hat{z} \\ b-\hat{b} \end{bmatrix} = \begin{bmatrix} l_{x1}l_{x2}l_{x3}l_{x4} \\ l_{y1}l_{y2}l_{y3}l_{y4} \\ l_{z1}l_{z2}l_{z3}l_{z4} \\ l_{b1}l_{b2}l_{b3}l_{b4} \end{bmatrix} \cdot \begin{bmatrix} R_1-\hat{R}_1 \\ R_2-\hat{R}_3 \\ R_3-\hat{R}_3 \\ R_4-\hat{R}_4 \end{bmatrix}$$
$$= \begin{bmatrix} l_{x1}(R_1-\hat{R}_1) \\ l_{y1}(R_1-\hat{R}_1) \\ l_{z1}(R_1-\hat{R}_1) \\ l_{b1}(R_1-\hat{R}_1) \end{bmatrix} + \cdots + \begin{bmatrix} l_{x4}(R_4-\hat{R}_4) \\ l_{y4}(R_4-\hat{R}_4) \\ l_{z4}(R_4-\hat{R}_4) \\ l_{b4}(R_4-\hat{R}_4) \end{bmatrix} \tag{54.97}$$

Equation (54.97) shows that the coefficient l_{x1} brings the differential range component $(R_1 - \hat{R}_1)$ into the filter to correct the x coordinate. Likewise l_{y1} uses $(R_1 - \hat{R}_1)$ to correct the y coordinate, and so on. The first column of H^{-1} thus corrects all four position coordinates from the information supplied by the $(R_1 - \hat{R}_1)$ differential range measurement. Likewise the second, third, and fourth columns of H^{-1} apply the position vector corrections due to the second, third, and fourth satellites, respectively.

An interesting question is, "What if these corrections are applied sequentially, rather than simultaneously?" That is, what if the filter is cycled every time a single new range measurement is obtained? The answer is that such a procedure works well, provided that the sampling rate is high

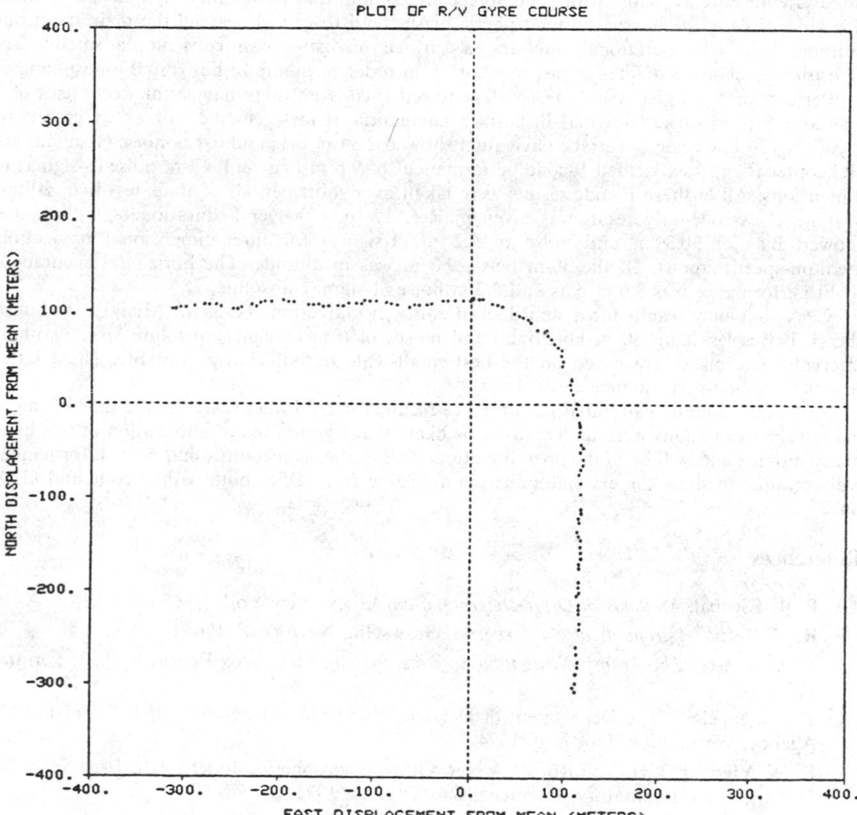

Fig. 54.19 Sequential receiver track.

compared with the maneuvering rate of the vehicle. As an example, Fig. 54.19 shows the actual output of a sequentially corrected Kalman filter operating with GPS (C/A only) on board a ship. The vessel initially steams east at 10 knots (5.16 m/s). It then turns south smoothly, executing the 90° turn in 55 seconds. The range sample rate is one range every 1.2 seconds. The track displays little random noise, roughly 6 m, 1-sigma, and no noticeable overshoot on the turn. The sequential correction of the filter has no noticeable effect for this case.

GPS Performance and Error Characterization

At the time of this writing not much had been published concerning the actual performance of GPS. What little had been released by the military or gotten into print about civil experiments showed the performance to be extraordinary. Actually the geodetic positioning accuracy of the system is so great that there is no standard against which to measure it. It seems likely that GPS itself will become the standard for measuring geodetic and navigation accuracies.

The article by Henderson and Strada[17] was the first widely available "official" release of GPS test data. Other sketchy data have appeared at various technical symposia. The author, for example, has been a coinvestigator on a civil GPS maritime experiment, from which interesting data are released from time to time.

The first really informative performance result was that the C/A-only performance is not 10 times worse than P-code performance, notwithstanding the factor of 10 difference in code resolution. Were the L2 frequency to be standardized for C/A code, so that a C/A-only user could have the ionospheric correction, it is likely that P- and C/A-code performances would be commensurate for some users.

For surface navigators, the ionospheric perturbations are not particularly troublesome, so long as three satellites are available with good spread in azimuth and reasonably high elevation angles. For this kind of case, the horizontal ionospheric biases tend to cancel, leaving the main contribution in altitude. With 18 operational satellites, a surface navigator can construct a satellite selection algorithm to choose satellites as just mentioned, in order to minimize horizontal ionospheric effects.

Performance data from Ref. 17 (p. 82) showed P-code absolute navigation accuracies of 2.84 m bias and 5.47 m noise (1-sigma) in surface navigation at low velocities for one user set. Another P-code set in low-velocity surface navigation showed 2.84 m bias and 6.4 m noise (1-sigma) absolute. Helicopter flight tests yielded P-code performances of 9.6 m bias and 4.7 m noise (1-sigma) in three dimensions. All of these P-code results were taken over short periods of time, less than 20 minutes.

One C/A-code-only result was given in Ref. 17 (p. 82) over a duration of six minutes. This showed bias of 30.3 m and noise of 9.2 m (1-sigma) for three-dimensional navigation of a medium-speed aircraft. Of the 30-m bias, 29.0 m was in altitude. The horizontal accuracy of that flight performance was 8.6 m bias and 8.3 m noise (1-sigma), absolute.

C/A-code-only results for a ship docked either at Galveston, Texas, or Miami have consistently shown biases less than 30 m horizontal and noises of 6 m (1-sigma) absolute, over two-hour time intervals. The biases are based on the best small-scale harbor charts available. These latter C/A results were best-performance cases.

Although carefully calibrated performance results for GPS were scarce at the time of this writing, many GPS evaluations were under way. It is likely that by the time of publication of this handbook, many more results will be in the open literature. This author conjectures that new performance results will continue to show the extreme precision available from GPS, both with P code and C/A-code-only.

References

1 P. R. Escobal, *Methods of Orbit Determination*, Wiley, New York, 1965.

2 R. H. Battin, *Astronautical Guidance*, McGraw-Hill, New York, 1964.

3 T. A. Stansell, *The Transit Navigation Satellite System*, Magnavox Report R-5933, Torrance, CA, 1978.

4 T. O. Seppelin, "The Department of Defense World Goedetic System 1972," Defense Mapping Agency, Washington, DC, May 1974.

5 K. W. Yip and O. H. Von Roos, "A New Global Ionospheric Model," *JPL Deep Space Network*, Jet Propulsion Laboratory, Pasadena, CA, Dec. 15, 1975, pp. 70–87.

6 M. D. Eggers and D. M. Bullock, *Computer Simulation of Ionospheric Wave Propagation for Detection of Range Error for Satellite Navigation Systems*, Research Memorandum #81-06, Telecommunication and Control Systems Laboratory, Dept. of Electrical Engineering, Texas A&M University, College Station, Texas, May 19, 1981, p. 44.

7 W. H. Guier and G. C. Weiffenbach, "A Satellite Doppler Navigation System," *Proc. IRE*, pp. 507–516 (Apr. 1960).

8 J. H. Painter, "Designing Pseudo-Random Coded Ranging Systems," *IEEE Trans. Aero. Electr. Syst.* **AES-3**(1), Jan. 1967, pp. 14–27.

9 A. Bartholomew, "Satellite Frequency Standards," in *Global Positioning System*, The Institute of Navigation, Washington, DC, 1980, pp. 21–28.

10 S. W. Golomb, *Shift Register Sequences*, Holden-Day, San Francisco, 1967, p. 33.

11 J. J. Spilker, "Signal Structure and Performance Characteristics," in *Global Positioning System*, Institute of Navigation, Washington, DC, 1980, pp. 29–54.

12 S. W. Golomb, L. D. Baumert, M. F. Easterling, et al., *Digital Communications With Space Applications*, Prentice-Hall, Englewood Cliffs, NJ, 1964, pp. 85–105.

13 H. W. Farris, M. P. Ristenbatt, et al., *An Introduction to Pseudo-Random Systems*, Vol. 1, *Basic Concepts and Techniques*, Tech. Report 104-1, Cooley Electronics Laboratory, University of Michigan, Ann Arbor, 1960 (classified confidential).

14 P. Ward, "An Advanced NAVSTAR GPS Multiplex Receiver," *Record of IEEE 1980 Position, Location, and Navigation Symposium*, Atlantic City, NJ, Dec. 8–11, 1980, pp. 51–58.

15 W. C. Lindsey and M. K. Simon, *Telecommunication Systems Engineering*, Prentice-Hall, Englewood Cliffs, NJ, 1973, pp. 62–64.

16 H. Parsiani, *A Navigation Algorithm for Single Channel Low-Cost GPS Receiver*, PhD Dissertation, Dept. of Electrical Engineering, Texas A & M University, Aug. 1979, pp. 28–45.

17 D. W. Henderson and J. A. Strada, "NAVSTAR Field Test Results," *Global Positioning System*, The Institute of Navigation, Washington, DC, 1980, pp. 234–246.

PART 12
COMPUTERS

CHAPTER 55
NUMBER SYSTEMS AND DATA REPRESENTATION

ALAN B. MARCOVITZ

Florida Atlantic University, Boca Raton, Florida

This chapter is concerned with the representation of numeric and alphanumeric information in a digital computer. In the computer, information is coded into standard length binary words consisting of ones and zeros. The primary emphasis is on the storage of and arithmetic on a wide range of numbers. For numerical data, it is necessary to convert between decimal and binary, store both unsigned and signed numbers, and do arithmetic computations on each. Methods of handling numbers beyond the normal range of the computer are presented. A brief discussion of error detection and correction is also included.

55.1 BINARY, OCTAL, AND HEXADECIMAL NUMBERS

Each of these number systems, like the decimal system, is *positional*, that is, the weight or value associated with a digit depends on the position of the digit as well as the value of the digit. For example, in decimal the 3 in the number 643 has value 3, whereas in the number 634 it has value 30. We will examine integers first, but will later expand the concepts to include fractions.

A decimal number can be expanded as a power series such that the integer

$$\ldots d_3 d_2 d_1 d_0$$

can be written as

$$\ldots + \quad d_3 \times 10^3 \quad + \quad d_2 \times 10^2 \quad + \quad d_1 \times 10 \quad + \quad d_0$$

The number 634 then becomes

$$6 \times 100 \quad + \quad 3 \times 10 \quad + \quad 4$$

In decimal, the base or radix is 10 and each of the digits in the series d_i is in the range $0 \leqslant d_1 < 10$.

Any positive integer (greater than 1) can be used as the radix r of a number system, yielding the expression

$$\ldots + \quad d_3 \times r^3 \quad + \quad d_2 \times r^2 \quad + \quad d_1 \times r \quad + \quad d_0$$

where each digit d_i must be in the range $0 \leqslant d_i < r$.

We are interested in four bases:

Binary	$r = 2$	$0 \leqslant d_i \leqslant 1$
Octal	$r = 8$	$0 \leqslant d_i \leqslant 7$
Decimal	$r = 10$	$0 \leqslant d_i \leqslant 9$
Hexadecimal (hex)	$r = 16$	$0 \leqslant d_i \leqslant 15$

In each of these systems the standard decimal digits are used, except that the digits 10 to 15 in hexadecimal are represented by the letters A to F (in order).

To convert from any base to decimal, the series is evaluated, with the arithmetic being done in decimal.

Examples.

Binary. $1011 = 1 \times 2^3 \quad + \quad 0 \times 2^2 \quad + \quad 1 \times 2 \quad + \quad 1 = 8 + 2 + 1 = 11$
$10010 = 16 + 2 = 18$
$11111111 = 128 + 64 + 32 + 16 + 8 + 4 + 2 + 1 = 255$
$111001011 = 256 + 128 + 64 + 8 + 2 + 1 = 459$

Octal. $471 = 4 \times 8^2 \quad + \quad 7 \times 8 \quad + \quad 1 = 313$
$77 = 7 \times 8 \quad + \quad 7 = 63$

Hexadecimal. $139 = 1 \times 16^2 \quad + \quad 3 \times 16 \quad + \quad 9 = 313$
$\text{ABC} = 10 \times 256 \quad + \quad 11 \times 16 \quad + \quad 12 = 2748$

If there is any doubt as to what base is being used, the base is often written as a subscript, for example,

$$471_8 = 139_{16} = 313_{10}$$

Table 55.1 shows the first 32 powers of 2 and of 1/2 (needed later for fractions). Note that the powers of 8 and of 16 are also included in the table, since $8^m = 2^{3m}$ and $16^k = 2^{4k}$.

Example (From Table 55.1).

$$8^3 = 2^9 = 512$$
$$16^2 = 2^8 = 256$$
$$8^4 = 16^3 = 2^{12} = 4096$$

Table 55.2 shows the first 32 positive integers in decimal, binary, octal, and hexadecimal (also binary coded decimal, which will be discussed later).

Decimal numbers are discussed here because these are the ones with which we are all familiar. Binary is included because the internal storage of all computers uses binary devices. Although some machines code decimal numbers in binary, an understanding of binary and of conversions between binary and decimal is essential. Octal and hexadecimal are included because they are a shorthand notation for binary, enabling us to more easily write and remember binary numbers.

To convert between binary and octal, one need only recognize that octal represents the grouping of binary digits three at a time. As an example, consider the binary number 110101 (corresponding to decimal 53). It is written in series form as

$$1 \times 2^5 \quad + \quad 1 \times 2^4 \quad + \quad 0 \times 2^3 \quad + \quad 1 \times 2^2 \quad + \quad 0 \times 2 \quad + \quad 1$$

or grouping terms in threes and factoring out $2^3 = 8$, as

$$8 \times (1 \times 4 \quad + \quad 1 \times 2 \quad + \quad 0) + (1 \times 4 \quad + \quad 0 \times 2 \quad + \quad 1)$$

or finally,

$$8 \times 6 \quad + \quad 5$$

Each term in parentheses is a legitimate octal digit (which can take on the values from 0, if the three binary digits are 000, to 7, if the binary digits are 111). If there are more bits,† the series will have

† Bit is a contraction of the words binary digit.

TABLE 55.1. POWERS OF 2

2^n	n	2^{-n}
1	0	1.0
2	1	0.5
4	2	0.25
8	3	0.125
16	4	0.062 5
32	5	0.031 25
64	6	0.015 625
128	7	0.007 812 5
256	8	0.003 906 25
512	9	0.001 953 125
1 024	10	0.000 976 562 5
2 048	11	0.000 488 281 25
4 096	12	0.000 244 140 625
8 192	13	0.000 122 070 312 5
16 384	14	0.000 061 035 156 25
32 768	15	0.000 030 517 578 125
65 536	16	0.000 015 258 789 062 5
131 072	17	0.000 007 629 394 531 25
262 144	18	0.000 003 814 697 265 625
524 288	19	0.000 001 907 348 632 812 5
1 048 576	20	0.000 000 953 674 316 406 25
2 097 152	21	0.000 000 476 837 158 203 125
4 194 304	22	0.000 000 238 418 579 101 562 5
8 388 608	23	0.000 000 119 209 289 550 781 25
16 777 216	24	0.000 000 059 604 644 775 390 625
33 554 432	25	0.000 000 029 802 322 387 695 312 5
67 108 864	26	0.000 000 014 901 161 193 847 656 25
134 217 728	27	0.000 000 007 450 580 596 923 828 125
268 435 456	28	0.000 000 003 725 290 298 461 914 062 5
536 870 912	29	0.000 000 001 862 645 149 230 957 031 25
1 073 741 824	30	0.000 000 000 931 322 574 615 478 515 625
2 147 483 648	31	0.000 000 000 465 661 287 307 739 257 812 5

more terms; the next term is 2^6, which will be factored out and produce the 8^2 term in the octal series. The basic method is to group the binary number in 3-bit groups, starting at the right (the least significant bit) and converting each group to a digit between 0 and 7.

Examples (Binary to Octal).

$$101110111 = 101\ \ 110\ \ 111 = 567$$
$$1101010 = 1\ \ 101\ \ 010 = 001\ \ 101\ \ 010 = 152$$

In the second series, leading zeros were added to complete the group of three. (This obviously does not change the value of the number.)

Conversion from octal to binary is just a digit-by-digit translation. Each octal digit is replaced by its 3-bit binary equivalent.

Examples (Octal to Binary).

$$713 = 111\ \ 001\ \ 011 = 111001011$$
$$177 = 001\ \ 111\ \ 111 = 1111111$$

TABLE 55.2. DECIMAL-BINARY EQUIVALENTS

Decimal	Binary	Octal	Hexadecimal	Binary Coded Decimal First Digit	Binary Coded Decimal Second Digit
0	000000	00	00	0000	0000
1	000001	01	01	0000	0001
2	000010	02	02	0000	0010
3	000011	03	03	0000	0011
4	000100	04	04	0000	0100
5	000101	05	05	0000	0101
6	000110	06	06	0000	0110
7	000111	07	07	0000	0111
8	001000	10	08	0000	1000
9	001001	11	09	0000	1001
10	001010	12	0A	0001	0000
11	001011	13	0B	0001	0001
12	001100	14	0C	0001	0010
13	001101	15	0D	0001	0011
14	001110	16	0E	0001	0100
15	001111	17	0F	0001	0101
16	010000	20	10	0001	0110
17	010001	21	11	0001	0111
18	010010	22	12	0001	1000
19	010011	23	13	0001	1001
20	010100	24	14	0010	0000
21	010101	25	15	0010	0001
22	010110	26	16	0010	0010
23	010111	27	17	0010	0011
24	011000	30	18	0010	0100
25	011001	31	19	0010	0101
26	011010	32	1A	0010	0110
27	011011	33	1B	0010	0111
28	011100	34	1C	0010	1000
29	011101	35	1D	0010	1001
30	011110	36	1E	0011	0000
31	011111	37	1F	0011	0001

Here the spaces are shown just to indicate the process and, in the second series, leading zeros are dropped.

To convert from binary to hexadecimal, bits are grouped in fours, each group translating to a hexadecimal digit. In the reverse process, each hex digit is replaced by 4 bits.

Examples (Binary to Hexadecimal).

10101001 = 1010 1001 = A9
101110111 = 0001 0111 0111 = 177
2B0 = 0010 1011 0000 = 1010110000

Finally, to convert from octal to hexadecimal (or vice versa), the easiest procedure is to first convert to binary, regroup the bits, and then convert to the other base.

Examples (Octal to Hexadecimal).

$$713_8 = 111\ 001\ 011 = 0001\ 1100\ 1011 = 1CB_{16}$$
$$3A0_{16} = 0011\ 1010\ 0000 = 001\ 110\ 100\ 000 = 1640_8$$

Conversion from decimal to any of the other number systems could also be carried out by evaluating a power series, but the arithmetic must be performed in the number system to which one is converting (e.g., binary or octal). Since most people are not comfortable working in bases other than decimal, other methods have been developed that require only decimal arithmetic. To convert to base r, divide the number by r. The remainder is the lowest order digit. The whole part is divided by r again; the remainder is the next digit. This process continues until the whole part goes to 0; one digit is produced by each division.

Examples (459_{10}).

	Remainder		Remainder
$459/2 = 229$	1	$459/8 = 57$	3
$229/2 = 114$	1	$57/8 = 7$	1
$114/2 = 57$	0	$7/8 = 0$	7
$57/2 = 28$	1	$459_{10} = 713_8$	
$28/2 = 14$	0		
$14/2 = 7$	0		
$7/2 = 3$	1	$459/16 = 28$	11 (B)
$3/2 = 1$	1	$28/16 = 1$	12 (C)
$1/2 = 0$	1	$1/16 = 0$	1
$459_{10} = 111001011_2$		$459_{10} = 1CB_{16}$	

Note that these results agree with the conversions in some of the earlier examples. Two shortcuts can be used. First, the division process can be terminated at any point when the whole part can be converted directly. In the decimal-to-binary conversion, for example, if one recognizes that $14_{10} = 1110_2$, the division can be stopped. It produced the bits 01011; the binary for 14, 1110, is then added to the beginning. Also you can convert from decimal to binary by first converting to octal and then from octal to binary. This requires one third the number of divisions and is thus usually faster.

Before looking at fractions, let us briefly examine why this method works. Consider the following number:

$$171_{10} = 253_8 = 2 \times 8^2 \quad + \quad 5 \times 8 \quad + \quad 3$$

When we divide 171 by 8, we get 21 with a remainder of 3. In the series of powers of 8, each of the terms except the first divides evenly; thus the lowest order term must be the remainder. The whole number resulting is

$$\frac{2 \times 8^2}{8} + \frac{5 \times 8}{8} = 2 \times 8 \quad + \quad 5$$

When the process is repeated, the second term, 5, is the remainder; the other terms divide evenly, leaving us with a whole part of 2. The final division gives the 2 as the third octal digit.

The series representation of numbers applies to fractions (nonintegral values) as well as integers. Thus the decimal

$$0.63 = 6 \times 10^{-1} \quad + \quad 3 \times 10^{-2}$$

The same expansion applies for other bases, with the terms being negative powers of the radix r (or equivalently, powers of $1/r$). To convert a number to decimal, the series is evaluated. (Table 55.1 can be utilized.)

Examples.

Binary. $.101 = 1 \times 2^{-1} \quad + \quad 0 \times 2^{-2} \quad + \quad 1 \times 2^{-3} = .5 \quad + \quad .125 = .625$
$.0110011 = .25 \quad + \quad .125 \quad + \quad .015625 \quad + \quad .0078125 = .3984375$

Octal. $.5 = 5 \times 8^{-1} = .625$
$.23 = 2 \times 8^{-1} \quad + \quad 3 \times 8^{-2} = .296875$

Hexadecimal. $.A = 10 \times 16^{-1} = .625$
$.18 = 16^{-1} \quad + \quad 8 \times 16^{-2} = .09375$

To convert from decimal to any radix, multiply by the radix. The whole part of the product is the first digit of the answer. The fractional part is multiplied again, producing the second digit. This process continues until the fraction is zero.

Examples.

$$.375 \times 2 = \underline{0}.75$$
$$.75 \times 2 = \underline{1}.50$$
$$.5 \times 2 = \underline{1}.00$$
$$.375_{10} = .011_2$$
$$.3828125 \times 8 = \underline{3}.0625$$
$$.0625 \times 8 = \underline{0}.50$$
$$.5 \times 8 = \underline{4}.0$$
$$.3828125_{10} = .304_8$$
$$.625 \times 16 = \underline{10}.0$$
$$.625_{10} = .A_{16}$$

To convert from binary to octal (hexadecimal), group bits in threes (fours), starting at the radix point. To convert from octal (hexadecimal) to binary, replace each digit by 3 (4) bits.

Examples.

$$.011_2 = .3_8$$
$$.011_2 = .0110_2 = .6_{16}$$
$$.304_8 = .011\ 000\ 100_2$$
$$= .0110\ 0010_2 = .62_{16}$$

In these examples, the process terminated because the fraction eventually went to 0. That does not always happen, as in the following case:

$$.6 \times 2 = \underline{1}.2$$
$$.2 \times 2 = \underline{0}.4$$
$$.4 \times 2 = \underline{0}.8$$
$$.8 \times 2 = \underline{1}.6$$
$$.6 \times 2 = \underline{1}.2$$
$$\vdots$$
$$.6_{10} = .1001100110011\ldots$$

This should not be surprising, since the conversion of 1/3 to decimal is 0.33333 . . . , a nonterminating fraction. The repetition also occurs in octal and hexadecimal; it may repeat each digit or group of digits.

Example.

$$.6_{10} = .999\ldots_{16} = .46314631\ldots_8$$

When mixed numbers are to be converted, the whole part and the fractional part are converted separately.

Examples.

To Hex. 25.625_{10}
$25/16 = 1$ Rem 9 thus integer = 19
$.625 \times 16 = 10.0$ thus fraction = .A
$25.625_{10} = 19.A$

To Octal. 212.630859375

$212/8 = 26$ Rem 4
$26/8 = 3$ Rem 2
$3/8 = 0$ Rem 3

$$.630859375 \times 8 = \underline{5}.046875000$$
$$.046875 \times 8 = \underline{0}.375$$
$$.375 \times 8 = \underline{3}.0$$
$$212.630859375_{10} = 324.503$$

55.2 BINARY ADDITION OF POSITIVE NUMBERS[1]

Positive binary integers are added in much the same way as positive decimal integers. The least significant digit of each number is added, producing a sum and a carry to the next highest order digit. Then that carry is added to those digits, producing the second digit of the sum and the carry to the third digit. A table for 1 bit of the sum is shown in Table 55.3, where the digit of the first number is referred to as a and that of the second number as b.

Example (1 Bit at a Time).

	00	00	00	10
00011100	00011100	00011100	00011100	00011100
10111010	10111010	10111010	10111010	10111010
	0	1	1	0

11	11	01	0 0	
00011100	00011100	00011100	00011100	00011100
10111010	10111010	10111010	10111010	10111010
1	0	1	1	0 11010110

In this example we added two 8-bit numbers and produced an 8-bit sum. Really the sum is 9 bits, but the extra bit is zero and can thus be ignored. That will happen whenever the sum is 255 or less (in this case we added 28 and 186 to produce a sum of 214). At each step, the carry from the previous step (always a 0 the first time) is added to the appropriate bit of each number, producing 1 bit of the sum plus a carry to the next step.

Let us now consider another example.

Example.

1111111	
10110111	183
01001001	73
1 00000000	256

Note that the sum requires 9 bits; it is larger than 255. (Actually we can get as high as 510 from eight-digit integers or 511 if there is a carry into the lowest order position.) What do we do with this extra bit? There is no place in the word to store it.

This condition is referred to as overflow. Overflow is defined as the production of a result that is out of the range of the machine. Most machines provide a carry bit to store that extra bit, so that the programmer can deal with the condition. (Overflow is not always an error. If multiple precision numbers are used, overflow is merely a carry from one group of 8 bits to the next. We will return to this in Sections 55.4 and 55.6.) Let us next look at a hardware diagram of an adder for 8-bit positive

TABLE 55.3. ONE-BIT BINARY ADDER

Carry In	a	b	Carry Out	Sum
0	0	0	0	0
0	0	1	0	1
0	1	0	0	1
0	1	1	1	0
1	0	0	0	1
1	0	1	1	0
1	1	0	1	0
1	1	1	1	1

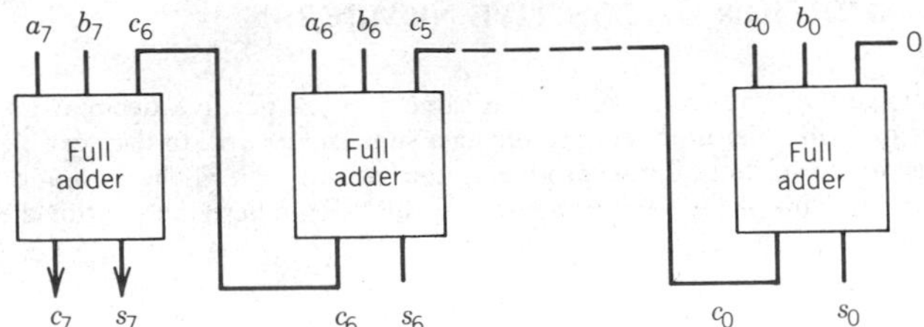

Fig. 55.1 Eight-bit binary adder.

integers. Figure 55.1 shows a block diagram of that hardware. Each block, labeled full adder, is capable of adding 1 bit. Its behavior is described by Table 55.3. The carry in to the low-order bit (bit 0) is always 0 and the carry out from the high-order bit is C_7, the extra bit of the sum.

55.3 SIGNED NUMBERS

When using a k-digit number in radix r, r^k values can be represented. For example, three decimal digits provide $10^3 = 1000$ representations and 8 bits provide $2^8 = 256$ representations. Up to this point, we used these numbers to represent the positive integers 0 to 999 for three-digit decimal numbers and 0 to 255 for 8-bit binary numbers.

When representing signed numbers, that is, numbers that can be either positive or negative, about half of the available representations are used for positive values and half for negative. This would give us a range for 8-bit numbers of approximately $-127 \leqslant n \leqslant 127$.†

Two common methods are used for storing signed integers: sign magnitude and two's complement. (The one's-complement scheme has also been used but is not very popular now.)

In sign magnitude, the leading bit is reserved for the sign of the number—0 for positive and 1 for negative. The remaining bits store the magnitude. The following examples illustrate this storage scheme using 8 bits.

Examples.

17	00010001
− 17	10010001
127	01111111
− 127	11111111
0	00000000
− 0	10000000

As can be seen from these examples, the range of numbers n that can be represented is $-127 \leqslant n \leqslant 127$. In general, a k-digit sign-magnitude representation has a range of $-(2^{k-1} - 1) \leqslant n \leqslant (2^{k-1} - 1)$.

Although sign magnitude is convenient, two drawbacks have limited its use in computers. First, there are two representations of zero, one with a positive sign bit and the other with a negative sign bit. This complicates the hardware used for testing whether a number is, for example, less than zero, since both zeros should be treated the same. A second problem occurs in the addition of sign-magnitude numbers; we will discuss that in the next section.

Two's complement is a modular number system designed to simplify arithmetic. At this point we will be content to examine the storage format. As in sign magnitude, the leading bit is usually used as a sign bit—0 for positive and 1 for negative. Positive numbers are stored directly; thus they have precisely the same format as in sign magnitude. For an 8-bit system, positive integers up to $2^7 - 1 = 127$ can be stored.

†This gives us 255 quantities, including 0. As we will see, sometimes the 256th is wasted; sometimes the range is extended. Also, it is not necessary to maintain this symmetry. We could represent $-55 \leqslant n \leqslant 200$ or any other range of 256 numbers. We will not pursue this approach, since it is rarely used.

To find the representation of a negative number in a two's-complement number system, several approaches are possible. Perhaps the simplest approach (and the one used by most computer hardware) is as follows.

1. Find the representation of the magnitude of the number.
2. Replace every bit by its complement (Comp).
3. Add 1 to the result.

Examples.

− 1.	+1	00000001	
	Comp	11111110	
	Add 1	1	
	Store	11111111	
− 17.	+17	00010001	
	Comp	11101110	
	Add 1	1	
	Store	11101111	
− 0.	+0	00000000	
	Comp	11111111	
	Add 1	1	
	Store	1 00000000	(ignore the ninth bit carry out; thus there is only one representation of 0)
− 128.	+128	10000000	
	Comp	01111111	
	Add 1	1	
	Store	10000000	

Note that the binary 10000000 could represent either +128 or −128. The negative value is usually chosen, since the leading bit is 1.

A second conversion technique for negative numbers is to add the number to 2^k (256 for $k = 8$) and then convert the result to binary.

Examples.

− 1.	256 − 1 = 255	
	255:	11111111
− 17.	256 − 17 = 239	
	239:	11101111
− 0.	256 − 0 = 256	
	256:	1 00000000 (ignore ninth bit)
− 128.	256 − 128 = 128	
	128:	10000000

As can be seen from these examples, there is only one representation of 0 in two's complement and the range of negative numbers is extended to −128, compared with −127 for sign magnitude. In general, a k-digit two's-complement representation has a range of $-2^{k-1} \leqslant n \leqslant (2^{k-1} - 1)$.

Interpreting numbers follows a similar pattern. If the leading digit is 0, the number is positive and is just the direct binary representation. If the leading bit is 1, the number is negative. We then complement each bit and *add* 1. (We could have subtracted 1 and then complemented each bit to do the exact reverse of our encoding procedure; the result would be the same.)

Examples.

01010110	+ 86
10101010	Negative number, complement each bit
01010101	
1	Add 1
01010110	Magnitude = 86; number = − 86
10000000	Negative number, complement each bit
01111111	
1	Add 1
10000000	Magnitude = 128; number = − 128

Table 55.4 lists the value for each representation (using 5-bit numbers) for positive integers and for sign-magnitude and two's-complement signed integers.

TABLE 55.4. FIVE-BIT NUMBERS AS POSITIVE, SIGN MAGNITUDE, AND TWO'S COMPLEMENT INTEGERS

Representation	Positive Integers	Sign Magnitude	Two's Complement
00000	0	+0	+0
00001	1	+1	+1
00010	2	+2	+2
00011	3	+3	+3
00100	4	+4	+4
00101	5	+5	+5
00110	6	+6	+6
00111	7	+7	+7
01000	8	+8	+8
01001	9	+9	+9
01010	10	+ 10	+ 10
01011	11	+ 11	+ 11
01100	12	+ 12	+ 12
01101	13	+ 13	+ 13
01110	14	+ 14	+ 14
01111	15	+ 15	+ 15
10000	16	− 0	− 16
10001	17	− 1	− 15
10010	18	− 2	− 14
10011	19	− 3	− 13
10100	20	− 4	− 12
10101	21	− 5	− 11
10110	22	− 6	− 10
10111	23	− 7	−9
11000	24	− 8	−8
11001	25	− 9	−7
11010	26	− 10	− 6
11011	27	− 11	− 5
11100	28	− 12	− 4
11101	29	− 13	− 3
11110	30	− 14	− 2
11111	31	− 15	− 1

55.4 BINARY ARITHMETIC

Let us now look at the addition of signed numbers. We indicated earlier that one of the disadvantages of sign magnitude is the difficulty of addition. To add two numbers in a sign-magnitude representation, say, a and b, we must first determine whether their signs agree. If so, we add the magnitudes and keep the same sign. If the signs are different, we must subtract the smaller magnitude from the larger and retain the sign of the larger. The following examples illustrate some of the different things that might happen in addition.

Examples.

+23	a	
+46	b	Add magnitudes, retain sign
+69		
−19	a	
−16	b	Add magnitudes, retain sign
−35		
+35	a	Calculate $\|a\| - \|b\|$
−19	b	Use sign of a
+16		
−19	a	Calculate $\|b\| - \|a\|$
+35	b	Use sign of b
+16		
−35	a	Calculate $\|a\| - \|b\|$
+19	b	Use sign of a
−16		

The implementation of this requires an adder, a subtractor (even if subtraction is not required as an operation), and a magnitude comparator (to determine which number is to be subtracted from which).

To add numbers in two's-complement representation we need just add their representations, ignoring the carry out of the last stage. Thus we can use the same 8-bit binary adder as we did for unsigned (positive) binary numbers. We then ignore the carry out of the most significant bit.

Examples.

30		00011110
21		00010101
51		00110011
3		00000011
−2		11111110
+1	(1)	00000001
−3		11111101
2		00000010
−1		11111111
−3		11111101
−2		11111110
−5	(1)	11111011

Note that in each of these cases the answer is correct; the carry out of the last stage is not an overflow indication. Overflow would result if the addition produced a result out of the range, that is, larger than 127 or smaller than −128. The following examples illustrate the two such cases.

Examples.

60		00111100
100		01100100
−96		10100000
−60		11000100
−100		10011100
+96	(1)	01100000

In each case we added two numbers of the same sign and produced a result of the opposite sign. This is the indication of overflow. (Overflow can also be computed from the carry in and carry out of the last stage.) Many computers provide hardware to detect overflow from two's-complement addition and store that in a special overflow bit.

Subtraction is most often done by first computing the negative of the subtrahend and then adding. The negative is obtained by complementing each bit of the subtrahend and then adding 1. The addition of 1 is usually accomplished as part of the addition of the two numbers by putting a 1 into the low-order carry-in bit. The two steps (complement and add) are performed by a single subtraction instruction.

Example. To compute 3 − 2:

```
                        1
  00000011       00000011
- 00000010       11111101   (bit-by-bit complement)
             (1) 00000001
```

A circuit for two's-complement addition and subtraction, using the same adder as for positive integers, is shown in Fig. 55.2, where sub = 1 if the operation is subtraction and sub = 0 if the operation is addition. For subtraction, the subtrahend must be complemented before addition. Notice that the carry produced by addition is the complement of the borrow. In this example, the borrow is 0, since 2 is less than 3 and the carry bit is 1. Some computers store this bit in the carry bit on a subtract instruction; others complement it and store the borrow there.

Multiplication[2] is accomplished by successive additions and shifting. The product is always double length, that is, the product of two 8-bit numbers is a 16-bit number. The multiplication process in binary is the same as in decimal. The multiplicand is multiplied successively by each bit of the multiplier, the results are positioned, and these products are added. Of course the 1-bit multiplication is simple, since multiplying by 1 leaves the multiplicand unchanged and multiplying by 0 results in all 0's.

Example.

```
          1 1 0 1    (13)
          1 0 1 1    (11)
          -------
          1 1 0 1
        1 1 0 1
      0 0 0 0
    1 1 0 1
    ---------------
    1 0 0 0 1 1 1 1  (143)
```

The addition involves computing the sum of four numbers in this example. It is normally done by adding each row into a partial product as it is computed. Thus a more descriptive version of the

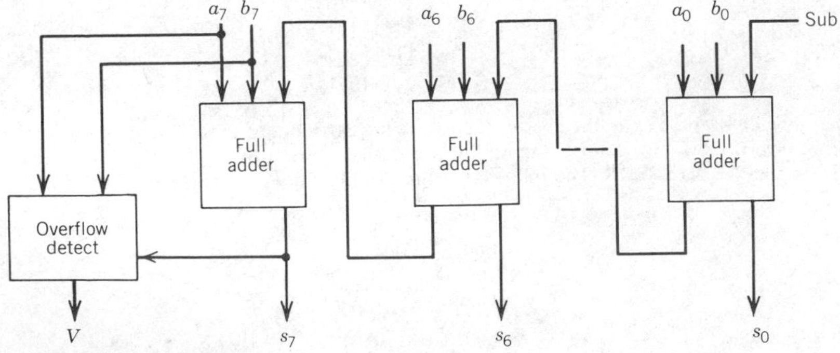

Fig. 55.2 Two's-complement adder–subtractor.

Dividend: 10001111 *Divisor*: 1011

Step 1. Subtract divisor from left 5 bits of dividend. (Subtraction may be accomplished by true subtraction or by complementing the division and then adding.)

```
10001
 1011
 0110    Subtraction successful (i.e., result nonnegative); high-order bit of quotient is 1
```

Step 2. Catenate next bit of dividend and subtract again.

```
01101
 1011
 0010    Subtraction successful; next bit is 1
```

Step 3. Catenate next bit and subtract.

```
00101
 1011
   *     Result negative; next bit is 0
```

Step 4. Restore dividend to former value (00101); catenate last bit of dividend; subtract.

```
001011
  1011
  0000   Subtraction successful; last bit is 1
```

Answer. 1011.

* The form of the answer will depend on the hardware used for subtraction. But the result will contain a negative indicator.

Fig. 55.3 Division example.

example is as follows:

```
         1 1 0 1
       × 1 0 1 1
         1 1 0 1
       1 1 0 1
     1 0 0 1 1 1
     0 0 0 0
   0 1 0 0 1 1 1
   1 1 0 1
 1 0 0 0 1 1 1 1
```

Note that at each addition only 4-bit numbers need be added. The computer algorithm is described in Chapter 59.

Division[2] is accomplished by a process of repeated trial subtractions. Consider the example shown in Fig. 55.3 in which the product of the multiplication example (10001111) is divided by 1011. This division terminates after four steps for these 4-bit numbers. The remainder is 0000 and thus the answer is an integer.

55.5 BINARY CODED DECIMAL (BCD)[3]

Up to this point we have assumed that any number that is to be stored in the computer is either entered in binary (or some shorthand for binary) or is converted from decimal to binary before it is stored or operated on. On output, either the display is binary or a conversion back to decimal must take place.

Working in binary is unnatural for the human, although many small systems (particularly microprocessors) are designed to operate in that mode. Conversion takes time (a nontrivial program) or a considerable amount of hardware.

Another approach that is employed in some machines is to code the decimal number in binary, one digit at a time. The most commonly used procedure is to replace each decimal digit by a 4-bit

TABLE 55.5. BINARY CODED DECIMAL

Digit	Binary Code
0	0000
1	0001
2	0010
3	0011
4	0100
5	0101
6	0110
7	0111
8	1000
9	1001

binary number (just the binary equivalent), as shown in Table 55.5. (Other codes are used occasionally but will not be discussed here.)

Examples.

Decimal	BCD
74	0111 0100
29	0010 1001

BCD has two advantages compared with binary. First, as we have just seen, the conversion process is much simpler. This is particularly important in such applications as calculators, where intermediate results are constantly being displayed in decimal. Second, it eliminates some errors introduced in the conversion process. Fractions that can be represented exactly in decimal can be represented exactly in BCD but not necessarily in binary.

Example.

Decimal. .24

BCD. .00100100

Binary. .001111010111 (12 places)

Exact decimal of 12-bit binary. .239990234375

If the 12-bit binary number is converted back to decimal (as on the last line) in this example, the result is off by 1 in the fifth decimal place. This could cause problems, particularly in financial computations.

BCD has two major disadvantages, both involving cost. First, BCD requires additional storage space. To represent a three-digit decimal number, 12 bits are required, 4 for each digit. This three-digit number provides 1000 values. But if binary were used, 1024 values could be represented with only 10 bits. Thus 20% more storage (12 bits compared with 10) is required for a slightly smaller range. Second, arithmetic is more complex, requiring additional hardware.

Before considering addition of BCD numbers, let us look at the storage of signed numbers. As in binary, a complement number system is commonly used in computers. The equivalent of two's complement is ten's complement (where the generic term is radix complement). With k digits ($4k$ bits), 10^k representations are available. Using approximately half of these for positive numbers, the range is $-5 \times 10^{k-1} \leqslant n \leqslant (5 \times 10^{k-1} - 1)$. For $k = 2$, we have $-50 \leqslant n \leqslant 49$ and for $k = 4$, $-5000 \leqslant n \leqslant 4999$.

Positive numbers are stored directly. As in binary, two computational approaches are available to compute the storage for negative numbers. First, the storage for the negative number n can be computed by finding $10^k + n$. Second, complement each digit with respect to 9 (i.e., subtract it from 9) and then add 1 to the number.

Examples (Four Digit).

Number	Decimal Storage	Binary Storage (16 Bits)
5	0005	0000 0000 0000 0101
1234	1234	0001 0010 0011 0100
− 1234	8766	1000 0111 0110 0110
− 5000	5000	0101 0000 0000 0000

In this method, the first digit of all positive numbers (and 0) is between 0 and 4, and the first digit of all negative numbers is in the range 5 to 9.

Sometimes it is undesirable to limit the range of the first digit. Instead an extra bit is added as a sign bit. (We now have $4k + 1$ bits, i.e., 17 bits. This is sometimes called 4 1/2 digits.) The range of numbers is thus extended to $-10000 \leqslant n \leqslant 9999$. In the first computation, the number to which n is added becomes 20000. The second computation remains the same except that the sign digit (only 1 bit) is complemented with respect to 1.

Examples.

Number	Decimal Storage	Binary Storage
5	00005	0 0000 0000 0000 0101
− 1234	18766	1 1000 0111 0110 0110
− 9321	10679	1 0000 0110 0111 1001
+ 9999	09999	0 1001 1001 1001 1001

Now let us look at the addition process for BCD numbers.[4] Addition is usually a two-step process. First, binary addition is performed and then, when necessary, a correction factor is added. When the codes for one digit of the addend and augend are added together and then added to the carry from the next lowest digit, the result is in the range 00000 through 10011 (0 to 19). As can be seen from Table 55.6, the results are correct when the sum is less than 10 but needs a correction when the sum is greater than or equal to 10. An inspection of the results show that if 6 (binary 00110) is added to the output of the binary adder in these cases, the answer is correct.

Examples.

```
 5    0101          6     0110     9     1001
 3    0011          4     0100     8     1000
 ---------         --     ----    --   ------
 8    1000         10     1010    17   1 0001
no correction      +6      110    +6      110
                        ------         ------
                        1 0000         1 0111
```

In some machines the correction is done by hardware. A special ADD DECIMAL instruction is used, which calls on the addition and correction circuitry.

In many 8-bit microprocessors, two decimal digits can be stored in a word. Decimal addition is accomplished by first performing binary addition (using the ADD instruction). The carry out of the low-order digit is saved in the half carry bit H, and the carry from the high-order digit is saved in C. The DECIMAL ADJUST instruction (DAA) then performs the correction (just shown) on each digit. For the right digit, H is considered the leading bit of the sum. For the left digit, C is considered the leading bit of the sum.

Example.

Decimal	Storage	Binary Addition
29	29	0010 1001
− 03	97	1001 0111
26	(1) 26	1100 0000
		C = 0 H = 1

TABLE 55.6. BINARY CODED DECIMAL (BCD) ADDER

Binary Adder					Number Represented	Correct Representation				
C_0^*	S_1^*	S_2^*	S_3^*	S_4^*		C_0	S_1	S_2	S_3	S_4
0	0	0	0	0	0 0	0	0	0	0	0
0	0	0	0	1	0 1	0	0	0	0	1
0	0	0	1	0	0 2	0	0	0	1	0
0	0	0	1	1	0 3	0	0	0	1	1
0	0	1	0	0	0 4	0	0	1	0	0
0	0	1	0	1	0 5	0	0	1	0	1
0	0	1	1	0	0 6	0	0	1	1	0
0	0	1	1	1	0 7	0	0	1	1	1
0	1	0	0	0	0 8	0	1	0	0	0
0	1	0	0	1	0 9	0	1	0	0	1
0	1	0	1	0	1 0	1	0	0	0	0
0	1	0	1	1	1 1	1	0	0	0	1
0	1	1	0	0	1 2	1	0	0	1	0
0	1	1	0	1	1 3	1	0	0	1	1
0	1	1	1	0	1 4	1	0	1	0	0
0	1	1	1	1	1 5	1	0	1	0	1
1	0	0	0	0	1 6	1	0	1	1	0
1	0	0	0	1	1 7	1	0	1	1	1
1	0	0	1	0	1 8	1	1	0	0	0
1	0	0	1	1	1 9	1	1	0	0	1
1	0	1	0	0	—			—		
1	0	1	0	1	—			—		
1	0	1	1	0	—			—		
1	0	1	1	1	—			—		
1	1	0	0	0	—			—		
1	1	0	0	1	—			—		
1	1	0	1	0	—			—		
1	1	0	1	1	—			—		
1	1	1	0	0	—			—		
1	1	1	0	1	—			—		
1	1	1	1	0	—			—		
1	1	1	1	1	—			—		

Correction 1. Right digit (sum 10000 = 16).

$$\begin{array}{r} 0000 \\ \underline{0110} \\ 0110 \end{array}$$

Correction 2. Left digit (sum 01100 = 12).

$$\begin{array}{r} 1100 \\ \underline{0110} \\ C = 1\ 0010 \end{array}$$

Answer. 0010 0110 (26).

Both corrections are done (in the order shown) by the hardware as a result of one DAA instruction. If the first correction produces a carry, it is added to the second digit before the correction.

Example.

Decimal	Storage	Binary Addition	
− 23	77	0111 0111	
26	26	0010 0110	
3	(1) 03	1001 1101	C = 0 H = 0
Right digit (01101 = 13), correct		0110	
		1010 0011	
Left digit (01010 = 10), correct		0110	
Decimal: 03		(1) 0000 0011	

55.6 MULTIPLE PRECISION ARITHMETIC

The standard size of data for arithmetic operations is the word. The number of bits in a word varies from as few as 4 in some microprocessors to 64 in some larger computers. Except for the largest word sizes, there is need to handle numbers of greater precision than is available in a single word. Two basic approaches are utilized. In some machines, instructions are provided for arithmetic on various data sizes. For example, in the Motorola 68000, three data types are permitted—byte (8 bits), word (16 bits), and long word (32 bits). The arithmetic instructions can operate on all three types of data.

In many machines[5] (whether or not the first approach is utilized), instructions are provided to extend the precision of addition and subtraction through software. For the ease of discussion, we will assume an 8-bit word size, although the concept applies equally well to other word sizes. When computing the sum of two 8-bit numbers, a 9-bit result is obtained, with the ninth bit usually stored as a carry bit. If the numbers were longer, that ninth bit would just be the carry to the next higher bit. To add two 16-bit numbers (in an 8-bit machine), first the low 8 bits of each number must be added. The sum is saved as the low 8 bits of the answer. Next the high 8 bits of the two words are added, along with the carry produced by the first addition. That sum is saved as the high 8 bits of the answer. A special instruction, ADD WITH CARRY (ADC), that adds the carry bit to the two numbers is provided.

Example (in Hexadecimal). To add

$$\begin{array}{r} 3456 \\ \underline{ABCD} \\ E023 \end{array}$$

first add

$$\begin{array}{r} 56 \\ \underline{CD} \\ 123 \end{array}$$

then add with carry

$$\begin{array}{r} 1 \\ 34 \\ \underline{AB} \\ 0E0 \end{array}$$

This same process can be applied to numbers requiring three or more words of precision. The lowest order words are added using ADD; then each of the higher order words are added, in turn, using ADC. Multiple precision can also be applied to decimal addition and to subtraction.

55.7 FLOATING-POINT ARITHMETIC[6,7]

We could design a computer with only fixed-point numbers. We could force everyone to scale their calculations so as to deal only with integers or only with fractions, or we could even put the binary point always in the middle. Thus if we had 32-bit words, we might use 1 for the sign, 16 to the left of the binary point, and 15 to the right of the binary point. But this allows us to have numbers as large as 2^{16}, or approximately 65,000, and gives us just over four decimal places to the right of the decimal point. Thus we could not represent (other than by zero) the number 0.00001 and we could not nearly represent 1 million. In scientific calculations, numbers as large as 10^{25} are not uncommon nor are numbers as small as 10^{-20}. To represent both quantities in a fixed-point number, we would need 150 bits, a much larger number than in any available computer. Consequently, floating-point notation has been introduced in most scientific computers to allow the representation of a larger range of numbers in a 32-bit word.

We will introduce floating point in decimal and then discuss how the numbers are stored in binary. We are doing this only because decimal numbers are more familiar. With five decimal digits plus two sign bits, we can now represent numbers in the range of magnitudes $.001 \times 10^{-99} \leqslant X \leqslant .999 \times 10^{99}$. Here three digits plus one sign bit are used for the *mantissa*, the significant digits, and two digits and one sign bit are used for the *exponent*, the power of 10. In this format (which is a common one), the mantissa is a fraction. This form gives us more than enough range for almost any calculation, whereas a five-digit fixed-point number could not even represent a million. This particular representation scheme is likely not adequate for many purposes because of lack of precision. Three places is not normally enough. A 32-bit word usually allots 8 or 9 bits to the exponent (including its sign) and the remaining 23 or 24 bits to the mantissa. This gives approximately seven significant digits, which is adequate for many calculations. (Some machines provide double-precision floating-point operations for situations where more precision is needed.)

To preserve precision, floating-point numbers are nearly always stored with the leading digit of the mantissa as nonzero (except when the number is exactly zero). Thus for example, 23 would be stored as $.230 \times 10^2$ rather than as $.023 \times 10^3$. This reduces the range slightly (down to $.100 \times 10^{-99}$), but this is not a problem.

Let us now see how we could add two floating-point numbers. As the first example, we will choose a simple one.

$$\begin{array}{r} -.231 \times 10^1 \\ \underline{.642 \times 10^1} \\ .411 \times 10^1 \end{array} \qquad \begin{array}{r} -2.31 \\ \underline{6.42} \\ 4.11 \end{array}$$

In this case we just added the mantissas; nothing else needed to be done. The first complication results when the two numbers to be added have different exponents. For example,

$$\begin{array}{r} .412 \times 10^1 \\ \underline{.264 \times 10^2} \\ \end{array} \qquad \begin{array}{l} 4.12 \\ \underline{26.4} \\ 30.52 \end{array}$$

We cannot just add the mantissas since the exponents are different. One of the mantissas must be shifted. In particular, we must adjust the numbers by shifting the mantissa of the number with the smaller exponent to the right by as many places as the exponents differ. The exponent of that number is then increased by the number of places shifted. This leaves the value unchanged, except for roundoff, that is, $.410 \times 10^1 = .041 \times 10^2$. Here since the exponents differ by 1, the first mantissa must be shifted one place to the right, as follows

$$\begin{array}{r} .041 \times 10^2 \\ \underline{.264 \times 10^2} \\ .305 \times 10^2 \end{array}$$

We can add two mantissas only when they have been shifted until the exponents are the same. We do have the possibility that the smaller number is so small that it disappears when shifted. For example

$$\begin{array}{r} .237 \times 10^{21} \\ \underline{+.641 \times 10^{16}} \end{array}$$

We need to shift the second number five places to the right; that leaves it as $.000 \times 10^{21}$ and the sum is just the first operand.

Let us now look at two other complications. First, consider

$$\begin{array}{r} -.581 \times 10^2 \\ \underline{-.723 \times 10^2} \\ -1.304 \times 10^2 \end{array}$$

The sum we produced is too large; it requires a digit to the left of the decimal point. It is *not* overflow, however. We can adjust the result by shifting one place to the right and adding one to the exponent, that is, the sum is

$$-.130 \times 10^3$$

We sometimes must adjust the sum, then, by shifting one place to the right. The only time we will get overflow is if the exponent was already at its maximum value before the adjustment, that is, if the result came out to be larger than 10^{99}. Then we truly have overflow.

The other adjustment that might be required occurs when we add two numbers that are nearly equal but of opposite sign. For example, consider

$$\begin{array}{r} .653 \times 10^2 \\ \underline{-.621 \times 10^2} \\ .032 \times 10^2 \end{array}$$

But this sum is not in the proper form; the leading digit is a zero. Consequently we must normalize the number by shifting it one place to the left and decreasing the exponent by one. The correct form is then

$$.320 \times 10^1$$

More than one place of adjustment might be needed. For example, we have

$$\begin{array}{r} -.511 \times 10^{-1} \\ \underline{.509 \times 10^{-1}} \\ -.002 \times 10^{-1} \\ .200 \times 10^{-3} \end{array}$$

A note of caution is in order here. If the sum is zero, we could shift indefinitely without finding a leading nonzero digit. We must check for that. Another problem occurs when we must make this correction and the exponent is already at its lower limit (at -99). We then get a number too small to represent; this is called *underflow*. This error condition can be treated in the same way as overflow, or it is often more meaningful to just set the result to 0. After all, the magnitude is closer to zero than we can represent.

Now let us look at how floating-point numbers are stored in a computer. All that we need is the mantissa (with its sign) and the exponent (and its sign). The number is stored as a binary fraction (the mantissa) and the proper power of 2 (the exponent) or as a hexadecimal fraction with the proper power of 16. As an example, consider the storage of 23.75. First we convert to binary.

$$\begin{aligned} 23.75 &= 1\ 0\ 1\ 1\ 1\ .1\ 1 \\ &= .1\ 0\ 1\ 1\ 1\ 1\ 1 \times 2^5 \end{aligned}$$

Thus in a binary scheme, the mantissa would be +.1 0 1 1 1 1 1 and the exponent +1 0 1 (binary 5).

In hexadecimal we would have

$$\begin{aligned} 23.75 &= 17.C_{16} \\ &= .17C \times 16^2 \end{aligned}$$

Since hexadecimal numbers are really stored as binary, we have a storage of

+.0 0 1 0 0 1 1 1 1 1 0 0 *and* +0 0 1 0.

What is the difference between the two schemes? Note that what eventually gets stored as the mantissa is exactly the same in both cases except for the leading zeros. In binary there is never a leading 0; in hex the leading *digit* cannot be 0, that is, there can never be four leading binary zeros. The savings in using hex is in the number of bits required to represent the exponent. To represent numbers up to about 2^{256}, we need 8 bits of exponent (plus sign). But $2^{256} = 16^{64}$ and we need only 6 bits of exponent (plus sign). Thus we save a little in the size of the exponent by going to hex. But at the same time we lose a little in precision because of the leading (nonsignificant) zeros. (The average number of leading zeros in hex, assuming all digits are equally likely, is .73. Thus we have a gain of slightly more than one bit of precision.)

The mantissa is almost always stored in the same number system as integers are stored, since the same kinds of operations (often using the same hardware) are required. The exponent may use a different format, or it may also be stored in the same system as the mantissa. The reasons for considering a different scheme for the exponent is that a more restricted set of operations is required and that the number 0 can be more effectively represented. A common scheme for storing the exponent in, say, a 9-bit format, including sign, is to store the exponent, e, as

$$2^8 + e$$

where e must be in the range $-256 = -2^8 \leqslant e \leqslant 2^8 - 1 = 255$. This is sometimes referred to as an

S	Exponent	Mantissa

Fig. 55.4 Floating-point format.

excess 256(2^8) format. Thus for example we have

	Storage
0	1 0 0 0 0 0 0 0 0
1	1 0 0 0 0 0 0 0 1
− 1	0 1 1 1 1 1 1 1 1
255	1 1 1 1 1 1 1 1 1
− 256	0 0 0 0 0 0 0 0 0

Note that all positive exponents (including 0) begin with a 1 and all negative exponents begin with a 0. Furthermore, if you treat the whole 9 bits as a positive binary integer, the larger that integer the larger the exponent is. Also, if you subtract using normal binary arithmetic (the smaller from the larger), you get the correct answer. The other key reason for using this scheme is that all zeros in both the mantissa and the exponent represents a true floating point zero. This is desirable since it makes floating point and integer zero the same and greatly simplifies tests for 0. On the other hand, if a complement scheme were used, we would have the choice of (two's complement)

$$.0\ 0\ 0 \times 2^{-256}$$

which would have an exponent of 1 0 0 0 0 0 0 0 0 (which differs from the integer representation of zero) or an all-zero exponent, which would be

$$.0\ 0\ 0 \times 2^{0}$$

That is not very satisfactory either, since the next number in magnitude that can be represented, $.100 \times 2^{-256}$, has a very different storage format.

A common format for floating-point numbers is shown in Fig. 55.4, where S is the sign bit of the mantissa. The sign, including the exponent, is usually 8 or 9 bits, with the rest for the mantissa.

Example—IBM 370. In the IBM 370, the floating-point number has the format of Fig. 55.4, where the exponent is 7-bit excess 64($2^6 + e = 64 + e$), the mantissa is a 24-bit fraction, and the S bit is the sign of the mantissa. The decimal value of the number is given by

$$\pm \text{mantissa} \times 16^{(e-64)}$$

The number +1 is stored as $16^{-1} \times 16^{65}$, or in hexadecimal (mantissa = .10000 and $e = 65 = 41_{16}$)

$$41100000$$

The 32-bit form has values ranging from

$$16^{-65} \text{ to } (1 - 16^{-6}) \times 16^{63}$$

55.8 ERROR DETECTION AND CORRECTION

If the storage or transmission medium is error prone, it may become necessary to add bits to the information to provide for the detection and/or correction of errors.[2,6,8]

The word can be extended by a single bit, often called a parity bit, so that any single error in that word (including the parity bit) can be detected. The parity bit is chosen so that the total number of 1's in the extended word is even. If in the processing of the information a single error is made, that is, if one 1 becomes a 0 or one 0 becomes a 1, then the total number of 1's is now odd. Thus an error exists if there is an odd number of 1's in the extended word. This method will detect one error in a word (or, in fact, any odd number of errors—3, 5, etc.) but it will not detect an even number of errors. Nevertheless, it provides for a great improvement in reliability if the probability of making a single error is small and errors are independent. For example, if a parity bit is added to an 8-bit word and the probability of an error in any bit is .01, then the probability of an undetected error in a word is .00336. Without error detection, the probability of an error is .0773 (23 times as high) and of course all errors are undetected. Note, however, that the addition of a check bit reduces the probability of correct transmission (from .923 to .913), since there are now 9 bits in which to make an error instead of 8.

Examples.

Data	Extended Word (Check Digit on Right)
0 1 1 0 1 0 1 1	0 1 1 0 1 0 1 1 1
0 0 0 0 0 0 0 0	0 0 0 0 0 0 0 0 0
1 1 1 1 1 1 1 1	1 1 1 1 1 1 1 1 0

Stored or Transmitted	Retrieved or Received	Error Detected
0 1 1 0 1 0 1 1 1	0 1 1 0 1 0 1 1 1	No
0 1 1 0 1 0 1 1 1	1 1 1 0 1 0 1 1 1	Yes
0 1 1 0 1 0 1 1 1	0 1 1 0 1 0 1 1 0	Yes (data correct; error in parity bit)
0 1 1 0 1 0 1 1 1	1 1 1 1 1 0 1 1 1	No (double error not detected)

The description of parity here provides for an even number of 1's in the extended word. It would work just as well if the parity bit were chosen so that the total number of bits is odd. A word of caution about terminology is in order. The terms "odd parity" and "even parity" are used inconsistently in the literature. Even parity is sometimes used to mean that the total number of 1's in the extended word is even (as in our examples). But it is also sometimes used to mean that the parity bit is to be 1 if there is an even number of 1's in the data (which is exactly the opposite).

An error detection code can only tell that an error has been made somewhere in the word; it does not tell which bit is wrong. An error-correcting code automatically corrects the error. A code that provides correction of all single errors, called a Hamming code, is illustrated in Tables 55.7 and 55.8.

TABLE 55.7. HAMMING CODE PARITY CHECKS FOR 8 DATA BITS

	Data Bit							
Check Bit	7	6	5	4	3	2	1	0
3	X	X		X	X		X	
2	X		X	X		X	X	
1		X	X	X				X
0					X	X	X	X

TABLE 55.8. HAMMING CODE CORRECTION MAP

Check Failed				Bit in Error
3	2	1	0	
	None			None
			X	C0
		X		C1
		X	X	D0
	X			C2
	X		X	D2
	X	X		D5
X				C3
X			X	D3
X		X		D6
X	X			D7
X	X		X	D1
X	X	X		D4

The 4 check bits are appended to the data bits. Each check bit is computed to create a parity check over the bits indicated by an X in Table 55.7. (It could be even or odd.) To find errors, the four parity checks are performed. Each possible single error produces a unique pattern of failures. Table 55.8 contains a correction map, relating the failed tests with the error.

Examples (Even Number of 1's Including Parity Bit).

Data: 0 0 1 0 1 1 0 1 Check: 1 0 0 1
Total word: 0 0 1 0 1 1 0 1 1 0 0 1

Error 1: 0 1 1 0 1 1 0 1 1 0 0 1 (in data bit 6)
Checks: Fail, pass, fail, pass D6
Corrected data: 0 $\underline{0}$ 1 0 1 1 0 1

Error 2: 0 0 1 0 1 1 0 1 0 0 0 1 (in check bit 3)
Checks: Fail, pass, pass, pass C3
Corrected data: 0 0 1 0 1 1 0 1

Error 3: 0 0 1 0 0 0 0 1 1 0 0 1 (in data bits 3 and 2)
Checks: Fail, fail, pass, pass D7
Corrected data: $\underline{1}$ 0 1 0 0 0 0 1 (now has 3 errors)

Note that in the last example a double error results in a pattern that points to an error in a correct bit. Because 12 bits are used instead of 8, the probability of no error is reduced to .886. But all single errors are correctable. Thus the probability of an uncorrected error is only .0062. Further, if a thirteen bit (fifth check bit) is added to provide overall parity, double errors can be detected, since all double errors will result in one or more discrepancies in the first four checks but a correct fifth bit. In that case the probability of an uncorrected error is .00725 and of an undetected error is .00027. Of course, there is a price paid for this reliability, namely the use of 13 bits instead of 8.

The same error detection and error correction concepts can be used for longer words. Table 55.9 shows the number of check bits required for single error detection for various word lengths. (One additional check bit is needed if double error detection is also required.)

55.9 ALPHANUMERIC DATA REPRESENTATIONS[5,8,9]

Textual information is often entered into a computer and processed. It must be coded in binary. Although many codes have been utilized, the most commonly used code today is the 7-bit ASCII code (American Standard Code for Information Interchange), which is shown in Table 55.10. In addition to uppercase and lowercase letters, numbers, and special arithmetic and punctuation symbols, a number of codes are provided to control the transmission or printing of information. From the table, the letter H would be represented as $1001000 = 48_{16}$.

Typically each character is stored in 1 byte (8 bits) of memory. A parity bit (see Section 55.8) is usually added to the 7-bit code.

TABLE 55.9. CHECK BITS REQUIRED FOR SINGLE ERROR DETECTION WITH GIVEN WORD LENGTH

Bits of Data	Check Bits
1	2
2–4	3
5–11	4
12–26	5
27–57	6
58–120	7

TABLE 55.10. SEVEN-BIT ASCII CODE

	$b_6b_5b_4$							
$b_3b_2b_1b_0$	000	001	010	011	100	101	110	111
0 0 0 0	NUL[a]	DLE	SP	0	@	P		p
0 0 0 1	SOH	DC1	!	1	A	Q	a	q
0 0 1 0	STX	DC2	"	2	B	R	b	r
0 0 1 1	ETX	DC3	#	3	C	S	c	s
0 1 0 0	EOT	DC4	$	4	D	T	d	t
0 1 0 1	ENQ	NAK	%	5	E	U	e	u
0 1 1 0	ACK	SYN	&	6	F	V	f	v
0 1 1 1	BEL	ETB	'	7	G	W	g	w
1 0 0 0	BS	CAN	(	8	H	X	h	x
1 0 0 1	HT	EM	)	9	I	Y	i	y
1 0 1 0	LF	SUB	*	:	J	Z	j	z
1 0 1 1	VT	ESC	+	;	K	[	k	{
1 1 0 0	FF	FS	,	<	L	\	l	|
1 1 0 1	CR	GS	-	=	M	]	m	}
1 1 1 0	SO	RS	.	>	N	^	n	~
1 1 1 1	SI	US	/	?	O	_	o	DEL

[a] Control characters: NUL, null; SOH, start of heading; STX, start of text; ETX, end of text; EOT, end of transmission; ENQ, enquiry; ACK, acknowledge; BEL, bell; BS, backspace; HT, horizontal tab; LF, line feed; VT, vertical tab; FF, form feed; CR, carriage return; SO, shift out; SI, shift in; DLE, data link escape; DC1, device control 1; DC2, device control 2; DC3, device control 3; DC4, device control 4; NAK, negative acknowledge; SYN, synchronous idle; ETB, end of transmission block; CAN, cancel; EM, end of medium; SUB, substitute; ESC, escape; FS, file separator; GS, group separator; RS, record separator; US, unit separator; SP, space; DEL, delete.

References

1 H. Taub, *Digital Circuits and Microprocessors*, McGraw-Hill, New York, 1982.

2 R. K. Richards, *Arithmetic Operations in Digital Computers*, D. Van Nostrand, New York, 1955.

3 V. Rajaraman and T. Radhakrishnan, *An Introduction to Digital Computer Design*, 2nd ed., Prentice-Hall, Englewood Cliffs, NJ, 1983.

4 R. L. Krutz, *Microprocessors and Logic Design*, Wiley, New York, 1980.

5 S. B. Newell, *Introduction to Microcomputing*, Harper & Row, New York, 1982.

6 T. C. Bartee, *Digital Computer Fundamentals*, 6th ed., McGraw-Hill, New York, 1985.

7 M. E. Sloan, *Computer Hardware and Organization*, 2nd ed., Science Research Associates, Chicago, 1983.

8 D. D. Givone and R. P. Roesser, *Microprocessors/Microcomputers: An Introduction*, McGraw-Hill, New York, 1980.

9 J. W. Gault and R. L. Pimmel, *Introduction to Microcomputer-Based Digital System*, McGraw-Hill, New York, 1982.

CHAPTER 56

COMPUTER ORGANIZATION AND ARCHITECTURE

EDWARD J. LANCEVICH

Robotic Vision Systems Inc., Hauppauge, New York

56.1 GENERAL ORGANIZATION—BLOCK DIAGRAM

The functional organization of a general computer system is shown in block-diagram form in Fig. 56.1. The basic functional blocks are the central processing unit (CPU), the main memory, and the input–output (I/O) processor.

The CPU is responsible for the interpretation and execution of instructions held within the main memory. Communications between the CPU and main memory are carried out over two functionally distinct busses: the address bus and the data bus.

To access a particular instruction in the memory, the CPU sends the address of the instruction over the address bus to the memory and receives the instruction at that address via the data bus. Part of the instruction is used by the CPU to identify the operation to be performed; this part is called the operation code of the instruction. The remaining information is utilized to determine the location(s) of the data on which the operation is to be performed.

The action of reading an instruction into the CPU and preparing for its execution is called the fetch cycle of the computer. To complete an instruction, the CPU decodes the operation code, generates the control signals necessary to access the required operands, and then controls the execution of the instruction.

For example, suppose the operation specified is to add two numbers held in two CPU registers and store the result into a third CPU register. To execute this instruction, the CPU would identify the two registers and generate the appropriate control signals to connect the registers to the arithmetic–logic unit (ALU) in the CPU. The CPU would also set up the ALU to function as an adder and direct the ALU's output to the third register.

The process of carrying out the operations specified by an instruction is called the execution[1,2] cycle of the computer.

The names "fetch cycle" and "execution cycle" are derived from the cyclic nature of the computer's operation; once the machine starts running, it repeats the fetch and execute cycles continuously. To refer to both cycles collectively, the term "machine[1,2] cycle" is used.

Implicit in the example is the notion that the CPU may be functionally divided into three subunits: the control unit, which is responsible for carrying out the fetch and execute cycles; the ALU, which carries out both arithmetic (e.g., addition and subtraction) and logic functions (e.g.,

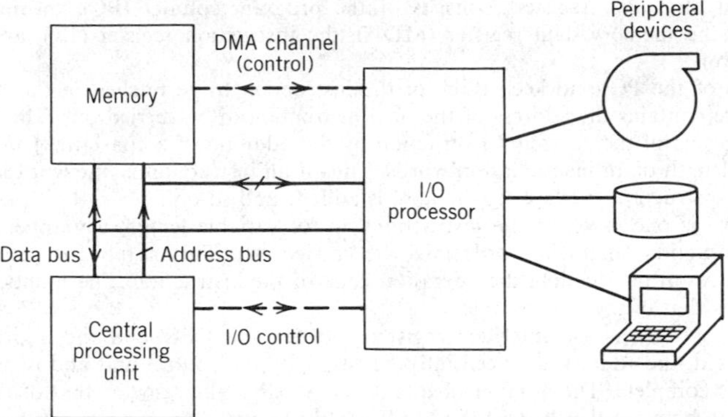

Fig. 56.1 Block diagram of general computer system. DMA, direct memory access; I/O, input–output.

AND, OR); and a set of registers, which provides storage for data in the CPU as well as certain control functions. These three units are detailed in later sections.

The main memory of the computer consists of a set of bit† strings. One or more of these bit strings constitutes a computer instruction. The length (number of bits) of a string is called the word[1,2] length of the computer and is the smallest distinctly addressable unit within the machine's memory. The maximum number of strings that can be accessed by the computer is called the address space of the machine.

Since addresses are binary numbers, the address space is an integral power of two, that is, 2^k, where k represents the number of lines in the address bus.

Typically, the address space of a computer is designated by the notation X K words; in this notation X represents the number of 1K ($1\text{K} = 2^{10} = 1024$) subsets of words in the address space. For example, a PDP-11[3,4] has an address space of 64-K: 8-bit words or 64-K bytes. A byte is commonly used to refer to a group of 8 bits.

The main memory is a random access memory. Independent of the order of the addresses appearing on the address bus, the time taken to access the data for any random address is constant. The random access characteristic distinguishes main memory from secondary storage devices such as magnetic tapes, disks, and drums.

The I/O processor of the computer is responsible for controlling the flow of information between I/O devices, the CPU, and memory sections of the machine. As shown in Fig. 56.1, the I/O processor has two channels of communication to the other devices (the dotted lines in the figure).

The channel between the memory and the I/O processor is called a direct memory access (DMA) channel and is used by devices that can transfer data at high rates of speed (speeds approaching the processor clock speed). The DMA channel speeds up the overall computer operation, since the CPU does not have to become directly involved in the transaction, allowing the concurrent execution of programs and data transfer.

The DMA channel is set up physically by enabling the I/O processor to gain control over the address and data busses during the time that the CPU is not accessing memory. This technique is called cycle stealing.

For devices whose data transfer rates are low with respect to the processor clock speed, data transfer is generally controlled by the CPU. In this case the I/O processor controls the inputting (outputting) of data from (to) the peripheral device, temporarily stores the data, and signals the CPU when data are available (or when the peripheral device is ready for new data). In this mode of data transfer, the I/O processor never gains control of the address and data busses.

In both modes of transfer, special instructions are interpreted by the I/O processor to carry out the process reliably. The instructions are detailed in Section 56.3.

56.2 CPU REGISTERS

The CPU contains a set of high-speed temporary data storage locations called registers. Some of the registers are dedicated to control and are accessible only by the control unit. The remaining registers are general-purpose registers and are accessible by the programmer. The following discussion will clarify the distinction.

† A bit is a binary digit whose value is 0 or 1.

The basic set of *control* registers consists of the program counter (PC), the memory address register (MAR), the memory data register (MDR), the instruction register (IR), and the program status word (PSW).

The function of the PC is to keep track of the instruction to be fetched in the "next" machine cycle; it therefore contains the address of the next instruction to be carried out. The PC is modified within the fetch cycle of the "current" instruction by the addition of a constant. The number added to the PC is the length of an instruction in words. Thus if an instruction is one word long, 1 is added to the PC; if an instruction is two words long, 2 is added; and so on.

If the machine is one in which the instructions are of variable length, a number of fetch cycles equal to the instruction length in words must be carried out. This number is determined by the control unit as a result of decoding the operation code of the instruction. The number of bits in the PC is the same as that of the address bus.

The MAR functions as the interface register between the CPU and the address bus. When memory is accessed, the address is placed into the MAR by the control unit and remains there until the transaction is complete. The number of bits in the MAR is the same as that of the address bus.

The distinction between the PC and the MAR should be noted at this point. If memory need *not* be referenced during the execution cycle of an instruction, the PC and MAR serve the same purpose. However, many of the machine instructions do reference memory and operate on the data there. Since the data address is generally different from the address of the next instruction, the MAR is necessary.

The function of the MDR is to provide a temporary (buffer) storage area for data that are exchanged between the CPU and memory. The data can be instructions (obtained in the fetch cycle) or operand data (obtained in the execution cycle). Because of its direct connection to the data bus, the MDR contains the same number of bits as the bus.

The IR is a register that holds the operation code of the instruction throughout the machine cycle. The code is used by the control unit in the CPU to generate the appropriate signals that control execution of the instruction. The length of the IR is the bit length of the operation code.

The program status word (PSW) stores information pertinent to the currently executing program. For example, on completion of an ALU function, a set of bits called condition codes (or flags) is altered. These bits specify whether the result of an arithmetic operation was zero or negative (or positive) or whether the result overflowed, in the case of two's complement arithmetic. The program can test these bits in ensuing instructions and conditionally change its flow of control depending on their value.

Additionally, the PSW contains bits that enable the computer to respond to asynchronous requests for service generated by I/O devices or internal error conditions. The signals are called interrupt signals and are detailed in Section 56.5.

The remaining set of registers are the general-purpose registers. They are used to store information temporarily and they also hold operands that take part in ALU operations. Sometimes the instruction set of the computer and the addressing scheme of the architecture constrain the use of some of these registers. For example, in the DEC PDP-11, which supports stack addressing, a specific register is assumed to hold the stack address, while in the IBM 370, which supports base-relative addressing, at least one of the registers must be used to hold the base (first) address of the currently executing program. While each machine allows the information within the register to be manipulated as ordinary data, during some instructions the data are used explicitly to derive a memory address. The advantage of using registers to hold computational data is that of speed. If operands need not be fetched from memory, the average two-operand computation is substantially speeded up.

56.3 INSTRUCTION TYPES

In a general-purpose computer, instructions may be broadly classified into five categories: arithmetic and logic instructions, data movement instructions, block data operations, program control instructions, and I/O instructions.

Arithmetic and logic instructions include both binary and unary operations. Binary operations require two operands and produce a single result. Addition and subtraction, as well as multiplication and division, are standard operations on most machines, with the exception of some minicomputers and microprocessors.

The logic operations included within most instruction sets are the AND operation, the INCLUSIVE OR operation, and the EXCLUSIVE OR operation.

The AND operation is carried out on a bit-by-bit basis between two operands and produces a logical one only if both bits are equal to 1. A major use of the AND function is to isolate a subset of bits from a longer operand. Fig. 56.2 illustrates the procedure graphically. Suppose that bits 4 through 7 are to be isolated from an 8-bit operand. Applying the AND function between the original operand and the bit pattern 11110000 forces the least significant bits of the result to zero (bits 3 through 0);

7 6 5 4 3 2 1 0	Bit position
1 0 1 1 0 1 0 1	Original operand—A
1 1 1 1 0 0 0 0	Mask bit pattern—B
1 0 1 1 0 0 0 0	A and B

Fig. 56.2 Illustration of the AND function.

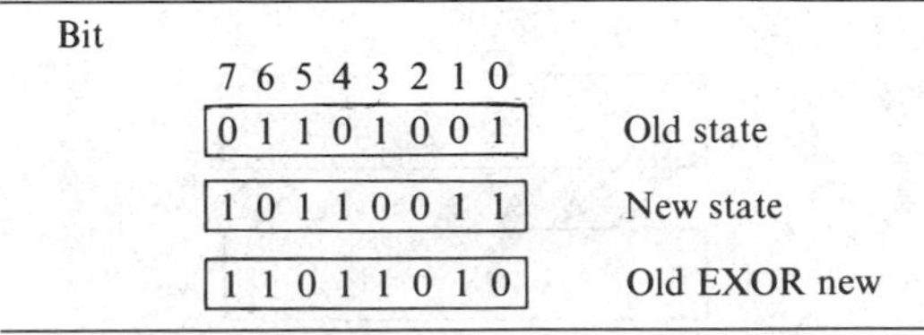

Fig. 56.3 Illustration of the EXCLUSIVE OR function. Ones indicate changes.

the most significant bits of the result are equal to the original bit pattern in bits 4 through 7. If the original bit was a 1, ANDing the bit with a 1 does not change it; if the original bit was a 0, the AND operation similarly retains the value. This filtering operation is called masking.

The INCLUSIVE OR[3,5] operation may be used to detect whether an operand is equal to zero. If the original operand is INCLUSIVE ORed with a set of zeros, the only time the result will be zero is when the original operand contained zeros; otherwise the result will be the original operand. (The INCLUSIVE OR of 2 bits is zero only if both bits are zero.)

The EXCLUSIVE OR[3,5] operation produces a 1 between two bits only if both bits differ. This operation might be used to detect changes in I/O devices. Assume that each bit of an operand corresponds to an on-off indicator for a set of devices and that the bits are set as a result of reading the state of the devices (via the I/O processor). If the scan is carried out periodically by EXCLUSIVE ORing the operand (previous scanned state) with the new scanned state, 1's appear only on those device indicators that changed. This is illustrated in Fig. 56.3.

In addition to their primary functions, the arithmetic and logic instructions cause the condition flags or codes to be set (see Section 56.2). Thus if the arithmetic system is a two's-complement system, the addition operation hardware will normally detect an overflow condition, a carry out of the most significant bit, and whether the result was equal to zero or was positive or negative. The condition codes are set accordingly.

Unary operations require a single operand and produce a single result.

Changing the sign of an operand is done by the negate (NEG) operation. In most machines this results in the two's complement of the operand being taken.

Increment (INC) and decrement (DEC) operations provide a counting capability for controlling program loops as well as an expedient method for changing addresses in an index register (see Section 56.4 for a complete discussion of indexed addressing). A clear[3] operation in which the operand is set to zero is also commonly found.

Shift and rotate[6] operations are other unary operations that are included in most instruction sets. Shifts may be to the right or to the left and either logical or arithmetic.

A logical shift[7] (left or right) moves the original bits of the operand one position (left or right). On a left shift, the most significant bit of the original operand is lost, while on a right shift, the least significant bit is lost. The bit vacated is filled with zero in both cases. An example is shown in Fig. 56.4. The arithmetic shift left is identical to the logical shift left. However, in an arithmetic shift right, the most significant bit is treated as a sign bit and the sign bit is retained in the most significant bit. Shift left corresponds to multiplication by 2, and shift right corresponds to division by 2.

The rotate operations are similar to the shift operations except that on a left rotate the most significant bit becomes the least significant bit, while on a right rotate the least significant bit becomes the most significant bit. This is illustrated in Fig. 56.5. Rotate operations provide a convenient way to check whether a bit is 1 or 0 without loss of information. The bit to be examined may be rotated left into the sign bit of the operand, with the appropriate condition code set as a result. In some architectures, rotation and shift instructions include the carry bit of the ALU.[3,4,8–10]

Data movement operations result in copying data from one operand location to another operand location. In addition to the operation code, these instructions require address information identifying

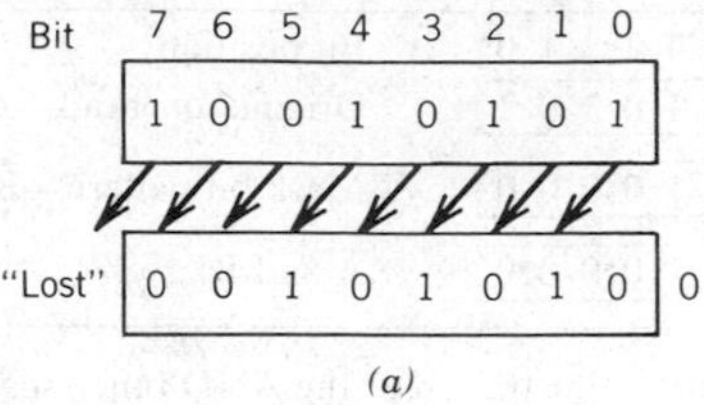

(a)

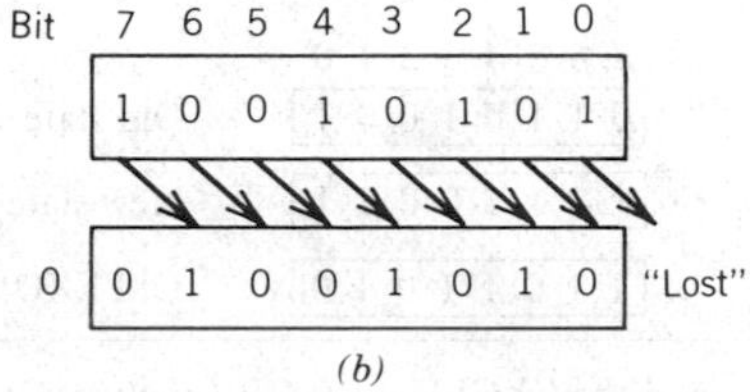

(b)

Fig. 56.4 Shift operations: (*a*) left shift, (*b*) right shift.

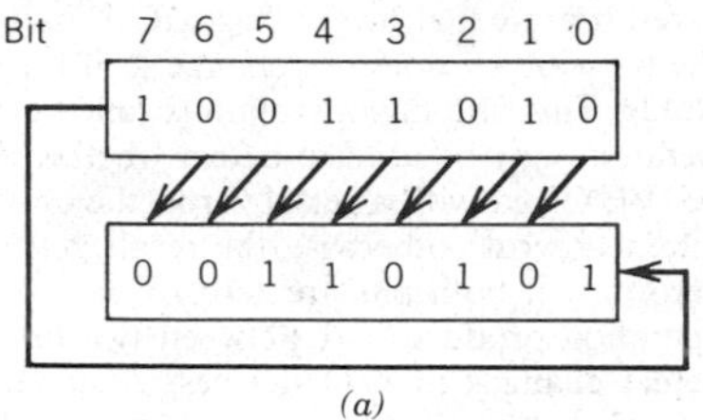

(a)

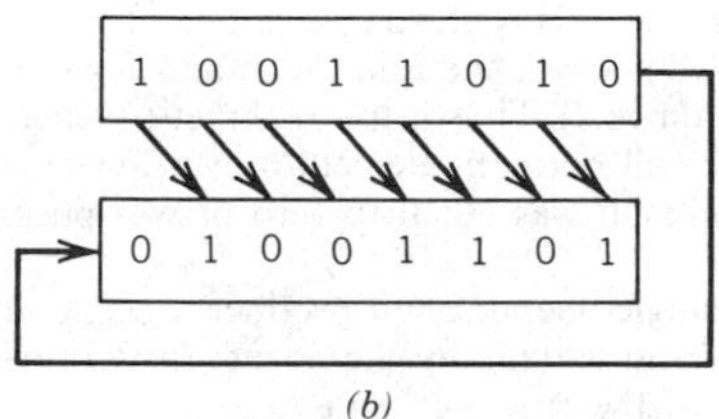

(b)

Fig. 56.5 Rotate operations: (*a*) left rotate, (*b*) right rotate.

both the source and destination operands. In a general-purpose computer, data may be moved from (1) register to register, (2) register to memory, (3) memory to register, and (4) memory to memory.

Most modern machines also provide a block data movement capability. For example, the IBM 370[6] and the Zilog Z-8000[11] instruction sets contain register-to-memory and memory-to-register moves that move all the register operands to or from main memory. These functions normally are performed when a machine is interrupted and when returning from an interrupt service program. The same machines also have a memory-to-memory move for a block of data. These instructions require, in addition to source and address information, the number of bytes to be moved. The latter information must be encoded as part of the instruction.

Block data operations are operations performed on a set of operands rather than on a single operand. (While block data movement instructions could be considered within this category, it was felt that they are conceptually better placed within the data movement category.) These operations are becoming more commonplace in recent computers, for example, the Z-8000 and the Intel 8086.[11]

Searching a set of operands for the first occurrence of a particular bit pattern is accomplished by a compare-and-repeat operation. Specified within the instruction is the test operand, the starting address of the set of operands, and the operand containing the length of the set. This instruction is useful in an editing operation, for example, where the input string might represent text information input from a keyboard and the objective is to remove characters representing blanks.

Comparing two sets of operands in memory on an operand-by-operand basis is another typical block data instruction. This instruction is useful when searching a table of symbols (composed of a set of operands) for the occurrence or nonoccurrence of a particular symbol in the table. This instruction requires the starting address of both operands as well as of the length operand.

"Translate" is an instruction that enables a set of data to be replaced on a one-to-one, operand-for-operand basis from a translation table. The instruction operates in the following manner and is illustrated in Fig. 56.6. As part of the instruction, the starting address of the source operand set (the data to be translated) is specified, along with the starting address of the translation table. In addition, the length of the source operand is specified. Each operand is then numerically added to the starting address of the translation table and is replaced by the operand found at the resultant address. This process is continued until the entire source operand is replaced.

One application of this instruction would be in a communications processing environment. Suppose a terminal inputs a message in EBCDIC (Extended Binary Coded Decimal Interchange Code), but that in passing messages over communications links in the network, ASCII (American Standard Code for Information Interchange) is used. Using this one instruction, the message can be translated in its entirety.

Program control instructions enable a program to depart from the inherent sequential nature of the machine cycle of the computer. In other words, sections of instructions may be passed over either as the result of a condition code being set or as a direct result of program design. The instruction corresponding to the former case is the conditional jump instruction, while in the latter case an unconditional jump instruction would be used. Both types of jump instruction require specification of the address in memory at which the next instruction will be found. Execution takes place simply by setting the PC to that address. The conditional jump instruction must additionally contain the condition code setting corresponding to the execution of the jump. In case the condition(s) is not satisfied, the next sequential instruction is executed.

Also included in the class of program control instructions are the subroutine "call" instructions. These may be conditional or unconditional. Their operation is similar to the jump instructions, with one important difference: The address of the next sequential instruction is saved in the machine's memory or registers.

The address transferred to by a call instruction is the first address of the subroutine. The subroutine then proceeds to execute. The last instruction of the subroutine is a "return" instruction, which executes by recovering the address saved by the "call" and placing it in the PC.

Subroutines are used for two major reasons in structuring a program. First, if the same instructions appear repeatedly in different areas of a program, they may be made into a subroutine to save memory space. Second, a complex program may be conceptually split into a set of smaller units, each of which is conveniently coded into a subroutine. The original program might then consist primarily of calls to subroutines. A common location for the return address storage is in a *stack*[4,12,13] (see Section 56.4), a last-in first-out data structure controlled by the CPU. If this approach is used, the return address is placed on the top of the stack by the "call" instruction and taken from the top of the stack by the "return" instruction.

I/O instructions[2,6,8,14–16] cannot be readily generalized, since they are highly dependent on the

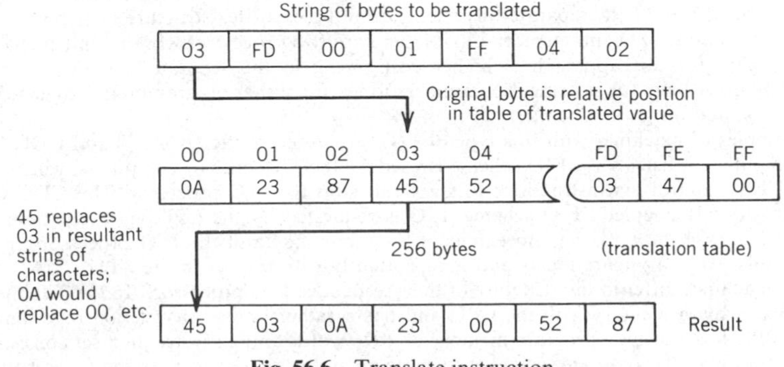

Fig. 56.6 Translate instruction.

computer system of interest as well as the particular devices to which the system interfaces. A rough classification can be made, however, on the basis of the bussing structure of the computer system.

Some CPUs time multiplex their data and address busses among all the attached peripheral equipment. In particular, all[2,4] models of the PDP-11 possess such an architecture (the bus system is called the UNIBUS). In addition, most microprocessors are constructed with a similar bussing arrangement.[11] The CPU does not generate any special control signals to the devices that distinguish data bound for memory from those bound for a peripheral device controller. Neither does the CPU distinguish between data read from the memory or from a peripheral device.

From a programming point of view, accessing either the memory or a peripheral simply requires the same set of instructions. Systems that operate in the manner described are called memory-mapped I/O systems.

Programming a device in memory-mapped I/O systems does require knowledge of the device and its characteristics, even though no special instructions are necessary. The device is characterized as a set of memory locations that are further broken up into two subcategories: a set of status and control registers and a set of information registers.

The status and control registers generally contain information about the state of the device, for example, whether it is idle or busy, whether maintenance is being carried out on it, and so on. Control information is also stored in these registers such as the type of parity (even or odd) and the rate of data transmission. The information in the control and status registers is primarily used to provide an overall picture of the hardware as it is being programmed.

The information registers, on the other hand, provide a buffer for information being transferred between the CPU and the peripheral. In the case of a device transferring data on a character-by-character basis, there may be only two registers, one holding data from CPU to device and one holding data from device to CPU. If programmed I/O is carried out on a unidirectional device (transmit only or receive only), then only one register is necessary.

In contrast to program-controlled I/O, a DMA device generally requires more information registers. In this case the CPU must provide a starting address within main memory for the data and the number of bytes to be transferred. These two registers, in addition to the data buffers, are necessary for DMA devices. Figure 56.7 helps to illustrate these concepts more concretely. The figure shows the layout of the receiver status register and the receiver buffer register for a DZ-11 interface that connects the PDP-11 CPU to an asynchronous communications line via the UNIBUS. The line can be used for transmission and reception, but only the receiver is considered here. Details on the transmitter may be found in Ref. 17.

The receiver status and buffer registers are located at addresses 174000 and 174004, respectively. These are standard addresses reserved for the unit. The function of the receiver buffer register is to buffer the character read from the line. In the status register, there is a set of bits (15–12) devoted to detection of various error conditions occurring within the unit itself. For example, bits 14 and 13 are set when an unexpected event occurs within the device (e.g., if the modem carrier turns off on the line, bit 14 is set), while bit 12 is set if the receiver buffer has not been read by the program and another character is entering the peripheral. Bits 10 and 9 enable the program to select the character length via software, while the baud rate of the device may be similarly programmed by appropriately setting bits 3 and 4. Bit 6, when set, enables the device to interrupt the CPU when a character is received, while bit 7 indicates whether the receiver buffer contains a character. The remaining bits are control bits used when connecting the DZ-11 to a modem.

The alternative approach to carrying out I/O is to provide a separate bus and a stored program capability, distinct from the CPU. The IBM 370[6,16] series of computers and the CDC 6600/7600[1] computers take this approach. The physical bus structures for these systems are shown in Fig. 56.8.

The chief advantage of the separate bus over the multiplexed bus structure is increased throughput, since with separate I/O and memory busses there are two processors working in true parallelism. The penalty paid for this approach is higher cost owing to the separate bussing structure, and additionally, the necessity for systems programmers to learn another programming language (the I/O processor is a stored program machine).

Two examples of machines with this type of I/O structure are the IBM 370 and CDC Cyber 70 series of computers. Master-slave I/O scheme is used by the 370 series of computers, where the CPU has specific I/O control via instructions to start and stop the I/O processor. The CDC Cyber 70 series uses a "loosely coupled" I/O scheme. I/O is requested by the CPU via a common memory location. In this case the CPU has no explicit I/O commands and the I/O processors (peripheral processing units) work asynchronously and independently with respect to the CPU.

The two machines differ in the structure of their respective I/O processors. In the 370, the I/O[6,16] processor shares main memory with the CPU and has a rather limited instruction repertoire. In the CDC, each PPU has a distinct program memory of 4096 words and an instruction set consisting of 66 different operations. The only area shared[1] in main memory is used to pass messages between the processors.

	Register	15 (MSB)	14	13	12	11	10	09	08	07	06	05	04	03	02	01	00 (LSB)
		BYTES HIGH								BYTES LOW							
DR0	CONTROL & STATUS (CSR)	RO TRDY	RW TIE	RO SA	RW SAE	NOT USED	RO TLINE C	RO TLINE B	RO TLINE A	RO RDONE	RW RIE	RW MSE	RW CLR	RW MAINT	NOT USED	NOT USED	NOT USED
DR2	RECEIVER BUFFER (RBUF)	RO DATA VALID	RO OVRN	RO FRAM ERR	RO PAR ERR	NOT USED	RO RX LINE C	RO RX LINE B	RO RX LINE A	RO RBUF D7	RO RBUF D6	RO RBUF D5	RO RBUF D4	RO RBUF D3	RO RBUF D2	RO RBUF D1	RO RBUF D0
DR2	LINE PARAMETER (LPR)	NOT USED	NOT USED	NOT USED	WO RX ON	WO FREQ D	WO FREQ C	WO FREQ B	WO FREQ A	WO ODD PAR	WO PAR ENAB	WO STOP CODE	WO CHAR LGTH B	WO CHAR LGTH A	WO LINE C	WO LINE B	WO LINE A

Fig. 56.7 Status and buffer register layout of the DZ11. From Digital Equipment Corp.,[17] with permission.

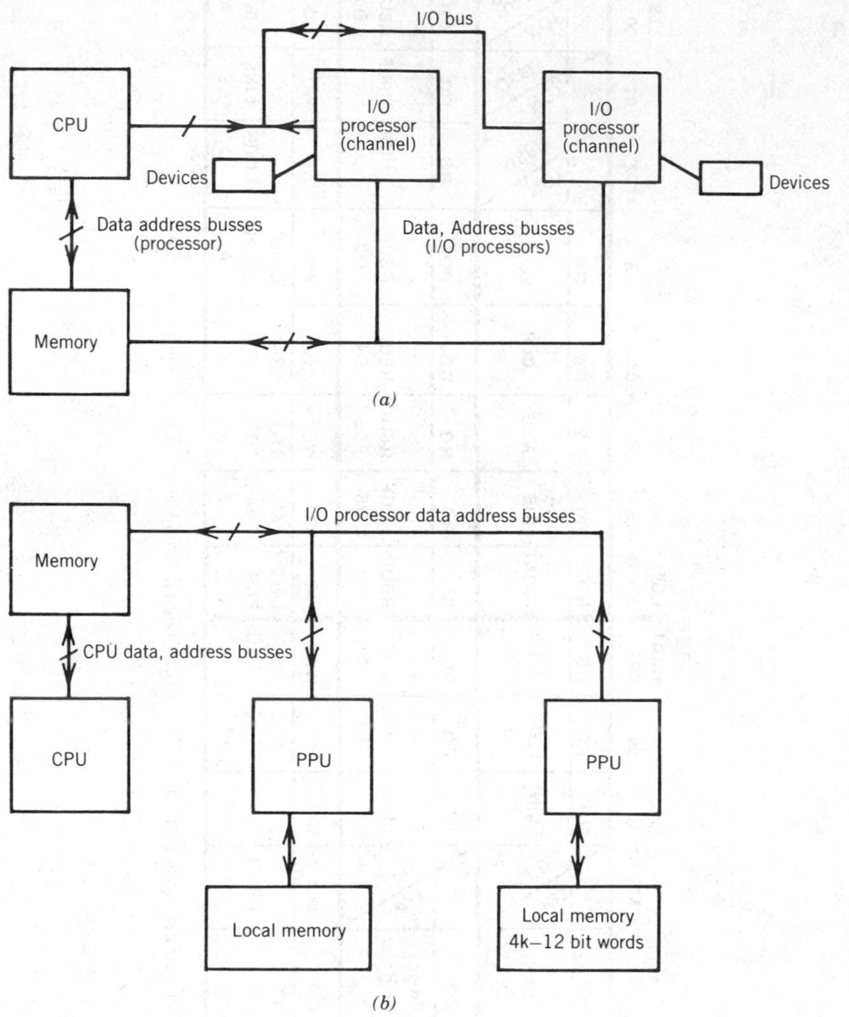

Fig. 56.8 Bussing structures of computer systems: (*a*) IBM 370, (*b*) CDC 6600. CPU, central processing unit; I/O, input–output; PPU, peripheral processing unit.

56.4 INSTRUCTION FORMATS AND ADDRESSING MODES

To implement a particular instruction set on a computer, a design choice must be made as to the layout of the bits of the instructions. The layout is called the instruction format. Generally an instruction consists of an operation code and other data pertinent to the successful completion of the operation. For example, if the operation code specifies an arithmetic operation, the location of the operands must be explicitly (or implicitly) specified as well as the location of the resultant operand. For block data operations, a block data memory-to-memory move operation, for example, the starting address of both blocks and the total count of the words to be moved should be included within the instruction.

The most general form of an instruction is shown in Fig. 56.9. The instruction is broken up into a set of fields. The first field is called the operation code and the remaining fields data fields. The data fields are discussed in the latter part of this section; the focus here is on the operation code.

The major purpose of the operation code is specification of an action by the computer. In addition the code also contains information as to the instruction length. Most modern computers have variable-length instructions for maximum flexibility and minimum memory utilization. More

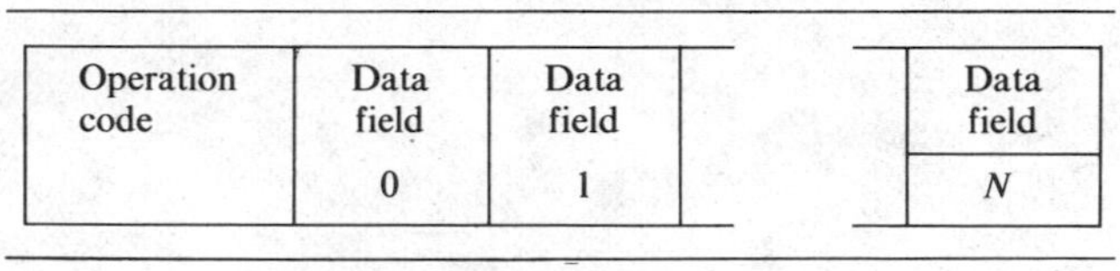

Operation code	Data field	Data field	…	Data field
	0	1		N

Fig. 56.9 General instruction format.

complex instructions (block data types) require the longest lengths, while simple instructions (transferring data from one general purpose register to another) require the shortest lengths. The only constraint invoked on the instruction lengths is that they be an integral multiple of the word length of the machine. Figure 56.10 shows the various instruction formats for the IBM 370 computer. There are five formats into which an instruction is placed.

The RR format (register-register) has three fields. The operation code specifies the action and the remaining fields designate 1 of 16 general-purpose registers as source and destination operands. The length of this instruction is 16 bits. The 32-bit instructions are classified as RX (register-memory), SI (storage-immediate), and RS (register-storage) instructions.

In the RX format, the R1 field specifies a general-purpose register as one operand, while the X (index), B (base), and D (displacement) fields specify a memory address using the index-base-displacement addressing scheme common to these computers (discussed later). The SI format specifies an 8-bit constant as one operand (constant field), and the B and D fields specify the address of another operand. The RS format is similar to the RX format, except that the R3 field in this case is used as a third operand address. The final format is the 48-bit SS (storage-storage) format. In this case the B and D fields specify the starting addresses of two block operands and the length field, the lengths of the blocks taking part in the operation.

In the 370 machines, the operation code is fixed at 8 bits regardless of the length of the instruction.

In contrast to having a fixed operation code, some machines have variable-length operation codes. These codes are referred to as expanding operation codes.

Most general-purpose computers support instruction sets that contain zero, one, two, and three address fields within them. For example, a machine architecture supporting stack addressing may contain a "push" operation (zero address), a complement operation (one address instruction), an addition operation (two addresses), and a data movement operation (three addresses). Consider such an architecture and assume that the number of three-address instructions for the machine is 10, the number of two-address instructions is 12, the number of one-address instructions is 13, and the number of zero-address instructions is 10. Further assume that the address field information is 4 bits long. (While on the surface the address constraint appears to be restrictive, if the machine holds all its address information in one of 16 general purpose registers it is justified.) The expanding operation code scheme for this machine is illustrated in Fig. 56.11. The instructions requiring the maximum number of address fields are assigned 4-bit op codes, with the code 1111 reserved as a prefix to all the two-address op codes. The 8-bit double prefix 1111 1111 is assigned to all the single-address operation codes, and finally the 12-bit prefix 1111 1111 1111 is assigned to all the zero-address operation codes. Because of the symmetry of the example, the basic instruction size is 16 bits, independent of the addressing modes of the machine. The transition to each new op code level as described in the example results in only 15 op codes being available at each level,† since each op code is defined as being prefixed by a fixed number of ones. Suppose the machine required 30 two-address instructions (8-bit op codes). Obviously the illustrated scheme would not work. One approach to solving this problem is illustrated in Fig. 56.12. In this case two 4-bit prefixes, 1110 and 1111, are used to make the transition from 4-bit to 8-bit op codes. The tradeoff here is a reduction of the number of three-address instructions from 15 to 14. The extra prefix allows 16 additional 8-bit operation codes to be defined. Table 56.1 illustrates the expanding operation code idea as implemented on a specific processor, the PDP-11/70.

An alternative to the prefix method of varying the length of the operation codes is to assign a field within the instruction that when decoded specifies the length. This approach allows full decoding of the number of bits at each level, greatly increasing the number of op codes available for use. To illustrate, if 2 bits were assigned to define the length of the instructions in the foregoing example, 16 three-address codes would be available, 256 (2^8) two-address codes, 4096 (2^{12}) one-address codes, and 65,356 (2^{16}) zero-address codes. The cost would be 2 bits per instruction word (Fig. 56.13).

†Except at the zero-address level.

RR format

Operation Code	Register 1	Register 2
8	4	4

RS format

Operation Code	Register 1	Register 3	Base Register 2	Displacement 2
8	4	4	4	12

RX format

Operation Code	Register 1	Index Register 2	Base Register 2	Displacement 2
8	4	4	4	12

SI format

Operation Code	Byte Constant	Base Register 1	Displacement 1
8	8	4	12

SS format

Operation Code	Operand Length	Base Register 1	Displacement 1	Base Register 2	Displacement 2
8	8	4	12	4	12

Fig. 56.10 Instruction formats, system 370. RR, register-register; RS, register-storage; RX, register-memory; SI, storage-immediate; SS, storage-storage.

4 bits—three-address op codes

XXXX	A1	A2	A3	$XXXX \neq 1111$ 15 op codes

8 bits—two-address op codes

1111	XXXX	A1	A2	$XXXX \neq 1111$ 15 op codes

12 bits—one-address op codes

1111	1111	XXXX	A1	$XXXX \neq 1111$ 15 op codes

16 bits—zero-address op codes

1111	1111	1111	XXXX	16 op codes

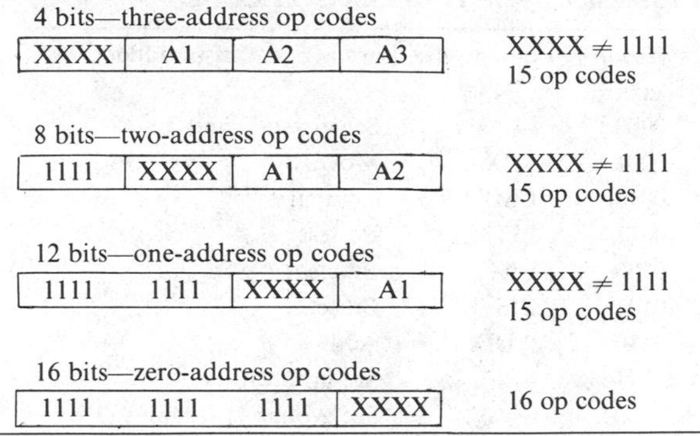

Fig. 56.11 Expanding operation (op) codes.

XXXX	A1	A2	A3	$XXXX \neq 1111$ or 1110 14 op codes
1111 or 1110	XXXX	A2	A3	$XXXX \neq 1111$ 30 op codes

Fig. 56.12 Increasing the number of available operation (op) codes (double prefixing).

L	Operation codes + data fields
2	16

00⇒ three-address 4-bit op codes, 16 instructions
01⇒ two-address 8-bit op codes, 256 instructions
10⇒ one-address 12-bit op codes, 4,096 instructions
11⇒ zero-address of 16-bit op codes, 65,536 instructions

Fig. 56.13 Increasing the number of variable-length operation (op) codes (length field). L, length of instruction.

The *addressing modes* of a computer refer to the methods utilized in the machine to generate an operand address. All instructions require at least one operand for their execution; many require two or more. The operands are stored in either a CPU general-purpose register or in memory. If the operand is in a general-purpose register of a CPU that has k such registers, then a minimum $\log_2 k$ bits are necessary to specify the register location within the CPU. For example, referring back to Fig. 56.10, the RR, RX, and RS specifications involve an operand stored in a CPU register. In this case $k = 16$, and 4 bits are necessary to encode the address information. In the case of a memory operand, however, a complete address must be specified. This address is called the *effective address* of the operand and may be specified in a number of ways. For the remainder of the discussion, it will be assumed that the length of an effective address is in bits.

One approach used to generate memory addresses is to include them in the instruction. This method is called direct[5] addressing. No calculations are necessary in the CPU to derive an address. As part of the execution sequence, the address is read into the CPU (during the fetch cycle), then the operand at that address is read into the CPU to take part in the action dictated by the instruction. In a computer that supports direct addressing, at least two memory accesses (fetch instruction, fetch operand) are required for execution of the instruction.

A major drawback to direct addressing is the large number of bits required when more than one operand is included in an instruction. For example, in a block data movement instruction on the 370,[6] two address operands are required. One address specifies the source, and the other the destination. Since addresses are 24 bits in this machine, a total of 48 bits would be required for both operands, as well as 8 bits for the operation code and additional bits for the length of the block.

TABLE 56.1. PDP-11 OPERATION CODES

Group 1, 4-bit opcodes. Format:[a] xxxxssssssdddddd

xxxx		
0000	—	See groups 2 and 3
0001	MOV	Move
0010	CMP	Compare
0011	BIT	Bit test
0100	BIC	Bit clear
0101	BIS	Bit set
0110	ADD	Add
0111	—	See group 10
1000	—	See groups 11, 12, and 13
1001	MOVB	Move byte
1010	CMPB	Compare byte
1011	BITB	Bit test byte
1100	BICB	Bit clear byte
1101	BISB	Bit set byte
1110	SUB	Subtract
1111	—	(Floating-point instructions)

Group 2, 8-bit opcodes. Format: 00000xxxkkkkkkkk

xxx		
000	—	See group 4
001	BR	Branch
010	BNE	Branch not equal
011	BEQ	Branch equal
100	BGE	Branch greater than or equal
101	BLT	Branch less than
110	BGT	Branch greater than
111	BLE	Branch less than or equal

Group 3, 7-bit opcodes. Format: 00001xxrrrdddddd

xx		
00	JSR	Jump to subroutine
01	—	See group 5
10	—	See group 6
11	—	Spare

Group 4, 10-bit opcodes. Format: 00000000xxdddddd

xx		
00	—	See group 7
01	JMP	Jump
10	—	See group 8
11	SWAB	Swap bytes

TABLE 56.1 (*Continued*)

Group 5, 10-bit opcodes. Format: 0000101xxxdddddd

xxx		
000	CLR	Clear
001	COM	Complement (one's complement)
010	INC	Increment
011	DEC	Decrement
100	NEG	Negate (two's complement)
101	ADC	Add carry
110	SBC	Subtract carry
111	TST	Test

Group 6, 10-bit opcodes. Format: 0000110xxxdddddd

xxx		
000	ROR	Rotate right 1 bit
001	ROL	Rotate left 1 bit
010	ASR	Arithmetic shift right 1 bit
011	ASL	Arithmetic shift left 1 bit
100	(MARK)	Clean up stack. dddddd = count
101	(MFPI)	Move from previous instruction space
110	(MTPI)	Move to previous instruction space
111	(SXT)	Sign extend

Group 7, 12-bit opcodes. Format: 0000000000xxcccc

xx		
00	—	See group 9
01	—	Spare
10	CCC	Clear condition codes
11	SCC	Set condition codes

Group 8, 13-bit opcodes. Format: 0000000010xxxrrr

xxx		
000	RTS	Return from subroutine
011	(SPL)	Set priority level

Group 9, 16-bit opcodes. Format: 0000000000000xxx

xxx		
000	HALT	Halt
001	WAIT	Wait
010	RTI	Return from interrupt
011	BPT	Breakpoint
100	IOT	I/O trap
101	RESET	Reset
110	(RTT)	Return from trap

TABLE 56.1 (*Continued*)

Group 10, 7-bit opcodes. Format: 0111xxxrrrdddddd

xxx		
000	(MUL)	Multiply
001	(DIV)	Divide
010	(ASH)	Arithmetic shift
011	(ASHC)	Arithmetic shift combined
100	(XOR)	Exclusive or
111	(SOB)	Subtract one and branch

Group 11, 8-bit opcodes. Format: 10000xxxkkkkkkkk

xxx		
000	BPL	Branch on plus
001	BMI	Branch on minus
010	BHI	Branch high
011	BLOS	Branch low or same
100	BVC	Branch on overflow clear
101	BVS	Branch on overflow set
110	BCC	Branch on carry clear
111	BCS	Branch on carry set

Group 12, 10-bit opcodes. Format: 1000101xxxdddddd

xxx		
000	CLRB	Clear byte
001	COMB	Complement byte
010	INCB	Increment byte
011	DECB	Decrement byte
100	NEGB	Negate byte
101	ADCB	Add carry byte
110	SBCB	Subtract carry byte
111	TSTB	Test byte

Group 13, 10-bit opcodes. Format: 1000110xxxdddddd

xxx		
000	RORB	Rotate byte right 1 bit
001	ROLB	Rotate byte left 1 bit
010	ASRB	Shift byte right 1 bit
011	ASLB	Shift byte left 1 bit
100	—	Spare
101	(MFPD)	Move from previous data space
110	(MTPD)	Move to previous data space
111	—	Spare

Source. A. S. Tanenbaum,[1] reprinted with permission.
[a] ssssss specifies a source; dddddd specifies a destination; rrr specifies a register; x specifies opcode bits; kkkkkkkk specifies an offset or constant; cccc specifies condition code bits.

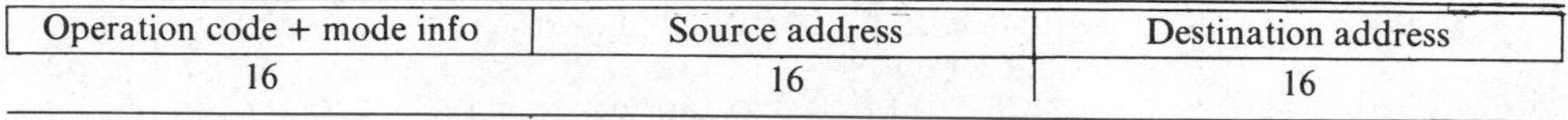

Operation code + mode info	Source address	Destination address
16	16	16

Fig. 56.14 Direct addressing example.

Address field of PDP-11 instruction

Mode bits	Register
3	3

Mode = 001 ⇒ register indirect addressing

Example

Address	Field
001	010

Operand address found in register 2, 3000 (hexadecimal). R2 [3000]

Fig. 56.15 Register indirect addressing.

When direct addressing is disallowed, this instruction is carried out using only 32 bits of address information. A striking example of the difference in instruction lengths when using direct addressing as opposed to register addressing can be seen in the following example. The ADD instruction format for a PDP-11[4,18] with direct addressing is shown in Fig. 56.14. Note that the instruction is 6 bytes long. The 16-bit source address follows the first 2 bytes of the operation code and the destination address follows the source address. If the operands are stored in registers, the instruction requires only 2 bytes, and the source and destination registers are effectively part of the operation code.

Another approach to addressing is to use a CPU general-purpose register to store the information. In this case the technique is called register indirect[18] addressing. Figure 56.15 shows an example. In this case the operand address is contained within a general-purpose register. To specify this form of addressing, the instruction needs three fields: the operation code, a field specifying that the addressing is indirect, and a field specifying the register. Since the number of general-purpose registers is far less than the number of memory locations, the instruction length (relative to direct addressing) is short. The tradeoff is that the instruction requires another information field. In addition, the number of memory accesses for the same operation using register indirect addressing is smaller than when using direct addressing. To support this addressing mode, the general-purpose registers within the machine must be long enough to hold a full address.

An extension to the register indirect addressing technique is the idea of indexed addressing. In an indexed addressing architecture, an operand address is derived by adding a constant stored in the instruction to the contents of a register to generate an address. (If the constant is equal to zero, the effect is the same as register indirect addressing.) An example of this form of addressing is shown in Fig. 56.16 for the PDP-11. The source operand for the instruction is in one register (R0), while the

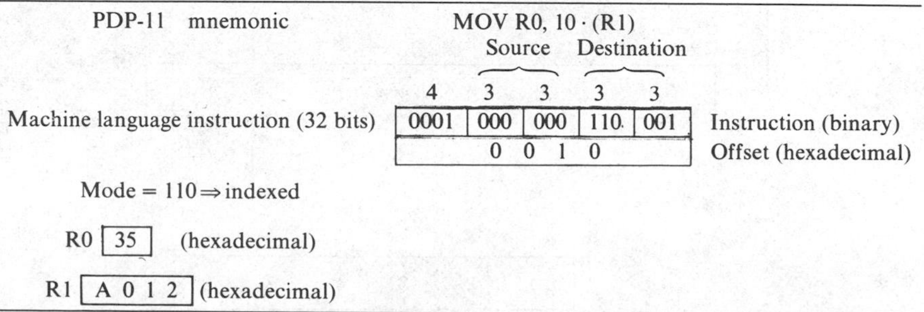

Fig. 56.16 Indexed addressing. The contents of R0 are moved to the location specified by the contents of R1 plus 10 (A022). R1 is *not* altered.

```
          MOV #10., R0
          CLR R1
LOOP:     INCB TABLE (R1); Add 1 to byte at Table +c(R1)
          INC R1; Add 1 to the index register
          DEC R0
          BNE LOOP; Branch back if count not zero
```

Fig. 56.17 Indexed addressing, example program. This PDP-11 program segment increments the first 10 bytes in locations TABLE + 0 through TABLE + 9 by address modification using index register R1. R0 is the loop counter.

16-bit offset (constant) is added to register R1 to generate the 16-bit effective address. The index technique is useful when generating consecutive memory addresses, as in a program that performs repetitive operations on a set of operands one after the other. In this case the repeated instruction would store the start address of the operand set, and at each iteration the index register would be adjusted to effect the proper address modification. An example of the technique is illustrated in Fig. 56.17. In this case each operand is addressed sequentially by adding 1 to the index register at each iteration of the loop. Some machines provide a mechanism for automatically[4,11] incrementing the index register. This feature is called autoindexing and results in shortening an iterative program by one step. When provided on a machine, the incrementing may be done prior to computing the effective address (preindexing) or after computing the effective address (postindexing).

A family of machines that uses the indexed addressing technique exclusively in computing addresses is the IBM 360/370[6,16] computer systems. The effective address is generated by adding the contents of one of the general-purpose registers (called a base register) to a 12-bit offset. Depending on the instruction type, another register's contents may be added to the base register and offset to form the final address. The technique is illustrated in Fig. 56.18. It allows programs in the system to be moved from one location to another without modification of the machine code. Note that if the first instruction in the program loads the starting address of the program into the base register, address information within each instruction need not be recomputed, since it is an offset relative to the base register. At assembly time, the offsets are easily computed relative to starting address zero.

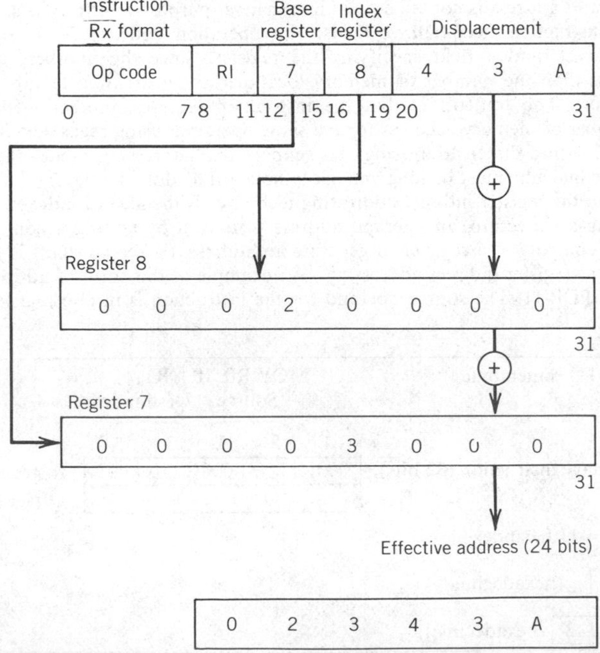

Fig. 56.18 Base-indexed addressing, IBM 370. All values in hexadecimal.

Program address		Source		Destination		
		Reg	Mode	Reg	Mode	
1000	Op code	1	0	7	6	
1002	4	5	6	2		Offset
1004						

Fig. 56.19 Relative addressing, PDP-11. During instruction execution, the PC contains 1004 effective destination address = 1004 + 4562 = 5566 (values in hexadecimal).

Symbolic instruction encoding
ADD @A, R0

Memory Address		Source Reg	Source Mode	Destination Reg	Destination Mode	
1000	Op code	7	7	0	0	
1002	4	5	6	2		Offset
1004						
⋮						
5566	4	0	0	0		

Fig. 56.20 Indirect addressing, PDP-11. Effective source address = 1004 + 4562 = 5566, but since indirect addressing is specified (source mode = 7), operand is found at location 4000.

Code possessing this attribute is called position independent[1,18] or relocatable code. This type of code is well suited to a multiprogramming environment, in which memory is shared among many users.

In modern computer architectures, the idea of relocation was extended by defining the base register to be the program counter (PC).[4] Since the program counter contains the address of the next instruction, if the offsets are computed relative to the PC, relocatable code automatically results. This form of addressing is commonly referred to as relative addressing. The technique is illustrated in Fig. 56.19 for the PDP-11.

Memory indirect addressing is conceptually similar to register indirect addressing (the technique is also known as deferred addressing). Indirect addressing[4,11,18] is best described by an example. Consider the instruction shown in Fig. 56.20. The interpretation of the instruction is to add the operand found at the address specified by the operand at A to R0, then store the result in R0. Note that the data at address A are interpreted as an address; the operand added to R0 is found at that address. This technique is generally used in machines with a limited number of general-purpose registers as a replacement for register indirect addressing. As an example, consider the PDP-8.[2] This machine has only one accumulator, and all instructions are 12 bits long. Excluding the operation code, 9 bits of the instruction are used for addressing. Since an effective address is 12 bits long, indirect addressing must be used to access the entire address space. The 9 address bits are used as follows.

Memory is logically divided on the machine into 32 128-word pages (see Fig. 56.21). Two of the nine address bits are used as control bits by the hardware. Bit 3 of the instruction is used as a page indicator. If the bit is 0, the effective address is derived by appending the 7 remaining bits in the instruction to 5 leading zeros, specifying an address on page 0. If the bit is a 1, the effective address is derived by appending the 7 bits to the 5 most significant bits of the program counter (the current page the program is executing on). Bit 4 is the indirect bit. If it is a 1, the data at the effective address are used as a 12-bit address, otherwise as an operand.

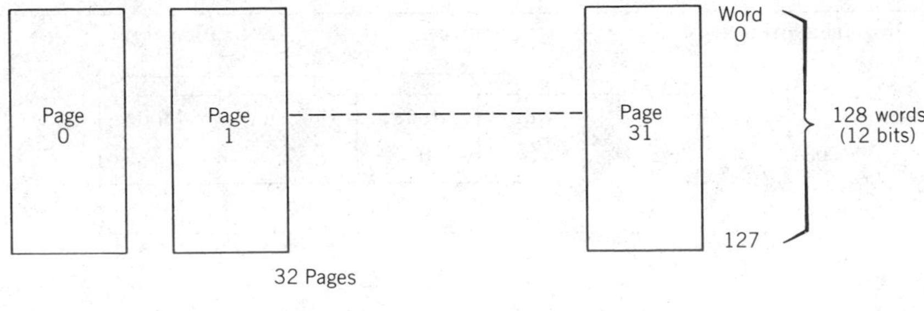

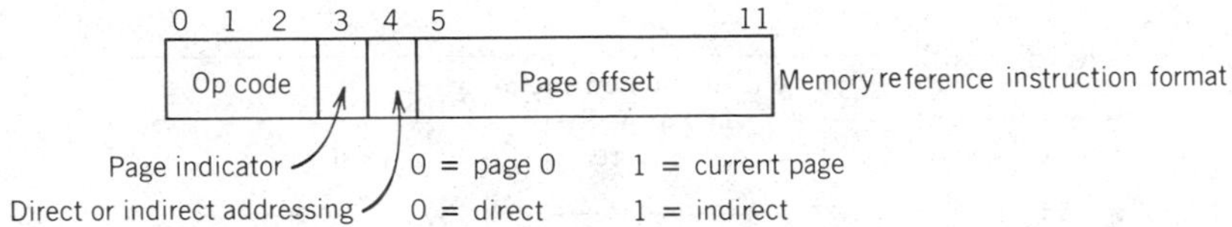

Fig. 56.21 Memory layout, PDP-8.

56.5 OPERATING MODES

Large modern computing systems generally service a great many users. In this section the architectural features necessary to facilitate this type of operation are discussed.

An operating system is a set of programs that allows the user of a computer system controlled access to its resources. The resources include the CPU, memory, and I/O devices. The operating system assigns CPU time, allocates memory space, and assigns and controls I/O devices for each user. These functions are carried out transparently; that is, the programmer writes the program as if the entire computer system is dedicated to his or her job. To provide operating system features, the architecture of the machine must possess certain properties. To begin with, the machine must have at least two[19] different modes of operation. One is called the supervisor (privileged) mode and the other the user (problem) mode.

When the machine is in the supervisor mode, the CPU can execute all machine instructions. This is the mode of operation in which the operating systems programs run, giving them total control of the computer system. All peripheral device service requests go through the operating system, since I/O instructions are generally issued only when the machine is in the supervisor mode.

In the problem mode, the CPU is restricted from executing the entire instruction set. In particular, machine control and I/O instructions are not allowed to be executed. The restriction on the I/O instructions is made clear by an example. Suppose that there is one printer to service all the users of the computer system, and the CPU is being shared on a round-robin[2,20,21] basis. If a line of output was printed by one user in his or her time slot and then the next user printed a line in his time slot, the printout would be a set of interlaced lines from the set of users of the system. Quite clearly it would be difficult to sort one user's lines from another. To resolve this situation, when I/O is required by a program the program issues a special instruction to the operating systems program. This instruction is called a supervisor call[19] (trap)[4] instruction. The effect of the instruction within the CPU is to generate an internal signal at the next machine cycle that causes the machine to switch its mode to the supervisor mode. In addition, the signal causes one of the operating systems programs to start execution. Because execution of the special instruction causes the program to be interrupted, the signal is called an interrupt signal.

When the interrupt occurs, the operating systems program proceeds to set up the mechanism for data transfer with the problem program. This involves allocating buffer space in memory for the data, suspending the operation of the interrupted program until I/O is completed, and allocating the CPU to another program. A convenient tool for describing the progression of a problem program in a multiuser environment is a state[19,22,23] diagram. Figure 56.22 illustrates the state diagram for a typical problem program. In the RUN state, the program is executing; in the HOLD state, the program is waiting for I/O to complete; in the READY state, the program is ready to use the CPU. The events causing the state transitions are also shown in the figure.

As illustrated in the state diagram, once I/O is complete, the job progresses to the READY state,

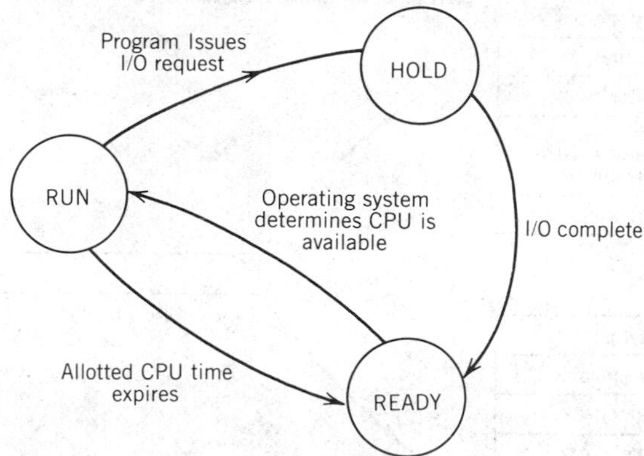

Fig. 56.22 State diagram of a program in a resource-sharing environment. CPU, central processing unit; I/O, input–output.

and when the CPU becomes available, the job goes back into execution. To start the problem program at its correct address, the operating system must retain that information at the time the interrupt occurs. The address is available in the PC at the time of the interrupt, since the PC contains the address of the next instruction to be executed. The situation is entirely analogous to a subroutine branch. In the interrupt case, the supervisor call instruction transfers control to a hardwired address. In machines[2,6] that support stack instructions, the PC contents are normally stored in the stack when the interrupt occurs. In machines that do not support stacks directly, the return address is stored in a fixed set of memory locations. The interrupt actions for a PDP-11 (stack) and a 370 (nonstack) are shown in Fig. 56.23. In addition to the return address, the state of the machine must also be stored when an interrupt occurs. The machine state is the contents of all the CPU registers as well as the condition codes when the interrupt occurs. In some processors,[11] all the registers are saved automatically, while in others, the machine state must be saved by software. I/O devices also use the interrupt mechanism to communicate with the CPU. The only difference between a peripheral and an internal interrupt is that the latter is synchronous with the processor's clock, while the peripheral interrupt occurs when the device is ready to accept or to transmit information.

From a system point of view, the mechanics of both are the same. When a peripheral generates the interrupt, however, the starting address of the service program to handle the interrupt may be generated externally.

Two basic methods are used to identify the interrupting source: the hardwired method and the vectored method. In the hardwired method, a set of interrupt signal lines is available to the peripheral system designer to use as interrupt request lines to the processor. Each line has a particular memory address associated with it to which control is transferred on receipt of a signal on the line. The interrupt service program must be placed at the address if the line is used by the device. In the vectored method, the peripheral device provides the location of the address of the interrupt service program as well as a new program status word (PSW) value. In the PDP-11, for example, these entities are hardwired into a pair of addresses assigned to the peripheral. When an interrupt request is received at the CPU, the CPU reads the address of the interrupt service program from the device. The new PSW is assumed to be located at the next word in memory. Figure 56.24 illustrates the procedure.

In a multiuser system there are many sources of interrupts. For example, there is a real-time clock generating interrupts periodically which allows the operating system to perform scheduling. There are many I/O devices, ranging from high-speed devices (disks, drums, and tapes) to low-speed devices (terminals and printers). In addition, there may be circuitry that detects serious system faults, such as power failures. In light of the number of devices, the operating system must have the capability to prioritize the sources of the interrupts as well as the ability to ignore them.

The priority structure imposed on the various sources depends on the use of the system. In a real-time control system, an analog-to-digital (A/D) converter may have priority over all other interrupts so as not to lose any data. In a computer center environment, the real-time clock would have priority over most other interrupts so that the throughput of the system is maximized. Naturally, a power failure or other catastrophic type of interrupt takes precedence over all others.

On many machines the priority structure is built into the hardware; if two or more interrupts occur simultaneously, the interrupt with the highest priority is serviced. The remaining interrupts are

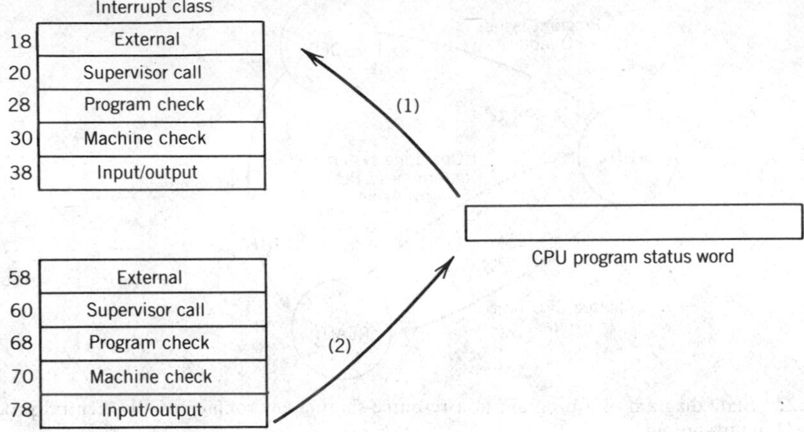

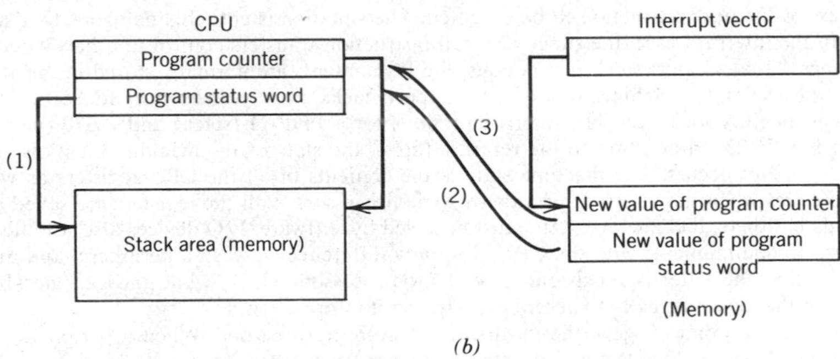

Fig. 56.23 Interrupt action: (*a*) IBM 370. On interrupt, the central processing unit (CPU) program status word (PSW) is stored in the old PSW area (1) and then the new PSW replaces it (2). One word in each area is assigned for the five distinct interrupts handled by the 370. (*b*) PDP-11. On interrupt, the CPU program counter and PSW are placed into the stack area (1, 2). The new values are loaded from consecutive memory locations whose address is determined by the interrupt vector address. The address is returned by the device to the CPU (3) as part of acknowledging the interrupt.

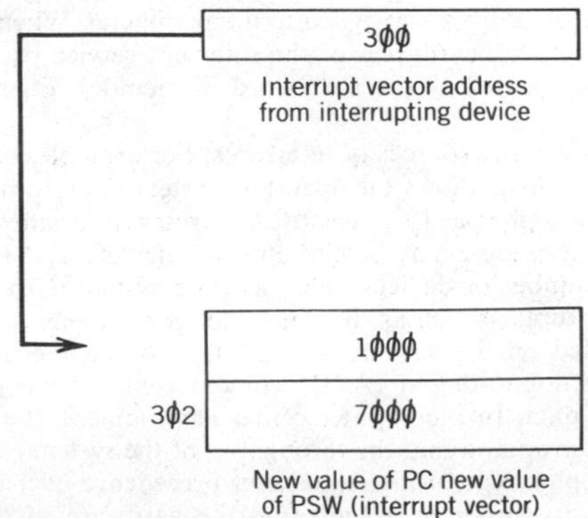

Fig. 56.24 Vectored interrupt, PDP-11. Interrupt service routine starts at location 1000. PC, program counter; PSW, program status word.

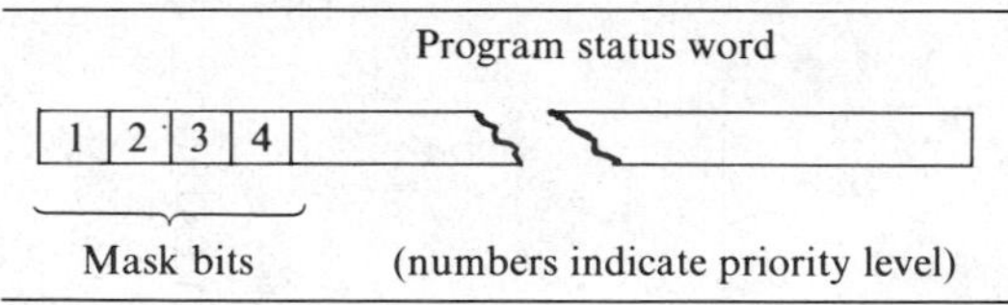

Fig. 56.25 Interrupt masking. When the mask bit is a 1, the interrupt at that level is recognized by the processor; when it is 0, the request is deferred.

deferred. A common technique used to implement the deferral of interrupts is shown in Fig. 56.25. Part of the PSW functions as a set of "mask" bits. There is 1 bit for each priority level. If the mask bit is a 1, the interrupt requests at that level are passed to the CPU, while if the bit is a 0, the requests are ignored. When an interrupt occurs at a particular level, the new value of the PSW is loaded as part of the acknowledge sequence. The mask bits in the new PSW generally are set to 0 for interrupt requests at a lower priority (of the same priority) and to 1 for higher priority devices.

If more than one device shares the same interrupt line to the processor, external hardware can be built to arbitrate the requests in case of simultaneous inputs. An alternative is to use a software polling approach. A 370 assembly language program using polling is shown in Fig. 56.26. The program is entered when the I/O processor generates an interrupt. It is assumed that the I/O processor has a card reader and line printer attached to it. When one of these devices completes I/O, it generates a device-end interrupt. The code for this interrupt is left in the channel status word[6,16] (CSW). If the interrupt was a device end, the service program determines the source by examining the interrupt code, found in the I/O PSW. The service program then restarts the device that completed by writing its I/O control program address into the I/O buffer area. More details on I/O programming on the 370 and other CPUs are found in Refs. 1 through 4, 11, 13, 16, 18, 19, 24, and 25.

56.6 CONTROL UNIT DESIGN

Proper machine operation[5,26] requires information transfer and processing to take place in a particular time sequence. The sequencing function is carried out by the control unit (CU) of the CPU. This section describes the design techniques applicable to the CU.

A typical CPU block diagram, showing the registers and interconnecting busses, is shown in Fig. 56.27. Since actual machines vary greatly in architectural detail, this is only loosely related to a real machine. However, the structure illustrated suffices as a vehicle for the discussion of general principles of CU design.

The diagram shows a register file, whose output is attached to the inputs of two multiplexers (LMUX and RMUX). The multiplexer outputs form the inputs to a multifunction ALU, where the FSEL (function select) input leads determine the function. The ALU output goes through a shifter that can perform a 1-bit shift left or right or alternatively no shift. The output of the shifter is fed into demultiplexing circuitry (DEMUX) so that it may be loaded into any one of the registers. The register is selected by the DSEL (destination select) leads.

The register file contains the MAR, MDR, and PC, while the IR is a separate register. The IR outputs are connected to the CU inputs.

In addition to the IR outputs, the CU inputs include the flag bits, interrupt lines, and a clock. The outputs of the CU control the flow of data in the register–ALU section as well as between the CPU and memory.

To illustrate the CU function, consider the operations required to implement a fetch cycle. Assume that the PC is contained within the register file. The first task to be performed is to transfer the PC contents to the MAR. This is carried out as follows.

1. The PC is selected as one of the inputs to the ALU.
2. A register containing zeros is selected as the other ALU input.
3. The ALU operation selected (via FSEL) is addition (of the two ALU inputs). No shift is the selected shifter operation.
4. The output of the shifter is directed to the MAR (via the DSEL) lines.

The CU generates the appropriate selection and control signals that close the gates in the data paths simultaneously so that the operation is carried out. The next step is to gate the contents of the MAR onto the address bus and generate a read signal to the memory. Afterward the PC is incremented. These operations are carried out by a set of substeps analogous to the transfer of the PC contents to the MAR.

```
                                    EXTERNAL SYMBOL DICTIONARY
SYMBOL  TYPE  ID  ADDR    LENGTH  LD  ID
LIST     SD   01  001000  000099

LOC     OBJECT CODE  ADDR1  ADDR2  STMT   SOURCE STATEMENT
001000                               1    LIST   START  X'1000'
001000  05F0                         2           BALR   15,0      ESTABLISH BASE REGISTER
001002                               3           USING  *,15      TELL THE ASSEMBLER
001002  8000  F096   01098           4           SSM    =X'00'    TURN ALL I/O INTERRUPTS OFF
                                     5    *
001006  4110  F036          01038    6    LOOP   LA     1,READ    SET CAW TO CHANNEL PROGRAM FOR READING
00100A  5010  0048          00048    7           ST     1,CAW
00100E  9C00  000C   0000C           8           SIO    X'00C'    START READER
001012  9D00  000C   0000C           9           TIO    X'00C'    WAIT UNTIL READER IS FINISHED
001016  4770  F010          01012   10           BNZ    *-4
                                    11    *
00101A  4110  F03E          01040   12           LA     1,PRINT   SET CAW TO CHANNEL PROGRAM FOR PRINTING
00101E  5010  0048          00048   13           ST     1,CAW
001022  9C00  000E   0000E          14           SIO    X'00E'    START PRINTER
001026  9D00  000E   0000E          15           TIO    X'00E'    WAIT UNTIL PRINTER IS FINISHED
00102A  4770  F024          01026   16           BNZ    *-4
00102E  47F0  F004          01006   17           B      LOOP      PROCESS NEXT CARD
                                    18    *
```

(Statements 1–17 bracketed: executed by CPU)

```
001032  000000000000
001038  0200104800000050       [19  READ     CCW   X'02', BUFFER, X'00', 80   READ A CARD
001040  0900104800000050  exe- |20  PRINT    CCW   X'09', BUFFER, X'00', 80   PRINT A CARD
001048                    cuted|21  BUFFER   DS    CL80                       80-BYTE BUFFER AREA
                          by   |22  *
000048                 channel |23  CAW      EQU   72                         CAW = LOCATION 72
                               [24           END
001098  00                      25                 = X'00'

                                         RELOCATION DICTIONARY
   POS.ID        REL.ID       FLAGS      ADDRESS
     01            01           08       001039
     01            01           08       001041

                                         CROSS REFERENCE
SYMBOL      LEN    VALUE   DEFN      REFERENCES
BUFFER     00080   001048  00021     0019  0020
CAW        00001   000048  00023     0007  0013
LIST       00001   001000  00001
LOOP       00004   001006  00006     0017
PRINT      00008   001040  00020     0012
READ       00008   001038  00019     0006
```

Fig. 56.26 Software polling, IBM 370. From S. Madnick and J. Donovan,[19] reprinted with permission.

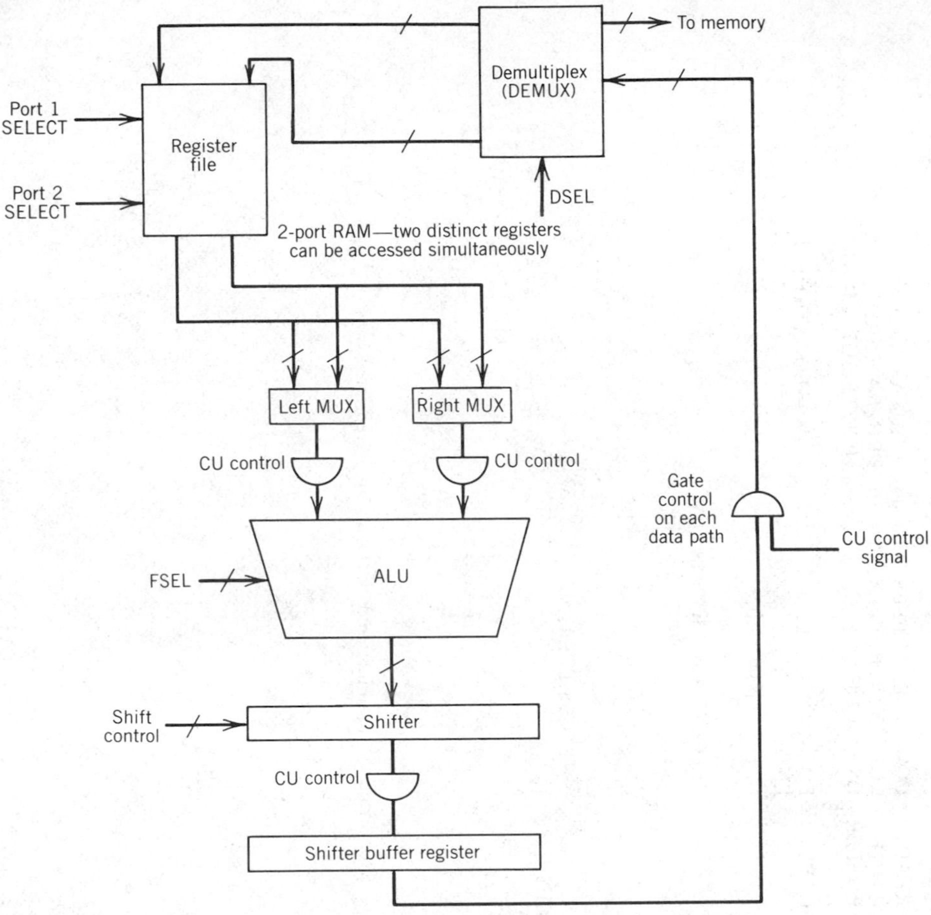

Fig. 56.27 Central processing unit (CPU) block diagram. CU, control unit; ALU, arithmetic–logic unit; DSEL, destination select; FSEL, function select.

When the operation code is read, it is transferred directly to the MDR and then to the IR, where it becomes one of the inputs to the CU. The control unit decodes the instruction and sequences the ALU–register sections of the CPU to carry out the instruction.

In the fetch cycle described, four steps must be sequentially executed. Assuming that each of the steps takes one clock pulse to execute, the entire sequence requires four clocks. Within each step, further subdivision of time must occur. Consider the data transfer between the PC and MAR. To carry the operation out successfully, the select lines on the various elements must first be set up. Then the output of the ALU can be steered to the MAR. The point is that the selection and transfer of information cannot take place simultaneously, since the ALU requires time to perform the addition. Thus the operation requires two subcycles (refer to Fig. 56.27).

The actual gating may cause even more problems. Consider the updating of the PC. In this operation, the PC is the destination as well as the source register. If the ALU is not carefully designed, its output may change two or more times within the subcycle, causing erroneous PC readings. Control unit design requires careful timing analysis to avoid race conditions and usually requires multiphase clock outputs along with the gate control signals.

Hardwired CU Design

In the hardwired approach, the sequencing is implemented by a control gate network similar to that of Fig. 56.27. Each control signal is formed by ANDing† together a multiphase clock output and one

†The symbol "·" represents AND.

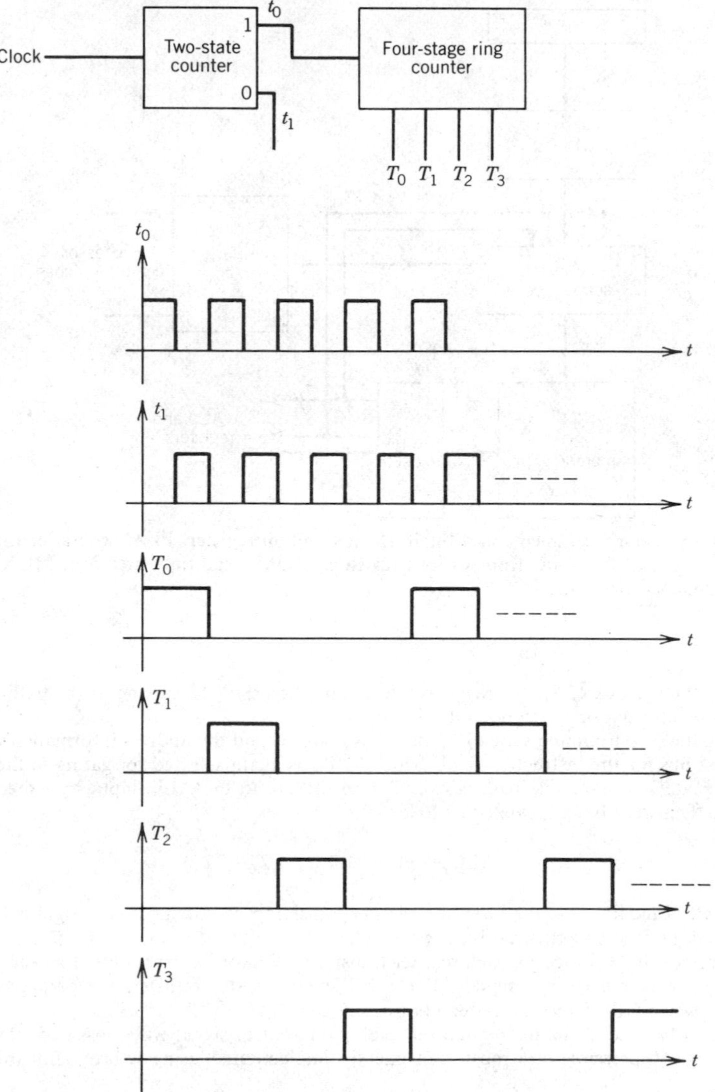

Fig. 56.28 Generation of timing signals for control unit sequencing.

or more nonclock control signals. Figure 56.28 shows a possible configuration for the generation of the timing signals.

The two-stage ring counter is driven at the clock frequency. The divide-by-2 counter drives a four-stage ring counter. The outputs of the two-state counter are the multiphase clock signals, and those of the four-stage ring counter the timing state signals. Suppose the timing state signals are called T_0, T_1, T_2, and T_3 and the subclock signals are t_0 and t_1. Then $T_0 \cdot t_1$ represents the second timing state in master clock cycle T_0. Now further assume that a single flip-flop distinguishes between the fetch and execute cycles in the CPU. Then $F \cdot T_0 \cdot t_0$ represents the first subcycle of the first timing state within the fetch cycle. This control signal is used for the following gating (Fig. 56.27):

1. Gates the PC onto the right inputs of the ALU.
2. Gates select information to the ALU and shifter.
3. Gates 0 onto the left inputs of the ALU.

On the trailing edge of the t_0 clock, the shifter outputs are gated into the SBR (shifter buffer register).

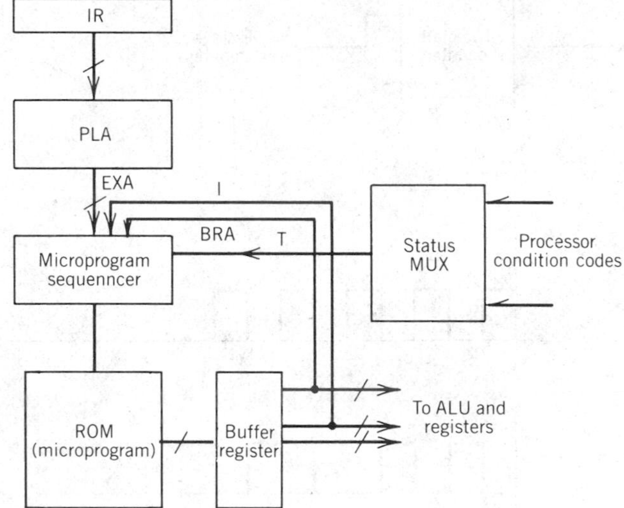

Fig. 56.29 Microprogrammed control unit. IR, instruction register; PLA, programmable logic array; EXA, I, BRA, and T, inputs from various registers; ROM, read-only memory; MUX, multiplexer; ALU, arithmetic–logic unit.

During the t_1 clock of T_0, the MAR is selected on the DSEL leads, and at the trailing edge of the t_1 clock, the information is strobed into the register.

During the second timing state (T_1), the PC is updated and the address information is placed onto the address bus for the instruction fetch. Thus the PC is again selected for gating to the ALU by the control signal $F \cdot T_1 \cdot t_0$. The path is closed from the PC to the ALU input by either $F \cdot T_0 \cdot t_0$ or $F \cdot T_1 \cdot t_0$. This may be abbreviated as follows:

$$\mathrm{ALU_r} \leftarrow \mathrm{PC} \qquad F \cdot t_0 \cdot (T_0 + {}^{\dagger}T_1)$$

The left half of the line is in the form of a register transfer[5,22,27] statement. The right half of the line is the control signal causing the transfer to be enabled.

To complete the CU design, each register transfer necessary for execution of all the instructions is listed, along with the control signals. For a particular register transfer, the path control signal is generated by the OR of the associated control signals.

Owing to the size of the instruction set, each path control signal will consist of many terms. The resulting logic expressions are most conveniently implemented using a programmable logic array (PLA).

Microprogrammed CU Design

In this design technique, the set of control signals required to perform the sequence of register transfer operations (microoperations) is stored in a read only memory (ROM).

The block diagram of a microprogrammed control unit is shown in Fig. 56.29. The CPU control registers are shown as a separate block. The output of the IR in this diagram is applied to a PLA. The PLA outputs are input to the microprogram sequencer. These outputs represent the starting address of the microprogram associated with the operation code in the IR.

The microprogram sequencer is connected to the address leads of a control ROM. This ROM contains the microinstructions, the control signals that permit the proper execution of the machine's macro instruction set. The ROM data outputs are buffered in a register.

In addition to the outputs of the PLA, the sequencer has three other inputs: the I inputs, coming from the I field of the buffer register; the BRA inputs, coming from the branch field of the buffer

† "+" represents logical OR.

register; and the T (test) input, coming from the status multiplexer. The inputs to the status multiplexer are the condition flags of the processor. The overall operation of the control unit is as follows.

An instruction is fetched from memory and its operation code is placed into the IR. The PLA derives the starting address of the microprogram to carry out the operation. The EXA inputs of the sequencer are internally buffered in a register called the microprogram counter (μPC). The outputs of the μPC are connected to the ROM address bus.

The microprogram is then executed sequentially and terminates with a transfer instruction that results in the fetch cycle microcode being executed.

Each microinstruction is composed of a set of fields. The fields are used to perform definitive functions with respect to the controlled architecture. To illustrate the point, refer back to Fig. 56.27. One field is necessary to control the flow of data to each side of the ALU. The size of the field is determined by the number of registers to be switched to the ALU. For example, if eight registers are required to be switched, 3 bits are necessary for each field.

In addition to the fields for the ALU selection, a field is necessary for the destination select and also for the shifter function. In addition, the memory control functions are also field controlled.

When the microoperation is read into the buffer register, the outputs select the appropriate functions to be performed within the clock cycle. The falling edge of the clock causes the execution of the operations.

As an example of microprogramming, reconsider the fetch cycle of the CPU. A symbolic encoding of the microprogram is shown in Table 56.2. The comments should be self-explanatory.

Complex operations require lengthy microcode. In designing the algorithm for the execution of an instruction, it may be convenient to have a branching capability within the microprogrammed control. For example, the last instruction in the execution cycle of a microprogrammed machine should be an unconditional branch to the microcode for the fetch cycle. In addition, conditional branching may aid in the design of complex algorithms. The branch capability is implemented on the model CPU by a combination of the status multiplexer, the status select bits (a field of the microinstruction), and the branch field of the microinstruction.

Internally, the microprogram sequencer is structured as shown in Fig. 56.30. From the block diagram, the ROM address is selected from one of three sources: the μPC, an internal stack memory, or the branch field of the microinstruction. The source selection is a function of the I bus and the SEL lines, two of the microinstruction fields.

The SEL lines switch one of the condition code outputs (e.g., the zero flag) to the input of the PLA. The PLA is used to determine the source (via the multiplexer) of the next microinstruction address. The other PLA inputs come from the I bus.

The I-bus leads are used to determine the microinstruction type. Suppose that there are three types of microinstructions: (1) an execute type, where the next microinstruction resides at the next location in the ROM; (2) an unconditional branch type, where the next microinstruction address is specified by the branch field of the current microinstruction; and (3) a conditional branch type, where the branch address in the current microinstruction is used if the condition is met (if the condition is not met the program proceeds sequentially). With these three types, the selected source address as a function of the instruction type and the status multiplexer output are shown in Table 56.3. The I bus consists of two bits (I_1 and I_0). The code 00 represents an execute-type microinstruction. In this case the selected input to the ROM is the μPC, which contains the address of the next

TABLE 56.2. MICROPROGRAM EXAMPLE

Instruction format

Control fields—Each field corresponds to a function in the microprogram control unit (Fig. 56.29)

	LMUX[a]	RMUX	LSEL	RSEL	SHIFTER	FSEL[b]	DSEL	Comment
1)	Port 1	Port 2	PC	0	0	L + R	MAR	MAR ← PC
2)	Port 1	Port 2	PC	1	0	L + R	PC	PC ← PC + 1
3)	Port 1	Port 2	MAR	0	0	L + R	Memory	Memory ← MAR

[a] LMUX, left multiplexer, RMUX, right MUX; LSEL, left select; RSEL, right SEL; FSEL, function SEL; DSEL, destination SEL; MAR, memory address register; PC, program counter.
[b] FSEL: Encodes ALU function, that is, L + R, means add left input to right input.

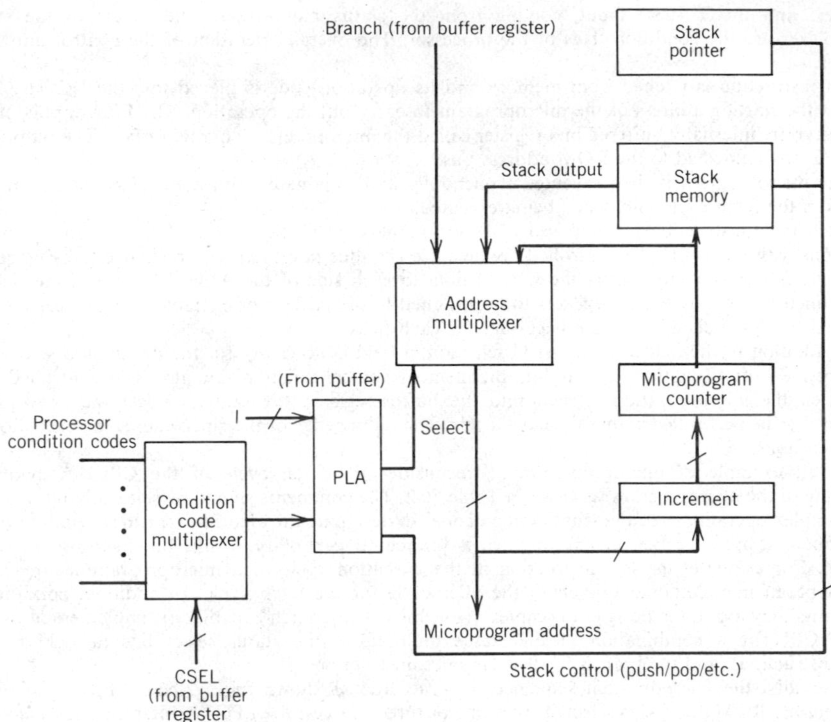

Fig. 56.30 Microprogram sequencer block diagram. PLA, programmable logic array; CSEL, count select.

TABLE 56.3. ADDRESS SELECTION AS FUNCTION OF INSTRUCTION TYPE AND CONDITION CODES

I_1	I_0	Condition[a]	Address of Next Microinstruction[b]
0	0	X	μPC (next sequential address)
0	1	X	BRA input (unconditional branch)
1	0	1	BRA input } conditional branch
1	0	0	μPC input } conditional branch

[a] The CSEL inputs select the particular condition to be tested.
[b] μPC, microprogram counter; BRA, branch.

sequential instruction. (Note that the address outputs have 1 added to them via the increment circuit and are fed back into the μPC). The selected input in this case is not a function of the T input.

Code 01 represents an unconditional branch. Independent of T, the selected source is the BRA field of the current microinstruction.

Code 10 represents a conditional branch. The condition tested is programmed by the SEL field of the current microinstruction. If the condition is true (T = 1), then the BRA field is output, while if the condition is false, the μPC is output.

The stack permits the microcode to use subroutines. (Return addresses may be stored on the stack.) Now assuming a subroutine branch instruction type within the microinstruction set, the stack inputs to the address multiplexer become clear. For a subroutine branch, the BRA field of the microinstruction will be selected, and simultaneously the μPC is placed in the stack. For a subroutine return type of instruction, the stack inputs are selected as the source of the ROM address (the equivalent of a pop operation from the stack).

To implement the stack features, additional hardware must be added to the sequencer structure. A stack pointer register is necessary to keep track of the current top-of-stack address. Another lead on the I bus is also needed. Conditional subroutine branches and returns are implemented in a manner similar to the conditional branch instructions.

Microprogram sequencers of this type are available as single-unit large-scale integrated circuits. They are made using TTL (transistor-transistor logic) and ECL (emitter-coupled logic) technology, to obtain high-speed operation. Chapter 59 deals with microprogramming in detail and the references contain details of the circuitry used in the design of the systems.[8,9,22,27–30]

56.7 CLASSIFICATION OF COMPUTER SYSTEMS

Computer system architectures have been classified by Flynn,[31] and to a large degree this classification has been used extensively in the literature. In this section, the classification is reviewed and examples of systems are given for each class.

The classification is based on two ideas: the concept of an instruction stream and the concept of a data stream. An instruction stream may be thought of as a program in execution and a data stream as a sequence of data operated on by a (set of) program(s). Single processors execute a single instruction stream, the machine language of the processor. Multiple processors may execute multiple instruction streams or they may execute the same instruction stream (identical machines). Data streams may also be classified as single data sets or multiple data sets.

Forming all possible combinations of instruction stream and data stream leads to the following categories of computer systems:

1. Single instruction, single data (SISD).
2. Multiple instruction, single data (MISD).
3. Single instruction, multiple data (SIMD).
4. Multiple instruction, multiple data (MIMD).

The general-purpose architectures described thus far fall into the categories SISD and MISD. The PDP-11, for example, is an SISD-type machine, since all instructions, including I/O instructions, are contained within the basic processor instruction set. For a particular job, the necessary data are kept within one data set, consistent with the definition. Processing of a program on this machine is sequential (in time).

The IBM-370 system is an example of the MISD machine type. In this case the I/O processor may be considered a separate, limited computer. The data on which the CPU and I/O processor operate are, for a particular job, contained within one data set. Processing of a job on this machine is also sequential.

Sequential processors of these types are generally well suited for implementation of sequential algorithms and represent a majority of the systems presently in operation. Speed improvement has been gained within these systems by hardware advances rather than by structural changes in the algorithms they execute. The basic speed-up techniques involve overlapping the fetch and execute[22] cycles of the machine and the use of semiconductor memory technology. The most common approach in overlapping the fetch and execute cycles is pipelining. In such an architecture, the functional steps involved in the two machine cycles are allocated to different units. The result is a sequential assembly line of hardware units, as shown in Fig. 56.31. While one instruction is being executed, another is being fetched (in the same time). The concurrency of operations speeds up program execution time on the average. Examples of pipeline processors are the IBM 360/195[1] and the INTEL 8086.[11]

The SIMD and the MIMD both are, from a practical standpoint, multiprocessor systems. The SIMD classification[31] relates to multiprocessor systems with identical processors operating on physically different data sets. This type of processor is also referred to as an array processor, which is structurally shown in Fig. 56.32. This structure is most useful for algorithms that exhibit a high degree of parallelism, for example, matrix multiplication and weather prediction problems. Examples of this type of machine are the ILLIAC IV,[1,32] produced by the Burroughs Corporation and designed at the University of Illinois, and the CRAY-1.[12,14,33]

The final category of machine is the MIMD class. While these systems are also multiprocessor systems, they differ from the MISD type in that they operate on diverse data bases. An important example of this classification is the computer–communications network.[1,34–36] In particular, the ARPA network is one in which diverse computers, each having separate data bases and local operating systems, are interconnected via communications channels to each other. The communications channels consist of telephone lines and radio links (Fig. 56.33). Communications are handled by minicomputers devoted to that task. The communications computers are called interface message

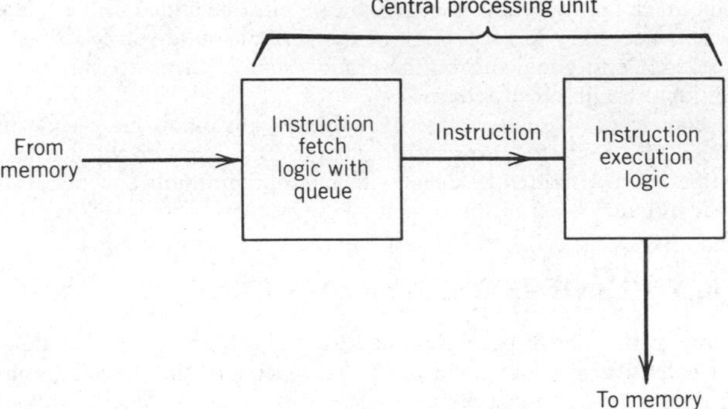

Fig. 56.31 Pipelined architecture.

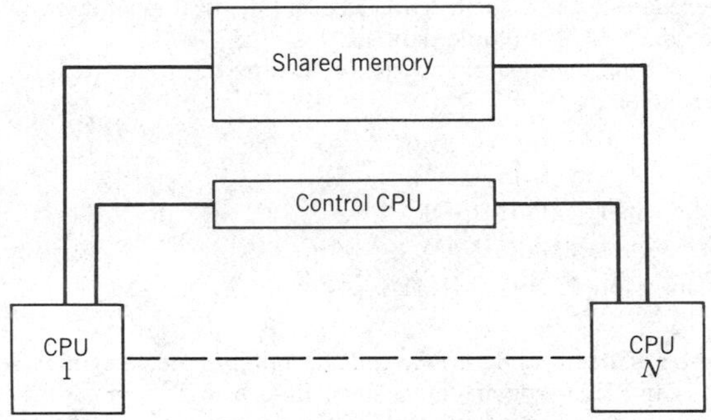

Fig. 56.32 Parallel processor functional diagram. CPU, central processing unit.

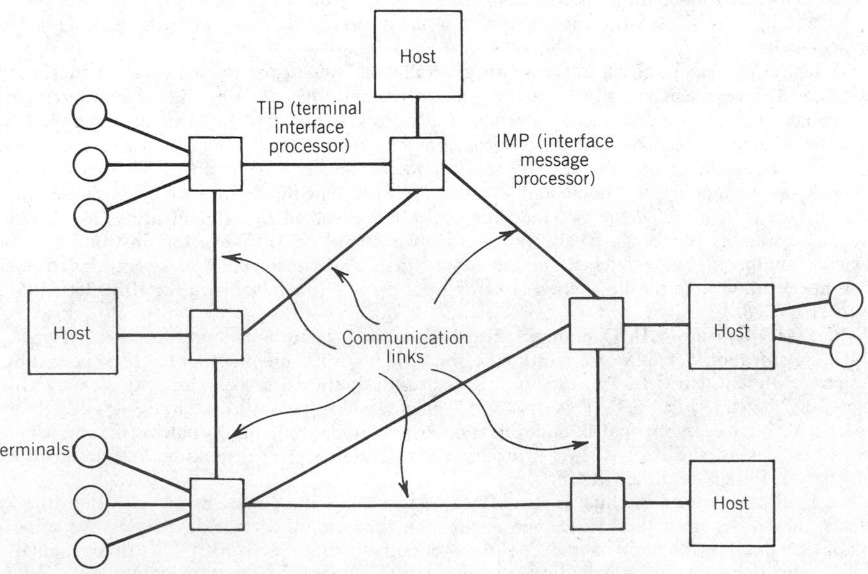

Fig. 56.33 Distributed computer network.

processors (IMPs). The network structure makes the entire set of resources of each of the machines available to the network subscriber. Other examples of computer networking include Local Area Networks (LANs), electronic mail service, off-track betting, and financial networks. Computer networks are discussed in Chapter 65.

In addition to geographically distributed systems, there are local computer networks where the computational functions are distributed among a set of processors. For example, a time-sharing system may use a microcomputer to off load a host machine of the task of terminal communication. The microcomputer accepts input from a set of keyboards on a character-by-character basis, checks the protocol, and forms messages to be sent to the host for processing.

Because of the hardware costs, multiprocessing was rarely used in the past. Presently, however, the rapid decline in hardware costs due to VLSI (very large-scale integration) technology is making these systems feasible and cost effective. The design considerations for the MIMD systems include such issues as bussing structure, functional decomposition, and operating system. Some multimicroprocessor systems have been designed and built. Reference 37 contains a detailed description of the implementation of the HYDRA system and Ref. 36 explores the design issues facing future systems in depth.

56.8 COMPARISON OF COMPUTERS

Table 56.4 shows a comparison of a large computer (IBM 370/168), two minicomputers, and a microprocessor. The comparison parameters are typical. With the ever-increasing use of LSI microprocessors, in addition to implementation of large word length machines using bit-slice logic, the trend will be toward drastically lowered costs in the high-end computers and more utilization of low-end microprocessors in small-business applications.

TABLE 56.4. COMPARISON OF COMPUTERS

	IBM 370/168	DEC PDP 11/45	Computer Automation NAKED MINI	Intel MCS-80
Cost	$4.5 million	$50,000	$2500	$250
Word length (bits)	32	16	16	8
Memory capacity (8-bit bytes)	8.4 million	256K[a]	64K	64K
Processor Add time	0.13 μs	0.9 μs	3.2 μs	2.0 μs
Maximum 1/0 data rate (bytes/second)	16 million	4 million	1,400,000	500,000
Number of general purpose registers	64	16	3	7
Peripherals (from manufacturer)	All types	Wide variety	Disk, tape, card, line printer, CRT, cassette	Paper tape reader, floppy disk, PROM programmer
Software	All types	Wide variety	Operating system, assembler, FORTRAN, BASIC	Assembler, monitor, PL/M, editor

Source. L. Leventhal,[38] reprinted with permission.
[a] 1K = 1024 bits.

Software and hardware support is increasing for the microprocessors. Several high-level languages are now available for them, including BASIC, FORTRAN, Pascal, and COBOL. In addition, disk-based operating systems are commonplace. LSI and VLSI technology will cause the comparison parameters to merge together such that one stand-alone system will be indistinguishable from another.

References

1 A. Tanenbaum, *Structured Computer Organization*, Prentice-Hall, Englewood Cliffs, NJ, 1976.

2 C. Bell and A. Newell, *Computer Structures: Readings and Examples*, McGraw-Hill, New York, 1971.

3 R. Eckhouse, Jr. and L. Morris, *Minicomputer Systems*, 2nd ed., Prentice-Hall, Englewood Cliffs, NJ, 1979.

4 Digital Equipment Corporation, *PDP-11/70 Processor Handbook*, Maynard, MA, 1975.

5 D. L. Dietmeyer, *Logic Design of Digital Systems*, 2nd ed., Allyn & Bacon, Boston, MA, 1978.

6 G. Amdahl, G. Blaauw, and F. Brooks, Jr., "Architecture of the IBM System/360," *IBM J. Res. Develop*. **8**:87–101 (Apr. 1964).

7 W. A. Wulf, M. Shaw, P. N. Hilfinger, and L. Flon, *Fundamental Structures of Computer Science*, Addison-Wesley, Reading, MA, 1980.

8 J. R. Mick and J. Brick, *Bit-Slice Microprocessor Design*, McGraw-Hill, New York, 1980.

9 AMD Corporation, *The 2900 Family Data Book*, Sunnyvale, CA, 1980.

10 A. M. Abd-alla and A. C. Meltzer, *Principles of Digital Computer Design*, Vol. 1, Prentice-Hall, Englewood Cliffs, NJ, 1976.

11 J. Wakerly, *Microcomputer Architecture and Programming*, Wiley, New York, 1981.

12 J. Iliffe, *Advanced Computer Design*, Prentice-Hall International, London, 1982.

13 Burroughs Corp., *B1700 Reference Manual*, Detroit, MI, 1973.

14 R. M. Russell, "The CRAY-1 Computer System," *Comm. ACM* **21**(1):63–72 (Jan. 1978).

15 U. W. Pooch and R. Chattergy, *Minicomputers: Hardware, Software and Selection*, West, St. Paul, MN, 1980.

16 A. Padegs, "Channel Design Considerations," *IBM Sys. J.* **3**(2):165–180 (1964).

17 Digital Equipment Corp., *DZ-11 Users Manual*, Maynard, MA, 1979.

18 H. S. Stone and D. P. Siewiorek, *Introduction to Computer Organization and Data Structures: PDP-11 Edition*, McGraw-Hill, New York, 1972.

19 S. Madnick and J. Donovan, *Operating Systems*, McGraw-Hill, New York, 1974.

20 A. C. Shaw, *The Logical Design of Operating Systems*, Prentice-Hall, Englewood Cliffs, NJ, 1974.

21 E. G. Coffman, Jr. and P. Denning, *Operating Systems Theory*, Prentice-Hall, Englewood Cliffs, NJ, 1973.

22 M. M. Mano, *Digital Logic and Computer Design*, Prentice-Hall, Englewood Cliffs, NJ, 1979.

23 V. T. Rhyne, *Fundamentals of Digital Systems Design*, Prentice-Hall, Englewood Cliffs, NJ, 1973.

24 G. J. Myers, *Advances in Computer Architecture*, Wiley, New York, 1978.

25 R. W. Kline, *Digital Computer Design*, 2nd ed., Prentice-Hall, Englewood Cliffs, NJ, 1982.

26 J. Donovan, *Systems Programming*, McGraw-Hill, New York, 1972.

27 Y. Chu, *Computer Organization and Microprogramming*, Prentice-Hall, Englewood Cliffs, NJ, 1972.

28 S. Husson, *Microprogramming: Principles and Practice*, Prentice-Hall, Englewood Cliffs, NJ, 1970.

29 M. M. Mano, *Computer System Architecture*, 2nd ed., Prentice-Hall, Englewood Cliffs, NJ, 1982.

30 F. J. Hill and G. R. Peterson, *Digital Systems: Hardware Organization and Design*, 2nd ed., Wiley, New York, 1978.

31 M. J. Flynn, "Very High Speed Computing Systems," *Proc. IEEE* **54**(12) (Dec. 1966).

32 R. Davis, "The ILLIAC-IV Processing Element," *IEEE Trans. Comp.* **C-18**(9):800–816 (Sep. 1969).

33 E. W. Kozdrowicki and D. J. Theis, "Second Generation of Vector Supercomputers," *Computer* **13**(11):71–83 (Nov. 1980).

34 P. H. Enslow, Jr., "Multiprocessor Organization—A Survey," *ACM Comp. Sur.* **9**:103–129 (Mar. 1977).

35 G. Adams and T. Rolander, "Design Motivation for Multiple Processor Microcomputer Systems," *Comp. Des.* **17**:81–89 (Mar. 1978).

36 B. A. Bowen and R. J. A. Brown, *The Logical Design of Multiple-Microprocessor Systems*, Prentice-Hall, Englewood Cliffs, NJ, 1980.

37 W. A. Wulf, R. Levin, and S. Harbison, *HYDRA / C.mmp An Experimental Computer System*, McGraw-Hill, New York, 1981.

38 L. Leventhal, *Introduction to Microprocessors: Software, Hardware, Programming*, Prentice-Hall, Englewood Cliffs, NJ, 1978.

CHAPTER 57
DATA STRUCTURES

JAMES A. M. McHUGH

New Jersey Institute of Technology, Newark, New Jersey

57.1 INTRODUCTION

Computers process information. The structure and the efficiency of the processing algorithms used are highly dependent on how this information is organized. At the machine level, information is represented using bit strings, bytes, and words. At the software level, a variety of general-purpose problem-oriented techniques for arranging information, called data structures, have been developed. A data structure is a table of data characterized by the methods for accessing and manipulating its information, its storage organization, and the kind of structural information it contains.

The building blocks of data structures are called components. Components are usually fixed-size blocks of storage. They are also called cells, nodes, atoms, elements, and (especially for data structures stored on external storage devices) records. Components may be divided into parts called fields. Data structures fall into two broad categories, sequential and linked, depending on the storage organization of their components.

Sequential Organization

The one-dimensional array is typically organized as a sequential data structure. The components of the array are stored sequentially and contiguously. This storage order facilitates the indexed access that characterizes an array. The Ith component of an array, where each component consists of N bytes, is at an easily calculated position. Assuming A is the starting address of the array, the Ith component starts at $A + N*(I - 1)$.

Linked Organization

In linked data structures, on the other hand, the components of the data structure do not lie in any particular storage order and need not even occupy contiguous areas of memory. Instead the organization is determined by storage addresses, or pointers, that are contained in components and that link each component to its structurally associated components. For example, in trees, a type of linked structure, the components form a hierarchy similar to a family tree, with parent components pointing to children components, and so on.

For most of the data structures we describe, the linked organization is typical. Some, such as queues and stacks, have alternative sequential representations. Others, such as hash tables, have sequential storage representations but nonsequential forms of access. Specifics on the access characteristics and procedures, storage representation, and areas of applicability of the standard major data structures follow.

57.2 LINKED LINEAR LISTS

The most fundamental linked data structure is the linked linear list. Each component has a single pointer that gives the address of a unique successor component. Thus the list is like a chain. To locate a component, one follows the trail of pointers from a first component to the target component. Maintenance of the list structure as components are inserted and deleted from the chain is accomplished by a few pointer adjustments and a minimum of data movement. On the other hand, a list does not support the kind of direct indexed access allowed in an array.

A diagram of a linked linear list is shown in Fig. 57.1. The large boxes represent the components, which consist of one or more data fields and a single pointer field. Entry to the list is via a separate head pointer, which gives the address of the first component on the list. A special nil pointer flags the end of the list. Figure 57.2 shows an implementation using arrays. Each component consists of a row from DATA and a corresponding row from PTR (the pointer row). The pointer values are row numbers rather than absolute storage addresses. A nil pointer is denoted by 0. The diagram actually contains two linked lists; one a list of spare or unused components with entry point "free"; the other a list of in-use or active components with entry point "head." Following the pointer trail gives rows 5, 2, and 3 as the list-order active components, and rows 4 and 1 as the list-order free components. To insert a new component in the active list (entry point head), one deletes a spare component from the free list (entry point free), enters the appropriate data into it, and inserts the component on the active list. The converse procedure is used to delete an active component.

This kind of array implementation, with direct programmer control over storage allocation, is typical in FORTRAN. Some programming languages, such as Pascal and PL/I, provide built-in features that simplify the programmer's role in storage management. A discussion of lists and other data structures in the context of programming languages can be found in the literature.[1]

List Operations

The basic operations are traversal, search, insertion, and deletion. We assume the array representation for the list, DATA, POINTER, HEAD, and FREE having the same meaning as before.

Traversal. The traversal procedure enters the list at its head, follows the trail of pointers, and processes components along the way. The pseudocode for the procedure is shown in Fig. 57.3*a*.

Search. The search procedure advances the pointer NEXT until a component matching a given searchkey is found, or returns a NIL pointer in NEXT if the search is unsuccessful. The pseudocode for the procedure is shown in Fig. 57.3*b*.

Insertion. Insertion requires testing for the availability of a free component, locating the point of insertion, and performing the insert. If no free space is available when an insertion is attempted, an overflow is said to occur. This usually indicates an error condition and causes an exit from the procedure. The point for insertion is context dependent. For example, for an ordered list arranged in ascending order of data values, the new component is inserted immediately after the last component whose data value is less than the data value of the insert. Pointers PREV and NEXT are moved through the list in tandem until PREV points to before the insert position and NEXT to after it. The new component is then inserted after component PREV and before component NEXT. The specific

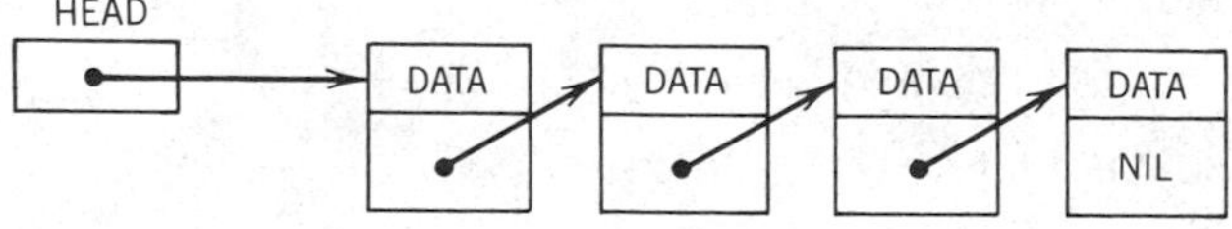

Fig. 57.1 Linear linked list.

	DATA	PTR	
1	—	0	HEAD
2	12.3	3	5
3	42.5	0	FREE
4	—	1	4
5	83.6	2	

Fig. 57.2 Array implementation of a linear linked list.

```
Set NEXT := HEAD.
WHILE NEXT ≠ 0 DO:
    Process DATA(NEXT).
    Set NEXT := PTR(NEXT).
END DO.
```

(*a*)

```
Set NEXT := HEAD and FOUND := FALSE.
WHILE NEXT ≠ 0 AND FOUND = FALSE DO:
    IF DATA (NEXT) ≠ SEARCHKEY
        set NEXT := PTR(NEXT),
    ELSE
        set FOUND := TRUE.
END DO.
```

(*b*)

```
IF FREE ≠ 0 (* Space available? *)
    Set NEW := FREE and FREE := PTR(FREE).
    (* Find insertion position. *)
    Set PREV := 0, NEXT := HEAD, and FOUND := FALSE.
    WHILE NEXT ≠ 0 and FOUND = FALSE DO:
        IF DATA(NEXT) < INSERTKEY
            set PREV := NEXT and NEXT := PTR(NEXT),
        ELSE
            set FOUND := TRUE.
    END DO.
    (* Make insert. *)
    IF PREV ≠ 0
        set PTR(PREV) := NEW and PTR(NEW) := NEXT,
    ELSE
        set PTR(NEW) := HEAD and HEAD := NEW.
ELSE (* Free list empty *)
    set an overflow message.
```

(*c*)

```
IF PTR(P) ≠ 0
    set OLD := PTR(P) and PTR(P) := PTR(OLD).
```

(*d*)

Fig. 57.3 Linked list algorithms: (*a*) Traversal, (*b*) search, (*c*) insertion, (*d*) deletion.

pointer adjustments made depend on whether the insertion is done at the head of the list or not. Let INSERTKEY denote the DATA value of the insert. The pseudocode for the insert procedure is shown in Fig. 57.3*c*.

Delete. The pseudocode for the procedure that deletes the component after the one pointed to by *P* is shown in Fig. 57.3*d*. The deleted component OLD can be inserted on the free list or another list, as appropriate.

Doubly Linked Lists

In a doubly linked list, each component points forward to its successor as well as backward to its predecessor, with separate head and tail pointers addressing the first and last components. Such a list can be traversed in either direction from any component. This simplifies such operations as insertion. A standard representation implements the double links with a pair of pointers in each component. An interesting alternative representation merges the successor and predecessor addresses into a single pointer field by taking the exclusive-or of their bit-string representations. The price of using a single pointer is a reduction in the total number of components that can be addressed and a slightly more complex traversal procedure.[2]

Inverted Lists

Searching a list is more efficient if the list is kept in order on the search key. Thus a directory of telephone customers might be kept in order by customer name, since name-driven searches are most common. But it might also be necessary to access customers given an address or phone number. One solution is to keep multiple lists, each in order on a different key. A more storage-efficient solution is to keep a single list and use multiple pointer fields to encode the different key orders. Components could have fields: NAME, ADDRESS, TELEPHONE, NAMEPTR, ADDRPTR, TELEPTR, with each pointer field pointing to the alphanumeric successor with respect to that field. NAMEPTR gives the address of the component with the alphabetically next name, and so on. Separate pointers to the head of each list are also required. The maintenance procedures for such inverted lists (or multilinked lists) are similar to those for a singly linked list.[3,4]

Appropriate Use of Linked Lists

Use of linked lists is appropriate if components are frequently inserted and deleted, if the storage requirements and shape of the data table are expected to vary greatly during program execution, or if the pointer capabilities of a list capture important structural features of a problem. For example, consider an airline reservations system involving the booking of passengers on flights. A possible data structure consists of a linked list with separate flight information components for each flight. Attached to each flight component are two sublists, a list of the passengers for the flight and a list of its standbys. The lists rapidly grow and shrink as flights are created, reservations canceled, customers changed from standby to scheduled status, and so on. The maintenance procedures of a linked organization allow efficient insertion and deletion of components with use of a few pointer changes. The shape and storage requirements of the data table are volatile, since both the number and size of flights and sublists vary dramatically. Once again, the linked structure is appropriate because it supports efficient use of storage. Storage released as components are deleted from one part of the structure can be recycled for newly created components added to another part. Finally, the natural correspondence between the data structure and the problem should simplify program development.[5,6]

When the choice of data structure is between a linear array ordered on some key and a linear linked list ordered on the same key, more specific guidelines can be given. Basically, the array is superior in search and sorting performance, while the list is better able to accommodate dynamic maintenance (i.e., insertion and deletion of components). Thus an array with N components can be sorted in $N \log_2(N)$ operations, and a binary search locates a component with a given key in at most $\log_2(N)$ steps.[7] However, inserting a new component in an ordered array requires moving an average of $N/2$ components to make room for the new component. The search performance of a linked list suffers from the strictly sequential nature of list access. To locate a component with a randomly chosen key requires accessing an average of half the components in the list. But inserting a component, once the point of insertion has been identified, can be done in a few pointer adjustments. A general rule is that for a static population kept in order on some key and subject to frequent search but with few insertions or deletions expected, an ordered array is appropriate. In a dynamic environment, however, the list may be superior.[8]

Applications of Linked Lists

Managing the allocation and de-allocation of variable size blocks of storage is called dynamic memory management. It is a problem that occurs both for operating systems requesting space for new jobs and for programs building linked data structures. Initially one starts with a single large block of free space. As storage requests arise, they are carved out of this free space. Blocks allocated at one time are returned to the free space at arbitrary later times. As allocations and de-allocations progress, the original free area becomes a checkerboard of free and in-use space. To manage this configuration, one takes the storage blocks, whether allocated or free, as components of a doubly linked list. Blocks appear on the list in order of their starting address. A small amount of space is reserved in each block for pointers to its successor and predecessor blocks as well as for a status bit that indicates if the block is free or in use. To allocate a block of M bytes requires searching the storage list for a free block of size $N(\geqslant M)$. If no such block is available, the request is rejected. Otherwise the located block is split into two blocks of sizes M and $N - M$. One is allocated, the other remains free; the blocks are linked and replace the single component or block they were derived from. To de-allocate a block, its predecessor and successor blocks—which are bordering storage areas—are accessed. If either of these blocks are free, they are merged with the block being de-allocated into a single larger free block. This simplifies the subsequent allocation of space.[9]

Some storage management policies do not immediately return de-allocated blocks to a free list. Instead a so-called garbage collection procedure is invoked when the amount of space on the free list becomes precariously low. The garbage collector scans memory for blocks that are currently pointed to by some data structure and marks them as in use. Unmarked blocks are then reclaimed by being returned to a list of free blocks.[2,8]

An application area that has simultaneously served as a major stimulus to the theory of data structures is interactive graphics.[3,10–12] The data structures used are highly dependent on the particular graphics system, but the general idea is to represent a graphics display by a data structure, each pictorial element of the display being defined by the parameters in a corresponding component of the data structure. For example, assuming line-drawing commands are primitive instructions of the display system, a particular component might indicate that a line of a certain length and angle should be drawn from a current position. A software display procedure scans the data structure and uses the contents of the components to identify, set the parameters of, and execute the appropriate hardware display commands.

Linked lists can be used to efficiently represent sparse arrays and symbolic expressions such as polynomials. An array is sparse if most of its entries are zero. A storage-efficient representation of a sparse array uses a linked list, each component of which corresponds to a nonzero entry of the array and contains that entry's row, column, value, and pointers to the next nonzero components in its row and column. Access to individual components is slower than for the standard sequential representation of an array, but space is conserved and array operations can be implemented efficiently.[3] The symbolic manipulation of polynomials can also be based on a list representation. Each component of the list contains the coefficient and exponent of a term of the polynomial and a pointer to the component for the next term. The standard polynomial operations are efficiently supported by typical list operations.[9]

Another application of lists is to text editing, in which text is considered as a linked list of characters (or blocks of characters) and such text-processing operations as inserting, deleting, and replacing lines or strings of text are facilitated by the list representation. For this and other applications, see Refs. 13 through 16.

57.3 QUEUES AND STACKS

Definitions

Several important kinds of lists have controlled access points where insertions and deletions can be made only at particular points on the lists. These include:

Queues. Insertions are made at one end of the list and deletions are made at the other.

Stacks. Insertions and deletions are made at one end of the list.

Deques. Insertions and deletions are made at either end of the list.

The algorithms for deques are simple extensions of those for queues, and so deques will not be considered further. For more information see Knuth.[8]

Queues

A queue is like a waiting line at a checkout counter. Arrivals are inserted at one end, called the tail of the queue, and components are processed or deleted from the other end, the head of the queue.

Appropriate Use. Queues are used when the rate of demand for a service can exceed the capacity of a system to process service requests, when the excess service requests are backlogged rather than rejected, and when the arrival order of the requests must be maintained. Situations of this kind are common in systems programming. For example, operating systems maintain queues of jobs awaiting scheduling; input–output (I/O) request queues, in which programs put requests for transfers of data to or from I/O devices; and task queues, in which ready programs await access to a central processing unit (CPU).

It is often appropriate to keep multiple queues, each one corresponding to a group of backlogged requests of a given priority. A processor serving such queues would handle all requests from a higher priority queue before any request from a lower priority queue. A set of queues, ordered by priority, can also be realized using a data structure called a heap. Heaps have the advantage that insertions and deletions can be done in $\log_2(N)$ operations, where N is the number of entries on the heap.[2]

Perhaps the classic application of queues is to input–output buffering. The problem that occurs is how to interface a high-speed processing device, such as a CPU, with slower speed I/O devices, such as disks or printers. Specifically, a program executing on a CPU may generate output for a printer more rapidly than the printer can handle the requests. Consequently, output records must be temporarily stored and eventually transmitted in their order of arrival. This is precisely the circumstance under which a queue is appropriate. Typically, the queue of output records resides in a storage area called a buffer, which corresponds in size to the exact amount of information that the I/O device can transfer at a given time. A collection of such buffers can be linked together into a circular list called a buffer pool. At any time, some of these buffers are full and awaiting transmission. Others are empty and waiting to be filled by the CPU. In between, a partly filled buffer is still being filled with output. Since the CPU adds at the tail of this output queue and the I/O device transmits from its head, the CPU and I/O device effectively race each other around the circular list of I/O buffers. Whenever the CPU runs out of empty buffers, because records for output are being produced faster than the I/O device can consume them, the CPU is blocked—until such time as the I/O device frees up a buffer by transmitting its data. Conversely, if the I/O device catches up with the CPU, it awaits a signal that an additional buffer has been filled before continuing transmission, so as to avoid transmitting empty or partially filled buffers.[8]

Queue Representations and Access Algorithms. Queues can be implemented using either linked or sequential representations. The linked representations have the advantage that components can be inserted and deleted by pointer changes only, a minimum of data movement. This is important in applications where components migrate from one queue to another. On the other hand, the sequential implementation does not require explicitly maintaining a free list of backup storage space for future components, and also allows faster sequential traversal of the queue. Two linked representations are standard. These are the linear representation and the circular queue.

In the linear linked representation, separate head and tail pointers identify the first and last components of the queue and each component points to its queue-order successor, the last component having a nil pointer.

In a circular linked representation, all access is via a single tail pointer. Each component points to its queue-order successor, except for the last component which points to the first component on the queue. Access to the head of the queue is indirect: first the last component is accessed via TAIL, then that component's pointer is followed to locate the head. The cost of this indirect access to the head of the queue is compensated for by the resulting simplification in the access algorithms.

The insertion and deletion algorithms for the circular linked representation of a queue follow. The algorithms for the linear linked representation are similar, and so are omitted.

1. Insert (circular linked representation). Let TAIL, DATA, and PTR denote the tail, data, and pointer fields of a circular queue. Let NEW be a free component obtained from a separately maintained free list. Let VALUE denote the data to be inserted. The pseudocode for the procedure to INSERT NEW on the queue is shown in Fig. 57.4*a*.

2. Delete (circular linked representation). To delete a component from the (head of the) queue, use the pseudocode shown in Fig. 57.4*b*.

Queues can also be implemented using a sequential storage representation as opposed to either the linear or circular linked representations. The algorithms appropriate to this representation follow. Let Q be a linear array of size N and initialize HEAD and TAIL to N.

```
Set DATA(NEW) := VALUE.

IF TAIL = NIL                                  (* Empty queue? *)

    set TAIL := NEW and PTR(NEW) := NEW,

ELSE

    Set PTR(NEW) := PTR(TAIL),                 (* Make NEW
    set PTR(TAIL) := NEW, and                     the new
    set TAIL := NEW.                              tail. *)
```

(*a*)

```
IF TAIL = NIL                                  (* Empty queue? *)

    set P := NIL,

ELSE

    Set P := PTR(TAIL).                        (* Identify head. *)

    IF PTR(TAIL) = TAIL                        (* Remove head. *)

        set TAIL := NIL,

    ELSE

        set PTR(TAIL) := PTR(P).
```

(*b*)

Fig. 57.4 Circular queue algorithms: (*a*) Insert, (*b*) delete.

1. **Insert (sequential representation).** The pseudocode for the procedure for inserting an element on a sequentially represented queue is shown in Fig. 57.5*a*. Overflow occurs when no space is available for the insertion.
2. **Delete (sequential representation).** The pseudocode for the delete procedure is shown in Fig. 57.5*b*. Underflow occurs when an attempt is made to delete an element from an empty queue.

The sequential representation algorithms in Fig. 57.5 are brief but subtle. TAIL points to the last component, while HEAD points to the position before the first component. At most, $N - 1$ components can be stored at a time, even though the underlying array is of size N. The array Q acts like a circular list. INSERT moves TAIL clockwise, and DELETE moves HEAD clockwise. Overflow occurs when TAIL catches up with HEAD. Underflow occurs when HEAD catches up with TAIL. Overflows are usually considered error conditions that should normally cause program termination, while underflow is usually a meaningful condition reflecting completion of queue processing.

Stacks

Stacks are lists for which insertions and deletions are allowed at only one end. The insertion operation is called PUSH and the deletion operation is called POP. The distinguished end of the stack is called the TOP of the stack. A stack behaves much like an ordinary stack of plates. Plates are easily added to or removed from the top of such a stack, but can be inserted or deleted at the middle or bottom only with difficulty.

Appropriate Use. Stacks occur whenever data are processed in the reverse of their order of generation. For example, consider subroutine calls. When one subroutine calls another, status information about the caller has to be saved. This includes the return address, the point in the calling routine to which control should return on completion of the called routine. If a nested sequence of subroutine calls occurs where routine S1 calls routine S2, which in turn calls routine S3, and so on, a

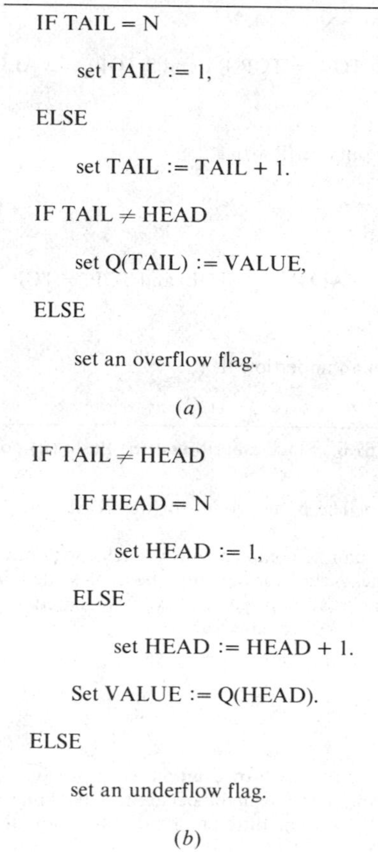

```
IF TAIL = N
    set TAIL := 1,
ELSE
    set TAIL := TAIL + 1.
IF TAIL ≠ HEAD
    set Q(TAIL) := VALUE,
ELSE
    set an overflow flag.
```

(*a*)

```
IF TAIL ≠ HEAD
    IF HEAD = N
        set HEAD := 1,
    ELSE
        set HEAD := HEAD + 1.
    Set VALUE := Q(HEAD).
ELSE
    set an underflow flag.
```

(*b*)

Fig. 57.5 Sequential queue algorithms: (*a*) Insert, (*b*) delete.

whole sequence of return addresses has to be saved—one address for each of the invocations. Since subroutines are returned to in precisely the reverse of their calling order, a stack represents the appropriate storage mechanism. The values of general registers are also usually stacked. This allows the called routine free access to the registers, whose original values can be reestablished from the stacked copy on return. Thus subroutine calls require pushing parameters on a stack, while returning from a subroutine entails popping parameters in order to reestablish the original environment.[17] The same general idea can be used to implement recursive routines, wherein a subroutine is allowed to call itself.[1]

Stacks are also used when interrupts occur. An interrupt is a signal to the CPU, by a device or program, for operating system service. Interrrupts cause an automatic transfer of control from whatever program is currently executing to an operating system routine. Thus interrupts have an effect similar to a subroutine call, and stacks are used here for the same reasons.[18]

Stacks play a pervasive role in the compilation of higher level languages. A very simple illustration of their use is in the evaluation of arithmetic expressions written in postfix form, where operators follow their operands. For example, the ordinary algebraic expression $(A + B)*(C - D)$ would be written in postfix as: $AB + CD - *$. To evaluate such an expression, the expression is scanned left to right and operands are pushed on a stack as they are encountered. When an operator is scanned, the top two operands are popped, the operator applied, and the result pushed on the stack. On completion, the top of the stack is the result of the evaluation of the expression.[17]

Some other applications of stacks are sorting, dynamic storage allocation in block-structured languages such as PASCAL, and tree traversal algorithms.[1] Horowitz and Sahni[9] give an instructive example of the use of stacks to solve a maze traversal problem.

Stack Representations and Access Algorithms. Stacks, like queues, have linked and sequential representations. For stacks, the sequential representation is the more common. Let S be a linear array

```
IF TOP < N
    set TOP := TOP + 1 and S(TOP) := VALUE,
ELSE
    set an overflow flag.
```
(*a*)

```
IF TOP > 0
    set VALUE := S(TOP) and TOP := TOP − 1,
ELSE
    set an underflow flag.
```
(*b*)

Fig. 57.6 Stack algorithms: (*a*) Push, (*b*) pop.

of size N. Initialize TOP to zero. The pseudocodes for the PUSH and POP procedures are shown in Fig. 57.6.

Some applications require a pair of complementary stacks, with one growing as the other shrinks and conversely. Two such stacks can conveniently be represented in a shared storage area by anchoring the bases of the stacks at opposite ends of the shared region. Techniques for storage sharing between three or more stacks are more subtle.[2]

57.4 TREES

Definition

A tree is a type of linked data structure. An example is shown in Fig. 57.7. The figure depicts a hierarchical organization of data such as might occur in a banking environment. Components at different levels in the hierarchy contain different kinds of information. The topmost component

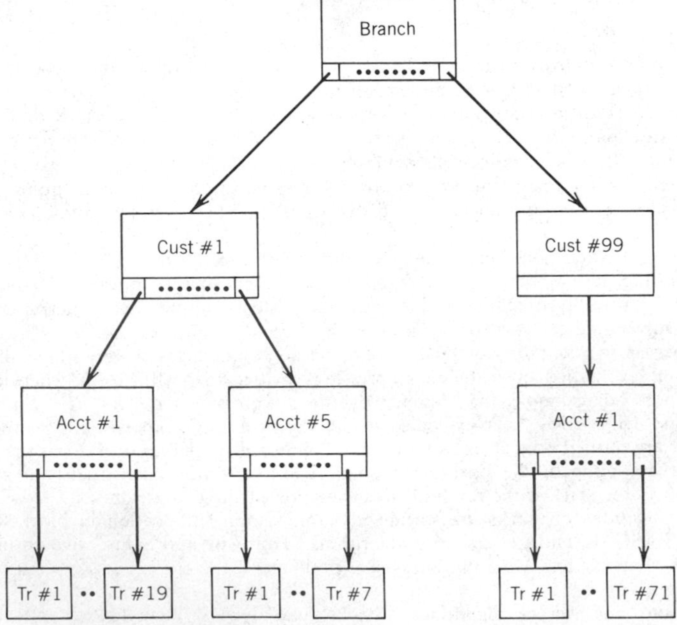

Fig. 57.7 Banking data base. Cust, customers; Acct, account; Tr, trial.

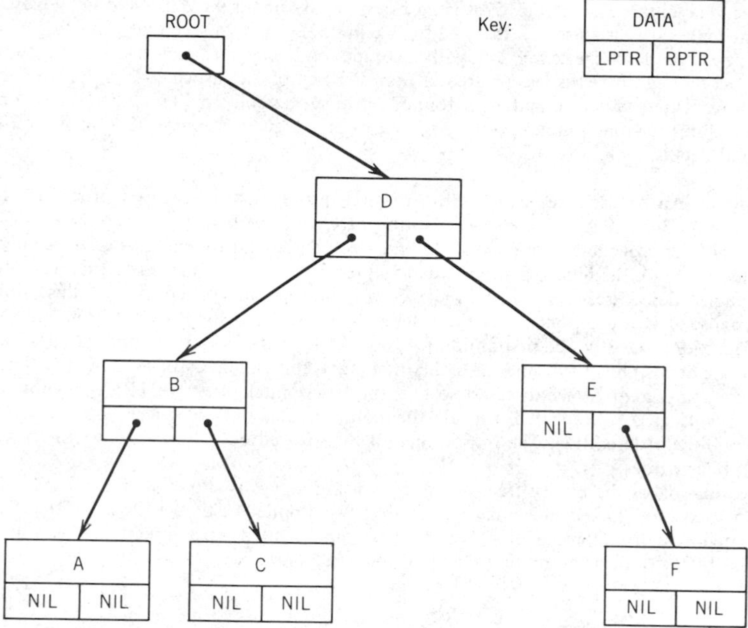

Fig. 57.8 Binary tree. LPTR, left pointer; RPTR, right pointer.

contains information on a bank branch. Lower levels contain data on customers, customer accounts, and account transactions, respectively. Each component owns or points to a variable number of components at the next, lower level in the hierarchy. For example, each customer component points to the account components associated with that customer. The defining characteristics of such a tree are:

1. There is a unique entry point called the root.
2. Each component has zero or more immediate successor components.
3. There is a unique access path or pointer trail from the root to any component.
4. No access path cycles or backtracks on itself.

The terminology of family trees is typically used. If a component A points to a component B, then A is called the parent of B and B is called the child of A. Components with a common parent are called siblings and together comprise a family. Ancestors and descendants of components are defined in the obvious way. Together with its descendants, each component determines a subtree of the original tree. Components with no children are called end points. If the children of each component are rank ordered, the tree is called an ordered tree.[8]

A binary tree is shown in Fig. 57.8. The single entry pointer is called ROOT. Each component has exactly two pointer fields, LPTR (left pointer) and RPTR (right pointer), either or both of which may be nil. The descendants to the left of a component form that component's left subtree. Its right subtree is defined similarly. Binary trees are important because of their use as search trees and for representing arbitrary ordered trees.

Appropriate Use. Trees are appropriate when a data collection is naturally hierarchical or in a dynamic environment when searches are frequent. In the banking example of Fig. 57.7, the hierarchical relationship between components is an intrinsic feature of the data. Using a data structure that models this organization tends to simplify programming. Standard algorithms for trees can efficiently implement standard hierarchical actions on the data, such as inserting a new transaction under an account. The linked organization of a tree also accommodates the unpredictable storage requirements caused by the highly variable number of components at each level of the tree.[19]

A binary tree is appropriate when a data collection is dynamic, that is, when components are frequently inserted, deleted, and searched for on the basis of some key. A key is a datum that uniquely identifies a component. An example is a social security number in a personnel record, or an account number for a bank account. A binary search tree is defined by the property that the key in each component is greater than all keys in that component's left subtree and is less than all keys in its right subtree. The binary tree of Fig. 57.8 is an example, assuming the data field is the key. Such an arrangement makes it easy to search the tree for a component with a given key. Starting at the entry component, one compares the search key with a component's key, and moves left or right in the tree, accordingly as the search key is less or greater than the key of the inspected component. The search terminates successfully when a match is found, or unsuccessfully on moving past an end point. Formal procedures for manipulating binary search trees are given hereafter. See Refs. 7, 20, and 21 for additional details.

Representation. An ordered tree can be transformed into a binary tree by a procedure called the Natural Correspondence.[8] An ordered tree and its corresponding binary tree are shown in Fig. 57.9. The first step of the procedure is to create a binary tree component for each ordered tree component. The next step is to link the binary components together in such a way that the structural information of the original ordered tree is implicit in the corresponding binary tree. As an illustration of the process, suppose A is a component of the ordered tree and A′ is its associated binary component. (Refer to Fig. 57.9.) Let B be the first child of A, and let B′ be the binary component associated with B. Let C be the next sibling of A, and let C′ be the binary component associated with C. The pointers of the binary component A′ are directed so that the left pointer of A′ [LPTR(A′)] points to B′ and RPTR(A′) points to C′. Repeated for all the components, this produces the binary tree that corresponds to the ordered tree. The importance of this procedure is that it allows us to restrict our attention to binary trees.

A representation of the binary tree of Fig. 57.8 using arrays is shown in Fig. 57.10. The pattern follows that used for a linked linear list except that two pointer fields, LPTR and RPTR, are used. The associated free list (components 7 and 5) is implemented as a linked list with its pointers embedded in LPTR (the corresponding RPTR values are ignored).

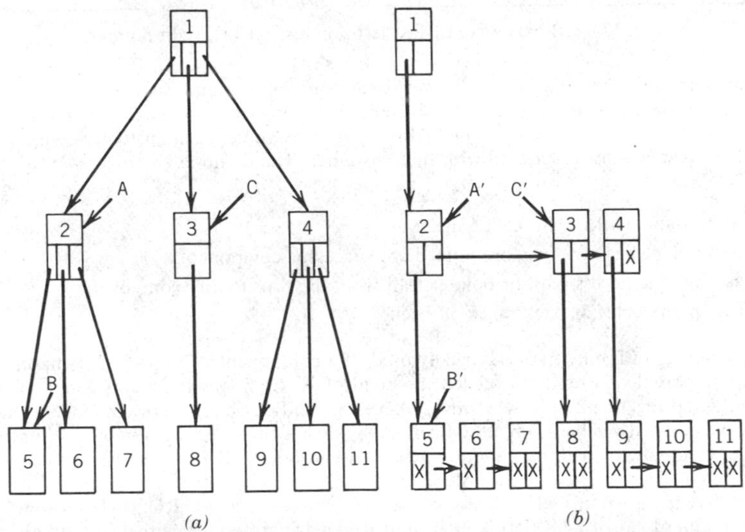

Fig. 57.9 Natural Correspondence: (*a*) ordered tree; (*b*) corresponding binary tree.

	DATA	LPTR	RPTR	
1	F	0	0	ROOT
2	D	3	4	2
3	B	6	8	
4	E	0	1	
5	—	0	—	
6	A	0	0	FREE
7	—	5	—	7
8	C	0	0	

Fig. 57.10 Array representation of a binary tree.

Trees are typically represented as linked data structures. But more compact, sequential representations are sometimes used. These are appropriate if storage is at a premium or the tree is often traversed sequentially, and as long as the shape of the tree is static. For example, an arbitrary ordered tree can be represented compactly using two bits per component to store structural information.[2]

Operations on Binary Search Trees. Typically the key is a field in the data area of a component. But, for simplicity, we assume the data area itself is the key. The algorithms use the array representation for trees illustrated in Fig. 57.10.

Search. The pseudocode for the search procedure is shown in Fig. 57.11*a*. On exit, P is NIL or points to where the key was found.

Insert. Before adding a member to the search tree, we first have to search for a component with a duplicate key. If one exists we reject the attempted insertion. Otherwise we insert the new member at

```
Set P := ROOT and FOUND := FALSE.

WHILE P ≠ 0 AND FOUND = FALSE DO:

    IF DATA(P) = SEARCHKEY

        set FOUND := TRUE,                    (* Exit. *)

    ELSE

        IF DATA(P) > SEARCHKEY

            set P := LPTR(P),                 (* Move left. *)

        ELSE
            set P := RPTR(P).                 (* Move right. *)

END DO.
```

(*a*)

```
Set P := ROOT, Q := 0, and FOUND := FALSE.

WHILE P ≠ 0 AND FOUND = FALSE DO:             (* Perform search. *)

    IF DATA(P) = SEARCHKEY                    (* Test for duplicate. *)

        set FOUND := TRUE,

    ELSE                                      (* Advance pointers. *)

        set Q := P.

        IF DATA(P) > SEARCHKEY

            set P := LPTR(P),

        ELSE

            set P := RPTR(P).

END DO.

IF FOUND = FALSE                              (* Perform Insert. *)

  Set NEW := FREE and FREE := LPTR(FREE).     (* Allocate space. *)
```

Fig. 57.11 Binary tree algorithms: (*a*) Search, (*b*) insert, (*c*) delete.

```
        Set DATA(NEW) := SEARCHKEY.                          (* Enter data *)
        Set LPTR(NEW) := 0 and RPTR(NEW) := 0.               (* Make endpoint. *)
        IF Q = 0                                             (* Link Q to P. *)
            set ROOT := NEW,
        ELSE
            IF DATA(Q) < SEARCHKEY
                set RPTR(Q) := NEW,
            ELSE
                set LPTR(Q) := NEW.
    ELSE                                                     (* Reject insert. *)
        Set duplicate key message.
```

(*b*)

```
    IF LPTR(P) = 0 OR RPTR(P) = 0                            (* P an endpoint? *)
        IF LPTR(P) = 0
            set LINK(Q) := RPTR(P),
        ELSE
            set LINK(Q) := LPTR(P).
    ELSE                                                     (* P not an endpoint. *)
        Set C := LPTR(P).
        IF RPTR(C) = 0
            Set LINK(Q) := C,
            set RPTR(C) := RPTR(P),
        ELSE
            Set R′ := C and R := RPTR(C).                    (* Find R, rightmost
            WHILE RPTR(R) ≠ 0 DO:                                component in left
                Set R′ := R and R := RPTR(R).                    subtree of P, and
            END DO.                                              parent R′ of R. *)
            Set DATA(P) := DATA(R).                          (* Replace P by R. *)
            Set RPTR(R′) := LPTR(R).                         (* Delete R instead of P. *)
```

(*c*)

Fig. 57.11 (*Continued*)

the end of the failed search path, as a child of the last component on that path. The required search is a slight modification of the basic search procedure. It moves a parent-child pair of pointers (Q and P) through the tree, so on exit either P points to the duplicate key or Q points to the end point where the search fails. The pseudocode for the insert procedure is shown in Fig. 57.11*b*.

Deletion. The deletion procedure is intricate. The actions required are highly dependent on the circumstances, such as deleting a component that is a root, deleting a component with empty left or right subtrees, and so on. The first step is to apply the search procedure from INSERT to locate the candidate for deletion (P) and its parent (Q). If the search fails, no further action is required. Otherwise we let LINK(Q) refer to LPTR(Q), RPTR(Q), or ROOT, accordingly, as P is the left or right child of Q or the entry component of the tree. The pseudocode for the procedure to delete the component P is shown in Fig. 57.11*c*. The deleted component can be added to a free list for subsequent reuse.

Traversal. Traversing a tree means visiting all of its components. Traversal procedures are used to display the contents of a tree in some suitable order. They are also required to search for a component when the search parameter is not the key on which the tree is organized. In that case the binary search procedure is inapplicable, and a scan of potentially all the components is required.

The components of a tree can be visited in different orders. For binary search trees, "inorder" traversal is most appropriate because it allows the components to be traversed according to key order. According to inorder traversal, the left subtree of a root is displayed first, then the root, and finally the right subtree. For example, referring to Fig. 57.8, the components of the subtree rooted at B displayed in inorder are A, B, C. The components of the subtree rooted at E displayed in inorder are E, F. The whole tree, displayed according to inorder, is A, B, C, D, E, F. The pseudocode for an inorder traversal procedure is shown in Fig. 57.12. The procedure uses a stack, initially empty, for temporary storage of pointers. For other traversal orders and additional applications, see Knuth.[8]

Trees can also be traversed without stacks, with consequent storage savings but at the cost of increased algorithmic complexity. Threaded trees are one standard device for facilitating stackless tree traversal.[9]

Performance. The height of a search tree determines its performance. The height is defined as the length or number of components in the longest search path. The height of a well-balanced binary tree with N components is on the order of $\log_2(N)$. The number of steps to search, insert, or delete is largely determined by the height of the tree. Consequently, these procedures take an average of $\log_2(N)$ steps in a well-balanced tree.[7] Performance deteriorates if the tree grows in a lopsided

```
Set P := ROOT and EXIT := FALSE.

REPEAT UNTIL EXIT = TRUE:

    WHILE P ≠ 0 DO:                  (* Advance to leftmost

        Push P on stack.                component under P,

        Set P := LPTR(P).               storing pointers to bypassed

    END DO.                             components on a stack. *)

    IF stack = empty                 (* Any components unvisited? *)

        set EXIT := TRUE,

    ELSE

        Pop stack top to P.          (* Backup to bypassed component. *)

        Visit DATA(P).               (* Visit that component. *)

        Set P := RPTR(P).            (* Prepare to traverse its

END REPEAT.                             right subtree. *)
```

Fig. 57.12 "In order" tree traversal.

fashion. If this occurs, the tree can be reorganized into a balanced tree. Alternatively, AVL trees[2] (after their developers Adelson-Velskii and Landis) can be used. AVL trees use insert and delete procedures that also perform balancing operations on the tree to keep its height close to the minimum.

Trees can also be organized to take advantage of prior information about search frequencies. Specifically, given a completely stable population of components with known search frequencies, an optimum search tree can be constructed that minimizes the expected length of search paths. The construction requires N^2 steps for a population with N components.[7]

An m-ary tree allows m children under each component of the tree. A type of m-ary tree called a B-tree is useful for storing large tree-structured indices on rotating storage devices such as disks. A binary tree is inappropriate on such a device because the N components of the tree may be scattered over many tracks. If a binary tree is used, an average of $\log_2(N)$ components lie on the search path. These components can lie on different tracks, and accessing each track requires an additional disk-seek operation. For example, if $N = 16 \times 10^6$, then $\log_2(N)$ is approximately 24. Therefore an average of 24 disk seeks would be required to find a component. For a B-tree with m children under each component, the expected search path length is only $\log_m(N)$. If we use a B-tree wtih $m = 24$, $\log_{24}(N)$ is approximately 5. Therefore only three seeks are required to locate a typical component. The maintenance operations are more complicated than for a binary tree, but this is compensated for by the savings in I/O time because of the reduced number of seeks.[7]

57.5 HASH TABLES

Definition

A hash table consists of (1) a block of storage divided into equal size components and (2) a function that maps keys into addresses of components in the block.

If K is a key and H is a hash function, then $H(K)$ gives the address of a position in the table that is called the home address of K. A component with that key is inserted at the position and found there on retrieval. This represents the best possible insertion and search performance attainable by a data structure.

For example, consider a table consisting of space for 1000 components each 100 bytes in length. Suppose that the keys of entries to be inserted in the table are 10 character identifiers. Define a hash function H of a key K by the formula

$$H(K) = T + 100^* V(K)$$

where T is the starting address of the table and $V(K)$ is a function that assigns a numeric value to K. For $H(K)$ to represent the starting address of a component in the table, V must lie between 0 and 999. A naive choice for V, but one that illustrates some basic points, follows. Assume characters are internally represented in 8 bits, as for EBCDIC (Extended Binary Coded Decimal Interchange Code). Define $V(K)$ to be the decimal value of, say, the first character of K. If K is "SMITH, JOHN," then the first character is S. Its EBCDIC representation is 11000101, which has a decimal value of 197. Therefore H would associate the key with the 197th position in the table. The corresponding entry would be inserted at that position.

Actually the insertion procedure is not quite so simple. Complications arise because distinct keys do not necessarily hash to distinct home addresses. Indeed that is usually both undesirable and unattainable. It is unattainable because the number of possible keys is typically enormous. If distinct keys mapped to distinct positions, the storage area would have to be huge. It is undesirable because most of the a priori possible keys would not occur in a given problem, and so most of that huge table would never be used. Consequently, hash functions are expected to map some keys to identical home positions.

When the key of a new entry hashes to an already occupied position, a collision is said to occur. In the example, this happens to any identifiers wtih the same first character. A well-designed hash function should minimize the chance of collision. The hash function in the example performs especially poorly in this respect, since it addresses only about one fourth of the available table addresses (256 out of 1000 possible locations). This forces new entries into a small portion of the available space and so increases the chance of collision.

Collisions are to be expected, however, and so there must be a policy for resolving them when they do occur. One simple expedient is to put a colliding entry in the next, sequential, available component. Once again, the hash function of the example interfaces poorly with such a collision policy. For this policy not only does not address most of the table but the addressed components are all contiguous (lying in the first 256 locations). The next-available-space collision policy uses precisely this same portion of the table to handle collisions. This increases the chance of subsequent collisions.

Clearly, for use of a hash table major consideration must be given to the choice of hash function and the collision resolution policy.[2,22]

Appropriate Use

A hash table is the data structure of choice for problems requiring primarily insertions and key-driven retrievals but few deletions or key-ordered traversals. Access performance can be excellent. With an appropriate choice of hash function and collision procedure, only one to two collisions are expected per insertion or retrieval, even when the table is 90% full.[2] A disadvantage is that hash tables cannot be easily extended. If the hash table becomes full and a larger table is required, all the existing entries must be rehashed.

A typical application of hash tables is as symbol tables for assemblers or compilers.[23] Assemblers and compilers must be able to identify the symbols that occur in programs and to store those symbols and their attributes for later reference. The symbols can be names of variables, mnemonics for operation codes in an assembly language, and so on. Attributes include such characteristics of symbols as their data type (for a variable name) or their machine code representation (for an assembly operation code). A collection of such entries, each consisting of a symbol and its attributes, constitutes a symbol table. The symbol itself is the key of the entry. Usually insertions and retrievals predominate, and so a hash table is appropriate. On the other hand, symbol tables for variable names in block-structured languages such as Pascal and Algol are subject to deletions as well. In that case a balanced binary search tree may be the most suitable data structure. If there is a real-time requirement, an AVL tree may be preferred because of the bad worst-case behavior of a hash table when it becomes very full (more than 95%) and because of the difficulty of extending the table.

Functions

A good hash function should be easy to compute and randomly scatter the expected keys around the table, minimizing the risk of collision. If the key is small enough to be operated on by the arithmetic instructions of the machine, the division method is often appropriate:

$$H(K) = T + S^*(K \text{ MOD } P)$$

where T is the starting address of the table; S is the component size, in bytes; K is the key; and P is a large prime number.[22] The corresponding table naturally should have space for P components. Both theory and experience indicate that choosing P as a large prime minimizes the risk of collision. If the key is too large to be manipulated directly by arithmetic instructions, then fold shifting can be applied.[22] This breaks an N-bit key into smaller M-bit subgroups and then sums or exclusive-or's these M-bit groups into a final M-bit result. This effectively provides an index into a table with 2^M components. Fold shifting is sometimes used merely to precondition or reduce the size of the key to the point where an arithmetic hashing function such as the division method can be applied. If the expected keys are known to exhibit some regular pattern, digit analysis can be used.[24] This method selects characters from fixed positions in a key in such a way that the resulting subkey is both small enough to be used as a table index and random enough to minimize collisions.

Collision Resolution

Chaining and open addressing are the basic techniques for handling collisions. A technique called bucketing is also used for hash tables on external storage devices but is not considered here.[2]

Chaining. In chaining, a colliding entry is inserted on a list. The list is called an overflow list. A separate overflow list is maintained for each position in the table. Space for components on the overflow lists is allocated from a free list, which is separate from the main hash table.

Open addressing. In open addressing, the collision procedure systematically searches the hash table for a free position. If one is found, the insertion is made there. Otherwise the request for insertion is rejected. The search sequence can be generated in different ways.[2]

There is a tradeoff between the time it takes to compute a hash function and the degree of risk of collision. In an internally stored table it may be more costly to compute a complex hash function with superior collision avoidance properties than to merely handle collisions when they occur. On the other hand, for an externally stored table the cost of the extra I/O incurred during collision resolution may far exceed the cost of computing a better hash function.

Access Procedures

The procedures for inserting an entry into a hash table, for searching for an entry with a given key, and for deleting an entry follow.

Insert. Insertion entails hashing and perhaps collision resolution. The key of the entry to be inserted is hashed to a target address. If the target component is available, the entry is inserted there. Otherwise a collision resolution procedure is used. If the procedure succeeds in identifying an alternative available position, the insertion is made there. Otherwise the table is flagged as full and the request to insert is rejected.

Search. The search procedure first hashes the search key. If a matching key is found at that home position, the search stops, successfully. If the home position is empty, the search stops, unsuccessfully, because the key is not in the table. Otherwise, the search procedure starts scanning the collision resolution trail that starts at the home position. This scan stops when a matching key is found or if the trail ends before a match occurs. Under chaining, the collision resolution trail is merely an overflow list, and the scan stops at the end of this list, or before if a match is found. Under open addressing, the scan stops if an empty position is encountered (which indicates the end of the trail), if the trail returns to the home position (which indicates the entry is not in the table), or if a match is found.

Delete. Deletion requires some care because a position from which an entry is deleted may lie in the middle of an open addressing collision sequence. If made available without also being specially flagged, such a position would incorrectly act as a stopping position for an open address search. Therefore positions from which entries are deleted must be marked as such. Although they are distinguished from free positions for the benefit of the search procedure, deleted positions are still available for insertions.[2]

References

1 M. C. Harrison, *Data Structures and Programming*, Scott, Foresman, Glenview, IL, 1973.

2 T. A. Standish, *Data Structure Techniques*, Addison-Wesley, Reading, MA, 1980.

3 J. L. Pfaltz, *Computer Data Structures*, McGraw-Hill, New York, 1977.

4 C. C. Gotlieb and L. R. Gotlieb, *Data Types and Structures*, Prentice-Hall, Englewood Cliffs, NJ, 1978.

5 M. V. Wilkes, "Lists and Why They Are Useful," *Comp. J.*, pp. 278–281 (Jan. 1965).

6 W. L. Honig and C. R. Carlson, "Toward an Understanding of (Actual) Data Structures," *Comp. J.*, pp. 98–104 (May 1978).

7 N. Wirth, *Algorithms + Data Structures = Programs*, Prentice-Hall, Englewood Cliffs, NJ, 1976.

8 D. Knuth, *The Art of Computer Programming*, Vol. 1, *Fundamental Algorithms*, Addison-Wesley, Reading, MA, 1968.

9 E. Horowitz and S. Sahni, *Fundamentals of Data Structures*, Computer Science, Potomac, MD, 1976.

10 R. Williams, "A Survey of Data Structures for Computer Graphics Systems," *Comp. Surv.*, pp. 1–23 (Mar. 1971).

11 W. Newman and R. Sproull, *Principles of Interactive Graphics*, McGraw-Hill, New York, 1973.

12 J. H. Sexton, "An Introduction to Data Structures with Some Emphasis on Graphics," *Comp. J.*, pp. 444–447 (Sep. 1972).

13 T. G. Lewis and M. Z. Smith, *Applying Data Structures*, Houghton Mifflin, Boston, MA, 1976.

14 J. P. Tremblay and P. G. Sorenson, *An Introduction to Data Structures with Applications*, McGraw-Hill, New York, 1976.

15 K. J. Thurber and P. C. Patton, *Data Structures and Computer Architecture*, Lexington Books, Lexington, MA, 1977.

16 A. T. Berztiss, *Data Structures: Theory and Practice*, 2nd ed., Academic, New York, 1975.

17 W. M. McKeeman, "Stack Computers," in H. S. Stone, ed., *Introduction to Computer Architecture*, 2nd ed., Science Research Associates, Chicago, 1980.

18 K. B. Magleby, "Introduction to Minicomputers," in H. S. Stone, ed., *Introduction to Computer Architecture*, 2nd ed., Science Research Associates, Chicago, 1980.

19 G. Wiederhold, *Database Design*, McGraw-Hill, New York, 1977.

20 E. M. Reingold, J. Nievergelt, and N. Deo, *Combinatorial Algorithms: Theory and Practice*, Prentice-Hall, Englewood Cliffs, NJ, 1977.

21 J. Nievergelt, "Binary Search Trees and File Organization," *Comp. Surv.*, pp. 195–207 (Sep. 1974).

22 W. D. Maurer and T. G. Lewis, "Hash Table Methods," *Comp. Surv.*, pp. 5–20 (Mar. 1975).

23 J. Donovan, *Systems Programming*, McGraw-Hill, New York, 1972.

24 C. D. Knott, "Hashing Functions," *Comp. J.*, pp. 265–278 (Aug. 1975).

CHAPTER 58
PROGRAMMING

PETER G. ANDERSON

JOHN A. BILES

JAMES R. CARBIN

WARREN R. CARITHERS

JAMES A. CHMURA

CHRIS COMTE

LAWRENCE A. COON

MARY ANN DVONCH

HENRY A. ETLINGER

JAMES HAMMERTON

JACK HOLLINGSWORTH

GUY JOHNSON

PETER H. LUTZ

RAYNO D. NIEMI

School of Computer Science and Technology
Rochester Institute of Technology

58.1 STRUCTURED PROGRAMMING

"Structured programming" is a term that means different things to different people. While it is difficult to pinpoint the exact start of the structured programming "revolution," we can say that structured programming has come to embody a discipline. This discipline includes a collection of ideas that can be applied when the solution to a problem means producing a program that can run on a computer. Faithfully following the tenets of structured programming leads to the development of a product—a program possessing certain characteristics. In addition, however, structured programming attempts to address the process used to achieve the desired outcome. E. W. Dijkstra's name surfaces time and time again when looking back at the roots of structured programming. In "Programming Considered as a Human Activity,"[1] Dijkstra argued convincingly for breaking down complex problems into a series of smaller problems and then solving the smaller problems (the divide-and-conquer approach). He also recounted his experiments with "goto-less" programs, describing the increase in clarity brought about when his programs were written without GOTO statements, and the flow of program control described in terms of IF statements (for conditional program paths) and WHILE statements (for repetitive paths, i.e., loops).

The theoretical basis for structured programming was provided first by Bohm and Jacopini and later by Ashcroft and Manna; they showed[1] the possibility of converting any flowchart into an equivalent one that uses only combinations of standard "structured" flowchart components (corresponding to IF and WHILE statements).

Dijkstra's paper, "Structured Programming,"[1] argued for "GOTO-free programs" principally so that the programs would lend themselves to proofs of correctness (i.e., programmers would be able to reason about their programs).

From the mid-seventies until the present, structured programming has been incorporated into a larger structured methodology that addresses all aspects of program development. This new appreciation of programming represents a growing sophistication by those who work with computers.

The structured programming revolution has brought a wealth of new tools for expression that programmers must deal with. One particularly useful method is that of "pseudocode," or "program design language," where the structured programming control structures (IF and WHILE statements and procedure calls) are combined with ordinary English text to describe a program at the highest (i.e., least detailed) level. This method, which replaces the classical flowchart as a design and documentation technique, is exploited in the following section, for a "bubble-sort" program.

58.2 OVERVIEW OF THE PROGRAMMING PROCESS

Writing a Program

This section gives an overview of how one creates a program. To aid in the discussion we will develop a small program using a language known as pseudocode or program design language (PDL), which is an informally defined mixture of English text and Pascal- or Ada-like syntax. The advantages of pseudocode are that it can be easier to understand than formal programming languages and it does not encumber the programmer with syntax details that are unimportant during algorithm design. However, since pseudocode is similar to the eventual implementation language, the first approximation to a solution can evolve naturally into a finished product in a stepwise, orderly fashion. The basic pseudocode control structures are shown in Fig. 58.1.

The program to be written will sort a list of N integers into increasing magnitude order. The method we will use will be to search the input list for the largest value, move it to the end of the list, and repeat the process for the remaining $N - 1$ values. The process continues until only one value remains to be sorted (a list of length 1 is by definition sorted). Our method of moving the largest element to the end of the list will be to compare adjacent elements, swapping them if they are out of order. This method is called a bubble sort because the smallest element is allowed to "bubble" to the top of the list.[2]

The list of numbers is denoted by a name, and individual elements of the list are denoted by the name of the list and a subscript. Thus the first element in the list A is denoted by A[1] and the twenty-first element of list B is denoted by B[21].

For our program we will call the list of numbers LIST and declare our intention of using the list in Fig. 58.2. This declaration names LIST as a vector (or one-dimensional array) of at most 100 integer values, named LIST[1], LIST[2], . . . , LIST[100]. Thus our program will be able to handle up to 100 elements in the list. The declaration of LIMIT as a single integer value will be used to indicate how many of the 100 possible elements in the list are actually being used at any time.

A method for determining when all but one of the original values have been sorted uses the integer variable LAST (which must be declared with LIMIT) as a place marker. Elements from LIST[1] through LIST[LAST] will be unsorted, and elements from LIST[LAST + 1] through LIST[LIMIT] will be sorted and larger than all the unsorted elements. Initially LAST will be LIMIT, to indicate that all the elements are unsorted. The pseudocode for an initial bubble sort is shown in Fig. 58.3.

The notation " := " means to set the variable on the left equal to the contents of the variable on the right or to the value of the constant on the right. The "while" statement is a repetitive command that will execute all statements between the beginning "while" and the terminating "end while" as long as the condition is "true." The "true" condition corresponds to LAST > 1. The condition will eventually become "false," since LAST is being decreased by 1 each time the body of the "while" is executed. The decrease in LAST by 1 is given symbolically as LAST := LAST − 1 (see Fig. 58.4).

```
while "condition" do
    —
    —
    statements to be executed as long as "condition"
    is true
    —
    —
end while

for counter := "initial value" to "final value" do
    —
    —
    statements to be executed with the counter set to
    the values in the range
    —
    —
end for

if "condition" then
    —
    —
    statements to be executed in case "condition" is true
    —
    —
else
    —
    —
    statements to be executed in case "condition" is false
    —
    —
end if
```

Fig. 58.1 Basic pseudocode control structures.

```
Var
    LIST : array [1..100] of integer;
    LIMIT : integer;
```

Fig. 58.2 Declarations.

```
LAST := LIMIT;
while LAST > 1 do
    Bubble the largest of the elements from LIST[1] through
    LIST[LAST] to the end of the list (i.e., LIST[LAST]).
    Decrease LAST by 1.
end while
```

Fig. 58.3 Initial bubble sort.

```
LAST := LIMIT;

while LAST > 1 do

    for "each element between LIST[1] and LIST[LAST − 1]" do
        compare the element with the next and, if they are
        out of order, swap them.

    end for

    LAST := LAST − 1;

end while
```

Fig. 58.4 Refinement of the bubble-sort method.

```
LAST := LIMIT;

while LAST > 1 do

    for I := 1 to LAST − 1 do
        if LIST[I] > LIST[I + 1] then
            TEMP := LIST[I];
            LIST[I] := LIST[I + 1];
            LIST[I + 1] := TEMP;

        end if

    end for

    LAST := LAST − 1;

end while
```

Fig. 58.5 Final refinement of the bubble sort.

```
for I := 1 to LIMIT do

    write (LIST[I]);

end for
```

Fig. 58.6 The "for" statement.

The portion of the while statement that "bubbles" the largest element to the end involves comparing adjacent elements and swapping them when they are out of order. Thus the while statement in Fig. 58.3 is refined to the form in Fig. 58.4.

The "stepping through" the array LIST from LIST[1] to LIST[LAST − 1] is accomplished by a "for" statement, which allows an integer variable, I, to take on successive values from 1 through LAST − 1, inclusive. This variable is used as a subscript to LIST. It is incremented by 1 at the "end for" terminator. Swapping is done in three steps using a temporary integer variable, TEMP. Both TEMP and I must be declared with LIMIT and LAST. The final refinement is shown in Fig. 58.5.

The body of the "then" portion of the "if" is executed between the "if" and "end if" terminator only when the condition is "true." Otherwise the body of the "then" portion is not executed. The items in the list can be printed out by use of a "for" statement, as shown in Fig. 58.6.

Finally, all the pieces are put together to obtain the complete pseudocode program, as shown in Fig. 58.7. Note the start of the begin-end pair to delimit the entire program, and the matching of the control structures "if," "while," and "for" with their own terminators.

```
Var

        LIST : array [1..100] of integer;
        LIMIT, LAST, I, TEMP : integer;

begin
        LIMIT := 0;

        while "not end of input file" do
            LIMIT := LIMIT + 1;
            if LIMIT > 100 then
                write ("Too many elements in list.");
                halt;
            end if
            read (LIST[LIMIT]);
        end while

        if LIMIT = 0 then
            halt;
        end if

        LAST := LIMIT;
        while LAST > 1 do
            for I := 1 to LAST - 1 do
                if LIST[I] > LIST[I + 1] then
                    TEMP := LIST[I];
                    LIST[I] := LIST[I + 1];
                    LIST[I + 1] := TEMP;
                end if
            end for
        end while

        for I := 1 to LIMIT do
            write (LIST[I]);
        end for
end.
```

Fig. 58.7 Final version of bubble-sort pseudocode program.

Running a Program

After we have designed and written our program in a high-level language, we probably would like to run it. This is done in four steps:

1. Compile the program into "machine code."
2. Link the pieces of the program together.
3. Load the resulting "object code" into the computer's memory.
4. Run the program.

Compilation. The compilation process is done by a program called a compiler, which translates the high-level language "source" program into machine-readable "object code." The basic goal is to translate all the symbolic parts of the source program (variable names, statement labels, operators like + or −, etc.) into numbers for the machine (addresses for variables and labels, machine instruction "op-codes" for operators, etc.).

In translating a source program, a compiler will read through it completely one or more times. In each such "pass" the compiler will carry out some of the following tasks:[3] lexical analysis, syntactic analysis, symbol table management, code generation, and optimization. Some languages, most notably Pascal, were designed around a "one-pass" compiler, which means that a Pascal compiler will read through a Pascal program only once and perform all the mentioned tasks at once. Other languages, such as FORTRAN, usually have "two-pass" compilers. In the first pass the goal is to do the lexical analysis and build the symbol table. In the second pass the code is generated.

The lexical analysis breaks up the program into indivisible "tokens." For instance, in the Pascal statement

PRICE := COST − DISCOUNT + TAX

PRICE would be treated as a single token, := would be another token, COST another, and so on, seven tokens in all.

After doing the lexical analysis, the compiler performs a "syntactic analysis," often called parsing. In parsing the tokens, the compiler makes sure that they are arranged in sequences that are "legal" for the language being compiled. If a particular statement does not parse correctly, the compiler prints an error message that is meant to point the programmer to the mistake.

The parser also groups the tokens of complex statements into "subexpressions" that reflect the order in which the operations will be done. In the given Pascal statement, for instance, a standard Pascal compiler would generate code that would do the subtraction of DISCOUNT from COST before the addition of that difference to TAX. On the other hand, it would be possible to build a parser that did the addition first and subtracted the sum from COST. This would yield different results from those produced by the standard parser.

This order of "precedence" for operations is defined in an unambiguous way for a particular language. This means that a given legal statement in a particular language can be parsed in only one way, as determined by the "syntactic rules" of the language. The compiler is written to recognize and translate all programs that adhere to those rules. Programs that do not adhere to the rules will generate error messages.

The next task the compiler must perform is generation of the object code that, when executed, does the things intended by the program. This object code is in the same format as that produced by an assembler and may be thought of as a sequence of numbers that will be loaded into the machine's memory.

A single statement is a high-level language typically compiles into a sequence of several machine instructions. This is because most computers have instructions to do only very simple things, while high-level languages usually provide powerful conceptual tools for solving problems. It is the compiler's job to translate these complex tools into simple instructions, that is, to map the complex operation into many smaller steps.

The code generated by a compiler for a program written in a high-level language usually is somewhat less efficient than the code from an assembly language program written to do the same thing. The code generated by early compilers was notoriously inefficient, but newer compilers come much closer to "optimal" code. The refinement of code to eliminate redundant instructions is called optimization and may be done in an extra pass through the program.

Linking and Loading. After a source program has been compiled into object code, it may need to be "linked" together and "loaded" into memory. These two operations, linking and loading, are often done by the same program, a "linking loader," but we will discuss the operations separately.

The linker ties together the separate modules that may have been created by the compiler. In languages like FORTRAN, for instance, each subroutine or function ends up as a separate "object module." The major problem in linking these object modules together into a single load module involves modifying the addresses of the local variables in each object module to reflect where that module appears relative to the other object modules.

Most compilers facilitate this modification by creating "zero-relative" addresses for variables and statement labels whose actual addresses depend on where the module in which they appear is loaded. The linker, then, modifies each of the zero-relative addresses in an object module by adding to them the sum of the lengths of the object modules that already have been loaded.

Loading a Program. The loader simply copies the linked load module into the main memory of the machine so that the program finally can be executed. Its major responsibility is to tell the operating system the address where the load module starts and the length of the module.

Executing a Program. After a program has been compiled, linked, and loaded it is finally ready to be "executed." The only thing necessary for the operating system to know is the "start address" of the program. This is the physical address in memory of the first executable instruction in the main procedure of the program. The program will "terminate" when the execution of an instruction generates a "trap," at which time the control of the machine transfers back to the operating system.

58.3 SYSTEMS SOFTWARE

The concept of systems software includes all of the support programs that are usually bought from the computer vendor at the time that the computer itself is bought (this software may either be offered "bundled" in a package deal with the computer hardware or "unbundled," that is, to be purchased separately). A bare computer is an extremely intractable tool for the typical applications programmer, and the following components are usually considered necessary to extend the hardware into a generally useful system:

Assembler.
Compilers and interpreters.
Linker.
Debugger.
File management systems and editors.
Input–output drivers.
Operating system.
Control language system.

Assemblers, compilers, and interpreters translate symbolic computer programs into the hardware binary language of the computer itself. These programs are probably the greatest labor-saving tools in the entire computer repertoire. In addition to providing a notation to use at the human or application level, they are very often equipped with run-time libraries of programs for such common procedures as mathematical function evaluation, sorting, and input–output formatting.

A linker is a computer program that combines program components, such as a main program, subroutines, and components from the system library, into a single unit for running. This yields an automated use of libraries that have been provided by the vendor, the local computer support facility, or the individual project team programmers. Such libraries allow for the easy construction and use of off-the-shelf software components and the maturity of software construction into an engineeringlike discipline.

A debugger is a specialized program-testing system that allows a programmer to test a program under development, to put it through a selected subset of its paces, to interrogate and modify its variables, to measure its performance, and to detect and remove the "bugs" (errors) in it.

Input–output drivers are used to control the peripherals of a computer: card readers and punches, magnetic and paper tape drives, printers, hard and floppy disk units, users' terminals, and any equipment that interfaces the computer with the outside world. These complicated processes are developed once and used by all the applications.

File management systems and editors provide a convenient method of storing, retrieving, and modifying programs and data. The computer system's on-line storage of disks and off-line storage of magnetic tapes and disks make up a far more convenient and flexible filing system than are file drawers for paper systems. File management systems extend in power to elaborately cross-referenced "data base systems" (described later), and file editors extend to become "word processors," which allow document composers to produce up-to-date finished works as easily as they might form rough drafts or marked-up typed drafts using paper systems.

Operating systems serve computer users by managing the system's resources for them. These resources usually include the input–output (I/O) devices (so multiple users can print without having their output scrambled), the main and auxiliary memory, and the central processing unit (CPU) of the computer itself. The operating system shares these resources to make the computer simulate several computers to serve several users concurrently, so that dozens of users on time-sharing terminals have the impression of being the sole users of the system.

The control language system is the processor that interprets users' requests and passes them on to the operating system. Users specify, in the control language, what they intend to do on the computer: compile a program, run a program with certain data, print a file, and so on.

58.4 PROGRAMMING LANGUAGES

Overview

A programming language allows the user to represent an algorithm in a way that is meaningful to the computer. Problems, and therefore, algorithms to solve the problems, vary greatly. From the need to express greatly different algorithms has grown a potpourri of programming languages, varying greatly in type and capability. The well-informed professional defines the problem, determines an algorithm to solve the problem, and then weighs the merits of the various programming languages available. He or she then chooses the language best suited to the problem. To make a decision, various aspects of a language need to be considered.

One major division in programming languages is low- vs. high-level languages. Low-level languages tend to be very close to machine code. This causes them to be difficult to understand and to require much programming effort to perform minimal tasks. The most commonly used low-level language is assembler language, which is specific to each machine. Programs written in assembler language are not usually portable to other machines.

High-level languages tend to be user friendly and reasonably easy to read. A single high-level statement can do the work of many assembler statements. Some well-known high level languages are FORTRAN, COBOL, BASIC, PL/I, SNOBOL, and Pascal. One might even consider the commands accepted by the operating system of a computer as a high-level language. In this respect, the UNIX† operating system has a set of extremely powerful commands.

Programming languages can be generally categorized as translated or interpreted. Typically, a program written in a translated language is submitted to a compiler or an assembler, and another version of the program is produced. The new version, called an object file, is written in machine code and is nonintelligible to all but the computer. The object file is loaded and run when the program is needed. FORTRAN, COBOL, C, PL/I, and Ada are some of the more notable translated languages. In general, compiled languages require declarations for variables and do static-type checking. However, many compiled languages allow some type of dynamic storage allocation.

A program written in an interpreted language is not sent through a compiler or assembler. Rather, the interpreter is loaded into main memory along with the program. The program is then interpreted line by line as it runs, with appropriate machine code being fed to the computer by the interpreter. Typically, an interpreted language carries run-time descriptors but does not require declaration of variables. Dynamic type checking is done by most interpreted languages, and generally all storage is dynamically allocated. Some of the most well-known interpreted languages are BASIC, APL, LISP, and SNOBOL.

The efficiency of a programming language is expressed in the speed at run time, the amount of space needed to operate the program, and the amount of programming effort required. Translated languages are very efficient at run time. However, this run-time efficiency is paid for at the price of loss of flexibility. Interpreted languages are designed to offer maximum program flexibility, but generally take more time and space to run.

Most languages are not "pure" compiled or "pure" interpreted but rather are a slight crossbreeding between the two. Although FORTRAN 77 is considered a compiled language, frequently its format statements are implemented by I/O interpreters. On the other hand, in most implementations of SNOBOL, the SNOBOL program is translated into more machine-friendly code and then the new code is interpreted. Pascal is the real hybrid between translated and interpreted. Some versions of Pascal generate and interpret "P code" (a programming system for a mythical Pascal computer), and others continue to translate P code into the machine language of the executing machine.

Other properties of a language that should be considered when judging usability of the language include:

The data structures it offers.

Its portability.

The amount of documentation available.

The conciseness of language definition.

Its suitability to the problem.

The availability of a compiler or interpreter.

Specific Languages

Several programming languages will be described in terms of their history, strong points, and references. An example program will be given, the bubble-sort program, to provide a Rosetta stone. A second example program is also given to show the particular strengths of each language, particularly for the more specialized languages.

Algol. Algol is the product of an international group of computer scientists. Initially called the International Algebraic Language, it was made public in the Algol 58 and Algol 60 Reports. Among the features for which Algol became famous is the fact that it was initially defined in a metalanguage, later to be known as BNF, and was the vehicle for making this metalanguage widely known and accepted.[4,5] The language was simple—it had only six statement types and few restrictions. For example, identifiers could be arbitrarily long, and an array could have any number of dimensions. Yet the language was powerful in that it included block structure, recursive functions and procedures, pass by value, and pass by name. Originally the language did not include input and output facilities,

†UNIX is a trademark of Bell Laboratories.

```
procedure demo begin
real array X[1 : 100];

procedure bubble (A, FIRST, LAST); real array A[FIRST : LAST];
begin integer I; real TEMP; boolean SWAPPED;
    LOOP : SWAPPED := false;

    for I := FIRST step 1 until LAST − 1 do

        if A[I] > A[I + 1] then begin
            TEMP := A[I];
            A[I] := A[I + 1];
            A[I + 1] := TEMP;
            SWAPPED := true; comment: record that a swap was needed;
        end;

    if SWAPPED then go to LOOP; comment: repeat loop;
end;

read(X); bubble (X, 1, 100); write (X)
end
```

Fig. 58.8 Bubble-sort program in Algol.

although these were provided later by procedures. All user-defined functions and procedures were expected to be internally defined within the calling program. With this feature Algol could be called a strongly typed language. The language was designed for scientific calculation and was the publication language for algorithms within the ACM (Association for Computing Machinery) community. Although popular in Europe, Algol never supplanted FORTRAN in the United States. Nevertheless, Algol, its metalanguage, and the research efforts made in implementing Algol have had a profound effect on computer science and on later languages such as PL/I, Pascal, and Ada, unmatched perhaps by any other single language. The Algol version of the bubble-sort program is given in Fig. 58.8.

COBOL. During the late 1950s the Department of Defense and other government agencies were awarding contracts. The programming languages used for these contracts varied from high to low level. This made the job of contract monitoring and program maintenance extremely costly, as auditors and programmers had to be trained to work with a variety of programming languages. Consequently, the idea of one common language for all government contracts was adopted. The Committee On Data Systems Languages was created in 1959. CODASYL, composed of representatives of U.S. government agencies, computer manufacturers, computer users, and universities, decided to create a new language because most of the current languages at that time were ill suited to the task of data processing. Differences in hardware from computer manufacturers also contributed to the difficulty of adopting any current language of that time. The new language, common business-oriented language (COBOL), was to be understandable by people with very little computer background.[6,7]

The initial specifications were published in April, 1960, with revisions in 1963 and 1965. To preserve the common features of COBOL among the different manufacturers, the American National Standards Institute (ANSI) published in 1968 suggestions for making COBOL a standard programming language. Thus, one standard was made available for the construction of COBOL compilers. COBOL compilers that conform to the standard are referred to as ANSI COBOL compilers. In 1974, ANSI revised the standard COBOL specifications, and new standards are currently under review by the computer community.

COBOL is the most widely used computer language on the market today. Some estimates are that as many as 80% of new business applications use COBOL. COBOL has achieved this high use because of its adoption by many contractors at the insistence of the government: its standardization by ANSI; its "Englishlike" sentences, which are easier to read than those of other languages; its extensive data-editing features; its file management facilities; and finally its evolving nature as the needs of the business community change. The main detractions of COBOL are its verbosity, its lack of mathematical capabilities, and its lack of structured constructs, which are found in Algol or PL/I. However, with the proper discipline, COBOL programs can be reasonably well structured.

The bubble-sort program implemented in COBOL is shown in Fig. 58.9. Every COBOL program

```
IDENTIFICATION DIVISION.
PROGRAM-ID. BUBBLE-SORT.

ENVIRONMENT DIVISION.
INPUT-OUTPUT SECTION.
FILE-CONTROL.
    SELECT UNSORTED-FILE ASSIGN TO "DATAFILE".
    SELECT SORTED-REPORT ASSIGN TO PRINTER.

DATA DIVISION.
FILE SECTION.
FD  UNSORTED-FILE.
01  DATA-VALUE-RECORD.
    05  DATA-VALUE-IN       PIC S9999V99.
FD  SORTED-REPORT.
01  REPORT-LINE PIC X(133).
WORKING-STORAGE SECTION.

01  FLAGS.
    05  MORE-DATA-REMAINS-FLAG   PIC X VALUE "Y".
        88 NO-MORE-DATA-REMAINS VALUE "N".
    05  SWITCH-FLAG   PIC X VALUE "Y".
        88 NO-VALUES-SWITCHED    VALUE "N".
01  NUMBER-TABLE.
    05  NO-VALUES   PIC S999 USAGE IS COMPUTATIONAL.
    05  DATA-VALUE   PIC S9999V99 OCCURS 100 TIMES
            USAGE IS COMPUTATIONAL-3.
    05  DATA-SUB       PIC S999 USAGE IS COMPUTATIONAL.
    05  DATA-TEMP        PIC S9999V99 USAGE IS COMPUTATIONAL.
01  HEADINGS.
    05  FILLER   PIC X(14) VALUE "SORTED VALUES".
01  DETAIL-LINE.
    05  FILLER   PIC XXX.
    05  DATA-VALUE-OUT   PIC -ZZZ9.99.


PROCEDURE DIVISION.
MAIN-LINE-ROUTINE.
    PERFORM READ-ROUTINE.
    PERFORM SORT-ROUTINE UNTIL NO-VALUES-SWITCHED.
    PERFORM WRITE-ROUTINE.
    STOP RUN.

READ-ROUTINE.
    OPEN INPUT UNSORTED-FILE.
    MOVE ZERO TO DATA-SUB.
    READ UNSORTED-FILE AT END MOVE "N" TO MORE-DATA-REMAINS-FLAG.
    PERFORM READ-A-VALUE UNTIL NO-MORE-DATA-REMAINS.
    MOVE DATA-SUB TO NO-VALUES.
    CLOSE UNSORTED-FILE.
READ-A-VALUE.
    ADD 1 TO DATA-SUB.
    MOVE DATA-VALUE-IN TO DATA-VALUE (DATA-SUB).
    READ UNSORTED-FILE AT END MOVE "N" TO MORE-DATA-REMAINS-FLAG.

SORT-ROUTINE.
    MOVE "N" TO SWITCH-FLAG.
    PERFORM SORT-PASS VARYING DATA-SUB FROM 1 BY 1
        UNTIL DATA-SUB > NO-VALUES - 1.
SORT-PASS.
    IF DATA-VALUE (DATA-SUB) > DATA-VALUE (DATA-SUB + 1)
        MOVE DATA-VALUE (DATA-SUB) TO DATA-TEMP
        MOVE DATA-VALUE (DATA-SUB + 1) TO DATA-VALUE (DATA-SUB)
        MOVE DATA-TEMP TO DATA-VALUE (DATA-SUB + 1)
        MOVE "Y" TO SWITCH-FLAG.
```

Fig. 58.9 Bubble-sort program in COBOL.

```
WRITE-ROUTINE.
    OPEN OUTPUT SORTED-REPORT.
    WRITE REPORT-LINE FROM HEADINGS AFTER ADVANCING PAGE.
    PERFORM WRITE-A-VALUE VARYING DATA-SUB FROM 1 BY 1
        UNTIL DATA-SUB > NO-VALUES.
    CLOSE SORTED-REPORT.
WRITE-A-LINE.
    MOVE SPACES TO DETAIL-LINE.
    MOVE DATA-VALUE (DATA-SUB) TO DATA-VALUE-OUT.
    WRITE REPORT-LINE FROM DETAIL-LINE AFTER ADVANCING 1 LINES.
```

Fig. 58.9 (*Continued*)

must have the four divisions in the order shown. The identification division gives the name of the program and other identification information. (A number of optional items have been omitted in the figure for the sake of brevity and clarity.) The environment division contains features based on one computer manufacturer's hardware, especially the names of the physical devices and files to be associated with the logical COBOL names (SELECT-ASSIGN statement).

The data division defines the structure of all files and the type and size of all identifiers (variables) used in the program. The PIC (or picture) clause defines the size and type [the type by the symbols 9 (numeric) and X (alphanumeric)] and the length by the number of these symbols. The symbol "S" indicates to store the sign of the number, while "V" indicates the implied decimal point location as the data do not contain a decimal point. An item used for arithmetic can contain only S, V, and 9. The hierarchical structure of data is implemented with the level number. A record must begin with the level 01. A subordinate field uses a higher number. Thus the DETAIL-LINE is composed of two items: FILLER and DATA-VALUE-OUT. The word "filler" is a reserved word, that is, a word in the syntax of COBOL that has a special meaning. It is used when a field does not need to be referenced directly by name. This field is referenced implicitly when DETAIL-LINE is referenced. DATA-VALUE-OUT is an edited field, that is, a field that is used to edit a number into a form for printing. This field has a minus (−) if negative, or a blank if positive, followed by four digits with zeros suppressed, a decimal point, and two decimal places. The 88 level has a special use and indicates a "condition name." The condition name has a true/false value depending on whether the preceding identifier contains the value given in the 88 level entry.

The procedure division contains the algorithm for using the data. Statements are grouped into named paragraphs that can be referenced by the PERFORM statements. The PERFORM statement causes the statements in the specified paragraph to be executed once or repeatedly until a condition becomes true. The AT END phrase in the READ statement is executed only when an end-of-file condition is encountered.

A typical COBOL application program would not have a sort coded with loops as shown in Fig. 58.9 but rather the SORT statement would be utilized.[6,7] The SORT statement uses an unsorted file as input to a sort routine. The output file would contain the sorted records, which would be read in and processed. A sort description (SD) entry in the file section specifies the length of the record to be sorted and whether the file is to be sorted in ascending or descending order.

FORTRAN IV and FORTRAN 77. FORTRAN (Formula Translator),[8] developed in the mid-fifties by IBM primarily for scientific and engineering applications, was the first widely used high-level language. Its major strengths are very efficient execution and the ease with which modules can be linked together. This provision for separate compilation allows the development of a library of routines that can be used in developing new software.

The basic structure of a FORTRAN program is quite simply the main module followed by the internal subprograms. External subprograms need not be recompiled but can be linked in, provided they are referenced by an EXTERNAL statement or by an explicit CALL.

The character set is limited to capital letters, digits, and the nine special characters

, . ( ) ' + − / =

In the source code, columns 1 through 5 form a numeric label field, the sixth position indicates a continuation line, and columns 7 through 72 are for actual statements. Some compilers relax this (card-oriented) convention and allow a more free format (for use with modern equipment).

Since the design emphasis was on efficient execution, there are severe limitations on the number of data types and control structures. There are five data types: integer, real, complex, double precision, and logical (Boolean). Data structures are limited to simple variables and arrays of up to three dimensions. Numeric variables are typed implicitly by their first letter, I to N for integer and the others for real; or explicitly by declaration statements such as REAL I, J, or INTEGER A. Logical variables must be typed explicitly with a LOGICAL declaration statement. Arrays are

```
          READ (5, 500)A

500       FORMAT (I10)
          CALL BUBBLE (FIRST, LAST)
          WRITE (6, 600)A

600       FORMAT (1H, 100(I5, 2X))
          STOP
          END

          SUBROUTINE BUBBLE (FIRST, LAST)
          INTEGER FIRST, I, LAST, TEMP, A(10)
          COMMON A
          LOGICAL SWAPPD

10        SWAPPD = .FALSE.
          DO 100 I = FIRST, LAST - 1
          IF (.NOT.(A(I).GT.A(I + 1))) GOTO 100
          TEMP = A(I)
          A(I) = A(I + 1)
          A(I + 1) = TEMP
          SWAPPD = .TRUE.

100       CONTINUE
          IF (SWAPPD) GOTO 10
          RETURN
          END
```

Fig. 58.10 Bubble-sort program in FORTRAN IV.

defined in a declaration statement or in DIMENSION statement by giving the number of entries in each dimension. For example, DIMENSION A(3, 4) and REAL I(10) define real arrays of two and one dimension; A is an array whose subscripts run from 1 to 3 and 1 to 4, and I is an array whose (single) subscript runs from 1 to 10.

There are three basic control structures: GOTO, an unconditional branch; IF, a conditional branch; and DO, a loop construct. Provision is made for internal and external subprograms. These are separate modules and are headed by a FUNCTION or SUBROUTINE statement. Access is by a CALL statement for subroutines and by use of the name for functions. External subprograms must be listed in an EXTERNAL statement. Arguments can be passed by means of an argument list or by giving the variable global scope by means of a COMMON statement in both the calling module and subprogram. To improve efficiency, the use of recursion (i.e., a subroutine calling itself) is prohibited.

Input and output are handled by READ and WRITE statements, respectively. With either of these statements, the programmer specifies a device by number (a printer, card reader, tape, or disk file), a format number, and a list of program variables to be read in or written out. The FORMAT statement specifies the way the programmer wants the conversion between internal computer binary and external readable characters to be done.

The bubble-sort program written in FORTRAN IV[8] is shown in Fig. 58.10.

The limitations of FORTRAN IV led to the development of FORTRAN 77,[9–11] which is now the sanctioned version of the language. Major new features include the addition of the IF-THEN-ELSE construct, the provision for dynamic dimensioning of arrays, and the addition of the CHARACTER data type. The bubble-sort program written in FORTRAN 77 is shown in Fig. 58.11.

PL/I. The development of PL/I began in 1963 when an IBM users group formed a subcommittee called the Advanced Language Development Committee. The original purpose of this committee was to design a language to succeed FORTRAN. Its work was strongly backed by IBM, which was looking for a new language to make better use of its yet unannounced new computers.

The committee invited many experts from around the world to suggest features that might be made part of this new language. In addition, the committee members examined existing languages to determine which of their features might be incorporated into the new language. Thus what started as an extension to FORTRAN quickly became a completely different language.

This new language was initially called NPL, for new programming language. However, this name was dropped owing to a conflict with the National Physics Laboratory in England. The name PL/I,

```
         INTEGER A(10)
         READ (5, 500) A
500      FORMAT (I10)
         CALL BUBBLE (1, 10, A, 10)
         WRITE (6, 600) A
600      FORMAT (1H, 100(I5, 2X))
         STOP
         END

         SUBROUTINE BUBBLE (FIRST, LAST, A, SIZE)
         INTEGER FIRST, I, LAST, TEMP, A(SIZE)
         LOGICAL SWAPPD
 10      SWAPPD = .FALSE.
         DO 100 I = FIRST, LAST - 1
             IF (A(I).GT.A(I + 1)) THEN
                 TEMP = A(I)
                 A(I) = A(I + 1)
                 A(I + 1) = TEMP
                 SWAPPD = .TRUE.
             END IF
100      CONTINUE
         IF (SWAPPD) GOTO 10
         RETURN
         END
```

Fig. 58.11 Bubble-sort program in FORTRAN 77.

for programming language I, was adopted. Implementation was begun at IBM in England in 1964. The first manual was published in 1965 and the first compiler was released in 1966.

As might be expected from a language designed by committee, PL/I is a combination of many different constructions and features. Up to this time, high-level languages had been designed to function most effectively in specific application areas (e.g., COBOL for commercial environments, FORTRAN for science and engineering). PL/I was designed with many features from these and other languages. It was thought that PL/I was general enough to be used to solve problems in any environment.[12]

Perhaps the most prominent feature of PL/I is its liberal use of defaults. Virtually every type of programming construction that might be presented to the compiler has a corresponding default value. The developers reasoned that since compiler technology had advanced significantly since the early days of FORTRAN and COBOL, and machines were becoming more and more powerful, this power should be harnessed to make life easier for programmers. What they did not appreciate was the way in which defaults can actually complicate the process of program development. PL/I programmers have to learn all of the defaults so that errors are not introduced into their code because of an assumption made by the compiler.

PL/I was one of the first high-level languages to offer liberal access to run-time facilities of the operating system. It provides ways for programmers to handle most run-time conditions gracefully. Most other high-level languages handle all run-time errors and conditions by terminating the user program. While PL/I is designed to be machine independent, it has a tendency to be operating system dependent. User programs coded to use features of a specific operating system will not run under a different system.

Other interesting features of PL/I include an elaborate compile time macro facility, string data types (both character and bit), and variable precision numeric data types.

When PL/I was first proposed it was thought to be the language of the future, and the first compilers were eagerly awaited. As the drawbacks of such a powerful language became apparent, users became reluctant to turn to PL/I. Unlike FORTRAN and COBOL, which are standardized and supported on many vendors' hardware, PL/I is supported by few vendors other than IBM. Users are not eager to invest large amounts of capital into software that would tie them to one vendor's hardware. The bubble-sort program in PL/I is shown in Fig. 58.12.

```
SORTER: PROCEDURE OPTIONS (MAIN);
DECLARE I, B(1 : 10) FIXED (31);

BUBBLE: PROCEDURE (A, LO, HI);
DECLARE I, TEMP, A( * ), LO, HI FIXED (31),
    SWAPPED BIT (1);

    SWAPPED = '0'B;   /*INITIALIZE SWAPPED TO FALSE.*/
    DO WHILE (~ SWAPPED);
        DO I = LO TO HI - 1;
            IF A(I) > A(I + 1) THEN DO;
                TEMP = A(I);
                A(I) = A(I + 1);
                A(I + 1) = TEMP;
                SWAPPED = '1'B; /*RECORD THE SWAP.*/
            END;
        END;
    END;
END BUBBLE;

/*MAIN PROGRAM BEGINS HERE.*/
DO I = 1 TO 10; /*INPUT*/
    READ(B(I));
END;
CALL BUBBLE (B, 1, 10); /*SORT B*/

DO I = 1 TO 10; /*OUTPUT.*/
    WRITE(B(I)):
END;

END SORTER;
```

Fig. 58.12 Bubble-sort program in PL/I.

BASIC. The computer language BASIC (beginner's all-purpose symbolic instruction code) grew from the environment of Dartmouth College in the mid-1960s under the guidance of John G. Kemeny and Thomas E. Kurtz.[13] Their intent was to design a language that was simple for nonexpert programmers to learn and use to solve simple programming problems. The original language was designed with a minimal number of control structures and data structures. The first three characters of each line form a unique keyword, the syntax rules for all types of statements are as identical as possible, and the variable names are short. These factors make BASIC easy to use. The original versions of basic were quite minimal in nature and were intended as a first language for students. After they had mastered BASIC, they would use other computer languages, notably FORTRAN, for producing faster and more efficient programs. The bubble-sort program in BASIC is shown in Fig. 58.13.

Each statement begins with a line number that is in ascending order. The first item after the line number is the keyword, which specifies the type of command. The first line has the keyword REM, which specifies that this line is a remark and does not contain any command. The DIM statement specifies that the one-dimensional array named A (called a list in BASIC) has 50 elements. BASIC also has two-dimensional arrays called tables. The READ statement takes the next available number of the data area and stores it in variable N (equals 8). The data area is defined by the DATA statements, which specify an ordered set of numbers to be stored in the internal data area. The numbers are stored in the order they are typed, with all the numbers for the lowest numbered DATA statement appearing before any numbers on the next lowest numbered DATA statement. The FOR and NEXT statements specify that the statements between them are to be carried out a specified number of times with the loop counter (J and K) to be incremented by 1 each time through the loop. The IF statement tests the specified condition [is A(K) less than or equal to A(K + 1)?] and if it is true, control is transferred to line number 140 (skips the instruction to interchange the numbers). The LET statement assigns the value of an arithmetic expression to the variable named to the left of the "=" symbol. The MAT statements are used to perform matrix operations, that is, operations on an entire list or table. The MAT READ statement causes the next N (equals 8) values in the data area to

```
10   REM SORT A LIST OF NUMBERS USING THE BUBBLE SORT
20   DIM A(50)
30   REM READ THE NUMBER OF VALUES (UP TO 50) TO BE SORTED
40   REM AND THE VALUES FROM THE INTERNAL DATA LIST
50   READ N
60   MAT READ A(N)
70   REM PERFORM BUBBLE SORT ON LIST A
80   LET S = 0
90   FOR I = 1 TO N - 1
100      IF A(I) < = A(I + 1) THEN 140
110      LET T = A(I)
120      LET A(I) = A(I + 1)
130      LET A(I + 1) = T
135      LET S = 1
140  NEXT I
150  IF S = 1 THEN 80
160  REM PRINT LIST
170  MAT PRINT A
180  DATA 8
190  DATA 34.5, 23.6, 0, 5.7, -87.5, 90, -0.7, 44.2
200  END
```

Fig. 58.13 Bubble-sort program in BASIC.

be stored in elements 1 through 8 of list A [A(1) is 34.5, A(2) is 23.6, . . . , A(8) is 44.2]. The MAT PRINT statement prints all elements of A that have been assigned a value.

Pascal. Pascal was designed and developed by Niklaus Wirth. It is in the Algol family of languages and includes a rich set of data structures and data-structure-building facilities. A preliminary version was developed in 1968, and the standard version was defined in 1974 in the *Pascal User Manual and Report*.[14] The IEEE Computer Standards Committee has also published a standard.[15]

Pascal was designed primarily as a teaching language. As such it contains features that easily support structured programming techniques such as modularization and top-down design. These include strong typing, limitations on the scope of variables and subprograms, sophisticated control and data structures, and subprograms that can be called recursively. These features and the fact that Pascal is concise yet easy to learn and use, constitute its major strengths. Pascal gained immediate acceptance in the academic community, and the success of the structured programming approach has led to its increased popularity in the workplace.

The basic structure of a Pascal program consists of a declaration section followed by a block of action statements that make up the algorithm proper. The declaration section is where constants, user-defined data types, variables, and data structures are defined and typed and where user-defined subprograms are placed. There are two types of subprograms in Pascal, functions and procedures. Functions are used like variables, but their values are determined by the execution of the subprogram. Procedures are invoked by a separate statement and have the effect of replacing that statement by the procedure body. The structure of subprograms parallels that of main programs. Each subprogram has a declaration section in which the programmer can define elements local to that subprogram. The declaration section is followed by a block of executable statements that specify the subprogram's actions.

Input and output is most often handled in terms of "text files," where the external file (generally card reader, printer, or interactive terminal) is human readable and the internal program variables are computer binary. The built-in Pascal procedures that perform this I/O are *read* and *write*, with variations *readln* and *writeln*, which read to the end of an input line and write an entire output line, including carriage return, respectively. The bubble-sort program in Pascal is shown in Fig. 58.14.

Statements are separated by semicolons. More than one statement can be on a line, and a statement can span several lines. All variables must be declared and their data types specified. An important concept is that of scope. The scope of a program element is the module in which it is defined and all the modules textually contained in it. It is not visible to program elements that textually include it. For instance, in this example, since the array A is declared in the main program, its scope includes the procedure "bubble," and references to A cause no problems so it need not be declared there. The same would be true of the variable I in the main program, except for the declaration for I in the procedure. This creates another variable that is local to "bubble" and prevents

```
program bubble (input, output);
const    FIRST = 1;
         LAST = 10;
var A : array[FIRST..LAST] of integer;
    I : integer;

procedure bubble;
var I, temp : integer;
    SWAPPED : boolean;
begin
    repeat
        SWAPPED := false; {assume that no swaps were done}
        for I := FIRST to LAST - 1 do
            if A[I] > A[I + 1] then
            begin {interchange two out of order items}
                temp := A[I];
                A[I] := A[I + 1];
                A[I + 1] := temp;
                SWAPPED := true {record that a swap was needed}
            end
    until not SWAPPED {sorting is done?}
end;

begin {The main program begins here.}
    for I := FIRST to LAST do {input elements to be sorted}
        read (A[I]);
    bubble;   {call sort procedure}

    for I := FIRST to LAST do {print the sorted array}
        writeln (A[I])
end.
```

Fig. 58.14 Bubble-sort program in Pascal.

the value in the main program from being referenced. The result is two distinct variables with the same name.

Two of the most important features of Pascal that are not used in the previous example are records and pointers.[16] The record structure can be thought of as an extension of the concept of an array, where the entries can be of different types. The entries are referenced by the name of the record, and the entry name is separated by a dot rather than by position number. Consider the record defined in Fig. 58.15. If a variable widgit is assigned the type *partrecord*, then *widgit.partno* and *widgit.price* refer to the part number and price.

A pointer is a variable that references (or points to) a storage location; its value is the address of that location. To create a pointer, one must declare a user-defined data type to be a pointer to the element that will occupy the storage area, then declare variables of that type. To assign values to the pointer variable, allocate storage locations by using the standard procedure *new* with the variable as an argument. For example, to access part records as just defined, by pointers rather than by names, use the declarations shown in Fig. 58.16. Then the statements *new*(*widget*) and *new*(*gizmo*) create two storage locations that will contain the information on those parts and place the addresses in widget and gizmo. The constructions *widget^.partno* and *gizmo^.price* refer to the widget's part number and the gizmo's price.

```
part = record

            partno, amount : integer;
            name : packed array [1..9] of char;
            price : real

       end
```

Fig. 58.15 Record declarations in Pascal.

```
type item = ^node;
     {variables of this type point to elements of type node}
var widget, gizmo : item;
```

Fig. 58.16 Declarations of pointer variables in Pascal.

C. C is a general-purpose programming language developed originally by Dennis Ritchie for the DEC PDP-11 running the UNIX† operating system. C is not tied to any particular system. However, compilers exist for several other machines, both large and small. The language has its origins in the language BCPL, by way of an intermediate language called B (which was developed in 1970 for the first UNIX system running on the PDP-7). Both BCPL and B are "typeless" languages, in that the only data type is the machine word (much like an assembly or machine language). C, however, is a typed language, where the fundamental data objects are characters, integers, and floating-point numbers. It is not, however, strongly typed (in the sense of Pascal and Ada): It is usually quite easy to perform conversion from one data type to another, although C does not allow conversion from all types to all other data types. C also provides facilities for creating "composite" data types through the use of pointers, arrays, structures, and functions. Some of the flexibility of the language is shown by the fact that although it is a block-structured language, most of the UNIX operating system, the C compiler itself, and most other UNIX applications programs are written in C.

C is a relatively low-level language in that it deals with many of the same kinds of objects that assembler languages deal with—characters, numbers, and addresses. C provides operations that work directly with these data types. It does not provide operations for such composite data types as character strings, lists, and entire arrays, although such operations are relatively easy to put together out of the simpler operations that are supported. C itself does not have any storage allocation facility other than static declaration and the ability to declare "local" variables in functions, although dynamic storage allocation is provided in most implementations. Input and output are done by way of implementation-defined functions rather than by statements in the language. C offers no complex flow-of-control constructs (such as coroutines or parallel operations) but rather only such straightforward constructs as tests, loops, grouping, and subprograms.

Statements in C are terminated with a semicolon (as opposed to Pascal, where statements are separated by semicolons). In terms of control structures, C provides looping statements with the controlling test at the beginning (for, while) and at the end (do); decision making (if); and selection of one of a set of possible cases (switch).[17] The forms of these statements are shown in Fig. 58.17.

```
for (expr1; expr2; expr3)
     stmt

while (condition)
     stmt

do
     stmt
while (condition);

if (condition)
     stmt1
else
     stmt2

switch (expr)
   {
     case selector_1
     case selector_2
           .
           .
     case selector_n
   }
```

Fig. 58.17 Control statements in the C language.

†UNIX is a trademark of Bell Laboratories.

```
case    const:
        stmt1;
        stmt2;
          .
          .
        stmtn;
```

Fig. 58.18 C's case selector.

The term "stmt" in this figure is any single C statement, and "expr" and "condition" are expressions to be evaluated. A block statement, which is a collection of one or more single statements enclosed in braces ("{ }," grouping operators similar to Pascal's begin and end), may be used in place of a single statement whenever execution of more than one statement is desired (as the body of a loop, for example). "Case selector" is the specification of a constant value and a series of C statements to be executed, as shown in Fig. 58.18.

C provides a macro definition capability through the use of the #define statement. This statement has the following syntax:

```
# define identifier definition_string
```

where "identifier" is any legal C identifier, and "definition_string" is the string of characters that will be substituted for "identifier" wherever it occurs (from the point of the definition onward) in the program. The simplest form of macro definition is something like

```
# define TRUE 1
```

which has the effect of associating the name TRUE with the value 1 (this is much like the const feature of Pascal). It is possible to define macros that take arguments, as in

```
# define swap(A, B){x = A; A = B; B = x; }
```

which exchanges the contents of the arguments A and B. Macro expansion is done prior to the actual compilation of a statement; the reference

```
swap(alpha, gamma)
```

will be expanded into the block statement

```
{x = alpha; alpha = gamma; gamma = x; }
```

which will then be compiled.

```
#define TRUE 1                  /*boolean true value*/
#define FALSE 0                 /*boolean false value*/

bubble (a, size)                /*function header*/
    int size, a[ ];             /*argument type definitions*/
{
    int swapped, i, temp;       /*local (auto) variable definitions*/

     do
     {
          swapped = FALSE; /*assume no swaps are needed*/
          for (i = 0; i < = size - 2; i + +)
               if (a[i] > a[i + 1]) /*two items out of order?*/
               {
                    temp = a[i]; /*so swap them*/
                    a[i] = a[i + 1];
                    a[i + 1] = temp;
                    swapped = TRUE; /*and record the fact*/
               } /*endif*/
     }
     while (swapped); /*continue as long as a change was made*/

} /*end bubble*/
```

Fig. 58.19 Bubble-sort program in C language.

Input and output are not an integral part of the C language but are handled through subroutine calls to installation specific routines. Neither the names of I/O functions nor their syntax are known directly to the C compiler.

The bubble-sort program in C is shown in Fig. 58.19. This function is called with two arguments: the array to be sorted (which is assumed to be an array of integers) and the size of the array. Arrays in C are based at zero; thus an array of size 10 has indices 0 through 9.

Ada. Ada (named after Ada Augusta, Countess of Lovelace, Charles Babbage's associate—a woman considered the world's first programmer) is a trademark of the U.S. Department of Defense. The definitive document for the Ada programming language is available as a military standard.[18] A series of language requirements documents, culminating with *STEELMAN*,[19] were issued by the U.S. Department of Defense. Several contractors then designed programming languages according to these requirements (all the competing languages were more or less based on the Pascal language).

Ada differs from other programming in that it addresses a much wider range of computer software life cycle, in particular the activity of software system design. Ada is further enlarged in its scope by having a specifically defined operating environment (run-time system).[20]

Pascal's advantage over its predecessors was mainly in its incorporation of support for structured programming in the program control flow and a coherent approach to data structuring. Ada incorporated those features of Pascal and added features to support modularity, specifically packages and tasks.[21,22]

Packages. Packages are collections of logically related units, such as special data structures and their associated operations or a family of subroutines. Packages may have portions of implementation details that are private or hidden from the user, with a public part giving the interface details needed to use the package. Packages may be "generic" or parameterized by the data types needed for a given application. For example, there could be a generic sorting package that could be used to sort any data type (this is similar to "macro" facilities that are available in some other language systems). The sample bubble-sort program is shown in Fig. 58.20. It uses a system-supplied package known as text_io, which contains all the routines needed to do input and output, where the external medium is for human consumption, that is, it is text (the internal representation is generally binary and unreadable).

```
with text_io; use text_io; use integer_io;
procedure sorter is

SWAPPED : BOOLEAN;
I, TEMP : INTEGER;
A    : array (1..10) of INTEGER;

for I in A'FIRST..A'LAST loop
     get (A(I));
end loop;

loop
     SWAPPED := false;
     for I in A'FIRST..A'LAST - 1 loop
          if A(I) > A(I + 1) then
               TEMP := A(I);
               A(I) := A(I + 1);
               A(I + 1) := TEMP;
               SWAPPED := true;          —record that a swap was needed.
          end if;
     end loop;
     exit when not SWAPPED;              —sorting is done?
end loop;

for I in A'FIRST..A'LAST loop
     put (A(I));
     put (', ');
end loop;

end sorter;
```

Fig. 58.20 Bubble-sort program in Ada.

A concept supported in Ada and not generally available in other popular programming languages is that of "overloading," where names and symbols may stand for more than one thing. The Ada system resolves which meaning is intended by the context. In the sample program, the procedure "put" is overloaded as a procedure for printing integers (the first use) and as a procedure for printing character strings (the second use, where the character string "comma-blank-blank" is used to separate the numbers in the output). The "write" procedure and the arithmetic operators are overloaded in Pascal, but in Ada, programmers can define overloaded procedures and operators. For another example, a programmer can define "+" and "*" as matrix addition and multiplication.

Tasks. Tasks are units for specifying sequences of actions that may be executed in parallel with other similar units. The parallel tasks may synchronize and pass information by means of an Ada feature known as task rendezvous. Parallel tasks may be run on a multiple processor system with each task on a separate computer, or on a single processor system with the tasks taking turns to run by time slicing.

The bubble-sort procedure in Ada is shown in Fig. 58.20. Note that comments are begun by two hyphens and ended by the end of line (much less error prone than the comment conventions "{ . . . }" of Pascal or "/* . . . */" of C, where the absence of the end-of-comment delimiter can cause a large portion of the program to be lost, often with no indication of the error). Ada's control structures are more regular than Pascal's. For example, in Pascal one writes

```
for i := j to k do
    write (A[i])
```

if there is only one statement within the loop body, but

```
for i := j to k do
    begin
        read (A[i]);
        write (A[i])
    end
```

since the two statements within the loop body need to be grouped. In Ada, on the other hand, one writes

```
for i in j..k loop
    put (A(i));
end loop;
```

and

```
for i in j..k loop
    get (A(i));
    put (A(i));
end loop;
```

Note also the differences in the use of the semicolon; in Pascal semicolons *separate* statements, but in Ada semicolons *terminate* statements. The Ada choice is much simpler to explain and to follow, so it is less error prone.

SNOBOL4. The programming language SNOBOL4 comes from Bell Laboratories and has been available since the mid-1960s. As the name indicates, the language has gone through several versions as it matured to its present state.[23,24]

Although SNOBOL4 contains data types, operations, and control structures suitable for a general-purpose programming language, its strong points and areas of application are in the area of character string processing. A character string is simply a sequence of letters, digits, punctuators, and other symbols; for example, words, sentences, books, computer programs, and lines in a telephone directory are all character strings. SNOBOL4 directly supports several operations on character strings, such as concatenation (combining two strings to form one longer string) and substring matching (finding an occurrence of a string within another string, such as finding a keyword in a book).

The most interesting and powerful feature of SNOBOL4 is that of the "pattern" for matching substrings of longer strings. With patterns, one can compose very short powerful programs, such as conversion of computer programs in one manufacturer's dialect of FORTRAN to another's, modification of record structures in a data file, or performance of symbolic algebra.

```
*       SORT THE ARRAY OF NUMBERS, A.
        SWAPPED = 'FALSE'
GOAGAIN I = 1
TEST GT (A⟨I⟩, A⟨I + 1⟩) :S(SWAP)F(NOSWAP)
SWAP    TEMP = A⟨I⟩
        A⟨I⟩ = A⟨I + 1⟩
        A⟨I + 1⟩ = TEMP
        SWAPPED = 'TRUE'
NOSWAP
        I = I + 1
        A⟨I + 1⟩    :S(TEST)
        EQ(SWAPPED, 'TRUE') :S(GOAGAIN)
```

Fig. 58.21 Bubble-sort program in SNOBOL4.

Some very complicated programs, such as natural language (i.e., English or French) processing or chemical compound structure manipulation may have their underlying character handling done by SNOBOL4, which makes that aspect of the problem, at least, very simply handled.

The bubble-sort program SNOBOL4 is shown in Fig. 58.21. This program does not use any of the distinctive features of SNOBOL4, but it does indicate the basic syntax of the language.

Each statement in the program is on a separate line (originally, like FORTRAN, SNOBOL4 was for punched-card oriented systems), and begins with an optional label (GOAGAIN, TEST, SWAP, NOSWAP). The labels are referred to by the conditional go-to section, which is the final (and optional) part of each statement. For example, the statement

```
TEST   GT(A⟨I⟩, A⟨I + 1⟩)   :S(SWAP) F(NOSWAP)
```

evaluates the function GT (called a predicate) which, if true (when A⟨I⟩ is greater than A⟨I + 1⟩), "succeeds" and control jumps to SWAP (the S stands for succeed); if GT is not true, the statement is said to fail and control jumps to NOSWAP.

SNOBOL4 does not directly support through its control structures the ideas of structured programming; most control flow is done by means of conditional and unconditional go-to's. However, SNOBOL4 does have a subroutine facility (not discussed here). Most SNOBOL4 programs or subroutines are sufficiently short to avoid the plague of spaghetti code).

Input and output operations in SNOBOL4 are particularly simple. To obtain input from the standard input device (a file, the card reader, or a user's terminal), one simply evaluates the variable INPUT:

```
NEWDATA = INPUT   :F(ALLGONE)
```

Control is transferred to the label ALLGONE in case end of input has been reached. To produce output (to a file, a printer, or a terminal), one assigns a new value to the variable OUTPUT:

```
OUTPUT = ANSWER
```

An example of SNOBOL4 pattern matching is shown in Fig. 58.22. This program segment converts an English word to Pig Latin (the initial consonant substring of a word is moved from the front of the word to the end and followed by "AY"; if the word begins with a vowel, simply attach "WAY" to the end; and note "QU" is a consonant string for these purposes). The reader should try to follow this and verify that DOG becomes OGDAY, STREET becomes EETSTRAY, APPLE becomes APPLEWAY, SQUIRT becomes IRTSQUAY.

Like many of the other programming languages, we see that SNOBOL4 has the usual sort of assignment statement,

```
variable = value
```

where the right-hand side *value* is the new value for the left-hand side *variable*. SNOBOL4 also has

```
variable pattern = value
```

where *value* is the new value for the substring of *variable* matched by *pattern*. In Fig. 58.22 we

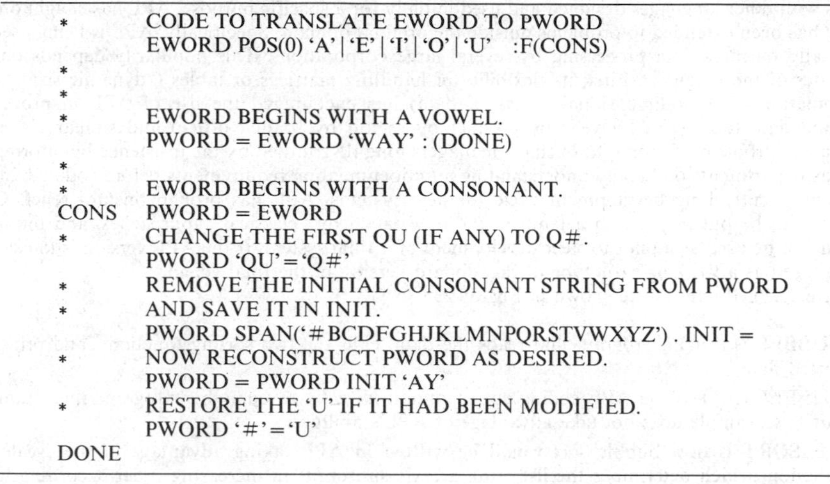

```
*       CODE TO TRANSLATE EWORD TO PWORD
        EWORD POS(0) 'A' | 'E' | 'I' | 'O' | 'U'    :F(CONS)
*
*
*       EWORD BEGINS WITH A VOWEL.
        PWORD = EWORD 'WAY' : (DONE)
*
*       EWORD BEGINS WITH A CONSONANT.
CONS    PWORD = EWORD
*       CHANGE THE FIRST QU (IF ANY) TO Q#.
        PWORD 'QU' = 'Q#'
*       REMOVE THE INITIAL CONSONANT STRING FROM PWORD
*       AND SAVE IT IN INIT.
        PWORD SPAN('#BCDFGHJKLMNPQRSTVWXYZ') . INIT =
*       NOW RECONSTRUCT PWORD AS DESIRED.
        PWORD = PWORD INIT 'AY'
*       RESTORE THE 'U' IF IT HAD BEEN MODIFIED.
        PWORD '#' = 'U'
DONE
```

Fig. 58.22 Pig Latin procedure in SNOBOL4.

changed the 'QU' substring of PWORD to 'Q#'; we changed the initial span of consonants to null (there was no right-hand side value); and we changed the '#' in PWORD to 'U'.

The statement

$$\text{variable}_1 \text{ pattern . variable}_2 = \text{value}$$

replaces the matched substring of $\underline{\text{variable}_1}$ with the value and also saves the original matched substring in $\underline{\text{variable}_2}$.

SNOBOL4 has a rich repertoire of patterns, and we have only seen a small sample of these:

'QU' matches the string 'QU'.
'A' | 'E' | 'I' | 'O' | 'U' matches any vowel.
POS(0) 'A' | 'E' | 'I' | 'O' | 'U' matches any vowel at the beginning of a string.
SPAN('#BCDFGHJKLMNPQRSTVWXYZ') matches the longest possible string of consonants and #s.

Patterns can be built on top of other patterns using concatenation and alternation ("|") to get almost any pattern matching one can imagine.

APL. APL ("a programming language")[25] was invented by Dr. Kenneth Iverson while he was at Harvard University. In 1966 the first experimental version of the language was made available for internal use at IBM. Since then the language has become an official product of IBM. The development of the language has been promoted most vigorously by I. P. Sharp Company of Toronto, Canada. The company spearheaded the development and promotion of a version called APL + ("APL Plus"), the principal extensions of which are (1) the ability to handle data files and (2) a report-formatting capability that owes much to FORTRAN and little to APL.

APL was conceived as a tool for mathematicians, statisticians, and engineers. The constructs and the notation are mathematical. A rich selection of primitive operations is focused on solution techniques of interest to these classes of professionals; for example, the "domino" operator, which specifies matrix inversion. APL relies on interpretive execution of the source program which, like BASIC, is fundamental to the encouragement of programming by the end user rather than reliance on people whose specialization is the programming itself rather than the application area. As with BASIC, Iverson's idea was to provide users with a flexible interface to computing power: flexible with respect to its capabilities, to the types of problems addressable, to the expansions and contractions of the solutions to those problems, and to the implementation of them. The key to the last of these is interpretive execution, which enables users to interrupt running programs and then to print program variables, to change their values, and to change the program during the course of execution and then to reinitiate execution from the point of interruption.

As with other languages designed and used initially for a specific purpose, APL has caught on and its use has been extended to problems outside the original domain. Specifically, APL is being used for essentially business data processing by several large corporations. This popularity depends on two attributes of the language. First, its flexibility of handling matrices or tables ("dynamic sizing" and manipulation of multidimensional arrays of data) has encouraged the use of APL in processing planning data (e.g., sales by year by product by region by branch office) and similar classes of large-array problems. Second, to business managers long disgruntled by the insistence by information systems departments on "really understanding and documenting requirements before coding" and by the consequently long development cycle for new systems, APL has brought instant relief. Good systems can be put in place in a tenth of the time taken by "classical" approaches, and the initial system can be used as a pilot to steer development of a final system. If the APL version is too slow, it can be used as a working prototype or breadboard version of the final version.

Three APL functions are shown in Fig. 58.23:

1. BUBBLE is a main program that reads the data, calls the two sorting functions, and prints the sorted data.
2. BUBBLESORT is the APL coding corresponding directly to our other programming examples, but this example does not take advantage of APL's abilities.
3. APLSORT is how bubble sort would be written in APL taking advantage of the "scale up" function, which determines the list of an array's subscripts in increasing order according to the values in the array.

LISP. The language LISP (list processor) was designed at MIT as a language for use in symbol processing.[26] LISP provides facilities for manipulation of nonnumeric as well as numeric data; the ease with which symbolic data can be handled makes LISP the language of choice for many applications where such data are used. Another interesting feature of LISP is the fact that all information, both programs and data, is represented in the same manner, allowing such uncommon actions as the dynamic creation and evaluation of functions.

```
      ∇ BUBBLE
 [1]    'INPUT VECTOR OF NUMBERS TO BE SORTED'
 [2]    VEC←⎕
 [3]    RES←APLSORT VEC
 [4]    'OUTPUT OF APLSORT'
 [5]    RES
 [6]    FIN←BUBBLESORT VEC
 [7]    'OUTPUT OF BUBBLESORT'
 [8]    FIN
 [9]    'END'

      ∇ RR←BUBBLESORT A
 [1]    AGAIN : SWAP←0
 [2]    I←1
 [3]    LOOP:
 [4]    →(I = ρA)/EXIT
 [5]    →(A[I] ≤ A[I + 1])/TEST
 [6]    TEMP←A[I]
 [7]    A[I]←A[I + 1]
 [8]    A[I + 1]←TEMP
 [9]    SWAP←1
[10]    TEST : I←I + 1
[11]    →LOOP
[12]    EXIT:→((SWAP = 1), SWAP = 0)/AGAIN, FINISH
[13]    'ERROR'; →0
[14]    FINISH:'END OF BUBBLESORT'
[15]    RR←A

      ∇ R←APLSORT NUMS
 [1]    R←NUMS[⍋NUMS]
```

Fig. 58.23 How to sort an array in APL.

Assembler Languages. The name "assembly language" does not stand for any individual, standardized programming language but rather the symbolic form for specifying machine language programs for any given computer. Since the instruction and data formats of various models of computers differ widely, the assembler languages also vary. But like other aspects of this realm, although the details are never the same the underlying concepts are largely invariant.

Assembly languages provide the following labor-saving services for programmers:

1. Mnemonics for the machine operations (such as "L" for "load register" and "MVC" for "move characters").
2. Symbolic names ("labels") for the locations of instructions and data.
3. Symbolic names for constants.
4. Expressions involving symbolic names and constants.
5. Data initialization in human-readable forms (i.e., decimal constants rather than binary patterns).
6. Commentary on separate comment lines and comments attached to each statement.

Assembly language programming will be illustrated in terms of the IBM System 360/370 basic assembly language (also known as BAL) to express the bubble-sort procedure.[27] The syntax of BAL is as follows:

1. Comment lines contain an asterisk ("*") in position one.
2. Labels, if present, start in position one.
3. The operation code ("opcode") follows the label, preceded by one or more blanks. If the opcode is a machine instruction mnemonic, then the current line corresponds to a single machine instruction; otherwise the line is used to reserve storage for program variables (possibly initializing them) or providing directives to the assembler.
4. Following the opcode, separated by one or more blanks, is the operand field, which consists of a list of specifiers of program variables, machine registers, and parameters that affect the interpretation of the opcode.

```
* 10  SWAPPD = .FALSE.
L10   MVI SWAPPD, 0set the byte SWAPPD to 0.
*     DO 100 I = FIRST, LAST-1
      L  1, FIRST   put the value of FIRST into general register 1.
      ST  1, I  store the contents of general register 1 in I.
DO100  L  1, I
      C  1, LAST   compare the contents of register 1 with LAST.
      BGT DO100X   go to DO100X if I is greater than or equal to LAST.
*     IF (.NOT. (A(I) .GT. A(I + 1))) GOTO 100
      SLL 1, 2 shift general register left two bits
      LE  0, A(1)   load the Ith element of A into floating point register 0.
      CE  0, A + 4(1) compare floating point register 0 with the
            I + 1st element of A
      BLE L100 go to L100 if compare is "less than or equal."
*     TEMP = A(I)
      STE 0, TEMP   store floating point register 0 in TEMP.
*     A(I) = A(I + 1)
      LE  0, A + 4(1)
      STE  0, A(1)
*     A(I + 1) = TEMP
      LE  0, TEMP
      STE  0, A + 4(1)
*     SWAPPD = .TRUE.
      MVI SWAPPD, 1
* 100   CONTINUE
L100 B  DO100   go to label DO100
*     IF (SWAPPD) GOTO 10
DO100X  CLI SWAPPD, 1 compare SWAPPD with 1.
      BE  L10 go to L10 if compare was "equal."
```

Fig. 58.24 Bubble-sort program in assembler language.

5. Following the operand list, separated by one or more blanks, is the comment field, which consists of any annotation the programmer may wish to make about the statement (comments are optional).

The code of the FORTRAN bubble-sort subroutine is used to annotate groups of assembly language statements in Fig. 58.24.

As the reader can see from this example, assembly language is very difficult to read and write and it lacks the features that are taken for granted in high-level languages, such as IF statements, loops, and subroutines. Everything must be done as a sequence of tiny steps; each step is very simple, but there are so many of them that the programs become overpoweringly complicated. It has been noted by many observers that programmers' productivity, measured in terms of lines of code produced per day, is independent of the programming language chosen (reported numbers are in the range of 7 to 50 lines per day, with the variation depending only on individual ability). In spite of this, there are several reasons put forward for the use of assembly language:

1. High-level language compilers are not as efficient as assembly language programmers. Compiler-generated code is longer and slower. (This argument becomes increasingly less valid as compiler technology advances. The compiler-generated code is more likely to be correct and maintainable.)

2. New machines may not have working compilers, and assemblers are easier to build than compilers. (This argument is overwhelming for the application programmer, but it applies to very few machines and only for a short period of time.)

58.5 SYSTEM SIMULATION

Systems that involve simulation models represent a computer application area of increasing interest and importance. Simply stated, simulation involves a methodology for the study of the dynamics of a system, or alternatively, the study of the interrelationship of the components that make up the system. The system can be defined as consisting of the set of parts that are functionally organized to make up the whole. For example, a harbor consists of piers, tugs, ships, channels, waterways, and so on. These all combine to make up a system that can be defined as a harbor. In fact, without the components the harbor is nothing more than just an inlet of water. It is the interrelationship of the components of the harbor that is of interest to the modeler.

A model is a representation of the system. It describes those components that make up the system in sufficient detail for us to be able to study the behavior of the system over valid ranges of operational constraints and, more importantly, to be able to make either valid predictions about the behavior of the system or select alternative plans of action. Therefore it is important that components that have a significant effect on the behavior of the system be included in the model. Components to which the model is most sensitive are of particular interest, although the simulation modeler is always faced with justifying (or rationalizing) both the inclusion (or exclusion) and the behavior of each component included in the model, regardless of its impact.[28] Two types of activities occur in the typical model: endogenous activities (parameters) that describe the activities that occur within the system and exogenous activities (input) that describe those activities that have an effect on the system.

While a model attempts to replicate to some degree a real process, the simulation model is not an attempt to mimic the real world. Some of the many reasons for this philosophy include: It is not uncommon to have the performance of a system depend only on a relatively few components; it would often be impractical to make the model very detailed, as opposed to a prototype of the system; the interrelations within the system are sometimes not clear or are beyond the scope of analysis; the collection of large amounts of empirical data used to determine either parameters, input values, or interrelationships may be time consuming, costly, and/or impractical; both cost and time constraints may limit the scope of the model.

It has been said[29] that "computer simulation is the court of last resort." If there was any other way to analyze the system, it would be utilized. Unfortunately, this is sometimes interpreted to mean that there always is a court of last resort, but alas, some problems cannot be analyzed, since an in-depth systems analysis is not plausible.

Computer models are most likely to be stochastic, as opposed to deterministic. A deterministic process is one in which every input value produces in a reproducible fashion some output value. A mathematical equation such as $a = b + 4$ represents a deterministic process, since for each input value for b the value of a is always uniquely determined without any irregularity. In contrast, a stochastic process involves the concept of randomness. Given sufficient empirical data, predictions about the behavior of a stochastic model can be made assuming that the systems analysis has determined the interrelationships between the components of the model. Stochastic models typically are made up of subsystems that are most likely stochastic themselves. It is the interaction of the subsystems of a system that is of interest to the modeler.

Computer models can also be classified as either discrete-event or continuous models.[28] Continuous models are utilized to represent the dynamics of processes, and more specifically the behavior of continuous physical devices or phenomena that are best represented by mathematical techniques such as differential equations or numerical integration. Rather than attempt to replicate the continuous behavior of a system, discrete-event models utilize discrete intervals of time in which to observe the behavior of the system. The underlying assumption is that the selected unit of time is short enough to allow the capture of adequate performance statistics without the possibility that extreme performance variance would have been observed had a smaller period of time been selected. Sometimes a discrete-event model may have a continuous model as one of its subsystems. For example, this would occur when a subsystem could be well described by the laws of physics. Unfortunately, the inclusion of a continuous subsystem within a discrete-event model is not always feasible if one utilizes a discrete-event simulation language such as GPSS or Simscript II.5.[28–30] It should be noted, however, that not all discrete event simulation is accomplished utilizing a special-purpose discrete-event simulation language. Models can be implemented in FORTRAN, PL/I, Pascal, or Ada. Even BASIC and COBOL are possible languages. However, to support the stochastic nature of the model, a pseudorandom number must be made available.

A simulation study has several phases, and success in each of them is essential if a rewarding outcome is expected. First comes the system analysis phase, to define the model itself. This includes determination of the components to be included in the model. Observation of interrelationships, and the gathering of empirical data that will be used both for guidance in the design of the model and for parameter values and input data for the model. Second is the computer programming, to code the model for computer processing. This includes the selection of a suitable programming language.

The third phase involves debugging the computer program that represents the model. In modeling, this involves a number of verification steps. Verification refers to determining whether the program produces output consistent with how the model is expected to behave. A common technique for this process is "boxing,"[29] in which one portion of the program is "boxed," with the input and output from the boxed portion analyzed. Since we are dealing with nonprocedural-type processing in which several events can occur concurrently, verification of the program is not only more difficult than usual but even more essential.

The fourth phase is validation. In this phase we strive to increase the user's confidence in the model; that is, how well does the output from the model correspond to the expected behavior of the system? Not unlike the first three phases, validation is essential before predictions can be made or alternatives can be chosen. The fifth phase involves iteration. Iterative processing of the model is done for three reasons: (1) to continue the validation process, (2) to determine the sensitivity of the model to its various components, which may suggest alternate strategies or the addition or deletion of components to the model before subsequent computer runs are made, (3) to determine whether the model has reached a steady-state condition—is the variance in the output due to the stochastic processes involved in the model (a steady-state condition) or is the variance due to fluctuations that would be eliminated if the model was executed for a longer period of time or was executed for n additional iterations before the output from the model was analyzed for the purpose of making predictions or selecting alternatives? The cost, time, and effort involved in a successful computer simulation is typically nontrivial.

In addition to general-purpose programming languages, several special-purpose discrete-event simulation languages are available.

GPSS

GPSS (circa 1960) was originally intended only for internal use by IBM for the analysis of engineering projects.[29,31] It was released publicly in 1961 as GPSS. (Originally it was known as GPS.) GPSS is an acronym for general purpose simulation system and today is supported by IBM as GPSS-V. Several other computer manufacturers and software houses support GPSS dialects for Burroughs, Honeywell, UNIVAC, Xerox, and DEC equipment. GPSS is still probably the most commonly utilized special-purpose discrete-event simulation language. GPSS is a block-oriented language in which each block of the block diagram corresponds to a GPSS instruction. Inherent within a GPSS processor is at least one pseudorandom number generator as well as a "clock" that is automatically incremented for the user. GPSS is a transaction-oriented language. Transactions are generated and flow through the model from block to block until they are eventually terminated. Standard statistics accumulated by GPSS reflect the cumulative effect of all transactions that have passed through the model during the elapsed time period. GPSS best describes the global effect on the model of all transactions, compared with an event-oriented language in which it is easier to capture statistics about each individual entity. GPSS programs are typically made up of several subprograms; it is the interaction between the subprograms that reflects the stochastic nature of the system being modeled.

Simscript II.5

Simscript II.5 (circa 1961) was developed by the Rand Corporation.[30] It has Englishlike syntax, a high-level command description as opposed to the more assembler-oriented syntax of GPSS, and it encourages the logical formation of modules. Simscript focuses on the concept of an event, which is defined as any process that causes the system to change. In effect, an event is a routine. Unlike in GPSS, in Simscript time elapses between events rather than within events. In a Simscript program, an entity passes from one event to the next as the system clock advances. As with GPSS, implementations of Simscript always provide for at least one pseudorandom number generator and a built-in system clock that is automatically incremented. While it is definitely possible to capture global statistics concerning the performance of the model, it is much easier to capture statistics about each event in Simscript as compared with GPSS. Simscript also overcomes one of the greatest weaknesses of GPSS by allowing FORTRAN-like arithmetic statements within a Simscript program.

SIMULA

SIMULA (circa 1960) is based on Algol-60; a common implementation is SIMULA-67.[31] One of its strongest features is that it is embedded within a host language (Algol) so that all of the processing capabilities of a general-purpose programming language are available to the user. SIMULA views the model of the system as a collection of processes in which the state of a process can be changed as the data structure associated with that process changes. SIMULA is more closely aligned with GPSS and its transaction orientation.

GASP

GASP (circa 1974) was developed by Pritsker Associates of West Lafayette, Indiana.[32] GASP utilizes an event-scheduling approach, similar to Simscript. It includes facilities for hybrid models, which include both discrete-event and continuous components.

SLAM

SLAM (circa 1979) is another high-level simulation language that allows for both discrete-event and continuous-event simulation. It has been implemented on the VAX/780, CDC-6000, IBM 360/370, UNIVAC 1108, PRIME 400 and 700, and Harris 550. SLAM employs a network structure with specialized symbols for nodes and branches. With its syntax, SLAM attempts to pictorially represent a system.

58.6 DATA BASE SYSTEMS

A data base is defined as a collection of stored operational data used by the application systems of some enterprise.[33] A data base system is then a system whose purpose is to store, maintain, and provide information from the data base.

The basic building blocks of the data base are entities and relationships between entities. Entities are those objects in which an enterprise has interest. An example of an entity is the collection of machines that an enterprise owns. Entities are described by characteristics called attributes. In a data base, an entity is composed of one or more attributes. For example, the entity *machine* may be composed of the machine's name, number, and location.

Entities are associated with one another through relationships. Relationships may be implicit or explicit depending on the model used to design the data base. For example, the entity *machine* may be related to an entity describing parts that make up the machine. Machines and parts would be related to one another because a machine is composed of parts. The relationship between machine and parts would be that which indicates that a machine is composed of parts.

An advantage of data base systems is the modeling capability they provide. A data base system may be modeled so that the organization of the data itself conveys meaning about the enterprise. There are currently three well-known models for data base management systems: relational, network, and hierarchical.[33]

The relational model is derived from set theory in mathematics. Relations represent entities and are defined in a tabular form. Each relation (table) is composed of attributes (columns) that describe the relation. Each row of a table, called a tuple, represents one occurrence of the entity. Relationships between entities are not specified in the data base, but are specified by the user through the query language. Information is retrieved from a relational data base using queries based on relational algebra.

In the network data model, relationships between entities are expressed explicitly by ownership relations. An entity may be an owner in many sets and may be a member of many sets. The hierarchical data model is a special case of the network data model, in which a general graph structure is constrained to be a tree structure.

A data base management system is a collection of software designed to store, maintain, and retrieve information from the data base. Data base management systems contain data definition facilities that allow the user to design and describe the data base. The data definition facility is supported by underlying software that directs the physical storage of the data. The storage of the actual data or information in the data base is completely transparent to the user. A data base system offers each user a "view" of the data consistent with the user's particular needs. A data manipulation facility is also available within the data base system. This manipulation language allows the user to add and delete information from the data base as well as update and retrieve data as necessary. Retrieval and updating may be performed on line with a nonprocedural query language or via calls from application programs written in COBOL, FORTRAN, or assembly language.

The use of a data base system by an enterprise provides many advantages. One such advantage is the centralized control of data. Many functional areas in an enterprise use the same data. Consequently, centralized control of the data facilitates sharing the data. Since only one copy of the data is stored, a single data base operation executed on the data in the data base performs the operation for all users. As a result, not only is storage kept to a minimum but all data used by the enterprise are kept current and consistent, which ensures data integrity. However, this data integrity must be paid for by increased security precautions. Through centralized control of the access to the data, a level of security is provided in that only those who need the data may have access to it.

Another advantage to data base systems is that they provide physical data independence. Physical data independence exists between application programs and the data base when changes in one have no effect on the other. A high degree of physical data independence allows the data base application programs to be changed with minimal data base maintenance effort required. Physical data independence also allows the data base to be changed with minimal effect on the application programs.

The future of data base technology is promising, but supporters fall into two camps. There are those who espouse the relational model for its simplicity and theoretical foundation. However, the relational model has not been successfully implemented to effectively handle very large data bases. Those who espouse the network model favor a model that has been implemented but that has considerable complexity and little flexibility.[34]

References

1 E. N. Yourdon, ed., *Classics in Software Engineering*, Yourdon Press, New York, 1979.

2 D. E. Knuth, *The Art of Computer Programming*, Vol. 3, *Sorting and Searching*, Addison-Wesley, Reading, MA, 1975.

3 A. V. Aho and J. D. Ullman, *Principles of Compiler Design*, Addison-Wesley, Reading, MA, 1975.

4 P. Naur et al., "Revised Report on the Algorithmic Language Algol-60," *Commun. ACM* **6**(1): 1–17 (1963).

5 A. van Wijngaarten, "Revised Report on the Algorithmic Language Algol-68," *Acta Inf.* **5**: 1–236 (1975).

6 D. D. McCracken, *A Simplified Guide to Structured COBOL Programming*, Wiley, New York, 1976.

7 L. S. Cohn, *Effective Use of ANS COBOL Computer Programming Language*, Wiley, New York, 1975.

8 *ANSI X3.9-1966*, USA Standard FORTRAN, Mar. 7, 1966.

9 *ANSI X3.9-1978*, American National Standard Programming Language FORTRAN, Apr. 3, 1978.

10 M. Boillot, *Understanding FORTRAN*, 2nd ed., West, St. Paul, MN, 1978.

11 L. Meissner and E. Organick, *FORTRAN 77: Featuring Structured Programming*, Addison-Wesley, Reading, MA, 1979.

12 C. T. Fike, *PL/I for Scientific Programmers*, Prentice-Hall, Englewood Cliffs, NJ, 1970.

13 J. G. Kemeny and T. E. Kurtz, *BASIC Programming*, Wiley, New York, 1980.

14 K. Jensen and N. Wirth, *Pascal User Manual and Report*, 2nd ed., Springer-Verlag, New York, 1976.

15 *Computer Programming Language, Pascal*, IEEE, Los Alamitos, CA.

16 G. M. Schneider, S. W. Weingart, and D. M. Perlman, *An Introduction to Programming and Problem Solving with Pascal*, 2nd ed., Wiley, New York, 1982.

17 B. W. Kernighan and D. M. Ritchie, *The C Programming Language*, Prentice-Hall, Englewood Cliffs, NJ, 1978.

18 *Ada Programming Language*, MIL-STD-1815, 10 Dec. 1980.

19 Department of Defense, *STEELMAN Requirements for High Order Computer Programming Language*, Jun. 1978.

20 Department of Defense, *Requirements for Ada Programming Support Environments—STONEMAN*, Feb. 1980.

21 I. C. Pyle, *The Ada Programming Language*, Prentice-Hall, Englewood Cliffs, NJ, 1981.

22 P. Wegner, *Programming with Ada: An Introduction by Means of Graded Examples*, Prentice-Hall, Englewood Cliffs, NJ, 1980.

23 R. Griswold et al., *The SNOBOL4 Programming Language*, Prentice-Hall, Englewood Cliffs, NJ, 1971.

24 R. Griswold and M. Griswold, *A SNOBOL4 Primer*, Prentice-Hall, Englewood Cliffs, NJ, 1973.

25 L. Gilman and A. J. Rose, *APL—An Interactive Approach*, Wiley, New York, 1974.

26 J. McCarthy et al., *LISP 1.5 Programmer's Manual*, 2nd ed., MIT, Cambridge, MA, 1965.

27 G. Struble, *Assembly Language Programming: The IBM System 360/370*, 2nd ed., Addison-Wesley, Reading, MA, 1975.

28 J. A. Payne, *Introduction to Simulation*, McGraw-Hill, New York, 1982.

29 T. J. Schreiber, *Simulation Using GPSS*, Wiley, New York, 1974.

30 T. P. Wyman, *Simulation Modeling: A Guide to Using Simscript*, Wiley, New York, 1970.

31 P. A. Bobillier, B. C. Kahan, and A. R. Probst, *Simulation with GPSS and GPSS-V*, Prentice-Hall, Englewood Cliffs, NJ, 1976.

32 A. A. Pritsker, *The GASP-IV Simulation Language*, Wiley, New York, 1974.

33 C. J. Date, *An Introduction to Database Systems*, 2nd ed., Addison-Wesley, Reading, MA, 1981.

34 E. Wong (moderator) et al., *ComputerWorld*, p. 78 (Sep. 13, 1982).

CHAPTER 59
CENTRAL PROCESSOR ORGANIZATION

MELVYN M. DROSSMAN

New York Institute of Technology
Old Westbury, New York

59.1 STRUCTURE AND FUNCTION

The central processing unit (CPU)[1] of a digital computer is the principal functional element of the computer system. It consists of two functional subunits: the control unit (CU) and the arithmetic–logic unit (ALU). The control unit interprets instructions, causes the other units of the computer to perform the functions required to execute the instructions, and synchronizes their operation. The ALU performs the arithmetic and logic operations that, along with input and output instructions, implement the computer's instruction set. The arithmetic instructions include addition and subtraction and, in some computers, multiplication and division. The logic operations include comparison, and some or all of the boolean operations AND, OR, XOR, and complement. A variety of shift operations are also performed. The CPU operates in conjunction with the main memory unit and input and output devices. It is necessary to describe the interfaces between the CPU and these devices to explain the operation of the CPU.

The main memory system[2] consists of the memory itself, a memory address register (MAR), a memory buffer register (MBR), and a memory control, as shown in Fig. 59.1. This memory contains N words of W bits per word. The memory address register contains the address of the word to be accessed. This register contains K bits, where $2^{**}K = N$, the number of words in the memory. ($2^{**}K$ represents 2 raised to the power K). The memory buffer register serves as an interface between the memory and the remainder of the computer. Data are read from the memory into the MBR and written from the MBR into the memory. The size of the MBR is equal to the word size of the memory. The memory control determines whether a word of data is transferred into or out of memory and when the transfer occurs. Three inputs are used in Fig. 59.1: the first is the read/write (RW) line, which is high for a memory–read operation and low for a memory–write operation; the second is the memory enable (ME) line, which indicates that a memory access operation is occurring; and the third is the clock input (Cl), which provides synchronization with the rest of the computer.

The interfaces with the input and output devices[3] are typified by the output device and controller shown in Fig. 59.2. The data register, DR, is loaded with L bits of data via the data bus by the CPU,

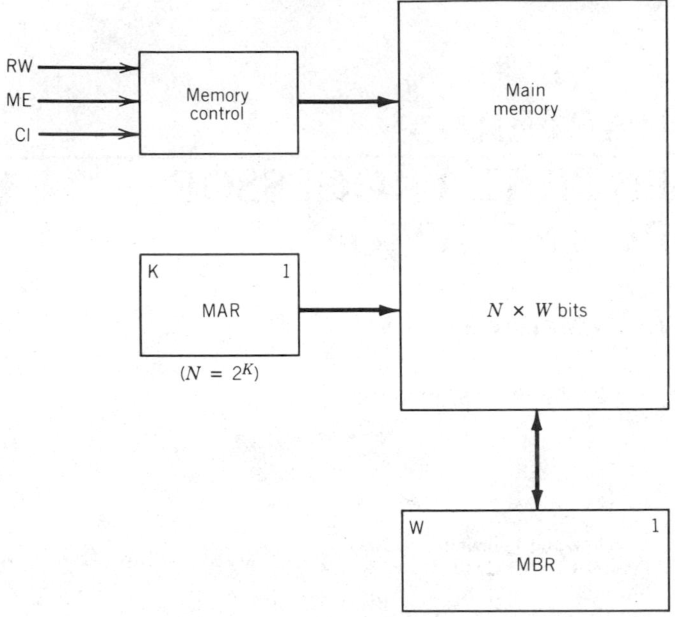

Fig. 59.1 Main memory system. RW, read–write; ME, memory enable; Cl, clock input; MAR, memory address register; MBR, memory buffer register.

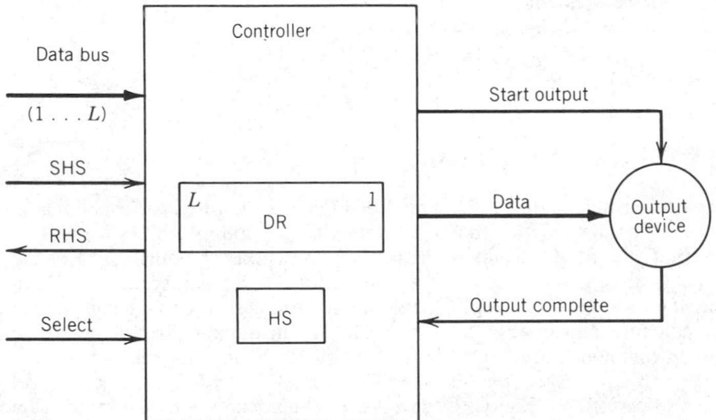

Fig. 59.2 Output system. DR, data register; HS, handshaking register; SHS, set HS; RHS, read HS.

which then sets the single-bit handshaking register, HS, high using the SHS control line. This causes the controller to send a "start output" control signal to the device to initiate the output operation. Data are transferred from DR to the device. When the output operation is complete the controller resets the HS register bit low. The CPU senses completion of the output operation by reading the HS register status via the RHS control line. The SELECT input line is used to select a particular controller; if it is low, the controller does not respond to any of the input lines and its RHS line is in a high-impedance state.

59.2 CONTROL UNIT

The control unit[4] manages the operation of the computer. It fetches and decodes instructions, generates the timing signals, and establishes the sequences of events that occur during the computer's operation. It contains a number of registers that store information required by the computer during its operation and it controls the transfer of information between these registers and other units of the computer.

Timing

The operation of a typical computer consists of the cyclical repetition of a basic sequence of four phases: the fetch phase, the indirect phase, the execution phase, and the interrupt phase. One instruction is processed per cycle. The instruction is fetched from memory during the fetch phase. During the indirect phase the actual address of the data is obtained. The instruction is executed during the execution phase. The interrupt phase permits stopping the normal instruction execution cycle to permit processing to be performed in response to an interrupt request caused by special circumstances. Once the interrupt processing is completed, the normal instruction cycle resumes with the fetch phase for the next instruction.

The timing signal for the central processing unit is derived from a clock generator, a circuit that generates a series of pulses at regular intervals of time. Sequencing and synchronization of the operation of different elements of the computer are achieved by directing clock pulses to different lines at different times during the instruction cycle by use of either a counter–decoder combination or a ring counter.

Registers

Operation of the control unit is based on information stored in the following registers, which are part of the control unit:

IC (Instruction Counter). This is also called the PC (program counter). This register is a combination storage register-counter that stores the address of the next instruction to be executed.

IR (Instruction Register). This is a storage register that is loaded with the instruction fetched from memory during the fetch phase of the instruction cycle. The instruction register is often replaced by a number of separate registers that store different parts of the instruction, that is, the operation code, the memory address, the indirect bit, and others.

Index Registers. These are generally combination storage registers-counters used for indexed addressing, in which the effective address of a datum is determined as the sum of the address field contained in the instruction and the content of an index register. This permits the same sequence of instructions to be repeated and to reference different memory locations on each iteration by incrementing the content of an index register.

Base Registers. These are storage registers used for base-displacement addressing in which the effective address of a datum is computed as the sum of the displacement, which is contained in the instruction address field, and the content of the base register. This is generally done to save space in the instruction by using a displacement field that is shorter than the total address field would be.

The following registers, while not part of the control unit, are part of the CPU and are involved with the operation of the control unit:

DR (Data Register). This is a storage register used to store the operand fetched from memory for use by the ALU. Sometimes the MBR is used in place of a DR.

Accumulators. These are used to store operands and partial results for computations. Some computers use a single accumulator, while others use multiple accumulators so that partial results do not have to be continually swapped in and out of memory. Instead, a different accumulator is used for each partial result and then partial results from different accumulators are combined. The accumulators operate in conjunction with the arithmetic–logic unit and are directly connected to it. The accumulator is generally a rather sophisticated register that is a combination of storage register, shift register, and possibly counter as well.

MQ (Multiplier-Quotient). This serves as an extension of the accumulator for multiplication and division operations that involve operands or results that are double the size of the other operands, that is, the product of two n bit numbers is $2n$ bits long. The MQ register is similar to, but somewhat more limited in its function than the accumulator.

General Registers. These are storage registers that are part of the CPU. They are used in conjunction with bus systems that typically permit them to function as base registers, index registers, and accumulators.

CC (Condition Code). This is a storage register that is part of the CPU. It stores a small number of bits, typically 4, that reflect the nature of the most recently computed result, for example, whether there was an overflow; whether the result was positive, negative, or zero. Conditional branch instructions do or do not branch, depending on the value of the condition code.

Operation

During the fetch phase, the instruction is fetched from the memory location whose address is in the instruction counter (IC) and stored in the instruction register (IR). The instruction counter is incremented by the number of memory words per instruction after each instruction fetch, causing it to contain the address of the next memory location at the beginning of the next fetch cycle. This results in the normal sequential execution of instructions from consecutive memory locations. If a branch operation occurs, the IC is loaded with the address of the alternate instruction during the instruction execution phase. Then, when it reaches the next fetch operation, it will contain the address of the instruction at the branch location rather than the next sequential instruction.

Indirect addressing permits an indirect address to be used in a computer instruction. The indirect address, rather than specifying the location of the datum, specifies the address in which the direct address, which is the address of the datum, is located. The indirect phase is the time during which the control unit fetches the direct address from the indirect address in memory. There is a special bit in the instruction that indicates whether the address field contains an indirect address or a direct address.

During the instruction execution phase, operands are fetched from memory (if necessary) into the DR, the operation specified by the operation code of the instruction is performed by the ALU using operands from the accumulator and DR and putting the result into the accumulator, and results are stored in memory (if necessary).

An interrupt[5] is a procedure that permits the computer to respond to a high-priority processing requirement that is distinct from the ongoing program and asynchronous with respect to it. The interrupt is initiated by a logic signal on a particular line becoming active; this is called an interrupt request. In response to this request, the computer ceases execution of the current program and transfers control of the CPU to another program in main memory called the interrupt handler. The transfer of control from the executing program to the interrupt handler takes place during the interrupt phase of the instruction cycle. When the interrupt handler completes the interrupt processing, the computer continues with the fetch of the next instruction in the regular program.

59.3 INSTRUCTIONS

The instruction set[6] is divided into groups on the basis of the addressing mode used. The similarity of the instructions within a group makes it possible to use a common description for the entire group. The instruction types described here represent those that are widely available on present-day computers.

Register Reference Instructions. These are instructions in which the operand or operands are all located in CPU registers. If there are general registers or multiple registers of a particular type, the instruction will contain fields that identify the particular register(s) to use as well as the operation to be performed, as specified by the operation code. Instructions in this category perform arithmetic and logical operations on data that are in registers, perform unconditional and conditional branch operations based on data in registers, and handle interrupt-related operations.

Memory Reference Instructions. These are instructions in which one or both operands are in memory. If there is a single operand in memory, usually a second operand is located in a register. This type of instruction is generally called a fixed-word-length instruction because the length of both operands is equal to the register size, that is, the size of the datum processed by a single instruction. The computer word length is generally, although not necessarily, equal to the memory word size. The two operands in this instruction type are combined according to the operation code and the result generally replaces the operand in the register. This type of operation is used to perform arithmetic and logical operations on data that are in memory.

An instruction that contains both operands in memory is generally a variable-word-length instruction because the size of the operands is not fixed by the computer architecture. Additional fields are used in the instruction to specify the length of the operands. The result of applying the operation code to the operands generally replace one of the operands in memory. Instructions of this type are usually used to move data from one place in memory to another. This type of instruction is used to perform variable-length arithmetic operations; addition of binary coded decimal values is an example of such an instruction.

Input–Output Instructions. These are instructions that cause the transfer of data between an input or output device and either a CPU register, generally the accumulator, or memory. Common units of data transferred in a single I/O operation include the bit; the byte, which is 8 bits; and the word, which is generally 4 bytes in a mainframe computer.

TABLE 59.1. PDP-11 INSTRUCTION SET

Operation	Description	Operation	Description
ADC	Add carry	CLC	Clear carry condition code
ADD	Add	CLN	Clear negative condition code
ASL	Arithmetic shift left	CLR	Clear
ASR	Arithmetic shift right	CLV	Clear overflow condition code
BCC	Branch on carry clear	CLZ	Clear zero condition code
BCS	Branch on carry set	CMP	Compare
BEQ	Branch on equal	COM	Complement
BGE	Branch on greater or equal	DEC	Decrement
BGT	Branch on greater than	HALT	Halt
BHI	Branch on higher	INC	Increment
BHIS	Branch on higher or same	JMP	Jump
BIC	Bit clear	JSR	Jump subroutine
BICB	Bit clear byte	MOV	Move
BIS	Bit set	MOVB	Move byte
BISB	Bit set byte	NEG	Negate
BIT	Bit test	ROL	Rotate left
BLE	Branch on less than or equal	ROR	Rotate right
BLO	Branch on lower	RTI	Return from interrupt
BLOS	Branch on lower or same	RTS	Return from subroutine
BLT	Branch on less than	SEC	Set carry condition code
BMI	Branch on minus	SEN	Set negative condition code
BNE	Branch on not equal	SEV	Set overflow condition code
BPL	Branch on plus	SEZ	Set zero condition code
BR	Branch	SUB	Subtract
BVC	Branch on overflow clear	TST	Test
BVS	Branch on overflow set	TSTB	Test byte

The instructions for a PDP-11 minicomputer[7] given in Table 59.1 are typical of the instruction set for a modern, medium-sized computer.

59.4 BUS ORGANIZATION

The busses[8] in a computer are the electrical paths over which data are transferred between registers and other units of the computer. A bus is a set of parallel lines over which a number of bits representing a datum, for example, a word, may be simultaneously transferred. Two kinds of bus structure are described here: the hard-wired bus and the central bus structure.

In the hard-wired bus structure, connections are made between registers to permit transfer of data between them as required for execution of those operations that must be performed. Figure 59.3 shows a hard-wired bus structure for a typical computer. Only the data lines and busses are shown in this figure; the control lines are not shown.

The central bus structure is generally associated with relatively sophisticated computers; it results in a more flexible architecture than does the hard-wired bus. A computer that uses multiple general-purpose registers generally uses this type of architecture. Figure 59.4 shows such a bus structure. The A SELECT and B SELECT busses carry control signals from the control unit into the multiplexers that determine the two operands for the ALU. The DESTINATION SELECT bus and ENABLE line into the decoder determine which register, if any, the result will be loaded into. Other control lines required for the ALU are not shown in this figure. It is obvious from the figure that it is possible to combine the content of any two registers using any of the operations that the ALU can perform and to load the result in any of the registers. The hard-wired CPU, on the other hand, permits performance of only certain data transfers.

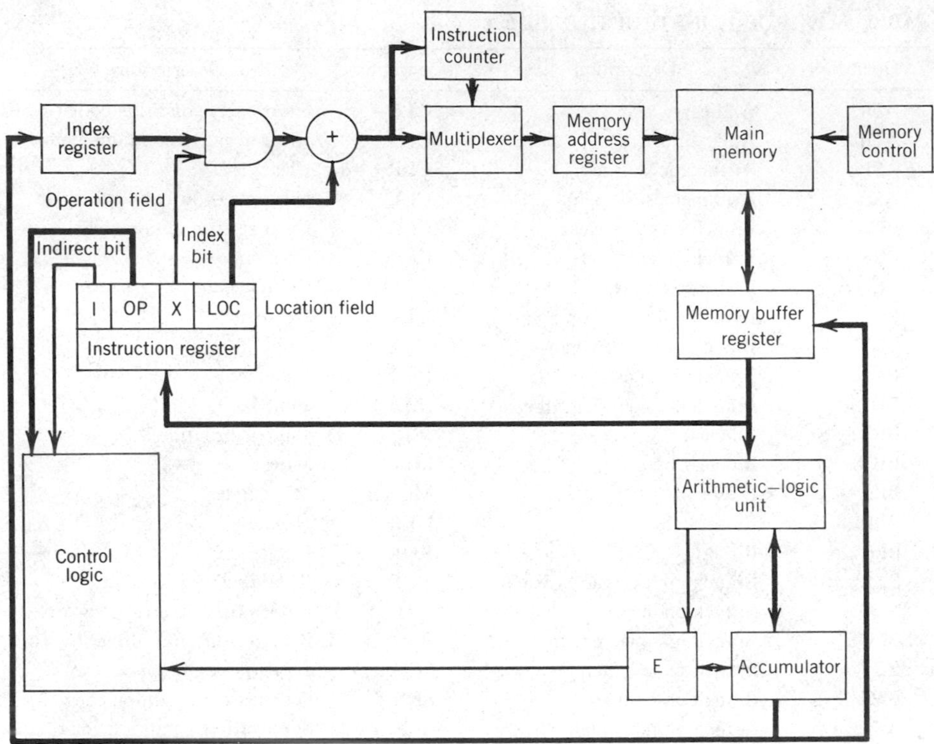

Fig. 59.3 Hard-wired bus structure. I, indirect bit; OP, operation field; X, index bit; LOC, location field; E, extension bit.

59.5 CPU OPERATION

The operation of the hard-wired CPU shown in Fig. 59.5 is typical of modern computers. The single lines entering registers, gates, or the ALU whose sources are not shown in the figure represent control lines from the control unit. The functions of these lines are described in the following.

The first phase in the operating cycle is the instruction fetch phase. This phase consists of four consecutive clock pulses during which the following actions occur:

Clock Pulse 1. SEL (select) is set low so that the output of the multiplexer (MUX) is the content of the IC, and LMAR (load MAR) is set high causing the MAR to be loaded with the content of the IC.

Clock Pulse 2. INCIC (increment IC) is set high causing the content of the IC to be increased by 1. At the same time, RW (read/write) is set low to indicate a memory read operation, and MOP (memory operate) is set high causing the instruction at the memory location whose address is stored in the MAR to be loaded into the MBR.

Clock Pulse 3. LIR (load IR) is set high causing the content of the MBR to be loaded into the IR.

Clock Pulse 4. During this clock pulse the control unit sets flip-flops that determine the next phase. If the I bit of the IR is a 1, the flip-flops are set to transfer control to the indirect phase; if the I bit is a 0, the flip-flops are set to transfer control to the execution phase.

In summary, the instruction at the memory location whose address is in the IC is fetched into the IR and the content of the IC is incremented by 1. The control unit transfers control to the indirect phase or the execution phase, depending on the value of the I bit.

The indirect phase consists of four consecutive clock pulses during which the following actions

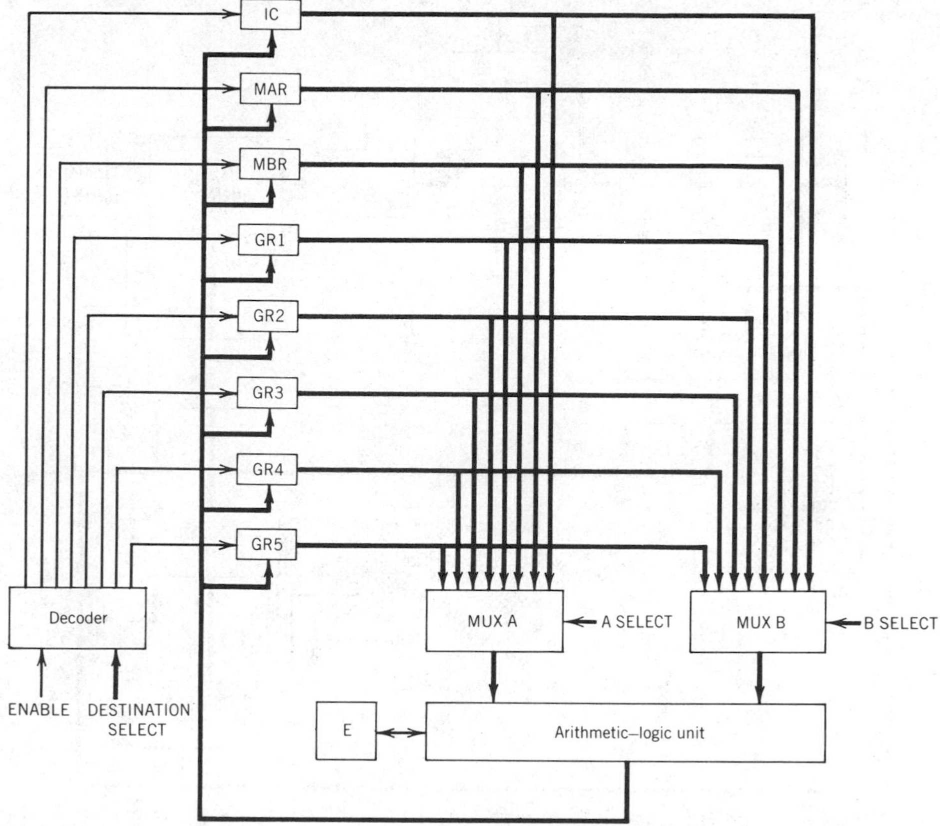

Fig. 59.4 Central bus structure. E, extension bit; MUX, multiplexer; IC, instruction counter; MAR, memory address register; MBR, memory buffer register; GR, general register.

occur:

Clock Pulse 1. ENX (enable indexing) is set low so that the output of the adder feeding the MUX is equal to the content of the location (LOC) field of the IR. SEL is set high so that the output of the MUX is equal to the input from the adder (the content of the LOC field). LMAR is set high causing the output of the MUX to be loaded into the MAR (the address field of the instruction in the IR is loaded into the MAR).

Clock Pulse 2. RW is set low causing a memory read, and MOP is set high causing the word at the addressed memory location to be loaded into the MBR.

Clock Pulse 3. LLOC (load location) is set high causing the location field bits of the word in the MBR to be loaded into the LOC field of the IR, that is, the direct address replaces the indirect address in the LOC field of the IR. The remainder of the IR is unchanged.

Clock Pulse 4. During this clock pulse the control unit sets the phase control flip-flops to transfer control to the execution phase.

In summary, the indirect phase occurs only if indirect addressing is called for (I = 1). It results in the indirect address (content of the location field of the instruction) being replaced by the direct address (content of the location field of the word at the indirect address in memory) in the LOC field of the instruction register. Control is then transferred to the execution phase.

The execution phase consists of four or fewer clock pulses during which the instruction is executed. The action during each of the clock pulses depends on which particular instruction is being executed. The operations required to execute an ADD instruction are described as an example:

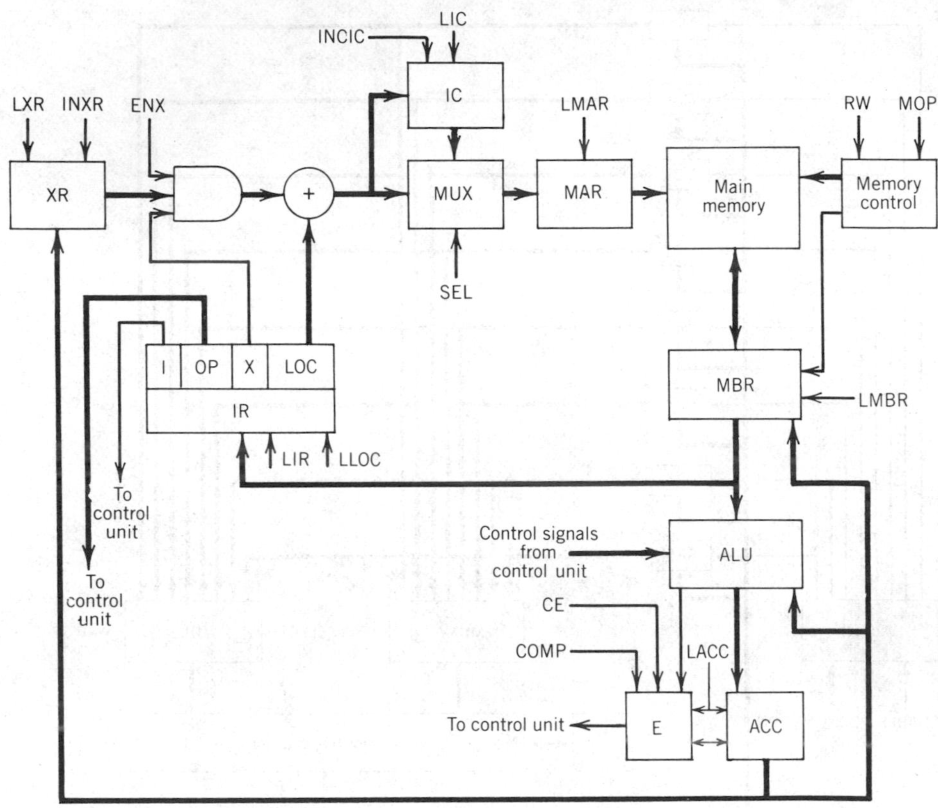

Fig. 59.5 Central processing unit. XR, index register; LXR, load XR; INXR, increment XR; ENX, enable indexing; IC, instruction counter; INCIC, increment IC; LIC, load IC; MUX, multiplexer; SEL, select; MAR, memory address register; LMAR, load MAR; R/W, read–write; MOP, memory operate; MBR, memory buffer register; LMBR, load MBR; ALU, arithmetic–logic unit; ACC, accumulator; E, extension bit; CE, clear E; COMP, complement; IR, instruction register; I, indirect bit; OP, operation field; X, index bit; LOC, location field.

Clock Pulse 1. ENX is set high so that the output of the adder feeding the MUX is equal to the content of the IR LOC field (the direct address) plus the content of the index register (XR) if the index bit, X, is 1. Otherwise the output of the adder is simply the content of the IR LOC field. SEL is set high so that the output of the MUX is equal to the adder output, and LMAR is set high causing the MUX output to be loaded into the MAR. This causes the address of the memory operand to be loaded into the MAR.

Clock Pulse 2. RW is set low to indicate a memory read operation is to be performed. MOP is set high causing the memory operand to be loaded into the MBR.

Clock Pulse 3. A control signal from the control unit is set, which causes the ALU to generate the sum of its two inputs, one from the MBR and the other from the accumulator (ACC), at its output. LACC (load ACC) is set high causing this output to be loaded into the accumulator.

Clock Pulse 4. The control unit checks whether an interrupt request is pending if interrupts are enabled. If there is, the phase control flip-flops are set to transfer control to the interrupt phase next; otherwise they are set to transfer control to the instruction fetch phase.

As another example of an instruction execution, the steps required for performing a STORE instruction are:

Clock Pulse 1. LMBR (load MBR) is set high causing the content of the ACC to be loaded into the MBR.

Clock Pulse 2. ENX is set high causing the sum of the IR LOC field and XR to appear at the output of the adder feeding the MUX if the X field of the IR contains a 1 bit; otherwise the output of the adder is simply the content of the LOC field. SEL is set high so that the output of the MUX is equal to the output of the adder, and LMAR is set high causing the output of the MUX to be loaded into the MAR.

Clock Pulse 3. RW is set high causing a memory write operation to take place, and MOP is set high causing the content of the MBR to be written into the memory location whose address is in the MAR.

Clock Pulse 4. The control unit checks the interrupt conditions and sets the phase control flip-flops to transfer control to the interrupt phase or to the instruction fetch phase, depending on the conditions that exist.

The transition from one phase to another is controlled by two phase control flip-flops. They are set–reset (SR) flip-flops whose outputs define the phase of the instruction cycle that the CPU is currently in. These values, denoted P and H, are shown in Table 59.2. On the basis of this table, the

TABLE 59.2. INSTRUCTION CYCLE PHASES

P	H	Phase
0	0	Fetch
0	1	Indirect
1	0	Execution
1	1	Interrupt

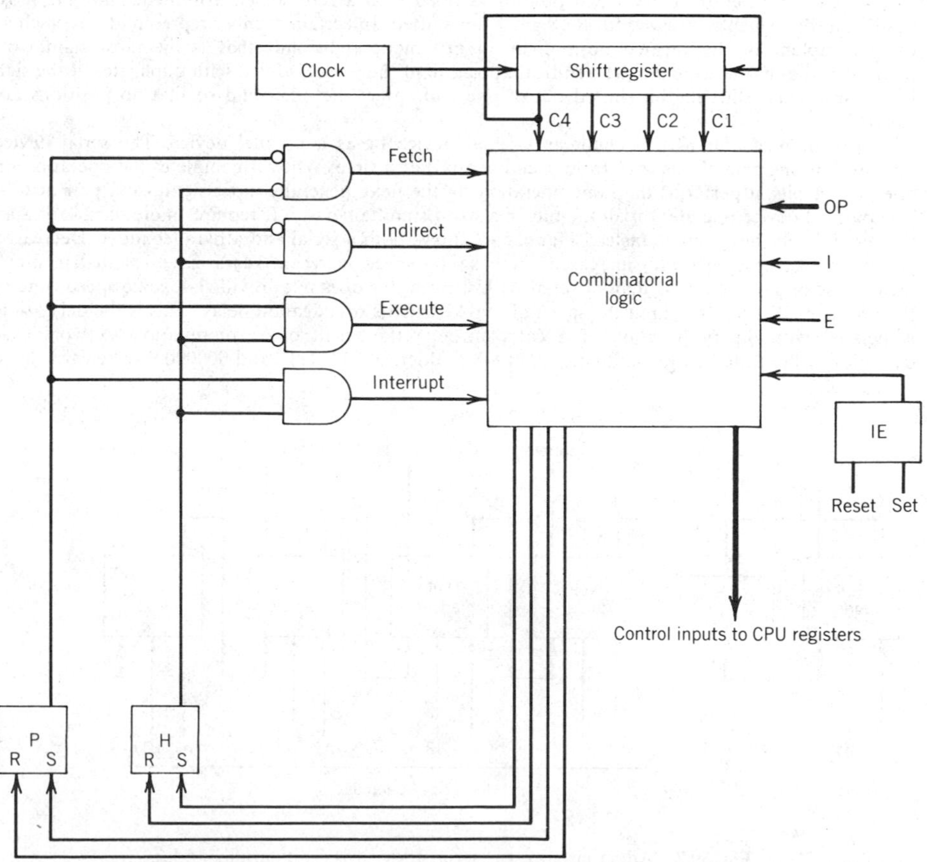

Fig. 59.6 Control unit timing circuit. CPU, central processing unit; IE, interrupt enable.

actions performed during the fourth clock pulse in the fetch phase are to set the H flip-flop if I = 1 or to set the P flip-flop if I = 0. During the fourth clock pulse of the indirect phase, the P flip-flop is set and the H flip-flop is reset. During the fourth clock pulse of the execution phase, the H flip-flop is set if there is an interrupt request pending and the interrupt enable (IE in Fig. 59.6) bit is set; otherwise the P flip-flop is reset. The IE flip-flop is controlled by instructions that enable and disable the interrupt system.

The control unit timing circuit is shown in Fig. 59.6. Timing is handled by a ring counter that is implemented using a shift register. This register is loaded with a single 1 bit that is shifted to the left; when it reaches the left end it is fed back to the right input. This produces the sequence of four sequential clock pulses required for each of the phases.[4] The combinatorial circuit is composed of logic gates, which generate the control inputs shown in Fig. 59.5.

59.6 ARITHMETIC–LOGIC UNIT

The arithmetic–logic unit performs the arithmetic and logic functions of the CPU. The arithmetic functions generally include addition and subtraction of binary numbers and binary coded decimal (BCD) numbers. In some minicomputers and in all mainframe computers, the ALU also performs multiplication and division of binary and, possibly, BCD numbers; in the smaller minicomputers and in most microcomputers, software subroutines are used to perform these operations. Addition, subtraction, multiplication, and division of floating-point numbers are also done by the ALU in some minicomputers and mainframe computers; this capability is often available as an option. The logical operations comprise three types: boolean, shift, and comparison operations. The boolean operations include some or all of the following operations performed on corresponding bits of the operands: logical complement, AND, INCLUSIVE OR, and EXCLUSIVE OR. The shift operations include a subset of arithmetic and logical, left and right, linear and circular shifts. In a logical shift all the bits are shifted one position; the vacated position is filled with a zero. In an arithmetic shift the most significant bit, which is reserved for the sign, is not shifted. In certain number representations, such as two's complement, the vacated position in an arithmetic right shift that is the most significant numeric position (the second most significant position in the word) is filled with duplicates of the sign bit. In a circular shift the bit shifted out of one end enters the other end so that no positions are vacated.

An arithmetic unit may be built as a serial device or as a parallel device. The serial device operates on one pair of bits (one from each operand) at a time. When it completes the operation on one pair of bits, it performs the same operation on the next, generally more significant, pair of bits. The parallel device operates on all the bits in a word simultaneously. It requires more hardware than the serial device but is much faster.[9] Figure 59.7 shows both a serial and a parallel adder. Decreased hardware costs, together with increased emphasis on speed in recent years, have resulted in much greater use of parallel ALUs than of serial ALUs. Even this does not provide adequate speed; one of the problems that tends to limit the speed of the ALU is its propagation delay. This is the delay time associated with the propagation of a carry through the result of a computation. A worst-case condition is illustrated by the addition, in an 8-bit adder, of 11111111 and 00000001. The carry must

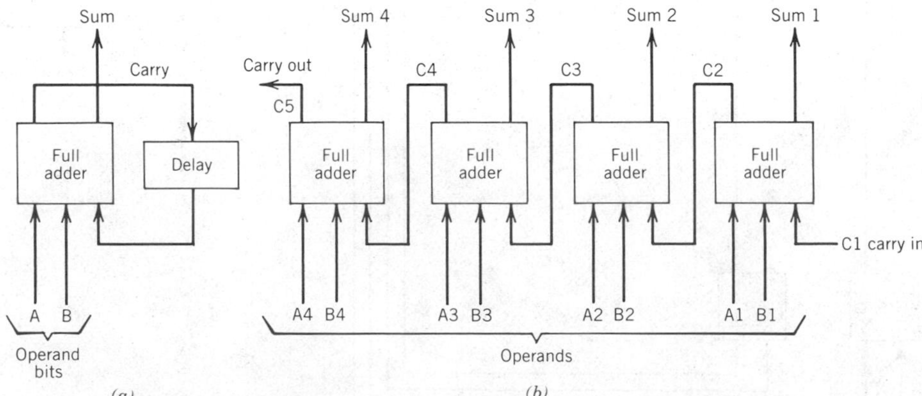

Fig. 59.7 Adder circuits: (*a*) serial adder, (*b*) 4-bit parallel adder.

then propagate, one stage at a time, from the least significant bit position to the most significant bit position. Sufficient time must be left between clock pulses for this process to reach completion before the start of the next clock pulse. This requires about $(n - 1)$ times the delay time of a single stage for an n-stage adder, in addition to the time it takes to generate a sum with no carries. Special circuits, called carry look-ahead generators,[10] are used when maximum speed is very important. These circuits check multiple stages of the ALU and adjust the carries in all of them simultaneously. This increases the speed of the ALU, but also increases the complexity and cost of the circuit.

ALU Design

The ALU may be partitioned into a set of identical modules, one for each bit position in the operands. Each stage of the ALU, in addition to requiring data inputs, also requires control inputs, which determine the function it is to perform. The block diagram of an n-stage ALU of this type is shown in Fig. 59.8. A_i is the ith bit of the A operand, B_i is the ith bit of the B operand, and F_i is the ith bit of the result, F. C_1 is the input carry and C_0 is the output carry. Many designs for a single stage of the ALU are possible.[11] They differ in the number and choice of elementary functions that the ALU can perform. A typical design for an ALU stage is shown in Fig. 59.9. This design uses four FUNCTION SELECT lines. One of these lines, the one labeled P, determines whether the carry bit will be propagated or not. For arithmetic functions, $P = 1$ causing the carry bit to be propagated; for logical functions, $P = 0$ and the carry bit is not propagated. The other three FUNCTION SELECT lines, S1, S2, and S3, determine the particular arithmetic or logical function that is to be performed. The operation of the ALU stage may be investigated by an analysis of its logic circuit (see Fig. 59.9). The logic functions are found by setting $P = 0$; they are listed and described in Table 59.3, in which the complement of a variable, Q, is denoted by Q'.

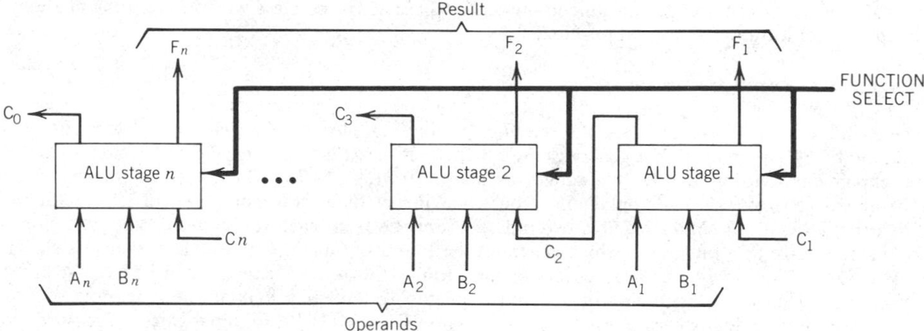

Fig. 59.8 n-stage arithmetic–logic unit (ALU). C_0, output carry; C_1, input carry.

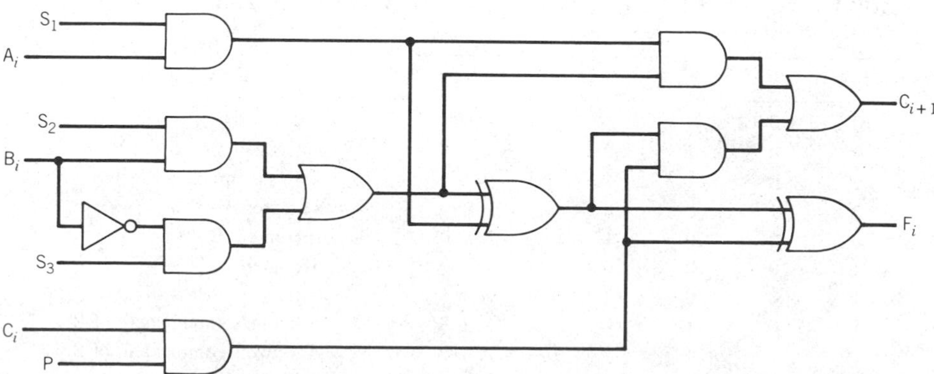

Fig. 59.9 Typical arithmetic–logic unit (ALU) stage. P, S_1, S_2, and S_3 are function select lines.

TABLE 59.3. LOGIC FUNCTIONS OF THE ARITHMETIC–LOGIC UNIT (ALU) STAGE (P = 0)

S_1	S_2	S_3	Function	Description
0	0	0	F = 0	Clear all bits
0	0	1	F = B′	Complement B
0	1	0	F = B	Transfer B
0	1	1	F = 1	Set all bits
1	0	0	F = A	Transfer A
1	0	1	F = (A XOR B)′	Exclusive NOR
1	1	0	F = A XOR B	Exclusive OR
1	1	1	F = A′	Complement A

The arithmetic functions are found by setting P = 1. Not all of the function select values result in useful arithmetic functions. Some duplicate logical functions, while others produce results that are not useful. The useful functions are tabulated and described in Table 59.4.

The basic ALU design just described does not provide for shifting operations. These operations are frequently implemented by means of a combinatorial circuit whose input is connected to the basic ALU and whose output is a shifted version of the input. The circuit for a four-stage left–right shifter is shown in Fig. 59.10. There are three control signals: SHIFT LEFT, SHIFT RIGHT, and NO SHIFT. If all of them are low, the output of the shifter is all 0 bits. There are two additional inputs to fill the vacated positions for the left and right shifts; these may be used to implement circular shifts. This shifter cannot be used for arithmetic shifts. For these shifts, changes would have to be made for the most significant (left-most) bit position.

Carry Look-Ahead Generator

The use of a carry look-ahead generator requires a more complex ALU in which the adder circuit is separate from the circuitry that generates the other functions. Furthermore, the adder circuit cannot be implemented using simple full adders as shown in Fig. 59.7*b*. The operation of the carry look-ahead generator depends on simultaneous checking of the output conditions of all the adders. The carry look-ahead circuit requires two inputs, P and G, from each full adder circuit. The circuit for a single stage of a full adder, which generates the P and G functions as well as the sum, is shown in Fig. 59.11. The carry look-ahead generator for a four-stage adder is shown in Fig. 59.12.[4] A block diagram showing the construction of an 8-bit adder using the carry look-ahead generators of Fig. 59.12 and the full adder stages of Fig. 59.11 is shown in Fig. 59.13.[10] Additional gates are required to generate the final carry for each four-stage section of the adder circuit as shown in the figure.

TABLE 59.4. ARITHMETIC FUNCTIONS OF ARITHMETIC–LOGIC UNIT (ALU) STAGE (P = 1)

S_1	S_2	S_3	C_i	Function	Description
0	0	0		F = C_i	Input carry
0	0	1	0	F = B′	One's complement of B
0	0	1	1	F = B′ + 1	Two's complement of B
0	1	0	0	F = B	Transfer B
0	1	0	1	F = B + 1	Increment B
1	0	0	0	F = A	Transfer A
1	0	0	1	F = A + 1	Increment A
1	0	1	0	F + A + B′	A + one's complement of B
1	0	1	1	F = A + B′ + 1	A + two's complement of B
1	1	0		F = A + B + Ci	Add A, B, input carry
1	1	1	0	F = A − 1	Decrement A

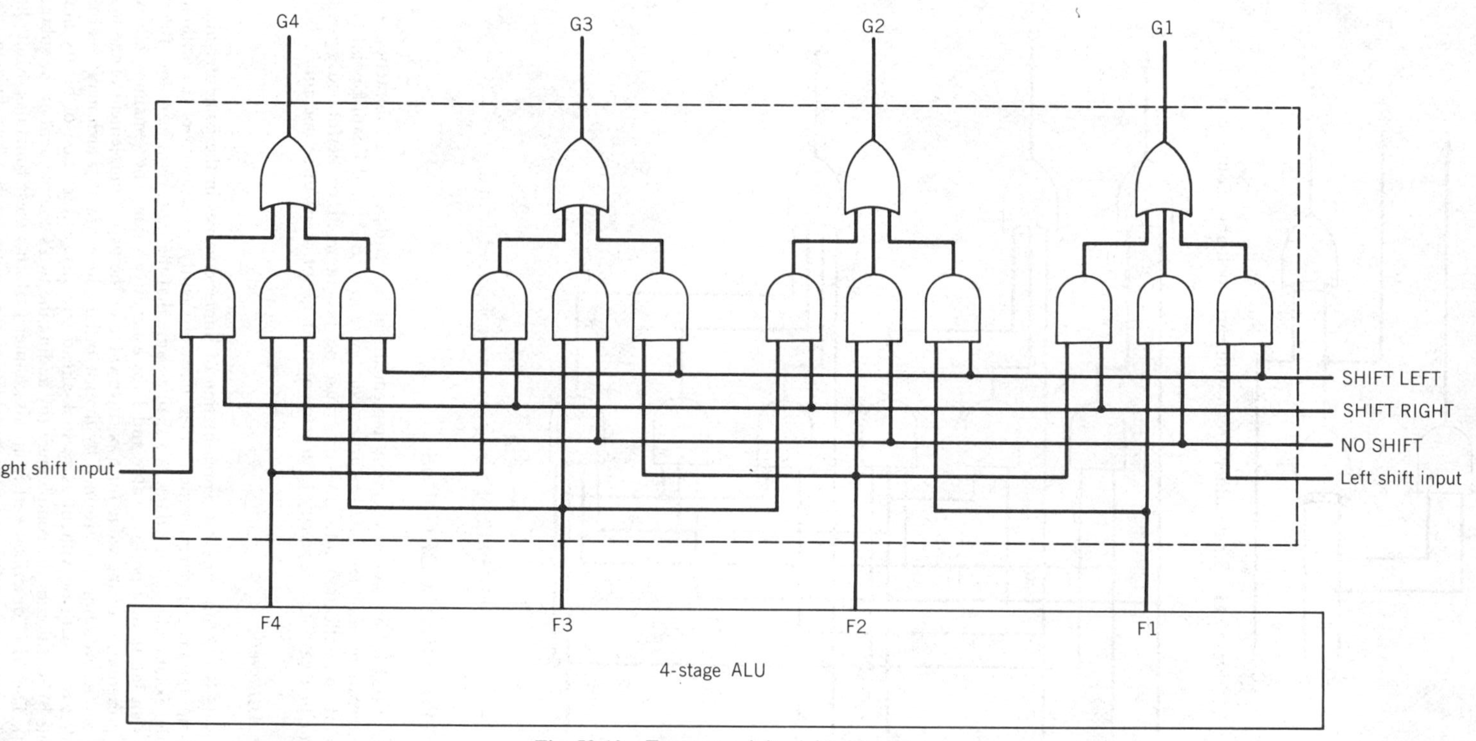

Fig. 59.10 Four-stage left–right shifter.

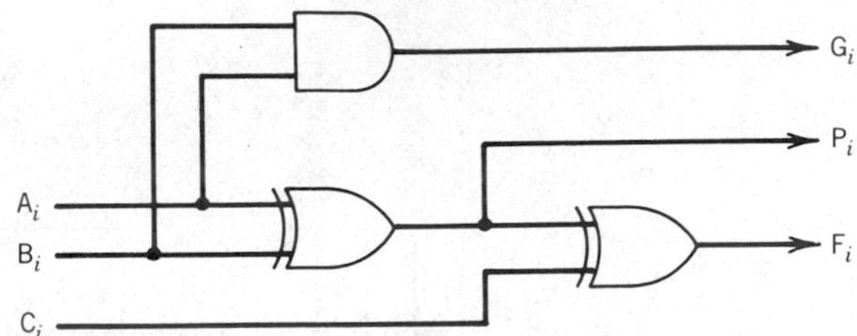

Fig. 59.11 Single adder circuit with carry look-ahead outputs.

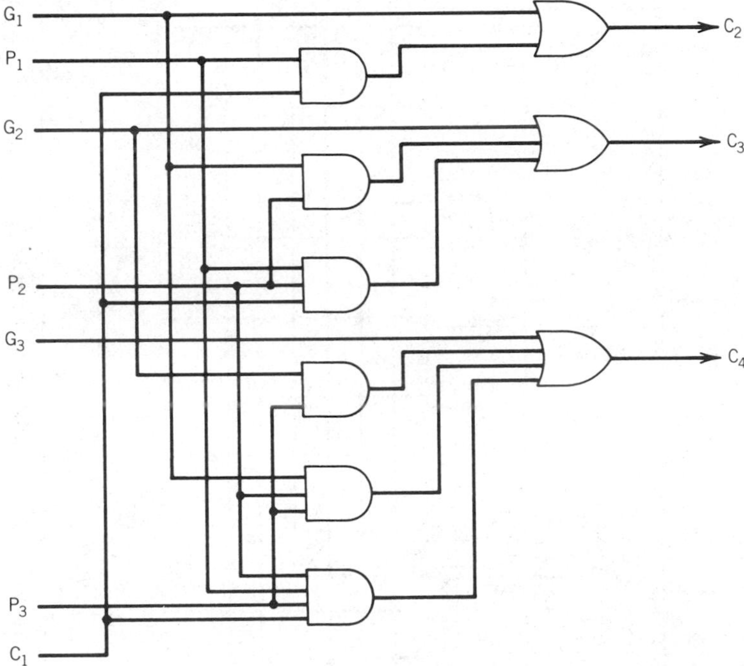

Fig. 59.12 Four-stage carry look-ahead circuit.

59.7 ARITHMETIC ALGORITHMS

The ALU described in the previous section implements binary addition and subtraction, certain of the boolean operations, and shifts. Other arithmetic operations, BCD and floating-point arithmetic, and fixed-point multiplication and division, must be programmed or performed by additional hardware. In either case it is necessary to develop algorithms for doing these operations.

Fixed-Point Arithmetic

Fixed-point code is the fundamental representation of numeric values in the computer and the one for which computations are generally performed most efficiently. The design of the arithmetic unit depends on the exact coding of the data. We shall assume that the most significant bit of the word is used for the sign bit, 0 for a positive value and 1 for a negative value. The remaining bits will store the binary representation of the value for a positive value or the two's complement of the magnitude of a negative value. Addition is performed by the use of an adder circuit, which may be of the serial type or parallel type, with or without carry look-ahead circuitry. Subtraction of binary numbers is generally done by addition of the complement (by adding the two's complement of the subtrahend to the minuend). The ALU described earlier in this chapter performs these functions. Multiplication is generally performed by repeated addition and shifts of the multiplicand, while division is generally performed by repeated subtractions and shifts.

One problem arises in fixed-point addition and subtraction that has not been discussed: the

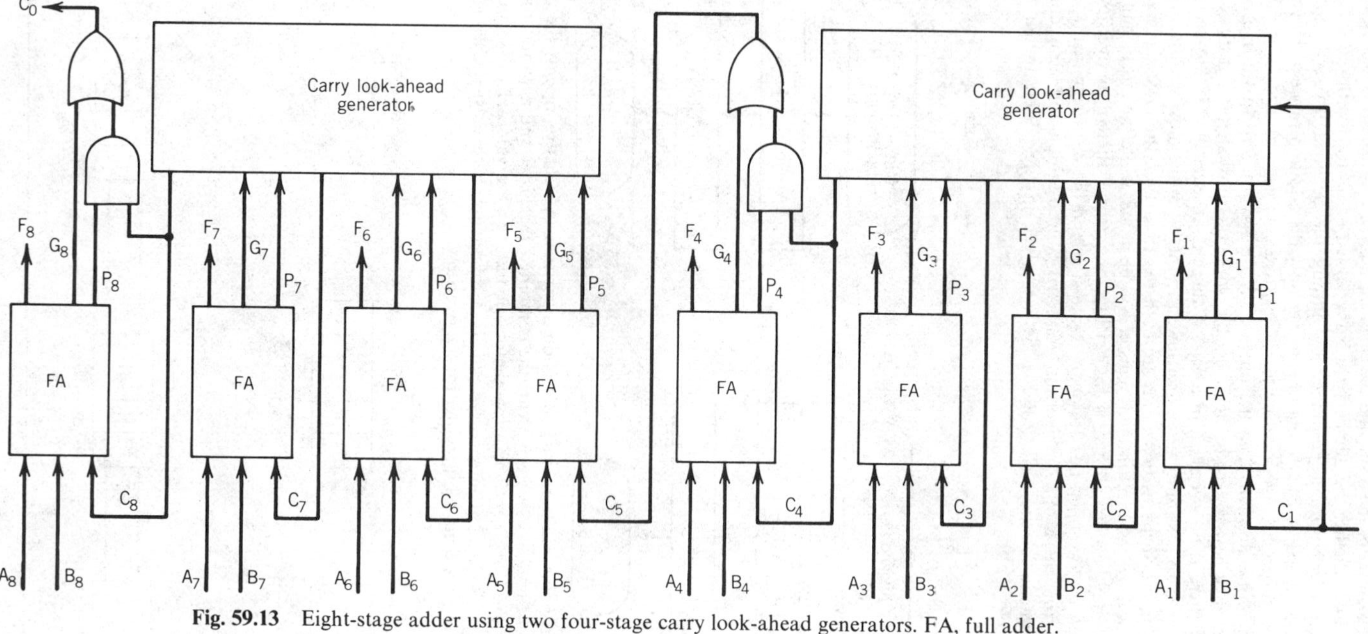

Fig. 59.13 Eight-stage adder using two four-stage carry look-ahead generators. FA, full adder.

handling of the carry and overflow. In performing either of these operations, the carry bit generated by the ALU is loaded into the E register, as shown in Fig. 59.5. This bit is not the overflow, however, since the most significant bit in the ALU is not a numeric bit but a sign bit. The overflow bit[4] is determined by the sign bit of the result and the carry bit into E. It is necessary to consider the signs of the two operands and whether the numbers are being added or subtracted in order to determine the relationships. Since subtraction is performed by adding the negative (two's complement) of the subtrahend to the minuend, it is sufficient to consider only the addition process. Three cases must be considered: both numbers positive, both numbers negative, and each number of opposite sign.

If both numbers are positive, then the operands each have a zero in the most significant bit position. The sum must have a zero in the E register, while the most significant bit of the ALU may contain a zero, if there is no overflow, or a one, if there is an overflow. If the two operands have different signs there is no possibility of an overflow, and the bit in the most significant position of the ALU is a valid sign bit. If the two operands are both negative, then the most significant bit of each operand is a one and the result bit in the E register will be a one, while the most significant bit of the result in the ALU may be a one, if there is no overflow, or a zero, if there is an overflow. An overflow has occurred when the two operands are of the same sign and the most significant (sign) bit of the ALU is different from the E bit. An algorithm for fixed-point addition and subtraction, using two's complement arithmetic for the system of Fig. 59.5, is given in Fig. 59.14. In this figure, ACC(1)

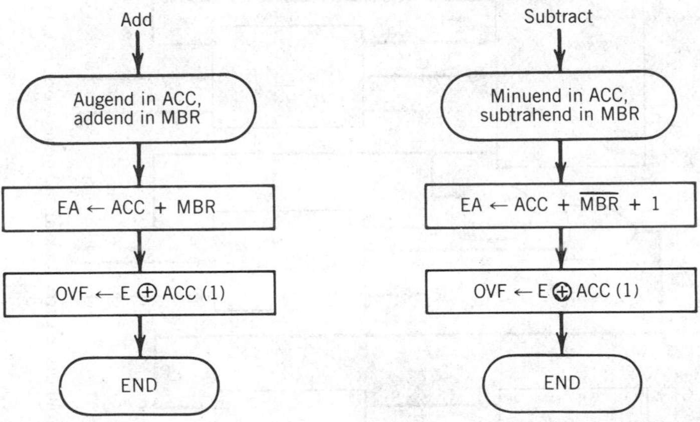

Fig. 59.14 Fixed-point addition and subtraction algorithm. ACC(1) represents the most significant (sign) bit of the accumulator. EA represents the combined E register and accumulator with the most significant bit in the E register. The arrow denotes loading of the value of the expression at its tail into the location at its head. The encircled plus sign represents the exclusive OR or modulo-2 sum. OVF represents an additional 1-bit register for the overflow.

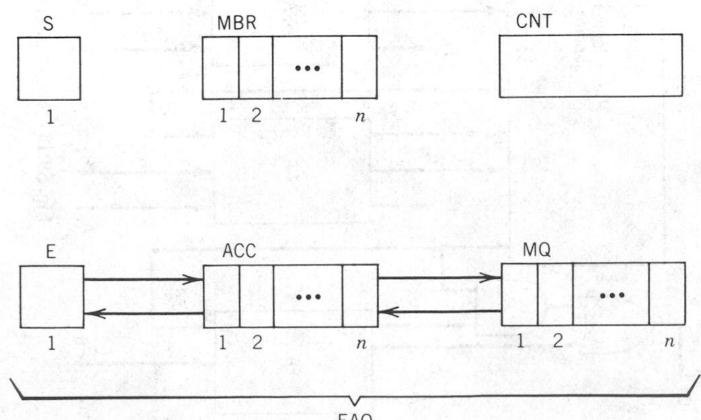

Fig. 59.15 Registers for multiplication and division. MBR, memory buffer register; CNT, count register; ACC, accumulator; MQ, multiplier–quotient register; EAQ, combined E, ACC, and MQ registers.

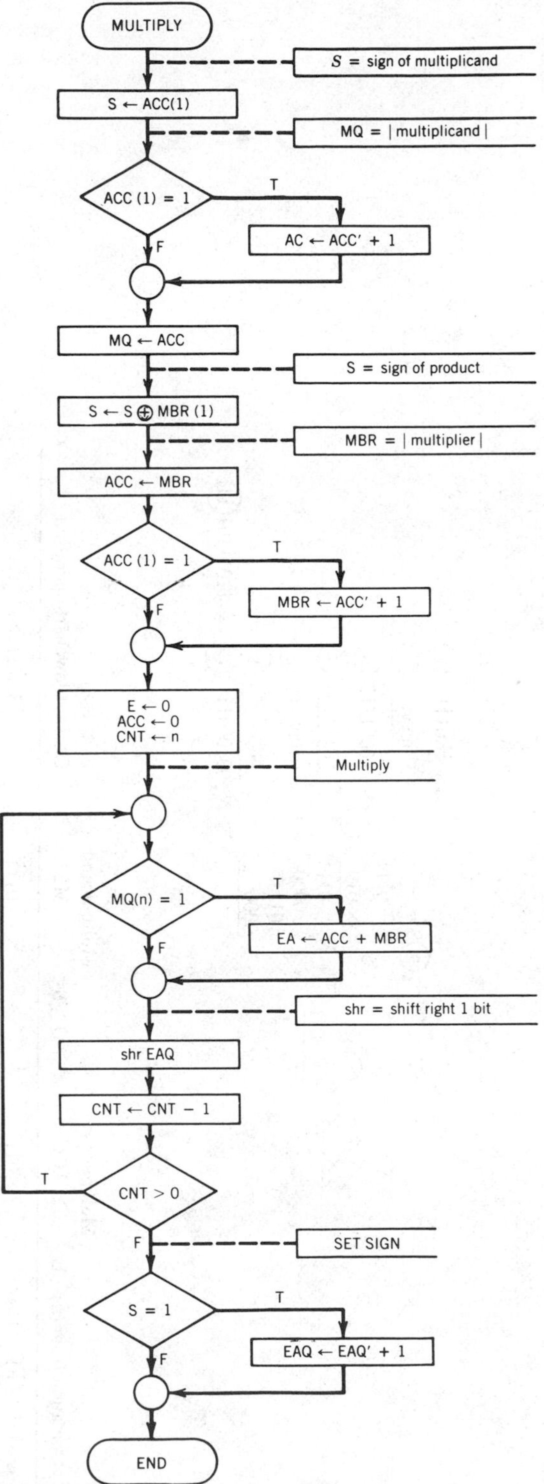

Fig. 59.16 Fixed-point multiplication algorithm. ACC, accumulator; MQ, multiplier–quotient register; MBR, memory buffer register; CNT, count register; EA, combined E register and accumulator; EAQ, combined E register, accumulator, and MQ register.

```
                             (2) S = -            0111      (4)
                                  0111            0000      (5)
 +7         (1) 0111    =>      x 0101   (3) =>  ------
x -5   =>     x 1011              0111           000111     (6)
----                             0000             0111      (7)
-35                             0111             -------
                               0000              0100011    (8)
                                                  0000      (9)
                                                 --------
                                                 00100011  (10)  => 11011101 (11)
```

(*a*)

Operation from (*a*)	E	Multiplier = −5 ACC	MQ	S	Multiplicand = +7 MBR	CNT	Flowchart operation
(1)		1011			0111		Initial
(2)		1011		1	0111		S ← ACC(1)
(3)		0101		1	0111		ACC ← ACC′ + 1
		0101	0101	1	0111		MQ ← ACC
(2)		0101	0101	1	0111		S ← S ⊕ MBR(1)
		0111	0101	1	0111		ACC ← MBR
	0	0000	0101	1	0111	100	ACC ← 0; E ← 0; CNT ← n = 4

(4)	0	0000 0111					MQ(4) = 1 ⇒	iteration	(1)
	0	0111	0101	1	0111	100	EA ← ACC + MBR		
	0	0011	1010	1	0111	011	shr EAQ; CNT ← CNT − 1; CNT > 0		
(5)	0	0011	1010	1	0111	011	MQ(4) = 0 ⇒ no operation	iteration	(2)
(6)	0	0001	1101	1	0111	010	shr EAQ; CNT ← CNT − 1; CNT > 0		
(7)	0	0001 0111					MQ(4) = 1 ⇒	iteration	(3)
	0	1000	1101	1	0111	010	EA ← ACC + MBR		
(8)	0	0100	0110	1	0111	001	shr EAQ; CNT ← CNT − 1; CNT > 0		
(9)	0	0100	0110	1	0111	001	MQ(4) = 0 ⇒ no operation	iteration	(4)
(10)	0	0010	0011	1	0111	000	shr EAQ; CNT ← CNT − 1; CNT = 0		
(11)	1	1101	1101	1	0111	000	S = 1 ⇒ EAQ ← EAQ′ + 1		

Product = −35

(*b*)

Fig. 59.17 Fixed-point multiplication computation: (*a*) Binary multiplication, (*b*) implementation. ACC, accumulator; E, register; MQ, multiplier–quotient register; S, single-bit sign register; MBR, memory buffer register; CNT, count register; EAQ, combination of E, ACC, and MQ registers.

represents the most significant (sign) bit of the accumulator and EA represents the combined E register and accumulator treated as a single register, with the most significant bit in the E register. The arrow denotes loading of the value of the expression at its tail into the location at its head. For example, $A \leftarrow 5$ indicates that location A is loaded with the value 5. The encircled plus operator represents the exclusive OR or modulo-2 sum, and OVF represents an additional 1-bit register for the overflow. At the end, the signed result is in the accumulator and the overflow is in OVF.

The product of two numbers, each one word long, will be a number two words in length. To accommodate the two-word product an additional register, called the MQ (multiplier–quotient) register, is used. This register is one word long and is treated as an extension of the accumulator. The most significant word of the product will be in the accumulator and the least significant word in the MQ register. In performing a division, the dividend, which will be two words long, is loaded into the combined ACC and MQ registers. The registers required for multiplication and division are shown in Fig. 59.15. The E, ACC, and MQ registers are connected so that left and right shifts may be performed treating the three of them as a single register. The combined register is denoted as EAQ. A single-bit sign register, S, and a count register, CNT, together with the MBR complete the register set required for these operations. The count register is used to keep track of the number of shifts that have been performed. It must be able to store any integer value from zero to n, the word length of the computer.

The algorithm for fixed-point multiplication[12] is shown in Fig. 59.16. The process is started with the multiplicand in ACC and the multiplier in MBR and ends with the product in the AQ (combined accumulator and MQ register) register, since the E register will contain a nonsignificant bit equal to the sign bit of the accumulator. The steps in multiplying $+7$ by -5 are shown in Fig. 59.17 for a 4-bit word length. The steps in the binary computation are shown in Fig. 59.17*a*. Figure 59.17*b* illustrates the contents of the various registers as the steps in the flowchart are executed. The computer steps are also cross referenced to the steps in the manual computation.

The fixed-point division algorithm[12] is shown in Fig. 59.18. The technique is essentially the inverse of the multiplication algorithm. The principal activity is the successive subtraction of the divisor from the most significant bits of the dividend, if the dividend bits are large enough, followed by a shift each time. The subtraction is performed by adding the two's complement of the divisor to the appropriate dividend bits. An additional problem that arises in division, which does not exist in multiplication, is the possibility of overflow in the event that the divisor is very small. In this flowchart the dividend is assumed to have been loaded into the combined AQ register and the divisor is assumed to have been loaded into the MBR. The initial part of the flowchart generates the sign of the quotient in the S register and places the magnitude of the dividend into the EAQ register and the magnitude of the divisor into the MBR register. The CNT register is then loaded with the word length of the computer.

The next part of the flowchart determines whether there will be an overflow. The dividend is two full words in length, while the divisor is one full word in length. If the most significant half of the dividend (the part in the ACC) is greater than or equal to the divisor (in the MBR), there will be an overflow. This situation is detected by subtracting the content of the MBR from the content of the ACC (subtraction is performed by addition of two's complement of subtrahend).[4] If the result is positive, there is an overflow. If there is an overflow, the dividend is restored by adding the divisor back to it and the OVF flip-flop is set to 1. If there is no overflow, the dividend is restored by adding the divisor back to it, the OVF flip-flop is reset to 0, and the division process is continued. The division process is a cyclical one that commences with a shift left of the dividend to correctly align the divisor for subtraction. A check is then made of the E bit; if it is a 1, the corresponding quotient bit is a 1 and the divisor is subtracted by adding its two's complement. If the E bit is a 0, the divisor is subtracted by adding its two's complement; if the resulting E bit is a 1, the result is positive and the quotient bit is set to 1, while a 0 value of the E bit indicates a negative result, in which case the dividend is restored by adding back the divisor and the quotient bit is left as a 0 (the value shifted in by the initial SHIFT LEFT instruction). The loop is concluded by decrementing the count and is terminated by a 0 value of the count.

The algorithm is completed by replacing the quotient and remainder by their two's complements if the sign is negative. The quotient is in the MQ register and the remainder is in the accumulator register.

Figure 59.19 illustrates the division of -37 by $+7$ using a 4-bit word length. The binary division process is shown in Fig. 59.19*a*, while the computer steps are shown in Fig. 59.19*b*, which references the steps in the manual procedure and the flowchart statements.

Floating-Point Arithmetic

Floating-point numbers are assumed to consist of a sign bit (0 for +, 1 for −), a true binary fraction with the binary point assumed at the left of the most significant bit, and an exponent of the base 2.

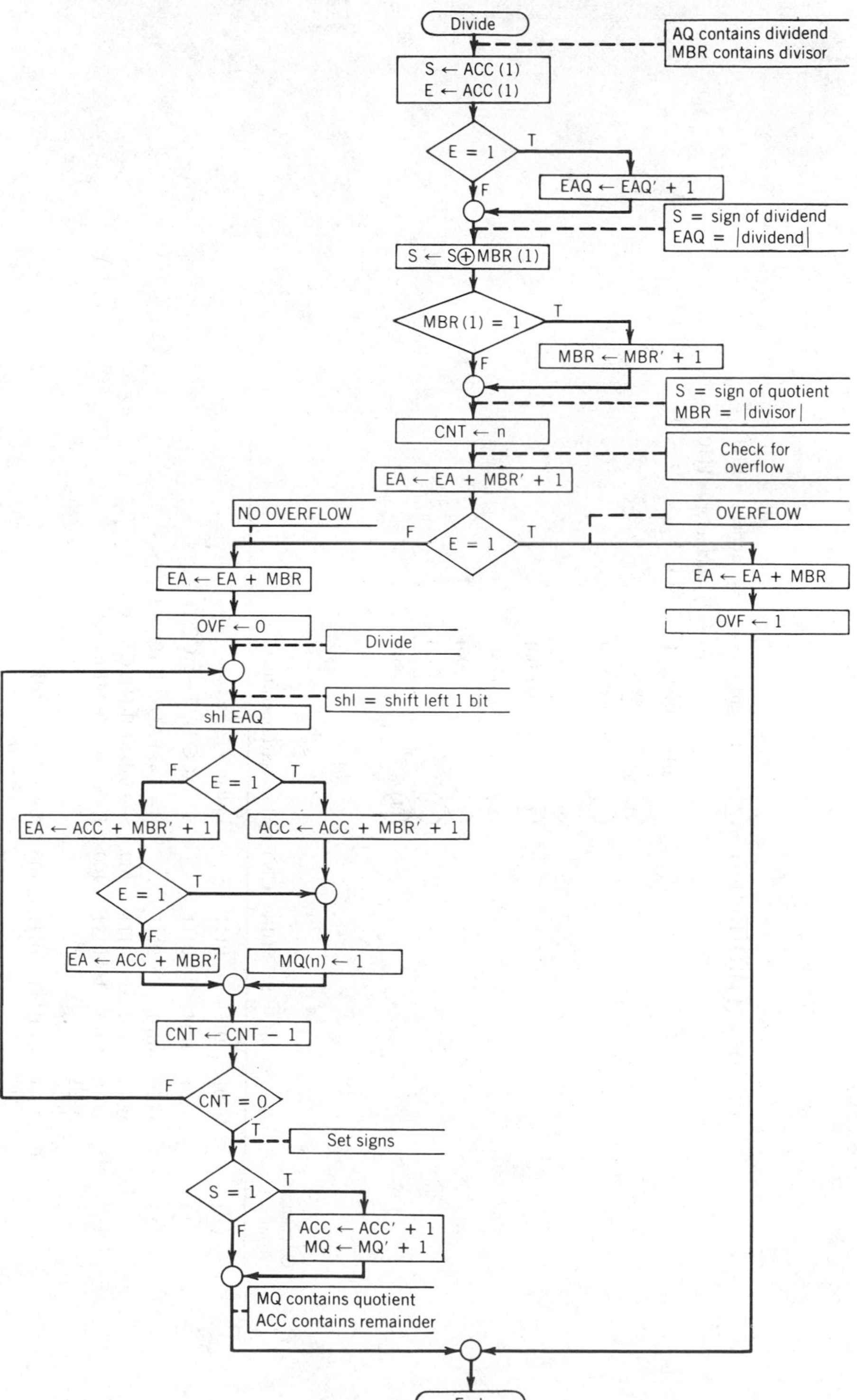

Fig. 59.18 Fixed-point division algorithm. AQ, combined ACC and MQ register; ACC, accumulator; MBR, memory buffer register; EAQ, combined E register, ACC, and MQ register; CNT, count register; MQ, multiplier–quotient register.

+7) − 37 ⇒ (1) 0111) 11011011 ⇒ (2) S(sign) = − ⇒ quotient = 1011

(3) 0101 (12) remainder = 1110

0111) 00100101

(4) −0000

(5) 0100101

(6) −0111

(7) 001001

(8) −0000

(9) 01001

(10) − 0111

(11) 0010

(*a*)

Operation from (*a*)	E	ACC	MQ	S	MBR	CNT	Flowchart operation
(1)		1101	1011		0111		Initial
(2)	1	1101	1011	1	0111		S ← ACC(1); E ← ACC(1)
(3)	1	0010	0101	1	0111		E = 1 ⇒ EAQ ← EAQ′ + 1
	1	0010	0101	1	0111	100	S ← S ⊕ MBR(1); CNT ← n = 4
		0010 1001	0101	1	0111	100	EA ← ACC + MBR′ + 1
	0	1011					
	0	1011 0111	0101	1	0111	100	EA ← EA + MBR OVF ← 0
	1	0010					

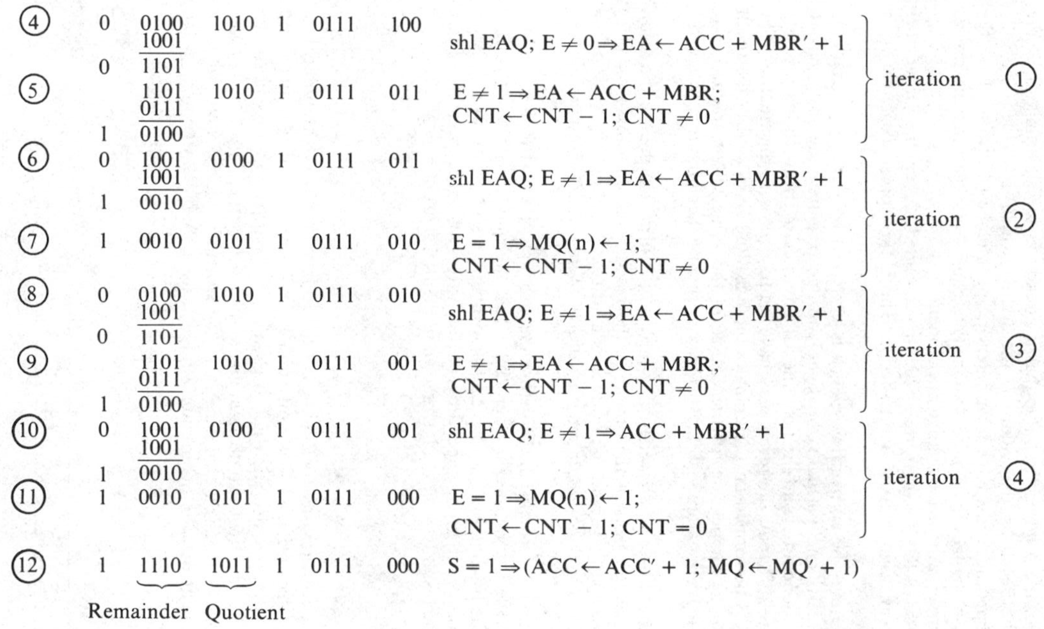

(*b*)

Fig. 59.19 Fixed-point division computation: (*a*) Binary division, (*b*) implementation.

The value of the number is the signed fraction, represented in true magnitude form, multiplied by 2 raised to the exponent.

Floating-point numbers that are to be added or subtracted are initially in the accumulator and MBR. The sign bits are denoted AS and BS, respectively, the fractions are denoted A and B, respectively, and the exponents are denoted a and b, respectively. The storage of these values in the registers is shown in Fig. 59.20. A1 is the most significant bit of the A field.

The addition and subtraction routines[4] are shown in the flowchart of Fig. 59.21. The result is located in the accumulator on completion of the routine. The first step determines whether the operation is a subtraction; if it is, the sign of the subtrahend is changed and an addition is performed. If the operand in the MBR is zero, the result is the value that is already in the accumulator. If the accumulator operand is zero, the result is the value in the MBR, which is loaded into the accumulator. Otherwise, the two operands must have their binary points aligned (they must have the same exponent). The next operation changes the smaller exponent to equal the larger one and inserts leading zeros in the fraction to compensate for the change in exponent.

Once the binary points are aligned, the fractions are added or subtracted, depending on whether their signs are the same or different, using a fixed-point routine. The E flip-flop is set if there is a carry out of the fraction.

If the signs are the same there may be an overflow, which is corrected by shifting the fraction, augmented by the E bit on its left, one place to the right and incrementing the exponent to compensate. If this results in an exponent beyond the machine capacity, the overflow flip-flop is set.

If the signs of the operands are different, the result may have leading zeros in the fraction. The normalization of the result, which is performed next, eliminates these leading zeros (unless the result is zero) and changes the exponent to compensate for the change in the value of the fraction. The normalization process is terminated prematurely if the exponent reaches the minimum value the computer is capable of representing.

Floating-point multiplication and division[4] are relatively simple to perform. The sign of the result is the EXCLUSIVE OR function of the signs of the operands, the result fraction is the product or quotient of the fractions in the two operands, and the result exponent is the sum or difference of the operand exponents. The only additional considerations are those of normalizing the result and handling overflows and underflows.

Both operands are assumed to be normalized. Therefore the product of two fractions cannot contain an integer bit but may have a single leading zero bit, while the quotient of two fractions cannot contain a leading zero in the fraction but may have a single integer bit. Overflow or underflow may occur in either multiplication or division, depending on the values of the exponents. These conditions are checked for after the result is computed, as shown in the flowchart of Fig. 59.22.

BCD Computations

Some computers provide arithmetic operations for binary coded decimal numbers. Multiplication and division are performed by successive additions and subtractions, respectively. Addition and subtraction may be performed serially, one decimal digit at a time, or in parallel. In either case the essential circuit is one that will add or subtract two decimal digits, each represented in the BCD code. Negative values are represented by their nine's complement in the system described here. In most practical systems, a ten's complement or sign–magnitude would be used. A block diagram for an

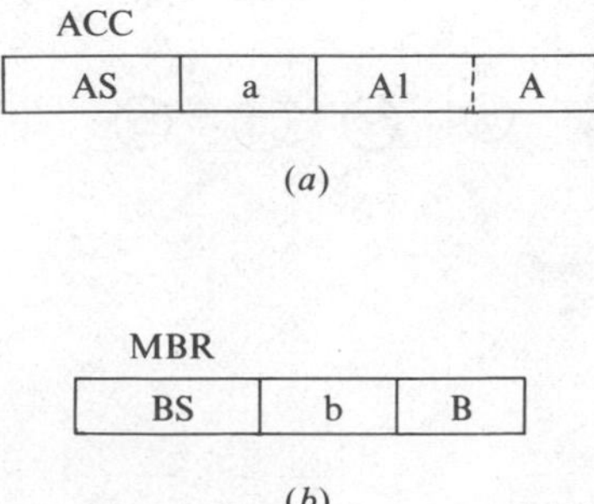

Fig. 59.20 Floating-point values: (*a*) Accumulator (ACC), (*b*) memory buffer register (MBR). AS, BS denote sign bits; A, B denote fractions; and a, b denote exponents.

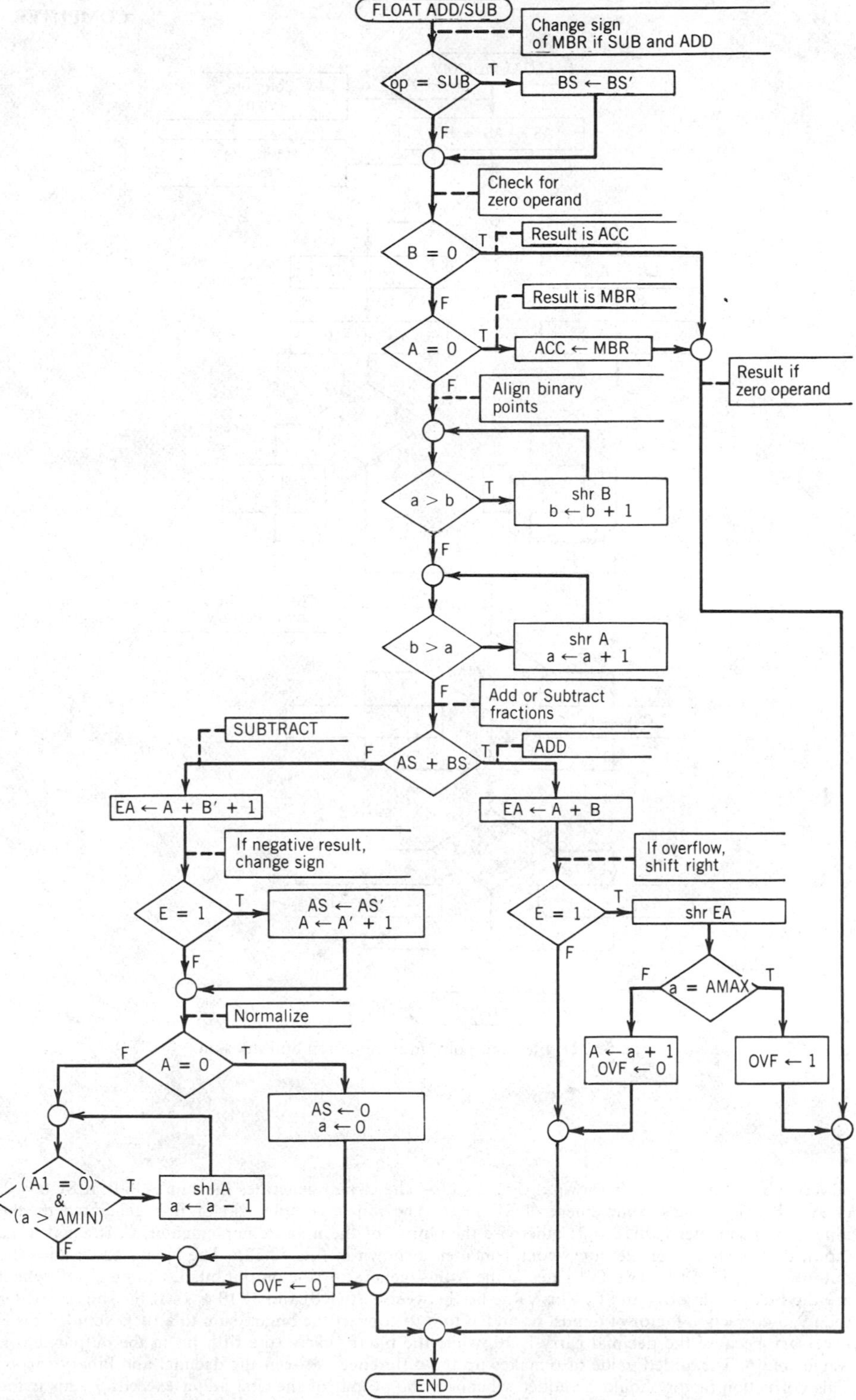

Fig. 59.21 Floating-point addition and subtraction. MBR, memory buffer register; ACC, accumulator; SHR, shift right; OVF, overflow.

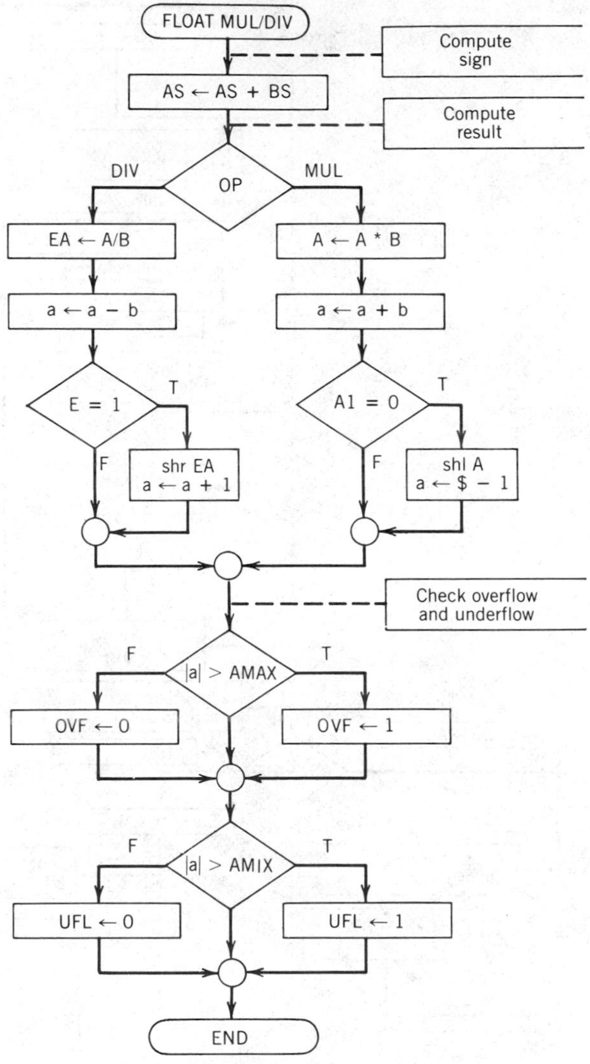

Fig. 59.22 Floating-point multiplication and division.

adder–subtractor circuit[11] is shown in Fig. 59.23*a*. The circuit generates the sum A + B, if SUB = 0, or A − B using nine's complement, if SUB = 1. The nine's complement of B is generated by the nine's complementer if SUB = 1; otherwise the output of the nine's complementer, C, is equal to its input, B. The circuit for the nine's complementer is shown in Fig. 59.23*b*. The binary adder adds the outputs of the EXCLUSIVE OR gates to the A bits, generating the output bits D1 through D5, which are the binary sum of A and C. This value lies between 0 (0 + 0) and 18 (9 + 9). If the sum is greater than 9, a correction factor of 6 must be added to it to convert the binary sum to a BCD sum.[12] This is necessary because the decimal carry is 10, while the binary carry (the fifth bit in the output) has a weight of 16. The added value of 6 makes up the difference between the decimal and binary carrys. This correction factor should be added whenever the output of the first adder exceeds 9; this is the case if S5 = 1 or if S4 = 1 and S3 or S2 = 1. The logic circuitry shown between the two adders adds the value 6 in these cases. The final carry, D5, is the output of the OR gate, as shown in Fig. 59.23*a*.

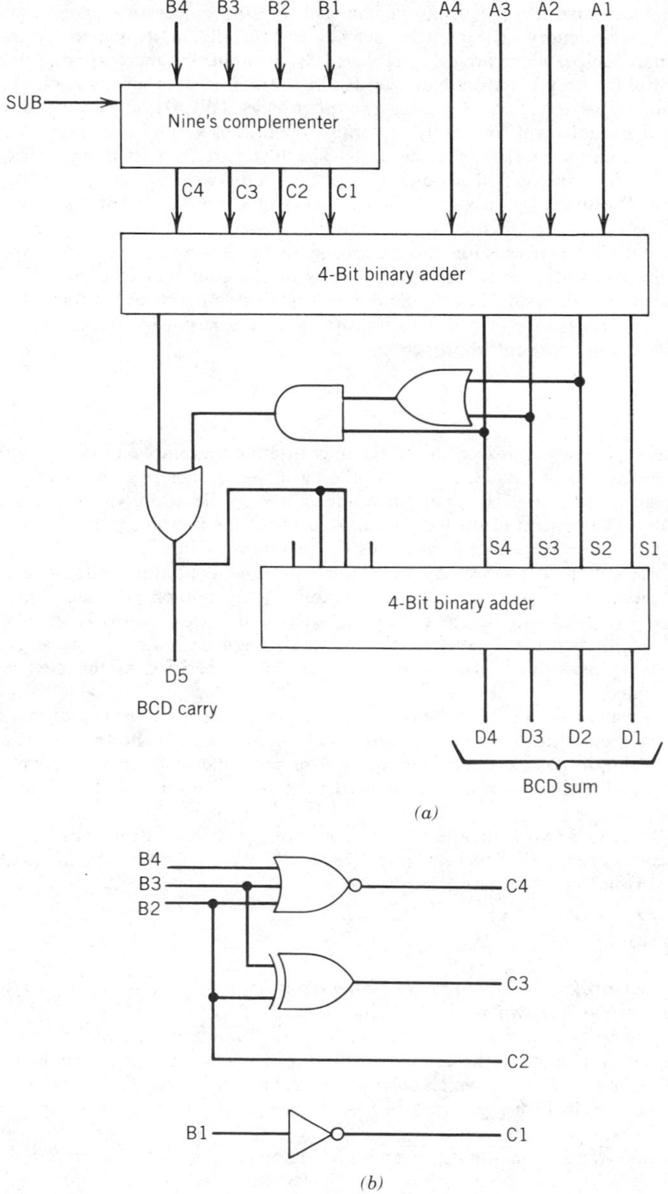

Fig. 59.23 Binary coded decimal (BCD) Adder–subtractor, (*a*) and nine's complementer (*b*).

59.8 MICROPROGRAMMING

The operation of the control units discussed thus far is determined by the design of their logic circuits. This is called hard-wired logic. A more recent approach to the design of central processing units is the use of a microprogrammed control unit.[13] This approach uses a universal controller, or microprogrammed control unit, in conjunction with a *control memory* that contains a sequence of microinstructions that implement each regular computer instruction, called a macroinstruction. (This use of the word "macroinstruction" is different from that in the context of an assembly language.) The advantage of a microprogrammed control unit over a hard-wired control unit is the ease with which the computer's instruction set may be changed. It is only necessary to change the microprogram in the control memory. The control memory is generally implemented by means of ROM (read-only memory) chips, which are easily replaced. In dynamically microprogrammable computers, part of the control memory is implemented by RAM (random access memory) chips, which can be user programmed. This is called a *writable control memory* (WCM); it permits the user to write microprograms that implement frequently performed operations very efficiently.

Each microinstruction consists of three parts. The first part consists of branch conditions that specify when this microinstruction should be executed; otherwise the controller skips to another microinstruction. The second part consists of the action that indicates what microoperations should be performed. The third part contains the next address, the address of the next microinstruction to be executed. The controller that executes the microprogram fetches microinstructions from the control memory and also has status inputs from other parts of the computer (the condition code), which allows it to determine whether the branch conditions specified in the microinstructions are satisfied or not. There are numerous ways in which to implement such microprogrammed control units.[14] The one described here illustrates one approach.

Operation

The first problem to be solved is location of the microroutine that implements a given macroinstruction. This is done by use of a small memory, called a mapping memory, that has a single word for each computer operation code. The operation code is used as the address to select a word from the mapping memory. The content of the word is the address of the first microinstruction (located in the control memory) of the routine that implements the macroinstruction.

This microinstruction is fetched by the controller. The condition code values input to the controller are checked against the branch conditions. If the branch conditions are satisfied, this microinstruction is skipped and the one in the next control memory location is fetched and executed. If the branch conditions are not satisfied, the microoperations specified in the activity part of the microinstruction are performed, after which the instruction specified in the next address field is fetched and executed.

Subroutines are used in microprograms, as they are in ordinary programs, to implement instruction sequences that occur multiple times. Nested subroutines are not necessary for microprograms because they are simpler than ordinary programs. The hardware required to support the subroutine capability consists of a single register that is used to store the return address when a subroutine is called.

Figure 59.24 shows a block diagram for a microprogrammed control unit. The control signals output by the controller are similar to, or might be the same as, the ones used as inputs to the gates, registers, and ALU in Fig. 59.5.

Microinstructions

The following description of microprogramming relates to the CPU shown in Fig. 59.5. The microinstruction format is shown in Fig. 59.25.

The 4-bit branch conditions field is composed of two 2-bit subfields. Bits 1 and 2 comprise the condition field, which determines the conditions under which a branch action will be taken, while bits 3 and 4 comprise the branch field, which determines the kind of branch that occurs. The functions of these bits are described in Tables 59.5 and 59.6. ACC(S) refers to the sign (most significant) bit of the accumulator.

Each bit position in the action field specifies a microoperation that is executed if it has a 1 bit. The microoperations are defined in Table 59.7. The branch condition and operation are evaluated and performed after execution of the microoperations specified in the action field.

Microroutines

The microroutines shown in Table 59.8 illustrate use of the microinstructions in implementing various macroinstructions and performing operations required for the instruction cycle. The 28-bit microinstructions are represented by seven hexadecimal digits. The address is 8 bits; it is represented by two hexadecimal digits with values that range from 00 to FF. The first seven microinstructions implement the instruction fetch phase and indirect addressing.

The next series of microinstructions, shown in Table 59.9, implement the load accumulator (LDA) macroinstruction, the add (ADD) macroinstruction, and the store accumulator (STA) macroinstruction. Post indexing is used in this CPU. This means that the direct address is first obtained from the

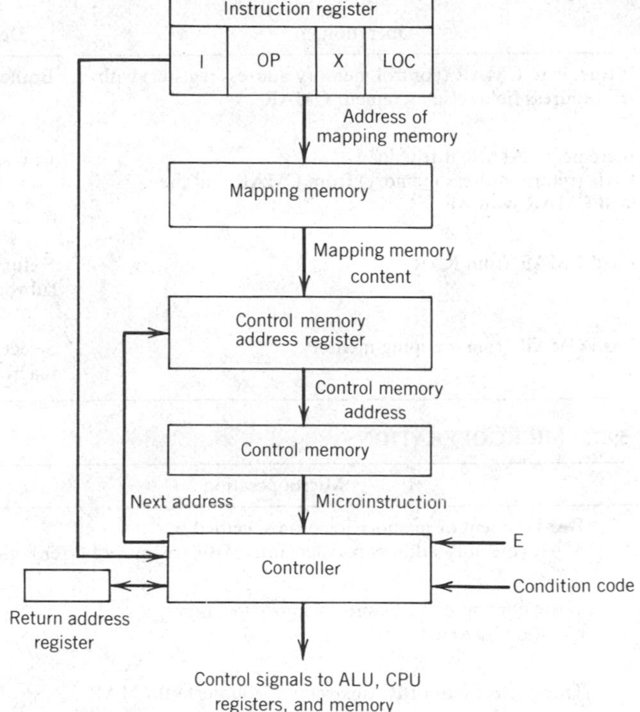

Fig. 59.24 Microprogrammed control unit. ALU, arithmetic–logic unit; CPU, central processing unit.

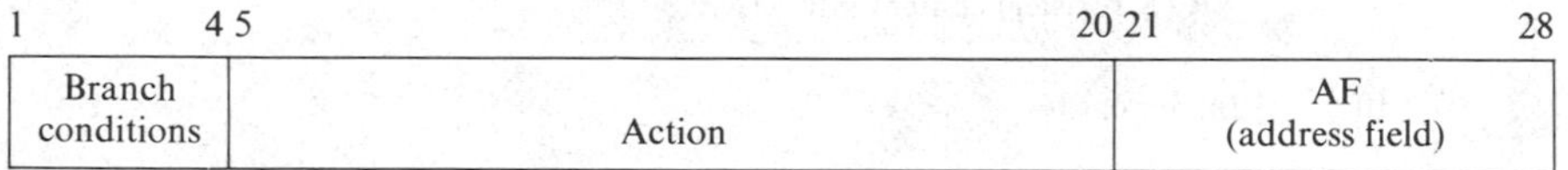

Fig. 59.25 Microinstruction format.

TABLE 59.5. BRANCH CONDITION FIELD

Bits 1	2	Condition	Description
0	0	Always true	Unconditional branch
0	1	I = 1	Branch for indirect addressing
1	0	ACC(S) = 1	Branch if ACC (accumulator) is negative
1	1	ACC = 0	Branch if ACC is zero

memory location specified by the indirect address (the indirect address is the address specified in the instruction) when indirect addressing is used. If indirect addressing is not used, the direct address is the address specified in the instruction. The operand is then fetched from the effective address, which is the sum of the direct address and the content of the index register when indexing is used. If indexing is not used, the effective address is the direct address.

These microinstructions illustrate the microprogrammed implementation of some of the

TABLE 59.6. BRANCH OPERATION FIELD

Bits 3	4	Operation	Description
0	0	If true, load CMAR (control memory address register) with AF (address field) else increment CMAR	Branch operation
0	1	Increment CMAR; if true load RAR (return address memory) from CMAR and then load CMAR with AF	Call subroutine
1	0	Load CMAR from RAR	Return from subroutine
1	1	Load CMAR from mapping memory	Select routine for macrooperation

TABLE 59.7. MICROOPERATIONS

Bit	Microoperation
5	Read content of memory location specified by MAR (memory address register) into MBR (memory buffer register)
6	Write content of MBR into memory location specified by MAR
7	Transfer content of IC (instruction counter) into MAR
8	Transfer LOC (location) field into MAR
9	Transfer sum of LOC field and product of X and XR (X register) content into MAR
10	Increment IC
11	Transfer content of MBR into IR (instruction register)
12	Transfer address field (bits 21–28) from MBR into LOC field of IR
13	Load content of ACC (accumulator) into MBR
14	Load content of MBR into ACC
15	Load sum of contents of MBR and ACC into ACC
16	Complement ACC
17	Clear ACC
18	Increment ACC
19	Load content of ACC into XR
20	Increment XR

TABLE 59.8. MICROROUTINES FOR INSTRUCTION FETCH AND INDIRECT ADDRESSING

Address	Microinstruction	Description
00	0200001	Begin instruction fetch: Load IC (instruction counter) into MAR (memory address register) and branch to 01
01	0840002	Read instruction from memory into MBR (memory buffer register), increment IC, and branch to 02
02	5020004	Transfer instruction from MBR into IR (instruction register) and branch to subroutine at 04 if indirect bit is 1
03	3000000	Load address of microroutine for given operation code. This completes instruction fetch
04	0100005	Begin indirect subroutine: Load indirect address from LOC (location) field into MAR and branch to 05
05	0800006	Read content of memory location specified by MAR into MBR and branch to 06
06	2010000	Load direct address into LOC field of IR and return from subroutine

macroinstructions. Each macroinstruction has its own microroutine. The multiply and divide algorithms, as well as the floating-point operations, can be implemented by microprograms. The relatively small capacity of the control memory makes it feasible to use more expensive, faster memory elements than are used for the main memory. This and related factors make it possible to execute these operations much more quickly when they are microprogrammed than when they are programmed using macroinstructions stored in the main memory. Microprograms for multiplication, division, and the floating-point operations require additional registers; these are not described in this chapter.[14]

Horizontal and Vertical Microprogramming

In some microprogrammed computers, each bit in the action field is connected to a single control line in the CPU, for example, LIC and INXR in Fig. 59.5. Such a computer generally has a very large microinstruction word length with no simple relationship between action bits and the function to be performed by the microinstruction. A computer of this type is said to use horizontal microprogramming.[1] Such a computer provides a high degree of flexibility in terms of permitting the simultaneous execution of several microoperations.

Some microprogrammed computers are designed to minimize the word length of the microinstructions. Each bit of the action field represents a complete microoperation and a decoder is necessary to translate these bits into control signals for the CPU. Such computers are said to use vertical microprogramming. These computers are generally much less flexible in terms of permitting simultaneous microoperations.[1]

A diagonally microprogrammed microcomputer[4] is one that lies between these two extremes. In such a computer, groups of control signals that would not be activated simultaneously are encoded into a small number of action field bits. This results in a computer with an intermediate microinstruction word length, the flexibility to permit many (though not all possible) microoperations to be performed simultaneously, and the need for a much simpler decoder than that required for a vertically microprogrammed computer.

TABLE 59.9. MICROROUTINES FOR LOAD ACCUMULATOR (LDA), ADD (ADD), AND STORE ACCUMULATOR (STA) MACROINSTRUCTIONS

Address	Microinstruction	Description
07	0080008	Start LDA microroutine: Transfer effective address into MAR (memory address register) and branch to 08
08	0800009	Read content of memory location specified by MAR into MBR (memory buffer register) and branch to 09
09	0004000	Transfer content of MBR into ACC (accumulator) and branch to 00 (fetch). This ends LDA microroutine
0A	008000B	Start ADD microroutine: Transfer effective address into MAR and branch to 0B
0B	080000C	Read content of memory location specified by MAR into MBR and branch to 0C
0C	0002000	Load sum of contents of MBR and ACC into ACC and branch to 00 (fetch). This ends ADD microroutine
0D	008000E	Start STA microroutine: Transfer effective address into MAR and branch to 0E
0E	000800F	Transfer content of ACC into MBR and branch to 0F
0F	0400000	Write content of MBR to memory location specified by MAR and branch to 00 (fetch). This ends STA microroutine

59.9 BIT-SLICE MICROPROCESSORS

Fixed-word-length microprocessors are mainly used in applications where small size and/or economy are important considerations. The initial microprocessors had 4-bit word lengths and were used for hand calculators and controllers. Eight-bit microprocessors made microcomputers possible, while 16-bit microprocessors have made higher performance microcomputers possible. There are even some 32-bit microprocessors available. Most of these use metal oxide semiconductor (MOS) technology to achieve high circuit density and low power dissipation; it also results in substantially slower operation than bipolar semiconductors.

Bit-slice microprocessors are designed to permit simple implementation of custom designed high-performance digital computers.[10] ECL (emitter-coupled logic) bipolar semiconductor chips are widely used to provide very high speed. The bit-slice microprocessor consists of a family of three kinds of building blocks, or chips, which are combined to form a microcomputer that is inherently a microprogrammed computer. The three kinds of chips are control memory chips, a controller, and the arithmetic–logic unit (ALU) slice.

A typical CPU contains a variety of registers. Some of these are data registers, which store information that is being processed. All of these registers store data having a given size, called the

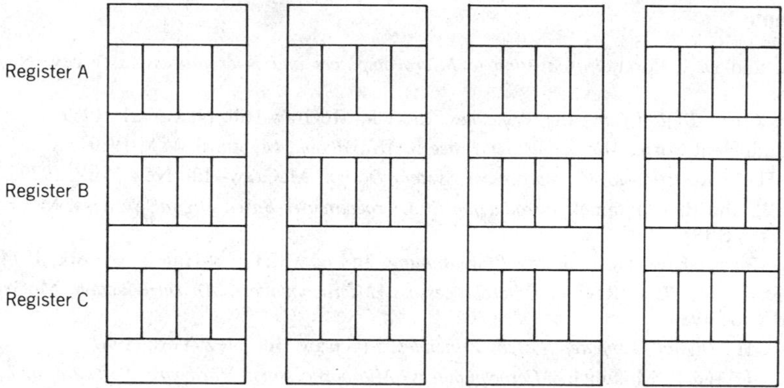

Fig. 59.26 Arithmetic–logic unit (ALU) slices.

word length, which is determined by the design of the computer. Other registers, of various lengths, are used to store various kinds of information, for example, address and status.

The registers that are not data registers are incorporated in the controller. The controller and control memory perform the functions described for the microprogrammed CPU as well as provide various registers required of the CPU. The data registers are implemented by connecting together a number of ALU slices, each of which contains a fixed number of bits, most frequently 2 or 4, of all the data registers used in the CPU. Figure 59.26 illustrates the use of four 4-bit ALU slices to form a CPU with a 16-bit word length. All the ALU slices are identical and have a part of every data register. The number of registers in the computer and their function is fixed by the design of the ALU slices. The word length of the computer is determined by the number of slices that the designer connects together.

References†

1 A. S. Tanenbaum, *Structured Computer Organization*, Prentice-Hall, Englewood Cliffs, NJ, 1976.

2 T. L. Booth, *Digital Networks and Computer Systems*, Wiley, New York, 1971.

3 B. Soucek, *Minicomputers in Data Processing and Simulation*, Wiley-Interscience, New York, 1972.

4 M. M. Mano, *Computer System Architecture*, 2nd ed., Prentice-Hall, Englewood Cliffs, NJ, 1982.

5 R. H. Eckhouse, Jr. and L. R. Morris, *Minicomputer Systems Organization, Programming, and Applications (PDP-11)*, 2nd ed., Prentice-Hall, Englewood Cliffs, NJ, 1979.

6 A. Osborne, *An Introduction to Microcomputers*, Vol. 1, *Basic Concepts*, 2nd ed., Osborne/McGraw-Hill, Berkeley, CA, 1980.

7 H. S. Stone, and D. P. Siewiorek, *Introduction to Computer Organization and Data Structures: PDP-11 Edition*, McGraw-Hill, New York, 1975.

8 G. D. Kraft and N. T. Wing, *Mini/Microcomputer Hardware Design*, Prentice-Hall, Englewood Cliffs, NJ, 1979.

9 L. Nashelsky, *Introduction to Digital Computer Technology*, Wiley, New York, 1972.

10 D. P. Siewiorek, C. G. Bell, and A. Newell, *Computer Structures: Principles and Examples*, McGraw-Hill, New York, 1982.

11 T. C. Bartee, *Digital Computer Fundamentals*, 4th ed., McGraw-Hill, New York, 1977.

12 J. O'Malley, *Introduction to the Digital Computer*, Holt, Rinehart and Winston, New York, 1972.

13 *Microprogramming Handbook*, 2nd ed., Microdata Corp., Santa Ana, CA, 1971.

14 Y. Chu, *Computer Organization and Microprogramming*, Prentice-Hall, Englewood Cliffs, NJ, 1972.

†The variety of topics covered in this chapter and the rate of technological growth in this area make it impossible to provide information on the latest developments. A list of periodicals that contain information on current work in the computer field is included at the end of the bibliography.

Bibliography

Barna, A. and D. I. Porat, *Introduction to Microcomputers and Microprocessors*, Wiley, New York, 1976.

Burroughs Corp., *Digital Computer Principles*, 2nd ed., McGraw-Hill, New York, 1969.

Digital Equipment Corp., *Microcomputer Processor Handbook*, Maynard, MA, 1980.

Garland, H., *Introduction to Microprocessor System Design*, McGraw-Hill, New York, 1979.

Gault, J. W. and R. L. Pimmel, *Introduction to Microcomputer-Based Digital Systems*, McGraw-Hill, New York, 1982.

Gear, W., *Computer Organization and Programming*, 2nd ed., McGraw-Hill, New York, 1974.

Givone, D. D. and R. P. Roesser, *Microprocessors/Microcomputers: An Introduction*, McGraw-Hill, New York, 1980.

Hellerman, H., *Digital Computer System Principles*, McGraw-Hill, New York, 1967.

Hilburn, J. L. and P. M. Julich, *Microcomputers/Microprocessors: Hardware, Software, and Applications*, Prentice-Hall, Englewood Cliffs, NJ, 1976.

Intel Corp., *MCS-48 Family of Single Chip Microcomputers: User's Manual*, Santa Clara, CA, 1978.

Intel Corp., *The 8086 Family User's Manual*, Santa Clara, CA, 1979.

Katzan, H., Jr., *Computer Systems Organization and Programming*, Science Research Associates, Chicago, 1976.

Korn, G. A., *Microprocessors and Small Digital Computer Systems for Engineers and Scientists*, McGraw-Hill, New York, 1977.

Leventhal, L. A., *Introduction to Microprocessors: Software, Hardware, and Programming*, Prentice-Hall, Englewood Cliffs, 1978.

Motorola Semiconductor Products, Inc., *Microprocessor Applications Manual*, McGraw-Hill, New York, 1975.

Osborne, A., S. Jacobson, and J. Kane, *An Introduction to Microcomputers,* Vol. 2, *Some Real Products,* Jun. 1977 revision, Osborne/McGraw-Hill, Berkeley, CA, 1977.

Peatman, J. B., *Microcomputer-Based System Design*, McGraw-Hill, New York, 1977.

Richards, R. K., *Arithmetic Operations in Digital Computers*, D. Van Nostrand, Princeton, NJ, 1955.

Simpson, W. D., G. Luecke, D. L. Cannon, and D. H. Clemens, *9900 Family Systems Design and Data Book*, Texas Instruments Inc., Houston, 1978.

Soucek, B., *Microprocessors and Microcomputers*, Wiley, New York, 1976.

Texas Instruments, *990 Computer Family Systems Handbook*, 2nd ed., Austin, TX, 1975.

Turner, J. F., *Digital Computer Analysis*, Merrill, Columbus, OH, 1968.

Zaks, R., *Microprocessors from Chips to Systems*, 2nd ed., Sybex, Berkeley, CA, 1977.

Computer, published monthly by IEEE Computer Society, Los Alamitos, CA.

Computer Design, published monthly by Computer Design, Littleton, MA.

IEEE Micro, published quarterly by IEEE Computer Society, Los Alamitos, CA.

IEEE Transactions on Computers, published bimonthly by IEEE, New York.

Mini-Micro Systems, published monthly by Cahners, Boston, MA.

CHAPTER 60
MEMORY SYSTEMS

JAY MICȞLIN

Exxon Corporation
Florham Park, New Jersey

60.1 INTRODUCTION

Memory is at the center of any computing system.[1,2] The processor(s), input–output (I/O), and peripherals all must somehow attach to memory. The size, speed, and basic architecture of memory often determine the character and performance of the entire system.

The last few years have brought remarkable improvements in memory technology, and they have given rise to a number of trends. First, memory prices have decreased dramatically. A megabyte (1,048,576 8-bit bytes) cost about $300, 000 in the early 1970s but costs only about $15,000 in the early 1980s.

The size of memory has also decreased. A megabyte of memory used to occupy an entire room but is now small enough to fit in the hand. If the average book published today contains about one half to 1 million characters, one can store the book in electronic memory that is smaller than the book itself.

With memory cheaper and smaller, designers are using more of it. Figure 60.1 shows the amount of memory coupled to the largest commercial processors, by year. As late as 1972 the largest processors had about 2 to 3 megabytes of memory. Today in the 1980s, they have 10 times that, and they will have 20 times it within two to three years. In fact, they would have 20 times it now if only they could address that much.

The ability to address large amounts of memory is perhaps the key bottleneck today. For reasons to be discussed later, many of today's processors cannot address more than 16 or 32 megabytes. Most new processors are designed to address on the order of 4 gigabytes (billion bytes), and hence will be able to use as much memory as can be justified.

The large increase in available memory has also had an effect on programming technology. In the past one went to great lengths to squeeze programs into memory. Today almost any program fits in memory without difficulty, so programs are easier to understand and maintain. Indeed the emphasis is now on containing programming and software costs instead of limiting the use of hardware. However, because of the addressing limitation, a program's *data* usually cannot fit in memory, so programmers must manage the data using I/O. There is little doubt that this will change in the next five years as memory continues to become cheaper and programmers become scarce and expensive.

Manufacturers have also taken advantage of large, cheap memories by adding capabilities to their operating systems. A large system used to reside in about 1/4 megabyte. Today it occupies a full megabyte and offers many sophisticated facilities. This has not been the case with the smaller microcomputers, but even their average memory size has increased by a factor of 2 in less than two years.

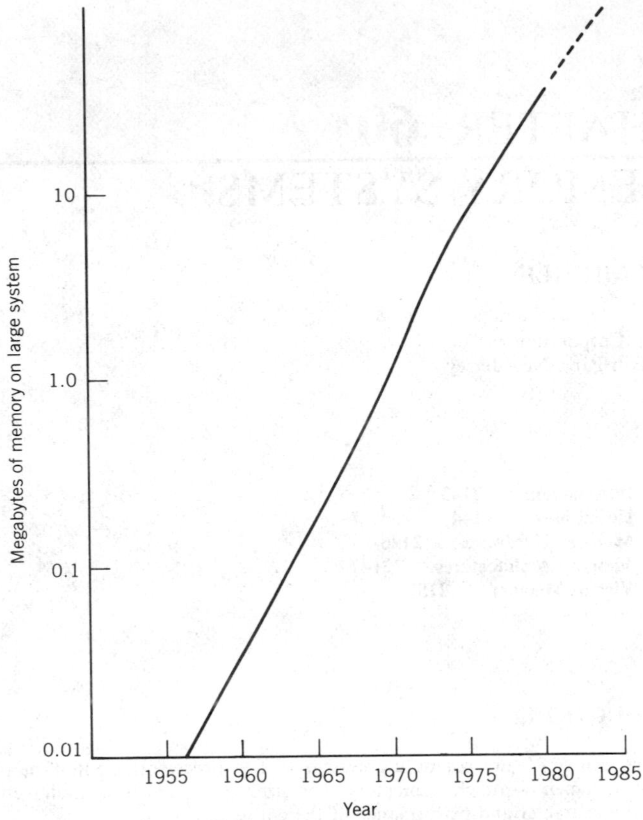

Fig. 60.1 Memory coupled to large commercial processors by year.

Finally, the low cost of memory has encouraged the use of error detecting and correcting facilities. Nearly all memory systems now include at least a parity bit on each word to detect errors. Medium and large systems almost always go a step further by adding enough extra bits to actually *correct* any single-bit error and detect any double-bit error. This has contributed appreciably to the reliability of memory systems.

The remainder of this chapter is organized as follows: Section 60.2 defines the terms that are commonly used to discuss memory systems. Section 60.3 briefly describes the actual electronic hardware that makes up modern memories. Section 60.4 describes memory architectures, that is, ways of organizing memories for maximum speed and efficiency. Section 60.5 discusses *virtual memory*, a technology that combines hardware and software to make memory more useful.

60.2 DEFINITIONS

The size of a memory is measured in bits, bytes, or words. A "byte" is by definition 8 bits, and a word may be 16, 24, 32, or 60 bits, depending on the manufacturer of the equipment. For convenience, we will discuss memory sizes in bytes. A "kilobyte" is, by convention, 1024 bytes (2^{10}), and a "megabyte" is, by convention, 1,048,576 bytes (2^{20}). A "gigabyte" is 1024 megabytes.

RAM. The memory with which we will be primarily concerned is called *main memory or random access memory* (RAM). It is the aggregate of memory on which the processor calls for programs and data. The term "memory," unmodified, usually refers to main memory.

ROM. The basic operations that programs perform on memory are reading and writing. *Read-only memory* (*ROM*) is a special kind of memory that cannot be written except, of course, when it is first initialized. The advantage of ROM is that it is cheap and fast. It also continues to hold data even when the power is shut off. ROM is used in large computers to store the microprograms that define the architecture of a machine. In microcomputers, ROM is commonly used to store primitive operating systems or language interpreters. In effect, the ROM provides an initial program for the computer to execute at power-up or, perhaps, at the push of a button. Without an initial program in ROM, a computer would be unable to perform any useful work until a program were manually loaded.

PROM. PROM, or programmable ROM, might also be called write-once memory. ROM is usually programmed during one of the masking steps in fabrication, while PROM is programmed individually by "burning in" ones and zeros for each unit. PROM chips are used whenever small quantities or prototypes are needed and mass production of a mask-programmed ROM is unjustified. PROMs are sometimes used in production equipment if the programming is expected to change so that it is inappropriate to commit to mass production of ROMs.

EPROM. An EPROM is an erasable PROM, and it is usually constructed using metal nitride on silicon (MNOS) technology. An EPROM might be called a write rarely memory. It cannot be written in the same way as conventional RAM, even though it is erasable. Rather, EPROM is written by a special process and is erased by exposure to ultraviolet light.

Writable Control Storage. Another special kind of memory that is related to ROM is writable control storage, or WCS. In any machine that can be microprogrammed, the microprogram is kept in a control store that is usually a ROM. To allow alteration of the microprogram in the field, some or all of the ROM in the control store is replaced with RAM. The resulting control store is writable, and it is referred to as WCS. The RAM used is usually bipolar, since speed is of the essence. The current trend is to use WCS for all microprogramming. The WCS is typically loaded from a floppy disk using a bootstrap loader in ROM.

Static and Dynamic RAMs. Two kinds of RAM chips are in common use. One kind, dynamic RAM, requires additional, outboard circuitry to generate a periodic refresh pulse. Without such circuitry, the bits stored in the RAM would decay and become unreliable. By contrast, static RAM retains information as long as power is applied, and does not require a refresh pulse. The tradeoff between the two is that the dynamic RAMs are cheaper but the static RAMs require less circuitry, since a refresh pulse generator is not needed. However, many microprocessors now include an on-board circuit to generate refresh pulses for dynamic RAMs, so the advantage of static RAMs is less pronounced.

Volatile and Nonvolatile Memories. A memory is volatile if it cannot reliably retain data when power is shut off. A memory is nonvolatile if it retains data across a power shutdown. Most modern RAM is volatile, and ROM is, of course, nonvolatile.

Modes of Access. There are three modes in which a memory is commonly accessed, and a given technology typically supports one of the modes. The modes are as follows:

Random Access. This means that any byte of the memory can be addressed and retrieved. Furthermore, the same amount of time is needed to retrieve each byte. Random access is the technology of main memories, and this is why they are made up of circuits called RAMs.

Direct Access. This means that any byte of the memory can be addressed and retrieved but that more time is needed for these operations with some bytes than with others. Bubble and charge-coupled technologies usually support direct access.

Sequential Access. This means that to access the *n*th byte one must first access the first $(n - 1)$ bytes. Magnetic tape is the classic example. Bubble and charge-coupled devices are also inherently sequential, but they are fast enough to be treated as direct access given the proper support circuits.

Addressability. In theory one can add as much memory to a computer as fits in the cabinet. However, in practice another factor limits how large a memory can be. That factor is *addressability*, or *addressing range.* All computer programs are eventually executed as machine instructions, and only a certain number of bits are reserved in the instructions to hold address values. If the number of address bits is N, then the maximum number of bytes that a program can refer to is 2^N. This number of bytes is called the program's *address space*.

For example, 24 bits has been a common width for the address portion of a machine instruction ever since the IBM System/360 pioneered 24-bit addressing in the early 1960s. At the time, this was considered much more than adequate. The maximum address space a program can have on a machine with a 24-bit address width is 16 megabytes (2^{24}). In a similar vein, many of today's microcomputers offer a 16-bit address width that supports only a 64 kilobyte (2^{16}) address space.

Now, even on the largest machines, few programs are larger than perhaps a megabyte, and the average program is much smaller. But many programs have to deal with huge *data structures* that exceed 16 megabytes. Addressing support for such large structures is becoming increasingly important.[3] The most modern computers are being designed with address widths that are 32 (Digital Equipment VAX series)[4] or even 48 (IBM System 38)[5] bits wide.

Hierarchical Memory. Any modern computer system contains many types of memory. The reason for the different types is a cost-size-speed tradeoff. There is ultra-high-speed cache memory to act as a buffer between the processor and the main memory. There is the main memory itself. There is virtual memory, which is a combination of hardware and software. And there is long-term file memory that is implemented using I/O devices.

As shown in Fig. 60.2, all of this memory can be thought of as a hierarchy, with the fastest and most expensive components at the top and the slowest and cheapest components at the bottom. The trend is toward *automatic management of memory hierarchies*.[3] Programs will "see" only a huge main memory. Combinations of hardware and software will move data among devices so that data is actually in main memory when needed.

60.3 MEMORY HARDWARE

This section describes the basic hardware that makes up the components of a memory system. The electronics of the components is a subject covered extensively elsewhere in this handbook and will not be considered here.

Magnetic Core

Magnetic core memory[6] was the prevalent technology from the late 1950s until 1972. It was developed to a high art with access times of as little as 400 ns and sizes to 3 megabytes. However, it was an incredibly labor-intensive technology that was unable to produce memory below about 4 to 5 cents per bit. Semiconductors, which are the standard today, are much less labor intensive and have achieved prices as low as 0.1 to 0.2 cents per bit. Many people, however, still use the term "core" to refer to memory.

Semiconductor Memory

Field-effect transistors (FETs) and metal-oxide-semiconductor (MOS) technology[7] are what brought about the revolutionary switch to semiconductors for main memory. Bipolar transistors evolved too but were limited to ultra-high-speed applications such as cache (Section 60.4), because of their cost.

MOSFET chips contained 1 to 2 kilobits in the early 1970s, rising to 64 kilobits today. So far, densities have been increasing by a factor of almost 4 every two years,[8] although the rate will probably slow as the limit of the technology is approached. Still, with newer etching methods such as electron beams, higher densities will be reached. There is some speculation that up to three or even four orders of magnitude more density is possible.

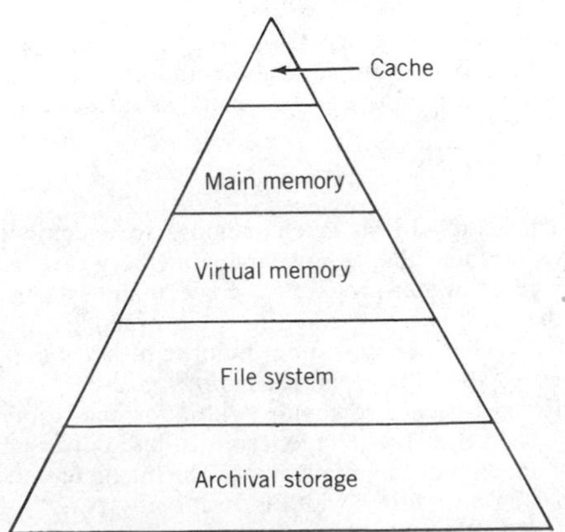

Fig. 60.2 Memory as a hierarchy.

MOSFET memory has the same advantages as other MOSFET technology. It is small and highly suitable for integration requiring as little as one equivalent device per bit stored. Also it consumes little power and dissipates little heat. MOSFET memories can easily achieve access times on the order of 500 ns, and access times of 200 ns are not uncommon.

Bipolar memories make less efficient use of a chip than do MOSFETs, if only because the bipolars require two or three transistors per cell. Also bipolars dissipate more heat and require more power. But a bipolar memory can be accessed in as little as 10 to 20 ns.

Other Technologies

Two new memory technologies may play a major role in the next few years. These are magnetic bubbles[9] and electrostatic charge-coupled devices.[10]

Average access time to a bit in a bubble store is on the order of 10 ms, so bubbles are viewed as a possible replacement for disks and drums. The problem up to now has been that rotating magnetic technologies have so advanced that bubbles are not economically competitive with them.

Bubbles have the great advantage that they are useful and economic in small quantities. Commercial bubble devices are available in the 256-kilobit range, while hard disks are not economic below about 10 megabytes. One can speculate that bubbles will see increasing use as the market for small, personal computers expands.

Bubbles may also become more attractive if so-called bubble array devices become available. In these devices many bubbles are moved en masse and densities are greater while access times are shorter. As this is written, a number of bubble devices are commercially available and in use. The best known use is as on-board memory in certain terminals. There are no bubble array devices on the market yet.

Charge-coupled devices, or CCDs, are degenerate forms of the MOS RAMs discussed earlier. The major difference between the two is that CCDs are accessed in a shift register mode, while RAMs are accessed randomly. This simplifies the decoding circuits on a CCD chip and results in greater densities. It also results in millisecond access times, compared with the nanosecond access times of the RAMs. Therefore CCDs, like bubbles, are viewed as competitors for rotating magnetic devices, not for main memories.

CCDs are marginally less expensive than bubbles and have appeared in at least one commercial product as a drum replacement. However, bubbles have the advantage of nonvolatility. That is, bubble chips, which are packaged with permanent magnets, retain information when power is disconnected. CCDs can be made nominally nonvolatile in a system by including a small emergency power supply.

60.4 MEMORY ARCHITECTURES

Basic Memory

On the highest level, any memory system can be thought of as shown in Fig. 60.3. There is a memory address register (MAR) into which the central processing unit (CPU) or other equipment deposits an address. The READ line is then set to 1, and some time later the data at the address appears in the memory data register (MDR). (The term "memory buffer register" is often used in place of MDR.) For storing into memory, the data to be stored are placed in the MDR while the address is placed in the MAR. The WRITE line is set to 1, and some time later the operation is complete.

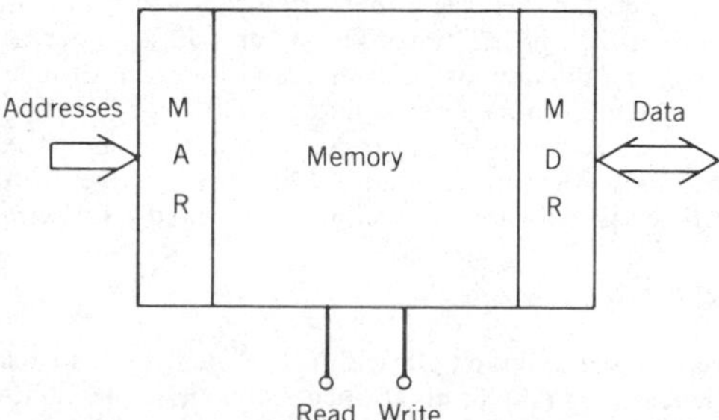

Fig. 60.3 Basic memory system. MAR, memory address register; MDR, memory data register.

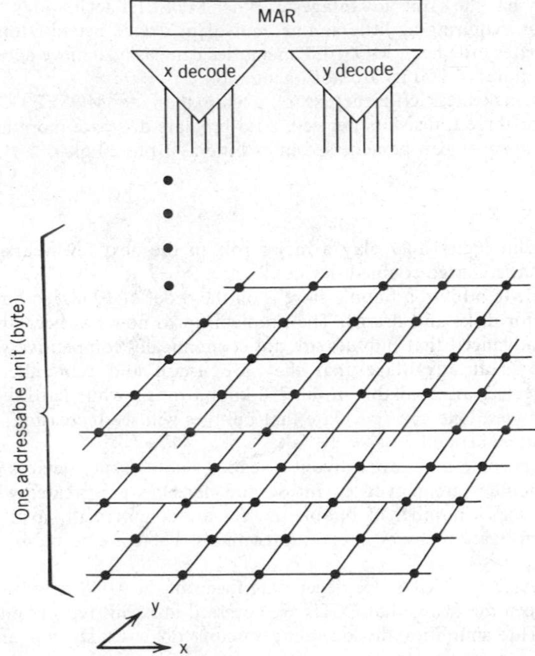

Fig. 60.4 Memory array bits are stored at the x, y crossings. MAR, memory address register.

On the lowest level, the memory is made up of a three-dimensional array of storage elements, as shown in Fig. 60.4. The elements may be magnetic cores, bipolar flip-flops, or various forms of FET circuits. Two dimensions of the array, conventionally referred to as the x and y directions, make up an addressing matrix. When a binary address is deposited in the MAR, the address is decoded into x and y values.

The third dimension, or depth of the array, corresponds to the word size of a machine. When an address selects an x, y point, it simultaneously selects all elements with that x, y value. Thus in a byte-oriented machine where everything is organized into 8-bit bytes, the array is eight deep, and a given address selects 8 bits. If, as is usually the case, the memory includes parity or error-correcting bits, the memory is deeper in the vertical dimension to account for them.

Effect of Path Width

One measure of the memory's speed is how many bytes per second it can deliver. This measure is sometimes called the *bandwidth of the memory*, and values on the order of 100 megabytes per second are not uncommon in large computers. The term "bandwidth," as used here, is analogous to but not exactly the same as the "bandwidth" specification used in communications. Here, bandwidth means aggregate bytes per second. One can increase a memory's bandwidth in two ways. One way is to use faster hardware that delivers a byte in less time. The other way is to access more bytes in the same amount of time, that is, to make the memory deliver many bytes in a single access.

The efficacy of a multibyte path to memory is that the CPU (or I/O, for that matter) often needs more than a single byte. Therefore the overhead cycles needed to start off requests for the additional bytes are saved. It is possible to organize a memory so that with very little additional circuitry it delivers many bytes at a time. This organizing technique is called *interleaving*.

Interleaving

Interleaving is always coupled with a large path width. If a memory is to access, for example, 8 bytes at a time, it can access them sequentially or all at once. Obviously, the all-at-once case is much faster. Interleaving is a technique that allows such parallel access. The cost of interleaving is the additional hardware needed to make many things go on at the same time.

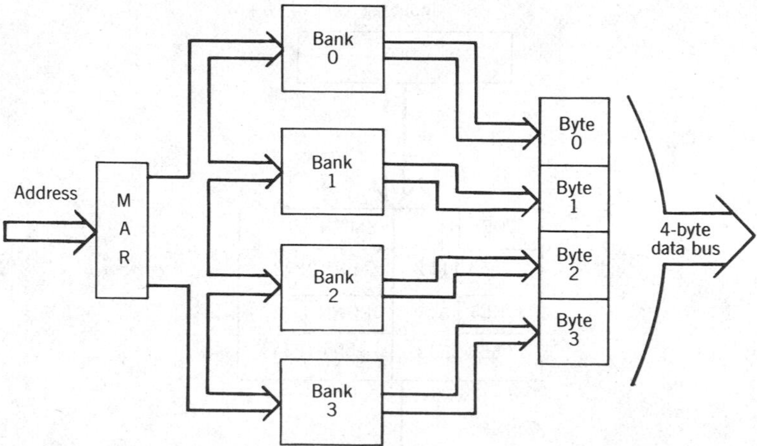

Fig. 60.5 Four-way interleaving. MAR, memory address register.

Consider a 1024-byte memory as an example. Think of it as organized into four independent *memory banks*, each with its own read, write, and decoding electronics, as shown in Fig. 60.5. The four banks operate as follows:

Bank 0 contains the bytes whose addresses are 0 modulo-4, that is, bytes 0, 4, 8, and so on.

Bank 1 contains the bytes whose addresses are 1 modulo-4, that is, bytes 1, 5, 9, and so on.

Bank 2 contains the bytes whose addresses are 2 modulo-4.

In general, bank *i* contains bytes with addresses *i* modulo-4.

If a request is made for byte 0, all four banks can operate simultaneously, in parallel, to produce bytes 0 through 3, that is, 4 bytes in the time it takes one bank to produce a single byte. The memory's effective speed actually increases, in this case by a factor of 4. This is a classic use of *parallelism* to increase the effective speed of a system.

Figure 60.5 shows 4-way interleaving. Other possibilities are 8-way, 16-way, and occasionally 2-way. The greater the interleaving, the more the parallelism and the greater the effective speed of the memory system. Of course, the path width must be large enough to pass all of the data that the memory is accessing in parallel. Also the programs being executed must have reasonably frequent use for all the extra data, and that is usually the case.

Associative Memories

Up to now, memories that are substantially linear have been discussed. One specifies an address, which is just a number, and the memory accesses the byte with that number. There is another kind of memory, associative, that works in an entirely different manner.[11]

A synonym for *associative memory* is *content addressable memory*. Each item in the memory has a *tag* or "key" associated with it. Instead of specifying an address, one specifies a tag value. If that tag exists anywhere in the memory, the associated item is retrieved.

Figure 60.6 illustrates the concept of an associative memory for looking up numbers in a telephone book. Suppose a program is searching for "John Doe's" telephone number. With a conventional memory, the program selects an entry, retrieves it, and tests whether the name (i.e., tag) associated with it is "John Doe." If not, the program selects another entry and continues searching. With an associative memory, the program has only to supply the tag "John Doe." If it is in the memory, the hardware will select it and return the associated telephone number.

So far, associative memories have been too expensive for general use except in two cases, caches and address translators.

Cache Memories. *Cache*, or *buffer*, memories[12] are another way to increase the apparent speed of a memory system without resorting to faster electronics on a wholesale basis. Figure 60.7 illustrates a memory system with a cache. The cache is much smaller but much faster than the basic memory. The idea is that whenever the CPU requests data from memory, those data plus a block of data "near"

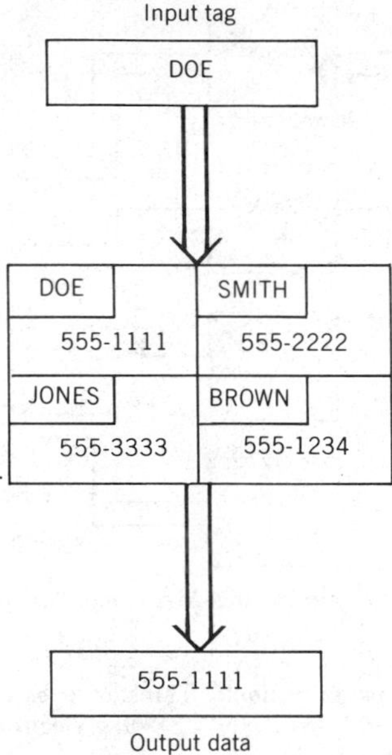

Fig. 60.6 Associative memory.

them are loaded into the cache. The next time the CPU needs data, there is a great likelihood that the data will already be in the cache. In that case, the transfer will proceed at cache speed, which is much faster than memory speed.

The basic parameter that governs the performance of a memory with a cache is the *cache hit ratio*. It is the fraction of memory requests that can be satisfied with data already in the cache, and is usually expressed as a percentage. If the cache hit ratio is r, the time to access the cache is t, and the time to access the main memory is T, then the effective access time of the entire memory system is given by the formula

$$\text{Effective access time} = r \times t + (1 - r) \times T$$

Consider as an example a memory that can deliver 64 bytes of data to the cache in 500 ns. Assume that the cache can deliver 8 bytes of data to the CPU in 50 ns. If all fetches had to go to

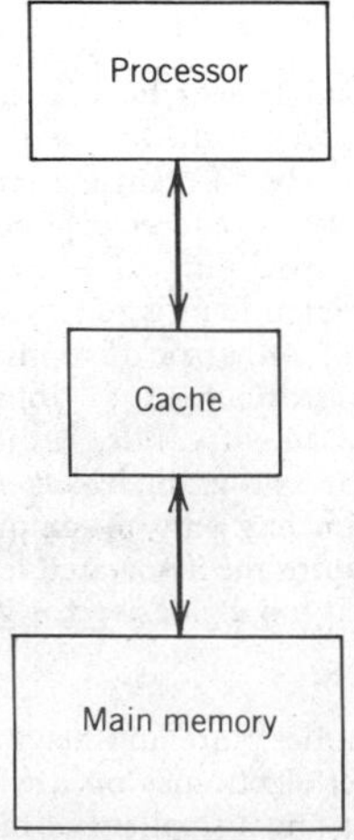

Fig. 60.7 Memory system with cache.

memory, access time would always be 500 ns. But assume that the cache hit ratio is 90%, so that 90% of the time the data requested are already in the cache. Then 90% of the time the access time is only 50 ns. The effective access time of the memory system is then

$$\text{Effective access time} = 0.90 \times 50 \text{ ns} + 0.10 \times 500 \text{ ns} = 95 \text{ ns}$$

Because the data were found in the cache 90% of the time, the effective speed of the entire memory system increased by a factor of more than 5.

Cache hit ratio is affected by a number of things, most notably the size of the cache and the nature of the program being executed. The size of the cache is essentially a cost vs. speed tradeoff. The cache must be built using extremely fast bipolar elements. Bipolar memory is also quite expensive, requires a good deal of power, and generates a good deal of heat. Typically, one uses 4 to 8 kilobytes of cache memory in small systems with 1/2 to 1 megabyte of memory, as in the DEC PDP-11 and VAX series of machines. One uses 32 to 64 kilobytes of cache memory in large systems to buffer 4 to 32 megabytes of memory, as in the IBM 3033 and 3081, or the Amdahl 470 and 580 series.

The nature of the program affects cache hit ratio through a property called *locality of reference*, and we shall see that this same phenomenon affects virtual memory (Section 60.5). Locality of reference is the phenomenon that if a program refers to an item of data, the likelihood is great that the program will soon refer to more data that are "nearby." Similarly, if a program executes an instruction, the probability is great that the program will next execute another instruction that is "nearby."

A program with good locality almost always accesses instructions and data close to places in memory that were recently accessed. The likelihood is high that such a program will find what it needs in the cache. A program with poor locality randomly accesses data and jumps repeatedly among groups of instructions. Such a program will repeatedly encounter "cache faults" and be forced to access memory at relatively slow speeds.

Still, caches improve the speed of even poor candidate programs. It is common for a program with good locality of reference to achieve a 98% to 99% cache hit ratio. It is uncommon to see even a poorly organized program do worse than about 80%, and even this increases the effective memory speed by a factor of perhaps 3. It is hard to overestimate the effect of a cache on an overall computer system. Overall speed increases of 50%, 100%, or more are common when a cache is installed.

Organization of Cache Memories. The memory system "knows" whether an item of data is in the cache or not, since in one way or another the cache has properties of an associative memory. The memory system is presented an address, and it uses that address as a tag. If the tag matches a tag stored in the cache, the bytes of memory that have been requested are associated with that tag. If there is no tag presently in the cache that matches the requested address, the memory system must bypass the cache and revert to main memory.

The most popular cache architecture today is called *set direct mapped* and is shown in Fig. 60.8. In this design the cache is broken into small blocks, typically 64 bytes long. The blocks are grouped into *lines* according to the parts of the memory they are permitted to hold. In a typical case there are 16 lines organized such that 64-byte blocks of memory whose addresses end in the bits "0000" are mapped into one of the blocks in line 0; memory blocks whose addresses end in "0001" are mapped into one of the blocks in line 1; and in general, memory blocks whose addresses end in any value, *i*, are mapped into the cache line with the same value.

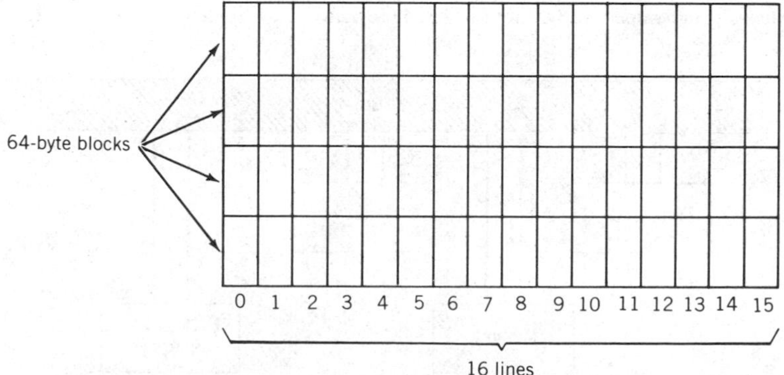

Fig. 60.8 Set direct map cache with 16 lines and four-way associativity.

When a request is made to the cache, the cache can determine which line might contain the requested data by examining the last few bits of the address supplied. Once a line is selected, the blocks in it are examined associatively.

If there are N blocks in each line, the cache is said to be N-way associative. If any of the N blocks have a tag that matches the address requested, the cache can complete the memory operation using the data in that block. If none of the N blocks has a tag matching the address requested, a cache fault occurs and the operation reverts to main memory.

Bus Architectures

Figure 60.9 illustrates a bus architecture. It is common in small computers, and has more to do with organization and flexibility than with speed. The architecture is such that any equipment, whether a CPU or an addition to memory or a peripheral device, simply plugs into the bus.

The flexibility inherent in this approach is excellent. One can add memory to a system by plugging in a memory board. One can attach peripherals with similar ease. And *each item on the bus is addressed as if it were memory*. Thus storing into some address range might be equivalent to writing a block to disk. Multiple CPUs can use the same mechanism to communicate with each other, although few multi-CPU systems have emerged.

The price one pays for a common bus is speed, since there may be contention among the many devices connected together. However, the speed penalty does not seem to matter much unless there is appreciable multiprogramming going on, so bus organizations are ideal for small and medium-sized machines.

60.5 VIRTUAL MEMORY

"Virtual memory"[13] is a catchall term for a wide variety of technologies. In a generic sense it refers to a peculiar way in which some operating systems manage memory. They allow programs to behave as if there were more memory than is actually present in hardware.

Most operating systems on large computers, and many on medium and small computers, implement some sort of virtual memory. Some of the latest microprocessors now have hardware facilities to assist an operating system in supporting virtual memory. Whatever the size of computer, virtual memory has a profound effect on programming technology. The following paragraphs discuss the motivation for virtual memory, how it is implemented, and some performance considerations.

Motivation

There are two underlying reasons for virtual memory. The first is efficient memory management. Early attempts at memory management allocated contiguous regions (i.e., address ranges) of memory to each competing program. But this often resulted in fragmentation of the memory in such a way that major parts of it were temporarily lost. Consider an example in which two programs are running, each with a 100-kilobyte region, as shown in Fig. 60.10*a*. When the two programs end, there are two 100-kilobyte unused holes that can accommodate two new 100-kilobyte programs. But unless the holes are contiguous, they cannot accommodate a single 200-kilobyte program, as shown in Fig. 60.10*b*. In computer systems that support several programs running at a time, it is quite common for the memory to become fragmented and develop inaccessible holes.

The second and more important reason for virtual memory is programming efficiency. Virtual memory allows programmers to be more or less ignorant of how the memory hardware, that is, the

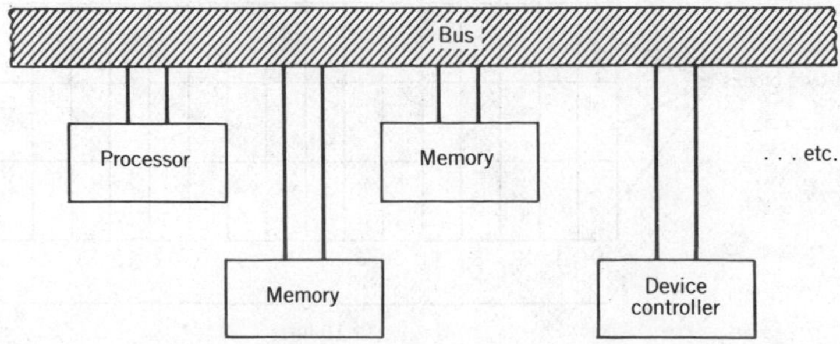

Fig. 60.9 Bus architecture.

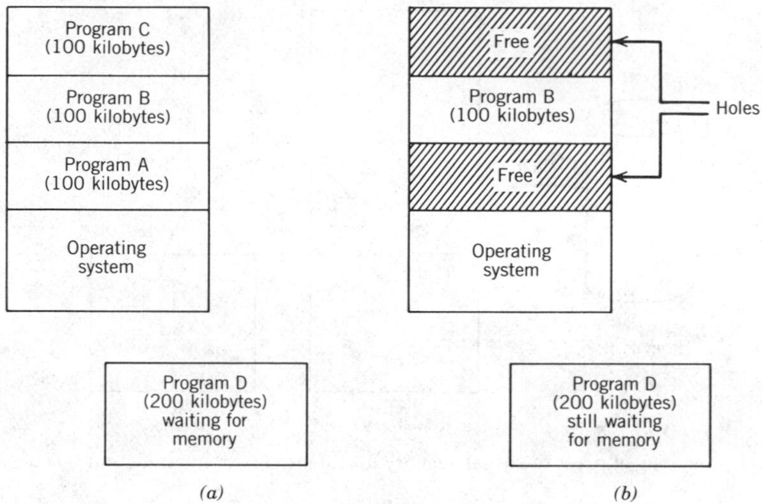

Fig. 60.10 Memory fragmentation: (*a*) Programs A and C running, (*b*) two free 100-kilobyte holes that are not contiguous and thus cannot accept the waiting 200-kilobyte program.

real memory, is configured. It allows programmers to write programs that run on many differently configured systems. It allows a system to be reconfigured without need for modifying any programs. It allows programmers to spend their time assuring that the logic of a program is correct instead of trying to squeeze the program into real memory. All of these advantages are taking on increased importance, since programmers have become relatively scarce and expensive compared with hardware.

Paging

Paging,[14] which is more properly called *demand paging*, is one of two technologies for implementing virtual memories. The other technology is called *segmentation*.[15] Today segmentation is less common than paging, and the discussion of virtual memory here will concentrate on paging.

The basic idea behind paging is that the operating system automatically keeps active parts of a program in memory and inactive parts out on a drum or disk. When an inactive part is needed, the operating system moves it into memory, and in exchange moves some other part of the program out. To ease the task of managing this continuous swapping of parts of a program in and out of memory, the system conceptually divides all of memory into fixed size units called *pages*.

Consider an example. Suppose a computer has a real memory of 1 megabyte, while it is desirable for programs to behave as if there were 16 megabytes. That is, we seek to implement a 16-megabyte virtual memory. The operating system conceptually divides the 1-megabyte real memory into pages. In a typical case, each page is a 4096-byte unit (4 kilobytes), so there are 256 pages in the 1-megabyte real memory.

The operating system also divides the 16-megabyte virtual memory into 4-kilobyte pages, and there are 4096 such pages (16×256) in the virtual memory. Obviously, all 4096 pages of virtual memory will not fit in real memory at one time. Therefore the operating system reserves space on a disk or drum to hold pages from the virtual memory that do not fit in real memory at any given moment. The situation is depicted in Fig. 60.11.

Now suppose that a program that is larger than 1 megabyte begins execution. Although the program is large, in *any given interval* it uses only a relatively small amount of memory. The part of its memory that a program actually uses in an interval is called the program's *working set*.[16] The operating system will arrange things so that the program's working set, which is a number of pages, is loaded into real memory. The rest of the memory that the program "thinks" it has is kept out on the paging disk or drum.

At some point the program will address an area of virtual memory that is not loaded into real memory. When that happens, the hardware generates a *page fault interrupt*. The operating system gains control and suspends the program while reading the required page into real memory. When the page has arrived, the operating system returns control to the program and processing continues. The

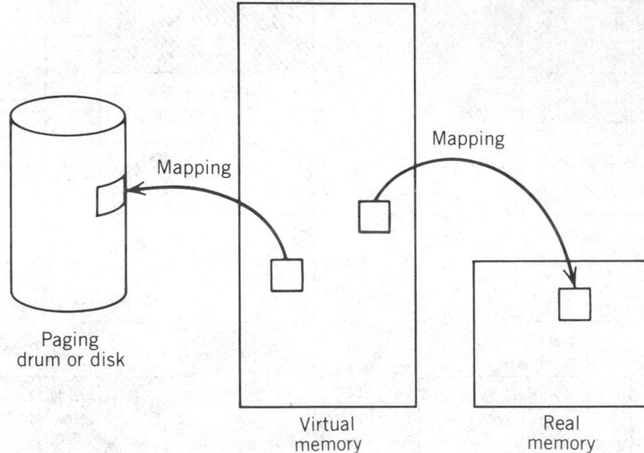

Fig. 60.11 Mapping of pages of virtual memory to real memory or input–output (I/O) devices.

program remains unaware that anything unusual has occurred (unless, of course, the program is sensitive to real time and can sense that it has been interrupted).

At some later point, the program will generate another virtual address that is not represented in real memory. But this time it may occur that the entire megabyte of real memory is already committed to various pages. In that case, the operating system will make a decision on which page to eject from real memory to make room. The usual choice is the page that has been *least recently used* (LRU), on the grounds that it is the least likely to be missed. That page is copied out to the disk or drum, and thus space is freed for a page in real memory. The required page is then read in. Then the operating system redispatches the program to proceed, still unaware that an underlying paging mechanism is allowing it to address more memory than is physically present.

The generalization to a situation where many programs share a computer is straightforward. At any moment the program that is executing may generate a page fault. When that occurs, the operating system will read into real memory whatever page is required. To do so the operating system may have to choose a page, which may belong to another program, to eject. Furthermore, while a program is suspended waiting for its page to come in, other programs can be dispatched to use the CPU.

Obviously there are many parameters and complexities to a paging mechanism. There are various algorithms for choosing which page to eject when space is required in real memory. The algorithm described here is called global LRU. There are many possible choices of page size; the 4-kilobyte size used in the example is only one such size.[13]

When a program refers to an address in virtual memory, that address may be mapped anywhere in real memory using the paging mechanism. Indeed, two addresses that the program "thinks" are contiguous could easily be mapped into places that are far apart in real memory. Yet the real memory takes addresses literally. If a program requests address "x," the memory will presumably retrieve address "x," even though the operating system has read the real data in a page at address "y."

What happens is that the memory subsystem is augmented with an *address translator*. Whenever the program generates an address in virtual memory, the translator consults a set of page tables maintained by the operating system and converts the address to the proper one in real memory. Only then is the address passed on to the MAR and decoders.

Address translation can be a slow process, since the hardware must perform a table look-up for every address that every program generates. That kind of overhead would be unacceptable, so the translator usually has an array of *associative registers* sometimes called a *translation lookaside buffer (TLB)*. When a program generates an address, the translator first consults the TLB. If an address in the same page was generated recently, the TLB already contains the translation. Otherwise the translator resorts to the page tables. Because the TLB is an associative memory, the look-up is almost instantaneous. In fact, it can be overlapped with other addressing operations so that no time is lost at all in the case of translations found in the buffer.

Similarity to Cache Memory

The description of paging given here should be familiar because it is the same sort of process that was discussed for a cache memory (Section 60.4). The cache is a high-speed store that at any moment contains the most likely to be used sections of real memory. Similarly, real memory is a high-speed store containing the most likely to be used sections of virtual memory. And in both cases, addresses are intercepted and translated associatively so that the memory read or write is directed to the fastest possible hardware.

Continuing the analogy to a cache, virtual memory works because most programs exhibit *locality of reference*. If they refer to a location in memory, it is highly likely that they will soon refer to another location that is nearby. In the case of the cache, nearby means in the same block. In virtual memory, nearby means in the same page.

Performance of Virtual Memory Systems

With either paging or segmentation, the performance of a virtual memory system depends on the operating system making "good guesses" as to what should or should not be kept in real memory at any moment.[17] Performance also depends on having enough real memory. Otherwise the operating system will constantly be moving things out to make room, and many of those moves will be against its "better judgment." If something is moved out of real memory too soon, the program that uses it will encounter an additional page fault sooner than desirable. In the extreme case, programs will encounter page faults almost as soon as they are dispatched, and the operating system will spend all of its time moving things into and out of real memory. Such an extreme case is called *thrashing*. The entire system essentially comes to a halt because the real memory, that is, the hardware, is overcommitted.

References

1 D. J. Theis, "An Overview of Memory Technologies," *Datamation* (Jan. 1978).

2 H. J. Gray, *High Speed Digital Memories and Circuits*, Addison-Wesley, Reading, MA, 1976.

3 D. N. Becker et al., *Toward More Usable Systems—The LSRAD Report*, SHARE Inc., California, 1980.

4 Digital Equipment Corp., *VAX11/780 Technical Summary*, Massachusetts, 1978.

5 IBM Corp., *The IBM System/38*, IBM General Systems Division, New York, 1978.

6 H. Hellerman, *Digital Computer System Principles*, McGraw-Hill, New York, 1973.

7 Intel Corp., *The Semiconductor Memory Book*, Wiley, New York, 1978.

8 M. Pfister, *Data Processing Technology and Economics*, Santa Monica Publishing, Santa Monica, CA, 1976.

9 A. H. Bobeck and H. E. D. Scovil, "Magnetic Bubbles," *Scient. Am.*, pp. 78–91 (Jun. 1971).

10 D. Toombs, "CCD and Bubble Memories," *IEEE Spect.* **15**(4) and **15**(5), April and May, 1978.

11 J. Minker, "An Overview of Associative Memory or Content-Addressable Memory Systems and a KWIC Index to the Literature: 1956–1970," *Comp. Rev.* **10**:453–504 (Oct. 1971).

12 A. V. Pohm et al., "The Cost and Performance Tradeoffs of Buffered Memory," *Proc. IEEE* **63**(8) (Aug. 1975).

13 P. J. Denning, "Virtual Memory," *Comp. Sur.* **2**(3) (Sep. 1970).

14 IBM Corp., *Introduction to Virtual Storage in System/370—Student Text*, IBM manual GR20-4260, New York.

15 J. H. Saltzer, *Traffic Control in a Multiplexed Computer System*, Ph.D. thesis, MIT, 1966.

16 P. J. Denning, *Resource Allocation in a Multiprocess Computer System*, Ph.D. thesis, MIT, 1968.

17 Y. Bard, "Performance Analysis of Virtual Memory Time-Sharing Systems," *IBM Syst. J.* **14**(4): 366–385 (1975).

CHAPTER 61
COMPUTER SYSTEM INPUT–OUTPUT (I/O)

SAM GOLDWASSER

University of Pennsylvania
Philadelphia, Pennsylvania

61.1 INTRODUCTION

The transfer of information between digital systems—whether for the purpose of communications, storage, control, or human operator interaction—represents one of the most challenging and diverse areas of electrical and computer engineering. From the point of view of design, as well as analysis and understanding, input–output (I/O) represents a rapidly changing field with both evolutionary (improved performance) and revolutionary (fundamentally new technologies) developments occurring regularly.

The punched card reader, magnetic tape unit, and line printer no longer dominate the computer room environment. Today devices as diverse as gigabit mass storage systems; interprocessor communication links; analog transducers; process controllers; image and graphics scanner, recognition, and display equipment; speech and music analyzers and synthesizers; and much more, represent the rule rather than the exception. The advent of the microprocessor has had a substantial impact by opening up a wide variety of applications areas requiring the development of specialized I/O devices and interfacing techniques.

From the computer system standpoint (Fig. 61.1), I/O comprises only one of the four principal components of a computing machine. However, the ramifications of I/O extend this simple representation and may be dealt with on several different levels, including the following:

The physical and electrical properties of the actual I/O device.

The interface or controller with which the central processor interacts.

The software and operating system support of the I/O environment.

The I/O unit may in actuality consist of many individual I/O *interfaces* or *controllers* and/or may

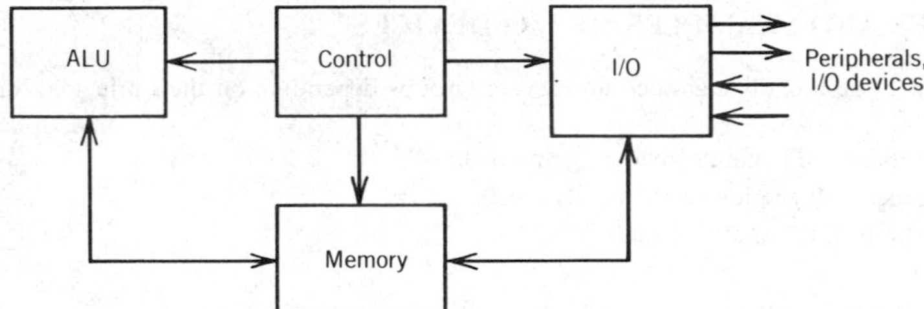

Fig. 61.1 Digital computer organization. ALU, arithmetic–logic unit; I/O, input–output unit.

include a separate special-purpose *I/O processor*—distinct and independent—but managed by the main CPU.

The I/O problem in general can be distinguished from the other logical and electrical considerations in a data processing system because it necessitates a *conversion* in one or more of the following areas:

Speed. Data transfer rate, digital sampling, synchronous and asynchronous requests.

Logical. Data format, coding, protocol.

Electrical. Signal levels, modality, analog-to-digital (A/D) and digital-to-analog (D/A).

Physical. Electromechanical, optical, audio, or other functions.

I/O operations almost always require a change of speed or *synchronization* between the central processing unit (CPU) and the I/O device. This imposes restrictions for both the hardware and software. For example, the most familiar peripheral, the video display terminal (VDT), can typically be updated over a serial communications line at a rate not exceeding 960 characters per second (cps); a line printer (which involves electromechanical components) at 2000 cps; or an electric motor controller at 10 cps. However, a typical CPU can process instructions hundreds or thousands of times faster than this and should be able to manage multiple I/O devices (simultaneously) and perform other computational functions as well, instead of waiting for each I/O operation to complete.

I/O transactions always involve cooperation between the CPU and peripheral (hardware and software) and enforcement of a logical *protocol* often involving interlocked (request–acknowledge or master–slave) data transfer communications. Overall specification and design of the system will influence logical data format (e.g., 80 columns per record on a Hollerith punched card) or coded representation (e.g., 12-bit left-justified signed numbers from an A/D converter). In addition, nearly all I/O eventually involves a fundamental change in the electrical or physical representation of information. Within the computer, data and control signals are usually in the form of voltage levels or pulses consistent with a given logic family (i.e., transistor-transistor logic or emitter-coupled logic). The I/O operation often involves a conversion from or to analog, mechanical, magnetic, audio, or other form for storage, transmission, or display.

In this chapter we will deal with the I/O field from three primary points of view:

Hardware. Interface characteristics, behavior, and limitations of typical I/O devices and their controllers.

Software. Support and control of I/O at both the device interface (machine) and operating system levels of implementation.

Design. Overall guidelines in the specification and selection of I/O implementation techniques for effectively matching applications requirements with system capabilities, particularly in relation to minicomputer and microcomputer interfacing.

We will emphasize general principles rather than the details of all available I/O devices and peripherals currently on the market. Additional information may be found in references dealing with computer system architecture and organization. Manufacturers' literature contains the theory, operation, and programming of their specific devices.

61.2 TYPES AND EXAMPLES OF I/O DEVICES

I/O devices may be broadly classified into several groups depending on their principal functions:

Communications. Terminals, readers, printers, links.
Mass Storage. Magnetic disks, tape, data cell.
Analog. A/D, D/A, audio, visual.
Interactive. Digitizers, sensors.
Display. Imagery, graphics, process control.

In many cases a particular peripheral spans several classes in its principal function. For example, a terminal includes aspects of communications, interaction, and display technology. In addition, the actual implementation of a physical I/O device usually requires multiple types of technology.

Some of the devices most frequently encountered are described in subsequent sections with respect to user characteristics and implementation.

Communications Devices

Table 61.1 lists characteristics of the most common peripherals that are heavily communications oriented. Note that we include traditional I/O devices such as punched card readers and line printers in this classification.

The *terminal* is probably the most common and visible peripheral representing the primary means of interacting with modern computer systems. The terminal consists of a typewriterlike keyboard for data input and a printer or video display monitor for data output. The printer version is usually referred to as a hard-copy unit, while the cathode ray tube (CRT) is designated soft copy. Depressing a key on the terminal generates a unique digital code which is transmitted to the computer to which it is attached. In addition to the normal alphanumeric keys (upper and lower case), special function and control keys facilitate user interaction with the computer program. On a printing terminal, codes sent by the computer cause their associated characters to be printed or to perform such functions as carriage return, linefeed, backspace, and so on.

In a video display terminal (VDT) a blinking cursor signifies the physical location at which the next received character will be displayed. This cursor, usually an underscore or a box, moves as characters are received but it may also be positioned with special code sequences and special keys. The most common VDTs can display 16 to 24 lines of text. More advanced versions include additional off-screen memory and may implement scrolling, programmable windowing, special-purpose character sets for limited graphics, and multiple color capability.

Modern VDTs generally employ a raster scan display format similar to that of commercial TV, forming characters out of a 5×7 or 7×9 dot matrix, as shown in Fig. 61.2. More sophisticated character fonts are utilized when display image quality is important, such as in the graphic arts or publishing fields.

Printing terminals are utilized when hard copy is desired. The original teletype machine—the workhorse of the industry for many years—has been replaced by faster, quieter devices. Impact printers generally employ either a dot matrix or a daisy wheel mechanism. The daisy wheel printer produces good type quality and is often used as the output device for word processing systems.

The *punched card*, long symbolic of the computer age, is an example of a batch or noninteractive medium. The common Hollerith format card actually predates the invention of the programmed digital computer by a substantial margin. *Punched cards*, *punched paper tape*, and *magnetic strip cards* are generally of a fixed format. The required physical organization of punched holes or the magnetic pattern is standardized and determined by the electromechanical or optical design of the handling devices.

Modern punched media readers are of photoelectric design with light-emitting diode-photodiode assemblies used to sense the pattern in a single column while the media is moved through the mechanism. Magnetic card readers sense the flux transitions representing the coded information.

Magnetic cards are most useful when a small amount of information is adequate and the need may arise to modify it. The most familiar examples of such information is the coded stripe on most credit cards, the program cards for some handheld calculators, and the automatic ticket and fare collection systems on modern mass transit.

A wide variety of printers are available, characterized by speed, performance, and implementation technology. Low-speed printers or character printers (which are often associated with keyboards as terminals) are generally one of three types:

TABLE 61.1. TYPICAL COMMUNICATIONS DEVICES

Device	Speed[a]	Capacity	Technology	Interface	Comments
Terminals					
TTY	10 cps	80 c/line	Electromech	Serial-intr[b]	ASR33 type of teletype machine
Printer	30 cps	132 c/line	Dot matrix	Serial-intr	5×7 or 7×9 impact or thermal
Cathode ray tube	960 cps	80 c, 24 lines	Raster scan	Serial-intr	Video display terminal
Readers					
Card	600 card/min	80 c/card	Mech-photo	Parallel-Intr, DMA[c]	Hollerith or similar format
Paper tape	500 cps	10 c/in.	Mech-photo	Parallel-Intr	1 in., 8-column paper tape
OCR	1000 c/min	Typed pages	Optical	Serial-Intr	Optical character recognition
Magnetic card	1 card/s	200 c/card	Magnetic	Parallel-Intr	Magnetic stripe credit cards
Printers					
Dot Matrix	200 cps	132 c/line	Impact	Parallel-Intr, DMA	Low-cost moving carriage
Line	200 line/min	132 c/line	Impact	Parallel-DMA	Disk, drum, or chain printer
Page	1 page/s	Variable	Photo-xerox	Parallel-DMA	High-performance page printer
Daisy wheel	60 cps	132 c/line	Impact	Serial-Intr	High-quality text processor
Print/plot	6 p/min	2048 dots/line	Electrostatic	Parallel-DMA	Combination printer and plotter
Facsimile					
Graphics	1 page/min	2048×2048	Photo-thermal	Parallel-DMA	Text and graphics transmission
Photo	1 page/min	145 dots/in.	Laser-thermal	Parallel-DMA	Newsphoto distribution

[a] c, characters; cps, c per second.
[b] intr, interrupt.
[c] DMA, direct memory access.

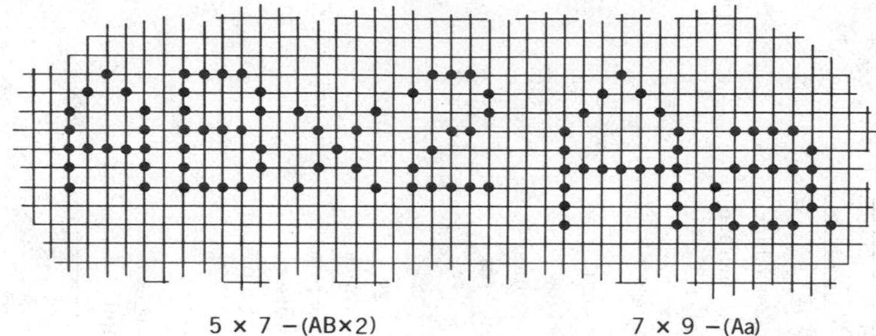

Fig. 61.2 Dot matrix characters.

Dot Matrix Impact. These form characters from a 5 × 7 or 7 × 9 dot matrix. Although the actual print head contains only seven (or nine) pins and solenoids, each character is formed through five (or seven) successive impacts through an inked ribbon. The typical speed of these devices is about 30 cps.

Dot Matrix Thermal. These use electrically heated pins to activate thermally sensitive paper. They are otherwise similar to the impact printer.

Type Ball or Daisy Wheel. In these a molded plastic, chromium plated, or metal type element provides all characters and symbols for a given font. The element is electromechanically positioned and caused to strike an inked ribbon. Typically, speeds are in the range of 15 to 100 cps. The typing elements are easily changed to enable the use of special symbols (such as math or engineering, etc.) or of different type styles.

High speed *line* or *page* printers are used as batch output devices. An entire line or page of text, symbols, and/or graphics is printed in a single operation. High-speed printers are available in the following types:

Drum or Disk. In a *drum* or *disk* printer, a continuously rotating cylinder provides an individual complete character set for each of 132 print positions. For each line, solenoid-driven hammers behind the paper are activated at just the right instant so that the desired character is passing in front of the paper and the inked ribbon.

Chain. *Chain* printers utilize a continuously moving horizontal chain instead of a cylinder. Otherwise, the principles are similar with the entire character set being repeated several times on the continuous length of chain.

Optical. *Optical* printers form characters from stored film images with appropriate positioning and sizing. CRT printers and typesetters form characters on the face of a special cathode ray tube while laser printers use a high resolution raster scanned laser beam. The development of the image in either case is usually based on a xerographic process. Optical printers are usually referred to as page printers.

Dot Matrix Printer/Plotters. These use an entire row of electrostatically activated stylii (1024 or 2048) and can print from an internally stored character set or can plot arbitrary graphics on a point-by-point (but a line at a time) basis.

Facsimile. These devices enable the input and output of arbitrary graphics and/or photographic image data. A typical unit might support 145 lines per inch of resolution and a 32-level tone scale. Some devices are designed exclusively for efficient transmission of documents and include sophisticated internal coding and bandwidth compression capabilities.

Communications Codes

For the devices just described to be generally useful, standardized coding definitions and formats must be employed. Three such codes are Hollerith, EBCDIC (Extended Binary Coded Decimal Interchange Code), and ASCII (American Standard Code for Information Interchange).

The original punched card used (and still uses) the Hollerith coding format, with letters and numerals represented by a 2-out-of-12 pattern of holes and control and punctuation represented by other patterns of two, three, or more punched-hole patterns.

The most popular code for use with terminals and printers today is *ASCII*, unless the devices are IBM compatible, in which case EBCDIC still reigns supreme. The ASCII code uses 8 bits to represent up to 256 possible characters and control combinations, although subsets of 64 or 128 (6 or 7 bits) are very common.

Slow and medium-speed terminals and printers are interfaced via a serial (bit sequential) asynchronous data transmission mechanism. A single character frame, shown in Fig. 61.3, typically consists of a start bit, a fixed number of data bits (typically 8), an optional parity bit for error checking, and 1 or 2 stop bits (1 for a start bit and 0's for stop bits). Each of these "bits" is of fixed and accurately controlled duration known to both the transmitter and receiver. The *start* bit signals the receiver, which lines up or synchronizes its internal timing with the center of each bit period. The *data bits* follow and are sampled close to their midpoints. The *parity bit* (if used) permits detection of single bit errors. The *stop bits* signify the end of frame and enforce a separation between successive characters. On the actual communications, line and logic levels are inverted with 0 corresponding to a positive voltage and 1 corresponding to a negative voltage ($\approx \pm 8$ to ± 12 V). Also shown in Fig. 61.3 are examples: for the character A $= 101_8 = 01000001_2$ and the character 2 $= 62_8 = 00110010$. Since the least significant bit (LSB) appears first (note that the line inverts the logic levels), A is represented as 10000010 and 2 is shown as 01001100. The start and stop bits are also shown inverted.

Typical common data rates in bits per second or *baud* range from 110 or less for mechanical teletypes to 9600 or more for video display CRT terminals. The actual maximum character transmission speed corresponding to these values would thus be 10 to 873 cps (for 8 data bits, no parity, and 2 stop bits).

Mass Storage Devices

The three principal applications areas for mass storage technology relate to (1) *temporary* or *swapping* storage for implementation of virtual memory management systems; (2) *on-line* file, program, and data access; and (3) *archival* storage and transfer of information. The characteristics of some of the most common mass storage devices are summarized in Table 61.2.

Fixed-head magnetic disks (and drums) are rotating media that store binary information encoded as regions of magnetization in a ferrous or metallic coating. An electromagnetic read–write (R/W) head is associated with each *track*, as shown in Fig. 61.4. Each track is further divided into *sectors* by angular position. The smallest addressable unit is usually a disk *block*, corresponding to the physical sector. A typical fixed-head disk may have 100 tracks, 32 sectors per track, and 256 16-bit words per sector, for a total capacity of 819,200 words. In a fixed-head device, the *access time* is simply the delay from the initiation of the data request to the time when the start of the desired block passes under the read–write head, since head selection is done electronically and instantaneously.

The cost per bit for fixed-head devices is relatively high because of the large number of read–write heads and associated electronics. However, the access time is relatively short, since there is no head movement. The current role of the fixed-head disk or drum has largely been assumed by solid-state technology owing to its competitive cost and faster access. Several such technologies are in use or under development.

Bulk memory (either core or solid state) is simply a large conventional RAM which, through the use of a special controller, is interfaced as a block-addressable device indistinguishable from a disk by the software. Access time (to the start of the data transfer) is zero and transfer time is determined by bus bandwidth.

Magnetic bubble devices store information as small (typically 1 μm) circular regions of reverse magnetization in a monolithic magnetic garnet crystal.[1] These bubbles (because of their appearance

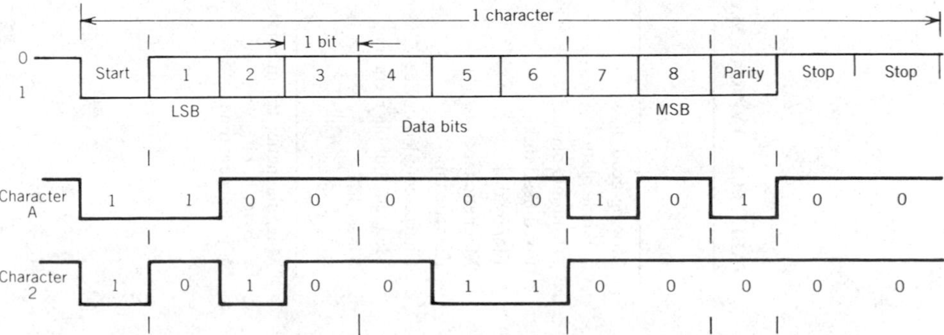

Fig. 61.3 Serial communications code format and examples. LSB, least significant bit; MSB, most significant bit.

TABLE 61.2. TYPICAL MASS STORAGE DEVICES

Device	Medium[a]	Capacity[b]	Access Time		Transfer	Application
			Seek	Latency[c]		
Moving-head disk						
Single platter	Mag	10 MB/pack	50 ms	8.3 ms	4 μs/word	General file and program storage
Multiple platter	Mag	300 MB/pack	50 ms	8.3 ms	2 μs/word	File storage and large data bases
Fixed-head disk	Mag	1 MB/drive	0	8.3 ms	10 μs/word	Temporary storage/swapping disk
Floppy disk	Mag	256 kB/diskette	0.5 s	83 ms	40 μs/byte	General storage for microsystems
Magnetic tape	Mag	40 MB/reel	30s*		5μs/byte	Archival storage/standard transfer media
Cassette tape	Mag	256 kB/cassette	30s*		300 μs/byte	Low-cost storage device
Charge-coupled device	IC	64 kb/chip	0.1 ms*		0.2 μs/bit	High-speed store/special applications
Magnetic bubble memory	IC	92 kb/chip	2 ms*		10 μs/bit	Temporary storage (nonvolatile)
Video disk	Opt	1.2 GB/disk	1 s	16.6 ms	3 μs/word	Emerging very large capacity storage device
Bulk memory	IC	4 MB/unit	0	0	2 μs/word	Fixed-head disk replacement

[a] Mag, magnetic; IC, integrated circuit; Opt, optical/laser.
[b] B, byte; b, bit.
[c] *, just access time.

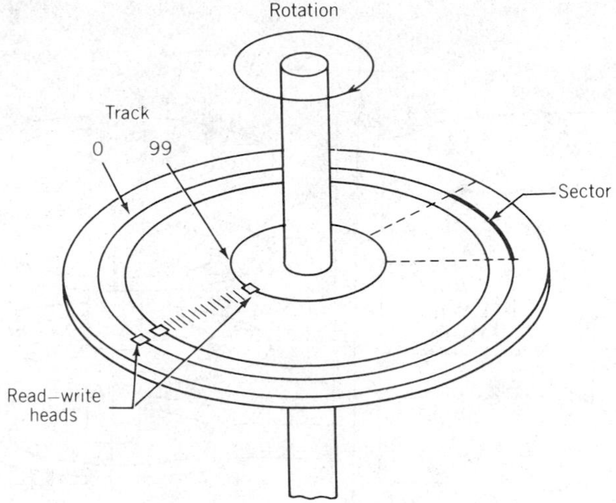

Fig. 61.4 Fixed-head disk drive.

when viewed under a microscope in polarized light) may be created, manipulated, and sensed through a combination of external rotating magnetic fields and permalloy patterns photolithographically produced on the crystal surface. A typical device can store several hundred thousand bits with the presence or absence of a bubble in a given position in a series-parallel arrangement of shift registers indicating a binary 1 or 0, respectively. Typical access time is a few milliseconds with a transfer rate of several hundred kilobits per second.

Charge-coupled devices (CCDs) store data as charge packets that may be moved through a configuration of shift registerlike arrays.[2] CCDs may be used to store finely quantized as well as analog data in addition to binary bits, since the magnitude of the charge is closely preserved as the packets are transferred through the array. Unlike the magnetically based devices, the information in a CCD is volatile—it disappears when power is turned off.

On-Line File Storage

The most common file system device is the moving-head disk drive (MHDD), which is illustrated schematically in Fig. 61.5.[3,4] As with the fixed-head disk, information is stored as regions of magnetization on the surface of a rotating medium. However, instead of individual R/W heads for each track, a single head is used for the entire surface of each side of each *platter*. The R/W heads, which float a few microinches above the disk surface on a layer of air (noncontact air bearing), are all mounted on a movable carriage that can position them simultaneously over any desired radial track location using a linear (usually voice-coil-actuated) motor. A single drive unit may use a fixed (permanently mounted) and/or removable *disk pack* consisting of from 1 to 10 or more platters (2–20 + surfaces). The collection of tracks (one above the other on each surface) simultaneously accessible without head movement is called a logical *cylinder*.†

The actual time required to access a particular quantity of data has three major components:

$$T_A = T_S + T_L + T_T$$

where T_S = *seek time* (time required to physically position heads over desired cylinder)

T_L = *rotational latency* (time needed for disk to rotate so that desired starting sector is passing under R/W heads)

T_T = *transfer time* (time required for actual number of words requested to be transferred to/from main memory)

Switching between tracks on the same cylinder is performed electronically and is instantaneous. However, if head movement is required during a data transfer, the cylinder-to-cylinder seek and subsequent rotational latency must also be considered.

†Note that the tracks of the fixed-head disk are actually logical cylinders for a disk pack with only one surface.

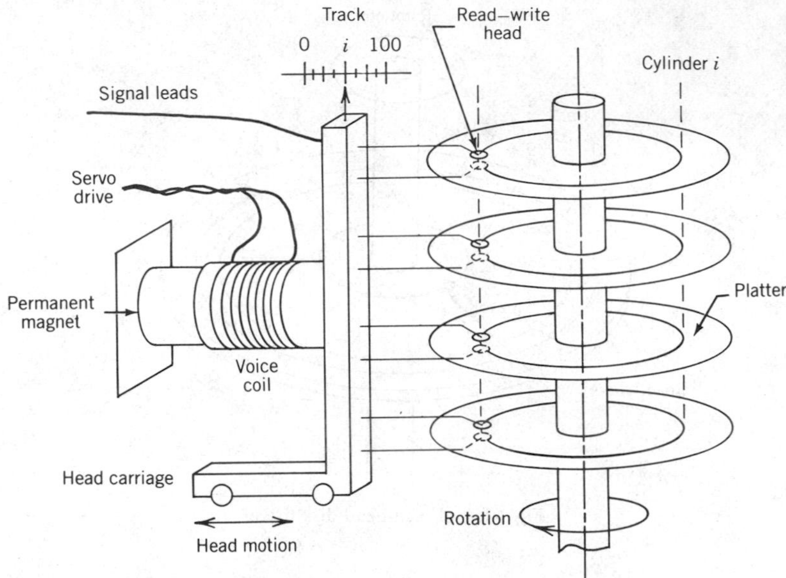

Fig. 61.5 Moving-head disk drive.

A typical disk file of this type may consist of 10 platters with 19 data surfaces (one for timing and control), 500 cylinders, 32 sectors per track, and 256 16-bit words per sector, for a total capacity of approximately 78 *million* words. Typical performance includes an average seek time of 50 ms and an average rotational latency of 8.33 ms (3600 rpm).

Floppy disk systems represent the low-cost cousins of the MHDDs discussed. The flexible diskette consists of a material similar to magnetic tape that rotates inside a protective sleeve. The speed of rotation is about one tenth that of a hard disk with the single R/W head in physical contact with the medium during actual data transfers. A stepping motor positioner controls head movement. As a result of their low cost, floppy disks are extremely popular in microprocessor-based systems and for personal computers.

Magnetic tape represents a medium widely used for archival storage as well as a common, standardized vehicle for transfer of data between computer installations. The most common tape medium is 1/2 in.-wide oxide coated with up to 2400 ft or more on a single reel. Either seven or nine tracks of information are recorded across the width of the tape in parallel.

Unlike the disks already described, tape is predominantly a sequential access storage device with no fixed block addressing structure. The smallest addressable unit of information on tape is the *record*, which may range in size from a few bytes to many kilobytes. With recording densities as high as 6250 bytes/in., over 150 megabytes can be stored on a single reel at low cost. Access to any given piece of data may require substantial time (i.e., 30 seconds or more).

Other variations of the basic magnetic tape concept are used for special-purpose and military applications. Tape cassettes are popular with microcomputers because of their convenience and low cost.

Data cell memories permit random access to data stored on individual magnetic strips, which are mechanically located and loaded onto a high-speed rotating drum. These systems provide very large on-line capacity (10^{12} bits or more) and fairly rapid access times—a few seconds or less.

Magnetic Coding Techniques

Each type of magnetic recording medium requires its own coding format to ensure data integrity and to optimize performance.

Every sector (block) recorded on a magnetic disk generally consists of three parts: header, data block, and error control, as shown in Fig. 61.6. The header includes synchronization information and

Header	Data block (256 words)	Error control

Fig. 61.6 Disk sector format.

special control such as access protection. Thus a block may be marked as read only, for example. The data block consists of the actual stored information. The error control area implements error detecting and, perhaps, correcting capability. A modulo-N ($N = 2^{16}$) checksum will reliably detect all single errors and virtually all multiple errors. Hamming or other parity check codes and more sophisticated convolutional codes are often used to permit detection *and* correction of single or multiple bit errors. The actual location of the start of the sector is usually determined by (or derived from) physical references (notches) on the disk pack through a photooptic or magnetic sensor.

The recording format on magnetic tape (Fig. 61.7) consists of physical beginning-of-tape and end-of-tape markers (reflective strips), a series of data records of arbitrary length separated by interrecord gaps, and a logical end-of-tape marker following the last record. Protective leader and trailer (unrecorded, about 10 ft) regions precede and follow the actual data area. Logical end-of-file markers follow each file. The interrecord gaps are of sufficient size to permit the tape motion to be stopped within these regions, usually about 1/2 in. Actual file directories are rarely kept on the tape itself. A specific file is located by skipping forward or backward over intervening files rapidly to locate the desired data.

Coding techniques employed in the actual magnetic recording of information on disks and tapes are usually one of two types. The first of these, typified by nonreturn-to-zero (NRZ) requires a separate clock track to recover timing information. Moreover, the bandwidth of the recorded data can vary over a wide range, depending on the pattern of 1's and 0's being accessed.

Self-clocking codes such as double frequency modulation (DFM) were developed to solve both of these problems. The timing information is derived from the data by ensuring that transitions always occur at regular intervals (at least once per bit cell), permitting a phase-locked loop to lock to the data stream. Further, the required bandwidth is now restricted to a 2 : 1 range instead of having to extend down near DC. More sophisticated techniques restrict the bandwidth requirements even further.

The two most common formats currently in general use are 800 bits per inch (bpi) NRZ and 1600 bpi phase encoded. Typical tape speeds range from 12 inches per second (ips) to greater than 150 ips.

Graphics and Image Displays

The development of the technology for the presentation of graphical or pictorial information represents one of the most active areas of I/O system-related research.[5–7]

A typical graphics display system, shown in Fig. 61.8, consists of three major components: (1) a hardware interface to permit direct access to the contents of the host computer memory; (2) a display

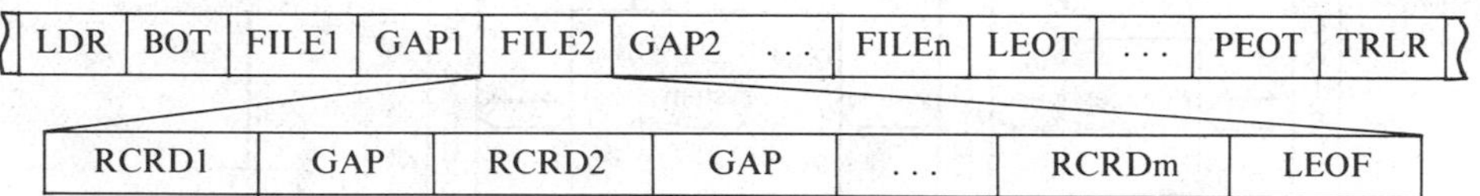

Fig. 61.7 Magnetic tape organization. LDR, unrecorded leader; BOT, beginning of tape marker; LEOT, logical end-of-tape; PEOT, physical end-of-tape marker; TRLR, unrecorded trailer; LEOF, logical end-of-file.

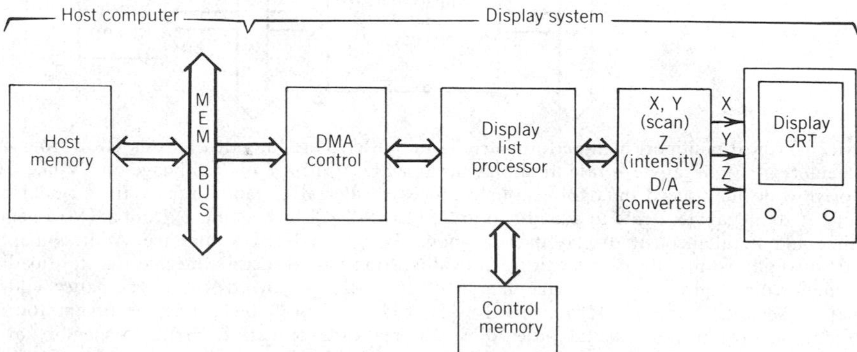

Fig. 61.8 Typical random scan display processor organization. DMA, direct memory access; D/A, digital to analog; CRT, cathode ray tube.

list processor for interpreting a graphics data structure representing vectors, conic sections, and so on; and (3) a scan converter or similar device for interfacing to the actual CRT output device.

A *display list* or *display program* residing in the main memory of the host computer is interpreted to produce a display on the screen that is refreshed many times per second. Typical display instructions generate arbitrary vectors, conic sections, or other more complex geometric figures. Graphics subroutine capability permits the construction of complex, modular structures from simpler primitives.

To minimize objectionable flicker on the CRT screen, the entire display must be refreshed 30 or more times per second. However, if a storage CRT or other form of intermediate buffering is employed, update is only required when a particular region of the display needs to be modified.

The type of system just described is usually implemented with *random scan* techniques where the CRT electron beam can be explicitly directed to any point on the screen. *Raster scan* systems (of which commercial TV and the VDT are the most common examples) refresh the CRT screen in a fixed pattern, a constant number of times (typically 30 or 60) per second. Typical *frame buffers* display the contents of a *dual port memory* with each point on the screen corresponding to a single bit or bit field in memory whose address is a function of instantaneous beam position. More advanced systems support hardware zoom, pan, arbitrary rotation of the image, and the formatting of the image to permit the display of multiple segments or windows of independent data.

A typical multiport system is shown in Fig. 61.9. In addition to the capability of accessing data from the computer and displaying it on the CRT, real-time input from a TV camera or some other video source is possible and a feedback loop (from display output to refresh memory input) permits high-speed image processing to be performed and viewed as it takes place.

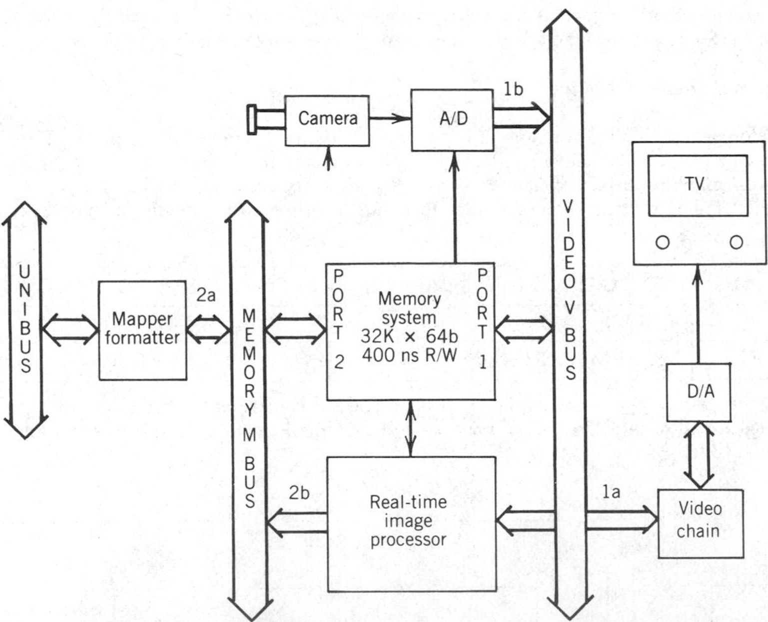

Fig. 61.9 Typical multiport high-performance frame buffer. Port 1a, video refresh. Read from dual port memory at 10 megabyte/s rate to display up to a 512 × 512 × 8 bit/pel image on TV-like CRT and/or provide high-speed data for real-time processor. Port 1b, frame grab. Write to dual port memory at 10 megabyte/s rate to acquire up to a 512 × 512 × 8 bit/pel image from a TV camera in real time and simultaneously display data as above. Port 2a, UNIBUS interface. Address mapped formatted window into display memory, providing transparent access mechanism for loading, retrieving, or direct processing of image data at up to a 1.25 megaword/s rate. Formatter/address translator effectively supports 128/256/512 × 256/512 × 1/2/4/8 bit/pel image organization in either normal or transpose spatial order. Port 2b, real-time processor. Write to memory at 10 megabyte/s rate after processing of data read-out of Port 1 either from memory or camera. Examples of desired computations: spatial filtering (convolution), tone scale transforms, edge location, segment movement, and feature extraction. A/D, analog-to-digital; D/A, digital-to-analog.

TABLE 61.3. TYPICAL SPECIFICATIONS FOR RASTER SCAN FRAME BUFFERS

	Low-Cost	Image Displays			High Performance
	Video Games	Monochrome	Color		Text Processing Multiple Format
Horizontal resolution	64	512	512	800	512–1280
Vertical Resolution	64	512	512	1000	512–1024
Tone scale	3 bit	8 bit	3×8 bit	1 bit	3×8 bit + 4 overlay
Independent segments	1	2	2	1	Many, list driven
Tone scale maps	None	1 + PC[a]	3	None	One/color/segment
Image dynamics	None	Limited	Limited	None	Zoom, pan, rotation, etc.

Video output: Standard TV, 30 frames/s; high bandwidth TV, 60 frames/s

[a] Usually provides pseudocolor (PC) map as option. PC map translates each tone of gray scale into program-specified arbitrary color.

Since the frame buffer is essentially a digital storage tube, a special-purpose microprocessor front end can be used to implement intelligent graphics capability. Frame buffers are most effective for the processing and display of multitoned or continuous-toned gray-scale imagery by use of several bits in memory at each point to represent an arbitrary brightness level. False or pseudocolor mappings from gray level and full color capability are easily achieved.

Some typical specifications for common raster scan frame buffers are summarized in Table 61.3.

Analog Devices

Digital-to-analog (*D/A*) *converters*, such as that illustrated in Fig. 61.10, output a voltage, current, frequency, or other quantity that is a function of a digital number input. In its simplest form, a voltage output D/A interface consists of a flip-flop register, which can be loaded by the CPU, and a buffered D/A converter module.

For high-speed real-time applications where uniform sampling is desired, a double buffer register or first-in first-out (FIFO) can be used, loaded from the CPU at its convenience and read.

Analog-to-digital (A/D) converters enable input of sampled data signals. A typical A/D subsystem, as shown in Fig. 61.11, permits multiple channels of analog data to be sampled and converted in

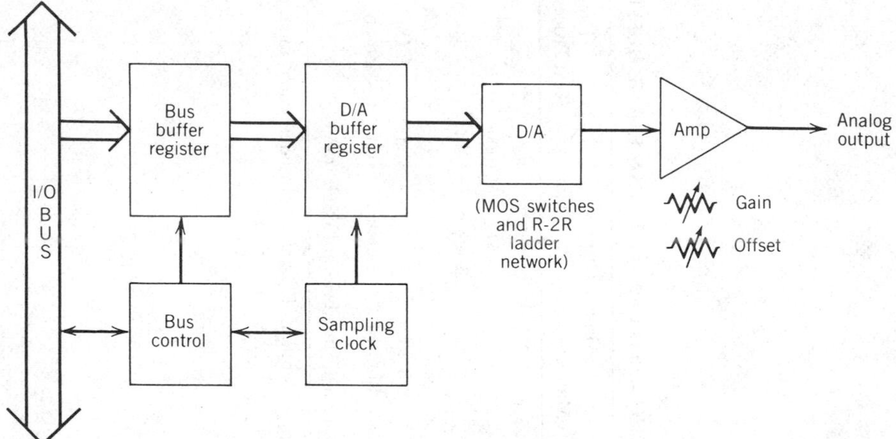

Fig. 61.10 Digital-to-analog (D/A) convertor interface. I/O, input–output; MOS, metal oxide semiconductor.

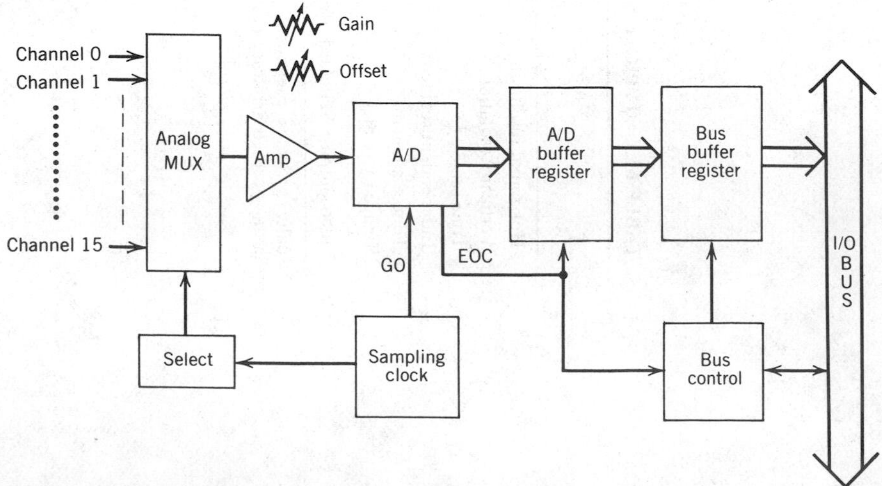

Fig. 61.11 Analog-to-digital (A/D) data acquisition system interface. I/O, input–output; MUX, multiplexer.

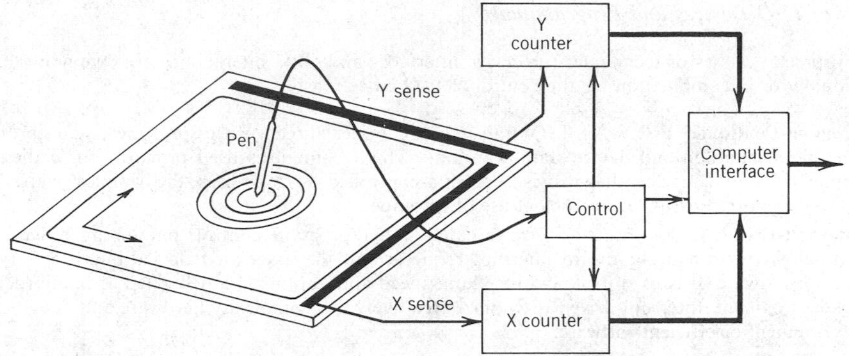

Fig. 61.12 Graphics digitizing tablet.

sequence or random order. In addition to linear quantization, floating point, logarithmic, or other companding functions are often utilized.

Interactive controls such as joysticks, trackballs, light pens, and digitizing tablets are all devices that convert mechanical position into digital coordinates. *Joysticks* and trackballs are mechanically operated devices that usually consist of a mechanism linked to potentiometers (one for each axis) and a multiplexed A/D. A *light pen* is a photosensitive wand that permits an operator to make a selection or point to an object on a CRT screen. The position of contact is easily determined, since the beam X, Y coordinates are known at the time when light output is detected. With a *graphics digitizing tablet*, the position of a probe on an X, Y rectangular tablet surface (see Fig. 61.12) is sensed and made available to the CPU. In the simplest form of the tablet, a probe emits an acoustic or electromagnetic pulse. The delay time of arrival of this pulse at the X and Y linear transducers is determined and made available to the CPU. More sophisticated systems include a microprocessor to perform coordinate conversion and correction for transducer nonlinearity.

61.3 GENERAL SYSTEMS CONSIDERATIONS

Many considerations influence the selection and specification of an I/O device for a particular application. The ultimate choice of the hardware interface and its associated software programming techniques is also determined by many factors, including:

Data Rate. How quickly must information be transferred, formatted, or processed? Must the system operate in real time? Will the requests occur synchronously (periodic) or asynchronously (random)? Can the CPU determine the timing of data transfers? For example, output to a terminal or printer is generally non-real time and asynchronous—the CPU can determine the rate of data transfers as long as some specified rate is not exceeded. A magnetic tape or disk unit is synchronous for the duration of the transfer of a block or record—the CPU must maintain a fixed rate or risk losing data as a result of timing errors.

Mode of Access. Is the information bit, byte, word, sequential, or block oriented (like a disk), or is nonsequential or random access required? Must the value of each data item be monitored as it is transferred (i.e., testing for control characters like end of line or escape sequences in the case of a terminal)?

Applications Environment. Will this be a dedicated processor or a multiprocessing system? Are the requirements very specific or is generality for future expansion or modification desirable?

Electrical/Physical. How far away is the actual I/O device to be located from the CPU? Is it an electrically noisy environment (as in an industrial plant with large machinery)? Factors like these will influence the choice of a data transmission technique: serial vs. parallel, fiber optic or coax, and so on. What physical conditions are required by the device (temperature, humidity, vibration)?

Cost Tradeoffs. What will the budget support in terms of equipment expenses, custom hardware design engineering, support software development, testing, and maintenance?

The hardware implementation and the method of software control to be employed will depend to a large extent on the particular application and the answers to questions such as these. For simple low-speed noncritical I/O, almost any technique will suffice. However, for the high-data-rate, high-reliability application, cost–performance tradeoffs will be of paramount importance.

Types of I/O Devices and Programming

Four principal classes of techniques (or device interfaces and CPU interaction) are commonly used individually or in combination for the control of I/O data transfers.

Program-Controlled I/O. All I/O transfers are under direct processor supervision through command initiation and device status checking. This technique is used primarily for dedicated applications and device diagnostics. Program-controlled I/O utilizes the simplest hardware interface, but represents an inefficient use of resources.

Interrupt-Driven I/O. Functions are initiated under program control, but timing is handled through asynchronous hardware interrupt requests and the associated device interrupt service routine software. It is used for low- to medium-speed applications in a dedicated or multiprocessing environment. Interrupt-driven I/O utilizes relatively low complexity hardware and permits the development of efficient software.

Direct Memory Access (DMA). I/O transfer of an entire block of data is set up under program control and implemented through the use of special-purpose hardware that transfers data directly to or from main memory. This occurs without further CPU intervention, concurrently with program execution. DMA supports medium to high-speed devices very efficiently and is quite efficient, but requires a more complex hardware interface. Common uses include magnetic tape or disk controllers, interprocessor links, and high-speed graphics processors.

Address-Mapped I/O. Data transfers are performed between the CPU and the logical region of the program address space, which is actually part of an I/O device, such as an image display or a bus window rather than main memory. This technique supports both program manipulation of data and direct transfers between the I/O device and mass storage. Address-mapped I/O represents an efficient, transparent mechanism for use with structured zero-latency devices. Note that this is *not* an extension of DMA but rather an alternative technique available for certain applications.

Note that most (nontrivial) devices can (must) be controlled using a combination of the techniques just summarized. A disk controller, for example, may be initialized and interrogated using mapped program access to its internal registers. It performs the actual block transfers using direct memory access and signifies function completion or an error condition by issuing an interrupt request. For diagnostic and maintenance purposes, program-controlled access is often utilized.

For most commercial devices, the manufacturer selects or recommends the primary method of control. However, the engineer faced with the task of designing a special-purpose interface must evaluate the alternatives and tradeoffs. With the present (and future) trend in the declining cost of digital hardware, factors such as the required software overhead and flexibility may prove more significant than the initial cost of physical hardware. For example, DMA I/O, while quite versatile and applicable to a wide range of speed requirements, necessitates a substantial software overhead in terms of operation and error recovery. In later sections, we will deal with each of the four techniques in more detail from the perspectives of hardware implementation, software support, and engineering design.

61.4 I/O SYSTEM BUS STRUCTURES

The overall I/O configuration is generally the least standardized section of a computer system. Although only a small number of processor options is usually available, the selection of the type and number of I/O devices is largely applications dependent. Furthermore, the I/O configuration is likely to change as the system is developed and expanded. It is highly desirable to be able to add or remove devices without rewiring the basic machine while at the same time minimizing cost and complexity. The fundamental method for achieving these goals in both hardware and software design is through *modularity*. The common vehicle for achieving modularity in the I/O system is the common *I/O bus* shown in Fig. 61.13, which interconnects I/O devices, the CPU, and main memory and supports flexible communications between these units. The I/O bus enables multiple devices to:

1. Transfer information over common shared data paths, reducing hardware complexity.
2. Be added or removed by simple plugging or unplugging of their cables.
3. Interface via a mechanism (the I/O bus) employing standardized logical, electrical, and physical specifications.

This standardization enables entire families of computers with various levels of performance (e.g., PDP-11's) to utilize the same peripherals regardless of the internal processor organization. In

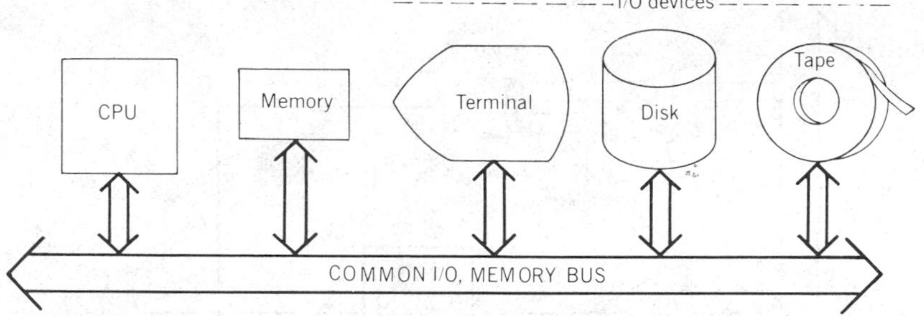

Fig. 61.13 Typical shared bus architecture. CPU, central processing unit; I/O, input–output.

addition, the designer of an I/O device interface need not be familiar with the CPU design or even know which particular machine the interface is to work with. He or she only must understand the bus characteristics.

The necessary specifications for a particular I/O bus can be classified into three categories:

1. **Logical.** Definition and grouping of related signals, logical polarity. For example, address, data, control, synchronization, and so on.
2. **Electrical, physical.** Signal levels, characteristic impedance and termination, allowable length and loading rules, cable and connector types.
3. **Protocol.** Rules for use of the bus such as timing, synchronization, handshaking, and arbitration.

These items are discussed in more detail as follows.

Logical Specifications: Data Transfer Signals

Address. Every physical device register or memory cell is assigned a unique logical location specified by a binary *address*. During a data transfer (initiated by the device in control of the bus, or *bus master*), the address selects the source (for reads) or destination (for writes) location, or *slave* device. For example, the CPU (as master) may request a read from location 1000 in main memory (the slave).

Data. Parallel paths over which actual digital information flows, typically 8 to 32 signal lines or more.

Control. Specify type and direction of data transfer. They include functions such as READ, WRITE, READ/PAUSE, WRITE/BYTE, and so on.

These signals (and those described hereafter) are usually *bidirectional*. Thus for any given transaction or cycle, information may flow from master to slave or from slave to master over the *same physical wires*, depending on the type and direction of the data transfer. Some busses multiplex more than one set of signals on the same wires (i.e., address and data on the LSI-11 Q bus) to save on hardware. Others time multiplex all signals onto a single UHF coax.

Logical Specifications: Synchronization Signals

Nearly all I/O busses utilize separate synchronization signals that implement request–acknowledge handshaking to ensure reliable data communications and support variable-speed devices such as memory of various types.

Request. Examples are MSYN, CYCREQ, and so on. The master device initiates data transfer.

Acknowledge. Examples are SSYN, CYCACK, and so on. The slave device responds, indicating completion of transfer.

A basic UNIBUS slave interface is illustrated schematically in Fig. 61.14.[8–10] The I/O "device" in this case is a read–write data register (single word memory cell). A simplified UNIBUS timing diagram is shown in Fig. 61.15. Referring to the figures, to store (write) into the data register the master (CPU) applies the address and data to be stored and asserts the R/W line, C1. It then waits 150 ns before asserting MSYN. If the addresses match (address comparison), the logical AND of C1,

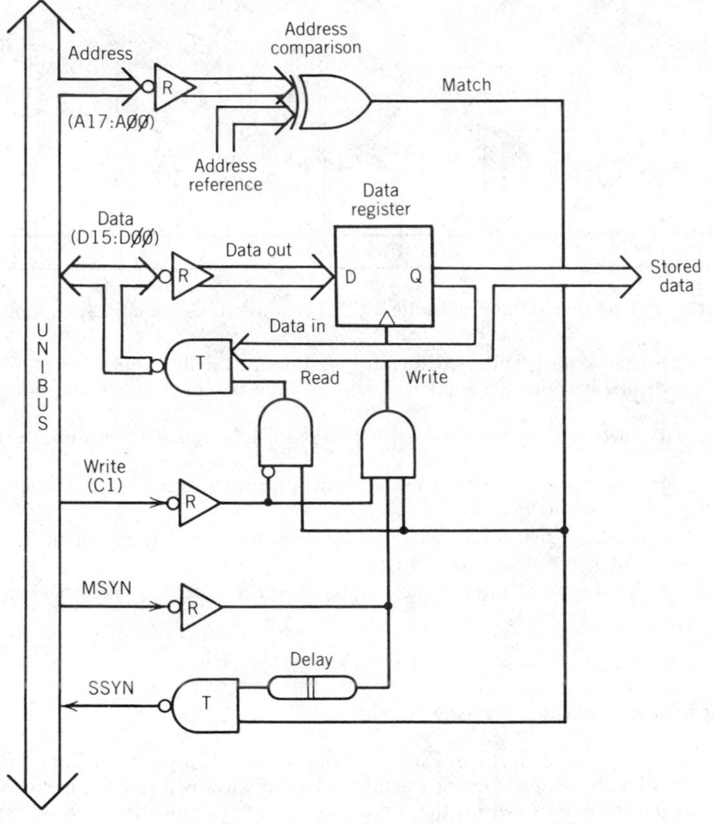

Fig. 61.14 Simplified read–write data register UNIBUS interface. MSYN, master sync (data request); SSYN, slave sync (data acknowledge).

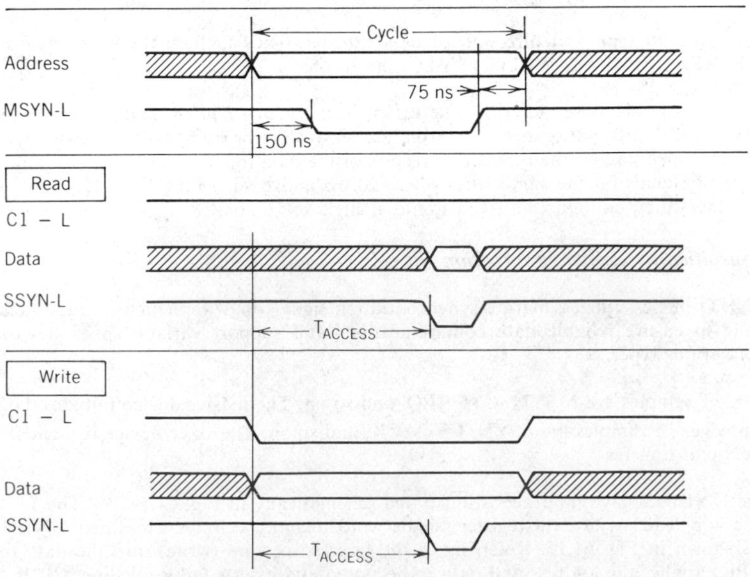

Fig. 61.15 PDP-11 data transfer timing.

MSYN (master sync-data request), and MATCH clocks the data register, loading it with the information on the data lines. In addition, the device responds with the acknowledge signal, SSYN (slave sync). To read back the stored data, the CPU applies the address, waits 150 ns, and asserts MSYN. The logical AND of negated C1 and MATCH enable the stored data to be put onto the bus by the interface, which then responds with SSYN. In either case the cycle is terminated by the negation of MSYN by the CPU and the removal a short time later of address (data) and control. The total time for a typical cycle is approximately 400 ns.

Bus Arbitration

Since an I/O bus is shared by multiple devices—CPU and I/O controllers—a mechanism is required to assign use of the bus on a time-shared priority basis. A *bus master* (the most common being the CPU) is always in control of the bus. The most common reason for devices other than the CPU to become bus master is for the purpose of performing DMA transfers to or from memory without processor intervention. Some machines (like the PDP-11 or LSI-11 series) also use the bus to transfer device vector addresses during interrupt transactions.

The arbitration is accomplished with a request-acknowledge protocol also. The device requests use of the bus (BUSREQ). At the end of the current bus cycle (or when the current bus master relinquishes control), the highest priority requesting device is granted bus mastership (BUSGNT). Bus master may now use the full capabilities of the bus to perform data transfers to or from memory or registers, and so on.

Vectored interrupt processing is similar, but the master (device) provides the address of a vector to the CPU. Note that DMA cycles may be requested at any time (except between read and write of the READ MODIFY WRITE instruction) but interrupt processing can only take place between instructions.

Implementation of Bidirectional Bus Structure

Logically a bus must provide a mechanism to permit a device in control (master) to communicate with any other addressable device (slave) over shared bus wiring. The implementation of bidirectional signals is usually accomplished in one of two ways: with an open collector bus or with a tristate bus.

Open Collector or Wired-ORed Bus. The device interface or *transceiver* for each signal consists of an open collector driver (bus transmitter) for placing information on the bus and a high-impedance input buffer (bus receiver). Refer to Fig. 61.16. Normally only one transmitter will be active at any

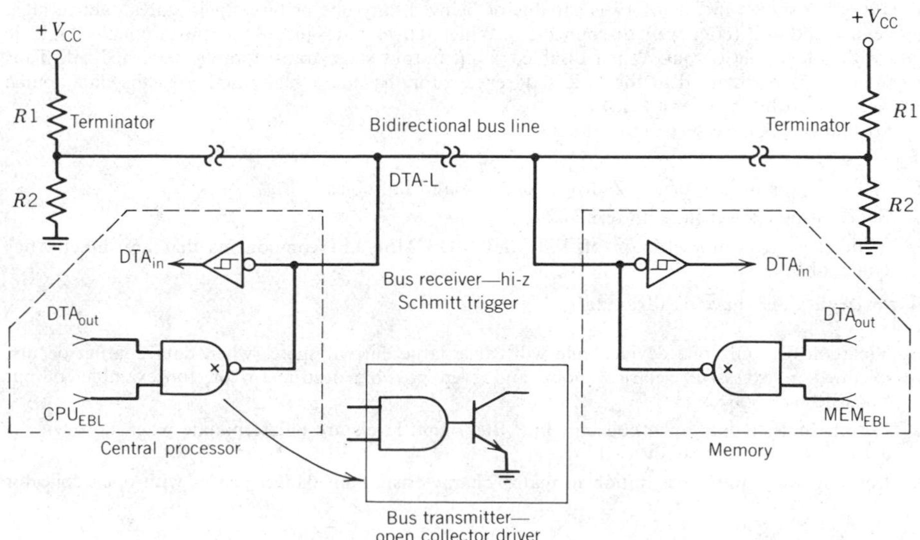

Fig. 61.16 Typical open collector bus configuration. DTA, data; CPU, central processing unit; MEM, memory.

given time but multiple devices will be listening. By convention, negative logic polarity is used for most bus signals. Thus a logical zero will be a high-voltage level and a logical one will correspond to a signal near ground potential.

Now if the CPU as master wants to read from memory (slave), CPU_{EBL} will be low and MEM_{EBL} will be high, setting up a data path from the memory data register to the CPU over the common bus data lines.

The termination serves two primary purposes: It provides the proper high signal level on the bus (typically +3 V) and it matches bus characteristic impedance to minimize reflections, ringing, and cross talk. Refer to Fig. 61.16:

$$R_1 \| R_2 \approx Z_0$$

$$V_{CC} \cdot \frac{R_2}{R_1 + R_2} \geqslant \text{logic high level}$$

Note that the impedance matching is only approximate, since most real bus configurations using ribbon cable are poor compared with ideal transmission lines.

Bus Loading. Each bus transceiver constitutes both a static (DC resistive) and a dynamic (AC capacitive) load, which limits the maximum number of devices that can be plugged into the bus without buffering. The shunt R decreases the high signal level. The shunt C increases rise and fall times and introduces discontinuities along the bus length which are sources of reflections. Bus specifications generally include loading rules based on standard bus receivers, transmitters, and cable type. For the PDP-11 UNIBUS, these are:

Maximum load	20 (load = one receiver and one transmitter)
Maximum length	50 ft total (termination $R1/R2 = 180/390$)
Transmitter gate type	7438 or equivalent
Receiver gate type	8T37 (hi-Z Schmitt trigger) or equivalent

When these restrictions are unacceptable, modules known as *bus repeaters* are available which permit additional devices to be installed on the bus and enable its physical length to be extended.

Tristate Bus. One disadvantage of the open collector bus is that the high (signal)-level drive is supplied entirely by the termination network and is purely resistive.[11] In addition, even short lengths of bus require termination at least at one end to maintain proper signal levels. An alternative technique is to use active pull-up devices for the bus transmitters (the receivers are unchanged). A *tristate* driver shown in Fig. 61.17 is capable of being in any one of three logic states—active high, active low, and hi-Z (effectively disconnected). When active, it acts just like a conventional TTL logic gate with totem-pole output. When disabled, both output stage transistors are switched off. Thus every device is maintained in the hi-Z state except for the one actually acting as the data source (master for a write, slave for a read).

Tristate busses have several advantages:

1. Active high and low for low-Z drive, good rise and fall times.
2. Short bus lengths require no termination.
3. Wide (and growing) selection of TTL and MOS MSI/LSI components that are directly bus compatible.

However, they also have disadvantages:

1. Violation of "only one device" rule will cause large current spikes when data conflict occurs, resulting in excessive electrical noise and even possible destruction of, for example, output transistors.
2. May be noisier than open collector bus. Bus signal levels are indeterminate when no device is active if termination is not used.
3. Long busses require termination to match characteristic impedance, just as with open collector bus.

Tristate busses are extremely popular with microprocessor systems, especially for internal data communications, while open collector types are often used in interfacing minicomputers.

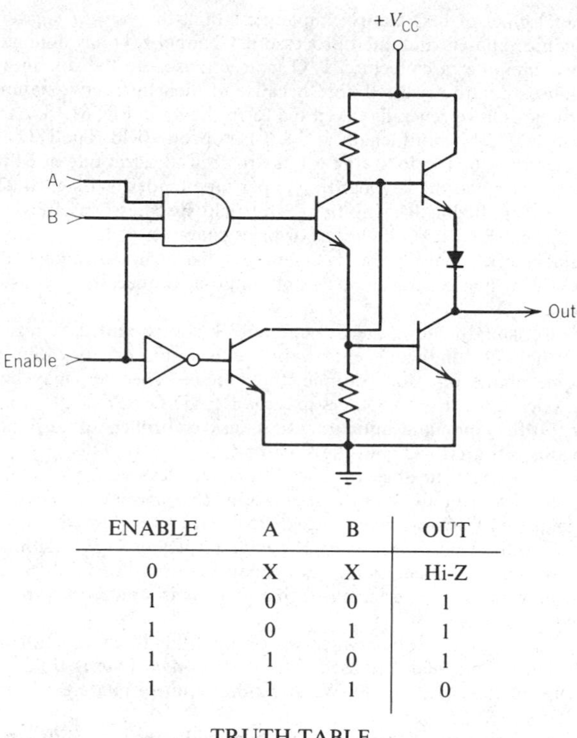

ENABLE	A	B	OUT
0	X	X	Hi-Z
1	0	0	1
1	0	1	1
1	1	0	1
1	1	1	0

TRUTH TABLE

Fig. 61.17 Simplifed tristate bus driver.

61.5 HARDWARE-SOFTWARE INTERFACE

Overall control of the actual hardware device at the most fundamental software level is generally accomplished in one of three ways:

1. *Special I/O instructions* are provided in the architecture of the CPU, which perform the basic operations required of each device.

2. *Mapped registers* in reserved address space permit all normal memory reference instructions to be used.

3. *Common I/O channel* utilizing a special peripheral processor provides a unified interface mechanism.

All I/O devices require certain classes of operations to implement their control:

Command Initiation. Set up of parameters and issuance of operation request. For a terminal, this might be "load the ASCII code for the letter A and print it"; for a magnetic disk, "write out the 256 word block starting at memory location 1000 to disk address 23450."

Status Checking. Interrogation of the state of the device to determine if the requested operation has completed, if the device is available, or if an error has occurred (and what type of error). Testing to see if a character has been received from a terminal or what type of error caused a premature abort of a magnetic tape operation are typical examples.

Interrupts. Most devices have the capability to suspend the current program running on the CPU and execute a section of code that services the immediate needs of the device. A terminal can interrupt whenever a new character (receive or transmit) needs to be processed; a disk will interrupt when a data transfer has completed or an error has occurred. Once the peripherals' needs have been satisfied, control is returned to the interrupted program exactly where it left off.

Of the three general types of I/O control implementation, the *mapped register* approach is most popular in modern minicomputers and microprocessors. Channel I/O has dominated the mainframe and high-performance computer architecture. I/O instructions were the sole means of control used on many early machines and are discussed here because of their instructive nature.

Special I/O instructions have generally taken the form shown in Fig. 61.18. As a specific example, consider the PDP-8 with a 12-bit word length and a 3-bit opcode field. The I/O class of instructions is identified by opcode group 6. The following 6 bits are used to select one of 64 possible devices. In the case of the teletype terminal, the keyboard or paper tape reader is assigned 03, while the printer or paper tape punch is identified as 04. The final 3-bit field specifies one of eight possible functions such as CLEAR, TEST, or SET FLAGS; or read or print character code.

As devices became more sophisticated, I/O instructions came to impose a severe burden on device control: They were not general enough and imposed unnecessary constraints on both the hardware and software.

Mapped I/O control simply allows the device control, status, and data registers to appear like memory locations to the CPU. Individual bits or bit fields within each location control or monitor device functions. Some of the bits, for example, the function specifier, may be accessed by both READ and WRITE operations (R/W). Others may be READ ONLY (R/O) for status checking or WRITE ONLY (W/O) for command initiation. In a disk controller, the disk address registers are usually R/W, the status bits are R/O, and the start or GO bit is W/O.

Mapped control is extremely flexible. In a well-designed device interface, all CPU instructions may be utilized to access and monitor device operation. (No special I/O opcodes are necessary.) Arithmetic and/or logical instructions may be used freely wherever desired. Only as many addresses as are actually needed will be occupied by each device (typically 1–32). Although these represent (logical) space on the system I/O memory bus, the penalty is almost certainly insignificant. From the hardware designer's point of view, interfacing in this manner is straightforward and almost entirely machine independent.

The register map for a communications interface on the PDP-11 computer is shown in Fig. 61.19. In this case several kinds of bit fields are used. The *receiver data* byte is R/O, the *transmitter data* byte is W/O, interrupt enables (IE) are R/W, and other control bits are a combination of access types.

One of the objectives of using *channel I/O* is to provide a more unified software (as well as, perhaps, hardware) interface for various I/O devices. The control required of a channel is generally more involved than that required of any single device. However, it is very flexible and essentially entirely device independent, a claim that is rarely made for interfaces using mapped registers. Nevertheless, the techniques available for setting up and controlling channel operations include those already discussed.

Operating System and Device Driver

The user of the modern computer system rarely comes in direct contact with the low-level hardware registers of I/O devices. Almost invariably, in anything larger than a very dedicated microprocessor the details of control and synchronization of I/O transactions are handled by the operating system

OPCODE	Device Specification	Function

Fig. 61.18 Input–output (I/O) instruction format. OPCODE identifies this as an I/O instruction. Device Specification Selects the target device type and unit. Function specifies the operation to be performed.

ADDRESS	NAME	15 8	7 0		FUNCTION
777560	RDS	– – – –	DONE	IE	Reader CSR
777562	RDD	– – – –	Receiver data		Reader data
777564	PRS	– – – –	READY	IE	Printer CSR
777566	PRD	– – – –	Transmitter data		Printer data

Fig. 61.19 PDP-11 communications interface register map. IE, interrupt enables; CSR, control status register.

(OS) through the use of high-level (more or less) *device-independent* system calls. The intricacies of the individual device characteristics are managed by the software *device driver* associated with each type of hardware device. The driver program interprets parameters passed to it by the OS from the user program and performs all operations needed to manipulate the hardware registers of the device and interact with it to accomplish the requested data transfer.

The use of a device driver provides several key advantages over direct control of I/O:

Device Independence. Devices from different manufacturers as well as devices with similar operational characteristics can present an identical software interface to the high level program.

Simplification of User Programs. The applications programmer is freed from worrying about most of the details of the hardware implementation, timing, synchronization, and control.

Isolation and Protection. Through determining what can and cannot be done by the user, inadvertent I/O-induced system crashes can be avoided or at least minimized.

In its simplest form, a device driver program consists of four sections:

1. Initialization. When the device is first loaded, that is, when its services are first requested, the hardware is reset to a known state. This code, invoked by the OS, is usually executed only once.

2. Command. Execution of actual I/O functions originate from here. Such operations as READ, WRITE, SEEK, and so on, which will later require attention when they are completed, are initiated from the command section. After a function is invoked, the calling program may be resumed or suspended awaiting completion of the command.

3. Interrupt. Following completion of the requested function (or detection of an error condition), most devices will interrupt the CPU, causing the driver to be entered in the interrupt section of code to perform any post processing and/or error checking required. Completion of function will generally signal or restart the user program.

4. Special Function. Specific manipulation of the device hardware or other services not requiring completion interrupt processing is performed here. This may include operations such as setting a specific flag or loading a control table.

As an example, consider the control of a simple moving-head disk unit. The initiation section of the driver, section 1, is invoked to set the controller to a known state and reset any previous error conditions. Commands such as READ, WRITE, SEEK, or VERIFY are handled by section 2. This includes loading the actual disk controller registers with the memory (bus) address, disk address (cylinder, track, sector), words count, and function type and then issuing the GO command to initiate the actual device operation. When the operation is completed, the driver is again entered at section 3.

Guidelines for Hardware-Software Interface Design

The design engineer involved with the development of peripheral interfaces can greatly ease the burden of the programmer through thoughtful specification and implementation. Here are some considerations for mapped register-type interfaces:

Read–Write Registers. Where possible, all control functions (bits or bit fields relating to control —function select, address, data port, etc.) should possess full read and write capability. Not only does this make initial program development and debugging more straightforward, but it also greatly aids in the ease of writing meaningful, comprehensive diagnostic routines for the devices.

Relevance. Related bits or functions should be grouped together. If the target machine supports byte addressing, byte accessing should be made available to modify or test meaningful groups of bits.

Free Access. If possible, checking and, perhaps, modification of registers should be as unrestricted as timing and mode selection permit.

Consistency. The data available in the mapped registers should *always* reflect the actual internal state of the device.

Transparency. The hardware designer should try to think in terms of the software designer when specifying the architecture for a peripheral interface. Functions and operations should behave naturally, logically, and consistently. The hardware design is done only once. The I/O programmer must contend with its characteristics for much longer.

To a very great extent, effective interface design is more of an art than a science. The good design will be a pleasure to program. The poorly thought-out design, even if it is completely functional and reliable, will continue to present frustrations for the software designer. Above all, the importance of clear, concise, and complete documentation cannot be overemphasized. The functions and interactions of all of the logical bit fields must be described in detail, including usage examples. Any behavior that is at all unusual should be specifically cited.

61.6 PROGRAM-CONTROLLED I/O

The simplest type of I/O interaction between a hardware device and the CPU software is known as program-controlled input–output.[12,13] Contained within the device interface are a set of *flags* or *status bits* whose state may be interrogated by the I/O program (driver). Through this mechanism, synchronization of the software-initiated data transfers with the timing requirements of the device may be achieved. Flags generally provide the following types of status information:

READY. The I/O device is idle and awaiting a command. By testing this bit, the program can determine when it is possible to issue another command after the completion of a previous operation. Other similar flags are DONE, BUSY, and IDLE. In a communications interface, the READY bit indicates that a character may be transmitted. BUSY indicates that the device is in the process of sending a character. Note that BUSY and READY (or DONE) are not necessarily complements of one another if an error has occurred.

ATTENTION. An external event has occurred requiring action by the CPU. Depressing a key on a terminal or some other external event could set this bit.

ERROR. The device has detected a fault condition either from an invalidly specified or executed command or from a hardware error. Error conditions are sometimes classified as *hard* (invalid command) or *soft* (data checksum error or other transient failure). The specific errors are indicated by additional status information bits.

ACKNOWLEDGE. In a communications interface, the device at the other end of the line has responded to a data transmission or request.

Mode or State Information. Instantaneous condition indication to aid in program optimization. As an example, the current angular position of disk drive may usually be sensed in a disk controller to permit optimal ordering of disk requests to minimize access times.

Any single I/O device may have more than one of these flag types.

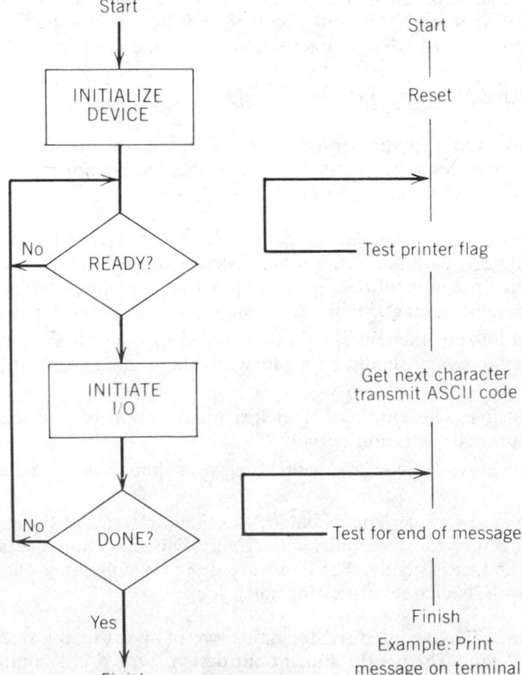

Fig. 61.20 Program-controlled input–output (I/O) operation.

Figure 61.20 illustrates the basic procedure for I/O transfer using this technique. In the specific case of a terminal, only one actual flag (READY) is necessary.

Program-controlled I/O possesses two basic advantages when compared with the techniques to be described in later sections. First, it is extremely easy to implement in hardware. Second, the principles of operation are easy to understand. However, since the processor may have to spend a great deal of time in wait loops testing flags and branching on conditions such as device readiness or operations completion, it is extremely inefficient in utilization of CPU time. In all but the simplest dedicated systems, useful processing can almost always be performed during these periods. In addition, to handle more than one device at a time necessitates *polling*—checking the flags of all active devices to ascertain requests and monitor their operation.

Program-controlled status-driven I/O techniques are rarely used in modern general-purpose computer systems. They are employed primarily for implementation of special-purpose or single-user systems where concurrency is not required, for use in the testing and diagnostics of I/O devices and their interfaces, and for use in dedicated applications where efficiency is of secondary importance. Note, however, that most or all of the functions of well-designed I/O interfaces can be exercised without resorting to more advanced programming techniques (such as interrupt or DMA). This proves extremely useful for diagnostic purposes to isolate a failure to a particular hardware function.

61.7 INTERRUPT-DRIVEN I/O

Interrupt techniques provide the capability for I/O operations and computation to proceed concurrently. Instead of checking status flags and waiting for an I/O device to accept or provide data or for some other I/O event to take place, the I/O device will signal the CPU asynchronously with the running program. The interrupt mechanism permits the I/O device essentially to force execution of a special service routine when the device requires attention, temporarily suspending the main program. When the device's needs have been met, the original program running at the time of the request is resumed.

Consider the following situation: A particular process computes a buffer containing a string of 256 characters to be sent to a 30-cps printing terminal. The actual computation requires approximately 10 seconds of CPU time. Using a simple programmed I/O approach, the total time required to compute and print would be 10 + 256/30 seconds. (See Fig. 61.21.) Processing and I/O are performed sequentially. Any increase in efficiency with programmed I/O techniques would necessitate a substantial increase in program complexity—checking flags after each character was computed, and so on.

Now (referring to Fig. 61.22) consider the same problem programmed using interrupt techniques. Instead of computing all 256 characters and then initiating the I/O, after each character has been generated it is placed in a buffer accessible to both the main program and to a low-level *interrupt service routine*. Whenever the printer is ready to accept another character, it issues an *INTERRUPT request*, which causes the main program to be temporarily suspended while saving its complete *state*. An *interrupt service routine* is entered (in the driver), which checks to see if any newly computed or previous data are in the buffer awaiting output. If there are some, commands are issued to the hardware to print another character and the buffer pointer (OUTPTR) is updated. If the buffer is empty, signifying that there are no data to print, an EMPTY flag is set. The main program is then resumed exactly where it left off.

Now observe the timing in Fig. 61.23. Instead of the processing and I/O having to take place sequentially as before, these functions can be interleaved and proceed in *parallel*. If the results of the computation occur at a faster rate than the printer can accept them, the I/O becomes the limiting factor. If the opposite situation exists, then the processing will determine the time. For this type of I/O, if the appropriate buffering is used, the total time will be very close to the maximum of *processing time* or *input–output time*. Note that we are making the implicit assumption that the program support of I/O incurs a small overhead compared with the main processing and with device latency and transfer time.

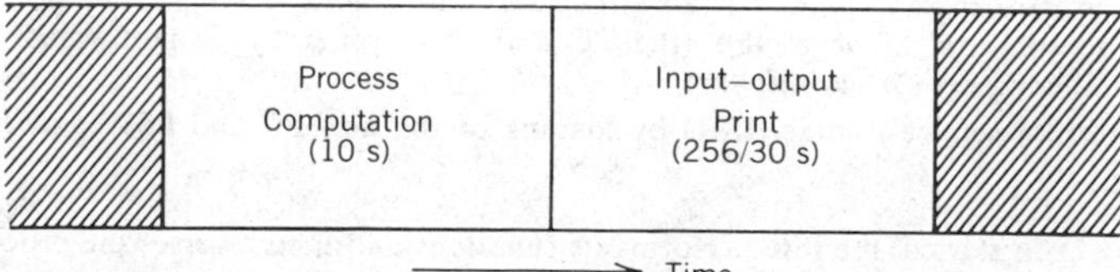

Fig. 61.21 Sequential timing using programmed input–output (I/O).

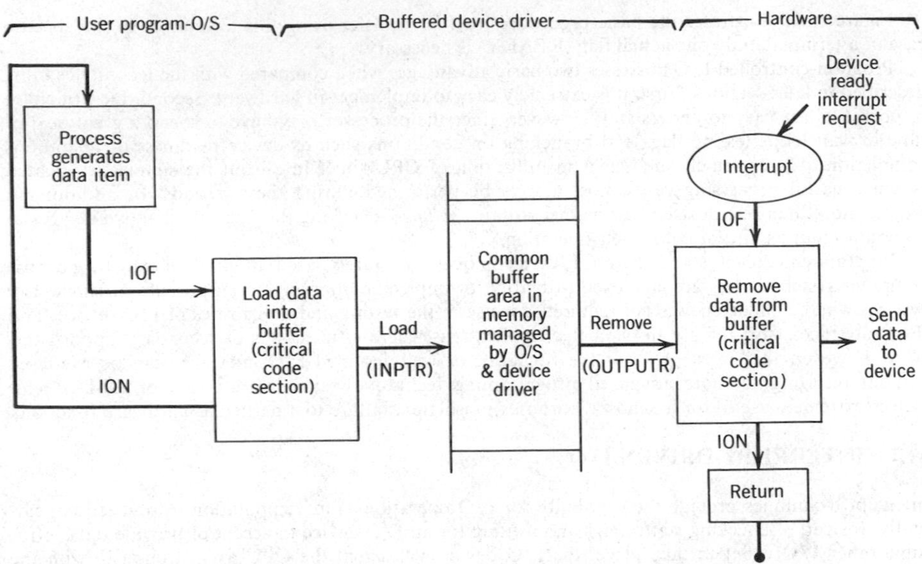

Fig. 61.22 Interaction of hardware and software for simple data transfer. IOF, interrupts disabled (off); ION, interrupts enabled (on) (this and lower priority devices). O/S, operating system.

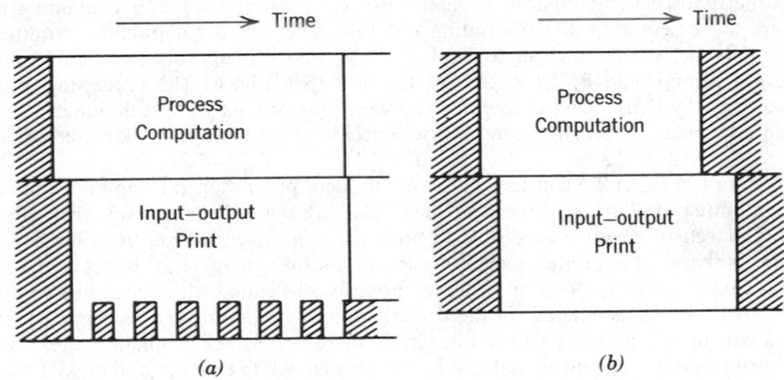

Fig. 61.23 Improved timing using interrupt-driven I/O (input–output): (*a*) $T_P > T_I$, (*b*) $T_P < T_I$. T_P = process time; T_I = I/O time (per character).

Interrupts

An *interrupt* is an *asynchronous* signal, with respect to the currently running program, by an I/O device that an event (perhaps requiring immediate attention) has occurred. The acceptance of the interrupt request by the CPU consists of the following steps:

1. *Suspension* of current program execution *in between* instructions.
2. *Saving* of the complete *state* of the interrupted program so that it can be (transparently) restored. This includes the program counter (PC), processor status word (PSW), contents of the general registers, and so forth. Some of this (the PC and PSW generally) is performed by the CPU hardware and some by system software.
3. *Start* of interrupt service routine (ISR) by loading of the new PC and PSW and continuing of processing.

After having been started, the ISR performs its functions and then reverses the process by which it was entered, thus restoring the original state of the main program at the exact point at which it was interrupted. Thus the processing of an interrupt is entirely transparent and invisible to the main program.

Interrupts are usually used to signify:

READY, DONE, ATTENTION. A single operation has been completed (like printing a character) or an I/O device wants attention.

FUNCTION COMPLETE. I/O processing consisting of several steps or data transfers (such as the writing on to a disk of a block of data) has been completed.

ERROR. A device controller has detected an error condition (such as a magtape read error) requiring immediate attention:

Interrupt Polling and Vectored Interrupts

Whenever an interrupt request is processed, there must be a method for identifying the device that caused the interrupt. In older machines (e.g., PDP-8) and primitive microprocessors, only a single interrupt request signal is available. For these machines it is necessary for the ISR to *poll* the various possible devices (as in Fig. 61.24*a*) by checking the status flags of each one to determine where the interrupt originated. This operation represents a substantial overhead, especially if a large number of devices are active at the same time.

Vectored interrupts solve this problem by providing hardware that automatically identifies the requesting device, saves the processor state, and branches directly to the appropriate device service routine (Fig. 61.24*b*). Generally a specific set of memory addresses assigned to each hardware device contains the necessary new state information. The starting addresses of this *vector* are either provided by the requesting device at the time of its interrupt request or are a direct mapping from the device (line) number.

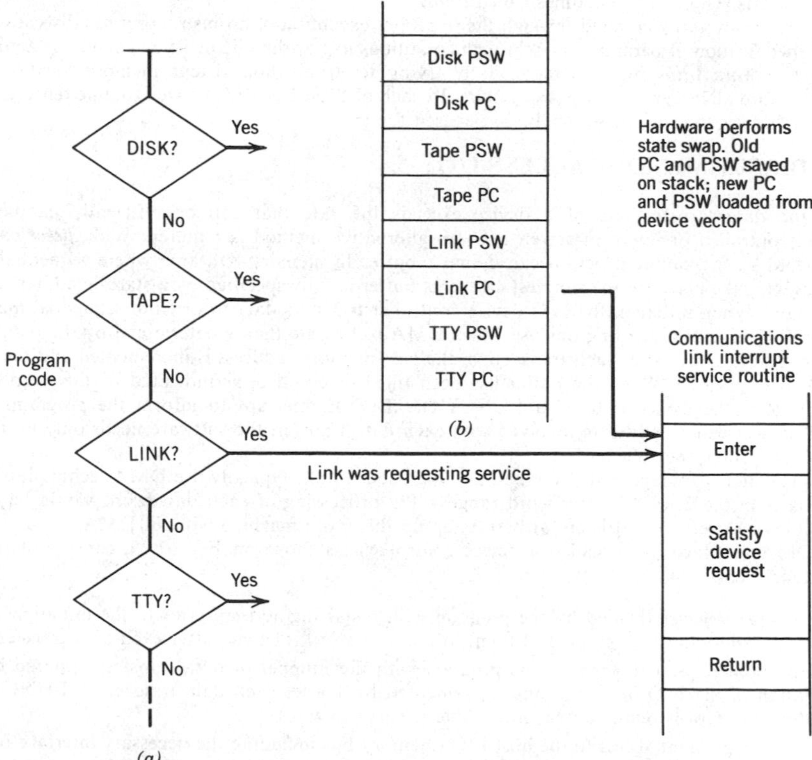

Fig. 61.24 Comparison of polled and vectored interrupt system implementation: (*a*) Interrupt request (polled system), (*b*) automatic dispatch (vectored system). PSW, processor status word; PC, program counter.

Interrupt Priorities

In general, certain devices or particular kinds of requests require more immediate attention than others.[14] For example, a high-speed device or one operating in a real-time environment will have more critical needs than a slow-speed terminal. Furthermore, under certain conditions interrupts for some devices may need to be locked out or temporarily inhibited. Therefore, devices capable of requesting interrupts usually have associated with them hardware interrupt ENABLE bit and a hardware priority for each interrupt function.

The ENABLE bit can be used to turn interrupt requests for the device on and off unconditionally. The priority of the device determines whether its request will be honored when compared with the current processor priority and which device will be serviced first when multiple physical devices request attention at the same time. Since the processor priority is software controlled, low-priority devices may be inhibited from requesting interrupts while high-priority devices are being serviced. Furthermore, to guarantee hazard-free program execution, control of processor priority may be necessary to lock out interrupts during critical sections of system or driver code.

The interrupt system of the PDP-11 family of minicomputers and microcomputers includes control flip-flops within each device to individually enable or disable interrupt requests. The (hardware) priority of each device is determined by the set of *request–grant* lines on the UNIBUS that it uses and by its physical position on the bus. The UNIBUS arbitration scheme resolves multiple simultaneous requests on a given priority by granting requests on those devices that are physically closer to the processor first. The required vector memory address for each device interrupt function is provided by the device interrupt controller during the processing of the interrupt bus cycle. The vector in memory contains the (PC) address of the service routine and the new PSW (including new priority).

Software interrupts implement the same type of behavior as device interrupts but are initiated through the use of a special CPU instruction or other similar means. They are used for various purposes, including interprocess communications, for operating system calls, and as diagnostic and debugging aids. These are sometimes called *traps*.

Exceptions are traps initiated through the improper execution of an instruction usually caused by invalid specification of parameters or through conditions like arithmetic overflow, divide by zero, and so on. *Hardware* traps may be induced by trying to access nonexistent memory locations, by improper word alignment, or by power failure. In each of these cases the service routine represents an error handler for the specific anomaly that caused the trap.

61.8 DIRECT MEMORY ACCESS I/O

When the data transfer rate of a device exceeds the rate that can be efficiently handled by program-controlled or interrupt-driven I/O, an alternative method is required. With *direct memory access* (DMA), the calling program can define a buffer in memory—an area where sequential data transfers will take place. In the simplest case, this buffer is fully specified by a starting address and a word count. When initiated, the I/O device controller transfers data to or from sequential memory locations by directly requesting bus use—thus DMA—at a rate that is determined by the peripheral device characteristics. After each transaction, the *bus* (memory) address is incremented and the word count is decremented. When the buffer has been filled or emptied, as indicated by the word count equaling zero, the device initiates a FUNCTION DONE interrupt to inform the program. Thus instead of overseeing and being involved with each data transfer, the software needs only to initiate the I/O operation once per buffer.

Devices such as magnetic disks, tapes, and high-speed links typically use DMA techniques. Their data rate is in the 2 to 20 μs per word range. CPU processing of each data word would be either impossible or extremely inefficient at best, tying up the entire machine without DMA.

A DMA interface (often called a device controller), as shown in Fig. 61.25, consists of several basic components:

Bus address register (loaded by the program with a starting address prior to the initiation of the I/O operation and incremented to point to the next word (or byte) after each data transfer).

Word count register (loaded by the program with the number of data transfers required before initiation of the I/O operation and decremented by 1 after each data transfer. A FUNCTION DONE interrupt is requested when its value is equal to zero).

Logic to implement access to the host I/O (memory bus including the necessary interface buffers and circuitry for bus request and interrupt control).

Logic to interface to the external peripheral device (includes functions such as data buffering, serial-parallel conversion, and the reception or generation of timing signals).

Because of the additional and more demanding functions required of DMA device controllers, the

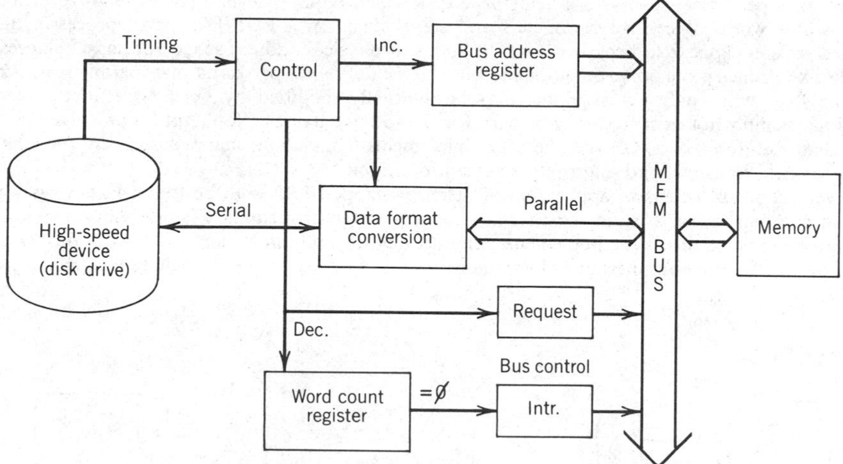

Fig. 61.25 Typical DMA (direct memory access) interface organization.

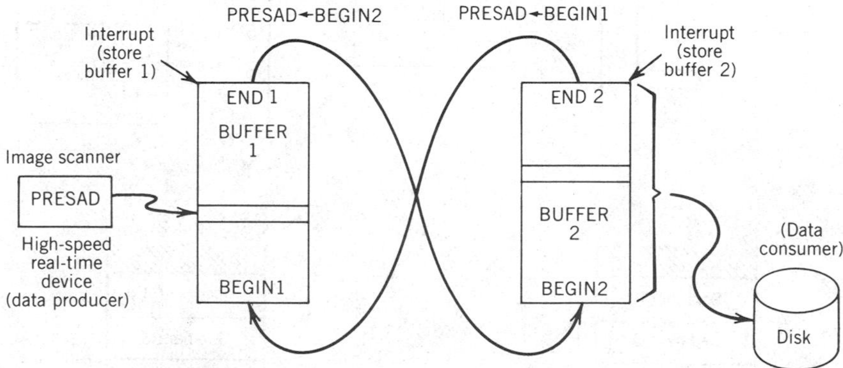

Fig. 61.26 Use of double buffering to implement real time input–output (I/O).

hardware interfaces tend to be considerably more complex. The additional cost is only justified when other techniques would impose a significant penalty in terms of efficiency or flexibility.

For applications requiring synchronous and/or real-time data transfers, such as graphics, music, or speech I/O, DMA with double buffering (refer to Fig. 61.26) provides an excellent compromise between interrupt-driven techniques and a specialized I/O processor. In this application class, double buffering implies that the device DMA controller fills (or empties) one buffer while the other is stored (or replenished) by the disk. The actual buffer switching and disk operation is managed by the software initiated by the DMA FUNCTION DONE interrupts (word count = 0) if there is sufficient time between data transfers. Otherwise a slightly more sophisticated DMA controller may include the capability to automatically handle double-buffered data structures by providing duplicate sets of bus address and word count registers in the hardware.

61.9 MEMORY-MAPPED I/O

Memory-mapped (address-mapped) or other similar direct access technique permits the most natural transparent mechanism of communicating with certain types of I/O devices. Controller registers are nearly always accessible in this manner in modern minicomputers and microprocessors. The most important major specific I/O applications of memory-mapped techniques are with respect to graphics displays, image frame buffers, and interprocessor bus links. (That is, if we are to exclude actual random access main memory as an instance of this technique.)

Briefly, the I/O device hardware interface is designed such that program interaction takes place as though the CPU "thinks" that it is accessing ordinary main memory—mapped into a region of physical address space on the bus. The simplest case is represented by the device control–status registers of a peripheral such as a moving-head disk which requires several (typically 4–16 or more) 16-bit words worth of control parameters and status flags. In a PDP-11 type of processor, these mapped registers appear to occupy several locations in physical address space and may be accessed exactly like ordinary memory. Some of the bits, bit fields, bytes, or words may permit read–write, read–only, or write–only access. Some may be internally modified by device operation, whereas others may simply not exist (i.e., may be hard wired to a logic 0 or 1). Note that for any given device, more than one form of I/O access is often implemented, including mapped registers for control, DMA for data transfers, and interrupts for synchronization.

A very common extensive application of memory-mapped I/O is to be found in devices where random program access to an internal memory or an isolated or secondary memory bus is desirable. A prime example of this is the implementation of a dual or multiprocessor system with bus window communications. Such an interbus link (as shown in Fig. 61.27*a*) will often include the capability for

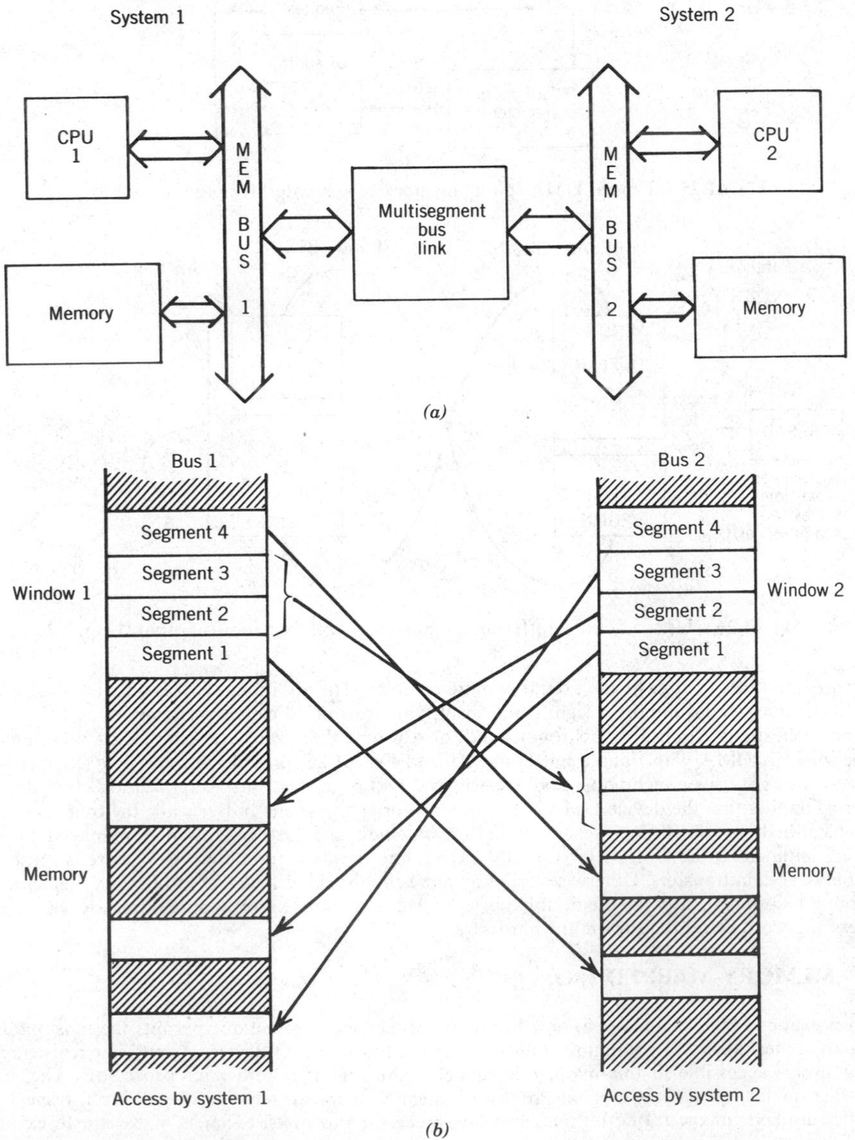

Fig. 61.27 Interprocessor bus link and possible segment map: (*a*) Configuration, (*b*) mapping determined by contents of relocation registers for each system. CPU, central processing unit.

multisegment mapping, relocation, and protection (Fig. 61.27*b*) so that a master CPU can freely access all locations on a slave bus. Internal registers (themselves accessed as a mapped region of bus I/O space) determine segment number, size, relocation offset, protection codes, and others.

In the example processor 1 (in system 1), through address translation logic, can access any portion of the bus for processor 2 (in system 2) or vice versa. There is a speed penalty, as the target (slave) bus must be requested for each data transfer.

It is important to consider certain issues which arise in systems of this type, including shared resources. Cycles and deadlocks can result if both processors attempt to access the other's bus simultaneously or, for example, if processor 1 attempts to access address space on processor 2's bus, which is currently mapped into processor 1. Therefore close cooperation between the machines is mandatory. The problems occur because separate resources are required for separate simultaneous transactions. A unidirectional link with processor 2 incapable of requesting bus 1 use or a third shared memory bus scheme (as shown in Fig. 61.28) would be acceptable.

Through the use of multiple processors, bus links, and shared memory, computer networks can be constructed that provide for efficient access to a large distributed data base.

The random access frame buffer represents another common example of a memory-mapped I/O device interface. Frame buffers are used in the image processing, graphics field to display and permit manipulation of pictures, diagrams, or complex designs on a TV-like screen.

In a typical case (see Fig. 61.29), the refresh memory contains an image, 1 byte (8 bits) for each displayed point starting at the upper lefthand corner of the screen. The data are read out at high speed through the *video chain* to produce a standard video signal and refresh the display 30 or 60 times per second.

It is highly desirable to be able to freely access the contents of the display memory at the same time that the display refresh is being performed. Free access by the CPU is not only desirable from the programmer's point of view but also permits the effects of image-processing algorithms to be observed as the algorithms progress. This capability facilitates the development of interactive systems.

These capabilities are readily accomplished by mapping the *refresh memory* into the address space of the host CPU through the *formatter*, which acts as special type of bus window. The timing is set up so that requests from the host have priority and "steal cycles" from the display. If the refresh memory is fast enough, host accesses may be interspersed between display cycles or a first-in first-out (FIFO) buffer may be used to prefetch the required display data and thus entirely eliminate any degradation of the displayed image caused by simultaneous access.

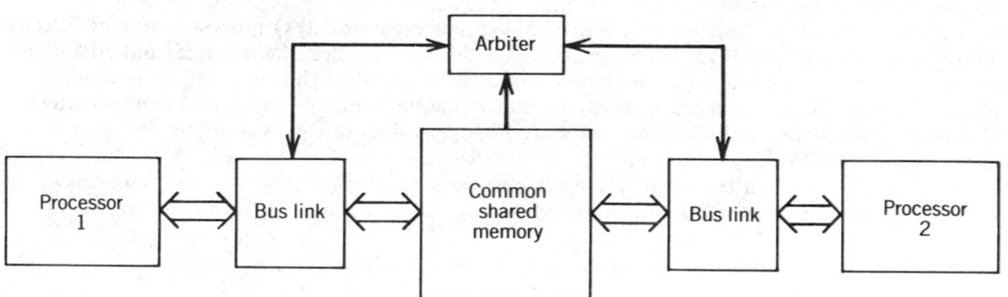

Fig. 61.28 Acceptable shared memory configuration.

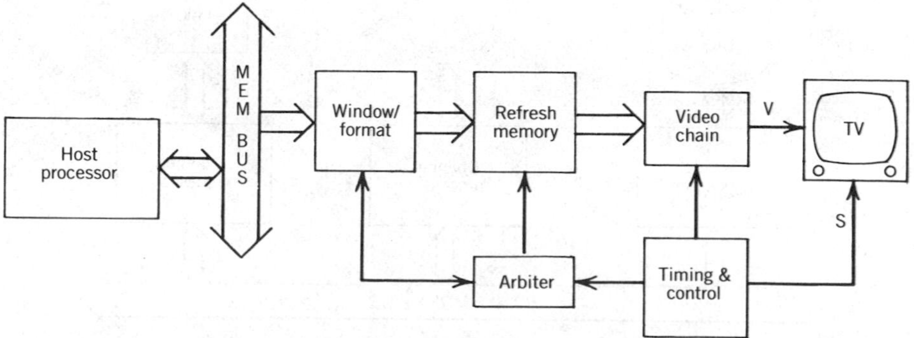

Fig. 61.29 Basic memory-mapped frame buffer.

The formatter may also perform address and data translation to make the various possible display resolutions or configurations transparent to host system software. For example, this permits direct access to either a horizontal or vertical (transpose) strip of an image.

Memory-mapped techniques are only applicable where the timing requirements of the source and destination devices are similar, since any READ access necessitates an explicit transfer of data from the slave, and the master must wait for the response. Thus it would *not* be appropriate to consider mapped access to a moving-head disk, where the seek times and rotational latency effects are significant.

61.10 MORE ADVANCED I/O TECHNIQUES

We can extend and generalize the I/O processes previously described in several ways:

Device interface uniformity permits various devices from different manufacturers and with dissimilar functions to share the same programming techniques.

Command chaining provides for the specification of multiple simultaneous or sequential I/O operations without CPU intervention.

Data chaining enables multiple noncontiguous, circular, or interleaved data buffers to be automatically linked together.

Data formatting and preprocessing incorporates automatic functions directly into the I/O hardware.

The term "*channel*" is usually used to signify a special-purpose peripheral controller designed to implement some or all of these functions.[15] (Some manufacturers use the term to simply indicate a dedicated DMA interface.) A *channel processor* is thus a high-speed controller that incorporates an instruction set optimized for I/O functions. In the simplest case, the channel processor presents an interface to the software that is identical, but fixed, for each device. In the most advanced systems, the interface format, I/O command sequence, and data buffer arrangement are fully programmable, being specified by an I/O *program* that is interpreted by the channel processor.

A "*dedicated*" channel is one in which the entire hardware interface is used for only one device or group of devices (e.g., a multiunit disk file). Figure 61.30 illustrates a system with several dedicated channels. Although there may be more than one such channel, the hardware in this case is like a very sophisticated DMA controller.

In a *selector channel* (as shown in Fig. 61.31), a single physical I/O processor may be selected among multiple (typically 16) I/O devices. Once the device is selected, the resources and attention of the channel are assigned to it for the duration of a block transfer. Thus it would be undesirable to place slow-speed devices on a selector channel that would unnecessarily tie up the channel hardware. However, this mechanism does permit the same high peak data rate as is possible with a dedicated channel or simple DMA.

Multiple devices, each being handled by its own I/O program, compete for the resources of the I/O processor in a *multiplexer channel* (Fig. 61.32). Access through the channel to memory is

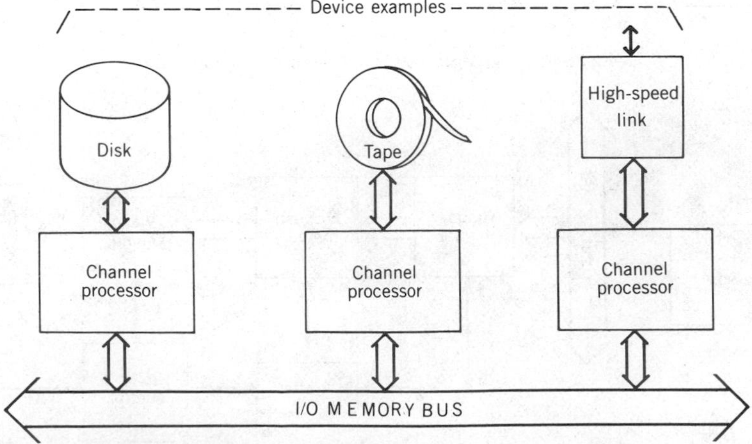

Fig. 61.30 Multiple dedicated channels. I/O, input–output.

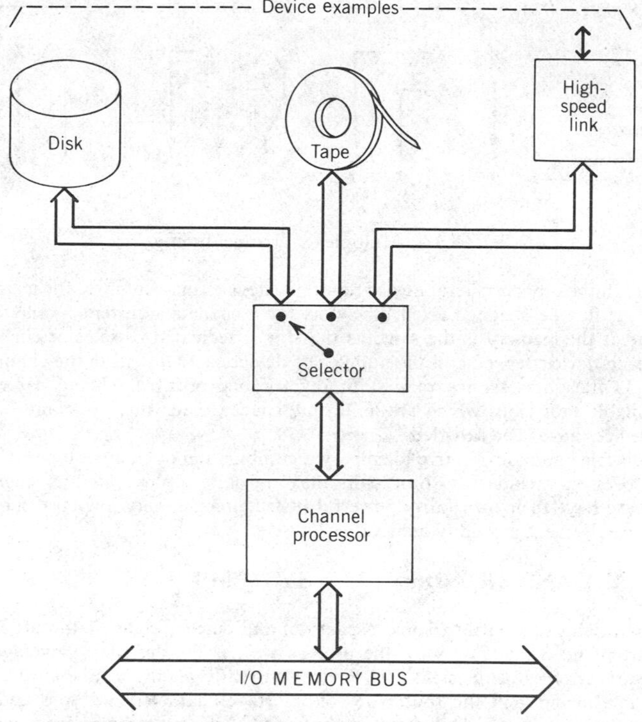

Fig. 61.31 Selector channel organization. I/O, input–output.

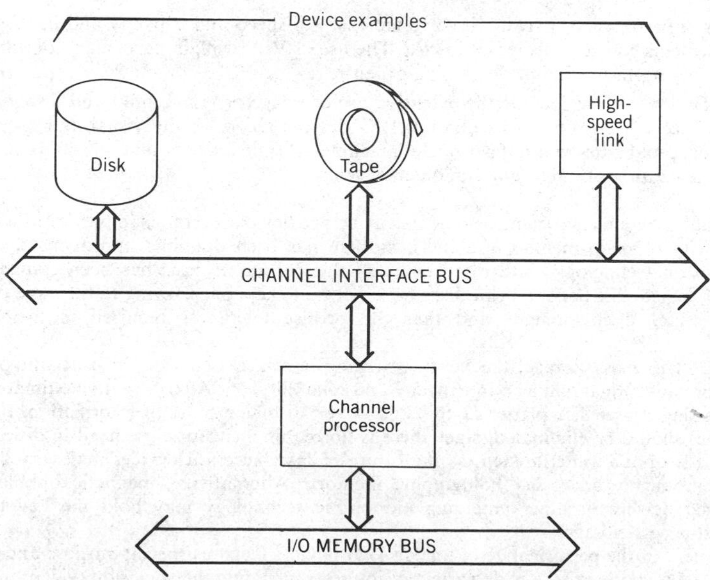

Fig. 61.32 Multiplexer channel organization. I/O, input–output.

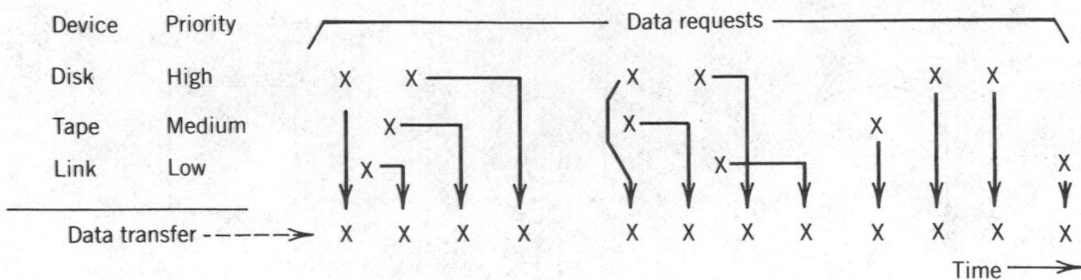

Fig. 61.33 Multiplexer channel timing.

interleaved as determined by peripheral device requests. Devices may compete for memory use on the basis of priority or may be assigned fixed time slots for their data transfers. While the total (peak) data transfer rate at the memory is the same as that for a dedicated or selector channel, the actual throughput for each device depends on the number of devices actually using the channel at the same time. Figure 61.33 illustrates service request timing for one possible scheme. Here three devices compete for available time slots, which are uniformly spaced. Each must wait until requests for all higher priority devices have been satisfied.

It is often desirable and cost effective to employ a combination of channel types in a large system. Modern microprocessors include I/O processors that are quite sophisticated as single-chip support devices. Since these have their own fairly powerful instruction set, very low-cost implementation of complex buffer management is readily achievable.

61.11 SUMMARY AND TRENDS IN I/O SYSTEMS

This chapter has briefly summarized some aspects of the current state of the art with respect to computer I/O from the point of view of the user as well as the designer. Coverage was given to identifying the various types of physical I/O devices currently readily available, the overall system aspects of I/O interfacing, and the four types of interface characteristic types and programming techniques commonly employed. The devices mentioned include examples from communications, mass storage, analog I/O, image and graphics, and control applications. Each one of these may be programmed with status-driven, interrupt, DMA, or mapped I/O techniques or a combination of these. Finally, some consideration was given to the general-purpose channel or emulator as a unified I/O implementation technique.

The state of the I/O "art" is advancing at an extraordinary rate. We can identify two broad classes of change:

1. **Advances in performance.** These involve increases in speed and capacity and in ease of design and implementation, and decreases in cost. The use of VLSI will, in particular, contribute greatly to increasing sophistication at decreasing expense.
2. **New modalities.** These include the extensive use of image or visual input and display as well as audio or speech recognition and control. I/O devices incorporating visual and speech pattern recognition capabilities will impart to the machine extended senses that will be fundamental to the next generation of intelligent computer systems.

The evolutionary advances can to some extent be predicted. Several examples can be cited.

The capacity of main memory and mass storage has been doubling approximately every two years. For silicon technology—integrated circuits—the decline in costs has been comparable. The economics of devices like moving-head disks have not quite kept pace, owing to the large dependence on electromechanical components and thus the precise tolerances required in their machined assemblies.

With respect to mass storage, the newly emerging laser-accessed video disk technology has the potential to provide a quantum leap in capacity and reliability.[16,17] A conservative estimate places the capacity of a single laser disk platter at 10^{16} bits—over 10 billion bytes of information. Failures due to disk crashes should be eliminated, since there is no requirement for close head-to-disk separation. However, this is only an interim step, as the future of laser-accessed devices may very likely reside with the electronically addressed holographic memory. Alternatively, magnetic bubble lattice or charge-coupled devices or other emerging monolithic technology may hold the key to virtually unlimited fast-access storage.

Advancement in the performance of analog I/O devices (such as special-purpose and high speed A/Ds and D/As) is driven by a wide variety of interests including the television (video) industry and the military. Routine digital-to-analog conversion at rates well in excess of 100 million 12-bit samples per second, and analog-to-digital conversion at 50 million 8-bit samples per second, are commonplace. The TV industry, in particular, is converting to digital processing to achieve the spectacular special effects that are made possible through the use of state-of-the-art digital techniques. This real-time processing requires a combined I/O rate of over 60 million bytes per second to achieve broadcast-quality color.

Digitally stored audio is advancing to provide quality in terms of low distortion and dynamic range previously unheard of with analog techniques. Here the challenge is not so much in terms of speed as in quantization resolution. Nonlinear A/D and D/A will provide a perceived 100 dB or more in dynamic range and signal-to-noise ratio (S/N).

Communication between processing systems separated by a few feet, or even thousands of miles, is being performed with advanced I/O techniques using optical fibers as well as microwave links, permitting data rates of hundreds of millions of bits per second.

Display technology is advancing in terms of both performance and new capabilities. Graphics peripherals with very sophisticated dedicated processors are becoming increasingly common interfaced to general-purpose minicomputers. Real-time interactive display of multisegment object-oriented three-dimensional imagery is possible through the exploitation of VLSI technology. Graphical communication is becoming increasingly important in all phases of scientific research and industrial design.

References

1 R. Bernhard, "Bubbles Take on Disks," *IEEE Spect.*, pp. 30–33 (May 1980).

2 D. Toombs, "Charge Coupled Memories," *IEEE Spect.*, pp. 22–30 (Apr. 1978).

3 R. M. White, "Disk Storage Technology," *Scient. Am.*, pp. 138–149 (Aug. 1980).

4 C. S. Chi, "Higher Densities for Disk Memories," *IEEE Spect.*, pp. 39–43 (Mar. 1981).

5 E. J. Lerner, "Fast Graphics Use Parallel Techniques," *IEEE Spect.*, pp. 34–39 (Mar. 1981).

6 W. M. Newman and R. F. Sproull, *Principles of Interactive Computer Graphics*, McGraw-Hill, New York, 1979 (Ch. 3, "Line Drawing Displays"; Ch. 15, "Raster Graphics Fundamentals").

7 H. C. Andrews, "Digital Image Processing," *IEEE Spect.*, pp. 38–49 (Apr. 1979).

8 H. C. Andrews, "PDP-11 Unibus Design Description," Digital Equipment Corp., *Logical, Physical and Electrical Specifications* (1979).

9 H. C. Andrews, "Microcomputer Processors (LSI-11)," Ch. 4 in Digital Equipment Corp., *The LSI-11 Bus* (1978).

10 H. C. Andrews, "Memories and Peripherals (LSI-11)," Ch. 2 in Digital Equipment Corp., *Devices and Interfacing* (1978).

11 B. A. Artwick, *Microcomputer Interfacing*, Prentice-Hall, Englewood Cliffs, NJ, 1980 (Ch. 5).

12 B. A. Artwick, "I/O Programming," in Digital Equipment Corp., *Introduction to Programming (PDP-8)*, PDP-8 Handbook Series, 1972.

13 B. A. Artwick, "I/O," in Digital Equipment Corp., *PDP-8/e Small Computer Handbook*, PDP-8 Handbook Series, 1973.

14 B. A. Artwick, *PDP-11 Peripherals Handbook*, Digital Equipment Corp., Massachusetts, 1976.

15 S. E. Madnick and J. J. Donovan, *Operating Systems*, McGraw-Hill, New York, 1974 (Chs. 2 and 5).

16 K. Bulthuis et al., "Ten Billion Bits on a Disk," *IEEE Spect.*, pp. 26–33 (Aug. 1979).

17 G. C. Keeney, "Digital Recording on Disk; 10 Billion Bits per Side on 30-cm Disk," *IEEE Spectr.*, pp. 33–38 (Feb. 1979).

Bibliography

Andrews, M., *Programming Microprocessor Interfaces for Control and Instrumentation*, Prentice-Hall, Englewood Cliffs, NJ, 1982.

Bywater, R. E. H., *Hardware/Software Design of Digital Systems*, Prentice-Hall International, London, 1981.

Doty, K. L., *Fundamental Principles of Microcomputer Architecture*, Matrix, Champaign, IL, 1979.

Garret, P. H., *Analog I/O Design—Acquisition, Conversion, and Recovery*, Reston, Reston, VA, 1981.

Kraft, G. D. and W. N. Toy, *Mini/Microcomputer Hardware Design*, Prentice-Hall, Englewood Cliffs, NJ, 1979.

Tananebaum, A. S., *Structured Computer Organization*, Prentice-Hall, Englewood Cliffs, NJ, 1976.

Tocci, R. J. and L. P. Laskowski, *Microprocessors and Microcomputers—Hardware and Software*, Prentice-Hall, Englewood Cliffs, NJ, 1982.

CHAPTER 62
MICROCOMPUTERS AND MICROPROCESSORS

EDWARD J. LANCEVICH

Robotic Vision Systems Inc.
Hauppauge, New York

62.1 PROCESSOR TYPES AND TECHNOLOGIES

Microprocessors are complete central processing units (CPUs) packaged as a single large-scale integration (LSI) circuit chip or a set of LSI chips. To function as computer systems, support circuits usually must be interfaced to the microprocessor [memory and input–output (I/O) devices]. Microcomputers, on the other hand, generally are single LSI chips that function as complete computer systems. They contain a CPU, memory, and I/O capabilities in one package. Examples of microcomputers are the General Instruments PIC 1650,[1] the Intel 8048,[2] and the Texas Instruments TMS 1000[3–5] series of chips. Examples of microprocessors are the Intel 8085[5,6] and 8086,[7] the Signetics 2650,[8] and the Motorola MC6800.[9]

The term "microprocessor"[10] was coined to indicate that while the circuits performed their operations under stored program control, they were something less than minicomputers and maxicomputers. In particular, microprocessors were slower than minicomputers, had limited instruction sets, and could access only limited amounts of memory. While indeed this was the case, with the onset of the new technology it has become far less true, as evidenced by the newest microprocessors. For example, the Zilog Z-8001[11] has the capability of directly addressing 2^{23} bytes[12] (8 megabytes) of main memory, has a 16-bit arithmetic–logic unit (ALU), and has instructions to operate on 32- and 64-bit operands. In addition, the architecture is designed to support multiprogramming. The machine has a basic architecture akin to the third generation maxicomputers of the late 1960s.[13]

One way microprocessors may be classified is with respect to the basic size of a data operand handled by the processors ALU.[6,14] The most common designations are 4-bit, 8-bit, and 16-bit microprocessors. In addition, there are bit-slice microprocessors.[5,15–17] A bit-slice processor is constructed from a set of ALU devices interconnected in parallel. This structure affords a designer the means of constructing a microprocessor of size nk bits, where n is the number of ALU chips utilized in the design and k is the number of bits per ALU chip. In the bit-slice microprocessor, the control function is implemented by a special sequencer chip[6,16] that is directed by a stored program in read-only memory (ROM). Bit-slice architectures are generally customized by microprogramming.[18,19] A typical example of a family of chips designed for slice architectures is the AMD 2900[16]

TABLE 62.1. MICROPROCESSOR CHARACTERISTICS

Processor	Technology[a]	MIN-MAX Execution Time[b]	Supply Voltage (V)	Power Dissipation (Typical)
Fairchild[22] 9445	IIL	0.3–1.3 μs	5	2 W
TI 9900[23]	IIL	2.7–20 μs	5	525 mW
Motorola[5] 10800 (slice), ECL 4-bit	ECL	5–50 ns, microoperation	−5, −2	1.3 W
AMD 2901 (slice), 4-bit	TTL	55–110 ns, microoperation	5	1 W
TMS 1000[5]	PMOS	3.3 μs	−15	105 mW
Intel 8085[5]	NMOS	1.3–5.75 μs	5	850 mW
Motorola[5] MC6800	NMOS	2–12 μs	5	600 mW
Zilog Z-80[5]	NMOS	2–11.5 μs	5	1 W
Signetics[8] 2650	NMOS	4.8–9.6 μs	5	525 mW
RCA CDP[5] 1802	CMOS	2.5–3.75 μs	10	40 mW
Intel 8086[7]	HMOS	0.4–37.8 μs	5	1.25 W
Motorola[24] MC6800L10	HMOS	0.4–17.0 μs	5	1.5 W

[a] IIL, integrated injection logic; ECL, emitter-coupled logic; TTL, transistor–transistor logic; PMOS, *p*-channel metal oxide semiconductor; NMOS, *n*-channel MOS; CMOS, complementary MOS; HMOS, high-density, short-channel MOS.

[b] Shortest instruction execution time, longest instruction execution time, assuming maximum clock rate for device.

series of chips. The ALU chips (2901 and 2903) are 4-bits wide and, in conjunction with a 2910 microsequencer, form the nucleus of the bit-slice CPU. An 8-bit processor would require the interconnection of two ALU chips as well as a ROM and microprogram sequencer.[16]

Another method for classifying microprocessors is with respect to the semiconductor technology used in implementing the devices. In broad terms, two basic technologies are used in the production of microprocessors: bipolar technology, which includes transistor–transistor logic (TTL), integrated injection logic (IIL), and emitter-coupled logic (ECL), and metal oxide semiconductor (MOS)[20] technology, including *p*-channel MOS (PMOS), *n*-channel MOS (NMOS), complementary MOS (CMOS), and high-density, short-channel MOS (HMOS). The general characteristics of bipolar technologies are high-speed operation, relatively high power dissipation, low supply voltages, and relatively poor noise immunity. They are further characterized by a low chip density. Examples of bipolar microprocessors are the AMD 2900[16] series of chips (TTL), the TI 9900[2,21] and Fairchild Microflame II,[22] both made in IIL, and the Motorola 10800,[5] an ECL[20] processor. Some bipolar processors are quantitatively characterized in Table 62.1.

The MOS processors generally outperform the bipolar processors in almost all categories, except for speed. Indeed the largest noise immunity is provided by the CMOS[20] processors, owing to the wide range of voltages over which they can operate. The tradeoff in this case is speed; the higher the applied voltage the higher the speed. NMOS[20] has replaced the earlier PMOS[20] technology as the predominant semiconductor in processor construction. This technology affords single power supply operation, better chip densities, and compatibility with TTL support devices. While TTL compatibility was an important factor in many early microprocessor systems, such compatibility is becoming less important owing to the availability of replacement devices within the confines of MOS. Examples of processors made in PMOS are the Intel 8008[12] and 4040,[12] while typical NMOS processors are the Intel 8080,[25] Motorola's MC6800,[9] and the Signetics 2650.[8] CMOS processors include the RCA CDP 1802[5] and the Intersil 1610.[5] Table 62.1 also provides quantitative characterization of some MOS processors.

62.2 ARCHITECTURAL FEATURES AND ADDRESSING MODES

In order to describe the architectural features of microprocessors, it is convenient to classify them into two groups: 8-bit microprocessors and 16-bit microprocessors. While 4-bit microprocessors are found in some applications, they are not nearly as commonplace as their 8- and 16-bit counterparts. Detailed descriptions of these processors may be found in Refs. 2, 5, 7, 10 through 12, and 26 through 28.

Eight-bit microprocessors generally are single-address[12,29] architectures in which an implicit register is used to hold the result of ALU operations. In the case of binary operations, one operand is specified explicitly, while the remaining operand is assumed to be in the implicit result register. In some microprocessors, more than one register may be used as the result register, in which case the processors are more appropriately classified as two-address machines. For discussion purposes, the focus will be on four 8-bit microprocessors: the Intel 8085, Motorola MC6800, Zilog Z-80, and Signetics 2650. Most other 8-bit processors have architectural features similar to those possessed by these processors.

Each of the model processors is packaged in a single 40-pin dual inline package[20,30] and is powered from a single 5-V power supply. External clock signals must be supplied to the MC6800, 2650, and Z-80. The 8085[5] needs only a crystal or an *RC* network to provide clock signals from an on-board oscillator. The remaining leads may be subdivided into three general categories: an address bus (a set of lines devoted to addressing information), a data bus, and a set of control lines. Most of these signal lines are tristate[20]; that is, in addition to providing the high-low voltage levels, they may be placed into a third state. The third state is the high-impedance state that in effect electrically disconnects the internal circuit connected to the signal line from the external pin. This feature is useful in the design of microprocessor-based logic systems, since it helps to minimize external circuitry (see Section 62.5).

The address bus is utilized primarily to address the memory chips associated with the computer system. Memory refers both to program memory and data memory. Most often, the program resides in a ROM that preserves its integrity when power is removed and, as implied by its name, cannot be changed by a program in execution. Data typically reside in read–write memory, normally made of semiconductor material, which loses its information on removal of power. This memory is commonly referred to as RAM.† The fundamental addressable unit of information in these processors is the 8-bit

†This is a misnomer, since RAM stands for random-access memory, which is defined as a memory in which any random location addressed is accessed in the same amount of time. This definition also applies to ROMs.

byte and the maximum size of the memory for the 8085, Z-80, and MC6800 is 2^{16} bytes (64 kilobytes) (16-bit address bus), while the 2650 can access 2^{15} bytes (15-bit address bus).

The address bus is also utilized for interfacing peripheral devices to the microprocessor. Information on the bus (or some part of the bus) is decoded by the peripherals involved.[10,12,14,29,31] To distinguish peripheral addressing data from memory addressing information, some processors generate a single control signal. When the signal is a 1, the address information pertains to I/O devices; when it is 0, it pertains to memory devices. The processors that contain the control lead (2650, 8085, Z-80)[5] have one or more of their instructions dedicated to I/O. The signal is generated in response to the I/O instruction execution. In the MC6800,[9] this lead is missing, indicating that no special instructions are dedicated to I/O. With this processor and others similar to it, peripheral devices must be "memory mapped."[9,10,12,31] That is, a portion of the total address space must be devoted to peripheral devices and not to memory locations. Memory-mapped devices have the advantage of making available to the I/O programmer all the memory reference instructions existing in the processor instruction set. Note that the memory mapping of I/O devices can be carried out using the other processors too, giving them an extra degree of flexibility.

In two of the model processors (Z-80 and MC6800), all the address bus leads are directly available at external pins, while in the other two processors, a subset of the address bus pins are functionally shared (multiplexed) in time. On the 8085, the least significant eight address leads (A0–A7)[5] are multiplexed with the processor data bus, while on the 2650,[8] address lines 14 and 13 (the most significant address bits) are multiplexed with I/O control signals. During use of the latter two processors, to enable addressing over the maximum address range of the machine external hardware may be required to demultiplex the address line. This would be the case, for example, if a member of the 27xx[10] (see Section 62.6) family of ROMs were to be used for program storage. On the other hand, if an 8085-compatible ROM (e.g., Intel 8755)[5] were used for program storage, the ROM itself would provide the demultiplexing.

The data bus is an 8-bit bidirectional communications link between the processor, the memory, and the peripherals. Bus direction is determined internally by the type of instruction being executed (read or write). A read pulse and a write pulse are supplied over a pair of separate control leads in the Z-80, 8085, and 2650 microprocessors. These pulses are generally used to strobe the data onto the bus (read) or from the bus (write). For the 6800 microprocessor, bus direction is defined by a single $R/\overline{W}$ line.[9,10] This signal, logically combined with one of the processor clock signals,[9,10] is equivalent to the separate read–write strobes described for the other machines. The control signals for accessing memory, as well as those necessary for peripheral control, are found on all microprocessors. Other common control signals include one or more interrupt inputs and interrupt acknowledge outputs.

An interrupt[10,12] is an asynchronous signal generated by a peripheral as a response to a change in its state. For example, an interrupt might be generated in response to a power failure in the system or in response to the reading of a character from a keyboard. Typically, the interrupt mode of communication is used when the peripheral speed is slow relative to the processor's instruction execution time. The interrupt[13] causes the program in execution to be suspended and an I/O program to be started to determine the action of the microprocessors. On detection of a power failure, for example, the interrupt processing program might switch all system RAM to a battery backup to preserve its integrity until power can be restored.

The address of the interrupt processing program is determined in a number of ways, depending on the internal structure of the processor. One approach is to require that the peripheral supply the address of the program. This approach is implemented on the 8085, Z-80, and 2650. In addition, a signal must be generated by the processor that is used by the peripheral as a gating strobe for the address information. This signal is called the interrupt acknowledge signal (INTA). For example, when the INTA signal is output on the 8085, 1 byte of the data will be "jammed" onto the data bus by the peripheral. The data that is normally placed on the bus in the case of the 8085 is a 1-byte, JUMP TO SUBROUTINE instruction. This instruction transfers to one of eight addresses, dependent on the setting of 3 bits within it. The situation is depicted in Fig. 62.1. Bits 3 to 5, depending on their value, define the correct starting address of the interrupt program as shown.

A variant of this approach is used on both the 2650 and the Z-80† microprocessors. In the Z-80 processors, the 8 bits are used as the least significant bits of a pointer to (the address of) the processing program's address. The upper 8 bits are determined by the 8-bit contents of a CPU register called the interrupt vector (IV) register. The starting address is then taken from 2 bytes as already defined. The process is illustrated in Fig. 62.2.

In the 2650, the byte of data supplied by the peripheral is used as an address of the starting address of the interrupt routine, as with the Z-80. In this machine, however, the most significant

†In the Z-80, this is an alternate operational mode for handling interrupts, since the Z-80 is designed to be an upward compatible version of the 8085.

D7	D6	D5	D4	D3	D2	D1	D0
1	1	X	X	X	1	1	1

D5	D4	D3	Interrupt Program Address
0	0	0	0
0	0	1	8
0	1	0	16
0	1	1	24
1	0	0	32
1	0	1	40
1	1	0	48
1	1	1	56

Fig. 62.1 Interrupt addressing for the Intel 8085 microprocessor.

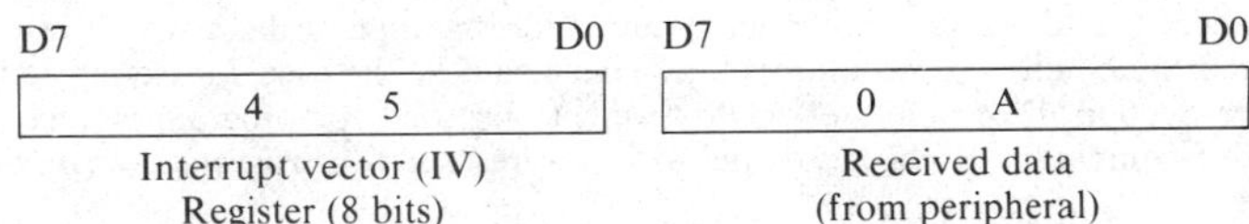

Fig. 62.2 Interrupt address for the Zilog Z-80 microprocessor. The subroutine address is found at locations $450A_{16}$ and $450B_{16}$. 450B contains the most significant address bits and 450A the least significant bits. The IV register is set by the program.

address bits are set to zero, and because of additional constraints,[8] the table must be in the first 64 bytes of the processor's address space.

The alternate approach to interrupt handling is to have the CPU transfer to a fixed address within the memory. This approach is implemented in all the processors discussed except the 2650, and is the only mode of operation of the 6800. Multiple interrupt capability in this case must be handled by multiple input pins on the processor. For example, in the 6800 two interrupt pins are provided ($\overline{IRQ}$ and $\overline{NMI}$).[9] When a request is made, the processor causes a jump to an address associated with the line. Simultaneous multiple requests are prioritized by internal arbitration.

The model CPUs illustrate a wide breadth of design features. Each of the processors contains a set of registers. In some processors, registers are "general purpose," that is, they may be used both for address information and for data. The Z-80, 8080, and 2650 possess registers of this type. The 6800, on the other hand, has separate registers for data and addresses. Each processor has a set of conditional flag bits (flip-flops) which generally are set by the execution of an arithmetic[5,10,12] or logic instruction. These bits are automatically tested by conditional jump instructions (see Section 62.3) to enable a programmer to break up the normal sequential flow of control of the program. Figure 62.3 illustrates the register configuration for each of the processors.

Each processor has a variety of methods for generating operand addresses. The methods are collectively referred to as the addressing modes of the machine. Addressing information is necessary on instructions that operate on data in memory locations or in CPU registers and must be encoded as part of the instruction.

Direct addressing means that the entire memory address is encoded as part of the instruction. For example, suppose that on a processor there is an instruction that causes 0 to be written into a memory location. Assume further that the mnemonic for the instruction is CLR. To clear location 1000 using direct addressing, the programmer would write CLR 1000. The actual machine language encoding would require 3 bytes. One byte would be used to specify the operation code and 2 bytes would specify the address. This is illustrated in Fig. 62.4. The first byte indicates the operation and an additional code that tells the processor how the address is to be formed. For example, the 2 might indicate the operation and the 0 the addressing mode.

Register indirect addressing is another way of specifying a memory location. This addressing mode is supported on the Z-80 and 8085[5] microprocessors. In this case the data in two of the processor's general-purpose registers are interpreted as a 16-bit address. The H and L registers are most commonly used, but for a class of instructions the D and E or the B and C registers may also be used. An example is shown in Fig. 62.5. In this case the mnemonic is CLR M and the machine code

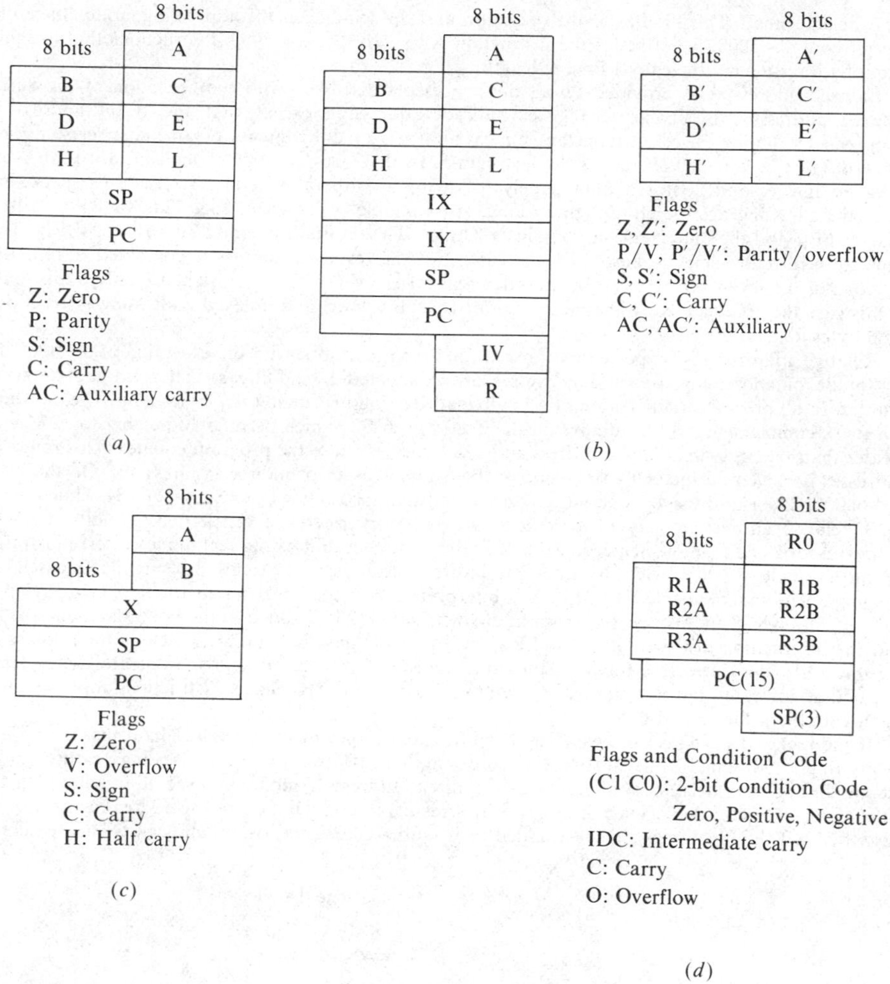

8 bits
8 bits
A
B
C
D
E
H
L
SP
PC
Flags
Z: Zero
P: Parity
S: Sign
C: Carry
AC: Auxiliary carry
(a)
8 bits
8 bits
A
B
C
D
E
H
L
IX
IY
SP
PC
IV
R
8 bits
8 bits
A′
B′
C′
D′
E′
H′
L′
Flags
Z, Z′: Zero
P/V, P′/V′: Parity/overflow
S, S′: Sign
C, C′: Carry
AC, AC′: Auxiliary
(b)
8 bits
8 bits
A
B
X
SP
PC
Flags
Z: Zero
V: Overflow
S: Sign
C: Carry
H: Half carry
(c)
8 bits
8 bits
R0
R1A
R2A
R1B
R2B
R3A
R3B
PC(15)
SP(3)
Flags and Condition Code
(C1 C0): 2-bit Condition Code
Zero, Positive, Negative
IDC: Intermediate carry
C: Carry
O: Overflow
(d)

Fig. 62.3 Microprocessor register configurations: (*a*) Intel 8085, (*b*) Zilog Z-80, (*c*) Motorola 6800, (*d*) Signetics 2650.

Instruction
CLR 1000
Machine code (hexadecimal)
2
0
10 00
Operation code
Mode code
Address

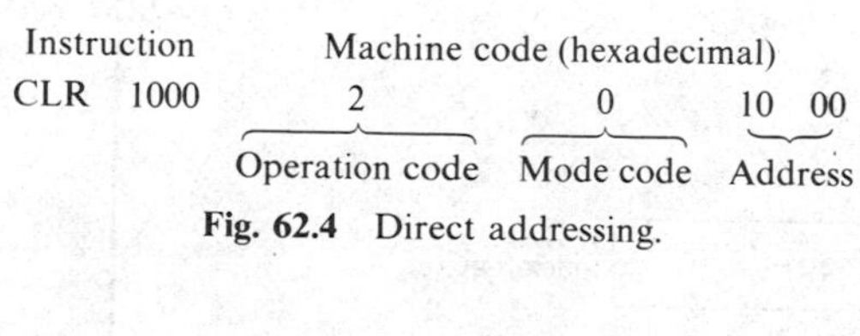

Fig. 62.4 Direct addressing.

Instruction
CLR M
Machine code (hexadecimal)
2
1
Operation code
Mode code (register indirect)
Indirect address register
(16 bits)
1 0 0 0

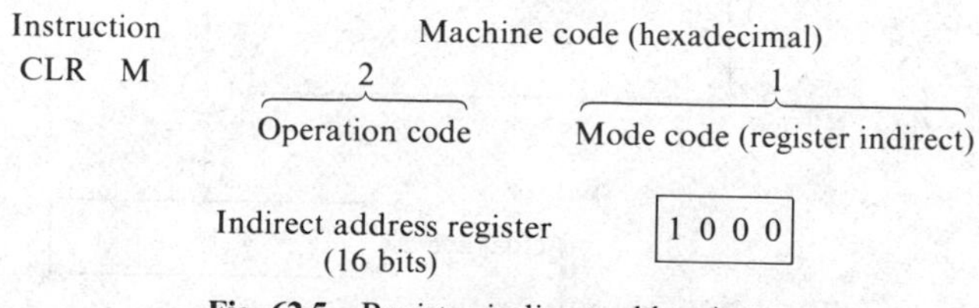

Fig. 62.5 Register indirect addressing.

is 21 (hexadecimal). The 2 indicates the operation and the 1 indicates the addressing mode. Since the register contains 1000, the effect of the instruction is to zero location 1000. Instructions that use this form of addressing are typically 1 byte in length.

Indexed addressing is provided by all the processors discussed with the exception of the 8085. Indexed addressing is similar to register indirect addressing except that the 16-bit address is computed by adding an offset (constant binary number) to the contents of an index register. The constant must be encoded as part of the instruction. In the Z-80[5] and 6800,[9] the constant is an 8-bit unsigned integer and is stored immediately following the operation code. Instructions specifying indexed addressing are therefore 2 bytes long. An example is shown in Fig. 62.6. In this case the offset is $6F_{16}$ and the index register contains a 2450_{16}. The resultant operand address is $24BF_{16}$. The same form of addressing is allowed on the 2650, but the resultant address is computed differently. Any one of the seven registers can be an index register; however, these registers are only 8 bits long. In this case the offset added to the index register is 15 bits long, and indexed addressing instructions are 3 bytes long on the 2650.

Relative addressing is supported by three of the four processors, with the exception of the 8085. In this mode of addressing, operand addresses are calculated by adding an offset to the program counter. In all processors, the length of a relative addressing instruction is 2 bytes long. The constant is a two's complement[30] 8-bit number on the Z-80 and 6800, which permits addressing to 127 bytes greater than the program counter address and 128 bytes less than the program counter. This range of addresses is halved on the 2650, since one of the 8 bits is used for indirect addressing.[8] On the Z-80 and 6800, relative addressing is permitted only with jump instructions (see Section 62.3), while on the 2650, relative addressing may be used with any memory reference instruction. Finally, memory indirect addressing is implemented on the 2650. The indirect addressing technique is best illustrated by an example. Figure 62.7 illustrates an indirect addressing instruction using 2650 assembly language. The mnemonic LODR, 1 * 25 is interpreted as a load register instruction using relative indirect addressing. In this case the offset in the instruction (25) is added to the program counter (PC) and the information at the resultant address (125) and the next byte (126) is used as the address of the operand (6345), which is finally loaded into register 1. The asterisk in the operand field of the instruction indicates indirect addressing within the machine language instruction and the most significant bit of the second byte is set to a 1.

If indirect addressing is used with an absolute address specification, the resultant instruction is 3 bytes long. Additionally, the 2650 architecture implements two unique addressing features, auto increment and auto decrement addressing. In a direct addressing mode, indexed addressing with the auto increment (auto decrement) option specifies results in computation of the address as previously described but the index register has 1 added to it (subtracted from it) automatically. This feature is

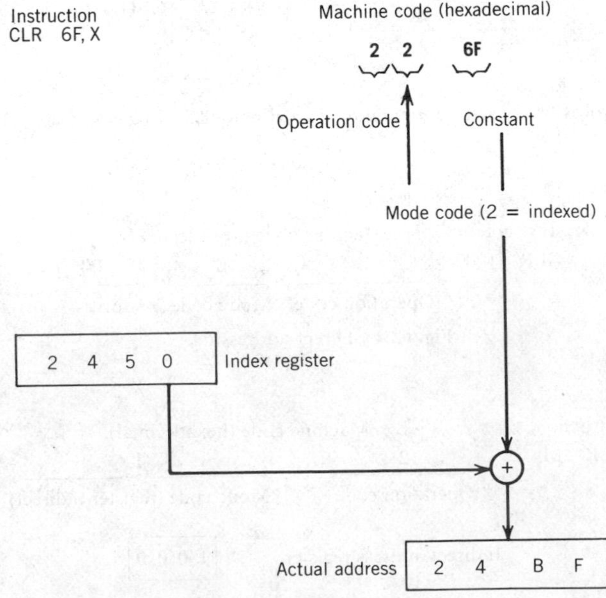

Fig. 62.6 Indexed addressing.

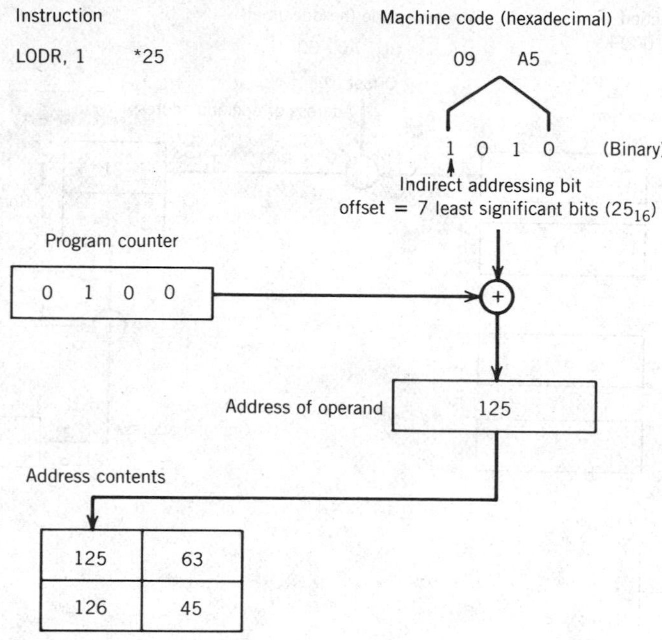

Fig. 62.7 Indirect addressing. Register 1 is loaded from memory location 6345. "*" implies indirect addressing.

useful in writing programs that loop sequentially through a table of data. When used in an indirect addressing mode, the incrementing is not done on the register contents, as is the case, for example, in the PDP-11 minicomputer[32] (see Chapter 56), but it is done on the address operand. Figures 62.8 and 62.9 illustrate both direct and indirect auto increment addressing.

Each of the processors includes a "stack"[33] mechanism for storing return addresses for subroutine jump instructions. In the 2650, the stack memory is 16 bytes of temporary storage contained directly in the CPU, while in the other processors the stack memory is in the RAM attached to the processor.

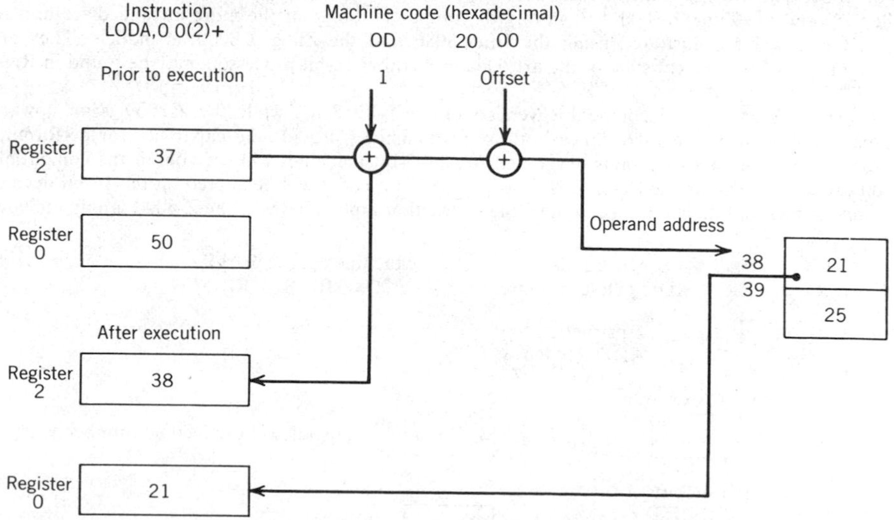

Fig. 62.8 Autoincrement addressing mode (direct). The instruction loads register 0 with the byte found at the address formed by the 8-bit constant in register 2 plus the 15-bit constant 0. Prior to execution, register 2 has the constant 1 added to it. the "+" means auto increment.

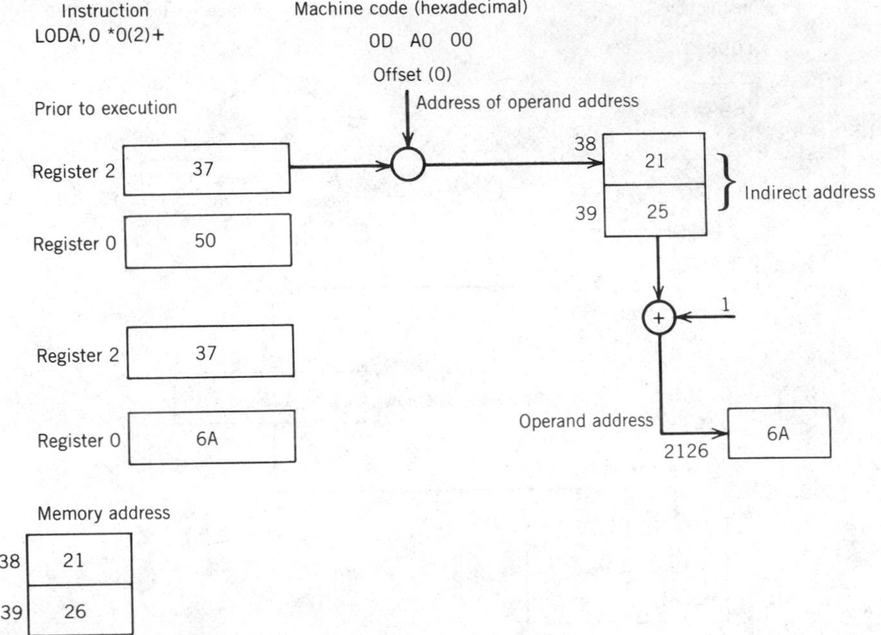

Fig. 62.9 Autoincrement addressing mode (indirect). The instruction loads register 0 with the byte found at the indirect address specified by the 8-bit constant in register 2 plus the 15-bit constant 0. The indirect address has 1 added to it prior to loading.

The latter approach allows the use of the stack for storing an extensive amount of return addresses. Use of the stack is facilitated by a special set of stack instructions (see Section 62.3).

Sixteen-bit processors are enhanced versions of the 8-bit processors, providing both capability for 8-bit and 16-bit ALU operations while generally extending the address space of the machine. In addition, most 16-bit processors have more flexible addressing modes, equivalent to or in some cases extending those provided on larger computers such as the PDP-11[32] and IBM-370 series. Examples of 16-bit processors are the General Instruments CP 1610,[34] the Texas Instruments 9900,[2,21] the Intel 8086,[2,7,35] and the Zilog Z-8000[2,11,36] family of computers. Some of the more advanced features of 16-bit CPUs will be illustrated using the Intel 8086 and the Zilog Z-8000 as models. They are representative of the current state of the art. Details of other 16-bit processors may be found in Refs. 2, 5, 22, 24, 31, and 35.

The 8086 is an upward compatible version of the 8-bit 8085, while the Z-8000 is an upward compatible version of the Z-80. Since both 16-bit machines provide the capability for performing operations on byte data, programs developed for the 8-bit machines will execute on the compatible 16-bit processors. The only difference is that the operation codes will be altered on the 16-bit devices (the mnemonics and machine execution of the instruction are unaltered). The Z-8000 family includes

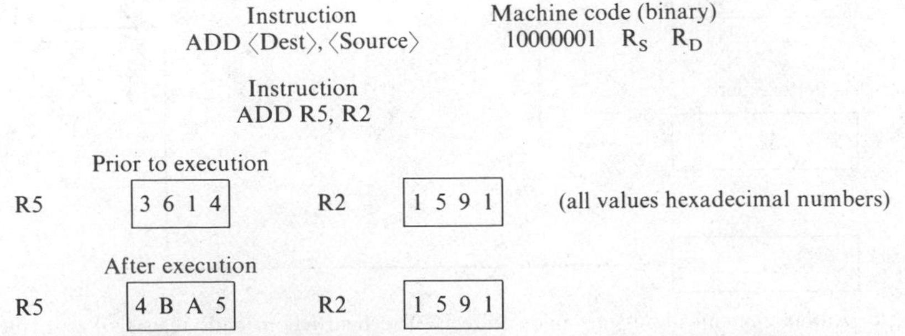

Fig. 62.10 Coding of Zilog Z-8000 add instruction. R_S and R_D are 4-bit numbers referring to the source and destination registers, respectively. The operation causes the contents of the source and destination registers to be added together and the result to be placed in the destination.

two processors, the Z-8001 and the Z-8002. The Z-8001 is a 48-pin device, with the capability of directly addressing 2^{23} bytes (8 megabytes) of memory, while the Z-8002 is a 40-pin device capable of accessing 2^{16} bytes (64 kilobytes). This constitutes the only essential difference between the devices. The 8086 can address 2^{20} bytes (1 megabyte) of memory and is packaged in a standard 40-pin chip. Both processors are two-address[37] machines in which two operand instructions (e.g., ADD) necessitate the specification of two addresses (a source and a destination address) as part of the instruction. The result of the binary operation is placed into the destination operand address. An example of a Z-8000 register to register ADD instruction is shown in Fig. 62.10.

Each processor contains a set of general-purpose registers. The 8086 additionally contains registers dedicated solely to addressing information. The registers of both CPUs are shown in Fig. 62.11. In the Z-8001 there are 18 16-bit registers. However, only 16 of the registers are accessible to the programmer at any one time.

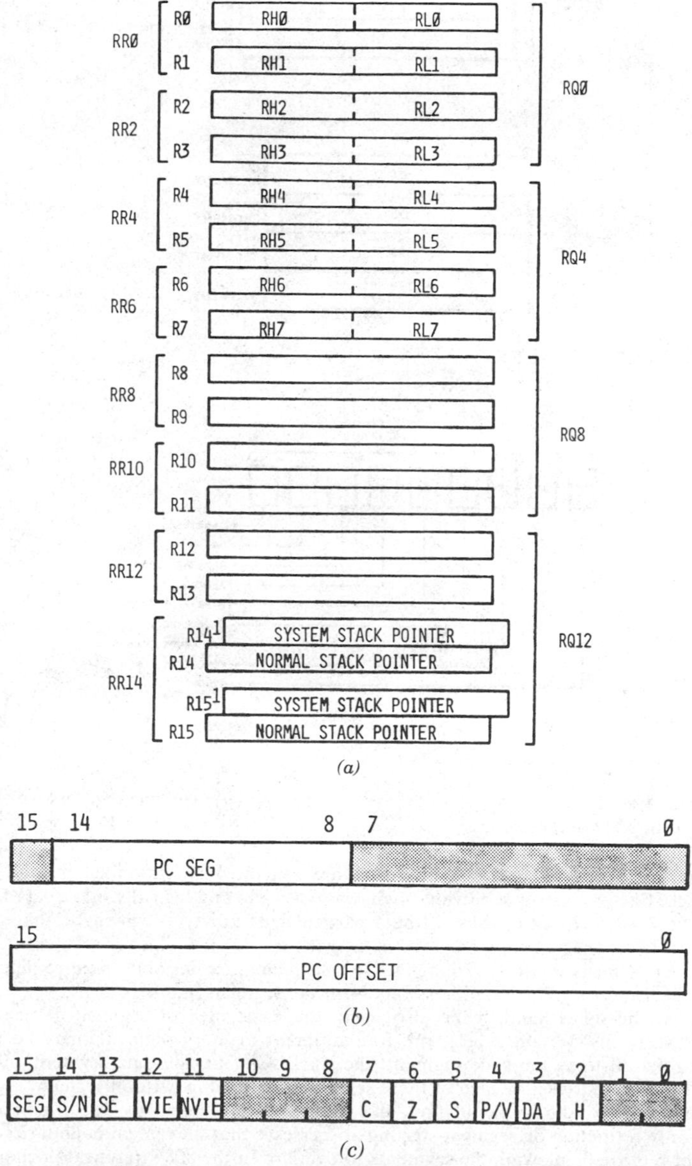

Fig. 62.11 Zilog Z-8001 register layouts: (*a*) General registers, (*b*) program counter, (*c*) flag and control word. From AMD Corp.,[11] reprinted with permission.

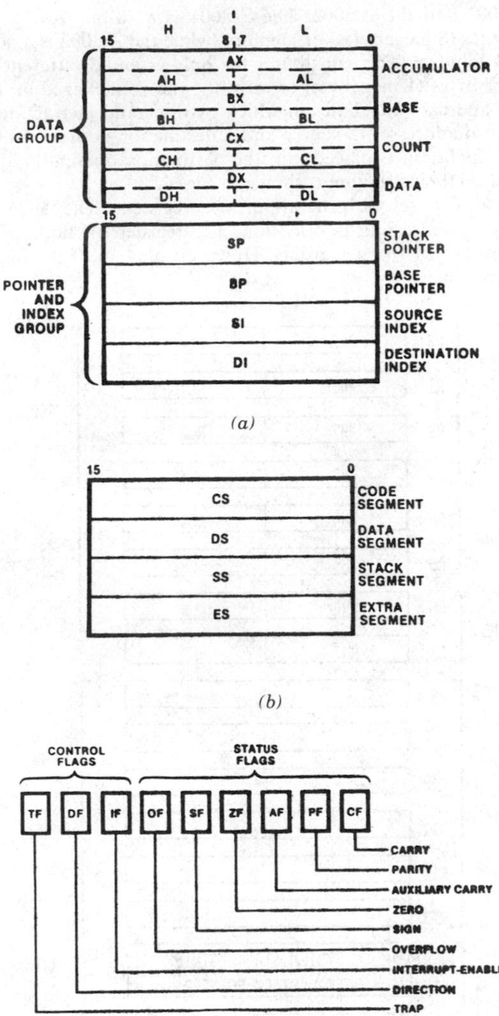

Fig. 62.12 Intel 8086 register layouts: (*a*) General registers, (*b*) segment registers, (*c*) flags. From Intel Corp.,[7] reprinted with permission.

In both machines, full 23- or 20-bit addresses are generated in a number of ways but with a similar approach. Each processor subdivides memory physically and logically into a set of 64 kilobyte segments. In the Z-8001, 7 address bits define a particular 64 kilobyte segment so that the machine physically could be connected to a memory of megabytes: 128 64-kilobyte segments. The Z-8002 provides only for a memory of 64 kilobytes. The segments are disjoint (nonoverlapping) in the Z-8001, since the 7 most significant address bits form the segment number.

The 8086, on the other hand, while also using the basic idea of segmented memory, allows overlapping of segments. Within a program, four different memory segments may be used. These segments[7] are referred to as the code segment, the stack segment, the data segment, and the extra segment. The starting addresses of each of these segments is stored in a 16-bit segment register on the CPU (when used in an address calculation, this address is multiplied by 16 to generate a 20-bit address). The only restriction on segment starting addresses is that they lie on boundaries where the 4 least significant bits are 0, allowing the segments to overlap. In the 8086, physical memory addresses are generated by adding a 16-bit offset to one of the 20-bit segment starting addresses; the resultant address is used to access the operand. The method of computing the offset defines the addressing

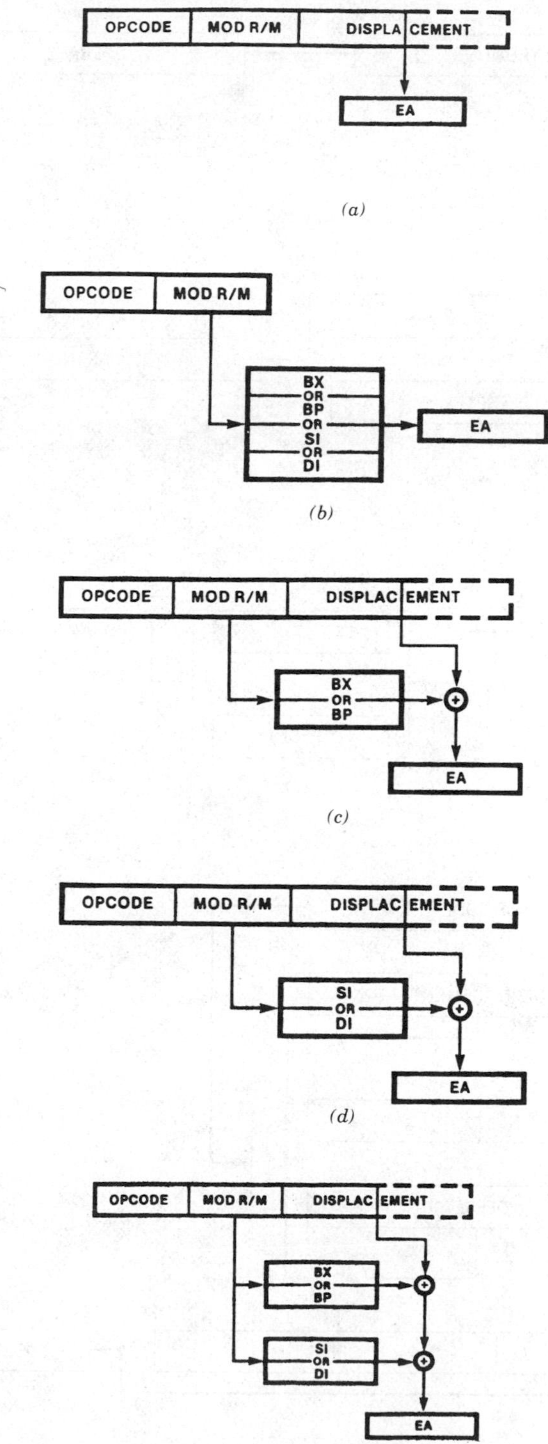

Fig. 62.13 Addressing modes for the Intel 8086 microprocessor: (*a*) Direct, (*b*) register indirect, (*c*) base, (*d*) indexed, (*e*) base indexed. From Intel Corp.,[7] reprinted with permission.

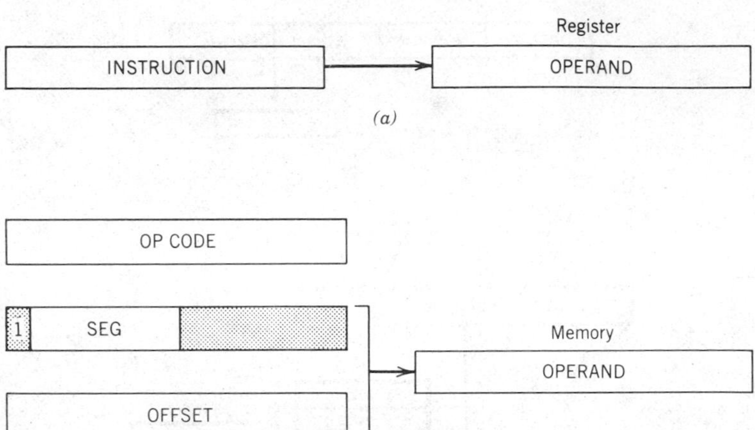

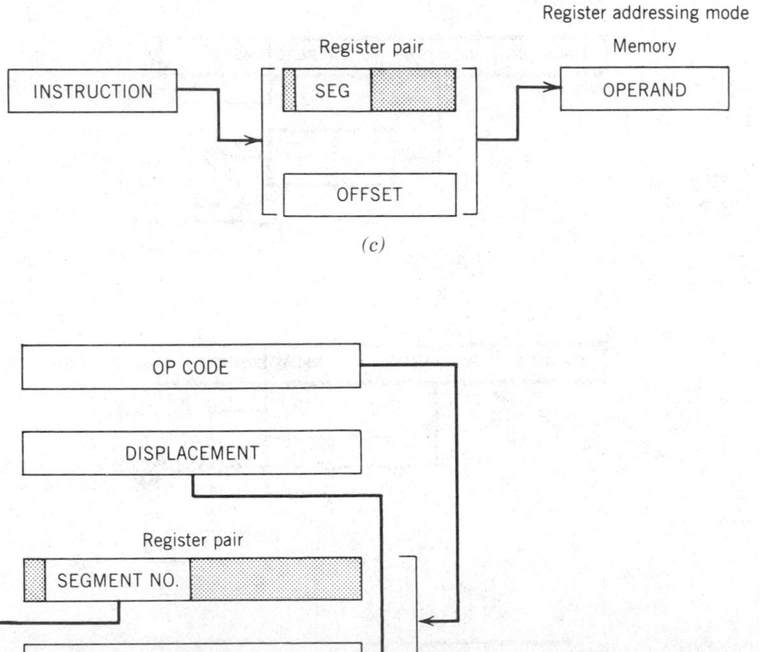

Fig. 62.14 Addressing modes for the Zilog Z-8000 microprocessor: (*a*) Register, (*b*) Direct, (*c*) Indirect Register, (*d*) Base, (*e*) Indexed, (*f*) Base Indexed. From AMD Corp.,[11] reprinted with permission.

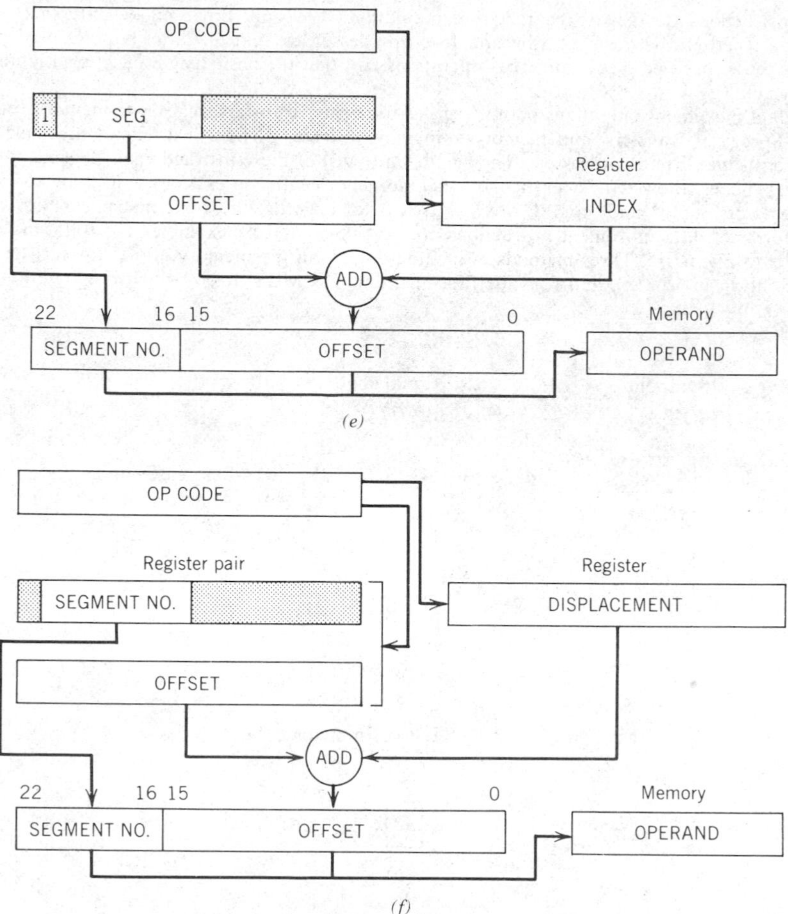

Fig. 62.14 (*Continued*)

mode of the processor. First the 16-bit offset may be stored with the instruction itself, the direct addressing mode of operation. Alternatively the address of the operand may be specified by one of the base or index registers, in which case the mode is called register indirect addressing. The remaining addressing modes all constitute a form of indexed addressing. In "based" addressing, the offset is derived by adding to registers BP or BX a constant value stored within the instruction. Indexed addressing is similar, except that the registers used are SI and DI. In based index addressing, both an index and base register can be used as well as a displacement to specify an address. The addressing modes are shown in Fig. 62.13. If only register operands are specified in an instruction, they are encoded as part of the instruction, while "immediate" constants (see Section 62.3) are also stored as part of the instruction. The segment registers are generally initialized once within a program. The Z-8000 has equivalent addressing modes, except that within the Z-8001 system, a segment number is explicitly required to be specified in the direct addressing modes. In addition, the PC in the Z-8001 consists of two registers; a 16-bit register to hold the offset information and a 7-bit segment base address. Segment addresses in the Z-8000 must be on 64 kilobyte boundaries. Since the Z-8002 supports only a single segment, the segment number is not necessary. The address modes of the Z-8000 are summarized in Fig. 62.14.

62.3 INSTRUCTION SETS

Microprocessor instruction sets may be generally classified into four major categories[10,25]: data movement, arithmetic and logic operations, transfer of control instructions, and I/O and machine control instructions. In addition, the 16-bit processor instruction sets have been expanded to include special instructions devoted to character manipulation. The only other major differences in the instruction repertoire between the 8-bit and 16-bit processor are that the latter requires two operand addresses to be specified on certain ALU operations and that the basic length of an instruction is 16 bits rather than 8.

The data movement operations permit register-to-register transfers, register-to-memory transfers, memory-to-register transfers, and memory-to-memory transfers. These instructions are included in various forms on all microprocessors. The length (in bytes) of these instructions is dependent on the addressing modes allowed in a particular machine. The length varies from 1 to 3 bytes on 8-bit microprocessors[8,25,28,38] and from 1 word (2 bytes) to 4 words on 16-bit machines.[4,7,11,24,35] The various forms of data movement instructions are best illustrated by example. The following discussion refers to Fig. 62.15. Three methods of loading data from the memory into the A register of the 8085 are shown in (*a*), (*b*), and (*c*). The first instruction is MOV A, M, which is 1 byte long. The

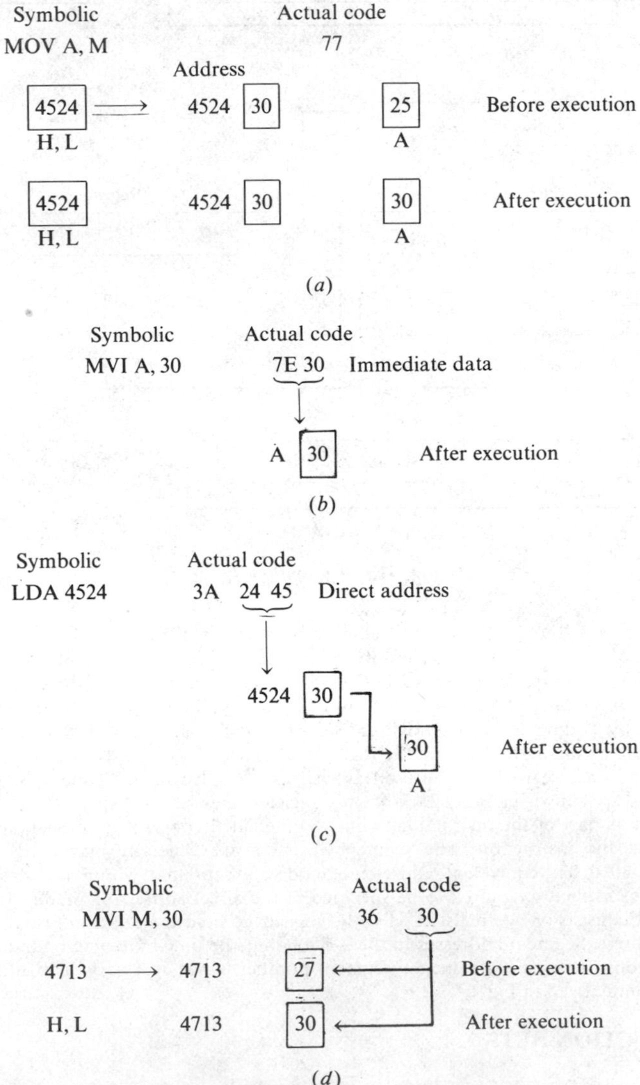

Fig. 62.15 Data movement instructions for the Intel 8085 microprocessor: (*a*) Register indirect addressing, (*b*) immediate addressing, (*c*) direct addressing, (*d*) memory-to-memory move.

addressing mode is register indirect addressing, in which the memory address is assumed to be in the H and L registers. The second instruction is MVI A, 30, where the data to be moved from memory is encoded as part of the instruction itself (immediately following the operation code). The instruction occupies 2 bytes. The final example is the LDA 4524 instruction, which causes data to be moved from memory location 4524 into the A register. It is a direct addressing instruction. The address is encoded as part of the instruction, making the instruction 3 bytes in length.

The other predominant addressing modes included in microprocessor instruction sets (indexed addressing and relative addressing) require an offset to be added to the index register, where the offset is encoded as part of the instruction. In this case the instruction length will be 2 bytes, if the offset is 8 bits, or 3 bytes, if the offset is 16 bits. An example of this instruction appears in Fig. 62.16.

On 8-bit processors, memory-to-memory operations are generally excluded from the instruction set, with two exceptions. In the 8085 and Z-80, immediate addressing can be used with the destination address specified by register indirect addressing. An example is shown in Fig. 62.15*d*. In this case the MVI M, 30 instruction causes 30 to be written into the address specified by H and L (4713).

The other exception occurs in the instruction set of the Z-80. This instruction allows a block of data to be moved from one place in memory to another. Figure 62.17 illustrates the instruction symbolically designated LDIR. When the instruction is executed, the H and L registers must contain the starting address of the data block, the D and E registers the starting address of where the data are to be moved, and the B and C registers the number of bytes to be transferred. Data are moved, byte by byte, from one location in the source area to the corresponding location in the destination area. The count is automatically decremented and when it reaches 0, the operation ends. In 16-bit processors, memory-to-memory operations are the rule rather than the exception.

The arithmetic and logic operations can be categorized as one- and two-operand types. Two-operand (binary) instructions include addition and subtraction, universally available on all 8-bit microprocessors, as well as the logic operations AND, OR, and EXCLUSIVE OR. The arithmetic operations are carried out using two's complement arithmetic. Addition and subtraction also result in altering of the condition codes (or flags) in the CPU. For example, a zero result would set the zero flag, a positive result (most significant bit = 0) would set the sign flag, and a carryout of the most

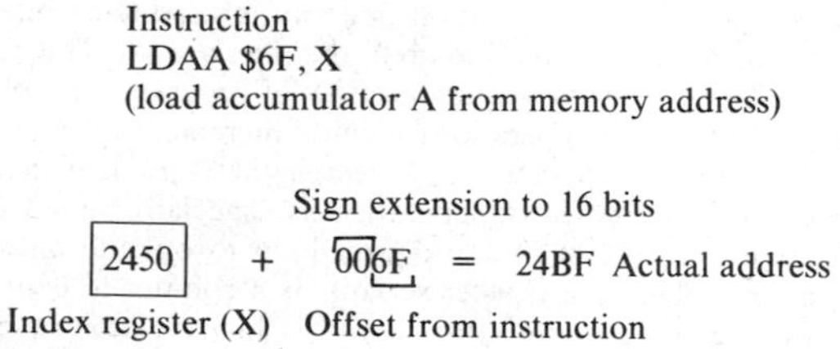

Instruction
LDAA \$6F, X
(load accumulator A from memory address)

Sign extension to 16 bits

2450 + 006F = 24BF Actual address

Index register (X) Offset from instruction

Fig. 62.16 Indexed addressing for the Motorola 6800 microprocessor.

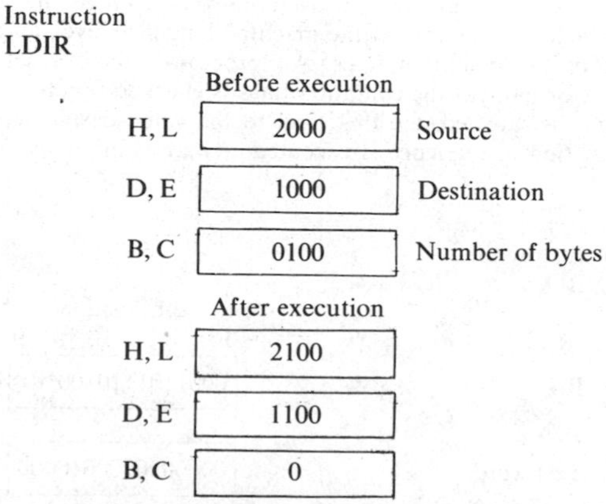

Instruction
LDIR

Before execution

H, L	2000	Source
D, E	1000	Destination
B, C	0100	Number of bytes

After execution

H, L	2100
D, E	1100
B, C	0

Fig. 62.17 Block data movement instruction for the Zilog Z-80 microprocessor. Bytes in sequential locations 2000–20FF are transferred into 1000–10FF, respectively.

significant bit would set the carry flag. This latter property is especially useful in the performance of multibyte (or multiword) additions or subtractions, since in these cases the carry represents the carry between consecutive bytes (words). A programming example of multibyte addition is included in Section 62.4. If the processor has an overflow flag, it is also affected by the arithmetic operations. Of the 8-bit processors, the 8085 and Z-80 are the only two that include 16-bit addition instructions, and the Z-80 includes a 16-bit subtraction instruction as well.[5] In both of the processors, the H, L register pair (Fig. 62.15) is used as a 16-bit accumulator for the result.[5,10,25] The second operand is data in any one of the 16-bit register pairs (or registers).

The Z-8000[11] and 8086[7] processors include two's complement multiplication and division[37] instructions as well as the fundamental addition and subtraction operations in their instruction sets. On both machines, 8- or 16-bit operands may be multiplied together, generating 16- or 32-bit results. The Z-8000 has the extended capability of multiplying 32-bit operands together, generating a 64-bit result. Divisions are performed using 32-bit dividends and 16-bit divisors, yielding a 16-bit quotient and 16-bit remainder, or using 16-bit dividends with 8-bit divisors. The Z-8000 includes the capability of dividing 64-bit operands by 32-bit operands. When a division operation is performed and the resulting quotient is too large to be placed into the appropriate number of bits, the overflow flag is set in the processor (e.g., division of a 64-bit number by a 16-bit number would result in a quotient exceeding 32 bits, setting the overflow flag). During performance of the multiplication and division operations, one operand is assumed to be in a register on the CPU and the other is specified by one of the addressing modes of the machine.

The binary logic operations included in the instruction sets of the machine are the AND operation, the OR (inclusive) operation, and the XOR (exclusive or) operation. These operations are carried out on a bit-by-bit basis between the source and destination operands, with the result being stored in the destination operand. The addressing modes of the machine dictate the specification of the operands. Some processors include instructions that can be used to detect the presence of a 1 in a single bit position. This instruction causes the flags to be set according to the logical AND of a source operand and a test byte (word) in which the position of the bit to be tested is set equal to 1. Figure 62.18 illustrates the operation, using the Z-8002 mnemonic (BIT). In this case the test word has a 1 in bit position 5 and the corresponding word (in register R4) has a 0 in it. On instruction execution, the zero flag is set. Note that the source operand is not changed during this operation.

The compare operation is similar to the bit test operation, except that two operands are compared on an arithmetic basis. In this case the destination operand is subtracted internally from the source operand, and while neither operand location is altered, the flags are set. The condition flags may then be tested by jump instructions to alter the sequential flow of program control.

Single-operand arithmetic and logic operations include increment (INC) (a word or byte), which results in a 1 being added to the operand, and decrement (DEC), which results in a 1 being subtracted from the operand. The increment operation is especially suited for generating a set of sequential memory addresses during use of a program loop to operate on a data array, while the decrement operation can be applied to a register serving as a counter to control the number of loop iterations. Forming the two's complement of an operand is also available, as well as shift and rotate operations. The latter are particularly useful when multiplying or dividing an operand by 2. As a final example of unary logic operations, the complement (bit by bit) of a single operand is included in most instruction sets.

Transfer of control operations involves instructions that change the address in the program counter, the jump or branch instructions. At the machine language level, all control is transferred on the basis of the value of the condition flags. All processors have a set of jump on condition instructions, in which one or more of the condition flags is checked by the hardware. If the condition implied by the instruction is satisfied, the PC is set to the value designated within the instruction; otherwise the next instruction in sequence is executed. As an example, consider the code shown in

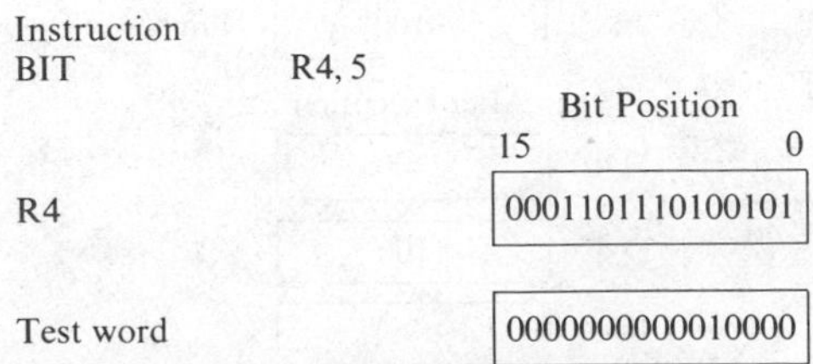

Fig. 62.18 Bit test instruction for the Zilog Z-8000 microprocessor. Flags set according to R4 are ANDed with test word, the result of which is zero.

```
        ADDA    #$A120
        BEQ     RS
        STAA    #$A230
          .
          .
          .
          .

RS      LDAB    # - 1
          .
          .
          .
          .
```

Fig. 62.19 Code illustrating conditional branching for the Motorola 6800 microprocessor. Control transfers to location RS if the result of the ADDA instruction on the first line is 0.

Fig. 62.19, which applies to a 6800 processor. The mnemonic BEQ means "branch if the last operation to affect the zero flag set it," or in other words, branch if the last ALU operation resulted in a zero. The instruction specifies the next address if the condition is satisfied. In the 6800, all branch operations, with one exception, use relative addressing. An 8-bit offset is part of the instruction. The exception is due to the fact that only 256 addresses relative to the PC can be reached. The single direct addressing transfer instruction permits the range to be extended to the full processor address space.

In the 8085, all jump instructions use direct addressing, while the other processors' jump instructions include a mixture of both relative and direct addressing. In addition, the 16-bit[7,11,24] processors allow indexed jump instructions. Unconditional branch instructions are also found in all machines.

A subroutine is a sequence of code that generally is entered from many places within a program to perform a specific task. It then returns control to the program section that initially caused its execution.

An example of a subroutine would be a program segment to read a character from a keyboard and place it in a register. This set of codes may be used repetitively within the framework of a larger program. Since the starting address of the subroutine is known, a jump instruction is sufficient to cause the subroutine to execute. But since the subroutine is started from programs in many different locations, the problem is to correctly return control to the program. To effect a correct return, the machine must store the address of the instruction immediately following the jump-to-subroutine instruction. In all of the model microprocessors, the return address is placed in the stack memory. The stack memory is an area of RAM (or of the CPU in the case of the 2650). The stack memory has the property that the last piece of data stored in it is the first piece removed. The address of the next available† location in the stack is stored in the stack pointer (SP) register. When a jump-to-subroutine instruction is executed, the PC contents are placed into the next 2 byte addresses in the stack, and the SP register is set to the next available location. This is generally done by decrementing the SP register. Consider the example in Fig. 62.20. The next available location in the stack memory is 100, so the PC contents are stored in memory locations 100 and 99. When the operation is finished, the SP contains a 98.

When the subroutine terminates, the program executes a return-from-subroutine instruction. Execution of the instruction results in the current top of stack information being put into the program counter, and in the value of the SP being incremented by 2. The correct return address is restored and the program continues.

The last-in first-out structure of the stack is particularly powerful for implementing subroutine linkage, since it allows subroutines to execute other subroutines. As long as each of the subroutines returns control using the special return instruction, correct transfer of control is assured. In some processors, conditional jump-to-subroutine instructions as well as conditional return instructions are allowed. The only computer in the set of model processors not having this capability is the 6800.

Finally, some processors have special instructions that automatically decrement a register and test it for zero. If the result is nonzero, a program control is transferred to the address specified within the instruction. As an example, consider the code (Z-8000) in Fig. 62.21. The DJNZ R4, 100 mnemonic causes the register R4 to be decremented and if the result is nonzero, control is passed to the

†In some processors, the current top of stack address is in the memory, but conceptually the operation is the same.

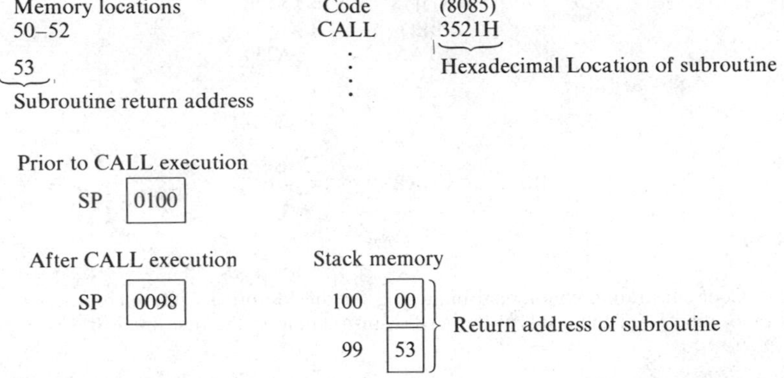

Fig. 62.20 Use of the stack memory for holding subroutine return addresses.

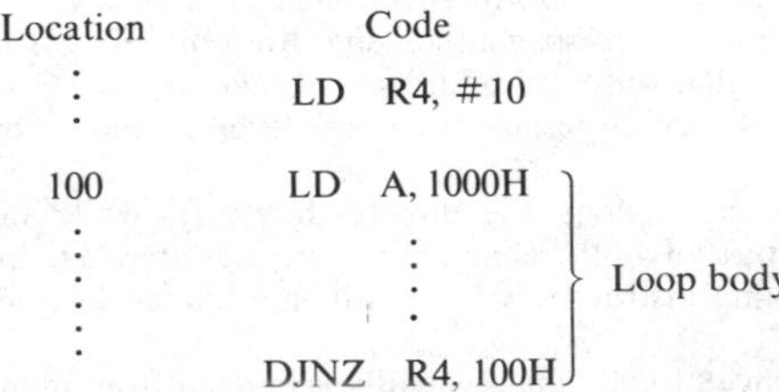

Fig. 62.21 Loop control instruction for the Zilog Z-8000 microprocessor. In this example, R4 is loaded with the constant 10 prior to execution. Each time the set of instructions marked loop body executes, R4 is decremented and tested for 0. If it is not equal to 0, control is passed back to location 100. The loop executes exactly 10 times.

instruction at location 100. This instruction is especially useful in counting the number of iterations in a loop.

Machine control and input–output instructions are highly machine dependent. These instructions generally include means for enabling or disabling interrupts (see Section 62.2), setting or resetting the flag bits, and performing I/O operations. In the section on interfacing (62.5), the examples will illustrate some of the instructions for particular microprocessors. The references contain detailed descriptions for all the microprocessors.[2,3,5,7,9–12,14,16,24,25,29,31,38–40]

62.4 PROGRAMMING EXAMPLES

In this section, some examples of microprocessor programs are presented along with a description of the algorithm that is implemented. The objective is to solidify the conceptual material presented in the previous sections with concrete examples. Each instruction is presented using the manufacturer's[24,26,28,38] recommendation as to the format. Generally, each of the instructions contains an operation field and a mnemonic representation of the operation code. These two fields are followed by one or more operands. If the operands are registers, they are designated symbolically. If the operands are numeric (8-bit or 16-bit constants), they are listed as hexadecimal numbers. On the left side of each program is the address of the instruction. These numbers are also in hexadecimal. The codes for multibyte instructions are listed with a blank between each byte. The format shown is similar to the listings most assembler programs produce.

Multibyte Addition, Z-80 Processor

The program in Fig. 62.22 is an example of multibyte[10] addition on the Z-80 microprocessor. The program adds two 32-bit integer numbers assumed to reside at memory locations 200-203 (the addend) and 204-207 (the augend). The sum is placed in locations 200-203. The H, L registers and the D, E registers contain the address of the augend and addend byte, respectively.

Location	Code	Mnemonic	Comments
0000:	21 00 02	LD HL, 200	SET HL to 200 – AUGEND ADD
0003:	11 03 02	LD DE, 203	SET DE to 203 – ADDEND ADD
0006:	06 04	LD B, 4	SET B to = 4 (Count)
0008:	B6	OR A	CLEAR CARRY
000A:	1A	LD A, (DE)	LOAD A WITH ADDEND BYTE
000B:	8E	ADC (HL)	ADD AUGEND BYTE
000C:	77	LD (HL), A	STORE RESULT
000D:	23	INC HL	ADJUST BYTE
000E:	13	INC DE	ADDRESSES
000F:	10 F9	DJNZ 000A	REPEAT IF NOT DONE

Fig. 62.22 Multibyte addition program, Zilog Z-80 microprocessor.

The algorithm used in the 32-bit addition procedure is a byte-by-byte addition of respective addend and augend locations using the ADC instruction. This causes the carry to be added into the next byte. The ADC instruction is done four times with the carry bit initially set to 0. Loop control is done using the DJNZ instruction, which decrements the B register and returns to the first instruction in the loop if the B register is not equal to zero.

Block Data Transfer Program, 8080, 8085 Microprocessor

The program in Fig. 62.23 illustrates the concept of register indirect addressing on the 8085. The H, L register pair is used for the address of the source data and the D, E register pair is used for the address of the destination. Register C counts the number of bytes.[28] The starting source address is 100, the starting destination address is 200, and the number of bytes transferred is 50.

Subroutine, MC6800 Processor

The example in Fig. 62.24 is a subroutine in 6800 code that finds the arithmetic maximum of a set of 100 8-bit numbers and leaves the result in accumulator A. The subroutine saves the value of the B register in the stack prior to execution and restores it before returning to the main program. The index register X is assumed to hold the starting address of the array when the subroutine is entered.

The program uses the indexed address approach to compare each byte in the array to the current maximum value, stored in the A register. If the value in A is larger, the index register is incremented, generating the address of the next byte to be compared. Otherwise the new maximum is loaded into A and the index register is incremented to form the address of the next comparison byte. The B register keeps track of the number of comparisons. The branch instructions (BLE and BNE) are relative addressing instructions; the second byte of the instruction, when sign extended and added to the program counter, generates the correct transfer address. For example, in the BNE 106 operation, the PC would contain 0110 and the sign-extended offset is $FFF6_{16}$. These two numbers, when added (neglecting the extra carry), yield 106.

Many more programming examples may be found in the references.[2,3,10,26,27,39,40]

Location	Code	Mnemonic	Comments
0300:	21 01 91	LXI H, 100	Initialize H, L
0303:	11 00 02	LXI D, 200	Initialize D, E
0306:	0E 50	MVI C, 50	Initialize count
0308:	7E	MOV A, M	Read source data to A
0309:	12	STAX D	Write source data to destination
030A:	23	INX H	Generate next source address
030B:	13	INX D	Generate next destination address
030C:	0D	DCR C	Decrement count
030D:	C2 08 03	JNZ 308	Jump if not equal to zero

Fig. 62.23 Block data movement, Intel 8085 microprocessor code.

Location	Main Program Operation	Mnemonic
0200-0203:	BD 01 00	JSR MAX

Location	Subroutine Operation	Mnemonic	Comments
0100:	37	PSHB	Save B in stack
0101:	C6 FF	LDAB #$FF	B contains number of comparisons
0103:	A6 00	LDAA X	A contains first number
0105:	08	INX	Add 1 to index register
0106:	A1 00	CMPA X	Compare current max to new number
0108:	2F 02	BLE $10C	
010A:	A6 00	LDAA X	
010C:	08	INX	
010D:	5A	DECB	
010E:	26 F6	BNE $106	
0110:	33	PULB	
0111:	39	RTS	

Fig. 62.24 Comparison subroutine, Motorola 6800 microprocessor code.

62.5 INTERFACING PRINCIPLES

Most microcomputer systems involve interconnection between a microprocessor, memory devices, and various I/O devices. The I/O devices may be as simple as a set of switches or as complex as a multiple disk drive system. The I/O devices generally operate at a different speed than the processor, asynchronous to the processor's clock, and use analog as well as digital driving signals. While the broad concept of interfacing implies connection of any of these devices to a microcomputer system, the focus of this section is more restrictive. The purpose is to develop some fundamental principles that apply to a wide range of devices.

The functional model of an I/O device that will be used throughout the discussion is shown in Fig. 62.25. The block marked I/O port(s) represents the electronic circuit interfacing directly to the microcomputer system. The circuits generally consist of decoding logic, registers, and some line drivers and receivers. They are the focal point of this discussion. The block marked device controller represents the circuitry that actually operates the device. For example, this may be a set of circuits that drive a stepper motor in a disk drive, operate solenoids in a printer, and so on. Finally, the block marked device represents the physical unit, for example, a disk drive or a printer. While the device controller and I/O ports are functionally separated for modeling purposes, it should be pointed out that these two functions may be integrated in an actual design. In particular, there are many microprocessor support circuits available with the dual role.[5,14,40] Data flow direction is defined with respect to the microprocessor: An input port delivers data from the device to the processor and an output port receives data from the microprocessor.

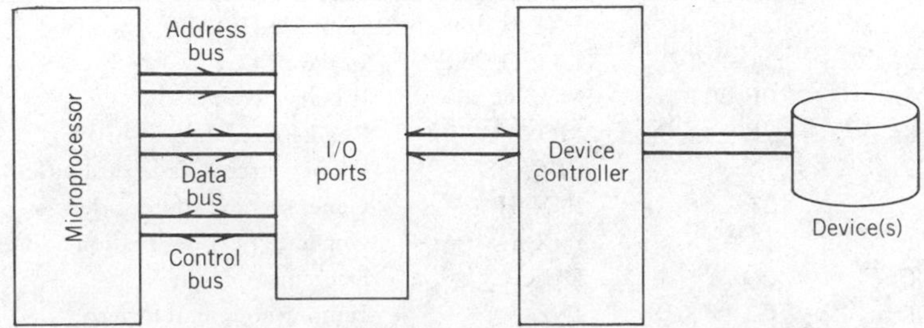

Fig. 62.25 Functional block diagram of input–output (I/O) device interfaced to a microprocessor.

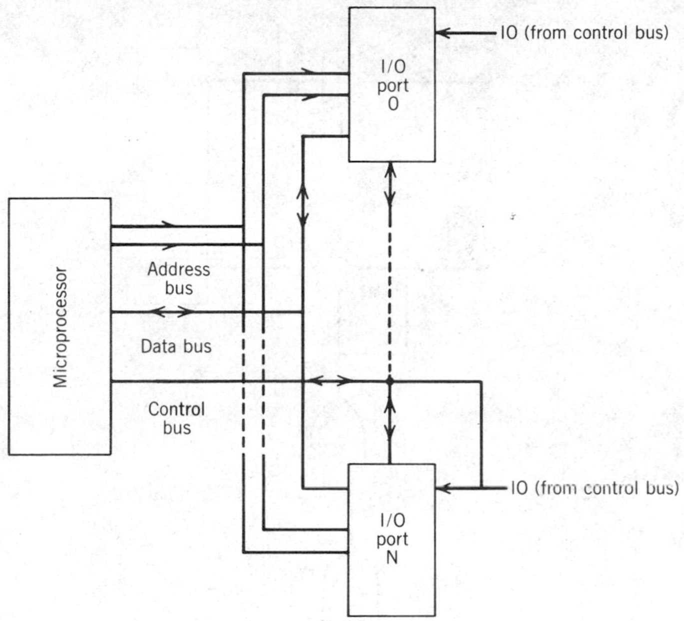

Fig. 62.26 Bus connection for multiple input–output (I/O) ports. Port selection is based on address information plus IO control signal.

Figure 62.26 illustrates the connection of a set of I/O ports within a typical microcomputer system. Each port is attached to the processor over a set of three functionally distinct busses: the address bus, the data bus, and the control bus. Since each port is connected in parallel to the busses, it is clear that the busses are time shared among them. If two or more ports attempt communication with the processor simultaneously, a problem known as bus contention arises. The result of this problem is erroneous communication. The address and control busses are used to eliminate bus contention.

Ports are distinguished from one another to the system by different addresses, analogous to memory locations. To communicate with an I/O port, the processor sends the port address to the address bus. Simultaneously, the processor activates a control signal (IO) indicating that the address information is an I/O port address rather than a memory address. Each port receives the information and decodes it. The selected device then conditions itself to be attached to the data bus and to either send or receive information (the port is either an input or output port). After a short delay, the microprocessor sends one of two control pulses, a READ or WRITE strobe, depending on the direction of data transfer. Figure 62.27 illustrates the timing for an input port. If the data to be transferred to the microprocessor are assumed to be in register A and the address of the input port is FF_{16} (assuming an 8-bit address bus), the NAND gate decodes the address and control line, enabling the condition lead C to the OR gate. When the READ strobe appears, the OR gate causes the inputs of the register to be driven onto the data bus from the tristate buffers. The only time that the port is electrically connected to the data bus is for the duration of the READ strobe due to the tristate device.

Tristate[30] input ports are commonly utilized in microprocessor-based systems design because of the simplicity of interfacing. Wired-OR[30] capability is built into the interface and a minimal amount of external components are required. The only components that may be required are terminating networks to minimize noise on the bus lines due to transmission line effects. A simple output port is shown in Fig. 62.28. In this case the output port address is also FF and the NAND gate is activated with the occurrence of the IO control signal and the correct address. The OR gate is enabled as before but the WRITE strobe latches the data into the register.

The device conditioning operation is performed by decoding an address and the IO control signal. The IO control signal is generally available only on those microprocessors that contain special input–output operations in their instruction sets. Examples of such processors are the Z-80, 8085, and 2650. On the other hand, the 6800 processor does not contain any special input–output instructions. In this case one of the address lines is usually used to distinguish between I/O ports and memory

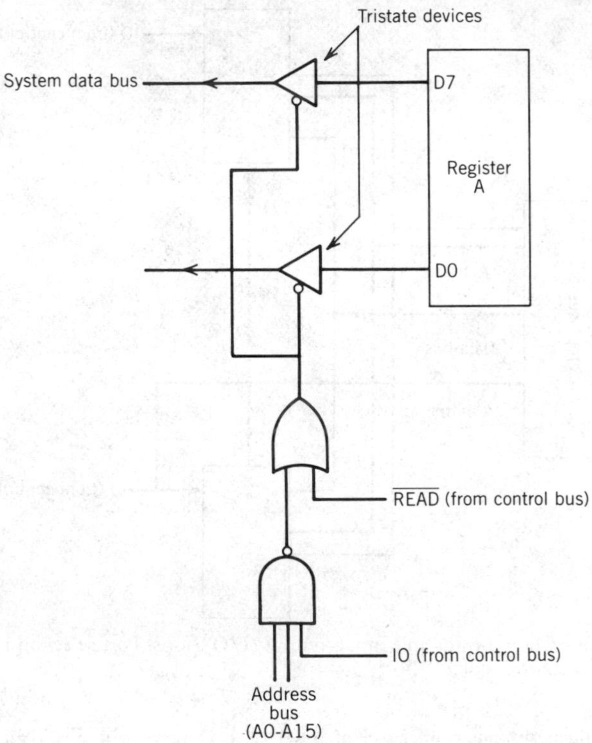

Fig. 62.27 Control signals, input port (all components are part of the input port).

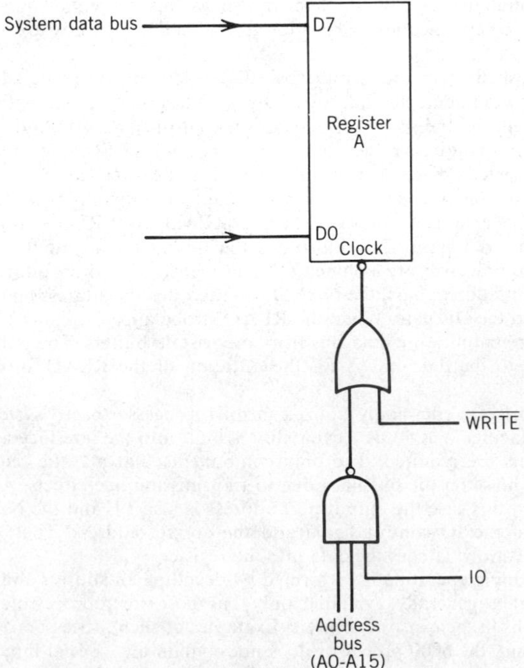

Fig. 62.28 Control signals, output port (all components are port of the output port).

addresses. For example, if address line 15 (A15) is selected for this function, then A15 being high could represent an I/O instruction and A15 being low would represent a memory operation. (See Fig. 62.28.)

The latter I/O scheme is called memory-mapped input–output. The advantage of this system is that all memory reference instructions may be used to gain access to the I/O ports. A disadvantage is that some memory address space is lost. In the example, only one half (32K) of the total machine addresses are available for storage. To improve the space for memory available under a memory-mapped I/O approach, more than one address line can be decoded to derive the IO control signal. To illustrate, suppose that only 16 devices are interfaced to the system. If the devices are assigned the hexadecimal addresses FFFO-FFFF, then the 12 most significant address bits could be ANDed together to give the IO control signal; the 4 least significant bits would then be decoded at each device. The remaining 65,520 addresses would be available for RAM and ROM.[10,24,38]

For processors possessing special I/O instructions, a memory-mapped I/O scheme could also be implemented using the same techniques. Indeed if it were prudent to do so in a designer's judgment, both could be combined in systems containing these processors.[5,14]

Processors possessing special I/O instructions normally have a single control line that distinguishes I/O operations and memory operations, since the two are disjoint. For example, in the 8085 the control line is designated IO/M (high implies I/O operation, low implies memory operation).

The control signals described are included in the design of virtually all I/O ports and are necessary for reliable data transfers. However, the questions of protocol and timing remain. That is, how is communication between the processor and an I/O device initiated? Once initiated how is the data transfer carried out reliably, that is, what are the timing considerations that must be observed when carrying out the data transfer?

Data communication between the processor and an I/O port can be initiated by either device, depending on the system design. If the processor initiates the data transfer, the procedure is called *polling*.[14,41] This procedure is most easily described by an example. Consider a single-user computer system that has a cathode ray tube (CRT) output and a keyboard input. Collectively, the two devices are called the terminal. The keyboard is attached to an input port and the CRT is attached to an output port. A monitor program[24] is utilized to control the operation of the system. The monitor program accepts keyboard commands that result in actions by the processor (e.g., display of a memory location). To make the situation somewhat more realistic, assume that the CRT controller operates more slowly than the processor. That is, assuming the processor was just programmed to transfer one character after another, the device could not empty the buffer and display the information as fast as it was loaded. Under these circumstances, a simple handshake[25] protocol is necessary for reliable data transfer. The controller must provide a 1-bit handshake signal that can be monitored by the microprocessor (on an input port). If the signal is high, the device is assumed to be busy. If it is low, the device is idle and ready to accept a new character. The flowchart[33,40] associated with the character transfer is shown in Fig. 62.29. The program monitors the handshake bit of the device and loops until the CRT is ready for a new character; when it is ready, the processor transfers the data. This constitutes an asynchronous data transfer, since the processor must wait until the controller is free before transferring data. The processor cannot transfer the data at its own clock rate.

In this system assume that the program responds to single-character commands from the keyboard (the letter D could mean to display the current register contents, for example). The basic functional flowchart[33,40] for the system is shown in Fig. 62.30. The input program receives a command, interprets it (causing some response and display), and returns to wait for another command. The monitor program then polls the keyboard for additional inputs.

If the system were extended to two terminals, the monitor program could be written so as to alternately share the processor's time between the devices. It could poll one terminal and then the other until a new command[24] was entered, interpret and execute the command, then return to polling the terminals. The polling approach would work provided that the interpretation of a command could be carried out within the duration of a keypress signal. Under these circumstances, if the two terminals demanded servicing simultaneously, the program could service one and start the second prior to the disappearance of the second keypress. If enough terminals were added to the system, however, the polling approach would eventually break down, since coupled with the response problem described for two terminals there would be more and more overhead program time required for switching between terminals. This would result in less time to do useful work (i.e., command interpretation). In addition, during the time the keyboards are scanned there is a low probability of an input command (the keypress signal is asynchronous with respect to the processor), and a large percentage of time is wasted in doing nothing.

In cases such as that suggested by the example, where a large number of inputs must be observed by the processor and where the inputs are random arrivals[5] with respect to the processor's clock, the controller is designed so that it initiates communication with the processor. Initiation of communication by the controller is provided by the microprocessor's interrupt facilities.

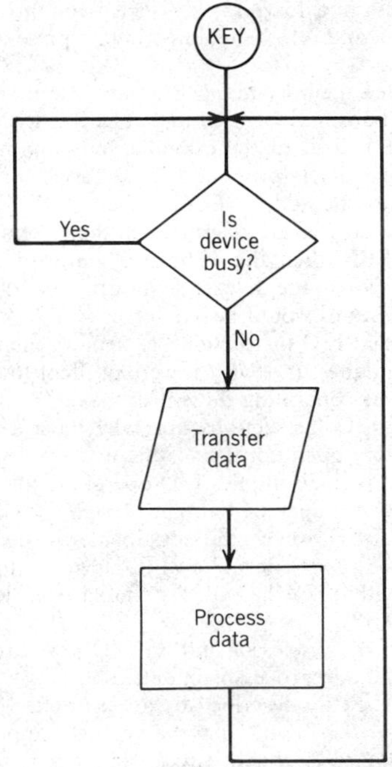

Fig. 62.29 Flowchart for keyboard polling.

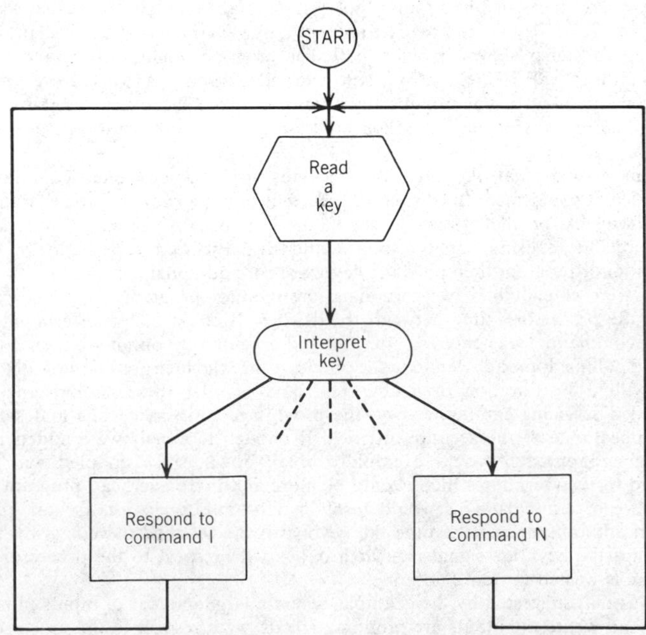

Fig. 62.30 Functional flowchart for keyboard monitor program.

Interrupt signals have been described in detail in Section 62.1. Recall that there are two major methods for processing interrupts, the fixed address (vector) method and the variable address (vector).† In the first case an interrupt[5] request entering the processor causes the processor to transfer to a fixed address in memory, where a program exists to service the I/O device. On completion the program executes a return from interrupt instruction, causing the processor to continue with the program that was executing when it was interrupted.

Acknowledgment of the interrupt request should cause the interrupt request to be cleared. As an example, consider the circuit shown in Fig. 62.31. This circuit is set up as an interrupting input port for a keyboard. In this case the keypress signal from the keyboard is latched in a flip-flop whose output is attached to the interrupt request line. When the interrupt is serviced by the processor and the I/O program reads the character from the input port, the flip-flop is cleared. The flip-flop is now ready for a new keypress signal.

In the variable vector approach, the interrupting[10,24,31] device must provide part (or all) of the address of the service routine. This requires that the processor return a strobe signal in response to the interrupt request, which is used by the device to gate the required address onto the data bus. The hardware required to do this for the keyboard input port is shown in Fig. 62.32. It is assumed that 8 bits of address information are supplied as part of the vectored address. The upper tristate has its inputs set up to place FF onto the bus. When the interrupt acknowledge signal (INTA) is received, the signal untristates the upper device, causing the address of the service routine to be placed on the bus. As before, when the service routine reads the input port, the request flip-flop is cleared.

The two hardware design examples illustrate how interrupts may be handled from single devices. To handle multiple devices, the external port hardware generally increases in complexity. This is primarily due to the fact that the number of interrupt inputs on microprocessors is limited. To illustrate, consider the 8085 microprocessor. The device has four interrupt request leads vectored internally and one interrupt request line on which a vector must be supplied externally. In addition, the processor has one interrupt acknowledge signal. Suppose two interrupting devices are to be connected to this processor using the latter technique (each device supplying its own vector). The basic problem is that there is only one interrupt input line and one acknowledge signal to handle both interrupts. A solution to this problem and the necessary additional hardware is shown in Fig. 62.33. The technique is called daisy[10,12] chaining. The idea is that the interrupt acknowledge signal is passed to each device in the "chain" depending on whether the interrupt request is enabled. For example, if both INTR2 and INTR1 are on, then INTA gates the vector information from device 1 onto the bus. The service program clears the request (INTR1) when the data are read. Since INTR2 is still on, the INT line is still held high via the OR gate and the next INTA signal is passed to the second port. INTR2 is then cleared by its service routine once the data are read. This approach to multiple interrupt handling has a built-in priority structure. The first device in the chain is the first one serviced.

There are many other approaches to the problem of multiple interrupting devices. Most of these schemes are highly dependent on a particular microprocessor's interrupt structure, and the details will be found in the references.[4,7,8,10,12,17,21,25,30,31,40]

Once communications have been established between the microprocessor and an I/O device, data must be transferred reliably. For single data transfers, establishment of communications usually implies that the device is ready to send or receive information, and transfer takes place on one of the

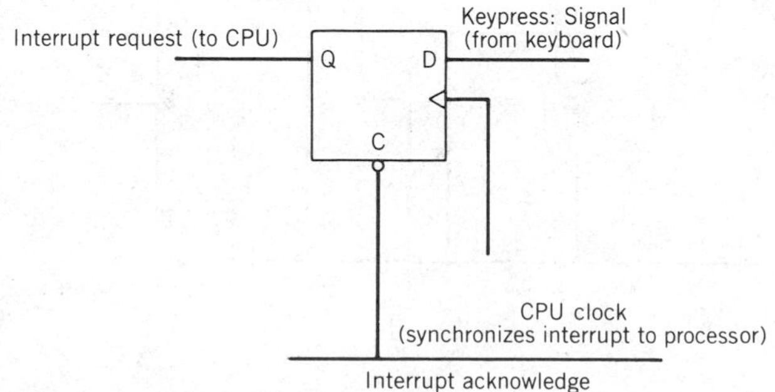

Fig. 62.31 Simple interrupt structure.

†The term "vector" is commonly used to refer to the address of the interrupt I/O service program.

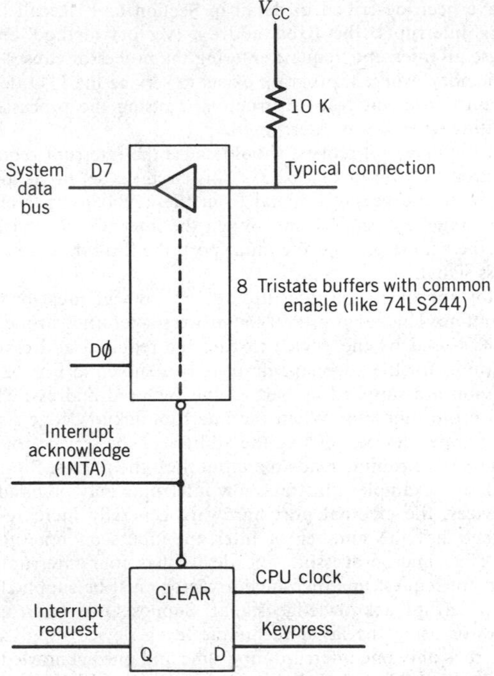

Fig. 62.32 Simple vectored interrupt port.

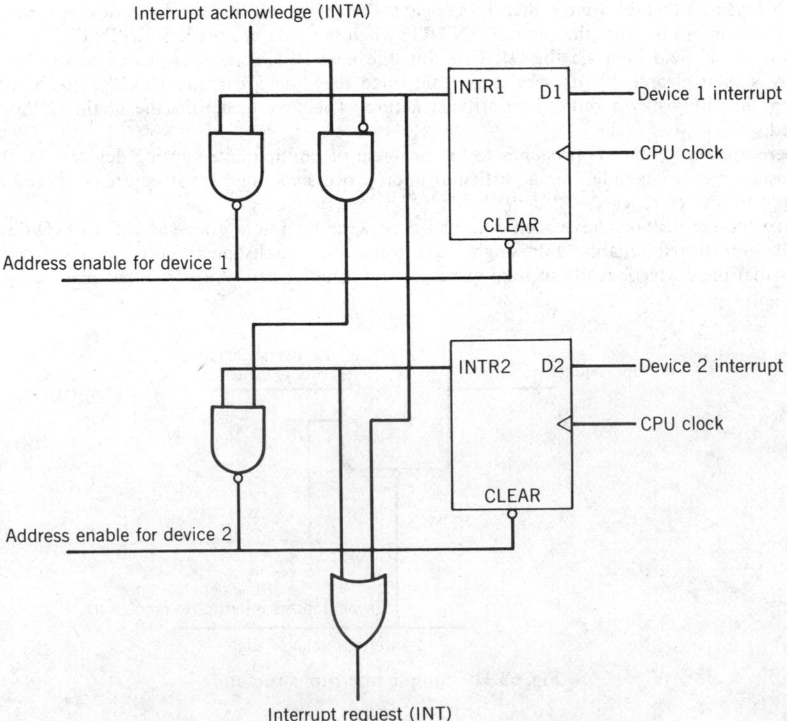

Fig. 62.33 Daisy-chained priority interrupt system. Address enables connected to tristate buffers similar to those in Fig. 62.32.

edges of the read or write pulse. On the other hand, if a block of information is to be transferred, other factors come into consideration.

In synchronous block data transfer, the device receives or transmits data from or to the processor at a rate determined by the processor's clock. For example, if in the monitor system described, a message were to be printed to the screen from a set of continuous memory locations, the flowchart in Fig. 62.34 would dictate a synchronous data transfer. In this case the characters are sent to the output port by the program one after the other until the "end of message" character is detected. The important point to note is that there is no test to see if the port is ready to accept new data, since it is assumed to be always ready. The rate at which the output device accepts the data is the time between write pulses. This is a function of the processor's execution time and the amount of program instructions necessary to read a character from memory and output it to the controller.

In asynchronous data transfer, the character is not sent until the device is ready for it. A modified flowchart for the data transfer is shown in Fig. 62.35. In this case, prior to output the processor checks the "ready" bit. Only when the port is ready will the next piece of data be sent. This handshaking approach to transferring more than one piece of data sequentially to a device is similar to the use of the ready bit in establishment of communications via polling, as was discussed.

In a synchronous block data transfer, where the peripheral can reliably accept data as fast as the processor can supply it from memory, the extra steps of the data having to be input to the CPU and

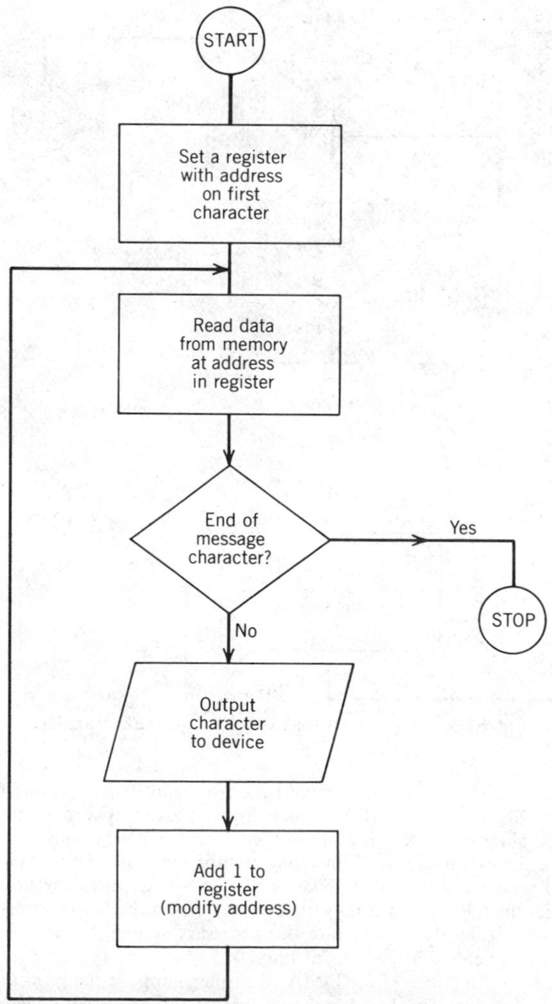

Fig. 62.34 Flowchart for synchronous data transfer.

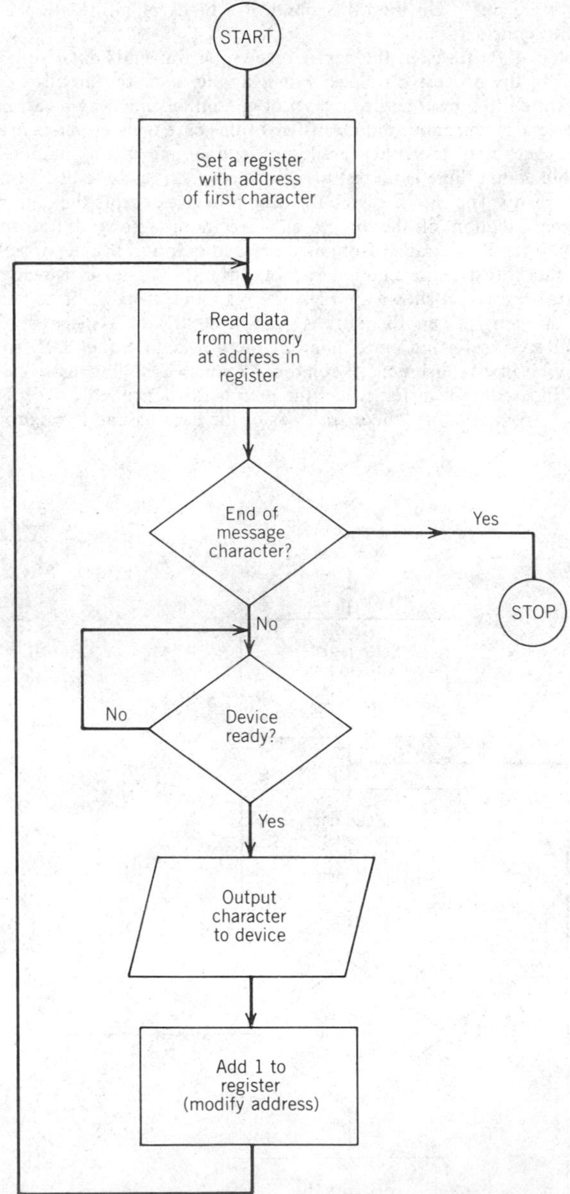

Fig. 62.35 Flowchart for asynchronous data transfer.

output from it cause a system slowdown. To avoid the overhead caused by the CPU intervention, a technique called direct memory access (DMA) may be employed. The devices[14,31] to which DMA applies are disk drives, magnetic tapes, and high-speed communication lines.

DMA involves the direct transfer of data from memory to a device without going through the CPU. Thus data read onto the bus should go directly to the I/O port. The approach used to carry this out is to disconnect the CPU (electrically) from all the system busses (address, data, and control), and to have special-purpose hardware generate the necessary signals to carry out the function. The special-purpose hardware is called a DMA controller.

To illustrate the general functions of the DMA controller, suppose the processor wishes to transfer a block of data to a disk drive. The DMA controller from the processor's viewpoint appears as an

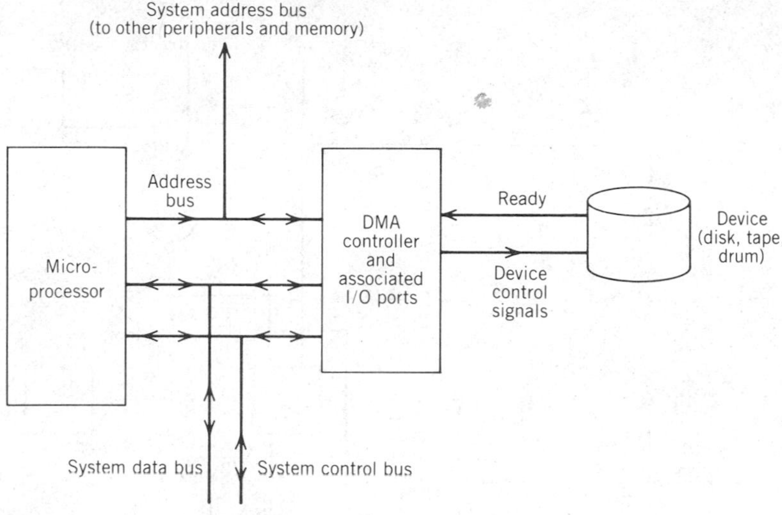

Fig. 62.36 Functional block diagram of a system with direct memory access (DMA) control. I/O, input–output.

I/O port. The processor transfers the starting memory address of the block to be transferred along with the number of words to the controller. The device ready line, instead of being monitored by the processor, is now monitored by the DMA controller. When the controller senses the ready line, it interprets the signal as a DMA request and sends a "hold" signal to the microprocessor. The hold signal causes the processor to tristate all of its bus lines. The DMA controller then takes over the address data and control busses and causes the transfer of information directly. The block diagram of a system including the DMA controller is shown in Fig. 62.36.

The design of DMA systems is rather complex, since the DMA controller in effect must generate most of the bus signals associated with the processor. Owing to the advances and lowered costs of LSI hardware, integrated circuits that perform the required functions are currently available.

62.6 MICROPROCESSOR MEMORY DEVICES

External memory is necessary in most microprocessor systems for program and data storage. Primarily, LSI semiconductor memory devices are used in microprocessor systems and the chips are categorized as ROMs or RAMs.

A block diagram of a generalized memory circuit is shown in Fig. 62.37. The dotted lines indicate the physical boundary of the chip. Address bus information is applied to the address inputs where it is internally decoded by an $n - 2^n$ [6,30] decoder. Each decoder output selects one of the rows in the memory chip. The k cells within the selected row are then available to read from or written to.

If the $R/\overline{W}$ line is high, the information from the data cell is gated through the lower AND gate and switched to the input of the tristate buffer. The chip select line is then used to untristate the device and make the information stored available to the data output pins. Information is written by making $R/\overline{W}$ low and the chip select high. This combination enables the data input lines to be gated to the memory cell, where the W signal strobes in the information.

The chip select signal is generated by a logical combination of address lines and the IO/M signal (IO/M must be low to signal a memory operation). For example, suppose that a memory chip has 12 address inputs and the microprocessor address bus is 16 bits. If the least significant address bus lines are connected to the chip, any memory reference instruction using address $X000_{16}$-$X\text{FFF}_{16}$, $0 \leqslant X \leqslant \text{F}$, will enable a row in the device. If the device should respond to a particular value of X, say $X = 0$, then the chip select should be high when all the most significant 4 bits of the address bus are low and IO/M is low. (The IO/M line is a single lead that distinguishes between IO and memory operations. IO/M = 1 implies IO, IO/M = 0 memory.)

The size of a memory device is usually specified by the number of rows and the number of bits per row. The number of rows is specified as X K, where 1K = 1024 (2^{10}). A 2716 ROM, for example, will be classified as a 2K × 8 ROM.

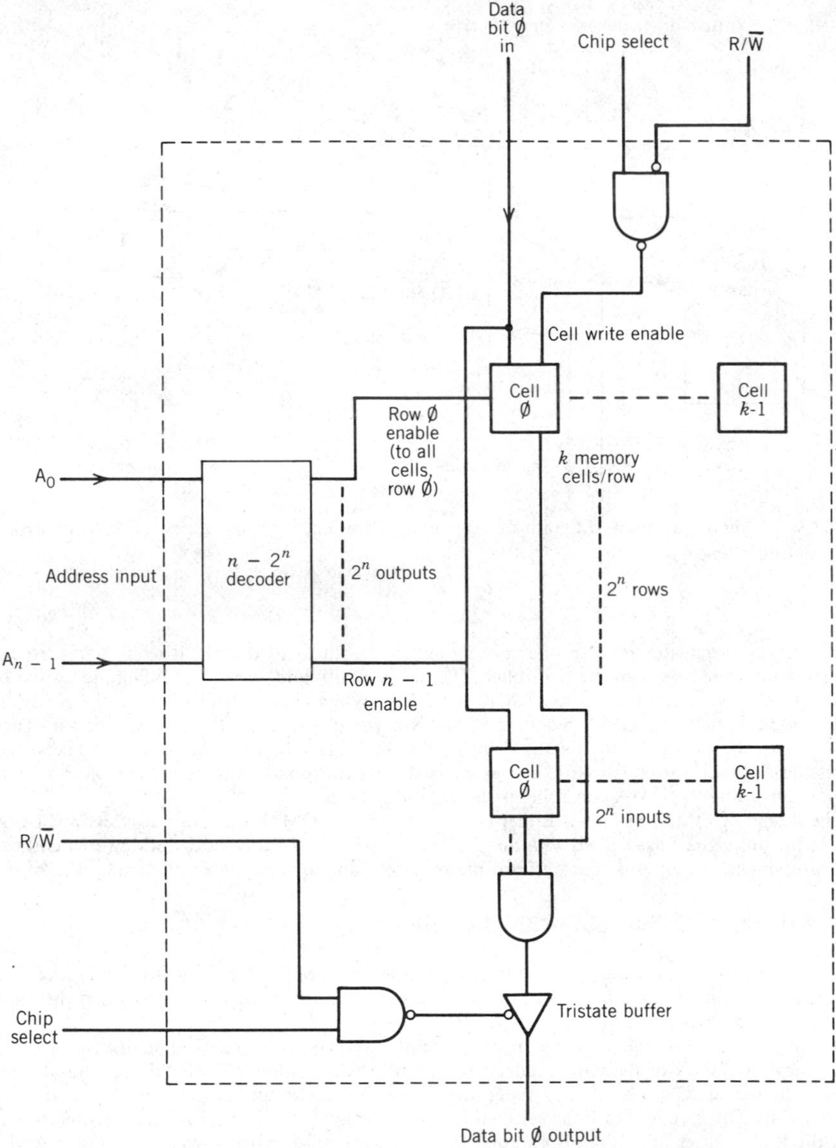

Fig. 62.37 Memory-functional block diagram. R/W, read–write.

In some memory chips, the data input and output lines are combined on one lead. Internal circuitry is used to demultiplex the line. For simplification of internal address decoding circuitry, most memory devices use a two-dimensional decoding strategy.[6,30]

There are two types of semiconductor RAMs, static and dynamic. The static RAM uses an S-R (set-reset) flip-flop as its basic storage cell, while the dynamic RAM uses a capacitor as its storage element. The flip-flop is a bistable device and remains in one of its two states indefinitely until directly altered by a data signal. The capacitor in the dynamic RAM, on the other hand, is a quasistable device. It also has two states, charged and discharged. However, due to leakage in its charged state, the device must periodically be refreshed. This refresh function generally necessitates external circuitry not required by the static devices. The tradeoff is greater density in the dynamic devices owing to the simplicity of the storage cell.

Programs for small systems applications, as well as for permanent data tables, are most conveniently stored in ROM. In addition to retaining its information when power is removed, ROM has a cell structure that achieves greater memory densities than does that of RAM. The cell is generally made up of diode junctions (closed or open circuited). These devices are detailed in Chapter 60.

A major consideration in choosing a set of memory devices for a microprocessor system is the access time. The access time of the memory is defined as the time from which address inputs become stable until the time the data are available at the output of the chip. To obtain this specification, manufacturers test with the chip untristated. A typical timing diagram for a ROM is shown in Fig. 62.38. The diagram also applies to a RAM being read. For a particular processor to work with the ROM (RAM) at the processor's clock rate, the (worst case) minimum "access" time of the processor must be less than the maximum access time of the memory device. The minimum processor access

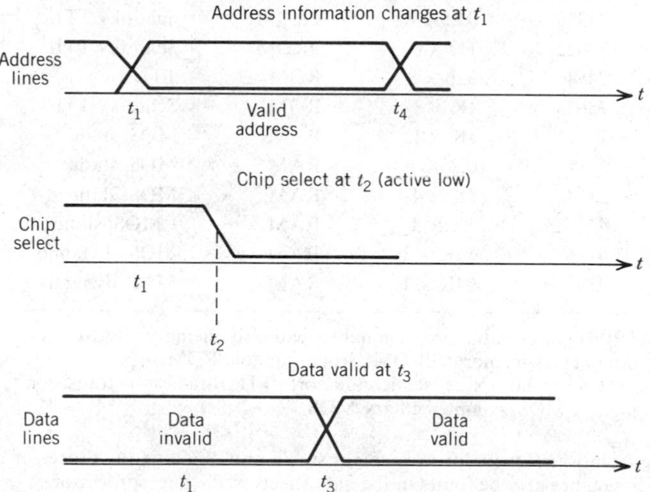

Fig. 62.38 Timing diagram for ROM (read-only memory)–RAM (random access memory) operation. $t_{31} \equiv t_3 - t_1$ = read access time (memory). $t_{41} = t_4 - t_1$ = processor access time (address must be valid for t_{31}).

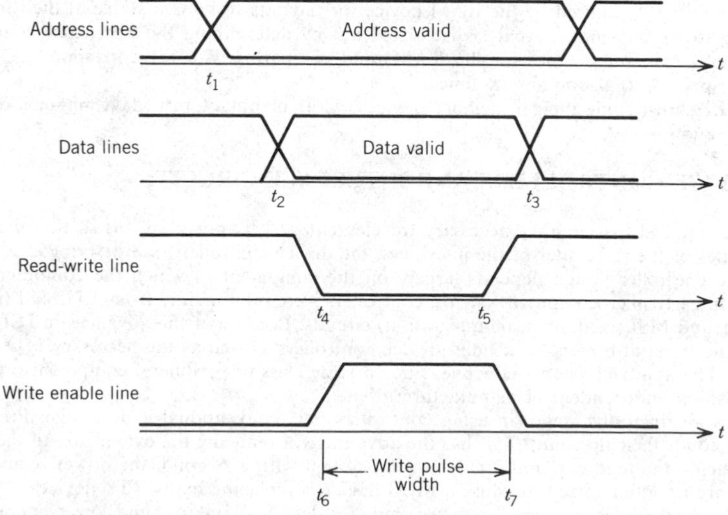

Fig. 62.39 Timing for RAM (read-only memory) write operation. $t_6 - t_2 = t_{62}$ = data set-up time. $t_3 - t_7 = t_{37}$ = data hold time. $t_6 - t_1 = t_{61}$ = address set-up time. $t_7 - t_6 = t_{76}$ = write pulse time.

TABLE 62.2. MEMORY DEVICES (PARTIAL LISTING)

Device	Size	Type[a]	Technology[b]
2764	8K × 8	EPROM	MOS
2732	4K × 8	EPROM	MOS
2716	2K × 8	EPROM	MOS
2708	1K × 8	EPROM	MOS
2704	0.5K × 8	EPROM	MOS
2316	2K × 8	ROM	MOS
1742	0.25K × 8	ROM	MOS
4210	0.25K × 4	ROM	MOS
74S287	0.25K × 4	PROM	Schottky TTL
77S228	1K × 8	PROM	Schottky TTL
7488	32b × 8	ROM	TTL
75S28	1K × 8	ROM	Schottky TTL
2102	1K × 1	RAM	MOS, static
2111	0.25K × 4	RAM	MOS, static
2114	1K × 4	RAM	MOS, static
74C929	1K × 8	RAM	CMOS, static
4116	16K × 1	RAM	MOS, dynamic
4164	64K × 1	RAM	MOS, dynamic

[a] EPROM, erasable programmable read-only memory; RAM, random access memory; PROM, programmable ROM.
[b] MOS, metal oxide semiconductor; TTL, transistor–transistor logic; CMOS, complementary MOS.

time is the time from which the address becomes stable until the time the processor strobes the data from the bus. These times may be found in the data sheets of the microprocessor.

For a RAM, the write cycle of the device must be considered as well (read cycle analysis is the same as for the ROM). While specific devices differ slightly, the RAM write cycle generally depends on four times, as shown in Fig. 62.39: the address set-up time, the time from which the address changes until the time the WRITE strobe goes low; the data set-up time, the time the data are stable on the bus until the time the WRITE strobe goes low; the width of the write pulse; and the data hold time, the time[5,10,17,31] required by the RAM device for the data to remain stable at their inputs once the WRITE strobe has ended. Analysis is carried out by determining the worst-case conditions for each parameter (minimum times for the RAM) and comparing with the corresponding worst-case times for the processor (also minimum times).

Table 62.2 shows some typical memory devices, while manufacturers' data manuals provide the latest information.

62.7 DEVICE CONTROLLERS AND INTERFACE CIRCUITS

Section 62.5 detailed the circuitry necessary for elementary I/O ports. In this section the functions and properties of the remainder of the interfaces, the device controllers, are described.

A device controller design depends largely on the equipment to which the computer system is interfaced. In early microcomputer systems, controllers were built mainly from TTL, SSI (small-scale integration), and MSI (medium-scale integration) circuits. Because of the advances in LSI and VLSI hardware, the trend at present is to build device controllers as well as the necessary I/O ports on a single chip. The approach is cost effective, since a large class of peripheral equipment operates in a standard fashion, independent of manufacturer.

There are distinct advantages in using controllers over conventional logic design. Primarily, the controllers reduce the chip count. This has the advantage of reducing the overall size of the system as well as reducing the time required for debugging system wiring. Second, the power requirements of the system are usually reduced, because a MOS device is replacing many TTL devices. The reduced power demand translates into smaller power supplies, less heat sinking, and lower operational and production costs. Finally, the cost of the part is generally lower than the cost of the equivalent MSI and SSI circuits.

The only disadvantage in using a controller is that it may not exactly fit the specifications of a particular piece of equipment. In most cases when this occurs, additional logic circuitry can be added external to the controller to customize the design.

Controllers are produced by microprocessor manufacturers to be compatible with their particular computer. This means, for example, that an Intel 8257 DMA controller would operate on the system busses of an 8085 computer with minimum internal circuitry. However, because of the structural similarity of many of the microprocessors, it is relatively easy to derive the necessary control signals of one computer from another. In many cases the controller manufacturer provides application notes for interfacing the chip to the system busses of a number of processors.

Device controllers possess some common features. Usually they are packaged in a 40-pin dip. Inputs to the system busses are generally driven by tristate devices, internal to the chip. As in memory devices, the controllers' lines become untristated by the occurrence of an active high (or low) signal on a single chip select lead. Some of the controllers use a stored program to operate the peripheral. They are actually special-purpose microprocessors. These controllers generally require an external clocking signal.

A functional block diagram of a controller is shown in Fig. 62.40. Internally the controller contains a status register and a control register as well as a set of data registers. The device control unit interfaces directly with the peripheral unit, sending and receiving the required signals. The internal registers are also accessible to the device control unit.

The microprocessor communicates with the device controller via the internal registers. From the processor's viewpoint, the registers are simply input–output ports. The selection mechanism for the internal I/O ports is similar to that described in Section 62.5.

As shown in Fig. 62.40, the controller connects directly to the READ, WRITE, and I/O lines and the decoding takes place internally. If more than one input port is in the controller, there generally are leads that connect to the low-order bits of the address bus. The address lines are decoded internally to select a specific port. As an example, consider a controller with four input ports and an active low chip select line. Assume that the input ports of the controller are designated by addresses FC, FD, FE, and FF. The internal circuitry of the controller is shown in Fig. 62.41. The chip select lead is connected to the output of the NAND gate, which detects the occurrence of highs on address lines 7-2; the chip select enables the 2-4 decoder. Address lines 1 and 0 are the inputs to the decoder

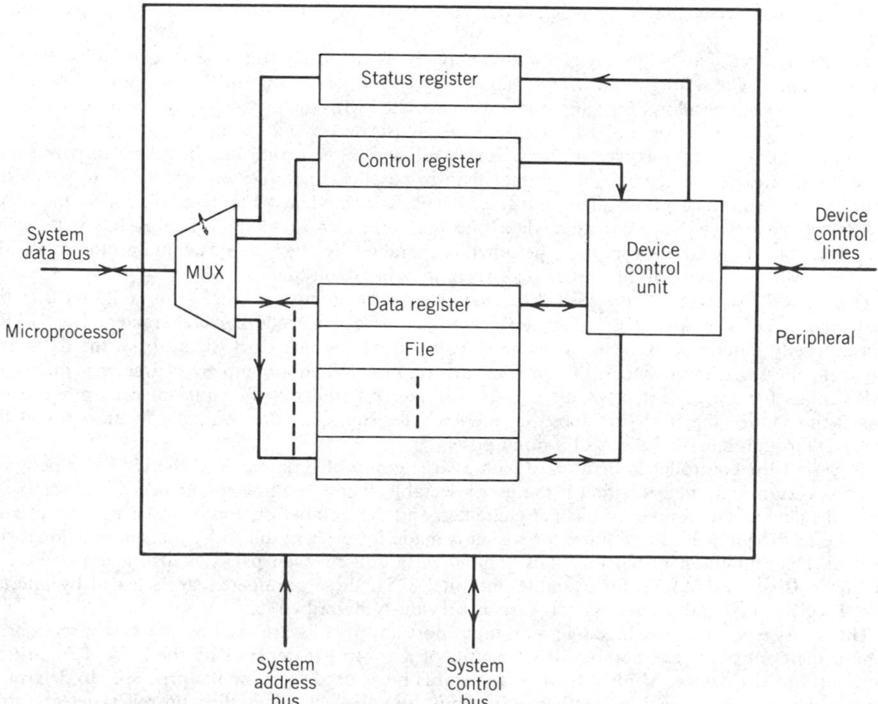

Fig. 62.40 Device controller block diagram.

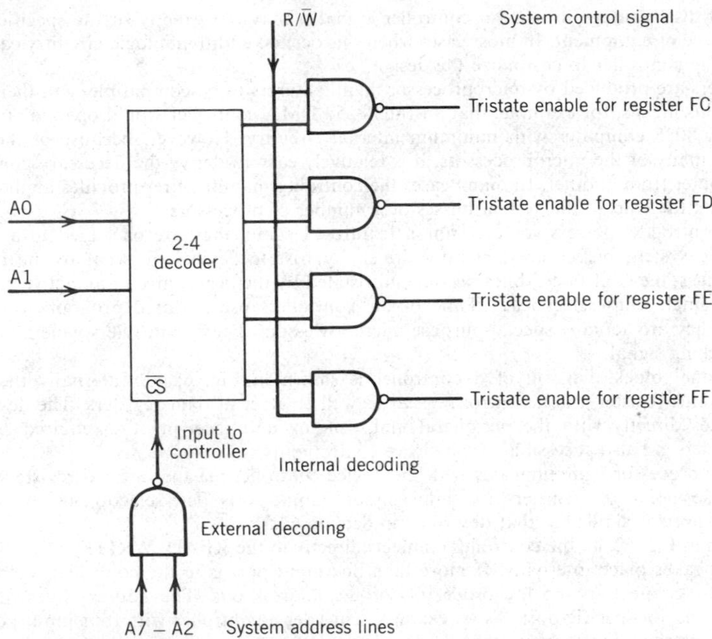

Fig. 62.41 Decoding circuitry for multiregister device controller. R/W, read–write; CS, chip select.

and the outputs enable one of the four NAND gates. The READ pulse completes the selection by untristating the selected buffer. Writing to an output port in the controller is handled in a similar manner.

The control register in the device controller provides (in some controllers) a primitive form of "programming." The setting or clearing of single bits within the register, or the encoding of groups of bits within the register allows certain parameters associated with the peripheral device to be set by the microprocessor control program. This notion is best illustrated by an example.

A universal asynchronous receive transmit (UART) circuit is a controller designed for peripherals that transmit and receive data serially rather than in parallel. Examples of peripherals to which the controller may interface are printers, CRT terminals, and modems. Figure 62.42 shows the block diagram of the device. The maximum data rate the controller can handle is 30K bits per second (b/s). Data are transmitted from these peripherals character by character. The line protocol calls for a continuous logic 1 signal to be transmitted when no data are on the line.

The UART distinguishes the start of a character by detecting a change from a 1 to a 0 at the serial input line. The first 0 is called a start bit and must precede each character. The UART algorithmically determines the center of the start bit (it derives the clock signal from the data) and then scans the serial input line at the bit transmission rate. When a complete character is input, the UART scans for a 1. The 1 bit is the end-of-character bit (the stop bit) and must always end the transmission. Once the data are stored in a UART register, they are available to be input to the processor through one of the UART's input ports.

To permit the controller to be utilized with a large group of serial devices, the UART is designed such that certain parameters within it are user selectable. These parameters include the transmission rate of the device, the number of bits per character, and the number of stop bits (this number usually is related to the bit rate). In addition, provision is made for a parity bit to be transmitted along with the data. Determination of whether it is computed as odd or even parity is also a user-selectable parameter. In some UARTs, for example, the Intel 8251, these parameters are selected by internal control register bits and are set via software as previously described.

The status register is associated with an input port and its bits are used by the microprocessor to monitor the controller. The number of status bits of a controller depends on the level of sophistication built into the device. Minimally at least 1 status bit is used to allow the processor to determine whether the peripheral is busy. Other status bits may be set when the controller detects error

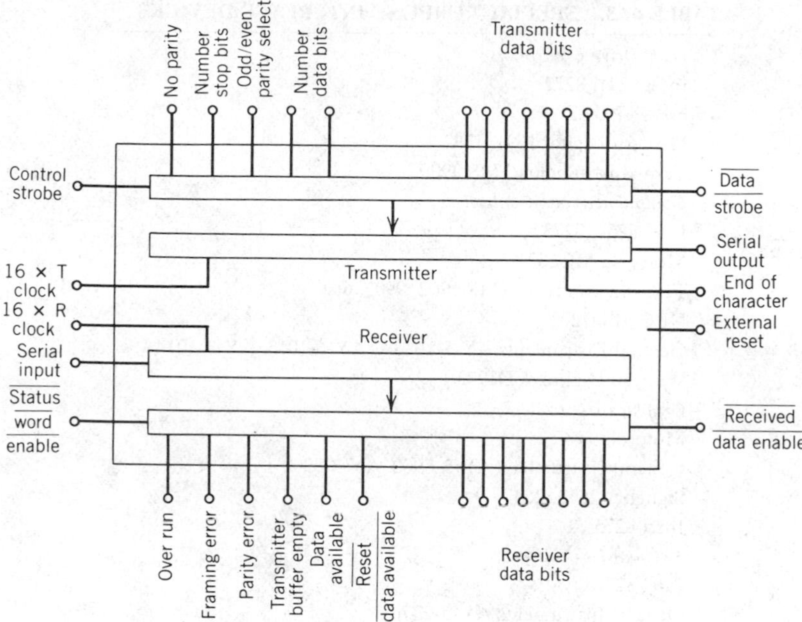

Fig. 62.42 Universal asynchronous receive transmit (UART) block diagram. From General Instrument Corp. *General Instrument Microelectronics*, 1982, reprinted with permission.

conditions. The larger the number of error conditions checked by the controller, the more status bits required in the register.

The UART is designed to detect and indicate certain error conditions arising in the data reception as well as to indicate the busy–idle condition of the receiver and transmitter. The error conditions checked by the UART are:

Parity Error. The received character had an incorrect parity bit.

Framing Error. The received character had a missing stop bit.

Receiver Overrun. The received character was not read from the UART by the microprocessor prior to the start of a new character bit.

The busy–idle condition for the transmitter section of the device is on a lead called TBMT (transmit buffer empty) and the busy–idle bit for the receiver section is on a lead designated DAV. The transmit buffer register and the receiver buffer registers are the data registers associated with the UART.

From the discussion it should be clear that the interface to the UART requires four ports: two input ports, the status port and the receiver port, and two output ports, the control register port and the transmitter register port. The UART can be made into an interrupting input or output port by generation of the interrupt request signal from the busy–idle signals of the device. This requires additional external hardware. While it is true that most device controllers contain control, status, and data registers, it must not be concluded that all of them do. Prior to use of a device controller in a system application, the data sheets for the device should be carefully studied.

Because of the number of device controller chips available, it is useful to classify them into categories organized with respect to applications areas. The available circuits generally fall into the following areas:

1. Disk drive controllers.
2. Communications controllers.
3. Graphics controllers.
4. Keyboard encoders.
5. Signal processing controllers.

TABLE 62.3. SPECIAL-PURPOSE INTERFACE DEVICES

Disk Drive Controllers
Intel 8271, 8272
NEC μPD765
Western Digital 1791, 1792
Texas Instruments TMS 9909
Communications Controllers
Intel 8251, 8273
Motorola MC6821
Texas Instruments TMS 9902, 9903, 6011
NEC μPD7201
General Instruments AY-5-1013A, AY-6-1013, AY-3-1014A
Western Digital WD1933
Graphics Controllers
Motorola 6847
General Instruments AY-3-8900, AY-3-8900-1
Signetics 2636, 2637
Intel 8275
Keyboard Encoders
Intel 8278
General Instruments AY-5-2376
Signal Processing
Intel
NEC μPD7720

Table 62.3 summarizes the devices and the functions performed by one or more of the controllers listed. The semiconductor manufacturers involved in the production of the controllers are also listed. The author apologizes if any devices have not been mentioned.

In addition to being applied to device controllers, LSI techniques have also been applied to "general-purpose" hardware, for example, circuitry for controlling DMA transfer. The functions of a DMA controller have been described in Section 62.5. The functions have been incorporated into a

TABLE 62.4. GENERAL-PURPOSE INTERFACE DEVICES

Identification	Device
TI SBP 9961	Interrupt controller timer
TI TMS 9914	General-purpose interface bus adapter
TI TMS 9911	DMA (direct memory access) controller
TI TMS 9901	Programmable systems interface
Intel 8254	Interval timer
Intel 8255	Programmable peripheral interface
Intel 8232	Floating-point processor
Intel 8231A	Arithmetic processing unit
Intel 8257	DMA controller
AMD Am 9511	Arithmetic processor
Motorola MC6820	Peripheral interface adapter
Motorola MC6846	ROM (read-only memory) I/O (input–output) timer
Motorola MC6840	Programmable timer module
Motorola MC6828	Priority interrupt controller
Zilog Z-80 DMA	DMA controller
Zilog Z-80 CTC	Clock timer circuit
Zilog Z-80 PIO	Peripheral I/O circuit

DMA controller that can handle up to four DMA input devices simultaneously. In addition to the DMA controller, another useful device is a programmable peripheral interface (PPI). These circuits are produced by many manufacturers and vary slightly from one to the other; however, they all perform similar functions.

PPIs contain two or more I/O ports. The direction of each port is programmable through bit settings in a control register vis-à-vis the device controller. Additionally, some of the external pins can be used to provide handshake signals with an internal device controller. This device functions as a single-chip multiport device. One or more can be used to interface an external device controller. The low cost of the devices makes them practical in systems not requiring all the port capabilities, leaving room for the expansion of the system.

Other general-purpose devices are floating-point and fixed-point arithmetic processors, interrupt controllers, and timers (real-time clocks with internal timers). Table 62.4 summarizes the general-purpose device hardware.

62.8 MICROPROCESSOR DEVELOPMENT SYSTEMS AND DESIGN AIDS

The design and implementation of a microprocessor-based system involves the development of both hardware and software. To efficiently carry out the development process, a number of software- and hardware-oriented aids are available to the designer. They include:

1. Programs, compilers, and assemblers, which help in program development and documentation.
2. Simulation programs, which aid in functional hardware testing without prototypes.
3. Debugging programs, which allow register contents and memory contents to be displayed and altered while the test system is in operation.
4. Logic analyzers, which allow sequences of bus information to be stored while the test system is in operation.
5. Prototyping systems for which sets of standard functional units are available.
6. Microprocessor development systems, which incorporate many of the design aids just enumerated.

The end result of program development is the choice of a computer language in which to implement the system algorithms. The choice depends on the application and the program complexity. The three language categories available to the designer are machine language, assembly language, and high-level language.

Languages

Microprocessor languages[3] are described within manufacturers' data manuals[2,7,11,16] in terms of a symbolic representation for each instruction. These symbolic representations are called mnemonics and are generally chosen to indicate to a user the nature of the operation that the instruction performs. Thus the mnemonics ADD, SUB, MUL, and INC are meaningful in that they imply the operations of addition, subtraction, multiplication, and incrementation. Along with each mnemonic, the binary code representing the operation is listed. The binary operation codes constitute the machine language of the processor.

In addition to the operation code, other information must be included to completely specify an instruction. For example, in a "move immediate" instruction on the 8085 (or any other processor), the destination register as well as the immediate operand must be specified. This information is symbolically written in a format as shown in Fig. 62.43.

Within an instruction line, three fields are defined. The mnemonic occupies the operation field and the other information occupies the operands field. If more than one operand is necessary, they are separated by commas within the operands field. The left-most field is the label field. This field is optionally used to specify a symbolic address for the instruction. To the left of the instruction in

Address	Code	Label	Operation	Operands	Comments
0004	0E 80	HERE	MVI	C, 80 H	Load 80 to C register

Fig. 62.43 Illustration of the fields in a line of code. HERE represents (symbolically) the address of the instruction in memory. MVI is the mnemonic (8085 processor) for the instruction. C is the register name. 80 H is the constant 80 H(exadecimal). 0004 is the starting address of the instruction. 0E 80 is the hexadecimal machine code.

columns marked address and code, is the machine language instruction. The address (or location) is the (hexadecimal) starting address of the instruction and the remaining digits represent the instruction.

For short programs (50–100 bytes), where operands are readily computed, machine language is a reasonable method of development. The mnemonic codes for each instruction can be written out, along with the associated numeric values of the operands. The operation code associated with each instruction can then be looked up (in the manufacturer's data sheets) and the entire instruction encoded. Instruction addresses are determined by counting the number of bytes of code starting from the program's initial address. The addresses of each instruction are necessary for encoding jump instructions that reference them. An example of an 8085 machine language program is given in Fig. 62.44. This program contains two jump instructions, one to address 0002 and one to address 0015. These were determined relative to a starting address of 0000.

For a larger size program, the machine language approach becomes unreliable. The probability of an error in translation (due to an address miscalculation) grows. In addition, if nondirect addressing modes such as relative addressing are used, the computation of the offsets becomes tedious, since it involves not only the absolute address of the distinction of the jump instruction but also the value in the program counter at the time the instruction executes.

An assembler program automates the tasks of computing offsets and keeping track of addresses. In addition, it allows the programmer the means to specify operands symbolically. An example of an assembly language program is shown in Fig. 62.45. Note that in the assembly language program, jump instructions reference symbolic addresses, not numeric addresses. For each symbolic address in an operand field there is a line of code with the same symbolic address in the address field. The symbolic address is generally called a label. The assembler program assigns an actual address to each label. The procedure is as follows.

The first statement of the program,[38] ORG $3000, is a command that informs the assembler that the program starts at location 3000 in memory. Since the assembler knows the length of each instruction from its operation code, it can maintain the actual address of each instruction it processes.

Address	Code	Operation
0000	OE 80	MVI C,80H
0002	DB FF	IN FF
0004	A1	ANA C
0005	CA 02 00	JZ 2
0008	C3 15 00	JMP 15H

Fig. 62.44 Machine language program for Intel 8085 microprocessor. Codes and addresses are in hexadecimal.

Number	Address	Code	Label	Operation	Operand	Comments
0001	067A			ORG	$3000	
0002	3000	F6FF02	OVER	LDAB	$FF02	$ = > HEX
0003	3003	2AFB		BPL	OVER	
0004	3005	8D02		BSR	COMP	
0005	3007	20F7		BRA	OVER	
0006	3009	C47F	COMP	ANDB	#$7F	
0007	300B	810A		CMPA	#10	
0008	300D	39		RTS		
0009	300E			END		

Fig. 62.45 Assembly language program for the Motorola 6809 microprocessor. The code is 1 to 4 bytes, depending on the instruction and addressing mode. OVER, COMP are symbolic addresses. The # symbol implies immediate addressing. The $ symbol prior to a number means hexadecimal.

Once it processes an instruction with a label, it assigns the label the numeric code and stores the value in a table of symbols.

Generally, because jumps can be to an address greater than that of the jump instruction, the assembler program cannot build the entire instruction when initially scanned (the label in the operand field has not yet been defined). Thus another "pass" over the data is required to generate code once all symbols have been defined. Assembler programs operating in this manner are called two-pass assemblers (one pass to assign numeric values to all symbols, the second to generate code).

In addition to their symbolic addressing capabilities, assembler programs incorporate a number of other useful features. For example, suppose it is desired to move the 8-bit ASCII equivalent of the letter A into the A register on an 8085, using a move immediate instruction. The immediate operand is allowed to be specified in a number of ways by the assembler. First it can be specified as a numerical entity in a number of different bases: as a binary number, 01001010B, the "B" denoting base 2; as an octal number, 112O; as a decimal number, 74; and as a hexadecimal number, 4AH. In addition, the number can be specified literally as "A"; the quote marks inform the assembler program to replace the value of the operand by its ASCII equivalent. Other features of assemblers include macro and conditional assembly facilities.[2,3,10,38,40,42]

Assembler programs may be obtained from microprocessor manufacturers in two forms. The cross assembler, which is an assembler program that runs on another computer (usually any machine supporting FORTRAN, the most popular language used for the task), and a "self" assembler, usually contained within a microprocessor development system. The latter program executes on a processor with the same language.

High-level languages imply programming languages that exhibit machine independence. For microprocessors (and other computers), the languages fall into two categories. The first kind is a compiled language where machine language instructions are generated for the processor. The second is interpreted code, in which a translation program transforms the high-level language into an intermediate code recognized by a software-defined machine, and the software machine (a program written in the assembly language of the processor on which it is actually executing) "executes" the intermediate code.

Examples of popular interpretive languages are Pascal and BASIC. They find their major application in microprocessor-based personal and small business systems with large memory spaces. The large memory requirement exists because the interpreter must always be resident in memory.

Examples of compiled languages are PL/M and FORTRAN. PL/M was developed for Intel's 8080 and 8008 microprocessors and constitutes a subset of PL/I instructions. It is distinctively different from PL/I in the sense that it only supports two data types, an 8-bit BYTE or a 16-bit ADDRESS. Memory locations are referenced in PL/M symbolically, and declaration statements are used to associate a data type with the symbol. Additionally, the language supports array data structures. PL/M is a procedure-oriented language, which allows the programmer to break up a large program into a set of smaller modules (called procedures). Links between procedures are established through special CALL instructions.

As in PL/I, assignment statements utilizing arithmetic operators may be employed of the general form ⟨Address⟩ = ⟨Expression⟩. The compiler generates the 8085 code for the expression evaluation and then the code to place the result into the address. PL/I-like flow[38] of control structures is also supported within the language.

The major advantage in creating a program using a language such as PL/M is the relief from "bit" pushing, that is, data movement instructions to keep tabs on all variables in a program. In addition, no direct knowledge of the machine architecture is necessary. The penalty paid is generally inefficiency in machine language coding (as generated by the compiler). This may be costly in memory space and, more importantly, in execution time, but offset by the savings in program development time.

Simulators

Another program, useful in the initial stages of system integration is the simulator. The simulator is a program that can run on any machine and can be written in any programming language. As the name implies, the simulator mimics the operations of the CPU, the CPU registers, and the memory of the system. Since input–output devices are almost all application dependent, most simulator programs do not include them. Generally, a provision is made for the user to define his or her own I/O devices within the framework of the program. The simulator can then be used to interpret the design software and provides a means of allowing the user to monitor the register contents at every step of the program. The results can be displayed on a terminal or printed on a hard-copy device.

Continuous monitoring such as this is called a trace. If the register contents are only to be observed at the occurrence of a particular event (e.g., the program reaching a specific address), this is easily incorporated within the simulator. The event causing the output is called a break point. Obviously (given enough memory within the computer system on which the simulator executes), multiple break points can be accommodated, as well as partial traces. In addition to the register contents, memory contents may be observed.

There are a number of drawbacks to the simulator. To begin with, the simulator must run on a processor other than the one used in the system. The size of the simulator calls for a minicomputer or larger machine. Second, because of the interpretive nature of the program (the simulator mimics the operation of the target processor), simulations usually require a rather large amount of computer time. Finally, because of the nature of the processes, interrupt-driven programs and time-dependent programs are difficult to simulate. Real-time interfaces cannot be simulated at all. The major advantage the simulator affords the designer is the ability to easily examine register contents, modify registers and memory at various break points, and test programs independent of building the hardware.

Prototyping Systems

A single-board computer is a system mounted on a printed circuit board that contains the elements essential to a microprocessor-based system. The system usually contains a processor, a ROM containing a monitor program, sockets for additional ROM, a small amount of RAM, and a few parallel I/O ports. Additionally, a keyboard and a set of seven segment displays are included. There is also room on the printed circuit board for customized circuitry.

The monitor program responds to single key commands. The commands provide such functions as:

1. Single stepping of the computer.
2. Examination of register and memory contents.
3. Modification of memory contents.
4. Execution of a program starting at a keyed-in address.

The single-board computer is excellent for small systems programmed in machine language.

A bus-oriented prototyping system consists of a set of printed circuit cards connected into a common bus through edge connectors. The bus signal lines are specified by the proto system manufacturer or alternatively may be user defined. The number of bus lines range from 56 to 100.

The hardware is generally mounted in a card cage and the plugs are commoned at the back of the cage using a PC board or wire wrapping. Each card is modularized according to a specific hardware function. For example, there would be one or more CPU cards (each containing a different processor), memory cards with switch selectable addressing, I/O cards containing serial and parallel I/O ports, and others. Standard hardware modules such as these simplify system design, since the majority of any microprocessor system requires the core devices. Customizing is done on a separate board or boards.

Because of the modularization, the system can be made as flexible as desired in terms of development aids. For example, a disk and a terminal may be connected to the system and development software purchased from the manufacturer.

Microprocessor Development Systems

An integrated hardware–software system oriented toward the development of microcomputer systems is a microprocessor development system (MDS).[5]

The basic MDS consists of a ROM-based monitor program, along with RAM for program development and general-purpose I/O boards which allow terminals (CRT keyboard) and secondary storage device (floppy disks) to be connected to the system. The monitor program allows access to disk-based files which at minimum contain an assembler for the microprocessor of interest. A text editor program, allowing both entry and modification of program files, completes the basic system. This core structure allows entry and assembly of new programs.

To extend the capabilities of the MDS, a number of optional modules are available. For example, a PROM programmer allows newly developed programs to be burned directly into ROM. This device can be interfaced into the I/O space of the MDS. The most useful device is a "universal" programmer, which has the capability of programming an assortment of ROMs. Control of the operation is by the user via the keyboard, and the programs necessary reside as part of the MDS operating system.

To perform system integration efficiently, an optional in-circuit emulation (ICE)[†7] system may be

[†]More devices and capabilities included in a basic MDS usually require more than just a simple monitor program. Generally the additional software resides in programs stored on disk. These programs will be referred to as a monitor system.

attached to the MDS. The ICE system consists of a logic analyzer and a processor that executes the user (development) software in real time. The ICE processor is the same as the one around which the user system is designed. The ICE processor interfaces the user system over a 40-pin cable and dip pin connector, allowing it access to the user's bus. With this connection the ICE processor can access the I/O devices in the user's system as well as the memory space. ICE is a trademark of Intel Corporation. The ICE processor serves not only to execute user's system programs but to interact with the development system as well. This dual role enables the MDS to monitor the execution of the user's program and to communicate commands to the ICE, resulting in a controlled execution.

62.9 MICROPROCESSOR SELECTION

For a product application,[43] a number of factors must be considered in choosing a microprocessor. Among the most important of these are

1. Technical merit.
2. Cost.
3. Availability.
4. Support tools.

Technical Merit

The microprocessor must have adequate memory address space and input–output capability to satisfy the application requirements. In addition, it must have ample speed. For example, in an application involving the use of floating-point arithmetic, benchmark programs can be run to compare the speed of two or more devices. Memory address space is generally determined by the size of the address bus. However, if an application requires the use of a high-level language, the code generated by the compiler may be significantly larger than if the application is written in assembly language, enough so that it may overflow a small memory.

In addition, the instruction set and the addressing modes of machines are important. A machine that does not support binary coded decimal (BCD) arithmetic in an application that requires it would necessitate the writing of subroutines for the application. This would affect the overall speed and the size of the program.

Cost

The cost of a system generally translates into minimizing the overall chip count, when the design is done at that level, or in minimizing the overall board count, when the design is at the modular level. In the latter case, many manufacturers market systems consisting of functional units that plug into standard busses. In both cases the choice of a device depends on the application and the support devices (or units) available to the designer for the particular application. A further consideration is the expandability of the system, in case the requirements change. The choice of board vs. chip design depends on the recovery of the one-time engineering development costs.

Availability

It is of the utmost importance that parts are readily available to the production department when the need arises, as well as to the engineering department when initial development is started. This may dictate selection of a device that is multiple sourced. An additional benefit to this approach is that with competition, unit costs may be minimized. On the other hand, a new device may ultimately suit the product much better than current ones, and it may be worthwhile to delay production until it becomes available. The choice of available software also becomes an important factor in the choice of one system over another.

Support Tools

The tools described in Section 62.8 are vitally important in getting a production job off the ground. If the application calls for programming in a high-level language, the processor chosen must have the appropriate compiler available to it as well as a development system that supports software development. In addition, a good macroassembler should be available to the user and either an in-circuit emulator or a single-board computer with a built-in debugging monitor. The latter should also provide a convenient means of loading the program into its RAM from the development system.

References

1 General Instrument Corp., *Microelectronics Data Catalog*, New York, 1980.

2 J. Wakerly, *Micro-Computer Architecture and Programming*, Wiley, New York, 1981.

3 J. G. Webster and W. D. Simpson, *Software Design for Microprocessors*, Texas Instruments, Dallas, 1976.

4 Texas Instruments, Inc., *Microprocessors/Microcomputers/System Design*, McGraw-Hill, New York, 1980.

5 A. Osborne, S. Jacobson, and J. Kane, *An Introduction to Microcomputers*, Vol. 1, *Basic Concepts*, Vol. 2, *Some Real Products*, Osborne/McGraw-Hill, New York, 1977, 1980.

6 M. Mano, *Digital Logic and Computer Design*, Prentice-Hall, Englewood Cliffs, NJ, 1979.

7 Intel Corp., *The 8086 Family User's Manual*, Santa Clara, CA, 1980.

8 Signetics Corp., *2650 Operations Manual*, CA, 1976.

9 Motorola Corp., *MC6800 Applications Manual*, Phoenix, AZ, 1975.

10 L. Leventhal, *Introduction to Microprocessors: Software, Hardware, and Programming*, Prentice-Hall, Englewood Cliffs, NJ, 1978.

11 AMD Corp., *AM Z8001/2 Processor Instruction Set*, Sunnyvale, CA, 1980.

12 B. Soucek, *Microprocessors and Microcomputers*, Wiley, New York, 1976.

13 A. G. Lippiatt, *The Architecture of Small Computer Systems*, Prentice-Hall International, London, 1980.

14 J. Peatman, *Microcomputer-Based System Design*, McGraw-Hill, New York, 1979.

15 G. Myers, *Digital System Design with Bit-Slice Logic*, Wiley, New York, 1980.

16 AMD Corp., *The 2900 Family Data Book*, Sunnyvale, CA, 1980.

17 J. Mick and J. Brick, *Bit-Slice Microprocessor Design*, McGraw-Hill, New York, 1979.

18 Advanced Micro Devices Corp., *Microprogramming Handbook*, Sunnyvale, CA, 1977.

19 H. Katzan, Jr., *Microprogramming Primer*, McGraw-Hill, New York, 1977.

20 H. Taub and D. Schilling, *Digital Integrated Electronics*, McGraw-Hill, New York, 1977.

21 Texas Instruments, Inc., *TMS 9900 Microprocessor Data Manual*, Austin, TX, 1980.

22 Fairchild Corp., *9445 Microflame II*, New York, 1980.

23 Texas Instruments, *9900 Family Systems Design and Data Book*, Austin, TX, 1978.

24 Motorola Corp., *MC68000 User's Manual*, 3rd ed., Prentice-Hall, Englewood Cliffs, NJ, 1982.

25 Intel Corp., *Intel 8080 Microcomputer System User's Manual*, Santa Clara, CA, 1975.

26 H. Garland, *Introduction to Microprocessor System Design*, McGraw-Hill, New York, 1978.

27 R. Goody, *Microcomputer Fundamentals*, SRA, Chicago, IL, 1979.

28 Intel Corp., *Component Data Catalog*, Santa Clara, CA, 1982.

29 D. Givone and R. Roesser, *Microprocessors/Microcomputers: An Introduction*, McGraw-Hill, New York, 1980.

30 H. Taub, *Digital Circuits and Microprocessors*, McGraw-Hill, New York, 1982.

31 G. Craft and W. Toy, *Mini/Microcomputer Hardware Design*, Prentice-Hall, Englewood Cliffs, NJ, 1980.

32 Digital Equipment Corp., *The PDP 11/70 Processor Handbook*, Maynard, MA, 1980.

33 D. Knuth, *The Art of Computer Programming*, Vol. 1, *Fundamental Algorithms*, 2nd ed., Addison-Wesley, Reading, MA, 1973.

34 General Instrument Corp., *CP1600/1610 16-Bit Single-Chip Microprocessor Data Manual*, New York, 1979.

35 T. Cantrell, "An 8088 Processor for the S-100 Bus," Part 3, *Byte* **3**(11) (Nov. 1980).

36 Zilog Corp., *Z8000 PLZ/ASM Assembly Language Programming Manual*, CA, 1979.

37 M. M. Mano, *Computer System Architecture*, 2nd ed., Prentice-Hall, Englewood Cliffs, NJ, 1982.

38 Motorola Corporation, *MC6809-MC6809E Microprocessor Programmer's Manual*, Phoenix, AZ, 1980.

39 R. Krutz, *Microprocessors and Logic Design*, Wiley, New York, 1980.

40 M. Andrews, *Programming Microprocessor Interfaces for Instrumentation and Control*, Prentice-Hall, Englewood Cliffs, NJ, 1982.

41 S. Leibson, "The Input/Output Primer, Part 4: The BCD and Serial Interfaces," *Byte* **7**(5) (May 1982).

42 R. Zaks, *The CP/M Handbook*, Sybex, Berkeley, CA, 1980.

43 J. Wakerly, "How to Choose a Microcomputer," *Computer* **12**(2):24 (Feb. 1979).

Bibliography

Artwick, B. A., *Microcomputer Interfacing*, Prentice-Hall, Englewood Cliffs, NJ, 1980.

Blakeslee, T., *Digital Design with Standard MSI and LSI*, Wiley, New York, 1975.

Bowen, B. and Buhr, R., *The Logical Design of Multiple-Microprocessor Systems*, Prentice-Hall, Englewood Cliffs, NJ, 1980.

Buzen, J. P., "I/O Subsystem Architecture," *Proc. IEEE*, **63**:871–879 (Jun. 1975).

Digital Equipment Corp., *PDP11 Peripherals and Interfacing Handbook*, Maynard, MA, 1972.

Electronic Industries Assoc., *EIA Standard RS-232-C: Interface Between Data Terminal Equipment and Data Communications Equipment Employing Serial Binary Data Interchange*, Washington, DC, 1969.

Garrett, P., *Analog Systems for Microprocessors and Minicomputers*, Reston, Reston, VA, 1978.

Illfe, J. K., *Advanced Computer Design*, Prentice-Hall International, London, 1982.

Intel Corp., *MCS-86 Assembly Language Reference Manual*, Santa Clara, CA, 1980.

Intersil Corp., *Data Acquisition Handbook*, CA, 1980.

Lofthus, A. and Ogden, D., "16-Bit Microprocessor Performs Like a Minicomputer," *Electronics* **49**(12) (May 20, 1976).

NEC Microcomputers, Inc., *1981 Catalog*, MA, 1981.

Ogdin, C. A., *Microcomputer Design*, Prentice-Hall, Englewood Cliffs, NJ, 1978.

Signetics Corp., *TTL Data Manual*, CA, 1980.

Weitzman, C., *Distributed Micro/Minicomputer Systems*, Prentice-Hall, Englewood Cliffs, NJ, 1980.

Weitzman, C., *Minicomputer Systems*, Prentice-Hall, Englewood Cliffs, NJ, 1974.

Western Digital Corp., *1981 Catalog*, CA, 1981.

CHAPTER 63
SOFTWARE ENGINEERING

FREDERIC L. SWERN

Stevens Institute of Technology
Hoboken, New Jersey

63.1 INTRODUCTION

The declining cost of computer hardware has resulted in an increase in both the number and complexity of new applications. To control rising costs, many of the ideas and practices of the established engineering disciplines have been applied to software development. The term "software engineering" was chosen in 1968 to describe techniques related to constructing high-quality software.[1]

Presently, software engineering consists of a collection of techniques covering a broad range of topics, including the following:

1. Systematic techniques for programming.
2. Portability and adaptability of software.
3. Program testing and validation.
4. Proof of correctness.
5. Program analysis and performance evaluation.
6. Security and protection.
7. Management of software development.
8. Data base management.
9. Documentation.

The use of software engineering practices has been shown to significantly reduce program development costs on large projects. A recent study[2] at Boeing Computer Services showed an average cost reduction of 73% over forecast costs for three projects. However, it is estimated that the techniques are not being widely used, underlining a need for education of both programmers and managers in this area.

63.2 PROGRAMMING METHODOLOGY

Systematic Techniques for Programming

While early programs were written in an ad hoc manner, modern programming practice dictates that they should be well organized and structured. Some of the principles that may be applied to achieve systematic programming include the following:

1. **Modular decomposition or structured design.** The problem is broken down into smaller pieces, or modules. A module is a disjoint unit that is constructed independently of other modules, possibly along functional boundaries. The size of each module is such that it is easily maintained by a programmer, and there are well-defined interfaces between modules to facilitate program integration and testing.

2. **Top-down design.** A complex system is decomposed into modules in a hierarchical manner, starting with the highest level and working down. First, the high level modules are laid out and coded. Testing may begin using dummy modules or stubs in place of lower level modules.

3. **Bottom-up design.** An alternative scheme to top-down design is bottom-up design, where the lowest level modules are coded first and tested with dummy drivers while design proceeds hierarchically upward.

4. **Structured programming.** Program modules are written in such a manner that logic flow is from beginning to end without explicit JUMP or GOTO statements. More information on structured programming is given in Section 58.2. Such a program is easier to validate, test, and maintain.

5. **Localization.** Portions of a system that are related are brought together in modular proximity. This may be the basis for a particular decomposition. For instance, a single program module may handle all references to a particular device. The advantages of this technique lie in improved resource utilization, readability, and maintainability.

6. **Hiding.** Details of a module that are not essential to its understanding are made inaccessible. The idea here is to amplify the important information in a module by eliminating the unimportant. An example is the assigning of symbolic names to certain resources, the resources to be defined later.

Program Testing and Validation

Software testing consists of a number of distinct phases to be carried out both during the development process of the software product and as a final validation of its integrity. The software is always tested against its specifications. It is apparent that the ability of a validation scheme to ensure correct functionality is inherently related to complete, correct, and unambiguous specifications. In addition, revalidation of software after maintenance requires a set of tests that demonstrates not only the ability of the modifications to produce the desired result, but also that the integrity of the system as a whole is not adversely affected.

Structured Walkthroughs. During the programming phase, structured walkthroughs are used in the same manner as the old engineering design review. The programmer presents to a group of peers his or her current project efforts in a structured manner. The group then aids in detecting errors and problems which the programmer may later correct.

Static Analysis. Many errors may also be detected prior to program execution by static analysis. This technique finds errors that are related to the structure of the code. Some examples of errors that might be detected include variables that are never initialized, code segments that are never executed, errors in variable name spellings, and subroutine argument correspondence. Much of this information can be obtained from the output of a good compiler. It is often desirable to use program flow graphs (an example is shown in Fig. 63.1 along with the pseudocode describing it) to aid in the analysis. Many structural errors become more apparent on the graph.[3]

Dynamic Testing. Dynamic testing of a software system often proceeds in a bottom-up manner. Each of the lowest level modules is tested individually with use of a dummy driver program to exercise it through its specification set. Subsystems are then built from a group of modules and tested in a similar manner. Finally, system integration testing is concerned with testing the interfaces between subsystems and validating the overall system.

An alternate scheme of dynamic testing is the top-down approach in which the higher level modules are tested with use of program stubs for the lower level routines. This is particularly attractive when coupled with a top-down coding scheme.

In some situations, neither top-down or bottom-up testing can be followed and a mixed testing strategy is pursued. The availability of program modules and the feasibility of constructing either program stubs or dummy drivers govern the choice of method.

Proof of Program Correctness. Certain applications require proving that a program is correct with respect to its specifications. The specifications imply assertions about the input and output conditions of the program. Another set of assertions, called a predicate, is constructed from the computational state at some particular point in the program. Many predicates can be associated with each program. If it can then be rigorously shown that each predicate is correct given that all preceding assertions are

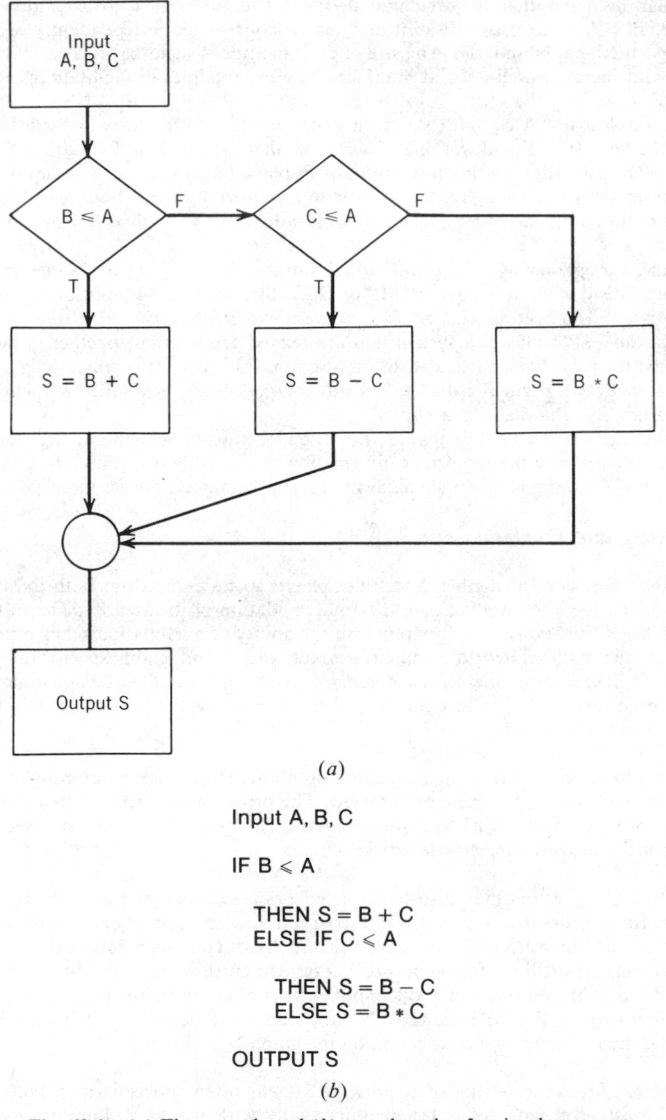
Input
A, B, C
B ≤ A
F
C ≤ A
F
T
T
S = B + C
S = B − C
S = B * C
Output S
(a)
Input A, B, C
IF B ≤ A
THEN S = B + C
ELSE IF C ≤ A
THEN S = B − C
ELSE S = B * C
OUTPUT S
(b)

Fig. 63.1 (*a*) Flow graph and (*b*) pseudocode of a simple program.

true, the program has been proved correct. Currently, proof of correctness has only been employed successfully in small systems.[4]

Monitoring Program Execution

Two reasons for monitoring program execution are to validate the program and to evaluate its performance. In the first case it is desired to ensure that all paths in a program have been exercised and all statements executed at least once. It may not be practical to generate a set of test data that covers completely all possible input conditions. However, a statistical sample of inputs coupled with judicious program monitoring may verify that a given test sequence is reasonably complete. In the second case it is desired to improve resource utilization of a program [e.g., time, memory, input–output (I/O) devices, etc.].

Two monitoring methods may be used:

1. Event driven. The event-driven monitor consists of software routines activated from different points in the program code, called hooks. These hooks may be placed in the program by subroutine calls, compiler options, or run time options. They might cause a counter location in storage to be incremented each time a particular path in the program is taken. By placing these counters at strategic points (see Fig. 63.2), it is possible to obtain a picture of the execution characteristics of the program. In addition to counting, these hooks may perform other functions, such as checking and recording the sequence of execution of program paths. In some cases pertinent data may be written to a dataset at the occurrence of each event. This results in a trace of program execution which may later be analyzed. A hook or program segment inserted for any of these purposes is called a software probe or software monitor.[5]

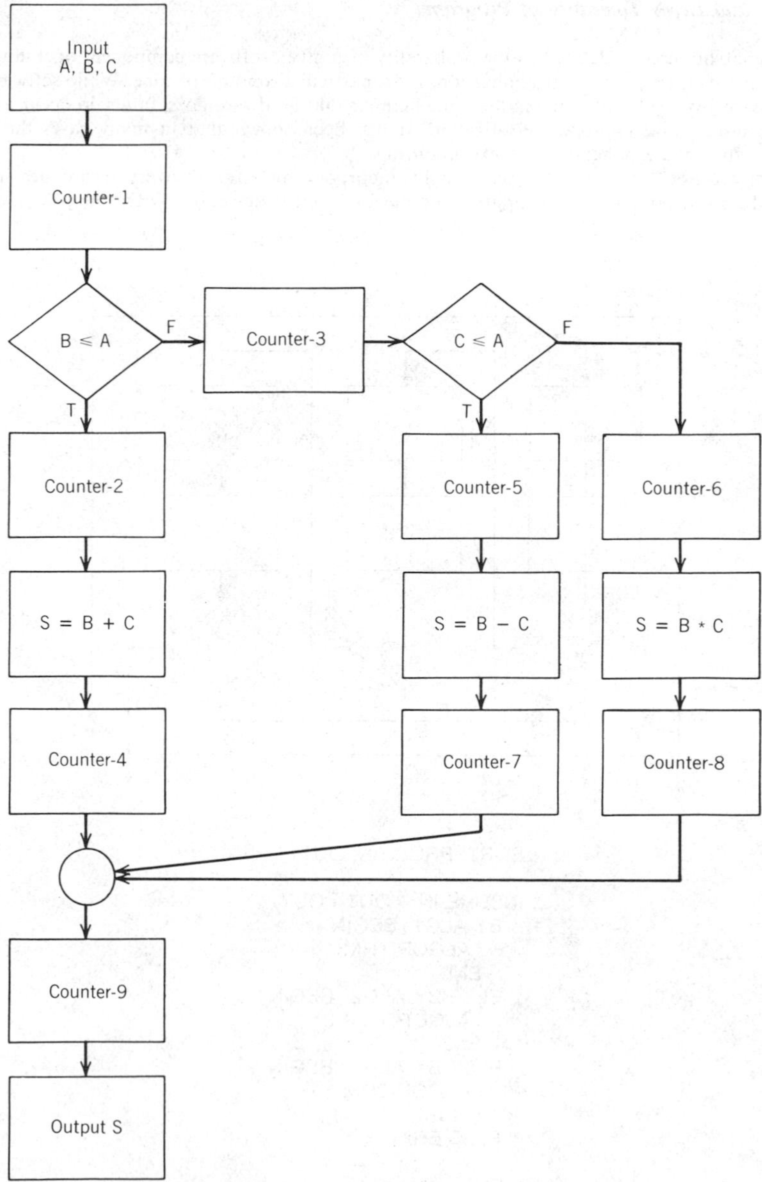

Fig. 63.2 Example of counters placed in a program.

2. Statistical. Statistical monitoring is implemented by sampling program data at given time intervals and recording it. When the sample is large enough, statistical inferences may be drawn as to the execution characteristics of the program. For instance, recording of the instruction address at each sample allows a graph of the frequency of execution of each module or of each source statement to be constructed. This may be helpful in identifying inefficient code in critical applications.

Regardless of the monitoring method, the code added may significantly corrupt the sampled environment. It is possible to avoid the problem by using a dedicated piece of hardware, called a hardware monitor, to take and store the samples. Obviously, monitoring is limited to those quantities available in the hardware of the machine to be monitored, such as memory address, I/O device address, and so on.

Reliability and Error Tolerance of Programs

Software reliability can be defined as the probability of a given software component operating in an acceptable manner, in a particular application, for a particular amount of time. While software does not deteriorate physically with time in the same manner that hardware does, faults do occur owing to incomplete testing and improper specification. It has been shown that in many cases the failure model for software is exponential with execution time.[6]

Reliability can be significantly increased by incorporating error recovery techniques into the software. However, error recovery implies some means of error detection. When the software errors

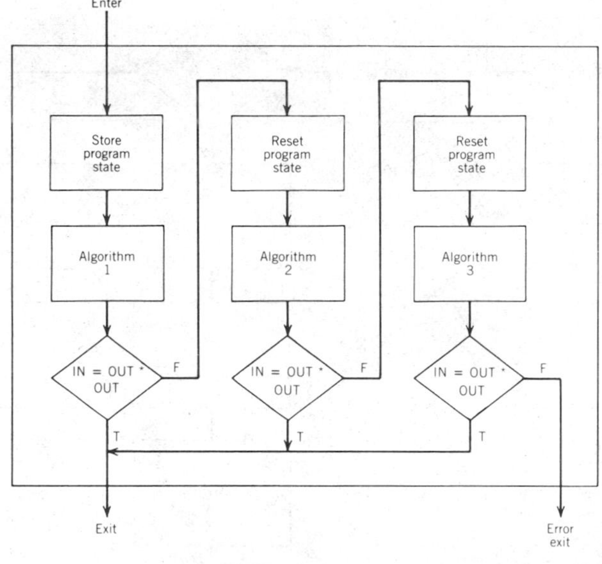


```
SQRT: PROC (IN, OUT)

INSURE IN = OUT * OUT
  BY ALG1 : BEGIN
    ⟨ALGORITHM 1⟩
    END
  ELSE BY ALG2 : BEGIN
    ⟨ALGORITHM 2⟩
    END
  ELSE BY ALG3 : BEGIN
    ⟨ALGORITHM 3⟩
    END
  ELSE ERROR

END
```

Fig. 63.3 Recovery block structure for a square root routine using Randell's notation.[7]

are induced by hardware errors, any of the standard detection techniques (such as redundancy, error correcting codes, etc.) may be used to trigger the recovery procedure. However, coding errors are harder to detect in that any attempt at redundancy would be unsuccessful. Often some type of acceptance check on the input data at crucial modules in the program is added to detect possible faults.

One recovery scheme employs blocks of data associated with each level of a multilevel process recording pertinent information on the state of the process on entry to that level. The blocks are hierarchically structured and protected from alteration by lower levels. When a failure is detected, control may be diverted to a higher level routine, which may retry that level with a different routine, abort the computation, or execute some user-supplied error recovery procedure. An example of a square root routine with three error alternatives is given in Fig. 63.3. The form of pseudocode used in the figure for representing the recovery process was suggested by Randall,[1] and a flow diagram is also shown.

63.3 MANAGEMENT OF SOFTWARE PROJECTS

Chief Programmer Teams

The concept of a chief programmer team was developed to reduce the problems of coordinating programmers on large software projects. A small group of programming professionals, usually three to six people, make up a team, including a program librarian, a chief programmer, and his or her backup programmer. Administrators and clerical personnel may be added to the group depending on the size of the project.

The chief programmer is the group manager and is also responsible for overall design of the software and coordination of the effort. He splits the problem into modules and assigns them to the programmers in the team, possibly programming some of the highest level modules himself. The backup programmer reviews the design with him and serves as a second in command. All of the keypunch and secretarial services are supplied by the librarian.

The chief-programmer-team concept tries to abandon the idea of a programmer being solely responsible for his or her own work and replaces it with the idea of software as a group effort. The introduction of this type of organization is often met by hostility on the part of programmers, as they are afraid of losing control over their own work. However, after a suitable training and adjustment period, both the productivity of the programmer and the quality of the program product are significantly increased.

Projects that are too large for a single chief programmer team might be organized into a number of such teams. A project manager might then break the system down into pieces suitable for each team. There has not been a significant amount of experience with large projects to assess the effectiveness of this method.

Documentation Techniques

Computer program documentation should specify requirements for program design, development, and testing. In addition, it should provide a detailed and complete description of the program for operation and maintenance. This includes the following:

1. A program performance specification.
2. A program design specification.
3. The computer program itself.
4. A data base of variable names and a cross reference of where they are used in the various program modules (sometimes called a traceability matrix).
5. An operator's manual.
6. A set of test procedures.
7. A set of test results.

The generation of good documentation results in proper coordination of the software team effort. It also gives a history of the design process to be used through the life of the software. While early programmers considered flowcharting and the inclusion of a large number of comments sufficient to satisfy documentation requirements, modern programming practice recognizes a need for additional items. Techniques such as the use of descriptive variable names and comments in the prolog to each program segment, in addition to comments interspersed throughout the program, aid in understanding. Each module requires a clear description of the algorithms performed and a list of the input–output conditions. Diagrams for module flow, function flow, and data flow are helpful in documenting system operation (see, for example, Refs. 8 and 9).

Source Code Management

Most program development projects involve large amounts of source code that is constantly undergoing update. For ease of access it is usually stored in programming libraries on direct access devices.

During the software life cycle, a program may undergo a number of changes as both individual versions to satisfy different users and maintenance changes to correct programming errors. It becomes necessary to identify each version so that, should the need arise, access may always be gained to the proper source. One method involves associating with each program module a group of deltas, where each delta represents a set of source code updates for a specific purpose. The delta datasets contain only the changed lines with line numbers and have unique identifying names. A typical sequence of deltas is shown in Fig. 63.4. Once written, the program modules and delta datasets are never changed; maintenance is applied by creating a new delta.[10]

The keeping of careful records of which deltas were used to create a given object module allows re-creation of the corresponding set of program source files. This is especially helpful when multiple versions of a program exist in the field. The list of deltas can be noted somewhere in the executable code so that, by examination of a system dump, the correct source can be found.

Increasing Productivity

The consideration of human factors in software engineering can lead to a significant increase in programmer productivity. While the availability of software tools can greatly simplify a given job, the usefulness of such tools is greatly enhanced by a well-designed user interface. This is especially true in the design of interactive aids. Any interactive system should be friendly, easy to use, and foolproof. A normal dialogue should not be confusing, threatening, or boring to the user, and error messages should be informative, indicating the corrective steps to be taken if possible. User inputs should be simple for the beginner, yet not cumbersome for the experienced user. This last requirement is often met by allowing abbreviations to be used for common keywords, and also by incorporating extra messages that may be called up by beginners. Sufficient checking should be incorporated so that most common errors are diagnosed and handled properly.[11]

While modern programming practices such as structured programming and modular design help reduce programming errors (which increases productivity), so does the use of programming languages

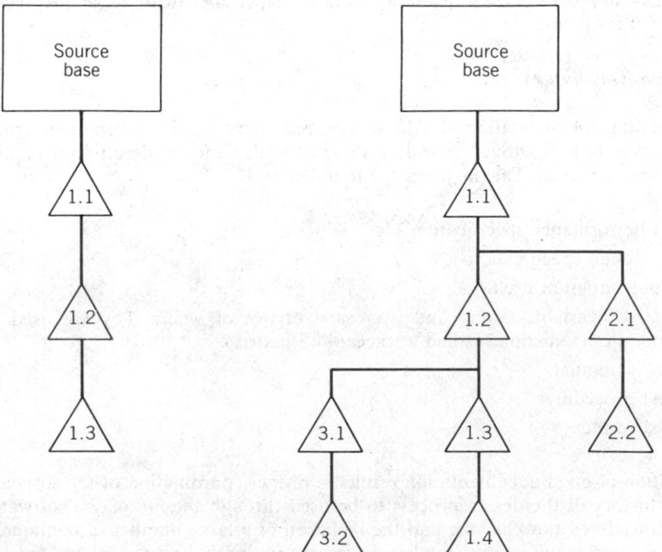

Fig. 63.4 A simple delta structure and a complex delta structure.

that contain inherent fault checking. Most implementations of FORTRAN and COBOL are very lax in this respect, since they allow compilation of any syntactically valid statements regardless of the programming usage. It would be desirable, for instance, to check the bounds of subscripts, the agreement of arguments in subroutine calls, and the like, to minimize errors. Newer languages like Pascal have been designed with these considerations in mind. However, at this time most programming efforts are continuing with either FORTRAN or COBOL.

Requirements Generation

Before any software system can be designed, the programmer must carefully assess exactly what the system is going to do and how it is going to do it. Too often the lack of a solid approach to requirements definition results in a system that is extremely costly, wasteful, and incapable of doing whatever it is supposed to do.

The requirements definition process usually results in the generation of a document to be used in the software design process. The document must be complete and unambiguous. Many techniques have been proposed, and a number of computer interactive systems have been devised.[9,12,13] A model of an interactive scheme is shown in Fig. 63.5.

One view of requirements definition divides it into three subjects that form boundary conditions on the system design that follows. The reasons why the system is needed and how it interacts with its environment must be considered. This is called context analysis. A functional specification of the system tells what functions it must accomplish, and design constraints tell how it will be built. A good set of requirements is consistent, testable, feasible, and flexible.[12]

Among the programs available for requirements generation is structured analysis and design techniques [SADT ®] marketed by SofTech. This is a graphics-based language that provides a hierarchy of diagrams to structurally present the information. The organization of a team of individuals to interact with the SADT system and to review the ongoing work allows achievement of a coordinated method of documenting the results in a clear and precise manner. A SADT model consists of an organized sequence of diagrams representing the requirements of the system in a top-down manner.[9,12]

Another set of programs was developed by TRW Defense and Space Systems Group called requirements engineering and validation system (REVS). This includes a requirements statement language (RSL) that provides a means for stating requirements in terms of structures of flow graphs. These are translated into a centralized data base which is checked for consistency. Finally, a model may be simulated from the requirements and checked for validity.[13]

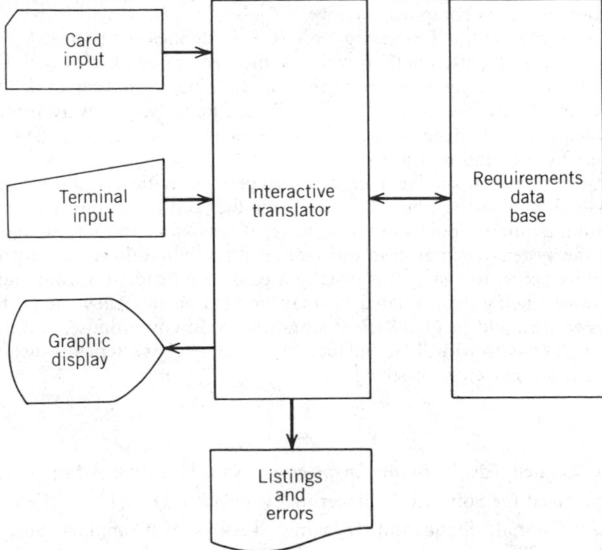

Fig. 63.5 Block diagram of an interactive requirements generation system.

Maintenance and the Software Cycle

The software development process may be characterized by a number of key steps collectively called the software life cycle:

1. Requirements analysis.
2. Software specification.
3. Program design.
4. Verification and testing.
5. Performance evaluation.
6. Maintenance and configuration management.

These six steps may be separated into three phases: definition, implementation, and maintenance.

The staggering cost of the maintenance phase is exemplified by studies showing that up to 70% of software costs during the life cycle are spent on maintenance compared with the other two phases.[14]

Software maintenance can be compared to its hardware counterpart, where much experience has been gained in efficient organization. Both require the correction of design errors. In software, these design errors are referred to as bugs. Hardware is also prone to failure due to deterioration, whereas software has no counterpart since it does not wear out. A large percentage of software maintenance involves changes and improvements to meet a new set of specifications. New capabilities are often added to an existing system by altering, deleting, and extending existing code rather than by redesigning the system as a whole.

It can be seen that the high cost of software maintenance is not indicative of poor-quality workmanship during the definition and implementation phases, but rather of a high level of change in requirements after the program has been developed. However, the design and documentation techniques that were applied during the first two phases are often absent or at a reduced level during the maintenance phase. This results in poorer programs at higher cost. In addition, the testing of modifications is often not as thorough as during the development phase, resulting in serious exposure to system failure.

Security

The security of a computer system from use by unauthorized persons and maintenance of the system from compromise by unscrupulous users have gained significant interest as a large number of applications are becoming available interactively via teleprocessing. While many computer crimes have involved employees with access to sensitive portions of the system, the malicious outsider represents a significant threat to computer security.

The architecture of the central processing unit (CPU) defines a privileged state in which all instructions are operative and a user state in which some instructions are denied. Only with proper authorization will the machine enter the privileged state. The restriction of I/O instructions as privileged denies the malicious user access to the devices. Similarly, hardware protection of storage blocks may be implemented, or page segmentation tables in virtual memory systems are used, to protect storage access by unauthorized users.

Physical security and administrative control of access to critical resources is crucial to the prevention of abuse. Independent and objective periodic audits of the system can review its adequacy. By embedding monitor points in the software it is possible to form an audit trail of critical events occurring in the system (such as cash transactions, data alterations, etc.) for later scrutiny.

The key to enabling access to a system is usually a password or identification number. To reduce the chances of passwords being disseminated to unauthorized parties, they should be changed from time to time. A password should be of sufficient length to avoid compromise, and random in nature. Perhaps in the future the password will be replaced by personal characteristics such as voiceprints or fingerprints, but this is not currently feasible.[15]

References

1 P. Naur and B. Randell, eds., *Software Engineering*, Nato Scientific Affairs Div., Brussels, 1969.

2 W. Myers, "The Need for Software Engineering," *Computer* **11**(2):12–26 (Feb. 1978).

3 R. E. Fairley, "Tutorial: Static and Dynamic Testing of Computer Software," *Computer* **11**(4):14–23 (Apr. 1978).

4 D. Gries, "An Illustration of Current Ideas on the Derivation of Correctness Proofs and Correct Programs," *IEEE Trans. Soft. Eng.* **SE-2**(4):238–243 (Dec. 1976).

5 U. C. Huang, "Program Instruction and Software Testing," *Computer* **11**(4):25–32 (Apr. 1978).

6 J. D. Musa, "A Theory of Software Reliability and Its Application," *IEEE Trans. Soft. Eng.* **SE-1**(3):312–327 (Sep. 1975).

7 B. Randell, "System Structure for Software Fault Tolerance," *IEEE Trans. Soft. Eng.* **SE-1**(2): 220–232 (Jun. 1975).

8 E. Yourdon and L. Constantine, *Structured Design*, Prentice-Hall, Englewood Cliffs, NJ, 1979.

9 T. DeMarco, *Structured Analysis and System Specification*, Prentice-Hall, Englewood Cliffs, NJ, 1979.

10 M. J. Rochkind, "The Source Code Control System," *Proceed. First National Conf. Software Engineering*, IEEE, New York, 1975, pp. 37–43.

11 B. Schneiderman, "Human Factors Experiments in Designing Interactive Systems," *Computer* **12**(12):9–19 (Dec. 1979).

12 D. T. Ross and K. E. Schoman, Jr., "Structured Analysis for Requirements Definition," *IEEE Trans. Soft. Eng.* **SE-3**(1):6–15 (Jan. 1977).

13 T. E. Bell, D. C. Bixler, and M. E. Dyer, "An Extendable Approach to Computer-Aided Software Requirements Engineering," *IEEE Trans. Soft. Eng.* **SE-3**(1):49–59 (Jan. 1977).

14 M. M. Lehman, "Programs, Life Cycles, and Laws of Software Evolution," *Proceed. IEEE* **68**(9):1060–1076 (Sep. 1980).

15 P. S. Brown, "Computer Security—A Survey," National Computer Conference, 1976, pp. 53–63.

Bibliography

Cooper, J. D., "Corporate Level Software Management," *IEEE Trans. Soft. Eng.* **SE-4**(4):319–325 (Jul. 1978).

Gane, C. and Sarson, T., *Structured System Analysis: Tools and Techniques*, Prentice-Hall, Englewood Cliffs, NJ, 1979.

Jackson, M. A., *Principles of Program Design*, Academic, New York, 1975.

Myers, G. J., *Reliable Software Through Composite Design*, Petrocelli/Charter, New York, 1975.

Ross, D. T., "Structured Analysis (SA): A Language for Communicating Ideas," *IEEE Trans. Soft. Eng.* **SE-3**(1):16–33 (Jan. 1977).

Ross, D. T., Goodenough, J. B., Irvine, C. A., "Software Engineering: Process, Principles, and Goals," *Computer* **8**(5):17–27 (May 1975).

Shooman, M. L., *Software Engineering: Design/Reliability/Management*, McGraw-Hill, New York, 1983.

Tausworthe, R. C., *Standardized Development of Computer Software*, Prentice-Hall, Englewood Cliffs, NJ, 1977.

Wasserman, A. I., "A Top-Down View of Software Engineering," *Proceed. First National Conference on Software Engineering*, IEEE, New York, 1975, pp. 1–7.

Zelkowitz, M. V., Shaw, A. C., Gannon, J. D., *Principles of Software Engineering and Design*, Prentice-Hall, Englewood Cliffs, NJ, 1979.

CHAPTER 64
COMPUTER GRAPHICS

GUY JOHNSON

Rochester Institute of Technology
Rochester, New York

64.1 INTRODUCTION

The term "computer graphics" refers to a body of knowledge and techniques that can be used to program a computer to generate geometric images. These are usually formed on the viewing surface of a cathode ray tube (CRT) or drawn on paper with the use of a computer-driven plotting device. The primary emphasis of the research done in this field centers on the generation of these images by the computer. There is a related field, usually called image processing, that deals with the analysis and enhancement of images acquired in a computer-readable format (the results of space probes, etc.). At the present time, the software used in these two fields has little in common. The fields do, however, make use of much of the same computer hardware and peripheral devices.

Applications of computer graphics include generation of artwork on a two-dimensional surface, business graphics that attempt to display data in a more visual manner, interactive design of products by engineers, and interactive games in a video arcade. Reference 1 contains a series of photographic plates illustrating the graphic output of many of these systems. References 2 and 3 contain state-of-the-art articles on topics in computer graphics. For many of these activities the computer display is constantly changed until the result is finalized and then reproduced on a more permanent device (the so-called hardcopy output). The common factor in all these applications is the use of the computer by an individual sitting at some form of a computer terminal manipulating and displaying geometric objects.

The areas of research by individuals in the computing sciences often center on current problems that restrict the use of computer graphics. These include:

1. Design of appropriate languages for manipulating geometric objects. See Ref. 4 for examples.
2. Design of efficient algorithms to perform tasks on the computer in a time period short enough to be useful to the computer user. This includes special algorithms for three-dimensional geometry, vector drawing, and raster display devices. References 1, 4, and 5 discuss these issues in detail.
3. Design of peripheral devices to allow a comfortable and safe working environment.
4. Creation of the appropriate mathematics to model geometric objects and thus have a firm foundation for their implementation in the computer. See Refs. 1, 4, 5, and 6.
5. Design of programming tools (graphics packages) that enable a competent programmer to build a program incorporating geometric output. Reference 7 contains several references on one such package called the CORE system.

6. Design of graphics tools for use in an embedded computer. An individual may be using a device and not know a computer is generating an image. The traditional computer input devices may not be attached.

64.2 GRAPHICS HARDWARE

Any number of peripheral devices attached to a computer may be used for the presentation of geometric images. Early users of graphics output used ordinary line printers to produce their images. Production was accomplished by having the computer overstrike characters at a given position until the desired level of black color was achieved. Viewed from a distance the images were sometimes remarkably realistic. Since that time, however, numerous devices have appeared on the market with improved capabilities. Some of these create a permanent image, while others simply display the image on a screen. Some include interactive capabilities, while others are used for output only. The major divisions in the technology are described in the following. Of course any given device on the market may include technology from several of these classes.

Storage Tubes

The storage tube was the earliest graphic output form to appear in quantity as a computer output device. The image was generated on the screen by an electron beam controlled by output from the computer. The beam would excite the phosphor on the face of the screen and it would emit light for a fairly long period of time. An image would thus be drawn by the computer and the screen itself would preserve that image. The main drawback of this device is the inability to erase a portion of a figure. Many interactive applications of a computer are based on the ability of the user to manipulate the image on the screen. Once a line is drawn on the screen of a storage tube, it remains until it fades with time or the entire screen is erased. The storage tube is excellent for the presentation of detailed drawings composed of lines, but does not allow the user to manipulate the image selectively.

Vector Refresh Devices

These devices use the cathode ray tube (CRT) as an output screen and allow selective erasure. They are referred to as refresh devices, since the image is in fact generated approximately 30 times a second on the face of the screen. The image fades rapidly and must constantly be redrawn to remain visible. Thus if a line must be erased, it may simply be omitted in the next refresh of the screen. Usually the computer sends commands to these devices to draw a sequence of lines. As long as no more information is received, the lines are refreshed constantly. However, if more lines are added or some are deleted, the image changes on the next refresh operation. Since refreshing happens 30 times a second, the result appears to be instantaneous.

These devices are useful for interactive applications in which the image changes according to user actions. The resolution of the device is usually not quite as fine, however, as that of the storage tube and the expense is usually greater.

Raster Devices

This class of devices is becoming more prevalent as the price of computer memory continues to decline. This class also uses the CRT as an output screen, as do the vector refresh devices. However, the screen is refreshed from a description of the image in the computer's memory. Corresponding to a dot on the screen there will be some number of bits allocated in the computer's memory for the description of the color to be displayed. This varies from 1 bit to 8 bits or more, usually depending on expense. The computer need only change the bit pattern in its own memory to change the image on the screen. Thus the device allows selective updating of screen areas. The major drawback is the use of the dots or "pixels" on the screen to draw lines. In many cases the line appears to be jagged, since it is composed of dots. The previous devices did not have this appearance. However, solid areas composed of pixels of the same color are easily displayed.

Hardcopy Devices

To provide a permanent copy of the image, several different computer-controlled devices have appeared. The earliest was the pen and ink plotter, in which the motion of the pen was controlled by commands from the computer. The plotter produces sharp images, and pens of different color may be used. The major drawback is the relatively slow speed of the device.

More recently a device called the electrostatic plotter is becoming available. It creates an image in a manner similar to that of the raster display in that the image is composed of dots. The main advantage is the increased speed of this device in the generation of the output plot.

64.3 GRAPHICS SOFTWARE

Interactive Picture Languages

A conversational graphics editor is a computer program that provides users with the ability to draw on the screen in an arbitrary fashion. The users do not need to know anything about programming a computer. They usually need to know about terminal usage (which keys cause what actions on the screen) and to learn a graphics command language. This language usually includes the primitive actions of drawing lines, circles, boxes, and so on. Various colors may be chosen, depending on the capability of the graphics peripheral. The users respond to prompts from the computer by typing commands such as those listed in Fig. 64.1.

In the figure the computer prompts the user with a "*" character and the user types in a command. The first response is a command to draw a red line between certain points on the screen. These points may be specified by giving the row and column position on the screen. The second response directs the graphics editor to draw a rectangle at a certain position again. This is an example of a more elaborate primitive object available in the editor. The third command is the creation of a more complex figure by the user. This figure is composed of three lines, all green, drawn between certain points. The final command is a directive to display the object created in step 3. This capability allows one to define and place more complex objects on the screen.

These programs usually allow the user to create special objects from the primitives and then manipulate the objects. It becomes quite easy to construct complex images, since the same objects may be repeated many times in many places with changes in size, color, and angles. These packages are used quite often for the creation of figures in presentation graphics. Text may usually be included in the picture in a variety of fonts and angle of presentation.

```
* DRAWLINE here there IN red
* DRAWBOX there IN green
* CREATE-AN-object BY
        DRAWLINE here there IN green
        DRAWLINE here there IN green
        DRAWLINE here there IN green
  END-CREATE
* DRAW-object
```

Fig. 64.1 Sample graphic editor commands in response to the "*" prompt character.

Computer-Aided Design of Two- and Three-Dimensional Objects

A common application of the use of computer graphics software is the design of complex two- and three-dimensional objects by interactive manipulation of shapes. Such a system enables an engineer with no programming expertise to create drawings of these objects for later manufacture. The systems usually function by providing the engineer with a number of primitive shapes that may be combined to produce more complex objects. The advantage to the systems is the precision of the output and the small effort needed to make changes on the design. The engineer usually manipulates the design on a CRT screen until the desired form is obtained and then he or she produces a final hardcopy. In this case the engineer uses the computer graphics system to visualize the object. The same data used in the production of the drawing may be used in later production steps in the manufacture of the part. These packages tend to be somewhat specialized to serve a particular industry (e.g., aerospace or automobile) and a library of common parts soon accumulate.

Data Plotting

Another field of application of computer graphics is in the area of data plotting. Acquisition of data by computers has become so easy that inevitably more data are acquired than can be comprehended by a human being. Analysis and presentation of those data is the real problem. A number of graphics programs have been written to prepare plots of data in one form or another, following some mathematical analysis. Examples include computer-generated images such as bar graphs, pie charts, x, y coordinate plots, and three-dimensional plots of multivariate data projected to two dimensions on the display surface. In many cases the user is expected to prepare the data ahead of time in some computer-accessible format (magnetic tape, diskette, etc.) and then present this data to the program.

The program analyzes the data according to specifications laid down by the programmer and then outputs the results in graphic formats. Again users need not know anything about programming or generation of the image on the output peripheral. They do need to collect data and present them according to the specifications of the program.

More exotic programs may be tied to data acquisition systems, in which data are collected in real time and the output is constantly updated according to the new data as they accumulate. Real-time applications may take this approach, since the output is more readily appreciated by the user. Certainly a bar chart, pie chart, or three-dimensional plot is more readily understood by a human being and thus more appropriate output for these programs. Tables of numbers or specific numbers (averages, standard deviations, etc.) sometimes have less impact.

These images are often presented on a graphics peripheral and then, depending on the result, transferred to a hardcopy device.

Support for Programmer

The creation of programs such as those just described demands some unique tools for image construction in the hands of the programmer. Most frequently, graphics support is provided to the programmer in the form of primitive graphic subroutines or procedures that he or she may use in the construction of his own program. Many varieties of such packages exist. Most are collections of commonly used subroutines for performance of drawing actions on the local graphics peripheral at a given computer location. Proliferation of these special packages has led to the development of a standardized specification called the CORE system. This was proposed by the ACM SIGGRAPH Graphics Standards Planning Committee (GSPC77). See Ref. 7 for the specification and further discussion.

Design of these packages is difficult because of the multitude of graphics peripherals that may be controlled by a computer, ranging from pen and ink plotters over sophisticated printers, storage tube displays and raster displays to vector refresh displays. The capabilities of these devices can and do affect the sturcture of the application program that is being created. The CORE system is an attempt to provide a standardized set of subroutines that can minimize the problems in transporting a program among computers controlling different output devices.

The functional capabilities of the CORE system include:

1. **Output primitives.** These are subroutines that allow the programmer to specify what objects he or she wishes to draw. For example, the LINE ABS 2 and LINE REL 2 subroutines are used to draw lines on the display surface.
2. **View of object.** These subroutines allow the programmer to specify a selection region for viewing and a selected point for viewing from. Two-dimensional and three-dimensional viewing are allowed. For example, the WINDOW and VIEWPORT subroutines are used to perform automatic scaling and projection of images onto the screen.
3. **Segments.** Any complex image must be broken into manageable portions, and subroutines are provided to allow the programmer to provide segments of the total image. For example, the CREATE SEGMENT and CLOSE SEGMENT routines allow the programmer to group many primitive graphics commands as one object and then to manipulate the object.
4. **Logical input devices.** This convention allows the programmer to receive input from logically defined devices rather than from physical devices. The program may be transported more easily, since physical devices characteristics are not built into the high-level program. For example, the READ LOCATOR subroutine is used to input a screen location from the user. The actual device may vary. It may be a joystick or a graphics tablet, but the programmer does not need to be concerned. The logical properties of the device are all that is assumed in the use of the subroutine. It must be capable of delivering the coordinates of a point on the display surface.

The following examples of pseudocode programs use the CORE system to perform simple graphics display.

Two-Dimensional Graphics. The first example, shown in Fig. 64.2, is a pseudocode program to draw a picture of two houses on the display device. This is done by invoking output primitives of the CORE system from the program. The display surface has an imaginary coordinate system overlaid. This enables the programmer to specify end points of lines, points for placement of symbols, and so on. The graphics package is responsible for scaling the image to fit on the particular device being used. The package is also responsible for sending the appropriate commands to draw the desired geometric symbols. This may involve sending a device-dependent graphics command to the display device; this command is stored in the memory of the device. This is the usual technique when using a refresh vector device. Alternatively, the package may modify a segment of memory (usually called the frame buffer) from which the device obtains information for drawing the image. Algorithms used in these packages are discussed in Refs. 1, 4, and 5. The mathematics that underlie many of the algorithms in the graphics packages and the higher level systems are discussed in Ref. 6.

Specifically, this program makes use of two primitives from the CORE system, MOVE ABS 2 and LINE ABS 2. The locations are given in terms of the traditional x and y coordinates. The

```
program house
procedure rectangle(x0,y0,height,width);
begin
(* procedure to draw a rectangle of arbitrary height and width *)
MOVE ABS 2(x0,y0);
LINE ABS 2(x0+width,y0);
LINE ABS 2(x0+width,y0+height);
LINE ABS 2(x0,y0+height);
LINE ABS 2(x0,y0);
end;

procedure roof(x0,y0,width);
begin
(* procedure to draw a flat roof on the house *)
MOVE ABS 2(x0,y0);
LINE ABS 2(x0+width,y0);
end;

begin (* main program *)

(* Draw the first house. *)
rectangle(25,25,50,75);
roof(20,75,85);

(* Draw the second house. *)
rectangle(150,150,50,75);
roof(145,200,85);

end. (* main program *)
```

Fig. 64.2 Pseudocode to draw two houses on the screen.

first primitive is used to position the start of a new line at a given point on the screen. Thus MOV ABS 2($x0$, $y0$) establishes the current position at the point with coordinates ($x0$, $y0$). The second primitive is used to actually draw the line to another point. Hence LINE ABS 2($x0$ + width, $y0$) draws a line from the point with coordinates ($x0$, $y0$) to the point with coordinates ($x0$ + width, $y0$).

There is an internal procedure called RECTANGLE in the program that, when executed, causes the line-drawing primitives to be executed sequentially and produces a box on the screen. The location and size of the box are given as parameters to the procedure when called from the main program. In Fig. 64.2, the procedure called RECTANGLE(25, 25, 50, 70) places the current position at point (25, 25) and draws a horizontal line to point (100, 25). It then draws a vertical line to point (100, 75). Next it draws a horizontal line to point (25, 75). Finally it draws a vertical line back down to point (25, 25).

There is also an internal procedure called ROOF in the program that adds a flat roof to the house when called from the main program. In Fig. 64.2, the procedure call ROOF(20, 75, 85) places the current position at (20, 75) and draws a horizontal line to (105, 75). This will, incidentally, overwrite the previous line generated by the RECTANGLE procedure. By adding more primitive line-drawing commands to the procedure, perhaps a more elaborate roof might be drawn. This program is meant to be simple.

Finally the main program uses the defined procedures to draw two houses on the screen. They might appear as in Fig. 64.3. The imaginary coordinate screen is placed over the display surface. Certainly a more elaborate main program might draw many houses of various sizes if that was the requirement. The program itself would need to be modified but not the RECTANGLE or ROOF procedures.

Algorithm for Line Drawing. An example of one of the possible line-drawing algorithms is given here. This algorithm could be used to implement the LINE ABS 2 primitive used in the CORE subroutine package. It will calculate for a raster display device which pixels must be set when given the end points of the line segment to be displayed. For such a device, drawing a line segment corresponds to setting the intensity of these pixels.

Given the coordinates of the end points ($x1$, $y1$) and ($x2$, $y2$), all points that lie on the line must satisfy the equation

$$y = \frac{(y2 - y1)}{(x2 - x1)}(x - x1) + y1$$

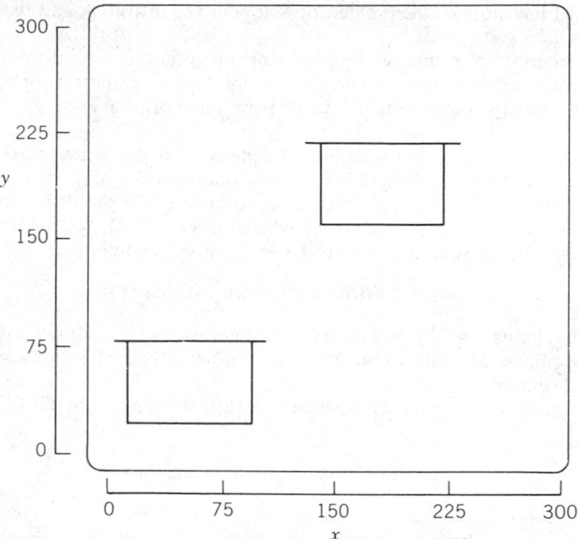

Fig. 64.3 Two houses drawn on the screen from example program of Fig. 64.2.

See Ref. 5 for further derivations. Expressed in the usual slope intercept form

$$y = mx + b$$

where

$$m = \frac{y2 - y1}{x2 - x1}$$

$$b = y1 - mx1$$

For a given point to lie on the line within the end points the following inequalities must also hold:

$$\min(x1, x2) <= x3 <= \max(x1, x2)$$

$$\min(y1, y2) <= y3 <= \max(y1, y2)$$

A raster display device can only display pixels at regular intervals in the screen coordinate system. Given the two points to connect with a line, one must determine which pixels lie on the line between the two end points.

If the slope of the line, the value m, lies between 0 and 1, one may increment the value of x starting at $x1$ and determine the y value of the pixel that lies on the line. This will allow the values of all points ($x3$, $y3$) to be calculated and displayed. This is exactly the problem that is solved by the LINE ABS 2 primitive of the CORE system. Expressed in pseudocode this would be

```
X3 := x1;
While X3 < = x2 do
    begin
    (* Calculate y value for given x value *)
    Y3 := Trunc(m * X3 + b);

    (* Display pixel on surface *)
    Set Pixel(X, Y, Intensity);

    (* Increment x value *)
    X3 := X3 + 1
    end;
end While
```

If the line has slope between 1 and infinity (a vertical line), one must increment the y value to avoid displaying gaps in the line. Other slopes (negative) require similar pseudocode.

A great deal of effort has gone into the development of algorithms for primitive graphic display such as the line-drawing algorithm. There are many drawbacks to the simple algorithm presented, including the use of real numbers, the use of multiplication, and the special code for lines of differing slopes. More detailed but more efficient algorithms are derived in Ref. 8, and these would be used in any production graphics system.

Algorithms of interest in computer graphics include primitives to display circles, arcs, and polygons. If these are implemented efficiently, an interactive graphics application can be programmed to quickly display complex figures made from these primitives.

Windows and Viewports. The program in Fig. 64.2 performed the drawing actions by referencing absolute coordinates on the face of the CRT screen. Since devices may vary and the programmer may wish to use numbers in the program that are appropriate to the problem domain, the concept of a real world and a window on that world are introduced in the CORE system. The programmer may define a rectangular window by using the WINDOW primitive as follows:

```
WINDOW(minX, maxX, minY, maxY)
```

This establishes the limits on the imaginary coordinate system assumed previously. Following execution of that primitive, any calls to the MOVE ABS 2 or LINE ABS 2 procedures may reference coordinates in that window.

To map this window to the actual screen there is a further primitive by the name VIEWPORT. It is used as follows:

```
VIEWPORT(minX, maxX, minY, maxY)
```

The values of the parameters are used to specify a portion of the actual screen. These values are usually specified as numbers between 0 and 1, and this selects a rectangular region of the screen for viewing the image. Thus this may or may not include the full screen area. In any case, any actions the programmer takes to draw in the window will actually appear in the viewport specified on the screen. After the use of the WINDOW primitive, this VIEWPORT primitive must be used to define a rectangular area of the screen that will be used for presentation of the figure. Any actions that the program takes to draw lines in the window defined on the "real world" will be automatically scaled by the graphics package to fit in the coordinates of the viewport on the screen. See the example in Fig. 64.4 of a program to draw a house twice on different areas of the screen. The main program first defines the minimum and maximum coordinates of the window. All future calls to the LINE ABS 2

```
program house2
procedure rectangle(x0,y0,height,width);
(* Procedure to draw a rectangle of arbitrary width and height. *)
begin
MOVE ABS 2(x0,y0);
LINE ABS 2(x0+width,y0);
LINE ABS 2(x0+width,y0+height);
LINE ABS 2(x0,y0+height);
LINE ABS 2(x0,y0);
end;

procedure roof(x0,y0,width);
(* Procedure to add a flat roof to the house *)
begin
MOVE ABS 2(x0,y0);
LINE ABS 2(x0+width,y0);
end;

begin (* main program *)
(* Set the window for the programmers world *)
window(0,300,0,300)

(* Set the viewport on the display surface *)
viewport(0,0.5,0.5,1.0);

(* Draw the first house *)
rectangle(25,25,50,75);
roof(20,75,85);

(* Draw the second house *)
rectangle(150,150,50,75);
roof(145,200,85);
end. (* main program *)
```

Fig. 64.4 Pseudocode program to draw two houses using viewports and windows.

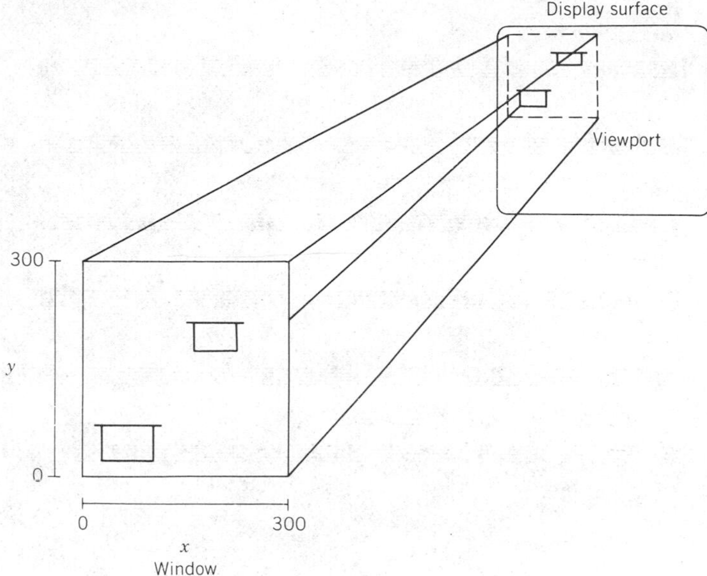

Fig. 64.5 Mapping of world window to screen viewport for program of Fig. 64.4.

primitive should reference points within this rectangular space. This is accomplished with the WINDOW(0, 300, 0, 300) procedure call. Next the region on the screen is selected using the VIEWPORT(0, 0.5, 0.5, 1.0) procedure call. This selects the upper left-hand quadrant of the screen for actual display. Now the houses are drawn by the same procedures as before, and the drawing is automatically scaled to fit in the upper left quadrant of the display screen. Figure 64.5 illustrates the mapping of the programmer's real world or window onto the screen viewport.

Interaction. The main advantage to using the computer to generate geometric images lies in the interactive capabilities of the modern machine. Many devices have been developed to allow a user sitting at the terminal to control the execution of a program. These include devices such as the keyboard itself, the joystick, the game paddles, the touch tablets, the light pens, and others. A well-written application that uses graphic output usually allows the user to change that output in some manner while it is being created. For example, a program may be written to paint the screen with a certain color depending on what keyboard key is depressed. The position of the paint may be determined by the current reading on the game paddles or the position of a joystick. When the user has more freedom in the execution of the program, many application systems become possible. A program that always paints the same picture or always displays the same data is only marginally useful. Interaction allows the user's own creativity to control the direction of the output.

The CORE system provides primitives for the interaction. These are based on the concept of logical devices, such as

Keyboard. Devices for typing character strings.

Locator. Devices for inputting screen coordinates.

Valuator. Devices for inputting a single real number.

Button. Devices for selecting among alternatives.

Pick. Devices for picking a symbol displayed on the display surface.

An actual peripheral device will be placed in one of these categories and the programmer need not depend on particular characteristics of a given manufacturer. Only the logical characteristics of the device are used. Thus one actual device may be substituted for another. Most interactive algorithms may be expressed as a polling loop waiting for input from one or more of the devices attached to the computer. For more discussion see Refs. 1, 4, and 5.

64.4 CONCLUSION

At the present time, computer users, regardless of their background, may accomplish many applications using graphic input and output. Many application systems have been developed that require no

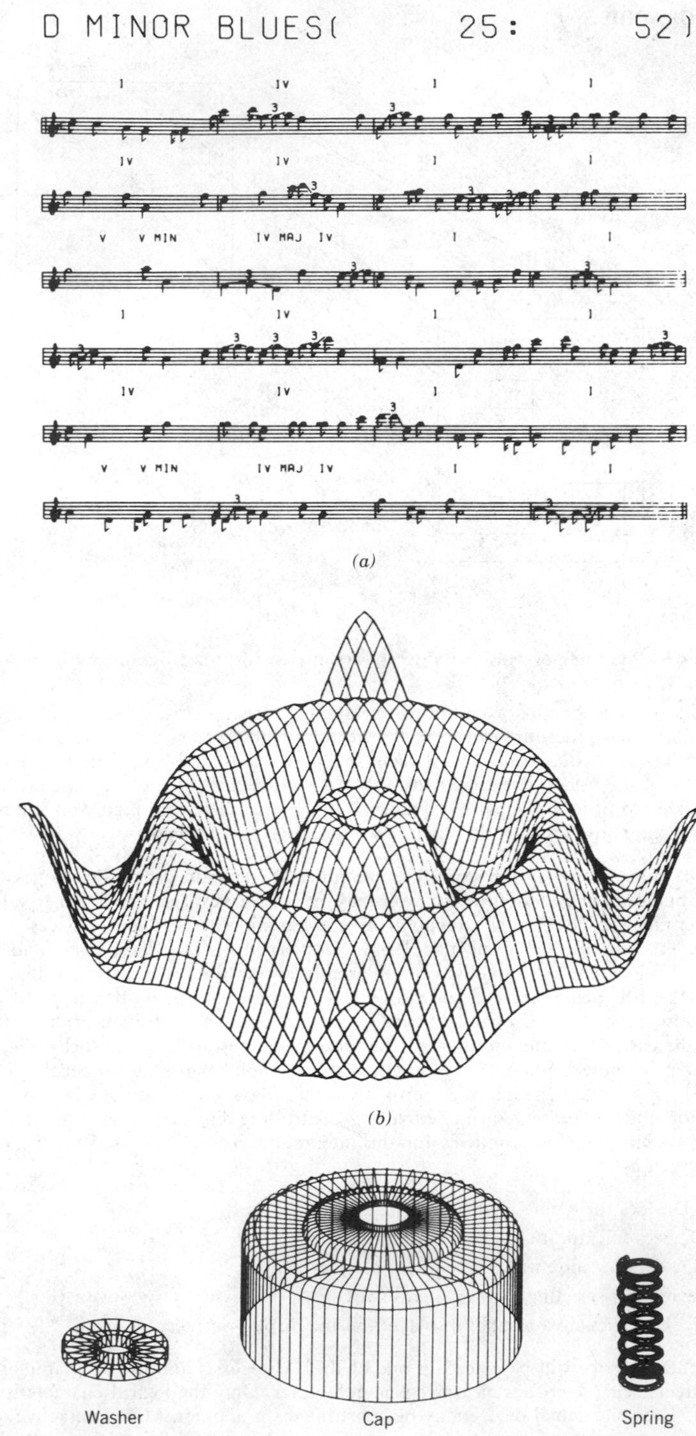

Fig. 64.6 Examples of computer graphics applications: (*a*) Musical notation, (*b*) three-dimensional surface, (*c*) washer, cap, and spring, (*d*) skeleton picture of an aerosol cap incorporating the washer, cap, and spring from Fig. 64.6*c*. From I. O. Angell, *A Practical Introduction to Computer Graphics*, Wiley, New York, 1981, reprinted with permission.

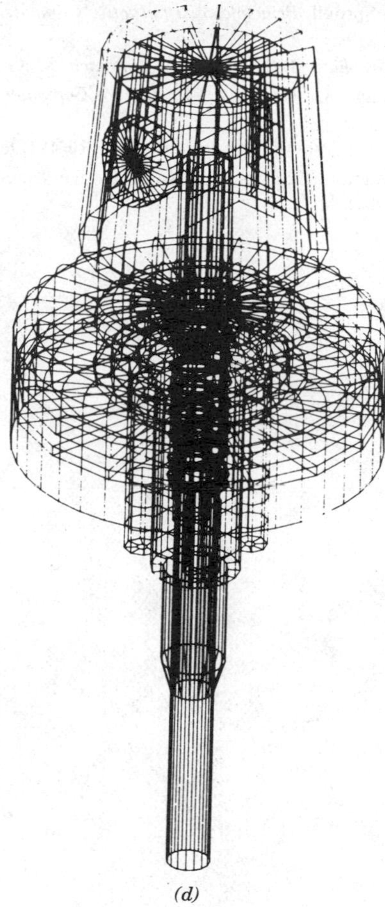

(d)

Fig. 64.6 (*Continued*)

more than the pressing of a button or a key at the appropriate time. Because of these application systems, many tasks are performed more easily or with better results than in the past. Some examples are shown in Fig. 64.6. For example, computer drafting by engineers rather than by programmers has many advantages. Engineers are usually not computer programmers. The systems that have been developed allow individuals trained in some field to use the computer with its interactive graphic operation and remain within the terminology of their field.

The results in the field of computer graphics are having an immediate impact on the creation of many new devices. With the proliferation of inexpensive microprocessors and the decline in price of graphic output devices, many applications are being developed around interactive graphic output. Development will undoubtedly continue and expand. More graphic output devices are being manufactured with increased capability in terms of higher quality output and faster response. As programmers become more familiar with this form of output, these devices will become the cornerstones of the application systems.

References

1 J. D. Foley and A. Van Dam, *Fundamentals of Interactive Computer Graphics*, Addison-Wesley, Reading, MA, 1982.

2 *ACM Transactions on Graphics*, Association for Computing Machinery.

3 *IEEE Computer Graphics and Applications*, IEEE.

4 W. N. Newman and R. F. Sproull, *Principles of Interactive Computer Graphics*, 2nd ed., McGraw-Hill, New York, 1979.

5 S. Harrington, *Computer Graphics: A Programming Approach*, McGraw-Hill, New York, 1983.

6 D. F. Rogers and J. A. Adams, *Mathematical Elements for Computer Graphics*, McGraw-Hill, New York, 1976.

7 "Graphics Standards," special issue of *ACM Comp. Surv.* **10**(4) (Dec. 1978).

8 R. F. Sproull, "Using Program Transformations to Derive Line-Drawing Algorithms", *ACM Trans. Graph.* **1**(4) (Oct. 1982).

CHAPTER 65
COMPUTER COMMUNICATION NETWORKS

RICHARD VAN SLYKE

Polytechnic Institute of New York
Brooklyn, New York

Computer communication networks perform one or both of two functions: (1) They connect users (terminal devices) to one or more shared computers and possibly to other users and (2) they connect computers to one another. The scale of these networks ranges from worldwide networks over networks connecting devices in a single building or building complex (called *local area networks*) to bus networks connecting many microprocessors in a single device (*multiprocessors*).

65.1 NETWORK ELEMENTS

Communication networks, being networks, consist of nodes and links connecting them. Nodes, in turn, are of two types: *termination devices*, in which data leave or enter the network for transfer, and *communication devices*, which provide the network functions. Termination devices include terminals, host computers, and gateways to other networks. Communication devices include switches, concentrators, multiplexers, front-end processors, and network controllers. The links connecting these nodes correspond to transmission channels.

Termination Devices

Terminals. Terminal devices range from low-speed sensors such as burglar alarms operating at a few bits per second to full-motion digital TV that operates in the neighborhood of 90 million bits per second (b/s). Common terminal devices are teletypewriters, which operate in the range of 50 to 1200 b/s; cathode ray tube (CRT) and graphics terminals, which operate from 110 to about 9600 b/s; and remote job entry terminals, which operate in the range of 2000 to 50,000 b/s.

Hosts. Hosts are computers connected to the network that provide information-processing services to other network termination devices.

Gateways. Gateways are special terminal devices that serve as interfaces between computer communication networks. A gateway can simultaneously be a node in more than one network or the gateway function can be implemented by gateway "halves," where a half in each network converts data passing between networks into a form that the corresponding gateway half in the next network can understand using channels outside both networks.[1]

Communication Devices

Communication devices perform a wide variety of functions, the most important of which are multiplexing, concentrating, switching, polling, and interfacing. Multiplexing, concentrating, polling, and switching are all ways of using transmission facilities more efficiently.

Multiplexing. In multiplexing, transmission channels are partitioned into fixed subchannels for different users (Fig. 65.1*a*). There are several ways in which the subchannels can be kept apart. In space division, the signals are propagated by separate transmission and receiving antennas or are separated by being in different wires in a bundle of wires. In frequency division, subchannels use different parts of the channel frequencies. In time division multiplexing, the use of the channel is allocated to different subchannels at different times. Less common methods are polarization and code division. Polarization was used to virtually double the satellite transponder capacity of the RCA SATCOM satellite by the carrying of channels on each of two different polarizations.[2] Code division is used with spread spectrum and is beyond the scope of this discussion.[3]

Concentrating. In multiplexing, a channel is divided into *fixed* subchannels. Therefore the sum of the capacities of the subchannels cannot exceed the capacity of the carrying channel. In many cases the subchannels are only occasionally used at full capacity. In such cases *concentration* is used to *dynamically* allocate access to the carrying channel. With concentration, the sum of the *peak* data rates of the users can exceed the capacity of the carrying channel (see Fig. 65.1*b*). Clearly if all the carried channels are active at their peak data rates at once, data back up. If the sum of the data rates is only temporarily higher than the capacity of the carrying line, the overflow can be stored

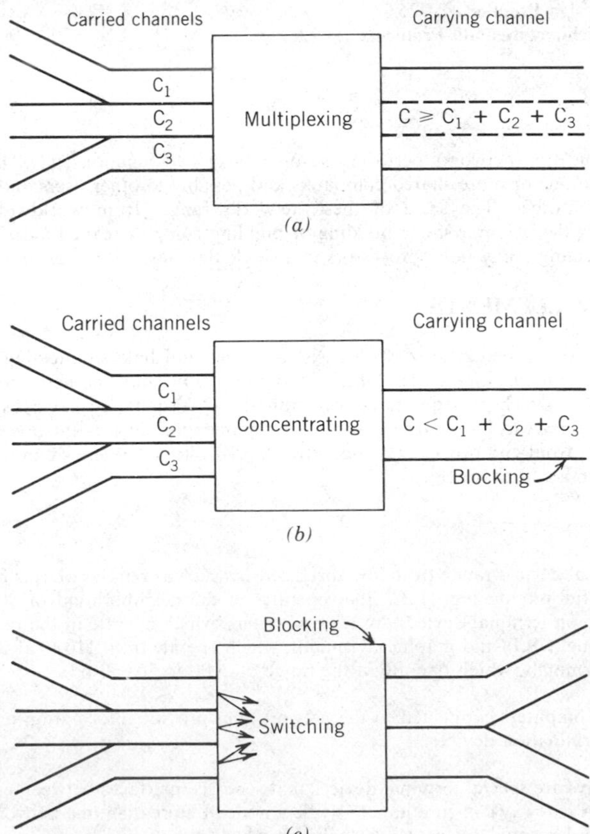

Fig. 65.1 Examples of ways to use transmission facilities efficiently: (*a*) Multiplexing, (*b*) concentrating, (*c*) switching.

temporarily in buffers in the concentrating device. Of course if the sum of the *average* data rates of the carried subchannels is more than the capacity of the carrying line for long periods of time, information will eventually be lost. Thus there can be a bottleneck in the channel serving the concentrator and data on the carried channels can be "blocked" by the congestion in the carrying channel. One should not be confused by the term "statistical multiplexing," which refers to a concentration technique rather than to multiplexing.

Switching. A switch makes it possible for data coming in on one channel to leave on any of a number of other channels depending on the state of the switch. In general, for a network with n nodes to have the ability to communicate between each pair of nodes in the absence of switching, a channel would be needed between each of the $n(n-1)/2$ node pairs. Switching can reduce the number of communication lines needed in the network to as few as $n-1$. Often switches can support paths between each input port and each output port individually but not for all possible simultaneous connections. Thus congestion can occur in the switch and a desired path can be "blocked" in the switch (Fig. 65.1*c*).

Polling. Polling is another method of sharing a channel among users. In this case a controlling device such as a terminal controller, front-end processor, or computer asks each user of the channel in sequence if it has something to transmit. Thus if the user does not need the channel it uses no more of it than the amount required to support the polling process itself.

Interfacing. Interface functions include code and speed conversion, analog-to-digital (A/D) signal conversion, conversion between synchronous and asynchronous communication, serial–parallel conversion, and changing data formats.

Transmission Channels

The categories of underlying channels that can be used to provide transmission are illustrated in Fig. 65.2. The first element in the description of the channel is the frequencies used in the transmission. These can range from audio up through high, very high, ultra high, and microwave frequencies to millimeter waves and light and fiber-optic systems. The transmission can be analog or digital. The propagation can be into free space either on a point-to-point basis, as in the microwave system used for telephone traffic, or in a broadcast mode, as is used by satellite communication. The signal can also be constrained or guided by physical boundaries. Guided channels include waveguides, fiber-optic channels, coaxial cable, and wire pairs.

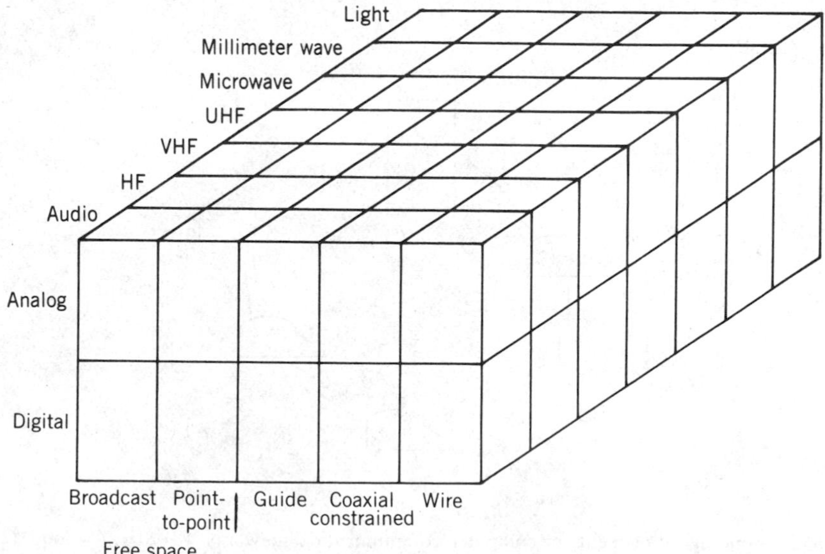

Fig. 65.2 Kinds of channels.

65.2 NETWORK TOPOLOGIES

Star, Loop, Tree, and Mesh Networks

The basic topological forms of computer communication networks that are connected by point-to-point link channels (as distinguished from broadcast channels) are stars, loops, trees, and mesh networks (Fig. 65.3). Star, loop, and tree networks are relatively simple in that the routing is uniquely defined and switching is relatively simple except possibly at the central site. These three topologies are frequently used for connecting terminals to a central computer. Mesh networks are significantly more complex, since there are alternate routes between source–destination pairs and because switching is required. Mesh networks are often used to connect computers together on a peer basis.

Hierarchical Networks

Networks can be connected hierarchically (Fig. 65.4)—in theory, in a number of levels. In the vast majority of implementations there are at most two levels: *local access subnetworks*, which connect end users (terminals, computers) into the network, and a *backbone subnetwork*, to provide switching and efficient long-distance communication. Typically local access is accomplished by star, loop, or tree structures, while the backbone subnetwork is often a mesh network.

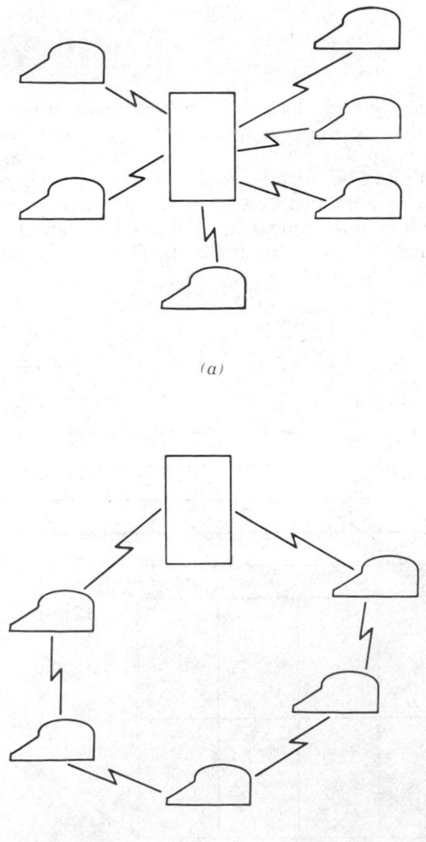

Fig. 65.3 Topological elements of computer communication networks: (*a*) Star, (*b*) loop, (*c*) tree, (*d*) mesh.

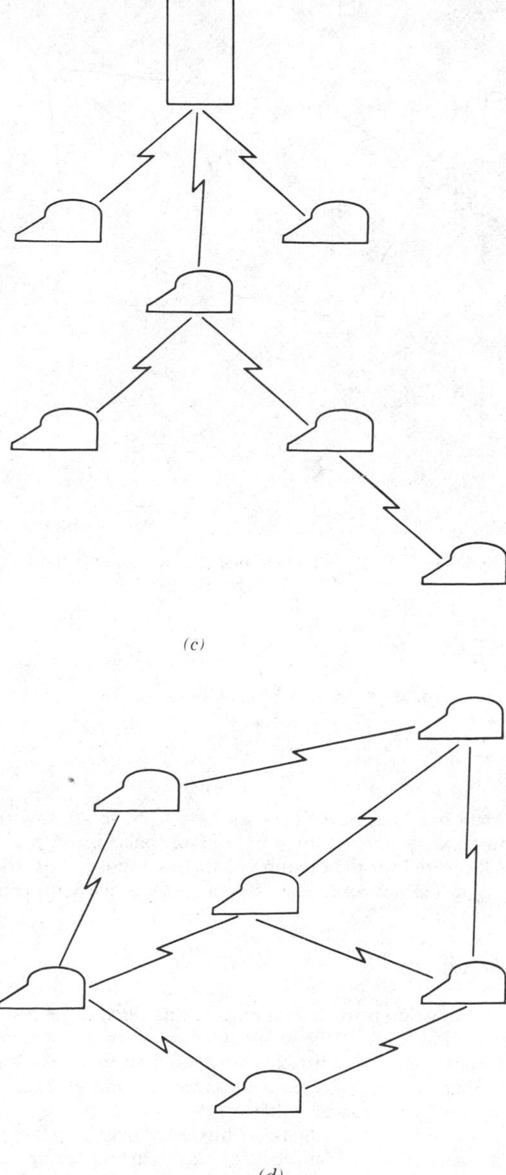

(c)

(d)

Fig. 65.3 (*Continued*)

Broadcast Channels

The use of broadcast channels such as satellite channels or radio channels operated in a broadcast mode gives rise to new topological variations. Satellite channels, because their costs do not strongly depend on distance, are usually used in the backbone. Only recently has the intrinsically broadcast nature of satellite communication been exploited. Previously these channels have been used point to point rather than as broadcast channels. Microwave radio is being developed for local access as an alternative to telephone company local loops. In 1979 the Federal Communications Commission allocated microwave frequencies for use in local communication for common carrier data networks.

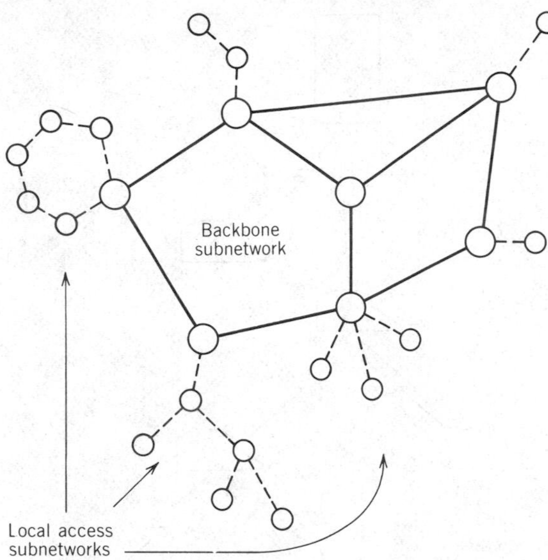

Fig. 65.4 Two-level hierarchical network connection.

65.3 SWITCHING MODES

Data can be switched through networks in two basic modes: circuit switched or store-and-forward switched.

Circuit Switching

In circuit switching, sufficient free lines in the network have to be found to form a path between the data origination point and the data termination point before transmission can begin. However, once this path is identified it is dedicated for the duration of the transaction, and transmission proceeds in real time with no buffering in the network. This, for example, is the kind of switching used in the telephone network for voice traffic.

Store-and-Forward Switching

When store-and-forward transmission is used, data units are buffered at each switch and transmission can proceed as long as a free link is available to the next switch on some transmission path toward the destination. There are two major approaches to store-and-forward switching—message and packet switching. With message switching, entire messages are sent at once and buffered at each switch. More recently (since 1969), in packet-switched networks the messages have been broken into "packets" that are limited in length to a few thousand bits. Moreover, in packet switching the packets are ordinarily buffered in the high-speed memory of the switching computers, while in message switching the messages are often buffered in slower peripherals such as disk, drum, and/or tapes. The major motivation for use of packet switching instead of message switching is to reduce delay. Each time a unit of data is buffered in a switch it is delayed by the transmission time of the data unit. Thus if the data units are small, the delay is reduced. In addition, since the units are smaller they can be buffered in high-speed memory, further reducing delays by shortening processing times in the switches. As a consequence, while real-time interaction using message switching is not usually feasible, this kind of traffic can be supported using packet switching.

Comparison of Switching Modes

Circuit switching is most appropriate for traffic that is characterized by long and relatively steady transmission. Message switching is most appropriate for occasional messages that are quite long and when delay constraints are not critical. Packet switching is indicated when the traffic is "bursty"; that is, there are many short transmissions each at a relatively high transmission rate. Store-and-forward networks of both types also offer natural speed conversion because of the buffering. In circuit switching the data rates at each end of the circuit must be the same.

Hybrid networks are currently being developed which support packet switching for bursty traffic and circuit switching for longer transmissions such as file transfers. It is hoped that these networks can support a wider range of traffic conditions more cost effectively.[4]

65.4 DESIGN PROCESS

Computer communication networks are designed to carry given traffic subject to constraints on performance, cost, and implementation time.

Performance Measures

Some important performance criteria are transfer rate, availability, reliability, accuracy, channel establishment time, network delay, line turnaround delay, transparency, and security.[5]

Transfer Rate. The number of information bits transferred through the communications network divided by the time required for their transfer and acceptance.

Availability. The fraction of a time interval of interest that the network is capable of performing its assigned functions.

Reliability. Measures the ability of the network to maintain its functions until an information transfer has been successfully completed.

Accuracy. Addresses the correctness and completeness of the received information.

Channel Establishment Time. Measures the time required to establish a circuit connection between calling and called terminals.

Network Delay. Measures the time required for information to travel through the network to the receiving communication device.

Line Turnaround Delay. If transmission can take place in only one direction at a time, *line turnaround delay* measures the time required to reverse directions.

Transparency. The lack of code or procedural constraints imposed by the network.

Security. The ability of the network to transfer information without loss, disclosure, or modification—whether inadvertent or deliberate.

Traffic Characteristics and Network Configuration

Some important traffic characteristics are the average traffic volume from network sources, the distribution of this traffic on a source–destination basis, transmission rates of the terminal devices, the message length distribution, the distribution intervals between message arrivals, and the duty cycle as measured by the ratio of the average data rate to the peak transmission rate.

For example, in time-sharing systems usually most of the traffic is going to and from a central site. This traffic pattern indicates a tree or star topology for the network rather than a mesh configuration. A traffic pattern in which most traffic goes to a small number of sites suggests a hierarchical network, perhaps with a mesh network connecting the small number of highly active sites and with tree, star, or perhaps loop local-access subnetworks connecting each of the remaining sites to one of the backbone sites. The data rate of the traffic sources obviously determines the channel capacity required.

The factor most affecting network architecture may be the duty cycle or the ratio of the average data rate to the peak transmission rate. This ratio is always less than or equal to unity. When this ratio is unity, the data source is transmitting continually at a constant rate. For smaller values of the ratio, the source transmits only on occasion. For values near unity, the traffic source can essentially use a channel with capacity equal to the transmission rate. Multiplexing for channel sharing and circuit switching for the network-switching mode are often appropriate in this case. However, if the ratio is near zero, dedicated facilities would often be idle. This suggests the use of concentration, statistical multiplexing, or random access techniques for channel sharing. Packet or message switching would be appropriate as the switching mode.

Design Steps

Even given an accurate representation of the traffic requirements, the overall design of an information network is complex. Therefore it is essential to break the design into more manageable parts. The

TABLE 65.1. TYPICAL DESIGN SUBPROBLEMS WITH QUANTITATIVE SOLUTION METHODS

Given	Determine	To	References
Traffic requirements, network topology, routing of traffic networks	Capacity of network transmission channels	Optimize tradeoff between channel costs and traffic delay	6, pp. 329–340; 7, pp. 58–102
Traffic requirements, network topology, capacity of network transmission channels	Routing of traffic in network	Minimize traffic delay	6, pp. 340–348; 7, pp. 212–241
Traffic requirements, network topology	Capacity of network transmission channels, routing of traffic in network	Optimize tradeoff between channel costs and traffic delay	6, pp. 348–351
Traffic requirements	Network topology, routing of traffic in network, Capacity of network transmission channels	Optimize tradeoff between channel costs and traffic delay	6, pp. 351–360; 7, pp. 201–208; 8, pp. 39–47; 9, pp. 32–34, 67–80
Terminal locations, traffic requirements	Location of multiplexers, concentrators, and/or switches	Minimize channel costs	7, pp. 195–201; 8, pp. 30–35; 9, pp. 83–84
Terminal locations; traffic requirements; location of multiplexers, concentrators, and/or switches	Assignment of terminals to multiplexers, concentrators, and/or switches	Minimize channel costs	7, pp. 171–188; 8, pp. 35–38; 9, pp. 80–83, 84–87

design process is then also partitioned into a number of design steps. The general design process for two-level hierarchical networks can be characterized by the following steps:

1. Determine potential candidates for network service points.
2. Locate switches for local access.
3. Determine sites to be included in the network that can be cost-effectively served.
4. Assign on-net sites to switches (more generally, design local-access topology).
5. Determine capacity for local-access channel.
6. Determine backbone network topology.
7. Determine backbone routing of traffic in the backbone network.
8. Determine the capacity of backbone channels.

This list is not meant to imply that the design process occurs in a linear fashion. There is much iteration and looping back and forth between the various steps. For example, the sites that can be cost-effectively served by the network are, in part, determined by the location of the local access switches. On the other hand, the cost of local-access switches is amortized by the number of sites that they will serve. References 6 through 9 give good descriptions of available network design algorithms. Some examples of problems for which methods have been developed are shown in Table 65.1. Exact algorithms are practicable only for the simplest and smallest problems. Most of the algorithms used in practical design work are heuristic.

65.5 ARCHITECTURES AND PROTOCOLS

Data communication networks are extremely complex. Therefore it is useful to layer the functions performed by the network hierarchically. The layers range from the functions of simple physical interaction to sophisticated high-level functional support for user applications. The layering concept is based on each layer making use of and adding value to the lower layers. The specification of the

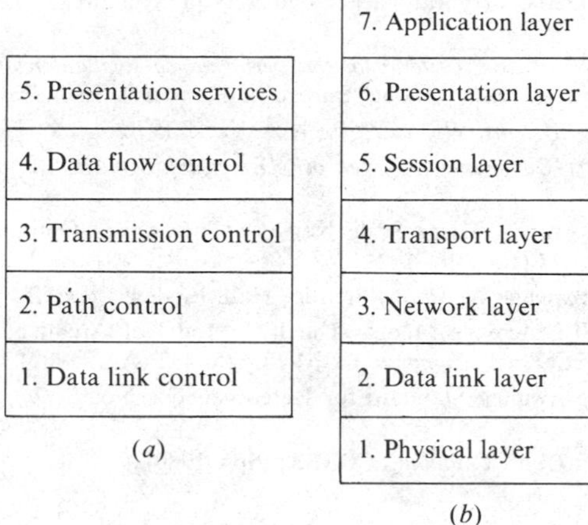

Fig. 65.5 Layered architectures: (*a*) IBM's systems network architecture (SNA), (*b*) ISO's (International Standards Organization's) Open System Interconnection (OSI) Standard.

functions in each layer and of the interfaces between the layers define the *architecture* of the network. A network architecture is *closed* or *open*, depending on whether it is based on equipment from one manufacturer or is designed to allow the use of equipment from different manufacturers. IBM's System Network Architecture (SNA) is one example of a closed architecture. The International Standards Organization's Open System Interconnection (OSI) standard outlines an open architecture. Figure 65.5 shows five levels of SNA and seven levels of OSI.

Architecture Layers[10,11,12]

The highest layer in the OSI architecture is the *applications layer*, which directly serves the end user. Included are the functions required to initiate, maintain, terminate, and record data concerning establishment of connections for data transfer among applications processes. All the other layers support this layer. An application can be thought of as composed of cooperating applications processes which communicate using application layer protocols. The next lower level, the *presentation layer*, provides services to the application layer to enable it to interpret the meaning of the data exchanged. These services manage entry, exchange, display, and control of structured data including translation, format conversion, and code conversion. The *session layer* assists in the support of interaction between cooperating presentation entities. It establishes a binding relationship between two presentation entities and unbinds them when the relationship is terminated. It also controls data exchange between the entities by delimiting and synchronizing data operations between them.

The *transport layer* provides transparent transfer of data between session entities. This layer provides reliable and cost-effective data transfer to the session layer. This requires that the transport layer be responsible for optimizing the use of available communications services to provide required performance at minimum cost. The *network layer* isolates transport entities from routing and switching considerations by providing the functional and procedural means to exchange data between two transport entities over the network.

The *data link layer* is responsible for establishing, maintaining, and releasing data links between network entities. This provides for the transfer of data across a data link. Finally the *physical layer* deals with the mechanical, electrical, functional, and procedural elements required to establish, maintain, and release physical connections between data link entities.

Protocols define the means of interaction between functions at the same level in the architecture. Standards exist for protocols at the physical layer (e.g., X.21, V.24, V.35) and for the data link layer (e.g., HDLC). The most likely basis for protocols for the network layer is level 3 of the X.25 protocol for connections between users and packet-switched networks. Protocol standards are in the process of development for the higher levels.

References

1 V. C. Cerf and R. E. Kahn, "A Protocol for Packet Network Interconnection," *IEEE Trans. Commun.* **COM-22**:205–216 (May 1974).

2 J. Martin, *Communication Satellite Systems*, Prentice-Hall, Englewood Cliffs, NJ, 1978.

3 R. C. Dixon, *Spread Spectrum Systems*, 2nd. ed., Wiley, New York, 1984.

4 N. Keynes and M. Gerla, "Hybrid Packet and Circuit Switching," *Telecommunications*, pp. 65–72 (Jul. 1978).

5 D. S. Grubb and I. W. Cotton, *Criteria for the Performance Evaluation of Data Communication Services for Computer Networks*, National Bureau of Standards Technical Note 882, Sep. 1975.

6 L. Kleinrock, *Queuing Systems*, Vol. 2, Wiley, New York, 1976.

7 M. Schwartz, *Computer-Communication Network Design and Analysis*, Prentice-Hall, Englewood Cliffs, NJ, 1977.

8 R. Boorstyn and H. Frank, "Large Scale Network Topological Optimization," *IEEE Trans. Commun.* **COM-25**:29–47 (Jan. 1977).

9 A. S. Tannenbaum, *Computer Networks*, Prentice-Hall, Englewood Cliffs, NJ, 1981.

10 H. Zimmerman, "OSI Reference Model—The ISO Model of Architecture for Open Systems Interconnection," *IEEE Trans. Commun.* **COM-28**:425–432 (Apr. 1980).

11 H. C. Folts, "A Long-Awaited Standard for Heterogeneous Nets," *Data Commun.*, pp. 63–73 (Jan. 1981).

12 Special issue of *IEEE Trans. Commun.*, **COM-28** (Spr. 1980).

Bibliography

Abramson, N. and F. Kuo, eds., *Computer Networks*, Prentice-Hall, Englewood Cliffs, NJ, 1973.

Bingham, J. E. and G. W. P. Davies, *Planning for Data Communications*, Halsted, New York, 1977.

Cypser, R. J., *Communications Architecture for Distributed Systems*, Addison-Wesley, Reading, MA, 1978.

Davies, D. W., Barber, D. L. A., Price, W. L., and Solomonides, C. M., *Computer Networks and Their Protocols*, Wiley, New York, 1979.

Doll, D. R., *Data Communications*, Wiley, New York, 1978.

McNamara, J. E., *Technical Aspects of Data Communications*, 2nd ed., Digital, Bedford, MA, 1982.

Martin, J., *Telecommunications and the Computer*, 2nd ed., Prentice-Hall, Englewood Cliffs, NJ, 1976.

Martin, J., *Computer Networks and Distributed Processing*, Prentice-Hall, Englewood Cliffs, NJ, 1981.

Martin, J., *Design and Strategy for Distributed Data Processing*, Prentice-Hall, Englewood Cliffs, NJ, 1981.

PART 13

ENERGY ENGINEERING

CHAPTER 66
ENERGY CONVERSION

JERALD D. PARKER

Oklahoma State University
Stillwater, Oklahoma

66.1 THERMOELECTRIC CONVERSION

Thermoelectric Effects

Three important physical phenomena are utilized in thermoelectric energy conversion devices. They are

1. Seebeck effect.
2. Peltier effect.
3. Thompson effect.

The *Seebeck effect*, based on the discovery of Thomas Johann Seebeck in 1821, is the creation of an electric current in the junction of two dissimilar materials on which a temperature difference has been imposed.[1a] It is the basis on which the thermocouple is used to measure temperatures. A typical thermocouple circuit, shown in Fig. 66.1, consists of a hot junction T_H of two dissimilar materials A and B and cold junctions T_C where the two metals are usually converted to copper lead wires. The cold junction is usually located in an ice bath of known temperature. A potentiometer uses a measurable voltage to balance the current created in the wires by the temperature difference between the hot and cold junctions. For a particular combination of materials making up the thermocouple junctions, the voltage is proportional in a known way to the temperature difference between the hot and cold junction. The Seebeck coefficient of the couple is given by the change in voltage V of the couple with temperature T

$$\frac{dV}{dT} = (\alpha_A - \alpha_B)$$

where α_A and α_B are the Seebeck coefficients for the junction materials A and B, respectively.

The *Peltier effect* was discovered by Charles Athanase Peltier in 1834 and clarified by Emil Levy in 1838. Because of this effect, heat is given up or absorbed by a junction of two dissimilar materials

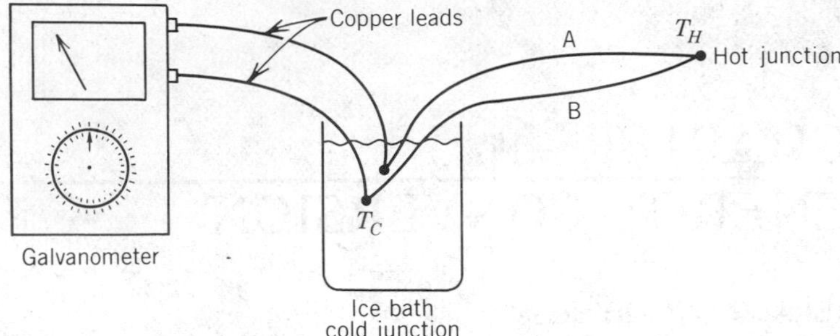

Fig. 66.1 Typical thermocouple circuit.

through which a current is passing. This heat absorbed or evolved is distinguished from any I^2R joule heating that would normally exist in the materials due to current flow.

An energy balance on a junction gives

$$q_j = I^2R_j + I(\pi_{T(A)} - \pi_{T(B)})$$

where q_j = energy given up or absorbed by the junction
I = current flow through junction
R_j = resistance of junction
$\pi_{T(A)}$ = Peltier coefficient of material A at temperature T
$\pi_{T(B)}$ = Peltier coefficient of material B at temperature T

By using Kelvin's second relation from irreversible thermodynamics, this balance can also be written in terms of the Seebeck coefficients of the junction material

$$q_j = I^2R_j + IT(\alpha_A - \alpha_B)$$

Notice that q_j can be positive or negative depending on the relative magnitudes of the two terms and the direction of current flow that fixes the sign of the last term.

The *Thompson effect* was discovered by William Thompson (Lord Kelvin) in 1855 as he applied thermodynamic reasoning to show that a relation existed between the Peltier and Seebeck effects. This effect occurs in a rod in which both a temperature and voltage gradient exist concurrently, that is, in a rod in which both heat and current are flowing in the same or opposite directions. The Thompson effect causes the heat developed to be greater or less than the I^2R joule heat that would normally exist without the pressure of the temperature gradient. This rate is given by

$$q = I^2R + I\left(T\frac{d\alpha}{dT}\Delta T\right)$$

where the last term is the *Thompson heat*. The sign of the last term depends on the direction of the current and the temperature gradient, and either adds to or subtracts from the joule heat given by the first term. The *Thompson* coefficient τ' is given by

$$\tau' = -T\frac{d\alpha}{dT}$$

where the negative sign is necessary to make the value of τ' positive if heat must be added to maintain the temperature constant.

The three thermoelectric effects can each be studied from three viewpoints: classical thermodynamics, irreversible thermodynamics, and solid-state theory,[1a,b,c] the latter being particularly useful in applications involving semiconductors.

Thermoelectric Generators

Thermoelectric generators are now widely used for generating electrical power in space vehicles, in automatic weather information transmitters, in navigation aids, and in portable military field generators. Energy sources for these devices include solar energy, fossil fuels, radioisotopes, and nuclear reactors. The source chosen and the configuration of the device depend on the function required and where the device is to be located. Several device configurations are given in Angrist.[1a]

The performance of a thermoelectric device is given in terms of its *thermal efficiency*, η_t, the ratio of the electrical power output P_o to the thermal power input Q_H to the hot junction:

$$\eta_t = \frac{P_o}{Q_H}$$

The important parameters in design of a thermoelectric generator are the summed thermal conductance K and electrical resistance R of the two junction elements and the combined Seebeck coefficient α for the junction. They can be grouped into a property called the figure of merit, Z:

$$Z = \frac{\alpha^2}{RK}$$

Semiconductor materials are usually superior to metals in terms of having higher values of Z. Assuming that the junction is made of a combination of n- and p-type semiconductor material whose geometry has been optimized, the maximum value of the figure of merit Z^* is given by

$$Z^* = \frac{(|\alpha_n| + |\alpha_p|)^2}{\left[(\rho_n k_n)^{1/2} + (\rho_p k_p)^{1/2}\right]^2}$$

where k is the thermal conductivity and ρ is the electrical resistivity of the junction material.

The geometry is optimized where the product RK is a minimum or when $RK = [(\rho_n k_n)^{1/2} + (\rho_p k_p)^{1/2}]^2$.

The thermal efficiency of the device is optimized when the ratio of the load resistance R_0 across the device to the internal resistance R is given by

$$L_0 = \left(\frac{R_0}{R}\right)_{\text{OPT}} = \left[1 + Z^* \frac{(T_C + T_H)}{2}\right]^{1/2}$$

where T_C and T_H are the temperatures of the cold and hot junctions.

For a device with optimum geometry RK_{OPT} and optimum load-to-internal resistance ratio L_0, the thermal efficiency depends on the optimum figure of merit Z^* and the temperatures of the hot and cold junctions T_H and T_C:

$$\eta_t(\text{MAX}) = \frac{(L_0 - 1)[(T_H - T_C)/T_H]}{L_0 + (T_C/T_H)}$$

A plot of thermal efficiency of a thermoelectric generator as a function of hot and cold junction temperatures and optimum figure of merit is given in Fig. 66.2.

The maximum power output of a thermoelectric generator occurs when the load ratio L_0 equals 1.0. The thermal efficiency for maximum power output is then

$$\eta_t(\text{MAX POWER}) = \frac{(T_H - T_C)/T_H}{4/Z^* T_H + 2 - \frac{1}{2}[(T_H - T_C)/T_H]}$$

Often in design of a generator one wishes to have optimum power with minimum size or weight. A lower limit is imposed on the length of the element to keep contact resistance small compared with resistance of the element, so a minimum volume in practice means minimum cross-sectional area. Maximum power production per unit cross-sectional area of a device occurs when

$$\frac{A_u}{A_p} = \left(\frac{\rho_u}{\rho_p}\right)^{1/2}$$

where A_u and A_p are the cross-sectional areas of the n- and p-type elements making up the junction.

Thermoelectric devices are usually staged to provide suitable voltage and power outputs. A simple electrical series stage is shown in Fig. 66.3. Its output voltage is N times the output voltage of a single stage where N is the number of stages in series. The efficiency of the series stage device is the same as that of a single couple.

Because of the large temperature gradient that usually exists across a thermoelectric device, and because thermoelectric materials are very temperature dependent, an improvement in performance can be obtained by placing a number of generators in thermal series. This device, called a *cascaded generator*, is illustrated in Fig. 66.4. The thermal analysis of this device is similar to that for any cascaded heat engine. The thermal efficiency η_t of a cascaded device with N stages, each having an efficiency of $\eta_t(i)$, is

$$\eta_t = 1 - \prod_{i=1}^{i=N} [1 - \eta_t(i)]$$

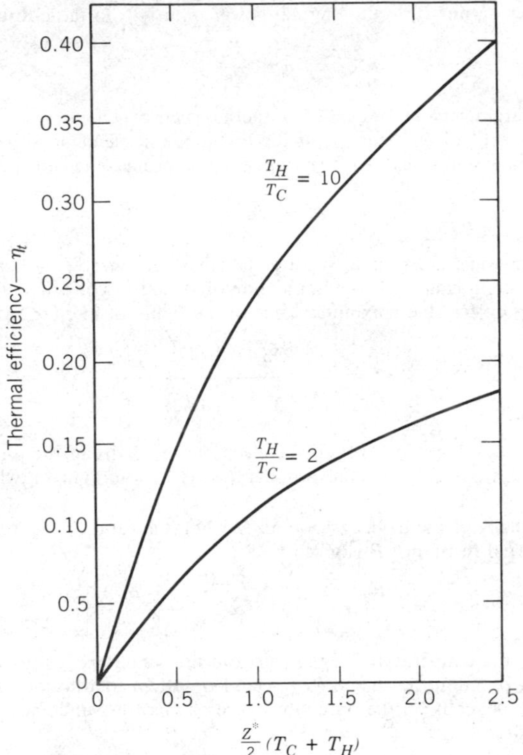

Fig. 66.2 Thermal efficiency of a thermoelectric generator as a function of hot and cold junction temperatures and optimum figure of merit Z^*.

where $\prod$ represents the product of N terms. Electrical resistance required between stages creates thermal resistances that usually limit the number of stages to three or less.

Another approach that attempts to permit materials to operate in a temperature range where their figure of merit is highest is use of the segmented arm generator. In this device each arm of the generator is made up of several different materials in series. The efficiency of this device is apparently

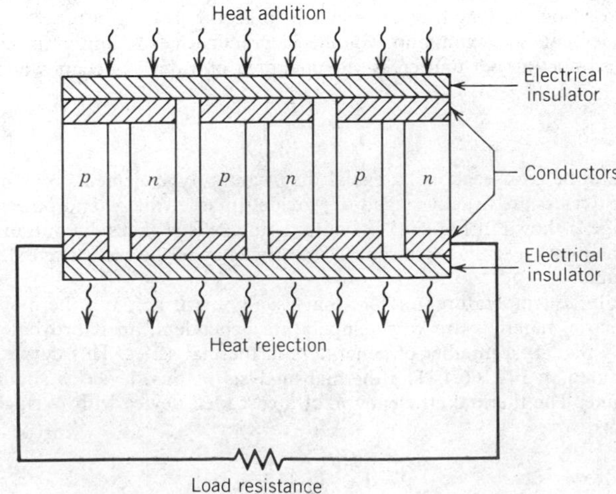

Fig. 66.3 Simple electrical series single-stage thermoelectric device.

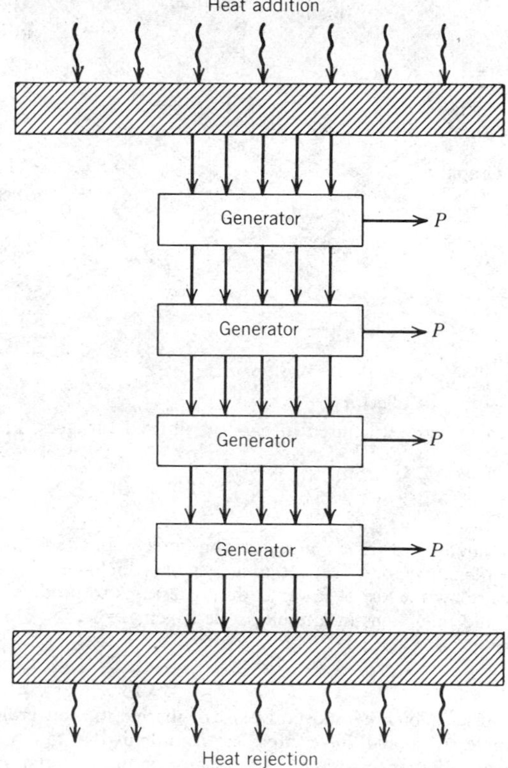

Fig. 66.4 Cascaded thermoelectric generator with four stages, each producing power P.

less than the efficiency of a corresponding multistage device using the same materials over the same temperature ranges.

66.2 THERMIONIC CONVERSION

Basic Principles

The thermionic converter is a device that converts thermal or heat energy into electric energy by the phenomenon of thermionic emission. In the simplest form, this device consists of a high-temperature surface, called the emitter, separated from a lower temperature surface, called a collector, by a vacuum or a plasma. Electrons on the hot surface have a high level of energy and some of them escape, pass across the vacuum or plasma gap to the colder collector surface, and flow through an external circuit to the emitter, producing power in the process. The electron flow also carries some thermal energy from the emitter to the collector, which must be removed. Thus the device operates much like a heat engine, using an electron gas as the working substance. Heat is added at the emitter and some of it is rejected at the collector, with the difference appearing as work in the external electric circuit.

The efficiency of the converter is limited severely by the buildup of charge in the space between the emitter and converter. The most common way of controlling this space charge buildup is by the introduction of positive ions, usually cesium, in the interelectrode space. This can be done in one of three modes:

1. Plasma mode, with low cesium pressure and high emitter temperatures.
2. High-pressure mode, which causes the emitter surface to be cesium coated and which requires a small interelectrode spacing to limit electron collision losses.
3. Ignited mode at low temperatures, where ions are generated mainly by impact. If heat is furnished from an external power source, the mode is described as "ball of fire." If the heating is done internally by the emission current, the term "arc" or "ignited mode" is used.

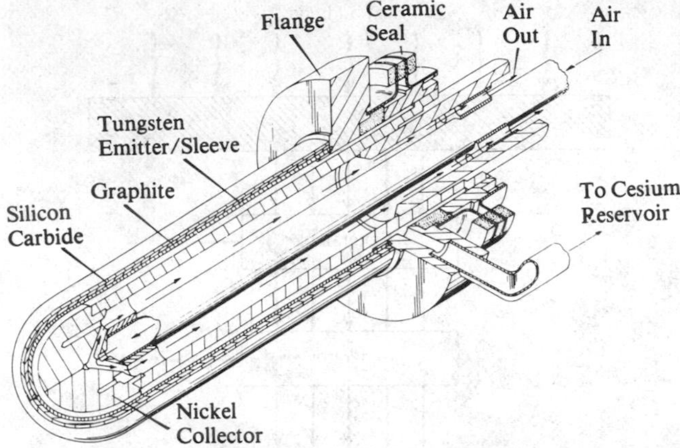

Fig. 66.5 Thermionic energy converter used in gas- or oil-fired furnaces. From Ref. 1b, reprinted with permission.

The majority of all thermionic converters in operation today operate in the ignited mode, with volume ionization of the cesium vapor.[1a] Recent advances have included the use of additives such as oxygen or barium, which allow the use of lower cesium pressures to produce a given value of work function on the emitter, and reductions in interelectrode spacing.

Design of Converters

Material development and selection are important parts of thermionic converter design. In the case of emitter surfaces, the material should have high electron emission rate capability, low rate of deterioration, low thermal radiation emittance, and if any of the material vaporizes it should not poison the collector.

Collector material should have as low a work function as possible so that the electrons will give up only a small amount of energy as they enter the collector surface.

Because high temperatures create many problems, some research efforts have been toward development of converters to operate at lower temperatures. Heat pipes have also been utilized as a means of bringing large heat fluxes to the emitter surfaces and to remove heat from the collector surfaces. Heat pipes are also useful as a thermal transformer, allowing heat flux reductions (or increases) as large as 10 to 1. Heat pipes have been proposed for use with thermionic devices having heat sources such as nuclear reactors, concentrating solar collectors, and radioisotopes.

Thermionic devices show particular promise for space propulsion systems powered by nuclear reactors. They also may be used some day as topping cycles for conventional fossil power plants and as a means of removing power from fusion reactors.

A thermionic energy converter that has been producing electricity for more than 11,000 hours is shown in Fig. 66.5.[2] It is designed for use in gas- or coal-fired furnaces and has a silicon carbide-coated hot shell, formed by vapor deposition that protects the diodes from corrosion by furnace gases. Incremental power generated by arrays of these devices is expected to cost about $500 per kilowatt in a 1982 estimate.

66.3 PHOTOVOLTAIC CONVERSION

Basic Principles

Photovoltaic devices convert the energy of sunlight directly into electric energy. The bypassing of the usual step involving thermal energy means that the Carnot principle limitation on conversion efficiency does not apply. This factor makes these devices very interesting in the search for new conversion methods.

Photovoltaic devices are usually made of semiconductor materials in small cells, and these small cells are then combined into large arrays. By connecting the cells in proper series and parallel arrangements, the desired voltages and power levels can be obtained. The fundamentals of photovoltaic devices and descriptions of individual solar components have been given in Chapter 15. This section will discuss some of the requirements and characteristics of arrays and systems and the solar energy that furnishes their energy input.

Solar Availability

The radiant energy emitted by the sun closely resembles the energy that would be emitted by a black body (an ideal radiator) at about 10,800°F or 5982°C. Figure 66.6 shows the spectral distribution of the radiation from the sun as it arrives at the outer edge of the earth's atmosphere (the upper curve). The peak radiation occurs at a wavelength of about 0.48×10^{-6} m in the green portion of the visible spectrum. Forty percent of the total energy emitted by the sun occurs in the visible portion of the spectrum, between 0.4 and 0.7×10^{-6} m. Fifty-one percent is in the near infrared region between 0.7 and 3.5×10^{-6}. About 9% is in the ultraviolet region below 0.4×10^{-6} m.

Solar energy arrives at the outer edge of the earth's atmosphere at a rate of about 1353 W/m^2, a value referred to as the solar constant. Part of this radiation is reflected back to space, part is absorbed by the atmosphere and reemitted, and part is scattered by atmospheric particles. As a result, only about two thirds of the sun's energy reaches the surface of the earth. At 40° north latitude, for example, the noontime radiation rate on a flat surface normal to the sun's rays is about 950 W/m^2 on a clear day. This would be the approximate maximum rate at which solar energy could be collected at that latitude. A photovoltaic device tracking the sun so as to always be normal to the sun's rays could gather approximately 40 MJ/m^2/day as an upper limit. This will vary somewhat with latitude and season.

Since a photovoltaic device may convert only 10% of the energy striking it, and since the percent sunshine might be about 70%, then the device would have to have about 0.780 m^2 to provide 1 kW-hr

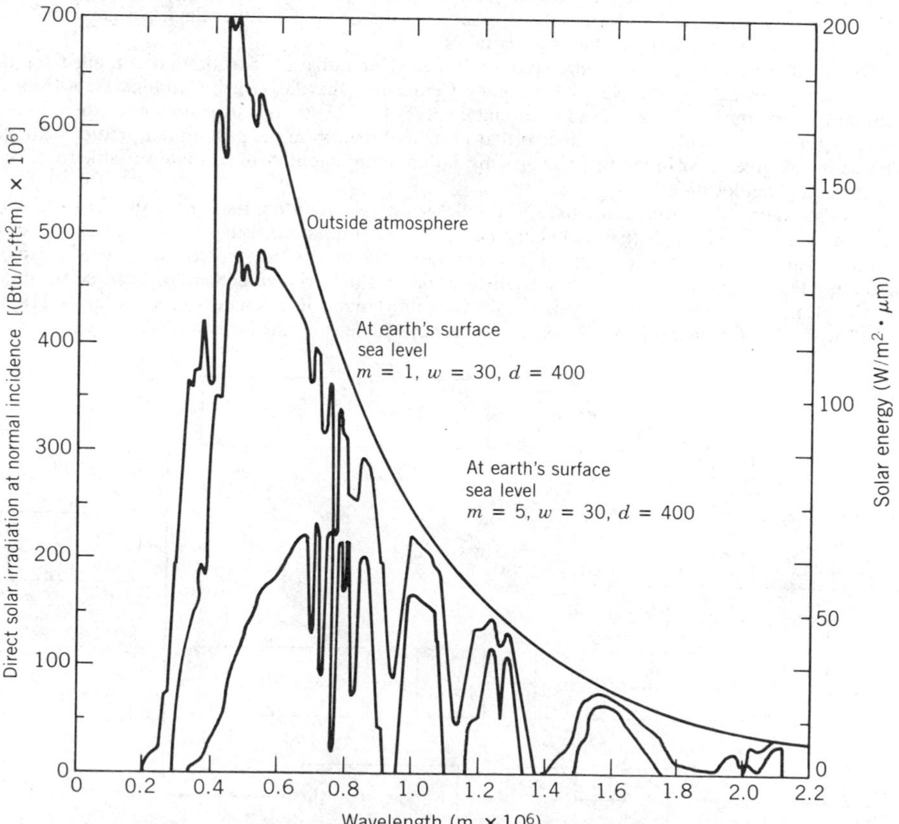

Fig. 66.6 Clear day spectral distribution of direct insolation at normal incidence. (Adapted from J. L. Threlkelb and R. C. Jordan, *ASHRAE Trans.* **64**, p. 50, with permission.)

of energy per day. The important question usually is: Would the cost of constructing, operating, and maintaining a solar system consisting of 0.780 m^2 of tracking photovoltaic devices justify a reduction in conventional energy usage of 1 kW-hr per day? Fixed devices would be expected to deliver considerably less energy for each square meter of surface. A most important consideration concerns the ability to use the electricity produced from solar energy when it is available. In utility or private systems, energy demand will rarely correlate with solar energy availability. Electricity is difficult to store economically in large quantities and some energy may be lost if storage is not available. The development of suitable storage systems will be important to photovoltaic development.

The amount of solar energy available to collect in a system depends on whether the devices move to follow or partially follow the sun or whether they are fixed. In the case of fixed devices, the tilt from horizontal and the orientation of the devices may be significant to the amount of energy collected.

Massive amounts of solar insolation data have been collected over the years by various government and private agencies. The majority of these data are hourly or daily solar insolation on a horizontal surface, and the data vary considerably in reliability. Fixed solar devices are usually tilted at some angle from the horizontal so as to provide a maximum amount of total solar energy collected over the year. In some cases the tilt angle is changed on a regular basis, such as over a month. What one needs in preliminary economic studies is the rate of solar insolations on tilted surfaces.

Figure 66.7 shows the procedure for the conversion of measured horizontal insolation to give values of insolation on a tilted surface. The measured insolation data on a horizontal surface consist of direct radiation from the sun and diffuse radiation from the sky. The total radiation must be split into these two components (step a) and each component analyzed separately (steps b and c). In addition, the solar energy reflected from the ground and other surroundings must be added into the total (step d). Procedures for doing this are given in References 3 through 6.

A very useful table of solar insolation values for 122 cities in the United States and Canada is given in Ref. 7. These data were developed from measured weather data using the methods of Liu and Jordan[4] and of Klein,[5] and are only as reliable as the original weather data, perhaps $\pm 10\%$. A summary of the data for several cities is given in Table 66.1.

One of the more exhaustive compilations of U.S. solar radiation data is that compiled for the Department of Energy by the National Climatic Center in Asheville, North Carolina. Data from 26 sites were rehabilitated and then used to estimate values for 222 stations. A summary of these data is given by Lunde.[8] It should be remembered that measured data from the past do not predict what will happen in any given year in the future. Solar insolation in any month can be quite variable from year to year at a given location.

Another approach is commonly used to predict solar insolation on a specified surface at a given location. This method is to first calculate the clear day insolation, using knowledge of the sun's location in the sky at the given time. This clear day insolation is then corrected by use of factors describing the clearness of the sky at a given location and the average percent of possible sunshine.

The clear sky insolation on a given surface is readily found in references such as the ASHRAE *Handbook of Fundamentals*. Table 66.2 shows percent possible sunshine for several cities.

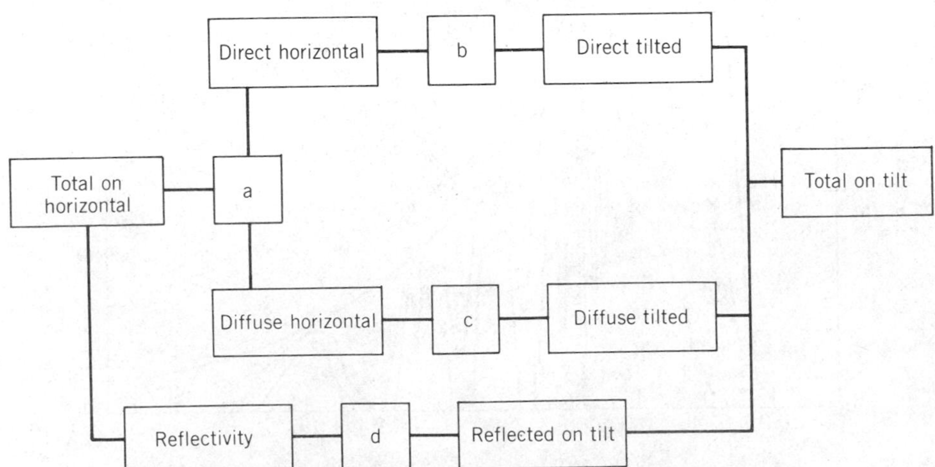

Fig. 66.7 Conversion of horizontal insolation to insolation on tilted surface.

Photovoltaic Arrays

Arrays of photovoltaic cells are used to provide electrical energy at the power levels required. These arrays may be fixed or movable (for tracking the sun); if movable they may utilize optical concentrators to increase the power output per cell.[9]

Flat, nonconcentrating arrays will probably be preferred in the temperate zones of the world, since much of the solar energy there is diffuse.[10] Flat arrays can utilize the diffuse energy that concentrating arrays cannot use, and although a flat array may be fixed it can gather as much or more energy than concentrating, tracking collectors in these zones. Flat arrays may also be designed to track the sun, even when concentrating devices are not used.

Fixed flat-plate arrays are easier to install, need less maintenance, and blend in better with buildings compared with tracking systems. The arrays are made up of modules, each of which is a collection of several solar cells connected together in series-parallel arrangements. To prevent failure of an entire module in the case where one cell in a series arrangement fails, parallel interconnecting tabs are placed at points of equal potential. These tabs usually carry no current, but should a cell fail, the tabs bypass the defective cell.

The modules are encapsulated, usually with glass covers to provide protection against weather and damage by handling and to reduce corrosion. A typical module, made by Arco, is about 1 by 4 ft and generates about 120 to 200 W-hr per day. The module is constructed by creating a permanent banding of the glass cover, the cells, and a back cover, using layers of polyvinylbutyral (PVB). Each module has electrical terminals for connection to other modules in series and/or parallel arrangement. Some modules are dual voltage, providing, for example, 6 or 12 V, depending on the terminal connection selected on the module. Most commercial solar modules on the market today are made with silicon solar cells. Efficiencies are in the range of 14%.

Fresnel lenses or reflecting parabolic troughs may be used to concentrate the sun's rays, reducing the total number or area of cells required. Concentrating the sunlight onto solar cells permits the use of the more expensive, complex photovoltaic devices with higher efficiencies.[11] With concentrating devices, the supporting structures must move to track the sun, and yet they must be strong and rigid enough to withstand wind loads. The supporting structures, tracking and drive mechanisms, and controls make these systems much more expensive than the fixed flat-plate systems. Since the concentrating systems can utilize only the direct component of sunlight, their use may be more common in the arid regions where there is little cloudiness or haze. It is also important to recognize that perhaps only 25% of the direct sunlight incident on the optical system will reach the cells.

With high sunlight levels focused onto the solar cells, it is important that the cells be designed with low internal resistance to carry the high currents generated. This can be done by the addition of conductive grids, closely spaced on the surface of the cell. Fins or a coolant may be used to remove thermal energy from the cell and to maintain the cell at some suitable operating temperature. The heat removed by a coolant may be used for space or water heating, possibly improving the system economics.

In concentrating systems, gallium-arsenide cells or other more expensive types may be preferred to the silicon cell.

Maycock and Stirewalt[12] have described several functioning photovoltaic systems. At that time (September 1981), the largest solar photovoltaic system in the world was the 100-kW system at Natural Bridges Monument in Utah. The system's accomplishments and problems are discussed. Other systems described include the first stand-alone unit designed specifically to supply power to a small voltage (Schuchuli, Arizona), a 3.5-kW system, and the 25-kW array at Mead, Nebraska, used for pumping irrigation water. Residential and centralized photovoltaic power systems are described by Green,[13] an excellent source of information on all aspects of photovoltaic devices and systems. An interesting proposal is the development of solar cell "shingles" that provide a protective roof cover.

A large system planned at the time of writing of this material is rated at 1 mW, with Arco photovoltaic panels occupying 20 acres, for Southern California Edison Company. The flat modules will be mounted on 100 double-axis trackers controlled by computers. A large-scale inverter will be used to convert the DC electricity generated by the panels to standard AC.

A lightweight solar array as tall as a 10-story building was built for a 1984 launch from a space-shuttle orbiter. It was made of a lightweight, flexible polymide film, with wraparound contact cells welded directly to the array blanket. The 105-ft × 13.5-ft array was folded accordionlike in the orbiter's cargo bay into a package less than 4 in. thick. It was folded and unfolded several times in orbit to test its characteristics. The panel was designed to generate 66 W/kg, over three times the ratio of conventional systems.

Large size (300 GW) photovoltaic systems designed for generation of electricity in orbital space and transmission by microwave to earth have recently been criticized in a National Research Council Report.[14]

TABLE 66.1. AVERAGE DAILY RADIATION ON TILTED SURFACES FOR SELECTED CITIES

City	Slope	Average Daily Radiation (Btu/day · ft^2)											
		Jan.	Feb.	Mar.	Apr.	May	Jun.	Jul.	Aug.	Sept.	Oct.	Nov.	Dec.
Albuquerque, NM	hor.	1134	1436	1885	2319	2533	2721	2540	2342	2084	1646	1244	1034
	30	1872	2041	2295	2411	2346	2390	2289	2318	2387	2251	1994	1780
	40	2027	2144	2319	2325	2181	2182	2109	2194	2369	2341	2146	1942
	50	2127	2190	2283	2183	1972	1932	1889	2028	2291	2369	2240	2052
	vert.	1950	1815	1599	1182	868	754	795	1011	1455	1878	2011	1927
Atlanta, GA	hor.	839	1045	1388	1782	1970	2040	1981	1848	1517	1288	975	740
	30	1232	1359	1594	1805	1814	1801	1782	1795	1656	1638	1415	1113
	40	1308	1403	1591	1732	1689	1653	1647	1701	1627	1679	1496	1188
	50	1351	1413	1551	1622	1532	1478	1482	1571	1562	1679	1540	1233
	vert.	1189	1130	1068	899	725	659	680	811	990	1292	1332	1107
Boston, MA	hor.	511	729	1078	1340	1738	1837	1826	1565	1255	876	533	438
	30	830	1021	1313	1414	1677	1701	1722	1593	1449	1184	818	736
	40	900	1074	1333	1379	1592	1595	1623	1536	1450	1234	878	803
	50	947	1101	1322	1316	1477	1461	1494	1448	1417	1254	916	850
	vert.	895	950	996	831	810	759	791	857	993	1044	842	820
Chicago, IL	hor.	353	541	836	1220	1563	1688	1743	1485	1153	763	442	280
	30	492	693	970	1273	1502	1561	1639	1503	1311	990	626	384
	40	519	716	975	1239	1425	1563	1544	1447	1307	1024	662	403
	50	535	723	959	1180	1322	1341	1421	1363	1274	1034	682	415
	vert.	479	602	712	746	734	707	754	806	887	846	610	373

Ft. Worth, TX	hor.	927	1182	1565	1078	2065	2364	2253	2165	1841	1450	1097	898
	30	1368	1550	1807	1065	1891	2060	2007	2097	2029	1859	1604	1388
	40	1452	1601	1803	1020	1755	1878	1845	1979	1995	1907	1698	1488
	50	1500	1614	1758	957	1586	1663	1648	1820	1914	1908	1749	1549
	vert.	1315	1286	1196	569	728	679	705	890	1185	1459	1509	1396
Lincoln, NB	hor.	629	950	1340	1752	2121	2286	2268	2054	1808	1329	865	629
	30	958	1304	1605	1829	2004	2063	2088	2060	2092	1818	1351	1027
	40	1026	1363	1620	1774	1882	1909	1944	1971	2087	1894	1450	1113
	50	1068	1389	1597	1679	1724	1720	1763	1838	2030	1922	1512	1170
	vert.	972	1162	1156	989	856	788	828	992	1350	1561	1371	1100
Los Angeles, CA	hor.	946	1266	1690	1907	2121	2272	2389	2168	1855	1355	1078	905
	30	1434	1709	1990	1940	1952	1997	2138	2115	2066	1741	1605	1439
	40	1530	1776	1996	1862	1816	1828	1966	2002	2037	1788	1706	1550
	50	1587	1799	1953	1744	1644	1628	1758	1845	1959	1791	1762	1620
	vert.	1411	1455	1344	958	760	692	744	918	1230	1383	1537	1479
New Orleans, LA	hor.	788	954	1235	1518	1655	1633	1537	1533	1411	1316	1024	729
	30	1061	1162	1356	1495	1499	1428	1369	1456	1490	1604	1402	1009
	40	1106	1182	1339	1424	1389	1309	1263	1371	1451	1626	1464	1058
	50	1125	1174	1292	1324	1256	1170	1137	1259	1381	1610	1490	1082
	vert.	944	899	847	719	599	546	548	647	843	1189	1240	929
Portland, OR	hor.	578	872	1321	1495	1889	1992	2065	1774	1410	1005	578	508
	30	1015	1308	1684	1602	1836	1853	1959	1830	1670	1427	941	941
	40	1114	1393	1727	1569	1746	1739	1848	1771	1680	1502	1020	1042
	50	1184	1442	1727	1502	1622	1594	1702	1673	1651	1539	1073	1116
	vert.	1149	1279	1326	953	889	824	890	989	1172	1309	1010	1109

Source. Department of Energy,[5] reprinted with permission.

TABLE 66.2. MEAN PERCENTAGE OF POSSIBLE SUNSHINE FOR SELECTED U.S. CITIES

Station	Jan.	Feb.	Mar.	Apr.	May	June	July	Aug.	Sept.	Oct.	Nov.	Dec.	Annual
Albuquerque, NM	70	72	72	76	79	84	76	75	81	80	79	70	76
Atlanta, GA	48	53	57	65	68	68	62	63	64	67	60	47	60
Boston, MA	47	56	57	56	59	62	64	63	61	58	48	48	57
Chicago, IL	44	49	53	56	63	69	73	70	65	61	47	41	59
Ft. Worth, TX	56	57	65	66	67	75	78	78	74	70	63	58	68
Lincoln, NB	57	59	60	60	63	69	76	71	67	66	59	55	64
Los Angeles, CA	70	69	70	67	68	69	80	81	80	76	79	72	73
New Orleans, LA	49	50	57	63	66	64	58	60	64	70	60	46	59
Portland, OR	27	34	41	49	52	55	70	65	55	42	28	23	48

Source. Original data from *Climatic Atlas of the United States*, U.S. Gov. Printing Off., 1968, reprinted from Lunde,[8] with permission.

66.4 FUEL CELLS

Basic Principles—Fuels

Fuel cells convert the energy of a fuel directly to electricity by an electrochemical process. Fuel cells and batteries operate on a similar principle except that a fuel cell operates with a continuous supply of fuel. Fuel cell efficiency is not limited by the Carnot principle, and therefore these devices have potential for much higher conversion efficiencies than do conventional power plants. Because they are produced in modules they offer great flexibility in use and installation and are factory produced with relatively short lead times. They can operate on a variety of fuels, are quiet in operation, and create almost no emissions to foul the environment. Because of their ability to generate a high ratio of electrical energy to thermal energy, they often are able to match building loads very well and are considered by some the "ultimate cogeneration package."[15]

A fuel cell consists of electrodes (a cathode and anode) separated by an electrolyte, Fig. 66.8. The electrolyte may be aqueous acid, aqueous alkaline, molten salt, or solid.

A hydrogen-rich fuel is supplied to a chamber behind the anode, and oxygen (or air) is supplied to a chamber behind the cathode. The cathodes are porous and permit the interfacing of the gases with the electrolyte. The hydrogen at the anode is split into hydrogen ions and electrons. With an acid electrolyte, the positive hydrogen ions move toward the cathode as the electrons flow through the external circuit, as shown in Fig. 66.8. With an alkaline electrolyte, negative ions move from the cathode to the electrode as the electrons flow externally, again from the anode to the cathode.

If the fuel is something other than hydrogen, for example, methane, a fuel processor or reformer is usually added to produce a hydrogen-rich gas. A common type is the catalytic steam reformer.[16]

Since the power produced by a fuel cell is DC, a power conditioner is usually added to produce AC power compatible with that of the utility. A complete fuel cell plant might have the arrangement shown in Fig. 66.9.[17]

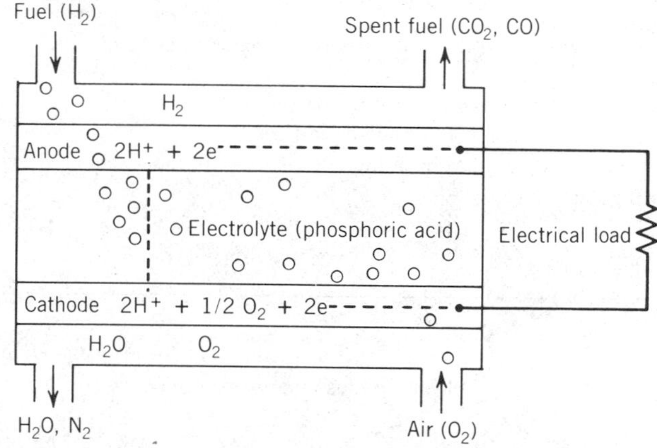

Fig. 66.8 Simplified diagram of an acid-type fuel cell.

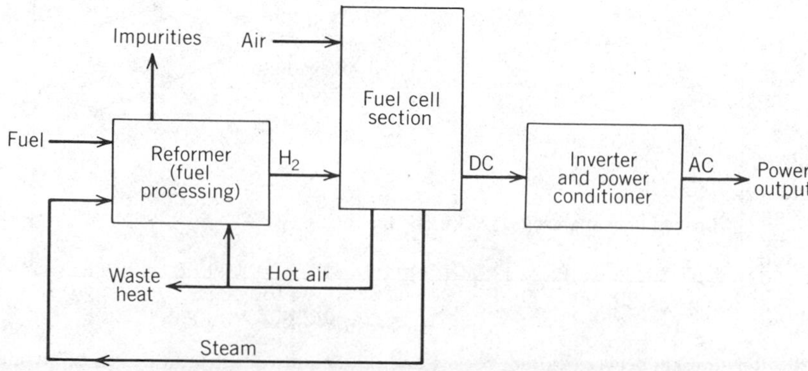

Fig. 66.9 Simplified schematic of a fuel cell plant.

Cell Performance

Angrist[1a] has a detailed discussion of fuel cells. He defines the ideal efficiency of a simple fuel cell (without regeneration) by

$$\eta_i = \frac{\Delta G}{\Delta H} = 1 - \frac{T\Delta S}{\Delta H}$$

where ΔG = change in Gibbs free energy
ΔH = heat of reaction or heat of combustion
T = absolute temperature

This ideal efficiency is shown to be

$$\eta_i = \frac{ItV}{\Delta H}$$

where I = current
t = time of current flow
V = reversible electromotive force of the cell

In operation under load, the electromotive force of the cell will decrease to the actual voltage V_{act}. In this case the voltage efficiency will be

$$\eta_v = \frac{V_{act}}{V}$$

The effect of current density on cell output voltage for a molten carbonate fuel cell is given in Fig. 66.10.[18]

Values of the heat of formation ΔH and the change in the Gibbs function are given in Table 66.3 at several temperatures, along with the computed values of ideal efficiency of the fuel cell and the Carnot efficiencies.[19] Note the superiority of the fuel cell at lower temperatures, with the opposite true at high temperatures.

Fuel cell performance may be less than the theoretical value because of

1. Undesirable reactions taking place.
2. Hindrance of the desired reaction at a node or cathode.
3. Formation of a concentration gradient in the electrolyte or in the reactants.
4. I^2R heating in the electrolyte.

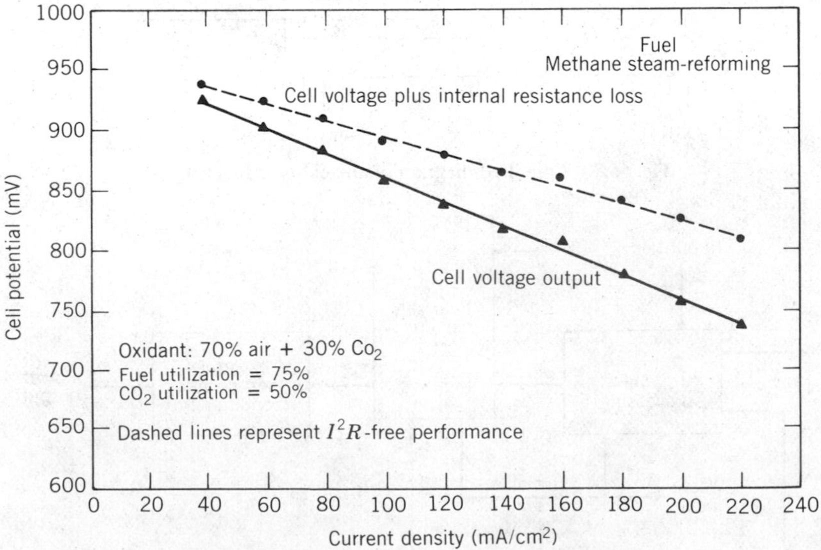

Fig. 66.10 Relationship between current density and cell output voltage for a molten carbonate fuel cell. From Marianowski et al.,[18] with permission.

TABLE 66.3. RELEVANT THERMODYNAMIC PROPERTIES[a] FOR THE HYDROGEN-OXYGEN FUEL CELL

T (K)	$\Delta H°$ (kcal/mole)	$\Delta G°$ (kcal/mole)	η_i ($\Delta G°/\Delta H°$)	Carnot Efficiency ($T_H - T_C/T_H$)
298	− 57.80	− 54.64	0.94	0
400	− 58.04	− 53.52	0.92	0.26
500	− 58.27	− 52.36	0.90	0.40
1000	− 59.21	− 46.03	0.78	0.70
2000	− 60.26	− 32.31	0.54	0.85

[a] The values of $\Delta G°$ and $\Delta H°$ listed here are for the reaction that produces H_2O from hydrogen and oxygen. The sign on these values must be reversed if hydrogen and oxygen are produced from H_2O.

Design of Fuel Cells

Attempts to improve the performance of fuel cells have led to designs that are both compact and reliable. Fuel cells have been used to furnish electricity in space vehicles and in addition to provide drinking water for astronauts. Since all of the energy input is not converted to electricity, there is a need to remove heat from the fuel cells. Sometimes this heat has useful application in the fuel processing (see Fig. 66.10) or in furnishing space heat to a building. A typical 2000-W hydrogen–oxygen module used in a space application consists of 32 series-connected cell sections with each section containing two cells connected in parallel. Each module contains its own control assembly and valves, condenser, water pumps, and inverters. The electrolyte in these devices is KOH, immobilized in an asbestos matrix.

In a typical cogeneration package for use with electric utilities, the module produces 40 kW. The matrix of cells uses phosphoric acid as the electrolyte and natural gas as fuel. The electrolyte operates at approximately 200°C. The units weigh 8000 lb and are $9 \times 5 \times 6.5$ ft in size. A larger version module has been used to make up a 4.8-MW generating package for experimental operation by a utility in New York City. The efficiency of this system is expected to be about 42%.

Not quite so far advanced as the phosphoric acid cell is the molten carbonate cell, which will operate with electrolyte at 650°C. At that temperature the electrolyte is a two-phase mixture of liquid carbonate salt and small particles of lithium aluminate. The design must prevent or minimize the evaporation of the molten carbonate and must accommodate the thermal expansion differences between the tile and the adjacent metal parts. The cell is poisoned by sulfur above 1 ppm. This type of cell will probably find its greatest application in base loading service.[20]

Fuel cells have also been described as showing promise as a vehicle power source.[21,22] Further basic information on fuel cells can be found in a book edited by K. R. Williams[23] and in an article by A. P. Fickett.[24]

66.5 MAGNETOHYDRODYNAMICS

Basic Principles

Faraday discovered that whenever a conductor moves through a magnetic field, a current is induced in the conductor. When the conductor is a fluid, we have the basis for an MHD (magnetohydrodynamic) generator. Figure 66.11 illustrates the principle of an MHD generator. A duct is arranged so that a magnetic field is induced across the channel. When a conductive fluid passes through the duct, an electric field results. If the upper and lower walls of the duct are electrically insulated from each other, these walls will have a difference in electrical potential. Current will flow through any external electrical circuit connecting these two walls. This direct current can be converted to AC by an inverter.

Because the gases flowing through a nozzle can be delivered at extremely high temperatures (2200°C) (4600°F), the MHD generator has the potential of producing electricity at very high thermal efficiencies, perhaps 45% to 50%. This is enough above the thermal efficiency of a conventional power plant ($\simeq$35%) to create a large amount of interest in MHD development. In addition, the MHD generator has the advantage of no moving parts such as exist in a turbine. Coal appears to be the most economical choice as a fuel, although the use of nuclear plants shows some promise.

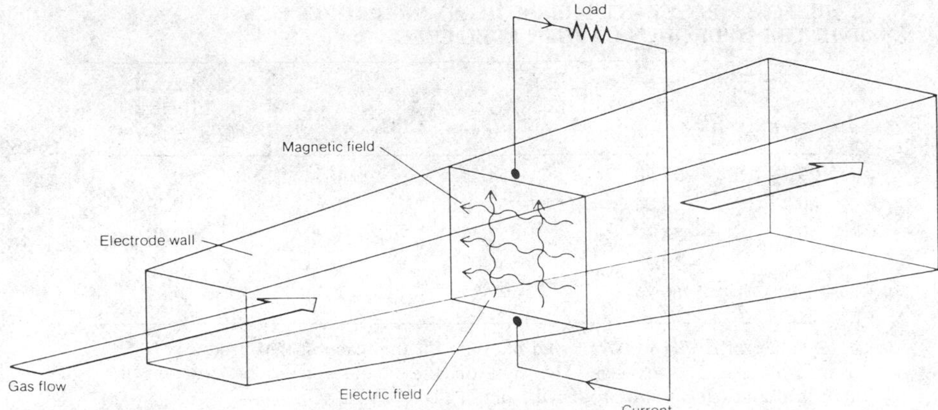

Fig. 66.11 Simplified schematic of a magnetohydrodynamic (MHD) generator.

Most gases are not electrically conductive at ordinary temperatures and are only fair conductors at the temperatures that would be created in combustion processes. To improve the conductivity of the flowing fluid and thus the performance of the generator, most designs propose the use of some seeding compound such as potassium carbonate added upstream of the nozzle, as in Fig. 66.12. This seed material would also combine with the sulfur that might be released in the combustion of the coal and would form potassium sulfate, which could be removed with comparative ease downstream of the MHD channel and steam generator.[25]

The steam generated downstream of the MHD channel is used to drive a conventional steam turbine to produce electricity in addition to that produced in the MHD channel. This conventional steam generator turbine system is called a *bottoming cycle*; the MHD part of the system is called a *topping cycle*. It is the combination of these two systems that results in the comparatively high thermal efficiencies quoted.

To attain the extremely high gas temperatures desired for efficient operation, two approaches are utilized:

1. The air is preheated in a fossil fuel-fired heat exchanger upstream of the combustor. Exhaust gases may be used for this task, although this creates some serious heat exchanger problems owing to the potassium seed and coal slag.

2. The combustion air is enriched with oxygen. This is the approach diagrammed in Fig. 66.12 and the one that appears to have the most promise.[26]

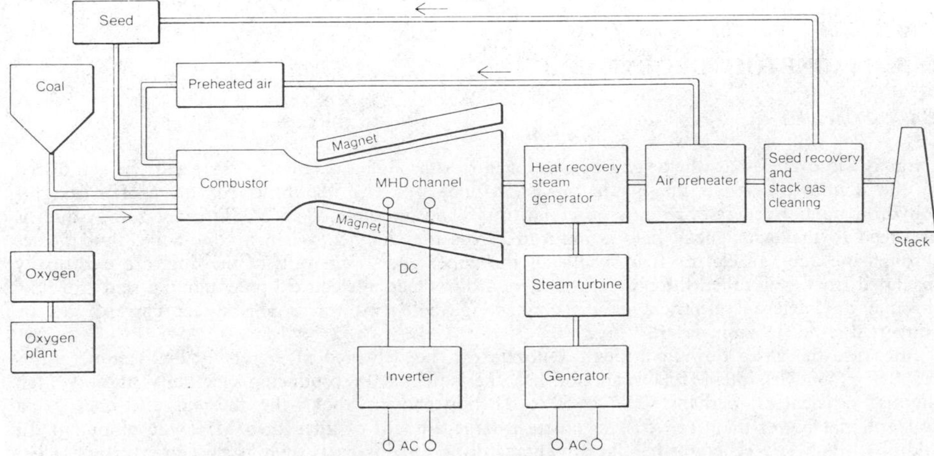

Fig. 66.12 Simplified schematic of a magnetohydrodynamic (MHD) power plant with seeding.

The high gas temperatures in the nozzle of the MHD channel create some unusual material requirements, and heat transfer computations are very critical. An overview of these problems is given by Postlethwaite and Shuyter.[27] Electrode lifetimes of 8000 hours now seem possible using water-cooled copper with platinum and/or stainless steel cladding in critical areas.[2(p. 10)]

To furnish the magnetic field at suitable flux levels and efficiencies, cryogenically cooled superconducting magnets are usually proposed. This will create some new and interesting design problems.

An excellent analysis of an MHD generator and other useful information are given in a recent text by Angrist.[1a]

66.6 SOLAR THERMAL CONVERSION

Concentrating Collectors

The major components of a typical solar thermal electric system, shown in Fig. 66.13, include the solar collectors, a thermal storage system, a heat engine, and an electric generator. In keeping with the second law of thermodynamics, heat will always be rejected to a sink, usually directly to the ambient air or indirectly to the air through some source of cooling water such as a lake, a river, or a cooling tower. Accounting for losses, the amount of energy produced as electricity will be equal to the difference between the energy collected and the energy rejected. The fraction of the sun's available energy that can be converted to electricity is improved by reducing the losses and the fraction that is rejected to the sink. The losses from each component can be reduced by good design; the remaining fraction of the available solar energy that is rejected to the sink will depend on the temperature level at which that solar energy is collected. Again using the second law but stating it in terms of the Carnot principle—"No heat engine, operating continuously between a given fixed source temperature T_H and a given fixed sink temperature T_C can have a thermal efficiency greater than a Carnot engine operating between these temperatures." The thermal efficiency η of a Carnot engine, the fraction of the heat input from the source that can be converted into useful work, is given by

$$\eta_C = 1 - \frac{T_C}{T_H} \tag{66.1}$$

If the sink temperature T_C is assumed to be the ambient temperature and a fixed value, it can be seen from Eq. (66.1) that increasing the temperature of the source T_H increases the thermal efficiency of the ideal heat engine. We would expect the same trend to be true for an actual heat engine, since its limiting performance is the ideal engine. As a result, we see that it is desirable to operate the solar collectors (the heat source) at high temperatures relative to ambient temperature if good conversion efficiencies are to be possible. This is why we must utilize concentrating collectors in solar thermal electric systems. Nonconcentrating (flat-plate) collectors cannot operate at sufficiently high temperatures to give overall system conversion efficiencies that are economical.

The importance of collector concentration ratio to solar thermal electric systems can be understood by making a simplified energy balance on a solar collector and putting the results in terms of a collector efficiency. The useful energy q_u furnished by a collector is, assuming that all heat loss is due

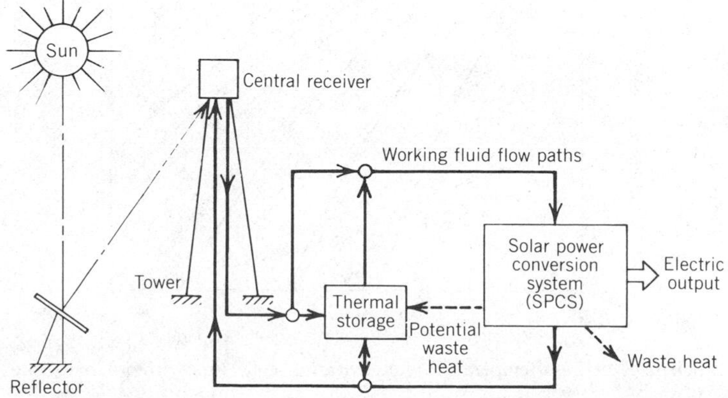

Fig. 66.13 Typical solar thermal electric power plant of the central receiver type.

to thermal radiation to an infinite environment at T_R

$\dot{q}_u$ = rate solar energy absorbed by receiver − net rate energy radiated away from receiver

$$\dot{q}_u = A_a G_s \rho(\tau\alpha)_r = \epsilon_r \sigma A_r (T_r^4 - T_a^4) \tag{66.2}$$

where A_a = area of aperture of collector, amount of area normal to and intercepting sun's rays
A_r = area of collector receiver, area on which sun's rays are concentrated, A_a/A_r = concentration ratio
G_s = irradiation rate of sun's direct rays normal to aperture area
ρ = reflectance of concentrating surface
$(\tau\alpha)_2$ = transmittance absorptance product of receiver for sunlight
ϵ_r = emittance of receiver surface
σ = Stefan–Boltzmann constant
T_r = absolute temperature of receiver
T_a = ambient absolute temperature

Defining the collector efficiency as

$$\eta_{COL} = \frac{\dot{q}u}{A_s G_s} \tag{66.3}$$

we can use Eq. (66.2) to derive the following

$$\eta_{COL} = \rho(\tau\alpha)_r = \frac{\sigma}{G_S} \frac{\epsilon_r A_r}{A_a} (T_r^4 - T_a^4) \tag{66.4}$$

$A_a/\epsilon_r A_r$ is the emittance-modified concentration ratio, which is a useful parameter with which to study concentrating collector systems.

Figure 66.14[28] is a graphic presentation of Eq. (66.4), assuming a $\rho(\tau\alpha)$ value of about 0.77 (which fixes the maximum efficiency) and a solar insolation rate of 0.8 kW/m^2. The collector efficiency drops off rapidly with collection temperature at the lower concentration ratios. Higher concentration ratios permit relatively high collection efficiencies at the higher collector temperature.

The maximum theoretical efficiency of a solar thermal electric plant can be determined by combining Eqs. (66.1) and (66.4), assuming $T_r = T_H$ and $T_C = T_a$. The overall plant efficiency is the product of the collector efficiency and the thermal efficiency of the ideal Carnot cycle. The result is shown in Fig. 66.15.[28] The advantage of a high concentration ratio is apparent in Fig. 66.15.

In addition to the higher temperatures and the corresponding higher efficiencies obtained with concentrating collectors, such systems generally have two other advantages. The reduced absorber area allows a reduction in mass flow rate of coolant, and this gives an increasing reduction in pumping power required. In addition, the weight of concentrating collectors per unit of collection area is usually less than that for flat-plate collectors.

There are some rather serious basic disadvantages to most concentrating solar collectors, however.

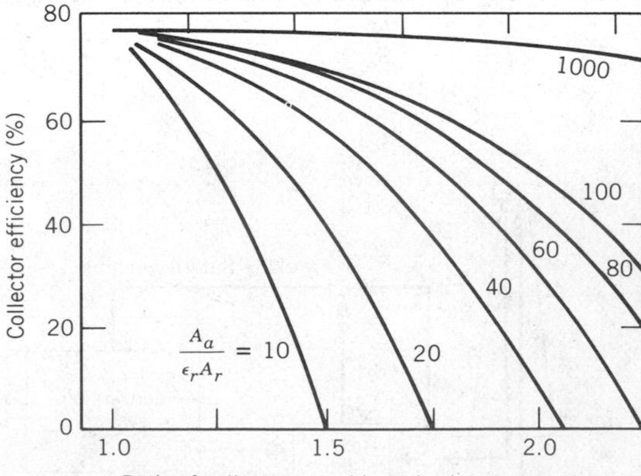

Fig. 66.14 Efficiencies of high-temperature concentrating collectors. From Kreith and Kreider,[28] with permission.

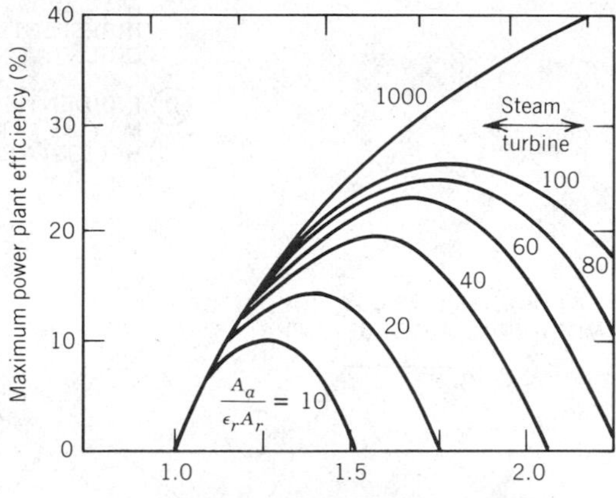

Fig. 66.15 Theoretical maximum efficiency of a solar power plant. From Kreith and Kreider,[28] with permission.

These are:

1. Only the direct (or beam) component of solar radiation is used.
2. The collectors must track the sun, which makes the control more complex and increases maintenance.
3. The optical system must be kept clean.

Types of Concentrators

A text that emphasizes concentrating collector systems is by Meinel and Meinel.[29] Concentrating collectors consist of an optical system and a receiver system. It is natural to classify concentrating collectors in terms of the kinds of optical systems and/or receiver systems utilized. An important classification involves whether there is a single receiver accepting energy (central receiver system) or numerous receivers (distributed receiver system). Generally, the larger (> 5 $MW_{electric}$) electric generating systems utilize the central receiver system to best advantage. The central receiver system almost always involves the use of a tower on which to locate the receiver, hence the name "power tower" is frequently used. In the central receiver system, a large number of mirrors (heliostats) are individually moved so as to concentrate the sun's rays onto the receiver. A system of this type is shown in Fig. 66.13. Central receiver systems are described in Refs. 30 and 31.

The main element of the collector subsystem is the heliostat. Fortunately, heliostat design does not need to be unique for each central receiver system, making the economics of mass production more probable. Most heliostats are made with second surface mirrors of relatively thin (1.5–3.0 mm) low-iron glass to minimize absorption losses. A typical heliostat consists of 10 to 14 panels with a total of 40 to 60 m^2 of area.

Individual heliostats are moved independently in two axes every 5 to 15 seconds and a central computer oversees the motion of the entire array and feeds information to several field controllers, which in turn oversee control of individual heliostat motors. The heliostats are staggered and arranged most densely near the tower.

A typical central receiver system plant in the southwestern United States of 100 $MW_{electric}$ size and a capacity factor of 0.42 would require approximately 12,500 heliostats (50 m^2 each) and would be placed over about 700 acres (2.4 km^2) of land. The losses that might occur with the use of heliostats are shown in Fig. 66.16.

The receiver, mounted at the top of the tower, absorbs the concentrated solar rays and transfers the heat to the transfer or working fluid. Two typical types of receivers are external and cavity. The choice of receiver type depends on the transfer fluid, design operating temperatures, heliostat layout, and cost and economics desired. The external receiver is generally less costly but less efficient. Thermal losses from an external receiver are shown in Fig. 66.17.

Maximum heat flux on a receiver can be about 0.5 million Btu/hr ft^2 (1.5 MW/m^2), which corresponds to a concentration ratio of about 1600. Tower height depends primarily on the size of the

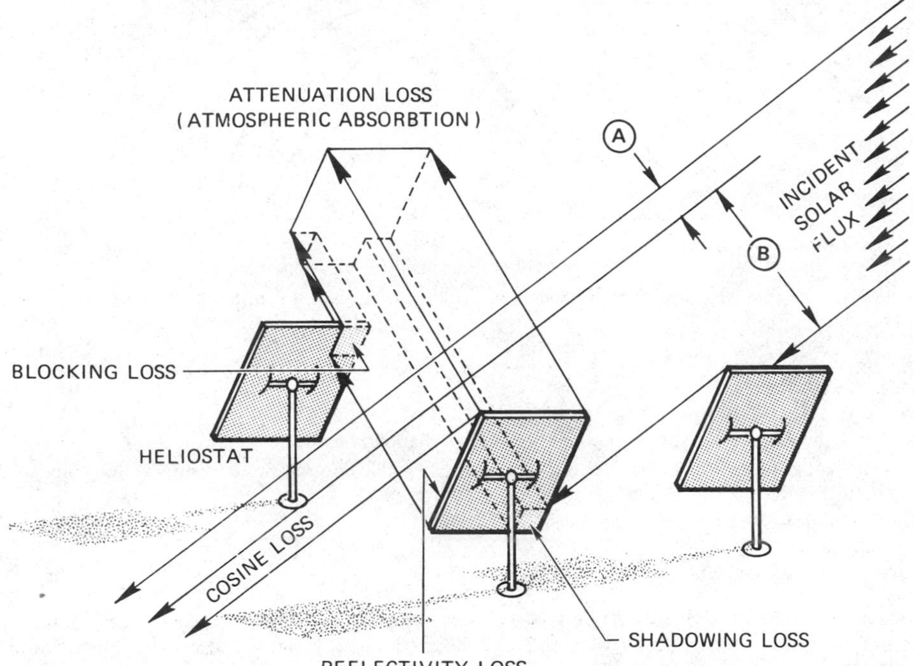

Fig. 66.16 Optical losses in central receiver systems.

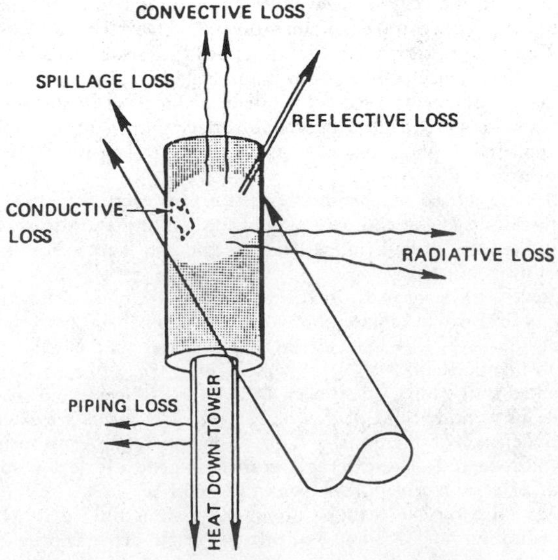

Fig. 66.17 Thermal losses from an external receiver.

heliostat field, which of course depends on plant size. For a surrounding field of heliostats, a 380-$MW_{thermal}$ plant would have a tower height of about 140 to 170 m.

In the distributed collector system (distributed receiver system), energy is collected at many receivers that are dispersed over the collection area. The energy from individual receivers is then usually piped to a central location for conversion to electrical energy. The distributed collectors are usually of the one-axis tracking type. The most common type is the line focus, parabolic trough collector, available in several commerçial models. This type of collector system may be arranged in one of three possible tracking modes.

1. Horizontal north-south axis, east-west tracking.
2. Horizontal east-west axis, north-south tracking.
3. Tilted axis parallel to polar axis of earth, east-west tracking.

The seasonal and total yields of these three modes would, of course, be different. The maximum possible hourly solar fluxes for collectors operating in each of these three modes are given in Meinel and Meinel.[29]

One way of concentrating sunlight on a linear receiver is by means of a Fresnel lens. Such a system is shown in Fig. 66.18.

The lens works well only if it is normal to the sun's rays. The collector must therefore be tilted at an angle of latitude plus declination, and this tilt angle will change continually throughout the year. The system is rotated east-west for daily tracking of the sun.

Some of the more recent efforts in developing line focus concentrating collectors are described in Ref. 30. Included is a description of the fixed-mirror distributed-focus system shown in Fig. 66.19. These systems can operate up to 750°C, but have a lower energy yield than do tracking systems.

The point-focusing parabolic concentrator is considered by some the ultimate type of solar collector. An excellent discussion of these devices is given in Ref. 32. In a point-focusing device, a heat engine may be located at the focal point of each collector. Collectors of this type are shown in Fig. 66.20. The ultimate measure of a concentrating collector is the figure of merit, the British thermal units collected per dollar per square foot of cost. The figure of merit depends on the reflectivity of the mirror surfaces, the surface quality of the substrate, and the pointing errors caused by controls and the flexibility of the structure.

A comparison of the maximum possible annual energy yields for various types of tracking modes is given in Table 66.4.[29]

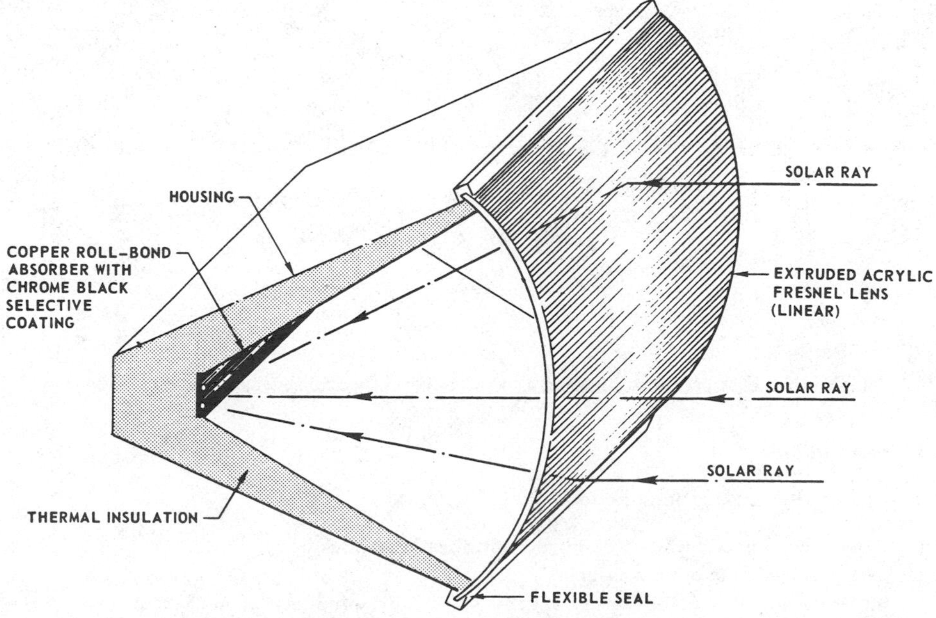

Fig. 66.18 Fresnel lens solar system.

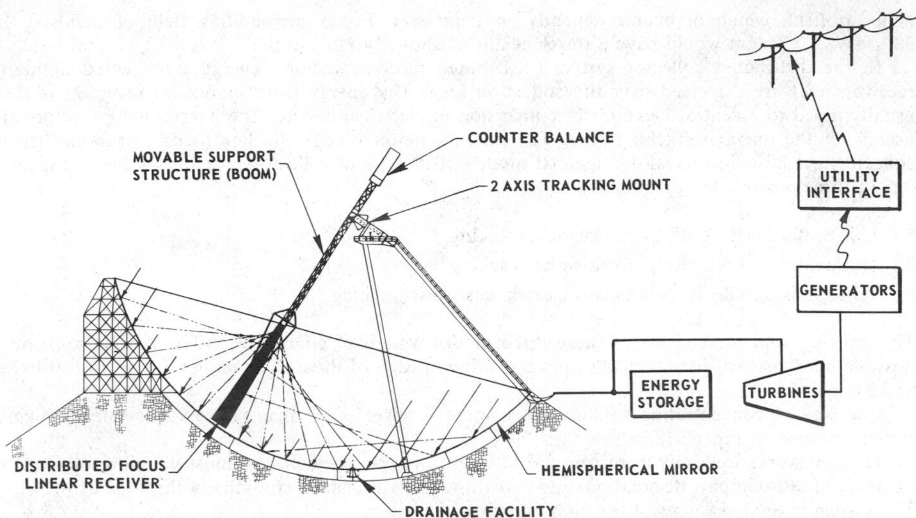

Fig. 66.19 A 20-m-diameter bowl collector of the fixed-mirror distributed-focus type. From Ref. 30, reprinted with permission.

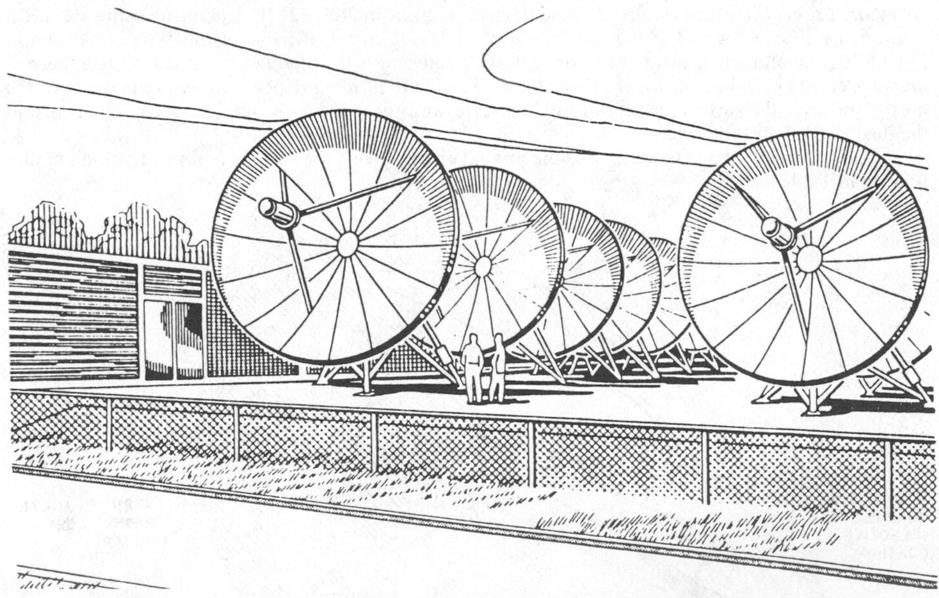

Fig. 66.20 Disk parabolic collectors, for the General Electric installation at Shenandoah.

Thermal Storage

Solar thermal systems require some form of energy storage to

1. Provide continuous operation during periods of variable insolation.
2. Extend operation into nonsolar hours.
3. Buffer potentially harmful system transients induced by abrupt insolation changes.
4. Assume availability of productive capacity during emergency periods.
5. Optimize the dispatching of energy into the grid.

TABLE 66.4. MAXIMUM POSSIBLE ANNUAL ENERGY YIELDS

Configuration	Latitude (°)	Energy Yields 10^3kWh/m^{2}yr	10^6 kJ/m^{2}yr	kBtu/ft^{2}yr
Fully tracking (*D*)	45	2.85	10.3	910
	30	3.11	11.3	985
Horizontal plate (*D*)	45	1.64	5.9	520
	30	1.92	6.9	610
Horizontal plate ($D + S$)	45	1.92	6.9	610
	30	2.26	8.1	710
Fixed (+15°) ($D + S$)	45	2.03	7.3	642
	30	2.25	8.1	713
Polar EW tracking (*D*)	45	2.80	10.0	886
	30	3.00	11.0	952
Horizontal EW:NS tracking (*D*)	45	2.30	8.3	728
	30	2.45	8.8	775
Horizontal NS:EW tracking (*D*)	45	1.70	6.1	537
	30	2.51	9.0	790

Source. Meinel and Meinel,[29] with permission.

Of the several forms of energy storage that might be used, thermal storage seems to be the most promising for solar system backup. This thermal storage may be either sensible or latent. Thermal storage may be accomplished by the use of two or more tanks (hot tank–cold tank storage) or by the use of a single tank in which there is temperature stratification (thermocline). The use of solid materials such as rock in the thermocline system is a means of increasing capacity and reducing mixing, and it allows the use of a lesser amount of expensive fluid.

To study the effect of storage size on solar plant performance, the term "solar multiple" is convenient. The solar multiple is the ratio, defined for one instant during the year, of the peak power absorbed by the receiver to the thermal power required to drive the power unit.

Capacity factor is dependent solar multiple and hours of storage built into a system. The term "oversized storage" means no additional capacity factor can be obtained by increasing plant storage size.

Power Cycles

There are three classes of heat engine cycles that have been most frequently mentioned for use with solar thermal electric systems. They are the Rankine cycle, the Brayton cycle, and the Stirling cycle. Each cycle has its particular advantages and drawbacks.

The Rankine cycle has been a popular choice because of its common use in conventional electrical generation systems. In larger solar systems that might utilize the central receiver system (power tower), the actual components utilized in the end-use subsystem could be the common, commercially available equipment used in conventional fossil fuel plants. In smaller solar thermal power systems where the Rankine cycle is selected, the working fluid might be an organic fluid such as R-11 refrigerant, particularly useful where lower collector temperatures (< 500°F) may be utilized. An organic Rankine cycle (ORC) power conversion assembly is suitable for use with point source concentrators.[30] Where high collector temperatures (> 1100°F) can be attained, gas power cycles such as the Brayton cycle or the Stirling cycle would most likely be employed.

Relatively high temperatures and low pressures exist for gas turbine compared with conventional superheated steam turbines. A great deal of interest has been shown in the use of gas turbines (Brayton cycle) in large central receiver systems. This interest is due primarily to the reduced need for cooling water and the promise of high thermal efficiencies. The closed Brayton cycle can use small high-pressure components.

The open Brayton cycle attains high efficiencies because of high operating temperatures. Recuperators improve the performance of the Brayton cycle, as does their operation with fossil augmentation (hybrid system) and as a topping unit over a steam (Rankine) cycle.

High-temperature turbines using ceramic vanes and blades and hybrid and combined cycles show promise for future improvement. The Stirling cycle offers particular promise in solar applications because of the very high thermal efficiencies attainable. It appears to be especially adaptable to point-focus devices, which can be electrically connected. One of the problems with a conventional Stirling engine is that it is difficult to seal due to the cranking mechanism required and the need to operate at high pressure. This can be overcome by the use of the free-piston Stirling engine. Additional advantages are the simplicity and the fact that no lubricant is needed. The device has been used with a linear generator.

66.7 WIND ENERGY CONVERSION SYSTEMS

Wind energy has not been utilized extensively by industry in the United States because of economics. Wind energy seems to be most promising in the generation of electricity at specific sites where wind velocities are high and reasonably steady. These sites tend to be remote from most industrial locations. Noise generation, safety considerations, and television interference also tend to discourage the use of wind devices in congested areas. Some object to wind devices on the basis that they mar the beauty of the landscape.

Since the excess electricity generated by a wind energy system would likely be put into the electrical distribution system of a utility, the electrical utility industry may become deeply involved in the use of these systems in the future.

Availability

It has been estimated that the wind power available across the continental United States, including offshore sites and the Aleutian arc, is equivalent to approximately 10^5 GW of electricity. This is about 100 times the present electrical generating capacity of the United States. Figure 66.21 shows the areas in the United States where the average wind velocities exceed 18 mph (6 m/s) at 150 ft (45.7 m) above ground level. The wind velocity varies approximately as the 1/7 power of distance from the ground.

The power in a moving air stream per unit area normal to the flow is proportional to the cube of the wind velocity. Thus small changes in wind velocity lead to much larger changes in power available. The equation for calculating the power density of the wind is

$$\frac{P}{A} = \frac{1}{2}\rho V^3 \tag{66.5}$$

where P = power contained in wind
A = area normal to wind velocity
ρ = density of air (about 0.07654 lb/ft^3 or 1.23 kg/m^3)
V = velocity of air stream

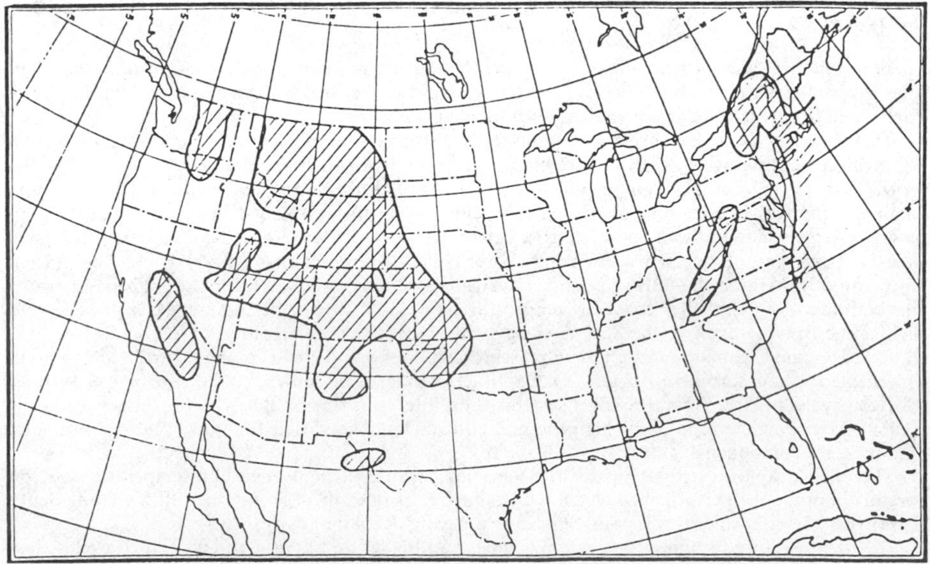

Fig. 66.21 Areas in the United States where average wind speeds exceed 18 mph (8 mps) at 150 ft (45.7 m) elevation above ground level. From Eldridge,[33] reprinted with permission.

Consistent units should be selected for use in Eq. (66.5). It is convenient to rewrite it as

$$\frac{P}{A} = KV^3 \tag{66.6}$$

If the power density P/A is desired in watts per square foot, the value of K depends on the units selected for the velocity V. Values of K for various units of velocity are given in Table 66.5.

The fraction of the power in a wind stream that is converted to mechanical shaft power by a wind device is given by the *power coefficient* C_p.

Only 16/27 or 0.5926 of the power in a wind stream can be extracted by a wind machine, since there must be some flow velocity downstream from the device for the air to move out of the way. This upper limit is called the *Betz coefficient* (or Glauert's limit). No wind device can extract this theoretical maximum. More typically, a device might extract some fraction, such as 70%, of the theoretical limit. Thus a real device might extract approximately (0.5926)(0.70) = 41% of the power available. Such a device would have an aerodynamic efficiency of 0.70 and a power coefficient of 0.41. The power conversion capability could be determined by using Eq. (66.6) and Table 66.5. Assume a 20-mph wind. Then

$$\left(\frac{P}{A}\right)_{\text{actual}} = (5.08 \times 10^{-3})(20)^3(0.41) = 16.7\ \text{W/ft}^2$$

Notice that for a 30-mph wind the power conversion capability would be 56.2 W/ft^2, or more than three times as much.

Because the power conversion capability of a wind device varies as the cube of the wind velocity, one cannot predict the annual energy production from a wind device using mean wind velocity. Such a prediction would tend to underestimate the actual energy available.

Wind Devices

Wind conversion devices have been proposed and built in a very wide variety of types. The most general types are shown in Fig. 66.22.[33] The most common type is the horizontal-axis head-on type, typical of conventional farm windmills. The axis of rotation is parallel to the direction of the wind stream. Where the wind direction is variable, the device must be turned into the wind, either by a tail vane or, in the case of larger systems, by a servo device. The rotational speed of the single-, double-, or three-bladed devices can be controlled by feathering of the blades or by flap devices or varying the load.

In most horizontal-axis wind turbines, the generator is directly coupled to the turbine shaft, sometimes through a gear drive. In the case of the bicycle multibladed type, the generator may be belt driven off the rim, or the generator hub may be driven directly off the rim by friction. In the latter case there is no rotational speed control except that imposed by the load.

In the case of a vertical-axis wind turbine (VAWT) such as the Savonius or Darrieus types, the direction of the wind is not important, which is a tremendous advantage. The system is simple and no stresses are created by yawing or turning into the wind, as occurs on horizontal-axis devices. A VAWT is also lighter in weight, requires only a short tower base, and can have the generator near the ground. VAWT enthusiasts claim much lower costs than those for comparable horizontal-axis systems.

TABLE 66.5. VALUES OF K TO GIVE P/A (W/FT^2) IN EQ. (66.6)[a]

Units of V	K
ft/s	1.61×10^{-3}
mph	5.08×10^{-3}
km/hr	1.22×10^{-3}
m/s	5.69×10^{-2}
Knots	7.74×10^{-3}

[a] To convert watts per square foot to watts per square meter, multiply by 10.76.

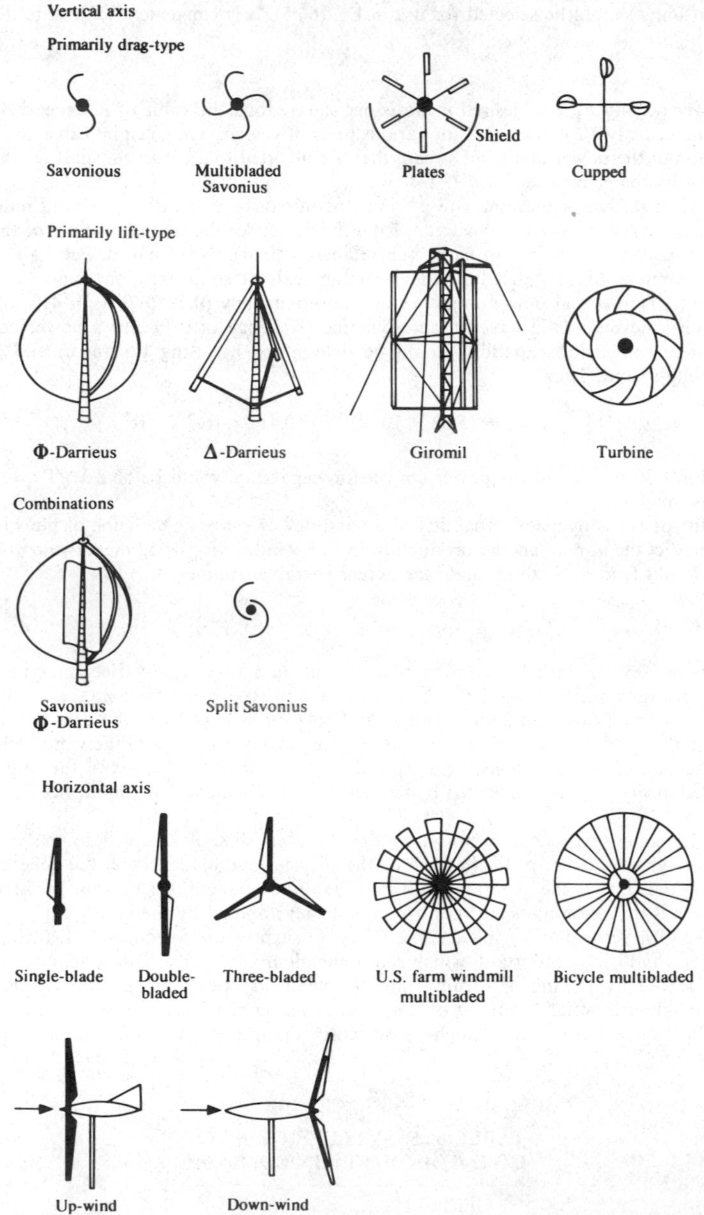

Fig. 66.22 Types of wind-conversion devices. From Eldridge,[33] reprinted with permission.

The side wind loads on a VAWT are accommodated by guy wires or cables stretched from the ground to the upper bearing fixture.

In the Darrieus-type VAWT, two or three blades are most common. The curved blades have an airfoil cross section with very low starting torque and a high tip-to-wind speed.

The Savonius-type turbine has a very high starting torque but a relatively low tip-to-wind speed. It is primarily a drag-type device, whereas the Darrieus type is primarily a lift-type device. The Savonius and the Darrieus types are sometimes combined in a single turbine to give good starting torque and maintenance of good performance at high rotational speeds.

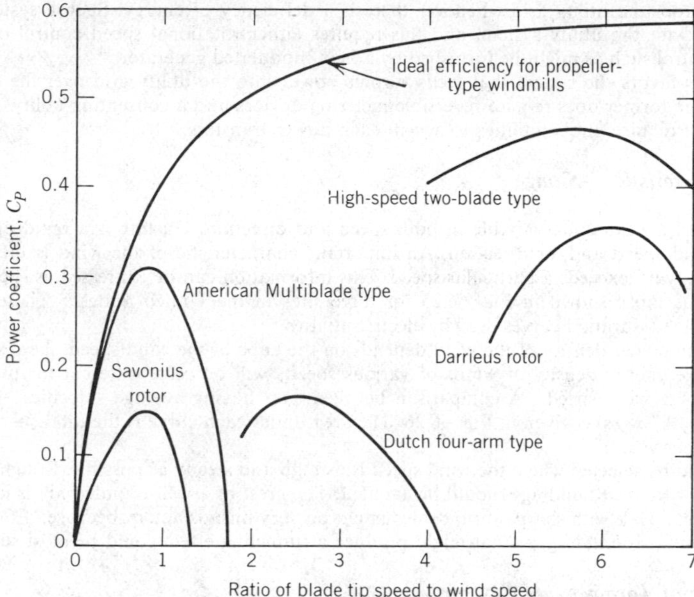

Fig. 66.23 Typical pressure coefficients of several wind turbine devices. From Eldridge,[33] with permission.

Figure 66.23 shows the variation of the power coefficient C_p as the ratio of blade tip speed to wind speed varies for different types of wind devices.[33] Two-blade types operating at relatively high speed ratios have the highest value of C_p, in the range of 0.45, which is fairly close to the limiting value of the Betz coefficient (0.593). The Darrieus rotor has a slightly lower maximum value, but like the two-blade type performs best at high rotational speeds. The American (bicycle) multiblade type, which has a high starting torque, is seen to perform best at lower ratios of tip to wind speed, as does the Savonius.

For comparison, in a 17-mph (7.6-m/s) wind, a 2000-kW horizontal-axis wind turbine would have a diameter of 220 ft (67 m), and a 2000-kW Darrieus type would have a diameter of 256 ft (78 m) and would stand about 312 ft (95 m) tall.[33]

Wind Systems

Because the typical wind device cannot furnish energy to exactly match the demand, a storage system and a backup conventional energy source may be made a part of the total wind energy system (Fig. 66.24). The storage system might be batteries and the backup system might be an electric utility. The system may be designed to put electrical power into the utility grid whenever there is a surplus and to

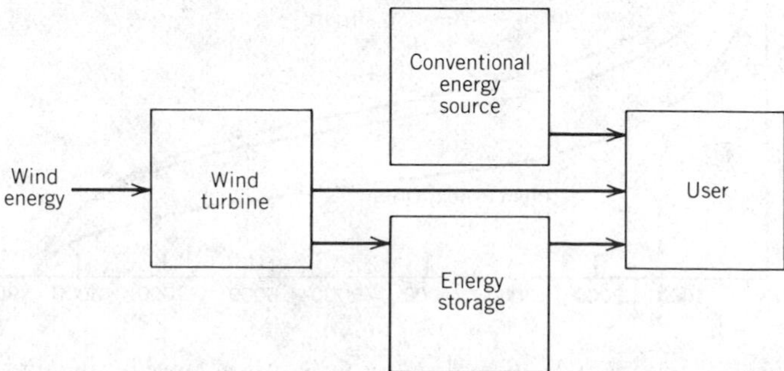

Fig. 66.24 Typical wind energy conversion system with storage.

draw power from the utility grid whenever there is a deficiency of energy. Such a system must be synchronized with the utility system, and this requires either rotational speed control or electronic frequency control such as might be furnished by a field-modulated generator.[34]

Economics favors the system that feeds surplus power into the utility grid over the system with storage, but the former does require reversible metering devices and a consenting utility. Some states have laws that require public utilities to accept such power transfers.

Wind Characteristics—Siting

Wind is almost always quite variable in both speed and direction. Gusting is a rapid up-and-down change in wind speed and/or direction. An important characteristic of the wind is the number of hours that the wind exceeds a particular speed. This information can be expressed as speed-duration curves, such as those shown in Fig. 66.25 for three sites in the United States.[33] These curves are similar to the load-duration curves used by electric utilities.

Because the power density of the wind depends on the cube of the wind speed, the distribution of annual average energy density of winds of various speeds will be quite different for two sites with different average wind speeds. A comparison between sites having average velocities of 13 and 24 mph (5.8 and 10.7 m/s) is given in Fig. 66.26. The area under each curve is the total energy available per unit area per year.

Sites should be selected where the wind speed is as high and steady as possible. Rough terrain and the presence of trees or buildings should be avoided. The crest of a well-rounded hill is ideal in most cases, whereas a peak with sharp, abrupt sides might be very unsatisfactory because of flow reversals near the ground. Mountain gaps that might produce a funneling effect could be most suitable.

Performance of Turbines and Systems

Three important wind speeds that might be selected in designing a wind energy conversion system (WECS) are (1) cut-in wind speed, (2) rated wind speed, and (3) cut-off wind speed. The names are descriptive in each case. The wind turbine is kept from turning at all by some type of brake as long as the wind speed is below the cut-in value. The wind turbine is shut off completely at the cut-off wind speed to prevent damage to the turbine. The rated wind speed is the lowest speed at which the system can generate its rated power. If frequency control were not important, a wind turbine would be permitted to rotate at a variable speed as the wind speed changed. In practice, however, since frequency control must be maintained, the wind turbine rotational speed might be controlled by varying the load on the generator when the wind speed is between the cut-in and rated speed. When the wind speed is greater than the rated speed but less than cut-out speed, the spin can be controlled by changing the blade pitch on the turbine. This is shown in Fig. 66.27 for the 100-kW DOE/NASA system at Sandusky, Ohio. A system such as that shown in Fig. 66.27 does not result in large losses of available wind power if the average energy content of the wind at that site is low for speeds below the cut-in speed and somewhat above the rated speed.

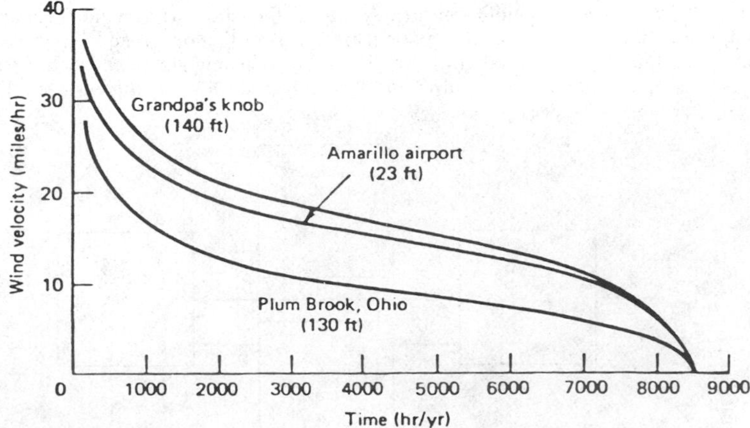

Fig. 66.25 Annual average speed-duration curves for three sites in United States. From Eldridge,[33] with permission.

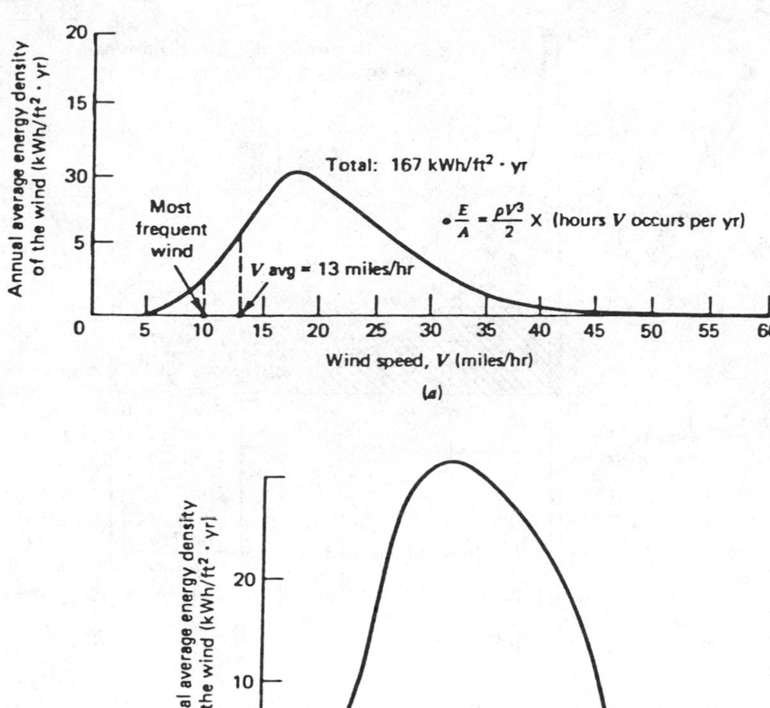

Fig. 66.26 Comparison of distribution of annual average energy density at two sites: (*a*) V_{avg} = 13 mph, (*b*) V_{avg} = 24 mph. From Eldridge,[33] reprinted with permission.

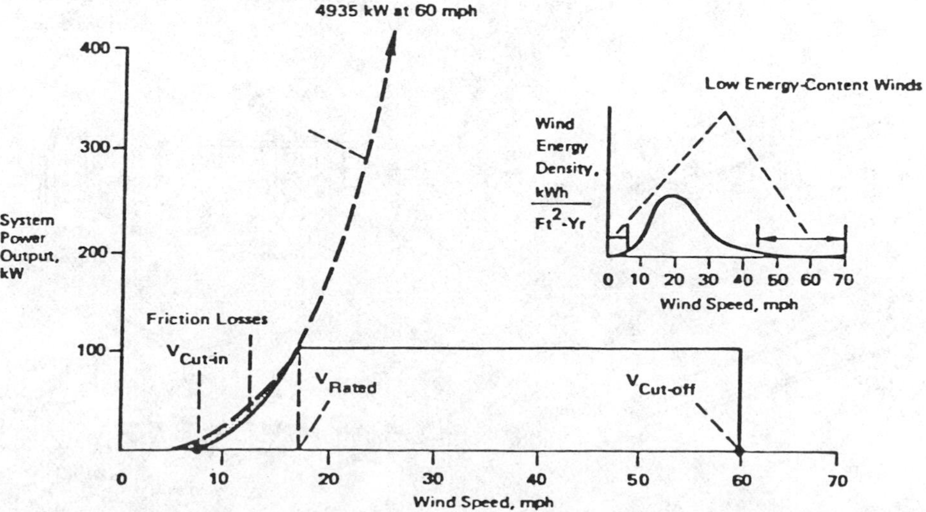

Fig. 66.27 Power output of a 100-kW wind energy conversion system at various wind speeds. From Eldridge,[33] reprinted with permission.

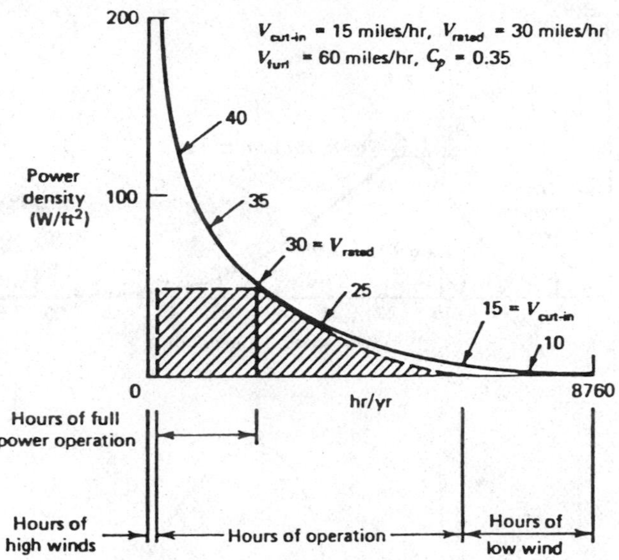

Fig. 66.28 Actual annual power density output of a wind energy conversion system. From Eldridge,[33] reprinted with permission.

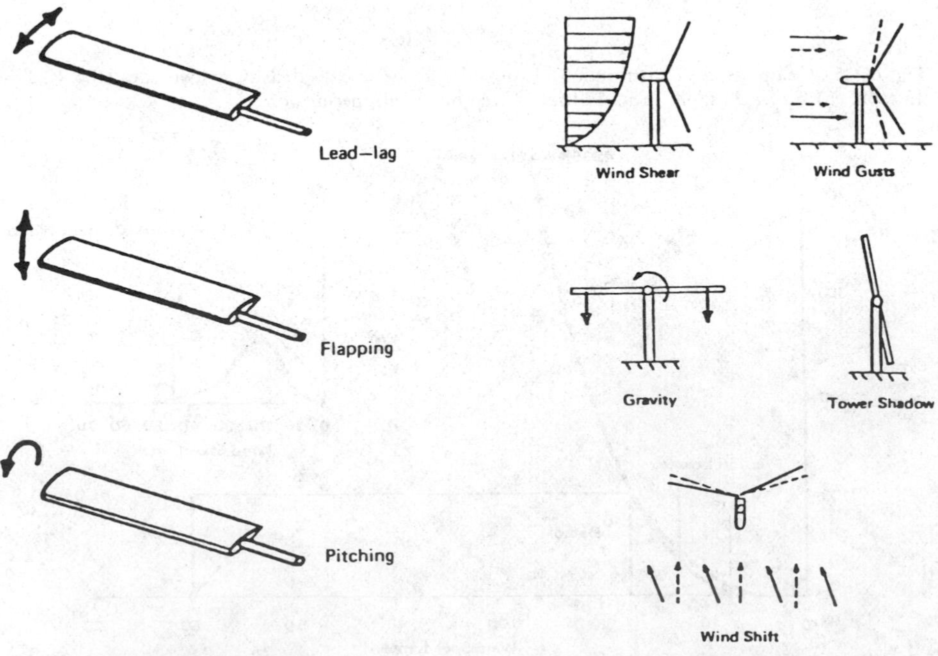

Fig. 66.29 Rotor blade motions and their causes. From Eldridge,[33] reprinted with permission.

Another useful curve is the actual annual power density output of a WECS (Fig. 66.28).[33] The curve shows the hours that the device would actually operate and the hours of operation at full rated power. The curve is for a system with a rated wind speed of 30 mph (13.4 m/s), a cut-in velocity of 15 mph (6.7 m/s), and a cut-off velocity of 60 mph (26.8 m/s) with constant output above 30 mph.

Loadings and Acoustics

Blades on wind turbine devices have a variety of extraneous loads imposed on them. Rotor blades may be subject to lead-lag motions, flapping, and pitching. These motions and some of their causes are shown in Fig. 66.29.[33] These loads can have a serious effect on the system performance, reliability, and lifetime.

Acoustics can be a serious problem with wind devices, especially in populated areas. The DOE/NASA device at Boone, North Carolina, caused some very serious low-frequency ($\simeq$ 1-Hz) noises that caused the system to be abandoned.

66.8 GEOTHERMAL CONVERSION

Resource Type and Availability

Geothermal energy is that obtained by heat transfer from the earth. Temperature increases with depth below the surface of the earth, except for the daily and seasonal variations that occur very near the surface. The average temperature gradient near the surface but below the level of seasonal variation is about 1° increase per 100 m, or about 85°F per mile. For these average conditions a depth of about 20,000 ft would have to be attained to find temperatures of 300 to 400°F. For most applications these conditions do not lead to economically viable geothermal systems for either heating or electrical power generation.

Fortunately, there are geothermal regions where the temperature gradients may be up to ten times the average. It is in these regions, such as in the Geysers region of northern California, where commercially feasible geothermal systems are possible. Geothermal sites are usually classified into one of four types:[35] (1) dry steam, (2) wet steam, (3) hot dry rock, and (4) geopressured zones.

In dry steam reservoirs, since no liquid water is present the steam can be taken directly to a turbine to produce electricity. The steam is usually filtered to remove solid material prior to its entry into the turbine. A mixture of water and steam is present in wet steam reservoirs. The hot water usually contains a fairly high level of dissolved solids. The steam and water are usually separated so that the steam can be utilized in a turbine. In some cases the steam and/or water is used in a heat exchanger to heat and possibly evaporate a second fluid. Disposal of the hot water with its high level of dissolved solids is often a critical problem.

In hot dry-rock systems, water is present due to natural causes. Steam is generated by pumping water down to the hot dry rock and the resulting steam is brought to the surface in a second pipe.

The usual procedure proposed for such systems is to drill two separate holes and to fracture the hot rock between them with high pressure fluid or with explosives. Hot dry rocks are the most abundant and widely distributed geothermal resource.

Geopressured zones contain high-pressure high-temperature water, usually saturated with natural gas. These zones exist along a great deal of the Gulf Coast of Texas and Louisiana, at depths of approximately 18,000 ft. The presence of large quantities of natural gas makes these formations attractive, but many technical problems are associated with their development. One estimate has been made that between 4300 and 4400×10^{15} Btu (4080 to 4171×10^{18} J) of the available 170×10^{18} Btu (161×10^{21} J) of energy in the geopressured resources can be recovered with existing technology.

An assessment of all types of geothermal resources in the United States found that "the currently identified resources could supply 23,000 $\pm$ 3400 MW electric power and $40 \pm 13 \times 10^{15}$ Btu ($42 \pm 13 \times 10^{18}$ J) of heat for space heating or process heat with existing technology for 30 years. The energy potentially available in unidentified resources has been estimated as 72,000 to 127,000 MW electric and 144 to 294×10^{15} Btu (218 to 332×10^{18} J) of heat for space heat for 30 years."[36,37]

Nunz[38] has stated that "the total energy content of the formations underlying the 50 states, to a depth of 10 km and at 'commercially interesting' temperatures above 150°C is about 13.2 million quad, or about 170 thousand times the present total annual consumption of energy in the U.S. Of this practically infinite geothermal resource base virtually all—more than 99 percent—exists in hot dry rock." If only 2% were recoverable it would be sufficient to provide the entire nontransportation energy requirements of the United States over 2000 years at the present rate of consumption.[39] It is important to know that a geothermal resource is not an inexhaustible supply of energy. Continued removal of energy at a given source will often result in a reduction of local temperatures with time, resulting eventually in abandonment.

Basic System Types

Geothermal energy will have wide application in space heating and in providing process heat. In these cases the problems will be primarily ones of heat exchange, corrosion, and cost reduction. Some of these problems have been discussed by Wehlage.[40] The important problems of heat transfer in the earth will need solution if we are to accurately predict system life. Heat transfer in the earth surrounding geothermal systems has been described by Cheng.[41]

The typical system for the generation of electricity where dry steam is available is shown schematically in Fig. 66.30. This is the Rankine cycle commonly used in conventional fossil-fired plants. The dry steam, after filtering, is expanded through a turbine and then condensed into water. This water is then usually reinjected into a suitable ground formation.

If hot water is available from the geothermal formation, the binary cycle may be utilized (Fig. 66.31). In this case, a working fluid, such as a refrigerant or volatile hydrocarbon is used in a closed circuit Rankine cycle. The hot water is used to evaporate the working fluid in heat exchangers. The water is then pumped into a suitable ground formation. The solids content of the water is usually too high to permit its dumping into a surface stream or lake.

Where wet steam or high-pressure hot water is available, the system shown in Fig. 66.32 might be utilized. In this case the steam and water are separated and the steam directed to a turbine. The water is partially flashed into steam by reducing its pressure. This steam is then separated and used simultaneously with the high-pressure turbine exhaust to drive a low-pressure turbine. The exhaust from the low-pressure turbine is then condensed and used as fresh water or mixed with the unflashed geothermal water and sent to a disposal well.

Hot dry-rock systems may be designed to produce dry steam, wet steam, or hot water and then utilized with some version of the three systems described.

Some of the factors limiting geothermal development in the United States have been discussed by Rex.[42] The cost of drilling geothermal wells is a major cost factor, representing 40% to 80% of the cost

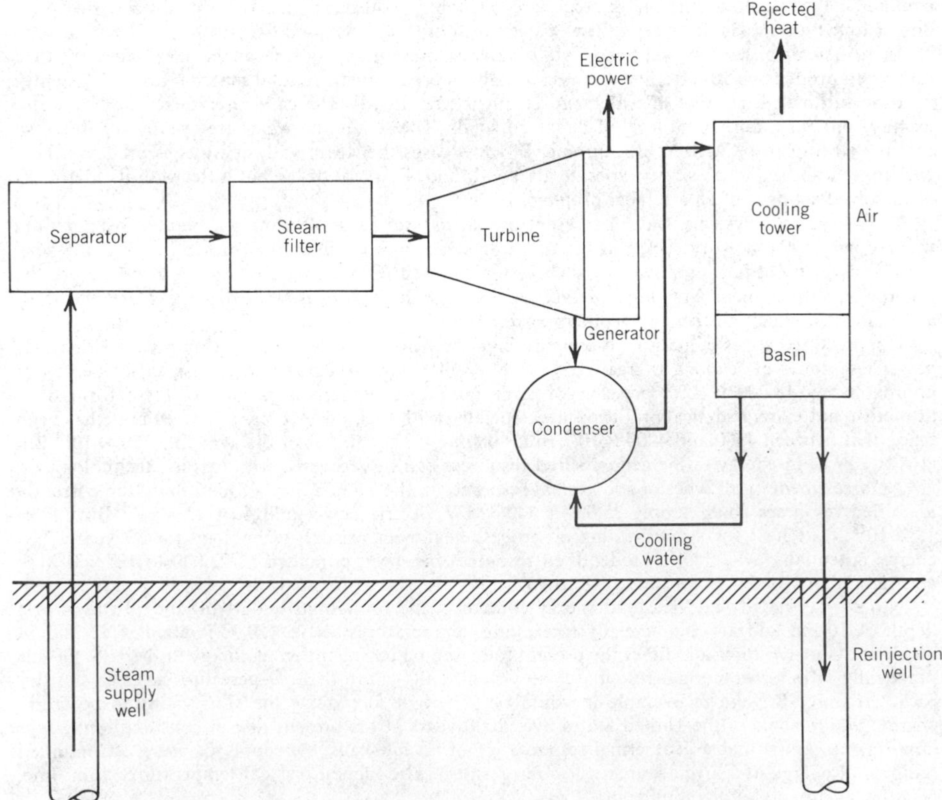

Fig. 66.30 Schematic of a dry steam geothermal power plant using a cooling tower.

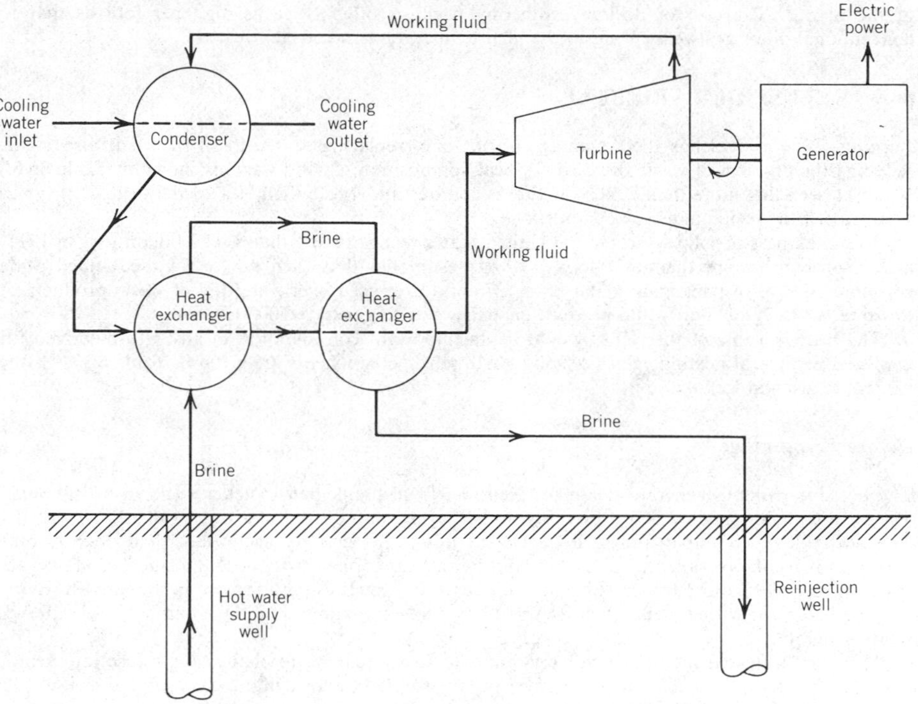

Fig. 66.31 Schematic of a binary geothermal power plant.

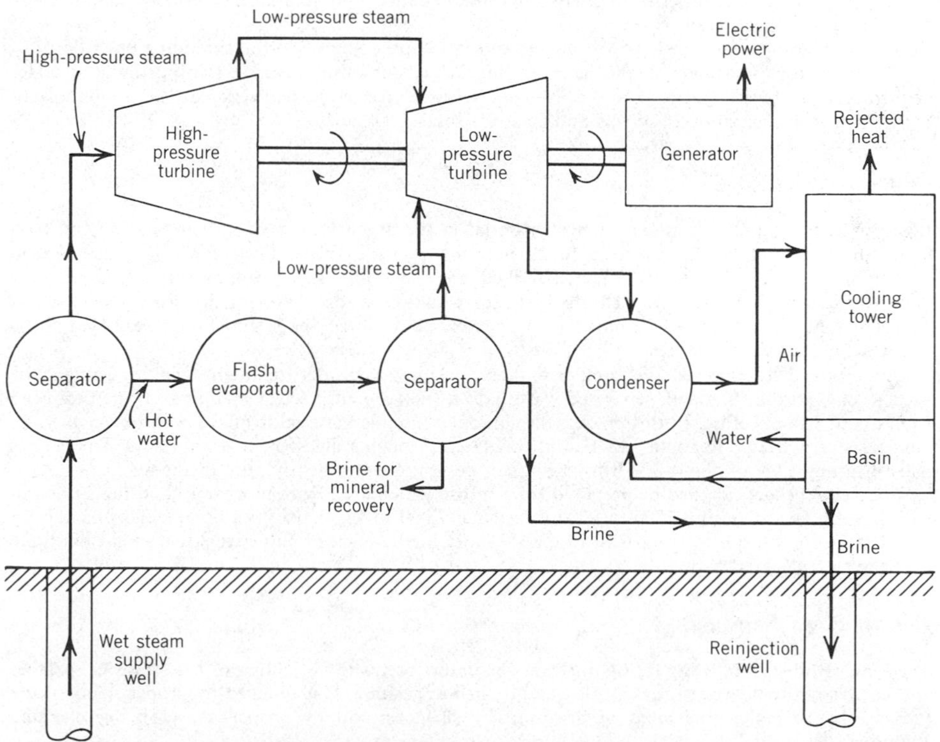

Fig. 66.32 Schematic of a wet steam geothermal power plant.

of the plant.[43] The cost of drilling geothermal wells is often twice as high per foot as that for conventional oil or gas wells owing to the high temperatures involved.

66.9 REFUSE-DERIVED FUEL

Burning waste as fuel has the advantage of not only replacing scarce fossil fuels but also greatly reducing the problem of waste disposal. Typical composition of solid waste is shown in Table 66.6.[44] It can be seen that more than 70% by weight is combustible. More than 90% of the volume of typical solid waste can be eliminated by combustion.

The total mass of solid wastes in the United States reached more than 4 1/2 billion tons in 1971[44] and is probably more than that today. It was estimated that each person in the United States consumed 600 lb of packaging material in 1976 and produced over 5 lb/day of waste products. At this rate, a family of four would produce enough waste to generate 2000 kW/yr.

The heating value of the refuse would be an important consideration in any refuse-derived fuel application. Typical heating values of solid waste refuse components are given in Table 66.7.[44] Other values are given in Ref. 45.

Refuse Preparation

The possible paths to generate steam or electricity from municipal wastes are shown in Fig. 66.33. The most common method is for the refuse to be burned unprepared in a waterwall steam generator. This technology is simple and well developed and costs can be accurately predicted. Another approach is for the refuse to be placed in landfill and gas formed from decomposition of the organic material recovered and burned. This is not efficient and requires land for use in the landfill. Refuse may be given some treatment, such as shredding and separation, and then burned in a waterwall steam generator.

More sophisticated methods involve treatment after shredding to change the refuse into a more desirable fuel form. This may involve converting the shredded refuse into a gas or liquid or into solid pellets. Shredding, which can be done wet or dry (as received), converts the refuse into a relatively homogeneous mixture. The shredding is usually done by hammermills or crushers. This shredding operation is costly both in terms of energy and maintenance. One reference gives 1977 maintenance costs of 60 cents per ton of waste.[45]

Problems with fire, explosions, vibrations, and noise are common in the shredding operation.

Density separation increases the fuel's heating value, minimizes wear on transporting and boiler heat-transfer surfaces, and makes the ash more usable. Resource recovery can be an important byproduct, with separation of metals and glass for resale.

Basic Processes

Pyrolysis is the thermal decomposition of material in the absence of oxygen. The product can be a liquid or a gas suitable for use as a fuel. There are many pyrolysis projects in the research and development stage. Morgenroth[46] has described an apparently successful pyrolytic heat-recovery system. The new plant saved $53,000 the first year while disposing of 90% of the firm's waste. It was expected that the system would pay out in approximately three years. Emissions were said to be below standards set by the EPA.

Anaerobic digestion processes, similar to those used in wastewater treatment facilities, can also be used to convert the shredded, separated waste into a fuel. About 3 scf of methane can be produced from about 1 lb of refuse. In this process the shredded organic material is mixed with nutrients in an aqueous slurry, heated to about 140°F, and circulated through a digester for several days. The off-gas has a heating value of about 600 Btu/cf but can be upgraded to nearly pure methane.

Solid fuel pellets can also be prepared from refuse which are low in inorganics and moisture and with heating values around 7500 to 8000 Btu/lb_m (17,000 to 19,000 kJ/kg). Some pellets have been found to be too fibrous to be ground in the low- and medium-speed pulverizers that might normally be found in coal-fired plants.

Corrosion and Fouling

The major problems in firing refuse in steam generators seems to be fouling of heat-transfer surfaces and corrosion. Fouling is caused by slag and fly-ash deposition. It is reduced by proper sizing of the furnace, by proper arrangement of heat-transfer surfaces, and by proper use of boiler cleaning equipment.[46]

TABLE 66.6. TYPICAL COMPOSITION OF SOLID WASTE

Food wastes (12% by weight)
- Garbage (10%)
- Fats (2%)

Noncombustibles (24% by weight)
- Ashes (10%)
- Metals (8%): cans, wire, and foil
- Glass and ceramics (6%): bottles primarily

Rubbish (64% by weight)
- Paper (42%): various types, some with fillers
- Leaves (5%)
- Grass (4%)
- Street sweepings (3%)
- Wood (2.4%): packaging, furniture, logs, twigs
- Brush (1.5%)
- Greens (1.5%)
- Dirt (1%)
- Oil, paints (0.8%)
- Plastics (0.7%): polyvinyl chloride, polyethylene, styrene, etc., as found in packaging, housewares, furniture, toys, and nonwoven synthetics
- Rubber (0.6%): shoes, tires, toys, etc.
- Rags (0.6%): cellulose, protein, and woven synthetics
- Leather (0.3%): shoes, tires, toys, etc.
- Unclassified (0.6%)

Source. Rolstein et al.,[44] used with permission.

TABLE 66.7. TYPICAL HEATING VALUES OF SOLID WASTE REFUSE COMPONENTS[a]

Component	MJ/kg	Btu/lb
Domestic Refuse		
Garbage	4.23	1,820
Grass	8.88	3,820
Leaves	11.39	4,900
Rags (cotton, linen)	14.97	6,440
Brush, branches	16.60	7,140
Paper, cardboard, cartons, bags	17.81	7,660
Wood, crates, boxes, scrap	18.19	7,825
Industrial Scrap and Plastic Refuse		
Boot, shoe trim, and scrap	19.76	8,500
Leather scrap	23.24	10,000
Cellophane	27.89	12,000
Waxed paper	27.89	12,000
Rubber	28.45	12,240
Polyvinyl chloride	40.68	17,500
Tires	41.84	18,000
Oil, waste, fuel-oil residue	41.84	18,000
Polyethylene	46.12	19,840
Agricultural		
Bagasse	8.37–15.11	3,600–6,500
Bark	10.46–12.09	4,500–5,200
Rice hulls	12.15–15.11	5,225–6,500
Corncobs	18.60–19.29	8,000–8,300
Composite		
Municipal	10.46–15.11	4,500–6,500
Industrial	15.34–16.97	6,600–7,300
Agricultural	6.97–13.95	3,300–6,000

Source. Rolstein et al.,[44] used with permission.

[a] Calorific value in megajoule per kilogram (British thermal unit per pound) as fired.

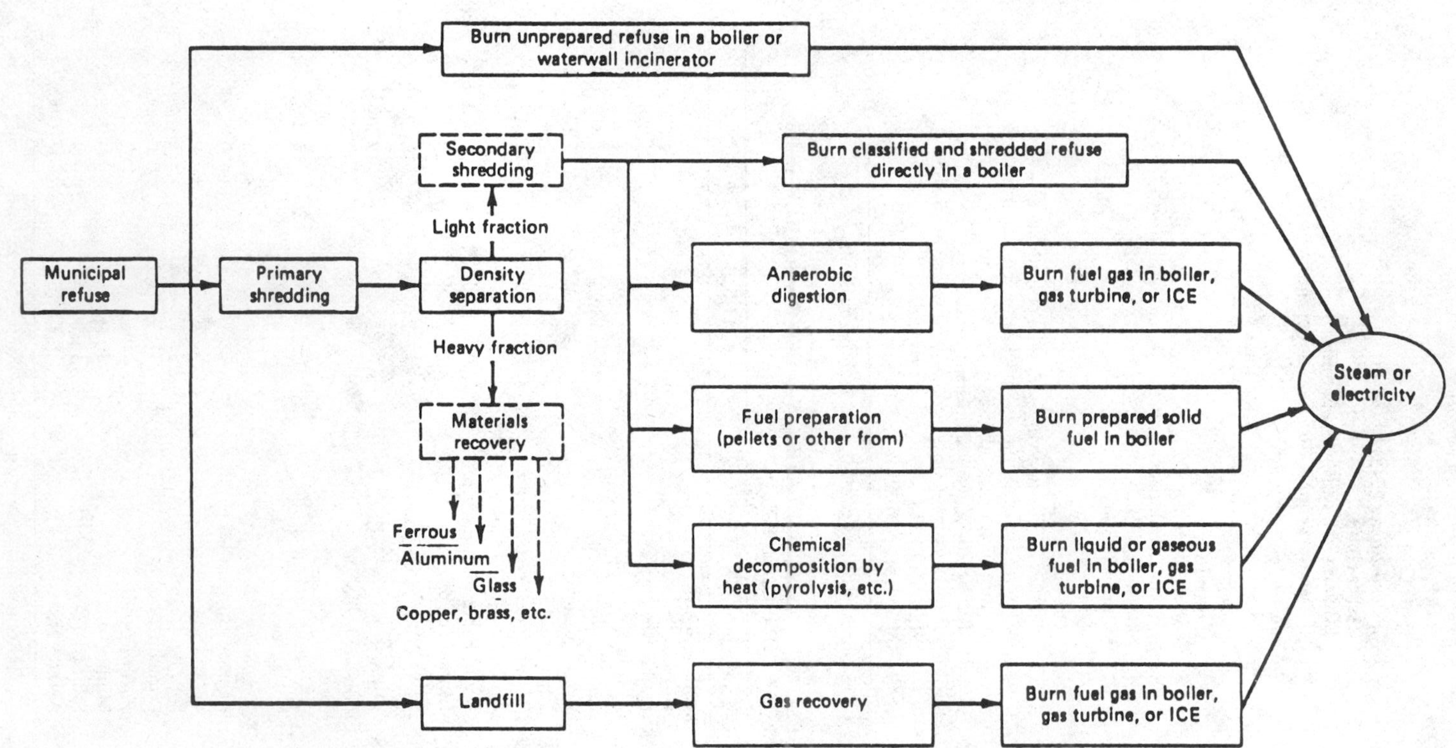

Fig. 66.33 Possible paths to convert municipal refuse to fuel. From Ref. 45, with permission.

Corrosion in refuse-derived fuel systems is usually due to:

1. Reducing environment caused by stratification or improper distribution of fuel and air.
2. Halogen corrosion caused by the presence of polyvinyl chloride (PVC) in the refuse.
3. Low temperatures, caused when some surface in contact with the combustion gases is below the dewpoint temperature of the gas.

It appears that many existing coal-fired boilers can be modified to use suitably prepared refuse as a fuel.

66.10 NUCLEAR CONVERSION

Nuclear energy has become a significant factor in the production of electricity. At the end of 1982 a total of 294 nuclear power plants were operating in 25 nations and producing more than 8% of the world's electricity.

All existing nuclear power plants operate on a sustained fission process. In the fission process an atom of a heavy element, usually uranium 235 or plutonium 239, absorbs a neutron and then splits to form lighter elements. The energy released is due to the loss of mass during the process. Neutrons released during the disintegration may strike other fuel atoms, which then fission in turn, leading to a sustained or chain reaction.

Since plutonium does not exist in significant quantities in nature, uranium 235 is the most common nuclear fuel. Natural uranium consists primarily of a mixture of two isotopes, uranium 235 (0.711 wt%) and uranium 238 (99.3 wt%). Most nuclear power plants can utilize the U-235 but not the U-238. For this reason the fuel for a power plant is usually enriched by separation of the isotope. Because of the difficulty and expense of the enrichment process, a typical enriched fuel for a commercial power plant would be about 2% to 4% U-235 by weight. The enriched uranium is converted into uranium dioxide (for water-cooled reactors) and formed into fuel pellets. These pellets are placed in fuel rods, which make up a fuel assembly, which in turn is placed in the reactor vessel. A coolant flowing around the fuel rods removes the heat of fission and provides energy for the turbine that generates the electricity. For typical low-enrichment fuels, the coolant is also the moderator for fission reaction, slowing down the released neutrons to thermal levels so that they will more readily react with the fuel and continue a sustained reaction. If ordinary water is used, the reactor is called a light-water reactor, or LWR. If deuterium (heavy water) is used as a moderator and coolant, the reactor is called a heavy-water reactor, or HWR. The reactor may be a pressurized-water reactor (PWR) if no phase change occurs in the reactor. It is called a boiling-water reactor, or BWR, if steam is allowed to form within the reactor. Reactors operating with highly enriched fuel do not require that neutron energies be reduced to thermal levels for sustained fission, and they can operate as fast reactors. Such reactors can use liquid metals as coolants, for example, since no moderator is needed.

Nuclear reactors in the United States are contained within a large building, called the containment building, that is built strong enough and tight enough to contain all radioactive material that might be released in an accident. Fission-type reactors produce radioactive material, and the fuel rods must be changed regularly to remove these waste products and to replenish the spent fissile material. Storage of these waste products has been a center of controversy between proponents and opponents of nuclear power. Agreement on suitable long-term storage methods and sites must be reached if the industry is to continue to expand.

Nuclear plants operate most efficiently as base-loaded plants, since they have relatively high capital costs, have relatively low fuel cost, and cannot be designed to follow rapidly changing loads. Although operating at total costs that are competitive with those of fossil-fuel plants, the typical light-water nuclear power plant operates at a lower overall thermodynamic efficiency. This is because the steam temperatures are lower than in conventional plants. Because of this, a nuclear plant will dissipate a larger quantity of waste heat to the cooling towers or other sinks than will a conventional power plant having the same electrical capacity.

A large amount of research is under way throughout the world to develop breeder reactors. These reactors can produce fissile material, such as plutonium 239, by allowing fast neutrons in the reactors to bombard blankets of uranium 238. The breeder is capable of producing new fuel at a rate that is greater than the burn-up of fuel in the reactor. Breeder reactors would thus be capable of expanding the usefulness of the world's uranium resources manyfold. The breeder will be necessary if the nuclear industry is to be an important factor in electrical power production well into the next century. The Clinch River liquid metal fast breeder reactor (LMFBR) has been the United States' main effort in the development of breeder technology. Funds for this project have been an item of controversy in Congress for several years.

Fusion energy, obtained by the sustained reaction of combining deuterium or tritium into helium, is still in a very early development stage. It is a promising source of power, since the fuels are abundant in nature or are easily produced, there are little or no radioactive wastes, and the reactor vessel contains little stored energy, which eliminates the emergency cooling systems required in a fission reactor. Fusion systems will require extensive research and development before they can begin to produce fuel on a commercial scale.

References

1a S. W. Angrist, *Direct Energy Conversion*, 4th ed., Allyn & Bacon, Boston, 1982.

1b S. W. Angrist, *Direct Energy Conversion*, 3rd ed., Allyn & Bacon, Boston, 1976.

1c M. A. Kettani, *Direct Energy Conversion*, Addison-Wesley, Reading, MA, 1970.

2 "News Trends," *Mach. Des.* (Mar. 25, 1982).

3 S. E. Fuller, "The Relationship Between Diffuse, Total and Extraterrestrial Solar Radiation," *J. Sol. Ener. Technol.* **18**(3):259–263 (1976).

4 B. V. H. Liu and R. C. Jordan, "Availability of Solar Energy for Flat-Plate Solar Heat Collectors," *Application of Solar Energy for Heating and Cooling of Buildings*, ASHRAE GRP-170, American Society of Heating, Refrigerating and Air Conditioning Engineers, New York, 1977.

5 S. A. Klein, "Calculations of Monthly Average Insolation on Tilted Surfaces," Vol. 1, *Proceed. ISES and SESC Joint Conference*, Winnipeg, 1976.

6 D. W. Ruth and R. E. Chant, "The Relationship of Diffuse Radiation to Total Radiation in Canada," *J. Sol. Ener. Technol.* **18**(2):153–154 (1976).

7 *Introduction to Solar Heating and Cooling Designing and Sizing*, DOE/CS-0011 UC-59a, b, c, Aug. 1978.

8 P. J. Lunde, *Solar Thermal Engineering*, Wiley, New York, 1980.

9 J. J. Loferski, "Photovoltaics I: Solar Cell Arrays," *IEEE Spectr.*, p. 26 (Feb. 1980).

10 M. Wolf, "Photovoltaics II: Flat Panels," *IEEE Spectr.*, p. 32 (Feb. 1980).

11 C. E. Backus, "Photovoltaics III: Concentrators," *IEEE Spectr.*, p. 34 (Feb. 1980).

12 P. D. Maycock and E. N. Stirewalt, "Solar Cell Systems that Work," *IEEE Spectr.*, p. 40 (Sep. 1981).

13 M. A. Green, *Solar Cells*, Prentice-Hall, Englewood Cliffs, NJ, 1982.

14 "News Trends," *Mach. Des.*, p. 12 (Aug. 20, 1981).

15 "The Ultimate Cogeneration Package," *Ener. Manage.*, p. 26 (Apr. 1983).

16 E. A. Gillis, "Fuel Cells for Electric Utilities," *Chem. Eng. Prog.*, pp. 88–93 (Oct. 1980).

17 L. A. Kilar, "Fuel Cells: In Pursuit of the Ideal Power Plant," *Power*, pp. 37–39 (May 1979).

18 L. G. Marianowski et al., "Fuel Cell Research on Second Generation Molten Carbonate Fuel Cells," National Fuel Cell Seminar, Boston, Jun. 21–23, 1974.

19 H. A. Liebhofsky, "The Fuel Cell and the Carnot Cycle," *J. Electrochem. Soc.* **106**:1068 (1959).

20 I. M. Berman and P. S. Schmidt, "Fuel Cells and Coal Derived Fuels," *Power Eng.*, pp. 62–68 (Oct. 1980).

21 "Fuel Cells Show Promise as a Vehicle Power Source," *Automot. Eng.* **88**(4):7884 (Apr. 1980).

22 J. B. McCormick, "Fuel Cells for Transportation," *Res. Devel.*, p. 88 (Apr. 1980).

23 K. R. Williams, ed., *An Introduction to Fuel Cells*, Elsevier, Amsterdam, 1966.

24 A. P. Fickett, "Fuel Cell Power Plants," *Scient. Amer.*, p. 70 (Dec. 1978).

25 J. N. Chapman, S. S. Strom, Y. C. L. Wu, "MHD Steam Power—Promises, Progress and Problems," *Mech. Eng.*, p. 30 (Sep. 1981).

26 "MHD: Direct Channel from Heat to Electricity," *EPRI J.* **5**(3):21–25 (Apr. 1980).

27 R. W. Postlethwaite and M. M. Sluyter, "MHD Heat Transfer Problems—An Overview," *Mech. Eng.*, p. 32 (Mar. 1978).

28 F. Kreith and J. F. Kreider, *Principles of Solar Engineering*, McGraw-Hill, New York, 1978.

29 A. P. Meinel and M. P. Meinel, *Applied Solar Energy*, Addison-Wesley, Reading, MA, 1976.

30 "Solar Thermal Energy Systems," Annual Technical Progress Report, FY 1980, DOE/CS/4042-2, Solar Energy Research Institute, Golden, CO, Jul. 1981.

31 K. W. Bettleson, "Solar Power Tower Design Guide: Solar Thermal Central Receiver Systems, a Source of Electricity and/or Process Heat," SAND81-2005, Sandia National Laboratories, Livermore, CA, Apr. 1981.

32 *Proceedings of the Solar Thermal Concentrating Collector Technology Symposium*, SERI/TP-34-048, Solar Energy Research Institute, Golden, CO, Aug. 1978.

33 F. R. Eldridge, *Wind Machines*, NSF-RA-N-75-051, prepared for NSF by Mitre Corp., Oct. 1975, U.S. Government Printing Office, Washington, DC (stock No. 038-000-00272-4).

34 R. Ramakumar, "Wind Power," *IEEE Handbook*, McGraw-Hill, New York, 1979.

35 *Geothermal Energy and Our Environment*, DOE/EV-0088.

36 R. W. Potter II, "Geothermal Energy: An Assessment," *Mech. Eng.*, pp. 20–23 (May 1981).

37 C. A. Brook et al., "Hydrothermal Convection Systems with Reservoir Temperatures −90°C," *U.S. Geol. Sur. Circ. 790*, pp. 18–85 (1979).

38 G. J. Nunz, "Hot Dry Rock Geothermal Energy," *Mech. Eng.*, pp. 26–31 (Nov. 1980).

39 "Definition Report: Geothermal Energy Research, Development and Demonstration Program," ERDA-86, ERDA, Division of Geothermal Energy, Oct. 1975.

40 E. F. Wehlage, "Geothermal Energy Needed: Effective Heat Transfer Equipment," *Mech. Eng.*, pp. 27–33 (Aug. 1976).

41 P. Cheng, "Heat Transfer in Geothermal Systems," in *Advances in Heat Transfer*, Vol. 14, Academic, New York, 1978.

42 R. W. Rex, "Factors Limiting Geothermal Development," *Geotherm. Ener.* **7**(12):30–33 (Dec. 1979).

43 J. C. Rowley, "Geothermal Energy," *Physics Today*, p. 36 (Jan. 1977).

44 R. F. Rolstein et al., "Solid Waste as Refuse Derived Fuel," *Nucl. Technol.* **36**:314–327 (mid-Dec. 1977).

45 "Power from Wastes—A Special Report," *Power* (Feb. 1975).

46 A. W. Morgenroth, "Solid Waste as a Replacement Energy Source," *ASHRAE J.*, pp. 44–46 (Jun. 1978).

Bibliography

Carlson, D. E., "Photovoltaics V: Amorphous Silicon Cells," *IEEE Spectr.*, p. 39 (Feb. 1980).

Conversion of Refuse to Energy, First International Conference, Montreux, Switzerland, Nov. 3–5, 1975.

Hirschfield, F., "Wind Power—Pipe Dream or Reality?," *Mech. Eng.*, pp. 20–26 (Sep. 1977).

Jackson, F. R., *Energy from Solid Waste*, Noyes Data Corp., Park Ridge, NJ, 1974.

Kuo, S. C., T. L. O. Horton and K.-T. Shu, "Parametric Analysis of Power Conversion Systems for Central Receiver Power Generation," ASME Paper 78-WA/Sol-2, presented at ASME Meeting, San Francisco, CA, Dec. 10–15, 1978.

Loferski, J. J., "Photovoltaics IV: Advanced Materials," *IEEE Spectr.*, p. 37 (Feb. 1980).

Present Status and Research Needs in Energy Recovery from Wastes, 1976 ASME Conference, Oxford, Ohio.

The Solar Thermal Report, Vol. 2, No. 6, Jet Propulsion Lab., Pasadena, CA, Aug. 1981.

Wallace, R. H., et al., "Assessment of Geopressured-Geothermal Resources in the Northern Gulf of Mexico Basin," *U.S. Geol. Sur. Circ. 790*, pp. 132–155, 1979.

CHAPTER 67
ELECTRIC POWER SYSTEMS

DANIEL D. LINGELBACH

Oklahoma State University
Stillwater, Oklahoma

Electric power systems consist of generators, transmission lines, power substations, subtransmission lines, distribution substations, and distribution lines or feeder circuits, which are interconnected to provide electrical energy to the customer's load.

67.1 POWER SYSTEM DISTRIBUTION

Loads in distribution systems are conveniently classified as (1) industrial, (2) commercial, and (3) residential and rural in decreasing levels of power demand. That part of the power system from the generator to and including the distribution lines is usually owned and operated by the electric utility.

Commencing with the electrical load, the planning, design, and operation are the responsibility of the individual customer.

The design of the distribution system that serves the individual components of the load involves certain basic considerations. These are (1) safety, (2) reliability, (3) cost, (4) voltage quality, (5) ease of maintenance, and (6) flexibility.

Safety entails two aspects: protection of life and protection of equipment and property. Protection of life is paramount and this can be enhanced by utilizing

1. Adequate high-quality electrical components.
2. Proper and easy-to-operate system arrangements.
3. Metal-enclosed construction with proper interlocks.
4. Proper insulation or guarding of energized conductors.
5. Proper electrical system and equipment grounding.
6. Proper installation practices and adequate maintenance programs.
7. Adequate short-circuit interrupting capabilities of switching and protective devices properly coordinated to ensure selective removal of faulty system components.

Reliability involves not only the use of high-quality components with low failure rates for critical areas but in some cases provision of alternate sources of power in the event of equipment failure or for equipment servicing. A decision to provide an alternate source or redundancy of equipment is an engineering one based on an evaluation of all aspects, including costs and the consequences of loss of electrical power. Also simple system designs are less susceptible to operational errors during an emergency than are complicated designs. Because of expected increases in the cost of energy, not only must first costs of a system be considered but the efficiency of system components and circuit arrangements must be investigated.

Voltage quality involves the magnitude of the voltage, the frequency, the harmonics, and in case of three phase, the balance of the voltage magnitudes between phases. The ideal is a pure sine wave with constant magnitude and frequency. Design considerations should be given to voltage spread, voltage regulation, and load devices that produce harmonics in the voltage and current wave forms and transients in the system. Such devices are electronic inverters, some speed and voltage control devices, and switching operations under load.

Proper maintenance is necessary for improved safety and reliability, and the system designer should incorporate features in the system to facilitate routine maintenance and inspection.

Flexibility means the design should contain features that make it easy to accommodate changes in load location, magnitude, or character. Sufficient space should be planned in the original design to allow for expansion without undue circuit rearrangement.

Because of the diverse nature of different customers' electrical requirements, several basic circuit arrangements should be investigated to arrive at the most suitable one. However, certain basic factors need to be considered before a decision is made.

First, the loads must be analyzed as to their magnitude, location within the facility, variation during the day, and type, such as resistive, inductive, capacitive, motor, and so on. Loads can be described in terms of several factors: demand factor, load factor, diversity factor, maximum demand, peak, and average. The first three are used in determining the effective load of a group of devices, while the last three apply to a given load or device. The following is a description of these load characteristics or factors.

Demand factor is the ratio of the maximum demand on a system to the total connected load (sum of the continuous rating of the connected devices on the system).

Load factor is the ratio of the average load over a specified period of time to the peak load occurring during that period.

Diversity factor is the ratio of the sum of the maximum demands on the various components of the system to the maximum demand of the system (diversity factor is $\geqslant 1$).

Maximum demand is the integrated maximum load that occurs for a specified period of time, usually 15-minute periods or 30-minute periods.

The maximum demand is determined by applying demand and diversity factors to the connected load. The demand factor varies considerably with different loads and is based on experience obtained from similar applications.[†] The demand on the system then is the total connected load multiplied by the demand factor assuming a diversity factor of 1. If the diversity factor is known, the demand thus obtained should be divided by the diversity factor to obtain the actual demand. In many cases it is desirable to use a diversity factor of 1 to allow for the addition of loads and for expansion.

Basic Circuit Arrangements—Industrial Plants

Of the many possible variations of substation and circuit arrangements, a few basic designs have emerged and are applicable to most power distribution systems. These basic circuit arrangements utilize the philosophy of supplying power to the load center substation at the primary voltage and distributing it on relatively short low-voltage circuits to the utilization devices. These basic arrangements are classified as radial feeders, secondary selective feeders, secondary network, primary selective feeders, and looped primary. The various circuit arrangements are the result of a tradeoff between cost and continuity or reliability of service during failure of certain system components.

The radial feeder arrangement supplies power at primary voltages of 2.4 to 13.8 kV by cable to substations located close to the centers of electrical load. This is illustrated by the one-line diagram shown in Fig. 67.1. This arrangement will be the least expensive in the majority of installations, since there is no duplication of equipment. However, failure of a primary cable or transformer will result in failure of service to the area supplied by the faulty equipment. Also during maintenance, the area served by this equipment will be completely deenergized.

With sufficient substation capacity installed, the radial arrangement can adequately care for practically any diversity that will be encountered due to shifting loads. The system is simple and with adequate properly installed equipment it is safe and easy to operate and expand. The short secondary feeders result in good voltage regulation.

To provide a higher level of reliability, the secondary selective circuit arrangement can be used. Here the system utilizes two transformers and two primary feeders to supply each load center area.

[†]Tables of some demand factors, load densities, and the energy needed to produce certain products are given in Ref. 1.

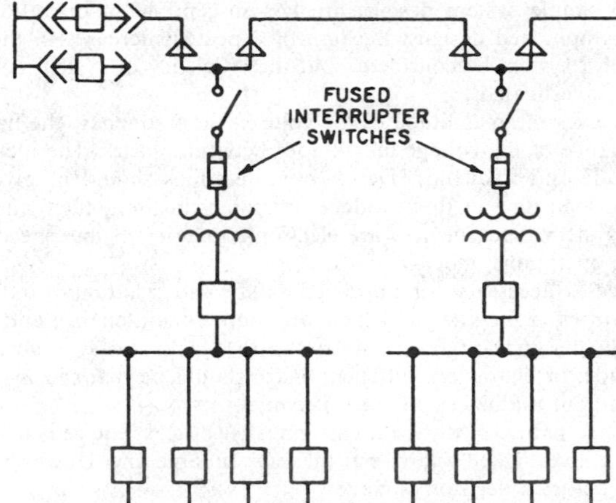

Fig. 67.1 Typical radial circuit arrangement.

Duplicate paths of supply are available from the source to supply each secondary bus; this makes it possible to provide power at all secondary busses when a transformer or primary feeder circuit is out of service. These duplicate paths can be achieved by a tie between two single-transformer substations through a secondary feeder or the use of a double-ended system with a tie breaker between the secondary busses. This is illustrated in Fig. 67.2, where the tie breakers B are normally interlocked with the two transformer breakers A so that B cannot be closed unless one of the transformer breakers A is open. This arrangement minimizes the short circuit duty imposed on the low-voltage secondary circuit breakers. When a primary cable or transformer fault occurs, service can be restored to the interrupted loads by opening the transformer breaker associated with the faulted circuit and closing tie breakers at all faulted locations. The remaining primary feeder and transformer must be capable of carrying the additional load for the time the other units are out of service.

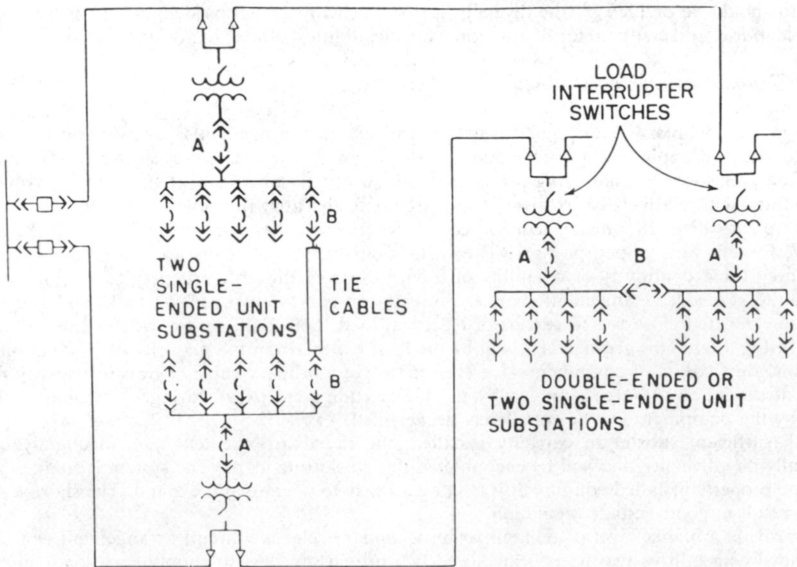

Fig. 67.2 Typical secondary selective circuit arrangement with two possible secondary tie arrangements.

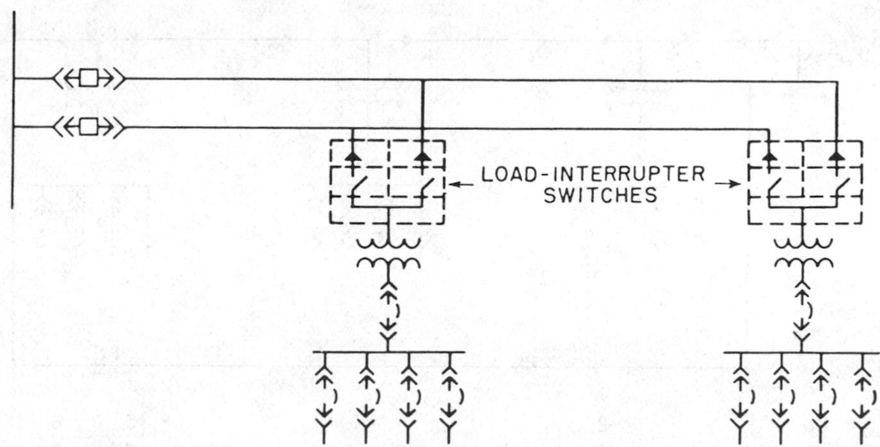

Fig. 67.3 Typical primary selective circuit arrangement with two single-throw interlocked interrupter switches.

For the primary selective arrangement, two primary feeders are brought to each substation transformer, as shown in Fig. 67.3. Half of the transformers are normally connected to each of the two primary feeders. The system must be designed so that the remaining feeder has sufficient capacity to carry the entire load.

For a feeder failure, and when switching is to be done while one primary feeder circuit is energized, the safest approach is use of adequate power circuit breakers to make the transfer. However, the cost is relatively high for such an arrangement and this approach is not used unless automatic transfer is desired. The usual practice is to use two load-interrupter switches interlocked so that only one can be closed at a time to make the transfer from one feeder to the other, as shown in Fig. 67.3.

For a transformer fault, the preferred procedure for determining which transformer is involved is to deenergize the good feeder and switch each transformer one at a time to the good feeder, energizing the feeder after connecting each transformer. This procedure is followed until all transformers have been connected to the good feeder or until the feeder breaker trips, which indicates the last transformer connected was faulted.

The primary selective arrangement provides about the same degree of service as the secondary selective system. This depends on the reserve transformer capacity of the secondary selective system.

So far the systems described have had radial primary feeders. When load centers are relatively far apart, the use of looped primary may offer some advantages. The looped primary system may utilize a single primary feeder breaker and one sectionalizing load interrupter switch at each transformer, or two primary feeder breakers and two sectionalizing switches at each transformer, as shown in Fig. 67.4.

For a fault in a transformer or primary feeder, in either arrangement the primary feeder breaker or breakers will open and interrupt the service to all loads on that loop. To locate the fault, all load interrupter switches are opened and then closed one at a time in sequence. It is safer to close only the load interrupter switches when the primary breakers are open. This will eliminate the problem of closing the switch in on a fault. When the fault is located, it can be isolated by leaving the appropriate load interrupter switches open. In the upper arrangement in Fig. 67.4, a transformer or feeder loop fault will result in that load being out of service. For the lower arrangement in Fig. 67.4, only a transformer fault will result in its load being out of service, since the two switches at each transformer make it possible to isolate a fault at any place in the loop.

The upper loop arrangement in Fig. 67.4 costs a little more than the radial arrangement but can provide service to the remainder of the system for a feeder loop fault that has been isolated (Ref. 2, p. 2). For a fault on the primary feeder in the radial system, service to all loads is interrupted until the fault is corrected.

The lower loop arrangement in Fig. 67.4 costs less than the primary selective arrangement and will provide, as does the primary selective, service to all loads for a primary feeder that has been isolated.[2(p. 21)] However, for transformer faults, service will be interrupted to associated loads in both loop primary and primary selective arrangements. The main disadvantage of the loop primary

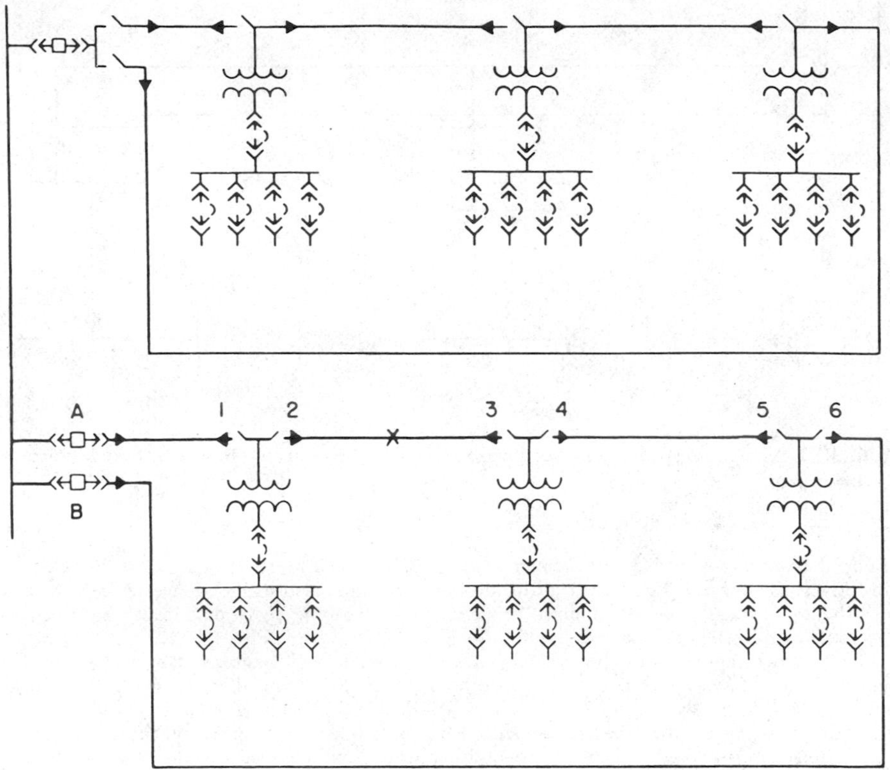

Fig. 67.4 Two possible looped primary circuit arrangements with sectionalizing switches.

arrangement is that a transformer or primary feeder fault, until it is isolated, causes an interruption in service to all loads.

The form of network system most frequently used in industrial plants is the primary selective-secondary network arrangement shown in Fig. 67.5. This system arrangement differs from the previously described ones in that a transformer or primary feeder fault will not cause even a momentary interruption of power to any of the loads. As seen from Fig. 67.5, this is because the transformer secondaries are interconnected and operated in parallel and two or more primary feeder circuits are used to supply the system. This provides more than one parallel path from the power supply to any load.

Parallel operation of the primary feeder circuits is possible through the use of the network protectors, as shown in Fig. 67.5. The network protector consists basically of an electrically operated air circuit breaker that is controlled by a directional-power relay and by a phasing voltage relay. For a primary feeder fault, power will flow from the secondary to the fault through all of the network protectors associated with the fault. This reversed power flow causes the network protector to operate and isolate the fault from the secondary. During this time the primary feeder breaker has tripped to isolate the fault from the primary supply. When the fault is eliminated and voltage is restored on the feeder by closing of the feeder breaker, the network relays on all associated network protectors allow the protectors to close automatically when conditions of power flow are from primary to secondary.

For normal operation the transformers are interconnected to the primary feeders in such a manner that the same number are on each feeder. For two primary feeders, half of the transformers would be on each feeder. For only a primary feeder fault following the normal tripping operation, the transformers that have been disconnected by the circuit breaker and network protector operation can be manually switched to a remaining energized feeder. The network protectors associated with those transformers will automatically close when the transformer is energized.

In addition to providing a high degree of service to the loads, this network arrangement offers great flexibility to meet shifting and growing loads because of the interconnected secondaries. The tie circuits between transformers allow adjacent transformers to share load and thereby permit loads on

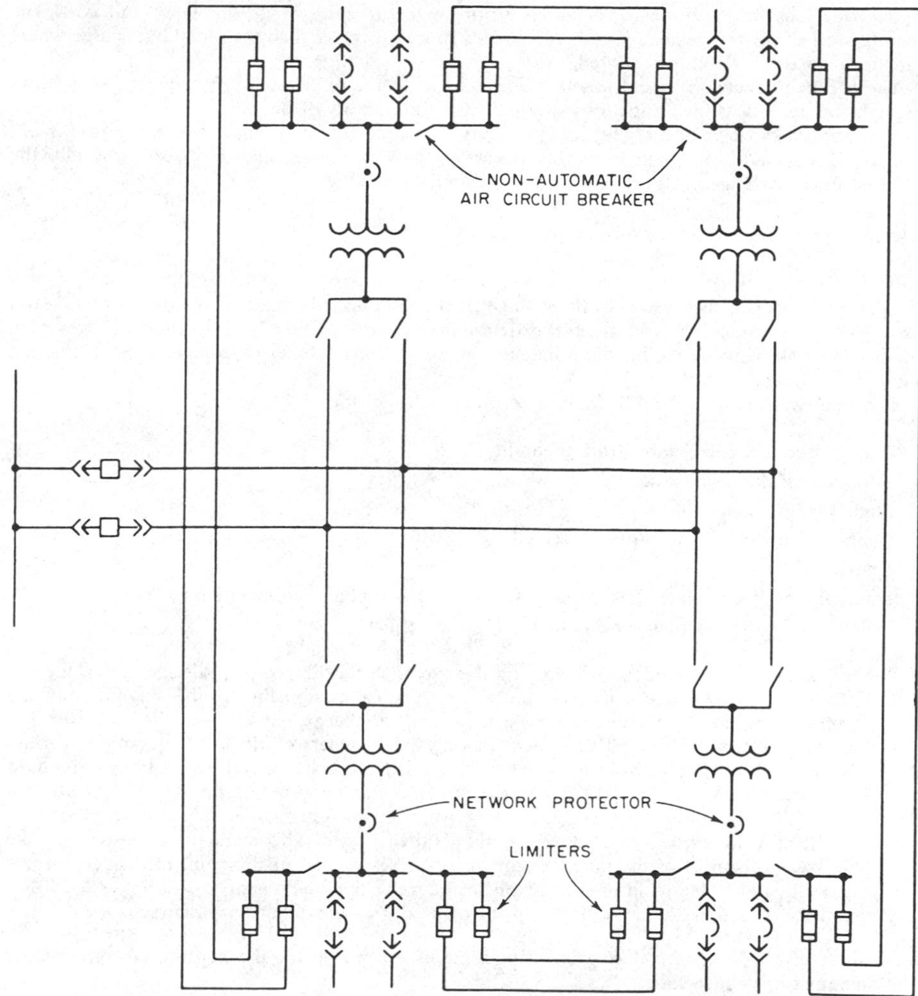

Fig. 67.5 Typical primary selective-secondary network with network protection.

some busses that are in excess of the transformer rating at that bus. The amount of power that can be transferred between transformer busses depends on the tie-circuit impedance, the transformer impedance, and the load characteristics.

Basic Circuit Arrangements—Commercial Buildings

Commercial buildings utilize circuit arrangements that are similar to those described for industrial systems but with certain significant differences. The geometric orientation of commercial buildings will probably differ from that of industrial plants. Industrial plants tend to be low and long, while commercial buildings are high and narrow. However, the volumetric measurements and electrical loads may be the same.

The differences that have to be considered are (1) the extremely high value of floor and volumetric space in commercial buildings, (2) the limited floor-to-ceiling space, and (3) the fact that many commercial buildings in urban areas are supplied at utilization voltage 208 Y/120 V or 480 Y/277 V from the electric utility's secondary low-voltage network system.

Four of the basic circuit arrangements described for industrial systems are used in commercial buildings. These are the radial, the secondary selective, the secondary network, and the primary selective.

The one most widely used is the radial arrangement, because of its simplicity and cost. The secondary selective arrangement is from 10% to 30% more expensive than the radial but is used where a greater degree of reliability is desired.[1]

The secondary network is only used where a high degree of reliability is required, such as in hospitals, because of its high initial cost and increased complexity of operation.

In special cases where primary feeder faults may be a problem, the primary selective arrangement that provides an alternate path from the source to the transformer may be used. Normally the preferred circuit arrangements are the radial or secondary selective.

Selection of System Voltages—Industrial Plants

Probably the most important single factor in the design of a power system is the selection of the voltage level. This is so because once the voltage level is established, changing it is very difficult and costly. Broadly speaking, the voltage level is determined from economic considerations. However, in practice the level may be modified by industry standards, availability of equipment, construction materials, and other factors.

The economic analysis should take into account the following:

1. Class of the service available from the utility.
2. Total size of the installation.
3. Plans for future growth.
4. Characteristics of equipment being served.
5. Density of the load.
6. Safety considerations including the qualifications of operating and maintenance personnel.
7. Whether this is a new plant or one that is to be enlarged.

Factors that have a tremendous influence on the overall cost of the voltage selected are (1) the size of the feeder circuits, switchgear, motors, and other electrical equipment, (2) the magnitude of the system fault current, and (3) the circuit arrangement and number of phases.

Over the years, through mutual agreement among the designers of electrical equipment, certain standard voltage levels have evolved. These are published in the American National Standard Institute publication ANSI C84.1-1970 entitled *Voltage Ratings for Electric Power Systems and Equipment (60 Hz)*. This lists all the preferred standard nominal system voltages and other associated nominal systems voltages generally in use in the United States. This standard also specifies the acceptable voltages spread or the tolerance limits for these voltages at the point of delivery by the supplying utility and at the point of connection to the utilization equipment.

Terms that are used in describing various voltage conditions are defined as follows.

Nominal Voltage. Nominal value assigned to a circuit or system for the purpose of conveniently designating its voltage class.

Service Voltage. The voltage at the point where the electric system of the supplier and user interconnect.

Utilization Voltage. The voltage at the line terminals of the utilization equipment.

Voltage Spread. The difference between the maximum and minimum steady-state voltages existing at a given voltage level. Transient or momentary fluctuations in voltage, such as those that occur during motor starting, switching, or faults, are not covered in voltage spread.

Voltage Drop. The difference at any instant between the source end and load end of various system components such as transformers, feeders, branch circuits, and so on.

Voltage Regulation. The change in voltage between no load and full load in terms of the full-load voltage at the load end of a given component. The source-end voltage is usually assumed to be constant for this calculation.

Acceptable voltage spread or tolerance limits were determined from a consideration of the effects of voltage variations on the performance and life of various types of utilization equipment.

Since in various voltage ranges, different factors are affected differently by the voltage level, the selection of the voltage value is usually divided into (1) voltages of 600 V and less, (2) voltages of 601 to 15,000 V, (3) voltages of 15,001 to 34,500 V, and (4) voltages of 34,500 V and higher.

Voltages of 600 V or Less. Since the majority of loads in industrial plants are integral horsepower polyphase motors and welders, the choice of nominal system voltage in this class for serving these loads is 208, 240, 480 (or 480 Y/277), or 600 V. For a system of a given kilovolt-ampere (kVA) size, the cost of a 240-V system is approximately twice that of the 480-V or 600-V size system. Reasons other than cost sometimes determine the voltage level selected. In comparing 480-V vs. 600-V systems, the cost is about the same (600 V being a few percent less), so availability of standard equipment is the determining factor (Ref. 3, p. 45). In ordering machine tools or other utilization equipment, it is often difficult to obtain them with 575 (or 550)-V rated motors; likewise, pumps and other equipment that are stocked by manufacturers usually have 220 (or 230)-V and 440 (or 460)-V ratings: Another advantage of 480 V is the possibility of using 480 Y/277 V distribution, with 277 V for fluorescent lighting.

When comparing 480- with 240-V systems, economics is the predominant factor if the systems are to be comparable in performance. In industries where there is considerable dampness, such as in dairies and slaughterhouses, it is felt that 240 V is safer than 480 V. Also, during work on energized circuits there is a greater chance for injury from electric shock with higher potentials to ground or phase to phase. Operating records show that the biggest factor in safety is to properly and securely ground all noncurrent-carrying parts so that insulation breakdowns do not place dangerous potential on the noncurrent-carrying parts. However, any voltage above 50 V can be lethal.[3] In all locations, current-carrying conductors should be enclosed in securely and properly grounded enclosures; when work is required on them they should be deenergized.

There are some applications where 208 Y/120 V is more economical than a 480-V system because the types of utilization equipment involved must operate at 120 V. When such equipment comprises more than 50% or 60% of the total load, then 208 Y/120 V may be more economical than a 480-V system. Examples are industries that use hand tools, such as a clothing manufacturing facility that uses motor-operated hand shears or an electronic assembly plant that uses soldering irons, electric drills, and other hand tools.

As a general practice, 480-V systems are recommended for industrial plants. Where lower voltages are required, the 208 Y/120 V may be used. Where a significant fluorescent lighting load exists, the 480 Y/277-V system should be considered.

Voltage of 601 V to 15 kV. For industrial plants, voltages in this class are mainly used for primary power distribution. Exceptions are the very large chemical plants and steel mills, among others.

The National Electrical Code (NEC) has no special restrictions on voltages of 15,000 V or below, so there is generally no reason for transforming to 2400 or 4100 V for distribution to the load-center substation. The voltage is transformed to the utilization voltages at the substation.

Above 15,000 V, the NEC requires that service enter either a metal-enclosed switchgear or a transformer vault (Sec. 230–202h, 1978, NEC). Therefore, for utility supplies above 15,000 V, transformation to some lower voltage is both economically and technically desirable for distributing the power to the plant load-center substations. Studies have shown that either of two voltages, 4160 or 13,200 V, will suffice in the majority of cases. Generally, for supply transformer or generation capacity below 10 MVA, 4160 V is the more economical distribution voltage. For capacity above 20 MVA, 13,200 V is the more economical, but for capacity between 10 MVA and 20 MVA, the more economical voltage depends on other factors such as expected plant growth (Ref. 3, p. 46).

The size of motors in a plant influences which system voltage is more economical. In general, for motors less than 200 hp, a 480-V system is more economical in terms of initial cost. For motors greater than 200 hp, the 2400-V system is more economical, provided the concentrated loads are less than 7500 kVA per bus. For loads above 7500 kVA, a 4160-V system may be necessary to obtain equipment to handle the short-circuit current available.

Selection of System Voltages—Commercial Buildings

For commercial buildings, generally the preferred utilization voltage is 480 Y/277 V. Here the three-phase loads are connected directly to the 480-V line conductors, while the ceiling fluorescent lighting is connected phase to neutral at 277 V. To provide 120 V for convenience outlets and 208 V for office machinery, a 480-208 Y/120-V transformer is used.

For many existing buildings in downtown areas, only 208 Y/120 V was available at the time of construction; but in recent years most utilities will provide spot network installations at 480 Y/277 V.

For large motors used for air conditioning and other purposes, consideration should be given to connecting these to a separate transformer, to reduce the effect of the starting voltage drop on other connected equipment. For motors over 200 hp, a 2400-V system may be more economical than the 480-V system. It should be realized that maintenance electricians in commercial buildings are rarely qualified to service equipment above 600 V. Therefore, contracted maintenance may be required.

67.2 VOLTAGE SPREAD AND REGULATION

For any specific nominal voltage, the voltages actually existing will vary depending on the location of the measurement and the time at which it is made. This is illustrated by the one-line diagram shown in Fig. 67.6 for a typical utility system. Each section of the system has a voltage drop associated with

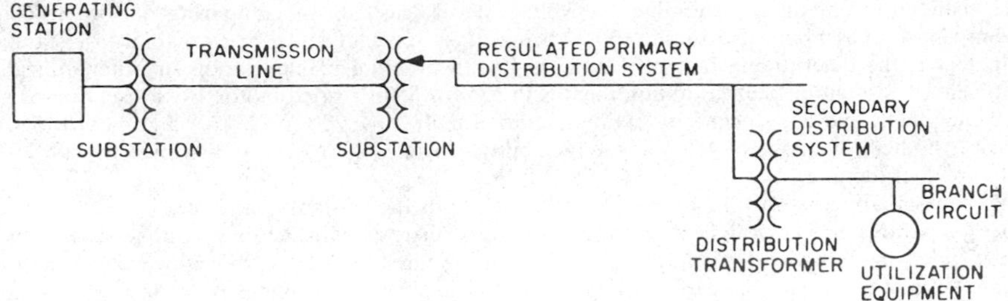

Fig. 67.6 Typical utility power generation, transmission, and distribution system.

it depending on the load (current) and power factor, both of which may change with time. The transformers at the substation supplying the primary distribution system generally are equipped with tap-changing under-load equipment. This equipment changes the transformer turns ratio, which makes it possible to maintain the primary distribution source-end voltage within a narrow range for different load conditions.

The ANSI C84.1-1970 standard lists two tolerance voltage ranges or voltage spreads that are considered acceptable under specified conditions. The ranges are range A, which specifies the limits under most operating conditions, and range B, which allows minor excursions outside the range-A limitations.

These ranges were determined from a consideration of the effects of voltage variations on the performance and life of various types of utilization equipment. The responsibility of the equipment manufacturer and the supplier for these ranges A and B are: "Utilization equipment shall be designed and rated to give satisfactory performance throughout range A, and insofar as practicable, utilization equipment shall be designed to give acceptable performance in the extreme limits of range B although not necessarily as good a performance as in range A. For the supplier, his system shall be designed and operated such that most of the service voltages are within the service voltage range A. Excursions of the service voltage into the extreme limits of service voltage range B are to be limited in extent, frequency, and duration. When they occur, corrective action should be undertaken within a reasonable time to improve the voltage to within the range A limits."

The range of voltage tolerances defined in ANSI C84.1-1970 standard for range A, expressed in terms of a nominal 120-V system are a maximum of 126 V and a minimum of 110 V for lighting equipment and 108 V for other than lighting equipment. The profile of these limits allows a 9-V drop in the primary distribution feeder, a 3-V drop in the distribution transformer, a 4-V drop in the building wiring for lighting equipment and a 6-V drop in the building wiring for other than lighting equipment. This assumes that the substation transformer is provided with tap-changing under-load equipment to hold the voltage at the source end of the primary distribution system essentially constant.

For commercial buildings, the responsibility of maintaining the primary distribution voltage drop and probably the distribution transformer voltage drop within limits rests with the supplying utility.

For industrial plants, depending on the supply voltage and size of the plant, the responsibility for maintaining the voltage drop within limits for all three parts of the system generally rests with the user.

Owing to the phasor relationship between voltage and currents, the resistance and reactance voltage drop, the total voltage drop in a given portion of the system not only depends on the current but also the load power factor. Figure 67.7 shows a phasor diagram of the voltage and current relations for calculating a voltage drop in a given part of the system. For the usual power factors encountered, an approximate formula can be derived for calculating the voltage drop V_d.

The formula is

$$V_d = IR\cos\theta + IX\sin\theta \tag{67.1}$$

where V_d = voltage drop in circuit (one conductor) line to neutral in volts (this assumes balanced three-phase load)

I = current flowing in conductor or transformer, A

R = resistance for one conductor or phase of transformer, Ω

X = reactance for one conductor or phase of transformer, Ω

θ = angle between load voltage and load current, which is also angle whose cosine is power factor of load

$\cos\theta$ = load power factor, decimals

$\sin\theta$ = load reactive factor, decimals

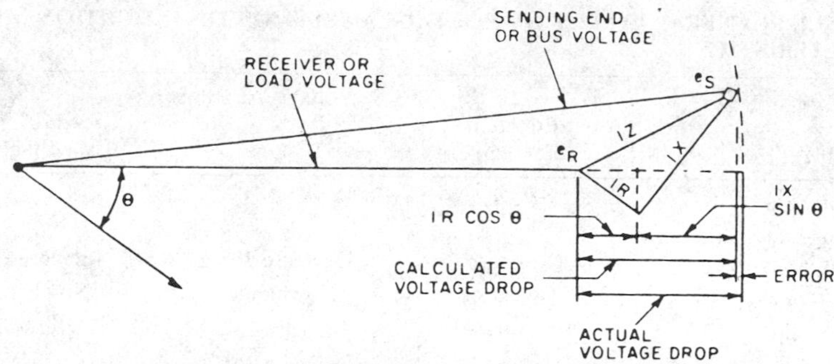

Fig. 67.7 Phasor diagram of voltage relations for voltage drop calculations.

To obtain the line-to-line voltage drop in a three-phase system, multiply the line-to-neutral voltage by $\sqrt{3}$. For single-phase loads, using a two-wire no-neutral arrangement, or for balanced single-phase loads on a three-wire single-phase system, multiply the line to neutral voltage drop by 2 to obtain the line-to-line voltage drop. For unbalanced loads on either system, the voltage drop in the neutral should be calculated and included in the total drop.

Effect of Voltage Variations on Utilization Equipment. Even though the utilization equipment is to be designed to give satisfactory performance within the allowed voltage spread of the utilization voltage, any deviation of the voltage from the equipment's rated value will result in a reduction in the life or performance of the equipment. The effect may be minor or serious, depending on the characteristics of the device and on how long and how large a deviation exists.

The effects of voltage variations on some of the usual utilization equipment are shown in Table 67.1 and Figs. 67.8, 67.9, and 67.10. For induction motors the most significant effects are: for low voltage, reduced starting torque and increased full-load temperature rise; for high voltage, increased starting current and torque and decreased power factor. Increased torque may cause coupling damage or damage to the driven equipment. For incandescent lamps, the greatest effect is that high voltage decreases the lamp life drastically, as shown in Fig. 67.8. Similar life reduction occurs in electron tubes, as shown in Fig. 67.9. Fluorescent lamps are less affected by voltage variations than are incandescent.

Synchronous machines are affected similarly to induction motors, except the pull-out torque varies directly with the voltage, unless the DC field excitation varies with the supply voltage.

High-intensity discharge lamp lumen output is affected by voltage, similar to incandescent lamps; however, at about 20% under voltage the arc will be extinguished. The lamp cannot then be restarted until the mercury condenses, which takes four to eight minutes unless special cooling equipment is used. Repeated starting can decrease the life of the lamp, since its life is inversely related to the number of starts.

For capacitors, the kilovar input varies as the square of the voltage, so reduced capacity exists at reduced voltages.

If the voltage spread or voltage conditions are not within acceptable limits, the following basic changes should be considered to correct the poor condition: (1) carry the power at a higher voltage closer to the load and lessen the distance traveled at the lower voltage, (2) reduce system impedance, (3) use regulating equipment to compensate for the voltage drops, and (4) use switched shunt or possibly series capacitors.

Since voltage drop is a function of current and impedance, a reduction in these will reduce the drop. Some possible changes to reduce these are:

1. Use closely spaced conductors, such as cables instead of open wiring, to reduce the reactance.
2. Use a low-voltage busway of interleaved construction to give more uniform current through the bus.
3. Sometimes use of two or more smaller cables in parallel instead of one large one reduces the voltage drop.
4. Use lower impedance transformers (but the increased short-circuit current will need to be investigated).
5. Correct the power factor at the utilization equipment with shunt capacitors.

TABLE 67.1. GENERAL EFFECT OF VOLTAGE VARIATION ON INDUCTION MOTOR CHARACTERISTICS

Variable	Function of Voltage	Voltage Variation 90% Voltage	110% Voltage
Characteristic			
Starting and maximum running torque	$(\text{Voltage})^2$	Decrease 19%	Increase 21%
Synchronous speed	Constant	No change	No change
Percent slip	$1/(\text{Voltage})^2$	Increase 23%	Decrease 17%
Full-load speed		Decrease $1\frac{1}{2}$%	Increase 1%
Efficiency			
Full load	—	Decrease 2%	Increase $\frac{1}{2}$–1%
$\frac{3}{4}$ Load	—	Practically no change	Practically no change
$\frac{1}{2}$ Load	—	Increase 1–2%	Decrease 1–2%
Power Factor			
Full load	—	Increase 1%	Decrease 3%
$\frac{3}{4}$ Load	—	Increase 2–3%	Decrease 4%
$\frac{1}{2}$ Load	—	Increase 4–5%	Decrease 5–6%
Full-load current	—	Increase 11%	Decrease 7%
Starting current	Voltage	Decrease 10–12%	Increase 10–12%
Temperature rise,[a] full load	—	Increase 6–7° C	Decrease 1–2° C
Maximum overload capacity	$(\text{Voltage})^2$	Decrease 19%	Increase 21%
Magnetic noise, no load in particular	—	Decrease slightly	Increase slightly

[a] These data apply to motors of over 25 hp.

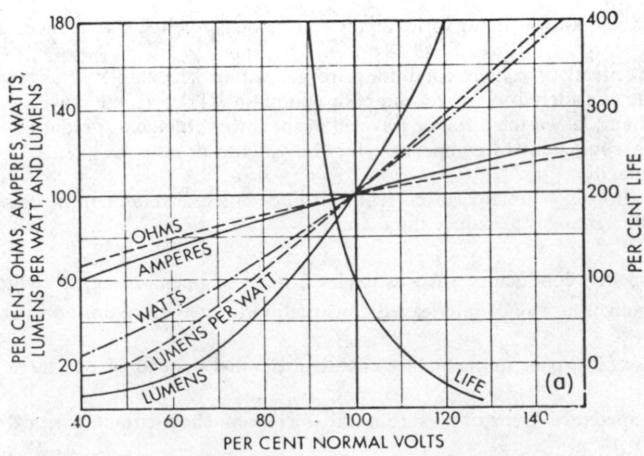

Fig. 67.8 Characteristics of incandescent lamps as a function of voltages.

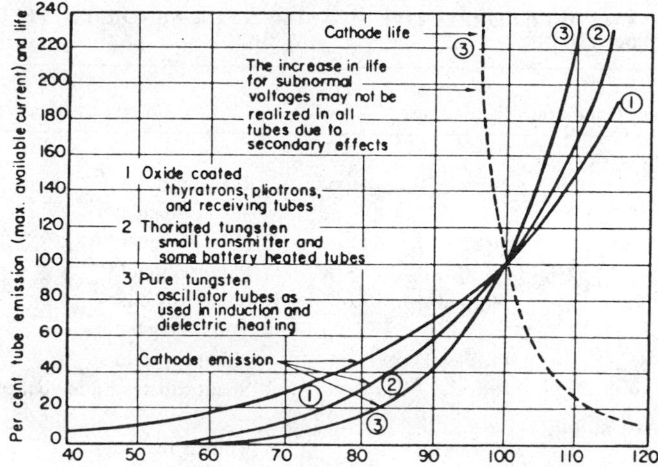

Fig. 67.9 Calculated values of electronic tube emission and life.

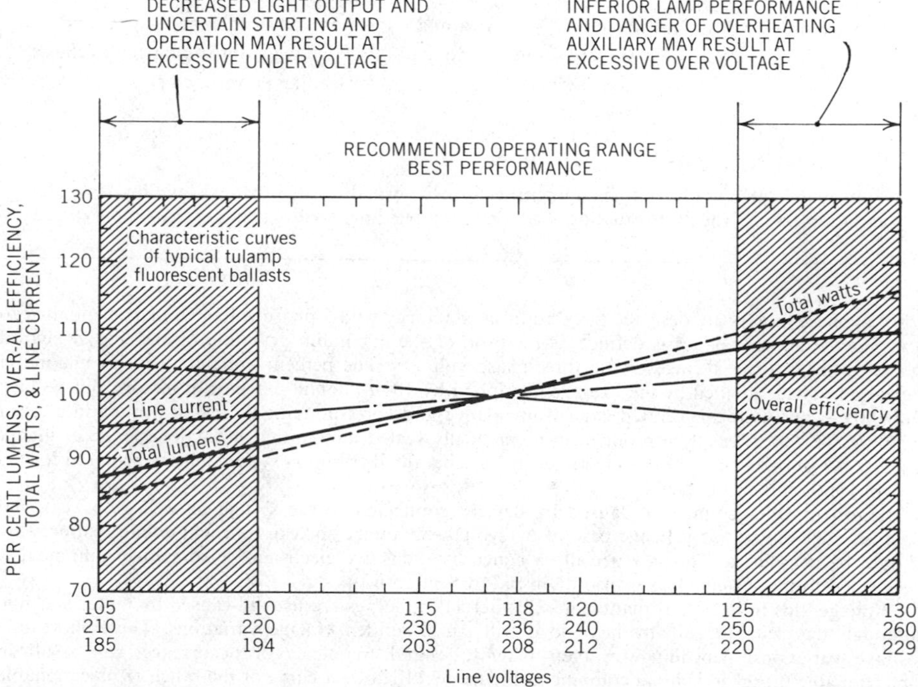

Fig. 67.10 Characteristics of fluorescent lamps as a function of voltage applied to the ballast.

Most modern transformers have taps in the winding to change the turns ratio. Changing taps will not change the voltage spread but only the average voltage level. Even when such a change is implemented, it may not be possible to achieve the required voltage spread, and a voltage regulator may have to be used. Table 67.2 summarizes and gives some techniques to improve poor voltage conditions existing in a system.

Besides voltage spread, three other voltage characteristics should be investigated in the distribution system of industrial plants or commercial buildings. These are phase voltage unbalance, transient voltage variations, and harmonics. Since it is somewhat common in four-wire grounded-wye distribution systems to serve single-phase loads line to neutral, unequal voltage drops can occur in the phases.

TABLE 67.2. POSSIBLE CORRECTIVE MEASURES FOR ABNORMAL FEEDER VOLTAGE CONDITIONS

Feeder Voltage Condition	Circuit Loading	Power Factor	Correct by Means of
Low-Load Voltage			
Low	Normal	Good	Transformer tap setting Voltage regulator
High feeder drop	Normal	Good	Voltage regulator Parallel circuit
High feeder drop	Normal	Low	Shunt capacitors Voltage regulator
High feeder drop	Overload	Low	Shunt capacitors Shunt capacitor and voltage regulator
High-Load Voltage			
Normal drop	Normal		Voltage regulator Transformer tap setting
Normal drop	No load		Transformer tap setting Voltage regulator
Voltage rise	"No load" (except shunt capacitors are on)	Leading at no load	Automatic switching of capacitors Voltage regulator (if no penalty clause for leading power factor)
Variable-Supply Voltage			

Voltage regulator or load-tap-changing transformer normally only practical solution. If voltage variation due to widely fluctuating loads, notably welding, series capacitors commonly used.

Polyphase motors are designed to perform satisfactorily with a phase voltage unbalance limited to 1%. The voltage unbalance is defined as the ratio of the maximum deviation of any line to neutral phase voltage to the average of the three-phase voltages. The percent unbalance is this maximum deviation voltage divided by the average multiplied by 100 to obtain percent. In motors, this voltage unbalance causes an exaggerated current unbalance and corresponding increased heating in a given winding. This is especially a problem in hermetically sealed motors that normally operate at higher current densities. Because of this characteristic, a separate distribution system may be required for the polyphase motors in the facility.

Transient voltages may be caused by devices connected to the system as well as by external sources, such as lightning. In the case of overvoltages (voltage spikes), transient voltage suppressors should be considered. This is especially a concern if sensitive electronic or computer equipment is connected to the system. This is also discussed in Section 67.4.

Voltage dips in industrial plants and commercial buildings are usually caused by motor starting, although the switching of any large load will cause sudden voltage variations. These short-time voltage variations, depending on their magnitude and frequency of occurrence, can result in objectionable flicker in lighting equipment. Figure 67.11 shows a curve of the range of objectionable as well as perceptible flicker. These are average values from a series of tests, but if voltage dips exist in the regions defined in this figure, corrective action should be considered. Sometimes a separate lighting feeder is required to correct the problem. Reduced voltage starting, reactor or resistance starting, or switching in shunt capacitors at start will reduce the starting current drawn by motors.

Harmonics are integral multiples of the system frequency (60 Hz) and are caused by nonlinear devices being connected to the system. All ferromagnetic devices have the potential for causing harmonic currents and corresponding harmonic voltages. These devices include transformers, motors, and iron core reactors. Also gas discharge lamps, arc equipment, and rectifiers and switching devices such as phase-controlled rectifiers (variable-speed control devices and light dimmers) can cause harmonic voltages and currents.

Harmonic effects vary widely in different parts of the system and affect devices differently. Adjacent communication systems are particularly sensitive to harmonics, and special precautions have to be taken to isolate these systems or to suppress the harmonics.

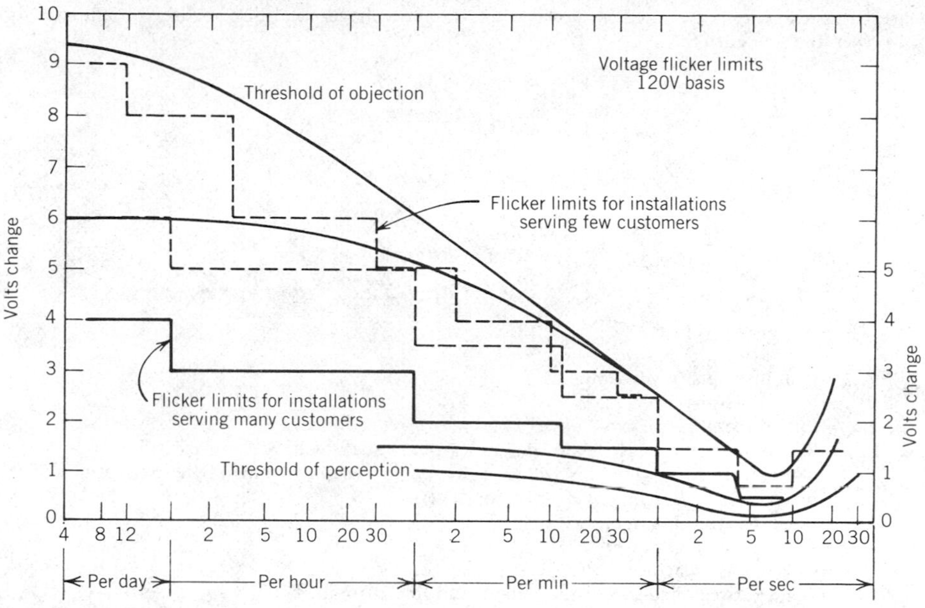

Fig. 67.11 Voltage flicker limits.

67.3 POWER FACTOR IMPROVEMENT AND KILOVAR SUPPLY

The use of shunt capacitors has previously been suggested to improve the voltage regulation of a given component or system in which the power factor is low. Correcting the power factor may be justified for other reasons as well. These reasons are: (1) to lower the cost of electrical energy where the electric utility rates vary with power factor at the metering point, (2) to reduce energy losses in conductors and transformers, and (3) to make available the full capacity of transformers, switches, circuit breakers, busses, and conductors for real power only, thereby reducing the capital investment and annual charges.

Most utilization equipment requires both real and reactive power for satisfactory performance. Incandescent lamps and resistive heaters are two exceptions. Only real power develops real work, and supplying the reactive power at the load (instead of from a long distance away) reduces the circuit current required to supply the real power.

The relevant equations involved in power factor correction calculations are repeated here for convenience. For single phase circuits:

$$\begin{aligned} P &= VI\cos\Theta = VI(\text{pf}) \\ Q &= VI\sin\Theta \\ VI &= \sqrt{P^2+Q^2} \\ I_a &= I\cos\Theta \\ I_r &= I\sin\Theta \\ I &= \sqrt{I_a^2+I_r^2} \end{aligned} \tag{67.2}$$

where P = real (active) power, W
Q = reactive power, var
VI = volt-amperes, VA
Θ = angle between voltage and current at location of power computation
I = line current, A
V = line-to-neutral voltage, V
I_a = active power component of line current, A
I_r = reactive component of current, A
pf = power factor, equal to cos Θ

For three-phase circuits, the following equations are used in power factor calculations. Three-phase balanced systems are rated in terms of line-to-line voltage, total three-phase volt-amperes, or real power and line current.

$$
\begin{aligned}
VA_{3\phi} &= \sqrt{3}\, V_L I_L = 3VA_\phi = 3V_{LN}I \\
P_{3\phi} &= \sqrt{3}\, V_L I_L \cos\Theta_p = 3P_\phi = 3V_{LN}I_L \cos\Theta_p \\
Q_{3\phi} &= \sqrt{3}\, V_L I_L \sin\Theta_p = 3Q_\Theta = 3V_{LN}I_L \sin\Theta_p \\
V_L &= \sqrt{3}\, V_{LN} \\
\text{pf} &= \cos\Theta_p \\
VA_\phi &= V_{LN}I
\end{aligned}
\qquad (67.3)
$$

where V_L = line-to-line voltage
I_L = line current
$VA_{3\phi}$ = total three-phase volt-amperes
$P_{3\phi}$ = total three-phase power, W
$Q_{3\phi}$ = total three-phase reactive power, var
V_{LN} = line-to-neutral voltage, where neutral is point equal voltage from each line (For 4-wire wye-connected systems this voltage is available to supply single-phase loads and is usual type of system existing at distribution level)
VA_ϕ = single-phase volt-amperes
P_ϕ = single-phase real power
Q_ϕ = single-phase reactive power
ϕ_p = angle between line-to-neutral voltage and line current
pf = power factor per phase, equal to $\cos\phi_p$

For unbalanced loads, the real power P_ϕ and the reactive power Q_ϕ for all the loads on a given phase are added (algebraically for reactive loads) to obtain the power per phase. For a balanced three-phase load, each phase has one third of the three-phase values of power. With unbalanced loads, each line current will be different, and the line-to-neutral as well as line-to-line voltages may be different. This voltage unbalance must be limited to 1% for satisfactory motor performance.

Because a low power factor requires an additional component of current, 90° out of phase with the active (power) component of current, to supply the reactive power, additional voltage drops and power losses exist in the circuit. In addition, extra generator, transformer, and conductor capacity is required to supply this reactive current. Since most motor loads are inductive, lagging reactive power is required. Capacitors supply leading reactive power, so when they are placed in the system in parallel with the inductive loads they can relieve the generators from supplying such power. This makes available additional capacity for additional loads, as illustrated in Fig. 67.12. The reduction in current and the additional real power capacity available can be calculated by using the previously listed equations.

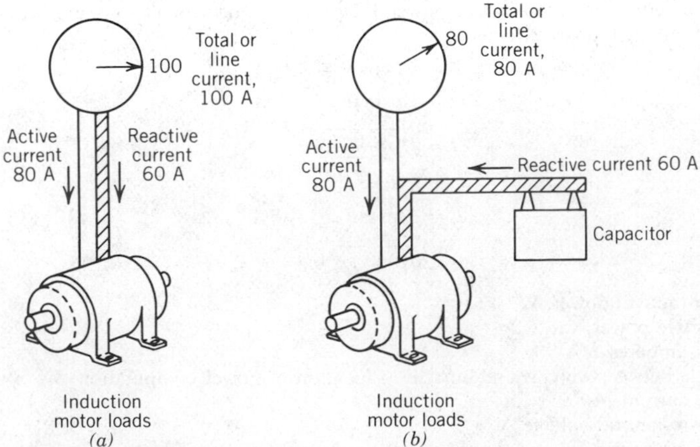

Fig. 67.12 (*a*) Current components before addition of capacitors; (*b*) current components after addition of capacitors.

Example. Assume a 100-kW induction motor is receiving 100 kW at 0.8 pf (power factor) lagging. For this system, determine the additional capacity available from improving the power factor to unity.

The volt-ampere capacity required of the system to supply this motor is

$$\text{VA} = \frac{P}{\text{pf}} = \frac{100}{0.8} = 125\ \text{kVA}$$

Of the 125 kVA, 75 kvar is reactive power, since

$$Q = \text{VA} \sin \Theta = 125 \sin[\cos^{-1} 0.8] = 125(0.6) = 75\ \text{kvar}$$

or

$$Q = P \tan \Theta = 100\ (\tan \Theta) = 75\ \text{kvar}$$

If a capacitor bank of 75 kvar is connected to the motor, the system need only supply

$$\text{VA} = \sqrt{P^2 + Q^2} = \sqrt{100^2 + (75 - 75)^2} = 100\ \text{kVA}$$

Therefore, by correcting the power factor to unity at the load ($Q = 0$), 25 kVA is available from the system to supply another load. Because of the relationships involving the volt-ampere calculations, this 25 kVA is the value of the load that can be added at the "improved" power factor. If the power factor of the added load is different, then more than 25 kVA can be added, depending on what the power factor is.

The additional load that can be added to the system due to power factor improvement is termed amount of capacity released. The incremental increase in released capacity for a given amount of reactive power supplied by a capacitor decreases as the power factor becomes larger. Therefore it may not be cost effective for the user to increase the power factor to unity.

Since system capacity can be increased by use of additional substation and distribution facilities as well as by addition of capacitors to improve the power factor, the installed costs of the various alternatives should be compared. In many cases the addition of capacitors is the cost-favorable alternative.

In the previous example, if the power factor had been increased to only 0.95 lagging, the amount of released capacity would be 19.7 kVA at the improved power factor. This calculation is shown as follows.

Example. Assume again a 100-kW induction motor being supplied 100 kW at 0.8 pf lagging from the system. Determine the additional capacity available from correcting the power factor to 0.95 lagging at the load.

Again the system capacity required for this uncorrected motor is

$$\text{VA} = \frac{P}{\text{pf}} = \frac{100}{0.8} = 125\ \text{kVA}$$

The reactive power required is $Q = \text{VA} \sin \Theta$, where $\cos \Theta = \text{pf}$. $Q = 125(\cos^{-1} 0.8) = 75$ kvar. The desired power factor is 0.95; therefore Q from the system is $Q = P \tan \Theta$, where now $\cos \Theta = 0.95$. $Q = 100(0.329) = 32.9$ kvar from the system. Since 75 kvar is required by the motor, the capacitors must supply 42.1 kvar. The new volt-ampere required from the system is $\text{VA} = \sqrt{p^2 + Q^2} = \sqrt{100^2 + (32.9)^2} = 105.3$ kVA. The released capacity is $125 - 105.3 = 19.7$ kVA.

The kilovar of capacitors per kilovolt-ampere of released capacity for the second case was $42.1/19.7 = 2.1$, while for the first case it was $75/25 = 3$. To correct from 0.95 to unity requires $19.7/5.3 = 3.7$ kvar of capacitors per kilovolt-ampere of released capacity. This shows why it may not be desirable for the user to correct to unity. Approximate values of released capacity can be conveniently obtained by the use of the curves in Fig. 67.13.

As previously indicated, increased capacity is not the only benefit from correction of the power factor. In some cases, power factor correction can improve the voltage conditions existing in a system, especially if the system power factor is low. The possibility of improvement also depends on the system or circuit reactance to resistance ratio, called X/R ratio. Using the approximate Eq. (67.1) one can determine the improvement in the voltage condition by power factor improvement. These calculations assume the capacitor is connected across the load component and is switched with the load.

For a given allowable percentage voltage drop in a feeder cable, improvement in the power factor can given a significant increase in the allowable load.

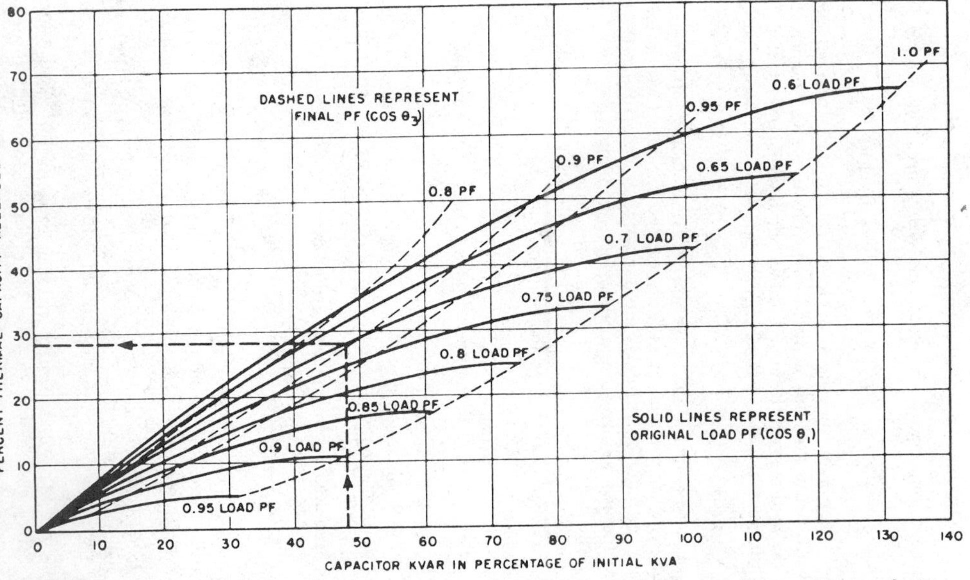

Fig. 67.13 Curves showing percent released capacity as a function of the percentage of power factor correcting capacitors. Example: For a 100-kVA load at 70% power factor (pf) lagging, the addition of shunt capacitors of 48 kvar will reduce the net system load to 73.8 kVA, thereby making available (releasing) 26.2 kVA that can be added to the system at the power factor of approximately 90%, without exceeding the original 100 kVA of the system load. If the added load is at a power factor of 70%, then 28.5 kVA can be added, as shown by the dashed line in the figure.

Example. Assume a 4% voltage drop at rated current in a cable with an X/R ratio of 2 and a load pf = 0.8 lag. Determine the additional capacity available from improving the power factor.

From Eq. (67.1) for the approximate voltage drop, $V_d = I(R\cos\Theta + X\sin\Theta)$, which in terms of per unit becomes

$$\frac{V_d}{V_r} = I_0(R\cos\Theta + X\sin\Theta) = 0.04 \text{ per unit}$$

where V_r is the rated voltage and I is the line current. For rated load these are equal to 1 per unit (100%).

With $X/R = 2$, then

$$R(\cos\Theta + 2\sin\Theta) = 0.04$$

and with $\cos\Theta = 0.8 = \text{pf}$, then $\sin\Theta = 0.6$ and

$$R = \frac{0.04}{0.8 + 2(0.6)} = 0.02 \text{ per unit}$$

Improving the power factor not only changes Θ in the equation for V_d but also reduces I.

Assume the power factor is increased to unity by adding 0.6 per unit of capacitor at the load. (See per-unit calculations in Section 67.4.) The current is now 0.8 per unit and Θ is 0^0. Therefore $V_d = 0.8(0.02)(1 + 0) = 0.016$ per unit.

If 0.04 per-unit voltage drop is still allowable, then the current at unity power can be increased to $0.04/0.016 = 2.5$ times its initial value.

If the power factor were improved to only 0.95, then $\Theta = \cos^{-1}0.95 = 18.19^0$ and $I = 0.8/0.95 = 0.842$ per unit, so $V_d = 0.842(0.02)[0.95 + 2(0.33)] = 0.0271$ per unit. For an allowable voltage drop of 0.04 per unit, the current at pf = 0.95 can be increased $0.04/0.0271 = 1.48$ times its initial value. This assumes that the voltage drop is the limiting factor and that the system capacity is sufficient to handle the increased current.

An additional benefit to improved power factor is reduced losses. At any given power factor the total current is the power component of the current divided by the power factor. Since losses vary as the square of the total current, the losses will vary as the square of the power factor. Therefore

$$\text{kW losses} = \left(\frac{\text{original pf}}{\text{improved pf}}\right)^2 (\text{original kW losses})$$

and loss reduction is

$$\text{Loss reduction kW} = \left[1 - \left(\frac{\text{original pf}}{\text{improved pf}}\right)^2\right] (\text{original kW losses})$$

Loss reduction alone may not be enough to justify the use of capacitors for power factor improvement, but with the additional benefits of voltage regulation improvement, increased capacity, and reduction of the power factor penalty in the utility bill, the use of capacitors to improve the power factor to closer to unity can usually be justified.

The greatest benefit obtained from capacitors for power factor improvement occurs when they are located as close to the low power factor load as possible. In some cases this is not possible or advisable. Other possible locations are along or at the beginning of a feeder circuit, substation bus, or transformer primary bus.

The reactive power supplied by a capacitor varies as the square of the voltage and directly as the frequency, and these may have to be considered when applying capacitors. Capacitors manufactured according to NEMA (National Electrical Manufacturers Association) standards are capable of supplying 135% of rated kilovars. This includes the effects of harmonics, overvoltage, overfrequency, and manufacturing tolerances.

The National Electrical Code (NEC) requires that capacitors have built-in discharge resistors or be used so that when deenergized, the voltage drops to 50 V or less in one minute for those on 600-V circuits or less. For application on over 600-V circuits, the voltage is required to drop to 50 V or less in five minutes.

Capacitors applied to induction motors are restricted in size when connected on the load side of motor controllers. The NEC limits the rated kilovars to be no greater than the induction motor no-load reactive power.† These restrictions are required to protect the motor from developing an overvoltage due to self-excitation when capacitor and motor are disconnected from the line as a unit

†Article 460, *Capacitors*, National Electrical Code, National Fire Protection Association, Boston, MA.

and from excessive transient torque if the motor is restarted before coming to a standstill. It is recommended that the starting controller be designed so that a restart is not initiated before the motor terminal voltage has collapsed to zero.

If the capacitors are switched separately, special circuit breakers or switches are required to handle the surge currents and reactive current of the capacitors.

67.4 POWER SYSTEM PROTECTION

Even well-designed power systems, with reasonable precautions to provide sufficient insulation clearances, and so on, occasionally experience short circuits resulting in abnormally high currents. Such system high currents (system faults) can result in injury to personnel; extensive property damage, either directly or as a result of a fire or explosion; and loss of production. These possibilities, plus consideration for other users of the utility service, should be added to the other engineering requirements when developing the protective system.

The main objective of the protective system is to isolate any faults safely with a minimum of damage to circuits and equipment and with a minimum amount of shutdown of plant operation. This entails isolating only that portion of the system involved with the fault and arranging to place that part of the system back in service as soon as is safely possible. The protective system consists of relays, circuit breakers, and fuses. The relays are used to detect the kind of system fault and the location of the fault and to initiate action to isolate a minimum amount of the system containing the fault.

The types of faults to be detected by the relays are three-phase, phase-to-phase (line-to-line), double line-to-ground (line-to-line-to-ground), and single phase-to-ground short circuits. Relays sense the magnitude of the voltage and current in various portions of the system and initiate action to isolate the fault and to provide back-up protection (redundancy) in case of failure to some portion of the protective system.

Isolation of the faulted part of the system is accomplished by circuit breakers that open and interrupt the short-circuit current and possibly reclose, if the fault is of a temporary nature. The command for the circuit breaker to open or close is given by the relays controlling that breaker's action.

Fuses also interrupt short-circuit currents and are actuated by the magnitude and length of time the current has passed through the fuse. A given fuse can only open a circuit; it cannot reclose after having interrupted the current. Certain mechanical arrangements exist where another unblown fuse can be mechanically used to close the circuit. Generally, fuses only open the circuit and function only on the current.

Since protective relays can sense current and/or voltage, they can provide a wide range of functions. Some of these functions are overcurrent, time-delayed overcurrent, voltage-constrained overcurrent, directional overcurrent, directional power, voltage, frequency, differential current, current balance, phase sequence, and distance relaying.

Large industrial systems, and those with internal power generation in parallel with the utility supply, may require almost all of the relays mentioned to control only one or two critical circuit breakers. Because of the special knowledge required of relay characteristics and of the proper design of a protective system, only the basics of system protection can be presented here.

Any power system should have, as a minimum, overcurrent protection. Overcurrent, whether occurring gradually from overloading or suddenly from short circuits, should result in some action to remove the cause of the overcurrent or to isolate (deenergize) that part of the system having the fault. Owing to the extreme hazard presented by short-circuit currents, it is mandatory that they be removed quickly. This can normally be accomplished by fuse or circuit-breaker operation at one or more locations. However, interruption at only one location will usually result in the least system disturbance. Thus overcurrent protective equipment should be selective in its operation. Overcurrent protection, like any kind of system protection, should have two components: the primary component, which should function first in removing the faulted equipment, and the "back-up" component, which functions only when the primary protective component has failed. Proper coordination of the short-circuit protective devices isolates only the smallest possible part of the system when a fault occurs. For this to be accomplished, these protective devices must be selective in operation so that the one closest to the fault (on the source side) will operate first. If any device should fail to function, the next closest device (on the source side) should open the circuit. This requires that at least two protective devices be in series between any fault or overload and the power source.

The fuse or circuit breaker (interrupting device) or a combination of fuses and circuit breakers must be capable of withstanding the voltages and currents it or they will be subjected to as well as being able to interrupt safely any current that the system is capable of delivering at that location during a fault. Therefore, for proper selection of the needed interrupting devices, the fault current existing within the system due to faults at various locations must be known. Other parts of the system, such as the cables, bus ducts, transformers, disconnect switches, and local generators, must also be capable of carrying the fault currents until the fault is cleared or isolated by the protective equipment. Thus the real starting point of planning system overcurrent protection is calculating the short-circuit (fault) currents.

Short-Circuit Calculations

The current that flows during a short circuit or fault is determined by the impedance existing between the point of the fault and the power source or sources supplying the system, and is not directly a function of the load on the system.

During a fault, the sources of the fault current are not only generators but also motors or synchronous condensers† (synchronous capacitors) operating on the system at the time of the fault. All rotating machines deliver fault current, which decreases with time after the initiation of the fault. To account for this decrease in fault current, each machine, especially the generators, are represented electrically by a constant voltage source in series with a reactance having different values. These values depend on the time from the start of the fault.

For fault calculations, the reactance of the generator is represented by three different values. These values are called subtransient, transient, and steady-state synchronous reactance and are symbolized by X_d'', X_d', and X_d, respectively. For fault calculations involving the current within the first few cycles after the fault, the subtransient reactance X_d'' is used for the generator circuit model. For fault calculations involving currents after the first few cycles up to one half second after the fault, the transient reactance X_d' is used for the generator circuit model; for fault calculations several seconds or longer after the fault, steady-state synchronous reactance X_d is used in the generator circuit mode.[4] This is illustrated in Fig. 67.14*a*.

In addition to the variable reactance characteristic of a generator subjected to a short circuit, the instantaneous variation of the current is not solely sinusoidal alternating current but contains a decreasing DC component. This is illustrated in Fig. 67.14*b*. This DC component or offset exists whenever an inductive current is subjected to a change in AC excitation. However, the amount of the DC component is a function of the resistance–inductance ratio of the circuit and the point on the AC cycle when the switching occurs. This DC component decays rapidly but should be considered when determining the withstand current required of a device. Also when certain fast-acting interrupting devices are used, the DC component should be included in selecting the rating.[5]

The different reactance values used for the generator equivalent circuit are also used for synchronous motors but will in general be different numerical values. However, the motor cannot sustain a steady-state current, since only its inertia and that of its load provide the source for the current and the current decreases as the motor slows down.

In the case of the induction motor, excitation for the rotor comes from the AC supply, and under fault conditions this excitation dies out rapidly and disappears in a few cycles. Therefore induction motor contributions are associated with the subtransient synchronous reactance X_d''; for calculating the interrupting duty, $1.5X_d''$ is used in the circuit model.[6] The fault current contribution then will be about two thirds of locked motor starting current at full voltage.

Capacitor discharge currents (except synchronous condensers which are treated as synchronous motors) are generally not included in fault current calculations, as this is of a higher frequency nature (Ref. 2, p. 100).

Because of the special transformation of the electrical variables required to calculate unbalanced faults, only balanced or symmetrical fault calculations will be explained here. For distribution system faults, generally the greatest fault current occurs for the balanced three-phase fault (Ref. 2, p. 100). Therefore values obtained from these calculations will be on the safe side when selecting the proper equipment. However, problems may arise when using balanced fault current calculations for setting the sensitivity of protective relays.

All fault calculations are essentially based on the application of a circuit theorem called Thévenin's theorem (see Chapter 19). Essentially, the theorem states that at any location in an electrical network, the network from that location back to the sources can be represented by one equivalent voltage source in series with one equivalent impedance. The equivalent voltage source is the open-circuit voltage that appears at the location of interest when the remainder of the network is disconnected. The equivalent series impedance is the impedance looking into the network at the location of interest with all sources replaced by their internal impedances. Figure 67.15 illustrates this.

†A synchronous condenser is essentially an overexcited synchronous motor with no load connected to the shaft, which supplies reactive power to the system.

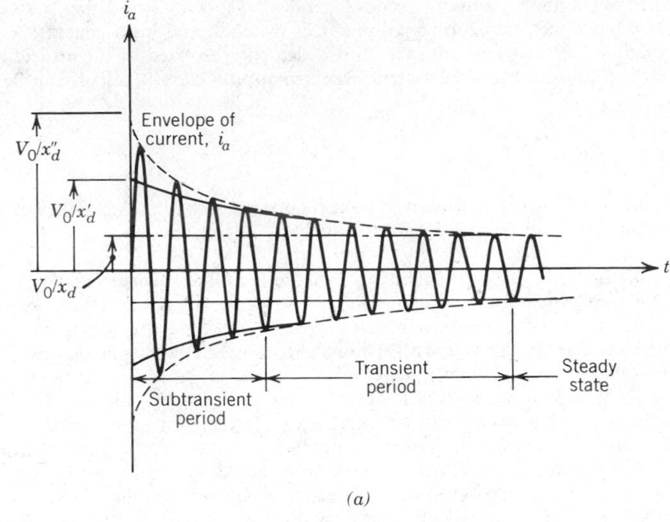

(a)

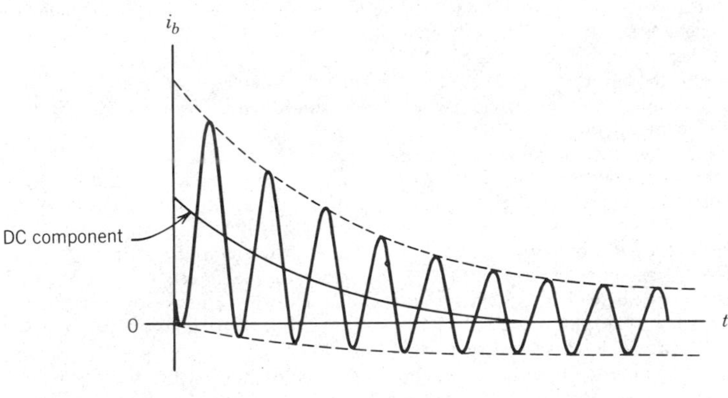

(b)

Fig. 67.14 Armature current vs. time (a) in a short-circuited generator and (b) showing DC offset in a short-circuited generator.

Per-Unit Calculations. Calculation of system electrical quantities under normal and fault conditions is greatly simplified by use of the per-unit technique. This involves selecting a base value for two of the system quantities and then calculating the base values for the remaining electrical quantities. The per-unit value of a given quantity is the ratio of that quantity to the base value of that same quantity. Of the basic electrical quantities involved in the calculations, such as voltage, current, power, and impedance, only two can have arbitrary values assigned to them, since all the others are related to these two. These basic relationships are

$$\begin{aligned} \mathrm{VA} &= VI \\ Z &= V/I \\ P &= VI \cos \Theta \\ Q &= VI \sin \Theta \\ R &= Z \cos \Theta_z \\ X &= Z \sin \Theta_z \end{aligned} \qquad (67.4)$$

The base values as well as the per-unit quantities must satisfy the same basic relationships as the

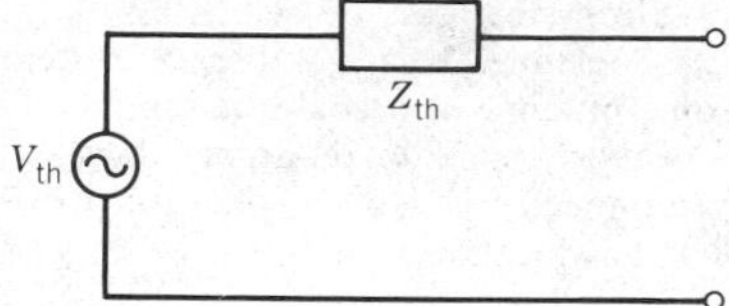

Fig. 67.15 Thévenin's equivalent circuit of an electrical system at a fault.

original quantities. That is, for single-phase circuits

$$\mathrm{VA}_{\mathrm{base}} = V_{\mathrm{base}} I_{\mathrm{base}}$$

$$Z_{\mathrm{base}} = \frac{V_{\mathrm{base}}}{I_{\mathrm{base}}} = \frac{(V_{\mathrm{base}})^2}{\mathrm{VA}_{\mathrm{base}}} \tag{67.5}$$

$$\mathrm{VA}_{\mathrm{pu}} = V_{\mathrm{pu}} I_{\mathrm{pu}}, \qquad Z_{\mathrm{pu}} = \frac{V_{\mathrm{pu}}}{I_{\mathrm{pu}}}, \qquad \text{etc.}$$

Since real power P and reactive power Q are respectively the real part and the imaginary part (j part) of the complex volt-amperes, the base for VA, P, and Q is the same. Likewise R and X, being the real and imaginary parts of complex Z, the base for Z, R, and X is the same. From the definition of per unit, the following relations exist.

$$V_{\mathrm{pu}} = \frac{V_{\mathrm{actual}}}{V_{\mathrm{base}}} \qquad I_{\mathrm{pu}} = \frac{I_{\mathrm{actual}}}{I_{\mathrm{base}}} \qquad Z_{\mathrm{pu}} = \frac{Z_{\mathrm{actual}}}{Z_{\mathrm{base}}}$$

$$\mathrm{VA}_{\mathrm{pu}} = \frac{\mathrm{VA}_{\mathrm{actual}}}{\mathrm{VA}_{\mathrm{base}}} \qquad P_{\mathrm{pu}} = \frac{P_{\mathrm{actual}}}{\mathrm{VA}_{\mathrm{base}}} \qquad Q_{\mathrm{pu}} = \frac{Q_{\mathrm{actual}}}{\mathrm{VA}_{\mathrm{base}}} \tag{67.6}$$

$$R_{\mathrm{pu}} = \frac{R_{\mathrm{actual}}}{Z_{\mathrm{base}}} \qquad X_{\mathrm{pu}} = \frac{X_{\mathrm{actual}}}{Z_{\mathrm{base}}}$$

For these equations, the actual and base values of any given quantity must be in the same units. For equipment such as transformers and generators, the base values used in determining a given per-unit impedance are the rated nameplate values. That is, for a single-phase 100-kVA 480 Y/120-V transformer with 4% (0.04 per-unit) impedance, the base values to determine Z_{base} are the rated voltage and rated kilovolt-ampere. The base ohms referred to the high side is calculated as

$$Z_{\mathrm{base}\,H} = \frac{V_{\mathrm{base}\,H}}{I_{\mathrm{base}\,H}} = \frac{(V_{\mathrm{base}\,H})^2}{\mathrm{VA}_{\mathrm{base}}} = \frac{(480)^2}{100 \times 10^3} = 2.304\ \Omega \tag{67.7}$$

Therefore from the definition of per unit, the actual value of the transformer impedance, when referred to the 480-V side, is $0.04(2.304) = 0.0922\ \Omega$. One big advantage of the per-unit or percent form is that the transformer impedance in per unit is the same when referred to either side of the transformer. This is so because the base values must satisfy the same relationships as the actual values.

For example, the 100-kVA 480 Y/120-V transformer has $0.0922\ \Omega$ when referred to the high side but will have $0.0922/(4)^2 = 0.00576\ \Omega$ when referred to the low side. Now calculating the impedance referred to the low side from the per-unit definition we get

$$Z_{\mathrm{base}\,L} = \frac{V_{\mathrm{base}\,L}}{I_{\mathrm{base}\,L}} = \frac{(V_{\mathrm{base}\,L})^2}{\mathrm{VA}_{\mathrm{base}}} = \frac{(120)^2}{100 \times 10^3} = 0.0144\ \Omega \tag{67.8}$$

$Z_{\mathrm{ohms}\,L} = 0.04(0.0144) = 5.76 \times 10^{-3}\ \Omega$, which is the same value calculated previously. Another advantage of per-unit calculations is that per-unit impedance value is essentially the same for equipment of a given class, even with a wide range of ratings.[7] This makes it easier to detect mistakes in the calculations.

Since the impedance of open-wire lines and cables is given in ohms, it is necessary to convert these to per unit. To take full advantage of the convenience associated with per-unit calculations it is necessary that the ratio of base voltages of each section of the system be the same as the transformer turns ratio interconnecting each section of the system. Unlike the base voltage, the kilovolt-ampere base of the system is the same throughout; therefore the base currents vary inversely as the base voltages.

The kilovolt-ampere base of the system can be selected arbitrarily but is usually selected to simplify calculations as much as possible. If a number of system components have the same kilovolt-ampere rating, this or some multiple of it could be used as the base.

Regarding cable calculations, if the system kilovolt-ampere base is 1000 kVA and the voltage base of the section where the cable is connected is 2400 V, then for a cable having 1 – Ω resistance the per-unit resistance for the cable at this location would be

$$R_{\mathrm{pu}} = \frac{1\,\Omega}{Z_{\mathrm{base}}}, \qquad \text{where} \quad Z_{\mathrm{base}} = \frac{(2400)^2}{1000 \times 10^3} = 5.76\,\Omega \tag{67.9}$$

Thus $R_{\mathrm{pu}} = 1/5.76 = 0.174$ per unit on a 1000-kVA and 2400-V base. If the cable were used at a location in the same system where the voltage was 7200 V, then the per-unit resistance would be

$$R_{\mathrm{pu}} = \frac{1(1000 \times 10^3)}{(7200)^2} = \frac{1}{51.84} = 0.0193 \text{ per unit} \tag{67.10}$$

which is much smaller, since the base voltage is larger.

Since the base voltage and base kilovolt-ampere can be arbitrarily selected, it is necessary to convert per-unit impedance calculated on one base to per-unit values calculated on the system base. This is accomplished by the following equation:

$$\frac{Z_{\mathrm{pu}\,1}}{Z_{\mathrm{pu}\,2}} = \left(\frac{\text{kVA base 1}}{\text{kVA base 2}}\right)\left(\frac{\text{kV base 2}}{\text{kV base 1}}\right)^2 \tag{67.11}$$

where the subscript 2 could represent the given values and 1 the system values. This means the per-unit impedance value for a 4%-Z 100-kVA 2400-V transformer in the equivalent circuit of a system with 200 kVA, 2400 V base values would be

$$Z_{\mathrm{pu}\,1} = 0.04\left(\frac{200}{100}\right)\left(\frac{2400}{2400}\right)^2 = 0.08 \text{ pu}$$

For balanced faults, all calculations are performed on a per-phase basis, that is, line-current and line-to-neutral voltages are used. For three-phase calculations, the relationships that relate the voltage, current, and volt-amperes must also be used with the base quantities. However, here the per-unit calculations are made as if they were single phase. The basic three-phase relationships are

$$\begin{aligned} V_{\mathrm{L\text{-}L}} &= \sqrt{3}\, V_{\mathrm{L\text{-}N}} \\ \mathrm{VA}_{3\phi} &= \sqrt{3}\, V_{\mathrm{L\text{-}L}} I_L = 3 V_{\mathrm{L\text{-}N}} I_L \end{aligned} \tag{67.12}$$

where $V_{\mathrm{L\text{-}L}}$ = line-to-line voltage
$V_{\mathrm{L\text{-}N}}$ = line-to-neutral voltage
I_L = line current.

As an aid in performing the short-circuit calculations, the following procedure is suggested:

1. Draw a one-line diagram of the system showing the major components such as transformers (both power and instrument), circuit breakers, distribution lines, and cables; all motors 50 hp and over and generators within the facility, if any; and the utility supply connection. See Fig. 67.16.

2. Select a kilovolt-ampere base and voltage base and calculate the per-unit impedance of each significant system component referred to this base. The voltage base of each section of the system must have the same ratio as the line-to-line voltage ratio of the transformer interconnecting them.

3. Using the arrangement on the one-line diagram, draw an impedance diagram indicating the per-unit impedance value on the system base of each component. The impedance values used for generators and motors are determined by the time range for which the currents are to be calculated. For the first few cycles, X_d'' is used; for 3 to 30 cycles, X_d' is used; and after a second or two, X_d is used.[4] See Fig. 67.17. The utility is represented by a voltage source, usually 1 per unit if the open-circuit voltage is not known, and a series impedance, which has a magnitude that is the reciprocal of the short-circuit capability when expressed in per unit on the system kilovolt-ampere base.

4. Select the location where the fault current is desired and, using network reduction equations, calculate the impedance between the fault location and the reference terminal of the sources. Here sources are all generators and motors 50 hp or larger.[5] The fault current in per unit then is the open-circuit voltage in per unit before the fault, divided by the total impedance from the fault to the reference terminal. The fault current at various locations in the system is determined by using the proper equations for currents in various branches of the network. Likewise, the voltages at various points can also be determined.

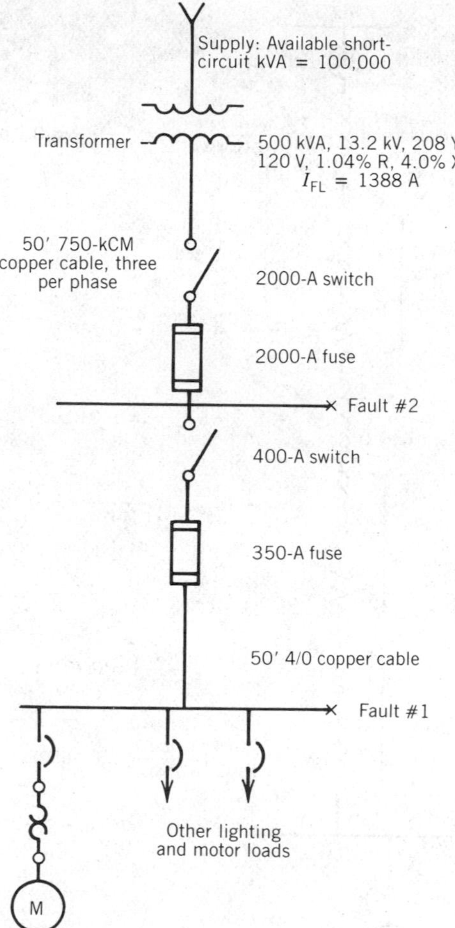

Fig. 67.16 One-line diagram of a feeder circuit.

5. Depending on the particular interrupting device, various factors are recommended for determining the momentary and interrupting currents. These factors are defined in Ref. 5.

The preceding procedure is best illustrated by a simplified example.

Example. The one-line diagram of a hypothetical power distribution system is shown in Fig. 67.16.

The next step is to express all of the impedances of each component in per unit on a common system kilovolt-ampere base and voltage base level for each system part. Here, choose the system kilovolt-ampere base as 10,000 and the voltage on the utility side as 13.2 kV. Then the current base on the 208-V side is $I_{\text{base}} = 10{,}000/\sqrt{3}\,(0.208) = 27{,}760$ A and the impedance base is

$$Z_{\text{base }L} = \frac{(0.208)^2}{10} = \frac{(\text{kV})^2}{\text{MVA}} = 0.00433\ \Omega \tag{67.13}$$

Calculations. Equivalent short-circuit impedance of the utility, assuming all reactance, is

$$X_u = \frac{10{,}000}{100{,}000} = 0.1\text{ pu} = \frac{\text{system kVA base}}{\text{short-circuit kVA}} \tag{67.14}$$

For the transformers with 4% X and 1.04% R, the per-unit reactance and resistance on the system

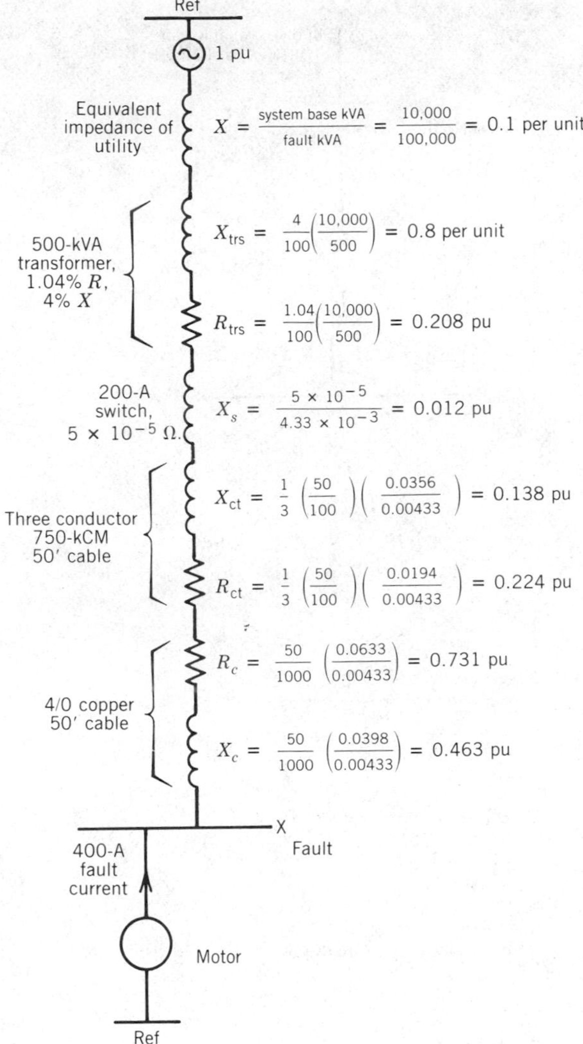

Fig. 67.17 Impedance diagram and summary of calculations for a short-circuited feeder. pu, per unit; trs, transformer; R, resistance; X, reactance; ct, identifies cable resistance and reactance.

base are

$$X_{\text{trs}} = X_{\text{tr}} = \left(\frac{\text{system kVA}}{\text{transformer kVA}} \right) = \left(\frac{4}{100} \right) \left(\frac{10{,}000}{500} \right) = 0.8 \text{ per unit} \tag{67.15}$$

$$R_{\text{trs}} = \left(\frac{1.04}{100} \right) \left(\frac{10{,}000}{500} \right) = 0.208 \text{ pu} \tag{67.16}$$

For a 2000-A switch with 5.10^{-5} Ω reactance,[8] the per-unit value on the system base is

$$X_s = \frac{5 \times 10^{-5}}{4.33 \times 10^{-3}} = 0.012 \text{ pu}$$

For 50 ft of 3 conductor 750-kCM conductor cable with 0.0194 Ω resistance and 0.0356 Ω reactance per 1000 ft,[8] the per-unit values on the system base are:

$$R_{cc} = \frac{l}{1000}\left(\frac{R}{Z_{\text{base}}}\right) = \left(\frac{50}{1000}\right)\left(\frac{0.0194}{0.00433}\right) = 0.244 \text{ pu each conductor} \tag{67.17}$$

$$R_{ct} = \frac{0.224}{3} = 0.075 \text{ pu total for cable} \tag{67.18}$$

$$X_{ct} = \frac{1}{3}\left(\frac{50}{1000}\right)\left(\frac{0.0365}{0.00433}\right) = 0.138 \text{ pu} \tag{67.19}$$

For the 400-A switch with 8×10^{-5} Ω reactance,[8] the per-unit value on the system base is

$$X_s = \frac{8 \times 10^{-5}}{4.33 \times 10^{-3}} = \frac{X_s\ \Omega}{Z_{\text{base}}} = 0.018 \text{ pu} \tag{67.20}$$

For 50 ft of 4/0 conductor cable with 0.0633 Ω resistance and 0.0398 Ω reactance,[8] the per-unit values on the system base are

$$R_{ct} = \frac{50}{1000}\left(\frac{0.0633}{0.00433}\right) = 0.731 \text{ pu} \tag{67.21}$$

$$X_{ct} = \frac{50}{1000}\left(\frac{0.0398}{0.00433}\right) = 0.463 \text{ pu} \tag{67.22}$$

For the motor load, assume that 50% of the feeder rating serving the lighting and motor loads are induction motors. For a 4/0 cable, the feeder rating is approximately 200 A. The motor contribution will be about equal to two thirds of the locked rotor current at full voltage and is assumed to be about four times the full load current in this example.[5] This gives $I_m = 0.5(4)(200) = 400$ A.

In calculating the short-circuit current, the contribution from the utility in per unit is

$$\begin{aligned} I_{\text{scu}} &= \frac{1}{Z_{\text{total}}} \\ Z_{\text{total}} &= R_{\text{total}} + jX_{\text{total}} \\ R_{\text{total}} &= 0.208 + 0.224 + 0.463 = 0.895 \text{ pu} \\ X_{\text{total}} &= 0.1 + 0.8 + 0.012 + 0.138 + 0.731 = 1.781 \\ |Z_{\text{total}}| &= |0.895 + j1.781| = 1.993 \text{ pu} \\ I_{\text{pu}} &= 1/Z_{\text{total}} = 0.502 \text{ pu} \end{aligned} \tag{67.23}$$

The utility contribution in amperes is $I_{\text{pu}}(I_{\text{base}}) = 1/1.993(27{,}760) = 13{,}929$ A. Motor load contribution is 400 A. Total short-circuit current to the fault is

$$13{,}929 + 400 = 14{,}329 \text{ A} \tag{67.24}$$

The current to point 1 (fault 1) in Fig. 67.16 is greatly limited by the impedance of the 4/0 cable. If the fault had occurred at point 2 (fault 2), the current contribution from the utility would have been

$$I_{\text{scu}} = \frac{1}{1.120}(27{,}760) = 24{,}690 \text{ A} \tag{67.25}$$

If only the reactance values had been used in calculating the short-circuit current, the fault current at point 1 would have been

$$I_{\text{sc}1} = 400 + \frac{1}{1.781}(27{,}760) = 15{,}587 + 400 = 15{,}978 \text{ A}$$

and at point 2

$$I_{\text{sc}2} = 400 + \frac{1}{1.05}(27{,}760) = 26{,}838 \text{ A}$$

To account for the DC offset of the instantaneous short-circuit current, certain multiplying factors are applied to the current calculated from the impedance. The multiplying factors are applied in the selection of protective devices such as circuit breakers and fuses. These factors and an explanation and discussion of a simplified E/X method are given in Refs. 5 and 6.

67.5 APPLICATION OF OVERCURRENT PROTECTIVE DEVICES

An adequate protective device should provide the capability of both the normal switching functions of opening and closing a circuit under normal loads and the function of interrupting safely the short-circuit currents.

Circuit-opening devices that provide one or both of these capabilities are (1) circuit breakers, (2) fuses, (3) load break switches, (4) interruptor switches, (5) disconnect switches, and (6) contactors. Only the first two provide automatic operation to clear fault currents, and only the circuit breaker provides both normal switching function and fault interrupting capability. Interruptor switches and disconnect switches are manually operated devices used for circuit and equipment isolation on deenergized circuits. Interruptor switches are suitable for opening and closing normal load currents. The load break switch is normally designed to carry 1.5 times its rated current, but interrupting only its rated current. Within its rating, a circuit-opening device should be capable of carrying and opening its normal current, safely interrupting any current, and being safely closed on any current.

Circuit breakers function by separating a pair of electrical contacts in some medium to help interrupt the arc. Circuit breakers are capable of being remotely operated, with relays being the detecting device in medium and high-voltage systems, while low-voltage power circuit breakers incorporate a built-in direct-acting series trip mechanism.

Certain parameters must be satisfied in selecting the proper low-voltage power circuit breaker. These parameters are rated voltage, rated maximum design voltage, rated maximum voltage, rated frequency, rated continuous current, rated short-time circuit, rated short-circuit interrupting current, and rated control coil (if one exists) voltage. This ensures that the circuit breaker is capable of closing, carrying, and interrupting the largest possible fault current at its circuit location. Its rating must be equal to or greater than the available short circuit currents at that location.

In fuses, the fusible element provides the detecting function and the interrupting function, but not the switching function. Fuses may be used instead of circuit breakers in certain applications. Fuses are classified as either current limiting or noncurrent limiting, with the higher interrupting ratings in the current-limiting class. Current-limiting fuses act to limit the current in less than one quarter cycle to a value much less than the available peak current at that location. The degree of current limiting depends on the relative magnitude of the available short-circuit current and the continuous rating of the fuse. Current-limiting fuses are also very fast in operation at high values of fault currents. In comparing fuses and circuit breakers, there are some advantages as well as disadvantages. The main advantages of fuses are:

1. Interrupting ratings are greater than those of circuit breakers.
2. They are inexpensive.
3. They are mechanically simple.
4. They are small.
5. Current-limiting reduces mechanical and thermal stresses on components.
6. They are easy to coordinate, since their melting and clearing characteristics are more consistent.
7. They are fail-safe devices, while deterioration of a breaker may cause unsafe operation (Ref. 2, p. 136).

Some disadvantages of fuses are:

1. Failure of a fuse in a three-phase circuit may result in single-phase operation of AC motors with subsequent damage to the motor.
2. They are not capable of remote-controlled operation.
3. They are hazardous to personnel during replacement of fuse elements, unless draw-out fused switches are used.
4. It is possible to replace blown-out fuses with the wrong-size rated fuse element.
5. One fuse can blow in a three-phase circuit, reducing the magnitude of the current in the other phases to the extent that those fuses will not clear.
6. Replacement fuses must be stocked (Ref. 2, p. 136).

Accomplishment of the objective of isolating only the smallest part of the system containing the fault requires coordination of the characteristics of the interrupting devices. Selective coordinating of protective devices is obtained by selecting the protective device closest to the fault to operate the fastest, with progressively slower operating times at the same current for protective devices further from the fault (closer to the source). Which device is the closest is probably best determined by plotting the time-vs.-current characteristic curves of all protective devices in series toward the source on the same sheet of graph paper, usually log-log coordinate paper, as shown in Fig. 67.18. The curves must be such that transformer inrush, normal load swings, and motor starting currents do not trip the breakers or cause melting of the fuse elements. A necessary condition for selective coordination is that the time-vs.-current characteristics do not cross over within the available

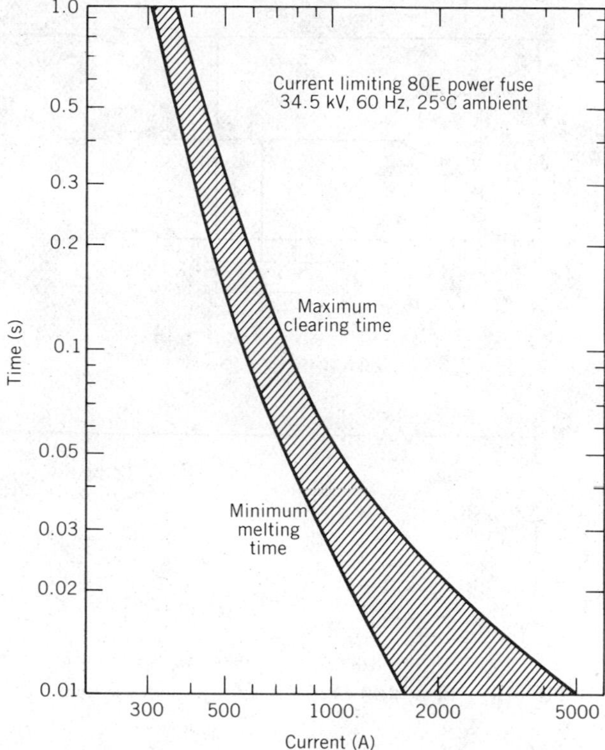

Fig. 67.18 Time-current characteristic of a power fuse.

short-circuit current range. To allow for the difference between minimum and maximum clearing time of circuit breaker and the difference between the minimum melting current and maximum total clearing time of fuses, some clear space must exist between the time-vs.-current characteristic plots of adjacent protective devices.

For fuse coordination, the maximum total clearing time of the downstream fuse must fall below the minimum melt characteristic of the next upstream fuse for all currents up to the available fault current at that location.

For current-limiting fuses, selective coordination is obtained when the *total* clearing energy of the load fuse is less than the *melting* energy of the next fuse toward the source. Extensive coverage of these procedures is given in Refs. 8 through 10.

67.6 OVERVOLTAGE PROTECTION

Except for lightning, most causes of overvoltages are within the system itself. The most prominent of these are (1) static charges, (2) contact with other high-voltage systems, (3) resonance effects of series inductive and capacitive circuits, (4) intermittent grounds and repetitive restrikes, (5) switching surges, (6) forced current zero interruptions, (7) autotransformers under abnormal conditions, and (8) lightning.

The problem of overvoltages due to resonance cannot be discussed without some discussion of system grounding. A system ground is a connection to ground of one of the *current-carrying* conductors of a distribution system. This connection can be an intentional one, such as a solid conductor or an impedance, or an unintentional one consisting of the distributed capacitance of a conductor to ground. In the latter case, the system is called ungrounded, although it is in reality grounded by the distributed capacitance.

Resonance can occur in the so-called ungrounded system whenever an inductive reactance is connected to ground. Resonance exists between the inductive reactance and the distributed capacitance. If the total distributed capacitance line to neutral is X_{c0}, then a circuit that is equivalent to

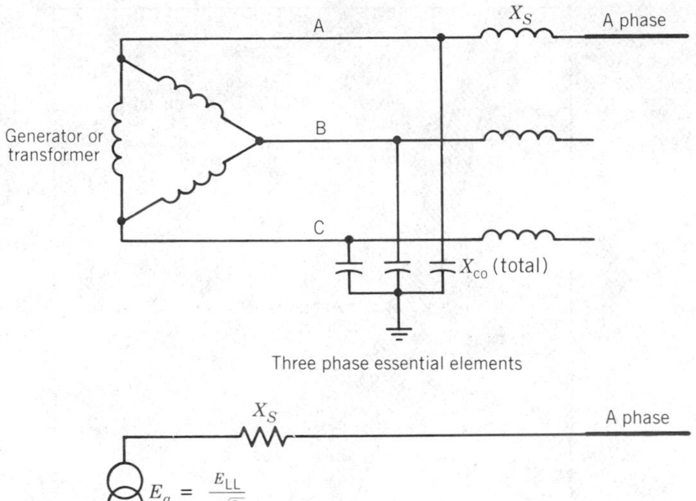

Three phase essential elements

X_S

A phase

$E_a = \frac{E_{LL}}{\sqrt{3}}$

$X_{co/3}$

Equivalent circuit referred to a phase conductor

Fig. 67.19 Equivalent circuit of an ungrounded system.

such an ungrounded system is given in Fig. 67.19. When an external impedance such as a fault Z_f occurs, if Z_f is inductive the voltage E_f is approximately that given by the following equation:[1]

$$E_f = \frac{Z_f}{X_{c0/3} - Z_f}\left(\frac{V_{\text{L-L}}}{\sqrt{3}}\right) \tag{67.27}$$

When Z_f approaches $X_{c0/3}$ in value, high overvoltages can exist in the system. This will frequently cause multiple or simultaneous equipment failures on ungrounded systems. Proper system grounding will reduce the overvoltages significantly to tolerable values.

An "arcing ground," or repetitive intermittent ground, can result in overvoltages. Again this results from the existence of the distributed capacitance X_{c0}, and the overvoltage phenomenon is similar to that associated with restrikes of switches interrupting capacitive circuits (Ref. 11, p. 286). Proper grounding can provide a very effective method of controlling these overvoltages.

Switching surges can result from energizing of transmission lines, disconnection of two systems that have become unsynchronized, interruption of system faults, and open-cycle autotransformer starting of motors. Certain abnormal conditions associated with switching may produce moderate overvoltages. These overvoltages are due to the resonance of the circuit effective inductance L and capacitance C and oscillate at a frequency given by the equation $f = 1/2\pi\sqrt{LC}$ hertz. These oscillations are normally much higher than the system frequency and are superimposed on the system voltage. To reduce these overvoltages, surge protective capacitors or metal oxide varistors (MOVs) may be utilized in the system (Ref. 1, p. 43, Sec. 2).

Current-limiting fuses result in forcing current zero interruption and can lead to overvoltages associated with the circuit inductance. Proper application of such current-limiting fuses can control this problem.

The unfaulted phases of a three-phase system subjected to a single line-to-ground fault can experience voltages from line to ground of not more than $\sqrt{3}$ times the normal system line-to-neutral voltage on grounded neutral systems. For the "effectively grounded" system, this is limited to 1.4 times. Even ungrounded systems, except in the rarest of cases, experience a voltage on the unfaulted phases of only $\sqrt{3}$ times the normal line-to-neutral voltage (Ref. 1, p. 43, Sec. 2).

The most destructive overvoltage is caused by lightning. The overvoltage may be caused by a direct stroke to the line or by electrostatic induction from a stroke to the earth in the vicinity of the line. This can result in a large quantity of charge existing on the line in the vicinity of the stroke. The action of these charged particles is to repel each other, resulting in a traveling wave of voltage and current in opposite directions along the line. This wave of charge travels at a speed determined by the line characteristics, but is generally slightly less than the speed of light. As with all wave phenomena, when the traveling wave encounters a change in line impedance such as a transformer, a reflected wave results. This is superimposed on the original voltage wave, resulting in increased voltage at that point. If the surge impedance at the junction is low, the resulting current decreases the voltage. This is the basic function of lightning arrestors which, when selected properly, protect the equipment from damaging voltage magnitudes.

Power system equipment experiences various types of overvoltages, and these can be divided as to the length of time they exist. To determine the ability of the equipment to withstand these overvoltages, two types of dielectric (insulation) tests are performed. One is the so-called low-frequency (60-Hz) test, usually of one minute duration, and the other is the "impulse" test of a few microseconds duration. The impulse test is used to determine the ability of a piece of equipment to survive lightning overvoltages.

The waveform of the voltage or current for impulse testing is a sawtooth form characterized by a 1.5-μs rise time to its crest value and a 40-μs decay to 50% of the crest value. The crest value of this wave is called the basic insulation level (BIL) of the equipment. To simplify the design and application, a series of standard BILs has been established for various types of electrical system equipment. Tables showing the BILs for some system components are given in Ref. 1.

To protect equipment from overvoltages of various kinds, especially lightning overvoltages, devices such as lightning arrestors, surge protective capacitors, and transient voltage suppressors are utilized. The lower the voltage rating, the better will be the protection provided the insulation of the equipment.

Lightning arrestors are generally available in two types, valve and expulsion. The valve arrestor consists of a spark gap in series with a "valve" element. The valve element has the characteristic of having low resistance for the higher overvoltages across it, and a high resistance at the normal system voltages. The ideal arrestor is one that presents an open circuit for the normal voltage and a short circuit for voltages above its rating. The arrestor must be capable of draining the lightning charge to ground rapidly but interrupting the "follow current" available from the system supply voltage. The spark gap and valve element are designed to interrupt the follow current and quickly restore the high resistance of the arrestor to the normal voltage. If the system voltage is too high, the arrestor gap will not reseal and arrestor failure will occur owing to the follow current. Therefore it is important that arrestors are not applied where the system line-to-ground voltage exceeds the voltage rating of the arrestor.

The expulsion arrestor utilizes the out-gasing characteristics of certain materials which, when exposed to an electric arc, interrupt the follow current. The arrestor is then in series with a spark gap to provide the necessary voltage rating of the arrestor. Expulsion arrestors must be used in the open, where the gaseous discharge will not present a safety problem.

The problem of selecting the arrestor voltage rating is one of determining the maximum sustained line-to-ground voltage that will occur at a location and then choosing the closest arrestor rating that is still greater than the sustained voltage value determined. The lower the arrestor rating, the lower the sparkover potential and *IR* voltage drop and thus the better the protection of the insulation of the equipment. Determination of the maximum sustained overvoltage requires special analytical techniques, utilizing symmetrical components that are beyond the development here. Such calculating methods are given in Refs. 7 and 12 through 14.

Since lightning involves very rapid rates of rise of voltage (high voltage gradients), arrestors should be located close to the equipment being protected. This requires adequate grounding of the arrestors utilizing solid and direct interconnection of arrestor, system, and apparatus grounds. Conductors connecting arrestors to the line and to grounds should be as straight as possible with no or as few bends as possible, because of the traveling-wave phenomenon of lightning.

Rotating machine windings have low impulse voltage strength, and with the steep wavefront of the voltage surge, the front end turns experience greater voltages stress. Therefore surge protective capacitors should be used in connection with arrestors protecting rotating machines, to reduce the steepness of the voltage wavefront. In general, each application must be analyzed individually to determine the proper selection of protective equipment.

67.7 EQUIPMENT AND SYSTEM GROUNDING

As mentioned previously, every system is grounded, either intentionally through a connection to ground or unintentionally through the distributed capacitance to ground of the various components. Grounding usually refers to the intentional ground. A system ground is the connection to ground of one of the current-carrying conductors of a distribution system or of an interior wiring system.

The type of impedance involved in the ground connection designates the type of system ground. The main types of system grounding are (1) solid, (2) low resistance, (3) high resistance, (4) reactance, (5) resonant, and (6) ungrounded (distributed capacitance). The nature of the grounding has an effect on the magnitudes of the line-to-ground voltages that exist during steady and transient conditions.

In addition to the control of system overvoltages, intentional grounding of system neutrals provides for sensitive and quicker fault protection based on detecting ground currents. Circuit protection devices on grounded systems are arranged to remove the faulty part from the system regardless of the type of fault.

For low-voltage systems, the National Electrical Code requires the system to be solidly grounded except for a few cases. At one time, low-voltage systems were generally ungrounded, as some felt that certain advantages existed for such arrangement. Generally experience has indicated that grounded systems result in less stress on the insulation and less equipment failures. Some higher voltage systems (4160Y/2400 V) and above are operated ungrounded, but the recommended practice is to ground all systems either solidly or through reactances or resistances.[2] Resistance and reactance grounding is usually considered for generators and motors for reducing and limiting the line-to-neutral fault currents. If not limited, these currents could result in burning and melting of the faulted equipment and circuits and increased mechanical stresses. Additional discussions of these may be obtained from Refs. 1 and 15.

Equipment grounding is the solid connecting to ground of one or more *non*current-carrying metal parts of the wiring system or apparatus, such as metal conduits, outlet boxes, motor frames, and control enclosures. The main purpose of equipment grounding is to provide safety to personnel. The solid connections to ground prevent dangerous potentials from existing between the noncurrent-carrying parts and ground. The grounding means should limit these potentials to a maximum of 50 V and have sufficient capacity to carry the maximum ground fault currents without thermal distress or attendant fire hazard.

The grounding system consists of (1) grounding electrodes, (2) ground bus, and (3) grounding conductors. The grounding electrode is a conductor embedded in the earth, used to maintain the ground potential on the ground bus connected to it and to conduct the current to earth. The ground bus is used to obtain a uniform potential in a given area of the system and is tied to the grounding conductors, which are connected to the equipment frames and metallic enclosures. Only by intentionally grounding metallic enclosures in such a manner as to ensure adequate current-carrying capability and low value of ground fault circuit impedance can both electric shock hazard and fire hazard be avoided. Investigations by insurance companies and others bring out the necessity of making good electrical junctions between sections of conduit or cable raceways and enclosures. The only sure way is to use a grounding conductor *inside* the conduit or raceway. Any grounding conductor outside is not effective in shunting the short-circuit current away from the conduit (Ref. 2, p. 12).

Grounding electrodes may be existing underground conductors such as bare piping systems, metal building frameworks, well casings, steel piling, or other underground metallic structures. The other ground electrodes are "made electrodes" specifically designed for grounding purposes. Since the resistance to ground is so variable, depending on soil conditions and metallic surface conditions, it is important that ground resistance measurements be made initially and periodically thereafter. The NEC states that grounding electrodes must have less than 25 Ω resistance to earth. A method of measuring and calculating ground resistances is given in Ref. 15.

References

1 D. Brereton, ed., *Electric Equipment Specifications Manual*, Part III, AIA File No. 31-R, General Electric Co., 1959.

2 IEEE Standard 141-1969, *Electric Power Distribution for Industrial Plants*, IEEE Standards Board, New York, 1969.

3 IEEE Standard 141-1976, *Electric Power Distribution for Industrial Plants*, IEEE Standards Board, New York, 1976.

4 IEEE Standard 241-1974, *Recommended Practice for Electric Power Systems in Commercial Buildings*, IEEE Standards Board, New York, 1974.

5 ANSI/IEEE C37.010-1979, *Application Guide for AC High-Voltage Circuit Breakers Rated on a Symmetrical Current Basis*, IEEE Standards Board, New York, 1979.

6 ANSI/IEEE C37.5-1979, *Guide for Calculation of Fault Currents for Application of AC High-Voltage Circuit Breakers Rated on a Total Current Basis*, IEEE Standards Board, New York, 1979.

7 O. I. Elgerd, *Electric Energy Systems Theory, An Introduction*, McGraw-Hill, New York, 1971.

8 *Engineering Dependable Protection for an Electrical Distribution System*, Part I-A, "Simple

Approach to Short Circuit Calculations," AIA File No. 31 d6, Bussman Mfg., Div. McGraw-Edison Co., St. Louis, MO, 1968.

9 *Engineering Dependable Protection for an Electrical Distribution System*, Part II, "Selective Coordination of Overcurrent Protective Devices for Low Voltage Systems," AIA File No. 31 d6, Bussman Mfg., Div. McGraw-Edison Co., St. Louis, MO, 1969.

10 *Engineering Dependable Protection for an Electrical Distribution System*, Part III, "Component Protection for Electrical Systems," AIA File No. 31 d6, Bussman Mfg., Div. McGraw-Edison Co., St. Louis, MO, 1972.

11 D. Beeman, ed., *Industrial Power Systems Handbook*, McGraw-Hill, New York, 1955.

12 W. D. Stevenson, Jr., *Elements of Power System Analysis*, 3rd ed., McGraw-Hill, New York, 1975.

13 G. W. Stagg and A. H. El-Abiad, *Computer Methods in Power Systems*, McGraw-Hill, New York, 1968.

14 C. A. Gross, *Power System Analysis*, Wiley, New York, 1979.

15 IEEE Standard 142-1972, *Grounding of Industrial and Commercial Power Systems*, IEEE Standards Board, New York, 1972.

Bibliography

AIEE Committee Report, "Bibliography of Industrial System Coordination and Protection Literature," *IEEE Trans. Appl. Indus.* **82**:1–2 (Mar. 1963).

ANSI C114.1-1973/IEEE Standard 142-1972, *Recommended Practice for Grounding of Industrial and Commercial Power Systems*, IEEE Standards Board, New York, 1973.

ANSI/IEEE C37.04-1979, *American National Standard Rating Structure for AC High-Voltage Circuit Breakers Rated on a Symmetrical Current Basis*, IEEE Standards Board, New York, 1979.

Edison Electric Institute, *Underground Systems Reference Book*, Edison Electric Institute, New York, 1957.

Electric Power Research Institute, *Transmission Line Reference Book—345 KV and Above*, 2nd ed., Fred Weidner and Sons, 1982.

Fink, D. G., ed., *Standard Handbook for Electrical Engineers*, 11th ed., McGraw-Hill, New York, 1978.

General Electric, *Industrial Power Systems Data Book*, Engineering Planning and Development Section, Apparatus Sales Div., General Electric Co., Schenectady, NY.

IEEE Committee Report, *IEEE Trans. Pow. Appar. Syst.* **PAS-99**(1):99–107 (Jan. 1980).

IEEE JH2112-1, *Protection Fundamentals for Low-Voltage Electric Distribution Systems in Commercial Buildings*, IEEE Industry Applications Society, 1974.

IEEE Standard 242-1975, *Recommended Practice for Protection and Coordination of Industrial and Commercial Power Systems*, IEEE Standards Board, New York, 1975.

IEEE Standard 446-1980, *Recommended Practice for Emergency and Standby Power Systems for Industrial and Commercial Applications*, IEEE Standards Board, New York, 1980.

IEEE Standard 493-1980, *Recommended Practice for the Design of Reliable Industrial and Commercial Power Systems*, IEEE Standards Board, New York, 1980.

Mason, C. R., *The Art and Science of Protective Relaying*, Wiley, New York, 1956.

Peters, O. S., "Ground Connections for Electric Systems," National Bureau of Standards, Boulder, CO, Technologic Papers 108, 1918.

Proceedings of Conference for Protective Relay Engineers, Texas A&M University, 1947 to 1980.

Rubenberg, R., *Transient Performance of Electric Power Systems*, McGraw-Hill, New York, 1950.

Soares, E. C., *Grounding Electrical Distribution Systems for Safety*, Marsh, Wayne, NJ, 1966.

Stevenson, W. D., Jr., *Elements of Power System Analysis*, 4th ed., McGraw-Hill, New York, 1982.

Weeks, W. L., *Transmission and Distribution of Electrical Energy*, Harper & Row, New York, 1981.

Westinghouse Electric, *Electrical Transmission and Distribution Reference Book*, Westinghouse Electric Corp., East Pittsburgh, PA, 1950, 1964.

Westinghouse Electric, *Applied Protective Relaying, A New "Silent Sentinels" Publication*, Westinghouse Electric Corp., Relay Instrument Div., Newark, NJ, 1970.

CHAPTER 68

MOTORS, GENERATORS, AND ILLUMINATION

DANIEL D. LINGELBACH

Oklahoma State University
Stillwater, Oklahoma

Electric power was first generated on a commercial basis using direct current (DC) generators driven by steam engines or water power. Consequently, DC motors were used first in the power utilization equipment to replace horses for powering streetcars. These DC traction motors possessed inherent characteristics that made them suitable for this type of application.

The invention of the AC motor accelerated the development of alternating current systems. Later it became evident that AC electrical systems were more economical for generation and distribution than were DC systems.

Even though today essentially all electrical energy is generated in AC form, DC motors are being used in increasing numbers because of their advantages and flexibility in adjustable speed drives.

68.1 BASIC TYPES OF AC AND DC MACHINES

Both DC generators and motors are divided into essentially four types, depending on the manner in which the main magnetic field is excited. These types are separately excited, shunt, series, and compound. In the case of generators, the shunt, series, and compound are usually self-excited, that is, the armature voltage available from the residual flux in the field is used to energize the field in such a manner as to add to the residual flux. The generator is said to build up when certain speed and field resistance values are satisfied and the voltage builds up to near rated. Each of the different types of excitation produce different speed-torque and voltage-current characteristics in motors and generators, respectively.

AC machines are classified as to the manner in which the rotor receives its excitation. If the rotor excitation is DC, the machine is a synchronous machine, since to develop torque the rotor must run at the same speed as the armature or stator magnetomotive force (mmf) (called synchronous speed).

Unlike that of the synchronous motor, the rotor of the induction motor receives its excitation (induced voltage) from the armature (stator) magnetomotive force by relative motion between the rotor conductors and the air gap flux set up by the stator.

For this to occur requires that the rotor turn at a speed less than synchronous speed for motor operation and greater than synchronous speed for generator operation. The stator windings of the induction motor are similar to those of the synchronous machine, although synchronous machines can tolerate much larger air gaps between rotor and stator iron.

68.2 PERFORMANCE CHARACTERISTICS OF AC MACHINES

Polyphase Induction Motors

Polyphase induction motors have two basic types of rotor windings. One is called squirrel cage because the rotor conductors are shorted together at the ends. The other is wound like the stator but with slip rings and brushes to provide for connections to external components. Induction machines can operate as generators but have not found much application except in special cases, since the excitation has to come from the stator supply.

Squirrel-cage induction motors are the most common of the two types and are essentially constant-speed machines. The change in speed from no load to full load is only a few percent, depending on the motor design. Variations in the rotor design produce different torque, current, and power factor values as a function of speed as well as different efficiencies as a function of load. The designs for polyphase squirrel-cage motors with ratings from $\frac{1}{2}$ to 500 hp have been standardized and classified by the National Electrical Manufacturers Association (NEMA), and are known as NEMA design A, B, C, and D.

The essential differences in the four designs are illustrated in Fig. 68.1. The NEMA design B is the general-purpose single-speed AC motor with normal starting torque and normal starting current. The normal starting torque ranges from 70% to 275% of the full-load torque, depending on size and speed. Generally motors rated at lower speeds have smaller percent starting torques than those rated at higher speeds. Likewise, for a given speed motors rated at a higher horsepower have lower percent starting torques than the smaller horsepower motors (see Ref. 1, parts 1, 12). Therefore a 100-hp 1800-rpm motor has a smaller percent starting torque than a 10-hp 1800-rpm motor. Breakdown torque for the design B motor ranges from 175% to about 300% of full load for normal starting current. Normal starting current is about six times the full-load current. The design B motors are the lowest cost of all induction motors. They are used for essentially constant-speed applications that do not require frequent starting or reversing operations. Also they are not suitable for driving high-inertia loads. Typical applications are motor–generator sets, fan blowers, and centrifugal pumps.

The NEMA design A motor has the same starting, pull-up, and breakdown torque of the design B but with higher starting currents.

NEMA design C motors are designed for loads having high starting torques, such as in conveyors, crushers, reciprocating pumps, and compressors. For a given starting torque, the design C operates at a higher efficiency and power factor than the larger horsepower design B having the same starting torque (see Ref. 1, Sec. 3, p. 4). For design C, the starting torque ranges from 200% to 250%, with breakdown torques from 190% to 225%. The NEMA design D motor has a starting torque of not less than 275% for motors rated up to 175 hp. It is designed for accelerating high-inertia loads and for

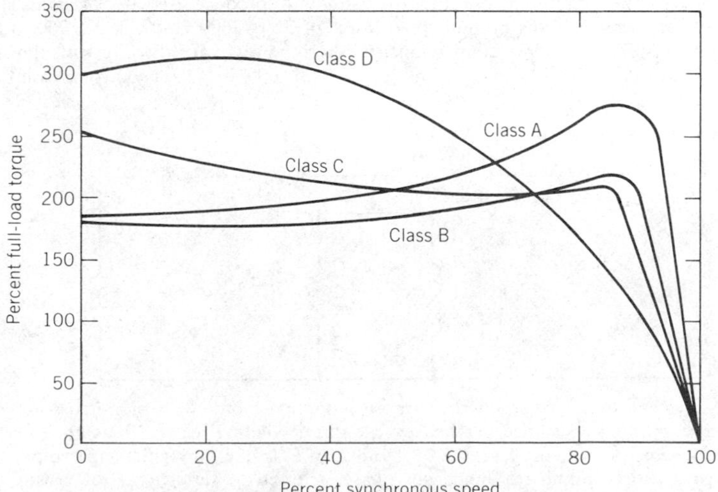

Fig. 68.1 Typical torque–speed characteristics of NEMA (National Electrical Manufacturers Association) design class squirrel-cage motors.

frequently starting and reversing operations. Typical applications are in conveyors, punch presses, elevators, small cranes, and hoists.

The performance of induction motors is computed from an equivalent circuit that is similar to that of a transformer but with the coupling between windings being a function of the speed. This speed effect is most conveniently given in terms of the relative speed normalized by rated speed and called slip. Actually it is slip in per unit. Slip is given by

$$s = \frac{n_s - n_r}{n_s} \tag{68.1}$$

where n_s is the synchronous speed of the armature stator magnetomotive force given in revolutions per minute, as $n_s = 120f_s/P$, with f_s the stator frequency in hertz and P the number of stator poles, and n_r is the actual rotor speed in revolutions per minute. The equivalent circuit in terms of slip for a general-purpose induction motor is given in Fig. 68.2. Voltages in the figure are all per phase (line to neutral), currents are line currents, and impedances are all referred to the stator. The subscript s refers to the stator and r refers to the rotor.

To obtain the mechanical power output, P_0, the mechanical friction and windage loss must be subtracted from the mechanical power P_m developed by the rotor $[P_m = I_r^2(1 - s)r_r/s]$. Torque is obtained from the output power by dividing by the rotor speed in radians per second. The output torque T_0 is given by

$$T_0 = \frac{P_0}{\omega_r} = \frac{P_0}{(1 - s)4\pi f_s/P_s} \tag{68.2}$$

One characteristic of induction motors, and a result also obtained from the equivalent circuit, is that the value of the maximum torque is independent of the rotor resistance, while the slip is directly proportional to the rotor resistance. This phenomenon is utilized in the wound-rotor induction motor. For the wound-rotor motor with slip rings and brushes, the rotor resistance can be increased by connecting external resistors across the brushes. This external resistance is in series with the internal rotor resistance, therefore making it possible to produce maximum torque at higher values of slip (lower speed). Therefore, with the proper external resistance, maximum torque can be obtained at start. By shorting out the external resistance, a lower slip and higher efficiency can be obtained during full-load condition. With the increased external resistance at start, the starting current is reduced for a given torque. With variable external resistance, the wound rotor becomes an adjustable varying-speed motor. For loads with constant torque–speed characteristics or loads with torque that increase with speed, such as fans or pumps, the wound-rotor induction motor can be satisfactorily used as an adjustable varying-speed motor. Wound-rotor induction motors have been successfully applied to conveyors, cranes, hoists, and elevators.

Squirrel-cage induction motors are available in multispeed ranges, with the speed being essentially constant in each speed range. Multispeed motors are available in two, three, or four definite operating speeds. This is accomplished by reconnecting the winding to produce a greater or smaller number of poles. The simplest reconnections produce pole ratios of 2 : 1 (3600 : 1800, 1800 : 900, or 1200 : 600 rpm). For constant-torque winding connections, the horsepowers vary directly with the speed. Two winding motors can give speed ratios of 1800 : 1200, 1800 : 600, or 1200 : 900 rpm. Combinations of

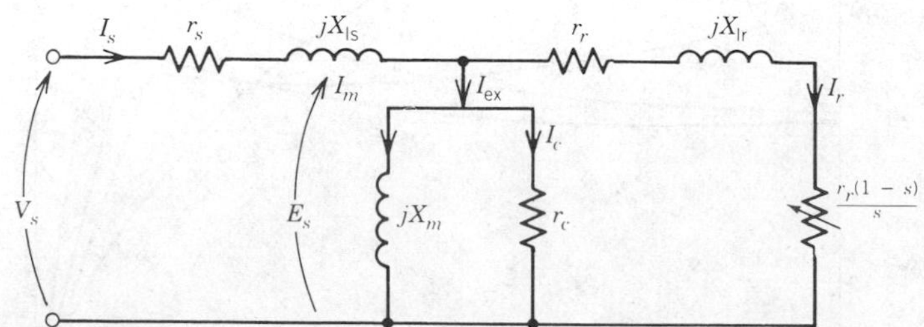

Fig. 68.2 Equivalent circuit of a general-purpose induction motor with parameters referred to the stator. r_s, stator winding resistance per phase; x_{ls}, stator winding leakage reactance per phase; r_c, core loss component, so that core loss is E_s^2/r_c and $I_c = E_s/r_c$; x_m, magnetizing component, so that $I_m = E_s/x_m$; r_r, rotor winding resistance per phase, referred to stator; x_{lr}, rotor winding leakage reactance per phase, referred to stator; $r_r(1 - s)/s$, equivalence resistance representing mechanical power developed by rotor, $P_m = I_r^2(1 - s)r_r/s$; V_s, stator applied voltage per phase; I_s, stator line current; E_s, stator induced voltage; I_r, rotor line current, referred to stator; I_c, core loss component of stator current; I_m, magnetizing component of stator current; I_{ex}, stator exciting current.

windings and winding reconnections can give speed ranges of 1800 : 1200 : 900 : 600 rpm. Even here, adjustable-speed drives should be considered. Multispeed motors may also be used to accelerate, drive, and decelerate high-inertia loads by using the proper winding connection sequence and an external brake.

Synchronous Machines

The synchronous machine has a stator and stator winding similar to those of the induction motor. However, the rotor of the synchronous machine receives DC excitation to set up a constant magnetic field. The rotor then must be turning at synchronous speed in order for the rotor DC field to interact with the rotating field set up by the stator excited from alternating current.

The synchronous machine can be modeled by the equivalent circuit of Fig. 68.3. Synchronous machine operation is better understood in terms of a phasor diagram developed from the equivalent circuit. For generator operation the phasor E_f must lead the terminal voltage phasor V_t, while for motor operation E_f must lag V_t. Examples of generator and motor phasor diagrams of the machine during system operation are shown in Figs. 68.4, 68.5, 68.6, and 68.7.

Synchronous machine performance, when operating connected to a system at constant V_t, can be predicted by the phasor diagrams. One important observation about synchronous machine operation is that the power factor can be changed by changing E_f. Since E_f is a function of speed and DC rotor field current, E_f can be changed by changing the rotor field current. For motor operation the current

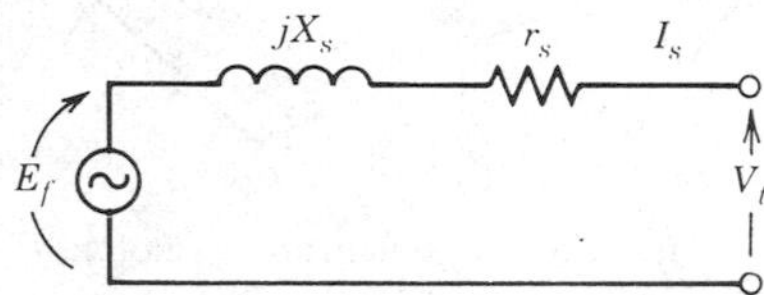

Fig. 68.3 Equivalent circuit of a synchronous machine with generator reference for the current. E_f, voltage generated by flux set up by rotor, a nonlinear function of rotor field current i_f times the synchronous speed n_s; x_s, synchronous reactance, includes voltage generated in stator due to revolving stator magnetomotive force as well as stator leakage reactance; r_s, stator resistance, much smaller than x_s; I_s, stator current (for motor reference, reference current is directed in at voltage reference); V_t, terminal voltage of machine which for system operation is fixed by system.

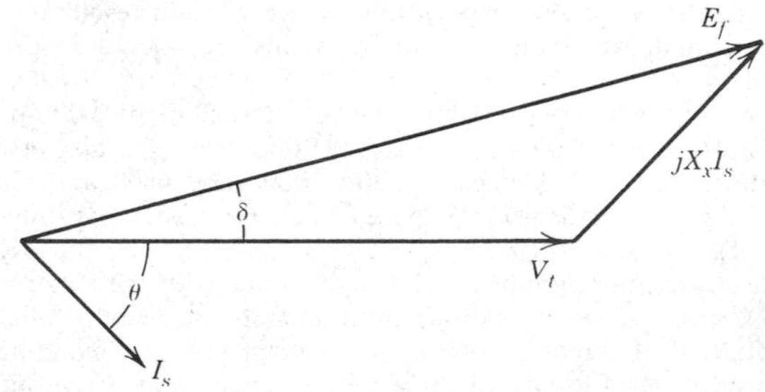

Fig. 68.4 Phasor diagram of an overexcited synchronous generator. E_f, voltage generated by flux set up by rotor; x_s, synchronous reactance; I_s, stator current; V_t, terminal voltage.

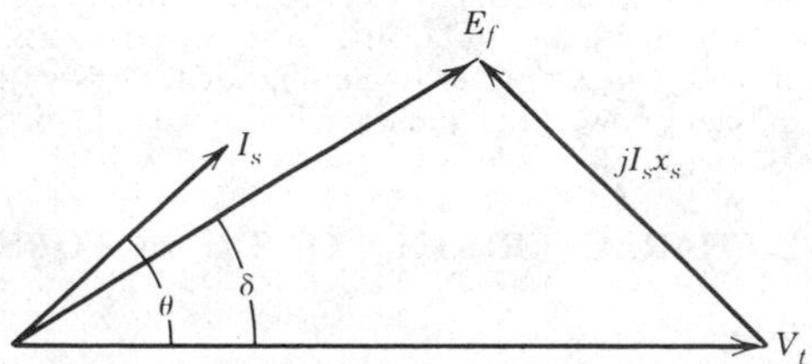

Fig. 68.5 Phasor diagram of an underexcited synchronous generator. E_f, voltage generated by flux set up by rotor; I_s, stator current; x_s, synchronous reactance; V_t, terminal voltage.

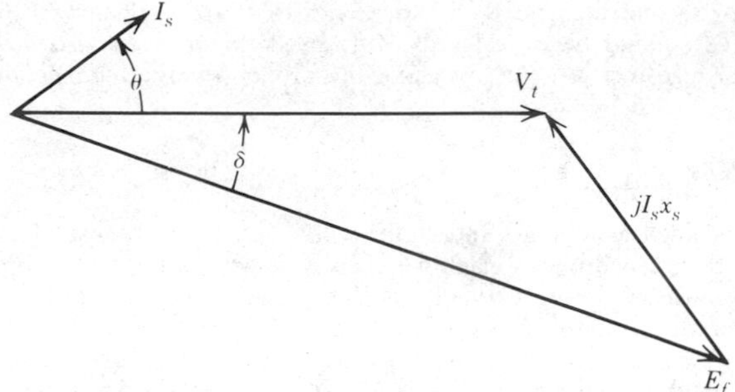

Fig. 68.6 Phasor diagram of an overexcited synchronous motor. I_s, stator current; V_t, terminal voltage; x_s, synchronous reactance; E_f, voltage generated by flux set up by rotor.

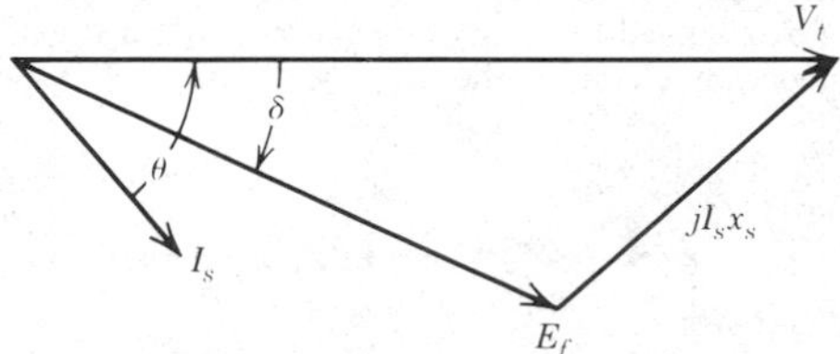

Fig. 68.7 Phasor diagram of an underexcited synchronous motor. V_t, terminal voltage; I_s, stator current; x_s, synchronous reactance; E_f, voltage generated by flux set up by rotor.

is leading when $|E_f \cos\delta| > |V_t|$ and is lagging when $|E_f \cos\delta| < |V_t|$, assuming $r_s \ll x_s$. Leading motor current delivers reactive power to the system. Therefore the power factor can be adjusted to provide some of the necessary reactive power required by the inductive loads on the system. This is one advantage of use of the synchronous motor. The same is true of the synchronous generator, although this generator delivers reactive power to the system when its current is lagging its terminal voltage. In either case, motor or generator operation, overexcitation results in reactive power being delivered to the system. Inductive (induction motors) loads are said to absorb or require reactive power from the system.

Synchronous motors are available in standard ratings from 20 to 100,000 hp with speeds from 80 to 1800 rpm for 60 Hz. However, it is not practical to build motors at all horsepower ratings at all speeds. The voltage ratings are 200, 230, 460, 575, 2300, 4000, 4600, 6600, and 13,200 V for 60 Hz and are rated at unity or 0.8 power factor leading. Again not all horsepower ratings are available in all voltages (Ref. 1, Part 21).

Since the DC field of the rotor cannot develop any torque with the stator magnetomotive force except at synchronous speed, a synchronous motor must be started by other means. The most convenient way is to utilize the induction motor action developed by the damping windings placed on the rotor to damp out mechanical oscillations due to load changes at synchronous speed. Therefore starting currents are similar to those of the induction motor and vary from 500% to 650% for normal starting torques.

Although synchronous motors are sometimes used because of their synchronous speed characteristic, speed may not be the primary reason for their selection. One advantage is their ability to change the power factor continuously from leading to lagging. Another advantage is that for unity power factor motors, efficiency is from 1% to 3% higher (including exciter losses) than that of an induction motor of the same speed and horsepower. For some types of loads, the better efficiency can be the deciding factor, especially as energy costs increase.

68.3 PERFORMANCE CHARACTERISTICS OF DC MOTORS AND GENERATORS

Even though essentially all the electrical energy produced is in the AC form, DC motors are utilized in significant amounts when variable or adjustable speed control is required.

DC motors are extremely flexible in their performance characteristics, and a wide variety of speed–torque characteristics can be easily obtained. The one main advantage of DC motors over AC motors is the wide adjustable speed range available with easy adjustment. Also DC motors have higher efficiency compared with AC motors in applications requiring frequent acceleration and deceleration of large-inertia loads.

The disadvantages associated with DC motors are that they are larger for a given horsepower and speed than AC motors and also cost more per horsepower. They also involve more maintenance and are more complicated. In spite of these disadvantages, DC motors find increasingly more applications as more processes become automated.

The two basic types are shunt and series, with the compound having some characteristics of both types. The shunt-wound motor is the most popular for modern industrial applications, although the shunt field may be excited from a source separate from the armature.

The equivalent circuit for predicting the performance of the shunt DC motor is given in Fig. 68.8. From the figure, the performance equations are

$$\begin{aligned} V_t &= E_g + I_a r_a \\ I_f &= \frac{V_t}{R_f + R_e} \\ E_g &= g(I_f)\omega_m \qquad \text{nonlinear in } I_f \\ I_L &= I_a + I_f \end{aligned} \tag{68.3}$$

where V_t = applied voltage from DC supply
E_g = generated voltage due to armature rotation and is function of I_f
I_a = armature current
I_L = supply current
I_f = field current
R_f = shunt field resistance
R_e = external resistance in shunt field circuit to control speed
r_a = armature resistance

Since the power input to the armature circuit is $V_t I_a$ and the losses are $I_a^2 r_a$, the remaining power $E_g I_a$ is the electrical power converted to mechanical energy. The mechanical power developed is given by $T_{em}\omega_m$; therefore the torque T_{em} is computed by

$$T_{em} = \frac{E_g I_a}{\omega_m} = \frac{V_t - I_a r_a}{\omega_m} \qquad \text{Nm} \tag{68.4}$$

where ω_m is the speed in radians per second. Since $E_g = k\phi\omega_m$, then ω_m is given by

$$\omega_m = \frac{E_g}{K\phi} = \frac{V_t - I_a r_a}{K\phi} \tag{68.5}$$

where $K\phi$ is the term proportional to the flux ϕ and is obtained from the no-load saturation curve. The no-load saturation curve is a plot of E_g at a given constant speed ω_{mo} vs. the field ampere turns. The $K\phi$ term is obtained by dividing E_g by ω_{mo} to obtain $K\phi$ as a function of the net field ampere turns (mmf) due to both series and shunt fields.

The performance equations of the compound motor of Fig. 68.9 are similar to those of the shunt motor except for the following

$$\begin{aligned} V_t &= E_g + I_a(r_a + r_s) \\ E_g &= g(I_f, I_a)\omega_m \qquad \text{nonlinear} \\ E_g &= K\phi\omega_m \end{aligned} \tag{68.6}$$

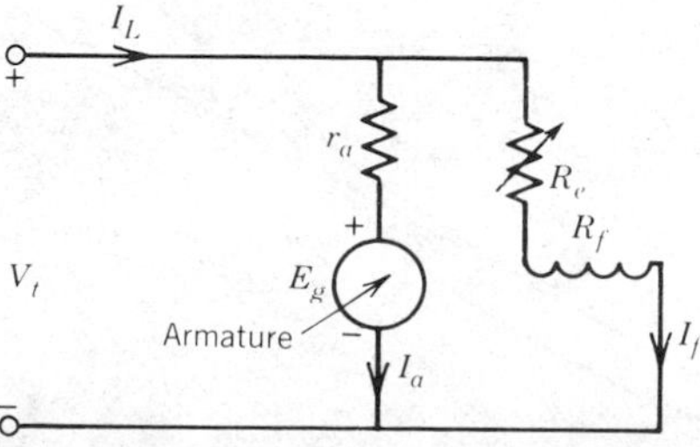

Fig. 68.8 Equivalent circuit of a DC shunt motor. I_L, supply current; V_t, applied voltage from DC supply; r_a, armature resistance; E_g, generated voltage due to armature rotation; I_a, armature current; R_e, external resistance in shunt field circuit; R_f, shunt field resistance; I_f, field current.

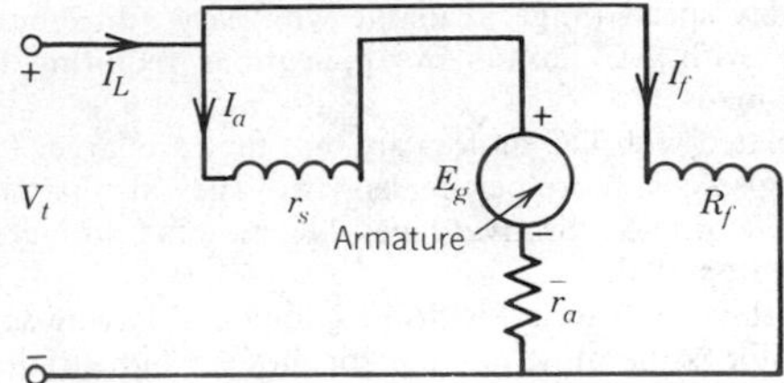

Fig. 68.9 Equivalent circuit of a DC compound motor. I_L, supply current; V_t, applied voltage from DC supply; I_a, armature current; r_s, series field stator resistance; E_g, generated voltage due to armature rotation; I_f, field current; R_f, shunt field resistance; r_a, armature resistance.

where r_s is the series field resistance. The changes involve an additional resistance voltage drop due to the series field and the additional magnetomotive force from the series field.

Using these equations, the speed–armature current and torque–armature current characteristics can be obtained. These are illustrated in Figs. 68.10 and 68.11.

Separate excitation of the shunt field allows the value of the no-load speed of a shunt or compound motor to be varied without materially changing the shape of the curve showing speed as a function of the armature current (load). Where adjustable constant speed is required, the shunt motor

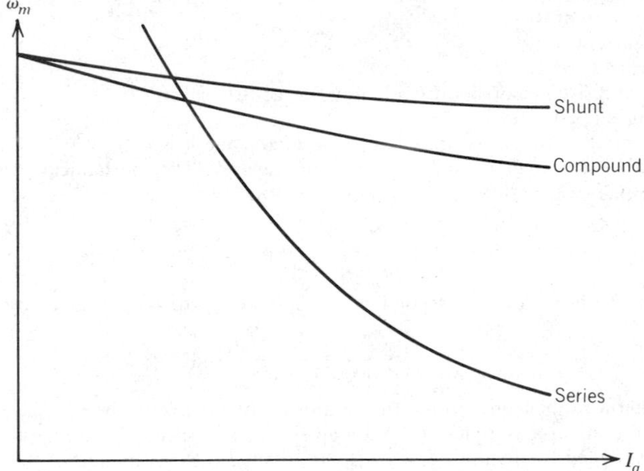

Fig. 68.10 Speed (ω_m) vs. armature current (I_a) of typical DC motors.

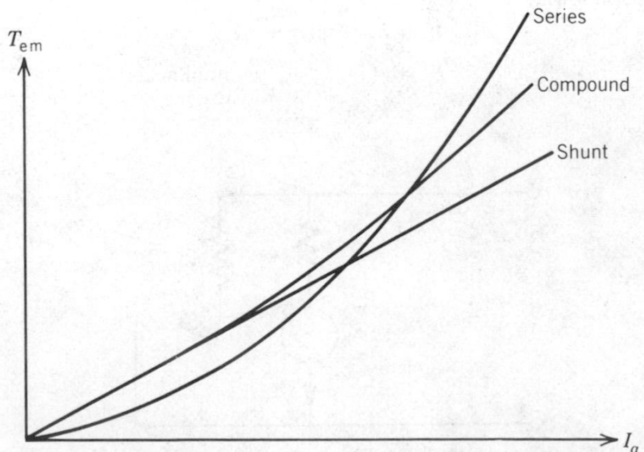

Fig. 68.11 Torque (T_{em}) vs. armature current (I_a) of typical DC motors.

should be a first consideration. Such motor applications include fans, blowers, centrifugal pumps, conveyors, and woodworking and metal-working machinery. Starting torque varies directly with the armature current and is limited to 125% to 200% because of commutation problems.

For loads requiring widely varying speed with load and high starting torque, the series motor is the choice. Since the field is connected in series, the flux at light loads is very small and the speed is high, as indicated by Eq. (68.5), which also applies to series motors. In fact, if there is no external load on the motor the speed is only limited by friction and windage, and the maximum safe speed will likely be exceeded. Therefore series motors are directly connected to their loads. Loads requiring high starting torques with widely varying speeds are those of transportation and hoisting equipment, winches, and capstans. The starting torque is limited to 300% to 375% by commutation and heating constraints.

The cumulative compound motor has the series field aiding the shunt field and consequently has a larger speed regulation than the shunt motor. Compound motors are used when the starting torque required is larger than that available from the shunt of the same horsepower but lower than that of the series. Because of its larger drop in speed with load, the compound motor is suited for driving punch presses, shears, crushers, reciprocating compressors, and plunger pumps. The maximum running torque varies from 175% to 250%, again limited by commutation and heating.

DC motors can usually meet the average needs of industry regarding size and speed. However, there are limits for both. The general rule of thumb is that machines are available in sizes and speeds for applications where the product of the speed in revolutions per minute and horsepower is less than 2×10^6.

DC generators can be either separately excited or self-excited. In separate excitation, the current through the field, whether shunt or series, is independent of the voltage or current delivered by the generator. Therefore the change in the terminal voltage of the generator is influenced only by the armature resistance drop and some reaction of the armature magnetomotive force with the field through magnetic saturation of the pole tips to cause an additional drop in terminal voltage.

For self-excited shunt, the field is in parallel (shunt) with the terminal voltage, and anything that decreases the terminal voltage decreases the field current. This effect can be best described in terms of a circuit model of the DC generator shown in Fig. 68.12. From the figure the following equations can be written for generator operation:

$$V_t = E_g - I_a r_a$$
$$I_f = \frac{V_t}{R_f + R_e} \tag{68.7}$$
$$I_a = I_f + I_L$$

where E_g, I_a, I_f, I_L, R_f, R_e, and r_a are defined in Eq. (68.3). As the load current I_L increases, the terminal voltage decreases, causing less field current and consequently less flux. Less flux ϕ results in less E_g and less V_t. This feedback effect is stable but results in a greater decrease in terminal voltage in the shunt-connected generator than with a separately excited generator. The terminal voltage V_t vs. load current I_L is called the external characteristic of a DC generator. The general shape of the external characteristic for the four basic types of generators is shown in Fig. 68.13.

The series generator has its field connected in series with the armature, so the flux varies with the load current. Consequently its voltage increases with increasing load until the magnetic circuit is well into saturation. After that the voltage will decrease with load.

The compound generator contains both a shunt and series field, so its external characteristic exhibits the characteristic of both the shunt and series. For the compound generator, the following

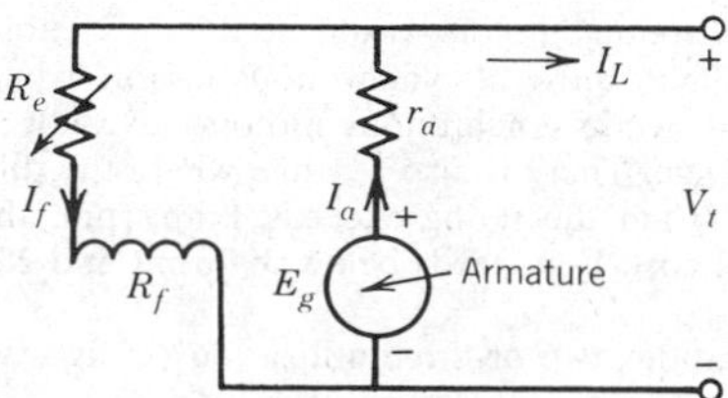

Fig. 68.12 Equivalent circuit of a self-excited shunt DC generator. R_e, external resistance in shunt field circuit; R_f, shunt field resistance; I_a, armature current; r_a, armature resistance; E_g, generated voltage due to armature rotation; I_L, supply current; V_t, terminal voltage.

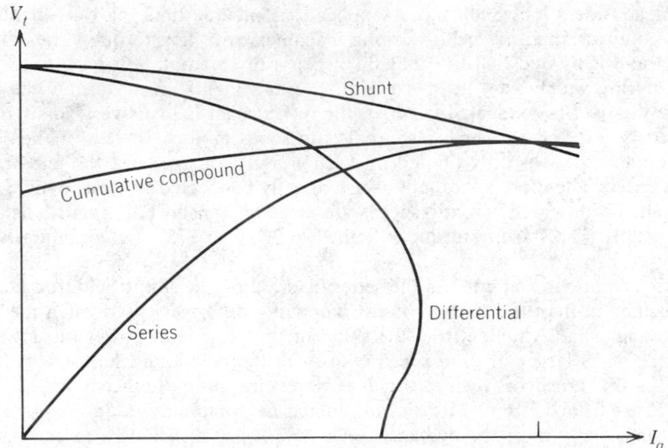

Fig. 68.13 Terminal voltage (V_t) vs. armature current (I_a) for DC generators.

equations obtained from Fig. 68.9 can be written:

$$\begin{aligned} V_t &= E_g - I_a(r_a + r_s) \\ I_f &= \frac{V_t}{R_f + R_e} \\ I_a &= I_f + I_L \\ \phi &= g(I_f N_f \pm I_a N_s) \end{aligned} \tag{68.8}$$

where r_s = series field resistance
N_f = shunt field turns
N_s = series field turns
and the other symbols are defined in Eq. (68.3).

If the series field-ampere-turns aid the shunt field, the generator is cumulatively compounded, while if they oppose, it is differentially compounded.

68.4 SELECTION AND APPLICATION OF AC AND DC MOTORS

Many factors should be considered in selecting the proper motor for a given application. Among these are the availability of motors in a given speed, horsepower, and voltage range; the values of the utilization voltage available; the type of load; the frequency of starting; ambient conditions; the type of enclosure required; the efficiency of operation; maintenance; and the initial cost and cost of operation.

AC Motors

In comparing induction motors with synchronous motors, the cost of the starting and control equipment must be considered as well as efficiency and operating costs. There are horsepower sizes and speeds at which induction motors would be the only choice, and likewise there are horsepower sizes and speeds at which synchronous motors would be the only choice. In general, the initial costs of induction motors are lower than those of synchronous motors when higher speeds and relatively lower horsepowers are required, while synchronous motors have lower initial cost in the low-speed relatively higher horsepower range. There is also a range where the difference is not significant. The general rule of thumb is that when the rating exceeds 1 hp/rpm, the synchronous motor and its control will have a lower initial cost. For speeds below 300 rpm and 200 hp, a gear motor may result in the lower initial cost.[1]

Generally, in industrial locations two or three utilization voltages will be available, ranging from 13.6 kV downward.

The design and consequently the ability of a given machine to withstand the mechanical stresses due to starting are directly related to the voltage rating. In addition, the number of poles and consequently the speed influence the rigidity of the stator windings and the reactance of the windings.

In general, higher-speed (fewer poles) machines have lower reactance than low-speed machines and thus higher starting currents. Therefore voltage rating may be influenced by the speed rating.

The voltage rating also affects the efficiency and, for induction motors, the power factor. For a given horsepower and speed, a point is reached where a higher voltage level results in a decrease in efficiency. For induction motors the power factor drops off appreciably as the voltage rating is increased.

For motors subjected to greater than the normal number of starts per day, the voltage ratings should be lower, since the mechanical winding strength will be greater. This is also true for fluctuating loads.[1]

Any cost comparison at different voltage levels must take into consideration the cost of starting equipment and the additional costs associated with the utilization voltages being considered.

The bar graph shown in Fig. 68.14 is an attempt to include the effect of these various factors on the initial cost. The heavy vertical lines that extend across the two voltage bars indicate the horsepower at which the costs are the same at either voltage.

The minimum recommended horsepower indicated in Fig. 68.14 in each voltage is based on a rating that is consistent with good motor design. The maximum horsepower indicated for a given voltage is based primarily on considerations other than good design. These considerations are usually economic and the availability of starting equipment. These minima and maxima recommended are not usually the limits of available ratings. Selecting either a motor of too high or too low a voltage rating for a given horsepower invites a shortening of motor life, increased maintenance, and motor winding failures.

To meet the various environmental conditions, AC motors are available with various types of enclosures. NEMA has defined various enclosures with eight types of open motors and six types of totally enclosed motors (Ref. 1, Sec. 1.25). In general, motors operated in explosive atmospheres must be totally enclosed, explosion proof.

The life of the coil insulation is directly related to the kind of material and the operating temperature. Therefore motors applied where the ambient temperature is high must have higher temperature-rated insulation. NEMA has set standards for insulating materials and has defined "hot spot" temperature ratings for each. Class A material is rated for a maximum "hot spot" temperature of 105°C, class B at 130°C, class F at 155°C, and class H at 180°C (Ref. 1, Sec. 12.41).

Speed and horsepower selection involves matching the speed–torque characteristic of the motor with the speed–torque characteristic of the load such that motor torque is always greater than that of the load until the curves intersect. For the induction motor, the torque at the point of intersection should be below the maximum torque, otherwise excessive slip would exist and overheating could result because of the low efficiency at such an operating point. Since the speed–torque characteristic

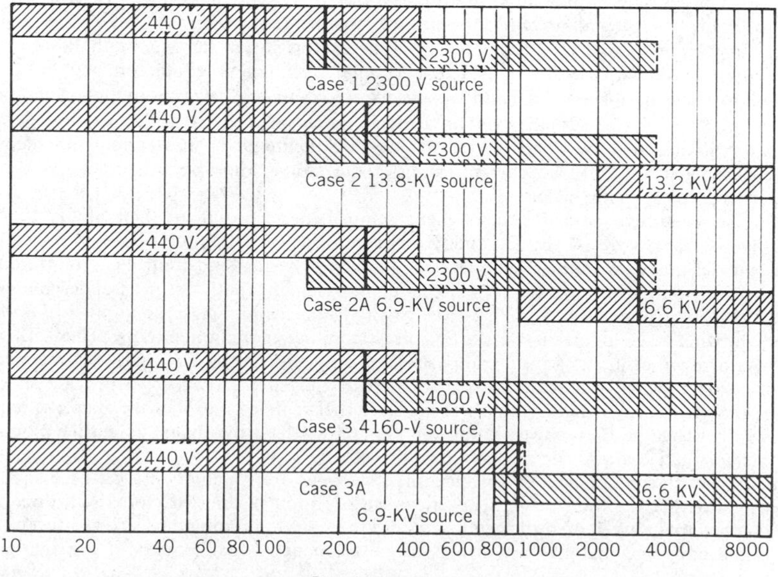

Fig. 68.14 Suggested motor rating voltage vs. horsepower on the basis of comparable costs.

is only for steady-state conditions, the effect of the inertia of the motor or load is not included. During starting, the inertia of the load and motor needs to be considered. The amount of torque needed to accelerate a given load is directly proportional to the inertia constant and the angular acceleration. Inertia constants in the English system are given in terms of the weight in pounds (W) and the radius of gyration (k) or Wk^2, which for the motor and load may be available from the manufacturer.†

DC Motors

Because of an operating limitation caused by commutation problems, DC motors are usually only available in speed–horsepower ratings where the product of the speed in revolutions per minute times the rated horsepower is less than 2×10^6. In addition, for speeds below 400 rpm and 100 hp, standard DC motors are not available. For the larger sizes, the lower speed limit is about 100 rpm. For the lower speed and horsepower applications, gear motors should be considered.

Unlike AC motors, DC motors are only available in a few voltage levels, and the selection of the voltage rating is probably an economic one. Whenever there is a choice, generally the higher voltage rating is more economical. Above 3 hp the 250-V rated motor is more economical, but below 3 hp there is little difference in the costs between the 125- or 250-V rated motors and controls.

The general dividing line for voltage selection between 250 and 600 V is where the product of speed in revolutions per minute times the horsepower rating is about 45×10^4 for speeds from 400 to 1150 rpm. That is, when the speed times horsepower is greater than 45×10^4, the 600-V rated motor would be more economical. However, there is a range of ± 100 hp where there is little difference in costs between the 250- and 600-V rated motor. This is a general rule, and the final selection should be determined after consideration of all cost factors.

68.5 FUNDAMENTALS OF MOTOR CONTROL

Controls for motors involve some means of safely starting and stopping them and of providing overcurrent protection for the motor and also short-circuit protection at or beyond the motor controller.

Modern induction motors are designed to withstand full-voltage starting from standstill without damage to the motor windings. Since full-voltage starting is the most economical, it should always be considered as the first choice. However, in some cases the motor-starting current, which can be as high as six times rated, will produce an objectionable light flicker where power and lighting circuits are the same. Electric utilities usually have specific regulations regarding the allowable voltage dips, depending on how often they occur (see Section 67.2). Even when such a problem exists, it is often desirable to still consider full-voltage starting as the first approach, since it gives the highest equivalent torque efficiency along with that of autotransformer starting.

In selecting either a manual or magnetic contactor for motor starting, both the current and voltage ratings should be carefully considered. Manual or magnetic contactors suitable for motor starting will be rated in horsepower. Manual starters are commonly used on induction motors of $7\frac{1}{2}$ hp and below. These devices usually contain a toggle mechanism that provides a set of quick-make quick-break contacts. Also overload relays are integrally mounted in these units with their heaters selected according to the full-load current rating of the motor. Short-circuit protection is normally provided by separately mounted fuses.

NEMA has classified manual and magnetic controllers according to their ability to interrupt currents. These classes are A, B, D, and E. The horsepower or continuous-current ratings are classified into various sizes, 00 through 9 (Ref. 2, Part 2-110). Magnetic controllers for various duties such as jogging, plugging, reversing, nonjogging, nonplugging and so on are designated in an ANSI/NEMA publication (Ref. 2, Part 2-231). Owing to their long life (large number of operations), contactors provide a good means of starting and stopping induction motors. They also provide thermal overload protection as an integral part of the contactor. They are used in connection with a fusable disconnect switch, a circuit breaker, or a fuse and circuit breaker to provide short-circuit protection at or beyond the load terminals of the starter. If remote control of the starter is required or the motor is beyond the horsepower rating for manual starters, the magnetic starter (contactor) is used. Contactors can also provide undervoltage protection.

To reduce the starting or in-rush current, three methods are usually considered for squirrel-cage induction motors: (1) autotransformer, (2) series resistor, and (3) series reactor. In all three methods additional contactors are required to switch in or out the current limiting device. The autotransformer method is the only one that has a starting torque efficiency equal to that of the full-voltage or

†Radii of gyration for rotating bodies are available in most Mechanics textbooks.

line-starting arrangement. The starting torque efficiency is the ratio of the starting torque to the starting current required from the supply. Since torque varies as the square of the voltage across the motor, and the autotransformer supply line current varies inversely as the voltage applied to the motor, this ratio is unchanged. For series resistor or reactor starting, part of the supply voltage is dropped across the resistor or reactor with no current transformation. Since the torque varies as the square of the motor voltage, but the starting line current varies directly with the motor voltage, the torque efficiency is less than with full-voltage starting. The reduced line current does reduce the voltage drop in other parts of the system, but the reduction in the critical areas may not justify the additional cost of a reduced-voltage starting arrangement.

Another form of reduced in-rush current starting is part-winding starting. Portions of the motor winding, such as one half of a two-part parallel winding, are connected in steps so that the starting current is limited to smaller increments. However, the torque is also correspondingly limited. Some simplified diagrams of the various starting methods are shown in Figs. 68.15, 68.16, and 68.17.

Stopping or braking of induction motors can be accomplished by mechanical brakes or electrical braking. Mechanical brakes are required for holding or braking the motor after it has reached reduced speed.

Electrical braking can be classified as dynamic braking, plug braking (plugging), or capacitor braking. Another form of electrical braking is called regenerative braking. This involves the motor

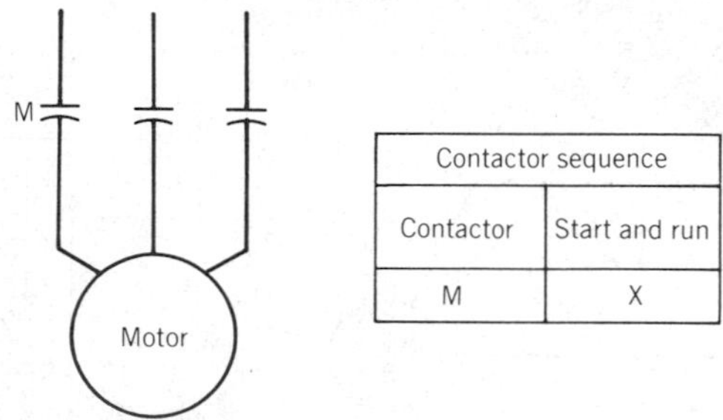

Contactor sequence	
Contactor	Start and run
M	X

Fig. 68.15 Full-voltage controller, line starting. M, main contacts; X, closed contacts.

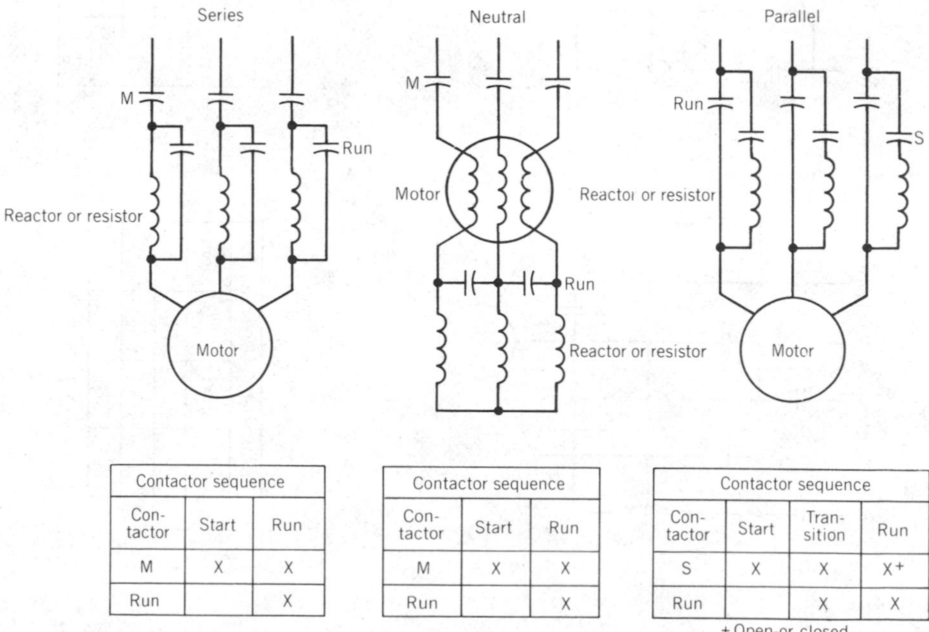

Contactor sequence		
Con-tactor	Start	Run
M	X	X
Run		X

Contactor sequence		
Con-tactor	Start	Run
M	X	X
Run		X

Contactor sequence			
Con-tactor	Start	Tran-sition	Run
S	X	X	X+
Run		X	X

+Open-or closed.

Fig. 68.16 Reduced-voltage controller reactor or resistor. M, main contacts; X, closed contacts; S start contacts.

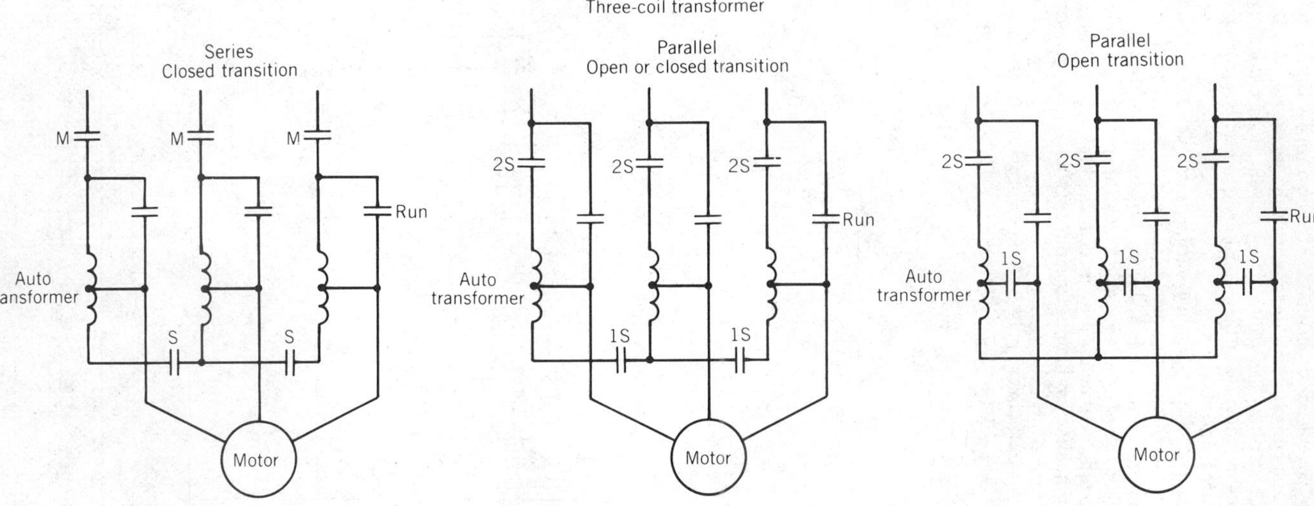

A

Contactor sequence			
Con-tactor	Start	Tran-sition	Run
M	X	X	X
S	X		
Run			X

Closed transition

B

Contactor sequence					
Con-tactor	Start	Open Tran-sition	Closed tran-sition		Run
			1	2	
1S	X				
2S	X		X	X	
Run				X	X

For open transition, 1S and 2S may be contacts of same contactor

C

Contactor sequence		
Contactor	Start	Run
1S	X	
2S	X	
Run		X

Open transition 1S and 2S may be contacts of same contactor

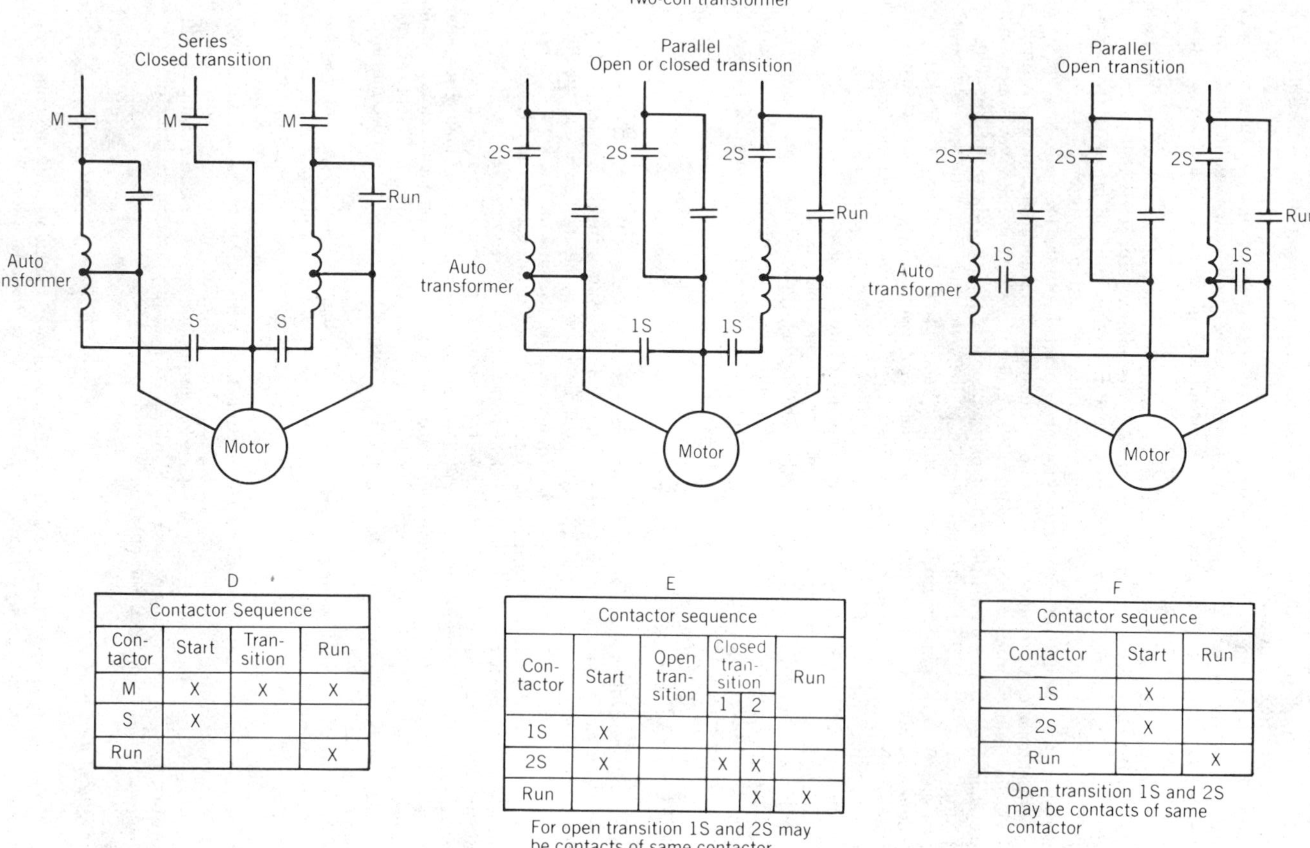

D

Contactor Sequence			
Con-tactor	Start	Tran-sition	Run
M	X	X	X
S	X		
Run			X

E

Contactor sequence					
Con-tactor	Start	Open tran-sition	Closed tran-sition		Run
			1	2	
1S	X				
2S	X		X	X	
Run				X	X

For open transition 1S and 2S may be contacts of same contactor

F

Contactor sequence		
Contactor	Start	Run
1S	X	
2S	X	
Run		X

Open transition 1S and 2S may be contacts of same contactor

Fig. 68.17 Reduced-voltage autotransformer controllers. M, main contacts; X, closed contacts; S, start contacts.

overspeeding and can be accomplished for special machines by changing the number of poles to produce a lower rated speed. This provides partial stopping with completion of the stopping cycle performed by a mechanical brake.

Dynamic braking is accomplished by removing the polyphase voltage from the stator and applying a DC or single-phase AC voltage. This produces an eddy current effect in the rotor, where the amount of braking is controlled by the amount of the excitation.

Capacitor braking involves using capacitors connected across the terminals of the stator, which when selected properly produce a resonance with the inductance of the motor winding and a kind of self-excitation. This braking effect goes to zero at about 25% of speed, so a mechanical brake is required to produce final braking.

Plug braking or plugging involves reversing the phase sequence of the voltage applied to the stator, thereby reversing the direction of the stator magnetomotive force. Here the voltage must be removed when the speed reaches zero; otherwise it will start to turn the motor in the opposite direction.[3] The inherent disadvantage is the large heat loss produced in the rotor.

Since synchronous motors have no starting torque associated with the DC field, they are started by induction motor action produced by the damper windings. Even though the starting is similar to that of induction motors, more elaborate controls are needed to synchronize the synchronous motor with the system and to remove the field if the motor pulls out of step. For the motor to pull its load into synchronism, the DC field must be energized at the right time and with the right polarity. Additional sensing devices and controls are required to accomplish this.[2]

DC motor starters, like those of AC, can be manual or automatic; however, full-voltage starting cannot be performed on DC motors. Because of commutation limitations, the starting current has to be limited to a maximum of 200%. This is accomplished by inserting additional resistance in the armature circuit while full field excitation is maintained and by shorting out portions of this resistance to obtain the desired torque. The amount of resistance shorted out is such as to limit the maximum armature current to 200%. For individual DC motors, variable-voltage power supplies can be used to provide power to the armature, thereby controlling the armature current during starting.

Regenerative braking and dynamic braking can be used effectively in stopping DC motors. For holding operations, mechanical brakes are required.

68.6 ILLUMINATION

Good illumination involves attention to a number of elements, such as the class of room and the service for which the illumination is required; the luminaire best suited for the service; the effect of the color and the reflectance of ceilings, walls, and floors; and the light intensity, distribution, diffusion, color, and intrinsic brightness of the luminaire as well as glare and shadows.

Recommended values for the intensities of illumination desirable for accurate and easy seeing for the room class and service are given in the *Illuminating Engineering Society* (*IES*) *Lighting Handbook*.[4] Some of these values are shown in Table 68.1. Luminaires are generally classified as providing (1) direct, (2) semidirect, (3) semi-indirect, and (4) indirect light. A wide variety of luminaires provide light within each of these four classes, and their selection would be based on such factors as surface brightness, efficiency, and maintenance. Color and reflectance of walls, ceilings, and floors affect the overall utilization factor used in calculating the light output necessary to obtain the level of illumination desired.

The light intensity should be ample to enable clear and distinct vision. The light distribution should be such that the illumination over a given part of the room or work area is at least nearly uniform. Except for special lighting effects, the color of light approaching daylight or that inclined toward the yellow in general will be found the most satisfactory. The light sources should be placed so that the rays do not pass directly into the eye and their intrinsic brightness is not reduced by diffusing enclosures. Also objects capable of high specular reflections should be removed from the range of vision. Glare, which is the result of intense brightness in concentrated areas within the direct line of vision, must be completely eliminated. Shadows are necessary for distinguishing outlines, but such shadows should be diffused and not too abrupt or dense.

Basic Lighting Terms

The knowledge that light is composed of photons that travel in all directions helps in comprehending the terminology used in lighting design. Luminous intensity (candlepower) is the index of the light-giving power of a source in a given direction; the unit of luminous intensity is the candela (cd).

A uniform point source of light having an intensity (candlepower) of 1 cd, by definition, produces a luminous flux density of 1 lm/sr. A steradian (sr) is a unit of solid angle; for a sphere this angle is 4Π. It is obtained by dividing the surface area of a sphere by the square of the radius. Thus a solid angle of 1 sr will sweep out an area of 1 ft^2 on the sphere having a radius of 1 ft. Since a 1-cd source

TABLE 68.1. TYPICAL RECOMMENDED LEVELS OF ILLUMINATION

Area	Footcandles on Tasks	Lux[a] on Tasks
Auditoriums		
Assembly only	15	160
Exhibitions	30	320
Social activities	5	54
Clothing manufacture (men's)		
Receiving, opening, storing, shipping	30	320
Examining (perching)	2000[b]	21,500[b]
Sponging, decating, winding, measuring	30	320
Piling up and marking	100	1,100
Cutting	300[b]	3,200[b]
Patternmaking, preparation of trimming, piping, canvas and shoulder pads	50	540
Fitting, bundling, shading, stitching	30	320
Shops	100	1,100
Inspection	500[b]	5,400[b]
Pressing	300[b]	3,200[b]
Sewing	500[b]	5,400[b]
Offices		
Drafting rooms		
Detailed drafting and designing, cartography	200[c]	2,200[c]
Rough layout drafting	150[c]	1,600[c]
Accounting offices		
Auditing, tabulating, bookkeeping, business-machine operation, computer operation	150[c]	1,600[c]
General offices		
Reading poor reproductions, business-machine operation, computer operation	150[c]	1,600[c]
Reading handwriting in hard pencil or on poor paper, reading fair reproductions, active filing, mail sorting	100[c]	1,100[c]
Reading handwriting in ink or medium pencil on good-quality paper, intermittent filing	70[c]	750[c]
Private offices		
Reading poor reproductions, business-machine operation	150[c]	1,600[c]
Reading handwriting in hard pencil or on poor paper, reading fair reproductions	100[c]	1,100[c]
Reading handwriting in ink or medium pencil on good-quality paper	70[c]	750[c]
Reading high-contrast or well-printed materials	30[c]	330[c]
Conferring and interviewing	30	330
Conference rooms		
Critical seeing tasks	100[c]	1,100[c]
Conferring	30	330
Note taking during projection (variable)	30[c]	330[c]

Source. IES Lighting Handbook. For complete listing see Handbook.

[a] Lux is SI unit equal to 9.29 fc.

[b] Obtained with combination of general lighting and specialized supplementary lighting.

[c] Equivalent sphere illumination.

emits light flux at the rate of 1 lm/sr, the resultant illumination on 1 m^2 is naturally 1 lm/m^2 or 1 lx. From this, then, the total luminous flux produced by a source of 1 cd is 4Π lm. The *intensity* of illumination is the density of the light flux falling on a surface in lumens per square meter, or lux. (Lumens per square foot = footcandle.) Luminous intensity (candlepower) is the angular density of the flux, while illumination intensity is the surface density of the flux.

Illumination lux (footcandles) is independent of the character or color of the surface on which it falls and is different from the brightness of a surface. The surface brightness (luminance), whether the surface is a luminous source or a reflecting surface, depends on the character and color of the surface in addition to the value of the illumination emitted or reflected by it. A perfectly diffusing surface emitting or reflecting light at the rate of 1 lm/m^2 has a brightness of 1 lx (lumen per square foot has a brightness of 1 footlambert). The average brightness of any reflecting surface in lux (footlambert) is the reflection factor of the surface times the illumination in lux (in footcandles) received by the surface.

The efficiency of a light source is the ratio of the total luminous flux to the total power consumed, expressed as lumens per watt.

The coefficient of utilization of an illumination installation on a given plane is the total flux received by that plane divided by the total flux from the lamps illuminating it. When not otherwise specified, the plane of reference is the working plane, a horizontal plane 30 in. from the floor.

Lamp Characteristics

For commercial buildings and industrial plants, five types of lamps are generally used. These are incandescent, fluorescent, mercury vapor, metal halide, and high-pressure sodium vapor. These lamps vary in luminous efficiency, with the high-pressure sodium vapor being the most efficient. The colors are distorted in the case of the vapor lamps, with mercury being mainly in the violet range and sodium in the orange. Fluorescent lamps have two to three times the efficiency of incandescent lamps.

The performance of luminaires consisting of lamps and their various reflectors, louvers, diffusers, housing, and support is given by a candlepower distribution curve. Such a curve is available from the manufacturer of the luminaire. Candlepower is determined from the definition of illumination intensity in lux (footcandles); the flux Ψ and E are related by

$$E = \frac{\psi}{\text{area}}$$

The total flux ψ from a point source I of 1 cd is $\psi = 4\Pi I$. A sphere surrounding the point source I has an area of $4\Pi r^2$; therefore the illumination E is

$$E = \frac{4\Pi I}{4\Pi r^2} = \frac{I}{r^2}$$

By measuring the illumination E at various vertical angles and known distance around a luminaire having cylindrical symmetry, we can compute the intensity (candlepower) I at each angular position. Plotting this intensity (candlepower) vs. angle on a polar plot produces the luminous intensity (candlepower) distribution curve. A typical luminous intensity (candlepower) distribution curve is shown in Fig. 68.18. The total lumen output of a given luminaire or lamp is not only a function of the intensity (candlepower) existing in a given direction and the surface area having that light intensity but also of the angular location. The intensity (candlepower) distribution curve can be misleading, since the total lumens falling on the surface area of a sphere included in a 10° zone at ±5° from the 0°-angle position is less than one tenth that on the area of a 10° zone at ±5° from the 90°-angle position. The area in the 10° zone about the 0°-angle position is a circle, while the area at the 90°-angle position is a circular band.

The zonal multiplying factors of each 10° zone at different angular locations are given in Table 68.2. Figure 68.19 shows the polar plot of intensity (candlepower) vs. angle for two different candlepower distributions, where B has a greater lumen output than does A. The difference is the zonal area of a spherical surface changing with the angle. Polar distribution curve A provides more light directly under the unit, while B provides more light out to the sides.

Lighting Calculations

There are two basic methods for calculating size, number, and possible spacing of luminaires to produce a given level (footcandle) of illumination on a given plane or surface. One, the point-by-point method, utilizes the inverse square law for a point source of light to determine what is needed to produce a given level of illumination on a given area. This method is somewhat laborious, since it

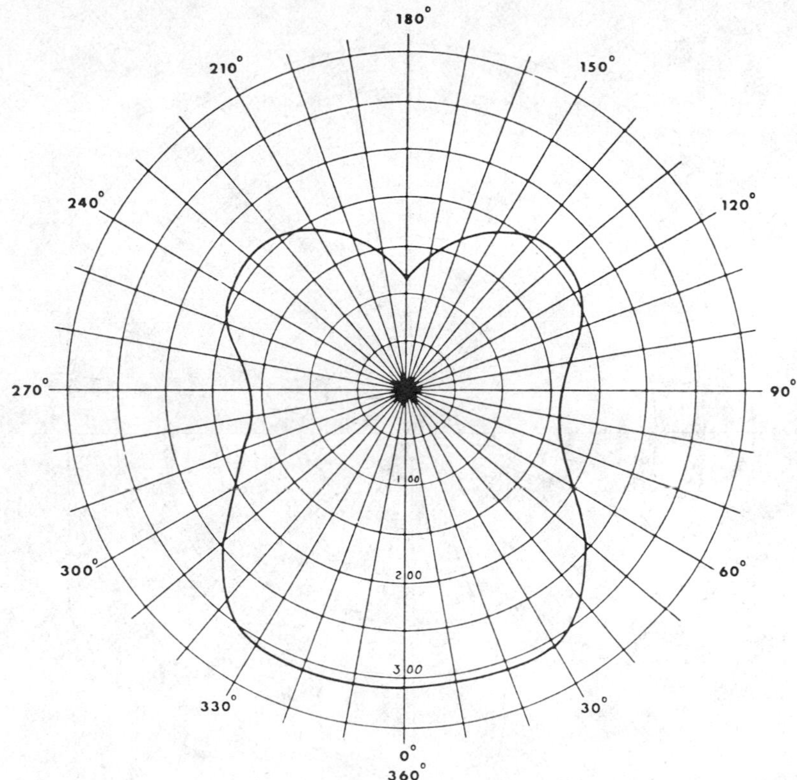

Fig. 68.18 Candlepower distribution curve for an incandescent fixture enclosing globe.

involves computing the contribution from each luminaire in a system to the illumination of a given area. This method is useful in determining illumination levels produced by single or multiple fixtures for spotlighting and floodlighting. One disadvantage of this method is that it does not generally account for interreflections from room surfaces, when used for determining illumination levels for general lighting.

The lumen method is based on the assumption that, for similar conditions as to type of luminaire;

TABLE 68.2. ZONAL MULTIPLYING FACTORS FOR COMPUTATION OF FLUX FROM CANDLEPOWER CURVES

Midzone Angle α From Vertical Degrees			Zonal Multiplier, 10° Zones
5	or	175	0.0954
15	or	165	0.2835
25	or	155	0.4629
35	or	145	0.6282
45	or	135	0.7744
55	or	125	0.8972
65	or	115	0.9926
75	or	105	1.0579
85	or	95	1.0911

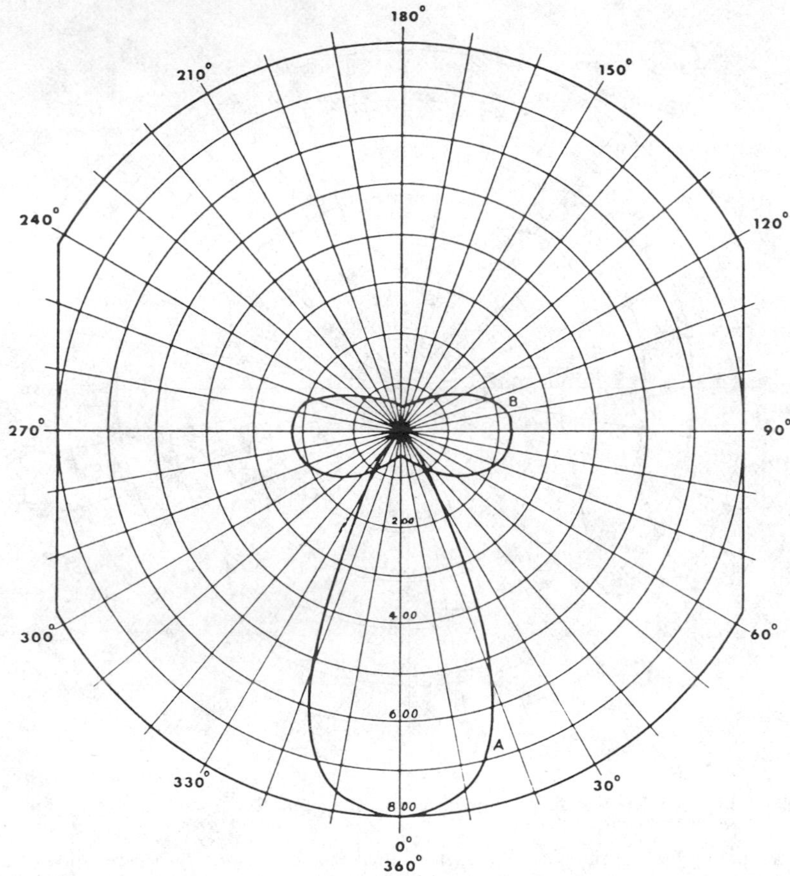

Fig. 68.19 Comparison of candlepower distribution curves. Fixture B has a greater lumen output than does fixture A.

color and texture of walls, ceiling, and floor; room dimensions; and so on, the lumens on the working plane will be a definite percentage of the total lumens emitted by the light sources.

Point-by-Point Method. The basic relationships for the point-by-point method are derived from the geometry of the diagram shown in Fig. 68.20 and the fact that the illumination from a concentrated source varies as the square of the distance, when the distance is more than five times the maximum dimension of the source. The terms used in the equations are defined as

I_a = intensity (candlepower) at angle α from vertical obtained from intensity (candlepower) distribution
E_n = illumination on normal (perpendicular) surface
E_h = illumination on horizontal surface
E_v = illumination on vertical surface
d = horizontal distance from vertical through source to point of interest (as shown in figure)
h = vertical distance from horizontal plane to source
l = total distance from source to point of interest
α = angle that light rays make with vertical

From the inverse square law

$$E_n = \frac{I_a}{l^2} \quad \text{lx (fc)} \tag{68.9}$$

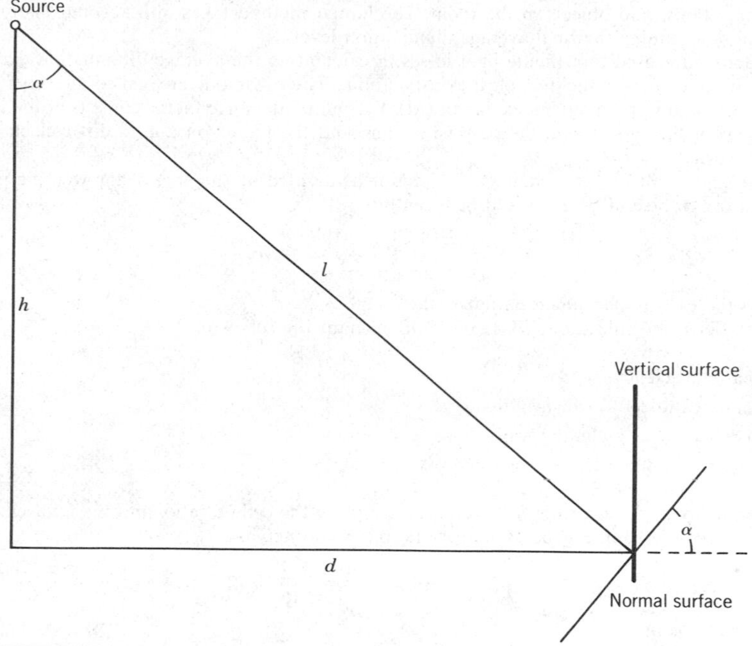

Fig. 68.20 Diagram used to define variables used in the equations for the point-by-point method of calculating size, number, and placement of luminaires to give a required level of illumination.

From geometry

$$E_n = \frac{I_a}{h^2 + d^2} = I_a \frac{\cos^2\alpha}{h^2} \tag{68.10}$$

as $l = h/\cos\alpha$.

$$E_h = E_n \cos\alpha = \frac{I_a \cos\alpha}{l^2} = \frac{I_a \cos^3\alpha}{h^2} = I_a h/l^3 \quad \text{lx (fc)} \tag{68.11}$$

Likewise

$$E_v = E_n \sin\alpha = I_a \frac{\sin^3\alpha}{d^2} = I_a d/l^3 \quad \text{lx (fc)}$$

Since for given tasks the lux (footcandles) are known or given, I_a is the variable of interest. Solving for I_a when E_h is known gives

$$I_a = E_h \frac{h^2}{\cos^3\alpha} = \frac{E_h l^3}{h} \quad \text{cd}$$

If tube sources such as fluorescent lights are used, then for long strings of these the illumination varies only inversely as the first power of the distance, so different equations would have to be developed when using these as a source.

If the light source is a large surface (infinite surface), such as a ceiling covered with fluorescent fixtures with diffusing troffers, then the illumination is constant with distance when the point of interest is away from the wall.

Lumen Method. The design of general lighting systems is determined by room dimensions; structural features; reflection characteristics of walls and ceiling; and mounting height, intensity (candlepower), distribution, and maintenance characteristics of the luminaires. The basic goal of general lighting design is to deliver a specified average lux (footcandle) level of illumination to a working plane or other plane of reference in a room. However, the light emitted by the sources is variously affected and reduced by the reflection, diffusion, and absorption of the parts of the luminaire and the walls, ceiling, floor, and objects in the room. The lumen method takes into account many of these variables in determining the final average illumination level.

Two factors are used to estimate light losses in calculating the average illumination that reaches the work surface during the life of the installation. These factors are called the coefficient of utilization (CU) and the maintenance factor (MU). The maintenance factor consists of the reduction in the output of the source with the passage of time and the reduction due to dirt collecting on the luminaires' surfaces.

Since the lux (footcandles) required are the lumens divided by the area of the working plane, the maintained lux (footcandles) E_n would be given by

$$E_h = \frac{I_a \times \text{CU} \times \text{MU}}{\text{area in m}^2} \quad \text{lx}$$

where I_a is the total initial lumen output of the lamp.

The coefficient of utilization (CU) takes into account the following:

1. Luminaire efficiency.
2. Luminaire candlepower distribution.
3. Room reflectances including walls, floor, and ceiling.
4. Room proportions in terms of a zonal-cavity ratio.

The zonal-cavity ratios consist of a room cavity ratio, the ceiling ratio, and the floor cavity ratio. The cavities are defined in Fig. 68.21, and the ratio is computed as

$$\text{Cavity ratio} = \frac{5h(L + W)}{L \times W}$$

where L = room length
W = room width
h = height of ceiling, room, or floor cavity
as shown in Fig. 68.21 for the particular cavity ratio being computed.

In the design of installations, a usual requirement is to determine how many lamps and fixtures are required to produce the desired lux (footcandles). Therefore the total initial lamp lumens is given by

$$I_a = \frac{E_h \times \text{working area}}{\text{CU} \times \text{MU}}$$

Coefficient of utilization (CU) values for a range of room ratios and wall, floor, and ceiling reflectance values for each type of luminaire are available from the manufacturer. Typical values for a few types are given in Fig. 68.22. Also included are several values of maintenance factors for different environmental conditions. Usually provided are maximum spacing factors as a ratio of the height of the source from the working plane.

The maintenance factor (MU) is equal to the product of the lamp lumen depreciation factor (LLD) and the luminaire dirt depreciation factor (LDD):

$$\text{MU} = \text{LLD} \times \text{LDD}$$

Lamp manufacturers can provide data on lumen depreciation, but a typical curve of lumens vs. life is given in Fig. 68.23. Values of the luminaire dirt depreciation factor are indicated in Fig. 68.22 by the

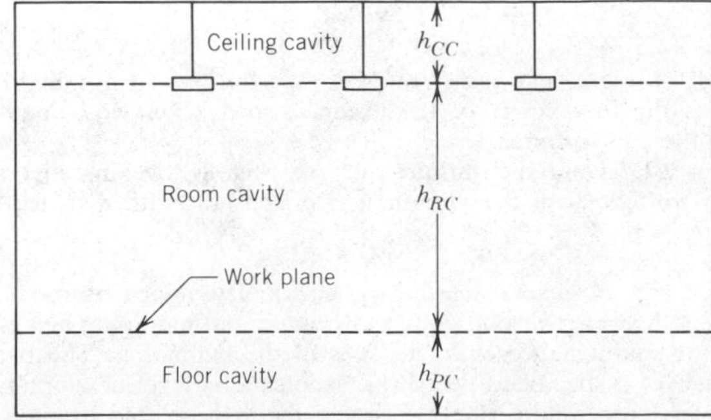

Fig. 68.21 Zonal cavities used in the lumen method of calculating size, number, and placement of luminaires to give a required level of illumination.

Typical Distribution and Maximum Spacing[a]	ρ_{CC}[c] →	80			70			50			30			10			0	Typical Luminaires and Luminaire Maintenance Category[e]
	ρ_W[d] →	50	30	10	50	30	10	50	30	10	50	30	10	50	30	10	0	
	RCR[b] ↓	Coefficients of Utilization for 20 Per Cent Effective Floor Cavity Reflectance, ρ_{FC}																
7																		
0% / 70%	1	.75	.72	.70	.73	.71	.69	.70	.68	.67	.68	.66	.65	.65	.64	.63	.62	
	2	.67	.63	.59	.65	.62	.59	.63	.60	.57	.61	.58	.56	.59	.57	.55	.54	
	3	.60	.55	.51	.59	.54	.51	.57	.53	.50	.55	.52	.49	.53	.50	.48	.47	
	4	.54	.48	.44	.53	.48	.44	.51	.47	.43	.50	.46	.43	.48	.45	.42	.41	
	5	.48	.42	.38	.47	.42	.38	.46	.41	.37	.44	.40	.37	.43	.39	.36	.35	
	6	.43	.37	.33	.42	.37	.33	.41	.36	.33	.40	.36	.32	.39	.35	.32	.31	
	7	.39	.33	.29	.38	.33	.29	.37	.32	.28	.36	.31	.28	.35	.31	.28	.27	2-lamp, 2′-wide white troffer
	8	.35	.29	.25	.34	.29	.25	.33	.28	.25	.32	.28	.25	.32	.28	.24	.23	with prismatic lens. (Multi-
	9	.31	.25	.21	.31	.25	.21	.30	.24	.21	.29	.24	.21	.28	.24	.21	.20	ply 0.9 for 4-lamp)
Max. S/MH_{wp} = 1.3	10	.28	.23	.19	.28	.22	.19	.27	.22	.19	.26	.22	.19	.26	.22	.19	.17	LDD Maint. Category V
30																		
40% / 35%	1	.70	.68	.65	.65	.63	.61	.55	.53	.52	.45	.45	.44	.37	.36	.36	.32	
	2	.62	.58	.55	.58	.54	.51	.49	.47	.44	.41	.39	.38	.34	.33	.32	.28	
	3	.56	.50	.47	.52	.47	.44	.44	.41	.38	.37	.35	.33	.30	.29	.28	.25	
	4	.49	.44	.40	.46	.41	.38	.40	.36	.33	.33	.31	.29	.28	.26	.25	.22	
	5	.44	.39	.35	.41	.36	.33	.36	.32	.29	.30	.27	.25	.25	.23	.22	.20	
	6	.40	.34	.30	.37	.32	.29	.32	.28	.25	.27	.24	.22	.23	.21	.19	.17	
	7	.36	.30	.26	.34	.28	.25	.29	.25	.22	.25	.22	.19	.21	.19	.17	.15	
	8	.33	.27	.23	.30	.25	.22	.26	.22	.20	.22	.20	.17	.19	.16	.15	.13	2-lamp prismatic lens bottom
	9	.29	.24	.20	.27	.22	.19	.24	.20	.17	.20	.17	.15	.17	.15	.13	.11	unit with open top
Max. S/MH_{wp} = 1.2	10	.27	.21	.18	.25	.20	.17	.22	.18	.15	.19	.15	.13	.16	.13	.12	.11	LDD Maint. Category VI
31																		
45% / 40%	1	.79	.76	.73	.73	.70	.68	.02	.00	.58	.51	.50	.49	.42	.41	.40	.36	
	2	.70	.65	.61	.65	.61	.57	.55	.52	.50	.46	.44	.42	.38	.36	.35	.32	
	3	.62	.56	.52	.58	.53	.49	.49	.46	.43	.41	.39	.36	.34	.33	.31	.28	
	4	.55	.49	.44	.51	.46	.42	.44	.40	.37	.37	.34	.32	.31	.29	.27	.24	
	5	.49	.43	.38	.46	.40	.36	.40	.35	.32	.34	.30	.27	.28	.25	.24	.21	
	6	.44	.38	.33	.41	.35	.31	.36	.31	.28	.30	.27	.24	.25	.23	.21	.19	
	7	.40	.34	.29	.37	.31	.27	.32	.28	.24	.27	.24	.21	.23	.20	.19	.18	Direct-indirect with metal or
	8	.35	.29	.25	.33	.27	.24	.29	.24	.21	.24	.21	.18	.20	.18	.16	.14	dense diffusing sides and
	9	.32	.25	.21	.30	.24	.20	.26	.21	.18	.22	.19	.16	.19	.16	.14	.12	35° x 45° louver shielding
Max. S/MH_{wp} = 1.3	10	.30	.23	.19	.28	.22	.19	.24	.20	.16	.20	.17	.14	.17	.14	.12	.12	LDD Maint. Category II
15																		
0% / 65%	1	.74	.72	.71	.72	.71	.70	.70	.69	.68	.67	.67	.66	.65	.65	.64	.63	
	2	.69	.67	.65	.68	.66	.65	.66	.65	.63	.64	.63	.62	.63	.62	.61	.60	
	3	.66	.63	.61	.65	.63	.61	.64	.61	.60	.62	.60	.59	.61	.59	.58	.57	
	4	.63	.60	.58	.62	.59	.57	.60	.58	.57	.59	.58	.56	.58	.57	.55	.55	
	5	.60	.57	.54	.59	.56	.54	.58	.56	.54	.57	.55	.53	.56	.54	.53	.52	
	6	.57	.54	.52	.57	.54	.52	.56	.54	.52	.55	.53	.51	.55	.53	.51	.50	
	7	.55	.52	.50	.55	.52	.50	.54	.51	.50	.53	.51	.49	.53	.51	.49	.48	
	8	.53	.50	.48	.52	.50	.48	.52	.49	.47	.51	.49	.47	.51	.49	.47	.46	
	9	.50	.47	.45	.50	.47	.45	.49	.47	.45	.49	.47	.45	.48	.46	.45	.44	R-40 flood with reflector skirt
Max. S/MH_{wp} = 0.6	10	.48	.45	.43	.48	.45	.43	.47	.44	.43	.47	.44	.42	.46	.44	.42	.42	LDD Maint. Category IV

[a] Ratio of maximum spacing between luminaire centers to mounting (or ceiling) height above the work plane. See "Luminaire Spacing" on page 9-16.
[b] RCR = Room Cavity Ratio.
[c] ρ_{CC} = Per cent effective ceiling cavity reflectance.
[d] ρ_W = Per cent wall reflectance.
[e] See pages 9-16 and 9-17.

Fig. 68.22 Coefficients of utilization (CU) for 20% effective floor cavity reflectance. LLD, lamp lumen depreciation. First column is the ratio of maximum spacing between luminaire centers to mounting (or ceiling) height above work plane. From *IES Lighting Handbook*,[4] reprinted with permission.

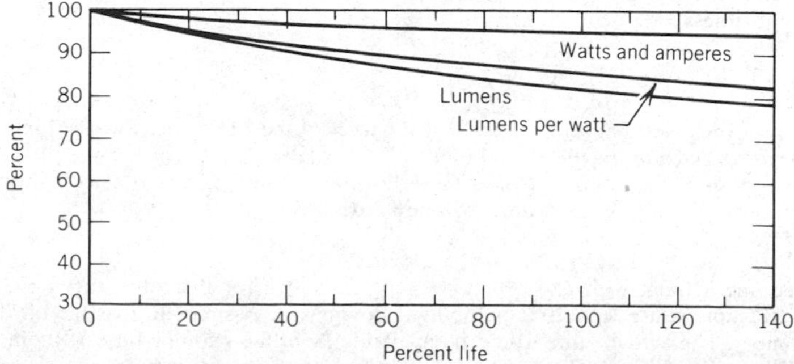

Fig. 68.23 Typical lamp lumen depreciation (LLD) values as a function of lamp life.

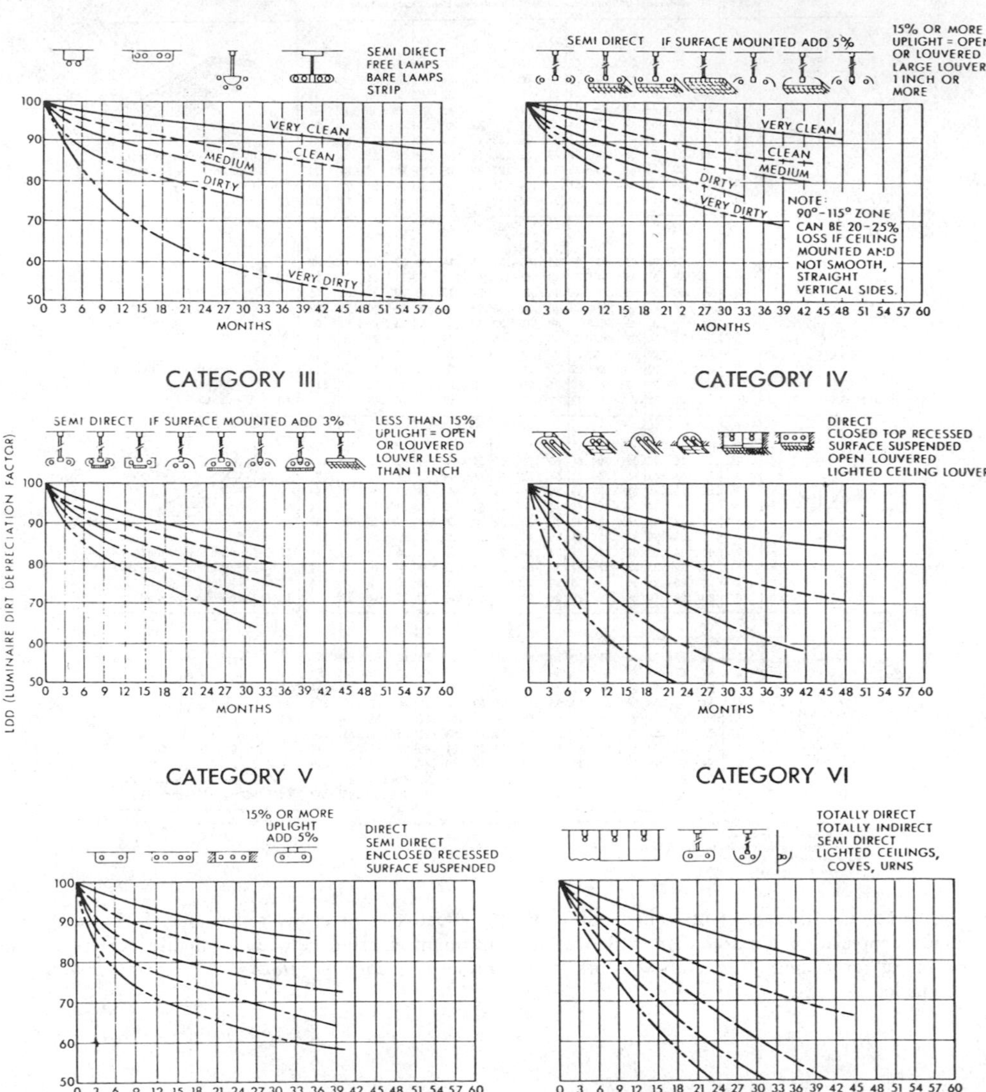

Fig. 68.24 Luminaire dirt depreciation factors (LLD) for six luminaire categories (I to VI) and for five degrees of dirtiness.

LDD category for the particular luminaire. Typical curves of the LDD are shown in Fig. 68.24 for the luminaire categories given in Fig. 68.22 and others.

As an illustration of the lumen method, the following example is given (since lighting levels are still given in footcandles, the English units will be used here).

Example. Assume a room with $L = 100$ ft, $W = 60$ ft, and h for the room cavity = 7 ft. Use the two-lamp fluorescent fixture with 70% of the light downward. Assume an average of 70 lm/W for fluorescent lamps. (The actual value would be available from the manufacturer.) Room reflectances are: ceiling 80%, walls 50%, and floor 20%. The required illumination is 35 fc.

The cavity ratio (h_{RC}) for the room zone is

$$h_{RC} = \frac{5h(L + W)}{L \times W} = \frac{5(7)(100 + 60)}{100(60)} = 0.93$$

Use $h_{RC} = 1$.

From the coefficient of utilization (CU) in Fig. 68.22, CU = 0.75.

A maintenance factor (MU) of 0.7 is used, obtained by multiplying an LLD factor of 0.8 from Fig. 68.23 for a lamp life of 120% by the LDD of 0.85 for category V from Fig. 68.24. This assumes a "clean" luminaire maintained every 18 months.

$$\text{Total lumens required} = \frac{\text{fc} \times \text{floor area}}{\text{CU} \times \text{MU}} = \frac{35(100)(60)}{0.75(0.7)} = 400{,}000 \text{ lm}$$

$$\text{Total wattage} = \frac{400{,}000}{70 \text{ lm/W}} = 5714 \text{ W}$$

With two 4-ft 40-W fluorescent tubes per fixture, each fixture produces approximately 5600 lm.

$$\text{Number of fixtures} = \frac{400{,}000}{5600} = 72$$

The maximum space ratio is $S = 1.3$ (from Fig. 68.22). Therefore the maximum space between fixtures is 1.3(7) = 9.1 ft. Because of the spacing limitation, seven rows of 11 fixtures each would be required, giving 37.4 fc average illumination.

Only the fundamentals for calculating the illumination have been presented. Many other factors need to be determined to arrive at the final design. Some of these are color of the light, glare, uniformity of illumination, shadows, sound produced by lamp ballasts, initial cost, efficiency, operating costs, maintenance, and heat dissipation. A review of some of these factors is given in IEEE Standard 241-1974, *Recommended Practice for Electric Power Systems in Commercial Buildings*, Chapter 10, and in Ref. 5.

References

1 ANSI/NEMA Standards Publ. No. MG1-1978, *Motors and Generators*, National Electrical Manufacturers Association, Washington, DC, 1978, with revisions.

2 ANSI/NEMA Standards Publ. No. ICS2-1978, *Industrial Control Devices, Controllers and Assemblies*, National Electrical Manufacturers Association, Washington, DC, 1978, with revisions.

3 R. A. Millermaster, *Harwood's Control of Electric Motors*, 4th ed., Wiley, New York, 1970.

4 *IES Lighting Handbook*, 5th ed., Illuminating Engineering Society, New York, 1972.

5 J. O. Kraehenbuehl, *Electric Illumination*, Wiley, New York, 1951.

Bibliography

Heumann, G. W., *Magnetic Control of Industrial Motors*, 2nd ed., Wiley, New York, 1954.

Kaufman, J. E., ed., *IES Lighting Handbook*, 6th ed., Illuminating Engineering Society, New York, 1981.

Kosow, I. L., *Control of Electric Machines*, Prentice-Hall, Englewood Cliffs, NJ, 1973.

McPartland, J. F., ed., *National Electrical Code^{R} Handbook*, McGraw-Hill, New York, 1981.

CHAPTER 69
ENERGY MANAGEMENT

WAYNE C. TURNER

Oklahoma State University
Stillwater, Oklahoma

69.1 WHAT IS ENERGY MANAGEMENT?[†]

Energy management is:

> *The judicious and effective use of energy to maximize profits (minimize costs) and enhance competitive positions.*

Therefore energy management is energy conservation, but it is also much more. In fact, any technique or study that encourages the wise use of energy is energy management.

Designing equipment to have dual fire capability and standby fuel storage is energy management, for it may mean the difference between continued operation of a plant or complete shutdown. It is certainly not energy conservation, since no Btus (British thermal units) are saved. Demand leveling for minimization of electrical utility bills is another example of energy management that is not energy conservation. Of course any energy conservation effort that leads to less costs and more profit is energy management. *In the author's opinion, any energy conservation effort that leads to less overall profit (more total cost) is poor energy management.*

The primary objective of energy management is, again, profit improvement (cost reduction) and enhancement of competitive position, but leading to these overall goals are some subsidiary objectives such as

- Energy conservation.
- Good communications on energy matters.
- Effective energy monitoring and reporting mechanisms.
- Maintenance of employee comfort, safety, and morale at satisfactory levels
- Incorporation of new products or services that may be marketable to others in energy management.
- Continued steady supplies of goods and services during curtailments or brownouts.

[†]Much of the material presented herein is drawn from W. C. Turner, ed., *Energy Management Handbook* and G. Salvendy, ed., *Industrial Engineering Handbook* both Wiley-Interscience, New York, 1982.

TABLE 69.1 TYPICAL RESULTS OF ENERGY MANAGEMENT PROGRAMS[a]

Level	Action	Accumulated Savings
1	Low cost–no cost activities only	5–10%
2	Capital intensive, engineering design	25–30%
3	Dedicated long-range programs	40–50%

[a] Savings are cumulative for each level and for all succeeding levels. For example, level 2 savings are for level 1 and level 2 activities.

TABLE 69.2 REASONS FOR INVOLVEMENT IN ENERGY MANAGEMENT

Large savings opportunities
Improvement in balance of payments, as less energy needs to be imported
Energy prices for various sources rose up to eight-fold in the period 1971–1979
Usually reduces pollution and improves the environment

Since this is a handbook on electrical and electronic engineering, we can add the following:

Maintenance of voltage level and consistency for peak performance of electrical machinery.
Maximization of plant power factor for minimum billing cost and peak equipment performance.
Minimization of demand during "peak" periods to reduce billing costs.

It is difficult to predict results from good energy management, since results vary widely owing to the nature of the business, geographic location, local utility billing procedures, and other factors. However, savings in energy consumed have run as high as 70% over original costs and seem to run as demonstrated in Table 69.1.

The literature shows many case studies with savings of 40% to 60%. General Services Administration (GSA) recently constructed an office building utilizing only 20% the energy normally used. With savings like this, it is no wonder energy management is popular. However, if more reasons are needed, Table 69.2 presents some.

69.2 DESIGNING, INITIATING, AND MANAGING ENERGY MANAGEMENT PROGRAMS

Program Design

Energy management programs are often "made" or "broken" in the design phase before they are ever implemented. Several ingredients seem to be necessary for success. Management commitment is the most important single factor. It is vital that management be dedicated to the concept of energy management and, more importantly, the dedication must be demonstrated. For example, some managers show their dedication by personally giving presentations at plant-wide or departmental meetings. In these meetings, management can show why the program is necessary, what results are expected, and what the program means to the employees. Some managers send a personal letter to all employees explaining the same. Some use billboards and posters, and some do all this plus more. Whatever method is chosen, management must realize that *funding cost-effective proposals* is the most important single way of demonstrating. Management everywhere faces capital budgeting problems and, unfortunately, energy projects can often be postponed easily. Postponing is a sure way, however, of destroying enthusiasm.

Organizationally, there should be one person who has complete responsibility for the energy management program. This person should be talented and should report high in the organization. More than likely, the person will be an engineer with several years of plant level experience. This person is often called the energy management coordinator (EMC).

If the plant utilizes a large amount of electricity, an electrical or electronics engineer is a good candidate for this position. Industrial, mechanical, and chemical engineers also are good candidates. It is most important that this person be dedicated to the principles of energy management.

To date, no one discipline covers all the techniques necessary for a dynamic energy management program, so the EMC needs a backup crew that can be called on as needed. This group can be a formal committee or an informal group.

Finally, employee involvement is necessary for a successful program. Employees know their job best and can often initiate ideas with substantial savings. They also like to be kept informed and usually will not cooperate (like most of us) if not informed. Their involvement can be obtained through suggestion systems, newsletters, membership on energy committees, and other creative ideas applicable to unique situations.

Program Start-Up

Early activities are the most critical ones for energy management programs. Everyone is curious and quite a few are skeptical. Early failures will reinforce skepticism and dampen excitement, but early success will go a long way toward winning over skeptics and further motivating supporters.

Consequently, early projects should be chosen that

1. Have a high probability of success.
2. Have few if any negative consequences.
3. Have large savings with favorable economic consequences.

For example, one company replaced mercury vapor lights in one of two refrigerated warehouses with high-pressure sodium lights. The system was carefully designed to improve visibility. Savings were large in that high-pressure sodium illumination requires less energy to operate and also produces less waste heat that must be offset by the refrigeration. The project was a success, to the point that the employees in the other refrigerated warehouse asked for high-pressure sodium lights in their area.

In another organization, over $10,000 per year was saved by shifting the operation of a large electric motor to a time not covered by peak demand.

Projects like these can be found in any plant if they are sought.

Program Management

Creativity is vitally important for a successful energy management program. Real progress usually occurs through a combination of engineering design and creative concept development. Management of the energy management program should encourage rather than stifle creative thought. Assuming the right person was chosen for the EMC, management should let that person run the program and judge by results, not by procedures.

Also management must encourage motivation through recognition of accomplishments. Bonuses, letters in employee files, or even a simple pat on the back all help cultivate enthusiasm.

Energy accountability through good monitoring and control procedures is vitally important. Ideally, each cost center in the plant should have its energy flow measured and monitored. This way, management of each cost center can be held accountable for the energy consumed. Unfortunately, the previous practice has been to meter energy at only one or a few locations in the plant, so the desired level of metering is almost always not available and retrofit is not practical. Management should strive, however, to evolve toward this ideal, so that someday energy will be a part of the budgetary process instead of overhead, as it is today.

Several companies have made significant strides toward good energy accounting systems. General Motors, 3M, and Carborundum are three of the better known ones. Seidman and Seidman, a consulting firm, has also done significant work in this area.

69.3 ENERGY AUDITS

The words "energy audits" mean different things to different people. In fact the literature shows entirely different meanings depending on the author and the situation. Basically, however, an energy audit is designed to answer one, two, or all three of the following questions:

1. How much energy is being consumed of each type and what is the cost?
2. Where is the energy being consumed (major users)?
3. What changes can be made to improve operations that
   (a) require little to no capital investment?
   (b) require significant amounts of capital investment?

While it may be desirable to answer these questions in the order they are given here, many companies skip either question 1 or question 2 or both to be able to get to question 3 quickly. Certainly this can be done with some success, but eventually most companies do answer at least question 1. These questions will be analyzed here one at a time.

1. How much energy is being consumed of each type and what is the cost? This question can be easily answered through what is often called a gross audit. The gross audit is a simple audit designed to tell a company how much energy is being consumed and how much it costs. While doing this, the audit may demonstrate how efficiently the energy is being used. For example, power factor and demand billing components sometimes offer dramatic cost reduction opportunity with little change in the amount of energy consumed.

A sample form is presented in Fig. 69.1. This form contains substantial information, perhaps more or less than may be needed for a particular study. The form should be designed for the organization's specific needs. Other sources provide a number of different forms, whose total contents are approximately Fig. 69.1.

Before going through the form, it is advisable to examine the utility billing schedules. Most sources are simple and involve a rate per unit of consumption only. For example, gasoline is billed at so much per gallon, coal per ton, and so on. Electricity, however, is more complicated and will be examined here.

The first level of electrical billing involves a charge per kilowatt hour (kWh) (measure of electrical usage equal to 3.6×10^5 cal or 3412 Btu for one hour) of electricity consumed. This is normally on a declining-block basis, in that the rate per kilowatt hour decreases as consumption increases. This rate schedule is normally used for residential and small commercial establishments.

The next schedule (larger commercial and small industry) involves a similar rate schedule per kilowatt hour *and* a demand charge. To understand demand charging, consider Fig. 69.2. In this figure, the demand on the power company in kilowatts is plotted against time for the due month for two companies, A and B. For company A, the demand fluctuates dramatically both within a day and between days. Customer B's demand stays relatively constant. Note that the utility company has to provide the peak demand capability to the companies and that A's peak demand is considerably higher than B's. Since the kilowatt hours of electricity consumed is the area under the curves and B's peak demand equals A's average demand, the same total number of kilowatt hours is consumed by each.

Obviously, the utility would rather serve customer B, so the utility charges a rate per kilowatt for the peak demand in addition to the consumption charge. This encourages A to flatten the profile toward B's, since the cost would be reduced (less demand charge) for the same total consumption.

Peak demand is usually measured as the average demand over a certain period (5, 15, or 30 minutes). Therefore, equipment requiring warm-up, such as electrical furnaces, contributes a large amount toward demand, while quick-start equipment, such as electric motors, does not add considerably to demand since the start-up surge lasts only a few seconds.

The next level of billing maturity and complexity includes consumption and demand billing as well as a new item—power factor. Inductive loads such as inductive motors, furnaces, and fluorescent lights develop a reactive current that is out of phase with the working current. This reactive current does no real work, yet the distribution system must carry it and the utility supply must be sized for it. This is shown in the phasor diagram, Fig. 69.3.

The company can supply its own reactive power through capacitors (which are simple devices capable of storing energy), synchronous motors, or synchronous condensors. Consequently, large industry is usually on a rate schedule penalizing it (in dollar cost) for low power factors (often 75% or lower) and sometimes rewarding it for high power factors (often 85% or higher). The actual method of billing varies widely, but the net result is encouragement to raise the power factor.

Required calculations are summarized as follows:

$$\text{kilowatts} = (\text{horsepower})\left(0.746\,\frac{\text{kW}}{\text{hp}}\right)$$

$$\text{kilowatts} = (\text{volts})(\text{amps})\left(\sqrt{3}\right)^{\dagger}(\text{power factor})\left(\frac{1}{1000}\right)$$

Volts and amps are measured quantities.

$$\text{kilovolt amperes} = (\text{volts})(\text{amps})\left(\sqrt{3}\right)^{\dagger}\left(\frac{1}{1000}\right)$$

†For a three-phase motor. For a single phase motor, simply drop $\sqrt{3}$ and substitute a 1.

Gross Audit of All Energy Resources

Date of audit ________

Date of last audit ________

			Electricity					Natural Gas		
Year / Month	Units of Production (1)	kWh (2)	kW Demand Actual (3)	kW Demand Billed (4)	Power Factor (5)	Total Cost (6)	$(2) \times A$[a] Btu/mo. (7)	kcf (8)	Total Cost (9)	$(8) \times 8$ Btu/mo. (10)
Jan										
Feb										
Mar										
Apr										
May										
Jun										
Jul										
Aug										
Sep										
Oct										
Nov										
Dec										
Total										
Conversion factors		A = 10,000 Btu/kWh or A = 3412 Btu/kWh						B = 1,000,000 Btu/kcf		

[a] 10,000 Btu/kWh is conversion to electricity at source, 3412 Btu/kWh is the Btu equivalent of 1 kWh. Whichever one is used, be sure to be consistent when calculating consumption and savings.

Fig. 69.1 Gross energy audit form.

Typically an electric utility has three or more rate schedules progressing as just discussed. The choice of rate schedule determines the gross audit form section for electricity in that the schedule dictates what columns are necessary.

Often a company can save money by going to a higher or lower rate schedule. Also utilities can make mistakes (but not very often); the energy management coordinator or someone should check each bill.

To fill in the gross audit form or forms, utility records for all energy supplies should be made available for one or two years past. Two is best, but one is acceptable. Then the data can be entered. The form is self-explanatory except for two additional comments.

1. Billed demand may differ from actual demand if the company is on a ratchet clause, where previous peaks can override actual peaks, or when power factor billing involves a modification of demand.
2. The kilowatt hour is 3.6×10^5 cal, or 3412 Btu, but the form lists 10,000 as the conversion to British thermal unit. This simply takes into account generating efficiency.

2. Where is the energy being consumed (major users)? This question can best be answered through a comprehensive engineering study involving heat balance equations or lots of metering.

Gasoline			Other Fuels			Monthly Totals			
Gallon (11)	Cost (12)	(11) × C[a] Btu/mo. (13)	# Units (gal, ft[a]) ton, etc.) Specify (14)	Cost (15)	Btu # Units × Btu/Unit (16)	(6) + (9) + (12) + (15) Cost (17)	(7) + (10) + (13) + (16) Btu (18)	(18) + (1) Btu/Unit (19)	Comments
C = 130,000 Btu/gal			D = $\overline{\text{specify}}$ Btu/Unit						

Fig. 69.1 (*Continued*)

Consequently, it is a time-consuming and expensive endeavor. Figures 69.4 and 69.5 present typical forms that might be used for this type of audit.

The analyst should not attempt to account for 100% of the energy consumed in the plant. This would entail examining all clocks, water coolers, lights, small electric motors, and so on. These really contribute little. Instead the analyst should attempt to account for 75% to 80% of the total energy consumed. By examining only the major consumers (few in number), 80% or more of the energy consumption can usually be found (ABC principle).

Many companies do not do this type of audit for the following reasons:

1. They are expensive and time consuming and really save little to no energy directly.
2. Energy management programs can be conducted without them (although not as comprehensively).
3. Most good plant engineers can "ballpark" these figures without doing the actual metering.

3. What changes can be made to improve operations? This audit is the first real step toward saving energy dollars. The audit is a tour (or a number of tours) of the facility with checklists and an open mind and the delineation of a number of energy management ideas.

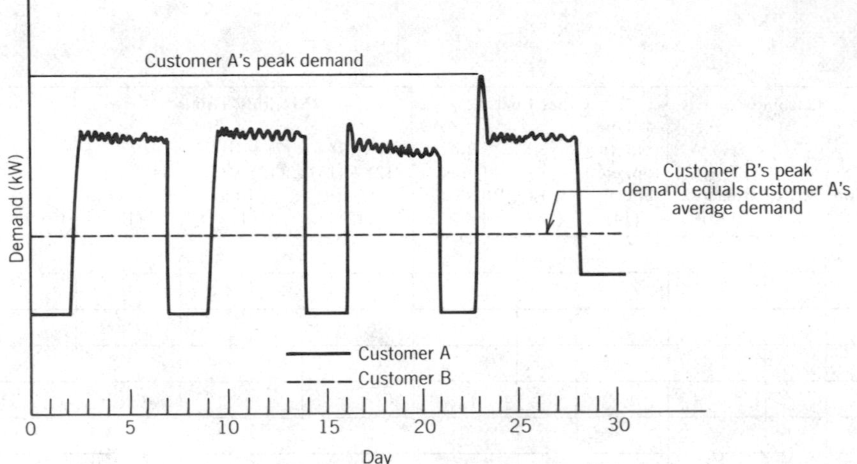

Fig. 69.2 Demand billing.

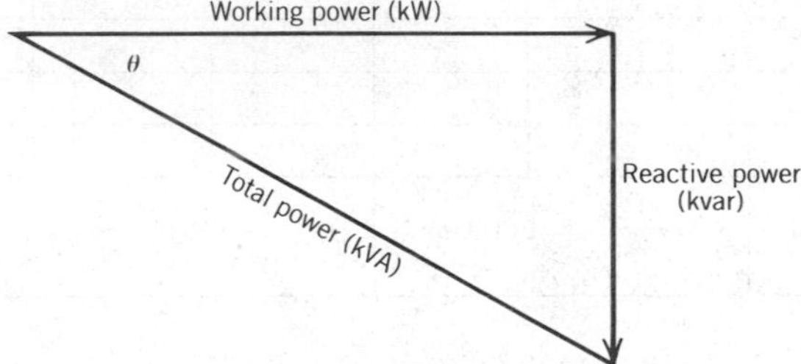

Fig. 69.3 Phasor diagram explaining power factor. Power factor = working power/total power $= \cos\theta$.

There are basically two types of energy management ideas. The first is operating and maintenance changes (O&M) (often called low-cost no-cost ideas) that involve little or no investment or engineering design. The second is energy management opportunities (EMOs) that require investment and/or engineering design. Naturally, the system should be fine tuned by implementing all applicable O&M changes before instituting many EMOs. O&M changes, however, are long-lived possibilities and will always be around, so their implementation never stops. Frequently, systems tend to deteriorate, so the same O&M idea may be found useful each time an audit is done.

Some believe the checklist should only include O&M ideas, while others believe it should contain EMO ideas also. Probably some combination is desirable, since problems and ideas do not separate according to this classification and an audit will likely yield numerous ideas in both categories.

The audit is not a one-time occurrence. It should be repeated periodically and at different times. Likely times are:

1. During operating hours alone.
2. During operating hours with others (management, employees, plant engineer, etc.).
3. During nonoperating hours (night, 2 or 3 AM) to look for air handlers or other equipment left running, lights left on, air or steam leaks (they can be heard), and so on.
4. During nonoperating hours (weekend) for essentially the same reasons as in 3.

There is no shortage of forms or checklists. Several are referenced and a sample is provided in the next section.

Date of audit ________
Date of last audit ________

Description of Unit	Rated Horsepower[a] (hp) (1)	Wattage[b] (W) (2)	Operating Hours (h) (3)	"Avg"[c] (Average) Load (4)	(2) × (3) × (4) (Wh) (5)	(5) ÷ 1000 (kWh) (6)	(6) × 10,000 (Btu) (7)	Comments
Total								

[a] 1 hp = 746 W.
[b] This can be taken off nameplate of equipment.
[c] For motors, average load is usually 70%, therefore insert 0.70. If wattage has been measured use 1.0; for fluorescent tubes, add 20% for ballast, that is, insert 1.20.

Fig. 69.4 Detailed audit for electrical units.

	Natural Gas				Gasoline			
Description of Unit	kcf/hr (1)	(1) × 10^6 (Btu/hr) (2)	Operating Time (3)	(2) × (3) (Btu) (4)	Miles or Hours (5)	Gallons (6)	(6) × (1.3 × 10^5) (Btu) (7)	Comments
Total								

Fig. 69.5 Detailed audit for natural gas and/or gasoline units.

69.4 CHECKLISTS OF ENERGY MANAGEMENT PROJECTS

Operating and Maintenance Changes

Most buildings in use today were designed when energy was inexpensive and comfort was paramount. Consequently, architects and engineers designed a great deal of flexibility and maximum provisions into the systems. This means that a large amount of control is available, and that *without careful monitoring* energy-consuming systems will tend to move toward overenergy consumption. Consequently, many operation and maintenance changes are available that cost practically nothing to implement but save considerable energy. The following is a sample list drawn from *Instructions for Energy Auditors*, US Dept. of Energy, p. 63 on.

Maintenance and Operational Changes

Space Heating

1. Lower thermostats during the heating season and raise thermostats during the cooling season. If your building is like most, you can save about 8% of your heating fuel bill by lowering the thermostat(s) a mere 5°.
2. Use night setback on heating systems with zone controls. Maintaining 55°F or 60°F at night will reduce energy consumption by 5% to 6% during the night hours.
3. Check burner firing period. If it is improper, it could be a sign of faulty controls.
4. Check automatic temperature-control system and related control valves and accessory equipment to ensure that they are regulating the system properly.
5. Check boiler stack temperature. If it is too high (more than 150°F above steam or water temperature), clean tubes and adjust fuel burner.
6. Check flue gas analysis on a periodic basis: The efficient combustion of fuel in a boiler requires burner adjustment to achieve proper stack temperature of no more than 150° above steam or water temperature. There should be no carbon monoxide. For a gas-fired unit, CO_2 should be present at 9% or 10%. For #2 oil, 11.5% to 12.8%, for #6 oil, 13% to 13.8%.
7. Adjust air–fuel ratios of firing equipment; the air to fuel ratio must be maintained properly. If there is insufficient air the fire will smoke and cause tubes to become covered with soot and carbon and thus to lose heat, wasting energy. Most fuel services companies will test your units for a token fee and provide you with specific recommendations.
8. Examine operating procedures when more than one boiler is involved. It is far better to operate one boiler at 90% capacity than two at 45% capacity each. The more boilers used, the greater the heat loss.
9. Lower steam pressure to the minimum pressure that will satisfy needs.
10. Turn off the boiler natural gas standing pilot during the summer months when the boiler is off.
11. Institute an operating procedure for multiple boiler plants to ensure maximum loading of one boiler before a second boiler is put into service.
12. Do not heat parking garages, docks, and platform areas.
13. Shut off boilers in the spring and fall when the air conditioning machine is on and temperature control is not needed.
14. Feel the pipe on the downstream side of steam traps. If it is excessively hot, the trap is probably passing steam. This can be caused by dirt in the trap, excessive steam pressure, or worn trap parts (especially valve and seats). If it is moderately hot, as hot as a hot water pipe, for example, it probably is passing condensate, which it should do. If it is cold, the trap is not working at all. Replace or repair it.
15. Keep a daily log of pressure, temperature, and other data obtained from instrumentation. This is the best method available to determine the need for tube and nozzle cleaning, pressure or linkage adjustments, and related measures. Variations from normal can be spotted quickly, enabling immediate action to avoid serious trouble. On an oil-fired unit, indications of problems include an oil pressure drop, which may indicate a plugged strainer; faulty regulator valve; or an air leak in the suction line. An oil temperature drop can indicate temperature control malfunction or a fouled heating element. On a gas-fired unit, a drop in gas supply pressure can indicate a drop in gas supply pressure or malfunctioning regulator.
16. Reduce blowdown losses.

Ventilation

1. Reduce fresh air to legal limits. Often ventilation systems draw excess fresh air. This requires warming the air to the inside temperature at the cost of energy.
2. Reduce cubic feet per minute per occupant of outdoor air requirements to the minimum, considering the tasks the occupants are performing, the room volume, and periods of occupancy.
3. Reduce exhaust air quantities to practical limits.
4. Establish a ventilation operation schedule so exhaust system operates only when it is needed.
5. Reduce outdoor air to the minimum required to balance the exhaust requirements and maintain a slight positive pressure to retard infiltration.
6. If possible, use permanently sealed windows to reduce infiltration in climatic zones where this is a large energy user.
7. Operate the ventilation system only when the building is occupied. Also consider shutting off the air handling units on normal heating days before the building is emptied. If the radiators are located properly, they should be able to maintain space temperature above freezing.
8. Increase the mixed air temperature setting on units to 65°F. If this is not practical, consider adjusting the fresh-air linkage so that the mixed air temperature cannot go below 55°F.
9. In summer when outdoor air temperature at night is lower than indoor temperature, use full outdoor air ventilation to remove excess heat and precool the structure, to reduce the air-conditioning load.
10. Adjust the time clock day–night settings to operate ventilation units fewer hours during the day cycle.
11. Adjust all VAV (variable air volume) boxes so they operate precisely. This will prevent overheating or overloading, both of which waste energy.
12. In noncritical areas, reduce the fresh air drawn into the ventilation system to 25%. If a 50% situation is reduced to 25%, there will be a 50% saving in the heat required to raise the incoming air to indoor conditions.

Air Conditioning

1. Set room temperature at 78°F in summer.
2. Use water-cooled refrigeration units rather than air-cooled ones since the former are up to 20% more efficient.
3. Turn off the cooling system during the night. Use ventilation air to cool the building at night.

Central Air Handling Equipment

Operating Procedures

1. Whenever possible, use outdoor air for cooling rather than mechanical refrigeration. Use the economizer cycle, where installed.

Hot Water

1. Reduce generating and storage-temperature levels to the minimum required for washing hands, usually about 110°F. Boost hot water temperature locally for kitchens and other areas where it is needed, rather than provide higher than necessary temperatures for the entire building.
2. Deenergize booster heaters in kitchens at night.
3. Consider replacing existing hot water faucets with spray-type faucets with flow restrictors where practical.
4. If water pressure exceeds 40 to 50 lb, install a pressure-reducing valve on the main service to restrict the amount of hot water that flows from the tap.
5. Operate only one of the domestic hot water heaters. If one unit carries the load, leave the other off for standby.
6. Inspect insulation on storage tanks and piping. Repair or replace as needed.
7. Reduce laundry-room hot water temperature to 160°F. This temperature will achieve the removal of soils, blood, and so on, which is the need of a hospital laundry. Sterilization should be done on

a separate basis. Hot water temperature reduction from 180°F to 160°F will save 160 Btu per gallon of water heated or 2 million Btu/yr per patient.

8. Use improved cakepan coatings to minimize amount of washing water needed. If the length of the washing cycle can be reduced to 25%, the energy and dollars for washing go down accordingly.

Domestic Water

1. If water pressure is in excess of 40 psi, use a pressure-reducing valve to lower the pressure.
2. Test hot water heater controls.

Lighting

1. Remove unnecessary lamps and/or replace present lamps with higher efficiency units.
2. Use energy-conserving fluorescent lamps. These lamps save about 15% of the energy for the same illumination.
3. Remove lamps or fixtures. If only lamps are removed, disconnect ballasts, since a ballast accounts for 10% to 30% of the lamp's power drain.
4. Control exterior lighting. If a photocell is used to turn on the lamps and a timer to turn them off, there may be savings of as much as a third of present consumption.
5. Shut off lights in unoccupied rooms.
6. Reduce area lighting in all intense lighting areas and substitute task lighting.
7. Use color-coded light switches to avoid nonessential lights being put on during nonselling hours; designate those needed for cleaning, register reading, and security.
8. Set a limit of 10 minutes after closing for the bulk of floor lights to be on; full night schedule to be in effect within 30 minutes.

General Buildings

1. Prepare an energy profile of the building in as much detail as possible.
2. Conduct a survey of the total building on a space-by-space basis to determine actual user needs.
3. A low power factor on an electrical system within a building will increase the losses in the electric utility system and reduce the system's capacity. Many electric utility companies have a penalty charge for low power factor. Correcting the power factor can provide for more efficient use of energy as well as a reduction in the cost of electricity. Electrical devices known as capacitors can be installed to correct low power factor.
4. Put each apartment on separate electric meters. Lease without electric utility or pay rewards for usage below experiential standards.
5. Make the monthly energy consumption and cost data available to the manager and chief operating engineer so that they can evaluate and compare against months and normal budget.
6. Involve total building staff with energy conservation measures so that each individual has responsibility.
7. Provide a temperature control training program for operating engineers that will give them a thorough understanding of how the heating and ventilating system was designed to operate. Include optimization of energy via temperature control.
8. Install storm windows or double-glaze windows throughout. A single 36 ft^2 window will save about 3.5 million Btu per year with storm windows added.
9. Examine the entire building for air leaks, around windows, doors, and any other place that leaks might occur. Seal up the leaks. Open windows or outside doors during the heating or cooling seasons are criminal if the heaters or air conditioners are running.
10. Provide proper insulation for all equipment that is heated or refrigerated. This includes tanks, ovens, dryers, washers, steam lines, boilers, and refrigerators.
11. Add insulation to the roof whenever the roof is going to be resurfaced or repaired. If no insulation is already in place, do not wait—put insulation on immediately.
12. Close off unused areas and rooms. Where possible, be certain that blinds or other shading devices are drawn, registers closed, and so on.
13. Utilize solid-state motor drives instead of motor generator sets for elevators.

Laundry Department

1. Use cold water detergents where permitted and satisfactory for your purposes.
2. Repair all steam leaks promptly.
3. Do not take make-up air from air-conditioned spaces. Laundry make-up air should be taken from the outdoors. Stealing air from the conditioned spaces results in infiltration into them of outside air. Such air has to be cooled and dehumidified, a needless waste of energy.
4. Set schedules to reduce peak use and demand charges. Whenever possible, make arrangements to do laundry work during periods when the least amount of energy is being used at the property (nonpeak demand hours). If practical, try to limit the amount of equipment being used at the same time.

Food Preparation and Storage Equipment

1. Cook in largest volumes possible.
2. Cook meat slowly at low temperatures. Cooking on roast for five hours at 250°F could save 25% to 50% of the energy that would be used in cooking for three hours at 350°F.
3. Use the minimum amount of compressed air at the minimum required pressure. Have an outside air intake for the compressor. Compressed air is very expensive and should be avoided if at all possible.
4. Provide ovens, fryers, and washers with loads all of the time they are heated and on. An oven not baking one hour out of seven is an oven wasting 14% of its energy.

69.5 ENERGY MANAGEMENT OPPORTUNITIES

The really big savings in energy management usually require some engineering design and capital expenditure. Consequently, it is best to fine tune the system with O&M changes before attempting these energy management opportunities (EMOs). The following is a sample list, also drawn from *Instructions for Energy Auditors*, US Dept. of Energy, p. 83 on.

Energy Measures

Space Heating

1. Preheat oil to increase efficiency.
2. Replace existing boilers with modular boilers.
3. Use waste heat from pan driers to preheat make-up water. Each gallon of water each degree Fahrenheit raised saves 8 Btus.
4. Preheat combustion air to increase boiler efficiency.
5. Reduce blowdown losses. Blowing down a boiler has two purposes: (1) to maintain a low concentration of dissolved and suspended solids in the boiler water and (2) to remove sludge in the boiler to avoid priming and carryover. There are two principal types of blowdown: intermittent manual blowdown and continuous blowdown. Manual blowdown (or sludge blowdown) is necessary whether or not continous blowdown is installed, with frequency depending on the amount of solids in the boiler make-up water and the type of water treatment used. Continuous blowdown results in a steady energy drain because make-up water must be heated. In either case, blowdown energy losses can be minimized by installing automatic blowdown controls and heat recovery systems. Automatic blowdown controls monitor the conductivity and pH of the boiler water periodically and blow down the boiler only when required to maintain acceptable water quality. Further savings can be realized if the blowdown water is piped through a heat exchanger or a flash tank with a heat exchanger. In this way, for example, heat from the boiler blowdown flash tank can be used for feed water heaters.
6. Replace old, inefficient burners with new, efficient ones. Changing a 60% unit for a 75% one with a 25,000 gal/yr experience will save 5000 gal or more than $2000/yr. Payback is about one year.
7. Install loading dock door seals. While it may not be possible to completely seal off such an opening, the effort to do so is rewarding in energy savings.
8. Insulate hot, bare heating pipes. A 160°F bare hot water pipe, $1\frac{1}{2}$ in. in diameter, not insulated, will lose 13 million Btu/yr for each 10-ft length. Economic thicknesses can be supplied by contractors.

9. Replace all damaged insulation on heating pipes, including those in steam, condensate, hot water supply, and hot water return systems.

Ventilation

1. Consider installing economizer–enthalpy controls on air-handling units in noncritical areas to minimize cooling energy required, by using proper amounts of outdoor and return air to permit "freeze cooling" by outside air when possible.
2. When more than 10,000 cfm are involved, and when building configuration permits, consider installation of heat recovery devices such as a rotary heat exchanger.
3. Consider utilizing revolving doors for main access, in addition to the swinging doors needed by those in wheelchairs or on crutches.
4. In locations where strong winds occur for long duration, consider installing wind screens to protect external doors from direct blast of prevailing winds.
5. Install an automation system to operate the ventilation units so that supply air temperature and return air–fresh air dampers can be adjusted to maintain the desired space temperature in the room.
6. Consider installation of an air curtain, especially in delivery areas.

Air Conditioning

1. Replace inefficient air conditioners. Newer units may save as much as 25% or more on the energy consumed for the same cooling.

Domestic Hot Water

1. Install shower head restrictors. This may save up to 50% of the hot water consumed.
2. Change the shower heads from the existing gallons per minute rating to a 2-GPM rating.
3. Install a small domestic hot water heater to maintain the desired temperature in the water storage tank to eliminate the need for running one of the large space heater boilers at a very low efficiency during the summer months.

Lighting

1. Remove unnecessary lamps, if removal will still provide illumination levels required. When lamps are removed from a fluorescent luminaire, all lamps controlled by a given ballast should be removed to prevent ballast failure or reduced lamp life. Except circuit-interrupting lamp holders, also consider disconnecting ballasts that otherwise would continue to consume energy.
2. Consider replacing present lamps with those of lower wattage which provide the same amount of illumination or (if acceptable in light of tasks involved) a lower level of illumination.
3. Replace all incandescent parking lighting with mercury vapor HID (high-intensity discharge) or sodium lamps.
4. When existing circuitry makes it impossible to utilize less than 25% of the light in a given large space whenever light is needed, and when persons work during normally unoccupied periods, consider development of a desk lamp issuance program, which enables persons working during unoccupied periods to use a simple desk lamp or two instead of a large bank of luminaires.
5. When natural light is available in a building, consider the use of photocell switching to turn off banks of lighting in areas where the natural light is sufficient for the task.
6. Move desks and other work surfaces to a position and orientation that will use installed luminaires to their greatest advantage (instead of moving the luminaires).

General Building

1. Add storm windows. Consider adding one to every other window of the building. Storms on 50% of the windows would still leave enough windows for ventilation during spring and fall schedules.
2. Insulate all roofs, walls, and floors having exterior exposures.
3. Use heat pumps in place of electrical resistance heating and take advantage of the favorable coefficient of performance.

4. Install a small domestic hot water heater to maintain the desired temperature in the water storage tank to eliminate the need for running one of the large space heater boilers at a very low efficiency during the summer months.

Bibliography

ASHRAE, *ASHRAE Standard 90-75*, American Society of Heating, Refrigerating, and Air Conditioning Engineers, New York, 1975.

Byrer, T. G., "Energy Conservation in the Metalworking Industry," SME Publ. #MM75-129, Dearborn, MI (1975).

Cassel, R. T., "Energy Conservation at the Rocketdyne Division of Rockwell International," SME Publ. #E76-109, Dearborn, MI (1976).

Clark, G. W., "Good Lighting with Energy Conservation," SME Publ. #M76-105, Dearborn, MI (1976).

Coggins, J. L., "Techniques of Energy Management," SME Publ. #EM76-114, Dearborn, MI (1976).

Connelly, R. R., "Conserve Energy by Controlling the Demand for Electric Power," SME Publ. #EM76-101, Dearborn, MI (1976).

Dale, J. C., "The Energy Audit—First Step in an Energy Conservation Audit Program," SME Publ. #EM76-104, Dearborn, MI (1976).

Dekoker, N., "GM Energy Management—Organization and Results," SME Publ. #MM75-132, Dearborn, MI (1975).

Doolittle, J. S., *Energy*, Matrix, Champaign, IL, 1977.

Dubin, F. S., H. L. Mindell, S. Bloome, *How to Save Energy and Cut Costs in Existing Industrial and Commercial Buildings*, Noyes Data Corp., Park Ridge, NJ, 1976.

Eckler, N. H., "Energy Conservation in a Research and Development Facility," SME Publ. #MM75-916, Dearborn, MI (1975).

Energy Management Section, General Motors Corporation, *Industrial Energy Conservation—101 Ideas at Work*.

Energy Task Force of American Council on Education, *Notes on Energy Management*, National Assoc. College & University Business Offices, Washington, DC, 1976.

Hardy, R. D., "Energy Savings with Powder Coatings," SME Publ. #FC74-189, Dearborn, MI (1974).

Hauser, L. B., "New Energy Usage Patterns in Manufacturing," SME Publ. #76-100, Dearborn, MI (1976).

IES Lighting Handbook, Illuminating Engineering Society, New York, 1980.

Increasing Energy Efficiency—A Program of Industrial Workshops, Participant's Workbook, Federal Energy Administration, Dallas, TX, 1976.

Industrial Ventilation, American Conf. Governmental Industrial Hygienists, Lansing, MI, 1976.

Institute of Real Estate Management of Natl. Assoc. Realtors, "Energy Cost Reduction for Apartment Owners and Managers," Dept. of Energy, Washington, DC (1977).

Jones, T., "Conserving Energy in Pretreatment Processes," SME Publ. #FC74-190, Dearborn, MI (1974).

Klisiewicz, P. E., "Electrical Supply Capacities and Limitations," SME Publ. #EM76-112, Dearborn, MI (1976).

Linder, E. E., "The Consultant's Role in Energy Management for Production," SME Publ. #EM76-113, Dearborn, MI (1976).

Managing the Energy Dilemma—Technical Reference Manual, Participant's Workbook, Federal Energy Administration, Dallas, TX, 1976.

Neal, G. W., "Evaluating On-Site Power Generation for Industrial Plants," SME Publ. #EM76-110, MI (1976).

NECA and NEMA, *Total Energy Management—A Practical Handbook on Energy Conservation and Management*, National Electrical Contractors Assoc. and National Electrical Manufacturers Assoc., Washington, DC, 1976.

NECA and NEMA, *Total Energy Management in Existing Buildings*, National Electrical Contractors Assoc. and National Electrical Manufacturers Assoc., Washington, DC, 1977.

NICA, "Principles of Heat Transfer and Introduction to ETI," National Insulation Contractors Assoc. (Jun. 1976).

Ostrander, B. W., "Fuel Savings Electronic Air Cleaners Applied to Welding Smoke," SME Publ. #76-102, Dearborn, MI (1976).

Ozarks Regional Commission, *Regional Energy Alternative Study*, Mathtech, Inc., Princeton, 1977.

The Potential for Energy Conservation in Nine Selected Industries, Vol. 2, FEA, Washington, DC, 1975.

Powers, H. R., "Economic & Energy Savings Waterborne Coatings," SME Publ. #FC75-561, Dearborn, MI (1975).

Presley, M. and W. Turner, *Save $$$—Conserve Energy*, School of Industrial Engineering & Management, Oklahoma State University, Stillwater, OK, 1977.

Relick, W. J., "Survey Your Plant's Electrical Power Needs and Save," SME Publ. #MM75-130, Dearborn, MI (1975).

Robnett, J. D., "Energy Conservation at DuPont," SME Publ. #EM76-106, Dearborn, MI (1976).

Shaffer, E. W., "Computerized Approach to Energy Conservation," SME Publ. #EM76-107, Dearborn, MI (1976).

Spencer, R. S. and G. L. Decker, "Energy and Capital Conservation Through Exploitation of the Industrial Steam Base," SME Publ. #EM76-103, Dearborn, MI (1976).

Stafford, J., "Energy Conservation in Appliance Manufacturing," SME Publ. #EM76-108, Dearborn, MI (1976).

Thekdi, A. C., "Airless Paint Drying for Energy Savings in Finishing Industry," SME Publ. #FC76-221, Dearborn, MI (1976).

Turner, W. C., et al., eds., *Energy Management Handbook*, Wiley, New York, 1982.

US Dept. of Commerce, *Building Energy Handbook*, Vol. 1, *Methodology for Energy Survey and Appraisal*, US Government Printing Office, ERDA-76/163/1, Washington, DC, 1976.

US Dept. of Commerce, *Building Energy Handbook*, Vol. 2, ERDA-76/163/2, US Government Printing Office, Washington, DC, 1976.

US Dept. of Commerce, *Energy Conservation Program Guide for Industry and Commerce*, US Government Printing Office, NBS Handbook 115, Washington, DC, 1974.

US Dept. of Energy, *Instructions for Energy Auditors*, US Government Printing Office, Washington, DC, 1979.

INDEX